W9-BBV-994

CRC Handbook of Chemistry and Physics

76th Edition

EDITORIAL ADVISORY BOARD

CRC Handbook of Chemistry and Physics

A Ready-Reference Book of Chemical and Physical Data

Editor-in-Chief

David R. Lide, Ph.D.
Former Director, Standard Reference Data
National Institute of Standards and Technology

Associate Editor

H. P. R. Frederikse, Ph.D.
(Retired)
Ceramics Division
National Institute of Standards and Technology

CRC Press
Boca Raton New York London Tokyo

Direct all inquiries to CRC Press, Inc., 2000 Corporate Blvd., N.W., Boca Raton, Florida 33431.

International Standard Book Number 0-8493-0596-9
Library of Congress Card Number 13
Printed in the United States of America 1 2 3 4 5 6 7 8 9 0
Printed on acid-free paper

To Mary, David, and Grace Lide

PREFACE
76th Edition

The highlight of the 76th Edition of the *Handbook of Chemistry and Physics* is a completely revised table of "Physical Constants of Inorganic Compounds". Judging from user inquiries, this table and its companion table for organic compounds are the most widely used parts of the *Handbook*. In addition to incorporating updated values of the physical constants, the table includes new information such as Chemical Abstracts Service Registry Numbers and pointers to other tables in the *Handbook* where additional properties are covered. The compounds in the table have been selected to include the most important inorganic substances encountered in research, industry, and environmental studies. An effort has been made to include inorganic compounds that are important in modern materials science. The table appears in a more readable format, and indexes by synonym and CAS Registry Number are given to facilitate location of compounds.

Several other parts of the *Handbook* have also been expanded or revised in this edition. The information on vapor pressure is enhanced by addition of a table covering values for less volatile substances at elevated temperatures. The table on solvents has been updated and expanded to include 200 common laboratory solvents. A new listing of carcinogenic chemicals, derived from the 1994 report of the National Toxicology Program, is given. On the physics side, the "Table of the Isotopes" and the "Summary Table of Particle Properties" have been completely updated, as have the tables on "Electron Affinities" and "Atomic and Molecular Polarizabilities". New tables appear on "Techniques for Materials Characterization", "Nonlinear Optical Properties", and "Organic Superconductors", and the tables on "High Temperature Superconductors" and "Physical and Optical Properties of Minerals" have been updated. Finally, several older tables have been converted to SI units and reset in a more readable format.

In order to make room for new material in the last few editions, some of the older tables have been dropped. The Editor will appreciate comments on the removal of these tables; if there is sufficient interest, an effort will be made to restore them in abbreviated form.

Suggestions and criticisms from users are very important to the success of this book. Suggestions on new topics of current interest (and on older tables whose usefulness is questionable) are particularly helpful. Address all comments to Editor, Handbook of Chemistry and Physics, CRC Press, Inc., 2000 Corporate Blvd. N.W., Boca Raton, FL 33431.

The *Handbook of Chemistry and Physics* is dependent on the efforts of many contributors throughout the world. The list of current contributors follows this Preface. Dr. H.P.R. Frederikse, Associate Editor, has made a major contribution during the last few years in expanding and improving the tables in solid state physics and materials science. I am indebted to Dr. S.L. Phillips and Dr. D.L. Perry for sending me information in advance from their forthcoming CRC Press publication *Handbook of Inorganic Compounds*, which was very helpful in planning the new inorganic table in this edition. I should finally like to express my gratitude to those at CRC Press who are responsible for the production of the *Handbook of Chemistry and Physics*, especially Paul Gottehrer, Barbara Caras, and James Brody. It is a great pleasure to work with them.

David R. Lide
February 10, 1995

Note on the Ordering of Chemical Compounds: The decision on the order in which to list chemical compounds in a table is always difficult. An alphabetical list by name has the disadvantage that several different synonyms are often in common use, with the result that a reader may conclude incorrectly that a compound is not present if he looks it up under the wrong name. Listing by line formula is satisfactory for simple inorganic compounds, but is cumbersome for organics. A listing by molecular formula is attractive because clear rules can be given for locating a compound, but the user may have to go to some effort to determine the molecular formula. In this book the choice is made on pragmatic grounds. The long tables, "Physical Constants of Organic Compounds" and "Physical Constants of Inorganic Compounds", are ordered by systematic name, but indexes to synonyms, formulas, and CAS Registry Numbers are provided. If the table is very short and includes only common substances, the listing is usually alphabetical by common name or formula. The remaining tables are ordered by molecular formula using a modification of the Hill convention. In this convention the molecular formula is written with C first, H second, and then all other elements in alphabetical order of their chemical symbols. For tables with organic compounds only, the sequence of entries then follows the alphabetical order of elements in the molecular formula and the number of atoms of each element, in ascending order, e.g., C_3H_7Cl, C_3H_7N, C_3H_7NO, $C_3H_7NO_2$, etc. In tables containing non-carbon compounds, those are listed first, followed by a separate listing of compounds that do contain carbon. This is in contrast to the strict Hill convention as followed by Chemical Abstracts Service, where the molecular formulas beginning with A and B precede the formulas for carbon-containing compounds, while those beginning with D... Z follow. For tabular displays, as opposed to an index, it appears more convenient to the user if the non-carbon compounds are listed as a block, rather than being split by the longer list of carbon compounds.

For organic compounds, a quick way to determine the molecular formula is to use the "Physical Constants of Organic Compounds" table, which starts on Page **3**-1, and its synonym index on Page **3**-586.

CURRENT CONTRIBUTORS

Douglas Ambrose
Department of Chemistry
University College London
London WC1H OAJ
England

Donald L. Baulch
School of Chemistry
University of Leeds
Leeds L52 9JT
England

Lev I. Berger
California Institute of Electronics
and Materials Science
2115 Flame Tree Way
Hemet, California 92343

Martin J. Berger
5011 Elm St.
Bethesda, Maryland 20814

George W. Burns
National Institute of Standards and
Technology
Gaithersburg, Maryland 20899

A. K. Covington
Department of Chemistry
University of Newcastle
Newcastle upon Tyne NE1 7RU
England

Hans Dolezalek
1812 Drury Lane
Alexandria, Virginia 22307

Howard T. Evans
U.S. Geological Survey
Reston, Virginia 22092

Lev R. Fokin
Institute for High Temperatures
Academy of Sciences
Izorskaye St. 13/19
Moscow 127412, Russia

H. P. R. Frederikse
9625 Dewmar Lane
Kensington, Maryland 20895

Karl A. Gschneidner
Ames Laboratory
Energy and Mineral Resources
Research Institute
Iowa State University
Ames, Iowa 50011

Lev V. Gurvich
Institute for High Temperatures
Academy of Sciences
Izorskaye St. 13/19
Moscow 127412, Russia

C. R. Hammond
17 Greystone
West Hartford, Connecticut 06107

Robert F. Hampson
Chemical Kinetics Division
National Institute of Standards and
Technology
Gaithersburg, Maryland 20899

Norman E. Holden
National Nuclear Data Center
Brookhaven National Laboratory
Upton, New York 11973

H. D. B. Jenkins
Department of Molecular Sciences
University of Warwick
Coventry CV4 7AL
England

Henry V. Kehiaian
ITODYS
1 rue Guy de la Brosse
75005 Paris
France

J. A. Kerr
Atmospheric Chemistry Group
EAWAG
CH-8600 Dubendorf-Zurich
Switzerland

William F. Krupke
Lawrence Livermore Laboratory
Livermore, California 94550

Kozo Kuchitsu
Department of Chemistry
Josai University
1-1 Keyakidai
Sakado 350-02
Japan

John Latham
Department of Physics
University of Manchester
Manchester M60 1QD
England

Sharon G. Lias
Chemical Kinetics and
Thermodynamics Division
National Institute of Standards and
Technology
Gaithersburg, Maryland 20899

Reiner Luckenbach
Beilstein Institute
Carl-Bosch-Haus
Varrentrapstrasse 40-42
D-6000 Frandfurt/Main 90
Germany

Arthur Maki
15012 24 Ave., S. E.
Mill Creek, Washington 98012

B. W. Mangum
Temperature and Pressure Division
National Insitute of Standards and
Technology
Gaithersburg, Maryland 20899

Thomas M. Miller
Geophysics Laboratory
Hanscom AFB, Massachusetts
01731

Dale L. Perry
Lawrence Berkeley Laboratory
1 Cyclotron Road
Berkeley, California 94720

Sidney L. Phillips
Camatx/Basic Chemistry
171 El Toyonal
Orinda, California 94563

CURRENT CONTRIBUTORS (continued)

C. N. R. Rao
Solid State and Structural Chemistry Unit
Indian Institute of Science
Bangalore 560-012
India

A. K. Raychaudhuri
Solid State and Structural Chemistry Unit
Indian Institute of Science
Bangalore 560-012
India

Hannes Tammet
Air Electricity Laboratory
18 Ulikooli St.
Tartu 202400
Estonia

B. N. Taylor
National Institute of Standards and Technology
Gaithersburg, Maryland 20899

D. Thorburn Burns
Department of Chemistry
Queen's University of Belfast
Belfast BT9 5AG
Northern Ireland

Thomas G. Trippe
Particle Data Group
Lawrence Berkeley Laboratory
1 Cyclotron Road
Berkeley, California 94720

Martin A. Uman
Department of Electrical Engineering
University of Florida
Gainesville, Florida 32611

Petr Vanýsek
Department of Chemistry
Northern Illinois University
DeKalb, Illinois 60115

John H. Weaver
Department of Chemical Engineering and Materials Science
University of Minnesota
Minneapolis, Minnesota 55425

Anthony Wexler
Department of Mechanical Engineering
University of Delaware
Newark, Delaware 19716

Wolfgang L. Wiese
Atomic Plasma and Radiation Division
National Institute of Standards and Technology
Gaithersburg, Maryland 20899

Gwyn P. Williams
National Synchrotron Light Source
Brookhaven National Laboratory
Upton, New York 11973

Christian Wohlfarth
Institut für Physikalische Chemie
Martin Luther University
D-06217 Merseburg
Germany

TABLE OF CONTENTS

SECTION 1: BASIC CONSTANTS, UNITS, AND CONVERSION FACTORS

SECTION 2: SYMBOLS, TERMINOLOGY, AND NOMENCLATURE

SECTION 3: PHYSICAL CONSTANTS OF ORGANIC COMPOUNDS

SECTION 4: PROPERTIES OF THE ELEMENTS AND INORGANIC COMPOUNDS

SECTION 5: THERMOCHEMISTRY, ELECTROCHEMISTRY, AND KINETICS

SECTION 6: FLUID PROPERTIES

SECTION 7: BIOCHEMISTRY

SECTION 8: ANALYTICAL CHEMISTRY

SECTION 9: MOLECULAR STRUCTURE AND SPECTROSCOPY

SECTION 10: ATOMIC, MOLECULAR, AND OPTICAL PHYSICS

SECTION 11: NUCLEAR AND PARTICLE PHYSICS

SECTION 12: PROPERTIES OF SOLIDS

SECTION 13: POLYMER PROPERTIES

SECTION 14: GEOPHYSICS, ASTRONOMY, AND ACOUSTICS

SECTION 15: PRACTICAL LABORATORY DATA

SECTION 16: HEALTH AND SAFETY INFORMATION

APPENDIX A: MATHEMATICAL TABLES

Section 1
Basic Constants, Units, and Conversion Factors

FUNDAMENTAL PHYSICAL CONSTANTS

The 1986 CODATA Recommended Values

By E. Richard Cohen and Barry N. Taylor as published in the *Journal of Research of the National Bureau of Standards,* 92, 85, 1987. Discussions of the background, data selection and evaluation procedures are presented in CODATA Bulletin Number 63, November 1986, "The 1986 Adjustment of the Fundamental Physical Constants", a Report of the CODATA Task Group on Fundamental Physical Constants (36 pages) published by Pergamon Press.

The 1986 recommended values of the fundamental physical constants are given in five tables. Table 1 is an abbreviated list containing the quantities which should be of greatest interest to most users. Table 2 is a more complete compilation. Table 3 is a list of related "maintained units and standard values." Table 4 contains a number of scientifically, technologically, and metrologically useful energy conversion factors. Table 5 is an extended covariance matrix containing the variances, covariances, and correlation coefficients of the unknowns and a number of different constants (included for convenience) from which the like quantities of other constants may be calculated. (B. N. Taylor, W. H. Parker, and D. N. Langenberg, *Rev. Mod. Phys.*, 41, 375, 1969. Such a matrix is necessary because the variables in a least-square adjustment are correlated.

Table 1
SUMMARY OF THE 1986 RECOMMENDED VALUES OF THE FUNDAMENTAL PHYSICAL CONSTANTS

Quantity	Symbol	Value	Units	Relative Uncertainty (ppm)
speed of light in vacuum	c	299 792 458	$\mathrm{m\,s^{-1}}$	(exact)
permeability of vacuum	μ_0	$4\pi \times 10^{-7}$	$\mathrm{N\,A^{-2}}$	
		=12.566 370 614...	$10^{-7}\,\mathrm{N\,A^{-2}}$	(exact)
permittivity of vacuum	ϵ_0	$1/\mu_0 c^2$		
		=8.854 187 817...	$10^{-12}\,\mathrm{F\,m^{-1}}$	(exact)
Newtonian constant of gravitation	G	6.672 59(85)	$10^{-11}\,\mathrm{m^3\,kg^{-1}\,s^{-2}}$	128
Planck constant	h	6.626 0755(40)	$10^{-34}\,\mathrm{J\,s}$	0.60
$h/2\pi$	$\hbar$	1.054 572 66(63)	$10^{-34}\,\mathrm{J\,s}$	0.60
elementary charge	e	1.602 177 33(49)	$10^{-19}\,\mathrm{C}$	0.30
magnetic flux quantum, $h/2e$	Φ_0	2.067 834 61(61)	$10^{-15}\,\mathrm{Wb}$	0.30
electron mass	m_e	9.109 3897(54)	$10^{-31}\,\mathrm{kg}$	0.59
proton mass	m_p	1.672 6231(10)	$10^{-27}\,\mathrm{kg}$	0.59
proton-electron mass ratio	m_p/m_e	1836.152 701(37)		0.020
fine-structure constant, $\mu_0 ce^2/2h$	α	7.297 353 08(33)	10^{-3}	0.045
inverse fine-structure constant	α^{-1}	137.035 9895(61)		0.045
Rydberg constant, $m_e c\alpha^2/2h$	R_∞	10 973 731.534(13)	$\mathrm{m^{-1}}$	0.0012
Avogadro constant	N_A, L	6.022 1367(36)	$10^{23}\,\mathrm{mol^{-1}}$	0.59
Faraday constant, $N_A e$	F	96 485.309(29)	$\mathrm{C\,mol^{-1}}$	0.30
molar gas constant	R	8.314 510(70)	$\mathrm{J\,mol^{-1}\,K^{-1}}$	8.4
Boltzmann constant, R/N_A	k	1.380 658(12)	$10^{-23}\,\mathrm{J\,K^{-1}}$	8.5
Stefan-Boltzmann constant, $(\pi^2/60)k^4/\hbar^3c^2$	σ	5.670 51(19)	$10^{-8}\,\mathrm{W\,m^{-2}\,K^{-4}}$	34
		Non-SI units used with SI		
electron volt, $(e/\mathrm{C})\,\mathrm{J} = \{e\}\,\mathrm{J}$	eV	1.602 177 33(49)	$10^{-19}\,\mathrm{J}$	0.30
(unified) atomic mass unit, $1\,\mathrm{u} = m_u = \frac{1}{12}m(^{12}\mathrm{C})$	u	1.660 5402(10)	$10^{-27}\,\mathrm{kg}$	0.60

NOTE: An abbreviated list of the fundamental constants of physics and chemistry based on a least-squares adjustment with 17 degrees of freedom. The digits in parentheses are the one-standard-deviation uncertainty in the last digits of the given value. Since the uncertainties of many entries are correlated, the full covariance matrix must be used in evaluating the uncertainties of quantities computed from them.

Table 2
THE 1986 RECOMMENDED VALUES OF THE FUNDAMENTAL PHYSICAL CONSTANTS

Quantity	Symbol	Value	Units	Relative Uncertainty (ppm)
		GENERAL CONSTANTS		
		Universal Constants		
speed of light in vacuum	c	299 792 458	$\mathrm{m\,s^{-1}}$	(exact)
permeability of vacuum	μ_0	$4\pi \times 10^{-7}$	$\mathrm{N\,A^{-2}}$	
		=12.566 370 614...	$10^{-7}\,\mathrm{N\,A^{-2}}$	(exact)
permittivity of vacuum	ϵ_0	$1/\mu_0 c^2$		
		=8.854 187 817...	$10^{-12}\,\mathrm{F\,m^{-1}}$	(exact)
Newtonian constant of gravitation	G	6.672 59(85)	$10^{-11}\,\mathrm{m^3\,kg^{-1}\,s^{-2}}$	128
Planck constant	h	6.626 0755(40)	$10^{-34}\,\mathrm{J\,s}$	0.60
in electron volts, $h/\{e\}$		4.135 6692(12)	$10^{-15}\,\mathrm{eV\,s}$	0.30
$h/2\pi$	$\hbar$	1.054 572 66(63)	$10^{-34}\,\mathrm{J\,s}$	0.60
in electron volts, $\hbar/\{e\}$		6.582 1220(20)	$10^{-16}\,\mathrm{eV\,s}$	0.30

Table 2
THE 1986 RECOMMENDED VALUES OF THE FUNDAMENTAL PHYSICAL CONSTANTS (continued)

Quantity	Symbol	Value	Units	Relative Uncertainty (ppm)
Planck mass, $(\hbar c/G)^{\frac{1}{2}}$	m_P	2.176 71(14)	10^{-8} kg	64
Planck length, $\hbar/m_Pc = (\hbar G/c^3)^{\frac{1}{2}}$	l_P	1.616 05(10)	10^{-35} m	64
Planck time, $l_P/c = (\hbar G/c^5)^{\frac{1}{2}}$	t_P	5.390 56(34)	10^{-44} s	64
		Electromagnetic Constants		
elementary charge	e	1.602 177 33(49)	10^{-19} C	0.30
	e/h	2.417 988 36(72)	10^{14} A J^{-1}	0.30
magnetic flux quantum, $h/2e$	Φ_o	2.067 834 61(61)	10^{-15} Wb	0.30
Josephson frequency-voltage ratio	$2e/h$	4.835 9767(14)	10^{14} Hz V^{-1}	0.30
quantized Hall conductance	e^2/h	3.874 046 14(17)	10^{-5} S	0.045
quantized Hall resistance, $h/e^2 = \mu_o c/2\alpha$	R_H	25 812.8056(12)	Ω	0.045
Bohr magneton, $e\hbar/2m_e$	μ_B	9.274 0154(31)	10^{-24} J T^{-1}	0.34
in electron volts, $\mu_B/\{e\}$		5.788 382 63(52)	10^{-5} eV T^{-1}	0.089
in hertz, μ_B/h		1.399 624 18(42)	10^{10} Hz T^{-1}	0.30
in wavenumbers, μ_B/hc		46.686 437(14)	m^{-1} T^{-1}	0.30
in kelvins, μ_B/k		0.671 7099(57)	K T^{-1}	8.5
nuclear magneton, $e\hbar/2m_p$	μ_N	5.050 7866(17)	10^{-27}J T^{-1}	0.34
in electron volts, $\mu_N/\{e\}$		3.152 451 66(28)	10^{-8} eV T^{-1}	0.089
in hertz, μ_N/h		7.622 5914(23)	MHz T^{-1}	0.30
in wavenumbers, μ_N/hc		2.542 622 81(77)	10^{-2} m^{-1} T^{-1}	0.30
in kelvins, μ_N/k		3.658 246(31)	10^{-4} K T^{-1}	8.5
		ATOMIC CONSTANTS		
fine-structure constant, $\mu_o ce^2/2h$	α	7.297 353 08(33)	10^{-3}	0.045
inverse fine-structure constant	α^{-1}	137.035 9895(61)		0.045
Rydberg constant, $m_e c\alpha^2/2h$	R_∞	10 973 731.534(13)	m^{-1}	0.0012
in hertz, $R_\infty c$		3.289 841 9499(39)	10^{15} Hz	0.0012
in joules, $R_\infty hc$		2.179 8741(13)	10^{-18} J	0.60
in eV, $R_\infty hc/\{e\}$		13.605 6981(40)	eV	0.30
Bohr radius, $\alpha/4\pi R_\infty$	a_o	0.529 177 249(24)	10^{-10} m	0.045
Hartree energy, $e^2/4\pi\epsilon_o a_o = 2R_\infty hc$	E_h	4.359 7482(26)	10^{-18} J	0.60
in eV, $E_h/\{e\}$		27.211 3961(81)	eV	0.30
quantum of circulation	$h/2m_e$	3.636 948 07(33)	10^{-4} m^2 s^{-1}	0.089
	h/m_e	7.273 896 14(65)	10^{-4} m^2 s^{-1}	0.089
		Electron		
electron mass	m_e	9.109 3897(54)	10^{-31} kg	0.59
		5.485 799 03(13)	10^{-4} u	0.023
in electron volts, $m_e c^2/\{e\}$		0.510 999 06(15)	MeV	0.30
electron–muon mass ratio	m_e/m_μ	4.836 332 18(71)	10^{-3}	0.15
electron–proton mass ratio	m_e/m_p	5.446 170 13(11)	10^{-4}	0.020
electron–deuteron mass ratio	m_e/m_d	2.724 437 07(6)	10^{-4}	0.020
electron–α-particle mass ratio	m_e/m_α	1.370 933 54(3)	10^{-4}	0.021
electron specific charge	$-e/m_e$	$-1.758\,819\,62(53)$	10^{11} C kg^{-1}	0.30
electron molar mass	$M(e)$, M_e	5.485 799 03(13)	10^{-7} kg/mol	0.023
Compton wavelength, $h/m_e c$	λ_C	2.426 310 58(22)	10^{-12} m	0.089
$\lambda_C/2\pi = \alpha a_o = \alpha^2/4\pi R_\infty$	λ_C	3.861 593 23(35)	10^{-13} m	0.089
classical electron radius, $\alpha^2 a_o$	r_e	2.817 940 92(38)	10^{-15} m	0.13
Thomson cross section, $(8\pi/3)r_e^2$	σ_e	0.665 246 16(18)	10^{-28} m^2	0.27
electron magnetic moment	μ_e	928.477 01(31)	10^{-26} J T^{-1}	0.34
in Bohr magnetons	μ_e/μ_B	1.001 159 652 193(10)		1×10^{-5}
in nuclear magnetons	μ_e/μ_N	1838.282 000(37)		0.020
electron magnetic moment anomaly, $\mu_e/\mu_B - 1$	a_e	1.159 652 193(10)	10^{-3}	0.0086
electron g-factor, $2(1 + a_e)$	g_e	2.002 319 304 386(20)		1×10^{-5}
electron–muon magnetic moment ratio	μ_e/μ_μ	206.766 967(30)		0.15
electron–proton magnetic moment ratio	μ_e/μ_p	658.210 6881(66)		0.010
		Muon		
muon mass	m_μ	1.883 5327(11)	10^{-28} kg	0.61
		0.113 428 913(17)	u	0.15
in electron volts, $m_\mu c^2/\{e\}$		105.658 389(34)	MeV	0.32
muon–electron mass ratio	m_μ/m_e	206.768 262(30)		0.15
muon molar mass	$M(\mu)$, M_μ	1.134 289 13(17)	10^{-4} kg/mol	0.15
muon magnetic moment	μ_μ	4.490 4514(15)	10^{-26} J T^{-1}	0.33
in Bohr magnetons,	μ_μ/μ_B	4.841 970 97(71)	10^{-3}	0.15
in nuclear magnetons,	μ_μ/μ_N	8.890 5981(13)		0.15

Table 2
THE 1986 RECOMMENDED VALUES OF THE FUNDAMENTAL PHYSICAL CONSTANTS (continued)

Quantity	Symbol	Value	Units	Relative Uncertainty (ppm)
muon magnetic moment anomaly, $[\mu_\mu/(e\hbar/2m_\mu)] - 1$	a_μ	1.1659230(84)	10^{-3}	7.2
muon g-factor, $2(1 + a_\mu)$	g_μ	2.002331846(17)		0.0084
muon–proton magnetic moment ratio	μ_μ/μ_p	3.18334547(47)		0.15
		Proton		
proton mass	m_p	1.6726231(10)	10^{-27} kg	0.59
		1.007276470(12)	u	0.012
in electron volts, $m_pc^2/\{e\}$		938.27231(28)	MeV	0.30
proton–electron mass ratio	m_p/m_e	1836.152701(37)		0.020
proton–muon mass ratio	m_p/m_μ	8.8802444(13)		0.15
proton specific charge	e/m_p	9.5788309(29)	10^7 C kg^{-1}	0.30
proton molar mass	$M(\mathrm{p}), M_p$	1.007276470(12)	10^{-3} kg/mol	0.012
proton Compton wavelength, h/m_pc	$\lambda_{C,p}$	1.32141002(12)	10^{-15} m	0.089
$\lambda_{C,p}/2\pi$	$\lambda_{C,p}$	2.10308937(19)	10^{-16} m	0.089
proton magnetic moment	μ_p	1.41060761(47)	10^{-26} J T^{-1}	0.34
in Bohr magnetons	μ_p/μ_B	1.521032202(15)	10^{-3}	0.010
in nuclear magnetons	μ_p/μ_N	2.792847386(63)		0.023
diamagnetic shielding correction for protons in pure water, spherical sample, 25 °C, $1 - \mu'_p/\mu_p$	σ_{H_2O}	25.689(15)	10^{-6}	
shielded proton moment (H_2O, sph., 25 °C)	μ'_p	1.41057138(47)	10^{-26} J T^{-1}	0.34
in Bohr magnetons	μ'_p/μ_B	1.520993129(17)	10^{-3}	0.011
in nuclear magnetons	μ'_p/μ_N	2.792775642(64)		0.023
proton gyromagnetic ratio	γ_p	26752.2128(81)	10^4 s^{-1} T^{-1}	0.30
	$\gamma_p/2\pi$	42.577469(13)	MHz T^{-1}	0.30
uncorrected (H_2O, sph., 25 °C)	γ'_p	26751.5255(81)	10^4 s^{-1} T^{-1}	0.30
	$\gamma'_p/2\pi$	42.576375(13)	MHz T^{-1}	0.30
		Neutron		
neutron mass	m_n	1.6749286(10)	10^{-27} kg	0.59
		1.008664904(14)	u	0.014
in electron volts, $m_nc^2/\{e\}$		939.56563(28)	Mev	0.30
neutron–electron mass ratio	m_n/m_e	1838.683662(40)		0.022
neutron–proton mass ratio	m_n/m_p	1.001378404(9)		0.009
neutron molar mass	$M(\mathrm{n}), M_n$	1.008664904(14)	10^{-3} kg/mol	0.014
neutron Compton wavelength, h/m_nc	$\lambda_{C,n}$	1.31959110(12)	10^{-15} m	0.089
$\lambda_{C,n}/2\pi$	$\lambda_{C,n}$	2.10019445(19)	10^{-16} m	0.089
neutron magnetic moment *	μ_n	0.96623707(40)	10^{-26} J T^{-1}	0.41
in Bohr magnetons	μ_n/μ_B	1.04187563(25)	10^{-3}	0.24
in nuclear magnetons	μ_n/μ_N	1.91304275(45)		0.24
neutron–electron magnetic moment ratio	μ_n/μ_e	1.04066882(25)	10^{-3}	0.24
neutron–proton magnetic moment ratio	μ_n/μ_p	0.68497934(16)		0.24
		Deuteron		
deuteron mass	m_d	3.3435860(20)	10^{-27} kg	0.59
		2.013553214(24)	u	0.012
in electron volts, $m_dc^2/\{e\}$		1875.61339(57)	MeV	0.30
deuteron–electron mass ratio	m_d/m_e	3670.483014(75)		0.020
deuteron–proton mass ratio	m_d/m_p	1.999007496(6)		0.003
deuteron molar mass	$M(\mathrm{d}), M_d$	2.013553214(24)	10^{-3} kg/mol	0.012
deuteron magnetic moment *	μ_d	0.43307375(15)	10^{-26} J T^{-1}	0.34
in Bohr magnetons,	μ_d/μ_B	0.4669754479(91)	10^{-3}	0.019
in nuclear magnetons,	μ_d/μ_N	0.857438230(24)		0.028
deuteron–electron magnetic moment ratio	μ_d/μ_e	0.4664345460(91)	10^{-3}	0.019
deuteron–proton magnetic moment ratio	μ_d/μ_p	0.3070122035(51)		0.017
		PHYSICO-CHEMICAL CONSTANTS		
Avogadro constant	N_A, L	6.0221367(36)	10^{23} mol^{-1}	0.59
atomic mass constant, $\frac{1}{12}m(^{12}C)$	m_u	1.6605402(10)	10^{-27} kg	0.59
in electron volts, $m_uc^2/\{e\}$		931.49432(28)	MeV	0.30
Faraday constant	F	96485.309(29)	C mol^{-1}	0.30
molar Planck constant	N_Ah	3.99031323(36)	10^{-10} J s mol^{-1}	0.089
	N_Ahc	0.11962658(11)	J m mol^{-1}	0.089
molar gas constant	R	8.314510(70)	J mol^{-1} K^{-1}	8.4

Table 2
THE 1986 RECOMMENDED VALUES OF THE FUNDAMENTAL PHYSICAL CONSTANTS (continued)

Quantity	Symbol	Value	Units	Relative Uncertainty (ppm)
Boltzmann constant, R/N_A	k	1.380658(12)	10^{-23} J K^{-1}	8.5
in electron volts, $k/\{e\}$		8.617385(73)	10^{-5} eV K^{-1}	8.4
in hertz, k/h		2.083674(18)	10^{10} Hz K^{-1}	8.4
in wavenumbers, k/hc		69.50387(59)	m^{-1} K^{-1}	8.4
molar volume (ideal gas), RT/p				
$T = 273.15$ K, $p = 101\,325$ Pa	V_m	22.41410(19)	L/mol	8.4
Loschmidt constant, N_A/V_m	n_o	2.686763(23)	10^{25} m^{-3}	8.5
$T = 273.15$ K, $p = 100$ kPa	V_m	22.71108(19)	L/mol	8.4
Sackur-Tetrode constant (absolute entropy constant), **				
$\frac{5}{2} + \ln\{(2\pi m_u kT_1/h^2)^{\frac{3}{2}} kT_1/p_o\}$				
$T_1 = 1$ K, $p_o = 100$ kPa	S_o/R	−1.151693(21)		18
$p_o = 101\,325$ Pa		−1.164856(21)		18
Stefan-Boltzmann constant, $(\pi^2/60)k^4/\hbar^3c^2$	σ	5.67051(19)	10^{-8} W m^{-2} K^{-4}	34
first radiation constant, $2\pi hc^2$	c_1	3.7417749(22)	10^{-16} W m^2	0.60
second radiation constant, hc/k	c_2	0.01438769(12)	m K	8.4
Wien displacement law constant, $b = \lambda_{max}T = c_2/4.965\,114\,23\ldots$	b	2.897756(24)	10^{-3} m K	8.4

NOTE: This list of the fundamental constants of physics and chemistry is based on a least-squares adjustment with 17 degrees of freedom. The digits in parentheses are the one-standard-deviation uncertainty in the last digits of the given value. Since the uncertainties of many of these entries are correlated, the full covariance matrix must be used in evaluating the uncertainties of quantities computed from them.

* The scalar magnitude of the neutron moment is listed here. The neutron magnetic dipole is directed oppositely to that of the proton, and corresponds to the dipole associated with a spinning negative charge distribution. The vector sum, $\mu_d = \mu_p + \mu_n$, is approximately satisfied.

** The entropy of an ideal monatomic gas of relative atomic weight A_r is given by $S = S_o + \frac{1}{2}R$ in $A_r - R$ in $(p/p_o) + \frac{5}{2}R$ in(T/K).

Table 3
MAINTAINED UNITS AND STANDARD VALUES

Quantity	Symbol	Value	Units	Relative Uncertainty (ppm)
electron volt, (e/C) J $= \{e\}$ J	eV	1.60217733(49)	10^{-19} J	0.30
(unified) atomic mass unit, 1 u $= m_u = \frac{1}{12}m(^{12}\mathrm{C})$	u	1.6605402(10)	10^{-27} kg	0.59
standard atmosphere	atm	101 325	Pa	(exact)
standard acceleration of gravity	g_n	9.80665	m s^{-2}	(exact)
'As-Maintained' Electrical Units				
BIPM maintained ohm, Ω_{69-BI}				
$\Omega_{BI85} \equiv \Omega_{69-BI}$(1 Jan 1985)	Ω_{BI85}	$1 - 1.563(50)\times10^{-6}$	Ω	
		$= 0.999998437(50)$	Ω	0.050
Drift rate of Ω_{69-BI}	$\frac{d\Omega_{69-BI}}{dt}$	−0.0566(15)	μΩ/a	—
BIPM maintained volt,	V_{76-BI}	$1 - 7.59(30)\times10^{-6}$	V	
$V_{76-BI} \equiv 483\,594$ GHz$(h/2e)$		$= 0.99999241(30)$	V	0.30
BIPM maintained ampere,	A_{BI85}	$1 - 6.03(30)\times10^{-6}$	A	
$A_{BIPM} = V_{76-BI}/\Omega_{69-BI}$		$= 0.99999397(30)$	A	0.30
X-Ray Standards				
Cu x-unit : $\lambda(\mathrm{CuK}\alpha_1) \equiv 1537.400$ xu	xu(CuKα_1)	1.00207789(70)	10^{-13} m	0.70
Mo x-unit : $\lambda(\mathrm{MoK}\alpha_1) \equiv 707.831$ xu	xu(MoKα_1)	1.00209938(45)	10^{-13} m	0.45
Å* : $\lambda(\mathrm{WK}\alpha_1) \equiv 0.209100$ Å*	Å*	1.00001481(92)	10^{-10} m	0.92

Table 3
MAINTAINED UNITS AND STANDARD VALUES (continued)

Quantity	Symbol	Value	Units	Relative Uncertainty (ppm)
lattice spacing of Si (in vacuum, 22.5 °C), +	a	0.543 101 96(11)	nm	0.21
$d_{220} = a/\sqrt{8}$	d_{220}	0.192 015 540(40)	nm	0.21
molar volume of Si, $M(\mathrm{Si})/\rho(\mathrm{Si}) = N_A a^3/8$	$V_m(\mathrm{Si})$	12.058 8179(89)	$\mathrm{cm^3/mol}$	0.74

NOTE: A summary of "maintained" units and "standard" values and their relationship to SI units, based on a least-squares adjustment with 17 degrees of freedom. The digits in parentheses are the one-standard-deviation uncertainty in the last digits of the given value. Since the uncertainties of many of these entries are correlated, the full covariance matrix must be used in evaluating the uncertainties of quantities computed from them.

\+ The lattice spacing of single-crystal Si can vary by parts in 10^7 depending on the preparation process. Measurements at PTB indicate also the possibility of distortions from exact cubic symmetry of the order of 0.2 ppm.

Table 4
ENERGY CONVERSION FACTORS

	J	kg	$\mathrm{m^{-1}}$	Hz
1 J =	1	$1/\{c^2\}$ 1.112 650 06 $\times 10^{-17}$	$1/\{hc\}$ 5.034 1125(30) $\times 10^{24}$	$1/\{h\}$ 1.509 188 97(90) $\times 10^{33}$
1 kg =	$\{c^2\}$ 8.987 551 787 $\times 10^{16}$	1	$\{c/h\}$ 4.524 4347(27) $\times 10^{41}$	$\{c^2/h\}$ 1.356 391 40(81) $\times 10^{50}$
1 $\mathrm{m^{-1}}$ =	$\{hc\}$ 1.986 4475(12) $\times 10^{-25}$	$\{h/c\}$ 2.210 2209(13) $\times 10^{-42}$	1	$\{c\}$ 299 792 458
1 Hz =	$\{h\}$ 6.626 0755(40) $\times 10^{-34}$	$\{h/c^2\}$ 7.372 5032(44) $\times 10^{-51}$	$1/\{c\}$ 3.335 640 952 $\times 10^{-9}$	1
1 K =	$\{k\}$ 1.380 658(12) $\times 10^{-23}$	$\{k/c^2\}$ 1.536 189(13) $\times 10^{-40}$	$\{k/hc\}$ 69.503 87(59)	$\{k/h\}$ 2.083 674(18) $\times 10^{10}$
1 eV =	$\{e\}$ 1.602 177 33(49) $\times 10^{-19}$	$\{e/c^2\}$ 1.782 662 70(54) $\times 10^{-36}$	$\{e/hc\}$ 806 554.10(24)	$\{e/h\}$ 2.417 988 36(72) $\times 10^{14}$
1 u =	$\{m_u c^2\}$ 1.492 419 09(88) $\times 10^{-10}$	$\{m_u\}$ 1.660 5402(10) $\times 10^{-27}$	$\{m_u c/h\}$ 7.513 005 63(67) $\times 10^{14}$	$\{m_u c^2/h\}$ 2.252 342 42(20) $\times 10^{23}$
1 hartree =	$\{2R_\infty hc\}$ 4.359 7482(26) $\times 10^{-18}$	$\{2R_\infty h/c\}$ 4.850 8741(29) $\times 10^{-35}$	$\{2R_\infty\}$ 21 947 463.067(26)	$\{2R_\infty c\}$ 6.579 683 8999(78) $\times 10^{15}$

	K	eV	u	hartree
1 J =	$1/\{k\}$ 7.242 924(61) $\times 10^{22}$	$1/\{e\}$ 6.241 5064(19) $\times 10^{18}$	$1/\{m_u c^2\}$ 6.700 5308(40) $\times 10^{9}$	$1/\{2R_\infty hc\}$ 2.293 7104(14) $\times 10^{17}$
1 kg =	$\{c^2/k\}$ 6.509 616(55) $\times 10^{39}$	$\{c^2/e\}$ 5.609 5862(17) $\times 10^{35}$	$1/\{m_u\}$ 6.022 1367(36) $\times 10^{26}$	$\{c/2R_\infty h\}$ 2.061 4841(12) $\times 10^{34}$
1 $\mathrm{m^{-1}}$ =	$\{hc/k\}$ 0.014 387 69(12)	$\{hc/e\}$ 1.239 842 44(37) $\times 10^{-6}$	$\{h/m_u c\}$ 1.331 025 22(12) $\times 10^{-15}$	$1/\{2R_\infty\}$ 4.556 335 2672(54) $\times 10^{-8}$
1 Hz =	$\{h/k\}$ 4.799 216(41) $\times 10^{-11}$	$\{h/e\}$ 4.135 6692(12) $\times 10^{-15}$	$\{h/m_u c^2\}$ 4.439 822 24(40) $\times 10^{-24}$	$1/\{2R_\infty c\}$ 1.519 829 8508(18) $\times 10^{-16}$
1 K =	1	$\{k/e\}$ 8.617 385(73) $\times 10^{-5}$	$\{k/m_u c^2\}$ 9.251 140(78) $\times 10^{-14}$	$\{k/2R_\infty hc\}$ 3.166 829(27) $\times 10^{-6}$
1 eV =	$\{e/k\}$ 11 604.45(10)	1	$\{e/m_u c^2\}$ 1.073 543 85(33) $\times 10^{-9}$	$\{e/2R_\infty hc\}$ 0.036 749 309(11)
1 u =	$\{m_u c^2/k\}$ 1.080 9478(91) $\times 10^{13}$	$\{m_u c^2/e\}$ 931.494 32(28) $\times 10^{6}$	1	$\{m_u c/2R_\infty h\}$ 3.423 177 25(31) $\times 10^{7}$
1 hartree =	$\{2R_\infty hc/k\}$ 3.157 733(27) $\times 10^{5}$	$\{2R_\infty hc/e\}$ 27.211 3961(81)	$\{2R_\infty h/m_u c\}$ 2.921 262 69(26) $\times 10^{-8}$	1

NOTE: To use this table note that all entries on the same line are equal; the unit at the top of a column applies to all of the values beneath it. Example: 1 eV = 806544.10 $\mathrm{m^{-1}}$.

Table 5
EXPANDED COVARIANCE AND CORRELATION COEFFICIENT MATRIX FOR THE 1986 RECOMMENDED SET OF FUNDAMENTAL PHYSICAL CONSTANTS

	α^{-1}	K_V	K_Ω	μ_μ/μ_p	e	h	m_e	N_A	F
α^{-1}	1997	−1062	925	3267	−3059	−4121	−127	127	−2932
K_V	*−0.080*	87988	90	−1737	89050	177038	174914	−174914	−85864
K_Ω	*0.416*	*0.006*	2477	1513	−835	−744	1105	−1105	−1939
μ_μ/μ_p	*0.498*	*−0.040*	*0.207*	21523	−5004	−6742	−208	208	−4796
e	*−0.226*	*0.989*	*−0.055*	*−0.112*	92109	181159	175042	−175042	−82933
h	*−0.154*	*0.997*	*−0.025*	*−0.077*	*0.997*	358197	349956	−349956	−168797
m_e	*−0.005*	*0.997*	*0.038*	*−0.002*	*0.975*	*0.989*	349702	−349702	−174660
N_A	*0.005*	*−0.997*	*−0.038*	*0.002*	*−0.975*	*−0.989*	*−1.000*	349702	174660
F	*−0.217*	*−0.956*	*−0.129*	*−0.108*	*−0.902*	*−0.931*	*−0.975*	*0.975*	91727

The elements of the covariance matrix appear on and above the major diagonal in (parts in $10^9)^2$; correlation coefficients appear in *italics* below the diagonal. The values are given to as many as six digits only as a matter of consistency. The correlation coefficient between m_e and N_A appears as -1.000 in this table because the auxiliary constants were considered to be exact in carrying out the least-squares adjustment. When the uncertainties of m_p/m_e and M_p are properly taken into account, the correlation coefficient is -0.999 and the variances of m_e and N_A are slightly increased.

STANDARD ATOMIC WEIGHTS (1993)

This table of atomic weights is reprinted from the 1993 report of the IUPAC Commission on Atomic Weights and Isotopic Abundances, in which new values are given for Ir, Sb, Fe, and Ti. The Standard Atomic Weights apply to the elements as they exist naturally on Earth, and the uncertainties take into account the isotopic variation found in most laboratory samples. Further comments on the variability are given in the footnotes.

The number in parentheses following the atomic weight value gives the uncertainty in the last digit. An entry in brackets indicates the mass number of the longest-lived isotope of an element that has no stable isotopes and for which a Standard Atomic Weight cannot be defined because of wide variability in isotopic composition (or complete absence) in nature.

REFERENCE

IUPAC Commission on Atomic Weights and Isotopic Abundances, Atomic Weights of the Elements, 1993, *Pure Appl. Chem.*, 66, 2423, 1994.

Name	Symbol	At. no.	Atomic Weight	Footnotes		
Actinium	Ac	89	[227]			
Aluminum	Al	13	26.981539(5)			
Americium	Am	95	[243]			
Antimony	Sb	51	121.760(1)	g		
Argon	Ar	18	39.948(1)	g		r
Arsenic	As	33	74.92159(2)			
Astatine	At	85	[210]			
Barium	Ba	56	137.327(7)			
Berkelium	Bk	97	[247]			
Beryllium	Be	4	9.012182(3)			
Bismuth	Bi	83	208.98037(3)			
Boron	B	5	10.811(5)	g	m	r
Bromine	Br	35	79.904(1)			
Cadmium	Cd	48	112.411(8)	g		
Calcium	Ca	20	40.078(4)	g		
Californium	Cf	98	[251]			
Carbon	C	6	12.011(1)	g		r
Cerium	Ce	58	140.115(4)	g		
Cesium	Cs	55	132.90543(5)			
Chlorine	Cl	17	35.4527(9)		m	
Chromium	Cr	24	51.9961(6)			
Cobalt	Co	27	58.93320(1)			
Copper	Cu	29	63.546(3)			r
Curium	Cm	96	[247]			
Dysprosium	Dy	66	162.50(3)	g		
Einsteinium	Es	99	[252]			
Erbium	Er	68	167.26(3)	g		
Europium	Eu	63	151.965(9)	g		
Fermium	Fm	100	[257]			
Fluorine	F	9	18.9984032(9)			
Francium	Fr	87	[223]			
Gadolinium	Gd	64	157.25(3)	g		
Gallium	Ga	31	69.723(1)			
Germanium	Ge	32	72.61(2)			
Gold	Au	79	196.96654(3)			
Hafnium	Hf	72	178.49(2)			
Hahnium*	Ha	105	[262]			
Helium	He	2	4.002602(2)	g		r
Holmium	Ho	67	164.93032(3)			
Hydrogen	H	1	1.00794(7)	g	m	r
Indium	In	49	114.818(3)			
Iodine	I	53	126.90447(3)			
Iridium	Ir	77	192.217(3)			
Iron	Fe	26	55.845(2)			

STANDARD ATOMIC WEIGHTS (1993) (continued)

Name	Symbol	At. no.	Atomic Weight	Footnotes		
Krypton	Kr	36	83.80(1)	g	m	
Lanthanum	La	57	138.9055(2)	g		
Lawrencium	Lr	103	[262]			
Lead	Pb	82	207.2(1)	g		r
Lithium	Li	3	6.941(2)	g	m	r
Lutetium	Lu	71	174.967(1)	g		
Magnesium	Mg	12	24.3050(6)			
Manganese	Mn	25	54.93805(1)			
Mendelevium	Md	101	[258]			
Mercury	Hg	80	200.59(2)			
Molybdenum	Mo	42	95.94(1)	g		
Neodymium	Nd	60	144.24(3)	g		
Neon	Ne	10	20.1797(6)	g	m	
Neptunium	Np	93	[237]			
Nickel	Ni	28	58.6934(2)			
Niobium	Nb	41	92.90638(2)			
Nitrogen	N	7	14.00674(7)	g		r
Nobelium	No	102	[259]			
Osmium	Os	76	190.23(3)	g		
Oxygen	O	8	15.9994(3)	g		r
Palladium	Pd	46	106.42(1)	g		
Phosphorus	P	15	30.973762(4)			
Platinum	Pt	78	195.08(3)			
Plutonium	Pu	94	[244]			
Polonium	Po	84	[209]			
Potassium	K	19	39.0983(1)	g		
Praseodymium	Pr	59	140.90765(3)			
Promethium	Pm	61	[145]			
Protactinium	Pa	91	231.03588(2)			
Radium	Ra	88	[226]			
Radon	Rn	86	[222]			
Rhenium	Re	75	186.207(1)			
Rhodium	Rh	45	102.90550(3)			
Rubidium	Rb	37	85.4678(3)	g		
Ruthenium	Ru	44	101.07(2)	g		
Rutherfordium**	Rf	104	[261]			
Samarium	Sm	62	150.36(3)	g		
Scandium	Sc	21	44.955910(9)			
Selenium	Se	34	78.96(3)			
Silicon	Si	14	28.0855(3)			r
Silver	Ag	47	107.8682(2)	g		
Sodium	Na	11	22.989768(6)			
Strontium	Sr	38	87.62(1)	g		r
Sulfur	S	16	32.066(6)	g		r
Tantalum	Ta	73	180.9479(1)			
Technetium	Tc	43	[98]			
Tellurium	Te	52	127.60(3)	g		
Terbium	Tb	65	158.92534(3)			
Thallium	Tl	81	204.3833(2)			
Thorium	Th	90	232.0381(1)	g		
Thulium	Tm	69	168.93421(3)			
Tin	Sn	50	118.710(7)	g		
Titanium	Ti	22	47.867(1)			
Tungsten	W	74	183.84(1)			
Uranium	U	92	238.0289(1)	g	m	
Vanadium	V	23	50.9415(1)			
Xenon	Xe	54	131.29(2)	g	m	
Ytterbium	Yb	70	173.04(3)	g		

STANDARD ATOMIC WEIGHTS (1993) (continued)

Name	Symbol	At. no.	Atomic Weight	Footnotes
Yttrium	Y	39	88.90585(2)	
Zinc	Zn	30	65.39(2)	
Zirconium	Zr	40	91.224(2)	g

* Current IUPAC name is unnilpentium, symbol Unp.

** Current IUPAC name is unnilquadium, symbol Unq.

g geological specimens are known in which the element has an isotopic composition outside the limits for normal material. The difference between the atomic weight of the element in such specimens and that given in the table may exceed the stated uncertainty.

m modified isotopic compositions may be found in commercially available material because it has been subjected to an undisclosed or inadvertent isotopic fractionation. Substantial deviations in atomic weight of the element from that given the table can occur.

r range in isotopic composition of normal terrestrial material prevents a more precise atomic weight being given; the tabulated atomic weight value should be applicable to any normal material.

ATOMIC MASSES AND ABUNDANCES

This table lists the mass (in atomic mass units, symbol u) and the natural abundance (in percent) of the stable nuclides and a few important radioactive nuclides. A complete table of all nuclides may be found in Section 10 ("Table of the Isotopes").

Numbers in parentheses give the uncertainty in the last digit(s) of the stated values. The uncertainty in the natural abundance includes both the estimated measurement uncertainty and the reported range of variation in different terrestrial sources of the element (see Reference 3 for more details). A * in the Abundance column indicates a radioactive nuclide not present in nature or which varies so widely that a meaningful natural abundance cannot be defined.

REFERENCES

1. Holden, N. E., "Table of the Isotopes", in Lide, D. R., Ed., *CRC Handbook of Chemistry and Physics*, 74th Ed., CRC Press, Boca Raton FL, 1993.
2. Wapstra, A. H., and Audi, G., *Nucl. Phys.*, A432, 1, 1985.
3. IUPAC Commission on Atomic Weights and Isotopic Abundances, *Pure Appl. Chem.*, 63, 991, 1991.

Z	Isotope	Mass in u	Abundance in %
1	^{1}H	1.007825035(12)	99.985(1)
1	^{2}H	2.014101779(24)	0.015(1)
1	^{3}H	3.01604927(4)	*
2	^{3}He	3.01602931(4)	0.000137(3)
2	^{4}He	4.00260324(5)	99.999863(3)
3	^{6}Li	6.0151214(7)	7.5(2)
3	^{7}Li	7.0160030(9)	92.5(2)
4	^{9}Be	9.0121822(4)	100
5	^{10}B	10.0129369(3)	19.9(2)
5	^{11}B	11.0093054(4)	80.1(2)
6	^{12}C	12 (by definition)	98.90(3)
6	^{13}C	13.003354826(17)	1.10(3)
6	^{14}C	14.003241982(27)	*
7	^{14}N	14.003074002(26)	99.634(9)
7	^{15}N	15.00010897(4)	0.366(9)
8	^{16}O	15.99491463(5)	99.762(15)
8	^{17}O	16.9991312(4)	0.038(3)
8	^{18}O	17.9991603(9)	0.200(12)
9	^{19}F	18.99840322(15)	100
10	^{20}Ne	19.9924356(22)	90.48(3)
10	^{21}Ne	20.9938428(21)	0.27(1)
10	^{22}Ne	21.9913831(18)	9.25(3)
11	^{23}Na	22.9897677(10)	100
12	^{24}Mg	23.9850423(3)	78.99(3)
12	^{25}Mg	24.9858374(8)	10.00(1)
12	^{26}Mg	25.9825937(8)	11.01(2)
13	^{27}Al	26.9815386(8)	100
14	^{28}Si	27.9769271(7)	92.23(1)
14	^{29}Si	28.9764949(7)	4.67(1)
14	^{30}Si	29.9737707(7)	3.10(1)
15	^{31}P	30.9737620(6)	100
16	^{32}S	31.97207070(25)	95.02(9)
16	^{33}S	32.97145854(23)	0.75(4)
16	^{34}S	33.96786665(22)	4.21(8)
16	^{36}S	35.96708062(27)	0.02(1)
17	^{35}Cl	34.968852721(69)	75.77(7)
17	^{37}Cl	36.96590262(11)	24.23(7)
18	^{36}Ar	35.96754552(29)	0.337(3)
18	^{38}Ar	37.9627325(9)	0.063(1)
18	^{40}Ar	39.9623837(14)	99.600(3)
19	^{39}K	38.9637074(12)	93.2581(44)
19	^{40}K	39.9639992(12)	0.0117(1)
19	^{41}K	40.9618254(12)	6.7302(44)
20	^{40}Ca	39.9625906(13)	96.941(18)
20	^{42}Ca	41.9586176(13)	0.647(9)
20	^{43}Ca	42.9587662(13)	0.135(6)
20	^{44}Ca	43.9554806(14)	2.086(12)
20	^{46}Ca	45.953689(4)	0.004(3)
20	^{48}Ca	47.952533(4)	0.187(4)
21	^{45}Sc	44.9559100(14)	100
22	^{46}Ti	45.9526294(14)	8.0(1)
22	^{47}Ti	46.9517640(11)	7.3(1)
22	^{48}Ti	47.9479473(11)	73.8(1)
22	^{49}Ti	48.9478711(11)	5.5(1)
22	^{50}Ti	49.9447921(12)	5.4(1)
23	^{50}V	49.9471609(17)	0.250(2)
23	^{51}V	50.9439617(17)	99.750(2)
24	^{50}Cr	49.9460464(17)	4.345(13)
24	^{52}Cr	51.9405098(17)	83.789(18)
24	^{53}Cr	52.9406513(17)	9.501(17)
24	^{54}Cr	53.9388825(17)	2.365(7)
25	^{55}Mn	54.9380471(16)	100
26	^{54}Fe	53.9396127(15)	5.8(1)
26	^{56}Fe	55.9349393(16)	91.72(30)
26	^{57}Fe	56.9353958(16)	2.1(1)
26	^{58}Fe	57.9332773(16)	0.28(1)
27	^{59}Co	58.9331976(16)	100
28	^{58}Ni	57.9353462(16)	68.077(9)
28	^{60}Ni	59.9307884(16)	26.223(8)
28	^{61}Ni	60.9310579(16)	1.140(1)
28	^{62}Ni	61.9283461(16)	3.634(2)
28	^{64}Ni	63.9279679(17)	0.926(1)
29	^{63}Cu	62.9295989(16)	69.17(3)
29	^{65}Cu	64.9277929(20)	30.83(3)
30	^{64}Zn	63.9291448(19)	48.6(3)
30	^{66}Zn	65.9260347(17)	27.9(2)
30	^{67}Zn	66.9271291(17)	4.1(1)
30	^{68}Zn	67.9248459(18)	18.8(4)
30	^{70}Zn	69.925325(4)	0.6(1)
31	^{69}Ga	68.925580(3)	60.108(9)
31	^{71}Ga	70.9247005(25)	39.892(9)
32	^{70}Ge	69.9242497(16)	21.23(4)
32	^{72}Ge	71.9220789(16)	27.66(3)
32	^{73}Ge	72.9234626(16)	7.73(1)
32	^{74}Ge	73.9211774(15)	35.94(2)
32	^{76}Ge	75.9214016(17)	7.44(2)
33	^{75}As	74.9215942(17)	100
34	^{74}Se	73.9224746(16)	0.89(2)

Z	Isotope	Mass in u	Abundance in %
34	^{76}Se	75.9192120(16)	9.36(11)
34	^{77}Se	76.9199125(16)	7.63(6)
34	^{78}Se	77.9173076(16)	23.78(9)
34	^{80}Se	79.9165196(19)	49.61(10)
34	^{82}Se	81.9166978(23)	8.73(6)
35	^{79}Br	78.9183361(26)	50.69(7)
35	^{81}Br	80.916289(6)	49.31(7)
36	^{78}Kr	77.920396(9)	0.35(2)
36	^{80}Kr	79.916380(9)	2.25(2)
36	^{82}Kr	81.913482(6)	11.6(1)
36	^{83}Kr	82.914135(4)	11.5(1)
36	^{84}Kr	83.911507(4)	57.0(3)
36	^{86}Kr	85.910616(5)	17.3(2)
37	^{85}Rb	84.911794(3)	72.165(20)
37	^{87}Rb	86.909187(3)	27.835(20)
38	^{84}Sr	83.913430(4)	0.56(1)
38	^{86}Sr	85.9092672(28)	9.86(1)
38	^{87}Sr	86.9088841(28)	7.00(1)
38	^{88}Sr	87.9056188(28)	82.58(1)
39	^{89}Y	88.905849(3)	100
40	^{90}Zr	89.9047026(26)	51.45(3)
40	^{91}Zr	90.9056439(26)	11.22(4)
40	^{92}Zr	91.9050386(26)	17.15(2)
40	^{94}Zr	93.9063148(28)	17.38(4)
40	^{96}Zr	95.908275(4)	2.80(2)
41	^{93}Nb	92.9063772(27)	100
42	^{92}Mo	91.906808(4)	14.84(4)
42	^{94}Mo	93.9050853(26)	9.25(3)
42	^{95}Mo	94.9058411(22)	15.92(5)
42	^{96}Mo	95.9046785(22)	16.68(5)
42	^{97}Mo	96.9060205(22)	9.55(3)
42	^{98}Mo	97.9054073(22)	24.13(7)
42	^{100}Mo	99.907477(6)	9.63(3)
43	^{98}Tc	97.907215(4)	*
44	^{96}Ru	95.907599(8)	5.52(6)
44	^{98}Ru	97.905287(7)	1.88(6)
44	^{99}Ru	98.9059389(23)	12.7(1)
44	^{100}Ru	99.9042192(24)	12.6(1)
44	^{101}Ru	100.9055819(24)	17.0(1)
44	^{102}Ru	101.9043485(25)	31.6(2)
44	^{104}Ru	103.905424(6)	18.7(2)
45	^{103}Rh	102.905500(4)	100
46	^{102}Pd	101.905634(5)	1.02(1)
46	^{104}Pd	103.904029(6)	11.14(8)
46	^{105}Pd	104.905079(6)	22.33(8)
46	^{106}Pd	105.903478(6)	27.33(3)
46	^{108}Pd	107.903895(4)	26.46(9)
46	^{110}Pd	109.905167(20)	11.72(9)
47	^{107}Ag	106.905092(6)	51.839(7)
47	^{109}Ag	108.904757(4)	48.161(7)
48	^{106}Cd	105.906461(7)	1.25(4)
48	^{108}Cd	107.904176(6)	0.89(2)
48	^{110}Cd	109.903005(4)	12.49(12)
48	^{111}Cd	110.904182(3)	12.80(8)
48	^{112}Cd	111.902758(3)	24.13(14)
48	^{113}Cd	112.904400(3)	12.22(8)
48	^{114}Cd	113.903357(3)	28.73(28)
48	^{116}Cd	115.904754(4)	7.49(12)
49	^{113}In	112.904061(4)	4.3(2)
49	^{115}In	114.903880(4)	95.7(2)
50	^{112}Sn	111.904826(5)	0.97(1)
50	^{114}Sn	113.902784(4)	0.65(1)
50	^{115}Sn	114.903348(3)	0.34(1)
50	^{116}Sn	115.901747(3)	14.53(1)
50	^{117}Sn	116.902956(3)	7.68(7)
50	^{118}Sn	117.901609(3)	24.23(11)
50	^{119}Sn	118.903310(3)	8.59(4)
50	^{120}Sn	119.9021991(29)	32.59(10)
50	^{122}Sn	121.9034404(30)	4.63(3)
50	^{124}Sn	123.9052743(17)	5.79(5)
51	^{121}Sb	120.9038212(29)	57.36(8)
51	^{123}Sb	122.9042160(24)	42.64(8)
52	^{120}Te	119.904048(21)	0.096(2)
52	^{122}Te	121.903054(3)	2.603(4)
52	^{123}Te	122.9042710(22)	0.908(2)
52	^{124}Te	123.902823(2)	4.816(6)
52	^{125}Te	124.904433(3)	7.139(6)
52	^{126}Te	125.903314(3)	18.95(1)
52	^{128}Te	127.904463(4)	31.69(1)
52	^{130}Te	129.906229(5)	33.80(1)
53	^{127}I	126.904473(5)	100
54	^{124}Xe	123.9058942(22)	0.10(1)
54	^{126}Xe	125.904281(8)	0.09(1)
54	^{128}Xe	127.9035312(17)	1.91(3)
54	^{129}Xe	128.9047801(21)	26.4(6)
54	^{130}Xe	129.9035094(17)	4.1(1)
54	^{131}Xe	130.905072(5)	21.2(4)
54	^{132}Xe	131.904144(5)	26.9(5)
54	^{134}Xe	133.905395(8)	10.4(2)
54	^{136}Xe	135.907214(8)	8.9(1)
55	^{133}Cs	132.905429(7)	100
56	^{130}Ba	129.906282(8)	0.106(2)
56	^{132}Ba	131.905042(9)	0.101(2)
56	^{134}Ba	133.904486(7)	2.417(27)
56	^{135}Ba	134.905665(7)	6.592(18)
56	^{136}Ba	135.904553(7)	7.854(36)
56	^{137}Ba	136.905812(6)	11.23(4)
56	^{138}Ba	137.905232(6)	71.70(7)
57	^{138}La	137.907105(6)	0.0902(2)
57	^{139}La	138.906347(5)	99.9098(2)
58	^{136}Ce	135.907140(50)	0.19(1)
58	^{138}Ce	137.905985(12)	0.25(1)
58	^{140}Ce	139.905433(4)	88.48(10)
58	^{142}Ce	141.909241(4)	11.08(10)
59	^{141}Pr	140.907647(4)	100
60	^{142}Nd	141.907719(4)	27.13(12)
60	^{143}Nd	142.909810(4)	12.18(6)
60	^{144}Nd	143.910083(4)	23.80(12)
60	^{145}Nd	144.912570(4)	8.30(6)
60	^{146}Nd	145.913113(4)	17.19(9)
60	^{148}Nd	147.916889(4)	5.76(3)
60	^{150}Nd	149.920887(4)	5.64(3)
61	^{145}Pm	144.912743(4)	*
62	^{144}Sm	143.911998(4)	3.1(1)
62	^{147}Sm	146.914895(4)	15.0(2)
62	^{148}Sm	147.914819(4)	11.3(1)

ATOMIC MASSES AND ABUNDANCES (continued)

Z	Isotope	Mass in u	Abundance in %
62	^{149}Sm	148.917181(4)	13.8(1)
62	^{150}Sm	149.917273(4)	7.4(1)
62	^{152}Sm	151.919729(4)	26.7(2)
62	^{154}Sm	153.922206(4)	22.7(2)
63	^{151}Eu	150.919847(8)	47.8(15)
63	^{153}Eu	152.921225(4)	52.2(15)
64	^{152}Gd	151.919786(4)	0.20(1)
64	^{154}Gd	153.920861(4)	2.18(3)
64	^{155}Gd	154.922618(4)	14.80(5)
64	^{156}Gd	155.922118(4)	20.47(4)
64	^{157}Gd	156.923956(4)	15.65(3)
64	^{158}Gd	157.924019(4)	24.84(12)
64	^{160}Gd	159.927049(4)	21.86(4)
65	^{159}Tb	158.925342(4)	100
66	^{156}Dy	155.924277(8)	0.06(1)
66	^{158}Dy	157.924403(5)	0.10(1)
66	^{160}Dy	159.925193(4)	2.34(6)
66	^{161}Dy	160.926930(4)	18.9(2)
66	^{162}Dy	161.926795(4)	25.5(2)
66	^{163}Dy	162.928728(4)	24.9(2)
66	^{164}Dy	163.929171(4)	28.2(2)
67	^{165}Ho	164.930319(4)	100
68	^{162}Er	161.928775(4)	0.14(1)
68	^{164}Er	163.929198(4)	1.61(2)
68	^{166}Er	165.930290(4)	33.6(2)
68	^{167}Er	166.932046(4)	22.95(15)
68	^{168}Er	167.932368(4)	26.8(2)
68	^{170}Er	169.935461(4)	14.9(2)
69	^{169}Tm	168.934212(4)	100
70	^{168}Yb	167.933894(5)	0.13(1)
70	^{170}Yb	169.934759(4)	3.05(6)
70	^{171}Yb	170.936323(3)	14.3(2)
70	^{172}Yb	171.936378(3)	21.9(3)
70	^{173}Yb	172.938208(3)	16.12(21)
70	^{174}Yb	173.938859(3)	31.8(4)
70	^{176}Yb	175.942564(4)	12.7(2)
71	^{175}Lu	174.940770(3)	97.41(2)
71	^{176}Lu	175.942679(3)	2.59(2)
72	^{174}Hf	173.940044(4)	0.162(3)
72	^{176}Hf	175.941406(4)	5.206(5)
72	^{177}Hf	176.943217(3)	18.606(4)
72	^{178}Hf	177.943696(3)	27.297(4)
72	^{179}Hf	178.9458122(29)	13.629(6)
72	^{180}Hf	179.9465457(30)	35.100(7)
73	^{180}Ta	179.947462(4)	0.012(2)
73	^{181}Ta	180.947992(3)	99.988(2)
74	^{180}W	179.946701(5)	0.13(4)
74	^{182}W	181.948202(3)	26.3(2)
74	^{183}W	182.950220(3)	14.3(1)
74	^{184}W	183.950928(3)	30.67(15)
74	^{186}W	185.954357(4)	28.6(2)
75	^{185}Re	184.952951(3)	37.40(2)
75	^{187}Re	186.955744(3)	62.60(2)
76	^{184}Os	183.952488(4)	0.02(1)
76	^{186}Os	185.953830(4)	1.58(30)
76	^{187}Os	186.955741(3)	1.6(3)
76	^{188}Os	187.955860(3)	13.3(7)
76	^{189}Os	188.958137(4)	16.1(8)
76	^{190}Os	189.958436(4)	26.4(12)
76	^{192}Os	191.961467(4)	41.0(8)
77	^{191}Ir	190.960584(4)	37.3(5)
77	^{193}Ir	192.962917(4)	62.7(5)
78	^{190}Pt	189.959917(7)	0.01(1)
78	^{192}Pt	191.961019(5)	0.79(6)
78	^{194}Pt	193.962655(4)	32.9(6)
78	^{195}Pt	194.964766(4)	33.8(6)
78	^{196}Pt	195.964926(4)	25.3(6)
78	^{198}Pt	197.967869(6)	7.2(2)
79	^{197}Au	196.966543(4)	100
80	^{196}Hg	195.965807(5)	0.15(1)
80	^{198}Hg	197.966743(4)	9.97(8)
80	^{199}Hg	198.968254(4)	16.87(10)
80	^{200}Hg	199.968300(4)	23.10(16)
80	^{201}Hg	200.970277(4)	13.18(8)
80	^{202}Hg	201.970617(4)	29.86(20)
80	^{204}Hg	203.973467(5)	6.87(4)
81	^{203}Tl	202.972320(5)	29.524(14)
81	^{205}Tl	204.974401(5)	70.476(14)
82	^{204}Pb	203.973020(5)	1.4(1)
82	^{206}Pb	205.974440(4)	24.1(1)
82	^{207}Pb	206.975872(4)	22.1(1)
82	^{208}Pb	207.976627(4)	52.4(1)
83	^{209}Bi	208.980374(5)	100
84	^{209}Po	208.982404(5)	*
85	^{210}At	209.987126(12)	*
86	^{222}Rn	222.017570(3)	*
87	^{223}Fr	223.019733(4)	*
88	^{226}Ra	226.025402(3)	*
89	^{227}Ac	227.027750(3)	*
90	^{232}Th	232.038054(2)	100
91	^{231}Pa	231.035880(3)	*
92	^{234}U	234.0409468(24)	0.0055(5)
92	^{235}U	235.0439242(24)	0.7200(12)
92	^{238}U	238.0507847(23)	99.2745(60)
93	^{237}Np	237.0481678(23)	*
94	^{239}Pu	239.052157(2)	*
94	^{244}Pu	244.064199(5)	*

ELECTRON CONFIGURATION OF NEUTRAL ATOMS IN THE GROUND STATE

Atomic	n =	K	L		M			N				O				P			Q
		1	2		3			4				5				6			7
no.	Element	s	s	p	s	p	d	s	p	d	f	s	p	d	f	s	p	d	s
1	H	1																	
2	He	2																	
3	Li	2	1																
4	Be	2	2																
5	B	2	2	1															
6	C	2	2	2															
7	N	2	2	3															
8	O	2	2	4															
9	F	2	2	5															
10	Ne	2	2	6															
11	Na	2	2	6	1														
12	Mg	2	2	6	2														
13	Al	2	2	6	2	1													
14	Si	2	2	6	2	2													
15	P	2	2	6	2	3													
16	S	2	2	6	2	4													
17	Cl	2	2	6	2	5													
18	Ar	2	2	6	2	6													
19	K	2	2	6	2	6		1											
20	Ca	2	2	6	2	6		2											
21	Sc	2	2	6	2	6	1	2											
22	Ti	2	2	6	2	6	2	2											
23	V	2	2	6	2	6	3	2											
24	Cr	2	2	6	2	6	5*	1											
25	Mn	2	2	6	2	6	5	2											
26	Fe	2	2	6	2	6	6	2											
27	Co	2	2	6	2	6	7	2											
28	Ni	2	2	6	2	6	8	2											
29	Cu	2	2	6	2	6	10*	1											
30	Zn	2	2	6	2	6	10	2											
31	Ga	2	2	6	2	6	10	2	1										
32	Ge	2	2	6	2	6	10	2	2										
33	As	2	2	6	2	6	10	2	3										
34	Se	2	2	6	2	6	10	2	4										
35	Br	2	2	6	2	6	10	2	5										
36	Kr	2	2	6	2	6	10	2	6										
37	Rb	2	2	6	2	6	10	2	6			1							
38	Sr	2	2	6	2	6	10	2	6			2							
39	Y	2	2	6	2	6	10	2	6	1		2							
40	Zr	2	2	6	2	6	10	2	6	2		2							
41	Nb	2	2	6	2	6	10	2	6	4*		1							
42	Mo	2	2	6	2	6	10	2	6	5		1							
43	Tc	2	2	6	2	6	10	2	6	5		2							
44	Ru	2	2	6	2	6	10	2	6	7		1							
45	Rh	2	2	6	2	6	10	2	6	8		1							
46	Pd	2	2	6	2	6	10	2	6	10*									
47	Ag	2	2	6	2	6	10	2	6	10		1							
48	Cd	2	2	6	2	6	10	2	6	10		2							
49	In	2	2	6	2	6	10	2	6	10		2	1						
50	Sn	2	2	6	2	6	10	2	6	10		2	2						
51	Sb	2	2	6	2	6	10	2	6	10		2	3						
52	Te	2	2	6	2	6	10	2	6	10		2	4						
53	I	2	2	6	2	6	10	2	6	10		2	5						
54	Xe	2	2	6	2	6	10	2	6	10		2	6						
55	Cs	2	2	6	2	6	10	2	6	10		2	6			1			
56	Ba	2	2	6	2	6	10	2	6	10		2	6			2			

ELECTRON CONFIGURATION OF NEUTRAL ATOMS IN THE GROUND STATE (continued)

Atomic		K	L		M			N				O				P			Q
	n =	1	2		3			4				5				6			7
no.	Element	s	s	p	s	p	d	s	p	d	f	s	p	d	f	s	p	d	s
57	La	2	2	6	2	6	10	2	6	10		2	6	1		2			
58	Ce	2	2	6	2	6	10	2	6	10	1*	2	6	1		2			
59	Pr	2	2	6	2	6	10	2	6	10	3	2	6			2			
60	Nd	2	2	6	2	6	10	2	6	10	4	2	6			2			
61	Pm	2	2	6	2	6	10	2	6	10	5	2	6			2			
62	Sm	2	2	6	2	6	10	2	6	10	6	2	6			2			
63	Eu	2	2	6	2	6	10	2	6	10	7	2	6			2			
64	Gd	2	2	6	2	6	10	2	6	10	7	2	6	1		2			
65	Tb	2	2	6	2	6	10	2	6	10	9*	2	6			2			
66	Dy	2	2	6	2	6	10	2	6	10	10	2	6			2			
67	Ho	2	2	6	2	6	10	2	6	10	11	2	6			2			
68	Er	2	2	6	2	6	10	2	6	10	12	2	6			2			
69	Tm	2	2	6	2	6	10	2	6	10	13	2	6			2			
70	Yb	2	2	6	2	6	10	2	6	10	14	2	6			2			
71	Lu	2	2	6	2	6	10	2	6	10	14	2	6	1		2			
72	Hf	2	2	6	2	6	10	2	6	10	14	2	6	2		2			
73	Ta	2	2	6	2	6	10	2	6	10	14	2	6	3		2			
74	W	2	2	6	2	6	10	2	6	10	14	2	6	4		2			
75	Re	2	2	6	2	6	10	2	6	10	14	2	6	5		2			
76	Os	2	2	6	2	6	10	2	6	10	14	2	6	6		2			
77	Ir	2	2	6	2	6	10	2	6	10	14	2	6	7		2			
78	Pt	2	2	6	2	6	10	2	6	10	14	2	6	9		1			
79	Au	2	2	6	2	6	10	2	6	10	14	2	6	10		1			
80	Hg	2	2	6	2	6	10	2	6	10	14	2	6	10		2			
81	Tl	2	2	6	2	6	10	2	6	10	14	2	6	10		2	1		
82	Pb	2	2	6	2	6	10	2	6	10	14	2	6	10		2	2		
83	Bi	2	2	6	2	6	10	2	6	10	14	2	6	10		2	3		
84	Po	2	2	6	2	6	10	2	6	10	14	2	6	10		2	4		
85	At	2	2	6	2	6	10	2	6	10	14	2	6	10		2	5		
86	Rn	2	2	6	2	6	10	2	6	10	14	2	6	10		2	6		
87	Fr	2	2	6	2	6	10	2	6	10	14	2	6	10		2	6		1
88	Ra	2	2	6	2	6	10	2	6	10	14	2	6	10		2	6		2
89	Ac	2	2	6	2	6	10	2	6	10	14	2	6	10		2	6	1	2
90	Th	2	2	6	2	6	10	2	6	10	14	2	6	10		2	6	2	2
91	Pa	2	2	6	2	6	10	2	6	10	14	2	6	10	2*	2	6	1	2
92	U	2	2	6	2	6	10	2	6	10	14	2	6	10	3	2	6	1	2
93	Np	2	2	6	2	6	10	2	6	10	14	2	6	10	4	2	6	1	2
94	Pu	2	2	6	2	6	10	2	6	10	14	2	6	10	6*	2	6		2
95	Am	2	2	6	2	6	10	2	6	10	14	2	6	10	7	2	6		2
96	Cm	2	2	6	2	6	10	2	6	10	14	2	6	10	7*	2	6	1	2
97	Bk	2	2	6	2	6	10	2	6	10	14	2	6	10	9	2	6		2
98	Cf	2	2	6	2	6	10	2	6	10	14	2	6	10	10	2	6		2
99	Es	2	2	6	2	6	10	2	6	10	14	2	6	10	11	2	6		2
100	Fm	2	2	6	2	6	10	2	6	10	14	2	6	10	12	2	6		2
101	Md	2	2	6	2	6	10	2	6	10	14	2	6	10	13	2	6		2
102	No	2	2	6	2	6	10	2	6	10	14	2	6	10	14	2	6		2
103	Lr	2	2	6	2	6	10	2	6	10	14	2	6	10	14	2	6	1	2
104	Rf	2	2	6	2	6	10	2	6	10	14	2	6	10	14	2	6	2	2

* Note irregularity.

REFERENCE

W. L. Wiese and G. A. Martin, in *A Physicist's Desk Reference*, American Institute of Physics, New York, 1989, 94.

PERIODIC TABLE OF THE ELEMENTS

New notation / Previous IUPAC form / CAS version

KEY TO CHART

Atomic Number →	50 +2	← Oxidation States
Symbol →	Sn +4	
1993 Atomic Weight →	118.71	
	18 18 4	← Electron Configuration

1 Group IA	2 IIA	3 IIIA IIIB	4 IVA IVB	5 VA VB	6 VIA VIB	7 VIIA VIIB	8 (VIIIA VIII)	9 VIIIA VIII	10 (VIIIA VIII)	11 IB	12 IIB	13 IIIB IIIA	14 IVB IVA	15 VB VA	16 VIB VIA	17 VIIB VIIA	18 VIIIA	Shell
1 H +1 −1 1.00794 1																	2 He 0 4.0020602 2	K
3 Li +1 6.941 2-1	4 Be +2 9.012182 2-2											5 B +3 10.811 2-3	6 C +2 +4 −4 12.011 2-4	7 N +1 +2 +3 +4 +5 −1 −2 −3 14.00674 2-5	8 O −2 15.9994 2-6	9 F −1 18.9984032 2-7	10 Ne 0 20.1797 2-8	K-L
11 Na +1 22.989768 2-8-1	12 Mg +2 24.3050 2-8-2											13 Al +3 26.981539 2-8-3	14 Si +2 +4 −4 28.0855 2-8-4	15 P +3 +5 −3 30.97362 2-8-5	16 S +4 +6 −2 32.066 2-8-6	17 Cl +1 +5 +7 −1 35.4527 2-8-7	18 Ar 0 39.948 2-8-8	K-L-M
19 K +1 39.0983 -8-8-1	20 Ca +2 40.078 -8-8-2	21 Sc +3 44.955910 -8-9-2	22 Ti +2 +3 +4 47.867 -8-10-2	23 V +2 +3 +4 +5 50.9415 -8-11-2	24 Cr +2 +3 +6 51.9961 -8-13-1	25 Mn +2 +3 +4 +7 54.93085 -8-13-2	26 Fe +2 +3 55.845 -8-14-2	27 Co +2 +3 58.93320 -8-15-2	28 Ni +2 +3 58.6934 -8-16-2	29 Cu +1 +2 63.546 -8-18-1	30 Zn +2 65.39 -8-18-2	31 Ga +3 69.723 -8-18-3	32 Ge +2 +4 72.61 -8-18-4	33 As +3 +5 −3 74.92159 -8-18-5	34 Se +4 +6 −2 78.96 -8-18-6	35 Br +1 +5 −1 79.904 -8-18-7	36 Kr 0 83.80 -8-18-8	-L-M-N
37 Rb +1 85.4678 -18-8-1	38 Sr +2 87.62 -18-8-2	39 Y +3 88.90585 -18-9-2	40 Zr +4 91.224 -18-10-2	41 Nb +3 +5 92.90638 -18-12-1	42 Mo +6 95.94 -18-13-1	43 Tc +4 +6 +7 (98) -18-13-2	44 Ru +3 101.07 -18-15-1	45 Rh +3 102.90550 -18-16-1	46 Pd +2 +4 106.42 -18-18-0	47 Ag +1 107.8682 -18-18-1	48 Cd +2 112.411 -18-18-2	49 In +3 114.818 -18-18-3	50 Sn +2 +4 118.710 -18-18-4	51 Sb +3 +5 −3 121.760 -18-18-5	52 Te +4 +6 −2 127.60 -18-18-6	53 I +1 +5 +7 −1 126.90447 -18-18-7	54 Xe 0 131.29 -18-18-8	-M-N-O
55 Cs +1 132.90543 -18-8-1	56 Ba +2 137.327 -18-8-2	57* La +3 138.9055 -18-9-2	72 Hf +4 178.49 -32-10-2	73 Ta +5 180.9479 -32-11-2	74 W +6 183.84 -32-12-2	75 Re +4 +6 +7 186.207 -32-13-2	76 Os +3 +4 190.23 -32-14-2	77 Ir +3 +4 192.217 -32-15-2	78 Pt +2 +4 195.08 -32-17-1	79 Au +1 +3 196.96654 -32-18-1	80 Hg +1 +2 200.59 -32-18-2	81 Tl +1 +3 204.3833 -32-18-3	82 Pb +2 +4 207.2 -32-18-4	83 Bi +3 +5 208.98037 -32-18-5	84 Po +2 +4 (209) -32-18-6	85 At (210) -32-18-7	86 Rn 0 (222) -32-18-8	-N-O-P
87 Fr +1 (223) -18-8-1	88 Ra +2 226.025 -18-8-2	89** Ac +3 227.028 -18-9-2	104 Unq +4 (261) -32-10-2	105 Unp (262) -32-11-2	106 Unh (263) -32-12-2	107 Uns (262) -32-13-2	108 Uno (265) -32-14-2	109 Une (266) -32-15-2	110 Uun (269) -32-16-2									O P Q

*Lanthanides	58 Ce +3 +4 140.115 -19-9-2	59 Pr +3 140.90765 -21-8-2	60 Nd +3 144.24 -22-8-2	61 Pm +3 (145) -23-8-2	62 Sm +2 +3 150.36 -24-8-2	63 Eu +2 +3 151.965 -25-8-2	64 Gd +3 157.25 -25-9-2	65 Tb +3 158.92534 -27-8-2	66 Dy +3 162.50 -28-8-2	67 Ho +3 164.93032 -29-8-2	68 Er +3 167.26 -30-8-2	69 Tm +3 168.93421 -31-8-2	70 Yb +2 +3 173.04 -32-8-2	71 Lu +3 174.967 -32-9-2	N O P
**Actinides	90 Th +4 232.0381 -18-10-2	91 Pa +5 +4 231.03588 -20-9-2	92 U +3 +4 +5 +6 238.0289 -21-9-2	93 Np +3 +4 +5 +6 237.048 -22-9-2	94 Pu +3 +4 +5 +6 (244) -24-8-2	95 Am +3 +4 +5 +6 (243) -25-8-2	96 Cm +3 (247) -25-9-2	97 Bk +3 +4 (247) -27-8-2	98 Cf +3 (251) -28-8-2	99 Es +3 (252) -29-8-2	100 Fm +3 (257) -30-8-2	101 Md +2 +3 (258) -31-8-2	102 No +2 +3 (259) -32-8-2	103 Lr +3 (260) -32-9-2	O P Q

The new IUPAC format numbers the groups from 1 to 18. The previous IUPAC numbering system and the system used by Chemical Abstracts Service (CAS) are also shown. For radioactive elements that do not occur in nature, the mass number of the most stable isotope is given in parentheses.

REFERENCES

1. G. J. Leigh, Editor, *Nomenclature of Inorganic Chemistry*, Blackwell Scientific Publications, Oxford, 1990.
2. *Chemical and Engineering News*, 63(5), 27, 1985.
3. Atomic Weights of the Elements, 1993, *Pure & Appl. Chem.*, 66, 2423, 1994

INTERNATIONAL TEMPERATURE SCALE OF 1990 (ITS-90)

B. W. Mangum

A new temperature scale, the International Temperature Scale of 1990 (ITS-90), was officially adopted by the Comité International des Poids et Mesures (CIPM), meeting 26—28 September 1989 at the Bureau International des Poids et Mesures (BIPM). The ITS-90 was recommended to the CIPM for its adoption following the completion of the final details of the new scale by the Comité Consultatif de Thermométrie (CCT), meeting 12—14 September 1989 at the BIPM in its 17th Session. The ITS-90 became the official international temperature scale on 1 January 1990. The ITS-90 supersedes the present scales, the International Practical Temperature Scale of 1968 (IPTS-68) and the 1976 Provisional 0.5 to 30 K Temperature Scale (EPT-76).

The ITS-90 extends upward from 0.65 K, and temperatures on this scale are in much better agreement with thermodynamic values that are those on the IPTS-68 and the EPT-76. The new scale has subranges and alternative definitions in certain ranges that greatly facilitate its use. Furthermore, its continuity, precision, and reproducibility throughout its ranges are much improved over that of the present scales. The replacement of the thermocouple with the platinum resistance thermometer at temperatures below 961.78°C resulted in the biggest improvement in reproducibility.

The ITS-90 is divided into four primary ranges:

1. Between 0.65 and 3.2 K, the ITS-90 is defined by the vapor pressure-temperature relation of ^{3}He, and between 1.25 and 2.1768 K (the λ point) and between 2.1768 and 5.0 K by the vapor pressure-temperature relations of ^{4}He. T_{90} is defined by the vapor pressure equations of the form:

$$T_{90}/\mathrm{K} = A_0 + \sum_{i=1}^{9} A_i\left[\left(\ln(p/\mathrm{Pa}) - B\right)/C\right]^i$$

   The values of the coefficients A_i, and of the constants A_o, B, and C of the equations are given below.
2. Between 3.0 and 24.5561 K, the ITS-90 is defined in terms of a ^{3}He or ^{4}He constant volume gas thermometer (CVGT). The thermometer is calibrated at three temperatures — at the triple point of neon (24.5561 K), at the triple point of equilibrium hydrogen (13.8033 K), and at a temperature between 3.0 and 5.0 K, the value of which is determined by using either ^{3}He or ^{4}He vapor pressure thermometry.
3. Between 13.8033 K (–259.3467°C) and 1234.93 K (961.78°C), the ITS-90 is defined in terms of the specified fixed points given below, by resistance ratios of platinum resistance thermometers obtained by calibration at specified sets of the fixed points, and by reference functions and deviation functions of resistance ratios which relate to T_{90} between the fixed points.
4. Above 1234.93 K, the ITS-90 is defined in terms of Planck's radiation law, using the freezing-point temperature of either silver, gold, or copper as the reference temperature.

Full details of the calibration procedures and reference functions for various subranges are given in:

The International Temperature Scale of 1990, *Metrologia*, 27, 3, 1990; errata in *Metrologia*, 27, 107, 1990.

Defining Fixed Points of the ITS-90

Material[a]	Equilibrium state[b]	Temperature	
		T_{90} (K)	t_{90} (°C)
He	VP	3 to 5	–270.15 to –268.15
e-H_2	TP	13.8033	–259.3467
e-H_2 (or He)	VP (or CVGT)	≈17	≈ –256.15
e-H_2 (or He)	VP (or CVGT)	≈20.3	≈ –252.85
Ne[c]	TP	24.5561	–248.5939
O_2	TP	54.3584	–218.7916
Ar	TP	83.8058	–189.3442
Hg[c]	TP	234.3156	–38.8344
H_2O	TP	273.16	0.01
Ga[c]	MP	302.9146	29.7646
In[c]	FP	429.7485	156.5985
Sn	FP	505.078	231.928
Zn	FP	692.677	419.527
Al[c]	FP	933.473	660.323
Ag	FP	1234.93	961.78
Au	FP	1337.33	1064.18
Cu[c]	FP	1357.77	1084.62

Defining Fixed Points of the ITS-90 (continued)

[a] e-H_2 indicates equilibrium hydrogen, that is, hydrogen with the equilibrium distribution of its ortho and para states. Normal hydrogen at room temperature contains 25% para hydrogen and 75% ortho hydrogen.

[b] VP indicates vapor pressure point; CVGT indicates constant volume gas thermometer point; TP indicates triple point (equilibrium temperature at which the solid, liquid, and vapor phases coexist); FP indicates freezing point, and MP indicates melting point (the equilibrium temperatures at which the solid and liquid phases coexist under a pressure of 101 325 Pa, one standard atmosphere). The isotopic composition is that naturally occurring.

[c] Previously, these were secondary fixed points.

Values of Coefficients in the Vapor Pressure Equations for Helium

Coef.or constant	^{3}He 0.65—3.2 K	^{4}He 1.25—2.1768 K	^{4}He 2.1768—5.0 K
A_0	1.053 447	1.392 408	3.146 631
A_1	0.980 106	0.527 153	1.357 655
A_2	0.676 380	0.166 756	0.413 923
A_3	0.372 692	0.050 988	0.091 159
A_4	0.151 656	0.026 514	0.016 349
A_5	–0.002 263	0.001 975	0.001 826
A_6	0.006 596	–0.017 976	–0.004 325
A_7	0.088 966	0.005 409	–0.004 973
A_8	–0.004 770	0.013 259	0
A_9	–0.054 943	0	0
B	7.3	5.6	10.3
C	4.3	2.9	1.9

CONVERSION OF TEMPERATURES FROM THE 1948 AND 1968 SCALES TO ITS-90

This table gives temperature corrections from older scales to the current International Temperature Scale of 1990 (see the preceding table for details on ITS-90). The first part of the table may be used for converting Celsius temperatures in the range -180 to 4000°C from IPTS-68 or IPTS-48 to ITS-90. Within the accuracy of the corrections, the temperature in the first column may be identified with either t_{68}, t_{48}, or t_{90}. The second part of the table is designed for use at lower temperatures to convert values expressed in kelvins from EPT-76 or IPTS-68 to ITS-90.

The references give analytical equations for expressing these relations. Note that Reference 1 supersedes Reference 2 with respect to corrections in the 630 to 1064°C range.

REFERENCES

1. Burns, G. W. et al., in *Temperature: Its Measurement and Control in Science and Industry,* Vol. 6, Schooley, J. F., Ed., American Institute of Physics, New York, 1993.
2. Goldberg, R. N. and Weir, R. D., *Pure and Appl. Chem.*, 1545, 1992.

t/°C	t_{90}-t_{68}	t_{90}-t_{48}
-180	0.008	0.020
-170	0.010	0.017
-160	0.012	0.007
-150	0.013	0.000
-140	0.014	0.001
-130	0.014	0.008
-120	0.014	0.017
-110	0.013	0.026
-100	0.013	0.035
-90	0.012	0.041
-80	0.012	0.045
-70	0.011	0.045
-60	0.010	0.042
-50	0.009	0.038
-40	0.008	0.032
-30	0.006	0.024
-20	0.004	0.016
-10	0.002	0.008
0	0.000	0.000
10	-0.002	-0.006
20	-0.005	-0.012
30	-0.007	-0.016
40	-0.010	-0.020
50	-0.013	-0.023
60	-0.016	-0.026
70	-0.018	-0.026
80	-0.021	-0.027
90	-0.024	-0.027
100	-0.026	-0.026
110	-0.028	-0.024
120	-0.030	-0.023
130	-0.032	-0.020
140	-0.034	-0.018
150	-0.036	-0.016
160	-0.037	-0.012
170	-0.038	-0.009
180	-0.039	-0.005
190	-0.039	-0.001
200	-0.040	0.003
210	-0.040	0.007
220	-0.040	0.011
230	-0.040	0.014
240	-0.040	0.018
250	-0.040	0.021
260	-0.040	0.024
270	-0.039	0.028
280	-0.039	0.030
290	-0.039	0.032
300	-0.039	0.034
310	-0.039	0.035
320	-0.039	0.036
330	-0.040	0.036
340	-0.040	0.037
350	-0.041	0.036
360	-0.042	0.035
370	-0.043	0.034
380	-0.045	0.032
390	-0.046	0.030
400	-0.048	0.028
410	-0.051	0.024
420	-0.053	0.022
430	-0.056	0.019
440	-0.059	0.015
450	-0.062	0.012
460	-0.065	0.009
470	-0.068	0.007
480	-0.072	0.004
490	-0.075	0.002
500	-0.079	0.000
510	-0.083	-0.001
520	-0.087	-0.002
530	-0.090	-0.001
540	-0.094	0.000
550	-0.098	0.002
560	-0.101	0.007
570	-0.105	0.011
580	-0.108	0.018
590	-0.112	0.025
600	-0.115	0.035
610	-0.118	0.047
620	-0.122	0.060
630	-0.125	0.075
640	-0.11	0.12
650	-0.10	0.15
660	-0.09	0.19
670	-0.07	0.24
680	-0.05	0.29
690	-0.04	0.32
700	-0.02	0.37
710	-0.01	0.41
720	0.00	0.45
730	0.02	0.49
740	0.03	0.53
750	0.03	0.56
760	0.04	0.60
770	0.05	0.63
780	0.05	0.66
790	0.05	0.69
800	0.05	0.72
810	0.05	0.75
820	0.04	0.76
830	0.04	0.79
840	0.03	0.81
850	0.02	0.83
860	0.01	0.85
870	0.00	0.87
880	-0.02	0.87
890	-0.03	0.89
900	-0.05	0.90
910	-0.06	0.92
920	-0.08	0.93
930	-0.10	0.94
940	-0.11	0.96
950	-0.13	0.97
960	-0.15	0.97
970	-0.16	0.99
980	-0.18	1.00
990	-0.19	1.02
1000	-0.20	1.04
1010	-0.22	1.05
1020	-0.23	1.07
1030	-0.23	1.10
1040	-0.24	1.12
1050	-0.25	1.14
1060	-0.25	1.17
1070	-0.25	1.19
1080	-0.26	1.20
1090	-0.26	1.20
1100	-0.26	1.2
1200	-0.30	1.4
1300	-0.35	1.5
1400	-0.39	1.6
1500	-0.44	1.8
1600	-0.49	1.9
1700	-0.54	2.1

CONVERSION OF TEMPERATURES FROM THE 1948 AND 1968 SCALES TO ITS-90 (continued)

$t/°C$	$t_{90}-t_{68}$	$t_{90}-t_{48}$
1800	-0.60	2.2
1900	-0.66	2.3
2000	-0.72	2.5
2100	-0.79	2.7
2200	-0.85	2.9
2300	-0.93	3.1
2400	-1.00	3.2
2500	-1.07	3.4
2600	-1.15	3.7
2700	-1.24	3.8
2800	-1.32	4.0
2900	-1.41	4.2
3000	-1.50	4.4
3100	-1.59	4.6
3200	-1.69	4.8
3300	-1.78	5.1
3400	-1.89	5.3
3500	-1.99	5.5
3600	-2.10	5.8
3700	-2.21	6.0
3800	-2.32	6.3
3900	-2.43	6.6
4000	-2.55	6.8

T/K	$T_{90}-T_{76}$	$T_{90}-T_{68}$
5	-0.0001	
6	-0.0002	
7	-0.0003	
8	-0.0004	
9	-0.0005	
10	-0.0006	
11	-0.0007	
12	-0.0008	
13	-0.0010	
14	-0.0011	-0.006
15	-0.0013	-0.003
16	-0.0014	-0.004
17	-0.0016	-0.006
18	-0.0018	-0.008
19	-0.0020	-0.009
20	-0.0022	-0.009
21	-0.0025	-0.008
22	-0.0027	-0.007
23	-0.0030	-0.007
24	-0.0032	-0.006
25	-0.0035	-0.005
26	-0.0038	-0.004
27	-0.0041	-0.004

T/K	$T_{90}-T_{76}$	$T_{90}-T_{68}$
28		-0.005
29		-0.006
30		-0.006
31		-0.007
32		-0.008
33		-0.008
34		-0.008
35		-0.007
36		-0.007
37		-0.007
38		-0.006
39		-0.006
40		-0.006
41		-0.006
42		-0.006
43		-0.006
44		-0.006
45		-0.007
46		-0.007
47		-0.007
48		-0.006
49		-0.006
50		-0.006
51		-0.005
52		-0.005
53		-0.004
54		-0.003
55		-0.002
56		-0.001
57		0.000
58		0.001
59		0.002
60		0.003
61		0.003
62		0.004
63		0.004
64		0.005
65		0.005
66		0.006
67		0.006
68		0.007
69		0.007
70		0.007
71		0.007
72		0.007
73		0.007
74		0.007
75		0.008
76		0.008

T/K	$T_{90}-T_{76}$	$T_{90}-T_{68}$
77		0.008
78		0.008
79		0.008
80		0.008
81		0.008
82		0.008
83		0.008
84		0.008
85		0.008
86		0.008
87		0.008
88		0.008
89		0.008
90		0.008
91		0.008
92		0.008
93		0.008
94		0.008
95		0.008
96		0.008
97		0.009
98		0.009
99		0.009
100		0.009
110		0.011
120		0.013
130		0.014
140		0.014
150		0.014
160		0.014
170		0.013
180		0.012
190		0.012
200		0.011
210		0.010
220		0.009
230		0.008
240		0.007
250		0.005
260		0.003
270		0.001
273.16		0.000
300		-0.006
400		-0.031
500		-0.040
600		-0.040
700		-0.055
800		-0.089
900		-0.124

INTERNATIONAL SYSTEM OF UNITS (SI)

The International System of units (SI) was adopted by the 11th General Conference on Weights and Measures (CGPM) in 1960. It is a coherent system of units built from seven *SI base units*, one for each of the seven dimensionally independent base quantities: they are the meter, kilogram, second, ampere, kelvin, mole, and candela, for the dimensions length, mass, time, electric current, thermodynamic temperature, amount of substance, and luminous intensity, respectively, The definitions of the SI base units are given below. The *SI derived units* are expressed as products of powers of the base units, analogous to the corresponding relations between physical quantities but with numerical factors equal to unity.

In the International System there is only one SI unit for each physical quanitity. This is either the appropriate SI base unit itself or the appropriate SI derived unit. However, any of the approved decimal prefixes, called *SI prefixes*, may be used to construct decimal multiples or submultiples of SI units.

It is recommended that only SI units be used in science and technology (with SI prefixes where appropriate). Where there are special reasons for making an exception to this rule, it is recommended always to define the units used in terms of SI units. This section was reprinted with the permission of IUPAC.

Definitions of SI Base Units

Meter — the meter is the length of path travelled by light in vacuum during a time interval of 1/299 792 458 of a second (17th CGPM, 1983).

Kilogram — The kilogram is the unit of mass; it is equal to the mass of the international prototype of the kilogram (3rd CGPM, 1901).

Second — The second is the duration of 9 192 631 770 periods of the radiation corresponding to the transition between the two hyperfine levels of the ground state of the cesium-133 atom (13th CGPM, 1967).

Ampere — The ampere is that constant current which, if maintained in two straight parallel conductors of infinite length, of negligible circular cross-section, and placed 1 meter apart in vacuum, would produce between these conductors a force equal to 2×10^{-7} newton per meter of length (9th CGPM, 1948).

Kelvin — The kelvin, unit of thermodynamic temperature, is the fraction 1/273.16 of the thermodynamic temperature of the triple point of water (13th CGPM, 1967)

Mole — The mole is the amount of substance of a system which contains as many elementary entities as there are atoms in 0.012 kilogram of carbon-12. When the mole is used, the elementary entities must be specified and may be atoms, molecules, ions, electrons, other particles, or specified groups of such particles (14th CGPM, 1971).

Examples of the use of the mole:

1 mol of H_2 contains about 6.022×10^{23} H_2 molecules, or 12.044×10^{23} H atoms
1 mol of HgCl has a mass of 236.04 g
1 mol of Hg_2Cl_2 has a mass of 472.08 g
1 mol of Hg_2^{2+} has a mass of 401.18 g and a charge of 192.97 kC
1 mol of $Fe_{0.91}S$ has a mass of 82.88 g
1 mol of e^- has a mass of 548.60 μg and a charge of –96.49 kC
1 mol of photons whose frequency is 10^{14} Hz has energy of about 39.90 kJ

Candela — The candela is the luminous intensity, in a given direction, of a source that emits monochromatic radiation of frequency 540×10^{12} hertz and that has a radiant intensity in that direction of (1/683) watt per steradian (16th CGPM, 1979).

Names and Symbols for the SI Base Units

Physical quantity	*Name of SI unit*	*Symbol for SI unit*
length	meter	m
mass	kilogram	kg
time	second	s
electric current	ampere	A
thermodynamic temperature	kelvin	K
amount of substance	mole	mol
luminous intensity	candela	cd

SI Derived Units with Special Names and Symbols

Physical quantity	*Name of SI unit*	*Symbol for SI unit*	*Expression in terms of SI base units*	
frequency[1]	hertz	Hz	s^{-1}	
force	newton	N	$m\ kg\ s^{-2}$	
pressure, stress	pascal	Pa	$N\ m^{-2}$	$= m^{-1}\ kg\ s^{-2}$
energy, work, heat	joule	J	$N\ m$	$= m^2\ kg\ s^{-2}$
power, radiant flux	watt	W	$J\ s^{-1}$	$= m^2\ kg\ s^{-3}$
electric charge	coulomb	C	$A\ s$	
electric potential, electromotive force	volt	V	$J\ C^{-1}$	$= m^2\ kg\ s^{-3}\ A^{-1}$
electric resistance	ohm	Ω	$V\ A^{-1}$	$= m^2\ kg\ s^{-3}\ A^{-2}$
electric conductance	siemens	S	Ω^{-1}	$= m^{-2}\ kg^{-1}\ s^3\ A^2$
electric capacitance	farad	F	$C\ V^{-1}$	$= m^{-2}\ kg^{-1}\ s^4\ A^2$
magnetic flux density	tesla	T	$V\ s\ m^{-2}$	$= kg\ s^{-2}\ A^{-1}$
magnetic flux	weber	Wb	$V\ s$	$= m^2\ kg\ s^{-2}\ A^{-1}$
inductance	henry	H	$V\ A^{-1}\ s$	$= m^2\ kg\ s^{-2}\ A^{-2}$
Celsius temperature[2]	degree Celsius	°C	K	
luminous flux	lumen	lm	cd sr	
illuminance	lux	lx	$cd\ sr\ m^{-2}$	
activity[3] (radioactive)	becquerel	Bq	s^{-1}	
absorbed dose[3] (of radiation)	gray	Gy	$J\ kg^{-1}$	$= m^2\ s^{-2}$
dose equivalent[3] (dose equivalent index)	sievert	Sv	$J\ kg^{-1}$	$= m^2\ s^{-2}$
plane angle[4]	radian	rad	1	$= m\ m^{-1}$
solid angle[4]	steradian	sr	1	$= m^2\ m^{-2}$

(1) For radial (circular) frequency and for angular velocity the unit rad s^{-1}, or simply s^{-1}, should be used, and this may not be simplified to Hz. The unit Hz should be used only for frequency in the sense of cycles per second.

(2) The Celsius temperature θ is defined by the equation:

$$\theta/°C = T/K - 273.15$$

The SI unit of Celsius temperature interval is the degree Celsius, °C, which is equal to the kelvin, K. °C should be treated as a single symbol, with no space between the ° sign and the letter C. (The symbol °K, and the symbol °, should no longer be used.)

(3) The units gray and sievert are admitted for reasons of safeguarding human health.

(4) The units radian and steradian are described as 'SI supplementary units'. However, in chemistry, as well as in physics, they are usually treated as dimensionless derived units, and this was recognized by CIPM in 1980. Since they are then of dimension 1, this leaves open the possibility of including them or omitting them in expressions of SI derived units, In practice this means that rad and sr may be used when appropriate and may be omitted if clarity is not lost thereby.

SI Prefixes

To signify decimal multiples and submultiples of SI units the following prefixes may be used.

Factor	Prefix	Symbol	Factor	Prefix	Symbol
10^{24}	yotta	Y	10^{-1}	deci	d
10^{21}	zetta	Z	10^{-2}	centi	c
10^{18}	exa	E	10^{-3}	milli	m
10^{15}	peta	P	10^{-6}	micro	µ
10^{12}	tera	T	10^{-9}	nano	n
10^{9}	giga	G	10^{-12}	pico	p
10^{6}	mega	M	10^{-15}	femto	f
10^{3}	kilo	k	10^{-18}	atto	a
10^{2}	hecto	h	10^{-21}	zepto	z
10^{1}	deka	da	10^{-24}	yocto	y

Prefix symbols should be printed in roman (upright) type with no space between the prefix and the unit symbol.

*Example*kilometer, km

When a prefix is used with a unit symbol, the combination is taken as a new symbol that can be raised to any power without the use of parentheses.

Examples $1\ \text{cm}^3 = (0.01\ \text{m})^3 = 10^{-6}\ \text{m}^3$
$1\ \mu\text{s}^{-1} = (10^{-6}\ \text{s})^{-1} = 10^{6}\ \text{s}^{-1}$
$1\ \text{V/cm} = 100\ \text{V/m}$
$1\ \text{mmol/dm}^3 = \text{mol m}^{-3}$

A prefix should never be used on its own, and prefixes are not to be combined into compound prefixes.

Example pm, not µµm

The names and symbols of decimal multiples and sub-multiples of the SI base unit of mass, the kg, which already contains a prefix, are constructed by adding the appropriate prefix to the word gram and symbol g.

Examples mg, not µkg; Mg, not kkg

The SI prefixes are not to be used with °C.

Units in Use Together with the SI

These units are not part of the SI, but it is recognized that they will continue to be used in appropriate contexts. SI prefixes may be attached to some of these units, such as milliliter, ml; millibar, mbar; megaelectronvolt, MeV; kilotonne, ktonne.

Physical quantity	*Name of unit*	*Symbol for unit*	*Value in SI units*
time	minute	min	60 s
time	hour	h	3600 s
time	day	d	86 400 s
plane angle	degree	°	$(\pi/180)$ rad
plane angle	minute	′	$(\pi/10\,800)$ rad
plane angle	second	″	$(\pi/648\,000)$ rad
length	ångström[1]	Å	10^{-10} m
area	barn	b	10^{-28} m^2
volume	litre	l, L	dm^3 $= 10^{-3}$ m^3
mass	tonne	t	Mg $= 10^3$ kg
pressure	bar[1]	bar	10^5 Pa $= 10^5$ N m^{-2}
energy	electronvolt[2]	eV $(= e \times V)$	$\approx 1.60218 \times 10^{-19}$ J
mass	unified atomic mass unit[2,3]	u $(= m_a(^{12}C)/12)$	$\approx 1.66054 \times 10^{-27}$ kg

(1) The ångström and the bar are approved by CIPM for 'temporary use with SI units', until CIPM makes a further recommendation, However, they should not be introduced where they are not used at present.

(2) The values of these units in terms of the corresponding SI units are not exact, since they depend on the values of the physical constants e (for the electronvolt) and N_A (for the unified atomic mass unit), which are determined by experiment.

(3) The unified atomic mass unit is also sometimes called the dalton, with symbol Da, although the name and symbol have not been approved by CGPM.

Atomic Units

For the purposes of quantum mechanical calculations of electronic wavefunctions, it is convenient to regard certain fundamental constants (and combinations of such constants) as though they were units. They are customarily called *atomic units* (abbreviated: au), and they may be regarded as forming a coherent system of units for the calculation of electronic properties in theoretical chemistry, although there is no authority from CGPM for treating them as units. The first five atomic units in the table below have special names and symbols. Only four of these are independent; all others may be derived by multiplication and division in the usual way, and the table includes a number of examples.

The relation of atomic units to the corresponding SI units involves the values of the fundamental physical constants, and is therefore not exact. The numerical values in the table are based on the 1986 CODATA values of the fundamental constants. The numerical results of calculations in theoretical chemistry are frequently quoted in atomic units, or as numerical values in the form (*physical quantity*)/(*atomic unit*), so that the reader may make the conversion using the current best estimates of the physical constants.

Physical quantity	*Name of unit*	*Symbol for unit*	*Definition and value of unit in SI*
mass	electron rest mass	m_e	$m_e \approx 9.1095 \times 10^{-31}$ kg
charge	elementary charge	e	$e \approx 1.6022 \times 10^{-19}$ C
action	Planck constant/2π	$\hbar$	$\hbar = h/2\pi \approx 1.0546 \times 10^{-34}$ J s
length	bohr	a_0	$4\pi\varepsilon_0\hbar^2/m_e e^2 \approx 5.2918 \times 10^{-11}$ m
energy	hartree	E_h	$\hbar^2/m_e a_0^2 \approx 4.3598 \times 10^{-18}$ J
time	au of time	$\hbar/E_h$	$\approx 2.4189 \times 10^{-17}$ s
velocity[1]	au of velocity	$a_0 E_h/\hbar$	$\approx 2.1877 \times 10^{6}$ m s^{-1}
force	au of force	E_h/a_0	$\approx 8.2389 \times 10^{-8}$ N
momentum, linear	au of momentum	$\hbar/a_0$	$\approx 1.9929 \times 10^{-24}$ N s
electric current	au of current	$eE_h/\hbar$	$\approx 6.6236 \times 10^{-3}$ A
electric field	au of electric field	E_h/ea_0	$\approx 5.1422 \times 10^{11}$ V m^{-1}
electric dipole moment	au of electric dipole	ea_0	$\approx 8.4784 \times 10^{-30}$ C m
magnetic flux density	au of magnetic flux density	$\hbar/ea_0^2$	$\approx 2.3505 \times 10^{5}$ T
magnetic dipole moment[2]	au of magnetic dipole	$e\hbar/m_e$	$= 2\mu_B \approx 1.8548 \times 10^{-23}$ J T^{-1}

(1) The numerical value of the speed of light, when expressed in atomic units, is equal to the reciprocal of the fine structure constant α; c/(au of velocity) = $c\hbar/a_0E_h = \alpha^{-1}$ 137.04.
(2) The atomic unit of magnetic dipole moment is twice the Bohr magneton, μ_B.

REFERENCES

Quantities, Units and Symbols in Physical Chemistry, Ian Mills, Editor, Blackwell Scientific Publications, Oxford, 1987.
Symbols, Units, Nomenclature and Fundamental Constants in Physics, E. R. Cohen and P. Giacomo, Editors, International Union of Pure and Applied Physics Document 25, 1987; also appears in Physica 146A, 1—8, 1987.
ISO Standards Handbook 2, Units of Measurement, International Organization for Standardization, Geneva, 1982.

CONVERSION FACTORS

The following table gives conversion factors from various units of measure to SI units. It is reproduced from NIST Special Publication 811, *Guide for the Use of the International System of Units* (Superintendent of Documents, U.S. Government Printing Office, 1991), which in turn was derived from IEEE Std 268—1982, *IEEE Standard Metric Practice* (© 1982 by the Institute of Electrical and Electronics Engineers, Inc.).

The SI values are expressed in terms of the base, supplementary, and derived units of SI in order to provide a coherent presentation of the conversion factors and facilitate computations (see the table "International System of Units" in this Section). Powers of ten can be avoided by using SI prefixes and shifting the decimal point if necessary. Conversion from a non-SI unit to a different non-SI unit may be carried out by using this table in two stages, e.g.,

1 cal (thermochemical) = 4.184 J
1 Btu (mean) = 1.05587 E+03 J

Thus, 1 Btu (mean) = (1.05587 E+03/4.184) cal (thermochemical)
= 252.359 cal (thermochemical).

Conversion factors are presented for ready adaptation to computer readout and electronic data transmission. The factors are written as a number equal to or greater than one and less than ten with six or less decimal places. This number is followed by the letter E (for exponent), plus or minus symbol, and two digits which indicate the power of 10 by which the number must be multiplied to obtain the correct value. For example:

3.523 907 E–02 is $3.523\,907 \times 10^{-2}$

or

0.035 239 07

Similarly:

3.386 389 E+03 is $3.386\,389 \times 10^{3}$

or

3 386.389

An asterisk (*) after the sixth decimal place indicates that the conversion factor is exact and that all subsequent digits are zero. All other conversion factors have been rounded to the figures given in accordance with accepted practice. Where less than six decimal places are shown, more precision is not warranted.

To convert from	to	Multiply by
abampere	ampere (A)	1.000 000*E+01
abcoulomb	coulomb (C)	1.000 000*E+01
abfarad	farad (F)	1.000 000*E+09
abhenry	henry (H)	1.000 000*E−09
abmho	siemens (S)	1.000 000*E+09
abohm	ohm (Ω)	1.000 000*E−09
abvolt	volt (V)	1.000 000*E−08
acre foot	meter^3 (m^3)	1.233 5 E+03
acre	meter^2 (m^2)	4.046 873 E+03
ampere hour	coulomb (C)	3.600 000*E+03
ångström	meter (m)	1.000 000*E−10
are	meter^2 (m^2)	1.000 000*E+02
astronomical unit	meter (m)	1.495 979 E+11
atmosphere (standard)	pascal (Pa)	1.013 250*E+05
atmosphere (technical = 1 kgf/cm^2)	pascal (Pa)	9.806 650*E+04
bar	pascal (Pa)	1.000 000*E+05
barn	meter^2 (m^2)	1.000 000*E−28
barrel (for petroleum, 42 gal)	meter^3 (m^3)	1.589 873 E−01
board foot	meter^3 (m^3)	2.359 737 E−03
British thermal unit (International Table)	joule (J)	1.055 056 E+03
British thermal unit (mean)	joule (J)	1.055 87 E+03

CONVERSION FACTORS (continued)

To convert from	to	Multiply by
British thermal unit (thermochemical)	joule (J)	1.054 350 E+03
British thermal unit (39°F)	joule (J)	1.059 67 E+03
British thermal unit (59°)	joule (J)	1.054 80 E+03
British thermal unit (60°)	joule (J)	1.054 68 E+03
Btu (International Table) · ft/(h · ft² · °F) (thermal conductivity)	watt per meter kelvin [W/(m · K)]	1.730 735 E+00
Btu (thermochemical) · ft/(h · ft² · °F) (thermal conductivity)	watt per meter kelvin [W/(m · K)]	1.729 577 E+00
Btu (International Table) · in/(h · ft² · °F) (thermal conductivity)	watt per meter kelvin [W/(m · K)]	1.442 279 E−01
Btu (thermochemical) · in/(h · ft² · °F) (thermal conductivity)	watt per meter kelvin [W/(m · K)]	1.441 314 E−01
Btu (International Table) · in/(s · ft² · °F) (thermal conductivity)	watt per meter kelvin [W/(m · K)]	5.192 204 E+02
Btu (thermochemical) · in/(s · ft² · °F) (thermal conductivity)	watt per meter kelvin [W/(m · K)]	5.188 732 E+02
Btu (International Table)/h	watt (W)	2.930 711 E−01
Btu (International Table)/s	watt (W)	1.055 056 E+03
Btu (thermochemical)/h	watt (W)	2.928 751 E−01
Btu (thermochemical)/min	watt (W)	1.757 250 E+01
Btu (thermochemical)/s	watt (W)	1.054 350 E+03
Btu (International Table)/ft²	joule per meter² (J/m²)	1.135 653 E+04
Btu (thermochemical)/ft²	joule per meter² (J/m²)	1.134 893 E+04
Btu (International Table)/(ft² · h)	watt per meter² (W/m²)	3.154 591 E+00
Btu (International Table)/(ft² · s)	watt per meter² (W/m²)	1.135 653 E+04
Btu (thermochemical)/(ft² · h)	watt per meter² (W/m²)	3.152 481 E+00
Btu (thermochemical)/(ft² · min)	watt per meter² (W/m²)	1.891 489 E+02
Btu (thermochemical)/(ft² · s)	watt per meter² (W/m²)	1.134 893 E+04
Btu (thermochemical)/(in² · s)	watt per meter² (W/m²)	1.634 246 E+06
Btu (International Table)/(h · ft² · °F)	watt per meter² kelvin [W/(m² · K)]	5.678 263 E+00
Btu (thermochemical)/(h · ft² · °F)	watt per meter² kelvin [W/m² · K)]	5.674 466 E+00
Btu (International Table)/(s · ft² · °F)	watt per meter² kelvin [W/(m² · K)]	2.044 175 E+04
Btu (thermochemical)/(s · ft² · °F)	watt per meter² kelvin [W/(m² · K)]	2.042 808 E+04
Btu (International Table)/lb	joule per kilogram (J/kg)	2.326 000*E+03
Btu (thermochemical)/lb	joule per kilogram (J/kg)	2.324 444 E+03
Btu (International Table)/(lb · °F) (specific heat capacity)	joule per kilogram kelvin (J/(kg · K)]	4.186 800*E+03
Btu (thermochemical)/(lb · °F) (specific heat capacity)	joule per kilogram kelvin [J/(kg · K)]	4.184 000*E+03
Btu (International Table)/ft³	joule per meter³ (J/m³)	3.725 895 E+04
Btu (thermochemical)/ft³	joule per meter³ (J/m³)	3.723 402 E+04
bushel	meter³ (m³)	3.523 907 E−02
calorie (International Table)	joule (J)	4.186 800*E+00
calorie (mean)	joule (J)	4.190 02 E+00
calorie (thermochemical)	joule (J)	4.184 000*E+00
calorie (15 °C)	joule (J)	4.185 80 E+00
calorie (20 °C)	joule (J)	4.181 90 E+00
calorie (kilogram, International Table)	joule (J)	4.186 800*E+03
calorie (kilogram, mean)	joule (J)	4.190 02 E+03
calorie (kilogram, thermochemical)	joule (J)	4.184 000*E+03
cal (thermochemical)/cm²	joule per meter² (J/m²)	4.184 000*E+04
cal (International Table)/g	joule per kilogram (J/kg)	4.186 800*E+03
cal (thermochemical)/g	joule per kilogram (J/kg)	4.184 000*E+03
cal (International Table)/(g · °C)	joule per kilogram kelvin [J/(kg · K)]	4.186 800*E+03
cal (thermochemical)/(g · °C)	joule per kilogram kelvin [J/(kg · K)]	4.184 000*E+03

CONVERSION FACTORS (continued)

To convert from	to	Multiply by
cal (thermochemical)/min	watt (W)	6.973 333 E−02
cal (thermochemical)/s	watt (W)	4.184 000*E+00
cal (thermochemical)/(cm^2 · min)	watt per meter^2 (W/m^2)	6.973 333 E−02
cal (thermochemical)/(cm^2 · s)	watt per meter^2 (W/m^2)	4.184 000 E+04
cal (thermochemical)/(cm · s · °C)	watt per meter kelvin [W/(m · K)]	4.184 000 E+02
cd/in^2	candela per meter^2 (cd/m^2)	1.550 003 E+03
carat (metric)	kilogram (kg)	2.000 000*E−04
centimeter of mercury (0 °C)	pascal (Pa)	1.333 22 E+03
centimeter of water (4 °C)	pascal (Pa)	9.806 38 E+01
centipoise	pascal second (Pa . s)	1.000 000*E−03
centistokes	meter^2 per second (m^2/s)	1.000 000*E−06
chain	meter^2 (m^2)	5.067 075 E+01
circular mil	meter^2 (m^2)	5.067 075 E−10
clo	kelvin meter^2 per watt (K · m^2/W)	2.003 712 E−01
cup	milliliter (mL)	2.366 E+02
curie	becquerel (Bq)	3.700 000*E+10
darcy[a]	meter^2 (m^2)	9.869 233 E−13
day	second (s)	8.640 000*E+04
day (sidereal)	second (s)	8.616 409 E+04
degree (angle)	radian (rad)	1.745 329 E−02
degree Celsius	kelvin (K)	$T_K = t_{°C} + 273.15$
degree centigrade	[see note below]	
degree Fahrenheit	degree Celsius (°C)	$t_{°C} = (t_{°F} - 32)/1.8$
degree Fahrenheit	kelvin (K)	$T_K = (t_{°F} + 459.67)/1.8$
degree Rankine	kelvin (K)	$T_K = T_{°R}/1.8$
°F · h · ft^2/Btu (International Table)	kelvin meter^2 per watt (K · m^2/W)	1.761 102 E−01
°F · h · ft^2/Btu (thermochemical)	kelvin meter^2 per watt (K · m^2/W)	1.762 280 E−01
°F · h · ft^2/[Btu (International Table) · in] (thermal resistivity)	kelvin meter^2 per watt (K · m^2/W)	6.933 472 E+00
°F · h · ft^2/[Btu (thermochemical) · in] (thermal resistivity)	kelvin meter^2 per watt (K · m^2/W)	6.938 112 E+00
denier	kilogram per meter (kg/m)	1.111 111 E−07
dyne	newton (N)	1.000 000*E−05
dyne · cm	newton meter (N · m)	1.000 000*E−07
dyne/cm^2	pascal (Pa)	1.000 000*E−01
electronvolt	joule (J)	1.602 19 E−19
EMU of capacitance	farad (F)	1.000 000*E+09
EMU of current	ampere (A)	1.000 000*E+01
EMU of electric potential	volt (V)	1.000 000*E−08
EMU of inductance	henry (H)	1.000 000*E−09
EMU of resistance	ohm (Ω)	1.000 000*E−09
ESU of capacitance	farad (F)	1.112 650 E−12
ESU of current	ampere (A)	3.335 641 E−10
ESU of electric potential	volt (V)	2.997 925 E+02
ESU of inductance	henry (H)	8.987 552 E+11
ESU of resistance	ohm (Ω)	8.987 552 E+11
erg	joule (J)	1.000 000*E−07
erg/cm^2 · s	watt per meter^2 (W/m^2)	1.000 000*E−03
erg/s	watt (W)	1.000 000*E−07
faraday (based on carbon-12)	coulomb (C)	9.648 70 E+04
faraday (chemical)	coulomb (C)	9.649 57 E+04
faraday (physical)	coulomb (C)	9.652 19 E+04

CONVERSION FACTORS (continued)

To convert from	to	Multiply by
fathom	meter (m)	1.828 8 E+00
fermi (femtometer)	meter (m)	1.000 000*E−15
fluid ounce (US)	meter^3 (m^3)	2.957 353 E−05
foot	meter (m)	3.048 000*E−01
foot (US survey)	meter (m)	3.048 006 E−01
foot of water (39.2 °F)	pascal (Pa)	2.988 98 E+03
ft^2	meter^2 (m^2)	9.290 304*E−02
ft^2/h (thermal diffusivity)	meter^2 per second (m^2/s)	2.580 640*E−05
ft^2/s	meter^2 per second (m^2/s)	9.290 340*E−02
ft^3 (volume; section modulus)	meter^3 (m^3)	2.831 685 E−02
ft^3/min	meter^3 per second (m^3/s)	4.719 474 E−04
ft^3/s	meter^3 per second (m^3/s)	2.831 685 E−02
ft^4 (second moment of area)[b]	meter^4 (m^4)	8.630 975 E−03
ft/h	meter per second (m/s)	8.466 667 E−05
ft/min	meter per second (m/s)	5.080 000*E−03
ft/s	meter per second (m/s)	3.048 000*E−01
ft/s^2	meter per second^2 (m/s^2)	3.048 000*E−01
footcandle	lux (lx)	1.076 391 E+01
footlambert	candela per meter^2 (cd/m^2)	3.426 259 E+00
ft · lbf	joule (J)	1.355 818 E+00
ft · lbf/h	watt (W)	3.766 161 E−04
ft · lbf/min	watt (W)	2.259 697 E−02
ft · lbf/s	watt (W)	1.355 818 E+00
ft · poundal	joule (J)	4.214 011 E−02
g, standard acceleration of free fall	meter per second^2 (m/s^2)	9.806 650*E+00
gal	meter per second^2 (m/s^2)	1.000 000*E−02
gallon (Canadian liquid)	meter^3 (m^3)	4.546 090*E−03
gallon (UK liquid)	meter^3 (m^3)	4.546 090*E−03
gallon (US liquid)	meter^3 (m^3)	3.785 412 E−03
gallon (US liquid) per day	meter^3 per second (m^3/s)	4.381 264 E−08
gallon (US liquid) per minute	meter^3 per second (m^3/s)	6.309 020 E−05
gallon (US liquid) per (hp · h) (SFC, specific fuel consumption)	meter^3 per joule (m^3/J)	1.410 089 E−09
gamma	tesla (T)	1.000 000*E−09
gauss	tesla (T)	1.000 000*E−04
gilbert	ampere (A)	7.957 747 E−01
gill (UK)	meter^3 (m^3)	1.420 654 E−04
gill (US)	meter^3 (m^3)	1.182 941 E−04
grade	degree (angular)	9.000 000*E−01
grade	radian (rad)	1.570 796 E−02
grain	kilogram (kg)	6.479 891*E−05
grain/gal (US liquid)	kilogram per meter^3 (kg/m^3)	1.711 806 E−02
gram	kilogram (kg)	1.000 000*E−03
g/cm^3	kilogram per meter^3 (kg/m^3)	1.000 000*E+03
gram-force/cm^2	pascal (PA)	9.806 650*E+01
hectare	meter^2 (m^2)	1.000 000*E+04
horsepower (550 ft · lbf/s)	watt (W)	7.456 999 E+02
horsepower (boiler)	watt (W)	9.809 50 E+03
horsepower (electric)	watt (W)	7.460 000*E+02
horsepower (metric)	watt (W)	7.354 99 E+02
horsepower (water)	watt (W)	7.460 43 E+02
horsepower (UK)	watt (W)	7.457 0 E+02
hour	second (s)	3.600 000*E+03
hour (sidereal)	second (s)	3.590 170 E+03
hundredweight (long)	kilogram (kg)	5.080 235 E+01
hundredweight (short)	kilogram (kg)	4.535 924 E+01

CONVERSION FACTORS (continued)

To convert from	to	Multiply by
inch	meter (m)	2.540 000*E−02
inch of mercury (32 °F) [c]	pascal (Pa)	3.386 38 E+03
inch of mercury (60 °F) [c]	pascal (Pa)	3.376 85 E+03
inch of water (39.2 °F)	pascal (Pa)	2.490 82 E+02
inch of water (60 °F)	pascal (Pa)	2.488 4 E+02
in^2	$meter^2$ (m^2)	6.451 600*E−04
in^3 (volume; section modulus) [d]	$meter^3$ (m^3)	1.638 706 E−05
in^3/min	$meter^3$ per second (m^3/s)	2.731 177 E−07
in^4 (second moment of area) [b]	$meter^4$ (m^4)	4.162 314 E−07
in/s	meter per second (m/s)	2.540 000*E−02
in/s^2	meter per $second^2$ (m/s^2)	2.540 000*E−02
kayser	1 per meter (1/m)	1.000 000*E+02
kelvin	degree Celsius	$t_{°C} = T_K - 273.15$
kilocalorie (International Table)	joule (J)	4.186 800*E+03
kilocalorie (mean)	joule (J)	4.190 02 E+03
kilocalorie (thermochemical)	joule (J)	4.184 000*E+03
kilocalorie (thermochemical)/min	watt (W)	6.973 333 E+01
kilocalorie (thermochemical)/s	watt (W)	4.184 000*E+03
kilogram-force (kgf)	newton (N)	9.806 650*E+00
kgf · m	newton meter (N · m)	9.806 650*E+00
kgf · s^2/m (mass)	kilogram (kg)	9.806 650*E+00
kgf/cm^2	pascal (Pa)	9.806 650*E+04
kgf/m^2	pascal (Pa)	9.806 650*E+00
kgf/mm^2	pascal (Pa)	9.806 650*E+06
km/h	meter per second (m/s)	2.777 778 E−01
kilopond (1 kp = 1 kgf)	newton (N)	9.806 650*E+00
kW · h	joule (J)	3.600 000*E+06
kip (1000 lbf)	newton (N)	4.448 222 E+03
kip/in^2 (ksi)	pascal (Pa)	6.894 757 E+06
knot (international)	meter per second (m/s)	5.144 444 E−01
lambert	candela per $meter^2$ (cd/m^2)	1/π *E+04
lambert	candela per $meter^2$ (cd/m^2)	3.183 099 E+03
langley	joule per $meter^2$ (J/m^2)	4.184 000*E+04
light year [e]	meter (m)	9.460 73 E+15
liter [f]	$meter^3$ (m^3)	1.000 000*E−03
lumen per ft^2	lumen per $meter^2$ (lm/m^2)	1.076 391 E+01
maxwell	weber (Wb)	1.000 000*E−08
mho	siemens (S)	1.000 000*E+00
microinch	meter (m)	2.540 000*E−08
micron	meter (m)	1.000 000*E−06
mil	meter (m)	2.540 000*E−05
mile (international)	meter (m)	1.609 344*E+03
mile (US statute)	meter (m)	1.609 3 E+03
mile (international nautical)	meter (m)	1.852 000*E+03
mile (US nautical)	meter (m)	1.852 000*E+03
m^2 (international)	$meter^2$ (m^2)	2.589 988 E+06
mi^2 (US statute)	$meter^2$ (m^2)	2.589 998 E+06
mi/h (international)	meter per second (m/s)	4.470 400*E−01
mi/h (international)	kilometer per hour (km/h)	1.609 344*E+00
mi/min (international)	meter per second (m/s)	2.682 240*E+01
mi/s (international)	meter per second (m/s)	1.609 344*E+03
millibar	pascal (Pa)	1.000 000*E+02
millimeter of mercury (0 °C) [c]	pascal (Pa)	1.333 22 E+02
minute (angle)	radian (rad)	2.908 882 E−04
minute	second (s)	6.000 000*E+01
minute (sidereal)	second (s)	5.983 617 E+01
oersted	ampere per meter (A/m)	7.957 747 E+01
ohm centimeter	ohm meter (Ω · m)	1.000 000*E−02

CONVERSION FACTORS (continued)

To convert from	to	Multiply by
ohm circular-mil per ft	ohm meter (Ω · m)	1.662 426 E−09
ounce (avoirdupois)	kilogram (kg)	2.834 952 E−02
ounce (troy or apothecary)	kilogram (kg)	3.110 348 E−02
ounce (UK fluid)	meter^3 (m^3)	2.841 307 E−05
ounce (US fluid)	meter^3 (m^3)	2.957 353 E−05
ounce-force	newton (N)	2.780 139 E−01
ozf · in	newton meter (N · m)	7.061 552 E−03
oz (avoirdupois)/gal (UK liquid)	kilogram per meter^3 (kg/m^3)	6.236 023 E+00
oz (avoirdupois)/gal (US liquid)	kilogram per meter^3 (kg/m^3)	7.489 152 E+00
oz (avoirdupois)/in^3	kilogram per meter^3 (kg/m^3)	1.729 994 E+03
oz (avoirdupois)/ft^2	kilogram per meter^2 (kg/m^2)	3.051 517 E−01
oz (avoirdupois)/yd^2	kilogram per meter^2 (kg/m^2)	3.390 575 E−02
parsec	meter (m)	3.085 678 E+16
peck (US)	meter^3 (m^3)	8.809 768 E−03
pennyweight	kilogram (kg)	1.555 174 E−03
perm (0 °C)	kilogram per pascal second meter^2 [kg/Pa · s · m^2)]	5.721 35 E−11
perm (23 °C)	kilogram per pascal second meter^2 [kg/(Pa · s · m^2)]	5.745 25 E−11
perm · in (0 °C)	kilogram per pascal second meter [kg/(Pa · s · m)]	1.453 22 E−12
perm · in (23 °C)	kilogram per pascal second meter [kg/(Pa · s · m)]	1.459 29 E−12
phot	lumen per meter^2 (lm/m^2)	1.000 000*E+04
pica (printer's)	meter (m)	4.217 518 E−03
pint (US dry)	meter^3 (m^3)	5.506 105 E−04
pint (US liquid)	meter^3 (m^3)	4.731 765 E−04
point (printer's)	meter (m)	3.514 598*E−04
poise (absolute viscosity)	pascal second (Pa · s)	1.000 000*E−01
pound (avoirdupois)[g]	kilogram (kg)	4.535 924 E−01
pound (troy or apothecary)	kilogram (kg)	3.732 417 E−01
lb/ft	kilogram per meter (kg/m)	1.488 164 E+00
lb · ft^2 (moment of inertia)	kilogram meter^2 (kg · m^2)	4.214 011 E−02
lb · in^2 (moment of inertia)	kilogram meter^2 (kg · m^2)	2.926 397 E−04
lb/ft · h	pascal second (Pa · s)	4.133 789 E−04
lb/ft · s	pascal second (Pa · s)	1.488 164 E+00
lb/ft^2	kilogram per meter^2 (kg/m^2)	4.882 428 E+00
lb/ft^3	kilogram per meter^3 (kg/m^3)	1.601 846 E+01
lb/gal (UK liquid)	kilogram per meter^3 (kg/m^3)	9.977 633 E+01
lb/gal (US liquid)	kilogram per meter^3 (kg/m^3)	1.198 264 E+02
lb/h	kilogram per second (kg/s)	1.259 979 E−04
lb/hp · h (SFC, specific fuel consumption	kilogram per joule (kg/J)	1.689 659 E−07
lb/in	kilogram per meter (kg/m)	1.785 797 E+01
lb/in^3	kilogram per meter^3 (kg/m^3)	2.767 990 E+04
lb/min	kilogram per second (kg/s)	7.559 873 E−03
lb/s	kilogram per second (kg/s)	4.535 924 E−01
lb/yd^3	kilogram per meter^3 (kg/m^3)	5.932 764 E−01
poundal	newton (N)	1.382 550 E−01
poundal/ft^2	pascal (Pa)	1.488 164 E+00
poundal · s/ft^2	pascal second (Pa · s)	1.488 164 E+00
pound-force (lbf)[h]	newton (N)	4.448 222 E+00
lbf · ft	newton meter (N · m)	1.355 818 E+00
lbf · ft/in	newton meter per meter (N · m/m)	5.337 866 E+01
lbf · in	newton meter (N · m)	1.129 848 E−01
lbf · in/in	newton meter per meter (N · m/m)	4.448 222 E+00
lbf · s/ft^2	pascal second (Pa · s)	4.788 026 E+01

CONVERSION FACTORS (continued)

To convert from	to	Multiply by
lbf · s/in^2	pascal second (Pa · s)	6.894 757 E+03
lbf/ft	newton per meter (N/m)	1.459 390 E+01
lbf/ft^2	pascal (Pa)	4.788 026 E+01
lbf/in.	newton per meter (N/m)	1.751 268 E+02
lbf/in^2 (psi)	pascal (Pa)	6.894 757 E+03
lbf/lb (thrust/weight [mass] ratio)	newton per kilogram (N/kg)	9.806 650 E+00
quad	joule (J)	1.055 E+18
quart (US dry)	meter^3 (m^3)	1.101 221 E−03
quart (US liquid)	meter^3 (m^3)	9.463 529 E−04
rad (absorbed dose)	gray (Gy)	1.000 000*E−02
rem (dose equivalent)	sievert (Sv)	1.000 000*E−02
rhe	1 per pascal second [1/Pa · s)]	1.000 000*E+01
rod	meter (m)	5.029 210 E+00
roentgen	coulomb per kilogram (C/kg)	2.58 E−04
second (angle)	radian (rad)	4.848 137 E−06
second (sidereal)	second (s)	9.972 696 E−01
shake	second (s)	1.000 000*E−08
slug	kilogram (kg)	1.459 390 E+01
slug/ft · s	pascal second (Pa · s)	4.788 026 E+01
slug/ft^3	kilogram per meter^3 (kg/m^3)	5.153 788 E+02
statampere	ampere (A)	3.335 641 E−10
statcoulomb	coulomb (C)	3.335 641 E−10
statfarad	farad (F)	1.112 650 E−12
stathenry	henry (H)	8.987 552 E+11
statmho	siemens (S)	1.112 650 E−12
statohm	ohm (Ω)	8.987 552 E+11
statvolt	volt (V)	2.997 925 E+02
stere	meter^3 (m^3)	1.000 000*E+00
stilb	candela per meter^2 (cd/m^2)	1.000 000*E+04
stokes (kinematic viscosity)	meter^2 per second (m^2/s)	1.000 000*E−04
tablespoon	milliliter (mL)	1.479 E+01
teaspoon	meter^3 (m^3) milliliter (mL)	4.929 E+00
tex	kilogram per meter (kg/m)	1.000 000*E−06
therm (EEG)[i]	joule (J)	1.055 060*E+08
therm (US)[i]	joule (J)	1.054 804*E+08
ton (assay)	kilogram (kg)	2.916 667 E−02
ton (long, 2240 lb)	kilogram (kg)	1.016 047 E+03
ton (metric)	kilogram (kg)	1.000 000*E+03
ton (explosive energy of one ton of TNT)	joule (J)	4.184 E+09[j]
ton of refrigeration (12 000 Btu/h)	watt (W)	3.517 E+03
ton (register)	meter^3 (m^3)	2.831 685 E+00
ton (short, 2000 lb)	kilogram (kg)	9.071 847 E+02
ton (long)/yd^3	kilogram per meter^3 (kg/m^3)	1.328 939 E+03
ton (short)/yd^3	kilogram per meter^3 (kg/m^3)	1.186 553 E+03
ton (short)/h	kilogram per second (kg/s)	2.519 958 E−01
ton-force (2000 lbf)	newton (N)	8.896 443 E+03
tonne	kilogram (kg)	1.000 000*E+03
torr (mmHg, 0 °C)[c]	pascal (Pa)	1.333 22 E+02
unit pole	weber (Wb)	1.256 637 E−07
W · h	joule (J)	3.600 000*E+03
W · s	joule (J)	1.000 000*E+00
W/cm^2	watt per meter^2 (W/m^2)	1.000 000*E+04
W/in^2	watt per meter^2 (W/m^2)	1.550 003 E+03

CONVERSION FACTORS (continued)

To convert from	to	Multiply by
yard	meter (m)	9.144 000*E−01
yd^2	$meter^2$ (m^2)	8.361 274 E−01
yd^3	$meter^3$ (m^3)	7.645 549 E−01
yd^3/min.	$meter^3$ per second (m^3/s)	1.274 258 E−02
year (365 days)	second (s)	3.153 600*E+07
year (sidereal)	second (s)	3.155 815 E+07
year (tropical)	second (s)	3.155 693 E+07

Note: The centigrade temperature scale is obsolete. The unit, degree centigrade, is only approximately equal to the degree Celsius.

[a] The darcy is a unit for measuring permeability of porous solids.

[b] This is sometimes called the moment of section or area moment of ineria of a plane section about a specified axis.

[c] Conversion factors for mercury manometer pressure units are calculated using the standard value for the acceleration of gravity and the density of mercury at the stated temperature. Higher levels of precision are not justified because the definitions of the units do not take into account the compressibility of mercury or the density value change caused by the revised practical temperature scale, ITS-90.

[d] The exact conversion factor is 1.638 706 4*E–05.

[e] This conversion factor is based on the astronomical unit of time of one day (86 400 seconds); an interval of 36 525 days is one Julian century. (See the Astronomical Almanac for the Year 1991, page K6, U.S. Government Printing Office, Washington, D.C.)

[f] In 1964 the General Conference on Weights and Measures reestablished the name liter as a special name for the cubic decimeter. Between 1901 and 1964, the liter was slightly larger (1.000 028 dm^3); in the use of high-accuracy volume data of that time interval, this fact must be kept in mind.

[g] The exact conversion factor is 4.535 923 7*E–01.

[h] The exact conversion factor is 4.448 221 615 260 5*E+00.

[i] The therm (EEC) is legally defined in the Council Directive of 20 December 1979, Council of the European Communities. The therm (US) is legally defined in the Federal Register of July 27, 1968. Although the therm (EEC), which is based on the International Table Btu, is frequently used by engineers in the US, the therm (US) is the legal unit used by the US natural gas industry.

[j] Defined (not measured) value.

Note concerning the foot:

The U.S. Metric Law of 1866 gave the relationship, 1 meter equals 39.37 inches. Since 1893 the U.S. yard has been derived from the meter. In 1959 a refinement was made in the definition of the yard to bring the U.S. yard and the yard used in other countries into agreement. The U.S. yard was changed from 3600/3937 m to 0.9144 m exactly. The new length is shorter by exactly two parts in a million. At the same time it was decided that any data in feet derived from and published as a result of geodetic surveys within the United States would remain with the old standard (1 ft = 1200/3937 m) until further decision This foot is named the U.S. survey foot and has the following relationships:

1 rod (pole or perch) = 16 $^1/_2$ feet
1 chain = 66 feet
1 mile (U.S. statute) = 5280 feet

CONVERSION OF TEMPERATURES

From	To	
°Celsius	°Fahrenheit	$t_F = (t_C \times 1.8) + 32$
	Kelvin	$T_K = t_C + 273.15$
	°Rankine	$T_R = (t_C + 273.15) \times 1.8$
°Fahrenheit	°Celsius	$t_C = \dfrac{t_F - 32}{1.8}$
	Kelvin	$T_k = \dfrac{t_F - 32}{1.8} + 273.15$
	°Rankine	$T_R = t_F + 459.67$
Kelvin	°Celsius	$t_C = T_K - 273.15$
	°Rankine	$T_R = T_K \times 1.8$
°Rankine	°Fahrenheit	$t_F = T_R - 459.67$
	Kelvin	$T_K = \dfrac{T_R}{1.8}$

DESIGNATION OF LARGE NUMBERS

	U.S.A.	Other countries
10^6	million	million
10^9	billion	milliard
10^{12}	trillion	billion
10^{15}	quadrillion	billiard
10^{18}	quintillion	trillion

CONVERSION FACTORS FOR ENERGY UNITS

$E = h\nu = hc\tilde{\nu} = kT;\ E_m = LE$

		wavenumber $\tilde{\nu}$	frequency ν	energy E			molar energy E_m		temperature T
		cm^{-1}	MHz	aJ	eV	E_h	kJ/mol	kcal/mol	K
$\tilde{\nu}$:	1 cm^{-1}	≙ 1	2.997925×10^4	1.986447×10^{-5}	1.239842×10^{-4}	4.556335×10^{-6}	11.96266×10^{-3}	2.85914×10^{-3}	1.438769
ν:	1 MHz	≙ 3.33564×10^{-5}	1	6.626076×10^{-10}	4.135669×10^{-9}	1.519830×10^{-10}	3.990313×10^{-7}	9.53708×10^{-8}	4.79922×10^{-5}
	1 aJ	≙ 50341.1	1.509189×10^9	1	6.241506	0.2293710	602.2137	143.9325	7.24292×10^4
E:	1 eV	≙ 8065.54	2.417988×10^8	0.1602177	1	3.674931×10^{-2}	96.4853	23.0605	1.16045×10^4
	1 E_h	≙ 219474.63	6.579684×10^9	4.359748	27.2114	1	2625.500	627.510	3.15773×10^5
E_m	1 kJ/mol	≙ 83.5935	2.506069×10^6	1.660540×10^{-3}	1.036427×10^{-2}	3.808798×10^{-4}	1	0.239006	120.272
	1 kcal/mol	≙ 349.755	1.048539×10^7	6.947700×10^{-3}	4.336411×10^{-2}	1.593601×10^{-3}	4.184	1	503.217
T:	1 K	≙ 0.695039	2.08367×10^4	1.380658×10^{-5}	8.61738×10^{-5}	3.16683×10^{-6}	8.31451×10^{-3}	1.98722×10^{-3}	1

Examples of the use of this table: 1 aJ ≙ 50341 cm^{-1}
1 eV ≙ 96.4853 kJ mol^{-1}

The symbol ≙ should be read as meaning "corresponds to" or "is equivalent to".

CONVERSION FACTORS FOR PRESSURE UNITS

	Pa	kPa	MPa	bar	atmos	Torr	μmHg	psi
Pa	1	0.001	0.000001	0.00001	9.8692×10^{-6}	0.0075006	7.5006	0.0001450377
kPa	1000	[illegible]	0.001	0.01	0.0098692	7.5006	7500.6	0.1450377
MPa	1000000	000	1	10	9.8692	7500.6	7500600	145.0377
bar	100000	00	0.1	1	0.98692	750.06	750060	14.50377
atmos	101325	01.325	0.101325	1.01325	1	760	760000	14.69594
Torr	133.322	0.133322	0.000133322	0.00133322	0.00131579	1	1000	0.01933672
μmHg	0.133322	0.000133322	1.33322×10^{-7}	1.33322×10^{-6}	1.31579×10^{-6}	0.001	1	1.933672×10^{-5}
psi	6894.757	6.894757	0.006894757	0.06894757	0.068046	51.7151	51715.1	1

To convert a pressure value from a unit in the left hand column to a new unit, multiply the value by the factor appearing in the column for the new unit. For example:

1 kPa = 9.8692×10^{-3} atmos
1 Torr = 1.33322×10^{-4} MPa

Notes: μmHg is often referred to as "micron"
Torr is essentially identical to mmHg
psi is pounds per square inch

CONVERSION BETWEEN TEMPERATURE SCALES

$$T/°C = T/K - 273.15 = 5/9(T/°F - 32)$$

T/K	*T*/°C	*T*/°F	*T*/K	*T*/°C	*T*/°F
0	–273.15	–459.67	256.15	–17	1.4
3.15	–270	–454	257.15	–16	3.2
13.15	–260	–436	258.15	–15	5
23.15	–250	–418	259.15	–14	6.8
33.15	–240	–400	260.15	–13	8.6
43.15	–230	–382	261.15	–12	10.4
53.15	–220	–364	262.15	–11	12.2
63.15	–210	–346	263.15	–10	14
73.15	–200	–328	264.15	–9	15.8
83.15	–190	–310	265.15	–8	17.6
93.15	–180	–292	266.15	–7	19.4
103.15	–170	–274	267.15	–6	21.2
113.15	–160	–256	268.15	–5	23
123.15	–150	–238	269.15	–4	24.8
133.15	–140	–220	270.15	–3	26.6
143.15	–130	–202	271.15	–2	28.4
153.15	–120	–184	272.15	–1	30.2
163.15	–110	–166	273.15	0	32
173.15	–100	–148	274.15	1	33.8
183.15	–90	–130	275.15	2	35.6
193.15	–80	–112	276.15	3	37.4
203.15	–70	–94	277.15	4	39.2
213.15	–60	–76	278.15	5	41
223.15	–50	–58	279.15	6	42.8
224.15	–49	–56.2	280.15	7	44.6
225.15	–48	–54.4	281.15	8	46.4
226.15	–47	–52.6	282.15	9	48.2
227.15	–46	–50.8	283.15	10	50
228.15	–45	–49	284.15	11	51.8
229.15	–44	–47.2	285.15	12	53.6
230.15	–43	–45.4	286.15	13	55.4
231.15	–42	–43.6	287.15	14	57.2
232.15	–41	–41.8	288.15	15	59
233.15	–40	–40	289.15	16	60.8
234.15	–39	–38.2	290.15	17	62.6
235.15	–38	–36.4	291.15	18	64.4
236.15	–37	–34.6	292.15	19	66.2
237.15	–36	–32.8	293.15	20	68
238.15	–35	–31	294.15	21	69.8
239.15	–34	–29.2	295.15	22	71.6
240.15	–33	–27.4	296.15	23	73.4
241.15	–32	–25.6	297.15	24	75.2
242.15	–31	–23.8	298.15	25	77
243.15	–30	–22	299.15	26	78.8
244.15	–29	–20.2	300.15	27	80.6
245.15	–28	–18.4	301.15	28	82.4
246.15	–27	–16.6	302.15	29	84.2
247.15	–26	–14.8	303.15	30	86
248.15	–25	–13	304.15	31	87.8
249.15	–24	–11.2	305.15	32	89.6
250.15	–23	–9.4	306.15	33	91.4
251.15	–22	–7.6	307.15	34	93.2
252.15	–21	–5.8	308.15	35	95
253.15	–20	–4	309.15	36	96.8
254.15	–19	–2.2	310.15	37	98.6
255.15	–18	–0.4	311.15	38	100.4

CONVERSION BETWEEN TEMPERATURE SCALES (continued)

T/K	*T*/°C	*T*/°F	*T*/K	*T*/°C	*T*/°F
312.15	39	102.2	353.15	80	176
313.15	40	104	354.15	81	177.8
314.15	41	105.8	355.15	82	179.6
315.15	42	107.6	356.15	83	181.4
316.15	43	109.4	357.15	84	183.2
317.15	44	111.2	358.15	85	185
318.15	45	113	359.15	86	186.8
319.15	46	114.8	360.15	87	188.6
320.15	47	116.6	361.15	88	190.4
321.15	48	118.4	362.15	89	192.2
322.15	49	120.2	363.15	90	194
323.15	50	122	364.15	91	195.8
324.15	51	123.8	365.15	92	197.6
325.15	52	125.6	366.15	93	199.4
326.15	53	127.4	367.15	94	201.2
327.15	54	129.2	368.15	95	203
328.15	55	131	369.15	96	204.8
329.15	56	132.8	370.15	97	206.6
330.15	57	134.6	371.15	98	208.4
331.15	58	136.4	372.15	99	210.2
332.15	59	138.2	373.15	100	212
333.15	60	140	383.15	110	230
334.15	61	141.8	393.15	120	248
335.15	62	143.6	403.15	130	266
336.15	63	145.4	413.15	140	284
337.15	64	147.2	423.15	150	302
338.15	65	149	433.15	160	320
339.15	66	150.8	443.15	170	338
340.15	67	152.6	453.15	180	356
341.15	68	154.4	463.15	190	374
342.15	69	156.2	473.15	200	392
343.15	70	158	483.15	210	410
344.15	71	159.8	493.15	220	428
345.15	72	161.6	503.15	230	446
346.15	73	163.4	513.15	240	464
347.15	74	165.2	523.15	250	482
348.15	75	167	533.15	260	500
349.15	76	168.8	543.15	270	518
350.15	77	170.6	553.15	280	536
351.15	78	172.4	563.15	290	554
352.15	79	174.2	573.15	300	572

CONVERSION FACTORS FOR THERMAL CONDUCTIVITY UNITS

MULTIPLY ↓ by appropriate factor to OBTAIN →	Btu_{IT} h^{-1} ft^{-1} F^{-1}	Btu_{IT} in. h^{-1} ft^{-2} F^{-1}	Btu_{th} h^{-1} ft^{-1} F^{-1}	Btu_{th} in. h^{-1} ft^{-2} F^{-1}	cal_{IT} s^{-1} cm^{-1} C^{-1}	cal_{th} s^{-1} cm^{-1} C^{-1}	$kcal_{th}$ h^{-1} m^{-1} C^{-1}	J s^{-1} cm^{-1} K^{-1}	W cm^{-1} K^{-1}	W m^{-1} K^{-1}	mW cm^{-1} K^{-1}
Btu_{IT} h^{-1} ft^{-1} F^{-1}	1	12	1.00067	12.0080	4.13379×10^{-3}	4.13656×10^{-3}	1.48916	1.73073×10^{-2}	1.73073×10^{-2}	1.73073	17.3073
Btu_{IT} in. h^{-1} ft^{-2} F^{-1}	8.33333×10^{-2}	1	8.33891×10^{-2}	1.00067	3.44482×10^{-4}	3.44713×10^{-4}	0.124097	1.44228×10^{-3}	1.44228×10^{-3}	0.144228	1.44228
Btu_{th} h^{-1} ft^{-1} F^{-1}	0.999331	11.9920	1	12	4.13102×10^{-3}	4.13379×10^{-3}	1.48816	1.72958×10^{-2}	1.72958×10^{-2}	1.72958	17.2958
Btu_{th} in. h^{-1} ft^{-2} F^{-1}	8.32776×10^{-2}	0.999331	8.33333×10^{-2}	1	3.44252×10^{-4}	3.44482×10^{-4}	0.124014	1.44131×10^{-3}	1.44131×10^{-3}	0.144131	1.44131
cal_{IT} s^{-1} cm^{-1} C^{-1}	2.41909×10^{2}	2.90291×10^{3}	2.42071×10^{2}	2.90485×10^{3}	1	1.00067	3.60241×10^{2}	4.1868	4.1868	4.1868×10^{2}	4.1868×10^{-3}
cal_{th} s^{-1} cm^{-1} C^{-1}	2.41747×10^{2}	2.90096×10^{3}	2.41909×10^{2}	2.90291×10^{3}	0.999331	1	3.6×10^{2}	4.184	4.184	4.184×10^{2}	4.184×10^{-3}
$kcal_{th}$ h^{-1} m^{-1} C^{-1}	0.671520	8.05824	0.671969	8.06363	2.77592×10^{-3}	2.77778×10^{-3}	1	1.16222×10^{-2}	1.16222×10^{-2}	1.16222	11.6222
J s^{-1} cm^{-1} K^{-1}	57.7789	6.93347×10^{2}	57.8176	6.93811×10^{2}	0.238846	0.239006	86.0421	1	1	1×10^{2}	1×10^{3}
W cm^{-1} K^{-1}	57.7789	6.93347×10^{2}	57.8176	6.93811×10^{2}	0.238846	0.239006	86.0421	1	1	1×10^{2}	1×10^{3}
W m^{-1} K^{-1}	0.577789	6.93347	0.578176	6.93811	2.38846×10^{-3}	2.39006×10^{-3}	0.860421	1×10^{-2}	1×10^{-2}	1	10
mW cm^{-1} K^{-1}	5.77789×10^{-2}	0.693347	5.78176×10^{-2}	0.693811	2.38846×10^{-4}	2.39006×10^{-4}	8.60421×10^{-2}	1×10^{-3}	1×10^{-3}	0.1	1

CONVERSION FACTORS FOR ELECTRICAL RESISTIVITY UNITS

To convert from ↓ multiply by appropriate factor to Obtain →	abΩ cm	μΩ cm	Ω cm	StatΩ cm	Ω m	Ω cir. mil ft^{-1}	Ω in.	Ω ft
abohm centimeter	1	1×10^{-3}	10^{-9}	1.113×10^{-21}	10^{-11}	6.015×10^{-3}	3.937×10^{-10}	3.281×10^{-11}
microohm centimeter	10^{3}	1	10^{-6}	1.113×10^{-18}	10^{-6}	6.015	3.937×10^{-7}	3.281×10^{-6}
ohm centimeter	10^{8}	10^{5}	1	1.113×10^{-12}	1×10^{-2}	6.015×10^{6}	3.937×10^{-1}	3.281×10^{-2}
statohm centimeter (esu)	8.987×10^{20}	8.987×10^{17}	8.987×10^{11}	1	8.987×10^{9}	5.406×10^{18}	3.538×10^{11}	2.949×10^{10}
ohm meter	10^{11}	10^{8}	10^{2}	1.113×10^{-10}	1	6.015×10^{8}	3.937×10^{1}	3.281
ohm circular mil per foot	1.662×10^{2}	1.662×10^{-1}	1.662×10^{-7}	1.850×10^{-19}	1.662×10^{-9}	1	6.54×10^{-6}	5.45×10^{-9}
ohm inch	2.54×10^{9}	2.54×10^{6}	2.54	2.827×10^{-12}	2.54×10^{-2}	1.528×10^{7}	1	8.3×10^{-2}
ohm foot	3.048×10^{10}	3.048×10^{7}	3.048×10^{-1}	3.3924×10^{-11}	3.048×10^{-1}	1.833×10^{8}	12	1

CONVERSION FACTORS FOR CHEMICAL KINETICS

Equivalent second order rate constants

A \ B	$cm^3\ mol^{-1}\ s^{-1}$	$dm^3\ mol^{-1}\ s^{-1}$	$m^3\ mol^{-1}\ s^{-1}$	$cm^3\ molecule^{-1}\ s^{-1}$	$(mm\ Hg)^{-1}\ s^{-1}$	$atm^{-1}\ s^{-1}$	$ppm^{-1}\ min^{-1}$	$m^2\ kN^{-1}\ s^{-1}$
$1\ cm^3\ mol^{-1}\ s^{-1} =$	1	10^{-3}	10^{-6}	1.66×10^{-24}	$1.604 \times 10^{-5}\ T^{-1}$	$1.219 \times 10^{-2}\ T^{-1}$	2.453×10^{-9}	$1.203 \times 10^{-4}\ T^{-1}$
$1\ dm^3\ mol^{-1}\ s^{-1} =$	10^3	1	10^{-3}	1.66×10^{-21}	$1.604 \times 10^{-2}\ T^{-1}$	$12.19\ T^{-1}$	2.453×10^{-6}	$1.203 \times 10^{-1}\ T^{-1}$
$1\ m^3\ mol^{-1}\ s^{-1} =$	10^6	10^3	1	1.66×10^{-18}	$16.04\ T^{-1}$	$1.219 \times 10^4\ T^{-1}$	2.453×10^{-3}	$120.3\ T^{-1}$
$1\ cm^3\ molecule^{-1}\ s^{-1} =$	6.023×10^{23}	6.023×10^{20}	6.023×10^{17}	1	$9.658 \times 10^{18}\ T^{-1}$	$7.34 \times 10^{21}\ T^{-1}$	1.478×10^{15}	$7.244 \times 10^{19}\ T^{-1}$
$1\ (mm\ Hg)^{-1}\ s^{-1} =$	$6.236 \times 10^4\ T$	$62.36\ T$	$6.236 \times 10^{-2}\ T$	$1.035 \times 10^{-19}\ T$	1	760	4.56×10^{-2}	7.500
$1\ atm^{-1}\ s^{-1}$	$82.06\ T$	$8.206 \times 10^{-2}\ T$	$8.206 \times 10^{-5}\ T$	$1.362 \times 10^{-22}\ T$	1.316×10^{-3}	1	6×10^{-5}	9.869×10^{-3}
$1\ ppm^{-1}\ min^{-1} =$ at 298 K, 1 atm total pressure	4.077×10^8	4.077×10^5	407.7	6.76×10^{-16}	21.93	1.667×10^4	1	164.5
$1\ m^2\ kN^{-1}\ s^{-1} =$	$8314\ T$	$8.314\ T$	$8.314 \times 10^{-3}\ T$	$1.38 \times 10^{-20}\ T$	0.1333	101.325	6.079×10^{-3}	1

To convert a rate constant from one set of units A to a new set B find the conversion factor for the row A under column B and multiply the old value by it, e.g. to convert $cm^3\ molecule^{-1}\ s^{-1}$ to $m^3\ mol^{-1}\ s^{-1}$ multiply by 6.023×10^{17}.

Table adapted from High Temperature Reaction Rate Data No. 5, The University, Leeds (1970).

Equivalent third order rate constants

A \ B	$cm^6\ mol^{-2}\ s^{-1}$	$dm^6\ mol^{-2}\ s^{-1}$	$m^6\ mol^{-2}\ s^{-1}$	$cm^6\ molecule^{-2}\ s^{-1}$	$(mm\ Hg)^{-2}\ s^{-1}$	$atm^{-2}\ s^{-1}$	$ppm^{-2}\ min^{-1}$	$m^4\ kN^{-2}\ s^{-1}$
$1\ cm^6\ mol^{-2}\ s^{-1} =$	1	10^{-6}	10^{-12}	2.76×10^{-48}	$2.57 \times 10^{-10}\ T^{-2}$	$1.48 \times 10^{-4}\ T^{-2}$	1.003×10^{-19}	$1.447 \times 10^{-8}\ T^{-2}$
$1\ dm^6\ mol^{-2}\ s^{-1} =$	10^6	1	10^{-6}	2.76×10^{-42}	$2.57 \times 10^{-4}\ T^{-2}$	$148\ T^{-2}$	1.003×10^{-13}	$1.447 \times 10^{-2}\ T^{-2}$
$1\ m^6\ mol^{-2}\ s^{-1} =$	10^{12}	10^6	1	2.76×10^{-36}	$257\ T^{-2}$	$1.48 \times 10^8\ T^{-2}$	1.003×10^{-7}	$1.447 \times 10^4\ T^{-2}$
$1\ cm^6 molecule^{-2}\ s^{-1} =$	3.628×10^{47}	3.628×10^{41}	3.628×10^{35}	1	$9.328 \times 10^{37}\ T^{-2}$	$5.388 \times 10^{43}\ T^{-2}$	3.64×10^{28}	$5.248 \times 10^{39}\ T^{-2}$
$1\ (mm\ Hg)^{-2}\ s^{-1} =$	$3.89 \times 10^9\ T^2$	$3.89 \times 10^3\ T^2$	$3.89 \times 10^{-3}\ T^2$	$1.07 \times 10^{-38}\ T^2$	1	5.776×10^5	3.46×10^{-5}	56.25
$1\ atm^{-2}\ s^{-1} =$	$6.733 \times 10^3\ T^2$	$6.733 \times 10^{-3}\ T^2$	$6.733 \times 10^{-9}\ T^2$	$1.86 \times 10^{-44}\ T^2$	1.73×10^{-6}	1	6×10^{-11}	9.74×10^{-5}
$1\ ppm^{-2}\ min^{-1} =$ at 298 K, 1 atm total pressure	9.97×10^{18}	9.97×10^{12}	9.97×10^6	2.75×10^{-29}	2.89×10^4	1.667×10^{10}	1	1.623×10^6
$1\ m^4\ kN^{-2}\ s^{-1} =$	$6.91 \times 10^7\ T^2$	$6.91\ T^2$	$69.1 \times 10^{-5}\ T^2$	$1.904 \times 10^{-40}\ T^2$	0.0178	1.027×10^4	6.16×10^{-7}	1

From *J. Phys. Chem. Ref. Data*, 9, 470, 1980, by permission of the authors and the copyright owner, the American Institute of Physics.

CONVERSION FACTORS FOR IONIZING RADIATION

CONVERSION BETWEEN SI AND OTHER UNITS

Quantity	Symbol for quantity	Expression in SI units	Expression in symbols for SI units	Special name for SI units	Symbols using special names	Conventional units	Symbol for conventional unit	Value of conventional unit in SI units
Activity	A	1 per second	s^{-1}	becquerel	Bq	curie	Ci	3.7×10^{10} Bq
Absorbed dose	D	joule per kilogram	$J\ kg^{-1}$	gray	Gy	rad	rad	0.01 Gy
Absorbed dose rate	$\dot{D}$	joule per kilogram second	$J\ kg^{-1}\ s^{-1}$		$Gy\ s^{-1}$	rad	$rad\ s^{-1}$	$0.01\ Gy\ s^{-1}$
Average energy per ion pair	W	joule	J			electronvolt	eV	1.602×10^{-19} J
Dose equivalent	H	joule per kilogram	$J\ kg^{-1}$	sievert	Sv	rem	rem	0.01 Sv
Dose equivalent rate	$\dot{H}$	joule per kilogram second	$J\ kg^{-1}\ s^{-1}$		$Sv\ s^{-1}$	rem per second	$rem\ s^{-1}$	$0.01\ Sv\ s^{-1}$
Electric current	I	ampere	A			ampere	A	1.0 A
Electric potential difference	U, V	watt per ampere	$W\ A^{-1}$	volt	V	volt	V	1.0 A
Exposure	X	coulomb per kilogram	$C\ kg^{-1}$			roentgen	R	$2.58 \times 10^{-4}\ C\ kg^{-1}$
Exposure rate	$\dot{X}$	coulomb per kilogram second	$C\ kg^{-1}\ s^{-1}$			roentgen	$R\ s^{-1}$	$2.58 \times 10^{-4}\ C\ kg^{-1}\ s^{-1}$
Fluence	ϕ	1 per meter squared	m^{-2}			1 per centimeter squared	cm^{-2}	$1.0 \times 10^4\ m^{-2}$
Fluence rate	Φ	1 per meter squared second	$m^{-2}\ s^{-1}$			1 per centimeter squared second	$cm^{-2}\ s^{-1}$	$1.0 \times 10^4\ m^{-2}\ s^{-1}$
Kerma	K	joule per kilogram	$J\ kg^{-1}$	gray	Gy	rad	rad	0.01 Gy
Kerma rate	$\dot{K}$	joule per kilogram second	$J\ kg^{-1}\ s^{-1}$		$Gy\ s^{-1}$	rad per second	$rad\ s^{-1}$	$0.01\ Gy\ s^{-1}$
Lineal energy	y	joule per meter	$J\ m^{-1}$			kiloelectron volt per micrometer	$keV\ \mu m^{-1}$	$1.602 \times 10^{-10}\ J\ m^{-1}$
Linear energy transfer	L	joule per meter	$J\ m^{-1}$			kiloelectron volt per micrometer	$keV\ \mu m^{-1}$	$1.602 \times 10^{-10}\ J\ m^{-1}$
Mass attenuation coefficient	μ/ρ	meter squared per kilogram	$m^2\ kg^{-1}$			centimeter squared per gram	$cm^2\ g^{-1}$	$0.1\ m^2\ kg^{-1}$
Mass energy transfer coefficient	μ_{tr}/ρ	meter squared per kilogram	$m^2\ kg^{-1}$			centimeter squared per gram	$cm^2\ g^{-1}$	$0.1\ m^2\ kg^{-1}$
Mass energy absorption coefficient	μ_{en}/ρ	meter squared per kilogram	$m^2\ kg^{-1}$			centimeter squared per gram	$cm^2\ g^{-1}$	$0.1\ m^2\ kg^{-1}$
Mass stopping power	S/ρ	joule meter squared per kilogram	$J\ m^2\ kg^{-1}$			MeV centimeter squared per gram	$MeV\ cm^2\ g^{-1}$	$1.602 \times 10^{-14}\ J\ m^2\ kg^{-1}$
Power	P	joule per second	$J\ s^{-1}$	watt	W	watt	W	1.0 W
Pressure	p	newton per meter squared	$N\ m^{-2}$	pascal	Pa	torr	torr	(101325/760)Pa
Radiation chemical yield	G	mole per joule	$mol\ J^{-1}$			molecules per 100 electron volts	molecules $(100\ eV)^{-1}$	$1.04 \times 10^{-7}\ mol\ J^{-1}$
Specific energy	z	joule per kilogram	$J\ kg^{-1}$	gray	Gy	rad	rad	0.01 Gy

CONVERSION FACTORS FOR IONIZING RADIATION (continued)

CONVERSION OF RADIOACTIVITY UNITS FROM MBq TO mCi AND μCi

MBq	mCi	MBq	mCi	MBq	μCi	MBq	μCi
7000	189.	500	13.5	30	810	1	27
6000	162.	400	10.8	20	540	0.9	24
5000	135.	300	8.1	10	270	0.8	21.6
4000	108.	200	5.4	9	240	0.7	18.9
3000	81.	100	2.7	8	220	0.6	16.2
2000	54.	90	2.4	7	189	0.5	13.5
1000	27.	80	2.16	6	162	0.4	10.8
900	24.	70	1.89	5	135	0.3	8.1
800	21.6	60	1.62	4	108	0.2	5.4
700	18.9	50	1.35	3	81	0.1	2.7
600	16.2	40	1.08	2	54		

CONVERSION OF RADIOACTIVITY UNITS FROM mCi AND μCi TO MBq

mCi	MBq	mCi	MBq	μCi	MBq	μCi	MBq
200	7400	10	370	1000	37.0	80	2.96
150	5550	9	333	900	33.3	70	2.59
100	3700	8	296	800	29.6	60	2.22
90	3330	7	259	700	25.9	50	1.85
80	2960	6	222	600	22.2	40	1.48
70	2590	5	185	500	18.5	30	1.11
60	2220	4	148	400	14.8	20	0.74
50	1850	3	111	300	11.1	10	0.37
40	1480	2	74.0	200	7.4	5	0.185
30	1110	1	37.0	100	3.7	2	0.074
20	740			90	3.33	1	0.037

CONVERSION OF RADIOACTIVITY UNITS

100 TBq (10^{14} Bq)	=	2.7 kCi (2.7×10^{3} Ci)	100 kBq (10^{5} Bq)	=	2.7 μCi (2.7×10^{-6}Ci)
10 TBq (10^{13} Bq)	=	270 Ci (2.7×10^{2} Ci)	10 kBq (10^{4} Bq)	=	270 nCi (2.7×10^{-7} Ci)
1 TBq (10^{12} Bq)	=	27 Ci (2.7×10^{1} Ci)	1 kBq (10^{3} Bq)	=	27 nCi (2.7×10^{-8} Ci)
100 GBq (10^{11} Bq)	=	2.7 Ci (2.7×10^{0} Ci)	100 Bq (10^{2} Bq)	=	2.7 nCi (2.7×10^{-9} Ci)
10 GBq (10^{10} Bq)	=	270 mCi (2.7×10^{-1} Ci)	10 Bq (10^{1} Bq)	=	270 pCi (2.7×10^{-10} Ci)
1 GBq (10^{9} Bq)	=	27 mCi (2.7×10^{-2} Ci)	1 Bq (10^{0} Bq)	=	27 pCi (2.7×10^{-11} Ci)
100 MBq (10^{8} Bq)	=	2.7 mCi (2.7×10^{-3} Ci)	100 mBq (10^{-1} Bq)	=	2.7 pCi (2.7×10^{-12} Ci)
10 MBq (10^{7} Bq)	=	270 μCi (2.7×10^{-4} Ci)	10 mBq (10^{-2} Bq)	=	270 fCi (2.7×10^{-13} Ci)
1 MBq (10^{6} Bq)	=	27 μCi (2.7×10^{-5} Ci)	1 mBq (10^{-3} Bq)	=	27 fCi (2.7×10^{-14} Ci)

CONVERSION OF ABSORBED DOSE UNITS

SI Units		Conventional
100 Gy (10^{2} Gy)	=	10,000 rad (10^{4} rad)
10 Gy (10^{1} Gy)	=	1,000 rad (10^{3} rad)
1 Gy (10^{0} Gy)	=	100 rad (10^{2} rad)
100 mGy (10^{-1} Gy)	=	10 rad (10^{1} rad)
10 mGy (10^{-2} Gy)	=	1 rad (10^{0} rad)
1 mGy (10^{-3} Gy)	=	100 mrad (10^{-1} rad)
100 μGy (10^{-4} Gy)	=	10 mrad (10^{-2} rad)
10 μGy (10^{-5} Gy)	=	1 mrad (10^{-3} rad)
1 μGy (10^{-6} Gy)	=	100 μrad (10^{-4} rad)
100 nGy (10^{-7} Gy)	=	10 μrad (10^{-5} rad)
10 nGy (10^{-8} Gy)	=	1 μrad (10^{-6} rad)
1 nGy (10^{-9} Gy)	=	100 nrad (10^{-7} rad)

CONVERSION OF DOSE EQUIVALENT UNITS

100 Sv (10^{2} Sv)	=	10,000 rem (10^{4} rem)
10 Sv (10^{1} Sv)	=	1,000 rem (10^{3} rem)
1 Sv (10^{0} Sv)	=	100 rem (10^{2} rem)
100 mSv (10^{-1} Sv)	=	10 rem (10^{1} rem)
10 mSv (10^{-2} Sv)	=	1 rem (10^{0} rem)
1 mSv (10^{-3} Sv)	=	100 mrem (10^{-1} rem)
100 μSv (10^{-4} Sv)	=	10 mrem (10^{-2} rem)
10 μSv (10^{-5} Sv)	=	1 mrem (10^{-3} rem)
1 μSv (10^{-6} Sv)	=	100 μrem (10^{-4} rem)
100 nSv (10^{-7} Sv)	=	10 μrem (10^{-5} rem)
10 nSv (10^{-8} Sv)	=	1 μrem (10^{-6} rem)
1 nSv (10^{-9} Sv)	=	100 nrem (10^{-7} rem)

VALUES OF THE GAS CONSTANT IN DIFFERENT UNIT SYSTEMS

Coleman J. Major

In SI units the value of the gas constant, R, is:

R = 8.314510 Pa m^3 K^{-1} mol^{-1}
= 8314.510 Pa L K^{-1} mol^{-1}
= 0.08314510 bar L K^{-1} mol^{-1}

This table gives the appropriate value of R for use in the ideal gas equation, $PV = nRT$, when the variables are expressed in other units.

Units of V, T, n			Units of P						
V	T	n	atm	psi	mm Hg	cm Hg	in Hg	in H_2O	ft H_20
ft^3	K	mol	0.00290	0.0426	2.20	0.220	0.0867	1.18	0.0982
		lb-mol	1.31	19.31	999	99.9	39.3	535	44.6
	°R	mol	0.00161	0.02366	1.22	0.122	0.0482	0.655	0.0546
		lb-mol	0.730	10.73	555	55.5	21.8	297	24.8
cm^3	K	mol	82.05	1206	62 400	6240	2450	33 400	2780
		lb-mol	37 200	547 000	2.83×10^7	2.83×10^6	1.11×10^6	1.51×10^7	1.26×10^6
	°R	mol	45.6	670	34 600	3460	1360	18 500	1550
		lb-mol	20 700	304 000	1.57×10^7	1.57×10^6	619 000	8.41×10^6	701 000
L	K	mol	0.08205	1.206	62.4	6.24	2.45	33.4	2.78
		lb-mol	37.2	547	28 300	2830	1113	15 140	1262
	°R	mol	0.0456	0.670	34.6	3.46	1.36	18.5	1.55
		lb-mol	20.7	304	15 700	1570	619	8410	701

Section 2
Symbols, Terminology, and Nomencalture

SYMBOLS AND TERMINOLOGY FOR PHYSICAL AND CHEMICAL QUANTITIES

The International Organization for Standardization (ISO), International Union of Pure and Applied Chemistry (IUPAC), and the International Union of Pure and Applied Physics (IUPAP) have jointly developed a set of recommended symbols for physical and chemical quantities. Consistent use of these recommended symbols helps assure unambiguous scientific communication. The list below is reprinted from Reference 1 with permission from IUPAC. Full details may be found in the following references:

1. Ian Mills, Ed., *Quantities, Units, and Symbols in Physical Chemistry*, Blackwell Scientific Publications, Oxford, 1988.
2. E. R. Cohen and P. Giacomo, *Symbols, Units, Nomenclature, and Fundamental Constants in Physics*, Document IUPAP-25, 1987; also published in *Physica*, 146A, 1—68, 1987.
3. *ISO Standards Handbook 2: Units of Measurement*, International Organization of Standardization, Geneva, 1982.

GENERAL RULES

The value of a physical quantity is expressed as the product of a numerical value and a unit, e.g.:

$T = 300$ K
$V = 26.2$ cm^3
$C_p = 45.3$ J mol^{-1} K^{-1}

The symbol for a physical quantity is always given in italic (sloping) type, while symbols for units are given in roman type. Column headings in tables and axis labels on graphs may conveniently be written as the physical quantity symbol divided by the unit symbol, e.g.:

T/K
V/cm^3
C_p/J mol K^{-1}

The values in the table or graph axis are then pure numbers.

Subscripts to symbols for physical quantities should be italic if the subscript refers to another physical quantity or to a number, e.g.:

C_p — heat capacity at constant pressure
B_n — nth virial coefficient

Subscripts which have other meanings should be in roman type:

m_{p} — mass of the proton
E_{k} — kinetic energy

The following tables give the recommended symbols for the major classes of physical and chemical quantities. The expression in the Definition column is given as an aid in identifying the quantity but is not necessarily the complete or unique definition. The SI Unit gives one (not necessarily unique) expression for the coherent SI unit for the quantity. Other equivalent unit expressions, including those which involve SI prefixes, may be used.

Name	*Symbol*	*Definition*	*SI unit*
SPACE AND TIME			
cartesian space coordinates	x, y, z		m
spherical polar coordinates	r, θ, ϕ		m, 1, 1
generalized coordinate	q, q_i		(varies)
position vector	$\boldsymbol{r}$	$\boldsymbol{r} = x\boldsymbol{i} + y\boldsymbol{j} + z\boldsymbol{k}$	m
length	l		m
special symbols:			
height	h		
breadth	b		
thickness	d, δ		
distance	d		

Name	*Symbol*	*Definition*	*SI unit*
radius	r		
diameter	d		
path length	s		
length of arc	s		
area	A, A_s, S		m^2
volume	$V, (v)$		m^3
plane angle	$\alpha, \beta, \gamma, \theta, \phi \ldots$	$\alpha = s/r$	rad, 1
solid angle	ω, Ω	$\omega = A/r^2$	sr, 1
time	t		s
period	T	$T = t/N$	s
frequency	ν, f	$\nu = 1/T$	Hz
circular frequency, angular frequency	ω	$\omega = 2\pi\nu$	rad s^{-1}, s^{-1}
characteristic time interval, relaxation time, time constant	τ, T	$\tau = \lvert dt/d\ln x \rvert$	s
angular velocity	ω	$\omega = d\phi/dt$	rad s^{-1}, s^{-1}
velocity	$\boldsymbol{v}, \boldsymbol{u}, \boldsymbol{w}, \boldsymbol{c}, \dot{\boldsymbol{r}}$	$\boldsymbol{v} = d\boldsymbol{r}/dt$	m s^{-1}
speed	v, u, w, c	$v = \lvert\boldsymbol{v}\rvert$	m s^{-1}
acceleration	$\boldsymbol{a}, (\boldsymbol{g})$	$\boldsymbol{a} = d\boldsymbol{v}/dt$	m s^{-2}

CLASSICAL MECHANICS

Name	*Symbol*	*Definition*	*SI unit*
mass	m		kg
reduced mass	μ	$\mu = m_1 m_2/(m_1 + m_2)$	kg
density, mass density	ρ	$\rho = m/V$	kg m^{-3}
relative density	d	$d = \rho/\rho^{\circ}$	1
surface density	ρ_A, ρ_S	$\rho_A = m/A$	kg m^{-2}
specific volume	v	$v = V/m = 1/\rho$	m^3 kg^{-1}
momentum	$\boldsymbol{p}$	$\boldsymbol{p} = m\boldsymbol{v}$	kg m s^{-1}
angular momentum, action	$\boldsymbol{L}$	$\boldsymbol{L} = \boldsymbol{r} \times \boldsymbol{p}$	J s
moment of inertia	I, J	$I = \sum m_i r_i^2$	kg m^2
force	$\boldsymbol{F}$	$\boldsymbol{F} = d\boldsymbol{p}/dt = m\boldsymbol{a}$	N
torque, moment of a force	$\boldsymbol{T}, (\boldsymbol{M})$	$\boldsymbol{T} = \boldsymbol{r} \times \boldsymbol{F}$	N m
energy	E		J
potential energy	E_p, V, Φ	$E_p = -\int \boldsymbol{F} \cdot d\boldsymbol{s}$	J
kinetic energy	E_k, T, K	$E_k = \frac{1}{2}mv^2$	J
work	W, w	$W = \int \boldsymbol{F} \cdot d\boldsymbol{s}$	J
Hamilton function	H	$H(q, p) = T(q, p) + V(q)$	J
Lagrange function	L	$L(q, \dot{q}) = T(q, \dot{q}) - V(q)$	J

Name	*Symbol*	*Definition*	*SI unit*
pressure	p, P	$p=F/A$	Pa, N m^{-2}
surface tension	γ, σ	$\gamma=\mathrm{d}W/\mathrm{d}A$	N m^{-1}, J m^{-2}
weight	$G, (W, P)$	$G=mg$	N
gravitational constant	G	$F=Gm_1 m_2/r^2$	N m^2 kg^{-2}
normal stress	σ	$\sigma=F/A$	Pa
shear stress	τ	$\tau=F/A$	Pa
linear strain, relative elongation	ε, e	$\varepsilon=\Delta l/l$	1
modulus of elasticity, Young's modulus	E	$E=\sigma/\varepsilon$	Pa
shear strain	γ	$\gamma=\Delta x/d$	1
shear modulus	G	$G=\tau/\gamma$	Pa
volume strain, bulk strain	θ	$\theta=\Delta V/V_0$	1
bulk modulus, compression modulus	K	$K=-V_0(\mathrm{d}p/\mathrm{d}V)$	Pa
viscosity, dynamic viscosity	η, μ	$\tau_{x,z}=\eta(\mathrm{d}v_x/\mathrm{d}z)$	Pa s
fluidity	ϕ	$\phi=1/\eta$	m kg^{-1} s
kinematic viscosity	ν	$\nu=\eta/\rho$	m^2 s^{-1}
friction coefficient	$\mu, (f)$	$F_{\mathrm{frict}}=\mu F_{\mathrm{norm}}$	1
power	P	$P=\mathrm{d}W/\mathrm{d}t$	W
sound energy flux	P, P_a	$P=\mathrm{d}E/\mathrm{d}t$	W
acoustic factors,			
reflection factor	ρ	$\rho=P_r/P_0$	1
acoustic absorption factor	$\alpha_a, (\alpha)$	$\alpha_a=1-\rho$	1
transmission factor	τ	$\tau=P_{tr}/P_0$	1
dissipation factor	δ	$\delta=\alpha_a-\tau$	1

ELECTRICITY AND MAGNETISM

Name	*Symbol*	*Definition*	*SI unit*
quantity of electricity, electric charge	Q		C
charge density	ρ	$\rho=Q/V$	C m^{-3}
surface charge density	σ	$\sigma=Q/A$	C m^{-2}
electric potential	V, ϕ	$V=\mathrm{d}W/\mathrm{d}Q$	V, J C^{-1}
electric potential difference	$U, \Delta V, \Delta\phi$	$U=V_2-V_1$	V
electromotive force	E	$E=\int(F/Q)\cdot \mathrm{d}s$	V
electric field strength	E	$E=F/Q=-\mathrm{grad}\ V$	V m^{-1}
electric flux	Ψ	$\Psi=\int D\cdot \mathrm{d}A$	C
electric displacement	D	$D=\varepsilon E$	C m^{-2}
capacitance	C	$C=Q/U$	F, C V^{-1}
permittivity	ε	$D=\varepsilon E$	F m^{-1}

Name	*Symbol*	*Definition*	*SI unit*
permittivity of vacuum	ε_0	$\varepsilon_0 = \mu_0^{-1} c_0^{-2}$	$\mathrm{F\,m^{-1}}$
relative permittivity	ε_r	$\varepsilon_r = \varepsilon/\varepsilon_0$	1
dielectric polarization (dipole moment per volume)	$\boldsymbol{P}$	$\boldsymbol{P} = \boldsymbol{D} - \varepsilon_0 \boldsymbol{E}$	$\mathrm{C\,m^{-2}}$
electric susceptibility	χ_e	$\chi_e = \varepsilon_r - 1$	1
electric dipole moment	$\boldsymbol{p}, \boldsymbol{\mu}$	$\boldsymbol{p} = Q\boldsymbol{r}$	C m
electric current	I	$I = \mathrm{d}Q/\mathrm{d}t$	A
electric current density	$\boldsymbol{j}, \boldsymbol{J}$	$I = \int \boldsymbol{j} \cdot \mathrm{d}\boldsymbol{A}$	$\mathrm{A\,m^{-2}}$
magnetic flux density, magnetic induction	$\boldsymbol{B}$	$\boldsymbol{F} = Q\boldsymbol{v} \times \boldsymbol{B}$	T
magnetic flux	Φ	$\Phi = \int \boldsymbol{B} \cdot \mathrm{d}\boldsymbol{A}$	Wb
magnetic field strength	$\boldsymbol{H}$	$\boldsymbol{B} = \mu \boldsymbol{H}$	$\mathrm{A\,m^{-1}}$
permeability	μ	$\boldsymbol{B} = \mu \boldsymbol{H}$	$\mathrm{N\,A^{-2}}$, $\mathrm{H\,m^{-1}}$
permeability of vacuum	μ_0		$\mathrm{H\,m^{-1}}$
relative permeability	μ_r	$\mu_r = \mu/\mu_0$	1
magnetization (magnetic dipole moment per volume)	$\boldsymbol{M}$	$\boldsymbol{M} = \boldsymbol{B}/\mu_0 - \boldsymbol{H}$	$\mathrm{A\,m^{-1}}$
magnetic susceptibility	$\chi, \kappa, (\chi_m)$	$\chi = \mu_r - 1$	1
molar magnetic susceptibility	χ_m	$\chi_m = V_m \chi$	$\mathrm{m^3\,mol^{-1}}$
magnetic dipole moment	$\boldsymbol{m}, \boldsymbol{\mu}$	$E_p = -\boldsymbol{m} \cdot \boldsymbol{B}$	$\mathrm{A\,m^2}$, $\mathrm{J\,T^{-1}}$
electrical resistance	R	$R = U/I$	Ω
conductance	G	$G = 1/R$	S
loss angle	δ	$\delta = (\pi/2) + \phi_I - \phi_U$	1, rad
reactance	X	$X = (U/I)\sin\delta$	Ω
impedance (complex impedance)	Z	$Z = R + \mathrm{i}X$	Ω
admittance (complex admittance)	Y	$Y = 1/Z$	S
susceptance	B	$Y = G + \mathrm{i}B$	S
resistivity	ρ	$\rho = E/j$	$\Omega\,\mathrm{m}$
conductivity	κ, γ, σ	$\kappa = 1/\rho$	$\mathrm{S\,m^{-1}}$
self-inductance	L	$E = -L(\mathrm{d}I/\mathrm{d}t)$	H
mutual inductance	M, L_{12}	$E_1 = L_{12}(\mathrm{d}I_2/\mathrm{d}t)$	H
magnetic vector potential	$\boldsymbol{A}$	$\boldsymbol{B} = \nabla \times \boldsymbol{A}$	$\mathrm{Wb\,m^{-1}}$
Poynting vector	$\boldsymbol{S}$	$\boldsymbol{S} = \boldsymbol{E} \times \boldsymbol{H}$	$\mathrm{W\,m^{-2}}$

QUANTUM MECHANICS

Name	*Symbol*	*Definition*	*SI unit*
momentum operator	$\hat{p}$	$\hat{p} = -\mathrm{i}\hbar\nabla$	$\mathrm{m^{-1}\,J\,s}$
kinetic energy operator	$\hat{T}$	$\hat{T} = -(\hbar^2/2m)\nabla^2$	J

Name	*Symbol*	*Definition*	*SI unit*
hamiltonian operator	$\hat{H}$	$\hat{H}=\hat{T}+V$	J
wavefunction, state function	Ψ, ψ, ϕ	$\hat{H}\psi=E\psi$	$(m^{-3/2})$
probability density	P	$P=\psi^*\psi$	(m^{-3})
charge density of electrons	ρ	$\rho=-eP$	$(C\ m^{-3})$
probability current density	S	$S=-i\hbar(\psi^*\nabla\psi - \psi\nabla\psi^*)/2m_e$	$(m^{-2}\ s^{-1})$
electric current density of electrons	j	$j=-eS$	$(A\ m^{-2})$
matrix element of operator $\hat{A}$	$A_{ij}, \langle i\|\hat{A}\|j\rangle$	$A_{ij}=\int\psi_i^*\hat{A}\psi_j\,d\tau$	(varies)
expectation value of operator $\hat{A}$	$\langle A\rangle, \bar{A}$	$\langle A\rangle=\int\psi^*\hat{A}\psi\,d\tau$	(varies)
hermitian conjugate of $\hat{A}$	$\hat{A}^\dagger$	$(\hat{A}^\dagger)_{ij}=(A_{ji})^*$	(varies)
commutator of $\hat{A}$ and $\hat{B}$	$[\hat{A},\hat{B}], [\hat{A},\hat{B}]_-$	$[\hat{A},\hat{B}]=\hat{A}\hat{B}-\hat{B}\hat{A}$	(varies)
anticommutator	$[\hat{A},\hat{B}]_+$	$[\hat{A},\hat{B}]_+=\hat{A}\hat{B}+\hat{B}\hat{A}$	(varies)
spin wavefunction	$\alpha; \beta$		1
coulomb integral	H_{AA}	$H_{AA}=\int\psi_A{}^*\hat{H}\psi_A\,d\tau$	J
resonance integral	H_{AB}	$H_{AB}=\int\psi_A{}^*\hat{H}\psi_B\,d\tau$	J
overlap integral	S_{AB}	$S_{AB}=\int\psi_A{}^*\psi_B\,d\tau$	1

ATOMS AND MOLECULES

Name	*Symbol*	*Definition*	*SI unit*
nucleon number, mass number	A		1
proton number, atomic number	Z		1
neutron number	N	$N=A-Z$	1
electron rest mass	m_e		kg
mass of atom, atomic mass	m_a, m		kg
atomic mass constant	m_u	$m_u=m_a(^{12}C)/12$	kg
mass excess	Δ	$\Delta=m_a-Am_u$	kg
elementary charge, proton charge	e		C
Planck constant	h		J s
Planck constant/2π	$\hbar$	$\hbar=h/2\pi$	J s
Bohr radius	a_0	$a_0=4\pi\varepsilon_0\hbar^2/m_e e^2$	m
Hartree energy	E_h	$E_h=\hbar^2/m_e a_0{}^2$	J
Rydberg constant	R_∞	$R_\infty=E_h/2hc$	m^{-1}
fine structure constant	α	$\alpha=e^2/4\pi\varepsilon_0\hbar c$	1
ionization energy	E_i		J

Name	*Symbol*	*Definition*	*SI unit*
electron affinity	E_{ea}		J
dissociation energy	E_d, D		J
from the ground state	D_0		J
from the potential minimum	D_e		J
principal quantum number (H atom)	n	$E = -hcR/n^2$	1
angular momentum quantum numbers	see under Spectroscopy		
magnetic dipole moment of a molecule	$\boldsymbol{m}$, $\boldsymbol{\mu}$	$E_p = -\boldsymbol{m}\cdot\boldsymbol{B}$	$J\,T^{-1}$
magnetizability of a molecule	ξ	$\boldsymbol{m} = \xi\boldsymbol{B}$	$J\,T^{-2}$
Bohr magneton	μ_B	$\mu_B = e\hbar/2m_e$	$J\,T^{-1}$
nuclear magneton	μ_N	$\mu_N = (m_e/m_p)\mu_B$	$J\,T^{-1}$
magnetogyric ratio (gyromagnetic ratio)	γ	$\gamma = \mu/L$	$C\,kg^{-1}$
g factor	g		1
Larmor circular frequency	ω_L	$\omega_L = (e/2m)B$	s^{-1}
Larmor frequency	ν_L	$\nu_L = \omega_L/2\pi$	Hz
longitudinal relaxation time	T_1		s
transverse relaxation time	T_2		s
electric dipole moment of a molecule	$\boldsymbol{p}$, $\boldsymbol{\mu}$	$E_p = -\boldsymbol{p}\cdot\boldsymbol{E}$	C m
quadrupole moment of a molecule	$\boldsymbol{Q}$; $\boldsymbol{\Theta}$	$E_p = \frac{1}{2}\boldsymbol{Q}:\boldsymbol{V}'' = \frac{1}{3}\boldsymbol{\Theta}:\boldsymbol{V}''$	$C\,m^2$
quadrupole moment of a nucleus	eQ	$eQ = 2\langle\Theta_{zz}\rangle$	$C\,m^2$
electric field gradient tensor	$\boldsymbol{q}$	$q_{\alpha\beta} = -\partial^2 V/\partial\alpha\partial\beta$	$V\,m^{-2}$
quadrupole interaction energy tensor	$\boldsymbol{\chi}$	$\chi_{\alpha\beta} = eQq_{\alpha\beta}$	J
electric polarizability of a molecule	α	p (induced) $= \alpha E$	$C\,m^2\,V^{-1}$
activity (of a radioactive substance)	A	$A = -\mathrm{d}N_B/\mathrm{d}t$	Bq
decay (rate) constant, disintegration (rate) constant	λ	$A = \lambda N_B$	s^{-1}
half life	$t_{\frac{1}{2}}$, $T_{\frac{1}{2}}$		s
mean life	τ		s
level width	Γ	$\Gamma = \hbar/\tau$	J

Name	*Symbol*	*Definition*	*SI unit*
disintegration energy	Q		J
cross section (of a nuclear reaction)	σ		m^2

SPECTROSCOPY

Name	*Symbol*	*Definition*	*SI unit*
total term	T	$T = E_{tot}/hc$	m^{-1}
transition wavenumber	$\tilde{\nu}$, (ν)	$\tilde{\nu} = T' - T''$	m^{-1}
transition frequency	ν	$\nu = (E' - E'')/h$	Hz
electronic term	T_e	$T_e = E_e/hc$	m^{-1}
vibrational term	G	$G = E_{vib}/hc$	m^{-1}
rotational term	F	$F = E_{rot}/hc$	m^{-1}
spin orbit coupling constant	A	$T_{s.o.} = A\langle \hat{L} \cdot \hat{S} \rangle$	m^{-1}
principal moments of inertia	$I_A; I_B; I_C$	$I_A \leq I_B \leq I_C$	kg m^2
rotational constants,			
in wavenumber	$\tilde{A}; \tilde{B}; \tilde{C}$	$\tilde{A} = h/8\pi^2 c I_A$	m^{-1}
in frequency	A; B; C	$A = h/8\pi^2 I_A$	Hz
inertial defect	Δ	$\Delta = I_C - I_A - I_B$	kg m^2
asymmetry parameter	κ	$\kappa = \frac{(2B - A - C)}{(A - C)}$	1
centrifugal distortion constants,			
S reduction	$D_J; D_{JK}; D_K; d_1; d_2$		m^{-1}
A reduction	$\Delta_J; \Delta_{JK}; \Delta_K; \delta_J; \delta_K$		m^{-1}
harmonic vibration wavenumber	ω_e; ω_r		m^{-1}
vibrational anharmonicity constant	$\omega_e x_e$; x_{rs}; $g_{tt'}$		m^{-1}
vibrational quantum numbers	v_r; l_t		1
Coriolis zeta constant	ζ_{rs}^{α}		1
angular momentum quantum numbers	see additional information below		
degeneracy, statistical weight	g, d, β		1
electric dipole moment of a molecule	p, μ	$E_p = -p \cdot E$	C m
transition dipole moment of a molecule	M, R	$M = \int \psi' p \psi'' d\tau$	C m
molecular geometry, interatomic distances,			
equilibrium distance	r_e		m
zero-point average distance	r_z		m
ground state distance	r_0		m

Name	*Symbol*	*Definition*	*SI unit*
substitution structure distance	r_s		m
vibrational coordinates,			
internal coordinates	R_i, r_i, θ_j, etc.		(varies)
symmetry coordinates	S_i		(varies)
normal coordinates			
mass adjusted	Q_r		$kg^{\frac{1}{2}}$ m
dimensionless	q_r		1
vibrational force constants,			
diatomic	f, (k)	$f=\partial^2 V/\partial r^2$	$J\ m^{-2}$
polyatomic,			
internal coordinates	f_{ij}	$f_{ij}=\partial^2 V/\partial r_i \partial r_j$	(varies)
symmetry coordinates	F_{ij}	$F_{ij}=\partial^2 V/\partial S_i \partial S_j$	(varies)
dimensionless normal coordinates	$\phi_{rst\ldots}$, $k_{rst\ldots}$		m^{-1}
nuclear magnetic resonance (NMR),			
magnetogyric ratio	γ	$\gamma=\mu/I\hbar$	$C\ kg^{-1}$
shielding constant	σ_A	$B_A=(1-\sigma_A)B$	1
chemical shift, δ scale	δ	$\delta=10^6(\nu-\nu_0)/\nu_0$	1
(indirect) spin–spin coupling constant	J_{AB}	$\hat{H}/h=J_{AB}\hat{I}_A \cdot \hat{I}_B$	Hz
direct (dipolar) coupling constant	D_{AB}		Hz
longitudinal relaxation time	T_1		s
transverse relaxation time	T_2		s
electron spin resonance, electron paramagnetic resonance (ESR, EPR),			
magnetogyric ratio	γ	$\gamma=\mu/s\hbar$	$C\ kg^{-1}$
g factor	g	$h\nu=g\mu_B B$	1
hyperfine coupling constant,			
in liquids	a, A	$\hat{H}_{hfs}/h=a\hat{S}\cdot\hat{I}$	Hz
in solids	T	$\hat{H}_{hfs}/h=\hat{S}\cdot T\cdot\hat{I}$	Hz

Angular momentum	*Operator symbol*	*Quantum number symbol* *Total*	*Z-axis*	*z-axis*
electron orbital	$\hat{L}$	L	M_L	Λ
one electron only	$\hat{l}$	l	m_l	λ

Angular momentum	*Operator symbol*	*Quantum number symbol* Total	*Z-axis*	*z-axis*
electron spin	$\hat{S}$	S	M_S	Σ
one electron only	$\hat{s}$	s	m_s	σ
electron orbital + spin	$\hat{L}+\hat{S}$			$\Omega=\Lambda+\Sigma$
nuclear orbital (rotational)	$\hat{R}$	R		K_R, k_R
nuclear spin	$\hat{I}$	I	M_I	
internal vibrational				
spherical top	$\hat{l}$	$l(l\zeta)$		K_l
other	$\hat{j}, \hat{\pi}$			$l(l\zeta)$
sum of $R+L(+j)$	$\hat{N}$	N		K, k
sum of $N+S$	$\hat{J}$	J	M_J	K, k
sum of $J+I$	$\hat{F}$	F	M_F	

ELECTROMAGNETIC RADIATION

Name	*Symbol*	*Definition*	*SI unit*
wavelength	λ		m
speed of light			
in vacuum	c_0		$\mathrm{m\,s^{-1}}$
in a medium	c	$c=c_0/n$	$\mathrm{m\,s^{-1}}$
wavenumber in vacuum	$\tilde{\nu}$	$\tilde{\nu}=\nu/c_0=1/n\lambda$	$\mathrm{m^{-1}}$
wavenumber (in a medium)	σ	$\sigma=1/\lambda$	$\mathrm{m^{-1}}$
frequency	ν	$\nu=c/\lambda$	Hz
circular frequency, pulsatance	ω	$\omega=2\pi\nu$	$\mathrm{s^{-1}}$, $\mathrm{rad\,s^{-1}}$
refractive index	n	$n=c_0/c$	1
Planck constant	h		J s
Planck constant/2π	$\hbar$	$\hbar=h/2\pi$	J s
radiant energy	Q, W		J
radiant energy density	ρ, w	$\rho=Q/V$	$\mathrm{J\,m^{-3}}$
spectral radiant energy density			
in terms of frequency	ρ_ν, w_ν	$\rho_\nu=\mathrm{d}\rho/\mathrm{d}\nu$	$\mathrm{J\,m^{-3}\,Hz^{-1}}$
in terms of wavenumber	$\rho_{\tilde{\nu}}, w_{\tilde{\nu}}$	$\rho_{\tilde{\nu}}=\mathrm{d}\rho/\mathrm{d}\tilde{\nu}$	$\mathrm{J\,m^{-2}}$
in terms of wavelenglth	ρ_λ, w_λ	$\rho_\lambda=\mathrm{d}\rho/\mathrm{d}\lambda$	$\mathrm{J\,m^{-4}}$
Einstein transition probabilities			
spontaneous emission	A_{nm}	$\mathrm{d}N_n/\mathrm{d}t=-A_{nm}N_n$	$\mathrm{s^{-1}}$
stimulated emission	B_{nm}	$\mathrm{d}N_n/\mathrm{d}t=-\rho_{\tilde{\nu}}(\tilde{\nu}_{nm})\times B_{nm}N_n$	$\mathrm{s\,kg^{-1}}$
stimulated absorption	B_{mn}	$\mathrm{d}N_n/\mathrm{d}t=\rho_{\tilde{\nu}}(\tilde{\nu}_{nm})B_{mn}N_m$	$\mathrm{s\,kg^{-1}}$
radiant power, radiant energy per time	Φ, P	$\Phi=\mathrm{d}Q/\mathrm{d}t$	W

Name	*Symbol*	*Definition*	*SI unit*
radiant intensity	I	$I = \mathrm{d}\Phi/\mathrm{d}\Omega$	$\mathrm{W\ sr^{-1}}$
radiant exitance, (emitted radiant flux)	M	$M = \mathrm{d}\Phi/\mathrm{d}A_{\mathrm{source}}$	$\mathrm{W\ m^{-2}}$
irradiance, (radiant flux received)	$E, (I)$	$E = \mathrm{d}\Phi/\mathrm{d}A$	$\mathrm{W\ m^{-2}}$
emittance	ε	$\varepsilon = M/M_{\mathrm{bb}}$	1
Stefan–Boltzmann constant	σ	$M_{\mathrm{bb}} = \sigma T^4$	$\mathrm{W\ m^{-2}\ K^{-4}}$
first radiation constant	c_1	$c_1 = 2\pi h c_0^2$	$\mathrm{W\ m^2}$
second radiation constant	c_2	$c_2 = hc_0/k$	K m
transmittance, transmission factor	τ, T	$\tau = \Phi_{\mathrm{tr}}/\Phi_0$	1
absorptance, absorption factor	α	$\alpha = \Phi_{\mathrm{abs}}/\Phi_0$	1
reflectance, reflection factor	ρ	$\rho = \Phi_{\mathrm{refl}}/\Phi_0$	1
(decadic) absorbance	A	$A = -\lg(1-\alpha_{\mathrm{i}})$	1
napierian absorbance	B	$B = -\ln(1-\alpha_{\mathrm{i}})$	1
absorption coefficient			
(linear) decadic	a, K	$a = A/l$	$\mathrm{m^{-1}}$
(linear) napierian	α	$\alpha = B/l$	$\mathrm{m^{-1}}$
molar (decadic)	ε	$\varepsilon = a/c = A/cl$	$\mathrm{m^2\ mol^{-1}}$
molar napierian	κ	$\kappa = \alpha/c = B/cl$	$\mathrm{m^2\ mol^{-1}}$
absorption index	k	$k = \alpha/4\pi\tilde{\nu}$	1
complex refractive index	$\hat{n}$	$\hat{n} = n + ik$	1
molar refraction	R, R_{m}	$R = \dfrac{(n^2-1)}{(n^2+2)} V_{\mathrm{m}}$	$\mathrm{m^3\ mol^{-1}}$
angle of optical rotation	α		1, rad

SOLID STATE

Name	*Symbol*	*Definition*	*SI unit*
lattice vector	$\boldsymbol{R}, \boldsymbol{R}_0$		m
fundamental translation vectors for the crystal lattice	$\boldsymbol{a}_1; \boldsymbol{a}_2; \boldsymbol{a}_3,$ $\boldsymbol{a}; \boldsymbol{b}; \boldsymbol{c}$	$\boldsymbol{R} = n_1\boldsymbol{a}_1 + n_2\boldsymbol{a}_2 + n_3\boldsymbol{a}_3$	m
(circular) reciprocal lattice vector	$\boldsymbol{G}$	$\boldsymbol{G} \cdot \boldsymbol{R} = 2\pi m$	$\mathrm{m^{-1}}$
(circular) fundamental translation vectors for the reciprocal lattice	$\boldsymbol{b}_1; \boldsymbol{b}_2; \boldsymbol{b}_3,$ $\boldsymbol{a}^*; \boldsymbol{b}^*; \boldsymbol{c}^*$	$\boldsymbol{a}_i \cdot \boldsymbol{b}_k = 2\pi\delta_{ik}$	$\mathrm{m^{-1}}$
lattice plane spacing	d		m
Bragg angle	θ	$n\lambda = 2d \sin\theta$	1, rad

Name	*Symbol*	*Definition*	*SI unit*
order of reflection	n		1
order parameters			
short range	σ		1
long range	s		1
Burgers vector	$\boldsymbol{b}$		m
particle position vector	$\boldsymbol{r}, \boldsymbol{R}_j$		m
equilibrium position vector of an ion	$\boldsymbol{R}_0$		m
displacement vector of an ion	$\boldsymbol{u}$	$\boldsymbol{u}=\boldsymbol{R}-\boldsymbol{R}_0$	m
Debye–Waller factor	B, D		1
Debye circular wavenumber	q_D		m^{-1}
Debye circular frequency	ω_D		s^{-1}
Grüneisen parameter	γ, Γ	$\gamma=\alpha V/\kappa C_V$	1
Madelung constant	$\alpha, \mathscr{M}$	$E_{coul}=\dfrac{\alpha N_A z_+ z_- e^2}{4\pi\varepsilon_0 R_0}$	1
density of states	N_E	$N_E=dN(E)/dE$	$J^{-1}\,m^{-3}$
(spectral) density of vibrational modes	N_ω, g	$N_\omega=dN(\omega)/d\omega$	$s\,m^{-3}$
resistivity tensor	ρ_{ik}	$\boldsymbol{E}=\boldsymbol{\rho}\cdot\boldsymbol{j}$	$\Omega\,m$
conductivity tensor	σ_{ik}	$\boldsymbol{\sigma}=\boldsymbol{\rho}^{-1}$	$S\,m^{-1}$
thermal conductivity tensor	λ_{ik}	$\boldsymbol{J}_q=-\boldsymbol{\lambda}\cdot \text{grad}\ T$	$W\,m^{-1}\,K^{-1}$
residual resistivity	ρ_R		$\Omega\,m$
relaxation time	τ	$\tau=l/v_F$	s
Lorenz coefficient	L	$L=\lambda/\sigma T$	$V^2\,K^{-2}$
Hall coefficient	A_H, R_H	$\boldsymbol{E}=\boldsymbol{\rho}\cdot\boldsymbol{j}+R_H(\boldsymbol{B}\times\boldsymbol{j})$	$m^3\,C^{-1}$
thermoelectric force	E		V
Peltier coefficient	Π		V
Thomson coefficient	$\mu, (\tau)$		$V\,K^{-1}$
work function	Φ	$\Phi=E_\infty-E_F$	J
number density, number concentration	$n, (p)$		m^{-3}
gap energy	E_g		J
donor ionization energy	E_d		J
acceptor ionization energy	E_a		J
Fermi energy	E_F, ε_F		J
circular wave vector, propagation vector	$\boldsymbol{k}, \boldsymbol{q}$	$k=2\pi/\lambda$	m^{-1}
Bloch function	$u_k(\boldsymbol{r})$	$\psi(\boldsymbol{r})=u_k(\boldsymbol{r})\exp(i\boldsymbol{k}\cdot\boldsymbol{r})$	$m^{-3/2}$
charge density of electrons	ρ	$\rho(\boldsymbol{r})=-e\psi^*(\boldsymbol{r})\psi(\boldsymbol{r})$	$C\,m^{-3}$

Name	Symbol	Definition	SI unit
effective mass	m^*		kg
mobility	μ	$\mu = v_{\mathrm{drift}}/E$	$\mathrm{m^2\,V^{-1}\,s^{-1}}$
mobility ratio	b	$b = \mu_n/\mu_p$	1
diffusion coefficient	D	$\mathrm{d}N/\mathrm{d}t = -DA(\mathrm{d}n/\mathrm{d}x)$	$\mathrm{m^2\,s^{-1}}$
diffusion length	L	$L = \sqrt{D\tau}$	m
characteristic (Weiss) temperature	θ, θ_W		K
Curie temperature	T_C		K
Néel temperature	T_N		K

STATISTICAL THERMODYNAMICS

Name	Symbol	Definition	SI unit
number of entities	N		1
number density of entities, number concentration	n, C	$n = N/V$	$\mathrm{m^{-3}}$
Avogadro constant	L, N_A		$\mathrm{mol^{-1}}$
Boltzmann constant	k, k_B		$\mathrm{J\,K^{-1}}$
gas constant (molar)	R	$R = Lk$	$\mathrm{J\,K^{-1}\,mol^{-1}}$
molecular position vector	$\boldsymbol{r}$ (x, y, z)		m
molecular velocity vector	$\boldsymbol{c}(c_x, c_y, c_z)$, $\boldsymbol{u}(u_x, u_y, u_z)$	$\boldsymbol{c} = \mathrm{d}\boldsymbol{r}/\mathrm{d}t$	$\mathrm{m\,s^{-1}}$
molecular momentum vector	$\boldsymbol{p}(p_x, p_y, p_z)$	$\boldsymbol{p} = m\boldsymbol{c}$	$\mathrm{kg\,m\,s^{-1}}$
velocity distribution function (Maxwell)	$f(c_x)$	$f(c_x) = (m/2\pi kT)^{\frac{1}{2}} \times \exp(-mc_x{}^2/2kT)$	$\mathrm{m^{-1}\,s}$
speed distribution function (Maxwell–Boltzmann)	$F(c)$	$F(c) = (m/2\pi kT)^{3/2} \times 4\pi c^2 \exp(-mc^2/2kT)$	$\mathrm{m^{-1}\,s}$
average speed	$\bar{c}$, $\bar{u}$, $\langle c\rangle$, $\langle u\rangle$	$\bar{c} = \int cF(c)\mathrm{d}c$	$\mathrm{m\,s^{-1}}$
generalized coordinate	q		(m)
generalized momentum	p	$p = \partial L/\partial \dot{q}$	$(\mathrm{kg\,m\,s^{-1}})$
volume in phase space	Ω	$\Omega = (1/h)\int p\,\mathrm{d}q$	1
probability	P		1
statistical weight, degeneracy	g, d, W, ω, β		1
density of states	$\rho(E)$	$\rho(E) = \mathrm{d}N/\mathrm{d}E$	$\mathrm{J^{-1}}$
partition function, sum over states,			
for a single molecule	q, z	$q = \sum_i g_i \exp(-\varepsilon_i/kT)$	1
for a canonical ensemble (system, or assembly)	Q, Z		1

Name	*Symbol*	*Definition*	*SI unit*
microcanonical ensemble	Ω		1
grand (canonical ensemble)	Ξ		1
symmetry number	σ, s		1
reciprocal temperature parameter	β	$\beta = 1/kT$	J^{-1}
characteristic temperature	Θ		K

GENERAL CHEMISTRY

Name	*Symbol*	*Definition*	*SI unit*
number of entities (e.g. molecules, atoms, ions, formula units)	N		1
amount (of substance)	n	$n_B = N_B/L$	mol
Avogadro constant	L, N_A		mol^{-1}
mass of atom, atomic mass	m_a, m		kg
mass of entity (molecule, or formula unit)	m_f, m		kg
atomic mass constant	m_u	$m_u = m_a(^{12}C)/12$	kg
molar mass	M	$M_B = m/n_B$	kg mol^{-1}
relative molecular mass (relative molar mass, molecular weight)	M_r	$M_{r,B} = m_B/m_u$	1
molar volume	V_m	$V_{m,B} = V/n_B$	m^3 mol^{-1}
mass fraction	w	$w_B = m_B/\Sigma m_i$	1
volume fraction	ϕ	$\phi_B = V_B/\Sigma V_i$	1
mole fraction, amount fraction, number fraction	x, y	$x_B = n_B/\Sigma n_i$	1
(total) pressure	p, P		Pa
partial pressure	p_B	$p_B = y_B p$	Pa
mass concentration (mass density)	γ, ρ	$\gamma_B = m_B/V$	kg m^{-3}
number concentration, number density of entities	C, n	$C_B = N_B/V$	m^{-3}
amount concentration, concentration	c	$c_B = n_B/V$	mol m^{-3}
solubility	s	$s_B = c_B$ (saturated solution)	mol m^{-3}
molality (of a solute)	m, (b)	$m_B = n_B/m_A$	mol kg^{-1}

Name	*Symbol*	*Definition*	*SI unit*
surface concentration	Γ	$\Gamma_B = n_B/A$	mol m^{-2}
stoichiometric number	ν		1
extent of reaction, advancement	ξ	$\Delta\xi = \Delta n_B/\nu_B$	mol
degree of dissociation	α		1

CHEMICAL THERMODYNAMICS

Name	*Symbol*	*Definition*	*SI unit*
heat	q, Q		J
work	w, W		J
internal energy	U	$\Delta U = q + w$	J
enthalpy	H	$H = U + pV$	J
thermodynamic temperature	T		K
Celsius temperature	θ, t	$\theta/°\mathrm{C} = T/\mathrm{K} - 273.15$	°C
entropy	S	$\mathrm{d}S \geqslant \mathrm{d}q/T$	J K^{-1}
Helmholtz energy, (Helmholtz function)	A	$A = U - TS$	J
Gibbs energy, (Gibbs function)	G	$G = H - TS$	J
Massieu function	J	$J = -A/T$	J K^{-1}
Planck function	Y	$Y = -G/T$	J K^{-1}
surface tension	γ, σ	$\gamma = (\partial G/\partial A_s)_{T,p}$	J m^{-2}, N m^{-1}
molar quantity X	X_m	$X_m = X/n$	(varies)
specific quantity X	x	$x = X/m$	(varies)
pressure coefficient	β	$\beta = (\partial p/\partial T)_V$	Pa K^{-1}
relative pressure coefficient	α_p	$\alpha_p = (1/p)(\partial p/\partial T)_V$	K^{-1}
compressibility,			
isothermal	κ_T	$\kappa_T = -(1/V)(\partial V/\partial p)_T$	Pa^{-1}
isentropic	κ_S	$\kappa_S = -(1/V)(\partial V/\partial p)_S$	Pa^{-1}
linear expansion coefficient	α_l	$\alpha_l = (1/l)(\partial l/\partial T)$	K^{-1}
cubic expansion coefficient	α, α_V, γ	$\alpha = (1/V)(\partial V/\partial T)_p$	K^{-1}
heat capacity,			
at constant pressure	C_p	$C_p = (\partial H/\partial T)_p$	J K^{-1}
at constant volume	C_V	$C_V = (\partial U/\partial T)_V$	J K^{-1}
ratio of heat capacities	γ, (κ)	$\gamma = C_p/C_V$	1
Joule–Thomson coefficient	μ, μ_{JT}	$\mu = (\partial T/\partial p)_H$	K Pa^{-1}
second virial coefficient	B	$pV_m = RT(1 + B/V_m + \cdots)$	m^3 mol^{-1}
compression factor (compressibility factor)	Z	$Z = pV_m/RT$	1

Name	*Symbol*	*Definition*	*SI unit*
partial molar quantity X	X_B, (X'_B)	$X_B = (\partial X/\partial n_B)_{T,p,n_{j\neq B}}$	(varies)
chemical potential (partial molar Gibbs energy)	μ	$\mu_B = (\partial G/\partial n_B)_{T,p,n_{j\neq B}}$	$J\,mol^{-1}$
absolute activity	λ	$\lambda_B = \exp(\mu_B/RT)$	1
standard chemical potential	μ°, μ^{o}		$J\,mol^{-1}$
standard partial molar enthalpy	H_B°	$H_B^{\circ} = \mu_B^{\circ} + TS_B^{\circ}$	$J\,mol^{-1}$
standard partial molar entropy	S_B°	$S_B^{\circ} = -(\partial \mu_B^{\circ}/\partial T)_p$	$J\,mol^{-1}\,K^{-1}$
standard reaction Gibbs energy (function)	$\Delta_r G^{\circ}$	$\Delta_r G^{\circ} = \sum_B \nu_B \mu_B^{\circ}$	$J\,mol^{-1}$
affinity of reaction	A, $(\mathcal{A})$	$A = -(\partial G/\partial \xi)_{p,T}$ $= -\sum_B \nu_B \mu_B$	$J\,mol^{-1}$
standard reaction enthalpy	$\Delta_r H^{\circ}$	$\Delta_r H^{\circ} = \sum_B \nu_B H_B^{\circ}$	$J\,mol^{-1}$
standard reaction entropy	$\Delta_r S^{\circ}$	$\Delta_r S^{\circ} = \sum_B \nu_B S_B^{\circ}$	$J\,mol^{-1}\,K^{-1}$
equilibrium constant	K°, K	$K^{\circ} = \exp(-\Delta_r G^{\circ}/RT)$	1
equilibrium constant, pressure basis	K_p	$K_p = \prod_B p_B^{\nu_B}$	$Pa^{\Sigma\nu}$
concentration basis	K_c	$K_c = \prod_B c_B^{\nu_B}$	$(mol\,m^{-3})^{\Sigma\nu}$
molality basis	K_m	$K_m = \prod_B m_B^{\nu_B}$	$(mol\,kg^{-1})^{\Sigma\nu}$
fugacity	f, $\tilde{p}$	$f_B = \lambda_B \lim_{p\to 0} (p_B/\lambda_B)_T$	Pa
fugacity coefficient	ϕ	$\phi_B = f_B/p_B$	1
activity and activity coefficient referenced to Raoult's law, (relative) activity	a	$a_B = \exp\left[\dfrac{\mu_B - \mu_B^{*}}{RT}\right]$	1
activity coefficient	f	$f_B = a_B/x_B$	1
activities and activity coefficients referenced to Henry's law, (relative) activity, molality basis	a_m	$a_{m,B} = \exp\left[\dfrac{\mu_B - \mu_B^{\circ}}{RT}\right]$	1
concentration basis	a_c	$a_{c,B} = \exp\left[\dfrac{\mu_B - \mu_B^{\circ}}{RT}\right]$	1

Name	*Symbol*	*Definition*	*SI Unit*
mole fraction basis	a_x	$a_{x,\mathrm{B}}=\exp\left[\dfrac{\mu_\mathrm{B}-\mu_\mathrm{B}^*}{RT}\right]$	1
activity coefficient,			
molality basis	γ_m	$a_{m,\mathrm{B}}=\gamma_{m,\mathrm{B}}m_\mathrm{B}/m^\circ$	1
concentration basis	γ_c	$a_{c,\mathrm{B}}=\gamma_{c,\mathrm{B}}c_\mathrm{B}/c^\circ$	1
mole fraction basis	γ_x	$a_{x,\mathrm{B}}=\gamma_{x,\mathrm{B}}x_\mathrm{B}$	1
ionic strength,			
molality basis	I_m, I	$I_m=\frac{1}{2}\Sigma m_\mathrm{B}z_\mathrm{B}^2$	mol kg^{-1}
concentration basis	I_c, I	$I_c=\frac{1}{2}\Sigma c_\mathrm{B}z_\mathrm{B}^2$	mol m^{-3}
osmotic coefficient,			
molality basis	ϕ_m	$\phi_m=(\mu_\mathrm{A}^*-\mu_\mathrm{A})/(RTM_\mathrm{A}\Sigma m_\mathrm{B})$	1
mole fraction basis	ϕ_x	$\phi_x=(\mu_\mathrm{A}-\mu_\mathrm{A}^*)/(RT\ln x_\mathrm{A})$	1
osmotic pressure	Π	$\Pi=c_\mathrm{B}RT$ (ideal dilute solution)	Pa

(i) *Symbols used as subscripts to denote a chemical process or reaction*

These symbols should be printed in roman (upright) type, without a full stop (period).

vaporization, evaporation (liquid→gas)	vap
sublimation (solid→gas)	sub
melting, fusion (solid→liquid)	fus
transition (between two phases)	trs
mixing of fluids	mix
solution (of solute in solvent)	sol
dilution (of a solution)	dil
adsorption	ads
displacement	dpl
immersion	imm
reaction in general	r
atomization	at
combustion reaction	c
formation reaction	f

(ii) *Recommended superscripts*

standard	$\ominus$, o
pure substance	*
infinite dilution	∞
ideal	id
activated complex, transition state	‡
excess quantity	E

Name	*Symbol*	*Definition*	*SI unit*
CHEMICAL KINETICS			
rate of change of quantity X	$\dot{X}$	$\dot{X}=\mathrm{d}X/dt$	(varies)
rate of conversion	$\dot{\xi}$	$\dot{\xi}=\mathrm{d}\xi/dt$	$\mathrm{mol\ s^{-1}}$
rate of concentration change (due to chemical reaction)	r_B, v_B	$r_B=\mathrm{d}c_B/dt$	$\mathrm{mol\ m^{-3}\ s^{-1}}$
rate of reaction (based on amount concentration)	v	$v=\dot{\xi}/V$ $=v_B{}^{-1}\mathrm{d}c_B/dt$	$\mathrm{mol\ m^{-3}\ s^{-1}}$
partial order of reaction	n_B	$v=k\Pi c_B{}^{n_B}$	1
overall order of reaction	n	$n=\Sigma n_B$	1
rate constant, rate coefficient	k	$v=k\Pi c_B{}^{n_B}$	$(\mathrm{mol^{-1}\ m^3})^{n-1}\mathrm{s^{-1}}$
Boltzmann constant	k, k_B		$\mathrm{J\ K^{-1}}$
half life	$t_{\frac{1}{2}}$	$c(t_{\frac{1}{2}})=c_0/2$	s
relaxation time	τ	$\tau=1/(k_1+k_{-1})$	s
energy of activation, activation energy	E_a, E	$E_a=RT^2\,\mathrm{d}\ln k/\mathrm{d}T$	$\mathrm{J\,mol^{-1}}$
pre-exponential factor	A	$k=A\exp(-E_a/RT)$	$(\mathrm{mol^{-1}\ m^3})^{n-1}\mathrm{s^{-1}}$
volume of activation	$\Delta^{\ddagger}V$	$\Delta^{\ddagger}V=-RT\times(\partial\ln k/\partial p)_T$	$\mathrm{m^3\ mol^{-1}}$
collision diameter	d	$d_{AB}=r_A+r_B$	m
collision cross-section	σ	$\sigma_{AB}=\pi d_{AB}{}^2$	$\mathrm{m^2}$
collision frequency	Z_A		$\mathrm{s^{-1}}$
collision number	Z_{AB}, Z_{AA}		$\mathrm{m^{-3}\,s^{-1}}$
collision frequency factor	z_{AB}, z_{AA}	$z_{AB}=Z_{AB}/Lc_Ac_B$	$\mathrm{m^3\ mol^{-1}\ s^{-1}}$
standard enthalpy of activation	$\Delta^{\ddagger}H^{\circ}, \Delta H^{\ddagger}$		$\mathrm{J\,mol^{-1}}$
standard entropy of activation	$\Delta^{\ddagger}S^{\circ}, \Delta S^{\ddagger}$		$\mathrm{J\,mol^{-1}\ K^{-1}}$
standard Gibbs energy of activation	$\Delta^{\ddagger}G^{\circ}, \Delta G^{\ddagger}$		$\mathrm{J\,mol^{-1}}$
quantum yield, photochemical yield	ϕ		1
ELECTROCHEMISTRY			
elementary charge (proton charge)	e		C
Faraday constant	F	$F=eL$	$\mathrm{C\,mol^{-1}}$

Name	*Symbol*	*Definition*	*SI unit*
charge number of an ion	z	$z_B = Q_B/e$	1
ionic strength	I_c, I	$I_c = \frac{1}{2}\sum c_i z_i^2$	mol m^{-3}
mean ionic activity	$a_\pm$	$a_\pm = m_\pm \gamma_\pm / m^\circ$	1
mean ionic molality	$m_\pm$	$m_\pm^{(\nu_+ + \nu_-)} = m_+^{\nu_+} m_-^{\nu_-}$	mol kg^{-1}
mean ionic activity coefficient	$\gamma_\pm$	$\gamma_\pm^{(\nu_+ + \nu_-)} = \gamma_+^{\nu_+} \gamma_-^{\nu_-}$	1
charge number of electrochemical cell reaction	n, (z)		1
electric potential difference (of a galvanic cell)	$\Delta V, E, U$	$\Delta V = V_R - V_L$	V
emf, electromotive force	E	$E = \lim_{I \to 0} \Delta V$	V
standard emf, standard potential of the electrochemical cell reaction	E°	$E^\circ = -\Delta_r G^\circ / nF$ $= (RT/nF) \ln K^\circ$	V
standard electrode potential	E°		V
emf of the cell, potential of the electrochemical cell reaction	E	$E = E^\circ - (RT/nF) \times \sum \nu_i \ln a_i$	V
pH	pH	$\mathrm{pH} \approx -\lg\left[\frac{c(\mathrm{H}^+)}{\mathrm{mol\,dm}^{-3}}\right]$	1
inner electric potential	ϕ	$\nabla\phi = -E$	V
outer electric potential	ψ	$\psi = Q/4\pi\varepsilon_0 r$	V
surface electric potential	χ	$\chi = \phi - \psi$	V
Galvani potential difference	$\Delta\phi$	$\Delta_\alpha^\beta \phi = \phi^\beta - \phi^\alpha$	V
volta potential difference	$\Delta\psi$	$\Delta_\alpha^\beta \psi = \psi^\beta - \psi^\alpha$	V
electrochemical potential	$\tilde{\mu}$	$\tilde{\mu}_B{}^\alpha = (\partial G / \partial n_B{}^\alpha)$	J mol^{-1}
electric current	I	$I = \mathrm{d}Q/\mathrm{d}t$	A
(electric) current density	j	$j = I/A$	A m^{-2}
(surface) charge density	σ	$\sigma = Q/A$	C m^{-2}
electrode reaction rate constant	k	$k_{ox} = I_a / (nFA \prod_i c_i^{n_i})$	(varies)
mass transfer coefficient, diffusion rate constant	k_d	$k_{d,B} = \|\nu_B\| I_{l,B} / nFcA$	m s^{-1}
thickness of diffusion layer	δ	$\delta_B = D_B / k_{d,B}$	m

Name	*Symbol*	*Definition*	*SI unit*
transfer coefficient (electrochemical)	α	$\alpha_c = \dfrac{-\lvert\nu\rvert RT}{nF} \dfrac{\partial \ln \lvert I_c \rvert}{\partial E}$	1
overpotential	η	$\eta = E_I - E_{I=0} - IR_u$	V
electrokinetic potential (zeta potential)	ζ		V
conductivity	$\kappa, (\sigma)$	$\kappa = j/E$	$\mathrm{S\,m^{-1}}$
conductivity cell constant	K_{cell}	$K_{\mathrm{cell}} = \kappa R$	$\mathrm{m^{-1}}$
molar conductivity (of an electrolyte)	Λ	$\Lambda_B = \kappa/c_B$	$\mathrm{S\,m^2\,mol^{-1}}$
ionic conductivity, molar conductivity of an ion	λ	$\lambda_B = \lvert z_B \rvert F u_B$	$\mathrm{S\,m^2\,mol^{-1}}$
electric mobility	$u, (\mu)$	$u_B = v_B/E$	$\mathrm{m^2\,V^{-1}\,s^{-1}}$
transport number	t	$t_B = j_B/\Sigma j_i$	1
reciprocal radius of ionic atmosphere	κ	$\kappa = (2F^2 I/\varepsilon RT)^{\frac{1}{2}}$	$\mathrm{m^{-1}}$

COLLOID AND SURFACE CHEMISTRY

Name	*Symbol*	*Definition*	*SI unit*
specific surface area	a, a_s, s	$a = A/m$	$\mathrm{m^2\,kg^{-1}}$
surface amount of B, adsorbed amount of B	n_B^s, n_B^a		mol
surface excess of B	n_B^σ		mol
surface excess concentration of B	Γ_B, (Γ_B^σ)	$\Gamma_B = n_B^\sigma/A$	$\mathrm{mol\,m^{-2}}$
total surface excess concentration	Γ, (Γ^σ)	$\Gamma = \sum_i \Gamma_i$	$\mathrm{mol\,m^{-2}}$
area per molecule	a, σ	$a_B = A/N_B^\sigma$	$\mathrm{m^2}$
area per molecule in a filled monolayer	a_m, σ_m	$a_{m,B} = A/N_{m,B}$	$\mathrm{m^2}$
surface coverage	θ	$\theta = N_B^\sigma/N_{m,B}$	1
contact angle	θ		1, rad
film thickness	t, h, δ		m
thickness of (surface or interfacial) layer	τ, δ, t		m
surface tension, interfacial tension	γ, σ	$\gamma = (\partial G/\partial A_s)_{T,p}$	$\mathrm{N\,m^{-1}}$, $\mathrm{J\,m^{-2}}$
film tension	Σ_f	$\Sigma_f = 2\gamma_f$	$\mathrm{N\,m^{-1}}$
reciprocal thickness of the double layer	κ	$\kappa = [2F^2 I_c/\varepsilon RT]^{\frac{1}{2}}$	$\mathrm{m^{-1}}$
average molar masses			
number-average	M_n	$M_n = \Sigma n_i M_i/\Sigma n_i$	$\mathrm{kg\,mol^{-1}}$
mass-average	M_m	$M_m = \Sigma n_i M_i^2/\Sigma n_i M_i$	$\mathrm{kg\,mol^{-1}}$

Name	Symbol	Definition	SI unit
Z-average	M_Z	$M_Z = \Sigma n_i M_i^3 / \Sigma n_i M_i^2$	kg mol^{-1}
sedimentation coefficient	s	$s = v/a$	s
van der Waals constant	λ		J
retarded van der Waals constant	β, B		J
van der Waals–Hamaker constant	A_H		J
surface pressure	π^s, π	$\pi^s = \gamma^0 - \gamma$	N m^{-1}

TRANSPORT PROPERTIES

Name	Symbol	Definition	SI unit
flux (of a quantity X)	J_X, J	$J_X = A^{-1}\,dX/dt$	(varies)
volume flow rate	q_V, $\dot{V}$	$q_v = dV/dt$	m^3 s^{-1}
mass flow rate	q_m, $\dot{m}$	$q_m = dm/dt$	kg s^{-1}
mass transfer coefficient	k_d		m s^{-1}
heat flow rate	ϕ	$\phi = dq/dt$	W
heat flux	J_q	$J_q = \phi/A$	W m^{-2}
thermal conductance	G	$G = \phi/\Delta T$	W K^{-1}
thermal resistance	R	$R = 1/G$	K W^{-1}
thermal conductivity	λ, k	$\lambda = J_q/(dT/dl)$	W m^{-1} K^{-1}
coefficient of heat transfer	h, (k, K, α)	$h = J_q/\Delta T$	W m^{-2} K^{-1}
thermal diffusivity	a	$a = \lambda/\rho c_p$	m^2 s^{-1}
diffusion coefficient	D	$D = J_n/(dc/dl)$	m^2 s^{-1}

The following symbols are used in the definitions of the dimensionless quantities: mass (m), time (t), volume (V), area (A), density (ρ), speed (v), length (l), viscosity (η), pressure (p), acceleration of free fall (g), cubic expansion coefficient (α), temperature (T), surface tension (γ), speed of sound (c), mean free path (λ), frequency (f), thermal diffusivity (a), coefficient of heat transfer (h), thermal conductivity (k), specific heat capacity at constant pressure (c_p), diffusion coefficient (D), mole fraction (x), mass transfer coefficient (k_d), permeability (μ), electric conductivity (κ), and magnetic flux density (B).

Name	Symbol	Definition	SI unit
Reynolds number	Re	$Re = \rho v l/\eta$	1
Euler number	Eu	$Eu = \Delta p/\rho v^2$	1
Froude number	Fr	$Fr = v/(lg)^{\frac{1}{2}}$	1
Grashof number	Gr	$Gr = l^3 g \alpha \Delta T \rho^2/\eta^2$	1
Weber number	We	$We = \rho v^2 l/\gamma$	1
Mach number	Ma	$Ma = v/c$	1
Knudsen number	Kn	$Kn = \lambda/l$	1
Strouhal number	Sr	$Sr = lf/v$	1
Fourier number	Fo	$Fo = at/l^2$	1
Péclet number	Pe	$Pe = vl/a$	1
Rayleigh number	Ra	$Ra = l^3 g \alpha \Delta T \rho/\eta a$	1
Nusselt number	Nu	$Nu = hl/k$	1
Stanton number	St	$St = h/\rho v c_p$	1

Name	*Symbol*	*Definition*	*SI units*
Fourier number for mass transfer	Fo^*	$Fo^* = Dt/l^2$	1
Péclet number for mass transfer	Pe^*	$Pe^* = vl/D$	1
Grashof number for mass transfer	Gr^*	$Gr^* = l^3 g\left(\frac{\partial \rho}{\partial x}\right)_{T,p}\left(\frac{\Delta x \rho}{\eta}\right)$	1
Nusselt number for mass transfer	Nu^*	$Nu^* = k_d l/D$	1
Stanton number for mass transfer	St^*	$St^* = k_d/v$	1
Prandtl number	Pr	$Pr = \eta/\rho a$	1
Schmidt number	Sc	$Sc = \eta/\rho D$	1
Lewis number	Le	$Le = a/D$	1
magnetic Reynolds number	Rm, Re_m	$Rm = v\mu\kappa l$	1
Alfvén number	Al	$Al = v(\rho\mu)^{\frac{1}{2}}/B$	1
Hartmann number	Ha	$Ha = Bl(\kappa/\eta)^{\frac{1}{2}}$	1
Cowling number	Co	$Co = B^2/\mu\rho v^2$	1

NOMENCLATURE FOR INORGANIC IONS AND RADICALS

Full details on inorganic chemical nomenclature may be found in:

1. Leigh, G. J., Editor, *Nomenclature of Inorganic Chemistry (Recommendations 1990)*, Blackwell Scientific Publications, Oxford, 1990.
2. Block, B. P., Powell, W. H., and Fernelius, W. C., *Inorganic Chemical Nomenclature*, American Chemical Society, Washington, D.C., 1990.

	Name				
Atom or group	as neutral molecule	as cation or cationic radical[1]	as anion	as ligand	as prefix for substituent in organic compounds
H	monohydrogen	hydrogen	hydride	hydrido	
F	monofluorine		fluoride	fluoro	fluoro
Cl	monochlorine	chlorine	chloride	chloro	chloro
Br	monobromine	bromine	bromide	bromo	bromo
I	monoiodine	iodine	iodide	iodo	iodo
I_3			triiodide		
ClO		chlorosyl	hypochlorite	hypochlorito	
ClO_2	chlorine dioxide	chloryl	chlorite	chlorito	
ClO_3		perchloryl	chlorate	chlorato	
ClO_4			perchlorate		
IO		iodosyl	hypoiodite		iodoso
IO_2		iodyl			iodyl or iodoxy
O	monoöxygen		oxide	oxo	oxo or keto
O_2	dioxygen		O_2^{2-}: peroxide O_2^{-}: hyperoxide	peroxo	peroxy
HO	hydroxyl		hydroxide	hydroxo	hydroxy
HO_2	(perhydroxyl)		hydrogen peroxide	hydrogen peroxo	hydroperoxy
S	monosulfur		sulfide	thio (sulfido)	thio
HS	(sulfhydryl)		hydrogen sulfide	thiolo	thiol or mercapto
S_2	disulfur		disulfide	disulfido	
SO	sulfur monoxide	sulfinyl (thionyl)			sulfinyl
SO_2	sulfur dioxide	sulfonyl (sulfuryl)	sulfoxylate		sulfonyl
SO_3	sulfur trioxide		sulfite	sulfito	
HSO_3			hydrogen sulfite	hydrogen sulfito	
S_2O_3			thiosulfate	thiosulfato	
SO_4			sulfate	sulfato	
Se	selenium		selenide	seleno	seleno
SeO		seleninyl			seleninyl
SeO_2		selenonyl			selenonyl
SeO_3	selenium trioxide		selenite	selenito	
SeO_4			selenate	selenato	
Te	tellurium		telluride	telluro	telluro
CrO_2		chromyl			
UO_2		uranyl			
NpO_2		neptunyl			
PuO_2		plutonyl			
AmO_2		americyl			
N	mononitrogen		nitride	nitrido	
N_3			azide	azido	
NH			imide	imido	imino
NH_2			amide	amido	amino
NHOH			hydroxylamide	hydroxylamido	hydroxylamino
N_2H_3			hydrazide	hydrazido	hydrazino
NO	nitrogen oxide	nitrosyl		nitrosyl	nitroso
NO_2	nitrogen dioxide	nitryl		nitro	nitro
ONO			nitrite	nitrito	
NS		thionitrosyl			
$(NS)_n$		thiazyl (*e.g.*, trithiazyl)			
NO_3			nitrate	nitrato	
N_2O_3			hyponitrite	hyponitrito	
P	phosphorus		phosphide	phosphido	
PO		phosphoryl			phosphoroso
PS		thiophosphoryl			
PH_2O_3			hypophosphite	hypophosphito	
PHO_3			phosphite	phosphito	
PO_4			phosphate	phosphato	
AsO_4			arsenate	arsenato	
VO		vanadyl			
CO	carbon monoxide	carbonyl		carbonyl	carbonyl
CS		thiocarbonyl			
CH_3O	methoxyl		methanolate	methoxo	methoxy
C_2H_5O	ethoxyl		ethanolate	ethoxo	ethoxy
CH_3S			methanethiolate	methanethiolato	methylthio
C_2H_5S			ethanethiolate	ethanethiolato	ethylthio
CN		cyanogen	cyanide	cyano	cyano
OCN			cyanate	cyanato	cyanato
SCN			thiocyanate	thiocyanato and isothiocyanato	thiocyanato and isothiocyanato
SeCN			selenocyanate	selenocyanato	selenocyanato
TeCN			tellurocyanate	tellurocyanato	
CO_3			carbonate	carbonato	
HCO_3			hydrogen carbonate	hydrogen carbonato	
CH_3CO_2			acetate	acetato	acetoxy
CH_3CO	acetyl	acetyl			acetyl
C_2O_4			oxalate	oxalato	

[1]If necessary, oxidation state is to be given by Stock notation.

ORGANIC RADICALS AND RING SYSTEMS

Listed below are the names of the major organic radicals and side chains, as well as the most important ring systems with their recommended numbering schemes. Full details on the nomenclature of organic chemistry may be found in the following references:

1. Richer, J. C., Panico, R., and Powell, W. H., *A Guide to IUPAC Nomenclature of Organic Compounds*, Blackwell Scientific Publications, Oxford, 1993.
2. *Chemical Abstracts Service Index Guide*, Chemical Abstracts Service, Columbus, OH.
3. Weast, R. C., and Grasselli, J. C., Editors, *Handbook of Data on Organic Compounds, Second Edition,* CRC Press, Boca Raton, FL, 1989.

Names of Radicals and Side-Chains

Name	Formula
acetamido (acetylamino)	CH_3CONH-
acetimido (acetylimino)	$CH_3C(=NH)-$
acetoacetamido	CH_3COCH_2CONH-
acetoacetyl	CH_3COCH_2CO-
acetonyl	CH_3COCH_2
acetonylidene	$CH_3COCH=$
acetyl	CH_3CO-
acrylyl	$CH_2=CHCO-$
adipyl (from adipic acid)	$-OC(CH_2)_4CO-$
alanyl (from alanine)	$CH_3CH(NH_2)CO-$
β-alanyl	$H N(CH_2)_2CO-$
allophanoyl	$H_2NCONHCO-$
allyl (2-propenyl)	$CH_2=CHCH_2-$
allylidene (2-propenylidene)	$CH_2=CHCH=$
amidino (aminoiminomethyl)	$H_2NC(=NH)-$
amino	H_2N-
amyl (pentyl)	$CH_3(CH_2)_4-$
anilino (phenylamino)	C_6H_5NH-
anisidino	$CH_3OC_6H_4NH-$
anisyl (from anisic acid)	$CH_3OC_6H_4CO-$
anthranoyl (2-aminobenzoyl)	$2-H_2NC_6H_4CO-$
arsino	AsH_2-
azelaoyl (from azelaic acid)	$-OC(CH_2)_7CO-$
azido	N_3-
azino	$=NN=$
azo	$-N=N-$
azoxy	$-N(O)N-$
benzal	$C_6H_5CH=$
benzamido (benzylamino)	C_6H_5CONH-
benzhydryl (diphenylmethyl)	$(C_6H_5)_2CH-$
benzimido (benzylimino)	$C_6H_5C(=NH)-$
benzoxy (benzoyloxy)	C_6H_5COO-
benzoyl	C_6H_5CO-
benzyl	$C_6H_5CH_2-$
benzylidine	$C_6H_5CH=$
benzyldyne	$C_6H_5C\equiv$
biphenylyl	$C_6H_5C_6H_5-$
biphenylene	$-C_6H_4C_6H_4-$
butoxy	C_4H_9O-
sec-butoxy	$C_2H_5CH(CH_3)O-$
tert-butoxy	$(CH_3)_3CO-$
butyl	$CH_3(CH_2)_3-$
iso-butyl (3-methylpropyl)	$(CH_3)_2(CH_2)_2-$
sec-butyl (1-methylpropyl)	$C_2H_5CH(CH_3)-$
tert-butyl (1,1, dimethylethyl)	$(CH_3)_3C-$
butyryl	C_3H_7CO-
caproyl (from caproic acid)	$CH_3(CH_2)_4CO-$
capryl (from capric acid)	$CH_3(CH_2)_8CO-$
caprylyl (from caprylic acid)	$CH_3(CH_2)_6CO-$
carbamido	$H_2NCONH-$
carbamoyl (aminocarbonyl)	H_2NCO-
carbamyl (aminocarbonyl)	H_2NCO-
carbazoyl (hydrazinocarbonyl)	$H_2NNHCO-$
carbethoxy	$C_2H_5O_2C-$
carbobenzoxy	$C_6H_5CH_2O_2C-$
carbonyl	$-C=O-$
carboxy	$HOOC-$
cetyl	$CH_3(CH_2)_{15}-$
chloroformyl (chlorocarbonyl)	$ClCO-$
cinnamyl (3-phenyl-2-propenyl)	$C_6H_5CH=CHCH_2-$
cinnamoyl	$C_6H_5CH=CHCO-$
cinnamylidene	$C_6H_5CH=CHCH=$
cresyl (hydroxymethylphenyl)	$HO(CH_3)C_6H_4-$
crotoxyl	$CH_3CH=CHCO-$
crotyl (2-butenyl)	$CH_3CH=CHCH_2$
cyanamido (cyanoamino)	$NCNH-$
cyanato	$NCO-$
cyano	$NC-$
decanedioyl	$-OC(CH_2)_8CO-$
decanoly	$CH_3(CH_2)_8CO-$
diazo	$N_2=$
diazoamino	$-NHN=N-$
disilanyl	H_2SiSiH_2-
disiloxanoxy	$H_3SiOSiH_2O-$
disulfinyl	$-S(O)S(O)-$
dithio	$-SS-$
enanthyl	$CH_3(CH_2)_5CO-$
epoxy	$-O-$
ethenyl (vinyl)	$CH_2=CH-$
ethinyl	$HC\equiv C-$
ethoxy	C_2H_5O-
ethyl	CH_3CH_2-
ethylthio	C_2H_5S-
formamido (formylamino)	$HCONH-$
formyl	$HCO-$
fumaroyl (from fumaric acid)	$-OCCH=CHCO-$
furfuryl (2-furanylmethyl)	$OC_4H_3CH_2-$
furfurylidene (2-furanylmethylene)	$OC_4H_3CH=$
furyl (furanyl)	OC_4H_3-
glutamyl (from glutamic acid)	$-OC(CH_2)_2CH(NH_2)CO-$
glutaryl (from glutaric acid)	$-OC(CH_2)_3CO-$
glycidyl (oxiranylmethyl)	CH_2-CHCH_2-
glycinamido	H_2NCH_2CONH-
glycolyl (hydroxyacetyl)	$HOCH_2CO-$
glycyl (aminoacetyl)	H_2NCH_2CO-
glyoxylyl (oxoacetyl)	$HCOCO-$
guanidino	$H_2NC(=NH)NH-$
guanyl	$H_2NC(=NH)-$
heptadecanoyl	$CH_3(CH_2)_{15}CO-$
heptanamido	$CH_3(CH_2)_{15}CONH-$
heptanedioyl	$-OC(CH_2)_5CO-$
heptanoyl	$CH_3(CH_2)_5CO-$
hexadecanoyl	$CH_3(CH_2)_4CO-$
hexamethylene	$(CH_2)_6=$
hexanedioyl	$-OC(CH_2)_4CO-$
hippuryl (N-benzoylglycyl)	$C_6H_5CONHCH_2CO-$
hydantoyl	$H_2NCONHCH_2CO-$
hydrazino	N_2NNH-
hydrazo	$-HNNH-$
hydrocinnamoyl	$C_6H_5(CH_2)_2CO-$

Radical	Formula
hydroperoxy	HOO−
hydroxamino	HONH−
hydroxy	HO−
imino	HN=
iodoso	OI−
isoamyl (isopentyl)	$(CH_3)_2CH(CH_2)_2-$
isobutenyl (2-methyl-1-propenyl)	$(CH_3)_2C{=}CH-$
isobutoxy	$(CH_3)_2CHCH\ O-$
isobutyl	$(CH_3)_2CHCH_2-$
isobutylidene	$(CH_3)_2CHCH{=}$
isobutyryl	$(CH_3)_2CHCO-$
isocyanato	OCN−
isocyano	CN−
isohexyl	$(CH_3)_2CH(CH_2)_3-$
isoleucyl (from isoleucine)	$C_2H_5CH(CH_3)CH(NH_2)CO-$
isonitroso	HON=
isopentyl	$(CH_3)_2CH(CH_2)_2-$
isopentylidene	$(CH_3)_2CHCH_2CH{=}$
isopropenyl	$H_2C{=}C(CH_3)-$
isopropoxy	$(CH_3)_2CHO-$
isopropyl	$(CH_3)_2CH-$
isopropylidene	$(CH_3)_2C{=}$
isothiocyanato (isothiocyano)	SCN−
isovaleryl (from isovaleric acid)	$(CH_3)_2CHCH_2CO-$
keto (oxo)	O=
lactyl (from lactic acid)	$CH_3CH(OH)CO-$
lauroyl (from lauric acid)	$CH_3(CH_2)_{10}CO-$
leucyl (from leucine)	$(CH_3)_2CHCH_2CH(NH_2)CO-$
levulinyl (From levulinic acid)	$CH_3CO(CH_2)_2CO-$
malonyl (from malonic acid)	$-OCCH_2CO-$
mandelyl (from mandelic acid)	$C_6H_5CH(OH)CO-$
mercapto	HS−
methacrylyl (from methacrylic acid)	$CH_2{=}C(CH_3)CO-$
methallyl	$CH_2{=}C(CH_3)CH_2-$
methionyl (from methionine)	$CH_3SCH_2CH_2CH(NH_2)CO-$
methoxy	CH_3O-
methyl	H_3C-
methylene	$H_2C{=}$
methylenedioxy	$-OCH_2O-$
methylenedisulfonyl	$-O_2SCH_2SO_2-$
methylol	$HOCH_2-$
methylthio	CH_3S-
myristyl (from myristic acid)	$CH_3(CH_2)_{12}CO-$
naphthal	$(C_{10}H_7)CH{=}$
naphthobenzyl	$(C_{10}H_7)CH_2-$
naphthoxy	$(C_{10}H_7)O-$
naphthyl	$(C_{10}H_7)-$
naphthylidene	$(C_{10}H_6){=}$
neopentyl	$(CH_3)_3CCH_2-$
nitramino	O_2NNH-
nitro	O_2N-
nitrosamino	ONNH−
nitrosimino	ONN=
nitroso	ON−
nonanoyl (from nonanoic acid)	$CH_3(CH_2)_7CO-$
oleyl (from oleic acid)	$CH_3(CH_2)_7CH{=}CH(CH_2)_7CO-$
oxalyl (from oxalic acid)	−OCCO−
oxamido	$H_2NCOCONH-$
oxo (keto)	O=
palmityl (from palmitic acid)	$CH_3CH_2)_{14}CO-$
pelargonyl (from pelargonic acid)	$CH_3(CH_2)_7CO-$
pentamethylene	$-(CH_2)_5-$
pentyl	$CH_3(CH_2)_4-$
phenacyl	$C_6H_5COCH_2-$
phenacylidene	$C_6H_5COCH{=}$
phenanthryl	$(C_{14}H_9)-$
phenethyl	$C_6H_5CH_2CH_2-$
phenoxy	C_6H_5O-
phenyl	C_6H_5-
phenylene	$-C_6H_4-$
phenylenedioxy	$-OC_6H_4O-$
phosphino	H_2P-

Radical	Formula
phosphinyl	$H_2P(O)-$
phospho	O_2P-
phosphono	$(HO)_2P(O)-$
phthalyl (from phthalic acid)	$1,2-C_6H_4(CO-)_2$
picryl (2,4,6-trinitrophenyl)	$2,4,6-(NO_2)_3C_6H_2-$
pimelyl (from pimelic acid)	$-OC(CH_2)_5CO-$
piperidino	$C_5H_{10}N-$
piperidyl (piperidinyl)	$(C_5H_{10}N)-$
piperonyl	$3,4-(CH_2O_2)C_6H_3CH_2-$
pivalyl (from pivalic acid)	$(CH_3)_3CCO-$
prenyl (3-methyl-2-butenyl)	$(CH_3)_2C{=}CHCH_2-$
propargyl (2-propynyl)	$HC{\equiv}CCH_2-$
propenyl	$CH_2{=}CHCH_2-$
iso-propenyl	$(CH_3)_2C{=}$
propionyl	CH_3CH_2CO-
propoxy	$CH_3CH_2CH_2O-$
propyl	$CH_3CH_2CH_2-$
iso-propyl	$(CH_3)_2CH-$
propylidene	$CH_3CH_2CH{=}$
pyridino	C_5H_5N-
pyridyl (pyridinyl)	$(C_5H_4N)-$
pyrryl (pyrrolyl)	$(C_4H_4N)-$
salicyl (2-hydroxybenzoyl)	$2-HOC_6H_4CO-$
selenyl	HSe−
seryl (from serine)	$HOCH_2CH(NH_2)CO-$
siloxy	H_3SiO-
silyl	H_3Si-
silylene	$H_2Si{=}$
sorbyl (from sorbic acid)	$CH_3CH{=}CHCH{=}CHCO-$
stearyl (from stearic acid)	$CH_3(CH_2)_{16}CO-$
styryl	$C_6H_5CH{=}CH-$
suberyl (from suberic acid)	$-OC(CH_2)_6CO-$
succinamyl	$H_2NCOCH_2CH_2CO-$
succinyl (from succinic acid)	$-OCCH_2CH_2CO-$
sulfamino	$HOSO_2NH-$
sulfamyl	H_2NSO-
sulfanilyl	$4-H_2NC_6H_4SO_2-$
sulfeno	HOS−
sulfhydryl (mercapto)	HS−
sulfinyl	OS=
sulfo	HO_3S-
sulfonyl	$-SO_2-$
terephthalyl	$1,4-C_6H_4(CO-)_2$
tetramethylene	$-(CH_2)_4-$
thenyl	$(C_4H_3S)CH-$
thienyl	$(C_4H_3S)-$
thiobenzoyl	C_6H_5CS-
thiocarbamyl	H_2NCS-
thiocarbonyl	−CS−
thiocarboxy	HOSC−
thiocyanato	NCS−
thionyl (sulfinyl)	−SO−
thiophenacyl	$C_6H_5CSCH_2-$
thiuram (aminothioxomethyl)	H_2NCS-
threonyl (from threonine)	$CH_3CH(OH)CH(NH_2)CO-$
toluidino	$CH_3C_6H_4NH-$
toluyl	$CH_3C_6H_4CO-$
tolyl (methylphenyl)	$CH_3C_6H_4-$
α-tolyl	$C_6H_5CH_2-$
tolylene (methylphenylene)	$(CH_3C_6H_3){=}$
α-tolylene	$C_6H_5CH{=}$
tosyl [(4-methylphenyl) sulfonyl)]	$4-CH_3C_6H_4SO_2-$
triazano	$H_2NNHNH-$
trimethylene	$-(CH_2)_3-$
triphenylmethyl (trityl)	$(C_6H_5)_3C-$
tyrosyl (from tyrosine)	$4-HOC_6H_4CH_2CH(NH_2)CO-$
ureido	$H_2NCONH-$
valeryl (from valeric acid)	C_4H_9CO
valyl (from valine)	$(CH_3)_2CHCH(NH_2)CO-$
vinyl	$CH_2{=}CH-$
vinylidene	$CH_2{=}C{=}$
xenyl (biphenylyl)	$C_6H_5C_6H_4-$
xylidino	$(CH_3)_2C_6H_3NH-$
xylyl (dimethylphenyl)	$(CH_3)_2C_6H_3-$
xylylene	$-CH_2C_6H_4CH_2-$

ORGANIC RING COMPOUNDS

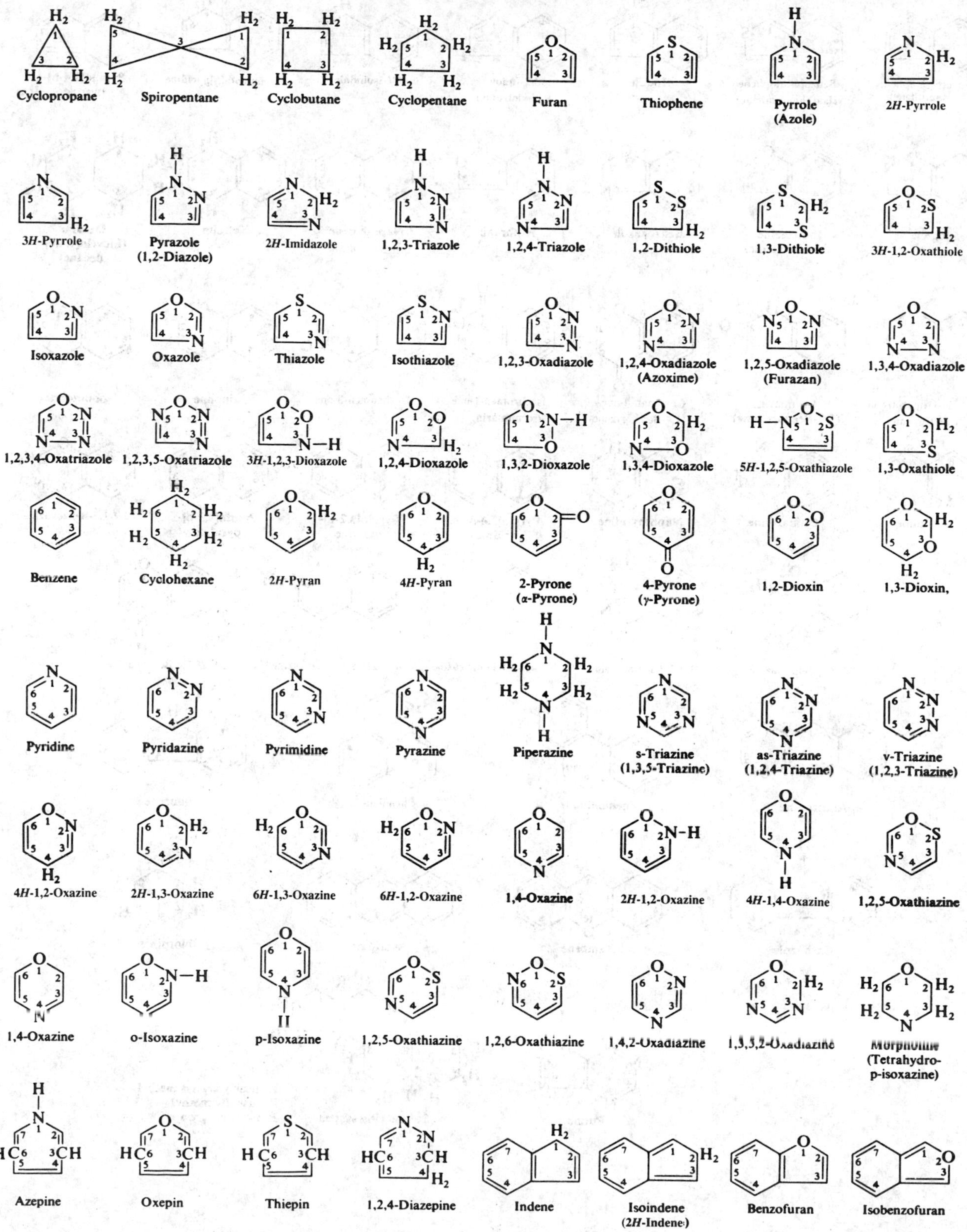
Cyclopropane
Spiropentane
Cyclobutane
Cyclopentane
Furan
Thiophene
Pyrrole
(Azole)
2H-Pyrrole
3H-Pyrrole
Pyrazole
(1,2-Diazole)
2H-Imidazole
1,2,3-Triazole
1,2,4-Triazole
1,2-Dithiole
1,3-Dithiole
3H-1,2-Oxathiole
Isoxazole
Oxazole
Thiazole
Isothiazole
1,2,3-Oxadiazole
1,2,4-Oxadiazole
(Azoxime)
1,2,5-Oxadiazole
(Furazan)
1,3,4-Oxadiazole
1,2,3,4-Oxatriazole
1,2,3,5-Oxatriazole
3H-1,2,3-Dioxazole
1,2,4-Dioxazole
1,3,2-Dioxazole
1,3,4-Dioxazole
5H-1,2,5-Oxathiazole
1,3-Oxathiole
Benzene
Cyclohexane
2H-Pyran
4H-Pyran
2-Pyrone
(α-Pyrone)
4-Pyrone
(γ-Pyrone)
1,2-Dioxin
1,3-Dioxin,
Pyridine
Pyridazine
Pyrimidine
Pyrazine
Piperazine
s-Triazine
(1,3,5-Triazine)
as-Triazine
(1,2,4-Triazine)
v-Triazine
(1,2,3-Triazine)
4H-1,2-Oxazine
2H-1,3-Oxazine
6H-1,3-Oxazine
6H-1,2-Oxazine
1,4-Oxazine
2H-1,2-Oxazine
4H-1,4-Oxazine
1,2,5-Oxathiazine
1,4-Oxazine
o-Isoxazine
p-Isoxazine
1,2,5-Oxathiazine
1,2,6-Oxathiazine
(Tetrahydro-
p-isoxazine)
Azepine
Oxepin
Thiepin
1,2,4-Diazepine
Indene
Isoindene
(2H-Indene)
Benzofuran
Isobenzofuran

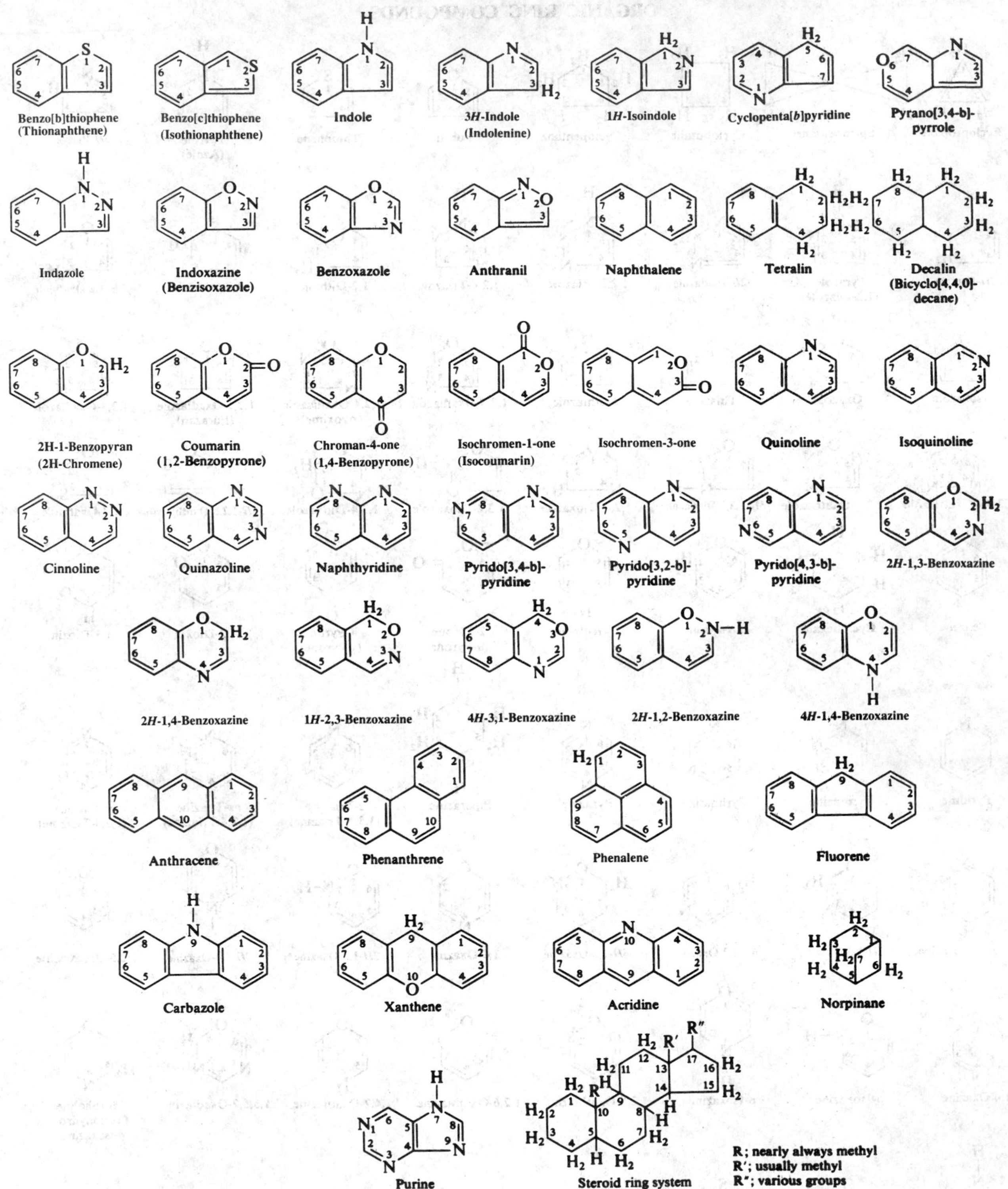
Benzo[b]thiophene (Thionaphthene)
Benzo[c]thiophene (Isothionaphthene)
Indole
3H-Indole (Indolenine)
1H-Isoindole
Cyclopenta[b]pyridine
Pyrano[3,4-b]-pyrrole
Indazole
Indoxazine (Benzisoxazole)
Benzoxazole
Anthranil
Naphthalene
Tetralin
Decalin (Bicyclo[4,4,0]-decane)
2H-1-Benzopyran (2H-Chromene)
Coumarin (1,2-Benzopyrone)
Chroman-4-one (1,4-Benzopyrone)
Isochromen-1-one (Isocoumarin)
Isochromen-3-one
Quinoline
Isoquinoline
Cinnoline
Quinazoline
Naphthyridine
Pyrido[3,4-b]-pyridine
Pyrido[3,2-b]-pyridine
Pyrido[4,3-b]-pyridine
2H-1,3-Benzoxazine
2H-1,4-Benzoxazine
1H-2,3-Benzoxazine
4H-3,1-Benzoxazine
2H-1,2-Benzoxazine
4H-1,4-Benzoxazine
Anthracene
Phenanthrene
Phenalene
Fluorene
Carbazole
Xanthene
Acridine
Norpinane
Purine
Steroid ring system
R; nearly always methyl
R′; usually methyl
R″; various groups

BIOCHEMICAL SYMBOLS AND ABBREVIATIONS

Prepared by the Office of Biochemical Nomenclature (OBN) of the NAS-NRC (Waldo E. Cohn, Director). Recent references on this topic are given at the end.

I. GENERAL

The symbols and abbreviations discussed here fall into two distinct classes.

A. *Symbols* for the monomeric units in macromolecules, to make up abbreviated structural formulas, *e.g.* Gly-Val-Thr for the tripeptide glycylvalylthreonine. These are fairly systematic. (See Sections II, III and V.) Symbols are not used for the monomers themselves, except in tables, lists, and figures.

B. *Abbreviations* for semisystematic or trivial names, *e.g.* ATP for adenosine triphosphate; FAD for flavin-adenine dinucleotide (see Sections IV, VI).

Structural analogues of a given compound are *not* abbreviated as derivatives of that compound.

II. POLYPEPTIDES AND PROTEINS

The symbols are derived from the trivial names or chemical names of the amino acids and of chemicals reacting with amino acids and polypeptides. For the sake of clarity, brevity, and listing in tables, the symbols have been, wherever possible, restricted to three letters, usually the first letters of the trivial names.

The symbols represent not only the names of the compounds but also their structural formulas.

Heteroatoms of amino acid residues (*e.g.* O^{β} and S^{β} of serine and cysteine, respectively, N^{ϵ} of lysine, N^{α} of glycine, etc.) do not explicitly appear in the symbol; such features are understood to be encompassed by the abbreviation.

Structural formulas of complicated features are used along with the abbreviated notation wherever necessary for clarity.

A. SYMBOLS FOR AMINO ACIDS

1. *Common Amino Acids.*

Alanine	Ala	Glycine	Gly	Phenylalanine	Phe
Arginine	Arg	Histidine	His	Proline	Pro
Asparagine*	Asn*	Isoleucine	Ile	Serine	Ser
Aspartic acid	Asp	Leucine	Leu	Threonine	Thr
Cysteine	Cys	Lysine	Lys	Tryptophan	Trp
Glutamic acid	Glu	Methionine	Met	Tyrosine	Tyr
Glutamine*	Gln*	Ornithine	Orn	Valine	Val

2. *Less Common Amino Acids.* The following principles and notations are used.

Hydroxyamino Acids.

Hydroxylysine	Hyl
3-Hydroxyproline	3Hyp
4-Hydroxyproline	4Hyp

allo-Amino Acids.

allo-Isoleucine	*a*Ile
allo-Hydroxylysine	*a*Hyl

"Nor" Amino Acids—"Nor" (*e.g. in norvaline*) is treated as part of the trivial name without special emphasis.

Norvaline	Nva
Norleucine	Nle

Higher Unbranched Amino Acids. The trivial name of the parent acid is abbreviated to leave no more than two or three letters, as convenient and necessary for clarity. The word "acid" ("*säure*" etc.) is omitted from the symbol as carrying no significant information. Unless otherwise indicated (see following paragraph), single groups are in the α position, two amino groups in the α,ω (monocarboxylic acids) or α, α′ positions (dicarboxylic acids). The location of amino acids in positions other than α and ω is shown by the appropriate Greek letter prefix.

* Asparagine and glutamine may also be denoted as Asp(NH_2) or Asp (with NH_2 attached below) and Glu(NH_2) or Glu (with NH_2 attached below), respectively.

Uncertainty as between Glu and Gln; and Asp and Asn, may be designated by Glx and Asx, respectively. (Added in 1967.)

Examples:

α-Aminobutyric acid	Abu	α, α'-Diaminopimelic	Dpm
α-Aminoadipic acid	Aad	β-Alanine	βAla
α-Aminopimelic acid	Apm	$\sigma\varepsilon$-Aminocaproic acid	εAcp
α, γ-Diaminobutyric acid	Dbu	β-Aminoadipic acid	βAad
α, β-Diaminopropionic acid	Dpr		

3. *N^{α}-Alkylated Amino Acids.*

Examples:

N-Methylglycine (sarcosine)	MeGly or Sar	*N*-Methylvaline, etc.	MeVal, etc.
N-Methylisoleucine	MeIle	*N*-Ethylglycine, etc.	EtGly, etc.

B. SEQUENCE AND DIRECTION

Where the sequence of residues in a peptide or protein is known, the symbols for the residues are written in order and joined by *hyphens*. Where the sequence is *not* known, the group of symbols, separated by *commas*, is enclosed in parentheses.

In the formulation of linear polypeptides or proteins, the symbol written at the left-hand end of a known sequence is that of the amino acid carrying the free amino group, and the symbol written at the right-hand end is that of the residue of the amino acid carrying the free carboxyl group.

Example: The condensed formula

Gly-Glu-Arg-Gly-Phe-(Phe, Tyr, Thr, Pro)-Lys-Ala-Trp-Tyr-Val-Ile-Hyp-Cys-Ala

is that of a polypeptide in which the sequence of the first five amino acids has been established, the glycine at the left carrying the free amino group. The sequence of the next four amino acids is unknown, but the last nine amino acids are in known order with alanine carrying the free carboxyl group.

The simple usage by which Gly-Gly-Gly stands for glycylglycylglycine appears to involve the employment of the *same* three letters Gly for *three* different residues or radicals—(b), (c), (d) below. However, if the hyphens are considered as part of each symbol, we have four distinct forms, for the free amino acid and the three residues, *viz.*:

(a) Gly = $NH_2 \cdot CH_2 \cdot CO_2H$; the free amino acid
(b) Gly- = $NH_2 \cdot CH_2 \cdot CO$—; the left hand unit
(c) -Gly- = --$NH \cdot CH_2 \cdot CO$—; the middle unit
(d) -Gly = --$NH \cdot CH_2 \cdot CO_2H$; the right hand unit

For peptides, a distinction may be made between the *peptide* itself, *e.g.* Gly-Glu (shown *without* hyphens at the ends of the symbols), and the *sequence*, *e.g.* -Gly-Glu- (shown *with* hyphens at the ends of the symbols).

If the direction of the link is to be specified, this may be done with an arrow thus (⟶), the point of the arrow indicating the nitrogen of the peptide bond . . . CO⟶NH . . .

Example: Gly⟶Ala⟶Val.

Unless otherwise indicated, polyfunctional amino acids, such as glutamic acid, aspartic acid, and lysine, are joined by normal α-peptide bonds.

Abnormal links, *e.g.* γ-peptide bonds or links formed through other functional groups, are indicated by the methods described below (Section IIC):

The amino acid symbols represent the natural (L) form. Other forms are indicated by the appropriate symbols (D or DL) immediately preceding the amino acid symbol.

Example: Leu-DPhe-Gly.

To show that a peptide is acting as a cation or anion, the amino-terminal and carboxyl-terminal ends of the peptide are marked with H and OH, respectively (I); these may be modified to show the appropriate state of ionization (II or III).

H-Gly-Val-Thr-OH (I)
$^{+}H_2$-Gly-Val-Thr-OH (II)
H-Gly-Val-Thr-O^{-} (III)

C. SUBSTITUTION

1. *Substitution in the α-Amino and α-Carboxyl Groups.* Examples:

N-Acetylglycine	Ac-Gly
Glycine ethyl ester	Gly-OEt
N^{α}-Acetyllysine	Ac-Lys
Serine methyl ester	Ser-OMe
α-Ethyl-*N*-acetylglutamate	Ac-Glu-OEt
Isoglutamine	$Glu\text{-}NH_2$

N-Ethyl-*N*-methylglycine EtMeGly, Et\Gly/Me (Et and Me on Gly), Et–Gly with Me above

2. *Substitution in the Side Chain.* Side chain substituents are portrayed above or below the amino acid symbol, or by placing the symbol for the substituent in parentheses immediately after the amino acid symbol. The use of parentheses is reserved for a *single* symbol denoting a side chain substituent. Where a more complex substituent is involved, the vertical stroke and a two-line abbreviation are used.

Aspartic acid β-methyl ester	Asp (OMe above) or Asp (OMe below) or Asp(OMe)
O-Methyltyrosine	Tyr (Me above) or Tyr (Me below) or Tyr(Me)
N^{ε}-Acetyllysine	Lys (Ac above) or Lys (Ac below) or Lys(Ac)
S-Ethylcysteine	Cys (Et above) or Cys (Et below) or Cys(Et)
O-Acetylserine	Ser (Ac above) or Ser (Ac below) or Ser(Ac)

Where the substituent may be attached to more than one position (as in histidine), the appropriate numeral should be beside the vertical bond.

D. POLYPEPTIDES

1. *Polypeptide Chains.* Polypeptides are dealt with in the same manner as substituted amino acids, *e.g.*

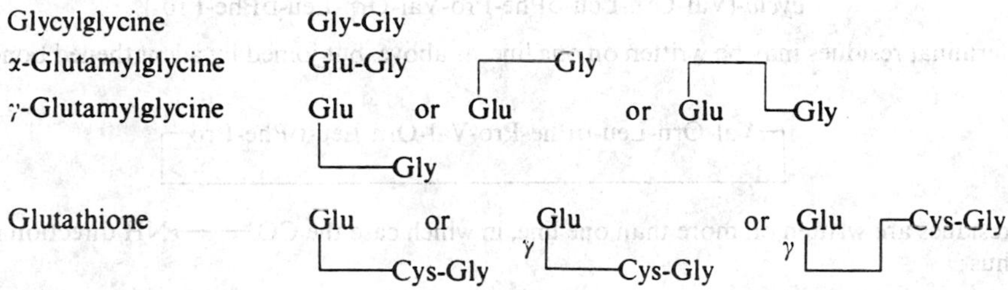

(Note that Glu (with Cys-Gly below) or Glu ⌐ Cys-Gly represents the corresponding thiolester with a bond between the γ-carboxyl of glutamic acid and the thiol group of cysteine.)

N^{ε}-α-Glutamyllysine	Glu (bond to Lys below) or Glu (bond to Lys above)
N^{ε}-γ-Glutamyllysine	Glu (Lys below) or Glu Lys (bond below) or Glu Lys (bond above)

The presence of free, substituted, or ionized functional groups can be represented (or stressed) as follows:

Glycyllysylglycine	H—Gly-Lys(H)-Gly—OH
Its dihydrochloride	$^+H_2$—Gly-Lys($^+H_2$)-Gly—OH·$2Cl^-$
Its sodium salt	Gly-Lys-Gly-O^-Na^+
Its N^ε-formylderivative	Gly-Lys(CHO)-Gly (CHO attached below Lys) or Gly-Lys(CHO)-Gly

2. *Peptides Substituted at N^α.*

Examples:

Glycylnitrosoglycine	Gly—Gly with NO attached below the bond
Glycylsarcosine	Gly—Gly with Me attached below the bond, or Gly-MeGly or Gly-Sar
N^α-Glycyl-N^α-acetylglycine	Gly—Gly with Ac attached above the bond
N^α-Glycyl-N^α-glycylglycine	Gly—Gly with a second Gly— attached above the bond

Etc.

E. CYCLIC POLYPEPTIDES

1. *Homodetic* cyclic polypeptides (the ring consists of amino acid residues in peptide linkage only). Three representations are possible:

a. The sequence is formulated in the usual manner but placed in parentheses and preceded by (an italic) *cyclo*.

Example: Gramicidin S =

cyclo-(Val-Orn-Leu-DPhe-Pro-Val-Orn-Leu-DPhe-Pro-)

b. The terminal residues may be written on one line, as above, but joined by a lengthened bond. Using the same example:

—Val-Orn-Leu-DPhe-Pro-Val-Orn-Leu-DPhe-Pro— (first and last residues joined by a lengthened bond below the line)

c. The residues are written on more than one line, in which case the CO ⟶ NH direction must be indicated by arrows, thus:

→Val → Orn → Leu → DPhe → Pro ┐
└ Pro ← DPhe ← Leu ← Orn ← Val ←

2. *Heterodetic* cyclic polypeptides (the ring consists of other residues in addition to amino acid residues in peptide linkage): These follow logically from the formulation of substituted amino acids.

Example:

Oxytocin Cys-Tyr-Ile-Asn-Gln-Cys-Pro-Leu-Gly-NH_2 (the two Cys residues joined by a bond below the line)

Cyclic ester of threonylglycylglycylglycine

Thr-Gly-Gly-Gly or H—Thr-Gly-Gly-Gly

F. SUBSTITUENTS

Groups substituted for hydrogen or for hydroxyl may be indicated either by their structural formulas or by accepted abbreviations, *e.g.*

Benzoylglycine (hippuric acid) —CO—Gly or C_6H_5—CO—Gly or Bz-Gly

Glycine methyl ester Gly—OCH_3 or Gly-OMe

1. *N-Protecting Groups of the Urethane Type.*

Benzyloxycarbonyl	Z-	*p*-Methoxyphenylazobenzyloxycarbonyl	Mz-
p-Nitrobenzyloxycarbonyl	Z(NO_2)-	*p*-Phenylazobenzyloxycarbonyl	Pz-
p-Bromobenzyloxycarbonyl	Z(Br)-	*t*-Butyloxycarbonyl	Boc-
p-Methoxybenzyloxycarbonyl	Z(OMe)-	Cyclopentyloxycarbonyl	Poc-

2. *Other N-Protecting Groups.*

Acetyl	Ac-	Phthalyl	Pht-	Dinitrophenyl	Dnp-
Benzoyl	Bz-	Benzyl	Bzl-	Benzylthiomethyl	Btm-
Tosyl	Tos-	Trityl	Trt-	*o*-Nitrophenylsulfenyl	Nps-
Trifluoroacetyl	Tfa-	Tetrahydropyranyl	Thp-	Dansyl	Dan-
				Phenylthiohydantoyl	Pth-

3. *Carboxyl-protecting Groups.*

Methoxy (methyl ester)	-OMe	*p*-Nitrophenoxy (*p*-nitrophenyl ester)	-ONp
Ethoxy (ethyl ester)	-OEt	Phenylthio (phenylthiolester)	-SPh
Tertiary butoxy (*tert*-butyl ester)	-OBu^t	*p*-Nitrophenylthio	-SNp
Benzyloxy (benzyl ester)	-OBzl	Cyanomethoxy	-OCH_2CN
Diphenylmethoxy (benzhydryl ester)	-OBzh		

Note: Contrary to the symbols for amino acid residues, the position of the dashes in the symbols for substituents carries no significant information.

III. CARBOHYDRATES

A. SYMBOLS

The following symbols are used to indicate monosaccharide units and their residues in oligosaccharides and polysaccharides.

Glucose	Glc	Fructose	Fru
Galactose	Gal	Ribose	Rib
Mannose	Man	etc.	

Other monosaccharides are represented similarly by the first three letters of their names, unless this leads to confusion with an existing symbol (*e.g.* Gly and Thr in the amino acid series).

A 2-deoxy sugar is designated by the symbol for its most common parent sugar with the prefix "d." Other deoxy sugars may be designated similarly with a positional numeral. Examples: 2-deoxyribose, dRib; 3-deoxyglucose, 3-dGlc.

B. PREFIXES AND SUFFIXES

Pyranose and furanose forms are designated where necessary by the suffixes *p* and *f*.

Configurational symbols D and L (small Roman capital letters) and anomeric prefixes are shown where necessary as prefixes. Examples: (i) an α-D-glucopyranose unit, αDGlc*p* *or* Glc*p*; (ii) α β-D-fructofuranose unit, βDFru*f* *or* Fru*f*.

C. SEQUENCE AND DIRECTION

Symbols thus formed are joined by short rules to indicate the links between units. The position and nature of the links are shown by numerals and the anomeric symbols α and β. Examples:

Maltose, αGlc*p*(1-4)Glc
Lactose, βGal*p*(1-4)Glc
Stachyose, αGal*p*(1-6)αGal*p*(1-6)αGal*p*(1-2)βFru*f*

A branched chain tetrasaccharide:

```
βGlc(1-3)βGal(1-4)Glc
               (2
                |
               1)αFuc
```

Arrows may be used to indicate the direction of the glycoside link, the arrow pointing away from the hemiacetal carbon of the link; *e.g.* lactose may be represented as βGal*p*(1 →4)Glc.

D. DERIVATIVES

Glyconic acids, glycuronic acids, 2-amino-2-deoxysaccharides, and their *N*-acetyl derivatives are designated by modified symbols. Examples, all in the glucose, Glc, series, are:

Gluconic acid	GlcA	Glucosamine	GlcN
Glucuronic acid	GlcUA	*N*-Acetylglucosamine	GlcNAc

IV. PHOSPHORYLATED COMPOUNDS: GENERAL

Phosphorylated compounds are designated by the name (or abbreviation) of the parent compound with a capital P as a prefix or suffix.

P is used as prefix where it symbolizes "phospho-" at the beginning of a name. P is used as a suffix where it symbolizes "phosphoric acid" or "phosphate" at the end of a name.

For compounds containing more than one position available for phosphorylation, the position of the phosphate group is indicated by number or Greek letter.

The capital P when linked to one radical indicates $-PO(OH)_2$ or any ion derived from it; when linked to two radicals it indicates —PO(OH)—, or the ion derived from it.

The pyrophosphate group (I) is represented by -P-P:

```
   O     O
   ↑     ↑
 —P—O—P—          (I)
   |     |
   OH    OH
```

Two separate phosphate groups, attached at different points to the same molecule, are represented by P_2.

Glucose 6-phosphate	Glucose-6-P or Glc-6-P
{ Glycerol 3-phosphate or α-phosphoglycerol	{ Glycerol-3-P / α-P-Glycerol
{ 3-Phosphoglyceric acid / Glycerate 3-phosphate	{ 3-P-Glyceric acid / Glycerate-3-P
Phosphoenolpyruvate	P-Enolpyruvate
Fructose 1,6-bisphosphate	Fructose-1,6-P_2 or Fru-1,6-P_2
{ Creatine phosphate / Phosphocreatine	{ Creatine-P / P-Creatine

V. NUCLEOTIDES AND NUCLEIC ACIDS

Two systems are recognized, one using three-letter symbols for the more common nucleosides (like those used for amino acids and monosaccharides in Sections II and III and a capital P for phosphate (Section IV), the other using single capital letters for the more common nucleosides and a lower-case p for the phosphate residue.

In either system, glycosyl linkages are assumed to be β and to involve only D-ribose or 2-deoxyribose, and the phosphodiester linkage is assumed to be 3′-5′ from left to right unless otherwise specified by appropriate *ad hoc* symbols or numerals.

BIOCHEMICAL SYMBOLS AND ABBREVIATIONS (continued)

A. THREE-LETTER SYMBOLS

1. The phosphate group is designated by a capital P.

2. The (ribo) nucleosides are designated by the following three-letter symbols, chosen to avoid confusion with the corresponding bases:

Ado	adenosine	Ino	inosine	Thd	ribosylthymine*	Urd	Uridine
Guo	guanosine	Xao	xanthosine	Cyd	cytidine	Ψrd	pseudouridine (5-ribosyluracil)

Ribosylnicotinamide may be designated by Nir.

3. The 2′-deoxyribonucleosides are designated by the symbols for the corresponding ribose derivatives with the prefix d. Examples are dAdo for 2′-deoxyadenosine, dThd for 2′-(deoxy)thymidine (2′-deoxyribosylthymine).

The letter d may also be used as a prefix to a series (an oligonucleotide) to indicate that *all* the sugars in the series are 2′-deoxyribosyl units.

4. The points of attachment of phosphate residues to a sugar, if other than 3′-P-5′, are designated by the appropriate primed numerals, separated by hyphens.

Examples:

(i) Adenosine 2′-phosphate: Ado-2′-P
Adenosine 5′-phosphate: Ado-5′-P or P-5′-Ado

(ii) 5′-*O*-Phosphoryldeoxyadenylyl-(3′-5′)-thymidine: P-5′-dAdo-3′-P-5′-dThd or P-dAdo-P-dThd.

The positional numerals may precede a series, as in 2′-5′-(Ado-P-Guo-P-Urd-P), which specifies Ado-2′-P-5′-Guo-2′-P-5′-Urd-2′-P. When the series in the left-to-right direction is 3′-5′-, as in example ii above, they may be omitted.

5. A cyclic phosphate group is designated by the two positional numerals for the points of attachment of a single P, as in Cyd-2′:3′-P. (The corresponding bisphosphate would be Cyd-2′,3′-P_2, or P-2′-Cyd-3′-P).

6. The so-called nucleoside diphosphate sugars, which are pyrophosphates, may be represented as follows: Urd-5′-P-P-Glc for uridine diphosphate glucose [*i.e.* uridine 5′-(α-D-glucopyranosyl diphosphate)]: Urd-5′-P-P-Gal.

B. ONE-LETTER SYMBOLS

The phosphate group is designated by a lower case p (to separate what would otherwise be a solid mass of capital letters).

The common (ribo)nucleosides are designated by single capital letters, thus:

A	adenosine	I	inosine	T	ribosylthymine*	U	uridine
G	guanosine	X	xanthosine	C	cytidine	Ψ	pseudouridine (5-ribosyluracil)

The following general symbols are also used: Pu, unspecified purine nucleoside; Py, unspecified pyrimidine nucleoside; N, unspecified nucleoside (not X). (R and Y may be used for Pu and Px, respectively.)

All other nucleosides should be represented by single capital letters, insofar as possible, defined as introduced (*e.g.* B for BrU). Where a nucleoside symbol must include more than one character, it should contain neither hyphens nor commas (*e.g.* -2MeG-, -6diMeA-, -BrU-).

The 2′-deoxyribonucleosides are designated by the same symbols preceded by d. Thus, dA = 2′-deoxyribosyladenine, dT = 2′-deoxyribosylthymine (= thymidine). (The use of r before ribonucleosides is optional; see below).

The points of attachment of phosphate residues, if other than 3′p5′, may be indicated as in A.4. A regular 3′-5′ sequence (read left to right), as in the natural nucleic acids, need not be specified by positional numerals.

Linkages other than 3′-5′, or sugars other than ribose or deoxyribose, should be indicated by special *ad hoc* symbols. d and r may precede whole chains, groups within chains, or individual nucleoside residues, as appropriate.

In this system, the substances in A.4 become: A2′p; pA; d(pApT).

Other examples are:

ApGpUp (3′-5′ trinucleotide, ending at right in a 3′-phosphate).
ApGpU-cyclic p (the same, but ending in a 2′ : 3′-cyclic phosphate).

The symbol > for "cyclic" is useful in the one-letter system. Thus this example can be represented as ApGpU > p. Unless otherwise stated, this symbol indicates a 2′ : 3′-cyclic phosphate residue.

pApApA (3′-5′ trinucleotide, starting with a 5′-phosphate at left, ending with unsubstituted 2′ and 3′ hydroxyl groups at right).

C. SEQUENCE

For more complex structures (large oligonucleotides or polynucleotides) in which known and unknown sequences may be involved, the p for phosphate between two nucleosides may be replaced by a *hyphen* (for a *known* sequence) or a *comma* (*unknown* sequence) to give a system identical with that used for amino acid sequences (see Section II.B).

* Not thymidine (deoxyribosylthymine). For this use dThd or T.

Regular 3′-5′ linking is assumed unless indicated otherwise. Thus GpApUp(CpCpUp)Gp—a 3′-ended 3′-5′ heptanucleotide of partially known sequence—becomes G-A-U(C,C,U)Gp or G-A-U(C_2,U)Gp; d-pTpTpCpTpTpC becomes d(pT-T-C-T-T-C).

D. HIGHER POLYMERS

For sequences too long or repetitive or obscure for detailed exposition, the prefix "poly" may be used in conjunction with the comma and hyphen. For example, the alternating regular sequence dA-dT-dA-dT... is poly d(A-T) or poly (dA-dT); the random copolymer of equal amounts of U and A is poly (U,A). The prefix "poly" may be omitted when the number (or proportions) of nucleoside residues is specified by subscript numerals, if known, or by the subscript *n* for molecules of indefinite size. Thus, poly d(A-T) may be expressed as $d(A\text{-}T)_n$ or $(dA\text{-}dT)_n$, and poly (U_2,A) as $(U_2,A)_n$. Composition and size are shown by appropriate numerical subscripts, as in $(U_2,A)_{50}$, which contains 100 U's and 50 A's in random sequence, and in $d(A\text{-}T)_{50}$, which contains 50 dA's and 50 dT's in regular alternating sequence. Multiple parentheses or brackets are used as in organic nomenclature for blocks within polymers, side chains, etc.

E. ASSOCIATION BETWEEN CHAINS

Associations between two or more nucleotide chains may be indicated by the *center dot*—as in $(A)_n \cdot (U)_n$ or $(dG)_n \cdot (dC)_n$ or $[(A)_n \cdot (U)_n \cdot (U)_n]$. The *absence of association* is indicated by the *plus* sign, as in poly (A) + poly (C). The *absence of definite information* on association is indicated by the comma (again meaning "unknown"), as in poly (A), poly (A-U) or $(A)_n$, $(A\text{-}U)_n$.

F. NUCLEIC ACIDS

1. The two types of nucleic acid are designated by their customary abbreviations:

RNA, ribonucleic acid or ribonucleate
DNA, deoxyribonucleic acid or deoxyribonucleate

2. Fractions or functions of RNAs are designated by prefixes: mRNA for "messenger" RNA, tRNA for "transfer" RNA, rRNA for "ribosomal" RNA, nRNA for "nuclear" RNA. (sRNA is specifically omitted and should not be used without definition. It should not be used to designate transfer RNA).

3. Transfer RNAs that accept specific amino acids are designated as, for example, $tRNA^{Ala}$ for the tRNA that accepts alanine (*i.e.* "alanine tRNA"); in the case of more than one such species, they are distinguished by subscripts, as $tRNA_1^{Ala}$, $tRNA_2^{Ala}$, etc. When a tRNA is bound to an amino acid, it is designated as, for example, alanyl-tRNA or alanyl-$tRNA^{Ala}$. Specification of its source is in parentheses before or after, as for example, alanyl-$tRNA_2^{Ala}$ (*E. coli*) or (*E. coli*) alanyl-$tRNA_2^{Ala}$.

VI. COENZYMES AND SPECIAL MONONUCLEOTIDES

a. The 5′-mono-, di-, and triphosphates of the common nucleosides are designated by the customary special abbreviations, *e.g.* AMP, ADP, and ATP for the derivatives of adenosine. The corresponding derivatives of cytidine, guanosine, inosine, xanthosine, uridine, and pseudouridine are designated by similar abbreviations in which the initial letters are C, G, I, X, U, and Ψ, respectively. As above, N means unknown (or "any") nucleoside. Thus, for example, IMP = inosine 5′-monophosphate; UDP = uridine 5′-diphosphate. Uridine diphosphate glucose may be designated by UDPG. NTP indicates an unspecified nucleoside diphosphate, or a mixture of them.

These compounds may also be designated by systematic symbols as indicated in paragraph V.A.4, *e.g.* Ado-5′-*P*, Ado-5′-*P*-*P*, Ado-5′-*P*-*P*-*P*, when required for consistency with the other nucleotides.

b. The dinucleotide coenzymes are designated by the following abbreviations:

Nicotinamide-adenine dinucleotide (formerly DPN, CoI)	NAD
Nicotinamide-adenine dinucleotide phosphate (formerly TPN, CoII)	NADP

The oxidized and reduced forms are designated by NAD^+ ($NADP^+$) and NADH (NADPH), respectively. They may be used in an equation as follows:

$$NAD^+ + XH_2 \rightleftarrows NADH + H^+ + X$$

FAD, $FADH_2$	Flavin-adenine dinucleotide, and its reduced form
FMN, $FMNH_2$	Flavin mononucleotide, and its reduced form
GSH, GSSG	Glutathione, and its oxidized form

CoA, acetyl-CoA *or* CoASH, CoASAc — Coenzyme A and its acetyl derivative (alternative forms)

c. Systematic symbols may be built up for some of these coenzymes as shown in paragraphs V.A.4 and V.A.6. *Examples:*

NAD = Nir-5′-*P*-*P*-5′-Ado
NADP = Nir-5′-*P*-*P*-5′-Ado-2′-*P*

VII. MISCELLANEOUS COMPOUNDS

ACTH	Adrenocorticotropin, adrenocorticotropic hormone, or corticotropin
BAL	2,3-dimercaptopropanol
CM-cellulose	*O*-(carboxymethyl)cellulose
DDT	1,1,1-trichloro-2,2-bis(*p*-chloro-phenyl)-ethane
DEAE-cellulose	*O*-(diethylaminoethyl)cellulose
DFP	di-isopropyl phosphorofluoridate
DNP-	2,4-dinitrophenyl-
DOC	11-deoxycorticosterone
DOCA	11-deoxycorticosterone acetate
DOPA	3,4-dihydroxyphenylalanine
DPT	diphosphothiamine (thiamine pyrophosphate, cocarboxylase)
EDTA	ethylenediaminetetraacetic acid (or -acetate)
FDNB	1-fluoro-2,4-dinitrobenzene
Hb	hemoglobin (deoxygenated)
HbCO	"carboxy" hemoglobin—*i.e.* hemoglobin *plus* carbon monoxide
HbO_2	oxyhemoglobin
MetHb	methemoglobin
Mb	deoxygenated myoglobin (may be modified in the same way as Hb)
MSH	melanocyte-stimulating hormone
P_i	orthophosphate (inorganic)
PP_i	pyrophosphate (inorganic)
TEAE-cellulose	*O*-(triethylaminoethyl)cellulose
Tris	tris(hydroxymethyl)aminomethane; 2-amino-2-hydroxymethylpropane-1,3-diol

VIII. ALPHABETICAL LISTS

For convenience the symbols and abbreviations are collected in the following two alphabetical lists.

Table I. Symbols for Common Monomeric Units in Macromolecules (or in Phosphorylated Compounds)

Symbol	Monomeric unit in macromolecule	Symbol	Monomeric unit in macromolecule
A, Ado	Adenosine	Hyl	Hydroxylysine
Ala	Alanine	Hyp	Hydroxyproline
Arg	Arginine	I, Ino	Inosine
Asp	Aspartic acid	Ile	Isoleucine
$Asp(NH_2)$, Asn	Asparagine	Leu	Leucine
C, Cyd	Cytidine	Lys	Lysine
Cys[a]	Cystine (half)	Man	Mannose
Cys	Cysteine	Met	Methionine
d	(Indicates "deoxy" in carbohydrates and nucleotides)	Nir	Ribosylnicotinamide
		Orn	Ornithine
f	(suffix) Furanose	P, p	Phosphate
Fru	Fructose	*p*	(suffix) Pyranose
Gal	Galactose	Phe	Phenylalanine
G, Glc	Glucose	Pro	Proline
G, Guo	Guanosine	Rib	Ribose
GlcA	Gluconic acid	Ser	Serine
GlcN	Glucosamine	Thr	Threonine
GlcNAc	*N*-Acetylglucosamine	Trp	Tryptophan
GlcUA	Glucuronic acid	T, Thd	Ribosylthymine[b]
Glu	Glutamic acid	Tyr	Tyrosine
$Glu(NH_2)$, Gln	Glutamine	U, Urd	Uridine
Gly	Glycine	Val	Valine
His	Histidine	Xao	Xanthosine

[a] With vertical bond above or below (see Section II).

[b] Not thymidine (deoxyribosylthymine). For this use dT or dThd.

Table II. Abbreviations for Semisystematic or Trivial Names

ACTH	Adrenocorticotropin, adrenocorticotropic hormone, or corticotropin	GSSG	oxidized glutathione
ADP	Adenosine 5′-diphosphate (pyro)	GTP	guanosine 5′-triphosphate (pyro)
AMP	Adenosine 5′-phosphate	Hb, HbCO, HbO_2	hemoglobin, carbon monoxide hemoglobin, oxyhemoglobin
ATP	Adenosine 5′-triphosphate (pyro)	IDP	inosine 5′-diphosphate (pyro)
BAL	2,3-Dimercaptopropanol	IMP	inosine 5′-phosphate
CDP	Cytidine 5′-diphosphate (pyro)	ITP	inosine 5′-triphosphate (pyro)
CM-cellulose	*O*-(carboxymethyl)cellulose	Mb, MbCO, MbO_2	myoglobin, carbon monoxide myoglobin, oxymyoglobin
CMP	Cytidine 5′-phosphate	MetHb, MetMb	methemoglobin, metmyoglobin
CoA (or CoASH)	Coenzyme A	MSH	melanocyte-stimulating hormone
CoASAc	Acetyl coenzyme A	NAD	nicotinamide-adenine dinucleotide (cozymase, Coenzyme I, diphosphopyridine nucleotide)
CTP	Cytidine 5′-triphosphate (pyro)	NADP	nicotinamide-adenine dinucleotide phosphate (Coenzyme II, triphosphopyridine nucleotide)
DEAE-cellulose	*O*-(diethylaminoethyl)cellulose	NMN	nicotinamide mononucleotide
DDT	1,1,1-trichloro-2,2-bis(*p*-chlorophenyl)-ethane	P_i	inorganic orthophosphate
DFP	di-isopropyl phosphorofluoridate	PP_i	inorganic pyrophosphate
DNA	deoxyribonucleic acid	RNA	ribonucleic acid
DNP-	2,4-dinitrophenyl-	TEAE-cellulose	*O*-(triethylaminoethyl)cellulose
DOPA	3,4-dihydroxyphenylalanine	TPN[b]	triphosphopyridine nucleotide
DPN[a]	diphosphopyridine nucleotide	Tris	tris(hydroxymethyl)aminomethane (2-amino-2-hydroxymethylpropane-1,3-diol)
DPT	diphosphothiamine (thiamine pyrophosphate, cocarboxylase)	UDP	uridine diphosphate (pyro)
EDTA	ethylenediaminetetraacetate	UDPG	uridine diphosphate glucose
FAD	flavin-adenine dinucleotide	UMP	uridine monophosphate
FDNB	1-fluoro-2,4-dinitrobenzene	UTP	uridine triphosphate (pyro)
FMN	riboflavin 5′-phosphate		
GDP	guanosine 5′-diphosphate (pyro)		
GMP	guanosine 5′-phosphate		
GSH	glutathione		

[a] Replaced by NAD (see Section VI).

[b] Replaced by NADP (see Section VI).

RECENT REFERENCES ON BIOCHEMICAL NOMENCLATURE

Biochemical Nomenclature and Related Documents. The Biochemical Society, London, 1978, 223 pp.

Enzyme Nomenclature 1984. Academic Press, Orlando, FL, 1984, 646 pp.
Supplement 1: Corrections and additions, *Eur. J. Biochem.*, 157(1), 1—26, 1986.
Supplement 2: Corrections and additions, *Eur. J. Biochem.*, 179, 489—533, 1989.
Supplement 3: Corrections and additions, *Eur. J. Biochem.*, 187(2), 263—281, 1990.

"Nomenclature of iron-sulfur proteins." Recommendations 1978. *Eur. J. Biochem.*, 93, 427—430, 1979; corrections, *Eur. J. Biochem.*, 95, 369, 1979 and 102, 315, 1979.

"Nomenclature for multienzymes." Recommendations 1989. *Eur. J. Biochem.*, 185(3), 485—486, 1989.

"Symbolism and terminology in enzyme kinetics." Recommendations 1981. *Eur. J. Biochem.*, 128(1), 281—291, 1982.

"Numbering of atoms in myo-inositol." Recommendations 1988. *Eur. J. Biochem.*, 180, 485—486, 1989.

"Nomenclature for incompletely specified bases in nucleic acid sequences." Recommendations 1984. *Eur. J. Biochem.*, 150(1), 1—5, 1985; corrections, *Eur. J. Biochem.*, 157, 1, 1986.

"Nomenclature of initiation, elongation and termination factors for translations in eukaryotes." Recommendations 1988. *Eur. J. Biochem.*, 186(1), 1—3, 1989.

"Abbreviations and symbols for the description of conformations of polynucleotide chains." Recommendation 1982. *Eur. J. Biochem.*, 131(1), 9—15, 1983.

"Nomenclature and symbolism for amino acids and peptides." Recommendations 1983. *Eur. J. Biochem.*, 138, 9—37, 1984; corrections, *Eur. J. Biochem.*, 152, 1, 1985.

"Nomenclature of glycoproteins, glycopeptides and peptidoglycans." Recommendations 1985. *Eur. J. Biochem.*, 159, 1—6, 1986; corrections, *Eur. J. Biochem.*, 185, 485—486, 1989.

"Conformational nomenclature for five and six-membered ring forms of monosaccharides and their derivatives." Recommendations 1980. *Eur. J. Biochem.*, 111, 295—298, 1980.

"Nomenclature of branched-chain monosaccharides." Recommendations 1980. *Eur. J. Biochem.*, 119(1), 5—8, 1981; corrections, *Eur. J. Biochem.*, 125(1), 1, 1982.

"Nomenclature of unsaturated monosaccharides." Recommendations 1980. *Eur. J. Biochem.*, 119(1), 1—3, 1981; corrections, *Eur. J. Biochem.*, 125(1), 1, 1982.

"Abbreviated terminology of oligosaccharide chains." Recommendations 1980. *J. Biol. Chem.*, 257(7), 3347—3351, 1982.

"Polysaccharide nomenclature." Recommendations 1980. *Eur. J. Biochem.*, 126(3), 439—441, 1982.

"Symbols for specifying the conformation of polysaccharide chains." Recommendations 1981. *Eur. J. Biochem.*, 131(1), 5—7, 1983.

"Nomenclature of tocopherols and related compounds." Recommendations 1981. *Eur. J. Biochem.*, 123(3), 473—475, 1982.

"Nomenclature of vitamin D." Recommendations 1981. *Eur. J. Biochem.*, 124(2), 223—227, 1982.

"Nomenclature of retinoids." Recommendations 1981. *Eur. J. Biochem.*, 129(1), 1—5, 1982.

"Nomenclature and symbols for folic acid and related compounds." Recommendations 1986. *Eur. J. Biochem.*, 168(2), 251—253, 1987.

"Prenol nomenclature." Recommendations 1986. *Eur. J. Biochem.*, 167(2), 181—184; 59(5), 683—689, 1987.

"Nomenclature of tetrapyrroles." Recommendations 1986. *Eur. J. Biochem.*, 178(2), 277—328, 1988.

"Nomenclature of steroids." Recommendations 1989. *Eur. J. Biochem.*, 186(3), 429—458, 1990.

"Eicosanoid nomenclature." *Prostaglandins*, 38(1), 125—133, 1989.

"List of symbols with units recommended for use in biotechnology." Provisional. *Pure Appl. Chem.*, 54(9), 1743—1749, 1982.

"Recommendations for the presentation of thermodynamic and related data in biology." Recommendations 1985. *Eur. J. Biochem.*, 153, 429—434, 1985.

SCIENTIFIC ABBREVIATIONS AND SYMBOLS

This table lists some symbols, abbreviations, and acronyms commonly encountered in the physical sciences. Entries in italic type are symbols for physical quantities; for more details see the table "Symbols and Terminology for Physical and Chemical Quantities" in this section. Additional information on units may be found in the table "International System of Units" in Section 1. Abbreviations are listed in alphabetical order without regard to case. Entries beginning with Greek letters fall at the end of the table.

Publication practices vary with regard to the use of capital letters or lower case for many abbreviations. An effort has been made to follow the most common practices in this table, but there are many exceptions. Likewise, policies on the use of periods in an abbreviation vary considerably. Periods are generally omitted in this table unless they are necessary for clarity. Periods should never appear in SI units.

a	absorption coefficient, acceleration, activity
a_0	Bohr radius
A	ampere
Å	ångstrom
A	absorbance, area, Helmholtz energy, mass number
AAS	atomic absorption spectroscopy
abs	absolute
ac	alternating current
ADP	adenosine diphosphate
ae	eon (10^9 years)
AES	atomic emission spectroscopy, Auger electron spectroscopy
AF	audio frequency
A_H	Hall coefficient
AI	artificial intelligence
Al	Alfen number
AM	amplitude modulation
AMP	adenosine monophosphate
amu	atomic mass unit (recommended symbol is u)
antilog	antilogarithm
AO	atomic orbital
APW	augmented plane wave
aq	aqueous
A_r	atomic weight (relative atomic mass)
ASCII	American National Standard Code for Information Interchange
atm	standard atmosphere
ATP	adenosine triphosphate
ATR	attenuated total internal reflection
at.wt.	atomic weight
AU	astronomical unit
av	average
avdp	avoirdupois
b	barn
B	magnetic flux density, second virial coefficient, susceptance
bar	bar
bbl	barrel
bcc	body centered cubic
BCS	Bardeen-Cooper-Schrieffer
Bé	Baumé
BeV	billion electronvolt
Bhn	Brinell hardness number
Bi	biot
BOD	biochemical oxygen demand
bp	boiling point
Bq	becquerel
Btu	British thermal unit
bu	bushel
c	amount concentration, specific heat, velocity
c_0	speed of light in vacuum
C	coulomb
°C	degree Celsius
C	capacitance, heat capacity, number concentration
cal	calorie
calc	calculated
cc	cubic centimeter
CCD	charge-coupled device
cd	candela
CD	circular dichroism
cfm	cubic feet per minute
cgs	centimeter-gram-second system
Ci	curie
CI	configuration interaction
cir	circular
cm	centimeter
c.m.	center of mass
CDNO	complete neglect of differential overlap
Co	Cowling number
conc	concentrated, concentration
const	constant
cos	cosine
cosh	hyperbolic cosine
cot	cotangent
coth	hyperbolic cotangent
cp	candle power
cP	centipoise
CP	chemically pure
cpd	contact potential difference
cps	cycles per second
CPU	central processing unit
csc	cosecant
ct	carat
cu	cubic
cw	continuous wave
cwt	hundredweight (112 pounds)
cyl	cylinder
d	day, deuteron
D	debye unit
D	diffusion coefficient, dissociation energy, electric displacement
Da	dalton
dB	decibel
dc	direct current
deg	degree
den	density
dev	deviation
diam	diameter
dil	dilute
dm	decimeter
DNA	deoxyribonucleic acid
doz	dozen
dpm	disintegrations per minute

dps	disintegrations per second
dr	dram
DSC	differential scanning calorimetry
DTA	differential thermal analysis
dyn	dyne
e	electron, base of natural logarithms
e	elementary charge, linear strain
E	electric field strength, electromotive force, energy, modulus of elasticity
E_h	Hartree energy
emf	electromotive force
emu	electromagnetic unit system
ENDOR	electron-nuclear double resonance
EPR	electron paramagnetic resonance
eq,eqn	equation
eqQ	quadrupole coupling constant
erg	erg
ESCA	electron spectroscopy for chemical analysis
e.s.d.	estimated standard deviation
ESR	electron spin resonance
esu	electrostatic unit system
ET	ephemeris time
e.u.	entropy unit
Eu	Euler number
eV	electronvolt
exp	exponential function
expt	experimental
ext	external
f	activity coefficient, aperture ratio, focal length, force constant, frequency, fugacity
F	farad
°F	degree Fahrenheit
F	Faraday constant, force, angular momentum
fcc	face centered cubic
FET	field effect transistor
FIR	far infrared
fl	fluid
FM	frequency modulation
Fo	Fourier number
fp	freezing point
fpm	feet per minute
fps	feet per second, foot-pound-second system
Fr	franklin
Fr	Froude number
ft	foot
FT	Fourier transform
ft-lb	foot pound
FTIR	Fourier transform infrared spectroscopy
g	gram
g	acceleration due to gravity, degeneracy, Landé g-factor
G	gauss
G	electrical conductance, Gibbs energy, gravitational constant, sheer modulus
gal	gallon
Gal	gal, galileo
GC	gas chromatography
GeV	gigaelectronvolt
GLC	gas-liquid chromatography
GMT	Greenwich mean time
gpm	gallons per minute
gps	gallons per second
gr	grain
Gr	Grashof number
GTO	gaussian type atomic orbital
GUT	grand unified theory
Gy	gray, gigayear
h	helion, hour
h	Planck constant
H	henry
H	enthalpy, hamiltonian function, magnetic field
H_0	Hubble constant
ha	hectare
Ha	Hartmann number
hav	haversine
hcp	hexagonal closed packed
HF	high frequency
hfs	hyperfine structure
HMO	Hückel molecular orbital
HOMO	highest occupied molecular orbital
hp	horsepower
HPLC	high-performance liquid chromatography
hr	hour
Hz	hertz
i	square root of minus one
I	electric current, ionic strength, moment of inertia, nuclear spin angular momentum, radiant intensity
IAT	international atomic time
IC	integrated circuit
ICR	ion cyclotron resonance
ID	inside diameter
IF	intermediate frequency
Im	imaginary part
IMPATT	impact ionization avalanche transit time
in.	inch
int	internal
I/O	input/output
IPTS	International Practical Temperature Scale
IR	infrared
ITS	International Temperature Scale (1990)
j	angular momentum, electric current density
J	joule
J	angular momentum, electric current density, flux, Massieu function
k	absorption index, Boltzmann constant, rate constant, thermal conductivity, wave vector
K	kelvin
K	absorption coefficient, bulk modulus, equilibrium constant, kinetic energy
kb	kilobar, kilobases (DNA or RNA)
kcal	kilocalorie
keV	kiloelectronvolt
kg	kilogram
kgf	kilogram force
kJ	kilojoule
km	kilometer
Kn	Knudsen number
kPa	kilopascal
kt	karat

kV	kilovolt
kva	kilovolt ampere
kW	kilowatt
kwh	kilowatt hour
l	liter
l	angular momentum, length
L	liter, lambert
L	Avogadro constant, inductance, Lagrange function
lat.	latitude
lb	pound
lbf	pound force
LCAO	linear combination of atomic orbitals
Le	Lewis function
LED	light emitter diode
LEED	low-energy electron diffraction
liq	liquid
lm	lumen
ln	logarithm (natural)
log	logarithm (common)
long.	longitude
LST	local sidereal time
LT	local time
lx	lux
ly	langley
l.y.	light year
m	meter, molal (as in 0.1 m solution)
m	magnetic dipole moment, mass, molality, angular momentum component
M	molar (as in 0.1 M solution)
M	magnetization, molar mass, mutual inductance, torque, angular momentum component
Ma	Mach number
max	maximum
MESFET	metal-semiconductor field-effect transistor
MeV	megaelectronvolt
MF	molecular formula
mg	milligram
MHD	magnetohydrodynamics
mi	mile
min	minimum, minute
MIR	mid-infrared
MKS	meter-kilogram-second system
MKSA	meter-kilogram-second-ampere system
ml,mL	milliliter
mm	millimeter
mmf	magnetomotive force
mmHg	millimeter of mercury
MO	molecular orbital
mol	mole
mol.wt.	molecular weight
MOS	metal-oxide semiconductor
MOSFET	metal-oxide semiconductor field-effect transistor
mp	melting point
MPa	megapascal
Mpc	megaparsec
M_r	molecular weight (relative molar mass)
mRNA	messenger RNA
ms	millisecond
MS	mass spectroscopy
MSL	mean sea level
mV	millivolt
mW	milliwatt
MW	megawatt, microwave, molecular weight
Mx	maxwell
n	neutron
n	amount of substance, number density, principal quantum number, refractive index
N	newton
N	angular momentum, neutron number
N_A	Avogadro constant
NAD	nicotinamide adenine dinucleotide
NADP	nicotinamide adenine dinucleotide phosphate
nbp	normal boiling point
N_E	density of states
ng	nanogram
NIR	near infrared
nm	nanometer
NMR	nuclear magnetic resonance
NQR	nuclear quadrupole resonance
ns	nanosecond
NTP	normal temperature and pressure
Nu	Nusselt number
obs	observed
OD	outside diameter
Oe	oersted
ORD	optical rotatory dispersion
oz	ounce
p	proton
p	dielectric polarization, electric dipole moment, momentum, pressure
P	poise
P	power, pressure, probability, sound energy flux
Pa	pascal
pc	parsec
PD	potential difference
pdl	poundal
Pe	Péclet number
pe	probable error
PES	photoelectron spectroscopy
pf	power factor
pg	picogram
pH	negative log of hydrogen ion concentration
pI	isoelectric point
p*K*	negative log of ionization constant
pm	picometer
ppb	parts per billion
ppm	parts per million
ppt	parts per thousand
Pr	Prandtl number
ps	picosecond
PS	photoelectron spectroscopy
psi	pounds per square inch
psia	pounds per square inch absolute
psig	pounds per square inch gage
pt	pint
PVT	pressure-volume-temperature
q	electric field gradient, flow rate, heat, wave vector (phonons)

Q	electric charge, heat, partition function, quadrupole moment, radiant energy, vibrational normal coordinate
QCD	quantum chromodynamics
QED	quantum electrodynamics
Q.E.D.	quod erat demonstrandum (which was to be proved)
QSO	quasi-stellar object
qt	quart
quad	quadrillion Btu (= $1.055 \cdot 10^{18}$ J)
q.v.	quod vide (which you should see)
r	position vector, radius
°R	degree Rankine
R	electrical resistance, gas constant, molar refraction, Rydberg constant
RA	right ascension
rad	radian
RAM	random access memory
Re	real part
rem	roentgen-equivalent-man
RF	radiofrequency
rms	root-mean-square
RNA	ribonucleic acid
ROM	read only memory
rpm	revolutions per minute
rps	revolutions per second
RRS	resonance Raman spectroscopy
RS	Raman spectroscopy
Ry	rydberg
s	second
s	path length, solubility, spin angular momentum, symmetry number
S	siemens
S	area, entropy, probability current density, Poynting vector, symmetry coordinate, spin angular momentum
Sc	Schmidt number
SCF	self-consistent field
SCR	silicon-controlled rectifier
sd	standard deviation
sec	secant, second
SEM	scanning electron microscope
SERS	surface-enhanced Raman spectroscopy
Sh	Sherwood number
SI	International System of Units
sin	sine
sinh	hyperbolic sine
SMOW	Standard Mean Ocean Water
SNU	solar neutrino unit
sp gr	specific gravity
sq	square
sr	steradian
Sr	Strouhal number
St	stoke
St	Stanton number
std	standard
STEM	scanning transmission electron microscope
STO	Slater type orbital
STP	standard temperature and pressure
Sv	sievert
t	metric tonne, triton
t	Celsius temperature, thickness, time, transport number
T	tesla
T	kinetic energy, period, term value, temperature (thermodynamic), torque, transmittance
tan	tangent
tanh	hyperbolic tangent
TE	transverse electric
TED	transferred electron device
TEM	transverse electromagnetic
temp	temperature
TFD	Thomas-Fermi-Dirac method
TL	thermoluminescence
TLC	thin-layer chromatography
TM	transverse magnetic
Torr	torr
tRNA	transfer RNA
u	unified atomic mass unit
u	Bloch function, electric mobility, velocity
U	electric potential difference, internal energy
UHF	ultrahigh frequency
UPS	ultraviolet photoelectron spectroscopy
UT	universal time
UV	ultraviolet
v	reaction rate, specific volume, velocity, vibrational quantum number
V	volt
V	electric potential, potential energy, volume
VB	valence band, valence bond
VHF	very high frequency
VIS	visible region of the spectrum
VSLI	very large scale integrated (circuit)
VUV	vacuum ultraviolet
v/v	volume for volume
w	energy density, mass fraction, velocity, work
W	watt
W	radiant energy, statistical weight, work
Wb	weber
We	Weber number
WKB	Wentzel-Kramers-Brillouin method
wt	weight
w/v	weight for volume
w/w	weight for weight
x	mole fraction
X	X unit
X	reactance
XPS	x-ray photoelectron spectroscopy
XRF	x-ray fluorescence
XRS	x-ray spectroscopy
Y	admittance, Planck function, Young's modulus
yd	yard
yr	year
z	charge number (of an ion), collision frequency factor
Z	atomic number, compression factor, collision number, impedance, partition function

SCIENTIFIC ABBREVIATIONS AND SYMBOLS (CONTINUED)

ZULU	Greenwich mean time
α	alpha particle
α	absorption coefficient, degree of dissociation, electric polarizability, expansion coefficient, fine structure constant
β	beta particle
γ	photon
γ	activity coefficient, conductivity, magnetogyric ratio, mass concentration, ratio of heat capacities, surface tension
Γ	Gruneisin parameter, level width, surface concentration
δ	chemical shift, Dirac delta function, Kronecker delta, loss angle
Δ	inertia defect, mass defect
ε	emittance, Levi-Civita symbol, linear strain, molar absorption coefficient, permittivity
η	overpotential, viscosity
κ	compressibility, conductivity, magnetic susceptibility, molar absorption coefficient
λ	absolute activity, radioactive decay constant, thermal conductivity, wavelength
Λ	angular momentum, ionic conductivity
μ	muon
μ	chemical potential, electric dipole moment, electric mobility, friction coefficient, Joule-Thompson coefficient, magnetic dipole moment, mobility, permeability
μF	microfarad
μg	microgram
μm	micrometer
μs	microsecond
ν	frequency, kinematic velocity, stoichiometric number
$\tilde{\nu}$	wavenumber
ν_e	neutrino
π	pion
Π	osmotic pressure, Peltier coefficient
ρ	density, reflectance, resistivity
σ	electrical conductivity, cross section, normal stress, shielding constant (NMR), Stefan-Boltzmann constant, surface tension
τ	transmittance, chemical shift, shear stress, relaxation time
ϕ	electrical potential, fugacity coefficient, osmotic coefficient, quantum yield, wavefunction
Φ	magnetic flux, potential energy, radiant power, work function
χ	magnetic susceptibility
χ_e	electric susceptibility
ψ	wavefunction
ω	circular frequency, angular velocity, harmonic vibration wavenumber, statistical weight
Ω	ohm
Ω	axial angular momentum, solid angle

GREEK, RUSSIAN, AND HEBREW ALPHABETS

The following table presents the Hebrew, Greek, and Russian alphabets, their letters, the names of the letters, the English equivalents.

HEBREW[1,4]			GREEK[7]			RUSSIAN	
א	aleph	’ [2]	Α α	alpha	a	А а	a
ב	beth	b, bh	Β β	beta	b	Б б	b
ג	gimel	g, gh	Γ γ	gamma	g, n	В в	v
ד	daleth	d, dh	Δ δ	delta	d	Г г	g
ה	he	h	Ε ϵ	epsilon	e	Д д	d
ו	waw	w	Ζ ζ	zeta	z	Е е	e
ז	zayin	z	Η η	eta	ē	Ж ж	zh
ח	heth	ḥ	Θ θ	theta	th	З з	z
ט	teth	ṭ	Ι ι	iota	i	И и Й й	i, ĭ
י	yodh	y	Κ κ	kappa	k	К к	k
כ ך	kaph	k, kh	Λ λ	lambda	l	Л л	l
ל	lamedh	l	Μ μ	mu	m	М м	m
מ ם	mem	m	Ν ν	nu	n	Н н	n
נ ן	nun	n	Ξ ξ	xi	x	О о	o
ס	samekh	s	Ο ο	omicron	o	П п	p
ע	ayin	‘	Π π	pi	p	Р р	r
פ ף	pe	p, ph	Ρ ρ	rho	r, rh	С с	s
צ ץ	sadhe	ṣ	Σ σ ς	sigma	s	Т т	t
ק	qoph	q	Τ τ	tau	t	У у	u
ר	resh	r	Υ υ	upsilon	y, u	Ф ф	f
שׂ	sin	ś	Φ φ	phi	ph	Х х	kh
שׁ	shin	sh	Χ χ	chi	ch	Ц ц	ts
ת	taw	t, th	Ψ ψ	psi	ps	Ч ч	ch
			Ω ω	omega	ō	Ш ш	sh
						Щ щ	shch
						Ъ ъ[9]	”
						Ы ы	y
						Ь ь[10]	’
						Э э	e
						Ю ю	yu
						Я я	ya

1 Where two forms of a letter are given, the second one is the form used at the end of a word.
2 Not represented in transliteration when initial.
4 The Hebrew letters are primarily consonants; a few of them are also used secondarily to represent certain vowels, when provided at all, is by means of a system of dots or strokes adjacent to the consonated characters.
7 The letter gamma is tranliterated ‘‘n’’ only before velars; the letter upsilon is transliterated ‘‘u’’ only as the final element in diphthongs.
9 This sign indicates that the immediately preceding consonant is not palatized even though immediately followed by a palatized vowel.
10 This sign indicates that the immediately preceding consonant is palatized even though not immediately followed by a palatized vowel.

Section 3
Physical Constants of Organic Compounds

PHYSICAL CONSTANTS OF ORGANIC COMPOUNDS

The basic physical constants for over 12,000 organic compounds are presented in this table, along with structures and references to other sources of information. An effort has been made to include the compounds most frequently encountered in the laboratory, the workplace, and the environment. The selection was based mainly on the appearance of the compounds in various specialized tables in this *Handbook* and in other widely used reference sources, such as the *Merck Index* and the *DIPPR Database of Pure Compound Properties*. The occurrence of a compound on regulatory lists of hazardous chemicals was also taken into consideration, as was the availability of reliable physical constant data. Clearly, criteria of this type are somewhat subjective, and compounds considered important by some users have undoubtedly been omitted. Suggestions for additional compounds or other improvements are welcomed.

The data in the table have been taken from many sources, including both compilations and the primary literature. Where conflicts were found, the value deemed most reliable was chosen. Some of the useful compilations of physical property data are listed at the end of this introduction.

The table is arranged alphabetically by the primary name, which is generally the Index Name from the 8th or 9th Collective Index of Chemical Abstracts Service (CAS). In a few cases, especially pesticides and pharmaceuticals, the common name is used rather than the more complex systematic name. By convention, CAS Index Names are written in inverted order, e.g., chloromethane is listed as methane, chloro and ethyl acetate as acetic acid, ethyl ester. Furthermore, certain important compounds are listed under Index Names which differ from the names by which they are commonly known (e.g. aniline appears as benzenamine and acetone as 2-propanone). In order to facilitate the location of compounds in the table, three indexes are provided:

- **Synonym Index:** Includes common synonyms, but not the primary name by which the table is arranged.
- **Molecular Formula Index:** Lists compounds by molecular formula in the Hill order (see Preface to this *Handbook*).
- **CAS Registry Number Index:** Lists compounds by Chemical Abstracts Service Registry Number.

Two lines of data appear for each compound. The explanation of the data fields follows.

Top Line:

- **No.:** An identification number used in the indexes and to identify the structure diagrams.
- **Name:** Primary name, generally the CAS Index Name.
- **Mol. Form.:** The molecular formula written in the Hill convention.
- **CAS RN:** The Chemical Abstracts Service Registry Number assigned by CAS as a unique identifier for the compound.
- **Merck No:** Monograph Number in *The Merck Index, Eleventh Edition.* It should be noted that this is not a unique identifier for a single compound, since several derivatives or isomers of a compound may be included in the same Monograph.
- **Beil. Ref:** Citation to the *Beilstein Handbook of Organic Chemistry*. An entry of 5-18-11-01234, for example, indicates that the compound may be found in the 5th Series, Volume 18, Subvolume 11, page 1234.
- **Solubility:** Solubility in common solvents on a relative scale: 1 = insoluble; 2 = slightly soluble; 3 = soluble; 4 = very soluble; 5 = miscible; 6 = decomposes. See List of Abbreviations for the solvent abbreviations.

Bottom line:

- **Synonym:** A synonym in common use. When the primary name is non-systematic, the systematic name appears here.
- **Mol. Wt.:** Molecular weight (relative molar mass) as calculated with the 1991 IUPAC Standard Atomic Weights.
- **mp/°C:** Normal melting point in °C. Although some values are quoted to 0.1°C, uncertainties are typically several degrees Celsius. A value is sometimes followed by "dec", indicating decomposition is observed at the stated temperature (so that it is probably not a true melting point). See the List of Abbreviations for other abbreviations.
- **bp/°C:** Boiling point in °C. When available, the normal boiling point is given first, without a superscript. This is the temperature at which the liquid phase is in equilibrium with the vapor at a pressure of 760 mmHg (101.325 kPa). Boiling point values at reduced pressure are also given in many cases; here the superscript indicates the pressure in mmHg. A "dec" or "exp" following the value indicates decomposition or explosion has been observed at the boiling point. A simple entry of "exp" (sometimes followed by a temperature) indicates explosion may occur on heating, even below the boiling point. An entry of "sub" indicates that no boiling point is available, but measurable vapor (sublimation) pressure has been observed upon heating the solid. A temperature may be given, but no precise meaning can be attached because the pressure is not specified.
- **den/g cm^{-3}:** density (mass per unit volume) in g/cm^{-3}. The superscript indicates the temperature in °C. Values are given only for the liquid and solid phases, and all values are true densities, not specific gravities. The number of decimal places gives a rough estimate of the accuracy of the value.
- **n_D:** Refractive index at the temperature indicated by the superscript. Unless otherwise indicated, all values refer to a wavelength of 589 nm (sodium D line). Values are given only for liquids and solids.

Structures are given, when available, in the section following the main table, using the **No.** in the first column as the linking identifier.

LIST OF ABBREVIATIONS

Ac	Acetyl	EtOH	Ethanol
ace	Acetone	exp	Explodes
acid	Acids	gl	Glacial
AcOEt	Ethyl acetate	HOAc	Acetic acid
alk	Alkalis	hp	Heptane
anh	Anhydrous	hx	Hexane
aq	Aqueous	lig	Ligroin
bz	Benzene	MeOH	Methanol
chl	Chloroform	os	Organic solvents
con	Concentrated	peth	Petroleum ether
ctc	Carbon tetrachloride	$PhNO_2$	Nitrobenzene
cyhex	Cyclohexane	PrOH	1-Propanol
dec	Decomposes	py	Pyridine
dil	Dilute	sub	Sublimes
diox	Dioxane	sulf	Sulfuric acid
DMF	Dimethyl formamide	temp.	Temperature
DMSO	Dimethyl sulfoxide	tfa	Trifluoroacetic acid
DMSO-d_6	Deuterated dimethyl sulfoxide	tol	Toluene
eth	Ethyl ether	xyl	Xylene

REFERENCES

1. Lide, D. R., and Milne, G. W. A., Editors, *Handbook of Data on Organic Compounds, Third Edition*, CRC Press, Boca Raton, FL, 1993.
2. Luckenbach, R., Editor, *Beilstein Handbook of Organic Chemistry*, Springer-Verlag, Heidelberg. Data can also be accessed through the on-line Beilstein Database.
3. Budavari, S., Editor, *The Merck Index, Eleventh Edition*, Merck & Co., Rahway, NJ, 1989.
4. Daubert, T. E., Danner, R. P., Sibul, H. M., and Stebbins, C. C., *Physical and Thermodynamic Properties of Pure Compounds: Data Compilation*, extant 1994 (core with 4 supplements), Taylor & Francis, Bristol, PA (also available as database).
5. *TRC Thermodynamic Tables*, Thermodynamic Research Center, Texas A & M University, College Station, TX.
6. Riddick, J. A., Bunger, W. B., and Sakano, T. K., *Organic Solvents, Fourth Edition*, John Wiley & Sons, New York, 1986.
7. *Physical Constants of Hydrocarbon and Non-Hydrocarbon Compounds,* ASTM Data Series DS 4B, ASTM, Philadelphia, 1988.
8. Stevenson, R. M., and Malanowski, S., *Handbook of the Thermodynamics of Organic Compounds*, Elsevier, New York, 1987.
9. Buckingham, J., Editor, *Dictionary of Organic Compounds, Fifth Edition* (and Supplements), Chapman and Hall, New York, 1982-1993.
10. Lide, D. R., and Kehiaian, H. V., *Handbook of Thermophysical and Thermochemical Data*, CRC Press, Boca Raton, FL, 1994.

PHYSICAL CONSTANTS OF ORGANIC COMPOUNDS

No.	Name Synonym	Mol. Form. Mol. Wt.	CAS RN mp/°C	Merck No. bp/°C	Beil. Ref. den/g cm^{-3}	Solubility n_D
1	Abate Phosphorothioic acid, *O,O'*-(thiodi-4,1-phenylene) *O,O,O',O'*-tetramethyl ester	$C_{16}H_{20}O_6P_2S_3$ 466.48	3383-96-8 30	9075	 1.32	
2	1-Acenaphthylenamine, 1,2-dihydro- 1-Acenaphthenamine	$C_{12}H_{11}N$ 169.23	40745-44-6 135	 sub	2-12-00-00764	bz 4; EtOH 4
3	Acenaphthylene Acenaphthalene	$C_{12}H_8$ 152.20	208-96-8 92.5	 280; 150^{28}	4-05-00-02138 0.8987^{16}	H_2O 1; EtOH 4; eth 4; bz 4
4	Acenaphthylene, 5-bromo-1,2-dihydro- 5-Bromoacenaphthene	$C_{12}H_9Br$ 233.11	2051-98-1 52	 335	4-05-00-01839 1.4392^{52}	H_2O 1; EtOH 2; chl 2 1.6565^{54}
5	Acenaphthylene, 5-chloro-1,2-dihydro- 5-Chloroacenaphthene	$C_{12}H_9Cl$ 188.66	5209-33-6 70.5	 319; 163^{13}	4-05-00-01837 1.1954^{70}	 1.6169^{100}
6	Acenaphthylene, 1,2-dihydro- Acenaphthene	$C_{12}H_{10}$ 154.21	83-32-9 93.4	23 279	4-05-00-01834 1.0242^{99}	H_2O 1; EtOH 2; bz 4; chl 2 1.6048^{95}
7	Acenaphthylene, 1,2-dihydro-5-iodo- 5-Iodoacenaphthene	$C_{12}H_9I$ 280.11	6861-64-9 64.3		2-05-00-00497 1.5000^{20}	bz 4; EtOH 4 1.6909^{65}
8	1,2-Acenaphthylenedione	$C_{12}H_6O_2$ 182.18	82-86-0 261	 sub	4-07-00-02498 1.4800^{20}	H_2O 1; EtOH 2; bz 2; lig 3
9	Acenaphthylene, 1,2,2a,3,4,5-hexahydro-	$C_{12}H_{14}$ 158.24	480-72-8 12.8	 249	4-05-00-01566 1.0290^{20}	chl 3
10	Acephate Phosphoramidothioic acid, acetyl-, *O,S*-dimethyl ester	$C_4H_{10}NO_3PS$ 183.17	30560-19-1 88		 1.35^{20}	
11	Acetaldehyde Ethanal	C_2H_4O 44.05	75-07-0 -123	32 20.1	4-01-00-03094 0.7834^{18}	H_2O 5; EtOH 5; eth 5; bz 5 1.3316^{20}
12	Acetaldehyde, chloro- Chloroacetaldehyde	C_2H_3ClO 78.50	107-20-0 -16.3	2108 85.5	4-01-00-03134	eth 3
13	Acetaldehyde, chloro-, trimer Chloroacetaldehyde, trimer	$C_6H_9Cl_3O_3$ 235.49	65438-35-9 87	 142^{10}	4-19-00-04718	eth 4
14	Acetaldehyde, ethoxy- α-Ethoxyacetaldehyde	$C_4H_8O_2$ 88.11	22056-82-2		3-01-00-03181 0.942^{20}	H_2O 4; ace 4; EtOH 4 1.3956^{20}
15	Acetaldehyde, hydroxy- Glycolaldehyde	$C_2H_4O_2$ 60.05	141-46-8 97		4-01-00-03955 1.366^{100}	chl 3 1.4772^{19}
16	Acetaldehyde, methoxy- Methoxyacetaldehyde	$C_3H_6O_2$ 74.08	10312-83-1	 92	4-01-00-03955 1.005^{25}	H_2O 4; ace 4; eth 4; EtOH 4 1.3950^{20}
17	Acetaldehyde, oxime Acetaldoxime	C_2H_5NO 59.07	107-29-9 45	35 115	4-01-00-03121 0.9656^{20}	H_2O 3; EtOH 5; eth 5; chl 3 1.4264^{20}
18	Acetaldehyde, phenylhydrazone	$C_8H_{10}N_2$ 134.18	935-07-9 99.5	 150^{40}; 135^{21}	3-15-00-00079	EtOH 4
19	Acetaldehyde, tetramer Metaldehyde II (tetramer)	$C_4H_{12}O_4$ 124.14	66056-06-2 47	 110	4-19-00-05643	ace 4; bz 4; eth 4; EtOH 4
20	Acetaldehyde, tribromo- Bromal	C_2HBr_3O 280.74	115-17-3	1372 174	4-01-00-03155 2.6649^{25}	ace 4; eth 4; EtOH 4 1.5939^{20}
21	Acetaldehyde, trichloro- Chloral	C_2HCl_3O 147.39	75-87-6 -57.5	9538 97.8	4-01-00-03142 1.512^{20}	H_2O 4; EtOH 3; eth 3 1.4580^{20}
22	Acetamide Ethanamide	C_2H_5NO 59.07	60-35-5 81	36 222.0	4-02-00-00399 0.9986^{85}	H_2O 4; EtOH 4 1.4278
23	Acetamide, *N*-acetyl-	$C_4H_7NO_2$ 101.11	625-77-4 79	 223.5	4-02-00-00416	H_2O 3; EtOH 3; eth 3; chl 3
24	Acetamide, *N*-acetyl-*N*-ethyl-	$C_6H_{11}NO_2$ 129.16	1563-83-3	 197	4-04-00-00352 1.0092^{20}	EtOH 4 1.4513^{20}
25	Acetamide, *N*-acetyl-*N*-methyl-	$C_5H_9NO_2$ 115.13	1113-68-4 -25	 195; 114.5^{61}	4-04-00-00183 1.0663^{25}	H_2O 5; eth 1 1.4502^{25}
26	Acetamide, *N*-acetyl-*N*-phenyl-	$C_{10}H_{11}NO_2$ 177.20	1563-87-7 37.5	 200^{100}; 142^{11}	4-12-00-00385	H_2O 2; EtOH 3; bz 3; tol 3
27	Acetamide, *N*-(aminocarbonyl)-	$C_3H_6N_2O_2$ 102.09	591-07-1 218	 sub 180	3-03-00-00119	H_2O 2; EtOH 3; eth 2
28	Acetamide, *N*-(2-aminoethyl)-	$C_4H_{10}N_2O$ 102.14	1001-53-2 51	 136^{27}	4-04-00-01193	H_2O 3; EtOH 3; eth 1; bz 3
29	Acetamide, *N*-(4-aminophenyl)- *p*-Aminoacetanilide	$C_8H_{10}N_2O$ 150.18	122-80-5 166.5	422 267	4-13-00-00137	H_2O 3; EtOH 4; eth 4
30	Acetamide, *N*-[(4-aminophenyl)sulfonyl]- Sulfacetamide	$C_8H_{10}N_2O_3S$ 214.25	144-80-9 183	8870	4-14-00-02662	H_2O 2; EtOH 3; eth 1; ace 4
31	Acetamide, *N*-[4-(aminosulfonyl)phenyl]- Acetylsulfanilamide	$C_8H_{10}N_2O_3S$ 214.25	121-61-9 219.5	87	4-14-00-02701	H_2O 3; EtOH 3; ace 3
32	Acetamide, *N*-[5-(aminosulfonyl)-1,3,4-thiadiazol-2-yl]- Acetazolamide	$C_4H_6N_4O_3S_2$ 222.25	59-66-5 260.5	45	4-27-00-08219	H_2O 2
33	Acetamide, *N*-(aminothioxomethyl)- Acetylthiourea	$C_3H_6N_2OS$ 118.16	591-08-2 165		4-03-00-00354	H_2O 2; EtOH 3; eth 2
34	Acetamide, *N*-[4-[[(aminothioxomethyl)hydrazono]methyl]phenyl]- Thiacetazone	$C_{10}H_{12}N_4OS$ 236.30	104-06-3 225 dec	9218	4-14-00-00075	H_2O 1; os 1; CS_2 1
35	Acetamide, *N*-[1,1'-biphenyl]-2-yl-	$C_{14}H_{13}NO$ 211.26	2113-47-5 121	 355	4-12-00-03224	H_2O 1; EtOH 4; eth 4
36	Acetamide, *N*-bromo- *N*-Bromoacetamide	C_2H_4BrNO 137.96	79-15-2 103.5	1386	4-02-00-00417	eth 4

No.	Name	Synonym	Mol. Form.	Mol. Wt.	CAS RN	mp/°C	Merck No.	bp/°C	Beil. Ref.	den/g cm^{-3}	Solubility	n_D
37	Acetamide, *N*-(4-bromophenyl)-	*p*-Bromoacetanilide	C_8H_8BrNO	214.06	103-88-8	168	1387		4-12-00-01504	1.717^{25}	H_2O 1; EtOH 3; eth 2; bz 2	
38	Acetamide, *N*-butyl-		$C_6H_{13}NO$	115.18	1119-49-9			229	4-04-00-00565	0.8960^{25}		1.4388^{25}
39	Acetamide, *N*-butyl-*N*-phenyl-		$C_{12}H_{17}NO$	191.27	91-49-6	24.5		281	4-12-00-00379	0.9912^{20}	chl 2	1.5146^{20}
40	Acetamide, *N,N'*-carbonylbis-		$C_5H_8N_2O_3$	144.13	638-20-0	154.5		sub	3-03-00-00121		H_2O 2; EtOH 2; eth 3; ace 3	
41	Acetamide, 2-chloro-	Chloroacetamide	C_2H_4ClNO	93.51	79-07-2	121	2109	225	4-02-00-00490		H_2O 3; EtOH 4; eth 2	
42	Acetamide, 2-chloro-*N,N*-bis(1-methylethyl)-		$C_8H_{16}ClNO$	177.67	7403-66-9	48.5		163^{55}; $86^{2.7}$	4-04-00-00516			1.4619^{25}
43	Acetamide, 2-chloro-*N*-3-chloroallyl-	Chloroacetamide, *N*-3-chloroallyl	$C_5H_7Cl_2NO$	168.02	100130-25-4	52.5		$112^{0.5}$	4-04-00-01087		peth 4	
44	Acetamide, 2-chloro-*N,N*-di-2-propenyl-	Allidochlor	$C_8H_{12}ClNO$	173.64	93-71-0		250	116^{1}; $92^{0.7}$	4-04-00-01064		H_2O 2; EtOH 3	1.4932^{25}
45	Acetamide, 2-chloro-*N*-hexyl-	α-Chloro-*N*-hexylacetamide	$C_8H_{16}ClNO$	177.67	5326-81-8	108.5		$100^{0.2}$	4-04-00-00713		peth 4	
46	Acetamide, 2-chloro-*N*-(3-methoxypropyl)-	α-Chloro-*N*-(3-methoxypropyl)acetamide	$C_6H_{12}ClNO_2$	165.62	1709-03-1	30		$88^{0.5}$	4-04-00-01645			1.4712^{25}
47	Acetamide, 2-chloro-*N*-(1-methylpropyl)-		$C_6H_{12}ClNO$	149.62	32322-73-9	45.5		$68^{0.7}$	4-04-00-00621		peth 4	
48	Acetamide, *N*-(4-chlorophenyl)-		C_8H_8ClNO	169.61	539-03-7	179		333	4-12-00-01178	1.385^{22}	H_2O 1; EtOH 3; eth 4; ctc 2	
49	Acetamide, 2-cyano-	Cyanoacetamide	$C_3H_4N_2O$	84.08	107-91-5	121.5	2695		4-02-00-01891		H_2O 4	
50	Acetamide, *N,N*-diacetyl-	Triacetamide	$C_6H_9NO_3$	143.14	641-06-5	79			4-02-00-00416		eth 4	
51	Acetamide, 2,2-dichloro-		$C_2H_3Cl_2NO$	127.96	683-72-7	99.4		234	4-02-00-00505		H_2O 3; EtOH 3; eth 3; ace 2	
52	Acetamide, 2,2-dichloro-*N*-[2-hydroxy-1-(hydroxymethyl)-2-(4-nitrophenyl)ethyl]-, [*R*-(*R**,*R**)]-	Chloramphenicol	$C_{11}H_{12}Cl_2N_2O_5$	323.13	56-75-7	150.5	2068	sub	3-13-00-02268		ace 4; EtOH 4; chl 4	
53	Acetamide, *N,N*-diethyl-		$C_6H_{13}NO$	115.18	685-91-6			185.5	4-04-00-00349	0.9130^{17}	H_2O 3; EtOH 3; eth 5; ace 5	1.4374^{17}
54	Acetamide, 2-(diethylamino)-*N*-(2,6-dimethylphenyl)-	Lidocaine	$C_{14}H_{22}N_2O$	234.34	137-58-6	68.5	5359	$180\text{-}2^{4}$	4-12-00-02538		bz 4; eth 4; EtOH 4; chl 4	
55	Acetamide, *N,N*-dimethyl-	*N,N*-Dimethylethanamide	C_4H_9NO	87.12	127-19-5	-20	3216	165	4-04-00-00180	0.9366^{25}	H_2O 5; EtOH 5; eth 5; ace 5	1.4380^{20}
56	Acetamide, *N*-(2,4-dimethylphenyl)-		$C_{10}H_{13}NO$	163.22	2050-43-3	129.3		170^{10}	4-12-00-02548		EtOH 4; chl 4	
57	Acetamide, *N,N*-diphenyl-	Diphenylacetamide	$C_{14}H_{13}NO$	211.26	519-87-9	103	3315	sub	4-12-00-00381		H_2O 2; EtOH 3; eth 2; chl 2	
58	Acetamide, *N,N*-dipropyl-		$C_8H_{17}NO$	143.23	1116-24-1			209.5	4-04-00-00476	0.8992^{17}	EtOH 4	1.4419^{17}
59	Acetamide, *N*-(2-ethoxyphenyl)-		$C_{10}H_{13}NO_2$	179.22	581-08-8	79		>240	3-13-00-00779		H_2O 1; EtOH 3; eth 3; chl 3	
60	Acetamide, *N*-(4-ethoxyphenyl)-	Phenacetin	$C_{10}H_{13}NO_2$	179.22	62-44-2	137.5	7155		4-13-00-01092		H_2O 2; EtOH 3; eth 2; ace 3	1.571
61	Acetamide, *N*-ethyl-	*N*-Ethylacetamide	C_4H_9NO	87.12	625-50-3			205; 104^{18}	4-04-00-00347	0.942^{4}	H_2O 5; EtOH 5; chl 3; HOAc 3	1.4338^{20}
62	Acetamide, *N*-ethyl-*N*-phenyl-		$C_{10}H_{13}NO$	163.22	529-65-7	55		260	4-12-00-00379	0.9938^{60}	H_2O 3; eth 3; ctc 3	
63	Acetamide, 2-fluoro-	Fluoroacetic acid amide	C_2H_4FNO	77.06	640-19-7	108	4095	sub	4-02-00-00454		H_2O 3; ace 3; chl 2	
64	Acetamide, *N*-9H-fluoren-2-yl-	2-(Acetylamino)fluorene	$C_{15}H_{13}NO$	223.27	53-96-3	193	4083		4-12-00-03373		H_2O 1; EtOH 3; eth 3; HOAc 3	
65	Acetamide, *N*-(2-hydroxyethyl)-		$C_4H_9NO_2$	103.12	142-26-7	63.5		$166\text{-}7^{8}$	4-04-00-01535	1.1079^{25}	H_2O 5; ace 3; bz 2; lig 2	1.4674^{20}
66	Acetamide, *N*-(2-hydroxy-1-naphthalenyl)-		$C_{12}H_{11}NO_2$	201.22	117-93-1	235 dec		sub	4-13-00-02095		ace 4; bz 4; eth 4; EtOH 4	
67	Acetamide, *N*-(4-hydroxyphenyl)-	Acetaminophen	$C_8H_9NO_2$	151.17	103-90-2	170	40		4-13-00-01091	1.293^{21}	H_2O 1; EtOH 4	
68	Acetamide, 2-mercapto-*N*-2-naphthalenyl-	Thionalide	$C_{12}H_{11}NOS$	217.29	93-42-5	111.5	9273		3-12-00-03052		H_2O 1; EtOH 4; os 4	
69	Acetamide, *N*-[2-(5-methoxy-1H-indol-3-yl)ethyl]-	Melatonin	$C_{13}H_{16}N_2O_2$	232.28	73-31-4	117	5695		5-22-12-00042			
70	Acetamide, *N*-(2-methoxyphenyl)-		$C_9H_{11}NO_2$	165.19	93-26-5	87.5		304	4-13-00-00833		H_2O 4; EtOH 4; eth 3; ace 3	
71	Acetamide, *N*-(4-methoxyphenyl)-	*p*-Acetanisidine	$C_9H_{11}NO_2$	165.19	51-66-1	131	43		4-13-00-01092		ace 4; EtOH 4; chl 4	
72	Acetamide, *N*-methyl-		C_3H_7NO	73.09	79-16-3	28		205	4-04-00-00176	0.9371^{25}	ace 4; bz 4; eth 4; EtOH 4	1.4301^{20}
73	Acetamide, *N*-[(methylamino)carbonyl]-		$C_4H_8N_2O_2$	116.12	623-59-6	180.5		dec	4-04-00-00207		H_2O 3; EtOH 2; eth 2; chl 3	

PHYSICAL CONSTANTS OF ORGANIC COMPOUNDS (continued)

No.	Name Synonym	Mol. Form. Mol. Wt.	CAS RN mp/°C	Merck No. bp/°C	Beil. Ref. den/g cm^{-3}	Solubility n_D
74	Acetamide, *N*-(1-methylethyl)-*N*-phenyl- *N*-Isopropylacetanilide	$C_{11}H_{15}NO$ 177.25	5461-51-8 40.5	 264	3-12-00-00466	lig 4
75	Acetamide, *N*-methyl-*N*-(2-methylphenyl)-	$C_{10}H_{13}NO$ 163.22	573-26-2 55.5	 260	3-12-00-01854	EtOH 3; chl 3
76	Acetamide, *N*-(2-methylphenyl)-	$C_9H_{11}NO$ 149.19	120-66-1 110	 296	4-12-00-01755 1.168^{15}	H_2O 2; EtOH 3; eth 3; ace 3
77	Acetamide, *N*-(3-methylphenyl)-	$C_9H_{11}NO$ 149.19	537-92-8 65.5	 303	4-12-00-01823 1.141^{15}	H_2O 2; EtOH 4; eth 4; chl 3
78	Acetamide, *N*-(4-methylphenyl)-	$C_9H_{11}NO$ 149.19	103-89-9 152	 307	4-12-00-01902 1.2120^{15}	eth 4; EtOH 4
79	Acetamide, *N*-methyl-*N*-phenyl- *N*-Methylacetanilide	$C_9H_{11}NO$ 149.19	579-10-2 103	5931 256	4-12-00-00378 1.0036^{105}	H_2O 3; EtOH 3; eth 3; chl 3 1.576
80	Acetamide, 2-nitro-*N*-phenyl- α-Nitroacetanilide	$C_8H_8N_2O_3$ 180.16	10151-95-8 94	 $100^{0.1}$	1-12-00-00193 1.419^{15}	bz 4; EtOH 4; lig 4
81	Acetamide, *N*-(2-nitrophenyl)-	$C_8H_8N_2O_3$ 180.16	552-32-9 94	 $100^{0.1}$	4-12-00-01574 1.419^{15}	H_2O 3; EtOH 3; eth 4; bz 3
82	Acetamide, *N*-(3-nitrophenyl)-	$C_8H_8N_2O_3$ 180.16	122-28-1 155	 $100^{0.0074}$	4-12-00-01595	H_2O 3; EtOH 3; eth 1; chl 3
83	Acetamide, *N*-(4-nitrophenyl)-	$C_8H_8N_2O_3$ 180.16	104-04-1 216	 $100^{0.008}$	4-12-00-01632	H_2O 2; EtOH 3; eth 2; chl 2
84	Acetamide, *N*-(5-nitro-2-propoxyphenyl)- 5'-Nitro-2'-propoxyacetanilide	$C_{11}H_{14}N_2O_4$ 238.24	553-20-8 102.5	6550	3-13-00-00881	
85	Acetamide, *N*-(5-nitro-2-thiazolyl)- Aminitrozole	$C_5H_5N_3O_3S$ 187.18	140-40-9 264.5	421	4-27-00-04676	alk 3
86	Acetamide, *N*-phenyl- Acetanilide	C_8H_9NO 135.17	103-84-4 114.3	42 304	4-12-00-00373 1.2190^{15}	H_2O 2; EtOH 4; eth 3; ace 4
87	Acetamide, *N*-(phenylmethyl)-	$C_9H_{11}NO$ 149.19	588-46-5 61	 157^{2}	4-12-00-02230	EtOH 4; eth 4
88	Acetamide, *N*-phenyl-*N*-propyl- *N*-Propylacetanilide	$C_{11}H_{15}NO$ 177.25	2437-98-1 49	 268	4-12-00-00379	eth 4; EtOH 4; lig 4
89	Acetamide, *N*-3-pyridinyl-	$C_7H_8N_2O$ 136.15	5867-45-8 133	 326.5	5-22-09-00017	H_2O 3; EtOH 3; eth 2; bz 2
90	Acetamide, *N*-4-quinolinyl-	$C_{11}H_{10}N_2O$ 186.21	32433-28-6 176	 sub	5-22-10-00250	EtOH 4
91	Acetamide, *N,N*'-(sulfonyldi-4,1-phenylene)bis- Acedapsone	$C_{16}H_{16}N_2O_4S$ 332.38	77-46-3 290	16	4-13-00-01361	EtOH 4
92	Acetamide, *N*-(5,6,7,9-tetrahydro-10-hydroxy-1,2,3-trimethoxy-9-oxobenzo[a]heptalen-7-yl)-, (*S*)- Colchiceine	$C_{21}H_{23}NO_6$ 385.42	477-27-0 178.5	2469	4-14-00-00943 1.24^{25}	H_2O 2; EtOH 4; eth 1; bz 1
93	Acetamide, *N*-(5,6,7,9-tetrahydro-1,2,3,10-tetramethoxy-9-oxobenzo[a]heptalen-7-yl)-, (*S*)- Colchicine	$C_{22}H_{25}NO_6$ 399.44	64-86-8 156	2470	4-14-00-00946	H_2O 4; EtOH 4
94	Acetamide, 2,2,2-tribromo- Tribromoacetamide	$C_2H_2Br_3NO$ 295.76	594-47-8 121.5	 sub	3-02-00-00485	H_2O 4; eth 4; EtOH 4
95	Acetamide, 2,2,2-trichloro-	$C_2H_2Cl_3NO$ 162.40	594-65-0 142	 240	4-02-00-00520	H_2O 2; EtOH 4; eth 4
96	Acetamide, 2,2,2-trichloro-*N,N*-diethyl-	$C_6H_{10}Cl_3NO$ 218.51	2430-00-4 27	 112^{12}	4-04-00-00351	 1.4900^{24}
97	Acetamide, 2,2,2-trichloro-*N,N*-dimethyl-	$C_4H_6Cl_3NO$ 190.46	7291-33-0 12	 231 dec	4-04-00-00182 1.390^{20}	bz 4; chl 4 1.5017^{25}
98	Acetamide, 2,2,2-trichloro-*N*-phenyl-	$C_8H_6Cl_3NO$ 238.50	2563-97-5 96	 169^{19}	4-12-00-00377	EtOH 4
99	Acetamide, 2,2,2-trifluoro-*N,N*-dimethyl-	$C_4H_6F_3NO$ 141.09	1547-87-1	 133	4-04-00-00182 1.232^{21}	 1.3611^{25}
100	Acetic acid Ethanoic acid	$C_2H_4O_2$ 60.05	64-19-7 16.6	47 117.9	4-02-00-00094 1.0492^{20}	H_2O 5; EtOH 5; eth 5; ace 5 1.3720^{20}
101	Acetic acid, (acetyloxy)-	$C_4H_6O_4$ 118.09	13831-30-6 67.5	 145^{12}	4-03-00-00576	H_2O 4; EtOH 4; eth 4; ace 4
102	Acetic acid, (acetyloxy)-, ethyl ester	$C_6H_{10}O_4$ 146.14	623-86-9	 179	4-03-00-00584 1.0880^{20}	eth 4; EtOH 4; HOAc 4 1.4112^{20}
103	Acetic acid, [2-(aminocarbonyl)phenoxy]- Salicylamide O-acetic acid	$C_9H_9NO_4$ 195.17	25395-22-6	8298	4-10-00-00184	
104	Acetic acid, aminooxo- Oxamic acid	$C_2H_3NO_3$ 89.05	471-47-6 210 dec	6870	4-02-00-01857	H_2O 2; EtOH 1; eth 1
105	Acetic acid, aminooxo-, hydrazide Semioxamazide	$C_2H_5N_3O_2$ 103.08	515-96-8 221 dec	8397	4-02-00-01868	H_2O 2; EtOH 1; eth 1; alk 4
106	Acetic acid, ammonium salt Ammonium, acetate	$C_2H_7NO_2$ 77.08	631-61-8 114	514 dec	 1.17^{20}	H_2O 4; EtOH 3; ace 2
107	Acetic acid, anhydride Acetic anhydride	$C_4H_6O_3$ 102.09	108-24-7 -73	48 139.5	4-02-00-00386 1.082^{20}	H_2O 4; EtOH 3; eth 5; bz 3 1.3901^{20}
108	Acetic acid, anhydride with nitric acid Acetyl nitrate	$C_2H_3NO_4$ 105.05	591-09-3	92 22^{70}; exp 60	4-02-00-00435 1.24^{15}	
109	Acetic acid, azido-	$C_2H_3N_3O_2$ 101.06	18523-48-3 16	 116^{12}; 92^{3}	4-02-00-00541 1.354^{33}	
110	Acetic acid, barium salt	$C_4H_6BaO_4$ 255.42	543-80-6	975	4-02-00-00094 2.468^{25}	H_2O 3

PHYSICAL CONSTANTS OF ORGANIC COMPOUNDS (continued)

No.	Name	Synonym	Mol. Form.	Mol. Wt.	CAS RN	mp/°C	Merck No.	bp/°C	Beil. Ref.	den/g cm^{-3}	Solubility	n_D
111	Acetic acid, bromo-	Bromoacetic acid	$C_2H_3BrO_2$	138.95	79-08-3	50	1388	208	4-02-00-00526	1.9335^{50}	H_2O 5; EtOH 5; eth 5; ace 3	1.4804^{50}
112	Acetic acid, bromochloro-	Bromochloroacetic acid	$C_2H_2BrClO_2$	173.39	5589-96-8	31.5		215; 104^{11}	4-02-00-00532	1.9848^{31}		
113	Acetic acid, bromochloro-, ethyl ester	Ethyl bromochloroacetate	$C_4H_6BrClO_2$	201.45	22524-32-9			174 dec	4-02-00-00532	1.5890^{22}	eth 4; EtOH 4	1.4639^{24}
114	Acetic acid, bromo-, 1,1-dimethylethyl ester		$C_6H_{11}BrO_2$	195.06	5292-43-3			$73\text{-}4^{25}$	3-02-00-00482		eth 4; EtOH 4	1.4430^{20}
115	Acetic acid, bromo-, ethyl ester	Ethyl bromacetate	$C_4H_7BrO_2$	167.00	105-36-2			168.5	4-02-00-00527	1.5032^{20}	H_2O 1; EtOH 5; eth 5; ace 3	1.4489^{20}
116	Acetic acid, bromofluoro-	Bromofluoroacetic acid	$C_2H_2BrFO_2$	156.94	359-25-1	50		183	4-02-00-00531		H_2O 4; EtOH 4; chl 4	
117	Acetic acid, bromo-, methyl ester	Methyl 2-bromoacetate	$C_3H_5BrO_2$	152.98	96-32-2			132	4-02-00-00527	1.6350^{20}	H_2O 1; EtOH 3; eth 3; ace 3	1.4520^{20}
118	Acetic acid, bromo-, 2-methylpropyl ester		$C_6H_{11}BrO_2$	195.06	59956-48-8			188; 74.5^{10}	1-02-00-00096	1.3269^{20}	ace 4; eth 4; EtOH 4	
119	Acetic acid, bromo-, phenyl ester		$C_8H_7BrO_2$	215.05	620-72-4	32		140^{20}	3-06-00-00599		eth 4; EtOH 4	
120	Acetic acid, bromo-, propyl ester	Propyl bromoacetate	$C_5H_9BrO_2$	181.03	35223-80-4			176	4-02-00-00528	1.4099^{20}	ace 4; eth 4; EtOH 4	1.4518^{20}
121	Acetic acid, butyl ester	Butyl acetate	$C_6H_{12}O_2$	116.16	123-86-4	-78	1535	126.1	4-02-00-00143	0.8825^{20}	H_2O 2; EtOH 5; eth 5; ace 3	1.3941^{20}
122	Acetic acid, calcium salt		$C_4H_6CaO_4$	158.17	62-54-4	dec	1643		4-02-00-00094		H_2O 3; EtOH 2	1.55
123	Acetic acid, chloro-	Chloroacetic acid (β)	$C_2H_3ClO_2$	94.50	79-11-8	63	2111	189.3	4-02-00-00474	1.4043^{40}	H_2O 4; EtOH 3; eth 3; bz 3	1.4351^{55}
124	Acetic acid, chloro-, anhydride	Chloroacetic anhydride	$C_4H_4Cl_2O_3$	170.98	541-88-8	46	2112	203	4-02-00-00487	1.5497^{20}		
125	Acetic acid, chloro-, butyl ester		$C_6H_{11}ClO_2$	150.60	590-02-3			183	4-02-00-00482	1.0704^{20}	eth 4; EtOH 4	1.4297^{20}
126	Acetic acid, chloro-, 2-chloroethyl ester		$C_4H_6Cl_2O_2$	157.00	3848-12-2			203; 102^{20}	4-02-00-00481	1.3600^{25}	eth 4	1.4619^{25}
127	Acetic acid, chlorodifluoro-	Chlorodifluoroacetic acid	$C_2HClF_2O_2$	130.48	76-04-0	25		122	4-02-00-00497		chl 3	1.3559^{20}
128	Acetic acid, chloro-, 1,1-dimethylethyl ester	*tert*-Butyl chloroacetate	$C_6H_{11}ClO_2$	150.60	107-59-5		1563	150; 50^{10}	3-02-00-00444		H_2O 6	1.4260^{20}
129	Acetic acid, chloro-, 1,2-ethanediyl ester		$C_6H_8Cl_2O_4$	215.03	6941-69-1	45.5		157^{8}	3-02-00-00448		H_2O 1; eth 4	
130	Acetic acid, chloro-, ethyl ester	Ethyl chloroacetate	$C_4H_7ClO_2$	122.55	105-39-5	-21	3741	144.3	4-02-00-00481	1.1585^{20}	H_2O 1; EtOH 5; eth 5; ace 5	1.4215^{20}
131	Acetic acid, chlorofluoro-	Chlorofluoroacetic acid	$C_2H_2ClFO_2$	112.49	471-44-3			162	4-02-00-00493	1.532^{25}		1.4085^{25}
132	Acetic acid, chlorofluoro-, butyl ester	Chlorofluoroacetic acid, butyl ester	$C_6H_{10}ClFO_2$	168.60	368-34-3			165.5	3-02-00-00453	1.124^{25}		1.4067^{25}
133	Acetic acid, chlorofluoro-, ethyl ester		$C_4H_6ClFO_2$	140.54	401-56-9			129	4-02-00-00493	1.225^{20}		1.3927^{20}
134	Acetic acid, chlorofluoro-, methyl ester	Methyl chlorofluoroacetate	$C_3H_4ClFO_2$	126.51	433-52-3			116	4-02-00-00493	1.323^{25}		1.3903^{25}
135	Acetic acid, chlorofluoro-, propyl ester	Chlorofluoroacetic acid, propyl ester	$C_5H_8ClFO_2$	154.57	348-70-9			147	3-02-00-00453	1.170^{25}		1.3994^{25}
136	Acetic acid, chloro-, 2-hydroxyethyl ester		$C_4H_7ClO_3$	138.55	35280-53-6			240 dec; $86^{1.5}$	3-02-00-00447	1.3300^{20}	H_2O 4; EtOH 4	1.4609^{20}
137	Acetic acid, chloro-, 2-methoxyethyl ester		$C_5H_9ClO_3$	152.58	13361-36-9			$86\text{-}6^{9}$	3-02-00-00447	1.2015^{20}	eth 4	1.4382^{20}
138	Acetic acid, chloro-, methyl ester	Chloroacetic acid, methyl ester	$C_3H_5ClO_2$	108.52	96-34-4	-32.1	5965	129.5	4-02-00-00480	1.236^{20}	ace 4; bz 4; eth 4; EtOH 4	1.4218^{20}
139	Acetic acid, chloro-, 1-methylethyl ester	Isopropyl chloroacetate	$C_5H_9ClO_2$	136.58	105-48-6			150.5	4-02-00-00482	1.0888^{20}	eth 4	1.4382^{20}
140	Acetic acid, chloro-, 4-methylphenyl ester		$C_9H_9ClO_2$	184.62	15150-39-7	32		254; 162^{41}	4-06-00-02112	1.1840^{35}	eth 4; EtOH 4	
141	Acetic acid, (4-chloro-2-methylphenoxy)-	MCPA	$C_9H_9ClO_3$	200.62	94-74-6	120	5645		4-06-00-01991		H_2O 2; EtOH 4; eth 4; bz 3	
142	Acetic acid, chloro-, 1-methylpropyl ester		$C_6H_{11}ClO_2$	150.60	17696-64-9			163.5	3-02-00-00443	1.060^{20}	eth 4; EtOH 4	1.4251^{19}
143	Acetic acid, chloro-, 2-methylpropyl ester		$C_6H_{11}ClO_2$	150.60	13361-35-8			170	4-02-00-00483	1.0612^{20}	ace 4; eth 4	1.4255^{20}
144	Acetic acid, chlorooxo-, ethyl ester		$C_4H_5ClO_3$	136.53	4755-77-5			137	4-02-00-01853	1.2226^{20}	bz 4; eth 4	
145	Acetic acid, chlorooxo-, methyl ester		$C_3H_3ClO_3$	122.51	5781-53-3			119	3-02-00-01583	1.3316^{20}		1.4189^{20}
146	Acetic acid, (4-chlorophenoxy)-	(*p*-Chlorophenoxy)acetic acid	$C_8H_7ClO_3$	186.59	122-88-3	156.5			4-06-00-00845		H_2O 4; chl 2	
147	Acetic acid, chloro-, phenyl ester	Phenyl chloroacetate	$C_8H_7ClO_2$	170.60	620-73-5	44.5		232.5	3-06-00-00598	1.2202^{44}	H_2O 1; EtOH 4; eth 4	1.5146^{44}
148	Acetic acid, chloro-, phenylmethyl ester		$C_9H_9ClO_2$	184.62	140-18-1			147.5^{9}; $85^{0.4}$	4-06-00-02264	1.1223^{4}	eth 4; EtOH 4	1.5426^{18}
149	Acetic acid, chloro-, propyl ester	Propyl chloroacetate	$C_5H_9ClO_2$	136.58	5396-24-7			161	4-02-00-00482	1.1022^{20}	eth 4	1.4261^{20}

No.	Name Synonym	Mol. Form. Mol. Wt.	CAS RN mp/°C	Merck No. bp/°C	Beil. Ref. den/g cm^{-3}	Solubility n_D
150	Acetic acid, cyano- Cyanoacetic acid	$C_3H_3NO_2$ 85.06	372-09-8 66	2696 $108^{0.15}$; 160 dec	4-02-00-01888	H_2O 3; EtOH 3; eth 3; chl 2
151	Acetic acid, cyano-, butyl ester	$C_7H_{11}NO_2$ 141.17	5459-58-5	 231; 115^{15}	4-02-00-01891 1.0010^{20}	 1.4200^{20}
152	Acetic acid, cyano-, ethyl ester Ethyl cyanoacetate	$C_5H_7NO_2$ 113.12	105-56-6 -22.5	3744 205	4-02-00-01889 1.0654^{20}	eth 4; EtOH 4 1.4175^{20}
153	Acetic acid, cyano-, hydrazide Cyacetacide	$C_3H_5N_3O$ 99.09	140-87-4 114.5	2688	4-02-00-01895	H_2O 4; EtOH 4
154	Acetic acid, cyano-, methyl ester Methyl cyanoacetate	$C_4H_5NO_2$ 99.09	105-34-0 -22.5	 200.5	4-02-00-01889 1.1225^{25}	eth 4; EtOH 4 1.4176^{20}
155	Acetic acid, cyclohexyl ester Cyclohexyl acetate	$C_8H_{14}O_2$ 142.20	622-45-7	 173; 96^{75}	4-06-00-00036 0.968^{20}	eth 4; EtOH 4 1.442^{20}
156	Acetic acid, cyclohexylidene-, methyl ester Methyl cyclohexylideneacetate	$C_9H_{14}O_2$ 154.21	40203-74-5	 99^{20}	4-09-00-00124 1.0021^{19}	 1.4838^{19}
157	Acetic acid, decyl ester Decyl acetate	$C_{12}H_{24}O_2$ 200.32	112-17-4 -15	 244	4-02-00-00168 0.8671^{20}	H_2O 1; EtOH 3; eth 3; bz 3 1.4273^{20}
158	Acetic acid, diazo-, ethyl ester Diazoacetic ester	$C_4H_6N_2O_2$ 114.10	623-73-4 -22	2980 140 dec	4-03-00-01495 1.0852^{18}	H_2O 2; EtOH 5; eth 5; bz 5 1.4605^{20}
159	Acetic acid, dibromo- Dibromoacetic acid	$C_2H_2Br_2O_2$ 217.84	631-64-1 49	 195^{250}, 130^{16}	4-02-00-00533	H_2O 4; EtOH 4; eth 3
160	Acetic acid, dibromo-, ethyl ester Ethyl dibromoacetate	$C_4H_6Br_2O_2$ 245.90	617-33-4	 194	4-02-00-00533 1.8991^{20}	H_2O 1; EtOH 5; eth 5 1.5017^{13}
161	Acetic acid, dibromofluoro-, ethyl ester Ethyl dibromofluoroacetate	$C_4H_5Br_2FO_2$ 263.89	565-53-7	 83^{33}	4-02-00-00534 1.894^{25}	 1.4633^{25}
162	Acetic acid, dichloro- Dichloroacetic acid	$C_2H_2Cl_2O_2$ 128.94	79-43-6 13.5	3037 194; 102^{20}	4-02-00-00498 1.5634^{20}	H_2O 5; EtOH 5; eth 5; ace 3 1.4658^{20}
163	Acetic acid, dichloro-, butyl ester Butyl dichloroacetate	$C_6H_{10}Cl_2O_2$ 185.05	29003-73-4	 193.5	4-02-00-00502 1.1820^{20}	eth 4; EtOH 4 1.4420^{20}
164	Acetic acid, dichloro-, ethyl ester Ethyl dichloroacetate	$C_4H_6Cl_2O_2$ 157.00	535-15-9	 155; 56^{10}	4-02-00-00501 1.2827^{20}	H_2O 2; EtOH 5; eth 5; ace 3 1.4386^{20}
165	Acetic acid, dichloro-, 2-hydroxyethyl ester	$C_4H_6Cl_2O_3$ 173.00	17704-30-2	 81-$2^{0.5}$	3-02-00-00460 1.438^{20}	EtOH 4 1.4735^{20}
166	Acetic acid, dichloro-, methyl ester Methyl dichloroacetate	$C_3H_4Cl_2O_2$ 142.97	116-54-1 -51.9	 142.9	4-02-00-00501 1.3774^{20}	H_2O 1; EtOH 3; ctc 3 1.4429^{20}
167	Acetic acid, dichloro-, 1-methylethyl ester	$C_5H_8Cl_2O_2$ 171.02	25006-60-4	 163.5	4-02-00-00502 1.2053^{20}	eth 4; EtOH 4 1.4328^{20}
168	Acetic acid, [2,3-dichloro-4-(2-methylene-1-oxobutyl)phenoxy]- Ethacrynic acid	$C_{13}H_{12}Cl_2O_4$ 303.14	58-54-8 122.5	3669		
169	Acetic acid, (2,4-dichlorophenoxy)- 2,4-D	$C_8H_6Cl_2O_3$ 221.04	94-75-7 140.5	2802 $160^{0.4}$	4-06-00-00908	H_2O 1; EtOH 3; bz 2
170	Acetic acid, (2,4-dichlorophenoxy)-, 2-butoxyethyl ester 2,4-D 2-Butoxyethyl ester	$C_{14}H_{18}Cl_2O_4$ 321.20	1929-73-3	 156-62^{1}	4-06-00-00911 1.232^{20}	
171	Acetic acid, (2,4-dichlorophenoxy)-, butyl ester 2,4-D Butyl ester	$C_{12}H_{14}Cl_2O_3$ 277.15	94-80-4	2802	3-06-00-00706	
172	Acetic acid, (2,4-dichlorophenoxy)-, methyl ester 2,4-D Methyl ester	$C_9H_8Cl_2O_3$ 235.07	1928-38-7 119	 141^{18}	4-06-00-00909	
173	Acetic acid, (2,4-dichlorophenoxy)-, 1-methylethyl ester 2,4-D Isopropyl ester	$C_{11}H_{12}Cl_2O_3$ 263.12	94-11-1 5	 140^{1}	4-06-00-00910 1.26^{25}	 1.5209^{25}
174	Acetic acid, diethoxy-, ethyl ester	$C_8H_{16}O_4$ 176.21	6065-82-3	 199	4-03-00-01494 0.985^{25}	 1.4100^{20}
175	Acetic acid, difluoro- Difluoroacetic acid	$C_2H_2F_2O_2$ 96.03	381-73-7 -1	 133	4-02-00-00455 1.526^{25}	 1.3470^{20}
176	Acetic acid, difluoro-, ethyl ester Ethyl difluoroacetate	$C_4H_6F_2O_2$ 124.09	454-31-9	 100	4-02-00-00455 1.1893^{9}	H_2O 1
177	Acetic acid, 1,1-dimethylethyl ester *tert*-Butyl acetate	$C_6H_{12}O_2$ 116.16	540-88-5	1537 95.1	4-02-00-00151 0.8665^{20}	EtOH 3; eth 3; chl 3; HOAc 3 1.3855^{20}
178	Acetic acid, dodecyl ester	$C_{14}H_{28}O_2$ 228.38	112-66-3 1.3	 265; 180^{40}	4-02-00-00170 0.8652^{22}	 1.4439^{20}
179	Acetic acid ethenyl ester Vinyl acetate	$C_4H_6O_2$ 86.09	108-05-4 -93.2	9896 72.5	4-02-00-00176 0.9317^{20}	H_2O 1; EtOH 5; eth 3; ace 3 1.3959^{20}
180	Acetic acid, ethoxy- 2-Ethoxyacetic acid	$C_4H_8O_3$ 104.11	627-03-2	 206.5	4-03-00-00574 1.1021^{20}	H_2O 4; EtOH 4; eth 4; chl 3 1.4194^{20}
181	Acetic acid, ethoxy-, ethyl ester	$C_6H_{12}O_3$ 132.16	817-95-8	 158	4-03-00-00581 0.9702^{20}	EtOH 3; eth 3; ace 3 1.4039^{20}
182	Acetic acid, ethoxy-, methyl ester Methyl ethoxyacetate	$C_5H_{10}O_3$ 118.13	17640-26-5	 147	4-03-00-00578 1.0112^{15}	ace 4; eth 4; EtOH 4
183	Acetic acid, ethoxy-, 5-methyl-2-(1-methylethyl)cyclohexyl ester, (1α,2β,5α)-	$C_{14}H_{26}O_3$ 242.36	579-94-2	 155^{20}	2-06-00-00047 0.9545^{20}	eth 4; EtOH 4; chl 4
184	Acetic acid, (ethoxysulfonyl)-, ethyl ester Acetic acid, sulfo, diethyl ester	$C_6H_{12}O_5S$ 196.22	59376-54-4	 $116^{0.5}$	3-04-00-00058 1.232^{16}	 1.436^{21}
185	Acetic acid, 2-ethylbutyl ester 2-Ethylbutyl acetate	$C_8H_{16}O_2$ 144.21	10031-87-5 <-100	 162.5	4-02-00-00161 0.8790^{20}	H_2O 1; EtOH 3; eth 3; ctc 3 1.4109^{20}
186	Acetic acid ethyl ester Ethyl acetate	$C_4H_8O_2$ 88.11	141-78-6 -83.6	3713 77.1	4-02-00-00127 0.9003^{20}	H_2O 3; EtOH 5; eth 5; ace 4 1.3723^{20}

No.	Name Synonym	Mol. Form. Mol. Wt.	CAS RN mp/°C	Merck No. bp/°C	Beil. Ref. den/g cm^{-3}	Solubility n_D
187	Acetic acid, 2-ethylhexyl ester 2-Ethylhexyl acetate	$C_{10}H_{20}O_2$ 172.27	103-09-3 -80	6683 199	4-02-00-00166 0.8718^{20}	H_2O 1; EtOH 3; eth 3 1.4204^{20}
188	Acetic acid, (ethylthio)-	$C_4H_8O_2S$ 120.17	627-04-3 -8.5	 164^{83}; 109^{5}	4-03-00-00603 1.1497^{20}	H_2O 4; EtOH 4; eth 4
189	Acetic acid, fluoro- Fluoroethanoic acid	$C_2H_3FO_2$ 78.04	144-49-0 35.2	4096 168	4-02-00-00446 1.3693^{36}	H_2O 3; EtOH 3
190	Acetic acid, fluoro-, 1,1-dimethylethyl ester	$C_6H_{11}FO_2$ 134.15	406-74-6	 132	4-02-00-00450 0.9904^{20}	 1.386^{20}
191	Acetic acid, fluoro-, ethyl ester	$C_4H_7FO_2$ 106.10	459-72-3	 120	4-02-00-00447 1.098^{25}	H_2O 4 1.3755^{20}
192	Acetic acid, fluoro-, methyl ester Methyl fluoroacetate	$C_3H_5FO_2$ 92.07	453-18-9 -35	 104.5	4-02-00-00447 1.1613^{15}	
193	Acetic acid, heptyl ester Heptyl acetate	$C_9H_{18}O_2$ 158.24	112-06-1 -50.2	 193	4-02-00-00162 0.8750^{15}	H_2O 1; EtOH 3; eth 3; ctc 3 1.4150^{20}
194	Acetic acid, hexyl ester	$C_8H_{16}O_2$ 144.21	142-92-7 -80.9	 171.5	4-02-00-00159 0.8779^{15}	eth 4; EtOH 4 1.4092^{20}
195	Acetic acid, hydrazide	$C_2H_6N_2O$ 74.08	1068-57-1 67	 137^{25}	4-02-00-00435	H_2O 3; EtOH 3; eth 2
196	Acetic acid, hydroxy- Glycolic acid	$C_2H_4O_3$ 76.05	79-14-1 79.5	4394 100	4-03-00-00571	H_2O 3; EtOH 3; eth 3
197	Acetic acid, hydroxy-, ethyl ester	$C_4H_8O_3$ 104.11	623-50-7	 160	4-03-00-00580 1.0826^{23}	eth 4; EtOH 4 1.4180^{20}
198	Acetic acid, hydroxy-, methyl ester	$C_3H_6O_3$ 90.08	96-35-5	 149; 52^{17}	4-03-00-00578 1.1677^{18}	H_2O 3; EtOH 5; eth 5
199	Acetic acid, hydroxy-, propyl ester Hydroxyacetic acid, propyl ester	$C_5H_{10}O_3$ 118.13	90357-58-7	 170.5	4-03-00-00588 1.0631^{18}	 1.4231^{18}
200	Acetic acid, iodo- Iodoacetic acid	$C_2H_3IO_2$ 185.95	64-69-7 82.5	4918 dec	4-02-00-00534	H_2O 3; EtOH 3; eth 2; chl 2
201	Acetic acid, iodo-, ethyl ester	$C_4H_7IO_2$ 214.00	623-48-3	 179	4-02-00-00535 1.8173^{13}	EtOH 3; eth 3 1.5079^{13}
202	Acetic acid, mercapto- Thioglycolic acid	$C_2H_4O_2S$ 92.12	68-11-1 -16.5	9265 120^{20}	4-03-00-00600 1.3253^{20}	H_2O 5; EtOH 5; eth 5; chl 2 1.5080^{20}
203	Acetic acid, mercapto-, ethyl ester	$C_4H_8O_2S$ 120.17	623-51-8	 157	4-03-00-00616 1.0964^{15}	EtOH 3; eth 3; ctc 2 1.4582^{20}
204	Acetic acid, mercapto-, methyl ester	$C_3H_6O_2S$ 106.15	2365-48-2	 $42\text{-}3^{10}$	4-03-00-00614	eth 4; EtOH 4 1.4657^{20}
205	Acetic acid, methoxy-	$C_3H_6O_3$ 90.08	625-45-6	 203.5	4-03-00-00574 1.1768^{20}	H_2O 3; EtOH 3; eth 3 1.4168^{20}
206	Acetic acid, methoxy-, ethyl ester Ethyl methoxyacetate	$C_5H_{10}O_3$ 118.13	3938-96-3	 142	3-03-00-00381 1.0118^{15}	H_2O 2; EtOH 4; eth 4; ctc 3 1.4050^{20}
207	Acetic acid, methoxy-, methyl ester	$C_4H_8O_3$ 104.11	6290-49-9	 131	4-03-00-00578 1.0511^{20}	H_2O 2; EtOH 4; eth 4; ace 3 1.3962^{20}
208	Acetic acid, methyl ester Methyl acetate	$C_3H_6O_2$ 74.08	79-20-9 -98	5932 56.8	4-02-00-00122 0.9342^{20}	H_2O 4; eth 4; EtOH 4 1.3614^{20}
209	Acetic acid, 1-methylethyl ester Isopropyl acetate	$C_5H_{10}O_2$ 102.13	108-21-4 -73.4	5093 88.6	4-02-00-00141 0.8718^{20}	H_2O 3; EtOH 3; eth 5; ace 3 1.3773^{20}
210	Acetic acid, [[5-methyl-2-(1-methylethyl)cyclohexyl]oxy]-, [1*R*-(1α,2β,5α)]- *l*-Methoxyacetic acid	$C_{12}H_{22}O_3$ 214.30	40248-63-3 54	 171^{11}	4-06-00-00157	H_2O 2; EtOH 4; eth 3; ace 4
211	Acetic acid, 2-methylphenyl ester *o*-Cresyl acetate	$C_9H_{10}O_2$ 150.18	533-18-6	2588 208	4-06-00-01960 1.0533^{15}	eth 4; EtOH 4 1.5002^{20}
212	Acetic acid, 3-methylphenyl ester *m*-Cresyl acetate	$C_9H_{10}O_2$ 150.18	122-46-3 12	2587 212	4-06-00-02047 1.043^{20}	bz 4; eth 4; EtOH 4 1.4978^{20}
213	Acetic acid, 4-methylphenyl ester	$C_9H_{10}O_2$ 150.18	140-39-6	 212.5	4-06-00-02112 1.0512^{17}	H_2O 2; EtOH 3; eth 3; ctc 2 1.5163^{22}
214	Acetic acid, 1-methylpropyl ester *sec*-Butyl acetate	$C_6H_{12}O_2$ 116.16	105-46-4 -98.9	1536 112	4-02-00-00148 0.8748^{20}	H_2O 1; EtOH 3; eth 3; ctc 2 1.3888^{20}
215	Acetic acid, 2-methylpropyl ester Isobutyl acetate	$C_6H_{12}O_2$ 116.16	110-19-0 -98.8	5014 116.5	4-02-00-00149 0.8712^{20}	H_2O 2; EtOH 5; eth 5; ace 3 1.3902^{20}
216	Acetic acid, (methylthio)-	$C_3H_6O_2S$ 106.15	2444-37-3	 $130\text{-}1^{27}$	4-03-00-00603 1.221^{20}	 1.495^{20}
217	Acetic acid, (2-naphthalenyloxy)- 2-Naphthoxyacetic acid	$C_{12}H_{10}O_3$ 202.21	120-23-0 156	6317	4-06-00-04274	H_2O 3; EtOH 3; eth 3; HOAc 3
218	Acetic acid, nitro-, ethyl ester Ethyl nitroacetate	$C_4H_7NO_4$ 133.10	626-35-7	 106^{25}; 83^{6}	4-02-00-00537 1.1953^{20}	H_2O 2; EtOH 5; eth 4; dil alk 3 1.4250^{20}
219	Acetic acid, 2-nitrophenyl ester	$C_8H_7NO_4$ 181.15	610-69-5 40.5	 253 dec; 141^{11}	4-06-00-01256	H_2O 3; EtOH 4; eth 4; ace 4
220	Acetic acid, nonyl ester Nonyl acetate	$C_{11}H_{22}O_2$ 186.29	143-13-5 -26	6597 210	4-02-00-00167 0.8785^{15}	 1.426^{20}
221	Acetic acid, octadecyl ester	$C_{20}H_{40}O_2$ 312.54	822-23-1 34.5	 208^{9}	4-02-00-00171 0.8510^{30}	EtOH 4
222	Acetic acid, octyl ester	$C_{10}H_{20}O_2$ 172.27	112-14-1 -38.5	 210	4-02-00-00165 0.8705^{20}	H_2O 1; EtOH 3; eth 3; ctc 2 1.4150^{20}
223	Acetic acid, oxo- Glyoxylic acid	$C_2H_2O_3$ 74.04	298-12-4 98	4407	4-03-00-01489	H_2O 4; EtOH 2; eth 2; bz 2
224	Acetic acid, oxo(phenylamino)-, ethyl ester	$C_{10}H_{11}NO_3$ 193.20	1457-85-8 65.5	 280	3-12-00-00550	ace 4; bz 4; eth 4; EtOH 4

No.	Name Synonym	Mol. Form. Mol. Wt.	CAS RN mp/°C	Merck No. bp/°C	Beil. Ref. den/g cm^{-3}	Solubility n_D
225	Acetic acid, [(4-oxo-2-phenyl-4H-1-benzopyran-7-yl)oxy]-, ethyl ester Efloxate	$C_{19}H_{16}O_5$ 324.33	119-41-5 123.7	3490	4-18-00-00695	chl 3
226	Acetic acid, 2,2'-oxybis-	$C_4H_6O_5$ 134.09	110-99-6 148	 dec	4-03-00-00577	H_2O 4; eth 4; EtOH 4
227	Acetic acid, pentyl ester Pentyl acetate	$C_7H_{14}O_2$ 130.19	628-63-7 -70.8	 149.2	4-02-00-00152 0.8756^{20}	H_2O 2; EtOH 5, eth 5; ctc 3 1.4023^{20}
228	Acetic acid, phenoxy- Phenoxyacetic acid	$C_8H_8O_3$ 152.15	122-59-8 98.5	7223 285 dec	4-06-00-00634	H_2O 3; EtOH 4; eth 4; bz 4
229	Acetic acid, phenoxy-, ethyl ester Ethyl phenoxyacetate	$C_{10}H_{12}O_3$ 180.20	2555-49-9	 $109\text{-}12^{3}$	4-06-00-00635 1.1082^{20}	 1.5080^{20}
230	Acetic acid, phenoxy-, methyl ester Methyl phenoxyacetate	$C_9H_{10}O_3$ 166.18	2065-23-8	 245	4-06-00-00635 1.1493^{20}	eth 4; EtOH 4 1.5155^{20}
231	Acetic acid, phenyl ester Phenyl acetate	$C_8H_8O_2$ 136.15	122-79-2	7238 196; 75^{8}	4-06-00-00611 1.0780^{20}	H_2O 2; EtOH 5; eth 5; ctc 3 1.5035^{20}
232	Acetic acid, 2-phenylethyl ester	$C_{10}H_{12}O_2$ 164.20	103-45-7 -31.1	 232.6	4-06-00-03073 1.0883^{20}	eth 4; EtOH 4 1.5171^{20}
233	Acetic acid, phenylmethyl ester Benzyl acetate	$C_9H_{10}O_2$ 150.18	140-11-4 -51.3	1137 213	4-06-00-02262 1.0550^{20}	H_2O 2; EtOH 5; eth 3; ace 3 1.5232^{20}
234	Acetic acid, potassium salt	$C_2H_3KO_2$ 98.14	127-08-2 292	7580	4-02-00-00094 1.57^{25}	H_2O 4; EtOH 3; eth 1; ace 1
235	Acetic acid, 2-propenyl ester Allyl acetate	$C_5H_8O_2$ 100.12	591-87-7	 103.5	4-02-00-00180 0.9275^{20}	H_2O 2; EtOH 5; eth 5; ace 3 1.4049^{20}
236	Acetic acid, propyl ester Propyl acetate	$C_5H_{10}O_2$ 102.13	109-60-4 -93	7853 101.5	4-02-00-00138 0.8878^{20}	H_2O 2; EtOH 5; eth 5; ctc 3 1.3842^{20}
237	Acetic acid, sodium salt	$C_2H_3NaO_2$ 82.03	127-09-3 324	8513	4-02-00-00094 1.528^{25}	H_2O 4; EtOH 2 1.464
238	Acetic acid, sulfo- Sulfoacetic acid	$C_2H_4O_5S$ 140.12	123-43-3 85	8932 245 dec	4-04-00-00102	H_2O 4; ace 4; EtOH 4
239	Acetic acid, 2,2'-sulfonylbis- Sulfonyldiacetic acid	$C_4H_6O_6S$ 182.15	123-45-5 187	8940	3-03-00-00424	H_2O 4; EtOH 4; eth 3; sulf 3
240	Acetic acid, 2,2'-thiobis- Thiodiglycolic acid	$C_4H_6O_4S$ 150.16	123-93-3 129	9260	4-03-00-00612	H_2O 2; EtOH 4; bz 3
241	Acetic acid, tribromo- Tribromoacetic acid	$C_2HBr_3O_2$ 296.74	75-96-7 132	9521 245 dec	4-02-00-00534	H_2O 3; EtOH 3; eth 3
242	Acetic acid, tribromo-, ethyl ester Ethyl tribromoacetate	$C_4H_5Br_3O_2$ 324.79	599-99-5	 225	3-02-00-00485 2.2260^{20}	EtOH 3; eth 3 1.5438^{13}
243	Acetic acid, trichloro- Trichloroacetic acid	$C_2HCl_3O_2$ 163.39	76-03-9 57.5	9539 196.5	4-02-00-00508 1.6126^{64}	H_2O 4; EtOH 3; eth 3; ctc 2 1.4603^{61}
244	Acetic acid, trichloro-, anhydride Trichloroacetic anhydride	$C_4Cl_6O_3$ 308.76	4124-31-6	 223 dec	4-02-00-00518 1.6908^{20}	eth 4; HOAc 4
245	Acetic acid, trichloro-, butyl ester	$C_6H_9Cl_3O_2$ 219.49	3657-07-6	 204	4-02-00-00515 1.2778^{20}	ctc 3 1.4525^{25}
246	Acetic acid, trichloro-, 2-chloroethyl ester	$C_4H_4Cl_4O_2$ 225.89	4974-21-4	 217	4-02-00-00515 1.5357^{20}	eth 4 1.4813^{20}
247	Acetic acid, trichloro-, 1,1-dimethylethyl ester	$C_6H_9Cl_3O_2$ 219.49	1860-21-5 25.5	 $54\text{-}5^{7}$	4-02-00-00516 1.2363^{25}	eth 4; EtOH 4 1.4398^{25}
248	Acetic acid, trichloro-, ethyl ester Ethyl trichloroacetate	$C_4H_5Cl_3O_2$ 191.44	515-84-4	9539 167.5	4-02-00-00514 1.3836^{20}	H_2O 1; EtOH 3; eth 3; bz 3 1.4505^{20}
249	Acetic acid, trichloro-, 2-hydroxyethyl ester	$C_4H_5Cl_3O_3$ 207.44	33560-17-7	 $130\text{-}4^{12}$	3-02-00-00474 1.532^{20}	EtOH 4 1.4775^{20}
250	Acetic acid, trichloro-, 2-methoxyethyl ester 2-Methoxyethyl trichloroacetate	$C_5H_7Cl_3O_3$ 221.47	35449-34-4 14.5	 $98\text{-}9^{17}$	3-02-00-00474 1.3826^{20}	bz 4; eth 4; EtOH 4 1.4563^{20}
251	Acetic acid, trichloro-, 1-methylpropyl ester	$C_6H_9Cl_3O_2$ 219.49	4484-80-4	 $93\text{-}4^{24}$	3-02-00-00472 1.2636^{20}	bz 4; eth 4; EtOH 4 1.4483^{20}
252	Acetic acid, trichloro-, 2-methylpropyl ester	$C_6H_9Cl_3O_2$ 219.49	33560-15-5	 188	4-02-00-00515 1.2636^{20}	bz 4; eth 4; EtOH 4 1.4483^{20}
253	Acetic acid, trichloro-, methyl ester Methyl trichloroacetate	$C_3H_3Cl_3O_2$ 177.41	598-99-2 -17.5	 153.8	4-02-00-00513 1.4874^{20}	H_2O 1; EtOH 4; eth 4; ctc 3 1.4572^{20}
254	Acetic acid, trichloro-, 1-methylethyl ester	$C_5H_7Cl_3O_2$ 205.47	3974-99-0	 175; 66^{15}	4-02-00-00515 1.3034^{20}	bz 4; eth 4; EtOH 4 1.4428^{20}
255	Acetic acid, trichloro-, 3-methylbutyl ester	$C_7H_{11}Cl_3O_2$ 233.52	57392-55-9	 217	4-02-00-00516 1.2314^{20}	eth 4; EtOH 4 1.4521^{20}
256	Acetic acid, (2,4,5-trichlorophenoxy)- 2,4,5-T	$C_8H_5Cl_3O_3$ 255.48	93-76-5 153	8999	4-06-00-00973	H_2O 1; EtOH 3; bz 4
257	Acetic acid, (2,4,5-trichlorophenoxy)-, 2-butoxyethyl ester 2,4,5-T Butoxyethyl ester	$C_{14}H_{17}Cl_3O_4$ 355.64	2545-59-7	 $163\text{-}6^{1}$	4-06-00-00978 1.280^{20}	ctc 3
258	Acetic acid, (2,4,5-trichlorophenoxy)-, butyl ester 2,4,5-T Butyl ester	$C_{12}H_{13}Cl_3O_3$ 311.59	93-79-8 28.5		4-06-00-00975	
259	Acetic acid, (2,4,5-trichlorophenoxy)-, 1-methylethyl ester 2,4,5-T Isopropyl ester	$C_{11}H_{11}Cl_3O_3$ 297.56	93-78-7 45		4-06-00-00974	chl 4
260	Acetic acid, (2,4,5-trichlorophenoxy)-, 1-methylpropyl ester 2,4,5-T *sec*-Butyl ester	$C_{12}H_{13}Cl_3O_3$ 311.59	61792-07-2 39		4-06-00-00975	
261	Acetic acid, trichloro-, propyl ester	$C_5H_7Cl_3O_2$ 205.47	13313-91-2	 187	4-02-00-00515 1.3221^{20}	eth 4; EtOH 4 1.4501^{20}
262	Acetic acid, trichloro-, trichloromethyl ester	$C_3Cl_6O_2$ 280.75	6135-29-1 34	 191.5	2-03-00-00015 1.6733^{35}	bz 4; eth 4; chl 4; lig 4

PHYSICAL CONSTANTS OF ORGANIC COMPOUNDS (continued)

No.	Name	Synonym	Mol. Form.	Mol. Wt.	CAS RN	mp/°C	Merck No.	bp/°C	Beil. Ref.	den/g cm^{-3}	Solubility	n_D
263	Acetic acid, trifluoro-	Trifluoroacetic acid	$C_2HF_3O_2$	114.02	76-05-1	-15.2	9595	73	4-02-00-00458	1.5351^{25}	H_2O 3; EtOH 3; eth 3; ace 3	
264	Acetic acid, trifluoro-, anhydride		$C_4F_6O_3$	210.03	407-25-0	-65		39.5	4-02-00-00469	1.490^{25}		1.269^{25}
265	Acetic acid, trifluoro-, butyl ester	Butyl trifluoroacetate	$C_6H_9F_3O_2$	170.13	367-64-6			102	4-02-00-00464	1.0268^{22}	chl 3	1.353^{22}
266	Acetic acid, trifluoro-, 1,1-dimethylethyl ester		$C_6H_9F_3O_2$	170.13	400-52-2			83	4-02-00-00465		chl 3	1.3300^{25}
267	Acetic acid, trifluoro-, ethyl ester		$C_4H_5F_3O_2$	142.08	383-63-1			61	4-02-00-00463	1.194^{20}		1.308^{20}
268	Acetic acid, trifluoro-, 2-methylpentyl ester	2-Methylpentyl trifluoroacetate	$C_8H_{13}F_3O_2$	198.19	10042-29-2			140	4-02-00-00466	1.0504^{20}	chl 2	
269	Acetic acid, trifluoro-, propyl ester	Propyl trifluoroacetate	$C_5H_7F_3O_2$	156.10	383-66-4			82.5	4-02-00-00464	1.1285^{25}	chl 3	1.3233^{22}
270	Acetochlor	Acetamide, 2-chloro-*N*-(ethoxymethyl)-*N*-(2-ethyl-6-methylphenyl)-	$C_{14}H_{20}ClNO_2$	269.77	34256-82-1							
271	Acetonitrile	Methyl cyanide	C_2H_3N	41.05	75-05-8	-43.8	62	81.6	4-02-00-00419	0.7857^{20}	H_2O 5; EtOH 5; eth 5; ace 5	1.3442^{30}
272	Acetonitrile, amino-	Aminoacetonitrile	$C_2H_4N_2$	56.07	540-61-4		423	58^{15}	4-04-00-02363		EtOH 4	
273	Acetonitrile, (butylamino)-		$C_6H_{12}N_2$	112.17	3010-04-6			85^{9}	4-04-00-02382	0.891^{25}		1.4337^{20}
274	Acetonitrile, chloro-	Chloromethyl cyanide	C_2H_2ClN	75.50	107-14-2			126.5	4-02-00-00492	1.1930^{20}	eth 4; EtOH 4	1.4202^{25}
275	Acetonitrile, (cyclohexylamino)-		$C_8H_{14}N_2$	138.21	1074-58-4	18		76^{1}	4-12-00-00073	0.9657^{25}		
276	Acetonitrile, cyclohexylidene-	Cyclohexylideneacetonitrile	$C_8H_{11}N$	121.18	4435-18-1			$107\text{-}8^{22}$	4-09-00-00124	0.9483^{15}	eth 4; EtOH 4	1.4382^{25}
277	Acetonitrile, dibromo-	Dibromoacetonitrile	C_2HBr_2N	198.84	3252-43-5			$67\text{-}9^{24}$	4-02-00-00533	2.296^{25}		1.5393^{20}
278	Acetonitrile, dichloro-	Dichloroacetonitrile	C_2HCl_2N	109.94	3018-12-0			112.5	4-02-00-00506	1.369^{20}	MeOH 3	1.4391^{25}
279	Acetonitrile, dichlorofluoro-	Dichlorofluoroacetonitrile	C_2Cl_2FN	127.93	353-82-2	-110		33	4-02-00-00508	1.3909^{25}		1.3682^{20}
280	Acetonitrile, (diethylamino)-		$C_6H_{12}N_2$	112.17	3010-02-4			170	4-04-00-02378	0.8660^{20}	H_2O 3	1.4260^{20}
281	Acetonitrile, (dimethylamino)-	Glycinonitrile, *N,N*-dimethyl-	$C_4H_8N_2$	84.12	926-64-7			137.5	4-04-00-02368	0.8649^{20}	H_2O 4; EtOH 4	1.4095^{20}
282	Acetonitrile, hydroxy-	Glyconitrile	C_2H_3NO	57.05	107-16-4	<-72		183 dec; 119^{24}	4-03-00-00598		H_2O 4; EtOH 4; eth 4; bz 1	1.4117^{19}
283	Acetonitrile, iodo-	Iodoacetonitrile	C_2H_2IN	166.95	624-75-9			185	2-02-00-00206	2.307^{25}		1.5744^{20}
284	Acetonitrile, methoxy-		C_3H_5NO	71.08	1738-36-9			119	3-03-00-00399	0.9492^{20}	H_2O 2; EtOH 3; eth 3; ace 3	1.3831^{20}
285	Acetonitrile, (methyleneamino)-	Methyleneaminoacetonitrile	$C_3H_4N_2$	68.08	109-82-0	129	5976					
286	Acetonitrile, (methylthio)-	Cyanomethyl methyl sulfide	C_3H_5NS	87.15	35120-10-6			$61\text{-}3^{15}$	4-03-00-00630	1.039^{25}		1.4826^{20}
287	Acetonitrile, phenoxy-	Phenoxyacetonitrile	C_8H_7NO	133.15	3598-14-9			239.5	4-06-00-00640	1.0991^{20}	eth 4; EtOH 4	1.5246^{20}
288	Acetonitrile, trichloro-	Trichloroacetonitrile	C_2Cl_3N	144.39	545-06-2	-42	9540	85.7	4-02-00-00524	1.4403^{25}	H_2O 1	1.4409^{20}
289	Acetonitrile, trifluoro-	Trifluoroacetonitrile	C_2F_3N	95.02	353-85-5			-68.8	4-02-00-00472			
290	Acetonitrile, trinitro-	Trinitroacetonitrile	$C_2N_4O_6$	176.05	630-72-8	41.5		exp 220	0-02-00-00229		eth 4	
291	Acetyl bromide	Acetic acid, bromide	C_2H_3BrO	122.95	506-96-7	-96	76	76	4-02-00-00398	1.6625^{16}	eth 5; ace 3; bz 5; chl 5	1.4486^{20}
292	Acetyl bromide, bromo-	Bromoacetyl bromide	$C_2H_2Br_2O$	201.85	598-21-0			148.5	4-02-00-00530	2.312^{22}	ace 3; ctc 3	1.5449^{20}
293	Acetyl bromide, trichloro-	Trichloroacetyl bromide	C_2BrCl_3O	226.28	34069-94-8			143	3-02-00-00476	1.898^{15}	bz 4; eth 4	
294	Acetyl chloride	Ethanoyl chloride	C_2H_3ClO	78.50	75-36-5	-112.8	79	50.7	4-02-00-00395	1.1051^{20}	eth 5; ace 5; bz 5; ctc 3	1.3886^{20}
295	Acetyl chloride, chloro-	Chloroacetyl chloride	$C_2H_2Cl_2O$	112.94	79-04-9	-22	2054	106	4-02-00-00488	1.4202^{20}	eth 5; ace 3; ctc 3	1.4530^{20}
296	Acetyl chloride, chlorofluoro-	Chlorofluoroacetyl chloride	C_2HCl_2FO	130.93	359-32-0			69.5	3-02-00-00453	1.468^{25}		1.3992^{25}
297	Acetyl chloride, dichloro-	Dichloroacetyl chloride	C_2HCl_3O	147.39	79-36-7		3040	108	4-02-00-00504	1.5315^{16}	H_2O 6; EtOH 6; eth 5	1.4591^{20}
298	Acetyl chloride, ethoxy-		$C_4H_7ClO_2$	122.55	14077-58-8			125	3-03-00-00396	1.1170^{25}	ace 4; eth 4	1.4204^{20}
299	Acetyl chloride, methoxy-		$C_3H_5ClO_2$	108.52	38870-89-2			112.5	3-03-00-00396	1.1871^{20}	eth 3; ace 3; ctc 3; chl 4	1.4199^{20}
300	Acetyl chloride, trichloro-	Trichloroacetyl chloride	C_2Cl_4O	181.83	76-02-8			117.9	4-02-00-00519	1.6202^{20}	eth 5	1.4695^{20}
301	Acetyl fluoride	Acetic acid, fluoride	C_2H_3FO	62.04	557-99-3	-84		20.8	4-02-00-00393	1.032^{25}	EtOH 5; eth 5; bz 3; chl 3	

No.	Name Synonym	Mol. Form. Mol. Wt.	CAS RN mp/°C	Merck No. bp/°C	Beil. Ref. den/g cm^{-3}	Solubility n_D
302	Acetyl iodide	C_2H_3IO 169.95	507-02-8	88 108	4-02-00-00399 2.0673^{20}	eth 4 1.5491^{20}
303	Acetyl isothiocyanate Isothiocyanic acid, acetyl	C_3H_3NOS 101.13	13250-46-9	 132.5	3-03-00-00276 1.1523^{13}	eth 3; CS_2 3 1.5231^{18}
304	Acifluorfen Benzoic acid, 5-[2-chloro-4-(trifluoromethyl)phenoxy]-2-nitro-	$C_{14}H_7ClF_3NO_5$ 361.66	50594-66-6 150	105		
305	Aconitane-3,8,13,14,15-pentol, 20-ethyl-1,6,16-trimethoxy-4-(methoxymethyl)-, Aconine	$C_{25}H_{41}NO_9$ 499.60	509-20-6 132	110	5-21-06-00310	H_2O 3; EtOH 3; eth 2; chl 3
306	Aconitane-3,8,13,14,15-pentol, 20-ethyl-1,6,16-trimethoxy-4-(methoxymethyl)-, 8-acetate 14-benzoate, Aconitine	$C_{34}H_{47}NO_{11}$ 645.75	302-27-2 204	113	5-21-06-00310	bz 4; EtOH 4; chl 4
307	Aconitane-3,8,13,14-tetrol, 20-ethyl-1,6,16-trimethoxy-4-(methoxymethyl)-, 8-acetate 14-benzoate, (1α,3alp Indaconitine	$C_{34}H_{47}NO_{10}$ 629.75	4491-19-4 202-3 dec	4842	5-21-06-00308	eth 4; EtOH 4; chl 4
308	Aconitane-3,8,13,14-tetrol, 20-ethyl-1,6,16-trimethoxy-4-(methoxymethyl)-, 8-acetate 14-(3,4-dimethoxybenzoate),(1α,3α,6α,14α,16β)- Pseudoaconitine	$C_{36}H_{51}NO_{12}$ 689.80	127-29-7 214	7924	5-21-06-00309	eth 4; EtOH 4
309	Aconitane-4,8,9-triol, 20-ethyl-1,14,16-trimethoxy-, 4-[2-(acetylamino)benzoate], (1α,14α,16β)- Lappaconitine	$C_{32}H_{44}N_2O_8$ 584.71	32854-75-4 217.5	5237	5-21-06-00291	H_2O 1; EtOH 2; eth 2; bz 3
310	Aconitane-8,13,14-triol, 20-ethyl-1,6,16-trimethoxy-4-(methoxymethyl)-, 8-acetate 14-(3,4-dimethoxybenzoate), Bikhaconitine	$C_{36}H_{51}NO_{11}$ 673.80	6078-26-8 164	1234	5-21-06-00303	eth 4; EtOH 4; chl 4
311	Aconitane-8,13,14-triol, 1,6,16-trimethoxy-4-(methoxymethyl)-20-methyl-, 8-acetate 14-benzoate, Delphinine	$C_{33}H_{45}NO_9$ 599.72	561-07-9 199	2867	5-21-06-00302	H_2O 1; chl 3; ace 3; eth 3
312	4-Acridinamine	$C_{13}H_{10}N_2$ 194.24	578-07-4 108	 183.5	5-22-11-00008	ace 4; bz 4; eth 4; EtOH 4
313	9-Acridinamine Aminacrine	$C_{13}H_{10}N_2$ 194.24	90-45-9 241	418	5-22-11-00008	EtOH 3; ace 3; dil HCl 4
314	Acridine Dibenzo[b,e]pyridine	$C_{13}H_9N$ 179.22	260-94-6 106(form a); 110(form b)	117 345.5	5-20-08-00199 1.005^{20}	H_2O 2; EtOH 4; eth 4; bz 4
315	Acridine, 9-chloro- 9-Chloroacridine	$C_{13}H_8ClN$ 213.67	1207-69-8 121	 sub	5-20-08-00210	H_2O 4; EtOH 4
316	3,6-Acridinediamine Proflavine	$C_{13}H_{11}N_3$ 209.25	92-62-6 285	7780	5-22-11-00322	H_2O 3; EtOH 4; eth 2; bz 2
317	3,9-Acridinediamine, 7-ethoxy- Ethacridine	$C_{15}H_{15}N_3O$ 253.30	442-16-0 226	3668	5-22-12-00243	
318	Acridine, 9-phenyl- 9-Phenylacridine	$C_{19}H_{13}N$ 255.32	602-56-2 184	 404	5-20-08-00564	H_2O 1; EtOH 2; eth 3; bz 4
319	Adenosine β-*D*-Ribofuranoside, adenine-9	$C_{10}H_{13}N_5O_4$ 267.24	58-61-7 235.5	143	4-26-00-03598	H_2O 2; EtOH 1
320	5'-Adenylic acid Adenosine 5'-monophosphate	$C_{10}H_{14}N_5O_7P$ 347.22	61-19-8 195 dec	148	4-26-00-03615	H_2O 4; EtOH 3
321	Ajmalan-17,21-diol, (17*R*,21α)- Ajmaline	$C_{20}H_{26}N_2O_2$ 326.44	4360-12-7 206	184	4-23-00-03212	H_2O 1; EtOH 3; eth 2; bz 2
322	Alachlor Acetamide, 2-chloro-*N*-(2,6-diethylphenyl)-*N*-(methoxymethyl)-	$C_{14}H_{20}ClNO_2$ 269.77	15972-60-8 40	193 $100^{0.02}$	 1.133^{25}	
323	β-Alanine 3-Aminopropanoic acid	$C_3H_7NO_2$ 89.09	107-95-9 200 dec	196	4-04-00-02526 1.437^{19}	H_2O 3; EtOH 2; eth 1; ace 1
324	*DL*-Alanine *DL*-2-Aminopropanoic acid	$C_3H_7NO_2$ 89.09	302-72-7 300 dec	195 sub 250	4-04-00-02481 1.424^{25}	H_2O 3; EtOH 4
325	*L*-Alanine (*S*)-2-Aminopropanoic acid	$C_3H_7NO_2$ 89.09	56-41-7 300 dec	195 sub 250	4-04-00-02480 1.432^{22}	H_2O 3; EtOH 2; eth 1; ace 1
326	Alanine, 3-amino- 2,3-Diaminopropionic acid	$C_3H_8N_2O_2$ 104.11	515-94-6 110	2962	4-04-00-02581	H_2O 4
327	*DL*-Alanine, *N*-benzoyl-	$C_{10}H_{11}NO_3$ 193.20	1205-02-3 165.5	 dec	4-09-00-00794	H_2O 3; EtOH 3; eth 2; DMSO 2
328	β-Alanine, *N*,*N*-diethyl-, ethyl ester Ethyl 3-(diethylamino)propionate	$C_9H_{19}NO_2$ 173.26	5515-83-3	 $83\text{-}4^{12}$	4-04-00-02538 0.881^{25}	 1.4253^{20}
329	β-Alanine, *N*-(2,4-dihydroxy-3,3-dimethyl-1-oxobutyl)-, (*R*)- Pantothenic acid	$C_9H_{17}NO_5$ 219.24	79-83-4	6964	4-04-00-02569	H_2O 4; bz 4; eth 4
330	β-Alanine, *N*-(3-ethoxy-3-oxopropyl)-*N*-methyl-, ethyl ester Propionic acid, 3,3'-iminodi-*N*-methyl, diethyl ester	$C_{11}H_{21}NO_4$ 231.29	6315-60-2	 $136\text{-}8^{4}$	3-04-00-01278 1.0172^{20}	eth 4; EtOH 4 1.4421^{20}
331	Alanine, 2-methyl- α-Aminoisobutyric acid	$C_4H_9NO_2$ 103.12	62-57-7 335	457 sub 280	4-04-00-02616	H_2O 4; EtOH 2; eth 1

No.	Name Synonym	Mol. Form. Mol. Wt.	CAS RN mp/°C	Merck No. bp/°C	Beil. Ref. den/g cm^{-3}	Solubility n_D
332	β-Alanine, *N*-methyl-, ethyl ester	$C_6H_{13}NO_2$ 131.17	2213-08-3	 80^{21}	3-04-00-01264 1.0064^{20}	eth 4; EtOH 4 1.4443^{20}
333	*L*-Alanine, 3-(2-propenylsulfinyl)-, (*S*)- Alliin	$C_6H_{11}NO_3S$ 177.22	556-27-4 165	251	4-04-00-03147	H_2O 4
334	Aldicarb Propanal, 2-methyl-2-(methylthio)-, *O*-[(methylamino)carbonyl]oxime	$C_7H_{14}N_2O_2S$ 190.27	116-06-3 99	216	 1.195^{25}	
335	Aldoxycarb Propanal, 2-methyl-2-(methylsulfonyl)-, *O*-[(methylamino)carbonyl]oxime	$C_7H_{14}N_2O_4S$ 222.27	1646-88-4			
336	Allethrin 2-Methyl-4-oxo-3-(2-propenyl)-2-cyclopenten-1-yl 2,2-dimethyl-3-(2-methyl-1-propenyl) cyclopropane carboxylate	$C_{19}H_{26}O_3$ 302.41	584-79-2		 1.010^{20}	
337	*D*-Allose	$C_6H_{12}O_6$ 180.16	2595-97-3 128	279	4-01-00-04299	H_2O 4
338	*D*-Altrose	$C_6H_{12}O_6$ 180.16	1990-29-0 103.5	319	4-01-00-04300	H_2O 4
339	Aluminum, trimethyl Trimethyl aluminum	C_3H_9Al 72.09	75-24-1 15.4	 130; 20^8	 0.752^{20}	
340	Aluminum, tris(2-hydroxypropanoato-O1,O2)- Aluminum lactate	$C_9H_{15}AlO_9$ 294.19	18917-91-4	350		H_2O 3
341	Ametryn *N*-Ethyl-*N'*-(1-methylethyl)-6-(methylthio)1,3,5-triazine-2,4-diamine	$C_9H_{17}N_5S$ 227.33	834-12-8 88	402		
342	Amitraz *N*-Methylbis(2,4-xylyliminomethyl)amine	$C_{19}H_{23}N_3$ 293.41	33089-61-1 86	503	 1.128^{20}	
343	Androsta-1,4-dien-3-one, 17-hydroxy-, (17β)- Boldenone	$C_{19}H_{26}O_2$ 286.41	846-48-0 165	1327		
344	Androsta-1,4-dien-3-one, 17-hydroxy-17-methyl-, (17β)- Methandrostenolone	$C_{20}H_{28}O_2$ 300.44	72-63-9 166	5862	4-08-00-01119	
345	Androstane	$C_{19}H_{32}$ 260.46	24887-75-0 50	668 $60^{0.003}$	4-05-00-01211	ace 4; eth 4; EtOH 4; peth 4
346	Androstane-17-carboxylic acid, (5β,17β)- Etiocholanic acid	$C_{20}H_{32}O_2$ 304.47	438-08-4 228.5	3824 sub 160	3-09-00-02644	
347	Androstan-17-one, 3-hydroxy-, (3β,5α)- Epiandrosterone	$C_{19}H_{30}O_2$ 290.45	481-29-8 178	3562	4-08-00-00641	
348	Androstan-17-one, 3-hydroxy-, (3α,5α)- Androsterone	$C_{19}H_{30}O_2$ 290.45	53-41-8 185	673	4-08-00-00642	H_2O 2; EtOH 3; eth 3; ace 3
349	Androstan-3-one, 17-hydroxy-, (5α,17β)- Stanolone	$C_{19}H_{30}O_2$ 290.45	521-18-6 181	8753 sub 135	4-08-00-00634	
350	Androstan-3-one, 17-hydroxy-17-methyl-, (5α,17β)- Mestanolone	$C_{20}H_{32}O_2$ 304.47	521-11-9 192.5	5816	4-08-00-00656	AcOEt 2
351	Androstan-3-one, 2-methyl-17-(1-oxopropoxy)-, (2α,5α,17β)- Dromostanolone propionate	$C_{23}H_{36}O_3$ 360.54	521-12-0 128	3436	4-08-00-00654	
352	Androst-4-ene-3,17-dione 4-Androstene-3,17-dione	$C_{19}H_{26}O_2$ 286.41	63-05-8 143(form a); 173(form b)	671	4-07-00-02381	
353	Androst-4-ene-3,11,17-trione Adrenosterone	$C_{19}H_{24}O_3$ 300.40	382-45-6 222	165 sub	4-07-00-02796	H_2O 2; EtOH 3; eth 3; ace 3
354	Androst-4-en-3-one, 4-chloro-17-hydroxy-, (17β)- Clostebol	$C_{19}H_{27}ClO_2$ 322.87	1093-58-9 189	2409	4-08-00-00989	
355	Androst-4-en-3-one, 9-fluoro-11,17-dihydroxy-17-methyl-, (11β,17β)- Fluoxymesterone	$C_{20}H_{29}FO_3$ 336.45	76-43-7 270	4113	4-08-00-02057	
356	Androst-4-en-3-one, 17-hydroxy-, (17β)- Testosterone	$C_{19}H_{28}O_2$ 288.43	58-22-0 155	9109	4-08-00-00974	H_2O 1; EtOH 3; eth 3; ace 3
357	Androst-4-en-3-one, 17-hydroxy-7,17-dimethyl-, (7β,17β)- Calusterone	$C_{28}H_{48}O$ 400.69	17021-26-0 157.5	1730	4-08-00-01029	
358	Androst-4-en-3-one, 17-hydroxy-17-methyl-, (17β)- 17-Methyltestosterone	$C_{20}H_{30}O_2$ 302.46	58-18-4 163.5	6044	4-08-00-01010	eth 4; EtOH 4
359	Androst-4-en-3-one, 17-(1-oxopropoxy)-, (17β)- Testosterone-17-propionate	$C_{22}H_{32}O_3$ 344.49	57-85-2 120	9115	3-08-00-00897	eth 4; py 4; EtOH 4
360	Anilazine 2,4-Dichloro-6-(o-chloroanilino)-s-triazine	$C_9H_5Cl_3N_4$ 275.52	101-05-3 160	685	 1.8^{20}	
361	Aniline, 3,3'-azobis[*N*,*N*-dimethyl- Azobenzene, 3,3'-bis(dimethylamino)	$C_{16}H_{20}N_4$ 268.36	21232-53-1 118	 sub	0-16-00-00305	bz 4; EtOH 4; chl 4
362	Aniline, *p*-isopentyl-	$C_{11}H_{17}N$ 163.26	104177-72-2	 257	4-12-00-02858 0.928^{15}	 1.5305^{20}
363	Anisole, 3-isopropyl-2-methyl- *o*-Cresol, 3-isopropyl, methyl ether	$C_{11}H_{16}O$ 164.25	31202-12-7 -0.5	 67^3	4-06-00-03328 0.9540^{25}	 1.5148^{20}
364	Anisole, 2,3,5-trinitro- 2,3,5-Trinitroanisole	$C_7H_5N_3O_7$ 243.13	7539-25-5 107.4		2-06-00-00253 1.618^{15}	ace 4; bz 4; EtOH 4

No.	Name Synonym	Mol. Form. Mol. Wt.	CAS RN mp/°C	Merck No. bp/°C	Beil. Ref. den/g cm^{-3}	Solubility n_D
365	2-Anthracenamine	$C_{14}H_{11}N$ 193.25	613-13-8 238.8	 sub	4-12-00-03435	H_2O 1; EtOH 3; os 2; con sulf 1
366	Anthracene	$C_{14}H_{10}$ 178.23	120-12-7 215.0	712 339.9	4-05-00-02281 1.28^{25}	H_2O 1; EtOH 2; eth 2; ace 2
367	Anthracene, 9,10-bis(phenylmethyl)- 9,10-Dibenzylanthracene	$C_{28}H_{22}$ 358.48	3613-42-1 246.5		3-05-00-02640 1.1787^{16}	H_2O 1; eth 2; bz 2; lig 1
368	1-Anthracenecarboxylic acid 1-Anthroic acid	$C_{15}H_{10}O_2$ 222.24	607-42-1 251.5	 sub	4-09-00-02670	H_2O 1; EtOH 3; eth 3; bz 2
369	2-Anthracenecarboxylic acid 2-Anthroic acid	$C_{15}H_{10}O_2$ 222.24	613-08-1 281	 sub	4-09-00-02671	HOAc 4
370	9-Anthracenecarboxylic acid 9-Anthroic acid	$C_{15}H_{10}O_2$ 222.24	723-62-6 217 dec	 sub	4-09-00-02671	H_2O 1; EtOH 3
371	2-Anthracenecarboxylic acid, 9,10-dihydro-4,5-dihydroxy-9,10-dioxo- Rhein	$C_{15}H_8O_6$ 284.22	478-43-3 321	8175 sub	4-10-00-04088	H_2O 2; EtOH 2; eth 2; ace 2
372	2-Anthracenecarboxylic acid, 9,10-dihydro-9,10-dioxo-	$C_{15}H_8O_4$ 252.23	117-78-2 291	 sub	4-10-00-03200	EtOH 2; eth 1; ace 3; bz 1
373	2-Anthracenecarboxylic acid, 7-β-*D*-glucopyranosyl-9,10-dihydro-3,5,6,8-tetrahydroxy-1-methyl-9,10-dioxo- Carminic acid	$C_{22}H_{20}O_{13}$ 492.39	1260-17-9 136 dec	1850	3-10-00-04874	H_2O 3; EtOH 3; eth 2; bz 1
374	Anthracene, 1-chloro- 1-Chloroanthracene	$C_{14}H_9Cl$ 212.68	4985-70-0 83.5		4-05-00-02292 1.1707^{100}	H_2O 1; EtOH 3; eth 3; bz 3 1.6959^{100}
375	Anthracene, 9,10-dibromo- 9,10-Dibromoanthracene	$C_{14}H_8Br_2$ 336.03	523-27-3 226	3000 sub	4-05-00-02295	H_2O 1; EtOH 2; eth 2; bz 2
376	Anthracene, 9,10-dihydro- 9,10-Dihydroanthracene	$C_{14}H_{12}$ 180.25	613-31-0 111	 305	4-05-00-02182 0.8976^{11}	H_2O 1; EtOH 3; eth 3; bz 3
377	Anthracene, 1,3-dimethyl- 1,3-Dimethylanthracene	$C_{16}H_{14}$ 206.29	610-46-8 83	 143^2	4-05-00-02327	eth 4; EtOH 4
378	Anthracene, 2,10-dimethyl- Anthracene, 9,3-dimethyl	$C_{16}H_{14}$ 206.29	27532-76-9 84	 140-5^2	4-05-00-02328	H_2O 1; EtOH 4; eth 4
379	9,10-Anthracenedione Anthraquinone	$C_{14}H_8O_2$ 208.22	84-65-1 286	717 377	4-07-00-02556 1.438^{20}	H_2O 1; EtOH 2; eth 2; bz 2
380	9,10-Anthracenedione, 1-amino- 1-Aminoanthraquinone	$C_{14}H_9NO_2$ 223.23	82-45-1 253.5	428 sub	4-14-00-00429	ace 4; bz 4; EtOH 4; chl 4
381	9,10-Anthracenedione, 2-amino- 2-Aminoanthraquinone	$C_{14}H_9NO_2$ 223.23	117-79-3 304.5	 sub	4-14-00-00447	H_2O 1; EtOH 2; eth 1; ace 3
382	9,10-Anthracenedione, 3-amino-1,2-dihydroxy-	$C_{14}H_9NO_4$ 255.23	3963-78-8 >300	 sub	2-14-00-00185	aq NH_3 4
383	9,10-Anthracenedione, 2-amino-1-hydroxy- Anthraquinone, 2-amino-1-hydroxy	$C_{14}H_9NO_3$ 239.23	568-99-0 258	 sub	4-14-00-00891	bz 4; eth 4; EtOH 4
384	9,10-Anthracenedione, 1-amino-2-methyl- C.I. Solvent Orange 35	$C_{15}H_{11}NO_2$ 237.26	82-28-0 205.5		4-14-00-00496	H_2O 1; EtOH 3; eth 2; bz 3
385	9,10-Anthracenedione, 1-bromo-	$C_{14}H_7BrO_2$ 287.11	632-83-7 186.5	 sub	3-07-00-04076	bz 4; EtOH 4
386	9,10-Anthracenedione, 2-bromo-	$C_{14}H_7BrO_2$ 287.11	572-83-8 205.8	 sub	3-07-00-04076	bz 4
387	9,10-Anthracenedione, 1-chloro-	$C_{14}H_7ClO_2$ 242.66	82 44 0 163	 sub	4-07-00-02559	H_2O 1; EtOH 2; eth 5; bz 3
388	9,10-Anthracenedione, 2-chloro-	$C_{14}H_7ClO_2$ 242.66	131-09-9 211	 sub	4-07-00-02559	H_2O 1; EtOH 2; eth 1; bz 2
389	9,10-Anthracenedione, 1,5-diamino-	$C_{14}H_{10}N_2O_2$ 238.25	129-44-2 319	 sub	4-14-00-00479	H_2O 1; EtOH 2; eth 2; ace 2
390	9,10-Anthracenedione, 2,7-dibromo- 2,7-Dibromoanthraquinone	$C_{14}H_6Br_2O_2$ 366.01	605-42-5 249	 sub	2-07-00-00718	bz 4; HOAc 4
391	9,10-Anthracenedione, 1,2-dihydroxy- Alizarin	$C_{14}H_8O_4$ 240.22	72-48-0 289.5	237	4-08-00-03256	H_2O 2; EtOH 3; eth 3; ace 3
392	9,10-Anthracenedione, 1,4-dihydroxy- Quinizarin	$C_{14}H_8O_4$ 240.22	81-64-1 200	8094	4-08-00-03260	H_2O 3; EtOH 3; eth 3; bz 3
393	9,10-Anthracenedione, 1,5-dihydroxy- Anthrarufin	$C_{14}H_8O_4$ 240.22	117-12-4 280	719 sub	4-08-00-03268	H_2O 1; EtOH 2; eth 2; ace 2
394	9,10-Anthracenedione, 1,7-dihydroxy- 9,10-Anthraquinone, 1,7-dihydroxy	$C_{14}H_8O_4$ 240.22	569-08-4 [illegible]	 sub	4-08-00-03270	H_2O 1; EtOH 3; eth 3; bz 3
395	9,10-Anthracenedione, 1,8-dihydroxy- Danthron	$C_{14}H_8O_4$ 240.22	117-10-2 193	2813 sub	4-08-00-03271	H_2O 1; EtOH 3; eth 3; ace 3
396	9,10-Anthracenedione, 2,7-dihydroxy-	$C_{14}H_8O_4$ 240.22	572-93-0 353.8	 sub	4-08-00-03273	H_2O 1; EtOH 3; eth 2; bz 2
397	9,10-Anthracenedione, 1,8-dihydroxy-3-(hydroxymethyl)- Aloe-emodol	$C_{15}H_{10}O_5$ 270.24	481-72-1 223.5	303 sub	4-08-00-03578	bz 4; eth 4; EtOH 4
398	9,10-Anthracenedione, 1,8-dihydroxy-3-methyl- Chrysophanic acid	$C_{15}H_{10}O_4$ 254.24	481-74-3 196	2263 sub	4-08-00-03277 0.92^{25}	bz 4; HOAc 4
399	9,10-Anthracenedione, 1,2-dihydroxy-3-nitro- Alizarine Orange	$C_{14}H_7NO_6$ 285.21	568-93-4 244 dec	240 sub	4-08-00-03259	H_2O 2; EtOH 3; bz 3; chl 3
400	9,10-Anthracenedione, 1,2-dihydroxy-4-nitro-	$C_{14}H_7NO_6$ 285.21	2243-71-2 289 dec	 sub	4-08-00-03259	bz 4; EtOH 4; chl 4; HOAc 4

No.	Name	Synonym	Mol. Form.	Mol. Wt.	CAS RN	mp/°C	Merck No.	bp/°C	Beil. Ref.	den/g cm^{-3}	Solubility	n_D
401	9,10-Anthracenedione, 1,8-dihydroxy-2,4,5,7-tetranitro-	Chrysamminic acid	$C_{14}H_4N_4O_{12}$	420.21	517-92-0	exp	2253	dec	1-08-00-00723		eth 4; EtOH 4	
402	9,10-Anthracenedione, 1,4-dimethyl-		$C_{16}H_{12}O_2$	236.27	1519-36-4	140.5		sub	4-07-00-02583		H_2O 1; EtOH 2; bz 3; xyl 3	
403	9,10-Anthracenedione, 2,3-dimethyl-		$C_{16}H_{12}O_2$	236.27	6531-35-7	210.8		sub	3-07-00-04131		EtOH 3; bz 3; chl 2; xyl 3	
404	9,10-Anthracenedione, 2,6-dimethyl-		$C_{16}H_{12}O_2$	236.27	3837-38-5	242		sub	3-07-00-04132		EtOH 4; HOAc 4; tol 4	
405	9,10-Anthracenedione, 1,5-dinitro-		$C_{14}H_6N_2O_6$	298.21	82-35-9	385		sub	4-07-00-02562		H_2O 1; EtOH 2; eth 2; bz 2	
406	9,10-Anthracenedione, 1,2,3,5,6,7-hexahydroxy-	Rufigallol	$C_{14}H_8O_8$	304.21	82-12-2		8267	sub	4-08-00-03750		H_2O 1; EtOH 2; eth 2; ace 3	
407	9,10-Anthracenedione, 1-hydroxy-		$C_{14}H_8O_3$	224.22	129-43-1	193.8		sub	4-08-00-02586		H_2O 1; EtOH 3; eth 3; bz 3	
408	9,10-Anthracenedione, 2-hydroxy-		$C_{14}H_8O_3$	224.22	605-32-3	306		sub	4-08-00-02598		H_2O 1; EtOH 3; eth 3; KOH 3	
409	9,10-Anthracenedione, 2-methyl-	2-Methylanthraquinone	$C_{15}H_{10}O_2$	222.24	84-54-8	177	5943	sub	4-07-00-02574		bz 4; EtOH 4; HOAc 4	
410	9,10-Anthracenedione, 1-nitro-		$C_{14}H_7NO_4$	253.21	82-34-8	231.5		270^{7}	4-07-00-02561		H_2O 1; EtOH 2; eth 2; ace 3	
411	9,10-Anthracenedione, 2-nitro-		$C_{14}H_7NO_4$	253.21	605-27-6	184.5		sub	4-07-00-02561		H_2O 1; EtOH 2; eth 2; ace 3	
412	9,10-Anthracenedione, 1,2,5,6-tetrahydroxy-		$C_{14}H_8O_6$	272.21	632-77-9	340		sub	3-08-00-04288		py 4; EtOH 4; HOAc 4	
413	9,10-Anthracenedione, 1,2,5,8-tetrahydroxy-	Quinalizarin	$C_{14}H_8O_6$	272.21	81-61-8	>275	8059		4-08-00-03683		H_2O 2; EtOH 2; eth 2; ace 2	
414	9,10-Anthracenedione, 1,3,5,7-tetrahydroxy-		$C_{14}H_8O_6$	272.21	632-82-6	360		sub	4-08-00-03683		ace 4; bz 4; EtOH 4; HOAc 4	
415	9,10-Anthracenedione, 1,2,3-trihydroxy-	Anthragallol	$C_{14}H_8O_5$	256.21	602-64-2	313	713	sub 290	4-08-00-03567		H_2O 2; EtOH 3; eth 3; CS_2 3	
416	9,10-Anthracenedione, 1,2,4-trihydroxy-	Purpurin	$C_{14}H_8O_5$	256.21	81-54-9	259	7962	sub	4-08-00-03568		H_2O 2; EtOH 4; eth 3; bz 4	
417	9,10-Anthracenedione, 1,2,6-trihydroxy-		$C_{14}H_8O_5$	256.21	82-29-1	330		459	4-08-00-03568		H_2O 3; EtOH 3; eth 2; bz 3	
418	9,10-Anthracenedione, 1,2,7-trihydroxy-		$C_{14}H_8O_5$	256.21	602-65-3	374		462	4-08-00-03569		bz 4; EtOH 4; HOAc 4	
419	9,10-Anthracenedione, 1,3,8-trihydroxy-6-methyl-	Emodin	$C_{15}H_{10}O_5$	270.24	518-82-1	257	3518	sub	4-08-00-03575		eth 4; EtOH 4	
420	Anthracene, 9-ethyl-	9-Ethylanthracene	$C_{16}H_{14}$	206.29	605-83-4	60			4-05-00-02326	1.0413^{99}	H_2O 1; EtOH 3; eth 3	1.6767^{99}
421	Anthracene, 9-ethyl-9,10-dihydro-	9-Ethyl-9,10-dihydroanthracene	$C_{16}H_{16}$	208.30	605-82-3			321.5	0-05-00-00649	1.048^{18}	bz 4; eth 4; EtOH 4	
422	Anthracene, 1,2,3,4,5,6-hexahydro-	1,2,3,4,5,6-Hexahydroanthracene	$C_{14}H_{16}$	184.28	6109-22-4	68.5		160^{15}	3-05-00-01676		bz 4	
423	Anthracene, 1-methyl-	1-Methylanthracene	$C_{15}H_{12}$	192.26	610-48-0	85.5		199.5	4-05-00-02310	1.0471^{99}	H_2O 1; EtOH 3; eth 3; bz 3	1.6802^{99}
424	Anthracene, 2-methyl-	2-Methylanthracene	$C_{15}H_{12}$	192.26	613-12-7	209		sub	4-05-00-02311	1.80^{0}	H_2O 1; EtOH 2; eth 2; ace 1	
425	Anthracene, 9-methyl-	9-Methylanthracene	$C_{15}H_{12}$	192.26	779-02-2	81.5		$196\text{-}7^{12}$	4-05-00-02312	1.065^{99}	EtOH 3; eth 3; ace 3; bz 3	1.6959^{99}
426	Anthracene, 9-nitro-	9-Nitroanthracene	$C_{14}H_9NO_2$	223.23	602-60-8	146		275^{17}	4-05-00-02296		H_2O 1; EtOH 2; ace 4; chl 2	
427	Anthracene, 1,2,3,4,5,6,7,8-octahydro-		$C_{14}H_{18}$	186.30	1079-71-6	78		294	4-05-00-01584	0.9703^{80}	H_2O 1; EtOH 3; bz 4; ctc 2	1.5372^{80}
428	Anthracene, 9-phenyl-	9-Phenylanthracene	$C_{20}H_{14}$	254.33	602-55-1	156		417	4-05-00-02628		H_2O 1; EtOH 3; eth 3; bz 3	
429	1,2,10-Anthracenetriol	Anthrarobin	$C_{14}H_{10}O_3$	226.23	577-33-3	208	718		4-06-00-07601		H_2O 2; EtOH 4; eth 4; ace 4	
430	1,8,9-Anthracenetriol	Anthralin	$C_{14}H_{10}O_3$	226.23	1143-38-0	179	714		4-06-00-07602		H_2O 1; EtOH 3; eth 2; ace 3	
431	1-Anthracenol		$C_{14}H_{10}O$	194.23	610-50-4	158		234^{13}	4-06-00-04928		H_2O 1; EtOH 4; eth 4; NaOH 3	
432	9-Anthracenol	Anthranol	$C_{14}H_{10}O$	194.23	529-86-2	152	716		4-06-00-04930			
433	9(10H)-Anthracenone	Anthrone	$C_{14}H_{10}O$	194.23	90-44-8	155	721		4-06-00-04930		ace 3; bz 3; con sulf 3; dil alk 3	
434	5,9,14,18-Anthrazinetetrone, 6,15-dihydro-	Indanthrene	$C_{28}H_{14}N_2O_4$	442.43	81-77-6	470-500 dec	4846		5-24-09-00504		H_2O 1; EtOH 1; eth 1; ace 1	
435	8'-Apo-β,ψ-carotenal, 3-hydroxy-, (3*R*)-	β-Citraurin	$C_{30}H_{40}O_2$	432.65	650-69-1	147	2326		3-08-00-01599		H_2O 1; EtOH 4; eth 4; ace 4	
436	β-*D*-Arabinopyranose	Arabinose (*D*)(β)	$C_5H_{10}O_5$	150.13	6748-95-4	155.5			4-01-00-04215	1.625^{25}	H_2O 4	
437	*DL*-Arginine		$C_6H_{14}N_4O_2$	174.20	7200-25-1	217 dec	805		3-04-00-01359		H_2O 1; EtOH 1; eth 1; bz 1	
438	*L*-Arginine		$C_6H_{14}N_4O_2$	174.20	74-79-3	244 dec	805		4-04-00-02648		H_2O 3; EtOH 2; eth 1	

No.	Name	Synonym	Mol. Form.	Mol. Wt.	CAS RN	mp/°C	Merck No.	bp/°C	Beil. Ref.	den/g cm^{-3}	Solubility	n_D
439	Arsenic acid, triethyl ester	Ethyl arsenate	$C_6H_{15}AsO_4$	226.10	15606-95-8			236.5	4-01-00-01359	1.3021^{20}		1.4343^{20}
440	Arsenous acid, triethyl ester	Ethyl arsenite	$C_6H_{15}AsO_3$	210.10	3141-12-6			165.5	4-01-00-01358	1.2239^{20}		1.4369^{13}
441	Arsine, diethyl-	Diethylarsine	$C_4H_{11}As$	134.05	692-42-2			105	3-04-00-01797	1.1338^{24}	ace 4; bz 4; eth 4; EtOH 4	1.4709
442	Arsine, dimethyl-	Dimethylarsine	C_2H_7As	106.00	593-57-7	-136.1		36	4-04-00-03665	1.208^{29}	ace 4; bz 4; eth 4; EtOH 4	
443	Arsine, diphenyl-	Diphenylarsine	$C_{12}H_{11}As$	230.14	829-83-4			174^{25}	4-16-00-01138	1.30^{25}	eth 4; EtOH 4	
444	Arsine, ethyl-	Ethylarsine	C_2H_7As	106.00	593-59-9			36	3-04-00-01797	1.214^{22}	eth 4; EtOH 4	
445	Arsine, ethyldifluoro-		$C_2H_5AsF_2$	141.98	430-40-0	-38.7		94.3	3-04-00-01799	1.708^{17}		
446	Arsine, methyl-	Methylarsine	CH_5As	91.97	593-52-2	-143		2	3-04-00-01795		ace 4; eth 4; EtOH 4	
447	Arsine, oxophenyl-	Oxophenylarsine	C_6H_5AsO	168.03	637-03-6	145	6903		4-16-00-01160		H_2O 1; EtOH 2; eth 1; bz 4	
448	Arsine, triethyl-	Triethylarsine	$C_6H_{15}As$	162.11	617-75-4			138.5	4-04-00-03666	1.150^{20}	ace 4; eth 4; EtOH 4	1.467^{20}
449	Arsine, trimethyl-	Trimethylarsine	C_3H_9As	120.03	593-88-4	-87.3		52	4-04-00-03665	1.144^{15}	bz 4; eth 4; EtOH 4	
450	Arsine, triphenyl-	Triphenylarsine	$C_{18}H_{15}As$	306.24	603-32-7	61		360	4-16-00-01130	1.2634^{18}	H_2O 1; EtOH 2; eth 4; bz 4	1.6888^{21}
451	Arsinic acid, dimethyl-	Cacodylic acid	$C_2H_7AsO_2$	138.00	75-60-5	195	1603		4-04-00-03681		H_2O 4; EtOH 3; eth 1	
452	Arsinous acid, dimethyl-, anhydride	Cacodyl oxide	$C_4H_{12}As_2O$	225.98	503-80-0	-25		150	3-04-00-01814	1.4816^{15}	eth 4; EtOH 4	1.5225^{9}
453	Arsinous chloride, dimethyl-	Dimethylchloroarsine	C_2H_6AsCl	140.44	557-89-1	<-45		109; 77^{37}	4-04-00-03670	1.505^{12}	EtOH 4	1.5203^{12}
454	Arsinous chloride, diphenyl-	Chlorodiphenylarsine	$C_{12}H_{10}AsCl$	264.59	712-48-1	44		337	3-16-00-00944	1.4820^{16}	ace 4; bz 4; eth 4; EtOH 4	1.6332^{56}
455	Arsonic acid, [4-[(aminocarbonyl)amino]phenyl]-	Carbarsone	$C_7H_9AsN_2O_4$	260.08	121-59-5	174	1788		3-16-00-01112		H_2O 2; EtOH 2; eth 1; chl 1	
456	Arsonic acid, (4-aminophenyl)	Arsanilic acid	$C_6H_8AsNO_3$	217.06	98-50-0	232	818		4-16-00-01190	1.9571^{10}	H_2O 3; EtOH 2; eth 3; ace 1	
457	Arsonic acid, ethyl-	Ethanearsonic acid	$C_2H_7AsO_3$	154.00	507-32-4	99.5	3677	$209\text{-}11^{12}$	4-04-00-03682		H_2O 4; EtOH 4	
458	Arsonic acid, methyl-	Methanearsonic acid	CH_5AsO_3	139.97	124-58-3	160.5	5864		4-04-00-03682		H_2O 3; EtOH 3	
459	Arsonic acid, (4-nitrophenyl)-	Nitarsone	$C_6H_6AsNO_5$	247.04	98-72-6	>310 dec	6483		4-16-00-01185		H_2O 2; EtOH 2; DMSO-d_6 2	
460	Arsonic acid, phenyl-	Benzenearsonic acid	$C_6H_7AsO_3$	202.04	98-05-5	158- dec	1075		4-16-00-01183		H_2O 4; EtOH 4	
461	Arsonic acid, propyl-	1-Propanearsonic acid	$C_3H_9AsO_3$	168.02	107-34-6	134.5	7810		4-04-00-03682		H_2O 4; EtOH 4; eth 1	
462	Arsonous dichloride, phenyl-	Dichlorophenylarsine	$C_6H_5AsCl_2$	222.93	696-28-6	-19		255	4-16-00-01161	1.6516^{20}	bz 4, eth 4, EtOH 4	1.6386^{15}
463	Arsonous difluoride, methyl-	Difluoromethylarsine	CH_3AsF_2	127.95	420-24-6	-29.7		76.5	3-04-00-01796	1.924^{18}		
464	Arsonous diiodide, phenyl-		$C_6H_5AsI_2$	405.84	6380-34-3	15		205^{14} 185^{10}	3-16-00-00958	1.6264^{15}		
465	*DL*-Ascorbic acid		$C_6H_8O_6$	176.13	62624-30-0	168.5			5-18-05-00026		H_2O 4; EtOH 3; eth 1; bz 1	
466	*L*-Ascorbic acid	Vitamin C	$C_6H_8O_6$	176.13	50-81-7	190-2 dec	855		5-18-05-00026	1.65^{25}	H_2O 4; EtOH 3; eth 1; bz 1	
467	*L*-Ascorbic acid, 6-deoxy-	6-Desoxy-*L*-ascorbic acid	$C_6H_8O_5$	160.13	528-81-4	168	2880	sub 160	4-18-00-02301		H_2O 4; ace 4; EtOH 4	
468	*L*-Asparagine	α-Aminosuccinamic acid	$C_4H_8N_2O_3$	132.12	70-47-3		859		4-04-00-03004	1.543^{15}	H_2O 3; EtOH 1; eth 1; MeOH 1	
469	*DL*-Asparagine, monohydrate		$C_4H_{10}N_2O_4$	150.13	69833-18-7	182.5	859	214 dec	2-04-00-00900	1.4540^{15}	H_2O 2; EtOH 1; eth 1; bz 1	
470	*DL*-Aspartic acid		$C_4H_7NO_4$	133.10	617-45-8	275			2-04-00-00000	1.6622^{13}	H_2O 2; EtOH 1; eth 1; bz 1	
471	*L*-Aspartic acid	*L*-Aminosuccinic acid	$C_4H_7NO_4$	133.10	56-84-8	270	862		4-04-00-02998	1.6603^{13}	H_2O 2; EtOH 1; eth 1; bz 1	
472	*L*-Aspartic acid, *N*-[2-[(2-amino-2-oxoethyl)amino]-2-carboxyethyl]-	Lycomarasmine	$C_9H_{15}N_3O_7$	277.23	7611-43-0	227-9 dec	5491		3-04-00-01521			
473	Aspidospermidine, 1-acetyl-17-methoxy-	Aspidospermine	$C_{22}H_{30}N_2O_2$	354.49	466-49-9	208	872	220^{2}	5-23-12-00210		H_2O 2; EtOH 3; eth 2; bz 3	
474	Asulam	Methyl [(4-aminophenyl)sulfonyl]carbamate	$C_8H_{10}N_2O_4S$	230.24	3337-71-1	144						
475	Atisine	Anthorine	$C_{22}H_{33}NO_2$	343.51	466-43-3	58.5	881				eth 4; EtOH 4; chl 4	
476	Atrazine	6-Chloro-*N*-ethyl-*N*′-(1-methylethyl)-1,3,5-triazine-2,4-diamine	$C_8H_{14}ClN_5$	215.69	1912-24-9	173	886					

No.	Name Synonym	Mol. Form. Mol. Wt.	CAS RN mp/°C	Merck No. bp/°C	Beil. Ref. den/g cm^{-3}	Solubility n_D
477	Avermectin B1a Abamectin	$C_{48}H_{72}O_{14}$ 873.09	71751-41-2 152	1		
478	7-Azabicyclo[4.1.0]heptane 2,3-Tetramethyleneaziridine	$C_6H_{11}N$ 97.16	286-18-0 21	 150; 50^{22}	5-20-04-00304 0.9480^{27}	bz 4; eth 4; EtOH 4
479	9-Azabicyclo[3.3.1]nonan-3-one, 9-methyl- Pseudopelletierine	$C_9H_{15}NO$ 153.22	552-70-5 54	7936 246	5-21-07-00059 1.001^{100}	H_2O 4; eth 4; EtOH 4 1.4760^{100}
480	1-Azabicyclo[2.2.2]octane Quinuclidine	$C_7H_{13}N$ 111.19	100-76-5 158	8109	5-20-04-00335	H_2O 4; ace 4; eth 4; EtOH 4
481	8-Azabicyclo[3.2.1]octane-2-carboxylic acid, 3-(benzoyloxy)-8-methyl-, [1*R*-(*exo,exo*)]- Benzoylecgonine	$C_{16}H_{19}NO_4$ 289.33	519-09-5 195	1125	5-22-05-00053	bz 4; EtOH 4
482	8-Azabicyclo[3.2.1]octane-2-carboxylic acid, 3-(benzoyloxy)-8-methyl-, methyl ester, [1*R*-(*exo,exo*)]- Cocaine	$C_{17}H_{21}NO_4$ 303.36	50-36-2 98	2450 $187^{0.1}$	5-22-05-00054	H_2O 2; EtOH 4; eth 4; ace 3 1.5022^{98}
483	8-Azabicyclo[3.2.1]octane-2-carboxylic acid, 3-hydroxy-8-methyl-, [1*R*-(*exo,exo*)]- Ecgonine	$C_9H_{15}NO_3$ 185.22	481-37-8 205	3467	5-22-05-00053	H_2O 4; EtOH 4
484	8-Azabicyclo[3.2.1]octane, 8-methyl- Tropane	$C_8H_{15}N$ 125.21	529-17-9	9689 166	5-20-04-00331 0.9251^{15}	
485	1-Azabicyclo[2.2.2]octan-3-ol 3-Quinuclidinol	$C_7H_{13}NO$ 127.19	1619-34-7 221	8110 sub 120	5-21-01-00271	ace 3
486	8-Azabicyclo[3.2.1]octan-3-ol, 8-methyl-, *exo*- Pseudotropine	$C_8H_{15}NO$ 141.21	135-97-7 109	7937 241	5-21-01-00219	H_2O 4; EtOH 4; eth 2; bz 3
487	8-Azabicyclo[3.2.1]octan-3-ol, 8-methyl-, benzoate (ester), *endo*-	$C_{15}H_{19}NO_2$ 245.32	19145-60-9 41.5	 177.5	5-21-01-00226	H_2O 4; eth 4; EtOH 4
488	8-Azabicyclo[3.2.1]octan-3-ol, 8-methyl-, benzoate (ester), *exo*- Tropacocaine	$C_{15}H_{19}NO_2$ 245.32	537-26-8 49	9686 dec	5-21-01-00226 1.0426^{100}	bz 4; eth 4; EtOH 4; peth 4 1.5080^{100}
489	8-Azabicyclo[3.2.1]octan-3-ol, 8-methyl-, *endo*- Tropine	$C_8H_{15}NO$ 141.21	120-29-6 64	9693 233	5-21-01-00219 1.016^{100}	H_2O 4; eth 4; EtOH 4 1.4811^{100}
490	8-Azabicyclo[3.2.1]octan-3-one, 8-methyl-	$C_8H_{13}NO$ 139.20	532-24-1 43	 227; 113^{25}	5-21-07-00032 1.9872^{100}	EtOH 3; eth 3; ace 3; bz 3 1.4598^{100}
491	8-Azabicyclo[3.2.1]oct-2-ene-2-carboxylic acid, 8-methyl-, (1*R*)- Ecgonidine	$C_9H_{13}NO_2$ 167.21	484-93-5 226-30 dec	3466	5-22-01-00446	H_2O 4
492	1H-Azepine, hexahydro- Hexamethylenimine	$C_6H_{13}N$ 99.18	111-49-9	 138	5-20-04-00003 0.8643^{22}	H_2O 3; EtOH 4; eth 4 1.4631^{20}
493	2H-Azepin-2-one, hexahydro- Caprolactam	$C_6H_{11}NO$ 113.16	105-60-2 69.3	1762 270	5-21-06-00444	H_2O 4; bz 4; EtOH 4; chl 4
494	Azetidine	C_3H_7N 57.10	503-29-7	 63	5-20-01-00136 0.8436^{20}	ace 4; bz 4; eth 4; EtOH 4 1.4287^{25}
495	2-Azetidinecarboxylic acid	$C_4H_7NO_2$ 101.11	2517-04-6 217 dec	923	5-22-01-00015	
496	2-Azetidinone	C_3H_5NO 71.08	930-21-2 73.5	 106^{15}	5-21-06-00312	eth 4; EtOH 4; chl 4
497	Azinphos-methyl *O,O*-Dimethyl *S*-[(4-oxo-1,2,3-benzotriazin-3(4H)-yl)methyl]phosphorodithioate	$C_{10}H_{12}N_3O_3PS_2$ 317.33	86-50-0 73	926	 1.44^{20}	
498	Aziridine Ethyleneimine	C_2H_5N 43.07	151-56-4 -77.9	3760 56	5-20-01-00003 0.832^{25}	H_2O 5; EtOH 3; eth 4; chl 2
499	1-Aziridineethanol	C_4H_9NO 87.12	1072-52-2	 168	5-20-01-00014 1.088^{25}	 1.4560^{20}
500	Aziridine, 1-methyl-	C_3H_7N 57.10	1072-44-2	 27.5	5-20-01-00007 0.7572^{19}	H_2O 5 1.3885^{19}
501	Aziridine, 2-methyl- 1,2-Propylenimine	C_3H_7N 57.10	75-55-8	 67	5-20-01-00150 0.812^{16}	
502	Azocine, octahydro-	$C_7H_{15}N$ 113.20	1121-92-2 29	 51-3^{15}	5-20-04-00156 0.896^{25}	 1.4720^{20}
503	1H-Azonine, octahydro-	$C_8H_{17}N$ 127.23	5661-71-2 -19	 91^{24}	5-20-04-00203 0.83^{25}	 1.4760^{20}
504	Azulene Bicyclo[5.3.0]decapentaene	$C_{10}H_8$ 128.17	275-51-4 99	939 270 dec; 125^{10}	4-05-00-01636	H_2O 1; EtOH 3; eth 3; ace 3
505	Azulene, 1,4-dimethyl-7-(1-methylethyl)- Guaiazulene	$C_{15}H_{18}$ 198.31	489-84-9 31.5	4464 167-8^{12}	4-05-00-01751 0.973^{20}	EtOH 3; eth 3; AcOEt 3
506	Azulene, 4,8-dimethyl-2-(1-methylethyl)-	$C_{15}H_{18}$ 198.31	529-08-8 32.3	 $122^{1.0}$	4-05-00-01754 0.9735^{19}	eth 4; EtOH 4
507	5-Azulenemethanol, 1,2,3,4,5,6,7,8-octahydro-α,α,3,8-tetramethyl-, Guaiol	$C_{15}H_{26}O$ 222.37	489-86-1 91	4466 288 dec; 165^{17}	3-06-00-00412 0.9074^{100}	H_2O 1; EtOH 3; eth 3 1.4716^{100}
508	Bacitracin	$C_{66}H_{103}N_{17}O_{16}S$ 1422.71	1405-87-4 223	948		H_2O 4; EtOH 4
509	Balan Benzenamine, *N*-butyl-*N*-ethyl-2,6-dinitro-4-(trifluoromethyl)-	$C_{13}H_{16}F_3N_3O_4$ 335.28	1861-40-1 66	1048 $121^{0.5}$; 148^7		

No.	Name Synonym	Mol. Form. Mol. Wt.	CAS RN mp/°C	Merck No. bp/°C	Beil. Ref. den/g cm^{-3}	Solubility n_D
510	Barban Carbamic acid, (3-chlorophenyl)-, 4-chloro-2-butynyl ester	$C_{11}H_9Cl_2NO_2$ 258.10	101-27-9 75	969		
511	Basic copper carbonate Copper, [μ-[carbonato(2-)-O:O']]dihydroxydi-	$CH_2Cu_2O_5$ 221.12	12069-69-1 200 dec	2634	 3.8^{20}	
512	Bayleton 2-Butanone, 1-(4-chlorophenoxy)-3,3-dimethyl-1-(1H-1,2,4-triazol-1-yl)-	$C_{14}H_{16}ClN_3O_2$ 293.75	43121-43-3 82	9507	 1.22^{20}	
513	Benalaxyl Methyl *N*-(2,6-dimethylphenyl)-*N*-(phenylacetyl)-*DL*-alaninate	$C_{20}H_{23}NO_3$ 325.41	71626-11-4 79		 1.27^{25}	
514	Bendiocarb 1,3-Benzodioxol-4-ol, 2,2-dimethyl-, methylcarbamate	$C_{11}H_{13}NO_4$ 223.23	22781-23-3 130	1044	 1.25^{20}	
515	Benomyl Carbamic acid, [1-[(butylamino)carbonyl]-1H-benzimidazol-2-yl]-, methyl ester	$C_{14}H_{18}N_4O_3$ 290.32	17804-35-2 dec	1053		
516	Bensulfuron-methyl 2-[[[[[(4,6-Dimethoxypyrimidin-2-yl)amino]carbonyl]amino]sulfonyl]methyl]benzoic acid, methyl ester	$C_{16}H_{18}N_4O_7S$ 410.41	83055-99-6 187			
517	Bensulide Phosphorodithioic acid, *O,O*-bis(1-methylethyl) *S*-[2-[(phenylsulfonyl)amino]ethyl] ester	$C_{14}H_{244}NO_4PS_3$ 397.52	741-58-2 34.4		 1.224^{20}	
518	Bentazon 1H-2,1,3-Benzothiadiazin-4(3H)-one, 3-(1-methylethyl)-, 2,2-dioxide	$C_{10}H_{12}N_2O_3S$ 240.28	25057-89-0 138	1060		
519	Benz[a]anthracene 1,2-Benzanthracene	$C_{18}H_{12}$ 228.29	56-55-3 84	1069	4-05-00-02549	EtOH 4
520	Benz[a]anthracene, 7,12-dimethyl- 9,10-Dimethyl-1,2-benzanthracene	$C_{20}H_{16}$ 256.35	57-97-6 122.5	3224	4-05-00-02587	ace 4; bz 4
521	Benz[a]anthracene, 11-methyl-	$C_{19}H_{14}$ 242.32	6111-78-0 118	 360 dec	4-05-00-02568	H_2O 1; EtOH 3; HOAc 3
522	Benz[a]anthracene, 3-methyl- 3-Methylbenz[a]anthracene	$C_{19}H_{14}$ 242.32	2498-75-1 156.5	 $160^{0.1}$	4-05-00-02562	H_2O 1; EtOH 3; ace 3; bz 3
523	Benz[a]anthracene, 8-methyl-	$C_{19}H_{14}$ 242.32	2381-31-9 156.5	 272^3; $160^{0.1}$	4-05-00-02566 1.2310^0	H_2O 1; EtOH 3; eth 3; bz 3
524	Benzadox Acetic acid, [(benzoylamino)oxy]-	$C_9H_9NO_4$ 195.17	5251-93-4 140			
525	Benzaldehyde Benzenecarboxaldehyde	C_7H_6O 106.12	100-52-7 -26	1065 179.0	4-07-00-00505 1.0415^{10}	H_2O 2; EtOH 5; eth 5; ace 4 1.5463^{20}
526	Benzaldehyde, 2-bromo-	C_7H_5BrO 185.02	6630-33-7 21.5	 230	4-07-00-00578	H_2O 1; EtOH 4; bz 4; ctc 2 1.5925^{20}
527	Benzaldehyde, 3-bromo-	C_7H_5BrO 185.02	3132-99-8	 234.5	4-07-00-00579	H_2O 1; EtOH 4; eth 4; ctc 2 1.5935^{20}
528	Benzaldehyde, 4-bromo-	C_7H_5BrO 185.02	1122-91-4 67	 $66\text{-}8^2$	4-07-00-00579	H_2O 1; EtOH 4; bz 4; chl 2
529	Benzaldehyde, 2-chloro-	C_7H_5ClO 140.57	89-98-5 12.4	 211.9	4-07-00-00561 1.2483^{20}	H_2O 2; EtOH 3; eth 3; ace 3 1.5662^{20}
530	Benzaldehyde, 3-chloro-	C_7H_5ClO 140.57	587-04-2 17.5	 213.5	4-07-00-00566 1.2410^{20}	H_2O 2; EtOH 3; eth 3; ace 3 1.5650^{20}
531	Benzaldehyde, 4-chloro-	C_7H_5ClO 140.57	104-88-1 47.5	 213.5	4-07-00-00568 1.196^{61}	H_2O 3; EtOH 4; eth 4; ace 3 1.555^{61}
532	Benzaldehyde, 3-chloro-4-hydroxy-	$C_7H_5ClO_2$ 156.57	2420-16-8 139	 150^{14}	4-08-00-00296	eth 4; EtOH 4
533	Benzaldehyde, 5-chloro-2-hydroxy-	$C_7H_5ClO_2$ 156.57	635-93-8 100.3	 105^{12}	4-08-00-00224	H_2O 1; EtOH 4; eth 3; alk 3
534	Benzaldehyde, 3,5-dibromo-2-hydroxy- 3,5-Dibromosalicylaldehyde	$C_7H_4Br_2O_2$ 279.92	90-59-5 86	3011 sub	3-08-00-00188	bz 4; eth 4; chl 4
535	Benzaldehyde, 2,5-dichloro- 2,5-Dichlorobenzaldehyde	$C_7H_4Cl_2O$ 175.01	6361-23-5 58	 232	4-07-00-00576	bz 4; eth 4; EtOH 4; chl 4
536	Benzaldehyde, 3,4-dichloro- 3,4-Dichlorobenzaldehyde	$C_7H_4Cl_2O$ 175.01	6287-38-3 44	 247.5	4-07-00-00576	H_2O 1; EtOH 3; eth 3; ctc 2
537	Benzaldehyde, 3,5-dichloro- [illegible]	$C_7H_4Cl_2O$ [illegible]	10203-08-4 [illegible]	 [illegible]	4-07-00-00577	ace 4; bz 4; eth 4; EtOH 4
538	Benzaldehyde, 3,4-diethoxy- 3,4-Diethoxybenzaldehyde	$C_{11}H_{14}O_3$ 194.23	2029-94-9 22	 279; 200^{50}	4-08-00-01766 1.0100^{22}	EtOH 4
539	Benzaldehyde, 4-(diethylamino)-	$C_{11}H_{15}NO$ 177.25	120-21-8 41	 172^{10}	4-14-00-00066	H_2O 4; EtOH 3; eth 3; bz 3
540	Benzaldehyde, 2,3-dihydroxy-	$C_7H_6O_3$ 138.12	24677-78-9 108	 235; 120^{16}	3-08-00-01979	ace 4; EtOH 4; HOAc 4
541	Benzaldehyde, 2,4-dihydroxy- β-Resorcylaldehyde	$C_7H_6O_3$ 138.12	95-01-2 135	8159 226^{22}	4-08-00-01753	H_2O 3; EtOH 4; eth 4; bz 2
542	Benzaldehyde, 3,4-dihydroxy- Protocatechualdehyde	$C_7H_6O_3$ 138.12	139-85-5 153 dec	7908	4-08-00-01762	H_2O 3; EtOH 4; eth 4
543	Benzaldehyde, 2,4-dimethoxy- 2,4-Dimethoxybenzaldehyde	$C_9H_{10}O_3$ 166.18	613-45-6 72	 290; 165^{10}	4-08-00-01754	H_2O 1; EtOH 3; eth 3; bz 3
544	Benzaldehyde, 2,5-dimethoxy- 2,5-Dimethoxybenzaldehyde	$C_9H_{10}O_3$ 166.18	93-02-7 52	 270; 146^{10}	4-08-00-01759	H_2O 2; EtOH 3; eth 3

No.	Name Synonym	Mol. Form. Mol. Wt.	CAS RN mp/°C	Merck No. bp/°C	Beil. Ref. den/g cm^{-3}	Solubility n_D
545	Benzaldehyde, 3,4-dimethoxy- Veratraldehyde	$C_9H_{10}O_3$ 166.18	120-14-9 43	9852 281; 155^{10}	4-08-00-01765	H_2O 2; EtOH 4; eth 4; chl 2
546	Benzaldehyde, 3,5-dimethoxy- 3,5-Dimethoxybenzaldehyde	$C_9H_{10}O_3$ 166.18	7311-34-4 46.3	 151^{16}	4-08-00-01786	H_2O 2; EtOH 3; bz 3; peth 2
547	Benzaldehyde, 2,4-dimethyl- 2,4-Dimethylbenzaldehyde	$C_9H_{10}O$ 134.18	15764-16-6 -9	 218	4-07-00-00706	EtOH 3; eth 3; ace 3; bz 3
548	Benzaldehyde, 2,5-dimethyl- Isoxylaldehyde	$C_9H_{10}O$ 134.18	5779-94-2	 220	4-07-00-00707 0.9500^{20}	EtOH 4; eth 3; ace 3; bz 3
549	Benzaldehyde, 3,5-dimethyl-	$C_9H_{10}O$ 134.18	5779-95-3 9	 221	4-07-00-00707	ace 4; bz 4; eth 4; EtOH 4
550	Benzaldehyde, 4-(dimethylamino)- Ehrlich's reagent	$C_9H_{11}NO$ 149.19	100-10-7 74.5	3219 $176\text{-}7^{17}$	4-14-00-00051 1.0254^{100}	H_2O 2; EtOH 3; eth 3; ace 3
551	Benzaldehyde, 2,4-dinitro- 2,4-Dinitrobenzaldehyde	$C_7H_4N_2O_5$ 196.12	528-75-6 72	3265 200^{15}	3-07-00-00923	H_2O 2; EtOH 3; eth 3; bz 3
552	Benzaldehyde, 2-ethoxy-	$C_9H_{10}O_2$ 150.18	613-69-4 21	 248	4-08-00-00180	EtOH 5; eth 5; chl 2
553	Benzaldehyde, 3-ethoxy-	$C_9H_{10}O_2$ 150.18	22924-15-8	 245.5	0-08-00-00060 1.0768^{20}	EtOH 3; eth 3; bz 3 1.5408^{20}
554	Benzaldehyde, 4-ethoxy-	$C_9H_{10}O_2$ 150.18	10031-82-0 13.5	 249	4-08-00-00254 1.08^{21}	EtOH 4; eth 4; bz 4
555	Benzaldehyde, 3-ethoxy-4-hydroxy- Ethylvanillin	$C_9H_{10}O_3$ 166.18	121-32-4 77.5	3815	4-08-00-01765	H_2O 2; EtOH 3; eth 3; bz 3
556	Benzaldehyde, 5-ethoxy-2-hydroxy- 5-Ethoxy-2-hydroxybenzaldehyde	$C_9H_{10}O_3$ 166.18	80832-54-8 51.5	 230	3-08-00-02004	eth 4; EtOH 4; chl 4
557	Benzaldehyde, 4-ethoxy-3-methoxy-	$C_{10}H_{12}O_3$ 180.20	120-25-2 64.5	 168^{13}	4-08-00-01766	H_2O 2; EtOH 3; eth 3; bz 3
558	Benzaldehyde, 2-ethyl-	$C_9H_{10}O$ 134.18	22927-13-5	 209	4-07-00-00696 1.0216^{17}	 1.5425^{17}
559	Benzaldehyde, 2-fluoro-	C_7H_5FO 124.11	446-52-6 -44.5	 175	4-07-00-00559 1.178^{25}	 1.5234^{20}
560	Benzaldehyde, 3-fluoro-	C_7H_5FO 124.11	456-48-4	 173	4-07-00-00559 1.17^{25}	 1.5206^{20}
561	Benzaldehyde, 4-fluoro-	C_7H_5FO 124.11	459-57-4 -10	 181.5	4-07-00-00559 1.1810^{19}	
562	Benzaldehyde, 2-(β-*D*-glucopyranosyloxy)- Helicin	$C_{13}H_{16}O_7$ 284.27	618-65-5 175	4544	5-17-07-00150	H_2O 4; EtOH 4
563	Benzaldehyde, hydrazone Benzylidene hydrazine	$C_7H_8N_2$ 120.15	5281-18-5 16	 140^{14}	4-07-00-00530	EtOH 3
564	Benzaldehyde, 2-hydroxy- Salicylaldehyde	$C_7H_6O_2$ 122.12	90-02-8 -7	8295 197	4-08-00-00176 1.1674^{20}	H_2O 2; EtOH 5; eth 5; ace 4 1.5740^{20}
565	Benzaldehyde, 3-hydroxy-	$C_7H_6O_2$ 122.12	100-83-4 108	 240	4-08-00-00240 1.1179^{130}	H_2O 2; EtOH 3; eth 3; ace 3
566	Benzaldehyde, 4-hydroxy- *p*-Hydroxybenzaldehyde	$C_7H_6O_2$ 122.12	123-08-0 117	4741	4-08-00-00251 1.129^{130}	H_2O 2; EtOH 4; eth 4; ace 2 1.5705^{130}
567	Benzaldehyde, 2-hydroxy-4,6-dimethoxy-	$C_9H_{10}O_4$ 182.18	708-76-9 70	 193^{25}, 165^{10}	4-08-00-02717	H_2O 1; EtOH 4; eth 4; bz 4
568	Benzaldehyde, 3-hydroxy-4,5-dimethoxy-	$C_9H_{10}O_4$ 182.18	29865-90-5 62.5	 179^{12}	4-08-00-02718	bz 4; EtOH 4; lig 4
569	Benzaldehyde, 4-hydroxy-3,5-dimethoxy- Syringaldehyde	$C_9H_{10}O_4$ 182.18	134-96-3 113	8996 $192\text{-}3^{14}$	4-08-00-02718	H_2O 2; EtOH 4; eth 4; bz 4
570	Benzaldehyde, 2-hydroxy-3-methoxy-	$C_8H_8O_3$ 152.15	148-53-8 44.5	 265.5	4-08-00-01747	H_2O 2; EtOH 4; eth 4; ctc 4
571	Benzaldehyde, 2-hydroxy-5-methoxy-	$C_8H_8O_3$ 152.15	672-13-9 4	 247.5	4-08-00-01759	eth 4; EtOH 4
572	Benzaldehyde, 3-hydroxy-4-methoxy-	$C_8H_8O_3$ 152.15	621-59-0 114	 179^{15}	4-08-00-01764 1.196^{25}	H_2O 2; EtOH 3; eth 3; bz 3
573	Benzaldehyde, 3-hydroxy-5-methoxy- 3-Hydroxy-5-methoxybenzaldehyde	$C_8H_8O_3$ 152.15	57179-35-8 130.5	 sub 110	3-08-00-02073	bz 4; EtOH 4; HOAc 4
574	Benzaldehyde, 4-hydroxy-3-methoxy- Vanillin	$C_8H_8O_3$ 152.15	121-33-5 81.5	9839 285	4-08-00-01763 1.056^{25}	H_2O 2; EtOH 4; eth 4; ace 4
575	Benzaldehyde, 2-hydroxy-3-methyl- *m*-Tolualdehyde, 2-hydroxy	$C_8H_8O_2$ 136.15	824-42-0 17	 208.5	4-08-00-00430	eth 4; EtOH 4; chl 4
576	Benzaldehyde, 2-hydroxy-4-methyl-	$C_8H_8O_2$ 136.15	698-27-1 60.5	 223	4-08-00-00434	eth 4; EtOH 4; chl 4
577	Benzaldehyde, 2-hydroxy-5-methyl-	$C_8H_8O_2$ 136.15	613-84-3 56	 217.5	4-08-00-00432 1.0913^{59}	eth 4; EtOH 4; chl 4 1.547^{59}
578	Benzaldehyde, 2-hydroxy-6-methyl-	$C_8H_8O_2$ 136.15	18362-36-2 32.3	 230	4-08-00-00430	bz 4
579	Benzaldehyde, 2-hydroxy-, oxime Salicylaldoxime	$C_7H_7NO_2$ 137.14	94-67-7 57	8296	1-08-00-00520	H_2O 2; EtOH 4; eth 4; bz 4
580	Benzaldehyde, 2-iodo-	C_7H_5IO 232.02	26260-02-6 37	 129^{14}	4-07-00-00582	H_2O 2; ace 3
581	Benzaldehyde, 3-iodo-	C_7H_5IO 232.02	696-41-3 57	 125^{13}	4-07-00-00583	EtOH 3; ace 3; bz 3
582	Benzaldehyde, 4-iodo-	C_7H_5IO 232.02	15164-44-0 77.5	 265	4-07-00-00584	H_2O 2; EtOH 3; bz 3
583	Benzaldehyde, 2-methoxy-	$C_8H_8O_2$ 136.15	135-02-4 37.5	 243.5	4-08-00-00180 1.1326^{20}	H_2O 1; EtOH 3; eth 4; ace 4 1.5600^{20}
584	Benzaldehyde, 3-methoxy-	$C_8H_8O_2$ 136.15	591-31-1	 231	4-08-00-00241 1.1187^{20}	H_2O 1; EtOH 3; eth 4; ace 4 1.5530^{20}

PHYSICAL CONSTANTS OF ORGANIC COMPOUNDS (continued)

No.	Name Synonym	Mol. Form. Mol. Wt.	CAS RN mp/°C	Merck No. bp/°C	Beil. Ref. den/g cm^{-3}	Solubility n_D
585	Benzaldehyde, 4-methoxy- *p*-Anisaldehyde	$C_8H_8O_2$ 136.15	123-11-5 0	693 248; 134^{12}	4-08-00-00252 1.119^{15}	H_2O 1; EtOH 5; eth 5; ace 4 1.5730^{20}
586	Benzaldehyde, 4-methoxy-3-methyl-	$C_9H_{10}O_2$ 150.18	32723-67-4	 $80\text{-}5^{1}$	4-08-00-00432 1.025^{25}	 1.5670^{20}
587	Benzaldehyde, 2-methyl- *o*-Tolualdehyde	C_8H_8O 120.15	529-20-4	9453 200; 94^{10}	4-07-00-00667 1.0328^{20}	H_2O 2; EtOH 3; eth 3; ace 4 1.5462^{20}
588	Benzaldehyde, 3-methyl-	C_8H_8O 120.15	620-23-5	 199	4-07-00-00669 1.0189^{21}	H_2O 2; EtOH 5; eth 5; ace 4 1.5413^{21}
589	Benzaldehyde, 4-methyl- *p*-Tolualdehyde	C_8H_8O 120.15	104-87-0	 $204\text{-}5.106^{10}$	4-07-00-00672 1.0194^{17}	H_2O 2; EtOH 5; eth 5; ace 5 1.5454^{20}
590	Benzaldehyde, 4-(1-methylethyl)- Cuminaldehyde	$C_{10}H_{12}O$ 148.20	122-03-2	2623 235.5	4-07-00-00723 0.9755^{20}	H_2O I; EtOH 3; eth 3; ctc 2 1.5301^{20}
591	Benzaldehyde, 2-nitro-	$C_7H_5NO_3$ 151.12	552-89-6 43.5	 153^{23}	4-07-00-00584 1.2844^{20}	H_2O 2; EtOH 4; eth 4; ace 4
592	Benzaldehyde, 3-nitro-	$C_7H_5NO_3$ 151.12	99-61-6 58.5	 164^{23}	4-07-00-00591 1.2792^{20}	H_2O 2; EtOH 3; eth 3; ace 4
593	Benzaldehyde, 4-nitro-	$C_7H_5NO_3$ 151.12	555-16-8 107	 sub	4-07-00-00598 1.496^{25}	H_2O 2; EtOH 4; eth 2; bz 3
594	Benzaldehyde, oxime, (*E*)-	C_7H_7NO 121.14	622-31-1 35	 119^{10}	4-07-00-00527 1.145^{20}	H_2O 3; EtOH 4; eth 4
595	Benzaldehyde, oxime, (*Z*)-	C_7H_7NO 121.14	622-32-2 36.5	 200	4-07-00-00527 1.1111^{20}	bz 4; eth 4; EtOH 4 1.5908^{20}
596	Benzaldehyde, pentafluoro-	C_7HF_5O 196.08	653-37-2 20	 167	4-07-00-00561	 1.4506^{20}
597	Benzaldehyde, 3-phenoxy-	$C_{13}H_{10}O_2$ 198.22	39515-51-0	 $140^{0.1}$	4-08-00-00242 1.147^{25}	 1.5954^{20}
598	Benzaldehyde, 2,3,5,6-tetramethyl- Durene aldehyde	$C_{11}H_{14}O$ 162.23	17432-37-0 20	 135^{11}	3-07-00-01148	 1.5560^{30}
599	Benzaldehyde, 2,4,6-trimethyl-	$C_{10}H_{12}O$ 148.20	487-68-3 14	 238.5	4-07-00-00730 1.0154^{25}	H_2O 1; EtOH 3; eth 3; ace 3
600	Benzamide Benzoic acid amide	C_7H_7NO 121.14	55-21-0 129.1	1067 290	4-09-00-00725 1.0792^{130}	H_2O 2; EtOH 4; eth 2; bz 2
601	Benzamide, 2-amino-	$C_7H_8N_2O$ 136.15	88-68-6 110.5 dec	 300	4-14-00-01010	H_2O 3; EtOH 3; eth 2; bz 2
602	Benzamide, 4-amino-*N*-[2-(diethylamino)ethyl]-, monohydrochloride Procainamide hydrochloride	$C_{13}H_{22}ClN_3O$ 271.79	614-39-1 167	7762	4-14-00-01154	H_2O 4; EtOH 3; eth 1; bz 1
603	Benzamide, *N*-[(4-aminophenyl)sulfonyl]- Sulfabenzamide	$C_{13}H_{12}N_2O_3S$ 276.32	127-71-9 181.5	8868	4-14-00-02664	
604	Benzamide, 2-bromo-	C_7H_6BrNO 200.03	4001-73-4 160.5	 sub	4-09-00-01013	H_2O 2; EtOH 3; eth 2; tfa 2
605	Benzamide, 3-bromo-	C_7H_6BrNO 200.03	22726-00-7 155.3	 sub	4-09-00-01015	H_2O 2; EtOH 3
606	Benzamide, 5-bromo-*N*,2-dihydroxy- 5-Bromosalicylhydroxamic acid	$C_7H_6BrNO_3$ 232.03	5798-94-7 232 dec	1424	4-10-00-00221	
607	Benzamide, *N*-butyl-4-chloro-2-hydroxy- Buclosamide	$C_{11}H_{14}ClNO_2$ 227.69	575-74-6 91.5	1450		
608	Benzamide, 5-chloro-*N*-(3,4-dichlorophenyl)-2-hydroxy- 3',4',5-Trichlorosalicylanilide	$C_{13}H_8Cl_3NO_2$ 316.57	642-84-2 247	9558	4-12-00-01273	
609	Benzamide, 3,5-dibromo-*N*-(4-bromophenyl)-2-hydroxy- Tribromsalan	$C_{13}H_8Br_3NO_2$ 449.92	87-10-5 227	9529		
610	Benzamide, 3,5-dichloro-*N*-(3,4-dichlorophenyl)-2-hydroxy- 3,3',4',5-Tetrachlorosalicylanilide	$C_{13}H_7Cl_4NO_2$ 351.01	1154-59-2 161	9127		
611	Benzamide, *N*,*N*-diethyl-4-hydroxy-3-methoxy- Ethamivan	$C_{12}H_{17}NO_3$ 223.27	304-84-7 95	3673	4-10-00-01481	chl 3
612	Benzamide, *N*,*N*-diethyl-3-methyl- DEET	$C_{12}H_{17}NO$ 191.27	134-62-3	2848 160^{19}; 111^{1}	4-09-00-01716 0.996^{20}	H_2O 4; bz 4; eth 4; EtOH 4 1.5212^{20}
613	Benzamide, *N*,2-dihydroxy- Salicylhydroxamic acid	$C_7H_7NO_3$ 153.14	89-73-6 168	8300 sub	4-10-00-00192	H_2O 2; EtOH 4; eth 4; HOAc 3
614	Benzamide, 3,5-dinitro- [illegible]	$C_7H_5N_3O_5$ 211.1[illegible]	121-81-3 184	6533	4-09-00-01351	H_2O 4
615	Benzamide, 2-ethoxy- Ethenzamide	$C_9H_{11}NO_2$ 165.19	938-73-8 133	3685	4-10-00-00175	H_2O 2; EtOH 4; eth 4; chl 2
616	Benzamide, 2-hydroxy- Salicylamide	$C_7H_7NO_2$ 137.14	65-45-2 142	8297 181.5^{14}	4-10-00-00169 1.175^{140}	H_2O 2; EtOH 3; eth 2
617	Benzamide, 2-hydroxy-*N*-phenyl- Salicylanilide	$C_{13}H_{11}NO_2$ 213.24	87-17-2 136.5	8299	4-12-00-00906	H_2O 3; EtOH 2; eth 2; bz 2
618	Benzamide, 4-methoxy-	$C_8H_9NO_2$ 151.17	3424-93-9 166.5	 295	4-10-00-00433	H_2O 4; EtOH 4
619	Benzamide, 2-methyl- *o*-Toluamide	C_8H_9NO 135.17	527-85-5 147	9454	4-09-00-01701	H_2O 2; EtOH 4; eth 2; bz 2
620	Benzamide, *N*-methyl-	C_8H_9NO 135.17	613-93-4 82	 291; 167^{12}	4-09-00-00727	EtOH 3; ace 3
621	Benzamide, *N*-(2-methylphenyl)-	$C_{14}H_{13}NO$ 211.26	584-70-3 145.5		4-12-00-01757 1.205^{15}	H_2O 2; EtOH 3; eth 3

No.	Name Synonym	Mol. Form. Mol. Wt.	CAS RN mp/°C	Merck No. bp/°C	Beil. Ref. den/g cm^{-3}	Solubility n_D
622	Benzamide, *N*-(3-methylphenyl)- *m*-Benzotoluidide	$C_{14}H_{13}NO$ 211.26	582-77-4 125		4-12-00-01825 1.170^{15}	EtOH 4
623	Benzamide, *N*-(4-methylphenyl)-	$C_{14}H_{13}NO$ 211.26	582-78-5 158		4-12-00-01910 1.202^{15}	eth 4; EtOH 4
624	Benzamide, 2-nitro-	$C_7H_6N_2O_3$ 166.14	610-15-1 176.6	 317	4-09-00-01049	H_2O 3; EtOH 3; eth 3
625	Benzamide, 3-nitro-	$C_7H_6N_2O_3$ 166.14	645-09-0 142.7	 312.5	4-09-00-01061	H_2O 3; EtOH 3; eth 3
626	Benzamide, *N*-phenyl- Benzanilide	$C_{13}H_{11}NO$ 197.24	93-98-1 163	1068 sub 117	4-12-00-00417 1.315^{25}	H_2O 1; EtOH 2; eth 2; HOAc 2
627	Benzamide, *N*,*N*,4-trimethyl-	$C_{10}H_{13}NO$ 163.22	14062-78-3 41	 156^{10}	0-09-00-00486	H_2O 2; EtOH 2
628	Benz[b]indeno[1,2-d]pyran-9(6H)-one, 6a,7-dihydro-3,4,6a,10-tetrahydroxy- Hematein	$C_{16}H_{12}O_6$ 300.27	475-25-2 250 dec	4553	4-18-00-03343	H_2O 1; EtOH 2; eth 1; bz 1
629	Benz[b]indeno[1,2-d]pyran-3,4,6a,9,10(6H)-pentol, 7,11b-dihydro-, *cis*-(+)- Hematoxylin	$C_{16}H_{14}O_6$ 302.28	517-28-2 140	4556	2-17-00-00273	H_2O 2; EtOH 3; eth 2; alk 3
630	Benz[c]acridine 12-Azabenz[a]anthracene	$C_{17}H_{11}N$ 229.28	225-51-4 132		5-20-08-00519	bz 4; eth 4; EtOH 4
631	Benz[e]acephenanthrylene Benzo[b]fluoranthene	$C_{20}H_{12}$ 252.32	205-99-2 168		4-05-00-02686	H_2O 1; bz 5
632	1H-Benz[e]indene, 2,3-dihydro-	$C_{13}H_{12}$ 168.24	4944-94-9	 294.5	4-05-00-01865 1.066^{20}	H_2O 1; HOAc 3 1.6290^{20}
633	Benzenamine Aniline	C_6H_7N 93.13	62-53-3 -6.0	687 184.1	4-12-00-00223 1.0217^{20}	H_2O 3; EtOH 5; eth 5; ace 5 1.5863^{20}
634	Benzenamine, 4-[(4-aminophenyl)(4-imino-2,5-cyclohexadien-1-ylidene)methyl]-2-methyl-, monohydrochloride Magenta I	$C_{20}H_{20}ClN_3$ 337.85	632-99-5 200 dec	5528		H_2O 2; EtOH 2; eth 1
635	Benzenamine, 4,4'-azobis- *p*-Diaminoazobenzene	$C_{12}H_{12}N_4$ 212.25	538-41-0 250.5	2954	4-16-00-00515	H_2O 2; EtOH 3; bz 4; chl 4
636	Benzenamine, 2,6-bis(1-methylethyl)-	$C_{12}H_{19}N$ 177.29	24544-04-5 -45	 257	3-12-00-02764 0.94^{25}	 1.5332^{20}
637	Benzenamine, 3,5-bis(trifluoromethyl)-	$C_8H_5F_6N$ 229.12	328-74-5	 85^{15}	3-12-00-02498 1.487^{25}	 1.4335^{20}
638	Benzenamine, 2-bromo- *o*-Bromoaniline	C_6H_6BrN 172.02	615-36-1 32	 229	4-12-00-01487 1.578^{20}	H_2O 1; EtOH 3; eth 3 1.6113^{20}
639	Benzenamine, 3-bromo- *m*-Bromoaniline	C_6H_6BrN 172.02	591-19-5 18.5	 251	4-12-00-01491 1.5793^{20}	H_2O 2; EtOH 3; eth 3 1.6260^{20}
640	Benzenamine, 4-bromo- *p*-Bromoaniline	C_6H_6BrN 172.02	106-40-1 66.4	1392 dec	4-12-00-01497 1.4970^{100}	H_2O 1; EtOH 3; eth 3; chl 2
641	Benzenamine, 2-bromo-4,6-dichloro-	$C_6H_4BrCl_2N$ 240.91	697-86-9 83.5	 273	2-12-00-00355	bz 4; EtOH 4; chl 4
642	Benzenamine, 4-bromo-*N*,*N*-diethyl-	$C_{10}H_{14}BrN$ 228.13	2052-06-4 38	 270	4-12-00-01500	H_2O 1; EtOH 4; eth 4
643	Benzenamine, 2-bromo-*N*,*N*-dimethyl-	$C_8H_{10}BrN$ 200.08	698-00-0	 $107\text{-}8^{14}$	4-12-00-01487 1.3839^{25}	EtOH 4 1.5748^{25}
644	Benzenamine, 3-bromo-*N*,*N*-dimethyl-	$C_8H_{10}BrN$ 200.08	16518-62-0 11	 259	4-12-00-01492 1.3651^{20}	EtOH 4; HOAc 4; con acid 3
645	Benzenamine, 4-bromo-*N*,*N*-dimethyl-	$C_8H_{10}BrN$ 200.08	586-77-6 55	 264	4-12-00-01499 1.3220^{100}	H_2O 1; EtOH 3; eth 4; os 3
646	Benzenamine, 2-bromo-4,6-dinitro-	$C_6H_4BrN_3O_4$ 262.02	1817-73-8 153.5	 sub	4-12-00-01734	EtOH 4; ace 4; HOAc 3
647	Benzenamine, 2-bromo-4-methyl-	C_7H_8BrN 186.05	583-68-6 26	 240	4-12-00-01992 1.510^{20}	H_2O 1; EtOH 3; eth 3 1.5999^{20}
648	Benzenamine, 2-bromo-5-methyl- 2-Bromo-5-methylaniline	C_7H_8BrN 186.05	53078-85-6 46	 129^{15}	3-12-00-01998 1.470^{25}	eth 4; EtOH 4
649	Benzenamine, 3-bromo-4-methyl-	C_7H_8BrN 186.05	7745-91-7 26	 255.5	4-12-00-01991	eth 4
650	Benzenamine, 4-bromo-2-methyl-	C_7H_8BrN 186.05	583-75-5 59.5	 240	4-12-00-01804	H_2O 2; EtOH 3; eth 4; chl 2
651	Benzenamine, 4-bromo-3-methyl-	C_7H_8BrN 186.05	6933-10-4 81	 240	4-12-00-01856	H_2O 1; EtOH 4; chl 2
652	Benzenamine, 5-bromo-2-methyl-	C_7H_8BrN 186.05	39478-78-9 33	 255 dec; 165^{25}	3-12-00-01923	eth 4; EtOH 4
653	Benzenamine, 2-bromo-6-nitro- Aniline, 2-bromo-6-nitro-	$C_6H_5BrN_2O_2$ 217.02	59255-95-7 74.5		2-12-00-00402 1.988^{25}	EtOH 4
654	Benzenamine, 4-bromo-2-nitro-	$C_6H_5BrN_2O_2$ 217.02	875-51-4 111.5	 sub	3-12-00-01670	EtOH 4
655	Benzenamine, 4-butyl-	$C_{10}H_{15}N$ 149.24	104-13-2	 261	4-12-00-02807 0.945^{20}	ctc 2
656	Benzenamine, *N*-butyl- *N*-Butylaniline	$C_{10}H_{15}N$ 149.24	1126-78-9 -14.4	 243.5	4-12-00-00256 0.9323^{20}	eth 4; EtOH 4 1.5341^{20}
657	Benzenamine, 4,4'-carbonimidoylbis[*N*,*N*-dimethyl- 4,4'-Dimethylaminobenzophenonimide	$C_{17}H_{21}N_3$ 267.37	492-80-8 136		4-14-00-00256	H_2O 1; EtOH 3; eth 2

PHYSICAL CONSTANTS OF ORGANIC COMPOUNDS (continued)

No.	Name	Synonym	Mol. Form.	Mol. Wt.	CAS RN	mp/°C	Merck No.	bp/°C	Beil. Ref.	den/g cm^{-3}	Solubility	n_D
658	Benzenamine, 2-chloro-	*o*-Chloroaniline	C_6H_6ClN	127.57	95-51-2	-14	2118	208.8	4-12-00-01115		H_2O 1; EtOH 5; eth 3; ace 3	1.5895^{20}
659	Benzenamine, 3-chloro-	*m*-Chloroaniline	C_6H_6ClN	127.57	108-42-9	-10.4	2118	230.5	4-12-00-01137	1.2161^{20}	H_2O 1; EtOH 5; eth 5; ace 5	1.5941^{20}
660	Benzenamine, 4-chloro-	*p*-Chloroaniline	C_6H_6ClN	127.57	106-47-8	72.5	2118	232	4-12-00-01166	1.429^{19}	H_2O 3; EtOH 3; eth 3; chl 3	1.5546^{87}
661	Benzenamine, 4-chloro-*N*,*N*-diethyl-		$C_{10}H_{14}ClN$	183.68	2873-89-4	45.5		252; 113^{6}	4-12-00-01170		EtOH 3	
662	Benzenamine, 2-chloro-*N*,*N*-dimethyl-		$C_8H_{10}ClN$	155.63	698-01-1			205	4-12-00-01116	1.1067^{20}	bz 4; EtOH 4	1.5578^{20}
663	Benzenamine, 4-chloro-*N*,*N*-dimethyl-		$C_8H_{10}ClN$	155.63	698-69-1	35.5		231	4-12-00-01169	1.0480^{100}	EtOH 3	
664	Benzenamine, 2-chloro-6-fluoro-	Aniline, 2-chloro-6-fluoro	C_6H_5ClFN	145.56	363-51-9			91^{30}	4-12-00-01238	1.316^{23}		1.5511^{23}
665	Benzenamine, 2-chloro-, hydrochloride	Aniline, 2-chloro, hydrochloride	$C_6H_7Cl_2N$	164.03	137-04-2	235			3-12-00-01281	1.505^{18}	H_2O 4	
666	Benzenamine, 2-chloro-5-methoxy-		C_7H_8ClNO	157.60	2401-24-3	27		142^{18}	4-13-00-00999			1.5848^{25}
667	Benzenamine, 4-chloro-2-methoxy-	4-Chloro-*o*-anisidine	C_7H_8ClNO	157.60	93-50-5	52		260	4-13-00-00885		EtOH 3; eth 3; bz 3; chl 3	
668	Benzenamine, 2-chloro-4-methyl-	2-Chloro-*p*-toluidine	C_7H_8ClN	141.60	615-65-6	7		220	3-12-00-02152	1.151^{20}	EtOH 2; bz 2	1.5748^{22}
669	Benzenamine, 2-chloro-5-methyl-	6-Chloro-*m*-toluidine	C_7H_8ClN	141.60	95-81-8	29.5		229	4-12-00-01849		EtOH 4	
670	Benzenamine, 2-chloro-*N*-methyl-		C_7H_8ClN	141.60	932-32-1			218	4-12-00-01116	1.1735^{11}	EtOH 3; ace 3; bz 3	1.5780^{25}
671	Benzenamine, 3-chloro-2-methyl-	3-Chloro-*o*-toluidine	C_7H_8ClN	141.60	87-60-5	1		245	4-12-00-01800		H_2O 3; EtOH 3; eth 1; bz 1	1.5880^{20}
672	Benzenamine, 3-chloro-4-methyl-	3-Chloro-*p*-toluidine	C_7H_8ClN	141.60	95-74-9	26		243	4-12-00-01985		EtOH 3; ctc 2	
673	Benzenamine, 4-chloro-2-methyl-		C_7H_8ClN	141.60	95-69-2	30.3		244	4-12-00-01796		EtOH 3; ctc 2	
674	Benzenamine, 4-chloro-3-methyl-	4-Chloro-*m*-toluidine	C_7H_8ClN	141.60	7149-75-9	83.5		241	4-12-00-01850		ace 4; bz 4; EtOH 4	
675	Benzenamine, 4-chloro-*N*-methyl-		C_7H_8ClN	141.60	932-96-7			240	4-12-00-01168	1.169^{11}	EtOH 3; ace 3; bz 3	1.5835^{20}
676	Benzenamine, 5-chloro-2-methyl-	5-Chloro-*o*-toluidine	C_7H_8ClN	141.60	95-79-4	26		239; 140^{38}	4-12-00-01794		EtOH 4	
677	Benzenamine, 5-chloro-2-nitro-		$C_6H_5ClN_2O_2$	172.57	1635-61-6	127.8		sub	4-12-00-01673		eth 4; EtOH 4	
678	Benzenamine, 3-chloro-*N*-phenyl-		$C_{12}H_{10}ClN$	203.67	101-17-7			338	4-12-00-01141	1.2^{25}	H_2O 1; EtOH 3; eth 3; bz 3	1.6513^{20}
679	Benzenamine, 4-chloro-*N*-phenyl-		$C_{12}H_{10}ClN$	203.67	1205-71-6	74		335	4-12-00-01170		bz 4; eth 4; EtOH 4; lig 4	
680	Benzenamine, *N*-cyclohexyl-		$C_{12}H_{17}N$	175.27	1821-36-9	16		279; 192^{73}	4-12-00-00268	1.0155^{20}	H_2O 1; EtOH 3; eth 3; bz 3	1.5610^{20}
681	Benzenamine, 2,4-dibromo-		$C_6H_5Br_2N$	250.92	615-57-6	79.5		156^{74}	4-12-00-01532	2.260^{20}	EtOH 3; eth 3; chl 3; HOAc 3	
682	Benzenamine, 2,6-dibromo-		$C_6H_5Br_2N$	250.92	608-30-0	87.5		263	4-12-00-01535		EtOH 4; eth 4; bz 4; chl 4	
683	Benzenamine, 3,4-dibromo-	3,4-Dibromoaniline	$C_6H_5Br_2N$	250.92	615-55-4	80.5		100	4-12-00-01535		EtOH 4	
684	Benzenamine, *N*,*N*-dibutyl-		$C_{14}H_{23}N$	205.34	613-29-6	-32.2		274.8	4-12-00-00256	0.9037^{20}	H_2O 1; EtOH 5; eth 5; ace 4	1.5186^{20}
685	Benzenamine, 2,3-dichloro-	2,3-Dichloroaniline	$C_6H_5Cl_2N$	162.02	608-27-5	24		252	0-12-00-00621		EtOH 3; eth 4; ace 3; bz 2	
686	Benzenamine, 2,4-dichloro-	2,4-Dichloroaniline	$C_6H_5Cl_2N$	162.02	554-00-7	63.5		245	4-12-00-01241	1.567^{20}	H_2O 2; EtOH 3; eth 3; chl 2	
687	Benzenamine, 2,5-dichloro-	2,5-Dichloroaniline	$C_6H_5Cl_2N$	162.02	95-82-9	50		251	4-12-00-01250		H_2O 2; EtOH 3; eth 3; bz 3	
688	Benzenamine, 2,6-dichloro-	2,6-Dichloroaniline	$C_6H_5Cl_2N$	162.02	608-31-1	39			2-12-00-00337		H_2O 2; EtOH 3; eth 3	
689	Benzenamine, 3,4-dichloro-	3,4-Dichloroaniline	$C_6H_5Cl_2N$	162.02	95-76-1	72	3041	272	4-12-00-01257		EtOH 3; eth 3; bz 2; chl 2	
690	Benzenamine, 3,5-dichloro-	3,5-Dichloroaniline	$C_6H_5Cl_2N$	162.02	626-43-7	52		261	4-12-00-01274		H_2O 1; EtOH 3; eth 3; ctc 3	
691	Benzenamine, 2,6-dichloro-4-ethoxy-	Aniline, 2,6-dichloro-4-ethoxy	$C_8H_9Cl_2NO$	206.07	51225-20-8	46		275	2-13-00-00276		bz 4; eth 4; EtOH 4	
692	Benzenamine, 2,6-dichloro-4-nitro-	2,6-Dichloro-4-nitroaniline	$C_6H_4Cl_2N_2O_2$	207.02	99-30-9	191			4-12-00-01681		EtOH 3; DMSO-d_6 2; acid 3	
693	Benzenamine, 3,4-diethoxy-	3,4-Diethoxyaniline	$C_{10}H_{15}NO_2$	181.23	39052-12-5	48		132-$5^{0.45}$	4-13-00-02507		bz 4; eth 4; EtOH 4	
694	Benzenamine, 2,6-diethyl-		$C_{10}H_{15}N$	149.24	579-66-8	3		243	4-12-00-02841	0.906^{25}		1.5452^{20}
695	Benzenamine, *N*,*N*-diethyl-	*N*,*N*-Diethylaniline	$C_{10}H_{15}N$	149.24	91-66-7	-38.8	3102	216.3	4-12-00-00252	0.9307^{20}	H_2O 2; EtOH 3; eth 4; ace 3	1.5409^{20}
696	Benzenamine, *N*,*N*-diethyl-2-methyl-		$C_{11}H_{17}N$	163.26	606-46-2	-60		209	4-12-00-01748	0.9286^{20}	H_2O 2; EtOH 5; eth 5; ctc 3	1.5153^{20}
697	Benzenamine, *N*,*N*-diethyl-4-methyl-		$C_{11}H_{17}N$	163.26	613-48-9			229	4-12-00-01875	0.9242^{16}	H_2O 2; EtOH 5; eth 5	

PHYSICAL CONSTANTS OF ORGANIC COMPOUNDS (continued)

No.	Name Synonym	Mol. Form. Mol. Wt.	CAS RN mp/°C	Merck No. bp/°C	Beil. Ref. den/g cm^{-3}	Solubility n_D
698	Benzenamine, *N,N*-diethyl-4-nitro-	$C_{10}H_{14}N_2O_2$ 194.23	2216-15-1 77.5		4-12-00-01618 1.225^{25}	EtOH 3; lig 2
699	Benzenamine, *N,N*-diethyl-4-nitroso- Aniline, *N,N*-diethyl-4-nitroso	$C_{10}H_{14}N_2O$ 178.23	120-22-9 87.5		4-12-00-01559 1.24^{15}	H_2O 2; EtOH 3; eth 3; ace 3
700	Benzenamine, 2,4-difluoro- 2,4-Difluoroaniline	$C_6H_5F_2N$ 129.11	367-25-9 -7.5	3131 170	4-12-00-01112 1.268^{25}	 1.5063^{20}
701	Benzenamine, 2,4-diiodo- 2,4-Diiodoaniline	$C_6H_5I_2N$ 344.92	533-70-0 95.5	3173	3-12-00-01508 2.748^{25}	ace 4; bz 4; eth 4; EtOH 4
702	Benzenamine, 2,5-dimethoxy-	$C_8H_{11}NO_2$ 153.18	102-56-7 82.5	 270	4-13-00-02548	H_2O 3; EtOH 3; chl 3; lig 3
703	Benzenamine, 2,6-dimethoxy- 2,6-Dimethoxyaniline	$C_8H_{11}NO_2$ 153.18	2734-70-5 76.5	 146^{23}	4-13-00-02526	bz 4; eth 4; EtOH 4; lig 4
704	Benzenamine, 2,3-dimethyl- 2,3-Xylidine	$C_8H_{11}N$ 121.18	87-59-2 <-15	 221.5	4-12-00-02497 0.9931^{20}	H_2O 2; EtOH 4; eth 4; ctc 3 1.5684^{20}
705	Benzenamine, 2,4-dimethyl- 2,4-Xylidine	$C_8H_{11}N$ 121.18	95-68-1 -14.3	 214	4-12-00-02545 0.9723^{20}	H_2O 2; EtOH 3; eth 3; bz 3 1.5569^{20}
706	Benzenamine, 2,5-dimethyl- 2,5-Xylidine	$C_8H_{11}N$ 121.18	95-78-3 15.5	 214	4-12-00-02567 0.9790^{21}	H_2O 2; eth 3; ctc 3 1.5591^{21}
707	Benzenamine, 2,6-dimethyl- 2,6-Xylidine	$C_8H_{11}N$ 121.18	87-62-7 11.2	 215	4-12-00-02521 0.9842^{20}	eth 4; EtOH 4 1.5610^{20}
708	Benzenamine, 3,4-dimethyl- 3,4-Xylidine	$C_8H_{11}N$ 121.18	95-64-7 51	 228	4-12-00-02502 1.076^{18}	H_2O 2; eth 3; chl 2; lig 4
709	Benzenamine, 3,5-dimethyl- 3,5-Xylidine	$C_8H_{11}N$ 121.18	108-69-0 9.8	 220.5	4-12-00-02561 0.9706^{20}	H_2O 2; eth 3; ctc 3 1.5581^{20}
710	Benzenamine, *N*,2-dimethyl-	$C_8H_{11}N$ 121.18	611-21-2	 207.5	4-12-00-01746 0.9709^{20}	H_2O 1; EtOH 5; eth 5; ace 3 1.5649^{20}
711	Benzenamine, *N*,3-dimethyl-	$C_8H_{11}N$ 121.18	696-44-6	 206.5	4-12-00-01815	H_2O 1; EtOH 5; eth 5; ace 3 1.5557^{25}
712	Benzenamine, *N*,4-dimethyl-	$C_8H_{11}N$ 121.18	623-08-5	 210	4-12-00-01874 0.9348^{55}	H_2O 1; EtOH 5; eth 5; ace 3 1.5568^{20}
713	Benzenamine, *N,N*-dimethyl- *N,N*-Dimethylaniline	$C_8H_{11}N$ 121.18	121-69-7 2.4	3223 194.1	4-12-00-00243 0.9557^{20}	H_2O 2; EtOH 3; eth 3; ace 3 1.5582^{20}
714	Benzenamine, 2-(1,1-dimethylethyl)-	$C_{10}H_{15}N$ 149.24	6310-21-0	 234	4-12-00-02823 0.977^{15}	EtOH 4; eth 4; bz 4; acid 3 1.5453^{20}
715	Benzenamine, 4-(1,1-dimethylethyl)-	$C_{10}H_{15}N$ 149.24	769-92-6 17	 241	4-12-00-02828 0.9525^{15}	H_2O 2; EtOH 5; eth 5; bz 4 1.5380^{20}
716	Benzenamine, *N*-(1,1-dimethylethyl)- *N-tert*-Butylaniline	$C_{10}H_{15}N$ 149.24	937-33-7	 215; 95^{19}	4-12-00-00257	EtOH 3; ace 4; bz 4; chl 4 1.5270^{20}
717	Benzenamine, *N,N*-dimethyl-, hydrochloride	$C_8H_{12}ClN$ 157.64	5882-44-0 90		4-12-00-00243 1.1156^{19}	H_2O 4; EtOH 4; chl 4
718	Benzenamine, *N,N*-dimethyl-2-nitro-	$C_8H_{10}N_2O_2$ 166.18	610-17-3 -20	 146^{20}	4-12-00-01564 1.1794^{20}	H_2O 3; EtOH 4; eth 3; chl 4 1.6102^{20}
719	Benzenamine, *N,N*-dimethyl-3-nitro-	$C_8H_{10}N_2O_2$ 166.18	619-31-8 60.5	 282.5	4-12-00-01591 1.313^{17}	H_2O 1; EtOH 3; eth 3
720	Benzenamine, *N,N*-dimethyl-4-nitroso- *p*-Nitroso-*N,N*-dimethylaniline	$C_8H_{10}N_2O$ 150.18	138-89-6 92.5	6559	4-12-00-01558 1.145^{20}	H_2O 2; EtOH 3; eth 3; chl 3
721	Benzenamine, *N,N*-dimethyl-4-(phenylazo)- 4-(Dimethylamino)azobenzene	$C_{14}H_{15}N_3$ 225.29	60-11-7 117	3218 dec	3-16-00-00340	H_2O 1; EtOH 4; eth 3; chl 2
722	Benzenamine, *N,N*-dimethyl-4-(2-phenylethenyl)-, (*E*)-	$C_{16}H_{17}N$ 223.32	838-95-9 150	 $145^{0.1}$	4-12-00-03400	H_2O 1; EtOH 3; chl 3
723	Benzenamine, *N,N*-dimethyl-4-[[3-(trifluoromethyl)phenyl]azo]- Azobenzene, 4-dimethylamino-3'-trifluoromethyl	$C_{15}H_{14}F_3N_3$ 293.29	328-96-1 80	 76^8	4-16-00-00487	chl 3
724	Benzenamine, 2,3-dinitro- 2,3-Dinitroaniline	$C_6H_5N_3O_4$ 183.12	602-03-9 128		4-12-00-01689 1.646^{50}	H_2O 1; EtOH 3; eth 2
725	Benzenamine, 2,4-dinitro- 2,4-Dinitroaniline	$C_6H_5N_3O_4$ 183.12	97-02-9 180	3263	4-12-00-01689 1.615^{14}	H_2O 1; EtOH 2; ace 2; HCl 2
726	Benzenamine, 2,6-dinitro- 2,6-Dinitroaniline	$C_6H_5N_3O_4$ 183.12	606-22-4 141.5	3264	4-12-00-01729	H_2O 1; EtOH 2; eth 3; bz 3
727	Benzenamine, 3,5-dinitro- 3,5-Dinitroaniline	$C_6H_5N_3O_4$ 183.12	618-87-1 163		4-12-00-01729 1.601^{50}	H_2O 1; EtOH 3; eth 3; ace 2
728	Benzenamine, *N,N*-diphenyl- Triphenylamine	$C_{18}H_{15}N$ 245.32	603-34-9 127	 365	4-12-00-00276 0.774^0	H_2O 1; EtOH 2; eth 3; bz 3 1.353^{16}
729	Benzenamine, 4-(diphenylmethyl)-	$C_{19}H_{17}N$ 259.35	603-38-3 84.5	 248^{12}	4-12-00-03470	bz 4; eth 4; lig 4
730	Benzenamine, *N,N*-dipropyl-	$C_{12}H_{19}N$ 177.29	2217-07-4	 242	4-12-00-00255 0.9104^{20}	H_2O 1; EtOH 3; eth 3; ace 3 1.5271^{20}
731	Benzenamine, 4,4'-(1,2-ethanediyl)bis-	$C_{14}H_{16}N_2$ 212.29	621-95-4 137	 sub	4-13-00-00401	H_2O 1; EtOH 4
732	Benzenamine, 4,4'-(1,2-ethenediyl)bis-, (*E*)-	$C_{14}H_{14}N_2$ 210.28	7314-06-9 231	 sub	4-13-00-00453	H_2O 2; bz 2; MeOH 3; CS_2 2
733	Benzenamine, 2-ethenyl-	C_8H_9N 119.17	3867-18-3	 112^{20}; 97^8	4-12-00-02914 1.0181^{20}	ace 3; bz 3 1.6124^{20}
734	Benzenamine, 3-ethenyl-	C_8H_9N 119.17	15411-43-5	 $112\text{-}5^{12}$	4-12-00-02915 1.0198^{20}	ace 3; bz 3 1.6069^{26}
735	Benzenamine, 4-ethenyl-	C_8H_9N 119.17	1520-21-4 23.5	 116^9	4-12-00-02915 1.010^{20}	ace 3; bz 3 1.6250^{22}

No.	Name Synonym	Mol. Form. Mol. Wt.	CAS RN mp/°C	Merck No. bp/°C	Beil. Ref. den/g cm^{-3}	Solubility n_D
736	Benzenamine, 2-ethoxy- *o*-Phenetidine	$C_8H_{11}NO$ 137.18	94-70-2 <-21	7187 232.5	4-13-00-00807	H_2O 2; EtOH 3; eth 3; ctc 2 1.5560^{20}
737	Benzenamine, 4-ethoxy- *p*-Phenetidine	$C_8H_{11}NO$ 137.18	156-43-4 2.4	7188 254	4-13-00-01017 1.0652^{16}	H_2O 2; EtOH 3; eth 3; chl 3 1.5528^{20}
738	Benzenamine, 3-ethoxy-*N*,*N*-diethyl-	$C_{12}H_{19}NO$ 193.29	1864-92-2	 286; $97^{0.6}$	4-13-00-00970	EtOH 3; bz 3; HOAc 3 1.5325^{25}
739	Benzenamine, 2-ethoxy-5-nitro- *o*-Phenetidine, 5-nitro	$C_8H_{10}N_2O_3$ 182.18	136-79-8 96.5	6538 $205\text{-}6^{14}$	4-13-00-00896	eth 4; EtOH 4
740	Benzenamine, 2-ethyl- 2-Ethylaniline	$C_8H_{11}N$ 121.18	578-54-1 -43	 209.5	4-12-00-02411 0.983^{22}	H_2O 2; EtOH 4; eth 4; chl 2 1.5584^{22}
741	Benzenamine, 3-ethyl-	$C_8H_{11}N$ 121.18	587-02-0 -64	 214; 94^{6}	4-12-00-02417 0.9896^{25}	eth 4; EtOH 4
742	Benzenamine, 4-ethyl- *p*-Ethylaniline	$C_8H_{11}N$ 121.18	589-16-2 -4.8	 217.5	4-12-00-02419 0.9679^{20}	H_2O 2; EtOH 4; eth 4; ctc 2 1.5554^{20}
743	Benzenamine, *N*-ethyl- *N*-Ethylaniline	$C_8H_{11}N$ 121.18	103-69-5 -63.5	3722 203.0	4-12-00-00250 0.9625^{20}	H_2O 1; EtOH 5; eth 5; ace 4 1.5559^{20}
744	Benzenamine, 2-ethyl-*N*-(2-ethylphenyl)-	$C_{16}H_{19}N$ 225.33	64653-59-4 29	 336^{603}; 173^{10}	4-12-00-02413	H_2O 1; EtOH 4; eth 4; chl 2 1.5550^{25}
745	Benzenamine, 2-ethyl-6-methyl-	$C_9H_{13}N$ 135.21	24549-06-2 -33	 231	4-12-00-02638 0.968^{25}	 1.5525^{20}
746	Benzenamine, *N*-ethyl-2-methyl-	$C_9H_{13}N$ 135.21	94-68-8 <-15	 216	4-12-00-01747 0.948^{25}	EtOH 3; eth 3 1.5456^{20}
747	Benzenamine, *N*-ethyl-3-methyl-	$C_9H_{13}N$ 135.21	102-27-2	 221	4-12-00-01816	EtOH 3; eth 3 1.5451^{20}
748	Benzenamine, *N*-ethyl-4-methyl- *N*-Ethyl-4-toluidine	$C_9H_{13}N$ 135.21	622-57-1	 217	4-12-00-01875 0.9391^{16}	EtOH 3; eth 3
749	Benzenamine, *N*-ethyl-*N*-methyl-	$C_9H_{13}N$ 135.21	613-97-8	 204	4-12-00-00252 0.92^{55}	H_2O 1; EtOH 5; eth 5; ctc 3
750	Benzenamine, *N*-ethyl-*N*-nitroso-	$C_8H_{10}N_2O$ 150.18	612-64-6	 $119\text{-}20^{15}$	4-16-00-00865 1.0874^{20}	H_2O 1; HOAc 3
751	Benzenamine, *N*-ethyl-*N*-phenyl-	$C_{14}H_{15}N$ 197.28	606-99-5	 295.5; 147^{10}	4-12-00-00276 1.0377^{20}	eth 4; EtOH 4 1.6095^{20}
752	Benzenamine, 2-fluoro- *o*-Fluoroaniline	C_6H_6FN 111.12	348-54-9 -34.6	 175; 55^{12}	4-12-00-01099 1.1513^{21}	H_2O 1; EtOH 3; eth 3; ctc 2 1.5421^{20}
753	Benzenamine, 3-fluoro- *m*-Fluoroaniline	C_6H_6FN 111.12	372-19-0	 188	4-12-00-01101 1.1561^{19}	H_2O 2; EtOH 3; eth 3; chl 2 1.5436^{20}
754	Benzenamine, 4-fluoro- *p*-Fluoroaniline	C_6H_6FN 111.12	371-40-4 -0.8	4098 182; 85^{19}	4-12-00-01104 1.1725^{20}	H_2O 2; EtOH 3; eth 3; ctc 2 1.5195^{20}
755	Benzenamine, 4-fluoro-2-methyl-	C_7H_8FN 125.15	452-71-1 14.2	 94^{16}	4-12-00-01786 1.1263^{18}	eth 3; ace 3; bz 3; ctc 3 1.5363^{18}
756	Benzenamine, hydrochloride Aniline, hydrochloride	C_6H_8ClN 129.59	142-04-1 198	687	4-12-00-00223 1.2215^{4}	H_2O 4; EtOH 4; eth 1; chl 1
757	Benzenamine, *N*-hydroxy- Phenylhydroxylamine	C_6H_7NO 109.13	100-65-2 83.5	7266	4-15-00-00004	bz 4; eth 4; EtOH 4; chl 4
758	Benzenamine, *N*-hydroxy-4-methyl-	C_7H_9NO 123.15	623-10-9 96	 117 dec	4-15-00-00019	eth 4; EtOH 4; chl 4
759	Benzenamine, *N*-hydroxy-*N*-nitroso-, ammonium salt Cupferron	$C_6H_9N_3O_2$ 155.16	135-20-6 163.5	2625	4-16-00-00889	DMSO 2
760	Benzenamine, 3-iodo-	C_6H_6IN 219.02	626-01-7 33	 145^{15}	4-12-00-01543	H_2O 1; EtOH 3; chl 3 1.6811^{20}
761	Benzenamine, 4-iodo- *p*-Iodoaniline	C_6H_6IN 219.02	540-37-4 67.5	4920	4-12-00-01544	H_2O 2; EtOH 3; eth 3; os 3
762	Benzenamine, 2-iodo-4-methyl- 2-Iodo-4-methylaniline	C_7H_8IN 233.05	29289-13-2 40	 dec	4-12-00-01996	ace 4; bz 4; eth 4; EtOH 4
763	Benzenamine, 5-iodo-2-methyl- *o*-Toluidine, 5-iodo	C_7H_8IN 233.05	83863-33-6 48.5	 273 dec	1-12-00-00391	EtOH 4
764	Benzenamine, *N*,*N*'-methanetetraylbis-	$C_{13}H_{10}N_2$ 194.24	622-16-2 169	 331; 175^{20}	4-12-00-00866	H_2O 2; EtOH 2; eth 2; bz 3
765	Benzenamine, *N*,*N*'-methanetetraylbis[4-methyl-	$C_{15}H_{14}N_2$ 222.29	726-42-1 58.5	 221^{20}	4-12-00-01938 1.1500^{20}	
766	Benzenamine, 2-methoxy- *o*-Anisidine	C_7H_9NO [illegible]	90-04-0 [illegible]	 [illegible]	4-13-00-00806 [illegible]	H_2O 2; EtOH 3; eth 3; ace 3 [illegible]
767	Benzenamine, 3-methoxy- *m*-Anisidine	C_7H_9NO 123.15	536-90-3 -1	 251	4-13-00-00953 1.096^{20}	H_2O 2; EtOH 3; eth 3; ace 3 1.5794^{20}
768	Benzenamine, 4-methoxy- *p*-Anisidine	C_7H_9NO 123.15	104-94-9 57.2	 243	4-13-00-01015 1.071^{57}	H_2O 3; EtOH 4; eth 4; ace 3 1.5559^{60}
769	Benzenamine, 2-methoxy-*N*,*N*-dimethyl-	$C_9H_{13}NO$ 151.21	700-75-4	 210	4-13-00-00816 1.0160^{23}	chl 4
770	Benzenamine, 2-methoxy-5-methyl- 5-Methyl-*o*-anisidine	$C_8H_{11}NO$ 137.18	120-71-8 53	 235	3-13-00-01577	H_2O 2; EtOH 3; eth 3; bz 3
771	Benzenamine, 3-methoxy-4-methyl-	$C_8H_{11}NO$ 137.18	16452-01-0 56.5	 251	4-13-00-01666	bz 4; eth 4; EtOH 4; lig 4
772	Benzenamine, 4-methoxy-2-methyl-	$C_8H_{11}NO$ 137.18	102-50-1 29.5	 248.5	4-13-00-01698 1.065^{25}	EtOH 4 1.5647^{20}
773	Benzenamine, 5-methoxy-2-methyl-	$C_8H_{11}NO$ 137.18	50868-72-9 47	 253; 140^{20}	4-13-00-01711	eth 4

No.	Name Synonym	Mol. Form. Mol. Wt.	CAS RN mp/°C	Merck No. bp/°C	Beil. Ref. den/g cm^{-3}	Solubility n_D
774	Benzenamine, 2-methoxy-6-methyl-	$C_8H_{11}NO$ 137.18	50868-73-0 31	 119^{16}	3-13-00-01549	EtOH 4
775	Benzenamine, 2-methoxy-5-nitro- 5-Nitro-*o*-anisidine	$C_7H_8N_2O_3$ 168.15	99-59-2 118		4-13-00-00896 1.2068^{15}	H_2O 3; EtOH 4; eth 3; ace 4
776	Benzenamine, 3-methoxy-4-nitro- *m*-Anisidine, 4-nitro-	$C_7H_8N_2O_3$ 168.15	16292-88-9 169	 sub	4-13-00-01005	ace 4; EtOH 4
777	Benzenamine, 5-methoxy-2-nitro-	$C_7H_8N_2O_3$ 168.15	16133-49-6 131	 sub	4-13-00-01003	EtOH 4
778	Benzenamine, 2-methyl- *o*-Toluidine	C_7H_9N 107.16	95-53-4 -16.3	9462 200.3	4-12-00-01744 0.9984^{20}	H_2O 2; EtOH 5; eth 5; ctc 5 1.5725^{20}
779	Benzenamine, 3-methyl- *m*-Toluidine	C_7H_9N 107.16	108-44-1 -31.2	9462 203.3	4-12-00-01813 0.9889^{20}	ace 4; bz 4; eth 4; EtOH 4 1.5681^{20}
780	Benzenamine, 4-methyl- *p*-Toluidine	C_7H_9N 107.16	106-49-0 43.7	9462 200.4	4-12-00-01866 0.9619^{20}	H_2O 2; EtOH 4; eth 3; ace 3 1.5534^{45}
781	Benzenamine, *N*-methyl- *N*-Methylaniline	C_7H_9N 107.16	100-61-8 -57	5941 196.2	4-12-00-00241 0.9891^{20}	H_2O 1; EtOH 3; eth 3; ctc 3 1.5684^{20}
782	Benzenamine, 4-(6-methyl-2-benzothiazolyl)-	$C_{14}H_{12}N_2S$ 240.33	92-36-4 194.8	 434	4-27-00-05052	EtOH 2; eth 2; bz 2; HOAc 2
783	Benzenamine, *N*-(3-methylbutyl)-	$C_{11}H_{17}N$ 163.26	2051-84-5	 254.5	4-12-00-00259 0.8912^{55}	eth 4; EtOH 4 1.5305^{20}
784	Benzenamine, 4,4'-methylenebis- *p,p'*-Diaminodiphenylmethane	$C_{13}H_{14}N_2$ 198.27	101-77-9 92.5	2958 398; 257^{18}	4-13-00-00390	H_2O 2; EtOH 4; eth 4; bz 4
785	Benzenamine, 4,4'-methylenebis[2-chloro- 4,4'-Methylenebis[2-chloroaniline]	$C_{13}H_{12}Cl_2N_2$ 267.16	101-14-4	5978	1-13-00-00074	ctc 3
786	Benzenamine, 4,4'-methylenebis[*N,N*-dimethyl- Michler's Base	$C_{17}H_{22}N_2$ 254.38	101-61-1 91.5	6099 390 dec; 183^3	4-13-00-00390	H_2O 1; EtOH 2; eth 4; bz 4
787	Benzenamine, 2-(1-methylethenyl)-	$C_9H_{11}N$ 133.19	52562-19-3	 95^{13}	4-12-00-02925 0.976^{25}	 1.5722^{20}
788	Benzenamine, 2-(1-methylethyl)- 2-Isopropylaniline	$C_9H_{13}N$ 135.21	643-28-7	 221; 95^{13}	4-12-00-02630 0.9760^{12}	H_2O 1; eth 3; bz 3; ctc 3
789	Benzenamine, 4-(1-methylethyl)- Cumidine	$C_9H_{13}N$ 135.21	99-88-7	2622 225	4-12-00-02632 0.953^{20}	
790	Benzenamine, *N*-(1-methylethyl)- *N*-Isopropylaniline	$C_9H_{13}N$ 135.21	768-52-5	 203	4-12-00-00255 0.9526^{25}	EtOH 3; eth 3; ace 3; bz 3 1.5380^{20}
791	Benzenamine, 2-methyl-, hydrochloride *o*-Toluidine hydrochloride	$C_7H_{10}ClN$ 143.62	636-21-5 215		4-12-00-01744	H_2O 4; EtOH 4
792	Benzenamine, *N*-methyl-, hydrochloride	$C_7H_{10}ClN$ 143.62	2739-12-0 122.7		4-12-00-00241 1.0660^{131}	H_2O 4; EtOH 3; eth 1; bz 1
793	Benzenamine, 2-methyl-5-(1-methylethyl)-	$C_{10}H_{15}N$ 149.24	2051-53-8 -16	 241	3-12-00-02733 0.9942^{20}	ctc 3; CS_2 3 1.5387^{20}
794	Benzenamine, 4-methyl-*N*-(1-methylethyl)-	$C_{10}H_{15}N$ 149.24	10436-75-6	 220	4-12-00-01876 0.9226^{20}	 1.5332
795	Benzenamine, 2-methyl-*N*-(2-methylphenyl)-	$C_{14}H_{15}N$ 197.28	617-00-5 52.5	 316; 192^{23}	4-12-00-01750	H_2O 2
796	Benzenamine, 3-methyl-*N*-(3-methylphenyl)-	$C_{14}H_{15}N$ 197.28	626-13-1 -12	 319.5	2-12-00-00467	H_2O 2; EtOH 3; eth 3; peth 3
797	Benzenamine, 4-methyl-*N*-(4-methylphenyl)-	$C_{14}H_{15}N$ 197.28	620-93-9 79.8	 330.5	4-12-00-01879	eth 4; peth 4
798	Benzenamine, 2-methyl-4-[(2-methylphenyl)azo]- 2',3-Dimethyl-4-aminoazobenzene	$C_{14}H_{15}N_3$ 225.29	97-56-3 102		4-16-00-00528	eth 4; EtOH 4
799	Benzenamine, 2-methyl-3-nitro-	$C_7H_8N_2O_2$ 152.15	603-83-8 92	 305	4-12-00-01811 1.3780^{15}	H_2O 2; EtOH 3; eth 3; bz 3
800	Benzenamine, 2-methyl-4-nitro-	$C_7H_8N_2O_2$ 152.15	99-52-5 133.5		4-12-00-01809 1.1586^{140}	H_2O 2; EtOH 3; bz 3; HOAc 3
801	Benzenamine, 2-methyl-5-nitro- 5-Nitro-*o*-toluidine	$C_7H_8N_2O_2$ 152.15	99-55-8 105.5		4-12-00-01807	H_2O 2; EtOH 3; eth 3; ace 3
802	Benzenamine, 2-methyl-6-nitro-	$C_7H_8N_2O_2$ 152.15	570-24-1 96		4-12-00-01807 1.1900^{100}	H_2O 2; EtOH 3; eth 3; bz 3
803	Benzenamine, 4-methyl-2-nitro-	$C_7H_8N_2O_2$ 152.15	89-62-3 116.3		4-12-00-02000 1.16^{121}	H_2O 2; EtOH 3; chl 3
804	Benzenamine, *N*-methyl-2-nitro-	$C_7H_8N_2O_2$ 152.15	612-28-2 38	 158^{18}	4-12-00-01564	H_2O 2; EtOH 3; eth 3; ace 3
805	Benzenamine, *N*-methyl-4-nitro-	$C_7H_8N_2O_2$ 152.15	100-15-2 152	 dec	4-12-00-01616 1.201^{155}	H_2O 1; EtOH 3; eth 2; bz 3
806	Benzenamine, *N*-methyl-*N*-nitroso-	$C_7H_8N_2O$ 136.15	614-00-6 14.7	 225 dec; 121^{13}	4-16-00-00864 1.1240^{20}	H_2O 1; EtOH 3; eth 3 1.5769^{20}
807	Benzenamine, *N*-methyl-*N*-phenyl- Methyldiphenylamine	$C_{13}H_{13}N$ 183.25	552-82-9 -7.5	5973 293.5	4-12-00-00275 1.0476^{20}	H_2O 1; EtOH 2; ctc 3; MeOH 2 1.6193^{20}
808	Benzenamine, 3-methyl-*N*-phenyl-	$C_{13}H_{13}N$ 183.25	1205-64-7 30	 316; 183^{17}	4-12-00-01818	bz 4; eth 4; EtOH 4 1.6350^{20}
809	Benzenamine, 2-methyl-*N*-(phenylmethylene)-	$C_{14}H_{13}N$ 195.26	5877-55-4	 314	4-12-00-01751 1.041^{20}	H_2O 1; ace 3 1.6310^{25}
810	Benzenamine, 3-methyl-*N*-(phenylmethylene)-	$C_{14}H_{13}N$ 195.26	5877-58-7 31	 315	3-12-00-01958 1.0390^{25}	ace 3; os 3 1.6353^{25}

No.	Name Synonym	Mol. Form. Mol. Wt.	CAS RN mp/°C	Merck No. bp/°C	Beil. Ref. den/g cm^{-3}	Solubility n_D
811	Benzenamine, 4-methyl-*N*-(phenylmethylene)-	$C_{14}H_{13}N$ 195.26	2272-45-9 35	 318; 178^{11}	2-12-00-00496	ace 4
812	Benzenamine, 2-(1-methylpropyl)-	$C_{10}H_{15}N$ 149.24	55751-54-7	 120-2^{16}	4-12-00-02814 0.9574^{20}	EtOH 3; ace 3; bz 3; ctc 2
813	Benzenamine, 4-(1-methylpropyl)-	$C_{10}H_{15}N$ 149.24	30273-11-1	 238; 118^{15}	4-12-00-02814 0.949^{15}	bz 4; eth 4 1.5360^{29}
814	Benzenamine, 4-(2-methylpropyl)-	$C_{10}H_{15}N$ 149.24	30090-17-6	 238; 112^{11}	4-12-00-02818 0.949^{15}	ace 4; bz 4; EtOH 4
815	Benzenamine, 4-methyl-*N*-propyl- *N*-Propyl-*p*-toluidine	$C_{10}H_{15}N$ 149.24	54837-90-0	 235	4-12-00-01876 0.9243^{20}	 1.5367
816	Benzenamine, *N*-(1-methylpropyl)- *N*-*sec*-Butylaniline	$C_{10}H_{15}N$ 149.24	6068-69-5	 225	4-12-00-00257	chl 2; os 3 1.5333^{20}
817	Benzenamine, *N*-(2-methylpropyl)-	$C_{10}H_{15}N$ 149.24	588-47-6	 232; 113^{11}	4-12-00-00257 0.940^{18}	H_2O 1; eth 4; bz 4 1.5328^{20}
818	Benzenamine, *N*-methyl-*N*,2,4,6-tetranitro- Nitramine	$C_7H_5N_5O_8$ 287.15	479-45-8 131.5	6488 exp 180	4-16-00-00895 1.57^{10}	H_2O 1; EtOH 2; eth 2; ace 3
819	Benzenamine, 2-(methylthio)- Aniline, 2-methylthio	C_7H_9NS 139.22	2987-53-3	 234	4-13-00-00910 1.111^{25}	 1.6239^{20}
820	Benzenamine, 4-(methylthio)-	C_7H_9NS 139.22	104-96-1	 272.5	4-13-00-01289 1.1379^{20}	EtOH 3; eth 3; ace 3; bz 3 1.6395^{20}
821	Benzenamine, nitrate Aniline nitrate	$C_6H_8N_2O_3$ 156.14	542-15-4 190 dec	687	4-12-00-00223 1.356^{4}	H_2O 4; eth 4; EtOH 4
822	Benzenamine, 2-nitro- *o*-Nitroaniline	$C_6H_6N_2O_2$ 138.13	88-74-4 71.2	6504 284	4-12-00-01563 1.442^{15}	H_2O 2; EtOH 3; eth 4; ace 4
823	Benzenamine, 3-nitro- *m*-Nitroaniline	$C_6H_6N_2O_2$ 138.13	99-09-2 114	6503 306 dec	4-12-00-01589 0.991^{25}	H_2O 2; EtOH 3; eth 3; ace 3
824	Benzenamine, 4-nitro- *p*-Nitroaniline	$C_6H_6N_2O_2$ 138.13	100-01-6 147	6505 332	4-12-00-01613 1.424^{20}	H_2O 1; EtOH 3; eth 3; ace 3
825	Benzenamine, *N*-nitro-	$C_6H_6N_2O_2$ 138.13	645-55-6 47.3	 exp	4-16-00-00889	H_2O 4; bz 4; eth 4; EtOH 4
826	Benzenamine, 4-nitro-*N*-phenyl-	$C_{12}H_{10}N_2O_2$ 214.22	836-30-6 135.3	 211^{30}	4-12-00-01619	H_2O 1; EtOH 4; ace 2; HOAc 4
827	Benzenamine, 5-nitro-2-propoxy- Aniline, 5-nitro-2-propoxy	$C_9H_{12}N_2O_3$ 196.21	553-79-7 49	6551	4-13-00-00897	EtOH 4
828	Benzenamine, 4-nitroso-	$C_6H_6N_2O$ 122.13	659-49-4 173.5	 dec	4-12-00-01558	H_2O 3; EtOH 3; bz 2
829	Benzenamine, 4-nitroso-*N*-phenyl- *p*-Nitrosodiphenylamine	$C_{12}H_{10}N_2O$ 198.22	156-10-5 143	6560	4-12-00-01560	H_2O 2; EtOH 4; eth 4; bz 4
830	Benzenamine, *N*-nitroso-*N*-phenyl- *N*-Nitroso-*N*-diphenylamine	$C_{12}H_{10}N_2O$ 198.22	86-30-6 66.5		4-16-00-00865	EtOH 2; bz 3; chl 2
831	Benzenamine, 4-octyl-	$C_{14}H_{23}N$ 205.34	16245-79-7 20	 310; 138^{5}	4-12-00-02887 0.9128^{20}	eth 4
832	Benzenamine, 4,4'-oxybis- Bis(*p*-aminophenyl) ether	$C_{12}H_{12}N_2O$ 200.24	101-80-4 189 dec		4-13-00-01038	
833	Benzenamine, 2,3,4,5,6-pentamethyl-	$C_{11}H_{17}N$ 163.26	2243-30-3 152.5	 277.5	3-12-00-02758	eth 4; EtOH 4
834	Benzenamine, 2-phenoxy-	$C_{12}H_{11}NO$ 185.23	2688-84-8 45.8	 308; 172^{14}	4-13-00-00809	EtOH 3; eth 3; ace 3; bz 3
835	Benzenamine, 3-phenoxy-	$C_{12}H_{11}NO$ 185.23	3586-12-7 37	 315; 180^{10}	4-13-00-00955 1.1583^{25}	EtOH 3; eth 3; ace 3; bz 3
836	Benzenamine, *N*-phenyl- Diphenylamine	$C_{12}H_{11}N$ 169.23	122-39-4 52.9	3317 302	4-12-00-00271 1.158^{22}	H_2O 1; EtOH 4; eth 3; ace 4
837	Benzenamine, 4-(phenylazo)- *p*-Aminoazobenzene	$C_{12}H_{11}N_3$ 197.24	60-09-3 127	430 >360	2-16-00-00149	H_2O 2; EtOH 3; eth 3; bz 3
838	Benzenamine, 2-(2-phenylethyl)-	$C_{14}H_{15}N$ 197.28	5697-85-8 33	 193^{19}	4-12-00-03311 1.0430^{20}	
839	Benzenamine, 4-(phenylmethyl)-	$C_{13}H_{13}N$ 183.25	1135-12-2 34.5	 300	4-12-00-03281 1.038^{25}	eth 4; EtOH 4; lig 4
840	Benzenamine, *N*-(phenylmethylene)- Benzylideneaniline	$C_{13}H_{11}N$ 181.24	538-51-2 54	1152 310	2-12-00-00113 1.038^{55}	H_2O 1; EtOH 3; eth 3; chl 2 1.600^{100}
841	Benzenamine, *N*-2-propenyl-	$C_9H_{11}N$ 133.19	589-09-3	 219; 106^{12}	4-12-00-00265 0.982^{25}	H_2O 2; EtOH 3; eth 5; ace 3 1.563^{20}
842	Benzenamine, 2-propyl-	$C_9H_{13}N$ [illegible]	1821-39-2	 [illegible]; 110^{15}	4-12-00-02580 [illegible]	H_2O 1; EtOH 3; eth 3 [illegible]
843	Benzenamine, *N*-propyl-	$C_9H_{13}N$ 135.21	622-80-0	 222; 98^{11}	4-12-00-00254 0.9443^{20}	eth 4; EtOH 4 1.5428^{20}
844	Benzenamine, *N*-sulfinyl-	C_6H_5NOS 139.18	1122-83-4	 200	4-12-00-01075 1.236^{25}	 1.6270^{20}
845	Benzenamine, 4,4'-sulfinylbis- 4,4'-Sulfinyldianiline	$C_{12}H_{12}N_2OS$ 232.31	119-59-5 175 dec	8927	4-13-00-01306	H_2O 3; EtOH 3
846	Benzenamine, 4,4'-sulfonylbis- Dapsone	$C_{12}H_{12}N_2O_2S$ 248.31	80-08-0 175.5	2820	4-13-00-01306	EtOH 3; DMSO-d_6 2
847	Benzenamine, 2,3,4,5-tetramethyl- Aniline, 2,3,4,5-tetramethyl	$C_{10}H_{15}N$ 149.24	2217-45-0 69	 259.5	3-12-00-02743	eth 4; EtOH 4; lig 4
848	Benzenamine, 2,3,4,6-tetramethyl-	$C_{10}H_{15}N$ 149.24	488-71-1 23.5	 255	4-12-00-02849 0.978^{24}	EtOH 4
849	Benzenamine, *N*,*N*,2,6-tetramethyl-	$C_{10}H_{15}N$ 149.24	769-06-2 -36	 196; 88^{20}	4-12-00-02522 0.9147^{20}	

No.	Name	Synonym	Mol. Form.	Mol. Wt.	CAS RN	mp/°C	Merck No.	bp/°C	Beil. Ref.	den/g cm^{-3}	Solubility	n_D
850	Benzenamine, 4,4'-thiobis-	Bis(4-aminophenyl) sulfide	$C_{12}H_{12}N_2S$	216.31	139-65-1	108.5			4-13-00-01306		H_2O 2; EtOH 4; eth 4; bz 4	
851	Benzenamine, 2,4,6-tribromo-	2,4,6-Tribromoaniline	$C_6H_4Br_3N$	329.82	147-82-0	122	9522	300	4-12-00-01538	2.35^{20}	H_2O 1; EtOH 2; eth 3; chl 3	
852	Benzenamine, 2,3,4-trichloro-	2,3,4-Trichloroaniline	$C_6H_4Cl_3N$	196.46	634-67-3	73		292	0-12-00-00626		EtOH 4	
853	Benzenamine, 2,4,5-trichloro-		$C_6H_4Cl_3N$	196.46	636-30-6	96.5		270	4-12-00-01277		EtOH 3; eth 3; CS_2 4; lig 2	
854	Benzenamine, 2,4,6-trichloro-		$C_6H_4Cl_3N$	196.46	634-93-5	78.5		262	4-12-00-01281		H_2O 1; EtOH 3; eth 3; chl 3	
855	Benzenamine, 2-(trifluoromethyl)-	*o*-(Trifluoromethyl)aniline	$C_7H_6F_3N$	161.13	88-17-5	35.5		68^{15}	4-12-00-01786	1.282^{25}		1.4810^{20}
856	Benzenamine, 3-(trifluoromethyl)-	*m*-(Trifluoromethyl)aniline	$C_7H_6F_3N$	161.13	98-16-8	5.5		187; $74\text{-}5^{10}$	4-12-00-01843	1.3047^{12}	H_2O 2; EtOH 3; eth 3	1.4787^{20}
857	Benzenamine, 4-(trifluoromethyl)-		$C_7H_6F_3N$	161.13	455-14-1	38		117.5^{60}	4-12-00-01982	1.283^{27}		1.4815^{25}
858	Benzenamine, 2,4,5-trimethyl-		$C_9H_{13}N$	135.21	137-17-7	68		234.5	4-12-00-02645	0.957^{25}	EtOH 4	
859	Benzenamine, 2,4,6-trimethyl-	Mesitylamine	$C_9H_{13}N$	135.21	88-05-1	-5		232.5	4-12-00-02648	0.9633^{25}	ctc 2	1.5495^{20}
860	Benzenamine, *N,N*,2-trimethyl-	*N,N*-Dimethyl-*o*-toluidine	$C_9H_{13}N$	135.21	609-72-3	-60		194.1	4-12-00-01747	0.9286^{20}	eth 4; EtOH 4	1.5152^{20}
861	Benzenamine, *N,N*,3-trimethyl-		$C_9H_{13}N$	135.21	121-72-2			212	4-12-00-01815	0.9410^{20}	EtOH 5; eth 5	1.5492^{20}
862	Benzenamine, *N,N*,4-trimethyl-		$C_9H_{13}N$	135.21	99-97-8			211	4-12-00-01874	0.9366^{20}	H_2O 1; EtOH 5; eth 5; ctc 3	1.5366^{20}
863	Benzenamine, 2,4,6-trinitro-		$C_6H_4N_4O_6$	228.12	489-98-5	193.5		exp	4-12-00-01735	1.762^{10}	H_2O 1; EtOH 2; eth 2; ace 3	
864	Benzenamine, 2,4,6-trinitro-*N*-(2,4,6-trinitrophenyl)-	Dipicrylamine	$C_{12}H_5N_7O_{12}$	439.21	131-73-7	244 dec	3340		4-12-00-01737		H_2O 1; EtOH 1; eth 2; ace 2	
865	Benzenaminium, *N*-ethyl-3-hydroxy-*N,N*-dimethyl-, chloride	Ammonium, dimethyl ethyl 3-hydroxyphenyl,chloride	$C_{10}H_{16}ClNO$	201.70	116-38-1		3487		4-13-00-00969			
866	Benzenaminium, *N,N,N*-trimethyl-, iodide	Phenyl trimethyl ammonium iodide	$C_9H_{14}IN$	263.12	98-04-4	224	7289		4-12-00-00249		H_2O 4; EtOH 3; ace 2; chl 1	
867	Benzene	[6]Annulene	C_6H_6	78.11	71-43-2	5.5	1074	80.0	4-05-00-00583	0.8765^{20}	H_2O 2; EtOH 5; eth 5; ace 5	1.5011^{20}
868	Benzeneacetaldehyde	Phenylacetaldehyde	C_8H_8O	120.15	122-78-1	33.5	7236	195	4-07-00-00664	1.0272^{20}	H_2O 2; EtOH 5; eth 5; ace 3	1.5255^{20}
869	Benzeneacetaldehyde, 3,4-dimethoxy-		$C_{10}H_{12}O_3$	180.20	5703-21-9			$125^{1.5}$	4-08-00-01828	1.148^{20}		1.5426^{20}
870	Benzeneacetaldehyde, 2-methoxy-		$C_9H_{10}O_2$	150.18	33567-59-8			245; 117^{17}	1-08-00-00543	1.0897^{20}	ace 4; EtOH 4	1.5393^{20}
871	Benzeneacetaldehyde, 4-methoxy-		$C_9H_{10}O_2$	150.18	5703-26-4			255.5	4-08-00-00426	1.096^{20}		1.5359^{20}
872	Benzeneacetaldehyde, α-methyl-		$C_9H_{10}O$	134.18	93-53-8			203.5		1.0089^{20}	EtOH 4	1.5176^{20}
873	Benzeneacetaldehyde, 2-methyl-		$C_9H_{10}O$	134.18	10166-08-2			221; 92^{10}	1-07-00-00163	1.0241^{10}	eth 4; EtOH 4; chl 4	
874	Benzeneacetaldehyde, 4-methyl-		$C_9H_{10}O$	134.18	104-09-6	40		221.5	4-07-00-00705	1.0052^{20}	eth 4; EtOH 4; chl 4	1.5255^{20}
875	Benzeneacetaldehyde, 2-nitro-		$C_8H_7NO_3$	165.15	1969-73-9	22.5		133^5	2-07-00-00229		bz 4; eth 4; EtOH 4	
876	Benzeneacetaldehyde, α-oxo-, aldoxime	Isonitrosoacetophenone	$C_8H_7NO_2$	149.15	532-54-7	129	5077		4-07-00-02132		H_2O 2; chl 3	
877	Benzeneacetaldehyde, α-phenyl-		$C_{14}H_{12}O$	196.25	947-91-1			315 dec; 157^7	4-07-00-01400	1.1061^{21}	H_2O 1; EtOH 4; eth 4; bz 4	1.5920^{21}
878	Benzeneacetamide	α-Phenylacetamide	C_8H_9NO	135.17	103-81-1	157	7237		4-09-00-01632		H_2O 2; EtOH 3; eth 2; bz 2	
879	Benzeneacetamide, *N*-(aminocarbonyl)-	Phenacemide	$C_9H_{10}N_2O_2$	178.19	63-98-9	215	7154		4-09-00-01636		bz 4; eth 4; EtOH 4	
880	Benzeneacetamide, 4-butoxy-*N*-hydroxy-	Bufexamac	$C_{12}H_{17}NO_3$	223.27	2438-72-4	154	1462					
881	Benzeneacetamide, α-ethyl-	α-Phenylbutyramide	$C_{10}H_{13}NO$	163.22	90-26-6	86	7249	185^{16}	4-09-00-01818		H_2O 3; ace 2; ctc 3	
882	Benzeneacetamide, α-(1-methylethyl)-		$C_{11}H_{15}NO$	177.25	5470-47-3	111.5		180^{14}	3-09-00-02518		eth 4; EtOH 4	
883	Benzeneacetic acid	Phenylacetic acid	$C_8H_8O_2$	136.15	103-82-2	76.7	7239	265.5	4-09-00-01614	1.228^6	H_2O 2; EtOH 4; eth 4; ace 3	
884	Benzeneacetic acid, α-acetyl-, ethyl ester		$C_{12}H_{14}O_3$	206.24	5413-05-8			156^{22}	4-10-00-02776	1.0855^{20}	eth 4; EtOH 4	1.5176^{20}
885	Benzeneacetic acid, α-amino-	α-Phenylglycine	$C_8H_9NO_2$	151.17	69-91-0	292 dec	7263	sub 255	4-14-00-01317		alk 3; os 2	
886	Benzeneacetic acid, 4-amino-	*p*-Aminophenylacetic acid	$C_8H_9NO_2$	151.17	1197-55-3	200 dec	475		3-14-00-01182		H_2O 1; EtOH 2; DMSO-d_6 2	
887	Benzeneacetic acid, anhydride		$C_{16}H_{14}O_3$	254.29	1555-80-2	73.3		195^{12}	4-09-00-01629		eth 4; chl 4	

No.	Name Synonym	Mol. Form. Mol. Wt.	CAS RN mp/°C	Merck No. bp/°C	Beil. Ref. den/g cm^{-3}	Solubility n_D
888	Benzeneacetic acid, 4-bromo-	$C_8H_7BrO_2$ 215.05	1878-68-8 116	 sub	3-09-00-02275	H_2O 2; EtOH 4; eth 4; CS_2 4
889	Benzeneacetic acid, 4-bromo-α-hydroxy-, (±)- Acetic acid, (4-bromophenyl)-hydroxy(*DL*)	$C_8H_7BrO_3$ 231.05	7021-04-7 119	1412	3-10-00-00480	H_2O 4; EtOH 4; eth 4; bz 4
890	Benzeneacetic acid, α-butyl-	$C_{12}H_{16}O_2$ 192.26	24716-09-4	 182^{20}; $118^{0.1}$	4-09-00-01897 1.0225^{19}	 1.5071^{19}
891	Benzeneacetic acid, α-chloro-, ethyl ester, (*S*)- Acetic acid, chlorophenyl-, ethyl ester, *L*-(+)-	$C_{10}H_{11}ClO_2$ 198.65	10606-73-2	 162^{45}; 133^{12}	3-09-00-02266 1.1594^{20}	 1.5152^{20}
892	Benzeneacetic acid, α-chloro-α-phenyl-, ethyl ester	$C_{16}H_{15}ClO_2$ 274.75	52460-86-3 43.5	 185^{14}	4-09-00-02508	eth 4; EtOH 4
893	Benzeneacetic acid, α-cyano-, ethyl ester	$C_{11}H_{11}NO_2$ 189.21	4553-07-5	 275 dec; 165^{20}	3-09-00-04262 1.091^{20}	ace 4; bz 4; eth 4; EtOH 4 1.5012^{25}
894	Benzeneacetic acid, 4-[2-(diethylamino)-2-oxoethoxy]-3-methoxy-, propyl ester Acetic acid, amide,(4(carboxy(propyl ester)methyl)-2-methoxyphenoxy)-*N*,*N*-diethyl	$C_{18}H_{27}NO_5$ 337.42	1421-14-3	7813 $210\text{-}2^{0.7}$		H_2O 1; EtOH 3; chl 3
895	Benzeneacetic acid, 2,5-dihydroxy- Homogentisic acid	$C_8H_8O_4$ 168.15	451-13-8 153	4658	4-10-00-01506	H_2O 4; EtOH 4; eth 4; bz 1
896	Benzeneacetic acid, α,4-dihydroxy-3-methoxy- Vanilmandelic acid	$C_9H_{10}O_5$ 198.18	55-10-7 131-3 dec	9840	3-10-00-02100	H_2O 4; ace 4; eth 4
897	Benzeneacetic acid, α-ethyl-	$C_{10}H_{12}O_2$ 164.20	90-27-7 47.5	 271	2-09-00-00356	eth 3; bz 3; ctc 3
898	Benzeneacetic acid, α-ethyl-, 2-(diethylamino)ethyl ester Butethamate	$C_{16}H_{25}NO_2$ 263.38	14007-64-8	1516 $167\text{-}9^{11}$	3-09-00-02463	 1.4909^{20}
899	Benzeneacetic acid, ethyl ester Ethyl phenylacetate	$C_{10}H_{12}O_2$ 164.20	101-97-3 -29.4	3794 227	4-09-00-01618 1.0333^{20}	eth 4; EtOH 4 1.4980^{20}
900	Benzeneacetic acid, α-ethyl-, methyl ester	$C_{11}H_{14}O_2$ 178.23	2294-71-5 77.5	 228	3-09-00-02462	eth 4; EtOH 4
901	Benzeneacetic acid, 4-fluoro- (*p*-Fluorophenyl)acetic acid	$C_8H_7FO_2$ 154.14	405-50-5 86	4105 164^2	4-09-00-01672	
902	Benzeneacetic acid, α-(hydroxymethyl)- 8-methyl-8-azabicyclo[3.2.1]oct-3-yl ester *endo*-(±)- Atropine	$C_{17}H_{23}NO_3$ 289.37	51-55-8 118.5	891 sub 95	5-21-01-00235	H_2O 4; EtOH 4; eth 1, chl 2
903	Benzeneacetic acid, α-hydroxy-, (*S*)-	$C_8H_8O_3$ 152.15	17199-29-0 134		4-10-00-00564 1.341^{25}	H_2O 4; eth 4; EtOH 4; chl 4
904	Benzeneacetic acid, 2-hydroxy-	$C_8H_8O_3$ 152.15	614-75-5 148	 240	4-10-00-00536	H_2O 2; eth 3; chl 2
905	Benzeneacetic acid, 3-hydroxy-	$C_8H_8O_3$ 152.15	621-37-4 132	 190^{11}	4-10-00-00541	H_2O 4; EtOH 4; eth 4; bz 3
906	Benzeneacetic acid, 4-hydroxy-	$C_8H_8O_3$ 152.15	156-38-7 152	 sub	4-10-00-00543	H_2O 2, EtOH 4; eth 4
907	Benzeneacetic acid, α-hydroxy-, ethyl ester, (±)-	$C_{10}H_{12}O_3$ 180.20	4358-88-7 35.5	 254	4-10-00-00569 1.1258^{20}	EtOH 3; eth 3; ctc 2; CS_2 3
908	Benzeneacetic acid, α-(hydroxymethyl)-, 9-methyl-3-oxa-9-azatricyclo[3.3.1.02,4]non-7-yl ester, [(7S)-(1α,2β,4β,5α,7β)]- Scopolamine	$C_{17}H_{21}NO_4$ 303.36	51-34-3	8361	2-27-00-00064	
909	Benzeneacetic acid, α-hydroxy-, ethyl ester	$C_{10}H_{12}O_3$ 180.20	774-40-3 35	 150^{21}	2-10-00-00121 1.1270^{20}	eth 4; EtOH 4
910	Benzeneacetic acid, α-hydroxy-, methyl ester, (±)-	$C_9H_{10}O_3$ 166.18	4358-87-6 58	 250 dec; 144^{20}	2-10-00-00120 1.1756^{20}	EtOH 4; chl 4
911	Benzeneacetic acid, α-hydroxy-, 8-methyl-8-azabicyclo[3.2.1]oct-3-yl ester, *endo*-(±)- Homatropine	$C_{16}H_{21}NO_3$ 275.35	87-00-3 99.5	4649	5-21-01-00234	H_2O 2; EtOH 3; eth 3; ace 3
912	Benzeneacetic acid, α-hydroxy-, 3-methylbutyl ester Mandelic acid isoamyl ester	$C_{13}H_{18}O_3$ 222.28	[illegible]	5000 172^{11}	4-10-00-00571	
913	Benzeneacetic acid, 4-hydroxy-3-methoxy- Homovanillic acid	$C_9H_{10}O_4$ 182.18	306-08-1 143.5	4662	4-10-00-01509	
914	Benzeneacetic acid, α-hydroxy-α-methyl- Atrolactic acid	$C_9H_{10}O_3$ 166.18	515-30-0 116.5	889	2-10-00-00157	H_2O 4; ace 4; bz 4; EtOH 4
915	Benzeneacetic acid, α-(hydroxymethyl)- Tropic acid	$C_9H_{10}O_3$ 166.18	529-64-6 118	9691	3-10-00-00564	eth 4; EtOH 4
916	Benzeneacetic acid, α-(hydroxymethyl)-, 8-azabicyclo[3.2.1]oct-3-yl ester, [3(*S*)-*endo*]- Norhyoscyamine	$C_{16}H_{21}NO_3$ 275.35	537-29-1 140.5	6623	5-21-01-00219	EtOH 4; chl 4
917	Benzeneacetic acid, α-(hydroxymethyl)-, 8-methyl-8-azabicyclo[3.2.1]oct-3-yl ester, [3(*S*)-*endo*]- Hyoscyamine	$C_{17}H_{23}NO_3$ 289.37	101-31-5 108.5	4795	5-21-01-00235	H_2O 2; EtOH 4; eth 2; bz 2

PHYSICAL CONSTANTS OF ORGANIC COMPOUNDS (continued)

No.	Name Synonym	Mol. Form. Mol. Wt.	CAS RN mp/°C	Merck No. bp/°C	Beil. Ref. den/g cm^{-3}	Solubility n_D
918	Benzeneacetic acid, α-hydroxy-α-phenyl- Benzilic acid	$C_{14}H_{12}O_3$ 228.25	76-93-7 151	1089 180 dec	4-10-00-01256	H_2O 2; EtOH 4; eth 4; ace 2
919	Benzeneacetic acid, α-hydroxy-α-phenyl-, ethyl ester	$C_{16}H_{16}O_3$ 256.30	52182-15-7 34	 201^{21}	4-10-00-01258 1.5620^{20}	eth 4; EtOH 4
920	Benzeneacetic acid, α-hydroxy-α-phenyl-, methyl ester	$C_{15}H_{14}O_3$ 242.27	76-89-1 75.8	 187^{13}	4-10-00-01258	eth 4; EtOH 4
921	Benzeneacetic acid, 2-methoxy-	$C_9H_{10}O_3$ 166.18	93-25-4 124	 $100\text{-}1^{2}$	4-10-00-00536	H_2O 3; EtOH 4; eth 4; ace 4
922	Benzeneacetic acid, 4-methoxy-	$C_9H_{10}O_3$ 166.18	104-01-8 87	 138^{2}	4-10-00-00544	H_2O 1; EtOH 4; eth 3; bz 3
923	Benzeneacetic acid, 4-methoxy-, ethyl ester	$C_{11}H_{14}O_3$ 194.23	14062-18-1	 $138\text{-}40^{70}$	4-10-00-00549 1.097^{25}	 1.5075^{20}
924	Benzeneacetic acid, α-methyl-, (±)-	$C_9H_{10}O_2$ 150.18	2328-24-7 <-20	 263	4-09-00-01779 1.1^{0}	 1.5237^{20}
925	Benzeneacetic acid, 3-methyl-	$C_9H_{10}O_2$ 150.18	621-36-3 62	 $120\text{-}3^{26}$	4-09-00-01792	H_2O 3; chl 3
926	Benzeneacetic acid, 4-methyl-	$C_9H_{10}O_2$ 150.18	622-47-9 93	 265	4-09-00-01795	bz 4; eth 4; EtOH 4
927	Benzeneacetic acid, α-methylene- Atropic acid	$C_9H_8O_2$ 148.16	492-38-6 106.5	890 267 dec	4-09-00-02050	H_2O 2; EtOH 3; eth 3; bz 3
928	Benzeneacetic acid, α-methylene-, 8-methyl-8-azabicyclo[3.2.1]oct-3-yl ester, *endo*- Apoatropine	$C_{17}H_{21}NO_2$ 271.36	500-55-0 62	770	5-21-01-00228	H_2O 2; EtOH 4; eth 4; ace 4
929	Benzeneacetic acid, methyl ester	$C_9H_{10}O_2$ 150.18	101-41-7	 216.5	4-09-00-01617 1.0622^{16}	H_2O 1; EtOH 5; eth 5; ace 3 1.5075^{20}
930	Benzeneacetic acid, α-(1-methylethyl)-	$C_{11}H_{14}O_2$ 178.23	3508-94-9 63	 159^{14}	3-09-00-02518	eth 4; EtOH 4
931	Benzeneacetic acid, α-methyl-α-phenyl-	$C_{15}H_{14}O_2$ 226.27	5558-66-7 176	 sub	4-09-00-02535	H_2O 2; EtOH 3; eth 4; bz 3
932	Benzeneacetic acid, 2-methylpropyl ester	$C_{12}H_{16}O_2$ 192.26	102-13-6	 247	4-09-00-01619 0.999^{18}	H_2O 1; EtOH 3; eth 3
933	Benzeneacetic acid, 4-nitro- *p*-Nitrophenylacetic acid	$C_8H_7NO_4$ 181.15	104-03-0 154	6543	4-09-00-01687	H_2O 2; EtOH 3; eth 3; bz 3
934	Benzeneacetic acid, α-oxo-	$C_8H_6O_3$ 150.13	611-73-4 66	 163^{15}	4-10-00-02737	H_2O 4; EtOH 3; eth 3; ctc 2
935	Benzeneacetic acid, α-oxo-, ethyl ester	$C_{10}H_{10}O_3$ 178.19	1603-79-8	 256.5	4-10-00-02738 1.1222^{25}	 1.5190^{25}
936	Benzeneacetic acid, α-phenyl- Diphenylacetic acid	$C_{14}H_{12}O_2$ 212.25	117-34-0 147	3316 194^{25}	4-09-00-02492 1.257^{15}	H_2O 2; EtOH 4; eth 3; chl 3
937	Benzeneacetic acid, α-phenyl-, anhydride	$C_{28}H_{22}O_3$ 406.48	1760-46-9 98	 222^{15}	4-09-00-02497	bz 4; chl 4
938	Benzeneacetic acid, α-phenyl-, 2-(diethylamino)ethyl ester, hydrochloride Adiphenine hydrochloride	$C_{20}H_{26}ClNO_2$ 347.88	50-42-0 113.5	151	3-09-00-03297	H_2O 4; EtOH 2; eth 2
939	Benzeneacetic acid, α-phenyl-, ethyl ester	$C_{16}H_{16}O_2$ 240.30	3468-99-3 60	 195^{25}	4-09-00-02493 1.1860^{20}	H_2O 1; EtOH 4; eth 4; CS_2 4
940	Benzeneacetic acid, 2-phenylethyl ester	$C_{16}H_{16}O_2$ 240.30	102-20-5 26.5	 $177\text{-}8^{4.5}$	3-09-00-02183 1.077^{25}	EtOH 4
941	Benzeneacetic acid, α-(phenylmethylene)- α-Phenylcinnamic acid	$C_{15}H_{12}O_2$ 224.26	3368-16-9 172.5	7253 sub	3-09-00-03438	eth 4; EtOH 4
942	Benzeneacetic acid, α-(phenylmethylene)-, ethyl ester	$C_{17}H_{16}O_2$ 252.31	24446-63-7 33.5	 $214\text{-}5^{28}$	3-09-00-03416 1.0971^{18}	 1.5972^{18}
943	Benzeneacetic acid, α-propyl-, (±)- Valeric acid, 2-phenyl-, (±)-	$C_{11}H_{14}O_2$ 178.23	7782-21-0 58	 280	4-09-00-01868	eth 4; EtOH 4
944	Benzeneacetic acid, 2,3,6-trichloro- Chlorfenac	$C_8H_5Cl_3O_2$ 239.48	85-34-7 161	2085	4-09-00-01681	
945	Benzeneacetonitrile Benzyl cyanide	C_8H_7N 117.15	140-29-4 -23.8	1145 233.5	4-09-00-01663 1.0205^{15}	 1.5211^{25}
946	Benzeneacetonitrile, 4-amino-	$C_8H_8N_2$ 132.17	3544-25-0 46	 312	4-14-00-01312	H_2O 2; EtOH 3; CS_2 2; os 3
947	Benzeneacetonitrile, α-bromo- α-Bromobenzyl cyanide	C_8H_6BrN 196.05	5798-79-8 29	1400 242 dec; 133^{12}	3-09-00-02278 1.539^{29}	H_2O 1; EtOH 4; eth 4; ace 4
948	Benzeneacetonitrile, 2-bromo-	C_8H_6BrN 196.05	19472-74-3 1	 140^{13}	4-09-00-01682	H_2O 1; EtOH 4
949	Benzeneacetonitrile, 2-chloro-	C_8H_6ClN 151.60	2856-63-5 24	 251	4-09-00-01674 1.1737^{18}	
950	Benzeneacetonitrile, 3-chloro-	C_8H_6ClN 151.60	1529-41-5	 $134\text{-}6^{10}$	4-09-00-01675 1.283^{25}	 1.5437^{20}
951	Benzeneacetonitrile, α-ethyl-	$C_{10}H_{11}N$ 145.20	769-68-6	 241	4-09-00-01822 0.977^{14}	H_2O 1; EtOH 3; eth 3; bz 3
952	Benzeneacetonitrile, 2-fluoro-	C_8H_6FN 135.14	326-62-5	 232; 102^{10}	4-09-00-01671 1.059^{25}	 1.5009^{20}
953	Benzeneacetonitrile, 4-fluoro-	C_8H_6FN 135.14	459-22-3	 $119\text{-}20^{18}$	4-09-00-01673 1.126^{25}	 1.5002^{20}

PHYSICAL CONSTANTS OF ORGANIC COMPOUNDS (continued)

No.	Name Synonym	Mol. Form. Mol. Wt.	CAS RN mp/°C	Merck No. bp/°C	Beil. Ref. den/g cm^{-3}	Solubility n_D
954	Benzeneacetonitrile, α-[(6-*O*-β-*D*-glucopyranosyl-β-*D*-glucopyranosyl)oxy]-, (*R*)- Amygdalin	$C_{20}H_{27}NO_{11}$ 457.43	29883-15-6 224.5	629	5-17-08-00118	H_2O 4; EtOH 2; eth 1; chl 1
955	Benzeneacetonitrile, α-(β-*D*-glucopyranosyloxy)- Mandelonitrile glucoside	$C_{14}H_{17}NO_6$ 295.29	138-53-4 122	5602		H_2O 4; EtOH 4
956	Benzeneacetonitrile, α-hydroxy-, (±)-	C_8H_7NO 133.15	613-88-7 22	 170	3-10-00-00473 1.1165^{20}	eth 4; EtOH 4; chl 4
957	Benzeneacetonitrile, 2-methoxy-	C_9H_9NO 147.18	7035-03-2 69.8	 141^{15}	4-10-00-00538	bz 4
958	Benzeneacetonitrile, 4-methoxy-	C_9H_9NO 147.18	104-47-2	 286.5	4-10-00-00555 1.0845^{20}	EtOH 3; eth 3; chl 3 1.5309^{20}
959	Benzeneacetonitrile, α-methyl-	C_9H_9N 131.18	1823-91-2	 231	4-09-00-01783 0.9854^{20}	eth 4; EtOH 4 1.5095^{25}
960	Benzeneacetonitrile, 2-methyl-	C_9H_9N 131.18	22364-68-7	 244	4-09-00-01791 1.0156^{22}	bz 4; eth 4; EtOH 4 1.5252^{20}
961	Benzeneacetonitrile, 3-methyl-	C_9H_9N 131.18	2947-60-6	 248; 133^{15}	4-09-00-01792 1.0022^{22}	H_2O 1; EtOH 3; eth 3; bz 3 1.5233^{20}
962	Benzeneacetonitrile, 4-methyl-	C_9H_9N 131.18	2947-61-7 18	 242.5	4-09-00-01796 0.992^{25}	H_2O 1; EtOH 3; eth 3; bz 3 1.5190^{20}
963	Benzeneacetonitrile, α-(1-methylethyl)-	$C_{11}H_{13}N$ 159.23	5558-29-2	 247	4-09-00-01879 0.967^{15}	bz 4; EtOH 4 1.5038^{25}
964	Benzeneacetonitrile, α-(2-methylpropyl)- Valeronitrile, 4-methyl-2-phenyl-	$C_{12}H_{15}N$ 173.26	5558-31-6	 264; 137^{15}	4-09-00-01903 0.942^{16}	bz 4; eth 4; EtOH 4
965	Benzeneacetonitrile, 2-nitro-	$C_8H_6N_2O_2$ 162.15	610-66-2 84	 178^{12}; 138^{1}	3-09-00-02283	ace 4; bz 4; eth 4; EtOH 4
966	Benzeneacetonitrile, 3-nitro-	$C_8H_6N_2O_2$ 162.15	621-50-1 63	 180^{15}	4-09-00-01687	bz 4; eth 4; EtOH 4; chl 4
967	Benzeneacetonitrile, 4-nitro- *p*-Nitrobenzyl cyanide	$C_8H_6N_2O_2$ 162.15	555-21-5 117	6512 195-7^{12}	4-09-00-01689	H_2O 2; EtOH 3; eth 3; bz 3
968	Benzeneacetonitrile, α-oxo-	C_8H_5NO 131.13	613-90-1 32.5	 206	4-10-00-02743	H_2O 1; EtOH 4; eth 4; chl 2
969	Benzeneacetonitrile, α-phenyl-	$C_{14}H_{11}N$ 193.25	86-29-3 74.3	 184^{16}	4-09-00-02505	EtOH 3; eth 4; chl 3; lig 2
970	Benzeneacetonitrile, α-propyl-	$C_{11}H_{13}N$ 159.23	5558-78-1	 255; 126^{13}	4-09-00-01869 0.9425^{20}	bz 4; EtOH 4 1.5000^{20}
971	Benzeneacetyl chloride Phenylacetyl chloride	C_8H_7ClO 154.60	103-80-0	 170^{250}; 105^{24}	4-09-00-01631 1.1682^{20}	eth 4 1.5325^{20}
972	Benzeneacetyl chloride, α-(acetyloxy)-	$C_{10}H_9ClO_3$ 212.63	1638-63-7 36	 152^{33}; 125^{10}	1-10-00-00089	eth 4; bz 4; chl 4; lig 4
973	Benzeneacetyl chloride, α-chloro-	$C_8H_6Cl_2O$ 189.04	2912-62-1	 120^{23}	2-09-00-00308 1.196^{25}	 1.5440^{20}
974	Benzeneacetyl chloride, α-ethyl-	$C_{10}H_{11}ClO$ 182.65	36854-57-6	 110-2^{10}	4-09-00-01817	ctc 3 1.5169^{20}
975	Benzeneacetyl chloride, α-phenyl-	$C_{14}H_{11}ClO$ 230.69	1871-76-7 56.5	 170^{16}	3-09-00-03300	lig 3
976	Benzene, azido-	$C_6H_5N_3$ 119.13	622-37-7 -27.5	 70^{11}	4-05-00-00759 1.0860^{20}	H_2O 1; EtOH 2; eth 2 1.5589^{25}
977	Benzene, 1-azido-4-bromo-	$C_6H_4BrN_3$ 198.02	2101-88-4 20	 105^{10}; 69^{2}	4-05-00-00761 1.5827^{25}	H_2O 1; EtOH 2; eth 4; bz 4
978	Benzene, 1-azido-4-chloro-	$C_6H_4ClN_3$ 153.57	3296-05-7 20	 96^{20}	4-05-00-00760 1.2634^{25}	H_2O 1; eth 3
979	Benzene, 1-azido-2-iodo-	$C_6H_4IN_3$ 245.02	54467-95-7	 90-100$^{0.9}$	0-05-00-00278 1.8893^{25}	ace 4; EtOH 4 1.6631^{25}
980	Benzene, (azidomethyl)-	$C_7H_7N_3$ 133.15	622-79-7	 108^{23}; 74^{11}	4-05-00-00877 1.0655^{25}	H_2O 1; EtOH 5; eth 5 1.5341^{25}
981	Benzene, 1-azido-2-methyl-	$C_7H_7N_3$ 133.15	31656-92-5 <-10	 90.5^{30}; 76^{12}	4-05-00-00876 1.0648^{25}	eth 4
982	Benzene, 1-azido-4-methyl-	$C_7H_7N_3$ 133.15	2101-86-2	 180 dec; 80^{10}	4-05-00-00876 1.0527^{23}	eth 4; EtOH 4
983	Benzene, 1,2-bis(bromomethyl)-	$C_8H_8Br_2$ 263.96	91-13-4 95	 [illegible]	4-05-00-00929 [illegible]	H_2O 1; EtOH 3; eth 3; ctc 3
984	Benzene, 1,3-bis(bromomethyl)-	$C_8H_8Br_2$ 263.96	626-15-3 77	 135-40^{20}	4-05-00-00946 1.959^{25}	H_2O 1; EtOH 3; eth 3; chl 3
985	Benzene, 1,4-bis(bromomethyl)-	$C_8H_8Br_2$ 263.96	623-24-5 144.5	 245	4-05-00-00970 2.012^{25}	H_2O 1; EtOH 4; eth 2; bz 3
986	Benzene, 1,2-bis(chloromethyl)-	$C_8H_8Cl_2$ 175.06	612-12-4 55	 239.5	4-05-00-00927 1.393^{25}	H_2O 1; EtOH 4; eth 4; ctc 3
987	Benzene, 1,3-bis(chloromethyl)-	$C_8H_8Cl_2$ 175.06	626-16-4 34.2	 251.5	4-05-00-00944 1.302^{20}	H_2O 1; EtOH 4; eth 4; chl 2
988	Benzene, 1,4-bis(chloromethyl)-	$C_8H_8Cl_2$ 175.06	623-25-6 100	 245 dec; 135^{16}	4-05-00-00967 1.417^{25}	H_2O 1; EtOH 4; eth 4; ace 4
989	Benzene, 1,4-bis(dichloromethyl)-	$C_8H_6Cl_4$ 243.95	7398-82-5 9.5		3-05-00-00857 1.606^{25}	H_2O 1; EtOH 3; eth 3; bz 3
990	Benzene, 1,4-bis(1,1-dimethylethyl)- Benzene, 1,4-di-*tert*-butyl	$C_{14}H_{22}$ 190.33	1012-72-2 79.5	 238; 109^{15}	4-05-00-01163 0.9850^{20}	H_2O 1; EtOH 3; eth 3

No.	Name Synonym	Mol. Form. Mol. Wt.	CAS RN mp/°C	Merck No. bp/°C	Beil. Ref. den/g cm^{-3}	Solubility n_D
991	Benzene, 1,4-bis(2-hydroxybenzylideneamino)-2-nitro Ethionine	$C_{20}H_{15}N_3O_4$ 361.36	13073-35-3 195	3693		
992	Benzene, 1,2-bis(1-methylethyl)-	$C_{12}H_{18}$ 162.27	577-55-9 -57	 204	4-05-00-01125 0.8701^{20}	H_2O 1; EtOH 5; eth 5; ace 5 1.4960^{20}
993	Benzene, 1,3-bis(1-methylethyl)-	$C_{12}H_{18}$ 162.27	99-62-7 -63.1	 203.2	4-05-00-01125 0.8559^{20}	H_2O 1; EtOH 5; eth 5; ace 5 1.4883^{20}
994	Benzene, 1,4-bis(1-methylethyl)-	$C_{12}H_{18}$ 162.27	100-18-5 -17	 210.3	4-05-00-01126 0.8568^{20}	H_2O 1; EtOH 5; eth 5; ace 5 1.4898^{20}
995	Benzene, 1,4-bis(phenylmethyl)-	$C_{20}H_{18}$ 258.36	793-23-7 87.5	 225^{18}	4-05-00-02512	EtOH 3; eth 2; bz 3; chl 3
996	Benzene, 1,4-bis(trichloromethyl)-	$C_8H_4Cl_6$ 312.84	68-36-0 109	1313	4-05-00-00968	chl 3
997	Benzene, 1,3-bis(trifluoromethyl)-	$C_8H_4F_6$ 214.11	402-31-3	 116	4-05-00-00943 1.3790^{25}	 1.3916^{25}
998	Benzene, bromo- Bromobenzene	C_6H_5Br 157.01	108-86-1 -30.6	1394 156.0	4-05-00-00670 1.4950^{20}	H_2O 1; EtOH 4; eth 4; bz 4 1.5597^{20}
999	Benzene, 1-bromo-2-(bromomethyl)-	$C_7H_6Br_2$ 249.93	3433-80-5 31	 129^{19}	4-05-00-00836	eth 4; EtOH 4; HOAc 4
1000	Benzene, 1-bromo-4-(bromomethyl)- *p*-Bromobenzyl bromide	$C_7H_6Br_2$ 249.93	589-15-1 63	1397	4-05-00-00836	H_2O 2; EtOH 3; eth 4; bz 3
1001	Benzene, (3-bromo-3-butenyl)- 1-Butene, 2-bromo-4-phenyl	$C_{10}H_{11}Br$ 211.10	62692-40-4	 118^{21}; 90^{5}	3-05-00-01209 1.2907^{20}	ace 4 1.5450^{20}
1002	Benzene, (4-bromo-2-butenyl)- 2-Butene, 1-bromo-4-phenyl-	$C_{10}H_{11}Br$ 211.10	40734-75-6	 $126\text{-}30^{10.5}$	4-05-00-01377 1.2660^{20}	eth 4 1.5678^{20}
1003	Benzene, (4-bromobutoxy)-	$C_{10}H_{13}BrO$ 229.12	1200-03-9 41	 154^{18}	4-06-00-00558	EtOH 2; ctc 2
1004	Benzene, 1-bromo-2-chloro-	C_6H_4BrCl 191.45	694-80-4 -12.3	 204	4-05-00-00680 1.6387^{25}	H_2O 1; bz 4; ctc 2 1.5809^{20}
1005	Benzene, 1-bromo-3-chloro-	C_6H_4BrCl 191.45	108-37-2 -21.5	 196	4-05-00-00680 1.6302^{20}	H_2O 1; EtOH 4; eth 4 1.5771^{20}
1006	Benzene, 1-bromo-4-chloro-	C_6H_4BrCl 191.45	106-39-8 68	 196	4-05-00-00681 1.576^{71}	H_2O 1; EtOH 2; eth 3; bz 3 1.5531^{70}
1007	Benzene, 1-bromo-3-(chloromethyl)-	C_7H_6BrCl 205.48	932-77-4 32.5	 119^{18}	4-05-00-00832	EtOH 4
1008	Benzene, 1-bromo-4-(chloromethyl)- *p*-Bromobenzyl chloride	C_7H_6BrCl 205.48	589-17-3 50	1398 236	4-05-00-00832	H_2O 1; EtOH 4; eth 4; peth 3
1009	Benzene, 1-bromo-4-cyclohexyl-	$C_{12}H_{15}Br$ 239.16	25109-28-8	 160^{25}	3-05-00-01258 1.283^{25}	bz 4; chl 4 1.5584^{20}
1010	Benzene, 1-bromo-2,3-dichloro-	$C_6H_3BrCl_2$ 225.90	56961-77-4 60	 245	0-05-00-00209	bz 4; eth 4; chl 4
1011	Benzene, 1-bromo-2,4-dichloro-	$C_6H_3BrCl_2$ 225.90	1193-72-2 26	 235	3-05-00-00564	bz 4; eth 4; chl 4
1012	Benzene, 1-bromo-3,5-dichloro-	$C_6H_3BrCl_2$ 225.90	19752-55-7 83	 232	2-05-00-00162	H_2O 1; EtOH 3; eth 3; bz 4
1013	Benzene, 2-bromo-1,3-dichloro-	$C_6H_3BrCl_2$ 225.90	19393-92-1 65	 242	0-05-00-00210	bz 4; eth 4; chl 4
1014	Benzene, 2-bromo-1,4-dichloro-	$C_6H_3BrCl_2$ 225.90	1435-50-3 35	 226; 119^{20}	3-05-00-00564	bz 4; eth 4; EtOH 4; lig 4
1015	Benzene, 4-bromo-1,2-dichloro-	$C_6H_3BrCl_2$ 225.90	18282-59-2 25	 237	3-05-00-00564	H_2O 1; EtOH 2; eth 4; bz 4
1016	Benzene, 4-bromo-1,2-dimethoxy-	$C_8H_9BrO_2$ 217.06	2859-78-1	 254.5	4-06-00-05621 1.702^{25}	 1.5743^{20}
1017	Benzene, 1-bromo-2,3-dimethyl-	C_8H_9Br 185.06	576-23-8	 214	4-05-00-00928 1.365^{25}	ace 4; bz 4; eth 4 1.5587^{20}
1018	Benzene, 1-bromo-2,4-dimethyl-	C_8H_9Br 185.06	583-70-0 -17	 205	4-05-00-00945 1.3419^{20}	H_2O 1; EtOH 4; eth 4; ace 4 1.5501^{20}
1019	Benzene, 1-bromo-3,5-dimethyl-	C_8H_9Br 185.06	556-96-7	 204	4-05-00-00945 1.362^{20}	eth 4; ace 3; bz 3 1.5462^{22}
1020	Benzene, 2-bromo-1,3-dimethyl-	C_8H_9Br 185.06	576-22-7	 203.5; 100^{20}	4-05-00-00945	eth 4; ace 3; bz 3 1.5552^{20}
1021	Benzene, 2-bromo-1,4-dimethyl-	C_8H_9Br 185.06	553-94-6 9	 199; $88\text{-}9^{13}$	4-05-00-00969 1.3582^{18}	H_2O 1; EtOH 4; bz 3 1.5514^{18}
1022	Benzene, 4-bromo-1,2-dimethyl-	C_8H_9Br 185.06	583-71-1 -0.2	 214.5	4-05-00-00928 1.3708^{20}	H_2O 1; EtOH 4; eth 4 1.5530^{20}
1023	Benzene, 1-bromo-4-(1,1-dimethylethyl)-	$C_{10}H_{13}Br$ 213.12	3972-65-4 19	 231.5	4-05-00-01050 1.2286^{20}	H_2O 1; eth 3; bz 3; chl 3 1.5436^{20}
1024	Benzene, 1-bromo-2,3-dinitro- 1-Bromo-2,3-dinitrobenzene	$C_6H_3BrN_2O_4$ 247.00	19613-76-4 101.5	 320	4-05-00-00749	EtOH 4
1025	Benzene, (1-bromoethenyl)-	C_8H_7Br 183.05	98-81-7 -44	 86^{14}; 71^{3}	4-05-00-01349 1.4025^{23}	 1.5881^{20}
1026	Benzene, 1-bromo-2-ethenyl-	C_8H_7Br 183.05	2039-88-5 -52.8	 209.2; 98^{20}	4-05-00-01349 1.4160^{20}	 1.5927^{20}
1027	Benzene, 1-bromo-3-ethenyl-	C_8H_7Br 183.05	2039-86-3	 $90\text{-}4^{20}$	3-05-00-01176 1.4059^{20}	 1.5933^{20}
1028	Benzene, 1-bromo-4-ethenyl-	C_8H_7Br 183.05	2039-82-9 7.7	 212; 103^{20}	4-05-00-01349 1.3984^{20}	H_2O 1; chl 4; HOAc 3 1.5947^{20}
1029	Benzene, (2-bromoethenyl)-, (*E*)-	C_8H_7Br 183.05	588-72-7 7	 219 dec; 108^{20}	4-05-00-01349 1.4269^{16}	H_2O 1; EtOH 5; eth 5; chl 3 1.6093^{20}

PHYSICAL CONSTANTS OF ORGANIC COMPOUNDS (continued)

No.	Name Synonym	Mol. Form. Mol. Wt.	CAS RN mp/°C	Merck No. bp/°C	Beil. Ref. den/g cm^{-3}	Solubility n_D
1030	Benzene, 1,1'-(bromoethenylidene)bis-	$C_{14}H_{11}Br$ 259.15	13249-58-6 41.5	 $175^{11.5}$	4-05-00-02177	EtOH 2; eth 3; ace 3; CS_2 3
1031	Benzene, (2-bromoethenyl)-, (*Z*)-	C_8H_7Br 183.05	588-73-8 -7	 $55\text{-}6^{2}$	4-05-00-01349 1.4322^{10}	 1.5990^{22}
1032	Benzene, 1-bromo-4-ethoxy-	C_8H_9BrO 201.06	588-96-5 4	 231	4-06-00-01045 1.4071^{25}	H_2O 1; EtOH 4; eth 4; chl 3 1.5517^{20}
1033	Benzene, (2-bromoethoxy)-	C_8H_9BrO 201.06	589-10-6 39	 240 dec; 128^{20}	4-06-00-00556 1.3555^{20}	H_2O 1; EtOH 4; eth 4
1034	Benzene, (1-bromoethyl)-, (±)-	C_8H_9Br 185.06	38661-81-3	 202-3. 85^{13}	4-05-00-00906 1.3605^{20}	EtOH 3; eth 3; bz 3; ctc 2 1.5612^{20}
1035	Benzene, 1-bromo-2-ethyl-	C_8H_9Br 185.06	1973-22-4 -67.9	 199.3	4-05-00-00905 1.3548^{20}	ace 4; bz 4; eth 4; EtOH 4 1.5472^{20}
1036	Benzene, 1-bromo-3-ethyl-	C_8H_9Br 185.06	2725-82-8	 202	4-05-00-00906 1.3493^{20}	 1.5465^{20}
1037	Benzene, 1-bromo-4-ethyl-	C_8H_9Br 185.06	1585-07-5 -43.5	 204	4-05-00-00906 1.3423^{20}	ace 4; bz 4; eth 4; EtOH 4 1.5445^{20}
1038	Benzene, (2-bromoethyl)-	C_8H_9Br 185.06	103-63-9	 219	4-05-00-00907 1.3587^{20}	H_2O 1; eth 3; bz 3; ctc 2 1.5372^{20}
1039	Benzene, (1-bromoethyl)-, (*R*)-	C_8H_9Br 185.06	1459-14-9	 203	4-05-00-00906 1.3108^{23}	H_2O 1; EtOH 3; eth 3 1.5612^{20}
1040	Benzene, 1-bromo-2-ethynyl- *o*-Bromophenylacetylene	C_8H_5Br 181.03	766-46-1	 $92\text{-}3^{20}$	3-05-00-01351 1.4434^{25}	 1.5962^{25}
1041	Benzene, 1-bromo-3-ethynyl- *m*-Bromophenylacetylene	C_8H_5Br 181.03	766-81-4	 $85\text{-}6^{16}$	3-05-00-01351 1.4466^{25}	 1.5896^{25}
1042	Benzene, 1-bromo-4-ethynyl-	C_8H_5Br 181.03	766-96-1 64.5	 89^{16}	4-05-00-01529	chl 3
1043	Benzene, 1-bromo-3-fluoro-	C_6H_4BrF 175.00	1073-06-9	 150	4-05-00-00678 1.7081^{20}	ctc 3 1.5257^{20}
1044	Benzene, 1-bromo-4-fluoro-	C_6H_4BrF 175.00	460-00-4 -17.4	 151.5	4-05-00-00678 1.593^{15}	H_2O 1; EtOH 3; eth 3; chl 3 1.5310^{15}
1045	Benzene, 1-bromo-4-(hexyloxy)-	$C_{12}H_{17}BrO$ 257.17	30752-19-3	 $156\text{-}7^{13}$	4-06-00-01047 1.2306^{20}	 1.5262^{20}
1046	Benzene, 1-bromo-2-iodo-	C_6H_4BrI 282.91	583-55-1 9.5	 257; 120^{15}	4-05-00-00698 2.2570^{25}	H_2O 1; EtOH 2; ace 3; HOAc 2 1.6618^{25}
1047	Benzene, 1-bromo-3-iodo- *m*-Bromoiodobenzene	C_6H_4BrI 282.91	591-18-4 -9.3	 252; 120^{18}	4-05-00-00698	H_2O 1; EtOH 2; HOAc 2
1048	Benzene, 1-bromo-4-iodo-	C_6H_4BrI 282.91	589-87-7 92	 252	4-05-00-00698	H_2O 1; EtOH 2; eth 3; chl 2
1049	Benzene, 1-bromo-4-isocyanato- *p*-Bromophenyl isocyanate	C_7H_4BrNO 198.02	2493-02-9	1418 226	1-12-00-00321	eth 4
1050	Benzene, 1-bromo-2-methoxy-	C_7H_7BrO 187.04	578-57-4 2.5	 216	4-06-00-01037 1.5018^{20}	H_2O 1; EtOH 4; eth 4 1.5727^{20}
1051	Benzene, 1-bromo-3-methoxy-	C_7H_7BrO 187.04	2398-37-0	 211; 105^{16}	4-06-00-01043	H_2O 1; EtOH 3; eth 3; bz 3 1.5635^{20}
1052	Benzene, 1-bromo-4-methoxy-	C_7H_7BrO 187.04	104-92-7 13.5	 215	4-06-00-01044 1.4564^{20}	H_2O 2; EtOH 4; eth 4; ctc 3 1.5642^{20}
1053	Benzene, (bromomethyl)- Benzyl bromide	C_7H_7Br 171.04	100-39-0 -3	1142 201	4-05-00-00829 1.4380^{25}	H_2O 1; EtOH 5; eth 5; ctc 3 1.5752^{20}
1054	Benzene, 1-bromo-2-methyl- *o*-Bromotoluene	C_7H_7Br 171.04	95-46-5 -27.8	 181.7	4-05-00-00825 1.4232^{20}	H_2O 1; EtOH 4; eth 4; bz 4 1.5565^{20}
1055	Benzene, 1-bromo-3-methyl- *m*-Bromotoluene	C_7H_7Br 171.04	591-17-3 -39.8	 183.7	4-05-00-00827 1.4099^{20}	H_2O 1; EtOH 3; eth 5; ace 3 1.5510^{20}
1056	Benzene, 1-bromo-4-methyl- *p*-Bromotoluene	C_7H_7Br 171.04	106-38-7 28.5	1429 184.3	4-05-00-00827 1.3959^{35}	H_2O 1; EtOH 3; eth 3; ace 3 1.5477^{20}
1057	Benzene, 1-(bromomethyl)-3,5-dimethyl- Mesitylene, α-bromo-	$C_9H_{11}Br$ 199.09	27129-86-8 40	 230; 118^{22}	4-05-00-01027	bz 4; eth 4; EtOH 4; chl 4
1058	Benzene, 1,1'-(bromomethylene)bis- Bromodiphenylmethane	$C_{13}H_{11}Br$ 247.13	776-74-9 45	 184^{20}; 152^{2}	4-05-00-01850	EtOH 3; bz 4; chl 3
1059	Benzene, 1-bromo-4-(1-methylethyl)-	$C_9H_{11}Br$ 199.09	586-61-8 -22.5	 218.7	4-05-00-00994 1.3145^{20}	H_2O 1; eth 3; bz 3; ctc 2 1.5569^{20}
1060	Benzene, (2-bromo-1-methylethyl)-	$C_9H_{11}Br$ 199.09	1459-00-3 -59.3	 188.5	4-05-00-00995 1.3020^{20}	
1061	Benzene, 1,1',1''-(bromomethylidyne)tris-	$C_{19}H_{15}Br$ 323.23	596-43-0 153	 230^{15}	4-05-00-02600 1.5500^{20}	
1062	Benzene, 1-(bromomethyl)-2-methyl-	C_8H_9Br 185.06	89-92-9 21	 217; 108^{16}	4-05-00-00928 1.3811^{23}	H_2O 1; EtOH 3; eth 3; ace 3 1.5730^{20}
1063	Benzene, 1-(bromomethyl)-3-methyl-	C_8H_9Br 185.06	620-13-3	 212.5	4-05-00-00945 1.3711^{23}	H_2O 1; EtOH 4; eth 4; os 3 1.5660^{20}
1064	Benzene, 1-(bromomethyl)-4-methyl-	C_8H_9Br 185.06	104-81-4 35	 220	4-05-00-00969 1.324^{25}	H_2O 1; EtOH 3; eth 4; chl 4
1065	Benzene, 2-bromo-1-methyl-4-(1-methylethyl)-	$C_{10}H_{13}Br$ 213.12	2437-76-5	 234.3	4-05-00-01063 1.2689^{18}	eth 4; EtOH 4; chl 4 1.5360^{20}
1066	Benzene, 1-bromo-2-methyl-4-nitro-	$C_7H_6BrNO_2$ 216.03	7149-70-4 78	 141^{17}	4-05-00-00860	eth 4
1067	Benzene, 1-(bromomethyl)-3-nitro-	$C_7H_6BrNO_2$ 216.03	3958-57-4 59.3	 162^{13}	4-05-00-00860	H_2O 1; EtOH 3

PHYSICAL CONSTANTS OF ORGANIC COMPOUNDS (continued)

No.	Name Synonym	Mol. Form. Mol. Wt.	CAS RN mp/°C	Merck No. bp/°C	Beil. Ref. den/g cm^{-3}	Solubility n_D
1068	Benzene, 1-bromo-3-methyl-2-nitro- 3-Bromo-2-nitrotoluene	$C_7H_6BrNO_2$ 216.03	52414-97-8 28.5	 153^{30}; 129^{10}	4-05-00-00860	eth 4; EtOH 4
1069	Benzene, 1-bromo-3-methyl-5-nitro- Toluene, 3-bromo-5-nitro-	$C_7H_6BrNO_2$ 216.03	52488-28-5 83	 269.5	3-05-00-00752	eth 4; EtOH 4
1070	Benzene, 1-bromo-2-methyl-3-nitro- Toluene, 2-bromo-6-nitro	$C_7H_6BrNO_2$ 216.03	55289-35-5 42	 143^{22}	3-05-00-00751	EtOH 4
1071	Benzene, 2-bromo-1-methyl-3-nitro-	$C_7H_6BrNO_2$ 216.03	41085-43-2 41.5	 161^{30}	4-05-00-00860	EtOH 4
1072	Benzene, 2-bromo-1-methyl-4-nitro-	$C_7H_6BrNO_2$ 216.03	7745-93-9 78	 175^{38}; 150^{20}	4-05-00-00861	eth 4
1073	Benzene, 2-bromo-4-methyl-1-nitro- 2-Bromo-4-methyl-1-nitrobenzene	$C_7H_6BrNO_2$ 216.03	40385-54-4 37.8	 269; 154^{20}	4-05-00-00861	eth 4; EtOH 4
1074	Benzene, 4-bromo-1-methyl-2-nitro- 4-Bromo-2-nitrotoluene	$C_7H_6BrNO_2$ 216.03	60956-26-5 47	 256.5	4-05-00-00860	eth 4; EtOH 4
1075	Benzene, 1-bromo-2-nitro-	$C_6H_4BrNO_2$ 202.01	577-19-5 43	 258	4-05-00-00728 1.6245^{80}	H_2O 1; EtOH 4; eth 3; ace 3
1076	Benzene, 1-bromo-3-nitro-	$C_6H_4BrNO_2$ 202.01	585-79-5 56	 265	4-05-00-00729 1.7036^{20}	H_2O 2; EtOH 3; eth 3; bz 3 1.5979^{20}
1077	Benzene, 1-bromo-4-nitro- *p*-Nitrobromobenzene	$C_6H_4BrNO_2$ 202.01	586-78-7 127	 256	4-05-00-00729 1.948^{25}	H_2O 1; EtOH 3; eth 3; bz 3
1078	Benzene, bromopentafluoro- Bromopentafluorobenzene	C_6BrF_5 246.96	344-04-7 -31	 137	4-05-00-00680 1.981^{25}	 1.4490^{20}
1079	Benzene, 1-bromo-4-phenoxy- 4-Bromophenyl phenyl ether	$C_{12}H_9BrO$ 249.11	101-55-3 18.72	 $124\text{-}8^{3\text{-}4}$	4-06-00-01047 1.6088^{20}	H_2O 1; eth 3; ctc 3 1.6084^{20}
1080	Benzene, 1-bromo-2-(2-phenylethenyl)-, (*E*)- *trans*-2-Bromostilbene	$C_{14}H_{11}Br$ 259.15	54737-45-0 34	 $143^{0.2}$	4-05-00-02164	H_2O I; EtOH 3; peth 3 1.6822^{25}
1081	Benzene, 1-bromo-2-(2-phenylethenyl)-, (*Z*)- Stilbene, 2-bromo (*cis*)	$C_{14}H_{11}Br$ 259.15	4877-77-4	 $121^{0.5}$	4-05-00-02164	H_2O 1; EtOH 3; peth 3 1.6404^{25}
1082	Benzene, 1-bromo-4-(1-propenyl)-	C_9H_9Br 197.07	4489-23-0 35	 238.5	2-05-00-00372 1.3323^{36}	 1.5903^{36}
1083	Benzene, 1-bromo-4-(2-propenyl)-	C_9H_9Br 197.07	2294-43-1	 224; 96^{12}	4-05-00-01363 1.324^{15}	 1.559^{15}
1084	Benzene, (3-bromo-1-propenyl)-	C_9H_9Br 197.07	4392-24-9 34	 130^{10}	3-05-00-01188 1.3428^{30}	EtOH 4 1.613^{20}
1085	Benzene, (3-bromopropoxy)-	$C_9H_{11}BrO$ 215.09	588-63-6 10.7	 127^{18}	4-06-00-00557 1.364^{16}	eth 4
1086	Benzene, (1-bromopropyl)-, (±)- Benzene, (1-bromopropyl) (*DL*)	$C_9H_{11}Br$ 199.09	63790-14-7	 105^{17}	2-05-00-00305 1.3098^{19}	 1.5517^{19}
1087	Benzene, (2-bromopropyl)-	$C_9H_{11}Br$ 199.09	2114-39-8	 $107\text{-}9^{16}$	4-05-00-00981 1.2908^{16}	chl 4 1.5450^{20}
1088	Benzene, (3-bromopropyl)-	$C_9H_{11}Br$ 199.09	637-59-2	 219.5; 117^{25}	4-05-00-00982 1.3106^{25}	H_2O 1; eth 4 1.5440^{25}
1089	Benzene, 1-bromo-2-(trifluoromethyl)-	$C_7H_4BrF_3$ 225.01	392-83-6	 167.5	4-05-00-00831 1.652^{25}	 1.4817^{20}
1090	Benzene, 1-bromo-3-(trifluoromethyl)-	$C_7H_4BrF_3$ 225.01	401-78-5 1	 151.5	4-05-00-00831 1.613^{25}	 1.4716^{20}
1091	Benzene, 1-bromo-4-(trifluoromethyl)-	$C_7H_4BrF_3$ 225.01	402-43-7	 160	4-05-00-00831 1.607^{25}	 1.4705^{25}
1092	Benzene, 1-bromo-2,3,5-trimethyl- 1-Bromo-2,3,5-trimethylbenzene	$C_9H_{11}Br$ 199.09	31053-99-3 <-15	 238	4-05-00-01014	bz 4 1.5516^{20}
1093	Benzene, 1-bromo-2,4,5-trimethyl-	$C_9H_{11}Br$ 199.09	5469-19-2 73	 234	4-05-00-01014	H_2O 1; EtOH 4
1094	Benzene, 2-bromo-1,3,5-trimethyl-	$C_9H_{11}Br$ 199.09	576-83-0 -1	 225	4-05-00-01027 1.3191^{10}	H_2O 1; eth 4; bz 3; ctc 2 1.5510^{20}
1095	Benzene, 1,1'-(1,3-butadiene-1,4-diyl)bis-, (*E,E*)-	$C_{16}H_{14}$ 206.29	538-81-8 154.3	 352	4-05-00-02319	bz 4; eth 4; EtOH 4; peth 4
1096	Benzene, 1,1'-(1,3-butadiene-1,4-diyl)bis-, (*Z,Z*)-	$C_{16}H_{14}$ 206.29	5807-76-1 70.5		4-05-00-02318 0.9697^{100}	H_2O 1; EtOH 2; eth 4; bz 4 1.6183^{100}
1097	Benzene, 1,3-butadienyl-, (*E*)-	$C_{10}H_{10}$ 130.19	16939-57-4 4.5	 76^{11}	4-05-00-01536 0.9286^{20}	H_2O I; EtOH 3; eth 3; ace 3 1.6089^{25}
1098	Benzene, 1,3-butadienyl-, (*Z*)-	$C_{10}H_{10}$ 130.19	31915-94-3 -57	 86^{11}	 0.9334^{20}	 1.6095^{20}
1099	Benzene, 1,1'-(1,4-butanediyl)bis- 1,4-Diphenylbutane	$C_{16}H_{18}$ 210.32	1083-56-3 52.5	 317	4-05-00-01937 0.9880^{20}	H_2O 1; EtOH 3; eth 3; chl 3
1100	Benzenebutanoic acid	$C_{10}H_{12}O_2$ 164.20	1821-12-1 52	 290	4-09-00-01811	H_2O 3; EtOH 3; eth 3
1101	Benzenebutanoic acid, 4-[bis(2-chloroethyl)amino]- Chlorambucil	$C_{14}H_{19}Cl_2NO_2$ 304.22	305-03-3 65	2064	4-14-00-01715	
1102	Benzenebutanoic acid, α,2-diamino-γ-oxo- Kynurenine	$C_{10}H_{12}N_2O_3$ 208.22	343-65-7 191 dec	5207	4-14-00-02562	
1103	Benzenebutanoic acid, 4-hydroxy-γ-(4-hydroxyphenyl)-γ-methyl- Diphenolic acid	$C_{17}H_{18}O_4$ 286.33	126-00-1 171.5	3312	4-10-00-01890	H_2O 4; ace 4; EtOH 4

No.	Name Synonym	Mol. Form. Mol. Wt.	CAS RN mp/°C	Merck No. bp/°C	Beil. Ref. den/g cm^{-3}	Solubility n_D
1104	Benzenebutanoic acid, γ-methyl-	$C_{11}H_{14}O_2$ 178.23	16433-43-5 13	 210^{85}; 165^{12}	4-09-00-01866 1.0554^{15}	eth 4; EtOH 4 1.5167^{20}
1105	Benzenebutanol, 4-methoxy-	$C_{11}H_{16}O_2$ 180.25	52244-70-9 3.5	 160^{8}; $125^{1.5}$	4-06-00-06005	 1.5267^{20}
1106	Benzenebutanol, α-methyl-	$C_{11}H_{16}O$ 164.25	2344-71-0	 $134\text{-}5^{16}$	3-06-00-01954 0.9643^{25}	 1.518^{19}
1107	Benzenebutanol, β-methyl-, (±)- 1-Butanol, 2-methyl-4-phenyl (*DL*)	$C_{11}H_{16}O$ 164.25	116783-11-0	 135^{11}	4-06-00-03380 0.9719^{20}	ace 4; bz 4; eth 4; EtOH 4 1.5173^{16}
1108	Benzenebutanol, β-phenyl-, (*R*)- 1-Butanol, 2,4-diphenyl-, (-)-	$C_{16}H_{18}O$ 226.32	17297-04-0 51.5	 178^{1}; $145^{0.1}$		chl 3 1.5686^{25}
1109	Benzene, 1-butenyl-, (*E*)-	$C_{10}H_{12}$ 132.21	1005-64-7 -43.1	 198.7	4-05-00-01374 0.9019^{20}	H_2O 1; EtOH 3; eth 3; bz 3 1.5420^{20}
1110	Benzene, 1-butenyl-, (*Z*)-	$C_{10}H_{12}$ 132.21	1560-09-4	 189	3-05-00-01205 0.9065^{25}	H_2O 1; EtOH 3; eth 3; bz 3 1.5390^{20}
1111	Benzene, 2-butenyl-	$C_{10}H_{12}$ 132.21	1560-06-1	 176	3-05-00-01207 0.8831^{20}	 1.5101^{20}
1112	Benzene, 3-butenyl-	$C_{10}H_{12}$ 132.21	768-56-9 -70	 177	4-05-00-01378 0.8831^{20}	H_2O 1; eth 3; bz 3 1.5059^{20}
1113	Benzene, 1,1'-(1-butenylidene)bis-	$C_{16}H_{16}$ 208.30	1726-14-3 -1.2	 296	4-05-00-02209 0.9934^{20}	 1.5604^{20}
1114	Benzene, butoxy-	$C_{10}H_{14}O$ 150.22	1126-79-0 -19.4	 210	4-06-00-00558 0.9351^{20}	eth 3; ace 3 1.4909^{20}
1115	Benzene, (butoxymethyl)-	$C_{11}H_{16}O$ 164.25	588-67-0	 223	4-06-00-02231 0.9227^{20}	ace 4; eth 4; EtOH 4 1.4833^{20}
1116	Benzene, 1-butoxy-3-methyl-	$C_{11}H_{16}O$ 164.25	23079-65-4	 229.2	4-06-00-02040 0.9406^{70}	eth 4 1.4970^{20}
1117	Benzene, 1-butoxy-4-methyl-	$C_{11}H_{16}O$ 164.25	10519-06-9	 229.5	3-06-00-01354 0.9205^{25}	eth 3 1.4970^{20}
1118	Benzene, butyl- Butylbenzene	$C_{10}H_{14}$ 134.22	104-51-8 -87.9	1549 183.3	4-05-00-01033 0.8601^{20}	H_2O 1; EtOH 5; eth 5; ace 5 1.4898^{20}
1119	Benzene, (1-butylhexadecyl)- Eicosane, 5-phenyl-	$C_{26}H_{46}$ 358.65	2400-04-6 30.2	 197^{1}	4-05-00-01223 0.8549^{20}	 1.4796^{20}
1120	Benzene, 1,1'-butylidenebis- 1,1-Diphenylbutane	$C_{16}H_{18}$ 210.32	719-79-9 27	 287	4-05-00-01944 0.9928^{20}	H_2O 1; EtOH 3; eth 3; bz 3 1.5664^{20}
1121	Benzene, 1-butyl-2-methyl-	$C_{11}H_{16}$ 148.25	1595-11-5	 208	3-05-00-00999 0.8710^{20}	ace 4; bz 4; eth 4 1.4960^{20}
1122	Benzene, 1-butyl-3-methyl-	$C_{11}H_{16}$ 148.25	1595-04-6	 205	3-05-00-00999 0.8590^{20}	ace 4; bz 4; eth 4 1.4910^{20}
1123	Benzene, 1-butyl-4-methyl-	$C_{11}H_{16}$ 148.25	1595-05-7 -85	 207	4-05-00-01094 0.8586^{20}	ace 4; bz 4; eth 4 1.4916^{20}
1124	Benzene, 3-butynyl-	$C_{10}H_{10}$ 130.19	16520-62-0	 190	4-05-00-01535 0.9258^{20}	 1.5208^{20}
1125	Benzenecarbodithioic acid, 2-hydroxy- Dithiosalicylic acid	$C_7H_6OS_2$ 170.26	527-89-9 49	3381	2-10-00-00078	bz 4; eth 4; EtOH 4
1126	Benzenecarboperoxoic acid Perbenzoic acid	$C_7H_6O_3$ 138.12	93-59-4 42	7109 100^{14}	4-09-00-00715	ace 4; bz 4; eth 4; EtOH 4
1127	Benzenecarboperoxoic acid, 3-chloro- 3-Chloroperoxybenzoic acid	$C_7H_5ClO_3$ 172.57	937-14-4 92-4 dec		4-09-00-00972	
1128	Benzenecarboperoxoic acid, 1,1-dimethylethyl ester Benzoyl *tert*-butyl peroxide	$C_{11}H_{14}O_3$ 194.23	614-45-9	 $75\text{-}6^{0.2}$	4-09-00-00715 1.021^{25}	 1.4990^{20}
1129	Benzenecarbothioamide, *N*-phenyl-	$C_{13}H_{11}NS$ 213.30	636-04-4 102	 dec	4-12-00-00423	H_2O 1; EtOH 4; eth 3; bz 3
1130	Benzenecarbothioic acid	C_7H_6OS 138.19	98-91-9 24	 $85\text{-}7^{10}$	4-09-00-01364 1.28^{20}	ace 4; bz 4; eth 4; EtOH 4 1.6040^{20}
1131	Benzene, chloro- Chlorobenzene	C_6H_5Cl 112.56	108-90-7 -45.2	2121 131.7	4-05-00-00640 1.1058^{20}	H_2O 1; EtOH 5; eth 5; bz 4 1.5241^{20}
1132	Benzene, 1-chloro-2-(chloromethyl)-	$C_7H_6Cl_2$ 161.03	611-19-8 -17	 217	4-05-00-00816 1.2699^{0}	H_2O 1; EtOH 2; eth 4; bz 4 1.5530^{20}
1133	Benzene, 1-chloro-3-(chloromethyl)-	$C_7H_6Cl_2$ 161.03	620-20-2	 216; 110^{25}	4-05-00-00816 1.2695^{15}	EtOH 4 1.5554^{20}
1134	Benzene, 1-chloro-4-(chloromethyl)- 4-Chlorobenzyl chloride	$C_7H_6Cl_2$ 161.03	104-83-6 31	 223	4-05-00-00816	ctc 2
1135	Benzene, 1-chloro-4-(chloromethyl)-2-nitro-	$C_7H_5Cl_2NO_2$ 206.03	57403-35-7	 122^{10}	4-05-00-00850 1.32^{25}	 1.5855^{20}
1136	Benzene, 1-chloro-4-[(chloromethyl)thio]-	$C_7H_6Cl_2S$ 193.10	7205-90-5	 $128\text{-}9^{12}$	4-06-00-01593 1.346^{25}	 1.6055^{20}
1137	Benzene, 1-chloro-2,3-dimethyl-	C_8H_9Cl 140.61	608-23-1	 188	4-05-00-00926 1.053^{20}	bz 4; chl 4 1.526^{20}
1138	Benzene, 1-chloro-2,4-dimethyl-	C_8H_9Cl 140.61	95-66-9 -32.5	 184	4-05-00-00943 1.0579^{20}	H_2O 1; ace 3; bz 4 1.5230^{25}
1139	Benzene, 2-chloro-1,3-dimethyl-	C_8H_9Cl 140.61	6781-98-2 -35	 186	4-05-00-00943 1.053^{20}	H_2O 1; ace 3; bz 4; chl 3 1.526^{20}
1140	Benzene, 2-chloro-1,4-dimethyl-	C_8H_9Cl 140.61	95-72-7 1.6	 187	4-05-00-00965 1.0589^{15}	H_2O 1; ace 3; bz 4; ctc 3
1141	Benzene, 4-chloro-1,2-dimethyl-	C_8H_9Cl 140.61	615-60-1 -6	 194	3-05-00-00816 1.0682^{15}	H_2O 1; ace 3; bz 4; ctc 3
1142	Benzene, 1-chloro-4-(1,1-dimethylethyl)-	$C_{10}H_{13}Cl$ 168.67	3972-56-3	 213	4-05-00-01048 1.0075^{18}	 1.5123^{20}

PHYSICAL CONSTANTS OF ORGANIC COMPOUNDS (continued)

No.	Name Synonym	Mol. Form. Mol. Wt.	CAS RN mp/°C	Merck No. bp/°C	Beil. Ref. den/g cm^{-3}	Solubility n_D
1143	Benzene, (2-chloro-1,1-dimethylethyl)- Neophyl chloride	$C_{10}H_{13}Cl$ 168.67	515-40-2	6375 223; 105^{18}	4-05-00-01048 1.047^{20}	ace 4; bz 4; eth 4; EtOH 4 1.5247^{20}
1144	Benzene, 1-chloro-2,4-dinitro- 1-Chloro-2,4-dinitrobenzene	$C_6H_3ClN_2O_4$ 202.55	97-00-7 53	2135 315	4-05-00-00744 1.4982^{75}	H_2O 1; EtOH 2; eth 3; bz 3 1.5857^{60}
1145	Benzene, 2-chloro-1,3-dinitro- 2-Chloro-1,3-dinitrobenzene	$C_6H_3ClN_2O_4$ 202.55	606-21-3 88	2136 315	4-05-00-00744 1.6867^{16}	H_2O 1; EtOH 3; eth 3; chl 2
1146	Benzene, 1,1'-(1-chloro-1,2-ethenediyl)bis-, (*E*)-	$C_{14}H_{11}Cl$ 214.69	948-98-1 53.5	 322; 128^{2}	4-05-00-02161	H_2O 1; EtOH 3; eth 3; bz 3
1147	Benzene, 1,1'-(1-chloro-1,2-ethenediyl)bis-, (*Z*)-	$C_{14}H_{11}Cl$ 214.69	948-99-2	 $160-2^{12}$	4-05-00-02162	H_2O 1; EtOH 3; eth 3; bz 3 1.6281^{19}
1148	Benzene, (1-chloroethenyl)-	C_8H_7Cl 138.60	618-34-8 -23	 199	4-05-00-01345 1.1016^{18}	eth 4; EtOH 4 1.5612^{20}
1149	Benzene, 1-chloro-2-ethenyl- *o*-Chlorostyrene	C_8H_7Cl 138.60	2039-87-4 -63.1	 188.7	4-05-00-01345 1.1000^{20}	EtOH 3; eth 3; ace 3; ctc 3 1.5649^{20}
1150	Benzene, 1-chloro-3-ethenyl-	C_8H_7Cl 138.60	2039-85-2	 $62-3^{6}$	4-05-00-01345 1.1168^{20}	H_2O 1; EtOH 3; eth 3 1.5625^{20}
1151	Benzene, 1-chloro-4-ethenyl-	C_8H_7Cl 138.60	1073-67-2 15.9	 192	4-05-00-01345 1.0868^{20}	H_2O 1; EtOH 3; eth 3; ace 5 1.5660^{20}
1152	Benzene, 1-(1-chloroethenyl)-2,4-dimethyl- Styrene, α-chloro-2,4-dimethyl	$C_{10}H_{11}Cl$ 166.65	74346-30-8	 $104-5^{19}$	2-05-00-00381 1.044^{13}	 1.5446^{13}
1153	Benzene, (2-chloroethenyl)-, (*E*)-	C_8H_7Cl 138.60	4110-77-4	 199	4-05-00-01346 1.1095^{18}	H_2O 1; EtOH 3; eth 3; ace 3 1.5648^{20}
1154	Benzene, 1,1',1''-(1-chloro-1-ethenyl-2-ylidene)tris[4-methoxy- Chlorotrianisene	$C_{23}H_{21}ClO_3$ 380.87	569-57-3 115	2173	4-06-00-07650	
1155	Benzene, (2-chloroethenyl)-, (*Z*)-	C_8H_7Cl 138.60	4604-28-8	 $59^{3.5}$	4-05-00-01346 1.1046^{25}	 1.5762^{25}
1156	Benzene, 1-chloro-2-ethoxy-	C_8H_9ClO 156.61	614-72-2	 210	4-06-00-00785 1.1288^{15}	bz 4; eth 4; EtOH 4 1.5284^{25}
1157	Benzene, 1-chloro-3-ethoxy- *m*-Chlorophenyl ethyl ether	C_8H_9ClO 156.61	2655-83-6	 207	4-06-00-00811 1.1712^{20}	H_2O 1; EtOH 4; eth 4; bz 4
1158	Benzene, 1-chloro-4-ethoxy-	C_8H_9ClO 156.61	622-61-7 21	 213	4-06-00-00823 1.1254^{20}	EtOH 3; eth 3; bz 4; ctc 2 1.5252^{20}
1159	Benzene, (2-chloroethoxy)-	C_8H_9ClO 156.61	622-86-6 28	 218.5	4-06-00-00556	H_2O 1; EtOH 4; eth 4; ace 4
1160	Benzene, (1-chloroethyl)-, (±)-	C_8H_9Cl 140.61	38661-82-4	 81^{17}	4-05-00-00898 1.0620^{20}	EtOH 3; eth 4; bz 3 1.5276^{20}
1161	Benzene, 1-chloro-2-ethyl-	C_8H_9Cl 140.61	89-96-3 -82.7	 178.4	4-05-00-00897 1.0569^{20}	H_2O 1; ace 3; bz 3; ctc 3 1.5218^{20}
1162	Benzene, 1-chloro-3-ethyl-	C_8H_9Cl 140.61	620-16-6 -55	 183.8	4-05-00-00897 1.0529^{20}	ace 4; bz 4; eth 4; EtOH 4 1.5195^{20}
1163	Benzene, 1-chloro-4-ethyl-	C_8H_9Cl 140.61	622-98-0 -62.6	 184.4	4-05-00-00897 1.0455^{20}	H_2O 1; EtOH 5; eth 5; ace 5 1.5175^{20}
1164	Benzene, (2-chloroethyl)-	C_8H_9Cl 140.61	622-24-2	 197.5; 92^{20}	4-05-00-00899 1.069^{25}	H_2O 1; EtOH 3; eth 3; ace 3 1.5276^{20}
1165	Benzene, [(2-chloroethyl)thio]-	C_8H_9ClS 172.68	5535-49-9	 $117-8^{11}$	4-06-00-01469 1.1769^{25}	eth 4; chl 4 1.5828^{20}
1166	Benzene, 1-chloro-2-fluoro-	C_6H_4ClF 130.55	348-51-6 -43	 137.6	4-05-00-00652 1.2233^{30}	H_2O 1; ace 3; bz 3 1.4918^{30}
1167	Benzene, 1-chloro-3-fluoro-	C_6H_4ClF 130.55	625-98-9	 127.6	4-05-00-00652 1.221^{25}	 1.4911
1168	Benzene, 1-chloro-4-fluoro-	C_6H_4ClF 130.55	352-33-0 -26.8	 130	4-05-00-00652 1.4990^{15}	H_2O 1; EtOH 3; eth 3; bz 3 1.4990^{15}
1169	Benzene, 1-chloro-3-fluoro-2-methyl-	C_7H_6ClF 144.58	443-83-4	 154	3-05-00-00691 1.191^{25}	 1.5026^{20}
1170	Benzene, 2-chloro-4-fluoro-1-methyl-	C_7H_6ClF 144.58	452-73-3	 152.5	3-05-00-00691 1.1972^{20}	 1.4985^{25}
1171	Benzene, 1-chloro-2-iodo-	C_6H_4ClI 238.45	615-41-8 0.7	 234.5	4-05-00-00695 1.9515^{25}	H_2O 1; ace 3; ctc 2 1.6331^{25}
1172	Benzene, 1-chloro-4-iodo-	C_6H_4ClI 238.45	637-87-6 57	 227	4-05-00-00696 1.886^{27}	H_2O 1; EtOH 3; chl 2
1173	Benzene, 1-chloro-2-iodo-4-methyl- Toluene, 4-chloro-3-iodo-	C_7H_6ClI 252.48	2401-22-1 -10	 249	3-05-00-00726 1.7940^{15}	 1.614^{23}
1174	Benzene, 4-chloro-1-iodo-2-methyl- Toluene, 5-chloro-2-iodo	C_7H_6ClI 252.48	23399-70-4 -21	 240	3-05-00-00726 1.702^{17}	 1.616^{23}
1175	Benzene, 4-chloro-2-iodo-1-methyl-	C_7H_6ClI 252.48	33184-48-4 -25	 242.5	3-05-00-00726 1.8358^{15}	 1.620^{23}
1176	Benzene, 1-chloro-2-isocyanato-	C_7H_4ClNO 153.57	3320-83-0 30.5	 200; 115^{43}	4-12-00-01129	ctc 2
1177	Benzene, 1-chloro-4-isothiocyanato-	C_7H_4ClNS 169.63	2131-55-7 46	 249.5	4-12-00-01214	H_2O 1; EtOH 3
1178	Benzene, 1-chloro-2-methoxy-	C_7H_7ClO 142.58	766-51-8 -26.8	 198.5	4-06-00-00785 1.1911^{20}	H_2O 1; EtOH 3; eth 3; chl 2 1.5480^{20}
1179	Benzene, 1-chloro-3-methoxy-	C_7H_7ClO 142.58	2845-89-8	 193.5	4-06-00-00811 1.1759^{12}	H_2O 1; EtOH 3; eth 3 1.5365^{20}
1180	Benzene, 1-chloro-4-methoxy-	C_7H_7ClO 142.58	623-12-1 <-18	 197.5	4-06-00-00822 1.201^{20}	H_2O 1; EtOH 4; eth 4; ctc 3 1.5390^{20}
1181	Benzene, [(chloromethoxy)methyl]-	C_8H_9ClO 156.61	3587-60-8	 103^{13}	4-06-00-02252 1.1350^{20}	 1.5192^{20}

No.	Name Synonym	Mol. Form. Mol. Wt.	CAS RN mp/°C	Merck No. bp/°C	Beil. Ref. den/g cm^{-3}	Solubility n_D
1182	Benzene, (chloromethyl)- Benzyl chloride	C_7H_7Cl 126.59	100-44-7 -45	1143 179	4-05-00-00809 1.1004^{20}	H_2O 1; EtOH 5; eth 5; ctc 2 1.5391^{20}
1183	Benzene, 1-chloro-2-methyl- *o*-Chlorotoluene	C_7H_7Cl 126.59	95-49-8 -35.6	2172 159.0	4-05-00-00805 1.0825^{20}	H_2O 1; EtOH 3; eth 5; ace 5 1.5268^{20}
1184	Benzene, 1-chloro-3-methyl- *m*-Chlorotoluene	C_7H_7Cl 126.59	108-41-8 -47.8	2172 161.8	4-05-00-00806 1.075^{20}	H_2O 1; EtOH 3; eth 5; bz 3 1.5214^{19}
1185	Benzene, 1-chloro-4-methyl- *p*-Chlorotoluene	C_7H_7Cl 126.59	106-43-4 7.5	2172 162.4	4-05-00-00806 1.0697^{20}	H_2O 1; EtOH 3; eth 5; ctc 3 1.5150^{20}
1186	Benzene, 1-(chloromethyl)-2,4-dimethyl-	$C_9H_{11}Cl$ 154.64	824-55-5	 215.5; 110^{20}	4-05-00-01012 1.0580^{19}	bz 4; eth 4; EtOH 4
1187	Benzene, 1,1'-(chloromethylene)bis-	$C_{13}H_{11}Cl$ 202.68	90-99-3 16	 140^{3}	4-05-00-01847 1.140^{25}	chl 3 1.5951^{20}
1188	Benzene, 1-(2-chloro-1-methylethenyl)-4-methyl- Styrene, β-chloro, α-4-dimethyl	$C_{10}H_{11}Cl$ 166.65	30926-60-4	 107-8^{10}	2-05-00-00381 1.058^{20}	 1.5549^{23}
1189	Benzene, (1-chloro-1-methylethyl)-	$C_9H_{11}Cl$ 154.64	934-53-2	 98^{1}	4-05-00-00992 1.192^{25}	 1.5290^{25}
1190	Benzene, 1-chloro-2-(1-methylethyl)-	$C_9H_{11}Cl$ 154.64	2077-13-6 -74.4	 191.1	4-05-00-00992 1.0341^{20}	ace 4; bz 4; eth 4; EtOH 4 1.5168^{20}
1191	Benzene, 1-(chloromethyl)-4-ethyl-	$C_9H_{11}Cl$ 154.64	1467-05-6	 95-6^{15}	4-05-00-01004	bz 4; EtOH 4; chl 4 1.5290^{25}
1192	Benzene, 1-chloro-4-(1-methylethyl)-	$C_9H_{11}Cl$ 154.64	2621-46-7 -12.3	 198.3	4-05-00-00992 1.0208^{20}	H_2O 1; EtOH 5; eth 5; ace 5 1.5117^{20}
1193	Benzene, (2-chloro-1-methylethyl)-	$C_9H_{11}Cl$ 154.64	824-47-5	 85^{13}	4-05-00-00992 1.047^{19}	 1.5245^{19}
1194	Benzene, 1-(chloromethyl)-2-fluoro-	C_7H_6ClF 144.58	345-35-7	 86^{40}	4-05-00-00813 1.216^{25}	 1.5150^{20}
1195	Benzene, 1-(chloromethyl)-4-fluoro-	C_7H_6ClF 144.58	352-11-4	 82^{26}	4-05-00-00813 1.207^{25}	 1.5130
1196	Benzene, 1,1',1''-(chloromethylidyne)tris-	$C_{19}H_{15}Cl$ 278.78	76-83-5 113.5	 310	4-05-00-02497	H_2O 1; EtOH 2; eth 4; ace 3
1197	Benzene, 1-(chloromethyl)-4-methoxy-	C_8H_9ClO 156.61	824-94-2 24.5	 262.5	4-06-00-02137 1.261^{20}	ace 4; bz 4; eth 4 1.580^{20}
1198	Benzene, 1-(chloromethyl)-2-methyl-	C_8H_9Cl 140.61	552-45-4	 198; 90^{20}	4-05-00-00926 1.063^{25}	eth 4; EtOH 4 1.5410^{25}
1199	Benzene, 1-(chloromethyl)-3-methyl-	C_8H_9Cl 140.61	620-19-9	 195.5	4-05-00-00943 1.064^{20}	H_2O 1, EtOH 3, eth 3 1.5345^{20}
1200	Benzene, 1-(chloromethyl)-4-methyl-	C_8H_9Cl 140.61	104-82-5	 201; 90^{20}	4-05-00-00966 1.0512^{20}	H_2O 1; EtOH 3; eth 5 1.5380
1201	Benzene, 2-chloro-1-methyl-4-(1-methylethyl)-	$C_{10}H_{13}Cl$ 168.67	4395-79-3	 216.5	4-05-00-01062 1.0104^{25}	H_2O 1; ace 3; bz 3 1.5078^{20}
1202	Benzene, 2-chloro-4-methyl-1-(1-methylethyl)-	$C_{10}H_{13}Cl$ 168.67	4395-80-6	 215; 113^{35}	3-05-00-00956 1.018^{15}	 1.5180^{20}
1203	Benzene, 1-(chloromethyl)-2-nitro-	$C_7H_6ClNO_2$ 171.58	612-23-7 51		4-05-00-00854	H_2O 1; EtOH 3; eth 3; ace 4 1.5557^{62}
1204	Benzene, 1-chloro-2-methyl-3-nitro-	$C_7H_6ClNO_2$ 171.58	83-42-1 37.8	 238	4-05-00-00854	H_2O 1; EtOH 3 1.5377^{69}
1205	Benzene, 1-chloro-2-methyl-4-nitro-	$C_7H_6ClNO_2$ 171.58	13290-74-9 42.5	 249	4-05-00-00855	eth 4
1206	Benzene, 1-(chloromethyl)-3-nitro-	$C_7H_6ClNO_2$ 171.58	619-23-8 46	 173^{34}	4-05-00-00855	ace 4; bz 4; eth 4; EtOH 4 1.5577^{62}
1207	Benzene, 1-chloro-3-methyl-5-nitro- 3-Chloro-5-nitrotoluene	$C_7H_6ClNO_2$ 171.58	16582-38-0 61		4-05-00-00855	EtOH 4 1.5404^{69}
1208	Benzene, 1-(chloromethyl)-4-nitro- 4-Nitrobenzyl chloride	$C_7H_6ClNO_2$ 171.58	100-14-1 71		4-05-00-00856	H_2O 1; EtOH 3; eth 3; ace 4 1.5647^{62}
1209	Benzene, 1-chloro-4-methyl-2-nitro-	$C_7H_6ClNO_2$ 171.58	89-60-1 7	 261; 118^{11}	4-05-00-00855	H_2O 1; ctc 3 1.5572^{20}
1210	Benzene, 2-chloro-1-methyl-4-nitro-	$C_7H_6ClNO_2$ 171.58	121-86-8 66.5	 260	4-05-00-00855	H_2O 2; EtOH 3; eth 3; chl 2 1.5470^{69}
1211	Benzene, 2-chloro-4-methyl-1-nitro-	$C_7H_6ClNO_2$ 171.58	38939-88-7 24	 146^{15}	4-05-00-00856	H_2O 1; ctc 3
1212	Benzene, 4-chloro-1-methyl-2-nitro-	$C_7H_6ClNO_2$ 171.58	89-59-8 38	 242; 115.5^{11}	4-05-00-00853 1.2559^{80}	H_2O 1; EtOH 3; eth 3; chl 2
1213	Benzene, 4-chloro-2-methyl-1-nitro-	$C_7H_6ClNO_2$ 171.58	5367-28-2 25.5		4-05-00-00854	H_2O 1; ctc 3 1.5490[illegible]
1214	Benzene, 1-chloro-2-nitro- *o*-Chloronitrobenzene	$C_6H_4ClNO_2$ 157.56	88-73-3 32.5	2151 245.5	4-05-00-00721 1.368^{242}	H_2O 1; EtOH 3; eth 3; ace 4
1215	Benzene, 1-chloro-3-nitro- *m*-Chloronitrobenzene	$C_6H_4ClNO_2$ 157.56	121-73-3 44.4	2151 235.5	4-05-00-00722 1.343^{50}	H_2O 1; EtOH 3; eth 3; bz 3 1.5374^{80}
1216	Benzene, 1-chloro-4-nitro- *p*-Chloronitrobenzene	$C_6H_4ClNO_2$ 157.56	100-00-5 83.5	2151 242	4-05-00-00723 1.2979^{90}	H_2O l; EtOH 2; eth 3; chl 3 1.5376^{100}
1217	Benzene, 1-chloro-2-nitro-4-(trifluoromethyl)- 2-Nitro-4-(trifluoromethyl)-1-chlorobenzene	$C_7H_3ClF_3NO_2$ 225.55	121-17-5 -2.5	 222; 95^{10}	4-05-00-00857 1.511^{25}	 1.4893^{20}
1218	Benzene, 1-chloro-4-nitro-2-(trifluoromethyl)- 4-Chloro-3-(trifluoromethyl)nitrobenzene	$C_7H_3ClF_3NO_2$ 225.55	777-37-7 22	 232	4-05-00-00857 1.527^{25}	 1.5083^{26}
1219	Benzene, 4-chloro-1-nitro-2-(trifluoromethyl)- 4-Chloro-2-(trifluoromethyl)nitrobenzene	$C_7H_3ClF_3NO_2$ 225.55	118-83-2		3-05-00-00748 1.526^{25}	 1.4980^{20}
1220	Benzene, chloropentafluoro- Chloropentafluorobenzene	C_6ClF_5 202.51	344-07-0	 117.9	4-05-00-00654 1.568^{25}	 1.4256^{20}

No.	Name Synonym	Mol. Form. Mol. Wt.	CAS RN mp/°C	Merck No. bp/°C	Beil. Ref. den/g cm^{-3}	Solubility n_D
1221	Benzene, 1-chloro-4-phenoxy- 4-Chlorophenyl phenyl ether	$C_{12}H_9ClO$ 204.66	7005-72-3	 284.5	4-06-00-00826 1.2026^{15}	 1.599
1222	Benzene, 1-chloro-2-(2-phenylethenyl)-, (*E*)-	$C_{14}H_{11}Cl$ 214.69	1657-52-9 39.5	 209^{30}; 139^{2}	4-05-00-02161	H_2O 1; EtOH 3; chl 3
1223	Benzene, 1-chloro-3-(2-phenylethenyl)-, (*E*)-	$C_{14}H_{11}Cl$ 214.69	14064-43-8 73.5	 $177^{0.2}$	4-05-00-02161	H_2O 1; EtOH 3; chl 3
1224	Benzene, 1-chloro-4-(phenylmethyl)-	$C_{13}H_{11}Cl$ 202.68	831-81-2 7.5	 299; $147\text{-}8^{8}$	4-05-00-01847 1.1247^{20}	ace 4
1225	Benzene, 1-chloro-4-(phenylsulfonyl)- Sulphenone	$C_{12}H_9ClO_2S$ 252.72	80-00-2 94	8969	4-06-00-01587	H_2O 1; EtOH 2; eth 3; ace 4
1226	Benzene, (1-chloro-1-propenyl)-	C_9H_9Cl 152.62	35673-03-1	 $90.5^{0.9}$	2-05-00-00372 1.085^{20}	ace 4; bz 4 1.5635^{15}
1227	Benzene, (2-chloro-1-propenyl)-	C_9H_9Cl 152.62	13099-50-8	 120^{20}; 61^{1}	3-05-00-01186 1.0738^{19}	ace 4; bz 4; chl 4 1.5565^{19}
1228	Benzene, (3-chloro-2-propenyl)-	C_9H_9Cl 152.62	6268-37-7	 211	4-05-00-01363 1.073^{13}	H_2O 1; eth 3; ace 3; bz 3 1.545^{14}
1229	Benzene, 1-chloro-3-(1-propenyl)-, (*E*)-	C_9H_9Cl 152.62	23204-80-0 8.5	 106.7^{13}	 1.0926^{20}	H_2O 1; EtOH 5; eth 5; ace 4 1.5851^{20}
1230	Benzene, (3-chloro-1-propenyl)-, (*E*)-	C_9H_9Cl 152.62	21087-29-6 8.5	 $106\text{-}7^{13}$	4-05-00-01360 1.0926^{20}	ace 4; bz 4; eth 4; EtOH 4 1.5851^{20}
1231	Benzene, 1-(3-chloro-2-propenyl)-4-methoxy- Benzene, 1(3-chloroallyl)-4-methoxy	$C_{10}H_{11}ClO$ 182.65	54644-23-4	 120^{11}	3-06-00-02416 1.155^{8}	 1.553^{8}
1232	Benzene, 1-chloro-4-(2-propenyloxy)-	C_9H_9ClO 168.62	13997-70-1	 $106\text{-}7^{12}$	4-06-00-00825 1.131^{15}	bz 4; eth 4; EtOH 4 1.5348^{25}
1233	Benzene, (3-chloropropoxy)-	$C_9H_{11}ClO$ 170.64	3384-04-1 12	 250	4-06-00-00557 1.1167^{20}	ace 4; eth 4 1.5235^{25}
1234	Benzene, (2-chloropropyl)-, (*R*)- *R*-(-)-2-Chloro-1-phenylpropane	$C_9H_{11}Cl$ 154.64	55449-46-2	 94^{17}	4-05-00-00980 1.038^{19}	H_2O 1; ace 3; bz 3; chl 3 1.5198^{22}
1235	Benzene, (3-chloropropyl)-	$C_9H_{11}Cl$ 154.64	104-52-9	 219.5	4-05-00-00980 1.056^{21}	ctc 2 1.5160^{25}
1236	Benzene, [(3-chloropropyl)thio]-	$C_9H_{11}ClS$ 186.71	4911-65-3	 $116\text{-}7^{4}$	3-06-00-00981 1.1536^{20}	ace 4; py 4 1.5752^{20}
1237	Benzene, 1-chloro-2-(trichloromethyl)- 1-Chloro-2-(trichloromethyl)benzene	$C_7H_4Cl_4$ 229.92	2136-89-2 29.4	 264.3	4-05-00-00823 1.5187^{20}	H_2O 1; eth 3; ace 3; ctc 2 1.5836^{20}
1238	Benzene, 1-chloro-3-(trichloromethyl)- 1-Chloro-3-(trichloromethyl)benzene	$C_7H_4Cl_4$ 229.92	2136-81-4 0.6	 255	4-05-00-00823 1.495^{14}	H_2O 1; eth 3; ace 3; ctc 2 1.4461^{20}
1239	Benzene, 1-chloro-4-(trichloromethyl)- 1-Chloro-4-(trichloromethyl)benzene	$C_7H_4Cl_4$ 229.92	5216-25-1	 245	4-05-00-00823 1.4463^{20}	ace 4; eth 4
1240	Benzene, 1-chloro-2-(trifluoromethyl)-	$C_7H_4ClF_3$ 180.56	88-16-4 -6	 152.2	4-05-00-00814 1.2540^{30}	chl 3 1.4513^{25}
1241	Benzene, 1-chloro-3-(trifluoromethyl)- Benzotrifluoride, 3-chloro	$C_7H_4ClF_3$ 180.56	98-15-7 -56	 137.5	4-05-00-00814 1.3311^{25}	 1.4438^{25}
1242	Benzene, 1-chloro-4-(trifluoromethyl)-	$C_7H_4ClF_3$ 180.56	98-56-6 -33	 138.5	4-05-00-00815 1.3340^{25}	 1.4431^{30}
1243	Benzene, 2-chloro-1,3,5-trimethyl-	$C_9H_{11}Cl$ 154.64	1667-04-5 <-20	 205	4-05-00-01026 1.0337^{30}	eth 4; EtOH 4 1.5212^{30}
1244	Benzene, 2-chloro-1,3,5-trinitro- Picryl chloride	$C_6H_2ClN_3O_6$ 247.55	88-88-0 83	7390	4-05-00-00757 1.797^{20}	H_2O 1; EtOH 3; eth 2; ace 4
1245	Benzene, cyclobutyl-	$C_{10}H_{12}$ 132.21	4392-30-7	 190^{155}; 101^{41}	4-05-00-01387 0.9378^{20}	 1.5277^{20}
1246	Benzene, 1-cyclohexen-1-yl-	$C_{12}H_{14}$ 158.24	771-98-2 -11	 252	4-05-00-01557 0.9939^{20}	MeOH 4 1.5718^{20}
1247	Benzene, 2-cyclohexen-1-yl-	$C_{12}H_{14}$ 158.24	15232-96-9	 $115\text{-}7^{16}$	4-05-00-01558 0.9800^{20}	 1.5530^{20}
1248	Benzene, cyclohexyl- Cyclohexylbenzene	$C_{12}H_{16}$ 160.26	827-52-1 7.3	 240.1	4-05-00-01424 0.9427^{20}	H_2O 1; EtOH 4; eth 3; ctc 2 1.5329^{20}
1249	Benzene, (1-cyclohexylethyl)-	$C_{14}H_{20}$ 188.31	4413-16-5	 265	4-05-00-01454 0.9773^{17}	 1.549^{17}
1250	Benzene, 1-cyclohexyl-2-methoxy-	$C_{13}H_{18}O$ 190.29	2206-48-6	 268	4-06-00-03899 1.007^{18}	 1.5365^{18}
1251	Benzene, 1-cyclohexyl-2-nitro- Cyclohexane, 2-nitrophenyl	$C_{12}H_{15}NO_2$ 205.26	7137-56-6 45	 174^{16}; $113^{0.5}$	3-05-00-01259 1.111^{23}	 1.5472^{25}
1252	Benzene, 1-cyclohexyl-4-nitro-	$C_{12}H_{15}NO_2$ 205.26	5458-48-0 58.5	 210^{25}	4-05-00-01426	bz 4; eth 4
1253	Benzene, (cyclohexyloxy)-	$C_{12}H_{16}O$ 176.26	2206-38-4	 261	4-06-00-00565 1.0077^{20}	H_2O 1; ace 3; bz 3 1.520^{22}
1254	Benzene, cyclopentyl-	$C_{11}H_{14}$ 146.23	700-88-9	 219	4-05-00-01409 0.9462^{20}	eth 4 1.5280^{20}
1255	Benzene, (3-cyclopentylpropyl)-	$C_{14}H_{20}$ 188.31	2883-12-7	 272	4-05-00-01455 0.9233^{19}	 1.5130^{19}
1256	Benzene, cyclopropyl- Cyclopropylbenzene	C_9H_{10} 118.18	873-49-4 -31	 173.6; 80^{37}	4-05-00-01370 0.9317^{20}	H_2O 1; eth 3; ace 3; chl 3 1.5285^{20}
1257	Benzenedecanoic acid, ι-octyl-	$C_{24}H_{40}O_2$ 360.58	1938-17-6 40.5	 $199\text{-}205^{0.09}$	2-09-00-00376 0.9310^{25}	 1.4894^{20}
1258	Benzene, decyl- Decylbenzene	$C_{16}H_{26}$ 218.38	104-72-3 -14.4	 298	4-05-00-01193 0.8555^{20}	ace 4; bz 4; eth 4; EtOH 4 1.4832^{20}
1259	1,4-Benzenediacetic acid, dimethyl ester	$C_{12}H_{14}O_4$ 222.24	36076-25-2 51.5	 189^{15}	2-09-00-00624	EtOH 3; eth 3

No.	Name Synonym	Mol. Form. Mol. Wt.	CAS RN mp/°C	Merck No. bp/°C	Beil. Ref. den/g cm^{-3}	Solubility n_D
1260	1,2-Benzenediamine *o*-Phenylenediamine	$C_6H_8N_2$ 108.14	95-54-5 102.5	7255 257	4-13-00-00038	H_2O 3; EtOH 4; eth 3; bz 3
1261	1,3-Benzenediamine *m*-Phenylenediamine	$C_6H_8N_2$ 108.14	108-45-2 63.5	7254 285	4-13-00-00079 1.0096^{58}	H_2O 4; EtOH 3; eth 3; bz 3 1.6339^{58}
1262	1,4-Benzenediamine *p*-Phenylenediamine	$C_6H_8N_2$ 108.14	106-50-3 146	7256 267	4-13-00-00104	H_2O 2; EtOH 3; eth 3; bz 3
1263	1,4-Benzenediamine, *N*-(4-aminophenyl)- 4,4'-Diaminodiphenylamine	$C_{12}H_{13}N_3$ 199.26	537-65-5 158	2957 dec	4-13-00-00203	eth 4; EtOH 4
1264	1,2-Benzenediamine, 4-chloro-	$C_6H_7ClN_2$ 142.59	95-83-0 76		4-13-00-00068	H_2O 2; EtOH 4; eth 4; bz 3
1265	1,3-Benzenediamine, *N,N*-diethyl-	$C_{10}H_{16}N_2$ 164.25	26513-20-2 152	 277	2-13-00-00026	EtOH 4
1266	1,4-Benzenediamine, *N,N*-diethyl- *N,N*-Diethyl-*p*-phenylenediamine	$C_{10}H_{16}N_2$ 164.25	93-05-0	 261	4-13-00-00109	bz 4
1267	1,2-Benzenediamine, *N,N*-dimethyl-	$C_8H_{12}N_2$ 136.20	2836-03-5	 218; 90^{22}	4-13-00-00042 0.995^{22}	H_2O 2; EtOH 4; eth 4; ace 4
1268	1,3-Benzenediamine, *N,N*-dimethyl-	$C_8H_{12}N_2$ 136.20	2836-04-6 <-20	 270; 138^{10}	4-13-00-00080 0.995^{25}	H_2O 2; EtOH 4; eth 4
1269	1,2-Benzenediamine, *N,N'*-dimethyl- *o*-Phenylenediamine, *N,N'*-dimethyl	$C_8H_{12}N_2$ 136.20	3213-79-4 34.5	 247	4-13-00-00042	 1.5914^{25}
1270	1,4-Benzenediamine, *N,N*-dimethyl- Dimethyl-*p*-phenylenediamine	$C_8H_{12}N_2$ 136.20	99-98-9 53	3242 263	4-13-00-00106 1.036^{20}	H_2O 3; EtOH 4; eth 4; bz 4
1271	1,4-Benzenediamine, *N,N'*-diphenyl- *N,N'*-Diphenyl-*p*-phenylenediamine	$C_{18}H_{16}N_2$ 260.34	74 31 7 144	3331	4-13-00-00116	EtOH 2; eth 2; bz 2; chl 2
1272	1,2-Benzenediamine, 4-ethoxy-	$C_8H_{12}N_2O$ 152.20	1197-37-1 71.5	 295	3-13-00-01362	H_2O 4; EtOH 3; eth 3; chl 3
1273	1,2-Benzenediamine, 4-methoxy-	$C_7H_{10}N_2O$ 138.17	102-51-2 51	 200^{21}; 168^{11}	4-13-00-01448	eth 4
1274	1,3-Benzenediamine, 4-methoxy- 4-Methoxy-*m*-phenylenediamine	$C_7H_{10}N_2O$ 138.17	615-05-4 67.5	5920	4-13-00-01425	EtOH 3; eth 3; DMSO-d_6 2
1275	1,2-Benzenediamine, 3-methyl-	$C_7H_{10}N_2$ 122.17	2687-25-4 63.5	 255	3-13-00-00277	ace 4; bz 4; EtOH 4
1276	1,2-Benzenediamine, 4-methyl- Toluene-3,4-diamine	$C_7H_{10}N_2$ 122.17	496-72-0 89.5	 265	4-13-00-00260	H_2O 4; lig 3
1277	1,3-Benzenediamine, 2-methyl- 2,6-Diaminotoluene	$C_7H_{10}N_2$ 122.17	823-40 5 106		3-13-00-00291	H_2O 3, EtOH 3; bz 3
1278	1,3-Benzenediamine, 4-methyl- Toluene-2,4-diamine	$C_7H_{10}N_2$ 122.17	95-80-7 99	 292	4-13-00-00235	H_2O 4; EtOH 4; eth 4; bz 4
1279	1,4-Benzenediamine, 2-methyl-	$C_7H_{10}N_2$ 122.17	95-70-5 64	 273.5	4-13-00-00246	H_2O 3; EtOH 3; eth 3; bz 2
1280	1,4-Benzenediamine, *N*-methyl-	$C_7H_{10}N_2$ 122.17	623-09-6 36	 258	4-13-00-00106	H_2O 4; EtOH 4; eth 4; ace 4
1281	1,2-Benzenediamine, 4-nitro- 4-Nitro-*o*-phenylenediamine	$C_6H_7N_3O_2$ 153.14	99-56-9 199.5	6544	4-13-00-00075	acid 3
1282	1,2-Benzenediamine, *N*-phenyl-	$C_{12}H_{12}N_2$ 184.24	534 85-0 79.5	 313	4-13-00-00043	H_2O 2; ace 3; bz 3; chl 3
1283	1,4-Benzenediamine, *N*-phenyl- *p*-Aminodiphenylamine	$C_{12}H_{12}N_2$ 184.24	101-54-2 66	 354	4-13-00-00113	H_2O 2; EtOH 4; eth 3; chl 2
1284	1,3-Benzenediamine, 4-(phenylazo)-, monohydrochloride Chrysoidine hydrochloride	$C_{12}H_{13}ClN_4$ 248.71	532-82-1 118.5	2262	2-16-00-00203	ace 4
1285	1,3-Benzenediamine, *N,N,N',N'*-tetraethyl- *N,N,N',N'*-Tetraethyl-*m*-phenylenediamine	$C_{14}H_{24}N_2$ 220.36	64287-26-9	 148^9	2-13-00-00026 0.9522^{12}	eth 4; EtOH 4 1.5537^{12}
1286	1,2-Benzenediamine, *N,N,N',N'*-tetramethyl-	$C_{10}H_{16}N_2$ 164.25	704-01-8 8.9	 215.5	4-13-00-00042 0.9560^{20}	
1287	1,4-Benzenediamine, *N,N,N',N'*-tetramethyl- Tetramethyl-*p*-phenylenediamine	$C_{10}H_{16}N_2$ 164.25	100-22-1 51	9159 260	4-13-00-00107	H_2O 2; EtOH 4; eth 4; bz 4
1288	Benzene, 1,2-dibromo- *o*-Dibromobenzene	$C_6H_4Br_2$ 235.91	583-53-9 7.1	 225	4-05-00-00682 1.9843^{20}	H_2O 1; EtOH 3; eth 5; ace 5 1.6155^{20}
1289	Benzene, 1,3-dibromo- *m*-Dibromobenzene	$C_6H_4Br_2$ 235.91	108-36-1 -7	 218	4-05-00-00682 1.9523^{20}	H_2O 1; EtOH 3; eth 5 1.6083^{17}
1290	Benzene, 1,4-dibromo- *p*-Dibromobenzene	$C_6H_4Br_2$ 235.91	106-37-6 87.3	3001 218.5	4-05-00-00683 2.261^{17}	H_2O 1; EtOH 3; eth 4; ace 4 1.574?
1291	Benzene, 1,3-dibromo-5-(bromomethyl)- 3,5-Dibromobenzyl bromide	$C_7H_5Br_3$ 328.83	56908-88-4 96	 173^{19}	3-05-00-00719	EtOH 4
1292	Benzene, 1,2-dibromo-3-chloro- 1,2-Dibromo-3-chlorobenzene	$C_6H_3Br_2Cl$ 270.35	104514-49-0 73.5	 264	1-05-00-00117	bz 4; eth 4; chl 4
1293	Benzene, 1,2-dibromo-4-chloro- 1,2-Dibromo-4-chlorobenzene	$C_6H_3Br_2Cl$ 270.35	60956-24-3 35.5	 256	0-05-00-00212	bz 4; eth 4; chl 4
1294	Benzene, 1,3-dibromo-2-chloro- 1,3-Dibromo-2-chlorobenzene	$C_6H_3Br_2Cl$ 270.35	19230-27-4 72	 265	3-05-00-00569	bz 4; eth 4; chl 4
1295	Benzene, 1,4-dibromo-2-chloro- 1,4-Dibromo-2-chlorobenzene	$C_6H_3Br_2Cl$ 270.35	3460-24-0 40.5	 259	0-05-00-00212	bz 4; eth 4; chl 4
1296	Benzene, 2,4-dibromo-1-chloro-	$C_6H_3Br_2Cl$ 270.35	29604-75-9 27	 258	0-05-00-00212	bz 4; eth 4; chl 4
1297	Benzene, 1,2-dibromo-3,4-dimethyl- 3,4-Dibromo-*o*-xylene	$C_8H_8Br_2$ 263.96	24932-49-8 7	 277	2-05-00-00285 1.782^{15}	EtOH 4
1298	Benzene, 1,5-dibromo-2,4-dimethyl-	$C_8H_8Br_2$ 263.96	615-87-2 70	 255.5	3-05-00-00839	EtOH 4

No.	Name	Synonym	Mol. Form.	Mol. Wt.	CAS RN	mp/°C	Merck No.	bp/°C	Beil. Ref.	den/g cm^{-3}	Solubility	n_D
1299	Benzene, (1,2-dibromoethyl)-		$C_8H_8Br_2$	263.96	93-52-7	75		133^{19}	4-05-00-00909		EtOH 3; eth 3; bz 3; chl 3	
1300	Benzene, 2,4-dibromo-1-fluoro-		$C_6H_3Br_2F$	253.90	1435-53-6	69		$103\text{-}5^{32}$	3-05-00-00569	2.047^{20}		1.5840^{20}
1301	Benzene, (dibromomethyl)-		$C_7H_6Br_2$	249.93	618-31-5			156^{23}	4-05-00-00836	1.8365^{28}	H_2O 1; EtOH 5; eth 5	1.6147^{20}
1302	Benzene, 1,2-dibromo-3-methyl-	2,3-Dibromotoluene	$C_7H_6Br_2$	249.93	61563-25-5	30.5		$126\text{-}8^{19}$	4-05-00-00835	1.8234^{15}		1.5984^{25}
1303	Benzene, 1,2-dibromo-4-methyl-	3,4-Dibromotoluene	$C_7H_6Br_2$	249.93	60956-23-2	-10		240; 123^{19}	4-05-00-00836	1.8197^{25}		1.5979^{25}
1304	Benzene, 1,3-dibromo-2-methyl-	2,6-Dibromotoluene	$C_7H_6Br_2$	249.93	69321-60-4	5.5		246	4-05-00-00836	1.812^{22}		
1305	Benzene, 1,4-dibromo-2-methyl-	2,5-Dibromotoluene	$C_7H_6Br_2$	249.93	615-59-8	5.6		236	4-05-00-00835	1.8127^{17}	H_2O 1	1.5982^{18}
1306	Benzene, 2,4-dibromo-1-methyl-		$C_7H_6Br_2$	249.93	31543-75-6	-9.7		$103\text{-}4^{11}$	4-05-00-00835	1.8176^{25}		1.5964^{25}
1307	Benzene, 1,2-dibromo-4-nitro-	3,4-Dibromonitrobenzene	$C_6H_3Br_2NO_2$	280.90	5411-50-7	58.5		296	4-05-00-00732	2.354^{8}	bz 4; EtOH 4; HOAc 4	1.9835^{111}
1308	Benzene, 1,3-dibromo-2-nitro-	2,6-Dibromonitrobenzene	$C_6H_3Br_2NO_2$	280.90	13402-32-9	84		sub	3-05-00-00621	2.211^{8}	ace 4; bz 4; EtOH 4	
1309	Benzene, 1,3-dibromo-5-nitro-		$C_6H_3Br_2NO_2$	280.90	6311-60-0	106			3-05-00-00621	2.36^{8}	bz 4; eth 4; EtOH 4	
1310	Benzene, 1,4-dibromo-2-nitro-		$C_6H_3Br_2NO_2$	280.90	3460-18-2	85.5			4-05-00-00732	2.368^{8}	H_2O 1; ace 3; bz 3; chl 2	
1311	Benzene, 2,4-dibromo-1-nitro-		$C_6H_3Br_2NO_2$	280.90	51686-78-3	62			3-05-00-00620	2.356^{8}	ace 4; bz 4	
1312	Benzene, 1,2-dibromo-3,4,5,6-tetrafluoro-		$C_6Br_2F_4$	307.87	827-08-7	12		198		2.260^{20}		1.5151^{25}
1313	Benzene, 2,4-dibromo-1,3,5-trimethyl-	Mesitylene, 2,4-dibromo-	$C_9H_{10}Br_2$	277.99	6942-99-0	65.5		285	3-05-00-00921		ctc 2	
1314	Benzene, 1,4-dibutoxy-		$C_{14}H_{22}O_2$	222.33	104-36-9	45.5		158^{15}	4-06-00-05721		ctc 3	
1315	1,3-Benzenedicarbonitrile	1,3-Dicyanobenzene	$C_8H_4N_2$	128.13	626-17-5	162		sub	4-09-00-03297	0.992^{40}	H_2O 2; EtOH 4; eth 3; bz 3	
1316	1,4-Benzenedicarbonitrile		$C_8H_4N_2$	128.13	623-26-7	224		sub	4-09-00-03328		H_2O 1; EtOH 2; eth 2; bz 3	
1317	1,3-Benzenedicarbonitrile, 2,4,5,6-tetrachloro-	Chlorothalonil	$C_8Cl_4N_2$	265.91	1897-45-6	250	2167	350		1.7^{25}	H_2O 1; ace 2; cyhex 2	
1318	1,2-Benzenedicarbonyl dichloride	Phthaloyl chloride	$C_8H_4Cl_2O_2$	203.02	88-95-9	15.5	7349	281.1	4-09-00-03261	1.4089^{20}		1.5684^{20}
1319	1,3-Benzenedicarbonyl dichloride		$C_8H_4Cl_2O_2$	203.02	99-63-8	43.5		276	4-09-00-03295	1.3880^{17}	H_2O 2; EtOH 2; eth 3	1.570^{47}
1320	1,4-Benzenedicarbonyl dichloride		$C_8H_4Cl_2O_2$	203.02	100-20-9	83.5		258; 125^{9}	4-09-00-03318		eth 3	
1321	1,2-Benzenedicarbonyl difluoride		$C_8H_4F_2O_2$	170.12	445-69-2	42.5		227	4-09-00-03260	1.3066^{50}	H_2O 1; peth 3	
1322	1,3-Benzenedicarboxaldehyde		$C_8H_6O_2$	134.13	626-19-7	89.5		246; 136^{13}	4-07-00-02139		H_2O 2; EtOH 4; eth 2; ace 3	
1323	1,4-Benzenedicarboxaldehyde		$C_8H_6O_2$	134.13	623-27-8	117		246	4-07-00-02140		H_2O 2; EtOH 4; eth 3; chl 3	
1324	1,2-Benzenedicarboxamide	Phthalamide	$C_8H_8N_2O_2$	164.16	88-96-0	222	7343	dec	4-09-00-03265		H_2O 2; EtOH 2; eth 1	
1325	1,2-Benzenedicarboxamide, *N,N,N',N'*-tetraethyl-	*N,N,N',N'*-Tetraethylphthalamide	$C_{16}H_{24}N_2O_2$	276.38	83-81-8	36	9137	204^{16}	4-09-00-03266			
1326	1,3-Benzenedicarboxamide, *N,N,N',N'*-tetraethyl-	Isophthalamide, *N,N,N',N'*-tetraethyl	$C_{16}H_{24}N_2O_2$	276.38	13698-87-8	86		242^{12}	4-09-00-03296		ace 4; bz 4; eth 4; EtOH 4	
1327	1,2-Benzenedicarboxylic acid	Phthalic acid	$C_8H_6O_4$	166.13	88-99-3	230 dec	7345	dec	4-09-00-03167	2.18^{191}	H_2O 2; EtOH 3; eth 2; chl 1	
1328	1,3-Benzenedicarboxylic acid	Isophthalic acid	$C_8H_6O_4$	166.13	121-91-5	347	5083	sub	4-09-00-03292		H_2O 2; EtOH 3; eth 1; bz 1	
1329	1,4-Benzenedicarboxylic acid	Terephthalic acid	$C_8H_6O_4$	166.13	100-21-0		9093	sub 300	4-09-00-03301		H_2O 1; EtOH 1; eth 1; ctc 2	
1330	1,3-Benzenedicarboxylic acid, 2-amino-		$C_8H_7NO_4$	181.15	39622-79-2	340		sub 267	4-14-00-01899		eth 4; EtOH 4	
1331	1,3-Benzenedicarboxylic acid, 5-amino-		$C_8H_7NO_4$	181.15	99-31-0	360		sub	1-14-00-00636		H_2O 1; EtOH 2	
1332	1,2-Benzenedicarboxylic acid, 3-benzoyl-		$C_{15}H_{10}O_5$	270.24	602-82-4	140.5		dec	3-10-00-04007		bz 4; EtOH 4	
1333	1,2-Benzenedicarboxylic acid, bis(2-ethoxyethyl) ester		$C_{16}H_{22}O_6$	310.35	605-54-9	34		345	2-09-00-00597	1.1229^{21}		
1334	1,2-Benzenedicarboxylic acid, bis(2-ethylhexyl) ester	Bis(2-ethylhexyl) phthalate	$C_{24}H_{38}O_4$	390.56	117-81-7	-55	1262	384	4-09-00-03181	0.981^{25}	ctc 2	1.4853^{20}
1335	1,2-Benzenedicarboxylic acid, bis(3-methylbutyl) ester	Diisoamyl phthalate	$C_{18}H_{26}O_4$	306.40	605-50-5		5006	334 dec	4-09-00-03179	1.0209^{16}	EtOH 4	1.4871^{20}

PHYSICAL CONSTANTS OF ORGANIC COMPOUNDS (continued)

No.	Name	Synonym	Mol. Form.	Mol. Wt.	CAS RN	mp/°C	Merck No.	bp/°C	Beil. Ref.	den/g cm^{-3}	Solubility	n_D
1336	1,2-Benzenedicarboxylic acid, bis(2-methylpropyl) ester		$C_{16}H_{22}O_4$	278.35	84-69-5			296.5; 159^{4}	4-09-00-03177	1.0490^{15}	ctc 3	
1337	1,2-Benzenedicarboxylic acid, bis(phenylmethyl) ester		$C_{22}H_{18}O_4$	346.38	523-31-9	44.0		277^{15}	4-09-00-03219		H_2O 1; EtOH 4; eth 4; lig 1	
1338	1,4-Benzenedicarboxylic acid, bis(2-methylpropyl) ester		$C_{16}H_{22}O_4$	278.35	18699-48-4	55		180^{6}	4-09-00-03305		eth 3	
1339	1,2-Benzenedicarboxylic acid, dibutyl ester	Dibutyl phthalate	$C_{16}H_{22}O_4$	278.35	84-74-2	-35	1586	340	4-09-00-03175	1.0465^{20}	H_2O 1; EtOH 5; eth 5; bz 5	1.4911^{20}
1340	1,2-Benzenedicarboxylic acid, 3,5-dichloro-	Phthalic acid, 3,5-dichloro-	$C_8H_4Cl_2O_4$	235.02	25641-98-9	164 dec		sub	4-09-00-03270		ace 4; eth 4; EtOH 4	
1341	1,2-Benzenedicarboxylic acid, dicyclohexyl ester		$C_{20}H_{26}O_4$	330.42	84-61-7	66			4-09-00-03189	1.383^{20}	H_2O 1; EtOH 3; eth 3; chl 2	1.431^{20}
1342	1,2-Benzenedicarboxylic acid, diethyl ester	Diethyl phthalate	$C_{12}H_{14}O_4$	222.24	84-66-2	-40.5	7345	295	4-09-00-03172	1.232^{14}	H_2O l; EtOH 5; eth 5; ace 3	1.5000^{21}
1343	1,2-Benzenedicarboxylic acid, dimethyl ester	Dimethyl phthalate	$C_{10}H_{10}O_4$	194.19	131-11-3	5.5	3243	283.7	4-09-00-03170	1.1905^{20}	H_2O 1; EtOH 5; eth 5; bz 3	1.5138^{20}
1344	1,2-Benzenedicarboxylic acid, dioctyl ester	Dioctyl phthalate	$C_{24}H_{38}O_4$	390.56	117-84-0	25			4-09-00-03180			
1345	1,2-Benzenedicarboxylic acid, diphenyl ester	Phenyl phthalate	$C_{20}H_{14}O_4$	318.33	84-62-8	73	7278	253^{14}	4-09-00-03217		H_2O 1; EtOH 2; eth 2; ctc 2	
1346	1,3-Benzenedicarboxylic acid, diethyl ester	Diethyl isophthalate	$C_{12}H_{14}O_4$	222.24	636-53-3	11.5		302	4-09-00-03294	1.1239^{17}	H_2O 1	1.508^{18}
1347	1,3-Benzenedicarboxylic acid, dimethyl ester	Dimethyl isophthalate	$C_{10}H_{10}O_4$	194.19	1459-93-4	67.5		282	4-09-00-03293	1.194^{20}	H_2O 2	1.5168^{20}
1348	1,3-Benzenedicarboxylic acid, 4,6-dimethyl-		$C_{10}H_{10}O_4$	194.19	2790-09-2	200		sub	3-09-00-04298		EtOH 4	
1349	1,4-Benzenedicarboxylic acid, 2,5-dichloro-	2,5-Dichloroterephthalic acid	$C_8H_4Cl_2O_4$	235.02	13799-90-1	306		sub	4-09-00-03335		H_2O 2; EtOH 3; eth 4	
1350	1,4-Benzenedicarboxylic acid, diethyl ester	Diethyl terephthalate	$C_{12}H_{14}O_4$	222.24	636-09-9	44		302	4-09-00-03304	1.0989^{45}	H_2O 1; EtOH 4; eth 3	
1351	1,4-Benzenedicarboxylic acid, dimethyl ester	Dimethyl terephthalate	$C_{10}H_{10}O_4$	194.19	120-61-6	141		288	4-09-00-03303	1.075^{141}	H_2O 2; EtOH 2; eth 3; chl 3	
1352	1,3-Benzenedicarboxylic acid, 4-hydroxy-	4-Hydroxyisophthalic acid	$C_8H_6O_5$	182.13	636-46-4	310	4758		4-10-00-02091		H_2O 1; EtOH 4; eth 4; chl 1	
1353	1,2-Benzenedicarboxylic acid, 3,4,5,6-tetrachloro-, monoethyl ester	Phthalic acid, tetrachloro-, monoethyl ester	$C_{10}H_6Cl_4O_4$	331.97	602-21-1	94.5		250 dec	3-09-00-04206		eth 4; EtOH 4	
1354	Benzene, 1,2-dichloro-	*o*-Dichlorobenzene	$C_6H_4Cl_2$	147.00	95-50-1	-16.7	3044	180	4-05-00-00654	1.3059^{20}	H_2O 1; EtOH 3; eth 3; ace 5	1.5515^{20}
1355	Benzene, 1,3-dichloro-	*m*-Dichlorobenzene	$C_6H_4Cl_2$	147.00	541-73-1	-24.8	3043	173	4-05-00-00657	1.2884^{20}	H_2O 1; EtOH 3; eth 3; ace 5	1.5459^{20}
1356	Benzene, 1,4-dichloro-	*p*-Dichlorobenzene	$C_6H_4Cl_2$	147.00	106-46-7	52.7	3045	174	4-05-00-00658	1.2475^{55}	H_2O 1; EtOH 5; eth 3; ace 5	1.5285^{20}
1357	Benzene, 1,2-dichloro-4-(chloromethyl)-		$C_7H_5Cl_3$	195.47	102-47-6	37.5		241	4-05-00-00820		H_2O 1; EtOH 3; ctc 3; os 3	
1358	Benzene, 1,3-dichloro-5-(chloromethyl)-		$C_7H_5Cl_3$	195.47	3290-06-0	36		$60^{0.35}$	3-05-00-00699		EtOH 4	
1359	Benzene, 1,2-dichloro-4-(dichloromethyl)-	Benzylidene chloride, 3,4-dichloro	$C_7H_4Cl_4$	229.92	56961-84-3			257	3-05-00-00702	1.515^{22}	bz 4; eth 4; EtOH 4	
1360	Benzene, 1,2-dichloro-3,4-dimethyl-	*o*-Xylene, 3,4-dichloro	$C_8H_8Cl_2$	175.06	68266-67-1	9		231	4-05-00-00926	1.2240^{20}		
1361	Benzene, 1,3-dichloro-2,5-dimethyl-		$C_8H_8Cl_2$	175.06	38204-89-6	15		222	3-05-00-00855	1.2156^{20}		
1362	Benzene, 1,4-dichloro-2,5-dimethyl-		$C_8H_8Cl_2$	175.06	1124-05-6	71		222	3-05-00-00855		chl 3	
1363	Benzene, 1,5-dichloro-2,4-dimethyl-		$C_8H_8Cl_2$	175.06	2084-45-9	68.5		223	3-05-00-00835		bz 4; eth 4; chl 4	
1364	Benzene, 2,3-dichloro-1,4-dimethyl-		$C_8H_8Cl_2$	175.06	34840-79-4	-2		230	3-05-00-00855	1.2310^{20}		
1365	Benzene, 1,1'-(1,2-dichloro-1,2-ethenediyl) bis-, (*E*)-		$C_{14}H_{10}Cl_2$	249.14	951-86-0	143.5		183^{18}	1-05-00-00304		H_2O 1; EtOH 4; eth 4	
1366	Benzene, 1,1'-(1,2-dichloro-1,2-ethenediyl) bis-, (*Z*)-		$C_{14}H_{10}Cl_2$	249.14	5216-32-0	67.5		179^{18}	1-05-00-00304		eth 4; EtOH 4	
1367	Benzene, 1,2-dichloro-3-ethenyl-	2,3-Dichlorostyrene	$C_8H_6Cl_2$	173.04	2123-28-6			$92\text{-}4^{4.5}$	3-05-00-01174	1.2826^{20}		1.5848^{20}
1368	Benzene, 1,2-dichloro-4-ethenyl-		$C_8H_6Cl_2$	173.04	2039-83-0			95^{5}	4-05-00-01347	1.256^{20}		1.5857^{20}
1369	Benzene, 1,3-dichloro-2-ethenyl-	2,6-Dichlorostyrene	$C_8H_6Cl_2$	173.04	28469-92-3	89		$64\text{-}5^{3}$	3-05-00-01174	1.2631^{20}		1.5752^{20}
1370	Benzene, 1,3-dichloro-5-ethenyl-	3,5-Dichlorostyrene	$C_8H_6Cl_2$	173.04	2155-42-2			59^{1}	3-05-00-01175	1.225^{25}		1.5745^{25}
1371	Benzene, 1,4-dichloro-2-ethenyl-	2,5-Dichlorostyrene	$C_8H_6Cl_2$	173.04	1123-84-8			$72\text{-}3^{2}$	4-05-00-01347	1.4045^{20}		1.5798^{20}
1372	Benzene, (2,2-dichloroethenyl)-		$C_8H_6Cl_2$	173.04	698-88-4			225	4-05-00-01348	1.2531^{20}	ace 3; chl 3	1.5852^{20}

No.	Name	Synonym	Mol. Form.	Mol. Wt.	CAS RN	mp/°C	Merck No.	bp/°C	Beil. Ref.	den/g cm^{-3}	Solubility	n_D
1373	Benzene, 2,4-dichloro-1-ethenyl-	2,4-Dichlorostyrene	$C_8H_6Cl_2$	173.04	2123-27-5			81^{6}	4-05-00-01347	1.243^{25}		1.5828^{20}
1374	Benzene, 1,1'-(dichloroethenylidene)bis[4-chloro-	2,2-Dichloro-1,1-bis(4-chlorophenyl)ethylene	$C_{14}H_8Cl_4$	318.03	72-55-9	89			4-05-00-02177			
1375	Benzene, 2,4-dichloro-1-ethoxy-		$C_8H_8Cl_2O$	191.06	5392-86-9	32		237	4-06-00-00885		H_2O 1; EtOH 3; eth 3; bz 3	
1376	Benzene, (1,2-dichloroethyl)-		$C_8H_8Cl_2$	175.06	1074-11-9			233.5	4-05-00-00901	1.240^{15}	H_2O 1; ace 3; bz 3	1.5544^{15}
1377	Benzene, 1,4-dichloro-2-ethyl-	2,5-Dichloroethylbenzene	$C_8H_8Cl_2$	175.06	54484-63-8			213.5	4-05-00-00901	1.239^{25}	eth 2; bz 3; chl 2; peth 2	
1378	Benzene, 1,1'-(2,2-dichloroethylidene)bis[4-chloro-	1,1-Dichloro-2,2-bis(*p*-chlorophenyl)ethane	$C_{14}H_{10}Cl_4$	320.04	72-54-8	109.5	3049		4-05-00-01884		chl 2	
1379	Benzene, (dichlorofluoromethyl)-		$C_7H_5Cl_2F$	179.02	498-67-9	-26.8		179	4-05-00-00818	1.3138^{11}	EtOH 4	1.5180^{11}
1380	Benzene, 1,4-dichloro-2-iodo-		$C_6H_3Cl_2I$	272.90	29682-41-5	22		253	3-05-00-00580		bz 4; eth 4; EtOH 4; chl 4	
1381	Benzene, 2,4-dichloro-1-methoxy-		$C_7H_6Cl_2O$	177.03	553-82-2	28.5		232; 125^{10}	4-06-00-00885		chl 2	
1382	Benzene, (dichloromethyl)-	Benzal chloride	$C_7H_6Cl_2$	161.03	98-87-3	-17	1064	205	4-05-00-00817	1.26^{25}	eth 4; EtOH 4	1.5502^{20}
1383	Benzene, 1,2-dichloro-3-methyl-	2,3-Dichlorotoluene	$C_7H_6Cl_2$	161.03	32768-54-0	6		207.5	4-05-00-00815	1.2458^{20}	bz 4	1.5511^{20}
1384	Benzene, 1,2-dichloro-4-methyl-	3,4-Dichlorotoluene	$C_7H_6Cl_2$	161.03	95-75-0	-15.2		208.9	4-05-00-00815	1.2564^{20}	H_2O 1; EtOH 5; eth 5; ace 5	1.5471^{20}
1385	Benzene, 1,3-dichloro-2-methyl-	2,6-Dichlorotoluene	$C_7H_6Cl_2$	161.03	118-69-4	25.8		198	4-05-00-00815	1.2686^{20}	H_2O 1; chl 3	1.5507^{20}
1386	Benzene, 1,3-dichloro-5-methyl-	3,5-Dichlorotoluene	$C_7H_6Cl_2$	161.03	25186-47-4	26		201.5	4-05-00-00816		H_2O 1	
1387	Benzene, 1,4-dichloro-2-methyl-	2,5-Dichlorotoluene	$C_7H_6Cl_2$	161.03	19398-61-9	5		200	3-05-00-00694	1.2535^{20}	H_2O 1; bz 3	1.5449^{20}
1388	Benzene, 2,4-dichloro-1-methyl-	2,4-Dichlorotoluene	$C_7H_6Cl_2$	161.03	95-73-8	-13.5		201	4-05-00-00815	1.2476^{20}	H_2O 1; ctc 3	1.5511^{20}
1389	Benzene, 1,1'-(dichloromethylene)bis-		$C_{13}H_{10}Cl_2$	237.13	2051-90-3			305 dec; 190^{21}	4-05-00-01848	1.235^{18}	eth 3; bz 3; ctc 3	
1390	Benzene, 1-(dichloromethyl)-4-methyl-		$C_8H_8Cl_2$	175.06	23063-36-7	48.5		105^{18}	3-05-00-00855		EtOH 2; bz 2; chl 2	
1391	Benzene, 1,2-dichloro-3-nitro-		$C_6H_3Cl_2NO_2$	192.00	3209-22-1	61.5		257.5	4-05-00-00725	1.721^{14}	H_2O 1; EtOH 3; eth 3; ace 3	
1392	Benzene, 1,2-dichloro-4-nitro-		$C_6H_3Cl_2NO_2$	192.00	99-54-7	43		255.5	4-05-00-00726	1.4558^{75}	H_2O 1; EtOH 3; eth 3; ctc 2	
1393	Benzene, 1,3-dichloro-2-nitro-	2,6-Dichloronitrobenzene	$C_6H_3Cl_2NO_2$	192.00	601-88-7	72.5		130^{8}	4-05-00-00726	1.603^{17}	eth 4; EtOH 4	
1394	Benzene, 1,3-dichloro-5-nitro-		$C_6H_3Cl_2NO_2$	192.00	618-62-2	65.4			4-05-00-00727	1.4000^{100}	H_2O 1; EtOH 3; eth 3	1.4000^{100}
1395	Benzene, 1,4-dichloro-2-nitro-		$C_6H_3Cl_2NO_2$	192.00	89-61-2	56		267	4-05-00-00726	1.439^{75}	H_2O 1; EtOH 3; eth 3; bz 3	1.4390^{75}
1396	Benzene, 2,4-dichloro-1-nitro-		$C_6H_3Cl_2NO_2$	192.00	611-06-3	34		258.5	4-05-00-00726	1.4790^{80}	H_2O 1; EtOH 3; eth 3; chl 2	1.5512^{70}
1397	Benzene, (3,3-dichloro-2-propenyl)-		$C_9H_8Cl_2$	187.07	38862-78-1	59		242; 142^{30}	4-05-00-01363	1.2028^{20}	eth 3; bz 3; chl 3; peth 3	
1398	Benzene, 1,2-dichloro-3,4,5,6-tetrafluoro	1,2-Dichlorotetrafluorobenzene	$C_6Cl_2F_4$	218.97	1198-59-0							
1399	Benzene, 1,3-dichloro-2,4,5,6-tetrafluoro-	1,3-Dichlorotetrafluorobenzene	$C_6Cl_2F_4$	218.97	1198-61-4							
1400	Benzene, 1,4-dichloro-2,3,5,6-tetrafluoro-	1,4-Dichlorotetrafluorobenzene	$C_6Cl_2F_4$	218.97	1198-62-5							
1401	Benzene, 1,2-dichloro-4-(trichloromethyl)-		$C_7H_3Cl_5$	264.36	13014-24-9	25.8		283.1	4-05-00-00824	1.5913^{20}		1.5886^{20}
1402	Benzene, 2,4-dichloro-1,3,5-trimethyl-	Mesitylene, 2,4-dichloro-	$C_9H_{10}Cl_2$	189.08	57386-83-1	59		243.5	3-05-00-00919		EtOH 3; eth 2; bz 2	
1403	Benzene, 1,2-diethenyl-	*o*-Divinylbenzene	$C_{10}H_{10}$	130.19	91-14-5			82^{14}	4-05-00-01540	0.9325^{22}	ace 3; bz 3	1.5767^{20}
1404	Benzene, 1,3-diethenyl-	*m*-Divinylbenzene	$C_{10}H_{10}$	130.19	108-57-6	-52.3		121^{76}; 52^{3}	3-05-00-01367	0.9294^{20}	ace 3; bz 3	1.5760^{20}
1405	Benzene, 1,4-diethenyl-	*p*-Divinylbenzene	$C_{10}H_{10}$	130.19	105-06-6	31		95^{18}; $34^{0.2}$	4-05-00-01541	0.913^{40}	ace 3; bz 3	1.5835^{25}
1406	Benzene, 1,2-diethoxy-		$C_{10}H_{14}O_2$	166.22	2050-46-6	44		219	4-06-00-05565	1.0075^{20}	EtOH 3; eth 4; ctc 3	1.5083^{25}
1407	Benzene, 1,3-diethoxy-		$C_{10}H_{14}O_2$	166.22	2049-73-2	12.4		235	4-06-00-05664	1.0170^{15}	H_2O 1; EtOH 3; eth 3	
1408	Benzene, 1,4-diethoxy-		$C_{10}H_{14}O_2$	166.22	122-95-2	72		246	4-06-00-05720		EtOH 4; eth 3; bz 3; ctc 3	
1409	Benzene, (diethoxymethoxy)-		$C_{11}H_{16}O_3$	196.25	14444-77-0			$103\text{-}4^{10}$	4-06-00-00610	1.014^{25}		1.4822^{20}
1410	Benzene, 1,4-diethoxy-2-methyl-	2,5-Diethoxytoluene	$C_{11}H_{16}O_2$	180.25	41901-72-8	24.5		248	3-06-00-04499	1.0134^{15}	bz 4; eth 4; EtOH 4	

No.	Name Synonym	Mol. Form. Mol. Wt.	CAS RN mp/°C	Merck No. bp/°C	Beil. Ref. den/g cm^{-3}	Solubility n_D
1411	Benzene, 1-(diethoxymethyl)-4-(1-methylethyl)-	$C_{14}H_{22}O_2$ 222.33	35364-90-0	 265; 142$^{18.5}$	2-07-00-00247 0.944^{19}	 1.484^{19}
1412	Benzene, (3,3-diethoxy-1-propynyl)-	$C_{13}H_{16}O_2$ 204.27	6142-95-6	 99-100^{2}	4-07-00-01173 0.991^{25}	 1.5170^{20}
1413	Benzene, 1,2-diethyl- *o*-Diethylbenzene	$C_{10}H_{14}$ 134.22	135-01-3 -31.2	 184	4-05-00-01065 0.8800^{20}	H_2O 1; EtOH 5; eth 5; ace 5 1.5035^{20}
1414	Benzene, 1,3-diethyl- *m*-Diethylbenzene	$C_{10}H_{14}$ 134.22	141-93-5 -83.9	 181.1	4-05-00-01066 0.8602^{20}	H_2O 1; EtOH 5; eth 5; ace 5 1.4955^{20}
1415	Benzene, 1,4-diethyl- *p*-Diethylbenzene	$C_{10}H_{14}$ 134.22	105-05-5 -42.83	 183.7	4-05-00-01067 0.8620^{20}	H_2O 1; EtOH 5; eth 5; ace 5 1.4967^{20}
1416	Benzene, 1,2-diethyl-3-methyl-	$C_{11}H_{16}$ 148.25	13632-93-4	 206.6	4-05-00-01105 0.8910^{20}	ace 4; bz 4; eth 4; EtOH 4 1.5105^{20}
1417	Benzene, 1,2-diethyl-4-methyl- Toluene, 3,4-diethyl-	$C_{11}H_{16}$ 148.25	13732-80-4	 203.6	4-05-00-01106 0.8762^{20}	ace 4; bz 4; eth 4; EtOH 4 1.5039^{20}
1418	Benzene, 1,3-diethyl-2-methyl- Toluene, 2,6-diethyl-	$C_{11}H_{16}$ 148.25	13632-95-6	 208.8	4-05-00-01105 0.8907^{20}	ace 4; bz 4; eth 4; EtOH 4 1.5106^{20}
1419	Benzene, 1,3-diethyl-5-methyl-	$C_{11}H_{16}$ 148.25	2050-24-0 -74.1	 205	4-05-00-01106 0.8748^{20}	H_2O 1; EtOH 5; eth 5; ace 5 1.5027^{20}
1420	Benzene, 1,4-diethyl-2-methyl-	$C_{11}H_{16}$ 148.25	13632-94-5	 207.1	4-05-00-01106 0.8758^{20}	ace 4; bz 4; eth 4; EtOH 4 1.5034^{20}
1421	Benzene, 2,4-diethyl-1-methyl-	$C_{11}H_{16}$ 148.25	1758-85-6	 205	4-05-00-01105 0.8748^{20}	ace 4; bz 4; eth 4; EtOH 4 1.5027^{20}
1422	Benzene, 1,2-difluoro- *o*-Difluorobenzene	$C_6H_4F_2$ 114.09	367-11-3 -34	 94	4-05-00-00637 1.1599^{18}	H_2O 1; ace 3; bz 3; chl 3 1.4451^{18}
1423	Benzene, 1,3-difluoro- *m*-Difluorobenzene	$C_6H_4F_2$ 114.09	372-18-9 -69.0	 82.6	4-05-00-00637 1.1572^{20}	H_2O 1; ace 3; bz 3 1.4374^{20}
1424	Benzene, 1,4-difluoro- *p*-Difluorobenzene	$C_6H_4F_2$ 114.09	540-36-3 -13	3132 89	4-05-00-00637 1.1701^{20}	H_2O 1; ace 3; bz 3; ctc 2 1.4422^{20}
1425	Benzene, (difluoromethyl)-	$C_7H_6F_2$ 128.12	455-31-2	 140	4-05-00-00801 1.1370^{19}	H_2O 1; EtOH 3 1.4577^{20}
1426	Benzene, 2,4-difluoro-1-nitro-	$C_6H_3F_2NO_2$ 159.09	446-35-5 9.8	 207	4-05-00-00720 1.4571^{14}	chl 2 1.5149^{14}
1427	Benzene, 1,2-diiodo-	$C_6H_4I_2$ 329.91	615-42-9 27	 287; 100^{3}	4-05-00-00700 2.54^{20}	H_2O 2; EtOH 2 1.7179^{20}
1428	Benzene, 1,3-diiodo-	$C_6H_4I_2$ 329.91	626-00-6 40.4	 285	4-05-00-00700 2.47^{25}	eth 4; EtOH 4; chl 4
1429	Benzene, 1,4-diiodo-	$C_6H_4I_2$ 329.91	624-38-4 131.5	 285	4-05-00-00700	H_2O 1; EtOH 3; eth 4; chl 2
1430	Benzene, 1,3-diisocyanato-2-methyl- Toluene-2,6-diisocyanate	$C_9H_6N_2O_2$ 174.16	91-08-7 18.3			H_2O 6; ace 3; bz 3
1431	Benzene, 2,4-diisocyanato-1-methyl- Toluene 2,4-diisocyanate	$C_9H_6N_2O_2$ 174.16	584-84-9 20.5	9456 251	4-13-00-00243 1.2244^{20}	ace 4; bz 4; eth 4
1432	Benzene, 1,4-diisothiocyanato- Bitoscanate	$C_8H_4N_2S_2$ 192.27	4044-65-9 132	1318	4-13-00-00174	
1433	1,3-Benzenedimethanamine *m*-Xylenediamine	$C_8H_{12}N_2$ 136.20	1477-55-0	 247	4-13-00-00293 1.052^{20}	H_2O 4; eth 4; EtOH 4
1434	1,3-Benzenedimethanol	$C_8H_{10}O_2$ 138.17	626-18-6 57	 156^{13}	4-06-00-05966 1.1610^{18}	H_2O 4; eth 4; EtOH 4
1435	1,4-Benzenedimethanol	$C_8H_{10}O_2$ 138.17	589-29-7 117.5	 138-43^{1}	4-06-00-05971	H_2O 4; ace 4; eth 4; EtOH 4
1436	Benzene, 1,2-dimethoxy- Veratrole	$C_8H_{10}O_2$ 138.17	91-16-7 22.5	9857 206	4-06-00-05564 1.0810^{25}	H_2O 2; EtOH 3; eth 3; ctc 3 1.5827^{21}
1437	Benzene, 1,3-dimethoxy-	$C_8H_{10}O_2$ 138.17	151-10-0 -52	 217.5	4-06-00-05663 1.0521^{25}	H_2O 2; EtOH 3; eth 3; bz 3 1.5231^{20}
1438	Benzene, 1,4-dimethoxy-	$C_8H_{10}O_2$ 138.17	150-78-7 59	 212.6	4-06-00-05718 1.0375^{55}	H_2O 2; EtOH 3; eth 4; bz 4
1439	Benzene, 1,2-dimethoxy-3-methyl-	$C_9H_{12}O_2$ 152.19	4463-33-6	 204	4-06-00-05860 1.0335^{20}	eth 4; EtOH 4 1.5121^{25}
1440	Benzene, 1,2-dimethoxy-4-methyl-	$C_9H_{12}O_2$ 152.19	494-99-5 24	 220	4-06-00-05879 1.0509^{25}	H_2O 1; ctc 2; os 4 1.5257^{25}
1441	Benzene, 1,3-dimethoxy-5-methyl-	$C_9H_{12}O_2$ 152.19	4179-19-5	 244	3-06-00-04533 1.0478^{15}	bz 4; eth 4; EtOH 4 1.5234^{20}
1442	Benzene, 1,4-dimethoxy-2-(1-methylethyl)-	$C_{11}H_{16}O_2$ 180.25	4132-71-2	 114.0^{16}	4-06-00-05990 [illegible]	bz 4; eth 4 [illegible]
1443	Benzene, 1-(dimethoxymethyl)-3-nitro-	$C_9H_{11}NO_4$ 197.19	3395-79-7	 162-4^{19}	4-07-00-00592 1.209^{15}	bz 4
1444	Benzene, 1-(dimethoxymethyl)-4-nitro-	$C_9H_{11}NO_4$ 197.19	881-67-4 24	 294	4-07-00-00600	bz 4
1445	Benzene, 1,2-dimethoxy-4-nitro-	$C_8H_9NO_4$ 183.16	709-09-1 98	 230^{15}	4-06-00-05627 1.1888^{133}	H_2O 1; EtOH 4; eth 4; chl 3
1446	Benzene, 1,3-dimethoxy-5-nitro-	$C_8H_9NO_4$ 183.16	16147-07-2 89		3-06-00-04347 1.1693^{133}	bz 4; EtOH 4
1447	Benzene, 1,4-dimethoxy-2-nitro-	$C_8H_9NO_4$ 183.16	89-39-4 72.5		4-06-00-05786 1.1666^{132}	H_2O 1; EtOH 3; bz 3; chl 3
1448	Benzene, 2,4-dimethoxy-1-nitro- 1,3-Dimethoxy-4-nitrobenzene	$C_8H_9NO_4$ 183.16	4920-84-7 76.5		4-06-00-05691 1.1876^{132}	H_2O 1; EtOH 3; os 3; lig 1
1449	Benzene, 1,2-dimethoxy-4-(1-propenyl)-	$C_{11}H_{14}O_2$ 178.23	93-16-3 18	 270.5	2-06-00-00918 1.0521^{20}	 1.5616^{20}

No.	Name	Synonym	Mol. Form.	Mol. Wt.	CAS RN	mp/°C	Merck No.	bp/°C	Beil. Ref.	den/g cm^{-3}	Solubility	n_D
1450	Benzene, 1,2-dimethoxy-4-(2-propenyl)-		$C_{11}H_{14}O_2$	178.23	93-15-2	-4		254.7	4-06-00-06337	1.0396^{20}	H_2O 1; EtOH 3; eth 3	1.5340^{20}
1451	Benzene, 1,2-dimethoxy-4-(1-propenyl)-, (*Z*)-		$C_{11}H_{14}O_2$	178.23	6380-24-1	70		270.5	4-06-00-06325	1.0521^{20}	H_2O 1; ace 3; bz 3	1.5616^{20}
1452	Benzene, 1,2-dimethyl-	*o*-Xylene	C_8H_{10}	106.17	95-47-6	-25.2	9988	144.5	4-05-00-00917	0.8802^{10}	H_2O 1; EtOH 5; eth 5; ace 5	1.5055^{20}
1453	Benzene, 1,3-dimethyl-	*m*-Xylene	C_8H_{10}	106.17	108-38-3	-47.8	9988	139.1	4-05-00-00932	0.8642^{20}	H_2O 1; EtOH 5; eth 5; ace 5	1.4972^{10}
1454	Benzene, 1,4-dimethyl-	*p*-Xylene	C_8H_{10}	106.17	106-42-3	13.2	9988	138.3	4-05-00-00951	0.8611^{20}	H_2O 1; EtOH 5; eth 5; ace 5	1.4958^{20}
1455	Benzene, (1,1-dimethylbutyl)-	Pentane, 2-methyl-2-phenyl	$C_{12}H_{18}$	162.27	1985-57-5			205.5	4-05-00-01118	0.8796^{10}		1.4955^{16}
1456	Benzene, 1,1'-(1,2-dimethyl-1,2-ethenediyl)bis-, (*E*)-		$C_{16}H_{16}$	208.30	782-06-9	107		156^{16}	4-05-00-02207	0.987^{20}	H_2O 1; EtOH 2; eth 3; MeOH 2	
1457	Benzene, 1,1'-(1,2-dimethyl-1,2-ethenediyl)bis-, (*Z*)-		$C_{16}H_{16}$	208.30	782-05-8	67.5			4-05-00-02207	1.004^{20}	H_2O 1; EtOH 3; HOAc 3; MeOH 3	1.5612^{78}
1458	Benzene, (1,1-dimethylethoxy)-		$C_{10}H_{14}O$	150.22	6669-13-2	-24		185.5	4-06-00-00559	0.9214^{20}		
1459	Benzene, (1,1-dimethylethyl)-	*tert*-Butylbenzene	$C_{10}H_{14}$	134.22	98-06-6	-57.8	1551	169.1	4-05-00-01045	0.8665^{20}	H_2O 1; EtOH 4; eth 4; ace 5	1.4927^{20}
1460	Benzene, 1-(1,1-dimethylethyl)-3,5-dimethyl-		$C_{12}H_{18}$	162.27	98-19-1	-18		207	4-05-00-01130	0.8668^{20}	ctc 3	
1461	Benzene, 1-(1,1-dimethylethyl)-3,5-dimethyl-2,4,6-trinitro-	2,4,6-Trinitro-3,5-dimethyl-*tert*-butylbenzene	$C_{12}H_{15}N_3O_6$	297.27	81-15-2	110			4-05-00-01132		H_2O 1; EtOH 2; eth 3; chl 3	
1462	Benzene, 1-(1,1-dimethylethyl)-4-methoxy-		$C_{11}H_{16}O$	164.25	5396-38-3			238	4-06-00-03297	0.9436^{20}		1.5039^{20}
1463	Benzene, 1-(1,1-dimethylethyl)-2-methoxy-4-methyl-3,5-dinitro-		$C_{12}H_{16}N_2O_5$	268.27	83-66-9	85		185^{16}	4-06-00-03402		H_2O 1; EtOH 2; eth 3; chl 3	
1464	Benzene, 1-(1,1-dimethylethyl)-2-methyl-		$C_{11}H_{16}$	148.25	1074-92-6	-50.3		200.4	4-05-00-01096	0.8897^{20}	ace 4; bz 4; eth 4; EtOH 4	1.5076^{20}
1465	Benzene, 1-(1,1-dimethylethyl)-3-methyl-		$C_{11}H_{16}$	148.25	1075-38-3	-41.4		189.3	4-05-00-01096	0.8657^{20}	ace 4; bz 4; eth 4; EtOH 4	1.4944^{20}
1466	Benzene, 1-(1,1-dimethylethyl)-4-methyl-	*p-tert*-Butyltoluene	$C_{11}H_{16}$	148.25	98-51-1	-52		190	4-05-00-01097	0.8612^{20}	H_2O 1; EtOH 2; eth 4; ace 3	1.4918^{20}
1467	Benzene, 1-(1,1-dimethylethyl)-3-nitro-		$C_{10}H_{13}NO_2$	179.22	23132-52-7	2		254; 136^{16}	4-05-00-01052	1.0643^{25}		
1468	Benzene, 1-(1,5-dimethyl-4-hexenyl)-4-methyl-	ar-Curcumene	$C_{15}H_{22}$	202.34	644-30-4			140^{19}	4-05-00-01465	0.8805^{20}	H_2O 1; bz 3	1.4989^{20}
1469	Benzene, 1,2-dimethyl-3-(1-methylethyl)-		$C_{11}H_{16}$	148.25	22539-65-7			202.6	4-05-00-01103	0.888^{20}	ace 4; bz 4; eth 4; EtOH 4	1.508^{20}
1470	Benzene, 1,2-dimethyl-4-(1-methylethyl)-		$C_{11}H_{16}$	148.25	4132-77-8			200	4-05-00-01103	0.8710^{20}		1.4951^{20}
1471	Benzene, 1,3-dimethyl-2-(1-methylethyl)-		$C_{11}H_{16}$	148.25	14411-75-7			199	4-05-00-01103	0.890^{20}	ace 4; bz 4; eth 4; EtOH 4	1.509^{20}
1472	Benzene, 2,4-dimethyl-1-(1-methylethyl)-		$C_{11}H_{16}$	148.25	4706-89-2	-82		197	4-05-00-01104	0.8734^{20}		1.4998^{25}
1473	Benzene, 2,4-dimethyl-1-(1-methylpropyl)-		$C_{12}H_{18}$	162.27	1483-60-9			84^{8}	4-05-00-01129	0.8664^{20}		1.4939^{25}
1474	Benzene, 1,2-dimethyl-3-nitro-		$C_8H_9NO_2$	151.17	83-41-0	15		240	4-05-00-00930	1.1402^{20}	H_2O 1; EtOH 3; ctc 3	1.5441^{20}
1475	Benzene, 1,2-dimethyl-4-nitro-	4-Nitro-*o*-xylene	$C_8H_9NO_2$	151.17	99-51-4	30.5		251; 143^{21}	4-05-00-00930	1.112^{15}	H_2O 1; EtOH 5	1.5202^{20}
1476	Benzene, 1,3-dimethyl-2-nitro-		$C_8H_9NO_2$	151.17	81-20-9	15		226	4-05-00-00948	1.112^{15}	H_2O 1; EtOH 4; ctc 3	1.5202^{20}
1477	Benzene, 1,3-dimethyl-5-nitro-		$C_8H_9NO_2$	151.17	99-12-7	75		274	4-05-00-00948		H_2O 1; EtOH 4; eth 4	
1478	Benzene, 1,4-dimethyl-2-nitro-		$C_8H_9NO_2$	151.17	89-58-7	-25		240.5	4-05-00-00971	1.132^{15}	H_2O 1; EtOH 3	1.5413^{20}
1479	Benzene, 2,4-dimethyl-1-nitro-		$C_8H_9NO_2$	151.17	89-87-2	9		247; 122^{18}	4-05-00-00948	1.135^{15}	H_2O 1; eth 3; ace 3; bz 3	1.5473^{25}
1480	Benzene, (1,1-dimethylpropyl)-		$C_{11}H_{16}$	148.25	2049-95-8			192.4	4-05-00-01090	0.8748^{20}		1.4958^{20}
1481	Benzene, (1,2-dimethylpropyl)-		$C_{11}H_{16}$	148.25	4481-30-5			187	4-05-00-01092	0.8672^{16}		1.4972^{16}
1482	Benzene, 1,3-dimethyl-2-propyl-		$C_{11}H_{16}$	148.25	17059-45-9			207.6	4-05-00-01102	0.8856^{20}	ace 4; bz 4; eth 4; EtOH 4	1.5063^{20}
1483	Benzene, 1,3-dimethyl-5-propyl-		$C_{11}H_{16}$	148.25	3982-64-7	-59.1		202.2	4-05-00-01102	0.8607^{20}	ace 4; bz 4; eth 4; EtOH 4	1.4952^{20}
1484	Benzene, (2,2-dimethylpropyl)-		$C_{11}H_{16}$	148.25	1007-26-7			186	4-05-00-01092	0.8581^{18}		1.4884^{18}
1485	Benzene, 1,2-dimethyl-3-propyl-	*o*-Xylene, 3-propyl	$C_{11}H_{16}$	148.25	17059-44-8			210.7	4-05-00-01101	0.8864^{20}	ace 4; bz 4; eth 4; EtOH 4	1.5075^{20}
1486	Benzene, 1,4-dimethyl-2-propyl-		$C_{11}H_{16}$	148.25	3042-50-0			204.3	4-05-00-01102	0.8717^{20}	ace 4; bz 4; eth 4; EtOH 4	1.4999^{20}
1487	Benzene, 2,4-dimethyl-1-propyl-	*m*-Xylene, 4-propyl-	$C_{11}H_{16}$	148.25	61827-85-8			206.6	4-05-00-01102	0.8723^{20}	ace 4; bz 4; eth 4; EtOH 4	1.4998^{20}

PHYSICAL CONSTANTS OF ORGANIC COMPOUNDS (continued)

No.	Name Synonym	Mol. Form. Mol. Wt.	CAS RN mp/°C	Merck No. bp/°C	Beil. Ref. den/g cm^{-3}	Solubility n_D
1488	Benzene, 1-(1,1-dimethylpropyl)-4-methoxy-	$C_{12}H_{18}O$ 178.27	2050-03-5	 238	4-06-00-03383 0.9436^{20}	 1.5039^{20}
1489	Benzene, 1,4-dimethyl-2,3,5-trinitro- *p*-Xylene, 2,3,5-trinitro-	$C_8H_7N_3O_6$ 241.16	602-27-7 139.5	 exp 410	3-05-00-00864 1.59^{19}	EtOH 4
1490	Benzene, 2,4-dimethyl-1,3,5-trinitro-	$C_8H_7N_3O_6$ 241.16	632-92-8 184		4-05-00-00950 1.604^{19}	H_2O 1; EtOH 2; eth 2; ace 3
1491	Benzene, 1,2-dinitro- *o*-Dinitrobenzene	$C_6H_4N_2O_4$ 168.11	528-29-0 118.5	 318; 194^{30}	4-05-00-00738 1.3119^{120}	H_2O 1; EtOH 3; bz 3; chl 3 1.565^{17}
1492	Benzene, 1,3-dinitro- *m*-Dinitrobenzene	$C_6H_4N_2O_4$ 168.11	99-65-0 90	 291; 167^{14}	4-05-00-00739 1.5751^{18}	H_2O 2; EtOH 4; eth 3; ace 4
1493	Benzene, 1,4-dinitro- *p*-Dinitrobenzene	$C_6H_4N_2O_4$ 168.11	100-25-4 174	 297; 183^{34}	4-05-00-00741 1.625^{18}	H_2O 1; EtOH 2; ace 3; bz 3
1494	Benzene, 1,1'-(1,2-dinitro-1,2-ethenediyl)bis-, (*Z*)-	$C_{14}H_{10}N_2O_4$ 270.24	1796-05-0 108.5	 dec	4-05-00-02171	H_2O 1; EtOH 3; eth 3; ace 4
1495	Benzene, 2,4-dinitro-1-phenoxy-	$C_{12}H_8N_2O_5$ 260.21	2486-07-9 71	 230^{27}	4-06-00-01375	eth 4; EtOH 4
1496	Benzene, 2,4-dinitro-1-(2-phenylethenyl)-, (*E*)- Stilbene, 2,4-dinitro (*trans*)	$C_{14}H_{10}N_2O_4$ 270.24	56456-42-9 144	 exp 412	4-05-00-02170	EtOH 1; bz 2; chl 3; CS_2 3
1497	1,2-Benzenediol Pyrocatechol	$C_6H_6O_2$ 110.11	120-80-9 105	8009 245	4-06-00-05557 1.1493^{22}	H_2O 4; bz 4; eth 4; EtOH 4 1.604
1498	1,3-Benzenediol Resorcinol	$C_6H_6O_2$ 110.11	108-46-3 111	8158 178^{16}	4-06-00-05658 1.2717^{25}	H_2O 3; EtOH 3; eth 3; bz 2
1499	1,4-Benzenediol *p*-Hydroquinone	$C_6H_6O_2$ 110.11	123-31-9 172.3	4738 287	4-06-00-05712 1.328^{15}	H_2O 3; EtOH 4; eth 3; ace 4
1500	1,2-Benzenediol, 4-(2-amino-1-hydroxyethyl)-, (*R*)- Norepinephrine	$C_8H_{11}NO_3$ 169.18	51-41-2 217 dec	6612		H_2O 2; EtOH 2; eth 2; alk 4
1501	1,4-Benzenediol, 2,5-bis(1,1-dimethylpropyl)- 2,5-Di-*tert*-pentylhydroquinone	$C_{16}H_{26}O_2$ 250.38	79-74-3 180	3300	3-06-00-04748	
1502	1,3-Benzenediol, 4-bromo-	$C_6H_5BrO_2$ 189.01	6626-15-9 103	 140^{12}	4-06-00-05687	H_2O 4; EtOH 2; eth 4; bz 2
1503	1,4-Benzenediol, 2-bromo-	$C_6H_5BrO_2$ 189.01	583-69-7 111.5	 sub	4-06-00-05780	H_2O 4; EtOH 4; eth 4; bz 4
1504	1,2-Benzenediol, 3-chloro-	$C_6H_5ClO_2$ 144.56	4018-65-9 48.5	 110^{11}	4-06-00-05613	lig 4
1505	1,2-Benzenediol, 4-chloro-	$C_6H_5ClO_2$ 144.56	2138-22-9 90.5	 $139^{10.5}$	4-06-00-05614	H_2O 4; ace 4; eth 4; EtOH 4
1506	1,3-Benzenediol, 5-chloro-	$C_6H_5ClO_2$ 144.56	52780-23-1 118	 sub	4-06-00-05684	H_2O 3; eth 3; ace 3; os 3
1507	1,4-Benzenediol, 2-chloro-	$C_6H_5ClO_2$ 144.56	615-67-8 108	 263	4-06-00-05767	H_2O 4; EtOH 3; eth 3; bz 4
1508	1,2-Benzenediol, diacetate	$C_{10}H_{10}O_4$ 194.19	635-67-6 64.5	 $142\text{-}3^{9}$	4-06-00-05582	H_2O 1; EtOH 4; eth 4; chl 4
1509	1,4 Benzenediol, diacetate	$C_{10}H_{10}O_4$ 194.19	1205-91-0 123.5		4-06-00-05741 0.8731^{25}	H_2O 3; EtOH 4; eth 4; chl 4
1510	1,3-Benzenediol, 4,6-dichloro-	$C_6H_4Cl_2O_2$ 179.00	137-19-9 113	 254	4-06-00-05685	H_2O 4; EtOH 4; eth 4; ace 4
1511	1,2-Benzenediol, 4,5-dimethyl-	$C_8H_{10}O_2$ 138.17	2785-74-2 87.5	 160^{26}	4-06-00-05950	H_2O 4; EtOH 4; eth 4; bz 2
1512	1,3-Benzenediol, 2,5-dimethyl-	$C_8H_{10}O_2$ 138.17	488-87-9 163	 278.5	4-06-00-05970	H_2O 3; EtOH 3; eth 3
1513	1,3-Benzenediol, 4,5-dimethyl-	$C_8H_{10}O_2$ 138.17	527-55-9 136	 284	4-06-00-05948 1.1994^{20}	H_2O 3; EtOH 4; eth 3; bz 2
1514	1,3-Benzenediol, 4,6-dimethyl-	$C_8H_{10}O_2$ 138.17	615-89-4 126.5	 276	4-06-00-05963	H_2O 3; EtOH 3; eth 3
1515	1,2-Benzenediol, 4,4'-(2,3-dimethyl-1,4-butanediyl)bis- Nordihydroguaiaretic acid	$C_{18}H_{22}O_4$ 302.37	500-38-9 185.5	6610	4-06-00-07771	H_2O 2; EtOH 3; eth 3; ace 3
1516	1,2-Benzenediol, 4-(1,1-dimethylethyl)-	$C_{10}H_{14}O_2$ 166.22	98-29-3 54.3	 285; 160^{22}	4-06-00-06014	tfa 3
1517	1,3-Benzenediol, 2,4-dinitro- 2,4-Dinitroresorcinol	$C_6H_4N_2O_6$ 200.11	519-44-8 147.5	3278	4-06-00-05696	H_2O 2; EtOH 2
1518	1,3-Benzenediol, 4-ethyl-	$C_8H_{10}O_2$ 138.17	2896-60-8 98.5	 160^{24}; 131^{15}	4-06-00-05924	H_2O 2; EtOH 2; eth 2
1519	1,3-Benzenediol, 4-hexyl- 4-Hexylresorcinol	$C_{12}H_{18}O_2$ 194.27	136-77-6 68	4633 334	4-06-00-06048	ace 4; eth 4; EtOH 4; chl 4
1520	1,4-Benzenediol, 2-(hydroxymethyl)- Gentisyl alcohol	$C_7H_8O_3$ 140.14	495-08-9 100	4292 sub 75	4-06-00-07380	H_2O 4; EtOH 4; chl 4
1521	1,2-Benzenediol, 4-[1-hydroxy-2-(methylamino)ethyl]-, (*R*)- Epinephrine	$C_9H_{13}NO_3$ 183.21	51-43-4 211.5	3569	4-13-00-02927	H_2O 2; EtOH 1; HOAc 3; min acid 3
1522	1,2-Benzenediol, 4-[1-hydroxy-2-[(1-methylethyl)amino]ethyl]- Ethanol, 1(3,4-dihydroxyphenyl)-2(isopropylamino)	$C_{11}H_{17}NO_3$ 211.26	7683-59-2 170.5	5105	3-13-00-02387	
1523	1,2-Benzenediol, 4-iodo- 4-Iodocatechol	$C_6H_5IO_2$ 236.01	76149-14-9 92	 sub	4-06-00-05625	ace 4; bz 4; eth 4; EtOH 4
1524	1,3-Benzenediol, 5-iodo- Resorcinol, 5-iodo	$C_6H_5IO_2$ 236.01	64339-43-1 92.5	 sub	2-06-00-00821	EtOH 4

No.	Name Synonym	Mol. Form. Mol. Wt.	CAS RN mp/°C	Merck No. bp/°C	Beil. Ref. den/g cm^{-3}	Solubility n_D
1525	1,2-Benzenediol, 3-methoxy-	$C_7H_8O_3$ 140.14	934-00-9 42.8	 163^{48}; 129^{10}	4-06-00-07329	chl 3
1526	1,3-Benzenediol, 5-methoxy-	$C_7H_8O_3$ 140.14	2174-64-3 80.3	 213^{16}	4-06-00-07362	H_2O 2; EtOH 4; eth 4; bz 2
1527	1,2-Benzenediol, 3-methyl-	$C_7H_8O_2$ 124.14	488-17-5 68	 241	4-06-00-05860	H_2O 3; EtOH 3; bz 3; chl 3
1528	1,2-Benzenediol, 4-methyl-	$C_7H_8O_2$ 124.14	452-86-8 65	 251	4-06-00-05878 1.1287^{74}	H_2O 3; EtOH 3; eth 3; ace 3 1.5425^{74}
1529	1,3-Benzenediol, 2-methyl-	$C_7H_8O_2$ 124.14	608-25-3 120	 264	4-06-00-05877	H_2O 4; bz 4; eth 4; EtOH 4
1530	1,3-Benzenediol, 4-methyl-	$C_7H_8O_2$ 124.14	496-73-1 106	 268.5	4-06-00-05864	H_2O 3; EtOH 3; eth 3; bz 2
1531	1,3-Benzenediol, 5-methyl- Orcinol	$C_7H_8O_2$ 124.14	504-15-4 107.5	6819 289.5	4-06-00-05892 1.290^{4}	H_2O 3; EtOH 3; eth 3; bz 3
1532	1,4-Benzenediol, 2-methyl-	$C_7H_8O_2$ 124.14	95-71-6 128	 163^{11}	4-06-00-05866	H_2O 4; EtOH 4; eth 4; ace 3
1533	1,2-Benzenediol, 4-[2-(methylamino)ethyl]- Deoxyepinephrine	$C_9H_{13}NO_2$ 167.21	501-15-5 188.5	2885	4-13-00-02606	
1534	1,3-Benzenediol, 4-(3-methylbutyl)-	$C_{11}H_{16}O_2$ 180.25	15116-17-3 67	 177^{7}	3-06-00-04700	bz 4; eth 4; EtOH 4
1535	1,2-Benzenediol, 4-(1-methylethyl)-	$C_9H_{12}O_2$ 152.19	2138-43-4 78	 271	3-06-00-04632	peth 3
1536	1,2-Benzenediol, 3-methyl-6-(1-methylethyl)-	$C_{10}H_{14}O_2$ 166.22	490-06-2 48	 267	4-06-00-06019	ace 4; eth 4
1537	1,3-Benzenediol, 2-[3-methyl-6-(1-methylethenyl)-2-cyclohexen-1-yl]-5-pentyl-, (1*R*-*trans*)- Cannabidiol	$C_{21}H_{30}O_2$ 314.47	13956-29-1 67	1750 $187\text{-}90^{2}$	3-06-00-05362 1.040^{40}	H_2O 1; EtOH 3; eth 3; bz 3 1.5404^{20}
1538	1,4-Benzenediol, 2-methyl-5-(1-methylethyl)-	$C_{10}H_{14}O_2$ 166.22	2217-60-9 148	 290	4-06-00-06019	eth 4; EtOH 4
1539	1,3-Benzenediol, 4-(2-methylpropyl)- Resorcinol, 4-isobutyl-	$C_{10}H_{14}O_2$ 166.22	18979-62-9 62.5	 166^{6}	3-06-00-04667	eth 4; EtOH 4
1540	1,2-Benzenediol, monoacetate	$C_8H_8O_3$ 152.15	2848-25-1 57.5	 189^{102}; 148^{25}	4-06-00-05581	H_2O 4; ace 4; EtOH 4; chl 4
1541	1,3-Benzenediol, 4-nitro-	$C_6H_5NO_4$ 155.11	3163-07-3 122	 178^{11}	4-06-00-05691	EtOH 3; eth 3; bz 3; ctc 2
1542	1,3-Benzenediol, 4-[(4-nitrophenyl)azo]- Magneson	$C_{12}H_9N_3O_4$ 259.22	74-39-5 200	5578	4-16-00-00266	H_2O 1; EtOH 2; bz 2; HOAc 2
1543	1,2-Benzenediol, 3-pentadecyl- 3-Pentadecylcatechol	$C_{21}H_{36}O_2$ 320.52	492-89-7 59.5	7061	4-06-00-06109	bz 4; eth 4; EtOH 4
1544	1,3-Benzenediol, 4-pentyl-	$C_{11}H_{16}O_2$ 180.25	533-24-4 72.5	 168^{6}	3-06-00-04693	bz 4; eth 4; EtOH 4
1545	1,3-Benzenediol, 5-(2-phenylethenyl)-, (*E*)- Pinosylvin	$C_{14}H_{12}O_2$ 212.25	22139-77-1 156	7419	3-06-00-05577	ace 4; bz 4; chl 4; HOAc 4
1546	1,4-Benzenediol, 2-(phenylmethyl)-	$C_{13}H_{12}O_2$ 200.24	1706-73-6 105.8	 230^{13}	3-06-00-05404	eth 4; EtOH 4
1547	1,2-Benzenediol, 3-(2-propenyl)-	$C_9H_{10}O_2$ 150.18	1125-74-2	 $141\text{-}5^{16}$	4-06-00-06334 1.1241^{20}	 1.5656^{20}
1548	1,2-Benzenediol, 4-propyl-	$C_9H_{12}O_2$ 152.19	2525-02-2 60	 177^{30}; $111^{0.2}$	3-06-00-04613 1.100^{18}	 1.4440^{18}
1549	1,3-Benzenediol, 4-propyl-	$C_9H_{12}O_2$ 152.19	18979-60-7 82.5	 172^{14}	4-06-00-05975	H_2O 3; EtOH 2; eth 3
1550	1,3-Benzenediol, 5-propyl-	$C_9H_{12}O_2$ 152.19	500-49-2 82.8	 169^{8}, 148^{3}	3-06-00-04622	ace 4; bz 4; eth 4; EtOH 4
1551	1,4-Benzenediol, 2,3,5,6-tetrabromo-	$C_6H_2Br_4O_2$ 425.70	2641-89-6 244		3-06-00-04440 3.023^{21}	eth 4; EtOH 4
1552	1,4-Benzenediol, 2,3,5,6-tetramethyl- Durohydroquinone	$C_{10}H_{14}O_2$ 166.22	527-18-4 233	3451	4-06-00-06028	EtOH 3; eth 2
1553	1,3-Benzenediol, 2,4,6-trimethyl-	$C_9H_{12}O_2$ 152.19	608-98-0 150.8	 274.5	4-06-00-06000	bz 4; eth 4; EtOH 4
1554	1,3-Benzenediol, 2,4,6-trinitro- Styphnic acid	$C_6H_3N_3O_8$ 245.11	82-71-3 175.5	8828 sub	4-06-00-05699	eth 4; EtOH 4
1555	Benzene, 1,2-dipropoxy-	$C_{12}H_{18}O_2$ 194.27	6280-98-4	 235.5	4-06-00-05566 0.9554^{33}	chl 2 1.4950^{27}
1556	Benzene, 1,3-dipropoxy- Resorcinol dipropyl ether	$C_{12}H_{18}O_2$ 194.27	56106-37-7	 253	3-06-00-04308 1.033^{20}	 1.5138^{83}
1557	Benzene, 1,4-dipropyl-	$C_{12}H_{18}$ 162.27	4815-57-0	 109^{23}	4-05-00-01124 0.8563^{19}	 1.4917^{19}
1558	1,3-Benzenedisulfonamide, 4-amino-6-chloro- Chloraminophenamide	$C_6H_8ClN_3O_4S_2$ 285.73	121-30-2 254.5	2067	4-14-00-02810	
1559	1,3-Benzenedisulfonamide, 4,5-dichloro- 1,3-Benzenedisulfonic acid, diamide,4,5-dichloro	$C_6H_6ClN_2O_4$ 205.58	120-97-8 228.7	3067	4-11-00-00555	
1560	1,3-Benzenedisulfonic acid, 4-hydroxy- Phenoldisulfonic acid	$C_6H_6O_7S_2$ 254.24	96-77-5 >100 dec	7207	2-11-00-00139	H_2O 4; EtOH 4
1561	1,2-Benzenedithiol	$C_6H_6S_2$ 142.25	17534-15-5 28.5	 238.5	4-06-00-05651	EtOH 4; eth 4; bz 4; AcOEt 3

No.	Name Synonym	Mol. Form. Mol. Wt.	CAS RN mp/°C	Merck No. bp/°C	Beil. Ref. den/g cm^{-3}	Solubility n_D
1562	1,3-Benzenedithiol	$C_6H_6S_2$ 142.25	626-04-0 27	 245	4-06-00-05705	bz 4; eth 4; EtOH 4
1563	1,3-Benzenedithiol, 4-chloro-	$C_6H_5ClS_2$ 176.69	58593-78-5	 $145\text{-}6^{13}$	4-06-00-05708 1.393^{25}	 1.6704^{20}
1564	1,2-Benzenedithiol, 4-methyl- Toluene-3,4-dithiol	$C_7H_8S_2$ 156.27	496-74-2 29	9457	4-06-00-05890	chl 3
1565	Benzene, docosyl- Docosane, 1-phenyl-	$C_{28}H_{50}$ 386.71	5634-22-0 51	 448	3-05-00-01142 0.8544^{20}	
1566	Benzene, dodecyl- 1-Phenyldodecane	$C_{18}H_{30}$ 246.44	123-01-3 3	 328	4-05-00-01200 0.8551^{20}	H_2O 1 1.4824^{20}
1567	Benzene, 1,1'-dodecylidenebis[4-methyl- Dodecane, 1,1-di-(4-tolyl)	$C_{26}H_{38}$ 350.59	55268-62-7	 $197^{1.0}$	3-05-00-01925 0.9117^{20}	 1.5001^{20}
1568	Benzene, eicosyl-	$C_{26}H_{46}$ 358.65	2398-68-7 44	 429	4-05-00-01222 0.8545^{20}	 1.4805^{20}
1569	Benzeneethanamine Phenylethylamine	$C_8H_{11}N$ 121.18	64-04-0	7186 197.5	4-12-00-02453 0.9580^{24}	H_2O 3; eth 4; eth 4; ctc 3 1.5290^{25}
1570	Benzeneethanamine, 4-bromo- Phenethylamine, 4-bromo	$C_8H_{10}BrN$ 200.08	73918-56-6	 $63\text{-}72^{0.2}$	4-12-00-02492 1.29^{25}	 1.5750^{20}
1571	Benzeneethanamine, 3,4-dimethoxy-	$C_{10}H_{15}NO_2$ 181.23	120-20-7	 $163\text{-}5^{14}$	4-13-00-02604	ctc 3 1.5464^{20}
1572	Benzeneethanamine, α,α-dimethyl- Phentermine	$C_{10}H_{15}N$ 149.24	122-09-8	7232 205; 100^{21}	4-12-00-02820	
1573	Benzeneethanamine, *N*,α-dimethyl-, (*S*)- Methamphetamine	$C_{10}H_{15}N$ 149.24	537-46-2	5859	4-12-00-02589	
1574	Benzeneethanamine, *N*,β-dimethyl- Phenylpropylmethylamine	$C_{10}H_{15}N$ 149.24	93-88-9	7280 207.5	3-12-00-02689 0.915^{25}	bz 4; eth 4; EtOH 4
1575	Benzeneethanamine, *N*,α-dimethyl-*N*-(phenylmethyl)-, (+)- Benzphetamine	$C_{17}H_{21}N$ 239.36	156-08-1	1130 $127^{0.02}$	4-12-00-02594	eth 4; EtOH 4; MeOH 4; chl 4 1.5515^{19}
1576	Benzeneethanamine, *N*-ethyl-α-methyl- *N*-Ethylamphetamine	$C_{11}H_{17}N$ 163.26	457-87-4	3720 $105\text{-}6^{14}$	3-12-00-02668	 1.4986^{25}
1577	Benzeneethanamine, 2-methoxy- Phenethylamine, 2-methoxy	$C_9H_{13}NO$ 151.21	2045-79-6	 236.5	3-13-00-01625 1.0400^{25}	 1.5422^{20}
1578	Benzeneethanamine, α-methyl-, (±)- Amphetamine	$C_9H_{13}N$ 135.21	300-62-9	616 203	4-12-00-02587 0.9306^{25}	H_2O 2; EtOH 3; eth 2; chl 3 1.518^{26}
1579	Benzeneethanamine, β-methyl-	$C_9H_{13}N$ 135.21	582-22-9	 210	4-12-00-02634 0.9433^{4}	bz 4; eth 4; EtOH 4 1.5255^{20}
1580	Benzeneethanamine, 4-methyl-	$C_9H_{13}N$ 135.21	3261-62-9	 214	4-12-00-02641 0.93^{25}	 1.5257^{20}
1581	Benzeneethanamine, *N*-methyl-	$C_9H_{13}N$ 135.21	589-08-2	 206	4-12-00-02453 0.93^{25}	 1.5162^{20}
1582	Benzeneethanamine, α-methyl-, (*S*)-, sulfate (2:1) Dextroamphetamine sulfate	$C_{18}H_{28}N_2O_4S$ 368.50	51-63-8 >300	2932	3-12-00-02665 1.15^{25}	H_2O 4
1583	Benzeneethanamine, α-methyl-*N*-(2,2,2-trichloroethylidene)- Amphecloral	$C_{11}H_{12}Cl_3N$ 264.58	5581-35-1	613 $96^{0.5}$		 1.530
1584	Benzeneethanamine, α-phenyl-	$C_{14}H_{15}N$ 197.28	25611-78-3	 311; 175^{15}	4-12-00-03313 1.031^{15}	eth 4; EtOH 4
1585	Benzeneethanamine, *N*-(2-phenylethyl)-	$C_{16}H_{19}N$ 225.33	6308-98-1 29	 360; 190^{15}	4-12-00-02459	eth 4; EtOH 4 1.5550^{25}
1586	Benzeneethanamine, 3,4,5-trimethoxy- Mescaline	$C_{11}H_{17}NO_3$ 211.26	54-04-6 35.5	5808 180^{12}	4-13-00-02919	H_2O 3; EtOH 3; eth 1; bz 3
1587	Benzene, 1,1'-(1,2-ethanediyl)bis- Dibenzyl	$C_{14}H_{14}$ 182.27	103-29-7 52.5	1219 284	4-05-00-01868 0.9780^{25}	H_2O 1; EtOH 3; eth 3; CS_2 3 1.5476^{60}
1588	Benzene, 1,1'-[1,2-ethanediylbis(oxy)]bis-	$C_{14}H_{14}O_2$ 214.26	104-66-5 98	 182^{12}	4-06-00-00573	H_2O 1; EtOH 2; eth 3; chl 3
1589	Benzene, 1,1'-(1,2-ethanediyl)bis[3-methyl-	$C_{16}H_{18}$ 210.32	4662-96-8	 298	4-05-00-01943 0.9703^{22}	 1.5566^{22}
1590	Benzene, 1,1'-(1,2-ethanediyl)bis[4-methyl- 1,2-Di-*p*-tolylethane	$C_{16}H_{18}$ 210.32	538-39-6 85	3386 178^{18}	4-05-00-01943	H_2O 1; EtOH 2; bz 3; peth 3
1591	Benzene, 1,1',1'',1'''-(1,2-ethanediylidene)tetrakis-	$C_{26}H_{22}$ 334.46	632-50-8 214.5	 360	4-05-00-02746	EtOH 2; bz 3; HOAc 3
1592	Benzeneethanol Phenethyl alcohol	$C_8H_{10}O$ 122.17	60-12-8 -27	7183 218.2	4-06-00-03067 1.0202^{20}	H_2O 2; EtOH 5; eth 5 1.5325^{20}
1593	Benzeneethanol, β-amino-	$C_8H_{11}NO$ 137.18	7568-92-5 58.3	 261; 160^{17}	4-13-00-01835	H_2O 4; EtOH 4
1594	Benzeneethanol, 2-amino-	$C_8H_{11}NO$ 137.18	5339-85-5	 $147\text{-}8^{3.5}$	3-13-00-01679 1.045^{25}	H_2O 4 1.5849^{20}
1595	Benzeneethanol, 4-bromo-	C_8H_9BrO 201.06	4654-39-1	 138^{9}	4-06-00-03082 1.436^{25}	 1.5735^{20}
1596	Benzeneethanol, 3-chloro-	C_8H_9ClO 156.61	5182-44-5	 $133\text{-}7^{13}$	4-06-00-03079 1.181^{25}	 1.5491^{20}
1597	Benzeneethanol, 4-chloro-	C_8H_9ClO 156.61	1875-88-3	 259	4-06-00-03079 1.1804^{20}	 1.5487^{20}
1598	Benzeneethanol, 4-chloro-α,α-dimethyl-	$C_{10}H_{13}ClO$ 184.67	5468-97-3 34	 126^{12}	4-06-00-03291	 1.5310^{20}
1599	Benzeneethanol, 2,5-dimethoxy-	$C_{10}H_{14}O_3$ 182.22	7417-19-8	 176^{18}	4-06-00-07393 1.055^{25}	 1.5395^{20}

No.	Name Synonym	Mol. Form. Mol. Wt.	CAS RN mp/°C	Merck No. bp/°C	Beil. Ref. den/g cm^{-3}	Solubility n_D
1600	Benzeneethanol, α,α-dimethyl-	$C_{10}H_{14}O$ 150.22	100-86-7 24	215	4-06-00-03290 0.9787^{16}	1.5173^{16}
1601	Benzeneethanol, α,α-diphenyl-	$C_{20}H_{18}O$ 274.36	4428-13-1 89.5	222^{11}	4-06-00-05057	H_2O 1; EtOH 4; eth 2; chl 2
1602	Benzeneethanol, β-ethyl-	$C_{10}H_{14}O$ 150.22	2035-94-1	125^{25}; 81^{2}	2-06-00-00488 0.989^{16}	1.43^{16}
1603	Benzeneethanol, 2-ethyl- Phenethyl alcohol, 2-ethyl	$C_{10}H_{14}O$ 150.22	22545-12-6	249.7; 127^{20}	4-06-00-03351 0.9972^{20}	1.5304^{20}
1604	Benzeneethanol, 4-ethyl- Phenethyl alcohol, 4-ethyl	$C_{10}H_{14}O$ 150.22	22545-13-7 7.9	250	4-06-00-03355 0.9907^{20}	1.5229^{20}
1605	Benzeneethanol, α-ethyl-α-methyl-, (±)- 2-Butanol, 3-chloro (*erythro, DL*)	C_4H_9ClO 108.57	116783-12-1	235; 56.1^{30}	3-06-00-01958 1.0610^{25}	eth 4; EtOH 4; chl 4 1.4397^{25}
1606	Benzeneethanol, 3-hydroxy-	$C_8H_{10}O_2$ 138.17	13398-94-2	$168\text{-}73^{4}$	4-06-00-05936 1.082^{25}	1.5643^{20}
1607	Benzeneethanol, α-methyl-	$C_9H_{12}O$ 136.19	698-87-3	125^{25}	4-06-00-03192 0.991^{20}	1.5190^{20}
1608	Benzeneethanol, β-methyl-	$C_9H_{12}O$ 136.19	1123-85-9	$105\text{-}6^{11}$	2-06-00-00477 0.975^{25}	H_2O 1; EtOH 3 1.5582^{2}
1609	Benzeneethanol, 2-methyl-	$C_9H_{12}O$ 136.19	19819-98-8 2	243.5	4-06-00-03234 1.016^{25}	1.5355^{20}
1610	Benzeneethanol, 4-methyl-	$C_9H_{12}O$ 136.19	699-02-5	244.5; 94^{6}	4-06-00-03244 1.0028^{20}	1.5267^{20}
1611	Benzeneethanol, β-(1-methylethyl)- 1-Butanol, 3-methyl-2-phenyl	$C_{11}H_{16}O$ 164.25	90499-41-5	130^{15}	4-06-00-03390 0.9694^{25}	ace 4; bz 4; eth 4; EtOH 4 1.5137^{20}
1612	Benzeneethanol, 2-nitro-	$C_8H_9NO_3$ 167.16	15121-84-3 2	267	3-06-00-01714 1.19^{25}	1.5637^{20}
1613	Benzeneethanol, α-phenyl-	$C_{14}H_{14}O$ 198.26	614-29-9 67	177^{15}	2-06-00-00637 1.0360^{70}	
1614	Benzeneethanol, α-phenyl-, (*S*)-	$C_{14}H_{14}O$ 198.26	5773-56-8 67.5	168^{10}	4-06-00-04701 1.0358^{70}	H_2O 1; EtOH 4; eth 4
1615	Benzeneethanol, β-phenyl- 2,2-Diphenylethanol	$C_{14}H_{14}O$ 198.26	1883-32-5 64.5	193^{20}; 144^{1}	4-06-00-04724	EtOH 2; eth 3; ace 3; chl 3
1616	Benzeneethanol, α,α,β-trimethyl- 2-Butanol, 2-methyl-3-phenyl	$C_{11}H_{16}O$ 164.25	3280-08-8 46	197; 106^{12}	4-06-00-03389 0.9794^{20}	ace 4; bz 4; eth 4; EtOH 4 1.5193^{20}
1617	Benzeneethanol, α,β,β-trimethyl-	$C_{11}H_{16}O$ 164.25	2977-31-3	197	4-06-00-03387	ace 4; bz 4; eth 4; EtOH 4 1.5161^{13}
1618	Benzene, 1,1',1'',1'''-(1-ethanyl-2-ylidyne)tetrakis- 1,1,1,2-Tetraphenylethane	$C_{26}H_{22}$ 334.46	2294-94-2 144.8	278^{21}	4-05-00-02747	bz 4
1619	Benzene, 1,1'-(1,2-ethenediyl)bis-, (*E*)- *trans*-Stilbene	$C_{14}H_{12}$ 180.25	103-30-0 123	8774 307; 166^{12}	4-05-00-02156 0.9707^{20}	H_2O 1; EtOH 2; eth 4; bz 4 1.6264^{17}
1620	Benzene, 1,1'-(1,2-ethenediyl)bis-, (*Z*)- *cis*-Stilbene	$C_{14}H_{12}$ 180.25	645-49-8 -5	8774 141^{12}	4-05-00-02155 1.0143^{20}	H_2O 1; EtOH 3; eth 3; ace 3 1.6130^{20}
1621	Benzene, 1,1'-(1,2-ethenediyl)bis[4-methoxy-	$C_{16}H_{16}O_2$ 240.30	4705-34-4 214.5	sub	4-06-00-06823	H_2O 1; ace 3; bz 3; HOAc 3
1622	Benzene, 1,1'-(1,2-ethenediyl)bis[2-methyl- 2,2'-Dimethylstilbene	$C_{16}H_{16}$ 208.30	10311-74-7 83	$176\text{-}80^{10}$	4-05-00-02208	eth 4; EtOH 4; MeOH 4
1623	Benzene, 1,1',1'',1'''-(1,2-ethenediylidene)tetrakis-	$C_{26}H_{20}$ 332.44	632-51-9 225	420	4-05-00-02780 1.155^{0}	H_2O 1; EtOH 2; eth 2; bz 4
1624	Benzene, ethenyl- Styrene	C_8H_8 104.15	100-42-5 -31	8830 145	4-05-00-01334 0.9060^{20}	H_2O 1; EtOH 3; eth 3; ace 3 1.5468^{20}
1625	Benzene, (1-ethenyl-3-butenyl)- 1,5-Hexadiene, 3-phenyl	$C_{12}H_{14}$ 158.24	1076-66-0	78^{8}	4-05-00-01555 0.8911^{25}	1.5141^{25}
1626	Benzene, 1-ethenyl-2,4-dimethyl-	$C_{10}H_{12}$ 132.21	2234-20-0	$79\text{-}80^{12}$	4-05-00-01386 0.905^{20}	1.539^{20}
1627	Benzene, 1-ethenyl-3,5-dimethyl-	$C_{10}H_{12}$ 132.21	5379-20-4	$57\text{-}8^{4}$	4-05-00-01386 0.894^{25}	1.5382^{20}
1628	Benzene, 2-ethenyl-1,4-dimethyl-	$C_{10}H_{12}$ 132.21	2039-89-6	69^{10}	4-05-00-01385 0.9072^{17}	1.5236^{17}
1629	Benzene, 4-ethenyl-1,2-dimethyl-	$C_{10}H_{12}$ 132.21	27831-13-6	$94\text{-}6^{20}$	4-05-00-01385 0.906^{25}	1.5463^{20}
1630	Benzene, 1-ethenyl-2-ethyl-	$C_{10}H_{12}$ 132.21	7564-63-8 -75.5	187.3; 68^{12}	3-05-00-01216 0.9017^{25}	1.5380^{20}
1631	Benzene, 1-ethenyl-3-ethyl-	$C_{10}H_{12}$ 132.21	7525-62-4 -101	190.0	3-05-00-01217 0.8945^{20}	1.5351^{20}
1632	Benzene, 1-ethenyl-4-ethyl-	$C_{10}H_{12}$ 132.21	3454-07-7 -49.7	192.3; 86^{20}	4-05-00-01384 0.8884^{25}	1.5376^{20}
1633	Benzene, 1-ethenyl-2-fluoro-	C_8H_7F 122.14	394-46-7	46^{32}	4-05-00-01342 1.0282^{20}	bz 4; eth 4; EtOH 4 1.5200^{20}
1634	Benzene, 1-ethenyl-3-fluoro-	C_8H_7F 122.14	350-51-6	$30\text{-}1^{4}$	4-05-00-01343 1.0177^{20}	H_2O 1; EtOH 3; eth 3; bz 3 1.5170^{20}
1635	Benzene, 1-ethenyl-4-fluoro-	C_8H_7F 122.14	405-99-2 -34.5	67.4^{50}; 30^{4}	4-05-00-01343 1.0220^{20}	H_2O 1; EtOH 3; eth 3; bz 3 1.5150^{20}
1636	Benzene, 1,1'-ethenylidenebis- 1,1-Diphenylethene	$C_{14}H_{12}$ 180.25	530-48-3 8.2	3323 277	4-05-00-02173 1.0232^{20}	H_2O 1; eth 3; chl 3 1.6085^{20}
1637	Benzene, 1-ethenyl-2-methoxy-	$C_9H_{10}O$ 134.18	612-15-7 29	197; 83^{12}	4-06-00-03771 1.0049^{17}	ace 4; bz 4; eth 4; EtOH 4 1.5388^{20}
1638	Benzene, 1-ethenyl-3-methoxy-	$C_9H_{10}O$ 134.18	626-20-0	$90\text{-}3^{15}$	4-06-00-03773 0.999^{16}	H_2O 1; EtOH 3; eth 3; bz 3 1.5586^{23}

No.	Name Synonym	Mol. Form. Mol. Wt.	CAS RN mp/°C	Merck No. bp/°C	Beil. Ref. den/g cm^{-3}	Solubility n_D
1639	Benzene, 1-ethenyl-4-methoxy-	$C_9H_{10}O$ 134.18	637-69-4	 205; 91^{13}	4-06-00-03775 1.0001^{13}	H_2O 1; EtOH 3; eth 3; bz 3 1.5642^{13}
1640	Benzene, 1-ethenyl-2-methyl- *o*-Methylstyrene	C_9H_{10} 118.18	611-15-4 -68.5	 169.8	4-05-00-01367 0.9077^{25}	H_2O 1; bz 3; chl 3 1.5437^{20}
1641	Benzene, 1-ethenyl-3-methyl- *m*-Methylstyrene	C_9H_{10} 118.18	100-80-1 -86.3	 164	4-05-00-01367 0.9076^{25}	H_2O 1; EtOH 3; eth 3; bz 3 1.5411^{20}
1642	Benzene, 1-ethenyl-4-methyl- *p*-Methylstyrene	C_9H_{10} 118.18	622-97-9 -34.1	 172.8	4-05-00-01369 0.9173^{25}	H_2O 1; bz 3 1.5420
1643	Benzene, 1-ethenyl-4-(1-methylethyl)-	$C_{11}H_{14}$ 146.23	2055-40-5 -44.7	 204.1	4-05-00-01408 0.8850^{20}	ace 4; bz 4; eth 4; EtOH 4 1.5289^{20}
1644	Benzene, 1-ethenyl-3-nitro-	$C_8H_7NO_2$ 149.15	586-39-0 -10	 $120\text{-}1^{11}$	4-05-00-01351 1.1552^{32}	H_2O 1; EtOH 3; eth 3; bz 3 1.5836^{20}
1645	Benzene, 1-ethenyl-4-nitro-	$C_8H_7NO_2$ 149.15	100-13-0 29	 dec	4-05-00-01351	EtOH 4; eth 4; chl 3; HOAc 3
1646	Benzene, (ethenyloxy)-	C_8H_8O 120.15	766-94-9	 155.5	4-06-00-00561 0.9770^{20}	H_2O 1; eth 4 1.5224^{20}
1647	Benzene, 2-ethenyl-1,3,5-trimethyl-	$C_{11}H_{14}$ 146.23	769-25-5 -37	 209	4-05-00-01408 0.9057^{20}	ctc 3 1.5296^{20}
1648	Benzene, 1,1',1''-(1-ethenyl-2-ylidene)tris-	$C_{20}H_{16}$ 256.35	58-72-0 72.5	 $220\text{-}1^{14}$	4-05-00-02575 1.0373^{78}	H_2O 1; EtOH 3; eth 4; chl 3 1.6292^{78}
1649	Benzene, ethoxy- Phenetole	$C_8H_{10}O$ 122.17	103-73-1 -29.5	7189 169.8	4-06-00-00554 0.9651^{20}	H_2O 1; EtOH 3; eth 3; ctc 3 1.5076^{20}
1650	Benzene, 2-ethoxy-1,3-dinitro- Phenetole, 2,6-dinitro-	$C_8H_8N_2O_5$ 212.16	13027-43-5 60.5	 137^{3}	3 06 00 00868	eth 4
1651	Benzene, (2-ethoxyethenyl)-	$C_{10}H_{12}O$ 148.20	17655-74-2 0	 224.5	 0.979^{20}	 1.5496^{20}
1652	Benzene, (2-ethoxyethoxy)- Ethane, 1-ethoxy-2-phenoxy-	$C_{10}H_{14}O_2$ 166.22	19594-02-6	 230	4-06-00-00572 1.018^{11}	H_2O 1
1653	Benzene, 1-ethoxy-4-ethyl-	$C_{10}H_{14}O$ 150.22	1585-06-4	 210	3-06-00-01665 0.9385^{17}	ace 4; bz 4; EtOH 4
1654	Benzene, 1-ethoxy-2-fluoro-	C_8H_9FO 140.16	451-80-9 -16.7	 171.4	4-06-00-00771 1.0874^{17}	H_2O 1; ace 3; bz 3; ctc 2 1.4932^{17}
1655	Benzene, 1-ethoxy-3-fluoro-	C_8H_9FO 140.16	458-03-7 -27.5	 172; 65.2^{15}	3-06-00-00669 1.0716^{16}	H_2O 1; bz 3 1.4847^{17}
1656	Benzene, 1-ethoxy-4-fluoro-	C_8H_9FO 140.16	459-26-7 -8.5	 173; 54^{7}	4-06-00-00774 1.0715^{18}	H_2O 1; bz 3; ctc 2; chl 3 1.4826^{18}
1657	Benzene, 1-ethoxy-4-iodo-	C_8H_9IO 248.06	699-08-1 29	 250	4-06-00-01077	bz 4; eth 4; EtOH 4; chl 4
1658	Benzene, (ethoxymethyl)- Benzyl ethyl ether	$C_9H_{12}O$ 136.19	539-30-0	1147 186	4-06-00-02229 0.9478^{20}	H_2O 1; EtOH 5; eth 5 1.4955^{20}
1659	Benzene, 1-ethoxy-2-methyl-	$C_9H_{12}O$ 136.19	614-71-1	 184	4-06-00-01944 0.9592^{13}	eth 4; EtOH 4 1.508^{13}
1660	Benzene, 1-ethoxy-3-methyl-	$C_9H_{12}O$ 136.19	621-32-9	 192	4-06-00-02039 0.949^{20}	H_2O 1; EtOH 3; eth 3 1.513^{20}
1661	Benzene, 1-ethoxy-4-methyl-	$C_9H_{12}O$ 136.19	622-60-6	 188.5	4-06-00-02099 0.9509^{18}	H_2O 1; EtOH 3; eth 3; ctc 2 1.5058^{18}
1662	Benzene, 1-ethoxy-2-nitro-	$C_8H_9NO_3$ 167.16	610-67-3 2.1	 267	4-06-00-01250 1.1903^{15}	eth 4; EtOH 4 1.5425^{20}
1663	Benzene, 1-ethoxy-3-nitro-	$C_8H_9NO_3$ 167.16	621-52-3 36	 264; 169^{70}	4-06-00-01271	H_2O 1; EtOH 3; eth 3
1664	Benzene, 1-ethoxy-4-nitro-	$C_8H_9NO_3$ 167.16	100-29-8 60	 283	4-06-00-01283 1.1176^{100}	H_2O 2; EtOH 2; eth 4; ace 5
1665	Benzene, 1-ethoxy-2-(1-propenyl)- Phenetole, 2-propenyl	$C_{11}H_{14}O$ 162.23	67191-37-1	 231; 120^{17}	0-06-00-00565 0.9731^{24}	 1.544^{24}
1666	Benzene, (3-ethoxy-1-propenyl)-	$C_{11}H_{14}O$ 162.23	1476-07-9	 $127\text{-}8^{22}$	3-06-00-02404 0.970^{15}	 1.547^{15}
1667	Benzene, 1-ethoxy-2-propyl- Phenetole, 2-propyl	$C_{11}H_{16}O$ 164.25	101144-90-5	 213; 100^{16}	0-06-00-00499 0.9240^{20}	 1.494^{20}
1668	Benzene, ethyl- Ethylbenzene	C_8H_{10} 106.17	100-41-4 -94.9	3723 136.1	4-05-00-00885 0.8670^{20}	H_2O 1; EtOH 5; eth 5; chl 2 1.4959^{20}
1669	Benzene, (1-ethylbutyl)-	$C_{12}H_{18}$ 162.27	4468-42-2 -55.4	 209	4-05-00-01117 0.8239^{25}	 1.4859^{20}
1670	Benzene, 1-ethyl-2,3-dimethyl- 3-Ethyl-*o*-xylene	$C_{10}H_{14}$ 134.22	933-98-2 -49.5	 194	4-05-00-01069 0.8881^{25}	 1.5117^{20}
1671	Benzene, 1-ethyl-2,4-dimethyl-	$C_{10}H_{14}$ 134.22	874-41-9 -62.9	 188.4	4-05-00-01070 0.8763^{20}	ace 4; bz 4; eth 4; EtOH 4 1.5038^{20}
1672	Benzene, 1-ethyl-3,5-dimethyl-	$C_{10}H_{14}$ 134.22	934-74-7 -84.3	 183.6	4-05-00-01071 0.8608^{25}	H_2O 1; EtOH 5; eth 5; ace 5 1.4981^{20}
1673	Benzene, 2-ethyl-1,3-dimethyl-	$C_{10}H_{14}$ 134.22	2870-04-4 -16.2	 190	4-05-00-01069 0.8864^{25}	 1.5107^{20}
1674	Benzene, 2-ethyl-1,4-dimethyl-	$C_{10}H_{14}$ 134.22	1758-88-9 -53.7	 186.9	4-05-00-01070 0.8732^{25}	H_2O 1; EtOH 5; eth 5; ace 5 1.5043^{20}
1675	Benzene, 4-ethyl-1,2-dimethyl-	$C_{10}H_{14}$ 134.22	934-80-5 -66.9	 189.5	4-05-00-01069 0.8706^{25}	H_2O 1; EtOH 5; eth 5; ace 5 1.5031^{20}
1676	Benzene, 1,1'-(1-ethyl-1,2-ethanediyl)bis-	$C_{16}H_{18}$ 210.32	5223-59-6	 290; 152^{11}	4-05-00-01939 0.9777^{20}	H_2O 1; EtOH 3; eth 3; bz 3 1.5554^{20}
1677	Benzene, 1,1'-ethylidenebis- 1,1-Diphenylethane	$C_{14}H_{14}$ 182.27	612-00-0 -17.9	 272.6	4-05-00-01880 0.9997^{20}	H_2O 1; EtOH 5; eth 5; bz 3 1.5756^{20}
1678	Benzene, 1,1'-ethylidenebis[4-chloro- 1,1-Bis(4-chlorophenyl)ethane	$C_{14}H_{12}Cl_2$ 251.15	3547-04-4 56	 320	4-05-00-01881	

PHYSICAL CONSTANTS OF ORGANIC COMPOUNDS (continued)

No.	Name Synonym	Mol. Form. Mol. Wt.	CAS RN mp/°C	Merck No. bp/°C	Beil. Ref. den/g cm^{-3}	Solubility n_D
1679	Benzene, 1,1',1''-ethylidynetris-	$C_{20}H_{18}$ 258.36	5271-39-6 95	 207^{18}	4-05-00-02510	H_2O 1; EtOH 2; eth 3; bz 3
1680	Benzene, 1-ethyl-2-iodo- *o*-Ethyliodobenzene	C_8H_9I 232.06	18282-40-1	 226	4-05-00-00910 1.6189^{10}	ace 4; bz 4 1.5941^{22}
1681	Benzene, 1-ethyl-4-iodo-	C_8H_9I 232.06	25309-64-2 -17	 209	3-05-00-00801 1.6095^{10}	ace 4; bz 4 1.5909^{22}
1682	Benzene, 1-ethyl-2-methoxy-	$C_9H_{12}O$ 136.19	14804-32-1	 187; 80^{14}	3-06-00-01656 0.9636^{19}	bz 4; eth 4 1.5142^{20}
1683	Benzene, 1-ethyl-3-methoxy-	$C_9H_{12}O$ 136.19	10568-38-4	 197; 74^{10}	3-06-00-01662 0.9575^{18}	H_2O 1; eth 4; bz 3; ctc 3 1.5102
1684	Benzene, 1-ethyl-4-methoxy-	$C_9H_{12}O$ 136.19	1515-95-3	 198	4-06-00-03021 0.9624^{15}	bz 4; eth 4 1.5120^{20}
1685	Benzene, 1-ethyl-2-methyl- *o*-Ethyltoluene	C_9H_{12} 120.19	611-14-3 -80.8	 165.2	4-05-00-00999 0.8807^{20}	H_2O 1; EtOH 5; eth 5; ace 5 1.5046^{20}
1686	Benzene, 1-ethyl-3-methyl- *m*-Ethyltoluene	C_9H_{12} 120.19	620-14-4 -95.5	 161.3	4-05-00-01001 0.8645^{20}	H_2O 1; EtOH 4; eth 4; ace 5 1.4966^{20}
1687	Benzene, 1-ethyl-4-methyl- *p*-Ethyltoluene	C_9H_{12} 120.19	622-96-8 -62.3	 162	4-05-00-01003 0.8614^{20}	H_2O 1; EtOH 4; eth 4; ace 5 1.4959^{20}
1688	Benzene, 1-ethyl-2-(1-methylethyl)- Benzene, 1-ethyl-2-isopropyl	$C_{11}H_{16}$ 148.25	18970-44-0	 193	4-05-00-01100 0.888^{20}	ace 4; bz 4; eth 4; EtOH 4 1.508^{20}
1689	Benzene, 1-ethyl-3-(1-methylethyl)- Benzene, 1-ethyl-3-isopropyl	$C_{11}H_{16}$ 148.25	4920-99-4 <-20	 192	3-05-00-01005 0.859^{20}	H_2O 1; EtOH 3; eth 3; bz 4 1.4921^{20}
1690	Benzene, 1-ethyl-4-(1-methylethyl)- Benzene, 1-ethyl-4-isopropyl	$C_{11}H_{16}$ 148.25	4218-48-8 <-20	 196.6	4-05-00-01100 0.8585^{20}	H_2O 1; eth 3; bz 4 1.4923^{20}
1691	Benzene, (1-ethyl-1-methylpropyl)- Pentane, 3-methyl-3-phenyl	$C_{12}H_{18}$ 162.27	1985-97-3	 205	4-05-00-01120 0.8773^{15}	 1.4972^{16}
1692	Benzene, 1-ethyl-2-nitro-	$C_8H_9NO_2$ 151.17	612-22-6 -12.3	 232.5	4-05-00-00911 1.1207^{20}	H_2O 1; EtOH 4; eth 4; ace 3 1.5356^{20}
1693	Benzene, 1-ethyl-3-nitro-	$C_8H_9NO_2$ 151.17	7369-50-8 -37.9	 242.5	4-05-00-00911 1.1345^{25}	ace 4; eth 4; EtOH 4
1694	Benzene, 1-ethyl-4-nitro-	$C_8H_9NO_2$ 151.17	100-12-9 -12.3	 245.5	4-05-00-00912 1.1192^{20}	H_2O 1; EtOH 4; eth 4; ace 3 1.5455^{20}
1695	Benzene, (1-ethyloctadecyl)- 3-Phenyleicosane	$C_{26}H_{46}$ 358.65	2400-02-4 29.3	 202^{1}	4-05-00-01222 0.8546^{20}	 1.4796^{20}
1696	Benzene, 1-ethyl-4-(phenylmethyl)-	$C_{15}H_{16}$ 196.29	620-85-9 -24	 297	4-05-00-01926 0.9777^{20}	EtOH 3; eth 3; chl 3 1.5616^{20}
1697	Benzene, 1,1'-(1-ethyl-1,3-propanediyl)bis- Pentane, 1,3-diphenyl-	$C_{17}H_{20}$ 224.35	838-45-9	 304; $113^{0.8}$	4-05-00-01963 0.9734^{21}	 1.553^{21}
1698	Benzene, 1,1'-(2-ethyl-1,3-propanediyl)bis- Propane, 1,1-dibenzyl	$C_{17}H_{20}$ 224.35	1520-45-2	 304.5	4-05-00-01964 0.9734^{21}	 1.553^{21}
1699	Benzene, (1-ethyl-1-propenyl)- 3-Phenyl-2-pentene	$C_{11}H_{14}$ 146.23	4701-36-4	 198; 92^{18}	4-05-00-01404 0.9173^{14}	 1.5266^{15}
1700	Benzene, (1-ethyl-2-propenyl)-	$C_{11}H_{14}$ 146.23	19947-22-9	 191.5; 71^{12}	3-05-00-01240 0.8458^{23}	 1.5030^{21}
1701	Benzene, (1-ethylpropyl)-	$C_{11}H_{16}$ 148.25	1196-58-3	 187.5	4-05-00-01090 0.8649^{20}	 1.4880^{20}
1702	Benzene, 1-ethyl-2-propyl-	$C_{11}H_{16}$ 148.25	16021-20-8	 203	4-05-00-01099 0.8744^{20}	 1.4992^{20}
1703	Benzene, 1-ethyl-4-propyl- 1-Ethyl-4-propylbenzene	$C_{11}H_{16}$ 148.25	20024-90-2	 205	4-05-00-01100 0.8594^{20}	ace 4; bz 4; eth 4; EtOH 4 1.4921^{21}
1704	Benzene, (ethylsulfonyl)-	$C_8H_{10}O_2S$ 170.23	599-70-2 42	 160^{12}	4-06-00-01469 1.1410^{20}	bz 4; eth 4; EtOH 4; chl 4
1705	Benzene, (ethylthio)-	$C_8H_{10}S$ 138.23	622-38-8	 205	4-06-00-01468 1.0211^{20}	EtOH 3 1.5670^{20}
1706	Benzene, 1-(ethylthio)-3-methyl- Ethyl *m*-tolyl sulfide	$C_9H_{12}S$ 152.26	34786-24-8	 219	3-06-00-01332 0.9987^{20}	 1.5590^{20}
1707	Benzene, 1-(ethylthio)-4-methyl-	$C_9H_{12}S$ 152.26	622-63-9	 220	4-06-00-02156 0.9996^{20}	 1.555^{20}
1708	Benzene, 1-ethyl-2,3,4-trimethyl- Benzene, 4-ethyl-1,2,3-trimethyl	$C_{11}H_{16}$ 148.25	61827-86-9	 220.4	4-05-00-01107 0.9019^{20}	ace 4; bz 4; eth 4; EtOH 4 1.5180^{20}
1709	Benzene, 1-ethyl-2,3,5-trimethyl- 1-Ethyl-2,3,5-trimethylbenzene	$C_{11}H_{16}$ 148.25	18262-85-6	 213	4-05-00-01107 0.8897^{20}	ace 4; bz 4; eth 4; EtOH 4 1.5118^{20}
1710	Benzene, 1-ethyl-2,4,5-trimethyl- 1-Ethyl-2,4,5-trimethylbenzene	$C_{11}H_{16}$ 148.25	17851-27-3 -13.5	 213	4-05-00-01108 0.883^{20}	ace 4; bz 4; eth 4; EtOH 4 1.5075^{20}
1711	Benzene, 2-ethyl-1,3,4-trimethyl- Benzene, 3-ethyl-1,2,4-trimethyl	$C_{11}H_{16}$ 148.25	61827-87-0	 216.6	4-05-00-01107 0.895^{20}	ace 4; bz 4; eth 4; EtOH 4 1.5133^{20}
1712	Benzene, 2-ethyl-1,3,5-trimethyl-	$C_{11}H_{16}$ 148.25	3982-67-0 -15.5	 212.4	4-05-00-01107 0.883^{20}	ace 4; bz 4; eth 4; EtOH 4 1.5074^{20}
1713	Benzene, 5-ethyl-1,2,3-trimethyl- 5-Ethyl-1,2,3-trimethylbenzene	$C_{11}H_{16}$ 148.25	31366-00-4	 215.8	4-05-00-01107 0.8863^{20}	ace 4; eth 4; EtOH 4 1.5101^{20}
1714	Benzene, 1,1'-(1,2-ethynediyl)bis- Diphenylacetylene	$C_{14}H_{10}$ 178.23	501-65-5 62.5	9428 300	4-05-00-02276 0.9657^{100}	H_2O 1; EtOH 2; eth 4; ctc 2
1715	Benzene, ethynyl- Ethynylbenzene	C_8H_6 102.14	536-74-3 -44.8	3817 143	4-05-00-01525 0.9300^{20}	H_2O 1; EtOH 5; eth 5; ace 3 1.5470^{20}
1716	Benzene, (ethynyloxy)-	C_8H_6O 118.14	4279-76-9 -36	 61-2^{25}	4-06-00-00565 1.0614^{20}	eth 4; EtOH 4 1.5125^{20}
1717	Benzene, fluoro- Fluorobenzene	C_6H_5F 96.10	462-06-6 -42.2	4099 84.7	4-05-00-00632 1.0225^{20}	bz 4; eth 4; EtOH 4; lig 4 1.4684^{30}
1718	Benzene, 1-fluoro-2,4-dinitro- 2,4-Dinitrophenyl fluoride	$C_6H_3FN_2O_4$ 186.10	70-34-8 25.8	4101 296	4-05-00-00742 1.4718^{54}	EtOH 3; chl 2 1.5690^{20}

No.	Name Synonym	Mol. Form. Mol. Wt.	CAS RN mp/°C	Merck No. bp/°C	Beil. Ref. den/g cm^{-3}	Solubility n_D
1719	Benzene, 1-fluoro-2-iodo-	C_6H_4FI 222.00	348-52-7 -41.5	 188.6	4-05-00-00693	ace 3; bz 3; chl 3 1.5910^{20}
1720	Benzene, 1-fluoro-4-iodo-	C_6H_4FI 222.00	352-34-1 -27	 183	4-05-00-00694 1.9523^{15}	H_2O 1; EtOH 3; eth 3; ace 3 1.5270^{22}
1721	Benzene, 1-fluoro-3-isothiocyanato-	C_7H_4FNS 153.18	404-72-8	 227	3-12-00-01276 1.27^{25}	 1.6186^{20}
1722	Benzene, 1-fluoro-2-methoxy-	C_7H_7FO 126.13	321-28-8 -39	 154.5	4-06-00-00771 1.5489^{17}	H_2O 1; eth 3; ctc 3 1.4969^{17}
1723	Benzene, 1-fluoro-3-methoxy-	C_7H_7FO 126.13	456-49-5 -35	 159; 51^{14}	4-06-00-00772 1.104^{25}	 1.4876^{20}
1724	Benzene, 1-fluoro-4-methoxy-	C_7H_7FO 126.13	459-60-9 -45	 157	4-06-00-00773 1.1781^{18}	eth 3 1.4886^{18}
1725	Benzene, (fluoromethyl)-	C_7H_7F 110.13	350-50-5 -35	 140; 40^{14}	4-05-00-00800 1.0228^{25}	ctc 3 1.4892^{25}
1726	Benzene, 1-fluoro-2-methyl- *o*-Fluorotoluene	C_7H_7F 110.13	95-52-3 -62	4108 115	4-05-00-00799 1.0041^{13}	H_2O 1; EtOH 4; eth 4 1.4704^{20}
1727	Benzene, 1-fluoro-3-methyl- *m*-Fluorotoluene	C_7H_7F 110.13	352-70-5 -87	4108 115	4-05-00-00799 0.9974^{20}	H_2O 1; EtOH 4; eth 4 1.4691^{20}
1728	Benzene, 1-fluoro-4-methyl- *p*-Fluorotoluene	C_7H_7F 110.13	352-32-9 -56	4108 116.6	4-05-00-00800 0.9975^{20}	H_2O 1; EtOH 4; eth 4 1.4699^{20}
1729	Benzene, 1-fluoro-4-methyl-2-nitro- Toluene, 4-fluoro-3-nitro	$C_7H_6FNO_2$ 155.13	446-11-7 26.5	 241	3-05-00-00743 1.2619^{28}	 1.5237^{28}
1730	Benzene, 2-fluoro-4-methyl-1-nitro- 3-Fluoro-4-nitrotoluene	$C_7H_6FNO_2$ 155.13	446-34-4 53.2	 97^{3}	4-05-00-00852 1.4380^{25}	
1731	Benzene, 4-fluoro-1-methyl-2-nitro-	$C_7H_6FNO_2$ 155.13	446-10-6 27	 213; 138^{83}	4-05-00-00851 1.2686^{20}	 1.5218^{20}
1732	Benzene, 4-fluoro-2-methyl-1-nitro-	$C_7H_6FNO_2$ 155.13	446-33-3 27.5	 217; $97\text{-}8^{10}$	4-05-00-00851 1.272^{28}	 1.5271^{20}
1733	Benzene, 1-fluoro-2-nitro- *o*-Fluoronitrobenzene	$C_6H_4FNO_2$ 141.10	1493-27-2 -6	 215 dec	4-05-00-00718 1.3285^{18}	eth 4; EtOH 4 1.5489^{17}
1734	Benzene, 1-fluoro-3-nitro- *m*-Fluoronitrobenzene	$C_6H_4FNO_2$ 141.10	402-67-5 41	 199; 86^{19}	4-05-00-00719 1.3254^{19}	H_2O 1; EtOH 3; eth 3; bz 2 1.5262^{15}
1735	Benzene, 1-fluoro-4-nitro- *p*-Fluoronitrobenzene	$C_6H_4FNO_2$ 141.10	350-46-9 21	 205	4-05-00-00719 1.3300^{20}	H_2O 1; EtOH 3; eth 3; ctc 2 1.5316^{20}
1736	Benzene, 1-fluoro-2-(trichloromethyl)-	$C_7H_4Cl_3F$ 213.47	488-98-2	 75^{5}	3-05-00-00701 1.453^{25}	 1.5432^{20}
1737	Benzene, 1-fluoro-2-(trifluoromethyl)-	$C_7H_4F_4$ 164.10	392-85-8	 114.5	4-05-00-00804 1.293^{25}	 1.4040^{25}
1738	Benzene, 1-fluoro-3-(trifluoromethyl)-	$C_7H_4F_4$ 164.10	401-80-9 -81.5	 101.5	4-05-00-00804 1.3021^{17}	
1739	Benzene, 1-fluoro-4-(trifluoromethyl)-	$C_7H_4F_4$ 164.10	402-44-8 -41.7	 103.5	4-05-00-00804 1.293^{25}	 1.4025^{20}
1740	Benzene, 2-fluoro-1,3,5-trimethyl-	$C_9H_{11}F$ 138.18	392-69-8 -36.7	 171.5	4-05-00-01025 0.9745^{25}	 1.4809^{25}
1741	Benzene, heneicosyl Heneicosylbenzene	$C_{27}H_{48}$ 372.68	40775-09-5 48	 439	4-05-00-01226 0.8545^{20}	 1.4804^{20}
1742	Benzene, heptacosyl- Heptacosylbenzene	$C_{33}H_{60}$ 456.84	61828-25-9 64	 397^{100}; 262^{1}	4-05-00-01243 0.8543^{20}	 1.4797^{20}
1743	Benzene, heptadecyl- Heptadecane, 1-phenyl-	$C_{23}H_{40}$ 316.57	14752-75-1 32	 397	3-05-00-01123 0.8546^{20}	 1.4810^{20}
1744	Benzene, 1,1'-(1-heptenylidene)bis- 1,1-Diphenyl-1-heptene	$C_{19}H_{22}$ 250.38	1530-20-7		4-05-00-02252 0.9673^{18}	EtOH 2 1.5648^{18}
1745	Benzene, heptyl-	$C_{13}H_{20}$ 176.30	1078-71-3 -48	 240; 109^{10}	4-05-00-01143 0.8567^{20}	H_2O 1; bz 3; chl 3 1.4865^{20}
1746	Benzene, 1,1'-heptylidenebis- 1,1-Diphenylheptane	$C_{19}H_{24}$ 252.40	1530-05-8 14	 333.5; 137^{1}	4-05-00-01992 0.9497^{20}	
1747	Benzene, (heptyloxy)-	$C_{13}H_{20}O$ 192.30	32395-96-3 -33.5	 267	4-06-00-00560 0.9170^{15}	ace 4; eth 4; EtOH 4 1.4912^{20}
1748	Benzenehexacarboxylic acid Mellitic acid	$C_{12}H_6O_{12}$ 342.17	517-60-2 286-8 dec	5706	4-09-00-03825	H_2O 4; EtOH 3; sulf 3
1749	Benzene, hexachloro- Hexachlorobenzene	C_6Cl_6 284.78	118-74-1 231.8	4600 325	4-05-00-00670 2.044^{23}	H_2O 1; EtOH 2; eth 3; bz 4 1.5691^{23}
1750	Benzene, hexacosyl- Hexacosane, 1-phenyl-	$C_{32}H_{58}$ 442.81	13024-80-1 62	 388^{100}; 250^{1}	4-05-00-01242 [illegible]	 [illegible]
1751	Benzene, hexadecyl-	$C_{22}H_{38}$ 302.54	1459-09-2 27	 385	4-05-00-01216 0.8547^{20}	H_2O 1; EtOH 2; eth 4; bz 4 1.4813^{20}
1752	Benzene, (hexadecyloxy)-	$C_{22}H_{38}O$ 318.54	35021-70-6 41.8	 200^{1}	3-06-00-00555 0.8434^{82}	 1.4556^{82}
1753	Benzene, 1,3-hexadienyl- 1-Phenyl-1,3-hexadiene	$C_{12}H_{14}$ 158.24	41635-77-2	 128^{16}	3-05-00-01378 0.9253^{12}	 1.6025^{12}
1754	Benzene, 1,5-hexadienyl- 1,5-Hexadiene, 1-phenyl-	$C_{12}H_{14}$ 158.24	1009-81-0	 $102\text{-}3^{8}$	4-05-00-01554 0.9005^{25}	 1.5421^{25}
1755	Benzene, 3,5-hexadienyl- 6-Phenyl-1,3-hexadiene	$C_{12}H_{14}$ 158.24	39669-95-9	 100.5^{11}	3-05-00-01378 0.9304^{13}	 1.5446^{13}
1756	Benzene, hexaethyl- Hexaethylbenzene	$C_{18}H_{30}$ 246.44	604-88-6 129	 298	4-05-00-01208 0.8305^{130}	H_2O 1; EtOH 3; eth 4; bz 4 1.4736^{130}
1757	Benzene, hexafluoro- Perfluorobenzene	C_6F_6 186.06	392-56-3 5.3	 80.2	4-05-00-00640 1.6184^{20}	 1.3777^{20}

No.	Name	Synonym	Mol. Form.	Mol. Wt.	CAS RN	mp/°C	Merck No.	bp/°C	Beil. Ref.	den/g cm^{-3}	Solubility	n_D
1758	Benzene, hexamethyl-	Mellitene	$C_{12}H_{18}$	162.27	87-85-4	166.5		263.4	4-05-00-01137	1.0630^{25}	H_2O 1; EtOH 3; eth 3; ace 3	
1759	Benzenehexanoic acid		$C_{12}H_{16}O_2$	192.26	5581-75-9	23		206^{30}	4-09-00-01895			1.5164^{21}
1760	Benzene, 2-hexenyl-	2-Hexene, 1-phenyl	$C_{12}H_{16}$	160.26	67590-77-6			108^{16}	0-05-00-00501	0.8898^{16}		1.5058^{16}
1761	Benzene, 3-hexenyl-		$C_{12}H_{16}$	160.26	35008-86-7			222	4-05-00-01418	0.8790^{20}		1.5039^{20}
1762	Benzene, 5-hexenyl-		$C_{12}H_{16}$	160.26	1588-44-9			$94\text{-}5^{10}$	3-05-00-01251	0.8839^{20}		1.5033^{20}
1763	Benzene, hexyl-		$C_{12}H_{18}$	162.27	1077-16-3	-61		226.1	4-05-00-01115	0.8575^{20}	H_2O 1; eth 5; bz 3; peth 3	1.4864^{20}
1764	Benzene, (1-hexylheptyl)-	7-Phenyltridecane	$C_{19}H_{32}$	260.46	2400-01-3			$183\text{-}4^{20}$	4-05-00-01210	0.8723^{20}		1.4931^{18}
1765	Benzene, (hexyloxy)-		$C_{12}H_{18}O$	178.27	1132-66-7	-19		240	4-06-00-00560	0.9174^{20}	H_2O 1; eth 3	1.4921^{20}
1766	Benzene, iodo-	Iodobenzene	C_6H_5I	204.01	591-50-4	-31.3	4922	188.4	4-05-00-00688	1.8308^{20}	H_2O 1; EtOH 3; eth 5; ace 5	1.6200^{20}
1767	Benzene, 1-iodo-2,3-dimethoxy-	3-Iodoveratrole	$C_8H_9IO_2$	264.06	25245-33-4	45.5		$144\text{-}5^{14}$	3-06-00-04262	1.7799^{20}		1.6127^{20}
1768	Benzene, 4-iodo-1,2-dimethoxy-		$C_8H_9IO_2$	264.06	5460-32-2	35		174^{30}	4-06-00-05625		bz 4; EtOH 4	
1769	Benzene, 1-iodo-2,3-dimethyl-		C_8H_9I	232.06	31599-60-7			229	3-05-00-00820	1.6395^{20}	ace 4	1.6074^{20}
1770	Benzene, 1-iodo-2,4-dimethyl-		C_8H_9I	232.06	4214-28-2			231 dec; 111^{14}	4-05-00-00947	1.6282^{16}	H_2O 1; ace 3; bz 3	1.6008^{16}
1771	Benzene, 1-iodo-3,5-dimethyl-	*m*-Xylene, 5-iodo-	C_8H_9I	232.06	22445-41-6			230.5	3-05-00-00840	1.6085^{18}	ace 4	1.5967^{18}
1772	Benzene, 2-iodo-1,3-dimethyl-		C_8H_9I	232.06	608-28-6	11.2		229.5	4-05-00-00947	1.6158^{20}	H_2O 1; ace 3; bz 3	1.6035^{20}
1773	Benzene, 2-iodo-1,4-dimethyl-		C_8H_9I	232.06	1122-42-5			227 dec	4-05-00-00970	1.6168^{17}	H_2O 1; ace 3; bz 3	1.5992^{17}
1774	Benzene, 4-iodo-1,2-dimethyl-	4-Iodo-*o*-xylene	C_8H_9I	232.06	31599-61-8			231.5	3-05-00-00821	1.6334^{18}	ace 4	1.6049^{18}
1775	Benzene, 1-iodo-2-methoxy-	*o*-Iodoanisole	C_7H_7IO	234.04	529-28-2		4921	241; 91^{2}	4-06-00-01070	1.8^{20}	EtOH 4; eth 4; ace 4; bz 4	
1776	Benzene, 1-iodo-4-methoxy-		C_7H_7IO	234.04	696-62-8	53		238; 138^{25}	4-06-00-01075		EtOH 3; eth 3; chl 3	
1777	Benzene, (iodomethyl)-		C_7H_7I	218.04	620-05-3	24.5		93^{10}	4-05-00-00842	1.7335^{25}	bz 4; eth 4; EtOH 4	1.6334^{25}
1778	Benzene, 1-iodo-2-methyl-		C_7H_7I	218.04	615-37-2			211.5	4-05-00-00838	1.713^{20}	H_2O 1; EtOH 5; eth 5	1.6079^{20}
1779	Benzene, 1-iodo-3-methyl-		C_7H_7I	218.04	625-95-6	-27.2		213	4-05-00-00839	1.705^{20}	H_2O 1; EtOH 5; eth 5	1.6053^{20}
1780	Benzene, 1-iodo-4-methyl-		C_7H_7I	218.04	624-31-7	36.5		211	4-05-00-00840	1.678^{20}	H_2O 1; EtOH 3; eth 3; chl 2	
1781	Benzene, 2-iodo-1-methyl-4-(1-methylethyl)-	2-Iodo-*p*-cymene	$C_{10}H_{13}I$	260.12	56739-95-8			139^{23}	3-05-00-00960	1.4205^{17}		1.5800^{17}
1782	Benzene, 2-iodo-4-methyl-1-(1-methylethyl)-	*p*-Cymene, 3-iodo-	$C_{10}H_{13}I$	260.12	4395-81-7			123^{13}; 80^{5}	3-05-00-00960	1.4113^{17}		1.5690^{17}
1783	Benzene, 1-iodo-2-nitro-		$C_6H_4INO_2$	249.01	609-73-4	54		290; 162^{18}	4-05-00-00732	1.9186^{75}	H_2O 1; EtOH 3; eth 3	
1784	Benzene, 1-iodo-3-nitro-		$C_6H_4INO_2$	249.01	645-00-1	38.5		280	4-05-00-00733	1.9477^{50}	H_2O 1; EtOH 3; eth 3	
1785	Benzene, 1-iodo-4-nitro-		$C_6H_4INO_2$	249.01	636-98-6	174.7		288	4-05-00-00734	1.8090^{155}	H_2O 1; EtOH 3; HOAc 3	
1786	Benzene, isocyanato-	Phenyl isocyanate	C_7H_5NO	119.12	103-71-9		7267	163; 55^{13}	4-12-00-00864	1.0956^{20}	eth 4; chl 2	1.5368^{20}
1787	Benzene, 1-isocyanato-2-methyl-		C_8H_7NO	133.15	614-68-6			185	4-12-00-01768		H_2O 1; eth 3	1.5282^{20}
1788	Benzene, 1-isocyanato-2-nitro-		$C_7H_4N_2O_3$	164.12	3320-86-3	41		137^{18}	3-12-00-01535		bz 4; eth 4; chl 4	
1789	Benzene, 1-isocyanato-3-nitro-		$C_7H_4N_2O_3$	164.12	3320-87-4	51		130^{11}	4-12-00-01604		bz 4; eth 4; chl 4	
1790	Benzene, 1-isocyanato-4-nitro-		$C_7H_4N_2O_3$	164.12	100-28-7	57		162^{20}; 137^{11}	4-12-00-01653		bz 4; eth 4; chl 4	
1791	Benzene, (isocyanomethyl)-		C_8H_7N	117.15	10340-91-7			199 dec; 93^{55}	4-12-00-02219	0.972^{15}		1.5193^{20}
1792	Benzene, isothiocyanato-	Phenyl isothiocyanate	C_7H_5NS	135.19	103-72-0	-21	7268	221	4-12-00-00867	1.1303^{20}	H_2O 1; EtOH 3; eth 3; ctc 3	1.6492^{23}
1793	Benzene, 1-isothiocyanato-2-methoxy-		C_8H_7NOS	165.22	3288-04-8			$131\text{-}2^{11}$	4-13-00-00855	1.1878^{20}		1.6458^{20}
1794	Benzene, (isothiocyanatomethyl)-		C_8H_7NS	149.22	622-78-6			243	4-12-00-02276	1.1246^{16}	H_2O 1; EtOH 5; eth 3	1.6049^{15}
1795	Benzenemethanamine	Benzylamine	C_7H_9N	107.16	100-46-9		1139	185; 90^{12}	4-12-00-02155	0.9813^{20}	H_2O 5; EtOH 5; eth 5; ace 4	1.5401^{20}
1796	Benzenemethanamine, 2-amino-		$C_7H_{10}N_2$	122.17	4403-69-4	61		269	4-13-00-00265		EtOH 4	

No.	Name Synonym	Mol. Form. Mol. Wt.	CAS RN mp/°C	Merck No. bp/°C	Beil. Ref. den/g cm^{-3}	Solubility n_D
1797	Benzenemethanamine, *N*-[bis(4-methoxyphenyl)methylene]- Benzylimidobis(*p*-methoxyphenyl)methane	$C_{22}H_{21}NO_2$ 331.41	524-96-9 90	1153	3-12-00-02254	eth 4; chl 4
1798	Benzenemethanamine, *N*,*N*-bis(phenylmethyl)-	$C_{21}H_{21}N$ 287.40	620-40-6 91.5	 385	4-12-00-02183 0.9912^{95}	H_2O 2; EtOH 2; eth 3; ctc 3
1799	Benzenemethanamine, 4-bromo- Benzyl amine, 4-bromo	C_7H_8BrN 186.05	3959-07-7 20	 250	3-12-00-02350	eth 4
1800	Benzenemethanamine, *N*-3-butynyl-*N*-methyl-	$C_{12}H_{15}N$ 173.26	15240-91-2	 127^{16}	4-12-00-02171 0.9372^{20}	 1.5202^{20}
1801	Benzenemethanamine, *N*-(2-chloroethyl)-*N*-(phenylmethyl)-, hydrochloride *N*-(2-Chloroethyl)dibenzylamine hydrochloride	$C_{16}H_{19}Cl_2N$ 296.24	55-43-6	2138	4-12-00-02180	
1802	Benzenemethanamine, 3-chloro-*N*-methyl-	$C_8H_{10}ClN$ 155.63	39191-07-6	 88^{4}	3-12-00-02342	chl 3 1.5350^{25}
1803	Benzenemethanamine, 2,4-dichloro-	$C_7H_7Cl_2N$ 176.04	95-00-1	 $124\text{-}6^{13}$	4-12-00-02376	chl 3 1.5762^{25}
1804	Benzenemethanamine, *N*,*N*-diethyl-α-phenyl- *N*,*N*-Diethylbenzhydrylamine	$C_{17}H_{21}N$ 239.36	519-72-2 58.5	3103 170^{17}	4-12-00-03283	
1805	Benzenemethanamine, 3,4-dimethoxy-	$C_9H_{13}NO_2$ 167.21	5763-61-1	 156^{12}; 120^{3}	4-13-00-02582 1.143^{25}	chl 3
1806	Benzenemethanamine, α,α-dimethyl-	$C_9H_{13}N$ 135.21	585-32-0	 106.5	4-12-00-02633 0.9423^{20}	 1.6181^{25}
1807	Benzenemethanamine, *N*,*N*-dimethyl- Dimethylbenzylamine	$C_9H_{13}N$ 135.21	103-83-3	 181	4-12-00-02161 0.915^{0}	H_2O 2; EtOH 5; eth 5 1.5011^{20}
1808	Benzenemethanamine, 2-ethoxy- 2-Ethoxybenzylamine	$C_9H_{13}NO$ 151.21	37806-29-4	 $69\text{-}74^{0.2}$	4-13-00-01676 1.015^{25}	 1.5326^{20}
1809	Benzenemethanamine, α-ethyl-	$C_9H_{13}N$ 135.21	2941-20-0	 206; 100^{16}	4-12-00-02582 0.9346^{25}	 1.5173^{23}
1810	Benzenemethanamine, *N*-ethyl-	$C_9H_{13}N$ 135.21	14321-27-8	 194	4-12-00-02163 0.9342^{17}	H_2O 2; EtOH 3; eth 3; bz 3 1.5117^{20}
1811	Benzenemethanamine, *N*-ethyl-*N*-phenyl- Ethylbenzylaniline	$C_{15}H_{17}N$ 211.31	92-59-1 35	3728 288; 185^{22}	4-12-00-02176 1.001^{55}	H_2O 1; EtOH 3; eth 3; chl 3 1.5943^{23}
1812	Benzenemethanamine, 2-methoxy-	$C_8H_{11}NO$ 137.18	6850-57-3	 229	4-13-00-01675 1.051^{25}	 1.5475^{20}
1813	Benzenemethanamine, 4-methoxy-	$C_8H_{11}NO$ 137.18	2393-23-9	 236.5	4-13-00-01719 1.050^{15}	H_2O 2, EtOH 2, eth 2 1.5462^{20}
1814	Benzenemethanamine, α-methyl-, (±)-	$C_8H_{11}N$ 121.18	618-36-0 32	 187	4-12-00-02424 0.9395^{15}	H_2O 3; EtOH 5; eth 5; chl 3 1.5238^{25}
1815	Benzenemethanamine, 2-methyl-	$C_8H_{11}N$ 121.18	89-93-0 -30	 206; 81^{15}	4-12-00-02518 0.9766^{19}	 1.5436^{19}
1816	Benzenemethanamine, 3-methyl-	$C_8H_{11}N$ 121.18	100-81-2	 203.5	4-12-00-02565 0.966^{25}	 1.5360^{20}
1817	Benzenemethanamine, 4-methyl-	$C_8H_{11}N$ 121.18	104-84-7 12.5	 195	4-12-00-02574 0.952^{20}	 1.5340^{20}
1818	Benzenemethanamine, *N*-methyl-	$C_8H_{11}N$ 121.18	103-67-3	 180.5	4-12-00-02161 0.9442^{18}	H_2O 4
1819	Benzenemethanamine, 4-(1-methylethyl)-	$C_{10}H_{15}N$ 149.24	4395-73-7	 227	4-12-00-02836	eth 4; EtOH 4 1.5182^{17}
1820	Benzenemethanamine, *N*-(1-methylethyl)-	$C_{10}H_{15}N$ 149.24	102-97-6	 200; 93^{10}	4-12-00-02165 0.892^{25}	 1.5025^{20}
1821	Benzenemethanamine, *N*-(2-methylphenyl)-	$C_{14}H_{15}N$ 197.28	5405-13-0 60	 302.5	4-12-00-02177 1.0142^{65}	ace 4; EtOH 4; chl 4 1.5861^{65}
1822	Benzenemethanamine, α-methyl-*N*-(1-phenylethyl)-	$C_{16}H_{19}N$ 225.33	10024-74-5	 296.5	4-12-00-02428 1.018^{15}	 1.573
1823	Benzenemethanamine, *N*-methyl-*N*-2-propynyl- Pargyline	$C_{11}H_{13}N$ 159.23	555-57-7	6988 $96\text{-}7^{11}$	2-12-00-00548 0.944^{25}	 1.5213^{20}
1824	Benzenemethanamine, *N*-(3-methylphenyl)- *m*-Toluidine, *N*-benzyl	$C_{14}H_{15}N$ 197.28	5405-17-4	 312	4-12-00-02178 1.0083^{65}	ace 4; bz 4; eth 4 1.5845^{65}
1825	Benzenemethanamine, *N*-(4-methylphenyl)- *p*-Toluidine, *N*-benzyl	$C_{14}H_{15}N$ 197.28	5405-15-2 19.5	 319; 181^{10}	4-12-00-02178 1.0064^{65}	bz 4; eth 4; EtOH 4; chl 4 1.5832^{65}
1826	Benzenemethanamine, α-phenyl- Benzhydrylamine	$C_{13}H_{13}N$ 183.25	91-00-9 34	1085 304; 176^{23}	4-12-00-03282 1.0633^{20}	H_2O 2; ace 3 1.5963
1827	Benzenemethanamine, *N*-phenyl- Benzylaniline	$C_{13}H_{13}N$ 183.25	103-32-2 37.5	1140 306.5	4-12-00-02172 1.0298^{65}	eth 4; EtOH 4 1.6118^{25}
1828	Benzenemethanamine, *N*-(phenylmethyl)- Dibenzylamine	$C_{14}H_{15}N$ 197.28	103-49-1 -26	2993 300 dec; 270^{250}	4-12-00-02179 1.0256^{22}	H_2O 1; EtOH 4; eth 4; ctc 3 1.5781^{20}
1829	Benzenemethanamine, *N*-phenyl-*N*-(phenylmethyl)-	$C_{20}H_{19}N$ 273.38	91-73-6 69	 226^{10}	3-12-00-02225 1.0444^{80}	H_2O 1; EtOH 2; eth 3; bz 3 1.6065^{80}
1830	Benzenemethanamine, α-propyl-	$C_{10}H_{15}N$ 149.24	2941-19-7	 221; 108^{16}	3-12-00-02716 0.9366^{20}	EtOH 4
1831	Benzene, 1,1',1'',1'''-methanetetrayltetrakis- Tetraphenylmethane	$C_{25}H_{20}$ 320.43	630-76-2 282	 431	4-05-00-02741	H_2O 1; EtOH 1; eth 1; bz 3
1832	Benzenemethanethiol Thiobenzyl alcohol	C_7H_8S 124.21	100-53-8 -30	9249 194.5	4-06-00-02632 1.058^{20}	H_2O 1; EtOH 4; eth 4; ctc 2 1.5151^{20}

No.	Name Synonym	Mol. Form. Mol. Wt.	CAS RN mp/°C	Merck No. bp/°C	Beil. Ref. den/g cm^{-3}	Solubility n_D
1833	Benzenemethanethiol, 4-chloro-	C_7H_7ClS 158.65	6258-66-8 19.5	 113^{17}	4-06-00-02770 1.202^{25}	 1.5893^{20}
1834	Benzenemethanethiol, α-methyl-, (*S*)- Ethanethiol, 1-phenyl (1)	$C_8H_{10}S$ 138.23	33877-11-1	 199; 83^{10}	4-06-00-03062 1.022^{20}	bz 4; eth 4; EtOH 4 1.5593^{20}
1835	Benzenemethanimine, α-phenyl-	$C_{13}H_{11}N$ 181.24	1013-88-3	 282	4-07-00-01365 1.0847^{19}	eth 4 1.6191^{19}
1836	Benzenemethanol Benzyl alcohol	C_7H_8O 108.14	100-51-6 -15.2	1138 205.3	4-06-00-02222 1.0419^{24}	H_2O 3; EtOH 3; eth 3; ace 3 1.5396^{20}
1837	Benzenemethanol, 2-amino-	C_7H_9NO 123.15	5344-90-1 83.5	 273	3-13-00-01615	H_2O 3; EtOH 3; eth 3; bz 4
1838	Benzenemethanol, α-(1-aminoethyl)-, hydrochloride, (*R**,*S**)-(±)- Phenylpropanolamine hydrochloride	$C_9H_{14}ClNO$ 187.67	154-41-6 194	7279	2-13-00-00371	H_2O 4; EtOH 3; eth 1; bz 1
1839	Benzenemethanol, α-(aminomethyl)- Phenylethanolamine	$C_8H_{11}NO$ 137.18	7568-93-6 56.5	7258 160^{17}	4-13-00-01801	H_2O 4; EtOH 3
1840	Benzenemethanol, α-(1-aminopropyl)- α-(α-Aminopropyl)benzyl alcohol	$C_{10}H_{15}NO$ 165.24	5897-76-7 79.5	483	3-13-00-01791	
1841	Benzenemethanol, 2-[bis(4- hydroxyphenyl)methyl]- Phenolphthalol	$C_{20}H_{18}O_3$ 306.36	81-92-5 201.5	7211	4-06-00-07623	
1842	Benzenemethanol, α,α-bis(1-methylethyl)-	$C_{13}H_{20}O$ 192.30	4397-05-1	 229; 158^{60}	4-06-00-03462 0.9755^{20}	eth 4 1.5239^{20}
1843	Benzenemethanol, 5-bromo-2-hydroxy- Bromosaligenin	$C_7H_7BrO_2$ 203.04	2316-64-5 113	1426	3-06-00-04541	bz 4; eth 4; EtOH 4; chl 4
1844	Benzenemethanol, α-(bromomethyl)-	C_8H_9BrO 201.06	2425-28-7	 109-10^2	4-06-00-03053 1.4994^{20}	H_2O 1 1.5800^{17}
1845	Benzenemethanol, 2-bromo-α-methyl-	C_8H_9BrO 201.06	5411-56-3	 109-10^2	4-06-00-03053 1.4994^{20}	 1.5800^{17}
1846	Benzenemethanol, α-butyl- 1-Phenyl-1-pentanol	$C_{11}H_{16}O$ 164.25	583-03-9	3921 141^{25}; 102^3	2-06-00-00503 0.9655^{20}	ace 4; eth 4; EtOH 4 1.4086^{25}
1847	Benzenemethanol, α-butyl-α-methyl- 2-Hexanol, 2-phenyl-	$C_{12}H_{18}O$ 178.27	4396-98-9	 129-30^4	3-06-00-01995 0.9522^{20}	 1.5091^{20}
1848	Benzenemethanol, 2-chloro-	C_7H_7ClO 142.58	17849-38-6 73	 230	4-06-00-02589	H_2O 2; EtOH 4; eth 4; lig 4
1849	Benzenemethanol, 4-chloro-	C_7H_7ClO 142.58	873-76-7 75	 235	4-06-00-02593	bz 4; eth 4; EtOH 4
1850	Benzenemethanol, 4-chloro-α-(4- chlorophenyl)-α-methyl- 1,1-Bis(4-chlorophenyl)ethanol	$C_{14}H_{12}Cl_2O$ 267.15	80-06-8 70	2086	4-06-00-04718	H_2O 1; EtOH 1; eth 3; bz 3
1851	Benzenemethanol, 4-chloro-α-(4- chlorophenyl)-α-(trichloromethyl)- 1,1-Bis(*p*-chlorophenyl)-2,2,2-trichloroethanol	$C_{14}H_9Cl_5O$ 370.49	115-32-2 77.5	3075	4-06-00-04722	H_2O 1; os 1
1852	Benzenemethanol, α-(chloromethyl)-	C_8H_9ClO 156.61	1674-30-2	 128^{17}	4-06-00-03045 1.1926^{20}	EtOH 3; eth 4 1.5523^{20}
1853	Benzenemethanol, 2,4-dichloro- 2,4-Dichlorobenzyl alcohol	$C_7H_6Cl_2O$ 177.03	1777-82-8 59.5	3048 150^{25}	4-06-00-02597	chl 3
1854	Benzenemethanol, 2,4-dichloro-α-methyl-	$C_8H_8Cl_2O$ 191.06	1475-13-4	 125-6^7	4-06-00-03048 1.293^{25}	 1.5605^{20}
1855	Benzenemethanol, α,α-diethyl-	$C_{11}H_{16}O$ 164.25	1565-71-5 <-17	 223.5	4-06-00-03381 0.9831^{20}	EtOH 4 1.5165^{20}
1856	Benzenemethanol, 3,4-dimethoxy-	$C_9H_{12}O_3$ 168.19	93-03-8	 298; 172^{12}	4-06-00-07381 1.178^{17}	H_2O 3; EtOH 3 1.555^{17}
1857	Benzenemethanol, α,α-dimethyl- α-Cumyl alcohol	$C_9H_{12}O$ 136.19	617-94-7 36	 202	4-06-00-03219 0.9735^{20}	H_2O 1; EtOH 3; eth 3; bz 3 1.5325^{20}
1858	Benzenemethanol, α,3-dimethyl- 1-(3-Methylphenyl)ethyl alcohol	$C_9H_{12}O$ 136.19	7287-81-2	 112^{12}	3-06-00-01823 0.9974^{15}	eth 4; EtOH 4 1.5240^{20}
1859	Benzenemethanol, α,4-dimethyl- 1-(4-Methylphenyl)ethanol	$C_9H_{12}O$ 136.19	536-50-5	3226 219	4-06-00-03242 0.9668^{25}	H_2O 1; EtOH 4; eth 4 1.5246^{20}
1860	Benzenemethanol, 2,4-dimethyl-	$C_9H_{12}O$ 136.19	16308-92-2 22	 232; 150^{44}	4-06-00-03250 1.0310^{20}	 1.5339^{20}
1861	Benzenemethanol, 3,5-dimethyl-	$C_9H_{12}O$ 136.19	27129-87-9	 219.5	0-06-00-00521 0.927^{25}	 1.5312^{20}
1862	Benzenemethanol, 4-(dimethylamino)-	$C_9H_{13}NO$ 151.21	1703-46-4 69	 123^1	4-13-00-01771 1.059^{14}	 1.5727^{14}
1863	Benzenemethanol, α-[1- (dimethylamino)ethyl]-, [*R*-(*R**,*S**)]- (1*R*,2*S*)-*N*-Methylephedrine	$C_{11}H_{17}NO$ 179.26	552-79-4 87.5	5987		H_2O 1; EtOH 3; eth 3; MeOH 3
1864	Benzenemethanol, α-(1,1-dimethylethyl)-	$C_{11}H_{16}O$ 164.25	3835-64-1 45	 111^{15}	1-06-00-00270	eth 4; EtOH 4
1865	Benzenemethanol, 4-(1,1-dimethylethyl)-	$C_{11}H_{16}O$ 164.25	877-65-6	 140^{20}	4-06-00-03402 0.928^{25}	 1.5179^{20}
1866	Benzenemethanol, α,α-diphenyl- Triphenylmethanol	$C_{19}H_{16}O$ 260.34	76-84-6 164.2	9653 380	4-06-00-05014 1.199^0	H_2O 1; EtOH 4; eth 4; ace 3
1867	Benzenemethanol, α,α-dipropyl- 4-Heptanol, 4-phenyl-	$C_{13}H_{20}O$ 192.30	4436-96-8	 134^{20}	2-06-00-00513 0.9470^{15}	 1.516^{10}
1868	Benzenemethanol, α-ethenyl- 1-Phenylallyl alcohol	$C_9H_{10}O$ 134.18	4393-06-0		2-06-00-00529 1.0249^{21}	H_2O 2; EtOH 3; eth 3; bz 3 1.5406^{20}
1869	Benzenemethanol, α-ethenyl-α-methyl- 3-Buten-2-ol, 2-phenyl-	$C_{10}H_{12}O$ 148.20	6051-52-1	 92-5^7	4-06-00-03839 0.9965^{21}	 1.5277^{21}

No.	Name Synonym	Mol. Form. Mol. Wt.	CAS RN mp/°C	Merck No. bp/°C	Beil. Ref. den/g cm^{-3}	Solubility n_D
1870	Benzenemethanol, α-ethyl- α-Ethylbenzyl alcohol	$C_9H_{12}O$ 136.19	93-54-9	3727 219	4-06-00-03183 0.9915^{25}	bz 4; eth 4; EtOH 4; MeOH 4 1.5169^{23}
1871	Benzenemethanol, α-[1-(ethylamino)ethyl]-, (*R**,*S**)- *N*-Ethylnorephedrine	$C_{11}H_{17}NO$ 179.26	37025-57-3 51.5	 143^{18}	4-13-00-01887	bz 4
1872	Benzenemethanol, α-ethyl-3-hydroxy- 1-(*m*-Hydroxyphenyl)-1-propanol	$C_9H_{12}O_2$ 152.19	55789-02-1 107	 177^{13}	3-06-00-04623	eth 4; EtOH 4
1873	Benzenemethanol, α-ethyl-4-methoxy-	$C_{10}H_{14}O_2$ 166.22	5349-60-0	 143^{20}	3-06-00-04624	ctc 3 1.5277^{20}
1874	Benzenemethanol, α-ethyl-α-methyl-, (*R*)-	$C_{10}H_{14}O$ 150.22	1006-06-0 -13	 211; 113^{23}	 0.984^{25}	H_2O 1; EtOH 3; eth 3 1.5185^{20}
1875	Benzenemethanol, α-ethyl-α-propyl- 3-Hexanol, 3-phenyl	$C_{12}H_{18}O$ 178.27	20731-93-5	 134^{27}	3-06-00-02000 0.964^{20}	 1.5100^{20}
1876	Benzenemethanol, α-ethynyl- 1-Phenylpropargyl alcohol	C_9H_8O 132.16	4187-87-5 22	 114^{12}	4-06-00-04066 1.0655^{20}	 1.5508^{20}
1877	Benzenemethanol, α-ethynyl-, carbamate Carfimate	$C_{10}H_9NO_2$ 175.19	3567-38-2 86.5	1845	4-06-00-04067	
1878	Benzenemethanol, α-ethynyl-α-methyl-	$C_{10}H_{10}O$ 146.19	127-66-2 52.3	 217.5; 102^{12}	4-06-00-04073 1.0314^{20}	
1879	Benzenemethanol, 3-fluoro-	C_7H_7FO 126.13	456-47-3	 201; 104^{22}	4-06-00-02589 1.164^{25}	 1.5095^{20}
1880	Benzenemethanol, 4-fluoro-	C_7H_7FO 126.13	459-56-3 23	 210	4-06-00-02589	 1.5080^{20}
1881	Benzenemethanol, α-hexyl- 1-Phenyl-1-heptanol	$C_{13}H_{20}O$ 192.30	614-54-0	 275	4-06-00-03454 0.946^{25}	 1.5024^{20}
1882	Benzenemethanol, 2-hydroxy- Salicyl alcohol	$C_7H_8O_2$ 124.14	90-01-7 87	8294 sub	4-06-00-05896 1.1613^{25}	H_2O 3; EtOH 3; eth 3; bz 3
1883	Benzenemethanol, 3-hydroxy-	$C_7H_8O_2$ 124.14	620-24-6 73	 300 dec	4-06-00-05907 1.161^{25}	H_2O 4; EtOH 4; eth 4; chl 2
1884	Benzenemethanol, 4-hydroxy- *p*-Hydroxybenzyl alcohol	$C_7H_8O_2$ 124.14	623-05-2 124.5	 252	4-06-00-05909	H_2O 4; EtOH 4; eth 3; bz 4
1885	Benzenemethanol, 2-hydroxy-3-methoxy-	$C_8H_{10}O_3$ 154.17	4383-05-5 61.5	 162^{12}	4-06-00-07378	chl 2
1886	Benzenemethanol, 4-hydroxy-3-methoxy-	$C_8H_{10}O_3$ 154.17	498-00-0 115	 dec	4-06-00-07381	H_2O 3; EtOH 3; eth 3; bz 3
1887	Benzenemethanol, 4-hydroxy-α-[(methylamino)methyl]- Synephrine	$C_9H_{13}NO_2$ 167.21	94-07-5 184.5	8994	2-13-00-00491	
1888	Benzenemethanol, 2-iodo-	C_7H_7IO 234.04	5159-41-1 92	 147-9^{32}	4-06-00-02605	 1.6349^{20}
1889	Benzenemethanol, 2-methoxy-	$C_8H_{10}O_2$ 138.17	612-16-8	 249	4-06-00-05896 1.0386^{25}	H_2O 1; EtOH 3; eth 5 1.5455^{20}
1890	Benzenemethanol, 3-methoxy-	$C_8H_{10}O_2$ 138.17	6971-51-3	 252	4-06-00-05907 1.112^{25}	 1.5440^{20}
1891	Benzenemethanol, 4-methoxy- Anise alcohol	$C_8H_{10}O_2$ 138.17	105-13-5 25	695 259.1	4-06-00-05909 1.109^{26}	H_2O 3; EtOH 4; eth 4; ctc 3 1.5420^{25}
1892	Benzenemethanol, 4-methoxy-, acetate	$C_{10}H_{12}O_3$ 180.20	104-21-2 84	 270; 150^{23}	4-06-00-05914 1.105^{25}	ctc 3
1893	Benzenemethanol, 2-methoxy-α-methyl-	$C_9H_{12}O_2$ 152.19	13513-82-1 -85.1	 125	3-06-00-04563 0.9647^{20}	ace 4; bz 4; eth 4; EtOH 4 1.4024^{20}
1894	Benzenemethanol, 3-methoxy-α-methyl-	$C_9H_{12}O_2$ 152.19	23308-82-9	 133^{15}	4-06-00-05929 1.0781^{19}	eth 4; EtOH 4 1.5325^{20}
1895	Benzenemethanol, 4-methoxy-α-methyl-	$C_9H_{12}O_2$ 152.19	3319-15-1	 310 dec, 140^{17}	4-06-00-05930 1.0794^{20}	ctc 3 1.5310^{25}
1896	Benzenemethanol, α-methyl- α-Methylbenzyl alcohol	$C_8H_{10}O$ 122.17	98-85-1 20	 205	2-06-00-00444 1.013^{25}	H_2O 1; EtOH 4; eth 4 1.5265^{20}
1897	Benzenemethanol, 2-methyl- *o*-Methylbenzyl alcohol	$C_8H_{10}O$ 122.17	89-95-2 38	 224; 118^{20}	4-06-00-03109 1.023^{40}	eth 4; EtOH 4; chl 4
1898	Benzenemethanol, 3-methyl- *m*-Methylbenzyl alcohol	$C_8H_{10}O$ 122.17	587-03-1 <-20	 215.5	4-06-00-03162 0.9157^{17}	H_2O 2; EtOH 4; eth 4; chl 3
1899	Benzenemethanol, 4-methyl- *p*-Methylbenzyl alcohol	$C_8H_{10}O$ 122.17	589-18-4 61.5	 217	4-06-00-03171 0.978^{22}	eth 4; EtOH 4
1900	Benzenemethanol, α-[1-(methylamino)ethyl]-, (*R**,*S**)-(±)- (±)-Ephedrine	$C_{10}H_{15}$[illegible] 165.24	[illegible] 76.5	[illegible] 135^{12}	[illegible] 1.1220^{20}	[illegible]
1901	Benzenemethanol, α-(1-methylethyl)- 1-Phenyl-2-methylpropyl alcohol	$C_{10}H_{14}O$ 150.22	611-69-8	 223	4-06-00-03289 0.9869^{14}	H_2O 1; EtOH 3; ace 3 1.5193^{14}
1902	Benzenemethanol, 4-(1-methylethyl)- Cumic alcohol	$C_{10}H_{14}O$ 150.22	536-60-7 28	2621 249	4-06-00-03348 0.9818^{20}	H_2O 1; EtOH 5; eth 5; bz 4 1.5210^{20}
1903	Benzenemethanol, α-[1-[methyl(3-phenyl-2-propenyl)amino]ethyl]- Cinnamedrine	$C_{19}H_{23}NO$ 281.40	90-86-8 75	2299	3-13-00-01734	
1904	Benzenemethanol, α-methyl-α-propyl-	$C_{11}H_{16}O$ 164.25	4383-18-0	 216	2-06-00-00505 0.9723^{22}	EtOH 4
1905	Benzenemethanol, α-(2-methylpropyl)-	$C_{11}H_{16}O$ 164.25	1565-86-2	 236; 112^{9}	4-06-00-03380 0.9537^{19}	ace 4; bz 4; eth 4; EtOH 4 1.5080^{18}
1906	Benzenemethanol, 2-nitro-	$C_7H_7NO_3$ 153.14	612-25-9 74	 270; 168^{20}	4-06-00-02608	H_2O 2; EtOH 3; eth 3

No.	Name Synonym	Mol. Form. Mol. Wt.	CAS RN mp/°C	Merck No. bp/°C	Beil. Ref. den/g cm^{-3}	Solubility n_D
1907	Benzenemethanol, 3-nitro-	$C_7H_7NO_3$ 153.14	619-25-0 30.5	 $175\text{-}80^{3}$	4-06-00-02609 1.296^{19}	H_2O 3; EtOH 3; eth 3; chl 2
1908	Benzenemethanol, 4-nitro- *p*-Nitrobenzyl alcohol	$C_7H_7NO_3$ 153.14	619-73-8 96.5	 255 dec; 185^{12}	4-06-00-02611	H_2O 2; EtOH 3; eth 3; ace 2
1909	Benzenemethanol, α-pentyl-	$C_{12}H_{18}O$ 178.27	4471-05-0	 170^{50}; 128^{5}	4-06-00-03417 0.9497^{25}	EtOH 3; eth 3 1.5105^{20}
1910	Benzenemethanol, α-phenyl- Benzohydrol	$C_{13}H_{12}O$ 184.24	91-01-0 69	1100 298; 180^{20}	4-06-00-04648	H_2O 2; EtOH 4; eth 4; ctc 4
1911	Benzenemethanol, α-2-propenyl-	$C_{10}H_{12}O$ 148.20	936-58-3	 228.5	4-06-00-03838 1.004^{18}	 1.5289^{21}
1912	Benzenemethanol, α-propyl-, (*R*)-	$C_{10}H_{14}O$ 150.22	22144-60-1 16	 232	3-06-00-01845 0.9740^{20}	eth 4; EtOH 4 1.5139^{20}
1913	Benzenemethanol, α-[(2-pyridinylamino)methyl]- Phenyramidol	$C_{13}H_{14}N_2O$ 214.27	553-69-5 83.5	7292	5-22-08-00301	
1914	Benzenemethanol, 3,4,5-trimethoxy- 3,4,5-Trimethoxybenzyl alcohol	$C_{10}H_{14}O_4$ 198.22	3840-31-1	 228^{25}	4-06-00-07695 1.232^{25}	 1.5439^{20}
1915	Benzene, methoxy- Anisole	C_7H_8O 108.14	100-66-3 -37.5	699 153.7	4-06-00-00548 0.9940^{20}	H_2O 1; EtOH 3; eth 3; ace 4 1.5174^{20}
1916	Benzene, 1-methoxy-2,3-dimethyl-	$C_9H_{12}O$ 136.19	2944-49-2 29	 199	4-06-00-03097 0.9596^{40}	H_2O 1; EtOH 4; eth 4; ace 3 1.5120^{40}
1917	Benzene, 1-methoxy-2,4-dimethyl-	$C_9H_{12}O$ 136.19	6738-23-4	 192	4-06-00-03127 0.9740^{16}	H_2O 1; EtOH 3; eth 3; bz 3 1.5190^{16}
1918	Benzene, 1-methoxy-3,5-dimethyl-	$C_9H_{12}O$ 136.19	874-63-5	 194; 89^{15}	4-06-00-03142 0.9627^{15}	H_2O 1; EtOH 3; eth 3; bz 3 1.5110^{20}
1919	Benzene, 2-methoxy-1,3-dimethyl-	$C_9H_{12}O$ 136.19	1004-66-6	 182.5	4-06-00-03114 0.9619^{14}	H_2O 1; EtOH 3; eth 3; bz 3 1.5053^{14}
1920	Benzene, 2-methoxy-1,4-dimethyl-	$C_9H_{12}O$ 136.19	1706-11-2	 194	4-06-00-03165 0.9693^{13}	H_2O 1; EtOH 3; eth 3; bz 3 1.5182^{15}
1921	Benzene, 4-methoxy-1,2-dimethyl-	$C_9H_{12}O$ 136.19	4685-47-6	 202	4-06-00-03100 0.9744^{14}	H_2O 1; EtOH 3; eth 3; bz 3 1.5198^{14}
1922	Benzene, 1-methoxy-2,3-dinitro- Anisole, 2,3-dinitro	$C_7H_6N_2O_5$ 198.14	16315-07-4 119		4-06-00-01369 1.23^{137}	EtOH 4; lig 4
1923	Benzene, 1-methoxy-2,4-dinitro-	$C_7H_6N_2O_5$ 198.14	119-27-7 94.5	 206^{12}	4-06-00-01372 1.3364^{131}	H_2O 2; EtOH 3; eth 3; ace 3 1.546^{15}
1924	Benzene, 1-methoxy-3,5-dinitro- 3,5-Dinitroanisole	$C_7H_6N_2O_5$ 198.14	5327-44-6 105.3		4-06-00-01385 1.558^{12}	ace 4; bz 4; MeOH 4
1925	Benzene, 2-methoxy-1,3-dinitro-	$C_7H_6N_2O_5$ 198.14	3535-67-9 118		3-06-00-00868 1.3000^{128}	EtOH 4
1926	Benzene, 2-methoxy-1,4-dinitro- Anisole, 2,5-dinitro-	$C_7H_6N_2O_5$ 198.14	3962-77-4 97	 $136\text{-}8^{2}$	4-06-00-01383 1.476^{18}	H_2O 2; EtOH 3; ace 3; bz 3
1927	Benzene, 4-methoxy-1,2-dinitro-	$C_7H_6N_2O_5$ 198.14	4280-28-8 71	 168^{3}	4-06-00-01384 1.3332^{110}	EtOH 4
1928	Benzene, (2-methoxyethenyl)-	$C_9H_{10}O$ 134.18	4747-15-3	 211.5	2-06-00-00521 0.9894^{23}	 1.5620^{24}
1929	Benzene, 1-methoxy-2-(3-methoxyphenoxy)-	$C_{14}H_{14}O_3$ 230.26	1655-71-6 54.8	 327.5	4-06-00-05670	bz 4; eth 4; EtOH 4
1930	Benzene, (methoxymethyl)- Benzyl methyl ether	$C_8H_{10}O$ 122.17	538-86-3 -52.6	1154 170	4-06-00-02229 0.9634^{20}	H_2O 1; EtOH 4; eth 4; bz 3 1.5008^{20}
1931	Benzene, 1-methoxy-2-methyl-	$C_8H_{10}O$ 122.17	578-58-5 -34.1	 171	4-06-00-01943 0.985^{25}	H_2O 1; EtOH 3; eth 3; ace 3 1.5161^{20}
1932	Benzene, 1-methoxy-3-methyl-	$C_8H_{10}O$ 122.17	100-84-5 -47	 175.5	4-06-00-02039 0.969^{25}	H_2O 1; EtOH 3; eth 3; ace 3 1.5130^{20}
1933	Benzene, 1-methoxy-4-methyl-	$C_8H_{10}O$ 122.17	104-93-8 -32	 175.5	4-06-00-02098 0.969^{25}	H_2O 1; EtOH 3; eth 3; chl 3 1.5112^{20}
1934	Benzene, 4-methoxy-1-methyl-2-nitro-	$C_8H_9NO_3$ 167.16	17484-36-5 17	 266.5	3-06-00-01384 1.207^{25}	 1.5525^{20}
1935	Benzene, 1-methoxy-2-nitro- 2-Nitroanisole	$C_7H_7NO_3$ 153.14	91-23-6 10.5	 277; 144^{4}	4-06-00-01249 1.2540^{20}	H_2O 1; EtOH 5; eth 5; ctc 3 1.5161^{20}
1936	Benzene, 1-methoxy-3-nitro-	$C_7H_7NO_3$ 153.14	555-03-3 38.5	 258	4-06-00-01270 1.373^{18}	H_2O 1; EtOH 3; eth 4
1937	Benzene, 1-methoxy-4-nitro- Anisole, 4-nitro	$C_7H_7NO_3$ 153.14	100-17-4 54	 274	4-06-00-01282 1.2192^{60}	H_2O 1; EtOH 4; eth 4; ctc 3 1.5070^{60}
1938	Benzene, 1-methoxy-2-phenoxy-	$C_{13}H_{12}O_2$ 200.24	1695-04-1 79	 288; 91^{7}	4-06-00-05571	bz 4; eth 4; EtOH 4
1939	Benzene, 1-methoxy-4-(2-phenylethenyl)-, (*E*)-	$C_{15}H_{14}O$ 210.28	1694-19-5 136.5	 142.5^{15}	4-06-00-04855	H_2O 1; EtOH 4; eth 4; ace 4
1940	Benzene, 1-methoxy-4-(1-phenylethyl)-	$C_{15}H_{16}O$ 212.29	2605-18-7	 $180\text{-}2^{19}$	4-06-00-04711 1.0473^{20}	bz 4; eth 4; HOAc 4 1.5725^{20}
1941	Benzene, 1-methoxy-2-(1-propenyl)-	$C_{10}H_{12}O$ 148.20	10577-44-3	 224; 104^{13}	4-06-00-03793 0.9962^{15}	 1.560^{15}
1942	Benzene, 1-methoxy-4-(1-propenyl)- Anethole	$C_{10}H_{12}O$ 148.20	104-46-1 21.3	675 234; 115^{12}	2-06-00-00523 0.9882^{20}	ace 4; bz 4; eth 4; EtOH 4 1.5615^{20}
1943	Benzene, 1-methoxy-4-(2-propenyl)- Estragole	$C_{10}H_{12}O$ 148.20	140-67-0	3657 215.5	4-06-00-03817 0.965^{25}	EtOH 4; chl 4 1.5195^{20}
1944	Benzene, 1-methoxy-4-(1-propenyl)-, (*E*)-	$C_{10}H_{12}O$ 148.20	4180-23-8 21.35	 234; 115^{12}	4-06-00-03796 0.9882^{20}	H_2O 2; EtOH 5; eth 5; ace 3 1.5615^{20}
1945	Benzene, 1-methoxy-4-propyl-	$C_{10}H_{14}O$ 150.22	104-45-0	 211.5	4-06-00-03181 0.9472^{20}	H_2O 2; EtOH 3; eth 4; ace 3 1.5045^{20}

No.	Name Synonym	Mol. Form. Mol. Wt.	CAS RN mp/°C	Merck No. bp/°C	Beil. Ref. den/g cm^{-3}	Solubility n_D
1946	Benzene, 2-methoxy-1,3,5-trinitro- Methyl picrate	$C_7H_5N_3O_7$ 243.13	606-35-9 69		4-06-00-01456 1.4947^{80}	H_2O 1; EtOH 4; eth 3; bz 4
1947	Benzene, methyl- Toluene	C_7H_8 92.14	108-88-3 -94.9	9455 110.6	4-05-00-00766 0.8669^{20}	H_2O 1; EtOH 5; eth 5; ace 3 1.4961^{20}
1948	Benzene, 1-methyl-2,4-bis(1-methylethyl)- Toluene, 2,4-diisopropyl-	$C_{13}H_{20}$ 176.30	1460-98-6	 82^{7}	4-05-00-01152 0.8636^{20}	ctc 3; CS_2 3 1.4912^{20}
1949	Benzene, 1-methyl-3,5-bis(1-methylethyl)-	$C_{13}H_{20}$ 176.30	3055-14-9 -60.8	 216.5	4-05-00-01152 0.8668^{20}	 1.4950^{20}
1950	Benzene, (1-methyl-1-butenyl)- 2-Phenyl-2-pentene	$C_{11}H_{14}$ 146.23	53172-84-2	 202	4-05-00-01402 0.8950^{26}	 1.5196^{26}
1951	Benzene, [(3-methylbutoxy)methyl]-	$C_{12}H_{18}O$ 178.27	122-73-6	 237; 118^{19}	3-06-00-01457 0.9098^{20}	eth 4; EtOH 4 1.4792^{20}
1952	Benzene, (1-methylbutyl)-	$C_{11}H_{16}$ 148.25	2719-52-0	 199	4-05-00-01088 0.8594^{21}	H_2O 1; EtOH 3; eth 3 1.4875^{25}
1953	Benzene, (2-methylbutyl)-	$C_{11}H_{16}$ 148.25	3968-85-2	 195.4	2-05-00-00332 0.8584^{20}	 1.4873^{20}
1954	Benzene, (3-methylbutyl)-	$C_{11}H_{16}$ 148.25	2049-94-7	 195	4-05-00-01089 0.856^{20}	H_2O 1; EtOH 3; eth 3; bz 4 1.4867^{10}
1955	Benzene, (1-methylcyclopropyl)-	$C_{10}H_{12}$ 132.21	2214-14-4	 91^{50}	4-05-00-01388	ace 3; bz 3; chl 3 1.5160^{20}
1956	Benzene, 1-methyl-2,4-dinitro- 2,4-Dinitrotoluene	$C_7H_6N_2O_4$ 182.14	121-14-2 71	 300 dec	4-05-00-00865 1.3208^{71}	H_2O 1; EtOH 3; eth 3; ace 4 1.442
1957	Benzene, 1-methyl-3,5-dinitro-	$C_7H_6N_2O_4$ 182.14	618-85-9 03	 sub	4-05-00-00867 1.2772^{111}	H_2O 2; EtOH 3; eth 3; bz 3
1958	Benzene, 2-methyl-1,3-dinitro- 2,6-Dinitrotoluene	$C_7H_6N_2O_4$ 182.14	606-20-2 66		4-05-00-00866 1.2833^{111}	EtOH 3; chl 3 1.479
1959	Benzene, 2-methyl-1,4-dinitro-	$C_7H_6N_2O_4$ 182.14	619-15-8 52.5		4-05-00-00866 1.282^{111}	EtOH 3; bz 3; CS_2 4
1960	Benzene, 4-methyl-1,2-dinitro- 3,4-Dinitrotoluene	$C_7H_6N_2O_4$ 182.14	610-39-9 58.3		4-05-00-00866 1.2594^{111}	H_2O 1; EtOH 3; chl 2; CS_2 3
1961	Benzene, 1,1'-methylenebis- Diphenylmethane	$C_{13}H_{12}$ 168.24	101-81-5 25.2	3329 265.0	4-05-00-01841 1.001^{26}	H_2O 1; EtOH 3; eth 3; chl 3 1.5753^{20}
1962	Benzene, 1,1'-methylenebis[2-bromo- Diphenylmethane, 2,2'-dibromo	$C_{13}H_{10}Br_2$ 326.03	61592-89-0	 $234\text{-}5^{40}$	3-05-00-01795 1.6197^{20}	 1.6300^{20}
1963	Benzene, 1,1'-methylenebis[4-chloro-	$C_{13}H_{10}Cl_2$ 237.13	101-76-8 55.5	 $186\text{-}90^{18}$	4-05-00-01848 1.365^{17}	EtOH 3
1964	Benzene, 1,1'-methylenebis[4-methyl-	$C_{15}H_{16}$ 196.29	4957-14-6 28	 286	4-05-00-01929 0.9800^{20}	eth 4; EtOH 4
1965	Benzene, 1,1'-[methylenebis(oxy)]bis[4-chloro- Bis(*p*-chlorophenoxy)methane	$C_{13}H_{10}Cl_2O_2$ 269.13	555-89-5 70.5	1258 $189\text{-}94^{6}$	4-06-00-00833	ace 4; bz 4
1966	Benzene, 1,1'-(1-methylene-1,2-ethanediyl)bis-	$C_{15}H_{14}$ 194.28	948-97-0 48	 289	4-05-00-02191 1.1014^{20}	 1.5903^{20}
1967	Benzene, (1-methylene-2-propenyl)-	$C_{10}H_{10}$ 130.19	2288-18-8	 $60\text{-}1^{17}$	4-05-00-01539 0.9266^{20}	H_2O 1; eth 3; bz 3; chl 3 1.5489^{20}
1968	Benzene, (1-methylenepropyl)-	$C_{10}H_{12}$ 132.21	2039-93-2	 182	4-05-00-01380 0.887^{25}	 1.5288^{20}
1969	Benzene, 1,1'-(1-methyl-1,2-ethanediyl)bis-	$C_{15}H_{16}$ 196.29	5814-85-7 52	 280.5	4-05-00-01915 0.9807^{20}	 1.5700^{20}
1970	Benzene, 1,1'-[(1-methyl-1,2-ethanediyl)bis(oxy)]bis- Propane, 1,2-diphenoxy	$C_{15}H_{16}O_2$ 228.29	69813-63-4 32	 $175\text{-}8^{12}$	2-06-00-00151 1.0748^{33}	ace 4; bz 4; eth 4; EtOH 4 1.5542^{33}
1971	Benzene, 1,1'-(1-methyl-1,2-ethenediyl)bis-	$C_{15}H_{14}$ 194.28	779-51-1 82	 285.5	4-05-00-02189 0.9857^{17}	EtOH 3; eth 3; bz 3; lig 2 1.5635^{17}
1972	Benzene, (1-methylethenyl)- α-Methylstyrene	C_9H_{10} 118.18	98-83-9 -23.2	 165.4	4-05-00-01364 0.9106^{20}	H_2O 1; EtOH 3; eth 3; ace 5 1.5386^{20}
1973	Benzene, 1-(1-methylethenyl)-4-(1-methylethyl)-	$C_{12}H_{16}$ 160.26	2388-14-9 -30.6	 220.8	4-05-00-01422 0.8936^{20}	ace 4; bz 4; eth 4; EtOH 4 1.5238^{20}
1974	Benzene, (1-methylethoxy)-	$C_9H_{12}O$ 136.19	2741-16-4 -33	 176.8	4-06-00-00557 0.9408^{25}	H_2O 3; EtOH 3; ace 3; bz 3 1.4975^{20}
1975	Benzene, (1-methylethyl)- Cumene	C_9H_{12} 120.19	98-82-8 -96.0	2619 152.4	4-05-00-00985 0.8618^{20}	H_2O 1; EtOH 5; eth 5; ace 5 1.4915^{20}
1976	Benzene, 1,1'-(1-methylethylidene)bis-	$C_{15}H_{16}$ 196.29	778-22-3 29	 282.5	4-05-00-01925 0.9980^{20}	
1977	Benzene, 1-(1-methylethyl)-2-nitro-	$C_9H_{11}NO_2$ 165.19	6526-72-3	 103^{9}	4-05-00-00997 1.101^{12}	H_2O 1; ace 3; bz 3; HOAc 3 1.5259^{20}
1978	Benzene, 1-(1-methylethyl)-4-nitro-	$C_9H_{11}NO_2$ 165.19	1817-47-6	 122^{9}	4-05-00-00997 1.0830^{20}	H_2O 1; ace 3; bz 3; lig 3 1.5367^{20}
1979	Benzene, 1-(1-methylethyl)-4-(phenylmethyl)-	$C_{16}H_{18}$ 210.32	886-58-8 13.4	 310	4-05-00-01951 1.007^{18}	
1980	Benzene, [(1-methylethyl)thio]-	$C_9H_{12}S$ 152.26	3019-20-3	 208	4-06-00-01471 0.9852^{20}	 1.5464^{20}
1981	Benzene, (1-methylheptyl)- 2-Phenyloctane	$C_{14}H_{22}$ 190.33	777-22-0	 $123\text{-}5^{20}$	4-05-00-01157 0.8611^{20}	 1.4837^{20}
1982	Benzene, 1,1',1''-methylidynetris- Triphenylmethane	$C_{19}H_{16}$ 244.34	519-73-3 94	9655 359; 200^{10}	4-05-00-02495 1.014^{99}	H_2O 1; EtOH 2; eth 4; bz 3 1.5839^{99}
1983	Benzene, 1,1',1''-[methylidynetris(oxy)]tris-	$C_{19}H_{16}O_3$ 292.33	16737-44-3 76.5	 269^{50} dec	4-06-00-00611	eth 4; EtOH 4
1984	Benzene, 1-methyl-2-(1-methylethenyl)-	$C_{10}H_{12}$ 132.21	7399-49-7	 175	3-05-00-01214 0.9180^{15}	 1.5112^{30}

No.	Name Synonym	Mol. Form. Mol. Wt.	CAS RN mp/°C	Merck No. bp/°C	Beil. Ref. den/g cm^{-3}	Solubility n_D
1985	Benzene, 1-methyl-3-(1-methylethenyl)-	$C_{10}H_{12}$ 132.21	1124-20-5	185	4-05-00-01383 0.9010^{25}	1.5335^{20}
1986	Benzene, 1-methyl-4-(1-methylethenyl)-	$C_{10}H_{12}$ 132.21	1195-32-0 -20	185.3	4-05-00-01383 0.8936^{23}	1.5283^{23}
1987	Benzene, 1-methyl-3-(1-methylethoxy)-	$C_{10}H_{14}O$ 150.22	19177-04-9	195	4-06-00-02040 0.931^{20}	1.4959^{20}
1988	Benzene, 1-methyl-2-(1-methylethyl)- *o*-Cymene	$C_{10}H_{14}$ 134.22	527-84-4 -71.5	2770 178.1	4-05-00-01057 0.8766^{20}	H_2O 1; EtOH 5; eth 5; ace 5 1.5006^{20}
1989	Benzene, 1-methyl-3-(1-methylethyl)- *m*-Cymene	$C_{10}H_{14}$ 134.22	535-77-3 -63.7	2770 175.1	4-05-00-01058 0.8610^{20}	H_2O 1; EtOH 5; eth 5; ace 5 1.4930^{20}
1990	Benzene, 1-methyl-4-(1-methylethyl)- *p*-Cymene	$C_{10}H_{14}$ 134.22	99-87-6 -68.9	2770 177.1	4-05-00-01060 0.8573^{20}	H_2O 1; EtOH 5; eth 5; ace 5 1.4909^{20}
1991	Benzene, 1-methyl-4-(1-methylethyl)-2-nitro-	$C_{10}H_{13}NO_2$ 179.22	943-15-7	126^{10}	4-05-00-01064 1.0744^{20}	eth 4; EtOH 4 1.5301^{20}
1992	Benzene, [2-methyl-1-(1-methylethyl)propyl]-	$C_{13}H_{20}$ 176.30	21777-84-4 -24.5	220.7	4-05-00-01146 0.8821^{10}	1.512
1993	Benzene, 1-methyl-2-[(4-methylphenyl)thio]- 2,4'-Ditolyl sulfide	$C_{14}H_{14}S$ 214.33	4279-70-3	173^{11}	4-06-00-02172 1.0774^{15}	eth 4; EtOH 4
1994	Benzene, 1-methyl-2-(1-methylpropyl)-	$C_{11}H_{16}$ 148.25	1595-06-8	196	4-05-00-01094 0.873^{20}	ace 4; bz 4; eth 4; EtOH 4 1.497^{20}
1995	Benzene, 1-methyl-2-(2-methylpropyl)-	$C_{11}H_{16}$ 148.25	36301-29-8 -73.3	196	4-05-00-01095 0.8649^{20}	ace 4; bz 4; eth 4; EtOH 4 1.4935^{20}
1996	Benzene, 1-methyl-3-(1-methylpropyl)-	$C_{11}H_{16}$ 148.25	1772-10-7	194	4-05-00-01095 0.858^{20}	ace 4; bz 4; eth 4; EtOH 4 1.490^{20}
1997	Benzene, 1-methyl-3-(2-methylpropyl)-	$C_{11}H_{16}$ 148.25	5160-99-6	194	3-05-00-01001 0.8536^{20}	ace 4; bz 4; eth 4; EtOH 4 1.4888^{20}
1998	Benzene, 1-methyl-4-(1-methylpropyl)-	$C_{11}H_{16}$ 148.25	1595-16-0	197	4-05-00-01095 0.8640^{19}	bz 4; eth 4; chl 4 1.493^{20}
1999	Benzene, 1-methyl-4-(2-methylpropyl)-	$C_{11}H_{16}$ 148.25	5161-04-6	196	4-05-00-01096 0.8517^{20}	ace 4; bz 4; eth 4; EtOH 4 1.4874^{20}
2000	Benzene, 1-methyl-2-nitro- *o*-Nitrotoluene	$C_7H_7NO_2$ 137.14	88-72-2 -10	6572 222	4-05-00-00845 1.1611^{19}	H_2O 1; EtOH 5; eth 5; ctc 3 1.5450^{20}
2001	Benzene, 1-methyl-3-nitro- *m*-Nitrotoluene	$C_7H_7NO_2$ 137.14	99-08-1 15.5	6572 232	4-05-00-00847 1.1581^{20}	H_2O 1; EtOH 3; eth 5; bz 3 1.5466^{20}
2002	Benzene, 1-methyl-4-nitro- *p*-Nitrotoluene	$C_7H_7NO_2$ 137.14	99-99-0 51.6	6572 238.3	4-05-00-00848 1.1038^{75}	H_2O 1; EtOH 3; eth 4; ace 4
2003	Benzene, (1-methyl-1-nitroethyl)-	$C_9H_{11}NO_2$ 165.19	3457-58-7	224 dec; 126^{15}	4-05-00-00998 1.1025^{20}	ace 4; bz 4 1.5209^{20}
2004	Benzene 1-methyl-4-(nitromethyl)-	$C_8H_9NO_2$ 151.17	29559-27-1 11.5	$150\text{-}1^{35}$	4-05-00-00972 1.1234^{20}	H_2O 1; bz 3 1.5278^{20}
2005	Benzene, 1-methyl-2-(2-nitrophenoxy)- 2-Nitrophenyl 2-tolyl ether	$C_{13}H_{11}NO_3$ 229.24	54106-40-0 39.5	$194\text{-}6^{14}$	4-06-00-01946 1.195^{20}	
2006	Benzene, 1-methyl-4-(4-nitrophenoxy)-	$C_{13}H_{11}NO_3$ 229.24	3402-74-2 69	224^{25}	4-06-00-02102	bz 4; HOAc 4
2007	Benzene, (1-methylnonadecyl)- Eicosane, 2-phenyl-	$C_{26}H_{46}$ 358.65	2398-66-5 29.0	204.5^{1}	4-05-00-01222 0.8547^{20}	1.4795^{20}
2008	Benzene, (1-methyl-1-pentenyl)- 2-Hexene, 2-phenyl	$C_{12}H_{16}$ 160.26	20247-89-6	$66\text{-}8^{1}$	4-05-00-01419 0.910^{20}	1.5200^{20}
2009	Benzene, (1-methylpentyl)-	$C_{12}H_{18}$ 162.27	6031-02-3	208	4-05-00-01116 0.869^{15}	1.492^{15}
2010	Benzene, (3-methylpentyl)- (3-Methylpentyl)benzene	$C_{12}H_{18}$ 162.27	54410-69-4	220	0-05-00-00444 0.8644^{14}	1.4896
2011	Benzene, (4-methylpentyl)-	$C_{12}H_{18}$ 162.27	4215-86-5	215	4-05-00-01116 0.8568^{16}	bz 4; eth 4; EtOH 4
2012	Benzene, 1-methyl-2-phenoxy-	$C_{13}H_{12}O$ 184.24	3991-61-5 21.7	263.5	4-06-00-01946 1.0480^{25}	
2013	Benzene, 1-methyl-3-phenoxy-	$C_{13}H_{12}O$ 184.24	3586-14-9	272	4-06-00-02041 1.051^{25}	1.5727^{20}
2014	Benzene, 1-methyl-2-(phenylmethyl)-	$C_{14}H_{14}$ 182.27	713-36-0 6.6	280.5	4-05-00-01892 1.0020^{20}	1.5763^{20}
2015	Benzene, 1-methyl-3-(phenylmethyl)-	$C_{14}H_{14}$ 182.27	620-47-3 -28	279	4-05-00-01893 0.9913^{20}	H_2O 1; EtOH 3; eth 3; bz 3 1.5712^{20}
2016	Benzene, 1-methyl-4-(phenylmethyl)-	$C_{14}H_{14}$ 182.27	620-83-7 -30	286	4-05-00-01894 0.9976^{20}	eth 4; bz 4; EtOH 4; chl 4 1.5712^{20}
2017	Benzene, 1-methyl-2-(phenylthio)-	$C_{13}H_{12}S$ 200.30	13963-35-4	308; 164^{12}	4-06-00-02018 1.0893^{20}	H_2O 1; ace 3; bz 3
2018	Benzene, 1-methyl-3-(phenylthio)-	$C_{13}H_{12}S$ 200.30	13865-48-0 -6.5	309.5	4-06-00-02080 1.0937^{15}	H_2O 1; ace 3; bz 3
2019	Benzene, 1-methyl-4-(phenylthio)-	$C_{13}H_{12}S$ 200.30	3699-01-2 15.7	317	4-06-00-02169 1.0986^{25}	H_2O 1; ace 3; bz 3 1.6225^{25}
2020	Benzene, 1,1'-(1-methyl-1,3-propanediyl)bis-, (±)- Butane, 1,3-diphenyl (*DL*)	$C_{16}H_{18}$ 210.32	116783-21-2	293	4-05-00-01938 0.9722^{20}	1.5525^{20}
2021	Benzene, 1,1'-(2-methyl-1,3-propanediyl)bis-	$C_{16}H_{18}$ 210.32	1520-46-3 -33.7	303	4-05-00-01939 0.9669^{20}	
2022	Benzene, 1,1'-(3-methyl-1-propene-1,3-diyl)bis- 1,3-Diphenyl-1-butene	$C_{16}H_{16}$ 208.30	7614-93-9 47.5	311	4-05-00-02205 0.9996^{20}	1.590^{15}
2023	Benzene, 1-methyl-2-(2-propenyl)-	$C_{10}H_{12}$ 132.21	1587-04-8	182.5	3-05-00-01212 0.9005^{20}	1.5187^{20}

No.	Name Synonym	Mol. Form. Mol. Wt.	CAS RN mp/°C	Merck No. bp/°C	Beil. Ref. den/g cm^{-3}	Solubility n_D
2024	Benzene, (2-methyl-1-propenyl)-	$C_{10}H_{12}$ 132.21	768-49-0	 $99^{43.5}$	4-05-00-01380 0.9029^{20}	 1.5388^{20}
2025	Benzene, (1-methyl-1-propenyl)-, (*E*)-	$C_{10}H_{12}$ 132.21	768-00-3 -23.5	 194.7	4-05-00-01378 0.9138^{25}	H_2O 1; bz 3; chl 3 1.5425^{20}
2026	Benzene, 1-methyl-2-(2-propenyloxy)-	$C_{10}H_{12}O$ 148.20	936-72-1	 206.5	4-06-00-01946 0.9698^{15}	H_2O 1; ctc 2 1.5188^{15}
2027	Benzene, 1-methyl-3-(2-propenyloxy)-	$C_{10}H_{12}O$ 148.20	1758-10-7	 212.5	4-06-00-02041 0.9564^{20}	bz 4 1.5179^{20}
2028	Benzene, 1-methyl-4-(2-propenyloxy)-	$C_{10}H_{12}O$ 148.20	23431-48-3	 214.5	4-06-00-02101 0.9719^{15}	H_2O 1; ace 3; ctc 2 1.5157^{24}
2029	Benzene, (1-methyl-1-propenyl)-, (*Z*)-	$C_{10}H_{12}$ 132.21	767-99-7	 194.7; 177^{500}	3-05-00-01209 0.9191^{25}	H_2O 1; bz 3; chl 3 1.5402^{25}
2030	Benzene, (1-methylpropoxy)-	$C_{10}H_{14}O$ 150.22	10574-17-1	 194.5	4-06-00-00559 0.9415^{20}	H_2O 1; eth 3; ctc 2 1.4926^{25}
2031	Benzene, (2-methylpropoxy)-	$C_{10}H_{14}O$ 150.22	1126-75-6	 197.5	4-06-00-00559 0.9232^{24}	 1.4932^{14}
2032	Benzene, 1-methyl-4-propoxy- Toluene, 4-propoxy	$C_{10}H_{14}O$ 150.22	5349-18-8 0	 210.4	3-06-00-01354 0.9496^{0}	
2033	Benzene, [(2-methylpropoxy)methyl]- Benzyl isobutyl ether	$C_{11}H_{16}O$ 164.25	940-49-8	 212	3-06-00-01457 0.9233^{20}	H_2O 1; eth 4; chl 4 1.4826^{20}
2034	Benzene, 1-(2-methylpropoxy)-2-nitro- 1-Isobutoxy-2-nitrobenzene	$C_{10}H_{13}NO_3$ 195.22	56245-02-4	 277.5	0-06-00-00218 1.1361^{20}	eth 4
2035	Benzene, 1-methylpropyl *sec*-Butylbenzene	$C_{10}H_{14}$ 134.22	135-98-8 -82.7	1550 173.5	 0.8580^{25}	H_2O 1; EtOH 5; eth 5; ace 5 1.4895^{20}
2036	Benzene, 1-methyl-2-propyl-	$C_{10}H_{14}$ 134.22	1074-17-5 -60.3	 185	4-05-00-01056 0.8697^{25}	 1.4996^{20}
2037	Benzene, 1-methyl-3-propyl-	$C_{10}H_{14}$ 134.22	1074-43-7 -82.5	 182	4-05-00-01056 0.8569^{25}	 1.4935^{20}
2038	Benzene, 1-methyl-4-propyl-	$C_{10}H_{14}$ 134.22	1074-55-1 -63.6	 183.4	4-05-00-01057 0.8544^{25}	H_2O 1; EtOH 3; eth 3 1.4922^{20}
2039	Benzene, (2-methylpropyl)- Isobutylbenzene	$C_{10}H_{14}$ 134.22	538-93-2 -51.4	5018 172.7	4-05-00-01042 0.8532^{20}	H_2O 1; EtOH 5; eth 5; ace 5 1.4866^{20}
2040	Benzene, (methylsulfinyl)-	C_7H_8OS 140.21	1193-82-4 32.0	 263.5; 140^{13}	4-06-00-01467	 1.5885^{20}
2041	Benzene, (methylthio)-	C_7H_8S 124.21	100-68-5	 193	4-06-00-01466 1.0579^{20}	H_2O 1; EtOH 3; ace 4; os 3 1.5868^{20}
2042	Benzene, [(methylthio)methyl]-	$C_8H_{10}S$ 138.23	766-92-7 -30	 210; 120^{48}	4-06-00-02633 1.0274^{20}	 1.5620^{20}
2043	Benzene, 1-(methylthio)-2-nitro-	$C_7H_7NO_2S$ 169.20	3058-47-7 64.5		4-06-00-01661 1.2628^{78}	H_2O 3; EtOH 3; bz 3; chl 3 1.6246^{78}
2044	Benzene, 1-(methylthio)-4-nitro-	$C_7H_7NO_2S$ 169.20	701-57-5 72	 $135\text{-}40^{2}$	4-06-00-01687 1.2391^{80}	H_2O 1; ace 3; bz 3 1.6401^{20}
2045	Benzene, 1-methyl-4-(trifluoromethyl)-	$C_8H_7F_3$ 160.14	6140-17-6	 129	 1.4320^{26}	eth 3; ace 3; bz 3; chl 3 1.4320^{26}
2046	Benzene, 1-methyl-2,3,4-trinitro- 2,3,4-Trinitrotoluene	$C_7H_5N_3O_6$ 227.13	602-29-9 112	 301	4-05-00-00872 1.62^{25}	H_2O 1; EtOH 2; eth 3; ace 3
2047	Benzene, 1-methyl-2,4,5-trinitro- 2,4,5-Trinitrotoluene	$C_7H_5N_3O_6$ 227.13	610-25-3 104	 exp 300	4-05-00-00872	H_2O 1; EtOH 2; eth 3; ace 3
2048	Benzene, 2-methyl-1,3,5-trinitro- 2,4,6-Trinitrotoluene	$C_7H_5N_3O_6$ 227.13	118-96-7 80.1	9643 exp 240	4-05-00-00873 1.654^{25}	H_2O 1; EtOH 2; eth 3; ace 4
2049	Benzene, nitro- Nitrobenzene	$C_6H_5NO_2$ 123.11	98-95-3 5.7	6509 210.8	4-05-00-00708 1.2037^{20}	H_2O 2; EtOH 4; eth 4; ace 4 1.5562^{20}
2050	Benzene, (2-nitroethenyl)-, (*E*)-	$C_8H_7NO_2$ 149.15	5153-67-3 60	 255	4-05-00-01352	H_2O 1; EtOH 3; eth 4; ace 3
2051	Benzene, (2-nitroethyl)-	$C_8H_9NO_2$ 151.17	6125-24-2 -23	 250; 137^{16}	4-05-00-00912 1.126^{24}	 1.5407^{19}
2052	Benzene, (nitromethyl)-	$C_7H_7NO_2$ 137.14	622-42-4	 226; 135^{25}	4-05-00-00850 1.1596^{20}	ace 4; eth 4 1.5323^{20}
2053	Benzene, 1-nitro-2-phenoxy-	$C_{12}H_9NO_3$ 215.21	2216-12-8 <-20	 235^{60}; 184^{8}	4-06-00-01252 1.2539^{22}	bz 4; eth 4; EtOH 4; chl 4 1.575^{20}
2054	Benzene, 1-nitro-4-phenoxy-	$C_{12}H_9NO_3$ 215.21	620-88-2 61	 320; 225^{30}	4-06-00-01287	H_2O 1; EtOH 2; eth 3; bz 3
2055	Benzene, 1-nitro-2-(2-phenylethenyl)-, (*E*)-	$C_{14}H_{11}NO_2$ 225.25	4264-29-3 73	 209^{11}	4-05-00-02166	H_2O 1; EtOH 3; eth 4
2056	Benzene, 1-nitro-2-(2-phenylethenyl)-, (*Z*)-	$C_{14}H_{11}NO_2$ 225.25	52200-02-5 64.8	 187^{11}	4-05-00-02166	EtOH 3; eth 4; AcOEt 4; HOAc 4
2057	Benzene, 1-nitro-2-(phenylthio)-	$C_{12}H_9NO_2S$ 231.28	4171-83-9 82	 210^{15}	4-06-00-01663	eth 4; EtOH 4
2058	Benzene, 1-nitro-4-(phenylthio)-	$C_{12}H_9NO_2S$ 231.28	952-97-6 56	 288^{100}; 240^{25}	4-06-00-01694	eth 4; EtOH 4
2059	Benzene, nitroso- Nitrosobenzene	C_6H_5NO 107.11	586-96-9 68.5	 58^{18}	4-05-00-00702	H_2O 1; EtOH 3; eth 3; bz 3
2060	Benzene, 1-nitro-2-(trifluoromethyl)-	$C_7H_4F_3NO_2$ 191.11	384-22-5 32.5	 217; 105^{20}	4-05-00-00852	H_2O 1; EtOH 4; bz 4; ctc 2
2061	Benzene, 1-nitro-3-(trifluoromethyl)- 3-(Trifluoromethyl)-1-nitrobenzene	$C_7H_4F_3NO_2$ 191.11	98-46-4 -2.4	 202.8; 103^{40}	4-05-00-00853 1.4357^{15}	H_2O 1; EtOH 3; eth 3; ctc 2 1.4719^{20}
2062	Benzene, nonadecyl-	$C_{25}H_{44}$ 344.62	29136-19-4 40	 419	3-05-00-01126 0.8545^{20}	 1.4807^{20}

No.	Name Synonym	Mol. Form. Mol. Wt.	CAS RN mp/°C	Merck No. bp/°C	Beil. Ref. den/g cm^{-3}	Solubility n_D
2063	Benzenenonanoic acid, θ-nonyl-	$C_{24}H_{40}O_2$ 360.58	1938-22-3 37	$200\text{-}4^{0.09}$	2-09-00-00376 0.9312^{25}	1.4891^{20}
2064	Benzene, nonyl-	$C_{15}H_{24}$ 204.36	1081-77-2 -24	280.5	3-05-00-01075 0.8584^{20}	1.4816^{20}
2065	Benzene, octacosyl- Octacosylbenzene	$C_{34}H_{62}$ 470.87	61828-26-0 66	405^{100}; 268^{1}	4-05-00-01243 0.8543^{20}	1.4796^{20}
2066	Benzene, octadecyl- Octadecylbenzene	$C_{24}H_{42}$ 330.60	4445-07-2 29	408	4-05-00-01219 0.8546^{20}	1.4809^{20}
2067	Benzene, octyl-	$C_{14}H_{22}$ 190.33	2189-60-8 -36	264.5	4-05-00-01157 0.8562^{20}	H_2O 1; eth 5; bz 5 1.4845^{20}
2068	Benzene, (1-octyldodecyl)- Eicosane, 9-phenyl-	$C_{26}H_{46}$ 358.65	2398-65-4 17.9	$196^{1.0}$	4-05-00-01223 0.8534^{20}	1.4790^{20}
2069	Benzene, (octyloxy)-	$C_{14}H_{22}O$ 206.33	1818-07-1 8	285	4-06-00-00560 0.9131^{15}	H_2O 1; EtOH 3; eth 3 1.4875^{20}
2070	Benzene, (3-octylundecyl)- 9-(2-Phenylethyl)heptadecane	$C_{25}H_{44}$ 344.62	5637-96-7 -26.7	$189^{1.0}$	4-05-00-01221 0.8560^{20}	1.4806^{20}
2071	Benzene, 1,1'-oxybis- Diphenyl ether	$C_{12}H_{10}O$ 170.21	101-84-8 26.8	7259 258.0	4-06-00-00568 1.0661^{30}	H_2O 1; EtOH 3; eth 3; bz 3 1.5787^{25}
2072	Benzene, 1,1'-oxybis[4-bromo-	$C_{12}H_8Br_2O$ 328.00	2050-47-7 60.5	339	4-06-00-01048 1.8^{25}	H_2O 1; EtOH 3; eth 4; bz 3
2073	Benzene, 1,1'-oxybis[4-chloro-	$C_{12}H_8Cl_2O$ 239.10	2444-89-5 30	313	4-06-00-00826 1.1231^{20}	H_2O 1 1.611^{20}
2074	Benzene, 1,1'-oxybis[2-ethyl-	$C_{16}H_{18}O$ 226.32	56911-77-4	318.5	1.0141^{18}	H_2O 1; eth 3; chl 3 1.5488^{18}
2075	Benzene, 1,1'-oxybis[2-methoxy-	$C_{14}H_{14}O_3$ 230.26	1655-70-5 79.5	330.5	3-06-00-04223	H_2O 1; EtOH 4; eth 4; lig 1
2076	Benzene, 1,1'-oxybis[2-methyl-	$C_{14}H_{14}O$ 198.26	4731-34-4	271	4-06-00-01947 1.047^{24}	bz 4; eth 4; EtOH 4
2077	Benzene, 1,1'-oxybis[3-methyl-	$C_{14}H_{14}O$ 198.26	19814-71-2	284; 136^{14}	4-06-00-02043 1.0323^{21}	H_2O 1; EtOH 3; eth 3; bz 3
2078	Benzene, 1,1'-oxybis[4-methyl-	$C_{14}H_{14}O$ 198.26	1579-40-4 51	285	4-06-00-02103	bz 4; eth 4; EtOH 4
2079	Benzene, 1,1'-[oxybis(methylene)]bis- Benzyl ether	$C_{14}H_{14}O$ 198.26	103-50-4 3.6	1146 298	4-06-00-02240 1.0428^{20}	H_2O 1; EtOH 5; eth 5; ctc 3 1.5168^{20}
2080	Benzene, 1,1'-(oxydi-2,1-ethanediyl)bis-	$C_{16}H_{18}O$ 226.32	2396-53-4	318.5; $86^{0.5}$	4-06-00-03071 1.0141^{18}	eth 4; chl 4 1.5488^{18}
2081	Benzene, 1,1'-(oxydiethylidene)bis-, (*R**,*R**)-(±)- Bis-(1-phenylethyl)ether (*DL*)	$C_{16}H_{18}O$ 226.32	53776-69-5	280.2	4-06-00-03034 1.0058^{15}	H_2O 1; eth 3; chl 3 1.5454^{21}
2082	Benzene, pentabromo- Pentabromobenzene	C_6HBr_5 472.59	608-90-2 160.5	sub	4-05-00-00687	bz 4; chl 4
2083	Benzene, pentabromomethyl-	$C_7H_3Br_5$ 486.62	87-83-2 288		3-05-00-00720 2.97^{17}	H_2O 1; EtOH 2; bz 3; HOAc 2
2084	Benzene, pentachloro- Pentachlorobenzene	C_6HCl_5 250.34	608-93-5 86	277	4-05-00-00669 1.8342^{16}	H_2O 1; EtOH 1; eth 2; bz 2
2085	Benzene, pentachloro(dichloromethyl)-	C_7HCl_7 333.25	2136-95-0 119.5	334	4-05-00-00824	EtOH 2
2086	Benzene, pentachloromethyl-	$C_7H_3Cl_5$ 264.36	877-11-2 224.8	301	4-05-00-00823	EtOH 2; eth 2; bz 3; tol 3
2087	Benzene, pentachloronitro- Quintozene	$C_6Cl_5NO_2$ 295.34	82-68-8 144	8108 328 dec	4-05-00-00728 1.718^{25}	H_2O 1; EtOH 2; bz 3; chl 3
2088	Benzene, pentacosyl- Pentacosylbenzene	$C_{31}H_{56}$ 428.79	61828-06-6 59	380^{100}; 244^{1}	4-05-00-01241 0.8544^{20}	1.4799^{20}
2089	Benzene, pentadecyl-	$C_{21}H_{36}$ 288.52	2131-18-2 22	373	3-05-00-01119 0.8548^{20}	1.4815^{20}
2090	Benzene, pentaethyl- Pentaethylbenzene	$C_{16}H_{26}$ 218.38	605-01-6 <-20	277	4-05-00-01197 0.8971^{19}	1.5127^{20}
2091	Benzene, pentafluoro- Pentafluorobenzene	C_6HF_5 168.07	363-72-4 -47.3	85.7	4-05-00-00639 1.514^{25}	1.3905^{20}
2092	Benzene, pentafluoroiodo-	C_6F_5I 293.96	827-15-6 -29	166	4-05-00-00695 2.212^{20}	1.4950^{25}
2093	Benzene, pentafluoromethoxy-	$C_7H_3F_5O$ 198.09	389-40-2 -37	138.5	4-06-00-00782 1.493^{20}	1.4087^{20}
2094	Benzene, pentafluoromethyl- 2,3,4,5,6-Pentafluorotoluene	$C_7H_3F_5$ 182.09	771-56-2 -29.8	117.5	1.440^{20}	1.4016^{25}
2095	Benzene, pentafluoro(trifluoromethyl)- Perfluorotoluene	C_7F_8 236.06	434-64-0 -65.6	104.5	4-05-00-00804	1.3670^{20}
2096	Benzene, pentaiodo- Pentaiodobenzene	C_6HI_5 707.60	608-96-8 172	sub	4-05-00-00702	bz 4; chl 4
2097	Benzene, pentamethyl- Pentamethylbenzene	$C_{11}H_{16}$ 148.25	700-12-9 54.5	232	4-05-00-01109 0.917^{20}	H_2O 1; EtOH 4; bz 4; chl 3 1.527^{20}
2098	Benzene, 1,1'-(1,5-pentanediyl)bis- 1,5-Diphenylpentane	$C_{17}H_{20}$ 224.35	1718-50-9 -8	327	4-05-00-01963 0.9812^{19}	H_2O 1 1.559^{19}
2099	Benzenepentanoic acid 5-Phenylvaleric acid	$C_{11}H_{14}O_2$ 178.23	2270-20-4 57.5	190^{30}	4-09-00-01864	H_2O 2; EtOH 4; os 3
2100	Benzenepentanoic acid, δ-oxo-, methyl ester	$C_{12}H_{14}O_3$ 206.24	1501-04-8 -2	186^{19}; 147^{8}	4-10-00-02783	H_2O 4
2101	Benzenepentanol	$C_{11}H_{16}O$ 164.25	10521-91-2	155^{20}	4-06-00-03373 0.9725^{20}	eth 4; EtOH 4 1.5156^{20}

No.	Name Synonym	Mol. Form. Mol. Wt.	CAS RN mp/°C	Merck No. bp/°C	Beil. Ref. den/g cm^{-3}	Solubility n_D
2102	Benzenepentanol, α-ethyl- 7-Phenylheptan-3-ol	$C_{13}H_{20}O$ 192.30	60012-63-7	 277.5; 150^{20}	3-06-00-02027 0.968^{20}	 1.5115^{20}
2103	Benzenepentanol, α-methyl- 6-Phenyl-2-hexanol	$C_{12}H_{18}O$ 178.27	38487-94-4	 148^{18}	3-06-00-01995 0.9567^{20}	 1.5079^{20}
2104	Benzene, 1-pentenyl-	$C_{11}H_{14}$ 146.23	826-18-6	 217	4-05-00-01401 0.8782^{22}	 1.5158^{20}
2105	Benzene, 3-pentenyl-	$C_{11}H_{14}$ 146.23	1745-16-0	 201	4-05-00-01401 0.8884^{16}	 1.5089^{16}
2106	Benzene, 4-pentenyl- 1-Pentene, 5-phenyl	$C_{11}H_{14}$ 146.23	1075-74-7	 201; 77^{10}	3-05-00-01238 0.8889^{20}	 1.5065^{20}
2107	Benzene, pentyl- Amylbenzene	$C_{11}H_{16}$ 148.25	538-68-1 -75	635 205.4	4-05-00-01085 0.8585^{20}	H_2O 1; EtOH 5; eth 5; ace 5 1.4878^{20}
2108	Benzene, 1,1'-pentylidenebis- Pentane, 1,1-diphenyl-	$C_{17}H_{20}$ 224.35	1726-12-1 -12	 308	4-05-00-01966 0.9659^{20}	H_2O 1 1.5511^{20}
2109	Benzene, (pentyloxy)- Pentyl phenyl ether	$C_{11}H_{16}O$ 164.25	2050-04-6	 200; 111^{17}	3-06-00-00552 0.9270^{20}	ace 4; EtOH 4 1.4947^{20}
2110	Benzene, 1,1'-(phenylmethylene)bis[4-methyl-	$C_{21}H_{20}$ 272.39	603-39-4 56	 218^{12}	4-05-00-02520	H_2O 1; EtOH 3; eth 4; bz 4
2111	Benzene, 1-(phenylmethyl)-4-propyl-	$C_{16}H_{18}$ 210.32	62155-41-3	 $152\text{-}5^{10}$	4-05-00-01950 0.9739^{18}	 1.3552^{20}
2112	Benzene, [(phenylmethyl)sulfonyl]-	$C_{13}H_{12}O_2S$ 232.30	3112-88-7 146		4-06-00-02647 1.1261^{153}	H_2O 1; EtOH 2; eth 2; bz 2
2113	Benzene, [(phenylmethyl)thio]	$C_{13}H_{12}S$ 200.30	831-91-4 43.5	 197^{27}	4-06-00-02644	H_2O 1; EtOH 3; eth 3; con sulf 3
2114	Benzenepropanal Hydrocinnamic aldehyde	$C_9H_{10}O$ 134.18	104-53-0 47	 224; 117^{28}	4-07-00-00692 1.0190^{20}	H_2O 1; EtOH 4; eth 5
2115	Benzenepropanal, 2-methyl- Hydrocinnamaldehyde, *o*-methyl-	$C_{10}H_{12}O$ 148.20	19564-40-0	 120^{13}	2-07-00-00246 0.998^{19}	 1.522^{19}
2116	Benzenepropanal, 4-methyl-	$C_{10}H_{12}O$ 148.20	5406-12-2	 223	2-07-00-00247 0.999^{14}	 1.525^{14}
2117	Benzenepropanal, α-methyl-4-(1-methylethyl)- 3-*p*-Cumenyl-2-methylpropionaldehyde	$C_{13}H_{18}O$ 190.29	103-95-7	 135^{99}; 115^9	4-07-00-00788 0.951^{15}	bz 4; eth 4; EtOH 4 1.5068^{20}
2118	Benzenepropanal, α-methyl-β-oxo-	$C_{10}H_{10}O_2$ 162.19	16837-43-7 119.3	 155^{125}	3-07-00-03497	bz 4; EtOH 4; MeOH 4
2119	Benzenepropanamine, α-methyl- 1-Methyl-3-phenylpropylamine	$C_{10}H_{15}N$ 149.24	22374-89-6 143	 223; 101^{14}	4-12-00-02811 0.9289^{15}	EtOH 4 1.5152^{20}
2120	Benzenepropanaminium, γ-(aminocarbonyl)-*N*-methyl-*N*,*N*-bis(1-methylethyl)-γ-phenyl-, iod Ammonium, (3-carbamoyl-3,3-diphenylpropyl) diisopropyl methyl,iodide	$C_{23}H_{33}IN_2O$ 480.43	71-81-8	5090	4-14-00-01853	
2121	Benzene, 1,1'-(1,3-propanediyl)bis- 1,3-Diphenylpropane	$C_{15}H_{16}$ 196.29	1081-75-0 6	 300.3	4-05-00-01913 1.007^{20}	 1.5760^{20}
2122	Benzene, 1,1'-[1,3-propanediylbis(oxy)]bis-	$C_{15}H_{16}O_2$ 228.29	726-44-3 62	 339	4-06-00-00585	H_2O 1; EtOH 3; eth 3
2123	Benzenepropanenitrile Hydrocinnamonitrile	C_9H_9N 131.18	645-59-0 -1	 261; 141^{25}	4-09-00-01764 1.0016^{20}	EtOH 3; eth 3; chl 2 1.5266^{28}
2124	Benzenepropanenitrile, β-oxo- Benzoylacetonitrile	C_9H_7NO 145.16	614-16-4 80.5	 160^{10}	4-10-00-02757	H_2O 2; EtOH 3; eth 3; bz 3
2125	Benzenepropanethiol	$C_9H_{12}S$ 152.26	24734-68-7	 109^{10}	4-06-00-03206 1.01^{25}	 1.5494^{20}
2126	Benzenepropanoic acid Hydrocinnamic acid	$C_9H_{10}O_2$ 150.18	501-52-0 107	4707 279.8	4-09-00-01752 1.0712^{49}	H_2O 3; EtOH 3; eth 3; bz 4
2127	Benzenepropanoic acid, α-acetyl-α-(phenylmethyl)-, ethyl ester	$C_{20}H_{22}O_3$ 310.39	42597-26-2 57	 233	3-10-00-03350	EtOH 2; eth 2
2128	Benzenepropanoic acid, β-(aminomethyl)- 4-Amino-3-phenylbutyric acid	$C_{10}H_{13}NO_2$ 179.22	1078-21-3 250-5 dec	476	4-14-00-01723	
2129	Benzenepropanoic acid, α,α-dimethyl-, methyl ester Methyl α,α-dimethylhydrocinnamate	$C_{12}H_{16}O_2$ 192.26	14248-22-7	 $121\text{-}2^{23}$	4-09-00-01879 1.0043^{25}	 1.4945^{25}
2130	Benzenepropanoic acid, β,β-dimethyl-	$C_{11}H_{14}O_2$ 178.23	1010-48-6 60	 167^{10}	4-09-00-01874	H_2O 3; ace 3; peth 3 1.5182^{20}
2131	Benzenepropanoic acid, ethyl ester	$C_{11}H_{14}O_2$ 178.23	2021-28-5	 247.2	4-09-00-01753 1.0147^{20}	eth 4; EtOH 4 1.4954^{20}
2132	Benzenepropanoic acid, α-ethyl-β-oxo-, ethyl ester	$C_{13}H_{16}O_3$ 220.27	24346-56-3	 152^7	3-10-00-03065 1.0706^{15}	eth 4 1.509^{15}
2133	Benzenepropanoic acid, α-hydroxy-, (±)- (±)-3-Phenyllactic acid	$C_9H_{10}O_3$ 166.18	828-01-3 98	 149^{15}	2-10-00-00154	H_2O 4; ace 4; eth 4; EtOH 4
2134	Benzenepropanoic acid, 4-hydroxy- *p*-Hydroxyhydrocinnamic acid	$C_9H_{10}O_3$ 166.18	501-97-3 130.8	 $208\text{-}10^{14}$	4-10-00-00631	H_2O 3; EtOH 3; eth 3; bz 3
2135	Benzenepropanoic acid, 4-hydroxy-3,5-diiodo-α-phenyl- Iodoalphionic acid	$C_{15}H_{12}I_2O_3$ 494.07	577-91-3 164	4919	4-10-00-01291	H_2O 1; EtOH 3; eth 3; bz 2
2136	Benzenepropanoic acid, 4-(4-hydroxy-3-iodophenoxy)-3,5-diiodo- 3,3',5-Triiodothyropropionic acid	$C_{15}H_{11}I_3O_4$ 635.96	51-26-3	9346	4-10-00-00641	
2137	Benzenepropanoic acid, α-methyl- α-Methyldihydrocinnamic acid	$C_{10}H_{12}O_2$ 164.20	1009-67-2 36.5	 272; 152^{12}	2-09-00-00357	eth 4; EtOH 4

No.	Name	Synonym	Mol. Form.	Mol. Wt.	CAS RN	mp/°C	Merck No.	bp/°C	Beil. Ref.	den/g cm^{-3}	Solubility	n_D
2138	Benzenepropanoic acid, β-methyl-, (±)-		$C_{10}H_{12}O_2$	164.20	772-17-8	46.5		$140-5^{3}$	4-09-00-01814	1.0701^{20}	H_2O 2; peth 3	1.5155^{20}
2139	Benzenepropanoic acid, methyl ester	Methyl dihydrocinnamate	$C_{10}H_{12}O_2$	164.20	103-25-3			$238.5; 91^{4}$	4-09-00-01753	1.0455^{25}	H_2O 1; EtOH 3; eth 3; bz 3	
2140	Benzenepropanoic acid, 1-methylethyl ester		$C_{12}H_{16}O_2$	192.26	22767-95-9			126^{11}	2-09-00-00339	0.9860^{25}	eth 4; EtOH 4	
2141	Benzenepropanoic acid, β-oxo-, ethyl ester	Ethyl benzoylacetate	$C_{11}H_{12}O_3$	192.21	94-02-0	<0	3726	267 dec; 167^{20}	4-10-00-02756	1.1202^{15}	H_2O 2; EtOH 3; eth 3	1.5317^{15}
2142	Benzenepropanoic acid, β-oxo-, methyl ester		$C_{10}H_{10}O_3$	178.19	614-27-7			265 dec; 151^{12}	4-10-00-02755	1.158^{29}	ace 4; eth 4; EtOH 4	1.537^{20}
2143	Benzenepropanoic acid, α-phenyl-, (±)-	Propionic acid, 2,3-diphenyl (*DL*)	$C_{15}H_{14}O_2$	226.27	94942-89-9	88(form a); 95(form b)		337.5	4-09-00-02528	1.1495^{96}	bz 4; eth 4; EtOH 4	
2144	Benzenepropanoic acid, α-(phenylmethyl)-		$C_{16}H_{16}O_2$	240.30	618-68-8	90		235^{18}	4-09-00-02546		bz 4; eth 4; EtOH 4	
2145	Benzenepropanoic acid, phenylmethyl ester		$C_{16}H_{16}O_2$	240.30	22767-96-0			$325; 198^{20}$	4-09-00-01755	1.090^{15}	eth 4	
2146	Benzenepropanoic acid, β-propyl-		$C_{12}H_{16}O_2$	192.26	5703-52-6			152^{4}	4-09-00-01900	1.025^{25}		1.5078
2147	Benzenepropanoic-β-14C acid, α-bromo-	Hydrocinnamic acid, α-bromo-	$C_9H_9BrO_2$	229.07	16503-53-0	52		150^{13}	4-09-00-01771	1.4790^{25}	ace 4; bz 4; EtOH 4	
2148	Benzenepropanol	Hydrocinnamyl alcohol	$C_9H_{12}O$	136.19	122-97-4	<-18		235	4-06-00-03198	0.995^{25}	H_2O 3; EtOH 5; eth 5; ctc 3	1.5357^{25}
2149	Benzenepropanol, carbamate	Phenprobamate	$C_{10}H_{13}NO_2$	179.22	673-31-4	102	7229				H_2O 1; EtOH 3; chl 3	
2150	Benzenepropanol, α,α-dimethyl-	Benzyl-*tert*-butanol	$C_{11}H_{16}O$	164.25	103-05-9	24.5		121^{13}	4-06-00-03380	0.9626^{21}	H_2O 1; EtOH 4; eth 4; ace 4	1.5077^{21}
2151	Benzenepropanol, β,β-dimethyl-		$C_{11}H_{16}O$	164.25	13351-61-6	34.5		$125^{14.5}$	2-06-00-00507		eth 4; EtOH 4	
2152	Benzenepropanol, α-ethyl-, (*S*)-	3-Pentanol, 1-phenyl (*d*)	$C_{11}H_{16}O$	164.25	71747-37-0	38		143^{19}	3-06-00-01953	0.9687^{20}	EtOH 4	
2153	Benzenepropanol, 4-hydroxy-3-methoxy-		$C_{10}H_{14}O_3$	182.22	2305-13-7	65		197^{15}	4-06-00-07405		eth 4; EtOH 4	1.5545^{25}
2154	Benzenepropanol, 4-methoxy-		$C_{10}H_{14}O_2$	166.22	5406-18-8	26		166^{18}	4-06-00-05982			1.5315^{20}
2155	Benzenepropanol, α-methyl-		$C_{10}H_{14}O$	150.22	2344-70-9			$239; 123^{15}$	4-06-00-03275	0.9899^{16}		1.517^{16}
2156	Benzenepropanol, γ-methyl-		$C_{10}H_{14}O$	150.22	2722-36-3			$139^{33}; 110^{6}$	4-06-00-03286	0.9832^{20}		1.4965^{20}
2157	Benzenepropanol, γ-methylene-	3-Buten-1-ol, 3-phenyl-	$C_{10}H_{12}O$	148.20	3174-83-2			123^{10}	4-06-00-03840	1.0272^{20}		1.5577^{20}
2158	Benzenepropanol, β-phenyl-		$C_{15}H_{16}O$	212.29	3536-29-6	51		$310; 190^{15}$	4-06-00-04754	1.0585^{24}		1.5742^{19}
2159	Benzenepropanol, γ-propyl-	1-Hexanol, 3-phenyl	$C_{12}H_{18}O$	178.27	67700-22-5			127^{5}	3-06-00-02001	0.955^{25}		1.5101^{25}
2160	Benzenepropanoyl chloride		C_9H_9ClO	168.62	645-45-4			225 dec; 105^{10}	4-09-00-01762	1.135^{21}	eth 3; CS_2 3	
2161	Benzene, 1,1'-(1-propene-1,3-diyl)bis-		$C_{15}H_{14}$	194.28	5209-18-7	16		$302; 178^{15}$	4-05-00-02187	1.0120^{20}		
2162	Benzene, 1-propenyl-		C_9H_{10}	118.18	637-50-3	-27.1		175.5	3-05-00-01185	0.9019^{25}	ace 4; bz 4; eth 4; EtOH 4	1.5508^{20}
2163	Benzene, 1-propenyl-, (*E*)-		C_9H_{10}	118.18	873-66-5	-29.3		178.3	4-05-00-01359	0.9023^{25}	H_2O 1; EtOH 5; eth 5; ace 5	1.5506^{20}
2164	Benzene, 1-propenyl-, (*Z*)-		C_9H_{10}	118.18	766-90-5	-61.6		167.5	4-05-00-01359	0.9088^{20}	H_2O 1; EtOH 5; eth 5; ace 5	1.5420^{20}
2165	Benzene, 2-propenyl-		C_9H_{10}	118.18	300-57-2	-40		156	4-05-00-01362	0.8920^{20}	H_2O 1; EtOH 3; eth 3; bz 3	1.5131^{20}
2166	Benzene, 1,1'-(1-propenylidene)bis-		$C_{15}H_{14}$	194.28	778-66-5	52		280-1 149^{11}	4-05-00-02192	1.0250^{20}	H_2O 1; EtOH 3; bz 3	1.5880^{20}
2167	Benzene, (2-propenyloxy)-		$C_9H_{10}O$	134.18	1746-13-0			191.7	4-06-00-00562	0.9811^{20}	H_2O 1; EtOH 3; eth 5; ctc 2	1.5223^{20}
2168	Benzene, propoxy-		$C_9H_{12}O$	136.19	622-85-5	-27		189.9	4-06-00-00556	0.9474^{20}	EtOH 3; eth 3	1.5014^{20}
2169	Benzene, propyl-	Propylbenzene	C_9H_{12}	120.19	103-65-1	-99.5	7856	159.2	4-05-00-00977	0.8620^{20}	H_2O 1; EtOH 5; eth 5; ace 5	1.4920^{20}
2170	Benzene, (1-propylheptadecyl)-	Eicosane, 4-phenyl-	$C_{26}H_{46}$	358.65	2400-03-5	31.4		199^{1}	4-05-00-01223	0.8546^{20}		1.4794^{20}
2171	Benzene, 1,1'-propylidenebis-	1,1-Diphenylpropane	$C_{15}H_{16}$	196.29	1530-03-6	13.3		281.6	4-05-00-01920	0.9951^{14}		1.5681^{14}
2172	Benzene, (propylthio)-		$C_9H_{12}S$	152.26	874-79-3	-45		220	4-06-00-01470	0.9995^{20}		1.5571^{20}
2173	Benzene, 1-propynyl-		C_9H_8	116.16	673-32-5			183	4-05-00-01530	0.942^{15}		1.563^{15}
2174	Benzeneseleninic acid	Phenylseleninic acid	$C_6H_6O_2Se$	189.07	6996-92-5	124.5			4-11-00-00708	1.93^{20}	H_2O 2; bz 1; alk 4	
2175	Benzene, 1,1'-selenobis-		$C_{12}H_{10}Se$	233.17	1132-39-4	2.5		301.5	4-06-00-01779	1.351^{20}	H_2O 1; EtOH 5; eth 5; bz 3	1.5500^{20}

No.	Name / Synonym	Mol. Form.	Mol. Wt.	CAS RN	mp/°C	Merck No.	bp/°C	Beil. Ref.	den/g cm^{-3}	Solubility	n_D
2176	Benzene, 1,1'-selenobis[4-methyl-	$C_{14}H_{14}Se$	261.23	22077-55-0	69.5		196^{16}	4-06-00-02218		H_2O 1; EtOH 3; eth 2; chl 2	
2177	Benzeneselenol	C_6H_6Se	157.07	645-96-5			183.6; 84^{25}	4-06-00-01777	1.4865^{15}	H_2O 1; EtOH 3; eth 4; ctc 4	
2178	Benzenesulfenyl chloride, 2,4-dinitro- 2,4-Dinitrobenzenesulfenyl chloride	$C_6H_3ClN_2O_4S$	234.62	528-76-7	99	3267		4-06-00-01772		bz 4; chl 4; HOAc 4; peth 2	
2179	Benzenesulfenyl chloride, 4-nitro-	$C_6H_4ClNO_2S$	189.62	937-32-6	52		$125^{0.1}$	4-06-00-01717		bz 4	
2180	Benzenesulfinic acid	$C_6H_6O_2S$	142.18	618-41-7	84		dec	4-11-00-00003		H_2O 2; EtOH 3; eth 3; bz 3	
2181	Benzenesulfinic acid, 2-methyl- *o*-Toluenesulfinic acid	$C_7H_8O_2S$	156.21	13165-77-0	80		dec	4-11-00-00008		H_2O 4; ace 4; eth 4; EtOH 4	
2182	Benzenesulfinic acid, 4-methyl- *p*-Toluenesulfinic acid	$C_7H_8O_2S$	156.21	536-57-2	86.5	9458		4-11-00-00009		H_2O 3; EtOH 4; eth 4; bz 2	
2183	Benzene, 1,1'-sulfinylbis-	$C_{12}H_{10}OS$	202.28	945-51-7	71.2		340^{16}	4-06-00-01489		EtOH 4; eth 4; bz 4; chl 2	
2184	Benzene, 1,1'-[sulfinylbis(methylene)]bis-	$C_{14}H_{14}OS$	230.33	621-08-9	134		210 dec	4-06-00-02651		H_2O 1; EtOH 4; eth 3	
2185	Benzenesulfinyl chloride	C_6H_5ClOS	160.62	4972-29-6	38		$71\text{-}2^{1.5}$	4-11-00-00005	1.3469^{25}	eth 3; chl 3	1.3470^{25}
2186	Benzenesulfinyl chloride, 4-methyl-	C_7H_7ClOS	174.65	10439-23-3	57		$113^{3.5}$	4-11-00-00010		chl 4	
2187	Benzenesulfonamide, 4-amino- Sulfanilamide	$C_6H_8N_2O_2S$	172.21	63-74-1	165.5	8898		4-14-00-02658	1.08^{25}	H_2O 3; EtOH 3; eth 3; ace 3	
2188	Benzenesulfonamide, 4-amino-*N*-[4-(aminosulfonyl)phenyl]- *N*4-Sulfanilylsulfanilamide	$C_{12}H_{13}N_3O_4S_2$	327.39	547-52-4	137	8905		4-14-00-02760		H_2O 2; EtOH 3; eth 3; ace 3	
2189	Benzenesulfonamide, 4-amino-*N*-(aminothioxomethyl)- Sulfathiourea	$C_7H_9N_3O_2S_2$	231.30	515-49-1	182	8921		4-14-00-02671		H_2O 1; EtOH 2	
2190	Benzenesulfonamide, 4-amino-*N*-(aminocarbonyl)- Sulfanilylurea	$C_7H_9N_3O_3S$	215.23	547-44-4	147 dec	8906		4-14-00-02667			
2191	Benzenesulfonamide, 4-amino-*N*-(aminoiminomethyl)- Sulfaguanidine	$C_7H_{10}N_4O_2S$	214.25	57-67-0	191.5	8879		4-14-00-02668			
2192	Benzenesulfonamide, 4-amino-*N*-[(butylamino)carbonyl]- Carbutamide	$C_{11}H_{17}N_3O_3S$	271.34	339-43-5	144.5	1839		4-14-00-02667			
2193	Benzenesulfonamide, 4-amino-*N*-(6-chloro-3-pyridazinyl)- Sulfachlorpyridazine	$C_{10}H_9ClN_4O_2S$	284.73	80-32-0	187	8871		4-25-00-02069			
2194	Benzenesulfonamide, 4-amino-*N*-(2,6-dimethoxy-4-pyrimidinyl)- Sulfadimethoxine	$C_{12}H_{14}N_4O_4S$	310.33	122-11-2	203.5	8876		4-25-00-03483			
2195	Benzenesulfonamide, 4-amino-*N*-(3,4-dimethyl-5-isoxazolyl)- Sulfisoxazole	$C_{11}H_{13}N_3O_3S$	267.31	127-69-5	191	8930		4-27-00-04747			
2196	Benzenesulfonamide, 4-amino-*N*-(4,6-dimethyl-2-pyrimidinyl)- Sulfamethazine	$C_{12}H_{14}N_4O_2S$	278.33	57-68-1	198.5	8886		4-25-00-02215		H_2O 3; acid 3; alk 3	
2197	Benzenesulfonamide, 4-amino-*N*-(6-methoxy-3-pyridazinyl)- Sulfamethoxypyridazine	$C_{11}H_{12}N_4O_3S$	280.31	80-35-3	182.5	8890		4-25-00-03332			
2198	Benzenesulfonamide, 4-amino-*N*-(5-methyl-3-isoxazolyl)- Sulfamethoxazole	$C_{10}H_{11}N_3O_3S$	253.28	723-46-6		8889		4-27-00-04685			
2199	Benzenesulfonamide, 4-amino-*N*-(4-methyl-2-pyrimidinyl)- Sulfamerazine	$C_{11}H_{12}N_4O_2S$	264.31	127-79-7	236	8884		4-25-00-02159		H_2O 2; EtOH 2; eth 1; ace 2	
2200	Benzenesulfonamide, 4-amino-*N*-(4-methyl-2-thiazolyl)- Sulfamethylthiazole	$C_{10}H_{11}N_3O_2S_2$	269.35	515-59-3	237	8891		4-27-00-04715		EtOH 4	
2201	Benzenesulfonamide, 4-amino-*N*-(1-phenyl-1H-pyrazol-5-yl)- Sulfaphenazole	$C_{15}H_{14}N_4O_2S$	314.37	526-08-9		8910		4-25-00-02029			
2202	Benzenesulfonamide, 4-amino-*N*-pyrazinyl- Sulfapyrazine	$C_{10}H_{10}N_4O_2S$	250.28	116-44-9	251	8912		4-25-00-02143		H_2O 1; EtOH 1; eth 1; ace 2	
2203	Benzenesulfonamide, 4-amino-*N*-2-pyridinyl- Sulfapyridine	$C_{11}H_{11}N_3O_2S$	249.29	144-83-2	192	8913		5-22-08-00424		H_2O 1; EtOH 3; bz 1; ctc 1	
2204	Benzenesulfonamide, 4-amino-*N*-2-pyrimidinyl- Sulfadiazine	$C_{10}H_{10}N_4O_2S$	250.28	68-35-9	255-6 dec	8874		4-25-00-02097		H_2O 2; EtOH 2; ace 2	
2205	Benzenesulfonamide, 4-amino-*N*-2-quinoxalinyl- Sulfaquinoxaline	$C_{14}H_{12}N_4O_2S$	300.34	59-40-5	247.5	8914		4-25-00-02606		H_2O 2; EtOH 2; ace 2; aq alk 3	
2206	Benzenesulfonamide, 4-amino-*N*-2-thiazolyl- Sulfathiazole	$C_9H_9N_3O_2S_2$	255.32	72-14-0	175(form a); 202(form b)	8920		4-27-00-04623		H_2O 2; EtOH 2; DMSO-d_6 2	

No.	Name Synonym	Mol. Form. Mol. Wt.	CAS RN mp/°C	Merck No. bp/°C	Beil. Ref. den/g cm^{-3}	Solubility n_D
2207	Benzenesulfonamide, *N*-[(butylamino)carbonyl]-4-methyl- Tolbutamide	$C_{12}H_{18}N_2O_3S$ 270.35	64-77-7 128.5	9432	4-11-00-00396 1.245^{25}	H_2O 2; EtOH 3; eth 3; chl 3
2208	Benzenesulfonamide, 4-chloro-*N*-[(propylamino)carbonyl]- Chloropropamide	$C_{10}H_{13}ClN_2O_3S$ 276.74	94-20-2	2187	4-11-00-00119	
2209	Benzenesulfonamide, 4-[(2,4-diaminophenyl)azo]- Prontosil	$C_{12}H_{14}ClN_5O_2S$ 327.79	103-12-8 249.5	8895	4-16-00-00563	H_2O 2; EtOH 3; ace 3; oils 3
2210	Benzenesulfonamide, *N*,*N*-dichloro-4-methyl- Dichloramine-T	$C_7H_7Cl_2NO_2S$ 240.11	473-34-7 83	3034	3-11-00-00301	H_2O 1; EtOH 3; eth 3; bz 3
2211	Benzenesulfonamide, *N*,4-dimethyl-	$C_8H_{11}NO_2S$ 185.25	640-61-9 78.5		4-11-00-00377 1.340^{25}	eth 4; EtOH 4
2212	Benzenesulfonamide, *N*,4-dimethyl-*N*-nitroso- *p*-Tolylsulfonylmethylnitrosamide	$C_8H_{10}N_2O_3S$ 214.25	80-11-5 60	9469	4-11-00-00478	H_2O 1; EtOH 4; eth 4
2213	Benzenesulfonamide, 4-[(phenylmethyl)amino]- *N*4-Benzylsulfanilamide	$C_{13}H_{14}N_2O_2S$ 262.33	104-22-3 171	1161	3-14-00-02026	
2214	Benzenesulfonic acid Besylic acid	$C_6H_6O_3S$ 158.18	98-11-3 65	1078	4-11-00-00027	H_2O 4; EtOH 4; eth 1; bz 2
2215	Benzenesulfonic acid, 2-amino- Orthanilic acid	$C_6H_7NO_3S$ 173.19	88-21-1 >320 dec	6835	4-14-00-02638	H_2O 2; EtOH 1; eth 1
2216	Benzenesulfonic acid, 3-amino- Metanilic acid	$C_6H_7NO_3S$ 173.19	121-47-1 dec	5832	4-14-00-02647	H_2O 2; EtOH 2; eth 1
2217	Benzenesulfonic acid, 4-amino- Sulfanilic acid	$C_6H_7NO_3S$ 173.19	121-57-3 288	8901	4-14-00-02655 1.485^{25}	H_2O 2; EtOH 1; eth 1
2218	Benzenesulfonic acid, 2-amino-5-[(4-amino-3-sulfophenyl)(4-imino-3-sulfo-2,5-cyclohexadien-1-ylidene)methyl]-3 Fuchsin, acid	$C_{20}H_{17}N_3Na_2O_9S_3$ 585.55	3244-88-0	103		H_2O 2; EtOH 2
2219	Benzenesulfonic acid, *sec*-butyl ester Benzenesulfonic acid, chloride	$C_{10}H_{14}O_3S$ 214.29	103275-78-1 14.5	251 dec; 120^{10}	4-11-00-00032 1.3830^{15}	H_2O 1; EtOH 4; eth 3; ctc 3
2220	Benzenesulfonic acid, 4-chloro- *p*-Chlorobenzenesulfonic acid	$C_6H_5ClO_3S$ 192.62	98-66-8 67	2122 $147\text{-}8^{25}$	4-11-00-00107	H_2O 3; EtOH 3; eth 1; bz 1
2221	Benzenesulfonic acid, 4-chloro-, 4-chlorophenyl ester Ovex	$C_{12}H_8Cl_2O_3S$ 303.17	80-33-1 86.5	6856	4-11-00-00109	H_2O 1; EtOH 2; ace 3
2222	Benzenesulfonic acid, 4-[[4-(dimethylamino)phenyl]azo]-, sodium salt Methyl orange	$C_{14}H_{14}N_3NaO_3S$ 327.34	547-58-0 dec	6019		H_2O 2; EtOH 2; eth 1; py 2
2223	Benzenesulfonic acid, ethyl ester Ethyl benzenesulfonate	$C_8H_{10}O_3S$ 186.23	515-46-8	1078 156^{15}	4-11-00-00030 1.2167^{20}	H_2O 2; EtOH 3; eth 4; chl 4 1.5081^{20}
2224	Benzenesulfonic acid, 4-hydrazino- *p*-Sulfophenylhydrazine	$C_6H_8N_2O_3S$ 188.21	98-71-5 286	4695	3-15-00-00865	H_2O 2; EtOH 2
2225	Benzenesulfonic acid, 2-hydroxy- *o*-Phenolsulfonic acid	$C_6H_6O_4S$ 174.18	609-46-1 145 dec		4-11-00-00574	H_2O 4; EtOH 4
2226	Benzenesulfonic acid, 4-hydroxy- *p*-Phenolsulfonic acid	$C_6H_6O_4S$ 174.18	98-67-9	7212	4-11-00-00582	H_2O 4; EtOH 4
2227	Benzenesulfonic acid, methyl ester Methyl benzenesulfonate	$C_7H_8O_3S$ 172.20	80-18-2	154^{20}	4-11-00-00029 1.2730^{17}	H_2O 2; EtOH 4; eth 4; chl 4 1.5151^{20}
2228	Benzenesulfonic acid, 2-methyl-	$C_7H_8O_3S$ 172.20	88-20-0 67.5	128.8^{25}	4-11-00-00228	H_2O 4; EtOH 3; eth 1
2229	Benzenesulfonic acid, 4-methyl- *p*-Toluenesulfonic acid	$C_7H_8O_3S$ 172.20	104-15-4 104.5	9459 140^{20}	4-11-00-00241	H_2O 4; EtOH 3; eth 3
2230	Benzenesulfonic acid, 4-methyl-, butyl ester	$C_{11}H_{16}O_3S$ 228.31	778-28-9	$164\text{-}6^{6}$	4-11-00-00250 1.1319^{20}	H_2O 1; eth 3; ctc 2 1.5050^{20}
2231	Benzenesulfonic acid, 4-methyl-, ethyl ester Ethyl *p*-toluenesulfonate	$C_9H_{12}O_3S$ 200.26	80-40-0 34.5	3814 173^{15}	4-11-00-00248 1.166^{48}	H_2O 1; EtOH 3; eth 3; ctc 2
2232	Benzenesulfonic acid, 4-methyl-, methyl ester	$C_8H_{10}O_3S$ 186.23	80-48-8 28.5	292; 186^{22}	4-11-00-00247 1.2087^{40}	H_2O 1; EtOH 4; eth 3; bz 4
2233	Benzenesulfonic acid, 4-methyl-, propyl ester	$C_{10}H_{14}O_3S$ 214.29	599-91-7 <-20	189^{9}	4-11-00-00249 1.144^{20}	1.4998^{20}
2234	Benzenesulfonic acid, 4-(phenylamino)- *N*-Phenylsulfanilic acid	$C_{12}H_{11}NO_3S$ 249.29	101-57-5 206	7284	3-14-00-02025	H_2O 4; EtOH 4
2235	Benzenesulfonic acid, propyl ester Propyl benzenesulfonate	$C_9H_{12}O_3S$ 200.26	80-42-2	$162\text{-}3^{15}$	4-11-00-00031 1.1804^{17}	H_2O 2; EtOH 3; eth 4; chl 4 1.5035^{25}
2236	Benzenesulfonothioic acid, 4-chloro-, *S*-phenyl ester Benzenesulfonic acid, *p*-chlorothio-, *S*-phenyl ester	$C_{12}H_9ClO_2S_2$ 284.79	1142-97-8 -86.8	118.5	4-11-00-00223 0.8341^{20}	1.4087^{14}
2237	Benzene, 1,1'-sulfonylbis- Diphenyl sulfone	$C_{12}H_{10}O_2S$ 218.28	127-63-9 128.5	3336 379	4-06-00-01490 1.252^{20}	H_2O 1; EtOH 3; eth 3; bz 3
2238	Benzene, 1,1'-sulfonylbis[4-bromo-	$C_{12}H_8Br_2O_2S$ 376.07	2050-48-8 172		4-06-00-01652 1.88^{25}	H_2O 1; EtOH 2; HOAc 3
2239	Benzene, 1,1'-sulfonylbis[4-chloro-	$C_{12}H_8Cl_2O_2S$ 287.17	80-07-9 147.9	250^{10}	4-06-00-01587	H_2O 2; EtOH 3; chl 3
2240	Benzene, 1,1'-sulfonylbis[4-methyl-	$C_{14}H_{14}O_2S$ 246.33	599-66-6 159	406	4-06-00-02174	H_2O 2; EtOH 3; eth 2; bz 3
2241	Benzene, 1,1'-[sulfonylbis(methylene)]bis-	$C_{14}H_{14}O_2S$ 246.33	620-32-6 152	290 dec	4-06-00-02651	H_2O 1; EtOH 2; ace 4; bz 3

No.	Name Synonym	Mol. Form. Mol. Wt.	CAS RN mp/°C	Merck No. bp/°C	Beil. Ref. den/g cm^{-3}	Solubility n_D
2242	Benzenesulfonyl chloride Phenylsulfonyl chloride	$C_6H_5ClO_2S$ 176.62	98-09-9 14.5	1080 251 dec	4-11-00-00049 1.3470^{15}	H_2O 1; EtOH 4; eth 3; ctc 3
2243	Benzenesulfonyl chloride, 4-(acetylamino)- Acetylsulfanilyl chloride	$C_8H_8ClNO_3S$ 233.68	121-60-8 149	99	4-14-00-02703	EtOH 4; eth 4; bz 3; chl 3
2244	Benzenesulfonyl chloride, 4-bromo- *p*-Brosyl chloride	$C_6H_4BrClO_2S$ 255.52	98-58-8 76	1395 153^{15}	4-11-00-00162	H_2O 1; eth 4; chl 3
2245	Benzenesulfonyl chloride, 4-chloro-	$C_6H_4Cl_2O_2S$ 211.07	98-60-2 51	 141^{15}	4-11-00-00114	eth 4; bz 4
2246	Benzenesulfonyl chloride, 4-fluoro-	$C_6H_4ClFO_2S$ 194.61	349-88-2 30	 106^{9}	4-11-00-00102	bz 4; eth 4; chl 4
2247	Benzenesulfonyl chloride, 4-iodo- Pipsyl chloride	$C_6H_4ClIO_2S$ 302.52	98-61-3 85	7458	4-11-00-00169	
2248	Benzenesulfonyl chloride, 2-methoxy-	$C_7H_7ClO_3S$ 206.65	10130-87-7 56	 $126^{0.3}$	4-11-00-00575	eth 3; peth 3
2249	Benzenesulfonyl chloride, 2-methyl- *o*-Toluenesulfonyl chloride	$C_7H_7ClO_2S$ 190.65	133-59-5 10.2	 154^{36}	4-11-00-00229 1.3383^{20}	H_2O 1; EtOH 3; eth 3; bz 3 1.5565^{20}
2250	Benzenesulfonyl chloride, 4-methyl- *p*-Toluenesulfonyl chloride	$C_7H_7ClO_2S$ 190.65	98-59-9 71	9460 $145\text{-}6^{15}$	4-11-00-00375	H_2O 1; EtOH 3; eth 3; bz 4
2251	Benzenesulfonyl fluoride Phenylsulfonyl fluoride	$C_6H_5FO_2S$ 160.17	368-43-4	 203.5	4-11-00-00049 1.3286^{20}	EtOH 3; eth 3 1.4932^{18}
2252	Benzenesulfonyl fluoride, 4-amino- *p*-Sulfanilyl fluoride	$C_6H_6FNO_2S$ 175.18	98-62-4 68.5	8904	4-14-00-02658	
2253	Benzene, 1,2,3,5-tetrabromo- 1,2,3,5-Tetrabromobenzene	$C_6H_2Br_4$ 393.70	634-89-9 99.5	 329	4-05-00-00686	bz 4; eth 4; EtOH 4
2254	Benzene, 1,2,4,5-tetrabromo- 2,3,5,6-Tetrabromobenzene	$C_6H_2Br_4$ 393.70	636-28-2 182	 3.1^{20}	4-05-00-00687 3.072^{20}	eth 4
2255	Benzene, 1,2,3,4-tetrabromo-5,6-dimethyl-	$C_8H_6Br_4$ 421.75	2810-69-7 262	 374.5	3-05-00-00820	bz 4
2256	Benzene, 1,2,3,5-tetrabromo-4-methoxy- 2,3,4,6-Tetrabromoanisole	$C_7H_4Br_4O$ 423.72	95970-07-3 113.5	 340	3-06-00-00766	EtOH 4
2257	1,2,4,5-Benzenetetracarboxylic acid Pyromellitic acid	$C_{10}H_6O_8$ 254.15	89-05-4 276	8015	4-09-00-03804	H_2O 2; EtOH 3
2258	1,2,4,5-Benzenetetracarboxylic acid, tetramethyl ester	$C_{14}H_{14}O_8$ 310.26	635-10-9 144	 sub	4-09-00-03804	EtOH 4
2259	Benzene, 1,2,3,4-tetrachloro- 1,2,3,4-Tetrachlorobenzene	$C_6H_2Cl_4$ 215.89	634-66-2 47.5	 254	4-05-00-00667	H_2O 1; EtOH 2; eth 4; CS_2 4
2260	Benzene, 1,2,4,5-tetrachloro- s-Tetrachlorobenzene	$C_6H_2Cl_4$ 215.89	95-94-3 139.5	 244.5	4-05-00-00668 1.858^{22}	H_2O 1; EtOH 2; eth 3; bz 3
2261	Benzene, 1,2,3,5-tetrachloro-4,6-dimethyl-	$C_8H_6Cl_4$ 243.95	877-09-8 223		4-05-00-00944 1.703^{25}	H_2O 1; EtOH 1; eth 3; bz 3
2262	Benzene, 1,2,3,5-tetrachloro-4-methyl- 2,3,4,6-Tetrachlorotoluene	$C_7H_4Cl_4$ 229.92	875-40-1 94	 276.5	3-05-00-00702	bz 4; EtOH 4
2263	Benzene, 1,2,4,5-tetrachloro-3-methyl- 2,3,5,6-Tetrachlorotoluene	$C_7H_4Cl_4$ 229.92	1006-31-1 93.5	 sub	3-05-00-00702	eth 4; EtOH 4
2264	Benzene, 1,2,4,5-tetrachloro-3-nitro-	$C_6HCl_4NO_2$ 260.89	117-18-0 99.5	 304	4-05-00-00728 1.744^{25}	H_2O 1; EtOH 3; bz 3; chl 3
2265	Benzene, tetracosyl- Tetracosylbenzene	$C_{30}H_{54}$ 414.76	61828-05-5 57	 373^{100}; 238^{1}	4-05-00-01239 0.8544^{20}	 1.4800^{20}
2266	Benzene, tetradecyl-	$C_{20}H_{34}$ 274.49	1459-10-5 16	 359	4-05-00-01212 0.8549^{20}	 1.4818^{20}
2267	Benzene, 1,1'-tetradecylidenebis-	$C_{26}H_{38}$ 350.59	55268-63-8 17.9	 207^{1}	4-05-00-02017 0.9187^{20}	 1.5202^{20}
2268	Benzene, 1,2,3,4-tetraethyl- 1,2,3,4-Tetraethylbenzene	$C_{14}H_{22}$ 190.33	642-32-0 11.8	 252; 121.7^{14}	4-05-00-01168 0.8875^{20}	eth 4; EtOH 4 1.5125^{20}
2269	Benzene, 1,2,4,5-tetraethyl- 1,2,4,5-Tetraethylbenzene	$C_{14}H_{22}$ 190.33	635-81-4 10	 250	4-05-00-01168 0.8788^{20}	eth 4; EtOH 4 1.5054^{20}
2270	Benzene, 1,2,3,4-tetrafluoro- 1,2,3,4-Tetrafluorobenzene	$C_6H_2F_4$ 150.08	551-62-2	 94.3	4-05-00-00639	 1.4054^{20}
2271	Benzene, 1,2,3,5-tetrafluoro- 1,2,3,5-Tetrafluorobenzene	$C_6H_2F_4$ 150.08	2367-82-0 -48	 84.4	4-05-00-00639 1.319^{25}	 1.4035^{20}
2272	Benzene, 1,2,4,5-tetrafluoro- 1,2,4,5-Tetrafluorobenzene	$C_6H_2F_4$ 150.08	327-54-8 4.5	 90.2	4-05-00-00639 1.4255^{20}	 1.4075^{20}
2273	Benzene, 1,2,3,4-tetraiodo- 1,2,3,4-Tetraiodobenzene	$C_6H_2I_4$ 581.70	634-68-4 136	 sub	0-05-00-00229	eth 1; EtOH 4
2274	Benzene, 1,2,3,5-tetraiodo- 1,2,3,5-Tetraiodobenzene	$C_6H_2I_4$ 581.70	634-92-4 148	 sub	0-05-00-00229	HOAc 4
2275	Benzene, 1,2,4,5-tetraiodo- 1,2,4,5-Tetraiodobenzene	$C_6H_2I_4$ 581.70	636-31-7 254	 sub	0-05-00-00229	HOAc 4
2276	Benzene, 1,2,4,5-tetrakis(1-methylethyl)- 1,2,4,5-Tetraisopropylbenzene	$C_{18}H_{30}$ 246.44	635-11-0 118.4	 259; 133^{17}	4-05-00-01207 0.758^{150}	
2277	Benzene, 1,2,3,4-tetramethoxy-5-(2-propenyl)-	$C_{13}H_{18}O_4$ 238.28	15361-99-6 25	 145^{12}	3-06-00-06691 1.087^{25}	 1.5146^{25}
2278	Benzene, 1,2,3,4-tetramethyl-	$C_{10}H_{14}$ 134.22	488-23-3 -6.2	 205	4-05-00-01072 0.9052^{20}	H_2O 1; EtOH 5; eth 5; ace 5 1.5203^{20}
2279	Benzene, 1,2,3,5-tetramethyl- Isodurene	$C_{10}H_{14}$ 134.22	527-53-7 -23.7	5050 198	4-05-00-01073 0.8903^{20}	H_2O 1; EtOH 5; eth 5; ace 5 1.5130^{20}
2280	Benzene, 1,2,4,5-tetramethyl- Durene	$C_{10}H_{14}$ 134.22	95-93-2 79.3	3450 196.8	4-05-00-01076 0.8380^{81}	H_2O 1; EtOH 5; eth 5; ace 5 1.4790^{81}

No.	Name	Synonym	Mol. Form.	Mol. Wt.	CAS RN	mp/°C	Merck No.	bp/°C	Beil. Ref.	den/g cm^{-3}	Solubility	n_D
2281	Benzene, 1,1'-(1,1,4,4-tetramethyl-1,4-butanediyl)bis-	Bineophyl	$C_{20}H_{26}$	266.43	17648-05-4	61.8		$146-50^{0.5}$	4-05-00-01997		chl 3	
2282	Benzene, 1,2,4,5-tetramethyl-3,6-dinitro-		$C_{10}H_{12}N_2O_4$	224.22	5465-13-4	211.5		sub	4-05-00-01080		H_2O 1; EtOH 2; eth 4; bz 3	
2283	Benzene, 1,1'-thiobis-	Phenyl sulfide	$C_{12}H_{10}S$	186.28	139-66-2	-25.9	7285	296	4-06-00-01488	1.1136^{20}	H_2O 1; EtOH 3; eth 5; bz 5	1.6334^{20}
2284	Benzene, 1,1'-thiobis[4-bromo-		$C_{12}H_8Br_2S$	344.07	3393-78-0	115		268.5^{40}	4-06-00-01651	1.84^{25}	H_2O 1; EtOH 2; eth 3; ctc 3	
2285	Benzene, 1,1'-thiobis[2-methyl-		$C_{14}H_{14}S$	214.33	4537-05-7	64		285	4-06-00-02019		eth 4; EtOH 4; chl 4	
2286	Benzene, 1,1'-thiobis[4-methyl-		$C_{14}H_{14}S$	214.33	620-94-0	57.3		>300; 175^{16}	4-06-00-02173		H_2O 1; EtOH 3; eth 4; ace 3	
2287	Benzene, 1,1'-[thiobis(methylene)]bis-	Benzyl sulfide	$C_{14}H_{14}S$	214.33	538-74-9	49.5	1162	dec	4-06-00-02649	1.0583^{50}	H_2O 1; EtOH 3; eth 3; CS_2 3	
2288	Benzene, 1,1'-thiobis[2,4,6-trinitro-	Picryl sulfide	$C_{12}H_4N_6O_{12}S$	456.26	2217-06-3				4-06-00-01775			
2289	Benzenethiol	Phenyl mercaptan	C_6H_6S	110.18	108-98-5	-14.9	9285	169.1	4-06-00-01463	1.0775^{20}	H_2O 1; EtOH 3; eth 3; bz 3	1.5893^{20}
2290	Benzenethiol, 2-amino-		C_6H_7NS	125.19	137-07-5	26		234	4-13-00-00910		EtOH 3; eth 3	1.4606^{20}
2291	Benzenethiol, 4-amino-		C_6H_7NS	125.19	1193-02-8	46		143^{17}	4-13-00-01289		H_2O 3; EtOH 3	
2292	Benzenethiol, 4-bromo-		C_6H_5BrS	189.08	106-53-6	73		230.5	4-06-00-01650	1.5260^{83}	H_2O 2; EtOH 2; eth 4; ctc 4	
2293	Benzenethiol, 2-chloro-		C_6H_5ClS	144.62	6320-03-2			205.5	4-06-00-01570	1.2752^{10}	H_2O 2; EtOH 2	
2294	Benzenethiol, 3-chloro-		C_6H_5ClS	144.62	2037-31-2			206	4-06-00-01576	1.2637^{13}	H_2O 1; EtOH 3; eth 3; chl 3	
2295	Benzenethiol, 4-chloro-		C_6H_5ClS	144.62	106-54-7	61		206	4-06-00-01581	1.1911^{20}	H_2O 1; EtOH 4; eth 4; bz 4	1.5480^{20}
2296	Benzenethiol, 4-ethoxy-		$C_8H_{10}OS$	154.23	699-09-2	1.6		238	4-06-00-05792	1.1070^{17}	ace 4; bz 4; eth 4; EtOH 4	
2297	Benzenethiol, 3-ethyl-	3-Ethylthiophenol	$C_8H_{10}S$	138.23	62154-77-2			211	3-06-00-01663	1.038^{20}		1.572^{30}
2298	Benzenethiol, 4-ethyl-		$C_8H_{10}S$	138.23	4946-13-8			211	4-06-00-03027	1.038^{20}		1.572^{20}
2299	Benzenethiol, 3-methoxy-		C_7H_8OS	140.21	15570-12-4			224.5; 114^{20}	4-06-00-05701		chl 3	1.5874^{20}
2300	Benzenethiol, 4-methoxy-		C_7H_8OS	140.21	696-63-9			228	4-06-00-05790	1.1313^{25}	EtOH 3; eth 3; bz 3; chl 2	1.5801^{25}
2301	Benzenethiol, 2-methyl-		C_7H_8S	124.21	137-06-4	15		195	4-06-00-02014	1.041^{20}	H_2O 1; EtOH 3; eth 4	1.570^{20}
2302	Benzenethiol, 3-methyl-		C_7H_8S	124.21	108-40-7	-20		195	4-06-00-02079	1.044^{20}	H_2O 1; EtOH 3; eth 5	1.572^{20}
2303	Benzenethiol, 4-methyl-		C_7H_8S	124.21	106-45-6	43		195	4-06-00-02153	1.0220^{51}	H_2O 1; EtOH 3; eth 4; chl 3	
2304	Benzenethiol, pentafluoro-		C_6HF_5S	200.13	771-62-0	-24		143		1.501^{25}		1.4645^{20}
2305	1,2,3-Benzenetriamine	1,2,3-Triaminobenzene	$C_6H_9N_3$	123.16	608-32-2	103		336	3-13-00-00551		H_2O 4; eth 4; EtOH 4	
2306	1,2,4-Benzenetriamine		$C_6H_9N_3$	123.16	615-71-4	97.3		340	4-13-00-00506		H_2O 4; EtOH 4; chl 4	
2307	Benzene, 1,2,3-tribromo-	1,2,3-Tribromobenzene	$C_6H_3Br_3$	314.80	608-21-9	87.5			3-05-00-00569	2.658^{25}	eth 4	
2308	Benzene, 1,2,4-tribromo-	1,2,4-Tribromobenzene	$C_6H_3Br_3$	314.80	615-54-3	44.5		275	4-05-00-00685		H_2O 1; EtOH 3; eth 4; ace 4	
2309	Benzene, 1,3,5-tribromo-	1,3,5-Tribromobenzene	$C_6H_3Br_3$	314.80	626-39-1	122.8		271	4-05-00-00685		H_2O 1; EtOH 2; eth 3; bz 3	
2310	Benzene, 1,2,3-tribromo-5-methoxy-	Anisole, 3,4,5-tribromo	$C_7H_5Br_3O$	344.83	73557-60-5	93.3		303	2-06-00-00195		ace 4; bz 4; EtOH 4	
2311	Benzene, 1,2,4-tribromo-5-methoxy-	2,4,5-Tribromoanisole	$C_7H_5Br_3O$	344.83	95970-10-8	105		307	4-06-00-01067		ace 4; bz 4; EtOH 4	
2312	Benzene, 1,2,5-tribromo-3-methoxy-	Anisole, 2,3,5-tribromo	$C_7H_5Br_3O$	344.83	73931-44-9	82		308.5	2-06-00-00192		ace 4; bz 4; EtOH 4	
2313	Benzene, 1,3,5-tribromo-2-methoxy-		$C_7H_5Br_3O$	344.83	607-99-8	88		298	4-06-00-01067	2.491^{25}	H_2O 2; EtOH 2; ace 4; bz 4	
2314	Benzene, 1,3,5-tribromo-2-methyl-	2,4,6-Tribromotoluene	$C_7H_5Br_3$	328.83	6320-40-7	70		290	3-05-00-00719	2.479^{17}		
2315	Benzene, 1,2,3-tribromo-5-nitro-	3,4,5-Tribromonitrobenzene	$C_6H_2Br_3NO_2$	359.80	3460-20-6	112		sub	4-05-00-00732	2.645^{25}		
2316	Benzene, 1,3,5-tribromo-2-nitro-		$C_6H_2Br_3NO_2$	359.80	3463-40-9	125		177^{11}	4-05-00-00732		eth 4; chl 4	
2317	1,3,5-Benzenetricarbonyl trichloride		$C_9H_3Cl_3O_3$	265.48	4422-95-1	36.3		180^{16}	4-09-00-03748		chl 3	
2318	1,2,3-Benzenetricarboxylic acid	Hemimellitic acid	$C_9H_6O_6$	210.14	569-51-7	200			4-09-00-03745	1.546^{20}	eth 4; EtOH 4	
2319	1,2,4-Benzenetricarboxylic acid	Trimellitic acid	$C_9H_6O_6$	210.14	528-44-9	219	9616		4-09-00-03746		H_2O 4; eth 4; EtOH 4	
2320	1,3,5-Benzenetricarboxylic acid, 2-chloro-	2-Chloro-1,3,5-benzenetricarboxylic acid	$C_9H_5ClO_6$	244.59	56961-24-1	285		sub	4-09-00-03749		H_2O 4; eth 4; EtOH 4	

PHYSICAL CONSTANTS OF ORGANIC COMPOUNDS (continued)

No.	Name Synonym	Mol. Form. Mol. Wt.	CAS RN mp/°C	Merck No. bp/°C	Beil. Ref. den/g cm^{-3}	Solubility n_D
2321	Benzene, 1,2,3-trichloro- 1,2,3-Trichlorobenzene	$C_6H_3Cl_3$ 181.45	87-61-6 53.5	9542 218.5	4-05-00-00664 1.4533^{25}	H_2O 1; EtOH 2; eth 4; bz 4
2322	Benzene, 1,2,4-trichloro- 1,2,4-Trichlorobenzene	$C_6H_3Cl_3$ 181.45	120-82-1 17	9543 213.5	4-05-00-00664 1.459^{25}	H_2O 1; EtOH 2; eth 4; ctc 2 1.5717^{20}
2323	Benzene, 1,3,5-trichloro- 1,3,5-Trichlorobenzene	$C_6H_3Cl_3$ 181.45	108-70-3 63.5	9544 208	4-05-00-00666	H_2O 1; EtOH 2; eth 4; bz 4
2324	Benzene, 1,2,4-trichloro-5-(chloromethyl)-	$C_7H_4Cl_4$ 229.92	3955-26-8	 273	4-05-00-00822 1.547^{20}	ace 4; eth 4; EtOH 4
2325	Benzene, 1,2,4-trichloro-5-[(4-chlorophenyl)sulfonyl]- Tetradifon	$C_{12}H_6Cl_4O_2S$ 356.06	116-29-0 146	9132	4-06-00-01636 1.151^{20}	
2326	Benzene, 1,2,3-trichloro-4-(dichloromethyl)- Benzylidene chloride, 2,3,4-trichloro	$C_7H_3Cl_5$ 264.36	56961-82-1 84	 278	1-05-00-00153	bz 4
2327	Benzene, 1,2,4-trichloro-5-(dichloromethyl)- Benzylidene chloride, 2,4,5-trichloro	$C_7H_3Cl_5$ 264.36	33429-70-8	 280.5	4-05-00-00824 1.5956^{20}	bz 4 1.5992^{20}
2328	Benzene, 1,1'-(2,2,2-trichloroethylidene)bis[4-chloro- DDT	$C_{14}H_9Cl_5$ 354.49	50-29-3 108.5	2832 260; $186^{0.05}$	4-05-00-01885	H_2O 1; EtOH 2; eth 4; ace 4
2329	Benzene, 1,1'-(2,2,2-trichloroethylidene)bis[4-methoxy- Methoxychlor	$C_{16}H_{15}Cl_3O_2$ 345.65	72-43-5 87	5913	4-06-00-06691 1.41^{25}	H_2O 1; EtOH 3; eth 4; bz 4
2330	Benzene, 1,2,4-trichloro-3-methoxy-	$C_7H_5Cl_3O$ 211.47	50375-10-5 45	 227	3-06-00-00717	ace 4; EtOH 4
2331	Benzene, 1,2,4-trichloro-5-methoxy-	$C_7H_5Cl_3O$ 211.47	6130-75-2 77.5	 254	4-06-00-00063	EtOH 4; ace 4; os 3
2332	Benzene, 1,3,5-trichloro-2-methoxy- 2,4,6-Trichloroanisole	$C_7H_5Cl_3O$ 211.47	87-40-1 61.5	9541 241	4-06-00-01005 1.640^{25}	EtOH 3; ace 4; bz 3; chl 3
2333	Benzene, (trichloromethyl)- (Trichloromethyl)benzene	$C_7H_5Cl_3$ 195.47	98-07-7 -5	1120 221	4-05-00-00820 1.3723^{20}	H_2O 1; EtOH 3; eth 3; bz 3 1.5580^{20}
2334	Benzene, 1,2,3-trichloro-4-methyl- 2,3,4-Trichlorotoluene	$C_7H_5Cl_3$ 195.47	7359-72-0 43.5	 244	4-05-00-00819	H_2O 1; EtOH 3; eth 3; ace 3
2335	Benzene, 1,2,3-trichloro-5-methyl- Toluene, 3,4,5-trichloro-	$C_7H_5Cl_3$ 195.47	21472-86-6 45.5	 246	0-05-00-00299	H_2O 1; EtOH 3; ace 3; os 3
2336	Benzene, 1,2,4-trichloro-5-methyl- 2,4,5-Trichlorotoluene	$C_7H_5Cl_3$ 195.47	6639-30-1 82.4	 231	4-05-00-00819	H_2O 1; EtOH 3; ace 3; os 3
2337	Benzene, 1,2,5-trichloro-3-methyl- 2,3,5-Trichlorotoluene	$C_7H_5Cl_3$ 195.47	56961-86-5 45.5	 230	0-05-00-00299	H_2O 1; EtOH 3; ace 3; os 3
2338	Benzene, 1,2,3-trichloro-5-nitro-	$C_6H_2Cl_3NO_2$ 226.45	20098-48-0 72.5		2-05-00-00187 1.807^{25}	EtOH 4
2339	Benzene, 1,2,4-trichloro-5-nitro-	$C_6H_2Cl_3NO_2$ 226.45	89-69-0 57.5	 288	4-05-00-00728 1.790^{23}	H_2O 1; EtOH 2; eth 3; bz 3
2340	Benzene, 1,3,5-trichloro-2,4,6-trifluoro- 1,3,5-Trichloro-2,4,6-trifluorobenzene	$C_6Cl_3F_3$ 235.42	319-88-0	 198.4		
2341	Benzene, tricosyl- Tricosylbenzene	$C_{29}H_{52}$ 400.73	61828-04-4 54	 365^{100}; 231^{1}	4-05-00-01237 0.8544^{20}	 1.4801^{20}
2342	Benzene, tridecyl- 1-Phenyltridecane	$C_{19}H_{32}$ 260.46	123-02-4 10	9575 346	3-05-00-01108 0.8550^{20}	 1.4821^{20}
2343	Benzene, 1,3,5-triethoxy- 1,3,5-Triethoxybenzene	$C_{12}H_{18}O_3$ 210.27	2437-88-9 43.5	 170^{24}	4-06-00-07362	eth 4; EtOH 4
2344	Benzene, 1,2,4-triethyl- 1,2,4-Triethylbenzene	$C_{12}H_{18}$ 162.27	877-44-1	 218; 99^{15}	4-05-00-01133 0.8738^{20}	H_2O 1; EtOH 3; eth 3 1.5024^{20}
2345	Benzene, 1,3,5-triethyl- 1,3,5-Triethylbenzene	$C_{12}H_{18}$ 162.27	102-25-0 -66.5	 215.9	4-05-00-01133 0.8631^{20}	H_2O 1; EtOH 4; eth 4 1.4969^{20}
2346	Benzene, 1,2,4-trifluoro-	$C_6H_3F_3$ 132.09	367-23-7	 90	4-05-00-00638 1.264^{25}	 1.4171^{20}
2347	Benzene, 1,3,5-trifluoro- 1,3,5-Trifluorobenzene	$C_6H_3F_3$ 132.09	372-38-3 -5.5	 75.5	4-05-00-00638 1.277^{25}	 1.4140^{20}
2348	Benzene, (trifluoromethyl)- (Trifluoromethyl)benzene	$C_7H_5F_3$ 146.11	98-08-8 -29.1	1121 102.1	4-05-00-00802 1.1884^{20}	EtOH 5; eth 5; ace 5; bz 5 1.4146^{20}
2349	Benzene, 1,2,3-triiodo- 1,2,3-Triiodobenzene	$C_6H_3I_3$ 455.80	608-29-7 116	 sub	1-05-00-00122	eth 4; EtOH 4; chl 4
2350	Benzene, 1,2,4-triiodo- 1,2,4-Triiodobenzene	$C_6H_3I_3$ 455.80	615-68-9 91.5	 sub	3-05-00-00585	EtOH 4; chl 4
2351	Benzene, 1,3,5-triiodo- 1,3,5-Triiodobenzene	$C_6H_3I_3$ 455.80	626-44-8 184.2	 sub	4-05-00-00702	H_2O 1; EtOH 2; eth 2; bz 2
2352	Benzene, 1,3,5-triiodo-2-methyl- 2,4,6-Triiodotoluene	$C_7H_5I_3$ 469.83	36994-79-3 118.5	 300 dec	1-05-00-00158	EtOH 4
2353	Benzene, 1,2,3-triiodo-5-nitro-	$C_6H_2I_3NO_2$ 500.80	53663-23-3 167		3-05-00-00626 3.256^{25}	bz 4; eth 4; EtOH 4; chl 4
2354	Benzene, 1,2,3-trimethoxy-	$C_9H_{12}O_3$ 168.19	634-36-6 48.5	 235	4-06-00-07329 1.1009^{45}	H_2O 1; EtOH 3; eth 3; bz 3
2355	Benzene, 1,3,5-trimethoxy-	$C_9H_{12}O_3$ 168.19	621-23-8 54.5	 255.5	4-06-00-07362	H_2O 1; EtOH 3; eth 3; bz 3
2356	Benzene, 1,2,4-trimethoxy-5-(1-propenyl)-	$C_{12}H_{16}O_3$ 208.26	494-40-6 67	 296	4-06-00-07476 1.165^{20}	eth 4; EtOH 4; peth 4 1.5683^{20}
2357	Benzene, 1,2,3-trimethyl- 1,2,3-Trimethylbenzene	C_9H_{12} 120.19	526-73-8 -25.4	 176.1	4-05-00-01007 0.8944^{20}	H_2O 1; EtOH 5; eth 5; ace 5 1.5139^{20}
2358	Benzene, 1,2,4-trimethyl- Pseudocumene	C_9H_{12} 120.19	95-63-6 -43.8	7929 169.3	4-05-00-01010 0.8758^{20}	H_2O 1; EtOH 5; eth 5; ace 5 1.5048^{20}

No.	Name Synonym	Mol. Form. Mol. Wt.	CAS RN mp/°C	Merck No. bp/°C	Beil. Ref. den/g cm^{-3}	Solubility n_D
2359	Benzene, 1,3,5-trimethyl- Mesitylene	C_9H_{12} 120.19	108-67-8 -44.7	5810 164.7	4-05-00-01016 0.8652^{20}	H_2O 1; EtOH 5; eth 5; ace 5 1.4994^{20}
2360	Benzene, 1,3,5-trimethyl-2,4-dinitro-	$C_9H_{10}N_2O_4$ 210.19	608-50-4 85.3	 exp 418	4-05-00-01029	H_2O 1; EtOH 3
2361	Benzene, 1,2,4-trimethyl-5-(1-methylethyl)-	$C_{12}H_{18}$ 162.27	10222-95-4 21	 221	3-05-00-01038 0.8795^{21}	 1.5065^{21}
2362	Benzene, 1,2,4-trimethyl-5-nitro-	$C_9H_{11}NO_2$ 165.19	610-91-3 71	 265	4-05-00-01015	EtOH 3; chl 3; peth 3
2363	Benzene, 1,3,5-trimethyl-2-nitro-	$C_9H_{11}NO_2$ 165.19	603-71-4 44	 255	4-05-00-01028 1.51^{25}	EtOH 4
2364	Benzene, 1,3,5-trimethyl-2,4,6-trinitro-	$C_9H_9N_3O_6$ 255.19	602-96-0 238.2	 exp 415	4-05-00-01029	H_2O 1; EtOH 2; eth 2; ace 3
2365	Benzene, 1,3,5-trinitro- *sym*-Trinitrobenzene	$C_6H_3N_3O_6$ 213.11	99-35-4 121.5	9639 315	4-05-00-00755 1.4775^{152}	H_2O 2; EtOH 2; eth 2; ace 4
2366	1,2,3-Benzenetriol Pyrogallol	$C_6H_6O_3$ 126.11	87-66-1 133	8010 309	4-06-00-07327 1.453^4	H_2O 4; EtOH 4; eth 4; ace 3 1.561^{134}
2367	1,2,4-Benzenetriol Hydroxyhydroquinone	$C_6H_6O_3$ 126.11	533-73-3 140.5	1081	4-06-00-07338	H_2O 4; EtOH 4; eth 4; bz 1
2368	1,3,5-Benzenetriol Phloroglucinol	$C_6H_6O_3$ 126.11	108-73-6 218.5	7301 sub	4-06-00-07361 1.46^{25}	H_2O 2; EtOH 4; eth 4; ace 3
2369	1,2,3-Benzenetriol, 5-methyl-	$C_7H_8O_3$ 140.14	609-25-6 129	 sub	4-06-00-07376	bz 4
2370	1,2,4-Benzenetriol, triacetate	$C_{12}H_{12}O_6$ 252.22	613-03-6 99	 300	4-06-00-07342	EtOH 3; chl 3; MeOH 3
2371	1,3,5-Benzenetriol, 2,4,6-trichloro- Trichloro-1,3,5-trihydroxybenzene	$C_6H_3Cl_3O_3$ 229.45	56961-23-0 136	 sub	0-06-00-01104	EtOH 4
2372	Benzene, 1,2,4-tris(1-methylethyl)- 1,2,4-Triisopropylbenzene	$C_{15}H_{24}$ 204.36	948-32-3	 244	4-05-00-01178 0.8574^{25}	 1.4896^{25}
2373	Benzene, 1,3,5-tris(1-methylethyl)-	$C_{15}H_{24}$ 204.36	717-74-8 -7.4	 238	4-05-00-01178 0.8545^{20}	ace 3; bz 3; chl 3 1.4882^{20}
2374	Benzene, undecyl-	$C_{17}H_{28}$ 232.41	6742-54-7 -5	 316	3-05-00-01100 0.8553^{20}	 1.4828^{20}
2375	9,10[1',2']-Benzenoanthracene, 9,10-dihydro- Triptycene	$C_{20}H_{14}$ 254.33	477-75-8	9663	4-05-00-02639	
2376	1H-Benzimidazole *N,N'*-Methenyl-*o*-phenylenediamine	$C_7H_6N_2$ 118.14	51-17-2 170.5	1091 >360	5-23-06-00196	H_2O 2; EtOH 4; eth 2; bz 1
2377	1H-Benzimidazole, 4,5-dichloro-2-(trifluoromethyl)- Chloroflurazole	$C_8H_3Cl_2F_3N_2$ 255.03	3615-21-2 213.5			
2378	1H-Benzimidazole, 1,5-dimethyl- 1,5-Dimethylbenzimidazole	$C_9H_{10}N_2$ 146.19	10394-35-1 95	 300	5-23-06-00380	bz 4; eth 4; EtOH 4
2379	1H-Benzimidazole, 2,5-dimethyl-	$C_9H_{10}N_2$ 146.19	1792-41-2 203	 350	5-23-06-00448	eth 4; EtOH 4
2380	1H-Benzimidazole, 5,6-dimethyl- Dimedazole	$C_9H_{10}N_2$ 146.19	582-60-5 205.5	3225 sub	5-23-06-00454	H_2O 3; EtOH 3; eth 3; chl 3
2381	1H-Benzimidazole, 1-ethyl-2-methyl- 1-Ethyl-2-methylbenzimidazole	$C_{10}H_{12}N_2$ 160.22	5805-76-5 51	 296	5-23-06-00326 1.073^{25}	
2382	1H-Benzimidazole, 1-methyl-	$C_8H_8N_2$ 132.17	1632-83-3 66	 286	5-23-06-00215 1.1254^{20}	peth 3 1.6013^7
2383	1H-Benzimidazole, 1-phenyl-	$C_{13}H_{10}N_2$ 194.24	2622-60-8 97	 221^{25}	5-23-06-00222	H_2O 2; EtOH 3; MeOH 3
2384	1H-Benzimidazole, 2-phenyl- Phenzidole	$C_{13}H_{10}N_2$ 194.24	716-79-0 293	7245	5-23-08-00437	H_2O 2; EtOH 3; bz 2; chl 2
2385	1H-Benzimidazole, 2-(phenylmethyl)- Bendazol	$C_{14}H_{12}N_2$ 208.26	621-72-7 187	1043	5-23-08-00498	bz 4; EtOH 4; gl HOAc 4
2386	2H-Benzimidazole-2-thione, 1,3-dihydro- 2-Benzimidazolethiol	$C_7H_6N_2S$ 150.20	583-39-1 298	1092	5-24-02-00413	EtOH 4
2387	1,2-Benzisothiazole 1-Thia-2-azaindene	C_7H_5NS 135.19	272-16-2 38	 220	2-27-00-00016	ctc 3
2388	1,2-Benzisothiazol-3(2H)-one, 1,1-dioxide Saccharin	$C_7H_5NO_3S$ 183.19	81-07-2 228 dec	8282 sub	4-27-00-02649 0.828^{25}	H_2O 2; EtOH 3; eth 2; ace 3
2389	1,2-Benzisoxazole Indoxazene	C_7H_5NO 119.12	271-95-4	 100^{26}	4-27-00-01067 1.1727^{21}	eth 4 1.5570^{20}
2390	2,1-Benzisoxazole Anthranil	C_7H_5NO 119.12	271-58-9 <-18	 215	4-27-00-01068 1.1827^{20}	ace 4; EtOH 4 1.5845^{20}
2391	Benz[j]aceanthrylene, 1,2-dihydro- Cholanthrene	$C_{20}H_{14}$ 254.33	479-23-2 174	2200	4-05-00-02638	H_2O 1; EtOH 3; bz 3; HOAc 3
2392	Benz[j]aceanthrylene, 1,2-dihydro-3-methyl- 3-Methylcholanthrene	$C_{21}H_{16}$ 268.36	56-49-5 180	5967 280^{80}	4-05-00-02648 1.28^{20}	H_2O 1
2393	11H-Benzo[a]fluorene	$C_{17}H_{12}$ 216.28	238-84-6 189.5	 405	4-05-00-02473	EtOH 2; eth 3; bz 3; chl 3
2394	Benzo[a]phenazine 7,12-Diazabenz[a]anthracene	$C_{16}H_{10}N_2$ 230.27	225-61-6 142.5	 >360	5-23-09-00494	H_2O 1; EtOH 3; eth 3; ace 3
2395	Benzo[a]pyrene 2,3-Benzopyrene	$C_{20}H_{12}$ 252.32	50-32-8 176.5	1113	4-05-00-02687	chl 4
2396	2H-Benzo[a]quinolizin-2-one, 1,3,4,6,7,11b-hexahydro-9,10-dimethoxy-3-(2-methylpropyl)- 1,2-Benzoquinolizine-2-one, 9,10-dimethoxy-1,2,3,4,6,7,11b-heptahydro-3-isobutyl	$C_{19}H_{27}NO_3$ 317.43	58-46-8 128	9118	5-21-13-00178	chl 3

No.	Name Synonym	Mol. Form. Mol. Wt.	CAS RN mp/°C	Merck No. bp/°C	Beil. Ref. den/g cm^{-3}	Solubility n_D
2397	2H,8H-Benzo[1,2-b:5,4-b']dipyran-2-one, 8,8-dimethyl- Xanthyletin	$C_{14}H_{12}O_3$ 228.25	553-19-5 131.5	9978 140-5$^{0.1}$	5-19-04-00470	EtOH 3; peth 3
2398	2H,8H-Benzo[1,2-b:5,4-b']dipyran-2-one, 5-methoxy-8,8-dimethyl- Xanthoxyletin	$C_{15}H_{14}O_4$ 258.27	84-99-1 133	9975	5-19-06-00039	H_2O 1; EtOH 3; eth 2; ace 3
2399	Benzo[b]thiophene Thianaphthene	C_8H_6S 134.20	95-15-8 32	9232 221	5-17-02-00006 1.1484^{32}	H_2O 1; EtOH 4; eth 3; ace 3 1.6374^{37}
2400	Benzo[b]thiophene-2-carboxylic acid Thionaphthene-2-carboxylic acid	$C_9H_6O_2S$ 178.21	6314-28-9 240.5	9274	5-18-06-00427	eth 4
2401	Benzo[b]thiophene-2,3-dione	$C_8H_4O_2S$ 164.18	493-57-2 121	 247	5-17-11-00251	H_2O 1; EtOH 3; bz 3; HOAc 3
2402	Benzo[b]thiophene, 5-methyl-	C_9H_8S 148.23	14315-14-1 20.5	 111-5^{12}	5-17-02-00031 1.111^{22}	 1.615^{22}
2403	Benzo[b]thiophene-4-ol	C_8H_6OS 150.20	3610-02-4 79.8	 sub	5-17-04-00199	EtOH 4
2404	Benzo[b]thiophen-3(2H)-one, 2-(3-oxobenzo[b]thien-2(3H)-ylidene)-	$C_{16}H_8O_2S_2$ 296.37	522-75-8 359	 sub	5-19-05-00281	H_2O 1; EtOH 1; bz 3; chl 2
2405	7H-Benzo[c]carbazole 7-Aza-7H-benzo[c]fluorene	$C_{16}H_{11}N$ 217.27	205-25-4 134	 448	5-20-08-00455	bz 4; EtOH 4; HOAc 4
2406	Benzo[c]phenanthrene Tetrahelicene	$C_{18}H_{12}$ 228.29	195-19-7 68	1129	4-05-00-02552	H_2O 1; EtOH 2; lig 2
2407	5H-Benzocyclohepten-5-one, 6,7,8,9-tetrahydro-	$C_{11}H_{12}O$ 160.22	826-73-3	 124^{7}; 108^{1}	4-07-00-01029 1.0780^{20}	EtOH 3 1.5698^{20}
2408	5H-Benzocyclohepten-5-one, 2,3,4,6-tetrahydroxy- Purpurogallin	$C_{11}H_8O_5$ 220.18	569-77-7 274-5 dec	7963	4-08-00-03456	
2409	3H-1,4-Benzodiazepin-2-amine, 7-chloro-*N*-methyl-5-phenyl-, 4-oxide Chlorodiazepoxide	$C_{16}H_{14}ClN_3O$ 299.76	58-25-3 236.2	2082		
2410	2H-1,4-Benzodiazepin-2-one, 7-chloro-1,3-dihydro-1-methyl-5-phenyl- Diazepam	$C_{16}H_{13}ClN_2O$ 284.74	439-14-5 132	2977	5-24-04-00300	
2411	2H-1,4-Benzodiazepin-2-one, 7-chloro-1,3-dihydro-5-phenyl- Nordazepam	$C_{15}H_{11}ClN_2O$ 270.72	1088-11-5 216.5	6608	5-24-04-00291	
2412	1,3,2-Benzodioxaborole	$C_6H_5BO_2$ 119.92	274-07-7 12	 88^{156}; 50^{50}	 1.2700^{20}	 1.5070^{20}
2413	1,3,2-Benzodioxaphosphole, 2-chloro-	$C_6H_4ClO_2P$ 174.52	1641-40-3 30	 80^{20}	4-06-00-05598 1.4650^{20}	 1.5712^{20}
2414	1,4-Benzodioxin, 2,3-dihydro-	$C_8H_8O_2$ 136.15	493-09-4	 103^{6}	5-19-01-00439 1.142^{25}	 1.5485^{20}
2415	1,3-Benzodioxole	$C_7H_6O_2$ 122.12	274-09-9	 172.5; 77^{27}	5-19-01-00434 1.064^{25}	 1.5398^{20}
2416	1,3-Benzodioxole-5-acetic acid, methyl ester Phenylacetic acid, 3,4-methylenedioxy, methyl ester	$C_{10}H_{10}O_4$ 194.19	326-59-0	 153-5^{10}	2-19-00-00295 1.246^{20}	 1.534
2417	1,3-Benzodioxole-5-carboxaldehyde Piperonal	$C_8H_6O_3$ 150.13	120-57-0 37	7445 263	5-19-04-00225	H_2O 2; EtOH 4; eth 5; ace 3
2418	1,3-Benzodioxole-5-carboxylic acid Piperonylic acid	$C_8H_6O_4$ 166.13	94-53-1 229	7447	5-19-07-00300	
2419	1,3-Benzodioxole-5-carboxylic acid, ethyl ester	$C_{10}H_{10}O_4$ 194.19	6951-08-2 18.5	 285.5; 135^{6}	5-19-07-00300	eth 4; EtOH 4; peth 4
2420	1,3-Benzodioxole-5-carboxylic acid, methyl ester	$C_9H_8O_4$ 180.16	326-56-7 53	 273 dec	5-19-07-00300	eth 4; EtOH 4
2421	1,3-Benzodioxole, 5-(chloromethyl)-	$C_8H_7ClO_2$ 170.60	20850-43-5 20.5	 134-5^{14}	5-19-01-00443 1.312^{25}	 1.5660^{20}
2422	1,3-Benzodioxole, 4,7-dimethoxy-5-(1-propenyl)-, (*E*)-	$C_{12}H_{14}O_4$ 222.24	17672-88-7 56	 303.5	4-19-00-01033	ace 4; bz 4; eth 4; EtOH 4
2423	1,3-Benzodioxole, 4,7-dimethoxy-5-(2-propenyl)- Apiole	$C_{12}H_{14}O_4$ 222.24	523-80-8 29.5	767 294; 179^{30}	5-19-03-00307 1.015^{20}	ace 4; bz 4; EtOH 4; lig 4 1.5360^{20}
2424	1,3-Benzodioxole-5-ethanamine	$C_9H_{11}NO_2$ 165.19	1484-85-1	 166^{20}	5-19-08-00407 1.225^{20}	 1.5620^{20}
2425	1,3-Benzodioxole, 5-ethenyl-	$C_9H_8O_2$ 148.16	7315-32-4	 224	5-19-01-00544 1.1488^{18}	eth 4; EtOH 4; chl 4 1.5802
2426	1,3-Benzodioxole-5-methanamine	$C_8H_9NO_2$ 151.17	2620-50-0	 138-9^{13}	5-19-08-00395 1.214^{25}	 1.5635^{20}
2427	1,3-Benzodioxole-5-methanol	$C_8H_8O_3$ 152.15	495-76-1 58	 157^{16}	5-19-02-00543	H_2O 2; EtOH 3; eth 3; bz 3
2428	1,3-Benzodioxole, 4-methoxy-6-(2-propenyl)- Myristicin	$C_{11}H_{12}O_3$ 192.21	607-91-0 <-20	6247 276.5	5-19-02-00631 1.1416^{20}	H_2O 1; EtOH 2; eth 3; bz 3 1.5403^{20}
2429	1,3-Benzodioxole, 5-(2-propenyl)- Safrole	$C_{10}H_{10}O_2$ 162.19	94-59-7 11.2	8287 234.5	5-19-01-00553 1.1000^{20}	H_2O 1; EtOH 4; eth 5; chl 5 1.5381^{20}
2430	1,3-Benzodioxole, 5-(1-propenyl)-, (*E*)-	$C_{10}H_{10}O_2$ 162.19	4043-71-4 6.8	 253	5-19-01-00552 1.1224^{20}	H_2O 1; EtOH 5; eth 5; ace 4 1.5782^{20}

No.	Name Synonym	Mol. Form. Mol. Wt.	CAS RN mp/°C	Merck No. bp/°C	Beil. Ref. den/g cm^{-3}	Solubility n_D
2431	1,3-Benzodioxole, 5-propyl- Dihydrosafrole	$C_{10}H_{12}O_2$ 164.20	94-58-6		5-19-01-00473	ctc 3
2432	[1,3]Benzodioxolo[5,6-c]-1,3-dioxolo[4,5-i]phenanthridinium, 13-methyl- Sanguinarine	$C_{20}H_{15}NO_5$ 349.34	2447-54-3 266	8320		ace 4; bz 4; eth 4; EtOH 4
2433	[1,3]Benzodioxolo[5,6-c]phenanthridinium, 1,2-dimethoxy-12-methyl- Chelerythrine	$C_{21}H_{19}NO_5$ 365.39	34316-15-9	2041 207		chl 4
2434	1,4-Benzodithiin	$C_8H_6S_2$ 166.27	255-50-5	 $67-70^{0.1}$	5-19-01-00536 1.2799^{20}	 1.6754^{25}
2435	Benzo[e]-1,3-dioxolo[4,5-l][2]benzazecin-12(5H)-one, 4,6,7,13-tetrahydro-9,10-dimethoxy-5-methyl- Cryptopine	$C_{21}H_{23}NO_5$ 369.42	482-74-6 223	2611	4-27-00-06652 1.315^{20}	H_2O 1; EtOH 2; eth 2; bz 2
2436	Benzo[e]pyrene 1,2-Benzpyrene	$C_{20}H_{12}$ 252.32	192-97-2 177.5	1114 311	4-05-00-02689	
2437	Benzo[f]quinoline β-Naphthoquinoline	$C_{13}H_9N$ 179.22	85-02-9 94	1115 352; $202-5^{8}$	5-20-08-00220	H_2O 2; EtOH 4; eth 4; ace 3
2438	Benzofuran Coumarone	C_8H_6O 118.14	271-89-6 <-18	1098 174	5-17-02-00003 1.0913^{25}	H_2O 1; EtOH 3; eth 3 1.5615^{17}
2439	2-Benzofurancarboxylic acid Coumarilic acid	$C_9H_6O_3$ 162.14	496-41-3 192.5	2562 312.5	5-18-06-00419	EtOH 4
2440	2-Benzofurancarboxylic acid, ethyl ester	$C_{11}H_{10}O_3$ 190.20	3199-61-9 30.5	 276; 161^{15}	5-18-06-00420 1.1656^{28}	 1.564^{27}
2441	Benzofuran, 2-(chloromethyl)-2,3-dihydro- 2-(Chloromethyl)-2,3-dihydrobenzofuran	C_9H_9ClO 168.62	53491-32-0 41.5	 $118-9^{11}$	4-17-00-00421 1.2183^{7}	ace 4; eth 4 1.5620^{7}
2442	Benzofuran, 2,3-dihydro- Coumaran	C_8H_8O 120.15	496-16-2 -21.5	2560 188.5	5-17-01-00581 1.058^{25}	eth 4; EtOH 4; chl 4 1.5497^{20}
2443	Benzofuran, 2,3-dihydro-2-methyl-	$C_9H_{10}O$ 134.18	1746-11-8	 197.5	5-17-01-00597 1.061^{25}	 1.5308
2444	Benzofuran, 2,3-dihydro-5-methyl- 2,3-Dihydro-5-methylbenzofuran	$C_9H_{10}O$ 134.18	76429-68-0	 210.5	4-17-00-00423 1.0463^{19}	 1.54^{20}
2445	Benzofuran, 2,5-dimethyl- 2,5-Dimethylbenzofuran	$C_{10}H_{10}O$ 146.19	29040-46-8	 220	5-17-02-00045 1.031^{20}	eth 4 1.5534^{20}
2446	Benzofuran, 2,6-dimethyl- 2,6-Dimethylbenzofuran	$C_{10}H_{10}O$ 146.19	24410-51-3	 217.5	5-17-02-00046 1.051^{12}	eth 4 1.554^{15}
2447	Benzofuran, 2,7-dimethyl- 2,7-Dimethylfuran	$C_{10}H_{10}O$ 146.19	59020-74-5	 216	4-17-00-00512 1.036^{20}	eth 4 1.5546^{20}
2448	Benzofuran, 3,5-dimethyl- 3,5-Dimethylbenzofuran	$C_{10}H_{10}O$ 146.19	10410-35-2	 220.5	5-17-02-00047 1.036^{20}	eth 4 1.550^{20}
2449	Benzofuran, 3,6-dimethyl- 3,6-DimethylfuranBenzofuran	$C_{10}H_{10}O$ 146.19	24410-50-2	 222	4-17-00-00512 1.0456^{20}	 1.5505^{20}
2450	Benzofuran, 4,6-dimethyl-	$C_{10}H_{10}O$ 146.19	116668-34-9	 219	0-17-00-00063 1.037^{20}	eth 4 1.5485^{21}
2451	Benzofuran, 4,7-dimethyl- 4,7-Dimethylbenzofuran	$C_{10}H_{10}O$ 146.19	28715-26-6	 216	0-17-00-00063 1.041^{17}	eth 4 1.549^{17}
2452	Benzofuran, 5,6-dimethyl- 5,6-Dimethylbenzofuran	$C_{10}H_{10}O$ 146.19	24410-52-4	 221	5-17-02-00048 1.060^{15}	eth 4 1.5516^{15}
2453	Benzofuran, 5,7-dimethyl- 5,7-Dimethylbenzofuran	$C_{10}H_{10}O$ 146.19	64965-91-9	 222	0-17-00-00063 1.0262^{18}	eth 4 1.5358^{20}
2454	Benzofuran, 6,7-dimethyl- 6,7-Dimethylfuran	$C_{10}H_{10}O$ 146.19	35355-36-3	 218	0-17-00-00063 1.038^{20}	eth 4 1.5478^{20}
2455	2-Benzofuranmethanol 2-Hydroxymethylbenzofuran	$C_9H_8O_2$ 148.16	55038-01-2	 $149-50^{14}$	5-17-04-00224 1.177^{13}	 1.5550^{11}
2456	Benzofuran, 2-methyl-	C_9H_8O 132.16	4265-25-2	 197.5	5-17-02-00022 1.0540^{20}	eth 4; EtOH 4 1.5495^{22}
2457	Benzofuran, 3-methyl- 3-Methylbenzo[b]furan	C_9H_8O 132.16	21535-97-7	 197; 86^{20}	5-17-02-00025 1.0540^{25}	H_2O 1; EtOH 3; eth 3 1.5536^{16}
2458	Benzofuran, 5-methyl- 5-Methylbenzo[b]furan	C_9H_8O 132.16	18441-43-5	 198	5-17-02-00030 1.0603^{19}	H_2O 1; EtOH 3; eth 3; sulf 3 1.5570^{19}
2459	Benzofuran, 7-methyl- 7-Methylbenzo[b]furan	C_9H_8O 132.16	17059-52-8 109	 190.5	5-17-02-00032 1.0490^{19}	H_2O 1; EtOH 3; eth 3; sulf 3 1.5525^{19}
2460	Benzofuran, 4-methyl-7-(1-methylethyl)- Benzofuran, 7-isopropyl-4-methyl	$C_{12}H_{14}O$ 174.24	95835-77-1	 241.5	0-17-00-00065 1.0145^{18}	 1.5363^{16}
2461	2(3H)-Benzofuranone	$C_8H_6O_2$ 134.13	553-86-6 50	 249	5-17-10-00006 1.2236^{14}	
2462	3(2H)-Benzofuranone	$C_8H_6O_2$ 134.13	7169-34-8 102.5	 152^{15}	5-17-04-00185	bz 4
2463	2(3H)-Benzofuranone, 3-[2-(diethylamino)ethyl]-3-phenyl- Amolanone	$C_{20}H_{23}NO_2$ 309.41	76-65-3 43.4	603 $192-4^{2.0}$	4-18-00-07995	 1.5614^{25}
2464	5-Benzofuranpropanol, 2-(1,3-benzodioxol-5-yl)-7-methoxy- Egonol	$C_{19}H_{18}O_5$ 326.35	530-22-3 118	 $228-30^{15}$	5-19-09-00583	chl 4
2465	Benzofuran, 4,5,6,7-tetrahydro-3,6-dimethyl-	$C_{10}H_{14}O$ 150.22	494-90-6 86	 80^{18}	5-17-01-00518 0.972^{15}	
2466	6H-Benzofuro[3,2-c][1]benzopyran-6-one, 3,9-dihydroxy- Coumestrol	$C_{15}H_8O_5$ 268.23	479-13-0 385 dec	2565	5-19-06-00405	H_2O 1; EtOH 2; eth 1; ace 2

No.	Name Synonym	Mol. Form. Mol. Wt.	CAS RN mp/°C	Merck No. bp/°C	Beil. Ref. den/g cm^{-3}	Solubility n_D
2467	Benzo[g]-1,3-benzodioxolo[5,6-a]quinolizinium, 5,6-dihydro-9,10-dimethoxy- Berberine	$C_{20}H_{19}NO_5$ 353.37	2086-83-1 145	1169		eth 4; EtOH 4
2468	5H-Benzo[g]-1,3-benzodioxolo[6,5,4-de]quinoline, 6,7,7a,8-tetrahydro-10,11-dimethoxy-7-methyl-, (*S*)- Dicentrine	$C_{20}H_{21}NO_4$ 339.39	517-66-8	3028	4-27-00-06486	chl 3
2469	5H-Benzo[g]-1,3-benzodioxolo[6,5,4-de]quinoline, 6,7,7a,8-tetrahydro-11-methoxy-7-methyl-, (*R*)- Laureline	$C_{19}H_{19}NO_3$ 309.36	81-38-9 114	5251	1-27-00-00461	H_2O 1; EtOH 3; eth 3; dil acid 3
2470	5H-Benzo[g]-1,3-benzodioxolo[6,5,4-de]quinolin-12-ol, 6,7,7a,8-tetrahydro-11-methoxy-7-methyl-, (*S*)- Bulbocapnine	$C_{19}H_{19}NO_4$ 325.36	298-45-3 199.5	1471	2-27-00-00554	H_2O 1; EtOH 3; chl 4; os 3
2471	Benzo[ghi]perylene 1,12-Benzperylene	$C_{22}H_{12}$ 276.34	191-24-2		4-05-00-02766	
2472	Benzo[g]pteridine-2,4(1H,3H)-dione, 7,8-dimethyl- Lumichrome	$C_{12}H_{10}N_4O_2$ 242.24	1086-80-2	5468	4-26-00-02538	
2473	Benzo[g]quinoline 6,7-Benzoquinoline	$C_{13}H_9N$ 179.22	260-36-6 114	 202^{14}	5-20-08-00195	eth 4; EtOH 4
2474	Benzo[h]quinoline	$C_{13}H_9N$ 179.22	230-27-3 52	 339; 233^{47}	5-20-08-00215 1.2340^{20}	H_2O 2; EtOH 3; eth 3; ace 3
2475	Benzoic acid Benzenecarboxylic acid	$C_7H_6O_2$ 122.12	65-85-0 122.4	1101 249.2	4-09-00-00273 1.2659^{15}	H_2O 2; EtOH 4; eth 4; ace 3 1.504^{132}
2476	Benzoic acid, 2-acetyl-	$C_9H_8O_3$ 164.16	577-56-0 114.5	 110^{-2}2	4-10-00-02766	H_2O 4; eth 4; EtOH 4
2477	Benzoic acid, 3-acetyl-	$C_9H_8O_3$ 164.16	586-42-5 172	 110^{-2}2	4-10-00-02769	H_2O 3; EtOH 5
2478	Benzoic acid, 4-acetyl-	$C_9H_8O_3$ 164.16	586-89-0 208	 sub	4-10-00-02769	H_2O 4
2479	Benzoic acid, 4-acetyl-, methyl ester	$C_{10}H_{10}O_3$ 178.19	3609-53-8 95	 142^4	3-10-00-03030	H_2O 4
2480	Benzoic acid, 2-(acetyloxy)- Aspirin	$C_9H_8O_4$ 180.16	50-78-2 135	873	4-10-00-00138	H_2O 3; EtOH 4; eth 3; bz 2
2481	Benzoic acid, 2-(acetyloxy)-5-bromo- 5-Bromoacetylsalicylic acid	$C_9H_7BrO_4$ 259.06	1503-53-3 60	1425	2-10-00-00064	H_2O 1; EtOH 4; eth 3
2482	Benzoic acid, 2-(acetyloxy)-, methyl ester Methyl acetylsalicylate	$C_{10}H_{10}O_4$ 194.19	580-02-9 51.5	5934 $134\text{-}6^9$	4-10-00-00147	eth 4; EtOH 4, chl 4
2483	Benzoic acid, 2-(acetyloxy)-, phenyl ester Phenyl acetylsalicylate	$C_{15}H_{12}O_4$ 256.26	134-55-4 96	7241	4-10-00-00156	
2484	Benzoic acid, 2-amino- *o*-Anthranilic acid	$C_7H_7NO_2$ 137.14	118-92-3 146.5	433 sub	4-14-00-01004 1.412^{20}	H_2O 3; EtOH 3; eth 3; bz 2
2485	Benzoic acid, 3-amino- Aniline-3-carboxylic acid	$C_7H_7NO_2$ 137.14	99-05-8 173	432	4-14-00-01092 1.51^{25}	H_2O 2; EtOH 2; eth 3; ace 4
2486	Benzoic acid, 4-amino- Aniline-4-carboxylic acid	$C_7H_7NO_2$ 137.14	150-13-0 188.5	434	4-14-00-01126 1.374^{20}	H_2O 3; EtOH 3; eth 3; ace 2
2487	Benzoic acid, 2-amino-5-bromo- 5-Bromoanthranilic acid	$C_7H_6BrNO_2$ 216.03	5794-88-7 219.5	1393	4-14-00-01081	DMSO 3
2488	Benzoic acid, 4-amino-, butyl ester Butamben	$C_{11}H_{15}NO_2$ 193.25	94-25-7 58	1504 $173\ 4^8$	4-14-00-01130	H_2O 1; EtOH 3; eth 3; bz 3
2489	Benzoic acid, 4-(aminocarbonyl)-	$C_8H_7NO_3$ 165.15	6051-43-0 300	 sub	4-09-00-03318	H_2O 1; EtOH 1; eth 1; bz 1
2490	Benzoic acid, 5-amino-2-chloro- 5-Amino-2-chlorobenzoic acid	$C_7H_6ClNO_2$ 171.58	89-54-3 188		4-14-00-01116 1.519^{15}	EtOH 4
2491	Benzoic acid, 2-amino-3,6-dichloro-	$C_7H_5Cl_2NO_2$ 206.03	3032-32-4 155	 sub	4-14-00-01078	ace 4; bz 4; eth 4; EtOH 4
2492	Benzoic acid, 3-amino-2,5-dichloro- Chloramben	$C_7H_5Cl_2NO_2$ 206.03	133-90-4 200	2063		DMSO 2
2493	Benzoic acid, 4-amino-, 2-(diethylamino)ethyl ester Procaine	$C_{13}H_{20}N_2O_2$ 236.31	59-46-1 61	7763	4-14-00-01138	H_2O 2; EtOH 3; eth 3; bz 3
2494	Benzoic acid, 4-[[(2-amino-1,4-dihydro-4-oxo-6-pteridinyl)methyl]formylamino]- Rhizopterin	$C_{15}H_{12}N_6O_4$ 340.30	119-20-0 >300	8181	4-26-00-03949	H_2O 1; EtOH 1; eth 1; aq alk 3
2495	Benzoic acid, 2-amino-, ethyl ester	$C_9H_{11}NO_2$ 165.19	[illegible] 13	 268	4-14-00-01008 1.1174^{20}	eth 4, EtOH 4 1.5646^{20}
2496	Benzoic acid, 3-amino-, ethyl ester	$C_9H_{11}NO_2$ 165.19	582-33-2	 294; 160^5	4-14-00-01093 1.1248^{22}	H_2O 2; EtOH 4; eth 4; ctc 3 1.5600^{22}
2497	Benzoic acid, 4-amino-, ethyl ester Ethyl aminobenzoate	$C_9H_{11}NO_2$ 165.19	94-09-7 92	3719 310	4-14-00-01129	H_2O 1; EtOH 4; eth 4; chl 3
2498	Benzoic acid, 4-amino-2-hydroxy- *p*-Aminosalicylic acid	$C_7H_7NO_3$ 153.14	65-49-6 150-1 dec	491	4-14-00-01967	H_2O 3; EtOH 3; eth 3; ace 3
2499	Benzoic acid, 5-amino-2-hydroxy- Mesalamine	$C_7H_7NO_3$ 153.14	89-57-6 283	5807	4-14-00-02058	H_2O 2; EtOH 1
2500	Benzoic acid, 4-amino-2-hydroxy-, hydrazide *p*-Aminosalicylic acid hydrazide	$C_7H_9N_3O_2$ 167.17	6946-29-8 195	492	4-14-00-01995	EtOH 4
2501	Benzoic acid, 3-amino-4-hydroxy-, methyl ester Orthocaine	$C_8H_9NO_3$ 167.16	536-25-4 143	6836	3-14-00-01477	H_2O 1; EtOH 4; eth 3; bz 2

No.	Name Synonym	Mol. Form. Mol. Wt.	CAS RN mp/°C	Merck No. bp/°C	Beil. Ref. den/g cm^{-3}	Solubility n_D
2502	Benzoic acid, 4-amino-2-hydroxy-, phenyl ester Phenyl *p*-aminosalicylate	$C_{13}H_{11}NO_3$ 229.24	133-11-9 153	7243	4-14-00-01979	
2503	Benzoic acid, 2-amino-, methyl ester Methyl anthranilate	$C_8H_9NO_2$ 151.17	134-20-3 24.5	5942 256	4-14-00-01008 1.1682^{10}	H_2O 2; EtOH 4; eth 4 1.5810
2504	Benzoic acid, 3-amino-, methyl ester	$C_8H_9NO_2$ 151.17	4518-10-9 39	 $152\text{-}3^{11}$	4-14-00-01093 1.232^{20}	EtOH 4; eth 4; bz 4; chl 4
2505	Benzoic acid, 4-amino-, 2-methylpropyl ester Isobutyl *p*-aminobenzoate	$C_{11}H_{15}NO_2$ 193.25	94-14-4 64.5	5017	4-14-00-01130	
2506	Benzoic acid, 2-amino-3-nitro- 3-Nitroanthranilic acid	$C_7H_6N_2O_4$ 182.14	606-18-8 208.5		4-14-00-01087 1.558^{15}	eth 4; EtOH 4
2507	Benzoic acid, 4-amino-, propyl ester Risocaine	$C_{10}H_{13}NO_2$ 179.22	94-12-2 75	8226	4-14-00-01130	bz 4; eth 4; EtOH 4; chl 4
2508	Benzoic acid, 4-(aminosulfonyl)- Carzenide	$C_7H_7NO_4S$ 201.20	138-41-0 290-2 dec	1885	4-11-00-00690	H_2O 1; EtOH 4; eth 2; bz 1
2509	Benzoic acid, ammonium salt	$C_7H_9NO_2$ 139.15	1863-63-4 194.5	515	4-09-00-00273	H_2O 3
2510	Benzoic acid, anhydride Benzoic anhydride	$C_{14}H_{10}O_3$ 226.23	93-97-0 42.5	1102 360	4-09-00-00550 1.989^{15}	H_2O 1; EtOH 3; eth 3; chl 2 1.5767^{15}
2511	Benzoic acid, 3-benzoyl-	$C_{14}H_{10}O_3$ 226.23	579-18-0 163.3	 sub	4-10-00-02982	eth 4; EtOH 4
2512	Benzoic acid, 4-benzoyl-	$C_{14}H_{10}O_3$ 226.23	611-95-0 199	 sub	3-10-00-03305	H_2O 2; EtOH 3; eth 3; bz 2
2513	Benzoic acid, 3-(benzoylamino)-	$C_{14}H_{11}NO_3$ 241.25	587-54-2 252.5		4-14-00-01105 1.510^4	EtOH 4
2514	Benzoic acid, 4-(benzoylamino)-2-hydroxy- Benzoyl-PAS	$C_{14}H_{11}NO_4$ 257.25	13898-58-3 260.5	1127	4-14-00-02029	
2515	Benzoic acid, 2-benzoyl-, ethyl ester	$C_{16}H_{14}O_3$ 254.29	604-61-5 59.5		3-10-00-03291 1.221^{64}	H_2O 1; EtOH 4; eth 4; chl 2 1.560^{64}
2516	Benzoic acid, 2-benzoyl-, methyl ester	$C_{15}H_{12}O_3$ 240.26	606-28-0 52	 351	4-10-00-02977 1.1903^{19}	H_2O 1; EtOH 4; eth 4; sulf 3 1.591^{20}
2517	Benzoic acid, 2-[bis(4-hydroxyphenyl)methyl]- Phenolphthalin	$C_{20}H_{16}O_4$ 320.34	81-90-3 230.5	7210	4-10-00-01932	eth 4; EtOH 4
2518	Benzoic acid, 2-bromo-	$C_7H_5BrO_2$ 201.02	88-65-3 150	 sub	4-09-00-01011 1.929^{25}	H_2O 2; EtOH 3; eth 3; ace 3
2519	Benzoic acid, 3-bromo-	$C_7H_5BrO_2$ 201.02	585-76-2 155	 >280	4-09-00-01013 1.845^{20}	H_2O 1; EtOH 3; eth 3
2520	Benzoic acid, 4-bromo- *p*-Bromobenzoic acid	$C_7H_5BrO_2$ 201.02	586-76-5 254.5	1396	4-09-00-01017 1.894^{20}	H_2O 2; EtOH 3; eth 3
2521	Benzoic acid, 2-bromo-, ethyl ester	$C_9H_9BrO_2$ 229.07	6091-64-1	 254.5	4-09-00-01012 1.4438^{15}	ace 4; bz 4; eth 4; EtOH 4 1.5455^{15}
2522	Benzoic acid, 3-bromo-, ethyl ester	$C_9H_9BrO_2$ 229.07	24398-88-7	 261	4-09-00-01014 1.4308^{19}	ace 4; bz 4; eth 4; EtOH 4 1.5430^{19}
2523	Benzoic acid, 4-bromo-, ethyl ester	$C_9H_9BrO_2$ 229.07	5798-75-4	 263; 125^{15}	3-09-00-01405 1.4332^{17}	H_2O 2; EtOH 3; eth 3; ace 3 1.5438^{17}
2524	Benzoic acid, 5-bromo-2-hydroxy-	$C_7H_5BrO_3$ 217.02	89-55-4 169.8	 sub 100	4-10-00-00217	H_2O 2; EtOH 4; eth 4; ace 2
2525	Benzoic acid, 3-bromo-, methyl ester	$C_8H_7BrO_2$ 215.05	618-89-3 32	 125^{15}	4-09-00-01014	H_2O 2; EtOH 3; eth 3
2526	Benzoic acid, 4-bromo-, methyl ester	$C_8H_7BrO_2$ 215.05	619-42-1 81		4-09-00-01017 1.689^{25}	EtOH 3; eth 3; ace 3; bz 4
2527	Benzoic acid, 2-bromo-4-nitro- 2-Bromo-4-nitrobenzoic acid	$C_7H_4BrNO_4$ 246.02	16426-64-5 166.5	 sub 150	3-09-00-01771	eth 4; EtOH 4
2528	Benzoic acid, 2-bromo-5-nitro- 2-Bromo-5-nitrobenzoic acid	$C_7H_4BrNO_4$ 246.02	943-14-6 180.5	 sub	3-09-00-01771	eth 4; EtOH 4
2529	Benzoic acid, 4-bromo-3-nitro- 4-Bromo-3-nitrobenzoic acid	$C_7H_4BrNO_4$ 246.02	6319-40-0 203.5	 sub	4-09-00-01236	EtOH 4
2530	Benzoic acid, 5-bromo-2-nitro- 2-Nitro-5-bromobenzoic acid	$C_7H_4BrNO_4$ 246.02	6950-43-2 140		1-09-00-00165 1.920^{18}	bz 4; EtOH 4
2531	Benzoic acid, 4-(butylamino)-, 2-(dimethylamino)ethyl ester, monohydrochloride Tetracaine hydrochloride	$C_{15}H_{25}ClN_2O_2$ 300.83	136-47-0 147	9123	4-14-00-01172	
2532	Benzoic acid, butyl ester Butyl benzoate	$C_{11}H_{14}O_2$ 178.23	136-60-7 -22.4	1552 250.3	4-09-00-00290 1.000^{20}	H_2O 1; EtOH 5; eth 5; ace 3 1.4940^{25}
2533	Benzoic acid, 4,4'-carbonylbis-	$C_{15}H_{10}O_5$ 270.24	964-68-1 365	 sub	4-10-00-03497	HOAc 4
2534	Benzoic acid, 2-chloro- *o*-Chlorobenzoic acid	$C_7H_5ClO_2$ 156.57	118-91-2 140.2	2125 sub	4-09-00-00956 1.544^{20}	H_2O 3; EtOH 4; eth 4; ace 4
2535	Benzoic acid, 3-chloro- *m*-Chlorobenzoic acid	$C_7H_5ClO_2$ 156.57	535-80-8 158	2124 sub	4-09-00-00969 1.496^{25}	H_2O 4; EtOH 3; eth 4; bz 2
2536	Benzoic acid, 4-chloro- *p*-Chlorobenzoic acid	$C_7H_5ClO_2$ 156.57	74-11-3 243	2126	4-09-00-00973	H_2O 1; EtOH 4; eth 2; ace 2
2537	Benzoic acid, 2-chloro-, ethyl ester	$C_9H_9ClO_2$ 184.62	7335-25-3	 243	4-09-00-00957 1.1942^{15}	eth 4; EtOH 4 1.5242^{15}
2538	Benzoic acid, 3-chloro-, ethyl ester	$C_9H_9ClO_2$ 184.62	1128-76-3	 243	4-09-00-00970 1.1859^{15}	H_2O 1; EtOH 3; eth 3 1.5223^{20}
2539	Benzoic acid, 4-chloro-, ethyl ester	$C_9H_9ClO_2$ 184.62	7335-27-5	 237.5	4-09-00-00975 1.1873^{14}	EtOH 4
2540	Benzoic acid, 3-chloro-4-hydroxy-	$C_7H_5ClO_3$ 172.57	3964-58-7 171	 sub	4-10-00-00463	H_2O 2; EtOH 4; eth 4; ace 4

No.	Name Synonym	Mol. Form. Mol. Wt.	CAS RN mp/°C	Merck No. bp/°C	Beil. Ref. den/g cm^{-3}	Solubility n_D
2541	Benzoic acid, 3-chloro-2-hydroxy-, ethyl ester Salicylic acid, 3-chloro, ethyl ester	$C_9H_9ClO_3$ 200.62	56961-32-1 21	 269.5	0-10-00-00101	EtOH 4
2542	Benzoic acid, 5-chloro-2-hydroxy-, methyl ester	$C_8H_7ClO_3$ 186.59	4068-78-4 50	 249 dec; 120^{12}	4-10-00-00209	EtOH 4
2543	Benzoic acid, 4-chloro-, methyl ester	$C_8H_7ClO_2$ 170.60	1126-46-1 43.5		4-09-00-00974 1.382^{20}	EtOH 4
2544	Benzoic acid, 2-chloro-3-nitro- 2-Chloro-3-nitrobenzoic acid	$C_7H_4ClNO_4$ 201.57	3970-35-2 183.5		4-09-00-01225 1.662^{18}	H_2O 3; EtOH 3
2545	Benzoic acid, 2-chloro-5-nitro-	$C_7H_4ClNO_4$ 201.57	2516-96-3 166.5		4-09-00-01228 1.608^{18}	H_2O 2; EtOH 3; eth 3; ace 2
2546	Benzoic acid, 3-chloro-2-nitro- 2-Nitro-3-chlorobenzoic acid	$C_7H_4ClNO_4$ 201.57	4771-47-5 238		4-09-00-01225 1.566^{18}	eth 4; EtOH 4
2547	Benzoic acid, 4-chloro-3-nitro-	$C_7H_4ClNO_4$ 201.57	96-99-1 182.8		4-09-00-01226 1.645^{18}	H_2O 1; EtOH 2; ace 2
2548	Benzoic acid, 4-chloro-3-nitro-, methyl ester Methyl 4-chloro-3-nitrobenzoate	$C_8H_6ClNO_4$ 215.59	14719-83-6 83		4-09-00-01226 1.522^{18}	EtOH 4
2549	Benzoic acid, 5-chloro-2-nitro-, methyl ester Methyl 5-chloro-2-nitrobenzoate	$C_8H_6ClNO_4$ 215.59	51282-49-6 48.5		0-09-00-00401 1.453^{18}	MeOH 4
2550	Benzoic acid, 3-cyano-	$C_8H_5NO_2$ 147.13	1877-72-1 219	 sub	4-09-00-03296	H_2O 2; EtOH 3; eth 3
2551	Benzoic acid, cyclohexyl ester	$C_{13}H_{16}O_2$ 204.27	2412-73-9 <-10	 285	4-09-00-00296 1.0429^{20}	H_2O 1; EtOH 3; eth 3 1.5200^{20}
2552	Benzoic acid, 2,3 diamino 2,3-Diaminobenzoic acid	$C_7H_8N_2O_2$ 152.15	003-81-8 201	 dec	3-14-00-01172	EtOH 4; HOAc 4
2553	Benzoic acid, 2,4-diamino- 4-Aminoanthranilic acid	$C_7H_8N_2O_2$ 152.15	611-03-0 140	 200 dec	4-14-00-01299	EtOH 4; HOAc 4
2554	Benzoic acid, 3,5-diamino- 3,5-Diaminobenzoic acid	$C_7H_8N_2O_2$ 152.15	535-87-5 228	2955	4-14-00-01304	H_2O 2; EtOH 3; eth 4; tfa 2
2555	Benzoic acid, 2,4-dibromo- 2,4-Dibromobenzoic acid	$C_7H_4Br_2O_2$ 279.92	611-00-7 174	 sub	4-09-00-01027	eth 4; EtOH 4
2556	Benzoic acid, 2,5-dibromo- 2,5-Dibromobenzoic acid	$C_7H_4Br_2O_2$ 279.92	610-71-9 157	 sub	4-09-00-01027	eth 4; EtOH 4; chl 4
2557	Benzoic acid, 2,6-dibromo- 2,6-Dibromobenzoic acid	$C_7H_4Br_2O_2$ 279.92	601-84-3 150.5	 209^{16}	4-09-00-01028	H_2O 3; EtOH 4; eth 4; ace 3
2558	Benzoic acid, 3,5-dibromo-2-hydroxy- 3,5-Dibromosalicylic acid	$C_7H_4Br_2O_3$ [illegible]	3147-55-5 220	3012	4-10-00-00222	ace 3
2559	Benzoic acid, 3,5-dibromo-2-hydroxy-, methyl ester Methyl 3,5-dibromosalicylate	$C_8H_6Br_2O_3$ 309.94	21702-79-4 149	 181^{12}	4-10-00-00223	eth 4
2560	Benzoic acid, 2,6-dibromo-3,4,5-trihydroxy- Dibromogallic acid	$C_7H_4Br_2O_5$ 327.91	602-92-6 150	3006	2-10-00-00347	H_2O 4; eth 4; EtOH 4
2561	Benzoic acid, 2,4-dichloro- 2,4-Dichlorobenzoic acid	$C_7H_4Cl_2O_2$ 191.01	50-84-0 164.2	 sub	4-09-00-00998	H_2O 3; EtOH 3; eth 3; ace 2
2562	Benzoic acid, 2,5-dichloro- 2,5-Dichlorobenzoic acid	$C_7H_4Cl_2O_2$ 191.01	50-79-3 154.4	 301	4-09-00-01005	H_2O 2; EtOH 3; eth 3
2563	Benzoic acid, 2,6-dichloro- 2,6-Dichlorobenzoic acid	$C_7H_4Cl_2O_2$ 191.01	50-30-6 144	 sub	4-09-00-01005	H_2O 3; EtOH 3; eth 3; bz 3
2564	Benzoic acid, 3,5-dichloro- 3,5-Dichlorobenzoic acid	$C_7H_4Cl_2O_2$ 191.01	51-36-5 188	 sub	4-09-00-01008	H_2O 2; EtOH 3; eth 3; lig2
2565	Benzoic acid, 4-[(dichloroamino)sulfonyl]- Halazone	$C_7H_5Cl_2NO_4S$ 270.09	80-13-7 195 dec	4503	2-11-00-00220	H_2O 2; chl 2, HOAc 4, peth 1
2566	Benzoic acid, 3,5-dichloro-2-hydroxy-	$C_7H_4Cl_2O_3$ 207.01	320-72-9 220.5	 sub	4-10-00-00213	H_2O 2; EtOH 4; eth 3
2567	Benzoic acid, 3,6-dichloro-2-methoxy- Dicamba	$C_8H_6Cl_2O_3$ 221.04	1918-00-9 115	3026	 1.57^{25}	
2568	Benzoic acid, 2,3-dihydroxy-	$C_7H_6O_4$ 154.12	303-38-8 204		4-10-00-01414 1.542^{20}	H_2O 3; EtOH 3; eth 3; ace 2
2569	Benzoic acid, 2,4-dihydroxy- β-Resorcylic acid	$C_7H_6O_4$ 154.12	89-86-1 225-7 dec	8160	4-10-00-01420	H_2O 3; EtOH 3; eth 3; bz 3
2570	Benzoic acid, 2,5-dihydroxy- Gentisic acid	$C_7H_6O_4$ 154.12	490-79-9 199.5	4290	4-10-00-01441	H_2O 4; EtOH 4; eth 4; ace 3
2571	Benzoic acid, 3,4-dihydroxy- Protocatechuic acid	$C_7H_6O_4$ 154.12	99-50-3 200-2 dec	7909	4-10-00-01459 1.524^{4}	H_2O 2; EtOH 4; eth 3; bz 1
2572	Benzoic acid, 2-(3,6-dihydroxy-9H-xanthen-9-yl)- Fluorescin	$C_{20}H_{14}O_5$ 334.33	518-44-5 126	4087	2-18-00-00307	H_2O 1; EtOH 3; eth 3; ace 3
2573	Benzoic acid, 2,4-dihydroxy-6-methyl- *o*-Orsellinic acid	$C_8H_8O_4$ 168.15	480-64-8 176 dec	6834	4-10-00-01526	EtOH 3; eth 3
2574	Benzoic acid, 2,4-dihydroxy-6-methyl-, ethyl ester	$C_{10}H_{12}O_4$ 196.20	2524-37-0 132	 sub	4-10-00-01526	eth 4; EtOH 4
2575	Benzoic acid, 3,4-dihydroxy-5-[(3,4,5-trihydroxybenzoyl)oxy]- Digallic acid	$C_{14}H_{10}O_9$ 322.23	536-08-3 268-70 dec	3135	3-10-00-02086	ace 4; EtOH 4
2576	Benzoic acid, 3,4-dimethoxy- Veratric acid	$C_9H_{10}O_4$ 182.18	93-07-2 181	9854 sub	4-10-00-01460	H_2O 1; EtOH 4; eth 4; chl 2
2577	Benzoic acid, 3,5-dimethoxy- 3,5-Dimethoxybenzoic acid	$C_9H_{10}O_4$ 182.18	1132-21-4 185.5	 sub	4-10-00-01501	eth 4; EtOH 4

No.	Name	Synonym	Mol. Form.	Mol. Wt.	CAS RN	mp/°C	Merck No.	bp/°C	Beil. Ref.	den/g cm^{-3}	Solubility	n_D
2578	Benzoic acid, 3,4-dimethoxy-, ethyl ester		$C_{11}H_{14}O_4$	210.23	3943-77-9	43.5		295.5	4-10-00-01469		eth 4; EtOH 4	
2579	Benzoic acid, 3,4-dimethoxy-, methyl ester		$C_{10}H_{12}O_4$	196.20	2150-38-1	60.8		283	4-10-00-01467		bz 4; eth 4; EtOH 4	
2580	Benzoic acid, 2,4-dimethyl-	4-Carboxy-1,3-dimethylbenzene	$C_9H_{10}O_2$	150.18	611-01-8	90		268	4-09-00-01801		H_2O 2; EtOH 3; ace 3; bz 3	
2581	Benzoic acid, 2,5-dimethyl-		$C_9H_{10}O_2$	150.18	610-72-0	132		sub	4-09-00-01802	1.069^{21}	H_2O 1; EtOH 3; eth 3; ace 3	
2582	Benzoic acid, 2,6-dimethyl-	2,6-Dimethylbenzoic acid	$C_9H_{10}O_2$	150.18	632-46-2	116		274.5; 155^{17}	4-09-00-01798		H_2O 2; EtOH 3; eth 3; lig 2	
2583	Benzoic acid, 3,5-dimethyl-	Mesitylenic acid	$C_9H_{10}O_2$	150.18	499-06-9	171.1		sub	4-09-00-01806		H_2O 2; EtOH 4; eth 4	
2584	Benzoic acid, 2-(dimethylamino)-		$C_9H_{11}NO_2$	165.19	610-16-2	72		sub	4-14-00-01017		H_2O 4; eth 4; EtOH 4	
2585	Benzoic acid, 4-(dimethylamino)-	*p*-(Dimethylamino)benzoic acid	$C_9H_{11}NO_2$	165.19	619-84-1	242.5	3221		4-14-00-01164		EtOH 3; eth 2	
2586	Benzoic acid, 6-[[6-[2-(dimethylamino)ethyl]-4-methoxy-1,3-benzodioxol-5-yl]acetyl]-2,3-dimethoxy-	Narceine	$C_{23}H_{27}NO_8$	445.47	131-28-2	138	6341		5-19-09-00051			
2587	Benzoic acid, 2-[[4-(dimethylamino)phenyl]azo]-	Methyl red	$C_{15}H_{15}N_3O_2$	269.30	493-52-7	183	6037		4-16-00-00504		H_2O 2; EtOH 3; ace 4; bz 4	
2588	Benzoic acid, 4-(1,1-dimethylethyl)-	*p-tert*-Butylbenzoic acid	$C_{11}H_{14}O_2$	178.23	98-73-7	164.5			4-09-00-01884		H_2O 1; EtOH 4; bz 4; chl 3	
2589	Benzoic acid, 2,4-dimethyl-, methyl ester	Methyl 2,4-dimethylbenzoate	$C_{10}H_{12}O_2$	164.20	23617-71-2	-2		232.5	2-09-00-00350			1.5052^{20}
2590	Benzoic acid, 3,4-dinitro-	3,4-Dinitrobenzoic acid	$C_7H_4N_2O_6$	212.12	528-45-0	166	3268		4-09-00-01242		H_2O 2; EtOH 4; eth 4	
2591	Benzoic acid, 3,5-dinitro-	3,5-Dinitrobenzoic acid	$C_7H_4N_2O_6$	212.12	99-34-3	205	3269		4-09-00-01242		EtOH 4; HOAc 4	
2592	Benzoic acid, 3,5-dinitro-, butyl ester	Butyl 3,5-dinitrobenzoate	$C_{11}H_{12}N_2O_6$	268.23	10478-02-1	62.5			4-09-00-01246		EtOH 3; DMSO-d_6 2	1.488
2593	Benzoic acid, 3,5-dinitro-, ethyl ester	Ethyl 3,5-dinitrobenzoate	$C_9H_8N_2O_6$	240.17	618-71-3	94			4-09-00-01243	1.295^{111}	EtOH 3	1.560
2594	Benzoic acid, 3,3'-[(1,6-dioxo-1,6-hexanediyl)diimino]bis[2,4,6-triiodo-	Iodipamide	$C_{20}H_{14}I_6N_2O_6$	1139.77	606-17-7	307 dec	4916		4-14-00-01122		H_2O 1; EtOH 2; eth 2; ace 2	
2595	Benzoic acid, 4-[(dipropylamino)sulfonyl]-	Probenecid	$C_{13}H_{19}NO_4S$	285.36	57-66-9	195	7760		4-11-00-00691			
2596	Benzoic acid, 2-ethoxy-		$C_9H_{10}O_3$	166.18	134-11-2	20.7		$211\text{-}2^{39}$	3-10-00-00098		H_2O 2; EtOH 2; ctc 2	
2597	Benzoic acid, 3-ethoxy-		$C_9H_{10}O_3$	166.18	621-51-2	137		sub	3-10-00-00245		H_2O 3; EtOH 3; eth 3; bz 3	
2598	Benzoic acid, 2-ethoxy-, ethyl ester	Ethyl 2-ethoxybenzoate	$C_{11}H_{14}O_3$	194.23	6290-24-0			251	3-10-00-00117	1.1005^{20}	eth 4; EtOH 4	
2599	Benzoic acid, 4-ethoxy-, ethyl ester	Ethyl 4-ethoxybenzoate	$C_{11}H_{14}O_3$	194.23	23676-09-7			275	4-10-00-00368	1.076^{12}	H_2O 1; EtOH 4; eth 4; ctc 2	
2600	Benzoic acid, 2-ethyl-		$C_9H_{10}O_2$	150.18	612-19-1	68		259	4-09-00-01789	1.0431^{100}	eth 4; EtOH 4	1.5099^{100}
2601	Benzoic acid, 3-ethyl-		$C_9H_{10}O_2$	150.18	619-20-5	47			4-09-00-01791	1.042^{100}	eth 4; EtOH 4	1.5345^{100}
2602	Benzoic acid, ethyl ester	Ethyl benzoate	$C_9H_{10}O_2$	150.18	93-89-0	-34	3725	212	4-09-00-00285	1.0511^{15}	H_2O 1; EtOH 3; eth 5; ace 3	1.5007^{20}
2603	Benzoic acid, 2-fluoro-		$C_7H_5FO_2$	140.11	445-29-4	126.5			4-09-00-00950	1.460^{25}	H_2O 2; EtOH 4; eth 4; bz 1	
2604	Benzoic acid, 3-fluoro-		$C_7H_5FO_2$	140.11	455-38-9	124			4-09-00-00952	1.474^{25}	H_2O 2; eth 3	
2605	Benzoic acid, 4-fluoro-	*p*-Fluorobenzoic acid	$C_7H_5FO_2$	140.11	456-22-4	185	4100		4-09-00-00953	1.479^{25}	H_2O 2; EtOH 3; eth 3; ace 2	
2606	Benzoic acid, 4-fluoro-, ethyl ester		$C_9H_9FO_2$	168.17	451-46-7	26		210	4-09-00-00953	1.146^{25}	eth 4; EtOH 4	1.4864^{20}
2607	Benzoic acid, 2-formyl-		$C_8H_6O_3$	150.13	119-67-5	98			4-10-00-02748	1.404^{25}	H_2O 3; EtOH 4; eth 4	
2608	Benzoic acid, 6-formyl-2,3-dimethoxy-	Opianic acid	$C_{10}H_{10}O_5$	210.19	519-05-1	150	6806		4-10-00-03863		EtOH 3; eth 3	
2609	Benzoic acid, 3-formyl-4-hydroxy-		$C_8H_6O_4$	166.13	584-87-2	244		sub	4-10-00-03622		eth 4; EtOH 4	
2610	Benzoic acid, hexyl ester		$C_{13}H_{18}O_2$	206.28	6789-88-4			272; 139^8	4-09-00-00293	0.9793^{20}	H_2O 1; EtOH 3; ace 3	
2611	Benzoic acid, hydrazide		$C_7H_8N_2O$	136.15	613-94-5	115		267 dec	4-09-00-00922		H_2O 3; EtOH 3; eth 2; ace 2	
2612	Benzoic acid, 2-hydroxy-	Salicylic acid	$C_7H_6O_3$	138.12	69-72-7	158	8301	211^{20}	4-10-00-00125	1.443^{20}	H_2O 2; EtOH 4; eth 4; ace 4	1.565
2613	Benzoic acid, 4-hydroxy-	*p*-Salicylic acid	$C_7H_6O_3$	138.12	99-96-7	214.5	4742		4-10-00-00345	1.46^{25}	H_2O 2; EtOH 4; eth 3; ace 3	
2614	Benzoic acid, 2-hydroxy-, butyl ester		$C_{11}H_{14}O_3$	194.23	2052-14-4	-5.9		271	4-10-00-00153	1.0728^{20}	ctc 2	1.5115^{20}
2615	Benzoic acid, 4-hydroxy-, butyl ester	Butylparaben	$C_{11}H_{14}O_3$	194.23	94-26-8	68.5	1583		4-10-00-00375		H_2O 2; EtOH 3; ctc 2	

No.	Name Synonym	Mol. Form. Mol. Wt.	CAS RN mp/°C	Merck No. bp/°C	Beil. Ref. den/g cm^{-3}	Solubility n_D
2616	Benzoic acid, 2-hydroxy-, 2-carboxyphenyl ester Salsalate	$C_{14}H_{10}O_5$ 258.23	552-94-3 147	8307	4-10-00-00165	ace 2
2617	Benzoic acid, 2-hydroxy-3,5-diiodo- 3,5-Diiodosalicylic acid	$C_7H_4I_2O_3$ 389.92	133-91-5 235.5	3175	4-10-00-00226	H_2O 2; EtOH 4; eth 4; bz 1
2618	Benzoic acid, 4-hydroxy-3,5-diiodo-	$C_7H_4I_2O_3$ 389.92	618-76-8 237	 260 dec	4-10-00-00479	H_2O 1; EtOH 4; eth 4; bz 2
2619	Benzoic acid, 2-hydroxy-, ethyl ester Ethyl salicylate	$C_9H_{10}O_3$ 166.18	118-61-6 45	3804 150-1^{10}	4-10-00-00149 1.1326^{20}	H_2O 1; EtOH 5; eth 4; ctc 3 1.5296^{20}
2620	Benzoic acid, 3-hydroxy-, ethyl ester	$C_9H_{10}O_3$ 166.18	7781-98-8 74		4-10-00-00321 1.0680^{131}	H_2O 2; EtOH 3; eth 3; chl 2
2621	Benzoic acid, 4-hydroxy-, ethyl ester Ethylparaben	$C_9H_{10}O_3$ 166.18	120-47-8 117	3792 297.5	4-10-00-00367	H_2O 2; EtOH 4; eth 4; chl 2
2622	Benzoic acid, 2-hydroxy-, 2-hydroxyethyl ester Glycol salicylate	$C_9H_{10}O_4$ 182.18	87-28-5 37	4395 173^{15}	3-10-00-00138 1.2526^{15}	H_2O 2; EtOH 4; eth 4; bz 4
2623	Benzoic acid, 4-hydroxy-3-iodo- 4-Hydroxybenzoic acid, 3-iodo	$C_7H_5IO_3$ 264.02	37470-46-5 174.3	 sub	4-10-00-00478	eth 4; EtOH 4
2624	Benzoic acid, 5-hydroxy-2-iodo- 3-Hydroxybenzoic acid, 6-iodo	$C_7H_5IO_3$ 264.02	57772-57-3 198	 sub 160	3-10-00-00261	eth 4; EtOH 4
2625	Benzoic acid, 4-hydroxy-3-methoxy- Vanillic acid	$C_8H_8O_4$ 168.15	121-34-6 211.5	9838 sub	4-10-00-01459	H_2O 2; EtOH 4; eth 3; DMSO 3
2626	Benzoic acid, 4-hydroxy-3-methoxy-, ethyl ester	$C_{10}H_{12}O_4$ 196.20	617-05-0 44	 292	3-10-00-01413	H_2O 1; EtOH 4; eth 4; chl 3
2627	Benzoic acid, 4-hydroxy-3-methoxy-, methyl ester	$C_9H_{10}O_4$ 182.18	3943-74-6 64	 286	3-10-00-01410	EtOH 3; chl 2; peth 3
2628	Benzoic acid, 4-hydroxy-3-methoxy-, 2-methylpropyl ester	$C_{12}H_{16}O_4$ 224.26	7152-88-7 56.5	 125^{2}	3-10-00-01415	chl 2
2629	Benzoic acid, 2-hydroxy-3-methyl- *o*-Cresotic acid	$C_8H_8O_3$ 152.15	83-40-9 165.5	2585	4-10-00-00601	H_2O 2; EtOH 3; eth 3; bz 3
2630	Benzoic acid, 2-hydroxy-4-methyl- *m*-Cresotic acid	$C_8H_8O_3$ 152.15	50-85-1 177	2584	4-10-00-00617	H_2O 2; EtOH 3; eth 4; bz 3
2631	Benzoic acid, 2-hydroxy-5-methyl- *p*-Cresotic acid	$C_8H_8O_3$ 152.15	89-56-5 151	2586	4-10-00-00610	H_2O 2; EtOH 3; eth 3; bz 3
2632	Benzoic acid, 3-hydroxy-5-methyl-	$C_8H_8O_3$ 152.15	585-81-9 210	 sub	4-10-00-00609	H_2O 3; EtOH 4; eth 4; chl 2
2633	Benzoic acid, 2-hydroxy-, 3-methylbutyl ester Isopentyl salicylate	$C_{12}H_{16}O_3$ 208.26	87-20-7	5007 278; 151^{15}	4-10-00-00153 1.0535^{20}	H_2O 1; EtOH 4; eth 3; ctc 2 1.5080^{20}
2634	Benzoic acid, 2-hydroxy-, methyl ester Methyl salicylate	$C_8H_8O_3$ 152.15	119-36-8 -8	6038 222.9	4-10-00-00143 1.181^{25}	eth 4; EtOH 4; chl 4 1.535^{20}
2635	Benzoic acid, 2-hydroxy-, 1-methylethyl ester Isopropyl salicylate	$C_{10}H_{12}O_3$ 180.20	607-85-2	 238	4-10-00-00152 1.0729^{20}	H_2O 1; EtOH 5; eth 5 1.5065^{20}
2636	Benzoic acid, 3-hydroxy-, methyl ester	$C_8H_8O_3$ 152.15	19438-10-9 73	 281, 178^{17}	4-10-00-00319 1.1528^{100}	EtOH 3; bz 3; chl 2; peth 3
2637	Benzoic acid, 4-hydroxy-, methyl ester Methylparaben	$C_8H_8O_3$ 152.15	99-76-3 131	6021 275 dec	4-10-00-00360	H_2O 2; EtOH 4; eth 4; ace 4
2638	Benzoic acid, 2-hydroxy-3-methyl-, methyl ester	$C_9H_{10}O_3$ 166.18	23287-26-5 29	 235	4-10-00-00602 1.1683^{25}	 1.5354^{16}
2639	Benzoic acid, 2-hydroxy-4-methyl-, methyl ester	$C_9H_{10}O_3$ 166.18	4670-56-8 27.5	 243	2-10-00-00139 1.1483^{15}	 1.5378^{15}
2640	Benzoic acid, 2-hydroxy-5-methyl-, methyl ester	$C_9H_{10}O_3$ 166.18	22717-57-3 -1	 244.5	4-10-00-00611 1.1673^{25}	 1.5351^{15}
2641	Benzoic acid, 2-hydroxy-3-methyl-6-(1-methylethyl)-	$C_{11}H_{14}O_3$ 194.23	4389-53-1 136	 sub	4-10-00-00738	eth 4; EtOH 4
2642	Benzoic acid, 2-hydroxy-6-methyl-3-(1-methylethyl)- *o*-Thymotic acid	$C_{11}H_{14}O_3$ 194.23	548-51-6 127	9342 sub	3-10-00-00629	bz 4; eth 4; EtOH 4
2643	Benzoic acid, 2-hydroxy-, 2-methylpropyl ester Isobutyl salicylate	$C_{11}H_{14}O_3$ 194.23	87-19-4 5.9	 261	3-10-00-00121 1.0639^{20}	H_2O 1; EtOH 3; eth 3; ctc 3 1.5087^{20}
2644	Benzoic acid, 2-hydroxy-, 1-naphthalenyl ester 1-Naphthyl salicylate	$C_{17}H_{12}O_3$ 264.28	550-97-0 83	6334	[illegible]	[illegible]
2645	Benzoic acid, 2-hydroxy-, 2-naphthalenyl ester 2-Naphthyl salicylate	$C_{17}H_{12}O_3$ 264.28	613-78-5 95.5	6335	4-10-00-00158 1.11^{116}	H_2O 1; EtOH 2; eth 3; bz 3
2646	Benzoic acid, 2-hydroxy-3-nitro- 3-Nitrosalicylic acid	$C_7H_5NO_5$ 183.12	85-38-1 148	6553	4-10-00-00228	H_2O 2; EtOH 4; eth 4; ace 3
2647	Benzoic acid, 2-hydroxy-5-nitro- 5-Nitrosalicylic acid	$C_7H_5NO_5$ 183.12	96-97-9 229.5	6554	4-10-00-00255 1.650^{20}	H_2O 2; EtOH 4; eth 4; ace 4
2648	Benzoic acid, 2-hydroxy-5-[(4-nitrophenyl)azo]- Alizarine Yellow R	$C_{13}H_9N_3O_5$ 287.23	2243-76-7 253-4 dec	241	4-16-00-00372	H_2O 4; EtOH 4
2649	Benzoic acid, 2-hydroxy-, pentyl ester	$C_{12}H_{16}O_3$ 208.26	2050-08-0	 270	4-10-00-00153 1.064^{15}	H_2O 2; EtOH 5; eth 5 1.506^{20}
2650	Benzoic acid, 2-hydroxy-, phenyl ester Phenyl salicylate	$C_{13}H_{10}O_3$ 214.22	118-55-8 130.5	7282 173^{12}	4-10-00-00154 1.2614^{30}	H_2O 1; EtOH 4; eth 3; ace 4

No.	Name Synonym	Mol. Form. Mol. Wt.	CAS RN mp/°C	Merck No. bp/°C	Beil. Ref. den/g cm^{-3}	Solubility n_D
2651	Benzoic acid, 2-hydroxy-, phenylmethyl ester Benzyl salicylate	$C_{14}H_{12}O_3$ 228.25	118-58-1	1160 320	4-10-00-00157 1.1799^{20}	H_2O 2; EtOH 3; eth 3; ctc 3 1.5805^{20}
2652	Benzoic acid, 2-hydroxy-, propyl ester	$C_{10}H_{12}O_3$ 180.20	607-90-9 97	 239	4-10-00-00152 1.0979^{20}	ctc 3; CS_2 3 1.5161^{20}
2653	Benzoic acid, 4-hydroxy-, propyl ester Propylparaben	$C_{10}H_{12}O_3$ 180.20	94-13-3 97	7879	4-10-00-00374 1.0630^{102}	H_2O 1; EtOH 3; eth 3; chl 2 1.5050^{102}
2654	Benzoic acid, 2-hydroxy-5-[[4-[(2-pyridinylamino)sulfonyl]phenyl]azo]- Sulfasalazine	$C_{18}H_{14}N_4O_5S$ 398.40	599-79-1 220 dec	8917	5-22-08-00433	
2655	Benzoic acid, 2-hydroxy-5-sulfo- Sulfosalicylic acid	$C_7H_6O_6S$ 218.19	97-05-2 120	8944	4-11-00-00702	H_2O 5; EtOH 5; eth 4
2656	Benzoic acid, 2,2'-iminobis- Diphenylamine-2,2'-dicarboxylic acid	$C_{14}H_{11}NO_4$ 257.25	579-92-0 296-7 dec	3318	4-14-00-01058	
2657	Benzoic acid, 2-iodo- *o*-Iodobenzoic acid	$C_7H_5IO_2$ 248.02	88-67-5 163	4923 exp 233	4-09-00-01030 2.25^{25}	H_2O 2; EtOH 4; eth 4; ace 2
2658	Benzoic acid, 3-iodo-	$C_7H_5IO_2$ 248.02	618-51-9 188.3	 sub	4-09-00-01033	H_2O 2; EtOH 4; eth 2
2659	Benzoic acid, 4-iodo-	$C_7H_5IO_2$ 248.02	619-58-9 270	 sub	4-09-00-01035 2.184^{20}	H_2O 2; EtOH 2; eth 3
2660	Benzoic acid, 2-iodo-, methyl ester	$C_8H_7IO_2$ 262.05	610-97-9	 280; 146^{16}	4-09-00-01031	EtOH 3 1.6052^{20}
2661	Benzoic acid, 3-iodo-, methyl ester	$C_8H_7IO_2$ 262.05	618-91-7 54.5	 277; 150^{18}	4-09-00-01034	H_2O 1; EtOH 3; eth 4; ace 4
2662	Benzoic acid, 4-iodo-, methyl ester	$C_8H_7IO_2$ 262.05	619-44-3 114.8	 sub	4-09-00-01036 2.0200^{10}	EtOH 3; eth 3
2663	Benzoic acid, 2-mercapto- *o*-Thiosalicylic acid	$C_7H_6O_2S$ 154.19	147-93-3 168.5	9291 sub	4-10-00-00272	H_2O 3; EtOH 3; eth 3; lig 2
2664	Benzoic acid, 2-methoxy-	$C_8H_8O_3$ 152.15	579-75-9 101	 200	4-10-00-00130	H_2O 2; EtOH 4; eth 4; bz 3
2665	Benzoic acid, 3-methoxy-	$C_8H_8O_3$ 152.15	586-38-9 107	 170^{10}	4-10-00-00316	H_2O 2; EtOH 3; eth 3; bz 3
2666	Benzoic acid, 4-methoxy- *p*-Anisic acid	$C_8H_8O_3$ 152.15	100-09-4 185	696 276.5	4-10-00-00346	H_2O 1; EtOH 4; eth 4; chl 3
2667	Benzoic acid, 2-(4-methoxybenzoyl)- *o*-(*p*-Anisoyl)benzoic acid	$C_{15}H_{12}O_4$ 256.26	1151-15-1 146	702	3-10-00-04429	eth 4; EtOH 4; tol 4
2668	Benzoic acid, 4-methoxy-, butyl ester	$C_{12}H_{16}O_3$ 208.26	6946-35-6	 183^{40}	3-10-00-00307 1.054^{16}	 1.5141^{16}
2669	Benzoic acid, 2-methoxy-, ethyl ester	$C_{10}H_{12}O_3$ 180.20	7335-26-4	 261	4-10-00-00150 1.1124^{20}	eth 4; EtOH 4 1.5224^{20}
2670	Benzoic acid, 3-methoxy-, ethyl ester	$C_{10}H_{12}O_3$ 180.20	10259-22-0	 260.5	4-10-00-00321 1.0993^{20}	eth 4; EtOH 4 1.5161^{20}
2671	Benzoic acid, 4-methoxy-, ethyl ester	$C_{10}H_{12}O_3$ 180.20	94-30-4 7.5	 269.5	4-10-00-00368 1.1038^{20}	H_2O 1; EtOH 3; eth 3 1.5254^{20}
2672	Benzoic acid, 3-methoxy-2-(methylamino)-, methyl ester Damascenine	$C_{10}H_{13}NO_3$ 195.22	483-64-7 28	2809 271; 147^{10}	4-14-00-02073	bz 4; eth 4; EtOH 4; lig 4
2673	Benzoic acid, 2-methoxy-, methyl ester	$C_9H_{10}O_3$ 166.18	606-45-1	 246.5	4-10-00-00144 1.1571^{19}	H_2O 1; EtOH 3 1.534^{19}
2674	Benzoic acid, 3-methoxy-, methyl ester	$C_9H_{10}O_3$ 166.18	5368-81-0	 248	4-10-00-00319 1.1310^{20}	H_2O 1; EtOH 3 1.5224^{20}
2675	Benzoic acid, 4-methoxy-, methyl ester	$C_9H_{10}O_3$ 166.18	121-98-2 49	 244	4-10-00-00360	H_2O 1; EtOH 3; eth 3; chl 3
2676	Benzoic acid, 2-methyl- *o*-Toluic acid	$C_8H_8O_2$ 136.15	118-90-1 103.7	9461 259	4-09-00-01697 1.062^{115}	H_2O 1; EtOH 4; eth 4; chl 3 1.512^{115}
2677	Benzoic acid, 3-methyl- *m*-Toluic acid	$C_8H_8O_2$ 136.15	99-04-7 108.7	9461	4-09-00-01712 1.054^{112}	H_2O 2; EtOH 4; eth 4; chl 2 1.509
2678	Benzoic acid, 4-methyl- *p*-Toluic acid	$C_8H_8O_2$ 136.15	99-94-5 179.6	9461	4-09-00-01724	H_2O 1; EtOH 4; eth 4; tfa 2
2679	Benzoic acid, 2-(methylamino)-	$C_8H_9NO_2$ 151.17	119-68-6 180.5	 $80^{0.01}$	4-14-00-01015	H_2O 2; EtOH 4; eth 4; bz 4
2680	Benzoic acid, 2-(methylamino)-, ethyl ester Anthranilic acid, *N*-methyl-, ethyl ester	$C_{10}H_{13}NO_2$ 179.22	35472-56-1 39	 266; 172^{45}	2-14-00-00213	eth 4
2681	Benzoic acid, 2-(methylamino)-, methyl ester	$C_9H_{11}NO_2$ 165.19	85-91-6 19	 255	4-14-00-01016 1.120^{15}	H_2O 1; EtOH 3; eth 3 1.5839^{15}
2682	Benzoic acid, 2-methyl-, anhydride	$C_{16}H_{14}O_3$ 254.29	607-86-3 38.5	 >325; 220^{11}	4-09-00-01700	eth 4
2683	Benzoic acid, 3-methyl-, anhydride	$C_{16}H_{14}O_3$ 254.29	21436-44-2 71	 230^{17}	4-09-00-01715	ace 4; bz 4; eth 4; EtOH 4
2684	Benzoic acid, 2-(4-methylbenzoyl)- 2-(*p*-Toluyl)benzoic acid	$C_{15}H_{12}O_3$ 240.26	85-55-2 146	9465	4-10-00-02991	H_2O 2; EtOH 4; eth 4; ace 4
2685	Benzoic acid, 3-methyl-, butyl ester Butyl *m*-toluate	$C_{12}H_{16}O_2$ 192.26	6640-77-3	 262.3	3-09-00-00397 1.0040^{20}	H_2O 1; EtOH 3; eth 5 1.4950^{20}
2686	Benzoic acid, 3,3'-methylenebis[6-hydroxy- 5,5'-Methylenedisalicylic acid	$C_{15}H_{12}O_6$ 288.26	122-25-8 243.5	5984	4-10-00-02337	ace 4; eth 4; EtOH 4
2687	Benzoic acid, methyl ester Methyl benzoate	$C_8H_8O_2$ 136.15	93-58-3 -15	5947 199	4-09-00-00283 1.0933^{15}	H_2O 1; EtOH 3; eth 5; ctc 3 1.5164^{20}
2688	Benzoic acid, 2-(1-methylethyl)-	$C_{10}H_{12}O_2$ 164.20	2438-04-2 64	 $160\text{-}1^{25}$	4-09-00-01839	H_2O 2; EtOH 4; eth 4; bz 3
2689	Benzoic acid, 4-(1-methylethyl)- Cumic acid	$C_{10}H_{12}O_2$ 164.20	536-66-3 117.5	2620 sub	4-09-00-01843 1.162^4	H_2O 2; EtOH 4; eth 4; peth 3

No.	Name Synonym	Mol. Form. Mol. Wt.	CAS RN mp/°C	Merck No. bp/°C	Beil. Ref. den/g cm^{-3}	Solubility n_D
2690	Benzoic acid, 2-methyl-, ethyl ester	$C_{10}H_{12}O_2$ 164.20	87-24-1 <-10	 227; 113^{18}	4-09-00-01699 1.0325^{21}	H_2O 1; EtOH 5; eth 5 1.507^{22}
2691	Benzoic acid, 3-methyl-, ethyl ester	$C_{10}H_{12}O_2$ 164.20	120-33-2	 234	4-09-00-01713 1.0265^{21}	eth 4; EtOH 4 1.5052^{22}
2692	Benzoic acid, 4-methyl-, ethyl ester	$C_{10}H_{12}O_2$ 164.20	94-08-6	 232	4-09-00-01726 1.0269^{18}	H_2O 1; EtOH 5; eth 5 1.5089^{18}
2693	Benzoic acid, 1-methylethyl ester	$C_{10}H_{12}O_2$ 164.20	939-48-0	 216	4-09-00-00289 1.0163^{15}	H_2O 1; EtOH 3; eth 3; ace 3 1.4890^{20}
2694	Benzoic acid, 4-(1-methylethyl)-, methyl ester	$C_{11}H_{14}O_2$ 178.23	20185-55-1	 126^{14}	4-09-00-01843 1.018^{19}	 1.515^{19}
2695	Benzoic acid, 4-(1-methylethyl)-, 2-methylpropyl ester Benzoic acid, 4-isopropyl, isobutyl ester	$C_{14}H_{20}O_2$ 220.31	6315-03-3	 155^{14}	2-09-00-00359 0.966^{19}	 1.497^{19}
2696	Benzoic acid, 2-methyl-, methyl ester	$C_9H_{10}O_2$ 150.18	89-71-4 <-50	 215	4-09-00-01699 1.068^{20}	H_2O 1; EtOH 5; eth 5
2697	Benzoic acid, 3-methyl-, methyl ester	$C_9H_{10}O_2$ 150.18	99-36-5	 221	4-09-00-01713 1.061^{20}	H_2O 1; EtOH 3; ctc 2
2698	Benzoic acid, 4-methyl-, methyl ester	$C_9H_{10}O_2$ 150.18	99-75-2 33.2	 220	4-09-00-01726	H_2O 1; EtOH 4; eth 4
2699	Benzoic acid, 2-methylphenyl ester	$C_{14}H_{12}O_2$ 212.25	617-02-7	 309; $155^{8.5}$	4-09-00-00305 1.114^{19}	eth 4; EtOH 4
2700	Benzoic acid, 4-methylphenyl ester	$C_{14}H_{12}O_2$ 212.25	614-34-6 71.5	 316	4-09-00-00306	eth 4; EtOH 4
2701	Benzoic acid, 2-methylpropyl ester	$C_{11}H_{14}O_2$ 178.23	120-50-3	 242	4-09-00-00291 0.9990^{20}	H_2O 1; EtOH 5; eth 5; ace 3
2702	Benzoic acid, 2-[(1-naphthalenylamino)carbonyl]- Naptalam	$C_{18}H_{13}NO_3$ 291.31	132-66-1 185	6338	3-12-00-02876 1.4^{20}	H_2O 1; EtOH 2; ace 2; bz 2
2703	Benzoic acid, 2-nitro- *o*-Nitrobenzoic acid	$C_7H_5NO_4$ 167.12	552-16-9 147.5		4-09-00-01046 1.575^{20}	H_2O 3; EtOH 4; eth 3; ace 4
2704	Benzoic acid, 3-nitro- *m*-Nitrobenzoic acid	$C_7H_5NO_4$ 167.12	121-92-6 141		4-09-00-01055 1.494^{20}	H_2O 2; EtOH 4; eth 4; ace 4
2705	Benzoic acid, 4-nitro- *p*-Nitrobenzoic acid	$C_7H_5NO_4$ 167.12	62-23-7 242	 sub	4-09-00-01072 1.610^{20}	ace 4; eth 4; EtOH 4; chl 4
2706	Benzoic acid, 4-nitro-, butyl ester	$C_{11}H_{13}NO_4$ 223.23	120-48-9 35.3	 160^8	3-09-00-01544	bz 4; eth 4
2707	Benzoic acid, 2-nitro-, ethyl ester	$C_9H_9NO_4$ 195.17	610-34-4 30	 275; 178^{23}	4-09-00-01047	eth 4; EtOH 4
2708	Benzoic acid, 3-nitro-, ethyl ester	$C_9H_9NO_4$ 195.17	618-98-4 47	 297	4-09-00-01056	H_2O 1; EtOH 4; eth 4
2709	Benzoic acid, 4-nitro-, ethyl ester	$C_9H_9NO_4$ 195.17	99-77-4 57	 186.3; 153^8	4-09-00-01074	H_2O 1; EtOH 3; eth 3
2710	Benzoic acid, 2-nitro-, methyl ester	$C_8H_7NO_4$ 181.15	606-27-9 -13	 275	4-09-00-01047 1.2855^{20}	H_2O 1; EtOH 3; eth 3; bz 3
2711	Benzoic acid, 3-nitro-, methyl ester	$C_8H_7NO_4$ 181.15	618-95-1 78	 279^{60}	4-09-00-01056	H_2O 1; EtOH 2; eth 2; MeOH 2
2712	Benzoic acid, pentachloro- Perchlorobenzoic acid	$C_7HCl_5O_2$ 294.35	1012-84-6 209.3	 sub	4-09-00-01011	EtOH 4; tol 4; HOAc 4
2713	Benzoic acid, pentamethyl- Permethylbenzoic acid	$C_{12}H_{16}O_2$ 192.26	2243-32-5 210.5	 sub	4-09-00-01920	H_2O 1; EtOH 4
2714	Benzoic acid, 2-phenoxy-	$C_{13}H_{10}O_3$ 214.22	2243-42-7 113	 355	4-10-00-00132 1.1553^{50}	H_2O 1; EtOH 4; eth 4; chl 3
2715	Benzoic acid, 2-(phenylamino)- *N*-Phenylanthranilic acid	$C_{13}H_{11}NO_2$ 213.24	91-40-7 183.5	7244	4-14-00-01019	H_2O 1; EtOH 4; eth 2; bz 2
2716	Benzoic acid, phenyl ester Phenyl benzoate	$C_{13}H_{10}O_2$ 198.22	93-99-2 71	7246 314	4-09-00-00303 1.235^{20}	H_2O 1; EtOH 3; eth 3; chl 3
2717	Benzoic acid, 2-phenylhydrazide	$C_{13}H_{12}N_2O$ 212.25	532-96-7 168	 314	4-15-00-00169	H_2O 2; EtOH 3; eth 2; bz 3
2718	Benzoic acid, 2-(phenylmethyl)-	$C_{14}H_{12}O_2$ 212.25	612-35-1 118	 sub	4-09-00-02516	H_2O 2; EtOH 3; eth 3; bz 3
2719	Benzoic acid, 4-(phenylmethyl)-	$C_{14}H_{12}O_2$ 212.25	620-86-0 158.8	 sub	4-09-00-02518	bz 4; eth 4; EtOH 4; chl 4
2720	Benzoic acid, phenylmethyl ester Benzyl benzoate	$C_{14}H_{12}O_2$ 212.25	120-51-4 21	1141 323.5	4-09-00-00307 1.1121^{25}	H_2O 1; EtOH 3; eth 3; ace 3 1.5680^{20}
2721	Benzoic acid, 4-[[(phenylmethyl)sulfonyl]amino]- Carinamide	$C_{14}H_{13}NO_4S$ 291.33	536-95-8 229.5	1163	4-14-00-01264	EtOH 4
2722	Benzoic acid, 2-propenyl ester	$C_{10}H_{10}O_2$ 162.19	583-04-0		4-09-00-00296 1.0569^{15}	H_2O 1; EtOH 3; eth 3; ace 3 1.5178^{20}
2723	Benzoic acid, *o*-propyl- Benzoic acid, 2-propyl	$C_{10}H_{12}O_2$ 164.20	2438-03-1 58	 273; 164^{20}	4-09-00-01836 1.0020^{15}	eth 4; EtOH 4
2724	Benzoic acid, propyl ester	$C_{10}H_{12}O_2$ 164.20	2315-68-6 -51.6	 211	4-09-00-00288 1.0230^{20}	H_2O 1; EtOH 5; eth 5 1.5000^{20}
2725	Benzoic acid, sodium salt	$C_7H_5NaO_2$ 144.11	532-32-1 >300	8527	4-09-00-00273	H_2O 3
2726	Benzoic acid, 2-[[[4-[(2-thiazolylamino)sulfonyl]phenyl]amino]carbonyl]- Phthalylsulphathiazole	$C_{17}H_{13}N_3O_5S_2$ 403.44	85-73-4 273	7351	4-27-00-04639	H_2O 1; EtOH 2; eth 1; chl 1

No.	Name Synonym	Mol. Form. Mol. Wt.	CAS RN mp/°C	Merck No. bp/°C	Beil. Ref. den/g cm^{-3}	Solubility n_D
2727	Benzoic acid, 2,3,6-trichloro- 2,3,6-Trichlorobenzoic acid	$C_7H_3Cl_3O_2$ 225.46	50-31-7 124.5		4-09-00-01009	H_2O 2; eth 3
2728	Benzoic acid, 2,4,5-trichloro- 2,4,5-Trichlorobenzoic acid	$C_7H_3Cl_3O_2$ 225.46	50-82-8 166.5	 sub	4-09-00-01010	H_2O 1; EtOH 3; eth 3
2729	Benzoic acid, 2-[[3-(trifluoromethyl)phenyl]amino]- Flufenamic acid	$C_{14}H_{10}F_3NO_2$ 281.23	530-78-9 133.5	4060	3-14-00-00905	DMSO 3
2730	Benzoic acid, 2,3,4-trihydroxy-	$C_7H_6O_5$ 170.12	610-02-6 221	 sub	4-10-00-01971	H_2O 2; EtOH 3; eth 3; ace 3
2731	Benzoic acid, 3,4,5-trihydroxy- Gallic acid	$C_7H_6O_5$ 170.12	149-91-7 253 dec	4251	3-10-00-02070 1.694^{6}	H_2O 2; EtOH 4; eth 2; ace 3
2732	Benzoic acid, 3,4,5-trihydroxy-, propyl ester Propyl gallate	$C_{10}H_{12}O_5$ 212.20	121-79-9 130	7872	4-10-00-02003	H_2O 2
2733	Benzoic acid, 2,4,5-trimethoxy-	$C_{10}H_{12}O_5$ 212.20	490-64-2 145	 300	4-10-00-01980	H_2O 4; bz 4; EtOH 4; peth 4
2734	Benzoic acid, 3,4,5-trimethoxy-	$C_{10}H_{12}O_5$ 212.20	118-41-2 172.3	 $225\text{-}7^{10}$	4-10-00-01996	H_2O 2; EtOH 4; eth 4; chl 4
2735	Benzoic acid, 2,4,6-trinitro- 2,4,6-Trinitrobenzoic acid	$C_7H_3N_3O_8$ 257.12	129-66-8 228 dec	9640	4-09-00-01362	H_2O 2; EtOH 4; eth 3; ace 3
2736	1H,5H-Benzo[ij]quinolizine-1,6(7H)-dione, 2,3-dihydro- Julolidine, 1,6-dioxo	$C_{12}H_{11}NO_2$ 201.22	39052-57-8 145.5	 $190\text{-}210^{0.3}$	4-21-00-05525	EtOH 4
2737	1H,5H-Benzo[ij]quinolizine, 2,3,6,7-tetrahydro- Julolidine	$C_{12}H_{15}N$ 173.26	479-59-4 40	 280 dec; 155^{17}	5-20-07-00189 1.003^{20}	 1.568^{25}
2738	Benzo[j]fluoranthene Dibenzo[a,jk]fluorene	$C_{20}H_{12}$ 252.32	205-82-3 166		4-05-00-02687	H_2O 1; EtOH 2; HOAc 2
2739	Benzo[k]fluoranthene 2,3,1',8'-Binaphthylene	$C_{20}H_{12}$ 252.32	207-08-9 217	 480	4-05-00-02686	H_2O 1; EtOH 3; bz 3; HOAc 3
2740	Benzonitrile Phenyl cyanide	C_7H_5N 103.12	100-47-0 -12.7	1107 191.1	4-09-00-00892 1.0093^{15}	H_2O 2; EtOH 5; eth 5; ace 4 1.5289^{20}
2741	Benzonitrile, 2-amino-	$C_7H_6N_2$ 118.14	1885-29-6 51	 263	4-14-00-01013	H_2O 2; EtOH 4; eth 4; ace 4
2742	Benzonitrile, 3-amino-	$C_7H_6N_2$ 118.14	2237-30-1 54.3	 289	4-14-00-01095	H_2O 2; EtOH 4; eth 4; ace 4
2743	Benzonitrile, 5-amino-2-methyl-	$C_8H_8N_2$ 132.17	50670-64-9 88	 $100\text{-}10^{22}$	3-14-00-01201	EtOH 4
2744	Benzonitrile, 2-bromo-	C_7H_4BrN 182.02	2042-37-7 55.5	 252	4-09-00-01013	H_2O 3; EtOH 4; chl 2
2745	Benzonitrile, 3-bromo-	C_7H_4BrN 182.02	6952-59-6 39.5	 225	4-09-00-01016	EtOH 4; eth 4; chl 2
2746	Benzonitrile, 4-bromo-	C_7H_4BrN 182.02	623-00-7 114	 236	4-09-00-01025	H_2O 3; EtOH 3; eth 3; chl 3
2747	Benzonitrile, 2-chloro-	C_7H_4ClN 137.57	873-32-5 46.3	 232	4-09-00-00965	H_2O 2; EtOH 3; eth 3; chl 3
2748	Benzonitrile, 3-chloro-	C_7H_4ClN 137.57	766-84-7 41	 100^{15}	4-09-00-00973	H_2O 1; EtOH 3; eth 3
2749	Benzonitrile, 4-chloro-	C_7H_4ClN 137.57	623-03-0 95	 223; 95^{5}	4-09-00-00992 1.1133^{17}	H_2O 2; EtOH 3; eth 3; bz 3
2750	Benzonitrile, 2-(chloromethyl)-	C_8H_6ClN 151.60	612-13-5 61	 252	4-09-00-01707	EtOH 4
2751	Benzonitrile, 3,5-dibromo-4-hydroxy- Bromoxynil	$C_7H_3Br_2NO$ 276.91	1689-84-5 190	1431	4-10-00-00475	
2752	Benzonitrile, 2,6-dichloro- Dichlobenil	$C_7H_3Cl_2N$ 172.01	1194-65-6 144.5	3029 270	4-09-00-01006	
2753	Benzonitrile, 2,6-dimethoxy- 2,6-Dimethoxybenzonitrile	$C_9H_9NO_2$ 163.18	16932-49-3 119	 310	2-10-00-00260	H_2O 2; EtOH 4; eth 2; ace 3
2754	Benzonitrile, 2-ethoxy- *o*-Ethoxybenzonitrile	C_9H_9NO 147.18	6609-57-0 5	 260.5	0-10-00-00097 1.0650^{15}	eth 4; EtOH 4; lig 4
2755	Benzonitrile, 4-ethoxy-	C_9H_9NO 147.18	25117-74-2 61.5	 258	3-10-00-00344	eth 4; EtOH 4; lig 4
2756	Benzonitrile, 4-fluoro-	C_7H_4FN 121.11	1194-02-1 34.8	 188.8	4-09-00-00954 1.1070^{55}	chl 2; peth 3 1.4925^{55}
2757	Benzonitrile, 3-formyl-	C_8H_5NO 131.13	24964-64-5 76.5	 210	4-10-00-02752	H_2O 4; EtOH 4; eth 4; chl 4
2758	Benzonitrile, 4-formyl-	C_8H_5NO 131.13	105-07-7 100.5	 133^{12}	4-10-00-02754	H_2O 3; EtOH 4; eth 4; chl 4
2759	Benzonitrile, 2-hydroxy-	C_7H_5NO 119.12	611-20-1 98	 149^{14}	4-10-00-00191 1.1052^{100}	H_2O 2; EtOH 4; eth 4; bz 4 1.5372^{100}
2760	Benzonitrile, 2-methoxy-	C_8H_7NO 133.15	6609-56-9 24.5	 255.5	4-10-00-00191 1.1063^{20}	EtOH 3; eth 4
2761	Benzonitrile, 3-methoxy-	C_8H_7NO 133.15	1527-89-5	 $111\text{-}2^{13}$	4-10-00-00329 1.089^{25}	 1.5402^{20}
2762	Benzonitrile, 4-methoxy-	C_8H_7NO 133.15	874-90-8 61.5	 256.5	4-10-00-00441	H_2O 1; EtOH 4; eth 4; bz 3
2763	Benzonitrile, 2-methyl- *o*-Tolunitrile	C_8H_7N 117.15	529-19-1 -13.5	9463 205	4-09-00-01703 0.9955^{20}	H_2O 1; EtOH 5; eth 5; ctc 2 1.5279^{20}
2764	Benzonitrile, 3-methyl- *m*-Tolunitrile	C_8H_7N 117.15	620-22-4 -23	 213	4-09-00-01717 1.0316^{20}	H_2O 1; EtOH 5; eth 5; ctc 2 1.5252^{20}
2765	Benzonitrile, 4-methyl- *p*-Tolunitrile	C_8H_7N 117.15	104-85-8 29.5	9464 217.0	4-09-00-01738 0.9762^{30}	H_2O 1; EtOH 4; eth 4; ctc 2

No.	Name Synonym	Mol. Form. Mol. Wt.	CAS RN mp/°C	Merck No. bp/°C	Beil. Ref. den/g cm^{-3}	Solubility n_D
2766	Benzonitrile, 2-methyl-5-nitro-	$C_8H_6N_2O_2$ 162.15	939-83-3 106	 174^{18}	3-09-00-02314	ace 4; bz 4; eth 4; EtOH 4
2767	Benzonitrile, pentafluoro-	C_7F_5N 193.08	773-82-0 2.4	 162	4-09-00-00956 1.563^{20}	 1.4402^{25}
2768	Benzonitrile, 4-(2-phenylethenyl)-, (*Z*)- 4-Stilbenecarbonitrile (*cis*)	$C_{15}H_{11}N$ 205.26	14064-68-7 44.5	 $130^{0.1}$	4-09-00-02615	H_2O 2; EtOH 3; chl 3
2769	Benzonitrile, 2-(trifluoromethyl)-	$C_8H_4F_3N$ 171.12	447-60-9 18	 205	4-09-00-01705	ctc 3
2770	Benzonitrile, 3-(trifluoromethyl)-	$C_8H_4F_3N$ 171.12	368-77-4 14.5	 189	3-09-00-02327 1.2813^{20}	 1.4508^{20}
2771	Benzophenone, 3-(dimethylamino)- 3-(Dimethylamino)benzophenone	$C_{15}H_{15}NO$ 225.29	31766-07-1 47	 216^{15}	4-14-00-00247	EtOH 3
2772	2H-1-Benzopyran 1,2-Chromene	C_9H_8O 132.16	254-04-6	1112 132^{102}; 91^{13}	5-17-02-00019 1.0993^{16}	H_2O 1 1.5869^{24}
2773	1H-2-Benzopyran-3-carboxylic acid, 6-[2-(4,9-dihydro-8-hydroxy-5,7-dimethoxy-4,9-dioxonaphtho[2,3-b]furan-2-yl Collinomycin	$C_{27}H_{20}O_{12}$ 536.45	27267-69-2 281	2478	5-19-08-00331	ace 4; diox 4; chl 4
2774	3H-2-Benzopyran-7-carboxylic acid, 4,6-dihydro-8-hydroxy-3,4,5-trimethyl-6-oxo-, (3*R*-*trans*)- Citrinin	$C_{13}H_{14}O_5$ 250.25	518-75-2 178-9 dec	2329	5-18-09-00061	H_2O 1; EtOH 2; eth 2; ace 3
2775	2H-1-Benzopyran-3-carboxylic acid, 2-oxo- Coumarin-3-carboxylic acid	$C_9H_6O_3$ 162.14	531-81-7 190 dec	2564	5-18-08-00323	EtOH 4
2776	1H-2-Benzopyran, 3,4-dihydro- Isochroman	$C_9H_{10}O$ 134.18	493-05-0	 90^{12}	5-17-01-00595 1.067^{25}	 1.5444^{20}
2777	2H-1-Benzopyran, 3,4-dihydro-	$C_9H_{10}O$ 134.18	493-08-3	 215; 98^{18}	5-17-01-00593 1.0610^{20}	H_2O 3; os 5 1.5444^{20}
2778	2H-1-Benzopyran, 3,4-dihydro-2-methyl- 2-Methylchroman	$C_{10}H_{12}O$ 148.20	13030-26-7	 224.5	5-17-01-00609 1.0340^{20}	bz 4; eth 4; EtOH 4 1.532^{13}
2779	2H-1-Benzopyran, 3,4-dihydro-3-methyl- Chroman, 3-methyl	$C_{10}H_{12}O$ 148.20	70401-56-8	 102-4^{15}	4-17-00-00430 1.002^{22}	 1.5335^{20}
2780	2H-1-Benzopyran, 3,4-dihydro-6-methyl- Chroman, 6-methyl-	$C_{10}H_{12}O$ 148.20	3722-74-5	 234	4-17-00-00431 1.0374^{14}	 1.5392^{25}
2781	2H-1-Benzopyran, 3,4-dihydro-7-methyl- Chroman, 7-methyl-	$C_{10}H_{12}O$ 148.20	3722-73-4	 141-3^{60}	4-17-00-00431 1.028^{26}	 1.5380^{20}
2782	2H-1-Benzopyran, 3,4-dihydro-2-phenyl- Flavan	$C_{15}H_{14}O$ 210.28	494-12-2 46.3	 332; 190^{16}	5-17-02-00[illegible]	bz 4; chl 4; EtOH 4
2783	[1]Benzopyrano[3,4-b]furo[2,3-h][1]benzopyran-6(6aH)-one, 1,2,12,12a-tetrahydro-8,9-dimethoxy-2-(1-methylethen Rotenone	$C_{23}H_{22}O_6$ 394.42	83-79-4 176	8248 210-$20^{0.5}$	5-19-10-00581	H_2O 1; EtOH 3; eth 2; ace 3
2784	[2]Benzopyrano[6,5,4-def][2]benzopyran-1,3,6,8-tetrone	$C_{14}H_4O_6$ 268.18	81-30-1 450	 sub 320	5-19-05-00423	H_2O 1; Na2CO_3 3; HOAc 3
2785	2H-1-Benzopyran-6-ol, 3,4-dihydro-2,8-dimethyl-2-(4,8,12-trimethyltridecyl)-, [2*R*-[2*R**(4*R**,8*R**)]]- δ-Tocopherol	$C_{27}H_{46}O_2$ 402.66	119-13-1	9419 $150^{0.001}$		H_2O 1; EtOH 4; eth 4; ace 4
2786	2H-1-Benzopyran-7-ol, 3,4-dihydro-3-(4-hydroxyphenyl)-, (*S*)- Equol	$C_{14}H_{14}O_3$ 230.26	531-95-3 189.5	3583	5-17-05-00413	
2787	2H-1-Benzopyran-6-ol, 3,4-dihydro-2,5,7,8-tetramethyl-2-(4,8,12-trimethyltridecyl)-, [2*R*-[2*R**(4*R**,8*R**)]]- Vitamin E	$C_{29}H_{50}O_2$ 430.71	59-02-9 3.0	9931 $210^{0.1}$	5-17-04-00168 0.950^{25}	H_2O 1; EtOH 3; eth 3; ace 3 1.5045^{25}
2788	2H-1-Benzopyran-6-ol, 3,4-dihydro-2,5,7,8-tetramethyl-2-(4,8,12-trimethyltridecyl)-, acetate, [2*R*-[2*R**(4*R**,8*R** Vitamin E acetate	$C_{31}H_{52}O_3$ 472.75	58-95-7 -27.5	9932 $184^{0.01}$	5-17-04-00168 0.9533^{21}	H_2O 1; EtOH 2; eth 3; ace 3 1.497^{20}
2789	2H-1-Benzopyran-6-ol, 3,4-dihydro-2,5,8-trimethyl-2-(4,8,12-trimethyltridecyl)- β-Tocopherol	$C_{28}H_{48}O_2$ 416.69	148-03-8	9417 200-$10^{0.1}$		ace 4; eth 4; EtOH 4; chl 4
2790	2H-1-Benzopyran-6-ol, 3,4-dihydro-2,7,8-trimethyl-2-(4,8,12-trimethyltridecyl)- γ-Tocopherol	$C_{28}H_{48}O_2$ 416.69	7616-22-0 -3	9418 200-$10^{0.1}$		H_2O 1; EtOH 5; eth 5; ace 5
2791	1H-2-Benzopyran-1-one Isocoumarin	$C_9H_6O_2$ 146.15	491-31-6 47	 286	5-17-10-00149	H_2O 1; EtOH 4; eth 4; bz 4
2792	2H-1-Benzopyran-2-one Coumarin	$C_9H_6O_2$ 146.15	91-64-5 71	2563 301.7	5-17-10-00143 0.935^{20}	H_2O 3; EtOH 3; eth 4; chl 4
2793	4H-1-Benzopyran-4-one	$C_9H_6O_2$ 146.15	491-38-3 59	 sub	5-17-10-00139 1.2900^{20}	H_2O 2; EtOH 3; eth 3; bz 3
2794	4H-1-Benzopyran-4-one, 7-[[6-*O*-(6-deoxy-α-*L*-mannopyranosyl)-β-*D*-glucopyranosyl]oxy]-2,3-dih Hesperidin	$C_{28}H_{34}O_{15}$ 610.57	520-26-3 262	4591	5-18-05-00218	py 4; EtOH 4; HOAc 4
2795	4H-1-Benzopyran-4-one, 7-[[2-*O*-(6-deoxy-α-*L*-mannopyranosyl)-β-*D*-glucopyranosyl]oxy]-2,3-dihydro-5-hy Naringin	$C_{27}H_{32}O_{14}$ 580.54	10236-47-2	6345	5-18-04-00528	H_2O 2; EtOH 2; eth 1; bz 1

No.	Name Synonym	Mol. Form. Mol. Wt.	CAS RN mp/°C	Merck No. bp/°C	Beil. Ref. den/g cm^{-3}	Solubility n_D
2796	4H-1-Benzopyran-4-one, 3-[(6-deoxy-α-*L*-mannopyranosyl)oxy]-2-(3,4-dihydroxyphenyl)-5,7-dihydroxy- Quercitrin	$C_{21}H_{20}O_{11}$ 448.38	522-12-3 170	8047	5-18-05-00514	H_2O 1; EtOH 3; eth 1; HOAc 3
2797	1H-2-Benzopyran-1-one, 3,4-dihydro-	$C_9H_8O_2$ 148.16	4702-34-5	 176^{20}, $112^{0.4}$	5-17-10-00018 1.197^{25}	 1.5629^{25}
2798	2H-1-Benzopyran-2-one, 3,4-dihydro-	$C_9H_8O_2$ 148.16	119-84-6 25	 272	5-17-10-00013 1.169^{18}	H_2O 1; EtOH 2; eth 2; ctc 2 1.5563^{20}
2799	4H-1-Benzopyran-4-one, 2,3-dihydro- 4-Chromanone	$C_9H_8O_2$ 148.16	491-37-2 36.5	 160^{50}, 127^{13}	5-17-10-00014 1.1291^{100}	EtOH 3; eth 4; ace 4; bz 4 1.5750
2800	4H-1-Benzopyran-4-one, 2,3-dihydro-5,7-dihydroxy-2-(3-hydroxy-4-methoxyphenyl)-, (*S*)- Hesperetin	$C_{16}H_{14}O_6$ 302.28	520-33-2 227.5	4590 sub 205	4-18-00-03215	eth 4; EtOH 4
2801	4H-1-Benzopyran-4-one, 2,3-dihydro-5,7-dihydroxy-3-(3-hydroxy-4-methoxyphenyl)- Flavanone, 4'-methoxy-3',5,7-trihydroxy	$C_{16}H_{14}O_6$ 302.28	99365-26-1 227.5	 sub 205	2-18-00-00204	H_2O 1; EtOH 4; eth 3; bz 2
2802	4H-1-Benzopyran-4-one, 2,3-dihydro-5,7-dihydroxy-2-(4-hydroxyphenyl)-, (*S*)- Naringenin	$C_{15}H_{12}O_5$ 272.26	480-41-1 251	6344	5-18-04-00524	bz 4; eth 4; EtOH 4
2803	4H-1-Benzopyran-4-one, 2,3-dihydro-6-methyl-	$C_{10}H_{10}O_2$ 162.19	39513-75-2 35	 $141\text{-}3^{13.5}$	5-17-10-00036 1.1245^{57}	 1.555^{57}
2804	2H-1-Benzopyran-2-one, 6,7-dihydroxy- Esculetin	$C_9H_6O_4$ 178.14	305-01-1 276	3645 sub	5-18-03-00202	H_2O 2; EtOH 3; eth 2; ace 3
2805	2H-1-Benzopyran-2-one, 7,8-dihydroxy- Daphnetin	$C_9H_6O_4$ 178.14	486-35-1 262	2816 sub	5-18-03-00211	H_2O 3; EtOH 3; eth 2; bz 2
2806	4H-1-Benzopyran-4-one, 5,7-dihydroxy-2-(4-hydroxyphenyl)- Apigenin	$C_{15}H_{10}O_5$ 270.24	520-36-5 347.5	763	5-18-04-00574	H_2O 1; EtOH 3; py 3; dil alk 4
2807	4H-1-Benzopyran-4-one, 5,7-dihydroxy-3-(4-hydroxyphenyl)- Genistein	$C_{15}H_{10}O_5$ 270.24	446-72-0 301-2 dec	4281	5-18-04-00594	
2808	2H-1-Benzopyran-2-one, 7,8-dihydroxy-6-methoxy- Fraxetin	$C_{10}H_8O_5$ 208.17	574-84-5 231	4178	5-18-04-00332	EtOH 4
2809	4H-1-Benzopyran-4-one, 5,7-dihydroxy-2-(4-methoxyphenyl)- Acacetin	$C_{16}H_{12}O_5$ 284.27	480-44-4 263	9	5-18-04-00575	EtOH 4
2810	4H-1-Benzopyran-4-one, 5,7-dihydroxy-6-methoxy-2-phenyl- Oroxylin A	$C_{16}H_{12}O_5$ 284.27	480-11-5 231.5	6830	5-18-04-00569	ace 4; eth 4; EtOH 4
2811	4H-1-Benzopyran-4-one, 5,7-dihydroxy-2-phenyl- Chrysin	$C_{15}H_{10}O_4$ 254.24	480-40-0 285.5	2261	5-18-04-00076	H_2O 1; EtOH 3; eth 2; ace 3
2812	4H-1-Benzopyran-4-one, 2-(3,4-dihydroxyphenyl)-2,3-dihydro-5,7-dihydroxy-, (*S*)- Eriodictyol	$C_{15}H_{12}O_6$ 288.26	552-58-9 267 dec	3616	5-18-05-00214	EtOH 4; HOAc 4
2813	4H-1-Benzopyran-4-one, 2-(3,4-dihydroxyphenyl)-3,7-dihydroxy- Fisetin	$C_{15}H_{10}O_6$ 286.24	528-48-3 330	4026	5-18-05-00291	H_2O 1; EtOH 3; eth 2; ace 3
2814	4H-1-Benzopyran-4-one, 2-(3,4-dihydroxyphenyl)-5,7-dihydroxy- Luteolin	$C_{15}H_{10}O_6$ 286.24	491-70-3 329-30 dec	5483	5-18-05-00296	H_2O 2; EtOH 3; eth 3; alk 3
2815	4H-1-Benzopyran-4-one, 2-(3,4-dihydroxyphenyl)-3,5-dihydroxy-7-methoxy- Rhamnetin	$C_{16}H_{12}O_7$ 316.27	90-19-7 295	8170	5-18-05-00495	H_2O 2; EtOH 3; ace 3; dil alk 4
2816	4H-1-Benzopyran-4-one, 2-(2,4-dihydroxyphenyl)-3,5,7-trihydroxy- Morin	$C_{15}H_{10}O_7$ 302.24	480-16-0 303.5	6179	5-18-05-00492	H_2O 2; EtOH 4; eth 2; bz 3
2817	4H-1-Benzopyran-4-one, 2-(3,4-dihydroxyphenyl)-3,5,7-trihydroxy- Quercetin	$C_{15}H_{10}O_7$ 302.24	117-39-5 316.5	8044 sub	5-18-05-00494	H_2O 2; EtOH 3; eth 2; ace 3
2818	2H-1-Benzopyran-2-one, 5,7-dimethoxy- Limettin	$C_{11}H_{10}O_4$ 206.20	487-06-9 149	5370 200 dec	5-18-03-00199	H_2O 2; EtOH 4; eth 1; ace 4
2819	4H-1-Benzopyran-4-one, 8-[(dimethylamino)methyl]-7-methoxy-3-methyl-2-phenyl- Dimefline	$C_{20}H_{21}NO_3$ 323.39	1165-48-6 109.5	3190	5-18-12-00083	chl 3
2820	2H-1-Benzopyran-2-one, 3-ethyl- Coumarin, 3-ethyl-	$C_{11}H_{10}O_2$ 174.20	66898-39-3 72.3	 299	5-17-10-00185	eth 4; EtOH 4
2821	2H-1-Benzopyran-2-one, 7-(β-*D*-glucopyranosyloxy)- Skimmin	$C_{15}H_{16}O_8$ 324.29	93-39-0 220	8506	5-18-01-00409	H_2O 3; EtOH 3; eth 1; chl 1
2822	2H-1-Benzopyran-2-one, 6-(β-*D*-glucopyranosyloxy)-7-hydroxy- Esculin	$C_{15}H_{16}O_9$ 340.29	531-75-9 205 dec	3646 230 dec	5-18-03-00208	H_2O 2; EtOH 2; eth 2; chl 3

No.	Name Synonym	Mol. Form. Mol. Wt.	CAS RN mp/°C	Merck No. bp/°C	Beil. Ref. den/g cm^{-3}	Solubility n_D
2823	2H-1-Benzopyran-2-one, 8-(β-*D*-glucopyranosyloxy)-7-hydroxy-6-methoxy- Fraxin	$C_{16}H_{18}O_{10}$ 370.31	524-30-1 205	4179	5-18-04-00334	
2824	4H-1-Benzopyran-4-one, 3-[4-(β-*D*-glucopyranosyloxy)phenyl]-5,7-dihydroxy- Sophoricoside	$C_{21}H_{20}O_{10}$ 432.38	152-95-4 274	8675	5-18-04-00597	
2825	2H-1-Benzopyran-2-one, 6-hydroxy-	$C_9H_6O_3$ 162.14	6093-68-1 250		5-18-01-00385 1.25^{25}	EtOH 4
2826	2H-1-Benzopyran-2-one, 7-hydroxy- Umbelliferone	$C_9H_6O_3$ 162.14	93-35-6 230.5	9758 sub	5-18-01-00386	EtOH 4; HOAc 4; chl 4
2827	4H-1-Benzopyran-4-one, 7-hydroxy-3-(4-hydroxyphenyl)- Daidzein	$C_{15}H_{10}O_4$ 254.24	486-66-8 323 dec	2805 sub	5-18-04-00089	EtOH 3; eth 3
2828	4H-1-Benzopyran-4-one, 5-hydroxy-3-(4-hydroxyphenyl)-7-methoxy- Prunetin	$C_{16}H_{12}O_5$ 284.27	552-59-0 239.5	7923	5-18-04-00595	
2829	2H-1-Benzopyran-2-one, 7-hydroxy-6-methoxy- Scopoletin	$C_{10}H_8O_4$ 192.17	92-61-5 204	8363	5-18-03-00203	H_2O 2; EtOH 2; eth 2; bz 1
2830	4H-1-Benzopyran-4-one, 7-hydroxy-3-(4-methoxyphenyl)- Formononetin	$C_{16}H_{12}O_4$ 268.27	485-72-3 256.5	4157	5-18-04-00089	
2831	2H-1-Benzopyran-2-one, 7-hydroxy-4-methyl- Hymecromone	$C_{10}H_8O_3$ 176.17	90-33-5 195.5	4792	5-18-01-00439	H_2O 2; EtOH 3; eth 2; chl 2
2832	2H-1-Benzopyran-2-one, 4-hydroxy-3-(3-oxo-1-phenylbutyl)- Warfarin	$C_{19}H_{16}O_4$ 308.33	81-81-2 161	9950	5-18-04-00162	H_2O 1; EtOH 3; ace 3; diox 3
2833	2H-1-Benzopyran-2-one, 3-methyl- 3-Methylcoumarin	$C_{10}H_8O_2$ 160.17	2445-82-1 91	 292.5	5-17-10-00165	H_2O 2; EtOH 3; chl 3; KOH 1
2834	2H-1-Benzopyran-2-one, 5-methyl- 5-Methylcoumarin	$C_{10}H_8O_2$ 160.17	42286-84-0 65.8	 173^{12}	5-17-10-00167	H_2O 2; EtOH 4; eth 4; bz 4
2835	2H-1-Benzopyran-2-one, 6-methyl-	$C_{10}H_8O_2$ 160.17	92-48-8 76.5	 304; 174^{14}	5-17-10-00168	EtOH 4; eth 4; bz 4; chl 2
2836	2H-1-Benzopyran-2-one, 7-methyl- 7-Methylcoumarin	$C_{10}H_8O_2$ 160.17	2445-83-2 128	 171.5^{11}	5-17-10-00169	H_2O 2; EtOH 4; eth 3; HOAc 4
2837	2H-1-Benzopyran-2-one, 8-methyl- 8-Methylcoumarin	$C_{10}H_8O_2$ 160.17	1807-36-9 110.5	 178^{20}	5-17-10-00170	EtOH 4; eth 4; bz 4; ctc 4
2838	4H-1-Benzopyran-4-one, 3-methyl- Tricromyl	$C_{10}H_8O_2$ 160.17	85 90.5	9574	5-17-10-00164	chl 3
2839	2H-1-Benzopyran-2-one, 3,3'-methylenebis[4-hydroxy- Dicumarol	$C_{19}H_{12}O_6$ 336.30	66-76-2 290	3080	5-19-06-00682	
2840	4H-1-Benzopyran-4-one, 2-phenyl- Flavone	$C_{15}H_{10}O_2$ 222.24	525-82-6 100	4030	5-17-10-00552	H_2O 1; EtOH 3; eth 3; ace 3
2841	4H-1-Benzopyran-4-one, 3-phenyl- Isoflavone	$C_{15}H_{10}O_2$ 222.24	574-12-9 148	5057	5-17-10-00562	
2842	4H-1-Benzopyran 4-one, 3,5,7-trihydroxy-2-(2-hydroxyphenyl)- Datiscetin	$C_{15}H_{10}O_6$ 286.24	480-15-9 277.5	2822	4-18-00-03281	ace 4; eth 4; EtOH 4
2843	4H-1-Benzopyran-4-one, 3,5,7-trihydroxy-2-(4-hydroxyphenyl)- Kaempferol	$C_{15}H_{10}O_6$ 286.24	520-18-3 277	5156	5-18-05-00251	H_2O 2; EtOH 4; eth 4; ace 4
2844	4H-1-Benzopyran-4-one, 5,6,7-trihydroxy-2-phenyl- Baicalein	$C_{15}H_{10}O_5$ 270.24	491-67-8 264-5 dec	954	5-18-04-00569	H_2O 2; EtOH 3; eth 3; ace 3
2845	2H-1-Benzopyran-2-one, 6,7,8-trimethoxy-	$C_{12}H_{12}O_5$ 236.22	6035-49-0 104.2	 $90\text{-}100^{0.2}$	5-18-04-00333	EtOH 3; eth 3
2846	2H-1-Benzopyran-3,5,7-triol, 2-(3,4-dihydroxyphenyl)-3,4-dihydro-, (2*S*-*cis*)-	$C_{15}H_{14}O_6$ 290.27	35323-91-2 242		5-17-08-00447 1.344^4	ace 4; EtOH 4
2847	1-Benzopyrylium, 3,5,7-trihydroxy-2-(4-hydroxy-3,5-dimethoxyphenyl)-, chloride Malvidin chloride	$C_{17}H_{15}ClO_7$ 366.75	643-84-5 >300	5596	5-17-08-00514	H_2O 2; EtOH 3; MeOH 3
2848	1-Benzopyrylium, 3,5,7-trihydroxy-2-(4-hydroxyphenyl)-, chloride Pelargonidin chloride	$C_{15}H_{11}ClO_5$ 306.70	134-04-3 >350	7014	1-18-00-00410	H_2O 3; EtOH 4; chl 2; con sulf 3
2849	1-Benzopyrylium, 3,5,7-trihydroxy-2-(3,4,5-trihydroxyphenyl)-, chloride Delphinidin	[illegible] 338.70	[illegible] >350	[illegible]	[illegible]	H_2O 4; EtOH 4; MeOH 4; AcOEt 3
2850	Benzo[rst]pentaphene 1,2:7,8-Dibenzopyrene	$C_{24}H_{14}$ 302.38	189-55-9 281.5		4-05-00-02803	
2851	2H-1,2,4-Benzothiadiazine-7-sulfonamide, 3-bicyclo[2.2.1]hept-5-en-2-yl-6-chloro-3,4-dihydro-, 1,1-dioxide Cyclothiazide	$C_{14}H_{16}ClN_3O_4S_2$ 389.88	2259-96-3 234	2760		
2852	2H-1,2,4-Benzothiadiazine-7-sulfonamide, 6-chloro-, 1,1-dioxide Chlorothiazide	$C_7H_6ClN_3O_4S_2$ 295.73	58-94-6 350 dec	2169	4-27-00-08036	

No.	Name Synonym	Mol. Form. Mol. Wt.	CAS RN mp/°C	Merck No. bp/°C	Beil. Ref. den/g cm^{-3}	Solubility n_D
2853	2H-1,2,4-Benzothiadiazine-7-sulfonamide, 6-chloro-3-(cyclopentylmethyl)-3,4-dihydro-, 1,1-dioxide Cyclopenthiazide	$C_{13}H_{18}ClN_3O_4S_2$ 379.89	742-20-1 238	2750		
2854	2H-1,2,4-Benzothiadiazine-7-sulfonamide, 6-chloro-3-(dichloromethyl)-3,4-dihydro-, 1,1-dioxide Trichloromethiazide	$C_8H_8Cl_3N_3O_4S_2$ 380.66	133-67-5 270 dec	9537		H_2O 2; EtOH 3
2855	2H-1,2,4-Benzothiadiazine-7-sulfonamide, 6-chloro-3,4-dihydro-, 1,1-dioxide Hydrochlorothiazide	$C_7H_8ClN_3O_4S_2$ 297.74	58-93-5 274	4704		
2856	2H-1,2,4-Benzothiadiazine-7-sulfonamide, 6-chloro-3,4-dihydro-3-(2-methylpropyl)-, 1,1-dioxide Buthiazide	$C_{11}H_{16}ClN_3O_4S_2$ 353.85	2043-38-1 221.5	1519		
2857	2H-1,2,4-Benzothiadiazine-7-sulfonamide, 6-chloro-3,4-dihydro-2-methyl-3-[[(2,2,2-trifluoroethyl)thio]methyl]- Polythiazide	$C_{11}H_{13}ClF_3N_3O_4S_3$ 439.89	346-18-9 214	7561		
2858	2H-1,2,4-Benzothiadiazine-7-sulfonamide, 3,4-dihydro-6-(trifluoromethyl)-, 1,1-dioxide Hydroflumethiazide	$C_8H_8F_3N_3O_4S_2$ 331.30	135-09-1 270.5	4716	4-27-00-08035	
2859	2-Benzothiazolamine 2-Aminobenzothiazole	$C_7H_6N_2S$ 150.20	136-95-8 132	435	4-27-00-04824	H_2O 2; EtOH 3; eth 3; chl 3
2860	6-Benzothiazolamine 6-Aminobenzothiazole	$C_7H_6N_2S$ 150.20	533-30-2 87	436	4-27-00-04884	H_2O 1; EtOH 3; eth 1
2861	Benzothiazole Benzosulfonazole	C_7H_5NS 135.19	95-16-9 2	1118 231	4-27-00-01069 1.2460^{20}	H_2O 2; EtOH 4; eth 4; ace 3 1.6379^{20}
2862	Benzothiazole, 2-chloro- 2-Chlorobenzothiazole	C_7H_4ClNS 169.63	615-20-3 24	 248	4-27-00-01072 1.3715^{10}	ace 4; eth 4; EtOH 4 1.6338^{10}
2863	Benzothiazole, 2-methyl- 2-Methylbenzothiazole	C_8H_7NS 149.22	120-75-2 14	 238	4-27-00-01080 1.1763^{19}	H_2O 1; EtOH 3; chl 3 1.6092^{19}
2864	Benzothiazole, 2-phenyl- 2-Phenylbenzothiazole	$C_{13}H_9NS$ 211.29	883-93-2 115	 371	4-27-00-01385	H_2O 1; EtOH 3; eth 3; CS_2 3
2865	2-Benzothiazolesulfonamide, 6-ethoxy- Ethoxzolamide	$C_9H_{10}N_2O_3S_2$ 258.32	452-35-7 189	3711	4-27-00-04404	
2866	2(3H)-Benzothiazolethione 2-Mercaptobenzothiazole	$C_7H_5NS_2$ 167.26	149-30-4 181	5759	4-27-00-02709 1.42^{20}	H_2O 1; EtOH 3; eth 2; bz 2
2867	2(3H)-Benzothiazolethione, 3-methyl-	$C_8H_7NS_2$ 181.28	2254-94-6 90	 335	4-27-00-02714	H_2O 1; EtOH 2; eth 2; bz 4
2868	2(3H)-Benzothiazolone	C_7H_5NOS 151.19	934-34-9 139	 360	4-27-00-02693	H_2O 1; EtOH 4; eth 4
2869	4H-1-Benzothiopyran-4-one, 2,3-dihydro-	C_9H_8OS 164.23	3528-17-4 29	 154^{12}	5-17-10-00016 1.2487^{14}	 1.6395^{20}
2870	1,2,4-Benzotriazine 1,2,4-Triazanaphthalene	$C_7H_5N_3$ 131.14	254-87-5 75.8	 237.5	4-26-00-00166	H_2O 4; eth 4; EtOH 4
2871	1H-Benzotriazole 1,2,3-Triaza-1H-indene	$C_6H_5N_3$ 119.13	95-14-7 100	1119 204^{15}	4-26-00-00093	H_2O 2; EtOH 3; bz 3; chl 3
2872	1H-Benzotriazole, 1-methyl-	$C_7H_7N_3$ 133.15	13351-73-0 64.5	 270.5	4-26-00-00095	bz 4; EtOH 4; HOAc 4
2873	1H-Benzotriazole, 5-nitro- 5-Nitrobenzotriazole	$C_6H_4N_4O_2$ 164.12	2338-12-7 217		4-26-00-00129	
2874	1,2,3-Benzoxadiazole, 5,7-dinitro- 2,3-Benzoxadiazole, 5,7-dinitro	$C_6H_2N_4O_5$ 210.11	87-31-0 158			EtOH 4
2875	3H-2,1-Benzoxathiol-3-one, 1,1-dioxide	$C_7H_4O_4S$ 184.17	81-08-3 129.5	 184^{18}	5-19-04-00215	bz 4; chl 4
2876	2-Benzoxazolamine, 5-chloro- Zoxazolamine	$C_7H_5ClN_2O$ 168.58	61-80-3 184.5	10098	4-27-00-04819	EtOH 4
2877	Benzoxazole 1-Oxa-3-azaindene	C_7H_5NO 119.12	273-53-0 31	 182.5	4-27-00-01069 1.1754^{20}	H_2O 1; EtOH 3; sulf 3 1.5594^{20}
2878	Benzoxazole, 2-chloro- 2-Chlorobenzoxazole	C_7H_4ClNO 153.57	615-18-9 7	 201.5	2-27-00-00017 1.3453^{18}	 1.5678^{20}
2879	Benzoxazole, 2,5-dimethyl- 2,5-Dimethylbenzoxazole	C_9H_9NO 147.18	5676-58-4	 218.5	4-27-00-01101 1.0880^{18}	ctc 3 1.5412^{20}
2880	Benzoxazole, 2-methyl- 2-Methylbenzoxazole	C_8H_7NO 133.15	95-21-6 9.5	 200.5	4-27-00-01078 1.1211^{20}	H_2O 1; EtOH 4; eth 5 1.5497^{20}
2881	2-Benzoxazolol 2-Hydroxybenzoxazole	$C_7H_5NO_2$ 135.12	69564-68-7 141	 230^{30}	2-27-00-00233	eth 4; EtOH 4
2882	2(3H)-Benzoxazolone	$C_7H_5NO_2$ 135.12	59-49-4 138	 335; 230^{30}	4-27-00-02677	H_2O 2; EtOH 3; eth 3; tfa 3
2883	2(3H)-Benzoxazolone, 5-chloro- Chlorzoxazone	$C_7H_4ClNO_2$ 169.57	95-25-0 191.5	2197	4-27-00-02683	EtOH 4; MeOH 4
2884	Benzoyl azide Benzazide	$C_7H_5N_3O$ 147.14	582-61-6 32	 exp	4-09-00-00945 1.1680^{35}	eth 4; EtOH 4
2885	Benzoyl bromide Benzoic acid, bromide	C_7H_5BrO 185.02	618-32-6 -24	 218.5	4-09-00-00724 1.570^{15}	eth 5 1.5868^{25}
2886	Benzoyl chloride Benzenecarbonyl chloride	C_7H_5ClO 140.57	98-88-4 -1.0	1124 197.2; 71^{9}	4-09-00-00721 1.2120^{20}	eth 5; bz 3; ctc 3; CS_2 3 1.5537^{20}
2887	Benzoyl chloride, 2-bromo-	C_7H_4BrClO 219.46	7154-66-7 11	 243	4-09-00-01013	ctc 2 1.5963^{20}
2888	Benzoyl chloride, 4-bromo-	C_7H_4BrClO 219.46	586-75-4 42	 246; 181^{125}	4-09-00-01023	EtOH 4; eth 4; bz 4; lig 4

PHYSICAL CONSTANTS OF ORGANIC COMPOUNDS (continued)

No.	Name	Synonym	Mol. Form.	Mol. Wt.	CAS RN	mp/°C	Merck No.	bp/°C	Beil. Ref.	den/g cm^{-3}	Solubility	n_D
2889	Benzoyl chloride, 4-butyl-		$C_{11}H_{13}ClO$	196.68	28788-62-7			155-6^{22}	3-09-00-02520	1.051^{25}		1.5351^{20}
2890	Benzoyl chloride, 2-chloro-		$C_7H_4Cl_2O$	175.01	609-65-4	-4		238	4-09-00-00962		ctc 3	1.5726^{16}
2891	Benzoyl chloride, 3-chloro-	*m*-Chlorobenzoyl chloride	$C_7H_4Cl_2O$	175.01	618-46-2			225	4-09-00-00972			1.5677^{20}
2892	Benzoyl chloride, 4-chloro-		$C_7H_4Cl_2O$	175.01	122-01-0	16		222	4-09-00-00984	1.3770^{20}	chl 2	1.5756^{20}
2893	Benzoyl chloride, 2,4-dichloro-	2,4-Dichlorobenzoyl chloride	$C_7H_3Cl_3O$	209.46	89-75-8	16.5		150^{34}, 111$^{7.5}$	4-09-00-01002		ctc 3	1.5895^{20}
2894	Benzoyl chloride, 3,4-dichloro-	3,4-Dichlorobenzoyl chloride	$C_7H_3Cl_3O$	209.46	3024-72-4	25		242	3-09-00-01379		ctc 2	
2895	Benzoyl chloride, 4-(1,1-dimethylethyl)-		$C_{11}H_{13}ClO$	196.68	1710-98-1			135^{20}	4-09-00-01887	1.007^{25}		1.5364^{20}
2896	Benzoyl chloride, 3,5-dinitro-	3,5-Dinitrobenzoyl chloride	$C_7H_3ClN_2O_5$	230.56	99-33-2	74	3270	196^{12}	4-09-00-01350		eth 3; chl 3	
2897	Benzoyl chloride, 2-fluoro-		C_7H_4ClFO	158.56	393-52-2	4		90-2^{15}	4-09-00-00951	1.328^{25}		1.5365^{20}
2898	Benzoyl chloride, 3-fluoro-		C_7H_4ClFO	158.56	1711-07-5	-30		189	4-09-00-00953	1.304^{25}		1.5285^{20}
2899	Benzoyl chloride, 4-fluoro-		C_7H_4ClFO	158.56	403-43-0	9		82^{20}	4-09-00-00954	1.342^{25}		1.5296^{20}
2900	Benzoyl chloride, 2-hydroxy-		$C_7H_5ClO_2$	156.57	1441-87-8	19		92^{15}	3-10-00-00150	1.3112^{20}	eth 4	1.5812^{20}
2901	Benzoyl chloride, 4-methoxy-	*p*-Anisoyl chloride	$C_8H_7ClO_2$	170.60	100-07-2	24.5	703	262.5	4-10-00-00430	1.261^{20}	eth 3; ace 3; bz 4; ctc 2	1.580^{20}
2902	Benzoyl chloride, 2-methyl-		C_8H_7ClO	154.60	933-88-0			213.5	3-09-00-02304		eth 4; EtOH 4	1.5549^{20}
2903	Benzoyl chloride, 3-methyl-		C_8H_7ClO	154.60	1711-06-4	-23		219.5	4-09-00-01716	1.0265^{21}	eth 4; EtOH 4	1.505^{22}
2904	Benzoyl chloride, 4-methyl-		C_8H_7ClO	154.60	874-60-2	-3		226	4-09-00-01733	1.1686^{20}	ctc 3	1.5547^{20}
2905	Benzoyl chloride, 3-nitro-		$C_7H_4ClNO_3$	185.57	121-90-4	36		276.5	4-09-00-01061		eth 4	
2906	Benzoyl chloride, 4-nitro-	*p*-Nitrobenzoyl chloride	$C_7H_4ClNO_3$	185.57	122-04-3	75	6511	203^{105}, 151^{15}	4-09-00-01191		eth 3	
2907	Benzoyl chloride, 4-pentyl-		$C_{12}H_{15}ClO$	210.70	49763-65-7			121^{8}	4-09-00-01910	1.036^{25}		1.5300^{20}
2908	Benzoyl chloride, 4-(pentyloxy)-		$C_{12}H_{15}ClO_2$	226.70	36823-84-4			196-200^{30}	4-10-00-00431	1.087^{25}		1.5434^{20}
2909	Benzoyl fluoride	Benzoic acid, fluoride	C_7H_5FO	124.11	455-32-3	-28		154.5	4-09-00-00721	1.1400^{20}	EtOH 4; eth 4; ctc 3	
2910	Benzoyl iodide		C_7H_5IO	232.02	618-38-2	3		128^{20}	4-09-00-00725	1.746^{18}	eth 4; EtOH 4	
2911	Berbaman-12-ol, 6,6',7-trimethoxy-2,2'-dimethyl-	Berbamine	$C_{37}H_{40}N_2O_6$	608.73	478-61-5	198.5	1168		2-27-00-00891		H_2O 2; EtOH 3; eth 3; chl 3	
2912	Beryllium, bis(2,4-pentanedionato-O,O')-, (T-4)-	Beryllium acetylacetonate	$C_{10}H_{14}BeO_4$	207.23	10210-64-7	108	1180	270				
2913	Bicyclo[2.2.1]hepta-2,5-diene		C_7H_8	92.14	121-46-0	-19.1		89.5	4-05-00-00879	0.9064^{20}	H_2O 1; EtOH 3; eth 3; ace 3	1.4702^{20}
2914	Bicyclo[2.2.1]heptane		C_7H_{12}	96.17	279-23-2	87.5		105.3	4-05-00-00258		ace 4; bz 4; eth 4; EtOH 4	
2915	Bicyclo[4.1.0]heptane	Norcarane	C_7H_{12}	96.17	286-08-8		6605	116.5	4-05-00-00257	0.853^{25}		1.4564^{20}
2916	Bicyclo[2.2.1]heptane-2-carboxaldehyde	2-Norbornanecarboxaldehyde	$C_8H_{10}O$	122.17	19396-83-9			70-2^{22}	4-07-00-00157	1.0227^{19}	eth 4	1.4760^{25}
2917	Bicyclo[2.2.1]heptane-2-carboxylic acid, 4,7,7-trimethyl-3-oxo-	*d*-Camphocarboxylic acid	$C_{11}H_{16}O_3$	196.25	18530-30-8	127.5	1737		3-10-00-02925		bz 4; eth 4; EtOH 4	
2918	Bicyclo[2.2.1]heptane, 2-chloro-		$C_7H_{11}Cl$	130.62	29342-53-8	-5		162; 52^{11}	4-05-00-00259	1.0603^{20}	chl 4	1.4849^{20}
2919	Bicyclo[2.2.1]heptane, 2-chloro-1,7,7-trimethyl-	Bornane, 2-chloro-	$C_{10}H_{17}Cl$	172.70	6120-13-4	132.8		207	4-05-00-00319		eth 4; EtOH 4	
2920	Bicyclo[2.2.1]heptane, 2-chloro-1,7,7-trimethyl-, *endo*-	Bornyl chloride	$C_{10}H_{17}Cl$	172.70	464-41-5	132	1341	207.5	4-05-00-00319		bz 4; eth 4; EtOH 4; peth 4	
2921	Bicyclo[2.2.1]heptane, 2,2-dimethyl-3-methylene-, (±)-		$C_{10}H_{16}$	136.24	565-00-4	51.5		158.5	4-05-00-00462	0.879^{20}	H_2O 1; EtOH 4; eth 4	1.4551^{54}
2922	Bicyclo[2.2.1]heptane, 2,2-dimethyl-5-methylene-		$C_{10}H_{16}$	136.24	497-32-5			152	4-05-00-00465	0.8591^{20}		1.4645^{25}
2923	Bicyclo[2.2.1]heptane, 7,7-dimethyl-2-methylene-, (±)-	α-Fenchene (*DL*)	$C_{10}H_{16}$	136.24	2623-54-3			155	3-05-00-00389	0.8660^{20}	ace 4; eth 4; EtOH 4	1.4705^{20}

No.	Name Synonym	Mol. Form. Mol. Wt.	CAS RN mp/°C	Merck No. bp/°C	Beil. Ref. den/g cm^{-3}	Solubility n_D
2924	Bicyclo[3.1.1]heptane, 6,6-dimethyl-2-methylene-, (1*R*)-	$C_{10}H_{16}$ 136.24	19902-08-0 -50	7415 164.6	4-05-00-00456 0.8654^{20}	bz 4; eth 4; EtOH 4; chl 4 1.4789^{20}
2925	Bicyclo[3.1.1]heptane, 6,6-dimethyl-2-methylene- β-Pinene	$C_{10}H_{16}$ 136.24	127-91-3 -61.5	7415 166	 0.860^{25}	H_2O 1; bz 3; EtOH 3; eth 3 1.4768^{25}
2926	Bicyclo[2.2.1]heptane-2,3-dione, 1,7,7-trimethyl-, (1*R*)-	$C_{10}H_{14}O_2$ 166.22	10334-26-6 199	 sub	4-07-00-02039	H_2O 3; EtOH 4; eth 4; bz 3
2927	Bicyclo[2.2.1]heptane-1-methanamine, 7,7-dimethyl- α-Camphylamine	$C_{10}H_{19}N$ 153.27	54131-44-1	 195; 95^{12}	1-12-00-00129 0.8688^{20}	 1.4728^{18}
2928	Bicyclo[3.1.1]heptane-2-methanamine, 6,6-dimethyl-, [1*S*-(1α,2β,5α)]-	$C_{10}H_{19}N$ 153.27	73522-42-6	 $94\text{-}9^{27}$	4-06-00-00384 0.915^{25}	 1.4877^{20}
2929	Bicyclo[2.2.1]heptane-1-methanesulfonic acid, 7,7-dimethyl-2-oxo-, (1*S*)- *D*-Camphorsulfonic acid	$C_{10}H_{16}O_4S$ 232.30	3144-16-9 195 dec	1740	4-11-00-00642	H_2O 4; eth 1; HOAc 2
2930	Bicyclo[3.1.1]heptane-2-methanol, 6,6-dimethyl-, [1*S*-(1α,2β,5α)]-	$C_{10}H_{18}O$ 154.25	51152-12-6	 $70\text{-}2^{1}$	4-06-00-00278 0.986^{25}	 1.4890^{20}
2931	Bicyclo[2.2.1]heptane, 1,3,3-trimethyl-, (+)- Fenchane(*d*)	$C_{10}H_{18}$ 138.25	116435-30-4 <-15	 151	3-05-00-00256 0.8345^{20}	H_2O 1; EtOH 3; eth 3; HOAc 2 1.44714^{20}
2932	Bicyclo[2.2.1]heptane, 1,7,7-trimethyl-	$C_{10}H_{18}$ 138.25	464-15-3	 161	4-05-00-00319	H_2O 1; EtOH 3; eth 3; AcOEt 3
2933	Bicyclo[2.2.1]heptane, 2,2,3-trimethyl-	$C_{10}H_{18}$ 138.25	473-19-8 62.5	 166	3-05-00-00263 0.8276^{67}	ace 4; EtOH 4
2934	Bicyclo[3.1.1]heptane, 2,6,6-trimethyl-, (1α,2α,5α)-(±)- Pinane (*DL*)	$C_{10}H_{18}$ 138.25	112456-46-9	 164.5	1-05-00-00048 0.8551^{20}	bz 4; eth 4 1.4609^{20}
2935	Bicyclo[3.1.1]heptane, 2,6,6-trimethyl-, [1*R*-(1α,2β,5α)]-	$C_{10}H_{18}$ 138.25	4795-86-2 -53	 169	4-05-00-00318 0.8560^{20}	bz 4; eth 4 1.4629^{20}
2936	Bicyclo[4.1.0]heptane, 3,7,7-trimethyl-, [1*S*-(1α,3β,6α)]-	$C_{10}H_{18}$ 138.25	2778-68-9	 169	4-05-00-00316 0.8410^{20}	 1.4569
2937	Bicyclo[4.1.0]heptane, 3,7,7-trimethyl-, [1*S*-(1α,3α,6α)]-	$C_{10}H_{18}$ 138.25	6069-97-2	 169	4-05-00-00316 0.841^{20}	 1.456
2938	Bicyclo[2.2.1]heptan-2-ol, 1,7-dimethyl-, (*exo,syn*)- β-Santenol	$C_9H_{16}O$ 140.23	509-12-6 101.5	 192; 60^{15}	4-06-00-00242	eth 4
2939	Bicyclo[2.2.1]heptan-2-ol, 1-phenyl-, acetate	$C_{15}H_{18}O_2$ 230.31	71173-15-4 44	159^{13}, $128^{1.0}$	3-06-00-02755	ctc 2
2940	Bicyclo[2.2.1]heptan-2-ol, 1,3,3-trimethyl-, *endo*-(±)-	$C_{10}H_{18}O$ 154.25	36386-49-9 39	 199.5	3-06-00-00288 0.9420^{40}	eth 4; EtOH 4
2941	Bicyclo[2.2.1]heptan-2-ol, 1,3,3-trimethyl-, (1*S*-*endo*)- *l*-α-Fenchyl alcohol	$C_{10}H_{18}O$ 154.25	512-13-0 48	 94^{20}	4-06-00-00278 0.9034^{84}	
2942	Bicyclo[2.2.1]heptan-2-ol, 1,7,7-trimethyl-, *endo*-(±)-	$C_{10}H_{18}O$ 154.25	6627-72-1 210.5	 sub	4-06-00-00281 1.011^{20}	H_2O 1; EtOH 4; eth 4; bz 4
2943	Bicyclo[2.2.1]heptan-2-ol, 1,7,7-trimethyl-, *exo*-(±)-	$C_{10}H_{18}O$ 154.25	24393-70-2 212	 sub	4-06-00-00282 1.10^{20}	H_2O 1; EtOH 4; eth 4; bz 2
2944	Bicyclo[3.1.1]heptan-2-ol, 4,6,6-trimethyl-, [1*R*-(1α,2α,4α,5α)]-	$C_{10}H_{18}O$ 154.25	515-88-8 58	 218	3-06-00-00284 0.940^{20}	 1.4702^{20}
2945	Bicyclo[2.2.1]heptan-2-ol, 1,7,7-trimethyl-, acetate, *endo*-(±)-	$C_{12}H_{20}O_2$ 196.29	36386-52-4 <-17	 223.5	4-06-00-00283 1.4630^{20}	ctc 3 1.4630^{20}
2946	Bicyclo[2.2.1]heptan-2-ol, 1,7,7-trimethyl-, acetate, *exo*-(±)-	$C_{12}H_{20}O_2$ 196.29	17283-45-3	 116^{21}	4-06-00-00283 0.9841^{20}	ace 4; EtOH 4 1.4640^{20}
2947	Bicyclo[2.2.1]heptan-2-ol, 4,7,7-trimethyl-, acetate, (1*R*-*endo*)- Epiborneol, acetate	$C_{12}H_{20}O_2$ 196.29	22621-74-5 <-15	 101^{11}	2-06-00-00092 0.9872^{14}	 1.4651^{14}
2948	Bicyclo[2.2.1]heptan-2-ol, 1,7,7-trimethyl-, acetate, *endo*- Borneol, acetate	$C_{12}H_{22}O_2$ 198.31	76-49-3 29	1339 221	3-06-00-00303	
2949	Bicyclo[2.2.1]heptan-2-ol, 1,7,7-trimethyl-, formate, (1*R*-*endo*)- Bornyl formate (*d*)	$C_{11}H_{18}O_2$ 182.26	74219-20-8	 90^{10}	3-06-00-00301 1.009^{22}	 1.4708^{15}
2950	Bicyclo[2.2.1]heptan-2-one, 3-(acetyl-17*O*)-1,7,7-trimethyl- Camphor, 3-acetyl	$C_{12}H_{18}O_2$ 194.27	15068-90-3	 $127\text{-}8^{15}$	2-07-00-00563 1.0314^{19}	 1.4949^{17}

No.	Name Synonym	Mol. Form. Mol. Wt.	CAS RN mp/°C	Merck No. bp/°C	Beil. Ref. den/g cm^{-3}	Solubility n_D
2951	Bicyclo[2.2.1]heptan-2-one, 1-(bromomethyl)-7,7-dimethyl-, (1*S*)- *d*-Camphor, 10-bromo-	$C_{10}H_{15}BrO$ 231.13	64161-50-8 78	 265 dec	4-07-00-00218	ace 4; bz 4; eth 4; EtOH 4
2952	Bicyclo[2.2.1]heptan-2-one, 3-bromo-1,7,7-trimethyl-, (1*R*-*endo*)-	$C_{10}H_{15}BrO$ 231.13	10293-06-8 76	 274 dec	 1.449^{25}	bz 4; eth 4; EtOH 4
2953	Bicyclo[2.2.1]heptan-2-one, 3,3-dibromo-1,7,7-trimethyl-, (1*R*)- α,α'-Dibromo-*d*-camphor	$C_{10}H_{14}Br_2O$ 310.03	514-12-5 61	3002	4-07-00-00220 1.854^{21}	H_2O 1; EtOH 4; eth 4; bz 4
2954	Bicyclo[2.2.1]heptan-2-one, 3,3-dimethyl-, (1*S*)-	$C_9H_{14}O$ 138.21	6069-71-2 41.9	 193; 76^{12}	4-07-00-00177 0.807^{20}	eth 4
2955	Bicyclo[3.1.1]heptan-2-one, 6,6-dimethyl-, (1*R*)-	$C_9H_{14}O$ 138.21	38651-65-9 -1	 209	4-07-00-00176 0.9807^{20}	eth 4; EtOH 4 1.4787^{20}
2956	Bicyclo[3.1.1]heptan-3-one, 6,6-dimethyl-2-methylene-, (1*S*)-	$C_{10}H_{14}O$ 150.22	19890-00-7 -1.8	 $67\text{-}9^{4}$	 0.9875^{15}	 1.4950^{20}
2957	Bicyclo[2.2.1]heptan-2-one, 3-hydroxy-1,7,7-trimethyl- 3-Hydroxycamphor	$C_{10}H_{16}O_2$ 168.24	10373-81-6 205.5	4747	4-08-00-00081	eth 4; EtOH 4; chl 4
2958	Bicyclo[2.2.1]heptan-2-one, 3-methylene- 3-Methylene-2-norbornanone	$C_8H_{10}O$ 122.17	5597-27-3	 $69\text{-}71^{11}$	4-07-00-00301 0.997^{25}	 1.4891^{20}
2959	Bicyclo[2.2.1]heptan-2-one, 1,4,7,7-tetramethyl-	$C_{11}H_{18}O$ 166.26	10309-50-9 168	 213	2-07-00-00113	eth 4; EtOH 4
2960	Bicyclo[2.2.1]heptan-2-one, 1,3,3-trimethyl-, (±)-	$C_{10}H_{16}O$ 152.24	18492-37-0 -18	3911 193.5	3-07-00-00393 0.9492^{15}	H_2O 1; EtOH 4; eth 3; ace 3 1.4702^{20}
2961	Bicyclo[2.2.1]heptan-2-one, 1,7,7-trimethyl-, (1*R*)- (+)-Camphor	$C_{10}H_{16}O$ 152.24	464-49-3 178.8	1738 207.4	4-07-00-00213 0.990^{25}	H_2O 1; EtOH 4; eth 4; ace 3 1.5462
2962	Bicyclo[2.2.1]heptan-2-one, 1,7,7-trimethyl-, oxime, (1*S*)- (-)-Camphor oxime	$C_{10}H_{17}NO$ 167.25	36065-15-3 118	 252 dec	4-07-00-00215 1.01^{116}	eth 4; EtOH 4
2963	Bicyclo[2.2.1]heptan-2-one, 4,7,7-trimethyl-, (1*S*)-	$C_{10}H_{16}O$ 152.24	10292-98-5 186.5	 213	3-07-00-00421	H_2O 2; EtOH 4; eth 4; peth 4
2964	Bicyclo[3.1.1]heptan-2-one, 4,6,6-trimethyl-, [1*R*-(1α,4α,5α)]-	$C_{10}H_{16}O$ 152.24	515-90-2	 222	4-07-00-00210 0.961^{20}	 1.4752^{20}
2965	Bicyclo[3.1.1]heptan-3-one, 2,6,6-trimethyl-, [1*S*-(1α,2β,5α)]-	$C_{10}H_{16}O$ 152.24	14575-93-0	 212	4-07-00-00209 0.964^{25}	 1.4735^{25}
2966	Bicyclo[2.2.1]hept-5-ene-2-carbonitrile	C_8H_9N 119.17	95-11-4 13	 $82\text{-}6^{10}$	 0.999^{25}	 1.4885^{20}
2967	Bicyclo[2.2.1]hept-5-ene-2-carboxaldehyde	$C_8H_{10}O$ 122.17	5453-80-5	 $67\text{-}70^{12}$	2-07-00-00125 1.018^{25}	 1.4893^{20}
2968	Bicyclo[2.2.1]hept-2-ene, 2,3-dimethyl- 2-Norbornene, 2,3-dimethyl-	C_9H_{14} 122.21	529-16-8	 140.5	4-05-00-00427 0.8698^{17}	eth 3; ace 3; bz 3 1.4688^{17}
2969	Bicyclo[3.1.1]hept-2-ene-2-ethanol, 6,6-dimethyl-	$C_{11}H_{18}O$ 166.26	128-50-7	 235; 110^{10}	4-06-00-00398 0.973^{25}	chl 3 1.4930^{20}
2970	Bicyclo[2.2.1]hept-2-ene, 1,7,7-trimethyl-	$C_{10}H_{16}$ 136.24	464-17-5 113	 146	2-05-00-00105	bz 4; eth 4; EtOH 4
2971	Bicyclo[3.1.1]hept-2-ene, 2,6,6-trimethyl-, (1*S*)- (-)-α-Pinene	$C_{10}H_{16}$ 136.24	7785-26-4 -64	7414 155	4-05-00-00452 0.859^{20}	 1.4660^{20}
2972	Bicyclo[3.1.1]hept-2-ene, 2,6,6-trimethyl-, (±)- (±)-α-Pinene	$C_{10}H_{16}$ 136.24	2437-95-8 -55	7414 156.2	2-05-00-00097 0.8582^{20}	H_2O 1; EtOH 5; eth 5; chl 5 1.4658^{20}
2973	Bicyclo[3.1.1]hept-2-ene, 2,6,6-trimethyl 2-Pinene	$C_{10}H_{16}$ 136.24	80-56-8 -64	 155.9	 0.8539^{25}	 1.4632^{25}
2974	Bicyclo[4.1.0]hept-3-ene, 3,7,7-trimethyl-, (1*S*)-	$C_{10}H_{16}$ 136.24	498-15-7	 171; 123^{200}	4-05-00-00450 0.8549^{30}	ace 4; bz 4; eth 4 1.469^{3}
2975	Bicyclo[3.1.1]hept-3-en-2-ol, 4,6,6-trimethyl-, (1α,2β,5α)	$C_{10}H_{16}O$ 152.24	1845-30-3 15.5	 90^{10}	 0.9684^{25}	 1.4912^{25}
2976	Bicyclo[3.1.1]hept-3-en-2-ol, 4,6,6-trimethyl-, (1α,2α,5α)-	$C_{10}H_{16}O$ 152.24	1820-09-3 24	 92^{10}	4-06-00-00383 0.9657^{25}	 1.4908^{25}
2977	Bicyclo[3.1.1]hept-2-en-6-one, 2,7,7-trimethyl- Chrysanthenone	$C_{10}H_{14}O$ 150.22	473-06-3	2256 $88\text{-}9^{12}$	4-07-00-00327	EtOH 4 1.4720^{22}
2978	Bicyclo[3.1.1]hept-3-en-2-one, 4,6,6-trimethyl-, (1*R*)- *d*-Verbenone	$C_{10}H_{14}O$ 150.22	18309-32-5 9.8	9862 227.5	4-07-00-00327 0.9978^{20}	H_2O 3; EtOH 3; ace 3; bz 3 1.4993^{18}
2979	[Bi-1,4-cyclohexadien-1-yl]-3,3',6,6'-tetrone, 2,2'-dihydroxy-4,4'-dimethyl- Phenicin	$C_{14}H_{10}O_6$ 274.23	128-68-7 230.5	7194	4-08-00-03660	H_2O 2; EtOH 4; chl 4; HOAc 4
2980	Bicyclo[3,1,0]hexane, 1-methyl- 1-Methylbicyclo[3,1,0]hexane	C_7H_{12} 96.17	4625-24-5	 93.1		

No.	Name Synonym	Mol. Form. Mol. Wt.	CAS RN mp/°C	Merck No. bp/°C	Beil. Ref. den/g cm^{-3}	Solubility n_D
2981	Bicyclo[3.1.0]hexane, 4-methylene-1-(1-methylethyl)-, (±)- 4(10)-Thujene, (±)-	$C_{10}H_{16}$ 136.24	15826-80-9	 164; 69^{30}	3-05-00-00365 0.8437^{20}	H_2O 1; EtOH 4; eth 3; ace 3 1.4676^{20}
2982	Bicyclo[3.1.0]hexane, 4-methyl-1-(1-methylethyl)-	$C_{10}H_{18}$ 138.25	471-12-5	 157	4-05-00-00317 0.8139^{20}	 1.4376^{20}
2983	Bicyclo[3.1.0]hexan-3-ol, 4-methylene-1-(1-methylethyl)- 4(10)-Thujene-3-ol	$C_{10}H_{16}O$ 152.24	59905-55-4	 208	4-06-00-00382 0.9461^{25}	 1.4871^{25}
2984	Bicyclo[3.1.0]hexan-3-ol, 4-methylene-1-(1-methylethyl)-, [1*S*-(1α,3β,5α)]-	$C_{10}H_{16}O$ 152.24	471-16-9	 208	4-06-00-00382 0.9488^{19}	eth 3 1.4871^{25}
2985	Bicyclo[3.1.0]hexan-3-ol, 4-methyl-1-(1-methylethyl)-	$C_{10}H_{18}O$ 154.25	513-23-5	 209	3-06-00-00278 0.9210^{20}	 1.4621^{25}
2986	Bicyclo[3.1.0]hex-2-ene, 2-methyl-5-(1-methylethyl)-	$C_{10}H_{16}$ 136.24	2867-05-2	 151	4-05-00-00451 0.8301^{20}	 1.4515^{20}
2987	Bicyclo[3.1.0]hex-3-en-2-one, 4-methyl-1-(1-methylethyl)-	$C_{10}H_{14}O$ 150.22	24545-81-1	 220	0-07-00-00159 0.9572^{15}	 1.48325
2988	1,1'-Bicyclohexyl Cyclohexylcyclohexane	$C_{12}H_{22}$ 166.31	92-51-3 4	 238		H_2O 2; EtOH 3; eth 3
2989	[1,1'-Bicyclohexyl]-2-one	$C_{12}H_{20}O$ 180.29	90-42-6 -32	 264	4-07-00-00248 0.9696^{25}	 1.4877^{25}
2990	Bicyclo[3.3.1]nonane	C_9H_{16} 124.23	280-65-9 145.5	 169.5	4-05-00-00293	EtOH 4
2991	Bicyclo[4.2.0]octane	C_8H_{14} 110.20	278-30-8	 136	4-05-00-00276 0.8573^{20}	 1.4613^{20}
2992	Bicyclo[2.2.2]octane, 2-methyl- 2-Methylbicyclo[2.2.2]octane	C_9H_{16} 124.23	766-53-0 33.5	 159	4-05-00-00295 0.8664^{40}	 1.4608^{40}
2993	Bicyclo[4.2.0]oct-2-ene	C_8H_{12} 108.18	13367-29-8	 138	4-05-00-00413 0.8948^{20}	 1.4810^{20}
2994	[1,1'-Bicyclopentyl]-2-ol Bicyclopentyl, 2-hydroxy	$C_{10}H_{18}O$ 154.25	4884-25-7 20	 235.5	3-06-00-00261 0.9785^{15}	 1.4884^{17}
2995	[1,1'-Bicyclopentyl]-2-one	$C_{10}H_{16}O$ 152.24	4884-24-6 -13	 232.5	4-07-00-00198 0.9745^{21}	 1.4763
2996	Bicyclo[7.2.0]undec-4-ene, 4,11,11-trimethyl-8-methylene-, [1*R*-(1*R**,4E,9*S**)]- Caryophyllene	$C_{15}H_{24}$ 204.36	87-44-5	1884 $122^{13.5}$	4-05-00-01182 0.9075^{20}	bz 4 1.4986^{20}
2997	Bifenox Benzoic acid, 5-(2,4-dichlorophenoxy)-2-nitro-, methyl ester	$C_{14}H_9Cl_2NO_5$ 342.13	42576-02-3 85	1228		
2998	Bifenthrin 2-Methylbiphenyl-3-ylmethyl (*Z*)-(1RS,3RS)-3-(2-chloro-3,3,3-trifluoroprop-1-enyl)- 2,2-dimethylcyclopropanecarboxylate	$C_{23}H_{22}ClF_3O_2$ 422.87	82657-04-3 69	1229	 1.2^{125}	
2999	[2,2'-Bi-1H-indene]-1,1',3,3'(2H,2'H)-tetrone, 2,2'-dihydroxy- Hydrindantin	$C_{18}H_{10}O_6$ 322.27	5103-42-4 250 dec	4698	4-08-00-03701	
3000	21H-Biline-8,12-dipropanoic acid, 2,17-diethenyl-1,10,19,22,23,24-hexahydro-3,7,13,18-tetramethyl-1,19-dioxo- Bilirubin	$C_{33}H_{36}N_4O_6$ 584.67	635-65-4	1235	4-26-00-03268	H_2O 1; EtOH 2; eth 2; bz 3
3001	21H-Biline-8,12-dipropanoic acid, 3,18-diethenyl-1,19,22,24-tetrahydro-2,7,13,17-tetramethyl-1,19-dioxo- Biliverdine	$C_{33}H_{34}N_4O_6$ 582.66	114-25-0 >300	1236	4-26-00-03272	H_2O 1; EtOH 3; eth 2; bz 3
3002	[2,2'-Bimorphinan]-3,3',6,6'-tetrol, 7,7',8,8'-tetradehydro-4,5:4',5'-diepoxy-17,17'-dimethyl- Pseudomorphine	$C_{34}H_{36}N_2O_6$ 568.67	125-24-6 282.5	7934	4-27-00-08724	H_2O 1; EtOH 1; eth 1; chl 1
3003	Binapacryl 2-Butenoic acid, 3-methyl-, 2-(1-methylpropyl)-4,6-dinitrophenyl ester	$C_{15}H_{18}N_2O_6$ 322.32	485-31-4 70	1237	 1.27^{20}	
3004	1,1'-Binaphthalene 1,1'-Binaphthyl	$C_{20}H_{14}$ 254.33	604-53-5 160	 >360; 240^{12}	4-05-00-02634 1.3000^{20}	H_2O 1; EtOH 2; eth 3; ace 3
3005	2,2'-Binaphthalene	$C_{20}H_{14}$ 254.33	612-78-2 188.3	 452	4-05-00-02636	H_2O 1; EtOH 2; eth 3; bz 3
3006	[1,1'-Binaphthalene]-4,4'-diol	$C_{20}H_{14}O_2$ 286.33	1446-34-0 300	 sub	4-06-00-07021	eth 4; EtOH 4
3007	2,2'-Bioxirane, (*R**,*S**)- Erythritol anhydride	$C_4H_6O_2$ 86.09	564-00-1 -19	3621 138	5-19-01-00184 1.113^{18}	H_2O 5; EtOH 4 1.4330^{20}
3008	1,1'-Biphenyl Diphenyl	$C_{12}H_{10}$ 154.21	92-52-4 69	3314 256.1	4-05-00-01807 1.04^{20}	H_2O 1; EtOH 3; eth 3; bz 4 1.588^{77}
3009	[1,1'-Biphenyl]-4-acetic acid Felbinac	$C_{14}H_{12}O_2$ 212.25	5728-52-9 160.5	3893	4-09-00-02520	
3010	[1,1'-Biphenyl]-2-amine	$C_{12}H_{11}N$ 169.23	90-41-5 51	 299	4-12-00-03223	H_2O 1; EtOH 3; eth 3; bz 3
3011	[1,1'-Biphenyl]-4-amine *p*-Biphenylamine	$C_{12}H_{11}N$ 169.23	92-67-1 53.5	1248 302	4-12-00-03241	H_2O 2; EtOH 3; eth 3; ace 3
3012	[1,1'-Biphenyl]-4-amine, 4'-chloro- 4-Amino-4'-chlorodiphenyl	$C_{12}H_{10}ClN$ 203.67	135-68-2 134	445	4-12-00-03269	ace 4; bz 4; eth 4

PHYSICAL CONSTANTS OF ORGANIC COMPOUNDS (continued)

No.	Name Synonym	Mol. Form. Mol. Wt.	CAS RN mp/°C	Merck No. bp/°C	Beil. Ref. den/g cm^{-3}	Solubility n_D
3013	[1,1'-Biphenyl]-2-amine, 6-methyl- Biphenyl, 2-amino-6-methyl	$C_{13}H_{13}N$ 183.25	76472-83-8 43.5	 355; 144^{2}	3-12-00-03234 1.9060^{0}	bz 4; eth 4; EtOH 4
3014	[1,1'-Biphenyl]-4-amine, 3-methyl-	$C_{13}H_{13}N$ 183.25	63019-98-7 43	 190^{15}	3-12-00-03238	bz 4; eth 4; EtOH 4
3015	[1,1'-Biphenyl]-4-amine, 4'-methyl- Biphenyl, 4-amino-4'-methyl	$C_{13}H_{13}N$ 183.25	1204-78-0 99	 190^{18}	2-12-00-00771	ace 4; bz 4; eth 4; EtOH 4
3016	1,1'-Biphenyl, 2-bromo-	$C_{12}H_9Br$ 233.11	2052-07-5 1.5	 297	4-05-00-01818 1.2175^{26}	eth 4; EtOH 4 1.6248^{25}
3017	1,1'-Biphenyl, 3-bromo-	$C_{12}H_9Br$ 233.11	2113-57-7	 300; 171^{17}	4-05-00-01818	H_2O 1 1.6411^{20}
3018	1,1'-Biphenyl, 4-bromo-	$C_{12}H_9Br$ 233.11	92-66-0 91.5	 310	4-05-00-01819 0.9327^{25}	H_2O 1; EtOH 3; eth 3; bz 3
3019	[1,1'-Biphenyl]-4-butanoic acid	$C_{16}H_{16}O_2$ 240.30	6057-60-9 119.5	 $190^{0.3}$	4-09-00-02558	eth 4; EtOH 4
3020	[1,1'-Biphenyl]-2-carbonitrile	$C_{13}H_9N$ 179.22	24973-49-7 41	 175^{13}	3-09-00-03269	eth 4; EtOH 4
3021	[1,1'-Biphenyl]-2-carboxylic acid	$C_{13}H_{10}O_2$ 198.22	947-84-2 114.3	 343.5	4-09-00-02472	H_2O 1; EtOH 4; bz 4; HOAc 4
3022	[1,1'-Biphenyl]-4-carboxylic acid	$C_{13}H_{10}O_2$ 198.22	92-92-2 228	 sub	4-09-00-02479	H_2O 1; EtOH 3; eth 3; bz 3
3023	1,1'-Biphenyl, 2-chloro-	$C_{12}H_9Cl$ 188.66	2051-60-7 34	 274	4-05-00-01816 1.1499^{32}	eth 4; EtOH 4; lig 4
3024	1,1'-Biphenyl, 3-chloro-	$C_{12}H_9Cl$ 188.66	2051-61-8 16	 284.5	4-05-00-01816 1.1579^{25}	ace 4; eth 4; EtOH 4 1.6181^{25}
3025	1,1'-Biphenyl, 4-chloro-	$C_{12}H_9Cl$ 188.66	2051-62-9 78.8	 292.9; 146^{10}	4-05-00-01816	H_2O 1; EtOH 3; eth 3; lig 3
3026	Biphenyl, 2-chloro-2'-methyl- 2-Chloro-2'-methylbiphenyl	$C_{13}H_{11}Cl$ 202.68	19493-31-3 17	 276; 174^{50}	3-05-00-01799	 1.588^{25}
3027	1,1'-Biphenyl, 2,2',3,3',4,4',5,5',6,6'-decafluoro-	$C_{12}F_{10}$ 334.12	434-90-2 67.5	 206	4-05-00-01816 1.785^{20}	
3028	[1,1'-Biphenyl]-2,2'-diamine	$C_{12}H_{12}N_2$ 184.24	1454-80-4 81	 162^{4}	4-13-00-00356 1.3090^{20}	H_2O 3; ace 3; bz 3
3029	[1,1'-Biphenyl]-2,4'-diamine 2,4'-Biphenyldiamine	$C_{12}H_{12}N_2$ 184.24	492-17-1 54.5	1249 363	4-13-00-00360	H_2O 1; EtOH 3; eth 3
3030	[1,1'-Biphenyl]-3,3'-diamine	$C_{12}H_{12}N_2$ 184.24	2050-89-7 93.5	 $205^{0.001}$	4-13-00-00363	H_2O 2; eth 4; bz 3
3031	[1,1'-Biphenyl]-4,4'-diamine Benzidine	$C_{12}H_{12}N_2$ 184.24	92-87-5 120	1086 401	4-13-00-00364	H_2O 2; EtOH 3; eth 2; DMSO 2
3032	[1,1'-Biphenyl]-4,4'-diamine, 2,2'-dichloro- *o*-Dichlorobenzidine	$C_{12}H_{10}Cl_2N_2$ 253.13	84-68-4 165	3046	4-13-00-00384	eth 4; EtOH 4
3033	[1,1'-Biphenyl]-4,4'-diamine, 3,3'-dimethoxy- *o*-Dianisidine	$C_{14}H_{16}N_2O_2$ 244.29	119-90-4 137	2970	4-13-00-02834	H_2O 1; EtOH 3; eth 3; ace 3
3034	[1,1'-Biphenyl]-4,4'-diamine, 3,3'-dimethyl- *o* Tolidine	$C_{14}H_{16}N_2$ 212.29	119-93-7 131.5	9437	4-13-00-00419	H_2O 2; EtOH 4; eth 4; chl 2
3035	[1,1'-Biphenyl]-4,4'-diamine, *N,N*'-diphenyl- *N,N*'-Diphenylbenzidine	$C_{24}H_{20}N_2$ 336.44	531-91-9 247	3319	4-13-00-00368	H_2O 1; EtOH 2; eth 2; bz 2
3036	[1,1'-Biphenyl]-4,4'-diamine, tetramethyl-	$C_{16}H_{20}N_2$ 240.35	34314-06-2 198	 >360	3-13-00-00429	bz 4; chl 4
3037	1,1'-Biphenyl, 4,4'-dibromo-	$C_{12}H_8Br_2$ 312.00	92-86-4 164	 357.5	4-05-00-01820	H_2O 1; EtOH 2; bz 3
3038	[1,1'-Biphenyl]-2,2'-dicarboxylic acid *o,o*'-Diphenic acid	$C_{14}H_{10}O_4$ 242.23	482-05-3 233.5	3309 sub	4-09-00-03552	H_2O 1; EtOH 3; eth 3; os 3
3039	[1,1'-Biphenyl]-2,2'-dicarboxylic acid, dimethyl ester	$C_{16}H_{14}O_4$ 270.28	5807-64-7 74	 204-6^{14}	4-09-00-03553	H_2O 1; EtOH 4; eth 3; bz 3
3040	1,1'-Biphenyl, 3,3'-dichloro-	$C_{12}H_8Cl_2$ 223.10	2050-67-1 29	 320	3-05-00-01739	bz 4; eth 4; EtOH 4
3041	1,1'-Biphenyl, 4,4'-dichloro-	$C_{12}H_8Cl_2$ 223.10	2050-68-2 149.3	 317	4-05-00-01817 1.4420^{0}	H_2O 1; EtOH 2; bz 3; chl 2
3042	1,1'-Biphenyl, 2,2'-difluoro-	$C_{12}H_8F_2$ 190.19	388-82-9 118.5		4-05-00-01815 1.393^{20}	H_2O 1; EtOH 3; xyl 3
3043	1,1'-Biphenyl, 3,3'-difluoro-	$C_{12}H_8F_2$ 190.19	396-64-5 8	 130^{14}	4-05-00-01815 1.192^{25}	ctc 2 1.5678^{20}
3044	1,1'-Biphenyl, 4,4'-difluoro- 4,4'-Difluorodiphenyl	$C_{12}H_8F_2$ 190.19	398-23-2 94.5	3133 254.5	4-05-00-01816	H_2O 1; EtOH 4; eth 3; ace 3
3045	1,1'-Biphenyl, 2,2'-dimethoxy- 2,2'-Dimethoxybiphenyl	$C_{14}H_{14}O_2$ 214.26	4877-93-4 155	 307.5	4-06-00-06645 1.268^{25}	H_2O 1; EtOH 4; eth 2; bz 4
3046	1,1'-Biphenyl, 3,3'-dimethoxy- 3,3'-Dimethoxybiphenyl	$C_{14}H_{14}O_2$ 214.26	6161-50-8 36	 328	4-06-00-06650	H_2O 1; EtOH 4; eth 4; ace 3
3047	1,1'-Biphenyl, 4,4'-dimethoxy-	$C_{14}H_{14}O_2$ 214.26	2132-80-1 175	 sub	4-06-00-06651	H_2O 1; EtOH 4; eth 2; bz 4
3048	1,1'-Biphenyl, 2,2'-dimethoxy-5,5'-dimethyl- Biphenyl, 2,2'-dimethoxy-5,5'-dimethyl	$C_{16}H_{18}O_2$ 242.32	7168-55-0 71	 188^{12}	3-06-00-05447	ace 4; bz 4; eth 4; EtOH 4
3049	1,1'-Biphenyl, 2,5'-dimethoxy-2',5-dimethyl- Biphenyl, 2,5'-dimethoxy-2',5-dimethyl	$C_{16}H_{18}O_2$ 242.32	72935-12-7 66	 168^{4}	4-06-00-06704	bz 4; peth 4
3050	1,1'-Biphenyl, 2,2'-dimethyl-	$C_{14}H_{14}$ 182.27	605-39-0 19.5	 256	4-05-00-01897 0.9906^{20}	H_2O 1; EtOH 4; eth 4; ace 3 1.5752^{20}
3051	1,1'-Biphenyl, 2,3'-dimethyl-	$C_{14}H_{14}$ 182.27	611-43-8	 270	4-05-00-01902 0.9924^{20}	H_2O 1; EtOH 4; eth 4; ace 3 1.5810^{20}

No.	Name Synonym	Mol. Form. Mol. Wt.	CAS RN mp/°C	Merck No. bp/°C	Beil. Ref. den/g cm^{-3}	Solubility n_D
3052	1,1'-Biphenyl, 2,3-dimethyl- 2,3-Dimethylbiphenyl	$C_{14}H_{14}$ 182.27	3864-18-4 42	 141^{14}; 98^{2}	4-05-00-01897	H_2O 1; eth 3; peth 3 1.5845^{23}
3053	1,1'-Biphenyl, 2,4'-dimethyl-	$C_{14}H_{14}$ 182.27	611-61-0	 277.2	4-05-00-01903 0.9924^{20}	H_2O 1; EtOH 4; eth 4; ace 3 1.5826^{20}
3054	1,1'-Biphenyl, 2,4-dimethyl- 2,4-Dimethylbiphenyl	$C_{14}H_{14}$ 182.27	4433-10-7	 273; 144^{20}	4-05-00-01897 0.9947^{20}	 1.5844^{20}
3055	1,1'-Biphenyl, 2,5-dimethyl-	$C_{14}H_{14}$ 182.27	7372-85-2	 $140^{14.5}$	4-05-00-01897 0.9931^{20}	 1.5819^{20}
3056	1,1'-Biphenyl, 2,6-dimethyl-	$C_{14}H_{14}$ 182.27	3976-34-9 -5	 262.5	4-05-00-01897 0.9907^{20}	 1.5745^{20}
3057	1,1'-Biphenyl, 3,3'-dimethyl-	$C_{14}H_{14}$ 182.27	612-75-9 9	 280	4-05-00-01903 0.9995^{20}	H_2O 1; EtOH 4; eth 4; ace 3 1.5946^{20}
3058	1,1'-Biphenyl, 3,4'-dimethyl-	$C_{14}H_{14}$ 182.27	7383-90-6 14.5	 289; 153^{15}	4-05-00-01905 0.9978^{20}	H_2O 1; bz 3 1.5968^{20}
3059	1,1'-Biphenyl, 3,4-dimethyl- 3,4-Dimethylbiphenyl	$C_{14}H_{14}$ 182.27	4433-11-8 29.5	 282	4-05-00-01903 1.0087^{20}	H_2O 1; bz 3; chl 3 1.6036^{20}
3060	1,1'-Biphenyl, 3,5-dimethyl- 3,5-Dimethylbiphenyl	$C_{14}H_{14}$ 182.27	17057-88-4 22.5	 274.5	4-05-00-01903 0.9990^{20}	 1.5952^{20}
3061	1,1'-Biphenyl, 4,4'-dimethyl-	$C_{14}H_{14}$ 182.27	613-33-2 125	 295	4-05-00-01906 0.917^{121}	H_2O 1; EtOH 2; eth 3; ace 3
3062	1,1'-Biphenyl, 2,2'-dinitro-	$C_{12}H_8N_2O_4$ 244.21	2436-96-6 126	 305	4-05-00-01826 1.45^{25}	H_2O 1; EtOH 4; eth 3; ace 2
3063	1,1'-Biphenyl, 2,4'-dinitro-	$C_{12}H_8N_2O_4$ 244.21	606-81-5 93.5		4-05-00-01827 1.474^{25}	H_2O 1; EtOH 4; eth 3; bz 3
3064	[1,1'-Biphenyl]-2,2'-diol	$C_{12}H_{10}O_2$ 186.21	1806-29-7 109	 320	4-06-00-06645 1.3420^{20}	H_2O 3; EtOH 3; eth 3; ace 3
3065	[1,1'-Biphenyl]-2,4'-diol	$C_{12}H_{10}O_2$ 186.21	611-62-1 162.5	 342	3-06-00-05387	eth 4
3066	[1,1'-Biphenyl]-3,3'-diol	$C_{12}H_{10}O_2$ 186.21	612-76-0 124.8	 247^{18}	3-06-00-05388	H_2O 2; EtOH 3; eth 3; bz 3
3067	[1,1'-Biphenyl]-2,2'-diol, 3,3'-dimethyl-	$C_{14}H_{14}O_2$ 214.26	32750-14-4 113	 sub	3-06-00-05445	bz 4; eth 4; EtOH 4
3068	[1,1'-Biphenyl]-2,2'-diol, 5,5'-dimethyl-	$C_{14}H_{14}O_2$ 214.26	15519-73-0 153.5	 sub	4-06-00-06705	H_2O 2; EtOH 3; eth 4; ace 3
3069	[1,1'-Biphenyl]-4,4'-diol, 3,3',5,5'-tetramethyl-	$C_{16}H_{18}O_2$ 242.32	2417-04-1 221.8	 sub	4-06-00-06744	EtOH 2; bz 2; gl HOAc 2; tol 2
3070	[1,1'-Biphenyl]-4,4'-disulfonic acid 4,4'-Biphenyldisulfonic acid	$C_{12}H_{10}O_6S_2$ 314.34	5314-37-4 72.5	 >200	3-11-00-00472	H_2O 4
3071	1,1'-Biphenyl, 3-ethoxy- 3-Ethoxybiphenyl	$C_{14}H_{14}O$ 198.26	54852-73-2 35	 305	3-06-00-03313	EtOH 3; eth 3; ace 3; bz 3
3072	1,1'-Biphenyl, 2-ethyl-	$C_{14}H_{14}$ 182.27	1812-51-7 -6	 266; 128^{11}	4-05-00-01896 0.9966^{20}	 1.5758^{25}
3073	1,1'-Biphenyl, 3-ethyl-	$C_{14}H_{14}$ 182.27	5668-93-9 -27.5	 283.5	4-05-00-01896 1.048^{25}	 1.5859^{25}
3074	1,1'-Biphenyl, 2-fluoro-	$C_{12}H_9F$ 172.20	321-60-8 73.5	 248	4-05-00-01815 1.2452^{25}	EtOH 3; eth 3; chl 3; lig 2
3075	1,1'-Biphenyl, 4-fluoro-	$C_{12}H_9F$ 172.20	324-74-3 74.2	 253	4-05-00-01815 1.247^{25}	EtOH 2; eth 3; gl HOAc 3
3076	1,1'-Biphenyl, 2,2',4,4',6,6'-hexamethyl-	$C_{18}H_{22}$ 238.37	4482-03-5 103.5	 297	4-05-00-01989 1.023^{50}	H_2O 1; EtOH 2; eth 3; bz 3
3077	1,1'-Biphenyl, 2-iodo-	$C_{12}H_9I$ 280.11	2113-51-1	 190^{36}; 158^{6}	4-05-00-01820 1.5991^{25}	H_2O 1; EtOH 3; eth 3; bz 3 1.6620^{20}
3078	1,1'-Biphenyl, 3-iodo-	$C_{12}H_9I$ 280.11	20442-79-9 26.5	 188^{16}	4-05-00-01821 1.5967^{25}	
3079	1,1'-Biphenyl, 4-iodo-	$C_{12}H_9I$ 280.11	1591-31-7 113.5	 320; 183^{11}	4-05-00-01821	H_2O 1; EtOH 3; eth 3; bz 3
3080	1,1'-Biphenyl, 2-methoxy-	$C_{13}H_{12}O$ 184.24	86-26-0 29	 274	4-06-00-04580 1.0233^{99}	H_2O 1; EtOH 3; ctc 2; peth 3 1.5641^{99}
3081	1,1'-Biphenyl, 4-methoxy-	$C_{13}H_{12}O$ 184.24	613-37-6 90	 157^{10}	4-06-00-04600 1.0278^{100}	H_2O 1; EtOH 3; eth 3 1.5744^{100}
3082	1,1'-Biphenyl, 2-methyl-	$C_{13}H_{12}$ 168.24	643-58-3 -0.2	 255.5	4-05-00-01855 1.010^{22}	H_2O 1; EtOH 3; eth 3 1.5914^{20}
3083	1,1'-Biphenyl, 3-methyl-	$C_{13}H_{12}$ 168.24	643-93-6 4.5	 272.7	4-05-00-01858 1.0182^{17}	H_2O 1; EtOH 3; eth 3; ctc 3 1.5972^{20}
3084	1,1'-Biphenyl, 4-methyl-	$C_{13}H_{12}$ 168.24	644-08-6 49.5	 267.5	4-05-00-01860 1.015^{27}	H_2O 1; EtOH 3; eth 3; ctc 2
3085	1,1'-Biphenyl, 2-nitro- *o*-Nitrobiphenyl	$C_{12}H_9NO_2$ 199.21	86-00-0 37.2	6513 320	4-05-00-01823 1.44^{25}	H_2O 1; EtOH 3; eth 3; chl 3
3086	1,1'-Biphenyl, 3-nitro-	$C_{12}H_9NO_2$ 199.21	2113-58-8 62	 227^{35}; 143^{9}	4-05-00-01823	H_2O 1; EtOH 3; eth 3; HOAc 3
3087	1,1'-Biphenyl, 4-nitro- *p*-Nitrobiphenyl	$C_{12}H_9NO_2$ 199.21	92-93-3 114	6514 340	4-05-00-01823	H_2O 1; EtOH 2; eth 3; bz 3
3088	[1,1'-Biphenyl]-2-ol *o*-Phenylphenol	$C_{12}H_{10}O$ 170.21	90-43-7 59	7276 286	4-06-00-04579 1.213^{25}	H_2O 1; EtOH 3; eth 4; ace 3
3089	[1,1'-Biphenyl]-3-ol *m*-Phenylphenol	$C_{12}H_{10}O$ 170.21	580-51-8 78	 >300	4-06-00-04597	H_2O 2; EtOH 4; eth 4; bz 4
3090	[1,1'-Biphenyl]-4-ol *p*-Phenylphenol	$C_{12}H_{10}O$ 170.21	92-69-3 166	7277 305	4-06-00-04600	H_2O 2; EtOH 4; eth 4; chl 4

No.	Name Synonym	Mol. Form. Mol. Wt.	CAS RN mp/°C	Merck No. bp/°C	Beil. Ref. den/g cm^{-3}	Solubility n_D
3091	[1,1'-Biphenyl]-4-ol, acetate	$C_{14}H_{12}O_2$ 212.25	148-86-7 89	 196^{13}	3-06-00-03326	H_2O 2; EtOH 2; eth 3; bz 3
3092	[1,1'-Biphenyl]-2-ol, 3-chloro- 2-Phenyl-6-chlorophenol	$C_{12}H_9ClO$ 204.66	85-97-2 6	7251 317 dec	4-06-00-04586 1.24^{25}	H_2O 1; EtOH 3; eth 3; ace 3 1.6237^{30}
3093	1,1'-Biphenyl, 2-(phenylmethyl)-	$C_{19}H_{16}$ 244.34	606-97-3 55	 283^{110}	3-05-00-02323	bz 4; eth 4; EtOH 4
3094	1,1'-Biphenyl, 4-(phenylmethyl)-	$C_{19}H_{16}$ 244.34	613-42-3 85	 $285\text{-}6^{110}$	4-05-00-02502 1.171^{0}	H_2O 1; EtOH 3; eth 4; bz 4
3095	1,1'-Biphenyl, 2,2',4,4'-tetramethyl-	$C_{16}H_{18}$ 210.32	3976-36-1 41	 287	3-05-00-01891	EtOH 3; eth 2; bz 2; chl 3
3096	1,1'-Biphenyl, 2,2',5,5'-tetramethyl-	$C_{16}H_{18}$ 210.32	3075-84-1 51.5	 285	3-05-00-01892	eth 3; bz 3
3097	[1,1'-Biphenyl]-3,3',5,5'-tetrol Diresorcinol	$C_{12}H_{10}O_4$ 218.21	531-02-2 310	3361	2-06-00-01129	H_2O 4; eth 4; EtOH 4
3098	2,3'-Bipiperidine 2,3'-Bipiperidyl	$C_{10}H_{20}N_2$ 168.28	2467-09-6 68.5	 269.5	4-23-00-00512	H_2O 4; EtOH 4
3099	2,2'-Bipyridine α,α'-Dipyridyl	$C_{10}H_8N_2$ 156.19	366-18-7 72	3355 273.5	5-23-08-00016	H_2O 2; EtOH 4; eth 4; bz 4
3100	2,3'-Bipyridine 2,3'-Bipyridyl	$C_{10}H_8N_2$ 156.19	581-50-0	 295.5	5-23-08-00026 1.140^{20}	H_2O 1; EtOH 4; eth 4; bz 4 1.6223^{20}
3101	2,4'-Bipyridine	$C_{10}H_8N_2$ 156.19	581-47-5 61.5	 281	5-23-08-00027	H_2O 2; EtOH 4; eth 4; chl 4
3102	3,3'-Bipyridine 3,3'-Dipyridyl	$C_{10}H_8N_2$ 156.19	581-46-4 68	 291.5	5-23-08-00027 1.1614^{20}	H_2O 4; EtOH 4; eth 2
3103	3,4'-Bipyridine 3,4'-Bipyridyl	$C_{10}H_8N_2$ 156.19	4394-11-0 62	 297; 144^{15}	5 23 08 00028	EtOH 4; peth 4
3104	4,4'-Bipyridine γ,γ'-Dipyridyl	$C_{10}H_8N_2$ 156.19	553-26-4 111	3356 305	5-23-08-00028	H_2O 2; EtOH 4; eth 3; bz 4
3105	2,3'-Bipyridine, 1,2,3,6-tetrahydro-, (*S*)- Anatabine	$C_{10}H_{12}N_2$ 160.22	581-49-7	661 $145\text{-}6^{10}$	5-23-06-00507 1.091^{19}	H_2O 5; EtOH 3; eth 3; bz 3 1.5676^{20}
3106	[5,5'-Bipyrimidine]- 2,2',4,4',6,6'(1H,1'H,3H,3'H,5H,5'H)-hexone- 5,5'-dihydroxy Alloxantin	$C_8H_6N_4O_8$ 286.16	76-24-4 253-5 dec	282	4-26-00-02782	H_2O 2; EtOH 2; eth 2
3107	2,3'-Biquinoline 2,3'-Biquinolyl	$C_{18}H_{12}N_2$ 256.31	612-81-7 175.8	 >400	5-23-10-00009	bz 4; eth 4; EtOH 4; chl 4
3108	Bis[1,3]benzodioxolo[4,5-c:5',6'-g]azecin- 13(5H)-one, 4,6,7,14-tetrahydro-5,14- dimethyl-, (*R*)- Corycavamine	$C_{21}H_{21}NO_5$ 367.40	521-85-7 149	2540	2-27-00-00621	EtOH 4; chl 4
3109	Bis[1,3]benzodioxolo[4,5-c:5',6'-g]azecin- 13(5H)-one, 4,6,7,14-tetrahydro-5-methyl- Protopine	$C_{20}H_{19}NO_5$ 353.37	130-86-9 208	7912	4-27-00-06881	H_2O 1; EtOH 2; eth 2; bz 2
3110	3H-Bis[1]benzopyrano[3,4-b:6',5'-e]pyran- 7(7aH)-one, 13,13a-dihydro-7a-hydroxy-9,10- dimethoxy-3,3-dimethyl Tephrosin	$C_{23}H_{22}O_7$ 410.42	76-80-2 198	9082	4-19-00-05271	ace 4; eth 4; chl 4
3111	Bismuthine, triphenyl- Triphenylbismuth	$C_{18}H_{15}Bi$ 440.30	603-33-8 77.6	 242^{14}	4-16-00-01226 1.715^{75}	EtOH 2; eth 3; ace 3; bz 3 1.7040^{75}
3112	2,2'-Bithiophene	$C_8H_6S_2$ 166.27	492-97-7 33	 260	5-19-01-00537	H_2O 1; EtOH 4; eth 3; ctc 3
3113	Boranamine, *N,N*,1,1-tetramethyl-	$C_4H_{12}BN$ 84.96	1113-30-0 -92	 65	4-04-00-04367	eth 4; ace 4; os 5
3114	Boranediamine, 1-fluoro-*N,N,N',N'*- tetramethyl-	$C_4H_{12}BFN_2$ 117.96	383-90-4 -44.3	 106	4-04-00-00303	ace 4; eth 4
3115	Borane, difluoro(4-methylphenyl)-	$C_7H_7BF_2$ 139.94	768-39-8	 128	4-16-00-01674 1.055^{25}	bz 4; eth 4 1.4535^{25}
3116	Borane, difluorophenyl-	$C_6H_5BF_2$ 125.91	368-98-9 -36.2	 98	4-16-00-01661 1.087^{25}	bz 4; eth 4 1.4441^{25}
3117	Borane, dimethoxy- Dimethoxyborane	$C_2H_7BO_2$ 73.89	4542-61-4 -130.6	 25.9	4-01-00-01269	H_2O 6
3118	Borane, triethyl Triethylborane	$C_6H_{15}B$ 98.00	97-94-9 -93	 95	 0.70^{23}	EtOH 3; eth 3 1.3971
3119	Borane, trimethyl- Trimethylborane	C_3H_9B 55.92	593-90-8 -161.5	 -20.2		
3120	Borane, trimethyl-, monoammoniate Aminetrimethylboron	$C_3H_{12}BN$ 72.95	1030-95-1 73.5			
3121	Borane, tripropyl-	$C_9H_{21}B$ 140.08	1116-61-6 -56	 159	4-04-00-04360 0.7204^{25}	 1.4135^{22}
3122	Borane, tris(3-methylbutyl)- Borane, triisopentyl-	$C_{15}H_{33}B$ 224.24	3062-81-5	 119^{14}	2-04-00-01023 0.7600^{25}	ace 4; eth 4; EtOH 4 1.4321
3123	Borane, tris(2-methylpropyl)-	$C_{12}H_{27}B$ 182.16	1116-39-8	 188; 86^{20}	4-04-00-04361 0.7380^{25}	bz 4; eth 4; EtOH 4 1.4188^{23}
3124	Boric acid, tributyl ester	$C_{12}H_{27}BO_3$ 230.16	688-74-4	 234	4-01-00-01544 0.8567^{20}	EtOH 3; eth 4; bz 3; MeOH 4 1.4106^{18}
3125	Boric acid, triethyl ester	$C_6H_{15}BO_3$ 145.99	150-46-9 -84.8	 120	4-01-00-01365 0.8546^{20}	EtOH 5; eth 5 1.3749^{20}
3126	Boric acid, trimethyl ester Trimethyl borate	$C_3H_9BO_3$ 103.91	121-43-7 -29.3	9626 67.5	4-01-00-01269 0.915^{25}	eth 4; MeOH 4 1.3568^{20}

No.	Name Synonym	Mol. Form. Mol. Wt.	CAS RN mp/°C	Merck No. bp/°C	Beil. Ref. den/g cm^{-3}	Solubility n_D
3127	Boric acid, tripropyl ester	$C_9H_{21}BO_3$ 188.07	688-71-1	 179.5	4-01-00-01436 0.8576^{20}	EtOH 4; eth 5; PrOH 3 1.3948^{20}
3128	Boric acid, tris(1-methylethyl) ester Isopropyl borate	$C_9H_{21}BO_3$ 188.07	5419-55-6	 140; 75^{76}	4-01-00-01488 0.8251^{20}	EtOH 4; eth 4; bz 4; PrOH 4 1.3777^{20}
3129	Borinic acid, dimethylthio-, methyl ester	C_3H_9BS 87.98	19163-05-4 -84	 71	4-04-00-04368	ace 4; eth 4
3130	Borinic acid, diphenyl-	$C_{12}H_{11}BO$ 182.03	2622-89-1 57.5	 $150\text{-}5^{20}$	4-16-00-01639 1.0740^{20}	EtOH 3; bz 3; peth 3
3131	3-Bornanecarboxylic acid Camphane-3-carboxylic acid	$C_{11}H_{18}O_2$ 182.26	91965-23-0 90.5	 153^{13}	3-09-00-00244	EtOH 4; HOAc 4
3132	Boron, (*N,N*-dimethylmethanamine)trihydro-, (T-4)- Trimethylamineborane	$C_3H_{12}BN$ 72.95	75-22-9 94	 172	4-04-00-00134 0.792^{25}	eth 4; EtOH 4
3133	Boronic acid, ethyl-	$C_2H_7BO_2$ 73.89	4433-63-0 40	 dec	4-04-00-04379	H_2O 4; eth 4; EtOH 4
3134	Boronic acid, (2-methylphenyl)-	$C_7H_9BO_2$ 135.96	16419-60-6 166.5	 dec	3-16-00-01277	bz 4; eth 4; EtOH 4
3135	Boronic acid, phenyl- Benzeneboronic acid	$C_6H_7BO_2$ 121.93	98-80-6 219	1076	4-16-00-01654	H_2O 2; EtOH 3; eth 3; bz 3
3136	Boronic acid, propyl-	$C_3H_9BO_2$ 87.91	17745-45-8 107	 dec	3-04-00-01964	H_2O 4; eth 4; EtOH 4
3137	Boron, trifluoro[1,1'-oxybis[ethane]]-, (T-4)- Boron trifluoride etherate	$C_4H_{10}BF_3O$ 141.93	109-63-7 -60.4	1352 125.5	4-01-00-01314 1.3572^{20}	eth 4; EtOH 4 1.4447^{20}
3138	Boron, trifluoro[oxybis[methane]]-, (T-4)- Boron trifluoride compd. with methyl ether	$C_2H_6BF_3O$ 113.88	353-42-4 -14	 127 dec	4-01-00-01245 1.2410^{20}	 1.302^{20}
3139	Bromacil 2,4(1H,3H)-Pyrimidinedione, 5-bromo-6-methyl-3-(1-methylpropyl)-	$C_9H_{13}BrN_2O_2$ 261.12	314-40-9 158	1370	 1.55^{25}	
3140	Bromopropylate Benzeneacetic acid, 4-bromo-α-(4-bromophenyl)-α-hydroxy-, 1-methylethyl ester	$C_{17}H_{16}Br_2O_3$ 428.12	18181-80-1 77	1422	 1.59^{20}	
3141	Bufa-20,22-dienolide, 16-(acetyloxy)-3,14-dihydroxy-, (3β,5β,16β)- Bufotalin	$C_{26}H_{36}O_6$ 444.57	471-95-4 223 dec	1466	5-18-04-00443	H_2O 1; EtOH 3; chl 3
3142	Bufa-4,20,22-trienolide, 6-(acetyloxy)-3-(β-*D*-glucopyranosyloxy)-8,14-dihydroxy-, (3β,6β)- Scilliroside	$C_{32}H_{44}O_{12}$ 620.69	507-60-8 169	8356 dec	4-18-00-03178	H_2O 2; EtOH 4; eth 1; ace 2
3143	*sec*-Bumeton 1,3,5-Triazine-2,4-diamine, *N*-ethyl-6-methoxy-*N*'-(1-methylpropyl)-	$C_{10}H_{19}N_5O$ 225.29	26259-45-0 87			
3144	Butachlor Acetamide, *N*-(butoxymethyl)-2-chloro-*N*-(2,6-diethylphenyl)-	$C_{17}H_{26}ClNO_2$ 311.85	23184-66-9 <-5	1498 $156^{.5}$	 1.070^{25}	
3145	1,2-Butadiene Methylallene	C_4H_6 54.09	590-19-2 -136.2	 10.9	4-01-00-00975 0.676^{0}	H_2O 1; EtOH 5; eth 5; bz 4 1.4205^{1}
3146	1,3-Butadiene Divinyl	C_4H_6 54.09	106-99-0 -108.9	1500 -4.4	4-01-00-00976 $^*0.6149^{25}$	H_2O 1; EtOH 3; eth 3; ace 4 1.4292^{-25}
3147	1,2-Butadiene, 4-bromo- 4-Bromo-1,2-butadiene	C_4H_5Br 132.99	18668-68-3	 110	3-01-00-00929 1.4255^{20}	ace 4 1.5248^{20}
3148	1,3-Butadiene, 1-bromo- 1-Bromo-1,3-butadiene	C_4H_5Br 132.99	21890-35-7	 93	4-01-00-00989 1.416^{0}	eth 4; EtOH 4; peth 4
3149	1,3-Butadiene, 2-bromo-	C_4H_5Br 132.99	1822-86-2	 $42\text{-}3^{165}$	4-01-00-00989 1.397^{20}	eth 4; EtOH 4 1.4988^{20}
3150	1,2-Butadiene, 4-chloro-	C_4H_5Cl 88.54	25790-55-0	 88	4-01-00-00975 0.9891^{20}	ace 4; bz 4; eth 4 1.4775^{20}
3151	1,3-Butadiene, 1-chloro-	C_4H_5Cl 88.54	627-22-5	 68	3-01-00-00949 0.9606^{20}	eth 4; EtOH 4; chl 4 1.4712^{20}
3152	1,3-Butadiene, 2-chloro- 2-Chloro-1,3-butadiene	C_4H_5Cl 88.54	126-99-8 -130	 59.4	4-01-00-00984 0.956^{20}	H_2O 2; eth 5; ace 5; bz 5 1.4583^{20}
3153	1,3-Butadiene, 1-chloro-2-methyl- Chloromethylbutadiene	C_5H_7Cl 102.56	35383-51-8	 107	3-01-00-00974 0.9710^{20}	ace 4; EtOH 4 1.4702^{20}
3154	1,3-Butadiene, 1-chloro-3-methyl- 4-Chloroisoprene	C_5H_7Cl 102.56	51034-46-9	 100.7	4-01-00-01005 0.9543^{20}	ace 4; eth 4; EtOH 4; chl 4 1.4719^{20}
3155	1,3-Butadiene, 2-chloro-3-methyl-	C_5H_7Cl 102.56	1809-02-5	 93	4-01-00-01004 0.9593^{20}	H_2O 1; EtOH 3; eth 4; ace 3 1.4686^{20}
3156	1,3-Butadiene, 1,1-dichloro- 1,1-Dichloro-1,3-butadiene	$C_4H_4Cl_2$ 122.98	6061-06-9	 $42\text{-}3^{90}$	4-01-00-00985 1.1831^{20}	H_2O 1; eth 4; bz 4; chl 4 1.5022^{20}
3157	1,3-Butadiene, 1,2-dichloro- 1,2-Dichloro-1,3-butadiene	$C_4H_4Cl_2$ 122.98	3574-40-1	 62^{105}; 35^{40}	4-01-00-00985 1.1991^{20}	ctc 4 1.4960^{20}
3158	1,3-Butadiene, 2,3-dichloro-	$C_4H_4Cl_2$ 122.98	1653-19-6	 98	4-01-00-00986 1.1829^{20}	chl 4 1.4890^{20}
3159	1,3-Butadiene, 2,3-dimethyl- Diisopropenyl	C_6H_{10} 82.15	513-81-5 -76	3228 68.8	4-01-00-01023 0.7222^{25}	ctc 3 1.4394^{20}
3160	1,3-Butadiene, 1-ethoxy-	$C_6H_{10}O$ 98.14	5614-32-4	 110.5	4-01-00-02222 0.8154^{20}	ace 4; bz 4; eth 4; EtOH 4 1.4529^{20}
3161	1,3-Butadiene, 2-ethoxy-	$C_6H_{10}O$ 98.14	4747-05-1	 94.5	4-01-00-02223 0.8177^{20}	ace 4; bz 4; eth 4; EtOH 4 1.4400^{20}
3162	1,3-Butadiene, 2-fluoro- Fluoroprene	C_4H_5F 72.08	381-61-3	 12	4-01-00-00982 0.843^{4}	 1.400^{4}
3163	1,3-Butadiene, 1,1,2,3,4,4-hexachloro- Hexachloro-1,3-butadiene	C_4Cl_6 260.76	87-68-3 -21	 215	4-01-00-00988 1.556^{25}	H_2O 1; EtOH 3; eth 3 1.5542^{20}

No.	Name Synonym	Mol. Form. Mol. Wt.	CAS RN mp/°C	Merck No. bp/°C	Beil. Ref. den/g cm^{-3}	Solubility n_D
3164	1,3-Butadiene, 1,1,2,3,4,4-hexafluoro- 1,3-Butadiene, hexafluoro	C_4F_6 162.03	685-63-2 -132	 6	4-01-00-00983 1.553^{-20}	 1.378^{-20}
3165	1,3-Butadiene, 2-iodo-	C_4H_5I 179.99	19221-28-4	 112	3-01-00-00956 1.7278^{20}	 1.5616
3166	1,2-Butadiene, 4-iodo- 4-Iodo-1,2-butadiene	C_4H_5I 179.99	67885-08-9	 130	3-01-00-00929 1.7129^{20}	 1.5709^{20}
3167	1,2-Butadiene, 4-methoxy-	C_5H_8O 84.12	36678-06-5	 88	4-01-00-02221 0.8286^{20}	EtOH 4 1.435^{20}
3168	1,3-Butadiene, 1-methoxy-	C_5H_8O 84.12	3036-66-6	 91.5	3-01-00-01975 0.8296^{20}	H_2O 3; EtOH 3 1.4594^{20}
3169	1,3-Butadiene, 2-methoxy-	C_5H_8O 84.12	3588-30-5	 75	4-01-00-02223 0.8272^{20}	ace 4; bz 4; eth 4; EtOH 4 1.4442^{20}
3170	1,2-Butadiene, 3-methyl-	C_5H_8 68.12	598-25-4 -113.6	 40.83	4-01-00-01006 0.6806^{25}	ace 4; bz 4; eth 4; EtOH 4 1.4203^{20}
3171	1,3-Butadiene, 2-methyl- Isoprene	C_5H_8 68.12	78-79-5 -145.9	5087 34.0	4-01-00-01001 0.679^{20}	H_2O 1; EtOH 5; eth 5; ace 5 1.4219^{20}
3172	1,3-Butadiene, 1,1,2,4,4-pentafluoro-3-(trifluoromethyl)- 1,3-Butadiene, pentafluoro-2-(trifluoromethyl)-	C_5F_8 212.04	384-04-3	 39	 1.527^{0}	ace 4; bz 4; eth 4 1.3000^{0}
3173	1,3-Butadiene, 1,2,3,4-tetrachloro-	$C_4H_2Cl_4$ 191.87	1637-31-6 -4	 188	4-01-00-00987 1.515^{15}	H_2O 1; EtOH 3; eth 3; ace 3 1.5455^{20}
3174	1,3-Butadiene, 1,2,3-trichloro- 1,2,3-Trichloro-1,3-butadiene	$C_4H_3Cl_3$ 157.43	1573-58-6	 $33\text{-}4^{7}$	3-01-00-00954 1.4060^{20}	eth 4; chl 4 1.5262^{20}
3175	2,3-Butadien-1-ol	C_4H_6O 70.09	18913-31-0	 128.5	4-01-00-02221 0.9164^{20}	ace 4; bz 4; eth 4; EtOH 4 1.4760^{20}
3176	1,3-Butadien-1-ol, acetate	$C_6H_8O_2$ 112.13	1515-76-0	 $60\text{-}1^{40}$	3-02-00-00295 0.945^{25}	 1.4690^{20}
3177	1,3-Butadiyne	C_4H_2 50.06	460-12-8 -36.4	 10.3	4-01-00-01116 0.7364^{0}	H_2O 4; EtOH 3; eth 4; ace 4 1.4189^{5}
3178	Butanal Butyraldehyde	C_4H_8O 72.11	123-72-8 -99	1591 74.8	4-01-00-03229 0.8016^{20}	H_2O 3; EtOH 5; eth 5; ace 4 1.3843^{20}
3179	Butanal, 2-bromo-	C_4H_7BrO 151.00	24764-97-4	 33^{17}	4-01-00-03241 1.469^{20}	eth 3; ace 3; bz 3; chl 3 1.4683^{20}
3180	Butanal, 2-chloro-	C_4H_7ClO 106.55	28832-55-5	 107	4-01-00-03238 1.1072^{17}	 1.4262^{17}
3181	Butanal, 4-chloro-	C_4H_7ClO 106.55	6139-84-0	 $50\text{-}1^{13}$	4-01-00-03240 1.106^{8}	ace 4; eth 4; EtOH 4 1.4466^{8}
3182	Butanal, 2,3-dichloro-	$C_4H_6Cl_2O$ 141.00	55775-41-3	 $58\text{-}60^{20}$	4-01-00-0[illegible] 1.2666^{21}	ace 4; eth 4; EtOH 4; chl 4 1.4618^{21}
3183	Butanal, 2-ethyl- Diethylacetaldehyde	$C_6H_{12}O$ 100.16	97-96-1	 $117\text{-}9^{160}$	4-01-00-03310 0.8110^{20}	H_2O 2; EtOH 5; eth 5; ctc 2 1.4025^{20}
3184	Butanal, 3-hydroxy- Aldol	$C_4H_8O_2$ 88.11	107-89-1	217 83^{20}	4-01-00-03984 1.103^{20}	H_2O 5; EtOH 5; eth 3; ace 4 1.4238^{20}
3185	Butanal, 2-methyl-, (±)-	$C_5H_{10}O$ 86.13	57456-98-1	 92.5	4-01-00-03285 0.8029^{20}	ace 4; eth 4; EtOH 4 1.3869^{20}
3186	Butanal, 3-methyl- Isovaleraldehyde	$C_5H_{10}O$ 86.13	590-86-3 -51	5118 92.5	4-01-00-03291 0.7977^{20}	H_2O 2; EtOH 3; eth 3 1.3902^{20}
3187	Butanal, 3-methyl-, oxime Butyraldehyde, 3-methyl, oxime	$C_5H_{11}NO$ 101.15	626-90-4 48.5	 161.3	4-01-00-03293 0.8934^{20}	ace 4; eth 4; EtOH 4 1.4367^{20}
3188	Butanal, oxime Butyraldehyde, oxime	C_4H_9NO 87.12	110-69-0 -29.5	 154	4-01-00-03234 0.923^{20}	H_2O 4; EtOH 5; eth 5; ace 4
3189	Butanal, phenylhydrazone Butyraldehyde, phenylhydrazone	$C_{10}H_{14}N_2$ 162.23	940-54-5 94	 192^{80}; 152^{14}	3-15-00-00080	ctc 2
3190	Butanal, 2-(phenylmethylene)-	$C_{11}H_{12}O$ 160.22	28467-92-7	 $157\text{-}8^{5}$	4-07-00-01024 1.0201^{22}	 1.578^{20}
3191	Butanal, 2,2,3-trichloro- 2,2,3-Trichlorobutyraldehyde	$C_4H_5Cl_3O$ 175.44	76-36-8	9545 164	4-01-00-03241 1.3956^{20}	H_2O 4; eth 4; EtOH 4 1.4755^{20}
3192	Butanal, 2,3,4-trihydroxy-, [*R*-(*R**,*R**)]- *D*-Erythrose	$C_4H_8O_4$ 120.11	583-50-6	3637	4-01-00-04172	H_2O 3; EtOH 4
3193	Butanal, 2,3,4-trihydroxy-, [*S*-(*R**,*S**)]- Threose (*D*-)	$C_4H_8O_4$ 120.11	95-43-2 129	9317	4-01-00-04173	
3194	Butanamide Butyramide	C_4H_9NO 87.12	541-35-5 114.8	1592 216	4-02-00-00804 0.8850^{120}	H_2O 2; EtOH 3; eth 2; bz 1 1.4087^{130}
3195	Butanamide, *N*-(aminocarbonyl)-2-bromo-2-ethyl- Carbromal	$C_7H_{13}BrN_2O_2$ 237.10	77-65-6 110	1837	4-03-00-00117 [illegible]	H_2O 2; ace 3; bz 3; chl 2
3196	Butanamide, *N*-(aminocarbonyl)-2-bromo-3-methyl- Bromisovalum	$C_6H_{11}BrN_2O_2$ 223.07	496-67-3 154	1385 sub	3-03-00-00123 1.56^{15}	ace 4; bz 4; eth 4; EtOH 4
3197	Butanamide, *N*-[5-(aminosulfonyl)-1,3,4-thiadiazol-2-yl]- Butazolamide	$C_6H_{10}N_4O_3S_2$ 250.30	16790-49-1	1511	4-27-00-08223	
3198	Butanamide, 2-bromo-2-ethyl- Diethylbromoacetamide	$C_6H_{12}BrNO$ 194.07	511-70-6 67	3105	3-02-00-00755	H_2O 2; EtOH 4; eth 4; bz 4
3199	Butanamide, 2-bromo-2-ethyl-3-methyl- Ibrotamide	$C_7H_{14}BrNO$ 208.10	466-14-8 50.5	4809 sub	4-02-00-00979	ace 4; bz 4; eth 4; EtOH 4
3200	Butanamide, *N*,*N*-diethyl-	$C_8H_{17}NO$ 143.23	1114-76-7	 206	4-04-00-00354	H_2O 4; EtOH 4 1.4403^{25}
3201	Butanamide, *N*,*N*-diethyl-3-methyl- Isovaleryl diethylamide	$C_9H_{19}NO$ 157.26	533-32-4	5122 211	4-04-00-00356 0.8764^{20}	eth 4; EtOH 4 1.4422^{20}

No.	Name Synonym	Mol. Form. Mol. Wt.	CAS RN mp/°C	Merck No. bp/°C	Beil. Ref. den/g cm^{-3}	Solubility n_D
3202	Butanamide, 2,4-dihydroxy-*N*-(3-hydroxypropyl)-3,3-dimethyl-, (*R*)- Dexpanthenol	$C_9H_{19}NO_4$ 205.25	81-13-0	2924 dec	4-04-00-01652 1.20^{20}	H_2O 4; EtOH 4; eth 2; MeOH 4 1.497^{20}
3203	Butanamide, *N*,*N*-dimethyl-	$C_6H_{13}NO$ 115.18	760-79-2 -40	 186; 125^{100}	4-04-00-00185 0.9064^{25}	ace 4; bz 4; eth 4; EtOH 4 1.4391^{25}
3204	Butanamide, *N*,*N*'-(3,3'-dimethyl[1,1'-biphenyl]-4,4'-diyl)bis[3-oxo- *N*,*N*'-Bis(acetoacetyl)-3,3'-dimethylbenzidine	$C_{22}H_{24}N_2O_4$ 380.44	91-96-3 212		3-13-00-00490	DMSO 2
3205	Butanamide, *N*-(4-hydroxyphenyl)- 4'-Hydroxybutyranilide	$C_{10}H_{13}NO_2$ 179.22	101-91-7 139.5	4745	4-13-00-01109	H_2O 4; EtOH 4
3206	Butanamide, 3-methyl- Isovaleramide	$C_5H_{11}NO$ 101.15	541-46-8 137	5119 226	4-02-00-00902	H_2O 3; EtOH 3; eth 3; peth 4
3207	Butanamide, 3-oxo-*N*-phenyl- Acetoacetanilide	$C_{10}H_{11}NO_2$ 177.20	102-01-2 86	50	4-12-00-00955	H_2O 2; EtOH 3; eth 3; bz 3
3208	Butanamide, *N*-phenyl-	$C_{10}H_{13}NO$ 163.22	1129-50-6 97	 189^{15}	4-12-00-00387 1.134^{25}	H_2O 1; EtOH 4; eth 4; chl 2
3209	1-Butanamine Butylamine	$C_4H_{11}N$ 73.14	109-73-9 -49.1	1543 77.0	4-04-00-00540 0.7414^{20}	H_2O 5; EtOH 3; eth 3 1.4031^{20}
3210	2-Butanamine, (±)- *sec*-Butylamine, (±)-	$C_4H_{11}N$ 73.14	33966-50-6 <-72	1544 63.5	4-04-00-00618 0.7246^{20}	H_2O 3; EtOH 5; eth 5; ace 4 1.3932^{20}
3211	1-Butanamine, *N*-butyl- Dibutylamine	$C_8H_{19}N$ 129.25	111-92-2 -62	3019 159.6	4-04-00-00550 0.7670^{20}	H_2O 3; EtOH 4; eth 4; ace 3 1.4177^{20}
3212	1-Butanamine, *N*-butyl-*N*-nitroso- Dibutylnitrosamine	$C_8H_{18}N_2O$ 158.24	924-16-3	 105^{8}	4-04-00-03389	
3213	1-Butanamine, *N*,*N*-dibutyl- Tributylamine	$C_{12}H_{27}N$ 185.35	102-82-9 -70	9530 216.5	4-04-00-00554 0.7770^{20}	H_2O 2; EtOH 4; eth 4; ace 3 1.4299^{20}
3214	1-Butanamine, 4,4-diethoxy-	$C_8H_{19}NO_2$ 161.24	6346-09-4	 196	4-04-00-01928 0.933^{25}	 1.4275^{20}
3215	1-Butanamine, *N*,*N*-dimethyl-	$C_6H_{15}N$ 101.19	927-62-8	 95	4-04-00-00546 0.7206^{20}	H_2O 5; EtOH 5; eth 5; ace 5 1.3970^{20}
3216	2-Butanamine, 2,3-dimethyl-	$C_6H_{15}N$ 101.19	4358-75-2 -70.5	 104.5	4-04-00-00733 0.7683^{0}	 1.4096^{17}
3217	2-Butanamine, 3,3-dimethyl-	$C_6H_{15}N$ 101.19	3850-30-4 -20	 102	4-04-00-00730 0.7668^{20}	H_2O 4 1.4105^{25}
3218	1-Butanamine, *N*-ethyl- Butylethylamine	$C_6H_{15}N$ 101.19	13360-63-9	 107.5	4-04-00-00547 0.7398^{20}	EtOH 5; eth 5; ace 5; bz 5 1.4040^{20}
3219	2-Butanamine, *N*-ethyl-, (±)- *sec*-Butyl ethyl amine (*DL*)	$C_6H_{15}N$ 101.19	116724-10-8 -104.3	 98	2-04-00-00636 0.7358^{20}	ace 4; bz 4; eth 4; EtOH 4
3220	1-Butanamine, 3-methyl- Isoamylamine	$C_5H_{13}N$ 87.16	107-85-7	4994 96	4-04-00-00696 0.7505^{20}	H_2O 5; EtOH 5; eth 5; ace 3 1.4083^{20}
3221	1-Butanamine, *N*-methyl- Butylmethylamine	$C_5H_{13}N$ 87.16	110-68-9	 91	4-04-00-00546 0.7637^{15}	
3222	2-Butanamine, 2-methyl-	$C_5H_{13}N$ 87.16	594-39-8 -105	 77	4-04-00-00694 0.731^{25}	H_2O 4; ace 4; eth 4; EtOH 4 1.3954^{25}
3223	2-Butanamine, 3-methyl-	$C_5H_{13}N$ 87.16	598-74-3 -50	 85.5	2-04-00-00644 0.7574^{19}	H_2O 4; EtOH 3 1.4096^{18}
3224	1-Butanamine, 2-methyl-*N*,*N*-bis(2-methylbutyl)-	$C_{15}H_{33}N$ 227.43	620-43-9	 232	0-04-00-00179 0.9^{13}	ace 4; bz 4; eth 4; EtOH 4 1.4330^{20}
3225	1-Butanamine, 3-methyl-*N*,*N*-bis(3-methylbutyl)-	$C_{15}H_{33}N$ 227.43	645-41-0	 235	4-04-00-00700 0.7848^{20}	H_2O 1; EtOH 4; eth 5; bz 5 1.4331^{20}
3226	1-Butanamine, 3-methyl-*N*-(3-methylbutyl)- Diisoamylamine	$C_{10}H_{23}N$ 157.30	544-00-3 -44	3178 188	4-04-00-00699 0.7672^{21}	H_2O 1; EtOH 3; eth 5 1.4235^{20}
3227	2-Butanamine, *N*-(1-methylpropyl)-	$C_8H_{19}N$ 129.25	626-23-3	 135	4-04-00-00620 0.7534^{20}	H_2O 4; EtOH 3; os 3 1.4162^{20}
3228	1-Butanamine, 1,1,2,2,3,3,4,4,4-nonafluoro-*N*,*N*-bis(nonafluorobutyl)-	$C_{12}F_{27}N$ 671.10	311-89-7	 178	4-02-00-00819 1.884^{25}	ace 3 1.291^{25}
3229	Butane	C_4H_{10} 58.12	106-97-8 -138.2	1507 -0.5	4-01-00-00236 $^{*}0.573^{25}$	H_2O 3; EtOH 4; eth 4; chl 4 1.3326^{20}
3230	Butane, 2,2-bis(ethylsulfonyl)- Sulfonethylmethane	$C_8H_{18}O_4S_2$ 242.36	76-20-0 76	8937 dec	3-01-00-02790 1.199^{85}	chl 3
3231	Butane, 1-bromo- Butyl bromide	C_4H_9Br 137.02	109-65-9 -112.4	1553 101.6	4-01-00-00258 1.2758^{20}	H_2O 1; EtOH 5; eth 5; ace 5 1.4401^{20}
3232	Butane, 2-bromo-, (±)- (±)-*sec*-Butyl bromide	C_4H_9Br 137.02	5787-31-5 -111.9	1554 91.2	4-01-00-00261 1.2585^{20}	ace 4; eth 4; chl 4 1.4366^{20}
3233	Butane, 1-bromo-4-chloro-	C_4H_8BrCl 171.46	6940-78-9	 175; 63^{10}	4-01-00-00264 1.488^{20}	H_2O 1; EtOH 3; eth 3; ctc 2 1.4885^{20}
3234	Butane, 2-bromo-1-chloro- 2-Bromo-1-chlorobutane	C_4H_8BrCl 171.46	79504-01-1	 146.5	4-01-00-00264 1.468^{20}	bz 4; eth 4; EtOH 4; chl 4 1.4880^{20}
3235	Butane, 1-bromo-3,3-dimethyl-	$C_6H_{13}Br$ 165.07	1647-23-0	 138	3-01-00-00409 1.1556^{20}	eth 4; EtOH 4; chl 4 1.4440^{20}
3236	Butane, 2-bromo-2,3-dimethyl- 2-Bromo-2,3-dimethylbutane	$C_6H_{13}Br$ 165.07	594-52-5 24.5	 133; 87^{180}	4-01-00-00374 1.1772^{10}	eth 4; chl 4 1.4517
3237	Butane, 1-bromo-4-fluoro- 1-Bromo-4-fluorobutane	C_4H_8BrF 155.01	462-72-6	 135	4-01-00-00263	eth 4; EtOH 4 1.4370^{25}
3238	Butane, 1-bromo-2-methyl-, (*S*)- *d*-Amyl bromide	$C_5H_{11}Br$ 151.05	534-00-9	636 121.6	4-01-00-00327 1.2234^{20}	H_2O 1; EtOH 3; eth 3; chl 4 1.4451^{20}
3239	Butane, 1-bromo-3-methyl- Isoamyl bromide	$C_5H_{11}Br$ 151.05	107-82-4 -112	4996 120.4	4-01-00-00328 1.2071^{20}	H_2O 1; EtOH 3; eth 3; ctc 2 1.4420^{20}

No.	Name Synonym	Mol. Form. Mol. Wt.	CAS RN mp/°C	Merck No. bp/°C	Beil. Ref. den/g cm^{-3}	Solubility n_D
3240	Butane, 2-bromo-2-methyl- *tert*-Amyl bromide	$C_5H_{11}Br$ 151.05	507-36-8	638 108	4-01-00-00327 1.197^{18}	 1.4421
3241	Butane, 1-chloro- Butyl chloride	C_4H_9Cl 92.57	109-69-3 -123.1	1560 78.6	4-01-00-00246 0.8862^{20}	H_2O 1; EtOH 5; eth 5; ctc 2 1.4021^{20}
3242	Butane, 2-chloro-, (±)- *sec*-Butyl chloride (*DL*)	C_4H_9Cl 92.57	53178-20-4 -131.3	1561 68.2	4-01-00-00248 0.8732^{20}	bz 4; eth 4; EtOH 4; chl 4 1.3971^{20}
3243	Butane, 3-chloro-1,1-diethoxy- Butyraldehyde, 3-chloro-, diethyl acetal	$C_8H_{17}ClO_2$ 180.67	51786-70-0	 $70\text{-}1^{12}$	4-01-00-03239 0.9709^{20}	bz 4; eth 4; EtOH 4 1.4195^{25}
3244	Butane, 3-chloro-1,1-dimethoxy- Butyraldehyde, 3-chloro-, dimethyl acetal	$C_6H_{13}ClO_2$ 152.62	50710-40-2	 $40\text{-}4^{7}$	4-01-00-03239 1.0024^{20}	 1.4160^{25}
3245	Butane, 1-chloro-2,3-dimethyl- 1-Chloro-2,3-dimethylbutane	$C_6H_{13}Cl$ 120.62	600-06-6	 124	4-01-00-00373 0.887^{22}	eth 4; EtOH 4; chl 4
3246	Butane, 1-chloro-3,3-dimethyl-	$C_6H_{13}Cl$ 120.62	2855-08-5	 115	4-01-00-00369 0.8670^{20}	eth 4; EtOH 4; chl 4 1.4161^{20}
3247	Butane, 2-chloro-2,3-dimethyl-	$C_6H_{13}Cl$ 120.62	594-57-0 -10.4	 112	4-01-00-00373 0.8780^{20}	ace 4; EtOH 4 1.4191^{20}
3248	Butane, 3-chloro-2,2-dimethyl-	$C_6H_{13}Cl$ 120.62	5750-00-5 0.9	 111	4-01-00-00369 0.8767^{20}	eth 4 1.4182^{20}
3249	Butane, 1-(2-chloroethoxy)-	$C_6H_{13}ClO$ 136.62	10503-96-5	 154 dec; 50^{11}	4-01-00-01519 0.9335^{20}	eth 4 1.4155^{20}
3250	Butane, 1-chloro-4-fluoro- 1-Chloro-4-fluorobutane	C_4H_8ClF 110.56	462-73-7	 114.7	4-01-00-00249 1.0627^{25}	eth 4; EtOH 4 1.4020^{25}
3251	Butane, 1-chloro-2-methyl-, (±)- Butane, 1-chloro-2-methyl (*DL*)	$C_5H_{11}Cl$ 106.60	114180-21-1	 99.9	4-01-00-00324 0.8810^{15}	EtOH 3; eth 3 1.4102^{25}
3252	Butane, 1-chloro-3-methyl- Isoamyl chloride	$C_5H_{11}Cl$ 106.60	107-84-6 -104.4	4998 98.9	4-01-00-00325 0.8750^{20}	H_2O 2; EtOH 5; eth 5; chl 4 1.4084^{20}
3253	Butane, 2-chloro-2-methyl-	$C_5H_{11}Cl$ 106.60	594-36-5 -73.5	 85.6	4-01-00-00324 0.8653^{20}	H_2O 2; EtOH 3; eth 3; ctc 3 1.4055^{20}
3254	Butane, 2-chloro-2,3,3-trimethyl-	$C_7H_{15}Cl$ 134.65	918-07-0 136	 sub	4-01-00-00411	eth 4
3255	Butane, decafluoro- Perfluorobutane	C_4F_{10} 238.03	355-25-9 -128.2	 -1.9	4-01-00-00245 1.6484^{25}	bz 3; chl 3
3256	Butanedial	$C_4H_6O_2$ 86.09	638-37-9	 170 dec; 58^{9}	4-01-00-03642 1.064^{20}	H_2O 4; ace 4; eth 4; EtOH 4 1.4262^{18}
3257	Butanediamide Succinamide	$C_4H_8N_2O_2$ 116.12	110-14-5 268 dec	8837 sub 125	4-02-00-01922	
3258	1,4-Butanediamine Putrescine	$C_4H_{12}N_2$ 88.15	110-60-1 27.5	7964 158.5	4-04-00-01283 0.877^{25}	H_2O 3 1.4969^{20}
3259	1,4-Butanediamine, *N,N'*-bis(3-aminopropyl)- Spermine	$C_{10}H_{26}N_4$ 202.34	71-44-3 29	8699 150^{5}	4-04-00-01301	
3260	1,4-Butanediamine, dihydrochloride Butane, 1,4-diamino, dihydrochloride	$C_4H_{14}Cl_2N_2$ 161.07	333-93-7 280 dec	 sub	4-04-00-01283	H_2O 4; EtOH 4; eth 1; bz 1
3261	2,3-Butanediamine, (*R**,*R**)-(±)-	$C_4H_{12}N_2$ 88.15	20699-48-3 -22	 $57\text{-}8^{60}$	4-04-00-01302 0.8499^{25}	 1.4408^{25}
3262	1,4-Butanediamine, *N,N,N',N'*-tetraethyl- *N,N,N',N'*-Tetraethyl-1,4-butanediamine	$C_{12}H_{28}N_2$ 200.37	69704-44-5	 $85\text{-}6^{5}$	4-04-00-01286	H_2O 4; eth 4; EtOH 4 1.4383^{25}
3263	1,4-Butanediamine, *N,N,N',N'*-tetramethyl- *N,N,N',N'*-Tetramethyl-1,4-diaminobutane	$C_8H_{20}N_2$ 144.26	111-51-3	9157 168	4-04-00-01284 0.7942^{15}	H_2O 5; EtOH 3; eth 3 1.4621^{25}
3264	Butane, 1,1-dibromo- 1,1-Dibromobutane	$C_4H_8Br_2$ 215.92	62168-25-6	 91^{101}; 72^{43}	4-01-00-00266 1.80^{25}	 1.4980^{25}
3265	Butane, 1,2-dibromo- α-Butylene dibromide	$C_4H_8Br_2$ 215.92	533-98-2 -65.4	1565 166.3	4-01-00-00266 1.7915^{20}	H_2O 1; eth 3; chl 3 1.4025^{20}
3266	Butane, 1,3-dibromo- 1,3-Dibromobutane	$C_4H_8Br_2$ 215.92	107-80-2	 174	4-01-00-00266 1.800^{20}	H_2O 1; eth 3; ctc 2; chl 3 1.507^{20}
3267	Butane, 1,4-dibromo- 1,4-Dibromobutane	$C_4H_8Br_2$ 215.92	110-52-1 -16.5	 197	4-01-00-00267 1.7890^{20}	H_2O 1; ctc 2; chl 3 1.5190^{20}
3268	Butane, 2,3-dibromo- 2,3-Dibromobutane	$C_4H_8Br_2$ 215.92	5408-86-6 -24	 161	3-01-00-00296 1.7893^{22}	H_2O 1; eth 3 1.5133^{22}
3269	Butane, 1,2-dibromo-2,3-dimethyl- 1,2-Dibromo-2,3-dimethylbutane	$C_6H_{12}Br_2$ 243.97	29916-45-8	 80^{17}	3-01-00-00414 1.6033^{20}	 1.5105^{20}
3270	Butane, 1,4-dibromo-2,3-dimethyl- 1,4-Dibromo-2,3-dimethylbutane	$C_6H_{12}Br_2$ 243.97	54462-70-3	 $106\text{-}8^{22}$	4-01-00-00375 1.620^{20}	 1.5128^{20}
3271	Butane, 2,3-dibromo-2,3-dimethyl- 1,2-Dibromo-1,1,2,2-tetramethylethane	$C_6H_{12}Br_2$ 243.97	594-81-0 171 dec	 78^{10}	4-01-00-00375 1.8110^{21}	
3272	Butane, 1,2-dibromo-2-methyl- 1,2-Dibromo-2-methylbutane	$C_5H_{10}Br_2$ 229.94	10428-64-5	 $47\text{-}8^{9}$	4-01-00-00330 1.6711^{20}	 1.5088^{20}
3273	Butane, 1,4-dibromo-2-methyl-, (*R*)- Butane, 1,4-dibromo-2-methyl (*d*)	$C_5H_{10}Br_2$ 229.94	69498-28-8	 $78\text{-}9^{12}$	4-01-00-00330 1.695^{17}	 1.5128^{20}
3274	Butane, 2,3-dibromo-2-methyl- 1,2-Dibromo-1,1,2-trimethylethane	$C_5H_{10}Br_2$ 229.94	594-51-4 7	 $61\text{-}4^{17}$	4-01-00-00330 1.6717^{20}	 1.5729^{25}
3275	Butane, 1,4-dibromo-1,1,2,2,3,3,4,4-octafluoro- 1,4-Dibromooctafluorobutane	$C_4Br_2F_8$ 359.84	335-48-8	 97		
3276	Butane, 1,1-dichloro- Butylidene chloride	$C_4H_8Cl_2$ 127.01	541-33-3	1571 113.8	4-01-00-00250 1.0863^{20}	H_2O 1; chl 3 1.4355^{20}
3277	Butane, 1,2-dichloro- 1,2-Dichlorobutane	$C_4H_8Cl_2$ 127.01	616-21-7	 124.1	4-01-00-00250 1.1116^{25}	H_2O 1; eth 3; ctc 2; chl 3 1.4450^{20}
3278	Butane, 1,3-dichloro- 1,3-Dichlorobutane	$C_4H_8Cl_2$ 127.01	1190-22-3	 134	4-01-00-00250 1.1158^{20}	H_2O 1; eth 3; ctc 2; chl 3 1.4445^{20}

No.	Name	Synonym	Mol. Form.	Mol. Wt.	CAS RN	mp/°C	Merck No.	bp/°C	Beil. Ref.	den/g cm^{-3}	Solubility	n_D
3279	Butane, 1,4-dichloro-	1,4-Dichlorobutane	$C_4H_8Cl_2$	127.01	110-56-5	-37.3		161	4-01-00-00250	1.1408^{20}	chl 4	1.4542^{20}
3280	Butane, 2,2-dichloro-	2,2-Dichlorobutane	$C_4H_8Cl_2$	127.01	4279-22-5	-74		104	4-01-00-00251	1.1048^{25}	H_2O 1; chl 3	1.4295
3281	Butane, 2,3-dichloro-, (*R**,*R**)-(±)-		$C_4H_8Cl_2$	127.01	2211-67-8	-80		118; 53.2^{80}	4-01-00-00251	1.063^{25}		1.4409^{25}
3282	Butane, 1,1-dichloro-3-methyl-	1,1-Dichloro-3-methylbutane	$C_5H_{10}Cl_2$	141.04	625-66-1			130	4-01-00-00326	1.0473^{20}	eth 4; EtOH 4	1.4344^{20}
3283	Butane, 1,2-dichloro-2-methyl-	1,2-Dichloro-2-methylbutane	$C_5H_{10}Cl_2$	141.04	23010-04-0			134	3-01-00-00360	1.0785^{20}	eth 4; EtOH 4; chl 4	1.4432^{21}
3284	Butane, 1,3-dichloro-3-methyl-		$C_5H_{10}Cl_2$	141.04	624-96-4			146.7	4-01-00-00326	1.0654^{20}	eth 4; chl 4	1.4455^{20}
3285	Butane, 1,4-dichloro-2-methyl-	1,4-Dichloro-2-methylbutane	$C_5H_{10}Cl_2$	141.04	623-34-7			168.5	3-01-00-00360	1.1003^{25}	chl 4	1.4562^{21}
3286	Butane, 2,3-dichloro-2-methyl-	Amylene dichloride	$C_5H_{10}Cl_2$	141.04	507-45-9		645	129	4-01-00-00325	1.0696^{15}	eth 4; EtOH 4	1.4450^{18}
3287	Butane, 2,3-dichloro-1,1,1,2,3,4,4,4-octafluoro-		$C_4Cl_2F_8$	270.94	355-20-4	-68		63	4-01-00-00252	1.6801^{20}		1.3100^{20}
3288	Butane, 1,4-diiodo-		$C_4H_8I_2$	309.92	628-21-7	5.8		125^{15} dec	4-01-00-00276	2.3659^{15}	H_2O 1; ctc 2; os 3	1.6239^{15}
3289	Butane, 2,2-dimethyl-	Neohexane	C_6H_{14}	86.18	75-83-2	-99		49.7	4-01-00-00367	0.6444^{25}	H_2O 1; EtOH 3; eth 3; ace 4	1.3688^{20}
3290	Butane, 2,3-dimethyl-	2,3-Dimethylbutane	C_6H_{14}	86.18	79-29-8	-128.8		57.9	4-01-00-00371	0.6616^{20}	H_2O 1; EtOH 3; eth 3; ace 4	1.3750^{20}
3291	Butanedinitrile	Succinonitrile	$C_4H_4N_2$	80.09	110-61-2	54.5	8843	266	4-02-00-01923	0.9867^{60}	H_2O 4; EtOH 3; eth 2; ace 3	1.4173^{60}
3292	Butanedinitrile, tetramethyl-	Tetramethylsuccinonitrile	$C_8H_{12}N_2$	136.20	3333-52-6	170.5			4-02-00-02054	1.070^{25}	EtOH 3	
3293	Butane, 1,4-dinitro-	1,4-Dinitrobutane	$C_4H_8N_2O_4$	148.12	4286-49-1	33.5		176^{13}	4-01-00-00280		H_2O 1; EtOH 2; eth 3; bz 3	
3294	Butanedioic acid	Succinic acid	$C_4H_6O_4$	118.09	110-15-6	188	8840	235 dec	4-02-00-01908	1.572^{25}	H_2O 2; EtOH 3; eth 3; ace 3	1.450
3295	Butanedioic acid, acetyl-, diethyl ester		$C_{10}H_{16}O_5$	216.23	1115-30-6			255; 133^{17}	4-03-00-01826	1.081^{20}	H_2O 1; EtOH 3; eth 3; bz 3	1.4346^{20}
3296	Butanedioic acid, bis(1,1-dimethylethyl) ester	Di-*tert*-butyl succinate	$C_{12}H_{22}O_4$	230.30	926-26-1	36.5	3024	$109\text{-}10^9$	4-02-00-01916			
3297	Butanedioic acid, bis(1-methylpropyl) ester	Di-*sec*-butyl succinate	$C_{12}H_{22}O_4$	230.30	626-31-3			256	3-02-00-01665	0.9735^{20}	bz 4; eth 4; EtOH 4	1.4238^{25}
3298	Butanedioic acid, bis(phenylmethyl) ester		$C_{18}H_{18}O_4$	298.34	103-43-5	49.5		245^{15}	4-06-00-02271	1.256^{25}	H_2O 1; EtOH 3; eth 3; bz 3	1.596
3299	Butanedioic acid, bromo-, (±)-		$C_4H_5BrO_4$	196.99	584-98-5	161			4-02-00-01929	2.073^{25}	H_2O 3; EtOH 3; HOAc 2	
3300	Butanedioic acid, bromo-	Bromosuccinic acid	$C_4H_5BrO_4$	196.99	923-06-8	161	1427		1-02-00-00268	2.073^{25}	EtOH 4	
3301	Butanedioic acid, dibutyl ester	Dibutyl succinate	$C_{12}H_{22}O_4$	230.30	141-03-7	-29.2	3023	274.5	4-02-00-01916	0.9752^{20}	H_2O 1; EtOH 3; eth 3; bz 3	1.4299^{20}
3302	Butanedioic acid, 2,3-dichloro-, (±)-	Succinic acid, 2,3-dichloro-, (±)-	$C_4H_4Cl_2O_4$	186.98	1114-09-6	175 dec			4-02-00-01928	1.844^{15}	ace 4; eth 4; chl 4	
3303	Butanedioic acid, diethyl ester	Diethyl succinate	$C_8H_{14}O_4$	174.20	123-25-1	-21	8840	217.7	4-02-00-01914	1.0402^{20}	H_2O 1; EtOH 5; eth 5; ace 3	1.4201^{20}
3304	Butanedioic acid, 2,3-dihydroxy-, (*R**,*S**)-	*meso*-Tartaric acid	$C_4H_6O_6$	150.09	147-73-9	147	9040		4-03-00-01218	1.666^{20}	H_2O 4; EtOH 4	
3305	Butanedioic acid, 2,3-dihydroxy-, bis(1-methylethyl) ester, (*R**,*R**)-(±)-	Tartaric acid, diisopropyl ester (*DL*)	$C_{10}H_{18}O_6$	234.25	58167-01-4	34		275; 154^{12}	2-03-00-00337	1.1166^{20}	ace 4; eth 4; EtOH 4	
3306	Butanedioic acid, 2,3-dihydroxy- [*R*-(*R**,*R**)]-, bis(phenylmethyl) ester		$C_{18}H_{18}O_6$	330.34	622-00-4	50		$250\text{-}70^4$	3-06-00-01537	1.2036^{72}	py 4; EtOH 4	
3307	Butanedioic acid, 2,3-dihydroxy-, diethyl ester, (*R**,*R**)-(±)-	Diethyl *DL*-tartrate	$C_8H_{14}O_6$	206.20	57968-71-5	18.7		281; 158^{14}	4-03-00-01232	1.2046^{20}	H_2O 2; EtOH 5; eth 5; ace 3	1.4438^{20}
3308	Butanedioic acid, 2,3-dihydroxy-, dimethyl ester, (*R**,*R**)-(±)-		$C_6H_{10}O_6$	178.14	608-69-5	90		280	4-03-00-01232	1.2604^{90}	H_2O 4; ace 4; eth 4; EtOH 4	
3309	Butanedioic acid, 2,3-dihydroxy- [*R*-(*R**,*R**)]-	*L*-Tartaric acid	$C_4H_6O_6$	150.09	87-69-4	169	9039		4-03-00-01219			
3310	Butanedioic acid, 2,3-dihydroxy- [*R*-(*R**,*R**)]-, dipropyl ester	Tartaric acid, dipropyl ester, (+)-	$C_{10}H_{18}O_6$	234.25	2217-14-3			303	2-03-00-00337	1.1390^{20}	H_2O 4; ace 4; eth 4; EtOH 4	
3311	Butanedioic acid, 2,3-dihydroxy-, (*R**,*R**)-(±)-	*DL*-Tartaric acid	$C_4H_6O_6$	150.09	133-37-9	206	9038		4-03-00-01229	1.788^{25}	H_2O 3; EtOH 3; eth 2; bz 1	
3312	Butanedioic acid, 2,3-dihydroxy-, [*S*-(*R**,*R**)]-	*D*-Tartaric acid	$C_4H_6O_6$	150.09	147-71-7	172.5	9037		4-03-00-01229	1.7598^{20}	DMSO 2	1.4955^{20}
3313	Butanedioic acid, 2,3-dihydroxy- [*R*-(*R**,*R**)]-, dibutyl ester		$C_{12}H_{22}O_6$	262.30	87-92-3	22		320	4-03-00-01232	1.0909^{20}	H_2O 4; ace 4; EtOH 4	1.4451^{20}
3314	Butanedioic acid, 2,3-dihydroxy- [*R*-(*R**,*R**)]-, disodium salt	Tartaric acid, disodium salt	$C_4H_4Na_2O_6$	194.05	868-18-8		8640				H_2O 3	

No.	Name Synonym	Mol. Form. Mol. Wt.	CAS RN mp/°C	Merck No. bp/°C	Beil. Ref. den/g cm^{-3}	Solubility n_D
3315	Butanedioic acid, 2,3-dihydroxy- [*R*-(*R**,*R**)]-, monoethyl ester Tartaric acid, ethyl ester	$C_6H_{10}O_6$ 178.14	608-89-9 90	3812	3-03-00-01028	H_2O 4; EtOH 4
3316	Butanedioic acid, [(dimethoxyphosphinothioyl)thio]-, diethyl ester Malathion	$C_{10}H_{19}O_6PS_2$ 330.36	121-75-5 2.8	5582 $156^{0.7}$ dec	4-03-00-01136 1.2076^{20}	H_2O 2; EtOH 3; eth 3; bz 3 1.4960^{20}
3317	Butanedioic acid, 2,3-dimethyl-	$C_6H_{10}O_4$ 146.14	13545-04-5 132 dec	 $87\text{-}90^{2}$	4-02-00-01998	ace 4; eth 4
3318	Butanedioic acid, dimethyl ester	$C_6H_{10}O_4$ 146.14	106-65-0 19	 196.4	4-02-00-01913 1.1198^{20}	H_2O 2; EtOH 3; eth 4; ace 3 1.4197^{20}
3319	Butanedioic acid, dipentyl ester	$C_{14}H_{26}O_4$ 258.36	645-69-2 -10.8	 171.5^{16}	3-02-00-01665 0.9616^{20}	
3320	Butanedioic acid, diphenyl ester Diphenyl succinate	$C_{16}H_{14}O_4$ 270.28	621-14-7 121	 330; 222.5^{15}	3-06-00-00605	H_2O 1; EtOH 3; eth 3; ace 3
3321	Butanedioic acid, dipropyl ester	$C_{10}H_{18}O_4$ 202.25	925-15-5 -5.9	 250.8	4-02-00-01916 1.0020^{20}	ace 4; bz 4; eth 4 1.4250^{20}
3322	Butanedioic acid, ethyl-, (*R*)-	$C_6H_{10}O_4$ 146.14	4074-24-2 96	 181.5	4-02-00-01995 1.0017^{20}	H_2O 4; eth 4; EtOH 4
3323	Butanedioic acid, ethyl methyl ester	$C_7H_{12}O_4$ 160.17	627-73-6 <-20	 208.2	0-02-00-00609 1.076^{20}	eth 4; EtOH 4
3324	Butanedioic acid, formyl-, diethyl ester	$C_9H_{14}O_5$ 202.21	5472-38-8	 $130\text{-}4^{15}$	4-03-00-01819	eth 4; EtOH 4 1.4486^{25}
3325	Butanedioic acid, methyl-, (±)-	$C_5H_8O_4$ 132.12	636-60-2 115	 dec	4-02-00-01948 1.4200^{0}	H_2O 4; EtOH 4; eth 3; chl 2 1.4303
3326	Butanedioic acid, methyl-, diethyl ester	$C_9H_{16}O_4$ 188.22	4676-51-1	 217.5	3-02-00-01696 1.012^{25}	 1.4199^{20}
3327	Butanedioic acid, methyl-, dimethyl ester	$C_7H_{12}O_4$ 160.17	1604-11-1	 196	3-02-00-01696 1.076^{25}	 1.4200^{20}
3328	Butanedioic acid, methylene- Itaconic acid	$C_5H_6O_4$ 130.10	97-65-4 175	5130 dec	4-02-00-02228 1.632^{25}	H_2O 3; EtOH 3; eth 2; ace 3
3329	Butanedioic acid, methylene-, diethyl ester	$C_9H_{14}O_4$ 186.21	2409-52-1 58.5	 228	4-02-00-02230 1.0467^{20}	EtOH 5; eth 3; ace 4; bz 3 1.4377^{20}
3330	Butanedioic acid, methylene-, dimethyl ester	$C_7H_{10}O_4$ 158.15	617-52-7 38	 208	4-02-00-02229 1.1241^{18}	EtOH 3; eth 3; ace 4; MeOH 3 1.4457^{20}
3331	Butanedioic acid, (1-methylethylidene)-	$C_7H_{10}O_4$ 158.15	584-27-0 100.5	 174 dec	4-02-00-02251	eth 4; EtOH 4
3332	Butanedioic acid, (1-methylpropyl)-, diethyl ester Succinic acid, *sec*-butyl-, diethyl ester	$C_{12}H_{22}O_4$ 230.30	69248-35-7	 $75\text{-}6^{0.2}$	3-02-00-01779 0.9745^{25}	 1.4293^{25}
3333	Butanedioic acid, monobutyl ester	$C_8H_{14}O_4$ 174.20	5150-93-6 8.6	 136.5^{3}	3-02-00-01665 1.0732^{20}	 1.4360^{20}
3334	Butanedioic acid, mono(2,2-dimethylhydrazide) Daminozide	$C_6H_{12}N_2O_3$ 160.17	1596-84-5 154.5	2810		
3335	Butanedioic acid, monoethyl ester Succinic acid, ethyl ester	$C_6H_{10}O_4$ 146.14	1070-34-4 8	 172^{42}; 119^{3}	4-02-00-01914 1.1466^{20}	H_2O 4; eth 4; EtOH 4 1.4327^{20}
3336	Butanedioic acid, monomethyl ester	$C_5H_8O_4$ 132.12	3878-55-5 58	 151^{20}; 122^{4}	4-02-00-01913	H_2O 3
3337	Butanedioic acid, monopentyl ester Succinic acid, pentyl ester	$C_9H_{16}O_4$ 188.22	97479-78-2 17.2	 147^{3}	2-02-00-00551 1.0460^{20}	 1.4378^{20}
3338	Butanedioic acid, monopropyl ester Succinic acid, monopropyl ester	$C_7H_{12}O_4$ 160.17	6946-88-9 15	 126^{3}	3-02-00-01664 1.1071^{20}	 1.4343^{20}
3339	Butanedioic acid, oxo- Oxalacetic acid	$C_4H_4O_5$ 132.07	328-42-7 161 dec	6863	4-03-00-01808	
3340	Butanedioic acid, oxo-, diethyl ester Diethyl oxalacetate	$C_8H_{12}O_5$ 188.18	108-56-5	3791 $131\text{-}2^{24}$	4-03-00-01809 1.131^{20}	H_2O 1; EtOH 5; eth 5; ace 4 1.4561^{17}
3341	Butanedioic acid, phenyl-, (±)-	$C_{10}H_{10}O_4$ 194.19	10424-29-0 168	 dec	4-09-00-03352	H_2O 2; EtOH 4; eth 4; ace 4
3342	Butanedioic acid, tetrahydroxy- Dihydroxytartaric acid	$C_4H_6O_8$ 182.09	76-30-2 114.5	3172	3-03-00-01419	
3343	Butanedioic acid, tetramethyl-	$C_8H_{14}O_4$ 174.20	630-51-3 200	 sub	4-02-00-02054 1.30^{25}	bz 4; EtOH 4; chl 4
3344	1,2-Butanediol, (±)-	$C_4H_{10}O_2$ 90.12	26171-83-5	 190.5	4-01-00-02507 1.0024^{20}	H_2O 3; EtOH 3; ace 3 1.4378^{20}
3345	1,3-Butanediol 1,3-Butylene glycol	$C_4H_{10}O_2$ 90.12	107-88-0 <-50	1500 207.5	4-01-00-02477 1.0053^{20}	 1.4401^{20}
3346	1,4-Butanediol Tetramethylene glycol	$C_4H_{10}O_2$ 90.12	110-63-4 20.1	 235	4-01-00-02515 1.0171^{20}	H_2O 5; EtOH 3; eth 2; DMSO 2 1.4460^{20}
3347	2,3-Butanediol, (*R**,*R**)-(±)-	$C_4H_{10}O_2$ 90.12	6982-25-8 7.6	 182.5	4-01-00-02525 1.0033^{20}	H_2O 5; EtOH 5; eth 3; ace 3 1.4310^{25}
3348	1,4-Butanediol, diacetate	$C_8H_{14}O_4$ 174.20	628-67-1 12	 229	4-02-00-00224 1.0479^{15}	 1.4251^{15}
3349	2,3-Butanediol, 1,4-dichloro- 1,4-Dichloro-2,3-butanediol	$C_4H_8Cl_2O_2$ 159.01	2419-73-0 126.5	 150^{30}	4-01-00-02531	EtOH 4
3350	2,3-Butanediol, 2,3-dimethyl- Pinacol	$C_6H_{14}O_2$ 118.18	76-09-5 43	7408 174.4	4-01-00-02575	H_2O 2; EtOH 4; eth 4; CS_2 2
3351	2,3-Butanediol, 2,3-dimethyl-, hexahydrate	$C_6H_{26}O_8$ 226.27	6091-58-3 45.4	 172	2-01-00-00553 0.967^{15}	eth 4; EtOH 4

No.	Name	Synonym	Mol. Form.	Mol. Wt.	CAS RN	mp/°C	Merck No.	bp/°C	Beil. Ref.	den/g cm^{-3}	Solubility	n_D
3352	1,2-Butanediol, 3-methyl-	3-Methyl-1,2-butanediol	$C_5H_{12}O_2$	104.15	50468-22-9			206	4-01-00-02549	0.9987^{0}	eth 4; EtOH 4	
3353	1,3-Butanediol, 2-methyl-	2-Methyl-1,3-butanediol	$C_5H_{12}O_2$	104.15	684-84-4			200; 115^{20}	4-01-00-02546	0.9912^{20}	H_2O 3; EtOH 3; eth 3	
3354	1,3-Butanediol, 3-methyl-		$C_5H_{12}O_2$	104.15	2568-33-4			202.5	4-01-00-02549	0.9448^{20}	H_2O 3; EtOH 3	1.4452^{20}
3355	2,3-Butanediol, 2-methyl-	2,3-Dihydroxy-2-methylbutane	$C_5H_{12}O_2$	104.15	5396-58-7			175	4-01-00-02547	0.9920^{25}	H_2O 4; eth 4; EtOH 4	1.4375^{20}
3356	1,2-Butanediol, 2-phenyl-	2-Phenyl-1,2-butanediol	$C_{10}H_{14}O_2$	166.22	90925-48-7	56		165^{23}	3-06-00-04667		H_2O 4; eth 4; EtOH 4	
3357	1,3-Butanediol, 1-phenyl-	1-Phenylbutane-1,3-diol	$C_{10}H_{14}O_2$	166.22	65469-88-7	73.5		175^{21}; 130^{2}	4-06-00-06006	1.0720^{25}	bz 4; EtOH 4; lig 4	
3358	1,3-Butanediol, 3-phenyl-		$C_{10}H_{14}O_2$	166.22	7133-68-8			150^{2}; $109^{0.5}$	4-06-00-06009	1.0865^{20}		1.5341^{20}
3359	2,3-Butanedione	Diacetyl	$C_4H_6O_2$	86.09	431-03-8	-2.4	2946	88	4-01-00-03644	0.9808^{18}	H_2O 4; EtOH 5; eth 5; ace 4	1.3951^{20}
3360	2,3-Butanedione, dioxime	Dimethylglyoxime	$C_4H_8N_2O_2$	116.12	95-45-4	245.5	3235	sub 234	3-01-00-03105		H_2O 1; EtOH 4; eth 4; bz 2	
3361	2,3-Butanedione, monooxime		$C_4H_7NO_2$	101.11	57-71-6	76.8		185.5	4-01-00-03646		H_2O 2; EtOH 4; eth 4; chl 4	
3362	1,3-Butanedione, 1-phenyl-		$C_{10}H_{10}O_2$	162.19	93-91-4	56		261.5	4-07-00-02151	1.0599^{74}	H_2O 1; eth 3; chl 2	1.5678^{78}
3363	1,3-Butanedione, 1-(2-pyridinyl)-		$C_9H_9NO_2$	163.18	40614-52-6	50		140^{15}	4-21-00-04750		bz 4; eth 4; EtOH 4; chl 4	
3364	1,3-Butanedione, 1-(3-pyridinyl)-		$C_9H_9NO_2$	163.18	3594-37-4	85		171^{15}	5-21-10-00180		H_2O 2; EtOH 3; eth 3; ace 3	
3365	1,3-Butanedione, 1-(4-pyridinyl)-		$C_9H_9NO_2$	163.18	75055-73-1	62		145^{18}	5-21-10-00180		bz 4; eth 4; EtOH 4	
3366	Butanedioyl dichloride	Succinyl chloride	$C_4H_4Cl_2O_2$	154.98	543-20-4	20	8844	193.3	4-02-00-01921	1.3748^{20}	eth 3; ace 3; bz 3	1.4683^{20}
3367	Butanedioyl dichloride, methylene-		$C_5H_4Cl_2O_2$	166.99	1931-60-8			89^{17}	3-02-00-01934		ace 3	1.4919^{20}
3368	1,4-Butanedithiol	Tetramethylenedithiol	$C_4H_{10}S_2$	122.26	1191-08-8	-53.9		195.5	4-01-00-02523	1.0021^{0}	H_2O 1; EtOH 4; ctc 2	1.5290^{20}
3369	Butane, 1-(ethenyloxy)-	Butyl vinyl ether	$C_6H_{12}O$	100.16	111-34-2	-92		94	4-01-00-02052	0.7888^{20}	H_2O 1; EtOH 4; eth 5; ace 4	1.4026^{20}
3370	Butane, 2-(ethenyloxy)-		$C_6H_{12}O$	100.16	1888-85-3			81	4-01-00-02054	0.7715^{20}		1.3970^{20}
3371	Butane, 1-(ethenyloxy)-3-methyl-	Isopentyl vinyl ether	$C_7H_{14}O$	114.19	39782-38-2			112.5	4-01-00-02055	0.7826^{20}	eth 4; EtOH 4	1.4072^{20}
3372	Butane, 1-ethoxy-	Butyl ethyl ether	$C_6H_{14}O$	102.18	628-81-9	-124		92.3	4-01-00-01518	0.7495^{20}	H_2O 1; EtOH 5; eth 5; ace 4	1.3818^{20}
3373	Butane, 2-ethoxy-	*sec*-Butyl ethyl ether	$C_6H_{14}O$	102.18	2679-87-0			81	4-01-00-01572	0.7503^{20}	H_2O 1; EtOH 4; eth 4	1.3802^{20}
3374	Butane, 1-ethoxy-3-methyl-	Ethyl isopentyl ether	$C_7H_{16}O$	116.20	628-04-6			112.5	4-01-00-01681	0.7688^{21}	eth 4; EtOH 4	
3375	Butane, 2-ethoxy-2-methyl-	Ethyl *tert*-amyl ether	$C_7H_{16}O$	116.20	919-94-8			102	3-01-00-01626	0.751^{18}	eth 4; EtOH 4	
3376	Butane, 1-(ethylthio)-	Butyl ethyl sulfide	$C_6H_{14}S$	118.24	638-46-0	-95.1		144.3	4-01-00-01558	0.8376^{20}	EtOH 4; chl 3	1.4492^{10}
3377	Butane, 1-(ethynyloxy)-		$C_6H_{10}O$	98.14	3329-56-4			104	4-01-00-02213	0.8200^{20}	eth 4; EtOH 4	1.4067
3378	Butane, 1-fluoro-	Butyl fluoride	C_4H_9F	76.11	2366-52-1	-134		32.5	4-01-00-00244	0.7789^{20}	EtOH 4	1.3396^{20}
3379	Butane, 2-fluoro-	*sec*-Butyl fluoride	C_4H_9F	76.11	359-01-3	-121.4		25.1	4-01-00-00244	0.7559^{25}		
3380	Butane, 1,1,2,2,3,4,4-heptachloro-		$C_4H_3Cl_7$	299.24	34973-41-6			$137.5^{13.5}$	4-01-00-00257	1.739^{20}	ace 4; ctc 4	1.5407^{20}
3381	Butane, 1-iodo-	Butyl iodide	C_4H_9I	184.02	542-69-8	-103	1572	130.6	4-01-00-00271	1.6154^{20}	H_2O 1; EtOH 5; eth 5; chl 4	1.5001^{20}
3382	Butane, 2-iodo-, (±)-	Butane, 2-iodo (*DL*)	C_4H_9I	184.02	52152-71-3	-104.2	1573	120	4-01-00-00272	1.5920^{20}	H_2O 1; EtOH 5; eth 5; chl 4	1.4991^{20}
3383	Butane, 1-iodo-2-methyl-, (*S*)-	Butane, 1-iodo-2-methyl (*d*)	$C_5H_{11}I$	198.05	29394-58-9			148; 132.6^{500}	3-01-00-00366	1.5253^{20}	eth 4; EtOH 4	1.4977^{20}
3384	Butane, 1-iodo-3-methyl-	Isoamyl iodide	$C_5H_{11}I$	198.05	541-28-6		5002	147	4-01-00-00331	1.5118^{20}	H_2O 2; EtOH 5; eth 5; ctc 2	1.4939^{20}
3385	Butane, 2-iodo-2-methyl-		$C_5H_{11}I$	198.05	594-38-7			124.5	4-01-00-00331	1.4937^{20}	H_2O 1; EtOH 5; eth 5	1.4981^{20}
3386	Butane, 1-isocyanato-3-methyl-		$C_6H_{11}NO$	113.16	1611-65-0			137	0-04-00-00186	0.806^{20}	H_2O 1; EtOH 5; eth 5	1.406^{20}
3387	Butane, 1-isocyano-	Butyl isocyanide	C_5H_9N	83.13	2769-64-4			120	4-04-00-00562	0.78^{20}	eth 4; EtOH 4	
3388	Butane, 1-isothiocyanato-		C_5H_9NS	115.20	592-82-5			168	4-04-00-00596	0.9546^{20}	eth 4; EtOH 4	1.501^{20}
3389	Butane, 2-isothiocyanato-, (±)-	*sec*-Butyl isothiocyanate (*DL*)	C_5H_9NS	115.20	116724-11-9			159.5	4-04-00-00624	0.944^{12}	eth 4; EtOH 4	
3390	Butane, 1-isothiocyanato-3-methyl-		$C_6H_{11}NS$	129.23	628-03-5			183	4-04-00-00707	0.9419^{17}	eth 4; EtOH 4	
3391	Butane, 1-methoxy-	Butyl methyl ether	$C_5H_{12}O$	88.15	628-28-4	-115.5		70.1	4-01-00-01518	0.7443^{20}	H_2O 1; EtOH 5; eth 5; ace 3	1.3736^{20}

No.	Name Synonym	Mol. Form. Mol. Wt.	CAS RN mp/°C	Merck No. bp/°C	Beil. Ref. den/g cm^{-3}	Solubility n_D
3392	Butane, 2-methoxy-, (*S*)- *sec*-Butyl methyl ether (*d*)	$C_5H_{12}O$ 88.15	66610-39-7	 60.7	4-01-00-01571 0.7400^{25}	
3393	Butane, 2-methoxy-, (±)- *sec*-Butyl methyl ether (*DL*)	$C_5H_{12}O$ 88.15	116783-23-4	 60	4-01-00-01571 0.7415^{20}	ace 4; eth 4; EtOH 4 1.3680^{25}
3394	Butane, 1-methoxy-2-methyl- Methyl (2-methylbutyl) ether	$C_6H_{14}O$ 102.18	62016-48-2	 90	4-01-00-01667 0.754^{18}	 1.3849^{20}
3395	Butane, 1-methoxy-3-methyl-	$C_6H_{14}O$ 102.18	626-91-5	 91	4-01-00-01681 0.7517^{20}	eth 4; EtOH 4 1.3830^{20}
3396	Butane, 2-methoxy-2-methyl-	$C_6H_{14}O$ 102.18	994-05-8	 86.3	4-01-00-01671 0.7703^{20}	eth 4; EtOH 4 1.3855^{20}
3397	Butane, 2-methyl- Isopentane	C_5H_{12} 72.15	78-78-4 -159.9	 27.8	4-01-00-00320 0.6201^{20}	H_2O 1; EtOH 5; eth 5 1.3537^{20}
3398	Butane, 1,1'-[methylenebis(oxy)]bis-	$C_9H_{20}O_2$ 160.26	2568-90-3 -58.1	 179.2	4-01-00-03029 0.8339^{20}	 1.4072^{17}
3399	Butane, 1-(1-methylethoxy)- Butyl isopropyl ether	$C_7H_{16}O$ 116.20	1860-27-1	 109	4-01-00-01519 0.7594^{15}	H_2O 1; EtOH 3; eth 3; ace 3 1.3870^{15}
3400	Butane, 3-methyl-1-nitro- 3-Methyl-1-nitrobutane	$C_5H_{11}NO_2$ 117.15	627-67-8 97	 163.5; 72^{27}	4-01-00-00332 0.9458^{25}	
3401	Butane, 1-(2-methylpropoxy)- Butyl isobutyl ether	$C_8H_{18}O$ 130.23	17071-47-5	 151	4-01-00-01594 0.7980^{22}	ace 4; eth 4; EtOH 4 1.4077^{21}
3402	Butane, 1-(methylthio)- Butyl methyl sulfide	$C_5H_{12}S$ 104.22	628-29-5 -97.8	 123.5	4-01-00-01557 0.8426^{20}	EtOH 4; MeOH 4 1.4477^{20}
3403	Butanenitrile Propyl cyanide	C_4H_7N 69.11	109-74-0 -111.9	1597 117.6	4-02-00-00806 0.7936^{20}	H_2O 2; EtOH 5; eth 5; bz 3 1.3842^{20}
3404	Butanenitrile, 4-bromo-	C_4H_6BrN 148.00	5332-06-9	 206	4-02-00-00836 1.4967^{20}	EtOH 3; eth 3; chl 3 1.4818^{20}
3405	Butanenitrile, 4-chloro-	C_4H_6ClN 103.55	628-20-6	 192	4-02-00-00827 1.0934^{15}	H_2O 1; EtOH 3; eth 3; ctc 2 1.4413^{20}
3406	Butanenitrile, 4,4-diethoxy-	$C_8H_{15}NO_2$ 157.21	18381-45-8	 104-6^{10}	3-03-00-01210 0.937^{25}	H_2O 2; EtOH 5; eth 5 1.4186^{20}
3407	Butanenitrile, 2-ethyl-	$C_6H_{11}N$ 97.16	617-80-1	 145.5	4-02-00-00953	eth 4; EtOH 4 1.3891^{24}
3408	Butanenitrile, 2-hydroxy-3-methyl-2-(1-methylethyl)- Diisopropyl ketone, cyanohydrin	$C_8H_{15}NO$ 141.21	4390-75-4 59	 111^{18}	4-03-00-00884	ace 4; bz 4; eth 4; EtOH 4
3409	Butanenitrile, 2-methyl- 2-Methylbutanenitrile	C_5H_9N 83.13	18936-17-9	 125	4-02-00-00892 0.7913^{15}	eth 4; EtOH 4 1.3933^{20}
3410	Butanenitrile, 3-methyl- Isobutyl cyanide	C_5H_9N 83.13	625-28-5 -101	 127.5	4-02-00-00902 0.7914^{20}	H_2O 2; EtOH 5; eth 5; ace 4 1.3927^{20}
3411	Butanenitrile, 4-phenoxy-	$C_{10}H_{11}NO$ 161.20	2243-43-8 45.5	 288	4-06-00-00646	eth 3
3412	Butane, 1-nitro- 1-Nitrobutane	$C_4H_9NO_2$ 103.12	627-05-4	 153	4-01-00-00277 0.970^{25}	H_2O 2; EtOH 5; eth 5; alk 3 1.4303^{20}
3413	Butane, 2-nitro-, (±)-	$C_4H_9NO_2$ 103.12	116781-85-2 -132	 140	4-01-00-00278 0.9854^{17}	chl 3 1.4044^{20}
3414	Butane, 1,1,1,2,3,4,4,4-octafluoro-2,3-bis(trifluoromethyl)- Perfluoro-2,3-dimethylbutane	C_6F_{14} 338.04	354-96-1 -15	 59.8		
3415	Butane, 1,1'-oxybis- Dibutyl ether	$C_8H_{18}O$ 130.23	142-96-1 -95.2	1568 140.2	4-01-00-01520 0.7684^{20}	H_2O 1; EtOH 5; eth 5; ace 4 1.3992^{20}
3416	Butane, 2,2'-oxybis-, (±)- *sec*-Butyl ether, (±)-	$C_8H_{18}O$ 130.23	17226-28-7	 120.5	3-01-00-01533 0.756^{25}	ace 4; eth 4; EtOH 4 1.393^{25}
3417	Butane, 1,1'-[oxybis(2,1-ethanediyloxy)]bis-	$C_{12}H_{26}O_3$ 218.34	112-73-2 -60	 256	4-01-00-02395 0.885^{25}	 1.4235^{20}
3418	Butane, 1,1'-oxybis[3-methyl- Diisoamyl ether	$C_{10}H_{22}O$ 158.28	544-01-4	5000 172.5	4-01-00-01682 0.7777^{20}	ace 4; EtOH 4; chl 4 1.4085^{20}
3419	Butane, 1,1,2,3,4-pentachloro- Butane, 1,1,2,3,4-pentachloro (liquid)	$C_4H_5Cl_5$ 230.35	77753-24-3	 95.5^{11}	4-01-00-00256 1.561^{18}	EtOH 4; ctc 4 1.5141^{18}
3420	Butane, 1,2,2,3,4-pentachloro- 1,2,2,3,4-Pentachlorobutane	$C_4H_5Cl_5$ 230.35	2431-52-9	 85^{10}	4-01-00-00256 1.5543^{20}	ace 4; chl 4 1.5157^{20}
3421	Butane, 1-propoxy- Butyl propyl ether	$C_7H_{16}O$ 116.20	3073-92-5	 118.1	4-01-00-01519 0.7772^{0}	H_2O 1; EtOH 4; eth 4
3422	Butane, 1,1'-sulfinylbis-	$C_8H_{18}OS$ 162.30	2168-93-6 32.6	 dec	4-01-00-01561 0.8317^{23}	H_2O 1; EtOH 3; eth 3 1.4669^{20}
3423	Butane, 1,1'-sulfonylbis-	$C_8H_{18}O_2S$ 178.30	598-04-9 45	 291	4-01-00-01561 0.9885^{47}	H_2O 1; EtOH 3; eth 3
3424	Butane, 1,1,4,4-tetrabromo-	$C_4H_6Br_4$ 373.71	116779-77-2	 138-45^{10}	3-01-00-00298 2.529^{20}	bz 4; eth 4; chl 4 1.6077^{20}
3425	Butane, 1,2,2,3-tetrabromo-	$C_4H_6Br_4$ 373.71	116779-78-3 -2	 97.5^{7}	3-01-00-00298 2.510^{20}	
3426	Butane, 1,2,3,4-tetrabromo-, *DL*- Butane, 1,2,3,4-tetrabromo (*DL*)	$C_4H_6Br_4$ 373.71	2657-65-0 118	 260; 180^{60}	4-01-00-00271	ace 4; eth 4; EtOH 4; lig 4
3427	Butane, 1,1,1,2-tetrachloro- 1,1,1,2-Tetrachlorobutane	$C_4H_6Cl_4$ 195.90	39966-95-5	 135	4-01-00-00253 1.3907^{20}	 1.4920^{25}
3428	Butane, 1,2,2,3-tetrachloro- 1,2,2,3-Tetrachlorobutane	$C_4H_6Cl_4$ 195.90	79630-70-9 -48	 182	3-01-00-00286 1.4276^{18}	ace 4; chl 4 1.491^{20}
3429	Butane, 1,2,3,3-tetrachloro- 1,2,3,3-Tetrachlorobutane	$C_4H_6Cl_4$ 195.90	13138-51-7	 90^{32}; 56^{10}	4-01-00-00254 1.4204^{20}	ace 4; eth 4; EtOH 4 1.4958^{20}
3430	Butane, 2,2,3,3-tetramethyl- 2,2,3,3-Tetramethylbutane	C_8H_{18} 114.23	594-82-1 100.7	 106.4	4-01-00-00447 0.8242^{20}	H_2O 1; eth 3; chl 3 1.4695^{20}

No.	Name Synonym	Mol. Form. Mol. Wt.	CAS RN mp/°C	Merck No. bp/°C	Beil. Ref. den/g cm^{-3}	Solubility n_D
3431	1,2,3,4-Butanetetrol, (*R**,*S**)- Erythritol	$C_4H_{10}O_4$ 122.12	149-32-6 121.5	3620 330.5	4-01-00-02807 1.451^{20}	H_2O 3; eth 1; bz 1
3432	1,2,3,4-Butanetetrol, tetranitrate, (*R**,*S**)- Erythrityl tetranitrate	$C_4H_6N_4O_{12}$ 302.11	7297-25-8 61	3622	4-01-00-02809	EtOH 4
3433	Butane, 1,1'-thiobis- Dibutyl sulfide	$C_8H_{18}S$ 146.30	544-40-1 -79.7	1590 185	4-01-00-01559 0.8386^{20}	eth 4; EtOH 4; chl 4 1.4530^{20}
3434	Butane, 2,2'-thiobis-	$C_8H_{18}S$ 146.30	626-26-6	 165	4-01-00-01586 0.8348^{20}	H_2O 1; EtOH 4; eth 4 1.4506^{20}
3435	Butane, 1,1'-thiobis[2-methyl- Bis(2-methylbutyl) sulfide	$C_{10}H_{22}S$ 174.35	96034-00-3	 165; 95^{13}	0-01-00-00387 0.8348^{20}	eth 4; EtOH 4 1.4506^{20}
3436	Butane, 1,1'-thiobis[3-methyl- Isoamyl sulfide	$C_{10}H_{22}S$ 174.35	544-02-5 -74.6	 211	4-01-00-01689 0.8323^{20}	H_2O 1; EtOH 5; eth 4 1.4520^{20}
3437	1-Butanethiol Butyl mercaptan	$C_4H_{10}S$ 90.19	109-79-5 -115.7	1575 98.5	4-01-00-01555 0.8416^{20}	H_2O 2; EtOH 4; eth 4; chl 2 1.4440^{20}
3438	2-Butanethiol, (±)- 2-Butanethiol (*DL*)	$C_4H_{10}S$ 90.19	91840-99-2 -165	1576 85	4-01-00-01584 0.8295^{20}	EtOH 3; eth 3; bz 3; ctc 2 1.4366^{20}
3439	1-Butanethiol, 2-methyl-, (*S*)- (+)-2-Methyl-1-butanethiol	$C_5H_{12}S$ 104.22	20089-07-0	 118.2	4-01-00-01668 0.8420^{20}	 1.4440^{20}
3440	1-Butanethiol, 3-methyl- Isopentyl mercaptan	$C_5H_{12}S$ 104.22	541-31-1	 120	4-01-00-01688 0.8350^{20}	H_2O 1; EtOH 5; eth 5; ctc 3 1.4418^{20}
3441	2-Butanethiol, 2-methyl- 2-Methyl-2-butanethiol	$C_5H_{12}S$ 104.22	1679-09-0	 99.1	4-01-00-01674 0.8120^{20}	 1.4385^{20}
3442	Butane, 1,1,2-tribromo- 1,1,2-Tribromocyclobutane	$C_4H_7Br_3$ 294.81	3675-68-1	 216.2	2-01-00-00084 2.1835^{20}	eth 4; EtOH 4; chl 4 1.5626^{17}
3443	Butane, 1,2,2-tribromo- 1,2,2-Tribromobutane	$C_4H_7Br_3$ 294.81	3675-69-2	 213.8	2-01-00-00085 2.1691^{20}	eth 4; EtOH 4; chl 4 1.5624^{10}
3444	Butane, 1,2,3-tribromo- 1,2,3-Tribromobutane	$C_4H_7Br_3$ 294.81	632-05-3 -19	 220	4-01-00-00271 2.1907^{20}	eth 4; EtOH 4; chl 4 1.5680^{20}
3445	Butane, 1,2,4-tribromo- 1,2,4-Tribromobutane	$C_4H_7Br_3$ 294.81	38300-67-3 -18	 215	4-01-00-00271 2.170^{20}	eth 4; EtOH 4; chl 4 1.5608^{20}
3446	Butane, 1,3,3-tribromo- 1,3,3-Tribromobutane	$C_4H_7Br_3$ 294.81	62127-46-2	 202.5	3-01-00-00298 2.1445^{20}	eth 4; EtOH 4; chl 4 1.5564^{20}
3447	Butane, 2,2,3-tribromo- 2,2,3-Tribromobutane	$C_4H_7Br_3$ 294.81	62127-47-3 1.8	 206	3-01-00-00298 2.1723^{20}	H_2O 1; EtOH 3; eth 3; ctc 2 1.5602^{20}
3448	Butane, 1,1,3-trichloro- 1,1,3-Trichlorobutane	$C_4H_7Cl_3$ 161.46	13279-87-3	 152	4-01-00-00252 1.317^{15}	H_2O 1; EtOH 3; eth 3; chl 4 1.4600^{15}
3449	Butane, 1,2,3-trichloro- 1,2,3-Trichlorobutane	$C_4H_7Cl_3$ 161.46	18338-40-4	 168; 63^{20}	3-01-00-00285 1.3164^{20}	H_2O 1; EtOH 3; eth 3; chl 4 1.4790^{20}
3450	Butane, 1,2,4-trichloro- 1,2,4-Trichlorobutane	$C_4H_7Cl_3$ 161.46	1790-22-3	 $61\text{-}3^{10}$	4-01-00-00253 1.3175^{20}	H_2O 1; bz 3; ctc 3; chl 4 1.4820^{20}
3451	Butane, 2,2,3-trichloro- 2,2,3-Trichlorobutane	$C_4H_7Cl_3$ 161.46	10403-60-8	 144	3-01-00-00285 1.2699^{20}	H_2O 1; chl 4 1.4645^{20}
3452	Butane, 2,2,3-trichloro-1,1,1,3,4,4,4-heptafluoro-	$C_4Cl_3F_7$ 287.39	335-44-4 4	 98	4-01-00-00253 1.7484^{20}	 1.3530^{20}
3453	Butane, 1,2,3-trichloro-2-methyl- Butane, 2-methyl-1,2,3-trichloro	$C_5H_9Cl_3$ 175.48	62521-69-1	 184	3-01-00-00361 1.2527^{20}	chl 4
3454	Butane, 2,2,3-trichloro-3-methyl- Butane, 2-methyl-2,3,3-trichloro	$C_5H_9Cl_3$ 175.48	98070-91-8	 182.5	3-01-00-00361 1.215^{15}	chl 4; HOAc 4 1.473^{21}
3455	Butane, 1,1,3-trimethoxy-	$C_7H_{16}O_3$ 148.20	10138-89-3	 157	4-01-00-03985 0.935^{25}	 1.4032^{20}
3456	Butane, 1,3,3-trimethoxy-	$C_7H_{16}O_3$ 148.20	6607-66-5	 $61\text{-}3^{20}$	4-01-00-03995 0.94^{25}	 1.4096^{20}
3457	Butane, 2,2,3-trimethyl- 2,2,3-Trimethylbutane	C_7H_{16} 100.20	464-06-2 -25	 80.8	4-01-00-00410 0.6901^{20}	H_2O 1; EtOH 3; eth 3; ace 4 1.3864^{20}
3458	1,2,3-Butanetriol 1-Methylglycerol	$C_4H_{10}O_3$ 106.12	4435-50-1	 170^{20}	4-01-00-02774	H_2O 4; EtOH 4 1.4462^{20}
3459	1,2,4-Butanetriol	$C_4H_{10}O_3$ 106.12	3068-00-6	 $172\text{-}4^{12}$	2-01-00-00596 1.018^{20}	H_2O 4; EtOH 4 1.4688^{20}
3460	Butanoic acid Butyric acid	$C_4H_8O_2$ 88.11	107-92-6 -5.7	1593 163.7	4-02-00-00779 0.9577^{20}	H_2O 5; EtOH 5; eth 5; ctc 2 1.3980^{20}
3461	Butanoic acid, 2-acetyl-3-methyl-, ethyl ester	$C_9H_{16}O_3$ 172.22	1522-46-9	 201; $97\text{-}8^{20}$	4-03-00-01608 0.9648^{18}	H_2O 1; EtOH 5; eth 5 1.4256^{18}
3462	Butanoic acid, 2-acetyl-3-oxo-, ethyl ester	$C_8H_{12}O_4$ 172.18	603-69-0	 210	4-03-00-01781 1.1045^{20}	bz 4; eth 4; EtOH 4 1.4690^{20}
3463	Butanoic acid, 2-amino-, (±)-	$C_4H_9NO_2$ 103.12	2835-81-6 305 dec	 sub	3-04-00-01296 1.2300^{20}	H_2O 4; EtOH 2; eth 1; bz 1
3464	Butanoic acid, 4-amino- γ-Aminobutyric acid	$C_4H_9NO_2$ 103.12	56-12-2 203 dec	441	4-04-00-02600	H_2O 4; EtOH 2; eth 1; ace 2
3465	Butanoic acid, 3-amino-4-hydroxy- γ-Hydroxy-β-aminobutyric acid	$C_4H_9NO_3$ 119.12	589-44-6 216	455	4-04-00-03216	
3466	Butanoic acid, 4-amino-2-hydroxy-4-oxo-	$C_4H_7NO_4$ 133.10	66398-52-5 147.5		3-03-00-00918 1.577^{18}	H_2O 3; EtOH 2; eth 1; MeOH 2
3467	Butanoic acid, anhydride Butyric anhydride	$C_8H_{14}O_3$ 158.20	106-31-0 -75	1594 200	4-02-00-00802 0.9668^{20}	eth 3; ctc 2 1.4070^{20}
3468	Butanoic acid, 2-bromo-, (±)- *DL*-α-Bromobutyric acid	$C_4H_7BrO_2$ 167.00	2385-70-8 -4	1401 217 dec; 181^{250}	4-02-00-00833 1.5641^{20}	H_2O 3; EtOH 3; eth 3
3469	Butanoic acid, 2-bromo-, ethyl ester	$C_6H_{11}BrO_2$ 195.06	533-68-6	 177; 43^{5}	4-02-00-00834 1.3273^{20}	H_2O 1; EtOH 5; eth 5; chl 3 1.4475^{20}

No.	Name Synonym	Mol. Form. Mol. Wt.	CAS RN mp/°C	Merck No. bp/°C	Beil. Ref. den/g cm^{-3}	Solubility n_D
3470	Butanoic acid, 4-bromo-, ethyl ester	$C_6H_{11}BrO_2$ 195.06	2969-81-5	 82^{10}	4-02-00-00836 1.363^{25}	 1.4559^{20}
3471	Butanoic acid, 2-bromo-3-methyl-, (±)-	$C_5H_9BrO_2$ 181.03	10323-40-7 44	 230 dec; 138^{25}	3-02-00-00706 1.459^{20}	H_2O 2; EtOH 4; eth 3; ace 3
3472	Butanoic acid, 3-bromo-3-methyl- β-Bromoisovaleric acid	$C_5H_9BrO_2$ 181.03	5798-88-9 74	1410	4-02-00-00905	bz 4; eth 4; EtOH 4
3473	Butanoic acid, 2-bromo-, methyl ester	$C_5H_9BrO_2$ 181.03	3196-15-4	 168	4-02-00-00834 1.4528^{20}	EtOH 4 1.4029^{25}
3474	Butanoic acid, 4-bromo-, methyl ester	$C_5H_9BrO_2$ 181.03	4897-84-1	 186.5	4-02-00-00835 1.4^{25}	EtOH 4 1.4567^{25}
3475	Butanoic acid, 2-bromo-3-methyl-, ethyl ester	$C_7H_{13}BrO_2$ 209.08	609-12-1	 186	4-02-00-00906 1.2760^{20}	eth 4; EtOH 4 1.4496^{20}
3476	Butanoic acid, 2-bromo-3-methyl-, methyl ester	$C_6H_{11}BrO_2$ 195.06	26330-51-8	 177	3-02-00-00706 1.352^{13}	eth 4; EtOH 4 1.4530^{20}
3477	Butanoic acid, 2-bromo-3-oxo-, ethyl ester	$C_6H_9BrO_3$ 209.04	609-13-2	 210 dec; 107^{15}	4-03-00-01551 1.4294^{16}	eth 4; EtOH 4 1.463^{14}
3478	Butanoic acid, 4-bromo-3-oxo-, ethyl ester	$C_6H_9BrO_3$ 209.04	13176-46-0	 115^{14}	4-03-00-01551 1.4840^{20}	eth 4; EtOH 4 1.5281^{20}
3479	Butanoic acid, butyl ester Butyl butanoate	$C_8H_{16}O_2$ 144.21	109-21-7 -91.5	1556 166	4-02-00-00789 0.8700^{20}	H_2O 1; EtOH 5; eth 5; ctc 3 1.4075^{20}
3480	Butanoic acid, 2-chloro-	$C_4H_7ClO_2$ 122.55	4170-24-5	 189^{627}, 101^{15}	4-02-00-00821 1.1790^{20}	H_2O 2; EtOH 4; eth 4 1.441^{20}
3481	Butanoic acid, 3-chloro-, (±)-	$C_4H_7ClO_2$ 122.55	625-68-3 16	 116^{22}	4-02-00-00823 1.1898^{20}	EtOH 3; eth 4; ctc 2 1.4221^{20}
3482	Butanoic acid, 4-chloro-	$C_4H_7ClO_2$ 122.55	627-00-9 16	 196^{22}; $68^{0.2}$	4-02-00-00825 1.2236^{20}	EtOH 4 1.4642^{20}
3483	Butanoic acid, 2-chloro-, ethyl ester	$C_6H_{11}ClO_2$ 150.60	7425-45-8	 163.5	4-02-00-00822 1.056^{18}	eth 4; EtOH 4 1.4180^{25}
3484	Butanoic acid, 3-chloro-, ethyl ester	$C_6H_{11}ClO_2$ 150.60	7425-48-1		4-02-00-00824 1.0517^{20}	EtOH 4 1.4246^{20}
3485	Butanoic acid, 4-chloro-, ethyl ester	$C_6H_{11}ClO_2$ 150.60	3153-36-4	 184	4-02-00-00825 1.0756^{20}	ace 4; eth 4; EtOH 4 1.4311^{20}
3486	Butanoic acid, 2-chloro-2-methyl- Butyric acid, 2-chloro-2-methyl-	$C_5H_9ClO_2$ 136.58	73758-54-0	 200 dec	3-02-00-00688 1.1204^{20}	eth 4; EtOH 4 1.4445^{20}
3487	Butanoic acid, 2-chloro-3-methyl-	$C_5H_9ClO_2$ 136.58	921-08-4 21	 210; 126^{32}	4-02-00-00904 1.135^{13}	H_2O 1; EtOH 3; eth 3 1.4450^{11}
3488	Butanoic acid, 2-chloro-, methyl ester	$C_5H_9ClO_2$ 136.58	26464-32-4	 146	3-02-00-00622 1.0979^{14}	eth 4; EtOH 4 1.4247^{20}
3489	Butanoic acid, 4-chloro-, methyl ester	$C_5H_9ClO_2$ 136.58	3153-37-5	 175; 55^4	4-02-00-00825 1.1201^{20}	H_2O 1; EtOH 4; eth 4; ace 3 1.4321^{20}
3490	Butanoic acid, 2-chloro-3-methyl-, ethyl ester Ethyl 2-chloro-3-methylbutanoate	$C_7H_{13}ClO_2$ 164.63	91913-99-4	 179	0-02-00-00316 1.021^{13}	eth 4; EtOH 4
3491	Butanoic acid, 2-chloro-2-methyl-, ethyl ester Butyric acid, 2-chloro-2-methyl, ethyl ester	$C_7H_{13}ClO_2$ 164.63	58190-94-6	 176	3-02-00-00688 1.069^{14}	eth 4; EtOH 4 1.4388^{11}
3492	Butanoic acid, 2-chloro-2-methyl-3-oxo-, ethyl ester Butyric acid, 2-chloro-2-methyl-3-oxo, ethyl ester	$C_7H_{11}ClO_3$ 178.62	37935-39-0	 $116\text{-}7^{75}$	2-03-00-00433 1.157^{18}	EtOH 4 1.4382^{18}
3493	Butanoic acid, 4-(4-chloro-2-methylphenoxy)- 4-(4-Chloro-2-methylphenoxy)butyric acid	$C_{11}H_{13}ClO_3$ 228.68	94-81-5 100		4-06-00-01996	
3494	Butanoic acid, 4-chloro-3-oxo-, ethyl ester	$C_6H_9ClO_3$ 164.59	638-07-3 -8	 220 dec; 115^{14}	4-03-00-01550 1.218^{25}	 1.4520^{20}
3495	Butanoic acid, 2-cyano-2-ethyl-, ethyl ester Cyanoacetic acid, diethyl, ethyl ester	$C_9H_{15}NO_2$ 169.22	1619-56-3	 214.5	3-02-00-01761	eth 4; EtOH 4 1.4200^{27}
3496	Butanoic acid, cyclohexyl ester	$C_{10}H_{18}O_2$ 170.25	1551-44-6	 213	4-06-00-00037 0.9572^0	H_2O 1; EtOH 3; ctc 2; os 3
3497	Butanoic acid, 2,3-dibromo-	$C_4H_6Br_2O_2$ 245.90	600-30-6 87	 $100\text{-}10^{20}$	4-02-00-00837	H_2O 2; EtOH 4; eth 4; bz 4
3498	Butanoic acid, 2,3-dibromo-, ethyl ester	$C_6H_{10}Br_2O_2$ 273.95	609-11-0 58.5	 113^{30}	3-02-00-00634 1.6800^{20}	H_2O 2; EtOH 3; eth 3; ctc 2
3499	Butanoic acid, 2,4-dibromo-, ethyl ester	$C_6H_{10}Br_2O_2$ [illegible]	36847-51-5	 [illegible]	4-02-00-00838 [illegible]	H_2O 1; EtOH 3; eth 3 [illegible]
3500	Butanoic acid, 2,2-dichloro- Butyric acid, 2,2-dichloro-	$C_4H_6Cl_2O_2$ 157.00	13023-00-2	 $107\text{-}10^{14}$	2-02-00-00254 1.389^{20}	H_2O 2; EtOH 4; eth 4
3501	Butanoic acid, 3,4-dichloro-, butyl ester Butyric acid, 3,4-dichloro, butyl ester	$C_8H_{14}Cl_2O_2$ 213.10	116723-94-5	 $127\text{-}8^{15}$	4-02-00-00829 1.143^{25}	 1.454^{24}
3502	Butanoic acid, 3,4-dichloro-, methyl ester	$C_5H_8Cl_2O_2$ 171.02	819-93-2	 90^{15}	4-02-00-00829 1.278^{23}	 1.459^{23}
3503	Butanoic acid, 2,2-dichloro-3-oxo-, ethyl ester	$C_6H_8Cl_2O_3$ 199.03	6134-66-3	 206; 91^{12}	2-03-00-00427 1.291^{16}	EtOH 4 1.4492^{17}
3504	Butanoic acid, 2,2-diethyl-3-oxo-, ethyl ester	$C_{10}H_{18}O_3$ 186.25	1619-57-4	 216; 64^3	4-03-00-01625 0.9717^{18}	H_2O 1; EtOH 5; eth 5; ctc 3 1.4326^{17}
3505	Butanoic acid, 2,3-dihydroxypropyl ester	$C_7H_{14}O_4$ 162.19	557-25-5	 280; 117^{10}	4-02-00-00798 1.129^{18}	H_2O 4; EtOH 4 1.4531^{20}
3506	Butanoic acid, 2,2-dimethyl-	$C_6H_{12}O_2$ 116.16	595-37-9 -14	 186	4-02-00-00954 0.9276^{20}	H_2O 2; EtOH 3; eth 3 1.4145^{20}

No.	Name Synonym	Mol. Form. Mol. Wt.	CAS RN mp/°C	Merck No. bp/°C	Beil. Ref. den/g cm^{-3}	Solubility n_D
3507	Butanoic acid, 2,3-dimethyl- Butyric acid, 2,3-dimethyl	$C_6H_{12}O_2$ 116.16	14287-61-7 -1.5	 191.7	4-02-00-00958 0.9275^{20}	eth 4; EtOH 4 1.4146^{20}
3508	Butanoic acid, 3,3-dimethyl- *tert*-Butylacetic acid	$C_6H_{12}O_2$ 116.16	1070-83-3 6.5	1538 190	4-02-00-00955 0.9124^{20}	EtOH 3; eth 3 1.4096^{20}
3509	Butanoic acid, 1,1-dimethylethyl ester	$C_8H_{16}O_2$ 144.21	2308-38-5	 146	4-02-00-00790	ace 4; eth 4; EtOH 4 1.4007^{17}
3510	Butanoic acid, 3,3-dimethyl-2-oxo-	$C_6H_{10}O_3$ 130.14	815-17-8 90.5	 189; 80^{15}	4-03-00-01595	H_2O 2; eth 3; bz 3; chl 3
3511	Butanoic acid, 2,2-dimethyl-3-oxo-, ethyl ester	$C_8H_{14}O_3$ 158.20	597-04-6	 182	4-03-00-01594 0.9759^{20}	eth 4; EtOH 4 1.4180^{20}
3512	Butanoic acid, 1,2-ethanediyl ester	$C_{10}H_{18}O_4$ 202.25	105-72-6 -80	 240	4-02-00-00796 1.0005^{20}	eth 4; EtOH 4 1.4262^{20}
3513	Butanoic acid, ethenyl ester Vinyl butyrate	$C_6H_{10}O_2$ 114.14	123-20-6	 116.7; 64^{130}	4-02-00-00792 0.9006^{20}	
3514	Butanoic acid, 2-ethyl- Diethylacetic acid	$C_6H_{12}O_2$ 116.16	88-09-5 -31.8	3099 194	4-02-00-00950 0.9239^{20}	H_2O 2; EtOH 5; eth 5; ctc 2 1.4132^{20}
3515	Butanoic acid, ethyl ester Ethyl butanoate	$C_6H_{12}O_2$ 116.16	105-54-4 -98	3733 121.5	4-02-00-00787 0.8844^{15}	H_2O 2; EtOH 3; eth 3; ctc 2 1.4000^{20}
3516	Butanoic acid, 2-ethyl-2-methyl-	$C_7H_{14}O_2$ 130.19	19889-37-3 <-20	 207	4-02-00-00981	EtOH 4 1.4250^{20}
3517	Butanoic acid, 2-ethyl-, methyl ester	$C_7H_{14}O_2$ 130.19	816-11-5	 137	4-02-00-00950 0.8886^{12}	 1.4067^{12}
3518	Butanoic acid, 2-ethyl-3-oxo-, ethyl ester	$C_8H_{14}O_3$ 158.20	607-97-6	 198.0; 80^{10}	4-03-00-01592 0.9847^{16}	EtOH 5; eth 5 1.4214^{25}
3519	Butanoic acid, 2-ethyl-3-oxo-, methyl ester	$C_7H_{12}O_3$ 144.17	51756-08-2	 182	4-03-00-01592 0.995^{14}	ace 4; eth 4; EtOH 4
3520	Butanoic acid, 2-furanylmethyl ester	$C_9H_{12}O_3$ 168.19	623-21-2	 212.5	4-17-00-01247 1.0530^{20}	H_2O 2; EtOH 3; eth 5
3521	Butanoic acid, heptafluoro-	$C_4HF_7O_2$ 214.04	375-22-4 -17.5	 121	4-02-00-00810 1.651^{20}	H_2O 3; eth 3; tol 3; peth 1 1.295^{25}
3522	Butanoic acid, heptafluoro-, anhydride	$C_8F_{14}O_3$ 410.06	336-59-4 -43	 106.5	4-02-00-00817 1.665^{20}	 1.285^{20}
3523	Butanoic acid, heptafluoro-, ethenyl ester	$C_6H_3F_7O_2$ 240.08	356-28-5	 79	4-02-00-00815 1.418^{20}	 1.309
3524	Butanoic acid, heptafluoro-, ethyl ester	$C_6H_5F_7O_2$ 242.09	356-27-4	 95	4-02-00-00813 1.394^{20}	H_2O 2; eth 3; ace 3 1.3011^{20}
3525	Butanoic acid, heptafluoro-, methyl ester	$C_5H_3F_7O_2$ 228.07	356-24-1 -86	 80	4-02-00-00812 1.483^{20}	H_2O 2; eth 3; ace 3 1.295^{20}
3526	Butanoic acid, heptafluoro-, 1-methylethyl ester Isopropyl heptafluorobutyrate	$C_7H_7F_7O_2$ 256.12	425-23-0	 106	4-02-00-00813 1.324^{20}	 1.310^{20}
3527	Butanoic acid, heptyl ester	$C_{11}H_{22}O_2$ 186.29	5870-93-9 -57.5	 225.8	4-02-00-00791 0.8637^{20}	EtOH 4 1.4231^{20}
3528	Butanoic acid, hexyl ester	$C_{10}H_{20}O_2$ 172.27	2639-63-6 -78	 208	4-02-00-00791 0.8652^{20}	H_2O 1; EtOH 3; chl 2; os 3 1.4160^{15}
3529	Butanoic acid, hydrazide Butyrhydrazide	$C_4H_{10}N_2O$ 102.14	3538-65-6 45.5	 138^{20}; 120^{10}	4-02-00-00807	H_2O 4; eth 4; EtOH 4
3530	Butanoic acid, 2-hydroxy-, (±)-	$C_4H_8O_3$ 104.11	600-15-7 44.2	 260 dec; 140^{14}	3-03-00-00561 1.125^{20}	H_2O 3; EtOH 3; eth 3
3531	Butanoic acid, 3-hydroxy-, (±)-	$C_4H_8O_3$ 104.11	625-71-8 49	 130^{12}; $94^{0.1}$	4-03-00-00761	H_2O 4; EtOH 4; eth 4; bz 1 1.4424^{20}
3532	Butanoic acid, 2-hydroxy-, ethyl ester, (±)- Butyric acid, 2-hydroxy, ethyl ester (*DL*)	$C_6H_{12}O_3$ 132.16	68057-83-0	 167	4-03-00-00756 1.0069^{20}	EtOH 4 1.4179^{20}
3533	Butanoic acid, 3-hydroxy-, ethyl ester, (±)-	$C_6H_{12}O_3$ 132.16	35608-64-1	 185; 76^{15}	4-03-00-00762 1.017^{20}	H_2O 3; EtOH 3; ctc 2 1.4182^{20}
3534	Butanoic acid, 3-hydroxy-, ethyl ester	$C_6H_{12}O_3$ 132.16	5405-41-4	 185; 76^{15}	4-03-00-00762 1.017^{20}	H_2O 4; EtOH 4 1.4182^{20}
3535	Butanoic acid, 3-hydroxy-3-methyl-	$C_5H_{10}O_3$ 118.13	625-08-1 <-32	 162^{12}	4-03-00-00827 0.9384^{20}	H_2O 4; eth 4; EtOH 4 1.5081^{20}
3536	Butanoic acid, 2-hydroxy-, 2-methylpropyl ester Butyric acid, 2-hydroxy, isobutyl ester	$C_8H_{16}O_3$ 160.21	116723-95-6	 196	0-03-00-00302 0.944^{15}	 1.4182
3537	Butanoic acid, 3-mercapto-	$C_4H_8O_2S$ 120.17	26473-49-4	 110^{10}; $88^{2.5}$	4-03-00-00769 1.1371^{20}	 1.4782^{20}
3538	Butanoic acid, 2-methyl-, (±)- (±)-2-Methylbutyric acid	$C_5H_{10}O_2$ 102.13	600-07-7 <-80	 177	4-02-00-00889 0.934^{20}	H_2O 2; EtOH 5; eth 5; chl 3 1.4051^{20}
3539	Butanoic acid, 3-methyl- Isovaleric acid	$C_5H_{10}O_2$ 102.13	503-74-2 -29.3	5120 176.5	4-02-00-00895 0.931^{20}	H_2O 3; EtOH 5; eth 5; chl 5 1.4033^{20}
3540	Butanoic acid, 2-methyl-, 10-(acetyloxy)-9,10-dihydro-8,8-dimethyl-2-oxo-2H,8H-benzo[1,2-b:3,4-b']dipyran-9-yl Visnadine	$C_{21}H_{24}O_7$ 388.42	477-32-7 85.5	9915	5-19-06-00272	H_2O 1; EtOH 3; eth 3
3541	Butanoic acid, 3-(methylamino)-, ethyl ester Butyric acid, 3-methylamino-, ethyl ester	$C_7H_{15}NO_2$ 145.20	68384-70-3	 $72^{12.5}$	3-04-00-01313 0.9282^{20}	 1.4250^{20}
3542	Butanoic acid, 3-methyl-, anhydride	$C_{10}H_{18}O_3$ 186.25	1468-39-9	 215	4-02-00-00901 0.9327^{20}	eth 4 1.4043^{20}
3543	Butanoic acid, 3-methyl-, butyl ester Butyl isovalerate	$C_9H_{18}O_2$ 158.24	109-19-3		3-02-00-00698	 1.4058^{25}

No.	Name Synonym	Mol. Form. Mol. Wt.	CAS RN mp/°C	Merck No. bp/°C	Beil. Ref. den/g cm^{-3}	Solubility n_D
3544	Butanoic acid, 3-methylbutyl ester Isopentyl butanoate	$C_9H_{18}O_2$ 158.24	106-27-4	4997 179	4-02-00-00791 0.865^{19}	H_2O 1; EtOH 4; eth 4 1.4110^{20}
3545	Butanoic acid, 2-methyl-, butyl ester Butyric acid, (2-methylbutyl) ester	$C_9H_{18}O_2$ 158.24	15706-73-7	 179	0-02-00-00304 0.8620^{20}	 1.4135^{20}
3546	Butanoic acid, 3-methyl-, 1,1-dimethylpropyl ester *tert*-Amyl isovalerate	$C_{10}H_{20}O_2$ 172.27	542-37-0	647 173.5	0-02-00-00312 0.8729^{0}	EtOH 4
3547	Butanoic acid, methyl ester Methyl butanoate	$C_5H_{10}O_2$ 102.13	623-42-7 -85.8	5957 102.8	4-02-00-00786 0.8984^{20}	H_2O 2; EtOH 5; eth 5; ctc 2 1.3878^{20}
3548	Butanoic acid, 3-methyl-, ethyl ester Ethyl isovalerate	$C_7H_{14}O_2$ 130.19	108-64-5 -99.3	3772 135.0	4-02-00-00898 0.8656^{20}	EtOH 4; eth 4 1.3962^{20}
3549	Butanoic acid, 1-methylethyl ester Isopropyl butyrate	$C_7H_{14}O_2$ 130.19	638-11-9	 130.5	4-02-00-00789 0.8588^{20}	H_2O 1; EtOH 3 1.3936^{20}
3550	Butanoic acid, 2-methyl-, ethyl ester, (*S*)-	$C_7H_{14}O_2$ 130.19	10307-61-6	 133; 35^{16}	4-02-00-00890 0.8689^{25}	H_2O 1; EtOH 3; bz 3 1.3964^{20}
3551	Butanoic acid, 3-methyl-, 3-methylbutyl ester Isoamyl isovalerate	$C_{10}H_{20}O_2$ 172.27	659-70-1	5003 190.4	4-02-00-00899 0.8583^{19}	 1.4130^{19}
3552	Butanoic acid, 3-methyl-, methyl ester Methyl isovalerate	$C_6H_{12}O_2$ 116.16	556-24-1	6006 116.5	4-02-00-00897 0.8808^{20}	H_2O 1; EtOH 4; eth 4; ace 4 1.3927^{20}
3553	Butanoic acid, 3-methyl-, 1-methylethyl ester Butyric acid, 3-methyl, isopropyl ester	$C_8H_{16}O_2$ 144.21	32665-23-9	 142; 70^{55}	3-02-00-00698 0.8538^{17}	ace 4; eth 4; EtOH 4 1.3960^{20}
3554	Butanoic acid, 3-methyl-, 5-methyl-2-(1-methylethyl)cyclohexyl ester	$C_{15}H_{28}O_2$ 240.39	16409-46-4	 129^{9}	2-06-00-00043 0.9089^{15}	H_2O 1; EtOH 3; ace 3 1.4486^{20}
3555	Butanoic acid, 3-methyl-, 2-methylpropyl ester Isobutyl isovalerate	$C_9H_{18}O_2$ 158.24	589-59-3	5029 168.5	4-02-00-00898 0.853^{20}	H_2O 1; EtOH 5; eth 5; ace 4 1.4057^{20}
3556	Butanoic acid, 3-methyl-2-oxo-	$C_5H_8O_3$ 116.12	759-05-7 31.5	 170.5	4-03-00-01577 0.9968^{20}	H_2O 3; EtOH 3; eth 3 1.3850^{16}
3557	Butanoic acid, 2-methyl-3-oxo-, ethyl ester	$C_7H_{12}O_3$ 144.17	609-14-3	 187	4-03-00-01573 0.9941^{20}	H_2O 2; EtOH 3; eth 3; ace 4 1.4185^{20}
3558	Butanoic acid, 2-methyl-3-oxo-, methyl ester	$C_6H_{10}O_3$ 130.14	17094-21-2	 177.4	3-03-00-01225 1.0217^{25}	eth 4; EtOH 4 1.416^{24}
3559	Butanoic acid, 1-methylpentyl ester, (*S*)- Butyric acid (1-methylpentyl) ester (*d*)	$C_{10}H_{20}O_2$ 172.27	116723-96-7	 85^{20}	1-02-00-00120 0.8744^{21}	eth 4; EtOH 4
3560	Butanoic acid, 3-methyl-, phenylmethyl ester	$C_{12}H_{16}O_2$ 192.26	103-38-8	 245; 136^{25}	4-06-00-02266 0.9983^{15}	 1.4884^{20}
3561	Butanoic acid, 3-methyl-, 1,2,3-propanetriyl ester	$C_{18}H_{32}O_6$ 344.45	620-63-3	 332.5	4-02-00-00900 0.9984^{20}	eth 4; EtOH 4 1.4354^{20}
3562	Butanoic acid, 3-methyl-, propyl ester Propyl isovalerate	$C_8H_{16}O_2$ 144.21	557-00-6	 155.9	4-02-00-00898 0.8617^{20}	eth 4; EtOH 4 1.4031^{20}
3563	Butanoic acid, 2-methylpropyl ester Isobutyl butanoate	$C_8H_{16}O_2$ 144.21	539-90-2	5020 156.9	4-02-00-00790 0.8364^{18}	H_2O 2; EtOH 5; eth 5 1.4032^{20}
3564	Butanoic acid, 1-methylpropyl ester, (*S*)- Butyric acid, *sec*-butyl ester (*d*)	$C_8H_{16}O_2$ 144.21	116836-55-6 -91.5	 152; 54^{18}	3-02-00-00600 0.8737^{13}	EtOH 3; bz 3; CS_2 3; py 3 1.4011^{20}
3565	Butanoic acid, 3-methyl-, 1,7,7-trimethylbicyclo[2.2.1]hept-2-yl ester, (1*R*-*endo*)- *d*-Bornyl isovalerate	$C_{15}H_{26}O_2$ 238.37	53022-14-3	1342 257.5	2-06-00-08488 0.955^{25}	eth 4; EtOH 4
3566	Butanoic acid, octyl ester	$C_{12}H_{24}O_2$ 200.32	110-39-4 -55.6	 244.1	4-02-00-00791 0.8629^{20}	EtOH 4 1.4267^{15}
3567	Butanoic acid, 2-oxo-	$C_4H_6O_3$ 102.09	600-18-0 33	 $80\text{-}2^{16}$	4-03-00-01524 1.200^{17}	H_2O 4; EtOH 4; eth 2 1.3972^{20}
3568	Butanoic acid, 3-oxo- Acetoacetic acid	$C_4H_6O_3$ 102.09	541-50-4 36.5	51 100 dec	4-03-00-01527	H_2O 4; eth 4; EtOH 4
3569	Butanoic acid, 3-oxo-, butyl ester	$C_8H_{14}O_3$ 158.20	591-60-6 -35.6	 127^{50}; 85^{8}	4-03-00-01536 0.9671^{25}	H_2O 2; EtOH 5; bz 5; lig 5 1.4137^{20}
3570	Butanoic acid, 3-oxo-, 2-chloroethyl ester	$C_6H_9ClO_3$ 164.59	54527-68-3	 198; 107^{14}	4-03-00-01534 1.190^{20}	bz 4; eth 4; EtOH 4 1.4430^{20}
3571	Butanoic acid, 3-oxo-, ethyl ester Ethyl acetoacetate	$C_6H_{10}O_3$ 130.14	141-97-9 -45	3714 180.8	4-03-00-01528 1.0368^{10}	H_2O 4; EtOH 5; eth 5; bz 3 1.4171^{20}
3572	Butanoic acid, 4-oxo-, ethyl ester	$C_6H_{10}O_3$ 130.14	10138-10-0	 162; 85^{12}	4-03-00-01555 1.0490^{22}	chl 2 1.4287^{20}
3573	Butanoic acid, 3-oxo-, methyl ester Methyl acetoacetate	$C_5H_8O_3$ 116.12	105-45-3 27.5	5933 171.7	4-03-00-01527 1.0762^{20}	H_2O 4; EtOH 5; eth 5; ctc 3 1.4184^{20}
3574	Butanoic acid, 3-oxo-, 1-methylethyl ester Isopropyl acetoacetate	$C_7H_{12}O_3$ 144.17	542-08-5 -27.3	5094 186	4-03-00-01535 0.9835^{20}	eth 4; EtOH 4; lig 4 1.4173^{20}
3575	Butanoic acid, 4-oxo-4-(phenylamino)- Succinanilic acid	$C_{10}H_{11}NO_3$ 193.20	102-14-7 148.5	8839	4-12-00-00468	H_2O 2; EtOH 3; eth 4
3576	Butanoic acid, 3-oxo-2-(phenylmethylene)-, ethyl ester	$C_{13}H_{14}O_3$ 218.25	620-80-4 60.5	 296; 180^{17}	3-10-00-03158	H_2O 1; EtOH 2; eth 2; bz 2
3577	Butanoic acid, 3-oxo-, 2-propenyl ester	$C_7H_{10}O_3$ 142.15	1118-84-9 -85	 196; 66.5^{14}	4-03-00-01538 1.0366^{20}	H_2O 3; EtOH 5; bz 5; lig 3 1.4398^{20}
3578	Butanoic acid, 4-oxo-4-[[4-[(2-thiazolylamino)sulfonyl]phenyl]amino]- Succinylsulphathiazole	$C_{13}H_{13}N_3O_5S_2$ 355.40	116-43-8 193.5	8850	4-27-00-04637	H_2O 1; EtOH 2; eth 1; ace 2
3579	Butanoic acid, pentyl ester Amyl butyrate	$C_9H_{18}O_2$ 158.24	540-18-1 -73.2	639 186.4	4-02-00-00790 0.8713^{15}	H_2O 1; EtOH 4; eth 4 1.4123^{20}
3580	Butanoic acid, 2-phenoxy-	$C_{10}H_{12}O_3$ 180.20	13794-14-4 98	 258	3-06-00-00616	H_2O 2; EtOH 3; eth 3; ace 3

No.	Name Synonym	Mol. Form. Mol. Wt.	CAS RN mp/°C	Merck No. bp/°C	Beil. Ref. den/g cm^{-3}	Solubility n_D
3581	Butanoic acid, 4-phenoxy-, ethyl ester Ethyl 4-phenoxybutyrate	$C_{12}H_{16}O_3$ 208.26	2364-59-2	 $170\text{-}3^{25}$	2-06-00-00159 1.042^{35}	eth 4; EtOH 4 1.491^{35}
3582	Butanoic acid, 2-phenoxyethyl ester	$C_{12}H_{16}O_3$ 208.26	23511-70-8	 251; 88^{2}	3-06-00-00571 1.0388^{21}	ace 4; eth 4; EtOH 4
3583	Butanoic acid, 2-phenoxy-, phenyl ester 2-Phenoxybutyric acid, phenyl ester	$C_{16}H_{16}O_3$ 256.30	116836-19-2 48.5	 $202\text{-}3^{25}$	4-06-00-00645 1.135^{15}	
3584	Butanoic acid, phenyl ester Phenyl butyrate	$C_{10}H_{12}O_2$ 164.20	4346-18-3	 225	4-06-00-00615 1.0382^{15}	H_2O 1; EtOH 3; eth 3
3585	Butanoic acid, phenylmethyl ester	$C_{11}H_{14}O_2$ 178.23	103-37-7	 239	4-06-00-02266 1.0111^{20}	H_2O 1; EtOH 4; eth 4; ctc 3 1.4920^{20}
3586	Butanoic acid, 1,2,3-propanetriyl ester Tributyrin	$C_{15}H_{26}O_6$ 302.37	60-01-5 -75	9532 307.5	4-02-00-00799 1.0350^{20}	H_2O 1; EtOH 3; eth 4; ace 3 1.4359^{20}
3587	Butanoic acid, 2-propenyl ester	$C_7H_{12}O_2$ 128.17	2051-78-7	 142; 44.5^{15}	4-02-00-00793 0.9017^{20}	H_2O 1; EtOH 5; eth 5; ctc 2 1.4158^{20}
3588	Butanoic acid, propyl ester Propyl butanoate	$C_7H_{14}O_2$ 130.19	105-66-8 -95.2	7858 143.0	4-02-00-00788 0.8730^{20}	H_2O 2; EtOH 5; eth 5 1.4001^{20}
3589	Butanoic acid, 3-thioxo-, ethyl ester Butyric acid, 3-thioxo, ethyl ester	$C_6H_{10}O_2S$ 146.21	7740-33-2	 75^{15}	4-03-00-01553 1.0554^{31}	eth 4; EtOH 4 1.4712^{26}
3590	Butanoic acid, 2,2,3-trichloro- 2,2,3-Trichlorobutyric acid	$C_4H_5Cl_3O_2$ 191.44	5344-55-8 60	 237	3-02-00-00629	eth 4
3591	Butanoic acid, 2,2,3-trichloro-, ethyl ester Butyric acid, 2,2,3-trichloro, ethyl ester	$C_6H_9Cl_3O_2$ 219.49	116723-97-8	 212	0-02-00-00281 1.3114^{20}	eth 4; EtOH 4
3592	Butanoic acid, 4,4,4-trifluoro-3-oxo-, ethyl ester	$C_6H_7F_3O_3$ 184.11	372-31-6	 133	4-03-00-01548 1.2586^{15}	EtOH 3; eth 3 1.3783^{15}
3593	1-Butanol Butyl alcohol	$C_4H_{10}O$ 74.12	71-36-3 -89.8	1540 117.7	4-01-00-01506 0.8098^{20}	H_2O 3; EtOH 5; eth 5; ace 4 1.3993^{20}
3594	2-Butanol *sec*-Butyl alcohol	$C_4H_{10}O$ 74.12	78-92-2 -114.7	1541 99.5	2-01-00-00400 0.8063^{20}	H_2O 4; EtOH 5; eth 5; ace 4 1.3978^{20}
3595	1-Butanol, 2-amino-, (±)-	$C_4H_{11}NO$ 89.14	13054-87-0 -2	 178	4-04-00-01705 0.9162^{20}	H_2O 5; EtOH 5; eth 5; chl 2 1.4489^{25}
3596	1-Butanol, 3-amino-	$C_4H_{11}NO$ 89.14	2867-59-6	 $82\text{-}5^{19}$	4-04-00-01710	H_2O 4; EtOH 4 1.4534^{25}
3597	1-Butanol, 4-amino-	$C_4H_{11}NO$ 89.14	13325-10-5	 205; 125^{34}	4-04-00-01711 0.967^{12}	H_2O 3; EtOH 3; eth 1 1.4625^{20}
3598	2-Butanol, 3-amino-	$C_4H_{11}NO$ 89.14	42551-55-3 19	 160; 79^{20}	3-04-00-00779 0.9399^{20}	H_2O 4; eth 4; EtOH 4 1.4445^{25}
3599	1-Butanol, 2-amino-3-methyl-, (±)-	$C_5H_{13}NO$ 103.16	16369-05-4	 $75\text{-}7^{8}$	4-04-00-01777 0.93^{25}	 1.4543^{20}
3600	1-Butanol, 3-bromo-	C_4H_9BrO 153.02	6089-12-9	 $69\text{-}71^{20}$	4-01-00-01553 1.7107^{18}	 1.5010^{21}
3601	2-Butanol, 3-bromo-, (*R**,*S**)-(±)- 2-Butanol, 3-bromo-, *erythro*-(±)-	C_4H_9BrO 153.02	19246-39-0	 154; 53^{13}	4-01-00-01580 1.4550^{20}	eth 4; EtOH 4 1.4780^{20}
3602	1-Butanol, 2-chloro-	C_4H_9ClO 108.57	26106-95-6	 $74\text{-}6^{25}$	4-01-00-01549 1.062^{25}	eth 4; EtOH 4 1.4438^{20}
3603	1-Butanol, 3-chloro-	C_5H_9ClO 120.58	2203-35-2	 175; 66^{15}	3-01-00-01517 1.0883^{20}	EtOH 3; eth 3; ace 3; ctc 2 1.4518^{20}
3604	1-Butanol, 4-chloro-	C_4H_9ClO 108.57	928-51-8	 $84\text{-}5^{16}$	4-01-00-01550 1.0883^{20}	eth 4; EtOH 4 1.4518^{20}
3605	2-Butanol, 1-chloro- α-Butylene chlorohydrin	C_4H_9ClO 108.57	1873-25-2	 141	4-01-00-01578 1.068^{25}	EtOH 3; eth 3 1.4400^{20}
3606	2-Butanol, 3-chloro- β-Butylene chlorohydrin	C_4H_9ClO 108.57	563-84-8	 139	3-01-00-01538 1.0669^{20}	EtOH 4; eth 4 1.4432^{20}
3607	2-Butanol, 4-chloro-	C_4H_9ClO 108.57	2203-34-1	 67^{20}	4-01-00-01578	eth 4; EtOH 4 1.4408^{20}
3608	2-Butanol, 1-chloro-2,3-dimethyl- 1-Chloro-2,3-dimethyl-2-butanol	$C_6H_{13}ClO$ 136.62	66235-62-9	 163	4-01-00-01730 1.049^{20}	 1.4459^{20}
3609	2-Butanol, 1-chloro-3,3-dimethyl- 1-Chloro-3,3-dimethyl-2-butanol	$C_6H_{13}ClO$ 136.62	36402-31-0 13	 157.5	4-01-00-01728 1.0063^{25}	 1.4432^{25}
3610	2-Butanol, 1-chloro-2-methyl-	$C_5H_{11}ClO$ 122.59	74283-48-0	 151	4-01-00-01673 1.0161^{20}	EtOH 4 1.4469^{20}
3611	2-Butanol, 3-chloro-2-methyl-	$C_5H_{11}ClO$ 122.59	21326-62-5	 141.5	3-01-00-01627 1.0295^{20}	eth 4; EtOH 4 1.4436^{20}
3612	1-Butanol, 2,3-dibromo- 2,3-Dibromo-1-butanol	$C_4H_8Br_2O$ 231.91	4021-75-4 32	 99.5^{10}	4-01-00-01554 1.9475^{20}	 1.5442^{20}
3613	1-Butanol, 3,4-dibromo- 3,4-Dibromobutanol	$C_4H_8Br_2O$ 231.91	87018-30-2	 114^{11}	3-01-00-01518 1.98^{15}	 1.548^{15}
3614	2-Butanol, 1,4-dibromo- 2-Hydroxy-1,4-dibromobutane	$C_4H_8Br_2O$ 231.91	19398-47-1	 $114\text{-}5^{15}$	4-01-00-01581 1.953^{20}	 1.544^{20}
3615	1-Butanol, 2,2-dimethyl- 2,2-Dimethyl-1-butanol	$C_6H_{14}O$ 102.18	1185-33-7 <-15	 136.5	4-01-00-01726 0.8283^{20}	H_2O 2; EtOH 3; eth 3 1.4208^{20}
3616	1-Butanol, 2,3-dimethyl-, (±)-	$C_6H_{14}O$ 102.18	20281-85-0	 144.5	4-01-00-01729 0.8297^{20}	EtOH 3; eth 3; ace 3 1.4195^{20}
3617	1-Butanol, 3,3-dimethyl-	$C_6H_{14}O$ 102.18	624-95-3 -60	 143	4-01-00-01729 0.844^{15}	H_2O 2; EtOH 3; eth 3; ace 3 1.4323^{15}
3618	2-Butanol, 2,3-dimethyl-	$C_6H_{14}O$ 102.18	594-60-5 -14	 118.4	4-01-00-01729 0.8236^{20}	H_2O 3; EtOH 5; eth 5 1.4176^{20}
3619	2-Butanol, 3,3-dimethyl-, (±)- 3,3-Dimethylbutan-2-ol (*DL*-)	$C_6H_{14}O$ 102.18	20281-91-8 5.6	 120.4	4-01-00-01727 0.8122^{25}	H_2O 2; EtOH 3; eth 5 1.4148^{20}
3620	1-Butanol, 2-ethyl- 2-Ethyl-1-butanol	$C_6H_{14}O$ 102.18	97-95-0 <-15	 147	4-01-00-01725 0.8326^{20}	H_2O 2; EtOH 3; eth 3; chl 3 1.4220^{20}

No.	Name Synonym	Mol. Form. Mol. Wt.	CAS RN mp/°C	Merck No. bp/°C	Beil. Ref. den/g cm^{-3}	Solubility n_D
3621	1-Butanol, 4-fluoro- 4-Fluoro-1-butanol	C_4H_9FO 92.11	372-93-0	 58^{15}	4-01-00-01546	ace 4; eth 4; EtOH 4 1.3942^{15}
3622	1-Butanol, 2,2,3,3,4,4,4-heptafluoro-	$C_4H_3F_7O$ 200.06	375-01-9	 95	4-01-00-01547 1.600^{20}	EtOH 3; ace 3 1.294^{20}
3623	2-Butanol, 1-[(2-hydroxyethyl)amino]- 2-Hydroxyethyl 2-hydroxybutyl amine	$C_6H_{15}NO_2$ 133.19	6967-43-7	 137^9	4-04-00-01725 1.0310^{20}	H_2O 4; ace 4; EtOH 4 1.4690^{30}
3624	2-Butanol, 3,3'-iminobis- Bis-(3-hydroxy-2-butyl)amine	$C_8H_{19}NO_2$ 161.24	6959-06-4	 $112\text{-}5^3$	3-04-00-00781 0.9775^{20}	H_2O 4; ace 4; EtOH 4 1.4162^{20}
3625	1-Butanol, 3-methoxy- 3-Methoxy-1-butanol	$C_5H_{12}O_2$ 104.15	2517-43-3	 157	4-01-00-02509 0.923^{23}	EtOH 4; eth 3; ace 4; chl 2 1.4148^{25}
3626	1-Butanol, 2-methyl- 2-Methyl-1-butanol	$C_5H_{12}O$ 88.15	137-32-6	5952 128	4-01-00-01666 0.8150^{25}	H_2O 2; EtOH 5; eth 5; ace 4 1.4092^{20}
3627	1-Butanol, 3-methyl- Isopentyl alcohol	$C_5H_{12}O$ 88.15	123-51-3 -117.2	5081 131.1	4-01-00-01677 0.8104^{20}	ace 4; eth 4; EtOH 4 1.4053^{20}
3628	2-Butanol, 2-methyl- *tert*-Amyl alcohol	$C_5H_{12}O$ 88.15	75-85-4 -8.8	7096 102.4	4-01-00-01668 0.8096^{20}	H_2O 3; EtOH 5; eth 5; ace 4 1.4052^{20}
3629	2-Butanol, 3-methyl-, (±)- (±)-3-Methyl-2-butanol	$C_5H_{12}O$ 88.15	70116-68-6	5953 112.9	4-01-00-01675 0.8180^{20}	H_2O 2; EtOH 5; eth 5; ace 4 1.4089^{20}
3630	1-Butanol, 2-methyl-, acetate	$C_7H_{14}O_2$ 130.19	624-41-9	 140	4-02-00-00156 0.8740^{20}	ace 4; eth 4; EtOH 4 1.4040^{20}
3631	1-Butanol, 3-methyl-, acetate Isopentyl acetate	$C_7H_{14}O_2$ 130.19	123-92-2 -78.5	4993 142.5	4-02-00-00157 0.876^{15}	H_2O 2; EtOH 5; eth 5; ace 3 1.4000^{20}
3632	1-Butanol, 3-methyl-, benzoate Isoamyl benzoate	$C_{12}H_{16}O_2$ 192.26	94-46-2	4995 261	4-09-00-00292 0.993^{15}	EtOH 4
3633	1-Butanol, 3-methyl-, carbamate	$C_6H_{13}NO_2$ 131.17	543-86-2 59	 220	4-03-00-00058 0.9438^{71}	eth 4; EtOH 4 1.4175^{20}
3634	2-Butanol, 2-methyl-, carbamate *tert*-Amyl carbamate	$C_6H_{13}NO_2$ 131.17	590-60-3 86	641	1-03-00-00014	ace 4; bz 4
3635	1-Butanol, 3-methyl-, carbonate (2:1)	$C_{11}H_{22}O_3$ 202.29	2050-95-5	 234; 122^{16}	3-03-00-00011 0.9067^{20}	H_2O 1 1.4174^{20}
3636	1-Butanol, 3-methyl-, formate Isopentyl formate	$C_6H_{12}O_2$ 116.16	110-45-2 -93.5	5001 123.5	4-02-00-00030 0.877^{20}	H_2O 2; EtOH 3; eth 5; ctc 2 1.3967^{20}
3637	1-Butanol, 3-methyl-, nitrate Isoamyl nitrate	$C_5H_{11}NO_3$ 133.15	543-87-3	5004 148	4-01-00-01683 0.996^{22}	 1.4122^{21}
3638	1-Butanol, 3-methyl-2-nitro-	$C_5H_{11}NO_3$ 133.15	77392-54-2	 111^{10}	3-01-00-01645 1.0889^{25}	 1.4430^{20}
3639	1-Butanol, 3-methyl-, propanoate Isopentyl propanoate	$C_8H_{16}O_2$ 144.21	105-68-0	 160.2	4-02-00-00709 0.8697^{20}	eth 4; EtOH 4 1.4069^{20}
3640	1-Butanol, 2-nitro- 2-Nitro-1-butanol	$C_4H_9NO_3$ 119.12	609-31-4 -47	 105^{10}	4-01-00-01555 1.1332^{25}	H_2O 3; EtOH 5; eth 5; ace 3 1.4390^{20}
3641	2-Butanol, 1-nitro- 2-Hydroxy-1-nitrobutane	$C_4H_9NO_3$ 119.12	3156-74-9	 204; 75^2	4-01-00-01582 1.1353^{20}	H_2O 3; EtOH 4; eth 4; ace 3 1.4435^{20}
3642	2-Butanol, 3-nitro- 3-Nitro-2-butanol	$C_4H_9NO_3$ 119.12	6270-16-2	 112^{38}; $55^{0.5}$	4-01-00-01583 1.1166^7	 1.4414^{20}
3643	2-Butanol, 1,1,1-trichloro- Ethyl(trichloromethyl)carbinol	$C_4H_7Cl_3O$ 177.46	6111-61-1	 171; 83^{22}	4-01-00-01579 1.3630^{25}	ace 4; bz 4; eth 4; EtOH 4 1.4800^{20}
3644	2-Butanol, 2,3,3-trimethyl-	$C_7H_{16}O$ 116.20	594-83-2 17	 131	4-01-00-01755 0.8380^{25}	ace 4; eth 4; EtOH 4 1.4233^{22}
3645	2-Butanone Methyl ethyl ketone	C_4H_8O 72.11	78-93-3 -86.6	5991 79.5	4-01-00-03243 0.8054^{20}	H_2O 4; EtOH 5; eth 5; ace 5 1.3788^{20}
3646	2-Butanone, 1-chloro- Chloromethyl ethyl ketone	C_4H_7ClO 106.55	616-27-3	 137.5	4-01-00-03255 1.0850^{20}	MeOH 4 1.4372^{20}
3647	2-Butanone, 3-chloro-	C_4H_7ClO 106.55	4091-39-8	 115	4-01-00-03256 1.0554^{25}	 1.4219^{20}
3648	2-Butanone, 4-chloro- 1-Chloro-3-butanone	C_4H_7ClO 106.55	6322-49-2	 120 dec	4-01-00-03256 1.0680^{23}	eth 4; EtOH 4 1.4284^{23}
3649	2-Butanone, 3-chloro-3-methyl- 3-Chloro-3-methyl-2-butanone	C_5H_9ClO 120.58	5950-19-6	 117.2	3-01-00-02818 1.0083^{20}	eth 4; EtOH 4 1.4204^{20}
3650	1-Butanone, 4-chloro-1-phenyl-	$C_{10}H_{11}ClO$ 182.65	939-52-6 19.5	 $130\text{-}3^4$	4-07-00-00711 1.137^{25}	 1.5459^{20}
3651	1-Butanone, 4-[4-(4-chlorophenyl)-4-hydroxy-1-piperidinyl]-1-(4-fluorophenyl)- Haloperidol	$C_{21}H_{23}ClFNO_2$ 375.87	52-86-8 151.5	4511	5-21-02-00377	
3652	1-Butanone, 1-cyclohexyl-	$C_{10}H_{18}O$ 154.25	1462-27-7	 94^{13}	4-07-00-00080 0.903^{20}	 1.4537^{20}
3653	2-Butanone, 3,4-dibromo- 1,2-Dibromoethyl methyl ketone	$C_4H_6Br_2O$ [illegible]	25109-57-3	 80-1	3-01-00-02788 1.9002^{18}	 1.5314^{16}
3654	2-Butanone, 1,3-dichloro- 1,3-Dichloro-2-butanone	$C_4H_6Cl_2O$ 141.00	16714-77-5	 166.5	3-01-00-02786 1.3116^{20}	ace 4; bz 4; eth 4; EtOH 4 1.4686^{20}
3655	2-Butanone, 3,4-dichloro-	$C_4H_6Cl_2O$ 141.00	58625-77-7	 $65\text{-}70^{16}$	3-01-00-02787 1.293^{18}	 1.4628^{18}
3656	2-Butanone, 4-(diethylamino)-	$C_8H_{17}NO$ 143.23	3299-38-5	 84^{30}	4-04-00-01930 0.8630^{20}	ace 4; bz 4; eth 4; EtOH 4 1.4333^{24}
3657	2-Butanone, 3,3-dimethyl- Pinacolone	$C_6H_{12}O$ 100.16	75-97-8 -52.5	7409 106.1	4-01-00-03310 0.7229^{25}	H_2O 2; EtOH 3; eth 3; ace 3 1.3952^{20}
3658	2-Butanone, 3,3-diphenyl- 2,2-Diphenyl-3-butanone	$C_{16}H_{16}O$ 224.30	2575-20-4 41	 310.5	4-07-00-01451 1.069^{20}	H_2O 1; EtOH 3; eth 4; chl 4 1.5748^{20}
3659	2-Butanone, 4-(2-furanyl)-	$C_8H_{10}O_2$ 138.17	699-17-2	 203	4-17-00-04559 1.0361^{19}	 1.4696^{17}
3660	2-Butanone, 1-hydroxy-	$C_4H_8O_2$ 88.11	5077-67-8	 160; 78^{60}	4-01-00-03989 1.0272^{20}	H_2O 4; EtOH 4; eth 4 1.4189^{20}

No.	Name	Synonym	Mol. Form.	Mol. Wt.	CAS RN	mp/°C	Merck No.	bp/°C	Beil. Ref.	den/g cm^{-3}	Solubility	n_D
3661	2-Butanone, 3-hydroxy-	Acetoin	$C_4H_8O_2$	88.11	513-86-0	15	55	143	2-01-00-00870	1.0044^{20}	H_2O 4; ace 4; EtOH 4	1.4171^{20}
3662	2-Butanone, 4-hydroxy-		$C_4H_8O_2$	88.11	590-90-9			109^{30}; 90^{11}	4-01-00-03994	1.1089^{24}	H_2O 5; EtOH 5; eth 5; ace 4	1.4585^{14}
3663	2-Butanone, 4-(4-hydroxy-3-methoxyphenyl)-	Zingerone	$C_{11}H_{14}O_3$	194.23	122-48-5	40.5	10072	$187\text{-}8^{14}$	4-08-00-01866		eth 4	
3664	2-Butanone, 4-hydroxy-3-methyl-		$C_5H_{10}O_2$	102.13	3393-64-4			$90\text{-}5^{15}$	4-01-00-04009	0.993^{25}		1.4340^{20}
3665	1-Butanone, 1-(2-hydroxyphenyl)-		$C_{10}H_{12}O_2$	164.20	2887-61-8	10		$124\text{-}6^{14}$	4-08-00-00497	1.0683^{24}		1.5379^{25}
3666	1-Butanone, 2-hydroxy-1-phenyl-	Butyrophenone, 2-hydroxy-	$C_{10}H_{12}O_2$	164.20	16183-46-3			$132\text{-}3^{12}$	4-08-00-00502	1.077^{19}		1.529^{19}
3667	1-Butanone, 4-hydroxy-1-phenyl-		$C_{10}H_{12}O_2$	164.20	39755-03-8	30		$187\text{-}8^{9}$	3-08-00-00449		EtOH 4	
3668	2-Butanone, 3-methyl-	Methyl isopropyl ketone	$C_5H_{10}O$	86.13	563-80-4	-92		94.3	4-01-00-03287	0.8051^{20}	H_2O 2; EtOH 5; eth 5; ace 4	1.3880^{20}
3669	1-Butanone, 3-methyl-1-(2-methylphenyl)-	Isobutyl 2-tolyl ketone	$C_{12}H_{16}O$	176.26	58138-81-1			248; $85\text{-}6^{3}$	4-07-00-00765	0.9578^{20}		1.5104^{20}
3670	1-Butanone, 3-methyl-1-(4-methylphenyl)-	Isobutyl 4-tolyl ketone	$C_{12}H_{16}O$	176.26	61971-91-3			259; 137^{18}	4-07-00-00765	0.9557^{20}		1.5081^{20}
3671	1-Butanone, 3-methyl-1-phenyl-		$C_{11}H_{14}O$	162.23	582-62-7			236.5	4-07-00-00739	0.9701^{16}	H_2O 1; EtOH 5; eth 5; ace 4	1.5139^{15}
3672	1-Butanone, 1-(4-methylphenyl)-	Butyrophenone, 4-methyl	$C_{11}H_{14}O$	162.23	4160-52-5	12		251.5	4-07-00-00743	0.9745^{20}	eth 4; EtOH 4	1.5232^{20}
3673	2-Butanone, (1-methylpropylidene)hydrazone		$C_8H_{16}N_2$	140.23	5921-54-0			171.5	4-01-00-03252	0.8404^{20}		1.4511^{20}
3674	1-Butanone, 1-(1-naphthalenyl)-	1-Butyronaphthone	$C_{14}H_{14}O$	198.26	2876-62-2			317	4-07-00-01305	1.0861^{25}	bz 4; eth 4; EtOH 4	1.596^{27}
3675	2-Butanone, oxime		C_4H_9NO	87.12	96-29-7	-29.5		152.5	4-01-00-03250	0.9232^{20}	H_2O 3; EtOH 5; eth 5; chl 3	1.4410^{20}
3676	1-Butanone, 1-phenyl-		$C_{10}H_{12}O$	148.20	495-40-9	12		228.5	4-07-00-00709	0.988^{20}	H_2O 1; EtOH 5; eth 5; ace 4	1.5203^{20}
3677	2-Butanone, 1-phenyl-		$C_{10}H_{12}O$	148.20	1007-32-5			230; 111^{16}	4-07-00-00712	1.002^{20}	H_2O 1; EtOH 3; eth 5; ace 4	
3678	2-Butanone, 4-phenyl-		$C_{10}H_{12}O$	148.20	2550-26-7	-13		233.5	4-07-00-00713	0.9849^{22}	H_2O 1; EtOH 3; eth 3; ace 4	1.511^{22}
3679	2-Butanone, 1,3,4-trihydroxy-, (*S*)-	*L*-Erythrulose	$C_4H_8O_4$	120.11	533-50-6	dec	3641	dec	4-01-00-04176		H_2O 4; EtOH 4	
3680	Butanoyl bromide	Butyryl bromide	C_4H_7BrO	151.00	5856-82-6			129	4-02-00-00804	1.4162^{17}		1.596^{17}
3681	Butanoyl chloride	Butyryl chloride	C_4H_7ClO	106.55	141-75-3	-89	1598	102	4-02-00-00803	1.0277^{20}	eth 5	1.4121^{20}
3682	Butanoyl chloride, 2-bromo-		C_4H_6BrClO	185.45	22118-12-3			151	4-02-00-00834	1.5320^{20}	eth 4	1.5320^{20}
3683	Butanoyl chloride, 2-chloro-		$C_4H_6Cl_2O$	141.00	7623-11-2			130.5	3-02-00-00623	1.2360^{17}	eth 4	1.4475^{20}
3684	Butanoyl chloride, 3-chloro-	3-Chlorobutyryl chloride	$C_4H_6Cl_2O$	141.00	1951-11-7			$40\text{-}1^{12}$	4-02-00-00824	1.2163^{20}		1.4509^{20}
3685	Butanoyl chloride, 4-chloro-		$C_4H_6Cl_2O$	141.00	4635-59-0			173.5	4-02-00-00826	1.2581^{20}	eth 3	1.4616^{20}
3686	Butanoyl chloride, 2-chloro-3-methyl-	Butyryl chloride, 2-chloro-3-methyl	$C_5H_8Cl_2O$	155.02	35383-59-6			148.5; 56^{20}	4-02-00-00904	1.135^{13}	eth 4	
3687	Butanoyl chloride, 4-chloro-3-oxo-	Butyryl chloride, 4-chloro-3-oxo	$C_4H_4Cl_2O_2$	154.98	41295-64-1			$117\text{-}9^{17}$	3-03-00-01207	1.4397^{20}	bz 4	1.4860^{20}
3688	Butanoyl chloride, 2,2-dimethyl-		$C_6H_{11}ClO$	134.61	5856-77-9			132	4-02-00-00955	0.9810^{20}	eth 4	1.4245^{20}
3689	Butanoyl chloride, 2,3-dimethyl-	Butyryl chloride, 2,3-dimethyl	$C_6H_{11}ClO$	134.61	51760-90-8			136; 38^{18}	3-02-00-00761	0.9795^{20}	eth 4	
3690	Butanoyl chloride, 3,3-dimethyl-		$C_6H_{11}ClO$	134.61	7065-46-5			130; 68^{100}	4-02-00-00956	0.9696^{20}	eth 4	1.4210^{20}
3691	Butanoyl chloride, 2-ethyl-		$C_6H_{11}ClO$	134.61	2736-40-5			140	4-02-00-00952	0.9825^{20}	eth 4	1.4234^{20}
3692	Butanoyl chloride, heptafluoro-		C_4ClF_7O	232.48	375-16-6			38.5	4-02-00-00818	1.55^{20}		1.288^{20}
3693	Butanoyl chloride, 3-methyl-	Isovaleryl chloride	C_5H_9ClO	120.58	108-12-3		5121	114	4-02-00-00901	0.9844^{20}	eth 3	1.4149^{20}
3694	Butanoyl chloride, 2-methyl-, (±)-	Butyryl chloride, 2-methyl (*DL*)	C_5H_9ClO	120.58	57526-28-0			116	4-02-00-00891	0.9917^{20}		1.4170^{20}
3695	2-Butenal, (*E*)-	*trans*-Crotonaldehyde	C_4H_6O	70.09	123-73-9	-76	2599	102.2	4-01-00-03447	0.8516^{20}	H_2O 3; EtOH 4; eth 4; ace 4	1.4366^{20}
3696	2-Butenal, 2-bromo-		C_4H_5BrO	148.99	24247-53-8			63^{15}	3-01-00-02981	1.5797^{13}		1.5184^{20}
3697	2-Butenal, 2-chloro-	α-Chlorocrotonaldehyde	C_4H_5ClO	104.54	53175-28-3			147.5	3-01-00-02981	1.1404^{23}	H_2O 2; EtOH 3; eth 3; ctc 3	1.4780^{25}
3698	2-Butenal, 2-methyl-		C_5H_8O	84.12	1115-11-3			117; 64^{110}	3-01-00-02990	0.8710^{20}	H_2O 3; EtOH 5; eth 5	1.4475^{20}
3699	2-Butenal, 2-methyl-, (*E*)-		C_5H_8O	84.12	497-03-0			117; 64^{119}	4-01-00-03464	0.8710^{20}	H_2O 4; EtOH 4	1.4475^{20}
3700	2-Butenal, 3-methyl-	Senecialdehyde	C_5H_8O	84.12	107-86-8		8399	134	4-01-00-03464	0.8722^{20}	H_2O 3; EtOH 3; eth 3	1.4528^{20}

No.	Name	Synonym	Mol. Form.	Mol. Wt.	CAS RN	mp/°C	Merck No.	bp/°C	Beil. Ref.	den/g cm^{-3}	Solubility	n_D
3701	2-Butenal, 3-phenyl-		$C_{10}H_{10}O$	146.19	1196-67-4	95		124^{12}; 109^{4}	4-07-00-01011			1.5876^{23}
3702	2-Butenamide, (*E*)-		C_4H_7NO	85.11	625-37-6	161.5		sub 140	2-02-00-00392	0.9461^{120}	H_2O 2; EtOH 3; eth 2; bz 3	1.4420^{165}
3703	3-Butenamide, 3-methyl-4-phenyl-	β-Benzalbutyramide	$C_{11}H_{13}NO$	175.23	7236-47-7	133	1063					
3704	1-Butene	1-Butylene	C_4H_8	56.11	106-98-9	-185.3	1513	-6.2	4-01-00-00765	$*0.588^{25}$	H_2O 1; EtOH 4; eth 4; bz 3	1.3962^{20}
3705	2-Butene, (*E*)-	*trans*-2-Butene	C_4H_8	56.11	624-64-6	-105.5	1514	0.8	4-01-00-00781	$*0.599^{25}$	bz 3	1.3848^{-25}
3706	2-Butene, (*Z*)-	*cis*-2-Butene	C_4H_8	56.11	590-18-1	-138.9	1514	3.7	4-01-00-00778	$*0.616^{25}$	H_2O 1; EtOH 4; eth 4; bz 3	1.3931^{-25}
3707	1-Butene, 1-bromo-, (*E*)-	*trans*-1-Butenyl bromide	C_4H_7Br	135.00	32620-08-9	-100.3		94.7		1.3209^{15}	H_2O 1; eth 3; ace 3; bz 3	1.4527^{20}
3708	1-Butene, 1-bromo-, (*Z*)-		C_4H_7Br	135.00	31849-78-2			86.1		1.3265^{15}	H_2O 1; eth 3; ace 3; bz 3	1.4536^{20}
3709	1-Butene, 2-bromo-	2-Bromo-1-butene	C_4H_7Br	135.00	23074-36-4	-133.4		88	4-01-00-00775	1.3209^{15}	H_2O 1; eth 3; ace 3; bz 3	1.4527^{20}
3710	1-Butene, 3-bromo-	α-Methallyl bromide	C_4H_7Br	135.00	22037-73-6			7^{10}	4-01-00-00775	1.2998^{25}		1.4618^{20}
3711	1-Butene, 4-bromo-		C_4H_7Br	135.00	5162-44-7			98.5	4-01-00-00775	1.3230^{20}	bz 4; eth 4; EtOH 4	1.4622^{20}
3712	2-Butene, 1-bromo-		C_4H_7Br	135.00	4784-77-4			104.5	4-01-00-00789	1.3371^{25}	H_2O 1; EtOH 3; eth 3; bz 4	1.4822^{20}
3713	2-Butene, 2-bromo-, (*E*)-		C_4H_7Br	135.00	3017-71-8	-114.6		85.6	4-01-00-00790	1.3323^{15}	H_2O 1; EtOH 3; eth 3; bz 4	1.4602^{16}
3714	2-Butene, 2-bromo-, (*Z*)-		C_4H_7Br	135.00	3017-68-3	-111.5		93.9	4-01-00-00790	1.3416^{15}	H_2O 1; EtOH 3; eth 3; bz 4	1.4631^{19}
3715	1-Butene, 4-bromo-3-chloro-3,4,4-trifluoro-	4-Bromo-3-chloro-3,4,4-trifluoro-1-butene	$C_4H_3BrClF_3$	223.42	374-25-4			99.5	4-01-00-00776	1.678^{25}		1.4092^{25}
3716	1-Butene, 2-bromo-3-methyl-	2-Bromo-3-methyl-1-butene	C_5H_9Br	149.03	31844-96-9			105	3-01-00-00800	1.2328^{20}	bz 4; eth 4; chl 4	1.4504^{20}
3717	2-Butene, 1-bromo-3-methyl-		C_5H_9Br	149.03	870-63-3			131 dec; 50^{40}	4-01-00-00824	1.2817^{20}	ace 4; bz 4; eth 4; EtOH 4	1.4930^{15}
3718	2-Butene, 2-bromo-3-methyl-		C_5H_9Br	149.03	3017-70-7			119.5	4-01-00-00824	1.2773^{20}	ace 4; eth 4; chl 4	1.4738^{20}
3719	2-Butene, 2-bromo-3-phenyl-	2-Bromo-3-phenyl-2-butene	$C_{10}H_{11}Br$	211.10	90841-14-8			$120\text{-}30^{11}$	4-05-00-01379	1.3348^{20}	bz 4; eth 4; chl 4	1.5811^{20}
3720	1-Butene, 1-chloro-, (*E*)-	*trans*-1-Chloro-1-butene	C_4H_7Cl	90.55	7611-87-2			68	3-01-00-00723	0.9205^{15}	H_2O 1; EtOH 5; eth 4; ace 3	1.4223^{15}
3721	1-Butene, 1-chloro-, (*Z*)-	*cis*-1-Chloro-1-butene	C_4H_7Cl	90.55	7611-86-1			63.5	3-01-00-00723	0.9153^{15}	ace 4; eth 4; EtOH 4; chl 4	1.4194^{15}
3722	1-Butene, 2-chloro-	2-Chloro-1-butene	C_4H_7Cl	90.55	2211-70-3			58.5	4-01-00-00769	0.9107^{15}	ace 4; bz 4; eth 4; EtOH 4	1.4165^{21}
3723	1-Butene, 3-chloro-	3-Chloro-1-butene	C_4H_7Cl	90.55	563-52-0		2131	64.5	4-01-00-00769	0.8978^{20}	eth 4; ace 4; chl 3	1.4149^{20}
3724	1-Butene, 4-chloro-		C_4H_7Cl	90.55	927-73-1			75	4-01-00-00771	0.9211^{20}	ace 4; eth 4; chl 4	1.4233^{20}
3725	2-Butene, 1-chloro-, (*E*)-		C_4H_7Cl	90.55	4894-61-5			85	4-01-00-00784	0.9295^{20}	H_2O 1; ace 3; chl 3	1.4350^{20}
3726	2-Butene, 1-chloro-, (*Z*)-		C_4H_7Cl	90.55	4628-21-1			84.1	4-01-00-00783	0.9426^{20}	H_2O 1; EtOH 3; ace 3; chl 3	1.4390^{20}
3727	2-Butene, 2-chloro-, (*E*)-		C_4H_7Cl	90.55	2211-68-9	-105.8		62.8	4-01-00-00785	0.9138^{20}	H_2O 1; EtOH 5; ace 3; chl 3	1.4190^{20}
3728	2-Butene, 2-chloro-, (*Z*)-		C_4H_7Cl	90.55	2211-69-0	-117.3		70.6	4-01-00-00785	0.9239^{20}	H_2O 1; EtOH 5; ace 3; chl 3	1.4240^{20}
3729	1-Butene, 3-chloro-2-(chloromethyl)-	3-Chloro-2-(chloromethyl)-1-butene	$C_5H_8Cl_2$	139.02	69295-21-2			155	3-01-00-00787	1.1233^{20}	ace 4; chl 4	1.4724^{20}
3730	2-Butene, 2-chloro-1,1-dimethoxy-	Crotonaldehyde-2-chloro, dimethylacetal	$C_6H_{11}ClO_2$	150.60	108365-83-9			58^{13}	3-01-00-02981	1.074^{18}		1.4466^{18}
3731	2-Butene, 1-chloro-2,3-dimethyl-	1-Chloro-2,3-dimethyl-2-butene	$C_6H_{11}Cl$	118.61	37866-06-1			112	4-01-00-00856	0.9355^{20}	ace 4; eth 4; chl 4	1.4605^{20}
3732	1-Butene, 1-chloro-2-methyl-	1-Chloro-2-methyl-1-butene	C_5H_9Cl	104.58	23378-11-2			96.5	3-01-00-00787	0.9170^{20}	ace 4; eth 4	1.4141^{20}
3733	1-Butene, 1-chloro-3-methyl-	1-Chloro-3-methyl-1-butene	C_5H_9Cl	104.58	23010-00-6			87	[illegible]		ace 4; eth 4; chl 4	1.4229^{20}
3734	1-Butene, 3-chloro-2-methyl-	3-Chloro-2-methyl-1-butene	C_5H_9Cl	104.58	5166-35-8			94	4-01-00-00819	0.9088^{20}	ace 4; eth 4; chl 4	1.4304^{20}
3735	2-Butene, 1-chloro-2-methyl-	1-Chloro-2-methyl-2-butene	C_5H_9Cl	104.58	13417-43-1			110	4-01-00-00822	0.9327^{20}	ace 4; eth 4; EtOH 4	1.4481^{20}
3736	2-Butene, 1-chloro-3-methyl-		C_5H_9Cl	104.58	503-60-6			109	4-01-00-00823	0.9273^{20}	ace 4; eth 4; EtOH 4; chl 4	1.4485^{20}
3737	2-Butene, 2-chloro-3-methyl-	2-Chloro-3-methyl-2-butene	C_5H_9Cl	104.58	17773-65-8			96	3-01-00-00794	0.9324^{20}	ace 4; eth 4; EtOH 4; chl 4	1.4320^{20}
3738	1-Butene, 1,1-dibromo-	1,1-Dibromo-1-butene	$C_4H_6Br_2$	213.90	73383-24-1			$53\text{-}5^{22}$	3-01-00-00727	1.8315^{20}		1.5168^{20}
3739	1-Butene, 1,2-dibromo-	1,2-dibromo-1-butene	$C_4H_6Br_2$	213.90	55030-56-3	-49.5		150	0-01-00-00204	1.887^{25}	eth 4; EtOH 4	
3740	1-Butene, 3,4-dibromo-	1,2-Dibromo-3-butene	$C_4H_6Br_2$	213.90	10463-48-6			52^{10}	4-01-00-00776	1.865^{24}		1.541^{21}

No.	Name Synonym	Mol. Form. Mol. Wt.	CAS RN mp/°C	Merck No. bp/°C	Beil. Ref. den/g cm^{-3}	Solubility n_D
3741	2-Butene, 1,3-dibromo- 2,4-Dibromo-2-butene	$C_4H_6Br_2$ 213.90	64930-16-1	168 dec	3-01-00-00746 1.8768^{20}	1.5485^{20}
3742	2-Butene, 1,4-dibromo-, (*E*)-	$C_4H_6Br_2$ 213.90	821-06-7 53.4	203; 74^{14}	4-01-00-00791	H_2O 2; EtOH 4; ace 3; chl 2
3743	1-Butene, 1,3-dichloro- 1,3-Dichloro-1-butene	$C_4H_6Cl_2$ 125.00	52497-07-1	125	4-01-00-00772 1.1341^{24}	ace 4; eth 4; EtOH 4; chl 4 1.4647^{20}
3744	1-Butene, 2,3-dichloro- 2,3-Dichloro-1-butene	$C_4H_6Cl_2$ 125.00	7013-11-8	112	4-01-00-00772 1.1340^{20}	ace 4; eth 4; chl 4 1.4580^{20}
3745	1-Butene, 3,4-dichloro- 1,2-Dichloro-3-butene	$C_4H_6Cl_2$ 125.00	760-23-6 -61	116	4-01-00-00772 1.1170^{20}	H_2O 1; EtOH 3; eth 3; bz 4 1.4641^{20}
3746	2-Butene, 1,1-dichloro- 1,1-Dichloro-2-butene	$C_4H_6Cl_2$ 125.00	56800-09-0	126	3-01-00-00741 1.1310^{20}	1.466^{18}
3747	2-Butene, 1,2-dichloro- 1,2-Dichloro-2-butene	$C_4H_6Cl_2$ 125.00	13602-13-6	130.5	3-01-00-00741 1.1601^{20}	ace 4; eth 4; chl 4 1.4734^{20}
3748	2-Butene, 1,3-dichloro-, (*E*)-	$C_4H_6Cl_2$ 125.00	7415-31-8	131; 53^{50}	4-01-00-00786 1.1585^{20}	ace 4; bz 4; eth 4; EtOH 4 1.4719^{20}
3749	2-Butene, 1,3-dichloro-, (*Z*)-	$C_4H_6Cl_2$ 125.00	10075-38-4	131; 34^{20}	4-01-00-00786 1.1605^{20}	ace 4; bz 4; eth 4; EtOH 4 1.4735^{20}
3750	2-Butene, 1,4-dichloro-, (*E*)- 1,4-Dichloro-*trans*-2-butene	$C_4H_6Cl_2$ 125.00	110-57-6 2.0	155.4	4-01-00-00787 1.183^{25}	ace 4; bz 4; eth 4; EtOH 4 1.4871^{25}
3751	2-Butene, 1,4-dichloro-, (*Z*)-	$C_4H_6Cl_2$ 125.00	1476-11-5 -48	152.5	4-01-00-00787 1.188^{25}	ace 4; bz 4; eth 4; EtOH 4 1.4887^{25}
3752	2-Butene, 2,3-dichloro-, (*E*)-	$C_4H_6Cl_2$ 125.00	1587-29-7	102	4-01-00-00787 1.1416^{20}	H_2O 1; EtOH 3; eth 3; ace 3 1.4582^{20}
3753	2-Butene, 2,3-dichloro-, (*Z*)-	$C_4H_6Cl_2$ 125.00	1587-26-4	125.5	4-01-00-00787 1.1618^{20}	H_2O 1; EtOH 3; eth 3; ace 3 1.4590^{20}
3754	2-Butene, 2,3-dichloro-1,1,1,4,4,4-hexafluoro- 2,3-Dichlorohexafluoro-2-butene	$C_4Cl_2F_6$ 232.94	303-04-8 -67.3	68.5	4-01-00-00787 1.6233^{20}	1.3459^{20}
3755	1-Butene, 3,3-dichloro-2-methyl- 3,3-Dichloro-2-methyl-1-butene	$C_5H_8Cl_2$ 139.02	42101-38-2		3-01-00-00787 1.1276^{18}	ace 4; eth 4; chl 4 1.4737^{18}
3756	2-Butene, 1,3-dichloro-2-methyl- 1,3-Dichloro-2-methyl-2-butene	$C_5H_8Cl_2$ 139.02	25148-87-2	150	3-01-00-00795 1.1293^{20}	ace 4; chl 4
3757	2-Butene, 1,4-dichloro-2-methyl- 1,4-Dichloro-2-methyl-2-butene	$C_5H_8Cl_2$ 139.02	29843-58-1	93^{50}; 56^{10}	3-01-00-00795 1.1526^{20}	ace 4; chl 4 1.4932^{20}
3758	2-Butene, 1,1-diethoxy- Crotonaldehyde, diethyl acetal	$C_8H_{16}O_2$ 144.21	10602-34-3	147.5	3-01-00-02980 0.8473^{18}	H_2O 1; EtOH 5; eth 5; ace 5 1.4097^{20}
3759	1-Butene, 2,3-dimethyl- 2,3-Dimethyl-1-butene	C_6H_{12} 84.16	563-78-0 -157.3	55.6	4-01-00-00852 0.6803^{20}	H_2O 1; EtOH 3; eth 3; ace 3 1.3995^{20}
3760	1-Butene, 3,3-dimethyl-	C_6H_{12} 84.16	558-37-2 -115.2	41.2	4-01-00-00850 0.6529^{20}	H_2O 1; EtOH 3; eth 3; ctc 3 1.3763^{20}
3761	2-Butene, 2,3-dimethyl- 2,3-Dimethyl-2-butene	C_6H_{12} 84.16	563-79-1 -74.6	73.3	4-01-00-00853 0.7080^{20}	H_2O 1; EtOH 3; eth 3; ace 3 1.4122^{20}
3762	2-Butenedinitrile, (*E*)-	$C_4H_2N_2$ 78.07	764-42-1 96.8	186	4-02-00-02219 0.9416^{111}	H_2O 3; EtOH 3; eth 3; ace 3 1.4349^{111}
3763	2-Butenedinitrile, 2-chloro-, (*Z*)- Fumaronitrile, chloro-	C_4HClN_2 112.52	71200-79-8 60.5	172; 68^{25}	3-02-00-01909 1.2499^{20}	1.4957^{20}
3764	2-Butenedioic acid (*E*)- Fumaric acid	$C_4H_4O_4$ 116.07	110-17-8 287 dec	4200 sub 165	3-02-00-01890 1.635^{20}	H_2O 2; EtOH 3; eth 2; ace 2
3765	2-Butenedioic acid (*Z*)- Maleic acid	$C_4H_4O_4$ 116.07	110-16-7 130.5	5585	4-02-00-02199 1.590^{20}	H_2O 4; EtOH 4; eth 3; ace 4
3766	2-Butenedioic acid (*E*)-, bis(1-methylethyl) ester Diisopropyl fumarate	$C_{10}H_{16}O_4$ 200.23	7283-70-7 2.1	280	4-02-00-02209	ace 4; eth 4; EtOH 4
3767	2-Butenedioic acid (*E*)-, bis(2-methylpropyl) ester Diisobutyl fumarate	$C_{12}H_{20}O_4$ 228.29	7283-69-4	170^{160}; 122^5	4-02-00-02211 0.9760^{20}	H_2O 1; EtOH 3; eth 3; ace 3 1.4432^{20}
3768	2-Butenedioic acid (*E*)-, bis(phenylmethyl) ester Benzyl fumarate	$C_{18}H_{16}O_4$ 296.32	538-64-7 59	1149 210-11^5	4-06-00-02275	eth 4; EtOH 4; chl 4
3769	2-Butenedioic acid, 2-bromo-, (*E*)-	$C_4H_3BrO_4$ 194.97	584-99-6 141	dec	4-02-00-02224	H_2O 4; EtOH 4; eth 4
3770	2-Butenedioic acid, 2-bromo-, (*Z*)- 2-Bromofumaric acid	$C_4H_3BrO_4$ 194.97	644-80-4 184.3	200 dec	4-02-00-02224	H_2O 4; EtOH 4
3771	2-Butenedioic acid, 2-chloro-, diethyl ester, (*E*)- Diethyl chloromaleate	$C_8H_{11}ClO_4$ 206.63	626-10-8	235 dec; 125^{19}	3-02-00-01928 1.1741^{20}	ace 4; eth 4; EtOH 4
3772	2-Butenedioic acid, 2-chloro-, diethyl ester, (*Z*)-	$C_8H_{11}ClO_4$ 206.63	10302-94-0	250 dec; 127^{10}	3-02-00-01909 1.1880^{20}	eth 4; EtOH 4 1.4571^{20}
3773	2-Butenedioic acid, 2-chloro-, dimethyl ester, (*E*)- Dimethyl chloromaleate	$C_6H_7ClO_4$ 178.57	19393-45-4	224; 100^{17}	3-02-00-01928 1.2899^{25}	eth 4; EtOH 4 1.4720^{18}
3774	2-Butenedioic acid (*E*)-, dibutyl ester	$C_{12}H_{20}O_4$ 228.29	105-75-9	150^4	4-02-00-02210 0.9869^{20}	H_2O 1; ace 3; chl 3 1.4469^{20}
3775	2-Butenedioic acid (*E*)-, diethyl ester	$C_8H_{12}O_4$ 172.18	623-91-6 1.5	214	4-02-00-02207 1.0452^{20}	H_2O 1; ace 3; chl 3 1.4412^{20}
3776	2-Butenedioic acid (*Z*)-, diethyl ester Diethyl maleate	$C_8H_{12}O_4$ 172.18	141-05-9 -8.8	3113 223	4-02-00-02207 1.0662^{20}	H_2O 1; EtOH 3; eth 3; chl 2 1.4416^{20}

No.	Name Synonym	Mol. Form. Mol. Wt.	CAS RN mp/°C	Merck No. bp/°C	Beil. Ref. den/g cm^{-3}	Solubility n_D
3777	2-Butenedioic acid, 2,3-dihydroxy-, (*Z*)- Dihydroxymaleic acid	$C_4H_4O_6$ 148.07	526-84-1 155 dec	3170	4-03-00-01975	H_2O 2; EtOH 3; eth 2; MeOH 2
3778	2-Butenedioic acid (*E*)-, dimethyl ester	$C_6H_8O_4$ 144.13	624-49-7 103.5	 193	4-02-00-02205 1.37^{20}	H_2O 1; ace 3; chl 3 1.4062^{111}
3779	2-Butenedioic acid (*Z*)-, dimethyl ester	$C_6H_8O_4$ 144.13	624-48-6 -19	 202	4-02-00-02204 1.1606^{20}	H_2O 1; eth 3; ctc 3; lig 1 1.4416^{20}
3780	2-Butenedioic acid (*Z*)-, diphenyl ester	$C_{16}H_{12}O_4$ 268.27	7242-17-3 73	 226^{15}	4-06-00-00628	ace 4; bz 4; eth 4; EtOH 4
3781	2-Butenedioic acid (*E*)-, di-2-propenyl ester Diallyl fumarate	$C_{10}H_{12}O_4$ 196.20	2807-54-7	 140^3	3-02-00-01906 1.060^{25}	ace 4; bz 4; eth 4; EtOH 4 1.4670^{25}
3782	2-Butenedioic acid (*Z*)-, di-2-propenyl ester Diallyl maleate	$C_{10}H_{12}O_4$ 196.20	999-21-3	 $108\text{-}10^3$	4-02-00-02214 1.0773^{20}	chl 3 1.4699^{20}
3783	2-Butenedioic acid (*E*)-, dipropyl ester Dipropyl fumarate	$C_{10}H_{16}O_4$ 200.23	14595-35-8	 110^5	4-02-00-02209 1.0129^{20}	EtOH 3; eth 3 1.4435^{20}
3784	2-Butenedioic acid (*Z*)-, dipropyl ester	$C_{10}H_{16}O_4$ 200.23	2432-63-5	 126^{12}	4-02-00-02209 1.0245^{20}	H_2O 1; EtOH 3; eth 3; ace 3 1.4434^{20}
3785	2-Butenedioic acid, 2-methyl-, (*E*)- Mesaconic acid	$C_5H_6O_4$ 130.10	498-24-8 204.5	5806 sub	4-02-00-02231 1.466^{20}	H_2O 2; EtOH 4; eth 3; bz 2
3786	2-Butenedioic acid, 2-methyl-, (*Z*)- Citraconic acid	$C_5H_6O_4$ 130.10	498-23-7 93.5	2323	4-02-00-02230 1.617^{25}	H_2O 4; eth 2; bz 1; chl 2
3787	2-Butenedioic acid, 2-methyl-, diethyl ester, (*E*)- Diethyl mesaconate	$C_9H_{14}O_4$ 186.21	2418-31-7	 229	4-02-00-02232 1.0434^{20}	ace 4; bz 4; eth 4; EtOH 4 1.4488^{20}
3788	2-Butenedioic acid, 2-methyl-, diethyl ester, (*Z*)-	$C_9H_{14}O_4$ 186.21	691-83-8	 228; 120^{20}	4-02-00-02232 1.0491^{20}	eth 4; EtOH 4; HOAc 4 1.4467^{20}
3789	2-Butenedioic acid, 2-methyl-, dimethyl ester, (*E*)-	$C_7H_{10}O_4$ 158.15	617-53-8	 203.5	4-02-00-02232 1.0914^{20}	ace 4; eth 4; EtOH 4 1.4512^{20}
3790	2-Butenedioic acid, 2-methyl-, dimethyl ester, (*Z*)-	$C_7H_{10}O_4$ 158.15	617-54-9	 210.5	4-02-00-02232 1.1153^{20}	ace 4; eth 4; EtOH 4 1.4473^{20}
3791	2-Butenedioic acid (*E*)-, monoethyl ester	$C_6H_8O_4$ 144.13	2459-05-4 70	 147^{16}	4-02-00-02206 1.1109^{87}	EtOH 3; ace 3; chl 2
3792	2-Butene-1,4-diol, (*E*)- *trans*-2-Butene-1,4-diol	$C_4H_8O_2$ 88.11	821-11-4 25	 131^{13}	4-01-00-02660 1.0700^{20}	H_2O 4; EtOH 4 1.4755^{20}
3793	2-Butene-1,4-diol, (*Z*)- *cis*-2-Butene-1,4-diol	$C_4H_8O_2$ 88.11	6117-80-2 4	 235	4-01-00-02660 1.0698^{20}	H_2O 3; EtOH 4 1.4782^{20}
3794	3-Butene-1,2-diol	$C_4H_8O_2$ 88.11	497-06-3	 196.5	4-01-00-02658 1.0470^{20}	H_2O 3; EtOH 3 1.4628^{21}
3795	2-Butenedioyl dichloride, (*E*)- Fumaric acid dichloride	$C_4H_2Cl_2O_2$ 152.96	627-63-4	 159	4-02-00-02217 1.408^{20}	 1.5004^{18}
3796	2-Butenedioyl dichloride, 2-chloro-, (*Z*)-	$C_4HCl_3O_2$ 187.41	17096-37-6	 185 dec	3-02-00-01909 1.564^{20}	eth 4; HOAc 4 1.5206^{20}
3797	1-Butene, 2-ethyl-3-methyl- 2-Ethyl-3-methylbut-1-ene	C_7H_{14} 98.19	7357-93-9	 89	4-01-00-00871 0.7150^{20}	H_2O 1; eth 3; ace 3; bz 3 1.410^{20}
3798	2-Butene, 1,1,2,3,4,4-hexachloro- 2-Butene, 1,1,2,3,4,4-hexachloro (liquid)	$C_4H_2Cl_6$ 262.78	13045-99-3 -19	 $97\text{-}8^{10}$	4-01-00-00789 1.650^{15}	H_2O 1; chl 3 1.5331
3799	1-Butene, 2-methyl- 2-Methyl-1-butene	C_5H_{10} 70.13	563-46-2 -137.5	 31.2	4-01-00-00818 0.6504^{20}	H_2O 1; EtOH 3; eth 3; bz 3 1.3378^{20}
3800	1-Butene, 3-methyl- 3-Methyl-1-butene	C_5H_{10} 70.13	563-45-1 -168.5	 20.1	4-01-00-00825 0.6213^{25}	H_2O 1; EtOH 5; eth 5; bz 3 1.3643^{20}
3801	2-Butene, 2-methyl- 2-Methyl-2-butene	C_5H_{10} 70.13	513-35-9 -133.7	644 38.5	4-01-00-00820 0.6623^{20}	H_2O 1; EtOH 3; eth 3; bz 3 1.3874^{20}
3802	2-Butenenitrile, (*E*)-	C_4H_5N 67.09	627-26-9 -51.5	 120	4-02-00-01507 0.8239^{20}	eth 3; ace 3 1.4225^{20}
3803	3-Butenenitrile Allyl cyanide	C_4H_5N 67.09	109-75-1 -87	288 119	4-02-00-01491 0.8341^{20}	H_2O 2; EtOH 5; eth 5 1.4060^{20}
3804	2-Butene, 1,1,1,2,3,4,4,4-octafluoro- Perfluoro-2-butene	C_4F_8 200.03	360-89-4 -129	 1.5	4-01-00-00783 1.5297^{25}	
3805	2-Butene, 1,1,1,4,4-pentachloro- 1,1,1,4,4-Pentachloro-2-butene	$C_4H_3Cl_5$ 228.33	77753-21-0	 $78\text{-}80^{11}$	3-01-00-00744 1.609^{21}	EtOH 4; chl 4 1.5538^{21}
3806	1-Butene, 1,3,4,4-tetrachloro- 1,3,4,4-Tetrachloro-1-butene	$C_4H_4Cl_4$ 193.89	2984-40-9	 88^{20}	4-01-00-00774 1.476^{20}	chl 4 1.4773^{20}
3807	1-Butene, 2,3,3,4-tetrachloro- 2,3,3,4-Tetrachloro-1-butene	$C_4H_4Cl_4$ 193.89	[illegible]	 $41\text{-}2^7$	[illegible] 1.4602^{20}	ace 4; chl 4 1.5135^{20}
3808	1-Butene, 1,3,4,4-tetrachloro-1,2,3,4-tetrafluoro- 1-Butene, 1,3,4,4-tetrachloro-tetrafluoro	$C_4Cl_4F_4$ 265.85	357-20-0	 140.8	4-01-00-00774 1.6902^{25}	 1.429^{20}
3809	1-Butene, 2,3,4-trichloro- 2,3,4-Trichloro-1-butene	$C_4H_5Cl_3$ 159.44	2431-50-7	 60^{20}	3-01-00-00725 1.3430^{20}	ace 4; chl 4 1.4944^{20}
3810	2-Butene, 1,2,4-trichloro- 1,2,4-Trichloro-2-butene	$C_4H_5Cl_3$ 159.44	2431-54-1	 $67\text{-}9^{10}$	3-01-00-00744 1.3843^{20}	bz 4; eth 4; EtOH 4; chl 4 1.5175^{20}
3811	2-Butene, 1,1,3-triethoxy- 2-Butenal, 3-ethoxy, diethyl acetal	$C_{10}H_{20}O_3$ 188.27	69190-65-4	 192.5	4-01-00-04082 0.908^{21}	EtOH 4 1.430^{21}
3812	1-Butene, 2,3,3-trimethyl-	C_7H_{14} 98.19	594-56-9 -109.9	 77.9	4-01-00-00873 0.7050^{20}	H_2O 1; eth 3; bz 3; chl 3 1.4025^{20}
3813	3-Butenoic acid	$C_4H_6O_2$ 86.09	625-38-7 -35	 169	4-02-00-01491 1.0091^{20}	H_2O 3; EtOH 5; eth 5 1.4239^{20}

No.	Name	Synonym	Mol. Form.	Mol. Wt.	CAS RN	mp/°C	Merck No.	bp/°C	Beil. Ref.	den/g cm^{-3}	Solubility	n_D
3814	2-Butenoic acid, (*E*)-	*trans*-Crotonic acid	$C_4H_6O_2$	86.09	107-93-7	72	2600	184.7	4-02-00-01498	0.9604^{77}	H_2O 4; EtOH 4; eth 3; ace 3	1.4249^{77}
3815	2-Butenoic acid, (*Z*)-	*cis*-Crotonic acid	$C_4H_6O_2$	86.09	503-64-0	15	5048	169	4-02-00-01497	1.0267^{20}	H_2O 4; EtOH 3	1.4483^{14}
3816	2-Butenoic acid, 3-amino-, ethyl ester, (*E*)-		$C_6H_{11}NO_2$	129.16	41867-20-3	34		212 dec	3-03-00-01199	1.0219^{19}	H_2O 1; EtOH 3; eth 3; bz 3	1.4988^{22}
3817	2-Butenoic acid, 4-amino-4-oxo-, (*Z*)-	Maleamic acid	$C_4H_5NO_3$	115.09	557-24-4	172.5	5583		4-02-00-02218		H_2O 4; EtOH 4	
3818	2-Butenoic acid, anhydride	Crotonic acid anhydride	$C_8H_{10}O_3$	154.17	623-68-7			247; 129^{19}	4-02-00-01505	1.0397^{20}	eth 4	1.4745^{20}
3819	2-Butenoic acid, 4-bromo-, ethyl ester, (*E*)-		$C_6H_9BrO_2$	193.04	37746-78-4			$97\text{-}8^{15}$	4-02-00-01517	1.402^{16}	EtOH 4	1.4925^{20}
3820	2-Butenoic acid, 3-chloro-, (*E*)-		$C_4H_5ClO_2$	120.54	6214-28-4	61		195	4-02-00-01510	1.1969^{69}	H_2O 2; EtOH 4; chl 3; CS_2 4	
3821	2-Butenoic acid, 3-chloro-, (*Z*)-		$C_4H_5ClO_2$	120.54	6213-90-7	61			4-02-00-01510	1.1905^{66}	H_2O 2; EtOH 4; peth 4	1.4704^{66}
3822	2-Butenoic acid, 4-chloro-	γ-Chlorocrotonic acid	$C_4H_5ClO_2$	120.54	16197-90-3	83		117^{13}	4-02-00-01511		eth 4; HOAc 4	
3823	2-Butenoic acid, 2-chloro-, ethyl ester	Ethyl β-chlorocrotonate	$C_6H_9ClO_2$	148.59	56216-12-7			176	4-02-00-01510	1.1133^{20}		1.453^{20}
3824	2-Butenoic acid, 2-chloro-, ethyl ester, (*E*)-	2-Butenoic acid, 2-chloro, ethyl ester (*trans*)	$C_6H_9ClO_2$	148.59	77825-53-7			177; 75^{30}	4-02-00-01510	1.1135^{20}	eth 4; EtOH 4	1.4538^{20}
3825	2-Butenoic acid, 2-chloro-, ethyl ester, (*Z*)-	2-Butenoic acid, 2-chloro, ethyl ester (*cis*)	$C_6H_9ClO_2$	148.59	77825-54-8			75^{30}	4-02-00-01510	1.1021^{18}	eth 4; EtOH 4	
3826	2-Butenoic acid, 3-chloro-, ethyl ester	3-Chlorocrotonic acid ethyl ester	$C_6H_9ClO_2$	148.59	38624-62-3			180; 75^{14}	4-02-00-01511	1.1062^{20}		1.459^{20}
3827	2-Butenoic acid, 3-chloro-, ethyl ester, (*E*)-		$C_6H_9ClO_2$	148.59	6127-92-0			184	4-02-00-01511	1.1062^{20}	eth 4; EtOH 4	1.4592^{20}
3828	2-Butenoic acid, 3-chloro-, ethyl ester, (*Z*)-		$C_6H_9ClO_2$	148.59	6127-93-1			161.4	4-02-00-01511	1.0860^{20}	eth 4; EtOH 4	1.4542^{19}
3829	2-Butenoic acid, 4-chloro-, ethyl ester	Crotonic acid, 4-chloro-, ethyl ester	$C_6H_9ClO_2$	148.59	15333-22-9			93^{10}	4-02-00-01511	1.090^{18}		1.462^{18}
3830	2-Butenoic acid, 2-chloro-, methyl ester, (*E*)-		$C_5H_7ClO_2$	134.56	22038-57-9			161.5; 63^{17}	4-02-00-01510	1.160^{20}	eth 4	1.4569^{23}
3831	2-Butenoic acid, 3-chloro-, methyl ester, (*E*)-		$C_5H_7ClO_2$	134.56	6372-01-6			$64\text{-}7^{14}$	4-02-00-01510	1.157^{20}	H_2O 2; EtOH 4; eth 4	1.4630^{20}
3832	2-Butenoic acid, 3-chloro-, methyl ester, (*Z*)-		$C_5H_7ClO_2$	134.56	6214-25-1			142.4; 70^{24}	4-02-00-01500	1.138^{20}	H_2O 2; eth 4; MeOH 4	1.4573^{19}
3833	2-Butenoic acid, 4-chloro-, methyl ester	Methyl γ-chlorocrotonate	$C_5H_7ClO_2$	134.56	15320-72-6			68^{11}	4-02-00-01511	1.169^{18}		1.467^{18}
3834	2-Butenoic acid, 3-cyclohexyl-	Cicrotoic acid	$C_{10}H_{16}O_2$	168.24	25229-42-9	85.5	2272					
3835	3-Butenoic acid, 2,2-dimethyl-	2,2-Dimethyl-3-butenoic acid	$C_6H_{10}O_2$	114.14	10276-09-2	-6		185	4-02-00-01574	0.9567^{15}		1.4305^{15}
3836	2-Butenoic acid, 2,3-dimethyl-, ethyl ester	Ethyl 2,3-dimethyl-2-butenoate	$C_8H_{14}O_2$	142.20	13979-28-7			155	4-02-00-01576	0.9072^{19}		1.430^{19}
3837	2-Butenoic acid, 2-ethyl-, (*E*)-	2-Ethylcrotonic acid	$C_6H_{10}O_2$	114.14	1187-13-9	45.5		209	3-02-00-01329	0.9578^{50}	eth 4; EtOH 4	1.4475^{50}
3838	2-Butenoic acid, ethyl ester		$C_6H_{10}O_2$	114.14	10544-63-5			136.5	2-02-00-00394	0.9175^{20}	eth 4; EtOH 4	1.4243^{20}
3839	2-Butenoic acid, ethyl ester, (*E*)-	Ethyl *trans*-crotonate	$C_6H_{10}O_2$	114.14	623-70-1		2600	138	4-02-00-01500	0.9175^{20}	H_2O 1; EtOH 3; eth 3	1.4243^{20}
3840	2-Butenoic acid, ethyl ester, (*Z*)-	*cis*-Ethyl crotonate	$C_6H_{10}O_2$	114.14	6776-19-8		2600	136	4-02-00-01497	0.9182^{20}	ace 4; eth 4; EtOH 4	1.4242^{20}
3841	3-Butenoic acid, ethyl ester	Ethyl 3-butenoate	$C_6H_{10}O_2$	114.14	1617-18-1			119	4-02-00-01491	0.9122^{20}	EtOH 3	1.4105^{20}
3842	3-Butenoic acid, 2-hydroxy-	Vinylglycolic acid	$C_4H_6O_3$	102.09	600-17-9	43		124^{9}	3-03-00-00685	1.2020^{21}	H_2O 4; eth 4; EtOH 4; chl 4	
3843	3-Butenoic acid, 2-hydroxy-, ethyl ester	Ethyl 2-hydroxy-3-butenoate	$C_6H_{10}O_3$	130.14	91890-87-8			173 dec; 68^{15}	3-03-00-00685	1.0470^{15}	H_2O 4; eth 4; EtOH 4	1.436^{13}
3844	2-Butenoic acid, 2-methyl-, (*E*)-	Tiglic acid	$C_5H_8O_2$	100.12	80-59-1	64.5	9365	198.5	4-02-00-01552	0.9641^{76}	eth 4; EtOH 4	1.4330^{76}
3845	2-Butenoic acid, 2-methyl-, (*Z*)-	Angelic acid	$C_5H_8O_2$	100.12	565-63-9	45.5	678	185	4-02-00-01551	0.9834^{49}	H_2O 2; EtOH 3; eth 4	1.4434^{47}
3846	2-Butenoic acid, 3-methyl-		$C_5H_8O_2$	100.12	541-47-9	69.5		197	4-02-00-01555	1.0062^{24}		
3847	2-Butenoic acid, methyl ester, (*E*)-	Methyl *trans*-crotonate	$C_5H_8O_2$	100.12	623-43-8	-42	2600	121	4-02-00-01500	0.9444^{20}	H_2O 1; EtOH 4; eth 4	1.4242^{20}
3848	2-Butenoic acid, methyl ester, (*Z*)-	Methyl *cis*-crotonate	$C_5H_8O_2$	100.12	4358-59-2		5048	118	4-02-00-01500			1.4175^{20}
3849	2-Butenoic acid, 2-methyl-, ethyl ester, (*E*)-		$C_7H_{12}O_2$	128.17	5837-78-5			156	4-02-00-01553	0.9200^{20}		1.4340^{20}
3850	2-Butenoic acid, 3-methyl-, ethyl ester		$C_7H_{12}O_2$	128.17	638-10-8			153.5	4-02-00-01556	0.9199^{21}		1.4345^{20}
3851	2-Butenoic acid, 2-methyl-, methyl ester, (*E*)-		$C_6H_{10}O_2$	114.14	6622-76-0			139	4-02-00-01553	0.9498^{20}		1.4370^{20}
3852	2-Butenoic acid, 3-methyl-, methyl ester		$C_6H_{10}O_2$	114.14	924-50-5	114		136.5	3-02-00-01313	0.9337^{20}		1.432^{20}

No.	Name Synonym	Mol. Form. Mol. Wt.	CAS RN mp/°C	Merck No. bp/°C	Beil. Ref. den/g cm^{-3}	Solubility n_D
3853	2-Butenoic acid, 4-oxo-4-(phenylamino)-, (*Z*)- Maleanilic acid	$C_{10}H_9NO_3$ 191.19	555-59-9 192 dec	5584	3-12-00-00571 1.418^{30}	
3854	2-Butenoic acid, 3-phenyl-, (*Z*)-	$C_{10}H_{10}O_2$ 162.19	704-79-0 131.5	 170^{14}	4-09-00-02057	bz 3; CS_2 3; peth 2
3855	3-Butenoic acid, 4-phenyl- Styrylacetic acid	$C_{10}H_{10}O_2$ 162.19	2243-53-0 87	 302	4-09-00-02054	H_2O 1; EtOH 4; eth 4; CS_2 4
3856	2-Butenoic acid, 2-propenyl ester Allyl crotonate	$C_7H_{10}O_2$ 126.16	20474-93-5	 $88\text{-}9^{70}$	4-02-00-01503 0.9440^{20}	 1.4465^{20}
3857	2-Butenoic acid, 2-propenyl ester, (*E*)-	$C_7H_{10}O_2$ 126.16	5453-44-1	 $88\text{-}9^{70}$	4-02-00-01503 0.9440^{20}	 1.4465^{20}
3858	3-Buten-2-ol, (±)-	C_4H_8O 72.11	6118-14-5 <-100	 97.3	4-01-00-02102 0.8318^{20}	 1.4137^{20}
3859	2-Buten-1-ol, (*E*)- *trans*-Crotyl alcohol	C_4H_8O 72.11	504-61-0 <-30	 121.2	4-01-00-02107 0.8521^{20}	H_2O 4; EtOH 5; eth 5; chl 3 1.4288^{20}
3860	2-Buten-1-ol, (*Z*)- *cis*-Crotyl alcohol	C_4H_8O 72.11	4088-60-2	 123	4-01-00-02107 0.8662^{20}	 1.4342^{25}
3861	3-Buten-1-ol	C_4H_8O 72.11	627-27-0	 113.5	4-01-00-02105 0.8424^{20}	H_2O 3; EtOH 5; eth 5; ace 3 1.4224^{20}
3862	2-Buten-1-ol, acetate	$C_6H_{10}O_2$ 114.14	628-08-0	 132	3-02-00-00283 0.9192^{20}	bz 4; eth 4; EtOH 4 1.4181^{20}
3863	2-Buten-1-ol, 2-chloro-	C_4H_7ClO 106.55	116723-93-4	 159	3-01-00-01900 1.1180^{20}	H_2O 4; EtOH 4 1.4682^{20}
3864	2-Buten-1-ol, 3-chloro- γ-Chlorocrotyl alcohol	C_4H_7ClO 106.55	40605-42-3 -42.9	 161.5	4-01-00-02108 1.095^{23}	 1.4652^{20}
3865	2-Buten-1-ol, 4-chloro- 4-Chloro-2-buten-1-ol	C_4H_7ClO 106.55	7523-44-6	 $64\text{-}5^{2}$	4-01-00-02110	eth 4; EtOH 4 1.4845^{20}
3866	3-Buten-2-ol, 1-chloro- 1-Chloro-3-buten-2-ol	C_4H_7ClO 106.55	671-56-7	 145.5	3-01-00-01893 1.111^{20}	chl 4 1.4643^{20}
3867	3-Buten-1-ol, 2-chloro-	C_4H_7ClO 106.55	75455-41-3	 $66\text{-}7^{30}$	4-01-00-02106 1.1044^{20}	eth 4; EtOH 4 1.4665^{20}
3868	3-Buten-2-ol, 3-chloro- 3-Chloro-3-buten-2-ol	C_4H_7ClO 106.55	6408-47-1	 $53\text{-}7^{19}$	3-01-00-01893 1.1138^{23}	eth 4; EtOH 4
3869	2-Buten-1-ol, 3-methyl-	$C_5H_{10}O$ 86.13	556-82-1	 140	4-01-00-02129 0.848^{25}	 1.4412^{20}
3870	3-Buten-2-ol, 2-methyl-	$C_5H_{10}O$ 86.13	115-18-4 -28	 97	4-01-00-02132 0.82^{20}	
3871	3-Buten-2-ol, 3-methyl- Methyl isopropenyl carbinol	$C_5H_{10}O$ 86.13	10473-14-0	 114	4-01-00-02127 0.8531^{17}	 1.4288^{17}
3872	2-Buten-1-ol, 3-phenyl- 2-Butene-1-ol, 3-phenyl	$C_{10}H_{12}O$ 148.20	1504-54-7	 145^{15}	4-06-00-03839 1.035^{20}	ctc 3 1.5678^{20}
3873	3-Buten-2-ol, 4-(2,6,6-trimethyl-1-cyclohexen-1-yl)- β-Ionol	$C_{13}H_{22}O$ 194.32	22029-76-1	 131^{15}; $80^{0.7}$	3-06-00-00401 0.9243^{20}	EtOH 3; eth 3; ace 3 1.4969^{20}
3874	3-Buten-2-ol, 4-(2,6,6-trimethyl-2-cyclohexen-1-yl)- α-Ionol	$C_{13}H_{22}O$ 194.32	25312-34-9	 127^{15}	3-06-00-00402 0.9174^{20}	 1.4735^{20}
3875	3-Buten-2-one Methyl vinyl ketone	C_4H_6O 70.09	78-94-4	6052 81.4	4-01-00-03444 0.864^{20}	H_2O 3; EtOH 3; eth 4; ace 4 1.4081^{20}
3876	2-Buten-1-one, 1,3-diphenyl- Dypnone	$C_{16}H_{14}O$ 222.29	495-45-4	3460 342.5	4-07-00-01682 1.1080^{15}	eth 4; EtOH 4 1.6343^{20}
3877	1-Buten-1-one, 2-ethyl- Diethylketene	$C_6H_{10}O$ 98.14	24264-08-2	 88.5	4-01-00-03477 0.831^{20}	 1.4135^{11}
3878	3-Buten-2-one, 4-(2-furanyl)-	$C_8H_8O_2$ 136.15	623-15-4 39.5	 229 dec; 113^{10}	5-17-09-00578 1.0496^{57}	H_2O 1; EtOH 4; eth 4; chl 4 1.5788^{45}
3879	3-Buten-2-one, 3-methyl- Isopropenyl methyl ketone	C_5H_8O 84.12	814-78-8 -54	 98	4-01-00-03462 0.8527^{20}	EtOH 4 1.4220^{20}
3880	2-Buten-1-one, 1-phenyl-	$C_{10}H_{10}O$ 146.19	495-41-0 20.5	 $111\text{-}2^{9}$	3-07-00-01410 1.025^{15}	 1.5626^{18}
3881	3-Buten-2-one, 4-phenyl- Benzylideneacetone	$C_{10}H_{10}O$ 146.19	122-57-6 41.5	1151 261	2-07-00-00287 1.0368^{15}	bz 4; eth 4; EtOH 4; chl 4
3882	3-Buten-2-one, 4-phenyl-, (*E*)-	$C_{10}H_{10}O$ 146.19	1896-62-4 42	 262	4-07-00-01003 1.0097^{45}	H_2O 1; EtOH 4; eth 3; ace 3 1.5836^{45}
3883	3-Buten-2-one, 4-(2,5,6,6-tetramethyl-1-cyclohexen-1-yl)- β-Irone	$C_{14}H_{22}O$ 206.33	79-70-9	4978 $85\text{-}90^{0.1}$	4-07-00-00377 0.9434^{21}	H_2O 2; EtOH 4; eth 4; bz 4 1.5017^{20}
3884	3-Buten-2-one, 4-(2,5,6,6-tetramethyl-2-cyclohexen-1-yl)- α-Irone	$C_{14}H_{22}O$ 206.33	79-69-6	4977 $87\text{-}92^{0.4}$	 0.9362^{20}	 1.5002^{20}
3885	3-Buten-2-one, 4-(2,6,6-trimethyl-1-cyclohexen-1-yl)-, (*E*)-	$C_{13}H_{20}O$ 192.30	79-77-6	 140^{18}; $73^{0.1}$	4-07-00-00361 0.9462^{20}	H_2O 2; EtOH 5; eth 5; chl 3 1.5198^{20}
3886	3-Buten-2-one, 4-(2,6,6-trimethyl-2-cyclohexen-1-yl)-, (*E*)-(±)-	$C_{13}H_{20}O$ 192.30	30685-95-1	 146^{28}; $75^{0.15}$	4-07-00-00363 0.9298^{21}	ace 4; eth 4; EtOH 4 1.5041^{20}
3887	2-Butenoyl chloride	C_4H_5ClO 104.54	10487-71-5	 124.5	2-02-00-00394 1.0905^{20}	ace 4 1.460^{18}
3888	2-Butenoyl chloride, 3-methyl-	C_5H_7ClO 118.56	3350-78-5	 146	4-02-00-01558 1.065^{25}	 1.4770^{20}

No.	Name Synonym	Mol. Form. Mol. Wt.	CAS RN mp/°C	Merck No. bp/°C	Beil. Ref. den/g cm^{-3}	Solubility n_D
3889	1-Buten-3-yne Vinylacetylene	C_4H_4 52.08	689-97-4	 5.1	4-01-00-01083 0.7094^{0}	H_2O 1; bz 3 1.4161^{1}
3890	1-Buten-3-yne, 4-chloro-	C_4H_3Cl 86.52	40589-38-6	 56	4-01-00-01085 1.0022^{20}	chl 4 1.4656^{20}
3891	1-Buten-3-yne, 1-methoxy-	C_5H_6O 82.10	2798-73-4	 123 dec	4-01-00-02301 0.906^{20}	H_2O 1; chl 3; os 4 1.4818^{20}
3892	1-Buten-3-yne, 2-methyl- Isopropenylacetylene	C_5H_6 66.10	78-80-8 -113	 32	4-01-00-01089 0.6801^{11}	chl 3 1.4140^{20}
3893	Butralin Benzenamine, 4-(1,1-dimethylethyl)-*N*-(1-methylpropyl)-2,6-dinitro-	$C_{14}H_{21}N_3O_4$ 295.34	33629-47-9 60	1532 $135^{0.5}$		
3894	*N-sec*-Butyl-2,6-dinitrobenzene Benzeneamine, *N*-(1-methylpropyl)-2,6-dinitro	$C_{10}H_{13}N_3O_4$ 239.23	55702-49-3			
3895	2-Butynal Tetrolaldehyde	C_4H_4O 68.08	1119-19-3 -26	 106.5	4-01-00-03540 0.9264^{17}	eth 3; ace 3; chl 3 1.446^{19}
3896	3-Butyn-2-amine, 2-methyl-	C_5H_9N 83.13	2978-58-7 18	 79.5	4-04-00-01145 0.79^{25}	 1.4235^{20}
3897	1-Butyne Ethylacetylene	C_4H_6 54.09	107-00-6 -125.7	 8.0	4-01-00-00969 0.6783^{0}	H_2O 1; EtOH 3; eth 3 1.3962^{20}
3898	2-Butyne Dimethylacetylene	C_4H_6 54.09	503-17-3 -32.3	 26.9	4-01-00-00971 0.6910^{20}	H_2O 1; EtOH 3; eth 3; ctc 3 1.3921^{20}
3899	1-Butyne, 3-chloro-	C_4H_5Cl 88.54	21020-24-6	 68.5	4-01-00-00970 1.4218^{25}	 1.4218^{25}
3900	2-Butyne, 1-chloro-	C_4H_5Cl 88.54	3355-17-7	 103	4-01-00-00973 1.0152^{20}	ace 4; eth 4; EtOH 4 1.4581^{20}
3901	1-Butyne, 3-chloro-3-methyl-	C_5H_7Cl 102.56	1111-97-3 -61	 76	4-01-00-01000 0.9061^{20}	
3902	2-Butynediamide Cellocidin	$C_4H_4N_2O_2$ 112.09	543-21-5 217 dec	1958	4-02-00-02295	H_2O 2; EtOH 2; eth 2; chl 2
3903	2-Butyne-1,4-diamine	$C_4H_8N_2$ 84.12	53878-96-9 44	 96^{16}	4-04-00-01392	H_2O 4; EtOH 4; MeOH 4
3904	2-Butyne, 1,4-dibromo- 1,4-Dibromo-2-butyne	$C_4H_4Br_2$ 211.88	2219-66-1	 92^{15}	4-01-00-00974 2.014^{18}	eth 3; ace 3; chl 4 1.588^{18}
3905	2-Butyne, 1,4-dichloro-	$C_4H_4Cl_2$ 122.98	821-10-3	 165.5	4-01-00-00973 1.258^{20}	eth 3; ace 3; ctc 2; chl 4 1.5058^{20}
3906	2-Butyne, 1,4-diiodo-	$C_4H_4I_2$ 305.88	116529-73-8 54.5	 $70^{0.1}$	4-01-00-00975	H_2O 2; EtOH 3; eth 3; ace 3
3907	1-Butyne, 3,3-dimethyl- *tert*-Butylacetylene	C_6H_{10} 82.15	917-92-0 -78.2	 37.7	4-01-00-01022 0.6623^{25}	 1.3736^{20}
3908	2-Butynedinitrile	C_4N_2 76.06	1071-98-3 20.5	 76.5	4-02-00-02295 0.9708^{25}	 1.4647^{25}
3909	2-Butynedioic acid, diethyl ester	$C_8H_{10}O_4$ 170.17	762-21-0 1.5	 184^{200}	4-02-00-02294 1.0075^{20}	EtOH 3; eth 3; ctc 3 1.4425^{20}
3910	2-Butynedioic acid, dimethyl ester	$C_6H_6O_4$ 142.11	762-42-5	 197 dec; 98^{20}	4-02-00-02291 1.1564^{20}	EtOH 3; eth 3; ctc 3 1.4434^{20}
3911	2-Butyne-1,4-diol Bis(hydroxymethyl)acetylene	$C_4H_6O_2$ 86.09	110-65-6 50	 238	4-01-00-02687	H_2O 4; EtOH 4; eth 2; ace 4 1.4804^{20}
3912	3-Butyne-1,2-diol 1-Butyne-3,4-diol	$C_4H_6O_2$ 86.09	616-28-4 40	 $64^{0.2}$	4-01-00-02689	H_2O 4; EtOH 4
3913	2-Butyne-1,4-diol, diacetate 1,4-Diacetoxy-2-butyne	$C_8H_{10}O_4$ 170.17	1573-17-7	 $122\text{-}3^{10}$	4-02-00-00244	ctc 3 1.4611^{20}
3914	2-Butyne, 1,1,1,4,4,4-hexafluoro-	C_4F_6 162.03	692-50-2 -117.4	 -24.6	4-01-00-00972	EtOH 3; eth 3; ace 3; ctc 3
3915	2-Butyne, 1-methoxy-	C_5H_8O 84.12	2768-41-4	 99.5	3-01-00-01973 0.8496^{20}	bz 4; eth 4; EtOH 4 1.4262^{20}
3916	1-Butyne, 3-methyl-	C_5H_8 68.12	598-23-2 -89.7	 26.3	4-01-00-00999 0.6660^{20}	H_2O 1; EtOH 5; eth 5 1.3723^{20}
3917	2-Butynoic acid	$C_4H_4O_2$ 84.07	590-93-2 78	 203	4-02-00-01690 0.9641^{20}	H_2O 4; eth 4; EtOH 4; chl 4
3918	2-Butynoic acid, ethyl ester	$C_6H_8O_2$ 112.13	4341-76-8	 163	4-02-00-01691 0.9641^{20}	 1.4372^{20}
3919	2-Butyn-1-ol	C_4H_6O 70.09	764-01-2 -2.2	 148	4-01-00-02220 0.9370^{20}	eth 4; EtOH 4 1.4530^{20}
3920	3-Butyn-1-ol	C_4H_6O 70.09	927-74-2 -63.6	 129	4-01-00-02219 0.9257^{20}	H_2O 4; EtOH 4 1.4409^{20}
3921	3-Butyn-2-ol	C_4H_6O 70.09	2028-63-9 -3	 106.5	4-01-00-02218 0.8618^{20}	H_2O 4; eth 4; EtOH 4 1.4207^{20}
3922	3-Butyn-2-ol, 2-methyl- 1,1-Dimethylpropargyl alcohol	C_5H_8O 84.12	115-19-5 3	5956 104	4-01-00-02229 0.8618^{20}	 1.4207^{20}
3923	3-Butyn-2-one Ethynyl methyl ketone	C_4H_4O 68.08	1423-60-5	 84	4-01-00-03539 0.8860^{15}	 1.4070^{20}
3924	3-Butyn-2-one, 4-phenyl-	$C_{10}H_8O$ 144.17	1817-57-8	 $75\text{-}6^{0.8}$	4-07-00-01175 0.99^{25}	 1.5762^{20}
3925	Butyrac 118 Butanoic acid, 4-(2,4-dichlorophenoxy)-	$C_{10}H_{10}Cl_2O_3$ 249.09	94-82-6 118	2828		
3926	Butyric acid, 3,4-dichloro- 3,4-Dichlorobutyric acid	$C_4H_6Cl_2O_2$ 157.00	29653-38-1 49	 123^{8}	4-02-00-00828	eth 4; chl 4; HOAc 4
3927	Butyric acid, 4-fluoro- 4-Fluorobutyric acid	$C_4H_7FO_2$ 106.10	462-23-7	 $76\text{-}8^{5}$	4-02-00-00809	eth 4; EtOH 4 1.3993^{25}

No.	Name Synonym	Mol. Form. Mol. Wt.	CAS RN mp/°C	Merck No. bp/°C	Beil. Ref. den/g cm^{-3}	Solubility n_D
3928	Butyric acid, 2-iodo-, ethyl ester	$C_6H_{11}IO_2$ 242.06	7425-47-0	 100^{21}; 46^{1}	4-02-00-00839 1.570^{17}	 1.4923^{20}
3929	Cadmium, dimethyl- Dimethyl cadmium	C_2H_6Cd 142.48	506-82-1 -4.5	3229 105.5; exp 150	 1.9846^{18}	peth 3 1.5488
3930	Calcium cyanide Calcyan	C_2CaN_2 92.11	592-01-8	1664		
3931	Camphoric acid, *trans*-(±)- Isocamphoric acid, (*DL*)	$C_{10}H_{16}O_4$ 200.23	560-08-7 195		3-09-00-03879 1.249^{25}	eth 4
3932	Carbamic acid, (aminocarbonyl)-, ethyl ester	$C_4H_8N_2O_3$ 132.12	626-36-8 196.5	 dec	4-03-00-00127	H_2O 1; EtOH 2; eth 1; bz 2
3933	Carbamic acid, butyl-, 2-[[(aminocarbonyl)oxy]methyl]-2-methylpentyl ester 2-Methyl-2-propyltrimethylene butylcarbamate carbamate	$C_{13}H_{26}N_2O_4$ 274.36	4268-36-4	9737		
3934	Carbamic acid, butyl-, butyl ester	$C_9H_{19}NO_2$ 173.26	13105-52-7	 88^{3}	4-04-00-00577 0.9221^{20}	eth 4; EtOH 4 1.4359^{20}
3935	Carbamic acid, butyl ester Butyl carbamate	$C_5H_{11}NO_2$ 117.15	592-35-8 53	 204 dec; 108^{14}	4-03-00-00054	EtOH 4; chl 2
3936	Carbamic acid, butyl-, ethyl ester	$C_7H_{15}NO_2$ 145.20	591-62-8 -22	 202; 100^{15}	4-04-00-00577 0.9434^{26}	 1.4278^{26}
3937	Carbamic acid, butyl-, methyl ester	$C_6H_{13}NO_2$ 131.17	2594-21-0 -18	 92^{15}	0-04-00-00158 0.9689^{23}	 1.4289^{23}
3938	Carbamic acid, *N*-butyl-*N*-nitro,butyl ester	$C_9H_{18}N_2O_4$ 218.25	110795-27-2	 98^{3}	 1.046^{20}	EtOH 5; eth 5 1.448^{20}
3939	Carbamic acid, (3-chlorophenyl)-, 1-methylethyl ester Chloropropham	$C_{10}H_{12}ClNO_2$ 213.66	101-21-3 41	2188 149^{2}	4-12-00-01149 1.18^{30}	 1.5388^{20}
3940	Carbamic acid, dichloro-, ethyl ester	$C_3H_5Cl_2NO_2$ 157.98	13698-16-3	 $55\text{-}6^{15}$	3-03-00-00050 1.349^{25}	 1.4595^{20}
3941	Carbamic acid, diethyl- Carbamic acid, *N,N*-diethyl	$C_5H_{11}NO_2$ 117.15	24579-70-2 -15	 171 dec	3-04-00-00222 0.9276^{20}	H_2O 4; eth 4; EtOH 4 1.4206^{20}
3942	Carbamic acid, (1,1-dimethylethyl)-, ethyl ester	$C_7H_{15}NO_2$ 145.20	1611-50-3 21.5	 72^{16}	4-04-00-00665 0.943^{15}	
3943	Carbamic acid, (1,1-dimethylethyl)-, methyl ester	$C_6H_{13}NO_2$ 131.17	27701-01-5 27	 63.3^{17}	3-04-00-00325 0.966^{15}	
3944	Carbamic acid, (1,1-dimethylethyl)nitro-, ethyl ester	$C_7H_{14}N_2O_4$ 190.20	55696-02-1	 56^{2} dec	4-04-00-03409 1.049^{20}	eth 4; EtOH 4 1.4331^{20}
3945	Carbamic acid, dimethyl-, 3-methyl-1-phenyl-1H-pyrazol-5-yl ester Pyrolan	$C_{13}H_{15}N_3O_2$ 245.28	87-47-8 50	8013 $160\text{-}2^{0.2}$	5-23-10-00530	ctc 3; CS_2 3
3946	Carbamic acid, diphenyl-, ethyl ester	$C_{15}H_{15}NO_2$ 241.29	603-52-1 72	 360	4-12-00-00850	H_2O 4; eth 4, bz 4
3947	Carbamic acid, ethyl-, butyl ester Carbamic acid, *N*-ethyl, butyl ester	$C_7H_{15}NO_2$ 145.20	16246-07-4	 66^{3}	 0.9396^{20}	eth 4; EtOH 4 1.4301^{20}
3948	Carbamic acid, ethyl ester Urethan	$C_3H_7NO_2$ 89.09	51-79-6 49	9789 185	4-03-00-00040 0.9862^{21}	H_2O 4; EtOH 4; eth 4; bz 4 1.4144^{51}
3949	Carbamic acid, ethyl-, ethyl ester	$C_5H_{11}NO_2$ 117.15	623-78-9	 176	4-04-00-00365 0.9813^{20}	H_2O 4; eth 4; EtOH 4 1.4215^{20}
3950	Carbamic acid, ethylnitro-, ethyl ester Carbamic acid, *N*-ethyl-*N*-nitro, ethyl ester	$C_5H_{10}N_2O_4$ 162.15	6274-16-4	 107^{31}	4-04-00-03404 1.161^{20}	eth 4; EtOH 4 1.4432^{20}
3951	Carbamic acid, ethylnitro-, methyl ester Carbamic acid, *N*-ethyl-*N*-nitro, butyl ester	$C_7H_{14}N_2O_4$ 190.20	6162-79-4	 79^{3}	4-04-00-03403 1.089^{20}	H_2O 4; eth 4; EtOH 4 1.4455^{20}
3952	Carbamic acid, (3-methylbutyl)-, ethyl ester Urethane, isopentyl	$C_8H_{17}NO_2$ 159.23	1611-52-5	 218	2-04-00-00648 0.9322^{20}	 1.4326^{20}
3953	Carbamic acid, methyl ester Methyl carbamate	$C_2H_5NO_2$ 75.07	598-55-0 54	5958 177	4-03-00-00037 1.1361^{56}	H_2O 4; EtOH 4; eth 3 1.4125^{56}
3954	Carbamic acid, methyl-, ethyl ester	$C_4H_9NO_2$ 103.12	105-40-8	 170	4-04-00-00200 1.0115^{20}	H_2O 4; EtOH 4 1.4183^{20}
3955	Carbamic acid, 1-methylethyl ester Isopropyl carbamate	$C_4H_9NO_2$ 103.12	1746-77-6 93	 183	4-03-00-00053 0.9951^{66}	
3956	Carbamic acid, (1-methylethyl)-, ethyl ester	$C_6H_{13}NO_2$ 131.17	2594-20-9	 79^{15}	4-04-00-00520 0.9531^{20}	eth 4; EtOH 4 1.4229^{20}
3957	Carbamic acid, (1-methylethyl)nitro-, ethyl ester Carbamic acid, *N*-isopropyl-*N*-nitro, ethyl ester	$C_6H_{12}N_2O_4$ 176.17	62261-05-6	 72^{7}	4-04-00-03407 1.110^{20}	eth 4; EtOH 4 1.4381^{20}
3958	Carbamic acid, methylphenyl-, methyl ester	$C_9H_{11}NO_2$ 165.19	28685-60-1 44	 235	2-12-00-00235 1.2960^{19}	eth 4; lig 4
3959	Carbamic acid, 2-methylpropyl ester Isobutyl carbamate	$C_5H_{11}NO_2$ 117.15	543-28-2 67	5021 207	4-03-00-00056	eth 4; EtOH 4 1.4098^{76}
3960	Carbamic acid, (1-methylpropyl)-, ethyl ester Urethane, *sec*-butyl	$C_7H_{15}NO_2$ 145.20	10212-74-5 -14	 194	4-04-00-00622 0.9404^{26}	 1.4267^{26}
3961	Carbamic acid, (2-methylpropyl)-, ethyl ester Isobutyl urethane	$C_7H_{15}NO_2$ 145.20	539-89-9 <-65	5037 110^{30}	4-04-00-00647 0.9432^{20}	eth 4; EtOH 4 1.4288^{20}

No.	Name Synonym	Mol. Form. Mol. Wt.	CAS RN mp/°C	Merck No. bp/°C	Beil. Ref. den/g cm^{-3}	Solubility n_D
3962	Carbamic acid, (1-methylpropyl)-, methyl ester Carbamic acid, *N-sec*-butyl, methyl ester	$C_6H_{13}NO_2$ 131.17	39076-02-3	 83^{16}	0-04-00-00162 0.9651^{25}	 1.4263^{25}
3963	Carbamic acid, nitro-, ethyl ester	$C_3H_6N_2O_4$ 134.09	626-37-9 64	 140 dec	4-03-00-00247 1.0074^{20}	H_2O 4; EtOH 4; eth 4; lig 2
3964	Carbamic acid, nitropropyl-, ethyl ester Carbamic acid, *N*-propyl-*N*-nitro, ethyl ester	$C_6H_{12}N_2O_4$ 176.17	13855-77-1	 66^{3}	4-04-00-03406 1.121^{20}	eth 4; EtOH 4 1.4431^{20}
3965	Carbamic acid, phenyl-, ethyl ester Phenylurethane	$C_9H_{11}NO_2$ 165.19	101-99-5 53	7291 237 dec	4-12-00-00619 1.1064^{30}	H_2O 1; EtOH 4; eth 4; bz 3 1.5376^{30}
3966	Carbamic acid, phenyl-, 1-methylethyl ester Propham	$C_{10}H_{13}NO_2$ 179.22	122-42-9 90	7828	4-12-00-00620 1.09^{20}	bz 4; EtOH 4 1.4989^{91}
3967	Carbamic acid, (phenylmethyl)-, ethyl ester	$C_{10}H_{13}NO_2$ 179.22	2621-78-5 49	 230	4-12-00-02250	bz 4; eth 4; EtOH 4; chl 4
3968	Carbamic acid, phenyl-, 2-methylpropyl ester	$C_{11}H_{15}NO_2$ 193.25	2291-80-7 86	 216	3-12-00-00614	bz 4; eth 4; EtOH 4
3969	Carbamic acid, propyl ester	$C_4H_9NO_2$ 103.12	627-12-3 60	 196	4-03-00-00052	ace 4; eth 4; EtOH 4
3970	Carbamic acid, propyl-, ethyl ester	$C_6H_{13}NO_2$ 131.17	623-85-8	 192	4-04-00-00480 0.9921^{15}	EtOH 4
3971	Carbamic chloride Carbamyl chloride	CH_2ClNO 79.49	463-72-9	1785 62 dec	3-03-00-00065	
3972	Carbamic chloride, diethyl- Carbamoyl chloride, *N,N*-diethyl	$C_5H_{10}ClNO$ 135.59	88-10-8	 186	4-04-00-00379	
3973	Carbamic chloride, dimethyl- Dimethylcarbamoyl chloride	C_3H_6ClNO 107.54	79-44-7 -33	 167	4-04-00-00224 1.168^{25}	 1.4540^{20}
3974	Carbamic chloride, methylphenyl-	C_8H_8ClNO 169.61	4285-42-1 88.5	 280	4-12-00-00839	eth 4; EtOH 4
3975	Carbamodithioic acid, dimethyl-, 2,4-dinitrophenyl ester	$C_9H_9N_3O_4S_2$ 287.32	89-37-2 152.5		4-06-00-01760 1.54^{20}	H_2O 1; EtOH 3; ace 3; bz 3
3976	Carbamothioic acid, butylethyl-, *S*-propyl ester Pebulate	$C_{10}H_{21}NOS$ 203.35	1114-71-2	7007 142^{20}	 0.9458^{20}	ace 4; bz 4; MeOH 4 1.4752^{20}
3977	Carbamothioic acid, diphenyl-, *S*-[2-(diethylamino)ethyl] ester Phencarbamide	$C_{19}H_{24}N_2OS$ 328.48	3735-90-8 48.5	7178 $120\text{-}6^{0.01}$		eth 4; chl 4; MeOH 4; peth 4
3978	Carbamothioic acid, *O*-ethyl ester	C_3H_7NOS 105.16	625-57-0 41	 dec	4-03-00-00294 1.069^{20}	eth 4; EtOH 4; chl 4 1.520^{20}
3979	Carbamothioic acid, *S*-ethyl ester	C_3H_7NOS 105.16	637-98-9 109	 dec	4-03-00-00294	eth 4; EtOH 4
3980	Carbamothioic chloride, dimethyl-	C_3H_6ClNS 123.61	16420-13-6 42.5	 98^{10}	3-04-00-00147	eth 4; chl 3; peth 3
3981	9H-Carbazole Dibenzo[b,d]pyrrole	$C_{12}H_9N$ 167.21	86-74-8 246.2	1792 354.7	5-20-08-00009	H_2O 1; EtOH 2; eth 2; ace 3
3982	9H-Carbazole-9-acetic acid (*N*-Carbazolyl)acetic acid	$C_{14}H_{11}NO_2$ 225.25	524-80-1 215	1793	5-20-08-00033	eth 4; EtOH 4; chl 4; HOAc 4
3983	9H-Carbazole, 9-acetyl-	$C_{14}H_{11}NO$ 209.25	574-39-0 69	 190^{6}	5-20-08-00030 1.158^{100}	 1.640^{100}
3984	9H-Carbazole, 9-butyl-	$C_{16}H_{17}N$ 223.32	1484-08-8 58	 $218\text{-}9^{19}$	5-20-08-00018	eth 4
3985	9H-Carbazole, 9-ethyl-	$C_{14}H_{13}N$ 195.26	86-28-2 68	 190^{10}	5-20-08-00016 1.059^{80}	H_2O 1; EtOH 4; eth 4 1.6394^{80}
3986	9H-Carbazole, 3-methyl-	$C_{13}H_{11}N$ 181.24	4630-20-0 208.5	 365	5-20-08-00076	bz 4; eth 4
3987	9H-Carbazole, 9-methyl-	$C_{13}H_{11}N$ 181.24	1484-12-4 89.34	 195^{12}	5-20-08-00015	eth 4
3988	9H-Carbazole, 9-(phenylmethyl)-	$C_{19}H_{15}N$ 257.33	19402-87-0 119	 267^{24}	5-20-08-00023	bz 4
3989	1H-Carbazole, 2,3,4,9-tetrahydro-	$C_{12}H_{13}N$ 171.24	942-01-8 120	 327.5	5-20-07-00468	H_2O 1; EtOH 3; eth 4; bz 4
3990	Carbendazim Carbamic acid, 1H-benzimidazol-2-yl-, methyl ester	$C_9H_9N_3O_2$ 191.19	10605-21-7 300 dec	1794	 1.45	
3991	Carbofuran 7-Benzofuranol, 2,3-dihydro-2,2-dimethyl-, methylcarbamate	$C_{12}H_{15}NO_3$ 221.26	1563-66-2 151	1810	 1.18	
3992	Carbon dioxide Carbonic anhydride	CO_2 44.01	124-38-9 -56.57 (5.11 atm)	1816 sub -78.5 (1 atm)	0-03-00-00004 $^{*}0.720^{25}$	
3993	Carbon disulfide Carbon bisulfide	CS_2 76.14	75-15-0 -111.5	1818 46	4-03-00-00395 1.2632^{20}	H_2O 3; EtOH 5; eth 5; chl 3 1.6319^{20}
3994	Carbonic acid, bis(1,1-dimethylethyl) ester	$C_9H_{18}O_3$ 174.24	34619-03-9 40	 158	4-03-00-00009	EtOH 4
3995	Carbonic acid, bis(1-methylethyl) ester	$C_7H_{14}O_3$ 146.19	6482-34-4	 147	3-03-00-00009 0.9162^{20}	EtOH 4 1.3932^{20}
3996	Carbonic acid, bis(2-methylphenyl) ester	$C_{15}H_{14}O_3$ 242.27	617-09-4 60	 $144^{0.5}$	3-06-00-01256	HOAc 4
3997	Carbonic acid, bis(2-methylpropyl) ester	$C_9H_{18}O_3$ 174.24	539-92-4	 190	4-03-00-00009 0.9138^{20}	H_2O 1; EtOH 5; eth 5 1.4072^{20}
3998	Carbonic acid, dibutyl ester Dibutyl carbonate	$C_9H_{18}O_3$ 174.24	542-52-9	1558 207	4-03-00-00008 0.9251^{20}	H_2O 1; EtOH 3; eth 3 1.4117^{20}

PHYSICAL CONSTANTS OF ORGANIC COMPOUNDS (continued)

No.	Name	Synonym	Mol. Form.	Mol. Wt.	CAS RN	mp/°C	Merck No.	bp/°C	Beil. Ref.	den/g cm^{-3}	Solubility	n_D
3999	Carbonic acid, diethyl ester	Diethyl carbonate	$C_5H_{10}O_3$	118.13	105-58-8	-43	3738	126	4-03-00-00005	0.9752^{20}	H_2O 1; EtOH 3; eth 3; chl 3	1.3845^{20}
4000	Carbonic acid, dimethyl ester	Methyl carbonate	$C_3H_6O_3$	90.08	616-38-6	0.5	5960	90.5	4-03-00-00003	1.065^{17}	H_2O 1; EtOH 3; eth 3; ctc 2	1.3687^{20}
4001	Carbonic acid, diphenyl ester	Phenyl carbonate	$C_{13}H_{10}O_3$	214.22	102-09-0	83	7250	306	4-06-00-00629	1.1215^{87}	H_2O 1; EtOH 3; eth 3; ctc 3	
4002	Carbonic acid, dipropyl ester	Dipropyl carbonate	$C_7H_{14}O_3$	146.19	623-96-1			168	4-03-00-00006	0.9435^{20}	H_2O 2; EtOH 5; eth 5	1.4008^{20}
4003	Carbonic acid, ethyl methyl ester	Ethyl methyl carbonate	$C_4H_8O_3$	104.11	623-53-0	-14		107.5	4-03-00-00004	1.012^{20}	eth 4; EtOH 4	1.3778^{20}
4004	Carbonic acid monosodium salt		$CHNaO_3$	84.01	144-55-8		8528				H_2O 3; EtOH 1	
4005	Carbonic chloride fluoride	Carbonyl chloride fluoride	$CClFO$	82.46	353-49-1				4-03-00-00031			
4006	Carbonic dibromide	Carbonyl bromide	CBr_2O	187.82	593-95-3			64.5		2.52^{15}		
4007	Carbonic dichloride	Phosgene	CCl_2O	98.92	75-44-5	-127.9	7310	8	4-03-00-00031	$*1.3719^{25}$	bz 3; ctc 3; chl 3; tol 3	
4008	Carbonic difluoride	Carbonyl fluoride	CF_2O	66.01	353-50-4	-111.2	1826	-84.5	4-03-00-00021	1.139^{25}		
4009	Carbonic dihydrazide	Carbohydrazide	CH_6N_4O	90.09	497-18-7	154	1811		4-03-00-00240	1.616^{20}	H_2O 4; EtOH 4	
4010	Carbonic dihydrazide, 2,2'-diphenyl-	*sym*-Diphenylcarbazide	$C_{13}H_{14}N_4O$	242.28	140-22-7	170	3321	dec	4-15-00-00182		H_2O 2; EtOH 3; eth 2; ace 3	
4011	Carbon monoxide	Carbon oxide	CO	28.01	630-08-0	-205	1820	-191.5	0-01-00-00720	0.7909^{-19}	H_2O 2; bz 3; HOAc 3	
4012	Carbonochloridic acid, butyl ester	Butyl chloroformate	$C_5H_9ClO_2$	136.58	592-34-7			142	4-03-00-00025	1.074^{25}	eth 5; ace 3; ctc 2	1.4114^{20}
4013	Carbonochloridic acid, 2-chloroethyl ester	2-Chloroethyl chloroformate	$C_3H_4Cl_2O_2$	142.97	627-11-2			155	4-03-00-00024	1.3847^{20}	H_2O 1; EtOH 3; eth 3; ace 3	1.4483^{20}
4014	Carbonochloridic acid, chloromethyl ester	Chloromethyl chloroformate	$C_2H_2Cl_2O_2$	128.94	22128-62-7			107	4-03-00-00030	1.465^{15}	ace 4; eth 4	1.4286^{22}
4015	Carbonochloridic acid, 3-chloropropyl ester		$C_4H_6Cl_2O_2$	157.00	628-11-5			177	3-03-00-00025	1.2926^{25}	H_2O 1	1.4456^{20}
4016	Carbonochloridic acid, dichloromethyl ester		$C_2HCl_3O_2$	163.39	22128-63-8			110.5	4-03-00-00031	1.5600^{15}	ace 4	
4017	Carbonochloridic acid, 2-ethoxyethyl ester		$C_5H_9ClO_3$	152.58	628-64-8			67.2^{14}	3-03-00-00029	1.1341^{25}	EtOH 4	1.4169^{25}
4018	Carbonochloridic acid, ethyl ester	Ethyl chloroformate	$C_3H_5ClO_2$	108.52	541-41-3	-80.6	3742	95	4-03-00-00023	1.1352^{20}	bz 4; eth 4; chl 4	1.3974^{20}
4019	Carbonochloridic acid, 2-methoxyethyl ester		$C_4H_7ClO_3$	138.55	628-12-6			58.7^{13}	3-03-00-00029	1.1905^{25}	eth 4	1.4163^{20}
4020	Carbonochloridic acid, methyl ester	Methyl chlorocarbonate	$C_2H_3ClO_2$	94.50	79-22-1		5966	70.5	4-03-00-00023	1.2231^{20}	EtOH 5; eth 5; bz 3; ctc 3	1.3868^{20}
4021	Carbonochloridic acid, 3-methylbutyl ester		$C_6H_{11}ClO_2$	150.60	628-50-2			154.3	4-03-00-00026	1.0288^{17}	eth 4; EtOH 4	1.4176^{20}
4022	Carbonochloridic acid, 1-methylethenyl ester	Isopropenyl chloroformate	$C_4H_5ClO_2$	120.54	57933-83-2			102	3-03-00-00028	1.101^{20}	eth 4	
4023	Carbonochloridic acid, 1-methylethyl ester	Isopropyl chloroformate	$C_4H_7ClO_2$	122.55	108-23-6			105	4-03-00-00024		eth 4	1.4013^{20}
4024	Carbonochloridic acid, 2-methylpropyl ester	Isobutyl chlorocarbonate	$C_5H_9ClO_2$	136.58	543-27-1		5023	128.8	4-03-00-00026	1.0426^{18}	EtOH 3; eth 5; bz 3; chl 3	1.4071^{18}
4025	Carbonochloridic acid, pentyl ester		$C_6H_{11}ClO_2$	150.60	638-41-5			$60\text{-}2^{15}$	4-03-00-00026		eth 3	1.4181^{18}
4026	Carbonochloridic acid, phenyl ester	Phenyl chloroformate	$C_7H_5ClO_2$	156.57	1885-14-9			71^{9}	4-06-00-00629			
4027	Carbonochloridic acid, phenylmethyl ester	Carbobenzoxy chloride	$C_8H_7ClO_2$	170.60	501-53-1		1807	103^{20}	4-06-00-02278	1.195^{25}	eth 3; ace 3; bz 3	1.5190^{20}
4028	Carbonochloridic acid, propyl ester	Propyl chlorocarbonate	$C_4H_7ClO_2$	122.55	109-61-5		7860	115.2	4-03-00-00024	1.0901^{20}	EtOH 5; eth 5	1.4035^{20}
4029	Carbonochloridic acid, trichloromethyl ester	Diphosgene	$C_2Cl_4O_2$	197.83	503-38-8	-57	3339	128	4-03-00-00033	1.6525^{14}	eth 4; EtOH 4	1.4566^{22}
4030	Carbonocyanidic acid, ethyl ester		$C_4H_5NO_2$	99.09	623-49-4			115.5	4-02-00-01862	1.003^{25}	H_2O 1; EtOH 3; eth 3; ctc 3	1.3820^{20}
4031	Carbonodithioic acid, *S,S*-diethyl ester		$C_5H_{10}OS_2$	150.27	623-80-3			197	3-03-00-00339	1.0[illegible]^{20}	eth 4; EtOH 4	1.[illegible]07^{18}
4032	Carbonothioic dichloride	Thiophosgene	CCl_2S	114.98	463-71-8			73	4-03-00-00281	1.508^{15}	H_2O 1; EtOH 3; eth 3	1.5442^{20}
4033	Carbonothioic dihydrazide	1,3-Diamino-2-thiourea	CH_6N_4S	106.15	2231-57-4	170 dec			4-03-00-00388		H_2O 4	
4034	Carbonotrithioic acid	Trithiocarbonic acid	CH_2S_3	110.22	594-08-1	-30	9671	57 dec	4-03-00-00428	1.47^{17}	EtOH 4; tol 4; chl 4	
4035	Carbon oxide sulfide	Carbonyl sulfide	COS	60.08	463-58-1	-138.8		-50	4-03-00-00271	1.028^{17}	H_2O 2; EtOH 3; KOH 4	1.24^{-87}
4036	Carbon selenide	Carbon diselenide	CSe_2	169.93	506-80-9	-45.5	1817	125.5	4-03-00-00436	2.6823^{20}		1.8454^{20}
4037	Carbophenothion	Phosphorodithioic acid, *S*-[[(4-chlorophenyl)thio]methyl] *O,O*-diethyl ester	$C_{11}H_{16}ClO_2PS_3$	342.87	786-19-6		1827	$82^{0.01}$		1.271^{20}		

No.	Name Synonym	Mol. Form. Mol. Wt.	CAS RN mp/°C	Merck No. bp/°C	Beil. Ref. den/g cm^{-3}	Solubility n_D
4038	Carbosulfan 2,3-Dihydro-2,2-dimethyl-7-benzofuranyl (dibutylamino)thio methylcarbamate	$C_{20}H_{32}N_2O_3S$ 380.55	55285-14-8	 126	 1.056^{20}	
4039	Carboxin 1,4-OxathiiN-3-carboxamide, 5,6-dihydro-2-methyl-*N*-phenyl-	$C_{12}H_{13}NO_2S$ 235.31	5234-68-4 94	1832		
4040	Card-20(22)-enolide, 3-[(6-deoxy-α-*L*-mannopyranosyl)oxy]-1,5,11,14,19-pentahydroxy-, (1β,3β Ouabain	$C_{29}H_{44}O_{12}$ 584.66	630-60-4 200	6854	5-18-05-00625	H_2O 2; EtOH 4
4041	Card-20(22)-enolide, 3-[(*O*-2,6-dideoxy-β-*D*-ribo-hexopyranosyl-(1-4)-*O*-2,6-dideoxy-β-*D*-ribo-hexopyranosyl-(1-4)-2,6-dideoxy-β-*D*, 12,14-dihydroxy Digoxin	$C_{41}H_{64}O_{14}$ 780.95	20830-75-5 248-50 dec	3150	5-18-04-00381	EtOH 4
4042	Card-20(22)-enolide, 3-[(*O*-2,6-dideoxy-β-*D*-ribo-hexopyranosyl-(1-4)-*O*-2,6-dideoxy-β-*D*-ribo-hexopyranosyl-(1-4)-2,6-dideoxy-β-*D*, 14,16-dihydroxy Gitoxin	$C_{41}H_{64}O_{14}$ 780.95	4562-36-1 285 dec	4328	5-18-04-00388	
4043	Card-20(22)-enolide, 3-[(*O*-2,6-dideoxy-β-*D*-ribo-hexopyranosyl-(1-4)-*O*-2,6-dideoxy-β-*D*-ribo-hexopyranosyl-(1-4)-2,6-dideoxy-β-*D*, 14-dihydroxy Digitoxin	$C_{41}H_{64}O_{13}$ 764.95	71-63-6 255.5	3146	5-18-03-00354	H_2O 2; EtOH 4; eth 3; chl 3
4044	Card-20(22)-enolide, 3,14-dihydroxy-, (3β,5β)- Digitoxigenin	$C_{23}H_{34}O_4$ 374.52	143-62-4 253	3145	5-18-03-00348	EtOH 3; MeOH 4
4045	Card-20(22)-enolide, 3-[(6-*O*-β-*D*-glucopyranosyl-β-D-glucopyranosyl)oxy]-14-hydroxy-, Uzarin	$C_{35}H_{54}O_{14}$ 698.81	20231-81-6 269	9809	4-18-00-01497	
4046	Card-20(22)-enolide, 3,11,14-trihydroxy-, (3β,5β,11α)- Sarmentogenin	$C_{23}H_{34}O_5$ 390.52	76-28-8 280	8334	5-18-04-00378	H_2O 1; EtOH 3; eth 1; ace 2
4047	Card-20(22)-enolide, 3,12,14-trihydroxy-, (3β,5β,12β)- Digoxigenin	$C_{23}H_{34}O_5$ 390.52	1672-46-4 222	3149	5-18-04-00380	EtOH 4; chl 2; MeOH 4
4048	Card-20(22)-enolide, 3,14,16-trihydroxy-, (3β,5β,16β)- Gitoxigenin	$C_{23}H_{34}O_5$ 390.52	545-26-6 234	4327	5-18-04-00385	H_2O 1; eth 2; chl 3
4049	Card-20(22)-enolide, 3,5,14-trihydroxy-19-oxo-, (3β,5β)- Strophanthidin	$C_{23}H_{32}O_6$ 404.50	66-28-4 173 dec	8818	5-18-05-00126	H_2O 1; EtOH 3; eth 1; ace 3
4050	4-Carene, (1*S*,3*R*,6*R*)-(-)- Δ4-Carene (*l*)	$C_{10}H_{16}$ 136.24	5208-49-1	 169; 64^{20}	4-05-00-00451 0.8441^{30}	ace 4; bz 4; eth 4 1.4740^{30}
4051	β,β-Carotene β-Carotene, all-*trans*-	$C_{40}H_{56}$ 536.88	7235-40-7 183	1860	4-05-00-02617 1.00^{20}	H_2O 1; EtOH 2; eth 3; ace 3
4052	β,ψ-Carotene γ-Carotene	$C_{40}H_{56}$ 536.88	472-93-5 178	1861	4-05-00-02616	H_2O 1; EtOH 1; eth 2; bz 3
4053	ψ,ψ-Carotene *trans*-Lycopene	$C_{40}H_{56}$ 536.88	502-65-8 175	5492	4-01-00-01165	EtOH 2; eth 3; bz 4; chl 4
4054	β,ε-Carotene, (6'*R*)- α-Carotene	$C_{40}H_{36}$ 516.73	7488-99-5 187.5	1859	4-05-00-02620 1.00^{20}	bz 4; eth 4; chl 4
4055	β,β-Carotene, 3'-(acetyloxy)-6',7'-didehydro-5,6-epoxy-5,5',6,6',7,8-hexahydro-3,5'-dihydroxy-8-oxo-, Fucoxanthin	$C_{40}H_{60}O_6$ 636.91	3351-86-8 168	4195	5-18-04-00673	eth 4; EtOH 4
4056	β,β-Carotene-3,3'-diol, 5,6:5',6'-diepoxy-5,5',6,6'-tetrahydro- Violaxanthin	$C_{40}H_{56}O_4$ 600.88	126-29-4 208	9902	5-19-03-00464	EtOH 3; eth 3; CS_2 3; peth 1
4057	β,β-Carotene-3,3'-diol, (3*R*,3'*R*)- Zeaxanthin	$C_{40}H_{56}O_2$ 568.88	144-68-3 215.5	10019 $226\text{-}9^{0.06}$	4-06-00-07017	H_2O 1; EtOH 2; eth 3; ace 3
4058	β,ε-Carotene-3,3'-diol, (3*R*,3'*R*,6'*R*)- Xanthophyll	$C_{40}H_{56}O_2$ 568.88	127-40-2 196	9972	3-06-00-05871	bz 4; eth 4; EtOH 4; peth 4
4059	β,ψ-Caroten-16'-oic acid, 3',4'-didehydro- Torularhodin	$C_{40}H_{52}O_2$ 564.85	514-92-1 211	9476	4-09-00-02767	py 4; chl 4; CS_2 4
4060	ψ,ψ-Caroten-16-ol Lycoxanthin	$C_{40}H_{56}O$ 552.88	19891-74-8 168	5499	4-01-00-02368	H_2O 1; EtOH 2; bz 3; CS_2 3
4061	β,β-Caroten-3-ol, (3*R*)- Cryptoxanthin	$C_{40}H_{56}O$ 552.88	472-70-8 160	2612	4-06-00-05111	bz 4; chl 4
4062	β,ψ-Caroten-3-ol, (3*R*)- Rubixanthin	$C_{40}H_{56}O$ 552.88	3763-55-1 160	8265	4-06-00-05110	EtOH 2; bz 3; chl 3; peth 2
4063	β,κ-Caroten-6'-one, 3,3'-dihydroxy-, (3*R*,3'*S*,5'*R*)- Capsanthin	$C_{40}H_{56}O_3$ 584.88	465-42-9 176	1768	4-08-00-02657	
4064	Carpaine 2,15-Dioxa-12,25-diazatricyclo[22.2.2.211,14]triacontane, carpaine deriv.	$C_{28}H_{50}N_2O_4$ 478.72	3463-92-1 121	1866	2-27-00-00209	ace 4; bz 4; eth 4; EtOH 4

No.	Name Synonym	Mol. Form. Mol. Wt.	CAS RN mp/°C	Merck No. bp/°C	Beil. Ref. den/g cm^{-3}	Solubility n_D
4065	CDEC Carbamodithioic acid, diethyl-, 2-chloro-2-propenyl ester	$C_8H_{14}ClNS_2$ 223.79	95-06-7	8882 129^{1}	 1.088	
4066	Cedrene 1H-3a,7-Methanoazulene, cedrene deriv.	$C_{15}H_{24}$ 204.36	11028-42-5	 262.5	3-05-00-01095 0.9342^{20}	H_2O 1; bz 3; lig 3 1.5034^{20}
4067	Cevane-3,4,12,14,16,17,20-heptol, 3-acetate, (3β,4α,16β)- Sabadine	$C_{29}H_{47}NO_8$ 537.69	124-80-1 258	8281	4-21-00-02883	ace 4; EtOH 4
4068	Cevane-3,4,7,14,15,16,20-heptol, 4,9-epoxy-, (3β,4α,7α,15α,16β)- Germine	$C_{27}H_{43}NO_8$ 509.64	508-65-6 220	4307	5-21-13-00706	bz 3; MeOH 3; alk 3; acid 3
4069	Cevane-3,4,12,14,16,17,20-heptol, 4,9-epoxy-, 3-(2-methyl-2-butenoate), [3β(*Z*),4α,16β]- Cevadine	$C_{32}H_{49}NO_9$ 591.74	62-59-9 213-4 dec	2025	4-21-00-06820	
4070	Cevane-3,4,6,7,14,15,16,20-octol, 4,9-epoxy-, (3β,4α,6α,7α,15α,16β Protoverine	$C_{27}H_{43}NO_9$ 525.64	76-45-9 221	7916	4-21-00-06841	H_2O 1; EtOH 3; bz 3; aq acid 3
4071	Chelidonine Stylophorine	$C_{20}H_{19}NO_5$ 353.37	476-32-4 135.5	2043 $220^{0.002}$	2-27-00-00615	H_2O 1; EtOH 3; eth 3; chl 3
4072	Chinomethionat Carbonic acid, dithio-, cyclic *S,S*-(6-methyl-2,3-quinoxalinediyl) ester	$C_{10}H_6N_2OS_2$ 234.30	2439-01-2 170	6933		
4073	*DL*-chiro-Inositol	$C_6H_{12}O_6$ 180.16	18685-70-6 253		4-06-00-07920 1.752^{15}	H_2O 4; EtOH 2; eth 1; bz 1
4074	*D*-chiro-Inositol, 2-deoxy- *D*-Quercitol	$C_6H_{12}O_5$ 164.16	488-73-3 236	8046	4-06-00-07883 1.5845^{13}	H_2O 4
4075	Chlorbenside Benzene, 1-chloro-4-[[(4-chlorophenyl)methyl]thio]-	$C_{13}H_{10}Cl_2S$ 269.19	103-17-3 75	2074	 1.4210^{20}	
4076	Chlorbromuron Urea, *N*'-(4-bromo-3-chlorophenyl)-*N*-methoxy-*N*-methyl-	$C_9H_{10}BrClN_2O_2$ 293.55	13360-45-7 96		 1.69^{20}	
4077	Chlordane 4,7-Methano-1H-indene, 1,2,4,5,6,7,8,8-octachloro-2,3,3a,4,7,7a-hexahydro-	$C_{10}H_6Cl_8$ 409.78	57-74-9 106	2079 175^{1}	 1.60^{25}	
4078	Chlorfenvinphos Phosphoric acid, 2-chloro-1-(2,4-dichlorophenyl)ethenyl diethyl ester	$C_{12}H_{14}Cl_3O_4P$ 359.57	470-90-6	2087 $170^{0.05}$		
4079	Chlorflurecol 9H-Fluorene-9-carboxylic acid, 2-chloro-9-hydroxy-	$C_{14}H_9ClO_3$ 260.68	2464-37-1		 1.496^{20}	
4080	Chloridazon 3(2H)-Pyridazinone, 5-amino-4-chloro-2-phenyl-	$C_{10}H_8ClN_3O$ 221.65	1698-60-8 205			
4081	Chlorimuron 2-(4-Chloro-6-methoxypyrimidin-2-ylcarbamoylsulfamoyl) benzoic acid, ethyl ester	$C_{15}H_{15}ClN_4O_6S$ 414.83	90982-32-4 186	2092		
4082	Chlorobenzilate Benzeneacetic acid, 4-chloro-α-(4-chlorophenyl)-α-hydroxy-, ethyl ester	$C_{16}H_{14}Cl_2O_3$ 325.19	510-15-6 37	 $157^{0.07}$	 1.2816^{20}	
4083	5-Chloro-3-methyl-4-nitro-1H-pyrazole 1H-Pyrazole, 3-chloro-5-methyl-4-nitro-	$C_4H_4ClN_3O_2$ 161.55	6814-58-0 111			
4084	Chloroneb Benzene, 1,4-dichloro-2,5-dimethoxy-	$C_8H_8Cl_2O_2$ 207.06	2675-77-6 134	 268		
4085	Chlorosulfuric acid, ethyl ester	$C_2H_5ClO_3S$ 144.58	625-01-4	 152.5; 93^{100}	4-01-00-01326 1.3502^{25}	eth 4; chl 4; lig 4 1.416^{20}
4086	Chlorosulfuric acid, methyl ester	CH_3ClO_3S 130.55	812-01-1 -70	 134	4-01-00-01252 1.4805^{25}	ace 4; bz 4; eth 4 1.4138^{18}
4087	Chlorosulfurous acid, ethyl ester	$C_2H_5ClO_2S$ 128.58	6378-11-6	 52.5^{44}; 32^{16}	3-01-00-01316 1.2766^{25}	eth 4 1.4550^{25}
4088	Chloroxuron Urea, *N*'-[4-(4-chlorophenoxy)phenyl]-*N,N*-dimethyl-	$C_{15}H_{15}ClN_2O_2$ 290.75	1982-47-4 151			
4089	Chlorpyrifos Phosphorothioic acid, *O,O*-diethyl *O*-(3,5,6-trichloro-2-pyridinyl) ester	$C_9H_{11}Cl_3NO_3PS$ 350.59	2921-88-2 42	2190		
4090	Chlorpyrifos-methyl Phosphorothioic acid, *O,O*-dimethyl *O*-(3,5,6-trichloro-2-pyridinyl) ester	$C_7H_7Cl_3NO_3PS$ 322.54	5598-13-0 43			
4091	Chlorsulfuron 2-Chloro-*N*-[[(4-methoxy-6-methyl-1,3,5-triazi*N*-2-yl)amino]carbonyl]benzenesulfonamide	$C_{12}H_{12}ClN_5O_4S$ 357.78	64902-72-3 176	2192		
4092	Chlorthiophos Phosphorothioic acid, *O*-[2,5-dichloro-4-(methylthio)phenyl] *O,O*-diethyl ester	$C_{11}H_{15}Cl_2O_3PS_2$ 361.25	21923-23-9	 $150^{0.001}$		

No.	Name	Synonym	Mol. Form.	Mol. Wt.	CAS RN	mp/°C	Merck No.	bp/°C	Beil. Ref.	den/g cm^{-3}	Solubility	n_D
4093	Cholane	1H-Cyclopenta[a]phenanthrene, hexadecahydro-10,13-dimethyl-17-(1-methylbutyl)-, [8R-[8α,9β,10α,1	$C_{24}H_{42}$	330.60	548-98-1	90	2198	$190^{0.001}$	4-05-00-01220			
4094	Cholan-24-oic acid	Cholanic acid	$C_{24}H_{40}O_2$	360.58	25312-65-6	163.5	2199		4-09-00-01992		EtOH 3; chl 3; HOAc 3	
4095	Cholan-24-oic acid, 3,12-dihydroxy-, (3α,5β,12α)-	Deoxycholic acid	$C_{24}H_{40}O_4$	392.58	83-44-3	177	2881		4-10-00-01608			
4096	Cholan-24-oic acid, 3,6-dihydroxy-, (3α,5β,6α)-	Hyodeoxycholic acid	$C_{24}H_{40}O_4$	392.58	83-49-8	198.5	4794		4-10-00-01600		H_2O 2; EtOH 3; eth 2; ace 2	
4097	Cholan-24-oic acid, 3,7-dihydroxy-, (3α,5β,7α)-	Chenodiol	$C_{24}H_{40}O_4$	392.58	474-25-9	119	2044		4-10-00-01604		H_2O 1; EtOH 4; eth 3; ace 4	
4098	Cholan-24-oic acid, 3,7-dihydroxy-, (3α,5β,7β)-	Ursodiol	$C_{24}H_{40}O_4$	392.58	128-13-2	203	9801		4-10-00-01604		EtOH 4; eth 2	
4099	Cholan-24-oic acid, 3-hydroxy-, (3α,5β)-	Lithocholic acid	$C_{24}H_{40}O_3$	376.58	434-13-9	186	5423		4-10-00-00785		H_2O 1; EtOH 3; eth 2; chl 3	
4100	Cholan-24-oic acid, 3,7,12-trihydroxy-, (3α,5β,7α,12α)-	Cholic acid	$C_{24}H_{40}O_5$	408.58	81-25-4	198	2206		4-10-00-02072		H_2O 2; EtOH 3; eth 4; ace 3	
4101	Cholan-24-oic acid, 3,7,12-trioxo-, (5β)-	Dehydrocholic acid	$C_{24}H_{34}O_5$	402.53	81-23-2	237	2858		4-10-00-03478		H_2O 1; EtOH 2; eth 1; ace 3	
4102	Cholesta-3,5-diene		$C_{27}H_{44}$	368.65	747-90-0	80		260^{13}	4-05-00-01620	0.925^{100}	H_2O 1; EtOH 3; eth 5; bz 5	
4103	Cholesta-5,7-dien-3-ol, (3β)-	7-Dehydrocholesterol	$C_{27}H_{44}O$	384.65	434-16-2	150.5	2857		4-06-00-04153		H_2O 1; EtOH 2; eth 3; ace 3	
4104	Cholesta-8,24-dien-3-ol, (3β,5α)-		$C_{27}H_{44}O$	384.65	128-33-6	110		$160^{0.001}$	3-06-00-02828		ace 3; chl 3; MeOH 3	
4105	Cholestane, (5α)-	28,29,30-Trinorlanostane	$C_{27}H_{48}$	372.68	481-21-0	80	2202	250^{1}	4-05-00-01227	0.9090^{88}	H_2O 1; EtOH 2; eth 4; bz 4	1.4887^{88}
4106	Cholestane, (5β)-	Coprostane	$C_{27}H_{48}$	372.68	481-20-9	72	2518		4-05-00-01226	0.9119^{87}	eth 4; chl 4	1.4884^{88}
4107	Cholestanol	Dihydrocholesterol	$C_{27}H_{48}O$	388.68	80-97-7	141.5	2203		4-06-00-03577		eth 4; chl 4	
4108	Cholestan-3-ol, (3α,5α)-	Epicholestanol	$C_{27}H_{48}O$	388.68	516-95-0	185.5	3564		4-06-00-03577		chl 3	
4109	Cholestan-3-ol, (3β,5β)-	Coprosterol	$C_{27}H_{48}O$	388.68	360-68-9	102	2519		4-06-00-03576		H_2O 1; EtOH 4; eth 4; bz 4	
4110	Cholest-5-en-3-ol (3β)-	Cholesterol	$C_{27}H_{46}O$	386.66	57-88-5	148.5	2204	360 dec; $233^{0.5}$	4-06-00-04000	1.067^{20}	H_2O 1; EtOH 2; eth 4; ace 2	
4111	Cholest-4-en-3-ol, (3β)-	Allocholesterol	$C_{27}H_{46}O$	386.66	517-10-2	132	253		3-06-00-02604		H_2O 1; EtOH 3; eth 4; ace 4	
4112	Cholest-5-en-3-ol, (3α)-	Epicholesterol	$C_{27}H_{46}O$	386.66	474-77-1	141.5	3565		4-06-00-04001		EtOH 2	
4113	6-Chrysenamine	6-Aminochrysene	$C_{18}H_{13}N$	243.31	2642-98-0	210.5	2258		4-12-00-03492			
4114	Chrysene	1,2-Benzophenanthrene	$C_{18}H_{12}$	228.29	218-01-9	258.2	2259	448	4-05-00-02554	1.274^{20}	H_2O 1; EtOH 2; eth 2; ace 2	
4115	Chrysene, 5,6-dimethyl-	5,6-Dimethylchrysene	$C_{20}H_{16}$	256.35	3697-27-6	129.3		$200^{0.5}$	4-05-00-02590		H_2O 1; EtOH 3; CS_2 3; HOAc 3	
4116	5,6-Chrysenedione		$C_{18}H_{10}O_2$	258.28	2051-10-7	239.5		sub	4-07-00-02646		H_2O 1; EtOH 3; eth 2; bz 3	
4117	Chrysene, 1-methyl-	1-Methylchrysene	$C_{19}H_{14}$	242.32	3351-28-8	256.5		sub 140	4-05-00-02572		H_2O 1; EtOH 3	
4118	Chrysene, octadecahydro-		$C_{18}H_{30}$	246.44	2090-14-4	115		353	3-05-00-01107		EtOH 4	
4119	Cinchonan-6',9-diol, (8α,9*R*)-	Cupreine	$C_{19}H_{22}N_2O_2$	310.40	524-63-0	202	2626		4-23-00-03260		EtOH 4	
4120	Cinchonan-9-ol, (9*S*)-	Cinchonine	$C_{19}H_{22}N_2O$	294.40	118-10-5	265	2289		5-23-12-00406			
4121	Cinchonan-9-ol, 10,11-dihydro-, (8α,9*R*)-	Hydrocinchonidine	$C_{19}H_{24}N_2O$	296.41	485-64-3	229	4705				EtOH 4	
4122	Cinchonan-9-ol, 10,11-dihydro-, (9*S*)-	Hydrocinchonine	$C_{19}H_{24}N_2O$	296.41	485-65-4	268.5	4706				H_2O 3; EtOH 2; eth 1	
4123	Cinchonan-9-ol, 10,11-dihydro-6'-methoxy-, (8α,9*R*)-	Hydroquinine	$C_{20}H_{26}N_2O_2$	326.44	522-66-7	172.5	4737		5-23-13-00340		ace 4; eth 4; EtOH 4; chl 4	
4124	Cinchonan-9-ol, 10,11-dihydro-6'-methoxy-, (9*S*)-	Hydroquinidine	$C_{20}H_{26}N_2O_2$	326.44	1435-55-8	168.5	4736		5-23-13-00340		EtOH 3; eth 3; ace 3; chl 3	
4125	Cinchonan-9-ol, 6'-methoxy-, (9*R*)-	Epiquinidine	$C_{20}H_{24}N_2O_2$	324.42	572-59-8	113	3570		0-23-00-00505		EtOH 4; eth 3	
4126	Cinchonan-9-ol, 6'-methoxy-, (9*S*)-	Quinidine	$C_{20}H_{24}N_2O_2$	324.42	56-54-2	174	8072		5-23-13-00395		H_2O 2; EtOH 3; eth 2; bz 3	

No.	Name Synonym	Mol. Form. Mol. Wt.	CAS RN mp/°C	Merck No. bp/°C	Beil. Ref. den/g cm^{-3}	Solubility n_D
4127	Cinchonan-9-ol, 6'-methoxy-, (8α,9*R*)- Quinine	$C_{20}H_{24}N_2O_2$ 324.42	130-95-0 57	8075	5-23-13-00395	H_2O 2; EtOH 4; eth 3; ace 2 1.625^{15}
4128	Cinchonan-9-ol, 6'-methoxy-, carbonate (2:1) (este*R*), (8α,9*R*)- Quinine carbonate	$C_{41}H_{46}N_4O_5$ 674.84	146-06-5 189	8077	4-23-00-03282	
4129	Cinchonan-9-ol, 6'-methoxy-, (8α,9*R*)-, monoformate (salt) Quinine, formate	$C_{21}H_{26}N_2O_4$ 370.45	130-90-5 149.5	8081	4-23-00-03274	H_2O 4; EtOH 4; chl 4
4130	Cinchonan-9-ol, 6'-methoxy-, monohydrochloride, (8α,9*R*)- Quinine hydrochloride	$C_{20}H_{25}ClN_2O_2$ 360.88	130-89-2 159	8085		H_2O 4; EtOH 4; chl 4
4131	Cinchonan-9-ol, 6'-methoxy-, (8α,9*R*)-, mono(2-hydroxybenzoate) (salt) Quinine salicylate	$C_{27}H_{32}N_2O_6$ 480.56	750-90-3 195	8088	4-23-00-03277	EtOH 4; chl 4
4132	Cinchonan-9-ol, 6'-methoxy-, (8α,9*R*)-, sulfate (2:1) (salt) Quinine sulfate	$C_{40}H_{50}N_4O_8S$ 746.93	804-63-7 235.2	8089		EtOH 4
4133	Cinchonan-9-ol, (8α,9*R*)- Cinchonidine	$C_{19}H_{22}N_2O$ 294.40	485-71-2 210.5	2288 sub	5-23-12-00406	H_2O 1; EtOH 3; eth 2; bz 1
4134	Cinchonan-9-one, 6'-methoxy-, (8α)- Quininone	$C_{20}H_{22}N_2O_2$ 322.41	84-31-1 108	8093	4-25-00-00204	bz 4; eth 4; EtOH 4
4135	Cinnamic acid, β-propyl- 2-Hexenoic acid, 3-phenyl-	$C_{12}H_{14}O_2$ 190.24	4362-01-0 94	 $183\text{-}4^{14}$	3-09-00-02811	bz 4
4136	Cinnoline 1,2-Benzodiazine	$C_8H_6N_2$ 130.15	253-66-7 38	2309 $114^{0.3}$	5-23-07-00124	eth 4; EtOH 4
4137	Ciodrin 2-Butenoic acid, 3-[(dimethoxyphosphinyl)oxy]-, 1-phenylethyl ester, (*E*)-	$C_{14}H_{19}O_6P$ 314.28	7700-17-6	2603 $135^{0.03}$	 1.19^{25}	
4138	Clethodim	$C_{17}H_{26}ClNO_3S$ 359.92	99129-21-2		 1.14^{20}	
4139	Clofentezine 3,6-Bis(2-chlorophenyl)-1,2,4,5-tetrazine	$C_{14}H_8Cl_2N_4$ 303.15	74115-24-5 182	2373		
4140	Clomazone 2-(2-Chlorobenzyl)-4,4-dimethyl-1,2-oxazolidin-3-one	$C_{12}H_{14}ClNO_2$ 239.70	81777-89-1	3203	 1.192^{20}	
4141	Cloprop Propanoic acid, 2-(3-chlorophenoxy)-	$C_9H_9ClO_3$ 200.62	101-10-0			
4142	Clopyralid 3,6-Dichloro-2-pyridinecarboxylic acid	$C_6H_3Cl_2NO_2$ 192.00	1702-17-6 151	2398		
4143	Con-5-enin-3-amine, *N,N*-dimethyl-, (3β)- Conessine	$C_{24}H_{40}N_2$ 356.60	546-06-5 125.5	2492 $165\text{-}7^{0.1}$		H_2O 2; chl 3; HOAc 3
4144	Conhydrine, *N*-methyl- *N*-Methylconhydrine	$C_9H_{19}NO$ 157.26	94826-08-1	 $94\text{-}5^{11}$	2-21-00-00011 0.9400^{20}	 1.4708^{19}
4145	Coronene	$C_{24}H_{12}$ 300.36	191-07-1 437.3	 525	4-05-00-02830 1.371^{25}	H_2O 1; bz 2; con sulf 1
4146	Corynan-16-carboxylic acid, 16,17,18,19-tetradehydro-17-methoxy-, methyl ester, (16*E*)- Corynantheine	$C_{22}H_{28}N_2O_3$ 366.46	18904-54-6 165.5	2546	4-25-00-01255	EtOH 4
4147	Coumaphos Coumarin, 3-chloro-7-hydroxy-4-methyl, *O*-ester with *O,O*-diethylphosphorothioate	$C_{14}H_{16}ClO_5PS$ 362.77	56-72-4 93	2559	 1.474	
4148	*o*-Cresol, 4-(3-methylpentyl)- Phenol, 2-methyl-4-(3-methyl pentyl)	$C_{13}H_{20}O$ 192.30	882-25-7 35	 $145\text{-}6^{11}$		ace 4 1.5200^{25}
4149	Crotonaldehyde, 2-bromo-, diethyl acetal 2-Butenal, 2-bromo, diethyl acetal	$C_8H_{15}BrO_2$ 223.11	98551-14-5	 86^{15}	1-01-00-00380 1.2255^{21}	 1.4565^{21}
4150	Crotonic acid, 2-chloro- 2-Chlorocrotonic acid	$C_4H_5ClO_2$ 120.54	600-13-5 99.8	 212; 111^{14}	4-02-00-01509	eth 4; EtOH 4
4151	Crotonic acid, 4-chloro-, (*E*)- 2-Butenoic acid, 4-chloro (*trans*)	$C_4H_5ClO_2$ 120.54	26340-58-9 83	 $117\text{-}8^{13}$	4-02-00-01511	eth 4
4152	Crotonic acid, 4,4-dichloro- 4,4-Dichlorocrotonic acid	$C_4H_4Cl_2O_2$ 154.98	99980-00-4 100.5	 130^{18}	4-02-00-01512 1.3331^{99}	eth 4; EtOH 4; chl 4 1.4597^{99}
4153	Crufomate Phosphoramidic acid, methyl-, 2-chloro-4-(1,1-dimethylethyl)phenyl methyl ester	$C_{12}H_{19}ClNO_3P$ 291.71	299-86-5 60	2607 $118^{0.01}$		
4154	Curan-17-ol, (16α)- Geissoschizoline	$C_{19}H_{26}N_2O$ 298.43	18397-07-4 135 dec	4274	5-23-12-00211	H_2O 1; EtOH 4; eth 4; chl 4
4155	Cyanamide Cyanogenamide	CH_2N_2 42.04	420-04-2 44	2691 140^{19}	4-03-00-00145 1.282^{20}	H_2O 4; EtOH 4; eth 3; ace 3 1.4418^{48}
4156	Cyanamide, diethyl-	$C_5H_{10}N_2$ 98.15	617-83-4 -80.6	 188	4-04-00-00381 0.854^{20}	H_2O 1; EtOH 3; eth 3 1.4126^{25}
4157	Cyanamide, dimethyl-	$C_3H_6N_2$ 70.09	1467-79-4	 163.5	4-04-00-00226	ace 4; eth 4; EtOH 4 1.4089^{19}
4158	Cyanamide, diphenyl- Diphenylcyanamide	$C_{13}H_{10}N_2$ 194.24	27779-01-7 73.5	 237^{60}	4-12-00-00857	EtOH 4; lig 4
4159	Cyanamide, di-2-propenyl- Diallylcyanamide	$C_7H_{10}N_2$ 122.17	538-08-9	2952 142^{90}; 95^{9}	4-04-00-01078	EtOH 3; eth 2; ctc 2
4160	Cyanamide, methylphenyl-	$C_8H_8N_2$ 132.17	18773-77-8 32	 137^{12}	4-12-00-00840 1.0400^{55}	

No.	Name Synonym	Mol. Form. Mol. Wt.	CAS RN mp/°C	Merck No. bp/°C	Beil. Ref. den/g cm^{-3}	Solubility n_D
4161	Cyanazine Propanenitrile, 2-[[4-chloro-6-(ethylamino)-1,3,5-triazin-2-yl]amino]-2-methyl-	$C_9H_{13}ClN_6$ 240.70	21725-46-2 168	2692		
4162	Cyanic acid Hydrogen cyanate (HOCN)	CHNO 43.03	420-05-3 -86	2693 23	4-03-00-00080 1.1402^{20}	H_2O 4; bz 4; eth 4; chl 4
4163	Cyanic acid, ethyl ester Ethyl cyanate	C_3H_5NO 71.08	627-48-5	162 dec; 30^{12}	0.89^{20}	EtOH 4 1.3788^{25}
4164	Cyanic acid, sodium salt Sodium cyanate	CNNaO 65.01	917-61-3 550	8552	1.893^{20}	H_2O 3; EtOH 2; eth 1
4165	Cyanogen bromide Bromine cyanide	CBrN 105.92	506-68-3 52	2700 61.5	4-03-00-00092 2.015^{20}	H_2O 3; EtOH 3; eth 3
4166	Cyanogen chloride Chlorine cyanide	CClN 61.47	506-77-4 -6.5	2701 13	4-03-00-00090 1.186^{20}	H_2O 3; EtOH 3; eth 4
4167	Cyanogen fluoride Fluorine cyanide	CFN 45.02	1495-50-7 -82	-46		
4168	Cycloate Carbamothioic acid, cyclohexylethyl-, *S*-ethyl ester	$C_{11}H_{21}NOS$ 215.36	1134-23-2 11.5	145^{10}	1.0156^{30}	
4169	Cyclobutanamine Aminocyclobutane	C_4H_9N 71.12	2516-34-9	82	4-12-00-00003 0.8328^{20}	1.4363^{19}
4170	Cyclobutane Tetramethylene	C_4H_8 56.11	287-23-0 -90.6	2720 12.6	4-05-00-00006 0.6890^{25}	H_2O 1; EtOH 4; eth 5; ace 4 1.365^{290}
4171	Cyclobutaneacetic acid, 3-(ethoxycarbonyl)-2,2-dimethyl-, ethyl ester Pinic acid diethyl ester	$C_{13}H_{22}O_4$ 242.32	28664-03-1	$142\text{-}6^{10}$	2-09-00-00529 1.0104^{20}	1.4496^{70}
4172	Cyclobutane, 1,2-bis(methylene)-	C_6H_8 80.13	14296-80-1	64	4-05-00-00389 0.7698^{20}	1.4232^{20}
4173	Cyclobutanecarbonitrile Cyanocyclobutane	C_5H_7N 81.12	4426-11-3	149.6		
4174	Cyclobutanecarboxylic acid Cyclobutylcarboxylic acid	$C_5H_8O_2$ 100.12	3721-95-7 -2	190; $74\text{-}5^{2}$	4-09-00-00006 1.0599^{20}	H_2O 2; EtOH 5; eth 5 1.4400^{20}
4175	Cyclobutanecarboxylic acid, ethyl ester Ethyl cyclobutanecarboxylate	$C_7H_{12}O_2$ 128.17	14924-53-9	159	3-09-00-00007 0.928^{25}	ctc 2 1.4261^{20}
4176	Cyclobutanecarboxylic acid, methyl ester Methyl cyclobutanecarboxylate	$C_6H_{10}O_2$ 114.14	765-85-5	135.5		
4177	Cyclobutane, 1-chloro-1,2,2,3,3,4,4-heptafluoro- Chloroheptafluorocyclobutane	C_4ClF_7 216.49	377-41-3			
4178	Cyclobutane, 1,1-dibromo- 1,1-Dibromocyclobutane	$C_4H_6Br_2$ 213.90	33742-81-3	160	0-05-00-00017 1.930^{20}	bz 4; EtOH 4; lig 4 1.5362^{20}
4179	Cyclobutane, 1,2-dibromo- 1,2-Dibromocyclobutane	$C_4H_6Br_2$ 213.90	89033-70-5 2.5	172.5	0-05-00-00017 1.972^{0}	
4180	1,1-Cyclobutanedicarboxylic acid, diethyl ester Diethyl 1,1-cyclobutanedicarboxylate	$C_{10}H_{16}O_4$ 200.23	3779-29-1	224	4-09-00-02788 1.0456^{20}	EtOH 4; ctc 2 1.4330^{26}
4181	1,2-Cyclobutanedicarboxylic acid, dimethyl ester, *trans*- Dimethyl *trans*-1,2-cyclobutanedicarboxylate	$C_8H_{12}O_4$ 172.18	7371-67-7	225; 118^{24}	3-09-00-03799 1.1276^{20}	1.4450^{20}
4182	1,3-Cyclobutanedicarboxylic acid, *cis*-	$C_6H_8O_4$ 144.13	2398-16-5 143.5	252	3-09-00-03801	H_2O 4; EtOH 4
4183	1,3-Cyclobutanedicarboxylic acid, 2,2-dimethyl-, *cis*- *cis*-Norpinic acid	$C_8H_{12}O_4$ 172.18	3211-48-1 175	sub 100	4-09-00-02813	eth 4; chl 4
4184	1,3-Cyclobutanedicarboxylic acid, 2,2-dimethyl-, dimethyl ester Norpinic acid, dimethyl ester	$C_{10}H_{16}O_4$ 200.23	91057-79-3	229; 113^{11}	3-09-00-03829 1.0700^{14}	1.4459^{17}
4185	1,3-Cyclobutanedicarboxylic acid, *trans*-	$C_6H_8O_4$ 144.13	7439-33-0 171	sub	3-09-00-03802	H_2O 4; EtOH 4
4186	Cyclobutane, 1,2-diethenyl-, *cis*-	C_8H_{12} 108.18	16177-46-1	38^{38}	4-05-00-00412 0.8010^{20}	1.4563^{20}
4187	Cyclobutane, 1,2-diethenyl-, *trans*- *trans*-1,2-Divinylcyclobutane	C_8H_{12} 108.18	6553-48-6	112.5	4-05-00-00412 0.7817^{20}	1.4451^{20}
4188	Cyclobutane, 1,2-dimethyl-, *trans*- *trans*-1,2-Dimethylcyclobutane	C_6H_{12} 84.16	15679-02-4 -122.5	58	4-05-00-00088 0.703^{20}	1.395^{20}
4189	Cyclobutane, 1,3-dimethyl-, *cis*- *cis*-1,3-Dimethylcyclobutane	C_6H_{12} 84.16	2398-09-6	60.5	4-05-00-00089 0.7060^{25}	1.3933^{20}
4190	Cyclobutane, 1,3-dimethyl-, *trans*- *trans*-1,3-Dimethylcyclobutane	C_6H_{12} 84.16	2398-10-9	57.5	4-05-00-00089 0.6970^{25}	1.3896^{20}
4191	Cyclobutane, ethyl- Ethylcyclobutane	C_6H_{12} 84.16	4806-61-5 -142.9	70.8	4-05-00-00087 0.7284^{20}	H_2O 1; EtOH 5; eth 5; ace 3 1.4020^{20}
4192	Cyclobutane, methyl- Methylcyclobutane	C_5H_{10} 70.13	598-61-8 -161.5	36.3	4-05-00-00021 0.6884^{20}	H_2O 1; EtOH 5; eth 5; ace 3 1.3866^{20}
4193	Cyclobutane, methylene- *exo*-Methylenecyclobutane	C_5H_8 68.12	1120-56-5 -134.7	42.2	4-05-00-00216 0.7401^{20}	1.4210^{20}
4194	Cyclobutane, octafluoro- Perfluorocyclobutane	C_4F_8 200.03	115-25-3 -40.1	6666 -5.9	4-05-00-00008 $*1.500^{25}$	eth 3
4195	Cyclobutane, 1,2,3,4-tetrabromo- 1,2,3,4-Tetrabromocyclobutane	$C_4H_4Br_4$ 371.69	101257-79-8	$110^{1.5}$	4-05-00-00012 2.5672^{20}	1.6303^{20}
4196	Cyclobutanol Hydroxycyclobutane	C_4H_8O 72.11	2919-23-5	124	4-06-00-00003 0.9218^{15}	1.4371^{20}

No.	Name Synonym	Mol. Form. Mol. Wt.	CAS RN mp/°C	Merck No. bp/°C	Beil. Ref. den/g cm^{-3}	Solubility n_D
4197	Cyclobutanone	C_4H_6O 70.09	1191-95-3 -50.9	99	4-07-00-00003 0.9547^{0}	H_2O 3; EtOH 4; eth 3; bz 3 1.4215^{20}
4198	Cyclobutene	C_4H_6 54.09	822-35-5	2	4-05-00-00207 0.733^{0}	ace 4; bz 3; peth 3
4199	Cyclobutene, 1,2-dichloro-3,3,4,4-tetrafluoro-	$C_4Cl_2F_4$ 194.94	377-93-5 -43.4	67	4-05-00-00209 1.534^{25}	1.3699^{25}
4200	Cyclobutene, hexafluoro-	C_4F_6 162.03	697-11-0 -60	5.5	4-05-00-00208 1.602^{-20}	1.298^{-20}
4201	Cyclobutene, 1-methyl- 1-Methylcyclobutene	C_5H_8 68.12	1489-60-7	37	4-05-00-00216 0.7244^{20}	1.4088^{18}
4202	Cyclodecane	$C_{10}H_{20}$ 140.27	293-96-9 10	202	4-05-00-00144 0.8538^{25}	1.4716^{20}
4203	1,2-Cyclodecanedione Sebacil	$C_{10}H_{16}O_2$ 168.24	96-01-5 40.5	8370 $104\text{-}5^{10}$	4-07-00-02009	
4204	Cyclodecanol	$C_{10}H_{20}O$ 156.27	1502-05-2 40.5	125^{12}	4-06-00-00138 0.9606^{20}	EtOH 3 1.4926^{20}
4205	Cyclodecanone	$C_{10}H_{18}O$ 154.25	1502-06-3 28	$106\text{-}7^{13}$	4-07-00-00076 0.9654^{20}	bz 4; eth 4; chl 4 1.4806^{20}
4206	Cyclodecanone, 2-hydroxy- Sebacoin	$C_{10}H_{18}O_2$ 170.25	96-00-4 38.5	8371 $134\text{-}8^{14}$	4-08-00-00037	
4207	Cyclodecene, (*E*)-	$C_{10}H_{18}$ 138.25	2198-20-1	$68\text{-}70^{10}$	4-05-00-00296 0.8672^{20}	H_2O 1; bz 3; chl 3 1.4822^{20}
4208	Cyclodecene, (*Z*)-	$C_{10}H_{18}$ 138.25	935-31-9	195	4-05-00-00295 0.8770^{20}	1.4854^{20}
4209	Cyclodecyne	$C_{10}H_{16}$ 136.24	3022-41-1	204	4-05-00-00430 0.8975^{20}	1.4903^{20}
4210	Cyclododecane	$C_{12}H_{24}$ 168.32	294-62-2 60.4	247	4-05-00-00169 0.82^{80}	
4211	Cyclododecanone Cyclodoecanone	$C_{12}H_{22}O$ 182.31	830-13-7 59	$126\text{-}8^{12}$	4-07-00-00100 0.9059^{66}	1.4571^{60}
4212	1,5,9-Cyclododecatriene CDT	$C_{12}H_{18}$ 162.27	4904-61-4 -17	240	4-05-00-01114 0.84^{100}	
4213	1,5,9-Cyclododecatriene, (*Z,E,E*)- 1,5,9-Cyclododecatriene, (*cis, trans, trans*)	$C_{12}H_{18}$ 162.27	53859-78-2 -18	$100\text{-}1^{11}$	4-05-00-01115 0.8910^{20}	bz 4; chl 4 1.5058^{20}
4214	1,5,9-Cyclododecatriene, (*Z,Z,Z*)-	$C_{12}H_{18}$ 162.27	4736-48-5 -1.5	108^{17}		bz 4; chl 4 1.5100^{25}
4215	Cyclododecene, (*E*)-	$C_{12}H_{22}$ 166.31	1486-75-5	$112\text{-}5^{17}$	4-05-00-00332	bz 4; chl 4 1.4850^{20}
4216	Cyclododecene, (*Z*)-	$C_{12}H_{22}$ 166.31	1129-89-1	133^{35}; 71^{2}	4-05-00-00332	bz 4; chl 4 1.4840^{20}
4217	9-Cycloheptadecen-1-one 9-Cycloheptadecene-1-one	$C_{17}H_{30}O$ 250.42	74244-64-7 32.5	343; 159^{2}	3-07-00-00524 0.9170^{33}	H_2O 2; EtOH 3; bz 3 1.4830^{33}
4218	9-Cycloheptadecen-1-one, (*Z*)- Civetone	$C_{17}H_{30}O$ 250.42	542-46-1 32.5	2337 343; 159^{2}	3-07-00-00524	
4219	1,3-Cycloheptadiene	C_7H_{10} 94.16	4054-38-0 -110.4	120.5	4-05-00-00390 0.868^{25}	1.4978^{20}
4220	2,4-Cycloheptadien-1-one, 2,6,6-trimethyl- Eucarvone	$C_{10}H_{14}O$ 150.22	503-93-5	105^{22}; 88^{16}	4-07-00-00311 0.9490^{20}	eth 3; ace 3 1.5087^{20}
4221	Cycloheptane	C_7H_{14} 98.19	291-64-5 -8.0	118.4	4-05-00-00092 0.8098^{20}	H_2O 1; EtOH 4; eth 4; bz 3 1.4436^{20}
4222	Cycloheptane, bromo- Cycloheptyl bromide	$C_7H_{13}Br$ 177.08	2404-35-5	101.5^{40}; 75^{12}	4-05-00-00093 1.2887^{22}	H_2O 1; eth 4; chl 4 1.4996^{20}
4223	Cycloheptanecarboxylic acid Cycloheptylcarboxylic acid	$C_8H_{14}O_2$ 142.20	1460-16-8	259; 130^{8}	4-09-00-00038 1.0423^{20}	ace 3 1.4753^{20}
4224	Cycloheptanecarboxylic acid, ethyl ester Ethyl cycloheptanecarboxylate	$C_{10}H_{18}O_2$ 170.25	32777-26-7	74^{3}	4-09-00-00038 0.9515^{20}	1.4482^{20}
4225	1,2-Cycloheptanedione	$C_7H_{10}O_2$ 126.16	3008-39-7 -40	$107\text{-}9^{17}$	4-07-00-01989 1.0583^{22}	EtOH 3 1.4689^{22}
4226	Cycloheptane, methyl- Methylcycloheptane	C_8H_{16} 112.22	4126-78-7	134	4-05-00-00114 0.800^{20}	bz 4; eth 4; EtOH 4; peth 4 1.439^{18}
4227	Cycloheptane, 1,1,2-trimethyl- 1,1,2-Trimethylcycloheptane	$C_{10}H_{20}$ 140.27	35099-89-9	$104\text{-}5^{100}$	3-05-00-00127 0.8243^{20}	1.4527^{20}
4228	Cycloheptane, 1,1,4-trimethyl- 1,1,4-Trimethylcycloheptane	$C_{10}H_{20}$ 140.27	2158-55-6	164	4-05-00-00146 0.8011^{20}	1.4420^{20}
4229	Cycloheptanol	$C_7H_{14}O$ 114.19	502-41-0 2	185	4-06-00-00094 0.9554^{20}	H_2O 2; EtOH 4; eth 4 1.40705^{20}
4230	Cycloheptanol, 1-methyl- 1-Methyl-1-cycloheptanol	$C_8H_{16}O$ 128.21	3761-94-2	183.5	4-06-00-00114 0.9392^{21}	1.4677^{22}
4231	Cycloheptanol, 2-methyl- 2-Methylcycloheptanol	$C_8H_{16}O$ 128.21	59777-92-3	191	3-06-00-00083 0.9492^{15}	1.4762^{15}
4232	Cycloheptanol, 4-methyl-	$C_8H_{16}O$ 128.21	90200-61-6	$105\text{-}6^{39}$	3-06-00-00084 0.9134^{31}	1.4574
4233	Cycloheptanol, 2,6,6-trimethyl- 3,3,7-Trimethylcycloheptanol	$C_{10}H_{20}O$ 156.27	33515-82-1	216	1-06-00-00018 0.9096^{24}	1.4369
4234	Cycloheptanone Suberone	$C_7H_{12}O$ 112.17	502-42-1	2728 178.5	4-07-00-00039 0.9508^{20}	H_2O 1; EtOH 4; eth 3 1.4608^{20}
4235	Cycloheptanone, 2-methyl- 2-Methylcycloheptanone	$C_8H_{14}O$ 126.20	932-56-9	185	4-07-00-00051 0.9395^{18}	1.461^{18}
4236	Cycloheptanone, 2-phenyl- 2-Phenylcycloheptanone	$C_{13}H_{16}O$ 188.27	14996-78-2 22.2	141^{5}; 125^{2}	4-07-00-01061	1.5395^{20}

No.	Name Synonym	Mol. Form. Mol. Wt.	CAS RN mp/°C	Merck No. bp/°C	Beil. Ref. den/g cm^{-3}	Solubility n_D
4237	Cycloheptanone, 2,2,6-trimethyl-	$C_{10}H_{18}O$ 154.25	1686-41-5	191	3-07-00-00138 0.878^{14}	1.4421^{16}
4238	Cycloheptanone, 2,6,6-trimethyl- Tetrahydroeucarvone	$C_{10}H_{18}O$ 154.25	4436-59-3	208.5	4-07-00-00079 0.9095^{18}	1.4568^{18}
4239	Cycloheptasiloxane, tetradecamethyl- Tetradecamethylcycloheptasiloxane	$C_{14}H_{42}O_7Si_7$ 519.08	107-50-6 -26	154^{20}	4-04-00-04130 0.9703^{20}	1.4040^{20}
4240	1,3,5-Cycloheptatriene Tropilidene	C_7H_8 92.14	544-25-2 -79.5	117; 60.5^{122}	4-05-00-00765 0.8875^{19}	H_2O 1; EtOH 3; eth 3; bz 4 1.5343^{20}
4241	1,3,6-Cycloheptatriene-1-carboxylic acid, 5,5-dimethyl- Thujic acid	$C_{10}H_{12}O_2$ 164.20	499-89-8	9325	4-09-00-01810	
4242	2,4,6-Cycloheptatrien-1-one	C_7H_6O 106.12	539-80-0 -7	113^{15}; 84^{6}	4-07-00-00501 1.095^{22}	bz 4; chl 4 1.6172^{22}
4243	2,4,6-Cycloheptatrien-1-one, 2-hydroxy-	$C_7H_6O_2$ 122.12	533-75-5 50.8	sub 40	4-08-00-00159	H_2O 3; eth 3; ace 3; os 3
4244	Cycloheptene	C_7H_{12} 96.17	628-92-2 -56	115	4-05-00-00244 0.8228^{20}	H_2O 1; EtOH 3; eth 3; bz 3 1.4552^{20}
4245	Cycloheptene, 1-methyl- 1-Methylcycloheptene	C_8H_{14} 110.20	1453-25-4	136	4-05-00-00265 0.8243^{22}	1.4581^{20}
4246	Cycloheptene, 5-methyl- 5-Methylcycloheptene	C_8H_{14} 110.20	2505-06-8	$69\text{-}70^{38}$	3-05-00-00211 0.7606^{31}	1.4202^{31}
4247	Cycloheptene, 1,4,4-trimethyl- 1,4,4-Trimethyl-1-cycloheptene	$C_{10}H_{18}$ 138.25	4755-36-6	165	3-05-00-00232 0.8185^{20}	1.4561^{20}
4248	1,3-Cyclohexadiene	C_6H_8 80.13	592-57-4 -89	80.5	4-05-00-00382 0.8405^{20}	H_2O 1; EtOH 3; eth 4; bz 3 1.4755^{20}
4249	1,4-Cyclohexadiene 1,4-Dihydrobenzene	C_6H_8 80.13	628-41-1 -49.2	85.5	4-05-00-00385 0.8471^{20}	H_2O 1; EtOH 5; eth 5; bz 3 1.4725^{20}
4250	1,3-Cyclohexadiene-1-carboxaldehyde, 2,6,6-trimethyl- Safranal	$C_{10}H_{14}O$ 150.22	116-26-7	8286 70^{1}	4-07-00-00318 0.9734^{19}	EtOH 4; peth 4 1.5281^{19}
4251	1,4-Cyclohexadiene-1,2-dicarbonitrile, 4,5-dichloro-3,6-dioxo- 2,3-Dichloro-5,6-dicyanobenzoquinone	$C_8Cl_2N_2O_2$ 227.01	84-58-2 214.5	3052	0-10-00-00902	bz 4; HOAc 4; diox 4
4252	1,3-Cyclohexadiene, 1,3-dimethyl- 1,3-Dimethyl-1,3-cyclohexadiene	C_8H_{12} 108.18	4573-05-1	135.5	3-05-00-00325 0.8373^{20}	1.4776^{20}
4253	1,3-Cyclohexadiene, 1,4-dimethyl- 1,4-Dimethyl-1,3-cyclohexadiene	C_8H_{12} 108.18	26120-52-5	136.5	4-05-00-00409 0.830^{20}	1.4792^{19}
4254	1,3-Cyclohexadiene, 1,5-dimethyl- 1,5-Dimethyl-1,3-cyclohexadiene	C_8H_{12} 108.18	1453-17-4	130	1-05-00-00063 0.821^{20}	1.471^{20}
4255	1,3-Cyclohexadiene, 2,5-dimethyl- 2,5-Dimethyl-1,3-cyclohexadiene	C_8H_{12} 108.18	2050-33-1	127	3-05-00-00325 0.8245^{18}	1.4631^{18}
4256	1,3-Cyclohexadiene, 2,6-dimethyl- 1,3-Cyclohexadiene, 3,5-dimethyl	C_8H_{12} 108.18	2050-32-0	131	0-05-00-00119 0.8225^{20}	1.4675^{20}
4257	2,5-Cyclohexadiene-1,4-dione *p*-Benzoquinone	$C_6H_4O_2$ 108.10	106-51-4 115.7	8103 sub	4-07-00-02065 1.318^{20}	H_2O 2; EtOH 3; eth 3; chl 3
4258	2,5-Cyclohexadiene-1,4-dione, 2,5-dichloro-3,6-dihydroxy- Chloranilic acid	$C_6H_2Cl_2O_4$ 208.98	87-88-7 283.5	2072	4-08-00-02707	H_2O 3
4259	2,5-Cyclohexadiene-1,4-dione, 2,5-dihydroxy-3,6-dinitro- Nitranilic acid	$C_6H_2N_2O_8$ 230.09	479-22-1 100 dec	6489 exp 170	3-08-00-03351	H_2O 4; EtOH 4; eth 1
4260	2,5-Cyclohexadiene-1,4-dione, 2,5-dihydroxy-3-methoxy-6-methyl- Spinulosin	$C_8H_8O_5$ 184.15	85-23-4 202.5	8706 sub 120	3-08-00-03983	H_2O 2; alk 3
4261	2,5-Cyclohexadiene-1,4-dione, 2,5-dihydroxy-3-undecyl- Embelin	$C_{17}H_{26}O_4$ 294.39	550-24-3 142.5	3513	4-08-00-02769	bz 4; eth 4; EtOH 4
4262	2,5-Cyclohexadiene-1,4-dione, 2,6-dimethoxy- 2,6-Dimethoxy-*p*-quinone	$C_8H_8O_4$ 168.15	530-55-2 256	3214 sub	4-08-00-02710	H_2O 2; EtOH 2; eth 2; tfa 3
4263	2,5-Cyclohexadiene-1,4-dione, 2,3-dimethyl-	$C_8H_8O_2$ 136.15	526-86-3 55	sub	4-07-00-02093	H_2O 2; EtOH 3; eth 3; chl 3
4264	2,5-Cyclohexadiene-1,4-dione, 2,6-dimethyl-	$C_8H_8O_2$ 136.15	527-61-7 72.5	sub	4-07-00-02095 1.0479^{28}	chl 3
4265	2,5-Cyclohexadiene-1,4-dione, 3-hydroxy-2-methoxy-5-methyl- Fumigatin	$C_8H_8O_4$ 168.15	484-89-9 116	4201	3-08-00-03374	ace 4; bz 4; eth 4; EtOH 4
4266	2,5-Cyclohexadiene-1,4-dione, 2-methyl-	$C_7H_6O_2$ 122.12	553-97-9 69	sub	4-07-00-02088 1.08^{75}	H_2O 2; EtOH 3; eth 3
4267	2,5-Cyclohexadiene-1,4-dione, 2-methyl-5-(1-methylethyl)-	$C_{10}H_{12}O_2$ 164.20	490-91-5 45.5	232	4-07-00-02100	chl 3
4268	2,5-Cyclohexadiene-1,4-dione, 2,3,5,6-tetrachloro- Chloranil	$C_6Cl_4O_2$ 245.88	118-75-2 290	2071 sub	4-07-00-02083	H_2O 1; EtOH 2; eth 3; chl 2
4269	2,5-Cyclohexadiene-1,4-dione, 2,3,5,6-tetrahydroxy- Tetroquinone	$C_6H_4O_6$ 172.09	319-89-1	9177	4-08-00-03604	H_2O 2; EtOH 4; eth 2; ctc 2
4270	2,5-Cyclohexadiene-1,4-dione, 2,3,5,6-tetramethyl- Duroquinone	$C_{10}H_{12}O_2$ 164.20	527-17-3 111.5	3452	4-07-00-02102	H_2O 1; EtOH 3; eth 3; ace 3
4271	1,3-Cyclohexadiene, 2-methyl- 4,5-Dihydrotoluene	C_7H_{10} 94.16	1489-57-2	107.5	3-06-00-00211 0.8260^{18}	1.4662^{18}

No.	Name Synonym	Mol. Form. Mol. Wt.	CAS RN mp/°C	Merck No. bp/°C	Beil. Ref. den/g cm^{-3}	Solubility n_D
4272	1,3-Cyclohexadiene, 5-methyl-, (±)- 1,3-Cyclohexadiene, 5-methyl (*DL*)	C_7H_{10} 94.16	116781-86-3	 101.5	3-05-00-00318 0.8354^{20}	bz 4; eth 4; EtOH 4; lig 4 1.4763^{20}
4273	1,3-Cyclohexadiene, 1-methyl-4-(1-methylethyl)-	$C_{10}H_{16}$ 136.24	99-86-5	 175	4-05-00-00435 0.8375^{19}	H_2O 1; EtOH 5; eth 5 1.477^{19}
4274	1,3-Cyclohexadiene, 2-methyl-5-(1-methylethyl)-, (±)-	$C_{10}H_{16}$ 136.24	13811-01-3	 175.5	3-05-00-00342 0.8410^{22}	H_2O 1; EtOH 1; eth 3 1.4772^{19}
4275	1,4-Cyclohexadiene, 1-methyl-4-(1-methylethyl)-	$C_{10}H_{16}$ 136.24	99-85-4	 183	4-05-00-00436 0.849^{20}	 1.4765^{14}
4276	1,3-Cyclohexadiene, 1,2,3,4,5,5,6,6-octafluoro-	C_6F_8 224.05	377-70-8	 63.5	4-05-00-00384 1.601^{20}	 1.3149^{20}
4277	1,4-Cyclohexadiene, 1,2,3,3,4,5,6,6-octafluoro-	C_6F_8 224.05	775-51-9	 57.5	4-05-00-00386 1.6010^{25}	 1.318^{18}
4278	2,5-Cyclohexadien-1-one, 4-[bis(4-hydroxyphenyl)methylene]- Aurin	$C_{19}H_{14}O_3$ 290.32	603-45-2 308-10 dec	899	4-08-00-02646	H_2O 1; EtOH 3; eth 2; bz 1
4279	2,5-Cyclohexadien-1-one, 2,6-dibromo-4-(chloroimino)- 2,6-Dibromoquinone-4-chlorimide	$C_6H_2Br_2ClNO$ 299.35	537-45-1 83	3010	4-07-00-02086	EtOH 4
4280	2,5-Cyclohexadien-1-one, 2,6-dichloro-4-(chloroimino)- Gibbs' reagent	$C_6H_2Cl_3NO$ 210.45	101-38-2 66	4315	3-07-00-03377	
4281	2,4-Cyclohexadien-1-one, 3,5-dihydroxy-2,6,6-tris(3-methyl-2-butenyl)-4-(3-methyl-1-oxobutyl)- Lupulon	$C_{26}H_{38}O_4$ 414.59	468-28-0 93	5482	3-07-00-04753	H_2O 1; EtOH 3; peth 3; hx 3
4282	2,4-Cyclohexadien-1-one, 3,5,6-trihydroxy-4,6-bis(3-methyl-2-butenyl)-2-(3-methyl-1-oxobutyl)-, (*R*)- Humulon	$C_{21}H_{30}O_5$ 362.47	26472-41-3 66.5	4673	4-08-00-03410	H_2O 2; EtOH 3; eth 3; ace 3
4283	Cyclohexanamine Cyclohexylamine	$C_6H_{13}N$ 99.18	108-91-8 -17.7	2735 134	4-12-00-00008 0.8191^{20}	H_2O 3; EtOH 4; eth 5; ace 5 1.4372^{20}
4284	Cyclohexanamine, *N*-cyclohexyl- Dicyclohexylamine	$C_{12}H_{23}N$ 181.32	101-83-7 -0.1	3085 256 dec; 114^{9}	4-12-00-00022 0.9123^{20}	H_2O 2; EtOH 3; eth 3; bz 3 1.4842^{20}
4285	Cyclohexanamine, *N,N*-diethyl-	$C_{10}H_{21}N$ 155.28	91-65-6	 193; 85^{20}	4-12-00-00019 0.872^{0}	EtOH 3; ctc 2
4286	Cyclohexanamine, *N,N*-dimethyl- Cyclohexyldimethylamine	$C_8H_{17}N$ 127.23	98-94-2	 162	4-12-00-00018	
4287	Cyclohexanamine, 2-ethyl- 2-Ethylcyclohexylamine	$C_8H_{17}N$ 127.23	6850-36-8	 170.5	3-12-00-00093 0.8744^{20}	 1.4682^{20}
4288	Cyclohexanamine, *N*-ethyl-	$C_8H_{17}N$ 127.23	5459-93-8	 164	4-12-00-00018 0.868^{0}	H_2O 2; EtOH 5; eth 5; ctc 2
4289	Cyclohexanamine, *N*-(2-ethylhexyl)-	$C_{14}H_{29}N$ 211.39	5432-61-1	 270; 140^{14}	3-12-00-00018 0.8473^{20}	ctc 2
4290	Cyclohexanamine, 1-ethynyl-	$C_8H_{13}N$ 123.20	30389-18-5	 $65\text{-}6^{20}$	4-12-00-00214 0.913^{25}	 1.4817^{20}
4291	Cyclohexanamine, *N,N*'-methanetetraylbis- Dicyclohexylcarbodiimide	$C_{13}H_{22}N_2$ 206.33	538-75-0 34.5	3086 123^{6}; $99^{0.5}$	4-12-00-00072	
4292	Cyclohexanamine, 2-methyl-, *cis*-	$C_7H_{15}N$ 113.20	2164-19-4	 153.5	4-12-00-00118 0.8778^{20}	 1.4688^{20}
4293	Cyclohexanamine, 2-methyl-, *trans*- *trans*-2-Methylcyclohexylamine	$C_7H_{15}N$ 113.20	931-10-2	 150	4-12-00-00118 0.8685^{20}	 1.4650^{20}
4294	Cyclohexanamine, 3-methyl-, *cis*-	$C_7H_{15}N$ 113.20	1193-16-4	 153	4-12-00-00121 0.8552^{20}	 1.4538^{20}
4295	Cyclohexanamine, 3-methyl-, *trans*-	$C_7H_{15}N$ 113.20	1193-17-5	 151.5	4-12-00-00121 0.8572^{20}	 1.4547^{20}
4296	Cyclohexanamine, 4-methyl-, *cis*-	$C_7H_{15}N$ 113.20	2523-56-0	 153.5	4-12-00-00122 0.8567^{20}	 1.4559^{20}
4297	Cyclohexanamine, 4-methyl-, *trans*-	$C_7H_{15}N$ 113.20	2523-55-9	 152	4-12-00-00122 0.8543^{20}	 1.4550^{20}
4298	Cyclohexanamine, *N*-methyl-	$C_7H_{15}N$ 113.20	100-60-7	 147	4-12-00-00017 0.8660^{23}	H_2O 2; EtOH 4; eth 5; chl 3 1.4560^{20}
4299	Cyclohexanamine, 4,4'-methylenebis-	$C_{13}H_{26}N_2$ 210.36	1761-71-3 15	 320	 0.92^{75}	
4300	Cyclohexanamine, 4,4'-methylenebis-, [cis(*cis*)]-	$C_{13}H_{26}N_2$ 210.36	6693-31-8 61	 141^{2}	4-13-00-00031	H_2O 1 1.5014^{27}
4301	Cyclohexanamine, 4,4'-methylenebis-, [*trans*(*cis*)]-	$C_{13}H_{26}N_2$ 210.36	6693-30-7 36.5	 $127\text{-}8^{1.2}$	4-13-00-00031 0.9608^{25}	H_2O 1 1.5046^{25}
4302	Cyclohexanamine, 4,4'-methylenebis-, [trans(*trans*)]-	$C_{13}H_{26}N_2$ 210.36	6693-29-4 64.7	 $131^{0.8}$	4-13-00-00031	H_2O 1 1.5032^{25}
4303	Cyclohexanamine, *N*-(1-methylethyl)-	$C_9H_{19}N$ 141.26	1195-42-2	 $60\text{-}5^{12}$	4-12-00-00019 0.859^{25}	 1.4480^{20}

No.	Name Synonym	Mol. Form. Mol. Wt.	CAS RN mp/°C	Merck No. bp/°C	Beil. Ref. den/g cm^{-3}	Solubility n_D
4304	Cyclohexanamine, *N*-2-propenyl-	$C_9H_{17}N$ 139.24	6628-00-8	 $65\text{-}6^{12}$	4-12-00-00021 0.962^{25}	 1.4664^{20}
4305	Cyclohexane Hexahydrobenzene	C_6H_{12} 84.16	110-82-7 6.6	2729 80.7	4-05-00-00027 0.7785^{20}	H_2O 1; EtOH 5; eth 5; ace 5 1.4266^{20}
4306	Cyclohexaneacetic acid Cyclohexylacetic acid	$C_8H_{14}O_2$ 142.20	5292-21-7 33	 245	4-09-00-00038 1.0423^{18}	H_2O 2; eth 3; ace 3; os 3 1.4775^{20}
4307	Δ1,α-Cyclohexaneacetic acid, 3,3-dimethyl- Allocyclogeranic acid	$C_{10}H_{16}O_2$ 168.24	6790-47-2	 145^{14}	4-09-00-00161 0.9921^{17}	 1.4765^{17}
4308	Cyclohexaneacetic acid, ethyl ester	$C_{10}H_{18}O_2$ 170.25	5452-75-5	 211	4-09-00-00038 0.9537^{14}	 1.451^{14}
4309	Cyclohexaneacetic acid, α-ethyl-1-hydroxy- Cyclobutyrol	$C_{10}H_{18}O_3$ 186.25	512-16-3 81.5	2721 164^{24}	4-10-00-00049 1.0010^{18}	ace 4; eth 4; EtOH 4; chl 4 1.4680^{18}
4310	Cyclohexaneacetic acid, methyl ester Methyl cyclohexaneacetate	$C_9H_{16}O_2$ 156.22	14352-61-5	 201	4-09-00-00038 0.9896^{24}	 1.459^{14}
4311	Cyclohexaneacetic acid, 2-methyl- Cyclohexanacetic acid, 2-methyl	$C_9H_{16}O_2$ 156.22	6617-04-5	 $145\text{-}7^{13}$	2-09-00-00013 1.012^{25}	 1.4656^{25}
4312	Cyclohexaneacetic acid, 3-methyl- Cyclohexanacetic acid, 3-methyl	$C_9H_{16}O_2$ 156.22	67451-76-7	 148^{18}	2-09-00-00013 0.9847^{20}	 1.495^{20}
4313	Cyclohexaneacetonitrile	$C_8H_{13}N$ 123.20	4435-14-7	 105^{22}	4-09-00-00040 0.9473^{21}	eth 4; EtOH 4 1.4787^{21}
4314	Cyclohexane, 1,2-bis(methylene)-	C_8H_{12} 108.18	2819-48-9	 125; $60\text{-}1^{90}$	4-05-00-00409 0.8229^{25}	H_2O 1; EtOH 3; eth 3; ace 4 1.4718^{25}
4315	Cyclohexane, bromo- Cyclohexyl bromide	$C_6H_{11}Br$ 163.06	108-85-0 -56.5	2736 166.2	4-05-00-00067 1.3359^{20}	H_2O 1; EtOH 5; eth 5; ace 5 1.4957^{20}
4316	Cyclohexane, (2-bromoethyl)-	$C_8H_{15}Br$ 191.11	1647-26-3 -57	 212	4-05-00-00116 1.2357^{20}	 1.4899^{20}
4317	Cyclohexane, 1-bromo-2-methoxy-	$C_7H_{13}BrO$ 193.08	24618-31-3	 $72\text{-}4^{8}$	3-06-00-00043 1.3257^{20}	 1.4871^{20}
4318	Cyclohexane, (bromomethyl)-	$C_7H_{13}Br$ 177.08	2550-36-9	 $76\text{-}7^{26}$	4-05-00-00100 1.2763^{25}	bz 4; eth 4; chl 4 1.4907^{30}
4319	Cyclohexane, 1-bromo-1-methyl-	$C_7H_{13}Br$ 177.08	931-77-1	 158	4-05-00-00100 1.2510^{20}	H_2O 1; EtOH 3; chl 3 1.4866^{20}
4320	Cyclohexane, 1-bromo-3-methyl- 3-Methylcyclohexyl bromide	$C_7H_{13}Br$ 177.08	13905-48-1	 181; 60^{11}	3-05-00-00076 1.275^{25}	H_2O 1; eth 4; bz 3 1.4979^{20}
4321	Cyclohexane, 1-bromo-4-methyl-1-(1-methylethyl)- *p*-Menthane, 4-bromo	$C_{10}H_{19}Br$ 219.16	116836-10-3	 $98\text{-}9^{11}$	0-05-00-00051 1.165^{20}	 1.4872^{20}
4322	Cyclohexane, (2-bromo-2-propenyl)- Propylene, 2-bromo-3-cyclohexyl	$C_9H_{15}Br$ 203.12	53608-85-8	 90^{15}	3-05-00-00224 1.215^{17}	eth 4 1.495^{17}
4323	Cyclohexane, 1,1'-(1,4-butanediyl)bis- 1,4-Dicyclohexylbutane	$C_{16}H_{30}$ 222.41	6165-44-2 12	 305	4-05-00-00357 0.8771^{21}	EtOH 2; eth 2; bz 2; chl 2 1.475^{21}
4324	Cyclohexanebutanol	$C_{10}H_{20}O$ 156.27	4441-57-0	 $103\text{-}4^{4}$	4-06-00-00139 0.902^{25}	 1.4660^{20}
4325	Cyclohexane, butyl- Butylcyclohexane	$C_{10}H_{20}$ 140.27	1678-93-9 -74.7	 180.9	4-05-00-00146 0.7902^{20}	H_2O 1 1.4408^{20}
4326	Cyclohexane, 1,1'-butylidenebis- Butane, 1,1-dicyclohexyl-	$C_{16}H_{30}$ 222.41	54890-00-5 -10.5	 293	4-05-00-00359 0.8841^{16}	 1.485^{16}
4327	Cyclohexane, 2-butyl-1,1,3-trimethyl- 2-Butyl-1,1,3-trimethylcyclohexane	$C_{13}H_{26}$ 182.35	54676-39-0	 $94\text{-}5^{10}$	4-05-00-00176 0.8292^{19}	 1.4563^{19}
4328	Cyclohexanecarbonitrile	$C_7H_{11}N$ 109.17	766-05-2			
4329	Cyclohexanecarbonitrile, 1-hydroxy-	$C_7H_{11}NO$ 125.17	931-97-5 35	 132^{20}	4-10-00-00021 1.0172^{20}	H_2O 4; eth 4 1.4693^{20}
4330	Cyclohexanecarbonyl chloride	$C_7H_{11}ClO$ 146.62	2719-27-9	 180	4-09-00-00026 1.0962^{15}	 1.4711^{29}
4331	Cyclohexanecarboxaldehyde	$C_7H_{12}O$ 112.17	2043-61-0	 159.3	4-07-00-00045 0.9035^{20}	H_2O 3; eth 3 1.4496^{20}
4332	Cyclohexanecarboxylic acid Hexahydrobenzoic acid	$C_7H_{12}O_2$ 128.17	98-89-5 31.5	2730 232.5	4-09-00-00016 1.0334^{22}	H_2O 2; EtOH 4; bz 4; ctc 2 1.4530^{20}
4333	Cyclohexanecarboxylic acid, 4-(aminomethyl)-, *trans*- Tranexamic acid	$C_8H_{15}NO_2$ 157.21	1197-18-8 >300	9487	3-14-00-00868	H_2O 4
4334	Cyclohexanecarboxylic acid, ethyl ester	$C_9H_{16}O_2$ 156.22	3289-28-9	 196	4-09-00-00018 0.9362^{20}	ace 4; eth 4; EtOH 4; chl 4 1.4501^{15}
4335	Cyclohexanecarboxylic acid, methyl ester	$C_8H_{14}O_2$ 142.20	4630-82-4	 183	4-09-00-00017 0.9954^{15}	H_2O 1; EtOH 3; eth 3; ace 3 1.4433^{20}
4336	Cyclohexanecarboxylic acid, 3-methyl-4-oxo- Cyclohexanone-4-carboxylic acid, 2-methyl	$C_8H_{12}O_3$ 156.18	101567-36-6 93.5	 195^{20}	4-10-00-02617	bz 4; EtOH 4; chl 4
4337	Cyclohexanecarboxylic acid, propyl ester Propyl cyclohexanecarboxylate	$C_{10}H_{18}O_2$ 170.25	6739-34-0	 215.5	4-09-00-00018 0.9530^{15}	ace 4; eth 4; EtOH 4; chl 4 1.4486^{15}
4338	Cyclohexanecarboxylic acid, 1,3,4,5-tetrahydroxy-, [1*R*-(1α,3α,4α,5β)]- Quinic acid	$C_7H_{12}O_6$ 192.17	77-95-2 162.5	8071	4-10-00-02257 1.64^{25}	H_2O 4; EtOH 4; HOAc 4
4339	Cyclohexane, chloro- Cyclohexyl chloride	$C_6H_{11}Cl$ 118.61	542-18-7 -44	2738 142	4-05-00-00048 1.000^{20}	H_2O 1; EtOH 5; eth 5; ace 5 1.4626^{20}
4340	Cyclohexane, cyclopentyl-	$C_{11}H_{20}$ 152.28	1606-08-2	 215.1	4-05-00-00328 0.8758^{20}	 1.4725^{20}
4341	Cyclohexane, (3-cyclopentylpropyl)- Propane, 1-cyclohexyl-3-cyclopentyl-	$C_{14}H_{26}$ 194.36	2883-07-0	 270	4-05-00-00347 0.8751^{20}	 1.4765^{20}

No.	Name Synonym	Mol. Form. Mol. Wt.	CAS RN mp/°C	Merck No. bp/°C	Beil. Ref. den/g cm^{-3}	Solubility n_D
4342	Cyclohexane, 1,1,2,2,3,3,4,4,5,6-decafluoro-5,6-bis(trifluoromethyl)- Cyclohexane, 1,2-dimethyl, perfluoro	C_8F_{16} 400.06	306-98-9 -56	 101.5	3-05-00-00098 1.829^{25}	 1.283^{25}
4343	Cyclohexane, 1,1,2,2,3,4,4,5,5,6-decafluoro-3,6-bis(trifluoromethyl)-	C_8F_{16} 400.06	374-77-6	 100.5	4-05-00-00124 1.8503^{20}	 1.2897^{20}
4344	Cyclohexane, decyl-	$C_{16}H_{32}$ 224.43	1795-16-0 -1.7	 299	4-05-00-00187 0.8186^{20}	 1.4534^{20}
4345	Cyclohexane, 1,1-dibromo- 1,1-Dibromocyclohexane	$C_6H_{10}Br_2$ 241.95	10489-97-1	 87^{14}	4-05-00-00070 1.7349^{20}	 1.5392^{25}
4346	Cyclohexane, 1,2-dibromo-, *cis-* *cis*-1,2-Dibromocyclohexane	$C_6H_{10}Br_2$ 241.95	19246-38-9 9.7	 115^{14}	4-05-00-00070 1.798^{25}	ace 4; bz 4; eth 4; lig 4 1.5514^{25}
4347	Cyclohexane, 1,2-dibromo-, *trans*-(±)-	$C_6H_{10}Br_2$ 241.95	5183-77-7 -4	 145^{100}; 105^{20}	4-05-00-00071 1.7759^{20}	ace 4; bz 4; eth 4; EtOH 4 1.5445^{19}
4348	Cyclohexane, 1,3-dibromo-, *trans-* *trans*-1,3-Dibromocyclohexane	$C_6H_{10}Br_2$ 241.95	29624-17-7 1	 116^{16}	4-05-00-00072	bz 4; EtOH 4 1.5480^{20}
4349	Cyclohexane, 1,4-dibromo-, *cis-* *cis*-1,4-Dibromocyclohexane	$C_6H_{10}Br_2$ 241.95	16661-99-7	 $137\text{-}8^{25}$	4-05-00-00072 1.7834^{20}	eth 4 1.5531^{20}
4350	1,2-Cyclohexanedicarboxylic acid, diethyl ester, *trans*-(±)- Cyclohexane-1,2-dicarboxylic acid, diethyl ester (*trans, DL*)	$C_{12}H_{20}O_4$ 228.29	96836-97-4	 133^{11}	4-09-00-02803 1.040^{20}	eth 4 1.4522^{13}
4351	1,3-Cyclohexanedicarboxylic acid, diethyl ester, *cis-* Cyclohexane-1,3-dicarboxylic acid, diethyl ester (*cis*)	$C_{12}H_{20}O_4$ 228.29	62059-56-7	 288; 151^{15}	3-09-00-03817 1.045^{20}	 1.452^{20}
4352	1,3-Cyclohexanedicarboxylic acid, dimethyl ester, *cis-* Dimethyl *cis*-1,3-cyclohexanedicarboxylate	$C_{10}H_{16}O_4$ 200.23	6998-82-9	 263	4-09-00-02806 1.0997^{20}	 1.4568^{20}
4353	1,3-Cyclohexanedicarboxylic acid, dimethyl ester, *trans-* Cyclohexane-1,3-dicarboxylic acid, dimethyl ester (*trans*)	$C_{10}H_{16}O_4$ 200.23	10021-92-8	 139^{20}	4-09-00-02806 1.1095^{20}	 1.4577^{25}
4354	1,4-Cyclohexanedicarboxylic acid, *trans-*	$C_8H_{12}O_4$ 172.18	619-82-9 312.5	 sub 300	4-09-00-02807	H_2O 2; EtOH 4; eth 2; ace 3
4355	1,4-Cyclohexanedicarboxylic acid, diethyl ester, *cis-* Cyclohexane-1,4-dicarboxylic acid, diethyl ester (*cis*)	$C_{12}H_{20}O_4$ 228.29	116724-15-3	 151^{13}	2-09-00-00524 1.015^{20}	eth 4
4356	1,4-Cyclohexanedicarboxylic acid, diethyl ester, *trans-* Cyclohexane-1,4-dicarboxylic acid, diethyl ester (*trans*)	$C_{12}H_{20}O_4$ 228.29	19145-96-1 43.5		2-09-00-00524 1.0110^{20}	eth 4 1.4337^{64}
4357	1,4-Cyclohexanedicarboxylic acid, dimethyl ester, *cis-* *cis*-Dimethyl 1,4-cyclohexanedicarboxylate	$C_{10}H_{16}O_4$ 200.23	3399-21-1 14	 132.5^{10}	4-09-00-02808 1.1112^{20}	
4358	Cyclohexane, 1,1-dichloro- 1,1-Dichlorocyclohexane	$C_6H_{10}Cl_2$ 153.05	2108-92-1 -47	 171	4-05-00-00050 1.1559^{20}	 1.4803^{20}
4359	Cyclohexane, 1,2-dichloro-, *cis-* *cis*-1,2-Dichlorocyclohexane	$C_6H_{10}Cl_2$ 153.05	10498-35-8 -1.5	 206.9	4-05-00-00050 1.2021^{20}	bz 4 1.4967^{20}
4360	Cyclohexane, 1,2-dichloro-, *trans*-(±)-	$C_6H_{10}Cl_2$ 153.05	5183-79-9 -6.3	 189	4-05-00-00051 1.1839^{20}	 1.4902^{20}
4361	Cyclohexane, 1,4-dichloro-, *cis-* *cis*-1,4-Dichlorocyclohexane	$C_6H_{10}Cl_2$ 153.05	16749-11-4 18	 80.3^{25}	4-05-00-00051 1.1900^{20}	 1.4942^{20}
4362	1,3-Cyclohexanedimethanamine Cyclohexane, 1,3-bis(aminomethyl)	$C_8H_{18}N_2$ 142.24	2579-20-6 <-70	 220	 0.945^{20}	H_2O 4; eth 4; EtOH 4
4363	1,3-Cyclohexanedimethanol Cyclohexane, 1,3-bis-(hydroxymethyl)	$C_8H_{16}O_2$ 144.21	3971-28-6	 105.5	4-06-00-05236 1.036^{25}	 1.4912^{20}
4364	Cyclohexane, 1,1-dimethyl- 1,1-Dimethylcyclohexane	C_8H_{16} 112.22	590-66-9 -33.3	 119.6	4-05-00-00117 0.7809^{20}	H_2O 1; EtOH 3; eth 3; ace 3 1.4290^{20}
4365	Cyclohexane, 1,2-dimethyl-, *cis-* *cis*-1,2-Dimethylcyclohexane	C_8H_{16} 112.22	2207-01-4 -49.9	 129.8	4-05-00-00118 0.7963^{20}	H_2O 1; EtOH 3; eth 5; ace 5 1.4360^{20}
4366	Cyclohexane, 1,2-dimethyl-, *trans-* *trans*-1,2-Dimethylcyclohexane	C_8H_{16} 112.22	6876-23-9 -90	 123.5	4-05-00-00118 0.7760^{20}	H_2O 1; EtOH 3; eth 3; ace 5 1.4270^{20}
4367	Cyclohexane, 1,3-dimethyl-, *cis* *cis*-1,3-Dimethylcyclohexane	C_8H_{16} 112.22	[illegible] -75.6	 120.1	4-05-00-00121 0.7660^{20}	H_2O 1; EtOH 5; eth 5; ace 5 1.4229^{20}
4368	Cyclohexane, 1,4-dimethyl-, *cis-* *cis*-1,4-Dimethylcyclohexane	C_8H_{16} 112.22	624-29-3 -87.4	 124.4	4-05-00-00122 0.7829^{20}	H_2O 1; EtOH 5; eth 5; ace 5 1.4230^{20}
4369	Cyclohexane, (1,1-dimethylethyl)-	$C_{10}H_{20}$ 140.27	3178-22-1 -41.2	 171.5	4-05-00-00147 0.8127^{20}	H_2O 1 1.4469^{20}
4370	Cyclohexane, 1,3-dimethyl-, *trans-* *trans*-1,3-Dimethylcyclohexane	C_8H_{16} 112.22	2207-03-6 -90.1	 124.5	 0.79^{15}	 1.4284^{25}
4371	Cyclohexane, 1,4-dimethyl-, *trans-* *trans*-1,4-Dimethylcyclohexane	C_8H_{16} 112.22	2207-04-7 -36.9	 119.4	 0.77^{15}	 1.4185^{25}
4372	1,2-Cyclohexanediol, *cis-*	$C_6H_{12}O_2$ 116.16	1792-81-0 100	 120^{15}	4-06-00-05193 1.0297^{101}	EtOH 3; ace 3; bz 3; chl 2
4373	1,2-Cyclohexanediol, *trans*-(±)-	$C_6H_{12}O_2$ 116.16	54383-22-1 105	 117^{13}	4-06-00-05194 1.147^{24}	H_2O 4; bz 4; eth 4; EtOH 4
4374	1,4-Cyclohexanediol, *trans-* *trans*-1,4-Dihydroxycyclohexane	$C_6H_{12}O_2$ 116.16	6995-79-5 143		4-06-00-05209 1.18^{20}	H_2O 3; EtOH 3; eth 1; ace 2

PHYSICAL CONSTANTS OF ORGANIC COMPOUNDS (continued)

No.	Name Synonym	Mol. Form. Mol. Wt.	CAS RN mp/°C	Merck No. bp/°C	Beil. Ref. den/g cm^{-3}	Solubility n_D
4375	1,2-Cyclohexanedione 1,2-Dioxocyclohexane	$C_6H_8O_2$ 112.13	765-87-7 40	 194	4-07-00-01982 1.1187^{21}	H_2O 3; EtOH 3; eth 3; bz 3 1.4995^{20}
4376	1,3-Cyclohexanedione	$C_6H_8O_2$ 112.13	504-02-9 105.5	3161	4-07-00-01985 1.0861^{91}	H_2O 3; EtOH 3; eth 2; ace 3 1.4576^{102}
4377	1,4-Cyclohexanedione Tetrahydroquinone	$C_6H_8O_2$ 112.13	637-88-7 78	 132^{20}	4-07-00-01986 1.0861^{91}	H_2O 3; EtOH 3; eth 3; ace 3
4378	1,3-Cyclohexanedione, 5,5-dimethyl- 5,5-Dimethyldihydroresorcinol	$C_8H_{12}O_2$ 140.18	126-81-8 149-51 dec	3231	4-07-00-01999	H_2O 2; eth 2; ace 3; ctc 3
4379	1,2-Cyclohexanedione, dioxime Nioxime	$C_6H_{10}N_2O_2$ 142.16	492-99-9 192	6477	4-07-00-01982	H_2O 3; ace 3; chl 3; tfa 2
4380	Cyclohexane, docosyl- Docosane, 1-cyclohexyl-	$C_{28}H_{56}$ 392.75	61828-07-7 54.4	 449	3-05-00-00167 0.8334^{20}	 1.4632^{20}
4381	Cyclohexane, dodecafluoro- Perfluorocyclohexane	C_6F_{12} 300.05	355-68-0 62.5 (trip. pt.)	 52.8 (subl. pt.)	4-05-00-00048	
4382	Cyclohexane, dodecyl-	$C_{18}H_{36}$ 252.48	1795-17-1 12.5	 331	4-05-00-00191 0.8223^{20}	 1.4559^{20}
4383	Cyclohexane, dotriacontyl- Dotriacontylcyclohexane	$C_{38}H_{76}$ 533.02	61828-17-9 74.8	 527	4-05-00-00204 0.8388^{20}	 1.4668^{20}
4384	Cyclohexane, eicosyl- 1-Cyclohexyleicosane	$C_{26}H_{52}$ 364.70	4443-55-4 48.5	 430	4-05-00-00198 0.8318^{20}	 1.4622^{20}
4385	Cyclohexaneethanamine, *N*,α-dimethyl- Propylhexedrine	$C_{10}H_{21}N$ 155.28	101-40-6	7873 205; 82^{10}	3-12-00-00108 0.8501^{20}	EtOH 4 1.4600^{20}
4386	Cyclohexane, 1,1'-(1,2-ethanediyl)bis- 1,2-Dicyclohexylethane	$C_{14}H_{26}$ 194.36	3321-50-4 11.5	 272.5	4-05-00-00344 0.8728^{24}	 1.4745^{24}
4387	Cyclohexaneethanol	$C_8H_{16}O$ 128.21	4442-79-9	 208	4-06-00-00119 0.9229^{20}	EtOH 3; eth 3; bz 3 1.4641^{20}
4388	Cyclohexane, ethenyl-	C_8H_{14} 110.20	695-12-5	 128	4-05-00-00268 0.8166^{19}	 1.455^{19}
4389	Cyclohexane, 1-ethenyl-1-methyl-2,4-bis(1-methylethenyl)-, (1α,2β,4β)- β-Elemene	$C_{15}H_{24}$ 204.36	33880-83-0	 $117\text{-}24^{15.5}$	4-05-00-01180 0.8749^{20}	 1.4935^{20}
4390	Cyclohexane, ethyl- Ethylcyclohexane	C_8H_{16} 112.22	1678-91-7 -111.3	 131.9	4-05-00-00115 0.7880^{20}	H_2O 1; EtOH 3; eth 3; ace 3 1.4330^{20}
4391	Cyclohexane, ethylidene- Ethylidenecyclohexane	C_8H_{14} 110.20	1003-64-1	 136	4-05-00-00267 0.822^{25}	 1.4618^{20}
4392	Cyclohexane, 1,1'-ethylidenebis- 1,1-Dicyclohexylethane	$C_{14}H_{26}$ 194.36	2319-61-1 -20.9	 256; 112^{7}	4-05-00-00345 0.9070^{25}	 1.4887^{25}
4393	Cyclohexane, 1-ethylidene-2-methyl- Cyclohexane, 2-methyl-1-ethylidene	C_9H_{16} 124.23	40514-70-3	 158	0-05-00-00078 0.823^{25}	 1.47
4394	Cyclohexane, 1-ethyl-1-methyl-	C_9H_{18} 126.24	4926-90-3	 152.2	4-05-00-00135 0.8025^{25}	 1.4419^{20}
4395	Cyclohexane, 1-ethyl-2-methyl-, *cis*-	C_9H_{18} 126.24	4923-77-7	 156	4-05-00-00136 0.8059^{25}	 1.4456^{20}
4396	Cyclohexane, 1-ethyl-2-methyl-, *trans*- Cyclohexane, 1-methyl-2-ethyl (*trans*)	C_9H_{18} 126.24	4923-78-8	 151.7	4-05-00-00136 0.790^{25}	 1.4381^{20}
4397	Cyclohexane, 1-ethyl-3-methyl-, *cis*- *cis*-1-Ethyl-3-methylcyclohexane	C_9H_{18} 126.24	19489-10-2	 156	4-05-00-00136 0.8094^{20}	 1.4432^{20}
4398	Cyclohexane, 1-ethyl-3-methyl-, *trans*-	C_9H_{18} 126.24	4926-76-5	 150	4-05-00-00136 0.7972^{20}	 1.4382^{20}
4399	Cyclohexane, 1-ethyl-4-methyl-, *cis*- *cis*-1-Ethyl-4-methylcyclohexane	C_9H_{18} 126.24	4926-78-7	 152.6	4-05-00-00136 0.7969^{20}	 1.4374^{20}
4400	Cyclohexane, 1-ethyl-4-methyl-, *trans*- *trans*-1-Methyl-4-ethylcyclohexane	C_9H_{18} 126.24	6236-88-0 -80.8	 149	4-05-00-00136 0.7798^{20}	 1.4304^{20}
4401	Cyclohexane, 1-ethyl-1-methyl-2,4-bis(1-methylethyl)-, [1*R*-(1α,2β,4β)]-	$C_{15}H_{30}$ 210.40	515-12-8	 $115\text{-}9^{10}$	4-05-00-00185 0.8509^{20}	bz 4; eth 4; peth 4 1.4640^{20}
4402	Cyclohexane, 1,1'-(2-ethyl-1,3-propanediyl)bis- Propane, 1,3-dicyclohexyl-2-ethyl	$C_{17}H_{32}$ 236.44	54833-34-0	 306	4-05-00-00363 0.8845^{20}	 1.483^{21}
4403	Cyclohexane, fluoro- Cyclohexyl fluoride	$C_6H_{11}F$ 102.15	372-46-3 13	 101	4-05-00-00044 0.9279^{20}	H_2O 1; py 3 1.4146^{20}
4404	Cyclohexane, heptadecyl- Heptadecane, 1-cyclohexyl-	$C_{23}H_{46}$ 322.62	19781-73-8 38	 397	3-05-00-00161 0.8290^{20}	 1.4603^{20}
4405	Cyclohexane, heptyl-	$C_{13}H_{26}$ 182.35	5617-41-4 -30	 244	4-05-00-00174 0.8109^{20}	 1.4484^{20}
4406	Cyclohexane, 1,2,3,4,5,6-hexachloro-, (1α,2β,3α,4β,5α,6β)- β-Lindane	$C_6H_6Cl_6$ 290.83	319-85-7	 $60^{0.50}$	4-05-00-00061 1.89^{19}	H_2O 1; EtOH 2; bz 2; chl 2
4407	Cyclohexane, 1,2,3,4,5,6-hexachloro-, (1α,2α,3α,4β,5α,6β)- δ-Lindane	$C_6H_6Cl_6$ 290.83	319-86-8 141.5	 $60^{0.36}$	4-05-00-00056	
4408	Cyclohexane, 1,2,3,4,5,6-hexachloro-, (1α,2α,3β,4α,5β,6β)-(±)- Cyclohexane, 1,2,3,4,5,6-hexachloro, (α, *DL*)	$C_6H_6Cl_6$ 290.83	60291-32-9 159.5	 288	4-05-00-00060 1.87^{20}	H_2O 1; EtOH 4; bz 3; chl 3
4409	Cyclohexane, 1,2,3,4,5,6-hexachloro-, (1α,2α,3β,4α,5α,6β)- Lindane	$C_6H_6Cl_6$ 290.83	58-89-9 112.5	5379 323.4	4-05-00-00058	ace 4; bz 4
4410	Cyclohexane, hexacosyl- Hexacosylcyclohexane	$C_{32}H_{64}$ 448.86	61828-11-3 64	 484	4-05-00-00203 0.8359^{20}	 1.4649^{20}

No.	Name Synonym	Mol. Form. Mol. Wt.	CAS RN mp/°C	Merck No. bp/°C	Beil. Ref. den/g cm^{-3}	Solubility n_D
4411	Cyclohexane, hexadecyl- Hexadecane, 1-cyclohexyl-	$C_{22}H_{44}$ 308.59	6812-38-0 33.6	 385	4-05-00-00196 0.8279^{20}	 1.4596^{20}
4412	Cyclohexanehexanoic acid Caproic acid, 6-cyclohexyl	$C_{12}H_{22}O_2$ 198.31	4354-56-7 34	 180^{11}	4-09-00-00095 0.9626^{20}	eth 3; chl 2 1.4750^{20}
4413	Cyclohexane, hexatriacontyl- Hexatriacontylcyclohexane	$C_{42}H_{84}$ 589.13	61828-21-5 80	 551	4-05-00-00205 0.8402^{20}	 1.4678^{20}
4414	Cyclohexane, hexyl-	$C_{12}H_{24}$ 168.32	4292-75-5 -43	 224	4-05-00-00170 0.8076^{20}	 1.4462^{20}
4415	Cyclohexane, iodo- Cyclohexyl iodide	$C_6H_{11}I$ 210.06	626-62-0	 180 dec; 81^{20}	4-05-00-00078 1.6244^{20}	H_2O 1; EtOH 3; eth 3; ace 3 1.5477^{20}
4416	Cyclohexane, isocyanato- Cyclohexyl isocyanate	$C_7H_{11}NO$ 125.17	3173-53-3	 172	4-12-00-00071 0.98^{25}	 1.4551^{20}
4417	Cyclohexane, isothiocyanato-	$C_7H_{11}NS$ 141.24	1122-82-3	 220	4-12-00-00073 0.996^{25}	H_2O 1; EtOH 3; eth 3; ctc 2 1.5375^{20}
4418	Cyclohexanemethanamine	$C_7H_{15}N$ 113.20	3218-02-8	 160	4-12-00-00123 0.87^{25}	 1.4630^{20}
4419	Cyclohexanemethanol Cyclohexylcarbinol	$C_7H_{14}O$ 114.19	100-49-2 -43	2737 183	4-06-00-00106 0.9297^{20}	eth 4; EtOH 4 1.4644^{20}
4420	Cyclohexanemethanol, α-methyl-	$C_8H_{16}O$ 128.21	1193-81-3	 189	1-06-00-00012 0.928^{25}	EtOH 4; eth 4; ctc 2 1.4656^{20}
4421	Cyclohexanemethanol, α-methyl-, (*S*)-	$C_8H_{16}O$ 128.21	3113-98-2	 105^{35}; 82^{12}	4-06-00-00117 0.9254^{20}	 1.4635^{25}
4422	Cyclohexanemethanol, 2-methyl-, *cis*- Cyclohexane, 1-hydroxymethyl-2-methyl (*cis*)	$C_8H_{16}O$ 128.21	3937-45-9	 188.5	4-06-00-00122 0.9342^{20}	 1.4689^{20}
4423	Cyclohexanemethanol, 2-methyl-, *trans*- Cyclohexane, 1-hydroxymethyl-2-methyl (*trans*)	$C_8H_{16}O$ 128.21	3937-46-0	 192	4-06-00-00122 0.9224^{20}	 1.4665^{20}
4424	Cyclohexanemethanol, 4-methyl-	$C_8H_{16}O$ 128.21	34885-03-5	 $75^{2.5}$	2-06-00-00030 0.9074^{20}	 1.4617^{20}
4425	Cyclohexane, methoxy-	$C_7H_{14}O$ 114.19	931-56-6 -74.4	 133	4-06-00-00026 0.8756^{20}	eth 4; EtOH 4 1.4355^{20}
4426	Cyclohexane, methyl- Methylcyclohexane	C_7H_{14} 98.19	108-87-2 -126.6	 100.9	4-05-00-00094 0.7694^{20}	H_2O 1; EtOH 3; eth 3; ace 5 1.4231^{20}
4427	Cyclohexane, (3-methylbutyl)-	$C_{11}H_{22}$ 154.30	54105-76-9	 196.5	3-05-00-00143 0.8023^{20}	bz 4; lig 4 1.4420^{20}
4428	Cyclohexane, methylene- Methylenecyclohexane	C_7H_{12} 96.17	1192-37-6 -106.7	 102.5	4-05-00-00250 0.8074^{20}	H_2O 1; eth 3; bz 3; chl 3 1.4523^{20}
4429	Cyclohexane, 1,1'-methylenebis- Dicyclohexylmethane	$C_{13}H_{24}$ 180.33	3178-23-2	 $110\text{-}5^{18}$	4-05-00-00340 0.8750^{20}	 1.4752^{20}
4430	Cyclohexane, 1,1'-(1-methyl-1,2-ethanediyl)bis- 1,2-Dicyclohexylpropane	$C_{15}H_{28}$ 208.39	41851-34-7 -22	 284.5	4-05-00-00350 0.8724^{21}	 1.479^{21}
4431	Cyclohexane, (1-methylethyl)- Isopropylcyclohexane	C_9H_{18} 126.24	696-29-7 -89.4	 154.8	4-05-00-00134 0.8023^{20}	H_2O 1; EtOH 4; eth 4; ace 5 1.4410^{20}
4432	Cyclohexane, 1,1'-(1-methylethylidene)bis- Propane, 2,2-dicyclohexyl	$C_{15}H_{28}$ 208.39	54934-90-6 15.6	 286	4-05-00-00350 0.9001^{23}	 1.490^{23}
4433	Cyclohexane, 1,1',1''-methylidynetris- Methane, tricyclohexyl-	$C_{19}H_{34}$ 262.48	1610-24-8 48	 325.5	4-05-00-00504 0.9273^{50}	H_2O 1; EtOH 2; eth 3; bz 3 1.4986^{40}
4434	Cyclohexane, 1-methyl-4-methylene-	C_8H_{14} 110.20	2808-80-2	 122	4-05-00-00271 0.7910^{19}	 1.4465^{18}
4435	Cyclohexane, 1-methyl-3-(1-methylethenyl)- *m*-Menth-8-ene	$C_{10}H_{18}$ 138.25	16605-36-0	 170	3-05-00-00235 0.8178^{20}	 1.4546
4436	Cyclohexane, 1-methyl-2-(1-methylethyl)-	$C_{10}H_{20}$ 140.27	16580-23-7	 171	4-05-00-00150 0.8134^{22}	 1.447^{21}
4437	Cyclohexane, 1-methyl-3-(1-methylethyl)-	$C_{10}H_{20}$ 140.27	16580-24-8	 166.5	3-05-00-00132 0.7963^{20}	 1.44^{24}
4438	Cyclohexane, 1-methyl-4-(1-methylethyl)-	$C_{10}H_{20}$ 140.27	99-82-1 -87.6	 170.7	4-05-00-00151 0.797^{25}	 1.4373^{20}
4439	Cyclohexane, 1-methyl-4-(1-methylethyl)-, *cis*-	$C_{10}H_{20}$ 140.27	6069-98-3 -89.9	 172	4-05-00-00151 0.8039^{20}	H_2O 1; EtOH 4; eth 4; bz 3 1.4431^{20}
4440	Cyclohexane, 1-methyl-4-(1-methylethyl)-, *trans*-	$C_{10}H_{20}$ 140.27	1678-82-6 -86.3	 170.6	4-05-00-00151 0.7928^{20}	bz 4; eth 4; EtOH 4; lig 4 1.4366^{20}
4441	Cyclohexane, 1-methyl-3-(1-methylethylidene)-	$C_{10}H_{18}$ 138.25	13828-34-7	 174	1-05-00-00043 0.8214^{20}	 1.4670^{20}
4442	Cyclohexane, 1-methyl-4-(1-methylethylidene)-	$C_{10}H_{18}$ 138.25	1124-27-2	 173	4-05-00-00302 0.819^{21}	 1.4568^{21}
4443	Cyclohexane, 1-methyl-1-nitro- 1-Methyl-1-nitrocyclohexane	$C_7H_{13}NO_2$ 143.19	59368-15-9 -71	 $118\text{-}20^{50}$	4-05-00-00103 1.0384^{20}	 1.4598^{20}
4444	Cyclohexane, 1-methyl-2-nitro- 2-Methylnitrocyclohexane	$C_7H_{13}NO_2$ 143.19	74221-86-6	 100^{20}	4-05-00-00103 1.046^{25}	 1.4608^{25}
4445	Cyclohexane, 1-methyl-4-nitro- 1-Methyl-4-nitrocyclohexane	$C_7H_{13}NO_2$ 143.19	89895-45-4	 102^{20}	4-05-00-00103 1.037^{25}	 1.4567^{23}
4446	Cyclohexane, 1-methyl-2-pentyl- 1-Methyl-2-pentylcyclohexane	$C_{12}H_{24}$ 168.32	54411-01-7	 217.5	4-05-00-00171 0.815^{20}	 1.4487^{20}
4447	Cyclohexane, 1,1'-(2-methyl-1,3-propanediyl)bis- 1,3-Dicyclohexyl-2-methylpropane	$C_{16}H_{30}$ 222.41	2883-08-1 0.6	 295.2	4-05-00-00358 0.8715^{20}	 1.4756^{20}

No.	Name	Synonym	Mol. Form.	Mol. Wt.	CAS RN	mp/°C	Merck No.	bp/°C	Beil. Ref.	den/g cm^{-3}	Solubility	n_D
4448	Cyclohexane, (1-methylpropyl)-	*sec*-Butylcyclohexane	$C_{10}H_{20}$	140.27	7058-01-7			179.3	4-05-00-00146	0.8131^{20}	H_2O 1; ace 3	1.4467^{20}
4449	Cyclohexane, 1-methyl-1-propyl-	1-Methyl-1-propylcyclohexane	$C_{10}H_{20}$	140.27	4258-93-9			172	4-05-00-00148	0.8101^{20}		1.4440^{20}
4450	Cyclohexane, 1-methyl-2-propyl-	1-Methyl-2-propylcyclohexane	$C_{10}H_{20}$	140.27	4291-79-6			176; 56^{13}	4-05-00-00148	0.8130^{19}		1.4468^{19}
4451	Cyclohexane, 1-methyl-3-propyl-	1-Methyl-3-propylcyclohexane	$C_{10}H_{20}$	140.27	4291-80-9			164.5	4-05-00-00149	0.7895^{21}		1.4377^{20}
4452	Cyclohexane, 1-methyl-4-propyl-	1-Methyl-4-propylcyclohexane	$C_{10}H_{20}$	140.27	4291-81-0			175.5	4-05-00-00149	0.797^{20}		1.4393^{20}
4453	Cyclohexane, (2-methylpropyl)-		$C_{10}H_{20}$	140.27	1678-98-4	-95		171.3	4-05-00-00147	0.7952^{20}	H_2O 1; EtOH 3; eth 4; ace 3	1.4386^{20}
4454	Cyclohexane, nitro-	Nitrocyclohexane	$C_6H_{11}NO_2$	129.16	1122-60-7	-34		205; 95^{22}	4-05-00-00081	1.0610^{20}	H_2O 1; EtOH 3; lig 3	1.4612^{19}
4455	Cyclohexane, (nitromethyl)-	(Nitromethyl)cyclohexane	$C_7H_{13}NO_2$	143.19	2625-30-1			123^{40}; 98^{10}	4-05-00-00103	1.0482^{20}		1.4705^{20}
4456	Cyclohexane, nonadecyl-		$C_{25}H_{50}$	350.67	22349-03-7	45		420	4-05-00-00197	0.8310^{20}		1.4616^{20}
4457	Cyclohexane, nonyl-		$C_{15}H_{30}$	210.40	2883-02-5	-10		282	4-05-00-00183	0.8163^{20}		1.4519^{20}
4458	Cyclohexane, octacosyl-	Octacosylcyclohexane	$C_{34}H_{68}$	476.91	61828-13-5	68		499	4-05-00-00203	0.8370^{20}		1.4656^{20}
4459	Cyclohexane, octadecyl-	Octadecane, 1-cyclohexyl-	$C_{24}H_{48}$	336.65	4445-06-1	41.6		409; 175^{1}	4-05-00-00197	0.8300^{20}		1.4610^{20}
4460	Cyclohexane, octyl-		$C_{14}H_{28}$	196.38	1795-15-9	-20		264	4-05-00-00178	0.8138^{20}		1.4503^{20}
4461	Cyclohexane, 1,1'-oxybis-		$C_{12}H_{22}O$	182.31	4645-15-2	-36		242.5	3-06-00-00019	0.9227^{20}		1.4741^{20}
4462	Cyclohexane, pentacosyl-	Pentacosylcyclohexane	$C_{31}H_{62}$	434.83	61828-10-2	62		476	4-05-00-00202	0.8353^{20}		1.4645^{20}
4463	Cyclohexane, pentadecyl-		$C_{21}H_{42}$	294.56	6006-95-7	29		373	4-05-00-00195	0.8267^{20}		1.4588^{20}
4464	Cyclohexane, 1,1'-(1,5-pentanediyl)bis-	Pentane, 1,5-dicyclohexyl-	$C_{17}H_{32}$	236.44	54833-31-7	-13.6		325	4-05-00-00362	0.8718^{21}		1.479^{21}
4465	Cyclohexane, pentyl-	Pentylcyclohexane	$C_{11}H_{22}$	154.30	4292-92-6	-57.5		203.7	4-05-00-00164	0.8037^{20}	ace 4; bz 4; eth 4; EtOH 4	1.4437^{20}
4466	Cyclohexane, 1,1'-(1,3-propanediyl)bis-	1,3-Dicyclohexylpropane	$C_{15}H_{28}$	208.39	3178-24-3	-17		291.5	4-05-00-00349	0.8728^{24}		1.4736^{24}
4467	Cyclohexanepropanenitrile, 2-oxo-		$C_9H_{13}NO$	151.21	4594-78-9			$138\text{-}42^{10}$	4-10-00-02623	1.0181^{20}	ctc 2	1.4755^{20}
4468	Cyclohexanepropanoic acid		$C_9H_{16}O_2$	156.22	701-97-3	16		276.5	4-09-00-00056	0.912^{25}	H_2O 3; eth 3; ctc 2	1.4638^{20}
4469	Cyclohexanepropanol, α-methyl-	3-Cyclohexyl-1-methylpropanol	$C_{10}H_{20}O$	156.27	10528-67-3			112^{16}	3-06-00-00123	0.903^{21}		1.464^{21}
4470	Cyclohexane, 2-propenyl-		C_9H_{16}	124.23	2114-42-3			132	4-05-00-00283	0.8135^{20}	EtOH 4; eth 3; ace 3; bz 3	1.4500^{20}
4471	Cyclohexane, propyl-	Propylcyclohexane	C_9H_{18}	126.24	1678-92-8	-94.9		156.7	4-05-00-00134	0.7936^{20}	H_2O 1; EtOH 5; eth 3; ace 5	1.4370^{20}
4472	Cyclohexane, 1,1'-propylidenebis-	Propane, 1,1-dicyclohexyl-	$C_{15}H_{28}$	208.39	54934-91-7	-23.5		283.6	4-05-00-00350	0.8886^{23}		1.485^{23}
4473	Cyclohexane, 2-propynyl-		C_9H_{14}	122.21	17715-00-3			157.5	4-05-00-00419	0.8449^{20}	eth 3	1.4605^{20}
4474	Cyclohexane, tetracosyl-	Tetracosylcyclohexane	$C_{30}H_{60}$	420.81	61828-09-9	59.5		467	4-05-00-00202	0.8347^{20}		1.4641^{20}
4475	Cyclohexane, tetradecyl-		$C_{20}H_{40}$	280.54	1795-18-2	24		360	4-05-00-00194	0.8254^{20}		1.4579^{20}
4476	Cyclohexane, 1,1'-tetradecylidenebis-	1,1-Dicyclohexyltetradecane	$C_{26}H_{50}$	362.68	55334-08-2	37.6		406	4-05-00-00374	0.8735^{20}		1.4799^{20}
4477	Cyclohexane, 1,1,3,4-tetramethyl-	1,1,3,4-Tetramethylcyclohexane	$C_{10}H_{20}$	140.27	24612-75-7			160	4-05-00-00159	0.7976^{20}		1.4380^{20}
4478	Cyclohexane, 1,1,3,5-tetramethyl-, *cis*-	*cis*-1,1,3,5-Tetramethylcyclohexane	$C_{10}H_{20}$	140.27	50876-32-9			152.5	4-05-00-00159	0.7813^{20}		1.4319^{20}
4479	Cyclohexane, tetratriacontyl-	Tetratriacontylcyclohexane	$C_{40}H_{80}$	561.08	61828-19-1	78		540	4-05-00-00205	0.8395^{20}		1.4673^{20}
4480	Cyclohexanethiol	Cyclohexyl mercaptan	$C_6H_{12}S$	116.23	1569-69-3			158.9	4-06-00-00072	0.9782^{20}	ace 4; bz 4; eth 4; EtOH 4	1.4921^{20}
4481	Cyclohexanethione	Thiocyclohexanone	$C_6H_{10}S$	114.21	2720-41-4			82^{17}; 74^{11}	3-07-00-00039		EtOH 3; eth 3; ace 3	1.5375^{20}
4482	Cyclohexane, triacontyl-	Triacontylcyclohexane	$C_{36}H_{72}$	504.97	61828-15-7	71.6		514	4-05-00-00204	0.8379^{20}		1.4662^{20}
4483	Cyclohexane, tricosyl-	Tricosylcyclohexane	$C_{29}H_{58}$	406.78	61828-08-8	57		459	4-05-00-00202	0.8341^{20}		1.4637^{20}
4484	Cyclohexane, tridecyl-	Tridecane, 1-cyclohexyl-	$C_{19}H_{38}$	266.51	6006-33-3	18.5		346	4-05-00-00192	0.8239^{20}		1.4570^{20}
4485	Cyclohexane, 1,1,2-trimethyl-	1,1,2-Trimethylcyclohexane	C_9H_{18}	126.24	7094-26-0	-29		145.2	4-05-00-00137	0.7963^{25}		1.4382^{20}
4486	Cyclohexane, 1,1,3-trimethyl-	Cyclogeraniolane	C_9H_{18}	126.24	3073-66-3	-65.7		136.6	4-05-00-00137	0.7749^{25}		1.4295^{20}
4487	Cyclohexane, 1,1,4-trimethyl-	1,1,4-Trimethylcyclohexane	C_9H_{18}	126.24	7094-27-1			135	4-05-00-00137	0.7685^{25}		1.4251^{20}

No.	Name Synonym	Mol. Form. Mol. Wt.	CAS RN mp/°C	Merck No. bp/°C	Beil. Ref. den/g cm^{-3}	Solubility n_D
4488	Cyclohexane, 1,2,3-trimethyl- 1,2,3-Trimethylcyclohexane	C_9H_{18} 126.24	1678-97-3	142	4-05-00-00137 0.7898^{20}	1.4346^{20}
4489	Cyclohexane, 1,2,3-trimethyl-, (1α,2β,3α)-	C_9H_{18} 126.24	1678-81-5 -66.9	144	4-05-00-00137 0.777^{25}	1.430^{20}
4490	Cyclohexane, 1,2,3-trimethyl-, (1α,2α,3α)-	C_9H_{18} 126.24	1839-88-9 -85	151.7	4-05-00-00137 0.7989^{25}	1.4405^{20}
4491	Cyclohexane, 1,2,3-trimethyl-, (1α,2α,3β)-	C_9H_{18} 126.24	7667-55-2 -85.7	151.2	4-05-00-00137 0.7995^{25}	1.4399^{20}
4492	Cyclohexane, 1,2,4-trimethyl-, (1α,2α,4α)-	C_9H_{18} 126.24	1678-80-4 -77.5	146.6	4-05-00-00138 0.783^{25}	1.4340^{20}
4493	Cyclohexane, 1,2,4-trimethyl-, (1α,2β,4α)-	C_9H_{18} 126.24	7667-59-6 -86	142.9	4-05-00-00138 0.7683^{25}	1.4266^{20}
4494	Cyclohexane, 1,2,4-trimethyl-, (1α,2α,4β)-	C_9H_{18} 126.24	7667-58-5 -91.8	146.7	4-05-00-00138 0.7870^{25}	1.4345^{20}
4495	Cyclohexane, 1,2,4-trimethyl-, (1α,2β,4β)-	C_9H_{18} 126.24	7667-60-9 -83.5	142.9	4-05-00-00138 0.7870^{25}	1.4341^{20}
4496	Cyclohexane, 1,3,5-trimethyl-, (1α,3α,5α)-	C_9H_{18} 126.24	1795-27-3 -49.7	138.5	4-05-00-00138 0.7708^{20}	H_2O 1; eth 3; bz 3; ctc 3 1.4269^{20}
4497	Cyclohexane, 1,3,5-trimethyl-, (1α,3α,5β)- *trans*-1,3,5-Trimethylcyclohexane	C_9H_{18} 126.24	1795-26-2 -107.4	140.5	4-05-00-00138 0.7794^{20}	bz 4; eth 4; lig 4 1.4307^{20}
4498	Cyclohexane, undecafluoro- Undecafluorocyclohexane	C_6HF_{11} 282.06	308-24-7	62.0		
4499	Cyclohexane, undecafluoro(trifluoromethyl)- Perfluoro(methylcyclohexane)	C_7F_{14} 350.05	355-02-2 -44.7	76.3	4-05-00-00097 1.7878^{25}	ace 3; bz 3; ctc 3; tol 3 1.285^{17}
4500	Cyclohexanol Cyclohexyl alcohol	$C_6H_{12}O$ 100.16	108-93-0 25.4	2731 160.8	4-06-00-00020 0.9624^{20}	H_2O 3; EtOH 3; eth 3; ace 3 1.4641^{20}
4501	Cyclohexanol, 2-amino-, *trans*-(±)- Cyclohexanol, 2-amino (*trans*, *DL*)	$C_6H_{13}NO$ 115.18	33092-82-9 68	219; 105^{10}	4-13-00-00540	bz 4; chl 4; HOAc 4
4502	Cyclohexanol, 3-(aminomethyl)-3,5,5-trimethyl- 1-Hydroxy-3-aminomethyl-3,5,5-trimethylcyclohexane	$C_{10}H_{21}NO$ 171.28	15647-11-7 45.5	265	0.969^{25}	1.4904^{20}
4503	Cyclohexanol, 2-bromo- 2-Bromo-1-cyclohexanol	$C_6H_{11}BrO$ 179.06	24796-87-0	85^{10}; 60^{1}	4-06-00-00063 1.4604^{20}	1.5169^{20}
4504	Cyclohexanol, 2-butyl-, *trans*- *trans*-2-Butylcyclohexanol	$C_{10}H_{20}O$ 156.27	35242-05-8	$111\text{-}2^{16}$	3-06-00-00122 0.9020^{20}	H_2O 1; eth 3; ace 3; bz 3 1.4641^{20}
4505	Cyclohexanol, 2-chloro-, *trans*- *trans*-2-Chlorocyclohexanol	$C_6H_{11}ClO$ 134.61	6628-80-4 29	93^{20}	4-06-00-00064 1.146^{16}	bz 4; eth 4; EtOH 4; chl 4 1.4899^{20}
4506	Cyclohexanol, 2-chloro-, *cis*-(±)-	$C_6H_{11}ClO$ 134.61	116783-28-9 36.5	$93\text{-}4^{26}$	4-06-00-00064 1.1261^{25}	bz 4; EtOH 4; chl 4 1.4894^{25}
4507	Cyclohexanol, 4-chloro-, *trans*- *trans*-4-Chlorocyclohexanol	$C_6H_{11}ClO$ 134.61	29538-77-0 82.5	106^{14}	4-06-00-00068 1.1435^{13}	EtOH 3; eth 3; bz 3; chl 3 1.4930^{17}
4508	Cyclohexanol, 1,2-dimethyl-, *cis*- Cyclohexanol, 1,2-dimethyl-, stereoisomer	$C_8H_{16}O$ 128.21	19879-11-9 23.2	82.8^{25}	4-06-00-00121 0.9250^{20}	1.4625^{20}
4509	Cyclohexanol, 1,2-dimethyl-, *trans*- Cyclohexanol, 1,2-dimethyl-, stereoisomer	$C_8H_{16}O$ 128.21	19879-12-0 13.2	74^{25}	4-06-00-00121 0.9187^{20}	1.4590^{20}
4510	Cyclohexanol, 1,3-dimethyl-, *cis*- *cis*-1,3-Dimethylcyclohexanol	$C_8H_{16}O$ 128.21	15466-94-1 27.5	84^{25}	3-06-00-00092 0.9022^{20}	1.4575^{20}
4511	Cyclohexanol, 1,3-dimethyl-, *trans*- *trans*-1,3-Dimethylcyclohexanol	$C_8H_{16}O$ 128.21	15466-93-0 14.5	77.6^{25}	3-06-00-00093 0.8894^{30}	1.4507^{20}
4512	Cyclohexanol, 1,4-dimethyl-, *cis*-	$C_8H_{16}O$ 128.21	16980-61-3 24	83.5^{25}	3-06-00-00098 0.9011^{30}	1.4564^{20}
4513	Cyclohexanol, 2,2-dimethyl- 2,2-Dimethylcyclohexanol	$C_8H_{16}O$ 128.21	1193-46-0 8	177	0.9225^{20}	1.4648^{20}
4514	Cyclohexanol, 2,4-dimethyl- 2,4-Dimethylcyclohexanol	$C_8H_{16}O$ 128.21	69542-91-2	177.5	4-06-00-00124 0.900^{20}	1.4500^{20}
4515	Cyclohexanol, 2,6-dimethyl- 2,6-Dimethyl-1-cyclohexanol	$C_8H_{16}O$ 128.21	5337-72-4 32.5	176	2-06-00-00028 0.944^{25}	1.4600^{20}
4516	Cyclohexanol, 3,3-dimethyl- 3,3-Dimethylcyclohexanol	$C_8H_{16}O$ 128.21	767-12-4 11.5	185; 99.5^{35}	3-06-00-00089 0.9128^{14}	1.4606^{15}
4517	Cyclohexanol, 3,4-dimethyl- 3,4-Dimethylcyclohexyl alcohol	$C_8H_{16}O$ 128.21	5715-23-1	189	3-06-00-00091 0.9073^{16}	1.458^{16}
4518	Cyclohexanol, 3,5-dimethyl- 3,5-Dimethylcyclohexanol	$C_8H_{16}O$ 128.21	5441-52-1 11.6	187	2-06-00-00029 0.898^{20}	1.4550^{20}
4519	Cyclohexanol, 4,4-dimethyl- 4,4-Dimethylcyclohexanol	$C_8H_{16}O$ 128.21	932-01-4 16	186	4-06-00-00120 0.9250^{20}	1.4613^{20}
4520	Cyclohexanol, 3-(dimethylamino)-	$C_8H_{17}NO$ 143.23	6890-03-5 73	231	3-13-00-00719 0.9737^{25}	EtOH 4 1.4852^{20}
4521	Cyclohexanol, 2-(1,1-dimethylethyl)-	$C_{10}H_{20}O$ 156.27	13491-79-7 45	139^{95}	3-06-00-00125 0.902^{25}	
4522	Cyclohexanol, 1-ethyl- 1-Ethylcyclohexanol	$C_8H_{16}O$ 128.21	1940-18-7 34.5	166	4-06-00-00115 0.9227^{25}	H_2O 2; bz 3; ctc 2; peth 3 1.4633^{20}
4523	Cyclohexanol, 2-ethyl-, *trans*-(±)- Cyclohexanol, 2-ethyl (*trans*, *DL*)	$C_8H_{16}O$ 128.21	89886-25-9	79^{12}	4-06-00-00117 0.9193^{21}	ace 4; bz 4; eth 4; peth 4 1.4640^{21}
4524	Cyclohexanol, 2-ethyl-, *cis*-(±)-	$C_8H_{16}O$ 128.21	116697-35-9	181	3-06-00-00085 0.9274^{21}	ace 4; bz 4; eth 4; peth 4 1.4655^{21}
4525	Cyclohexanol, 4-ethyl- 4-Ethylcyclohexanol	$C_8H_{16}O$ 128.21	4534-74-1	84^{10}	2-06-00-00026 0.889^{25}	1.4625^{20}
4526	Cyclohexanol, 1-ethyl-2-methyl- 1-Ethyl-2-methylcyclohexanol	$C_9H_{18}O$ 142.24	102370-18-3	182; 82^{73}	4-06-00-00133 0.9235^{20}	1.458^{20}

No.	Name	Synonym	Mol. Form.	Mol. Wt.	CAS RN	mp/°C	Merck No.	bp/°C	Beil. Ref.	den/g cm^{-3}	Solubility	n_D
4527	Cyclohexanol, 1-ethyl-3-methyl-	1-Ethyl-3-methylcyclohexanol	$C_9H_{18}O$	142.24	62067-44-1			88^{20}	2-06-00-00035	0.9013^{20}		1.459
4528	Cyclohexanol, 1,1'-(1,2-ethynediyl)bis-		$C_{14}H_{22}O_2$	222.33	78-54-6	112.5		182^{13}	4-06-00-06076		H_2O 1; EtOH 4; eth 4; ace 4	
4529	Cyclohexanol, 1-ethynyl-		$C_8H_{12}O$	124.18	78-27-3	31.5		174	4-06-00-00348	0.9873^{20}	H_2O 1; EtOH 3; bz 3; chl 2	1.4822^{20}
4530	Cyclohexanol, 1-methyl-	1-Methylcyclohexanol	$C_7H_{14}O$	114.19	590-67-0	25		155; 70^{25}	4-06-00-00095	0.9194^{20}	H_2O 1; EtOH 3; bz 3; chl 3	1.4595^{20}
4531	Cyclohexanol, 2-methyl-, *cis*-(±)-	(±)-*cis*-2-Methylcyclohexanol	$C_7H_{14}O$	114.19	615-38-3	7		165	4-06-00-00100	0.9360^{20}	EtOH 4	1.4640^{20}
4532	Cyclohexanol, 2-methyl-, *trans*-(±)-	(±)-*trans*-2-Methylcyclohexanol	$C_7H_{14}O$	114.19	615-39-4	-4		167.5	4-06-00-00100	0.9247^{20}	eth 4; EtOH 4	1.4616^{20}
4533	Cyclohexanol, 3-methyl-, *cis*-(±)-	*cis*-3-Methylcyclohexanol	$C_7H_{14}O$	114.19	5454-79-5	-5.5		168; 94^{12}		0.9155^{20}	eth 4; EtOH 4	1.4752^{20}
4534	Cyclohexanol, 3-methyl-, *trans*-(±)-	*trans*-3-Methylcyclohexanol	$C_7H_{14}O$	114.19	7443-55-2	-0.5		167; 84^{13}		0.9214^{30}	eth 4; EtOH 4	1.4580^{20}
4535	Cyclohexanol, 4-methyl-, *cis*-	*cis*-4-Methylcyclohexanol	$C_7H_{14}O$	114.19	7731-28-4	-9.2		173	4-06-00-00105	0.9170^{20}	eth 4; EtOH 4	1.4614^{20}
4536	Cyclohexanol, 4-methyl-, *trans*-	*trans*-4-Methylcyclohexanol	$C_7H_{14}O$	114.19	7731-29-5			174	4-06-00-00105	0.9118^{21}	H_2O 2; EtOH 5; eth 3	1.4561^{20}
4537	Cyclohexanol, 2-methylene-		$C_7H_{12}O$	112.17	4065-80-9			$82\text{-}4^{13}$	4-06-00-00208	0.955^{20}		1.4843^{20}
4538	Cyclohexanol, 1-(1-methylethyl)-		$C_9H_{18}O$	142.24	3552-01-0			$76\text{-}9^{10}$	4-06-00-00131	0.928^{19}		1.4683^{19}
4539	Cyclohexanol, 2-(1-methylethyl)-, *cis*-	*cis*-2-Isopropylcyclohexanol	$C_9H_{18}O$	142.24	10488-25-2	52.5		77^{13}	4-06-00-00131	0.9223^{25}		1.4665^{25}
4540	Cyclohexanol, 4-(1-methylethyl)-, *cis*-	*cis*-4-Isopropylcyclohexanol	$C_9H_{18}O$	142.24	22900-08-9			$72^{1.9}$	3-06-00-00111	0.9195^{20}		1.4671^{20}
4541	Cyclohexanol, 4-(1-methylethyl)-, *trans*-	*trans*-4-Isopropylcyclohexanol	$C_9H_{18}O$	142.24	15890-36-5			$79.5^{1.8}$	3-06-00-00112	0.9140^{20}		1.4658^{20}
4542	Cyclohexanol, 1-methyl-4-(1-methylethenyl)-	β-Terpineol	$C_{10}H_{18}O$	154.25	138-87-4	32.5		210; 90^{10}	4-06-00-00254	0.917^{20}		1.4747^{20}
4543	Cyclohexanol, 2-methyl-5-(1-methylethenyl)-, [1*S*-(1α,2β,5α)]-		$C_{10}H_{18}O$	154.25	22567-21-1			225; 107^{14}	3-06-00-00256	0.9274^{20}		1.4780^{20}
4544	Cyclohexanol, 5-methyl-2-(1-methylethenyl)-, [1*R*-(1α,2β,5α)]-		$C_{10}H_{18}O$	154.25	89-79-2	78		93^{14}	3-06-00-00257	0.911^{20}	H_2O 2; EtOH 3; eth 3	1.4723^{20}
4545	Cyclohexanol, 5-methyl-2-(1-methylethenyl)-, acetate, [1*R*-(1α,2β,5α)]-		$C_{12}H_{20}O_2$	196.29	57576-09-7	85		$112\text{-}4^{8}$	4-06-00-00255	0.925^{25}		1.4566^{20}
4546	Cyclohexanol, 1-methyl-4-(1-methylethyl)-		$C_{10}H_{20}O$	156.27	21129-27-1			208.5	4-06-00-00147	0.90^{20}		1.4619^{20}
4547	Cyclohexanol, 2-methyl-5-(1-methylethyl)-, [1*R*-(1α,2α,5β)]-		$C_{10}H_{20}O$	156.27	5563-78-0			217.5; 102^{18}	4-06-00-00148	0.9012^{20}	eth 4; EtOH 4	1.4632^{20}
4548	Cyclohexanol, 5-methyl-2-(1-methylethyl)-, (1α,2β,5α)-(±)-	(±)-Menthol	$C_{10}H_{20}O$	156.27	15356-70-4	38	5723	216	4-06-00-00151	0.903^{15}	H_2O 1; EtOH 4; eth 4; ace 4	1.4615^{20}
4549	Cyclohexanol, 5-methyl-2-(1-methylethyl)-, acetate, [1*R*-(1α,2β,5α)]-		$C_{12}H_{22}O_2$	198.31	2623-23-6			109^{10}	4-06-00-00153	0.9185^{20}		1.4469^{20}
4550	Cyclohexanol, 2-methyl-1-propyl-	2-Methyl-1-propylcyclohexanol	$C_{10}H_{20}O$	156.27	24580-48-1			$97\text{-}8^{34}$	3-06-00-00127	0.919^{20}		1.48^{20}
4551	Cyclohexanol, 3-methyl-1-propyl-	3-Methyl-1-propylcyclohexanol	$C_{10}H_{20}O$	156.27	24580-52-7			199; $96\text{-}8^{20}$	1-06-00-00018	0.8903^{21}		1.4566^{24}
4552	Cyclohexanol, 1-phenyl-	1-Phenylcyclohexan-1-ol	$C_{12}H_{16}O$	176.26	1589-60-2	63.5		157^{28}; 112^{5}	4-06-00-03908	1.035^{16}	bz 4; eth 4; EtOH 4; lig 4	1.5415^{16}
4553	Cyclohexanol, 2-phenyl-, *cis*-(±)-	Cyclohexanol, 2-phenyl (*cis*, *DL*)	$C_{12}H_{16}O$	176.26	40960-73-4	41.5		$140\text{-}1^{16}$	4-06-00-03909	1.035^{16}		1.5415^{16}
4554	Cyclohexanol, 2-phenyl-, *trans*-(±)-	Cyclohexanol, 2-phenyl (*trans*, *DL*)	$C_{12}H_{16}O$	176.26	40960-69-8	56.5		153^{16}	4-06-00-03909		EtOH 4; chl 4	
4555	Cyclohexanol, 1-(2-propenyl)-		$C_9H_{16}O$	140.23	1123-34-8			190	4-06-00-00235	0.9341^{22}		1.4756^{22}
4556	Cyclohexanol, 2-(2-propenyl)-, *trans*-	Cyclohexanol, 2-allyl-, *trans*-	$C_9H_{16}O$	140.23	24844-28-8			$94\text{-}6^{15}$	4-06-00-00235	0.947^{20}	HOAc 4	1.4778^{20}
4557	Cyclohexanol, 1-propyl-	1-Hydroxy-1-propylcyclohexane	$C_9H_{18}O$	142.24	5445-24-9			180 dec; 85^{20}	4-06-00-00129	0.934^{12}		1.468^{12}
4558	Cyclohexanol, 2-propyl-, *cis*-	*cis*-2-Propylcyclohexanol	$C_9H_{18}O$	142.24	5857-86-3			84^{10}	4-06-00-00130	0.9247^{11}		1.4688^{11}
4559	Cyclohexanol, 2-propyl-, *trans*-	*trans*-2-Propylcyclohexanol	$C_9H_{18}O$	142.24	5846-43-5			90^{14}	4-06-00-00130	0.9162^{11}		1.4668^{11}
4560	Cyclohexanol, 4-propyl-, *trans*-	*trans*-4-Propylcyclohexanol	$C_9H_{18}O$	142.24	77866-58-1			208; 103^{14}	3-06-00-00106	0.9072^{17}		1.4613^{17}
4561	Cyclohexanol, 4-propyl-, *cis*-		$C_9H_{18}O$	142.24	98790-22-8			99^{17}	3-06-00-00105	0.9102^{16}		1.4652^{16}
4562	Cyclohexanol, 1,2,2-trimethyl-	1,2,2-Trimethylcyclohexanol	$C_9H_{18}O$	142.24	31720-69-1			81^{20}; 76^{16}	3-06-00-00114	0.9230^{20}	bz 4; eth 4; EtOH 4	1.4680^{20}

No.	Name Synonym	Mol. Form. Mol. Wt.	CAS RN mp/°C	Merck No. bp/°C	Beil. Ref. den/g cm^{-3}	Solubility n_D
4563	Cyclohexanol, 1,2,6-trimethyl- 1,2,6-Trimethylcyclohexanol	$C_9H_{18}O$ 142.24	96244-13-2	78^{22}	1-06-00-00017 0.9126^{15}	ace 4; bz 4; eth 4; EtOH 4 1.4598^{15}
4564	Cyclohexanol, 1,3,5-trimethyl- 1,3,5-Trimethylcyclohexanol	$C_9H_{18}O$ 142.24	90760-75-1	181	3-06-00-00117 0.8876^{17}	eth 4; EtOH 4; chl 4 1.454^{16}
4565	Cyclohexanol, 2,2,5-trimethyl- 2,2,5-Trimethylcyclohexanol	$C_9H_{18}O$ 142.24	73210-25-0	188	0-06-00-00022 0.8955^{22}	EtOH 4 1.4569^{20}
4566	Cyclohexanol, 2,2,6-trimethyl- 1,1,3-Trimethyl-2-cyclohexanol	$C_9H_{18}O$ 142.24	10130-91-3 51	187; 87^{28}	4-06-00-00135 0.9128^{20}	eth 4; EtOH 4; chl 4 1.4600^{20}
4567	Cyclohexanol, 2,3,3-trimethyl-	$C_9H_{18}O$ 142.24	116724-17-5 28	197	1-06-00-00016 0.9002^{25}	ace 4; EtOH 4 1.4572^{25}
4568	Cyclohexanol, 2,3,6-trimethyl- 2,3,6-Trimethylcyclohexanol	$C_9H_{18}O$ 142.24	58210-03-0	195	0-06-00-00022 0.9117^{17}	EtOH 4; chl 4
4569	Cyclohexanol, 3,3,5-trimethyl-, *cis-* *cis*-3,3,5-Trimethylcyclohexanol	$C_9H_{18}O$ 142.24	933-48-2 37.3	202; 92^{12}	4-06-00-00135 0.9006^{16}	H_2O 1; EtOH 3; eth 3; chl 3 1.4550^{16}
4570	Cyclohexanol, 3,3,5-trimethyl-, *trans-* *trans*-3,3,5-Trimethylcyclohexanol	$C_9H_{18}O$ 142.24	767-54-4 55.8	189.2	4-06-00-00135 0.8631^{60}	H_2O 1; EtOH 3; eth 3; chl 3
4571	Cyclohexanone Pimelic ketone	$C_6H_{10}O$ 98.14	108-94-1 -31	2732 155.4	4-07-00-00015 0.9478^{20}	H_2O 3; EtOH 3; eth 3; ace 3 1.4507^{20}
4572	Cyclohexanone, 2-acetyl-	$C_8H_{12}O_2$ 140.18	874-23-7	$111\text{-}2^{18}$	4-07-00-01997 1.0782^{25}	ctc 3 1.5138^{20}
4573	Cyclohexanone, 2,6-bis(phenylmethylene)-	$C_{20}H_{18}O$ 274.36	897-78-9 117.5	$185\text{-}95^{20}$	3-07-00-02661	EtOH 2; bz 3; HOAc 3
4574	Cyclohexanone, 2-bromo-	C_6H_9BrO 177.04	822-85-5	$89\text{-}90^{14}$	4-07-00-00034 1.340^{25}	1.5085^{25}
4575	Cyclohexanone, 2-butyl- 2-Butylcyclohexanone	$C_{10}H_{18}O$ 154.25	1126-18-7	70^{2}	4-07-00-00079 0.905^{20}	H_2O 1; ctc 2; os 4 1.4545^{20}
4576	Cyclohexanone, 2-butylidene- 2-Butylidenecyclohexanone	$C_{10}H_{16}O$ 152.24	7153-14-2	$98\text{-}100^{10}$	4-07-00-00181 0.935^{20}	ace 4; bz 4; eth 4; EtOH 4 1.4800^{20}
4577	Cyclohexanone, 2-chloro- 2-Chlorocyclohexanone	C_6H_9ClO 132.59	822-87-7 23	82^{15}	4-07-00-00032 1.160^{20}	eth 3; bz 3; ctc 2; diox 3 1.4825^{20}
4578	Cyclohexanone, 4-chloro- 4-Chlorocyclohexanone	C_6H_9ClO 132.59	21299-26-3	95^{17}	4-07-00-00032	eth 4 1.4867^{20}
4579	Cyclohexanone, 2-(2-chlorophenyl)-2-(methylamino)-, (±)- Ketamine	$C_{13}H_{16}ClNO$ 237.73	6740-88-1 92.5	5174		
4580	Cyclohexanone, 2,2-dimethyl- 2,2-Dimethyl-1-cyclohexanone	$C_8H_{14}O$ 126.20	1193-47-1 20.5	172	4-07-00-00055 0.9146^{20}	1.4486^{20}
4581	Cyclohexanone, 2,3-dimethyl-, *cis-* *cis*-2,3-Dimethylcyclohexanone	$C_8H_{14}O$ 126.20	1551-88-8	178.5; 99^{57}	3-07-00-00088 0.9159^{20}	1.4505^{20}
4582	Cyclohexanone, 2,3-dimethyl-, *trans-* *trans*-2,3-Dimethylcyclohexanone	$C_8H_{14}O$ 126.20	1551-89-9	96^{57}	4-07-00-00057 0.908^{22}	1.4511^{25}
4583	Cyclohexanone, 2,4-dimethyl-, (2*S*-*trans*)- Cyclohexanone, 2,4-dimethyl (*trans*, d)	$C_8H_{14}O$ 126.20	93921-42-7	179; 69^{17}	3-07-00-00093 0.9004^{16}	ace 4; bz 4; eth 4 1.4488^{22}
4584	Cyclohexanone, 2,4-dimethyl-, *cis-*	$C_8H_{14}O$ 126.20	116783-29-0	177	3-07-00-00093 0.910^{15}	1.4493^{15}
4585	Cyclohexanone, 2,5-dimethyl-, *trans*-(±)- *trans*-(±)-2,5-Dimethylcyclohexanone	$C_8H_{14}O$ 126.20	66395-23-1	172	4-07-00-00059 0.9025^{20}	eth 4; EtOH 4 1.4446^{20}
4586	Cyclohexanone, 2,6-dimethyl- 2,6-Dimethylcyclohexanone	$C_8H_{14}O$ 126.20	2816-57-1	175	2-07-00-00025 0.925^{25}	1.4460^{20}
4587	Cyclohexanone, 3,3-dimethyl- 3,3-Dimethylcyclohexanone	$C_8H_{14}O$ 126.20	2979-19-3	180; $72\text{-}3^{25}$	4-07-00-00056 0.909^{15}	1.4482^{17}
4588	Cyclohexanone, 3,4-dimethyl- 3,4-Dimethylcyclohexanone	$C_8H_{14}O$ 126.20	5465-09-8	187	3-07-00-00089 0.906^{20}	1.4507^{20}
4589	Cyclohexanone, 3,5-dimethyl- 3,5-Dimethylcyclohexanone	$C_8H_{14}O$ 126.20	2320-30-1	182.5	4-07-00-00058 0.896^{20}	1.4407^{20}
4590	Cyclohexanone, 4,4-dimethyl- 4,4-Dimethylcyclohexanone	$C_8H_{14}O$ 126.20	4255-62-3 39	73^{14}	4-07-00-00056 0.932^{20}	1.4537^{24}
4591	Cyclohexanone, 2-[(dimethylamino)methyl]- 2-[(Dimethylamino)methyl]cyclohexanone	$C_9H_{17}NO$ 155.24	15409-60-6	$92^{10.5}$	4-14-00-00008 0.9504^{20}	H_2O 2; EtOH 3; eth 3 1.4672^{20}
4592	Cyclohexanone, 4-(1,1-dimethylpropyl)-	$C_{11}H_{20}O$ 168.28	16587-71-6	$124\text{-}5^{16}$	4-07-00-00096 0.920^{25}	1.4677^{20}
4593	Cyclohexanone, 4-ethyl- 4-Ethylcyclohexanone	$C_8H_{14}O$ 126.20	5441-51-0	193	3-07-00-00083 0.895^{25}	1.4515^{20}
4594	Cyclohexanone, 2-ethylidene- 2-Ethylidenecyclohexanone	$C_8H_{12}O$ 124.18	1122-25-4	92^{20}	4-07-00-00146 0.962^{20}	H_2O 1; os 4 1.4882^{20}
4595	Cyclohexanone, 2-ethyl-2-methyl- 2-Ethyl-2-methylcyclohexanone	$C_9H_{16}O$ 140.23	17206-52-9	195	2-07-00-00051 0.9037^{17}	1.4515^{17}
4596	Cyclohexanone, 2-methyl-, (±)- (±)-2-Methylcyclohexanone	$C_7H_{12}O$ 112.17	24965-84-2 -13.9	165	4-07-00-00041 0.9250^{20}	H_2O 1; EtOH 3; eth 3 1.4483^{25}
4597	Cyclohexanone, 3-methyl-, (±)-	$C_7H_{12}O$ 112.17	625-96-7 -73.5	169; 65^{15}	4-07-00-00043 0.9136^{20}	H_2O 1; EtOH 3; eth 3 1.4456^{20}
4598	Cyclohexanone, 4-methyl- 4-Methylcyclohexanone	$C_7H_{12}O$ 112.17	589-92-4 -40.6	170	4-07-00-00044 0.9138^{20}	H_2O 1; EtOH 3; eth 3; ctc 2 1.4451^{20}
4599	Cyclohexanone, 2-(1-methylethyl)- 2-Isopropylcyclohexanone	$C_9H_{16}O$ 140.23	1004-77-9 72.5	83^{17}	4-07-00-00066 0.922^{16}	H_2O 1; EtOH 3; eth 3; ace 3 1.4564^{15}
4600	Cyclohexanone, 4-(1-methylethyl)- 4-Isopropylcyclohexanone	$C_9H_{16}O$ 140.23	5432-85-9	$139\text{-}40^{100}$	4-07-00-00066 0.9158^{25}	1.4552^{25}
4601	Cyclohexanone, 2-methyl-5-(1-methylethenyl)-, (2*S*-*trans*)-	$C_{10}H_{16}O$ 152.24	619-02-3 -11	221.5	0.9253^{20}	1.4717^{20}

No.	Name Synonym	Mol. Form. Mol. Wt.	CAS RN mp/°C	Merck No. bp/°C	Beil. Ref. den/g cm^{-3}	Solubility n_D
4602	Cyclohexanone, 5-methyl-2-(1-methylethenyl)-	$C_{10}H_{16}O$ 152.24	529-00-0	100^{18}	4-07-00-00190 0.9198^{20}	1.4675^{20}
4603	Cyclohexanone, 5-methyl-2-(1-methylethenyl)-, *trans*-	$C_{10}H_{16}O$ 152.24	29606-79-9	100^{18}	4-07-00-00190 0.9198^{20}	1.4675^{20}
4604	Cyclohexanone, 2-methyl-5-(1-methylethyl)-, (2*S*-*trans*)-	$C_{10}H_{18}O$ 154.25	13163-73-0	220; 96^{15}	4-07-00-00085 0.904^{20}	ace 4; EtOH 4; chl 4 1.4553^{20}
4605	Cyclohexanone, 5-methyl-2-(1-methylethyl)-, (2*R*-*cis*)- Menthone(*d*)	$C_{10}H_{18}O$ 154.25	1196-31-2	204; 85^{14}	4-07-00-00087 0.8947^{20}	H_2O 2; EtOH 3; eth 3; ace 3 1.4503^{20}
4606	Cyclohexanone, 5-methyl-2-(1-methylethyl)-, (2*S*-*trans*)- (-)-Menthone	$C_{10}H_{18}O$ 154.25	14073-97-3 -6	5724 207	0.8954^{20}	H_2O 2; EtOH 5; eth 5; ace 3 1.4505^{20}
4607	Cyclohexanone, 5-methyl-2-(1-methylethylidene)-	$C_{10}H_{16}O$ 152.24	15932-80-6	$95\text{-}6^{11}$	4-07-00-00188 0.9367^{20}	1.4869^{20}
4608	Cyclohexanone, 5-methyl-2-(1-methylethylidene)-, (*R*)- Pulegone	$C_{10}H_{16}O$ 152.24	89-82-7	7955 224	4-07-00-00188 0.9346^{45}	H_2O 1; EtOH 5; eth 5; ctc 3 1.4894^{20}
4609	Cyclohexanone, 4-(2-methylpropyl)-	$C_{10}H_{18}O$ 154.25	42061-72-3	218; 98^{12}	4-07-00-00082 0.8914^{24}	1.4502^{24}
4610	Cyclohexanone, oxime (Hydroxyimino)cyclohexane	$C_6H_{11}NO$ 113.16	100-64-1 90	206	4-07-00-00021	H_2O 3; EtOH 3; eth 3; chl 2
4611	Cyclohexanone, 2-(1-piperidinylmethyl)- Pimeclone	$C_{12}H_{21}NO$ 195.30	534-84-9	7399 $118\text{-}20^{14}$	5-20-02-00330	
4612	Cyclohexanone, 2-propyl- 2-Propylcyclohexanone	$C_9H_{16}O$ 140.23	94-65-5	197	4-07-00-00064 0.927^{20}	H_2O 1; EtOH 3; eth 4; ace 3 1.4538^{20}
4613	Cyclohexanone, 4-propyl-	$C_9H_{16}O$ 140.23	40649-36-3	213; 97.5^{15}	3-07-00-00116 0.9072^{19}	1.4530^{19}
4614	Cyclohexanone, 2,2,3-trimethyl- 2,2,3-Trimethylcyclohexanone	$C_9H_{16}O$ 140.23	39257-08-4	$66\text{-}7^{10}$	4-07-00-00069 0.9234^{15}	1.4569^{15}
4615	Cyclohexanone, 2,2,4-trimethyl- 2,2,4-Trimethylcyclohexanone	$C_9H_{16}O$ 140.23	35413-38-8	$86\text{-}7^{12}$	3-07-00-00125 0.8914^{20}	1.4520^{20}
4616	Cyclohexanone, 2,2,5-trimethyl- 2,2,5-Trimethylcyclohexanone	$C_9H_{16}O$ 140.23	933-36-8	183	3-07-00-00126 0.8871^{24}	1.4432^{24}
4617	Cyclohexanone, 2,2,6-trimethyl-	$C_9H_{16}O$ 140.23	2408-37-9 -31.8	178.5	4-07-00-00069 0.9043^{18}	1.4470^{20}
4618	Cyclohexanone, 2,3,6-trimethyl- 2,3,6-Trimethylcyclohexanone	$C_9H_{16}O$ 140.23	42185-47-7	190	3-07-00-00126 0.9129^{18}	1.4464^{21}
4619	Cyclohexanone, 2,4,4-trimethyl- 2,4,4-Trimethylcyclohexanone	$C_9H_{16}O$ 140.23	2230-70-8	191	4-07-00-00070 0.902^{20}	1.4493^{20}
4620	Cyclohexanone, 2,4,5-trimethyl- 2,4,5-Trimethylcyclohexanone	$C_9H_{16}O$ 140.23	51299-55-9	193	4-07-00-00072 0.897^{20}	1.4479^{20}
4621	Cyclohexanone, 2,4,6-trimethyl- 2,4,6-Trimethylcyclohexanone	$C_9H_{16}O$ 140.23	90645-54-8	185	3-07-00-00127 0.8992^{20}	1.4458^{20}
4622	Cyclohexanone, 3,3,5-trimethyl- Dihydroisophorone	$C_9H_{16}O$ 140.23	873-94-9	189	4-07-00-00070 0.8919^{19}	1.4454^{15}
4623	Cyclohexanone, 3,4,4-trimethyl- 3,4,4-Trimethylcyclohexanone	$C_9H_{16}O$ 140.23	40441-35-8	$80\text{-}1^{13}$	3-07-00-00122 0.911^{25}	1.4552^{20}
4624	Cyclohexasiloxane, dodecamethyl- Dodecamethylcyclohexasiloxane	$C_{12}H_{36}O_6Si_6$ 444.93	540-97-6 -3	3400 245	4-04-00-04129 0.9672^{25}	H_2O 1 1.4015^{20}
4625	Cyclohexasiloxane, 2,4,6,8,10,12-hexamethyl- Hexamethylcyclohexasiloxane	$C_6H_{24}O_6Si_6$ 360.77	6166-87-6 -79		4-04-00-04100 1.006^{20}	H_2O 1 1.3944^{20}
4626	Cyclohexene Tetrahydrobenzene	C_6H_{10} 82.15	110-83-8 -103.5	2733 82.9	4-05-00-00218 0.8110^{20}	H_2O 1; EtOH 5; eth 5; ace 5 1.4465^{20}
4627	Cyclohexene, 1-bromo-	C_6H_9Br 161.04	2044-08-8	165	4-05-00-00236 1.3901^{20}	eth 3; ace 4; bz 3; chl 4 1.5134^{20}
4628	Cyclohexene, 3-bromo-	C_6H_9Br 161.04	1521-51-3	$80\text{-}2^{40}$	4-05-00-00237 1.3890^{20}	H_2O 1; eth 3; bz 3; chl 3 1.5320^{20}
4629	3-Cyclohexene-1-carbonitrile	C_7H_9N 107.16	100-45-8	$39^{1.5}$	4-09-00-00116	ctc 3 1.4716^{20}
4630	1-Cyclohexenecarbonitrile 1-Cyanocyclohexene	C_7H_9N 107.16	1855-63-6	81^{12}		
4631	1-Cyclohexene-1-carboxaldehyde	$C_7H_{10}O$ 110.16	1192-88-7	72^{15}	4-07-00-00132 0.9694^{20}	EtOH 3; eth 3 1.5005^{20}
4632	3-Cyclohexene-1-carboxaldehyde	$C_7H_{10}O$ 110.16	100-50-5 2	105	4-07-00-00134 0.9692^{20}	ace 3; ctc 2; MeOH 3
4633	1-Cyclohexene-1-carboxaldehyde, 4-(1-methylethenyl)- Perillaldehyde	$C_{10}H_{14}O$ 150.22	2111-75-3	7119	2-07-00-00130	
4634	1-Cyclohexene-1-carboxaldehyde, 4-(1-methylethenyl)-, (*R*)-	$C_{10}H_{14}O$ 150.22	5503-12-8	238; 99^{7}	4-07-00-00316 0.953^{20}	1.5058^{20}
4635	1-Cyclohexene-1-carboxaldehyde, 4-(1-methylethyl)- Phellandral	$C_{10}H_{16}O$ 152.24	21391-98-0	223	4-07-00-00187 0.93^{20}	1.4911^{20}
4636	1-Cyclohexene-1-carboxaldehyde, 2,6,6-trimethyl- β-Cyclocitral	$C_{10}H_{16}O$ 152.24	432-25-7	$95\text{-}100^{15}$	4-07-00-00192 0.959^{15}	1.4971^{15}

No.	Name Synonym	Mol. Form. Mol. Wt.	CAS RN mp/°C	Merck No. bp/°C	Beil. Ref. den/g cm^{-3}	Solubility n_D
4637	1-Cyclohexene-1-carboxylic acid Cyclohexene-1-carboxylic acid	$C_7H_{10}O_2$ 126.16	636-82-8 38	 241	4-09-00-00111 1.1092^{20}	H_2O 2; EtOH 3; ace 3 1.4902^{20}
4638	3-Cyclohexene-1-carboxylic acid	$C_7H_{10}O_2$ 126.16	4771-80-6 17	 234.5	4-09-00-00114 1.0820^{20}	H_2O 4; EtOH 3; ace 3 1.4814^{20}
4639	3-Cyclohexene-1-carboxylic acid, ethyl ester	$C_9H_{14}O_2$ 154.21	15111-56-5	 194.5	4-09-00-00114 0.9688^{20}	 1.4578^{20}
4640	3-Cyclohexene-1-carboxylic acid, methyl ester	$C_8H_{12}O_2$ 140.18	6493-77-2	 80^{20}	4-09-00-00114 0.995^{25}	 1.4610^{20}
4641	2-Cyclohexene-1-carboxylic acid, 2,6-dimethyl-4-oxo-, ethyl ester	$C_{11}H_{16}O_3$ 196.25	6102-15-4	 $157\text{-}8^{18}$	4-10-00-02674 1.0493^{20}	ace 3 1.4773^{20}
4642	Cyclohexene, 1-chloro- 1-Chlorocyclohexene	C_6H_9Cl 116.59	930-66-5	 142.5	4-05-00-00230 1.0361^{19}	eth 3; ace 3; ctc 3; chl 3 1.4797^{20}
4643	Cyclohexene, decafluoro- Perfluorocyclohexene	C_6F_{10} 262.05	355-75-9	 52.0	4-05-00-00229 1.6650^{25}	 1.293^{20}
4644	1-Cyclohexene-1,2-dicarboxylic acid, diethyl ester Cyclohexene-1,2-dicarboxylic acid, diethyl ester	$C_{12}H_{18}O_4$ 226.27	92687-41-7	 160^{14}	3-09-00-03940 1.0803^{19}	 1.4747^{19}
4645	1-Cyclohexene-1,3-dicarboxylic acid, diethyl ester Diethyl 1-cyclohexene-1,3-dicarboxylate	$C_{12}H_{18}O_4$ 226.27	38511-07-8	 150^{12}	3-09-00-03950 1.0772^{20}	 1.4722^{20}
4646	1-Cyclohexene-1,4-dicarboxylic acid, dimethyl ester Cyclohexene-1,4-dicarboxylic acid, dimethyl ester	$C_{10}H_{14}O_4$ 198.22	22646-79-3 39	 153^{20}	4-09-00-02897	ace 4; eth 4; EtOH 4
4647	Cyclohexene, 1,2-dichloro-3,3,4,4,5,5,6,6-octafluoro-	$C_6Cl_2F_8$ 294.96	336-19-6 -70	 113	4-05-00-00231 1.719^{20}	 1.375^{20}
4648	Cyclohexene, 1,2-dimethyl-	C_8H_{14} 110.20	1674-10-8 -84.1	 138	4-05-00-00268 0.8220^{25}	 1.4620^{20}
4649	Cyclohexene, 1,3-dimethyl- 1,3-Dimethyl-1-cyclohexene	C_8H_{14} 110.20	2808-76-6	 127	4-05-00-00270 0.799^{25}	 1.449^{20}
4650	Cyclohexene, 1,4-dimethyl- 1,4-Dimethyl-1-cyclohexene	C_8H_{14} 110.20	2808-79-9 -59.4	 128.5	4-05-00-00270 0.8005^{20}	 1.4457^{20}
4651	Cyclohexene, 1,5-dimethyl- Cyclohexene, 2,4-dimethyl-	C_8H_{14} 110.20	2808-77-7	 126	3-05-00-00216 0.8009^{25}	 1.450^{20}
4652	Cyclohexene, 1,6-dimethyl-	C_8H_{14} 110.20	1759-64-4	 131	4-05-00-00269 0.815^{20}	 1.454^{20}
4653	Cyclohexene, 3,3-dimethyl- 3,3-Dimethylcyclohexene	C_8H_{14} 110.20	695-28-3	 119	4-05-00-00268 0.804^{20}	 1.445^{20}
4654	Cyclohexene, 3,5-dimethyl- 3,5-Dimethylcyclohexene	C_8H_{14} 110.20	823-17-6	 124.5	3-05-00-00216 0.8005^{18}	 1.443^{18}
4655	Cyclohexene, 3,6-dimethyl- 3,6-Dimethylcyclohexene	C_8H_{14} 110.20	19550-40-4	 125	4-05-00-00271 0.802^{25}	 1.4433^{25}
4656	Cyclohexene, 4,4-dimethyl-	C_8H_{14} 110.20	14072-86-7 -74.4	 117	4-05-00-00268 0.7996^{20}	 1.4420^{20}
4657	Cyclohexene, 1-ethenyl-	C_8H_{12} 108.18	2622-21-1	 145	4-05-00-00405 0.8623^{15}	H_2O 1; eth 3; bz 3; MeOH 4 1.4915^{20}
4658	Cyclohexene, 4-ethenyl-	C_8H_{12} 108.18	100-40-3 -108.9	 128	4-05-00-00406 0.8299^{20}	H_2O 1; eth 3; bz 3; peth 3 1.4639^{20}
4659	Cyclohexene, 6-ethenyl-6-methyl-1-(1-methylethyl)-3-(1-methylethylidene)-, (*S*)-	$C_{15}H_{24}$ 204.36	5951-67-7	 $120\text{-}30^{8}$	3-05-00-01083 0.8782^{20}	ace 4; bz 4 1.5130^{26}
4660	Cyclohexene, 1-ethyl-	C_8H_{14} 110.20	1453-24-3 -109.9	 137	4-05-00-00266 0.8176^{25}	 1.4567^{20}
4661	Cyclohexene, 3-ethyl- 3-Ethyl-1-cyclohexene	C_8H_{14} 110.20	2808-71-1	 134	4-05-00-00266 0.814^{20}	 1.451^{20}
4662	Cyclohexene, 4-ethyl- 4-Ethyl-1-cyclohexene	C_8H_{14} 110.20	3742-42-5	 133	4-05-00-00267 0.810^{20}	 1.449^{20}
4663	Cyclohexene, 1-ethyl-4-methyl- 1-Ethyl-4-methylcyclohexene	C_9H_{16} 124.23	62088-36-2	 149	4-05-00-00286 0.8169^{16}	 1.453^{16}
4664	1-Cyclohexene-1-methanol, 4-(1-methylethenyl)-	$C_{10}H_{16}O$ 152.24	536-59-4	 244; 12.5^{12}	2-06-00-00102 0.9690^{20}	 1.5005^{20}
4665	1-Cyclohexene-1-methanol, 2,4,4-trimethyl- β-Cyclolavandulol	$C_{10}H_{18}O$ 154.25	107085-40-6	 $105\text{-}6^{15}$	4-06-00-00250 0.9153^{20}	 1.4814^{20}
4666	1-Cyclohexene-1-methanol, 2,6,6-trimethyl- Cyclogeranyl alcohol	$C_{10}H_{18}O$ 154.25	472-20-8 43.5		3-06-00-00259 0.945^{19}	 1.487^{19}
4667	2-Cyclohexene-1-methanol, 2,6,6-trimethyl- α-Cyclogeraniol	$C_{10}H_{18}O$ 154.25	6627-74-3	 215	4-06-00-00257 0.9382^{20}	 1.4843^{20}
4668	3-Cyclohexene-1-methanol, α,α,4-trimethyl-, (±)-	$C_{10}H_{18}O$ 154.25	2438-12-2 40.5	 220	4-06-00-00252 0.9337^{20}	ace 4; bz 4; eth 4; EtOH 4 1.4831^{20}
4669	3-Cyclohexene-1-methanol, α,α,4-trimethyl-, acetate	$C_{12}H_{20}O_2$ 196.29	80-26-2	 140^{40}; 105^{11}	4-06-00-00253 0.9659^{21}	H_2O 1; EtOH 3; eth 3; bz 3 1.4689^{21}
4670	Cyclohexene, 1-methyl-	C_7H_{12} 96.17	591-49-1 -120.4	 110.3	4-05-00-00245 0.8102^{20}	H_2O 1; eth 3; bz 3; ctc 3 1.4503^{20}

No.	Name Synonym	Mol. Form. Mol. Wt.	CAS RN mp/°C	Merck No. bp/°C	Beil. Ref. den/g cm^{-3}	Solubility n_D
4671	Cyclohexene, 3-methyl-, (±)- (±)-3-Methylcyclohexene	C_7H_{12} 96.17	56688-75-6 -115.5	 104	4-05-00-00247 0.7990^{20}	bz 4; eth 4; chl 4; peth 4 1.4414^{20}
4672	Cyclohexene, 4-methyl-	C_7H_{12} 96.17	591-47-9 -115.5	 102.7	4-05-00-00248 0.7991^{20}	H_2O 1; EtOH 3; eth 3 1.4414^{20}
4673	Cyclohexene, 3-methylene-6-(1-methylethyl)- β-Phellandrene	$C_{10}H_{16}$ 136.24	555-10-2	7152 171.5	2-05-00-00087 0.8520^{20}	H_2O 1; EtOH 1; eth 3 1.4788^{20}
4674	Cyclohexene, 3-methylene-6-(1-methylethyl)-, (+)-	$C_{10}H_{16}$ 136.24	6153-16-8	 171.5	4-05-00-00436 0.8520^{20}	eth 4 1.4788^{20}
4675	Cyclohexene, 4-methylene-1-(1-methylethyl)-	$C_{10}H_{16}$ 136.24	99-84-3	 173.5	4-05-00-00437 0.838^{22}	 1.4754^{22}
4676	Cyclohexene, 1-methyl-3-(1-methylethenyl)-	$C_{10}H_{16}$ 136.24	38738-60-2	 175	4-05-00-00434 0.8479^{18}	 1.4760^{18}
4677	Cyclohexene, 1-methyl-4-(1-methylethenyl)- Limonene	$C_{10}H_{16}$ 136.24	138-86-3 -95	5371 176	4-05-00-00440 0.8402^{21}	H_2O 1; EtOH 5; eth 5; ctc 3 1.4727^{19}
4678	Cyclohexene, 1-methyl-5-(1-methylethenyl)-	$C_{10}H_{16}$ 136.24	13898-73-2	 176; 95^{20}	4-05-00-00435 0.8481^{20}	eth 4; EtOH 4 1.4804^{20}
4679	Cyclohexene, 5-methyl-1-(1-methylethenyl)-, (±)- Cyclohexene, 1-isopropenyl-5-methyl (*DL*)	$C_{10}H_{16}$ 136.24	105065-24-5	 184	3-05-00-00336 0.8594^{20}	 1.4975
4680	Cyclohexene, 1-methyl-4-(1-methylethyl)-	$C_{10}H_{18}$ 138.25	5502-88-5	 174.5	4-05-00-00300 0.8457^{15}	 1.4735^{20}
4681	Cyclohexene, 1-methyl-4-(1-methylethyl)-, (*R*)-	$C_{10}H_{18}$ 138.25	1195-31-9	 173	4-05-00-00299 0.8246^{18}	bz 4; EtOH 4; peth 4 1.4563^{18}
4682	Cyclohexene, 3-methyl-6-(1-methylethyl)-, (3*R*-*trans*)-	$C_{10}H_{18}$ 138.25	5113-93-9	 167; 55-6^{12}	4-05-00-00300 0.824^{20}	 1.461^{20}
4683	Cyclohexene, 4-methyl-1-(1-methylethyl)-, (*R*)-	$C_{10}H_{18}$ 138.25	619-52-3	 167.5	4-05-00-00301 0.8078^{20}	H_2O 1; EtOH 3; eth 3; bz 3 1.4503^{15}
4684	Cyclohexene, 1-methyl-4-(1-methylethylidene)- Terpinolene	$C_{10}H_{16}$ 136.24	586-62-9	 186	4-05-00-00437 0.8632^{15}	H_2O 1; EtOH 5; eth 5; bz 3 1.4883^{20}
4685	Cyclohexene, 1-methyl-4-(5-methyl-1-methylene-4-hexenyl)-, (*S*)-	$C_{15}H_{24}$ 204.36	495-61-4	 129-$30^{10.5}$	4-05-00-01174 0.8673^{20}	 1.4880^{20}
4686	5-Cyclohexene-1,2,3,4-tetrone, 5,6-dihydroxy- Rhodizonic acid	$C_6H_2O_6$ 170.08	118-76-3 248 dec	 sub	4-08-00-03609	EtOH 3
4687	Cyclohexene, 1,2,3-trimethyl- 1,2,3-Trimethylcyclohexene	C_9H_{16} 124.23	72312-48-2	 150	1-05-00-00040 0.8347^{11}	 1.463^{11}
4688	Cyclohexene, 1,3,5-trimethyl- 1,3,5-Trimethylcyclohexene	C_9H_{16} 124.23	3643-64-9	 141	3-05-00-00226 0.8025^{11}	 1.449^{13}
4689	Cyclohexene, 1,4,4-trimethyl- 1,4,4-Trimethylcyclohexene	C_9H_{16} 124.23	3419-71-4	 140	1-05-00-00039 0.8032^{19}	 1.444^{23}
4690	Cyclohexene, 1,4,5-trimethyl- 1,4,5-Trimethylcyclohexene	C_9H_{16} 124.23	20030-32-4	 145	2-05-00-00050 0.805^{20}	 1.4482^{20}
4691	Cyclohexene, 1,5,5-trimethyl- α-Cyclogeraniolene	C_9H_{16} 124.23	503-46-8	 140	4-05-00-00286 0.7981^{23}	 1.4461^{21}
4692	Cyclohexene, 1,6,6-trimethyl- 1,6,6-Trimethylcyclohexene	C_9H_{16} 124.23	69745-49-9	 147	3-05-00-00225 0.8217^{20}	 1.456^{20}
4693	Cyclohexene, 1,5,6-trimethyl-	C_9H_{16} 124.23	116724-18-6	 140	3-05-00-00225 0.829^{25}	 1.4572^{25}
4694	2-Cyclohexen-1-ol	$C_6H_{10}O$ 98.14	822-67-3	 164	4-06-00-00196 0.9923^{15}	EtOH 3; ace 3 1.4790^{25}
4695	3-Cyclohexen-1-ol	$C_6H_{10}O$ 98.14	822-66-2	 164; 68^{16}	4-06-00-00200 0.9845^{23}	H_2O 2; eth 3; ace 3 1.4851^{22}
4696	2-Cyclohexen-1-ol, 2-methyl-5-(1-methylethenyl)-	$C_{10}H_{16}O$ 152.24	99-48-9	 228	2-06-00-00102 0.9484^{25}	 1.4942^{25}
4697	2-Cyclohexen-1-ol, 3-methyl-6-(1-methylethyl)-	$C_{10}H_{18}O$ 154.25	491-04-3	 97-$8^{15.5}$	2-06-00-00065 0.9119^{25}	 1.4729^{25}
4698	2-Cyclohexen-1-ol, 6-methyl-3-(1-methylethyl)- Carvenol	$C_{10}H_{18}O$ 154.25	586-27-6	 220	4-06-00-00253 0.925^{22}	 1.479
4699	3-Cyclohexen-1-ol, 4-methyl-1-(1-methylethyl)-	$C_{10}H_{18}O$ 154.25	562-74-3	 209	4-06-00-00250 0.926^{20}	 1.4785^{19}
4700	2-Cyclohexen-1-ol, 1,3,5-trimethyl- Cyclohexen-3-ol, 1,3,5-trimethyl	$C_9H_{16}O$ 140.23	33843-55-9 46	 87-90^{17}	1-06-00-00036 0.9132^{20}	 1.4735^{19}
4701	2-Cyclohexen-1-ol, 2,4,4-trimethyl- 2,4,4-Trimethyl-2-cyclohexen-1-ol	$C_9H_{16}O$ 140.23	73741-61-4	 193	1-06-00-00036 0.9310^{25}	bz 4; EtOH 4
4702	2-Cyclohexen-1-ol, 3,5,5-trimethyl- Isophorol	$C_9H_{16}O$ 140.23	470-99-5	 69^{5}	3-06-00-00230 0.9128^{20}	 1.4717^{20}
4703	2-Cyclohexen-1-one	C_6H_8O 96.13	930-68-7 -53	 170	4-07-00-00124 0.9620^{25}	EtOH 4; ace 3 1.4883^{20}
4704	3-Cyclohexen-1-one	C_6H_8O 96.13	4096-34-8	 170	4-07-00-00126 0.9620^{25}	ace 4; EtOH 4 1.4842^{18}
4705	2-Cyclohexen-1-one, 2,3-dimethyl-	$C_8H_{12}O$ 124.18	1122-20-9	 93-6^{20}	4-07-00-00148 0.9695^{20}	EtOH 3; eth 3 1.4995^{20}
4706	2-Cyclohexen-1-one, 2,5-dimethyl- 2,5-Dimethyl-2-cyclohexenone	$C_8H_{12}O$ 124.18	14845-35-3	 189.5	1-07-00-00051 0.938^{22}	eth 4; EtOH 4 1.4753^{22}
4707	2-Cyclohexen-1-one, 3,5-dimethyl-	$C_8H_{12}O$ 124.18	1123-09-7	 208.5	4-07-00-00149 0.9400^{20}	EtOH 3; eth 3 1.4812^{20}

No.	Name Synonym	Mol. Form. Mol. Wt.	CAS RN mp/°C	Merck No. bp/°C	Beil. Ref. den/g cm^{-3}	Solubility n_D
4708	2-Cyclohexen-1-one, 3,6-dimethyl-	$C_8H_{12}O$ 124.18	15329-10-9	 75^{19}	4-07-00-00150 1.007^{18}	H_2O 1; EtOH 3; eth 3 1.4805^{18}
4709	2-Cyclohexen-1-one, 2-hydroxy-3-methyl-6-(1-methylethyl)- Diosphenol	$C_{10}H_{16}O_2$ 168.24	490-03-9 83	3292 109^{10}		
4710	2-Cyclohexen-1-one, 2-methyl-	$C_7H_{10}O$ 110.16	1121-18-2	 178.5	4-07-00-00131 0.966^{20}	bz 3 1.4833^{20}
4711	2-Cyclohexen-1-one, 3-methyl-	$C_7H_{10}O$ 110.16	1193-18-6 -21	 201	4-07-00-00130 0.9693^{20}	H_2O 5; bz 3 1.49475^{20}
4712	2-Cyclohexen-1-one, 5-methyl- 5-Methyl-2-cyclohexenone	$C_7H_{10}O$ 110.16	7214-50-8	 185	4-07-00-00133 0.947^{25}	ace 4; bz 4; EtOH 4 1.4739^{25}
4713	3-Cyclohexen-1-one, 3-methyl-	$C_7H_{10}O$ 110.16	31883-98-4 -21	 201	4-07-00-00131 0.9693^{20}	bz 4 1.4947^{20}
4714	3-Cyclohexen-1-one, 4-methyl-	$C_7H_{10}O$ 110.16	5259-65-4	 171; 74^{17}	4-07-00-00131 0.9551^{20}	EtOH 3; ace 3; bz 3 1.4652^{20}
4715	2-Cyclohexen-1-one, 2-methyl-5-(1-methylethenyl)-, (*S*)- (+)-Carvone	$C_{10}H_{14}O$ 150.22	2244-16-8 <15	1883 231	4-07-00-00315 0.965^{20}	H_2O 2; EtOH 4; eth 3; chl 3 1.4989^{20}
4716	2-Cyclohexen-1-one, 3-methyl-6-(1-methylethyl)-, (±)- (±)-Piperitone	$C_{10}H_{16}O$ 152.24	6091-52-7 -19	7443 232.5	4-07-00-00186 0.9331^{20}	ace 4; EtOH 4 1.4845^{20}
4717	2-Cyclohexen-1-one, 6-methyl-3-(1-methylethyl)-, (±)- Carvenone (*DL*)	$C_{10}H_{16}O$ 152.24	23733-68-8	 235.5	4-07-00-00187 0.9203^{20}	ace 4 1.4826^{20}
4718	2-Cyclohexen-1-one, 3-methyl-6-(1-methylethylidene)- Piperitenone	$C_{10}H_{14}O$ 150.22	491-09-8	 $120\text{-}2^{14}$	4-07-00-00314 0.9774^{20}	EtOH 4; eth 4 1.5294^{20}
4719	2-Cyclohexen-1-one, 3,4,4-trimethyl- Cyclohexen-3-one, 1,6,6-trimethyl	$C_9H_{14}O$ 138.21	17299-41-1	 217; 98^{13}	4-07-00-00164 0.944^{25}	 1.4889^{25}
4720	2-Cyclohexen-1-one, 3,4,6-trimethyl- Cyclohexen-3-one, 1,4,6 trimethyl	$C_9H_{14}O$ 138.21	7474-10-4	 213	4-07-00-00170 0.943^{17}	 1.484^{17}
4721	2-Cyclohexen-1-one, 3,5,5-trimethyl- Isophorone	$C_9H_{14}O$ 138.21	78-59-1 -8.1	 215.2	4-07-00-00165 0.9255^{20}	 1.4766^{18}
4722	2-Cyclohexen-1-one, 3,6,6-trimethyl- 3,6,6-Trimethyl-2-cyclohexenone	$C_9H_{14}O$ 138.21	23438-77-9	 208; $86\text{-}8^{15}$	3-07-00-00286 0.927^{24}	 1.4798^{20}
4723	Cyclohexylamine, *N*,4-dimethyl- Cyclohexane, 1-methylamino-4-methyl	$C_8H_{17}N$ 127.23	90226-23-6 -15	 196	 0.8188^{22}	 1.4175^{22}
4724	Cyclononane	C_9H_{18} 126.24	293-55-0 11	 178.4	4-05-00-00133 0.8463^{25}	 1.4666^{20}
4725	Cyclononanone	$C_9H_{16}O$ 140.23	3350-30-9 34	 148^{24}; 94^{12}	4-07-00-00062 0.9560^{20}	EtOH 3 1.4729^{20}
4726	Cyclononasiloxane, octadecamethyl- Octadecamethylcyclononasiloxane	$C_{18}H_{54}O_9Si_9$ 667.39	556-71-8	 188^{20}	3-04-00-01886	bz 4; lig 4 1.4070^{20}
4727	Cyclononene, (*E*)-	C_9H_{16} 124.23	3958-38-1	 $94\text{-}6^{30}$	4-05-00-00280 0.8615^{20}	 1.4799^{20}
4728	Cyclononene, (*Z*)-	C_9H_{16} 124.23	933-21-1	 168	4-05-00-00280 0.8671^{20}	bz 4 1.4805^{20}
4729	Cyclononyne	C_9H_{14} 122.21	6573-52-0	 178	4-05-00-00419 0.8972^{20}	 1.4890^{20}
4730	2,4(1H)-Cyclo-3,4-secoakuammilanium, 3,17-dihydroxy-16-(methoxycarbonyl)-4-methyl-, (3β,16*R*)- Echitamine	$C_{22}H_{30}N_2O_5$ 402.49	6871-44-9 206	3474		H_2O 3; EtOH 3; eth 3; chl 3
4731	1,3-Cyclooctadecadiene	$C_{18}H_{32}$ 248.45	6568-58-7	 115^{3}	4-05-00-00503 0.8814^{25}	 1.4899^{20}
4732	1,4-Cyclooctadiene	C_8H_{12} 108.18	1073-07-0 -53	 145	 0.8754^{20}	
4733	1,5-Cyclooctadiene	C_8H_{12} 108.18	111-78-4 -56.4	 150.8	3-05-00-00321 0.8818^{25}	bz 4 1.4905^{25}
4734	1,5-Cyclooctadiene, (*Z*,*Z*)-	C_8H_{12} 108.18	1552-12-1 -70	 151; $51\text{-}2^{25}$	4-05-00-00403 0.8818^{25}	bz 4 1.4905^{25}
4735	Cyclooctanamine Aminocyclooctane	$C_8H_{17}N$ 127.23	5452-37-9 -48	 190	4-12-00-00128 0.928^{25}	 1.4804^{20}
4736	Cyclooctane	C_8H_{16} 112.22	292-64-8 14.8	 149	4-05-00-00111 0.8349^{20}	H_2O 1; bz 3; lig 3 1.4586^{20}
4737	Cyclooctane, butyl- Butylcyclooctane	$C_{12}H_{24}$ 168.32	16538-93-6	 $90\text{-}1^{10}$	4-05-00-00170 0.8260^{25}	 1.4585^{25}
4738	Cyclooctanecarboxaldehyde Formylcyclooctane	$C_9H_{16}O$ 140.23	6688-11-5	 96^{15}	3-07-00-00112 0.94^{25}	 1.4748^{20}
4739	Cyclooctanol	$C_8H_{16}O$ 128.21	696-71-9 25.1	 99^{16}	4-06-00-00113 0.9740^{20}	EtOH 3 1.4871^{20}
4740	Cyclooctanone	$C_8H_{14}O$ 126.20	502-49-8 29	 196	4-07-00-00049 0.9581^{20}	H_2O 1; EtOH 3; ace 3; bz 3 1.4694^{20}
4741	Cyclooctasiloxane, hexadecamethyl- Hexadecamethylcyclooctasiloxane	$C_{16}H_{48}O_8Si_8$ 593.24	556-68-3 31.5	 290	3-04-00-01886 1.177^{25}	bz 4; lig 4 1.4060^{20}
4742	1,3,5,7-Cyclooctatetraene [8]Annulene	C_8H_8 104.15	629-20-9 -4.7	 140.5	4-05-00-01331 0.9206^{20}	EtOH 3; eth 3; ace 3; bz 3 1.5381^{20}
4743	1,3,5,7-Cyclooctatetraene, 1-bromo-	C_8H_7Br 183.05	7567-22-8	 $52\text{-}3^{18}$	4-05-00-01334 1.4206^{25}	ace 3 1.5870^{25}
4744	1,3,5,7-Cyclooctatetraene, 1-butyl-	$C_{12}H_{16}$ 160.26	13402-37-4	 98^{20}	4-05-00-01417 0.8876^{25}	 1.5083^{25}

No.	Name	Synonym	Mol. Form.	Mol. Wt.	CAS RN	mp/°C	Merck No.	bp/°C	Beil. Ref.	den/g cm^{-3}	Solubility	n_D
4745	1,3,5,7-Cyclooctatetraene, 1-chloro-	Chlorocyclooctatetraene	C_8H_7Cl	138.60	29554-49-2			$50-1^{5.5}$	4-05-00-01334	1.1199^{25}	H_2O 1; ace 3; bz 3; chl 3	1.5542^{25}
4746	1,3,5,7-Cyclooctatetraene, 1-ethyl-		$C_{10}H_{12}$	132.21	13402-35-2			81^{37}	4-05-00-01373	0.8996^{25}		1.5187^{25}
4747	1,3,5,7-Cyclooctatetraene, 1-methyl-		C_9H_{10}	118.18	2570-12-9			84.5^{67}	4-05-00-01358	0.8978^{25}	H_2O 1; EtOH 3; eth 3; bz 3	1.5249^{25}
4748	1,3,5,7-Cyclooctatetraene, propyl-	Cyclooctatetraene, propyl	$C_{11}H_{14}$	146.23	13402-36-3			73^{9}	4-05-00-01400	0.8870^{25}		1.5131^{25}
4749	1,3,5-Cyclooctatriene		C_8H_{10}	106.17	1871-52-9	-83		145.5	4-05-00-00882	0.8971^{25}		1.5035^{25}
4750	1,3,6-Cyclooctatriene		C_8H_{10}	106.17	3725-30-2	-62		68^{60}	4-05-00-00883	0.8940^{25}		
4751	Cyclooctene, (*E*)-		C_8H_{14}	110.20	931-89-5	-59		143	4-05-00-00263	0.8483^{20}	EtOH 3; ctc 2; chl 3	1.4741^{25}
4752	Cyclooctene, (*Z*)-		C_8H_{14}	110.20	931-87-3	-12		138	3-05-00-00209	0.8472^{20}	EtOH 3; eth 3; ctc 3	1.4698^{20}
4753	2-Cycloocten-1-ol		$C_8H_{14}O$	126.20	3212-75-7			74^{2}	4-06-00-00217	0.9756^{25}		1.4959^{25}
4754	3-Cycloocten-1-one		$C_8H_{12}O$	124.18	4734-90-1			200.5	4-07-00-00142	0.990^{20}	ace 4; eth 4; chl 4	1.4953^{25}
4755	Cyclooctyne		C_8H_{12}	108.18	1781-78-8			158	4-05-00-00401	0.868^{20}		1.4850^{20}
4756	15H-Cyclopenta[a]phenanthrene, 16,17-dihydro-	1,2-Cyclopentenophenanthrene	$C_{17}H_{14}$	218.30	482-66-6	135.5	2749		4-05-00-02423		H_2O 1; EtOH 3; peth 3	
4757	Cyclopenta[c]furo[3',2':4,5]furo[2,3-h][1]benzopyran-1,11-dione, 2,3,6a,8,9,9a-hexahydro-4-methoxy-, (6a*R*-	Aflatoxin B2	$C_{17}H_{14}O_6$	314.29	7220-81-7	287.5	168		5-19-10-00552			
4758	Cyclopenta[c]pyran-1(4aH)-one, 5,6,7,7a-tetrahydro-4,7-dimethyl-	Nepetalactone	$C_{10}H_{14}O_2$	166.22	490-10-8		6383	$71-2^{0.05}$	4-17-00-04622	1.0663^{25}		1.4859^{25}
4759	Cyclopentadecane		$C_{15}H_{30}$	210.40	295-48-7	61.3			4-05-00-00181	0.8364^{61}		1.4592^{61}
4760	Cyclopentadecanol	Exaltol	$C_{15}H_{30}O$	226.40	4727-17-7	80.5		177^{11}; $145^{0.3}$	4-06-00-00185	0.930^{20}		1.4555^{98}
4761	Cyclopentadecanone		$C_{15}H_{28}O$	224.39	502-72-7	63		$120^{0.3}$	4-07-00-00113	0.8895^{25}	H_2O 2; EtOH 3; ace 3	1.4637^{60}
4762	Cyclopentadecanone, 2-methyl-	2-Methylcyclopentadecanone	$C_{16}H_{30}O$	238.41	52914-66-6			$171-3^{12}$	3-07-00-00208	0.9213^{16}		1.4812^{16}
4763	Cyclopentadecanone, 3-methyl-	Muscone	$C_{16}H_{30}O$	238.41	541-91-3		6222	329; $130^{0.5}$	4-07-00-00118	0.9221^{17}	ace 4; eth 4; EtOH 4	1.4802^{17}
4764	1,3-Cyclopentadiene	Pyropentylene	C_5H_6	66.10	542-92-7	-85	2744	41	4-05-00-00377	0.8021^{20}	H_2O 1; EtOH 5; eth 5; ace 3	1.4440^{20}
4765	1,3-Cyclopentadiene, 5-ethylidene-		C_7H_8	92.14	3839-50-7			32^{17}	4-05-00-00878	0.865^{18}		1.526^{20}
4766	1,3-Cyclopentadiene, 1,2,3,4,5,5-hexachloro-	Perchlorocyclopentadiene	C_5Cl_6	272.77	77-47-4	-9		239; $48^{0.3}$	4-05-00-00381	1.7019^{25}		1.5658^{20}
4767	1,3-Cyclopentadiene, 5-methylene-	Fulvene	C_6H_6	78.11	497-20-1			$7-8^{56}$	4-05-00-00764	0.8241^{20}	H_2O 1; bz 3; chl 3	1.4920^{20}
4768	1,3-Cyclopentadiene, 5-(1-methylethylidene)-		C_8H_{10}	106.17	2175-91-9			$49-50^{11}$	4-05-00-00974	0.881^{20}		1.5474^{20}
4769	1,3-Cyclopentadiene, 5-(2-propenylidene)-, (*E*)-	Fulvene, 6-vinyl(*trans*)	C_8H_8	104.15	116862-65-8	-35		45^{12} $20^{0.15}$		0.898^{20}	H_2O 1; bz 3; chl 3	
4770	1,3-Cyclopentadiene, 1,2,3,4-tetrachloro-5,5-dimethoxy-		$C_7H_6Cl_4O_2$	263.93	2207-27-4			109	4-07-00-00286	1.501^{25}		1.5282^{20}
4771	Cyclopentanamine	Cyclopentylamine	$C_5H_{11}N$	85.15	1003-03-8	-82.7		108	4-12-00-00004	0.8689^{20}	ace 3; bz 3; chl 3	1.4728^{25}
4772	Cyclopentanamine, 1-methyl-		$C_6H_{13}N$	99.18	40571-45-7			121	4-12-00-00113	0.8025^{25}	H_2O 4	1.4408^{25}
4773	Cyclopentanamine, 2-methyl-		$C_6H_{13}N$	99.18	41223-14-7			123	0-12-00-00007	0.801^{20}	H_2O 4	
4774	Cyclopentane	Pentamethylene	C_5H_{10}	70.13	287-92-3	-93.8	2746	49.3	4-05-00-00014	0.7457^{20}	H_2O I; EtOH 5; eth 5; ace 5	1.4065^{20}
4775	Cyclopentaneacetic acid	2-Cyclopentylacetic acid	$C_7H_{12}O_2$	128.17	1123-00-8	13.5		228	4-09-00-00033	1.0216^{18}		1.4523^{18}
4776	Cyclopentaneacetic acid, 3-methyl-	3-Methylcyclopentaneacetic acid	$C_8H_{14}O_2$	142.20	63370-69-4			237	3-09-00-00060	0.9818^{20}		1.4504^{20}
4777	Cyclopentane, bromo-	Cyclopentyl bromide	C_5H_9Br	149.03	137-43-9			137.5	4-05-00-00019	1.3873^{20}	ctc 2	1.4886^{20}
4778	Cyclopentane, butyl-	Butylcyclopentane	C_9H_{18}	126.24	2040-95-1	-108		156.6	4-05-00-00139	0.7846^{20}	ace 4; bz 4; eth 4; EtOH 4	1.4316^{20}
4779	Cyclopentanecarbonitrile	Cyanocyclopentane	C_6H_9N	95.14	4254-02-8							
4780	Cyclopentanecarboxaldehyde		$C_6H_{10}O$	98.14	872-53-7			133.5	4-07-00-00037	0.9371^{20}	H_2O 4; eth 4; EtOH 4	1.4432^{20}
4781	Cyclopentanecarboxylic acid	Cyclopentanoic acid	$C_6H_{10}O_2$	114.14	3400-45-1	-7		212; 104^{11}	4-09-00-00011	1.0527^{20}	H_2O 2; ctc 2; MeOH 3	1.4532^{20}

No.	Name Synonym	Mol. Form. Mol. Wt.	CAS RN mp/°C	Merck No. bp/°C	Beil. Ref. den/g cm^{-3}	Solubility n_D
4782	Cyclopentanecarboxylic acid, 1-amino- Cycloleucine	$C_6H_{11}NO_2$ 129.16	52-52-8 330 dec	2740	4-14-00-00974	
4783	Cyclopentanecarboxylic acid, 1-methyl- 1-Methylcyclopentanecarboxylic acid	$C_7H_{12}O_2$ 128.17	5217-05-0	 219	4-09-00-00035 1.0218^{20}	eth 4; EtOH 4 1.4529^{20}
4784	Cyclopentanecarboxylic acid, 2-methyl-5-(1-methylethyl)- Puleganic acid	$C_{10}H_{18}O_2$ 170.25	528-23-4 -18	 152^{25}; 139^{12}	4-09-00-00080 0.9642^{20}	 1.4524^{24}
4785	Cyclopentanecarboxylic acid, 2-oxo-, ethyl ester	$C_8H_{12}O_3$ 156.18	611-10-9	 221; 110^{16}	4-10-00-02602 1.0781^{21}	eth 3; bz 3 1.4519^{20}
4786	Cyclopentanecarboxylic acid, 1,2,2,3-tetramethyl- Campholic acid	$C_{10}H_{18}O_2$ 170.25	464-88-0 106	 254; 146^{12}	2-09-00-00019	H_2O 2; EtOH 3; eth 3
4787	Cyclopentanecarboxylic acid, 3-isopropyl-1-methyl-	$C_{12}H_{22}O_2$ 198.31	104068-42-0 18.5	 257	4-09-00-00079 0.9698^{19}	 1.4563^{20}
4788	Cyclopentanecarboxylic acid, 2-methyl-5-(1-methylethyl)-, ethyl ester Cyclopentanecarboxylic acid, 2-isopropyl-5-methyl, ethyl ester	$C_{12}H_{22}O_2$ 198.31	116530-99-5	 145^{4}	1-09-00-00017 0.9178^{12}	 1.4405^{12}
4789	Cyclopentane, chloro- Cyclopentyl chloride	C_5H_9Cl 104.58	930-28-9	 114	4-05-00-00018 1.0051^{20}	H_2O 1; eth 3; ace 3; bz 3 1.4510^{20}
4790	Cyclopentane, decyl- Decane, 1-cyclopentyl-	$C_{15}H_{30}$ 210.40	1795-21-7 -22	 279	4-05-00-00186 0.8110^{20}	ace 4; bz 4; eth 4; EtOH 4 1.4486^{20}
4791	1,1-Cyclopentanedicarboxylic acid, diethyl ester	$C_{11}H_{18}O_4$ 214.26	4167-77-5	 210	4-09-00-02793 1.0186^{20}	 1.4370^{25}
4792	1,2-Cyclopentanedicarboxylic acid, dimethyl ester, *trans*-(±)- Cyclopentane-1,2-dicarboxylic acid, dimethyl ester (*DL*, *trans*)	$C_9H_{14}O_4$ 186.21	80656-12-8	 118-9^{17}	4-09-00-02794 1.1130^{19}	 1.4498^{17}
4793	1,3-Cyclopentanedicarboxylic acid, *cis*-	$C_7H_{10}O_4$ 158.15	876-05-1 121	 300 dec	4-09-00-02795	H_2O 3; EtOH 3; eth 3; ace 3
4794	1,3-Cyclopentanedicarboxylic acid, 1,2,2-trimethyl-, (1*R* *cis*)-	$C_{10}H_{16}O_4$ 200.23	124-83-4 189.1		4-09-00-02851 1.186^{20}	H_2O 2; EtOH 4; eth 4; ace 3
4795	1,3-Cyclopentanedicarboxylic acid, 1,2,2-trimethyl-, dimethyl ester, (1*R*-*cis*)- Camphoric acid, dimethyl ester, (+)-	$C_{12}H_{20}O_4$ 228.29	15797-21-4 <-16	 264	4-09-00-02851 1.015^{20}	eth 4; EtOH 4 1.4627^{19}
4796	Cyclopentane, 1,1-diethyl- 1,1-Diethylcyclopentane	C_9H_{18} 126.24	2721-38-2 -95.6	 150.5	4-05-00-00141 0.7988^{25}	 1.4388^{20}
4797	Cyclopentane, 1,2-diethyl-, *cis*- *cis*-1,2-Diethylcyclopentane	C_9H_{18} 126.24	932-39-8 -118	 153.6	4-05-00-00141 0.7920^{25}	 1.4355^{20}
4798	Cyclopentane, 1,2-diethyl-, *trans*- *trans*-1,2-Diethylcyclopentane	C_9H_{18} 126.24	932-40-1 -95	 147.5	4-05-00-00141 0.7792^{25}	bz 4; eth 4; peth 4 1.4295^{20}
4799	Cyclopentane, 1,3-diethyl- 1,3-Diethylcyclopentane	C_9H_{18} 126.24	19398-75-5	 150	0-05-00-00045 0.783^{25}	 1.430^{20}
4800	Cyclopentane, 1,1-dimethyl- 1,1-Dimethylcyclopentane	C_7H_{14} 98.19	1638-26-2 -69.8	 87.5	4-05-00-00105 0.7499^{25}	 1.4136^{20}
4801	Cyclopentane, 1,2-dimethyl-, *cis*- *cis*-1,2-Dimethylcyclopentane	C_7H_{14} 98.19	1192-18-3 -54	 99.5	4-05-00-00106 0.7680^{25}	 1.4222^{20}
4802	Cyclopentane, 1,2-dimethyl-, *trans*- *trans*-1,2-Dimethylcyclopentane	C_7H_{14} 98.19	822-50-4 -117.6	 91.9	4-05-00-00107 0.7468^{25}	 1.4120^{20}
4803	Cyclopentane, 1,3-dimethyl-, *cis*- *cis*-1,3-Dimethylcyclopentane	C_7H_{14} 98.19	2532-58-3 -133.7	 90.8	4-05-00-00107 0.7402^{25}	 1.4089^{20}
4804	Cyclopentane, 1,3-dimethyl-, *trans*- *trans*-1,3-Dimethylcyclopentane	C_7H_{14} 98.19	1759-58-6 -134	 91.7	4-05-00-00108 0.7443^{25}	 1.4107^{20}
4805	Cyclopentane, (1,1-dimethylethyl)-	C_9H_{18} 126.24	3875-52-3 -95.8	 144.9	3-05-00-00124 0.7870^{25}	 1.4338^{20}
4806	1,3-Cyclopentanediol, *cis*-	$C_5H_{10}O_2$ 102.13	16326-97-9 31	 105^{5}; 90^{1}	3-06-00-04057 1.099^{15}	 1.4792^{20}
4807	1,3-Cyclopentanediol, *trans*- Cyclopentane-1,3-diol (*trans*)	$C_5H_{10}O_2$ 102.13	16326-98-0 40	 80-5$^{0.1}$	3-06-00-04057 1.094^{25}	 1.4830^{20}
4808	Cyclopentane, docosyl- Docosylcyclopentane	$C_{27}H_{54}$ 378.73	62016-55-1 45	 100	4-05-00-00201 [illegible]	bz 4; eth 4 1.4600[illegible]
4809	Cyclopentane, dodecyl-	$C_{17}H_{34}$ 238.46	5634-30-0 -5	 312	3-05-00-00156 0.8158^{20}	ace 4; bz 4; eth 4; EtOH 4 1.4518^{20}
4810	Cyclopentaneethanamine, *N*,α-dimethyl- Cyclopentamine	$C_9H_{19}N$ 141.26	102-45-4	2745 171	4-12-00-00137	 1.4500^{20}
4811	Cyclopentaneethanol	$C_7H_{14}O$ 114.19	766-00-7	 183; 96^{24}	4-06-00-00110 0.9180^{20}	eth 4 1.4577^{20}
4812	Cyclopentane, ethyl- Ethylcyclopentane	C_7H_{14} 98.19	1640-89-7 -138.4	 103.5	4-05-00-00104 0.7665^{20}	H_2O 1; EtOH 5; eth 5; ace 5 1.4198^{20}
4813	Cyclopentane, ethylidene- Ethylidenecyclopentane	C_7H_{12} 96.17	2146-37-4 -129.5	 115	4-05-00-00253 0.8020^{20}	 1.4481^{20}
4814	Cyclopentane, 1-ethyl-1-methyl- 1-Ethyl-1-methylcyclopentane	C_8H_{16} 112.22	16747-50-5 -143.8	 121.6	4-05-00-00126 0.7767^{25}	ace 4; bz 4; eth 4; EtOH 4 1.4272^{20}
4815	Cyclopentane, 1-ethyl-2-methyl-, *cis*- *cis*-1-Methyl-2-ethylcyclopentane	C_8H_{16} 112.22	930-89-2 -106	 128	4-05-00-00126 0.7852^{20}	 1.4293^{20}

No.	Name Synonym	Mol. Form. Mol. Wt.	CAS RN mp/°C	Merck No. bp/°C	Beil. Ref. den/g cm^{-3}	Solubility n_D
4816	Cyclopentane, 1-ethyl-2-methyl-, *trans-*	C_8H_{16} 112.22	930-90-5 -105.9	 121.2	4-05-00-00126 0.7649^{25}	 1.4219^{20}
4817	Cyclopentane, 1-ethyl-3-methyl-, *cis-* 1-Methyl-*cis*-3-ethylcyclopentane	C_8H_{16} 112.22	2613-66-3	 121	4-05-00-00127 0.7724^{20}	ace 4; bz 4; eth 4; EtOH 4 1.4203^{20}
4818	Cyclopentane, 1-ethyl-3-methyl-, *trans-* 1-Methyl-*trans*-3-ethylcyclopentane	C_8H_{16} 112.22	2613-65-2 -108	 121	4-05-00-00127 0.7619^{20}	 1.4186^{20}
4819	Cyclopentane, heneicosyl-	$C_{26}H_{52}$ 364.70	6703-82-8 42	 420	4-05-00-00200 0.8286^{20}	 1.4602^{20}
4820	Cyclopentane, heptyl- Heptane, 1-cyclopentyl-	$C_{12}H_{24}$ 168.32	5617-42-5 -53	 224	4-05-00-00173 0.8010^{20}	ace 4; bz 4; eth 4; EtOH 4 1.4421^{20}
4821	Cyclopentane, hexacosyl- Hexacosylcyclopentane	$C_{31}H_{62}$ 434.83	62016-58-4 56	 468	4-05-00-00203 0.8326^{20}	 1.4628^{20}
4822	Cyclopentane, hexadecyl- Hexadecane, 1-cyclopentyl-	$C_{21}H_{42}$ 294.56	6812-39-1 21	 368	4-05-00-00195 0.8228^{20}	ace 4; bz 4; eth 4; EtOH 4 1.4564^{20}
4823	Cyclopentane, hexyl-	$C_{11}H_{22}$ 154.30	4457-00-5 -73	 203	4-05-00-00168 0.7965^{20}	ace 4; bz 4; eth 4; EtOH 4 1.4392^{20}
4824	Cyclopentane, iodo- Cyclopentyl iodide	C_5H_9I 196.03	1556-18-9	 166.5	4-05-00-00020 1.7096^{20}	H_2O 1; eth 3; bz 3; ctc 2 1.5447^{20}
4825	Cyclopentanemethanol	$C_6H_{12}O$ 100.16	3637-61-4	 163	4-06-00-00092 0.9332^{20}	 1.4579^{20}
4826	Cyclopentane, methyl- Methylcyclopentane	C_6H_{12} 84.16	96-37-7 -142.5	 71.8	4-05-00-00084 0.7486^{20}	H_2O 1; EtOH 5; eth 5; ace 5 1.4097^{20}
4827	Cyclopentane, methylene- Methylenecyclopentane	C_6H_{10} 82.15	1528-30-9	 75.5	4-05-00-00241 0.7787^{20}	bz 3; chl 3 1.4355^{20}
4828	Cyclopentane, (1-methylethyl)- Isopropylcyclopentane	C_8H_{16} 112.22	3875-51-2 -111.4	 126.5	4-05-00-00125 0.7765^{20}	H_2O 1; EtOH 5; eth 3; ace 5 1.4258^{20}
4829	Cyclopentane, 1-methyl-2-(1-methylethyl)- Cyclopentane, 1-isopropyl-2-methyl-	C_9H_{18} 126.24	89223-57-4	 143	4-05-00-00141 0.7832^{15}	 1.4279^{20}
4830	Cyclopentane, 1-methyl-1-nitro- 1-Methyl-1-nitrocyclopentane	$C_6H_{11}NO_2$ 129.16	30168-50-4	 181; 92^{40}	4-05-00-00087 1.0395^{20}	 1.4504^{20}
4831	Cyclopentane, 1-methyl-2-nitro- 1-Methyl-2-nitrocyclopentane	$C_6H_{11}NO_2$ 129.16	102153-88-8	 185 dec; 98^{40}	1-05-00-00011 1.0381^{22}	 1.4488^{22}
4832	Cyclopentane, (1-methylpropyl)-	C_9H_{18} 126.24	4850-32-2	 154.4	3-05-00-00124 0.7905^{25}	 1.4357^{20}
4833	Cyclopentane, 1-methyl-1-propyl- 1-Methyl-1-propylcyclopentane	C_9H_{18} 126.24	16631-63-3	 146	4-05-00-00140 0.795^{25}	 1.437^{20}
4834	Cyclopentane, 1-methyl-2-propyl-, *cis-* *cis*-1-Methyl-2-propylcyclopentane	C_9H_{18} 126.24	932-43-4 -104	 152.6	4-05-00-00140 0.7881^{25}	 1.4343^{20}
4835	Cyclopentane, 1-methyl-2-propyl-, *trans-* *trans*-1-Methyl-2-propylcyclopentane	C_9H_{18} 126.24	932-44-5 -123	 146.4	4-05-00-00140 0.7735^{25}	 1.4274^{20}
4836	Cyclopentane, 1-methyl-3-propyl-, *cis-* *cis*-1-Methyl-3-propylcyclopentane	C_9H_{18} 126.24	2443-04-1	 148	3-05-00-00125 0.776^{25}	 1.426^{20}
4837	Cyclopentane, 1-methyl-3-propyl-, *trans-* *trans*-1-Methyl-3-propylcyclopentane	C_9H_{18} 126.24	2443-03-0	 148	3-05-00-00125 0.776^{25}	 1.426^{20}
4838	Cyclopentane, (2-methylpropyl)- Isobutylcyclopentane	C_9H_{18} 126.24	3788-32-7 -115.2	 148	4-05-00-00140 0.7769^{25}	 1.4298^{20}
4839	Cyclopentane, nitro- Nitrocyclopentane	$C_5H_9NO_2$ 115.13	2562-38-1	 90-1^{40}	4-05-00-00020 1.0776^{23}	H_2O I; bz 3; chl 3 1.4538^{20}
4840	Cyclopentane, (nitromethyl)- (Nitromethyl)cyclopentane	$C_6H_{11}NO_2$ 129.16	2625-31-2	 110^{35}	1-05-00-00011 1.0713^{20}	 1.4587^{20}
4841	Cyclopentane, nonyl-	$C_{14}H_{28}$ 196.38	2882-98-6 -29	 262	4-05-00-00181 0.8081^{20}	ace 4; bz 4; eth 4; EtOH 4 1.4467^{20}
4842	Cyclopentane, octadecyl- Octadecane, 1-cyclopentyl-	$C_{23}H_{46}$ 322.62	62016-53-9 30	 391	4-05-00-00196 0.8254^{20}	ace 4; bz 4; eth 4; EtOH 4 1.4581^{20}
4843	Cyclopentane, octyl- Octane, 1-cyclopentyl-	$C_{13}H_{26}$ 182.35	1795-20-6 -44	 243	4-05-00-00176 0.8048^{20}	 1.4446^{20}
4844	Cyclopentane, pentadecyl- Pentadecane, 1-cyclopentyl-	$C_{20}H_{40}$ 280.54	4669-01-6 17	 355	4-05-00-00195 0.8213^{20}	ace 4; bz 4; eth 4; EtOH 4 1.4554^{20}
4845	Cyclopentane, pentatriacontyl- Pentatriacontylcyclopentane	$C_{40}H_{80}$ 561.08	61827-98-3 74	 531	4-05-00-00205 0.8374^{20}	 1.4660^{20}
4846	Cyclopentane, pentyl-	$C_{10}H_{20}$ 140.27	3741-00-2 -83	 180	4-05-00-00160 0.7912^{20}	ace 4; bz 4; eth 4; EtOH 4 1.4356^{20}
4847	Cyclopentanepropanoic acid	$C_8H_{14}O_2$ 142.20	140-77-2	 130-2^{12}	4-09-00-00046 0.996^{25}	 1.4570^{20}
4848	Cyclopentane, 2-propenyl-	C_8H_{14} 110.20	3524-75-2 -110.7	 125	4-05-00-00272 0.793^{25}	chl 3 1.4412^{20}
4849	Cyclopentane, propyl- Propylcyclopentane	C_8H_{16} 112.22	2040-96-2 -117.3	 131	4-05-00-00125 0.7763^{20}	H_2O 1; EtOH 5; eth 5; ace 5 1.4266^{20}
4850	Cyclopentane, tetradecyl- Tetradecane, 1-cyclopentyl-	$C_{19}H_{38}$ 266.51	1795-22-8 9	 341	3-05-00-00158 0.8196^{20}	 1.4543^{20}
4851	Cyclopentane, 1,1,3,3-tetramethyl- 1,1,3,3-Tetramethylcyclopentane	C_9H_{18} 126.24	50876-33-0 -88.4	 118	4-05-00-00143 0.7469^{25}	 1.4125^{20}
4852	Cyclopentanethiol Cyclopentyl mercaptan	$C_5H_{10}S$ 102.20	1679-07-8	 132.1	4-06-00-00015 0.9550^{20}	
4853	Cyclopentane, triacontyl- Triacontylcyclopentane	$C_{35}H_{70}$ 490.94	61827-93-8 65	 498	4-05-00-00204 0.8350^{20}	 1.4644^{20}
4854	Cyclopentane, tridecyl- Tridecane, 1-cyclopentyl-	$C_{18}H_{36}$ 252.48	6006-34-4 5	 327	4-05-00-00192 0.8178^{20}	ace 4; bz 4; eth 4; EtOH 4 1.4531^{20}
4855	Cyclopentane, 1,1,2-trimethyl- Camphocean	C_8H_{16} 112.22	4259-00-1 -21.6	 114; 53^{100}	4-05-00-00127 0.7660^{20}	 1.4199^{20}

No.	Name Synonym	Mol. Form. Mol. Wt.	CAS RN mp/°C	Merck No. bp/°C	Beil. Ref. den/g cm^{-3}	Solubility n_D
4856	Cyclopentane, 1,1,3-trimethyl- 1,1,3-Trimethylcyclopentane	C_8H_{16} 112.22	4516-69-2 -142.4	 104.9	3-05-00-00111 0.7439^{25}	 1.4112^{20}
4857	Cyclopentane, 1,2,3-trimethyl-, (1α,2β,3α)- Cyclopentane, 1,2,3-trimethyl-, *trans*-1,2,*cis*-1,3-	C_8H_{16} 112.22	19374-46-0 -112.7	 110.4	4-05-00-00128 0.7492^{25}	 1.4138^{20}
4858	Cyclopentane, 1,2,3-trimethyl-, (1α,2α,3α)- Cyclopentane, 1,2,3-trimethyl-, *cis*-1,2,*cis*-1,3-	C_8H_{16} 112.22	2613-69-6 -116.4	 123	4-05-00-00128 0.7751^{25}	 1.4262^{20}
4859	Cyclopentane, 1,2,3-trimethyl-, (1α,2α,3β)-	C_8H_{16} 112.22	15890-40-1 -112	 117.5	4-05-00-00128 0.7661^{25}	 1.4218^{20}
4860	Cyclopentane, 1,2,4-trimethyl-, (1α,2β,4α)-	C_8H_{16} 112.22	16883-48-0 -130.8	 109.3	4-05-00-00129 0.7430^{25}	 1.4106^{20}
4861	Cyclopentane, 1,2,4-trimethyl-, (1α,2α,4α)- Cyclopentane, 1,2,4-trimethyl-, *cis*-1,2,*cis*-1,4-	C_8H_{16} 112.22	2613-72-1 -132.3	 116.8	4-05-00-00129 0.758^{25}	 1.4186^{20}
4862	Cyclopentane, 1,2,4-trimethyl-, (1α,2α,4β)-	C_8H_{16} 112.22	4850-28-6 -132.6	 116.7	4-05-00-00129 0.7592^{25}	 1.4186^{20}
4863	Cyclopentane, undecyl-	$C_{16}H_{32}$ 224.43	6785-23-5 -10	 296	4-05-00-00189 0.8135^{20}	ace 4; bz 4; eth 4; EtOH 4 1.4503^{20}
4864	Cyclopentanol Cyclopentyl alcohol	$C_5H_{10}O$ 86.13	96-41-3 -19	2747 140.4	4-06-00-00005 0.9488^{20}	H_2O 2; EtOH 3; eth 3; ace 3 1.4530^{20}
4865	Cyclopentanol, acetate Cyclopentyl acetate	$C_7H_{12}O_2$ 128.17	933-05-1	 $51\text{-}2^{12}$	4-06-00-00007 0.9522^{16}	
4866	Cyclopentanol, 1-ethynyl- 1-Ethynyl-1-cyclopentanol	$C_7H_{10}O$ 110.16	17356-19-3 27	 157.5	4-06-00-00339 0.962^{25}	 1.4751^{20}
4867	Cyclopentanol, 1-methyl-	$C_6H_{12}O$ 100.16	1402-03-9 36	 136; 53^{30}	4-06-00-00086 0.9044^{23}	 1.4429^{23}
4868	Cyclopentanol, 2-methyl-, *cis*- *cis*-2-Methylcyclopentanol	$C_6H_{12}O$ 100.16	25144-05-2	 148.5	3-06-00-00055 0.9379^{16}	 1.4504^{16}
4869	Cyclopentanol, 2-methyl-, *trans*- *trans*-2-Methylcyclopentanol	$C_6H_{12}O$ 100.16	25144-04-1	 150.5	4-06-00-00091 0.9248^{16}	 1.4499^{16}
4870	Cyclopentanol, 3-methyl- 3-Methylcyclopentanol	$C_6H_{12}O$ 100.16	18729-48-1	 148.5	2-06-00-00015 0.9158^{16}	eth 4; EtOH 4 1.4487^{16}
4871	Cyclopentanol, 1-phenyl-	$C_{11}H_{14}O$ 162.23	10487-96-4	 $132\text{-}3^{18}$	4-06-00-03884 1.0530^{20}	 1.5479^{20}
4872	Cyclopentanol, 1-propyl-	$C_8H_{16}O$ 128.21	1604-02-0 -37.5	 173.5	3-06-00-00101 0.9040^{25}	 1.4502^{25}
4873	Cyclopentanol, 2-propyl-, *cis*- *cis*-2-Propylcyclopentanol	$C_8H_{16}O$ 128.21	25172-42-3	 $79\text{-}80^{12}$	3-06-00-00101 0.9163^{9}	 1.4600^{9}
4874	Cyclopentanol, 2-propyl-, *trans*- *trans*-2-Propylcyclopentanol	$C_8H_{16}O$ 128.21	25172-43-4	 $78\text{-}9^{10}$	3-06-00-00101 0.9016^{9}	 1.4565^{9}
4875	Cyclopentanol, 1,2,2-trimethyl- 1,2,2-Trimethylcyclopentanol	$C_8H_{16}O$ 128.21	1121-95-5	 156; 60^{15}	1-06-00-00014 0.9101^{20}	 1.4513^{20}
4876	Cyclopentanol, 1,2,4-trimethyl- 1,2,4-Trimethylcyclopentanol	$C_8H_{16}O$ 128.21	34103-97-4	 158	1-06-00-00014 0.8850^{21}	 1.4424^{21}
4877	Cyclopentanol, 1,2,5-trimethyl- 1,2,5-Trimethylcyclopentanol	$C_8H_{16}O$ 128.21	33840-38-9	 $56\text{-}60^{8}$	1-06-00-00014 0.9121^{15}	 1.4554^{16}
4878	Cyclopentanone Adipic ketone	C_5H_8O 84.12	120-92-3 -51.3	2748 130.5	4-07-00-00005 0.9487^{20}	H_2O 1; EtOH 3; eth 5; ace 3 1.4366^{20}
4879	Cyclopentanone, 2-acetyl- 2-Acetylcyclopentanone	$C_7H_{10}O_2$ 126.16	1670-46-8	 $72\text{-}5^{20}$	4-07-00-01993 1.0431^{25}	 1.4906^{20}
4880	Cyclopentanone, 2-chloro-	C_5H_7ClO 118.56	694-28-0	 $72\text{-}4^{12}$	4-07-00-00012 1.185^{25}	 1.4750^{20}
4881	Cyclopentanone, 2-cyclopentylidene-	$C_{10}H_{14}O$ 150.22	825-25-2	 135^{25}	4-07-00-00320 1.0179^{18}	 1.5215^{18}
4882	Cyclopentanone, 2,5-dimethyl- 2,5-Dimethylcyclopentanone	$C_7H_{12}O$ 112.17	4041-09-2	 146.5	4-07-00-00048 0.882^{25}	 1.4310^{20}
4883	Cyclopentanone, 2-ethyl-5-methyl- 2-Ethyl-5-methylcyclopentanone	$C_8H_{14}O$ 126.20	51686-60-3	 164.5	3-06-00-02552 0.900^{12}	 1.4360^{22}
4884	Cyclopentanone, 4-isopropyl-2-methyl- Cyclopentanone, 3-isopropyl-5-methyl	$C_9H_{16}O$ 140.23	90645-61-7	 191.5	2-07-00-00034 0.8862^{20}	 1.4413^{19}
4885	Cyclopentanone, 2-methyl-	$C_6H_{10}O$ 98.14	1120-72-5 -75	 139.5	4-07-00-00036 0.9139^{20}	H_2O 3; EtOH 4; eth 4; ace 4 1.4364^{20}
4886	Cyclopentanone, 3-methyl-, (±)- (±)-3-Methylcyclopentanone	$C_6H_{10}O$ 98.14	6195-92-2 -58.4	 144	4-07-00-00037 0.913^{22}	H_2O 3; EtOH 4; eth 4; ace 4 1.4329^{20}
4887	Cyclopentanone, 2-(1-methylethyl)- 2-Isopropylcyclopentanone	$C_8H_{14}O$ 126.20	14845-55-7	 175	4-07-00-00060 0.9105^{20}	 [illegible]^{15}
4888	Cyclopentanone, [illegible]methyl-[illegible]-methylbutyl) Cyclopentanone, 2-isopentyl-3-methyl	$C_{11}H_{20}O$ 168.28	52033-97-3	 $98\text{-}9^{8}$	3-07-00-00181 0.8938^{20}	 1.4537^{20}
4889	Cyclopentanone, 2-methyl-2-(1-methylethyl)- Cyclopentanone, 2-isopropyl-2-methyl-	$C_9H_{16}O$ 140.23	32116-65-7	 97.5^{45}	3-07-00-00130 0.9067^{16}	 1.4495^{16}
4890	Cyclopentanone, 2-methyl-5-(1-methylethyl)-	$C_9H_{16}O$ 140.23	6784-18-5	 182	3-07-00-00130 0.889^{20}	 1.4402^{20}
4891	Cyclopentanone, 4-methyl-2-(1-methylethyl)- Cyclopentanone, 2-isopropyl-4-methyl	$C_9H_{16}O$ 140.23	69770-98-5	 183	1-07-00-00026 0.8850^{20}	 1.4392^{20}
4892	Cyclopentanone, oxime	C_5H_9NO 99.13	1192-28-5 57.8	 196	4-07-00-00008	H_2O 4; bz 4
4893	Cyclopentanone, 2-(phenylmethyl)- 2-Benzylcyclopentanone	$C_{12}H_{14}O$ 174.24	2867-63-2	 152^{18}, $84^{0.3}$	4-07-00-01043 1.038^{20}	 1.534^{13}
4894	Cyclopentanone, 2-propyl- 2-Propylcyclopentanone	$C_8H_{14}O$ 126.20	1193-70-0 -68.3	 183	4-07-00-00060 0.9017^{20}	 1.4429^{20}

No.	Name Synonym	Mol. Form. Mol. Wt.	CAS RN mp/°C	Merck No. bp/°C	Beil. Ref. den/g cm^{-3}	Solubility n_D
4895	Cyclopentanone, 3-propyl- 3-Propylcyclopentanone	$C_8H_{14}O$ 126.20	82322-93-8	190.5	3-07-00-00099 0.9041^{20}	1.4456^{12}
4896	Cyclopentanone, 2,2,4-trimethyl- 2,2,4-Trimethylcyclopentanone	$C_8H_{14}O$ 126.20	28056-54-4 -40.6	158	4-07-00-00061 0.877^{25}	1.4300^{20}
4897	Cyclopentanone, 2,2,5-trimethyl- 2,2,5-Trimethylcyclopentanone	$C_8H_{14}O$ 126.20	4573-09-5	152	3-07-00-00107 0.8781^{20}	1.4306^{20}
4898	Cyclopentanone, 2,3,5-trimethyl- 2,3,5-Trimethylcyclopentanol	$C_8H_{14}O$ 126.20	90112-26-8	158	1-07-00-00021 0.8778^{19}	1.4316^{19}
4899	Cyclopentanone, 2,4,4-trimethyl- 2,4,4-Trimethylcyclopentanone	$C_8H_{14}O$ 126.20	4694-12-6 -25.6	160.5	4-07-00-00061 0.8785^{18}	1.433^{18}
4900	1H-Cyclopentapyrimidine-2,4(3H,5H)-dione, 3-cyclohexyl-6,7-dihydro- Lenacil	$C_{13}H_{18}N_2O_2$ 234.30	2164-08-1 290	5318	5-24-07-00375 1.32^{25}	py 4
4901	Cyclopentasiloxane, decamethyl- Decamethylcyclopentasiloxane	$C_{10}H_{30}O_5Si_5$ 370.77	541-02-6 -38	2841 210	4-04-00-04128 0.9593^{20}	H_2O 1 1.3982^{20}
4902	Cyclopentasiloxane, 2,4,6,8,10-pentamethyl-	$C_5H_{20}O_5Si_5$ 300.64	6166-86-5 -108	169	4-04-00-04099 0.9985^{20}	1.3912^{20}
4903	Cyclopentene	C_5H_8 68.12	142-29-0 -135.1	44.2	4-05-00-00209 0.7720^{20}	H_2O 1; EtOH 3; eth 3; bz 3 1.4225^{20}
4904	3-Cyclopentene-1-acetic acid	$C_7H_{10}O_2$ 126.16	767-03-3 19	$93\text{-}4^{2.5}$	1.047^{25}	1.4676^{20}
4905	1-Cyclopentene-1-acetonitrile	C_7H_9N 107.16	22734-04-9	124^{100}	3-09-00-00151 0.951^{25}	1.4670^{20}
4906	Cyclopentene, 1-butyl- 1-Butyl-1-cyclopentene	C_9H_{16} 124.23	2423-01-0 -88.4	157	4-05-00-00289 0.8035^{25}	1.4486^{20}
4907	1-Cyclopentenecarbonitrile 1-Cyanocyclopentene	C_6H_7N 93.13	3047-38-9			
4908	1-Cyclopentene-1-carboxaldehyde	C_6H_8O 96.13	6140-65-4 -32	146	4-07-00-00127 0.970^{21}	1.4872^{17}
4909	1-Cyclopentene-1-carboxylic acid	$C_6H_8O_2$ 112.13	1560-11-8 123.5	210	4-09-00-00109	H_2O 3; EtOH 3; eth 3; peth 3
4910	1-Cyclopentene-1-carboxylic acid, 2,3,3-trimethyl-	$C_9H_{14}O_2$ 154.21	5587-63-3 135	255.5	4-09-00-00150	eth 4; EtOH 4
4911	1-Cyclopentene-1-carboxylic acid, 2,3,3-trimethyl-, methyl ester	$C_{10}H_{16}O_2$ 168.24	1460-88-4	203.5	0-09-00-00058 0.9656^{20}	1.4571^{20}
4912	2-Cyclopentene-1-carboxylic acid, 1,2,3-trimethyl- Laurolenic acid	$C_9H_{14}O_2$ 154.21	6894-69-5 6.5	192^{100}	2-09-00-00040 1.0318^{25}	1.4766
4913	3-Cyclopentene-1-carboxylic acid, 2,2,3-trimethyl- α-Campholytic acid	$C_9H_{14}O_2$ 154.21	6709-22-4 40.5	241.5; 162^{45}	4-09-00-00150 1.0145^{18}	1.4712^{17}
4914	Cyclopentene, 3-chloro-	C_5H_7Cl 102.56	96-40-2	$25\text{-}31^{30}$	4-05-00-00212 1.0577^{15}	eth 4; EtOH 4; chl 4 1.4708^{26}
4915	1-Cyclopentene-1,2-dicarboxylic acid, diethyl ester Diethyl 1-cyclopentene-1,2-dicarboxylate	$C_{11}H_{16}O_4$ 212.25	70202-92-5	140^{16}	3-09-00-03938 1.0805^{20}	1.4652^{20}
4916	1-Cyclopentene-1,3-dicarboxylic acid, diethyl ester Cyclopentene-1,3-dicarboxylic acid, diethyl ester	$C_{11}H_{16}O_4$ 212.25	30689-41-9	168^{21}	3-09-00-03939 1.1121^{24}	1.4564
4917	Cyclopentene, 1,2-dichloro-3,3,4,4,5,5-hexafluoro-	$C_5Cl_2F_6$ 244.95	706-79-6 -105.8	90.7	4-05-00-00213 1.6546^{20}	1.3676^{20}
4918	Cyclopentene, 1,2-dimethyl-	C_7H_{12} 96.17	765-47-9 -90.4	105.8	4-05-00-00254 0.7928^{25}	1.4448^{20}
4919	Cyclopentene, 1,3-dimethyl- 1,3-Dimethylcyclopentene	C_7H_{12} 96.17	62184-82-1	92	4-05-00-00255 0.766^{20}	1.428^{20}
4920	Cyclopentene, 1,4-dimethyl- 1,4-Dimethylcyclopentene	C_7H_{12} 96.17	19550-48-2	93.2	4-05-00-00255 0.779^{20}	1.4283^{20}
4921	Cyclopentene, 1,5-dimethyl-	C_7H_{12} 96.17	16491-15-9 -118	99	4-05-00-00254 0.780^{20}	1.4331^{20}
4922	Cyclopentene, 4,4-dimethyl-	C_7H_{12} 96.17	19037-72-0	88	4-05-00-00254 0.766^{25}	1.423^{20}
4923	Cyclopentene, 1,5-dimethyl-4-(1-methylethyl)- Cyclopentene, 2,3-dimethyl-4-isopropyl	$C_{10}H_{18}$ 138.25	6912-05-6	163	1-05-00-00046 0.8085^{22}	1.4503^{22}
4924	Cyclopentene, 1-ethyl-	C_7H_{12} 96.17	2146-38-5 -118.5	106.3	4-05-00-00252 0.7936^{25}	1.4412^{20}
4925	Cyclopentene, 3-ethyl-	C_7H_{12} 96.17	694-35-9	97.8	4-05-00-00252 0.7784^{25}	1.4315^{20}
4926	Cyclopentene, 4-ethyl- 4-Ethylcyclopentene	C_7H_{12} 96.17	3742-38-9	98.3	4-05-00-00253 0.778^{25}	1.431^{20}
4927	Cyclopentene, 4-isopropyl-1-methyl- 4-Isopropyl-1-methylcyclopentene	C_9H_{16} 124.23	90769-70-3	142.5	1-05-00-00041 0.7945^{21}	1.4403^{21}
4928	Cyclopentene, 1-methyl-	C_6H_{10} 82.15	693-89-0 -126.5	75.5	4-05-00-00239 0.7748^{25}	1.4322^{20}
4929	Cyclopentene, 3-methyl- 3-Methylcyclopentene	C_6H_{10} 82.15	1120-62-3	64.9	4-05-00-00240 0.7572^{25}	1.4216^{20}
4930	Cyclopentene, 4-methyl-	C_6H_{10} 82.15	1759-81-5 -160.8	65.7	4-05-00-00240 0.7634^{25}	1.4209^{20}

No.	Name Synonym	Mol. Form. Mol. Wt.	CAS RN mp/°C	Merck No. bp/°C	Beil. Ref. den/g cm^{-3}	Solubility n_D
4931	Cyclopentene, 3-methyl-1-(1-methylethyl)- Pulegene	C_9H_{16} 124.23	51115-02-7	 138.5	1-05-00-00040 0.791^{22}	 1.4380^{23}
4932	Cyclopentene, octachloro- Perchlorocyclopentene	C_5Cl_8 343.68	706-78-5 40	 283	4-05-00-00214 1.8200^{50}	H_2O 1; EtOH 4 1.5660^{50}
4933	1-Cyclopentene-1-pentanoic acid, δ-hydroxy-3,5-bis(1-methylpropyl)-β-oxo-	$C_{18}H_{30}O_4$ 310.43	53109-18-5 183		3-10-00-04194 1.269^{20}	eth 4; EtOH 4
4934	Cyclopentene, 1-pentyl-	$C_{10}H_{18}$ 138.25	4291-98-9 -83	 179	4-05-00-00306 0.8085^{25}	 1.4516^{20}
4935	Cyclopentene, 1-propyl- 1-Propyl-1-cyclopentene	C_8H_{14} 110.20	3074-61-1 -100.3	 131.2	4-05-00-00271 0.7978^{25}	 1.4452^{20}
4936	2-Cyclopentene-1-tridecanoic acid, (*S*)- Chaulmoogric acid	$C_{18}H_{32}O_2$ 280.45	29106-32-9 68.5	2037 $247\text{-}8^{20}$	4-09-00-00216	eth 4; chl 4
4937	Cyclopentene, 1,2,3-trimethyl-	C_8H_{14} 110.20	473-91-6	 121.6	3-05-00-00219 0.8039^{15}	 1.4464^{16}
4938	Cyclopentene, 1,5,5-trimethyl- 1,5,5-Trimethylcyclopentene	C_8H_{14} 110.20	62184-83-2	 109	3-05-00-00219 0.7824^{20}	 1.4324^{20}
4939	4-Cyclopentene-1,2,3-trione, 4,5-dihydroxy- Croconic acid	$C_5H_2O_5$ 142.07	488-86-8 150	 dec	4-08-00-03342	H_2O 4; EtOH 3
4940	2-Cyclopentene-1-undecanoic acid, (*R*)- Hydnocarpic acid	$C_{16}H_{28}O_2$ 252.40	459-67-6 60.5	4679	4-09-00-00215	EtOH 4; chl 4; peth 4
4941	2-Cyclopenten-1-one Cyclopenten-3-one	C_5H_6O 82.10	930-30-3	 136; 40^{12}	4-07-00-00122 0.989^{15}	eth 4; EtOH 4 1.4629^{15}
4942	2-Cyclopenten-1-one, 2,3-dihydroxy- Reductic acid	$C_5H_6O_3$ 114.10	80 72 8 212	8134	4-00-00-01714	H_2O 3, EtOH 3, eth 2; ace 2
4943	2-Cyclopenten-1-one, 2,4-dimethyl-	$C_7H_{10}O$ 110.16	23048-13-7	 165	4-07-00-00138 0.9423^{20}	 1.4655^{20}
4944	2-Cyclopenten-1-one, 2-methyl-	C_6H_8O 96.13	1120-73-6	 157	4-07-00-00127 0.9808^{16}	 1.4762^{15}
4945	2-Cyclopenten-1-one, 3-methyl- 3-Methyl-2-cyclopentenone	C_6H_8O 96.13	2758-18-1	 157.5	4-07-00-00126 0.9712^{20}	 1.4714^{20}
4946	2-Cyclopenten-1-one, 5-methyl- Cyclopenten-3-one, 4-methyl	C_6H_8O 96.13	14963-40-7	 140	3-07-00-00229 0.942^{18}	 1.4460^{18}
4947	2-Cyclopenten-1-one, 3-(1-methylethyl)-	$C_8H_{12}O$ 124.18	1619-28-9	 214	2-07-00-00061 0.9378^{20}	 1.4788^{20}
4948	2-Cyclopenten-1-one, 5-methyl-2-(1-methylethyl)-	$C_9H_{14}O$ 138.21	5587-79-1	 189	3-07-00-00291 0.9143^{20}	 1.4660^{20}
4949	2-Cyclopenten-1-one, 3-methyl-2-(2-pentenyl)-, (*Z*)- Jasmone	$C_{11}H_{16}O$ 164.25	488-10-8	5141 258; 134^{12}	4-07-00-00337 0.9437^{22}	H_2O 2; EtOH 3; eth 3; ctc 3 1.4979^{22}
4950	2-Cyclopenten-1-one, 3-methyl-2-pentyl-	$C_{11}H_{18}O$ 166.26	1128-08-1	 143^{22}; 116^{12}	4-07-00-00230 0.9165^{18}	 1.4767^{20}
4951	2-Cyclopenten-1-one, 3-phenyl-	$C_{11}H_{10}O$ 158.20	3810-26-2 -23	 234.2	4-07-00-01180 0.9711^{20}	EtOH 3; eth 2; ace 3; chl 3 1.5440^{20}
4952	Cycloprate Cyclopropanecarboxylic acid, hexadecyl ester	$C_{20}H_{38}O_2$ 310.52	54460-46-7	 $154^{0.05}$		
4953	Cyclopropanamine Cyclopropylamine	C_3H_7N 57.10	765-30-0 -35.4	 50.5	4-12-00-00003 0.8240^{20}	H_2O 5; EtOH 3; eth 3; chl 3 1.4210^{20}
4954	Cyclopropa[d]naphthalene, 1,1a,4,4a,5,6,7,8-octahydro-2,4a,8,8-tetramethyl-, [1a*S*-(1aα,4aβ,8a*R**)]- Thujopsene	$C_{15}H_{24}$ 204.36	470-40-6	9327	4-05-00-01189	
4955	Cyclopropane Trimethylene	C_3H_6 42.08	75-19-4 -127.4	2755 -32.8	4-05-00-00003 $*0.617^{25}$	H_2O 3; EtOH 4; eth 4; bz 3 1.3799^{-42}
4956	Cyclopropanecarbonitrile Cyclopropyl cyanide	C_4H_5N 67.09	5500-21-0	 135.1	4-09-00-00005 0.8946^{20}	eth 3; ctc 2; hx 3 1.4229^{20}
4957	Cyclopropanecarbonyl chloride, 2-phenyl-, *trans*- *trans*-2-Phenylcyclopropanecarbonyl chloride	$C_{10}H_9ClO$ 180.63	939-87-7	 $108\text{-}10^{2}$	3-09-00-02774 1.163^{25}	 1.5560^{20}
4958	Cyclopropanecarboxylic acid	$C_4H_6O_2$ 86.09	1759-53-1 18.5	 183	4-09-00-00003 1.0885^{20}	H_2O 3; EtOH 3; eth 3; ctc 2 1.4390^{20}
4959	Cyclopropanecarboxylic acid, 2,2-dimethyl-3-(2-methyl-1-propenyl)-, (1*R*-*cis*)-	$C_{10}H_{16}O_2$ [illegible]	26771-11-9 [illegible]	 [illegible]	4-09-00-00168	eth 4; EtOH 4; chl 4
4960	Cyclopropanecarboxylic acid, 2,2-dimethyl-3-(2-methyl-1-propenyl)-, 2-methyl-4-oxo-3-(2,4-pentadienyl)-2-cyclopenten-1-yl ester, [1R-[1α[S*(Z)],3β]]- Pyrethrin 1	$C_{21}H_{28}O_3$ 328.45	121-21-1	7978 $170^{0.1}$ dec	3-09-00-00215 1.5192^{18}	H_2O 1; EtOH 3; eth 3; ctc 3 1.5192^{18}
4961	Cyclopropanecarboxylic acid, ethyl ester Ethyl cyclopropylcarboxylate	$C_6H_{10}O_2$ 114.14	4606-07-9	 134	4-09-00-00003 0.9608^{15}	 1.4190^{20}
4962	Cyclopropanecarboxylic acid, 3-(3-methoxy-2-methyl-3-oxo-1-propenyl)-2,2-dimethyl-, 2-methyl-4-oxo-3-(2,4-pentadienyl)-2-cyclopenten-1-yl ester [1R-[1α[S*(Z)],3β(E)]]- Pyrethrin 2	$C_{22}H_{28}O_5$ 372.46	121-29-9	7978 $200^{0.1}$ dec	3-09-00-03988	H_2O 1; EtOH 3; eth 3; ctc 3 1.5258^{20}
4963	Cyclopropanecarboxylic acid, methyl ester Methyl cyclopropanecarboxylate	$C_5H_8O_2$ 100.12	2868-37-3	 114.9	4-09-00-00003 0.9848^{20}	ace 3; chl 3 1.4144^{19}

No.	Name Synonym	Mol. Form. Mol. Wt.	CAS RN mp/°C	Merck No. bp/°C	Beil. Ref. den/g cm^{-3}	Solubility n_D
4964	Cyclopropanecarboxylic acid, 2-methyl- 2-Methylcyclopropanecarboxylic acid	$C_5H_8O_2$ 100.12	29555-02-0	$191; 96.5^{14}$	2-09-00-00006 1.0267^{20}	1.4441^{20}
4965	Cyclopropanecarboxylic acid, 2,2-dimethyl-3-(2-methyl-1-propenyl)-, methyl ester, *trans*-(±)- Chrysanthemum monocarboxylic acid, methyl ester (*DL*, *trans*)	$C_{11}H_{18}O_2$ 182.26	15543-70-1	$108^{20}; 95^{13}$	4-09-00-00169 0.9274^{25}	1.4614^{25}
4966	Cyclopropane, (chloromethyl)-	C_4H_7Cl 90.55	5911-08-0 -90.9	88	4-05-00-00013 0.98^{25}	1.4350^{20}
4967	1,1-Cyclopropanedicarboxylic acid, diethyl ester	$C_9H_{14}O_4$ 186.21	1559-02-0	$216; 100^{12}$	4-09-00-02786 1.0535^{25}	EtOH 4; eth 4 1.4345^{18}
4968	1,2-Cyclopropanedicarboxylic acid, diethyl ester, *cis*- Cyclopropane-1,2-dicarboxylic acid, diethyl ester (*cis*)	$C_9H_{14}O_4$ 186.21	710-43-0		2-09-00-00513 1.062^{12}	eth 4; EtOH 4 1.4450^{20}
4969	1,2-Cyclopropanedicarboxylic acid, dimethyl ester Cyclopropane-1,2-dicarboxylic acid, dimethyl ester	$C_7H_{10}O_4$ 158.15	702-28-3	$219; 110^{3}$	1.1584^{16}	eth 4; EtOH 4 1.4472^{14}
4970	1,2-Cyclopropanedicarboxylic acid, *trans*-(±)- Cyclopropane-1,2-dicarboxylic acid (*trans*, *DL*)	$C_5H_6O_4$ 130.10	58616-95-8 175	210^{30}	4-09-00-02786 1.4600^{20}	H_2O 4; eth 4; EtOH 4
4971	Cyclopropane, 1,1-dimethyl- gem-Dimethylcyclopropane	C_5H_{10} 70.13	1630-94-0 -109	20.6	4-05-00-00024 0.6604^{20}	H_2O 1; EtOH 3; eth 4; sulf 4 1.3668^{20}
4972	Cyclopropane, 1,2-dimethyl-, *cis*- *cis*-1,2-Dimethylcyclopropane	C_5H_{10} 70.13	930-18-7 -140.9	37.0	4-05-00-00025 0.6889^{25}	H_2O 1; EtOH 3; eth 4; ctc 2 1.3829^{20}
4973	Cyclopropane, 1,2-dimethyl-, *trans*- *trans*-1,2-Dimethylcyclopropane	C_5H_{10} 70.13	2402-06-4 -149.6	28.2	4-05-00-00026 0.6648^{25}	eth 4; EtOH 4 1.3713^{20}
4974	Cyclopropane, ethyl- Ethylcyclopropane	C_5H_{10} 70.13	1191-96-4 -149.2	35.9	4-05-00-00023 0.6790^{25}	1.3786^{20}
4975	Cyclopropane, 1-ethyl-1-methyl- Cyclopropane, 1-methyl-1-ethyl	C_6H_{12} 84.16	53778-43-1 -130.2	56.8	4-05-00-00091 0.6968^{25}	1.3887^{20}
4976	Cyclopropane, 1-ethyl-2-methyl-, *cis*- *cis*-1-Methyl-2-ethylcyclopropane	C_6H_{12} 84.16	19781-68-1	58.7	3-05-00-00063 0.6935^{20}	1.3846^{20}
4977	Cyclopropane, 1-ethyl-2-methyl-, *trans*- *trans*-1-Methyl-2-ethylcyclopropane	C_6H_{12} 84.16	19781-69-2	59	3-05-00-00063 0.7146^{20}	ctc 3 1.3953^{20}
4978	Cyclopropanemethanol, 2,2-dimethyl-3-(2-methyl-1-propenyl)-	$C_{10}H_{18}O$ 154.25	5617-92-5	$66\text{-}9^{0.07}$	4-06-00-00264 0.888^{25}	1.4757^{20}
4979	Cyclopropanemethanol, α-methyl-	$C_5H_{10}O$ 86.13	765-42-4 -32.1	123.5	4-06-00-00020 0.8805^{20}	1.4316^{20}
4980	Cyclopropane, methoxy- Cyclopropyl methyl ether	C_4H_8O 72.11	540-47-6 -119	2756 44.7	3-06-00-00003 0.8100^{20}	H_2O 4; bz 4; eth 4; EtOH 4 1.3802^{20}
4981	Cyclopropane, methyl- Methylcyclopropane	C_4H_8 56.11	594-11-6 -177.3	0.7	4-05-00-00013 0.6912^{-20}	eth 4; EtOH 4
4982	Cyclopropane, (1-methylethenyl)-	C_6H_{10} 82.15	4663-22-3	70	4-05-00-00243 0.7500^{20}	1.4252^{20}
4983	Cyclopropane, (1-methylethyl)-	C_6H_{12} 84.16	3638-35-5 -112.9	58.3	4-05-00-00090 0.6936^{25}	1.3865^{20}
4984	Cyclopropane, propyl- Propane, 1-cyclopropyl-	C_6H_{12} 84.16	2415-72-7	69.1	4-05-00-00090 0.7062^{25}	1.3930^{20}
4985	Cyclopropane, 1,1,2-trimethyl- 1,1,2-Trimethylcyclopropane	C_6H_{12} 84.16	4127-45-1 -138.2	54	4-05-00-00091 0.6897^{25}	1.3864^{20}
4986	Cyclopropane, 1,2,3-trimethyl-, (1α,2α,3α)-	C_6H_{12} 84.16	4806-58-0	66	3-05-00-00063 0.7130^{25}	1.3970^{20}
4987	Cyclopropane, 1,2,3-trimethyl-, (1α,2α,3β)-	C_6H_{12} 84.16	4806-59-1	59.7	3-05-00-00063 0.6929^{25}	1.3873^{20}
4988	Cyclopropanone	C_3H_4O 56.06	5009-27-8 stable only at low temp.			
4989	1H-Cycloprop[e]azulen-4-ol, decahydro-1,1,4,7-tetramethyl-, (1a*R*,4*R*,4a*S*,7*R*,7a*S*,7b*S*)- Ledol	$C_{15}H_{26}O$ 222.37	577-27-5 105	5313 292	0.9078^{100}	ace 4; eth 4; EtOH 4 1.4667^{110}
4990	Cyclopropene	C_3H_4 40.06	2781-85-3	-36 dec		
4991	Cyclotetrasiloxane, ethenylheptamethyl-	$C_9H_{24}O_4Si_4$ 308.63	3763-39-1	84^{20}	0.9505^{20}	ctc 3; CS_2 3 1.4034^{20}
4992	Cyclotetrasiloxane, heptamethyl- Heptamethylcyclotetrasiloxane	$C_7H_{22}O_4Si_4$ 282.59	15721-05-8 -27	165	4-04-00-04125 0.9583^{20}	H_2O I 1.3965^{20}
4993	Cyclotetrasiloxane, octamethyl- Octamethylcyclotetrasiloxane	$C_8H_{24}O_4Si_4$ 296.62	556-67-2 17.5	6668 175.8	4-04-00-04125 0.9561^{20}	H_2O I; ctc 3 1.3968^{20}
4994	Cyclotetrasiloxane, octaphenyl-	$C_{48}H_{40}O_4Si_4$ 793.19	546-56-5 200.5	330^{1}	4-16-00-01530	H_2O I; EtOH 2; bz 3; chl 3
4995	Cyclotetrasiloxane, 2,4,6,8-tetrabutyl-2,4,6,8-tetramethyl- Siloxane, butylmethyl(cyclic tetramer)	$C_{20}H_{48}O_4Si_4$ 464.94	14685-29-1	291	4-04-00-04173 0.9230^{20}	1.4300^{20}
4996	Cyclotetrasiloxane, 2,4,6,8-tetraethenyl-2,4,6,8-tetramethyl-	$C_{12}H_{24}O_4Si_4$ 344.66	2554-06-5 -43.5	$224; 111^{12}$	4-04-00-04184 0.9875^{20}	ctc 3; CS_2 3

No.	Name Synonym	Mol. Form. Mol. Wt.	CAS RN mp/°C	Merck No. bp/°C	Beil. Ref. den/g cm^{-3}	Solubility n_D
4997	Cyclotetrasiloxane, 2,4,6,8-tetraethyl-2,4,6,8-tetramethyl- Cyclotetrasiloxane, 2,4,6,8-tetramethyl-2,4,6,8-tetraethyl	$C_{12}H_{32}O_4Si_4$ 352.73	7623-01-0 -43.5	 245; 111^{12}	4-04-00-04154 0.9600^{20}	ctc 3; CS_2 3
4998	Cyclotetrasiloxane, 2,4,6,8-tetraethyl-2,4,6,8-tetraphenyl- Cyclotetrasiloxane, 1,3,5,7-tetraethyl-1,3,5,7-tetraphenyl	$C_{32}H_{40}O_4Si_4$ 601.01	18758-34-4 106	 $212^{0.1}$	3-16-00-01211 1.1000^{20}	 1.5430^{25}
4999	Cyclotetrasiloxane, 2,4,6,8-tetramethyl-	$C_4H_{16}O_4Si_4$ 240.51	2370-88-9 -65	 134.5	4-04-00-04099 0.9912^{20}	H_2O i 1.3870^{20}
5000	Cyclotetrasiloxane, 2,4,6,8-tetramethyl-2,4,6,8-tetraphenyl-	$C_{28}H_{32}O_4Si_4$ 544.90	77-63-4 99	 237^{115}	3-16-00-01211 1.1183^{20}	H_2O 1; ace 5; hp 5 1.5461^{20}
5001	Cyclotrisilazane, 2,2,4,4,6,6-hexamethyl- 1,1,3,3,5,5-Hexamethylcyclotrisilazane	$C_6H_{21}N_3Si_3$ 219.51	1009-93-4 -10	 188	4-04-00-04114 0.9196^{20}	 1.448^{20}
5002	Cyclotrisiloxane, hexamethyl- Dimethylsiloxane cyclic trimer	$C_6H_{18}O_3Si_3$ 222.46	541-05-9 64.5	 134	4-04-00-04123 1.1200^{20}	H_2O 1
5003	Cyclotrisiloxane, 2,4,6-triethyl-2,4,6-trimethyl- 2,4,6-Triethyl-2,4,6-trimethylcyclotrisiloxane	$C_9H_{24}O_3Si_3$ 264.54	15901-49-2 -3	 199	4-04-00-04154	ctc 3; CS_2 3
5004	Cyclotrisiloxane, 2,4,6-triethyl-2,4,6-triphenyl- 2,4,6-Triethyl-2,4,6-triphenylcyclotrisiloxane	$C_{24}H_{30}O_3Si_3$ 450.76	546-33-8 177.5	 $166^{0.025}$	3-16-00-01211 1.0952^{25}	H_2O 1 1.5402^{25}
5005	Cyclotrisiloxane, 2,4,6-trimethyl-2,4,6-triphenyl-	$C_{21}H_{24}O_3Si_3$ 408.68	546-45-2 100	 $190^{1.5}$	3-16-00-01211 1.1062^{20}	 1.5397^{20}
5006	1,4,8-Cycloundecatriene, 2,6,6,9-tetramethyl-, (*E,E,E*)- Humulene	$C_{15}H_{24}$ 204.36	6753-98-6	4672 123^{10}	4-05-00-01171 0.8905^{20}	 1.5038^{20}
5007	7,20-Cycloveatchane-1,12,15-triol, 21-ethyl-4-methyl-16-methylene-, (1α,12α,15β)- Luciculine	$C_{22}H_{35}NO_3$ 361.52	5008-52-6 149	6286 $165^{0.02}$	4-21-00-02584	EtOH 4
5008	Cyfluthrin Cyano(4-fluoro-3-phenoxyphenyl)methyl 3-(2,2-dichloroethenyl)-2,2-dimethylcyclopropanecarboxylate	$C_{22}H_{18}Cl_2FNO_3$ 434.29	68359-37-5 60	2764		
5009	Cygon Phosphorodithioic acid, *O,O*-dimethyl *S*-[2-(methylamino)-2-oxoethyl] ester	$C_5H_{12}NO_3PS_2$ 229.26	60-51-5 52	3209 $117^{0.1}$	 1.277^{65}	
5010	Cyhalothrin	$C_{23}H_{19}ClF_3NO_3$ 449.86	91465-08-6 49.2			
5011	Cyhexatin Stannane, tricyclohexylhydroxy-	$C_{18}H_{34}OSn$ 385.18	13121-70-5 196	2767		
5012	Cypermethrin Cyclopropanecarboxylic acid, 3-(2,2-dichloroethenyl)-2,2-dimethyl-, cyano(3-phenoxy-phenyl)methyl ester	$C_{22}H_{19}Cl_2NO_3$ 416.30	52315-07-8 70	2775	 1.25^{20}	
5013	Cyprazine 1,3,5-Triazine-2,4-diamine, 6-chloro-*N*-cyclopropyl-*N'*-(1-methylethyl)-	$C_9H_{14}ClN_5$ 227.70	22936-86-3 167			
5014	*L*-Cysteine Propanoic acid, 2-amino-3-mercapto-, (*R*)-	$C_3H_7NO_2S$ 121.16	52-90-4 240 dec	2787	4-04-00-03144	H_2O 4; ace 4; EtOH 4
5015	*L*-Cysteine, *N*-acetyl- Acetylcysteine	$C_5H_9NO_3S$ 163.20	616-91-1 109.5	82	4-04-00-03160	
5016	*L*-Cysteine, *S*-(2-amino-2-carboxyethyl)-, (*R*)- *L*-Lanthionine	$C_6H_{12}N_2O_4S$ 208.24	922-55-4 293-5 dec	5234	3-04-00-01593	
5017	*L*-Cysteine, *S*-(carboxymethyl)- Carbocysteine	$C_5H_9NO_4S$ 179.20	638-23-3 206	1809	4-04-00-03150	
5018	*L*-Cysteine, *S,S'*-methylenebis- Djenkolic acid	$C_7H_{14}N_2O_4S_2$ 254.33	498-59-9 300-50 dec	3393	3-04-00-01591	
5019	*L*-Cysteine, methyl ester, hydrochloride Methylcysteine hydrochloride	$C_4H_{10}ClNO_2S$ 171.65	18598-63-5 140.5	5668	3-04-00-01601	
5020	*L*-Cystine 3,3'-Dithiobis(2-aminopropanoic acid)	$C_6H_{12}N_2O_4S_2$ 240.30	56-89-3 260-1 dec	2788	4-04-00-03155 1.677^{25}	H_2O 2; EtOH 1; eth 1; bz 1
5021	Cytidine 4-Amino-1-β-*D*-ribofuranosyl-2(1H)-pyrimidinone	$C_9H_{13}N_3O_5$ 243.22	65-46-3 230-1 dec	2792	4-25-00-03667	H_2O 4; EtOH 2
5022	3'-Cytidylic acid Cytidine 3'-monophosphate	$C_9H_{14}N_3O_8P$ 323.20	[illegible] 233 dec	[illegible]	[illegible]	[illegible]
5023	1,3-Decadiene 1-Hexyl-1,3-butadiene	$C_{10}H_{18}$ 138.25	2051-25-4	 169	4-01-00-01056 0.752^{30}	bz 4
5024	1,9-Decadiene α,ω-Decadiene	$C_{10}H_{18}$ 138.25	1647-16-1	 167	4-01-00-01056 0.75^{25}	 1.4325^{20}
5025	2,4-Decadienoic acid, (*E,E*)- 2,4-Decadienoic acid (*trans*-2, *trans*-4)	$C_{10}H_{16}O_2$ 168.24	30361-33-2 49.5	 $121^{0.3}$	4-02-00-01730	 1.5058^{31}
5026	2,4-Decadienoic acid, methyl ester, (*E,E*)-	$C_{11}H_{18}O_2$ 182.26	7328-33-8	 $70^{0.2}$	4-02-00-01731 0.982^{22}	 1.4918^{22}
5027	2,4-Decadienoic acid, methyl ester, (*E,Z*)-	$C_{11}H_{18}O_2$ 182.26	4493-42-9	 $71^{0.15}$	4-02-00-01731 0.9128^{22}	 1.4874^{22}

No.	Name Synonym	Mol. Form. Mol. Wt.	CAS RN mp/°C	Merck No. bp/°C	Beil. Ref. den/g cm^{-3}	Solubility n_D
5028	2,4-Decadienoic acid, methyl ester, (*Z,E*)- 2,4-Decadienoic acid, methyl ester (*cis*-2, *trans*-4)	$C_{11}H_{18}O_2$ 182.26	108965-84-0	$76-80^{0.5}$	4-02-00-01731 0.9131^{23}	1.4876^{23}
5029	2,4-Decadienoic acid, methyl ester, (*Z,Z*)- 2,4-Decadienoic acid, methyl ester (*cis*-2, *cis*-4)	$C_{11}H_{18}O_2$ 182.26	108965-86-2	$78^{0.6}$	4-02-00-01731 0.9095^{23}	1.4830^{23}
5030	Decanal Capraldehyde	$C_{10}H_{20}O$ 156.27	112-31-2 -5	208.5	4-01-00-03366 0.830^{15}	H_2O 1; EtOH 3; eth 3; ace 3 1.4287^{20}
5031	Decanamide Capramide	$C_{10}H_{21}NO$ 171.28	2319-29-1 108		4-02-00-01050 0.822^{130}	ace 4; eth 4; EtOH 4 1.4621^{110}
5032	1-Decanamine	$C_{10}H_{23}N$ 157.30	2016-57-1 17	220.5	4-04-00-00783 0.7936^{20}	H_2O 2; EtOH 5; eth 5; ace 5 1.4369^{20}
5033	Decane *n*-Decane	$C_{10}H_{22}$ 142.28	124-18-5 -29.7	174.1	4-01-00-00464 0.7300^{20}	H_2O 1; EtOH 5; eth 3; ctc 2 1.4102^{20}
5034	Decane, 1-bromo-	$C_{10}H_{21}Br$ 221.18	112-29-8 -29.2	240.6	4-01-00-00470 1.0702^{20}	H_2O 1; eth 4; ctc 3; chl 4 1.4557^{20}
5035	Decane, 1-bromo-10-fluoro- 1-Bromo-10-fluorodecane	$C_{10}H_{20}BrF$ 239.17	334-61-2	$131-2^{11}$	4-01-00-00471 1.152^{20}	eth 4; EtOH 4 1.4512^{25}
5036	Decane, 1-chloro-	$C_{10}H_{21}Cl$ 176.73	1002-69-3 -31.3	223.4	4-01-00-00469 0.8705^{20}	H_2O 1; eth 4; ctc 3; chl 4 1.4379^{20}
5037	Decane, 1-chloro-10-fluoro- 1-Chloro-10-fluorodecane	$C_{10}H_{20}ClF$ 194.72	334-62-3	115^{9}	4-01-00-00469 0.957^{20}	eth 4; EtOH 4 1.4333^{25}
5038	Decane, 1,10-dibromo- Decamethylene dibromide	$C_{10}H_{20}Br_2$ 300.08	4101-68-2 28	161^{9}, 128^{4}	4-01-00-00471 1.335^{30}	H_2O 1; EtOH 2; eth 3 1.4927^{25}
5039	Decane, 5,6-dibromo- 5,6-Dibromodecane	$C_{10}H_{20}Br_2$ 300.08	77928-86-0	119^{9}	4-01-00-00472 1.3484^{20}	1.4912^{20}
5040	Decane, 1,10-dichloro- Decamethylene dichloride	$C_{10}H_{20}Cl_2$ 211.17	2162-98-3 15.6	$167-8^{28}$	4-01-00-00469 0.9945^{25}	1.4586^{25}
5041	Decanedinitrile	$C_{10}H_{16}N_2$ 164.25	1871-96-1	204^{16}	4-02-00-02089 0.9313^{20}	H_2O 1; chl 3 1.4474^{20}
5042	Decanedioic acid Sebacic acid	$C_{10}H_{18}O_4$ 202.25	111-20-6 130.8	8369 295^{100}, 232^{10}	4-02-00-02078 1.2705^{20}	H_2O 2; EtOH 3; eth 3; bz 1 1.422^{133}
5043	Decanedioic acid, bis(2-ethylhexyl) ester Bis(2-ethylhexyl) sebacate	$C_{26}H_{50}O_4$ 426.68	122-62-3 -48	1263 256^{5}	4-02-00-02083 0.912^{25}	ace 4; bz 4; EtOH 4 1.451^{25}
5044	Decanedioic acid, dibutyl ester Dibutyl sebacate	$C_{18}H_{34}O_4$ 314.47	109-43-3 -10	344.5	4-02-00-02081 0.9405^{15}	H_2O 1; eth 3; ctc 3 1.4433^{15}
5045	Decanedioic acid, diethyl ester	$C_{14}H_{26}O_4$ 258.36	110-40-7 5	305; 188^{19}	4-02-00-02080 0.9646^{20}	H_2O 2; EtOH 3; ace 3; bz 1 1.4306^{20}
5046	Decanedioic acid, dimethyl ester	$C_{12}H_{22}O_4$ 230.30	106-79-6 38	175^{20}, 144^{5}	4-02-00-02080 0.9882^{28}	H_2O 1; EtOH 3; eth 3; ace 3 1.4355^{28}
5047	1,10-Decanediol Decamethylene glycol	$C_{10}H_{22}O_2$ 174.28	112-47-0 74	2842 192^{20}	4-01-00-02613	H_2O 2; EtOH 4; eth 2; DMSO 3
5048	2,9-Decanediol	$C_{10}H_{22}O_2$ 174.28	14021-92-2 33	114^{5}	3-01-00-02232	1.4505^{21}
5049	Decanedioyl dichloride	$C_{10}H_{16}Cl_2O_2$ 239.14	111-19-3 -2.5	220^{75}, 165^{11}	4-02-00-02088 1.1212^{20}	1.4684^{18}
5050	Decane, docosafluoro- Perfluorodecane	$C_{10}F_{22}$ 538.07	307-45-9	144.2		
5051	Decane, 1-(ethenyloxy)- Decyl vinyl ether	$C_{12}H_{24}O$ 184.32	765-05-9 -41	101^{10}	4-01-00-02057 0.812^{20}	1.4346^{20}
5052	Decane, 1-fluoro- Decyl fluoride	$C_{10}H_{21}F$ 160.28	334-56-5 -35	186.2	4-01-00-00468 0.8194^{20}	eth 4 1.4085
5053	Decane, 1-iodo-	$C_{10}H_{21}I$ 268.18	2050-77-3 -16.3	132^{15}	4-01-00-00472 1.2546^{20}	H_2O 1; EtOH 3; eth 3; ctc 3 1.4858^{20}
5054	Decane, 2-methyl- 2-Methyldecane	$C_{11}H_{24}$ 156.31	6975-98-0 -48.9	189.3	4-01-00-00491 0.7368^{20}	1.4154^{20}
5055	Decane, 3-methyl- 3-Methyldecane	$C_{11}H_{24}$ 156.31	13151-34-3 -92.9	188.1	4-01-00-00491 0.7422^{20}	1.4177^{20}
5056	Decane, 4-methyl- 4-Methyldecane	$C_{11}H_{24}$ 156.31	2847-72-5 -77.5	187		1.4352^{20}
5057	Decanenitrile Caprinitrile	$C_{10}H_{19}N$ 153.27	1975-78-6 -17.9	243, 106^{10}	4-02-00-01051 0.8199^{20}	ace 4; eth 4; EtOH 4; chl 4 1.4296^{20}
5058	1-Decanethiol Decyl mercaptan	$C_{10}H_{22}S$ 174.35	143-10-2 -26	240.6	4-01-00-01821 0.8443^{20}	H_2O 1; EtOH 3; eth 3 1.4509^{20}
5059	Decanoic acid Capric acid	$C_{10}H_{20}O_2$ 172.27	334-48-5 31.9	1759 268.7	4-02-00-01041 0.8858^{40}	ace 4; bz 4; eth 4; EtOH 4 1.4288^{40}
5060	Decanoic acid, anhydride	$C_{20}H_{38}O_3$ 326.52	2082-76-0 24.7		4-02-00-01049 0.8865^{25}	eth 4; EtOH 4 1.400^{25}
5061	Decanoic acid, 2-bromo-	$C_{10}H_{19}BrO_2$ 251.16	2623-95-2 4	$140-1^{2}$	4-02-00-01054 1.1912^{24}	eth 4 1.4595^{24}
5062	Decanoic acid, decyl ester	$C_{20}H_{40}O_2$ 312.54	1654-86-0 9.7	219^{15}	4-02-00-01045 0.8586^{20}	eth 4 1.4423^{20}
5063	Decanoic acid, ethyl ester Ethyl caprate	$C_{12}H_{24}O_2$ 200.32	110-38-3 -20	3734 241.5	4-02-00-01044 0.8650^{20}	eth 4; EtOH 4; chl 4 1.4256^{20}
5064	Decanoic acid, 10-fluoro- Capric acid, 10-fluoro	$C_{10}H_{19}FO_2$ 190.26	334-59-8 49	$135-8^{10}$	4-02-00-01051	eth 4; EtOH 4; lig 4
5065	Decanoic acid, 2-hexyl- 2-Hexyldecanoic acid	$C_{16}H_{32}O_2$ 256.43	25354-97-6	4630 $140-50^{0.02}$	4-02-00-01189	1.4432^{24}

No.	Name Synonym	Mol. Form. Mol. Wt.	CAS RN mp/°C	Merck No. bp/°C	Beil. Ref. den/g cm^{-3}	Solubility n_D
5066	Decanoic acid, methyl ester	$C_{11}H_{22}O_2$ 186.29	110-42-9 -18	 224	4-02-00-01044 0.8730^{20}	H_2O 1; EtOH 4; eth 4; ctc 2 1.4259^{20}
5067	Decanoic acid, 1-methylethyl ester	$C_{13}H_{26}O_2$ 214.35	2311-59-3	 121^{10}	4-02-00-01045 0.8543^{20}	 1.4221^{25}
5068	Decanoic acid, 2-octyl-	$C_{18}H_{36}O_2$ 284.48	619-39-6 38.5	 215^{13}	4-02-00-01254 0.8447^{70}	eth 4; EtOH 4
5069	Decanoic acid, propyl ester	$C_{13}H_{26}O_2$ 214.35	30673-60-0	 128.5^{10}	4-02-00-01045 0.8623^{20}	 1.4280^{20}
5070	1-Decanol Capric alcohol	$C_{10}H_{22}O$ 158.28	112-30-1 6.9	2847 231.1	4-01-00-01815 0.8297^{20}	H_2O 1; EtOH 5; eth 5; ace 5 1.4372^{20}
5071	2-Decanol, (±)- 2-Decanol (*DL*)	$C_{10}H_{22}O$ 158.28	74742-10-2 -2.4	 211	4-01-00-01823 0.8250^{20}	EtOH 3; eth 5; ace 5; bz 3 1.4326^{25}
5072	4-Decanol 1-Propylheptyl alcohol	$C_{10}H_{22}O$ 158.28	2051-31-2 -11	 210.5	4-01-00-01824 0.8261^{20}	H_2O 1; EtOH 3; ctc 3 1.4320^{20}
5073	1-Decanol, 10-chloro-	$C_{10}H_{21}ClO$ 192.73	51309-10-5 12.5	 $185\text{-}9^{15}$	4-01-00-01821 0.9630^{25}	eth 4; EtOH 4 1.4578^{20}
5074	1-Decanol, 10-fluoro- 10-Fluoro-1-decanol	$C_{10}H_{21}FO$ 176.27	334-64-5 22	 $136\text{-}7^{15}$	4-01-00-01821 0.919^{20}	eth 4; EtOH 4 1.4322^{25}
5075	2-Decanone Methyl octyl ketone	$C_{10}H_{20}O$ 156.27	693-54-9 14	 210; 96^{12}	4-01-00-03367 0.8248^{20}	H_2O 1; EtOH 3; eth 3; ctc 2 1.4255^{20}
5076	3-Decanone Ethyl heptyl ketone	$C_{10}H_{20}O$ 156.27	928-80-3 2.5	 203	4-01-00-03368 0.8251^{20}	EtOH 3; eth 3; ctc 3 1.4252^{20}
5077	4-Decanone Hexyl propyl ketone	$C_{10}H_{20}O$ 156.27	624-16-8 -9	 206.5	4-01-00-03368 0.824^{20}	H_2O 1; EtOH 5; eth 5 1.4240^{21}
5078	Decanoyl chloride Caprinoyl chloride	$C_{10}H_{19}ClO$ 190.71	112-13-0 -34.5	 95	4-02-00-01050 0.919^{25}	eth 3; ctc 3 1.4410^{20}
5079	Decasiloxane, docosamethyl- Decasiloxane, dicosamethyl	$C_{22}H_{66}O_9Si_{10}$ 755.62	556-70-7	 $183.^{4}$	3-04-00-01881 0.925^{20}	bz 4; lig 4 1.3988^{20}
5080	2-Decenal	$C_{10}H_{18}O$ 154.25	3913-71-1	 230	4-01-00-03511 0.845^{17}	 1.4533^{17}
5081	3-Decenal	$C_{10}H_{18}O$ 154.25	58474-80-9	 $93\text{-}4^{14}$	4-01-00-03512 0.850^{15}	 1.4462^{15}
5082	1-Decene	$C_{10}H_{20}$ 140.27	872-05-9 -66.3	 170.5	3-01-00-00858 0.7408^{20}	H_2O 1; EtOH 5; eth 5 1.4215^{20}
5083	4-Decene	$C_{10}H_{20}$ 140.27	19689-18-0	 170.6	4-01-00-00902 0.7404^{20}	 1.4243^{20}
5084	5-Decene, (*E*)-	$C_{10}H_{20}$ 140.27	7433-56-9 -73	 171	4-01-00-00902 0.7401^{20}	H_2O 1; EtOH 5; eth 5; ctc 2 1.4243^{20}
5085	5-Decene, (*Z*)-	$C_{10}H_{20}$ 140.27	7433-78-5 -112	 171; 73^{20}	3-01-00-00859 0.7445^{20}	H_2O 1; EtOH 5; eth 5; ctc 2 1.4258^{20}
5086	1-Decene, 2-bromo- 2-Bromo-1-decene	$C_{10}H_{19}Br$ 219.16	3017-67-2	 $115\text{-}6^{22}$	3-01-00-00859 1.0844^{20}	 1.4629^{20}
5087	2-Decene, 1-bromo- 1-Bromo-2-decene	$C_{10}H_{19}Br$ 219.16	14304-30-4	 121^{17}	4-01-00-00902 1.074^{18}	lig 4 1.4716^{18}
5088	2-Decenoic acid Δ 2-Decenoic acid	$C_{10}H_{18}O_2$ 170.25	3913-85-7 12	 165^{15}	4-02-00-01606 0.9280^{18}	 1.4616^{20}
5089	3-Decenoic acid	$C_{10}H_{18}O_2$ 170.25	15469-77-9 18	 $154\text{-}63^{11}$	4-02-00-01606 0.914^{15}	 1.4510^{18}
5090	4-Decenoic acid Deconic acid telomer	$C_{10}H_{18}O_2$ 170.25	26303-90-2	 149^{13}	4-02-00-01607 0.9197^{20}	bz 4; eth 4 1.4497^{20}
5091	9-Decenoic acid Caproleic acid	$C_{10}H_{18}O_2$ 170.25	14436-32-9	 158^{21}; 142^{4}	4-02-00-01605 0.9238^{15}	eth 4; EtOH 4 1.4507^{15}
5092	9-Decen-1-ol Decylenic alcohol	$C_{10}H_{20}O$ 156.27	13019-22-2	 236	4-01-00-02184 0.876^{25}	 1.4480^{20}
5093	3-Decen-2-one Heptylidene acetone	$C_{10}H_{18}O$ 154.25	10519-33-2	 $102\text{-}3^{15.3}$	4-01-00-03512 0.8473^{20}	 1.4480^{20}
5094	1-Decen-3-yne	$C_{10}H_{16}$ 136.24	33622-26-3	 76^{20}	4-01-00-01105 0.7873^{20}	 1.4620^{20}
5095	1-Decen-4-yne	$C_{10}H_{16}$ 136.24	24948-66-1	 $73\text{-}4^{22}$	3-01-00-01049 0.7880^{20}	 1.445^{20}
5096	2-Decen-4-yne	$C_{10}H_{16}$ 136.24	116668-40-7	 55^{5}	3-01-00-01049 0.7850^{25}	 1.4609^{25}
5097	1-Decyne Octylacetylene	$C_{10}H_{18}$ 138.25	764-93-2 -44	 174	4-01-00-01054 0.7655^{20}	H_2O 1; EtOH 3; eth 3; os 3 1.4265^{20}
5098	3-Decyne	$C_{10}H_{18}$ [illegible]	2384-85-2	 [illegible]	4-01-00-01055 [illegible]	 [illegible]
5099	4-Decyne	$C_{10}H_{18}$ 138.25	2384-86-3	 74.5^{19}	3-01-00-01017 0.772^{17}	 1.436^{17}
5100	5-Decyne Dibutylacetylene	$C_{10}H_{18}$ 138.25	1942-46-7 -73	 177; 78.8^{25}	4-01-00-01055 0.7690^{20}	H_2O 1; EtOH 3; eth 3 1.4331^{20}
5101	4-Decyne, 3,3-dimethyl- 3,3-Dimethyl-4-decyne	$C_{12}H_{22}$ 166.31	70732-45-5	 86^{20}	3-01-00-01026 0.7731^{20}	 1.4399^{20}
5102	Deltamethrin Cyano(3-phenoxyphenyl)methyl-3-(2,2-dibromoethenyl)-2,2-dimethylcyclopropanecarboxylate	$C_{22}H_{19}Br_2NO_3$ 505.21	52918-63-5 99	2869		
5103	Demeton S methyl Phosphorothioic acid, *S*-[2-(ethylthio)ethyl] *O*,*O*-dimethyl ester	$C_6H_{15}O_3PS_2$ 230.29	919-86-8	 $89^{0.15}$; 118^{1}	 1.207^{20}	

No.	Name	Synonym	Mol. Form.	Mol. Wt.	CAS RN	mp/°C	Merck No.	bp/°C	Beil. Ref.	den/g cm^{-3}	Solubility	n_D
5104	Desmedipham	Carbamic acid, [3-[[(phenylamino)carbonyl]oxy]phenyl]-, ethyl ester	$C_{16}H_{16}N_2O_4$	300.31	13684-56-5	120						
5105	Dialifor	Phosphorodithioic acid, *S*-[2-chloro-1-(1,3-dihydro-1,3-dioxo-2H-isoindol-2-yl)ethyl]	$C_{14}H_{17}ClNO_4PS_2$	393.85	10311-84-9	68	2949					
5106	Diallate	Carbamothioic acid, bis(1-methylethyl)-, *S*-(2,3-dichloro-2-propenyl) ester	$C_{10}H_{17}Cl_2NOS$	270.22	2303-16-4		2950	150^{9}				
5107	8,8'-Diapo-ψ,ψ-carotenedioic acid	Crocetin (*trans*)	$C_{20}H_{24}O_4$	328.41	27876-94-4	286	2592		3-02-00-02018		H_2O 2; EtOH 2; eth 1; bz 1	
5108	6,6'-Diapo-ψ,ψ-carotenedioic acid, monomethyl ester, 9-*cis*-	Bixin	$C_{25}H_{30}O_4$	394.51	6983-79-5	198	1320		3-02-00-02021		H_2O 1; EtOH 3; eth 2; ace 3	
5109	Diarsine, tetraethyl-	Tetraethyldiarsenic	$C_8H_{20}As_2$	266.09	612-08-8			186	3-04-00-01832	1.1388^{24}	eth 4; EtOH 4	1.4709
5110	Diarsine, tetramethyl-	Cacodyl	$C_4H_{12}As_2$	209.98	471-35-2	-6	1602	165	4-04-00-03683	1.447^{15}	eth 4; EtOH 4	
5111	1,4-Diazabicyclo[2.2.2]octane	Triethylenediamine	$C_6H_{12}N_2$	112.17	280-57-9	159	9584		5-23-03-00487		chl 3	
5112	Diazene, bis(2-ethoxyphenyl)-	Azobenzene, 2,2'-diethoxy-	$C_{16}H_{18}N_2O_2$	270.33	613-43-4	131		240 dec	4-16-00-00138		H_2O 1; EtOH 3; HCl 3	
5113	Diazene, bis(4-ethoxyphenyl)-		$C_{16}H_{18}N_2O_2$	270.33	588-52-3	162		dec	4-16-00-00174		H_2O 1; EtOH 2; eth 3; bz 3	
5114	Diazene, bis(2-methylphenyl)-		$C_{14}H_{14}N_2$	210.28	584-90-7	55.5			2-16-00-00019	1.0215^{65}	H_2O 1; EtOH 3; eth 4; bz 4	1.6180^{65}
5115	Diazene, bis(3-methylphenyl)-, (*E*)-		$C_{14}H_{14}N_2$	210.28	51437-67-3	54			4-16-00-00070	1.0123^{66}	EtOH 4; eth 4; bz 4	1.6152^{66}
5116	Diazene, bis(2-methylphenyl)-, 1-oxide, (*Z*)-	Azoxybenzene, 2,2'-dimethyl (*trans*)	$C_{14}H_{14}N_2O$	226.28	51284-68-5	60			4-16-00-00064	1.0215^{65}	bz 4; eth 4; EtOH 4	1.6180^{65}
5117	Diazene, bis(2-methylphenyl)-, 1-oxide, (*E*)-	Azoxybenzene, 2,2'-dimethyl(*trans*)	$C_{14}H_{14}N_2O$	226.28	116723-89-8				2-16-00-00318	1.0215^{65}	H_2O 1; EtOH 3; eth 3; bz 3	1.6180^{65}
5118	Diazene, bis(3-methylphenyl)-, 1-oxide, (*Z*)-	Azoxybenzene, 3,3'-dimethyl (*trans*)	$C_{14}H_{14}N_2O$	226.28	71297-97-7	39			4-16-00-00071	1.0123^{66}	bz 4; eth 4; EtOH 4; lig 4	1.6152^{66}
5119	Diazene, bis(3-methylphenyl)-, 1-oxide, (*E*)-	Azoxybenzene, 3,3'-dimethyl(*trans*)	$C_{14}H_{14}N_2O$	226.28	116723-90-1	39			2-16-00-00319	1.0123^{66}	H_2O 1; EtOH 3; eth 3; bz 3	1.6152^{66}
5120	Diazene, (4-bromophenyl)phenyl-, 1-oxide	Azoxybenzene, 4-bromo-	$C_{12}H_9BrN_2O$	277.12	16109-68-5	93.5			3-16-00-00584	1.4138^{100}	H_2O 1; EtOH 3; bz 3; MeOH 3	
5121	Diazenecarbothioic acid, phenyl-, 2-phenylhydrazide	Dithizone	$C_{13}H_{12}N_4S$	256.33	60-10-6	165-9 dec	3383		4-16-00-00018		H_2O 1; EtOH 2; eth 2; chl 3	
5122	Diazenecarboxylic acid, phenyl-, 2-phenylhydrazide	Diphenylcarbazone	$C_{13}H_{12}N_4O$	240.26	538-62-5	157 dec	3322		4-16-00-00017		H_2O 1; EtOH 4; bz 4; chl 4	
5123	Diazene, dibutyl	Azobutane	$C_8H_{18}N_2$	142.24	2159-75-3			60^{18}				
5124	Diazenedicarboxamide	Azodicarbonamide	$C_2H_4N_4O_2$	116.08	123-77-3	212 dec	932		3-03-00-00234			
5125	Diazene, dimethyl-	Azomethane	$C_2H_6N_2$	58.08	503-28-6	-78		1.5	3-04-00-01747	0.743^{0}	ace 4; eth 4; EtOH 4	1.4199^{19}
5126	Diazene, dimethyl-, (*E*)-		$C_2H_6N_2$	58.08	4143-41-3	-78		1.5	4-04-00-03366	0.743^{0}	ctc 3; hp 3	
5127	Diazene, di-1-naphthalenyl-	1,1'-Azobisnaphthalene	$C_{20}H_{14}N_2$	282.34	487-10-5	190		sub	2-16-00-00026		H_2O 1; EtOH 2; ace 3; bz 4	
5128	Diazene, di-2-naphthalenyl-	2,2'-Azonaphthalene	$C_{20}H_{14}N_2$	282.34	582-08-1	208		sub	2-16-00-00026		H_2O 1; EtOH 2; eth 2; bz 3	
5129	Diazene, diphenyl-, (*E*)-	*trans*-Azobenzene	$C_{12}H_{10}N_2$	182.22	17082-12-1	68.5	930	293	4-16-00-00007	1.203^{20}	H_2O 2; EtOH 3; eth 3; bz 3	1.6266^{78}
5130	Diazene, diphenyl-, (*Z*)-	*cis*-Azobenzene	$C_{12}H_{10}N_2$	182.22	1080-16-6	71	930		4-16-00-00007		H_2O 2; EtOH 3; eth 3; bz 3	
5131	Diazene, diphenyl-, 1-oxide, (*Z*)-	*cis*-Azoxybenzene	$C_{12}H_{10}N_2O$	198.22	20972-43-4	87	937		4-16-00-00010	1.166^{20}		1.633^{20}
5132	Diazene, dipropyl	Azopropane	$C_6H_{14}N_2$	114.19	821-67-0			114				
5133	Diazene, (4-ethoxyphenyl)phenyl-		$C_{14}H_{14}N_2O$	226.28	7466-38-8	85		339.5	2-16-00-00040	1.0400^{100}	EtOH 4; eth 4; ace 3; bz 4	1.6419^{100}
5134	Diazene, ethylphenyl-		$C_8H_{10}N_2$	134.18	935-08-0			177	4-16-00-00007	0.9628^{22}	bz 4; eth 4; EtOH 4	1.5579
5135	Diazene, (2-methoxyphenyl)phenyl-		$C_{13}H_{12}N_2O$	212.25	6319-21-7	41		$195\text{-}7^{14}$	3-16-00-00081	1.0728^{100}	H_2O 1; EtOH 3; ace 3; os 4	
5136	Diazene, (3-methoxyphenyl)phenyl-		$C_{13}H_{12}N_2O$	212.25	34238-81-8	33.3		193^{15}	4-16-00-00155	1.1023^{53}	H_2O 1; EtOH 3; ace 3; os 4	
5137	Diazene, (4-methoxyphenyl)phenyl-		$C_{13}H_{12}N_2O$	212.25	2396-60-3	56		340	2-16-00-00040	1.12^{75}	H_2O 1; EtOH 3; eth 3; ace 3	
5138	Diazene, (3-methylphenyl)phenyl-	Azobenzene, 3-methyl	$C_{13}H_{12}N_2$	196.25	17478-66-9	18.5		175^{19}	3-16-00-00047	1.065^{20}		
5139	Diazene, (4-methylphenyl)phenyl-	*p*-Methylazobenzene	$C_{13}H_{12}N_2$	196.25	949-87-1	71.5		312	3-16-00-00049		bz 4; eth 4; chl 4	

No.	Name Synonym	Mol. Form. Mol. Wt.	CAS RN mp/°C	Merck No. bp/°C	Beil. Ref. den/g cm^{-3}	Solubility n_D
5140	1H-1,4-Diazepine, 1-[(4-chlorophenyl)phenylmethyl]hexahydro-4-methyl- Homochlorocyclizine	$C_{19}H_{23}ClN_2$ 314.86	848-53-3	4653	5-23-03-00248	
5141	1H-1,4-Diazepine, hexahydro-1-methyl-	$C_6H_{14}N_2$ 114.19	4318-37-0	 154	5-23-03-00241 0.9111^{20}	 1.4769^{20}
5142	Dibenz[a,h]anthracene 1,2:5,6-Dibenzanthracene	$C_{22}H_{14}$ 278.35	53-70-3 269.5	2989	4-05-00-02722	H_2O 1; EtOH 2; ace 3; bz 3
5143	Dibenz[a,j]acridine 7-Azadibenz[a,j]anthracene	$C_{21}H_{13}N$ 279.34	224-42-0 216		5-20-08-00656	
5144	5H-Dibenz[b,f]azepine-5-carboxamide Carbamazepine	$C_{15}H_{12}N_2O$ 236.27	298-46-4 190.2	1783	5-20-08-00247	
5145	Dibenz[c,e]oxepin-5,7-dione	$C_{14}H_8O_3$ 224.22	6050-13-1 217	 sub	5-17-11-00525	H_2O 1; eth 2
5146	10H-Dibenzo[a,d]cyclohepten-10-one, 5,11-dihydro-	$C_{15}H_{12}O$ 208.26	6374-70-5 30	 $203\text{-}4^{7}$	4-07-00-01675 1.1635^{20}	 1.6324^{20}
5147	5H-Dibenzo[a,d]cyclohepten-5-one, 10,11-dihydro-	$C_{15}H_{12}O$ 208.26	1210-35-1 30	 $203\text{-}4^{7}$	4-07-00-01674 1.1635^{20}	 1.6324^{20}
5148	13H-Dibenzo[a,g]fluorene	$C_{21}H_{14}$ 266.34	207-83-0 174.5	 $195\text{-}200^{0.5}$	4-05-00-02693	H_2O 1; EtOH 2; eth 2; bz 3
5149	6H-Dibenzo[a,g]quinolizine, 5,8,13,13a-tetrahydro-2,3,9,10-tetramethoxy-13-methyl-, (13*S*-*trans*)- Corydaline (*d*)	$C_{22}H_{27}NO_4$ 369.46	518-69-4 136	2543	5-21-06-00173	bz 4; eth 4; EtOH 4; chl 4
5150	6H-Dibenzo[a,g]quinolizin-2-ol, 5,8,13,13a-tetrahydro-3,9,10-trimethoxy-13-methyl-, (13*S*-*trans*)- Isocorybulbine	$C_{21}H_{25}NO_4$ 355.43	22672-74-8 187.5	5045	5-21-06-00173 1.045^{20}	H_2O 1; EtOH 3; chl 3; acid 3
5151	6H-Dibenzo[a,g]quinolizin-3-ol, 5,8,13,13a-tetrahydro-2,9,10-trimethoxy-13-methyl-, (13*S*-*trans*)- Corybulbine	$C_{21}H_{25}NO_4$ 355.43	518-77-4 237.5	2539	5-21-06-00173	H_2O 1, EtOH 2; eth 2; ace 3
5152	13H-Dibenzo[a,h]fluoren-13-one 1,2:6,7-Dibenzo-9-fluorenone	$C_{21}H_{12}O$ 280.33	4599-94-4 214	 sub 200	3-07-00-02897	
5153	11H-Dibenzo[b,e][1,4]dioxepin-11-one, 2,4,7,9-tetrachloro-3-hydroxy-8-methoxy-1,6-dimethyl- Diploicin	$C_{16}H_{10}Cl_4O_5$ 424.06	527-93-5 232	3345	5-19-06-00364	
5154	6H-Dibenzo[b,d]pyran-1-ol, 6,6,9-trimethyl-3-pentyl- Cannabinol	$C_{21}H_{26}O_2$ 310.44	521-35-7 77	1751 $185^{0.05}$	5-17-04-00567	H_2O 1; EtOH 3; eth 3; ace 3
5155	4H-Dibenzo[de,g]quinoline-10,11-diol, 5,6,6a,7-tetrahydro-6-methyl-, (*R*)- Apomorphine	$C_{17}H_{17}NO_2$ 267.33	58-00-4 195 dec	776	5-21-05-00333	H_2O 2; EtOH 3; eth 3; ace 3
5156	4H-Dibenzo[de,g]quinoline-2,9-diol, 5,6,6a,7-tetrahydro-1,10-dimethoxy-6-methyl-, (*S*)- Boldine	$C_{19}H_{21}NO_4$ 327.38	476-70-0 163	1328	5-21-06-00118	EtOH 4; chl 4
5157	4H-Dibenzo[de,g]quinoline, 5,6,6a,7-tetrahydro-1,2,9,10-tetramethoxy-6-methyl-, (*S*)- *d*-Glaucine	$C_{21}H_{25}NO_4$ 355.43	475-81-0 120	4332	5-21-06-00122	ace 4; EtOH 4; chl 4
5158	4H-Dibenzo[de,g]quinolin-1-ol, 5,6,6a,7-tetrahydro-2,11-dimethoxy-6-methyl-, (*S*)- Isothebaine	$C_{19}H_{21}NO_3$ 311.38	568-21-8 203.5	5116	5-21-05-00658	H_2O 1; EtOH 5; eth 2; chl 5
5159	4H-Dibenzo[de,g]quinolin-10-ol, 5,6,6a,7-tetrahydro-1,2-dimethoxy-, (*R*)- Tsuduranine	$C_{18}H_{19}NO_3$ 297.35	517-97-5 204	9710	5-21-05-00655	ace 4; eth 4; EtOH 4
5160	4H-Dibenzo[de,g]quinolin-11-ol, 5,6,6a,7-tetrahydro-10-methoxy-6-methyl-, (*R*)- Apocodeine	$C_{18}H_{19}NO_2$ 281.35	641-36-1 123.5	771	5-21-05-00334	EtOH 2; eth 3; ace 3; bz 3
5161	4H-Dibenzo[de,g]quinolin-11-ol, 5,6,6a,7-tetrahydro-1,2,10-trimethoxy-6-methyl-, (*S*)- Isocorydine	$C_{20}H_{23}NO_4$ 341.41	475-67-2 185	5046	5-21-06-00132	chl 4
5162	4H-Dibenzo[de,g]quinolin-1-ol, 5,6,6a,7- [illegible] Corydine	$C_{20}H_{23}NO_4$ 341.41	476-69-7 149	2545	5-21-06-00134	eth 4; EtOH 4; chl 4
5163	Dibenzofuran 2,2'-Biphenylene oxide	$C_{12}H_8O$ 168.19	132-64-9 86.5	 287	5-17-02-00234 1.0886^{99}	H_2O 2; EtOH 3; eth 4; ace 3 1.6079^{99}
5164	Dibenzofuran, 2-bromo- 2-Bromodibenzofuran	$C_{12}H_7BrO$ 247.09	86-76-0 110	 220^{40}	5-17-02-00238	EtOH 3; HOAc 2
5165	Dibenzofuran, 3-bromo- 3-Bromodibenzofuran	$C_{12}H_7BrO$ 247.09	26608-06-0 121	 220^{40}	4-17-00-00588	H_2O 1; EtOH 3; eth 3
5166	Dibenzofuran, 3-nitro- 3-Nitrodibenzofuran	$C_{12}H_7NO_3$ 213.19	5410-97-9 181.5	 $180\text{-}5^{3}$	5-17-02-00239	H_2O 1; EtOH 2; eth 2; HOAc 3
5167	Dibenzothiophene	$C_{12}H_8S$ 184.26	132-65-0 99.5	 332.5	5-17-02-00239	H_2O 3; EtOH 4; bz 4; chl 3
5168	Diborane(6), mu-(methylthio)- Diborane, methylthio	CH_8B_2S 73.76	91572-15-5 -101.5	 53	4-01-00-01287	ace 4; bz 4

No.	Name	Synonym	Mol. Form.	Mol. Wt.	CAS RN	mp/°C	Merck No.	bp/°C	Beil. Ref.	den/g cm^{-3}	Solubility	n_D
5169	Dicarbonic acid, diethyl ester	Pyrocarbonic acid diethyl ester	$C_8H_{10}O_5$	162.14	1609-47-8		8008	$93\text{-}4^{18}$	4-03-00-00018	1.120^{20}	ace 4; EtOH 4; lig 4	1.3960^{20}
5170	Dichlorvos	Phosphoric acid, 2,2-dichloroethenyl dimethyl ester	$C_4H_7Cl_2O_4P$	220.98	62-73-7		3069	140^{20}; 84^{1}		1.415^{25}		
5171	Diclofop-methyl	Propanoic acid, 2-[4-(2,4-dichlorophenoxy)phenoxy]-, methyl ester	$C_{16}H_{14}Cl_2O_4$	341.19	51338-27-3	40	3072	$176^{0.1}$				
5172	Dicrotophos	Phosphoric acid, 3-(dimethylamino)-1-methyl-3-oxo-1-propenyl dimethyl ester, (*E*)-	$C_8H_{16}NO_5P$	237.19	141-66-2		3077	400; $130^{0.1}$		1.216^{15}		
5173	Diethatyl-ethyl	Glycine, *N*-(chloroacetyl)-*N*-(2,6-diethylphenyl)-, ethyl ester	$C_{16}H_{22}ClNO_3$	311.81	38727-55-8							
5174	Difenoconazole	*cis,trans*-3-Chloro-4-[4-methyl-2-(1H-1,2,4-triazol-1-ylmethyl)-1,3-dioxolan-2-yl]phenyl 4-chlorophenyl ether	$C_{19}H_{17}Cl_2N_3O_3$	406.27	119446-68-3	76		$220^{0.03}$				
5175	Difenzoquat methyl sulfate	1H-Pyrazolium, 1,2-dimethyl-3,5-diphenyl-, methyl sulfate	$C_{17}H_{17}N_2 \cdot CH_3O_4S$	360.43	43222-48-6	157						
5176	Diflubenzuron	Benzamide, *N*-[[(4-chlorophenyl)amino]carbonyl]-2,6-difluoro-	$C_{14}H_9ClF_2N_2O_2$	310.69	35367-38-5	239	3128					
5177	1,4:5,8-Dimethanonaphthalene, 1,2,3,4,10,10-hexachloro-1,4,4a,5,8,8a-hexahydro-,	Aldrin	$C_{12}H_8Cl_6$	364.91	309-00-2	104	219		3-05-00-01385		H_2O 1; EtOH 3; eth 3; ace 3	
5178	2,7:3,6-Dimethanonaphth[2,3-b]oxirene, 3,4,5,6,9,9-hexachloro-1a,2,2a,3,6,6a,7,7a-octahydro-,(1aα,2β,2aα,3β,6β,6aα,7β,7aα)-	Dieldrin	$C_{12}H_8Cl_6O$	380.91	60-57-1	175.5	3093		5-17-02-00087	1.75^{25}	H_2O 1; EtOH 2; ace 3; bz 3	
5179	Dimethipin	2,3-Dihydro-5,6-dimethyl-1,4-dithiin, 1,1,4,4-tetraoxide	$C_6H_{10}O_4S_2$	210.28	55290-64-7	165						
5180	Dimethyl tetrachloroterephthalate	1,4-Benzenedicarboxylic acid, 2,3,5,6-tetrachloro-, dimethyl ester	$C_{10}H_6Cl_4O_4$	331.97	1861-32-1	155	2830					
5181	Dinitramine	1,3-Benzenediamine, *N*3,*N*3-diethyl-2,4-dinitro-6-(trifluoromethyl)-	$C_{11}H_{13}F_3N_4O_4$	322.24	29091-05-2	98						
5182	Dinocap	2-Butenoic acid, 2-(1-methylheptyl)-4,6-dinitrophenyl ester	$C_{18}H_{24}N_2O_6$	364.40	6119-92-2			$136^{0.01}$				
5183	Dinoseb	Phenol, 2-(1-methylpropyl)-4,6-dinitro-	$C_{10}H_{12}N_2O_5$	240.22	88-85-7	40	3282			1.265^{45}		
5184	2,3-Dioxabicyclo[2.2.2]oct-5-ene, 1-methyl-4-(1-methylethyl)-	Ascaridole	$C_{10}H_{16}O_2$	168.24	512-85-6	3.3	852	115^{15}; $39^{0.2}$	5-19-01-00319	1.0103^{20}	H_2O 1; EtOH 3; ace 3; bz 3	1.4769^{20}
5185	1,7-Dioxadispiro[4.0.4.2]dodeca-3,9-diene-2,8-dione, *trans*-	Anemonin	$C_{10}H_8O_4$	192.17	508-44-1	158	674		5-19-05-00101		chl 4	
5186	1,3-Dioxane	1,3-Dioxacyclohexane	$C_4H_8O_2$	88.11	505-22-6	-45		106.1	5-19-01-00011	1.0342^{20}	H_2O 5; EtOH 5; eth 5; ace 5	1.4165^{20}
5187	1,4-Dioxane	1,4-Dioxacyclohexane	$C_4H_8O_2$	88.11	123-91-1	11.8	3294	101.5	5-19-01-00016	1.0337^{20}	H_2O 5; EtOH 5; eth 5; ace 5	1.4224^{20}
5188	1,4-Dioxane, 2,3-dichloro-		$C_4H_6Cl_2O_2$	157.00	95-59-0	30		$80\text{-}2^{10}$	4-19-00-00030	1.468^{20}	H_2O 1; eth 4; ace 4; bz 4	1.4928^{20}
5189	1,3-Dioxane, 2,4-dimethyl-		$C_6H_{12}O_2$	116.16	766-20-1			116.5	5-19-01-00072	0.9392^{20}	H_2O 2; acid 4	1.4136^{20}
5190	1,3-Dioxane-4,6-dione, 2,2-dimethyl-	Meldrum's acid	$C_6H_8O_4$	144.13	2033-24-1	94	5696		5-19-05-00008			
5191	1,3-Dioxane-2-ethanol, 4-hydroxy-α,6-dimethyl-	Paraldol	$C_8H_{11}O_4$	171.17	19404-07-0	90		90^{15}	4-01-00-03987	1.116^{20}	H_2O 4; eth 4; EtOH 4	1.4610^{20}
5192	1,4-Dioxane, 2,2,3,3,5,5,6-heptachloro-	1,4-Dioxane, heptachloro	$C_4HCl_7O_2$	329.22	6629-96-5	55		$123\text{-}8^{8}$	4-19-00-00032		H_2O 1; eth 4; ace 4; bz 4	
5193	1,3-Dioxane, 4-methyl-		$C_5H_{10}O_2$	102.13	1120-97-4	-44.5		114	5-19-01-00054	0.9758^{20}	H_2O 2; os 4	1.4159^{20}
5194	1,3-Dioxane, 4-methyl-4-phenyl-	*m*-Dioxane, 4-methyl-4-phenyl-	$C_{11}H_{14}O_2$	178.23	1200-73-3	37.5		256	5-19-01-00481	1.0864^{20}	H_2O 1; chl 2; os 3	1.5240^{20}
5195	1,3-Dioxane, 2-phenyl-		$C_{10}H_{12}O_2$	164.20	772-01-0	41		253	5-19-01-00459	1.6053^{60}	EtOH 4; eth 4	
5196	1,3-Dioxane, 4-phenyl-		$C_{10}H_{12}O_2$	164.20	772-00-9			247	5-19-01-00464	1.1038^{20}	H_2O 1; os 3	1.5306^{18}
5197	1,3-Dioxan-5-ol, 2-methyl-		$C_5H_{10}O_3$	118.13	3774-03-6			176; 79^{18}	4-19-00-00624	1.0705^{17}	H_2O 4	1.4375^{17}
5198	1,3,2-Dioxathiane, 4-methyl-, 2-oxide	1,3-Butanediol, sulfite	$C_4H_8O_3S$	136.17	4426-51-1	5		185	4-01-00-02510	1.2352^{20}	EtOH 4; eth 4; ace 4; bz 4	1.4661^{20}

No.	Name Synonym	Mol. Form. Mol. Wt.	CAS RN mp/°C	Merck No. bp/°C	Beil. Ref. den/g cm^{-3}	Solubility n_D
5199	1,3,2-Dioxathiolane, 2,2-dioxide	$C_2H_4O_4S$ 124.12	1072-53-3 99	 sub	3-01-00-02110	ace 4; bz 4; eth 4; EtOH 4
5200	1,3,2-Dioxathiolane, 2-oxide	$C_2H_4O_3S$ 108.12	3741-38-6 -11	 173	4-01-00-02409 1.4402^{20}	H_2O 4; EtOH 4; eth 4; ace 4 1.4463^{20}
5201	Dioxathion Phosphorodithioic acid, *S,S'*-1,4-dioxane-2,3-diyl *O,O,O',O'*-tetraethyl ester	$C_{12}H_{26}O_6P_2S_4$ 456.55	78-34-2 -20	3296	 1.257^{26}	
5202	1,4-Dioxin	$C_4H_4O_2$ 84.07	290-67-5	 76	5-19-01-00275 1.115^{20}	eth 3; ace 3; bz 3 1.4350^{20}
5203	1,4-Dioxin, 2,3-dihydro-	$C_4H_6O_2$ 86.09	543-75-9	 94.1	5-19-01-00181 1.0836^{20}	ctc 3 1.4372^{20}
5204	1,3-Dioxolane 1,3-Dioxacyclopentane	$C_3H_6O_2$ 74.08	646-06-0 -95	 78	5-19-01-00006 1.060^{20}	H_2O 5; EtOH 3; eth 3; ace 3 1.3974^{20}
5205	1,3-Dioxolane, 2-(3-bromophenyl)-	$C_9H_9BrO_2$ 229.07	17789-14-9	 132.5	5-19-01-00448 1.514^{25}	 1.5627^{20}
5206	1,3-Dioxolane-4-carboxaldehyde, 2,2-dimethyl-	$C_6H_{10}O_3$ 130.14	5736-03-8	 74^{50}	5-19-04-00041	H_2O 4 1.4189^{25}
5207	1,3-Dioxolane-4-methanol, 2,2-dimethyl- Isopropylidene glycerol	$C_6H_{12}O_3$ 132.16	100-79-8	5101 82^{10}	5-19-02-00362 1.064^{20}	 1.4383^{20}
5208	1,3-Dioxolane-4-methanol, 2-methyl-	$C_5H_{10}O_3$ 118.13	3773-93-1	 187	5-19-02-00350 1.1243^{17}	EtOH 4 1.4413^{17}
5209	1,3-Dioxolane-4-methanol, 2-phenyl-	$C_{10}H_{12}O_3$ 180.20	1708-39-0	 143-4^{2}	5-19-02-00562 1.1916^{17}	 1.5389^{17}
5210	1,3-Dioxolane, 2-methyl-	$C_4H_8O_2$ 88.11	497-26-7	 81.5	5-19-01-00042 0.9811^{20}	H_2O 4; EtOH 5; eth 5 1.4035^{17}
5211	1,3-Dioxolan-2-one Ethylene glycol carbonate	$C_3H_4O_3$ 88.06	96-49-1 36.4	 248	5-19-04-00006 1.3214^{39}	H_2O 5; EtOH 5; eth 5; bz 5 1.4148^{50}
5212	1,3-Dioxolan-4-one, 2,5-bis(trichloromethyl)- Chloralide	$C_5H_2Cl_6O_3$ 322.79	554-21-2 116	 272.5	4-19-00-01571	eth 4; HOAc 4
5213	1,3-Dioxolan-2-one, 4-methyl-	$C_4H_6O_3$ 102.09	108-32-7 -48.8	 242	5-19-04-00021 1.2047^{20}	H_2O 4; EtOH 4; eth 4; ace 4 1.4189^{20}
5214	1,3-Dioxolo[4,5-g]isoquinoline, 5,6,7,8-tetrahydro-4-methoxy-6-methyl- Hydrocotarnine	$C_{12}H_{15}NO_3$ 221.26	550-10-7 56	4715	4-27-00-06395	H_2O 1; EtOH 3; eth 3; ace 3
5215	1,3-Dioxolo[4,5-g]isoquinoline, 5,6,7,8-tetrahydro-6-methyl- Hydrohydrastinine	$C_{11}H_{13}NO_2$ 191.23	494-55-3 66	4732 303	4-27-00-06306	ace 4; bz 4; eth 4; EtOH 4
5216	1,3-Dioxolo[4,5-g]isoquinolin-5-ol, 5,6,7,8-tetrahydro-4-methoxy-6-methyl- Cotarnine	$C_{12}H_{15}NO_4$ 237.26	82-54-2 132-3 dec	2551	4-27-00-06448	H_2O 2; EtOH 3; eth 3; bz 3
5217	1,3-Dioxolo[4,5-g]isoquinolin-5-ol, 5,6,7,8-tetrahydro-6-methyl- Hydrastinine	$C_{11}H_{13}NO_3$ 207.23	6592-85-4 116.5	4689	4-27-00-06396	H_2O 3; EtOH 4; eth 4; chl 4
5218	1,3-Dioxolo[4,5-h]quinolin-8(9H)-one, 6-methoxy-9-methyl- Casimiroin	$C_{12}H_{11}NO_4$ 233.22	477-89-4	1894	4-27-00-06618	chl 2
5219	Diphenamid Benzeneacetamide, *N,N* dimethyl-α-phenyl-	$C_{16}H_{17}NO$ 239.32	957-51-7 135	3305	 1.17$^{23.3}$	
5220	Diphosphoramide, octamethyl- Schradan	$C_8H_{24}N_4O_3P_2$ 286.25	152-16-9 17	8351 188-22$^{0.3}$	4-04-00-00288 1.1343^{25}	H_2O 4; EtOH 4; chl 4 1.462^{25}
5221	Diphosphoric acid, tetraethyl ester Tetraethyl pyrophosphate	$C_8H_{20}O_7P_2$ 290.19	107-49-3 170 dec	9138 155^{3}	4-01-00-01342 1.1847^{20}	H_2O 5; EtOH 5; eth 5; ace 5 1.4180^{20}
5222	Dipropetryn 1,3,5-Triazine-2,4-diamine, 6-(ethylthio)-*N,N'*-bis(1-methylethyl)-	$C_{11}H_{21}N_5S$ 255.39	4147-51-7 105	3349		
5223	5H,10H-Dipyrrolo[1,2-a:1',2'-d]pyrazine-5,10-dione Pyrocoll	$C_{10}H_6N_2O_2$ 186.17	484-73-1 268	 sub	5-24-08-00263	eth 4; EtOH 4; chl 4
5224	Diquat dibromide Dipyrido[1,2-a:2',1'-c]pyrazinediium, 6,7-dihydro-, dibromide	$C_{12}H_{12}N_22Br$ 344.05	85-00-7 337	3359	 1.24^{20}	
5225	Diselenide, diphenyl Diphenyl diselenide	$C_{12}H_{10}Se_2$ 312.13	1666-13-3 63.5	 202^{11}	4-06-00-01781 1.557^{80}	EtOH 3; eth 3; xyl 3; MeOH 3 1.743^{20}
5226	Disilane, 1,2-dichloro-1,1,2,2-tetramethyl-	$C_4H_{12}Cl_2Si_2$ 187.22	4342-61-4	 49-50^{18}	4-04-00-04280 1.010^{20}	 1.4548^{20}
5227	Disilane, 1,2-difluoro-1,1,2,2-tetramethyl- 1,2-Difluorotetramethyldisilane	$C_4H_{12}F_2Si_2$ 154.31	661-68-7	 94	4-04-00-04279 0.9120^{20}	 1.3837^{20}
5228	Disilane, hexamethyl- Permethyldisilane	$C_6H_{18}Si_2$ 146.38	1450-14-2 13.5	 113.5	4-04-00-04277 0.7247^{22}	H_2O 1; eth 3; ace 3; bz 3 1.4229^{20}
5229	Disiloxane, 1,3-bis(bromomethyl)-1,1,3,3-tetramethyl- 1,3-Bis(bromomethyl)tetramethyldisiloxane	$C_6H_{16}Br_2OSi_2$ 320.17	2351-13-5	 102-4^{15}	4-04-00-04027 1.3918^{25}	 1.4719^{25}
5230	Disiloxane, 1,3-bis(chloromethyl)-1,1,3,3-tetramethyl-	$C_6H_{16}Cl_2OSi_2$ 231.27	2362-10-9 -90	 204; 92^{21}	4-04-00-04026 1.0381^{20}	 1.4398^{20}
5231	Disiloxane, 1,3-bis(dichloromethyl)-1,1,3,3-tetramethyl- 1,3-Bis-(dichloromethyl)tetramethyl disiloxane	$C_6H_{14}Cl_4OSi_2$ 300.16	2943-70-6	 149-50^{50}	4-04-00-04031 1.2213^{20}	 1.4660^{20}
5232	Disiloxane, 1,3-dichloro-1,1,3,3-tetramethyl-	$C_4H_{12}Cl_2OSi_2$ 203.22	2401-73-2 -37.5	 138	4-04-00-04119 1.038^{20}	

No.	Name Synonym	Mol. Form. Mol. Wt.	CAS RN mp/°C	Merck No. bp/°C	Beil. Ref. den/g cm^{-3}	Solubility n_D
5233	Disiloxane, 1,3-diethenyl-1,1,3,3-tetramethyl-	$C_8H_{18}OSi_2$ 186.40	2627-95-4 -99.7	39	4-04-00-04080 0.811^{20}	1.4123^{20}
5234	Disiloxane, 1,3-dimethoxy-1,1,3,3-tetramethyl- Bis(methoxydimethylsilyl) oxide	$C_6H_{18}O_3Si_2$ 194.38	18187-24-1	139	4-04-00-04117 0.9048^{20}	1.3835^{20}
5235	Disiloxane, hexaethyl- 1,1,1,3,3,3-Hexaethyldisiloxane	$C_{12}H_{30}OSi_2$ 246.54	994-49-0	233; 129^{30}	4-04-00-04055 0.8589^{0}	1.4340^{20}
5236	Disiloxane, hexamethyl- Hexamethyldisiloxane	$C_6H_{18}OSi_2$ 162.38	107-46-0 -66	99	4-04-00-04018 0.7638^{20}	H_2O 1 1.3774^{20}
5237	Disiloxane, 1,1,3,3-tetramethyl-	$C_4H_{14}OSi_2$ 134.33	3277-26-7	72	4-04-00-03991 0.7572^{20}	1.3700^{20}
5238	Disiloxane, 1,1,3,3-tetramethyl-1,3-diphenyl-	$C_{16}H_{22}OSi_2$ 286.52	56-33-7	155-8^{13}	4-16-00-01475 0.9763^{20}	ctc 3 1.5176^{20}
5239	Disiloxane, 1,1,1-trimethyl-3,3,3-triphenyl- 1,1,1-Trimethyl-3,3,3-triphenyldisiloxane	$C_{21}H_{24}OSi_2$ 348.59	799-53-1 51	349	4-16-00-01486 1.0320^{25}	ctc 3; CS_2 3
5240	Dispiro[5.1.5.1]tetradecane Dispiro[5,1,5,1]tetradecane	$C_{14}H_{24}$ 192.34	184-97-4 10	131^{10}	4-05-00-00491 0.872^{26}	1.4735^{27}
5241	Distannoxane, hexabutyl- Bis(tributyltin) oxide	$C_{24}H_{54}OSn_2$ 596.11	56-35-9 45	180^{2}		
5242	Disulfide, bis(1,1-dimethylethyl)	$C_8H_{18}S_2$ 178.36	110-06-5 -5	88^{21}	4-01-00-01638 0.9226^{20}	1.4899^{20}
5243	Disulfide, bis(3-methylbutyl)	$C_{10}H_{22}S_2$ 206.42	2051-04-9	247	4-01-00-01689 0.9192^{20}	1.4864^{20}
5244	Disulfide, bis(1-methylethyl)	$C_6H_{14}S_2$ 150.31	4253-89-8 -69	177	4-01-00-01503 0.9435^{20}	1.4916^{20}
5245	Disulfide, bis(3-methylphenyl)	$C_{14}H_{14}S_2$ 246.40	20333-41-9 -21	177^{3}	4-06-00-02085	EtOH 3; eth 3; ace 3
5246	Disulfide, bis(4-methylphenyl)	$C_{14}H_{14}S_2$ 246.40	103-19-5 47.5	210-15^{20}	4-06-00-02206 1.114^{51}	H_2O 1; EtOH 3; eth 4; ace 3
5247	Disulfide, bis(3-nitrophenyl) Nitrophenide	$C_{12}H_8N_2O_4S_2$ 308.34	537-91-7 84	6539	4-06-00-01686	EtOH 2; eth 3; chl 2
5248	Disulfide, bis(phenylmethyl) Dibenzyl disulfide	$C_{14}H_{14}S_2$ 246.40	150-60-7 71.5	2995	4-06-00-02760	H_2O 2; EtOH 3; eth 3; bz 3
5249	Disulfide, bis(trifluoromethyl)	$C_2F_6S_2$ 202.14	372-64-5	34.6	4-03-00-00278	EtOH 4; peth 4
5250	Disulfide, diacetyl Acetyl disulfide	$C_4H_6O_2S_2$ 150.22	592-22-3 20	105^{18}	4-02-00-00564	eth 4; EtOH 4
5251	Disulfide, dibenzoyl	$C_{14}H_{10}O_2S_2$ 274.36	644-32-6 134.5	dec	3-09-00-01977	H_2O 1; EtOH 2; eth 2; CS_2 3
5252	Disulfide, dibutyl	$C_8H_{18}S_2$ 178.36	629-45-8	117^{20}; 85^{3}	4-01-00-01562 0.9371^{20}	H_2O 1; EtOH 5; eth 5 1.4923^{20}
5253	Disulfide, diethyl Diethyl disulfide	$C_4H_{10}S_2$ 122.26	110-81-6 -101.5	154.1	4-01-00-01397 0.9931^{20}	H_2O 2; EtOH 5; eth 5 1.5073^{20}
5254	Disulfide, dimethyl Dimethyl disulfide	$C_2H_6S_2$ 94.20	624-92-0 -85	109.8	4-01-00-01281 1.0625^{20}	H_2O 1; EtOH 5; eth 5 1.5289^{20}
5255	Disulfide, di-1-naphthalenyl	$C_{20}H_{14}S_2$ 318.46	39178-11-5 91		4-06-00-04245 1.144^{20}	H_2O 1; EtOH 2; eth 4; chl 2
5256	Disulfide, di-2-naphthalenyl	$C_{20}H_{14}S_2$ 318.46	5586-15-2 139.5		4-06-00-04320 1.144^{145}	H_2O 1; EtOH 4; eth 4; lig 1 1.4555^{20}
5257	Disulfide, dipentyl	$C_{10}H_{22}S_2$ 206.42	112-51-6	119^{7}	4-01-00-01654 0.9221^{20}	H_2O 1; eth 3 1.4889^{20}
5258	Disulfide, diphenyl	$C_{12}H_{10}S_2$ 218.34	882-33-7 62	310	4-06-00-01560 1.353^{20}	H_2O 1; EtOH 3; eth 3; bz 3
5259	Disulfide, dipropyl	$C_6H_{14}S_2$ 150.31	629-19-6 -85.6	193.5	4-01-00-01454 0.9599^{20}	1.4981^{20}
5260	Disulfoton Phosphorodithioic acid, *O,O*-diethyl *S*-[2-(ethylthio)ethyl] ester	$C_8H_{19}O_2PS_3$ 274.41	298-04-4 -25	3371 $108^{0.01}$; 128^{1}	1.144^{20}	
5261	1,2-Dithiane	$C_4H_8S_2$ 120.24	505-20-4 32.5	80^{14}; 60^{5}	5-19-01-00010	eth 3; bz 3; chl 3 1.5981^{25}
5262	1,3-Dithiane	$C_4H_8S_2$ 120.24	505-23-7 54	89^{14}	5-19-01-00013	bz 4; eth 4; chl 4 1.5981^{25}
5263	1,4-Dithiane	$C_4H_8S_2$ 120.24	505-29-3 112.3	199.5	5-19-01-00036	H_2O 2; EtOH 3; eth 3; ctc 3
5264	4H-1,3,5-Dithiazine, dihydro-5-methyl- 1,3,5-Dithiazine, 4,5-dihydro-5-methyl	$C_4H_9NS_2$ 135.25	6302-94-9 65	185	4-27-00-06285	eth 4; EtOH 4; HOAc 4
5265	4H-1,3,5-Dithiazine, dihydro-2,4,6-trimethyl-, (2α,4α,6α)-	$C_6H_{13}NS_2$ 163.31	638-17-5 46	dec	2-27-00-00525 1.0613^{50}	eth 4; EtOH 4
5266	1,3-Dithiolane 1,3-Dithiacyclopentane	$C_3H_6S_2$ 106.21	4829-04-3 -50	175	5-19-01-00008 1.259^{17}	EtOH 3; eth 3; xyl 3 1.5975^{15}
5267	Dithiopyr *S,S'*-Dimethyl-2-difluoromethyl-4-isobutyl-6-trifluoromethylpyridine-3,5-dicarbothioate	$C_{15}H_{16}F_5NO_2S_2$ 401.42	97886-45-8 65			
5268	Docosane *n*-Docosane	$C_{22}H_{46}$ 310.61	629-97-0 44.4	368.6	4-01-00-00572 0.7944^{20}	H_2O 1; EtOH 3; eth 4; chl 3 1.4455^{20}
5269	Docosane, 5-butyl- 5-*n*-Butyldocosane	$C_{26}H_{54}$ 366.71	55282-16-1 208	244^{10}	4-01-00-00584 0.8058^{20}	1.4503^{20}
5270	Docosane, 7-butyl- 7-Butyldocosane	$C_{26}H_{54}$ 366.71	55282-15-0 3.2	241^{10}	4-01-00-00584 0.8046^{20}	1.4499^{20}

No.	Name Synonym	Mol. Form. Mol. Wt.	CAS RN mp/°C	Merck No. bp/°C	Beil. Ref. den/g cm^{-3}	Solubility n_D
5271	Docosane, 9-butyl- 9-Butyldocosane	$C_{26}H_{54}$ 366.71	55282-14-9 1.3	 242.5^{10}	4-01-00-00584 0.8044^{20}	 1.4498^{20}
5272	Docosane, 11-butyl- 11-Butyldocosane	$C_{26}H_{54}$ 366.71	13475-76-8	 242.5^{10}	4-01-00-00584 0.8041^{20}	 1.4499^{20}
5273	Docosane, decyl- Docosane, 11-decyl	$C_{32}H_{66}$ 450.88	28261-97-4 1.0	 286^{10}	4-01-00-00597 0.8131^{20}	 1.4546^{20}
5274	Docosane, 7-hexyl- 7-Hexyldocosane	$C_{28}H_{58}$ 394.77	55373-86-9 19.3	 258^{10}	4-01-00-00590 0.8080^{20}	 1.4517^{20}
5275	Docosane, 9-octyl- 9-Octyldocosane	$C_{30}H_{62}$ 422.82	55319-83-0 8.6	 272.5^{10}	4-01-00-00594 0.8114^{20}	 1.4537^{20}
5276	Docosanoic acid Behenic acid	$C_{22}H_{44}O_2$ 340.59	112-85-6 81	1033 306^{60}	4-02-00-01290 0.8223^{90}	H_2O 2; EtOH 2; eth 2 1.4270^{100}
5277	Docosanoic acid, ethyl ester	$C_{24}H_{48}O_2$ 368.64	5908-87-2 50	 240^{10}	4-02-00-01292	eth 4; EtOH 4
5278	Docosanoic acid, methyl ester	$C_{23}H_{46}O_2$ 354.62	929-77-1 54		4-02-00-01291	eth 4; EtOH 4 1.4339^{60}
5279	1-Docosanol	$C_{22}H_{46}O$ 326.61	661-19-8 72.5	 $180^{0.22}$	4-01-00-01906	H_2O 2; EtOH 4; eth 2; chl 3
5280	1-Docosene	$C_{22}H_{44}$ 308.59	1599-67-3 38	 367	3-01-00-00883 0.794^{25}	
5281	13-Docosenoic acid	$C_{22}H_{42}O_2$ 338.57	1072-39-5 34.7	 242^{5}	4-02-00-01676 0.8532^{20}	 1.4444^{70}
5282	13-Docosenoic acid, (*E*)- Brassidic acid	$C_{22}H_{42}O_2$ 338.57	506-33-2 60	1359 202^{30}, 256^{10}	4-02-00-01677 0.8585^{57}	 1.4347^{100}
5283	13-Docosenoic acid, (*Z*)- Erucic acid	$C_{22}H_{42}O_2$ 338.57	112-86-7 33.5	3619 265^{15}	4-02-00-01676 0.860^{55}	H_2O 1; EtOH 3; eth 4; ctc 3 1.4758^{20}
5284	13-Docosen-1-ol, (*Z*)- Erucyl alcohol	$C_{22}H_{44}O$ 324.59	629-98-1 35	 240^{10}	4-01-00-02209 0.8416^{33}	bz 4; EtOH 4; peth 4
5285	1,11-Dodecadiene	$C_{12}H_{22}$ 166.31	5876-87-9	 208.5	4-01-00-01067 0.7702^{20}	 1.4400^{20}
5286	3,9-Dodecadiyne	$C_{12}H_{18}$ 162.27	61827-89-2			
5287	5,7-Dodecadiyne Dibutylbutadiyne	$C_{12}H_{18}$ 162.27	1120-29-2	 103^{8}	4-01-00-01130	
5288	Dodecanal Lauraldehyde	$C_{12}H_{24}O$ [illegible]	112-54-9 44.5	 [illegible], $100^{2.5}$	4-01-00-03380 [illegible]	H_2O 1; EtOH 2; eth 3 [illegible]
5289	Dodecanamide	$C_{12}H_{25}NO$ 199.34	1120-16-7 110	 199^{12}	4-02-00-01103	H_2O 1; EtOH 3; eth 2; ace 3 1.4287^{110}
5290	Dodecanamide, *N*,*N*-diethyl-	$C_{16}H_{33}NO$ 255.44	3352-87-2	 $166\text{-}7^{2}$	4-04-00-00357 0.847^{25}	chl 3 1.4545^{20}
5291	1-Dodecanamine	$C_{12}H_{27}N$ 185.35	124-22-1 28.3	 259	4-04-00-00794 0.8015^{20}	H_2O 2; EtOH 5; eth 5; bz 5 1.4421^{20}
5292	Dodecane	$C_{12}H_{26}$ 170.34	112-40-3 -9.6	 216.3	4-01-00-00498 0.7487^{20}	H_2O 1; EtOH 4; eth 4; ace 4 1.4216^{20}
5293	Dodecane, 1-bromo- Lauryl bromide	$C_{12}H_{25}Br$ 249.23	143-15-7 -9.5	5260 276	4-01-00-00502 1.0399^{20}	H_2O 1; EtOH 3; eth 3; ace 5 1.4583^{20}
5294	Dodecane, 1-bromo-12-fluoro- 1-Bromo-12-fluorododecane	$C_{12}H_{24}BrF$ 267.22	353-29-7	 $85.8^{0.15}$	4-01-00-00502	ace 4; eth EtOH 4 1.4524^{25}
5295	Dodecane, 1-chloro-	$C_{12}H_{25}Cl$ 204.78	112-52-7 -9.3	 260	4-01-00-00501 0.8687^{20}	H_2O 1; EtOH 4; ace 5; bz 3 1.4433^{20}
5296	Dodecane, 1,12-dibromo-	$C_{12}H_{24}Br_2$ 328.13	3344-70-5 41	 215^{15}	4-01-00-00503	H_2O 1; EtOH 4; eth 3; chl 4
5297	Dodecane, 1,1-dimethoxy- Lauraldehyde, dimethyl acetal	$C_{14}H_{30}O_2$ 230.39	14620-52-1	 $132\text{-}4^{5}$	4-01-00-03381	eth 4; EtOH 4 1.4310^{25}
5298	Dodecane, 1,12-dimethoxy- 1,12-Dodecanediol, dimethyl ether	$C_{14}H_{30}O_2$ 230.39	73120-52-2 11.5	 $155\text{-}6^{15}$	4-01-00-02627 0.8563^{22}	 1.436^{22}
5299	Dodecanedioic acid	$C_{12}H_{22}O_4$ 230.30	693-23-2 128	 222^{25}	4-02-00-02126 1.15^{25}	tfa 3
5300	Dodecanedioic acid, dimethyl ester	$C_{14}H_{26}O_4$ 258.36	1731-79-9 31.9	 170^{10}	4-02-00-02126	H_2O 1; AcOEt 3
5301	1,12-Dodecanediol 1,12-Dihydroxydodecane	$C_{12}H_{26}O_2$ 202.34	5675-51-4 81.3	 189^{12}	4-01-00-02627	tfa 3
5302	Dodecane, 1-iodo-	$C_{12}H_{25}I$ [illegible]	4292-19-7 0.3	 [illegible]	4-01-00-00503 [illegible]	H_2O 1; EtOH 3; eth 5; ace 5 [illegible]
5303	Dodecanenitrile Lauronitrile	$C_{12}H_{23}N$ 181.32	2437-25-4 4	 277; 198^{100}	4-02-00-01104 0.8240^{20}	H_2O 1; EtOH 5; eth 5; ace 5 1.4361^{20}
5304	1-Dodecanethiol	$C_{12}H_{26}S$ 202.40	112-55-0	 $142\text{-}5^{15}$	4-01-00-01851 0.8435^{20}	H_2O 1; EtOH 3; eth 3; chl 3 1.4589^{20}
5305	Dodecanoic acid Lauric acid	$C_{12}H_{24}O_2$ 200.32	143-07-7 43.2	5254 91.4	4-02-00-01082 0.8679^{50}	H_2O 1; EtOH 4; eth 4; ace 3 1.4183^{82}
5306	Dodecanoic acid, anhydride	$C_{24}H_{46}O_3$ 382.63	645-66-9 41.8		4-02-00-01101 0.8533^{70}	EtOH 4 1.4292^{70}
5307	Dodecanoic acid, 2-bromo-	$C_{12}H_{23}BrO_2$ 279.22	111-56-8 32	 $157\text{-}9^{2}$	4-02-00-01106 1.1474^{74}	bz 4; eth 4; EtOH 4; lig 4 1.4585^{24}
5308	Dodecanoic acid, 2-bromoethyl ester	$C_{14}H_{27}BrO_2$ 307.27	6309-50-8	 176^{12}; 145^{3}	4-02-00-01092 1.088^{25}	 1.4547^{25}
5309	Dodecanoic acid, 2-bromo-, methyl ester	$C_{13}H_{25}BrO_2$ 293.24	617-60-7	 $169\text{-}71^{12}$	4-02-00-01107 1.113^{15}	 1.4572^{25}

No.	Name Synonym	Mol. Form. Mol. Wt.	CAS RN mp/°C	Merck No. bp/°C	Beil. Ref. den/g cm^{-3}	Solubility n_D
5310	Dodecanoic acid, 2,3-dihydroxypropyl ester, (±)- Glycerol, 1-laurate (*DL*)	$C_{15}H_{30}O_4$ 274.40	40738-26-9 63	 186^{2}	4-02-00-01096 0.9248^{97}	EtOH 2; eth 4; ace 4; bz 3 1.4350^{86}
5311	Dodecanoic acid, 1,2-ethanediyl ester Ethylene glycol dilaurate	$C_{26}H_{50}O_4$ 426.68	624-04-4 56.6	4393 188^{20}	4-02-00-01094	eth 4; EtOH 4
5312	Dodecanoic acid, ethyl ester Ethyl laurate	$C_{14}H_{28}O_2$ 228.38	106-33-2 -10	3774 271; 154^{15}	4-02-00-01092 0.8618^{20}	H_2O 1; EtOH 4; eth 5; ctc 2 1.4311^{20}
5313	Dodecanoic acid, 2-(2-hydroxyethoxy)ethyl ester Diethylene glycol monolaurate	$C_{16}H_{32}O_4$ 288.43	141-20-8 17.5	3110 >270	 0.96^{25}	EtOH 5; eth 5; ace 5; bz 3
5314	Dodecanoic acid, 2-methyl-	$C_{13}H_{26}O_2$ 214.35	2874-74-0 22	 153^{1}	4-02-00-01121 0.890^{18}	
5315	Dodecanoic acid, methyl ester Methyl laurate	$C_{13}H_{26}O_2$ 214.35	111-82-0 5.2	 267	4-02-00-01090 0.8702^{20}	H_2O 1; EtOH 5; eth 5; ace 5 1.4319^{20}
5316	Dodecanoic acid, 1-methylethyl ester	$C_{15}H_{30}O_2$ 242.40	10233-13-3	 196^{60}; 117^{2}	4-02-00-01092 0.8536^{20}	eth 4; EtOH 4 1.4280^{25}
5317	Dodecanoic acid, phenyl ester Phenyl laurate	$C_{18}H_{28}O_2$ 276.42	4228-00-6 24.5	 210^{15}	4-06-00-00618 0.9354^{30}	ace 4; eth 4; EtOH 4
5318	Dodecanoic acid, phenylmethyl ester	$C_{19}H_{30}O_2$ 290.45	140-25-0 8.5	 $209\text{-}11^{12}$	4-06-00-02267 0.9429^{25}	bz 4; eth 4; EtOH 4; peth 4 1.4812^{24}
5319	Dodecanoic acid, 1,2,3-propanetriyl ester	$C_{39}H_{74}O_6$ 639.01	538-24-9		4-02-00-01098 0.8986^{55}	H_2O 1; EtOH 3; eth 3; ace 4 1.4404^{60}
5320	Dodecanoic acid, propyl ester Propyl laurate	$C_{15}H_{30}O_2$ 242.40	3681-78-5	 205^{60}; 124^{2}	4-02-00-01092 0.8600^{20}	 1.4335^{20}
5321	1-Dodecanol Lauryl alcohol	$C_{12}H_{26}O$ 186.34	112-53-8 24	3402 259	4-01-00-01844 0.8309^{24}	H_2O 1; EtOH 3; eth 3; bz 2
5322	2-Dodecanol	$C_{12}H_{26}O$ 186.34	10203-28-8 19	 252	3-01-00-01793 0.8286^{20}	 1.4400^{20}
5323	3-Dodecanol	$C_{12}H_{26}O$ 186.34	10203-30-2 25	 130^{15}	4-01-00-01854 0.8223^{32}	
5324	6-Dodecanol	$C_{12}H_{26}O$ 186.34	6836-38-0 30	 225; 119^{9}	3-01-00-01794 0.8201^{40}	eth 4; EtOH 4
5325	2-Dodecanone Decyl methyl ketone	$C_{12}H_{24}O$ 184.32	6175-49-1 21	 246.5	4-01-00-03382 0.8198^{20}	H_2O 1; EtOH 3; eth 3; ace 3 1.4330^{20}
5326	6-Dodecanone Amyl hexyl ketone	$C_{12}H_{24}O$ 184.32	6064-27-3 10	 112^{9}	4-01-00-03383	 1.4302^{20}
5327	1-Dodecanone, 1-phenyl-	$C_{18}H_{28}O$ 260.42	1674-38-0 47	 201^{9}; 181^{5}	4-07-00-00847 0.8794^{18}	H_2O 1; ace 3; ctc 2 1.4700^{18}
5328	Dodecanoyl chloride	$C_{12}H_{23}ClO$ 218.77	112-16-3 -17	 145^{18}	4-02-00-01103 0.9169^{25}	eth 4 1.4458^{20}
5329	1,3,6,10-Dodecatetraene, 3,7,11-trimethyl-, (*E,E*)- α-Farnesene	$C_{15}H_{24}$ 204.36	502-61-4	3883 $129\text{-}32^{12}$	3-01-00-01067 0.8410^{20}	H_2O 1; eth 3; ace 3; peth 5 1.4836^{20}
5330	2,6,10-Dodecatrienal, 3,7,11-trimethyl-	$C_{15}H_{24}O$ 220.35	19317-11-4	 $172\text{-}4^{14}$	4-01-00-03603 0.893^{18}	 1.4995
5331	1,6,10-Dodecatriene, 7,11-dimethyl-3-methylene-, (*E*)- β-Farnesene	$C_{15}H_{24}$ 204.36	18794-84-8	3884 $121\text{-}2^{9}$	4-01-00-01133 0.8363^{20}	ace 4; eth 4; chl 4 1.4899^{20}
5332	1,6,10-Dodecatrien-3-ol, 3,7,11-trimethyl-, [*S*-(*Z*)]-	$C_{15}H_{26}O$ 222.37	142-50-7	6388 276; $70^{0.1}$	4-01-00-02336 0.8778^{20}	EtOH 4; eth 3; ace 3; os 3 1.4898^{20}
5333	2,6,10-Dodecatrien-1-ol, 3,7,11-trimethyl-, (*E,E*)-	$C_{15}H_{26}O$ 222.37	106-28-5	 160^{10}	4-01-00-02335 0.8846^{20}	H_2O 1; EtOH 4; eth 3; ace 3 1.4877^{20}
5334	2,6,10-Dodecatrien-1-ol, 3,7,11-trimethyl-, (*Z,E*)-	$C_{15}H_{26}O$ 222.37	3790-71-4	 156^{12}; $120^{0.3}$	4-01-00-02335 0.8908^{20}	ace 4; eth 4; EtOH 4 1.4877^{20}
5335	1-Dodecene	$C_{12}H_{24}$ 168.32	112-41-4 -35.2	 213.8	4-01-00-00914 0.7584^{20}	H_2O 1; EtOH 3; eth 3; ace 3 1.4300^{20}
5336	2-Dodecenedioic acid, (*E*)- Traumatic acid	$C_{12}H_{20}O_4$ 228.29	6402-36-4 165.5	9493	4-02-00-02279	eth 4; EtOH 4; chl 4
5337	2-Dodecenoic acid	$C_{12}H_{22}O_2$ 198.31	4412-16-2 17.1	 155^{3}; $127^{0.15}$	4-02-00-01619 0.9265^{20}	 1.4629^{25}
5338	4-Dodecenoic acid Linderic acid	$C_{12}H_{22}O_2$ 198.31	505-92-0 1.3	 171^{13}	4-02-00-01619 0.9081^{15}	bz 4; eth 4; chl 4 1.4529^{20}
5339	5-Dodecenoic acid	$C_{12}H_{22}O_2$ 198.31	2761-84-4 1.3	 $170\text{-}2^{13}$	4-02-00-01619 0.9081^{20}	
5340	11-Dodecenoic acid	$C_{12}H_{22}O_2$ 198.31	65423-25-8 20	 171^{13}; 144^{3}	4-02-00-01618 0.9014^{20}	 1.4510^{20}
5341	11-Dodecenoic acid, methyl ester Methyl 11-dodecenoate	$C_{13}H_{24}O_2$ 212.33	29972-79-0	 $138\text{-}9^{13}$	4-02-00-01618 0.8789^{22}	 1.4414^{20}
5342	1-Dodecen-3-yne	$C_{12}H_{20}$ 164.29	74744-36-8	 78^{4}	4-01-00-01112 0.7858^{25}	 1.4510^{25}
5343	1-Dodecyne Decylacetylene	$C_{12}H_{22}$ 166.31	765-03-7 -19	 215	4-01-00-01066 0.7788^{20}	 1.4340^{20}
5344	2-Dodecyne	$C_{12}H_{22}$ 166.31	629-49-2 -9	 105^{15}	0-01-00-00261 0.7917^{15}	 1.4828^{20}
5345	3-Dodecyne	$C_{12}H_{22}$ 166.31	6790-27-8	 95^{12}	3-01-00-01025 0.7871^{20}	ace 4; eth 4 1.4442^{20}

No.	Name Synonym	Mol. Form. Mol. Wt.	CAS RN mp/°C	Merck No. bp/°C	Beil. Ref. den/g cm^{-3}	Solubility n_D
5346	6-Dodecyne	$C_{12}H_{22}$ 166.31	6975-99-1	 210; 100^{14}	4-01-00-01067 0.7871^{20}	ace 4; eth 4; EtOH 4 1.4442^{20}
5347	Dodine Guanidine, dodecyl-, monoacetate	$C_{13}H_{29}N_3{\cdot}C_2H_4O_2$ 287.45	2439-10-3 136	3406		
5348	Dotriacontane Bicetyl	$C_{32}H_{66}$ 450.88	544-85-4 69.7	 467	4-01-00-00595 0.8124^{20}	H_2O 1; EtOH 2; eth 3; bz 4 1.4550^{20}
5349	Eburnamenine-14-carboxylic acid, 14,15-dihydro-14-hydroxy-, methyl ester, (3α,14β,16α)- Vincamine	$C_{21}H_{26}N_2O_3$ 354.45	1617-90-9 231.5	9888	4-25-00-01252	
5350	Eicosane	$C_{20}H_{42}$ 282.55	112-95-8 36.8	 343	4-01-00-00563 0.7886^{20}	H_2O 1; eth 3; ace 4; bz 3 1.4425^{20}
5351	Eicosane, 1-cyclopentyl- Cyclopentane, eicosyl	$C_{25}H_{50}$ 350.67	22331-38-0 38	 413	4-05-00-00198 0.8276^{20}	ace 4; bz 4; eth 4; EtOH 4 1.4595^{20}
5352	Eicosanedioic acid 1,18-Octadecanedicarboxylic acid	$C_{20}H_{38}O_4$ 342.52	2424-92-2 125.5	 233-4^{2}	4-02-00-02185	eth 3
5353	Eicosanedioic acid, diethyl ester	$C_{24}H_{46}O_4$ 398.63	42235-39-2 54.5	 240^{12}; 230^{2}	3-02-00-01881	eth 4; EtOH 4
5354	Eicosane, 9-octyl- Eicosane, 9-ocytl	$C_{28}H_{58}$ 394.77	13475-77-9 0.5	 257^{10}; $199^{0.5}$	4-01-00-00590 0.8075^{20}	 1.4515^{20}
5355	Eicosanoic acid Arachidic acid	$C_{20}H_{40}O_2$ 312.54	506-30-9 75.4	791 328 dec; 204^{1}	4-02-00-01275 0.8240^{100}	H_2O 1; EtOH 2; eth 4; bz 3 1.425^{100}
5356	Eicosanoic acid, ethyl ester	$C_{22}H_{44}O_2$ 340.59	18281-05-5 50	 295^{100}; 186^{2}	3-02-00-01067	H_2O 1; EtOH 3; eth 3; bz 3
5357	Eicosanoic acid, methyl ester	$C_{21}H_{41}O_2$ 325.56	1120-28-1 54.5	 215^{10}	4-02-00-01276	bz 4; eth 4; EtOH 4; chl 4 1.4317^{60}
5358	1-Eicosanol Arachic alcohol	$C_{20}H_{42}O$ 298.55	629-96-9 66.1	 309; 222^{3}	4-01-00-01900 0.8405^{20}	H_2O 1; EtOH 2; ace 4; bz 3 1.4350^{20}
5359	2-Eicosanol 1-Methyl-1-nonadecanol	$C_{20}H_{42}O$ 298.55	4340-76-5 63.5	 357	4-01-00-01902 0.8378^{20}	ace 4; bz 4 1.4912^{80}
5360	5,8,11,14-Eicosatetraenoic acid 5,8,11,14-Eicosotetraenoic acid	$C_{20}H_{32}O_2$ 304.47	7771-44-0 -49.5	 $169^{0.15}$	4-02-00-01802 0.9219^{20}	H_2O 1; EtOH 3; eth 3; ace 3 1.4824^{20}
5361	5,8,11,14-Eicosatetraenoic acid, (all-*Z*)- Arachidonic acid	$C_{20}H_{32}O_2$ 304.47	506-32-1 -49.5	792 163^{1}	4-02-00-01802 0.9082^{20}	ace 4; eth 4; EtOH 4; peth 4 1.4824^{20}
5362	1-Eicosene	$C_{20}H_{40}$ 280.54	3452-07-1 28.5	 341; 151^{2}	4-01-00-00934 0.7882^{30}	H_2O 1; bz 3; peth 3 1.4440^{30}
5363	11-Eicosenoic acid Δ 11-Eicosenoic acid	$C_{20}H_{38}O_2$ 310.52	2462-94-4 24	 267^{15}	4-02-00-01673 0.8826^{25}	EtOH 4; MeOH 4
5364	9-Eicosenoic acid	$C_{20}H_{38}O_2$ 310.52	506-31-0 23	 220^{6}; $170^{0.1}$	4-02-00-01672 0.8882^{25}	 1.4597^{25}
5365	9-Eicosenoic acid, (*Z*)-	$C_{20}H_{38}O_2$ 310.52	29204-02-2 24.5	 220^{6}	4-02-00-01672 0.8882^{25}	
5366	1-Eicosyne	$C_{20}H_{38}$ 278.52	765-27-5 36	 340	4-01-00-01077 0.8073^{20}	bz 4; peth 4 1.4501^{20}
5367	Emetan-6'-ol, 7',10,11-trimethoxy- Cephaeline	$C_{28}H_{38}N_2O_4$ 466.62	483-17-0 115.5	1970	5-23-13-00611	ace 4; EtOH 4; MeOH 4; chl 4
5368	Emetan, 6',7',10,11-tetramethoxy Emetine	$C_{29}H_{40}N_2O_4$ 480.65	483-18-1 74	3517	4-23-00-03422	H_2O 1; EtOH 3; eth 3; ace 3
5369	Endosulfan 5,9-Methano-2,4,3-benzodioxathiepin, 6,7,8,9,10,10-hexachloro-1,5,5a,6,9,9a-hexahydro-, 3-oxide	$C_9H_6Cl_6O_3S$ 406.93	115-29-7 106	3529 $106^{0.7}$	 1.745^{20}	
5370	Endothall, disodium salt 7-Oxabicyclo[2.2.1]heptane-2,3-dicarboxylic acid, disodium salt	$C_8H_{10}Na_2O_5$ 232.14	145-73-3 144	3530	 1.431^{20}	
5371	4,7-Epoxyisobenzofuran-1,3-dione, hexahydro-3a,7a-dimethyl-, (3aα,4β,7β,7aα)- Cantharidin	$C_{10}H_{12}O_4$ 196.20	56-25-7 218	1755 sub 84	5-19-05-00051	H_2O 1; EtOH 2; eth 2; ace 2
5372	EPTC Carbamothioic acid, dipropyl-, *S*-ethyl ester	$C_9H_{19}NOS$ 189.32	759-94-4	3580 127^{20}	 0.9546^{30}	
5373	Ergoline-8-carboxamide, 9,10-didehydro-*N*-(2-hydroxy-1-methylethyl)-6-methyl-, [8α(*S*)]- Ergometrinine	$C_{19}H_{23}N_3O_2$ 325.41	479-00-5 195-7 dec	3599	4-25-00-00950	chl 4
5374	Ergoline-8-carboxamide, 9,10-didehydro-6-methyl-, (8β)- Lysergamide	$C_{16}H_{17}N_3O$ 267.33	478-94-4 137.5	5505	4-25-00-00937	EtOH 2; ace 2; os 2
5375	Ergoline-8-carboxylic acid, 9,10-didehydro-6-methyl-, (8α)- Isolysergic acid	$C_{16}H_{16}N_2O_2$ 268.32	478-95-5 218 dec	5065	4-25-00-00935	H_2O 2; EtOH 2; py 3
5376	Ergoline-8-carboxylic acid, 9,10-didehydro-6-methyl-, (8β)- Lysergic acid	$C_{16}H_{16}N_2O_2$ 268.32	82-58-6 240 dec	5506	4-25-00-00934	H_2O 2; EtOH 3; eth 2; bz 2
5377	Ergostane, (5α)-	$C_{28}H_{50}$ 386.71	511-20-6 85	3602	4-05-00-01234	ace 4; eth 4; chl 4
5378	Ergostane, (5β)- Coproergostane	$C_{28}H_{50}$ 386.71	511-21-7 64	2516	3-05-00-01143	eth 4; chl 4

No.	Name Synonym	Mol. Form. Mol. Wt.	CAS RN mp/°C	Merck No. bp/°C	Beil. Ref. den/g cm^{-3}	Solubility n_D
5379	Ergostan-3-ol, (3β,5α)- Ergostanol	$C_{28}H_{50}O$ 402.70	6538-02-9 144.5	3603	4-06-00-03602	H_2O 1; eth 3; chl 3
5380	Ergosta-5,7,9(11),22-tetraen-3-ol, (3β,22E)- Dehydroergosterol	$C_{28}H_{42}O$ 394.64	516-85-8 146	2861 $230^{0.5}$	4-06-00-04838	ace 4; bz 4; eth 4; EtOH 4
5381	Ergosta-5,7,22-trien-3-ol, (3β,10α,22E)- Pyrocalciferol	$C_{28}H_{44}O$ 396.66	128-27-8 94	8007	3-06-00-03098	H_2O 1; EtOH 3; chl 3; MeOH 3
5382	Ergosta-5,7,22-trien-3-ol, (3β,9β,10α,22E)- Lumisterol	$C_{28}H_{44}O$ 396.66	474-69-1 118	5471	4-06-00-04408	H_2O 1; EtOH 3; eth 4; ace 4
5383	Ergosta-5,7,22-trien-3-ol, (3β,22E)- Ergosterol	$C_{28}H_{44}O$ 396.66	57-87-4 170	3607 $250^{0.01}$	4-06-00-04407	H_2O 1; EtOH 2; eth 2; bz 3
5384	Ergost-7-en-3-ol, (3β,5α)- γ-Ergostenol	$C_{28}H_{48}O$ 400.69	516-78-9 146	3606	4-06-00-04029	eth 3
5385	Ergost-8(14)-en-3-ol, (3β,5α)- α-Ergostenol	$C_{28}H_{48}O$ 400.69	632-32-6 131	3604	4-06-00-04031	EtOH 2; eth 3; bz 3; chl 3
5386	Ergost-5-en-3-ol, (3β,24*R*)- Campesterol	$C_{28}H_{48}O$ 400.69	474-62-4 157.5	1735	3-06-00-02680	
5387	Ergotaman-3',6',18-trione, 12'-hydroxy-2',5'-bis(1-methylethyl)-, Ergocornine	$C_{31}H_{39}N_5O_5$ 561.68	564-36-3 182-4 dec	3591	4-25-00-00963	H_2O 1; EtOH 3; ace 3; bz 3
5388	Ergotaman-3',6',18-trione, 12'-hydroxy-2',5'-bis(1-methylethyl)-, Ergocorninine	$C_{31}H_{39}N_5O_5$ 561.68	564-37-4 228 dec	3592	4-25-00-00963	ace 4; bz 4; EtOH 4; chl 4
5389	Ergotaman-3',6',18-trione, 12'-hydroxy-2'-(1-methylethyl)-5'-(2-methylpropyl)- Ergocryptine	$C_{32}H_{41}N_5O_5$ 575.71	511-09-1 212-4 dec	3595	4-25-00-00964	H_2O 1; EtOH 3; chl 3
5390	Ergotaman-3',6',18-trione, 12'-hydroxy-2'-(1-methylethyl)-5'-(2-methylpropyl)-, (5'α,8α)- Ergocryptinine	$C_{33}H_{41}N_5O_5$ 587.72	511-10-4 245 dec	3596	4-25-00-00964	ace 4; chl 4
5391	Ergotaman-3',6',18-trione, 12'-hydroxy-2'-(1-methylethyl)-5'-(phenylmethyl)- Ergocristine	$C_{35}H_{39}N_5O_5$ 609.73	511-08-0 175 dec	3593	4-25-00-00966	H_2O 1; EtOH 3; ace 3; chl 3
5392	Ergotaman-3',6',18-trione, 12'-hydroxy-2'-(1-methylethyl)-5'-(phenylmethyl)- Ergocristinine	$C_{35}H_{39}N_5O_5$ 609.73	511-07-9 237-8 dec	3594	4-25-00-00967	H_2O 1; EtOH 2; ace 2; chl 2
5393	Ergotaman-3',6',18-trione, 12'-hydroxy-2'-methyl-5'-(2-methylpropyl)-, Ergosine	$C_{30}H_{37}N_5O_5$ 547.65	561-94-4 228 dec	3601	4-25-00-00961	ace 3; chl 3; MeOH 2
5394	Ergotaman-3',6',18-trione, 12'-hydroxy-2'-methyl-5'-(phenylmethyl)-, (5'α)- Ergotamine	$C_{33}H_{55}N_5O_5$ 601.83	113-15-5 213-4 dec	3609	4-25-00-00964	bz 4; eth 4; chl 4
5395	Ergotaman-3',6',18-trione, 12'-hydroxy-2'-methyl-5'-(phenylmethyl)-, Ergotaminine	$C_{33}H_{35}N_5O_5$ 581.67	639-81-6 252 dec	3610	4-25-00-00966	H_2O 1; EtOH 2; ace 2; bz 2
5396	Erythromycin Propiocine	$C_{37}H_{67}NO_{13}$ 733.94	114-07-8 191	3626	5-18-10-00398	ace 4; eth 4; EtOH 4; chl 4
5397	Estra-1,3,5,7,9-pentaen-17-one, 3-hydroxy- Equilenin	$C_{18}H_{18}O_2$ 266.34	517-09-9 258.5	3581 sub 170	4-08-00-01420	EtOH 2; ace 2; chl 2
5398	Estra-1,3,5(10),7-tetraen-17-one, 3-hydroxy- Equilin	$C_{18}H_{20}O_2$ 268.36	474-86-2 239	3582 sub 170	4-08-00-01366	H_2O 2; EtOH 3; ace 3; diox 3
5399	Estra-1,3,5(10)-triene-3,17-diol (17β)- Estradiol	$C_{18}H_{24}O_2$ 272.39	50-28-2 178.5	3653	4-06-00-06611	ace 4; EtOH 4; Diox 4
5400	Estra-1,3,5(10)-triene-3,17-diol, (17α)- α-Estradiol	$C_{18}H_{24}O_2$ 272.39	57-91-0 221.5	3654	4-06-00-06611	H_2O 1; EtOH 3; eth 2; ace 3
5401	Estra-1,3,5(10)-triene-3,17-diol, (8α,17β)- Isoestradiol	$C_{18}H_{24}O_2$ 272.39	517-04-4 181	5051	4-06-00-06611	EtOH 3; diox 3
5402	Estra-5,7,9-triene-3,17-diol, (3β,17β)- Hexahydroequilenin	$C_{18}H_{24}O_2$ 272.39	517-07-7 168.5	4607 sub 150	4-06-00-06618	
5403	Estra-1,3,5(10)-triene-3,17-diol (17β)-, 3-benzoate Estradiol benzoate	$C_{25}H_{28}O_3$ 376.50	50-50-0 196	3655	4-09-00-00406	
5404	Estra-1,3,5(10)-triene-3,16,17-triol, (16α,17β)- Estriol	$C_{18}H_{24}O_3$ 288.39	50-27-1 288 dec	3659	4-06-00-07550 1.27^{25}	EtOH 3; eth 2; bz 2; tfa 2
5405	Estra-1,3,5(10)-triene-3,16,17-triol, (16β,17β)- 16-Epiestriol	$C_{18}H_{24}O_3$ 288.39	547-81-9 290	3567	4-06-00-07549	
5406	Estra-1,3,5(10)-trien-17-one, 3-hydroxy- Estrone	$C_{18}H_{22}O_2$ 270.37	53-16-7 260.2	3660	3-08-00-01171 1.236^{25}	H_2O 1; EtOH 2; eth 2; ace 3
5407	Estra-1,3,5(10)-trien-17-one, 3-hydroxy-, (8α)- 8-Isoestrone	$C_{18}H_{22}O_2$ 270.37	517-06-6 247	5052	3-08-00-01170	eth 4; Diox 4
5408	Estr-4-en-3-one, 17-(1-oxo-3-phenylpropoxy)-, (17β)- Nandrolone phenpropionate	$C_{27}H_{34}O_3$ 406.57	62-90-8	6283	4-09-00-01758	
5409	Ethalfluralin Benzenamine, *N*-ethyl-*N*-(2-methyl-2-propenyl)-2,6-dinitro-4-(trifluoromethyl)-	$C_{13}H_{14}F_3N_3O_4$ 333.27	55283-68-6 57	3671 256 dec		
5410	Ethanamine Ethylamine	C_2H_7N 45.08	75-04-7 -80.5	3718 16.5	4-04-00-00307 •0.677^{25}	H_2O 5; EtOH 5; eth 5 1.3663^{20}
5411	Ethanamine, 2-[(3-butyl-1-isoquinolinyl)oxy]-*N*,*N*-dimethyl- Dimethisoquin	$C_{17}H_{24}N_2O$ 272.39	86-80-6 146	3207 $155\text{-}7^{3}$	5-21-03-00414	H_2O 3; EtOH 3 1.5486^{20}

No.	Name Synonym	Mol. Form. Mol. Wt.	CAS RN mp/°C	Merck No. bp/°C	Beil. Ref. den/g cm^{-3}	Solubility n_D
5412	Ethanamine, 2-chloro-*N*,*N*-bis(2-chloroethyl)- 2,2',2''-Trichlorotriethylamine	$C_6H_{12}Cl_3N$ 204.53	555-77-1 -4	9560 $143\text{-}4^{15}$	4-04-00-00447	bz 4; eth 4; EtOH 4
5413	Ethanamine, 2-chloro-*N*-(2-chloroethyl)-*N*-methyl- Mechlorethamine	$C_5H_{11}Cl_2N$ 156.05	51-75-2 -60	5655 87^{18}; 64^{5}		H_2O 2; ctc 5; DMF 5
5414	Ethanamine, 2,2-diethoxy-	$C_6H_{15}NO_2$ 133.19	645-36-3 -78	 163	4-04-00-01918 0.9159^{25}	H_2O 4; eth 4; EtOH 4; chl 4 1.4123^{25}
5415	Ethanamine, 2,2-diethoxy-*N*,*N*-diethyl-	$C_{10}H_{23}NO_2$ 189.30	3616-57-7	 194.5	4-04-00-01919 0.863^{25}	H_2O 3; EtOH 3; eth 3; ctc 2 1.4189^{20}
5416	Ethanamine, 2,2-diethoxy-*N*,*N*-dimethyl-	$C_8H_{19}NO_2$ 161.24	3616-56-6	 170.5	3-04-00-00870 0.885^{7}	H_2O 4; EtOH 3; eth 3; ace 3 1.4129^{20}
5417	Ethanamine, *N*,*N*-diethyl- Triethylamine	$C_6H_{15}N$ 101.19	121-44-8 -114.7	9582 89	4-04-00-00322 0.7275^{20}	H_2O 3; EtOH 3; eth 3; ace 4 1.4010^{20}
5418	Ethanamine, *N*,*N*-diethyl-, hydrochloride	$C_6H_{16}ClN$ 137.65	554-68-7 260 dec	 sub 245	4-04-00-00322 1.0689^{21}	H_2O 4; EtOH 4; chl 4
5419	Ethanamine, 2,2-dimethoxy-	$C_4H_{11}NO_2$ 105.14	22483-09-6	 $135\text{-}9^{95}$	4-04-00-01918 0.965^{25}	 1.4170^{20}
5420	Ethanamine, 2,2-dimethoxy-*N*-methyl-	$C_5H_{13}NO_2$ 119.16	122-07-6	 140	2-04-00-00759 0.928^{25}	 1.4115^{20}
5421	Ethanamine, *N*,*N*-dimethyl-	$C_4H_{11}N$ 73.14	598-56-1 -140	 36.5	4-04-00-00312 0.675^{20}	 1.3705^{25}
5422	Ethanamine, 2-(diphenylmethoxy)-*N*,*N*-dimethyl- Diphenhydramine	$C_{17}H_{21}NO$ 255.36	58-73-1	3308	4-06-00-04658	
5423	Ethanamine, 2-ethoxy-	$C_4H_{11}NO$ 89.14	110-76-9	 108	4-04-00-01411 0.8512^{20}	H_2O 5; EtOH 5; eth 5; ace 3 1.4101^{20}
5424	Ethanamine, 2-ethoxy-*N*,*N*-dimethyl-	$C_6H_{15}NO$ 117.19	26311-17-1	 121	4-04-00-01425 0.806^{20}	 1.406^{20}
5425	Ethanamine, 2-ethoxy-*N*-(2-ethoxyethyl)- Bis-(2-ethoxyethyl) amine	$C_8H_{19}NO_2$ 161.24	124-21-0 -50	 194	4-04-00-01516 0.883^{25}	 1.4213^{20}
5426	Ethanamine, 2-ethoxy-*N*-methyl- Methyl(2-ethoxyethyl)amine	$C_5H_{13}NO$ 103.16	38256-94-9	 115	3-04-00-00647 0.8363^{20}	ace 4; bz 4; eth 4; EtOH 4 1.4147^{20}
5427	Ethanamine, *N*-ethyl- Diethylamine	$C_4H_{11}N$ 73.14	109-89-7 -49.8	3100 55.5	4-04-00-00313 0.7056^{20}	H_2O 4; EtOH 5; eth 3; ctc 3 1.3864^{20}
5428	Ethanamine, *N*-ethyl-, hydrochloride	$C_4H_{12}ClN$ 109.60	660 68 4 228.5		4-04-00-00313 1.0477^{22}	H_2O 4; EtOH 4
5429	Ethanamine, *N*-ethyl-*N*-hydroxy-	$C_4H_{11}NO$ 89.14	3710-84-7 10	 133	4-04-00-03304 0.8669^{20}	 1.4195^{20}
5430	Ethanamine, *N*-ethyl-*N*-methyl-	$C_5H_{13}N$ 87.16	616-39-7 -196	 66	4-04-00-00321 0.703^{25}	H_2O 4; EtOH 4; eth 4 1.3879^{25}
5431	Ethanamine, *N*-ethyl-*N*-nitro-	$C_4H_{10}N_2O_2$ 118.14	7119-92-8	 206.5	4-04-00-03403 1.057^{15}	eth 4; EtOH 4
5432	Ethanamine, *N*-ethyl-*N*-nitroso- *N*-Nitrosodiethylamine	$C_4H_{10}N_2O$ 102.14	55-18-5	6557 176.9	4-04-00-03386 0.9422^{20}	H_2O 3; EtOH 3; eth 3; chl 2 1.4386^{20}
5433	Ethanamine, hydrochloride	C_2H_8ClN 81.54	557-66-4 109.5	3718	4-04-00-00307 1.2160^{20}	H_2O 4; EtOH 4
5434	Ethanamine, *N*-hydroxy- *N*-Ethylhydroxylamine	C_2H_7NO 61.08	624 81 7 59-60 dec		4-04-00-03304 0.9079^{20}	H_2O 4; EtOH 4 1.4152^{66}
5435	Ethanamine, *N*-methyl-, hydrochloride Ethyl methyl amine, hydrochloride	$C_3H_{10}ClN$ 95.57	624-60-2 128		2-04-00-00589 1.0874^{20}	H_2O 4; EtOH 4; eth 1; chl 3
5436	Ethanamine, 1,1,2,2,2-pentafluoro-*N*,*N*-bis(pentafluoroethyl)-	$C_6F_{15}N$ 371.05	359 70 6	 70.3	4-02-00-00471 1.736^{20}	 1.262^{25}
5437	Ethanamine, *N*-(phenylmethylene)-	$C_9H_{11}N$ 133.19	6852-54-6	 196	4-07-00-00518 0.937^{20}	H_2O 1; EtOH 3; eth 3 1.5378^{15}
5438	Ethanaminium, 2-(acetyloxy)-*N*,*N*,*N*-trimethyl-, bromide Acetyl choline bromide	$C_7H_{16}BrNO_2$ 226.11	66-23-9 146	80	4-04-00-01446	H_2O 4
5439	Ethanaminium, 2-(acetyloxy)-*N*,*N*,*N*-trimethyl-, chloride Acetyl choline chloride	$C_7H_{16}ClNO_2$ 181.66	60-31-1 150	81	4-04-00-01446	H_2O 3; EtOH 3; eth 1
5440	Ethanaminium, 2-[(aminocarbonyl)oxy]-*N*,*N*,*N*-trimethyl-, chloride Carbamoylcholine chloride	$C_6H_{15}ClN_2O_2$ 182.65	51-83-2 210 dec	1780	4-04-00-01455	H_2O 4; EtOH 2; eth 1; chl 1
5441	Ethanaminium, 2,2',2''-[1,2,3- [illegible] Gallamine triethiodide	$C_{30}H_{60}I_3N_3O_3$ 891.54	65-29-2 147.5	4249	4-06-00-07333	H_2O 4; EtOH 4; eth 2; ace 2
5442	Ethanaminium, 2-chloro-*N*,*N*,*N*-trimethyl-, chloride	$C_5H_{13}Cl_2N$ 158.07	999-81-5 239 dec	2103	4-04-00-00445	
5443	Ethanaminium, 2-[(cyclohexylhydroxyphenylacetyl)oxy]-*N*,*N*-diethyl-*N*-methyl-, bromide Oxyphenonium bromide	$C_{21}H_{34}BrNO_3$ 428.41	50-10-2 191.5	6928	3-10-00-00911	H_2O 4; EtOH 2
5444	Ethanaminium, *N*-[4-[[4-(diethylamino)phenyl]phenylmethylene]-2,5-cyclohexadien-1-ylidene]-*N*-ethyl-, sulfate (1 Brilliant green	$C_{27}H_{34}N_2O_4S$ 482.64	633-03-4	1367	4-13-00-02281	H_2O 4; EtOH 4
5445	Ethanaminium, *N*,*N*-diethyl-*N*-methyl-2-[(9H-xanthen-9-ylcarbonyl)oxy]-, bromide Methantheline bromide	$C_{21}H_{26}BrNO_3$ 420.35	53-46-3 174.5	5869	5-18-06-00590	H_2O 3; EtOH 3; eth 1; chl 3

PHYSICAL CONSTANTS OF ORGANIC COMPOUNDS (continued)

No.	Name Synonym	Mol. Form. Mol. Wt.	CAS RN mp/°C	Merck No. bp/°C	Beil. Ref. den/g cm^{-3}	Solubility n_D
5446	Ethanaminium, 2-hydroxy-*N,N,N*-trimethyl-, chloride Choline chloride	$C_5H_{14}ClNO$ 139.63	67-48-1 305-5 dec	2208	4-04-00-01443	H_2O 4; EtOH 4
5447	Ethanaminium, *N,N,N*-triethyl-, bromide Tetraethylammonium bromide	$C_8H_{20}BrN$ 210.16	71-91-0 285-7 dec	9133	4-04-00-00331 1.3970^{20}	H_2O 4; EtOH 4; chl 4; MeOH 4
5448	Ethane	C_2H_6 30.07	74-84-0 -182.8	3676 -88.6	4-01-00-00108 0.5446^{-89}	bz 4
5449	Ethane, 1,2-bis(2-chloroethoxy)-	$C_6H_{12}Cl_2O_2$ 187.07	112-26-5	232	4-01-00-02379 1.195^{20}	ctc 3 1.4592^{25}
5450	Ethane, 1,1-bis(ethylthio)-	$C_6H_{14}S_2$ 150.31	14252-42-7	184	4-01-00-03162 0.9706^{20}	eth 3; ace 3; chl 3 1.5025^{20}
5451	Ethane, 1,2-bis(ethylthio)-	$C_6H_{14}S_2$ 150.31	5395-75-5 180	217; 108^{27}	4-01-00-02452 0.9815^{20}	EtOH 3; eth 3 1.5118^{20}
5452	Ethane, 1,2-bis(methylthio)-	$C_4H_{10}S_2$ 122.26	6628-18-8	183; 79^{11}	4-01-00-02451 1.0371^{20}	H_2O 4; ace 4; eth 4; EtOH 4 1.5292^{20}
5453	Ethanebis(thioic) acid, *S,S*-diethyl ester	$C_6H_{10}O_2S_2$ 178.28	615-85-0 27	235	1-02-00-00244 1.0565^{21}	eth 4
5454	Ethane, bromo- Ethyl bromide	C_2H_5Br 108.97	74-96-4 -118.6	3730 38.5	4-01-00-00150 1.4604^{20}	H_2O 2; EtOH 5; eth 5; chl 5 1.4239^{20}
5455	Ethane, 1-bromo-1-chloro- 1-Bromo-1-chloroethane	C_2H_4BrCl 143.41	593-96-4	83	4-01-00-00155 1.6671^{10}	1.4660^{2-}
5456	Ethane, 1-bromo-2-chloro-	C_2H_4BrCl 143.41	107-04-0 -16.7	107	4-01-00-00155 1.7392^{20}	H_2O 2; EtOH 3; eth 3; chl 3 1.4908^{20}
5457	Ethane, 1-bromo-2-chloro-1,1,2-trifluoro- 1-Bromo-2-chloro-1,1,2-trifluoroethane	$C_2HBrClF_3$ 197.38	354-06-3	52.5	4-01-00-00156 1.8636^{25}	1.3738^{20}
5458	Ethane, 2-bromo-2-chloro-1,1,1-trifluoro- Halothane	$C_2HBrClF_3$ 197.38	151-67-7	4517 50.2; 20^{243}	1.871^{20}	H_2O 2; peth 3 1.3697^{0}
5459	Ethane, 2-bromo-1,1-diethoxy-	$C_6H_{13}BrO_2$ 197.07	2032-35-1	$66\text{-}7^{18}$	4-01-00-03151 1.310^{25}	EtOH 3; eth 3 1.4387^{20}
5460	Ethane, 2-bromo-1,1-dimethoxy-	$C_4H_9BrO_2$ 169.02	7252-83-7	149	3-01-00-02672 1.430^{20}	eth 3; ace 3; chl 3 1.4450^{20}
5461	Ethane, 1-bromo-2-ethoxy- 2-Bromoethyl ethyl ether	C_4H_9BrO 153.02	592-55-2	127.5	4-01-00-01386 1.3852^{0}	H_2O 2; EtOH 5; eth 5 1.4447^{20}
5462	Ethane, 1-bromo-2-fluoro-	C_2H_4BrF 126.96	762-49-2	71.5	4-01-00-00154 1.7044^{25}	eth 4; EtOH 4 1.4236^{20}
5463	Ethane, 1-bromo-2-iodo- 1-Bromo-2-iodoethane	C_2H_4BrI 234.86	590-16-9 28	163	4-01-00-00168 2.516^{29}	
5464	Ethane, (bromomethoxy)-	C_3H_7BrO 138.99	53588-92-4	110	3-01-00-02594 1.4402^{20}	eth 4 1.4515^{20}
5465	Ethane, 1-bromo-2-methoxy-	C_3H_7BrO 138.99	6482-24-2	110 dec	4-01-00-01386 1.4623^{20}	1.44753^{20}
5466	Ethane, 2-bromo-1,1,1-trifluoro-	$C_2H_2BrF_3$ 162.94	421-06-7 -93.9	26	4-01-00-00154 1.7881^{20}	1.3331^{20}
5467	Ethane, chloro- Ethyl chloride	C_2H_5Cl 64.51	75-00-3 -138.7	3740 12.3	4-01-00-00124 $*0.8902^{25}$	H_2O 2; EtOH 4; eth 5; chl 2 1.3676^{20}
5468	Ethane, 1-chloro-1-(2-chloroethoxy)- 1-Chloroethyl 2-chloroethyl ether	$C_4H_8Cl_2O$ 143.01	1462-34-6	56^{17}	3-01-00-02655 1.1867^{20}	eth 4; EtOH 4; chl 4 1.4473^{20}
5469	Ethane, 2-chloro-1,1-diethoxy-	$C_6H_{13}ClO_2$ 152.62	621-62-5	157.4	4-01-00-03134 1.0180^{20}	H_2O 2; EtOH 5; eth 5; ctc 2 1.4170^{20}
5470	Ethane, 1-chloro-1,1-difluoro 1-Chloro-1,1-difluoroethane	$C_2H_3ClF_2$ 100.50	75-68-3 -130.8	-9.7	1.107^{25}	H_2O 1; bz 3
5471	Ethane, 2-chloro-1,1-difluoro 1-Chloro-2,2-difluoroethane	$C_2H_3ClF_2$ 100.50	338-65-8	35.1		
5472	Ethane, 2-chloro-1,1-dimethoxy-	$C_4H_9ClO_2$ 124.57	97-97-2	127.5	4-01-00-03134 1.068^{20}	EtOH 2; eth 2; bz 2; ctc 2 1.4150^{20}
5473	Ethane, 1-chloro-1-ethoxy-	C_4H_9ClO 108.57	7081-78-9	93.5	4-01-00-03119 0.9655^{20}	1.4053^{20}
5474	Ethane, 1-chloro-2-ethoxy-	C_4H_9ClO 108.57	628-34-2	107.5	4-01-00-01375 0.9895^{20}	H_2O 2; eth 5; chl 3 1.4113^{20}
5475	Ethane, 1-chloro-1-fluoro- 1-Chloro-1-fluoroethane	C_2H_4ClF 82.50	1615-75-4	16.2	4-01-00-00127	
5476	Ethane, 1-chloro-2-fluoro-	C_2H_4ClF 82.50	762-50-5	59	4-01-00-00127 1.1747^{20}	eth 4; EtOH 4 1.3775^{20}
5477	Ethane, 1-chloro-2-iodo-	C_2H_4ClI 190.41	624-70-4 -15.6	140	4-01-00-00167 2.1644^{25}	
5478	Ethane, (chloromethoxy)- Chloromethyl ethyl ether	C_3H_7ClO 94.54	3188-13-4	83	4-01-00-03047 1.0188^{15}	1.4040^{20}
5479	Ethane, 1-chloro-1-methoxy-	C_3H_7ClO 94.54	1538-87-0	73	4-01-00-03119 0.9902^{20}	eth 4 1.4004^{20}
5480	Ethane, 1-chloro-2-methoxy-	C_3H_7ClO 94.54	627-42-9	92.5	4-01-00-01375 1.0345^{20}	H_2O 4; eth 4 1.4111^{20}
5481	Ethane, 1-chloro-2-(methylthio)-	C_3H_7ClS 110.61	542-81-4	60^{30}, 44^{20}	4-01-00-01406 1.1225^{20}	EtOH 3; eth 3; ace 3 1.4902^{20}
5482	Ethane, 1-chloro-1-nitro- 1-Chloro-1-nitroethane	$C_2H_4ClNO_2$ 109.51	598-92-5	124.5	4-01-00-00172 1.2837^{20}	H_2O 1; EtOH 3; ctc 3; alk 3 1.4224^{20}
5483	Ethane, 1-chloro-2-nitro- 2-Chloro-1-nitroethane	$C_2H_4ClNO_2$ 109.51	625-47-8	173.5	4-01-00-00173 1.3550^{20}	1.407^{7}
5484	Ethane, chloropentafluoro- Chloropentafluoroethane	C_2ClF_5 154.47	76-15-3 -99.4	-37.9	4-01-00-00129 1.5678^{-42}	H_2O 1; EtOH 3; eth 3 1.2678^{-42}

No.	Name Synonym	Mol. Form. Mol. Wt.	CAS RN mp/°C	Merck No. bp/°C	Beil. Ref. den/g cm^{-3}	Solubility n_D
5485	Ethane, 1-chloro-1,1,2,2-tetrafluoro- 1-Chloro-1,1,2,2-tetrafluoroethane	C_2HClF_4 136.48	354-25-6 -117	 -10	4-01-00-00129	
5486	Ethane, 1-chloro-1,2,2,2-tetrafluoro- 1-Chloro-1,2,2,2-tetrafluoroethane	C_2HClF_4 136.48	2837-89-0	 -12		
5487	Ethane, 1-chloro-1,1,2-trifluoro 1-Chloro-1,1,2-trifluoroethane	$C_2H_2ClF_3$ 118.49	421-04-5	 12		
5488	Ethane, 1-chloro-1,2,2-trifluoro 1-Chloro-1,2,2-trifluoroethane	$C_2H_2ClF_3$ 118.49	431-07-2	 17		
5489	Ethane, 2-chloro-1,1,1-trifluoro- 2-Chloro-1,1,1-trifluoroethane	$C_2H_2ClF_3$ 118.49	75-88-7 -105.5	 6.1	4-01-00-00128 1.389^{0}	 1.3090^{0}
5490	Ethane, 1-chloro-1,1,2-trifluoro-2-iodo- Ethane, 1-chloro-2-iodo-1,1,2-trifluoro	C_2HClF_3I 244.38	354-26-7	 77^{630}	4-01-00-00167 2.181^{20}	 1.4320^{20}
5491	Ethanedial Glyoxal	$C_2H_2O_2$ 58.04	107-22-2 15	4405 50.4	4-01-00-03625 1.14^{20}	H_2O 4; EtOH 3; eth 3 1.3826^{20}
5492	Ethanedial, dioxime	$C_2H_4N_2O_2$ 88.07	557-30-2 178 dec	 sub	3-01-00-03084	H_2O 4; EtOH 4; eth 4
5493	Ethanediamide Oxamide	$C_2H_4N_2O_2$ 88.07	471-46-5 350 dec	6871	4-02-00-01860 1.667^{20}	H_2O 2; EtOH 2; eth 1
5494	Ethanediamide, *N,N'*-diethyl-	$C_6H_{12}N_2O_2$ 144.17	615-84-9 175		4-04-00-00359 1.169^{4}	EtOH 4
5495	Ethanediamide, *N,N'*-diphenyl-	$C_{14}H_{12}N_2O_2$ 240.26	620-81-5 254	 >360	4-12-00-00463	bz 4
5496	1,2-Ethanediamine Ethylenediamine	$C_2H_8N_2$ 60.10	107-15-3 11.1	3752 117	4-04-00-01166 0.8979^{20}	H_2O 4; EtOH 5; eth 1; bz 1 1.4565^{20}
5497	1,2-Ethanediamine, *N*-(2-aminoethyl)- Diethylenetriamine	$C_4H_{13}N_3$ 103.17	111-40-0 -39	 207	4-04-00-01238 0.9569^{20}	H_2O 5; EtOH 5; eth 1; lig 3 1.4810^{25}
5498	1,2-Ethanediamine, *N*-(2-aminoethyl)-*N'*-[2-[(2-aminoethyl)amino]ethyl]- Tetraethylenepentamine	$C_8H_{23}N_5$ 189.30	112-57-2	 341.5	4-04-00-01244	H_2O 3 1.5042^{20}
5499	1,2-Ethanediamine, *N,N'*-bis(2-aminoethyl)- Trientine	$C_6H_{18}N_4$ 146.24	112-24-3 12	9579 266.5	4-04-00-01242	H_2O 3; EtOH 3; acid 3 1.4971^{20}
5500	1,2-Ethanediamine, *N,N'*-bis(phenylmethyl)- Benzathine	$C_{16}H_{20}N_2$ 240.35	140-28-3 26	1072 195^{4}	4-12-00-02321 1.024^{20}	bz 4; eth 4; EtOH 4 1.5635^{20}
5501	1,2-Ethanediamine, *N,N'*-diethyl-	$C_6H_{16}N_2$ 116.21	111-74-0	 146	4-04-00-01174 0.8280^{20}	H_2O 4; eth 4; EtOH 4; tol 4 1.4340^{20}
5502	1,2-Ethanediamine, *N,N*-diethyl- *N,N*-Diethylethylenediamine	$C_6H_{16}N_2$ 116.21	100-36-7	 144	4-04-00-01175 0.8280^{20}	H_2O 5; EtOH 3; eth 3; ctc 3 1.4340^{20}
5503	1,2-Ethanediamine, *N,N'*-dimethyl	$C_4H_{12}N_2$ 88.15	110-70-3	 120	4-04-00-01171 0.828^{15}	EtOH 3; eth 3; dil HCl 3
5504	1,2-Ethanediamine, *N,N*-dimethyl-*N'*-(phenylmethyl)-	$C_{11}H_{18}N_2$ 178.28	103-55-9	 $122\text{-}4^{11}$	4-12-00-02320 0.922^{25}	 1.5089^{20}
5505	1,2-Ethanediamine, *N,N*-dimethyl-*N'*-(phenylmethyl)-*N'*-2-pyridinyl- Tripelennamine	$C_{16}H_{21}N_3$ 255.36	91-81-6	9651 $167\text{-}72^{0.1}$	5-22-08-00378	
5506	1,2-Ethanediamine, *N,N*-dimethyl-*N'*-phenyl-*N'*-(phenylmethyl)- Phenbenzamine	$C_{17}H_{22}N_2$ 254.38	961-71-7	7176 $179\text{-}80^{7}$	4-12-00-02324 1.016^{25}	 1.5794^{25}
5507	1,2-Ethanediamine, *N,N*-dimethyl-*N'*-2-pyridinyl-*N'*-(2-thienylmethyl)- Methapyrilene	$C_{14}H_{19}N_3S$ 261.39	91-80-5	5871 $173\text{-}5^{3}$	5-22-08-00401	 1.5915^{20}
5508	1,2-Ethanediamine, *N,N'*-diphenyl- 1,2-Dianilinoethane	$C_{14}H_{16}N_2$ 212.29	150-61-8 74	2969 229^{12}; 178^{2}	4-12-00-00986	H_2O 1; EtOH 3; eth 3; tfa 2
5509	1,2-Ethanediamine, *N*-ethyl-	$C_4H_{12}N_2$ 88.15	110-72-5	 129	4-04-00-01174 0.837^{25}	 1.4385^{20}
5510	1,2-Ethanediamine, *N'*-ethyl-*N,N*-dimethyl- Ethylenediamine, *N,N*-dimethyl-*N'*-ethyl	$C_6H_{15}N_2$ 115.20	123-83-1	 134.5	4-04-00-01174 0.738^{25}	 1.4222^{20}
5511	1,2-Ethanediamine, *N*-[(4-methoxyphenyl)methyl]-*N',N'*-dimethyl-*N*-2-pyridinyl- Pyrilamine	$C_{17}H_{23}N_3O$ 285.39	91-84-9	7996 201^{5}	5-22-08-00381	
5512	1,2-Ethanediamine, *N*-methyl-	$C_3H_{10}N_2$ 74.13	109-81-9	 115	4-04-00-01170 0.841^{25}	 1.4395^{20}
5513	1,2-Ethanediamine, monohydrate	$C_2H_{10}N_2O$ 78.11	6780-13-8 10	 118	4-04-00-01166 0.964^{20}	H_2O 4 1.4500^{20}
5514	1,2-Ethanediamine, *N*-1-naphthalenyl- *N*-(1-Naphthyl)ethylenediamine	$C_{12}H_{14}N_2$ 186.26	551-09-7	6329 204^{9}	4-12-00-00101 1.114^{25}	 1.6648^{25}
5515	1,2-Ethanediamine, *N,N,N',N'*-tetraethyl-	$C_{10}H_{24}N_2$ 172.31	150-77-6	 192	4-04-00-01177 0.808^{25}	 1.4343^{20}
5516	1,2-Ethanediamine, *N,N,N',N'*-tetramethyl- *N,N,N',N'*-Tetramethyldiaminoethane	$C_6H_{16}N_2$ 116.21	110-18-9 -55	 121	4-04-00-01172 0.77^{25}	 1.4179^{20}
5517	Ethane, 1,1-dibromo- Ethylidene dibromide	$C_2H_4Br_2$ 187.86	557-91-5 -63	 109	4-01-00-00157 2.0554^{20}	H_2O 1; EtOH 3; eth 4; ace 3 1.5128^{20}
5518	Ethane, 1,2-dibromo- Ethylene dibromide	$C_2H_4Br_2$ 187.86	106-93-4 9.9	3753 131.6	4-01-00-00158 2.1791^{20}	ace 4; bz 4; eth 4; EtOH 4 1.5387^{20}
5519	Ethane, 1,2-dibromo-1-chloro-1,2,2-trifluoro- 1,2-Dibromo-1-chloro-1,2,2-trifluoroethane	$C_2Br_2ClF_3$ 276.28	354-51-8 50	 93		
5520	Ethane, 1,2-dibromo-1,1-dichloro-	$C_2H_2Br_2Cl_2$ 256.75	75-81-0 -26	 195	4-01-00-00161 2.135^{20}	ace 4; bz 4; eth 4; EtOH 4 1.5662^{20}
5521	Ethane, 1,2-dibromo-1,2-dichloro-	$C_2H_2Br_2Cl_2$ 256.75	683-68-1 -26	 195	4-01-00-00161 2.135^{20}	H_2O 1; EtOH 3; eth 3; ace 3 1.5662^{20}

No.	Name Synonym	Mol. Form. Mol. Wt.	CAS RN mp/°C	Merck No. bp/°C	Beil. Ref. den/g cm^{-3}	Solubility n_D
5522	Ethane, 1,2-dibromo-1,1-difluoro- Genetron 132b-B2	$C_2H_2Br_2F_2$ 223.84	75-82-1 -61.3	 92.5	4-01-00-00160 2.2238^{20}	 1.4456^{20}
5523	Ethane, 1,2-dibromo-1-ethoxy-	$C_4H_8Br_2O$ 231.91	2983-26-8	 80^{20}	4-01-00-03152 1.7320^{20}	EtOH 4; chl 4 1.5044^{20}
5524	Ethane, 1,2-dibromo-1,1,2,2-tetrachloro-	$C_2Br_2Cl_4$ 325.64	630-25-1 202.5		3-01-00-00190 2.713^{25}	
5525	Ethane, 1,2-dibromo-1,1,2,2-tetrafluoro- 1,2-Dibromotetrafluoroethane	$C_2Br_2F_4$ 259.82	124-73-2 -110.4	 47.3	4-01-00-00160 2.149^{25}	 1.361^{25}
5526	Ethane, 1,2-dibromo-1,1,2-trifluoro- Halon 2302	$C_2HBr_2F_3$ 241.83	354-04-1	 76	4-01-00-00160 2.274^{27}	 1.4191^{24}
5527	Ethane, 1,2-dibutoxy Ethylene glycol dibutyl ether	$C_{10}H_{22}O_2$ 174.28	112-48-1 -69.1	 203.3	 0.8319^{25}	
5528	Ethane, 1,1-dichloro- Ethylidene dichloride	$C_2H_4Cl_2$ 98.96	75-34-3 -96.9	3766 57.4	4-01-00-00130 1.1757^{20}	H_2O 2; EtOH 4; eth 4; ace 3 1.4164^{20}
5529	Ethane, 1,2-dichloro- Ethylene dichloride	$C_2H_4Cl_2$ 98.96	107-06-2 -35.5	3754 83.5	4-01-00-00131 1.2351^{20}	H_2O 2; EtOH 4; eth 5; ace 3 1.4448^{20}
5530	Ethane, 1,1-dichloro-2,2-diethoxy-	$C_6H_{12}Cl_2O_2$ 187.07	619-33-0	 183.5	4-01-00-03140 1.1383^{14}	EtOH 3; eth 3; ctc 2
5531	Ethane, 1,2-dichloro-1,1-difluoro-	$C_2H_2Cl_2F_2$ 134.94	1649-08-7 -101.2	 46.8	4-01-00-00135 1.4163^{20}	 1.36193^{20}
5532	Ethane, 1,2-dichloro-1,2-difluoro-	$C_2H_2Cl_2F_2$ 134.94	431-06-1 -101.2	 46.8	4-01-00-00135 1.4163^{20}	 1.3619^{20}
5533	Ethane, 1,2-dichloro-1-ethoxy-	$C_4H_8Cl_2O$ 143.01	623-46-1	 145	4-01-00-03136 1.1370^{20}	chl 2 1.4435^{20}
5534	Ethane, 1,1-dichloro-1-fluoro- 1,1-Dichloro-1-fluoroethane	$C_2H_3Cl_2F$ 116.95	1717-00-6 -103.5	 32	4-01-00-00134 1.250^{10}	 1.3600^{10}
5535	Ethane, 1,2-dichloro-1-fluoro-	$C_2H_3Cl_2F$ 116.95	430-57-9 -60	 73.7	4-01-00-00134 1.3814^{20}	 1.4132^{20}
5536	Ethane, 1,1-dichloro-1-nitro- Ethide	$C_2H_3Cl_2NO_2$ 143.96	594-72-9	 123.5	4-01-00-00173	ctc 3
5537	Ethane, 1,1-dichloro-1,2,2,2-tetrafluoro- 1,1-Dichlorotetrafluoroethane	$C_2Cl_2F_4$ 170.92	374-07-2 -56.6	 4	4-01-00-00136 $*1.455^{25}$	bz 4; eth 4; EtOH 4 1.3092^{0}
5538	Ethane, 1,2-dichloro-1,1,2,2-tetrafluoro- Refrigerant 114	$C_2Cl_2F_4$ 170.92	76-14-2 -94	2608 3.8	4-01-00-00137 $*1.455^{25}$	eth 4; EtOH 4 1.3092^{0}
5539	Ethane, 1,2-dichloro-1,1,2-trifluoro-	$C_2HCl_2F_3$ 152.93	354-23-4 -78	 28	4-01-00-00136 1.50^{25}	
5540	Ethane, 2,2-dichloro-1,1,1-trifluoro- 2,2-Dichloro-1,1,1-trifluoroethane	$C_2HCl_2F_3$ 152.93	306-83-2		4-01-00-00135	
5541	Ethane, 1,1-diethoxy- Acetal	$C_6H_{14}O_2$ 118.18	105-57-7 -100	31 102.2	4-01-00-03103 0.8254^{20}	H_2O 3; EtOH 5; eth 5; ace 4 1.3834^{20}
5542	Ethane, 1,2-diethoxy- Ethylene glycol diethyl ether	$C_6H_{14}O_2$ 118.18	629-14-1	 119.4	4-01-00-02379 0.8484^{20}	ace 4; bz 4; eth 4; EtOH 4 1.3860^{20}
5543	Ethane, 1,1-diethoxy-2-iodo-	$C_6H_{13}IO_2$ 244.07	51806-20-3	 132^{90}; 100^{10}	4-01-00-03156 1.489^{22}	 1.4734^{22}
5544	Ethane, 1,1-difluoro-	$C_2H_4F_2$ 66.05	75-37-6 -117	 -24.9	4-01-00-00120 $*0.896^{20}$	 1.3011^{-72}
5545	Ethane, 1,2-difluoro- Freon 152	$C_2H_4F_2$ 66.05	624-72-6	 30.7	4-01-00-00121	bz 4; eth 4; chl 4
5546	Ethane, 1,1-diiodo- 1,1-Diiodoethane	$C_2H_4I_2$ 281.86	594-02-5	 179.5	4-01-00-00169 2.84^{25}	ace 4; eth 4; EtOH 4; chl 4 1.673^{20}
5547	Ethane, 1,2-diiodo-	$C_2H_4I_2$ 281.86	624-73-7 83	 200	4-01-00-00169 3.325^{20}	H_2O 2; EtOH 3; eth 3; ace 3 1.871^{20}
5548	Ethane, 1,1-dimethoxy- Dimethyl acetal	$C_4H_{10}O_2$ 90.12	534-15-6 -113.2	3215 64.5	4-01-00-03103 0.8501^{20}	H_2O 3; EtOH 3; eth 3; ace 4 1.3668^{20}
5549	Ethane, 1,2-dimethoxy- Ethylene glycol dimethyl ether	$C_4H_{10}O_2$ 90.12	110-71-4 -58	3213 85	4-01-00-02376 0.8691^{20}	H_2O 3; EtOH 3; eth 3; ace 3 1.3796^{20}
5550	Ethanedinitrile Cyanogen	C_2N_2 52.04	460-19-5 -27.9	2698 -21.1	4-02-00-01863 0.9537^{-21}	H_2O 3; EtOH 3; eth 3
5551	Ethane, 1,1-dinitro- 1,1-Dinitroethane	$C_2H_4N_2O_4$ 120.06	600-40-8	 185.5	4-01-00-00174 1.349^{24}	H_2O 2; EtOH 3; eth 3
5552	Ethane, 1,2-dinitro- 1,2-Dinitroethane	$C_2H_4N_2O_4$ 120.06	7570-26-5 39.5	 $94\text{-}6^{5}$	4-01-00-00175 1.4597^{20}	eth 4; EtOH 4 1.4468^{20}
5553	Ethanedioic acid Oxalic acid	$C_2H_2O_4$ 90.04	144-62-7 189.5 dec	6865 sub 157	4-02-00-01819 1.900^{17}	H_2O 3; EtOH 4; eth 2; bz 1
5554	Ethanedioic acid, bis(2-chloroethyl) ester	$C_6H_8Cl_2O_4$ 215.03	7208-92-6 45	 150^{4}	4-02-00-01849	bz 4; EtOH 4
5555	Ethanedioic acid, bis(3-methylbutyl) ester	$C_{12}H_{22}O_4$ 230.30	2051-00-5 -9	 267.5	4-02-00-01851 0.968^{11}	eth 4; EtOH 4
5556	Ethanedioic acid, bis(1-methylethyl) ester	$C_8H_{14}O_4$ 174.20	615-81-6	 191	4-02-00-01850 1.001^{20}	eth 4; EtOH 4 1.4100^{20}
5557	Ethanedioic acid, bis(2-methylpropyl) ester Diisobutyl oxalate	$C_{10}H_{18}O_4$ 202.25	2050-61-5	 229	4-02-00-01850 0.9737^{20}	ace 4; eth 4; EtOH 4 1.4180^{20}
5558	Ethanedioic acid, dibutyl ester	$C_{10}H_{18}O_4$ 202.25	2050-60-4 -30.5	 241; 96^{2}	4-02-00-01850 0.9873^{20}	H_2O 1; EtOH 3; eth 3 1.4234^{20}
5559	Ethanedioic acid, dicyclohexyl ester Dicyclohexyl oxalate	$C_{14}H_{22}O_4$ 254.33	620-82-6 43.5	 190^{73}	3-06-00-00026	H_2O 1; EtOH 4; eth 4; MeOH 3
5560	Ethanedioic acid, diethyl ester Diethyl oxalate	$C_6H_{10}O_4$ 146.14	95-92-1 -40.6	3115 185.7	4-02-00-01848 1.0785^{20}	H_2O 2; EtOH 5; eth 5; ace 5 1.4101^{20}

No.	Name Synonym	Mol. Form. Mol. Wt.	CAS RN mp/°C	Merck No. bp/°C	Beil. Ref. den/g cm^{-3}	Solubility n_D
5561	Ethanedioic acid, dihydrate Oxalic acid dihydrate	$C_2H_6O_6$ 126.07	6153-56-6 101.5		4-02-00-01819 1.653^{18}	H_2O 3; EtOH 3; eth 2
5562	Ethanedioic acid, dihydrazide	$C_2H_6N_4O_2$ 118.10	996-98-5 243 dec		4-02-00-01868 1.458^{22}	H_2O 3; EtOH 2; eth 2; bz 2
5563	Ethanedioic acid, dimethyl ester Dimethyl oxalate	$C_4H_6O_4$ 118.09	553-90-2 54.3	6020 163.5	4-02-00-01847 1.1716^{60}	H_2O 2; EtOH 3; eth 3; ace 3 1.379^{82}
5564	Ethanedioic acid, di-2-propenyl ester Diallyl oxalate	$C_8H_{10}O_4$ 170.17	615-99-6	 217	4-02-00-01851 1.1582^{20}	H_2O 1; EtOH 3; ace 3; bz 3 1.4481^{20}
5565	Ethanedioic acid, dipropyl ester	$C_8H_{14}O_4$ 174.20	615-98-5 -44.3	 211	4-02-00-01849 1.0188^{20}	H_2O 2; EtOH 5; eth 3 1.4158^{20}
5566	Ethanedioic acid, disodium salt Oxalic acid, disodium salt	$C_2Na_2O_4$ 134.00	62-76-0 260 dec	8601	4-02-00-01819 2.34^{25}	H_2O 2; EtOH 1; eth 1
5567	Ethanedioic acid, ethyl methyl ester	$C_5H_8O_4$ 132.12	615-52-1	 173.7	0-02-00-00535 1.1555^{0}	eth 4; EtOH 4
5568	Ethanedioic acid, lead salt Oxalic acid, lead salt	C_2O_4Pb 295.22	15843-48-8 300 dec	5294	4-02-00-01819	H_2O i; EtOH 1; HNO_3 3
5569	1,2-Ethanediol Ethylene glycol	$C_2H_6O_2$ 62.07	107-21-1 -13	3755 197.3	4-01-00-02369 1.1088^{20}	H_2O 5; EtOH 5; eth 3; ace 5 1.4318^{20}
5570	1,1-Ethanediol, diacetate Ethylidene diacetate	$C_6H_{10}O_4$ 146.14	542-10-9 18.9	3767 169	4-02-00-00282 1.070^{25}	eth 4; EtOH 4 1.3985^{25}
5571	1,2-Ethanediol, diacetate Ethylene glycol diacetate	$C_6H_{10}O_4$ 146.14	111-55-7 -31	3756 190	4-02-00-00217 1.1043^{20}	H_2O 4; EtOH 5; eth 5; ace 5 1.4159^{20}
5572	1,2-Ethanediol, dibenzoate	$C_{16}H_{14}O_4$ 270.28	94-49-5 73.5	 360 dec	4-09-00-00356	H_2O 1; eth 3; chl 3
5573	1,2-Ethanediol, diformate	$C_4H_6O_4$ 118.09	029-15-2	 174	4-02-00-00037 1.193^{0}	H_2O 2; EtOH 3; eth 3 1.3580
5574	1,2-Ethanediol, dinitrate Ethylene glycol dinitrate	$C_2H_4N_2O_6$ 152.06	628-96-6 -22.3	 198.5	4-01-00-02413 1.4918^{20}	eth 4; EtOH 4
5575	1,2-Ethanediol, 1,2-diphenyl-, (*R**,*R**)-(±)-	$C_{14}H_{14}O_2$ 214.26	655-48-1 122.5	 >300	4-06-00-06682	H_2O 1; EtOH 4; eth 4; ace 3
5576	1,2-Ethanediol, dipropanoate	$C_8H_{14}O_4$ 174.20	123-80-8	 211	4-02-00-00715 1.020^{15}	H_2O 2; EtOH 5; eth 5; ace 4
5577	1,2-Ethanediol, monoacetate Ethylene glycol, monoacetate	$C_4H_8O_3$ 104.11	542-59-6	3757 188	4-02-00-00214 1.108^{15}	H_2O 5; EtOH 5; eth 5
5578	1,2-Ethanediol, monobenzoate	$C_9H_{10}O_3$ 166.18	94-33-7 45	 150^{10}	4-09-00-00354 1.1101^{30}	EtOH 4
5579	1,2-Ethanediol, 1-phenyl- Styrene glycol	$C_8H_{10}O_2$ 138.17	93-56-1 67.5	8831 273	4-06-00-05939	H_2O 4; eth 4; bz 4; EtOH 4
5580	1,2-Ethanediol, 1,1,2,2-tetraphenyl- Benzopinacol	$C_{26}H_{22}O_2$ 366.46	464-72-2 182	1110	4-06-00-07053	H_2O 1; EtOH 2; eth 3; ace 3
5581	1,1-Ethanediol, 2,2,2-tribromo- Acetaldehyde, tribromo, hydrate	$C_2H_3Br_3O_2$ 298.76	507-42-6 53.5	1373 dec	4-01-00-03155 2.5661^{40}	eth 4; EtOH 4
5582	1,1-Ethanediol, 2,2,2-trichloro- Chloral hydrate	$C_2H_3Cl_3O_2$ 165.40	302-17-0 57	2061 96 dec	4-01-00-03143 1.9081^{20}	H_2O 4; bz 4; eth 4; EtOH 4
5583	Ethanedione, di-2-furanyl-, dioxime α-Furildioxime	$C_{10}H_8N_2O_4$ 220.18	522-27-0 167	4218	4-19-00-02008	EtOH 2; eth 2; bz 2; lig 2
5584	Ethanedione, diphenyl- Benzil	$C_{14}H_{10}O_2$ 210.23	134-81-6 94.8	1087 347	4-07-00-02502 1.084^{102}	H_2O 1; EtOH 4; eth 4; ace 3
5585	Ethanedioyl dichloride Oxalyl chloride	$C_2Cl_2O_2$ 126.93	79-37-8 -16	6807 63.5	4-02-00-01853 1.4785^{20}	eth 3 1.4316^{20}
5586	Ethane, 1,1-dipropoxy- 1,2-Dipropoxyethane	$C_8H_{18}O_2$ 146.23	18854-56-3			
5587	1,2-Ethanedisulfonic acid Ethylenedisulfonic acid	$C_2H_6O_6S_2$ 190.20	110-04-3 173	3678	4-04-00-00078	Diox 4
5588	Ethanedithioamide Rubeanic acid	$C_2H_4N_2S_2$ 120.20	79-40-3 170 dec	8256	4-02-00-01871	H_2O 2; EtOH 2; con sulf 3
5589	Ethane(dithioic) acid	$C_2H_4S_2$ 92.19	594-03-6	 66^{85}; 37^{15}	4-02-00-00572 1.24^{20}	H_2O 3; EtOH 4; eth 4; ace 4
5590	1,2-Ethanedithiol Ethylenedimercaptan	$C_2H_6S_2$ 94.20	540-63-6 -41.2	3679 146	4-01-00-02450 1.234^{20}	H_2O 1; EtOH 3; eth 3; ace 3 1.5590^{20}
5591	Ethane, 1-ethoxy-2-methoxy- 1-Ethoxy-2-methoxyethane	$C_5H_{12}O_2$ 104.15	5137-45-1	 102.1	3-01-00-02078 0.8529^{20}	 1.3868^{20}
5592	Ethane, 1,1'-[ethylidenebis(oxy)]bis[2-chloro-	$C_6H_{12}Cl_2O_2$ 187.07	14689-97-5	 195	4-01-00-03104 1.1737^{20}	eth 4; EtOH 4 1.4527^{20}
5593	Ethane, fluoro- Ethyl fluoride	C_2H_5F 48.06	353-36-6 -143.2	 -37.6	4-01-00-00120 0.7182^{20}	H_2O 3; EtOH 4; eth 4 [illegible]
5594	Ethane, hexabromo- Hexabromoethane	C_2Br_6 503.45	594-73-0	 200 dec	3-01-00-00193 3.823^{20}	EtOH 2; eth 2; CS_2 2 1.863
5595	Ethane, hexachloro- Hexachloroethane	C_2Cl_6 236.74	67-72-1 187 (triple point)	4601	4-01-00-00148 2.091^{20}	H_2O 1; EtOH 4; eth 4; bz 3
5596	Ethane, hexafluoro- Perfluoroethane	C_2F_6 138.01	76-16-4 -100.7	 -78.1	4-01-00-00123 1.590^{-78}	H_2O 1; EtOH 2; eth 2
5597	Ethane, iodo- Ethyl iodide	C_2H_5I 155.97	75-03-6 -111.1	3769 72.5	4-01-00-00163 1.9358^{20}	H_2O 2; EtOH 5; eth 3; chl 3 1.5133^{20}
5598	Ethane, isocyanato- Ethyl isocyanate	C_3H_5NO 71.08	109-90-0	 60	4-04-00-00402 0.9031^{20}	H_2O 1; EtOH 5; eth 5 1.3808^{20}
5599	Ethane, isocyano-	C_3H_5N 55.08	624-79-3 <-66	 79	4-04-00-00342 0.7402^{20}	H_2O 4; EtOH 5; eth 5; ace 3 1.3622^{20}
5600	Ethane, isothiocyanato- Ethyl isothiocyanate	C_3H_5NS 87.15	542-85-8 -5.9	3771 131.5	4-04-00-00403 0.9990^{20}	H_2O 1; EtOH 5; eth 5 1.5130^{20}

No.	Name Synonym	Mol. Form. Mol. Wt.	CAS RN mp/°C	Merck No. bp/°C	Beil. Ref. den/g cm^{-3}	Solubility n_D
5601	Ethane, 1,1',1'',1'''-[methanetetrayltetrakis(oxy)]tetrakis-	$C_9H_{20}O_4$ 192.26	78-09-1	 159.5	4-03-00-00006 0.9186^{20}	EtOH 5; eth 5; ctc 3 1.3905^{25}
5602	Ethane, methoxy- Ethyl methyl ether	C_3H_8O 60.10	540-67-0 -113	3783 7.4	4-01-00-01314 0.7251^{0}	H_2O 3; EtOH 5; eth 5; ace 3 1.3420^{4}
5603	Ethane, 1,1'-[methylenebis(oxy)]bis- Diethoxymethane	$C_5H_{12}O_2$ 104.15	462-95-3 -66.5	 88	4-01-00-03027 0.8319^{20}	H_2O 3; EtOH 5; eth 5; ace 4 1.3748^{18}
5604	Ethane, 1,1',1''-[methylidynetris(oxy)]tris- Ethyl orthoformate	$C_7H_{16}O_3$ 148.20	122-51-0	 143; 60^{20}	4-02-00-00025 0.8909^{20}	EtOH 3; eth 3 1.3922^{20}
5605	Ethane, 1,1',1''-[methylidynetris(thio)]tris-	$C_7H_{16}S_3$ 196.40	6267-24-9	 235 dec; 127^{12}	4-02-00-00093 1.053^{20}	eth 4; EtOH 4 1.5410^{15}
5606	Ethane, (methylthio)- Ethyl methyl sulfide	C_3H_8S 76.16	624-89-5 -105.9	 66.7	4-01-00-01392 0.8422^{20}	H_2O 1; EtOH 5; eth 3; chl 3 1.4404^{20}
5607	Ethane, nitro- Nitroethane	$C_2H_5NO_2$ 75.07	79-24-3 -89.5	6518 114.0	4-01-00-00170 1.0448^{25}	H_2O 2; EtOH 5; eth 5; ace 3 1.3917^{20}
5608	Ethane, 1,1'-oxybis- Diethyl ether	$C_4H_{10}O$ 74.12	60-29-7 -116.3	3762 34.5	4-01-00-01314 0.7138^{20}	H_2O 2; EtOH 5; eth 5; ace 4 1.3526^{20}
5609	Ethane, 1,1'-oxybis[2-bromo-	$C_4H_8Br_2O$ 231.91	5414-19-7	 115^{32}	4-01-00-01386 1.8222^{27}	 1.5131^{27}
5610	Ethane, 1,1'-oxybis[1-chloro-	$C_4H_8Cl_2O$ 143.01	6986-48-7	 116.5	4-01-00-03120 1.1060^{25}	eth 4; EtOH 4; chl 4 1.4186^{25}
5611	Ethane, 1,1'-oxybis[2-chloro- Bis(2-chloroethyl) ether	$C_4H_8Cl_2O$ 143.01	111-44-4 -51.9	3055 178.5	4-01-00-01375 1.22^{20}	H_2O 1; EtOH 3; eth 3; ace 3 1.451^{20}
5612	Ethane, 1,1'-oxybis[2-ethoxy- Diethylene glycol diethyl ether	$C_8H_{18}O_3$ 162.23	112-36-7 -45	3108 188	4-01-00-02394 0.9063^{20}	H_2O 4; EtOH 4; eth 3; os 4 1.4115^{20}
5613	Ethane, 1,1'-oxybis[2-methoxy- Diethylene glycol dimethyl ether	$C_6H_{14}O_3$ 134.18	111-96-6 -68	3148 162	4-01-00-02393 0.9434^{20}	H_2O 5; EtOH 5; eth 5 1.4097^{20}
5614	Ethane, 1,1'-[oxybis(methyleneoxy)]bis- Bis(ethoxymethyl)ether	$C_6H_{14}O_3$ 134.18	5648-29-3	 140.6		
5615	Ethane, 1,1'-oxybis[1,2,2,2-tetrachloro- Bis(1,2,2,2-tetrachloroethyl) ether	$C_4H_2Cl_8O$ 349.68	41284-12-2 40.2	 188; 130^{11}	2-01-00-00681	bz 4; peth 4
5616	Ethane, pentabromo- 1,1,1,2,2-Pentabromoethane	C_2HBr_5 424.55	75-95-6 56.5	 210^{300}	3-01-00-00193 3.312^{20}	H_2O 1; EtOH 3; eth 4; chl 3
5617	Ethane, pentachloro- Pentachloroethane	C_2HCl_5 202.29	76-01-7 -29	7058 159.8	4-01-00-00147 1.6796^{20}	H_2O 1; EtOH 5; eth 5 1.5025^{20}
5618	Ethane, pentachlorofluoro-	C_2Cl_5F 220.28	354-56-3 101.3	 135	4-01-00-00148 1.74^{25}	H_2O 1; EtOH 3; eth 3
5619	Ethaneperoxoic acid Peroxyacetic acid	$C_2H_4O_3$ 76.05	79-21-0 -0.2	7107 110	4-02-00-00390 1.226^{15}	H_2O 4; EtOH 3; eth 4; sulf 4 1.3974^{20}
5620	Ethane, 1,1'-selenobis-	$C_4H_{10}Se$ 137.08	627-53-2 55	 108	4-01-00-01411 1.2300^{20}	 1.4768^{20}
5621	Ethane, 1,1'-sulfinylbis-	$C_4H_{10}OS$ 106.19	70-29-1 14	 104^{25}; 90^{15}	4-01-00-01395 1.0092^{22}	H_2O 4; eth 4; EtOH 4
5622	Ethanesulfonic acid Ethylsulfonic acid	$C_2H_6O_3S$ 110.13	594-45-6 -17	 123^{1}	4-04-00-00033 1.3341^{25}	H_2O 4; EtOH 4 1.4335^{20}
5623	Ethanesulfonic acid, 2-amino- Taurine	$C_2H_7NO_3S$ 125.15	107-35-7 328	9043	4-04-00-03289	H_2O 4
5624	Ethanesulfonic acid, 2-(methylamino)- *N*-Methyltaurine	$C_3H_9NO_3S$ 139.18	107-68-6 241.5	6043	3-04-00-01699	H_2O 4; EtOH 1; eth 1
5625	Ethanesulfonic acid, 2-[[(3α,5β,7α,12α)-3,7,12-trihydroxy-24-oxocholan-24-yl]amino]- Taurocholic acid	$C_{26}H_{45}NO_7S$ 515.71	81-24-3 125 dec	9044	4-10-00-02078	H_2O 4; EtOH 4; eth 2; AcOEt 2
5626	Ethane, 1,1'-sulfonylbis-	$C_4H_{10}O_2S$ 122.19	597-35-3 73.5	 248	4-01-00-01396 1.357^{20}	H_2O 3; eth 3; bz 4; peth 1
5627	Ethanesulfonyl chloride	$C_2H_5ClO_2S$ 128.58	594-44-5	 174	4-04-00-00034 1.357^{22}	eth 4; CS_2 3 1.4531^{20}
5628	Ethanesulfonyl chloride, 2-bromo- 2-Bromoethanesulfonyl chloride	$C_2H_4BrClO_2S$ 207.48	54429-56-0	 102^{13}	4-04-00-00037 1.921^{20}	 1.5242^{20}
5629	Ethanesulfonyl chloride, 2-chloro-	$C_2H_4Cl_2O_2S$ 163.02	1622-32-8	 201.5	4-04-00-00036 1.555^{20}	 1.4920^{20}
5630	Ethane, 1,1'-tellurobis-	$C_4H_{10}Te$ 185.72	627-54-3	 137.5	4-01-00-01412 1.599^{15}	EtOH 4 1.5182^{15}
5631	Ethane, 1,1,1,2-tetrabromo- 1,2,2,2-Tetrabromoethane	$C_2H_2Br_4$ 345.65	630-16-0 0	 112^{18} dec	4-01-00-00162 2.8747^{20}	ace 4; bz 4; eth 4; EtOH 4 1.6277^{20}
5632	Ethane, 1,1,2,2-tetrabromo- Acetylene tetrabromide	$C_2H_2Br_4$ 345.65	79-27-6 0	9121 243.5; 151^{54}	4-01-00-00162 2.9655^{20}	H_2O 1; EtOH 5; eth 5; ace 3 1.6353^{20}
5633	1,1,2,2-Ethanetetracarboxylic acid, tetraethyl ester	$C_{14}H_{22}O_8$ 318.32	632-56-4 77	 305 dec	4-02-00-02415 1.0648^{80}	EtOH 3 1.4105^{80}
5634	Ethane, 1,1,1,2-tetrachloro- 1,1,1,2-Tetrachloroethane	$C_2H_2Cl_4$ 167.85	630-20-6 -70.2	 130.5	4-01-00-00143 1.5406^{20}	H_2O 2; EtOH 5; eth 5; ace 3 1.4821^{20}
5635	Ethane, 1,1,2,2-tetrachloro- Acetylene tetrachloride	$C_2H_2Cl_4$ 167.85	79-34-5 -43.8	9125 146.5	4-01-00-00144 1.5953^{20}	H_2O 2; EtOH 5; eth 5; ace 3 1.4940^{20}
5636	Ethane, 1,1,1,2-tetrachloro-2,2-difluoro- 1,1-Difluoroperchloroethane	$C_2Cl_4F_2$ 203.83	76-11-9 40.6	 91.5	4-01-00-00146 1.649^{25}	H_2O 1; EtOH 3; eth 3; chl 3
5637	Ethane, 1,1,2,2-tetrachloro-1,2-difluoro- Tetrachloro-1,2-difluoroethane	$C_2Cl_4F_2$ 203.83	76-12-0 26	 93	4-01-00-00146 1.6447^{25}	H_2O 1; EtOH 3; eth 3; chl 3 1.4130^{25}
5638	Ethane, 1,1,2,2-tetrachloro-1-fluoro-	C_2HCl_4F 185.84	354-14-3 -82.6	 116	4-01-00-00145 1.5497^{17}	 1.4390^{20}

No.	Name Synonym	Mol. Form. Mol. Wt.	CAS RN mp/°C	Merck No. bp/°C	Beil. Ref. den/g cm^{-3}	Solubility n_D
5639	Ethane, 1,1,1,2-tetrachloro-2-fluoro- 1,1,1,2-Tetrachloro-2-fluoroethane	C_2HCl_4F 185.84	354-11-0 -95.3	 116.5		
5640	Ethane, 1,1,1,2-tetrafluoro- 1,1,1,2-Tetrafluoroethane	$C_2H_2F_4$ 102.03	811-97-2 -101	 -26	4-01-00-00123	H_2O 1; eth 3
5641	Ethane, 1,1,2,2-tetrafluoro- 1,1,2,2-Tetrafluoroethane	$C_2H_2F_4$ 102.03	359-35-3 -89	 -19.9		
5642	Ethane, 1,1,2,2-tetrafluoro-1,2-dinitro- Ethane, 1,2-dinitro-1,1,2,2-tetrafluoro	$C_2F_4N_2O_4$ 192.03	356-16-1 -41.5	 58.5	3-01-00-00203 1.6024^{25}	H_2O 1; ace 3 1.3265^{25}
5643	Ethanethioamide Thioacetamide	C_2H_5NS 75.13	62-55-5 115.5	9246	4-02-00-00565	H_2O 4; EtOH 4; eth 2; bz 2
5644	Ethane, 1,1'-thiobis- Diethyl sulfide	$C_4H_{10}S$ 90.19	352-93-2 -103.9	3809 92.1	4-01-00-01394 0.8362^{20}	H_2O 2; EtOH 3; eth 3; ctc 2 1.4430^{20}
5645	Ethane, 1,1'-thiobis[2-chloro- Mustard gas	$C_4H_8Cl_2S$ 159.08	505-60-2 13.5	6225 216	4-01-00-01407 1.2741^{20}	 1.5313^{20}
5646	Ethanethioic acid Thioacetic acid	C_2H_4OS 76.12	507-09-5 <-17	9247 93; 26^{35}	4-02-00-00542 1.064^{20}	H_2O 3; EtOH 4; eth 5; ace 4 1.4648^{20}
5647	Ethanethioic acid, *S*-ethyl ester	C_4H_8OS 104.17	625-60-5	 116.4	4-02-00-00543 0.9792^{20}	H_2O 1; EtOH 4; eth 4 1.4583^{21}
5648	Ethanethiol Ethyl mercaptan	C_2H_6S 62.14	75-08-1 -147.8	3680 35.1	4-01-00-01390 0.8315^{25}	H_2O 2; EtOH 3; eth 3; ace 3 1.4310^{20}
5649	Ethanethiol, 2-amino- Cysteamine	C_2H_7NS 77.15	60-23-1 99.5	2785 dec	4-04-00-01570	H_2O 4; EtOH 4
5650	Ethanethiol, 2-(butylamino)- 2-(Butylamino)ethanethiol	$C_6H_{15}NS$ 133.26	5842-00-2	 $112\text{-}5^{10}$	4-04-00-01602 0.901^{25}	 1.4711^{20}
5651	Ethanethiol, 2-chloro- 2-Chloroethanethiol	C_2H_5ClS 96.58	4325 07-7	 113	4-01-00-01406 1.1826^{20}	eth 4; EtOH 4; Diox 4 1.4929^{20}
5652	Ethanethiol, 2,2'-thiobis-	$C_4H_{10}S_3$ 154.32	3570-55-6	 $135\text{-}6^{10}$	4-01-00-02455 1.183^{25}	 1.5982^{20}
5653	Ethane, 1,1,2-tribromo- 1,1,2-Tribromoethane	$C_2H_3Br_3$ 266.76	78-74-0 -29.3	 188.93	4-01-00-00161 2.6210^{20}	H_2O 1; EtOH 3; eth 3; bz 3 1.5933^{20}
5654	1,1,2-Ethanetricarboxylic acid, triethyl ester	$C_{11}H_{18}O_6$ 246.26	7459-46-3	 99^{05}	4-02-00-02364 1.074^{25}	 1.4290^{20}
5655	Ethane, 1,1,1-trichloro- Methylchloroform	$C_2H_3Cl_3$ 133.40	71-55-6 -30.4	9549 74.0	4-01-00-00138 1.3390^{20}	H_2O 2; EtOH 3; eth 5; chl 3 1.4379^{20}
5656	Ethane, 1,1,2-trichloro- 1,1,2-Trichloroethane	$C_2H_3Cl_3$ 133.40	79-00-5 -36.6	9550 113.8	4-01-00-00139 1.4397^{20}	H_2O 1; EtOH 3; eth 3; chl 3 1.4714^{20}
5657	Ethane, 1,1,1-trichloro-2,2-diethoxy- Chloral, diethyl acetal	$C_6H_{11}Cl_3O_2$ 221.51	599-97-3	 197	4-01-00-03144 1.266^{20}	H_2O 2; EtOH 5; eth 5; glycerol 5 1.4586^{25}
5658	Ethane, 1,1,1-trichloro-2,2-difluoro 1,1,1-Trichloro-2,2-difluoroethane	$C_2HCl_3F_2$ 169.38	354-12-1	 73		
5659	Ethane, 1,2,2-trichloro-1,1-difluoro-	$C_2HCl_3F_2$ 169.38	354-21-2	 71.9	4-01-00-00141 1.5447^{20}	 1.3889^{20}
5660	Ethane, 1,2,2-trichloro-1,2-difluoro- 1,2,2-Trichloro-1,2-difluoroethane	$C_2HCl_3F_2$ 169.38	354-15-4	 72.5		
5661	Ethane, 1,1,2-trichloro-2-fluoro-	$C_2H_2Cl_3F$ 151.39	359-28-4	 102	4-01-00-00141 1.5393^{20}	H_2O 1 1.4390^{20}
5662	Ethane, 1,1,1-trichloro-2,2,2-trifluoro- 1,1,1-Trichlorotrifluoroethane	$C_2Cl_3F_3$ 187.38	354-58-5 14.2	 46.1	4-01-00-00142 1.5790^{20}	H_2O 1; EtOH 3; eth 3; chl 3 1.3610^{36}
5663	Ethane, 1,1,2-trichloro-1,2,2-trifluoro- 1,1,2-Trichlorotrifluoroethane	$C_2Cl_3F_3$ 187.38	76-13-1 -35	 47.7	4-01-00-00142 1.5635^{25}	H_2O 1; EtOH 3; eth 5; bz 5 1.3557^{25}
5664	Ethane, 1,1,1-triethoxy-	$C_8H_{18}O_3$ 162.23	78-39-7	 145	4-02-00-00137 0.8847^{25}	H_2O 1; EtOH 5; eth 5; ctc 5 1.3980^{20}
5665	Ethane, 1,1,1-trifluoro- Methylfluoroform	$C_2H_3F_3$ 84.04	420-46-2 -111.3	 -47.5	4-01-00-00122	eth 3; chl 3
5666	Ethane, 1,1,2-trifluoro- 1,1,2-Trifluoroethane	$C_2H_3F_3$ 84.04	430-66-0 -84	 5	4-01-00-00123	
5667	Ethane, 1,1,1-trifluoro-2-iodo-	$C_2H_2F_3I$ 209.94	353-83-3	 54.5	4-01-00-00166 2.13^{25}	 1.4009^{20}
5668	Ethane, 1,1,1-trifluoro-2-nitro-	$C_2H_2F_3NO_2$ 129.04	819-07-8	 96	4-01-00-00172 1.3914^{20}	eth 4 1.3394^{20}
5669	Ethane, 1,1,1-trimethoxy-	$C_5H_{12}O_3$ 120.15	1445-45-0	 108	4-02-00-00127 0.9438^{25}	eth 4; EtOH 4 1.3859^{25}
5670	Ethanimidamide, *N,N'*-bis(4-ethoxyphenyl)-, monohydrochloride Holocaine, hydrochloride	$C_{18}H_{23}ClN_2O_2$ 334.85	620-99-5 191	7153	3-13-00-01069	H_2O 4; EtOH 4; chl 4
5671	Ethanimidamide, monohydrochloride Acetamidine, hydrochloride	[illegible] 94.54	[illegible] 177.5	[illegible]	4-02-00-00428	H_2O 4; EtOH 4
5672	Ethanol Ethyl alcohol	C_2H_6O 46.07	64-17-5 -114.1	3716 78.2	4-01-00-01289 0.7893^{20}	H_2O 5; EtOH 5; eth 5; ace 5 1.3611^{20}
5673	Ethanol, 2-[2-(acetyloxy)ethoxy]-	$C_6H_{12}O_4$ 148.16	2093-20-1	 $114\text{-}5^{7}$	2-02-00-00155 1.1208^{20}	 1.4320^{20}
5674	Ethanol, 1-amino- Acetaldehyde ammonia	C_2H_7NO 61.08	75-39-8 97	33 110 dec		H_2O 3; eth 2
5675	Ethanol, 2-amino- Ethanolamine	C_2H_7NO 61.08	141-43-5 10.5	3681 171	4-04-00-01406 1.0180^{20}	H_2O 5; EtOH 5; eth 2; bz 2 1.4541^{20}
5676	Ethanol, 2-[(2-aminoethyl)amino]- *N*-(Aminoethyl)ethanolamine	$C_4H_{12}N_2O$ 104.15	111-41-1	 239; 105^{10}	4-04-00-01558 1.0286^{20}	H_2O 5; EtOH 5; ace 3; bz 2 1.4863^{20}
5677	Ethanol, 2-[(2-amino-2-methylpropyl)amino]-	$C_6H_{16}N_2O$ 132.21	68750-16-3 5	 236		chl 3 1.4698^{25}

No.	Name Synonym	Mol. Form. Mol. Wt.	CAS RN mp/°C	Merck No. bp/°C	Beil. Ref. den/g cm^{-3}	Solubility n_D
5678	Ethanol, 2-[2-[2-(3-aminopropoxy)ethoxy]ethoxy]-	$C_9H_{21}NO_4$ 207.27	49542-66-7 -50	 184^{10}	 1.0663^{20}	H_2O 4; EtOH 4 1.4668^{20}
5679	Ethanol, 2,2'-[(3-aminopropyl)imino]bis- *N*-(3-Aminopropyl)diethanolamine	$C_7H_{18}N_2O_2$ 162.23	4985-85-7	 160^{1}	3-04-00-00720	
5680	Ethanol, 1-amino-2,2,2-trichloro- Chloral ammonia	$C_2H_4Cl_3NO$ 164.42	507-47-1 73	2057 100 dec	4-01-00-03147	bz 4; eth 4; EtOH 4
5681	Ethanol, 2-azido-	$C_2H_5N_3O$ 87.08	1517-05-1	 75^{40}	4-01-00-01389 1.146^{24}	H_2O 4
5682	Ethanol, 2-[bis(3-methylbutyl)amino]- Diisopentyl 2-hydroxyethylamine	$C_{12}H_{27}NO$ 201.35	3574-43-4	 248	3-04-00-00685 0.8492^{20}	eth 4; EtOH 4; chl 4 1.4435^{20}
5683	Ethanol, 2-[bis(1-methylethyl)amino]- *N,N*-Diisopropyl-2-aminoethanol	$C_8H_{19}NO$ 145.25	96-80-0	 190	4-04-00-01504 0.826^{25}	 1.4417^{20}
5684	Ethanol, 2-[bis(2-methylpropyl)amino]- Ethanolamine, *N,N*-diisobutyl	$C_{10}H_{23}NO$ 173.30	4535-66-4	 214	4-04-00-01509 0.8407^{20}	eth 4; EtOH 4; lig 4; chl 4 1.4355^{20}
5685	Ethanol, 2-bromo- Ethylene bromohydrin	C_2H_5BrO 124.97	540-51-2	3749 150; 51^{4}	4-01-00-01385 1.7629^{20}	H_2O 5; EtOH 5; eth 5; os 4 1.4915^{20}
5686	Ethanol, 2-bromo-, acetate	$C_4H_7BrO_2$ 167.00	927-68-4 -13.8	 162.5	4-02-00-00136 1.514^{20}	H_2O 4; EtOH 5; eth 5; chl 4 1.457^{23}
5687	Ethanol, 2-butoxy- Ethylene glycol monobutyl ether	$C_6H_{14}O_2$ 118.18	111-76-2 -74.8	1559 168.4	4-01-00-02380 0.9015^{20}	H_2O 5; EtOH 5, eth 5; ctc 2 1.4198^{20}
5688	Ethanol, 2-(2-butoxyethoxy)- Diethylene glycol monobutyl ether	$C_8H_{18}O_3$ 162.23	112-34-5 -68	1557 231	4-01-00-02394 0.9553^{20}	H_2O 5; EtOH 4; eth 4; ace 4 1.4306^{20}
5689	Ethanol, 2-(2-butoxyethoxy)-, acetate	$C_{10}H_{20}O_4$ 204.27	124-17-4 -32	 245	3-02-00-00308 0.985^{20}	ace 4; eth 4; EtOH 4 1.4262^{20}
5690	Ethanol, 2-[2-(2-butoxyethoxy)ethoxy]-	$C_{10}H_{22}O_4$ 206.28	143-22-6	 278	4-01-00-02402 0.9890^{20}	EtOH 4; MeOH 4 1.4389^{20}
5691	Ethanol, 2-(butylamino)-	$C_6H_{15}NO$ 117.19	111-75-1	 199; $91\text{-}2^{11}$	3-04-00-00682 0.8907^{20}	H_2O 4; EtOH 4; eth 4 1.4437^{20}
5692	Ethanol, 2,2'-(butylimino)bis-	$C_8H_{19}NO_2$ 161.24	102-79-4	 275; 80^{35}	4-04-00-01520 0.9592^{20}	chl 3 1.4625^{20}
5693	Ethanol, 2-chloro- Ethylene chlorohydrin	C_2H_5ClO 80.51	107-07-3 -67.5	3750 128.6	4-01-00-01372 1.2019^{20}	H_2O 5; EtOH 5; eth 2; chl 3 1.4419^{20}
5694	Ethanol, 1-chloro-, acetate	$C_4H_7ClO_2$ 122.55	5912-58-3	 123	4-02-00-00282 1.110^{20}	eth 4 1.4092^{0}
5695	Ethanol, 2-chloro-, acetate β-Chloroethyl acetate	$C_4H_7ClO_2$ 122.55	542-58-5	2137 145	4-02-00-00136 1.178^{20}	H_2O 1; EtOH 5; eth 5; ctc 3 1.4234^{20}
5696	Ethanol, 2-chloro-, benzoate	$C_9H_9ClO_2$ 184.62	939-55-9 118	 255	4-09-00-00288 1.1789^{20}	eth 4; EtOH 4
5697	Ethanol, 2-chloro-, carbonate (2:1) Bis(2-chloroethyl) carbonate	$C_5H_8Cl_2O_3$ 187.02	623-97-2 8	 241	4-03-00-00006 1.3506^{20}	H_2O 1 1.461^{20}
5698	Ethanol, 2-(2-chloroethoxy)-	$C_4H_9ClO_2$ 124.57	628-89-7	 $79\text{-}81^{5}$	3-01-00-02078 1.18^{25}	H_2O 4; EtOH 5; eth 5 1.4529^{20}
5699	Ethanol, 2-chloro-, methanesulfonate	$C_3H_7ClO_3S$ 158.61	3570-58-9	 $125\text{-}6^{9}$	4-04-00-00012 1.39^{25}	chl 3 1.4562^{20}
5700	Ethanol, 2-[2-[4-[(4-chlorophenyl)phenylmethyl]-1-piperazinyl]ethoxy]- Hydroxyzine	$C_{21}H_{27}ClN_2O_2$ 374.91	68-88-2	4786	5-23-01-00462	
5701	Ethanol, 2-chloro-, phosphate (3:1)	$C_6H_{12}Cl_3O_4P$ 285.49	115-96-8	 330; 194^{10}	4-01-00-01379 1.39^{25}	ctc 3 1.4721^{20}
5702	Ethanol, 2-chloro-, phosphite (3:1)	$C_6H_{12}Cl_3O_3P$ 269.49	140-08-9	 $112\text{-}5^{2}$	4-01-00-01378 1.328^{25}	 1.4868^{20}
5703	Ethanol, 2-(dibutylamino)-	$C_{10}H_{23}NO$ 173.30	102-81-8	 114^{18}	4-04-00-01506	
5704	Ethanol, 2,2-dichloro- 2,2-Dichloroethanol	$C_2H_4Cl_2O$ 114.96	598-38-9	 146	4-01-00-01383 1.4040^{25}	H_2O 2; EtOH 3; eth 3; ctc 2 1.4626^{25}
5705	Ethanol, 2,2-diethoxy-	$C_6H_{14}O_3$ 134.18	621-63-6	 167	4-01-00-03958 0.9^{25}	eth 4; EtOH 4 1.4073^{20}
5706	Ethanol, 2-(diethylamino)- 2-(Diethylamino)ethanol	$C_6H_{15}NO$ 117.19	100-37-8	3101 163	4-04-00-01471 0.8921^{20}	H_2O 5; EtOH 3; eth 3; ace 3 1.4412^{20}
5707	Ethanol, 2-[2-(diethylamino)ethoxy]-	$C_8H_{19}NO_2$ 161.24	140-82-9	 221.5; 92^{7}	4-04-00-01474 0.9421^{25}	 1.4480^{20}
5708	Ethanol, 2,2-difluoro-	$C_2H_4F_2O$ 82.05	359-13-7 -28.2	 95.5	4-01-00-01369 1.3084^{17}	ace 4; eth 4; EtOH 4 1.3345^{11}
5709	Ethanol, 2-(dimethylamino)- Deanol	$C_4H_{11}NO$ 89.14	108-01-0 -59	2834 134	4-04-00-01424 0.8866^{20}	H_2O 5; EtOH 5; eth 5; chl 3 1.4300^{20}
5710	Ethanol, 2-[(1,1-dimethylethyl)amino]-	$C_6H_{15}NO$ 117.19	4620-70-6 44	 176.5; 72^{14}	4-04-00-01509 0.8818^{20}	
5711	Ethanol, 2,2',2'',2'''-[(4,8-di-1-piperidinylpyrimido[5,4-d]pyrimidine- Dipyridamole	$C_{24}H_{40}N_8O_4$ 504.63	58-32-2 163	3354	4-26-00-03840	
5712	Ethanol, 2-(dipropylamino)-	$C_8H_{19}NO$ 145.25	3238-75-3	 196	4-04-00-01502 0.8576^{20}	eth 4; EtOH 4 1.4402^{20}
5713	Ethanol, 2,2'-[1,2-ethanediylbis(oxy)]bis- Triethylene glycol	$C_6H_{14}O_4$ 150.17	112-27-6 -7	9585 285	4-01-00-02400 1.1274^{15}	H_2O 5; EtOH 5; eth 2; bz 5 1.4531^{20}
5714	Ethanol, 2,2'-[1,2-ethanediylbis(oxy)]bis-, diacetate Triethylene glycol, diacetate	$C_{10}H_{18}O_6$ 234.25	111-21-7 -50	 286	4-02-00-00215 1.1153^{20}	H_2O 4; eth 4; EtOH 4
5715	Ethanol, 2,2'-(1,2-ethanediyldiimino)bis-	$C_6H_{16}N_2O_2$ 148.21	4439-20-7 97.5	 $135\text{-}7^{1}$	4-04-00-01559	H_2O 3

No.	Name Synonym	Mol. Form. Mol. Wt.	CAS RN mp/°C	Merck No. bp/°C	Beil. Ref. den/g cm^{-3}	Solubility n_D
5716	Ethanol, 2-(ethenyloxy)-	$C_4H_8O_2$ 88.11	764-48-7	 141.6	4-01-00-02387 0.9821^{20}	H_2O 3; EtOH 3; eth 3; bz 3 1.4564^{17}
5717	Ethanol, 2-ethoxy- Ethylene glycol monoethyl ether	$C_4H_{10}O_2$ 90.12	110-80-5 -70	3707 135	4-01-00-02377 0.9297^{20}	H_2O 4; ace 4; eth 4; EtOH 4 1.4080^{20}
5718	Ethanol, 2-ethoxy-, acetate Ethylene glycol monoethyl ether acetate	$C_6H_{12}O_3$ 132.16	111-15-9 -61.7	3708 156.4	4-02-00-00214 0.9740^{20}	H_2O 4; ace 4; eth 4; EtOH 4 1.4054^{20}
5719	Ethanol, 2-ethoxy-, carbonate (2:1)	$C_9H_{18}O_5$ 206.24	2049-74-3	 245.5	3-03-00-00017 1.0439^{20}	ace 4; eth 4; EtOH 4 1.4227^{20}
5720	Ethanol, 2-(2-ethoxyethoxy)- Diethylene glycol monoethyl ether	$C_6H_{14}O_3$ 134.18	111-90-0	1806 196	4-01-00-02393 0.9885^{20}	H_2O 5; EtOH 5; eth 4; ace 5 1.4300^{20}
5721	Ethanol, 2-(2-ethoxyethoxy)-, acetate Diethylene glycol monoethyl ether acetate	$C_8H_{16}O_4$ 176.21	112-15-2 -25	 218.5	3-02-00-00308 1.0096^{20}	H_2O 4; ace 4; eth 4; EtOH 4 1.4213^{20}
5722	Ethanol, 2-(ethylamino)-	$C_4H_{11}NO$ 89.14	110-73-6	 169.5	4-04-00-01465 0.914^{20}	H_2O 4; EtOH 4; eth 4; chl 3 1.444^{20}
5723	Ethanol, 2,2'-(ethylimino)bis-	$C_6H_{15}NO_2$ 133.19	139-87-7 -50	 247	3-04-00-00693 1.0135^{20}	H_2O 4; EtOH 4; eth 2 1.4663^{20}
5724	Ethanol, 2-[ethyl(3-methylphenyl)amino]-	$C_{11}H_{17}NO$ 179.26	91-88-3	 $118^{1.5}$	4-12-00-01819	ctc 3 1.5540^{20}
5725	Ethanol, 2-(ethylsulfonyl)- Ethylsulfonylethyl alcohol	$C_4H_{10}O_3S$ 138.19	513-12-2	3810	4-01-00-02430	chl 2
5726	Ethanol, 2-(ethylthio)- 2-(Ethylthio)ethanol	$C_4H_{10}OS$ 106.19	110-77-0 -100	3813 184	4-01-00-02430 1.0166^{20}	H_2O 2; EtOH 3; ace 4 1.4867^{20}
5727	Ethanol, 2-fluoro- Ethylene fluorohydrin	C_2H_5FO 64.06	371-62-0 -26.4	 103.5	4-01-00-01366 1.1040^{20}	H_2O 5; EtOH 5; eth 5; ace 4 1.3647^{18}
5728	Ethanol, 2-(hexyloxy)-	$C_8H_{18}O_2$ 146.23	112-25-4 -45.1	 208	4-01-00-02383 0.8878^{20}	H_2O 2; EtOH 4; eth 4 1.4291^{20}
5729	Ethanol, 2-hydrazino- 2-Hydrazinoethanol	$C_2H_8N_2O$ 76.10	109-84-2 -70	4696 219; $120^{17.5}$	4-04-00-03348 1.119^{25}	H_2O 4; EtOH 4; MeOH 4
5730	Ethanol, 2,2'-iminobis- Diethanolamine	$C_4H_{11}NO_2$ 105.14	111-42-2 28	3097 268.8	4-04-00-01514 1.0966^{20}	H_2O 4; EtOH 4; eth 2; bz 2 1.4776^{20}
5731	Ethanol, 2-iodo-	C_2H_5IO 171.97	624-76-0	 176 dec	4-01-00-01387 2.1967^{20}	H_2O 4; eth 4; EtOH 4 1.5713^{20}
5732	Ethanol, 2-mercapto- 2-Mercaptoethanol	C_2H_6OS 78.14	60-24-2	5760 158; 55^{13}	4-01-00-02428 1.1143^{20}	H_2O 3; EtOH 3; eth 3; bz 3 1.4996^{20}
5733	Ethanol, 2-methoxy- Ethylene glycol monomethyl ether	$C_3H_8O_2$ 76.10	109-86-4 -85.1	5961 124.1	4-01-00-02375 0.9647^{20}	H_2O 5; EtOH 4; eth 5; ace 3 1.4024^{20}
5734	Ethanol, methoxy-, acetate Acetic acid, 2-methoxyethyl ester	$C_5H_{10}O_3$ 118.13	32718-56-2	 145; 40^{12}	2-02-00-00154 1.0074^{19}	H_2O 0, EtOH 3, eth 3; ctc 2 1.4002^{20}
5735	Ethanol, 2-methoxy-, acetate 2-Methoxyethyl acetate	$C_5H_{10}O_3$ 118.13	110-49-6 -70	5962 143	4-02-00-00214 1.0074^{19}	H_2O 4; EtOH 4 1.4002^{20}
5736	Ethanol, 2-methoxy-, carbonate (2:1) Bis(2-methcxyethyl) carbonate	$C_7H_{14}O_5$ 178.19	626-84-6	 231	3-03-00-00017 1.0988^{20}	EtOH 3; eth 3; ace 3 1.4204^{20}
5737	Ethanol, 2-(2-methoxyethoxy)- 2-(2-Methoxyethoxy)ethanol	$C_5H_{12}O_3$ 120.15	111-77-3	5959 193	4-01-00-02392 1.035^{20}	H_2O 5; EtOH 4; eth 4; ace 5 1.4264^{20}
5738	Ethanol, 2-(methylamino)- *N*-Methyl-2-ethanolamine	C_3H_9NO 75.11	109-83-1	5939 158	4-04-00-01422 0.937^{20}	H_2O 5; EtOH 5; eth 5 1.4385^{20}
5739	Ethanol, 2-[(3-methylbutyl)amino]- *N*-Isopentylethanolamine	$C_7H_{17}NO$ 131.22	34240-76-1	 210; $82^{5.5}$	2-04-00-00728 0.8822^{20}	H_2O 4; eth 4; EtOH 4 1.4447^{20}
5740	Ethanol, 2-(1-methylethoxy)- 2-Isopropoxyethanol	$C_5H_{12}O_2$ 104.15	109-59-1	 145	4-01-00-02380 0.9030^{20}	H_2O 5; EtOH 5; eth 5; ace 3 1.4095^{20}
5741	Ethanol, 2-[(1-methylethyl)amino]-	$C_5H_{13}NO$ 103.16	109-56-8 128.5	 173	3-04-00-00681 0.8970^{20}	H_2O 5; EtOH 5; eth 5 1.4395^{20}
5742	Ethanol, 2,2'-(methylimino)bis-	$C_5H_{13}NO_2$ 119.16	105-59-9 -21	 247	4-04-00-01517 1.038^{25}	H_2O 4 1.4685^{20}
5743	Ethanol, 2-(methylphenylamino)-	$C_9H_{13}NO$ 151.21	93-90-3	 150^{14}	4-12-00-00280 0.9993^{15}	H_2O 3; EtOH 4; eth 4; ace 4
5744	Ethanol, 2-[(2-methylphenyl)amino]-	$C_9H_{13}NO$ 151.21	136-80-1	 285.5	4-12-00-01750 1.0794^{20}	eth 4; EtOH 4 1.5675^{20}
5745	Ethanol, 2-[(4-methylphenyl)amino]-	$C_9H_{13}NO$ 151.21	2933-74-6 42.5	 287	4-12-00-01880	bz 4; eth 4; EtOH 4; chl 4
5746	Ethanol, 2-(2-methylpropoxy)-	$C_6H_{14}O_2$ 118.18	4439-24-1	 160	4-01-00-02382 0.8900^{20}	 1.4143^{20}
5747	Ethanol, 2-[(2-methylpropyl)amino]- Ethanolamine, *N*-isobutyl	$C_6H_{15}NO$ 117.19	17091-40-6	 199; 90^{16}	3-04-00-00683 0.8818^{20}	H_2O 4; EtOH 4; eth 4 1.4402^{20}
5748	Ethanol, 2-[(2-methylpropyl)amino]-, 4-aminobenzoate (ester) Butethamine	$C_{13}H_{20}N_2O_2$ 200.01	2090-89-3	1517		
5749	Ethanol, 2-(methylthio)-	C_3H_8OS 92.16	5271-38-5	 68-70^{20}	4-01-00-02429 1.063^{20}	H_2O 4; eth 4; EtOH 4 1.4861^{30}
5750	Ethanol, 2-(2-naphthalenylamino)- Ethanolamine, *N*-(β-naphthyl)	$C_{12}H_{13}NO$ 187.24	36190-77-9 52	 197-8^{3}	4-12-00-03131	eth 4
5751	Ethanol, 2,2',2''-nitrilotris- Triethanolamine	$C_6H_{15}NO_3$ 149.19	102-71-6 20.5	9581 335.4	4-04-00-01524 1.1242^{20}	H_2O 5; EtOH 5; eth 2; bz 2 1.4852^{20}
5752	Ethanol, 2-nitro- 2-Nitroethanol	$C_2H_5NO_3$ 91.07	625-48-9 -80	 194; 102^{10}	4-01-00-01388 1.270^{15}	H_2O 5; EtOH 5; eth 5; bz 1 1.4438^{19}
5753	Ethanol, 2,2'-oxybis- Diethylene glycol	$C_4H_{10}O_3$ 106.12	111-46-6 -10.4	3109 245.8	4-01-00-02390 1.1197^{15}	H_2O 3; EtOH 3; eth 3; chl 3 1.4472^{20}
5754	Ethanol, 2,2'-oxybis-, diacetate Diethylene glycol, diacetate	$C_8H_{14}O_5$ 190.20	628-68-2 18	 200	4-02-00-00216 1.1068^{15}	EtOH 4 1.4348^{20}
5755	Ethanol, 2,2'-oxybis-, dibenzoate	$C_{18}H_{18}O_5$ 314.34	120-55-8 33.5	 280^{24}; 250^{1}	4-09-00-00356 1.1690^{15}	H_2O 4; EtOH 4

No.	Name Synonym	Mol. Form. Mol. Wt.	CAS RN mp/°C	Merck No. bp/°C	Beil. Ref. den/g cm^{-3}	Solubility n_D
5756	Ethanol, 2,2'-[oxybis(2,1-ethanediyloxy)]bis- Tetraethylene glycol	$C_8H_{18}O_5$ 194.23	112-60-7 -6.2	 328	4-01-00-02403 1.1285^{15}	H_2O 4; EtOH 3; eth 3; ctc 3 1.4577^{20}
5757	Ethanol, 2-(pentyloxy)-	$C_7H_{16}O_2$ 132.20	6196-58-3	 182	4-01-00-02383 0.8918^{15}	 1.4213^{25}
5758	Ethanol, 2-phenoxy- 2-Phenoxyethanol	$C_8H_{10}O_2$ 138.17	122-99-6 14	7226 245	4-06-00-00571 1.102^{22}	H_2O 1; EtOH 3; eth 3; chl 3 1.534^{20}
5759	Ethanol, 2-(phenylamino)-	$C_8H_{11}NO$ 137.18	122-98-5	 279.5; 150^{10}	4-12-00-00277 1.0945^{20}	H_2O 2; EtOH 4; eth 4; chl 4 1.5760^{20}
5760	Ethanol, 2,2'-(phenylimino)bis-	$C_{10}H_{15}NO_2$ 181.23	120-07-0 57	 200^{10}	4-12-00-00285 1.201^{60}	ace 4; bz 4; eth 4; EtOH 4
5761	Ethanol, 2-(phenylmethoxy)-	$C_9H_{12}O_2$ 152.19	622-08-2 <-75	 256	4-06-00-02241 1.0640^{20}	H_2O 4; eth 4; EtOH 4 1.5233^{20}
5762	Ethanol, 2-[(phenylmethyl)amino]-	$C_9H_{13}NO$ 151.21	104-63-2	 $153\text{-}6^{12}$	4-12-00-02184 1.065^{25}	 1.5430^{20}
5763	Ethanol, 2-(2-propenyloxy)-	$C_5H_{10}O_2$ 102.13	111-45-5	 158.5	4-01-00-02388 0.9580^{20}	H_2O 5; EtOH 4; bz 3; ctc 3 1.4358^{20}
5764	Ethanol, 2-propoxy- 2-Propoxyethanol	$C_5H_{12}O_2$ 104.15	2807-30-9	 149.8	4-01-00-02379 0.9112^{20}	H_2O 3; EtOH 4; eth 4 1.4133^{20}
5765	Ethanol, 2-(propylamino)-	$C_5H_{13}NO$ 103.16	16369-21-4	 183	3-04-00-00681 0.9005^{20}	 1.4428^{20}
5766	Ethanol, 2,2'-thiobis- 2,2'-Thiodiethanol	$C_4H_{10}O_2S$ 122.19	111-48-8 -10.2	9259 282	4-01-00-02437 1.1793^{25}	H_2O 5; EtOH 5; eth 3; bz 2 1.5211^{20}
5767	Ethanol, 2,2,2-tribromo- Tribromoethanol	$C_2H_3Br_3O$ 282.76	75-80-9 81	9525 $92\text{-}3^{10}$	3-01-00-01362	bz 4; eth 4; EtOH 4
5768	Ethanol, 2,2,2-trichloro- 2,2,2-Trichloroethanol	$C_2H_3Cl_3O$ 149.40	115-20-8 19	9551 152; 52^{11}	4-01-00-01383	H_2O 2; EtOH 5; eth 5; ctc 2 1.4861^{20}
5769	Ethanol, 2,2,2-trichloro-1-ethoxy- Chloral alcoholate	$C_4H_7Cl_3O_2$ 193.46	515-83-3 56.5	2056 115.5	4-01-00-03144 1.143^{40}	H_2O 3; EtOH 3; eth 3
5770	Ethanol, 2,2,2-trifluoro-	$C_2H_3F_3O$ 100.04	75-89-8 -43.5	 74	4-01-00-01370 1.3842^{20}	EtOH 4; eth 3; ace 3; bz 3 1.2907^{22}
5771	Ethanone, 2-(acetyloxy)-1-phenyl-	$C_{10}H_{10}O_3$ 178.19	2243-35-8 49	 270	4-08-00-00368 1.1169^{65}	H_2O 1; EtOH 4; eth 4; bz 2 1.5036^{65}
5772	Ethanone, 1-(3-aminophenyl)-	C_8H_9NO 135.17	99-03-6 98.5	 289.5	4-14-00-00096	H_2O 2; EtOH 3
5773	Ethanone, 1-(4-aminophenyl)-	C_8H_9NO 135.17	99-92-3 106	 294; 195^{15}	4-14-00-00100	eth 4; EtOH 4
5774	Ethanone, 2-amino-1-phenyl- Phenacylamine	C_8H_9NO 135.17	613-89-8 20	7159 251	4-14-00-00114	H_2O 1; eth 3; ctc 2 1.6160^{20}
5775	Ethanone, 1-(2-benzofuranyl)-	$C_{10}H_8O_2$ 160.17	1646-26-0 76	 126^{11}	5-17-10-00172	H_2O 3
5776	Ethanone, 1-[1,1'-biphenyl]-4-yl-	$C_{14}H_{12}O$ 196.25	92-91-1 121	 326	4-07-00-01407 1.2510^{0}	H_2O 1; EtOH 4; ace 4; chl 2
5777	Ethanone, 2-bromo-1-(4-bromophenyl)- *p*-Bromophenacyl bromide	$C_8H_6Br_2O$ 277.94	99-73-0 111	1415	4-07-00-00652	H_2O 1; EtOH 3; eth 3; chl 3
5778	Ethanone, 2-bromo-1-(4-chlorophenyl)- *p*-Chlorophenacyl bromide	C_8H_6BrClO 233.49	536-38-9 96.5	2153	4-07-00-00651	
5779	Ethanone, 2-bromo-1-(4-methylphenyl)-	C_9H_9BrO 213.07	619-41-0 51	 157^{14}	4-07-00-00705	eth 4; EtOH 4
5780	Ethanone, 1-(3-bromophenyl)-	C_8H_7BrO 199.05	2142-63-4 7.5	 133^{19}	4-07-00-00646	H_2O 1; ace 3; bz 3 1.5755^{20}
5781	Ethanone, 1-(4-bromophenyl)- *p*-Bromoacetophenone	C_8H_7BrO 199.05	99-90-1 50.5	1390 257; 130^{11}	4-07-00-00647 1.647^{25}	H_2O 1; EtOH 3; eth 3; bz 3 1.647
5782	Ethanone, 2-bromo-1-phenyl- ω-Bromacetophenone	C_8H_7BrO 199.05	70-11-1 50.5	1391 135^{18}	4-07-00-00649 1.647^{20}	H_2O 1; EtOH 3; eth 4; bz 4
5783	Ethanone, 1-(5-bromo-2-thienyl)-	C_6H_5BrOS 205.08	5370-25-2 94.5	 103^{4}	5-17-09-00392	EtOH 2; ctc 3
5784	Ethanone, 2-chloro-1-(4-chlorophenyl)-	$C_8H_6Cl_2O$ 189.04	937-20-2 101.5	 270	4-07-00-00644	EtOH 3; bz 3; MeOH 3
5785	Ethanone, 2-chloro-1,2-diphenyl-	$C_{14}H_{11}ClO$ 230.69	447-31-4 68.5	 dec	3-07-00-02106	EtOH 3; chl 2; alk 1
5786	Ethanone, 2-chloro-1-(4-hydroxy-3-methoxyphenyl)-	$C_9H_9ClO_3$ 200.62	6344-28-1 102	 208^{22}	3-08-00-02114	H_2O 2; EtOH 3
5787	Ethanone, 2-chloro-1-(4-methylphenyl)-	C_9H_9ClO 168.62	4209-24-9 57.5	 261.5	4-07-00-00704	eth 4; EtOH 4
5788	Ethanone, 1-(3-chlorophenyl)-	C_8H_7ClO 154.60	99-02-5	 244; 129^{30}	4-07-00-00638 1.2130^{40}	EtOH 3; eth 3; ace 3 1.5494^{20}
5789	Ethanone, 1-(4-chlorophenyl)- *p*-Chloroacetophenone	C_8H_7ClO 154.60	99-91-2 20	2114 232	4-07-00-00639 1.1922^{20}	H_2O 1; EtOH 5; eth 5; chl 3 1.5550^{20}
5790	Ethanone, 2-chloro-1-phenyl- ω-Chloroacetophenone	C_8H_7ClO 154.60	532-27-4 56.5	2115 247	4-07-00-00641 1.324^{15}	H_2O 1; EtOH 4; eth 4; ace 3
5791	Ethanone, 1-(5-chloro-2-thienyl)-	C_6H_5ClOS 160.62	6310-09-4 52	 117^{17}; $88^{4.5}$	5-17-09-00391	H_2O 4; eth 4; EtOH 4
5792	Ethanone, 1-cyclobutyl-	$C_6H_{10}O$ 98.14	3019-25-8	 138	4-07-00-00037 0.9020^{20}	H_2O 2 1.4322^{19}
5793	Ethanone, 1-(1-cyclohexen-1-yl)-	$C_8H_{12}O$ 124.18	932-66-1 73	 201.5	4-07-00-00143 0.9655^{20}	EtOH 3; eth 3 1.4881^{20}
5794	Ethanone, 1-cyclohexyl-	$C_8H_{14}O$ 126.20	823-76-7	 180.5	4-07-00-00053 0.9176^{20}	H_2O 1; eth 3 1.4565^{16}
5795	Ethanone, 1-cyclopentyl-	$C_7H_{12}O$ 112.17	6004-60-0	 158.5	4-07-00-00047 0.916^{20}	eth 3 1.4409^{20}

No.	Name Synonym	Mol. Form. Mol. Wt.	CAS RN mp/°C	Merck No. bp/°C	Beil. Ref. den/g cm^{-3}	Solubility n_D
5796	Ethanone, 1-cyclopropyl- Cyclopropyl methyl ketone	C_5H_8O 84.12	765-43-5 -68.3	 111.3	4-07-00-00014 0.8984^{20}	H_2O 4; eth 4; EtOH 4 1.4251^{20}
5797	Ethanone, 1-(3,5-dibromophenyl)-	$C_8H_6Br_2O$ 277.94	14401-73-1 68	 198^{15}	3-07-00-00984	bz 4; eth 4; EtOH 4
5798	Ethanone, 2,2-dibromo-1-phenyl-	$C_8H_6Br_2O$ 277.94	13665-04-8 36.5	 159^{13}	4-07-00-00653	EtOH 3; eth 3; chl 3
5799	Ethanone, 1-(2,4-dichlorophenyl)-	$C_8H_6Cl_2O$ 189.04	2234-16-4 33.5		4-07-00-00643	H_2O 1 1.5640^{20}
5800	Ethanone, 1-(2,5-dichlorophenyl)-	$C_8H_6Cl_2O$ 189.04	2476-37-1 12	 118^{12}	4-07-00-00643 1.321^{30}	 1.5595^{30}
5801	Ethanone, 1-(3,4-dichlorophenyl)-	$C_8H_6Cl_2O$ 189.04	2642-63-9 76	 135^{12}	4-07-00-00643	H_2O 1; ctc 3; lig 3
5802	Ethanone, 2,2-dichloro-1-phenyl-	$C_8H_6Cl_2O$ 189.04	2648-61-5 20.5	 249	4-07-00-00644 1.340^{16}	EtOH 3; bz 3; ctc 3 1.5686^{20}
5803	Ethanone, 1-(2,4-dihydroxyphenyl)- Resacetophenone	$C_8H_8O_3$ 152.15	89-84-9 146	8143	4-08-00-01792 1.18^{141}	H_2O 1; EtOH 3; eth 2; bz 2
5804	Ethanone, 1-(3,4-dihydroxyphenyl)-2-(methylamino)- Adrenalone	$C_9H_{11}NO_3$ 181.19	99-45-6 235-6 dec	161	4-14-00-00832	H_2O 2; EtOH 2; eth 2
5805	Ethanone, 1-(3,4-dimethoxyphenyl)-	$C_{10}H_{12}O_3$ 180.20	1131-62-0 51	 287	4-08-00-01815	H_2O 4; bz 4; EtOH 4; chl 4
5806	Ethanone, 1-[3-(dimethylamino)phenyl]-	$C_{10}H_{13}NO$ 163.22	18992-80-8 43	 148^{13}	4-14-00-00096	ace 4; eth 4
5807	Ethanone, 1-[10-[3-(dimethylamino)propyl]-10H-phenothiazin-2-yl]- Acepromazine	$C_{19}H_{22}N_2OS$ 326.46	61-00-7	27 $230^{0.5}$	4-27-00-02911	
5808	Ethanone, 1-[4-(1,1-dimethylethyl)phenyl]-	$C_{12}H_{16}O$ 176.26	943-27-1	 $136\text{-}8^{20}$	4-07-00-00769 0.9705^{25}	 1.518^{15}
5809	Ethanone, 1-(2,4-dimethylphenyl)-	$C_{10}H_{12}O$ 148.20	89-74-7	 228	4-07-00-00729 1.0121^{15}	eth 4; EtOH 4 1.5340^{20}
5810	Ethanone, 1-(2,5-dimethylphenyl)-	$C_{10}H_{12}O$ 148.20	2142-73-6 -18.1	 232.5	4-07-00-00728 0.9963^{19}	H_2O 1; EtOH 4; eth 4; bz 4 1.5291^{20}
5811	Ethanone, 1-(3,4-dimethylphenyl)-	$C_{10}H_{12}O$ 148.20	3637-01-2 -3	 246.5	4-07-00-00728 1.0090^{14}	H_2O 1; EtOH 4; eth 4; bz 4 1.5413^{15}
5812	Ethanone, 1,2-diphenyl-	$C_{14}H_{12}O$ 196.25	451-40-1 60	 320	4-07-00-01393 1.201^{0}	H_2O 2; EtOH 3; eth 3; ctc 3
5813	Ethanone, 2-ethoxy-1,2-diphenyl-	$C_{16}H_{16}O_2$ 240.30	574-09-4 62	 $194\text{-}5^{20}$	3-08-00-01279 1.1010^{17}	bz 4; eth 4; EtOH 4; lig 4 1.5727^{17}
5814	Ethanone, 2-ethoxy-1-phenyl- 2-Ethoxyacetophenone	$C_{10}H_{12}O_2$ 164.20	14869-39-7 43	 243.5	4-08-00-00365 1.0036^{78}	eth 4; EtOH 4; lig 4
5815	Ethanone, 1-(4-fluorophenyl)-	C_8H_7FO 138.14	403-42-9 -45	 196	4-07-00-00635 1.1382^{25}	H_2O 1; bz 3; chl 3 1.5081^{25}
5816	Ethanone, 2-fluoro-1-phenyl-	C_8H_7FO 138.14	450-95-3 29	 $90\text{-}1^{12}$	4-07-00-00636 1.152^{20}	 1.5200^{20}
5817	Ethanone, 1-(2-furanyl)-	$C_6H_6O_2$ 110.11	1192-62-7 33	 175	5-17-09-00381 1.098^{20}	H_2O 1; EtOH 3; eth 3 1.5017^{20}
5818	Ethanone, 1-[4-(β-*D*-glucopyranosyloxy)phenyl]- Picein	$C_{14}H_{18}O_7$ 298.29	530-14-3 195.5	7367	5-17-07-00151	H_2O 2; EtOH 3; eth 3; chl 1
5819	Ethanone, 1-(1H-indol-3-yl)-	$C_{10}H_9NO$ 159.19	703-80-0 192.3	 144^{10}	5-21-08-00297	EtOH 4
5820	Ethanone, 1-(1H-pyrrol-2-yl)-	C_6H_7NO 109.13	1072-83-9 90	 220	5-21-07-00204	H_2O 3; EtOH 3; eth 3
5821	Ethanone, 1-(1-hydroxycyclohexyl)-	$C_8H_{14}O_2$ 142.20	1123-27-9	 125.5; 91^{11}	4-08-00-00023 1.0248^{25}	eth 4; EtOH 4 1.4670^{25}
5822	Ethanone, 2-hydroxy-1,2-diphenyl-, (±)- (±)-Benzoin	$C_{14}H_{12}O_2$ 212.25	579-44-2 137	1103 344; 194^{12}	4-08-00-01279 1.310^{20}	EtOH 4; chl 4
5823	Ethanone, 1-(2-hydroxy-4-methoxyphenyl)-	$C_9H_{10}O_3$ 166.18	552-41-0 52.5	 158^{20}	4-08-00-01793 1.3102^{81}	bz 4; eth 4; EtOH 4; chl 4 1.5452^{81}
5824	Ethanone, 1-(4-hydroxy-3-methoxyphenyl)- Apocynin	$C_9H_{10}O_3$ 166.18	498-02-2 115	772 297; 234^{15}	4-08-00-01814	H_2O 2; EtOH 3; eth 4; ace 3
5825	Ethanone, 1-[6-hydroxy-2-(1-methylethenyl)-5-benzofuranyl]- Euparin	$C_{13}H_{12}O_3$ 216.24	532-48-9 121.5	3857	5-18-02-00029	eth 3; bz 3; chl 3; NaOH 2
5826	Ethanone, 1-(2-hydroxy-4-methylphenyl)-	$C_9H_{10}O_2$ 150.18	6921-64-8 21	 245	4-08-00-00480 1.1012^{10}	 1.5527^{13}
5827	Ethanone, 1-(2-hydroxy-5-methylphenyl)-	$C_9H_{10}O_2$ 150.18	1450-72-2 50	 210; 120^{20}	[illegible] 1.0797^{53}	bz 4; eth 4; EtOH 4; chl 4
5828	Ethanone, 1-(4-hydroxy-2-methylphenyl)-	$C_9H_{10}O_2$ 150.18	875-59-2 128	 313	4-08-00-00469 1.0592^{135}	eth 4; EtOH 4 1.5369^{135}
5829	Ethanone, 1-(1-hydroxy-2-naphthalenyl)-	$C_{12}H_{10}O_2$ 186.21	711-79-5 101	 325 dec	4-08-00-01175	bz 4; HOAc 4
5830	Ethanone, 1-(2-hydroxyphenyl)-	$C_8H_8O_2$ 136.15	118-93-4 5	 218	4-08-00-00320 1.1307^{20}	eth 4; EtOH 4; HOAc 4 1.5584^{20}
5831	Ethanone, 1-(3-hydroxyphenyl)-	$C_8H_8O_2$ 136.15	121-71-1 96	 296; 153^{5}	4-08-00-00334 1.0992^{109}	H_2O 2; EtOH 4; eth 4; bz 4 1.5348^{109}
5832	Ethanone, 1-(4-hydroxyphenyl)-	$C_8H_8O_2$ 136.15	99-93-4 109.5	 $147\text{-}8^{3}$	4-08-00-00339 1.1090^{109}	H_2O 2; EtOH 4; eth 4 1.5577^{109}
5833	Ethanone, 2-hydroxy-1-phenyl-	$C_8H_8O_2$ 136.15	582-24-1 90	 125^{12}; 56^{1}	4-08-00-00365 1.0963^{99}	H_2O 3; EtOH 3; eth 3; chl 3

No.	Name	Synonym	Mol. Form.	Mol. Wt.	CAS RN	mp/°C	Merck No.	bp/°C	Beil. Ref.	den/g cm^{-3}	Solubility	n_D
5834	Ethanone, 2-[6-(2-hydroxy-2-phenylethyl)-1-methyl-2-piperidinyl]-1-phenyl-, [2*R*-[2α,6α(*S**)]]-	Lobeline	$C_{22}H_{27}NO_2$	337.46	90-69-7	130.5	5432		5-21-12-00627		H_2O 2; EtOH 3; eth 3; ace 4	
5835	Ethanone, 1-(3-hydroxy-2-thienyl)-		$C_6H_6O_2S$	142.18	5556-07-0	51.5		$47^{0.2}$	5-18-01-00121	1.5000^{20}	EtOH 3	1.5795^{20}
5836	Ethanone, 1-(3-iodophenyl)-		C_8H_7IO	246.05	14452-30-3			129^{8}; 117^{4}	4-07-00-00654		bz 3	1.622^{20}
5837	Ethanone, 1-(4-iodophenyl)-		C_8H_7IO	246.05	13329-40-3	86		153^{18}	4-07-00-00654		EtOH 3; eth 2; bz 3; HOAc 3	
5838	Ethanone, 2-iodo-1-phenyl-		C_8H_7IO	246.05	4636-16-2	34.4		154^{15}	4-07-00-00654		ace 3	
5839	Ethanone, 2-methoxy-1,2-diphenyl-		$C_{15}H_{14}O_2$	226.27	3524-62-7	49.5		$188\text{-}9^{15}$	4-08-00-01280	1.1278^{14}	bz 4; eth 4; EtOH 4	
5840	Ethanone, 1-(3-methoxyphenyl)-		$C_9H_{10}O_2$	150.18	586-37-8	95.5		240	4-08-00-00334	1.0343^{19}	H_2O 3; EtOH 3; ace 3; ctc 3	1.5410^{20}
5841	Ethanone, 1-(4-methoxyphenyl)-		$C_9H_{10}O_2$	150.18	100-06-1	38.5		258	4-08-00-00340	1.0818^{41}	H_2O 2; EtOH 3; eth 3; ace 3	1.547^{41}
5842	Ethanone, 2-methoxy-1-phenyl-		$C_9H_{10}O_2$	150.18	4079-52-1	8		245; 125^{19}	4-08-00-00365	1.0897^{20}	H_2O 2; EtOH 3; ace 3	1.5393^{20}
5843	Ethanone, 1-(1-methylcyclohexyl)-		$C_9H_{16}O$	140.23	2890-62-2			186.2	4-07-00-00067	0.9504^{20}		1.4543^{20}
5844	Ethanone, 1-(3-methylcyclohexyl)-		$C_9H_{16}O$	140.23	7193-78-4			197	3-07-00-00120	0.912^{19}		1.4517^{19}
5845	Ethanone, 1-(4-methylcyclohexyl)-		$C_9H_{16}O$	140.23	1879-06-7			196	3-07-00-00121	0.9055^{18}		1.4509^{18}
5846	Ethanone, 1-(1-methylcyclopentyl)-		$C_8H_{14}O$	126.20	13388-93-7			48.4^{10}	3-07-00-00102	0.9104^{20}		1.4430^{20}
5847	Ethanone, 1-(2-methylcyclopentyl)-		$C_8H_{14}O$	126.20	1601-00-9			170.5; 59.6^{15}	3-07-00-00103	0.9222^{4}		1.4434^{16}
5848	Ethanone, 1-[4-(1-methylethyl)phenyl]-		$C_{11}H_{14}O$	162.23	645-13-6			254	4-07-00-00747	0.9753^{15}		1.5235^{20}
5849	Ethanone, 1-(1-methyl-1H-pyrrol-2-yl)-		C_7H_9NO	123.15	932-16-1			201^{252}; 93^{22}	5-21-07-00205	1.0445^{15}	EtOH 3; bz 3; chl 3	1.5403^{15}
5850	Ethanone, 1-[2-methyl-5-(1-methylethyl)phenyl]-		$C_{12}H_{16}O$	176.26	1202-08-0	<-20		249.5	4-07-00-00772	0.956^{20}		1.5181^{20}
5851	Ethanone, 1-(2-methylphenyl)-		$C_9H_{10}O$	134.18	577-16-2			214	4-07-00-00697	1.026^{20}		1.5276^{20}
5852	Ethanone, 1-(3-methylphenyl)-		$C_9H_{10}O$	134.18	585-74-0			220	4-07-00-00699	1.0165^{0}	EtOH 3; eth 3; ace 3; ctc 2	1.533^{15}
5853	Ethanone, 1-(4-methylphenyl)-		$C_9H_{10}O$	134.18	122-00-9	28		226; 93.5^{7}	4-07-00-00701	1.0051^{20}	bz 4; eth 4; EtOH 4; chl 4	1.5335^{20}
5854	Ethanone, 2,2'-(1-methyl-2,6-piperidinediyl)bis[1-phenyl-, *cis*-	Lobelanine	$C_{22}H_{25}NO_2$	335.45	579-21-5	99	5430		5-21-11-00432		ace 4; bz 4; EtOH 4; chl 4	
5855	Ethanone, 1-(6-methyl-3-pyridinyl)-		C_8H_9NO	135.17	36357-38-7	17.6		144^{50}	5-21-07-00425	1.0168^{25}	H_2O 4	1.5302^{25}
5856	Ethanone, 1-(5-methyl-2-thienyl)-		C_7H_8OS	140.21	13679-74-8	27.5		232.5	5-17-09-00425	1.1185^{25}	eth 3; ace 3; bz 3; chl 2	1.5604
5857	Ethanone, 1-(1-naphthalenyl)-		$C_{12}H_{10}O$	170.21	941-98-0	34		297	4-07-00-01292	1.1171^{21}	H_2O 1; EtOH 3; eth 3; ace 3	1.6280^{22}
5858	Ethanone, 1-(2-naphthalenyl)-		$C_{12}H_{10}O$	170.21	93-08-3	56		302	4-07-00-01294		EtOH 2; ctc 2	
5859	Ethanone, 1-(1-naphthalenyl)-2-phenyl-	Benzyl 1-naphthyl ketone	$C_{18}H_{14}O$	246.31	605-85-6	66.5		$194\text{-}6^{0.05}$	4-07-00-01790		eth 4	
5860	Ethanone, 1-(3-nitrophenyl)-		$C_8H_7NO_3$	165.15	121-89-1	81		202; 167^{18}	4-07-00-00656		H_2O 4; EtOH 2; eth 4; chl 2	
5861	Ethanone, 1-(4-nitrophenyl)-		$C_8H_7NO_3$	165.15	100-19-6	81.8		165^{5}	4-07-00-00657		eth 4; EtOH 4	
5862	Ethanone, 2-nitro-1-phenyl-		$C_8H_7NO_3$	165.15	614-21-1	106		158^{16}; 142^{10}	4-07-00-00659		eth 4; EtOH 4	1.5468^{30}
5863	Ethanone, 1-(2-nitrophenyl)-	Acetophene,2-nitro	$C_8H_7N_3O$	161.16	577-59-3	28.5		178^{32}; 158^{16}	4-07-00-00655	1.2370^{25}	H_2O 1; EtOH 4; eth 4; chl 4	1.5468^{20}
5864	Ethanone, 1-phenyl-	Acetophenone	C_8H_8O	120.15	98-86-2	20	65	202	4-07-00-00619	1.0281^{20}	H_2O 1; EtOH 3; eth 3; ace 3	1.5372^{20}
5865	Ethanone, 1,1'-(1,3-phenylene)bis-		$C_{10}H_{10}O_2$	162.19	6781-42-6	32		152^{15}	4-07-00-02156		H_2O 2; EtOH 3; bz 3; chl 3	
5866	Ethanone, 1-phenyl-, oxime		C_8H_9NO	135.17	613-91-2	60		245	3-07-00-00954	1.0515^{78}	H_2O 2; EtOH 4; eth 4; ace 4	
5867	Ethanone, 1-(2-pyridinyl)-		C_7H_7NO	121.14	1122-62-9			192	5-21-07-00385	1.077^{25}	EtOH 3; eth 3; ctc 2; HOAc 3	1.5203^{20}
5868	Ethanone, 1-(3-pyridinyl)-	Methyl 3-pyridyl ketone	C_7H_7NO	121.14	350-03-8	13.5	6034	220	5-21-07-00394		H_2O 3; EtOH 3; eth 3; acid 3	1.5341^{20}
5869	Ethanone, 1-(4-pyridinyl)-		C_7H_7NO	121.14	1122-54-9	16		212	5-21-07-00400	1.097^{25}	EtOH 2; eth 2; acid 2	1.5282^{25}

PHYSICAL CONSTANTS OF ORGANIC COMPOUNDS (continued)

No.	Name	Synonym	Mol. Form.	Mol. Wt.	CAS RN	mp/°C	Merck No.	bp/°C	Beil. Ref.	den/g cm^{-3}	Solubility	n_D
5870	Ethanone, 1-(2-thienyl)-		C_6H_6OS	126.18	88-15-3	10.5		213.5	5-17-09-00387	1.1679^{20}	H_2O 2; EtOH 5; eth 5; ctc 3	1.5667^{20}
5871	Ethanone, 2,2,2-trichloro-1-phenyl-		$C_8H_5Cl_3O$	223.49	2902-69-4			256.5	4-07-00-00645	1.425^{16}	eth 4; EtOH 4	
5872	Ethanone, 2,2,2-trifluoro-1-phenyl-		$C_8H_5F_3O$	174.12	434-45-7	-40		153	4-07-00-00637	1.279^{20}		1.4583^{20}
5873	Ethanone, 1-(2,3,4-trihydroxyphenyl)-	Gallacetophenone	$C_8H_8O_4$	168.15	528-21-2	173	4248		4-08-00-02721		H_2O 3; EtOH 4; eth 3; ace 4	
5874	Ethanone, 1-(2,3,4-trimethoxyphenyl)-		$C_{11}H_{14}O_4$	210.23	13909-73-4	15.8		296	4-08-00-02721			1.5384^{20}
5875	Ethanone, 1-(2,4,5-trimethylphenyl)-		$C_{11}H_{14}O$	162.23	2040-07-5	10.5		246.5	4-07-00-00751	1.0039^{15}	H_2O 1; EtOH 4; eth 4; bz 4	1.541^{15}
5876	Ethanone, 1-(2,4,6-trimethylphenyl)-		$C_{11}H_{14}O$	162.23	1667-01-2			241; 120^{12}	4-07-00-00749	0.9754^{20}	H_2O 1; EtOH 3; eth 3; ace 3	1.5175^{20}
5877	Ethanone, 1-(3,4,5-trimethylphenyl)-		$C_{11}H_{14}O$	162.23	2047-21-4	4.7		101.5^{3}	4-07-00-00749	1.0037^{25}		1.5420^{25}
5878	Ethenaminium, *N,N,N*-trimethyl-, hydroxide	Neurine	$C_5H_{13}NO$	103.16	463-88-7		6393		3-04-00-00442		H_2O 4; eth 4; EtOH 4	
5879	Ethene	Ethylene	C_2H_4	28.05	74-85-1	-169	3748	-103.7	4-01-00-00677	0.5678^{-104}	H_2O 1; EtOH 2; eth 3; ace 2	1.363^{100}
5880	Ethene, bromo-	Vinyl bromide	C_2H_3Br	106.95	593-60-2	-137.8		15.8	4-01-00-00718	1.4933^{20}	H_2O 1; EtOH 3; eth 3; ace 3	1.4380^{20}
5881	Ethene, 1-bromo-2-chloro-	Ethylene, 1-bromo-2-chloro-	C_2H_2BrCl	141.39	3018-09-5	-86.7		84.6	3-01-00-00671	1.7972^{15}		1.4982
5882	Ethene, 2-bromo-1,1-dichloro-		C_2HBrCl_2	175.84	5870 61-1	-88.5		107.5	4-01-00-00720	1.9053^{15}		
5883	Ethene, 1-bromo-1,2-difluoro-	1-Bromo-1,2-difluoroethylene	C_2HBrF_2	142.93	358-99-6			19	0-01-00-00189	1.8434^{25}		1.3846^{0}
5884	Ethene, chloro-	Vinyl chloride	C_2H_3Cl	62.50	75-01-4	-153.7	9898	-13.3	4-01-00-00700	0.9106^{20}	H_2O 2; EtOH 3; eth 4	1.3700^{20}
5885	Ethene, 2-chloro-1,1-difluoro-	1,1-Difluoro-2-chloroethylene	C_2HClF_2	98.48	359-10-4	-138.5		-18.5	4-01-00-00703			
5886	Ethene, 1-chloro-2-ethoxy-	2-Chlorovinyl ethyl ether	C_4H_7ClO	106.55	928-56-3			120	4-01-00-02061	1.0386^{20}		1.4385^{20}
5887	Ethene, (2-chloroethoxy)-	2-Chloroethyl vinyl ether	C_4H_7ClO	106.55	110-75-8	-70	2139	108	4-01-00-02051	1.0495^{20}	EtOH 4; eth 4; chl 2	1.4378^{20}
5888	Ethene, 1-chloro-2-iodo-		C_2H_2ClI	188.40	20244-71-7	-38.2		119	3-01-00-00674	2.2298^{25}		
5889	Ethene, chlorotrifluoro-	Chlorotrifluoroethylene	C_2ClF_3	116.47	79-38-9	-158		-27.8	4-01-00-00704	1.54^{-60}	bz 3; chl 3	1.38^{0}
5890	Ethene, 1,1-dibromo-		$C_2H_2Br_2$	185.85	593-92-0			92	4-01-00-00720	2.1779^{21}	EtOH 3; eth 3; ace 3; bz 3	
5891	Ethene, 1,2-dibromo-, (*E*)-	*trans*-1,2-Dibromoethylene	$C_2H_2Br_2$	185.85	590-12-5	-6.5		108	4-01-00-00721	2.2308^{20}	H_2O 1; EtOH 4; eth 4; ace 3	1.5505^{18}
5892	Ethene, 1,2-dibromo-, (*Z*)-	*cis*-1,2-Dibromoethylene	$C_2H_2Br_2$	185.85	590-11-4	-53		112.5	4-01-00-00720	2.2464^{20}	H_2O 1; EtOH 4; eth 4; ace 3	1.5428^{20}
5893	Ethene, 1,1-dibromo-2-ethoxy-	Ethylene, 1,1-dibromo-2-ethoxy	$C_4H_6Br_2O$	229.90	77295-79-5			172, 74^{15}	2-01-00-00473	1.7697^{18}	eth 4	
5894	Ethene, 1,1-dichloro-	Vinylidene chloride	$C_2H_2Cl_2$	96.94	75-35-4	-122.5	9900	31.6	4-01-00-00706	1.213^{20}	H_2O 1; EtOH 3; eth 4; ace 3	1.4249^{20}
5895	Ethene, 1,2-dichloro-, (*E*)-	*trans*-1,2-Dichloroethylene	$C_2H_2Cl_2$	96.94	156-60-5	-49.8	86	48.7	4-01-00-00709	1.2565^{20}	H_2O 2; EtOH 5; eth 5; ace 5	1.4454^{20}
5896	Ethene, 1,2-dichloro-, (*Z*)-	*cis*-1,2-Dichloroethylene	$C_2H_2Cl_2$	96.94	156-59-2	-80	86	60.1	4-01-00-00707	1.2837^{20}	H_2O 2; EtOH 5; eth 5; ace 5	1.4490^{20}
5897	Ethene, 1,1-dichloro-2,2-difluoro-		$C_2Cl_2F_2$	132.92	79-35-6	-116		19	4-01-00-00711	1.555^{-20}		1.383^{-20}
5898	Ethene, 1,2-dichloro-1,2-difluoro-		$C_2Cl_2F_2$	132.92	598-88-9	-130.5		21.1		1.4950^{0}		1.3777^{0}
5899	Ethene, 1,2-dichloro-1-ethoxy-		$C_4H_6Cl_2O$	141.00	42345-82-4			128.2	3-01-00-02950	1.1972^{25}		1.4558^{17}
5900	Ethene, 1,1-dichloro-2-fluoro-	1,1-Dichloro-2-fluoroethylene	C_2HCl_2F	114.93	359-02-4	-108.8		37.5	4-01-00-00711	1.3732^{16}		1.4031^{16}
5901	Ethene, 1,1-diethoxy-		$C_6H_{12}O_2$	116.16	2678-54-8			68^{100}	4-01-00-03420	0.7932^{20}		1.3643^{21}
5902	Ethene, 1,1-difluoro-	Vinylidene fluoride	$C_2H_2F_2$	64.03	75-38-7	-144		-85.7	4-01-00-00696		eth 4; EtOH 4	
5903	Ethene, 1,2-difluoro-, (*Z*)-	*cis*-1,2-Difluoroethylene	$C_2H_2F_2$	64.03	1630-77-9							
5904	Ethene, 1,2-diiodo-, (*Z*)-		$C_2H_2I_2$	279.85	590-26-1	-14		72.5^{16}	4-01-00-00724	3.0625^{20}	eth 3; chl 3	
5905	Ethene, ethoxy-	Ethyl vinyl ether	C_4H_8O	72.11	109-92-2	-115.8		35.5	4-01-00-02049	0.7589^{20}	H_2O 2; EtOH 3; eth 5; ctc 2	1.3767^{20}
5906	Ethene, fluoro-	Vinyl fluoride	C_2H_3F	46.04	75-02-5	-160.5		-72	4-01-00-00694		H_2O 1; EtOH 3; ace 3	
5907	Ethene, iodo-	Iodoethylene	C_2H_3I	153.95	593-66-8			56	4-01-00-00722	2.037^{20}	eth 4; EtOH 4	1.5385^{20}
5908	Ethene, methoxy-	Methyl vinyl ether	C_3H_6O	58.08	107-25-5	-122		5.5	4-01-00-02049	0.7725^{0}	H_2O 2; EtOH 4; eth 4; ace 4	1.3730^{0}
5909	Ethene, (methylsulfonyl)-		$C_3H_6O_2S$	106.15	3680-02-2			$122\text{-}4^{24}$	4-01-00-02065	1.2117^{20}	eth 3; ace 3	1.4636^{20}

No.	Name	Synonym	Mol. Form.	Mol. Wt.	CAS RN	mp/°C	Merck No.	bp/°C	Beil. Ref.	den/g cm^{-3}	Solubility	n_D
5910	Ethene, (methylthio)-		C_3H_6S	74.15	1822-74-8			69.5	4-01-00-02065	0.9026^{20}	eth 3; ace 3; chl 3	1.4837^{20}
5911	Ethene, nitro-		$C_2H_3NO_2$	73.05	3638-64-0	-55.5		98.5	4-01-00-00725	1.2212^{14}	EtOH 4; eth 4; ace 4; bz 4	1.4282^{20}
5912	Ethene, 1,1'-oxybis-	Divinyl ether	C_4H_6O	70.09	109-93-3	-101	9899	28.3	4-01-00-02058	0.773^{20}	H_2O 1; EtOH 5; eth 5; ace 5	1.3989^{20}
5913	Ethene, 1,1'-sulfonylbis-		$C_4H_6O_2S$	118.16	77-77-0	-26		234.5	4-01-00-02068	1.177^{25}		1.4765^{20}
5914	Ethene, tetrabromo-	Tetrabromoethylene	C_2Br_4	343.64	79-28-7	56.5		226	4-01-00-00722		H_2O 1; EtOH 3; eth 3; ace 3	
5915	Ethenetetracarbonitrile	Tetracyanoethylene	C_6N_4	128.09	670-54-2	199	9129	223	4-02-00-02450	1.348^{25}	eth 2; ace 3; bz 2; ctc 2	1.560^{25}
5916	Ethenetetracarboxylic acid, tetraethyl ester		$C_{14}H_{20}O_8$	316.31	6174-95-4	58		328 dec; 203^{13}	4-02-00-02450		H_2O 1; EtOH 4; eth 4	
5917	Ethene, tetrachloro-	Tetrachloroethylene	C_2Cl_4	165.83	127-18-4	-22.3	9126	121.3	4-01-00-00715	1.6227^{20}	H_2O 1; EtOH 5; eth 5; bz 5	1.5053^{20}
5918	Ethene, tetrafluoro-	Tetrafluoroethylene	C_2F_4	100.02	116-14-3	-142.5		-75.9	4-01-00-00698	1.519^{-76}	H_2O 1	
5919	Ethene, tetraiodo-	Tetraiodoethylene	C_2I_4	531.64	513-92-8	187	9151	sub	4-01-00-00724	2.983^{20}	bz 4; chl 4	
5920	Ethene, 1,1'-thiobis-		C_4H_6S	86.16	627-51-0	20		84	4-01-00-02068	0.9174^{15}	H_2O 2; EtOH 5; eth 5; ace 3	
5921	Ethene, tribromo-		C_2HBr_3	264.74	598-16-3			164	4-01-00-00722	2.708^{20}	H_2O 2; EtOH 4; eth 3; ace 3	1.6045^{16}
5922	Ethene, trichloro-	Trichloroethylene	C_2HCl_3	131.39	79-01-6	-84.7	9552	87.2	4-01-00-00712	1.4642^{20}	H_2O 2; EtOH 5; eth 5; ace 3	1.4773^{20}
5923	Ethene, trichlorofluoro-		C_2Cl_3F	149.38	359-29-5	-108.9		71	4-01-00-00715	1.5460^{20}	chl 3	1.4379^{20}
5924	Ethene, trifluoro-	Trifluoroethylene	C_2HF_3	82.03	359-11-5			-51	4-01-00-00697	1.26^{-70}	H_2O 1; EtOH 2; eth 3	
5925	Ethene, trifluoroiodo-		C_2F_3I	207.92	359-37-5			30	4-01-00-00723	2.284^{25}		1.4143^{0}
5926	1,4-Etheno-3H,7H-benzo[1,2-c:3,4-c']dipyran-3,7-dione, 9-(3-furanyl)-1,4,4a,5,6,6a,9,10,10a,10b-decahydro-4-hydroxy-4a,10a-dimethyl-, [1R-(1α,4β,4aα,6aβ,9β,10aα,10b)]	Columbin	$C_{20}H_{22}O_6$	358.39	546-97-4		2489		5-19-10-00546			
5927	Ethenone	Ketene	C_2H_2O	42.04	463-51-4	-151	5177	-49.8	4-01-00-03418		eth 2; ace 2	
5928	Ethenone, diphenyl-	Diphenylketene	$C_{14}H_{10}O$	194.23	525-06-4		3327	267.5	4-07-00-01653	1.1107^{13}		1.615^{14}
5929	Ethephon	Phosphonic acid, (2-chloroethyl)-	$C_2H_6ClO_3P$	144.49	16672-87-0	74	3686			1.2		
5930	Ethiolate	Carbamothioic acid, diethyl-, *S*-ethyl ester	$C_7H_{15}NOS$	161.27	2941-55-1			$52^{0.5}$, 142^{87}		0.9791^{30}		
5931	Ethion	Phosphorodithioic acid, *S,S'*-methylene *O,O,O',O'*-tetraethyl ester	$C_9H_{22}O_4P_2S_4$	384.48	563-12-2	-13	3691	$165^{0.3}$		1.22^{20}		
5932	Ethirimol	4(1H)-Pyrimidinone, 5-butyl-2-(ethylamino)-6-methyl-	$C_{11}H_{19}N_3O$	209.29	23947-60-6	160	3695			1.21^{25}		
5933	Ethofumesate	2-Ethoxy-2,3-dihydro-3,3-dimethyl-5-benzofuranyl methanesulfonate	$C_{13}H_{18}O_5S$	286.35	26225-79-6	71	3697			1.14		
5934	Ethoprop	Phosphorodithioic acid, *O*-ethyl *S,S*-dipropyl ester	$C_8H_{19}O_2PS_2$	242.34	13194-48-4		3702	$88^{0.2}$		1.094^{20}		
5935	Ethylene, 1-bromo-1,2-dichloro-, *trans*-	Ethylene, 1-bromo-1,2-dichloro (dis)	C_2HBrCl_2	175.84	6795-75-1	-83.5		113.8	3-01-00-00671	1.9133^{15}		1.5218^{15}
5936	Ethyne	Acetylene	C_2H_2	26.04	74-86-2	-80.7 (trip. pt.)	84	-84.7 (subl. point)	4-01-00-00939	*0.377^{25}	H_2O 2; EtOH 2; ace 3; bz 3	
5937	Ethyne, bromo-	Bromoacetylene	C_2HBr	104.93	593-61-3			4.7	3-01-00-00919		eth 4	
5938	Ethyne, chloro-	Chloroacetylene	C_2HCl	60.48	593-63-5	-126		-30	4-01-00-00957		EtOH 2	
5939	Ethyne, dibromo-		C_2Br_2	183.83	624-61-3	-24		exp 76	3-01-00-00919		ace 4; bz 4; eth 4; EtOH 4	
5940	Ethyne, dichloro-	Dichloroacetylene	C_2Cl_2	94.93	7572-29-4	-66		33	4-01-00-00957	1.261^{20}	EtOH 3; eth 3; ace 3	1.42790^{20}
5941	Ethyne, diiodo-	Diiodoacetylene	C_2I_2	277.83	624-74-8	81.5		exp	4-01-00-00958		ace 4; bz 4; eth 4; EtOH 4	
5942	Ethyne, ethoxy-		C_4H_6O	70.09	927-80-0			50	4-01-00-02211	0.8000^{20}		1.3796^{20}
5943	Ethyne, fluoro-	Fluoroacetylene	C_2HF	44.03	2713-09-9				4-01-00-00957			
5944	Ethyne, methoxy-		C_3H_4O	56.06	6443-91-0				4-01-00-02211	0.8001^{20}	eth 4; EtOH 4	1.3812^{20}

No.	Name Synonym	Mol. Form. Mol. Wt.	CAS RN mp/°C	Merck No. bp/°C	Beil. Ref. den/g cm^{-3}	Solubility n_D
5945	Etrimfos Phosphorothioic acid, *O*-(6-ethoxy-2-ethyl-4-pyrimidinyl) *O,O*-dimethyl ester	$C_{10}H_{17}N_2O_4PS$ 292.30	38260-54-7 -3.35	3847	 1.195^{20}	
5946	Famphur Phosphorothioic acid, *O*-[4-[(dimethylamino)sulfonyl]phenyl] *O,O*-dimethyl ester	$C_{10}H_{16}NO_5PS_2$ 325.35	52-85-7 53	3882		
5947	Fenamiphos Phosphoramidic acid, (1-methylethyl)-, ethyl 3-methyl-4-(methylthio)phenyl ester	$C_{13}H_{22}NO_3PS$ 303.36	22224-92-6 49	3901	 1.15^{20}	
5948	Fenarimol 5-Pyrimidinemethanol, α-(2-chlorophenyl)-α-(4-chlorophenyl)-	$C_{17}H_{12}Cl_2N_2O$ 331.20	60168-88-9 118	3903		
5949	Fenbuconazole 4-(4-Chlorophenyl)-2-phenyl-2-(1H-1,2,4-triazol-1-ylmethyl)butyronitrile	$C_{20}H_{19}ClN_4$ 350.85	114369-43-6 125			
5950	Fenbutatin oxide Distannoxane, hexakis(2-methyl-2-phenylpropyl)-	$C_{60}H_{78}OSn_2$ 1052.70	13356-08-6 138	3907		
5951	Fenitrothion Phosphorothioic acid, *O,O*-dimethyl *O*-(3-methyl-4-nitrophenyl) ester	$C_9H_{12}NO_5PS$ 277.24	122-14-5	3922 $118^{0.05}$; 164^{1}	 1.3227^{25}	
5952	Fenoxycarb Ethyl 2-(4-phenoxyphenoxy)ethylcarbamate	$C_{17}H_{19}NO_4$ 301.34	79127-80-3 53			
5953	Fenpropathrin Cyano(3-phenoxyphenyl)methyl 2,2,3,3-tetramethylcyclopropanecarboxylate	$C_{22}H_{23}NO_3$ 349.43	64257-84-7 47		 1.15^{25}	
5954	Fensulfothion Phosphorothioic acid, *O,O*-diethyl *O*-[4-(methylsulfinyl)phenyl] ester	$C_{11}H_{17}O_4PS_2$ 308.36	115-90-2	3943 $140^{0.01}$	 1.202^{20}	
5955	Fenthion Phosphorothioic acid, *O,O*-dimethyl *O*-[3-methyl 4-(methylthio)phenyl] ester	$C_{10}H_{15}O_3PS_2$ 278.33	55-38-9 7.5	3945 $87^{0.01}$	 1.246^{20}	
5956	Fenvalerate Benzeneacetic acid, 4-chloro-α-(1-methylethyl)-, cyano(3-phenoxyphenyl)methyl ester	$C_{25}H_{22}ClNO_3$ 419.91	51630-58-1	3952 dec	 1.15^{25}	
5957	Ferbam Iron, tris(dimethylcarbamodithioato *O,O*')-, (OC-6-11)-	$C_9H_{18}FeN_3S_6$ 416.51	14484-64-1 180 dec	3954		
5958	Ferrate(2-), chloro[7,12-diethenyl-3,8,13,17-tetramethyl-21H,23H-porphine-2,18-dipropanoato(4-)-*N*21,*N*22,*N*23, Hemin	$C_{34}H_{32}ClFeN_4O_4$ 651.95	16009-13-5 >300	4563		
5959	Ferrate(2-), [7,12-diethenyl-3,8,13,17-tetramethyl-21H,23H-porphine-2,18-dipropanoato(4-)-*N*21,*N*22,*N*23,*N*24 Hematin	$C_{34}H_{33}FeN_4O_5$ 633.51	15489-90-4 >200	4554	4-26-00-03047	H_2O 1; EtOH 3; eth 1; alk 3
5960	Ferrocene Bis(cyclopentadienyl)iron	$C_{10}H_{10}Fe$ 186.04	102-54-5 172.5	3985 249		H_2O 1
5961	Ferrocene, 1,1'-bis(1,1,3,3-tetramethyl-3-phenyldisiloxanyl)- Disiloxane, 1-(cyclopentadienyl)-1,1,3,3-tetramethyl-3-phenyl-, Fe deriv.	$C_{30}H_{42}FeO_2Si_4$ 602.85	12321-04-9 18.7	 $200\text{-}5^{0.03}$	 1.1063^{25}	 1.5473^{25}
5962	Fluazipop-butyl (±)-2-[4-[[5-(Trifluoromethyl)-2-pyridinyl]oxy]phenoxy]propanoic acid, butyl ester	$C_{19}H_{20}F_3NO_4$ 383.37	69806-50-4 13	4049 $170^{0.5}$	 1.21^{20}	
5963	Fluazipop-P-butyl (*R*)-2-[4-(5-Trifluoromethyl-2-pyridyloxy)phenoxy]propionic acid, butyl ester	$C_{19}H_{20}F_3NO_4$ 383.37	79241-46-6 5			
5964	Fluchloralin Benzenamine, *N*-(2-chloroethyl)-2,6-dinitro-*N*-propyl-4-(trifluoromethyl)-	$C_{12}H_{13}ClF_3N_3O_4$ 355.70	33245-39-5 42	4052		
5965	Flucythrinate Cythrin	$C_{26}H_{23}F_2NO_4$ 451.47	70124-77-5	4055 $108^{0.35}$	 1.189^{22}	
5966	Fluoranthene 1,2-(1,8-Naphthylene)benzene	$C_{16}H_{10}$ [illegible]	206-44-0 107.8	 384	4-05-00-02163 1.252^{0}	H_2O 1; EtOH 3; eth 3; bz 3
5967	9H-Fluorene 2,2'-Methylenebiphenyl	$C_{13}H_{10}$ 166.22	86-73-7 114.8	4081 295	4-05-00-02142 1.203^{0}	H_2O 1; EtOH 2; eth 3; ace 3
5968	9H-Fluorene, 2-bromo- 2-Bromofluorene	$C_{13}H_9Br$ 245.12	1133-80-8 113.5	 185^{135}	4-05-00-02147	H_2O 1; EtOH 3; chl 4; HOAc 3
5969	9H-Fluorene-2-carboxylic acid, 9-oxo-	$C_{14}H_8O_3$ 224.22	784-50-9 341	 sub 340	4-10-00-03040	H_2O 1; EtOH 3; HOAc 3
5970	9H-Fluorene, 9-[(2-chlorophenyl)methylene]- Fluorene, 9-(2-chlorobenzylidene)	$C_{20}H_{13}Cl$ 288.78	1643-49-8 69.5	 $180^{0.7}$	4-05-00-02631	EtOH 4; HOAc 4
5971	9H-Fluorene-2,7-diamine 2,7-Diaminofluorene	$C_{13}H_{12}N_2$ 196.25	525-64-4 166	4082	4-13-00-00449	H_2O 1; EtOH 3; chl 3
5972	9H-Fluorene, 2,7-dichloro- 2,7-Dichlorofluorene	$C_{13}H_8Cl_2$ 235.11	7012-16-0 128	 sub	4-05-00-02147	bz 4; chl 4

No.	Name	Synonym	Mol. Form.	Mol. Wt.	CAS RN	mp/°C	Merck No.	bp/°C	Beil. Ref.	den/g cm^{-3}	Solubility	n_D
5973	1H-Fluorene, dodecahydro-		$C_{13}H_{22}$	178.32	5744-03-6			260	4-05-00-00489	0.92^{25}		1.5012^{20}
5974	9H-Fluorene, 9-methyl-	9-Methylfluorene	$C_{14}H_{12}$	180.25	2523-37-7	46.5		$154-6^{15}$	4-05-00-02185	1.0263^{66}	H_2O 1; EtOH 3; eth 3; ace 3	1.610^{66}
5975	9H-Fluoren-9-one		$C_{13}H_8O$	180.21	486-25-9	84		341.5	4-07-00-01629	1.1300^{99}	H_2O 1; EtOH 3; eth 4; ace 3	1.6309^{99}
5976	9H-Fluoren-9-one, 2-chloro-		$C_{13}H_7ClO$	214.65	3096-47-7	125.5		sub	4-07-00-01634		EtOH 4	
5977	9H-Fluoren-9-one, 2-nitro-		$C_{13}H_7NO_3$	225.20	3096-52-4	224.3		sub	4-07-00-01636		EtOH 2; ace 3; sulf 3; HOAc 3	
5978	9H-Fluoren-9-one, 1,2,3,4-tetrahydro-	Phentydrone	$C_{13}H_{12}O$	184.24	634-19-5	81.5	7235	$139-40^{0.05}$	4-07-00-01302			
5979	9H-Fluoren-9-one, 2,4,7-trinitro-	2,4,7-Trinitrofluorenone	$C_{13}H_5N_3O_7$	315.20	129-79-3	176	9641		4-07-00-01638		H_2O 2; ace 4; bz 4; chl 4	
5980	Fluorodifen	Benzene, 2-nitro-1-(4-nitrophenoxy)-4-(trifluoromethyl)-	$C_{13}H_7F_3N_2O_5$	328.20	15457-05-3	94						
5981	Fluridone	4(1H)-Pyridinone, 1-methyl-3-phenyl-5-[3-(trifluoromethyl)phenyl]-	$C_{19}H_{14}F_3NO$	329.32	59756-60-4	155						
5982	Fluroxypyr	[(4-Amino-3,5-dichloro-6-fluoro-2-pyridyl)oxy]acetic acid	$C_7H_5Cl_2FN_2O_3$	255.03	69377-81-7	232	4128					
5983	Fluvalinate	2-[[2-Chloro-4-(trifluoromethyl)phenyl]amino]-3-methylbutanoic acid, (±)-cyano(3-phenoxy phenyl)methyl ester	$C_{26}H_{22}ClF_3N_2O_3$	502.92	102851-06-9			>450		1.29^{25}		
5984	Folpet	1H-Isoindole-1,3(2H)-dione, 2-[(trichloromethyl)thio]-	$C_9H_4Cl_3NO_2S$	296.56	133-07-3	177	4142					
5985	Fomesafen	5-[2-Chloro-4-(trifluoromethyl)phenoxy]-*N*-(methylsulfonyl)-2-nitrobenzamide	$C_{15}H_{10}ClF_3N_2O_6S$	438.77	72178-02-0	220				1.28^{20}		
5986	Fonofos	Phosphonodithioic acid, ethyl-, *O*-ethyl *S*-phenyl ester	$C_{10}H_{15}OPS_2$	246.33	944-22-9		4147	$130^{0.1}$		1.16^{25}		
5987	Formaldehyde	Methanal	CH_2O	30.03	50-00-0	-92	4148	-19.1	4-01-00-03017	0.815^{-20}	H_2O 3; EtOH 3; eth 5; ace 5	
5988	Formaldehyde, oxime		CH_3NO	45.04	75-17-2	2.5		109^{15}	4-01-00-03055	1.133^{25}	H_2O 3; EtOH 4; eth 4	
5989	Formamide	Methanamide	CH_3NO	45.04	75-12-7	2.55	4151	220	4-02-00-00045	1.1334^{20}	H_2O 5; EtOH 5; eth 2; ace 3	1.4472^{20}
5990	Formamide, *N*-cyclohexyl-		$C_7H_{13}NO$	127.19	766-93-8	39.5		260; $113^{0.7}$	4-12-00-00038	1.0123^{17}		
5991	Formamide, *N,N*-diethyl-		$C_5H_{11}NO$	101.15	617-84-5			177.5	4-04-00-00346	0.9080^{19}	H_2O 5; EtOH 4; eth 4; ace 5	1.4321^{25}
5992	Formamide, *N,N*-dimethyl-	*N,N*-Dimethylformamide	C_3H_7NO	73.09	68-12-2	-60.4	3232	153	4-04-00-00171	0.944^{25}	H_2O 5; EtOH 5; eth 5; ace 5	1.4305^{20}
5993	Formamide, *N,N*-diphenyl-		$C_{13}H_{11}NO$	197.24	607-00-1	73.5		337.5; 189^{13}	4-12-00-00370		H_2O 1; EtOH 3; eth 3; bz 3	
5994	Formamide, *N*-ethyl-	*N*-Ethylformamide	C_3H_7NO	73.09	627-45-2			198	4-04-00-00346	0.9552^{20}	H_2O 5; EtOH 5; eth 5	1.4320^{20}
5995	Formamide, *N*-methyl-	*N*-Methylformamide	C_2H_5NO	59.07	123-39-7	-3.8		199.5	4-04-00-00170	1.011^{19}	H_2O 4; ace 4; EtOH 4	1.4319^{20}
5996	Formamide, *N*-(2-methylphenyl)-		C_8H_9NO	135.17	94-69-9	62		288	3-12-00-01852	1.086^{55}	H_2O 3; EtOH 4	
5997	Formamide, *N*-(3-methylphenyl)-		C_8H_9NO	135.17	3085-53-8	<-18		290 dec; 176^{17}	3-12-00-01962		H_2O 3; NaOH 3	
5998	Formamide, *N*-methyl-*N*-phenyl-		C_8H_9NO	135.17	93-61-8	14.5		243	4-12-00-00370	1.0948^{20}	H_2O 2; EtOH 3; ace 3; ctc 2	1.5589^{20}
5999	Formamide, *N*-phenyl-	Formanilide	C_7H_7NO	121.14	103-70-8	46	4152	271	4-12-00-00368	1.1186^{50}	H_2O 3; EtOH 4; eth 3; bz 3	
6000	Formamide, *N*-(2,2,2-trichloro-1-hydroxyethyl)-	Chloral formamide	$C_3H_4Cl_3NO_2$	192.43	515-82-2	120	2060		3-02-00-00059		ace 4; eth 4; EtOH 4	
6001	Formetanate hydrochloride	Methanimidamide, *N,N*-dimethyl-*N'*-[3-[[(methylamino)carbonyl]oxy]phenyl]-, monohydrochloride	$C_{11}H_{16}ClN_3O_2$	257.72	23422-53-9							
6002	Formic acid	Methanoic acid	CH_2O_2	46.03	64-18-6	8.3	4153	101	4-02-00-00003	1.220^{20}	H_2O 5; EtOH 5; eth 5; ace 4	1.3714^{20}
6003	Formic acid, ammonium salt	Ammonium formate	CH_5NO_2	63.06	540-69-2		548		4-02-00-00003			
6004	Formic acid, butyl ester	Butyl formate	$C_5H_{10}O_2$	102.13	592-84-7	-91.5		106.1	4-02-00-00028	0.8885^{20}	H_2O 2; EtOH 5; eth 5; ace 3	1.3912^{20}
6005	Formic acid, cyclohexyl ester		$C_7H_{12}O_2$	128.17	4351-54-6			163	4-06-00-00035	1.0057^{0}	H_2O 1; EtOH 3; eth 4; HCOOH 3	1.4430^{20}

No.	Name Synonym	Mol. Form. Mol. Wt.	CAS RN mp/°C	Merck No. bp/°C	Beil. Ref. den/g cm^{-3}	Solubility n_D
6006	Formic acid, ethenyl ester Vinyl formate	$C_3H_4O_2$ 72.06	692-45-5		4-02-00-00033	
6007	Formic acid, ethyl ester Ethyl formate	$C_3H_6O_2$ 74.08	109-94-4 -79.6	3763 54.4	4-02-00-00023 0.9168^{20}	H_2O 3; EtOH 5; eth 5; ace 4 1.3598^{10}
6008	Formic acid, heptyl ester Heptyl formate	$C_8H_{16}O_2$ 144.21	112-23-2	 178.1	4-02-00-00031 0.8784^{20}	H_2O 1; EtOH 5; eth 5 1.4140^{20}
6009	Formic acid, hexyl ester Hexyl formate	$C_7H_{14}O_2$ 130.19	629-33-4 -62.6	 155.5	4-02-00-00031 0.8813^{20}	H_2O 1; EtOH 5; eth 5 1.4071^{20}
6010	Formic acid, methyl ester Methyl formate	$C_2H_4O_2$ 60.05	107-31-3 -99	5994 31.7	4-02-00-00020 0.9742^{20}	H_2O 4; EtOH 5; eth 3; chl 3 1.3433^{20}
6011	Formic acid, 1-methylethyl ester Isopropyl formate	$C_4H_8O_2$ 88.11	625-55-8	 68.2	4-02-00-00027 0.8728^{20}	H_2O 2; EtOH 5; eth 5; ace 4 1.3678^{20}
6012	Formic acid, 1-methylpropyl ester	$C_5H_{10}O_2$ 102.13	589-40-2	 97	4-02-00-00029 0.8846^{20}	H_2O 2; EtOH 5; eth 5; ace 3 1.3865^{20}
6013	Formic acid, 2-methylpropyl ester Isobutyl formate	$C_5H_{10}O_2$ 102.13	542-55-2 -95.8	5026 98.2	4-02-00-00029 0.8776^{20}	H_2O 2; EtOH 5; eth 5; ace 4 1.3857^{20}
6014	Formic acid, octyl ester	$C_9H_{18}O_2$ 158.24	112-32-3 -39.1	 198.8	4-02-00-00031 0.8744^{20}	H_2O 1; EtOH 3; eth 5; ctc 2 1.4208^{15}
6015	Formic acid, pentyl ester Pentyl formate	$C_6H_{12}O_2$ 116.16	638-49-3 -73.5	 130.4	4-02-00-00030 0.8853^{20}	H_2O 2; EtOH 5; eth 5 1.3992^{20}
6016	Formic acid, phenylmethyl ester Benzyl formate	$C_8H_8O_2$ 136.15	104-57-4	1148 203; 84^{10}	4-06-00-02262 1.081^{20}	H_2O 1; EtOH 3; eth 5; ace 3 1.5154^{20}
6017	Formic acid, 2-propenyl ester Allyl formate	$C_4H_6O_2$ 86.09	1838-59-1	 83.6	3-02-00-00046 0.9460^{20}	H_2O 2; EtOH 3; eth 5
6018	Formic acid, propyl ester Propyl formate	$C_4H_8O_2$ 88.11	110-74-7 -92.9	7871 80.9	4-02-00-00026 0.9058^{20}	H_2O 2; EtOH 5; eth 5; ctc 2 1.3779^{20}
6019	Formyl fluoride Fluoroformaldehyde	CHFO 48.02	1493-02-3 -142.2	4162 -26.5	4-02-00-00042 1.1950^{-30}	
6020	Fosetyl-Al Ethyl hydrogen phosphonate, aluminum salt	$C_6H_{18}AlO_9P_3$ 354.11	39148-24-8 >300	4167		
6021	Fosthiazate (*RS*)-*S*-*sec*-Butyl *O*-ethyl 2-oxo-1,3-thiazolidin-3-ylphosphonothioate	$C_9H_{18}NO_3PS_2$ 283.35	98886-44-3			
6022	*D*-Fructopyranose, 3-*O*-α-*D*-glucopyranosyl- *D*-Turanose	$C_{12}H_{22}O_{11}$ 342.30	547-25-1 168	9728	5-17-07-00213	H_2O 4; EtOH 3; MeOH 3
6023	β-*D*-Fructose β-Levulose	$C_6H_{12}O_6$ 180.16	53188-23-1 103 dec	4185	4-01-00-04401 1.60^{20}	H_2O 4; EtOH 3; ace 4; MeOH 3
6024	Fulminic acid Carbyloxime	CHNO 43.03	506-85-4 unstable in pure form			eth 3
6025	Furan Oxacyclopentadiene	C_4H_4O 68.08	110-00-9 -85.6	4206 31.5	5-17-01-00291 0.9514^{20}	H_2O 1; EtOH 4; eth 4; ace 3 1.4214^{20}
6026	2-Furanacetic acid	$C_6H_6O_3$ 126.11	2745-26-8 68.5	 $102^{0.4}$	5-18-06-00206	H_2O 3; bz 3; MeOH 3; peth 3
6027	2-Furanacetic acid, butyl ester 2-Furylacetic acid, butyl ester	$C_{10}H_{14}O_3$ 182.22	4915-23-5	 $110\text{-}111^{13}$	4-18-00-04062 1.0232^{25}	 1.4558^{25}
6028	2-Furanacetic acid, ethyl ester	$C_8H_{10}O_3$ 154.17	4915-21-3	 88^{15}	5-18-06-00206 1.0763^{25}	 1.4571^{25}
6029	2-Furanacetic acid, methyl ester	$C_7H_8O_3$ 140.14	4915-22-4	 $87\text{-}8^{21}$	5-18-06-00206 1.1250^{20}	 1.4638^{25}
6030	3-Furanacetic acid, tetrahydro-2,2-dimethyl-5-oxo- Terpenylic acid	$C_8H_{12}O_4$ 172.18	26754-48-3 90	9100	5-18-08-00032	H_2O 4
6031	2-Furanacetonitrile 2-Furylacetonitrile	C_6H_5NO 107.11	2745-25-7	 $75\text{-}80^{20}$	5-18-06-00207 1.0854^{25}	eth 4; EtOH 4 1.4693^{20}
6032	Furan, 2,5-bis(1,1-dimethylethyl)-	$C_{12}H_{20}O$ 180.29	4789-40-6	 210	5-17-01-00439 0.837^{20}	H_2O 1; EtOH 3; eth 3 1.4369^{20}
6033	Furan, 2-bromo-	C_4H_3BrO 146.97	584-12-3	 103	5-17-01-00295 1.6500^{20}	H_2O 2; EtOH 3; eth 3; ace 3 1.4980^{20}
6034	Furan, 3-bromo- 3-Bromofuran	C_4H_3BrO 146.97	22037-28-1	 103	5-17-01-00295 1.6606^{20}	ace 4; bz 4; eth 4; EtOH 4 1.4958^{20}
6035	Furan, 2-(bromomethyl)-	C_5H_5BrO 161.00	4437-18-7	 $32\text{-}4^{2}$	4-17-00-00268 1.557^{20}	 1.5380^{20}
6036	Furan, 2-(bromomethyl)tetrahydro-	C_5H_9BrO 165.03	1192-30-9	 170; 70^{22}	5-17-01-00080 1.3653^{20}	EtOH 3; eth 3 1.4850^{20}
6037	Furan, 2-(butoxymethyl)-	[illegible] 154.21	[illegible]	 190	2-17-00-00114 0.9516^{20}	eth 4; EtOH 4 1.4522^{20}
6038	Furan, 2,2'-*sec*-butylidenedi- Butane, 2,2-di(2-furyl)	$C_{12}H_{14}O_2$ 190.24	100121-73-1	 $64\text{-}6^{1}$	4-19-00-00287 1.0330^{20}	eth 4; EtOH 4 1.4970^{20}
6039	2-Furancarbonitrile	C_5H_3NO 93.08	617-90-3	 147	5-18-06-00122 1.0822^{20}	EtOH 3; eth 3 1.4798^{20}
6040	2-Furancarbonyl chloride	$C_5H_3ClO_2$ 130.53	527-69-5 -2	 173	5-18-06-00110 1.324^{25}	H_2O 1; eth 3; ctc 2; chl 3 1.5310^{20}
6041	2-Furancarboxaldehyde Furfural	$C_5H_4O_2$ 96.09	98-01-1 -36.5	4214 161.7	5-17-09-00292 1.1594^{20}	H_2O 3; EtOH 4; eth 5; ace 4 1.5261^{20}
6042	3-Furancarboxaldehyde	$C_5H_4O_2$ 96.09	498-60-2	 145; $70\text{-}2^{43}$	5-17-09-00372 1.109^{20}	 1.4945^{20}
6043	2-Furancarboxaldehyde, 5-bromo-	$C_5H_3BrO_2$ 174.98	1899-24-7 83.5	 201; 112^{16}	5-17-09-00320	eth 4; EtOH 4

No.	Name Synonym	Mol. Form. Mol. Wt.	CAS RN mp/°C	Merck No. bp/°C	Beil. Ref. den/g cm^{-3}	Solubility n_D
6044	2-Furancarboxaldehyde, 5-(hydroxymethyl)- 5-(Hydroxymethyl)-2-furaldehyde	$C_6H_6O_3$ 126.11	67-47-0 31.5	4764 114-6^{1}	5-18-01-00130 1.2062^{25}	H_2O 3; EtOH 3; eth 2; bz 3 1.5627^{18}
6045	2-Furancarboxaldehyde, 5-methyl-	$C_6H_6O_2$ 110.11	620-02-0	187; 89^{26}	5-17-09-00404 1.1072^{18}	H_2O 3; EtOH 4; eth 5; ctc 2 1.5264^{20}
6046	2-Furancarboxaldehyde, 5-nitro-	$C_5H_3NO_4$ 141.08	698-63-5 35.5	128-32^{10}	5-17-09-00323	H_2O 2; peth 3
6047	2-Furancarboxaldehyde, oxime, (*E*)-	$C_5H_5NO_2$ 111.10	620-03-1	205 dec; 98^{9}	5-17-09-00306 1.1550^{80}	H_2O 3; EtOH 4; eth 4; bz 4
6048	2-Furancarboxaldehyde, oxime, (*Z*)-	$C_5H_5NO_2$ 111.10	1450-58-4 91	100^{12}	5-17-09-00306 1.3800^{20}	H_2O 2; EtOH 4; eth 4; bz 4
6049	2-Furancarboxaldehyde, tetrahydro-	$C_5H_8O_2$ 100.12	7681-84-7	142; 46^{29}	5-17-09-00027 1.0727^{20}	H_2O 4; eth 4 1.4366^{20}
6050	2-Furancarboxylic acid 2-Furoic acid	$C_5H_4O_3$ 112.08	88-14-2 133.5	4219 231	5-18-06-00102	H_2O 3; EtOH 3; eth 4; ace 2
6051	3-Furancarboxylic acid	$C_5H_4O_3$ 112.08	488-93-7 122.5	sub 105	5-18-06-00196	H_2O 2; EtOH 3; eth 4; AcOEt 3
6052	2-Furancarboxylic acid, butyl ester	$C_9H_{12}O_3$ 168.19	583-33-5	233	4-18-00-03918 1.0555^{20}	H_2O 1; EtOH 3; eth 3; bz 3 1.4740
6053	2-Furancarboxylic acid, 5-chloro-, ethyl ester Ethyl 5-chloro-2-furancarboxylate	$C_7H_7ClO_3$ 174.58	4301-39-7 2	217; 108^{23}	5-18-06-00131 1.2418^{25}	
6054	3-Furancarboxylic acid, 4,5-dihydro-5-oxo-	$C_5H_4O_4$ 128.08	585-68-2 164	dec	4-18-00-05333	H_2O 4; EtOH 3; MeOH 3
6055	3-Furancarboxylic acid, 2,5-dimethyl-, ethyl ester	$C_9H_{12}O_3$ 168.19	29113-63-1	208.5; 96^{19}	5-18-06-00233 1.0478^{23}	1.4686^{20}
6056	2-Furancarboxylic acid, ethyl ester	$C_7H_8O_3$ 140.14	614-99-3 34.5	196.8	5-18-06-00104 1.1174^{21}	H_2O 1; EtOH 5; eth 5; ace 5 1.4797^{21}
6057	2-Furancarboxylic acid, 2-furanylmethyl ester	$C_{10}H_8O_4$ 192.17	615-11-2 27.5	122^{2}	4-18-00-03936 1.2347^{25}	ace 4; bz 4; eth 4; EtOH 4 1.5280^{20}
6058	2-Furancarboxylic acid, hexyl ester α-Furoic acid, hexyl ester	$C_{11}H_{16}O_3$ 196.25	39251-86-0	105-7^{1}	2-18-00-00267 1.0170^{20}	H_2O 1; EtOH 3; eth 3; bz 5
6059	2-Furancarboxylic acid, methyl ester	$C_6H_6O_3$ 126.11	611-13-2	181.3	5-18-06-00103 1.1786^{21}	H_2O 1; EtOH 3; eth 3; bz 3 1.4860^{20}
6060	2-Furancarboxylic acid, 5-methyl-	$C_6H_6O_3$ 126.11	1917-15-3 109.5	105^{1}	5-18-06-00217	H_2O 2; EtOH 3; eth 3; bz 2
6061	2-Furancarboxylic acid, 1-methylethyl ester	$C_8H_{10}O_3$ 154.17	6270-34-4	198.5	0-18-00-00275 1.0655^{24}	H_2O 1; EtOH 3; eth 3; ace 3 1.4682^{24}
6062	2-Furancarboxylic acid, 2-methylpropyl ester	$C_9H_{12}O_3$ 168.19	20279-53-2	222	2-18-00-00266 1.0388^{20}	ace 4; bz 4; eth 4; EtOH 4 1.4676^{20}
6063	3-Furancarboxylic acid, methyl ester	$C_6H_6O_3$ 126.11	13129-23-2	160	5-18-06-00196 1.1733^{15}	H_2O 1; EtOH 3; ace 3 1.4676^{20}
6064	3-Furancarboxylic acid, 2-methyl-, ethyl ester	$C_8H_{10}O_3$ 154.17	28921-35-9	85-7^{20}	5-18-06-00214 1.0102^{25}	eth 4 1.4620^{25}
6065	3-Furancarboxylic acid, 5-methyl- 3-Furoic acid, 5-methyl-	$C_6H_6O_3$ 126.11	21984-93-0 119	sub	5-18-06-00216	H_2O 4; eth 4; EtOH 4
6066	2-Furancarboxylic acid, 5-nitro-	$C_5H_3NO_5$ 157.08	645-12-5 186	sub	5-18-06-00139	H_2O 3; EtOH 3; eth 3; ace 2
6067	2-Furancarboxylic acid, octyl ester α-Furoic acid, octyl ester	$C_{13}H_{20}O_3$ 224.30	39251-88-2	126-7^{1}	2-18-00-00267 0.9885^{20}	H_2O 1; EtOH 3
6068	2-Furancarboxylic acid, pentyl ester 2-Furoic acid, pentyl ester	$C_{10}H_{14}O_3$ 182.22	4996-48-9	95-7^{1}	2-18-00-00266 1.0335^{20}	H_2O 1; EtOH 5
6069	2-Furancarboxylic acid, phenylmethyl ester	$C_{12}H_{10}O_3$ 202.21	5380-40-5	179-81^{18}	4-18-00-03921 1.1623^{22}	ace 4; eth 4 1.5550^{20}
6070	2-Furancarboxylic acid, 2-propenyl ester	$C_8H_8O_3$ 152.15	4208-49-5	207.5	5-18-06-00105 1.115^{25}	eth 3; ace 3; ctc 2 1.4945^{20}
6071	2-Furancarboxylic acid, propyl ester	$C_8H_{10}O_3$ 154.17	615-10-1	210.9; 112^{20}	5-18-06-00104 1.0745^{20}	H_2O 1; EtOH 3; eth 3; ace 3 1.4787^{20}
6072	2-Furancarboxylic acid, tetrahydro-	$C_5H_8O_3$ 116.12	16874-33-2 21	145^{25}	5-18-06-00010 1.1912^{20}	H_2O 4 1.4612^{20}
6073	Furan, 2-chloro- 2-Chlorofuran	C_4H_3ClO 102.52	3187-94-8	78	5-17-01-00294 1.1923^{20}	ace 4; eth 4; EtOH 4 1.4569^{20}
6074	Furan, 3-chloro- 3-Chlorofuran	C_4H_3ClO 102.52	50689-17-3	80	5-17-01-00295 1.2094^{20}	ace 4; eth 4 1.4601^{20}
6075	Furan, 2-(chloromethyl)-	C_5H_5ClO 116.55	617-88-9	49^{26}	5-17-01-00323 1.1783^{20}	bz 4; eth 4; EtOH 4 1.4941^{20}
6076	Furan, 2-(cyclohexyloxy)- Cyclohexyl 2-furyl ether	$C_{10}H_{14}O_2$ 166.22	99172-57-3	118-9^{28}	4-17-00-01219 1.0200^{28}	ace 4; eth 4; EtOH 4 1.4861^{28}
6077	Furan, 2,3-dibromo- 2,3-Dibromofuran	$C_4H_2Br_2O$ 225.87	30544-34-4	167	5-17-01-00295 1.9938^{25}	1.5458^{25}
6078	Furan, 2,5-dibromo- 2,5-Dibromofuran	$C_4H_2Br_2O$ 225.87	32460-00-7 9.5	164.5	5-17-01-00295 2.27^{20}	1.5455^{20}
6079	2,3-Furandicarboxylic acid Isocitrate lactone	$C_6H_4O_5$ 156.09	4282-24-0 226	sub	5-18-07-00043	H_2O 4; EtOH 4; eth 2; AcOEt 4
6080	2,4-Furandicarboxylic acid	$C_6H_4O_5$ 156.09	4282-28-4 266	sub	5-18-07-00045	H_2O 2; EtOH 4; ace 4; chl 2
6081	2,5-Furandicarboxylic acid Dehydromucic acid	$C_6H_4O_5$ 156.09	3238-40-2 342	sub	5-18-07-00047 1.7400^{20}	H_2O 2; EtOH 2

No.	Name Synonym	Mol. Form. Mol. Wt.	CAS RN mp/°C	Merck No. bp/°C	Beil. Ref. den/g cm^{-3}	Solubility n_D
6082	3,4-Furandicarboxylic acid, diethyl ester 2,3-Bis(ethoxycarbonyl)furan	$C_{10}H_{12}O_5$ 212.20	30614-77-8	 155^{13}	5-18-07-00053 1.14^{25}	 1.4717^{20}
6083	2,5-Furandicarboxylic acid, dimethyl ester Dimethyl 2,5-furandicarboxylate	$C_8H_8O_5$ 184.15	4282-32-0 112	 154^{15}	5-18-07-00048	eth 4; EtOH 4; chl 4
6084	Furan, 2-(diethoxymethyl)-	$C_9H_{14}O_3$ 170.21	13529-27-6	 191.5	5-17-09-00299 0.9976^{20}	EtOH 4 1.4451^{20}
6085	Furan, 2,2-diethyltetrahydro- 2,2-Diethyltetrahydrofuran	$C_8H_{16}O$ 128.21	1193-35-7	 146	4-17-00-00107 0.8703^{20}	ace 4; bz 4; eth 4; EtOH 4 1.4317^{20}
6086	Furan, 2,3-dihydro- 2,3-Dihydrofuran	C_4H_6O 70.09	1191-99-7	 54.5	5-17-01-00172 0.927^{25}	 1.4239^{20}
6087	Furan, 2,5-dihydro-2,5-dimethoxy- 2,5-Dimethoxy-2,5-dihydrofuran	$C_6H_{10}O_3$ 130.14	332-77-4	 161	5-17-05-00073 1.073^{25}	 1.4339^{20}
6088	Furan, 2,4-dimethyl- 3,5-Dimethylfuran	C_6H_8O 96.13	3710-43-8	 94	5-17-01-00343 0.8993^{20}	 1.4371^{20}
6089	Furan, 2,5-dimethyl- 2,5-Dimethylfuran	C_6H_8O 96.13	625-86-5 -62.8	 93.5	5-17-01-00344 0.8883^{20}	H_2O 1; EtOH 3; eth 3; ace 3 1.4363^{20}
6090	Furan, 2-(1,1-dimethylethyl)- 2-*tert*-Butylfuran	$C_8H_{12}O$ 124.18	7040-43-9	 119.5	5-17-01-00371 0.869^{20}	ace 4; eth 4; EtOH 4 1.4373^{20}
6091	2,5-Furandione Maleic anhydride	$C_4H_2O_3$ 98.06	108-31-6 52.8	5586 202	5-17-11-00055 1.314^{60}	H_2O 3; eth 3; ace 3; chl 3
6092	2,5-Furandione, 3-chloro-	C_4HClO_3 132.50	96-02-6 33	 196	5-17-11-00063 1.5375^{25}	 1.4980^{20}
6093	2,5-Furandione, dihydro- Succinic anhydride	$C_4H_4O_3$ 100.07	108-30-5 119	8841 261	5-17-11-00006 1.2^{20}	H_2O 1; EtOH 3; eth 2; chl 3
6094	2,5-Furandione, dihydro-3-methyl-	$C_5H_6O_3$ 114.10	4100-80-5 34	 239	5-17-11-00011 1.22^{25}	
6095	2,5-Furandione, dihydro-3-methylene-	$C_5H_4O_3$ 112.08	2170-03-8 69	 139^{30}; 114^{18}	5-17-11-00066	eth 2; chl 4
6096	2,5-Furandione, dihydro-3-phenyl-, (±)- Succinic anhydride, phenyl (*DL*)	$C_{10}H_8O_3$ 176.17	112489-85-7 54	 204^{22}	5-17-11-00275	ace 4; bz 4; eth 4; EtOH 4
6097	2,5-Furandione, 3,4-dimethyl-	$C_6H_6O_3$ 126.11	766-39-2 96	 223	5-17-11-00069 1.107^{100}	H_2O 2; EtOH 4; eth 4; bz 4
6098	2,5-Furandione, 3-methyl-	$C_5H_4O_3$ 112.08	616-02-4 7.5	 213.5	5-17-11-00065 1.2469^{16}	ace 4; eth 4; EtOH 4 1.4710^{21}
6099	2,5-Furandione, 3,3,4,4-tetrafluorodihydro-	$C_4F_4O_3$ 172.04	699-30-9	 54.5	5-17-11-00007 1.6209^{20}	 1.3240^{20}
6100	Furan, 2,5-diphenyl-	[illegible] 220.27	[illegible] 91	 344	5-17-02-00398	H_2O 1; EtOH 4; eth 4; ace 3
6101	Furan, 2,2'-[dithiobis(methylene)]bis-	$C_{10}H_{10}O_2S_2$ 226.32	4437-20-1 10	 167^{13}; $112^{0.5}$	4-17-00-01258	EtOH 4
6102	Furan, 2-ethenyl-	C_6H_6O 94.11	1487-18-9 -94	 99.5	5-17-01-00477 0.9445^{19}	 1.4992^{19}
6103	Furan, 2-ethoxy- Ethyl 2-furyl ether	$C_6H_8O_2$ 112.13	5809-07-4	 125.5	4-17-00-01219 0.9849^{23}	 1.4500^{23}
6104	Furan, 2-(ethoxymethyl)-	$C_7H_{10}O_2$ 126.16	6270-56-0	 149	5-17-03-00340 0.9844^{20}	H_2O 1; EtOH 3; eth 3 1.4523^{20}
6105	Furan, 2-ethyl- 2-Ethylfuran	C_6H_8O 96.13	3208-16-0	 92	5-17-01-00338 0.911^{15}	EtOH 3; eth 3; ace 3; bz 3 1.4466^{23}
6106	Furan, 2-ethyl-5-methyl- 2-Methyl-5-ethylfuran	$C_7H_{10}O$ 110.16	1703-52-2	 118.5	5-17-01-00357 0.8883^{20}	bz 4; eth 4; EtOH 4 1.4473^{20}
6107	Furan, 2-ethyltetrahydro- 2-Ethyltetrahydrofuran	$C_6H_{12}O$ 100.16	1003-30-1	 109	5-17-01-00095 0.8570^{19}	ace 4; bz 4; eth 4; EtOH 4 1.4147^{19}
6108	Furan, 2-ethynyl-	C_6H_4O 92.10	18649-64-4	 105.5	4-17-00-00396 0.9919^{20}	 1.5055^{20}
6109	Furan, 2,2,3,3,4,4,5-heptafluorotetrahydro-5-(nonafluorobutyl)- Perfluoro-2-butyltetrahydrofuran	$C_8F_{16}O$ 416.06	335-36-4	 102.6		
6110	Furan, 2-iodo- 2-Furyl iodide	C_4H_3IO 193.97	54829-48-0	 $43\text{-}5^{15}$	5-17-01-00295 2.024^{20}	eth 4 1.5661^{20}
6111	Furan, 3-iodo- 3-Iodofuran	C_4H_3IO 193.97	29172-20-1	 133; $37\text{-}8^{22}$	5-17-01-00296 2.045^{20}	H_2O 1; eth 3 1.5610^{20}
6112	Furan, 2-isopropoxy- 2-Furyl isopropyl ether	$C_7H_{10}O_2$ 126.16	98272-34-5	 135.5	4-17-00-01219 0.9689^{20}	 1.4419^{20}
6113	2-Furanmethanamine Furfurylamine	C_5H_7NO 97.12	617-89-0	 [illegible]	5-18-09-00541 [illegible]	H_2O 5; EtOH 5; eth 3; chl 3 1.4900[illegible]
6114	2-Furanmethanamine, *N*-(2-furanylmethyl)- Difurfurylamine	$C_{10}H_{11}NO_2$ 177.20	18240-50-1	 $135\text{-}42^{15}$	4-18-00-07092 1.1045^{20}	eth 4 1.5168^{20}
6115	2-Furanmethanamine, *N*-methyl-	C_6H_9NO 111.14	4753-75-7	 149	5-18-09-00541 0.989^{25}	 1.4729^{20}
6116	2-Furanmethanamine, tetrahydro- Tetrahydrofurfurylamine	$C_5H_{11}NO$ 101.15	4795-29-3	 153	5-18-09-00488 0.9752^{20}	H_2O 4; eth 4; EtOH 4 1.4551^{20}
6117	2-Furanmethanethiol	C_5H_6OS 114.17	98-02-2	 157	5-17-03-00351 1.1319^{20}	H_2O 1; chl 2 1.5329^{20}
6118	2-Furanmethanol Furfuryl alcohol	$C_5H_6O_2$ 98.10	98-00-0 -31	4215 171	5-17-03-00338 1.1296^{20}	H_2O 5; EtOH 4; eth 4; chl 3 1.4869^{20}
6119	2-Furanmethanol, acetate	$C_7H_8O_3$ 140.14	623-17-6	 179	5-17-03-00344 1.1175^{20}	H_2O 1; EtOH 3; eth 3 1.4327^{20}
6120	2-Furanmethanol, 5-(aminomethyl)tetrahydro-	$C_6H_{13}NO_2$ 131.17	589-14-0 <-70	 $103^{0.3}$	 1.1021^{25}	 1.4870^{25}

No.	Name Synonym	Mol. Form. Mol. Wt.	CAS RN mp/°C	Merck No. bp/°C	Beil. Ref. den/g cm^{-3}	Solubility n_D
6121	2-Furanmethanol, α-methyl-	$C_6H_8O_2$ 112.13	4208-64-4	 162.5	5-17-03-00374 1.0739^{25}	 1.4827^{15}
6122	2-Furanmethanol, 5-methyl-	$C_6H_8O_2$ 112.13	3857-25-8	 195 dec; 81^{23}	5-17-03-00387 1.0769^{20}	eth 4; EtOH 4 1.4853^{20}
6123	2-Furanmethanol, propanoate	$C_8H_{10}O_3$ 154.17	623-19-8	 195	2-17-00-00115 1.1085^{20}	H_2O 2; EtOH 3; eth 5; ace 3
6124	2-Furanmethanol, tetrahydro- Tetrahydrofurfuryl alcohol	$C_5H_{10}O_2$ 102.13	97-99-4 <-80	9146 178	5-17-03-00115 1.0524^{20}	ace 4; eth 4 1.4520^{20}
6125	2-Furanmethanol, tetrahydro-, acetate	$C_7H_{12}O_3$ 144.17	637-64-9	 193; 89^{18}	5-17-03-00119 1.0624^{20}	H_2O 4; eth 4; EtOH 4; chl 4 1.4350^{25}
6126	2-Furanmethanol, tetrahydro-, propanoate	$C_8H_{14}O_3$ 158.20	637-65-0	 205.5	5-17-03-00119 1.044^{20}	eth 4; EtOH 4; chl 4
6127	Furan, 2-methoxy- 2-Methoxyfuran	$C_5H_6O_2$ 98.10	25414-22-6	 110.5	5-17-03-00283 1.0646^{25}	 1.4468^{25}
6128	Furan, 2-(methoxymethyl)-	$C_6H_8O_2$ 112.13	13679-46-4	 132	5-17-03-00340 1.0163^{20}	H_2O 1; EtOH 3; eth 4 1.4570^{20}
6129	Furan, 2-(methoxymethyl)-5-nitro- 2-(Methoxymethyl)-5-nitrofuran	$C_6H_7NO_4$ 157.13	586-84-5	5917 $104\text{-}5^{3}$	4-17-00-01254 1.281^{20}	EtOH 4 1.5325^{20}
6130	Furan, 2-methyl- 2-Methylfuran	C_5H_6O 82.10	534-22-5 -87.5	 65	5-17-01-00322 0.9132^{20}	H_2O 2; EtOH 3; eth 3; ctc 2 1.4342^{20}
6131	Furan, 3-methyl- 3-Methylfuran	C_5H_6O 82.10	930-27-8	 66	5-17-01-00330 0.923^{18}	H_2O 1; EtOH 3; eth 3 1.4330^{19}
6132	Furan, 2,2'-methylenebis-	$C_9H_8O_2$ 148.16	1197-40-6	 94^{22}	5-19-01-00545 1.102^{20}	H_2O 1; EtOH 4; eth 4; ace 4 1.5049^{20}
6133	Furan, 3-(4-methyl-3-pentenyl)-	$C_{10}H_{14}O$ 150.22	539-52-6	 185.5	5-17-01-00515 0.9017^{20}	 1.4705^{21}
6134	Furan, 2-nitro-	$C_4H_3NO_3$ 113.07	609-39-2 30	 134^{123}, 84^{13}	5-17-01-00296	H_2O 3; EtOH 3; eth 3
6135	Furan, 2-(octyloxy)- 2-Furyl octyl ether	$C_{12}H_{20}O_2$ 196.29	100314-90-7	 $129\text{-}30^{18}$	4-17-00-01219 0.9214^{28}	eth 4 1.4520^{20}
6136	2-Furanol, 3,4-bis(1,3-benzodioxol-5-ylmethyl)tetrahydro- Cubebin	$C_{20}H_{20}O_6$ 356.38	18423-69-3 131.5	2616	5-19-12-00048	eth 4; EtOH 4; chl 4
6137	3-Furanol, tetrahydro-	$C_4H_8O_2$ 88.11	453-20-3	 181	5-17-03-00064 1.09^{25}	 1.4500^{20}
6138	2-Furanol, tetrahydro-, benzoate	$C_{11}H_{12}O_3$ 192.21	3333-44-6	 302; 139^{2}	5-17-03-00060 1.137^{20}	eth 4; EtOH 4
6139	2(3H)-Furanone, 3-acetyldihydro- α-Acetylbutyrolactone	$C_6H_8O_3$ 128.13	517-23-7	77 $107\text{-}8^{5}$	5-17-11-00016 1.1846^{20}	H_2O 4 1.4585^{20}
6140	2(3H)-Furanone, 5-butyldihydro-	$C_8H_{14}O_2$ 142.20	104-50-7	 $132\text{-}3^{20}$	5-17-09-00073 0.9796^{19}	EtOH 3; ctc 2 1.4451^{19}
6141	2(3H)-Furanone, 5-(chloromethyl)dihydro- Valeric acid, 5-chloro-4-hydroxy-, γ-lactone	$C_5H_7ClO_2$ 134.56	39928-72-8	 $132\text{-}5^{12}$	5-17-09-00025 1.625^{18}	EtOH 2; eth 2; bz 2
6142	2(3H)-Furanone, dihydro- γ-Butyrolactone	$C_4H_6O_2$ 86.09	96-48-0 -43.3	1596 204	5-17-09-00007 1.1284^{16}	ace 4; bz 4; eth 4; EtOH 4 1.4341^{20}
6143	2(3H)-Furanone, dihydro-3-hydroxy-4,4-dimethyl-, (*R*)- Pantolactone	$C_6H_{10}O_3$ 130.14	599-04-2 92	6963	5-18-01-00022	
6144	2(3H)-Furanone, dihydro-3-(hydroxyphenylmethyl)-4-[(1-methyl-1H-imidazol-5-yl)methyl]-, Isopilosine	$C_{16}H_{18}N_2O_3$ 286.33	491-88-3 187	5085	1-27-00-00612	EtOH 4
6145	2(3H)-Furanone, dihydro-3-methyl-	$C_5H_8O_2$ 100.12	1679-47-6	 $78\text{-}81^{10}$	5-17-09-00028 1.063^{25}	 1.4325^{20}
6146	2(3H)-Furanone, dihydro-5-methyl-, (±)-	$C_5H_8O_2$ 100.12	57129-69-8 -31	 206	5-17-09-00024 1.0465^{25}	H_2O 5; EtOH 3; ace 3; ctc 2 1.4328^{20}
6147	2(3H)-Furanone, dihydro-3-methylene- α-Methylene butyrolactone	$C_5H_6O_2$ 98.10	547-65-9	5981 $85\text{-}6^{10}$	5-17-09-00127 1.1206^{20}	H_2O 3; EtOH 4; eth 3; ace 3 1.4650^{20}
6148	2(3H)-Furanone, dihydro-5-methylene-	$C_5H_6O_2$ 98.10	10008-73-8	 $85\text{-}6^{10}$	5-17-09-00124 1.1206^{20}	ace 4; bz 4; eth 4; EtOH 4 1.4650^{20}
6149	2(3H)-Furanone, dihydro-5-pentyl- 4-Hydroxynonanoic acid lactone	$C_9H_{16}O_2$ 156.22	104-61-0		5-17-09-00084	
6150	2(3H)-Furanone, 5-ethyldihydro-	$C_6H_{10}O_2$ 114.14	695-06-7 -18	 215.5	5-17-09-00040 1.0261^{20}	H_2O 4; EtOH 4 1.4495^{20}
6151	2(3H)-Furanone, 3-ethyldihydro-4-[(1-methyl-1H-imidazol-5-yl)methyl]-, (3*S*-*cis*)- Pilocarpine	$C_{11}H_{16}N_2O_2$ 208.26	92-13-7 34	7395 260^{5}	2-27-00-00694	H_2O 3; EtOH 3; eth 2; bz 2
6152	2(3H)-Furanone, 5-heptyldihydro- 4-Hydroxyundecanoic acid lactone	$C_{11}H_{20}O_2$ 184.28	104-67-6	 286	5-17-09-00098 0.9494^{20}	EtOH 4 1.4512^{20}
6153	2(3H)-Furanone, 5-methyl-	$C_5H_6O_2$ 98.10	591-12-8 18	 56^{12}	5-17-09-00120 1.084^{20}	H_2O 3; EtOH 3; eth 3; ctc 2 1.4476^{20}
6154	2(5H)-Furanone, 5-methyl-	$C_5H_6O_2$ 98.10	591-11-7 <-17	 209; 98^{15}	5-17-09-00121 1.0810^{20}	H_2O 5; EtOH 3; eth 3 1.4454^{20}
6155	2(5H)-Furanone, 5-methylene- Protoanemonin	$C_5H_4O_2$ 96.09	108-28-1	7907 73^{11}	5-17-09-00371	H_2O 2; chl 3
6156	Furan, 2,2'-[oxybis(methylene)]bis-	$C_{10}H_{10}O_3$ 178.19	4437-22-3	 101^{2}	5-17-03-00347 1.1405^{20}	H_2O 1 1.5088^{20}
6157	Furan, 2-phenoxy- 2-Furyl phenyl ether	$C_{10}H_8O_2$ 160.17	60698-31-9	 $105\text{-}6^{18}$	4-17-00-01219 1.1010^{23}	eth 3 1.5418^{23}

No.	Name Synonym	Mol. Form. Mol. Wt.	CAS RN mp/°C	Merck No. bp/°C	Beil. Ref. den/g cm^{-3}	Solubility n_D
6158	Furan, 2-(phenoxymethyl)- Furfuryl phenyl ether	$C_{11}H_{10}O_2$ 174.20	4437-23-4	 139-40^{20}	5-17-03-00340 1.123^{17}	 1.5535^{17}
6159	Furan, 2-phenyl- 2-Phenylfuran	$C_{10}H_8O$ 144.17	17113-33-6	 107-8^{18}	5-17-02-00139 1.083^{20}	ace 4; bz 4 1.5920^{20}
6160	Furan, 2-(phenylthio)- 2-Furyl phenyl sulfide	$C_{10}H_8OS$ 176.24	16003-14-8	 119-20^{8}	5-17-03-00285 1.1341^{26}	eth 4; EtOH 4 1.5976^{20}
6161	2-Furanpropanoic acid	$C_7H_8O_3$ 140.14	935-13-7 58	 229	5-18-06-00224	H_2O 4; eth 4; chl 4
6162	2-Furanpropanoic acid, methyl ester	$C_8H_{10}O_3$ 154.17	37493-31-5	 89^{15}	5-18-06-00224 1.088^{20}	 1.4662^{20}
6163	2-Furanpropanoic acid, β-oxo-, ethyl ester	$C_9H_{10}O_4$ 182.18	615-09-8	 142-3^{10}	5-18-08-00143 1.165^{17}	H_2O 1; EtOH 3; eth 3
6164	2-Furanpropanoic acid, tetrahydro- Propionic acid, 3-(2-tetrahydrofuryl)	$C_7H_{12}O_3$ 144.17	935-12-6	 263	5-18-06-00033 1.1135^{20}	ace 4; EtOH 4 1.4578^{25}
6165	2-Furanpropanoic acid, tetrahydro-, ethyl ester	$C_9H_{16}O_3$ 172.22	4525-36-4	 222; 73^{2}	5-18-06-00034 1.023^{7}	ace 4; EtOH 4 1.440^{20}
6166	2-Furanpropanol, tetrahydro-α-[2-(tetrahydro-2-furanyl)ethyl]- 1,5-Bis(2-tetrahydrofuryl)-3-pentanol	$C_{13}H_{24}O_3$ 228.33	6265-26-5	 116$^{0.05}$	4-19-00-00704	chl 2 1.4541^{25}
6167	Furan, 2-propyl- 2-Propylfuran	$C_7H_{10}O$ 110.16	4229-91-8	 115	5-17-01-00354 0.8876^{20}	EtOH 3; eth 3; ace 3 1.4549^{20}
6168	Furan, tetrahydro- Tetrahydrofuran	C_4H_8O 72.11	109-99-9 -108.3	9144 65	5-17-01-00027 0.8892^{20}	H_2O 3; EtOH 4; eth 4; ace 4 1.4050^{20}
6169	Furan, tetrahydro-2,5-dimethoxy-	$C_6H_{12}O_3$ 132.16	696-59-3	 145.7	5-17-05-00004 1.02^{25}	 1.4180^{20}
6170	Furan, tetrahydro-2-(methoxymethyl)-	$C_6H_{12}O_2$ 116.16	19354-27-9	 140	5-17-03-00116 0.952^{25}	 1.4270^{20}
6171	Furan, tetrahydro-2-methyl- 2-Methyltetrahydrofuran	$C_5H_{10}O$ 86.13	96-47-9	 78	5-17-01-00078 0.8552^{20}	H_2O 3; EtOH 4; eth 4; ace 4 1.4059^{21}
6172	Furan, tetrahydro-3-methyl- 3-Methyltetrahydrofuran	$C_5H_{10}O$ 86.13	13423-15-9	 86.5	5-17-01-00081 0.8642^{20}	ace 4; bz 4; eth 4; EtOH 4 1.4122^{20}
6173	Furan, tetrahydro-2-propyl-	$C_7H_{14}O$ 114.19	3208-22-8	 136	5-17-01-00116 0.8547^{20}	 1.4242^{20}
6174	Furan, tetrahydro-2,2,4,4-tetramethyl- 2,2,4,4-Tetramethyltetrahydrofuran	$C_8H_{16}O$ 128.21	3358-28-9	 121	4-17-00-00109	H_2O 1; EtOH 5; eth 5 1.4074^{20}
6175	Furan, tetraphenyl- 2,3,4,5-Tetraphenylfuran	$C_{28}H_{20}O$ 372.47	1056-77-5 175	 220	4-17-00-00810	H_2O 1; EtOH 3; eth 3; ace 3
6176	Furazan, dimethyl-	$C_4H_6N_2O$ 98.10	4975-21-7 -7	 156	4-27-00-07096 1.0528^{14}	eth 4; EtOH 4 1.4327^{20}
6177	3aH-Furo[2,3-b]indol-3a-ol, 8a-(5-ethenyl-1-azabicyclo[2.2.2]oct-2-yl)-2,3,8,8a-tetrahydro- Conquinamine	$C_{19}H_{24}N_2O_2$ 312.41	464-86-8 123	2505	2-27-00-00667	H_2O 2; EtOH 3; eth 3; chl 3
6178	3aH-Furo[2,3-b]indol-3a-ol, 8a-(5-ethenyl-1-azabicyclo[2.2.2]oct-2-yl)-2,3,8,8a-tetrahydro- Quinamine	$C_{19}H_{24}N_2O_2$ 312.41	464-85-7 185.5	8060	2-27-00-00667	H_2O 1; EtOH 4; eth 3; ace 3
6179	Furo[2,3-b]quinoline, 4,8-dimethoxy- Fagarine	$C_{13}H_{11}NO_3$ 229.24	524-15-2 142	3879	4-27-00-02211	H_2O 2; EtOH 3; eth 3; bz 3
6180	Furo[2,3-b]quinoline, 4-methoxy- Dictamnine	$C_{12}H_9NO_2$ 199.21	484-29-7 133.5	3079	4-27-00-02030	H_2O 2; EtOH 4; eth 3; chl 3
6181	Furo[2,3-b]quinoline, 4,7,8-trimethoxy- Skimmianine	$C_{14}H_{13}NO_4$ 259.26	83-95-4 177	8505	4-27-00-02296	H_2O 1; EtOH 3; eth 2; chl 3
6182	Furo[2,3-b]quinolin-4(2H)-one, 3,9-dihydro-8-methoxy-9-methyl-2-(1-methylethyl)-, (*R*)- Lunacrine	$C_{16}H_{19}NO_3$ 273.33	82-40-6	5473	4-27-00-03566	chl 3
6183	4H-Furo[3,2-c]pyran-2(6H)-one, 4-hydroxy- Patulin	$C_7H_6O_4$ 154.12	149-29-1 111	7002	5-18-03-00005	H_2O 3; EtOH 3; eth 3; ace 3
6184	1H,12H-Furo[3',2':4,5]furo[2,3-h]pyrano[3,4-c][1]benzopyran-1,12-dione, 3,4,7a,10a-tetrahydro-5-meth Aflatoxin G1	$C_{17}H_{12}O_7$ 328.28	1165-39-5 245		5-19-12-00081	
6185	5H-Furo[3,2-g][1]benzopyran-5-one, 4,9-dimethoxy-7-methyl- Khellin	$C_{14}H_{12}O_5$ 260.25	82-02-0 154-5 dec	5189 180-200$^{0.05}$	5-19-06-00320	H_2O 1; EtOH 3; eth 2; ace 3
6186	5H-Furo[3,2-g][1]benzopyran-5-one, 4-methoxy-7-methyl- Visnagin	$C_{13}H_{10}O_4$ 230.22	82-57-5 144.5	9916	5-19-06-00030	H_2O 2; EtOH 2; chl 4
6187	7H-Furo[3,2-g][1]benzopyran-7-one, 9-methoxy- Methoxsalen	$C_{12}H_8O_4$ 216.19	298-81-7 148	5911	5-19-06-00015	H_2O 2; EtOH 4; eth 2; ace 2
6188	7H-Furo[3,2-g][1]benzopyran-7-one, 3-methoxy-2-(1-methylethyl)- Peucedanin	$C_{15}H_{14}O_4$ 258.27	133-26-6 85	7144 276-81^{17}	5-19-06-00042	H_2O 2; EtOH 3; eth 3; bz 2
6189	7H-Furo[3,2-g][1]benzopyran-7-one, 9-[(3-methyl-2-butenyl)oxy]- Imperatorin	$C_{16}H_{14}O_4$ 270.28	482-44-0 102	4839	5-19-06-00016	H_2O 2; EtOH 3; eth 3; bz 3
6190	3-Furoic acid, 5-bromo-, ethyl ester β-Furoic acid, 5-bromo, ethyl ester	$C_7H_7BrO_3$ 219.03	32460-20-1 17	 235; 135^{34}	5-18-06-00198 1.528^{20}	eth 4; EtOH 4
6191	Furo[3',4':6,7]naphtho[2,3-d]-1,3-dioxol-6(5aH)-one, 5,8,8a,9-tetrahydro-9-hydroxy-5-(3,4,5-trimethoxyphenyl)- Picropodophyllin	$C_{22}H_{22}O_8$ 414.41	477-47-4 228	7385	5-19-10-00665	ace 4; bz 4; eth 4; EtOH 4

No.	Name Synonym	Mol. Form. Mol. Wt.	CAS RN mp/°C	Merck No. bp/°C	Beil. Ref. den/g cm^{-3}	Solubility n_D
6192	Furo[3',4':6,7]naphtho[2,3-d]-1,3-dioxol-6(5aH)-one, 5,8,8a,9-tetrahydro-9-hydroxy-5-(3,4,5-trimetho Podophyllotoxin	$C_{22}H_{22}O_8$ 414.41	518-28-5 183	7520	5-19-10-00666	H_2O 2; EtOH 4; eth 1; ace 3
6193	Galactaric acid Mucic acid	$C_6H_{10}O_8$ 210.14	526-99-8 255 dec	4237	4-03-00-01292	
6194	Galactitol Dulcose	$C_6H_{14}O_6$ 182.17	608-66-2 189.5	4238 $275\text{-}80^{1}$	4-01-00-02844 1.47^{20}	H_2O 3; EtOH 2; eth 1; bz 1
6195	β-*D*-Galactopyranoside, (2α,3β,5α,15β,25*R*)-2,15-dihydroxyspirostan-3-yl Digitonin	$C_{56}H_{92}O_{29}$ 1229.33	11024-24-1 237.5	3144	5-19-03-00603	
6196	β-*D*-Galactopyranoside, (3β,5α,22β,25*S*)-spirosolan-3-yl *O*-β- Tomatine	$C_{50}H_{83}NO_{21}$ 1034.20	17406-45-0 270	9471	4-27-00-01954	EtOH 4; diox 4
6197	*D*-Galactose	$C_6H_{12}O_6$ 180.16	59-23-4 170	4241	4-01-00-04336	H_2O 4; EtOH 2; eth 1; bz 1
6198	Galactose, 6-deoxy-3-*O*-methyl- Digitalose	$C_7H_{14}O_5$ 178.19	4481-08-7 119	3142	4-01-00-04270	H_2O 4
6199	*D*-Galacturonic acid	$C_6H_{10}O_7$ 194.14	685-73-4 166	4242	1-03-00-00306	H_2O 3; EtOH 3; eth 1
6200	Galanthamine Lycoremine	$C_{17}H_{21}NO_3$ 287.36	357-70-0 126.5	4245	4-27-00-02184	ace 4; EtOH 4; chl 4
6201	Galanthan-1,2-diol, 3,12-didehydro-9,10-[methylenebis(oxy)]-, (1α,2β)- Lycorine	$C_{16}H_{17}NO_4$ 287.32	476-28-8 280	5498 sub	2-27-00-00547	H_2O 1; EtOH 2; eth 2; chl 2
6202	Gardona Phosphoric acid, 2-chloro-1-(2,4,5-trichlorophenyl)ethenyl dimethyl ester	$C_{10}H_9Cl_4O_4P$ 365.96	961-11-5			
6203	Gelsemine Spiro[3,5,8-ethanylylidene-1H-pyrano[3,4-c]pyridine-10,3'-[3H]indol]-2'(1'H)-one, 5-ethenyl	$C_{20}H_{22}N_2O_2$ 322.41	509-15-9 178	4277	4-27-00-07526	ace 4; bz 4; eth 4; EtOH 4
6204	Gibb-3-ene-1,10-dicarboxylic acid, 2,4a,7-trihydroxy-1-methyl-8-methylene-, 1,4a-lactone, Gibberellic acid	$C_{19}H_{22}O_6$ 346.38	77-06-5 234	4313	5-18-09-00269	ace 4; EtOH 4; MeOH 4
6205	*D*-Glucaric acid *D*-Tetrahydroxyadipic acid	$C_6H_{10}O_8$ 210.14	87-73-0 125.5	4344	4-03-00-01291	H_2O 4; EtOH 4; eth 2; chl 2
6206	*D*-Glucitol Sorbitol	$C_6H_{14}O_6$ 182.17	50-70-4 111	8680 $295^{3.5}$	4-01-00-02839 1.489^{20}	H_2O 4; ace 4 1.3330^{20}
6207	*D*-Glucitol, 1-amino-1-deoxy- Glucamine	$C_6H_{15}NO_5$ 181.19	488-43-7 127	4343	4-04-00-01913	H_2O 4; EtOH 4
6208	*D*-Glucitol, 1-deoxy-1-(methylamino)- *N*-Methylglucamine	$C_7H_{17}NO_5$ 195.22	6284-40-8 128.5	5995	4-04-00-01914	H_2O 3
6209	*D*-Gluconic acid Gluconic acid	$C_6H_{12}O_7$ 196.16	526-95-4 131	4350	4-03-00-01255	H_2O 3; EtOH 2; eth 1; bz 1
6210	*D*-Gluconic acid, 4-*O*-β-*D*-galactopyranosyl- Lactobionic acid	$C_{12}H_{22}O_{12}$ 358.30	96-82-2	5219	5-17-07-00436	H_2O 4; EtOH 2; eth 1; HOAc 2
6211	Gluconic acid, β-lactone Oxetane, gluconic acid deriv.	$C_6H_{10}O_6$ 178.14	3087-62-5 153		2-18-00-00190 1.610^{-5}	H_2O 2; EtOH 2; eth 1; ace 1
6212	*D*-Gluconic acid, δ-lactone δ-*D*-Gluconolactone	$C_6H_{10}O_6$ 178.14	90-80-2	4351	5-18-05-00011	
6213	*D*-Gluconic acid, monosodium salt Gluconic acid, sodium salt	$C_6H_{11}NaO_7$ 218.14	527-07-1	8569	4-03-00-01255	H_2O 3
6214	β-*D*-Glucopyranose, 4-*O*-β-*D*-galactopyranosyl-	$C_{12}H_{22}O_{11}$ 342.30	5965-66-2 254		5-17-07-00196 1.59^{20}	H_2O 4; EtOH 2; eth 1; chl 1
6215	α-*L*-Glucopyranose, 2-(methylamino)-2-deoxy- *N*-Methyl-α-*L*-glucosamine	$C_7H_{15}NO_5$ 193.20	42852-95-9	5996		
6216	α-*D*-Glucopyranose, pentaacetate	$C_{16}H_{22}O_{11}$ 390.34	604-68-2 113.3	sub	5-17-07-00318	H_2O 2; EtOH 2; eth 3; chl 3
6217	β-*D*-Glucopyranose, pentaacetate	$C_{16}H_{22}O_{11}$ 390.34	604-69-3 134	sub	5-17-07-00319 1.2740^{20}	H_2O 1; EtOH 2; eth 2; bz 3
6218	α-*D*-Glucopyranoside, β-*D*-fructofuranosyl Sucrose	$C_{12}H_{22}O_{11}$ 342.30	57-50-1 185.5	8855	5-17-08-00399 1.5805^{17}	H_2O 3; EtOH 2; eth 1; py 3 1.5376
6219	α-*D*-Glucopyranoside, β-*D*-fructofuranosyl *O*-α-*D*-galactopyranosyl-(1 Raffinose	$C_{18}H_{32}O_{16}$ 504.44	512-69-6 80	8120	4-17-00-03801 1.465^{25}	H_2O 4; EtOH 2; eth 1; MeOH 4
6220	α-*D*-Glucopyranoside, α-*D*-glucopyranosyl Trehalose	$C_{12}H_{22}O_{11}$ 342.30	99-20-7 203	9496	1.58^{24}	H_2O 4; EtOH 3; eth 1; bz 1
6221	α-*D*-Glucopyranoside, *O*-α-*D*-glucopyranosyl-(13)-β-*D*-fructofuranosyl Melezitose	$C_{18}H_{32}O_{16}$ 504.44	597-12-6 153	5698	5-17-08-00414 1.5565^{25}	H_2O 4
6222	β-*D*-Glucopyranoside, 2-(hydroxymethyl)phenyl Salicin	$C_{13}H_{18}O_7$ 286.28	138-52-3 207	8293 240 dec	5-17-07-00113 1.434^{20}	H_2O 4; EtOH 4; HOAc 4
6223	β-*D*-Glucopyranoside, 2-(hydroxymethyl)phenyl, 6-benzoate Populin	$C_{20}H_{22}O_8$ 390.39	99-17-2 180	7571	5-17-07-00343	

No.	Name Synonym	Mol. Form. Mol. Wt.	CAS RN mp/°C	Merck No. bp/°C	Beil. Ref. den/g cm^{-3}	Solubility n_D
6224	β-*D*-Glucopyranoside, 4-hydroxyphenyl Arbutin	$C_{12}H_{16}O_7$ 272.25	497-76-7 199.5	799	5-17-07-00110	H_2O 4; EtOH 3; eth 2; bz 1
6225	β-*D*-Glucopyranoside, 4-(3-hydroxy-1-propenyl)-2,6-dimethoxyphenyl Syringin	$C_{17}H_{24}O_9$ 372.37	118-34-3 192	8997	4-17-00-03002	EtOH 4
6226	β-*D*-Glucopyranoside, 4-(3-hydroxy-1-propenyl)-2-methoxyphenyl Coniferin	$C_{16}H_{22}O_8$ 342.35	531-29-3 186	2498	4-17-00-02999	H_2O 3; EtOH 2; eth 1; py 3
6227	α-*D*-Glucopyranoside, methyl α-Methylglucoside	$C_7H_{14}O_6$ 194.18	97-30-3 168	5997 $200^{0.2}$	5-17-07-00010 1.46^{30}	H_2O 4
6228	α-*D*-Glucopyranoside, methyl 2,3,4,6-tetra-*O*-methyl- α-*D*-Glucose, pentamethyl ether	$C_{11}H_{22}O_6$ 250.29	605-81-2	 $180^{0.4}$	5-17-07-00029 1.0944^{20}	H_2O 4; ace 4; eth 4; EtOH 4 1.4466^{20}
6229	α-*D*-Glucopyranoside, 1,3,4,6-tetra-*O*-acetyl-β-*D*-fructofuranosyl, tetraacetate Octaacetyl sucrose	$C_{28}H_{38}O_{19}$ 678.60	126-14-7 86.5	8856 250^{1}	5-17-08-00410 1.27^{16}	H_2O 2; EtOH 3; eth 3; ace 3 1.4660
6230	α-*D*-Glucopyranosiduronic acid, (3β,20β)-20-carboxy-11-oxo-30-norolean-12-en-3-yl 2-*O*- Glycyrrhizic acid	$C_{42}H_{62}O_{16}$ 822.94	1405-86-3 220 dec	4401	4-18-00-05156	H_2O 4
6231	β-*D*-Glucopyranosiduronic acid, 2,2,2-trichloroethyl Urochloralic acid	$C_8H_{11}Cl_3O_7$ 325.53	97-25-6 142	9798	4-18-00-05113	H_2O 4; EtOH 4
6232	β-*D*-Glucose	$C_6H_{12}O_6$ 180.16	28905-12-6 150		4-01-00-04306 1.5620^{18}	H_2O 4; EtOH 2; eth 1; py 3
6233	*D*-Glucose, 2-amino-2-deoxy- *D*-Glucosamine	$C_6H_{13}NO_5$ 179.17	3416-24-8	4352	4-04-00-02017	H_2O 4
6234	*D*-Glucose, 6-*O*-α-*L*-arabinopyranosyl- Vicianose	$C_{11}H_{20}O_{10}$ 312.27	14116-69-9 210 dec	9879	5-17-06-00112	H_2O 4
6235	*D*-Glucose, 6-deoxy- Quinovose	$C_6H_{12}O_5$ 164.16	7658-08-4 139.5	8106	4-01-00-04260	H_2O 4; EtOH 4
6236	*D*-Glucose, 6-*O*-(6-deoxy-α-*L*-mannopyranosyl)- Rutinose	$C_{12}H_{22}O_{10}$ 326.30	90-74-4 189-92 dec	8277	5-17-06-00269	H_2O 4; EtOH 4
6237	*D*-Glucose, 6-*O*-β-*D*-xylopyranosyl- Primeverose	$C_{11}H_{20}O_{10}$ 312.27	26531-85-1 210	7752	4-17-00-02447	H_2O 4; MeOH 4
6238	*D*-Glucuronic acid	$C_6H_{10}O_7$ 194.14	6556-12-3 165	4360	4-03-00-01996	H_2O 4; EtOH 4
6239	*D*-Glucuronic acid, γ-lactone *D*-Glucuronolactone	$C_6H_8O_6$ 176.13	32449-92-6 177.5	4362	5-18-05-00033 1.76^{20}	H_2O 3; EtOH 2; bz 1; MeOH 2
6240	*DL*-Glutamic acid	$C_5H_9NO_4$ 147.13	617-65-2 199 dec		4-04-00-03028 1.4601^{20}	H_2O 2; EtOH 1; eth 2; CS_2 1
6241	*L*-Glutamic acid (*S*)-2-Aminopentanedioic acid	$C_5H_9NO_4$ 147.13	56-86-0 224 dec	4363 sub 175	4-04-00-03028 1.538^{20}	
6242	*L*-Glutamic acid, *N*-(4-aminobenzoyl)- *N*-(*p*-Aminobenzoyl)-L(+)-glutamic acid	$C_{12}H_{14}N_2O_5$ 266.25	4271-30-1 173	437	4-14-00-01153	
6243	*L*-Glutamic acid, *N*-[4-[[(2-amino-1,4-dihydro-4-oxo-6-pteridinyl)methyl]amino]benzoyl]- Folic acid	$C_{19}H_{19}N_7O_6$ 441.40	59-30-3 250 dec	4140	4-26-00-03944	py 4; EtOH 4; HOAc 4
6244	*L*-Glutamic acid, *N*-[4-[[(2-amino-5-formyl-1,4,5,6,7,8-hexahydro-4-oxo-6-pteridinyl)methyl]amino]benzoyl]- Folinic acid	$C_{20}H_{23}N_7O_7$ 473.45	58-05-9 245 dec	4141	4-26-00-03881	H_2O 2
6245	*L*-Glutamic acid, 5-ethyl ester γ-Ethyl *L*-glutamate	$C_7H_{13}NO_4$ 175.18	1119-33-1 191	4364	4-04-00-03033	H_2O 2
6246	*DL*-Glutamic acid, 3-hydroxy-	$C_5H_9NO_5$ 163.13	5985-23-9 209	 dec	4-04-00-03253	H_2O 4; EtOH 1; eth 1; bz 1
6247	*L*-Glutamic acid, 5-[2-[4-(hydroxymethyl)phenyl]hydrazide] Agaritine	$C_{12}H_{17}N_3O_4$ 267.28	2757-90-6 205-9 dec	175		H_2O 4
6248	*L*-Glutamic acid, monosodium salt Glutamic acid, monosodium salt	$C_5H_8NNaO_4$ 169.11	142-47-2	6165		H_2O 3
6249	*L*-Glutamine 2-Aminoglutaramic acid	$C_5H_{10}N_2O_3$ 146.15	56-85-9 185-6 dec	4365	4-04-00-03038	H_2O 3; EtOH 1; eth 1; bz 1
6250	Glycine [illegible]	$C_2H_5NO_2$ [illegible]	56-40-6 [illegible]	4386	4-04-00-02349 [illegible]	H_2O 4; EtOH 1; eth 1; ace 2
6251	Glycine, *N*-acetyl- Aceturic acid	$C_4H_7NO_3$ 117.10	543-24-8 206	74	4-04-00-02399	H_2O 4; ace 4; EtOH 4
6252	Glycine, *N*-(4-aminobenzoyl)- *p*-Aminohippuric acid	$C_9H_{10}N_2O_3$ 194.19	61-78-9 198.5	454	4-14-00-01152	ace 4; bz 4; EtOH 4
6253	Glycine, *N*-(aminoiminomethyl)- Glycocyamine	$C_3H_7N_3O_2$ 117.11	352-97-6 >300	4391	4-04-00-02414	H_2O 2; EtOH 2; eth 2
6254	Glycine, *N*-(aminoiminomethyl)-*N*-methyl- Creatine	$C_4H_9N_3O_2$ 131.13	57-00-1 303 dec	2570	4-04-00-02426 1.33^{25}	H_2O 3; EtOH 2; eth 1
6255	Glycine, *N*-benzoyl- Hippuric acid	$C_9H_9NO_3$ 179.18	495-69-2 191.5	4636	4-09-00-00778 1.371^{20}	H_2O 3; EtOH 3; eth 2; bz 2
6256	Glycine, *N*,*N*-bis(carboxymethyl)- Nitrilotriacetic acid	$C_6H_9NO_6$ 191.14	139-13-9 242 dec	6499	4-04-00-02441	H_2O 2; EtOH 3; DMSO-d_6 2

No.	Name Synonym	Mol. Form. Mol. Wt.	CAS RN mp/°C	Merck No. bp/°C	Beil. Ref. den/g cm^{-3}	Solubility n_D
6257	Glycine, *N,N*-bis[2-[bis(carboxymethyl)amino]ethyl]- Pentetic acid	$C_{14}H_{23}N_3O_{10}$ 393.35	67-43-6	7083	4-04-00-02454	
6258	Glycine, *N,N*-bis(ethoxycarbonyl)-, ethyl ester Aminoacetic acid, *N,N*-(dicarboethoxy), ethyl ester	$C_{10}H_{17}NO_6$ 247.25	127665-32-1 36.5	 152^{10}	0-04-00-00365	bz 4; eth 4; EtOH 4
6259	Glycine, *N,N*-bis(2-hydroxyethyl)- Bicine	$C_6H_{13}NO_4$ 163.17	150-25-4 193-5 dec	1221	4-04-00-02390	H_2O 4; EtOH 1
6260	Glycine, *N*-(carboxymethyl)- Iminodiacetic acid	$C_4H_7NO_4$ 133.10	142-73-4 247.5	4832	4-04-00-02428	H_2O 2; EtOH 1; eth 1
6261	Glycine, *N,N*-dimethyl- *N,N*-Dimethylglycine	$C_4H_9NO_2$ 103.12	1118-68-9 185.5	3233	4-04-00-02365	H_2O 4; EtOH 3; eth 3; ace 3
6262	Glycine, *N,N'*-1,2-ethanediylbis[*N*-(carboxymethyl)- Edetic acid	$C_{10}H_{16}N_2O_8$ 292.25	60-00-4 245 dec	3484	4-04-00-02449	
6263	Glycine, ethyl ester	$C_4H_9NO_2$ 103.12	459-73-4	 149; 58^{18}	4-04-00-02355 1.0275^{10}	H_2O 5; EtOH 5; eth 5; ace 5 1.4242^{10}
6264	Glycine, *N*-(*N*-*L*-γ-glutamyl-*L*-cysteinyl)- Glutatione	$C_{10}H_{17}N_3O_6S$ 307.33	70-18-8 195	4369	4-04-00-03165	H_2O 4; EtOH 1; eth 1; DMF 3
6265	Glycine, *N*-glycyl- 2-(Aminoacetamido)acetic acid	$C_4H_8N_2O_3$ 132.12	556-50-3 215 dec	4399	4-04-00-02459	H_2O 3
6266	Glycine, hydrazide Aminoacetic acid hydrazide	$C_2H_7N_3O$ 89.10	14379-80-7 80.5	 150 dec	3-04-00-01121	EtOH 2; eth 1; chl 3
6267	Glycine, *N*-(4-hydroxyphenyl)- *N*-(4-Hydroxyphenyl)glycine	$C_8H_9NO_3$ 167.16	122-87-2 245-7 dec	4771	4-13-00-01210	H_2O 2; EtOH 2; eth 1; ace 2
6268	Glycine, *N*-methyl- Sarcosine	$C_3H_7NO_2$ 89.09	107-97-1 212-3 dec	8331	4-04-00-02363	
6269	Glycine, *N*-methyl-*N*-(1-oxododecyl)-, sodium salt Gardol	$C_{15}H_{28}NNaO_3$ 293.38	137-16-6	4267		H_2O 2
6270	Glycine, *N*-phenyl- Phenylaminoacetic acid	$C_8H_9NO_2$ 151.17	103-01-5 127.5	7262	4-12-00-00872	H_2O 4; EtOH 4
6271	Glycine, *N*-phenyl-, ethyl ester	$C_{10}H_{13}NO_2$ 179.22	2216-92-4 58	 273.5	4-12-00-00873	eth 4; EtOH 4
6272	Glycine, *N*-(phenylmethyl)-, ethyl ester	$C_{11}H_{15}NO_2$ 193.25	6436-90-4	 $175\text{-}9^{50}$	4-12-00-02277	EtOH 4; eth 4; bz 4 1.5041^{20}
6273	Glycine, *N*-[(3α,5β,7α,12α)-3,7,12-trihydroxy-24-oxocholan-24-yl]- Glycocholic acid	$C_{26}H_{43}NO_6$ 465.63	475-31-0 166.5	4390	4-10-00-02077	H_2O 2; EtOH 4; eth 2
6274	Glycogen	$(C_6H_{10}O_5)_x$ 162.14	9005-79-2	4392		H_2O 4; EtOH 1; eth 1
6275	Glyodin 1H-Imidazole, 2-heptadecyl-4,5-dihydro-, monoacetate	$C_{20}H_{40}N_2 \cdot C_2H_4O_2$ 368.60	556-22-9	4404	 1.035^{20}	
6276	Glyphosate Glycine, *N*-(phosphonomethyl)-	$C_3H_8NO_5P$ 169.07	1071-83-6 230 dec	4408		
6277	Glyphosine Glycine, *N,N*-bis(phosphonomethyl)-	$C_4H_{11}NO_8P_2$ 263.08	2439-99-8	4409		
6278	Guanidine Aminomethanamidine	CH_5N_3 59.07	113-00-8 50	4475	4-03-00-00148	H_2O 4; EtOH 4
6279	Guanidine, *N,N'*-bis(2-methylphenyl)-	$C_{15}H_{17}N_3$ 239.32	97-39-2 179		4-12-00-01764 1.10^{20}	H_2O 2; EtOH 2; eth 4; chl 3
6280	Guanidine, cyano- Dicyanodiamide	$C_2H_4N_4$ 84.08	461-58-5 211	3082	4-03-00-00160 1.404^{14}	H_2O 3; EtOH 3; eth 1; ace 3
6281	Guanidine, *N,N'*-dimethyl-*N''*-(phenylmethyl)- Bethanidine	$C_{10}H_{15}N_3$ 177.25	55-73-2 196	1208		
6282	Guanidine, *N,N'*-diphenyl- 1,3-Diphenylguanidine	$C_{13}H_{13}N_3$ 211.27	102-06-7 150	3325 170 dec	4-12-00-00769 1.13^{20}	H_2O 2; EtOH 3; eth 4; ctc 3
6283	Guanidine, (3-methyl-2-butenyl)- Galegine	$C_6H_{13}N_3$ 127.19	543-83-9 62.5	4246 dec	3-04-00-00465	H_2O 4; EtOH 4
6284	Guanidine, *N*-methyl-*N'*-nitro-*N*-nitroso- *N*-Methyl-*N'*-nitro-*N*-nitrosoguanidine	$C_2H_5N_5O_3$ 147.09	70-25-7	6017	4-04-00-03386	DMSO 3
6285	Guanidine, monohydrochloride	CH_6ClN_3 95.53	50-01-1 182.3		4-03-00-00148 1.354^{20}	H_2O 4; EtOH 4
6286	Guanidine, mononitrate	$CH_6N_4O_3$ 122.08	506-93-4 217	 dec	4-03-00-00148	H_2O 4; EtOH 4
6287	Guanidine, nitro- Nitroguanidine	$CH_4N_4O_2$ 104.07	556-88-7 239 dec	6529	4-03-00-00249	H_2O 2; EtOH 2; eth 1; alk 4
6288	Guanidine, *N,N',N''*-triphenyl-	$C_{19}H_{17}N_3$ 287.36	101-01-9 146.5	 dec	4-12-00-00866 1.163^{20}	H_2O 2; EtOH 3
6289	Guanosine 2-Amino-1,9-dihydro-9-β-*D*-ribofuranosyl-6H-purin-6-one	$C_{10}H_{13}N_5O_5$ 283.24	118-00-3 239 dec	4480	4-26-00-03901	H_2O 2; EtOH 1; eth 1; HOAc 4
6290	*D*-Gulose Gulose, *D*-	$C_6H_{12}O_6$ 180.16	4205-23-6	4490 dec	4-01-00-04333	H_2O 4
6291	Haloxyfop-(2-ethoxyethyl) (*RS*)-2-[4-(3-Chloro-5-trifluoromethyl-2-pyridyloxy)phenoxy]propionic acid, ethyl ester	$C_{17}H_{15}ClF_3NO_4$ 389.76	87237-48-7 60		 1.34	

No.	Name Synonym	Mol. Form. Mol. Wt.	CAS RN mp/°C	Merck No. bp/°C	Beil. Ref. den/g cm^{-3}	Solubility n_D
6292	7H-Benz[de]anthracen-7-one Benzanthrone	$C_{17}H_{10}O$ 230.27	82-05-3 170	1070		bz 2
6293	Helenine	$C_{15}H_{20}O_2$ 232.32	1407-14-3 78	 275; 192^{10}	4-17-00-05030	bz 4; eth 4; EtOH 4
6294	Heneicosane	$C_{21}H_{44}$ 296.58	629-94-7 40.5	 356.5	4-01-00-00569 0.7919^{20}	H_2O 1; EtOH 2; peth 3 1.4441^{20}
6295	Heneicosane, 11-decyl-	$C_{31}H_{64}$ 436.85	55320-06-4 10.0	 282.0^{10}	4-01-00-00595 0.8116^{20}	 1.4540^{20}
6296	Heneicosane, 11-(2,2-dimethylpropyl)-	$C_{26}H_{54}$ 366.71	55282-10-5 -21	 238.0^{10}	4-01-00-00585 0.8031^{20}	 1.4491^{20}
6297	Heneicosane, 11-phenyl- Benzene, (1-decylundecyl)-	$C_{27}H_{48}$ 372.68	6703-80-6 20.8	 $205^{1.0}$	4-05-00-01226 0.8531^{20}	 1.4788^{20}
6298	10-Heneicosene, 11-phenyl-	$C_{27}H_{46}$ 370.66	6703-78-2 48.2	 $203^{1.0}$ $190^{0.5}$	4-05-00-01506 0.8638^{20}	 1.4922^{20}
6299	Hentriacontane Untriacontane	$C_{31}H_{64}$ 436.85	630-04-6 67.9	 458	4-01-00-00594 0.781^{68}	EtOH 2; eth 2; bz 2; chl 2 1.4278^{90}
6300	16-Hentriacontanone	$C_{31}H_{62}O$ 450.83	502-73-8 83.8	 $250^{0.1}$	4-01-00-03413 0.7947^{91}	H_2O 1; EtOH 2; eth 3; ace 2 1.4297^{94}
6301	Heptachlor epoxide 2,5-Methano-2H-indeno[1,2-b]oxirene, 2,3,4,5,6,7,7-heptachloro-1a,1b,5,5a,6,6a-hexahydro-, (1aα,1bβ,2α,5α,5aβ,6β,6aα)-	$C_{10}H_5Cl_7O$ 389.32	1024-57-3 160			
6302	Heptacosane	$C_{27}H_{56}$ 380.74	593-49-7 59.5	 442	4-01-00-00586 0.7796^{60}	H_2O 1; EtOH 1; eth 2 1.4345^{65}
6303	14-Heptacosanone	$C_{27}H_{54}O$ 394.73	542-50-7 77.5	 $238^{0.1}$	4-01-00-03411 0.7986^{81}	
6304	Heptadecanal Margaric aldehyde	$C_{17}H_{34}O$ 254.46	629-90-3 36	 204^{26}	4-01-00-03395	bz 4; eth 4
6305	1-Heptadecanamine	$C_{17}H_{37}N$ 255.49	4200-95-7 49	 336	4-04-00-00824 0.8510^{20}	H_2O 1; EtOH 3; eth 3 1.4510^{20}
6306	Heptadecane	$C_{17}H_{36}$ 240.47	629-78-7 22	 302.0	4-01-00-00548 0.7780^{20}	H_2O 1; EtOH 2; eth 3; ctc 2 1.4369^{20}
6307	Heptadecane, 1-bromo- 1-Bromoheptadecane	$C_{17}H_{35}Br$ 319.37	3508-00-7 32	 349	4-01-00-00549 0.9916^{20}	H_2O 1; chl 4 1.4625^{20}
6308	Heptadecane, 1,17-dibromo-	$C_{17}H_{34}Br_2$ 398.26	81726-82-1 38	 208^{3}	3-01-00-00564	chl 4
6309	Heptadecane, 9-hexyl- 9-Hexylheptadecane	$C_{23}H_{48}$ 324.63	55124-79-3 -19.4	 213^{10}; $157^{0.5}$	4-01-00-00578 0.7976^{20}	 1.4465^{20}
6310	Heptadecane, 2-methyl- 2-Methylheptadecane	$C_{18}H_{38}$ 254.50	1560-89-0 5.7	 311	4-01-00-00556 0.7838^{15}	 1.4394^{15}
6311	Heptadecane, 9-methyl- 9-Methylheptadecane	$C_{18}H_{38}$ 254.50	18869-72-2	 $172\text{-}4^{10}$	3-01-00-00568 0.7810^{20}	 1.4388^{20}
6312	Heptadecanenitrile	$C_{17}H_{33}N$ 251.46	5399-02-0 34	 349	4-02-00-01195 0.8315^{20}	H_2O 1; EtOH 2; eth 4; chl 2 1.4467^{20}
6313	Heptadecane, 9-octyl-	$C_{25}H_{52}$ 352.69	7225-64-1 -13.8	 231.5^{10}	4-01-00-00583 0.8020^{20}	 1.4487^{20}
6314	Heptadecanoic acid Margaric acid	$C_{17}H_{34}O_2$ 270.46	506-12-7 61.3	5634 227^{100}	4-02-00-01193 0.8532^{60}	EtOH 2; eth 3; ace 3; bz 3 1.4342^{60}
6315	Heptadecanoic acid, ethyl ester	$C_{19}H_{38}O_2$ 298.51	14010-23-2 28	 185^{5}	4-02-00-01194 0.8517^{30}	ace 4; bz 4; eth 4; EtOH 4
6316	Heptadecanoic acid, methyl ester	$C_{18}H_{36}O_2$ 284.48	1731-92-6 30	 185^{9}; $152^{0.05}$	4-02-00-01194	H_2O 1; EtOH 3; eth 4; ace 3
6317	1-Heptadecanol Heptadecyl alcohol	$C_{17}H_{36}O$ 256.47	1454-85-9 53.8	 333	4-01-00-01884 0.8475^{20}	H_2O 1; EtOH 3; eth 3
6318	2-Heptadecanol	$C_{17}H_{36}O$ 256.47	16813-18-6 54	 155^{1}	4-01-00-01885	eth 4; EtOH 4 1.4407^{37}
6319	9-Heptadecanol	$C_{17}H_{36}O$ 256.47	624-08-8 61	 174^{9}	4-01-00-01886	H_2O 1; EtOH 3; eth 3; ace 3 1.4262^{80}
6320	2-Heptadecanone Pentadecyl methyl ketone	$C_{17}H_{34}O$ 254.46	2922-51-2 48	 320	4-01-00-03395 0.8049^{48}	H_2O 1; EtOH 2; eth 4; ace 3
6321	9-Heptadecanone	$C_{17}H_{34}O$ 254.46	540-08-9 53	 251.5; $142^{1.5}$	4-01-00-03396 0.8140^{48}	EtOH 2; MeOH 3
6322	1-Heptadecene Hexahydroaplotaxene	$C_{17}H_{34}$ 238.46	6765-39-5 11.5	 300	4-01-00-00929 0.7852^{20}	H_2O 1; eth 4; bz 3; lig 5 1.4432^{20}
6323	8-Heptadecene, 9-octyl-	$C_{25}H_{50}$ 350.67	24306-18-1	 227^{10}	4-01-00-00936 0.8086^{20}	 1.4554^{20}
6324	9-Heptadecenoic acid	$C_{17}H_{32}O_2$ 268.44	10136-52-4 12.5	 $175^{0.5}$	4-02-00-01632 0.8942^{20}	 1.4598^{20}
6325	2,4-Heptadienal, (*E,E*)-	$C_7H_{10}O$ 110.16	4313-03-5	 84.5	4-01-00-03549 0.881^{25}	 1.5315^{20}
6326	1,2-Heptadiene Butylallene	C_7H_{12} 96.17	2384-90-9	 105.5	4-01-00-01027 0.7306^{18}	 1.432^{18}
6327	1,4-Heptadiene	C_7H_{12} 96.17	5675-22-9	 93	3-01-00-00999 0.7270^{20}	bz 4; eth 4; peth 4 1.4370^{20}
6328	1,5-Heptadiene 1-Methyldiallyl	C_7H_{12} 96.17	1541-23-7	 94	3-01-00-00999 0.7186^{20}	ace 4; bz 4; eth 4; EtOH 4 1.4200^{20}

No.	Name Synonym	Mol. Form. Mol. Wt.	CAS RN mp/°C	Merck No. bp/°C	Beil. Ref. den/g cm^{-3}	Solubility n_D
6329	2,4-Heptadiene	C_7H_{12} 96.17	628-72-8	108	4-01-00-01029 0.7384^{20}	H_2O 1; EtOH 3; eth 3; ace 3 1.4578^{20}
6330	1,3-Heptadiene, 2,6-dimethyl-	C_9H_{16} 124.23	2436-84-2	141	4-01-00-01051 0.7648^{10}	1.4606^{22}
6331	1,5-Heptadiene, 2,6-dimethyl-	C_9H_{16} 124.23	6709-39-3 -70	143	4-01-00-01051 0.7648^{25}	
6332	2,4-Heptadiene, 2,4-dimethyl-	C_9H_{16} 124.23	74421-05-9	138	4-01-00-01051 0.7750^{4}	1.4587^{4}
6333	2,4-Heptadiene, 3,5-dimethyl-	C_9H_{16} 124.23	101935-28-8	142	4-01-00-01052 0.7728^{20}	1.4487^{20}
6334	1,6-Heptadiene-3,5-dione, 1,7-bis(4-hydroxy-3-methoxyphenyl)-, (*E,E*)- Curcumin	$C_{21}H_{20}O_6$ 368.39	458-37-7 183	2681	4-08-00-03697	EtOH 4; HOAc 4
6335	1,5-Heptadiene, 3-methyl- 3-Methyl-1,5-heptadiene	C_8H_{14} 110.20	4894-62-6 -57	111.	4-01-00-01040 0.7250^{25}	
6336	1,5-Heptadien-4-ol	$C_7H_{12}O$ 112.17	5638-26-6	156; 68^{24}	4-01-00-02248 0.8598^{20}	ace 4 1.4510^{25}
6337	1,6-Heptadien-4-ol 4-Hydroxy-1,6-heptadiene	$C_7H_{12}O$ 112.17	2883-45-6	151	4-01-00-02249 0.864^{25}	1.4505^{20}
6338	1,6-Heptadien-4-ol, 4-methyl- 1,6-Heptadiene-4-ol, 4-methyl	$C_8H_{14}O$ 126.20	25201-40-5	158.4	4-01-00-02260 0.8611^{20}	ace 4; EtOH 4 1.4500^{23}
6339	3,5-Heptadien-2-one	$C_7H_{10}O$ 110.16	3916-64-1	88^{28}	4-01-00-03549 0.8946^{19}	eth 4 1.5177^{19}
6340	2,5-Heptadien-4-one, 2,6-dimethyl- Phorone	$C_9H_{14}O$ 138.21	504-20-1 28	7307 197.5	4-01-00-03564 0.8850^{20}	H_2O 2; EtOH 3; eth 3; ace 3 1.4998^{20}
6341	1,6-Heptadien-3-yne 1,6-Heptadiene-3-yne	C_7H_8 92.14	5150-80-1	110	3-01-00-01061 0.787^{25}	bz 3; peth 3 1.4694^{25}
6342	1,5-Heptadiyne	C_7H_8 92.14	764-56-7	26^{30}	2-01-00-00247 0.8100^{21}	bz 4; peth 4 1.4521^{21}
6343	1,6-Heptadiyne	C_7H_8 92.14	2396-63-6 -85	112	4-01-00-01121 0.8164^{17}	H_2O 1; bz 3; HOAc 3 1.451^{17}
6344	Heptanal Heptaldehyde	$C_7H_{14}O$ 114.19	111-71-7 -43.3	4578 152.8	4-01-00-03314 0.8132^{25}	H_2O 2; EtOH 5; eth 5; ctc 2 1.4113^{20}
6345	Heptanal, 5-methyl-	$C_8H_{16}O$ 128.21	75579-88-3	72^{25}	4-01-00-03345 0.8164^{25}	1.4144^{25}
6346	Heptanal, oxime Enanthaldoxime	$C_7H_{15}NO$ 129.20	629-31-2 57.5	195	4-01-00-03316 0.8583^{55}	H_2O 2; EtOH 3; eth 3 1.4210^{20}
6347	Heptanal, 2-(phenylmethylene)-	$C_{14}H_{18}O$ 202.30	122-40-7 80	$174\text{-}5^{20}$	2-07-00-00310 0.9711^{20}	H_2O 1; ace 3; ctc 3 1.5381^{20}
6348	Heptanamide	$C_7H_{15}NO$ 129.20	628-62-6 96	254	4-02-00-00963 0.8521^{110}	H_2O 3; EtOH 3; eth 3; chl 2 1.4217^{110}
6349	1-Heptanamine Heptylamine	$C_7H_{17}N$ 115.22	111-68-2 -18	156	4-04-00-00734 0.7754^{20}	H_2O 2; EtOH 5; eth 5; chl 2 1.4251^{20}
6350	2-Heptanamine Tuaminoheptane	$C_7H_{17}N$ 115.22	123-82-0	9712 142	3-04-00-00374 0.7665^{19}	H_2O 2; EtOH 3; eth 3; chl 2 1.4199^{19}
6351	4-Heptanamine	$C_7H_{17}N$ 115.22	16751-59-0	139.5	4-04-00-00745 0.767^{20}	1.4172^{20}
6352	1-Heptanamine, *N*-heptyl- Diheptylamine	$C_{14}H_{31}N$ 213.41	2470-68-0 31.5	271; 135^{9}	4-04-00-00736 0.7956^{21}	H_2O 2; EtOH 3; eth 4
6353	2-Heptanamine, 6-methyl-, (±)-	$C_8H_{19}N$ 129.25	5984-58-7	6678 155	4-04-00-00764 0.767^{25}	1.4209^{20}
6354	2-Heptanamine, 6-methyl-*N*-(3-methylbutyl)- Octamylamine	$C_{13}H_{29}N$ 199.38	502-59-0	6671 $100\text{-}1^{7}$	3-04-00-00387	
6355	Heptane	C_7H_{16} 100.20	142-82-5 -90.6	4580 98.5	4-01-00-00376 0.6837^{20}	H_2O 1; EtOH 4; eth 5; ace 5 1.3878^{20}
6356	Heptane, 1-bromo- Heptyl bromide	$C_7H_{15}Br$ 179.10	629-04-9 -58	179	4-01-00-00391 1.1400^{20}	H_2O 1; EtOH 4; eth 4; ctc 2 1.4502^{20}
6357	Heptane, 2-bromo- 2-Heptyl bromide	$C_7H_{15}Br$ 179.10	1974-04-5 47	166	3-01-00-00431 1.1277^{20}	H_2O 1; bz 4; ctc 3; chl 3 1.4503^{20}
6358	Heptane, 4-bromo- 4-Heptyl bromide	$C_7H_{15}Br$ 179.10	998-93-6	84.6^{72}; 60^{18}	4-01-00-00392 1.1351^{20}	H_2O 1; bz 3; ctc 3; chl 3 1.4495^{20}
6359	Heptane, 1-bromo-7-fluoro-	$C_7H_{14}BrF$ 197.09	334-42-9	85^{11}	4-01-00-00392	eth 4; EtOH 4 1.4463^{20}
6360	Heptane, 1-chloro- Heptyl chloride	$C_7H_{15}Cl$ 134.65	629-06-1 -69.5	159	4-01-00-00389 0.8758^{20}	H_2O 1; EtOH 5; eth 5; ctc 2 1.4256^{20}
6361	Heptane, 2-chloro- 2-Chloroheptane	$C_7H_{15}Cl$ 134.65	1001-89-4	46^{19}	4-01-00-00390 0.8672^{20}	H_2O 1; eth 4; bz 3; chl 3 1.4221^{20}
6362	Heptane, 3-chloro- 3-Chloroheptane	$C_7H_{15}Cl$ 134.65	999-52-0	144; 48.3^{20}	4-01-00-00390 0.8960^{20}	bz 4; eth 4 1.4228^{20}
6363	Heptane, 4-chloro- 4-Chloroheptane	$C_7H_{15}Cl$ 134.65	998-95-8	144	4-01-00-00390 0.8710^{20}	bz 4; eth 4 1.4237^{20}
6364	Heptane, 3-chloro-3-ethyl-	$C_9H_{19}Cl$ 162.70	28320-89-0	46^{3}	3-01-00-00510 0.8856^{20}	EtOH 4; chl 4 1.4400^{20}
6365	Heptane, 1-chloro-7-fluoro-	$C_7H_{14}ClF$ 152.64	334-43-0	70^{10}	4-01-00-00390 0.993^{20}	bz 4; eth 4; EtOH 4; chl 4 1.4222^{25}
6366	Heptane, 2-chloro-2-methyl- 2-Chloro-2-methylheptane	$C_8H_{17}Cl$ 148.68	4325-49-9	50^{15}	4-01-00-00428 0.8568^{25}	bz 4; eth 4; EtOH 4; chl 4 1.4240^{25}
6367	Heptane, 2-chloro-6-methyl- 2-Chloro-6-methylheptane	$C_8H_{17}Cl$ 148.68	2350-19-8	74^{35}	3-01-00-00472	bz 4; eth 4; EtOH 4; chl 4 1.4260^{15}
6368	Heptane, 3-(chloromethyl)-	$C_8H_{17}Cl$ 148.68	123-04-6	172	4-01-00-00430 0.8769^{20}	H_2O 1; EtOH 3; eth 3; ace 3 1.4319^{20}

No.	Name Synonym	Mol. Form. Mol. Wt.	CAS RN mp/°C	Merck No. bp/°C	Beil. Ref. den/g cm^{-3}	Solubility n_D
6369	Heptane, 3-chloro-3-methyl- 3-Methyl-3-chloroheptane	$C_8H_{17}Cl$ 148.68	5272-02-6	 64^{27}	4-01-00-00430 0.8764^{20}	bz 4; eth 4; EtOH 4; chl 4 1.4317^{20}
6370	Heptane, 4-chloro-4-methyl- 2-Chloro-2-propylpentane	$C_8H_{17}Cl$ 148.68	61764-94-1	 50^{12}	3-01-00-00477 0.8690^{20}	bz 4; eth 4; EtOH 4; chl 4 1.4310^{15}
6371	1,7-Heptanediamine	$C_7H_{18}N_2$ 130.23	646-19-5 28.9	 224	4-04-00-01354	EtOH 3; eth 3; ace 3
6372	Heptane, 1,2-dibromo- 1,2-Dibromoheptane	$C_7H_{14}Br_2$ 258.00	42474-21-5	 228	3-01-00-00432 1.5086^{20}	 1.4986^{20}
6373	Heptane, 1,5-dibromo- 1,5-Dibromoheptane	$C_7H_{14}Br_2$ 258.00	1622-10-2	 $113\text{-}5^{11}$	4-01-00-00393 1.536^{15}	 1.5041^{15}
6374	Heptane, 1,7-dibromo- Heptamethylene dibromide	$C_7H_{14}Br_2$ 258.00	4549-31-9 41.7	 263	4-01-00-00393 1.5306^{20}	H_2O 1; eth 3; ace 3; bz 3 1.5034^{20}
6375	Heptane, 2,3-dibromo- 2,3-Dibromoheptane	$C_7H_{14}Br_2$ 258.00	21266-88-6	 101^{17}	3-01-00-00432 1.5139^{20}	 1.4992^{20}
6376	Heptane, 3,4-dibromo- 3,4-Dibromoheptane	$C_7H_{14}Br_2$ 258.00	21266-90-0	 107^{24}	3-01-00-00433 1.5182^{20}	 1.5010^{20}
6377	Heptane, 1,1-dichloro- 1,1-Dichloroheptane	$C_7H_{14}Cl_2$ 169.09	821-25-0	 191	4-01-00-00390 1.011^{20}	H_2O 1; eth 3; bz 3; ctc 3 1.4440^{20}
6378	Heptane, 1,2-dichloro- 1,2-Dichloroheptane	$C_7H_{14}Cl_2$ 169.09	10575-87-8	 $68\text{-}72^{7}$	3-01-00-00430 1.064^{20}	bz 4; eth 4; chl 4 1.4490^{20}
6379	Heptane, 1,7-dichloro- 1,7-Dichloroheptane	$C_7H_{14}Cl_2$ 169.09	821-76-1	 $124\text{-}5^{35}$	4-01-00-00390 1.0408^{25}	 1.4565^{25}
6380	Heptane, 2,2-dichloro- 2,2-Dichloroheptane	$C_7H_{14}Cl_2$ 169.09	65786-09-6	 77^{25}	3-01-00-00430 1.012^{20}	bz 4; eth 4; chl 4 1.4440^{20}
6381	Heptane, 4,4-dichloro-	$C_7H_{14}Cl_2$ 169.09	89796-76-9	 86^{27}	2-01-00-00117 1.008^{17}	eth 4; chl 4 1.448^{17}
6382	Heptane, 2,6-dichloro-2,6-dimethyl-	$C_9H_{18}Cl_2$ 197.15	35951-36-1 43	 93^{16}	3-01-00-00513	bz 4; chl 4
6383	Heptane, 2,2-dimethyl- 2,2-Dimethylheptane	C_9H_{20} 128.26	1071-26-7 -113	 132.7	4-01-00-00457 0.7105^{20}	H_2O 1; eth 3; ace 4; bz 5 1.4016^{20}
6384	Heptane, 2,3-dimethyl- 2,3-Dimethylheptane	C_9H_{20} 128.26	3074-71-3 -116	 140.5	4-01-00-00457 0.7260^{20}	H_2O 1; EtOH 5; eth 5; ace 5 1.4088^{20}
6385	Heptane, 2,4-dimethyl- 2,4-Dimethylheptane	C_9H_{20} 128.26	2213-23-2	 132.9	4-01-00-00457 0.7115^{25}	H_2O 1; EtOH 5; eth 5; ace 5 1.4034^{20}
6386	Heptane, 2,5-dimethyl- 3,6-Dimethylheptane	C_9H_{20} 128.26	2216-30-0	 136	4-01-00-00457 0.7198^{20}	ace 4; bz 4; eth 4; EtOH 4 1.4033^{20}
6387	Heptane, 2,6-dimethyl- 2,6-Dimethylheptane	C_9H_{20} 128.26	1072-05-5 -102.9	 135.2	4-01-00-00458 0.7089^{20}	chl 2 1.4011^{20}
6388	Heptane, 3,3-dimethyl- 3,3-Dimethylheptane	C_9H_{20} 128.26	4032-86-4	 137.3	4-01-00-00458 0.7254^{20}	H_2O 1; EtOH 5; eth 3; ace 4 1.4087^{20}
6389	Heptane, 3,4-dimethyl- 3,4-Dimethylheptane	C_9H_{20} 128.26	922-28-1	 140.6	3-01-00-00514 0.7314^{20}	H_2O 1; eth 3; ace 4; bz 5 1.4108^{20}
6390	Heptane, 3,5-dimethyl- 3,5-Dimethylheptane	C_9H_{20} 128.26	926-82-9	 136	4-01-00-00458 0.7225^{20}	H_2O 1; eth 3; ace 4; bz 5 1.4083^{20}
6391	Heptane, 4,4-dimethyl- 4,4-Dimethylheptane	C_9H_{20} 128.26	1068-19-5	 135.2	4-01-00-00458 0.7221^{20}	H_2O 1; eth 3; ace 4; bz 5 1.4076^{20}
6392	Heptanedinitrile	$C_7H_{10}N_2$ 122.17	646-20-8 -31.4	 155^{14}	4-02-00-02006 0.949^{18}	H_2O 1; EtOH 5; eth 5; chl 5 1.4472^{20}
6393	Heptanedioic acid Pimelic acid	$C_7H_{12}O_4$ 160.17	111-16-0 106	7401 272^{100}; 212^{10}	4-02-00-02003 1.329^{15}	H_2O 3; EtOH 3; eth 3; bz 1
6394	Heptanedioic acid, diethyl ester Diethyl pimelate	$C_{11}H_{20}O_4$ 216.28	2050-20-6 -24	 254; 140^{15}	4-02-00-02004 0.9945^{20}	eth 4; EtOH 4 1.4305^{20}
6395	Heptanedioic acid, dimethyl ester	$C_9H_{16}O_4$ 188.22	1732-08-7 -21	 120^{10}; 80^{1}	4-02-00-02004 1.0625^{20}	H_2O 2; EtOH 3; eth 3; bz 3 1.4309^{20}
6396	Heptanedioic acid, monoethyl ester	$C_9H_{16}O_4$ 188.22	33018-91-6 10	 182^{18}; 160^{4}	3-02-00-01742	 1.4415^{20}
6397	1,4-Heptanediol	$C_7H_{16}O_2$ 132.20	40646-07-9	 243; 127^{4}	4-01-00-02579 0.9542^{20}	 1.4510^{25}
6398	1,5-Heptanediol	$C_7H_{16}O_2$ 132.20	60096-09-5	 $135\text{-}6^{11}$	3-01-00-02213 0.9705^{20}	eth 4; EtOH 4; chl 4 1.4571^{22}
6399	1,7-Heptanediol	$C_7H_{16}O_2$ 132.20	629-30-1 22.5	 262	4-01-00-02580 0.9569^{25}	eth 4; EtOH 4 1.4520^{25}
6400	2,4-Heptanediol	$C_7H_{16}O_2$ 132.20	20748-86-1	 $107\text{-}8^{8}$	4-01-00-02580 0.926^{20}	 1.4386^{25}
6401	2,4-Heptanediol, 3-methyl- 3-Methylheptan-2,4-diol	$C_8H_{18}O_2$ 146.23	6964-04-1	 115[illegible]	0-01-00-00491 0.920[illegible]	H_2O 1; EtOH 3; ctc 2 1.4460[illegible]
6402	2,3-Heptanedione Acetyl valeryl	$C_7H_{12}O_2$ 128.17	96-04-8	 46^{13}	4-01-00-03697 0.919^{18}	 1.4150^{18}
6403	2,6-Heptanedione	$C_7H_{12}O_2$ 128.17	13505-34-5 33.5	 222; 97^{11}	4-01-00-03699 0.9399^{37}	bz 4; eth 4; EtOH 4 1.4277^{37}
6404	3,5-Heptanedione, 2,2,6,6-tetramethyl- Dipivaloylmethane	$C_{11}H_{20}O_2$ 184.28	1118-71-4	 $72\text{-}3^{6}$	4-01-00-03729 0.883^{25}	ctc 2 1.4589^{20}
6405	Heptane, 3-[(ethenyloxy)methyl]-	$C_{10}H_{20}O$ 156.27	103-44-6 -100	 76^{32}	4-01-00-02056 0.8108^{20}	 1.4247^{25}
6406	Heptane, 1-ethoxy-	$C_9H_{20}O$ 144.26	1969-43-3 -68.3	 166	4-01-00-01733 0.790^{16}	
6407	Heptane, 3-ethyl- 3-Ethylheptane	C_9H_{20} 128.26	15869-80-4 -114.9	 143.0	4-01-00-00456 0.7225^{25}	 1.4093^{20}
6408	Heptane, 4-ethyl- 4-Ethylheptane	C_9H_{20} 128.26	2216-32-2	 141.2	4-01-00-00457 0.7241^{25}	H_2O 1; EtOH 5; eth 3; ace 5 1.4096^{20}

PHYSICAL CONSTANTS OF ORGANIC COMPOUNDS (continued)

No.	Name Synonym	Mol. Form. Mol. Wt.	CAS RN mp/°C	Merck No. bp/°C	Beil. Ref. den/g cm^{-3}	Solubility n_D
6409	Heptane, 3-ethyl-2-methyl- Heptane, 2-methyl-3-ethyl	$C_{10}H_{22}$ 142.28	14676-29-0	 163	4-01-00-00480 0.7398^{25}	 1.4174^{20}
6410	Heptane, 3-ethyl-3-methyl-	$C_{10}H_{22}$ 142.28	17302-01-1	 163.8	4-01-00-00481 0.7463^{25}	 1.4208^{20}
6411	Heptane, 3-ethyl-4-methyl- 3-Ethyl-4-methylheptane	$C_{10}H_{22}$ 142.28	52896-91-0	 165	4-01-00-00481 0.7468^{25}	 1.4207^{20}
6412	Heptane, 3-ethyl-5-methyl- 3-Ethyl-5-methylheptane	$C_{10}H_{22}$ 142.28	52896-90-9	 158.2	4-01-00-00481 0.7368^{25}	 1.4164^{20}
6413	Heptane, 4-ethyl-2-methyl- 4-Ethyl-2-methylheptane	$C_{10}H_{22}$ 142.28	52896-88-5	 158	4-01-00-00481 0.7322^{25}	 1.4137^{20}
6414	Heptane, 4-ethyl-3-methyl- 4-Ethyl-3-methylheptane	$C_{10}H_{22}$ 142.28	52896-89-6	 165	4-01-00-00481 0.7466^{25}	 1.4206^{20}
6415	Heptane, 4-ethyl-4-methyl- 4-Ethyl-4-methylheptane	$C_{10}H_{22}$ 142.28	17302-04-4	 160.8	4-01-00-00481 0.7472^{25}	 1.4210^{25}
6416	Heptane, 5-ethyl-2-methyl-	$C_{10}H_{22}$ 142.28	13475-78-0	 159.7	4-01-00-00481 0.7318^{25}	 1.4134^{20}
6417	Heptane, 1-fluoro-	$C_7H_{15}F$ 118.19	661-11-0 -73	 117.9	4-01-00-00387 0.8062^{20}	H_2O 1; eth 3; ace 3; bz 3 1.3854^{20}
6418	Heptane, hexadecafluoro- Perfluoroheptane	C_7F_{16} 388.05	335-57-9 -78	 82.5	4-01-00-00388 1.7333^{20}	ace 4; eth 4; EtOH 4; chl 4 1.2618^{20}
6419	Heptane, 1-iodo-	$C_7H_{15}I$ 226.10	4282-40-0 -48.2	 204	4-01-00-00393 1.3791^{20}	H_2O 1; EtOH 3; eth 3; ace 3 1.4904^{20}
6420	Heptane, 2-iodo- 2-Iodoheptane	$C_7H_{15}I$ 226.10	18589-29-2	 98^{50}	4-01-00-00393 1.304^{20}	H_2O 1; ace 3; bz 3; ctc 3 1.4826
6421	Heptane, 1-methoxy-	$C_8H_{18}O$ 130.23	629-32-3	 151	3-01-00-01682 0.7862^{15}	ace 4; eth 4; EtOH 4 1.4073^{20}
6422	Heptane, 2-methyl- 2-Methylheptane	C_8H_{18} 114.23	592-27-8 -108.9	 117.6	4-01-00-00428 0.6980^{20}	H_2O 1; EtOH 5; eth 3; ace 5 1.3949^{20}
6423	Heptane, 3-methyl-, (*S*)-	C_8H_{18} 114.23	6131-25-5 -120	 116.5	4-01-00-00429 0.7075^{16}	H_2O 1; EtOH 3; eth 3; ace 5 1.4002^{18}
6424	Heptane, 4-methyl- 4-Methylheptane	C_8H_{18} 114.23	589-53-7 -121	 117.7	4-01-00-00431 0.7046^{20}	H_2O 1; EtOH 5; eth 3; ace 5 1.3979^{20}
6425	Heptane, 3-methylene-	C_8H_{16} 112.22	1632-16-2	 120	4-01-00-00884 0.7270^{20}	H_2O 1; eth 4; bz 4; peth 4 1.4157^{20}
6426	Heptane, 4-methylene- 2-Propyl-1-pentene	C_8H_{16} 112.22	15918-08-8	 117.7	4-01-00-00885 0.7198^{25}	 1.4136^{20}
6427	Heptane, 4-(1-methylethyl)- 4-Isopropylheptane	$C_{10}H_{22}$ 142.28	52896-87-4	 158.9	4-01-00-00481 0.7354^{25}	 1.4153^{20}
6428	Heptanenitrile	$C_7H_{13}N$ 111.19	629-08-3	 183; $70\text{-}2^{10}$	4-02-00-00963 0.8106^{20}	H_2O 1; eth 3; ace 3; bz 3 1.4104^{30}
6429	Heptane, 1-nitro- 1-Nitroheptane	$C_7H_{15}NO_2$ 145.20	693-39-0	 194	4-01-00-00395 0.9476^{17}	eth 4; EtOH 4
6430	Heptane, 1,1'-oxybis-	$C_{14}H_{30}O$ 214.39	629-64-1	 258.5	4-01-00-01733 0.8008^{20}	eth 4; EtOH 4 1.4275^{20}
6431	Heptane, 3,3'-[oxybis(methylene)]bis- Ether, dihexyl,2,2'-diethyl	$C_{16}H_{34}O$ 242.45	10143-60-9	 269.4	4-01-00-01785	ctc 2 1.4325^{20}
6432	Heptane, 1,1,1,2,2,3,3,4,4,5,5,6,6,7,7-pentadecafluoro- 1H-Pentadecafluoroheptane	C_7HF_{15} 370.06	375-83-7	 96.0	4-01-00-00388 1.725^{25}	 1.2690^{25}
6433	Heptane, 2,2,4,6,6-pentamethyl- 2,2,4,6,6-Pentamethylheptane	$C_{12}H_{26}$ 170.34	13475-82-6 -67	 177.8	4-01-00-00510 0.7463^{20}	 1.4440^{20}
6434	Heptane, 4-propyl- 4-Propylheptane	$C_{10}H_{22}$ 142.28	3178-29-8	 157.5	4-01-00-00480 0.7321^{25}	 1.4135^{20}
6435	Heptane, 1,1'-thiobis-	$C_{14}H_{30}S$ 230.46	629-65-2 70	 298	4-01-00-01739 0.8416^{20}	H_2O 1; eth 3 1.4606^{20}
6436	1-Heptanethiol Heptyl mercaptan	$C_7H_{16}S$ 132.27	1639-09-4 -43	 177	4-01-00-01738 0.8427^{20}	H_2O 1; EtOH 5; eth 5; chl 3 1.4521^{20}
6437	Heptane, 2,2,3-trimethyl- 2,2,3-Trimethylheptane	$C_{10}H_{22}$ 142.28	52896-92-1	 157.8	4-01-00-00481 0.7385^{25}	 1.4168^{20}
6438	Heptane, 2,2,4-trimethyl- 2,2,4-Trimethylheptane	$C_{10}H_{22}$ 142.28	14720-74-2	 148.3	4-01-00-00481 0.7237^{25}	H_2O 1; bz 3; ctc 3; chl 3 1.4092^{20}
6439	Heptane, 2,2,5-trimethyl- 2,2,5-Trimethylheptane	$C_{10}H_{22}$ 142.28	20291-95-6	 150.8	4-01-00-00482 0.7243^{25}	 1.4101^{20}
6440	Heptane, 2,2,6-trimethyl- 2,2,6-Trimethylheptane	$C_{10}H_{22}$ 142.28	1190-83-6 -105	 148.9	4-01-00-00482 0.7200^{25}	 1.4078^{20}
6441	Heptane, 2,3,3-trimethyl- 2,3,3-Trimethylheptane	$C_{10}H_{22}$ 142.28	52896-93-2	 160.2	4-01-00-00482 0.7450^{25}	 1.4202^{20}
6442	Heptane, 2,3,4-trimethyl- 2,3,4-Trimethylheptane	$C_{10}H_{22}$ 142.28	52896-95-4	 161	4-01-00-00482 0.7447^{25}	 1.4195^{20}
6443	Heptane, 2,3,5-trimethyl- 2,3,5-Trimethylheptane	$C_{10}H_{22}$ 142.28	20278-85-7	 160.7	4-01-00-00482 0.7413^{25}	 1.4169^{20}
6444	Heptane, 2,3,6-trimethyl- 2,3,6-Trimethylheptane	$C_{10}H_{22}$ 142.28	4032-93-3	 156.0	4-01-00-00482 0.7305^{25}	 1.4131^{20}
6445	Heptane, 2,4,4-trimethyl- 2,4,4-Trimethylheptane	$C_{10}H_{22}$ 142.28	4032-92-2	 151.0	4-01-00-00482 0.7308^{25}	 1.4142^{20}
6446	Heptane, 2,4,5-trimethyl- 2,4,5-Trimethylheptane	$C_{10}H_{22}$ 142.28	20278-84-6	 156.5	4-01-00-00483 0.7373^{25}	 1.4160^{20}
6447	Heptane, 2,4,6-trimethyl- 2,4,6-Trimethylheptane	$C_{10}H_{22}$ 142.28	2613-61-8	 147.6	4-01-00-00483 0.7190^{25}	 1.4071^{20}
6448	Heptane, 2,5,5-trimethyl- 2,5,5-Trimethylheptane	$C_{10}H_{22}$ 142.28	1189-99-7	 152.8	4-01-00-00483 0.7362^{25}	 1.4149^{20}

No.	Name Synonym	Mol. Form. Mol. Wt.	CAS RN mp/°C	Merck No. bp/°C	Beil. Ref. den/g cm^{-3}	Solubility n_D
6449	Heptane, 3,3,4-trimethyl- 3,3,4-Trimethylheptane	$C_{10}H_{22}$ 142.28	20278-87-9	 161.9	4-01-00-00483 0.7527^{25}	 1.4236^{20}
6450	Heptane, 3,3,5-trimethyl- 3,3,5-Trimethylheptane	$C_{10}H_{22}$ 142.28	7154-80-5	 155.7	4-01-00-00483 0.7248^{20}	H_2O 1; bz 3; ctc 3; chl 3 1.4170^{20}
6451	Heptane, 3,4,4-trimethyl- 3,4,4-Trimethylheptane	$C_{10}H_{22}$ 142.28	20278-88-0	 161.1	4-01-00-00484 0.7535^{25}	 1.4235^{20}
6452	Heptane, 3,4,5-trimethyl- 3,4,5-Trimethylheptane	$C_{10}H_{22}$ 142.28	20278-89-1	 162.5	4-01-00-00484 0.7519^{25}	 1.4229^{20}
6453	1,4,7-Heptanetriol	$C_7H_{16}O_3$ 148.20	3920-53-4 -35	 231^{25}; 146^{1}	4-01-00-02787 1.075^{18}	H_2O 3; EtOH 3; ace 3 1.4725^{20}
6454	2,4,6-Heptanetrione	$C_7H_{10}O_3$ 142.15	626-53-9 49	 121^{10}	4-01-00-03783 1.0599^{40}	H_2O 4; eth 4; EtOH 4 1.4930^{20}
6455	Heptanoic acid Enanthylic acid	$C_7H_{14}O_2$ 130.19	111-14-8 -7.5	4581 222.2	4-02-00-00958 0.9181^{20}	H_2O 2; EtOH 3; eth 3; ace 3 1.4170^{20}
6456	Heptanoic acid, anhydride	$C_{14}H_{26}O_3$ 242.36	626-27-7 -12.4	 269.5	4-02-00-00962 0.9321^{20}	H_2O 1; EtOH 3; eth 3 1.4335^{15}
6457	Heptanoic acid, 2-bromo- 2-Bromoheptanoic acid	$C_7H_{13}BrO_2$ 209.08	2624-01-3	 250 dec; 147^{12}	4-02-00-00967 1.319^{15}	ace 4; eth 4 1.471^{18}
6458	Heptanoic acid, 7-bromo-	$C_7H_{13}BrO_2$ 209.08	30515-28-7 31	 280; 165^{12}	4-02-00-00968	ace 4; bz 4; eth 4; EtOH 4
6459	Heptanoic acid, butyl ester Butyl enanthate	$C_{11}H_{22}O_2$ 186.29	5454-28-4 -67.5	 226.2	3-02-00-00768 0.8638^{20}	ace 4; bz 4; eth 4; EtOH 4 1.4204^{20}
6460	Heptanoic acid, 2,2-dimethyl-6-oxo- Geronic acid	$C_9H_{16}O_3$ 172.22	461-11-0	 278; 169^{12}	3-03-00-01264 1.0211^{20}	eth 4; EtOH 4 1.4488^{20}
6461	Heptanoic acid, 4,4-dimethyl-6-oxo- Isogeronic acid	$C_9H_{16}O_3$ 172.22	471-04-5 50	 178^{20}; 162^{10}	3-03-00-01266	H_2O 4; bz 4; eth 4; EtOH 4
6462	Heptanoic acid, ethyl ester Ethyl oenanthate	$C_9H_{18}O_2$ 158.24	106-30-9 -66.1	3790 187	4-02-00-00960 0.8817^{20}	H_2O 2; EtOH 3; eth 3; ctc 2 1.4100^{20}
6463	Heptanoic acid, 7-fluoro-	$C_7H_{13}FO_2$ 148.18	334-28-1	 133^{10}	4-02-00-00964 1.039^{20}	 1.4207^{25}
6464	Heptanoic acid, heptyl ester Heptyl heptanoate	$C_{14}H_{28}O_2$ 228.38	624-09-9 -33	 277	4-02-00-00961 0.8649^{20}	H_2O 1; EtOH 3; eth 3 1.4320^{20}
6465	Heptanoic acid, hexyl ester Hexyl heptanoate	$C_{13}H_{26}O_2$ 214.35	1119-06-8 -48	 261	3-02-00-00768 0.8611^{20}	ace 4; bz 4; eth 4; EtOH 4 1.429^{15}
6466	Heptanoic acid, methyl ester Methyl heptanoate	$C_8H_{16}O_2$ 144.21	106-73-0 -56	4501 174	4-02-00-00960 0.8815^{20}	H_2O 2; EtOH 3; eth 3; ace 2 1.4152^{20}
6467	Heptanoic acid, 2-methylpropyl ester	$C_{11}H_{22}O_2$ 186.29	7779-80-8	 208	1-02-00-00145 0.8593^{20}	ace 4; bz 4; eth 4; EtOH 4
6468	Heptanoic acid, octyl ester Octyl enanthate	$C_{15}H_{30}O_2$ 242.40	5132-75-2 -22.5	 290	3-02-00-00768 0.8596^{20}	ace 4; bz 4; eth 4; EtOH 4 1.4349^{15}
6469	Heptanoic acid, 6-oxo-	$C_7H_{12}O_3$ 144.17	3128-07-2 40.2	 251^{280}; 135^{1}	4-03-00-01598	H_2O 4; ace 4; eth 4; EtOH 4 1.4306^{25}
6470	Heptanoic acid, pentyl ester Amyl heptanoate	$C_{12}H_{24}O_2$ 200.32	7493-82-5 -50	 245.4	4-02-00-00960 0.8623^{20}	ace 4; bz 4; eth 4; EtOH 4 1.4263^{15}
6471	Heptanoic acid, propyl ester Propyl heptanoate	$C_{10}H_{20}O_2$ 172.27	7778-87-2 -63.5	 208	3-02-00-00767 0.8641^{15}	ace 4; bz 4; eth 4; EtOH 4 1.4183^{15}
6472	1-Heptanol Heptyl alcohol	$C_7H_{16}O$ 116.20	111-70-6 -34	4582 176.4	4-01-00-01731 0.8219^{20}	H_2O 2; EtOH 5; eth 5; ctc 2 1.4249^{20}
6473	2-Heptanol, (±)-	$C_7H_{16}O$ 116.20	52390-72-4	4583 159	4-01-00-01740 0.8167^{20}	H_2O 2; EtOH 3; eth 3; ctc 2 1.4210^{20}
6474	3-Heptanol, (*S*)-	$C_7H_{16}O$ 116.20	26549-25-7 -70	 157; 66^{18}	2-01-00-00444 0.8227^{20}	H_2O 2; EtOH 3; eth 3; ctc 2 1.4201^{20}
6475	4-Heptanol Dipropylcarbinol	$C_7H_{16}O$ 116.20	589-55-9 -41.2	 156	4-01-00-01743 0.8183^{20}	H_2O 1; EtOH 3; eth 3 1.4205^{20}
6476	4-Heptanol, acetate	$C_9H_{18}O_2$ 158.24	5921-84-6	 171	4-02-00-00163 0.8741^{0}	eth 4 1.4105^{19}
6477	1-Heptanol, 7-chloro- Heptamethylene chlorohydrin	$C_7H_{15}ClO$ 150.65	55944-70-2 11	 150^{20}	4-01-00-01738 0.9998^{15}	EtOH 4; peth 4 1.4537^{25}
6478	2-Heptanol, 1-chloro- 1-Chloro-2-heptanol	$C_7H_{15}ClO$ 150.65	53660-21-2	 93^{13}	4-01-00-01741 0.9885^{20}	EtOH 3; eth 3; ace 3 1.4499^{20}
6479	2-Heptanol, 2,6-dimethyl- 2,6-Dimethyl-2-heptanol	$C_9H_{20}O$ 144.26	13254-34-7	 173	2-01-00-00457 0.8186^{20}	 1.4242^{20}
6480	3-Heptanol, 2,3-dimethyl- Butylisopropylmethylcarbinol	$C_9H_{20}O$ 144.26	19549-71-4	 173; 75^{16}	4-01-00-01752 0.8395^{20}	 1.4355^{20}
6481	3-Heptanol, 2,6-dimethyl- 2,6-Dimethyl-3-heptanol	$C_9H_{20}O$ 144.26	19549-73-6	 175	3-01-00-01753 0.8212^{20}	 1.4246^{20}
6482	3-Heptanol, 3,5-dimethyl-	$C_9H_{20}O$ 144.26	19549-74-7	 $95\text{-}7^{50}$	4-01-00-01811 0.8177^{25}	 1.4251^{20}
6483	3-Heptanol, 3,6-dimethyl- 3,6-Dimethylheptan-3-ol	$C_9H_{20}O$ 144.26	1573-28-0	 173	4-01-00-01810 0.8285^{16}	eth 4; EtOH 4; lig 4 1.4326^{16}
6484	4-Heptanol, 2,4-dimethyl- 2,4-Dimethyl-4-heptanol	$C_9H_{20}O$ 144.26	19549-77-0	 171; 70^{26}	4-01-00-01810 0.8215^{20}	 1.4292^{20}
6485	4-Heptanol, 2,6-dimethyl- Diisobutylcarbinol	$C_9H_{20}O$ 144.26	108-82-7	 174.5	4-01-00-01810 0.8114^{20}	H_2O 1; EtOH 3; eth 3; ctc 2 1.4242^{20}
6486	4-Heptanol, 3,5-dimethyl- 3,5-Dimethyl-4-heptanol	$C_9H_{20}O$ 144.26	19549-79-2	 186	4-01-00-01811 0.836^{18}	 1.4283^{20}
6487	4-Heptanol, 4-ethyl-	$C_9H_{20}O$ 144.26	597-90-0	 182	4-01-00-01809 0.8350^{20}	eth 4; EtOH 4 1.4332^{20}

No.	Name	Synonym	Mol. Form.	Mol. Wt.	CAS RN	mp/°C	Merck No.	bp/°C	Beil. Ref.	den/g cm^{-3}	Solubility	n_D
6488	3-Heptanol, 3-ethyl-2-methyl-		$C_{10}H_{22}O$	158.28	66719-37-7			193	1-01-00-00214	0.8455^{20}		1.4378
6489	3-Heptanol, 3-ethyl-6-methyl-		$C_{10}H_{22}O$	158.28	66719-40-2			$83\text{-}6^{15}$	0-01-00-00427	0.852^{25}		1.4409
6490	3-Heptanol, 5-ethyl-4-methyl-		$C_{10}H_{22}O$	158.28	66731-94-0			$86\text{-}8^{14}$	3-01-00-01770	0.865^{17}		1.4565^{17}
6491	1-Heptanol, 7-fluoro-		$C_7H_{15}FO$	134.19	408-16-2			$98\text{-}9^{12}$	4-01-00-01736	0.956^{20}	eth 4; EtOH 4	1.4197^{25}
6492	1-Heptanol, 3-methyl-	3-Methyl-1-heptanol	$C_8H_{18}O$	130.23	1070-32-2			101^{20}	4-01-00-01782	0.824^{24}		1.4295^{25}
6493	1-Heptanol, 4-methyl-	4-Methyl-1-heptanol	$C_8H_{18}O$	130.23	817-91-4			188	4-01-00-01789	0.8065^{25}	EtOH 4	1.4253^{25}
6494	1-Heptanol, 6-methyl-	6-Methyl-1-heptanol	$C_8H_{18}O$	130.23	1653-40-3	-106		$188; 95.8^{20}$	4-01-00-01782	0.8176^{25}	H_2O 1; EtOH 3; eth 3	1.4251^{25}
6495	2-Heptanol, 2-methyl-	2-Methyl-2-heptanol	$C_8H_{18}O$	130.23	625-25-2	-50.4		156	4-01-00-01780	0.8142^{20}	H_2O 1; EtOH 3; eth 3	1.4250^{20}
6496	2-Heptanol, 3-methyl-	3-Methyl-2-heptanol	$C_8H_{18}O$	130.23	31367-46-1			166.1		0.8177^{25}	H_2O 1; EtOH 3; eth 3; ctc 3	1.4199^{25}
6497	2-Heptanol, 5-methyl-	5-Methyl-2-heptanol	$C_8H_{18}O$	130.23	54630-50-1	-61		170	4-01-00-01783	0.8174^{21}		
6498	2-Heptanol, 6-methyl-		$C_8H_{18}O$	130.23	4730-22-7	-105		174	3-01-00-01728	0.8218^{20}		1.4238^{10}
6499	3-Heptanol, 2-methyl-, (±)-		$C_8H_{18}O$	130.23	100296-26-2	-85		167.5	4-01-00-01781	0.8235^{20}	H_2O 2; EtOH 3; eth 3; ctc 3	1.4265^{20}
6500	3-Heptanol, 3-methyl-	2-Ethyl-2-hexanol	$C_8H_{18}O$	130.23	5582-82-1	-83		163	4-01-00-01783	0.8282^{20}	H_2O 1; EtOH 3; eth 3; ctc 3	1.4279^{20}
6501	3-Heptanol, 4-methyl-	4-Methyl-3-heptanol	$C_8H_{18}O$	130.23	14979-39-6	-123		170	4-01-00-01789	0.827^{25}		1.4300^{20}
6502	3-Heptanol, 5-methyl-		$C_8H_{18}O$	130.23	18720-65-5	-91.2		172	4-01-00-01783	0.8425^{25}		1.433^{24}
6503	4-Heptanol, 2-methyl-	6-Methyl-4-heptanol	$C_8H_{18}O$	130.23	21570-35-4	-81		164	4-01-00-01781	0.8207^{20}	eth 4; EtOH 4	1.4203
6504	4-Heptanol, 3-methyl-, (*R**,*S**)-(±)-	4-Heptanol, 3-methyl(*DL*)	$C_8H_{18}O$	130.23	92737-91-2			162	3-01-00-01731	0.8335^{25}	H_2O 2; EtOH 3; eth 3; ctc 3	1.4211^{25}
6505	4-Heptanol, 4-methyl-	4-Hydroxy-4-methylheptane	$C_8H_{18}O$	130.23	598-01-6	-82		161	4-01-00-01789	0.8248^{20}	H_2O 1; EtOH 3; eth 3; ctc 3	1.4258^{20}
6506	2-Heptanol, 6-methyl-, acetate	Acetic acid, (6-methyl-2-heptyl) ester	$C_{10}H_{20}O_2$	172.27	67952-57-2			187	0-02-00-00135	0.8474^{20}	EtOH 4	1.413^{20}
6507	3-Heptanol, 6-methyl-, acetate	Acetic acid, (6-methyl-3-heptyl) ester	$C_{10}H_{20}O_2$	172.27	32764-34-4			184.5	0-02-00-00135	0.8554^{20}	EtOH 4	1.4160^{20}
6508	4-Heptanol, 4-propyl-		$C_{10}H_{22}O$	158.28	2198-72-3			191	4-01-00-01831	0.8337^{21}	bz 4; eth 4; EtOH 4	1.4355^{21}
6509	4-Heptanol, 2,4,6-trimethyl-		$C_{10}H_{22}O$	158.28	60836-07-9			$181; 79^{12}$	4-01-00-01833	0.823^{21}		1.4334^{18}
6510	2-Heptanone	Methyl pentyl ketone	$C_7H_{14}O$	114.19	110-43-0	-35	4584	151.0	4-01-00-03318	0.8111^{20}	H_2O 4; EtOH 3; eth 3	1.4088^{20}
6511	3-Heptanone	Ethyl butyl ketone	$C_7H_{14}O$	114.19	106-35-4	-39		147	4-01-00-03321	0.8183^{20}	H_2O 1; EtOH 5; eth 5; ctc 2	1.4057^{20}
6512	4-Heptanone	Dipropyl ketone	$C_7H_{14}O$	114.19	123-19-3	-33	3351	144	4-01-00-03323	0.8174^{20}	H_2O 1; EtOH 5; eth 5; ctc 3	1.4069^{20}
6513	2-Heptanone, 1-chloro-	1-Chloro-2-heptanone	$C_7H_{13}ClO$	148.63	41055-92-9			83^{16}	4-01-00-03320	0.802^{20}	eth 4; EtOH 4	1.4371^{20}
6514	4-Heptanone, 2,6-dimethyl-	Diisobutyl ketone	$C_9H_{18}O$	142.24	108-83-8	-41.5		169.4	4-01-00-03360	0.8062^{20}	H_2O 1; EtOH 5; eth 5; ctc 3	1.412^{21}
6515	4-Heptanone, 3,5-dimethyl-	Di-*sec*-butyl ketone	$C_9H_{18}O$	142.24	19549-84-9			162	3-01-00-02893	0.826^{14}		1.4193^{14}
6516	3-Heptanone, 6-(dimethylamino)-4,4-diphenyl-, hydrochloride	Methadone hydrochloride	$C_{21}H_{28}ClNO$	345.91	1095-90-5	235	5852		3-14-00-00279		H_2O 4; EtOH 4	
6517	4-Heptanone, 2-hydroxy-		$C_7H_{14}O_2$	130.19	54862-92-9			95^{12}	4-01-00-04034	0.9296^{19}		1.4357^{19}
6518	2-Heptanone, 4-hydroxy-3-methyl-	4-Heptanol-2-one, 3-methyl	$C_8H_{16}O_2$	144.21	56072-27-6			110^{16}	3-01-00-03249	0.9238^{20}		1.442^{20}
6519	3-Heptanone, 5-hydroxy-5-methyl-	3-Heptanol-5-one, 3-methyl	$C_8H_{16}O_2$	144.21	39121-37-4			85^{15}	4-01-00-04045	0.9315^{15}		1.4367^{15}
6520	2-Heptanone, 3-methyl-	3-Methyl-2-heptanone	$C_8H_{16}O$	128.21	2371-19-9			164	3-01-00-02878	0.8218^{20}	H_2O 2; EtOH 3; eth 3; ace 3	1.4172^{20}
6521	2-Heptanone, 6-methyl-		$C_8H_{16}O$	128.21	928-68-7			167	4-01-00-03344	0.8151^{20}	H_2O 2; EtOH 4; eth 4; ace 5	1.4162^{20}
6522	3-Heptanone, 2-methyl-	Butyl isopropyl ketone	$C_8H_{16}O$	128.21	13019-20-0			158	4-01-00-03343	0.8163^{20}	H_2O 1; EtOH 3; eth 3; ace 4	1.4115^{20}
6523	3-Heptanone, 6-methyl-		$C_8H_{16}O$	128.21	624-42-0			164	4-01-00-03344	0.8304^{20}	H_2O 1; EtOH 3; eth 3; bz 3	1.4209^{20}
6524	4-Heptanone, 2-methyl-	Isobutyl propyl ketone	$C_8H_{16}O$	128.21	626-33-5			155	4-01-00-03343	0.813^{22}	H_2O 1; EtOH 3; eth 3	
6525	4-Heptanone, 3-methyl-	3-Methyl-4-heptanone	$C_8H_{16}O$	128.21	15726-15-5			154	4-01-00-03344	0.817^{20}		1.4103^{25}
6526	1-Heptanone, 1-phenyl-		$C_{13}H_{18}O$	190.29	1671-75-6	16.4		283.3	4-07-00-00780	0.9516^{20}	ace 4; eth 4; EtOH 4	1.5060^{20}
6527	Heptanoyl chloride		$C_7H_{13}ClO$	148.63	2528-61-2	-83.8		125.2	4-02-00-00963	0.9590^{20}	eth 3; ctc 2; lig 4	1.4345^{18}

No.	Name	Synonym	Mol. Form.	Mol. Wt.	CAS RN	mp/°C	Merck No.	bp/°C	Beil. Ref.	den/g cm^{-3}	Solubility	n_D
6528	Heptasiloxane, hexadecamethyl-	Hexadecamethylheptasiloxane	$C_{16}H_{48}O_6Si_7$	533.15	541-01-5	-78		270	4-04-00-04129	0.9012^{20}	bz 4; lig 4	1.3965^{20}
6529	1,3,5-Heptatriene		C_7H_{10}	94.16	2196-23-8	23		113.5; 31^{25}	4-01-00-01097	0.764^{20}		1.5079
6530	2-Heptenal	Butylacrolein	$C_7H_{12}O$	112.17	2463-63-0			166	4-01-00-03478	0.864^{17}		1.4468^{17}
6531	3-Heptenal		$C_7H_{12}O$	112.17	89896-73-1			151	4-01-00-03480	0.851^{15}	eth 3; ace 3; bz 3	1.4348^{15}
6532	5-Hepten-2-amine, *N*,6-dimethyl-	Isometheptene	$C_9H_{19}N$	141.26	503-01-5		5070	177	3-04-00-00467		eth 4; EtOH 4	
6533	1-Heptene		C_7H_{14}	98.19	592-76-7	-119.7		93.6	4-01-00-00857	0.6970^{20}	H_2O 1; EtOH 3; eth 3; ctc 2	1.3998^{20}
6534	2-Heptene, (*E*)-		C_7H_{14}	98.19	14686-13-6	-109.5		98	4-01-00-00860	0.7012^{20}	H_2O 1; EtOH 3; eth 3; ace 3	1.4045^{20}
6535	2-Heptene, (*Z*)-	*cis*-2-Heptene	C_7H_{14}	98.19	6443-92-1			98.4	3-01-00-00824	0.708^{20}	H_2O 1; EtOH 3; eth 3; ace 3	1.406^{20}
6536	3-Heptene, (*E*)-	*trans*-3-Heptene	C_7H_{14}	98.19	14686-14-7	-136.6		95.7	4-01-00-00861	0.6981^{20}	H_2O 1; EtOH 3; eth 3; ace 3	1.4043^{20}
6537	3-Heptene, (*Z*)-	*cis*-3-Heptene	C_7H_{14}	98.19	7642-10-6	-136.6		95.8	4-01-00-00861	0.7030^{20}	H_2O 1; EtOH 3; eth 3; ace 3	1.4059^{20}
6538	1-Heptene, 1-chloro-	1-Chloro-1-heptene	$C_7H_{13}Cl$	132.63	2384-75-0			155	3-01-00-00823	0.8948^{20}	ace 4; bz 4; eth 4; chl 4	1.4380^{20}
6539	1-Heptene, 2-chloro-		$C_7H_{13}Cl$	132.63	65786-11-0			139	3-01-00-00823	0.8895^{20}	eth 4; EtOH 4; chl 4	1.4349^{20}
6540	3-Heptene, 4-chloro-		$C_7H_{13}Cl$	132.63	2431-24-5			139	3-01-00-00827	0.883^{14}	eth 4; EtOH 4; chl 4	1.437^{14}
6541	2-Heptene, 6-chloro-2-methyl-		$C_8H_{15}Cl$	146.66	80325-37-7			$60\text{-}1^{15}$	2-01-00-00200	0.8931^{18}	ace 4; bz 4; eth 4; EtOH 4	1.4458^{18}
6542	2-Heptene, 2,3-dimethyl-	2,3-Dimethyl-2-heptene	C_9H_{18}	126.24	3074-64-4	-108.5		145.2	4-01-00-00897	0.731^{25}		1.4519^{20}
6543	1-Heptene, 2-methyl-	2-Methyl-1-heptene	C_8H_{16}	112.22	15870-10-7	-90		119.3	4-01-00-00881	0.7104^{25}		1.4123^{20}
6544	1-Heptene, 3-methyl-	3-Methyl-1-heptene	C_8H_{16}	112.22	4810-09-7			111	4-01-00-00883	0.707^{25}		1.406^{20}
6545	1-Heptene, 4-methyl-	4-Methyl-1-heptene	C_8H_{16}	112.22	13151-05-8			112.8	4-01-00-00884	0.713^{25}		1.410^{20}
6546	1-Heptene, 5-methyl-	5-Methyl-1-heptene	C_8H_{16}	112.22	13151-04-7			113.3	4-01-00-00884	0.7125^{25}		1.4094^{20}
6547	1-Heptene, 6-methyl-	6-Methyl-1-heptene	C_8H_{16}	112.22	5026-76-6			113.2	4-01-00-00883	0.7079^{25}		1.4070^{20}
6548	2-Heptene, 2-methyl-	2-Methyl-2-heptene	C_8H_{16}	112.22	627-97-4			122.6	4-01-00-00882	0.7200^{25}	H_2O 1; eth 3; bz 3; ctc 3	1.4170^{20}
6549	2-Heptene, 3 methyl-, (*E*)-	3-Methyl-*trans*-2-heptene	C_8H_{16}	112.22	22768-20-3			122	4-01-00-00883	1.419^{20}		
6550	2-Heptene, 3-methyl-, (*Z*)-	2-Heptene, 3-methyl (*cis*)	C_8H_{16}	112.22	22768-19-0			122	3-01-00-00843	0.725^{25}		1.419^{20}
6551	2-Heptene, 4-methyl-, (*E*)-	4-Methyl-*trans*-2-heptene	C_8H_{16}	112.22	66225 17-0			114	4-01-00-00885	0.712^{25}		1.410^{20}
6552	2-Heptene, 4-methyl-, (*Z*)-	2-Heptene, 4-methyl (*cis*)	C_8H_{16}	112.22	66225-16-9			114	4-01-00-00885	0.712^{25}		1.408^{20}
6553	2-Heptene, 6-methyl-	2-Methyl-5-heptene	C_8H_{16}	112.22	73548-72-8			117	3-01-00-00843	0.714^{25}		1.410^{20}
6554	3-Heptene, 2-methyl-	2-Methyl-3-heptene	C_8H_{16}	112.22	17618-76-7	-107.5		112	4-01-00-00882	0.702^{25}		1.407^{20}
6555	3-Heptene, 3-methyl-	2-Ethyl-2-hexene	C_8H_{16}	112.22	7300-03-0			121	3-01-00-00843	0.724^{25}		1.418^{20}
6556	3-Heptene, 4-methyl-	4-Methyl-3-heptene	C_8H_{16}	112.22	4485-16-9			121	1-01-00-00094	0.721^{25}		1.417^{20}
6557	3-Heptene, 5-methyl-		C_8H_{16}	112.22	13172-91-3			112	4-01-00-00884	0.709^{25}		1.410^{20}
6558	3-Heptene, 6-methyl-	2-Methyl-4-heptene	C_8H_{16}	112.22	3404-57-7			115	4-01-00-00882	0.709^{25}		1.410^{20}
6559	3-Heptene, 4-propyl-	4-Propyl-3-heptene	$C_{10}H_{20}$	[illegible]	4485-13-6			160.5	4-01-00-00906	0.7518^{17}		1.4302^{18}
6560	1-Heptene, 1,1,2,3,3,4,4,5,5,6,6,7,7,7-tetradecafluoro-	Perfluoro-1-heptene	C_7F_{14}	350.05	355-63-5			81.0	[illegible]			
6561	2-Heptenoic acid		$C_7H_{12}O_2$	128.17	18999-28-5	-19		226.5	4-02-00-01577	0.9575^{20}		1.4488^{20}
6562	4-Heptenoic acid		$C_7H_{12}O_2$	128.17	35194-37-7			$98\text{-}100^{5}$	4-02-00-01578	0.949^{15}		1.4418^{15}
6563	5-Heptenoic acid		$C_7H_{12}O_2$	128.17	3593-00-8			223	4-02-00-01578	0.9496^{20}		1.4444^{20}
6564	6-Heptenoic acid		$C_7H_{12}O_2$	128.17	1119-60-4	-6.5		226	4-02-00-01576	0.9515^{14}		1.4404^{14}
6565	2-Heptenoic acid, 6-methyl-		$C_8H_{14}O_2$	142.20	90112-75-7	3		239.5	1-02-00-00194	0.938^{20}		1.4511^{10}
6566	1-Hepten-4-ol		$C_7H_{14}O$	114.19	3521-91-3			152.1	3-01-00-01936	0.8384^{22}		1.4347^{20}
6567	2-Hepten-1-ol, (*E*)-		$C_7H_{14}O$	114.19	33467-76-4			178; 75^{10}	3-01-00-01936	0.8516^{20}	EtOH 3; ace 3	1.4460^{20}

No.	Name Synonym	Mol. Form. Mol. Wt.	CAS RN mp/°C	Merck No. bp/°C	Beil. Ref. den/g cm^{-3}	Solubility n_D
6568	2-Hepten-1-ol, (*Z*)- *cis*-2-Hepten-1-ol	$C_7H_{14}O$ 114.19	55454-22-3	178; 82^{20}	4-01-00-02155 0.8421^{20}	1.4410^{20}
6569	2-Hepten-4-ol, (±)-	$C_7H_{14}O$ 114.19	115113-98-9	153	4-01-00-02155 0.8445^{20}	EtOH 3; eth 3 1.4373^{20}
6570	3-Hepten-2-ol	$C_7H_{14}O$ 114.19	98991-54-9	66-7^{16}	4-01-00-02156 0.834^{17}	1.4391^{18}
6571	6-Hepten-2-ol	$C_7H_{14}O$ 114.19	24395-10-6	64-5^{13}	2-01-00-00488 0.8484^{18}	1.4387^{20}
6572	6-Hepten-3-ol 1-Heptene-4-ol	$C_7H_{14}O$ 114.19	19781-77-2	60-2^{11}	4-01-00-02154 0.8447^{18}	1.4369^{18}
6573	1-Hepten-4-ol, 4-methyl- 1-Heptene-4-ol, 4-methyl	$C_8H_{16}O$ 128.21	1186-31-8	159.5	3-01-00-01945 0.8345^{20}	eth 4; EtOH 4 1.4479^{18}
6574	2-Hepten-4-ol, 5-methyl-	$C_8H_{16}O$ 128.21	4048-31-1	69-70^{18}	2-01-00-00490 0.8473^{20}	1.4411^{20}
6575	2-Hepten-4-ol, 6-methyl- 6-Methyl-2-hepten-4-ol	$C_8H_{16}O$ 128.21	4798-62-3	75^{21}; 67^{11}	2-01-00-00490 0.8354^{15}	1.4378^{20}
6576	3-Hepten-2-ol, 2-methyl-	$C_8H_{16}O$ 128.21	116668-45-2	62-3^{14}	2-01-00-00490 0.8398^{10}	1.4416^{10}
6577	4-Hepten-2-ol, 2-methyl-	$C_8H_{16}O$ 128.21	116668-44-1 -28	63-4^{20}	3-01-00-01944 0.8424^{14}	1.4407^{14}
6578	4-Hepten-3-ol, 3-methyl-	$C_8H_{16}O$ 128.21	4048-30-0	103^{75}; 62^{15}	2-01-00-00490 0.8477^{17}	1.4465^{20}
6579	4-Hepten-3-ol, 4-methyl-	$C_8H_{16}O$ 128.21	81280-12-8	66^{17}	4-01-00-02170 0.8525^{18}	1.4479^{18}
6580	5-Hepten-2-ol, 6-methyl-	$C_8H_{16}O$ 128.21	1569-60-4	175	1-01-00-00230 0.8545^{20}	1.4505^{20}
6581	6-Hepten-1-ol, 3-methyl-	$C_8H_{16}O$ 128.21	4048-32-2	97-9^{22}	2-01-00-00491 0.8562^{19}	1.4470^{14}
6582	6-Hepten-2-ol, 2-methyl- 1,1-Dimethyl-5-hexen-1-ol	$C_8H_{16}O$ 128.21	77437-98-0	65-6^{16}	3-01-00-01945 0.8393^{16}	1.4393^{16}
6583	1-Hepten-3-one Butyl vinyl ketone	$C_7H_{12}O$ 112.17	2918-13-0	42-3^{15}	4-01-00-03477 0.8434^{20}	1.4305^{19}
6584	4-Hepten-2-one	$C_7H_{12}O$ 112.17	24332-22-7	61-2^{20}	3-01-00-03004 0.8618^{21}	1.4290^{21}
6585	6-Hepten-2-one Methyl 4-pentenyl ketone	$C_7H_{12}O$ 112.17	21889-88-3	41-3^{10}	4-01-00-03478 0.8673^{18}	1.4350^{18}
6586	6-Hepten-3-one 1-Hepten-5-one	$C_7H_{12}O$ 112.17	2565-39-1	142	4-01-00-03478 0.8487^{18}	1.4254^{18}
6587	3-Hepten-2-one, (*E*)-	$C_7H_{12}O$ 112.17	5609-09-6	62^{15}	4-01-00-03481 0.8496^{20}	H_2O 1; EtOH 4; eth 4 1.4436^{20}
6588	3-Hepten-2-one, (*Z*)- 3-Hepten-2-one (*cis*)	$C_7H_{12}O$ 112.17	69668-88-8	34-9^{11}	4-01-00-03481 0.8440^{20}	1.4325^{20}
6589	5-Hepten-2-one, (*E*)-	$C_7H_{12}O$ 112.17	1071-94-9	153	4-01-00-03480 0.8445^{20}	1.4309^{20}
6590	3-Hepten-2-one, 3-ethyl-4-methyl-	$C_{10}H_{18}O$ 154.25	54244-90-5	190; 83^{14}	3-01-00-03028 0.8559^{20}	1.4535^{20}
6591	4-Hepten-2-one, 3-ethyl-4-methyl-	$C_{10}H_{18}O$ 154.25	53252-08-7	69^{11}	3-01-00-03028 0.8450^{20}	1.4405^{20}
6592	4-Hepten-3-one, 5-ethyl-4-methyl-	$C_{10}H_{18}O$ 154.25	22319-28-4	80^{10}	4-01-00-03517 0.8622^{10}	1.4545^{19}
6593	5-Hepten-2-one, 4-ethyl-6-methyl- 2-Hepten-6-one, 4-ethyl-2-methyl	$C_{10}H_{18}O$ 154.25	104035-66-7	74-5^{13}	4-01-00-03517 0.8430^{20}	1.4546^{20}
6594	5-Hepten-3-one, 5-ethyl-4-methyl-	$C_{10}H_{18}O$ 154.25	74764-56-0 74.5	135-9^{3}	4-01-00-03517 0.8564^{21}	1.4452^{21}
6595	1-Hepten-3-one, 5-methyl-	$C_8H_{14}O$ 126.20	44829-76-7	40-1^{7}	3-01-00-03014 0.8479^{19}	1.4360^{19}
6596	2-Hepten-4-one, 5-methyl- 5-Methyl-2-hepten-4-one	$C_8H_{14}O$ 126.20	81925-81-7	171	4-01-00-03496 0.848^{25}	1.4400^{25}
6597	2-Hepten-4-one, 6-methyl-	$C_8H_{14}O$ 126.20	49852-35-9	170	4-01-00-03494 0.841^{25}	1.4388^{25}
6598	3-Hepten-2-one, 6-methyl-	$C_8H_{14}O$ 126.20	2009-74-7	179	3-01-00-03011 0.8443^{17}	1.4427^{10}
6599	4-Hepten-2-one, 6-methyl- 2-Methyl-3-hepten-6-one	$C_8H_{14}O$ 126.20	39273-81-9	163	1-01-00-00385 0.8345^{20}	1.4315
6600	4-Hepten-3-one, 5-methyl- Homomesityl oxide	$C_8H_{14}O$ 126.20	1447-26-3	166.5	3-01-00-03012 0.8591^{20}	1.4476^{20}
6601	5-Hepten-2-one, 3-methyl- 3-Methyl-5-hepten-2-one	$C_8H_{14}O$ 126.20	38552-72-6	62-4^{20}	4-01-00-03496 0.8463^{18}	1.4345^{18}
6602	5-Hepten-2-one, 5-methyl-	$C_8H_{14}O$ 126.20	10339-67-0	84^{45}	4-01-00-03494 0.8647^{16}	1.4434^{16}
6603	5-Hepten-2-one, 6-methyl-	$C_8H_{14}O$ 126.20	110-93-0 173.5		4-01-00-03493 0.8546^{16}	eth 4; EtOH 4 1.4445^{20}
6604	5-Hepten-3-one, 2-methyl-	$C_8H_{14}O$ 126.20	77958-21-5	161.5	1-01-00-00385 0.842^{20}	1.4310^{20}
6605	5-Hepten-3-one, 5-methyl- 5-Methylhept-5-en-3-one	$C_8H_{14}O$ 126.20	1190-34-7	63^{19}	3-01-00-03011 0.8524^{21}	1.4367^{21}
6606	1-Hepten-3-yne	C_7H_{10} 94.16	2384-73-8	110	4-01-00-01097 0.7603^{20}	ace 4; bz 4; EtOH 4; peth 4 1.4520^{25}
6607	1-Hepten-4-yne, 6,6-dimethyl-	C_9H_{14} 122.21	31508-08-4	125	0.758^{20}	ace 4; bz 4; EtOH 4; peth 4 1.4312^{20}

PHYSICAL CONSTANTS OF ORGANIC COMPOUNDS (continued)

No.	Name Synonym	Mol. Form. Mol. Wt.	CAS RN mp/°C	Merck No. bp/°C	Beil. Ref. den/g cm^{-3}	Solubility n_D
6608	6-Hepten-4-yn-3-ol, 3-ethyl- Diethylvinylethynylcarbinol	$C_9H_{14}O$ 138.21	3142-84-5	 62^{4}	3-01-00-02034 0.8875^{20}	eth 4; EtOH 4 1.4800^{20}
6609	1-Heptyne	C_7H_{12} 96.17	628-71-7 -81	 99.7	4-01-00-01025 0.7328^{20}	H_2O 2; EtOH 5; eth 5; bz 3 1.4087^{20}
6610	2-Heptyne 1-Methyl-2-butylacetylene	C_7H_{12} 96.17	1119-65-9	 112	4-01-00-01026 0.744^{25}	H_2O 1; EtOH 5; eth 5; bz 3 1.4230^{20}
6611	3-Heptyne 1-Ethyl-2-propylacetylene	C_7H_{12} 96.17	2586-89-2 -130.5	 107.2	4-01-00-01027 0.7336^{25}	H_2O 1; EtOH 5; eth 5; bz 3 1.4189^{20}
6612	1-Heptyne, 1-bromo- 1-Bromo-1-heptyne	$C_7H_{11}Br$ 175.07	19821-84-2	 164; 69^{25}	3-01-00-00998 1.2120^{22}	ace 4; eth 4; EtOH 4; chl 4 1.4678^{22}
6613	2-Heptyne, 1-bromo- 2-Heptynyl bromide	$C_7H_{11}Br$ 175.07	18495-26-6	 104^{55}; 84^{20}	4-01-00-01026	ace 4; eth 4; EtOH 4 1.4878^{25}
6614	1-Heptyne, 1-chloro- 1-Chloro-1-heptyne	$C_7H_{11}Cl$ 130.62	51556-10-6	 141	3-01-00-00997 0.9250^{24}	eth 4; EtOH 4 1.4411^{24}
6615	2-Heptyne, 7-chloro- 1-Chloro-5-heptyne	$C_7H_{11}Cl$ 130.62	70396-13-3	 166	3-01-00-00998	eth 4; EtOH 4 1.4507^{25}
6616	3-Heptyne, 7-chloro- 4-Heptynyl chloride	$C_7H_{11}Cl$ 130.62	51575-85-0	 164	3-01-00-00999	eth 4; EtOH 4 1.4517^{20}
6617	3-Heptyne, 2,6-dimethyl-	C_9H_{16} 124.23	19549-97-4	 130.6	3-01-00-01015 0.785^{20}	ace 4; eth 4
6618	3-Heptyne, 5,5-dimethyl- 5,5-Dimethyl-3-heptyne	C_9H_{16} 124.23	23097-98-5	 69^{100}	3-01-00-01015 0.7610^{20}	ace 4; eth 4 1.4360^{20}
6619	3-Heptyne, 5-ethyl-5-methyl- 5-Ethyl-5-methyl-3-heptyne	$C_{10}H_{18}$ 138.25	61228-10-2	 88^{100}	3-01-00-01021 0.7714^{20}	ace 4; eth 4 1.4386^{20}
6620	1-Heptyne, 1-iodo- 1-Iodo-1-heptyne	$C_7H_{11}I$ 222.07	54573-13-6	 $90\text{-}2^{17}$	3-01-00-00998 1.4701^{19}	 1.5123^{19}
6621	Hexacene	$C_{26}H_{16}$ 328.41	258-31-1 380	 sub	4-05-00-02834	H_2O 1; EtOH 1
6622	Hexacosane	$C_{26}H_{54}$ 366.71	630-01-3 56.4	 412.2	4-01-00-00583 0.7783^{60}	bz 4; lig 4; chl 4 1.4357^{60}
6623	Hexacosane, 13-dodecyl-	$C_{38}H_{78}$ 535.04	55517-73-2 13.7	 272^{1}	4-01-00-00600 0.8188^{20}	 1.4577^{20}
6624	Hexacosanoic acid Cerotic acid	$C_{26}H_{52}O_2$ 396.70	506-46-7 88.5		4-02-00-01310 0.8198^{100}	H_2O 1; EtOH 4; eth 4 1.4301^{100}
6625	Hexacosanoic acid, methyl ester Methyl hexacosanoate	$C_{27}H_{54}O_2$ 410.72	5802-82-4 63.8	 286^{15}; 237^{2}	4-02-00-01311	chl 3 1.4301^{80}
6626	1-Hexacosanol	$C_{26}H_{54}O$ 382.71	506-52-5 80	 305^{20} dec	4-01-00-01912	H_2O 1; EtOH 3; eth 3
6627	6,10-Hexadecadiyne	$C_{16}H_{26}$ 218.38	10160-99-3	 157^{10}	3-01-00-01067 0.7907^{20}	 1.4523^{20}
6628	Hexadecanal	$C_{16}H_{32}O$ 240.43	629-80-1 35	 200^{29}	4-01-00-03393	H_2O 1; EtOH 3; eth 3; ace 3
6629	Hexadecanamide	$C_{16}H_{33}NO$ 255.44	629-54-9 107	 236^{12}	4-02-00-01182 1.0000^{20}	H_2O 1; EtOH 2; eth 2; ace 2
6630	Hexadecanamide, *N*-phenyl-	$C_{22}H_{37}NO$ 331.54	6832-98-0 91.8	 283^{17}; 132^{10}	4-12-00-00400	ace 4; bz 4; EtOH 4; chl 4
6631	1-Hexadecanamine	$C_{16}H_{35}N$ 241.46	143-27-1 46.8	 322.5	4-04-00-00818 0.8129^{20}	H_2O 1; EtOH 4; eth 4; ace 3 1.4496^{20}
6632	Hexadecane Cetane	$C_{16}H_{34}$ 226.45	544-76-3 18.1	 286.8	4-01-00-00537 0.7733^{20}	H_2O 1; EtOH 2; eth 5; ctc 3 1.4345
6633	Hexadecane, 1-bromo-	$C_{16}H_{33}Br$ 305.34	112-82-3 18	 336	4-01-00-00542 0.9991^{20}	H_2O 1; eth 3 1.4618^{25}
6634	Hexadecane, 1-chloro-	$C_{16}H_{33}Cl$ 260.89	4860-03-1 17.9	 322	4-01-00-00542 0.8652^{20}	H_2O 1 1.4505^{20}
6635	Hexadecane, 1,16-dibromo- Hexadecamethylene dibromide	$C_{16}H_{32}Br_2$ 384.24	45223-18-5 56	 204^{4}	2-01-00-00138	chl 4
6636	Hexadecane, 1,1-dimethoxy- Palmitaldehyde, dimethyl acetal	$C_{18}H_{38}O_2$ 286.50	2791-29-9 10	 144^{2}	4-01-00-03393 0.8542^{20}	ace 4; eth 4; EtOH 4 1.4382^{25}
6637	Hexadecane, 6,11-dipentyl-	$C_{26}H_{54}$ 366.71	15874-03-0 -16.2	 231^{10}	4-01-00-00586 0.8072^{20}	 1.4502^{20}
6638	Hexadecane, 1-(ethenyloxy)-	$C_{18}H_{36}O$ 268.48	822-28-6 16	 160^{2}	4-01-00-02057 0.821^{27}	 1.4444^{25}
6639	Hexadecane, 1-fluoro- 1-Fluorohexadecane	$C_{16}H_{33}F$ [illegible]	408-38-8 19	 289	4-01-00-00542 0.8321^{20}	eth 4; lig 4 1.4317^{20}
6640	Hexadecane, 1-iodo-	$C_{16}H_{33}I$ 352.34	544-77-4 22	 212^{15}	4-01-00-00543 1.123^{25}	H_2O 1; EtOH 2; eth 3; [illegible] 1.4806^{20}
6641	Hexadecanenitrile	$C_{16}H_{31}N$ 237.43	629-79-8 31	 333	4-02-00-01183 0.8303^{20}	H_2O 1; EtOH 4; eth 4; ace 4 1.4450^{20}
6642	Hexadecane, 1,1'-oxybis-	$C_{32}H_{66}O$ 466.88	4113-12-6 55	 270 dec	4-01-00-01878 0.978^{19}	H_2O 1; EtOH 3; eth 3; chl 3
6643	1-Hexadecanethiol Cetyl mercaptan	$C_{16}H_{34}S$ 258.51	2917-26-2 19	 $123\text{-}8^{0.5}$	4-01-00-01881	H_2O 1; EtOH 2; eth 3; ctc 2
6644	Hexadecanoic acid Palmitic acid	$C_{16}H_{32}O_2$ 256.43	57-10-3 61.8	6947 351.5	4-02-00-01157 0.8527^{62}	H_2O 1; EtOH 3; eth 5; ace 3 1.43345^{60}
6645	Hexadecanoic acid, aluminum salt	$C_{48}H_{93}AlO_6$ 793.24	555-35-1	361	4-02-00-01157	H_2O 1; EtOH 3; peth 3; bz 3
6646	Hexadecanoic acid, anhydride	$C_{32}H_{62}O_3$ 494.84	623-65-4 64		4-02-00-01181 0.8388^{83}	eth 4 1.4364^{68}
6647	Hexadecanoic acid, butyl ester	$C_{20}H_{40}O_2$ 312.54	111-06-8 16.9		4-02-00-01167	H_2O 1; EtOH 3; eth 3 1.4312^{50}

No.	Name Synonym	Mol. Form. Mol. Wt.	CAS RN mp/°C	Merck No. bp/°C	Beil. Ref. den/g cm^{-3}	Solubility n_D
6648	Hexadecanoic acid, 2-[(dichloroacetyl)amino]-3-hydroxy-3-(4-nitrophenyl)propyl ester, [*R*-(*R**,*R**)]- Chloramphenicol palmitate	$C_{27}H_{42}Cl_2N_2O_6$ 561.55	530-43-8 90	2069	4-13-00-02753	bz 4; eth 4; EtOH 4
6649	Hexadecanoic acid, 1,2-ethanediyl ester	$C_{34}H_{66}O_4$ 538.90	624-03-3 72		4-02-00-01169 0.8594^{78}	H_2O 1; EtOH 1; eth 3; ace 4
6650	Hexadecanoic acid, ethyl ester	$C_{18}H_{36}O_2$ 284.48	628-97-7 24	 191^{10}	4-02-00-01165 0.8577^{25}	H_2O 1; EtOH 3; eth 3; ace 3 1.4347^{34}
6651	Hexadecanoic acid, hexadecyl ester Cetyl palmitate	$C_{32}H_{64}O_2$ 480.86	540-10-3 54	2023	4-02-00-01168 0.989^{20}	eth 4; EtOH 4 1.4398^{70}
6652	Hexadecanoic acid, 2-hydroxyethyl ester	$C_{18}H_{36}O_3$ 300.48	4219-49-2 51	 $173\text{-}4^{3}$	4-02-00-01168 0.8768^{60}	EtOH 4
6653	Hexadecanoic acid, methyl ester	$C_{17}H_{34}O_2$ 270.46	112-39-0 30	 417; 148^{2}	4-02-00-01165 0.8247^{75}	H_2O 1; EtOH 4; eth 3; ace 4
6654	Hexadecanoic acid, 1-methylethyl ester	$C_{19}H_{38}O_2$ 298.51	142-91-6 13.5	 160^{2}	4-02-00-01167 0.8404^{38}	ace 4; bz 4; eth 4; EtOH 4 1.4364^{25}
6655	Hexadecanoic acid, phenylmethyl ester Benzyl palmitate	$C_{23}H_{38}O_2$ 346.55	41755-60-6 36		3-06-00-01481 0.9109^{35}	bz 4; eth 4; EtOH 4; chl 4 1.4689^{50}
6656	Hexadecanoic acid, 1,2,3-propanetriyl ester Tripalmitin	$C_{51}H_{98}O_6$ 807.34	555-44-2 66.5	9648 315	4-02-00-01176 0.8752^{70}	H_2O 1; EtOH 2; eth 4; bz 3 1.4381^{80}
6657	Hexadecanoic acid, propyl ester	$C_{19}H_{38}O_2$ 298.51	2239-78-3 20.4	 190^{12}	4-02-00-01167 0.8455^{33}	 1.4392^{25}
6658	Hexadecanoic acid, 3,7,11,15-tetramethyl- Phytanic acid	$C_{20}H_{40}O_2$ 312.54	14721-66-5 -65			
6659	1-Hexadecanol Cetyl alcohol	$C_{16}H_{34}O$ 242.45	36653-82-4 49.3	2020 334	4-01-00-01876 0.8187^{50}	H_2O 1; EtOH 2; eth 4; ace 3 1.4283^{79}
6660	2-Hexadecanol	$C_{16}H_{34}O$ 242.45	14852-31-4 44	 314	4-01-00-01882 0.8338^{20}	 1.4479^{20}
6661	3-Hexadecanol	$C_{16}H_{34}O$ 242.45	593-03-3 50	 152^{4}	4-01-00-01882 0.8000^{67}	
6662	1-Hexadecanol, acetate	$C_{18}H_{36}O_2$ 284.48	629-70-9 -18.5	 $220\text{-}5^{205}$	4-02-00-00171 0.8574^{25}	H_2O 1; EtOH 2; ctc 3 1.4438^{20}
6663	3-Hexadecanone	$C_{16}H_{32}O$ 240.43	18787-64-9 43	 184^{17}; 140^{2}	4-01-00-03394	chl 3
6664	1-Hexadecanone, 1-phenyl-	$C_{22}H_{36}O$ 316.53	6697-12-7 59	 251^{15}	4-07-00-00860 0.8692^{76}	 1.4675^{76}
6665	Hexadecanoyl chloride	$C_{16}H_{31}ClO$ 274.87	112-67-4 12	 199^{20}	4-02-00-01182 0.9016^{25}	eth 4 1.4514^{20}
6666	6,10,14-Hexadecatrienoic acid Hiragonic acid	$C_{16}H_{26}O_2$ 250.38	4444-12-6	 $180\text{-}90^{15}$	3-02-00-01505 0.9296^{20}	eth 4; EtOH 4 1.4850^{50}
6667	1-Hexadecene 1-Cetene	$C_{16}H_{32}$ 224.43	629-73-2 4.1	 284.9	4-01-00-00927 0.7811^{20}	H_2O 1; EtOH 3; eth 3; ctc 3 1.4412^{20}
6668	7-Hexadecenoic acid Dinoroleic acid	$C_{16}H_{30}O_2$ 254.41	1191-75-9 33	 230^{10}	0-02-00-00460	EtOH 4; eth 4
6669	9-Hexadecenoic acid	$C_{16}H_{30}O_2$ 254.41	2091-29-4 0.5	 181^{1}	4-02-00-01629	bz 4; eth 4; peth 4
6670	9-Hexadecenoic acid, (*Z*)- Palmitoleic acid	$C_{16}H_{30}O_2$ 254.41	373-49-9 -0.1	 $131^{0.06}$	4-02-00-01629	
6671	1-Hexadecen-3-ol, 3,7,11,15-tetramethyl- Isophytol	$C_{20}H_{40}O$ 296.54	505-32-8	5084 $107\text{-}10^{0.01}$	2-01-00-00504 0.8519^{20}	bz 4; eth 4; EtOH 4 1.4571^{20}
6672	2-Hexadecen-1-ol, 3,7,11,15-tetramethyl-, [*R*-[*R**,*R**-(*E*)]]- Phytol	$C_{20}H_{40}O$ 296.54	150-86-7	7362 203^{10}	4-01-00-02208 0.8497^{25}	 1.4595^{25}
6673	1-Hexadecyne	$C_{16}H_{30}$ 222.41	629-74-3 15	 284	4-01-00-01073 0.7965^{20}	bz 4 1.4440^{20}
6674	2-Hexadecyne	$C_{16}H_{30}$ 222.41	629-75-4 20	 160^{15}	3-01-00-01028 0.8039^{20}	
6675	2,4-Hexadienal, (*E*,*E*)-	C_6H_8O 96.13	142-83-6	 174; 76^{30}	4-01-00-03545 0.898^{20}	 1.5384^{20}
6676	1,2-Hexadiene Propylallene	C_6H_{10} 82.15	592-44-9	 76	4-01-00-01011 0.7149^{20}	eth 4; chl 4 1.4282^{20}
6677	1,4-Hexadiene	C_6H_{10} 82.15	592-45-0	 65	4-01-00-01013 0.7000^{20}	H_2O 1; eth 4 1.4150^{20}
6678	1,5-Hexadiene Biallyl	C_6H_{10} 82.15	592-42-7 -140.7	 59.4	4-01-00-01013 0.6878^{25}	H_2O 1; EtOH 3; eth 3; bz 3 1.4042^{20}
6679	2,3-Hexadiene Methylethylallene	C_6H_{10} 82.15	592-49-4	 78	4-01-00-01016 0.684^{15}	 1.395^{20}
6680	2,4-Hexadiene	C_6H_{10} 82.15	592-46-1 -79	 80	3-01-00-00985 0.7196^{20}	H_2O 1; EtOH 3; eth 3; chl 3 1.4500^{20}
6681	1,3-Hexadiene, (*E*)-	C_6H_{10} 82.15	20237-34-7 -102.4	 73.2	4-01-00-01011 0.6995^{25}	 1.4406^{20}
6682	1,3-Hexadiene, (*Z*)-	C_6H_{10} 82.15	14596-92-0	 73.1	4-01-00-01011 0.7033^{25}	 1.4379^{20}
6683	1,4-Hexadiene, (*E*)-	C_6H_{10} 82.15	7319-00-8 -138.7	 65.0	 0.695^{25}	 1.4104^{20}
6684	1,4-Hexadiene, (*Z*)-	C_6H_{10} 82.15	7318-67-4	 66.3	 0.695^{25}	eth 4 1.4049^{20}
6685	2,4-Hexadiene, (*E*,*E*)-	C_6H_{10} 82.15	5194-51-4 -44.9	 82.2	4-01-00-01016 0.7101^{25}	 1.4510^{20}
6686	2,4-Hexadiene, (*E*,*Z*)-	C_6H_{10} 82.15	5194-50-3 -96.1	 83.5	4-01-00-01016 0.7185^{25}	 1.4560^{20}

No.	Name Synonym	Mol. Form. Mol. Wt.	CAS RN mp/°C	Merck No. bp/°C	Beil. Ref. den/g cm^{-3}	Solubility n_D
6687	2,4-Hexadiene, (*Z*,*Z*)-	C_6H_{10} 82.15	6108-61-8	85	0.7298^{25}	1.4606^{20}
6688	1,3-Hexadiene, 3-chloro-	C_6H_9Cl 116.59	101870-06-8	68^{117}	4-01-00-01012 0.9390^{20}	eth 4; chl 4 1.4770^{20}
6689	1,5-Hexadiene, 1,1,2,3,3,4,4,5,6,6-decachloro-	C_6Cl_{10} 426.59	29030-84-0 49	121$^{0.03}$	4-01-00-01016 1.905^{52}	ace 4; eth 4; chl 4 1.6012^{51}
6690	2,4-Hexadiene, 1,3-dichloro-	$C_6H_8Cl_2$ 151.03	73454-83-8	80-2^{17}	4-01-00-01017 1.1456^{20}	bz 4; chl 4 1.5271^{20}
6691	1,5-Hexadiene, 2,5-dimethyl-	C_8H_{14} 110.20	627-58-7 -75.6	114.3	4-01-00-01042 0.743^{20}	H_2O 1; ace 3; chl 3 1.43995^{21}
6692	2,4-Hexadiene, 2,5-dimethyl-	C_8H_{14} 110.20	764-13-6 14	134.5	4-01-00-01043 0.7577^{25}	H_2O 1; EtOH 3; eth 3; bz 3 1.4785^{20}
6693	1,5-Hexadiene-3,4-diol, (*R**,*R**)-(±)-	$C_6H_{10}O_2$ 114.14	19700-97-1 21.7	90-1^{8}	1.017^{19}	H_2O 3; EtOH 3; eth 3; chl 3 1.4790^{19}
6694	1,5-Hexadiene, 3-ethyl-2-methyl- 1,5-Hexadiene, 2-methyl-3-ethyl	C_9H_{16} 124.23	73398-15-9 -70	145	0-01-00-00260 0.7594^{25}	
6695	1,5-Hexadiene, 2-methyl-	C_7H_{12} 96.17	4049-81-4 -128.8	88.1	4-01-00-01030 0.7153^{25}	1.4183^{20}
6696	2,4-Hexadiene, 2-methyl-	C_7H_{12} 96.17	28823-41-8 -74.2	111.5	4-01-00-01030 0.7411^{25}	
6697	2,4-Hexadiene, 1,3,4,6-tetrachloro-	$C_6H_6Cl_4$ 219.92	100367-45-1	84-9^{2}	3-01-00-00987 1.4013^{20}	MeOH 4; chl 4 1.5465^{20}
6698	2,4-Hexadienoic acid Sorbic acid	$C_6H_8O_2$ 112.13	110-44-1 134.5	8677 228 dec; 153^{50}	2-02-00-00452 1.204^{19}	H_2O 3; EtOH 3; eth 4; chl 3
6699	2,4-Hexadienoic acid, ethyl ester	$C_8H_{12}O_2$ 140.18	110318-09-7	195.5 85^{20}	2-02-00-00452 0.9506^{20}	H_2O 1; EtOH 3; eth 3; chl 3 1.4951^{20}
6700	2,4-Hexadienoic acid, ethyl ester, (*E*,*E*)-	$C_8H_{12}O_2$ 140.18	2396-84-1	195.5	4-02-00-01703 0.9506^{20}	eth 4; EtOH 4; chl 4 1.4951^{20}
6701	2,5-Hexadienoic acid, 3-methoxy-5-methyl-4-oxo- Penicillic acid	$C_8H_{10}O_4$ 170.17	90-65-3 83	7033	3-03-00-01467	H_2O 3; EtOH 4; eth 4; ace 3
6702	2,4-Hexadienoic acid, methyl ester Methyl 2,4-hexadienoate	$C_7H_{10}O_2$ 126.16	1515-80-6 15	180; 70^{20}	2-02-00-00452 0.9777^{20}	H_2O 1; EtOH 3; eth 3 1.5025^{22}
6703	2,4-Hexadienoic acid, methyl ester, (*E*,*E*)	$C_7H_{10}O_2$ 126.16	689-89-4 15	180	4-02-00-01703 0.9777^{20}	eth 4; EtOH 4 1.5025^{22}
6704	2,4-Hexadienoic acid, potassium salt, (*E*,*E*)-	$C_6H_7KO_2$ 150.22	24634-61-5 >270 dec	7661	1.361^{25}	H_2O 4; EtOH 3
6705	1,4-Hexadien-3-ol	$C_6H_{10}O$ 98.14	1070-14-0	50-2^{15}	4-01-00-02238 0.859^{23}	1.4502^{23}
6706	1,5-Hexadien-3-ol 3-Hydroxy-1,5-hexadiene	$C_6H_{10}O$ 98.14	924-41-4	133.5	4-01-00-02239 0.8596^{25}	1.4464^{25}
6707	2,4-Hexadien-1-ol 1-Hydroxy-2,4-hexadiene	$C_6H_{10}O$ 98.14	111-28-4 30.5	8678 76^{12}	4-01-00-02239 0.8967^{23}	H_2O 1; EtOH 3; eth 3 1.4981^{20}
6708	3,5-Hexadien-2-ol 1,3 Hexadien-5-ol	$C_6H_{10}O$ 98.14	3280-51-1	77-8^{26}	4-01-00-02237 0.8678^{20}	EtOH 4 1.4896^{20}
6709	2,4-Hexadien-1-ol, (*E*,*E*)-	$C_6H_{10}O$ 98.14	17102-64-6 30.5	76^{12}	4-01-00-02239 0.8967^{20}	eth 4; EtOH 4 1.4981^{20}
6710	3,5-Hexadien-2-one, 6-phenyl-	$C_{12}H_{12}O$ 172.23	4173-44-8 68	171^{15}, 155^{12}	4-07-00-01183	bz 4; eth 4; EtOH 4; chl 4
6711	2,4-Hexadienoyl chloride, (*E*,*E*)-	C_6H_7ClO 130.57	2614-88-2	78^{15}	3-02-00-01460 1.0666^{19}	ace 4 1.5545^{20}
6712	1,3-Hexadien-5-yne	C_6H_6 78.11	10420-90-3 -81	83.5	4-01-00-01120 0.7806^{20}	bz 4 1.5095^{20}
6713	1,5-Hexadien-3-yne Divinylacetylene	C_6H_6 78.11	821-08-9 -88	85	4-01-00-01120 0.7851^{20}	H_2O 1; bz 3 1.5035^{20}
6714	1,5-Hexadien-3-yne, 2,5-dimethyl- Diisopropenylacetylene	C_8H_{10} 106.17	3725-05-1	123	4-01-00-01124 0.7863^{25}	bz 4; chl 4 1.4845^{20}
6715	1,4-Hexadiyne	C_6H_6 78.11	10420-91-4 <-80	81.7	3-01-00-01057 0.825^{0}	bz 4; chl 4
6716	1,5-Hexadiyne Bipropargyl	C_6H_6 78.11	628-16-0 -6	86	4-01-00-01118 0.8049^{20}	H_2O 1; EtOH 3; eth 3; ace 3 1.4380^{23}
6717	2,4-Hexadiyne Dimethyldiacetylene	C_6H_6 78.11	[illegible] 67.8	129.5	4-01-00-01119	EtOH 4; eth 4
6718	Hexanal Caproaldehyde	$C_6H_{12}O$ 100.16	66-25-1 -56	1761 131	4-01-00-03296 0.8335^{20}	H_2O 2; EtOH 4; eth 4; ace 3 1.4039^{20}
6719	Hexanal, 2-ethyl- 2-Ethylhexanal	$C_8H_{16}O$ 128.21	123-05-7 <-100	163	4-01-00-03345 0.8540^{20}	H_2O 1; EtOH 3; eth 3; ctc 2 1.4142^{20}
6720	Hexanal, 3-methyl- Caproaldehyde, 3-methyl	$C_7H_{14}O$ 114.19	19269-28-4	143	2-01-00-00756 0.8203^{20}	H_2O 1; EtOH 3; eth 3 1.4122^{20}
6721	Hexanamide	$C_6H_{13}NO$ 115.18	628-02-4 101	255	4-02-00-00929 0.999^{20}	bz 4; eth 4; EtOH 4; chl 4 1.4200^{110}
6722	Hexanamide, *N*-phenyl-	$C_{12}H_{17}NO$ 191.27	621-15-8 95		4-12-00-00392 1.112^{25}	eth 4; EtOH 4
6723	1-Hexanamine Hexylamine	$C_6H_{15}N$ 101.19	111-26-2 -22.9	132.8	4-04-00-00709 0.7660^{20}	H_2O 2; EtOH 5; eth 5; chl 3 1.4180^{20}
6724	2-Hexanamine, (±)-	$C_6H_{15}N$ 101.19	68107-05-1 -19	117.5	4-04-00-00721 0.7533^{20}	EtOH 3; eth 3 1.4080^{25}

No.	Name	Mol. Form.	CAS RN	Merck No.	Beil. Ref.	Solubility
	Synonym	Mol. Wt.	mp/°C	bp/°C	den/g cm^{-3}	n_D
6725	1-Hexanamine, *N,N*-dihexyl-	$C_{18}H_{39}N$	102-86-3		4-04-00-00711	H_2O 1; EtOH 4; eth 4; acid 3
		269.51		261.7	0.7976^{21}	
6726	1-Hexanamine, 2-ethyl-	$C_8H_{19}N$	104-75-6		4-04-00-00766	
	2-Ethyl-1-hexylamine	129.25				
6727	1-Hexanamine, *N*-hexyl-	$C_{12}H_{27}N$	143-16-8		4-04-00-00711	EtOH 3; eth 3
		185.35	-13.1	236; 75^{1}	0.7889^{20}	1.4339^{20}
6728	1-Hexanamine, 3-methyl-	$C_7H_{17}N$	65530-93-0		2-04-00-00653	
	Hexylamine, 3-methyl-	115.22		149; 67^{45}	0.772^{26}	1.4249^{25}
6729	1-Hexanamine, 4-methyl-	$C_7H_{17}N$	34263-68-8		2-04-00-00654	
	Hexylamine, 4-methyl-	115.22		153	0.7802^{20}	1.4238^{20}
6730	2-Hexanamine, 4-methyl-	$C_7H_{17}N$	105-41-9	6000	4-04-00-00747	H_2O 2; EtOH 4; eth 4; chl 4
	Methylhexaneamine	115.22		132.5	0.7655^{20}	1.4150^{25}
6731	Hexane	C_6H_{14}	110-54-3	4613	4-01-00-00338	H_2O 1; EtOH 4; eth 3; chl 3
		86.18	-95.3	68.7	0.6548^{25}	1.3749^{20}
6732	Hexane, 1-bromo-	$C_6H_{13}Br$	111-25-1		4-01-00-00352	H_2O 1; EtOH 5; eth 5; ace 3
	Hexyl bromide	165.07	-84.7	155.3	1.1744^{20}	1.4478^{20}
6733	Hexane, 2-bromo-	$C_6H_{13}Br$	3377-86-4		4-01-00-00353	H_2O 1; EtOH 4; eth 3; ace 3
	2-Bromohexane	165.07		144; 78^{90}	1.1658^{20}	1.4832^{25}
6734	Hexane, 3-bromo-	$C_6H_{13}Br$	3377-87-5		4-01-00-00353	ace 4; eth 4; EtOH 4; chl 4
	3-Bromohexane	165.07		142	1.1799^{20}	1.4472^{20}
6735	Hexane, 1-bromo-6-fluoro-	$C_6H_{12}BrF$	373-28-4		4-01-00-00353	ace 4; eth 4; EtOH 4; chl 4
	1-Bromo-6-fluorohexane	183.06		$67\text{-}8^{11}$	1.293^{20}	1.4436^{25}
6736	Hexane, 1-bromo-6-methoxy-	$C_7H_{15}BrO$	50592-87-5		4-01-00-01704	
		195.10		112^{35}; 98^{19}	1.1887^{25}	1.4469^{25}
6737	Hexane, 1-chloro-	$C_6H_{13}Cl$	544-10-5	2144	4-01-00-00349	H_2O 1; EtOH 3; eth 3; ace 3
	Hexyl chloride	120.62	-94	135	0.8785^{20}	1.4199^{20}
6738	Hexane, 2-chloro-	$C_6H_{13}Cl$	638-28-8		4-01-00-00349	ace 4; bz 4; eth 4; EtOH 4
	2-Hexyl chloride	120.62		122.5	0.8694^{21}	1.4142^{22}
6739	Hexane, 3-chloro-	$C_6H_{13}Cl$	2346-81-8		4-01-00-00349	ace 4; bz 4; eth 4; EtOH 4
	3-Chlorohexane	120.62		123	0.8684^{20}	1.4163^{20}
6740	Hexane, 2-chloro-2,5-dimethyl-	$C_8H_{17}Cl$	29342-44-7		4-01-00-00434	ace 4; bz 4; eth 4; EtOH 4
	2-Chloro-2,5-dimethylhexane	148.68		86^{100}	0.8476^{18}	1.4232^{20}
6741	Hexane, 3-chloro-2,3-dimethyl-	$C_8H_{17}Cl$	101654-30-2		3-01-00-00481	EtOH 4
		148.68		$41\text{-}3^{12}$	0.8869^{20}	1.4333^{25}
6742	Hexane, 1-chloro-6-fluoro-	$C_6H_{12}ClF$	1550-09-0		4-01-00-00350	eth 4; EtOH 4; chl 4
	1-Chloro-6-fluorohexane	138.61		168; 62^{15}	1.015^{20}	1.4168^{25}
6743	Hexane, 1-chloro-3-methyl-	$C_7H_{15}Cl$	101257-63-0		2-01-00-00119	eth 4; EtOH 4; chl 4
		134.65		151	0.8766^{20}	1.4274^{207}
6744	Hexane, 2-chloro-2-methyl-	$C_7H_{15}Cl$	4398-65-6		4-01-00-00398	eth 4; EtOH 4; chl 4
		134.65		135 dec; 60^{52}	0.8635^{20}	1.4200^{20}
6745	Hexane, 2-chloro-5-methyl-	$C_7H_{15}Cl$	58766-17-9		3-01-00-00436	eth 4; EtOH 4; chl 4
	2-Chloro-5-methylhexane	134.65		138 dec	0.863^{20}	
6746	Hexane, 3-chloro-3-methyl-	$C_7H_{15}Cl$	43197-78-0		3-01-00-00438	eth 4; EtOH 4; chl 4
		134.65		135	0.8787^{20}	1.4250^{20}
6747	Hexane, 3-chloro-2,2,3-trimethyl-	$C_9H_{19}Cl$	102449-95-6		3-01-00-00515	eth 4; EtOH 4; chl 4
		162.70		$64\text{-}5^{13}$	0.9010^{20}	1.4465^{20}
6748	Hexanedial	$C_6H_{10}O_2$	1072-21-5		4-01-00-03686	bz 4; eth 4; EtOH 4
		114.14	-8	$92\text{-}4^{9}$	1.003^{19}	1.4350^{20}
6749	1,6-Hexanediamine	$C_6H_{16}N_2$	124-09-4	4614	4-04-00-01320	H_2O 4; EtOH 3; bz 3
	Hexamethylenediamine	116.21	41.5	205		
6750	1,6-Hexanediamine, *N,N'*-dibutyl-	$C_{14}H_{32}N_2$	4835-11-4		4-04-00-01327	
	N,N'-Dibutylhexamethylenediamine	228.42		$137\text{-}40^{3\text{-}4}$		1.4470^{25}
6751	2,5-Hexanediamine, 2,5-dimethyl-	$C_8H_{20}N_2$	23578-35-0		4-04-00-01363	
		144.26		$63\text{-}4^{8}$	0.832^{25}	1.4459^{20}
6752	1,6-Hexanediamine, *N,N,N',N'*-tetramethyl-	$C_{10}H_{24}N_2$	111-18-2		4-04-00-01322	
		172.31		209.5	0.806^{25}	1.4359^{20}
6753	1,6-Hexanediaminium, *N,N'*-di-9H-fluoren-9-yl-*N,N,N',N'*-tetramethyl-dibromide	$C_{36}H_{42}Br_2N_2$	317-52-2	4606	4-12-00-03397	
	Hexafluorenium bromide	662.55	153.5			
6754	Hexane, 1,2-dibromo-	$C_6H_{12}Br_2$	624-20-4		3-01-00-00392	bz 4; eth 4; chl 4
	1,2-Dibromohexane	243.97		104^{36}; 82^{12}	1.5774^{20}	1.5024^{20}
6755	Hexane, 1,4-dibromo-	$C_6H_{12}Br_2$	25118-28-9		3-01-00-00392	
	1,4-Dibromohexane	243.97		$106\text{-}8^{15}$	1.602^{15}	1.5084^{15}
6756	Hexane, 1,5-dibromo-	$C_6H_{12}Br_2$	627-96-3		4-01-00-00354	
	1,5-Dibromohexane	243.97		153^{100}; 103^{9}	1.5989^{20}	1.5072^{15}
6757	Hexane, 1,6-dibromo-	$C_6H_{12}Br_2$	629-03-8		4-01-00-00354	H_2O 1; eth 3; ace 3; ctc 2
		243.97	-2.3	245.5	1.5948^{15}	1.5037^{20}
6758	Hexane, 2,3-dibromo-	$C_6H_{12}Br_2$	6423-02-5		3-01-00-00393	
	2,3-Dibromohexane	243.97		90^{16}	1.5812^{20}	1.5025^{20}
6759	Hexane, 2,5-dibromo-, (*R**,*R**)-(±)-	$C_6H_{12}Br_2$	54462-68-9		3-01-00-00393	H_2O 1; eth 3; ace 3; ctc 3
	Hexane, 2,5-dibromo(*DL*)	243.97	-44.64	108^{30} 94^{13}	1.5788^{20}	1.5007^{20}
6760	Hexane, 3,4-dibromo-	$C_6H_{12}Br_2$	89583-12-0		3-01-00-00394	
		243.97		$80\text{-}1^{13}$	1.6027^{20}	1.5043^{20}
6761	Hexane, 1,2-dichloro-	$C_6H_{12}Cl_2$	2162-92-7		4-01-00-00350	eth 4; chl 4
	1,2-Dichlorohexane	155.07		173	1.085^{15}	
6762	Hexane, 1,6-dichloro-	$C_6H_{12}Cl_2$	2163-00-0		4-01-00-00350	H_2O 1; eth 3; ctc 3; chl 3
	1,6-Dichlorohexane	155.07		204	1.0677^{20}	1.4572^{20}
6763	Hexane, 2,2-dichloro-	$C_6H_{12}Cl_2$	42131-89-5		3-01-00-00390	eth 4; chl 4
	2,2-Dichlorohexane	155.07		68^{49}	1.0150^{25}	1.4353^{25}

No.	Name Synonym	Mol. Form. Mol. Wt.	CAS RN mp/°C	Merck No. bp/°C	Beil. Ref. den/g cm^{-3}	Solubility n_D
6764	Hexane, 2,3-dichloro- 2,3-Dichlorohexane	$C_6H_{12}Cl_2$ 155.07	54305-87-2	 163.5	4-01-00-00350 1.0527^{11}	eth 4; chl 4
6765	Hexane, 2,5-dichloro-2,5-dimethyl- 2,5-Dimethyl-2,5-dichlorohexane	$C_8H_{16}Cl_2$ 183.12	6223-78-5 67.5		4-01-00-00435 0.9543^{20}	bz 4; eth 4; EtOH 4; chl 4
6766	Hexane, 2,5-dichloro-, (R^*,R^*)-(±)- (±)-2,5-Dichlorohexane	$C_6H_{12}Cl_2$ 155.07	41761-12-0 -38	 177; 106^{91}	4-01-00-00350 1.0474^{20}	chl 4 1.4491^{20}
6767	Hexane, 3,3-diethyl-	$C_{10}H_{22}$ 142.28	17302-02-2	 166.3	4-01-00-00484 0.7575^{25}	 1.4258^{20}
6768	Hexane, 3,4-diethyl- 3,4-Diethylhexane	$C_{10}H_{22}$ 142.28	19398-77-7	 163.9	4-01-00-00484 0.7472^{25}	 1.4190^{20}
6769	Hexane, 1,6-diiodo- Hexamethylene diiodide	$C_6H_{12}I_2$ 337.97	629-09-4 9.5	 163^{17}; 141^{10}	4-01-00-00356 2.0416^{15}	H_2O 1; EtOH 4; eth 4 1.5899^{15}
6770	Hexane, 1,6-diisocyanato- Hexamethylene diisocyanate	$C_8H_{12}N_2O_2$ 168.20	822-06-0	 122^{10}; 94^{1}	4-04-00-01349 1.0528^{20}	 1.4585^{20}
6771	Hexane, 2,2-dimethyl- 2,2-Dimethylhexane	C_8H_{18} 114.23	590-73-8 -121.1	 106.8	4-01-00-00432 0.6953^{20}	ace 4; bz 4; eth 4; EtOH 4 1.3935^{20}
6772	Hexane, 2,3-dimethyl- 2,3-Dimethylhexane	C_8H_{18} 114.23	584-94-1	 115.6	4-01-00-00432 0.6912^{25}	ace 4; bz 4; EtOH 4; lig 4 1.4011^{20}
6773	Hexane, 2,4-dimethyl- 2,4-Dimethylhexane	C_8H_{18} 114.23	589-43-5	 109.5	 0.6962^{25}	 1.3929^{25}
6774	Hexane, 2,5-dimethyl- 2,5-Dimethylhexane	C_8H_{18} 114.23	592-13-2 -91	 109.1	4-01-00-00434 0.6901^{25}	H_2O 1; EtOH 5; eth 3; ace 5 1.3925^{20}
6775	Hexane, 3,3-dimethyl- 3,3-Dimethylhexane	C_8H_{18} 114.23	563-16-6 -126.1	 111.9	4-01-00-00435 0.7100^{20}	H_2O 1; EtOH 5; eth 4; ace 4 1.4001^{20}
6776	Hexane, 3,4-dimethyl- 3,4-Dimethylhexane	C_8H_{18} 114.23	583-48-2	 117.7	4-01-00-00436 0.7151^{25}	H_2O 1; EtOH 5; eth 3; ace 5 1.4041^{20}
6777	Hexanedinitrile Adiponitrile	$C_6H_8N_2$ 108.14	111-69-3 1	 295	4-02-00-01975 0.9676^{20}	H_2O 2; EtOH 3; eth 2; chl 3 1.4380^{20}
6778	Hexane, 1,6-dinitro- 1,6-Dinitrohexane	$C_6H_{12}N_2O_4$ 176.17	4286-47-9 37.5	 100-3$^{0.3}$	4-01-00-00357	H_2O 1; HOAc 3
6779	Hexanedioic acid Adipic acid	$C_6H_{10}O_4$ 146.14	124-04-9 153.2	152 337.5	4-02-00-01956 1.360^{25}	H_2O 2; EtOH 4; eth 3; HOAc 1
6780	Hexanedioic acid, bis(2-ethylbutyl) ester Adipic acid, di-2-ethylbutyl ester	$C_{18}H_{34}O_4$ 314.47	10022-60-3 -15	 200^{19}	4-02-00-01964 0.934^{25}	ace 4; EtOH 4 1.4434^{20}
6781	Hexanedioic acid, bis(2-ethylhexyl) ester Bis(2-ethylhexyl) adipate	$C_{22}H_{42}O_4$ 370.57	103-23-1 -67.8	 214^{5}	4-02-00-01964 0.922^{25}	ace 4; eth 4; EtOH 4 1.4474^{20}
6782	Hexanedioic acid, bis(1-methylethyl) ester	$C_{12}H_{22}O_4$ 230.30	6938-94-9 -1.1	 120$^{6.5}$	4-02-00-01961 0.9569^{20}	ace 4; eth 4; EtOH 4 1.4247^{20}
6783	Hexanedioic acid, bis(2-methylpropyl) ester	$C_{14}H_{26}O_4$ 258.36	141-04-8	 186-8^{15}	4-02-00-01962 0.9543^{19}	 1.4301^{20}
6784	Hexanedioic acid, dibutyl ester	$C_{14}H_{26}O_4$ 258.36	105-99-7 -32.4	 165^{10}	4-02-00-01961 0.9613^{20}	H_2O 1; EtOH 5; eth 5 1.4369^{20}
6785	Hexanedioic acid, diethyl ester	$C_{10}H_{18}O_4$ 202.25	141-28-6 -19.8	 245	4-02-00-01960 1.0076^{20}	H_2O 1; EtOH 3; eth 3 1.4272^{20}
6786	Hexanedioic acid, dimethyl ester	$C_8H_{14}O_4$ 174.20	627-93-0 10.3	 115^{13}	4-02-00-01959 1.0600^{20}	H_2O 1; EtOH 3; eth 3; ctc 3 1.4283^{20}
6787	Hexanedioic acid, dipropyl ester	$C_{12}H_{22}O_4$ 230.30	106-19-4 -15.7	 151^{11}	4-02-00-01961 0.9790^{20}	eth 4; EtOH 4; chl 4 1.4314^{20}
6788	Hexanedioic acid, 2-methyl-	$C_7H_{12}O_4$ 160.17	626-70-0 64	 209^{13}	4-02-00-02010	H_2O 4; eth 4; EtOH 4; chl 4
6789	Hexanedioic acid, 3-methyl-, (±)- Adipic acid, 3-methyl (*DL*)	$C_7H_{12}O_4$ 160.17	81177-02-8 97	 230^{30}	4-02-00-02014	H_2O 4; EtOH 4; eth 4; ace 4
6790	Hexanedioic acid, 3-methyl-, diethyl ester Diethyl 3-methyladipate	$C_{11}H_{20}O_4$ 216.28	55877-01-5	 258; 126.5^{10}	4-02-00-02015 0.9948^{20}	 1.4335
6791	Hexanedioic acid, monoethyl ester	$C_8H_{14}O_4$ 174.20	626-86-8 29	 285	4-02-00-01960 0.9796^{20}	EtOH 3; eth 3; peth 3 1.4311^{20}
6792	Hexanedioic acid, monomethyl ester	$C_7H_{12}O_4$ 160.17	627-91-8 9	 158^{10}	3-02-00-01711 1.0623^{20}	EtOH 3 1.4283^{20}
6793	1,3-Hexanediol	$C_6H_{14}O_2$ 118.18	21531-91-9	 123^{13}	4-01-00-02555 0.958^{22}	 1.4461^{22}
6794	1,4-Hexanediol	$C_6H_{14}O_2$ 118.18	16432-53-4	 134-5$^{18.5}$	4-01-00-02556 0.982^{16}	eth 4; EtOH 4 1.4530^{16}
6795	1,5-Hexanediol	$C_6H_{14}O_2$ [illegible]	928-40-5	 [illegible]	4-01-00-02556 [illegible]	 [illegible]
6796	1,6-Hexanediol Hexamethylene glycol	$C_6H_{14}O_2$ 118.18	629-11-8 42.8	4610 208	4-01-00-02556	H_2O 3; EtOH 3; eth 2; ace 3 1.4579^{25}
6797	2,3-Hexanediol	$C_6H_{14}O_2$ 118.18	617-30-1 60	 205	4-01-00-02561 0.9900^{15}	H_2O 4; eth 4; EtOH 4 1.4510^{15}
6798	2,4-Hexanediol	$C_6H_{14}O_2$ 118.18	19780-90-6	 211; 105^{9}	4-01-00-02562 0.9516^{21}	 1.4418^{21}
6799	2,5-Hexanediol Diisopropanol	$C_6H_{14}O_2$ 118.18	2935-44-6 43	 218; 86^{1}	4-01-00-02562 0.9610^{20}	H_2O 3; EtOH 3; eth 3; ctc 2 1.4475^{20}
6800	3,4-Hexanediol, 3,4-diethyl- 3,4-Diethyl-3,4-hexanediol	$C_{10}H_{22}O_2$ 174.28	6931-71-1 28	 230	4-01-00-02621 0.9602^{13}	eth 4; EtOH 4 1.467^{13}
6801	2,5-Hexanediol, 2,5-dimethyl- 1,1,4,4-Tetramethyl-1,4-butanediol	$C_8H_{18}O_2$ 146.23	110-03-2 92	 214	4-01-00-02600 0.898^{20}	H_2O 3; EtOH 4; bz 4; chl 4
6802	1,3-Hexanediol, 2-ethyl- Ethohexadiol	$C_8H_{18}O_2$ 146.23	94-96-2 -40	3699 244	4-01-00-02597 0.9325^{22}	H_2O 2; EtOH 3; eth 3 1.4497^{20}

No.	Name Synonym	Mol. Form. Mol. Wt.	CAS RN mp/°C	Merck No. bp/°C	Beil. Ref. den/g cm^{-3}	Solubility n_D
6803	2,4-Hexanediol, 2-methyl-	$C_7H_{16}O_2$ 132.20	66225-35-2	121^{30}	2-01-00-00554 0.9321^{18}	1.4407^{18}
6804	2,4-Hexanedione Propionylacetone	$C_6H_{10}O_2$ 114.14	3002-24-2	160	4-01-00-03687 0.959^{20}	1.4516^{20}
6805	2,5-Hexanedione Acetonylacetone	$C_6H_{10}O_2$ 114.14	110-13-4 -5.5	63 194	4-01-00-03688 0.7370^{20}	H_2O 4; bz 4; eth 4; EtOH 4 1.4232^{20}
6806	3,4-Hexanedione Bipropionyl	$C_6H_{10}O_2$ 114.14	4437-51-8 -10	130	4-01-00-03690 0.941^{21}	1.4130^{21}
6807	3,4-Hexanedione, 2,5-dimethyl- Biisobutyryl	$C_8H_{14}O_2$ 142.20	4388-87-8	144.5	4-01-00-03713 0.9232^{20}	1.4206^{20}
6808	2,5-Hexanedione, 3-hydroxy- 3-Hydroxy-2,5-hexanedione	$C_6H_{10}O_3$ 130.14	61892-85-1	$62\text{-}7^{0.5}$	3-01-00-03317	bz 4; eth 4 1.4497^{25}
6809	2,3-Hexanedione, 5-methyl- 2-Methylhexa-4,5-dione	$C_7H_{12}O_2$ 128.17	13706-86-0	138	4-01-00-03701 0.908^{22}	1.4119^{20}
6810	2,5-Hexanedione, 3-methyl- 3-Methyl-2,5-hexanedione	$C_7H_{12}O_2$ 128.17	4437-50-7	196; 71^{10}	3-01-00-03137 0.9527^{20}	eth 4; EtOH 4 1.4260^{20}
6811	3,4-Hexanedione, 2,2,5,5-tetramethyl- Pivalil	$C_{10}H_{18}O_2$ 170.25	4388-88-9 -2	169	4-01-00-03726 0.8776^{20}	H_2O 1; eth 5 1.4157^{20}
6812	1,6-Hexanedithiol	$C_6H_{14}S_2$ 150.31	1191-43-1	$118\text{-}9^{15}$	4-01-00-02559 0.983^{25}	1.5110^{20}
6813	Hexane, 1-(ethenyloxy)- Hexyl vinyl ether	$C_8H_{16}O$ 128.21	5363-64-4	143	4-01-00-02055 0.7966^{20}	1.4171^{20}
6814	Hexane, 1-ethoxy-	$C_8H_{18}O$ 130.23	5756-43-4	143	4-01-00-01697 0.7722^{20}	eth 4; EtOH 4 1.4008^{20}
6815	Hexane, 3-ethyl- 3-Ethylhexane	C_8H_{18} 114.23	619-99-8	118.6	4-01-00-00431 0.7136^{20}	H_2O 1; EtOH 5; eth 5; ace 5 1.4018^{20}
6816	Hexane, 3-ethyl-2,2-dimethyl- 3-Ethyl-2,2-dimethylhexane	$C_{10}H_{22}$ 142.28	20291-91-2	156.1	4-01-00-00484 0.7447^{25}	1.4197^{20}
6817	Hexane, 3-ethyl-2,3-dimethyl- 3-Ethyl-2,3-dimethylhexane	$C_{10}H_{22}$ 142.28	52897-00-4	166	4-01-00-00484 0.7599^{25}	1.4270^{207}
6818	Hexane, 3-ethyl-2,4-dimethyl- 3-Ethyl-2,4-dimethylhexane	$C_{10}H_{22}$ 142.28	7220-26-0	162	4-01-00-00484 0.7514^{25}	1.4225^{20}
6819	Hexane, 3-ethyl-2,5-dimethyl- 3-Ethyl-2,5-dimethylhexane	$C_{10}H_{22}$ 142.28	52897-04-8	154.1	4-01-00-00484 0.7368^{25}	1.4157^{20}
6820	Hexane, 3-ethyl-3,4-dimethyl- 3-Ethyl-3,4-dimethylhexane	$C_{10}H_{22}$ 142.28	52897-06-0	162.1	4-01-00-00485 0.7596^{25}	1.4267^{207}
6821	Hexane, 4-ethyl-2,2-dimethyl- Hexane, 2,2-dimethyl-4-ethyl	$C_{10}H_{22}$ 142.28	52896-99-8	147	4-01-00-00485 0.7330^{20}	1.4131^{20}
6822	Hexane, 4-ethyl-2,3-dimethyl- 4-Ethyl-2,3-dimethylhexane	$C_{10}H_{22}$ 142.28	52897-01-5	162	4-01-00-00485 0.7516^{25}	1.4226^{20}
6823	Hexane, 4-ethyl-2,4-dimethyl- 4-Ethyl-2,4-dimethylhexane	$C_{10}H_{22}$ 142.28	52897-03-7	161.1	4-01-00-00485 0.7525^{25}	1.4235^{20}
6824	Hexane, 4-ethyl-3,3-dimethyl- 4-Ethyl-3,3-dimethylhexane	$C_{10}H_{22}$ 142.28	52897-05-9	162.9	4-01-00-00485 0.7598^{25}	1.4269^{20}
6825	Hexane, 3-ethyl-2-methyl- 3-Ethyl-2-methylhexane	C_9H_{20} 128.26	16789-46-1	138	4-01-00-00459 0.7310^{20}	1.4106^{20}
6826	Hexane, 3-ethyl-3-methyl- 3-Ethyl-3-methylhexane	C_9H_{20} 128.26	3074-76-8	140.6	4-01-00-00459 0.7371^{25}	1.4140^{20}
6827	Hexane, 3-ethyl-4-methyl- 4-Ethyl-3-methylhexane	C_9H_{20} 128.26	3074-77-9	140	3-01-00-00514 0.7420^{20}	1.4134^{20}
6828	Hexane, 4-ethyl-2-methyl- 4-Ethyl-2-methylhexane	C_9H_{20} 128.26	3074-75-7	133.8	4-01-00-00459 0.7195^{25}	1.4063^{20}
6829	Hexane, 1-fluoro- Hexyl fluoride	$C_6H_{13}F$ 104.17	373-14-8 -103	91.5	4-01-00-00348 0.7995^{20}	eth 3; bz 3 1.3738^{20}
6830	Hexane, 2-fluoro- 2-Fluorohexane	$C_6H_{13}F$ 104.17	372-54-3	86	4-01-00-00348 0.7916^{20}	1.3693^{20}
6831	Hexane, 1-iodo- Hexyl iodide	$C_6H_{13}I$ 212.07	638-45-9 -74.2	181	4-01-00-00355 1.4397^{20}	H_2O 1 1.4929^{20}
6832	Hexane, 2-iodo-, (*R*)- Hexane, 2-iodo (*l*)	$C_6H_{13}I$ 212.07	59654-13-6	90^{70}; 45^{9}	4-01-00-00355 1.4354^{17}	H_2O 1; ace 3; chl 3 1.4878^{25}
6833	Hexane, 1-methoxy- Hexyl methyl ether	$C_7H_{16}O$ 116.20	4747-07-3	126.1		
6834	Hexane, 2-methyl- 2-Methylhexane	C_7H_{16} 100.20	591-76-4 -118.2	90.0	4-01-00-00397 0.6787^{20}	H_2O 1; EtOH 3; eth 5; ace 5 1.3848^{20}
6835	Hexane, 3-methyl-, (*S*)- (+)-3-Methylhexane	C_7H_{16} 100.20	6131-24-4 -119	92	4-01-00-00399 0.6860^{20}	H_2O 1; EtOH 3; eth 5; ace 5 1.3887^{20}
6836	Hexane, 2-methyl-3-methylene- 2-Isopropyl-1-pentene	C_8H_{16} 112.22	16746-02-4	113	4-01-00-00887 0.721^{25}	1.414^{20}
6837	Hexane, 2-methyl-4-methylene- 1-Pentene, 2-ethyl-4-methyl-	C_8H_{16} 112.22	3404-80-6	110.3	4-01-00-00888 0.7152^{25}	1.4105^{20}
6838	Hexane, 3-methyl-4-methylene-	C_8H_{16} 112.22	3404-67-9	112.5	4-01-00-00890 0.725^{25}	1.4142^{20}
6839	Hexane, 2-methyl-3-(1-methylethyl)- Hexane, 3-isopropyl-2-methyl-	$C_{10}H_{22}$ 142.28	62016-13-1	165	4-01-00-00484 0.7436^{25}	1.4195^{20}
6840	Hexanenitrile Capronitrile	$C_6H_{11}N$ 97.16	628-73-9 -80.3	163.6	4-02-00-00930 0.8051^{20}	H_2O 1; EtOH 3; eth 3; chl 2 1.4068^{20}
6841	Hexane, 1-nitro- 1-Nitrohexane	$C_6H_{13}NO_2$ 131.17	646-14-0	193; 84^{21}	4-01-00-00356 0.9396^{20}	H_2O 1; EtOH 3; eth 3; ace 3 1.4270^{20}
6842	Hexane, 1,1'-oxybis-	$C_{12}H_{26}O$ 186.34	112-58-3	223	4-01-00-01698 0.7936^{20}	H_2O 1; eth 3; ctc 2 1.4204^{20}

PHYSICAL CONSTANTS OF ORGANIC COMPOUNDS (continued)

No.	Name Synonym	Mol. Form. Mol. Wt.	CAS RN mp/°C	Merck No. bp/°C	Beil. Ref. den/g cm^{-3}	Solubility n_D
6843	Hexane, tetradecafluoro- Perfluorohexane	C_6F_{14} 338.04	355-42-0 -87.1	 56.6	4-01-00-00348 1.6995^{20}	H_2O 1; eth 3; bz 3; chl 3 1.2515^{20}
6844	Hexane, 2,2,3,3-tetramethyl- 2,2,3,3-Tetramethylhexane	$C_{10}H_{22}$ 142.28	13475-81-5 -54	 160.3	4-01-00-00485 0.7609^{25}	 1.4282^{20}
6845	Hexane, 2,2,3,4-tetramethyl- 2,2,3,4-Tetramethylhexane	$C_{10}H_{22}$ 142.28	52897-08-2	 157	4-01-00-00485 0.7513^{25}	 1.4216^{20}
6846	Hexane, 2,2,3,5-tetramethyl- 2,2,3,5-Tetramethylhexane	$C_{10}H_{22}$ 142.28	52897-09-3	 148.4	4-01-00-00485 0.7336^{25}	 1.4142^{20}
6847	Hexane, 2,2,4,4-tetramethyl- 2,2,4,4-Tetramethylhexane	$C_{10}H_{22}$ 142.28	51750-65-3	 153.8	4-01-00-00486 0.7424^{25}	 1.4208^{20}
6848	Hexane, 2,2,4,5-tetramethyl- 2,2,4,5-Tetramethylhexane	$C_{10}H_{22}$ 142.28	16747-42-5	 147.9	4-01-00-00486 0.7316^{25}	 1.4132^{20}
6849	Hexane, 2,2,5,5-tetramethyl- 2,2,5,5-Tetramethylhexane	$C_{10}H_{22}$ 142.28	1071-81-4 -12.6	 137.4	4-01-00-00486 0.7148^{25}	 1.4055^{20}
6850	Hexane, 2,3,3,4-tetramethyl- 2,3,3,4-Tetramethylhexane	$C_{10}H_{22}$ 142.28	52897-10-6	 164.6	4-01-00-00486 0.7656^{25}	 1.4298^{20}
6851	Hexane, 2,3,3,5-tetramethyl- 2,3,3,5-Tetramethylhexane	$C_{10}H_{22}$ 142.28	52897-11-7	 153.1	4-01-00-00486 0.7449^{25}	 1.4196^{20}
6852	Hexane, 2,3,4,4-tetramethyl- 2,3,4,4-Tetramethylhexane	$C_{10}H_{22}$ 142.28	52897-12-8	 161.6	4-01-00-00487 0.7586^{25}	 1.4267^{20}
6853	Hexane, 2,3,4,5-tetramethyl- 2,3,4,5-Tetramethylhexane	$C_{10}H_{22}$ 142.28	52897-15-1	 156.2	4-01-00-00487 0.7456^{25}	 1.4204^{20}
6854	Hexane, 3,3,4,4-tetramethyl- 3,3,4,4-Tetramethylhexane	$C_{10}H_{22}$ 142.28	5171-84-6	 170.0	3-01-00-00534 0.7789^{25}	 1.4368^{20}
6855	Hexane, 1,1'-thiobis-	$C_{12}H_{26}S$ 202.40	6294-31-1	 230; 136^{20}	4-01-00-01706 0.8411^{20}	 1.4586^{20}
6856	1-Hexanethiol Hexyl mercaptan	$C_6H_{14}S$ 118.24	111-31-9 -81	 151	4-01-00-01705 0.8424^{20}	H_2O 1; EtOH 4; eth 4 1.4496^{20}
6857	2-Hexanethiol	$C_6H_{14}S$ 118.24	1679-06-7 -147	 142	4-01-00-01711 0.8345^{20}	H_2O 1; EtOH 3; eth 3; bz 3 1.4451^{20}
6858	3-Hexanethiol 3-Mercaptohexane	$C_6H_{14}S$ 118.24	1633-90-5	 57^{25}	4-01-00-01713 0.9206^{20}	EtOH 3 1.4496^{20}
6859	Hexane, 2,2,3-trimethyl- 2,2,3-Trimethylhexane	C_9H_{20} 128.26	16747-25-4	 133.6	4-01-00-00459 0.7257^{25}	 1.4106^{20}
6860	Hexane, 2,2,4-trimethyl- 2,2,4-Trimethylhexane	C_9H_{20} 128.26	16747-26-5 -120	 126.5	4-01-00-00459 0.711^{20}	 1.4033^{20}
6861	Hexane, 2,2,5-trimethyl- 2,2,5-Trimethylhexane	C_9H_{20} 128.26	3522-94-9 -105.7	 124.0	4-01-00-00460 0.7072^{20}	H_2O 1; EtOH 4; eth 4; ace 4 1.3997^{20}
6862	Hexane, 2,3,3-trimethyl- 2,3,3-Trimethylhexane	C_9H_{20} 128.26	16747-28-7 -116.8	 137.7	4-01-00-00461 0.7345^{25}	 1.4141^{20}
6863	Hexane, 2,3,4-trimethyl- 2,3,4-Trimethylhexane	C_9H_{20} 128.26	921-47-1	 139.1	4-01-00-00461 0.7354^{25}	 1.4144^{20}
6864	Hexane, 2,3,5-trimethyl- 2,3,5-Trimethylhexane	C_9H_{20} 128.26	1069-53-0 -127.9	 131.4	4-01-00-00461 0.7218^{20}	 1.4051^{20}
6865	Hexane, 2,4,4-trimethyl- 2,4,4-Trimethylhexane	C_9H_{20} 128.26	16747-30-1 -113.4	 130.7	4-01-00-00461 0.7201^{25}	 1.4074^{20}
6866	Hexane, 3,3,4-trimethyl- 3,3,4-Trimethylhexane	C_9H_{20} 128.26	16747-31-2 -101.2	 140.5	4-01-00-00462 0.7414^{25}	 1.4178^{20}
6867	1,2,5-Hexanetriol	$C_6H_{14}O_3$ 134.18	10299-30-6	 181^{10}; 149^3	4-01-00-02784 1.1012^{20}	H_2O 4; EtOH 4
6868	1,2,6-Hexanetriol 1,2,6-Trihydroxyhexane	$C_6H_{14}O_3$ 134.18	106-69-4	 170^3	4-01-00-02784 1.03^{20}	 1.58^{20}
6869	2,3,4-Hexanetriol	$C_6H_{14}O_3$ 134.18	93972-93-1 47	 256.5	3-01-00-02349	H_2O 4; EtOH 4
6870	Hexanoic acid Caproic acid	$C_6H_{12}O_2$ 116.16	142-62-1 -3	1760 205.2	4-02-00-00917 0.9274^{20}	H_2O 1; EtOH 3; eth 3; chl 3 1.4163^{20}
6871	Hexanoic acid, 6-(acetylamino)- ε-Acetamidocaproic acid	$C_8H_{15}NO_3$ 173.21	57-08-9 104.5	38	4-04-00-02700	
6872	Hexanoic acid, 2-acetyl-, ethyl ester	$C_{10}H_{18}O_3$ 186.25	1540-29-0	 221.5	4-03-00-01616 0.9523^{20}	ace 4; eth 4 1.4301^{20}
6873	Hexanoic acid, 6-amino- ε-Aminocaproic acid	$C_6H_{13}NO_2$ 131.17	60-32-2 205	442	4-04-00-02695	H_2O 4; EtOH 1; MeOH 2
6874	Hexanoic acid, anhydride	$C_{12}H_{22}O_3$ 214.30	2051-49-2 -41	 255 dec	4-02-00-00928 0.9240^{15}	eth 4; EtOH 4 1.4297^{20}
6875	Hexanoic acid, 2-bromo-, (±)	$C_6H_{11}BrO_2$ 195.06	2681-83-6 4	 240; 140^{25}	3-02-00-00736 1.281[illegible]	EtOH 3; eth 3
6876	Hexanoic acid, 6-bromo-	$C_6H_{11}BrO_2$ 195.06	4224-70-8 35	 $165\text{-}70^{20}$	4-02-00-00940	peth 4
6877	Hexanoic acid, 2-bromo-, ethyl ester, (±)-	$C_8H_{15}BrO_2$ 223.11	63927-44-6	 208.7	4-02-00-00938 1.2210^{25}	EtOH 3
6878	Hexanoic acid, 5-bromo-, ethyl ester Caproic acid, 5-bromo,ethyl ester	$C_8H_{15}BrO_2$ 223.11	90202-08-7	 121^{21}; 70^3	3-02-00-00737 1.1908^{25}	 1.4525^{25}
6879	Hexanoic acid, 6-bromo-, ethyl ester	$C_8H_{15}BrO_2$ 223.11	25542-62-5 33	 $126\text{-}7^{21}$	4-02-00-00940 1.238^{23}	 1.4566^{21}
6880	Hexanoic acid, butyl ester	$C_{10}H_{20}O_2$ 172.27	626-82-4 -64.3	 208	4-02-00-00922 0.8653^{20}	H_2O 1; EtOH 3; eth 5 1.4152^{20}
6881	Hexanoic acid, 2-cyano-, ethyl ester 2-Cyanocaproic acid, ethyl ester	$C_9H_{15}NO_2$ 169.22	7391-39-1	 105^9	3-02-00-01746 0.988^{15}	 1.4248^{20}
6882	Hexanoic acid, 2-ethyl- 2-Ethylhexanoic acid	$C_8H_{16}O_2$ 144.21	149-57-5	 228; 120^{13}	3-02-00-00804 0.9031^{25}	H_2O 3; EtOH 2; eth 3; ctc 3 1.4241^{20}

No.	Name	Synonym	Mol. Form.	Mol. Wt.	CAS RN	mp/°C	Merck No.	bp/°C	Beil. Ref.	den/g cm^{-3}	Solubility	n_D
6883	Hexanoic acid, 3-ethyl-	3-Ethylhexanoic acid	$C_8H_{16}O_2$	144.21	41065-91-2			106^{5}	4-02-00-01008	0.911^{30}		1.4287^{25}
6884	Hexanoic acid, ethyl ester	Ethyl hexanoate	$C_8H_{16}O_2$	144.21	123-66-0	-67	3735	167	4-02-00-00921	0.873^{20}	eth 4; EtOH 4	1.4073^{20}
6885	Hexanoic acid, 2-ethyl-, ethyl ester		$C_{10}H_{20}O_2$	172.27	2983-37-1			90^{28}	4-02-00-01004	0.8586^{25}		1.4123^{25}
6886	Hexanoic acid, 2-ethyl-3-methyl-	2-Ethyl-3-methylhexanoic acid	$C_9H_{18}O_2$	158.24	74581-94-5			231	3-02-00-00832	0.9060^{25}		1.4302^{25}
6887	Hexanoic acid, 3-ethyl-3-methyl-	3-Ethyl-3-methylhexanoic acid	$C_9H_{18}O_2$	158.24	50902-82-4			$139\text{-}40^{22}$	4-02-00-01036	0.9205^{25}		1.4375^{25}
6888	Hexanoic acid, 6-fluoro-	Caproic acid, 6-fluoro	$C_6H_{11}FO_2$	134.15	373-05-7			138^{28}, $67^{0.6}$	4-02-00-00932		eth 4; EtOH 4	1.4166^{25}
6889	Hexanoic acid, heptyl ester		$C_{13}H_{26}O_2$	214.35	6976-72-3	-34.4		261	4-02-00-00923	0.8611^{20}	ace 4; bz 4; eth 4; EtOH 4	1.4293^{15}
6890	Hexanoic acid, hexyl ester		$C_{12}H_{24}O_2$	200.32	6378-65-0	-55		246	4-02-00-00922	0.865^{18}	ace 4; bz 4; eth 4; EtOH 4	1.4264^{15}
6891	Hexanoic acid, 2-hydroxy-, (±)-		$C_6H_{12}O_3$	132.16	636-36-2	61.3		270	4-03-00-00838		H_2O 4; EtOH 4; eth 3; chl 4	
6892	Hexanoic acid, 3-hydroxy-2,2-dimethyl-, β-lactone	Caproic acid, 2,2-dimethyl-3-hydroxy, β-lactone	$C_8H_{14}O_2$	142.20	90112-82-6	143.5		$93\text{-}100^{14}$			DMSO 2	1.4260^{20}
6893	Hexanoic acid, 2-methyl-, (±)-	Caproic acid, 2-methyl (*DL*)	$C_7H_{14}O_2$	130.19	22160-12-9			215.5	4-02-00-00969	0.9612^{20}	H_2O 5; EtOH 5; eth 5; ace 5	1.4195^{20}
6894	Hexanoic acid, 3-methyl-		$C_7H_{14}O_2$	130.19	3780-58-3			213; 116^{15}	2-02-00-00298	0.9187^{20}		1.4222^{20}
6895	Hexanoic acid, 4-methyl-, (±)-		$C_7H_{14}O_2$	130.19	22160-41-4	-80		217.5	4-02-00-00974	0.9215^{20}	H_2O 2; EtOH 3; eth 3; ace 3	1.4211^{20}
6896	Hexanoic acid, 5-methyl-		$C_7H_{14}O_2$	130.19	628-46-6	<-25		216	4-02-00-00970	0.9138^{21}	ace 4; bz 4; eth 4; EtOH 4	1.4220^{20}
6897	Hexanoic acid, 3-methylbutyl ester		$C_{11}H_{22}O_2$	186.29	2198-61-0			225.5	3-02-00-00727	0.861^{20}	H_2O 1; EtOH 3; eth 3	
6898	Hexanoic acid, methyl ester	Methyl caproate	$C_7H_{14}O_2$	130.19	106-70-7	-71		149.5	4-02-00-00921	0.8846^{20}	H_2O 1; EtOH 4; eth 4; ace 3	1.4049^{20}
6899	Hexanoic acid, 3-methyl-, ethyl ester	Ethyl 3-methylhexanoate	$C_9H_{18}O_2$	158.24	41692-47-1			177	2-02-00-00298	0.8679^{20}		1.4119^{20}
6900	Hexanoic acid, 4-methyl-, ethyl ester	Ethyl 4-methylhexanoate	$C_9H_{18}O_2$	158.24	1561-10-0			180	4-02-00-00974	0.8708^{20}		1.4051^{20}
6901	Hexanoic acid, 4-methyl-5-oxo-		$C_7H_{12}O_3$	144.17	6818-07-1			160^{15}; 140^{5}	4-03-00-01601	0.9803^{20}		1.4320^{20}
6902	Hexanoic acid, 4-methyl-5-oxo-, ethyl ester	Valeric acid, 4-acetyl, ethyl ester	$C_9H_{16}O_3$	172.22	53068-88-5			114^{15}	4-03-00-01601	0.9803^{20}		1.4282^{25}
6903	Hexanoic acid, 4-methyl-5-oxo-, methyl ester	Valeric acid, 4-acetyl, methyl ester	$C_8H_{14}O_3$	158.20	36045-56-4			$113\text{-}7^{16}$	4-03-00-01601	1.0988^{20}		1.4288^{20}
6904	Hexanoic acid, octyl ester		$C_{14}H_{28}O_2$	228.38	4887-30-3	-28		275	3-02-00-00727	0.8603^{20}	ace 4; bz 4; eth 4; EtOH 4	1.4326^{15}
6905	Hexanoic acid, 5-oxo-		$C_6H_{10}O_3$	130.14	3128-06-1	13.5		274.5	4-03-00-01583	1.09^{25}	H_2O 3; EtOH 3; eth 3; ctc 2	1.4451^{20}
6906	Hexanoic acid, 5-oxo-, ethyl ester	Butyric acid, 4-acetyl, ethyl ester	$C_8H_{14}O_3$	158.20	13984-57-1			221.5	4-03-00-01583	0.989^{25}		1.4277^{20}
6907	Hexanoic acid, pentyl ester	Amyl caproate	$C_{11}H_{22}O_2$	186.29	540-07-8	-47	640	226	4-02-00-00922	0.8612^{25}	EtOH 3; eth 3; ace 3; ctc 2	1.4202^{25}
6908	Hexanoic acid, 1,2,3-propanetriyl ester		$C_{21}H_{38}O_6$	386.53	621-70-5	-60		>200	4-02-00-00926	0.9867^{20}	H_2O 1; EtOH 5; eth 5; ace 4	1.4427^{20}
6909	Hexanoic acid, propyl ester		$C_9H_{18}O_2$	158.24	626-77-7	-68.7		187	4-02-00-00922	0.8672^{20}	eth 4; EtOH 4	1.4170^{20}
6910	Hexanoic acid, 3,5,6-trichloro-2,2,3,4,4,5,6,6-octafluoro-	Hexanoic acid, 2,2,3,4,4,5,6,6-octafluoro-3,5,6-trichloro	$C_6HCl_3F_8O_2$	363.42	2106-54-9			$105\text{-}6^{5}$	4-02-00-00937	1.860^{20}		1.3903^{20}
6911	1-Hexanol	Caproyl alcohol	$C_6H_{14}O$	102.18	111-27-3	-44.6	4615	157.6	4-01-00-01694	0.8136^{20}	H_2O 2; EtOH 3; eth 5; ace 3	1.4178^{20}
6912	2-Hexanol, (±)-		$C_6H_{14}O$	102.18	20281-86-1			140	4-01-00-01708	0.8159^{20}	H_2O 2; EtOH 3; eth 3; ctc 2	1.4144^{20}
6913	2-Hexanol, (*R*)-		$C_6H_{14}O$	102.18	26549-24-6			137	3-01-00-01663	0.8178^{18}	H_2O 2; EtOH 3; eth 3	
6914	3-Hexanol, (±)-		$C_6H_{14}O$	102.18	17015-11-1			135	4-01-00-01711	0.8182^{20}	H_2O 2; EtOH 3; eth 5; ace 3	1.4167^{20}
6915	3-Hexanol, (*R*)-	3-Hexanol (*l*)	$C_6H_{14}O$	102.18	13471-42-6			135		0.8213^{20}	H_2O 2; EtOH 3; eth 5; ace 3	1.4140^{20}
6916	1-Hexanol, 6-chloro-		$C_6H_{13}ClO$	136.62	2009-83-8			107^{12}	4-01-00-01704	1.0241^{20}	H_2O 2; EtOH 4; eth 4	1.4550^{20}
6917	2-Hexanol, 1-chloro-	1-Chloro-2-hexanol	$C_6H_{13}ClO$	136.62	52802-07-0			$73\text{-}5^{12}$	4-01-00-01710	1.0139^{20}	ace 4; eth 4; EtOH 4	1.4478^{20}
6918	3-Hexanol, 1-chloro-	1-Chloro-3-hexanol	$C_6H_{13}ClO$	136.62	52418-81-2			120^{35}; 78^{6}	4-01-00-01712	1.003^{25}	ace 4; bz 4; eth 4; EtOH 4	1.446^{25}
6919	3-Hexanol, 5-chloro-		$C_6H_{13}ClO$	136.62	58588-28-6			$78\text{-}9^{13}$	4-01-00-01712	1.0012^{15}	ace 4; bz 4; eth 4; EtOH 4	1.4433^{19}
6920	1-Hexanol, 3,5-dimethoxy-		$C_8H_{18}O_3$	162.23	90952-10-6			114^{13}	4-01-00-02785	0.9631^{25}		1.4329^{25}

No.	Name Synonym	Mol. Form. Mol. Wt.	CAS RN mp/°C	Merck No. bp/°C	Beil. Ref. den/g cm^{-3}	Solubility n_D
6921	1-Hexanol, 2,5-dimethyl-	$C_8H_{18}O$ 130.23	6886-16-4	 179.5	0-01-00-00422 0.8280^{20}	 1.5095^{17}
6922	2-Hexanol, 2,3-dimethyl- 2,3-Dimethyl-2-hexanol	$C_8H_{18}O$ 130.23	19550-03-9	 160; 62^{10}	3-01-00-01737 0.8365^{20}	 1.4335^{20}
6923	2-Hexanol, 2,5-dimethyl- 2,5-Dimethyl-2-hexanol	$C_8H_{18}O$ 130.23	3730-60-7	 154.5	4-01-00-01792 0.8227^{20}	 1.4208^{20}
6924	2-Hexanol, 3,4-dimethyl-	$C_8H_{18}O$ 130.23	19550-05-1	 165.5	3-01-00-01740 0.840^{15}	 1.4325^{15}
6925	3-Hexanol, 2,2-dimethyl- Propyl-*tert*-butylcarbinol	$C_8H_{18}O$ 130.23	4209-90-9	 156	3-01-00-01737 0.8342^{20}	 1.4275^{20}
6926	3-Hexanol, 2,3-dimethyl-	$C_8H_{18}O$ 130.23	4166-46-5	 $61\text{-}2^{18}$	4-01-00-01791 0.8371^{20}	 1.4309^{20}
6927	3-Hexanol, 2,5-dimethyl- 2,5-Dimethyl-3-hexanol	$C_8H_{18}O$ 130.23	19550-07-3	 157.5	4-01-00-01793 0.8212^{20}	 1.4246^{20}
6928	3-Hexanol, 3,5-dimethyl-, (±)- 3-Hexanol, 3,5-dimethyl (*DL*)	$C_8H_{18}O$ 130.23	19113-78-1	 152	4-01-00-01792 0.8373^{20}	bz 4; eth 4; EtOH 4 1.4300^{20}
6929	3-Hexanol, 4,4-dimethyl-	$C_8H_{18}O$ 130.23	19550-09-5	 160.4	4-01-00-01793 0.8341^{20}	 1.4345^{20}
6930	1-Hexanol, 2-ethyl- 2-Ethyl-1-hexanol	$C_8H_{18}O$ 130.23	104-76-7 -70	3764 184.6	4-01-00-01783 0.8319^{25}	H_2O 1; EtOH 3; eth 3; ace 3 1.4300^{20}
6931	3-Hexanol, 3-ethyl-	$C_8H_{18}O$ 130.23	597-76-2	 160	4-01-00-01790 0.8373^{20}	H_2O 1; EtOH 3; eth 3; ace 3 1.4300^{20}
6932	1-Hexanol, 2-ethyl-3-methyl-	$C_9H_{20}O$ 144.26	66794-04-5	 193; 83^{10}	4-02-00-02107 0.8486^{20}	 1.4455^{20}
6933	2-Hexanol, 3-ethyl-2-methyl-	$C_9H_{20}O$ 144.26	66794-02-3	 71.5^{14}	4-01-00-01812 0.8334^{25}	 1.4331^{25}
6934	3-Hexanol, 3-ethyl-2-methyl-	$C_9H_{20}O$ 144.26	66794-03-4	 177.5	4-01-00-01811 0.8445^{25}	 1.4369^{25}
6935	3-Hexanol, 3-ethyl-5-methyl-	$C_9H_{20}O$ 144.26	597-77-3	 172; 68^{18}	3-01-00-01755 0.8396^{22}	eth 4; EtOH 4 1.4346^{13}
6936	1-Hexanol, 6-fluoro-	$C_6H_{13}FO$ 120.17	373-32-0	 $85\text{-}6^{14}$	4-01-00-01703 0.975^{20}	eth 4; EtOH 4 1.4141^{25}
6937	1-Hexanol, 2-methyl-, (±)- 1-Hexanol, 2-methyl (*DL*)	$C_7H_{16}O$ 116.20	111768-04-8	 164; 71^{15}	4-01-00-01745 0.8270^{20}	eth 4; EtOH 4 1.4226^{20}
6938	1-Hexanol, 2-methyl-, (*R*)- (*R*)-2-Methyl-1-hexanol	$C_7H_{16}O$ 116.20	66050-98-4	 164; 71^{15}	 0.8313^{13}	eth 4; EtOH 4 1.4245^{17}
6939	1-Hexanol, 3-methyl-, (±)-	$C_7H_{16}O$ 116.20	111768-08-2	 169; 91^{13}	4-01-00-01748 0.8258^{20}	H_2O 1; EtOH 4; eth 4; ace [illegible] 1.4245^{20}
6940	1-Hexanol, 4-methyl-, (±)- 1-Hexanol, 4-methyl (*DL*)	$C_7H_{16}O$ 116.20	111768-05-9	 173	4-01-00-01749 0.8239^{20}	ace 4; bz 4; eth 4; EtOH 4 1.4219^{20}
6941	1-Hexanol, 5-methyl- 5-Methyl-1-hexanol	C_7H_{16} 100.20	627-98-5	 170; $53\text{-}5^{15}$	4-01-00-01748 0.8119^{20}	eth 4; EtOH 4 1.4175^{20}
6942	2-Hexanol, 2-methyl-	$C_7H_{16}O$ 116.20	625-23-0	 143	4-01-00-01745 0.8119^{20}	H_2O 2; EtOH 5; eth 5 1.4175^{20}
6943	2-Hexanol, 3-methyl- 3-Methyl-2-hexanol	$C_7H_{16}O$ 116.20	2313-65-7	 $79\text{-}81^{52}$	 0.8820^{25}	H_2O 1; EtOH 4; eth 4; ace 3 1.4198^{18}
6944	2-Hexanol, 5-methyl- 5-Methyl-2-hexanol	$C_7H_{16}O$ 116.20	627-59-8	 151; 78^{28}	4-01-00-01747 0.814^{20}	H_2O 1; EtOH 3; eth 3 1.4180^{20}
6945	3-Hexanol, 2-methyl-, (±)- (±)-2-Methyl-3-hexanol	$C_7H_{16}O$ 116.20	100295-82-7	 143	4-01-00-01746 0.8407^{20}	H_2O 1; EtOH 3; eth 3; ctc 3 1.4149^{20}
6946	3-Hexanol, 3-methyl- 2-Ethyl-2-pentanol	$C_7H_{16}O$ 116.20	597-96-8	 143	4-01-00-01749 0.8233^{20}	H_2O 2; EtOH 3; eth 3; ctc 3 1.4231^{20}
6947	3-Hexanol, 5-methyl-, (±)- (±)-5-Methyl-3-hexanol	$C_7H_{16}O$ 116.20	100295-83-8	 148	4-01-00-01747 0.827^{25}	H_2O 1; EtOH 3; eth 3; ctc 2 1.4128^{20}
6948	1-Hexanol, 5-methyl-2-(1-methylethyl)- 2-Isopropyl-5-methyl-1-hexanol	$C_{10}H_{22}O$ 158.28	2051-33-4	 211	4-01-00-01833 0.8345^{20}	H_2O 1; EtOH 3; eth 4 1.4369^{20}
6949	3-Hexanol, 1-phenyl-	$C_{12}H_{18}O$ 178.27	2180-43-0 34	 170^{50}, 146^{16}	4-06-00-03418 0.9525^{20}	
6950	3-Hexanol, 2,2,5,5-tetramethyl-	$C_{10}H_{22}O$ 158.28	55073-86-4 52.5	 170	3-01-00-01772	ace 4; eth 4; EtOH 4; peth 4
6951	1-Hexanol, 3,5,5-trimethyl-	$C_9H_{20}O$ 144.26	3452-97-9	 194	4-01-00-01813 0.8236^{25}	 1.4300^{25}
6952	2-Hexanol, 2,3,4-trimethyl-	$C_9H_{20}O$ 144.26	21102-13-6	 57^{5}	3-01-00-01756 0.853^{15}	ace 4; eth 4; EtOH 4; peth 4 1.4415^{15}
6953	3-Hexanol, 2,2,3-trimethyl-	$C_9H_{20}O$ 144.26	5340-41-0	 170	3-01-00-01755 0.8474^{20}	EtOH 3; eth 3; ace 3 1.4402^{20}
6954	3-Hexanol, 2,3,5-trimethyl-	$C_9H_{20}O$ 144.26	65927-60-8	 72^{21}	4-01-00-01813 0.8256^{20}	ace 4; eth 4; EtOH 4 1.4321^{20}
6955	3-Hexanol, 2,4,4-trimethyl-	$C_9H_{20}O$ 144.26	66793-92-8	 170	3-01-00-01756 0.8489^{20}	EtOH 3; eth 3; ace 3 1.4395^{20}
6956	3-Hexanol, 2,5,5-trimethyl-	$C_9H_{20}O$ 144.26	66793-72-4	 77^{32}	3-01-00-01756 0.8250^{20}	ace 4; eth 4; EtOH 4 1.4286^{20}
6957	3-Hexanol, 3,4,4-trimethyl-	$C_9H_{20}O$ 144.26	66793-74-6	 165.5	0-01-00-00425 0.8322^{21}	EtOH 3; eth 3; ace 3 1.4341^{21}
6958	3-Hexanol, 3,5,5-trimethyl- 3,5,5-Trimethylhexan-3-ol	$C_9H_{20}O$ 144.26	66810-87-5	 62^{14}	3-01-00-01755 0.8350^{20}	ace 4; eth 4; EtOH 4 1.4352^{20}
6959	2-Hexanone Butyl methyl ketone	$C_6H_{12}O$ 100.16	591-78-6 -55.5	5955 127.6	4-01-00-03298 0.8113^{20}	H_2O 2; EtOH 5; eth 5; ace 3 1.4007^{20}
6960	3-Hexanone Ethyl propyl ketone	$C_6H_{12}O$ 100.16	589-38-8 -55.5	 123.5	4-01-00-03301 0.8118^{20}	H_2O 2; EtOH 5; eth 5; ace 3 1.4004^{20}

No.	Name Synonym	Mol. Form. Mol. Wt.	CAS RN mp/°C	Merck No. bp/°C	Beil. Ref. den/g cm^{-3}	Solubility n_D
6961	2-Hexanone, 6-bromo-	$C_6H_{11}BrO$ 179.06	10226-29-6	216; 136^{90}	3-01-00-02829 1.3494^{0}	eth 4; EtOH 4
6962	1-Hexanone, 1-(2,4-dihydroxyphenyl)-	$C_{12}H_{16}O_3$ 208.26	3144-54-5 57	343	4-08-00-01888	H_2O 1; EtOH 3; eth 3; ace 3
6963	2-Hexanone, 3,3-dimethyl- 3,3-Dimethyl-2-hexanone	$C_8H_{16}O$ 128.21	26118-38-7	149	4-01-00-03350 0.838^{0}	ace 4; eth 4; EtOH 4 1.4098^{20}
6964	2-Hexanone, 3,4-dimethyl- 3,4-Dimethyl-2-hexanone	$C_8H_{16}O$ 128.21	19550-10-8	158	3-01-00-02882 0.8295^{22}	EtOH 3; eth 3; ace 3 1.4193^{20}
6965	3-Hexanone, 2,2-dimethyl- *tert*-Butyl propyl ketone	$C_8H_{16}O$ 128.21	5405-79-8	147	4-01-00-03347 0.8105^{25}	EtOH 3; eth 3; ace 3 1.4119^{20}
6966	3-Hexanone, 2,5-dimethyl-	$C_8H_{16}O$ 128.21	1888-57-9	147.5	4-01-00-03349 0.8269^{0}	ace 4; eth 4; EtOH 4 1.4049^{20}
6967	3-Hexanone, 4,4-dimethyl-	$C_8H_{16}O$ 128.21	19550-14-2	148	4-01-00-03350 0.8285^{20}	bz 4; EtOH 4; chl 4 1.4208^{25}
6968	3-Hexanone, 6-(dimethylamino)-4,4-diphenyl- Normethadone	$C_{20}H_{25}NO$ 295.42	467-85-6	6628 164-7^{3}	4-14-00-00297	
6969	2-Hexanone, 5-hydroxy- 5-Hydroxy-2-hexanone	$C_6H_{12}O_2$ 116.16	56745-61-0	203^{270}; 80^{10}	4-01-00-04019 0.9626^{25}	H_2O 4; eth 4; EtOH 4 1.4312^{25}
6970	2-Hexanone, 6-hydroxy-	$C_6H_{12}O_2$ 116.16	21856-89-3	227	4-01-00-04020 0.9886^{15}	eth 4; EtOH 4 1.4494
6971	3-Hexanone, 4-hydroxy-	$C_6H_{12}O_2$ 116.16	4984-85-4	133^{227}; 73^{20}	4-01-00-04021 0.956^{21}	ace 4; EtOH 4 1.4340^{21}
6972	3-Hexanone, 5-hydroxy- 5-Hydroxy-3-hexanone	$C_6H_{12}O_2$ 116.16	33683-44-2	75-8^{12}	4-01-00-04021 0.950^{20}	1.4280^{25}
6973	3-Hexanone, 2-hydroxy-2-methyl- 2-Hexanol-3-one, 2-methyl	$C_7H_{14}O_2$ 130.19	18905-91-4	60-2^{15}	4-01-00-04035 0.899^{18}	1.419^{18}
6974	3-Hexanone, 5-hydroxy-2-methyl- 2-Hexanol-4-one, 5-methyl	$C_7H_{14}O_2$ 130.19	59357-07-2	72-3^{9}	4-01-00-04035 0.929^{20}	1.4278^{20}
6975	3-Hexanone, 4-hydroxy-2,2,5,5-tetramethyl- Pivaloin	$C_{10}H_{20}O_2$ 172.27	815-66-7 81	80^{10}	4-01-00-04060	eth 4
6976	2-Hexanone, 3-methyl- 3-Methyl-2-hexanone	$C_7H_{14}O$ 114.19	2550-21-2	143.5	3-01-00-02863 0.828^{25}	ace 4; bz 4; eth 4; EtOH 4 1.4035^{20}
6977	2-Hexanone, 4-methyl- Methyl 2-methylbutyl ketone	$C_7H_{14}O$ 114.19	105-42-0	144.5	4-01-00-03329 0.8130^{20}	H_2O 2; EtOH 4; eth 4; ace 4 1.4081^{24}
6978	2-Hexanone, 5-methyl- Methyl isopentyl ketone	$C_7H_{14}O$ 114.19	110-12-3	144	4-01-00-03329 0.888^{20}	H_2O 2; EtOH 5; eth 5; ace 4 1.4062^{20}
6979	3-Hexanone, 2-methyl-	$C_7H_{14}O$ 114.19	7379-12-6	135	4-01-00-03328 0.8091^{20}	EtOH 3; eth 3; ace 4; chl 3 1.4042^{20}
6980	3-Hexanone, 4-methyl-	$C_7H_{14}O$ 114.19	17042-16-9	134.5	4-01-00-03329 0.8162^{20}	ace 4; bz 4; eth 4; EtOH 4 1.4069^{20}
6981	3-Hexanone, 5-methyl-	$C_7H_{14}O$ 114.19	623-56-3	135	4-01-00-03328 0.8090^{20}	H_2O 1; EtOH 5; eth 5; ctc 3 1.4047^{20}
6982	2-Hexanone, 5-methyl-, oxime	$C_7H_{15}NO$ 129.20	624-44-2	195.5	0-01-00-00701 0.8881^{20}	chl 2 1.4448^{20}
6983	1-Hexanone, 1-phenyl-	$C_{12}H_{16}O$ 176.26	942-92-7 27	265	4-07-00-00754 0.9576^{20}	H_2O 2; EtOH 3; eth 3; ace 3 1.5027^{25}
6984	2-Hexanone, 3-phenyl-	$C_{12}H_{16}O$ 176.26	6306-30-5	235.5; 114^{13}	3-07-00-01160 0.970^{0}	1.5020^{20}
6985	3-Hexanone, 2-phenyl- 1-Phenylethyl propyl ketone	$C_{12}H_{16}O$ 176.26	65248-43-3	241	2-07-00-00258 0.941^{25}	1.4988^{25}
6986	Hexanoyl chloride Caproyl chloride	$C_6H_{11}ClO$ 134.61	142-61-0 -87	1763 153	4-02-00-00928 0.9784^{20}	eth 3; ace 3 1.4264^{20}
6987	Hexanoyl chloride, 2-ethyl-	$C_8H_{15}ClO$ 162.66	760-67-8	67-8^{11}	3-02-00-00805 0.939^{25}	1.4335^{20}
6988	Hexanoyl chloride, 3-methyl- Caproyl chloride, 3-methyl	$C_7H_{13}ClO$ 148.63	57323-93-0	163; 82^{50}	3-02-00-00777 0.967^{20}	bz 4 1.4293^{25}
6989	Hexanoyl chloride, 4-methyl- Caproyl chloride, 4-methyl	$C_7H_{13}ClO$ 148.63	50599-73-0	167	2-02-00-00299 0.9677^{20}	eth 4
6990	Hexasiloxane, tetradecamethyl- Tetradecamethylhexasiloxane	$C_{14}H_{42}O_5Si_6$ 459.00	107-52-8 -59	9131 245.5	4-04-00-04127 0.8910^{20}	bz 4 1.3948^{20}
6991	Hexatriacontane	$C_{36}H_{74}$ 506.98	630-06-8 76.5	298.4^{3}	4-01-00-00599 0.7803^{80}	1.4397^{80}
6992	1,3,5-Hexatriene, (*E*)-	C_6H_8 80.13	821-07-8 -12	78.5	4-01-00-01093 0.7369^{15}	H_2O 1; EtOH 3; ace 3; chl 3 1.5135^{20}
6993	1,3,5-Hexatriene, (*Z*)-	C_6H_8 80.13	2612-46-6 -12	78	0.7175^{20}	H_2O 1; EtOH 3; ace 3; chl 3 1.4577^{20}
6994	1,3,5-Hexatriene, 2,5-dimethyl-	C_8H_{12} 108.18	4916-63-6 -9	146	4-01-00-01102 0.7822^{20}	H_2O 1; ace 3; MeOH 3; lig 3 1.5122^{20}
6995	Hexazinone 1,3,5-Triazine-2,4(1H,3H)-dione, 3-cyclohexyl-6-(dimethylamino)-1-methyl-	$C_{12}H_{20}N_4O_2$ 252.32	51235-04-2 99	4617 dec	1.25	
6996	2-Hexenal, (*E*)-	$C_6H_{10}O$ 98.14	6728-26-3	146.5; 50^{20}	4-01-00-03468 0.8491^{20}	1.4480^{20}
6997	3-Hexenal, (*E*)-	$C_6H_{10}O$ 98.14	69112-21-6	42-3^{28}	4-01-00-03469 0.8455^{22}	H_2O 1; eth 3; ace 3 1.4275^{21}
6998	3-Hexenal, (*Z*)-	$C_6H_{10}O$ 98.14	6789-80-6	121	4-01-00-03469 0.8533^{22}	1.4300^{21}
6999	1-Hexene	C_6H_{12} 84.16	592-41-6 -139.7	63.4	4-01-00-00828 0.6731^{20}	bz 4; eth 4; EtOH 4; peth 4 1.3837^{20}

No.	Name Synonym	Mol. Form. Mol. Wt.	CAS RN mp/°C	Merck No. bp/°C	Beil. Ref. den/g cm^{-3}	Solubility n_D
7000	2-Hexene, (*E*)- *trans*-2-Hexene	C_6H_{12} 84.16	4050-45-7 -133	 67.9	4-01-00-00834 0.6732^{25}	H_2O 1; EtOH 3; eth 3; bz 3 1.3936^{20}
7001	2-Hexene, (*Z*)- *cis*-2-Hexene	C_6H_{12} 84.16	7688-21-3 -141.1	 68.8	4-01-00-00833 0.6869^{20}	H_2O 1; EtOH 3; eth 3; bz 3 1.3979^{20}
7002	3-Hexene, (*E*)-	C_6H_{12} 84.16	13269-52-8 -115.4	 67.1	4-01-00-00837 0.6772^{20}	H_2O 1; EtOH 3; eth 3; bz 3 1.3943^{20}
7003	3-Hexene, (*Z*)-	C_6H_{12} 84.16	7642-09-3 -137.8	 66.4	4-01-00-00837 0.6796^{20}	H_2O 1; EtOH 3; eth 3; bz 3 1.3947^{20}
7004	1-Hexene, 1-chloro- 1-Chloro-1-hexene	$C_6H_{11}Cl$ 118.61	22922-67-4	 121	4-01-00-00831 0.8872^{22}	ace 4; bz 4; eth 4; chl 4 1.4300^{22}
7005	1-Hexene, 2-chloro- 2-Chloro-1-hexene	$C_6H_{11}Cl$ 118.61	10124-73-9	 113	3-01-00-00803 0.8886^{25}	ace 4; bz 4; chl 4 1.4278^{25}
7006	1-Hexene, 5-chloro- 5-Chloro-1-hexene	$C_6H_{11}Cl$ 118.61	927-54-8	 120.7	4-01-00-00832 0.8891^{25}	ace 4; bz 4; eth 4; chl 4 1.4305^{20}
7007	2-Hexene, 4-chloro- 4-Chloro-2-hexene	$C_6H_{11}Cl$ 118.61	6734-98-1	 123	4-01-00-00835 0.8934^{20}	ace 4; bz 4; eth 4; chl 4 1.4400^{20}
7008	3-Hexene, 1-chloro- 3-Hexen-1-yl chloride	$C_6H_{11}Cl$ 118.61	62706-16-5	 61^{60}	4-01-00-00838 0.900^{24}	ace 4; bz 4; eth 4; chl 4 1.435^{24}
7009	3-Hexene, 3-chloro-, (*Z*)- 3-Hexene, 3-chloro (*cis*)	$C_6H_{11}Cl$ 118.61	17226-34-5	 117	4-01-00-00838 0.9009^{20}	ace 4; bz 4; eth 4; chl 4 1.4360^{20}
7010	3-Hexene, 1-chloro-4-ethyl-	$C_8H_{15}Cl$ 146.66	82507-04-8	 173	2-01-00-00201 0.9102^{20}	bz 4; chl 4 1.4524^{20}
7011	1-Hexene, 1,2-dichloro-, (*E*)-	$C_6H_{10}Cl_2$ 153.05	59697-51-7	 $63\text{-}5^{22}$	3-01-00-00803 1.1167^{25}	bz 4; chl 4 1.4576^{25}
7012	1-Hexene, 1,2-dichloro-, (*Z*)-	$C_6H_{10}Cl_2$ 153.05	59697-55-1	 88^{30}	3-01-00-00803 1.0812^{25}	bz 4; chl 4 1.4631^{25}
7013	1-Hexene, 2,3-dimethyl- 2,3-Dimethyl-1-hexene	C_8H_{16} 112.22	16746-86-4	 110.5	3-01-00-00845 0.7172^{25}	 1.4113^{20}
7014	1-Hexene, 2,4-dimethyl- 2,4-Dimethyl-1-hexene	C_8H_{16} 112.22	16746-87-5	 111.2	4-01-00-00887 0.716^{25}	 1.409^{20}
7015	1-Hexene, 2,5-dimethyl- 2,5-Dimethyl-1-hexene	C_8H_{16} 112.22	6975-92-4	 111.6	3-01-00-00846 0.7129^{25}	 1.4105^{20}
7016	1-Hexene, 3,3-dimethyl- 3,3-Dimethyl-1-hexene	C_8H_{16} 112.22	3404-77-1	 104	4-01-00-00889 0.7099^{25}	 1.4070^{20}
7017	1-Hexene, 3,4-dimethyl- 3,4-Dimethyl-1-hexene	C_8H_{16} 112.22	16745-94-1	 112	4-01-00-00889 0.720^{25}	 1.413^{20}
7018	1-Hexene, 3,5-dimethyl-	C_8H_{16} 112.22	7423-69-0	 104	3-01-00-00846 0.712^{25}	 1.404^{20}
7019	1-Hexene, 4,4-dimethyl- 4,4-Dimethyl-1-hexene	C_8H_{16} 112.22	1647-08-1	 107.2	3-01-00-00847 0.7157^{25}	 1.4102^{20}
7020	1-Hexene, 4,5-dimethyl- 4,5-Dimethyl-1-hexene	C_8H_{16} 112.22	16106-59-5	 109	4-01-00-00887 0.724^{25}	 1.414^{20}
7021	1-Hexene, 5,5-dimethyl- 5,5-Dimethyl-1-hexene	C_8H_{16} 112.22	7116-86-1	 104	4-01-00-00886 0.705^{25}	 1.4049^{20}
7022	2-Hexene, 2,3-dimethyl- 2,3-Dimethyl-2-hexene	C_8H_{16} 112.22	7145-20-2 -115.1	 121.8	4-01-00-00887 0.7366^{25}	 1.4268^{20}
7023	2-Hexene, 2,4-dimethyl- 2,4-Dimethyl-2-hexene	C_8H_{16} 112.22	14255-23-3	 110.6	4-01-00-00887 0.7171^{25}	 1.4118^{20}
7024	2-Hexene, 2,5-dimethyl- 2,5-Dimethyl-2-hexene	C_8H_{16} 112.22	3404-78-2	 112.2	4-01-00-00888 0.7182^{20}	 1.4140^{20}
7025	2-Hexene, 4,5-dimethyl- 2,3-Dimethyl-4-hexene	C_8H_{16} 112.22	73548-71-7	 110	3-01-00-00846 0.721^{25}	 1.413^{20}
7026	2-Hexene, 5,5-dimethyl-, (*E*)-	C_8H_{16} 112.22	39782-43-9	 104.1	4-01-00-00886 0.7023^{25}	 1.4055^{20}
7027	2-Hexene, 5,5-dimethyl-, (*Z*)- *cis*-5,5-Dimethyl-2-hexene	C_8H_{16} 112.22	39761-61-0	 106.9	4-01-00-00886 0.7125^{25}	 1.4113^{20}
7028	3-Hexene, 2,2-dimethyl-, (*E*)-	C_8H_{16} 112.22	690-93-7	 100.8	4-01-00-00885 0.6995^{25}	 1.4063^{20}
7029	3-Hexene, 2,2-dimethyl-, (*Z*)-	C_8H_{16} 112.22	690-92-6 -137.4	 105.5	4-01-00-00885 0.7086^{25}	 1.4099^{20}
7030	3-Hexene, 2,4-dimethyl-, (*E*)- 2,4-Dimethyl-*trans*-3-hexene	C_8H_{16} 112.22	61847-78-7	 107.6	4-01-00-00887 0.7101^{25}	 1.4126^{20}
7031	3-Hexene, 2,4-dimethyl-, (*Z*)- 3-Hexene, 2,4-dimethyl (*cis*)	C_8H_{16} 112.22	37549-89-6	 109	4-01-00-00887 0.7135^{25}	 1.4140^{20}
7032	1-Hexene, 1,1,2,3,3,4,4,5,5,6,6,6- dodecafluoro- Perfluoro-1-hexene	C_6F_{12} 300.05	755-25-4	 57.0	4-01-00-[illegible]	chl 4
7033	1-Hexene, 3-ethyl- 3-Propyl-1-pentene	C_8H_{16} 112.22	3404-58-8	 110.3	4-01-00-00885 0.711^{25}	 1.407^{20}
7034	3-Hexene, 3-ethyl-	C_8H_{16} 112.22	16789-51-8	 116	3-01-00-00845 0.725^{25}	 1.418^{20}
7035	3-Hexene, 1,2,3,4,5,6-hexachloro- 1,2,3,4,5,6-Hexachloro-3-hexene	$C_6H_6Cl_6$ 290.83	1725-74-2 58.5	 112^{2}	3-01-00-00808	MeOH 4; chl 4
7036	1-Hexene, 2-methyl- 2-Methyl-1-hexene	C_7H_{14} 98.19	6094-02-6 -102.8	 92	4-01-00-00863 0.7000^{20}	 1.4035^{20}
7037	1-Hexene, 3-methyl- 3-Methyl-1-hexene	C_7H_{14} 98.19	3404-61-3	 83.9	4-01-00-00865 0.6871^{25}	 1.3965^{20}
7038	1-Hexene, 4-methyl- 4-Methyl-1-hexene	C_7H_{14} 98.19	3769-23-1 -141.5	 86.7	3-01-00-00830 0.6942^{25}	 1.4000^{20}
7039	1-Hexene, 5-methyl- 5-Methyl-1-hexene	C_7H_{14} 98.19	3524-73-0	 85.3	4-01-00-00865 0.6877^{25}	 1.3967^{20}

No.	Name Synonym	Mol. Form. Mol. Wt.	CAS RN mp/°C	Merck No. bp/°C	Beil. Ref. den/g cm^{-3}	Solubility n_D
7040	2-Hexene, 2-methyl- 2-Methyl-2-hexene	C_7H_{14} 98.19	2738-19-4 -130.4	 95.4	4-01-00-00863 0.7038^{25}	 1.4106^{20}
7041	2-Hexene, 3-methyl-, (*Z*)-	C_7H_{14} 98.19	10574-36-4 -118.5	 95.6	4-01-00-00866 0.712^{20}	 1.4126^{20}
7042	2-Hexene, 4-methyl-, (*E*)-	C_7H_{14} 98.19	3683-22-5 -125.7	 87.6	4-01-00-00867 0.6925^{25}	 1.4025^{20}
7043	2-Hexene, 4-methyl-, (*Z*)-	C_7H_{14} 98.19	3683-19-0	 86.3	4-01-00-00867 0.6952^{25}	 1.4026^{20}
7044	2-Hexene, 5-methyl-, (*E*)-	C_7H_{14} 98.19	7385-82-2 -124.3	 88.1	4-01-00-00865 0.6883^{25}	 1.4006^{20}
7045	2-Hexene, 5-methyl-, (*Z*)-	C_7H_{14} 98.19	13151-17-2	 89.5	4-01-00-00865 0.697^{25}	 1.404^{20}
7046	3-Hexene, 2-methyl-, (*E*)-	C_7H_{14} 98.19	692-24-0 -141.6	 85.9	4-01-00-00864 0.6853^{25}	 1.4001^{20}
7047	3-Hexene, 2-methyl-, (*Z*)-	C_7H_{14} 98.19	15840-60-5	 86	4-01-00-00864 0.690^{25}	 1.401^{20}
7048	1-Hexene, 1,1,2-trichloro-	$C_6H_9Cl_3$ 187.50	53977-99-4	 $90\text{-}3^{10}$	3-01-00-00803 1.125^{25}	eth 4 1.4760^{25}
7049	2-Hexenoic acid	$C_6H_{10}O_2$ 114.14	1191-04-4 36.5	 216.5	4-02-00-01563 0.965^{20}	eth 4 1.4460^{40}
7050	3-Hexenoic acid Hydrosorbic acid	$C_6H_{10}O_2$ 114.14	4219-24-3 12	 208	4-02-00-01566 0.9640^{23}	 1.4935^{20}
7051	5-Hexenoic acid 5-Hexanoic acid	$C_6H_{10}O_2$ 114.14	1577-22-6 -37	 203	4-02-00-01562 0.9610^{20}	eth 4; EtOH 4 1.4343^{20}
7052	4-Hexenoic acid, 6-(1,3-dihydro-4-hydroxy-6-methoxy-7-methyl-3-oxo-5-isobenzofuranyl)-4-methyl-, (*E*)- Mycophenolic acid	$C_{17}H_{20}O_6$ 320.34	24280-93-1 141	6238	4-18-00-06513	H_2O 1; EtOH 4; eth 4; bz 2
7053	2-Hexenoic acid, ethyl ester Ethyl 2-hexenoate	$C_8H_{14}O_2$ 142.20	1552-67-6	 174.5; 80^{22}	3-02-00-01318 0.8986^{20}	 1.4348^{20}
7054	3-Hexenoic acid, ethyl ester Hydrosorbic acid, ethyl ester	$C_8H_{14}O_2$ 142.20	2396-83-0	 166.5	4-02-00-01567 0.8957^{20}	 1.4255^{20}
7055	2-Hexenoic acid, 2-methyl-, (*E*)-	$C_7H_{12}O_2$ 128.17	97961-66-5	 205; 118^{11}	4-02-00-01581 0.9627^{20}	eth 3; ace 3 1.4601^{20}
7056	2-Hexenoic acid, 4-methyl- 4-Methyl-2-hexenoic acid	$C_7H_{12}O_2$ 128.17	37549-83-0	 125^{13}	3-02-00-01336 0.9441^{20}	 1.4526^{20}
7057	2-Hexenoic acid, 5-methyl- 5-Methyl-2-hexenoic acid	$C_7H_{12}O_2$ 128.17	41653-96-7 16.5	 226.5	2-02-00-00411 0.942^{20}	 1.4425^{17}
7058	3-Hexenoic acid, 2-methyl- Hydrosorbic acid, 2-methyl-	$C_7H_{12}O_2$ 128.17	73513-50-5	 122^{24}	4-02-00-01582 0.9353^{20}	 1.4379^{20}
7059	3-Hexenoic acid, 3-methyl-	$C_7H_{12}O_2$ 128.17	35205-71-1	 113^{10}	3-02-00-01335 0.9549^{20}	 1.4469^{20}
7060	3-Hexenoic acid, 4-methyl- 4-Ethyl-3-pentenoic acid	$C_7H_{12}O_2$ 128.17	55665-79-7	 118^{12}	3-02-00-01336 0.9644^{16}	 1.4512^{16}
7061	4-Hexenoic acid, 5-methyl-	$C_7H_{12}O_2$ 128.17	5636-65-7 -28	 217	4-02-00-01581 0.9862^{0}	 1.4504
7062	3-Hexenoic acid, methyl ester Methyl 3-hexenoate	$C_7H_{12}O_2$ 128.17	2396-78-3	 $67\text{-}8^{34}$	4-02-00-01566 0.9132^{25}	 1.4240^{23}
7063	2-Hexenoic acid, 2-methyl-, ethyl ester	$C_9H_{16}O_2$ 156.22	26311-33-1	 72^{10}	3-02-00-01333 0.9031^{20}	 1.4407^{20}
7064	3-Hexenoic acid, 2-methyl-, ethyl ester	$C_9H_{16}O_2$ 156.22	21994-78-5	 78^{25}	3-02-00-01334 0.8778^{20}	 1.4237^{20}
7065	3-Hexenoic acid, 3-methyl-, ethyl ester	$C_9H_{16}O_2$ 156.22	21994-77-4	 $84\text{-}5^{26}$	3-02-00-01335 0.8961^{20}	 1.4309^{20}
7066	1-Hexen-3-ol	$C_6H_{12}O$ 100.16	4798-44-1	 134	4-01-00-02136 0.834^{22}	ace 4; eth 4; EtOH 4 1.4297^{18}
7067	4-Hexen-2-ol	$C_6H_{12}O$ 100.16	52387-50-5	 137.5	4-01-00-02139 0.8405^{18}	 1.4392^{20}
7068	5-Hexen-2-ol	$C_6H_{12}O$ 100.16	626-94-8	 139	3-01-00-01926 0.842^{16}	H_2O 2
7069	2-Hexen-1-ol, (*E*)-	$C_6H_{12}O$ 100.16	928-95-0	 157	4-01-00-02138 0.8490^{16}	 1.4340^{20}
7070	2-Hexen-1-ol, (*Z*)-	$C_6H_{12}O$ 100.16	928-94-9	 157	4-01-00-02138 0.8472^{20}	H_2O 3; EtOH 4; eth 3; ace 3 1.4397^{20}
7071	3-Hexen-1-ol, (*Z*)-	$C_6H_{12}O$ 100.16	928-96-1	 156.5	4-01-00-02141 0.8478^{22}	H_2O 3; EtOH 4; eth 3 1.4380^{20}
7072	4-Hexen-1-ol, (*E*)-	$C_6H_{12}O$ 100.16	928-92-7	 159	4-01-00-02140 0.8513^{20}	 1.4402^{20}
7073	1-Hexen-3-ol, 3,5-dimethyl-	$C_8H_{16}O$ 128.21	3329-48-4	 146.5	4-01-00-02173 0.8382^{20}	 1.4342^{20}
7074	4-Hexen-3-ol, 2,2-dimethyl- 2-Hexen-4-ol, 5,5-dimethyl	$C_8H_{16}O$ 128.21	37674-67-2	 $75\text{-}6^{50}$	4-01-00-02171 0.8308^{25}	 1.4369^{25}
7075	4-Hexen-3-ol, 2,5-dimethyl- 2,5-Dimethyl-4-hexen-3-ol	$C_8H_{16}O$ 128.21	60703-31-3	 162	2-01-00-00491 0.8444^{20}	 1.4449^{20}
7076	4-Hexen-3-ol, 3,5-dimethyl- 2-Hexen-4-ol, 2,4-dimethyl	$C_8H_{16}O$ 128.21	1569-43-3	 $48\text{-}50^{3\text{-}6}$	4-01-00-02172 0.8600^{17}	 1.4460^{17}
7077	1-Hexen-3-ol, 5-methyl-	$C_7H_{14}O$ 114.19	4798-46-3	 125; 58^{15}	3-01-00-01939 0.8306^{15}	 1.4263^{23}
7078	2-Hexen-1-ol, 5-methyl- 5-Methyl-2-hexenol	$C_7H_{14}O$ 114.19	77053-92-0	 169	4-01-00-02158 0.8355^{20}	 1.4390^{20}

No.	Name Synonym	Mol. Form. Mol. Wt.	CAS RN mp/°C	Merck No. bp/°C	Beil. Ref. den/g cm^{-3}	Solubility n_D
7079	3-Hexen-2-ol, 2-methyl-	$C_7H_{14}O$ 114.19	18812-62-9	49^{11}	4-01-00-02157 0.8536^{18}	1.443^{18}
7080	3-Hexen-2-ol, 3-methyl-	$C_7H_{14}O$ 114.19	110383-31-8	89^{55}	0-01-00-00447 0.8678^{9}	1.4487^{9}
7081	4-Hexen-3-ol, 2-methyl-	$C_7H_{14}O$ 114.19	4798-60-1	139.5	3-01-00-01939 0.8426^{20}	1.438^{20}
7082	4-Hexen-3-ol, 3-methyl-	$C_7H_{14}O$ 114.19	60111-14-0	$72\text{-}3^{60}$	4-01-00-02160 0.8471^{25}	1.4268^{16}
7083	5-Hexen-2-ol, 2-methyl- Dimethyl(vinylethyl)carbinol	$C_7H_{14}O$ 114.19	16744-89-1 -22	142.5	4-01-00-02158 0.8397^{15}	1.4349^{17}
7084	5-Hexen-3-ol, 3-methyl- Methylethylallylcarbinol	$C_7H_{14}O$ 114.19	1569-44-4	140; 61^{35}	4-01-00-02160 0.8432^{20}	1.4370^{20}
7085	5-Hexen-3-ol, 4-methyl- 4-Methyl-5-hexen-3-ol	$C_7H_{14}O$ 114.19	1838-77-3	140.5	3-01-00-01940 0.8452^{22}	1.4365^{22}
7086	3-Hexen-2-one 1-Butenyl methyl ketone	$C_6H_{10}O$ 98.14	763-93-9	140	4-01-00-03468 0.8655^{20}	ace 4; eth 4; EtOH 4 1.4418^{20}
7087	4-Hexen-2-one 2-Hexen-5-one	$C_6H_{10}O$ 98.14	25659-22-7	127	3-01-00-02994 0.8520^{16}	1.4300^{16}
7088	4-Hexen-3-one	$C_6H_{10}O$ 98.14	2497-21-4	138.5	4-01-00-03468 0.8559^{20}	EtOH 3; eth 3; ace 4 1.4388^{20}
7089	5-Hexen-2-one	$C_6H_{10}O$ 98.14	109-49-9	129.5	4-01-00-03467 0.833^{27}	1.4178^{27}
7090	3-Hexen-2-one, (*E*)-	$C_6H_{10}O$ 98.14	4376-23-2	140	4-01-00-03469 0.8665^{20}	EtOH 3; eth 3; ace 4 1.4418^{20}
7091	3-Hexen-2-one, 3,4-dimethyl-	$C_8H_{14}O$ 126.20	1635-02-5	158; 65^{20}	4-01-00-03500 0.8585^{20}	1.4418^{20}
7092	4-Hexen-2-one, 3,4-dimethyl- 2-Hexen-5-one, 3,4-dimethyl	$C_8H_{14}O$ 126.20	53252-21-4	154	4-01-00-03500 0.8538^{19}	1.4377^{19}
7093	1-Hexen-3-one, 5-methyl- Isobutyl vinyl ketone	$C_7H_{12}O$ 112.17	2177-32-4	41^{22}; 32^{10}	4-01-00-03483 0.8400^{15}	1.4293^{15}
7094	3-Hexen-2-one, 5-methyl- 2-Oxo-5-methylhex-3-ene	$C_7H_{12}O$ 112.17	5166-53-0	77.5^{50}	3-01-00-03005 0.8407^{22}	1.4395^{22}
7095	4-Hexen-2-one, 5-methyl-	$C_7H_{12}O$ 112.17	28332-44-7	76^{50}	4-01-00-03482 0.9012^{21}	1.4385^{20}
7096	4-Hexen-3-one, 2-methyl-	$C_7H_{12}O$ 112.17	53252-19-0	148	4-01-00-03483 0.843^{20}	EtOH 3; eth 3; ace 4 1.4345^{20}
7097	5-Hexen-2-one, 3-methyl- 1-Methyl-1-allyl acetone	$C_7H_{12}O$ 112.17	2550-22-3	137	4-01-00-03484 0.845^{15}	1.4215^{25}
7098	5-Hexen-2-one, 4-methyl- 4-Methyl-5-hexen-2-one	$C_7H_{12}O$ 112.17	61675-14-7	137.5	3-01-00-03006 0.8273^{25}	1.4193^{22}
7099	5-Hexen-2-one, 5-methyl- 5-Hexene-2-one,5-methyl	$C_7H_{12}O$ 112.17	3240-09-3	150	4-01-00-03482 0.8460^{20}	ace 4; eth 4; EtOH 4 1.4348^{20}
7100	1-Hexen-3-one, 5-methyl-1-phenyl-	$C_{13}H_{16}O$ 188.27	2892-18-4 43	154^{25}	4-07-00-01058 0.9509^{46}	H_2O 2; EtOH 3; bz 3; chl 3 1.5523^{25}
7101	5-Hexen-2-one, 4-phenyl-	$C_{12}H_{14}O$ 174.24	50552-30-2	$85\text{-}6^{1}$	3-07-00-01460 0.9848^{25}	1.5193^{25}
7102	5-Hexen-2-one, 6-phenyl- 6-Phenyl-5-hexen-2-one	$C_{12}H_{14}O$ 174.24	69371-59-1	154^{10}; $98^{0.3}$	3-07-00-01455 0.9967^{25}	1.5458^{25}
7103	1-Hexen-3-yne Ethylvinylacetylene	C_6H_8 80.13	13721-54-5	85	4-01-00-01091 0.7492^{20}	bz 4; eth 4; peth 4; chl 4 1.4522^{20}
7104	1-Hexen-4-yne	C_6H_8 80.13	5009-11-0	87	4-01-00-01092 0.767^{14}	1.446^{14}
7105	1-Hexen-5-yne Hex-5-en-1-yne	C_6H_8 80.13	14548-31-3	70	4-01-00-01092 0.7650^{20}	bz 4; eth 4; peth 4; chl 4 1.4318^{20}
7106	2-Hexen-4-yne 2-Hexene-4-yne	C_6H_8 80.13	14092-20-7	88.5	4-01-00-01092 0.7710^{20}	1.4918^{20}
7107	1-Hexen-3-yne, 5-chloro-5-methyl- Dimethylvinylethynylchloromethane	C_7H_9Cl 128.60	819-44-3	48^{28}	4-01-00-01098 0.9375^{15}	ace 4; bz 4; eth 4; EtOH 4 1.4778^{20}
7108	3-Hexen-1-yne, 3-propyl-	C_9H_{14} 122.21	688-52-8	136	3-01-00-01049 0.7799^{25}	bz 4; peth 4; chl 4 1.4432^{25}
7109	Hexitol, 2,5-anhydro-3,4-dideoxy- Tetrahydro-2,5-furandimethanol	$C_6H_{12}O_3$ 132.16	104-80-3 <-50	9145 265	4-17-00-02052 1.154^{20}	H_2O 4; ace 4; bz 4; EtOH 4
7110	*D-arabino*-Hexose, 2-deoxy- 2-Deoxy-*D*-glucose	$C_6H_{12}O_5$ 164.16	154-17-6 146.5	2886	4-01-00-04282	
7111	D-*ribo*-Hexose, 2,6-dideoxy- Digitoxose	$C_6H_{12}O_4$ 148.16	527-52-6 112	3147	4-01-00-04191	H_2O 4; ace 4; py 3; AcOEt 3
7112	*ribo*-Hexose, 2,6-dideoxy-3-*O*-methyl- Cymarose	$C_7H_{14}O_4$ 162.19	579-04-4 101	2769	4-01-00-04193	H_2O 4; ace 4; EtOH 4
7113	3-Hexynal	C_6H_8O 96.13	89533-67-5	$42\text{-}3^{17}$	4-01-00-03544 0.9036^{23}	1.4498^{22}
7114	1-Hexyne Butylacetylene	C_6H_{10} 82.15	693-02-7 -131.9	71.3	4-01-00-01006 0.7155^{20}	H_2O 1; EtOH 3; eth 3; bz 3 1.3989^{20}
7115	2-Hexyne 1-Methyl-2-propylacetylene	C_6H_{10} 82.15	764-35-2 -89.6	84.5	4-01-00-01009 0.7315^{20}	H_2O 1; EtOH 5; eth 5; bz 3 1.4138^{20}
7116	3-Hexyne Diethylacetylene	C_6H_{10} 82.15	928-49-4 -103	81	4-01-00-01009 0.7231^{20}	H_2O 1; EtOH 3; eth 3; bz 3 1.4115^{20}
7117	3-Hexyne, 2,5-dichloro-2,5-dimethyl- 3-Hexyne, 2,5-dimethyl-2,5-dichloro	$C_8H_{12}Cl_2$ 179.09	2431-30-3 29	178; 62^{14}	4-01-00-01042 1.0118^{20}	chl 4
7118	3-Hexyne-2,5-diol	$C_6H_{10}O_2$ 114.14	3031-66-1	120^{11}	3-01-00-02271 1.0180^{20}	1.4691^{20}

No.	Name Synonym	Mol. Form. Mol. Wt.	CAS RN mp/°C	Merck No. bp/°C	Beil. Ref. den/g cm^{-3}	Solubility n_D
7119	3-Hexyne-2,5-diol, 2,5-dimethyl-	$C_8H_{14}O_2$ 142.20	142-30-3 95	205	4-01-00-02699 0.947^{20}	H_2O 3; EtOH 4; eth 4; ace 4
7120	1-Hexyne, 5-methyl- 5-Methyl-1-hexyne	C_7H_{12} 96.17	2203-80-7 -125	92	3-01-00-01000 0.7274^{20}	H_2O 1; EtOH 3; eth 3; bz 3 1.4059^{20}
7121	2-Hexyne, 5-methyl-	C_7H_{12} 96.17	53566-37-3 -92.9	102.5	4-01-00-01030 0.7378^{20}	H_2O 1; eth 3; ace 3; bz 3 1.4176^{20}
7122	3-Hexyne, 2-methyl- 1-Ethyl-2-isopropylacetylene	C_7H_{12} 96.17	36566-80-0 -116.7	95.2	4-01-00-01029 0.7263^{20}	bz 4; eth 4; chl 4; peth 4 1.4120^{20}
7123	2-Hexynoic acid Propiolic acid, propyl-	$C_6H_8O_2$ 112.13	764-33-0 27	126^{24}	4-02-00-01699 0.9820^{20}	eth 4; EtOH 4; lig 4
7124	1-Hexyn-3-ol	$C_6H_{10}O$ 98.14	105-31-7 -80	142	4-01-00-02234 0.8704^{20}	ctc 3 1.4340^{25}
7125	2-Hexyn-1-ol	$C_6H_{10}O$ 98.14	764-60-3	100^{53}; 80^{21}	4-01-00-02235 0.8472^{20}	1.4341^{20}
7126	3-Hexyn-1-ol 3-Hexynol	$C_6H_{10}O$ 98.14	1002-28-4	163; 65^{12}	4-01-00-02236 0.8982^{20}	1.4530^{20}
7127	5-Hexyn-2-ol	$C_6H_{10}O$ 98.14	23470-12-4	152	4-01-00-02235 0.899^{20}	1.4481^{20}
7128	1-Hexyn-3-ol, 3-methyl- Ethynylmethylpropylcarbinol	$C_7H_{12}O$ 112.17	4339-05-3	137	4-01-00-02252 0.8620^{20}	H_2O 4; eth 4; EtOH 4 1.4338^{20}
7129	3-Hexyn-2-ol, 2-methyl- 2-Methyl-3-hexyn-2-ol	$C_7H_{12}O$ 112.17	5075-33-2 -6	146	4-01-00-02250 0.962^{25}	eth 4; EtOH 4 1.4392^{25}
7130	3-Hexyn-2-one 3-Hexyne-2-one	C_6H_8O 96.13	1679-36-3	$77\text{-}8^{75}$	4-01-00-03544 0.8804^{20}	1.4400^{20}
7131	5-Hexyn-2-one	C_6H_8O 96.13	2550-28-9	149	4-01-00-03543 0.9065^{20}	1.4366^{20}
7132	*L*-Histidine Glyoxaline-5-alanine	$C_6H_9N_3O_2$ 155.16	71-00-1 287 dec	4642	4-25-00-04343	H_2O 3; EtOH 2; eth 1; ace 1
7133	*L*-Histidine, *N*-β-alanyl- Carnosine	$C_9H_{14}N_4O_3$ 226.24	305-84-0 260	1857	4-25-00-04381	H_2O 4
7134	Homoserine (*S*)-2-Amino-4-hydroxybutanoic acid	$C_4H_9NO_3$ 119.12	672-15-1 203 dec	4661	4-04-00-03187	H_2O 4; EtOH 2; eth 1; bz 1
7135	*L*-Homoserine, *O*-[(aminoiminomethyl)amino]- Canavanine	$C_5H_{12}N_4O_3$ 176.18	543-38-4	1745		H_2O 4
7136	Hydramethylnon Tetrahydro-5,5-dimethyl-2-(1H)-pyrimidinone[3[4-(trifluoromethyl-styryl)-cinnamylidene- hydrazone	$C_{25}H_{24}F_6N_4$ 494.48	67485-29-4 190	4684		
7137	Hydrazine, 1,2-bis(1-methylethyl)-	$C_6H_{16}N_2$ 116.21	3711-34-0	125	4-04-00-03339 0.7894^{20}	ace 4; bz 4; eth 4; EtOH 4 1.4173^{20}
7138	Hydrazine, 1,2-bis(3-methylphenyl)-	$C_{14}H_{16}N_2$ 212.29	621-26-1 38	224	4-15-00-01220	H_2O 1; EtOH 3; eth 3; bz 3
7139	Hydrazine, 1,2-bis(4-methylphenyl)-	$C_{14}H_{16}N_2$ 212.29	637-47-8 135		4-15-00-01230 0.957^{20}	bz 4; eth 4; EtOH 4
7140	Hydrazine, 1,2-bis(2-methylpropyl)-	$C_8H_{20}N_2$ 144.26	3711-37-3	171; 63^{18}	2-04-00-00962 0.8002^{20}	ace 4; bz 4; eth 4; EtOH 4 1.4276
7141	Hydrazine, (4-bromophenyl)- (*p*-Bromophenyl)hydrazine	$C_6H_7BrN_2$ 187.04	589-21-9 108	1417	4-15-00-00282	eth 4; EtOH 4; lig 4
7142	Hydrazinecarbothioamide Thiosemicarbazide	CH_5N_3S 91.14	79-19-6 183	9292	4-03-00-00374	H_2O 4; EtOH 4
7143	Hydrazinecarboxamide, monohydrochloride Semicarbazide hydrochloride	CH_6ClN_3O 111.53	563-41-7 175-7 dec	8396	4-03-00-00177	H_2O 4
7144	Hydrazinecarboxamide, 2-[(5-nitro-2-furanyl)methylene]- Nitrofurazone	$C_6H_6N_4O_4$ 198.14	59-87-0 236-40 dec	6521	5-17-09-00335	H_2O 1; EtOH 2; eth 1; DMSO 2
7145	Hydrazinecarboxamide, 2-phenyl- Phenicarbazide	$C_7H_9N_3O$ 151.17	103-03-7 172	7193	4-15-00-00180	H_2O 2; EtOH 3; eth 2; ace 3
7146	Hydrazinecarboxamide, *N*-phenyl- 4-Phenylsemicarbazide	$C_7H_9N_3O$ 151.17	537-47-3 128	7283	3-12-00-00822	H_2O 2; EtOH 4; eth 1; chl 4
7147	Hydrazinecarboximidamide Aminoguanidine	CH_6N_4 74.09	79-17-4 dec	453	4-03-00-00236	H_2O 4; EtOH 4
7148	Hydrazinecarboxylic acid, ethyl ester	$C_3H_8N_2O_2$ 104.11	4114-31-2 46	198 dec; 93^9	4-03-00-00174	EtOH 3; eth 3; chl 2
7149	Hydrazinecarboxylic acid, methyl ester	$C_2H_6N_2O_2$ 90.08	6294-89-9 73	108^{12}	2-03-00-00078	H_2O 3; EtOH 3; bz 2; peth 1
7150	1,1-Hydrazinedicarboxylic acid, diethyl ester	$C_6H_{12}N_2O_4$ 176.17	5311-96-6 30.5	138^{12}	2-03-00-00079	EtOH 4
7151	1,2-Hydrazinedicarboxylic acid, diethyl ester	$C_6H_{12}N_2O_4$ 176.17	4114-28-7 135	250 dec	4-03-00-00175 1.324^8	eth 4; EtOH 4
7152	Hydrazine, 1,1-diethyl- 1,1-Diethylhydrazine	$C_4H_{12}N_2$ 88.15	616-40-0	99	4-04-00-03335 0.8804^{20}	bz 4; eth 4; EtOH 4; chl 4 1.4214^{20}
7153	Hydrazine, 1,2-diethyl- 1,2-Diethylhydrazine	$C_4H_{12}N_2$ 88.15	1615-80-1	85.5	4-04-00-03336 0.797^{26}	bz 4; eth 4; EtOH 4 1.4204^{20}
7154	Hydrazine, 1,1-dimethyl- 1,1-Dimethylhydrazine	$C_2H_8N_2$ 60.10	57-14-7 -58	3236 63.9	4-04-00-03322 0.791^{22}	H_2O 4; EtOH 4; eth 4; MeOH 4 1.4075^{22}
7155	Hydrazine, 1,2-dimethyl- 1,2-Dimethylhydrazine	$C_2H_8N_2$ 60.10	540-73-8	3237 81	4-04-00-03323 0.8274^{20}	H_2O 5; EtOH 5; eth 5 1.4209^{20}
7156	Hydrazine, (2,4-dinitrophenyl)- (2,4-Dinitrophenyl)hydrazine	$C_6H_6N_4O_4$ 198.14	119-26-6 194	3277	4-15-00-00380	H_2O 1; EtOH 3; eth 2; bz 2

No.	Name	Synonym	Mol. Form.	Mol. Wt.	CAS RN	mp/°C	Merck No.	bp/°C	Beil. Ref.	den/g cm^{-3}	Solubility	n_D
7157	Hydrazine, 1,1-diphenyl-	1,1-Diphenylhydrazine	$C_{12}H_{12}N_2$	184.24	530-50-7	50.5	3326	220^{40}	4-15-00-00055	1.190^{16}	bz 4; eth 4; EtOH 4; chl 4	
7158	Hydrazine, 1,2-diphenyl-	*N,N'*-Diphenylhydrazine	$C_{12}H_{12}N_2$	184.24	122-66-7	131			4-15-00-00056	1.158^{16}	EtOH 4; bz 2; DMSO-d_6 2; HOAc 1	
7159	Hydrazine, 1-ethyl-1-phenyl-	1-Phenyl-1-ethylhydrazine	$C_8H_{12}N_2$	136.20	644-21-3			237	3-15-00-00074	1.0181^{21}	ace 4; bz 4; eth 4; EtOH 4	1.5711^{21}
7160	Hydrazine, 1-ethyl-2-phenyl-	1-Ethyl-2-phenylhydrazine	$C_8H_{12}N_2$	136.20	622-82-2			241; 110^{14}	4-15-00-00054	1.0150^{20}	bz 4; eth 4; EtOH 4; chl 4	1.5676^{20}
7161	Hydrazine, methyl-	Methylhydrazine	CH_6N_2	46.07	60-34-4	-52.4	6001	87.5	4-04-00-03322		H_2O 3; EtOH 5; eth 3; ctc 3	1.4325^{20}
7162	Hydrazine, 1-(3-methylbutyl)-1-phenyl-	Hydrazine, 1-isopentyl-1-phenyl	$C_{11}H_{18}N_2$	178.28	636-10-2	236		260	0-15-00-00121	0.9588^{15}		
7163	Hydrazine, 1-methyl-2-(3-methylphenyl)-	Hydrazine, 1-methyl-2-(3-tolyl)	$C_8H_{12}N_2$	136.20	116836-06-7	60			2-15-00-00229	1.0265^{100}	bz 4; EtOH 4	
7164	Hydrazine, (3-methylphenyl)-		$C_7H_{10}N_2$	122.17	536-89-0			244 dec	4-15-00-01220	1.057^{20}	H_2O 2; EtOH 4; eth 4; bz 4	
7165	Hydrazine, 1-methyl-1-phenyl-		$C_7H_{10}N_2$	122.17	618-40-6			228; 131^{35}	4-15-00-00053	1.0404^{20}	H_2O 2; EtOH 5; eth 5; bz 5	1.5691^{20}
7166	Hydrazine, 1-methyl-2-phenyl-	1-Methyl-2-phenylhydrazine	$C_7H_{10}N_2$	122.17	622-36-6			231; 112^{14}	4-15-00-00054	1.0320^{20}	bz 4; eth 4; EtOH 4; chl 4	1.5733^{20}
7167	Hydrazine, 1-naphthalenyl-		$C_{10}H_{10}N_2$	158.20	2243-55-2	117		203^{20}	3-15-00-00728		bz 4; eth 4; EtOH 4; chl 4	
7168	Hydrazine, (4-nitrophenyl)-	(4-Nitrophenyl)hydrazine	$C_6H_7N_3O_2$	153.14	100-16-3	158 dec	6545		4-15-00-00317		H_2O 2; EtOH 3; eth 3; bz 3	
7169	Hydrazine, phenyl-	Phenylhydrazine	$C_6H_8N_2$	108.14	100-63-0	19.6	7264	243.5	4-15-00-00050	1.0986^{20}	H_2O 3; EtOH 5; eth 5; ace 4	1.6084^{10}
7170	Hydrazine, (1-phenylethyl)-		$C_8H_{12}N_2$	136.20	65-64-5			$69^{0.65}$	4-15-00-01267	0.9672^{25}		1.5436^{25}
7171	Hydrazine, phenyl-, hemihydrate		$C_{12}H_{18}N_4O$	234.30	6152-31-4	24		120^{12}	3-15-00-00067	1.0938^{25}		1.6081^{20}
7172	Hydrazine, phenyl-, monohydrochloride		$C_6H_9ClN_2$	144.60	59-88-1	243-6 dec	7265	sub			H_2O 4; EtOH 4	
7173	Hydrazine, triphenyl-	Triphenylhydrazine	$C_{18}H_{16}N_2$	260.34	606-88-2	142 dec			4-15-00-00058	0.869^{70}	EtOH 2; bz 2	
7174	Hydrazobenzene, 3-methyl-	Hydrazine, 1-phenyl-2-(3-tolyl)	$C_{13}H_{14}N_2$	198.27	621-25-0	61			2-15-00-00229	1.0265^{100}	bz 4; EtOH 4; lig 4	
7175	Hydrocinnamic acid, β-ethyl-β-methyl-	Valeric acid, 3-methyl-3-phenyl	$C_{12}H_{16}O_2$	192.26	105401-59-0	284		174^{14}	4-09-00-01905	1.050^{25}	H_2O 3	1.5197^{25}
7176	Hydrocyanic acid	Hydrogen cyanide	CHN	27.03	74-90-8	-13.4	4722	26	4-02-00-00050	0.6876^{20}	H_2O 5; EtOH 5; eth 5	1.2614^{20}
7177	Hydroperoxide, cyclohexyl	Cyclohexyl hydroperoxide	$C_6H_{12}O_2$	116.16	766-07-4	-20		$42^{0.1}$	4-06-00-00053	1.019^{20}	eth 4; EtOH 4; HOAc 4	1.4645^{25}
7178	Hydroperoxide, 1,1-dimethylethyl	*tert*-Butyl hydroperoxide	$C_4H_{10}O_2$	90.12	75-91-2	6	1569	89 dec; 36^{17}	4-01-00-01616	0.8960^{20}	H_2O 3; EtOH 3; eth 3; ctc 3	1.4015^{20}
7179	Hydroperoxide, ethyl	Ethyl hydrogen peroxide	$C_2H_6O_2$	62.07	3031-74-1	-100		95	4-01-00-01323	0.9332^{20}	H_2O 4; bz 4; eth 4; EtOH 4	1.3800^{20}
7180	Hydroperoxide, methyl		CH_4O_2	48.04	3031-73-0			38-40^{65}	4-01-00-01249	1.9967^{15}	H_2O 4; bz 4; eth 4; EtOH 4	1.3641^{15}
7181	Hydroperoxide, 1-methyl-1-phenylethyl	Cumene hydroperoxide	$C_9H_{12}O_2$	152.19	80-15-9				4-06-00-03221			
7182	Hydroprene	2,4-Dodecadienoic acid, 3,7,11-trimethyl-, ethyl ester, (*E,E*)-	$C_{17}H_{30}O_2$	266.42	41096-46-2			174^{19}		0.8955^{20}		
7183	Hypochlorous acid, 1,1-dimethylethyl ester	*tert*-Butyl hypochlorite	C_4H_9ClO	108.57	507-40-4		1570	77.5	4-01-00-01621	0.9583^{18}	H_2O 1; eth 4; ace 3; bz 4	1.403^{20}
7184	Hypochlorous acid, ethyl ester	Ethyl hypochlorite	C_2H_5ClO	80.51	624-85-1			37	4-01-00-01324	1.013^{-6}	H_2O 1; EtOH 3; eth 5; bz 5	
7185	Hypophosphoric acid, tetraethyl ester		$C_8H_{20}O_6P_2$	274.19	679-37-8			116-7^{2}	4-01-00-01358	1.1283^{18}	bz 3; ctc 3	1.4284^{20}
7186	Ibogamine, 12-methoxy-	Ibogaine	$C_{20}H_{26}N_2O$	310.44	83-74-9	148	4806		5-23-12-00284		chl 3	
7187	Imazalil	1H-Imidazole, 1-[2-(2,4-dichlorophenyl)-2-(2-propenyloxy)ethyl]-	$C_{14}H_{14}Cl_2N_2O$	297.18	35554-44-0	50	3537	dec		1.243^{23}		
7188	Imazapyr	2-(4-Isopropyl-4-methyl-5-oxo-2-imidazolin-2-yl)nicotinic acid	$C_{13}H_{15}N_3O_3$	261.28	81334-34-1	171						
7189	Imazaquin	(*RS*)-2-(4-Isopropyl-4-methyl-5-oxo-2-imidazolin-2-yl)quinoline-3-carboxylic acid	$C_{17}H_{17}N_3O_3$	311.34	81335-37-7	221	4826					
7190	Imazethapyr	(*RS*)-5-Ethyl-2-(4-isopropyl-4-methyl-5-oxo-2-imidazolin-2-yl)nicotinic acid	$C_{15}H_{19}N_3O_3$	289.33	81335-77-5	173						
7191	Imidazo[4,5-d]imidazole-2,5(1H,3H)-dione, tetrahydro-	Acetyleneurea	$C_4H_6N_4O_2$	142.12	496-46-8	300 dec	87		2-26-00-00260		H_2O 2; EtOH 1; eth 3; HCl 3	
7192	1H-Imidazole	1,3-Diazole	$C_3H_4N_2$	68.08	288-32-4	90.5	4828	257	5-23-04-00191	1.0303^{101}	H_2O 4; EtOH 4; eth 3; ace 3	1.4801^{101}
7193	1H-Imidazole, 1,1'-carbonylbis-	*N,N'*-Carbonyldiimidazole	$C_7H_6N_4O$	162.15	530-62-1	119	1825		5-23-04-00245			

PHYSICAL CONSTANTS OF ORGANIC COMPOUNDS (continued)

No.	Name	Synonym	Mol. Form.	Mol. Wt.	CAS RN	mp/°C	Merck No.	bp/°C	Beil. Ref.	den/g cm^{-3}	Solubility	n_D
7194	1H-Imidazole-1-carboxylic acid, 2,3-dihydro-3-methyl-2-thioxo-, ethyl ester	Carbimazole	$C_7H_{10}N_2O_2S$	186.23	22232-54-8	123.5	1803		4-24-00-00064		ace 4; chl 4	
7195	1H-Imidazole-4,5-dicarboxylic acid		$C_5H_4N_2O_4$	156.10	570-22-9	290 dec			4-25-00-01051	1.749^{25}	H_2O 2; EtOH 1; eth 1; bz 1	
7196	1H-Imidazole, 1,2-diethyl-	1,2-Diethylimidazole	$C_7H_{12}N_2$	124.19	51807-53-5			219	5-23-05-00165	0.9813^{25}	H_2O 4; eth 4; EtOH 4	
7197	1H-Imidazole, 4,5-dihydro-2-methyl-	Lysidine	$C_4H_8N_2$	84.12	534-26-9	107	5508	196.5	5-23-03-00385		H_2O 4; EtOH 4; eth 1; chl 3	
7198	1H-Imidazole, 4,5-dihydro-2-(phenylmethyl)-	2-Imidazolidine, 2-benzyl	$C_{10}H_{12}N_2$	160.22	59-98-3		9430		5-23-06-00488			
7199	1H-Imidazole, 4,5-dihydro-2,4,5-triphenyl-, *cis*-		$C_{21}H_{18}N_2$	298.39	573-33-1	134.5		198 dec	5-23-10-00053		bz 4; eth 4; EtOH 4	
7200	1H-Imidazole, 1,2-dimethyl-		$C_5H_8N_2$	96.13	1739-84-0			206	5-23-05-00039	1.0051^{11}	H_2O 4; eth 4; EtOH 4	
7201	1H-Imidazole, 1,4-dimethyl-	1,4-Dimethylimidazole	$C_5H_8N_2$	96.13	6338-45-0	25.8		199	5-23-05-00094	0.9960^{16}		
7202	1H-Imidazole, 1,2-dimethyl-5-nitro-	Dimetridazole	$C_5H_7N_3O_2$	141.13	551-92-8	138.5	3255		5-23-05-00058		eth 4; EtOH 4	
7203	1H-Imidazole-4-ethanamine	Histamine	$C_5H_9N_3$	111.15	51-45-6	86	4640	209^{18}	4-25-00-02049		H_2O 3; EtOH 3; eth 2; chl 3	
7204	1H-Imidazole-4-ethanaminium, α-carboxy-2,3-dihydro-*N*,*N*,*N*-trimethyl-2-thioxo-, hydroxide, in	Ergothioneine	$C_9H_{15}N_3O_2S$	229.30	497-30-3	290 dec	3611		4-25-00-04487		H_2O 4; EtOH 2; eth 1; ace 2	
7205	1H-Imidazole-1-ethanol		$C_5H_8N_2O$	112.13	1615-14-1	39		203^{20}	5-23-04-00309		H_2O 4; EtOH 4	
7206	1H-Imidazole-1-ethanol, 2-methyl-5-nitro-	Metronidazole	$C_6H_9N_3O_3$	171.16	443-48-1	160.5	6079		5-23-05-00063			
7207	1H-Imidazole, 1-ethyl-		$C_5H_8N_2$	96.13	7098-07-9			208	5-23-04-00261	0.999^{25}	H_2O 5	
7208	1H-Imidazole, 1-methyl-		$C_4H_6N_2$	82.11	616-47-7	-6		195.5	5-23-04-00256	1.0325^{20}	H_2O 4; ace 4; eth 4; EtOH 4	1.4970^{20}
7209	1H-Imidazole, 2-methyl-		$C_4H_6N_2$	82.11	693-98-1	144		267	5-23-05-00035		H_2O 4; EtOH 4	
7210	1H-Imidazole, 4-methyl-		$C_4H_6N_2$	82.11	822-36-6	56		263	5-23-05-00089	1.0416^{14}	H_2O 4; EtOH 4	1.5037^{14}
7211	1H-Imidazole, 2-nitro-	Azomycin	$C_3H_3N_3O_2$	113.08	527-73-1	287 dec	934		5-23-04-00471			
7212	1H-Imidazole, 1-phenyl-		$C_9H_8N_2$	144.18	7164-98-9	13		276	5-23-04-00267	1.1397^{15}	H_2O 1; eth 4; ace 4; chl 4	1.6025^{25}
7213	1H-Imidazole, 2-phenyl-		$C_9H_8N_2$	144.18	670-96-2	149.3		340	5-23-07-00194		EtOH 4	
7214	2H-Imidazole-2-thione, 1,3-dihydro-1-methyl-	Methimazole	$C_4H_6N_2S$	114.17	60-56-0	146	5892	280 dec	5-24-01-00305		H_2O 4; EtOH 3; eth 2; bz 2	
7215	1H-Imidazole, 2,4,5-triphenyl-		$C_{21}H_{16}N_2$	296.37	484-47-9	275		sub	5-23-10-00107		H_2O 1; EtOH 3; eth 3	
7216	2,4-Imidazolidinedione	Hydantoin	$C_3H_4N_2O_2$	100.08	461-72-3	220	4678		5-24-05-00188		H_2O 4; EtOH 3; eth 2; alk 3	
7217	2,4-Imidazolidinedione, 1,3-dibromo-5,5-dimethyl-	Dibromantine	$C_5H_6Br_2N_2O_2$	285.92	77-48-5	197-9 dec			5-24-05-00374			
7218	2,4-Imidazolidinedione, 5,5-dimethyl-		$C_5H_8N_2O_2$	128.13	77-71-4	178		sub	5-24-05-00348		H_2O 4; EtOH 4; eth 4; ace 4	
7219	2,4-Imidazolidinedione, 5,5-diphenyl-	Phenytoin	$C_{15}H_{12}N_2O_2$	252.27	57-41-0	286	7293		5-24-08-00376		H_2O 1; EtOH 3; eth 2; ace 3	
7220	2,4-Imidazolidinedione, 5-(1-ethylpentyl)-3-[(trichloromethyl)thio]-	Chlordantoin	$C_{11}H_{17}Cl_3N_2O_2S$	347.69	5588-20-5		2080				CS_2 3	
7221	2,4-Imidazolidinedione, 3-methyl-5-phenyl-	3-Methyl-5-phenylhydantoin	$C_{10}H_{10}N_2O_2$	190.20	6846-11-3	164.5	6026		5-24-08-00005		chl 3	
7222	2,4-Imidazolidinedione, 1-[[(5-nitro-2-furanyl)methylene]amino]-	Nitrofurantoin	$C_8H_6N_4O_5$	238.16	67-20-9	263	6520		5-24-05-00224			
7223	4-Imidazolidinehexanoic acid, 5-methyl-2-oxo-, (4*R*-*cis*)-	Desthiobiotin	$C_{10}H_{18}N_2O_3$	214.26	533-48-2	157	2914		4-25-00-01543		H_2O 3	
7224	4-Imidazolidinehexanoic acid, 5-methyl-2-oxo-, methyl ester, (4*R*-*cis*)-	Desthiobiotin, methyl ester	$C_{11}H_{20}N_2O_3$	228.29	6020-51-5	69.5		$194\text{-}7^{0.03}$	4-25-00-01544		EtOH 4	
7225	2-Imidazolidinethione	Ethylenethiourea	$C_3H_6N_2S$	102.16	96-45-7	203	3759		5-24-01-00165		H_2O 4; EtOH 3; eth 1; bz 1	
7226	Imidazolidinetrione	Parabanic acid	$C_3H_2N_2O_3$	114.06	120-89-8	244 dec	6970	sub 100	5-24-09-00009		H_2O 3; EtOH 4	
7227	2-Imidazolidinone	Ethyleneurea	$C_3H_6N_2O$	86.09	120-93-4	131	4830		5-24-01-00022		H_2O 4; EtOH 4; eth 2; chl 2	
7228	4-Imidazolidinone, 5,5-diphenyl-	Doxenitoin	$C_{15}H_{14}N_2O$	238.29	3254-93-1	183	3424		5-24-04-00145			
7229	4-Imidazolidinone, 5-(2-methylpropyl)-3-(2-propenyl)-2-thioxo-	Albutoin	$C_{10}H_{16}N_2OS$	212.32	830-89-7	210.5	210		4-24-00-01139			

No.	Name Synonym	Mol. Form. Mol. Wt.	CAS RN mp/°C	Merck No. bp/°C	Beil. Ref. den/g cm^{-3}	Solubility n_D
7230	4H-Imidazol-4-one, 2-amino-1,5-dihydro-1-methyl- Creatinine	$C_4H_7N_3O$ 113.12	60-27-5 300 dec	2571	4-25-00-03543	H_2O 3; EtOH 2; eth 1; ace 1
7231	Imidodicarbonic acid, diethyl ester	$C_6H_{11}NO_4$ 161.16	19617-44-8 50	 226	3-03-00-00050	H_2O 4
7232	Imidodicarbonic diamide Biuret	$C_2H_5N_3O_2$ 103.08	108-19-0 190 dec	1319	4-03-00-00141	H_2O 2; EtOH 4; eth 1
7233	Imidodicarbonimidic diamide Biguanide	$C_2H_7N_5$ 101.11	56-03-1 136	1233 142 dec	4-03-00-00162	H_2O 4; EtOH 3; bz 1; chl 1
7234	Imidodicarbonimidic diamide, *N*-phenyl- Phenyl biguanide	$C_8H_{11}N_5$ 177.21	102-02-3 143	7247	4-12-00-00771	
7235	1H-Indazole 1H-Benzopyrazole	$C_7H_6N_2$ 118.14	271-44-3 148	4848 269	5-23-06-00156	H_2O 3; EtOH 3; eth 3
7236	1H-Indazole, 3-chloro- 3-Chloroindazole	$C_7H_5ClN_2$ 152.58	29110-74-5 149	 sub	5-23-06-00173	H_2O 3; EtOH 4; eth 4; bz 4
7237	1H-Indazole, 1,5-dimethyl- 1,5-Dimethylindazole	$C_9H_{10}N_2$ 146.19	70127-93-4 62.5	 260	2-23-00-00157	ace 4; eth 4; EtOH 4; MeOH 4
7238	2H-Indazole, 2,3-dimethyl- 2,3-Dimethylindazole	$C_9H_{10}N_2$ 146.19	50407-18-6 79.5	 286	5-23-06-00317	eth 4; EtOH 4
7239	1H-Indazole, 1-methyl- 1-Methylindazole	$C_8H_8N_2$ 132.17	13436-48-1 60.5	 231	5-23-06-00159 1.0315^{99}	H_2O 1; EtOH 3; ace 3; chl 3
7240	1H-Indazole, 3-methyl- 3-Methylindazole	$C_8H_8N_2$ 132.17	3176-62-3 113	 280.5	5-23-06-00312	H_2O 3; EtOH 3; eth 3; chl 3
7241	1H-Indazole, 5-methyl- 5-Methylindazole	$C_8H_8N_2$ 132.17	1776-37-0 117	 294	5-23-06-00319	H_2O 3; EtOH 3; eth 3; chl 3
7242	2H-Indazole, 2-methyl- 2-Methyl-2H-indazole	$C_8H_8N_2$ 132.17	4838-00-0 56	 261	5-23-06-00165 1.0450^{99}	H_2O 1; EtOH 3; eth 3; ace 3
7243	1H-Inden-1-amine, 2,3-dihydro-, (±)- (*RS*)-1-Aminoindan	$C_9H_{11}N$ 133.19	61949-83-5	 221; 96^{8}	3-12-00-02798 1.038^{15}	H_2O 2; eth 3; ace 3; bz 3 1.5613^{20}
7244	1H-Inden-5-amine, 2,3-dihydro-	$C_9H_{11}N$ 133.19	24425-40-9 37.5	 248; 131^{15}	4-12-00-02929	H_2O 2; eth 3; ace 3; bz 3
7245	1H-Indene Indonaphthene	C_9H_8 116.16	95-13-6 -1.8	4851 182	4-05-00-01532 0.9960^{25}	H_2O 1; EtOH 5; eth 5; ace 3 1.5768^{20}
7246	1H-Indene-1-carboxylic acid 1-Indenecarboxylic acid	$C_{10}H_8O_2$ 160.17	5020-21-3 71	 193.5^{12}	3-09-00-03068	H_2O 2; EtOH 4; eth 2; bz 2
7247	1H-Indene, 1,2-dibromo-2,3-dihydro- 1,2-Dibromoindan	$C_9H_8Br_2$ 275.97	20357-79-3 32	 140^{11}	2-05-00-00377 1.7470^{25}	
7248	2H-Indene-2,2-dicarboxylic acid, 1,3-dihydro-1,3-dioxo-, diethyl ester Phthaloylmalonic acid, diethyl ester	$C_{15}H_{14}O_6$ 290.27	116836-20-5		1-10-00-00441 1.1896^{83}	eth 4; EtOH 4 1.541^{84}
7249	1H-Indene, 1,2-dichloro-2,3-dihydro- 1,2-Dichloroindan	$C_9H_8Cl_2$ 187.07	74925-48-7	 $87\text{-}90^{2}$	3-05-00-01202 1.254^{25}	 1.5715^{23}
7250	1H-Indene, 2,3-dihydro- Indan	C_9H_{10} 118.18	496-11-7 -51.4	4844 177.9	4-05-00-01371 0.9639^{20}	H_2O 1; EtOH 5; eth 5; chl 2 1.5378^{20}
7251	1H-Indene, 2,3-dihydro-1,1-dimethyl- 1,1-Dimethylindan	$C_{11}H_{14}$ 146.23	4912-92-9	 191	4-05-00-01415 0.919^{20}	 1.5135^{25}
7252	1H-Indene, 2,3-dihydro-1,2-dimethyl- 1,2-Dimethylindan	$C_{11}H_{14}$ 146.23	17057-82-8	 $79\text{-}80^{10}$	4-05-00-01415 0.927^{20}	 1.5186^{20}
7253	1H-Indene, 2,3-dihydro-4,7-dimethyl- 4,7-Dimethylindan	$C_{11}H_{14}$ 146.23	6682-71-9	 $94\text{-}7^{10}$	4-05-00-01416 0.949^{20}	 1.5342^{20}
7254	1H-Indene, 2,3-dihydro-5,6-dimethyl- 5,6-Dimethylindan	$C_{11}H_{14}$ 146.23	1075-22-5	 94^{10}	4-05-00-01416 0.9449^{20}	 1.5360^{20}
7255	1H-Indene, 2,3-dihydro-1-methyl- 1-Methylindan	$C_{10}H_{12}$ 132.21	767-58-8	 190.6	4-05-00-01397 0.938^{25}	H_2O 1 1.5266^{20}
7256	1H-Indene, 2,3-dihydro-2-methyl- 2-Methylindan	$C_{10}H_{12}$ 132.21	824-63-5	 195	4-05-00-01397 0.940^{25}	H_2O 1 1.5220^{20}
7257	1H-Indene, 2,3-dihydro-4-methyl- 4-Methylindan	$C_{10}H_{12}$ 132.21	824-22-6	 205.3	4-05-00-01397 0.9577^{20}	 1.5356^{20}
7258	1H-Indene, 2,3-dihydro-5-methyl- 5-Methylindan	$C_{10}H_{12}$ 132.21	874-35-1	 202	4-05-00-01398 0.944^{25}	 1.5336^{20}
7259	1H-Indene, 2,3-dihydro-2-phenyl- 2-Phenylindan	$C_{15}H_{14}$ 194.28	22253-11-8	 $162\text{-}3^{10}$	4-05-00-02198 1.0821^{16}	 1.5955^{15}
7260	1H-Indene, 2,3-dihydro-1,1,4,7-tetramethyl- 1,1,4,7-Tetramethylindan	$C_{13}H_{18}$ 174.28	1078-04-2	 $114\text{-}2^{15}$	3-05-00-01282 0.934^{25}	 1.5216^{25}
7261	1H-Indene, 2,3-dihydro-1,1,5-trimethyl- 1,1,5-Trimethylindan	$C_{12}H_{16}$ 160.26	40650-41-7	 86^{10}	4-05-00-01433 0.9119^{20}	 1.5126^{20}
7262	1H-Indene, 2,3-dihydro-1,4,7-trimethyl- 1,4,7-Trimethylindan	$C_{12}H_{16}$ 160.26	54340-87-3	 95^{10}	4-05-00-01434 0.938^{20}	 1.5252^{20}
7263	1H-Indene, 2,3-dihydro-1,1,3-trimethyl-3-phenyl- Indan, 1-phenyl-1,3,3-trimethyl	$C_{18}H_{20}$ 236.36	3910-35-8 52.5	 308.5	4-05-00-02246 1.0009^{20}	H_2O 1; EtOH 3; bz 3; MeOH 3 1.5681^{20}
7264	1H-Indene-1,3(2H)-dione	$C_9H_6O_2$ 146.15	606-23-5 131-2 dec		4-07-00-02344 1.37^{21}	H_2O 2; EtOH 4; eth 3; bz 3
7265	1H-Indene-1,3(2H)-dione, 2-(4-bromophenyl)- Bromindione	$C_{15}H_9BrO_2$ 301.14	1146-98-1 138	1381	4-07-00-02572	
7266	1H-Indene-1,3(2H)-dione, 2-(4-chlorophenyl)- Clorindione	$C_{15}H_9ClO_2$ 256.69	1146-99-2 145.5	2402	4-07-00-02571	bz 4; eth 4; EtOH 4
7267	1H-Indene-1,3(2H)-dione, 2,2-dihydroxy- Ninhydrin	$C_9H_6O_4$ 178.14	485-47-2 241-3 dec	6470	4-07-00-02786	H_2O 4; EtOH 3; eth 2; alk 3

No.	Name	Synonym	Mol. Form.	Mol. Wt.	CAS RN	mp/°C	Merck No.	bp/°C	Beil. Ref.	den/g cm^{-3}	Solubility	n_D
7268	1H-Indene-1,3(2H)-dione, 2-(diphenylacetyl)-	Diphenadione	$C_{23}H_{16}O_3$	340.38	82-66-6	146.5	3304		4-07-00-02838		ace 4; HOAc 4	1.670
7269	1H-Indene-1,3(2H)-dione, 2-(4-methoxyphenyl)-	Anisindione	$C_{16}H_{12}O_3$	252.27	117-37-3	156.5	698		3-08-00-02931			
7270	1H-Indene-1,3(2H)-dione, 2-phenyl-	Phenindione	$C_{15}H_{10}O_2$	222.24	83-12-5	150	7196		4-07-00-02570		H_2O 1; EtOH 3; eth 3; ace 3	
7271	1H-Indene, 1,3-diphenyl-	1,3-Diphenylindene	$C_{21}H_{16}$	268.36	4467-88-3	85		230^{15}	3-05-00-02474		ace 4; eth 4	
7272	1H-Indene, 2,3-diphenyl-	2,3-Diphenylindene	$C_{21}H_{16}$	268.36	5324-00-5	109.3		237^{12}	4-05-00-02641		H_2O 1; EtOH 3; eth 3; ace 3	
7273	1H-Indene, 1-ethyl-2,3-dihydro-	1-Ethylindan	$C_{11}H_{14}$	146.23	4830-99-3			212	4-05-00-01414	0.9348^{25}		1.5121^{25}
7274	1H-Indene, 5-hexyl-2,3-dihydro-	5-Hexylindan	$C_{15}H_{22}$	202.34	54889-55-3	-34		292.1	4-05-00-01474	0.9114^{20}		1.5122^{20}
7275	1H-Indene, 1-methyl-	1-Methylindene	$C_{10}H_{10}$	130.19	767-59-9			199; 82^{15}	3-05-00-01371	0.970^{25}		1.5616^{20}
7276	1H-Indene, 2-methyl-	2-Methylindene	$C_{10}H_{10}$	130.19	2177-47-1	80		208	4-05-00-01545	0.974^{25}	H_2O 1; eth 3; ace 3; bz 3	1.5652^{20}
7277	1H-Indene, 3-methyl-	3-Methylindene	$C_{10}H_{10}$	130.19	767-60-2			198	4-05-00-01545	0.972^{25}	H_2O 1; eth 3; ace 3; bz 3	1.5621^{20}
7278	1H-Indene, 4-methyl-	4-Methylindene	$C_{10}H_{10}$	130.19	7344-34-5			209; 88^{13}	2-05-00-00418	0.989^{25}		1.568^{20}
7279	1H-Indene, 6-methyl-	6-Methylindene	$C_{10}H_{10}$	130.19	20232-11-5			207; 90^{17}	4-05-00-01545	0.977^{25}		1.566^{20}
7280	1H-Indene, 7-methyl-	7-Methylindene	$C_{10}H_{10}$	130.19	7372-92-1			209; 90^{15}	2-05-00-00418	0.989^{25}		1.568^{20}
7281	1H-Indene, octahydro-	Hexahydroindan	C_9H_{16}	124.23	496-10-6	-53		167	4-05-00-00292	0.876^{25}		1.4702^{20}
7282	6H-Indeno[1,2-b]oxirene, 1a,6a-dihydro-		C_9H_8O	132.16	768-22-9	24.5		113^{20}; 98^{6}	5-17-02-00033	1.1255^{24}	chl 3	
7283	1H-Inden-4-ol, 7-chloro-2,3-dihydro-	Chlorindanol	C_9H_9ClO	168.62	145-94-8	92	2094		4-06-00-03828			
7284	1H-Inden-1-ol, 2,3-dihydro-		$C_9H_{10}O$	134.18	6351-10-6	54.8		220; 128^{12}	4-06-00-03824		bz 4; EtOH 4; chl 4	
7285	1H-Inden-5-ol, 2,3-dihydro-		$C_9H_{10}O$	134.18	1470-94-6	58		253	4-06-00-03829		H_2O 2; EtOH 4; eth 4; sulf 3	
7286	1H-Inden-1-one, 2,3-dihydro-		C_9H_8O	132.16	83-33-0	42		243; 129^{12}	4-07-00-00999	1.0943^{40}	H_2O 2; EtOH 4; eth 4; ace 4	1.561^{25}
7287	2H-Inden-2-one, 1,3-dihydro-	2-Indanone	C_9H_8O	132.16	615-13-4	59		218 dec	4-07-00-01002	1.0712^{69}	H_2O 1; EtOH 4; eth 4; ace 4	1.538^{67}
7288	1H-Inden-1-one, 2,3-dihydro-3,3-dimethyl-		$C_{11}H_{12}O$	160.22	26465-81-6			$130\text{-}1^{18}$	4-07-00-01035	1.0320^{14}		1.5453^{14}
7289	1H-Inden-1-one, 2,3-dihydro-2-methyl-		$C_{10}H_{10}O$	146.19	17496-14-9			250; 125^{18}	4-07-00-01019	1.0651^{21}		1.553^{22}
7290	1H-Inden-1-one, 2,3-dihydro-4-methyl-	4-Methyl-1-indanone	$C_{10}H_{10}O$	146.19	24644-78-8	95		144^{13}; 120^{5}	4-07-00-01020		bz 4; eth 4; EtOH 4	
7291	1H-Inden-1-one, 2,3-dihydro-2-phenyl-	2-Phenyl-1-indanone	$C_{15}H_{12}O$	208.26	16619-12-8	77.5		344	4-07-00-01673		ace 4; bz 4; EtOH 4	
7292	2H-Inden-2-one, octahydro-, *cis*-		$C_9H_{14}O$	138.21	5689-04-3	10		225; 113^{25}	4-07-00-00174	0.9970^{20}	EtOH 3; bz 3; lig 3	1.4830^{20}
7293	2H-Inden-2-one, octahydro-, *trans*-		$C_9H_{14}O$	138.21	16484-17-6	-12		218	4-07-00-00174	0.9807^{17}	EtOH 3; bz 3; lig 3	1.4769^{17}
7294	1H-Indole	2,3-Benzopyrrole	C_8H_7N	117.15	120-72-9	52.5	4869	254	5-20-07-00005	1.22^{25}	H_2O 3; EtOH 4; eth 4; bz 3	
7295	1H-Indole-3-acetic acid	Indoleacetic acid	$C_{10}H_9NO_2$	175.19	87-51-4	168.5	4870		5-22-03-00065		H_2O 1; EtOH 4; eth 3; ace 3	
7296	1H-Indole-3-butanoic acid	Indolebutyric acid	$C_{12}H_{13}NO_2$	203.24	133-32-4	124.5	4871		5-22-03-00140		bz 4; DMSO 3; peth 1	
7297	1H-Indole, 2,3-dihydro-		C_8H_9N	119.17	496-15-1			229	5-20-06-00238	1.069^{20}	H_2O 2; eth 3; ace 3; bz 3	1.5923^{20}
7298	1H-Indole, 1,3-dimethyl-		$C_{10}H_{11}N$	145.20	875-30-9	142		258.5	5-20-07-00071		eth 3	
7299	1H-Indole, 2,5-dimethyl-	2,5-Dimethylindole	$C_{10}H_{11}N$	145.20	1196-79-8	115.3		188^{40}	5-20-07-00111		H_2O 1; EtOH 2; eth 3; bz 3	
7300	1H-Indole-2,3-dione	Isatin	$C_8H_5NO_2$	147.13	91-56-5	203 dec	4985		5-21-10-00221		H_2O 3; EtOH 4; eth 2; ace 3	
7301	1H-Indole-5,6-dione, 2,3-dihydro-3-hydroxy-1-methyl-	Adrenochrome	$C_9H_9NO_3$	179.18	54-06-8	125 dec	162		5-21-13-00079		H_2O 4; EtOH 4; eth 1; bz 1	
7302	1H-Indole-3-ethanamine	Tryptamine	$C_{10}H_{12}N_2$	160.22	61-54-1	118	9705	$137^{0.15}$	5-22-10-00045		H_2O 1; EtOH 3; eth 1; ace 3	
7303	1H-Indole-3-ethanamine, *N,N*-dimethyl-	*N,N*-Dimethyltryptamine	$C_{12}H_{16}N_2$	188.27	61-50-7	46	3251		5-22-10-00048			
7304	1H-Indole-3-ethanamine, 5-methoxy-	5-Methoxytryptamine	$C_{11}H_{14}N_2O$	190.25	608-07-1	121.5	5926		5-22-12-00018			
7305	1H-Indole-3-ethanol	Tryptophol	$C_{10}H_{11}NO$	161.20	526-55-6	59	9708	174^{2}	5-21-03-00061		ace 4; eth 4; EtOH 4; chl 4	
7306	1H-Indole-3-ethanol, 2-(5-ethenyl-1-azabicyclo[2.2.2]oct-2-yl)-	Cinchonamine	$C_{19}H_{24}N_2O$	296.41	482-28-0	186	2287		4-23-00-02735		H_2O 1; EtOH 4; eth 4; bz 3	

PHYSICAL CONSTANTS OF ORGANIC COMPOUNDS (continued)

No.	Name Synonym	Mol. Form. Mol. Wt.	CAS RN mp/°C	Merck No. bp/°C	Beil. Ref. den/g cm^{-3}	Solubility n_D
7307	1H-Indole, 1-ethyl-	$C_{10}H_{11}N$ 145.20	10604-59-8 105	252.5; 122^{14}	5-20-07-00014 1.2563^{15}	
7308	1H-Indole, 3-ethyl-	$C_{10}H_{11}N$ 145.20	1484-19-1 37	279; 138^{6}	5-20-07-00099	bz 4; eth 4; EtOH 4
7309	1H-Indole, 3-ethyl-2-methyl-	$C_{11}H_{13}N$ 159.23	35246-18-5 44.5	292; 156^{12}	5-20-07-00136 1.0414^{0}	eth 4; EtOH 4
7310	1H-Indole-3-methanamine, *N*,*N*-dimethyl- Gramine	$C_{11}H_{14}N_2$ 174.25	87-52-5 138.5	4440	5-22-10-00025	H_2O 1; EtOH 3; eth 3; chl 3
7311	1H-Indole, 7-methoxy-	C_9H_9NO 147.18	3189-22-8	$108\text{-}10^{0.3}$	5-21-03-00024 1.126^{25}	 1.6120^{20}
7312	1H-Indole, 1-methyl-	C_9H_9N 131.18	603-76-9	237	5-20-07-00012 1.0707^{25}	H_2O 1; EtOH 3; eth 3; bz 3
7313	1H-Indole, 2-methyl-	C_9H_9N 131.18	95-20-5 61	272	5-20-07-00059 1.07^{20}	H_2O 2; EtOH 4; eth 4; ace 3
7314	1H-Indole, 3-methyl- Skatole	C_9H_9N 131.18	83-34-1 97.5	8503 266	5-20-07-00069	H_2O 3; EtOH 3; eth 3; ace 3
7315	1H-Indole, 4-methyl-	C_9H_9N 131.18	16096-32-5 5	267	5-20-07-00079 1.062^{20}	
7316	1H-Indole, 5-methyl-	C_9H_9N 131.18	614-96-0 60	267	5-20-07-00080 1.0202^{78}	H_2O 3; EtOH 3; eth 3; bz 3
7317	1H-Indole, 7-methyl- 7-Methylindole	C_9H_9N 131.18	933-67-5 85	266	5-20-07-00081 1.0202^{100}	
7318	1H-Indole, 3-methyl-2-phenyl-	$C_{15}H_{13}N$ 207.27	10257-92-8 91.5	$280\text{-}90^{120}$	5-20-08-00274	H_2O 1; EtOH 3; bz 3
7319	1H-Indole, octahydro-	$C_8H_{15}N$ 125.21	4375-14-8 143	185.5	5-20-04-00358 0.9472^{20}	bz 4; eth 4; EtOH 4 1.4892^{20}
7320	1H-Indole, 2-phenyl-	$C_{14}H_{11}N$ 193.25	948-65-2 190.5	250^{10}	5-20-08-00232	H_2O 2; eth 3; bz 3; chl 3
7321	1H-Indole, 1-(phenylmethyl)-	$C_{15}H_{13}N$ 207.27	3377-71-7 45.8	172^{2}	5-20-07-00016	bz 4; eth 4; EtOH 4; lig 4
7322	1H-Indole, 1,2,3-trimethyl- 1,2,3-Trimethylindole	$C_{11}H_{13}N$ 159.23	1971-46-6 18	283	5-20-07-00102	H_2O 2; EtOH 2; eth 2; bz 2
7323	1H-Indole, 2,3,5-trimethyl- 2,3,5-Trimethylindole	$C_{11}H_{13}N$ 159.23	21296-92-4 121.5	297	5-20-07-00143	EtOH 4; chl 4; lig 4
7324	Indolizine	C_8H_7N 117.15	274-40-8 75	205	5-20-07-00047	H_2O 1; EtOH 3
7325	Indolizine, octahydro-	$C_8H_{15}N$ 125.21	13618-93-4	75^{43}	5-20-04-00360 0.9074^{10}	eth 4; EtOH 4 1.4748
7326	1H-Indol-5-ol, 3-[2-(dimethylamino)ethyl]- Bufotenine	$C_{12}H_{16}N_2O$ 204.27	487-93-4 146.5	1467 $320^{0.1}$	5-22-12-00026	eth 4; EtOH 4
7327	2H-Indol-2-one, 1,3-dihydro-	C_8H_7NO 133.15	59-48-3 128	227^{23}; 195^{17}	5-21-08-00007	H_2O 3; EtOH 3; eth 3
7328	2H-Indol-2-one, 1,3-dihydro-1,3-dimethyl-	$C_{10}H_{11}NO$ 161.20	24438-17-3 55	273; 137^{11}	5-21-08-00077	eth 4; EtOH 4
7329	2H-Indol-2-one, 1,3-dihydro-3,3-dimethyl-	$C_{10}H_{11}NO$ 161.20	19155-24-9 154.3	300	5-21-08-00113	bz 4; eth 4; EtOH 4
7330	2H-Indol-2-one, 1,3-dihydro-4,7-dimethyl- 4,7-Dimethyl-2-indolinone	$C_{10}H_{11}NO$ 161.20	59022-71-8 159	$205\text{-}10^{15}$	4-21-00-03676	ace 4; EtOH 4; chl 4
7331	3H-Indol-3-one, 2-(1,3-dihydro-3-oxo-2H-indol-2-ylidene)-1,2-dihydro- Indigo	$C_{16}H_{10}N_2O_2$ 262.27	482-89-3 390 dec	4855 sub 300	5-24-08-00503	
7332	2H-Indol-2-one, 1-ethyl-1,3-dihydro-3-methyl- 2-Indolinone, 1-ethyl-3-methyl-	$C_{11}H_{13}NO$ 175.23	84258-49-1	283; $105^{0.5}$	2-21-00-00258	eth 4; EtOH 4 1.557^{25}
7333	Indolo[2',3':3,4]pyrido[2,1-b]quinazolin-5(7H)-one, 8,13,13b,14-tetrahydro-14-methyl-, (*S*)- Evodiamine	$C_{19}H_{17}N_3O$ 303.36	518-17-2 28	3865	4-26-00-00525	
7334	Indolo[2',3':3,4]pyrido[2,1-b]quinazolin-5(7H)-one, 8,13-dihydro- Rutecarpine	$C_{18}H_{13}N_3O$ 287.32	84-26-4 259.5	8271	4-26-00-00528	EtOH 2; ace 2; bz 2
7335	Inosine Hypoxanthine riboside	$C_{10}H_{12}N_4O_5$ 268.23	58-63-9 218 dec	4880	4-26-00-02086	EtOH 4
7336	*scyllo*-Inositol	$C_6H_{12}O_6$ 180.16	488-59-5 353 dec		4-06-00-07920 1.659^{19}	H_2O 2; EtOH 1; eth 1; bz 1
7337	Iodine cyanide Cyanogen iodide	CIN 152.92	506-78-5 146.7	2702 sub 45	4-03-00-00093 2.84^{18}	eth 4; EtOH 4
7338	Iprodione 1-Imidazolidinecarboxamide, 3-(3,5-dichlorophenyl)-*N*-(1-methylethyl)-2,4-dioxo-	$C_{13}H_{13}Cl_2N_3O_3$ 330.17	36734-19-7 136	4964		
7339	Iron carbonyl Iron pentacarbonyl	C_5FeO_5 195.90	13463-40-6 -20	4980 103	 1.5^{20}	H_2O 1; EtOH 2; bz 3; ace 3 1.453^{22}
7340	Isazophos Phosphorothioic acid, *O*-[2-chloro-l-(l-methylethyl)-lH-imidazol-4-yl] *O*,*O*-diethyl ester	$C_9H_{17}ClN_3O_3PS$ 313.74	67329-04-8	4988 170; $100^{0.001}$	 1.22^{20}	
7341	5-Isobenzofurancarboxylic acid, 1,3-dihydro-1,3-dioxo- Trimellitic anhydride	$C_9H_4O_5$ 192.13	552-30-7 162	9617 $240\text{-}3^{14}$	5-18-08-00562	
7342	1,3-Isobenzofurandione Phthalic anhydride	$C_8H_4O_3$ 148.12	85-44-9 130.8	7346 295	5-17-11-00253 1.527^{4}	H_2O 2; EtOH 3; eth 2; ace 3

No.	Name Synonym	Mol. Form. Mol. Wt.	CAS RN mp/°C	Merck No. bp/°C	Beil. Ref. den/g cm^{-3}	Solubility n_D
7343	1,3-Isobenzofurandione, 4,5-dichloro- Phthalic anhydride, 3,4-dichloro	$C_8H_2Cl_2O_3$ 217.01	56962-07-3 121	 329	4-17-00-06142	EtOH 4; chl 4
7344	1,3-Isobenzofurandione, 5,6-dichloro- 4,5-Dichlorophthalic anhydride	$C_8H_2Cl_2O_3$ 217.01	942-06-3 188	 313	5-17-11-00260	eth 4; EtOH 4; tol 4
7345	1,3-Isobenzofurandione, hexahydro- Hexahydrophthalic anhydride	$C_8H_{10}O_3$ 154.17	85-42-7 32	 145^{18}	5-17-11-00076	
7346	1,3-Isobenzofurandione, 4,5,6,7-tetrachloro- Tetrachlorophthalic anhydride	$C_8Cl_4O_3$ 285.90	117-08-8 254.5	 sub	5-17-11-00260 1.49^{275}	eth 2
7347	1,3-Isobenzofurandione, 4,5,6,7-tetrahydro- 3,4,5,6-Tetrahydrophthalic anhydride	$C_8H_8O_3$ 152.15	2426-02-0 74		5-17-11-00133 1.2^{105}	EtOH 3; eth 4; ace 3; chl 3
7348	1,3-Isobenzofurandione, 4,5,6,7-tetraiodo-	$C_8I_4O_3$ 651.70	632-80-4 327.5	 sub	5-17-11-00266	H_2O 1; EtOH 1; bz 1; HOAc 2
7349	1(3H)-Isobenzofuranone	$C_8H_6O_2$ 134.13	87-41-2 75	 290	5-17-10-00007 1.1636^{99}	H_2O 3; EtOH 4; eth 4; chl 2 1.536^{99}
7350	1(3H)-Isobenzofuranone, 3,3-bis(3,5-dibromo-4-hydroxyphenyl)- 3',3'',5',5''-Tetrabromophenolphthalein	$C_{20}H_{10}Br_4O_4$ 633.91	76-62-0 296	9122	5-18-04-00190	H_2O 1; EtOH 2; eth 4; alk 3
7351	1(3H)-Isobenzofuranone, 3,3-bis(4-hydroxy-3,5-diiodophenyl)-	$C_{20}H_{10}I_4O_4$ 821.91	386-17-4 308		4-18-00-01949 2.0201^{22}	H_2O 1; EtOH 1; eth 1; chl 2
7352	1(3H)-Isobenzofuranone, 3,3-bis[4-hydroxy-2-methyl-5-(1-methylethyl)phenyl]- Thymolphthalein	$C_{28}H_{30}O_4$ 430.54	125-20-2 253	9336	5-18-04-00194	H_2O 1; EtOH 3; eth 3; ace 3
7353	1(3H)-Isobenzofuranone, 3,3-bis(4-hydroxy-3-methylphenyl)- *o*-Cresolphthalein	$C_{22}H_{18}O_4$ 346.38	596-27-0 223	2582	5-18-04-00193	EtOH 4
7354	1(3H)-Isobenzofuranone, 3,3-bis(4-hydroxyphenyl)- Phenolphthalein	$C_{20}H_{14}O_4$ 318.33	77-09-8 262.5	7208	5-18-04-00188 1.277^{32}	H_2O 1; EtOH 4; eth 3; ace 4
7355	1(3H)-Isobenzofuranone, 6,7-dimethoxy- Meconin	$C_{10}H_{10}O_4$ 194.19	569-31-3 102.5	5665	5-18-03-00068	H_2O 2; EtOH 3; eth 3; ace 3
7356	1(3H)-Isobenzofuranone, 6,7-dimethoxy-3-(5,6,7,8-tetrahydro-4-methoxy-6-methyl-1,3-dioxolo[4,5-g]isoquinolin-5 Noscapine	$C_{22}H_{23}NO_7$ 413.43	128-62-1 176	6638	4-27-00-06838	H_2O 1; EtOH 3; eth 2; ace 4
7357	1(3H)-Isobenzofuranone, 6,7-dimethoxy-3-(5,6,7,8-tetrahydro-6-methyl-1,3-dioxolo[4,5-g]isoquinolin-5-yl)-, [S- Hydrastine	$C_{21}H_{21}NO_6$ 383.40	118-08-1 132	4688	4-27-00-06833	H_2O 1; ace 3; bz 3
7358	1(3H)-Isobenzofuranone, 3,3-diphenyl-	$C_{20}H_{14}O_2$ 286.33	596-29-2 120	 235^{15}	5-17-10-00735	chl 2
7359	Isofenphos Benzoic acid, 2-[[ethoxy[(1-methylethyl)amino]phosphinothioyl]oxy]-, 1-methylethyl ester	$C_{15}H_{24}NO_4PS$ 345.40	25311-71-1 <-12	5055 $120^{0.01}$	 1.134^{20}	
7360	Isofulminic acid	CHNO 43.03	51060-05-0 stable only at low temp.			
7361	2H-Isoindole-2-acetic acid, 1,3-dihydro-1,3-dioxo-, ethyl ester	$C_{12}H_{11}NO_4$ 233.22	6974-10-3 112.5	 300	5-21-10-00431	bz 4; eth 4; EtOH 4; chl 4
7362	1H-Isoindole-1,3(2H)-dione Phthalimide	$C_8H_5NO_2$ 147.13	85-41-6 238	7347	5-21-10-00270	bz 4
7363	1H-Isoindole-1,3(2H)-dione, 2-(2,6-dioxo-3-piperidinyl)- Thalidomide	$C_{13}H_{10}N_2O_4$ 258.23	50-35-1 270	9182	5-22-13-00224	py 4; diox 4
7364	1H-Isoindole-1,3(2H)-dione, *N*-ethyl-	$C_{10}H_9NO_2$ 175.19	5022-29-7 79	 285.5	5-21-10-00274	EtOH 3; eth 3
7365	1H-Isoindole-1,3(2H)-dione, 2-methyl-	$C_9H_7NO_2$ 161.16	550-44-7 134	 286	5-21-10-00273	H_2O 1; EtOH 2
7366	1H-Isoindole-1,3(2H)-dione, 2-phenyl-	$C_{14}H_9NO_2$ 223.23	520-03-6 210	 sub	5-21-10-00292	H_2O 1; EtOH 2; chl 5
7367	1H-Isoindole-1,3(2H)-dione, 2-(phenylmethyl)-	$C_{15}H_{11}NO_2$ 237.26	2142-01-0 116		5-21-10-00299 1.343^{18}	EtOH 3; DMSO-d_6 2; HOAc 3
7368	1H-Isoindole-1,3(2H)-dione, 3a,4,7,7a-tetrahydro-2-[(trichloromethyl)thio]- Captan	$C_9H_8Cl_3NO_2S$ 300.59	133-06-2 172.5	1771	5-21-10-00136 1.74^{25}	chl 4
7369	1H-Isoindol-1-one, 2,3-dihydro-	C_8H_7NO 133.15	480-91-1 151	 338; 103^{18}	5-21-08-00018	eth 4; EtOH 4; chl 4
7370	*L*-Isoleucine 2-Amino-3-methylpentanoic acid	$C_6H_{13}NO_2$ 131.17	73-32-5 285-6 dec	5064	4-04-00-02775	H_2O 3; EtOH 1
7371	Isopropalin Benzenamine, 4-(1-methylethyl)-2,6-dinitro-*N,N*-dipropyl-	$C_{15}H_{23}N_3O_4$ 309.37	33820-53-0	5089		
7372	5-Isoquinolinamine 5-Aminoisoquinoline	$C_9H_8N_2$ 144.18	1125-60-6 128	 196^{18}	5-22-10-00342	chl 2; lig 3
7373	Isoquinoline Benzopyridine	C_9H_7N 129.16	119-65-3 26.47	5110 243.2	5-20-07-00333 1.0910^{30}	H_2O 1; EtOH 4; eth 5; ace 4 1.6148^{20}
7374	Isoquinoline, 4-bromo- 4-Bromoisoquinoline	C_9H_6BrN 208.06	1532-97-4 41.5	 282.5	5-20-07-00363	eth 4
7375	5-Isoquinolinecarbonitrile	$C_{10}H_6N_2$ 154.17	27655-41-0 137	 sub 100	4-22-00-01208	eth 4; EtOH 4

No.	Name Synonym	Mol. Form. Mol. Wt.	CAS RN mp/°C	Merck No. bp/°C	Beil. Ref. den/g cm^{-3}	Solubility n_D
7376	Isoquinoline, 1-chloro- 1-Chloroisoquinoline	C_9H_6ClN 163.61	19493-44-8 37.5	 274.5	5-20-07-00361	H_2O 1; bz 3
7377	Isoquinoline, 1-[(3,4-diethoxyphenyl)methyl]-6,7-diethoxy- Isoquinoline, 6,7-diethoxy-1(3,4-diethoxybenzyl)	$C_{24}H_{29}NO_4$ 395.50	486-47-5 100	3682	5-21-06-00183	H_2O 1; EtOH 3; eth 2; chl 2
7378	Isoquinoline, 1-[(3,4-dimethoxyphenyl)methyl]-6,7-dimethoxy- Papaverine	$C_{20}H_{21}NO_4$ 339.39	58-74-2 147.5	6968 sub 135	5-21-06-00182 1.337^{20}	H_2O 2; EtOH 4; eth 2; ace 3 1.625
7379	Isoquinoline, 1-[(3,4-dimethoxyphenyl)methyl]-1,2,3,4-tetrahydro-6,7-dimethoxy-2-methyl-, (*S*)- Laudanosine	$C_{21}H_{27}NO_4$ 357.45	2688-77-9 89	5249	5-21-06-00049	ace 4; eth 4; EtOH 4; chl 4
7380	Isoquinoline, 7-methoxy- 7-Methoxyisoquinoline	$C_{10}H_9NO$ 159.19	39989-39-4 49	 194^{50}; 184^{34}	5-21-03-00326	H_2O 1; EtOH 3; lig 3
7381	Isoquinoline, 1-methyl- Isoquinaldine	$C_{10}H_9N$ 143.19	1721-93-3 10	 248	5-20-07-00407 1.0777^{20}	H_2O 2; eth 3; ace 3; bz 3 1.6095^{20}
7382	Isoquinoline, 3-methyl- 3-Methylisoquinoline	$C_{10}H_9N$ 143.19	1125-80-0 68	 249	5-20-07-00410	H_2O 2; eth 3; ace 3; chl 2
7383	Isoquinoline, 4-methyl- 4-Methylisoquinoline	$C_{10}H_9N$ 143.19	1196-39-0 <-75	 256	5-20-07-00413	H_2O 2; eth 3; ace 3; bz 3
7384	Isoquinoline, 6-methyl- 6-Methylisoquinoline	$C_{10}H_9N$ 143.19	42398-73-2 85.5	 265.5	5-20-07-00414	ace 4; bz 4; eth 4; EtOH 4
7385	Isoquinoline, 8-methyl- 8-Methylisoquinoline	$C_{10}H_9N$ 143.19	62882-00-2 205	 258	5-20-07-00414	ace 4; bz 4; eth 4
7386	Isoquinoline, 5-nitro- 5-Nitroisoquinoline	$C_9H_6N_2O_2$ 174.16	607-32-9 110	 sub	5-20-07-00365	H_2O 2; EtOH 4; eth 4; chl 4
7387	Isoquinoline, 3-(phenylmethyl)- Isoquinoline, 3-benzyl	$C_{16}H_{13}N$ 219.29	90210-56-3 104	 311^{23}	5-20-08-00383	bz 4; chl 4
7388	Isoquinoline, 1,2,3,4-tetrahydro-	$C_9H_{11}N$ 133.19	91-21-4 <-15	 232.5	5-20-06-00320 1.0642^{24}	H_2O 1; EtOH 3; bz 3; acid 3 1.5668^{20}
7389	Isoquinoline, 5,6,7,8-tetrahydro- 5,6,7,8-Tetrahydroisoquinoline	$C_9H_{11}N$ 133.19	36556-06-6 146.5	 218; 106^{13}	5-20-06-00319 1.0504^{10}	H_2O 1; ace 3; bz 3; chl 3 1.5276^{10}
7390	Isoquinoline, 1,2,3,4-tetrahydro-6,7-dimethoxy-1,2-dimethyl-, (±)- Carnegine	$C_{13}H_{19}NO_2$ 221.30	490-53-9	1854 170^1	5-21-04-00501	H_2O 4; eth 4; EtOH 4
7391	Isoquinoline, 1,2,3,4-tetrahydro-6,7-dimethoxy-1,2-dimethyl- Isoquinoline, 6,7-dimethoxy-1,2-dimethyl-1,2,3,4-tetrahydro	$C_{13}H_{19}NO_2$ 221.30	71783-56-7 262.5	 170^1; $100^{0.01}$	5-21-04-00501	H_2O 3; EtOH 3; eth 3; chl 3
7392	7-Isoquinolinol, 1-[(3,4-dimethoxyphenyl)methyl]-1,2,3,4-tetrahydro-6-methoxy-2-methyl-, (*S*)- Codamine	$C_{20}H_{25}NO_4$ 343.42	21040-59-5 127	2458	4-21-00-02704	eth 4; EtOH 4; chl 4
7393	8-Isoquinolinol, 1,2,3,4-tetrahydro-6,7-dimethoxy- Anhalamine	$C_{11}H_{15}NO_3$ 209.25	643-60-7 187.5	682	5-21-05-00513	eth 4; EtOH 4
7394	8-Isoquinolinol, 1,2,3,4-tetrahydro-6,7-dimethoxy-1,2-dimethyl-, (*S*)- Pellotine	$C_{13}H_{19}NO_3$ 237.30	83-14-7 111.5	7018	4-21-00-02525	ace 4; eth 4; EtOH 4; peth 4
7395	8-Isoquinolinol, 1,2,3,4-tetrahydro-6,7-dimethoxy-1-methyl-, (*S*)- Anhalonidine	$C_{12}H_{17}NO_3$ 223.27	17627-77-9 160.5	683	4-21-00-02524	H_2O 4; EtOH 4
7396	7-Isoquinolinol, 1,2,3,4-tetrahydro-1-[(4-hydroxyphenyl)methyl]-6-methoxy-, (*S*)- Coclaurine	$C_{17}H_{19}NO_3$ 285.34	486-39-5 220.5	2455	4-21-00-02605	
7397	6-Isoquinolinol, 1,2,3,4-tetrahydro-7-methoxy-1-methyl-, (*S*)- Salsoline	$C_{11}H_{15}NO_2$ 193.25	89-31-6 221.5	8308	5-21-04-00499	H_2O 2; EtOH 2; eth 1; bz 2
7398	1(2H)-Isoquinolinone, 3,4-dihydro-6,7-dimethoxy- Corydaldine	$C_{11}H_{13}NO_3$ 207.23	493-49-2 175	2542	5-21-13-00086	H_2O 4; bz 4; eth 4; EtOH 4
7399	Isothiocyanic acid	CHNS 59.09	3129-90-6			
7400	*DL*-Isovaline	$C_5H_{11}NO_2$ 117.15	595-39-1 316.5	 sub 300	4-04-00-02655	H_2O 3; EtOH 3; eth 2; bz 1
7401	*L*-Isovaline 2-Amino-2-methylbutyric acid	$C_5H_{11}NO_2$ 117.15	595-40-4	5124	4-04-00-02655	
7402	Isoxazole 1-Oxa-2-azacyclopentadiene	C_3H_3NO 69.06	288-14-2	 95	4-27-00-00959 1.078^{20}	H_2O 3 1.4298^{17}
7403	Isoxazole, 3,5-dimethyl- 3,5-Dimethylisoxazole	C_5H_7NO 97.12	300-87-8	 143	4-27-00-00978 0.99^{25}	 1.4421^{20}
7404	5-Isoxazolemethanol, 3-methyl-	$C_5H_7NO_2$ 113.12	14716-89-3	 130^{20}	4-27-00-01745 1.132^{25}	 1.4790^{20}
7405	Isoxazole, 5-methyl- 5-Methylisoxazole	C_4H_5NO 83.09	5765-44-6	 122	4-27-00-00966 1.023^{20}	DMSO 3 1.4386^{20}
7406	3-Isoxazolidinone, 4-amino-, (*R*)- Cycloserine	$C_3H_6N_2O_2$ 102.09	68-41-7 155-6 dec	2758	4-27-00-05549	H_2O 3; MeOH 2
7407	Isoxazolium, 2-ethyl-5-(3-sulfophenyl)-, hydroxide, inner salt Woodward's Reagent K	$C_{11}H_{11}NO_4S$ 253.28	4156-16-5 dec 207	9962		

No.	Name Synonym	Mol. Form. Mol. Wt.	CAS RN mp/°C	Merck No. bp/°C	Beil. Ref. den/g cm^{-3}	Solubility n_D
7408	Kepone 1,3,4-Metheno-2H-cyclobuta[cd]pentalen-2-one, 1,1a,3,3a,4,5,5,5a,5b,6-decachloro-octahydro-	$C_{10}Cl_{10}O$ 490.64	143-50-0 350 dec	2081	 1.61^{25}	
7409	Ketone, 1-methyl-4-piperidyl phenyl Piperidine, 4-benzoyl-*N*-methyl	$C_{13}H_{17}NO$ 203.28	92040-00-1 36	 190^{21}; 133^{2}	4-21-00-03689	H_2O 3; EtOH 3; eth 4; ace 3 1.5430^{23}
7410	Lactofen (±)-2-Ethoxy-1-ethyl-2-oxoethyl 5-[2-chloro-4-(trifluoromethyl)phenoxy]-2-nitrobenzoate	$C_{19}H_{15}ClF_3NO_7$ 461.78	77501-63-4			
7411	Lanosta-8,24-dien-3-ol, (3β)- Lanosterol	$C_{30}H_{50}O$ 426.73	79-63-0 140.5	5232	4-06-00-04188	eth 4; EtOH 4; chl 4
7412	*DL*-Leucine	$C_6H_{13}NO_2$ 131.17	328-39-2 293	 sub	2-04-00-00870 1.293^{18}	H_2O 3; EtOH 2; eth 1
7413	*L*-Leucine 2-Amino-4-methylpentanoic acid	$C_6H_{13}NO_2$ 131.17	61-90-5 293	5331 sub	4-04-00-02738 1.293^{18}	H_2O 2; EtOH 1; eth 1
7414	*DL*-Leucine, *N*-glycyl-	$C_8H_{16}N_2O_3$ 188.23	688-14-2 242 dec		4-04-00-02766 1.181^{25}	H_2O 4; EtOH 1
7415	Linuron Urea, *N*'-(3,4-dichlorophenyl)-*N*-methoxy-*N*-methyl-	$C_9H_{10}Cl_2N_2O_2$ 249.10	330-55-2 93	5387		
7416	Lup-20(29)-ene-3,28-diol, (3β)- Betulin	$C_{30}H_{50}O_2$ 442.73	473-98-3 250	1212 sub 240	4-06-00-06534	H_2O 1; EtOH 2; eth 3; bz 2
7417	Lup-20(29)-en-3-ol, (3β)- Lupeol	$C_{30}H_{50}O$ 426.73	545-47-1 216	5478	4-06-00-04200 0.9457^{218}	H_2O 1; EtOH 4; eth 4; ace 4 1.4910^{218}
7418	*DL*-Lysine 2,6-Diaminohexanoic acid	$C_6H_{14}N_2O_2$ 146.19	70-54-2 224		2-04-00-00858	H_2O 2
7419	*L*-Lysine 2,6-Diaminohexanoic acid	$C_6H_{14}N_2O_2$ 146.19	56-87-1 224-5 dec	5509	4-04-00-02717	H_2O 4; EtOH 1; eth 1; ace 1
7420	Maneb Manganese, [[1,2-ethanediylbis[carbamodithioato]](2-)]-	$C_4H_6MnN_2S_4$ 265.31	12427-38-2 dec 200	5603		
7421	Manganous dimethyldithiocarbamate Carbamic acid, dimethyldithio-, manganese(2+) salt	$C_6H_{12}MnN_2S_4$ 295.38	15339-36-3			
7422	*D*-Mannitol Cordycepic acid	$C_6H_{14}O_6$ 182.17	69-65-8 168	5629 $295^{3.5}$	4-01-00-02841 1.489^{20}	H_2O 4; EtOH 2; eth 1; py 2 1.3330
7423	*D*-Mannitol, 1,4:3,6-dianhydro-	$C_6H_{10}O_4$ 146.14	641-74-7 88	 274 dec	5-19-03-00201	H_2O 4
7424	*D*-Mannitol, hexanitrate Mannitol, hexanitrate	$C_6H_8N_6O_{18}$ 452.16	15825-70-4 107	5630 exp	4-01-00-02849 1.8^{20}	bz 4; eth 4; EtOH 4
7425	*D*-Mannose Seminose	$C_6H_{12}O_6$ 180.16	3458-28-4 132 dec	5632	4-01-00-04328 1.539^{20}	H_2O 4; EtOH 2; eth 1; bz 1
7426	Matridin-15-one Matrine	$C_{15}H_{24}N_2O$ 248.37	519-02-8	5641 223^{6}	5-24-02-00301	H_2O 3; EtOH 4; eth 3; ace 3 1.5286^{25}
7427	Mefluidide Acetamide, *N*-[2,4-dimethyl-5-[[(trifluoromethyl)sulfonyl]amino]phenyl]-	$C_{11}H_{13}F_3N_2O_3S$ 310.30	53780-34-0 184	5684		
7428	*m*-Menthan-6-ol Cyclohexanol-4-isopropyl-2-methyl	$C_{10}H_{20}O$ 156.27	1490-05-7	 119.21^{28}	2-06-00-00038 0.9140^{20}	eth 4; EtOH 4 1.4666^{20}
7429	*p*-Menth-3-ene, (*S*)-(-)- Cyclohexene, 1-isopropyl-4-methyl (*l*)	$C_{10}H_{18}$ 138.25	22564-83-6	 167	4-05-00-00301 0.8099^{19}	H_2O 1; EtOH 3; eth 3; bz 3 1.4511^{20}
7430	Mepiquat chloride Piperidinium, 1,1-dimethyl-, chloride	$C_7H_{16}NCl$ 149.66	24307-26-4 223	5746		
7431	Mercury, (acetato-*O*)phenyl- Phenylmercuric acetate	$C_8H_8HgO_2$ 336.74	62-38-4 153	7271	4-16-00-01720	chl 3
7432	Mercury, bis(4-methylphenyl)- Bis(*p*-tolyl)mercury	$C_{14}H_{14}Hg$ 382.86	537-64-4 245.7	3387	4-16-00-01705	
7433	Mercury, dibutyl- Dibutylmercury	$C_8H_{18}Hg$ 314.82	629-35-6	 105^{10}	4-04-00-04429 1.7779^{20}	 1.5057^{20}
7434	Mercury, diethyl- Diethyl mercury	$C_4H_{10}Hg$ 258.71	627-44-1	 159; 57^{16}	 2.43^{20}	eth 3; EtOH 2
7435	Mercury, dimethyl- Dimethyl mercury	C_2H_6Hg 230.66	593-74-8	3238 93	 3.17^{25}	H_2O 1; EtOH 4; eth 4 1.5452^{20}
7436	Mercury, diphenyl- Benzene, mercuriodi-	$C_{12}H_{10}Hg$ 354.80	587-85-9	 204^{10}	4-16-00-01702 2.318^{25}	H_2O 1; EtOH 2; eth 2; bz 3
7437	Mercury, [2-methyl-5-nitrophenolato(2-)-C6,O1]- Nitromersol	$C_7H_5HgNO_3$ 351.71	133-58-4	6531		H_2O 1; ace 2; EtOH 2; alk 3
7438	Merphos Phosphorotrithious acid, *S,S,S*-tributyl ester	$C_{12}H_{27}PS_3$ 298.52	150-50-5 100	 $137^{0.7}$; 176^{15}	 1.02^{20}	
7439	Metalaxyl *N*-(2,6-Dimethylphenyl)-*N*-(methoxyacetyl)alanine, methyl ester	$C_{15}H_{21}NO_4$ 279.34	57837-19-1 71	5826		
7440	Metaldehyde Acetaldehyde polymer	$(C_2H_4O)_x$	37273-91-9 246	5827 sub 115	4-19-00-05643	H_2O 1; EtOH 2; eth 2; ace 1
7441	Methamidophos Phosphoramidothioic acid, *O,S*-dimethyl ester	$C_2H_8NO_2PS$ 141.13	10265-92-6 46	5858	 1.31^{20}	

No.	Name Synonym	Mol. Form. Mol. Wt.	CAS RN mp/°C	Merck No. bp/°C	Beil. Ref. den/g cm^{-3}	Solubility n_D
7442	Methanamine Methylamine	CH_5N 31.06	74-89-5 -93.4	5938 -6.3	4-04-00-00118 *0.656^{25}	H_2O 4; EtOH 3; eth 5; ace 3
7443	Methanamine, 1,1-diethoxy-*N*,*N*-dimethyl-	$C_7H_{17}NO_2$ 147.22	1188-33-6	 129	4-04-00-00174 0.859^{25}	 1.4007^{20}
7444	Methanamine, *N*,*N*-dimethyl- Trimethylamine	C_3H_9N 59.11	75-50-3 -117.1	9625 2.8	4-04-00-00134 *0.627^{25}	H_2O 4; EtOH 3; eth 3; bz 3 1.3631^{0}
7445	Methanamine, *N*,*N*-dimethyl-, hydrochloride	$C_3H_{10}ClN$ 95.57	593-81-7 277.5	 sub 200	4-04-00-00134	H_2O 4; EtOH 4; chl 4
7446	Methanamine, hydrochloride	CH_6ClN 67.52	593-51-1 227.5	5938 227^{15}	4-04-00-00118	H_2O 4; EtOH 4
7447	Methanamine, *N*-hydroxy- Hydroxylamine, *N*-methyl	CH_5NO 47.06	593-77-1 87.5	 62.5^{15}	4-04-00-03299 1.0003^{20}	H_2O 4; EtOH 4 1.4164^{20}
7448	Methanamine, *N*-methyl- Dimethylamine	C_2H_7N 45.08	124-40-3 -92.2	3217 6.8	4-04-00-00128 0.6804^{0}	H_2O 4; EtOH 3; eth 3 1.350^{17}
7449	Methanamine, *N*-methyl-*N*-nitro-	$C_2H_6N_2O_2$ 90.08	4164-28-7 58	 187	4-04-00-03401 1.1090^{72}	H_2O 4; ace 4; eth 4; EtOH 4 1.4462^{72}
7450	Methanamine, *N*-methyl-*N*-nitroso- *N*-Nitrosodimethylamine	$C_2H_6N_2O$ 74.08	62-75-9	6558 154	4-04-00-03384 1.0059^{20}	H_2O 3; EtOH 3; eth 3; chl 3 1.4358^{20}
7451	Methanamine, *N*-nitro-	$CH_4N_2O_2$ 76.06	598-57-2 38	 80-5^{10}	4-04-00-03400 1.2433^{49}	H_2O 4; EtOH 4; eth 3; bz 4 1.4616^{49}
7452	Methanamine, *N*-(phenylmethylene)-	C_8H_9N 119.17	622-29-7	 185; 92^{34}	4-07-00-00518 0.9671^{14}	EtOH 3; eth 3; ace 3; chl 3 1.5526^{20}
7453	Methanaminium, *N*-[4-[bis[4-(dimethylamino)phenyl]methylene]-2,5-cyclohexadien-1-ylidene]-*N*-methyl-, chloride Crystal violet	$C_{25}H_{30}ClN_3$ 407.99	548-62-9 215 dec	4287		H_2O 4; chl 4
7454	Methanaminium, 1-carboxy-*N*,*N*,*N*-trimethyl-, hydroxide, inner salt Betaine	$C_5H_{11}NO_2$ 117.15	107-43-7 293 dec	1201	4-04-00-02369	H_2O 4; EtOH 3; eth 2; chl 2
7455	Methanaminium, *N*-[4-[[4-(dimethylamino)phenyl]phenylmethylene]-2,5-cyclohexadien-1-ylidene]-*N*-methyl-, chlorid Malachite green	$C_{23}H_{25}ClN_2$ 364.92	569-64-2	5581		H_2O 4; EtOH 4; MeOH 4
7456	Methanaminium, *N*,*N*,*N*-trimethyl-, bromide Tetramethylammonium bromide	$C_4H_{12}BrN$ 154.05	64-20-0 230 dec		4-04-00-00145 1.56^{25}	H_2O 4; EtOH 2; eth 1; bz 1
7457	Methanaminium, *N*,*N*,*N*-trimethyl-, chloride Tetramethylammonium chloride	$C_4H_{12}ClN$ 109.60	75-57-0 420 dec		4-04-00-00145 1.169^{20}	H_2O 3; EtOH 2; eth 1; bz 1
7458	Methanaminium, *N*,*N*,*N*-trimethyl-, iodide Tetramethylammonium iodide	$C_4H_{12}IN$ 201.05	75-58-1 >230 dec	9156	4-04-00-00145 1.829^{25}	H_2O 2; EtOH 2; eth 1; ace 2
7459	Methane	CH_4 16.04	74-82-8 -182.4	5863 -161.5	4-01-00-00003 0.4228^{-162}	H_2O 3; EtOH 3; eth 3; ace 2
7460	Methane, azido Methyl azide	CH_3N_3 57.06	624-90-8	 20.5; exp 500	 0.869^{15}	
7461	Methane, bis(methylthio)- Bis[methylmercapto]methane	$C_3H_8S_2$ 108.23	1618-26-4	1267 148	4-01-00-03088	
7462	Methane, bromo- Methyl bromide	CH_3Br 94.94	74-83-9 -93.7	5951 3.5	4-01-00-00068 1.6755^{20}	H_2O 2; EtOH 5; eth 5; chl 5 1.4218^{20}
7463	Methane, bromochloro- Bromochloromethane	CH_2BrCl 129.38	74-97-5 -87.9	 68.0	4-01-00-00074 1.9344^{20}	H_2O 1; EtOH 3; eth 3; ace 3 1.4838^{20}
7464	Methane, bromochlorodifluoro- Bromochlorodifluoromethane	$CBrClF_2$ 165.36	353-59-3 -159.5	 -3.7	4-01-00-00075	
7465	Methane, bromochlorodinitro- Bromochlorodinitromethane	$CBrClN_2O_4$ 219.38	33829-48-0 9.3	 75-6^{15}	2-01-00-00044 2.0393^{20}	EtOH 4 1.4793
7466	Methane, bromochlorofluoro- Bromochlorofluoromethane	$CHBrClF$ 147.37	593-98-6 -115	 36	4-01-00-00075 1.9771^{0}	H_2O 1; eth 3; ace 3; chl 3 1.4144^{25}
7467	Methane, bromodichloro- Bromodichloromethane	$CHBrCl_2$ 163.83	75-27-4 -57	 90	4-01-00-00076 1.980^{20}	H_2O 1; EtOH 4; eth 4; ace 4 1.4964^{20}
7468	Methane, bromodichlorofluoro- Bromodichlorofluoromethane	$CBrCl_2F$ 181.82	353-58-2		4-01-00-00076	
7469	Methane, bromodifluoro- Bromodifluoromethane	$CHBrF_2$ 130.92	1511-62-2 -14.5		4-01-00-00072 1.55^{16}	H_2O 3; EtOH 4
7470	Methane, bromofluoro- Bromofluoromethane	CH_2BrF 112.93	373-52-4	 19	4-01-00-00072	EtOH 3; chl 4
7471	Methane, bromoiodo- Bromoiodomethane	CH_2BrI 220.84	[illegible]	 139.5	[illegible] 2.926^{17}	chl 1 1.6410^{20}
7472	Methane, bromomethoxy-	C_2H_5BrO 124.97	13057-17-5	 87	4-01-00-03051 1.5976^{20}	 1.4562^{20}
7473	Methane, bromonitro- Bromonitromethane	CH_2BrNO_2 139.94	563-70-2	 149; 71^{40}	4-01-00-00106	EtOH 4 1.4880^{20}
7474	Methane, bromotrichloro- Bromotrichloromethane	$CBrCl_3$ 198.27	75-62-7 -5.7	 105	4-01-00-00077 2.012^{25}	eth 4; EtOH 4 1.5065^{20}
7475	Methane, bromotrifluoro- Bromotrifluoromethane	$CBrF_3$ 148.91	75-63-8 -172	 -57.8	4-01-00-00073 1.5800^{20}	chl 4
7476	Methane, bromotrinitro- Bromotrinitromethane	$CBrN_3O_6$ 229.93	560-95-2 17.5	 56^{10}	3-01-00-00116 2.0312^{20}	EtOH 4; chl 4 1.4808^{20}
7477	Methane, chloro- Methyl chloride	CH_3Cl 50.49	74-87-3 -97.7	5964 -24.0	4-01-00-00028 *0.911^{25}	H_2O 3; EtOH 3; eth 5; ace 5 1.3389^{20}
7478	Methane, chlorodifluoro- Chlorodifluoromethane	$CHClF_2$ 86.47	75-45-6 -157.4	 -40.7	4-01-00-00032 1.4909^{-69}	H_2O 4; eth 3; ace 3; chl 3

No.	Name Synonym	Mol. Form. Mol. Wt.	CAS RN mp/°C	Merck No. bp/°C	Beil. Ref. den/g cm^{-3}	Solubility n_D
7479	Methane, chlorodiiodo-	$CHClI_2$ 302.28	638-73-3 -4	 200 dec; 101^{33}	4-01-00-00097 3.1700^{0}	ace 4; eth 4; chl 4
7480	Methane, chlorodinitro- Chlorodinitromethane	$CHClN_2O_4$ 140.48	921-13-1	 $34\text{-}6^{13}$	3-01-00-00115 1.6125^{20}	 1.4575^{20}
7481	Methane, chlorofluoro- Chlorofluoromethane	CH_2ClF 68.48	593-70-4 -133	 -9.1	4-01-00-00032	chl 4
7482	Methane, chloroiodo- Chloroiodomethane	CH_2ClI 176.38	593-71-5	 109	4-01-00-00094 2.422^{20}	ace 4; bz 4; eth 4; EtOH 4 1.5822^{20}
7483	Methane, chloromethoxy- Chloromethyl methyl ether	C_2H_5ClO 80.51	107-30-2 -103.5	2146 59.5	4-01-00-03046 1.063^{10}	EtOH 3; eth 3; ace 3; chl 3 1.397^{20}
7484	Methane, chloro(methylthio)-	C_2H_5ClS 96.58	2373-51-5	 105	4-01-00-03079 1.153^{25}	 1.4963^{20}
7485	Methane, chloronitro- Chloronitromethane	CH_2ClNO_2 95.49	1794-84-9	 122.5	4-01-00-00106 1.466^{15}	H_2O 4
7486	Methane, chlorotrifluoro- Chlorotrifluoromethane	$CClF_3$ 104.46	75-72-9 -181	 -81.4	4-01-00-00034	
7487	Methane, chlorotrinitro-	$CClN_3O_6$ 185.48	1943-16-4 4.5	 134 dec; 56^{40}	3-01-00-00116 1.6769^{20}	eth 4; EtOH 4; chl 4 1.4500^{20}
7488	Methanediamine, 1-(2-furanyl)-*N,N'*-bis(2-furanylmethylene)- Hydrofuramide	$C_{15}H_{12}N_2O_3$ 268.27	494-47-3 117	4718	5-17-09-00306	H_2O 1; EtOH 4; eth 4
7489	Methanediamine, *N,N,N',N'*-tetraethyl-	$C_9H_{22}N_2$ 158.29	102-53-4	 165.8	4-04-00-00338 0.8000^{20}	 1.4420^{25}
7490	Methanediamine, *N,N,N',N'*-tetramethyl- Tetramethylmethanediamine	$C_5H_{14}N_2$ 102.18	51-80-9	 83	4-04-00-00153 0.7491^{18}	H_2O 3
7491	Methane, diazo- Diazomethane	CH_2N_2 42.04	334-88-3 -145	2983 -23	4-01-00-03056	eth 4; diox 4
7492	Methane, dibromo- Methylene bromide	CH_2Br_2 173.83	74-95-3 -52.5	5980 97	4-01-00-00078 2.4969^{20}	H_2O 2; EtOH 5; eth 5; ace 5 1.5420^{20}
7493	Methane, dibromochloro- Chlorodibromomethane	$CHBr_2Cl$ 208.28	124-48-1 -20	 120	4-01-00-00081 2.451^{20}	H_2O 1; EtOH 3; eth 3; ace 3 1.5482^{20}
7494	Methane, dibromochlorofluoro- Dibromochlorofluoromethane	CBr_2ClF 226.27	353-55-9	 80.3	4-01-00-00082 2.3173^{22}	 1.4570^{20}
7495	Methane, dibromodichloro- Dibromodichloromethane	CBr_2Cl_2 242.72	594-18-3 38	 150.2	4-01-00-00082 2.42^{25}	H_2O 1; EtOH 3; eth 3; ace 3
7496	Methane, dibromodifluoro- Dibromodifluoromethane	CBr_2F_2 209.82	75-61-6 -110.1	 25	4-01-00-00080	H_2O 3; eth 3; ace 3; bz 3
7497	Methane, dibromodinitro- Dibromodinitromethane	$CBr_2N_2O_4$ 263.83	2973-00-4 5.5	 158 dec; 77^{21}	3-01-00-00115 2.4439^{20}	EtOH 4 1.5280^{25}
7498	Methane, dibromofluoro- Dibromofluoromethane	$CHBr_2F$ 191.83	1868-53-7 -78	 64.9	4-01-00-00080 2.421^{20}	H_2O 1; EtOH 3; eth 3; ace 3 1.4685^{20}
7499	Methane, dichloro- Methylene chloride	CH_2Cl_2 84.93	75-09-2 -95.1	5982 40	4-01-00-00035 1.3266^{20}	H_2O 2; EtOH 5; eth 5; ctc 3 1.4242^{20}
7500	Methane, dichlorodifluoro- Dichlorodifluoromethane	CCl_2F_2 120.91	75-71-8 -158	3053 -29.8	4-01-00-00040	H_2O 3; EtOH 3; eth 3; HOAc 3
7501	Methane, dichlorodinitro- Dichlorodinitromethane	$CCl_2N_2O_4$ 174.93	1587-41-3	 121.5	3-01-00-00115 1.6124^{20}	bz 4; eth 4; EtOH 4; chl 4 1.4575^{20}
7502	Methane, dichlorofluoro- Dichlorofluoromethane	$CHCl_2F$ 102.92	75-43-4 -135	 8.9	4-01-00-00039 1.405^{9}	H_2O 1; EtOH 3; eth 3; ctc 3 1.3724^{9}
7503	Methane, dichloroiodo- Dichloroiodomethane	$CHCl_2I$ 210.83	594-04-7	 132	4-01-00-00095 2.392^{20}	ace 4; bz 4; eth 4; EtOH 4 1.5840^{20}
7504	Methane, difluoro- Methylene difluoride	CH_2F_2 52.02	75-10-5 -136	 -51.6	4-01-00-00024 $*0.960^{25}$	H_2O 1; EtOH 3
7505	Methane, difluoroiodo- Difluoroiodomethane	CHF_2I 177.92	1493-03-4 -122	 21.6	4-01-00-00092 3.238^{-19}	
7506	Methane, diiodo- Methylene iodide	CH_2I_2 267.84	75-11-6 6.1	5985 182	4-01-00-00096 3.3212^{20}	H_2O 2; EtOH 3; eth 3; bz 3 1.7425^{20}
7507	Methane, dimethoxy- Methylal	$C_3H_8O_2$ 76.10	109-87-5 -104.8	5936 42	4-01-00-03026 0.8593^{20}	ace 4; bz 4; eth 4; EtOH 4 1.3513^{20}
7508	Methane, dinitro- Dinitromethane	$CH_2N_2O_4$ 106.04	625-76-3 <-15	 exp 100	4-01-00-00107	H_2O 1; EtOH 3; eth 3
7509	Methanediol, dibenzoate	$C_{15}H_{12}O_4$ 256.26	5342-31-4 99	 225 dec	4-09-00-00448 1.275^{22}	ace 4; bz 4; eth 4
7510	Methanediol, 2-furanyl-, diacetate	$C_9H_{10}O_5$ 198.18	613-75-2 53.3	 220	5-17-09-00300	bz 4; eth 4; EtOH 4
7511	Methanediol, (3-nitrophenyl)-, diacetate (ester) Benzaldehyde, 3-nitro, diacetate	$C_{11}H_{11}NO_6$ 253.21	29949-19-7 72		3-07-00-00900 1.393^{25}	ace 4; bz 4; eth 4; EtOH 4
7512	Methanediol, phenyl-, diacetate	$C_{11}H_{12}O_4$ 208.21	581-55-5 46	 220	4-07-00-00516 1.11^{20}	bz 4; eth 4; EtOH 4
7513	Methane, diphenyl-*m*-tolyl- Triphenylmethane, 3-methyl	$C_{20}H_{18}$ 258.36	603-26-9 62	 357	3-05-00-02332 1.07^{16}	bz 4; eth 4; chl 4; lig 4
7514	Methanedisulfonic acid Methionic acid	$CH_4O_6S_2$ 176.17	503-40-2 98	5895	4-01-00-03054	H_2O 1; HNO_3 3
7515	Methane, fluoro- Methyl fluoride	CH_3F 34.03	593-53-3 -141.8	4103 -78.4	4-01-00-00022 $*0.566^{25}$	H_2O 4; EtOH 4; eth 4; bz 3
7516	Methane, fluorodiiodo-	$CHFI_2$ 285.83	1493-01-2 -34.5	 100	4-01-00-00097 3.1969^{22}	eth 4; EtOH 4

PHYSICAL CONSTANTS OF ORGANIC COMPOUNDS (continued)

No.	Name Synonym	Mol. Form. Mol. Wt.	CAS RN mp/°C	Merck No. bp/°C	Beil. Ref. den/g cm^{-3}	Solubility n_D
7517	Methane, fluoroiodo- Fluoroiodomethane	CH_2FI 159.93	373-53-5	 53.4	4-01-00-00092 2.366^{20}	ace 4; bz 4; eth 4; chl 4 1.5256^{20}
7518	Methane, iodo- Methyl iodide	CH_3I 141.94	74-88-4 -66.4	6002 42.5	4-01-00-00087 2.279^{20}	H_2O 2; EtOH 5; eth 5; ace 3 1.5380^{20}
7519	Methane, iodomethoxy- (Iodomethyl) methyl ether	C_2H_5IO 171.97	13057-19-7	 122	4-01-00-03052 2.030^{20}	 1.5472^{20}
7520	Methane, isocyanato- Methyl isocyanate	C_2H_3NO 57.05	624-83-9 -45	6004 39.5	4-04-00-00247 0.9230^{27}	H_2O 4 1.3419^{18}
7521	Methane, isothiocyanato- Methyl isothiocyanate	C_2H_3NS 73.12	556-61-6 36	6005 119	4-04-00-00248 1.0691^{37}	H_2O 2; EtOH 5; eth 4 1.5258
7522	Methane, nitro- Nitromethane	CH_3NO_2 61.04	75-52-5 -28.5	6532 101.1	4-01-00-00100 1.1371^{20}	H_2O 3; EtOH 3; eth 3; ace 3 1.3817^{20}
7523	Methane, oxybis- Dimethyl ether	C_2H_6O 46.07	115-10-6 -141.5	5990 -24.8	4-01-00-01245	H_2O 3; EtOH 3, eth 3; ace 3
7524	Methane, oxybis[bromo-	$C_2H_4Br_2O$ 203.86	4497-29-4 -34	 154.5	4-01-00-03052 2.2013^{20}	
7525	Methane, oxybis[chloro- Bis(chloromethyl) ether	$C_2H_4Cl_2O$ 114.96	542-88-1 -41.5	3058 106	4-01-00-03051 1.323^{15}	EtOH 5; eth 5 1.435^{21}
7526	Methane, selenobis-	C_2H_6Se 109.03	593-79-3	 55	4-01-00-01288 1.4077^{15}	eth 4; EtOH 4; chl 4
7527	Methanesulfenyl chloride, trichloro- Perchloromethyl mercaptan	CCl_4S 185.89	594-42-3	 147.5	4-03-00-00290 1.6947^{20}	eth 3 1.5484^{20}
7528	Methanesulfenyl fluoride, trifluoro- Methyl mercaptan, perfluoro	CF_4S 120.07	17742-04-0	 147.5	 1.6947^{20}	eth 4 1.5484^{20}
7529	Methane, sulfinylbis- Dimethyl sulfoxide	C_2H_6OS 78.14	67-68-5 18.5	3247 189	4-01-00-01277 1.1014^{20}	H_2O 3; EtOH 3; eth 3; ace 3 1.4170^{20}
7530	Methanesulfonic acid Methylsulfonic acid	CH_4O_3S 96.11	75-75-2 20	5865 167^{10}	4-04-00-00010 1.4812^{18}	H_2O 3 1.4317^{18}
7531	Methanesulfonic acid, ethyl ester Ethyl methanesulfonate	$C_3H_8O_3S$ 124.16	62-50-0	3782 86^{10}	4-04-00-00012	
7532	Methanesulfonic acid, methyl ester Methyl methanesulfonate	$C_2H_6O_3S$ 110.13	66-27-3 20	6010 202.5	4-04-00-00011 1.2943^{20}	 1.4138^{20}
7533	Methanesulfonic acid, trifluoro-, ethyl ester	$C_3H_5F_3O_3S$ 178.13	425-75-2	 115	4-03-00-00034 1.3740^{0}	eth 3
7534	Methane, sulfonylbis- Dimethyl sulfone	$C_2H_6O_2S$ 94.13	67-71-0 109	3246 238	4-01-00-01279 1.1700^{110}	H_2O 3; EtOH 3; bz 3 1.4226
7535	Methanesulfonyl chloride Methanesulfonic acid chloride	CH_3ClO_2S 114.55	124-63-0	5866 162; 55^{11}	4-04-00-00027 1.4805^{18}	H_2O 1; EtOH 3; eth 3 1.4573^{20}
7536	Methanesulfonyl chloride, trichloro-	CCl_4O_2S 217.89	2547-61-7 140.5	 170	4-03-00-00036	H_2O 1; EtOH 3; eth 3; CS_2 3
7537	Methanesulfonyl chloride, trifluoro-	$CClF_3O_2S$ 168.52	421-83-0	 162; 62^{18}	4-03-00-00035	H_2O 1 1.3344^{20}
7538	Methanesulfonyl fluoride Methanesulfonic acid fluoride	CH_3FO_2S 98.10	558-25-8	 123.5	4-04-00-00027	
7539	Methane, tetrabromo- Carbon tetrabromide	CBr_4 331.63	558-13-4 90.1	 189.5	4-01-00-00085 2.9608^{100}	H_2O 1, EtOH 3, eth 3; chl 3 1.5942^{100}
7540	Methane, tetrachloro- Carbon tetrachloride	CCl_4 153.82	56-23-5 -23	1822 76.8	4-01-00-00056 1.5940^{20}	H_2O 1; EtOH 3; eth 5; ace 3 1.4601^{20}
7541	Methane, tetrafluoro- Carbon tetrafluoride	CF_4 88.00	75-73-0 -183.5	1823 -128.0	4-01-00-00026 3.034^{25}	H_2O 2; bz 3; chl 3
7542	Methane, tetraiodo- Carbon tetraiodide	CI_4 519.63	507-25-5 171	1824 $130\text{-}40^{1-2}$	4-01-00-00098 4.23^{20}	py 4; chl 4
7543	Methane, tetramethoxy-	$C_5H_{12}O_4$ 136.15	1850-14-2 -5	 114	4-03-00-00004 1.023^{25}	 1.3845^{20}
7544	Methane, tetranitro- Tetranitromethane	CN_4O_8 196.03	509-14-8 13.8	9164 126.1	4-01-00-00107 1.6380^{20}	H_2O 1; EtOH 3; eth 3 1.4384^{20}
7545	Methane, thiobis- Dimethyl sulfide	C_2H_6S 62.14	75-18-3 -98.3	6042 37.3	4-01-00-01275 0.8483^{20}	H_2O 1; EtOH 3; eth 3 1.4438^{20}
7546	Methanethiol Methyl mercaptan	CH_4S 48.11	74-93-1 -123	5867 5.9	4-01-00-01273 0.8665^{20}	H_2O 2; EtOH 4; eth 4; chl 2
7547	Methanethione, diphenyl-	$C_{13}H_{10}S$ 198.29	1450-31-3 53.5	 174^{14}	4-07-00-01385	EtOH 2; eth 2; bz 4; chl 4
7548	Methane, tribromo- Bromoform	$CHBr_3$ 252.73	75-25-2 8.0	1407 149.1	4-01-00-00082 2.899^{15}	H_2O 2; EtOH 5; eth 5; bz 3 1.6005^{15}
7549	Methane, tribromochloro- Tribromochloromethane	CBr_3Cl 287.18	594-15-0 55	 158.5	4-01-00-00085 2.71^{15}	eth 4
7550	Methane, tribromofluoro- Tribromofluoromethane	CBr_3F 270.72	353-54-8 -73.6	 108		EtOH 3
7551	Methane, tribromonitro-	CBr_3NO_2 297.73	464-10-8 10	 127^{18}	4-01-00-00106 2.811^{12}	H_2O 1; EtOH 3; eth 3; ace 4 1.5790^{20}
7552	Methanetricarboxylic acid, triethyl ester Triethyl methanetricarboxylate	$C_{10}H_{16}O_6$ 232.23	6279-86-3 29	 253	4-02-00-02360 1.1084^{16}	 1.4243^{20}
7553	Methanetricarboxylic acid, trimethyl ester Trimethyl methanetricarboxylate	$C_7H_{10}O_6$ 190.15	1186-73-8 46.5	 242.7; 128^{15}	3-02-00-02023	bz 4; eth 4; EtOH 4; chl 4
7554	Methane, trichloro- Chloroform	$CHCl_3$ 119.38	67-66-3 -63.6	2141 61.1	4-01-00-00042 1.4832^{20}	H_2O 2; EtOH 5; eth 5; ace 3 1.4459^{20}
7555	Methane, trichlorofluoro- Trichlorofluoromethane	CCl_3F 137.37	75-69-4 -111.1	9553 23.7	4-01-00-00054	
7556	Methane, trichloroiodo- Trichloroiodomethane	CCl_3I 245.27	594-22-9 -7.8	 142	4-01-00-00095 2.355^{20}	ace 4; bz 4; eth 4; chl 4 1.5854^{20}

No.	Name Synonym	Mol. Form. Mol. Wt.	CAS RN mp/°C	Merck No. bp/°C	Beil. Ref. den/g cm^{-3}	Solubility n_D
7557	Methane, trichloronitro- Chloropicrin	CCl_3NO_2 164.37	76-06-2 -64	2156 112	4-01-00-00106 1.6558^{20}	H_2O 3; EtOH 5; ace 5; bz 5 1.4611^{20}
7558	Methane, trifluoro- Fluoroform	CHF_3 70.01	75-46-7 -155.1	4102 -82.1	4-01-00-00024 $*0.673^{25}$	H_2O 3; EtOH 4; ace 3; bz 3
7559	Methane, trifluoroiodo- Trifluoroiodomethane	CF_3I 195.91	2314-97-8	 -22.5	4-01-00-00092 2.3607^{-32}	 1.3790^{-32}
7560	Methane, triiodo- Iodoform	CHI_3 393.73	75-47-8 119	4926 218	4-01-00-00097 4.008^{25}	H_2O 1; EtOH 3; eth 3; ace 3
7561	Methane, trimethoxy-	$C_4H_{10}O_3$ 106.12	149-73-5 15	 104	4-02-00-00022 0.9676^{20}	EtOH 3; eth 3 1.3793^{20}
7562	Methane, trinitro- Trinitromethane	CHN_3O_6 151.04	517-25-9 15	9642 exp	4-01-00-00107 1.479^{20}	ace 4; EtOH 4 1.4451^{24}
7563	Methanimidamide Formamidine	CH_4N_2 44.06	463-52-5 81	 dec	4-02-00-00082	H_2O 4; EtOH 4
7564	Methanimidamide, *N*'-(4-chloro-2-methylphenyl)-*N*,*N*-dimethyl- Chlordimeform	$C_{10}H_{13}ClN_2$ 196.68	6164-98-3 35	2083 $156\text{-}7^{0.4}$	 1.105^{25}	bz 4; eth 4; EtOH 4 1.5885^{25}
7565	Methanimidamide, *N*,*N*'-diphenyl-	$C_{13}H_{12}N_2$ 196.25	622-15-1 142	 >250	4-12-00-00372	H_2O 2; EtOH 3; eth 4; ace 3
7566	Methanimidamide, *N*-hydroxy-	CH_4N_2O 60.06	624-82-8 114.5	 dec	4-02-00-00084	EtOH 4
7567	Methanimidic acid, *N*-phenyl-, ethyl ester	$C_9H_{11}NO$ 149.19	6780-49-0	 214; 87^{10}	4-12-00-00371 1.0051^{20}	eth 3; bz 3 1.5279^{20}
7568	1,4-Methanoazulene, decahydro-4,8,8-trimethyl-9-methylene-, [1*S*-(1α,3aβ,4α,8aβ)]- Longifolene	$C_{15}H_{24}$ 204.36	475-20-7	5446 258; 126^{15}	4-05-00-01192 0.9319^{18}	H_2O 1; bz 3 1.5040^{20}
7569	1,4-Methanoazulen-9-ol, decahydro-1,5,5,8a-tetramethyl-, [1*R*-(1α,3aβ,4α,8aβ,9*S**)]	$C_{15}H_{26}O$ 222.37	465-24-7 112	 287 dec	3-06-00-00426 1.0441^{20}	 1.519
7570	1H-3a,7-Methanoazulen-6-ol, octahydro-3,6,8,8-tetramethyl-, Cedrol	$C_{15}H_{26}O$ 222.37	77-53-2 86	1919	3-06-00-00424 0.9479^{90}	 1.4824^{90}
7571	7,14-Methano-2H,6H-dipyrido[1,2-a:1',2'-e][1,5]diazocine, dodecahydro-, Sparteine	$C_{15}H_{26}N_2$ 234.38	90-39-1 30.5	8692 325; 173^8	5-23-05-00497 1.0196^{20}	eth 4; EtOH 4; chl 4 1.5312^{20}
7572	7,14-Methano-2H,6H-dipyrido[1,2-a:1',2'-e][1,5]diazocin-6-one, dodecahydro-, [7*R*-(7α,7aβ,14α,14aalp Aphylline	$C_{15}H_{24}N_2O$ 248.37	577-37-7 54.5	761 200 dec	5-24-02-00298	ace 4; bz 4; eth 4; EtOH 4
7573	7,14-Methano-4H,6H-dipyrido[1,2-a:1',2'-e][1,5]diazocin-4-one, 7,7a,8,9,10,11,13,14-octahydro-, Anagyrine	$C_{15}H_{20}N_2O$ 244.34	486-89-5	660 265^{12}; 212^4	5-24-03-00410	H_2O 3; EtOH 4; eth 3; bz 3
7574	7,14-Methano-4H,6H-dipyrido[1,2-a:1',2'-e][1,5]diazocin-4-one, dodecahydro-9-hydroxy-, Hydroxylupanine	$C_{15}H_{24}N_2O_2$ 264.37	15358-48-2 169.5	4760	4-25-00-00082	H_2O 4; EtOH 4; chl 4
7575	2,5-Methano-2H-furo[3,2-b]pyrrol-6-ol, hexahydro-4-methyl-, (2α,3aβ,5α,6β,6aβ Scopoline	$C_8H_{13}NO_2$ 155.20	487-27-4 108.5	8365 248	2-27-00-00061 1.0891^{134}	H_2O 3
7576	4,7-Methano-1H-indene, 1,4,5,6,7,8,8-heptachloro-3a,4,7,7a-tetrahydro- Heptachlor	$C_{10}H_5Cl_7$ 373.32	76-44-8 95.5	4576	 1.57^9	bz 4; eth 4; EtOH 4; lig 4
7577	4,7-Methano-1H-indene, 3a,4,7,7a-tetrahydro-, (3aα,4α,7α,7aα)-	$C_{10}H_{12}$ 132.21	1755-01-7 32	 170 dec; 65^{14}	4-05-00-01399 0.9302^{35}	eth 4; EtOH 4 1.5050^{35}
7578	1,5-Methano-8H-pyrido[1,2-a][1,5]diazocin-8-one, 1,2,3,4,5,6-hexahydro-3-methyl-, (1*R*)- Caulophylline	$C_{12}H_{16}N_2O$ 204.27	486-86-2 137	1914	5-24-02-00538	H_2O 4; ace 4; bz 4; EtOH 4
7579	3,6-Methano-8H-1,5,7-trioxacyclopenta[ij]cycloprop[a]azulene-4,8(3H)-dione, hexahydro-2a-hydroxy Picrotoxin	$C_{30}H_{34}O_{13}$ 602.59	124-87-8 203.5	7388	4-19-00-05245	py 4; EtOH 4
7580	4,7-Methanoisobenzofuran-1,3-dione, 3a,4,7,7a-tetrahydro-, (3aα,4α,7α,7aα)- Carbic anhydride	$C_9H_8O_3$ 164.16	129-64-6 164.5	1801	5-17-11-00192 1.417^{25}	ace 4; bz 4; EtOH 4; chl 4
7581	Methanol Methyl alcohol	CH_4O 32.04	67-56-1 -97.6	5868 64.6	4-01-00-01227 0.7914^{20}	H_2O 5; EtOH 5; eth 5; ace 5 1.3288^{20}
7582	Methanol, chloro-, acetate	$C_3H_5ClO_2$ 108.52	625-56-9	 116	4-02-00-00280 1.194^{20}	eth 4; EtOH 4 1.409^{20}
7583	Methanol, (1,3,5-triazine-2,4,6-triyltriimino)tris- Trimethylolmelamine	$C_6H_{12}N_6O_3$ 216.20	1017-56-7	9631	4-26-00-01273	
7584	Methanol, (1,3,5-triazine-2,4,6-triyltrinitrilo)hexakis- Hexamethylolmelamine	$C_9H_{18}N_6O_6$ 306.28	531-18-0 137	4611	4-26-00-01274	H_2O 4
7585	Methanol, trichloro-, carbonate (2:1)	$C_3Cl_6O_3$ 296.75	32315-10-9 79	 203	4-03-00-00033 1.6290^{80}	
7586	1,6-Methanonaphthalen-1(2H)-ol, octahydro-4,8a,9,9-tetramethyl-, Patchouli alcohol	$C_{15}H_{26}O$ 222.37	5986-55-0 56	7001	3-06-00-00426 0.9906^{65}	H_2O 1; EtOH 3; eth 3 1.5029^{65}

No.	Name Synonym	Mol. Form. Mol. Wt.	CAS RN mp/°C	Merck No. bp/°C	Beil. Ref. den/g cm^{-3}	Solubility n_D
7587	Methanone, 1,3,5-benzenetriyltris[phenyl- 1,3,5-Tribenzoylbenzene	$C_{27}H_{18}O_3$ 390.44	25871-69-6 120	 260^{1}	4-07-00-02847	chl 3
7588	Methanone, 2-benzofuranylphenyl-	$C_{15}H_{10}O_2$ 222.24	6272-40-8 91	 360	5-17-10-00568	H_2O 1
7589	Methanone, bis(3-aminophenyl)-	$C_{13}H_{12}N_2O$ 212.25	611-79-0 173.5	 285^{11}	4-14-00-00255	eth 4; EtOH 4
7590	Methanone, bis(4-bromophenyl)-	$C_{13}H_8Br_2O$ 340.01	3988-03-2 177	 395	4-07-00-01380	bz 4; HOAc 4; chl 4
7591	Methanone, bis(3-chlorophenyl)-	$C_{13}H_8Cl_2O$ 251.11	7094-34-0 124	 163^{2}	4-07-00-01376	H_2O 1; EtOH 3; eth 3
7592	Methanone, bis(4-chlorophenyl)-	$C_{13}H_8Cl_2O$ 251.11	90-98-2 147.5	 353	4-07-00-01376 1.4500^{20}	H_2O 1; EtOH 3; eth 4; ace 2
7593	Methanone, bis[4-(dimethylamino)phenyl]- Michler's ketone	$C_{17}H_{20}N_2O$ 268.36	90-94-8 179	6100 360 dec	4-14-00-00255	H_2O 1; EtOH 2; eth 1; bz 4
7594	Methanone, bis(2,4-dimethylphenyl)- 2,2',4,4'-Tetramethylbenzophenone	$C_{17}H_{18}O$ 238.33	3478-88-4	 190^{10}	4-07-00-01470 1.043^{15}	 1.5790^{25}
7595	Methanone, bis(2-hydroxy-4-methoxyphenyl)- Benzophenone 6	$C_{15}H_{14}O_5$ 274.27	131-54-4 139.5	1109	4-08-00-03505	
7596	Methanone, bis(2-hydroxyphenyl)-	$C_{13}H_{10}O_3$ 214.22	835-11-0 59.5	 333	4-08-00-02446	H_2O 1; EtOH 3; eth 3; chl 3
7597	Methanone, bis(4-hydroxyphenyl)-	$C_{13}H_{10}O_3$ 214.22	611-99-4 210		4-08-00-02452 1.133^{131}	H_2O 2; EtOH 3; eth 3; ace 3
7598	Methanone, bis(4-iodophenyl)- 4,4'-Diiodobenzophenone	$C_{13}H_8I_2O$ 434.01	5630-56-8 238.5	 281^{12}	4-07-00-01381 2.2600^{20}	
7599	Methanone, bis(4-methylphenyl)-	$C_{15}H_{14}O$ 210.28	611-97-2 96.5	 334	4-07-00-01434	ace 4; bz 4; eth 4; EtOH 4
7600	Methanone, (2-bromophenyl)phenyl- 2-Bromobenzophenone	$C_{13}H_9BrO$ 261.12	13047-06-8 42	 345	3-07-00-02079	ace 4; EtOH 4; lig 4
7601	Methanone, (3-bromophenyl)phenyl-	$C_{13}H_9BrO$ 261.12	1016-77-9 81	 186^{6}	4-07-00-01378	EtOH 4
7602	Methanone, (4-bromophenyl)phenyl-	$C_{13}H_9BrO$ 261.12	90-90-4 82.5	 350	4-07-00-01378	H_2O 1; EtOH 2; eth 2; bz 2
7603	Methanone, (2-chlorophenyl)(4-chlorophenyl)-	$C_{13}H_8Cl_2O$ 251.11	85-29-0 67	 $214\text{-}5^{22}$	4-07-00-01376 1.393^{14}	EtOH 3; chl 2
7604	Methanone, (4-chlorophenyl)phenyl-	$C_{13}H_9ClO$ 216.67	134-85-0 77.5	 332	4-07-00-01375	EtOH 3; eth 3; ace 3; ctc 2
7605	Methanone, cyclobutylphenyl-	$C_{11}H_{12}O$ 160.22	5407-98-7	 260	4-07-00-01027 1.0426^{25}	 1.5472^{20}
7606	Methanone, dicyclohexyl-	$C_{13}H_{22}O$ 194.32	119-60-8	 159^{20}	4-07-00-00262 0.986^{0}	eth 3; ace 3; ctc 3 1.4860^{20}
7607	Methanone, dicyclopropyl- Dicyclopropyl ketone	$C_7H_{10}O$ 110.16	1121-37-5	 161	4-07-00-00139 0.977^{25}	 1.4670^{20}
7608	Methanone, (2,6-dihydroxy-4-methoxyphenyl)phenyl- Cotoin	$C_{14}H_{12}O_4$ 244.25	479-21-0 130.5	2553	4-08-00-03163	ace 4; bz 4; eth 4; EtOH 4
7609	Methanone, (2,4-dihydroxyphenyl)phenyl- Benzoresorcinol	$C_{13}H_{10}O_3$ 214.22	131-56-6 144	1117	4-08-00-02442	H_2O 1; EtOH 3; eth 4; bz 2
7610	Methanone, (3,4-dihydroxyphenyl)(2,4,6-trihydroxyphenyl)- Maclurin	$C_{13}H_{10}O_6$ 262.22	519-34-6 222.5	5519	3-08-00-04226	eth 4; EtOH 4
7611	Methanone, (6,7-dimethoxy-1-isoquinolinyl)(3,4-dimethoxyphenyl)- Papaveraldine	$C_{20}H_{19}NO_5$ 353.37	522-57-6 210.5	6966	5-21-13-00658	H_2O 1; EtOH 2; eth 2; bz 3
7612	Methanone, (2,4-dimethoxyphenyl)phenyl- 2,4-Dimethoxybenzophenone	$C_{15}H_{14}O_3$ 242.27	3555-84-8 87.5	 218^{10}	4-08-00-02442	H_2O 1; EtOH 4; eth 2; chl 4
7613	Methanone, [4-(dimethylamino)phenyl]phenyl- 4-(Dimethylamino)benzophenone	$C_{15}H_{15}NO$ 225.29	530-44-9 92.5	3222	4-14-00-00248	H_2O 1; EtOH 2; eth 4; chl 3
7614	Methanone, (2,4-dimethylphenyl)phenyl-	$C_{15}H_{14}O$ 210.28	1140-14-3	 321.5	4-07-00-01429 1.071^{20}	bz 4; eth 4; EtOH 4; chl 4
7615	Methanone, (3,4-dimethylphenyl)phenyl- 3,4-Dimethylbenzophenone	$C_{15}H_{14}O$ 210.28	2571-39-3 47.5	 341	4-07-00-01433	bz 4; eth 4; EtOH 4; chl 4
7616	Methanone, diphenyl- Benzophenone	$C_{13}H_{10}O$ 182.22	119-61-9 47.8	1108 305.4	4-07-00-01357	H_2O 1; EtOH 4; eth 4; ace 4 1.6077^{19}
7617	Methanone, di-2-thienyl- Di(2-thienyl) ketone	$C_9H_6OS_2$ 194.28	704-38-1 90	 326	5-19-04-00318	ace 4; eth 4
7618	Methanone, 2-furanylphenyl-	$C_{11}H_8O_2$ 172.18	2689-59-0 <-15	 285	5-17-10-00289 1.1732^{20}	eth 4; EtOH 4 1.6055^{20}
7619	Methanone, (2-hydroxy-4-methoxyphenyl)(2-hydroxyphenyl)- Dioxybenzone	$C_{14}H_{12}O_4$ 244.25	131-53-3	3298 $170\text{-}5^{1}$	4-08-00-03163	
7620	Methanone, (2-hydroxy-4-methoxyphenyl)phenyl- Oxybenzone	$C_{14}H_{12}O_3$ 228.25	131-57-7 65.5	6907	4-08-00-02442	ctc 3
7621	Methanone, [2-hydroxy-4-(octyloxy)phenyl]phenyl- Octabenzone	$C_{21}H_{26}O_3$ 326.44	1843-05-6 48.5	6662		
7622	Methanone, (2-hydroxyphenyl)phenyl-	$C_{13}H_{10}O_2$ 198.22	117-99-7 40	 250^{560}	4-08-00-01246	H_2O 1; EtOH 4; eth 4; bz 4
7623	Methanone, (2-methoxyphenyl)phenyl-	$C_{14}H_{12}O_2$ 212.25	2553-04-0 41	 210^{27}; 195^{18}	4-08-00-01246	H_2O 1; EtOH 3; bz 3; HOAc 3

PHYSICAL CONSTANTS OF ORGANIC COMPOUNDS (continued)

No.	Name Synonym	Mol. Form. Mol. Wt.	CAS RN mp/°C	Merck No. bp/°C	Beil. Ref. den/g cm^{-3}	Solubility n_D
7624	Methanone, (3-methoxyphenyl)phenyl-	$C_{14}H_{12}O_2$ 212.25	6136-67-0 44	 343; 201^{17}	4-08-00-01262	H_2O 1; EtOH 4; bz 4; HOAc 4
7625	Methanone, (4-methoxyphenyl)phenyl-	$C_{14}H_{12}O_2$ 212.25	611-94-9 61.5	 355; 168^{12}	4-08-00-01263	H_2O 1; EtOH 4; eth 4; ace 3
7626	Methanone, [2-(methylamino)phenyl]phenyl-	$C_{14}H_{13}NO$ 211.26	1859-76-3 69	 282	4-14-00-00244	bz 4; EtOH 4; HOAc 4
7627	Methanone, (4-methyl-2-nitrophenyl)phenyl- 4-Methyl-2-nitrobenzophenone	$C_{14}H_{11}NO_3$ 241.25	100224-75-7 126.5	 sub	0-07-00-00442	bz 4; EtOH 4; chl 4
7628	Methanone, (2-methylphenyl)(4-methylphenyl)- 2,4'-Dimethylbenzophenone	$C_{15}H_{14}O$ 210.28	1140-16-5	 317	4-07-00-01432 1.074^{19}	eth 4; EtOH 4; HOAc 4
7629	Methanone, (2-methylphenyl)-1-naphthalenyl-	$C_{18}H_{14}O$ 246.31	68723-25-1 64	 365	3-07-00-02639	eth 4; EtOH 4
7630	Methanone, (4-methylphenyl)(4-nitrophenyl)-	$C_{14}H_{11}NO_3$ 241.25	5350-47-0 124	 sub	4-07-00-01406	bz 4; eth 4; EtOH 4; chl 4
7631	Methanone, (2-methylphenyl)phenyl-	$C_{14}H_{12}O$ 196.25	131-58-8 <-18	 308; 128^{12}	4-07-00-01401 1.1098^{20}	H_2O 1; EtOH 4; os 3
7632	Methanone, (3-methylphenyl)phenyl-	$C_{14}H_{12}O$ 196.25	643-65-2	 317; 170^{9}	4-07-00-01403 1.088^{17}	H_2O 1; EtOH 3; eth 3; bz 3
7633	Methanone, (4-methylphenyl)phenyl-	$C_{14}H_{12}O$ 196.25	134-84-9 59.5	 228^{70}	4-07-00-01403 0.9926^{0}	H_2O 1; EtOH 2; eth 3; bz 3
7634	Methanone, 1-naphthalenyl-2-naphthalenyl- 1,2'-Dinaphthyl ketone	$C_{21}H_{14}O$ 282.34	605-79-8 136.5	 $235^{0.6}$	4-07-00-01886	chl 4
7635	Methanone, (3-nitrophenyl)phenyl-	$C_{13}H_9NO_3$ 227.22	2243-80-3 95	 234^{18}	4-07-00-01382	H_2O 1; EtOH 3
7636	Methanone, (4-nitrophenyl)phenyl-	$C_{13}H_9NO_3$ 227.22	1144-74-7 138		4-07-00-01382 1.406^{9}	bz 4
7637	Methanone, phenyl-1H-pyrrol-2-yl- Pyrrole, 2-benzoyl	$C_{11}H_9NO$ 171.20	7697-46-3 79	 305	5-21-08-00461	bz 4; EtOH 4; peth 4; HOAc 4
7638	Methanone, phenyl-2-pyridinyl-	$C_{12}H_9NO$ 183.21	91-02-1 42	 317	5-21-08-00566 1.1556^{20}	chl 3
7639	Methanone, phenyl-4-pyridinyl-	$C_{12}H_9NO$ 183.21	14548-46-0 72	 315; 170^{10}	5-21-08-00576	H_2O 2; EtOH 3; eth 3; bz 3
7640	Methanone, phenyl-4-quinolinyl-	$C_{16}H_{11}NO$ 233.27	54885-00-6 60	 222^{15}	1-21-00-00320	bz 4; eth 4; EtOH 4
7641	Methanone, phenyl-2-thienyl-	$C_{11}H_8OS$ 188.25	135-00-2 56.5	 300	5-17-10-00291 1.1890^{54}	H_2O 1; EtOH 3; eth 3 1.6181^{54}
7642	Methanone, phenyl-3-thienyl-	$C_{11}H_8OS$ 188.25	6453-99-2 63.5	 129^{3}	5-17-10-00295	H_2O 1; EtOH 3; eth 3
7643	Methazole 1,2,4-Oxadiazolidine-3,5-dione, 2-(3,4-dichlorophenyl)-4-methyl-	$C_9H_6Cl_2N_2O_3$ 261.06	20354-26-1 123	5876	 1.24^{25}	
7644	1,3,4-Metheno-1H-cyclobuta[cd]pentalene, 1,1a,2,2,3,3a,4,5,5,5a,5b,6-dodecachlorooctahydro- Mirex	$C_{10}Cl_{12}$ 545.54	2385-85-5 485 dec	6126	3-05-00-01237	bz 4; diox 4
7645	Methidathion Phosphorodithioic acid, *S*-[(5-methoxy-2-oxo-1,3,4-thiadiazol-3(2H)-yl)methyl], *O,O*-dimethyl ester	$C_6H_{11}N_2O_4PS_3$ 302.34	950-37-8 39	5891		
7646	Methiocarb Phenol, 3,5-dimethyl-4-(methylthio)-, methylcarbamate	$C_{11}H_{15}NO_2S$ 225.31	2032-65-7 120	5893		
7647	*DL*-Methionine	$C_5H_{11}NO_2S$ 149.21	59-51-8 281 dec		4-04-00-03190 1.340^{25}	H_2O 4; EtOH 2; eth 1; bz 1
7648	*L*-Methionine 2-Amino-4-methylthiobutanoic acid	$C_5H_{11}NO_2S$ 149.21	63-68-3 283 dec	5896	4-04-00-03189	H_2O 3; EtOH 1; eth 1; ace 1
7649	*L*-Methionine, *N*-acetyl- Methionamine	$C_7H_{13}NO_3S$ 191.25	65-82-7 105.5	90	4-04-00-03205	
7650	Methomyl Acetimidic acid, *N*-[(methylcarbamoyl)oxy]thio-, methyl ester	$C_5H_{10}N_2O_2S$ 162.21	16752-77-5 78	5905	 1.2946^{24}	
7651	Methoprene 2,4-Dodecadienoic acid, 11-methoxy-3,7,11-trimethyl-, 1-methylethyl ester, (*E,E*)-	$C_{19}H_{34}O_3$ 310.48	40596-69-8	5906 $100^{0.05}$	 0.926^{20}	
7652	Methylamine, *N*-(*o*-chlorobenzylidene)- Toluene, 2-chloro-α-methylimino	C_8H_8ClN 153.61	17972-08-6 7	 78-80^{2}	3-07-00-00866	 1.5660^{20}
7653	Methyl 3,6-dichloro-*o*-anisate Benzoic acid, 3,6-dichloro-2-methoxy-, methyl ester	$C_9H_8Cl_2O_3$ 235.07	6597-78-0	 $115^{2.4}$		
7654	Methylene bispropionate Propanoic acid, methylene ester	$C_7H_{12}O_4$ 160.17	7044-96-4	 98^{1}; 191^{745}	 1.053^{20}	
7655	Methylisonitrile Methyl isocyanate	C_2H_3N 41.05	593-75-9 -45	 59.6 exp	 0.756^{4}	
7656	Metobromuron Urea, 3-(*p*-bromophenyl)-1-methoxy-1-methyl-	$C_9H_{11}BrN_2O_2$ 259.10	3060-89-7 95	6061	 1.60^{20}	
7657	Metolachlor Acetamide, 2-chloro-*N*-(2-ethyl-6-methylphenyl)-*N*-(2-methoxy-1-methylethyl)-	$C_{15}H_{22}ClNO_2$ 283.80	51218-45-2	6067 $100^{0.001}$	 1.12^{20}	

No.	Name Synonym	Mol. Form. Mol. Wt.	CAS RN mp/°C	Merck No. bp/°C	Beil. Ref. den/g cm^{-3}	Solubility n_D
7658	Metribuzin 1,2,4-Triazin-5(4H)-one, 4-amino-6-(1,1-dimethylethyl)-3-(methylthio)-	$C_8H_{14}N_4OS$ 214.29	21087-64-9 126	6076	 1.31^{20}	
7659	Mevinphos 2-Butenoic acid, 3-[(dimethoxyphosphinyl)oxy]-, methyl ester, (*E*)-	$C_7H_{13}O_6P$ 224.15	7786-34-7 21 (E), 6.9 (Z)	6089 $101^{0.3}$		
7660	MGK 264 4,7-Methano-1H-isoindole-1,3(2H)-dione, 2-(2-ethylhexyl)-3a,4,7,7a-tetrahydro-	$C_{17}H_{25}NO_2$ 275.39	113-48-4 <-20	 157	 1.04	
7661	Molinate 1H-Azepine-1-carbothioic acid, hexahydro-, *S*-ethyl ester	$C_9H_{17}NOS$ 187.31	2212-67-1	 202^{10}	 1.063^{20}	
7662	Molybdenum carbonyl Molybdenum hexacarbonyl	C_6MoO_6 264.00	13939-06-5 dec 150	 subl		os 3
7663	Monocrotophos (*E*)-Dimethyl 1-methyl-3-(methylamino)-3-oxo-1-propenyl phosphate	$C_7H_{14}NO_5P$ 223.17	6923-22-4 55	6161 $125^{0.0005}$	 1.33^{20}	
7664	Morphinan, 6,7-didehydro-4,5-epoxy-3,6-dimethoxy-17-methyl-, (5α)- Dihydrothebaine	$C_{19}H_{23}NO_3$ 313.40	561-25-1 162.5	3164	2-27-00-00110	H_2O 1; EtOH 3; bz 3; AcOEt 3
7665	Morphinan-3,6-diol, 7,8-didehydro-4,5-epoxy-, (5α,6α)- Normorphine	$C_{16}H_{17}NO_3$ 271.32	466-97-7 273	6630		
7666	Morphinan-3,6-diol, 7,8-didehydro-4,5-epoxy-17-methyl- (5α,6α)- Morphine	$C_{17}H_{19}NO_3$ 285.34	57-27-2 255	6186 sub 190	4-27-00-02223	H_2O 1; EtOH 2; eth 1; ace 1
7667	Morphinan-3,6-diol, 7,8-didehydro-4,5-epoxy-17-methyl- (5α,6α)-, diacetate (ester) Diacetylmorphine	$C_{21}H_{23}NO_5$ 369.42	561-27-3 173	2948 $272\text{-}4^{12}$	4-27-00-02236 1.56^{25}	bz 4; chl 4
7668	Morphinan-3,6-diol, 7,8-didehydro-4,5-epoxy-17-methyl- (5α,6α)-, 17-oxide Morphine, *N*-oxide	$C_{17}H_{19}NO_4$ 301.34	639-46-3 274.5	6192	4-27-00-02249	H_2O 2, EtOH 2; ace 1; bz 1
7669	Morphinan-6-ol, 7,8-didehydro-4,5-epoxy-3-methoxy-17-methyl-, (5α,6α)- Codeine	$C_{18}H_{21}NO_3$ 299.37	76-57-3 157.5	2459 250^{22}, $140^{1.5}$	2-27-00-00136 1.32^{25}	H_2O 3; EtOH 4; eth 3; bz 3
7670	Morphinan-6-ol, 7,8-didehydro-4,5-epoxy-3-methoxy-17-methyl-, (5α,6β)-	$C_{18}H_{21}NO_3$ 299.37	509-64-8 171.5	 dec	2-27-00-00175 1.87^{4}	 1.675
7671	Morphinan-6-ol, 8,14-didehydro-4,5-epoxy-3-methoxy-17-methyl-, (5α,6α)- Neopine	$C_{18}H_{21}NO_3$ 299.37	467-14-1 127.5	6376		H_2O 3; EtOH 3; eth 3; bz 3
7672	Morphinan-8-ol, 6,7-didehydro-4,5-epoxy-3-methoxy-17-methyl-, (5α,8β)- Pseudocodeine	$C_{18}H_{21}NO_3$ 299.37	466-96-6 181.5	7927	2-27-00-00112 1.290^{80}	 1.574
7673	Morphinan-6-ol, 7,8-didehydro-4,5-epoxy-3-methoxy-17-methyl-, (5α,6α)-, phosphate (1:1) (salt) Codeine, phosphate	$C_{18}H_{24}NO_7P$ 397.36	52-28-8 220-35 dec	2462	4-27-00-02228	EtOH 4; chl 4
7674	Morphinan-6-ol, 4,5-epoxy-3-methoxy-17-methyl-, (5α,6α)- Dihydrocodeine	$C_{18}H_{23}NO_3$ 301.39	125-28-0 112.5	3154 248^{15}	2-27-00-00103	
7675	Morphinan-6-one, 7,8-didehydro-4,5-epoxy-14-hydroxy-3-methoxy-17-methyl-, (5α)- Hydroxycodeinone	$C_{18}H_{19}NO_4$ 313.35	508-54-3 275 dec	4752		
7676	Morphinan-6-one, 7,8-didehydro-4-hydroxy-3,7-dimethoxy-17-methyl-, (9α,13α,14α)- Sinomenine	$C_{19}H_{23}NO_4$ 329.40	115-53-7 162	8496	5-21-13-00516	H_2O 2; EtOH 3; eth 2; ace 3
7677	Morphinan-6-one, 7,8-didehydro-4-hydroxy-3-methoxy-17-methyl- Thebainone	$C_{18}H_{21}NO_3$ 299.37	467-98-1 151.5	9203	5-21-13-00268	H_2O 2; EtOH 2; eth 2; ace 3
7678	Morphinan-6-one, 4,5-epoxy-3-methoxy-17-methyl-, (5α)- Hydrocodone	$C_{18}H_{21}NO_3$ 299.37	125-29-1 198	4708	4-27-00-03580	H_2O 1; EtOH 3
7679	Morphinan, 6,7,8,14-tetradehydro-4,5-epoxy-3,6-dimethoxy-17-methyl-, (5α)- Thebaine	$C_{19}H_{21}NO_3$ 311.38	115-37-7 193	9202 sub 91	[illegible] 1.305^{20}	[illegible]
7680	4-Morpholinamine	$C_4H_{10}N_2O$ 102.14	4319-49-7	 166	4-27-00-00623 1.059^{25}	 1.4772^{20}
7681	Morpholine Tetrahydro-1,4-oxazine	C_4H_9NO 87.12	110-91-8 -4.9	6194 128	4-27-00-00015 1.0005^{20}	H_2O 5; EtOH 3; eth 3; ace 3 1.4548^{20}
7682	Morpholine, 4-acetyl-	$C_6H_{11}NO_2$ 129.16	1696-20-4 14.5	 152^{50}, 118^{12}	4-27-00-00274 1.1145^{20}	H_2O 5; EtOH 3; ace 3; ctc 3 1.4827^{20}
7683	Morpholine, 4-[3-(4-butoxyphenoxy)propyl]- Pramoxine	$C_{17}H_{27}NO_3$ 293.41	140-65-8	7707 196^{6}	4-27-00-00127	
7684	Morpholine, 4-butyl-	$C_8H_{17}NO$ 143.23	1005-67-0 -57.1	 213.5	4-27-00-00026 0.9068^{20}	H_2O 4; ace 4; bz 4; EtOH 4 1.4451^{20}
7685	4-Morpholinecarboxaldehyde	$C_5H_9NO_2$ 115.13	4394-85-8 21	 239	4-27-00-00274 1.1520^{20}	 1.4845^{20}

No.	Name Synonym	Mol. Form. Mol. Wt.	CAS RN mp/°C	Merck No. bp/°C	Beil. Ref. den/g cm^{-3}	Solubility n_D
7686	4-Morpholinecarboxamide, *N,N'*-1,2-ethanediylbis[*N*-butyl- Dimorpholamine	$C_{20}H_{38}N_4O_4$ 398.55	119-48-2 41.5	3257 $229^{0.4}$	4-27-00-00298	H_2O 4
7687	Morpholine, 2,6-dimethyl- 2,6-Dimethylmorpholine	$C_6H_{13}NO$ 115.18	141-91-3 -88	 146.6	4-27-00-00674 0.9329^{20}	H_2O 5; EtOH 5; ace 3; bz 5 1.4460^{20}
7688	Morpholine, 4,4'-dithiobis- 4,4'-Dithiodimorpholine	$C_8H_{16}N_2O_2S_2$ 236.36	103-34-4 124.5	3379	4-27-00-00613	chl 3
7689	4-Morpholineethanamine	$C_6H_{14}N_2O$ 130.19	2038-03-1 25.6	 205	4-27-00-00370 0.9897^{20}	H_2O 5; EtOH 5; ace 3; bz 5 1.4715^{20}
7690	Morpholine, 4,4'-(1,2-ethanediyl)bis-	$C_{10}H_{20}N_2O_2$ 200.28	1723-94-0 75	 285; 160^{25}	4-27-00-00375	H_2O 4; ace 4; bz 4; EtOH 4
7691	4-Morpholineethanol	$C_6H_{13}NO_2$ 131.17	622-40-2 -1.6	 227	4-27-00-00056 1.0710^{20}	H_2O 3; EtOH 3; ctc 2 1.4763^{20}
7692	4-Morpholineethanol, α-methyl-	$C_7H_{15}NO_2$ 145.20	2109-66-2	 $92\text{-}4^{13}$	4-27-00-00119 1.0174^{20}	H_2O 4; ace 4; bz 4; EtOH 4 1.4638^{20}
7693	Morpholine, 4-(2-ethoxyethyl)- Morpholine, *N*-(2-ethoxyethyl)	$C_8H_{17}NO_2$ 159.23	622-09-3 -100	 206	4-27-00-00057 0.963^{20}	H_2O 4; ace 4; bz 4; eth 4
7694	Morpholine, 4-ethyl- *N*-Ethylmorpholine	$C_6H_{13}NO$ 115.18	100-74-3	 138.5	4-27-00-00023 0.8996^{20}	H_2O 5; EtOH 5; eth 5; ace 3 1.4400^{20}
7695	Morpholine, 4-(2-hydroxybenzoyl)-	$C_{11}H_{10}NO_3$ 204.21	3202-84-4	8302	4-27-00-00346	DMSO 3
7696	Morpholine, 4-methyl- *N*-Methylmorpholine	$C_5H_{11}NO$ 101.15	109-02-4	 116	4-27-00-00022 0.9051^{20}	H_2O 3; EtOH 3; eth 3 1.4332^{20}
7697	Morpholine, 3-methyl-2-phenyl- Phenmetrazine	$C_{11}H_{15}NO$ 177.25	134-49-6	7200 139^{12}; 104^{1}	4-27-00-01050	
7698	Morpholine, 4-(4-methylphenyl)-	$C_{11}H_{15}NO$ 177.25	3077-16-5 51	 167^{30}	2-27-00-00004	eth 4; EtOH 4
7699	Morpholine, 4-nitroso- *N*-Nitrosomorpholine	$C_4H_8N_2O_2$ 116.12	59-89-2 29	6561 225; 140^{25}		H_2O 3
7700	Morpholine, 4-(phenylmethyl)-	$C_{11}H_{15}NO$ 177.25	10316-00-4 194	 260.5	4-27-00-00039 1.0387^{20}	ace 4; bz 4 1.5302^{20}
7701	4-Morpholinepropanamine 4-(3-Aminopropyl)morpholine	$C_7H_{16}N_2O$ 144.22	123-00-2 -15	 220; 134^{50}	4-27-00-00411 0.9854^{20}	H_2O 5; EtOH 5; ace 3; bz 5 1.4762^{20}
7702	Naled Phosphoric acid, 1,2-dibromo-2,2-dichloroethyl dimethyl ester	$C_4H_7Br_2Cl_2O_4P$ 380.78	300-76-5 27	6272 $110^{0.5}$	 1.96^{20}	
7703	Naphthacene 2,3-Benzanthracene	$C_{18}H_{12}$ 228.29	92-24-0 357	6288 sub	4-05-00-02545	H_2O 2; bz 2; con sulf 3; os 2
7704	2-Naphthacenecarboxamide, 7-chloro-4-(dimethylamino)-1,4,4a,5,5a,6,11,12a-octahydro-3,6,10,12,12a-pentahydroxy-6-methyl-1,11-dioxo, [4S-(4α,4aα,5aα,6β,12aα)]- Chlortetracycline	$C_{22}H_{23}ClN_2O_8$ 478.89	57-62-5 168.5	2193	4-14-00-02629	H_2O 1; EtOH 2; eth 1; ace 2
7705	2-Naphthacenecarboxamide, 4-(dimethylamino)-1,4,4a,5,5a,6,11,12a-octahydro-3,5,6,10,12,12a-hexahydroxy-6-methyl-1,11-dioxo-,[4S-(4α,4aα,5aα,6β,12aα)]- Oxytetracycline	$C_{22}H_{24}N_2O_9$ 460.44	79-57-2 184.5	6931	4-14-00-02633 1.634^{20}	
7706	2-Naphthacenecarboxamide, 4-(dimethylamino)-1,4,4a,5,5a,6,11,12a-octahydro-3,6,10,12,12a-pentahydroxy-6-methyl-1,11-dioxo-,[4S-(4α,4aα,5aα,6β,12aα)]- Tetracycline	$C_{22}H_{24}N_2O_8$ 444.44	60-54-8 170-5 dec	9130	4-14-00-02625	
7707	1,2-Naphthacenedione 9,10-Naphthacenequinone	$C_{18}H_{10}O_2$ 258.28	29276-40-2 294	 sub	3-07-00-04273	EtOH 4
7708	1-Naphthalenamine 1-Naphthylamine	$C_{10}H_9N$ 143.19	134-32-7 49.2	6318 300.8	4-12-00-03009 1.0228^{20}	chl 3 1.6140^{20}
7709	2-Naphthalenamine 2-Naphthylamine	$C_{10}H_9N$ 143.19	91-59-8 113	6319 366	4-12-00-03122 1.6414^{98}	H_2O 3; EtOH 3; eth 3 1.6493^{98}
7710	2-Naphthalenamine, *N,N*-bis(2-chloroethyl)- Chlornaphazine	$C_{14}H_{15}Cl_2N$ 268.19	494-03-1 55	2107 210^{5}	4-12-00-03126	ace 4; bz 4; eth 4; EtOH 4
7711	1-Naphthalenamine, 5-bromo- 1-Amino-5-bromonaphthalene	$C_{10}H_8BrN$ 222.08	4766-33-0 70	 sub	4-12-00-03111	bz 4; eth 4; EtOH 4; lig 4
7712	1-Naphthalenamine, decahydro-	$C_{10}H_{19}N$ 153.27	7250-95-5 13.5	 227.6	3-12-00-00183 0.9508^{21}	
7713	1-Naphthalenamine, *N,N*-diethyl-	$C_{14}H_{17}N$ 199.30	84-95-7	 285	3-12-00-02855 1.013^{20}	EtOH 3; eth 3; bz 3; ctc 2 1.5961^{20}
7714	1-Naphthalenamine, 5,8-dihydro- 1-Naphthylamine, 5,8-dihydro-	$C_{10}H_{11}N$ 145.20	32666-56-1 37.5	 247^{408}	4-12-00-02995	EtOH 4; chl 4
7715	1-Naphthalenamine, *N,N*-dimethyl- *N,N*-Dimethyl-1-naphthylamine	$C_{12}H_{13}N$ 171.24	86-56-6	3240 250; 140^{13}	4-12-00-03013 1.0423^{20}	H_2O 1; EtOH 3; eth 3; ctc 3 1.624^{15}
7716	2-Naphthalenamine, *N,N*-dimethyl-	$C_{12}H_{13}N$ 171.24	2436-85-3 52.5	 305	4-12-00-03125 1.0279^{60}	H_2O 1; EtOH 3; eth 3 1.6443^{53}
7717	1-Naphthalenamine, *N,N*-diphenyl- 1-Naphthylamine, *N,N*-diphenyl-	$C_{22}H_{17}N$ 295.38	61231-45-6 142	 $335\text{-}40^{80}$	0-12-00-01225	ace 4; bz 4; eth 4; EtOH 4
7718	1-Naphthalenamine, *N*-ethyl-	$C_{12}H_{13}N$ 171.24	118-44-5	 305; 191^{16}	4-12-00-03013 1.060^{20}	eth 4; EtOH 4 1.6477^{15}

No.	Name Synonym	Mol. Form. Mol. Wt.	CAS RN mp/°C	Merck No. bp/°C	Beil. Ref. den/g cm^{-3}	Solubility n_D
7719	2-Naphthalenamine, *N*-ethyl-	$C_{12}H_{13}N$ 171.24	2437-03-8 236	 316.5	4-12-00-03126 1.0545^{21}	 1.6544^{21}
7720	1-Naphthalenamine, 4-methyl-	$C_{11}H_{11}N$ 157.22	4523-45-9 51.5	 176^{12}	4-12-00-03190	eth 4
7721	1-Naphthalenamine, *N*-methyl-	$C_{11}H_{11}N$ 157.22	2216-68-4 174	 294.5	4-12-00-03013	eth 4; EtOH 4 1.6722^{20}
7722	2-Naphthalenamine, *N*-methyl-	$C_{11}H_{11}N$ 157.22	2216-67-3	 317	4-12-00-03125	eth 4; EtOH 4 1.6722^{20}
7723	1-Naphthalenamine, *N*-(2-methylphenyl)- 1-Naphthyl(2-tolyl)amine	$C_{17}H_{15}N$ 233.31	634-41-3 94.5	 200^{9}	3-12-00-02857	bz 4; eth 4; EtOH 4
7724	1-Naphthalenamine, *N*-(4-methylphenyl)-	$C_{17}H_{15}N$ 233.31	634-43-5 79	 360^{528}; 230^{10}	4-12-00-03016	bz 4; eth 4
7725	2-Naphthalenamine, *N*-(2-methylphenyl)- 2-Naphthyl(2-tolyl)amine	$C_{17}H_{15}N$ 233.31	644-15-5 95.5	 402.5	1-12-00-00535	ace 4; bz 4; eth 4; EtOH 4
7726	1-Naphthalenamine, *N*-1-naphthalenyl-	$C_{20}H_{15}N$ 269.35	737-89-3 115	 311^{15}	4-12-00-03018	H_2O 1; EtOH 3; eth 3; ace 3
7727	2-Naphthalenamine, *N*-2-naphthalenyl- β,β'-Dinaphthylamine	$C_{20}H_{15}N$ 269.35	532-18-3 172.2	3260 471	4-12-00-03131	H_2O 1; EtOH 2; eth 3; bz 2
7728	2-Naphthalenamine, *N*-phenyl- *N*-Phenyl-β-naphthylamine	$C_{16}H_{13}N$ 219.29	135-88-6 108	 395.5	4-12-00-03128	H_2O 1; EtOH 3; eth 3; bz 3
7729	2-Naphthalenamine, 1-(phenylazo)- Yellow AB	$C_{16}H_{13}N_3$ 247.30	85-84-7 103	10003		EtOH 4; HOAc 4
7730	2-Naphthalenamine, *N*-(phenylmethyl)-	$C_{17}H_{15}N$ 233.31	13672-18-9 68	 405	4-12-00-03130	bz 4; lig 4
7731	1-Naphthalenamine, 5,6,7,8-tetrahydro-	$C_{10}H_{13}N$ 147.22	2217-41-6 38	 279	4-12-00-02934 1.0625^{16}	H_2O 2; EtOH 3; eth 3; acid 3 1.5900^{20}
7732	Naphthalene	$C_{10}H_8$ 128.17	91-20-3 80.2	6289 217.9	4-05-00-01640 1.0253^{20}	H_2O 1; EtOH 3; eth 4; ace 4 1.5898^{25}
7733	1-Naphthaleneacetamide Acetic acid, amide,1-naphthyl	$C_{12}H_{11}NO$ 185.23	86-86-2	 sub 180	4-09-00-02426	H_2O 1; eth 3; bz 3; CS_2 3
7734	1-Naphthaleneacetic acid 1-Naphthylacetic acid	$C_{12}H_{10}O_2$ 186.21	86-87-3 135	6290 dec	4-09-00-02424	H_2O 2; EtOH 2; eth 4; ace 4
7735	1-Naphthaleneacetic acid, ethyl ester Ethyl 1-naphthylacetate	$C_{14}H_{14}O_2$ 214.26	2122-70-5 88.5	 222^{20}; 118^{13}	4-09-00-02425	EtOH 3; eth 3
7736	1-Naphthaleneacetonitrile	$C_{12}H_9N$ 167.21	132-75-2 32.5	 192^{18}; 163^{12}	4-09-00-02428	EtOH 3 1.6192^{20}
7737	2-Naphthaleneacetonitrile	$C_{12}H_9N$ 167.21	7498-57-9 83	 303	3-09-00-03212 1.092^{25}	
7738	Naphthalene, 1-azido-	$C_{10}H_7N_3$ 169.19	6921-40-0 12	 75$^{0.001}$	 1.1713^{25}	EtOH 4; eth 5; ace 5 1.6550^{25}
7739	Naphthalene, 1-bromo- 1-Naphthyl bromide	$C_{10}H_7Br$ 207.07	90-11-9 -1.8	1413 281	4-05-00-01665 1.4785^{20}	H_2O 3; EtOH 5; eth 5; ace 3 1.658^{20}
7740	Naphthalene, 2-bromo- 2-Bromonaphthalene	$C_{10}H_7Br$ 207.07	580-13-2 59	1414 281.5	4-05-00-01667 1.605^{25}	H_2O 1; EtOH 3; eth 3; bz 3 1.6382^{60}
7741	Naphthalene, 1-bromo-4-chloro- 1-Bromo-4-chloronaphthalene	$C_{10}H_6BrCl$ 241.51	53220-82-9 66.5	 303	3-05-00-01585	eth 4; HOAc 4
7742	Naphthalene, 1-(bromomethyl)-	$C_{11}H_9Br$ 221.10	3163-27-7 56	 183^{18}; 167^{10}	4-05-00-01693	ace 4; bz 4; eth 4; EtOH 4
7743	Naphthalene, 2-(bromomethyl)-	$C_{11}H_9Br$ 221.10	939-26-4 56	 213^{100}; 167^{14}	4-05-00-01698	EtOH 3; eth 3; chl 3; HOAc 3
7744	Naphthalene, 1-bromo-3-nitro-	$C_{10}H_6BrNO_2$ 252.07	7499-65-2 132.3	 sub	4-05-00-01678	eth 4; EtOH 4
7745	Naphthalene, 2-bromo-1-nitro- 2-Bromo-1-nitronaphthalene	$C_{10}H_6BrNO_2$ 252.07	4185-62-0 104.3	 >360	4-05-00-01677	bz 4; eth 4; EtOH 4
7746	Naphthalene, 1-butyl-	$C_{14}H_{16}$ 184.28	1634-09-9 -19.8	 289.3	4-05-00-01737 0.9738^{20}	H_2O 1; EtOH 3; eth 3; ace 3 1.5819^{20}
7747	Naphthalene, 2-butyl- 2-Butylnaphthalene	$C_{14}H_{18}$ 186.30	1134-62-9 -5	 292	4-05-00-01738 0.9673^{20}	ace 4; bz 4; EtOH 4 1.5777^{20}
7748	1-Naphthalenecarbonitrile	$C_{11}H_7N$ 153.18	86-53-3 37.5	 299	4-09-00-02407 1.1000[illegible]	H_2O 1; EtOH 4; eth 4; lig 3 1.6208^{18}
7749	2-Naphthalenecarbonitrile	$C_{11}H_7N$ 153.18	613-46-7 66	 306.5	4-09-00-02419 1.0755^{60}	H_2O 2; EtOH 3; eth 3; chl 2
7750	2-Naphthalenecarbonyl chloride	$C_{11}H_7ClO$ 190.63	2243-83-6 51	 305	4-09-00-02417	bz 4; eth 4; chl 4
7751	1-Naphthalenecarboxaldehyde	$C_{11}H_8O$ 156.18	66-77-3 33.5	 292	4-07-00-01286 1.1503^{20}	H_2O 1; EtOH 3; eth 3; ace 3 1.6507^{20}
7752	2-Naphthalenecarboxaldehyde	$C_{11}H_8O$ 156.18	66-99-9 62	 160^{19}	4-07-00-01288 1.0775^{99}	H_2O 2; EtOH 4; eth 4; ace 3 1.6211^{99}
7753	1-Naphthalenecarboxaldehyde, 2-ethoxy-	$C_{13}H_{12}O_2$ 200.24	19523-57-0 115	 185-7^{25}	3-08-00-01110	EtOH 3; HOAc 3
7754	1-Naphthalenecarboxaldehyde, 2-hydroxy-	$C_{11}H_8O_2$ 172.18	708-06-5 83	 192^{27}	4-08-00-01160	H_2O 1; EtOH 3; eth 3; aq alk 3
7755	1-Naphthalenecarboxaldehyde, 4-methoxy-	$C_{12}H_{10}O_2$ 186.21	15971-29-6 34.8	 212^{40}	4-08-00-01170 1.1879^{39}	

No.	Name Synonym	Mol. Form. Mol. Wt.	CAS RN mp/°C	Merck No. bp/°C	Beil. Ref. den/g cm^{-3}	Solubility n_D
7756	1-Naphthalenecarboxaldehyde, 4-methyl- 1-Naphthaldehyde, 4-methyl-	$C_{12}H_{10}O$ 170.21	33738-48-6 33.5	 175^{13}	3-07-00-01972 1.1252^{38}	
7757	1-Naphthalenecarboxamide	$C_{11}H_9NO$ 171.20	2243-81-4 205.8	 sub	4-09-00-02405	H_2O 2; EtOH 2; con HCl 3; HOAc 3
7758	1-Naphthalenecarboxylic acid 1-Naphthoic acid	$C_{11}H_8O_2$ 172.18	86-55-5 161	6301 >300	4-09-00-02402 1.398^{25}	eth 4; EtOH 4; chl 4 1.46
7759	2-Naphthalenecarboxylic acid 2-Naphthoic acid	$C_{11}H_8O_2$ 172.13	93-09-4 185.5	6302 >300	4-09-00-02414 1.077^{100}	H_2O 2; EtOH 3; eth 3; chl 3
7760	1-Naphthalenecarboxylic acid, 5-amino-	$C_{11}H_9NO_2$ 187.20	32018-88-5 212.3	 sub	4-14-00-01812	EtOH 4; HOAc 4
7761	1-Naphthalenecarboxylic acid, 5-bromo-	$C_{11}H_7BrO_2$ 251.08	16726-67-3 261	 sub	3-09-00-03153	H_2O 2; EtOH 2; eth 2; bz 4
7762	1-Naphthalenecarboxylic acid, 5-chloro- 1-Naphthoic acid, 5-chloro-	$C_{11}H_7ClO_2$ 206.63	16650-52-5 245	 sub	3-09-00-03149	H_2O 2; EtOH 4; bz 2; HOAc 2
7763	1-Naphthalenecarboxylic acid, 8-chloro-	$C_{11}H_7ClO_2$ 206.63	4537-00-2 171.5	 sub	3-09-00-03151	H_2O 3; EtOH 3; ace 4; bz 3
7764	1-Naphthalenecarboxylic acid, 3,4-dihydro- 3,4-Dihydro-1-naphthoic acid	$C_{11}H_{10}O_2$ 174.20	3333-23-1 125	 306	4-09-00-02338	H_2O 2; EtOH 3; chl 3; MeOH 3
7765	1-Naphthalenecarboxylic acid, ethyl ester	$C_{13}H_{12}O_2$ 200.24	3007-97-4	 310	4-09-00-02403 1.1264^{15}	H_2O 1; EtOH 3 1.5966^{15}
7766	1-Naphthalenecarboxylic acid, methyl ester	$C_{12}H_{10}O_2$ 186.21	2459-24-7 59.5	 168^{20}; $101^{0.04}$	4-09-00-02402 1.1290^{20}	bz 4; EtOH 4 1.6086^{20}
7767	1-Naphthalenecarboxylic acid, 5-nitro-	$C_{11}H_7NO_4$ 217.18	1975-44-6 241.5	 sub	4-09-00-02411	H_2O 1; EtOH 3; eth 2; bz 2
7768	2-Naphthalenecarboxylic acid, 3-amino- 3-Amino-2-naphthoic acid	$C_{11}H_9NO_2$ 187.20	5959-52-4 216.5	464	3-14-00-01341	EtOH 3; eth 3
7769	2-Naphthalenecarboxylic acid, 5-bromo- 5-Bromo-2-naphthoic acid	$C_{11}H_7BrO_2$ 251.08	1013-83-8 270	 sub	3-09-00-03196	bz 4; eth 4; EtOH 4; HOAc 4
7770	2-Naphthalenecarboxylic acid, ethyl ester	$C_{13}H_{12}O_2$ 200.24	3007-91-8 32	 308.5	4-09-00-02415 1.1143^{23}	eth 4; EtOH 4; chl 4; HOAc 4 1.5951^{23}
7771	2-Naphthalenecarboxylic acid, 1-hydroxy- 1-Hydroxy-2-naphthoic acid	$C_{11}H_8O_3$ 188.18	86-48-6 195	4766	4-10-00-01181	H_2O 2; EtOH 4; eth 4; bz 3
7772	2-Naphthalenecarboxylic acid, 3-hydroxy- 3-Hydroxy-2-naphthoic acid	$C_{11}H_8O_3$ 188.18	92-70-6 222.5	4767	4-10-00-01184	H_2O 2; EtOH 4; eth 4; bz 3
7773	2-Naphthalenecarboxylic acid, 3-hydroxy-, ethyl ester	$C_{13}H_{12}O_3$ 216.24	7163-25-9 85	 291	4-10-00-01186	ace 4; chl 4
7774	2-Naphthalenecarboxylic acid, 3-hydroxy-, hexadecyl ester Hexadecyl 3-hydroxy-2-naphthoate	$C_{27}H_{40}O_3$ 412.61	531-84-0 72.5	4604	3-10-00-01087	bz 4; HOAc 4
7775	2-Naphthalenecarboxylic acid, 3-hydroxy-, methyl ester	$C_{12}H_{10}O_3$ 202.21	883-99-8 75.5	 206	4-10-00-01186	H_2O 1; EtOH 3; os 3
7776	2-Naphthalenecarboxylic acid, methyl ester	$C_{12}H_{10}O_2$ 186.21	2459-25-8 77	 290	4-09-00-02414	bz 4; eth 4; EtOH 4; chl 4
7777	2-Naphthalenecarboxylic acid, 4,4'-methylenebis[3-hydroxy- Pamoic acid	$C_{23}H_{16}O_6$ 388.38	130-85-8 315	6955	4-10-00-02362	
7778	2-Naphthalenecarboxylic acid, 1,2,3,4-tetrahydro- 1,2,3,4-Tetrahydro-2-naphthoic acid	$C_{11}H_{12}O_2$ 176.22	53440-12-3 97.7	 168^{15}	3-09-00-02805	bz 4; eth 4; EtOH 4; chl 4
7779	2-Naphthalenecarboxylic acid, 5,6,7,8-tetrahydro-	$C_{11}H_{12}O_2$ 176.22	1131-63-1 155.5	 216^{14}	4-09-00-02096	EtOH 4; bz 3; HOAc 3; peth 1
7780	Naphthalene, 1-chloro- 1-Naphthyl chloride	$C_{10}H_7Cl$ 162.62	90-13-1 -2.5	2149 259; 106.5^{5}	4-05-00-01658 1.1938^{20}	H_2O 1; EtOH 3; eth 3; bz 3 1.6326^{20}
7781	Naphthalene, 2-chloro- 2-Chloronaphthalene	$C_{10}H_7Cl$ 162.62	91-58-7 61	2150 256	4-05-00-01660 1.1377^{71}	H_2O 1; EtOH 3; eth 3; bz 3 1.6079^{13}
7782	Naphthalene, 2-chlorodecahydro- Decalin, 2-chloro	$C_{10}H_{17}Cl$ 172.70	5597-81-9	 112^{15}	4-05-00-00313 1.0421^{20}	 1.5020^{20}
7783	Naphthalene, 1-(chloromethyl)-	$C_{11}H_9Cl$ 176.65	86-52-2 32	 291.5	4-05-00-01692 1.1813^{20}	H_2O 1; EtOH 3; ctc 3; peth 3 1.6380^{20}
7784	Naphthalene, 1-chloro-2-methyl- 1-Chloro-2-methylnaphthalene	$C_{11}H_9Cl$ 176.65	5859-45-0	 $150\text{-}2^{10}$	3-05-00-01632	ctc 3 1.6265^{20}
7785	Naphthalene, 2-(chloromethyl)-	$C_{11}H_9Cl$ 176.65	2506-41-4 48.5	 169^{20}	4-05-00-01697	H_2O 1; EtOH 3; peth 3
7786	Naphthalene, 1-chloro-5-nitro- 1-Chloro-5-nitronaphthalene	$C_{10}H_6ClNO_2$ 207.62	605-63-0 111	 >360	3-05-00-01597	EtOH 4
7787	Naphthalene, 1-chloro-8-nitro- 1-Chloro-8-nitronaphthalene	$C_{10}H_6ClNO_2$ 207.62	602-37-9 95.8	 175^{2}	4-05-00-01676	EtOH 3; bz 3; chl 2; HOAc 3
7788	Naphthalene, 2-chloro-1-nitro- 2-Chloro-1-nitronaphthalene	$C_{10}H_6ClNO_2$ 207.62	4185-63-1 99.5	 >360	4-05-00-01676 1.4854^{25}	EtOH 4; eth 4; ace 4; bz 4
7789	Naphthalene, decahydro- Decalin	$C_{10}H_{18}$ 138.25	91-17-8 -43	2839 155.5	3-05-00-00245 0.8965^{22}	ace 4; bz 4; eth 4; EtOH 4 1.4810^{20}
7790	Naphthalene, decahydro-, *cis-* *cis*-Decalin	$C_{10}H_{18}$ 138.25	493-01-6 -42.9	2839 195.8	4-05-00-00310 0.8965^{20}	H_2O 1; EtOH 5; eth 4; ace 4 1.4810^{20}
7791	Naphthalene, decahydro-, *trans-* *trans*-Decalin	$C_{10}H_{18}$ 138.25	493-02-7 -30.3	2839 187.3	4-05-00-00311 0.8699^{20}	H_2O 1; EtOH 4; eth 4; ace 4 1.4695^{20}

No.	Name Synonym	Mol. Form. Mol. Wt.	CAS RN mp/°C	Merck No. bp/°C	Beil. Ref. den/g cm^{-3}	Solubility n_D
7792	Naphthalene, decahydro-2-methylene-, *trans*- Decalin, 2-methylene (*trans*)	$C_{11}H_{18}$ 150.26	7787-72-6	 201; 82^{10}	3-05-00-00397 0.8897^{20}	 1.4841^{22}
7793	Naphthalene, 1-decyl- Decane, 1-(1-naphthyl)-	$C_{20}H_{28}$ 268.44	26438-27-7 15	 379	4-05-00-01786 0.9322^{20}	 1.5435^{20}
7794	Naphthalene, 2-decyl-	$C_{20}H_{28}$ 268.44	14188-79-5 20	 387	4-05-00-01786 0.9253^{20}	 1.5413^{20}
7795	1,2-Naphthalenediamine 1,2-Diaminonaphthalene	$C_{10}H_{10}N_2$ 158.20	938-25-0 98.5	 214^{13}; $150^{0.5}$	4-13-00-00337	H_2O 2; EtOH 4; eth 4; chl 4
7796	1,4-Naphthalenediamine 1,4-Diaminonaphthalene	$C_{10}H_{10}N_2$ 158.20	2243-61-0 120		4-13-00-00338	ace 4; bz 4; EtOH 4; chl 4 1.6441^{18}
7797	1,5-Naphthalenediamine	$C_{10}H_{10}N_2$ 158.20	2243-62-1 190	 sub	4-13-00-00340 1.4^{25}	H_2O 3; EtOH 3; eth 3; chl 4
7798	1,8-Naphthalenediamine Naphthalene, 1,8-diamino	$C_{10}H_{10}N_2$ 158.20	479-27-6 66.5	6291 205^{12}	4-13-00-00344 1.1265^{90}	eth 4; EtOH 4 1.6828^{99}
7799	2,3-Naphthalenediamine 2,3-Diaminonaphthalene	$C_{10}H_{10}N_2$ 158.20	771-97-1 199		4-13-00-00346 1.0968^{26}	H_2O 2; EtOH 4; eth 3 1.6392^{26}
7800	1,4-Naphthalenediamine, 2-methyl- Vitamin K6	$C_{11}H_{12}N_2$ 172.23	83-68-1 113.5	9937	3-13-00-00406	
7801	Naphthalene, 1,3-dibromo- 1,3-Dibromonaphthalene	$C_{10}H_6Br_2$ 285.97	52358-73-3 62.8	 290	3-05-00-01586	EtOH 4
7802	Naphthalene, 1,4-dibromo- 1,4-Dibromonaphthalene	$C_{10}H_6Br_2$ 285.97	83-53-4 83	 310	4-05-00-01669	H_2O 1; EtOH 3; eth 3; HOAc 2
7803	Naphthalene, 1,5-dibromo- 1,5-Dibromonaphthalene	$C_{10}H_6Br_2$ 285.97	7351-74-8 131.3	 326	4-05-00-01669	eth 4; EtOH 4
7804	1,8-Naphthalenedicarbonyl dichloride Naphthaloyl chloride	$C_{12}H_6Cl_2O_2$ 253.08	6423-29-6 85	 $197^{0.2}$	3-09-00-04468	bz 4; chl 4
7805	1,8-Naphthalenedicarboxylic acid Naphthalic acid	$C_{12}H_8O_4$ 216.19	518-05-8 260	6299	4-09-00-03534	H_2O 1; EtOH 2; eth 2
7806	1,8-Naphthalenedicarboxylic acid, diethyl ester Naphthalic acid, diethyl ester	$C_{16}H_{16}O_4$ 272.30	58618-39-6 59.5	 238^{19}; 202^{11}	4-09-00-03534 1.1399^{70}	
7807	Naphthalene, 1,2-dichloro- 1,2-Dichloronaphthalene	$C_{10}H_6Cl_2$ 197.06	2050-69-3 36	 296.5	4-05-00-01661 1.3147^{49}	EtOH 3; eth 3 1.5338^{49}
7808	Naphthalene, 1,3-dichloro- 1,3-Dichloronaphthalene	$C_{10}H_6Cl_2$ 197.06	2198-75-6 62.3	 291	4-05-00-01661	EtOH 3
7809	Naphthalene, 1,4-dichloro- 1,4-Dichloronaphthalene	$C_{10}H_6Cl_2$ 197.06	1825-31-6 67.5	 288; 147^{12}	4-05-00-01661 1.2997^{76}	H_2O 1; EtOH 2; eth 3; ace 4 1.6228^{76}
7810	Naphthalene, 1,5-dichloro- 1,5-Dichloronaphthalene	$C_{10}H_6Cl_2$ 197.06	1825-30-5 107	 sub	4-05-00-01662 1.4900^{20}	H_2O 1; EtOH 2; eth 3
7811	Naphthalene, 1,7-dichloro- 1,7-Dichloronaphthalene	$C_{10}H_6Cl_2$ 197.06	2050-73-9 63.5	 285.5	4-05-00-01662 1.2611^{100}	EtOH 3; eth 3; bz 3; HOAc 3 1.6092^{100}
7812	Naphthalene, 1,8-dichloro- 1,8-Dichloronaphthalene	$C_{10}H_6Cl_2$ 197.06	2050-74-0 89	 sub	4-05-00-01662 1.2924^{100}	EtOH 3; peth 3 1.6236^{100}
7813	Naphthalene, 2,6-dichloro- 2,6-Dichloronaphthalene	$C_{10}H_6Cl_2$ 197.06	2065-70-5 140.5	 285	4-05-00-01662	EtOH 2; eth 3; bz 3; chl 3
7814	Naphthalene, 1,2-dihydro-	$C_{10}H_{10}$ 130.19	447-53-0 -8	 206.5	4-05-00-01543 0.9974^{20}	 1.5814^{20}
7815	Naphthalene, 1,4-dihydro- Δ 2-Dialin	$C_{10}H_{10}$ 130.19	612-17-9 25	 211.5	4-05-00-01544 0.9928^{33}	 1.5577^{20}
7816	Naphthalene, 1,2-dihydro-3-methyl-	$C_{11}H_{12}$ 144.22	2717-44-4	 105^{13}; $64^{0.6}$	4-05-00-01552 0.9819^{20}	H_2O 1; eth 3; bz 3 1.5751^{20}
7817	Naphthalene, 1,2-dihydro-4-methyl-	$C_{11}H_{12}$ 144.22	4373-13-1	 112^{18}; 84^5	4-05-00-01551 0.9895^{20}	H_2O 1; eth 3; bz 3 1.5758^{20}
7818	Naphthalene, 1,2-dimethyl- 1,2-Dimethylnaphthalene	$C_{12}H_{12}$ 156.23	573-98-8 1.6	 266.5	4-05-00-01708 1.0179^{20}	H_2O 1; eth 3; bz 3 1.6166^{20}
7819	Naphthalene, 1,3-dimethyl- 1,3-Dimethylnaphthalene	$C_{12}H_{12}$ 156.23	575-41-7 -6	 263	4-05-00-01708 1.0144^{20}	H_2O 1; eth 3; bz 3 1.6140^{20}
7820	Naphthalene, 1,4-dimethyl- 1,4-Dimethylnaphthalene	$C_{12}H_{12}$ 156.23	571-58-4 7.6	 268	4-05-00-01709 1.0166^{20}	H_2O 1; EtOH 4; eth 5; ace 5 1.6127^{20}
7821	Naphthalene, 1,5-dimethyl- 1,5-Dimethylnaphthalene	$C_{12}H_{12}$ 156.23	571-61-9 82	 265	4-05-00-01710	bz 4; eth 4
7822	Naphthalene, 1,6-dimethyl- 1,6-Dimethylnaphthalene	[illegible] 156.23	575-43-9 -16.9	 264	4-05-00-01711 1.0021[illegible]	H_2O 1; eth 3; bz 3 1.6100[illegible]
7823	Naphthalene, 1,7-dimethyl- 1,7-Dimethylnaphthalene	$C_{12}H_{12}$ 156.23	575-37-1 -13.9	 263	4-05-00-01711 1.0115^{20}	H_2O 1; eth 3; bz 3 1.6083^{20}
7824	Naphthalene, 1,8-dimethyl- 1,8-Dimethylnaphthalene	$C_{12}H_{12}$ 156.23	569-41-5 65	 270	4-05-00-01712 1.003^{20}	H_2O 1; eth 3; bz 3
7825	Naphthalene, 2,3-dimethyl- Guajen	$C_{12}H_{12}$ 156.23	581-40-8 105	 268	4-05-00-01713 1.003^{20}	bz 4; eth 4 1.5060^{20}
7826	Naphthalene, 2,6-dimethyl- 2,6-Dimethylnaphthalene	$C_{12}H_{12}$ 156.23	581-42-0 112	 262	4-05-00-01714 1.003^{20}	
7827	Naphthalene, 2,7-dimethyl- 2,7-Dimethylnaphthalene	$C_{12}H_{12}$ 156.23	582-16-1 97	 265	4-05-00-01715 1.003^{20}	
7828	Naphthalene, 1-(1,1-dimethylethyl)-	$C_{14}H_{16}$ 184.28	17085-91-5 -11.3	 279	4-05-00-01739 0.9629^{20}	 1.5726^{20}
7829	Naphthalene, 2-(1,1-dimethylethyl)-	$C_{14}H_{16}$ 184.28	2876-35-9 -4	 276; 125^4	4-05-00-01740 0.9674^{20}	ace 4; bz 4; EtOH 4 1.5685^{20}

No.	Name Synonym	Mol. Form. Mol. Wt.	CAS RN mp/°C	Merck No. bp/°C	Beil. Ref. den/g cm^{-3}	Solubility n_D
7830	Naphthalene, 1,6-dimethyl-4-(1-methylethyl)- Cadalene	$C_{15}H_{18}$ 198.31	483-78-3	1607 294; 149^{10}	4-05-00-01758 0.9667^{25}	oils 4 1.5785^{25}
7831	Naphthalene, 1,3-dinitro- 1,3-Dinitronaphthalene	$C_{10}H_6N_2O_4$ 218.17	606-37-1 148	 sub	4-05-00-01680	H_2O 1; EtOH 3; ace 3
7832	Naphthalene, 1,5-dinitro- 1,5-Dinitronaphthalene	$C_{10}H_6N_2O_4$ 218.17	605-71-0 219	 sub	4-05-00-01680 1.5860^{20}	H_2O 1; EtOH 2; eth 4; ace 2
7833	Naphthalene, 1,8-dinitro- 1,8-Dinitronaphthalene	$C_{10}H_6N_2O_4$ 218.17	602-38-0 173	 445 dec	4-05-00-01681	H_2O 1; EtOH 2; ace 3; bz 2
7834	1,3-Naphthalenediol Naphthoresorcinol	$C_{10}H_8O_2$ 160.17	132-86-5 123.5	6316	4-06-00-06543	H_2O 3; EtOH 3; eth 3; ace 2
7835	1,5-Naphthalenediol	$C_{10}H_8O_2$ 160.17	83-56-7 262 dec	 sub	4-06-00-06554	H_2O 2; EtOH 2; eth 4; ace 4
7836	1,6-Naphthalenediol	$C_{10}H_8O_2$ 160.17	575-44-0 138	 sub	4-06-00-06557	H_2O 2; EtOH 2; eth 3; ace 3
7837	1,7-Naphthalenediol	$C_{10}H_8O_2$ 160.17	575-38-2 180.5	 sub	4-06-00-06559	H_2O 2; EtOH 4; eth 4; bz 3
7838	2,6-Naphthalenediol	$C_{10}H_8O_2$ 160.17	581-43-1 220	 sub	4-06-00-06566	H_2O 2; EtOH 3; eth 3; ace 3
7839	2,7-Naphthalenediol	$C_{10}H_8O_2$ 160.17	582-17-2 193	 sub	4-06-00-06570	H_2O 3; EtOH 3; eth 3; ace 2
7840	1,8-Naphthalenediol, cyclic sulfite 1-Naphthalenesulfonic acid, 8-hydroxy, sulfone	$C_{10}H_6O_3S$ 206.22	21849-97-8 97	 >360	2-06-00-00954	bz 4; chl 4
7841	1,4-Naphthalenediol, 2-methyl-, diacetate Menadiol diacetate	$C_{15}H_{14}O_4$ 258.27	573-20-6 113	5710	4-06-00-06581	EtOH 4
7842	1,2-Naphthalenedione 1,2-Naphthoquinone	$C_{10}H_6O_2$ 158.16	524-42-5 146	6314	4-07-00-02417 1.450^{25}	H_2O 3; EtOH 3; eth 3; sulf 3
7843	1,4-Naphthalenedione 1,4-Naphthoquinone	$C_{10}H_6O_2$ 158.16	130-15-4 128.5	6315 sub	4-07-00-02422	H_2O 2; EtOH 4; eth 3; bz 3
7844	1,2-Naphthalenedione, 3-bromo- 3-Bromo-1,2-naphthoquinone	$C_{10}H_5BrO_2$ 237.05	7474-83-1 178	 sub	4-07-00-02422	bz 4; EtOH 4
7845	1,4-Naphthalenedione, 2-bromo-3-hydroxy-	$C_{10}H_5BrO_3$ 253.05	1203-39-0 202	 sub	4-08-00-02364	ace 4
7846	1,4-Naphthalenedione, 2-bromo-3-methyl-	$C_{11}H_7BrO_2$ 251.08	3129-39-3 152.8	 sub 100	4-07-00-02431	ace 4; bz 4; eth 4; EtOH 4
7847	1,4-Naphthalenedione, 5-chloro- 5-Chloro-1,4-naphthoquinone	$C_{10}H_5ClO_2$ 192.60	40242-15-7 163	 sub	4-07-00-02425	EtOH 4; lig 4
7848	1,4-Naphthalenedione, 2-chloro-3-hydroxy-	$C_{10}H_5ClO_3$ 208.60	1526-73-4 216.5	 sub	4-08-00-02363	bz 4; eth 4; EtOH 4
7849	1,2-Naphthalenedione, 3,4-dichloro-	$C_{10}H_4Cl_2O_2$ 227.05	18398-36-2 184	 sub	4-07-00-02422	bz 4; chl 4
7850	1,4-Naphthalenedione, 2,3-dichloro- Dichlone	$C_{10}H_4Cl_2O_2$ 227.05	117-80-6 195	3032	4-07-00-02426	H_2O 1; EtOH 2; eth 2; bz 2
7851	1,4-Naphthalenedione, 5,6-dichloro- 1,4-Naphthoquinone, 5,6-dichloro	$C_{10}H_4Cl_2O_2$ 227.05	56961-93-4 181	 sub	0-07-00-00730	eth 4
7852	1,4-Naphthalenedione, 2,3-dihydroxy-	$C_{10}H_6O_4$ 190.16	605-37-8 282	 sub	4-08-00-02953	H_2O 2; EtOH 2; eth 2; ace 3
7853	1,4-Naphthalenedione, 5,8-dihydroxy-	$C_{10}H_6O_4$ 190.16	475-38-7 232	 sub	4-08-00-02946	H_2O 2; EtOH 2; eth 2; HOAc 3
7854	1,4-Naphthalenedione, 5,8-dihydroxy-6-(1-hydroxy-4-methyl-3-pentenyl)-, (*S*)- Alkannin	$C_{16}H_{16}O_5$ 288.30	23444-65-7 149	243 sub 140	3-08-00-04089	EtOH 4
7855	1,4-Naphthalenedione, 5,8-dihydroxy-6-methoxy-2-methyl-3-(2-oxopropyl)- Javanicin	$C_{15}H_{14}O_6$ 290.27	476-45-9 208 dec	5143	4-08-00-03646	alk 3
7856	1,4-Naphthalenedione, 3,5-dihydroxy-2-methyl-	$C_{11}H_8O_4$ 204.18	478-40-0 181	 sub 100	4-08-00-02958	H_2O 3; EtOH 3; eth 3; peth 3
7857	1,4-Naphthalenedione, 2-ethyl-3,5,6,7,8-pentahydroxy- Echinochrome A	$C_{12}H_{10}O_7$ 266.21	517-82-8 220 dec	3470 sub 120	4-08-00-03714	H_2O 2; EtOH 3; eth 4; ace 3
7858	1,4-Naphthalenedione, 6-ethyl-2,3,5,7,8-pentahydroxy-	$C_{12}H_{10}O_7$ 266.21	1471-96-1 220 dec	 $120^{0.0004}$	4-08-00-03714	ace 4; eth 4; EtOH 4
7859	1,4-Naphthalenedione, 2-hydroxy- Lawsone	$C_{10}H_6O_3$ 174.16	83-72-7 195-6 dec	5263	4-08-00-02360	EtOH 4; eth 1; bz 1; chl 1
7860	1,4-Naphthalenedione, 5-hydroxy- Juglone	$C_{10}H_6O_3$ 174.16	481-39-0 155	5150 sub	4-08-00-02368	H_2O 1; EtOH 3; eth 3; bz 3
7861	1,4-Naphthalenedione, 2-hydroxy-3-methyl- Phthiocol	$C_{11}H_8O_3$ 188.18	483-55-6 173.5	7352 sub	4-08-00-02375	ace 4; eth 4
7862	1,4-Naphthalenedione, 2-hydroxy-3-(3-methyl-2-butenyl)- Lapachol	$C_{15}H_{14}O_3$ 242.27	84-79-7 139.5	5235	4-08-00-02487	H_2O 1; EtOH 3; eth 3; bz 3
7863	1,4-Naphthalenedione, 5-hydroxy-2-methyl- 1,4-Naphthoquinone, 5-hydroxy-2-methyl	$C_{11}H_8O_3$ 188.18	481-42-5 78.5	7511 sub	4-08-00-02376	ace 4; bz 4; eth 4; EtOH 4
7864	1,4-Naphthalenedione, 2-methyl- Menadione	$C_{11}H_8O_2$ 172.18	58-27-5 107	5714	4-07-00-02430	H_2O 1; EtOH 2; eth 3; bz 3
7865	1,4-Naphthalenedione, 2-methyl-3-(3,7,11,15-tetramethyl-2-hexadecenyl)-, [*R*-[*R**,*R**-(*E*)]]- Vitamin K1	$C_{31}H_{46}O_2$ 450.71	84-80-0 -20	9933 $140\text{-}5^{0.001}$	4-07-00-02496 0.964^{25}	ace 4; bz 4; eth 4; EtOH 4 1.5250^{25}

No.	Name Synonym	Mol. Form. Mol. Wt.	CAS RN mp/°C	Merck No. bp/°C	Beil. Ref. den/g cm^{-3}	Solubility n_D
7866	1,4-Naphthalenedione, 2-(phenylamino)-	$C_{16}H_{11}NO_2$ 249.27	6628-97-3 193	 sub	4-14-00-00399	bz 4; eth 4
7867	1,5-Naphthalenedisulfonic acid Armstrong's acid	$C_{10}H_8O_6S_2$ 288.30	81-04-9 240-5 dec	812	4-11-00-00561 1.493^{25}	H_2O 4; EtOH 3; eth 1
7868	1,6-Naphthalenedisulfonic acid Eiver-pick acid	$C_{10}H_8O_6S_2$ 288.30	525-37-1 125 dec	6292	4-11-00-00562	H_2O 4; EtOH 3; eth 1
7869	2,7-Naphthalenedisulfonic acid Naphthalene-2,7-disulfonic acid	$C_{10}H_8O_6S_2$ 288.30	92-41-1 199	6294	4-11-00-00563	H_2O 3; con HCl 2
7870	1,3-Naphthalenedisulfonic acid, 7-amino- Amido-G-Acid	$C_{10}H_9NO_6S_2$ 303.32	86-65-7 274	412	4-14-00-02811	H_2O 4; EtOH 4
7871	1,3-Naphthalenedisulfonic acid, 7-hydroxy- 2-Naphthol-6,8-disulfonic acid	$C_{10}H_8O_7S_2$ 304.30	118-32-1	6308	4-11-00-00625	H_2O 3
7872	1,6-Naphthalenedisulfonic acid, 4-amino- 1-Naphthylamine-4,7-disulfonic acid	$C_{10}H_9NO_6S_2$ 303.32	85-75-6	6322	2-14-00-00476	H_2O 4
7873	1,7-Naphthalenedisulfonic acid, 4-amino- 1-Naphthylamine-4,6-disulfonic acid	$C_{10}H_9NO_6S_2$ 303.32	85-74-5	6321	2-14-00-00477	H_2O 4; EtOH 4
7874	2,7-Naphthalenedisulfonic acid, 4-amino-5-hydroxy- 1-Amino-8-naphthol-3,6-disulfonic acid	$C_{10}H_9NO_7S_2$ 319.32	90-20-0	6305	4-14-00-02824	H_2O 2; EtOH 2; eth 2
7875	2,7-Naphthalenedisulfonic acid, 4,5-dihydroxy- Chromotropic acid	$C_{10}H_8O_8S_2$ 320.30	148-25-4	2243	4-11-00-00636	H_2O 3; EtOH 1; eth 1; alk 3
7876	2,7-Naphthalenedisulfonic acid, 3-hydroxy- 2-Naphthol-3,6-disulfonic acid	$C_{10}H_8O_7S_2$ 304.30	148-75-4 dec	6307	4-11-00-00624	H_2O 4; EtOH 4
7877	1,3-Naphthalenedisulfonic acid, 6,6'-[(3,3'-dimethyl[1,1'-biphenyl]-4,4'-diyl)bis(azo)]bis[4-amino-5-hydroxy-, C.I. Direct blue 53	$C_{34}H_{24}N_6Na_4O_{14}S_4$ 960.82	314-13-6	3863		H_2O 3; EtOH 3; acid 3
7878	2,7-Naphthalenedisulfonic acid, 4,5-dihydroxy-3-[(4-nitrophenyl)azo]-, disodium salt	$C_{16}H_9N_3Na_2O_{10}S_2$ 513.37	548-80-1	2242		
7879	Naphthalene, 1-dodecyl- Dodecane, 1-(1-naphthyl)-	$C_{22}H_{32}$ 296.50	26438-28-8 27	 415	4-05-00-01790 0.9240^{20}	 1.5364^{20}
7880	Naphthalene, 2-dodecyl- Dodecane, 1-(2-naphthyl)-	$C_{22}H_{32}$ 296.50	60899-39-0 26	 414	4-05-00-01791 0.9177^{20}	 1.5343^{20}
7881	Naphthalene, 1-ethenyl-	$C_{12}H_{10}$ 154.21	826-74-4	 $124\text{-}5^{15}$	4-05-00-01833 1.036^{18}	 1.644^{20}
7882	Naphthalene, 2-ethenyl- 2-Vinylnaphthalene	$C_{12}H_{10}$ 154.21	827-54-3 66	 135^{18}; 95^2	4-05-00-01833	H_2O 1; EtOH 3; ace 3; bz 3
7883	Naphthalene, 1-ethoxy-	$C_{12}H_{12}O$ 172.23	5328-01-8 5.5	 280.5	4-06-00-04212 1.060^{20}	H_2O 1; EtOH 4; eth 4 1.5953^{25}
7884	Naphthalene, 2-ethoxy- 2-Ethoxynaphthalene	$C_{12}H_{12}O$ 172.23	93-18-5 37.5	3709 282	4-06-00-04257 1.0640^{20}	H_2O 1; EtOH 3; eth 3; tol 3 1.5975^{36}
7885	Naphthalene, 1-ethyl- 1-Ethylnaphthalene	$C_{12}H_{12}$ 156.23	1127-76-0 -13.9	 258.6	4-05-00-01705 1.0082^{20}	H_2O 1; EtOH 5; eth 5 1.6062^{20}
7886	Naphthalene, 2-ethyl- 2-Ethylnaphthalene	$C_{12}H_{12}$ 156.23	939-27-5 -7.4	 258	4-05-00-01707 0.9922^{20}	H_2O 1; EtOH 5; eth 5; chl 2 1.5999^{20}
7887	Naphthalene, 1-ethyl-5-methyl- 1-Methyl-5-ethylnaphthalene	$C_{13}H_{14}$ 170.25	17057-92-0 40	 133^{10}	3-05-00-01659	 1.600^{30}
7888	Naphthalene, 1-ethyl-1,2,3,4-tetrahydro- 1-Ethyltetralin	$C_{12}H_{16}$ 160.26	13556-58-6	 241	4-05-00-01429 0.9528^{20}	 1.5318^{20}
7889	Naphthalene, 2-ethyl-1,2,3,4-tetrahydro- 2-Ethyltetralin	$C_{12}H_{16}$ 160.26	32367-54-7 -42.8	 235	4-05-00-01429 0.9401^{15}	 1.5250^{15}
7890	Naphthalene, 5-ethyl-1,2,3,4-tetrahydro-	$C_{12}H_{16}$ 160.26	42775-75-7 -44.5	 245	4-05-00-01428 0.973^{20}	 1.540^{20}
7891	Naphthalene, 6-ethyl-1,2,3,4-tetrahydro- 6-Ethyltetralin	$C_{12}H_{16}$ 160.26	22531-20-0 -70	 244	4-05-00-01429 0.9632^{17}	 1.5414^{16}
7892	Naphthalene, 1-ethynyl-	$C_{12}H_8$ 152.20	15727-65-8 2	 $143\text{-}4^{25}$	3-05-00-01935 1.0513^{20}	 1.6360^{20}
7893	Naphthalene, 1-fluoro- 1-Fluoronaphthalene	$C_{10}H_7F$ 146.16	321-38-0 -9	 215; 80^{11}	4-05-00-01657 1.1322^{20}	H_2O 1; EtOH 3; eth 3; bz 3 1.5939^{20}
7894	Naphthalene, 2-fluoro- 2-Fluoronaphthalene	$C_{10}H_7F$ 146.16	323-09-1 61	 212; 90^{16}	4-05-00-01658	H_2O 1; EtOH 3; eth 3; bz 3
7895	Naphthalene, 1-heptyl- 1-Heptylnaphthalene	$C_{17}H_{22}$ 226.36	2876-52-0 -8	 340	4-05-00-01775 0.9491^{20}	 1.5582^{20}
[illegible]	[illegible] 2-Heptylnaphthalene	[illegible] 226.36	[illegible] 1	 341	4 05 00 01775 0.9410^{20}	 1.5556^{20}
7897	Naphthalene, 1,2,3,4,4a,8a-hexahydro- Naphthalene, 1,2,3,4,9,10-hexahydro	$C_{10}H_{14}$ 134.22	62690-62-4	 197	4-05-00-01082 0.934^{23}	bz 4; eth 4 1.5260^{16}
7898	Naphthalene, 1,2,4a,5,8,8a-hexahydro-4,7-dimethyl-1-(1-methylethyl)-, [1*S*-(1α,4aβ,8aα)]-	$C_{15}H_{24}$ 204.36	523-47-7	 274; 136^{11}	3-05-00-01086 0.9230^{20}	eth 4; lig 4 1.5059^{20}
7899	Naphthalene, 1,2,3,5,6,8a-hexahydro-4,7-dimethyl-1-(1-methylethyl)-, (1*S*-*cis*)-	$C_{15}H_{24}$ 204.36	483-76-1	 $125\text{-}6^{12}$	2-05-00-00348 0.9160^{15}	 1.5089^{15}
7900	Naphthalene, 1-hexyl- 1-Hexylnaphthalene	$C_{16}H_{20}$ 212.33	2876-53-1 -18	 322	4-05-00-01767 0.9566^{20}	 1.5647^{20}
7901	Naphthalene, 2-hexyl-	$C_{16}H_{20}$ 212.33	2876-46-2 -3	 324	4-05-00-01767 0.9479^{20}	 1.5620^{20}
7902	Naphthalene, 1-iodo-	$C_{10}H_7I$ 254.07	90-14-2 4.2	 302	4-05-00-01670 1.7399^{20}	H_2O 1; EtOH 5; eth 5; bz 5 1.7026^{20}

No.	Name Synonym	Mol. Form. Mol. Wt.	CAS RN mp/°C	Merck No. bp/°C	Beil. Ref. den/g cm^{-3}	Solubility n_D
7903	Naphthalene, 2-iodo-	$C_{10}H_7I$	612-55-5		4-05-00-01671	H_2O 1; EtOH 4; eth 4; HOAc 4
		254.07	54.5	308	1.6319^{99}	1.6662^{99}
7904	Naphthalene, 1-isocyanato-	$C_{11}H_7NO$	86-84-0	6330	4-12-00-03094	eth 3; bz 3
	1-Naphthyl isocyanate	169.18		269	1.1774^{20}	
7905	Naphthalene, 1-isopropyl-2,4,7-trimethyl-	$C_{16}H_{20}$	6995-30-8		3-05-00-01689	
	Cadalene, 3-methyl	212.33		$129\text{-}31^{1.5}$	0.9721^{30}	1.5840^{30}
7906	Naphthalene, 1-isothiocyanato-	$C_{11}H_7NS$	551-06-4	6331	4-12-00-03094	bz 4; eth 4; EtOH 4; chl 4
	1-Naphthyl isothiocyanate	185.25	58			
7907	2-Naphthalenemethanamine	$C_{11}H_{11}N$	2018-90-8		4-12-00-03207	eth 4; EtOH 4
		157.22	59.5	180^{24}; 148^{12}		
7908	1-Naphthalenemethanol	$C_{11}H_{10}O$	4780-79-4		4-06-00-04332	H_2O 2; EtOH 4; eth 4
		158.20	64	304; 163^{12}	1.1039^{80}	
7909	2-Naphthalenemethanol	$C_{11}H_{10}O$	1592-38-7		4-06-00-04340	H_2O 2; EtOH 3; eth 3
		158.20	81.3	178^{12}		
7910	1-Naphthalenemethanol, α,α-diphenyl-	$C_{23}H_{18}O$	630-95-5		4-06-00-05134	bz 4; eth 4; EtOH 4
	Benzhydrol, α-(1-naphthyl)	310.40	136.5	dec		
7911	1-Naphthalenemethanol, α-methyl-, (±)-	$C_{12}H_{12}O$	57605-95-5		4-06-00-04346	ace 4; bz 4; EtOH 4; chl 4
	(±)-1-(α-Naphthyl)ethanol	172.23	66	178^{15}	1.1190^{14}	1.6188^{25}
7912	1-Naphthalenemethanol, α-phenyl-	$C_{17}H_{14}O$	642-28-4		4-06-00-04989	H_2O 1; EtOH 3; eth 3; bz 3
		234.30	87.8	360		
7913	Naphthalene, 1-methoxy-	$C_{11}H_{10}O$	2216-69-5		4-06-00-04211	H_2O 1; EtOH 3; eth 3; bz 3
		158.20	<-10	269	1.0963^{14}	1.6940^{25}
7914	Naphthalene, 2-methoxy-	$C_{11}H_{10}O$	93-04-9	5918	4-06-00-04257	bz 4; eth 4; chl 4
	2-Methoxynaphthalene	158.20	73.5	274		
7915	Naphthalene, 1-methyl-	$C_{11}H_{10}$	90-12-0		4-05-00-01687	H_2O 1; EtOH 4; eth 4; bz 3
	1-Methylnaphthalene	142.20	-30.4	244.7	1.0202^{20}	1.6170^{20}
7916	Naphthalene, 2-methyl-	$C_{11}H_{10}$	91-57-6		4-05-00-01693	H_2O 1; EtOH 4; eth 4; bz 3
	2-Methylnaphthalene	142.20	34.4	241.1	1.0058^{20}	1.6015^{40}
7917	Naphthalene, 1-(3-methylbutoxy)-	$C_{15}H_{18}O$	20213-30-3		3-06-00-02925	
		214.31		319; 150^3	1.0069^{14}	1.5705^{16}
7918	Naphthalene, 2-(3-methylbutoxy)-	$C_{15}H_{18}O$	635-88-1		4-06-00-04259	eth 4; EtOH 4
	Isopentyl β-Naphthyl ether	214.31	26.5	325 dec	1.0155^{12}	1.5768^{12}
7919	Naphthalene, 1,1'-methylenebis-	$C_{21}H_{16}$	607-50-1		4-05-00-02646	H_2O 2; EtOH 2; eth 3; bz 3
		268.36	109	>360; 270^{14}		
7920	Naphthalene, 1-(1-methylethyl)-	$C_{13}H_{14}$	6158-45-8		4-05-00-01722	
		170.25	-16	268	0.9956^{20}	1.5952^{20}
7921	Naphthalene, 2-(1-methylethyl)-	$C_{13}H_{14}$	2027-17-0		4-05-00-01723	H_2O 1; EtOH 4; eth 4; bz 3
		170.25	14.5	268.2	0.9753^{20}	1.5848^{20}
7922	Naphthalene, 1-methyl-4-(1-methylethyl)-	$C_{14}H_{16}$	1680-53-1		3-05-00-01671	
		184.28		$145\text{-}8^{12}$	0.9934^{14}	1.5907^{14}
7923	Naphthalene, 1-methyl-7-(1-methylethyl)-	$C_{14}H_{16}$	490-65-3		4-05-00-01741	H_2O 1; eth 3; ace 3; bz 3
		184.28		140^{11}	0.9740^{20}	1.5833^{20}
7924	Naphthalene, 7-methyl-1-(1-methylethyl)-	$C_{14}H_{16}$	66577-17-1		3-05-00-01672	
	Naphthalene, 1-isopropyl-7-methyl	184.28		282	0.9833^{20}	1.5884^{20}
7925	Naphthalene, 1-methyl-4-nitro-	$C_{11}H_9NO_2$	880-93-3		3-05-00-01624	ace 4; eth 4; EtOH 4
	4-Methyl-1-nitronaphthalene	187.20	71.5	182^{18}		
7926	Naphthalene, 2-methyl-1-nitro-	$C_{11}H_9NO_2$	881-03-8		4-05-00-01698	H_2O 1; EtOH 3; ace 4; os 4
	2-Methyl-1-nitronaphthalene	187.20	81.5	188^{20}		
7927	Naphthalene, 1-(1-methylpropyl)-	$C_{14}H_{16}$	1680-58-6		4-05-00-01738	
		184.28		273	0.9742^{20}	1.5701^{20}
7928	Naphthalene, 1-(2-naphthalenyloxy)-	$C_{20}H_{14}O$	611-49-4		2-06-00-00600	bz 4; eth 4
	1,2'-Bis-naphthyl ether	270.33	81	264^{15}		
7929	Naphthalene, 1-nitro-	$C_{10}H_7NO_2$	86-57-7	6535	4-05-00-01673	H_2O 1; EtOH 4; eth 4; bz 4
	1-Nitronaphthalene	173.17	61	180^{14}	1.332^{20}	
7930	Naphthalene, 2-nitro-	$C_{10}H_7NO_2$	581-89-5		4-05-00-01675	H_2O 1; EtOH 4; eth 4
	2-Nitronaphthalene	173.17	79	314; 165^{15}		
7931	Naphthalene, 1-nonyl-	$C_{19}H_{26}$	26438-26-6		4-05-00-01784	
		254.42	8	366	0.9371^{20}	1.5477^{20}
7932	Naphthalene, 2-nonyl-	$C_{19}H_{26}$	61886-67-7		4-05-00-01784	
		254.42	12	372	0.9298^{20}	1.5454^{20}
7933	Naphthalene, octachloro-	$C_{10}Cl_8$	2234-13-1		4-05-00-01665	EtOH 2; bz 4; chl 4; lig 4
	Perchloronaphthalene	403.73	197.5	441^7; $248^{0.5}$		
7934	Naphthalene, octadecafluorodecahydro-	$C_{10}F_{18}$	306-94-5		3-05-00-00249	
	Perfluorodecalin	462.08	-10	142		
7935	Naphthalene, octafluoro-	$C_{10}F_8$	313-72-4		4-05-00-01658	
	Perfluoronaphthalene	272.10	87.5	209		
7936	Naphthalene, 1,2,3,4,4a,5,6,8a-octahydro-7-methyl-4-methylene-1-(1-methylethyl)-, (1α,4aβ,8aα)-	$C_{15}H_{24}$	39029-41-9		3-05-00-01088	
	γ-Cadinene	204.36		$124\text{-}8^{12}$	0.9182^{15}	1.3166^{20}
7937	Naphthalene, 1-octyl-	$C_{18}H_{24}$	2876-51-9		4-05-00-01779	
	1-Octylnaphthalene	240.39	-2	356	0.9427^{20}	1.5526^{20}
7938	Naphthalene, 2-octyl-	$C_{18}H_{24}$	2876-44-0		4-05-00-01779	
		240.39	12	357	0.9350^{20}	1.5501^{20}
7939	Naphthalene, 1,1'-oxybis-	$C_{20}H_{14}O$	607-52-3		4-06-00-04215	bz 4; eth 4
	1-Naphthyl ether	270.33	110	$280\text{-}5^{22}$		
7940	Naphthalene, 2,2'-oxybis-	$C_{20}H_{14}O$	613-80-9		4-06-00-04260	H_2O 1; EtOH 2; eth 4; bz 4
		270.33	105	250^{19} dec		

No.	Name Synonym	Mol. Form. Mol. Wt.	CAS RN mp/°C	Merck No. bp/°C	Beil. Ref. den/g cm^{-3}	Solubility n_D
7941	Naphthalene, 1-pentyl-	$C_{15}H_{18}$ 198.31	86-89-5 -22	 307	4-05-00-01755 0.9656^{20}	 1.5725^{20}
7942	Naphthalene, 2-pentyl- 2-Pentylnaphthalene	$C_{15}H_{18}$ 198.31	93-22-1 -4	 310	4-05-00-01755 0.9561^{20}	 1.5694^{20}
7943	Naphthalene, 1-(pentyloxy)- 1-Naphthyl pentyl ether	$C_{15}H_{18}O$ 214.31	108438-40-0 30	 322	4-06-00-04213	bz 4; eth 4; EtOH 4
7944	Naphthalene, 2-(pentyloxy)- 2-Naphthyl pentyl ether	$C_{15}H_{18}O$ 214.31	31059-19-5 24.5	 335	3-06-00-02973	bz 4; eth 4; EtOH 4; chl 4 1.5587^{30}
7945	Naphthalene, 1-phenoxy-	$C_{16}H_{12}O$ 220.27	3402-76-4 55.5	 349.5	4-06-00-04214	H_2O 1; EtOH 2; eth 3; ace 3
7946	Naphthalene, 2-phenoxy-	$C_{16}H_{12}O$ 220.27	19420-29-2 46	 335	3-06-00-02975	H_2O 1; eth 3; chl 3; HOAc 3
7947	Naphthalene, 1-phenyl- 1-Phenylnaphthalene	$C_{16}H_{12}$ 204.27	605-02-7 45	 334	4-05-00-02411 1.096^{20}	H_2O 1; EtOH 4; eth 4; bz 4 1.6664^{20}
7948	Naphthalene, 2-phenyl-	$C_{16}H_{12}$ 204.27	612-94-2 103.5	 345.5	4-05-00-02412 1.2180^{20}	EtOH 3; eth 4; bz 3; chl 3
7949	Naphthalene, 1-(phenylmethyl)-	$C_{17}H_{14}$ 218.30	611-45-0 59.5	 350	4-05-00-02420 1.166^{17}	H_2O 1; EtOH 2; eth 4; bz 3
7950	Naphthalene, 2-(phenylmethyl)- 2-Benzylnaphthalene	$C_{17}H_{14}$ 218.30	613-59-2 58	 350	4-05-00-02420 1.176^{12}	bz 4; eth 4; chl 4
7951	Naphthalene, 1-(phenylthio)-	$C_{16}H_{12}S$ 236.34	7570-98-1 41.8	 255^{43}, 222^{11}	4-06-00-04242 1.167^{15}	eth 4; EtOH 4
7952	1-Naphthalenepropanoic acid	$C_{13}H_{12}O_2$ 200.24	3243-42-3 156.5	 179^{11}	4-09-00-02435	H_2O 3; EtOH 3; bz 3
7953	2-Naphthalenepropanoic acid, 6-hydroxy- Allenolic acid	$C_{13}H_{12}O_3$ 216.24	553-39-9 180.5	247	3-10-00-01113	py 4; EtOH 4; MeOH 4
7954	Naphthalene, 1-(2-propenyl)-	$C_{13}H_{12}$ 168.24	2489-86-3	 266	3-05-00-01805 1.0228^{20}	EtOH 3; bz 4; chl 3 1.6140^{20}
7955	Naphthalene, 1-propoxy-	$C_{13}H_{14}O$ 186.25	20009-26-1	 293.5	4-06-00-04213 1.0447^{18}	 1.5928^{18}
7956	Naphthalene, 2-propoxy-	$C_{13}H_{14}O$ 186.25	19718-45-7 41	 305	3-06-00-02973	EtOH 4
7957	Naphthalene, 1-propyl-	$C_{13}H_{14}$ 170.25	2765-18-6 -8.6	 274.5	4-05-00-01721 0.9897^{20}	 1.5923^{20}
7958	Naphthalene, 2-propyl-	$C_{13}H_{14}$ 170.25	2027-19-2 -3	 273.5	4-05-00-01722 0.9770^{20}	 1.5872^{20}
7959	1-Naphthalenesulfonic acid	$C_{10}H_8O_3S$ 208.24	85-47-2 140	6295	4-11-00-00521	H_2O 3; EtOH 3; eth 2
7960	2-Naphthalenesulfonic acid β-Naphthylsulfonic acid	$C_{10}H_8O_3S$ 208.24	120-18-3 91	6296 dec	4-11-00-00527 1.441^{25}	H_2O 4; EtOH 4; eth 3; bz 2
7961	1-Naphthalenesulfonic acid, 2-amino- 2-Naphthylamine-1-sulfonic acid	$C_{10}H_9NO_3S$ 223.25	81-16-3	6326	4-14-00-02792	DMSO 3
7962	1-Naphthalenesulfonic acid, 4-amino- 1-Naphthylamine-4-sulfonic acid	$C_{10}H_9NO_3S$ 223.25	84-86-6 dec	6323	4-14-00-02793 1.6703^{25}	H_2O 1; EtOH 2; MeOH 3; py 3
7963	1-Naphthalenesulfonic acid, 5-amino- 1-Naphthylamine-5-sulfonic acid	$C_{10}H_9NO_3S$ 223.25	84-89-9	6324	4-14-00-02800	H_2O 3; eth 1
7964	1-Naphthalenesulfonic acid, 7-amino- Badische acid	$C_{10}H_9NO_3S$ 223.25	86-60-2	952	4-14-00-02802	HOAc 4
7965	1-Naphthalenesulfonic acid, 8-amino- 1-Naphthylamine-8-sulfonic acid	$C_{10}H_9NO_3S$ 223.25	82-75-7	6325	4-14-00-02802	gl HOAc 4
7966	1-Naphthalenesulfonic acid, 4-amino-3-hydroxy- 1-Amino-2-naphthol-4-sulfonic acid	$C_{10}H_9NO_4S$ 239.25	116-63-2	466	4-14-00-02825	
7967	2-Naphthalenesulfonic acid, 8-amino- 1,7-Cleve's acid	$C_{10}H_9NO_3S$ 223.25	119-28-8	2349	4-14-00-02805	
7968	1-Naphthalenesulfonic acid, 3,6-bis(1,1-dimethylethyl)-, ethyl ester Ethyl dibunate	$C_{20}H_{28}O_3S$ 348.51	5560-69-0	3745		chl 3
7969	1-Naphthalenesulfonic acid, 3,3'-[[1,1'-biphenyl]-4,4'-diylbis(azo)]bis[4-amino-, disodium salt Congo red	$C_{32}H_{22}N_6Na_2O_6S_2$ 696.68	573-58-0 >360	2493	1-16-00-00342	H_2O 2; EtOH 3; eth 1
7970	1-Naphthalenesulfonic acid, 3,4-dihydro-3,4-dioxo-, sodium salt 1,2-Naphthoquinone-4-sulfonic acid,sodium salt	$C_{10}H_5NaO_5S$ 260.20	521-24-4 287 dec	6387		
7971	1-Naphthalenesulfonic acid, 4-hydroxy- 1-Naphthol-4-sulfonic acid	$C_{10}H_8O_4S$ 224.24	84-87-7 170 dec	6311	4-11-00-00614	H_2O 4; eth 1
7972	1-Naphthalenesulfonic acid, 7-hydroxy- Croceic acid	$C_{10}H_8O_4S$ 224.24	132-57-0	2591	4-11-00-00623	H_2O 3
7973	2-Naphthalenesulfonic acid, 1-hydroxy- 1-Naphthol-2-sulfonic acid	$C_{10}H_8O_4S$ 224.24	567-18-0 >250	6310	4-11-00-00613	H_2O 2; EtOH 3; eth 1; dil HCl 2
7974	2-Naphthalenesulfonic acid, 6-hydroxy- 2-Naphthol-6-sulfonic acid	$C_{10}H_8O_4S$ 224.24	93-01-6 125	6312	4-11-00-00622	H_2O 4; EtOH 4; eth 1; HOAc 3
7975	1-Naphthalenesulfonyl chloride	$C_{10}H_7ClO_2S$ 226.68	85-46-1 68	 209^{20}, $147^{0.9}$	3-11-00-00383	bz 4; eth 4; EtOH 4

No.	Name Synonym	Mol. Form. Mol. Wt.	CAS RN mp/°C	Merck No. bp/°C	Beil. Ref. den/g cm^{-3}	Solubility n_D
7976	2-Naphthalenesulfonyl chloride	$C_{10}H_7ClO_2S$ 226.68	93-11-8 81	 201^{13}; $148^{0.5}$	4-11-00-00529	H_2O 1; EtOH 3; eth 4; bz 3
7977	1-Naphthalenesulfonyl chloride, 5-(dimethylamino)- Dansyl chloride	$C_{12}H_{12}ClNO_2S$ 269.75	605-65-2 70	2812	4-14-00-02801	
7978	2-Naphthalenesulfonyl chloride, 5,6,7,8-tetrahydro- 6-Tetralinsulfonyl chloride	$C_{10}H_{11}ClO_2S$ 230.71	61551-49-3 58	 197^{18}	3-11-00-00375	eth 4
7979	Naphthalene, 1,2,3,4-tetrahydro- Tetralin	$C_{10}H_{12}$ 132.21	119-64-2 -35.7	9152 207.6	4-05-00-01388 0.9660^{25}	H_2O 1; EtOH 4; eth 4; chl 3 1.5413^{20}
7980	Naphthalene, 1,2,3,4-tetrahydro-1,1-dimethyl- 1,1-Dimethyltetralin	$C_{12}H_{16}$ 160.26	1985-59-7	 221	4-05-00-01430 0.950^{20}	 1.5292^{20}
7981	Naphthalene, 1,2,3,4-tetrahydro-1,2-dimethyl- 1,2,3,4-Tetrahydro-1,2-dimethylnaphthalene	$C_{12}H_{16}$ 160.26	5195-40-4	 235	3-05-00-01262 0.9470^{20}	 1.5286^{20}
7982	Naphthalene, 1,2,3,4-tetrahydro-1,3-dimethyl- Tetralin, 1,3-dimethyl	$C_{12}H_{16}$ 160.26	5195-37-9	 234; 78^{1}	3-05-00-01263 0.940^{20}	 1.525^{20}
7983	Naphthalene, 1,2,3,4-tetrahydro-1,4-dimethyl- Tetralin, 1,4-dimethyl	$C_{12}H_{16}$ 160.26	4175-54-6	 226	4-05-00-01430 0.940^{20}	 1.525^{20}
7984	Naphthalene, 1,2,3,4-tetrahydro-1,5-dimethyl-	$C_{12}H_{16}$ 160.26	21564-91-0	 239	3-05-00-01263 0.941^{20}	 1.526^{20}
7985	Naphthalene, 1,2,3,4-tetrahydro-2,2-dimethyl- Tetralin, 2,2-dimethyl	$C_{12}H_{16}$ 160.26	13556-55-3	 230	4-05-00-01431 0.935^{20}	 1.5200^{20}
7986	Naphthalene, 1,2,3,4-tetrahydro-2,3-dimethyl- Tetralin, 2,3-dimethyl	$C_{12}H_{16}$ 160.26	21564-92-1 -5		3-05-00-01265 0.940^{20}	 1.523^{20}
7987	Naphthalene, 1,2,3,4-tetrahydro-2,5-dimethyl- Tetralin, 2,5-dimethyl	$C_{12}H_{16}$ 160.26	25419-37-8	 236	3-05-00-01263 0.946^{20}	 1.526^{20}
7988	Naphthalene, 1,2,3,4-tetrahydro-2,6-dimethyl-	$C_{12}H_{16}$ 160.26	7524-63-2 20	 240; 117^{15}	4-05-00-01432 0.941^{20}	 1.526^{20}
7989	Naphthalene, 1,2,3,4-tetrahydro-2,7-dimethyl- Tetralin, 2,7-dimethyl	$C_{12}H_{16}$ 160.26	13065-07-1	 237	3-05-00-01265 0.941^{20}	 1.526^{20}
7990	Naphthalene, 1,2,3,4-tetrahydro-2,8-dimethyl- Tetralin, 2,8-dimethyl	$C_{12}H_{16}$ 160.26	25419-36-7	 236	3-05-00-01264 0.941^{20}	 1.526^{20}
7991	Naphthalene, 1,2,3,4-tetrahydro-5,6-dimethyl- Tetralin, 5,6-dimethyl	$C_{12}H_{16}$ 160.26	20027-77-4	 252	4-05-00-01430 0.975^{20}	 1.552^{20}
7992	Naphthalene, 1,2,3,4-tetrahydro-5,7-dimethyl- Tetralin, 5,7-dimethyl	$C_{12}H_{16}$ 160.26	21693-54-9 -6	 253	4-05-00-01430 0.9583^{20}	 1.5405^{20}
7993	Naphthalene, 1,2,3,4-tetrahydro-5,8-dimethyl-	$C_{12}H_{16}$ 160.26	14108-88-4 -4.1	 254	4-05-00-01430 0.967^{20}	 1.547^{20}
7994	Naphthalene, 1,2,3,4-tetrahydro-6,7-dimethyl-	$C_{12}H_{16}$ 160.26	1076-61-5 10	 252	4-05-00-01431 0.954^{20}	 1.538^{20}
7995	Naphthalene, 1,2,3,4-tetrahydro-1-methyl-	$C_{11}H_{14}$ 146.23	1559-81-5	 220.6	4-05-00-01412 0.9583^{20}	 1.5353^{20}
7996	Naphthalene, 1,2,3,4-tetrahydro-2-methyl-	$C_{11}H_{14}$ 146.23	3877-19-8 -43.1	 221	4-05-00-01414 0.952^{20}	 1.531^{20}
7997	Naphthalene, 1,2,3,4-tetrahydro-5-methyl-	$C_{11}H_{14}$ 146.23	2809-64-5 -23	 234	4-05-00-01412 0.9720^{20}	 1.5439^{20}
7998	Naphthalene, 1,2,3,4-tetrahydro-6-methyl-	$C_{11}H_{14}$ 146.23	1680-51-9 -40	 229	4-05-00-01413 0.9537^{20}	 1.5357^{20}
7999	Naphthalene, 1,2,3,4-tetrahydro-1,1,2,6-tetramethyl- Irene	$C_{14}H_{20}$ 188.31	1681-22-7	 $120\text{-}5^{10}$	4-05-00-01460 0.9332^{20}	H_2O 1; ace 3; bz 3 1.5217^{20}
8000	Naphthalene, 1,2,3,4-tetrahydro-1,1,6-trimethyl-	$C_{13}H_{18}$ 174.29	475-03-6	 240; 90^{4}	4-05-00-01445 0.9356^{20}	EtOH 3; eth 3; bz 3; chl 3 1.5257^{20}
8001	Naphthalene, 1,2,4,7-tetramethyl-	$C_{14}H_{16}$ 184.28	16020-17-0 -3	 $174\text{-}5^{12}$	4-05-00-01746 1.011^{20}	 1.6032^{20}
8002	Naphthalene, 1,2,5,6-tetramethyl-	$C_{14}H_{16}$ 184.28	2131-43-3 118	 152^{12}	4-05-00-01746	H_2O 1; EtOH 3; bz 3; peth 3
8003	Naphthalene, 1,3,6,8-tetramethyl- 1,3,6,8-Tetramethylnaphthalene	$C_{14}H_{16}$ 184.28	14558-14-6 84.5	 $115\text{-}6^{2}$	4-05-00-01746	H_2O 1; EtOH 3
8004	Naphthalene, 1,3,6,8-tetranitro-	$C_{10}H_4N_4O_8$ 308.16	28995-89-3 207	 exp	4-05-00-01683 1.6400^{0}	bz 4; HOAc 4
8005	Naphthalene, 1,1'-thiobis-	$C_{20}H_{14}S$ 286.40	607-53-4 110	 289^{15}	4-06-00-04243	bz 4; HOAc 4
8006	Naphthalene, 2,2'-thiobis-	$C_{20}H_{14}S$ 286.40	613-81-0 151	 296^{15}	4-06-00-04314	EtOH 1; bz 4; CS_2 4; HOAc 4
8007	1-Naphthalenethiol 1-Naphthyl mercaptan	$C_{10}H_8S$ 160.24	529-36-2	6297 285 dec; 161^{20}	4-06-00-04241 1.1607^{20}	H_2O 2; EtOH 4; eth 4; dil alk 2 1.6802^{20}
8008	2-Naphthalenethiol 2-Naphthyl mercaptan	$C_{10}H_8S$ 160.24	91-60-1 81	6298 288	4-06-00-04312 1.550^{25}	H_2O 2; EtOH 4; eth 4; lig 4
8009	Naphthalene, 1,2,4-trimethyl- 1,2,4-Trimethylnaphthalene	$C_{13}H_{14}$ 170.25	2717-42-2 55.5	 146^{12}	4-05-00-01725	EtOH 3; eth 3; bz 3; chl 3
8010	Naphthalene, 1,2,5-trimethyl- 1,2,5-Trimethylnaphthalene	$C_{13}H_{14}$ 170.25	641-91-8 33.5	 140^{12}	4-05-00-01726 1.0103^{22}	H_2O 1; EtOH 2; eth 4; bz 4 1.6093^{22}
8011	Naphthalene, 1,2,6-trimethyl- 1,2,6-Trimethylnaphthalene	$C_{13}H_{14}$ 170.25	3031-05-8 14	 146^{10}	4-05-00-01726	H_2O 1; eth 3; bz 3 1.6010^{20}
8012	Naphthalene, 1,2,7-trimethyl-	$C_{13}H_{14}$ 170.25	486-34-0	 $147\text{-}8^{16}$	4-05-00-01727 1.0087^{20}	H_2O 1; eth 3; bz 3 1.6097^{20}

PHYSICAL CONSTANTS OF ORGANIC COMPOUNDS (continued)

No.	Name Synonym	Mol. Form. Mol. Wt.	CAS RN mp/°C	Merck No. bp/°C	Beil. Ref. den/g cm^{-3}	Solubility n_D
8013	Naphthalene, 1,3,7-trimethyl- 2,4,6-Trimethylnaphthalene	$C_{13}H_{14}$ 170.25	2131-38-6 12	 $131-3^{9}$	4-05-00-01728 0.9801^{21}	 1.5972^{15}
8014	Naphthalene, 1,2,5-trimethyl-8-(1-methylethyl)- Cadalene, 5-methyl	$C_{16}H_{20}$ 212.33	6897-76-3	 $144-5^{1.5}$	3-05-00-01689 0.9847^{30}	 1.5926^{30}
8015	Naphthalene, 1,3,8-trimethyl-5-(1-methylethyl)- Cadalene, 8-methyl	$C_{16}H_{20}$ 212.33	6897-88-7	 $135-6^{1.5}$	4-05-00-01770 0.9800^{30}	 1.5876^{30}
8016	Naphthalene, 1,3,5-trinitro-	$C_{10}H_5N_3O_6$ 263.17	2243-94-9 124	 exp 364	4-05-00-01682	ace 4; EtOH 4; chl 4
8017	1-Naphthalenol 1-Naphthol	$C_{10}H_8O$ 144.17	90-15-3 95	6303 288; 184^{40}	4-06-00-04208 1.0989^{99}	H_2O 1; EtOH 4; eth 4; ace 3 1.6224^{99}
8018	2-Naphthalenol 2-Naphthol	$C_{10}H_8O$ 144.17	135-19-3 123	6304 285	4-06-00-04253 1.28^{20}	H_2O 1; EtOH 4; eth 4; bz 3
8019	2-Naphthalenol, 6-amino-	$C_{10}H_9NO$ 159.19	4363-04-6 193	 212 dec	3-13-00-01902	H_2O 4; EtOH 4
8020	2-Naphthalenol, 8-amino-	$C_{10}H_9NO$ 159.19	118-46-7 206	 sub	3-13-00-01907	H_2O 3; EtOH 4; eth 3; bz 2
8021	2-Naphthalenol, benzoate 2-Naphthyl benzoate	$C_{17}H_{12}O_2$ 248.28	93-44-7 107	6328	4-09-00-00339	H_2O 1; EtOH 3; eth 2; HOAc 2
8022	2-Naphthalenol, 1-bromo-	$C_{10}H_7BrO$ 223.07	573-97-7 84	 130	4-06-00-04301	H_2O 1; EtOH 3; eth 3; bz 3
8023	2-Naphthalenol, 6-chloro-	$C_{10}H_7ClO$ 178.62	40604-49-7 116.5	 sub	4-06-00-04294	bz 4; eth 4; EtOH 4; chl 4
8024	2-Naphthalenol, decahydro-	$C_{10}H_{18}O$ 154.25	825-51-4	 109^{14}	1-06-00-00044 0.996^{25}	 1.4992^{20}
8025	1-Naphthalenol, 2,3-dichloro- 2,3-Dichloro-1-naphthol	$C_{10}H_6Cl_2O$ 213.06	71284-96-3 101	 sub	0-06-00-00612	bz 4; eth 4; EtOH 4; chl 4
8026	1-Naphthalenol, 2,4-dichloro-	$C_{10}H_6Cl_2O$ 213.06	2050-76-2 107.5	 180	4-06-00-04233	bz 4; eth 4; EtOH 4
8027	2-Naphthalenol, 1,4-dimethyl- 2-Naphthol, 1,4-dimethyl-	$C_{12}H_{12}O$ 172.23	4705-94-6 135.5	 315.5	4-06-00-04351	eth 4; EtOH 4
8028	2-Naphthalenol, 1-[(2,4-dimethylphenyl)azo]- C.I. Solvent Orange 7	$C_{18}H_{16}N_2O$ 276.34	3118-97-6 166		4-16-00-00234	eth 4; EtOH 4
8029	2-Naphthalenol, 1-[(2,5-dimethylphenyl)azo]- 1-(2,5-Xylylazo)-2-naphthol	$C_{18}H_{16}N_2O$ 276.34	85-82-5 153	[illegible]	4-16-00-00234	
8030	1-Naphthalenol, 4-methyl-	$C_{11}H_{10}O$ 158.20	10240-08-1 85.8	 166^{13}	4-06-00-04330	H_2O 3; EtOH 3; eth 3; ace 3
8031	2-Naphthalenol, 1-methyl-	$C_{11}H_{10}O$ 158.20	1076-26-2 112	 160^{10}	4-06-00-04329	H_2O 2; EtOH 4; eth 4; ace 4
8032	1-Naphthalenol, methylcarbamate Carbaryl	$C_{12}H_{11}NO_2$ 201.22	63-25-2 145	1789	4-06-00-04219 1.228^{25}	ace 4; DMF 4
8033	2-Naphthalenol, 1-[[2-methyl-4-[(2-methylphenyl)azo]phenyl]azo]- Scarlet red	$C_{24}H_{20}N_4O$ 380.45	85-83-6 186 dec	8349 260 dec	4-16-00-00249	chl 4; peth 4
8034	2-Naphthalenol, 1-nitro- 1-Nitro-2-naphthol	$C_{10}H_7NO_3$ 189.17	550-60-7 104	6536 $115^{0.05}$	4-06-00-04307	H_2O 3; EtOH 3; eth 4; chl 2
8035	2-Naphthalenol, 1-nitroso- 1-Nitroso-2-naphthol	$C_{10}H_7NO_2$ 173.17	131-91-9 109.5	6562	4-07-00-02419	bz 4; eth 4
8036	1-Naphthalenol, 2-(phenylazo)-	$C_{16}H_{12}N_2O$ 248.28	3375-23-3 138	 sub	4-16-00-00219	EtOH 4; HOAc 4
8037	2-Naphthalenol, 1-(phenylazo)- Azo, benzene 1'-naphthalene,2'-hydroxy	$C_{16}H_{12}N_2O$ 248.28	842-07-9		4-16-00-00228	
8038	2-Naphthalenol, 1-[[4-(phenylazo)phenyl]azo]- Sudan III	$C_{22}H_{16}N_4O$ 352.40	85-86-9 195	8858	4-16-00-00248	H_2O 1; EtOH 3; eth 3; ace 3
8039	1-Naphthalenol, 2-(phenylmethyl)-	$C_{17}H_{14}O$ 234.30	36441-32-4 73.5	 237^{12}	4-06-00-04990	H_2O 1; bz 3; lig 2
8040	2-Naphthalenol, 1-(phenylmethyl)- 2-Naphthol, 1-benzyl	$C_{17}H_{14}O$ 234.30	36441-31-3 114	 247^{14}	4-06-00-04988	ace 4; bz 4; eth 4; EtOH 4
8041	1-Naphthalenol, 1,2,3,4-tetrahydro-	$C_{10}H_{12}O$ 148.20	529-33-9	 $102-4^{2}$	3-06-00-02457 1.09^{25}	 1.5638^{20}
8042	1-Naphthalenol, 5,6,7,8-tetrahydro-	$C_{10}H_{12}O$ 148.20	529-35-1 70	 266; 143^{11}	4-06-00-03851 1.0556^{75}	
8043	2-Naphthalenol, 1,2,3,4-tetrahydro- 2-Tetralol	$C_{10}H_{12}O$ 148.20	[illegible] 15.5	[illegible] 140^{12}	[illegible]	
8044	2-Naphthalenol, 5,6,7,8-tetrahydro-	$C_{10}H_{12}O$ 148.20	1125-78-6 57	 275.5	4-06-00-03854 1.0552^{65}	
8045	1(2H)-Naphthalenone, 2-acetyl-3,4-dihydro-	$C_{12}H_{12}O_2$ 188.23	17216-08-9 56	 $135-42^{1}$	4-07-00-02358	chl 3
8046	1(2H)-Naphthalenone, 3,4-dihydro-	$C_{10}H_{10}O$ 146.19	529-34-0 8	 115^{6}	4-07-00-01015 1.0988^{16}	 1.5672^{20}
8047	2(1H)-Naphthalenone, 3,4-dihydro-	$C_{10}H_{10}O$ 146.19	530-93-8 18	 237	4-07-00-01018 1.1055^{27}	H_2O 1; eth 3; bz 3 1.5598^{20}
8048	1(2H)-Naphthalenone, 3,4-dihydro-6-methoxy- 6-Methoxy-α-tetralone	$C_{11}H_{12}O_2$ 176.22	1078-19-9 78	5924 171^{11}	4-08-00-00904	
8049	1(2H)-Naphthalenone, 3,4-dihydro-2-methyl-	$C_{11}H_{12}O$ 160.22	1590-08-5	 $127-31^{12}$	4-07-00-01033 1.057^{25}	 1.5535^{20}
8050	1(2H)-Naphthalenone, 3,4-dihydro-4-methyl-	$C_{11}H_{12}O$ 160.22	19832-98-5	 $133-4^{12}$	4-07-00-01031 1.0779^{19}	 1.5620^{19}

No.	Name Synonym	Mol. Form. Mol. Wt.	CAS RN mp/°C	Merck No. bp/°C	Beil. Ref. den/g cm^{-3}	Solubility n_D
8051	1(2H)-Naphthalenone, 7-ethyl-3,4-dihydro- 1-Tetralone, 7-ethyl	$C_{12}H_{14}O$ 174.24	22531-06-2	 152-3^{12}	4-07-00-01049 1.0556^{17}	bz 4; eth 4; EtOH 4 1.5599^{17}
8052	2(3H)-Naphthalenone, 4,4a,5,6,7,8-hexahydro-4,4a-dimethyl-6-(1-methylethylidene)-, (4*R-cis*)-	$C_{15}H_{22}O$ 218.34	15764-04-2 51.5	 144^{2}	 1.0035^{20}	ace 4 1.5370^{20}
8053	1(2H)-Naphthalenone, octahydro-, *cis*- 1-Decalone (*cis*)	$C_{10}H_{16}O$ 152.24	32166-40-8 2	 126^{20}	4-07-00-00201 1.008^{20}	 1.4936^{20}
8054	1(2H)-Naphthalenone, octahydro-, *trans*-	$C_{10}H_{16}O$ 152.24	21370-71-8 33	 122^{20}	4-07-00-00201 0.986^{20}	 1.4849^{21}
8055	2(1H)-Naphthalenone, octahydro-	$C_{10}H_{16}O$ 152.24	4832-17-1	 96$^{2.5}$	4-07-00-00203 0.979^{25}	 1.4900^{20}
8056	Naphtho[2,3-b]-1,4-dithiin-2,3-dicarbonitrile, 5,10-dihydro-5,10-dioxo- Dithianone	$C_{14}H_4N_2O_2S_2$ 296.33	3347-22-6 220	3375	5-19-08-00189	
8057	Naphtho[1,2-b]furan-2,8(3H,4H)-dione, 3a,5,5a,9b-tetrahydro-3,5a,9-trimethyl-, α-Santonin	$C_{15}H_{18}O_3$ 246.31	481-06-1 175	8325	5-17-11-00314 1.590^{25}	H_2O 2; EtOH 2; eth 2; bz 3
8058	Naphtho[1,2-b]furan-2,8(3H,4H)-dione, 3a,5,5a,9b-tetrahydro-4,5-dihydroxy-9-methoxy-3,5a-dimethyl- Cedrin	$C_{15}H_{18}O_6$ 294.30	6040-62-6 266	1918		EtOH 4; MeOH 4; chl 4
8059	Naphtho[2,3-b]furan-2(3H)-one, 3a,5,6,7,8,8a,9,9a-octahydro-5,8a-dimethyl-3-methylene-, Alantolactone	$C_{15}H_{20}O_2$ 232.32	546-43-0 76	198 275	5-17-10-00097	bz 4; eth 4; EtOH 4; chl 4
8060	Naphtho[1,2-c]furan-1,3-dione, 4,5-dihydro-	$C_{12}H_8O_3$ 200.19	37845-14-0 126.5	 227-30^{23}	5-17-11-00444	bz 4; MeOH 4
8061	Naphtho[1,2,3,4-def]chrysene Dibenzo[a,e]pyrene	$C_{24}H_{14}$ 302.38	192-65-4 233.5		4-05-00-02803	EtOH 2; ace 2; bz 2; tol 3
8062	Naphtho[2,3-f]quinoline-7,12-dione, 5,6-dihydroxy- Alizarine Blue	$C_{17}H_9NO_4$ 291.26	568-02-5 269	239	5-21-13-00590	bz 4; gl HOAc 4
8063	2-Naphthol, 8-chloro-	$C_{10}H_7ClO$ 178.62	29921-50-4 101	 307.5	3-06-00-02992	bz 4; eth 4; EtOH 4; chl 4
8064	2-Naphthylamine, 1,4-dimethyl- Naphthalene, 2-amino-1,4-dimethyl	$C_{12}H_{13}N$ 171.24	878-93-3 75	 333	0-12-00-01317	eth 4; EtOH 4
8065	1,5-Naphthyridine 1,5-Diazanaphthalene	$C_8H_6N_2$ 130.15	254-79-5 75	 111^{12}	5-23-07-00156 1.2100^{20}	
8066	1,8-Naphthyridine-3-carboxylic acid, 1-ethyl-1,4-dihydro-7-methyl-4-oxo-	$C_{12}H_{12}N_2O_3$ 232.24	389-08-2 229.5	6273		EtOH 2; eth 2; chl 3
8067	Napropamide Propanamide, *N,N*-diethyl-2-(1-naphthalenyloxy)-	$C_{17}H_{21}NO_2$ 271.36	15299-99-7 75	6336		
8068	Nickel carbonyl Nickel tetracarbonyl	C_4NiO_4 170.74	13463-39-3 -19.3	6411 43	 1.31^{25}	
8069	Nicosulfuron 2-(4,6-Dimethoxypyrimidi*N*-2-ylcarbamoylsulfamoyl)-*N,N*-dimethylnicotinamide	$C_{15}H_{18}N_6O_6S$ 410.41	111991-09-4 172			
8070	Nitralin Benzenamine, 4-(methylsulfonyl)-2,6-dinitro-*N,N*-dipropyl-	$C_{13}H_{19}N_3O_6S$ 345.38	4726-14-1 150	6486		
8071	Nitrapyrin Pyridine, 2-chloro-6-(trichloromethyl)-	$C_6H_3Cl_4N$ 230.91	1929-82-4 63	6490 136^{11}		
8072	Nitric acid, butyl ester Butyl nitrate	$C_4H_9NO_3$ 119.12	928-45-0	 133	4-01-00-01524 1.0228^{30}	H_2O 1; EtOH 3; eth 3; ctc 2 1.4013^{23}
8073	Nitric acid, decyl ester Decyl nitrate	$C_{10}H_{21}NO_3$ 203.28	2050-78-4	 128^{11}; 88^{1}	4-01-00-01819 0.951^{0}	EtOH 3; eth 3
8074	Nitric acid, ethyl ester Ethyl nitrate	$C_2H_5NO_3$ 91.07	625-58-1 -94.6	 87.2	4-01-00-01327 1.1084^{20}	H_2O 3; EtOH 5; eth 5 1.3852^{20}
8075	Nitric acid, methyl ester Methyl nitrate	CH_3NO_3 77.04	598-58-3 -82.5	6015 64.6 exp	4-01-00-01254 1.2075^{20}	H_2O 2; EtOH 3; eth 3 1.3748^{20}
8076	Nitric acid, 1-methylethyl ester Isopropyl nitrate	$C_3H_7NO_3$ 105.09	1712-64-7	 100	4-01-00-01475 1.034^{19}	EtOH 3; eth 3 1.3912^{16}
8077	Nitric acid, 1-methylpropyl ester	$C_4H_9NO_3$ 119.12	924-52-7	 124	4-01-00-01573 1.0264^{20}	eth 4; EtOH 4 1.4015^{20}
8078	Nitric acid, 2-methylpropyl ester Isobutyl nitrate	$C_4H_9NO_3$ 119.12	543-29-3	5031 123.4	4-01-00-01595 1.0152^{20}	 1.4028^{20}
8079	Nitric acid, octyl ester	$C_8H_{17}NO_3$ 175.23	629-39-0	 110-2^{20}	4-01-00-01762 0.8409^{17}	H_2O 2; EtOH 3; eth 3
8080	Nitric acid, propyl ester Propyl nitrate	$C_3H_7NO_3$ 105.09	627-13-4	7877 110	4-01-00-01424 1.0538^{20}	H_2O 2; EtOH 3; eth 3; ctc 3 1.3973^{20}
8081	Nitrofen Benzene, 2,4-dichloro-1-(4-nitrophenoxy)-	$C_{12}H_7Cl_2NO_3$ 284.10	1836-75-5 70	6519		
8082	Nitrofluorfen Benzene, 2-chloro-1-(4-nitrophenoxy)-4-(trifluoromethyl)-	$C_{13}H_7ClF_3NO_3$ 317.65	42874-01-1 68			
8083	Nitrous acid, butyl ester Butyl nitrite	$C_4H_9NO_2$ 103.12	544-16-1	1581 78	4-01-00-01523 0.9114^{25}	EtOH 5; eth 5 1.3762^{20}

No.	Name Synonym	Mol. Form. Mol. Wt.	CAS RN mp/°C	Merck No. bp/°C	Beil. Ref. den/g cm^{-3}	Solubility n_D
8084	Nitrous acid, decyl ester Decyl nitrite	$C_{10}H_{21}NO_2$ 187.28	1653-57-2	 $105\text{-}8^{12}$	4-01-00-01819	EtOH 3; eth 3 1.4247^{20}
8085	Nitrous acid, 1,1-dimethylethyl ester *tert*-Butyl nitrite	$C_4H_9NO_2$ 103.12	540-80-7	1582 63	4-01-00-01622 0.8670^{20}	H_2O 2; EtOH 3; eth 3; chl 3 1.3682^{20}
8086	Nitrous acid, 1,2-ethanediyl ester Ethylene glycol, dinitrite	$C_2H_4N_2O_4$ 120.06	629-16-3 <-15	 98	4-01-00-02411 1.2156^{0}	eth 4; EtOH 4
8087	Nitrous acid, ethyl ester Ethyl nitrite	$C_2H_5NO_2$ 75.07	109-95-5	3786 18	4-01-00-01327 0.899^{15}	EtOH 5; eth 5 1.3418^{10}
8088	Nitrous acid, heptyl ester Heptyl nitrite	$C_7H_{15}NO_2$ 145.20	629-43-6	 155.8	4-01-00-01735 0.8939^{0}	eth 4 1.4032^{20}
8089	Nitrous acid, hexyl ester Hexyl nitrite	$C_6H_{13}NO_2$ 131.17	638-51-7	 129; 52^{44}	4-01-00-01699 0.8778^{20}	H_2O 1; EtOH 3; eth 3 1.3987^{20}
8090	Nitrous acid, 3-methylbutyl ester Isoamyl nitrite	$C_5H_{11}NO_2$ 117.15	110-46-3	5005 99.2	4-01-00-01683 0.8828^{20}	H_2O 2; EtOH 5; eth 5 1.3918^{20}
8091	Nitrous acid, methyl ester Methyl nitrite	CH_3NO_2 61.04	624-91-9 -16	 -12	4-01-00-01253 0.991^{15}	EtOH 3; eth 3
8092	Nitrous acid, 1-methylethyl ester Isopropyl nitrite	$C_3H_7NO_2$ 89.09	541-42-4	5104 40	4-01-00-01474 0.8684^{15}	H_2O 1; EtOH 3; eth 3
8093	Nitrous acid, 1-methylpropyl ester	$C_4H_9NO_2$ 103.12	924-43-6	 68.5	4-01-00-01573 0.8726^{20}	eth 4; EtOH 4; chl 4 1.3710^{20}
8094	Nitrous acid, 2-methylpropyl ester Isobutyl nitrite	$C_4H_9NO_2$ 103.12	542-56-3	5032 67	4-01-00-01595 0.8699^{22}	H_2O 2; EtOH 3; eth 3 1.3715^{22}
8095	Nitrous acid, octyl ester	$C_8H_{17}NO_2$ 159.23	629-46-9	 174.5	4-01-00-01762 0.862^{17}	H_2O 2; EtOH 4; eth 4 1.4127^{20}
8096	Nitrous acid, pentyl ester	$C_5H_{11}NO_2$ 117.15	463-04-7	 104.5	4-01-00-01644 0.8817^{20}	H_2O 2; EtOH 5; eth 5 1.3851^{20}
8097	Nitrous acid, phenylmethyl ester Benzyl nitrite	$C_7H_7NO_2$ 137.14	935-05-7	 $80\text{-}3^{35}$	4-06-00-02568 1.075^{25}	 1.4989^{25}
8098	Nitrous acid, propyl ester Propyl nitrite	$C_3H_7NO_2$ 89.09	543-67-9	7878 48	4-01-00-01424 0.886^{20}	H_2O 2; EtOH 3; eth 3 1.3604^{20}
8099	Nonacosane Celidoniol, deoxy-	$C_{29}H_{60}$ 408.80	630-03-5 63.7	 440.8	4-01-00-00591 0.8083^{20}	H_2O 1; EtOH 4; eth 4; ace 4 1.4529^{20}
8100	1-Nonacosanol Nonacosanyl alcohol	$C_{29}H_{60}O$ 424.79	6624-76-6 84.5	 $200^{1.0}$	4-01-00-01916	EtOH 3
8101	Nonadecane	$C_{19}H_{40}$ 268.53	629-92-5 32.1	 [illegible]	4-01-00-00560 [illegible]	H_2O 1; EtOH 2; eth 3; ace 3 [illegible]
8102	1,2,3-Nonadecanetricarboxylic acid, 2-hydroxy- Agaricic acid	$C_{22}H_{40}O_7$ 416.56	666-99-9 142 dec	174	4-03-00-01284	H_2O 3; EtOH 2; eth 2; bz 1
8103	Nonadecanoic acid	$C_{19}H_{38}O_2$ 298.51	646-30-0 69.4	 297^{100}; 228^{10}	4-02-00-01256 0.8468^{70}	H_2O 1; EtOH 4; eth 4; bz 4
8104	1-Nonadecanol	$C_{19}H_{40}O$ 284.53	1454-84-8 63.3	 $166^{0.3}$	4-01-00-01898	eth 3; ace 3 1.4328^{75}
8105	10-Nonadecanone	$C_{19}H_{38}O$ 282.51	504-57-4 65.5	 >350; $156^{1.1}$	4-01-00-03402	H_2O 1; EtOH 2; eth 3; ace 3
8106	2-Nonadecanone	$C_{19}H_{38}O$ 282.51	629-66-3 57	 266^{110}; 165^{2}	4-01-00-03400 0.8108^{56}	H_2O 1; EtOH 2; eth 4; ace 3
8107	1-Nonadecyne	$C_{19}H_{36}$ 264.49	26186-01-6 37.5	 327	3-01-00-01030 0.8054^{20}	ace 4; bz 4; eth 4 1.4488^{20}
8108	2,6-Nonadienal Leaf aldehyde violet	$C_9H_{14}O$ 138.21	26370-28-5	 $85\text{-}7^{11}$	4-01-00-03560 0.8678^{20}	 1.4460^{20}
8109	2,4-Nonadienal, (*E*,*E*)-	$C_9H_{14}O$ 138.21	5910-87-2	 $97\text{-}8^{10}$	4-01-00-03560 0.862^{25}	 1.5207^{20}
8110	1,8-Nonadiene	C_9H_{16} 124.23	4900-30-5	 142.5	4-01-00-01048 0.7511^{20}	 1.4302^{20}
8111	2,7-Nonadiene	C_9H_{16} 124.23	51333-70-1 -72.5	 151.5	3-01-00-01013 0.7499^{20}	 1.4358^{20}
8112	2,6-Nonadien-1-ol	$C_9H_{16}O$ 140.23	7786-44-9	 $96\text{-}100^{11}$	4-01-00-02266 0.8604^{25}	 1.4598^{25}
8113	6,8-Nonadien-2-one, 8-methyl-5-(1-methylethyl)-, [*S*-(*E*)]- Solanone	$C_{13}H_{22}O$ 194.32	1937-54-8	8663 60^{1}	 0.870^{20}	 1.4755^{20}
8114	1,4-Nonadiyne	C_9H_{12} 120.19	[illegible]	 $83\text{-}4^{41}$	[illegible] 0.8112^{25}	 1.4518^{25}
8115	1,8-Nonadiyne	C_9H_{12} 120.19	2396-65-8 -27.3	 162	4-01-00-01125 0.8158^{20}	eth 3; ace 3 1.4490^{20}
8116	2,7-Nonadiyne	C_9H_{12} 120.19	31699-35-1 4.3	 180	3-01-00-01063 0.8332^{20}	 1.4674^{20}
8117	Nonanal Nonaldehyde	$C_9H_{18}O$ 142.24	124-19-6	 191	4-01-00-03352 0.8264^{22}	eth 3; chl 3 1.4273^{20}
8118	Nonanamide	$C_9H_{19}NO$ 157.26	1120-07-6 99.5	 sub	4-02-00-01023 0.8394^{110}	H_2O 1; EtOH 2; eth 2 1.4248^{110}
8119	1-Nonanamine	$C_9H_{21}N$ 143.27	112-20-9 -1	 202.2	4-04-00-00777 0.7886^{20}	H_2O 2; EtOH 3; eth 3; chl 2 1.4336^{20}
8120	Nonane	C_9H_{20} 128.26	111-84-2 -53.5	 150.8	4-01-00-00447 0.7176^{20}	H_2O 1; EtOH 4; eth 4; ace 5 1.4054^{20}
8121	Nonane, 1-bromo-	$C_9H_{19}Br$ 207.15	693-58-3	 88^{4}	4-01-00-00451 1.0165^{20}	 1.4533^{20}

No.	Name	Synonym	Mol. Form.	Mol. Wt.	CAS RN	mp/°C	Merck No.	bp/°C	Beil. Ref.	den/g cm^{-3}	Solubility	n_D
8122	Nonane, 5-butyl-		$C_{13}H_{28}$	184.37	17312-63-9			217.5	4-01-00-00517	0.7635^{18}		1.4273^{18}
8123	Nonane, 1-chloro-		$C_9H_{19}Cl$	162.70	2473-01-0	-39.4		203.5	4-01-00-00450	0.8720^{20}	H_2O 1; eth 3; chl 3	1.4345^{20}
8124	Nonane, 2-chloro-	2-Chlorononane	$C_9H_{19}Cl$	162.70	2216-36-6			190	0-01-00-00166	0.8790^{20}	chl 4	1.4420^{20}
8125	Nonane, 5-chloro-	5-Chlorononane	$C_9H_{19}Cl$	162.70	28123-70-8			$85\text{-}7^{14}$	2-01-00-00128	0.8639^{15}	eth 4	1.4314^{15}
8126	Nonane, 1-chloro-9-fluoro-		$C_9H_{18}ClF$	180.69	463-23-0			102^{11}	4-01-00-00450	0.966^{20}	eth 4; EtOH 4	1.4301^{25}
8127	Nonane, 1,2-dibromo-	1,2-Dibromononane	$C_9H_{18}Br_2$	286.05	73642-91-8			141.5^{20}	3-01-00-00506	1.3980^{20}		1.4942^{20}
8128	Nonane, 1,9-dichloro-	1,9-Dichlorononane	$C_9H_{18}Cl_2$	197.15	821-99-8			260; 138^{17}	4-01-00-00451	1.0173^{25}		1.4586^{25}
8129	Nonanedinitrile		$C_9H_{14}N_2$	150.22	1675-69-0			$198\text{-}9^{25}$	4-02-00-02059	0.9200^{19}	eth 4; EtOH 4; bz 4	1.4518^{19}
8130	Nonanedioic acid	Azelaic acid	$C_9H_{16}O_4$	188.22	123-99-9	106.5	921	287^{100}; 225^{10}	4-02-00-02055	1.225^{25}	H_2O 2; EtOH 3; eth 2; bz 2	1.4303^{111}
8131	Nonanedioic acid, bis(2-ethylbutyl) ester		$C_{21}H_{40}O_4$	356.55	105-03-3	-45		230^5	4-02-00-02057	0.928^{25}	ace 4; bz 4; EtOH 4	1.443^{25}
8132	Nonanedioic acid, bis(2-ethylhexyl) ester		$C_{25}H_{48}O_4$	412.65	103-24-2	-78		237^5	3-02-00-01787	0.915^{25}	H_2O 1; EtOH 3; ace 3; bz 3	1.446^{25}
8133	Nonanedioic acid, diethyl ester		$C_{13}H_{24}O_4$	244.33	624-17-9	-18.5		291.5	4-02-00-02056	0.9729^{20}	H_2O 1; EtOH 3; eth 3	1.4351^{20}
8134	Nonanedioic acid, dimethyl ester		$C_{11}H_{20}O_4$	216.28	1732-10-1			156^{20}	4-02-00-02056	1.0082^{20}	H_2O 1; EtOH 3; ace 3; bz 3	1.4367^{20}
8135	Nonanedioic acid, 3-methyl-	Azelaic acid, 3-methyl	$C_{10}H_{18}O_4$	202.25	76078-85-8	44		$168^{0.2}$	3-02-00-01819			1.4670^{19}
8136	1,9-Nonanediol	1,9-Dihydroxynonane	$C_9H_{20}O_2$	160.26	3937-56-2	45.8		173^{20}; 150^3	4-01-00-02607		H_2O 2; EtOH 4; eth 4; bz 3	
8137	2,4-Nonanedione	Methyl β-oxoheptyl ketone	$C_9H_{16}O_2$	156.22	6175-23-1	-18		$105\text{-}7^{22}$	4-01-00-03715	0.938^{25}		
8138	Nonanedioyl dichloride		$C_9H_{14}Cl_2O_2$	225.11	123-98-8			166^{18}	3-02-00-01789	1.4680^{20}	eth 3; bz 4	1.4680^{20}
8139	Nonane, icosofluoro	Perfluorononane	C_9F_{20}	488.07	375-96-2			125.3		1.800^{25}		
8140	Nonane, 2-methyl-	2-Methylnonane	$C_{10}H_{22}$	142.28	871-83-0	-74.6		167.1	4-01-00-00473	0.7281^{20}	H_2O 1; eth 3; bz 3; chl 3	1.4099^{20}
8141	Nonane, 3-methyl-	3-Methylnonane	$C_{10}H_{22}$	142.28	5911-04-6	-84.8		167.9	2-01-00-00130	0.7354^{20}	bz 4; eth 4; chl 4	1.4125^{20}
8142	Nonane, 4-methyl-	4-Methylnonane	$C_{10}H_{22}$	142.28	17301-94-9	-99		165.7	4-01-00-00475	0.7323^{20}	bz 4; eth 4; chl 4	1.4123^{20}
8143	Nonane, 5-methyl-	5-Methylnonane	$C_{10}H_{22}$	142.28	15869-85-9	-87.7		165.1	4-01-00-00475	0.7326^{20}	H_2O 1; eth 3; bz 3; chl 3	1.4116^{20}
8144	Nonanenitrile		$C_9H_{17}N$	139.24	2243-27-8	-34.2		224.4	4-02-00-01024	0.8178^{20}	H_2O 1; EtOH 3; eth 3; ctc 2	1.4255^{20}
8145	1-Nonanethiol	Nonyl mercaptan	$C_9H_{20}S$	160.32	1455-21-6	-20.1		220	4-01-00-01802	0.842^{25}		1.4548^{20}
8146	Nonanoic acid	Pelargonic acid	$C_9H_{18}O_2$	158.24	112-05-0	12.3	7013	254.5	4-02-00-01018	0.9052^{20}	H_2O 1; EtOH 3; eth 3; chl 3	1.4343^{19}
8147	Nonanoic acid, butyl ester	Pelargonic acid, butyl ester	$C_{13}H_{26}O_2$	214.35	50623-57-9	-38		$122\text{-}4^{20}$	3-02-00-00820	0.8520^{25}		1.4262^{25}
8148	Nonanoic acid, 9-cyano-, methyl ester	Sebacic acid, monomitrile, methyl ester	$C_{11}H_{19}NO_2$	197.28	53663-26-6	3.5		178^{16}	4-02-00-02089	0.934^{20}	eth 4	1.4398^{25}
8149	Nonanoic acid, ethenyl ester		$C_{11}H_{20}O_2$	184.28	6280-03-1			133.5^{50}	4-02-00-01020	0.8689^{30}		1.4447^{30}
8150	Nonanoic acid, ethyl ester	Ethyl nonanoate	$C_{11}H_{22}O_2$	186.29	123-29-5	-36.7	3793	227.0	4-02-00-01019	0.8657^{20}	H_2O 1; EtOH 3; eth 3; ace 3	1.4220^{20}
8151	Nonanoic acid, 9-fluoro-	Pelargonic acid, 9-fluoro	$C_9H_{17}FO_2$	176.23	463-16-1	18		$89^{0.2}$	4-02-00-01024		eth 4; EtOH 4	1.4289^{25}
8152	Nonanoic acid, heptyl ester	Pelargonic acid, heptyl ester	$C_{16}H_{32}O_2$	256.43	71605-85-1	-15.5		210^{75}	3-02-00-00821	0.8553^{25}		1.4350^{20}
8153	Nonanoic acid, 3-methyl-	Pelargonic acid, 3-methyl	$C_{10}H_{20}O_2$	172.27	35205-79-9			$147\text{-}8^{12}$	4-02-00-01057	0.9012^{20}		1.4342^{20}
8154	Nonanoic acid, methyl ester		$C_{10}H_{20}O_2$	172.27	1731-84-6			213.5	4-02-00-01019	0.8799^{15}	H_2O 1; EtOH 3; eth 3; ctc 2	1.4214^{20}
8155	Nonanoic acid, 3-methyl-, ethyl ester	Pelargonic acid 3-methyl, ethyl ester	$C_{12}H_{24}O_2$	200.32	86051-37-8			115^{13}	3-02-00-00849	0.8653^{20}		1.4240^{20}
8156	Nonanoic acid, pentyl ester	Pelargonic acid, pentyl ester	$C_{14}H_{28}O_2$	228.38	61531-45-1	-27		$130\text{-}2^{20}$	3-02-00-00820	0.8506^{25}		1.4318^{20}
8157	1-Nonanol	Nonyl alcohol	$C_9H_{20}O$	144.26	143-08-8	-5	6598	213.3	4-01-00-01798	0.8273^{20}	H_2O 1; EtOH 3; eth 3; ctc 2	1.4333^{20}
8158	2-Nonanol, (±)-		$C_9H_{20}O$	144.26	74683-66-2	-35		193.5	4-01-00-01803	0.8471^{20}	eth 4; EtOH 4	1.4353^{20}
8159	3-Nonanol, (±)-		$C_9H_{20}O$	144.26	74742-08-8	22		195; 93^{18}	4-01-00-01803	0.8250^{20}	H_2O 1; EtOH 3; eth 3	1.4289^{20}
8160	4-Nonanol, (*S*)-		$C_9H_{20}O$	144.26	52708-03-9			192.5; 94^{18}	3-01-00-01747	0.8282^{20}	H_2O 1; EtOH 3; eth 3	1.4197^{20}
8161	5-Nonanol	Dibutylcarbinol	$C_9H_{20}O$	144.26	623-93-8			193; 97^{20}	4-01-00-01804	0.8356^{20}	H_2O 1; EtOH 3	1.4289^{20}

PHYSICAL CONSTANTS OF ORGANIC COMPOUNDS (continued)

No.	Name Synonym	Mol. Form. Mol. Wt.	CAS RN mp/°C	Merck No. bp/°C	Beil. Ref. den/g cm^{-3}	Solubility n_D
8162	5-Nonanol, 5-butyl-	$C_{13}H_{28}O$ 200.36	597-93-3 20	 232 dec	4-01-00-01863 0.8408^{20}	EtOH 4 1.4445^{20}
8163	1-Nonanol, 9-chloro- 9-Chlorononan-1-ol	$C_9H_{19}ClO$ 178.70	51308-99-7 28	 147^{14}	4-01-00-01802	eth 4; EtOH 4 1.4575^{20}
8164	5-Nonanol, 2,8-dimethyl- 2,8-Dimethylnonan-5-ol	$C_{11}H_{24}O$ 172.31	19780-96-2	 105^9	0-01-00-00428 0.8305^{12}	 1.4380^{12}
8165	1-Nonanol, 9-fluoro-	$C_9H_{19}FO$ 162.25	463-24-1	 $125\text{-}6^{15}$	4-01-00-01801 0.928^{20}	eth 4; EtOH 4 1.4279^{25}
8166	1-Nonanol, 3-methyl- 3-Hexylbutanol-1	$C_{10}H_{22}O$ 158.28	22663-64-5	 114^{14}; 103^9	4-01-00-01825 0.8342^{20}	 1.4361^{20}
8167	1-Nonanol, 4-methyl-	$C_{10}H_{22}O$ 158.28	1489-47-0	 120^{17}	4-01-00-01826 0.826^{27}	 1.4364^{20}
8168	1-Nonanol, 8-methyl 8-Methyl-1-nonanol	$C_{10}H_{22}O$ 158.28	55505-26-5	 108^{10}		
8169	3-Nonanol, 2-methyl-	$C_{10}H_{22}O$ 158.28	26533-33-5	 115^{28}	4-01-00-01825 0.8290^{20}	 1.4346^{20}
8170	4-Nonanol, 4-methyl- 1-Methyl-1-propylhexanol	$C_{10}H_{22}O$ 158.28	23418-38-4	 $92\text{-}3^{15}$	4-01-00-01826 0.8245^{25}	 1.4338^{20}
8171	5-Nonanol, 5-methyl- 2-Butyl-2-hexanol	$C_{10}H_{22}O$ 158.28	33933-78-7	 $84\text{-}5^{10}$	4-01-00-01827 0.8290^{20}	 1.4341^{20}
8172	2-Nonanone Heptyl methyl ketone	$C_9H_{18}O$ 142.24	821-55-6 -7.5	 195.3	4-01-00-03353 0.8208^{20}	H_2O 1; EtOH 3; eth 3; ace 4 1.4210^{20}
8173	3-Nonanone Ethyl hexyl ketone	$C_9H_{18}O$ 142.24	925-78-0 -8	 190; 86^{20}	4-01-00-03354 0.8241^{20}	H_2O 1; EtOH 3, eth 3, ace 4 1.4208^{20}
8174	4-Nonanone Propyl amyl ketone	$C_9H_{18}O$ 142.24	4485-09-0	 187.5	4-01-00-03354 0.8190^{25}	H_2O 1; EtOH 3; eth 3; ace 4 1.4189^{20}
8175	5-Nonanone Dibutyl ketone	$C_9H_{18}O$ 142.24	502-56-7 -5.9	 188.4	4-01-00-03355 0.8217^{20}	H_2O 1; EtOH 3; eth 4; chl 4 1.4195^{20}
8176	5-Nonanone, 2-methyl- 2-Methyl-5-nonanone	$C_{10}H_{20}O$ 156.27	22287-02-1	 203.5	3-01-00-02900 0.8213^{20}	 1.4239^{20}
8177	Nonanoyl chloride	$C_9H_{17}ClO$ 176.69	764-85-2 -60.5	 215.3	4-02-00-01023 0.9463^{15}	eth 3; ace 3
8178	Nonasiloxane, eicosamethyl- Eicosamethylnonasiloxane	$C_{20}H_{60}O_8Si_9$ 681.46	2652-13-3	3494 307.5; 198^{16}	4-04-00-04130 0.9173^{20}	bz 4 1.3980^{20}
8179	2-Nonenal 2-Nonen-1-al	$C_9H_{16}O$ 140.23	2463-53-8	 $56\text{-}8^{0.1}$	4-01-00-03501 0.8418[illegible]	 1.4502[illegible]
8180	2-Nonenal, (*E*)-	$C_9H_{16}O$ 140.23	18829-56-6	6595	4-01-00-03502	
8181	6-Nonenamide, *N*-[(4-hydroxy-3-methoxyphenyl)methyl]-8-methyl-, (*E*)- Capsaicin	$C_{18}H_{27}NO_3$ 305.42	404-86-4 65	1767 $210\text{-}20^{0.01}$	4-13-00-02588	H_2O 1; EtOH 4; eth 3; bz 3
8182	1-Nonene	C_9H_{18} 126.24	124-11-8 -81.3	 146.9	4-01-00-00894 0.7253^{25}	 1.4257^{20}
8183	3-Nonene, (*E*)-	C_9H_{18} 126.24	20063-92-7	 147.5	3-01-00-00853 0.732^{21}	H_2O 1; eth 3; bz 3; ctc 3 1.4181^{21}
8184	4-Nonene, 5-butyl- 5-Butyl-4-nonene	$C_{13}H_{26}$ 182.35	7367-38-6	 215.5	4-01-00-00922 0.7745^{20}	 1.4375^{20}
8185	1-Nonene, 2-methyl- 2-Methylnon-1-ene	$C_{10}H_{20}$ 140.27	2980-71-4 -64.2	 168.4	4-01-00-00903 0.7412^{25}	 1.4241^{20}
8186	3-Nonene, 2-methyl-	$C_{10}H_{20}$ 140.27	53966-53-3	 161.0	4-01-00-00903 0.7340^{20}	 1.4202^{20}
8187	3-Nonenoic acid	$C_9H_{16}O_2$ 156.22	4124-88-3	 106^1	4-02-00-01598 0.9254^{20}	 1.4454^{25}
8188	8-Nonenoic acid	$C_9H_{16}O_2$ 156.22	31642-67-8 5	 $116\text{-}8^1$	4-02-00-01597 0.9146^{16}	 1.4492^{15}
8189	1-Nonen-3-ol 1-Vinylheptanol	$C_9H_{18}O$ 142.24	21964-44-3	 193.5	3-01-00-01947 0.824^{21}	 1.4382^{15}
8190	8-Nonen-1-ol	$C_9H_{18}O$ 142.24	13038-21-6	 135^{20}	3-01-00-01947 0.8394^{25}	 1.4450^{23}
8191	5-Nonen-4-one, 6-methyl-	$C_{10}H_{18}O$ 154.25	7036-98-8	 $90\text{-}2^{16}$	3-01-00-03022 0.8608^{20}	 1.4518^{20}
8192	6-Nonen-4-one, 6-methyl-	$C_{10}H_{18}O$ 154.25	53252-20-3	 94^{18}	3-01-00-03022 0.8413^{21}	 1.4429^{21}
8193	1-Nonen-3-yne	C_9H_{14} 122.21	[illegible]	 $27\text{-}8^4$	4-01-00-01103 0.7602[illegible]	ace 4; eth 4 1.4487[illegible]
8194	1-Nonen-4-yne	C_9H_{14} 122.21	31508-12-0	 58^{22}	4-01-00-01103 0.777^{25}	ace 4; eth 4 1.4418^{25}
8195	2-Nonen-4-yne	C_9H_{14} 122.21	90644-60-3	 70^{20}	3-01-00-01048 0.7832^{25}	ace 4; eth 4 1.4590^{25}
8196	1-Nonyne Heptylacetylene	C_9H_{16} 124.23	3452-09-3 -50	 150.8	4-01-00-01047 0.7658^{20}	H_2O 1; eth 3; bz 3; ctc 3 1.4217^{20}
8197	2-Nonyne	C_9H_{16} 124.23	19447-29-1	 161.9	3-01-00-01013 0.7690^{20}	H_2O 1; eth 3; lig 3 1.4337^{20}
8198	3-Nonyne	C_9H_{16} 124.23	20184-89-8	 157.1	3-01-00-01013 0.7616^{20}	H_2O 1; eth 3; lig 3 1.4299^{20}
8199	4-Nonyne	C_9H_{16} 124.23	20184-91-2	 152	4-01-00-01047 0.757^{25}	H_2O 1; eth 3; ace 3 1.4296^{25}
8200	1-Nonyne, 1-chloro-	$C_9H_{15}Cl$ 158.67	90722-14-8	 $75\text{-}7^{15}$	3-01-00-01012 0.906^{20}	eth 4 1.450^{20}
8201	4-Nonyne, 3,3-dimethyl- 3,3-Dimethyl-4-nonyne	$C_{11}H_{20}$ 152.28	29022-31-9	 82^{40}	3-01-00-01024 0.7667^{20}	bz 4; eth 4 1.4317^{20}

No.	Name Synonym	Mol. Form. Mol. Wt.	CAS RN mp/°C	Merck No. bp/°C	Beil. Ref. den/g cm^{-3}	Solubility n_D
8202	4-Nonyne, 8-methyl- 8-Methyl-4-nonyne	$C_{10}H_{18}$ 138.25	70732-43-3	104.5^{97}	3-01-00-01018 0.7681^{20}	bz 4; eth 4 1.4311^{20}
8203	2-Nonynoic acid	$C_9H_{14}O_2$ 154.21	1846-70-4 -8	$154\text{-}6^{18}$	4-02-00-01721 0.9525^{12}	1.4605^{25}
8204	3-Nonynoic acid	$C_9H_{14}O_2$ 154.21	56630-33-2 17	118^{2}	4-02-00-01721	1.4603^{25}
8205	5-Nonynoic acid	$C_9H_{14}O_2$ 154.21	56630-34-3 -3	115^{1}	4-02-00-01723	1.4558^{25}
8206	6-Nonynoic acid	$C_9H_{14}O_2$ 154.21	56630-31-0 14	130^{3}	4-02-00-01723	1.4578^{25}
8207	8-Nonynoic acid	$C_9H_{14}O_2$ 154.21	30964-01-3 19	100^{6}	4-02-00-01721	1.4524^{25}
8208	1-Nonyn-3-ol	$C_9H_{16}O$ 140.23	7383-20-2	85^{10}; 63^{2}	4-01-00-02265 0.8627^{20}	1.4440^{20}
8209	2-Nonyn-1-ol	$C_9H_{16}O$ 140.23	5921-73-3	109^{11}	4-01-00-02265 0.878^{16}	1.4576^{16}
8210	24-Norcholan-23-oic acid, (5β)- Norcholanic acid	$C_{23}H_{38}O_2$ 346.55	511-18-2 177	6606	3-09-00-02652	
8211	29-Nordammara-1,17(20),24-trien-21-oic acid, 6,16-bis(acetyloxy)-3,7-dioxo-, Helvolic acid	$C_{33}H_{44}O_8$ 568.71	29400-42-8 212 dec	4552	4-10-00-04083	H_2O 2; EtOH 2; eth 3; ace 3
8212	Norea Urea, *N,N*-dimethyl-*N*'-(octahydro-4,7-methano-1H-inde*N*-5-yl)-, (3aα,4α,5α,7α,7aα)-	$C_{13}H_{22}N_2O$ 222.33	18530-56-8 177	6611		
8213	Norflurazon 3(2H)-Pyridazinone, 4-chloro-5-(methylamino)-2-[3-(trifluoromethyl)phenyl]-	$C_{12}H_9ClF_3N_3O$ 303.67	27314-13-2 184	6618		
8214	Norleucine 2-Aminohexanoic acid	$C_6H_{13}NO_2$ 131.17	327-57-1 298.5	6624	4-04-00-02686	H_2O 4
8215	*DL*-Norleucine Hexanoic acid, 2-amino(*DL*)	$C_6H_{13}NO_2$ 131.17	616-06-8 327 dec	6624	3-04-00-01386 1.172^{25}	H_2O 3; EtOH 2; eth 1
8216	19-Norpregna-1,3,5(10)-trien-20-yne-3,17-diol, (17α)- 17α-Ethinyl-17β-estradiol	$C_{20}H_{24}O_2$ 296.41	57-63-6	3689	4-06-00-06877	chl 2
8217	*DL*-Norvaline	$C_5H_{11}NO_2$ 117.15	760-78-1 303	6636 sub	4-04-00-02629	H_2O 3; EtOH 1; eth 1; chl 1
8218	*L*-Norvaline (*S*)-2-Aminopentanoic acid	$C_5H_{11}NO_2$ 117.15	6600-40-4 307	6636	4-04-00-02629	H_2O 4
8219	Nuarimol 5-Pyrimidinemethanol, α-(2-chlorophenyl)-α-(4-fluorophenyl)-	$C_{17}H_{12}ClFN_2O$ 314.75	63284-71-9 126			
8220	Octacosane	$C_{28}H_{58}$ 394.77	630-02-4 64.5	431.6	4-01-00-00588 0.8067^{20}	H_2O 1; ace 5; bz 3; chl 3 1.4330^{70}
8221	Octacosanoic acid Montanic acid	$C_{28}H_{56}O_2$ 424.75	506-48-9 90.9		4-02-00-01318 0.8191^{100}	bz 4; chl 4 1.4313^{100}
8222	1-Octacosanol Montanyl alcohol	$C_{28}H_{58}O$ 410.77	557-61-9 83.3	6664 200^{1}	4-01-00-01915	
8223	9,12-Octadecadienoic acid (*Z,Z*)- Linoleic acid	$C_{18}H_{32}O_2$ 280.45	60-33-3 -8.5	5382 $229\text{-}30^{16}$	4-02-00-01754 0.9022^{20}	ace 4; bz 4; eth 4; EtOH 4 1.4699^{20}
8224	10,12-Octadecadienoic acid, (*E,E*)-	$C_{18}H_{32}O_2$ 280.45	1072-36-2 56.5		4-02-00-01752 0.8686^{70}	1.4689^{40}
8225	9,12-Octadecadienoic acid	$C_{18}H_{32}I_2$ 502.26	2197-37-7 -5	$229\text{-}30^{16}$	4-02-00-01754 0.9022^{20}	ace 4; bz 4; eth 4; EtOH 4 1.4699^{20}
8226	9,12-Octadecadienoic acid (*Z,Z*)-, ethyl ester Ethyl linoleate	$C_{20}H_{36}O_2$ 308.50	544-35-4	3776 272^{180}; 212^{12}	4-02-00-01757 0.8865^{20}	eth 4; EtOH 4
8227	9,12-Octadecadienoic acid, ethyl ester	$C_{20}H_{36}O_2$ 308.50	7619-08-1	272^{180} 212^{12}	2-02-00-00461 0.8865^{20}	H_2O 1; EtOH 3; eth 3
8228	9,12-Octadecadienoic acid (*Z,Z*)-, methyl ester Methyl linoleate	$C_{19}H_{34}O_2$ 294.48	112-63-0 -35	6008 215^{20}	4-02-00-01756 0.8886^{10}	eth 4; EtOH 4 1.4638^{20}
8229	9,12-Octadecadienoic acid, methyl ester Methyl 9,12-octadecadienoate	$C_{19}H_{34}O_2$ 294.48	2462-85-3 -35	215^{20} 169^{1}	2-02-00-00461 0.8886^{18}	H_2O 1; EtOH 3; eth 3; ctc 2 1.4638^{20}
8230	7,11-Octadecadiyne	$C_{18}H_{30}$ 246.44	103697-14-9	$167\text{-}8^{7}$	3-01-00-01068 0.841^{19}	1.4698^{19}
8231	Octadecanamide	$C_{18}H_{37}NO$ 283.50	124-26-5 109	250^{12}	4-02-00-01240	eth 4; chl 4
8232	Octadecanamide, *N*-(4-hydroxyphenyl)-	$C_{24}H_{41}NO_2$ 375.60	103-99-1 133.8	239.5^{10}	4-13-00-01113	H_2O 1; eth 2; ace 3; bz 2
8233	1-Octadecanamine	$C_{18}H_{39}N$ 269.51	124-30-1 52.9	346.8	4-04-00-00825 0.8618^{20}	H_2O 1; EtOH 3; eth 3; ace 2 1.4522^{20}
8234	Octadecane	$C_{18}H_{38}$ 254.50	593-45-3 28.2	316.3	4-01-00-00553 0.7768^{28}	H_2O 1; EtOH 2; eth 3; ace 3 1.4390^{20}
8235	Octadecane, 1-bromo-	$C_{18}H_{37}Br$ 333.40	112-89-0 28.2	210^{10}	4-01-00-00555 0.9848^{20}	H_2O 1; EtOH 3; eth 3; ctc 2 1.4631^{20}
8236	Octadecane, 1-chloro-	$C_{18}H_{37}Cl$ 288.94	3386-33-2 28.6	348	4-01-00-00554 0.8641^{20}	H_2O 1; ctc 2 1.4531^{20}
8237	Octadecane, 1,1-dimethoxy- Octadecanal, dimethyl acetal	$C_{20}H_{42}O_2$ 314.55	14620-55-4	$168\text{-}70^{3}$	4-01-00-03397	eth 4; EtOH 4 1.4410^{25}

No.	Name Synonym	Mol. Form. Mol. Wt.	CAS RN mp/°C	Merck No. bp/°C	Beil. Ref. den/g cm^{-3}	Solubility n_D
8238	Octadecanedioic acid, diethyl ester	$C_{22}H_{42}O_4$ 370.57	1472-90-8 54.5	240^{12}	4-02-00-02176	eth 4; EtOH 4
8239	Octadecanedioic acid, 9,10-dihydroxy-, (*R**,*R**)-(±)- Phloionic acid	$C_{18}H_{34}O_6$ 346.46	23843-52-9 126	7297	4-03-00-01250	
8240	Octadecane, 1-(ethenyloxy)-	$C_{20}H_{40}O$ 296.54	930-02-9 30	182^{3}	4-01-00-02057 0.8138^{40}	chl 2
8241	Octadecane, 3-ethyl-5-(2-ethylbutyl)-	$C_{26}H_{54}$ 366.71	55282-12-7	229.5^{10}	4-01-00-00586 0.8115^{20}	1.45239^{20}
8242	Octadecane, 1-iodo-	$C_{18}H_{37}I$ 380.40	629-93-6 34	383	4-01-00-00556 1.0994^{20}	H_2O 1; EtOH 2; eth 2 1.4810^{20}
8243	Octadecanenitrile	$C_{18}H_{35}N$ 265.48	638-65-3 41	362	4-02-00-01242 0.8325^{20}	H_2O 1; EtOH 3; eth 4; ace 4 1.4389^{45}
8244	1-Octadecanethiol Stearyl mercaptan	$C_{18}H_{38}S$ 286.57	2885-00-9 30	$204\text{-}10^{11}$	4-01-00-01894 0.8475^{20}	eth 4 1.4645^{20}
8245	Octadecane, 9-*p*-tolyl- Toluene, *p*-(1-octyldecyl)-	$C_{25}H_{44}$ 344.62	4445-08-3	185^{10}	4-05-00-01221 0.8549^{20}	1.4811^{20}
8246	9,11,13-Octadecanetrienoic acid (*Z*,*Z*,*Z*) Eleostearic acid	$C_{18}H_{30}O_2$ 278.44	3884-88-6 48.5			
8247	Octadecanoic acid Stearic acid	$C_{18}H_{36}O_2$ 284.48	57-11-4 68.8	8761 350 dec; 232^{15}	4-02-00-01206 0.9408^{20}	H_2O 1; EtOH 2; eth 4; ace 3 1.4299^{80}
8248	Octadecanoic acid, aluminum salt	$C_{54}H_{105}AlO_6$ 877.41	637-12-7 118	370	4-02-00-01206	H_2O 1; EtOH 3; peth 3
8249	Octadecanoic acid, anhydride	$C_{36}H_{70}O_3$ 550.95	638-08-4 72		4-02-00-01239 0.8365^{82}	H_2O 1; EtOH 1; eth 2; bz 2 1.4362^{80}
8250	Octadecanoic acid, 18-bromo- Stearic acid, 18-bromo	$C_{18}H_{35}BrO_2$ 363.38	2536-38-1 75.5	240^{4}	2-02-00-00361	bz 4; eth 4; EtOH 4
8251	Octadecanoic acid, butyl ester Butyl stearate	$C_{22}H_{44}O_2$ 340.59	123-95-5 27	1589 343	4-02-00-01219 0.854^{25}	H_2O 1; EtOH 3; ace 4 1.4328^{50}
8252	Octadecanoic acid, calcium salt	$C_{36}H_{70}CaO_4$ 607.03	1592-23-0 179.5	1710	4-02-00-01206	H_2O 1; EtOH 1; eth 1
8253	Octadecanoic acid, cyclohexyl ester Stearic acid, cyclohexyl ester	$C_{24}H_{46}O_2$ 366.63	104-07-4 44		4-06-00-00038 0.889^{15}	eth 4
8254	Octadecanoic acid, 9,10-dihydroxy- 9,10-Dihydroxystearic acid	$C_{18}H_{36}O_4$ 316.48	120-87-6 90	3171	4-03-00-01092	H_2O 1; EtOH 2; eth 2
8255	Octadecanoic acid, 2,3-dihydroxypropyl ester, (±)-	$C_{21}H_{42}O_4$ 358.56	22610-63-5 74		4-02-00-01225 0.9841^{20}	H_2O 1; EtOH 2; eth 2; lig 3 1.4400^{86}
8256	Octadecanoic acid, 1,2-ethanediyl ester	$C_{38}H_{74}O_4$ 595.00	627-83-8 79	241^{20}	4-02-00-01223 0.8581^{78}	H_2O 1; EtOH 1; eth 4; ace 4
8257	Octadecanoic acid, ethyl ester	$C_{20}H_{40}O_2$ 312.54	111-61-5 33	199^{10}	4-02-00-01218 1.057^{20}	H_2O 1; EtOH 3; eth 3; ace 4 1.4349^{40}
8258	Octadecanoic acid, hexadecyl ester	$C_{34}H_{68}O_2$ 508.91	1190-63-2 57		4-02-00-01220	ace 4; eth 4; chl 4 1.4410^{70}
8259	Octadecanoic acid, 2-[2-[2-(2-hydroxyethoxy)ethoxy]ethoxy]ethyl ester	$C_{26}H_{52}O_6$ 460.70	106-07-0 40	328	1.1285^{15}	1.4593^{20}
8260	Octadecanoic acid, 2-hydroxyethyl ester	$C_{20}H_{40}O_3$ 328.54	111-60-4 60.5	$189\text{-}91^{3}$	4-02-00-01222 0.8780^{60}	EtOH 2; eth 3 1.4310^{60}
8261	Octadecanoic acid, lead salt	$C_{36}H_{70}O_4Pb$ 774.15	7428-48-0 115.7		4-02-00-01206	H_2O 2; EtOH 1; eth 1
8262	Octadecanoic acid, 14-methyl- Stearic acid, 14-methyl	$C_{19}H_{38}O_2$ 298.51	94434-64-7 37.5	$182^{0.4}$	4-02-00-01265 0.9400^{20}	
8263	Octadecanoic acid, 17-methyl-	$C_{19}H_{38}O_2$ 298.51	2724-59-6 67.5	$180^{0.3}$	4-02-00-01260 0.8420^{70}	1.4336^{70}
8264	Octadecanoic acid, 9-methyl- Stearic acid, 9-methyl	$C_{19}H_{38}O_2$ 298.51	86073-38-3 40	$171^{0.1}$	4-02-00-01271 0.9980^{20}	
8265	Octadecanoic acid, 3-methylbutyl ester Stearic acid, isopentyl ester	$C_{23}H_{46}O_2$ 354.62	627-88-3 25.5	192^{2}	2-02-00-00353 0.855^{20}	H_2O 1; EtOH 2; eth 3; ace 3 1.433^{50}
8266	Octadecanoic acid, methyl ester	$C_{19}H_{38}O_2$ 298.51	112-61-8 39.1	443; 215^{15}	4-02-00-01216 0.8498^{40}	eth 4; chl 4 1.4367^{40}
8267	Octadecanoic acid, 1-methylethyl ester	$C_{21}H_{42}O_2$ 326.56	112-10-7 28	207^{6}	4-02-00-01219 0.8403^{38}	ace 4; eth 4; EtOH 4; chl 4
8268	Octadecanoic acid, 2-methylpropyl ester Isobutyl stearate	$C_{22}H_{44}O_2$ 340.59	646-13-9 28.9	5034 223^{15}	3-02-00-01017 0.8498^{20}	eth 4
8269	Octadecanoic acid, 12-oxo-, ethyl ester Stearic acid, 12-oxo, ethyl ester	$C_{20}H_{38}O_3$ 326.52	88472-61-1 38	199^{3}	3-03-00-01294	EtOH 4
8270	Octadecanoic acid, pentyl ester Stearic acid, pentyl ester	$C_{23}H_{46}O_2$ 354.62	6382-13-4 30		4-02-00-01220	eth 4; EtOH 4 1.4342^{50}
8271	Octadecanoic acid, phenyl ester Stearic acid, phenyl ester	$C_{24}H_{40}O_2$ 360.58	637-55-8 52	267^{15}	4-06-00-00618	H_2O 1; EtOH 3; eth 3
8272	Octadecanoic acid, 1,2,3-propanetriyl ester Tristearin	$C_{57}H_{110}O_6$ 891.50	555-43-1	9669	4-02-00-01233 0.8559^{90}	H_2O 1; EtOH 1; ace 3; bz 2 1.4395^{80}
8273	Octadecanoic acid, propyl ester	$C_{21}H_{42}O_2$ 326.56	3634-92-2 28.9	186.8^{2}	4-02-00-01219 0.8452^{38}	ace 4; eth 4; EtOH 4 1.4400^{30}
8274	Octadecanoic acid, 9,10,12,13-tetrabromo-, methyl ester Stearic acid, 9,10,12,13-tetrabromo, methyl ester	$C_{19}H_{34}Br_4O_2$ 614.09	62080-86-8 63	215^{15}	3-02-00-01049	eth 4; EtOH 4; chl 4 1.4346^{45}

No.	Name Synonym	Mol. Form. Mol. Wt.	CAS RN mp/°C	Merck No. bp/°C	Beil. Ref. den/g cm^{-3}	Solubility n_D
8275	Octadecanoic acid, (tetrahydro-2-furanyl)methyl ester Stearic acid, tetrahydrofurfuryl ester	$C_{23}H_{44}O_3$ 368.60	6940-09-6 22		4-02-00-01248 0.914^{25}	eth 4; EtOH 4
8276	Octadecanoic acid, 9,10,18-trihydroxy-, (*R**,*R**)- Phloionolic acid	$C_{18}H_{36}O_5$ 332.48	583-86-8 101.5	7298	4-03-00-01119	
8277	Octadecanoic acid, zinc salt	$C_{36}H_{70}O_4Zn$ 632.34	557-05-1 130	10063	4-02-00-01206	H_2O 1; EtOH 1; eth 1
8278	1-Octadecanol Stearyl alcohol	$C_{18}H_{38}O$ 270.50	112-92-5 59.5	 210.5^{15}	4-01-00-01888 0.8124^{59}	H_2O 1; EtOH 3; eth 3; ace 2
8279	3-Octadecanol, (±)- 3-Octadecanol (*DL*)	$C_{18}H_{38}O$ 270.50	111897-18-8 51	 202^{13}	4-01-00-01896 0.7858^{80}	 1.4290^{80}
8280	3-Octadecanone	$C_{18}H_{36}O$ 268.48	18261-92-2 51	 197^{11}	4-01-00-03398	ctc 3
8281	Octadecanoyl chloride	$C_{18}H_{35}ClO$ 302.93	112-76-5 23	 215^{15}	4-02-00-01240 0.8969^{0}	EtOH 2 1.4523^{24}
8282	9,11,13-Octadecatrienoic acid, (*E,E,E*)-	$C_{18}H_{30}O_2$ 278.44	544-73-0 71.5	 188^{1}	4-02-00-01787 0.8839^{80}	EtOH 4 1.5000^{80}
8283	9,12,15-Octadecatrienoic acid, ethyl ester, (*Z,Z,Z*)-	$C_{20}H_{34}O_2$ 306.49	1191-41-9	 218^{15}	4-02-00-01782 0.8919^{20}	eth 4; EtOH 4 1.4694^{20}
8284	9,11,13-Octadecatrienoic acid, (*E,Z,E*)-	$C_{18}H_{30}O_2$ 278.44	506-23-0 49	 235^{12} dec; 170^{1}	4-02-00-01787 0.9028^{50}	eth 4; EtOH 4 1.5112^{50}
8285	9,12,15-Octadecatrienoic acid, methyl ester, (*Z,Z,Z*)-	$C_{19}H_{32}O_2$ 292.46	301-00-8	 182^{3}	4-02-00-01782 0.896^{25}	 1.4709^{20}
8286	9,11,13-Octadecatrienoic acid, (*Z,Z,E*)-	$C_{18}H_{30}O_2$ 278.44	544-72-9 62		4-02-00-01787 0.9027^{20}	EtOH 4; peth 4 1.5114^{50}
8287	9,12,15-Octadecatrienoic acid, (*Z,Z,Z*)- Linolenic acid	$C_{18}H_{30}O_2$ 278.44	463-40-1 -16.5	5383 231^{17}; $129^{0.05}$	4-02-00-01781 0.9164^{20}	H_2O 1; EtOH 3; eth 3; bz 2 1.4800^{20}
8288	9-Octadecenal Octadecenyl aldehyde	$C_{18}H_{34}O$ 266.47	5090-41-5	 $168\text{-}9^{3}$	4-01-00-03533 0.8509^{20}	 1.4558^{20}
8289	9-Octadecenal, (*Z*)-	$C_{18}H_{34}O$ 266.47	2423-10-1	 $168\text{-}9^{3}$	4-01-00-03533 0.8509^{20}	 1.4558^{20}
8290	9-Octadecenamide, *N*-phenyl-, (*Z*)-	$C_{24}H_{39}NO$ 357.58	5429-85-6 41	 143.5^{10}	4-12-00-00411	bz 4; eth 4; HOAc 4
8291	1-Octadecene	$C_{18}H_{36}$ 252.48	112-88-9 17.5	 179^{15}; 145^{8}	4-01-00-00930 0.7891^{20}	H_2O 1; ace 3; ctc 3 1.4448^{20}
8292	9-Octadecene	$C_{18}H_{36}$ 252.48	5557-31-3 -30.5	 162^{9}	4-01-00-00932 0.7916^{20}	 1.4470^{20}
8293	9-Octadecenenitrile, (*Z*)-	$C_{18}H_{33}N$ 263.47	112-91-4 -1	 332 dec	4-02-00-01668 0.847^{17}	EtOH 4 1.4566^{20}
8294	9-Octadecenoic acid (*Z*)- Oleic acid	$C_{18}H_{34}O_2$ 282.47	112-80-1 13.4	6788 286^{100}	4-02-00-01641 0.8935^{20}	H_2O 1; EtOH 5; eth 5; ace 5 1.4582^{20}
8295	11-Octadecenoic acid, (*E*)- Vaccenic acid	$C_{18}H_{34}O_2$ 282.47	693-72-1 44	9810	4-02-00-01640	ace 3 1.4499^{60}
8296	2-Octadecenoic acid, (*E*)-	$C_{18}H_{34}O_2$ 282.47	2825-79-8 58.5		4-02-00-01632 0.8484^{90}	H_2O 1; EtOH 2; eth 3; bz 3
8297	9-Octadecenoic acid, (*E*)- Elaidic acid	$C_{18}H_{34}O_2$ 282.47	112-79-8 45	3497 288^{100}; 234^{15}	4-02-00-01647 0.8734^{45}	H_2O 1; EtOH 3; eth 3; bz 3 1.4499^{45}
8298	9-Octadecenoic acid (*Z*)-, butyl ester	$C_{22}H_{42}O_2$ 338.57	142-77-8 -26.4	 $227\text{-}8^{15}$	4-02-00-01653 0.8704^{15}	EtOH 4 1.4480^{25}
8299	9-Octadecenoic acid (*Z*)-, 2,3-dihydroxypropyl ester	$C_{21}H_{40}O_4$ 356.55	111-03-5 35	 $238\text{-}40^{3}$	2-02-00-00439 0.9420^{20}	H_2O 1; EtOH 3; eth 3; chl 3 1.4626^{20}
8300	6-Octadecenoic acid, 6,7-diiodo- Diiodotariric acid	$C_{18}H_{32}I_2O_2$ 534.26	533-86-8 50.3	 dec	4-02-00-01638	bz 4; eth 4; chl 4
8301	9-Octadecenoic acid (*Z*)-, ethyl ester	$C_{20}H_{38}O_2$ 310.52	111-62-6	 207^{13}	4-02-00-01651 0.8720^{20}	eth 4; EtOH 4 1.4515^{20}
8302	9-Octadecenoic acid, ethyl ester, (*E*)-	$C_{20}H_{38}O_2$ 310.52	6114-18-7 5.8	 $217\text{-}9^{15}$	4-02-00-01652 0.8664^{25}	eth 4; EtOH 4 1.4480^{25}
8303	9-Octadecenoic acid, 12-hydroxy-, [*R*-(*Z*)]- Ricinoleic acid	$C_{18}H_{34}O_3$ 298.47	141-22-0	8213 227^{10}	4-03-00-01026 0.9450^{21}	eth 4; EtOH 4 1.4716^{21}
8304	9-Octadecenoic acid, 12-hydroxy-, butyl ester, [*R*-(*Z*)]-	$C_{22}H_{42}O_3$ 354.57	151-13-3	 278^{12}	0-03-00-00388 0.9058^{22}	eth 4 1.4566^{22}
8305	9-Octadecenoic acid, 12-hydroxy-, ethyl ester, [*R*-(*Z*)]-	$C_{20}H_{38}O_3$ 326.52	55066-53-0	 258^{13}	4-03-00-01029 0.9045^{22}	 1.4618^{22}
8306	9-Octadecenoic acid (*Z*)-, 3-methylbutyl ester Oleic acid, isopentyl ester	$C_{23}H_{44}O_2$ 352.60	627-89-4	 $223\text{-}4^{10}$	3-02-00-01414 0.897^{15}	eth 4; EtOH 4
8307	9-Octadecenoic acid (*Z*)-, methyl ester Methyl oleate	$C_{19}H_{36}O_2$ 296.49	112-62-9 -19.9	 218.5^{20}	4-02-00-01649 0.8739^{20}	H_2O 1; EtOH 5; eth 5; chl 3 1.4522^{20}
8308	9-Octadecenoic acid, methyl ester, (*E*)-	$C_{19}H_{36}O_2$ 296.49	1937-62-8	 $213\text{-}5^{15}$	4-02-00-01651 0.8730^{20}	eth 4; EtOH 4 1.4513^{20}
8309	9-Octadecenoic acid (*Z*)-, phenylmethyl ester Oleic acid, benzyl ester	$C_{25}H_{40}O_2$ 372.59	55130-16-0	 237^{1}	4-06-00-02269 0.9302^{25}	eth 4; EtOH 4 1.4875^{25}

No.	Name Synonym	Mol. Form. Mol. Wt.	CAS RN mp/°C	Merck No. bp/°C	Beil. Ref. den/g cm^{-3}	Solubility n_D
8310	9-Octadecenoic acid (*Z*)-, 1,2,3-propanetriyl ester Triolein	$C_{57}H_{104}O_6$ 885.45	122-32-7 -32	9644 235-40^{18}	4-02-00-01664 0.8988^{40}	H_2O 1; EtOH 2; eth 4; chl 3 1.4621^{40}
8311	9-Octadecen-1-ol, (*E*)-	$C_{18}H_{36}O$ 268.48	506-42-3 36.5	 333	4-01-00-02204 0.8338^{40}	ace 4; eth 4; EtOH 4 1.4552^{40}
8312	9-Octadecen-1-ol, (*Z*)- Oleyl alcohol	$C_{18}H_{36}O$ 268.48	143-28-2 6.5	6791 205-10^{15}	4-01-00-02204 0.8489^{20}	H_2O 1; EtOH 3; eth 3; ctc 2 1.4606^{20}
8313	1-Octadecyne	$C_{18}H_{34}$ 250.47	629-89-0 22.5	 313	4-01-00-01075 0.8025^{20}	 1.4774^{20}
8314	2-Octadecyne	$C_{18}H_{34}$ 250.47	61847-97-0 30	 184^{15}	4-01-00-01075 0.8016^{30}	
8315	9-Octadecyne	$C_{18}H_{34}$ 250.47	35365-59-4 3	 163-4^{7}	4-01-00-01076 0.8012^{20}	 1.4488^{25}
8316	9-Octadecynoic acid 9-Stearolic acid	$C_{18}H_{32}O_2$ 280.45	506-24-1 48	 189$^{1.8}$	4-02-00-01751	H_2O 1; EtOH 3; eth 4 1.4510^{54}
8317	2,6-Octadienal, 3,7-dimethyl-, (*E*)-	$C_{10}H_{16}O$ 152.24	141-27-5	 229	4-01-00-03569 0.8888^{20}	H_2O 1; EtOH 5; eth 5 1.4898^{20}
8318	2,6-Octadienal, 3,7-dimethyl-, (*Z*)-	$C_{10}H_{16}O$ 152.24	106-26-3	 120^{20}	4-01-00-03569 0.8869^{20}	H_2O 1; EtOH 5; eth 5 1.4869^{20}
8319	1,7-Octadiene α,ω-Octadiene	C_8H_{14} 110.20	3710-30-3	 115.5	4-01-00-01038 0.734^{20}	 1.4245^{20}
8320	2,6-Octadiene 1,6-Dimethyldiallyl	C_8H_{14} 110.20	4974-27-0 -70	 124.5	4-01-00-01039 0.7406^{25}	 1.4292^{17}
8321	2,6-Octadiene, 1-bromo-3,7-dimethyl-, (*E*)- *trans*-Geranyl bromide	$C_{10}H_{17}Br$ 217.15	6138-90-5	 101^{12}; 47$^{0.005}$	4-01-00-01059 1.0940^{22}	 1.5027^{20}
8322	2,4-Octadiene, 3,7-dimethyl- 3,7-Dimethylocta-2,4-diene	$C_{10}H_{18}$ 138.25	56523-26-3	 165.5	3-01-00-01020 0.7933^{20}	 1.456^{20}
8323	2,5-Octadiene, 2,6-dimethyl-	$C_{10}H_{18}$ 138.25	116668-48-5	 59.5^{12}	3-01-00-01018 0.773^{20}	 1.4500^{20}
8324	2,6-Octadiene, 2,6-dimethyl-, (*Z*)- *cis*-2,6-Dimethyl-2,6-octadiene	$C_{10}H_{18}$ 138.25	2492-22-0	 168	4-01-00-01058 0.775^{21}	ace 4; eth 4; EtOH 4 1.4498^{20}
8325	2,6-Octadiene, 2,7-dimethyl- 2,7-Dimethyl-2,6-octadiene	$C_{10}H_{18}$ 138.25	16736-42-8 -74.4	 166	4-01-00-01061 0.7849^{20}	 1.4481^{20}
8326	2,6-Octadiene, 3,6-dimethyl-	$C_{10}H_{18}$ 138.25	116668-49-6	 134	4-01-00-01061 0.7767^{20}	 1.4445^{20}
8327	2,6-Octadiene, 4,5-dimethyl- 4,5-Dimethyl-2,6-octadiene	$C_{10}H_{18}$ 138.25	18476-57-8 -64.8	 153.5	3-01-00-01021 0.7611^{25}	 1.4375^{25}
8328	2,4-Octadiene, 7-methyl-	C_9H_{16} 124.23	2216-70-8	 149	2-01-00-00239 0.7521^{18}	 1.4543^{18}
8329	3,5-Octadiene, 4-methyl- 4-Methyl-3,5-octadiene	C_9H_{16} 124.23	36903-95-4	 149.5	2-01-00-00239 0.7640^{25}	 1.4628^{25}
8330	1,6-Octadiene, 7-methyl-3-methylene- β-Myrcene	$C_{10}H_{16}$ 136.24	123-35-3	6243 167	4-01-00-01108 0.8013^{15}	H_2O 1; EtOH 3; eth 3; bz 3 1.4722^{20}
8331	2,4-Octadienoic acid, 3,7-dimethyl-	$C_{10}H_{16}O_2$ 168.24	101715-60-0	 151-4^{14}	4-02-00-01736 0.959^{17}	 1.4919
8332	1,6-Octadien-3-ol, 3,7-dimethyl-, (±)-	$C_{10}H_{18}O$ 154.25	22564-99-4	 198; 86^{13}	4-01-00-02278 0.870^{15}	 1.4627
8333	2,6-Octadien-1-ol, 3,7-dimethyl-, (*E*)- *trans*-Geraniol	$C_{10}H_{18}O$ 154.25	106-24-1 <-15	4298 230	4-01-00-02277 0.8894^{20}	H_2O 1; EtOH 3; eth 3; ace 3 1.4766^{20}
8334	2,6-Octadien-1-ol, 3,7-dimethyl-, (*Z*)- Nerol	$C_{10}H_{18}O$ 154.25	106-25-2 <-15	6387 225; 125^{25}	4-01-00-02276 0.8756^{20}	EtOH 4 1.4746^{20}
8335	1,6-Octadien-3-ol, 3,7-dimethyl-, acetate, (*R*)-	$C_{12}H_{20}O_2$ 196.29	16509-46-9	 220; 115^{25}	4-02-00-00204 0.8951^{20}	eth 4; EtOH 4 1.4544^{21}
8336	2,6-Octadien-1-ol, 3,7-dimethyl-, acetate, (*Z*)-	$C_{12}H_{20}O_2$ 196.29	141-12-8	 134^{25}; 93^{3}	2-02-00-00153 0.905^{15}	 1.452^{20}
8337	1,6-Octadien-3-ol, 3,7-dimethyl-, formate Linalyl formate	$C_{11}H_{18}O_2$ 182.26	115-99-1	 100-3^{10}	4-02-00-00035 0.915^{25}	EtOH 4 1.456^{20}
8338	2,6-Octadien-1-ol, 3,7-dimethyl-, formate, (*E*)-	$C_{11}H_{18}O_2$ 182.26	105-86-2	 229 dec; 113^{25}	4-02-00-00035 0.9086^{25}	H_2O 1; EtOH 4; eth 3; ace 3 1.4659^{20}
8339	2,6-Octadien-4-yne, 3,6-diethyl-	$C_{12}H_{18}$ 162.27	100319-48-0	 170	3-01-00-01066 0.8196^{20}	ace 4; eth 4 1.4965^{20}
8340	2,6-Octadien-4-yne, 3,6-dimethyl- 3,6-Dimethyl-2,6-octadiene-4-yne	$C_{10}H_{14}$ 134.22	3725-07-3 -45	 170	4-01-00-01129 0.8071^{22}	ace 4; eth 4 1.4998^{20}
8341	1,5-Octadien-3-yne, 5-propyl- 5-Propyl-1,5-octadien-3-yne	$C_{11}H_{16}$ 148.25	70732-44-4	 57-8^{6}	4-01-00-01129 0.8047^{20}	ace 4; eth 4 1.4949^{20}
8342	1,7-Octadiyne	C_8H_{10} 106.17	871-84-1	 135.5; 59^{35}	4-01-00-01122 0.8169^{21}	eth 3 1.4521^{18}
8343	2,6-Octadiyne	C_8H_{10} 106.17	764-73-8 27	 62^{19}	4-01-00-01123 0.828^{80}	eth 4 1.4658^{30}
8344	3,5-Octadiyne	C_8H_{10} 106.17	16387-70-5	 163.5	4-01-00-01123 0.826^{0}	eth 3 1.4968^{0}
8345	3,5-Octadiyne, 2,7-dimethyl- Diisopropylbutadiyne	$C_{10}H_{14}$ 134.22	14813-68-4	 74^{12}	1-01-00-00128 0.8090^{20}	eth 4
8346	Octanal Caprylic aldehyde	$C_8H_{16}O$ 128.21	124-13-0	1766 171	4-01-00-03337 0.8211^{20}	ace 4; bz 4; eth 4; EtOH 4 1.4217^{20}
8347	Octanal, 7-hydroxy-3,7-dimethyl-	$C_{10}H_{20}O_2$ 172.27	107-75-5	 103^{3}	4-01-00-04058 0.9220^{20}	H_2O 2; EtOH 3; ace 3 1.4494^{20}
8348	Octanal, oxime	$C_8H_{17}NO$ 143.23	929-55-5 60	 112^{9}	2-01-00-00758	ace 4; EtOH 4

No.	Name	Synonym	Mol. Form.	Mol. Wt.	CAS RN	mp/°C	Merck No.	bp/°C	Beil. Ref.	den/g cm^{-3}	Solubility	n_D
8349	Octanamide		$C_8H_{17}NO$	143.23	629-01-6	108		239	4-02-00-00992	0.8450^{110}	H_2O 2; EtOH 4; eth 3; ace 3	
8350	1-Octanamine		$C_8H_{19}N$	129.25	111-86-4	0		179.6	4-04-00-00751	0.7826^{20}	H_2O 2; EtOH 4; eth 4; ctc 3	1.4292^{20}
8351	2-Octanamine, (±)-		$C_8H_{19}N$	129.25	44855-57-4	97		164	4-04-00-00762	0.7744^{20}	eth 4; EtOH 4	1.4232^{25}
8352	2-Octanamine, *N*-(1-methylheptyl)-		$C_{16}H_{35}N$	241.46	5412-92-0			283	4-04-00-00764	0.7948^{20}	H_2O 1; EtOH 4; eth 4	
8353	1-Octanamine, *N*-methyl-*N*-octyl-		$C_{17}H_{37}N$	255.49	4455-26-9	-30.1		158^{10}	4-04-00-00753			1.4424^{20}
8354	1-Octanamine, *N*-octyl-		$C_{16}H_{35}N$	241.46	1120-48-5	35.5		297.5	4-04-00-00753	0.7963^{26}	eth 4; EtOH 4	1.4415^{26}
8355	Octane		C_8H_{18}	114.23	111-65-9	-56.8	6672	125.6	4-01-00-00412	0.6986^{25}	H_2O 1; EtOH 5; eth 3; ace 5	1.3974^{20}
8356	Octane, 1-bromo-	Octyl bromide	$C_8H_{17}Br$	193.13	111-83-1	-55	6684	200	4-01-00-00422	1.108^{25}	H_2O 1; EtOH 5; eth 5; ctc 2	1.4524^{20}
8357	Octane, 2-bromo-, (±)-	Octane, 2-bromo (*DL*)	$C_8H_{17}Br$	193.13	60251-57-2			188.5	4-01-00-00423	1.0878^{25}	H_2O 1; EtOH 5; eth 5	1.4442^{25}
8358	Octane, 1-bromo-8-fluoro-		$C_8H_{16}BrF$	211.12	593-12-4			$118\text{-}20^{22.5}$	4-01-00-00424		eth 4; EtOH 4	1.4500^{20}
8359	Octane, 1-chloro-	Octyl chloride	$C_8H_{17}Cl$	148.68	111-85-3	-57.8		181.5	4-01-00-00419	0.8738^{20}	H_2O 1; EtOH 4; eth 4; ctc 2	1.4305^{20}
8360	Octane, 2-chloro-		$C_8H_{17}Cl$	148.68	628-61-5			172; 75^{28}	1-01-00-00060	0.8658^{17}	H_2O 1; EtOH 4; eth 3	1.4273^{21}
8361	Octane, 1-chloro-8-fluoro-		$C_8H_{16}ClF$	166.67	593-14-6			87^{10}	4-01-00-00420	0.978^{20}	ace 4; eth 4; EtOH 4	1.4266^{25}
8362	Octane, 3-chloro-3-methyl-		$C_9H_{19}Cl$	162.70	28320-88-9			$73\text{-}4^{15}$	3-01-00-00508	0.8680^{25}	ace 4; bz 4; eth 4; chl 4	1.4351^{20}
8363	Octane, 4-chloro-4-methyl-	4-Chloro-4-methyloctane	$C_9H_{19}Cl$	162.70	36903-89-6			$71^{14.5}$	3-01-00-00509	0.8723^{20}	eth 4; chl 4	1.4360^{20}
8364	Octanedial		$C_8H_{14}O_2$	142.20	638-54-0			235 dec; 97^{3}	4-01-00-03706		H_2O 4; EtOH 4	1.4439^{20}
8365	1,8-Octanediamine		$C_8H_{20}N_2$	144.26	373-44-4	52		225	3-04-00-00612		H_2O 4; eth 4; EtOH 4	
8366	Octane, 1,2-dibromo-	1,2-Dibromooctane	$C_8H_{16}Br_2$	272.02	6269-92-7			241	3-01-00-00468	1.4580^{20}		1.4970^{20}
8367	Octane, 1,4-dibromo-		$C_8H_{16}Br_2$	272.02	70690-24-3			$125\text{-}6^{11}$	4-01-00-00424	1.466^{11}		1.5003^{11}
8368	Octane, 1,5-dibromo-	1,5-Dibromooctane	$C_8H_{16}Br_2$	272.02	17912-17-3			$127\text{-}8^{11}$	4-01-00-00424	1.447^{16}		1.4968^{16}
8369	Octane, 1,8-dibromo-	Octamethylene dibromide	$C_8H_{16}Br_2$	272.02	4549-32-0	15.5		271	4-01-00-00424	1.4594^{25}	H_2O 1; eth 3; ctc 3; chl 3	1.4971^{25}
8370	Octane, 4,5-dibromo-		$C_8H_{16}Br_2$	272.02	61539-75-1			$84^{4.3}$	3-01-00-00469	1.4569^{20}		1.4981^{20}
8371	Octane, 1,8-dichloro-	1,8-Dichlorooctane	$C_8H_{16}Cl_2$	183.12	2162-99-4			241	4-01-00-00420	1.0248^{25}		1.4572^{25}
8372	Octane, 2,3-dichloro-		$C_8H_{16}Cl_2$	183.12	21948-47-0	-55.1		206	4-01-00-00425	0.9712^{20}		1.4523^{20}
8373	Octane, 2,2-dimethyl-	2,2-Dimethyloctane	$C_{10}H_{22}$	142.28	15869-87-1			155	4-01-00-00476	0.7208^{25}		1.4082^{20}
8374	Octane, 2,3-dimethyl-	2,3-Dimethyloctane	$C_{10}H_{22}$	142.28	7146-60-3			164.3	4-01-00-00476	0.7377^{20}		1.4146^{20}
8375	Octane, 2,4-dimethyl-	2,4-Dimethyloctane	$C_{10}H_{22}$	142.28	4032-94-4			156	4-01-00-00477	0.7226^{25}		1.4091^{20}
8376	Octane, 2,5-dimethyl-	2,5-Dimethyloctane	$C_{10}H_{22}$	142.28	15869-89-3			158.5	4-01-00-00477	0.7264^{25}		1.4112^{20}
8377	Octane, 2,6-dimethyl-	2,6-Dimethyloctane	$C_{10}H_{22}$	142.28	2051-30-1			160.4	2-01-00-00131	0.7313^{20}		1.4097^{20}
8378	Octane, 2,7-dimethyl-		$C_{10}H_{22}$	142.28	1072-16-8	-54.9		159.9	4-01-00-00478	0.7202^{25}	eth 3; HOAc 3	1.4086^{20}
8379	Octane, 3,3-dimethyl-	3,3-Dimethyloctane	$C_{10}H_{22}$	142.28	4110-44-5			161.2	4-01-00-00479	0.7351^{25}		1.4165^{20}
8380	Octane, 3,4-dimethyl-	3,4-Dimethyloctane	$C_{10}H_{22}$	142.28	15869-92-8			163.4	4-01-00-00479	0.7410^{25}		1.4182^{20}
8381	Octane, 3,5-dimethyl-	3,5-Dimethyloctane	$C_{10}H_{22}$	142.28	15869-93-9			159.4	4-01-00-00479	0.7329^{25}		1.4139^{20}
8382	Octane, 3,6-dimethyl-	3,6-Dimethyloctane	$C_{10}H_{22}$	142.28	15869-94-0			160.8	4-01-00-00479	0.7324^{25}		1.4139^{20}
8383	Octane, 4,4-dimethyl-	4,4-Dimethyloctane	$C_{10}H_{22}$	142.28	15869-95-1			157.5	4-01-00-00480	0.7312^{25}		1.4144^{20}
8384	Octane, 4,5-dimethyl-	4,5-Dimethyloctane	$C_{10}H_{22}$	142.28	15869-96-2			162.15	4-01-00-00480	0.7432^{25}		1.4190^{20}
8385	Octanedinitrile	Suberonitrile	$C_8H_{12}N_2$	136.20	629-40-3	-3.5		185^{15}	3-02-00-01768	0.954^{25}		1.4436^{20}
8386	Octanedioic acid	Suberic acid	$C_8H_{14}O_4$	174.20	505-48-6	144	8833	219^{20}	4-02-00-02028		H_2O 1; eth 5; bz 5	
8387	Octanedioic acid, diethyl ester		$C_{12}H_{22}O_4$	230.30	2050-23-9	5.9		282.6	4-02-00-02029	0.9811^{20}	H_2O 1; EtOH 3; eth 3; ctc 2	1.4328^{20}
8388	Octanedioic acid, dimethyl ester		$C_{10}H_{18}O_4$	202.25	1732-09-8	-3.1		268	4-02-00-02029	1.0217^{20}	H_2O 1; EtOH 3; eth 3; ace 3	1.4341^{20}

No.	Name Synonym	Mol. Form. Mol. Wt.	CAS RN mp/°C	Merck No. bp/°C	Beil. Ref. den/g cm^{-3}	Solubility n_D
8389	1,8-Octanediol	$C_8H_{18}O_2$ 146.23	629-41-4 63	 172^{20}	4-01-00-02592	H_2O 2; EtOH 4; eth 2; bz 3
8390	2,4-Octanediol	$C_8H_{18}O_2$ 146.23	90162-24-6	 $117\text{-}8^{8}$	3-01-00-02219 0.918^{25}	 1.4422^{25}
8391	4,5-Octanediol, (±)-	$C_8H_{18}O_2$ 146.23	22520-40-7 30	 110^{8}	4-01-00-02593	ctc 2 1.4419^{25}
8392	1,7-Octanediol, 3,7-dimethyl-	$C_{10}H_{22}O_2$ 174.28	107-74-4	 265	3-01-00-02233 0.937^{20}	bz 2; tol 2 1.4599^{20}
8393	2,4-Octanedione Valerylacetone	$C_8H_{14}O_2$ 142.20	14090-87-0	 $79\text{-}83^{20}$	4-01-00-03706 0.9233^{20}	 1.4559^{20}
8394	2,7-Octanedione	$C_8H_{14}O_2$ 142.20	1626-09-1 44	 110^{13}	4-01-00-03707	EtOH 4
8395	3,6-Octanedione	$C_8H_{14}O_2$ 142.20	2955-65-9 35.5	 98^{14}	4-01-00-03708	EtOH 4
8396	4,5-Octanedione	$C_8H_{14}O_2$ 142.20	5455-24-3	 168	4-01-00-03708 0.934^{0}	EtOH 4; eth 4; ace 4
8397	4,5-Octanedione, dioxime Dipropylglyoxime	$C_8H_{16}N_2O_2$ 172.23	61050-68-8 186.5	 sub	2-01-00-00846	eth 4; EtOH 4
8398	3,5-Octanedione, 6,6,7,7,8,8,8-heptafluoro-2,2-dimethyl-	$C_{10}H_{11}F_7O_2$ 296.18	17587-22-3 38	 $46\text{-}7^{5}$	 1.273^{25}	 1.3766^{20}
8399	2,3-Octanedione, 3-oxime 3-Oximino-2-octanone	$C_8H_{15}NO_2$ 157.21	584-92-9 59	 133^{13}	0-01-00-00795	eth 4
8400	3,6-Octanedione, 2,2,7,7-tetramethyl- 2,2,7,7-Tetramethyl-3,6-octanedione	$C_{12}H_{22}O_2$ 198.31	27610-88-4 2.5	 116^{17}; $57^{0.5}$	4-01-00-03734 0.900^{27}	EtOH 4 1.4400^{20}
8401	Octane, 1-(ethenyloxy)-	$C_{10}H_{20}O$ 156.27	929-62-4	 58^{4}	4-01-00-02056 0.8020^{20}	 1.4268^{20}
8402	Octane, 1-ethoxy-	$C_{10}H_{22}O$ 158.28	929-61-3 12.5	 186.3	4-01-00-01759 0.7847^{20}	EtOH 4 1.4127^{20}
8403	Octane, 3 ethyl 3-Ethyloctane	$C_{10}H_{22}$ 142.28	5881-17-4	 166.5	4-01-00-00476 0.7359^{25}	 1.4156^{20}
8404	Octane, 4-ethyl- 4-Ethyloctane	$C_{10}H_{22}$ 142.28	15869-86-0	 163.7	4-01-00-00476 0.7343^{25}	 1.4151^{20}
8405	Octane, 1-fluoro- Octyl fluoride	$C_8H_{17}F$ 132.22	463-11-6	 142.4	4-01-00-00418 0.8108^{20}	 1.3935^{20}
8406	Octane, 1-iodo-	$C_8H_{17}I$ 240.13	629-27-6 -45.7	 225.5	4-01-00-00420 1.3297^{20}	EtOH [illegible]; eth [illegible] 1.4889^{20}
8407	Octane, 2-iodo-, (±)- (±)-2-Octyl iodide	$C_8H_{17}I$ 240.13	36049-78-2	 210; 95^{16}	4-01-00-00425 1.3251^{20}	H_2O 1; EtOH 3; eth 3; lig 3 1.4896^{20}
8408	Octane, 2-methyl- 2-Methyloctane	C_9H_{20} 128.26	3221-61-2 -80.3	 143.2	4-01-00-00454 0.7095^{25}	H_2O 1; EtOH 3; eth 3; ctc 2 1.4031^{20}
8409	Octane, 3-methyl- 3-Methyloctane	C_9H_{20} 128.26	2216-33-3 -107.6	 144.2	 0.717^{25}	 1.4040^{25}
8410	Octane, 4-methyl- 4 Methyloctane	C_9H_{20} 128.26	2216-34-4 -113.3	 142.4	 0.716^{25}	 1.4039^{25}
8411	Octanenitrile Caprylnitrile	$C_8H_{15}N$ 125.21	124-12-9 -45.6	 205.2	4-02-00-00993 0.8136^{20}	eth 4 1.4203^{20}
8412	Octane, 1-nitro- 1-Nitrooctane	$C_8H_{17}NO_2$ 159.23	629-37-8 15	 208.5	4-01-00-00426 0.9346^{20}	 1.4322^{20}
8413	Octane, 2-nitro-, (+)-	$C_8H_{17}NO_2$ 159.23	116836-12-5	 $102\text{-}5^{23}$	4-01-00-00426 0.9166^{20}	 1.4280^{20}
8414	Octane, 3-nitro-	$C_8H_7NO_2$ 149.15	4609-92-1	 91^{25}	4-01-00-00427 0.917^{25}	 1.4210^{25}
8415	Octane, octadecafluoro- Perfluorooctane	C_8F_{18} 438.06	307-34-6	 105.9	4-01-00-00418 1.73^{20}	 1.282^{20}
8416	Octane, 1,1'-oxybis-	$C_{16}H_{34}O$ 242.45	629-82-3 -7.6	 283	4-01-00-01760 0.8063^{20}	H_2O 2; EtOH 3; eth 3; ctc 3 1.4327^{20}
8417	Octane, 1,1'-thiobis-	$C_{16}H_{34}S$ 258.51	2690-08-6	 180^{10}	4-01-00-01768 0.842^{25}	 1.4610^{20}
8418	1-Octanethiol Octyl mercaptan	$C_8H_{18}S$ 146.30	111-88-6 -49.2	 199.1	4-01-00-01767 0.8433^{20}	EtOH 3; ctc 2 1.4540^{20}
8419	2-Octanethiol, (±)- 2-Octanethiol (*DL*)	$C_8H_{18}S$ 146.30	10435-81-1 -79	 186.4	4-01-00-01777 0.8366^{20}	bz 4; eth 4; EtOH 4 1.4504^{20}
8420	Octane, 2,2,4-trimethyl- Octane, 2,2,4-trimethyl	$C_{11}H_{24}$ 156.31	18932-14-4	 100.[illegible]	4-01-00-00494 [illegible]	 [illegible]
8421	Octane, 2,4,7-trimethyl 2,4,7-Trimethyloctane	$C_{11}H_{24}$ 156.31	62016-38-0	 168.1		
8422	Octanoic acid Caprylic acid	$C_8H_{16}O_2$ 144.21	124-07-2 16.3	1765 239	4-02-00-00982 0.9106^{20}	H_2O 2; EtOH 5; chl 5; CH_3CN 5 1.4285^{20}
8423	Octanoic acid, 2-amino-, (±)-	$C_8H_{17}NO_2$ 159.23	644-90-6 270	 sub	4-04-00-02799	H_2O 2; EtOH 2; eth 2; bz 2
8424	Octanoic acid, anhydride	$C_{16}H_{30}O_3$ 270.41	623-66-5 -1	 282.5	3-02-00-00796 0.9065^{18}	ace 4; eth 4; EtOH 4 1.4358^{18}
8425	Octanoic acid, 2-bromo-	$C_8H_{15}BrO_2$ 223.11	2623-82-7	 140^{5}; $118^{0.5}$	4-02-00-01000 1.2785^{24}	 1.4613^{24}
8426	Octanoic acid, 8-bromo-	$C_8H_{15}BrO_2$ 223.11	17696-11-6 38.5	 147^{2}	4-02-00-01000	bz 4; eth 4; EtOH 4
8427	Octanoic acid, butyl ester	$C_{12}H_{24}O_2$ 200.32	589-75-3 -42.9	 240.5	4-02-00-00987 0.8628^{20}	ace 4; eth 4; EtOH 4 1.4232^{25}

No.	Name Synonym	Mol. Form. Mol. Wt.	CAS RN mp/°C	Merck No. bp/°C	Beil. Ref. den/g cm^{-3}	Solubility n_D
8428	Octanoic acid, ethyl ester Ethyl octanoate	$C_{10}H_{20}O_2$ 172.27	106-32-1 -43.1	3736 208.5	4-02-00-00987 0.866^{18}	H_2O 1; EtOH 4; eth 4; ctc 2 1.4178^{20}
8429	Octanoic acid, 8-fluoro- Caprylic acid, 8-fluoro	$C_8H_{15}FO_2$ 162.20	353-25-3 35	132^{4}	4-02-00-00994	eth 4; EtOH 4
8430	Octanoic acid, heptyl ester	$C_{15}H_{30}O_2$ 242.40	4265-97-8 -10.6	290.5	3-02-00-00794 0.8596^{20}	ace 4; eth 4; EtOH 4 1.4340^{20}
8431	Octanoic acid, hexyl ester	$C_{14}H_{28}O_2$ 228.38	1117-55-1 -30.6	277.4	3-02-00-00794 0.8603^{20}	H_2O 1; EtOH 3; eth 3; ace 3 1.4323^{25}
8432	Octanoic acid, 2-hydroxy-	$C_8H_{16}O_3$ 160.21	617-73-2 70	162^{10}	3-03-00-00642	H_2O 2; EtOH 4; eth 4; chl 2
8433	Octanoic acid, 3-methyl- Caprylic acid, 3-methyl	$C_9H_{18}O_2$ 158.24	6061-10-5	135^{16}	3-02-00-00826 0.899^{23}	1.4298^{25}
8434	Octanoic acid, methyl ester Methyl caprylate	$C_9H_{18}O_2$ 158.24	111-11-5 -40	192.9	4-02-00-00986 0.8775^{20}	H_2O 1; EtOH 4; eth 4; ctc 2 1.4170^{20}
8435	Octanoic acid, 1-methylethyl ester	$C_{11}H_{22}O_2$ 186.29	5458-59-3	93.8^{10}	4-02-00-00987 0.8555^{20}	1.4147^{25}
8436	Octanoic acid, 2-methyl-3-oxo-, ethyl ester Caprylic acid, 2-methyl-3-oxo, ethyl ester	$C_{11}H_{20}O_3$ 200.28	10488-94-5	$128-9^{12}$	4-03-00-01629 0.963^{0}	EtOH 4
8437	Octanoic acid, octyl ester	$C_{16}H_{32}O_2$ 256.43	2306-88-9 -18.1	306.8	4-02-00-00988 0.8554^{20}	ace 4; eth 4; EtOH 4 1.4352^{20}
8438	Octanoic acid, pentadecafluoro-, methyl ester	$C_9H_3F_{15}O_2$ 428.10	376-27-2	158	4-02-00-00995 1.684^{20}	1.304^{27}
8439	Octanoic acid, pentyl ester	$C_{13}H_{26}O_2$ 214.35	638-25-5 -34.8	260.2	3-02-00-00794 0.8613^{20}	H_2O 1; EtOH 3; eth 3; ace 3 1.4262^{25}
8440	Octanoic acid, 1,2,3-propanetriyl ester	$C_{27}H_{50}O_6$ 470.69	538-23-8 10	233	4-02-00-00991 0.9540^{20}	H_2O 1; EtOH 5; eth 4; bz 4 1.4482^{20}
8441	Octanoic acid, propyl ester	$C_{11}H_{22}O_2$ 186.29	624-13-5 -46.2	226.4	4-02-00-00987 0.8659^{20}	ace 4; eth 4; EtOH 4 1.4191^{25}
8442	Octanoic acid, 3,5,7,8-tetrachloro-2,2,3,4,4,5,6,6,7,8,8-undecafluoro- Octanoic acid, 3,5,7,8-tetrachloro-2,2,3,4,4,5,6,6,7,8,8-undecafluoro	$C_8HCl_4F_{11}O_2$ 479.89	2923-68-4	$154.0-6.5^{10}$	4-02-00-00998 1.899^{20}	1.3980^{20}
8443	1-Octanol Caprylic alcohol	$C_8H_{18}O$ 130.23	111-87-5 -15.5	6674 195.1	4-01-00-01756 0.8262^{25}	H_2O 1; EtOH 5; eth 5; ctc 3 1.4295^{20}
8444	2-Octanol, (±)-	$C_8H_{18}O$ 130.23	4128-31-8 -31.6	6675 180	4-01-00-01770 0.8193^{20}	H_2O 2; EtOH 3; eth 3; ace 3 1.4203^{20}
8445	3-Octanol Ethylamylcarbinol	$C_8H_{18}O$ 130.23	589-98-0 -45	171	3-01-00-01723 0.8258^{20}	
8446	4-Octanol, (±)-	$C_8H_{18}O$ 130.23	74778-22-6 -40.7	176.3	4-01-00-01779 0.8186^{20}	H_2O 2; EtOH 3; ctc 2 1.4248^{20}
8447	2-Octanol, acetate, (±)- Acetic acid, 2-octyl ester(*DL*)	$C_{10}H_{20}O_2$ 172.27	74112-36-0	195	4-02-00-00165 0.8626^{14}	eth 4; EtOH 4 1.4146^{20}
8448	1-Octanol, 8-chloro-	$C_8H_{17}ClO$ 164.68	23144-52-7	139^{19}	4-01-00-01766	eth 4; EtOH 4 1.4563^{25}
8449	1-Octanol, 3,7-dimethyl-, (±)- Geraniol, tetrahydro (*DL*)	$C_{10}H_{22}O$ 158.28	59204-02-3	212.5; 106^{13}	4-01-00-01830 0.8308^{10}	H_2O 1; EtOH 3; eth 3; ace 3 1.4367^{20}
8450	2-Octanol, 2,6-dimethyl- Tetrahydromyrcenol	$C_{10}H_{22}O$ 158.28	18479-57-7	80.5^{10}	4-01-00-01829 0.8023^{25}	1.4220^{25}
8451	3-Octanol, 2,3-dimethyl- Amylisopropylmethylcarbinol	$C_{10}H_{22}O$ 158.28	19781-10-3	$69-70^{5}$	4-01-00-01828 0.8401^{20}	1.4380^{20}
8452	3-Octanol, 2,7-dimethyl-	$C_{10}H_{22}O$ 158.28	66719-55-9	193.5	1-01-00-00214 0.8152^{20}	1.4302^{20}
8453	3-Octanol, 3,6-dimethyl-	$C_{10}H_{22}O$ 158.28	151-19-9 -67.5	202.2	1-01-00-00214 0.8347^{22}	1.4370^{20}
8454	3-Octanol, 3,6-dimethyl-(+-)- Linalool, tetrahydro (*DL*)	$C_{10}H_{22}O$ 158.28	57706-88-4 31.5	196.5	3-01-00-01767 0.8280^{20}	EtOH 4 1.4335^{20}
8455	4-Octanol, 2,4-dimethyl-	$C_{10}H_{22}O$ 158.28	33933-79-8	97^{27}	4-01-00-01828 0.8238^{20}	1.4338^{20}
8456	4-Octanol, 2,5-dimethyl-	$C_{10}H_{22}O$ 158.28	66719-53-7	$102-4^{34}$	1-01-00-00213 0.8125^{25}	1.4260^{25}
8457	4-Octanol, 2,6-dimethyl-	$C_{10}H_{22}O$ 158.28	66719-54-8	195	2-01-00-00460 0.8114^{20}	1.4270^{20}
8458	4-Octanol, 2,7-dimethyl- 2,7-Dimethyloctan-4-ol	$C_{10}H_{22}O$ 158.28	19781-11-4	96^{18}	3-01-00-01769 0.8139^{20}	1.4280^{20}
8459	3-Octanol, 3-ethyl-	$C_{10}H_{22}O$ 158.28	2051-32-3	199	3-01-00-01766 0.8361^{25}	EtOH 4 1.4390^{20}
8460	1-Octanol, 8-fluoro-	$C_8H_{17}FO$ 148.22	408-27-5	$106-7^{10}$	4-01-00-01766 0.945^{20}	eth 4; EtOH 4 1.4248^{25}
8461	1-Octanol, 3-methyl- 3-Methyl-1-octanol	$C_9H_{20}O$ 144.26	38514-02-2	110^{25}	4-01-00-01806 0.827^{4}	1.4328^{25}
8462	1-Octanol, 4-methyl- 4-Methyl-1-octanol	$C_9H_{20}O$ 144.26	38514-03-3	106^{17}	4-01-00-01807 0.820^{27}	1.4335^{25}
8463	1-Octanol, 7-methyl-	$C_9H_{20}O$ 144.26	2430-22-0 64.5	206	4-01-00-01806 0.8260^{25}	1.4316^{25}
8464	2-Octanol, 2-methyl- 2-Methyl-2-octanol	$C_9H_{20}O$ 144.26	628-44-4	178	4-01-00-01805 0.8210^{20}	H_2O 1; EtOH 3; eth 3 1.4280^{20}
8465	2-Octanol, 3-methyl-	$C_9H_{20}O$ 144.26	27644-49-1	75^{15}	3-01-00-01748 0.831^{27}	1.437^{27}
8466	3-Octanol, 3-methyl- Amylethylmethylcarbinol	$C_9H_{20}O$ 144.26	5340-36-3	97.5^{50}; 36^{3}	4-01-00-01807 0.8258^{25}	1.4257^{25}

No.	Name Synonym	Mol. Form. Mol. Wt.	CAS RN mp/°C	Merck No. bp/°C	Beil. Ref. den/g cm^{-3}	Solubility n_D
8467	3-Octanol, 4-methyl-	$C_9H_{20}O$ 144.26	66793-80-4	132-3^{20}	3-01-00-01750 0.8437^{25}	1.4372^{20}
8468	3-Octanol, 6-methyl- 6-Methyl-3-octanol	$C_9H_{20}O$ 144.26	40225-75-0	81-3^{15}	3-01-00-01749 0.8320^{28}	1.4372^{20}
8469	4-Octanol, 4-methyl- 4-Methyl-4-octanol	$C_9H_{20}O$ 144.26	23418-37-3	180; 78^{15}	4-01-00-01807 0.8267^{20}	1.4327^{20}
8470	4-Octanol, 5-methyl-	$C_9H_{20}O$ 144.26	59734-23-5	74-6^{9}	1-01-00-00211 0.8156^{25}	1.4262^{25}
8471	2-Octanone Hexyl methyl ketone	$C_8H_{16}O$ 128.21	111-13-7 -16	4632 172.5	4-01-00-03339 0.820^{20}	H_2O 2; EtOH 5; eth 5 1.4151^{20}
8472	3-Octanone Ethyl amyl ketone	$C_8H_{16}O$ 128.21	106-68-3	3721 167.5	4-01-00-03341 0.822^{25}	H_2O 1; EtOH 5; eth 5 1.4150^{20}
8473	4-Octanone Butyl propyl ketone	$C_8H_{16}O$ 128.21	589-63-9	163	4-01-00-03342 0.8146^{25}	H_2O 1; EtOH 5; eth 5; ctc 3 1.4173^{14}
8474	4-Octanone, 5-hydroxy- Butyroin	$C_8H_{16}O_2$ 144.21	496-77-5 -10	1595 185	4-01-00-04042 0.9107^{16}	1.4345^{16}
8475	2-Octanone, 3-methyl- 3-Methyl-2-octanone	$C_9H_{18}O$ 142.24	6137-08-2	64-5^{18}	4-01-00-03356 0.832^{27}	1.424^{27}
8476	3-Octanone, 4-methyl- 4-Methyl-3-octanone	$C_9H_{18}O$ 142.24	6137-15-1	62-4^{10}	4-01-00-03357 0.820^{25}	1.4186^{25}
8477	3-Octanone, 7-methyl- 7-Methyl-3-octanone	$C_9H_{18}O$ 142.24	5408-57-1	182.5	3-01-00-02889 0.8353^{25}	1.4748
8478	4-Octanone, 3-methyl-	$C_9H_{18}O$ 142.24	20754-04-5	174	4-01-00-03357 0.829^{14}	1.4200^{14}
8479	4-Octanone, 7-methyl- 7-Methyl-4-octanone	$C_9H_{18}O$ 142.24	20809-46-5	178	4-01-00-03356 0.8239^{20}	H_2O 1; EtOH 3; eth 3 1.4210^{20}
8480	1-Octanone, 1-phenyl-	$C_{14}H_{20}O$ 204.31	1674-37-9 22.8	285; 164^{15}	4-07-00-00798 0.9360^{30}	EtOH 3; eth 3
8481	Octanoyl chloride	$C_8H_{15}ClO$ 162.66	111-64-8 -63	195.6	4-02-00-00992 0.9535^{15}	eth 3 1.4335^{20}
8482	Octasiloxane, octadecamethyl- Octadecamethyloctasiloxane	$C_{18}H_{54}O_7Si_8$ 607.31	556-69-4	153$^{5.1}$	4-04-00-04130 0.913^{25}	bz 4; peth 4; lig 4 1.3970^{20}
8483	1,3,5,7-Octatetraene Octatetraene	C_8H_{10} 106.17	1482-91-3 50	sub	4-01-00-01124	peth 3; HOAc 3
8484	2,4,6-Octatrienal Octatrienal	$C_8H_{10}O$ 122.17	17609-31-3 30.0	75^{1}	4-01-00-03591 0.0091^{60}	eth 4; EtOH 4
8485	1,3,6-Octatriene, 3,7-dimethyl- Ocimene	$C_{10}H_{16}$ 136.24	13877-91-3	6661 177 dec; 73^{21}	3-01-00-01052 0.800^{20}	bz 4; eth 4; EtOH 4; chl 4 1.4862^{20}
8486	1,3,7-Octatriene, 3,7-dimethyl-	$C_{10}H_{16}$ 136.24	502-99-8	177 dec	3-01-00-01050 0.8000^{20}	H_2O 1; EtOH 3; eth 3; chl 3 1.4862^{20}
8487	2,4,6-Octatriene, 2,6-dimethyl-, (*E,E*)-	$C_{10}H_{16}$ 136.24	3016-19-1 -35.4	188; 91^{20}	4-01-00-01106 0.8118^{20}	1.5446^{20}
8488	2,4,6-Octatriene, 2,6-dimethyl-, (*E,Z*)-	$C_{10}H_{16}$ 136.24	7216-56-0 -20.6	89^{20}	4-01-00-01106 0.8060^{20}	1.5446^{20}
8489	2,4,6-Octatriene, (*E,E,E*)-	C_8H_{12} 108.18	15192-80-0 52	147.5	4-01-00-01101 0.7961^{23}	EtOH 3; chl 3; lig 3 1.5131^{27}
8490	6-Octenal, 3,7-dimethyl- Resorcinol, 4-acetyl	$C_{10}H_{18}O$ 154.25	106-23-0	2331 47^{1}	1-01-00-00387 0.85^{20}	H_2O 2; EtOH 3
8491	1-Octene Caprylene	C_8H_{16} 112.22	111-66-0 -101.7	1764 121.2	4-01-00-00874 0.7149^{20}	H_2O 1; EtOH 5; eth 3; ace 3 1.4087^{20}
8492	2-Octene, (*E*)-	C_8H_{16} 112.22	13389-42-9 -87.7	125	4-01-00-00879 0.7199^{20}	H_2O 1; EtOH 3; eth 3; ace 3 1.4132^{20}
8493	2-Octene, (*Z*)-	C_8H_{16} 112.22	7642-04-8 -100.2	125.6	4-01-00-00878 0.7243^{20}	H_2O 1; EtOH 3; eth 3; ace 3 1.4150^{20}
8494	3-Octene, (*E*)-	C_8H_{16} 112.22	14919-01-8 -110	123.3	4-01-00-00880 0.7152^{20}	H_2O 1; EtOH 3; eth 3; ace 3 1.4126^{20}
8495	3-Octene, (*Z*)-	C_8H_{16} 112.22	14850-22-7 -126	122.9	4-01-00-00880 0.7159^{20}	ace 4; bz 4; eth 4; EtOH 4 1.4135^{20}
8496	4-Octene, (*E*)-	C_8H_{16} 112.22	14850-23-8 -93.8	122.3	4-01-00-00880 0.7141^{20}	H_2O 1; EtOH 3; eth 3; ace 3 1.4114^{20}
8497	4-Octene, (*Z*)-	C_8H_{16} 112.22	7642-15-1 -118.7	122.5	4-01-00-00880 0.7212^{20}	ace 4; bz 4; eth 4; EtOH 4 1.4148^{20}
8498	1-Octene, 2-chloro-	$C_8H_{15}Cl$ 146.66	31283-43-9	169	3-01-00-00840 0.9273^{0}	ace 4; bz 4; eth 4
8499	2-Octene, 2-chloro-	$C_8H_{15}Cl$ 146.66	90202-21-4	167.5; 82^{50}	4-01-00-00879 0.8914^{16}	ace 4; bz 4; eth 4; EtOH 4 1.4424^{16}
8500	4-Octene, 4-chloro-, (*Z*)- 4-Octene, 4-chloro (*cis*)	$C_8H_{15}Cl$ 146.66	7321-48-4	165.3; 75^{50}	4-01-00-00881 0.8912^{20}	eth 4 1.4447^{20}
8501	2-Octene, 4-chloro-	$C_8H_{15}Cl$ 146.66	116668-50-9	153	3-01-00-00840 0.8924^{20}	ace 4; bz 4; eth 4; chl 4 1.4952^{20}
8502	1-Octene, 3,7-dimethyl- 2,6-Dimethyl-7-octene	$C_{10}H_{20}$ 140.27	4984-01-4	154	3-01-00-00862 0.7396^{20}	1.4212^{20}
8503	2-Octene, 2,6-dimethyl- 2,6-Dimethyl-2-octene	$C_{10}H_{20}$ 140.27	4057-42-5	163	3-01-00-00861 0.746^{22}	1.425^{22}
8504	4-Octene-2,7-dione	$C_8H_{12}O_2$ 140.18	2130-23-6 30.5	72$^{0.03}$	4-01-00-03757	eth 4; EtOH 4
8505	1-Octene, 2-methyl- 2-Methyl-1-octene	C_9H_{18} 126.24	4588-18-5 -77.8	144.8	4-01-00-00896 0.7343^{20}	1.4184^{20}
8506	2-Octene, 3-methyl-	C_9H_{18} 126.24	113426-22-5	145	3-01-00-00854 0.7409^{25}	1.4247^{20}

PHYSICAL CONSTANTS OF ORGANIC COMPOUNDS (continued)

No.	Name Synonym	Mol. Form. Mol. Wt.	CAS RN mp/°C	Merck No. bp/°C	Beil. Ref. den/g cm^{-3}	Solubility n_D
8507	3-Octene, 7-methyl-	C_9H_{18} 126.24	86668-33-9	142	3-01-00-00854 0.7278^{20}	1.4168^{20}
8508	4-Octene, 2-methyl-	C_9H_{18} 126.24	64501-77-5	139	3-01-00-00854 0.7379^{20}	1.4181^{20}
8509	1-Octen-4-ol	$C_8H_{12}O$ 124.18	40575-42-6	172; 68^{10}	3-01-00-01942 0.8373^{20}	1.4383^{20}
8510	2-Octen-1-ol	$C_8H_{16}O$ 128.21	22104-78-5	$87\text{-}9^{11}$	4-01-00-02164 0.850^{20}	1.4470^{20}
8511	2-Octen-4-ol	$C_8H_{16}O$ 128.21	4798-61-2	175	3-01-00-01943 0.8393^{20}	1.4395
8512	4-Octen-1-ol	$C_8H_{16}O$ 128.21	67700-26-9	95^{17}	4-01-00-02166 0.842^{20}	1.4462^{20}
8513	5-Octen-1-ol	$C_8H_{16}O$ 128.21	90200-83-2	91.5^{14}	4-01-00-02166 0.8492^{21}	1.4478^{21}
8514	6-Octen-1-ol, 3,7-dimethyl-, (±)- (±)-Citronellol	$C_{10}H_{20}O$ 156.27	26489-01-0	2332 224; 99^{10}	4-01-00-02188 0.8560^{20}	H_2O 2; EtOH 5; eth 5 1.4543^{20}
8515	6-Octen-3-ol, 3,7-dimethyl-	$C_{10}H_{20}O$ 156.27	18479-51-1	94^{14}	2-01-00-00494 0.8695^{15}	1.4569^{15}
8516	7-Octen-1-ol, 3,7-dimethyl-, (*S*)- (-)-α-Citronellol	$C_{10}H_{20}O$ 156.27	6812-78-8	8186 $114\text{-}5^{12}$	2-01-00-00497 0.8549^{20}	eth 4; EtOH 4 1.4556^{20}
8517	1-Octen-4-ol, 4-methyl-	$C_9H_{18}O$ 142.24	62108-06-9	178	2-01-00-00492 0.8440^{20}	1.4399
8518	2-Octen-4-ol, 7-methyl-	$C_9H_{18}O$ 142.24	4798-64-5	85^{12}	2-01-00-00492 0.8402^{14}	1.4449^{14}
8519	3-Octen-2-ol, 2-methyl-, (*Z*)- 3-Octen-2-ol, 2-methyl (*cis*)	$C_9H_{18}O$ 142.24	18521-07-8	177	4-01-00-02177 0.8378^{20}	1.4426^{20}
8520	4-Octen-3-ol, 3-methyl-	$C_9H_{18}O$ 142.24	90676-55-4	$69\text{-}70^{10}$	4-01-00-02178 0.812^{14}	1.4414^{14}
8521	5-Octen-4-ol, 5-methyl- 3-Octen-5-ol, 4-methyl	$C_9H_{18}O$ 142.24	36903-92-1	89^{16}	2-01-00-00492 0.8468^{25}	1.4446^{25}
8522	1-Octen-3-one, 5-methyl-	$C_9H_{16}O$ 140.23	116836-13-6	$72\text{-}3^{16}$	3-01-00-03018 0.8450^{19}	1.4380^{19}
8523	5-Octen-4-one, 2-methyl- 2-Methyl-5-octen-4-one	$C_9H_{16}O$ 140.23	17577-93-4	183	4-01-00-03504 0.839^{25}	1.4387^{25}
8524	5-Octen-4-one, 7-methyl-	$C_9H_{16}O$ 140.23	32064-78-1	$68\text{-}78^{24}$	3-01-00-03018 0.9011^{12}	1.4748
8525	7-Octen-4-one, 2-methyl- 1-Octen-5-one, 7-methyl	$C_9H_{16}O$ 140.23	54298-97-4	$62\text{-}3^{14}$	2-01-00-00801 0.8361^{12}	1.4288^{12}
8526	1-Octen-3-yne	C_8H_{12} 108.18	17679-92-4	62^{60}	4-01-00-01100 0.7830^{20}	eth 4 1.4592^{20}
8527	7-Octen-5-yn-4-ol, 4-methyl- 1-Octen-3-yne-5-ol, 5-methyl	$C_9H_{14}O$ 138.21	39118-35-9	80^{13}	3-01-00-02034 0.8851^{15}	EtOH 4 1.4785^{20}
8528	Octhilinone 3(2H)-Isothiazolone, 2-octyl-	$C_{11}H_{19}NOS$ 213.34	26530-20-1	6677 $120^{0.01}$		
8529	1-Octyne Hexylacetylene	C_8H_{14} 110.20	629-05-0 -79.3	126.3	4-01-00-01034 0.7461^{20}	H_2O 1; EtOH 3; eth 3 1.4159^{20}
8530	2-Octyne Methylpentylacetylene	C_8H_{14} 110.20	2809-67-8 -61.6	137.6	4-01-00-01035 0.7596^{20}	H_2O 1; EtOH 3; eth 3 1.4278^{20}
8531	3-Octyne	C_8H_{14} 110.20	15232-76-5 -103.9	133.1	4-01-00-01036 0.7529^{20}	H_2O 1; EtOH 3; eth 3 1.4250^{20}
8532	4-Octyne Dipropylacetylene	C_8H_{14} 110.20	1942-45-6 -101	131.6	4-01-00-01037 0.7509^{20}	H_2O 1; EtOH 3; eth 3 1.4248^{20}
8533	1-Octyne, 1-chloro- 1-Chloro-1-octyne	$C_8H_{13}Cl$ 144.64	64531-26-6	$61\text{-}2^{17}$	3-01-00-01005 0.912^{20}	eth 4; EtOH 4 1.445^{20}
8534	3-Octyne, 2-chloro-2-methyl- 2-Chloro-2-methyloct-3-yne	$C_9H_{15}Cl$ 158.67	20599-21-7	68^{15}	4-01-00-01048 0.8929^{20}	H_2O 1; eth 4 1.4480^{20}
8535	3-Octyne, 2,2-dimethyl- 2,2-Dimethyl-3-octyne	$C_{10}H_{18}$ 138.25	19482-57-6	79^{60}	3-01-00-01018 0.7491^{20}	eth 4 1.4270^{20}
8536	2-Octyne, 1-methoxy- Methyl (2-octyn-1-yl) ether	$C_9H_{16}O$ 140.23	18495-23-3	77^{19}	3-01-00-01996 0.8370^{25}	eth 4 1.4380^{20}
8537	3-Octyne, 7-methyl- 7-Methyl-3-octyne	C_9H_{16} 124.23	37050-06-9	87^{99}	3-01-00-01014 0.7599^{20}	eth 4 1.4280^{20}
8538	2-Octynoic acid Pentylpropiolic acid	$C_8H_{12}O_2$ 140.18	5663-96-7 4	148^{19}, 114^{2}	4-02-00-01714 0.9623^{13}	1.4595^{20}
8539	3-Octynoic acid	$C_8H_{12}O_2$ 140.18	57074-96-1 18	110^{2}	4-02-00-01715	1.4577^{25}
8540	5-Octynoic acid	$C_8H_{12}O_2$ 140.18	76469-08-4 8	111^{2}	4-02-00-01715 0.9787^{27}	1.4540^{25}
8541	7-Octynoic acid	$C_8H_{12}O_2$ 140.18	10297-09-3 19	123^{2}	4-02-00-01714	1.4502^{25}
8542	2-Octynoic acid, methyl ester	$C_9H_{14}O_2$ 154.21	111-12-6	107^{20}; 94^{10}	4-02-00-01715 0.926^{20}	1.4464^{20}
8543	2-Octyn-1-ol 2-Octynol	$C_8H_{14}O$ 126.20	20739-58-6 -18	$98\text{-}9^{15}$	4-01-00-02256 0.8805^{20}	eth 4 1.4556^{20}
8544	1-Octyn-3-ol, 3-methyl- 1-Octyne-3-ol, 3-methyl	$C_9H_{16}O$ 140.23	23580-51-0	75^{10}	4-01-00-02268 0.863^{10}	1.443^{10}
8545	Olean-12-en-28-oic acid, 3,23-dihydroxy-, (3β,4α)- Hederagenin	$C_{30}H_{48}O_4$ 472.71	465-99-6 333	4540		

No.	Name Synonym	Mol. Form. Mol. Wt.	CAS RN mp/°C	Merck No. bp/°C	Beil. Ref. den/g cm^{-3}	Solubility n_D
8546	Olean-12-en-28-oic acid, 3,16-dihydroxy-23-oxo-, (3β,4α,16α)- Quillaic acid	$C_{30}H_{46}O_5$ 486.69	631-01-6 294	8049	3-10-00-04652	ace 4; eth 4; py 4; EtOH 4
8547	Olean-12-en-28-oic acid, 3-hydroxy-, (3β)- Oleanolic acid	$C_{30}H_{48}O_3$ 456.71	508-02-1 310 dec	6787 sub 280	4-10-00-01164	H_2O 1; EtOH 2; eth 2; ace 2
8548	Olean-12-en-3-ol, (3β)- β-Amyrin	$C_{30}H_{50}O$ 426.73	559-70-6 197	654 260^{05}	4-06-00-04195	H_2O 1; EtOH 2; eth 3; bz 3
8549	Orcein	$C_{28}H_{24}N_2O_7$ 500.51	1400-62-0	6818	2-06-00-00876	
8550	*L*-Ornithine (*S*)-2,5-Diaminopentanoic acid	$C_5H_{12}N_2O_2$ 132.16	70-26-8 140	6826	4-04-00-02644	H_2O 4; EtOH 4
8551	*L*-Ornithine, *N*5-(aminocarbonyl)- Citrulline	$C_6H_{13}N_3O_3$ 175.19	372-75-8 235.5	2333	4-04-00-02647	H_2O 3; EtOH 1; MeOH 1
8552	Orthoformic acid, triisopentyl ester Isopentyl alcohol, orthoformate (3:1)	$C_{16}H_{34}O_3$ 274.44	5337-70-2	 268 dec	4-02-00-00030 0.8628^{20}	eth 4; EtOH 4 1.4238^{20}
8553	Oryzalin Benzenesulfonamide, 4-(dipropylamino)-3,5-dinitro-	$C_{12}H_{18}N_4O_6S$ 346.36	19044-88-3 141	6840		
8554	7-Oxabicyclo[4.1.0]heptane	$C_6H_{10}O$ 98.14	286-20-4 <-10	 131.5	5-17-01-00203 0.9663^{20}	H_2O 1; EtOH 4; eth 4; ace 4 1.4519^{20}
8555	7-Oxabicyclo[4.1.0]heptane, 3-ethenyl- Cyclohexane, 1,2-epoxy-4-vinyl	$C_8H_{12}O$ 124.18	106-86-5 <-100	 169; 70^{20}	4-17-00-00314 0.9581^{20}	 1.4700^{20}
8556	7-Oxabicyclo[2.2.1]heptane, 1-methyl-4-(1-methylethyl)-	$C_{10}H_{18}O$ 154.25	470-67-7 1	 173.5	5-17-01-00273 0.8997^{20}	H_2O 2, EtOH 5, eth 5, bz 3 1.4562^{20}
8557	7-Oxabicyclo[4.1.0]heptane, 3-oxiranyl- 4-Vinyl-1-cyclohexene dioxide	$C_8H_{12}O_2$ 140.18	106-87-6 <-55	 227	5-19-01-00295 1.0966^{20}	H_2O 4 1.4787^{20}
8558	6-Oxabicyclo[3.1.0]hexane	C_5H_8O 84.12	285-67-6	 102	5-17-01-00190 0.964^{25}	 1.4336^{20}
8559	3-Oxabicyclo[3.2.1]octane-2,4-dione, 1,8,8-trimethyl-, (±)-	$C_{10}H_{14}O_3$ 182.22	595-30-2 221	 270	5-17-11-00089 1.194^{20}	bz 4
8560	2-Oxabicyclo[2.2.2]octane, 1,3,3-trimethyl- Eucalyptol	$C_{10}H_{18}O$ 154.25	470-82-6 1.5	3851 176.4	5-17-01-00273 0.9267^{20}	H_2O 1; EtOH 3; eth 3; ctc 2 1.4586^{20}
8561	6-Oxabicyclo[3.2.1]oct-3-ene, 4,7,7-trimethyl-, (±)- (±)-Pinol	$C_{10}H_{16}O$ 152.24	60761-04-8	 183.5	4-17-00-00327 0.9515^{20}	eth 4; EtOH 4 1.4695^{20}
8562	Oxacyclohexadecan-2-one Exaltolide	$C_{15}H_{28}O_2$ 240.39	106-02-5	3867	5-17-09-00106	
8563	1,3,4-Oxadiazole 1-Oxa-3,4-diazacyclopentadiene	$C_2H_2N_2O$ 70.05	288-99-3	6859 150		 1.4300^{25}
8564	1,2,4-Oxadiazole, 3,5-diphenyl- 3,5-Diphenyl-1,2,4-oxadiazole	$C_{14}H_{10}N_2O$ 222.25	888-71-1 109	 290; 210^{17}	4-27-00-07210	bz 4; eth 4; EtOH 4
8565	Oxadiazon 1,2,4-Oxadiazolidin-3,5-dione, 2-*tert*-butyl-4-(2,4-dichloro-5-isopropoxyphenyl)-	$C_{15}H_{18}Cl_2N_2O_3$ 345.23	19666-30-9 90	6860		
8566	Oxadixyl *N*-(2,6-Dimethylphenyl)-2-methoxy-*N*-(2-oxo-3-oxazolidinyl)acetamide	$C_{14}H_{18}N_2O_4$ 278.31	77732-09-3 104			
8567	16(15H)-Oxaerythrinan-15-one, 1,2,6,7-tetradehydro-14,17-dihydro-3-methoxy-, (3β)- β-Erythroidine	$C_{16}H_{19}NO_3$ 273.33	466-81-9 99.5	3625	4-27-00-03568	H_2O 3; EtOH 4; eth 3; bz 4
8568	Oxamyl Ethanimidothioic acid, 2-(dimethylamino)-*N*-[[(methylamino)carbonyl]oxy]-2-oxo-, methyl ester	$C_7H_{13}N_3O_3S$ 219.26	23135-22-0 109	6873 dec	 0.97^{25}	
8569	1,4-Oxathiane	C_4H_8OS 104.17	15980-15-1	 147	5-19-01-00033 1.1174^{20}	H_2O 2
8570	1,2-Oxathiolane, 2,2-dioxide Propane sultone	$C_3H_6O_3S$ 122.14	1120-71-4		5-19-01-00005	chl 3
8571	Oxayohimban-16-carboxylic acid, 16,17-didehydro-19-methyl-, methyl ester, (19α)- Raubasine	$C_{21}H_{24}N_2O_3$ 352.43	483-04-5 258 dec	8129	4-27-00-07927	H_2O 1; MeOH 3
8572	Oxayohimbanium, 3,4,5,6,16,17-hexadehydro-16-(methoxycarbonyl)-19-methyl-, (19α)- Serpentine	$C_{21}H_{20}N_2O_3$ 348.40	18786-24-8 175	8415		H_2O 1; EtOH 3; eth 3; ace 3
8573	2H-1,3-Oxazine-2,4(3H)-dione, 5,5-diethyldihydro- Diethadione	$C_8H_{13}NO_3$ 171.20	702-54-5 97.5	3096	4-27-00-03263	
8574	2-Oxazolamine, 4,5-dihydro-5-phenyl- Aminorex	$C_9H_{10}N_2O$ 162.19	2207-50-3 137	490		
8575	Oxazole 1,3-Oxazole	C_3H_3NO 69.06	288-42-6	 69.5	4-27-00-00960	 1.4285^{17}
8576	Oxazole, 4,5-dihydro-2-methyl-	C_4H_7NO 85.11	1120-64-5	 111	4-27-00-00921 1.005^{25}	 1.4340^{20}
8577	Oxazole, 2,4-dimethyl- 2,4-Dimethyloxazole	C_5H_7NO 97.12	7208-05-1	 108	4-27-00-00980 0.9352^{15}	H_2O 4; eth 4; EtOH 4 1.4166^{15}
8578	Oxazole, 2,5-dimethyl- 2,5-Dimethyloxazole	C_5H_7NO 97.12	23012-11-5	 117.5	4-27-00-00985 0.9958^{21}	H_2O 4 1.4385^{21}

No.	Name	Synonym	Mol. Form.	Mol. Wt.	CAS RN	mp/°C	Merck No.	bp/°C	Beil. Ref.	den/g cm^{-3}	Solubility	n_D
8579	Oxazole, 2,4-diphenyl-	2,4-Diphenyloxazole	$C_{15}H_{11}NO$	221.26	838-41-5	103		339	4-27-00-01435		EtOH 4; eth 4; bz 4	
8580	Oxazole, 2,5-diphenyl-		$C_{15}H_{11}NO$	221.26	92-71-7	74		360	4-27-00-01435	1.0940^{100}	H_2O 1; EtOH 4; eth 4; chl 2	1.6231^{100}
8581	Oxazole, 4,5-diphenyl-	4,5-Diphenyloxazole	$C_{15}H_{11}NO$	221.26	4675-18-7	44		193^{15}; $132^{0.05}$	4-27-00-01438		H_2O 2; EtOH 4; eth 4; con acid 3	1.6283^{100}
8582	Oxazole, 5-methoxy-2-phenyl-	5-Methoxy-2-phenyloxazole	$C_{10}H_9NO_2$	175.19	40527-16-0			140^{10}	4-27-00-01958		ctc 3	1.584^{23}
8583	Oxazole, 4-methyl-	4-Methyloxazole	C_4H_5NO	83.09	693-93-6			88	4-27-00-00968	1.015^{25}		1.4317^{20}
8584	2,4-Oxazolidinedione, 5,5-dimethyl-	Dimethadione	$C_5H_7NO_3$	129.12	695-53-4	76.5	3201		4-27-00-03237			
8585	2,4-Oxazolidinedione, 5,5-dipropyl-	5,5-Dipropyl-2,4-oxazolidinedione	$C_9H_{15}NO_3$	185.22	512-12-9	42.5	3352	$148\text{-}50^{3}$	4-27-00-03269			
8586	2,4-Oxazolidinedione, 5-methyl-3-(2-propenyl)-	Aloxidone	$C_7H_9NO_3$	155.15	526-35-2		305	138^{35}; $86^{0.5}$	4-27-00-03214			1.4688^{25}
8587	2,4-Oxazolidinedione, 3,5,5-trimethyl-	Trimethadione	$C_6H_9NO_3$	143.14	127-48-0	46	9620	$78\text{-}80^{5}$	4-27-00-03237		H_2O 3; EtOH 4; eth 4; ace 4	
8588	2-Oxazolidinone, 5-[(2-methoxyphenoxy)methyl]-	Mephenoxalone	$C_{11}H_{13}NO_4$	223.23	70-07-5	144	5739					
8589	2-Oxazolidinone, 3-[[(5-nitro-2-furanyl)methylene]amino]-	Furazolidone	$C_8H_7N_3O_5$	225.16	67-45-8	255	4210		4-27-00-02530			
8590	Oxepane		$C_6H_{12}O$	100.16	592-90-5			119	5-17-01-00090	0.89^{25}		1.4400^{20}
8591	2-Oxepanone	Caprolactone	$C_6H_{10}O_2$	114.14	502-44-3	-18		215	5-17-09-00034	1.0693^{20}	EtOH 3; eth 3; ace 3	1.4611^{20}
8592	Oxetane	Trimethylene oxide	C_3H_6O	58.08	503-30-0	-97	9630	47.6	5-17-01-00011	0.8930^{25}	H_2O 5; EtOH 5; eth 3; ace 4	1.3961^{20}
8593	Oxetane, 3,3-dimethyl-	3,3-Dimethyloxetane	$C_5H_{10}O$	86.13	6921-35-3			80.6	5-17-01-00085	0.834^{25}		1.3965^{20}
8594	Oxetane, 2-methyl-		C_4H_8O	72.11	2167-39-7			59	5-17-01-00054	0.841^{25}		1.3885^{20}
8595	2-Oxetanone	β-Propiolactone	$C_3H_4O_2$	72.06	57-57-8	-33.4	7832	162	5-17-09-00003	1.1460^{20}	eth 5; chl 3	1.4105^{20}
8596	3-Oxetanone		$C_3H_4O_2$	72.06	6704-31-0							
8597	2-Oxetanone, 3,3-dimethyl-4-(1-methylethylidene)-		$C_8H_{12}O_2$	140.18	3173-79-3	-18		170	5-17-09-00181	0.947^{25}		1.4380^{20}
8598	2-Oxetanone, 4-methylene-	Diketene	$C_4H_4O_2$	84.07	674-82-8	-6.5		126.1	5-17-09-00115	1.0877^{20}		1.4379^{20}
8599	Oxirane	Ethylene oxide	C_2H_4O	44.05	75-21-8	-111.7	3758	10.6	5-17-01-00003	0.8821^{10}	H_2O 3; EtOH 3; eth 3; ace 3	1.3597^{7}
8600	Oxirane, (bromomethyl)-, (±)-	Propane, 3-bromo-1,2-epoxy (*DL*)	C_3H_5BrO	136.98	82584-73-4	-40		137	5-17-01-00023	1.615^{14}	H_2O 1; EtOH 3; eth 3; bz 3	1.4841^{20}
8601	Oxirane, 2,2'-[1,4-butanediylbis(oxymethylene)]bis-		$C_{10}H_{18}O_4$	202.25	2425-79-8			266; 155^{11}	5-17-03-00023	1.1^{25}		1.4611^{20}
8602	Oxirane, (butoxymethyl)-	Butyl glycidyl ether	$C_7H_{14}O_2$	130.19	2426-08-6			169; 75^{26}	5-17-03-00011	0.918^{20}		
8603	Oxiranecarboxaldehyde	Glycidaldehyde	$C_3H_4O_2$	72.06	765-34-4	-62		112.5	5-17-09-00006			1.4265^{20}
8604	Oxiranecarboxylic acid, 3-methyl-3-phenyl-, ethyl ester		$C_{12}H_{14}O_3$	206.24	77-83-8			273.5	5-18-06-00385	1.044^{20}		1.5182^{20}
8605	Oxirane, (chloromethyl)-, (±)-	*DL*-α-Epichlorohydrin	C_3H_5ClO	92.52	13403-37-7	-26	3563	117; 62^{100}	5-17-01-00020	1.1812^{20}	H_2O 1; EtOH 5; eth 5; bz 3	1.4361^{20}
8606	Oxirane, 2-(chloromethyl)-2-methyl-		C_4H_7ClO	106.55	598-09-4			122	5-17-01-00059	1.1011^{20}	H_2O 4; eth 4	1.4310^{20}
8607	Oxirane, 2,2-dimethyl-		C_4H_8O	72.11	558-30-5			52	5-17-01-00058	0.8112^{20}	EtOH 3; eth 3	1.3712^{22}
8608	Oxirane, 2,3-dimethyl-, *cis*-		C_4H_8O	72.11	1758-33-4	-80		60	5-17-01-00061	0.8226^{25}	eth 4; ace 4; bz 4; os 4	1.3802^{20}
8609	Oxirane, 2,3-dimethyl-, *trans*-		C_4H_8O	72.11	21490-63-1	-85		56.5	5-17-01-00061	0.8010^{25}	eth 4; ace 4; bz 4; os 4	1.3736^{20}
8610	Oxirane, [(1,1-dimethylethoxy)methyl]-		$C_7H_{14}O_2$	130.19	7665-72-7	-70		152	5-17-03-00011	0.898^{20}		
8611	Oxirane, [[4-(1,1-dimethylethyl)phenoxy]methyl]-		$C_{13}H_{18}O_2$	206.28	3101-60-8			$165\text{-}70^{14}$	5-17-03-00017	1.038^{25}		1.5145^{20}
8612	Oxirane, 2-(2,2-dimethylpropyl)-2-methyl-		$C_8H_{16}O$	128.21	107-48-2	-64		140.9	4-17-00-00112	0.8272^{20}	bz 4; eth 4	1.4097^{20}
8613	Oxirane, ethenyl-		C_4H_6O	70.09	930-22-3			68	5-17-01-00180	0.9006^{25}	EtOH 3; eth 3; bz 3; os 3	1.4168^{20}
8614	Oxirane, (ethoxymethyl)-	2,3-Epoxypropyl ethyl ether	$C_5H_{10}O_2$	102.13	4016-11-9			128	5-17-03-00010	0.9700^{20}	H_2O 3; EtOH 3; eth 3; ctc 2	1.4320^{20}

No.	Name	Synonym	Mol. Form.	Mol. Wt.	CAS RN	mp/°C	Merck No.	bp/°C	Beil. Ref.	den/g cm^{-3}	Solubility	n_D
8615	Oxirane, ethyl-	1,2-Epoxybutane	$C_4H_8O_2$	88.11	106-88-7			63.3	5-17-01-00056	0.8297^{20}	EtOH 4; eth 5; ace 4; os 4	1.3851^{20}
8616	Oxirane, (fluoromethyl)-		C_3H_5FO	76.07	503-09-3			85.5	5-17-01-00019	1.067^{25}		1.3715^{20}
8617	Oxirane, (iodomethyl)-		C_3H_5IO	183.98	624-57-7			161; 62^{24}	5-17-01-00024	1.982^{24}	eth 4; EtOH 4	
8618	Oxiranemethanol	Glycidol	$C_3H_6O_2$	74.08	556-52-5		4385	167 dec; $66^{2.5}$	5-17-03-00009	1.1143^{25}	H_2O 5; EtOH 5; eth 3; ace 3	1.4287^{20}
8619	Oxiranemethanol, 3-phenyl-		$C_9H_{10}O_2$	150.18	21915-53-7	26.5		134^4	4-17-00-01349	1.1512^{27}		1.5427^{27}
8620	Oxirane, (methoxymethyl)-		$C_4H_8O_2$	88.11	930-37-0			113	5-17-03-00010	0.9890^{20}	H_2O 4; ace 4; eth 4; EtOH 4	1.4320^{20}
8621	Oxirane, methyl-	Propylene oxide	C_3H_6O	58.08	75-56-9	-111.9	7869	35	5-17-01-00017	0.859^0	H_2O 4; eth 4; EtOH 4	1.3670^{20}
8622	Oxirane, 2-methyl-2-phenyl-		$C_9H_{10}O$	134.18	2085-88-3			$83\text{-}5^{15}$	5-17-01-00591	1.0280^{20}		1.5232^{20}
8623	Oxirane, (phenoxymethyl)-	Phenyl glycidyl ether	$C_9H_{10}O_2$	150.18	122-60-1			243	5-17-03-00013	1.1109^{21}		1.5307^{21}
8624	Oxirane, phenyl-	Phenylethylene oxide	C_8H_8O	120.15	96-09-3	-35.6		194.1	5-17-01-00577	1.0523^{16}	H_2O 1; EtOH 3; eth 3; chl 3	1.5342^{20}
8625	Oxirane, 2,2'-[1,3-phenylenebis(oxymethylene)]bis-	Diglycidyl resorcinol ether	$C_{12}H_{14}O_4$	222.24	101-90-6	42.5		$143\text{-}51^{0.4}$	5-17-03-00024	1.2183^{30}		1.5408^{20}
8626	Oxirane, [(2-propenyloxy)methyl]-	Allyl glycidyl ether	$C_6H_{10}O_2$	114.14	106-92-3			154	5-17-03-00012	0.9698^{20}		1.4332^{20}
8627	Oxirane, tetramethyl-		$C_6H_{12}O$	100.16	5076-20-0			$90\text{-}4^{74.5}$	5-17-01-00108	0.8156^{16}	H_2O 3	1.3984^{16}
8628	Oxirane, (trichloromethyl)-		$C_3H_3Cl_3O$	161.41	3083-23-6			149; 44^{13}	5-17-01-00023	1.495^{20}	eth 4; chl 3	1.4737^{25}
8629	Oxirane, trifluoro(trifluoromethyl)-	Perfluoropropylene oxide	C_3F_6O	166.02	428-59-1				5-17-01-00020			
8630	Oxyacanthan, 6',12'-dimethoxy-2,2'-dimethyl-6,7-[methylenebis(oxy)]-	Cepharanthine	$C_{37}H_{38}N_2O_6$	606.72	481-49-2	150	1981		4-27-00-09061			
8631	Oxyacanthan-12'-ol, 6,6',7-trimethoxy-2,2'-dimethyl-	Oxyacanthine	$C_{37}H_{40}N_2O_6$	608.73	548-40-3	216.5	6906		2-27-00-00894		H_2O 1; EtOH 3; eth 3; bz 3	
8632	Oxydemeton-methyl	Phosphorothioic acid, *S*-[2-(ethylsulfinyl)ethyl] *O,O*-dimethyl ester	$C_6H_{15}O_4PS_2$	246.29	301-12-2	<-20		$106^{0.01}$		1.289^{20}		
8633	Oxyfluorfen	Benzene, 2-chloro-1-(3-ethoxy-4-nitrophenoxy)-4-(trifluoromethyl)-	$C_{15}H_{11}ClF_3NO_4$	361.70	42874-03-3	84	6916	358.2 dec		1.35^{73}		
8634	Paraformaldehyde	Formaldehyde polymer	$(CH_2O)_x$		30525-89-4	164 dec	6974		5-19-09-00103		H_2O 2; EtOH 1; eth 1; alk 3	
8635	Pendimethalin	Benzenamine, *N*-(1-ethylpropyl)-3,4-dimethyl-2,6-dinitro-	$C_{13}H_{19}N_3O_4$	281.31	40487-42-1	56	7026	dec		1.19^{25}		
8636	Pentacene	Benzo[b]naphthacene	$C_{22}H_{14}$	278.35	135-48-8	257	7057		4-05-00-02721		H_2O 1; bz 2; $PhNO_2$ 3; os 2	
8637	Pentacosane		$C_{25}H_{52}$	352.69	629-99-2	54		401.9; 282^{40}	4-01-00-00582	0.8012^{20}	bz 3; chl 3	1.4491^{20}
8638	Pentacosane, 13-phenyl-	Benzene, (1-dodecyltridecyl)-	$C_{31}H_{56}$	428.79	6006-90-2	31.7		235^{10}	4-05-00-01241	0.8537^{20}		1.4787^{20}
8639	Pentacosane, 13-undecyl-		$C_{36}H_{74}$	506.98	55517-89-0	9.7		307^{10}	4-01-00-00600	0.8168^{20}		1.4567^{20}
8640	Pentadecanal	Pentadecaldehyde	$C_{15}H_{30}O$	226.40	2765-11-9	24.5		185^{25}	4-01-00-03391		ace 4; eth 4; EtOH 4	
8641	1-Pentadecanamine		$C_{15}H_{33}N$	227.43	2570-26-5	37.3		307.6	4-04-00-00817	0.8104^{20}	eth 4; EtOH 4	1.4480^{20}
8642	Pentadecane		$C_{15}H_{32}$	212.42	629-62-9	9.9		270.6	4-01-00-00529	0.7685^{20}	H_2O 1; EtOH 4; eth 4	1.4315^{20}
8643	Pentadecane, 1-bromo-		$C_{15}H_{31}Br$	291.32	629-72-1	19		322	4-01-00-00531	1.0675^{20}	H_2O 1; ace 3; chl 4	1.4611^{20}
8644	Pentadecane, 2,6,10,14-tetramethyl-	Pristane	$C_{19}H_{40}$	268.53	1921-70-6		7757	296	3-01-00-00570	0.783^{20}	bz 4; eth 4; chl 4; peth 4	1.4379^{20}
8645	Pentadecanoic acid	Pentadecylic acid	$C_{15}H_{30}O_2$	242.40	1002-84-2	52.3		257^{100}; 158^1	4-02-00-01147	0.8423^{80}	H_2O 1; EtOH 4; eth 3; ace 4	1.4254^{80}
8646	Pentadecanoic acid, methyl ester	Methyl pentadecanoate	$C_{16}H_{32}O_2$	256.43	7132-64-1	18.5		153.5	3-02-00-00936	0.8618^{25}	EtOH 3; eth 3; os 3	1.4390^{25}
8647	2-Pentadecanol	*sec*-Pentadecyl alcohol	$C_{15}H_{32}O$	228.42	1653-34-5	35		299	4-01-00-01871	0.8328^{20}		1.4463^{20}
8648	3-Pentadecanol		$C_{15}H_{32}O$	228.42	53346-71-7	38.8		388; 163^{12}	4-01-00-01871			1.4227^{80}
8649	1-Pentadecanol, 2-methyl-	2-Methyl-1-pentadecanol	$C_{16}H_{34}O$	242.45	25354-98-7	12.5		$185\text{-}7^{15}$	3-01-00-01827	0.8320^{25}		1.4453^{25}
8650	6-Pentadecanol, 6-methyl-		$C_{16}H_{34}O$	242.45	108836-86-8			$199\text{-}200^{50}$	3-01-00-01828	0.8316^{25}		1.4446^{25}

No.	Name Synonym	Mol. Form. Mol. Wt.	CAS RN mp/°C	Merck No. bp/°C	Beil. Ref. den/g cm^{-3}	Solubility n_D
8651	2-Pentadecanone	$C_{15}H_{30}O$ 226.40	2345-28-0 39.5	294	4-01-00-03391 0.8182^{39}	
8652	8-Pentadecanone	$C_{15}H_{30}O$ 226.40	818-23-5 43	291	4-01-00-03392 0.8180^{39}	EtOH 3; eth 3; bz 3; ctc 3
8653	1-Pentadecene	$C_{15}H_{30}$ 210.40	13360-61-7 -2.8	268.2	4-01-00-00926 0.7764^{20}	H_2O 1; ace 3 1.4389^{20}
8654	7-Pentadecene	$C_{15}H_{30}$ 210.40	15430-98-5	114$^{3.2}$	3-01-00-00875 0.7765^{20}	1.4420^{20}
8655	1-Pentadecen-3-ol	$C_{15}H_{30}O$ 226.40	99814-65-0 27.5	150-2^{6}	3-01-00-01961 0.834^{32}	1.4481^{32}
8656	1-Pentadecyne	$C_{15}H_{28}$ 208.39	765-13-9 10	268	4-01-00-01072 0.7928^{20}	ace 4 1.4419^{20}
8657	2,4-Pentadienal, 5-phenyl- 5-Phenyl-2,4-pentadienal	$C_{11}H_{10}O$ 158.20	13466-40-5 42.5	160^{3}; 133$^{1.0}$	4-07-00-01179	H_2O 1; EtOH 5; eth 4; bz 5
8658	1,2-Pentadiene Ethylallene	C_5H_8 68.12	591-95-7 -137.3	44.9	4-01-00-00993 0.6926^{20}	EtOH 5; eth 5; ace 5; bz 5 1.4209^{20}
8659	1,4-Pentadiene	C_5H_8 68.12	591-93-5 -148.8	26	4-01-00-00998 0.6608^{20}	H_2O 1; EtOH 4; eth 4; ace 4 1.3888^{20}
8660	2,3-Pentadiene 1,3-Dimethylallene	C_5H_8 68.12	591-96-8 -125.6	48.2	4-01-00-00999 0.6950^{20}	H_2O 1; EtOH 5; eth 5; ace 5 1.4284^{20}
8661	1,3-Pentadiene, (*E*)- *trans*-Piperylene	C_5H_8 68.12	2004-70-8 -87.4	42	4-01-00-00994 0.6710^{25}	1.4301^{20}
8662	1,3-Pentadiene, (*Z*)- *cis*-Piperylene	C_5H_8 68.12	1574-41-0 -140.8	44.1	4-01-00-00994 0.6910^{20}	EtOH 5; eth 5; ace 5; bz 5 1.4363^{20}
8663	1,3-Pentadiene, 3-chloro- 3-Chloro-1,3-pentadiene	C_5H_7Cl 102.56	37710-49-9	100	3-01-00-00962 0.9576^{20}	ace 4; bz 4; eth 4 1.4785^{20}
8664	1,2-Pentadiene, 1-chloro-3-methyl-	C_6H_9Cl 116.59	32337-74-9	68-70^{100}	3-01-00-00990 0.9562^{20}	eth 4
8665	1,3-Pentadiene, 2,4-dimethyl-	C_7H_{12} 96.17	1000-86-8 -114	93.2	4-01-00-01034 0.7343^{23}	1.4390^{23}
8666	1,2-Pentadiene, 3-methyl- 1-Ethyl-1-methylallene	C_6H_{10} 82.15	7417-48-3	70	4-01-00-01021 0.710^{25}	1.425^{20}
8667	1,2-Pentadiene, 4-methyl-	C_6H_{10} 82.15	13643-05-5	70	2-01-00-00232 0.703^{25}	1.424^{20}
8668	1,3-Pentadiene, 2-methyl-, (*Z*)- *cis*-2-Methyl-1,3-pentadiene	C_6H_{10} 82.15	1501-60-6 -117.6	76	4-01-00-01019 0.714^{25}	1.446^{20}
8669	1,3-Pentadiene, 3-methyl-	C_6H_{10} 82.15	4549-74-0	77	4-01-00-01021 0.730^{25}	1.452^{20}
8670	1,3-Pentadiene, 4-methyl- 1,1-Dimethylbutadiene	C_6H_{10} 82.15	926-56-7	76.5	4-01-00-01020 0.7181^{20}	1.4532^{20}
8671	1,4-Pentadiene, 2-methyl- 2-Methyl-1,4-pentadiene	C_6H_{10} 82.15	763-30-4	56	4-01-00-01020 0.689^{25}	1.405^{20}
8672	2,3-Pentadiene, 2-methyl-	C_6H_{10} 82.15	3043-33-2	72	4-01-00-01020 0.706^{25}	1.425^{20}
8673	2,4-Pentadienenitrile	C_5H_5N 79.10	1615-70-9 -60	136.5	4-02-00-01692 0.8444^{20}	1.4880
8674	2,4-Pentadienenitrile, (*E*)-	C_5H_5N 79.10	2180-68-9 -43	41^{13}	4-02-00-01695 0.8576^{20}	eth 3; ace 3 1.4986^{20}
8675	2,4-Pentadienenitrile, (*Z*)-	C_5H_5N 79.10	2180-69-0 -64	49^{32}; 32^{13}	4-02-00-01695 0.8541^{26}	eth 3; ace 3 1.4855^{20}
8676	2,4-Pentadienoic acid	$C_5H_6O_2$ 98.10	626-99-3 79	112 dec	3-02-00-01451	H_2O 4; EtOH 4; eth 4; bz 4
8677	2,4-Pentadienoic acid, 5-(1,3-benzodioxol-5-yl)- 2,4-Pentadienoic acid, 5(3,4-metylenedioxyphenyl)	$C_{12}H_{10}O_4$ 218.21	5285-18-7 215	7437	2-19-00-00300	H_2O 1; EtOH 3; eth 2; bz 2
8678	2,4-Pentadienoic acid, 5-(1,3-benzodioxol-5-yl)-, (*E,E*)-	$C_{12}H_{10}O_4$ 218.21	136-72-1 215.8	sub	5-19-07-00382	EtOH 4
8679	2,4-Pentadienoic acid, 5-(1-hydroxy-2,6,6-trimethyl-4-oxo-2-cyclohexen-1-yl)-3-methyl-, [*S*-(*Z,E*)]- Abscisic acid	$C_{15}H_{20}O_4$ 264.32	21293-29-8 160	6 sub 120		ace 4; eth 4; chl 4
8680	2,4-Pentadienoic acid, 5-phenyl-, ethyl ester Ethyl cinnamylideneacetate	$C_{13}H_{14}O_2$ 202.25	1552-95-0 26	149-50^{4}	2-09-00-00441 1.0467^{20}	eth 4; EtOH 4 1.5768^{80}
8681	1,4-Pentadien-3-ol	C_5H_8O 84.12	922-65-6	115.5	4-01-00-02229 0.860^{23}	1.4400^{17}
8682	1,4-Pentadien-3-one, 1,5-di-2-furanyl-	$C_{13}H_{10}O_3$ 214.22	886-77-1 60.5	181^{4}	5-19-04-00459	eth 4; EtOH 4; chl 4
8683	1,4-Pentadien-3-one, 1,5-diphenyl- Dibenzalacetone	$C_{17}H_{14}O$ 234.30	538-58-9 113 dec	2988 dec	4-07-00-01747	H_2O 1; EtOH 2; eth 2; ace 3
8684	1,3-Pentadiyne Methyldiacetylene	C_5H_4 64.09	4911-55-1 -38.5	55	4-01-00-01117 0.7909^{20}	H_2O 1; eth 3; bz 3; chl 3 1.4431^{21}
8685	1,5-Pentalenedione, hexahydro- Bicyclo[3.3.0]octane-2,6-dione	$C_8H_{10}O_2$ 138.17	62353-69-9 45	86-8$^{0.2}$	3-07-00-03280 1.1290^{60}	1.4877^{54}
8686	Pentalene, octahydro-, *cis*-	C_8H_{14} 110.20	1755-05-1 <-80	139	4-05-00-00277 0.8638^{25}	H_2O 1; EtOH 3 1.4595^{25}
8687	Pentalene, octahydro-, *trans*-	C_8H_{14} 110.20	5597-89-7 -30	132	4-05-00-00277 0.8624^{18}	H_2O 1; EtOH 3 1.4625^{18}
8688	2(1H)-Pentalenone, hexahydro- Bicyclo[3.3.0]octane-2-one	$C_8H_{12}O$ 124.18	56180-61-1	72^{13}	4-07-00-00156 1.0097^{20}	ace 4; EtOH 4 1.4790^{20}

No.	Name	Synonym	Mol. Form.	Mol. Wt.	CAS RN	mp/°C	Merck No.	bp/°C	Beil. Ref.	den/g cm^{-3}	Solubility	n_D
8689	2(1H)-Pentalenone, hexahydro-, *cis-*	*cis*-Bicyclo[3.3.0]octan-3-one	$C_8H_{12}O$	124.18	19915-11-8			$50^{2.3}$	4-07-00-00156	1.0660^{25}		1.4766^{25}
8690	Pentanal	Valeraldehyde	$C_5H_{10}O$	86.13	110-62-3	-91.5	9813	103	4-01-00-03268	0.8095^{20}	H_2O 2; EtOH 3; eth 3	1.3944^{20}
8691	Pentanal, 3-hydroxy-2-methyl-		$C_6H_{12}O_2$	116.16	615-30-5			$94\text{-}6^{23}$	4-01-00-04022	0.986^{25}	H_2O 4; ace 4; eth 4; EtOH 4	1.4502^{20}
8692	Pentanal, 3-hydroxy-2,2,4-trimethyl-		$C_8H_{16}O_2$	144.21	918-79-6			$118\text{-}20^{14}$	4-01-00-04049	0.9482^{20}		1.4501^{20}
8693	Pentanal, 2-methyl-	2-Methylvaleraldehyde	$C_6H_{12}O$	100.16	123-15-9			117	4-01-00-03304		H_2O 3; eth 3; ace 3; ctc 2	
8694	Pentanal, 3-methyl-		$C_6H_{12}O$	100.16	15877-57-3			121	3-01-00-02836	0.811^{25}		1.4026^{20}
8695	Pentanal, 2-oxo-		$C_5H_8O_2$	100.12	7332-93-6			112	4-01-00-03658		H_2O 4; eth 4; EtOH 4	1.4043^{25}
8696	Pentanal, 4-oxo-		$C_5H_8O_2$	100.12	626-96-0	<-21		187 dec	4-01-00-03659	1.0134^{21}	H_2O 4; ace 4; eth 4; EtOH 4	1.4257^{22}
8697	Pentanal, 2-propyl-	Valeraldehyde, 2-propyl-	$C_8H_{16}O$	128.21	18295-59-5			160	3-01-00-02880	0.8347^{15}		1.4142^{15}
8698	Pentanamide		$C_5H_{11}NO$	101.15	626-97-1	106		225	4-02-00-00874	0.8735^{110}	H_2O 4; EtOH 4; eth 4; chl 2	1.4183^{110}
8699	Pentanamide, *N,N*-dimethyl-		$C_7H_{15}NO$	129.20	6225-06-5	-51		141^{100}	3-04-00-00127	0.8962^{25}	H_2O 4; eth 4; EtOH 4	1.4419^{25}
8700	1-Pentanamine	Amylamine	$C_5H_{13}N$	87.16	110-58-7	-55	630	104.3	4-04-00-00674	0.7544^{20}	H_2O 5; EtOH 5; eth 5; ace 4	1.448^{20}
8701	2-Pentanamine		$C_5H_{13}N$	87.16	625-30-9			92	4-04-00-00689	0.7384^{20}	ace 4; bz 4; eth 4; EtOH 4	1.4027^{20}
8702	3-Pentanamine		$C_5H_{13}N$	87.16	616-24-0			89	4-04-00-00692	0.7487^{20}	EtOH 3; chl 2	1.4063^{20}
8703	1-Pentanamine, *N,N*-dimethyl-		$C_7H_{17}N$	115.22	26153-88-8			123	4-04-00-00675	0.743^{20}	eth 4	1.4083^{20}
8704	2-Pentanamine, 2,4-dimethyl-		$C_7H_{17}N$	115.22	64379-30-2			122	4-04-00-00749	0.7719^{20}		1.4009^{20}
8705	1-Pentanamine, *N,N*-dipentyl-		$C_{15}H_{33}N$	227.43	621-77-2			242.5	4-04-00-00676	0.7907^{20}	H_2O 1; EtOH 3; eth 3; acid 3	1.4366^{20}
8706	1-Pentanamine, 3-methyl-	3-Methylpentylamine	$C_6H_{15}N$	101.19	42245-37-4			122	3-04-00-00366	0.767[illegible]		1.4190[illegible]
8707	1-Pentanamine, 4-methyl-		$C_6H_{15}N$	101.19	5344-20-7	-94.4		125	4-04-00-00727	0.758^{52}		
8708	1-Pentanamine, *N*-pentyl-	Diamylamine	$C_{10}H_{23}N$	157.30	2050-92-2			202.5	4-04-00-00676	0.7771^{20}	H_2O 2; EtOH 4; eth 5; ace 3	1.4272^{20}
8709	Pentane		C_5H_{12}	72.15	109-66-0	-129.7	7072	36.0	4-01-00-00303	0.6262^{20}	H_2O 2; EtOH 5; eth 5; ace 5	1.3575^{20}
8710	Pentane, 1-bromo-	Amyl bromide	$C_5H_{11}Br$	151.05	110-53-2	-95	637	129.8	4-01-00-00312	1.2182^{20}	H_2O 1; EtOH 3; eth 5; bz 3	1.4447^{20}
8711	Pentane, 2-bromo-	2-Pentyl bromide	$C_5H_{11}Br$	151.05	107-81-3	-95.5		117.4	4-01-00-00312	1.2075^{20}	bz 4; eth 4; EtOH 4; chl 4	1.4413^{20}
8712	Pentane, 3-bromo-	3-Pentyl bromide	$C_5H_{11}Br$	151.05	1809-10-5	-126.2		118.6	4-01-00-00313	1.214^{20}	H_2O 1; EtOH 3; eth 3; bz 3	1.4441^{20}
8713	Pentane, 1-bromo-5-fluoro-	1-Bromo-5-fluoropentane	$C_5H_{10}BrF$	169.04	407-97-6			162	4-01-00-00313	1.3604^{25}	eth 4; EtOH 4	1.4406^{25}
8714	Pentane, 1-bromo-2-methyl-	2-Methylpentyl bromide	$C_6H_{13}Br$	165.07	25346-33-2			141	4-01-00-00361	1.1624^{20}	eth 4; chl 4	1.4495^{20}
8715	Pentane, 1-bromo-3-methyl-	3-Methylpentyl bromide	$C_6H_{13}Br$	165.07	51116-73-5			148	4-01-00-00365	1.1829^{20}	H_2O 1; eth 3; chl 4	1.4496^{20}
8716	Pentane, 1-bromo-4-methyl-		$C_6H_{13}Br$	165.07	626-88-0			145	4-01-00-00361	1.1683^{20}	eth 4; chl 4	1.4490
8717	Pentane, 2-bromo-2-methyl-	1-Bromo-1,1-dimethylbutane	$C_6H_{13}Br$	165.07	4283-80-1			142.5	4-01-00-00361		eth 4; chl 4	1.442^{23}
8718	Pentane, 3-bromo-3-methyl-	3-Bromo-3-methylpentane	$C_6H_{13}Br$	165.07	25346-31-0			130; 76^{100}	4-01-00-00365	1.1835^{20}	eth 4; chl 4	1.4525^{20}
8719	Pentane, 1-chloro-	Amyl chloride	$C_5H_{11}Cl$	106.60	543-59-9	-99	642	107.8	4-01-00-00309	0.8820^{20}	H_2O 1; EtOH 5; eth 5; bz 3	1.4127^{20}
8720	Pentane, 2-chloro-, (+)-	Pentane, 2-chloro (*d*)	$C_5H_{11}Cl$	106.60	29882-57-3	-137		96.9	4-01-00-00309	0.8698^{20}	H_2O 1; EtOH 3; eth 3; bz 3	1.4069^{20}
8721	Pentane, 3-chloro-	1-Ethylpropyl chloride	$C_5H_{11}Cl$	106.60	616-20-6	-105		97.5	4-01-00-00309	0.8731^{20}	H_2O 1; EtOH 3; eth 3; ace 2	1.4082^{20}
8722	Pentane, 2-chloro-2,3-dimethyl-		$C_7H_{15}Cl$	134.65	59889-45-1			$38\text{-}9^{20}$	4-01-00-00406		bz 4; eth 4	1.4264^{20}
8723	Pentane, 2-chloro-2,4-dimethyl-		$C_7H_{15}Cl$	134.65	35951-33-8			129; 33^{20}	4-01-00-00408	0.861^{20}	eth 4	1.4180^{20}
8724	Pentane, 3-chloro-2,3-dimethyl-		$C_7H_{15}Cl$	134.65	595-38-0			136 dec; 41^{20}	4-01-00-00406	0.882^{22}	eth 4; chl 4	1.4318^{20}
8725	Pentane, 4-chloro-2,2-dimethyl-		$C_7H_{15}Cl$	134.65	33429-72-0			93^{250}	4-01-00-00404	0.855^{20}	eth 4; chl 4	1.4180^{20}
8726	Pentane, 3-chloro-3-ethyl-		$C_7H_{15}Cl$	134.65	994-25-2			143.5	4-01-00-00403	0.8856^{20}	eth 4	1.4400^{20}
8727	Pentane, 3-chloro-3-ethyl-2,2-dimethyl-	Pentane 3-chloro-2,2-dimethyl-3-ethyl	$C_9H_{19}Cl$	162.70	86661-53-2			53^{6} dec	4-01-00-00462		eth 4; chl 4	1.4528^{25}
8728	Pentane, 1-chloro-5-fluoro-		$C_5H_{10}ClF$	124.59	407-98-7			143.2	4-01-00-00309	1.0325^{25}	eth 4; EtOH 4	1.4120^{23}

No.	Name / Synonym	Mol. Form. / Mol. Wt.	CAS RN / mp/°C	Merck No. / bp/°C	Beil. Ref. / den/g cm^{-3}	Solubility / n_D
8729	Pentane, 2-chloro-2-methyl-	$C_6H_{13}Cl$ 120.62	4325-48-8	110 dec; 36^{15}	4-01-00-00360 0.863^{20}	eth 4 1.4126^{20}
8730	Pentane, 2-chloro-4-methyl- 2-Chloro-4-methylpentane	$C_6H_{13}Cl$ 120.62	25346-32-1	113	3-01-00-00399 0.8610^{20}	eth 4 1.4113^{20}
8731	Pentane, 3-(chloromethyl)-	$C_6H_{13}Cl$ 120.62	4737-41-1	126; 83^{202}	3-01-00-00403 0.8914^{20}	bz 4; eth 4; chl 4 1.4222^{20}
8732	Pentane, 3-chloro-3-methyl-	$C_6H_{13}Cl$ 120.62	918-84-3	116	4-01-00-00365 0.8900^{20}	bz 4; eth 4; chl 4 1.4210^{20}
8733	Pentane, 2-chloro-2,4,4-trimethyl-	$C_8H_{17}Cl$ 148.68	6111-88-2 -26	147 dec; 44^{16}	4-01-00-00444 0.8746^{20}	EtOH 4 1.4308^{20}
8734	Pentanedial Glutaraldehyde	$C_5H_8O_2$ 100.12	111-30-8	4366 188 dec	4-01-00-03659	H_2O 5; EtOH 5; bz 3
8735	1,5-Pentanediamine Cadaverine	$C_5H_{14}N_2$ 102.18	462-94-2 9	1608 179	4-04-00-01310 0.873^{25}	H_2O 3; EtOH 3; eth 2 1.463^{20}
8736	1,4-Pentanediamine, *N*1,*N*1-diethyl- Novoldiamine	$C_9H_{22}N_2$ 158.29	140-80-7	6642 201	4-04-00-01308 0.814^{20}	1.4429^{20}
8737	1,5-Pentanediamine, *N*-(6-methoxy-8-quinolinyl)-*N*'-(1-methylethyl)-	$C_{18}H_{27}N_3O$ 301.43	86-78-2	$165\text{-}70^{0.02}$	5-22-12-00126	dil HCl 4 1.5785^{25}
8738	1,5-Pentanediaminium, *N*,*N*,*N*,*N*',*N*',*N*'-hexamethyl-, dibromide Ammonium, pentamethylenebis trimethyl, bromide	$C_{11}H_{28}Br_2N_2$ 348.16	541-20-8 301	7070	4-04-00-01311	H_2O 2
8739	Pentane, 1,3-dibromo- 1,3-Dibromopentane	$C_5H_{10}Br_2$ 229.94	42474-20-4	193; 64^{10}	4-01-00-00314 1.6721^{20}	1.5048^{207}
8740	Pentane, 1,4-dibromo- 1,4-Dibromopentane	$C_5H_{10}Br_2$ 229.94	626-87-9 -34.4	146^{150}; 99^{14}	4-01-00-00314 1.6222^{20}	1.5086^{20}
8741	Pentane, 1,5-dibromo-	$C_5H_{10}Br_2$ 229.94	111-24-0 -39.5	222.3	4-01-00-00314 1.7018^{20}	H_2O 1; bz 3; ctc 2; chl 3 1.5126^{20}
8742	Pentane, 2,3-dibromo- 2,3-Dibromopentane	$C_5H_{10}Br_2$ 229.94	5398-25-4 -56	178; 91^{50}	3-01-00-00347 1.6865^{14}	1.5087^{20}
8743	Pentane, 2,4-dibromo- 2,4-Dibromopentane	$C_5H_{10}Br_2$ 229.94	19398-53-9	60^{12}	3-01-00-00348 1.6659^{20}	1.4987^{20}
8744	Pentane, 1,2-dichloro- 1,2-Dichloropentane	$C_5H_{10}Cl_2$ 141.04	1674-33-5	148.3	3-01-00-00341 1.0872^{20}	H_2O 1; EtOH 3; chl 4 1.4485^{20}
8745	Pentane, 1,3-dichloro- 1,3-Dichloropentane	$C_5H_{10}Cl_2$ 141.04	30122-12-4	80.4^{60}	3-01-00-00342 1.0834^{20}	ace 4; chl 4 1.4485^{20}
8746	Pentane, 1,4-dichloro- 1,4-Dichloropentane	$C_5H_{10}Cl_2$ 141.04	626-92-6	162; 88^{60}	4-01-00-00309 1.0840^{20}	ace 3; chl 3 1.4503^{20}
8747	Pentane, 1,5-dichloro- 1,5-Dichloropentane	$C_5H_{10}Cl_2$ 141.04	628-76-2 -72.8	179	4-01-00-00310 1.1006^{22}	H_2O 1; EtOH 3; eth 3; bz 3 1.4564^{20}
8748	Pentane, 2,2-dichloro- 2,2-Dichloropentane	$C_5H_{10}Cl_2$ 141.04	34887-14-4	128.5	2-01-00-00095 1.040^{20}	H_2O 1; eth 3; bz 3; chl 3 1.434^{20}
8749	Pentane, 2,3-dichloro- 2,3-Dichloropentane	$C_5H_{10}Cl_2$ 141.04	600-11-3 -77.3	139	4-01-00-00310 1.0789^{20}	1.4464^{20}
8750	Pentane, 2,4-dichloro- 2,4-Dichloropentane	$C_5H_{10}Cl_2$ 141.04	625-67-2	145	3-01-00-00343 1.0634^{15}	H_2O 1; eth 4; bz 3; chl 4 1.447^{18}
8751	Pentane, 3,3-dichloro- 3,3-Dichloropentane	$C_5H_{10}Cl_2$ 141.04	21571-91-5	132; 32^{14}	3-01-00-00343 1.053^{20}	bz 4; chl 4 1.442^{20}
8752	Pentane, 1,2-dichloro-4,4-dimethyl- 1,2-Dichloro-4,4-dimethylpentane	$C_7H_{14}Cl_2$ 169.09	6065-90-3	175; 58^{12}	3-01-00-00445 1.0259^{20}	bz 4; chl 4 1.4489^{20}
8753	Pentane, 1,5-dichloro-3,3-dimethyl- 1,5-Dichloro-3,3-dimethylpentane	$C_7H_{14}Cl_2$ 169.09	62496-53-1	135^{80}; 58^{8}	4-01-00-00409 1.0563^{20}	chl 4 1.4652^{20}
8754	Pentane, 2,4-dichloro-2,4-dimethyl- 2,4-Dichloro-2,4-dimethylpentane	$C_7H_{14}Cl_2$ 169.09	33553-93-4 23.5	$51\text{-}7^{8}$	4-01-00-00408 1.0292^{20}	1.4537^{20}
8755	Pentane, 1,1-diethoxy-	$C_9H_{20}O_2$ 160.26	3658-79-5	59^{12}	4-01-00-03269 0.829^{22}	1.4029^{22}
8756	Pentane, 3,3-diethyl- 3,3-Diethylpentane	C_9H_{20} 128.26	1067-20-5 -33.1	146.3	4-01-00-00462 0.7536^{20}	H_2O 1; eth 3; bz 3; os 3 1.4206^{20}
8757	Pentane, 3,3-diethyl-2-methyl- 3,3-Diethyl-2-methylpentane	$C_{10}H_{22}$ 142.28	52897-16-2	172	4-01-00-00487 0.7755^{25}	1.4343^{20}
8758	Pentane, 1,5-diiodo- Pentamethylene diiodide	$C_5H_{10}I_2$ 323.94	628-77-3 9	149^{20}; 101^{3}	4-01-00-00317 2.1903^{15}	H_2O 1; eth 3; chl 3 1.6046^{15}
8759	Pentane, 2,4-diiodo-	$C_5H_{10}I_2$ 323.94	66719-29-7	147^{34}	0-01-00-00133 2.195^{20}	1.600^{20}
8760	Pentane, 2,2-dimethyl- 2,2-Dimethylpentane	C_7H_{16} 100.20	590-35-2 -123.8	79.2	4-01-00-00403 0.6739^{20}	H_2O 1; EtOH 3; eth 3; ace 5 1.3822^{20}
8761	Pentane, 2,3-dimethyl- 2,3-Dimethylpentane	C_7H_{16} 100.20	565-59-3	89.7	3-01-00-00445 0.6951^{20}	H_2O 1; EtOH 3; eth 3; ace 5 1.3919^{20}
8762	Pentane, 2,4-dimethyl- 2,4-Dimethylpentane	C_7H_{16} 100.20	108-08-7 -119.9	80.4	4-01-00-00406 0.6727^{20}	H_2O 1; EtOH 3; eth 3; ace 5 1.3815^{20}
8763	Pentane, 3,3-dimethyl- 3,3-Dimethylpentane	C_7H_{16} 100.20	562-49-2 -134.9	86.0	4-01-00-00409 0.6936^{20}	H_2O 1; EtOH 3; eth 3; ace 5 1.3909^{20}
8764	Pentane, 2,2-dimethyl-3-methylene-	C_8H_{16} 112.22	18231-53-3	110	3-01-00-00848 0.724^{25}	1.4159^{20}
8765	Pentane, 2,4-dimethyl-3-methylene- 1-Butene, 2-isopropyl-3-methyl-	C_8H_{16} 112.22	111823-35-9	104	3-01-00-00851 0.718^{25}	1.4086^{20}

No.	Name Synonym	Mol. Form. Mol. Wt.	CAS RN mp/°C	Merck No. bp/°C	Beil. Ref. den/g cm^{-3}	Solubility n_D
8766	Pentane, 2,4-dimethyl-3-(1-methylethyl)-	$C_{10}H_{22}$	13475-79-1		3-01-00-00534	
		142.28	-81.7	157.1	0.7545^{25}	1.4246^{20}
8767	Pentanedinitrile	$C_5H_6N_2$	544-13-8	4368	4-02-00-01941	EtOH 4; chl 4
	Glutaronitrile	94.12	-29	286	0.9911^{15}	1.4295^{20}
8768	Pentanedinitrile, 2-bromo-2-(bromomethyl)-	$C_6H_6Br_2N_2$	35691-65-7	3004		H_2O 1; ace 4; bz 4; DMF 4
	1,2-Dibromo-2,4-dicyanobutane	265.94	52			
8769	Pentane, 1,5-dinitro-	$C_5H_{10}N_2O_4$	6848-84-6		4-01-00-00319	H_2O 1; bz 3
	1,5-Dinitropentane	162.15		$134^{1.2}$		1.461^{20}
8770	Pentanedioic acid	$C_5H_8O_4$	110-94-1	4367	4-02-00-01934	H_2O 4; EtOH 4; eth 4; bz 1
	Glutaric acid	132.12	97.8	303 dec	1.429^{15}	1.4188^{106}
8771	Pentanedioic acid, 2-acetyl-, diethyl ester	$C_{11}H_{18}O_5$	1501-06-0		4-03-00-01835	H_2O 2; EtOH 3; eth 3; ctc 2
		230.26		271 dec	1.0712^{20}	1.4420^{15}
8772	Pentanedioic acid, diethyl ester	$C_9H_{16}O_4$	818-38-2		4-02-00-01937	eth 4
		188.22	-24.1	236.5	1.0220^{20}	1.4241^{20}
8773	Pentanedioic acid, 3,3-dimethyl-	$C_7H_{12}O_4$	4839-46-7		4-02-00-02023	H_2O 4; EtOH 4; eth 4; bz 2
		160.17	103.5	126^{415}; 89^2	1.4278^{20}	
8774	Pentanedioic acid, 3,3-dimethyl-, dimethyl ester	$C_9H_{16}O_4$	19184-67-9		3-02-00-01758	ctc 3
		188.22		$103\text{-}4^{15}$	1.0366^{20}	
8775	Pentanedioic acid, dimethyl ester	$C_7H_{12}O_4$	1119-40-0		4-02-00-01937	EtOH 4; eth 4; chl 3
		160.17	-42.5	214; 109^{21}	1.0876^{20}	1.4242^{20}
8776	Pentanedioic acid, diphenyl ester	$C_{17}H_{16}O_4$	47172-89-4		4-06-00-00626	ace 4; EtOH 4; lig 4
	Diphenyl glutarate	284.31	54	236.5^{15}		
8777	Pentanedioic acid, 3-ethyl-3-methyl-	$C_8H_{14}O_4$	5345-01-7		3-02-00-01781	H_2O 3; EtOH 3; eth 3; bz 3
	3-Ethyl-3-methylglutaric acid	174.20	87	261		
8778	Pentanedioic acid, 3-hydroxy-, diethyl ester	$C_9H_{16}O_5$	32328-03-3		4-03-00-01149	
	Diethyl 3-hydroxyglutarate	204.22		$156\text{-}7^{23}$	1.103^{25}	1.4368^{20}
8779	Pentanedioic acid, 3-methyl-	$C_6H_{10}O_4$	626-51-7		4-02-00-01992	H_2O 3; EtOH 3; eth 3; bz 2
	Glutaric acid, 3-methyl	146.14	87	$165\text{-}7^{0.5}$		
8780	Pentanedioic acid, monomethyl ester	$C_6H_{10}O_4$	1501-27-5		4-02-00-01937	
		146.14		$150\text{-}1^{10}$	1.169^{25}	1.4381^{20}
8781	Pentanedioic acid, 2-oxo-	$C_5H_6O_5$	328-50-7	5182	4-03-00-01813	H_2O 4; EtOH 4; eth 4; ace 3
	α-Ketoglutaric acid	146.10	115.5			
8782	Pentanedioic acid, 3-oxo-	$C_5H_6O_5$	542-05-2	60	4-03-00-01816	H_2O 3; EtOH 3; eth 2; bz 1
	Acetonedicarboxylic acid	146.10	138 dec			
8783	Pentanedioic acid, 3-oxo-, diethyl ester	$C_9H_{14}O_5$	105-50-0		4-03-00-01817	H_2O 2; EtOH 5
		202.21		250	1.113^{20}	
8784	Pentanedioic acid, 3-oxo-, dimethyl ester	$C_7H_{10}O_5$	1830-54-2		4-03-00-01817	
		174.15		150^{25}	1.185^{25}	1.4434^{20}
8785	1,3-Pentanediol	$C_5H_{12}O_2$	3174-67-2		4-01-00-02538	
		104.15		$120\text{-}2^{16}$	0.9863^{20}	1.4659^{20}
8786	1,4-Pentanediol	$C_5H_{12}O_2$	626-95-9		3-01-00-02191	H_2O 4; EtOH 4; chl 4
	4-Hydroxy-1-pentanol	104.15		202; 125^{10}	0.9883^{20}	1.4452^{23}
8787	1,5-Pentanediol	$C_5H_{12}O_2$	111-29-5	7073	4-01-00-02540	H_2O 3; EtOH 3; eth 2; bz 2
	Pentamethylene glycol	104.15	-18	239	0.9914^{20}	1.4494^{20}
8788	2,3-Pentanediol	$C_5H_{12}O_2$	42027-23-6		4-01-00-02543	H_2O 3; EtOH 3; eth 2
		104.15		187.5; 100^{17}	0.9798^{19}	1.4412^{25}
8789	2,4-Pentanediol	$C_5H_{12}O_2$	625-69-4		3-01-00-02195	H_2O 4; EtOH 4
	2,4-Amylene glycol	104.15		199; $97\text{-}8^{13}$	0.9635^{20}	1.4349^{20}
8790	1,2-Pentanediol, (±)-	$C_5H_{12}O_2$	91049-43-3		4-01-00-02538	
	1,2-Pentanediol (*DL*)	104.15		211	0.9784^{20}	1.4412^{19}
8791	1,5-Pentanediol, diacetate	$C_9H_{16}O_4$	6963-44-6		4-02-00-00226	
	Pentamethylene acetate	188.22		$122\text{-}3^{3}$	1.0296^{20}	1.4261^{19}
8792	1,3-Pentanediol, 2,2-dimethyl-	$C_7H_{16}O_2$	2157-31-5		0-01-00-00490	EtOH 4
	2,2-Dimethyl-1,3-dihydroxypentane	132.20	62.3	213		
8793	2,4-Pentanediol, 2-methyl-	$C_6H_{14}O_2$	107-41-5	4631	4-01-00-02565	H_2O 3; EtOH 3; eth 3; ctc 2
	Hexylene glycol	118.18	-50	197.1	0.923^{15}	1.4276^{20}
8794	2,4-Pentanediol, 3-methyl-	$C_6H_{14}O_2$	5683-44-3		4-01-00-02572	H_2O 3; EtOH 3
	3-Methyl-2,4-pentanediol	118.18		211.5	0.9640^{20}	1.4433^{20}
8795	1,3-Pentanediol, 2,2,4-trimethyl-	$C_8H_{18}O_2$	144-19-4		4-01-00-02604	H_2O 2; EtOH 4; eth 4; bz 3
	2,2,4-Trimethyl-1,3-pentanediol	146.23	51.5	235; $81\text{-}2^{1}$	0.936^{15}	1.4513^{15}
8796	1,4-Pentanediol, 2,2,4-trimethyl-	$C_8H_{18}O_2$	80864-10-4		0-01-00-00493	eth 4; EtOH 4
		146.23	86	210		
8797	2,3-Pentanedione	$C_5H_8O_2$	600-14-6		4-01-00-03660	H_2O 3; EtOH 5; eth 5; ace 5
	Acetylpropionyl	100.12		108	0.9565^{19}	1.4014^{19}
8798	2,4-Pentanedione	$C_5H_8O_2$	123-54-6	75	4-01-00-03662	H_2O 4; EtOH 5; eth 5; ace 5
	Acetylacetone	100.12	-23	138	0.9721^{25}	1.4494^{20}
8799	2,4-Pentanedione, 3,3-dimethyl-	$C_7H_{12}O_2$	3142-58-3		4-01-00-03705	eth 4
	3,3-Dimethylacetylacetone	128.17	19	173	0.9575^{20}	1.4306^{20}
8800	2,3-Pentanedione, dioxime	$C_5H_{10}N_2O_2$	4775-86-4		4-01-00-03661	EtOH 4
		130.15	172.5	sub		
8801	2,4-Pentanedione, 3-ethyl-	$C_7H_{12}O_2$	1540-34-7		4-01-00-03703	eth 4; EtOH 4; chl 4
		128.17		178.5	0.9531^{19}	1.4408^{19}
8802	2,4-Pentanedione, 1,1,1,5,5,5-hexafluoro-	$C_5H_2F_6O_2$	1522-22-1		4-01-00-03681	
	Acetylacetone, hexafluoro	208.06		54.1	1.485^{20}	1.3333^{20}
8803	2,3-Pentanedione, 3-oxime	$C_5H_9NO_2$	609-29-0		4-01-00-03661	eth 4; EtOH 4; chl 4
	3-Oximino-2-pentanone	115.13	58.5	183		
8804	1,4-Pentanedione, 1-phenyl-	$C_{11}H_{12}O_2$	583-05-1		4-07-00-02160	ace 4
		176.22		162^{12}		1.5250^{30}
8805	2,4-Pentanedione, 3-(2-propenyl)-	$C_8H_{12}O_2$	3508-78-9		4-01-00-03759	
		140.18		199; 92^{16}	0.974^{15}	1.4698^{14}

No.	Name	Synonym	Mol. Form.	Mol. Wt.	CAS RN	mp/°C	Merck No.	bp/°C	Beil. Ref.	den/g cm^{-3}	Solubility	n_D
8806	Pentanedioyl dichloride		$C_5H_6Cl_2O_2$	169.01	2873-74-7			217	4-02-00-01939	1.324^{20}	eth 3; chl 2	1.4728^{20}
8807	Pentane, dodecafluoro-	Perfluoropentane	C_5F_{12}	288.04	678-26-2	-10		29.2	4-01-00-00308			
8808	Pentane, 1-ethoxy-	Ethyl pentyl ether	$C_7H_{16}O$	116.20	17952-11-3			117.6	3-01-00-01602	0.7622^{20}	eth 4; EtOH 4	1.3927^{20}
8809	Pentane, 3-ethyl-	3-Ethylpentane	C_7H_{16}	100.20	617-78-7	-118.6		93.5	4-01-00-00402	0.6982^{20}	H_2O 1; EtOH 3; eth 3; ace 5	1.3934^{20}
8810	Pentane, 3-ethyl-2,2-dimethyl-		C_9H_{20}	128.26	16747-32-3	-99.3		133.8	4-01-00-00462	0.7438^{20}		1.4123^{20}
8811	Pentane, 3-ethyl-2,3-dimethyl-		C_9H_{20}	128.26	16747-33-4			144.7	4-01-00-00462	0.7508^{25}		1.4221^{20}
8812	Pentane, 3-ethyl-2,4-dimethyl-		C_9H_{20}	128.26	1068-87-7	-122.4		136.7	4-01-00-00462	0.7365^{20}		1.4131^{20}
8813	Pentane, 3-ethyl-2-methyl-	3-Ethyl-2-methylpentane	C_8H_{18}	114.23	609-26-7	-114.9		115.6	4-01-00-00437	0.7193^{20}	H_2O 1; EtOH 5; eth 3; ace 5	1.4040^{20}
8814	Pentane, 3-ethyl-3-methyl-	3-Ethyl-3-methylpentane	C_8H_{18}	114.23	1067-08-9	-90.9		118.2	4-01-00-00438	0.7274^{20}	H_2O 1; EtOH 5; eth 3; ace 5	1.4078^{20}
8815	Pentane, 3-ethyl-2,2,3-trimethyl-	3-Ethyl-2,2,3-trimethylpentane	$C_{10}H_{22}$	142.28	52897-17-3	-42.5		169.5	4-01-00-00487	0.7780^{25}		1.4420^{20}
8816	Pentane, 3-ethyl-2,3,4-trimethyl-	3-Ethyl-2,3,4-trimethylpentane	$C_{10}H_{22}$	142.28	52897-19-5			169.47	4-01-00-00487	0.7735^{25}		1.4333^{20}
8817	Pentane, 1-fluoro-	Pentyl fluoride	$C_5H_{11}F$	90.14	592-50-7	-120		62.8	4-01-00-00308	0.7907^{20}	eth 4; EtOH 4	1.3591^{2-}
8818	Pentane, 1-iodo-	Amyl iodide	$C_5H_{11}I$	198.05	628-17-1	-85.6		155	4-01-00-00315	1.5161^{20}	chl 3	1.4959^{20}
8819	Pentane, 2-iodo-, (±)-	Pentane, 2-iodo (*DL*)	$C_5H_{11}I$	198.05	52152-72-4			141	4-01-00-00316	1.5096^{20}	H_2O 1; eth 3; ace 3; bz 3	1.4961^{20}
8820	Pentane, 3-iodo-	3-Iodopentane	$C_5H_{11}I$	198.05	1809-05-8			145.5	3-01-00-00349	1.5176^{20}	ace 4; bz 4; eth 4	1.4974^{20}
8821	Pentane, 1-isocyano-	Pentyl isocyanide	$C_6H_{11}N$	97.16	18971-59-0	-51.1		155.5	3-04-00-00331	0.806^{20}	H_2O 1; EtOH 3	
8822	Pentane, 1-methoxy-	Methyl pentyl ether	$C_6H_{14}O$	102.18	628-80-8			99	4-01-00-01643	0.759^{22}	ace 4; eth 4; EtOH 4	1.3862^{22}
8823	Pentane, 2-methyl-	Isohexane	C_6H_{14}	86.18	107-83-5	-153.7		60.2	4-01-00-00358	0.650^{25}	H_2O 1; EtOH 3; eth 3; ace 5	1.3715^{20}
8824	Pentane, 3-methyl-	3-Methylpentane	C_6H_{14}	86.18	96-14-0	-162.9		63.2	4-01-00-00363	0.6598^{25}	H_2O 1; EtOH 3; eth 5; ace 5	1.3765^{20}
8825	Pentane, 3-methylene-	2-Ethyl-1-butene	C_6H_{12}	84.16	760-21-4	-131.5		64.7	4-01-00-00850	0.6894^{20}	H_2O 1; eth 3; ace 3; bz 3	1.3969^{20}
8826	Pentane, 1-(methylthio)-	Methyl pentyl sulfide	$C_6H_{14}S$	118.24	1741-83-9	-94		145.1	3-01-00-01608	0.8431^{20}	EtOH 3; eth 3; ace 3; bz 3	1.4506^{20}
8827	Pentanenitrile	Valeronitrile	C_5H_9N	83.13	110-59-8	-96.2		141.3	4-02-00-00875	0.8008^{20}	eth 3; ace 3; bz 3; ctc 2	1.3971^{20}
8828	Pentanenitrile, 5-bromo-		C_5H_8BrN	162.03	5414-21-1			111^{12}	4-02-00-00884	1.386^{25}		1.4780^{20}
8829	Pentanenitrile, 2-hydroxy-2-propyl-		$C_8H_{15}NO$	141.21	5699-74-1			119-20^{21}	4-03-00-00880	0.9077^{18}	EtOH 4	1.4337^{18}
8830	Pentanenitrile, 3-imino-2-methyl-	Valeronitrile, 3-imino-2-methyl	$C_6H_{10}N_2$	110.16	95642-49-2	46.3		257.5		0.9525^{19}	eth 4; EtOH 4	
8831	Pentanenitrile, 4-methyl-	Isoamyl cyanide	$C_6H_{11}N$	97.16	542-54-1	-51	4999	156.5	4-02-00-00946	0.8030^{20}	H_2O 1; EtOH 3; eth 5; ctc 2	1.4059^{20}
8832	Pentane, 1-nitro-	1-Nitropentane	$C_5H_{11}NO_2$	117.15	628-05-7			172.5	4-01-00-00317	0.9525^{20}	EtOH 3; eth 3; bz 3	1.4175^{20}
8833	Pentane, 2-nitro-	2-Nitropentane	$C_5H_{11}NO_2$	117.15	4609-89-6			149; 88^{100}	4-01-00-00318	0.938^{25}		1.4092^{25}
8834	Pentane, 3-nitro-	3-Nitropentane	$C_5H_{11}NO_2$	117.15	551-88-2		6537	154	4-01-00-00318	0.957^{0}	ace 4; eth 4; EtOH 4	
8835	Pentane, 1,1'-oxybis-	Amyl ether	$C_{10}H_{22}O$	158.28	693-65-2	-69	646	190	4-01-00-01643	0.7833^{20}	H_2O 1; EtOH 5; eth 5; chl 3	1.4119^{20}
8836	Pentane, 2,2,3,3,4-pentamethyl-	2,2,3,3,4-Pentamethylpentane	$C_{10}H_{22}$	142.28	16747-44-7	-36.4		166.1	4-01-00-00487	0.7767^{25}		1.4361^{20}
8837	Pentane, 2,2,3,4,4-pentamethyl-	2,2,3,4,4-Pentamethylpentane	$C_{10}H_{22}$	142.28	16747-45-8	-38.7		159.3	4-01-00-00487	0.7636^{25}		1.4307^{20}
8838	Pentane, 1,1,1,5-tetrachloro-	1,1,1,5-Tetrachloropentane	$C_5H_8Cl_4$	209.93	2467-10-9			112^{24}	4-01-00-00311	1.3416^{25}		1.4859^{25}
8839	Pentane, 2,2,3,3-tetramethyl-	2,2,3,3-Tetramethylpentane	C_9H_{20}	128.26	7154-79-2	-9.8		140.2	4-01-00-00463	0.7530^{25}		1.4236^{20}
8840	Pentane, 2,2,3,4-tetramethyl-	2,2,3,4-Tetramethylpentane	C_9H_{20}	128.26	1186-53-4	-121.0		133.0	4-01-00-00463	0.7389^{20}		1.4147^{20}
8841	Pentane, 2,2,4,4-tetramethyl-	Di-*tert*-butylmethane	C_9H_{20}	128.26	1070-87-7	-66.5		122.2	4-01-00-00464	0.7195^{20}	H_2O 1; EtOH 4; bz 4; os 4	1.4069^{20}
8842	Pentane, 2,3,3,4-tetramethyl-	2,3,3,4-Tetramethylpentane	C_9H_{20}	128.26	16747-38-9	-102.1		141.5	4-01-00-00464	0.7547^{20}		1.4222^{20}
8843	Pentane, 1,1'-thiobis-		$C_{10}H_{22}S$	174.35	872-10-6	-51.3		86$^{3.7}$	4-01-00-01654	0.8407^{20}	H_2O 1; eth 3	1.4561^{20}
8844	1-Pentanethiol	Pentyl mercaptan	$C_5H_{12}S$	104.22	110-66-7	-75.7	648	126.6	4-01-00-01653	0.850^{20}	H_2O 1; EtOH 5; eth 5	1.4469^{20}
8845	2-Pentanethiol	*sec*-Amylmercaptan	$C_5H_{12}S$	104.22	2084-19-7	-169		112.9	4-01-00-01662	0.8327^{20}	EtOH 3; lig 3	1.4412^{20}

No.	Name Synonym	Mol. Form. Mol. Wt.	CAS RN mp/°C	Merck No. bp/°C	Beil. Ref. den/g cm^{-3}	Solubility n_D
8846	3-Pentanethiol 3-Pentyl mercaptan	$C_5H_{12}S$ 104.22	616-31-9 -110.8	 105	4-01-00-01665 0.8410^{20}	EtOH 3; DMSO-d_6 2 1.4447^{20}
8847	Pentane, 2,2,3-trimethyl- 2-*tert*-Butylbutane	C_8H_{18} 114.23	564-02-3 -112.2	 110	4-01-00-00438 0.7161^{20}	H_2O 1; EtOH 5; eth 5; ace 5 1.4030^{20}
8848	Pentane, 2,2,4-trimethyl- Isooctane	C_8H_{18} 114.23	540-84-1 -107.3	5079 99.2	4-01-00-00439 0.6877^{25}	H_2O 1; EtOH 5; eth 3; ace 5 1.3915^{20}
8849	Pentane, 2,3,3-trimethyl- 2,3,3-Trimethylpentane	C_8H_{18} 114.23	560-21-4 -100.9	 114.8	4-01-00-00445 0.7262^{20}	H_2O 1; EtOH 4; eth 5; ace 5 1.4075^{20}
8850	Pentane, 2,3,4-trimethyl- 2,3,4-Trimethylpentane	C_8H_{18} 114.23	565-75-3 -109.2	 113.5	4-01-00-00446 0.7191^{20}	H_2O 1; EtOH 4; eth 5; ace 5 1.4042^{20}
8851	1,2,5-Pentanetriol 1,4,5-Pentanetriol	$C_5H_{12}O_3$ 120.15	14697-46-2	 $190\text{-}1^{13}$	4-01-00-02779 1.136^{20}	H_2O 4; EtOH 4 1.4730^{20}
8852	1,3,5-Pentanetriol	$C_5H_{12}O_3$ 120.15	4328-94-3	 188^{11}; $139^{0.1}$	4-01-00-02779 1.1291^{20}	H_2O 4; EtOH 4; eth 3; ace 3 1.4785^{20}
8853	Pentane, 1,1,1,2,2,3,4,4,5,5,5-undecafluoro-3-(trifluoromethyl) Perfluoro-3-methylpentane	C_6F_{14} 338.04	865-71-4 -115	 58.4		bz 3
8854	Pentane, 1,1,1,2,2,3,3,4,5,5,5-undecafluoro-4-(trifluoromethyl)- Perfluoro(2-methylpentane)	C_6F_{14} 338.04	355-04-4	 57.6	4-01-00-00360 1.7326^{20}	H_2O 1; bz 3 1.2564^{22}
8855	Pentanoic acid Valeric acid	$C_5H_{10}O_2$ 102.13	109-52-4 -34	9815 186.1	4-02-00-00868 0.9391^{20}	H_2O 3; EtOH 3; eth 3; ctc 2 1.4085^{20}
8856	Pentanoic acid, 2-acetyl-, ethyl ester	$C_9H_{16}O_3$ 172.22	1540-28-9	 224; 90^{15}	4-03-00-01602 0.9661^{20}	eth 4; EtOH 4 1.4255^{20}
8857	Pentanoic acid, 5-amino-	$C_5H_{11}NO_2$ 117.15	660-88-8 157-8 dec	 dec	4-04-00-02636	H_2O 3; EtOH 2; eth 1; bz 1
8858	Pentanoic acid, anhydride	$C_{10}H_{18}O_3$ 186.25	2082-59-9 -56.1	 227	4-02-00-00874 0.924^{20}	eth 4; EtOH 4 1.4171^{26}
8859	Pentanoic acid, 2-bromo-, (±)-	$C_5H_9BrO_2$ 181.03	2681-92-7	 118^{12}	4-02-00-00883 1.381^{20}	eth 4; EtOH 4
8860	Pentanoic acid, 2-bromo-, ethyl ester	$C_7H_{13}BrO_2$ 209.08	615-83-8	 191	3-02-00-00681 1.226^{18}	H_2O 1; EtOH 3; eth 3 1.4496^{20}
8861	Pentanoic acid, 5-bromo-, ethyl ester	$C_7H_{13}BrO_2$ 209.08	14660-52-7	 $107\text{-}8^{20}$	4-02-00-00884 1.2804^{20}	ctc 2 1.4543^{20}
8862	Pentanoic acid, butyl ester	$C_9H_{18}O_2$ 158.24	591-68-4 -92.8	 185.8	3-02-00-00671 0.8710^{15}	H_2O 2; EtOH 3; eth 3 1.4128^{20}
8863	Pentanoic acid, 2-chloro-, (±)-	$C_5H_9ClO_2$ 136.58	94347-45-2 -15	 222	4-02-00-00878 1.141^{13}	eth 4; EtOH 4 1.4481^{11}
8864	Pentanoic acid, 3-chloro- Valeric acid, 3-chloro	$C_5H_9ClO_2$ 136.58	51637-47-9 33	 112^{10}	4-02-00-00879 1.1484^{20}	eth 4; EtOH 4 1.4462^{20}
8865	Pentanoic acid, 4-chloro- 4-Chlorovaleric acid	$C_5H_9ClO_2$ 136.58	32607-54-8	 116^{10}	4-02-00-00879 1.1514^{20}	eth 4; EtOH 4 1.4458^{20}
8866	Pentanoic acid, 5-chloro-	$C_5H_9ClO_2$ 136.58	1119-46-6 18	 230	4-02-00-00880 1.3416^{25}	eth 4; EtOH 4 1.4555^{20}
8867	Pentanoic acid, 4-chloro-, ethyl ester Ethyl 4-chlorovalerate	$C_7H_{13}ClO_2$ 164.63	70786-82-2	 196	4-02-00-00879 1.0393^{20}	eth 4; EtOH 4 1.4310^{20}
8868	Pentanoic acid, 5-chloro-, ethyl ester	$C_7H_{13}ClO_2$ 164.63	2323-81-1	 205.5	4-02-00-00880 1.0561^{20}	H_2O 1, EtOH 3, eth 3, chl 2 1.4355^{20}
8869	Pentanoic acid, 2-chloroethyl ester	$C_7H_{13}ClO_2$ 164.63	7735-33-3	 185	3-02-00-00670 1.040^{12}	eth 4 1.4307^{11}
8870	Pentanoic acid, 2,3-dibromo- Valeric acid, 2,3-dibromo-	$C_5H_8Br_2O_2$ 259.93	79912-57-5 57	 $104\text{-}6^{13}$	3-02-00-00683	bz 4; CS_2 4
8871	Pentanoic acid, 2,5-dibromo- Valeric acid, 2,5-dibromo-	$C_5H_8Br_2O_2$ 259.93	1450-81-3	 173^{15}; 151^{5}	3-02-00-00683 1.8629^{25}	 1.5347^{25}
8872	Pentanoic acid, 2,3-dibromo-, ethyl ester Valeric acid, 2,3-dibromo, ethyl ester	$C_7H_{12}Br_2O_2$ 287.98	79912-55-3	 117^{14}	2-02-00-00269 1.6199^{15}	 1.4953^{15}
8873	Pentanoic acid, 2,5-dibromo-, ethyl ester Valeric acid, 2,5-dibromo-, ethyl ester	$C_7H_{12}Br_2O_2$ 287.98	29823-16-3	 $133\text{-}5^{14}$	3-02-00-00683 1.6289^{25}	 1.4947^{25}
8874	Pentanoic acid, 2,2-dimethyl-4-oxo-	$C_7H_{12}O_3$ 144.17	470-49-5 76.8	 138^{15}	3-03-00-01247	H_2O 4; bz 4; eth 4; EtOH 4
8875	Pentanoic acid, 2,4-dioxo-	$C_5H_6O_4$ 130.10	5699-58-1 99.5	 130^{37}	3-03-00-01331	H_2O 3; EtOH 3; eth 3; ace 3
8876	Pentanoic acid, 2,4-dioxo-, ethyl ester	$C_7H_{10}O_4$ 158.15	615-79-2 18	 214	4-03-00-01777 1.1251^{20}	eth 4; EtOH 4 1.4757^{17}
8877	Pentanoic acid, 2-ethyl-, (±)- Valeric acid, 2-ethyl (*DL*)	$C_7H_{14}O_2$ 130.19	116908-83-9	 209.2	4-02-00-00975 0.9311^{33}	eth 4; EtOH 4
8878	Pentanoic acid, ethyl ester Ethyl valerate	$C_7H_{14}O_2$ 130.19	539-82-2 -91.2	 146.1	4-02-00-00872 0.8770^{20}	H_2O 1; EtOH 5; eth 5; ctc 2 1.4120^{20}
8879	Pentanoic acid, 5-fluoro- Valeric acid, 5-fluoro-	$C_5H_9FO_2$ 120.12	407-75-0	 83^{2}	4-02-00-00876	eth 4; EtOH 4 1.4080^{25}
8880	Pentanoic acid, 2-furanylmethyl ester Furfuryl valerate	$C_{10}H_{14}O_3$ 182.22	36701-01-6	 228; 82^{1}	2-17-00-00115 1.0284^{20}	eth 4; EtOH 4
8881	Pentanoic acid, heptyl ester	$C_{12}H_{24}O_2$ 200.32	5451-80-9 -46.4	 245.2	4-02-00-00872 0.8623^{20}	ace 4; eth 4; EtOH 4 1.4254^{15}
8882	Pentanoic acid, hexyl ester	$C_{11}H_{22}O_2$ 186.29	1117-59-5 -63.1	 226.3	3-02-00-00671 0.8635^{20}	ace 4; eth 4; EtOH 4 1.4228^{15}
8883	Pentanoic acid, 2-hydroxy-	$C_5H_{10}O_3$ 118.13	617-31-2 34	 sub	4-03-00-00807	H_2O 3; EtOH 3; eth 3
8884	Pentanoic acid, 4-hydroxy-, ethyl ester Valeric acid, 4-hydroxy-, ethyl ester	$C_7H_{14}O_3$ 146.19	6149-46-8	 $85\text{-}6^{2}$	3-03-00-00612 0.9504^{25}	eth 4; EtOH 4 1.4265^{25}

No.	Name Synonym	Mol. Form. Mol. Wt.	CAS RN mp/°C	Merck No. bp/°C	Beil. Ref. den/g cm^{-3}	Solubility n_D
8885	Pentanoic acid, 2-hydroxyethyl ester Valeric acid, 2-hydroxyethyl ester	$C_7H_{14}O_3$ 146.19	16179-36-5	 190	3-02-00-00672 0.9839^{20}	 1.4220^{20}
8886	Pentanoic acid, 2-methoxyphenyl ester Guaiacol valerate	$C_{12}H_{16}O_3$ 208.26	531-39-5	4461 265	 1.05^{25}	bz 4; eth 4; EtOH 4
8887	Pentanoic acid, 2-methyl-, (±)-	$C_6H_{12}O_2$ 116.16	22160-39-0	 195.6	4-02-00-00942 0.9230^{20}	H_2O 3; EtOH 3; eth 3; ctc 2 1.413^{20}
8888	Pentanoic acid, 3-methyl-, (±)- Valeric acid, 3-methyl (*DL*)	$C_6H_{12}O_2$ 116.16	22160-40-3 -41.6	 197.5	4-02-00-00948 0.9262^{20}	eth 4; EtOH 4 1.4159^{20}
8889	Pentanoic acid, 4-methyl-	$C_6H_{12}O_2$ 116.16	646-07-1 -33	 200.5	4-02-00-00944 0.9225^{20}	H_2O 2; EtOH 3; eth 3; chl 3 1.4144^{20}
8890	Pentanoic acid, 2-methylene- Valeric acid, 2-methylene	$C_6H_{10}O_2$ 114.14	5650-75-9 24.4	 214	0-02-00-00437 0.9783^{25}	bz 4; eth 4; chl 4
8891	Pentanoic acid, methyl ester Methyl valerate	$C_6H_{12}O_2$ 116.16	624-24-8	 127.4	4-02-00-00871 0.8947^{20}	H_2O 2; EtOH 5; eth 5; ace 3 1.4003^{20}
8892	Pentanoic acid, 4-methyl-, ethyl ester	$C_8H_{16}O_2$ 144.21	25415-67-2	 163; 52^{10}	4-02-00-00945 0.8705^{20}	 1.4050^{20}
8893	Pentanoic acid, 1-methylethyl ester	$C_8H_{16}O_2$ 144.21	18362-97-5		3-02-00-00671 0.8579^{20}	H_2O 1; EtOH 3; eth 3; ace 3 1.4061^{20}
8894	Pentanoic acid, 4-methyl-, 2-methylpropyl ester	$C_{10}H_{20}O_2$ 172.27	25415-70-7	 172; 76^{10}	4-02-00-00945 0.8536^{20}	 1.4135^{20}
8895	Pentanoic acid, 2-methyl-4-oxo-	$C_6H_{10}O_3$ 130.14	6641-83-4	 165^{40}	4-03-00-01587 1.11^{25}	 1.4431^{20}
8896	Pentanoic acid, 3-methyl-4-oxo-	$C_6H_{10}O_3$ 130.14	6628-79-1 31.5	 241.5	3-03-00-01238 1.0932^{17}	H_2O 4; EtOH 4 1.4443^{17}
8897	Pentanoic acid, 4-methyl-3-oxo-, ethyl ester	$C_8H_{14}O_3$ 158.20	7152-15-0 -9	 173	4-03-00-01589 0.98^{25}	 1.250^{20}
8898	Pentanoic acid, 3-methyl-4-oxo-, methyl ester Levulinic acid, 3-methyl-, methyl ester	$C_7H_{12}O_3$ 144.17	25234-83-7	 200; 82^{10}	1-03-00-00240 1.022^{20}	 1.4052^{73}
8899	Pentanoic acid, 1-methylpropyl ester, (+)- Valeric acid, *sec*-butyl ester (*d*)	$C_9H_{18}O_2$ 158.24	116836-32-9	 174.5	3-02-00-00671 0.8605^{20}	bz 4; eth 4; py 4; EtOH 4 1.4070^{20}
8900	Pentanoic acid, 2-methylpropyl ester	$C_9H_{18}O_2$ 158.24	10588-10-0	 179	3-02-00-00671 0.8625^{25}	H_2O 1; EtOH 5; eth 3; ace 3 1.4046^{20}
8901	Pentanoic acid, octyl ester Valeric acid, octyl ester	$C_{13}H_{26}O_2$ 214.35	5451-85-4 -42.3	 261.6	3-02-00-00672 0.8615^{20}	ace 4; eth 4; EtOH 4 1.4273^{15}
8902	Pentanoic acid, 2-oxo-	$C_5H_8O_3$ 116.12	1821-02-9 6.5	 179	4-03-00-01558 1.0970^{14}	H_2O 2; eth 3; bz 3; chl 3
8903	Pentanoic acid, 4-oxo- Levulinic acid	$C_5H_8O_3$ 116.12	123-76-2 33	5352 245 dec	4-03-00-01560 1.1335^{20}	H_2O 4; EtOH 4; eth 4; chl 3 1.4396^{20}
8904	Pentanoic acid, 4-oxo-, butyl ester	$C_9H_{16}O_3$ 172.22	2052-15-5	 237.5	4-03-00-01563 0.9735^{20}	chl 2 1.4290^{20}
8905	Pentanoic acid, 2-oxo-, ethyl ester Valeric acid, 2-oxo-, ethyl ester	$C_7H_{12}O_3$ 144.17	50461-74-0	 182.5; 70^{11}	4-03-00-01558 0.9985^{18}	 1.4170^{18}
8906	Pentanoic acid, 3-oxo-, ethyl ester	$C_7H_{12}O_3$ 144.17	4949-44-4	 191	4-03-00-01559 1.0120^{20}	bz 4; eth 4; EtOH 4 1.4230^{20}
8907	Pentanoic acid, 4-oxo-, ethyl ester Ethyl levulinate	$C_7H_{12}O_3$ 144.17	539-88-8	3775 205.8	4-03-00-01562 1.0111^{20}	H_2O 4; EtOH 4 1.4229^{20}
8908	Pentanoic acid, 4-oxo-, methyl ester	$C_6H_{10}O_3$ 130.14	624-45-3	 196	4-03-00-01562 1.0511^{20}	H_2O 2; EtOH 3; eth 5; ace 3 1.4233^{20}
8909	Pentanoic acid, 4-oxo-, 1-methylethyl ester Levulinic acid, isopropyl ester	$C_8H_{14}O_3$ 158.20	21884-26-4	 209.3	4-03-00-01563 0.9842^{20}	ace 4; bz 4; eth 4; EtOH 4 1.4420^{20}
8910	Pentanoic acid, 4-oxo-, 2-methylpropyl ester	$C_9H_{16}O_3$ 172.22	3757-32-2	 228	4-03-00-01563 0.9669^{20}	ace 4; eth 4; EtOH 4 1.4249^{20}
8911	Pentanoic acid, 4-oxo-, phenylmethyl ester Levulinic acid, benzyl ester	$C_{12}H_{14}O_3$ 206.24	6939-75-9	 $132\text{-}4^{2}$	4-06-00-02481 1.0935^{20}	tol 4 1.5090^{20}
8912	Pentanoic acid, 4-oxo-, propyl ester Levulinic acid, propyl ester	$C_8H_{14}O_3$ 158.20	645-67-0	 218	4-03-00-01563 0.9896^{20}	H_2O 2; EtOH 3; eth 3; ace 3 1.4258^{20}
8913	Pentanoic acid, pentyl ester	$C_{10}H_{20}O_2$ 172.27	2173-56-0 -78.8	 203.7	4-02-00-00872 0.8638^{20}	H_2O 2; EtOH 5; eth 5 1.4164^{20}
8914	Pentanoic acid, propyl ester	$C_8H_{16}O_2$ 144.21	141-06-0 -70.7	 167.5	4-02-00-00872 0.8699^{20}	H_2O 1; EtOH 3; eth 3; chl 3 1.4065^{20}
8915	1-Pentanol Amyl alcohol	$C_5H_{12}O$ 88.15	71-41-0 -78.9	7074 137.9	4-01-00-01640 0.8144^{20}	H_2O 1; EtOH 5; eth 5; ace 3 1.4101^{20}
8916	2-Pentanol *sec*-Amyl alcohol	$C_5H_{12}O$ 88.15	6032-29-7 -73	7075 119.3	3-01-00-01609 0.8094^{20}	H_2O 4; EtOH 3; eth 3; ctc 3 1.4053^{20}
8917	3-Pentanol Diethyl carbinol	$C_5H_{12}O$ 88.15	584-02-1 -69	7076 116.2	4-01-00-01662 0.8203^{20}	H_2O 2; EtOH 3; eth 3; ace 3 1.4104^{20}
8918	2-Pentanol, acetate, (*R*)- *sec*-Amyl acetate (*R*)	$C_7H_{14}O_2$ 130.19	54638-10-7	 142	0-02-00-00132 0.8803^{18}	eth 4; EtOH 4 1.4012^{20}
8919	3-Pentanol, acetate	$C_7H_{14}O_2$ 130.19	620-11-1	 133	4-02-00-00155 0.8712^{20}	eth 4; EtOH 4 1.4005^{20}
8920	1-Pentanol, 5-amino-	$C_5H_{13}NO$ 103.16	2508-29-4 38.5	 221.5	4-04-00-01750 0.9488^{17}	H_2O 5; EtOH 5; ace 5 1.4618^{17}
8921	3-Pentanol, 2-amino-	$C_5H_{13}NO$ 103.16	116836-16-9	 171	1-04-00-00442 0.9289^{23}	 1.4458
8922	1-Pentanol, 2-amino-4-methyl-, (±)-	$C_6H_{15}NO$ 117.19	16369-17-8	 199	2-04-00-00748 0.9173^{13}	H_2O 4; EtOH 4
8923	1-Pentanol, 5-chloro-	$C_5H_{11}ClO$ 122.59	5259-98-3	 112^{12}	4-01-00-01650	eth 4; EtOH 4 1.4518^{20}
8924	2-Pentanol, 1-chloro-	$C_5H_{11}ClO$ 122.59	17658-32-1	 160; 60^{14}	3-01-00-01613 1.035^{20}	eth 4; EtOH 4 1.4404^{25}

No.	Name Synonym	Mol. Form. Mol. Wt.	CAS RN mp/°C	Merck No. bp/°C	Beil. Ref. den/g cm^{-3}	Solubility n_D
8925	3-Pentanol, 1-chloro- 1-Chloro-3-hydroxypentane	$C_5H_{11}ClO$ 122.59	32541-33-6	 173	4-01-00-01665 1.0327^{25}	eth 4; EtOH 4 1.448^{25}
8926	1-Pentanol, 2-(diethylamino)-4-methyl-, 4-aminobenzoate (ester), monomethanesulfonate (salt) Panthesin	$C_{18}H_{32}N_2O_5S$ 388.53	135-44-4 158	5332	4-14-00-01143	H_2O 4; EtOH 4
8927	1-Pentanol, 2,4-dimethyl-, (±)- 1-Pentanol, 2,4-dimethyl (*DL*)	$C_7H_{16}O$ 116.20	111768-02-6	 159	4-01-00-01753 0.793^{20}	eth 4; EtOH 4 1.427^{20}
8928	2-Pentanol, 2,4-dimethyl- Isobutyldimethylcarbinol	$C_7H_{16}O$ 116.20	625-06-9 <-20	 133.1	4-01-00-01753 0.8103^{20}	H_2O 1; EtOH 3; eth 3; ctc 3 1.4172^{20}
8929	3-Pentanol, 2,2-dimethyl- 2,2-Dimethyl-3-pentanol	$C_7H_{16}O$ 116.20	3970-62-5 -5	 135	3-01-00-01697 0.8253^{20}	H_2O 1; EtOH 3; eth 3 1.4223^{20}
8930	3-Pentanol, 2,3-dimethyl- 2,3-Dimethyl-3-pentanol	$C_7H_{16}O$ 116.20	595-41-5 <-30	 139.7	4-01-00-01752 0.833^{20}	H_2O 1; EtOH 3; eth 3; bz 2 1.4287^{20}
8931	3-Pentanol, 2,4-dimethyl-	$C_7H_{16}O$ 116.20	600-36-2 <-70	 138.7	4-01-00-01754 0.8288^{20}	H_2O 2; EtOH 3; eth 3 1.4250^{20}
8932	3-Pentanol, 3-ethyl-	$C_7H_{16}O$ 116.20	597-49-9 -12.5	 142	4-01-00-01750 0.8407^{22}	H_2O 1; EtOH 3; eth 3 1.4294^{20}
8933	2-Pentanol, 3-ethyl-2-methyl-	$C_8H_{18}O$ 130.23	19780-63-3	 156	4-01-00-01794 0.8382^{20}	bz 4; eth 4; EtOH 4; lig 4 1.4325^{20}
8934	3-Pentanol, 3-ethyl-2-methyl- Diethylisopropylcarbinol	$C_8H_{18}O$ 130.23	597-05-7	 160; 56^{48}	4-01-00-01794 0.8280^{20}	H_2O 1; EtOH 3; eth 3 1.4372^{10}
8935	1-Pentanol, 5-fluoro-	$C_5H_{11}FO$ 106.14	592-80-3	 $70\text{-}1^{11}$	4-01-00-01647	eth 4; EtOH 4 1.4057^{25}
8936	2-Pentanol, 1,1'-iminobis[2-methyl- Dipentylamine, 2,2'-dimethyl,2,2'-dihydroxy	$C_{12}H_{27}NO_2$ 217.35	85733-97-7	 $165\text{-}70^{15}$	 0.9264^{20}	ace 4; eth 4; EtOH 4 1.4584^{20}
8937	1-Pentanol, 2-methyl- 2-Methyl-1-pentanol	$C_6H_{14}O$ 102.18	105-30-6	 149	3-01-00-01665 0.8263^{20}	EtOH 3; eth 3; ace 3; ctc 3 1.4182^{20}
8938	1-Pentanol, 3-methyl-, (±)- 1-Pentanol, 3 methyl (*DL*)	$C_6H_{14}O$ 102.18	20281-83-8	 153	4-01-00-01722 0.8242^{20}	H_2O 1; EtOH 3; eth 3 1.4112^{23}
8939	1-Pentanol, 4-methyl- Isohexyl alcohol	$C_6H_{14}O$ 102.18	626-89-1	 151.9	4-01-00-01721 0.8131^{20}	H_2O 1; EtOH 3; eth 3 1.4134^{25}
8940	2-Pentanol, 2-methyl- 2-Methyl-2-pentanol	$C_6H_{14}O$ 102.18	590-36-3 -103	 121.1	4-01-00-01714 0.8350^{16}	H_2O 2; EtOH 3; eth 3 1.4100^{20}
8941	2-Pentanol, 3-methyl- 3-Methyl-2-pentanol	$C_6H_{14}O$ 102.18	565-60-6	 134.3	3-01-00-01672 0.8307^{20}	H_2O 2; EtOH 3; eth 3 1.4182^{20}
8942	2-Pentanol, 4-methyl- 4-Methyl-2-pentanol	$C_6H_{14}O$ 102.18	108-11-2 -90	 131.6	4-01-00-01717 0.8075^{20}	H_2O 2; EtOH 3; eth 3; ctc 2 1.4100^{20}
8943	3-Pentanol, 2-methyl- 2-Methyl-3-pentanol	$C_6H_{14}O$ 102.18	565-67-3	 126.5	4-01-00-01716 0.8243^{20}	H_2O 2; EtOH 5; eth 5 1.4175^{20}
8944	3-Pentanol, 3-methyl-	$C_6H_{14}O$ 102.18	77-74-7 -23.6	 122.4	4-01-00-01723 0.8286^{20}	H_2O 2; EtOH 5; eth 5; ctc 2 1.4186^{20}
8945	1-Pentanol, 2-methyl-, acetate Acetic acid, (2-methylpentyl) ester	$C_8H_{16}O_2$ 144.21	7789-99-3	 163	4-02-00-00160 0.8691^{25}	eth 4; EtOH 4
8946	2-Pentanol, 2-methyl-, acetate 1,1-Dimethylbutyl acetate	$C_8H_{16}O_2$ 144.21	34859-98-8	 153; 34^{10}	3-02-00-00257 0.8798^{18}	eth 4; EtOH 4 1.4068^{10}
8947	2-Pentanol, 4-methyl-, acetate 1,3-Dimethylbutyl acetate	$C_8H_{16}O_2$ 144.21	108-84-9	 147.5	4-02-00-00161 0.8805^{25}	eth 4; EtOH 4 1.3980^{20}
8948	3-Pentanol, 2-methyl-, acetate Acetic acid, (2-methyl-3-pentyl) ester	$C_8H_{16}O_2$ 144.21	35897-16-6	 149	4-02-00-00160 0.8688^{20}	eth 4; EtOH 4
8949	3-Pentanol, 3-methyl-, acetate Acetic acid, (3-methyl-3-pentyl) ester	$C_8H_{16}O_2$ 144.21	10250-47-2	 148	4-02-00-00161 0.8818^{20}	eth 4; EtOH 4 1.4169^{18}
8950	1-Pentanol, 2-nitro- 2-Nitro-1-pentanol	$C_5H_{11}NO_3$ 133.15	2899-90-3	 133^{28}, 117^{10}	3-01-00-01606 1.0818^{25}	 1.4405^{25}
8951	2-Pentanol, 1-nitro-	$C_5H_{11}NO_3$ 133.15	2224-37-5	 $87\text{-}8^{1}$	4-01-00-01661 1.0968^{20}	 1.4493^{20}
8952	2-Pentanol, 3-nitro- 3-Nitro-2-pentanol	$C_5H_{11}NO_3$ 133.15	5447-99-4	 $60^{0.5}$	3-01-00-01614 1.075^{25}	 1.4430^{20}
8953	1-Pentanol, 2,2,3,3,4,4,5,5-octafluoro- 2,2,3,3,4,4,5,5-Octafluoro-1-pentanol	$C_5H_4F_8O$ 232.07	355-80-6	6667 140.5	4-01-00-01648 1.6647^{20}	 1.3178^{20}
8954	2-Pentanol, 2,4,4-trimethyl- 2,4,4-Trimethyl-2-pentanol	$C_8H_{18}O$ 130.23	690-37-9 -20	 147.5	4-01-00-01796 0.8225^{20}	H_2O 1; EtOH 2; eth 3 1.4284^{20}
8955	3-Pentanol, 2,2,4-trimethyl- Isopropyl tert butylcarbinol	$C_8H_{18}O$ 130.23	5162-48-1 12	 150.5	4-01-00-01795 [illegible]	 [illegible]
8956	3-Pentanol, 2,3,4-trimethyl-	$C_8H_{18}O$ 130.23	3054-92-0	 156.5	4-01-00-01798 0.8492^{20}	 1.4353^{20}
8957	2-Pentanone Methyl propyl ketone	$C_5H_{10}O$ 86.13	107-87-9 -76.9	6032 102.2	4-01-00-03271 0.809^{20}	H_2O 2; EtOH 5; eth 5; ctc 2 1.3895^{20}
8958	3-Pentanone Diethyl ketone	$C_5H_{10}O$ 86.13	96-22-0 -39	3111 101.9	4-01-00-03279 0.816^{19}	H_2O 4; EtOH 5; eth 5; ctc 3 1.3905^{25}
8959	2-Pentanone, 4-amino-4-methyl- Diacetonamine	$C_6H_{13}NO$ 115.18	625-04-7	2942 $25^{0.14}$	3-04-00-00894	H_2O 3; EtOH 5; eth 5
8960	1-Pentanone, 1-(4-aminophenyl)-	$C_{11}H_{15}NO$ 177.25	38237-74-0 74.5	 $160\text{-}3^{3}$	4-14-00-00178	H_2O 1; EtOH 3; eth 3
8961	2-Pentanone, 5-chloro-	C_5H_9ClO 120.58	5891-21-4	 76^{34}	4-01-00-03277 1.0523^{20}	eth 3; ace 3; ctc 2 1.4375^{20}
8962	3-Pentanone, 1-chloro-	C_5H_9ClO 120.58	32830-97-0	 68^{20}	4-01-00-03284	eth 4; EtOH 4 1.4361^{20}
8963	2-Pentanone, 3,5-dichloro- 3,5-Dichloro-2-pentanone	$C_5H_8Cl_2O$ 155.02	58371-98-5	 99^{15}; 58^{3}	4-01-00-03277 1.239^{15}	 1.4632^{20}

No.	Name	Synonym	Mol. Form.	Mol. Wt.	CAS RN	mp/°C	Merck No.	bp/°C	Beil. Ref.	den/g cm^{-3}	Solubility	n_D
8964	2-Pentanone, 5-(diethylamino)-		$C_9H_{19}NO$	157.26	105-14-6			$83-5^{15}$	4-04-00-01935	0.861^{25}		1.4350^{20}
8965	2-Pentanone, 4,4-dimethyl-		$C_7H_{14}O$	114.19	590-50-1	-64		126	4-01-00-03331	0.809^{25}		1.4036^{20}
8966	3-Pentanone, 2,2-dimethyl-	2,2-Dimethyl-3-pentanone	$C_7H_{14}O$	114.19	564-04-5	-45		125.6	4-01-00-03331	0.8125^{20}	H_2O 2; EtOH 3; eth 3; ace 3	1.4065^{20}
8967	3-Pentanone, 2,4-dimethyl-	Diisopropyl ketone	$C_7H_{14}O$	114.19	565-80-0	-69		125.4	4-01-00-03334	0.8108^{20}	H_2O 1; EtOH 5; eth 5; bz 3	1.3999^{20}
8968	3-Pentanone, 2,4-dimethyl-, oxime	Diisopropyl ketone oxime	$C_7H_{15}NO$	129.20	1113-74-2	34		$181; 91.5^{21}$	4-01-00-03335	0.9022^{17}		
8969	3-Pentanone, 1,5-diphenyl-	Diphenethyl ketone	$C_{17}H_{18}O$	238.33	5396-91-8	13.5		352	4-07-00-01462	1.0356^{0}		
8970	2-Pentanone, 3-ethyl-4-methyl-		$C_8H_{16}O$	128.21	71172-57-1			154.5	3-01-00-02883	0.812^{20}	H_2O 3; EtOH 5; eth 5; bz 5	1.4105^{20}
8971	2-Pentanone, 1-hydroxy-	1-Pentanol-2-one	$C_5H_{10}O_2$	102.13	64502-89-2			152	4-01-00-04004	0.9860^{20}	H_2O 4; eth 4; EtOH 4	1.4234^{12}
8972	2-Pentanone, 3-hydroxy-		$C_5H_{10}O_2$	102.13	3142-66-3			147.5	3-01-00-03220	0.9500^{20}	H_2O 4; ace 4; bz 4; eth 4	1.4350^{10}
8973	2-Pentanone, 4-hydroxy-		$C_5H_{10}O_2$	102.13	4161-60-8			177	4-01-00-04005	1.0071^{20}	H_2O 3; EtOH 3; eth 3	1.4415^{18}
8974	2-Pentanone, 5-hydroxy-		$C_5H_{10}O_2$	102.13	1071-73-4			$209; 117^{33}$	4-01-00-04006	1.0071^{20}	H_2O 5; EtOH 3; eth 3	1.4390^{20}
8975	3-Pentanone, 2-hydroxy-	2-Pentanol-3-one	$C_5H_{10}O_2$	102.13	5704-20-1			152.5	4-01-00-04008	0.9742^{20}	eth 4; EtOH 4	1.4128^{20}
8976	2-Pentanone, 4-hydroxy-4-methyl-	Diacetone alcohol	$C_6H_{12}O_2$	116.16	123-42-2	-44	2944	167.9	4-01-00-04023	0.9387^{20}	H_2O 5; EtOH 5; eth 5; chl 3	1.4213^{20}
8977	2-Pentanone, 4-imino-		C_5H_9NO	99.13	870-74-6	43		209	4-01-00-03678	0.9427^{0}	H_2O 4; eth 4	
8978	2-Pentanone, 4-methoxy-4-methyl-	Pentoxone	$C_7H_{14}O_2$	130.19	107-70-0			160	4-01-00-04024	0.8980^{25}		1.418^{20}
8979	1-Pentanone, 1-(3-methoxyphenyl)-		$C_{12}H_{16}O_2$	192.26	20359-55-1			196^{30}	4-08-00-00533	1.012^{20}		1.5242^{31}
8980	2-Pentanone, 3-methyl-, (±)-	2-Pentanone, 3-methyl (*DL*)	$C_6H_{12}O$	100.16	55156-16-6			118	4-01-00-03309	0.8130^{20}	H_2O 2; EtOH 5; eth 5; chl 3	1.4002^{20}
8981	2-Pentanone, 4-methyl-	Isobutyl methyl ketone	$C_6H_{12}O$	100.16	108-10-1	-84	5095	116.5	4-01-00-03305	0.7978^{20}	H_2O 2; EtOH 5; eth 5; ace 5	1.3962^{20}
8982	3-Pentanone, 2-methyl-	Ethyl isopropyl ketone	$C_6H_{12}O$	100.16	565-69-5			113.5	4-01-00-03304	0.814^{18}	H_2O 2; EtOH 4; eth 5; ace 5	1.3975^{20}
8983	1-Pentanone, 4-methyl-1-phenyl-		$C_{12}H_{16}O$	176.26	2050-07-9	-2		255.5	3-07-00-01158	0.9623^{15}	ace 4; bz 4; eth 4; EtOH 4	1.533^{20}
8984	2-Pentanone, 3-methyl-1-phenyl-	Benzyl *sec*-butyl ketone	$C_{12}H_{16}O$	176.26	27993-42-6			$127-8^{17}$	4-07-00-00758	0.964^{20}		1.5019^{20}
8985	2-Pentanone, oxime	Methyl propyl ketone oxime	$C_5H_{11}NO$	101.15	623-40-5			168	4-01-00-03274	0.9095^{20}	H_2O 4; eth 4; EtOH 4	1.4450^{20}
8986	1-Pentanone, 1-phenyl-		$C_{11}H_{14}O$	162.23	1009-14-9	-9.4		245	4-07-00-00733	0.986^{20}	H_2O 1; EtOH 4; eth 4; ctc 2	1.5158^{20}
8987	2-Pentanone, 3-(phenylmethylene)-	Methyl α-ethylstyryl ketone	$C_{12}H_{14}O$	174.24	3437-89-6			$136-8^{12}$	3-07-00-01461	1.0005^{22}		1.5650^{22}
8988	3-Pentanone, 2,2,4,4-tetramethyl-		$C_9H_{18}O$	142.24	815-24-7	-25.2		152	4-01-00-03365	0.8240^{18}	H_2O 1; EtOH 3; eth 3; ace 3	1.4194^{20}
8989	3-Pentanone, 2,2,4-trimethyl-	*tert*-Butyl isopropyl ketone	$C_8H_{16}O$	128.21	5857-36-3			135.1	4-01-00-03351	0.8065^{20}	H_2O 1; eth 3; ace 3	1.4060
8990	Pentanoyl chloride	Valeroyl chloride	C_5H_9ClO	120.58	638-29-9	-110		109	4-02-00-00874	1.0155^{15}		1.4200^{20}
8991	Pentanoyl chloride, 2-chloro-		$C_5H_8Cl_2O$	155.02	61589-68-2			156	4-02-00-00879	1.1765^{20}	eth 4	1.4465^{20}
8992	Pentanoyl chloride, 5-chloro-		$C_5H_8Cl_2O$	155.02	1575-61-7			83^{12}	4-02-00-00881	1.210^{18}	eth 4	1.4639^{20}
8993	Pentanoyl chloride, 2-methyl-, (±)-	Valeryl chloride, 2-methyl (*DL*)	$C_6H_{11}ClO$	134.61	116908-84-0			141	3-02-00-00741	0.9781^{20}	eth 4	1.4330^{27}
8994	Pentanoyl chloride, 3-methyl-, (±)-	Valeryl chloride, 3-methyl (*DL*)	$C_6H_{11}ClO$	134.61	116908-85-1			141	4-02-00-00949	0.9781^{20}	eth 4	
8995	2,5,8,11,14-Pentaoxapentadecane	Tetraethylene glycol dimethyl ether	$C_{10}H_{22}O_5$	222.28	143-24-8		9141	275.3	4-01-00-02404	1.0114^{20}	H_2O 5; EtOH 3; eth 3; ctc 3	
8996	Pentaphene	2,3:6,7-Dibenzphenanthrene	$C_{22}H_{14}$	278.35	222-93-5	257	2992		4-05-00-02721		H_2O 1; EtOH 2; eth 2; bz 3	
8997	Pentasiloxane, dodecamethyl-	Dodecamethylpentasiloxane	$C_{12}H_{36}O_4Si_5$	384.84	141-63-9	-80	3401	$232; 105^{12}$	4-04-00-04124	0.8755^{20}	ctc 3; CS_2 3	1.3925^{20}
8998	1,2,3,5,6-Pentathiepane	Lenthionine	$C_2H_4S_5$	188.38	292-46-6	60.5	5321		5-19-12-00251			
8999	Pentatriacontane		$C_{35}H_{72}$	492.96	630-07-9	75		490	4-01-00-00598	0.8157^{20}	H_2O 1; eth 2; ace 3	1.4568^{20}
9000	18-Pentatriacontanone		$C_{35}H_{70}O$	506.94	504-53-0	89.0		$270^{0.1}$	4-01-00-03413	0.793^{95}	H_2O 1; EtOH 2; eth 2; ace 2	
9001	4-Pentenal	Pent-4-en-1-al	C_5H_8O	84.12	2100-17-6			99	4-01-00-03459	0.852^{20}	H_2O 1; eth 3; ace 3	1.4191^{20}
9002	2-Pentenal, 2-methyl-		$C_6H_{10}O$	98.14	623-36-9			136.5	4-01-00-03471	0.8581^{20}	H_2O 1; EtOH 3; eth 3; bz 3	1.4488^{20}
9003	4-Pentenamide, 2,2-diethyl-	Novonal	$C_9H_{17}NO$	155.24	512-48-1	75.5	6643		3-02-00-01351		eth 4; EtOH 4	

No.	Name Synonym	Mol. Form. Mol. Wt.	CAS RN mp/°C	Merck No. bp/°C	Beil. Ref. den/g cm^{-3}	Solubility n_D
9004	1-Pentene α-Amylene	C_5H_{10} 70.13	109-67-1 -165.2	7079 29.9	4-01-00-00808 0.6405^{20}	H_2O 1; EtOH 5; eth 5; bz 3 1.3715^{20}
9005	2-Pentene, (*E*)- *trans*-2-Pentene	C_5H_{10} 70.13	646-04-8 -140.2	7080 36.3	4-01-00-00814 0.6431^{25}	H_2O 1; EtOH 5; eth 5; bz 3 1.3793^{20}
9006	2-Pentene, (*Z*)- *cis*-2-Pentene	C_5H_{10} 70.13	627-20-3 -151.4	7080 36.9	4-01-00-00814 0.6556^{20}	H_2O 1; EtOH 5; eth 5; bz 3 1.3830^{20}
9007	1-Pentene, 1-bromo- 1-Bromo-1-pentene	C_5H_9Br 149.03	60468-23-7	 119	3-01-00-00775 1.2606^{20}	bz 4; eth 4; chl 4 1.4572^{20}
9008	1-Pentene, 2-bromo- 2-Bromo-1-pentene	C_5H_9Br 149.03	31844-95-8	 107.5	3-01-00-00775 1.228^{20}	bz 4; eth 4; lig 4 1.4535^{20}
9009	1-Pentene, 3-bromo- 1-Ethylallyl bromide	C_5H_9Br 149.03	53045-71-9	 30.5^{30}	2-01-00-00183 1.2417^{25}	ace 4; bz 4; chl 4 1.4626^{25}
9010	1-Pentene, 5-bromo-	C_5H_9Br 149.03	1119-51-3	 125.5	3-01-00-00775 1.2581^{20}	 1.4640^{20}
9011	2-Pentene, 1-bromo-	C_5H_9Br 149.03	20599-27-3	 123.5	3-01-00-00783 1.2545^{20}	ace 4; bz 4; chl 4 1.4731^{20}
9012	2-Pentene, 2-bromo- 2-Bromo-2-pentene	C_5H_9Br 149.03	80204-19-9	 111	3-01-00-00784 1.275^{20}	ace 4; bz 4; chl 4 1.4580^{20}
9013	2-Pentene, 3-bromo- 3-Bromo-2-pentene	C_5H_9Br 149.03	21964-23-8	 116	3-01-00-00784 1.271^{20}	ace 4; bz 4; chl 4 1.4628^{20}
9014	2-Pentene, 4-bromo-	C_5H_9Br 149.03	1809-26-3	 117 dec; 22^{9}	4-01-00-00816 1.2312^{21}	ace 4; bz 4; chl 4 1.4752^{21}
9015	2-Pentene, 5-bromo- 3-Pentenyl bromide	C_5H_9Br 149.03	51952-42-2	 121.7^{621}	4-01-00-00816 1.2715^{20}	ace 4; bz 4; chl 4 1.4695^{20}
9016	1-Pentene, 2-chloro- 2-Chloro-1-pentene	C_5H_9Cl 104.58	42131-85-1	 96	3-01-00-00774 0.872^{5}	eth 4; EtOH 4
9017	1-Pentene, 3-chloro- 1-Ethylallyl chloride	C_5H_9Cl 104.58	24356-00-1	 93	3-01-00-00774 0.8978^{20}	ace 4; eth 4; EtOH 4 1.4254^{20}
9018	1-Pentene, 4-chloro- 4-Chloro-1-pentene	C_5H_9Cl 104.58	10524-08-0	 97	2-01-00-00183 0.934^{15}	H_2O 1; eth 4; chl 4 1.417^{15}
9019	1-Pentene, 5-chloro-	C_5H_9Cl 104.58	928-50-7	 105	4-01-00-00810 0.9125^{20}	ace 4; eth 4 1.4297^{20}
9020	2-Pentene, 1-chloro- γ-Ethylallyl chloride	C_5H_9Cl 104.58	10071-60-0	 108	3-01-00-00782 0.908^{22}	ace 4; eth 4; EtOH 4; chl 4 1.4352^{22}
9021	2-Pentene, 2-chloro- 2-Chloro-2-pentene	C_5H_9Cl 104.58	67747-70-0	 96	3-01-00-00782 0.9067^{20}	ace 4; eth 4 1.4261^{20}
9022	2-Pentene, 3-chloro- 3-Chloro-2-pentene	C_5H_9Cl 104.58	34238-52-3	 91	3-01-00-00782 0.9125^{20}	ace 4; eth 4 1.423^{24}
9023	2-Pentene, 4-chloro-	C_5H_9Cl 104.58	1458-99-7	 103; 47^{25}	4-01-00-00816 0.8988^{20}	ace 4; eth 4; chl 4 1.4322^{20}
9024	2-Pentene, 5-chloro-	C_5H_9Cl 104.58	16435-50-0	 108	4-01-00-00816 0.9043^{20}	ace 4; bz 4; chl 4 1.4310^{20}
9025	1-Pentene, 3-chloro-2-methyl-	$C_6H_{11}Cl$ 118.61	4104-01-2	 122.5	3-01-00-00809	bz 4; eth 4; chl 4 1.4422^{20}
9026	2-Pentene, 5-chloro-2-methyl- 5-Chloro-2-methyl-2-pentene	$C_6H_{11}Cl$ 118.61	7712-60-9	 133	4-01-00-00843 0.9135^{20}	ace 4; chl 4
9027	1-Pentene, 1,1,2,3,3,4,4,5,5,5-decafluoro-	C_5F_{10} 250.04	376-87-4	 30	4-01-00-00810	chl 4 1.2571^{25}
9028	2-Pentene, 2,5-dichloro- 1,4-Dichloro-3-pentene	$C_5H_8Cl_2$ 139.02	20177-02-0	 $40\text{-}1^{8}$	4-01-00-00816 1.1182^{15}	chl 4
9029	1-Pentene, 2,3-dimethyl- 2,3-Dimethyl-1-pentene	C_7H_{14} 98.19	3404-72-6 -134.3	 84.3	3-01-00-00832 0.7051^{20}	H_2O 1; EtOH 5; eth 5; dil sulf 4 1.4033^{20}
9030	1-Pentene, 2,4-dimethyl- 2,4-Dimethyl-1-pentene	C_7H_{14} 98.19	2213-32-3 -124.1	 81.6	4-01-00-00871 0.6943^{20}	H_2O 1; EtOH 5; eth 5; bz 3 1.3986^{20}
9031	1-Pentene, 3,3-dimethyl- 3,3-Dimethyl-1-pentene	C_7H_{14} 98.19	3404-73-7 -134.3	 77.5	4-01-00-00873 0.6974^{20}	H_2O 1; EtOH 5; eth 5; bz 3 1.3984^{20}
9032	1-Pentene, 3,4-dimethyl- 3,4-Dimethyl-1-pentene	C_7H_{14} 98.19	7385-78-6	 80.8	4-01-00-00871 0.6934^{25}	 1.3992^{20}
9033	1-Pentene, 4,4-dimethyl-	C_7H_{14} 98.19	762-62-9 -136.6	 72.5	4-01-00-00869 0.6827^{20}	H_2O 1; EtOH 5; eth 5; bz 3 1.3818^{20}
9034	2-Pentene, 2,3-dimethyl- Ethyltrimethylethylene	C_7H_{14} 98.19	10574-37-5 -118.3	 97.5	4-01-00-00870 0.7277^{20}	H_2O 1; EtOH 3; eth 3; bz 3 1.4208^{20}
9035	2-Pentene, 2,4-dimethyl- 2,4-Dimethyl-2-pentene	C_7H_{14} 98.19	625-65-0 -127.7	 83.4	4-01-00-00872 0.6954^{20}	H_2O 1; EtOH 3; eth 3; bz 3 1.4040^{20}
9036	2-Pentene, 3,4-dimethyl- 3,4-Dimethyl-2-pentene	C_7H_{14} 98.19	24910-63-2 -124.2	 89	3-01-00-00833 0.7126^{20}	H_2O 1; EtOH 3; eth 3; bz 3 1.4070^{20}
9037	2-Pentene, 3,4-dimethyl-, (*E*)-	C_7H_{14} 98.19	4914-92-5 -124.2	 91.5	4-01-00-00871 0.7124^{25}	 1.4128^{20}
9038	2-Pentene, 3,4-dimethyl-, (*Z*)-	C_7H_{14} 98.19	4914-91-4 -113.4	 89.3	4-01-00-00871 0.7092^{25}	 1.4104^{20}
9039	2-Pentene, 4,4-dimethyl-, (*E*)-	C_7H_{14} 98.19	690-08-4 -115.2	 76.7	4-01-00-00868 0.6889^{20}	H_2O 1; EtOH 3; eth 3; bz 3 1.3982^{20}
9040	2-Pentene, 4,4-dimethyl-, (*Z*)-	C_7H_{14} 98.19	762-63-0 -135.4	 80.4	4-01-00-00868 0.6951^{25}	 1.4026^{20}
9041	2-Pentenedioic acid, diethyl ester Diethyl glutaconate	$C_9H_{14}O_4$ 186.21	2049-67-4	 237	 1.0496^{20}	eth 4; EtOH 4 1.4411^{20}
9042	2-Pentenedioic acid, diethyl ester, (*E*)- *trans*-2-Pentenedioic acid diethyl ester	$C_9H_{14}O_4$ 186.21	73178-43-5	 237; 126^{14}	4-02-00-02227 1.0496^{20}	eth 4; EtOH 4 1.4411^{20}

No.	Name Synonym	Mol. Form. Mol. Wt.	CAS RN mp/°C	Merck No. bp/°C	Beil. Ref. den/g cm^{-3}	Solubility n_D
9043	1-Pentene, 2-ethyl- 2-Ethyl-1-pentene	C_7H_{14} 98.19	3404-71-5	94	4-01-00-00867 0.7079^{20}	bz 4; eth 4; EtOH 4 1.405^{20}
9044	1-Pentene, 3-ethyl- 3-Ethyl-1-pentene	C_7H_{14} 98.19	4038-04-4 -127.5	84.1	4-01-00-00867 0.6917^{25}	1.3982^{20}
9045	2-Pentene, 3-ethyl- 3-Ethyl-2-pentene	C_7H_{14} 98.19	816-79-5	96	4-01-00-00868 0.7204^{20}	H_2O 1; EtOH 3; eth 3; bz 3 1.4148^{20}
9046	1-Pentene, 3-ethyl-2-methyl-	C_8H_{16} 112.22	19780-66-6 -112.9	109.5	4-01-00-00891 0.7262^{20}	1.4140^{20}
9047	1-Pentene, 3-ethyl-3-methyl- 3-Ethyl-3-methyl-1-pentene	C_8H_{16} 112.22	6196-60-7	112	4-01-00-00891 0.7264^{20}	1.418^{20}
9048	1-Pentene, 3-ethyl-4-methyl- 1-Pentene, 4-methyl-3-ethyl	C_8H_{16} 112.22	61847-80-1	107.5	4-01-00-00890 0.7158^{25}	1.4097^{20}
9049	2-Pentene, 3-ethyl-2-methyl- 3-Ethyl-2-methyl-2-pentene	C_8H_{16} 112.22	19780-67-7	115	4-01-00-00890 0.735^{25}	1.4247^{20}
9050	2-Pentene, 3-ethyl-4-methyl-, (*E*)- 2-Pentene, 4-methyl-3-ethyl (*trans*)	C_8H_{16} 112.22	42067-49-2	114.3	4-01-00-00890 0.7308^{25}	1.4210^{20}
9051	2-Pentene, 3-ethyl-4-methyl-, (*Z*)- 2-Pentene, 4-methyl-3-ethyl (*cis*)	C_8H_{16} 112.22	42067-48-1	116	4-01-00-00890 0.735^{25}	1.424^{20}
9052	1-Pentene, 2-methyl- 2-Methyl-1-pentene	C_6H_{12} 84.16	763-29-1 -135.7	62.1	4-01-00-00841 0.6799^{20}	H_2O 1; EtOH 3; bz 3; ctc 2 1.3920^{20}
9053	1-Pentene, 3-methyl- *sec*-Butylethene	C_6H_{12} 84.16	760-20-3 -153	54.2	3-01-00-00812 0.6675^{20}	H_2O 1; EtOH 3; bz 3; chl 3 1.3841^{20}
9054	1-Pentene, 4-methyl- Isobutylethene	C_6H_{12} 84.16	691-37-2 -153.6	53.9	4-01-00-00846 0.6642^{20}	H_2O 1; EtOH 3; bz 3; chl 3 1.3828^{20}
9055	2-Pentene, 2-methyl- 2-Methyl-2-pentene	C_6H_{12} 84.16	625-27-4 -135	67.3	4-01-00-00842 0.6863^{20}	H_2O 1; EtOH 3; bz 3; ctc 3 1.4004^{20}
9056	2-Pentene, 3-methyl-, (*E*)-	C_6H_{12} 84.16	616-12-6 -138.5	70.4	4-01-00-00848 0.6930^{25}	H_2O 1; EtOH 3; bz 3; ctc 3 1.4045^{20}
9057	2-Pentene, 3-methyl-, (*Z*)- (*cis*)-3-Methyl-2-pentene	C_6H_{12} 84.16	922-62-3 -134.8	67.7	4-01-00-00848 0.6886^{25}	H_2O 1; EtOH 3; bz 3; chl 3 1.4016^{20}
9058	2-Pentene, 4-methyl-, (*E*)- 4-Methyl-*trans*-2-pentene	C_6H_{12} 84.16	674-76-0 -140.8	58.6	4-01-00-00844 0.6686^{20}	H_2O 1; EtOH 3; bz 3; ctc 2 1.3889^{20}
9059	2-Pentene, 4-methyl-, (*Z*)- *cis*-4-Methyl-2-pentene	C_6H_{12} 84.16	691-38-3 -134.8	56.3	4-01-00-00844 0.6690^{20}	H_2O 1; EtOH 3; bz 3; chl 3 1.3800^{20}
9060	1-Pentene, 3-methylene-	C_6H_{10} 82.15	3404-63-5	75	4-01-00-01022 0.712^{25}	1.445^{20}
9061	3-Pentenenitrile, (*E*) *trans*-3-Pentenenitrile	C_5H_7N 81.12	16545-78-1	142.6		
9062	4-Pentenenitrile	C_5H_7N 81.12	592-51-8	140	4-02-00-01543 0.8239^{24}	H_2O 1; EtOH 5; eth 5 1.4213^{14}
9063	3-Pentenenitrile, 2-hydroxy-, (*Z*)-	C_5H_7NO 97.12	116908-78-2	139^{70}	4-03-00-01001 0.9675^{15}	bz 4; eth 4; EtOH 4; chl 4 1.4460^{21}
9064	Pentene, 2,4,4-trimethyl- Diisobutylene	C_8H_{16} 112.22	25167-70-8	101	4-01-00-00892	
9065	1-Pentene, 2,3,3-trimethyl- 2,3,3-Trimethyl-1-pentene	C_8H_{16} 112.22	560-23-6 -69	108.3	4-01-00-00894 0.7308^{25}	1.4174^{20}
9066	1-Pentene, 2,3,4-trimethyl- 2,3,4-Trimethyl-1-pentene	C_8H_{16} 112.22	565-76-4	108	4-01-00-00894 0.725^{25}	1.415^{20}
9067	1-Pentene, 2,4,4-trimethyl- 2,2,4-Trimethyl-4-pentene	C_8H_{16} 112.22	107-39-1 -93.5	101.4	4-01-00-00892 0.7150^{20}	H_2O 1; eth 3; bz 3; ctc 3 1.4086^{20}
9068	1-Pentene, 3,3,4-trimethyl-	C_8H_{16} 112.22	560-22-5	105	3-01-00-00851 0.725^{25}	1.4144^{20}
9069	1-Pentene, 3,4,4-trimethyl- 3,4,4-Trimethyl-1-pentene	C_8H_{16} 112.22	564-03-4	104	4-01-00-00891 0.715^{25}	1.412^{20}
9070	2-Pentene, 2,3,4-trimethyl- 2,3,4-Trimethyl-2-pentene	C_8H_{16} 112.22	565-77-5 -113.4	116.5	4-01-00-00894 0.7434^{20}	1.4274^{20}
9071	2-Pentene, 2,4,4-trimethyl- 2,4,4-Trimethyl-2-pentene	C_8H_{16} 112.22	107-40-4 -106.3	104.9	4-01-00-00891 0.7218^{20}	H_2O 1; eth 3; bz 3; ctc 3 1.4160^{20}
9072	2-Pentene, 3,4,4-trimethyl- 3,4,4-Trimethyl-2-pentene	C_8H_{16} 112.22	598-96-9	112	4-01-00-00891 0.735^{25}	EtOH 4 1.423^{20}
9073	4-Pentenoic acid Allylacetic acid	$C_5H_8O_2$ 100.12	591-80-0 -22.5	188.5	4-02-00-01542 0.9809^{20}	H_2O 2; EtOH 4; eth 4 1.4281^{20}
9074	4-Pentenoic acid, 2-acetyl-, ethyl ester	$C_9H_{14}O_3$ 170.21	610-89-9	208	3-03-00-01314 0.9898^{20}	EtOH 5; eth 5; bz 5 1.4388^{18}
9075	4-Pentenoic acid, 2,2-dimethyl- 2,2-Dimethylpent-4-enoic acid	$C_7H_{12}O_2$ 128.17	16386-93-9	116^{30}	4-02-00-01585 0.933^{20}	1.4338^{20}
9076	2-Pentenoic acid, 2-methyl-, (*E*)- 2-Pentenoic acid, 2-methyl (*trans*)	$C_6H_{10}O_2$ 114.14	16957-70-3 24.4	214; 112^{12}	4-02-00-01568 0.9751^{20}	H_2O 2; eth 3; chl 3; CS_2 3 1.4513^{20}
9077	2-Pentenoic acid, 3-methyl-, (*Z*)- 2-Pentenoic acid, 3-methyl (*cis*)	$C_6H_{10}O_2$ 114.14	19866-51-4 12	96^{5}	4-02-00-01571 0.9830^{20}	1.4650
9078	2-Pentenoic acid, 4-methyl- 4,4-Dimethyl-2-butenoic acid	$C_6H_{10}O_2$ 114.14	10321-71-8 35	217	3-02-00-01324 0.9529^{21}	ace 4; eth 4; EtOH 4 1.4489^{21}
9079	3-Pentenoic acid, 2-methyl-	$C_6H_{10}O_2$ 114.14	37674-63-8	199	3-02-00-01324 0.966^{15}	1.4402^{25}
9080	2-Pentenoic acid, 4-methyl-, ethyl ester Ethyl 4,4-dimethylcrotonate	$C_8H_{14}O_2$ 142.20	2351-97-5	174; $76\text{-}7^{25}$	3-02-00-01324 0.9048^{21}	1.4328^{20}
9081	3-Pentenoic acid, 3-methyl-, ethyl ester Ethyl 3-methyl-3-pentenoate	$C_8H_{14}O_2$ 142.20	86623-80-5	61^{13}	3-02-00-01328 0.9183^{20}	1.4298^{20}
9082	3-Pentenoic acid, 4-methyl-, ethyl ester Ethyl 4-methyl-3-pentenoate	$C_8H_{14}O_2$ 142.20	6849-18-9	58^{11}	3-02-00-01324 0.9134^{17}	1.4329^{17}

No.	Name Synonym	Mol. Form. Mol. Wt.	CAS RN mp/°C	Merck No. bp/°C	Beil. Ref. den/g cm^{-3}	Solubility n_D
9083	2-Pentenoic acid, 2-methyl-, methyl ester	$C_7H_{12}O_2$ 128.17	10478-12-3	51^{11}	4-02-00-01568 0.920^{25}	1.4336
9084	2-Pentenoic acid, 3-methyl-, methyl ester, (*E*)-	$C_7H_{12}O_2$ 128.17	17447-01-7	79^{50}	4-02-00-01571 0.9302^{20}	1.4446^{20}
9085	2-Pentenoic acid, 3-methyl-, methyl ester, (*Z*)- *cis*-3-Methyl-2-pentenoic acid methyl ester	$C_7H_{12}O_2$ 128.17	17447-00-6	74^{50}	4-02-00-01571 0.9279^{20}	1.4420^{20}
9086	3-Pentenoic acid, 3-methyl-, methyl ester Methyl 3-methyl-3-pentenoate	$C_7H_{12}O_2$ 128.17	2258-58-4	74^{50}	4-02-00-01573 0.9949^{20}	1.4306^{20}
9087	1-Penten-3-ol	$C_5H_{10}O$ 86.13	616-25-1	115	1-01-00-00227 0.839^{20}	H_2O 2; EtOH 5; eth 5 1.4239^{20}
9088	4-Penten-1-ol	$C_5H_{10}O$ 86.13	821-09-0	141	4-01-00-02119 0.8457^{20}	H_2O 2; eth 3; ctc 2 1.4309^{20}
9089	4-Penten-2-ol	$C_5H_{10}O$ 86.13	625-31-0	116	1-01-00-00227 0.8367^{20}	H_2O 4; EtOH 5; eth 5 1.4225^{20}
9090	2-Penten-1-ol, (*E*)-	$C_5H_{10}O$ 86.13	1576-96-1	138	4-01-00-02121 0.8471^{20}	EtOH 3; eth 3; ace 3 1.4341^{20}
9091	2-Penten-1-ol, (*Z*)-	$C_5H_{10}O$ 86.13	1576-95-0	138	4-01-00-02121 0.8529^{20}	EtOH 3; eth 3; ace 3 1.4354^{20}
9092	3-Penten-2-ol, (±)-	$C_5H_{10}O$ 86.13	42569-16-4	121.6; 65^{70}	4-01-00-02122 0.8328^{25}	ace 4; eth 4; EtOH 4 1.4280^{20}
9093	2-Penten-1-ol, 5-(2,3-dimethyltricyclo[2.2.1.02,6]hept-3-yl)-2-methyl-, stereoisomer α-Santalol	$C_{15}H_{24}O$ 220.35	115-71-9	8321 301.5	4-06-00-03521 0.9679^{20}	H_2O 1; EtOH 3 1.5023^{20}
9094	2-Penten-1-ol, 2-methyl-	$C_6H_{12}O$ 100.16	1610-29-3	167.5	4-01-00-02144 0.8501^{24}	1.444^{24}
9095	2-Penten-1-ol, 4-methyl-	$C_6H_{12}O$ 100.16	5362-55-0	159	4-01-00-02146 0.8489^{16}	1.4403^{16}
9096	3-Penten-1-ol, 4-methyl- 4-Methyl-3-pentenol	$C_6H_{12}O$ 100.16	763-89-3	157	4-01-00-02145 0.8600^{20}	1.4432^{21}
9097	3-Penten-2-ol, 2-methyl- 2-Methyl-3-penten-2-ol	$C_6H_{12}O$ 100.16	63468-05-3	112; 65^{72}	4-01-00-02145 0.8347^{20}	1.4302^{20}
9098	3-Penten-2-ol, 3-methyl- 3-Methyl-3-penten-2-ol	$C_6H_{12}O$ 100.16	2747-53-7	140	4-01-00-02149 0.8792^{0}	1.4428^{17}
9099	3-Penten-2-ol, 4-methyl-	$C_6H_{12}O$ 100.16	4325-82-0	134	4-01-00-02144 0.840^{15}	1.9377^{15}
9100	4-Penten-2-ol, 2-methyl-	$C_6H_{12}O$ 100.16	624-97-5 -73	119.5	4-01-00-02146 0.8300^{20}	H_2O 2; EtOH 3; eth 3; ctc 2 1.4268^{20}
9101	4-Penten-2-ol, 3-methyl-	$C_6H_{12}O$ 100.16	1569-59-1	125.5	3-01-00-01933 0.8429^{22}	1.4326^{22}
9102	4-Penten-2-ol, 4-methyl-	$C_6H_{12}O$ 100.16	2004-67-3	129	4-01-00-02143 0.8436^{20}	1.4297^{17}
9103	2-Penten-1-ol, 2-methyl-5-(2-methyl-3-methylenebicyclo[2.2.1]hept-2-yl)-, [1*S*-[1α,2α(*Z*),4α] β-Santalol	$C_{15}H_{24}O$ 220.35	77-42-9	8322 $167\text{-}8^{10}$	4-06-00-03519 0.9750^{20}	1.5115^{20}
9104	1-Penten-3-one Ethyl vinyl ketone	C_5H_8O 84.12	1629-58-9	103; 44^{90}	4-01-00-03457 0.8468^{20}	H_2O 1; EtOH 3; eth 3; ace 3 1.4195^{20}
9105	3-Penten-2-one, (*E*)-	C_5H_8O 84.12	3102-33-8	122	4-01-00-03460 0.8624^{20}	H_2O 3; eth 3; ace 3; ctc 3 1.4350^{20}
9106	1-Penten-3-one, 4,4-dimethyl-1-phenyl-	$C_{13}H_{16}O$ 188.27	538-44-3 43	154^{25}	3-07-00-01490 0.9508^{46}	1.5523^{25}
9107	1-Penten-3-one, 2-methyl-	$C_6H_{10}O$ 98.14	25044-01-3 -69.5	118.5	4-01-00-03470 1.8530^{20}	ace 4; EtOH 4 1.4289^{20}
9108	3-Penten-2-one, 4-methyl- Mesityl oxide	$C_6H_{10}O$ 98.14	141-79-7 -59	5811 130	4-01-00-03471 0.8653^{20}	H_2O 3; EtOH 5; eth 5; ace 3 1.4440^{20}
9109	4-Penten-2-one, 4-methyl-	$C_6H_{10}O$ 98.14	3744-02-3 -72.6	124.2	4-01-00-03470 0.8411^{20}	
9110	1-Penten-3-one, 1-phenyl-	$C_{11}H_{12}O$ 160.22	3152-68-9 38.5	142^{12}	2-07-00-00298 0.8697^{20}	H_2O 2; EtOH 4; eth 4; bz 4 1.5684^{20}
9111	1-Penten-3-yne Methylvinylacetylene	C_5H_6 66.10	646-05-9	59.5	4-01-00-01087 0.7401^{20}	bz 4; eth 4 1.4496^{20}
9112	1-Penten-4-yne	C_5H_6 66.10	871-28-3	42.5	4-01-00-01088 0.738^{16}	H_2O 1; eth 3; bz 3 1.4125^{16}
9113	3-Penten-1-yne, (*E*)- *trans*-3-Penten-1-yne	C_5H_6 66.10	2004-69-5	52.2	4-01-00-01085	
9114	3-Penten-1-yne, (*Z*)- *cis*-3-Penten-1-yne	C_5H_6 66.10	1574-40-9	44.6		
9115	3-Penten-1-yne, 3-ethyl-	C_7H_{10} 94.16	14272-54-9	96.5	4-01-00-01099 0.7886^{25}	bz 4; eth 4 1.4338^{25}
9116	1-Penten-3-yne, 2-methyl- Methylisopropenylacetylene	C_6H_8 80.13	926-55-6	$81\text{-}2^{100}$	4-01-00-01095	bz 4; eth 4 1.4002^{20}
9117	3-Penten-1-yne, 3-methyl- 3-Methyl-3-penten-1-yne	C_6H_8 80.13	1574-33-0	66.5	4-01-00-01096 0.739^{20}	eth 3; bz 3 1.4332^{20}
9118	4-Penten-2-yn-1-ol 1-Hydroxypent-2-yn-4-ene	C_5H_6O 82.10	2919-05-3	57^{10}; $44^{2.5}$	4-01-00-02303 0.9581^{20}	1.4986^{20}
9119	Pentitol, 1,2-dideoxy- 1,2,3-Pentanetriol	$C_5H_{12}O_3$ 120.15	5371-48-2	192^{63}	4-01-00-02778 1.0849^{34}	H_2O 4; EtOH 4; eth 4
9120	Pentonic acid, 5-C-[3,5-bis(1-methylpropyl)-1-cyclopenten-1-yl]-4-deoxy-	$C_{18}H_{32}O_5$ 328.45	491-14-5 196		3-10-00-02045 1.292^{14}	EtOH 4

No.	Name Synonym	Mol. Form. Mol. Wt.	CAS RN mp/°C	Merck No. bp/°C	Beil. Ref. den/g cm^{-3}	Solubility n_D
9121	1-Pentyn-3-amine, 3-ethyl-	$C_7H_{13}N$ 111.19	3234-64-8	$71-2^{90}$	4-04-00-01154 0.828^{25}	1.4409^{20}
9122	1-Pentyne Propylacetylene	C_5H_8 68.12	627-19-0 -90	40.1	4-01-00-00990 0.6901^{20}	H_2O 1; EtOH 4; eth 5; bz 3 1.3852^{20}
9123	2-Pentyne	C_5H_8 68.12	627-21-4 -109.3	56.1	4-01-00-00992 0.7058^{25}	H_2O 1; EtOH 4; eth 5; bz 3 1.4039^{20}
9124	1-Pentyne, 1-bromo- 1-Pentynyl bromide	C_5H_7Br 147.01	14752-60-4	$44-6^{52}$	3-01-00-00958 1.281^{13}	1.4579^{13}
9125	1-Pentyne, 3-chloro-3-ethyl- 3-Chloro-3-ethyl-1-pentyne	$C_7H_{11}Cl$ 130.62	6080-79-1	$73-6^{100}$	4-01-00-01032 0.9230^{19}	bz 4; eth 4; chl 4 1.4437^{19}
9126	1-Pentyne, 3-chloro-3-methyl- 3-Chloro-3-methyl-1-pentyne	C_6H_9Cl 116.59	14179-94-3	102.3	4-01-00-01021 0.9163^{20}	H_2O 1; eth 3; bz 3; chl 3 1.4330^{20}
9127	2-Pentyne, 4-chloro-4-methyl-	C_6H_9Cl 116.59	999-79-1	55^{70}	4-01-00-01018	ace 4; bz 4; chl 4 1.4143^{20}
9128	1-Pentyne, 3,3-dimethyl- 3,3-Dimethyl-1-pentyne	C_7H_{12} 96.17	918-82-1	70	4-01-00-01034 0.7032^{25}	1.3934^{20}
9129	1-Pentyne, 4,4-dimethyl- 4,4-Dimethyl-1-pentyne	C_7H_{12} 96.17	13361-63-2 -75.7	76.1	4-01-00-01033 0.7142^{20}	bz 4; eth 4; chl 4 1.3983^{20}
9130	2-Pentyne, 4,4-dimethyl-	C_7H_{12} 96.17	999-78-0 -82.4	83	4-01-00-01032 0.7176^{20}	H_2O 1; eth 3; bz 3; ctc 2 1.4071^{20}
9131	1-Pentyne, 3-ethyl- 3-Ethyl-1-pentyne	C_7H_{12} 96.17	21020-26-8	84	3-01-00-01002 0.7246^{25}	bz 4; eth 4; chl 4 1.4043^{25}
9132	1-Pentyne, 3-ethyl-3-methyl- 3-Ethyl-3-methyl-1-pentyne	C_8H_{14} 110.20	919-12-0	101.5	4-01-00-01046 0.7422^{20}	bz 4; eth 4; chl 4 1.4110^{20}
9133	1-Pentyne, 1-iodo- 1-Iodo-1-pentyne	C_5H_7I 194.02	14752-61-5	54^{23}	4-01-00-00992 1.6127^{19}	1.5148^{19}
9134	1-Pentyne, 3-methyl- 3-Methyl-1-pentyne	C_6H_{10} 82.15	922-59-8	57.7	4-01-00-01021 0.6992^{25}	1.3916^{20}
9135	1-Pentyne, 4-methyl-	C_6H_{10} 82.15	7154-75-8 -104.6	61.2	4-01-00-01019 0.7000^{25}	H_2O 1; bz 3; chl 3 1.3936^{20}
9136	2-Pentyne, 4-methyl-	C_6H_{10} 82.15	21020-27-9 -110.3	73.1	4-01-00-01018 0.7112^{25}	bz 4; chl 4 1.4057^{20}
9137	2-Pentynoic acid Ethylpropiolic acid	$C_5H_6O_2$ 98.10	5963-77-9 50	122^{10}; 81^{3}	4-02-00-01693 0.978^{20}	H_2O 4 1.4619^{20}
9138	4-Pentynoic acid Propargylacetic acid	$C_5H_6O_2$ 98.10	6089-09-4 57.7	110^{30}; 102^{17}	4-02-00-01693	eth 4; EtOH 4
9139	2-Pentynoic acid, methyl ester Methyl 2-pentynoate	$C_6H_8O_2$ 112.13	24342-04-9	72^{10}	4-02-00-01694 0.937^{20}	1.4455^{20}
9140	1-Pentyn-3-ol Ethylethynylcarbinol	C_5H_8O 84.12	4187-86-4	125	4-01-00-02224 0.8926^{15}	1.4347^{15}
9141	2-Pentyn-1-ol	C_5H_8O 84.12	6261-22-9	$61-2^{15}$	4-01-00-02226 0.913^{17}	1.4518^{17}
9142	3-Pentyn-1-ol	C_5H_8O 84.12	10229-10-4	154	4-01-00-02227 0.9002^{20}	1.4454^{20}
9143	4-Pentyn-1-ol	C_5H_8O 84.12	5390-04-5	154	4-01-00-02225 0.913^{20}	1.4414^{20}
9144	1-Pentyn-3-ol, 3,4-dimethyl-	$C_7H_{12}O$ 112.17	1482-15-1	134	4-01-00-02255 0.8691^{20}	H_2O 4; eth 4; EtOH 4 1.4372^{20}
9145	1-Pentyn-3-ol, 3-ethyl-	$C_7H_{12}O$ 112.17	6285-06-9	138	4-01-00-02254 0.8691^{20}	H_2O 2
9146	1-Pentyn-3-ol, 3-methyl- Meparfynol	$C_6H_{10}O$ 98.14	77-75-8 30.5	5731 120.5	2-01-00-00506 0.8688^{20}	1.4310^{20}
9147	2-Pentynoyl chloride, 4,4-dimethyl- 4,4-Dimethyl-2-pentynoyl chloride	C_7H_9ClO 144.60	52324-03-5	52^{20}	2-02-00-00454 0.9743^{20}	1.4443^{20}
9148	Perchloric acid, trichloromethyl ester Trichloro perchlorate	CCl_4O_4 217.82	67632-66-0 -55	exp 40	3-03-00-00037	eth 4
9149	Perfluidone *o*-Toluidine, 4-(phenylsulfonyl)-*N*-[(trifluoromethyl)sulfono]-	$C_{14}H_{12}F_3NO_4S_2$ 379.38	37924-13-3 143	7113		
9150	Permethrin Cyclopropanecarboxylic acid, 3-(2,2-dichloroethenyl)-2,2-dimethyl-, (3-phenoxyphenyl), methyl ester	$C_{21}H_{20}Cl_2O_3$ 391.29	52645-53-1 34	7132 $200^{0.01}$	1.23^{20}	
9151	Peroxide, acetyl benzoyl Acetozone	$C_9H_8O_4$ 180.16	644-31-5 37	71 130^{19}	4-09-00-00717	eth 4
9152	Peroxide, bis(2,4-dichlorobenzoyl)	$C_{14}H_6Cl_4O_4$ 380.01	133-14-2 106		4-09-00-01001	
9153	Peroxide, bis(1,1-dimethylethyl) DTBP	$C_8H_{18}O_2$ 146.23	110-05-4 -40	3446 111	4-01-00-01619 0.704^{20}	H_2O 1; ace 5; ctc 3; os 3 1.3890^{20}
9154	Peroxide, bis(1,1-dimethylpropyl)	$C_{10}H_{22}O_2$ 174.28	10508-09-5	58.5^{14}; 44^{10}	4-01-00-01671 0.821^{20}	1.4095^{20}
9155	Peroxide, bis(1-oxododecyl) Lauroyl peroxide	$C_{24}H_{46}O_4$ 398.63	105-74-8 49		4-02-00-01102	H_2O 1; chl 3
9156	Peroxide, diacetyl Acetyl peroxide	$C_4H_6O_4$ 118.09	110-22-5 30	63^{21}	4-02-00-00392	eth 4; EtOH 4
9157	Peroxide, dibenzoyl Benzoyl peroxide	$C_{14}H_{10}O_4$ 242.23	94-36-0 105	1128 exp	4-09-00-00717	H_2O 2; EtOH 3; eth 3; ace 3 1.543
9158	Peroxide, diethyl	$C_4H_{10}O_2$ 90.12	628-37-5 -70	65	4-01-00-01323 0.8240^{19}	H_2O 2; EtOH 5; eth 5 1.3715^{17}

PHYSICAL CONSTANTS OF ORGANIC COMPOUNDS (continued)

No.	Name Synonym	Mol. Form. Mol. Wt.	CAS RN mp/°C	Merck No. bp/°C	Beil. Ref. den/g cm^{-3}	Solubility n_D
9159	Peroxide, dimethyl	$C_2H_6O_2$ 62.07	690-02-8	 14	4-01-00-01249 0.8677^{0}	EtOH 2; eth 2; tol 3; HOAc 3 1.3503^{0}
9160	Perthane Ethane, 1,1-dichloro-2,2-bis(*p*-ethylphenyl)-	$C_{18}H_{20}Cl_2$ 307.26	72-56-0 56	3050		
9161	Perylene Dibenz[de,kl]anthracene	$C_{20}H_{12}$ 252.32	198-55-0 274	7137	4-05-00-02689 1.35^{25}	H_2O 1; EtOH 2; eth 2; ace 4
9162	9-Phenanthrenamine	$C_{14}H_{11}N$ 193.25	947-73-9 138.3	 sub	4-12-00-03443	eth 2; bz 2; chl 2
9163	Phenanthrene	$C_{14}H_{10}$ 178.23	85-01-8 99.2	7167 340	4-05-00-02297 0.9800^{4}	H_2O 1; EtOH 3; eth 3; ace 3 1.5943
9164	Phenanthrene, 9-bromo- 9-Phenanthryl bromide	$C_{14}H_9Br$ 257.13	573-17-1 64.5	 >360	4-05-00-02303 1.4093^{10}	H_2O 1; EtOH 3; eth 3; chl 2
9165	9-Phenanthrenecarboxylic acid 9-Phenanthroic acid	$C_{15}H_{10}O_2$ 222.24	837-45-6 257.3	 sub	4-09-00-02674	H_2O 1; EtOH 3; eth 3; bz 3
9166	1-Phenanthrenecarboxylic acid, 1,2,3,4,4a,5,6,9,10,10a-decahydro-1,4a-dimethyl-7-(1-methylethyl)-, Palustric acid	$C_{20}H_{30}O_2$ 302.46	1945-53-5 164.5	6950	4-09-00-02174	
9167	1-Phenanthrenecarboxylic acid, 1,2,3,4,4a,4b,5,6,10,10a-decahydro-1,4a-dimethyl-7-(1-methylethyl)-, Abietic acid	$C_{20}H_{30}O_2$ 302.46	514-10-3 173.5	2 250^{9}	4-09-00-02175	ace 4; bz 4; eth 4; EtOH 4
9168	1-Phenanthrenecarboxylic acid, 1,2,3,4,4a,4b,5,6,10,10a-decahydro-1,4a-dimethyl-7-(1-methylethyl)-, methyl ester, [1R-(1α,4aβ,4bα,10aα)]- Methyl abietate	$C_{21}H_{32}O_2$ 316.48	127-25-3	5930 $225\text{-}6^{16}$	4-09-00-02176 1.049^{20}	H_2O 1; EtOH 3; HOAc 3 1.5344
9169	1-Phenanthrenecarboxylic acid, 1,2,3,4,4a,4b,5,9,10,10a-decahydro-1,4a-dimethyl-7-(1-methylethyl)-, Levopimaric acid	$C_{20}H_{30}O_2$ 302.46	79-54-9 150	5348	4-09-00-02173	
9170	1-Phenanthrenecarboxylic acid, 7-ethenyl-1,2,3,4,4a,4b,5,6,7,9,10,10a-dodecahydro-1,4a,7-trimethyl-, [1R-(1α,4aβ,4bα,7β,10aα)]- Pimaric acid	$C_{20}H_{30}O_2$ 302.46	127-27-5 218.5	7398 282^{18}	4-09-00-02179	eth 4; py 4; EtOH 4
9171	Phenanthrene, 9,10-dihydro- 9,10-Dihydrophenanthrene	$C_{14}H_{12}$ 180.25	776-35-2 34.5	 $168\text{-}9^{15}$	4-05-00-02183 1.0757^{40}	chl 3 1.6415^{20}
9172	Phenanthrene, 9,10-dimethyl- 9,10-Phenanthraquinodimethane	$C_{16}H_{14}$ 206.29	604-83-1 144	 sub	4-05-00-02335	EtOH 2; bz 3; chl 3; HOAc 3
9173	3,4-Phenanthrenediol	$C_{14}H_{10}O_2$ 210.23	478-71-7 143	 sub 130	3-06-00-05689	eth 4; EtOH 4
9174	9,10-Phenanthrenedione Phenanthrenequinone	$C_{14}H_8O_2$ 208.22	84-11-7 209	7168	4-07-00-02565 1.405^{22}	H_2O 1; EtOH 2; eth 3; bz 2
9175	9,10-Phenanthrenedione, 2-hydroxy-	$C_{14}H_8O_3$ 224.22	4088-81-7 284	 sub	3-08-00-02929	H_2O 4
9176	9,10-Phenanthrenedione, 3-hydroxy- 9,10-Phenanthraquinone, 3-hydroxy	$C_{14}H_8O_3$ 224.22	57404-54-3 330 dec	 sub	2-08-00-00398	H_2O 4; EtOH 4
9177	9,10-Phenanthrenedione, 1-methyl-7-(1-methylethyl)-	$C_{18}H_{16}O_2$ 264.32	5398-75-4 197.5	 sub	4-07-00-02595	eth 4; EtOH 4
9178	Phenanthrene, dodecahydro- Dodecahydrophenanthrene	$C_{14}H_{22}$ 190.33	1322-67-4	 $81\text{-}2^{1.5}$	3-05-00-01074 0.9674^{20}	ace 4; bz 4 1.5102^{20}
9179	Phenanthrene, 9-ethyl- 9-Ethylphenanthrene	$C_{16}H_{14}$ 206.29	3674-75-7 62.5	 $199\text{-}200^{1}$	4-05-00-02332 1.0603^{78}	H_2O 1; EtOH 3; bz 3; peth 2 1.6582^{78}
9180	Phenanthrene, 3-methyl- 3-Methylphenanthrene	$C_{15}H_{12}$ 192.26	832-71-3 65	 350; 145^{6}	4-05-00-02314	H_2O 1; EtOH 3; ace 3; chl 2
9181	Phenanthrene, 4-methyl- 4-Methylphenanthrene	$C_{15}H_{12}$ 192.26	832-64-4 53.5	 $175\text{-}80^{10}$	3-05-00-02154	H_2O 1; EtOH 3; ctc 3
9182	Phenanthrene, 9-(1-methylethenyl)- 9-Isopropenylphenanthrene	$C_{17}H_{14}$ 218.30	58873-44-2 38	 163^{20}	3-05-00-02240	 1.6765^{22}
9183	Phenanthrene, 1-methyl-7-(1-methylethyl)- Retene	$C_{18}H_{18}$ 234.34	483-65-8 101	8162 390	4-05-00-02368 1.035^{25}	H_2O 1; EtOH 3; eth 3; bz 3
9184	Phenanthrene, 1,2,3,4,4a,9,10,10a-octahydro- 1,2,3,4,4a,9,10,10-Octahydrophenanthrene	$C_{14}H_{18}$ 186.30	16306-39-1	 159^{15}	4-05-00-01585 0.9973^{32}	 1.5527^{19}
9185	Phenanthrene, 1,2,3,4,5,6,7,8-octahydro- 1,2,3,4,5,6,7,8-Octahydrophenanthrene	$C_{14}H_{18}$ 186.30	5325-97-3 16.7	 295	4-05-00-01585 1.026^{20}	H_2O 1; ace 3; bz 3; CS_2 3 1.5569^{17}
9186	Phenanthrene, tetradecahydro-	$C_{14}H_{24}$ 192.34	5743-97-5	 $86\text{-}9^{2}$	4-05-00-00492 0.9447^{20}	H_2O 1; eth 3; ace 3; bz 3 1.5011^{20}
9187	Phenanthrene, 1,2,3,4-tetrahydro- 1,2,3,4-Tetrahydrophenanthrene	$C_{14}H_{14}$ 182.27	1013-08-7 33.5	 173^{11}	4-05-00-01909 1.0601^{40}	H_2O 1; EtOH 3; eth 3; ace 3
9188	Phenanthridine	$C_{13}H_9N$ 179.22	229-87-8 107.4	 349	5-20-08-00223	H_2O 2; EtOH 4; eth 4; ace 3
9189	1,10-Phenanthroline *o*-Phenanthroline	$C_{12}H_8N_2$ 180.21	66-71-7 117	7169 >300	5-23-08-00419	H_2O 4; EtOH 3; ace 3; bz 3
9190	1,7-Phenanthroline	$C_{12}H_8N_2$ 180.21	230-46-6 78	 360	5-23-08-00417	H_2O 3; EtOH 4; eth 1; bz 1
9191	4,7-Phenanthroline	$C_{12}H_8N_2$ 180.21	230-07-9 177	 sub 100	5-23-08-00430	H_2O 3; EtOH 4; eth 2; bz 2
9192	1,10-Phenanthroline, 2,9-dimethyl- Neocuproine	$C_{14}H_{12}N_2$ 208.26	484-11-7 159.5	6364	5-23-08-00527	
9193	Phenazine Dibenzopyrazine	$C_{12}H_8N_2$ 180.21	92-82-0 176.5	7173	5-23-08-00389	H_2O 2; EtOH 3; eth 2; bz 3

No.	Name Synonym	Mol. Form. Mol. Wt.	CAS RN mp/°C	Merck No. bp/°C	Beil. Ref. den/g cm^{-3}	Solubility n_D
9194	2,3-Phenazinediamine 2,3-Diaminophenazine	$C_{12}H_{10}N_4$ 210.24	655-86-7 264	2960 sub	4-25-00-03028	bz 4; EtOH 4
9195	Phenazine, 2-methyl- Tolazine	$C_{13}H_{10}N_2$ 194.24	1016-94-0 118.5	 350	5-23-08-00475	eth 4; EtOH 4; chl 4
9196	Phenazinium, 1-hydroxy-5-methyl-, hydroxide, inner salt Pyocyanine	$C_{13}H_{10}N_2O$ 210.24	85-66-5 133 dec	7965	5-24-04-00100	H_2O 2; EtOH 3; eth 1; ace 3
9197	1-Phenazinol Hemipyocyanine	$C_{12}H_8N_2O$ 196.21	528-71-2 158	4564 sub	5-23-12-00297	H_2O 2; EtOH 2; bz 3; os 4
9198	Phenmedipham Carbamic acid, (3-methylphenyl)-, 3-[(methoxycarbonyl)amino]phenyl ester	$C_{16}H_{16}N_2O_4$ 300.31	13684-63-4 143	7199		
9199	Phenol Hydroxybenzene	C_6H_6O 94.11	108-95-2 40.9	7206 181.8	4-06-00-00531 1.0545^{45}	H_2O 3; EtOH 3; eth 4; ace 5 1.5408^{41}
9200	Phenol, amino- Hydroxyphenylamine	C_6H_7NO 109.13	27598-85-2 83.5		2-15-00-00004	H_2O 3; EtOH 4; eth 4; bz 4
9201	Phenol, 2-amino- *o*-Aminophenol	C_6H_7NO 109.13	95-55-6 174	473 sub 153	4-13-00-00805 1.328^{25}	H_2O 3; EtOH 4; eth 3; bz 2
9202	Phenol, 3-amino- *m*-Aminophenol	C_6H_7NO 109.13	591-27-5 123	472 164^{11}	4-13-00-00952	H_2O 3; EtOH 4; eth 4; bz 2
9203	Phenol, 4-amino- *p*-Aminophenol	C_6H_7NO 109.13	123-30-8 187.5	474 $110^{0.3}$	4-13-00-01014	H_2O 2; EtOH 4; bz 1; chl 1
9204	Phenol, 2-amino-4-chloro- 2-Hydroxy-5-chloroaniline	C_6H_6ClNO 143.57	95-85-2 140		4-13-00-00878	DMSO 2
9205	Phenol, 2-amino-4,6-dichloro- 2-Amino-4,6-dichlorophenol	$C_6H_5Cl_2NO$ 178.02	527-62-8 95.5	448 sub 70	4-13-00-00889	
9206	Phenol, 4-amino-2,6-dichloro-	$C_6H_5Cl_2NO$ 178.02	5930-28-9 168	 sub	2-13-00-00274	H_2O 1; EtOH 4; eth 4; ace 3
9207	Phenol, 2-amino-4,6-dinitro- Picramic acid	$C_6H_5N_3O_5$ 199.12	96-91-3 169	7379	4-13-00-00909	bz 4; EtOH 4
9208	Phenol, 4-(2-aminoethyl)- Tyramine	$C_8H_{11}NO$ 137.18	51-67-2 164.5	9743 $205\text{-}7^{25}$	4-13-00-01788	H_2O 2; EtOH 3; bz 2; xyl 3
9209	Phenol, 2-amino-3-methyl-	C_7H_9NO 123.15	2835-97-4 150	 sub	2-13-00-00324	eth 4
9210	Phenol, 2-amino-4-methyl-	C_7H_9NO 123.15	95-84-1 136	 sub	4-13-00-01712	H_2O 2; EtOH 3; eth 3; bz 2
9211	Phenol, 3-amino-4-methyl-	C_7H_9NO 123.15	2836-00-2 157.8	 sub	4-13-00-01710	eth 4
9212	Phenol, 3-amino-5-methyl- *m*-Cresol, 5-amino-	C_7H_9NO 123.15	76619-89-1 139	 245	4-13-00-01697	EtOH 4
9213	Phenol, 4-amino-2-methyl-	C_7H_9NO 123.15	2835-96-3 176.5	 sub	4-13-00-01672	H_2O 2; EtOH 3; eth 3; bz 2
9214	Phenol, 2-amino-3-nitro-	$C_6H_6N_2O_3$ 154.13	603-85-0 216.5	 sub	4-13-00-00895	H_2O 4
9215	Phenol, 4-(2-aminopropyl)-, (±)- Hydroxyamphetamine	$C_9H_{13}NO$ 151.21	1518-86-1 125.5	4740	4-13-00-01871	H_2O 3; EtOH 3; bz 3; chl 3
9216	Phenol, 2,2'-azobis-	$C_{12}H_{10}N_2O_2$ 214.22	2050-14-8 173	 $140^{0.001}$	2-16-00-00033	H_2O 1; EtOH 2; eth 4; bz 2
9217	Phenol, 2-(2-benzothiazolyl)-	$C_{13}H_9NOS$ 227.29	3411-95-8 131	 $175\text{-}93^{3}$	4-27-00-02081	EtOH 3
9218	Phenol, 2-(2-benzoxazolyl)-	$C_{13}H_9NO_2$ 211.22	835-64-3 123.5	 338	4-27-00-02080	H_2O 2; EtOH 4; eth 3; ace 3
9219	Phenol, 2,4-bis(1,1-dimethylethyl)-	$C_{14}H_{22}O$ 206.33	96-76-4 56.5	 263.5	4-06-00-03493	ctc 2; alk 1 1.5080^{20}
9220	Phenol, 2,6-bis(1,1-dimethylethyl)-	$C_{14}H_{22}O$ 206.33	128-39-2 39	 161^{50}; 133^{20}	4-06-00-03492	EtOH 2; ctc 3; alk 1 1.5001^{20}
9221	Phenol, 2,6-bis(1,1-dimethylethyl)-4-ethyl-	$C_{16}H_{26}O$ 234.38	4130-42-1 44	 272	4-06-00-03529	alk 1
9222	Phenol, 2,4-bis(1,1-dimethylethyl)-5-methyl- DBMC	$C_{15}H_{24}O$ 220.35	497-39-2 62.1	2829 282	4-06-00-03510 0.912^{80}	H_2O 1; EtOH 3; eth 3; ace 3
9223	Phenol, 2,4-bis(1,1-dimethylethyl)-6-methyl-	$C_{15}H_{24}O$ 220.35	616-55-7 51	 269	4-06-00-03510 0.891^{80}	alk 1
9224	Phenol, 2,6-bis(1,1-dimethylethyl)-4-methyl- 2,6-Di-*tert*-butyl-4-methylphenol	$C_{15}H_{24}O$ 220.35	128-37-0 71	1548 265	4-06-00-03511 0.8937^{75}	H_2O 1; EtOH 3; ace 3; bz 3 1.4859^{75}
9225	Phenol, 2,6-bis(1,1-dimethylpropyl)-4-methyl- *p*-Cresol, 2,6-di-*tert*-pentyl-	$C_{17}H_{28}O$ 248.41	56103-67-4	 283	4-06-00-03533 0.931^{25}	 1.4950^{20}
9226	Phenol, 2,6-bis(1-methylethyl)- Propofol	$C_{12}H_{18}O$ 178.27	2078-54-8 18	7847 256	4-06-00-03435 1.5140^{20}	 1.5140^{20}
9227	Phenol, 2,6-bis(1-methylpropyl)- 2,6-Di-*sec*-butylphenol	$C_{14}H_{22}O$ 206.33	5510-99-6 -42	 257.5	4-06-00-03490	 1.5080^{20}
9228	Phenol, 2-bromo-	C_6H_5BrO 173.01	95-56-7 5.6	 194.5	4-06-00-01037 1.4924^{20}	H_2O 2; EtOH 3; eth 3; chl 2 1.589^{20}
9229	Phenol, 3-bromo-	C_6H_5BrO 173.01	591-20-8 33	 236.5	4-06-00-01042	H_2O 2; EtOH 4; eth 4; ctc 2
9230	Phenol, 4-bromo-	C_6H_5BrO 173.01	106-41-2 66.4	 238	4-06-00-01043 1.840^{15}	H_2O 3; EtOH 4; eth 4; chl 3
9231	Phenol, 3-bromo-5-chloro-	C_6H_4BrClO 207.45	56962-04-0 70	 256	2-06-00-00187	EtOH 4; HOAc 4
9232	Phenol, 4-bromo-2-chloro- 4-Bromo-2-chlorophenol	C_6H_4BrClO 207.45	3964-56-5 50.5	 233.5	3-06-00-00750 1.6170^{20}	EtOH 4; eth 4; ace 4; bz 4 1.5859^{20}

No.	Name Synonym	Mol. Form. Mol. Wt.	CAS RN mp/°C	Merck No. bp/°C	Beil. Ref. den/g cm^{-3}	Solubility n_D
9233	Phenol, 2-bromo-4,6-dichloro-	$C_6H_3BrCl_2O$ 241.90	4524-77-0 68	 220^{200}	3-06-00-00751	bz 4; eth 4; chl 4
9234	Phenol, 2-bromo-4,6-dinitro-	$C_6H_3BrN_2O_5$ 263.00	2316-50-9 119.3	 sub	3-06-00-00871	H_2O 2; EtOH 3; eth 4; bz 4
9235	Phenol, 4-bromo-2,6-dinitro-	$C_6H_3BrN_2O_5$ 263.00	40466-95-3 78	 sub	3-06-00-00872	bz 4; eth 4; EtOH 4; chl 4
9236	Phenol, 2-bromo-4-methyl-	C_7H_7BrO 187.04	6627-55-0 56.5	 213.5	4-06-00-02143 1.5422^{25}	H_2O 2; EtOH 3; bz 3; chl 3 1.5772^{20}
9237	Phenol, 2-bromo-5-methyl- 6-Bromo-*m*-cresol	C_7H_7BrO 187.04	14847-51-9 38.8	 209	3-06-00-01320	ace 4; eth 4; EtOH 4
9238	Phenol, 3-bromo-2-methyl-	C_7H_7BrO 187.04	7766-23-6 95	 55^{4}	0-06-00-00360	ace 4; bz 4; eth 4; EtOH 4
9239	Phenol, 3-bromo-4-methyl-	C_7H_7BrO 187.04	60710-39-6 56	 246	4-06-00-02143	ace 4; bz 4; eth 4; EtOH 4
9240	Phenol, 3-bromo-5-methyl-	C_7H_7BrO 187.04	74204-00-5 56.5	 161^{28}	2-06-00-00357	eth 4; EtOH 4
9241	Phenol, 4-bromo-2-methyl-	C_7H_7BrO 187.04	2362-12-1 64.7	 235	4-06-00-02006	ace 4; eth 4; EtOH 4
9242	Phenol, 4-bromo-3-methyl-	C_7H_7BrO 187.04	14472-14-1 63.5	 142^{23}	4-06-00-02072	eth 4; py 4
9243	Phenol, 2-bromo-3-nitro-	$C_6H_4BrNO_3$ 218.01	101935-40-4 147.5	 sub	3-06-00-00844	eth 4; EtOH 4
9244	Phenol, 4-bromo-2-nitro- 4-Bromo-2-nitrophenol	$C_6H_4BrNO_3$ 218.01	7693-52-9 92	 sub	4-06-00-01363	bz 4; eth 4; EtOH 4; chl 4
9245	Phenol, 2-butoxy-	$C_{10}H_{14}O_2$ 166.22	39075-90-6	 232.5	4-06-00-05566 1.026^{25}	 1.5113^{25}
9246	Phenol, 2-butyl-	$C_{10}H_{14}O$ 150.22	3180-09-4 -20	 235	4-06-00-03267 0.975^{20}	H_2O 1; EtOH 3; eth 3; alk 3 1.5180^{25}
9247	Phenol, 3-butyl	$C_{10}H_{14}O$ 150.22	4074-43-5	 248	4-06-00-03269 0.974^{20}	eth 4; EtOH 4
9248	Phenol, 4-butyl-	$C_{10}H_{14}O$ 150.22	1638-22-8 22	 248	4-06-00-03269 0.976^{22}	H_2O 1; EtOH 3; eth 3; ctc 2 1.5165^{25}
9249	Phenol, 4,4'-butylidenebis-	$C_{16}H_{18}O_2$ 242.32	4731-84-4 137	 270^{12}	4-06-00-06736	bz 4; EtOH 4; tol 4
9250	Phenol, 2-butyl-4-methyl-	$C_{11}H_{16}O$ 164.25	6891-45-8 19	 220, 124^{15}	4-06-00-03392 0.9680^{20}	
9251	Phenol, 2-chloro- *o*-Chlorophenol	C_6H_5ClO 128.56	95-57-8 9.8	2154 174.9	4-06-00-00782 1.2634^{20}	H_2O 2; EtOH 3; eth 3; bz 4 1.5524^{20}
9252	Phenol, 3-chloro- *m*-Chlorophenol	C_6H_5ClO 128.56	108-43-0 32.6	2154 214	4-06-00-00810 1.245^{45}	H_2O 2; EtOH 3; eth 3; bz 4 1.5565^{40}
9253	Phenol, 4-chloro- *p*-Chlorophenol	C_6H_5ClO 128.56	106-48-9 42.7	2154 220	4-06-00-00820 1.2651^{40}	H_2O 1; EtOH 4; eth 4; bz 4 1.5579^{40}
9254	Phenol, 2-chloro-4-cyclohexyl- 2-Chloro-4-cyclohexylphenol	$C_{12}H_{15}ClO$ 210.70	3964-61-2 40.5	 160^{12}; 122^{3}	3-06-00-02508 1.1600^{15}	ctc 3; CS_2 3
9255	Phenol, 4-chloro-2-cyclopentyl- Dowicide 9	$C_{11}H_{13}ClO$ 196.68	13347-42-7	3420 $181\text{-}5^{18}$	3-06-00-02478	
9256	Phenol, 2-chloro-3,4-dimethyl-	C_8H_9ClO 156.61	10283-15-5 27	 188	2-06-00-00456 1.5538^{20}	peth 4
9257	Phenol, 4-chloro-3,5-dimethyl- Chloroxylenol	C_8H_9ClO 156.61	88-04-0 115	2176 246	4-06-00-03152	H_2O 2; EtOH 3; eth 3; bz 2
9258	Phenol, 4-chloro-2,6-dinitro- 4-Chloro-2,6-dinitrophenol	$C_6H_3ClN_2O_5$ 218.55	88-87-9 81		4-06-00-01386 1.74^{22}	eth 4; EtOH 4; chl 4
9259	Phenol, 5-chloro-2,4-dinitro- 3-Chloro-4,6-dinitrophenol	$C_6H_3ClN_2O_5$ 218.55	54715-57-0 92		4-06-00-01385 1.74^{22}	eth 4; EtOH 4; chl 4; peth 4
9260	Phenol, 2-chloro-3-methyl-	C_7H_7ClO 142.58	608-26-4 56	 230	4-06-00-02062	H_2O 2; HOAc 4; peth 3
9261	Phenol, 2-chloro-4-methyl-	C_7H_7ClO 142.58	6640-27-3	 195.5	4-06-00-02135 1.1785^{27}	bz 4; eth 4; EtOH 4 1.5200^{27}
9262	Phenol, 2-chloro-5-methyl- 6-Chloro-*m*-cresol	C_7H_7ClO 142.58	615-74-7 45.5	 196	3-06-00-01315 1.215^{15}	H_2O 4; EtOH 4
9263	Phenol, 2-chloro-6-methyl-	C_7H_7ClO 142.58	87-64-9	 189; 80^{20}	4-06-00-01984	eth 3 1.5449^{20}
9264	Phenol, 3-chloro-2-methyl-	C_7H_7ClO 142.58	3260-87-5 86	 225	4-06-00-02000	bz 4; eth 4; EtOH 4
9265	Phenol, 3-chloro-4-methyl-	C_7H_7ClO 142.58	615-62-3 55.5	 228	[illegible]-01971	bz 4; eth 4; EtOH 4
9266	Phenol, 4-chloro-2-methyl- 4-Chloro-*o*-cresol	C_7H_7ClO 142.58	1570-64-5 51	 223	4-06-00-01987	H_2O 2; peth 3
9267	Phenol, 4-chloro-3-methyl- 4-Chloro-*m*-cresol	C_7H_7ClO 142.58	59-50-7 67	2133 235	4-06-00-02064	H_2O 2; EtOH 3; eth 3; chl 2
9268	Phenol, 4-chloro-5-methyl-2-(1-methylethyl)- Chlorothymol	$C_{10}H_{13}ClO$ 184.67	89-68-9 63	2171 258.5	4-06-00-03344	H_2O 4; EtOH 3; eth 3; bz 3
9269	Phenol, 5-chloro-2-nitro- 5-Chloro-2-nitrophenol	$C_6H_4ClNO_3$ 173.56	611-07-4 41	 sub	3-06-00-00836	H_2O 2; EtOH 3; eth 3; HOAc 3
9270	Phenol, 4-chloro-2-(phenylmethyl)- Clorophene	$C_{13}H_{11}ClO$ 218.68	120-32-1 48.5	2403 $160\text{-}2^{3.5}$	4-06-00-04636 1.185^{58}	ctc 3; CS_2 3
9271	Phenol, 4-chloro-2-(2-propenyl)-	C_9H_9ClO 168.62	13997-73-4 48	 $130\text{-}2^{15}$	4-06-00-03813 1.171^{15}	bz 4; EtOH 4
9272	Phenol, 4-cyclohexyl-	$C_{12}H_{16}O$ 176.26	1131-60-8 133	 294; 133^{4}	4-06-00-03904	H_2O 1; EtOH 4; eth 4; bz 3

No.	Name Synonym	Mol. Form. Mol. Wt.	CAS RN mp/°C	Merck No. bp/°C	Beil. Ref. den/g cm^{-3}	Solubility n_D
9273	Phenol, 2-cyclohexyl-4,6-dinitro- 4,6-Dinitro-o-cyclohexylphenol	$C_{12}H_{14}N_2O_5$ 266.25	131-89-5	2739	4-06-00-03903	
9274	Phenol, 2,4-diamino- 2,4-Diaminophenol	$C_6H_8N_2O$ 124.14	95-86-3 78-80 dec	2961	4-13-00-01425	H_2O 4; ace 4; EtOH 4
9275	Phenol, 4,4'-(1,2-diarsenediyl)bis[2-amino-, dihydrochloride Salvarsan	$C_{12}H_{20}As_2Cl_2O_5$ 465.04	139-93-5 185-95 dec	839	2-16-00-00560	H_2O 4
9276	Phenol, 2,4-dibromo- 2,4-Dibromophenol	$C_6H_4Br_2O$ 251.91	615-58-7 38	 238.5	4-06-00-01061 2.0700^{20}	H_2O 2; EtOH 4; eth 4; bz 4
9277	Phenol, 2,6-dibromo- 2,6-Dibromophenol	$C_6H_4Br_2O$ 251.91	608-33-3 56.5	 255; 162^{21}	4-06-00-01064	H_2O 3; EtOH 4; eth 4
9278	Phenol, 3,5-dibromo- 3,5-Dibromophenol	$C_6H_4Br_2O$ 251.91	626-41-5 81	 274; 121^{3}	4-06-00-01065	H_2O 2; EtOH 3; eth 3; lig 2
9279	Phenol, 2,4-dibromo-6-methyl-	$C_7H_6Br_2O$ 265.93	609-22-3 58	 265 dec; 105^{4}	3-06-00-01271	chl 3
9280	Phenol, 3,5-dibromo-2-methyl-	$C_7H_6Br_2O$ 265.93	14122-00-0 100.3	 285	2-06-00-00334	peth 4
9281	Phenol, 2,4-dibromo-3-methyl-6-(1-methylethyl)- Thymol, 2,6-dibromo	$C_{10}H_{12}Br_2O$ 308.01	52262-38-1 3.7	 160-1^{16}	2-06-00-00500 1.6568^{17}	
9282	Phenol, 2,3-dichloro- 2,3-Dichlorophenol	$C_6H_4Cl_2O$ 163.00	576-24-9 58		4-06-00-00883	EtOH 3; eth 3; bz 3; lig 3
9283	Phenol, 2,4-dichloro- 2,4-Dichlorophenol	$C_6H_4Cl_2O$ 163.00	120-83-2 45	3061 210	4-06-00-00885	H_2O 2; EtOH 3; eth 3; bz 3
9284	Phenol, 2,5-dichloro- 2,5-Dichlorophenol	$C_6H_4Cl_2O$ 163.00	583-78-8 59	 211	4-06-00-00942	H_2O 2; EtOH 4; eth 4; bz 3
9285	Phenol, 2,6-dichloro- 2,6-Dichlorophenol	$C_6H_4Cl_2O$ 163.00	87-65-0 68.5	3062 220; 82^{4}	4-06-00-00949	EtOH 4; eth 4; bz 3; peth 3
9286	Phenol, 3,4-dichloro- 3,4-Dichlorophenol	$C_6H_4Cl_2O$ 163.00	95-77-2 68	 253	4-06-00-00952	H_2O 2; EtOH 4; eth 4; bz 3
9287	Phenol, 3,5-dichloro- 3,5-Dichlorophenol	$C_6H_4Cl_2O$ 163.00	591-35-5 68	 233	4-06-00-00957	H_2O 2; EtOH 4; eth 4; peth 3
9288	Phenol, 2,4-dichloro-, acetate 2,4-Dichlorophenyl acetate	$C_8H_6Cl_2O_2$ 205.04	6341-97-5	 244.5	3-06-00-00704	EtOH 3; eth 3; chl 3
9289	Phenol, 2,4-dichloro-, benzenesulfonate Genite	$C_{12}H_8Cl_2O_3S$ 303.17	97-16-5 45.5	4283	4-11-00-00034	ctc 3; CS_2 3
9290	Phenol, 2,4-dichloro-3,5-dimethyl- Dichloroxylenol	$C_8H_8Cl_2O$ 191.06	133-53-9 83	3066	4-06-00-03156	eth 4
9291	Phenol, 2,6-dichloro-3,5-dimethyl-	$C_8H_8Cl_2O$ 191.06	1943-54-0 87.5	 105-10^{1}	4-06-00-03157	eth 4; chl 4
9292	Phenol, 2,4-dichloro-3-methyl- *m*-Cresol, 2,4-dichloro	$C_7H_6Cl_2O$ 177.03	17788-00-0 58	 236; 77^{4}	3-06-00-01319	eth 4; chl 4
9293	Phenol, 2,4-dichloro-5-methyl-	$C_7H_6Cl_2O$ 177.03	1124-07-8 73	 234	4-06-00-02069	chl 4; peth 3 1.572^{20}
9294	Phenol, 2,6-dichloro-3-methyl- 2,6-Dichloro-*m*-cresol	$C_7H_6Cl_2O$ 177.03	13481-70-4 28	 237; 77^{4}	3-06-00-01319	eth 4; chl 4
9295	Phenol, 2,6-dichloro-4-methyl-	$C_7H_6Cl_2O$ 177.03	2432-12-4 39	 231; 138^{28}	4-06-00-02141	eth 4; EtOH 4; HOAc 4
9296	Phenol, 2,6-dichloro-4-nitro-	$C_6H_3ClNO_3$ 172.55	618-80-4 127 exp		4-06-00-01361 1.822^{25}	eth 4; chl 4
9297	Phenol, 3-(diethylamino)-	$C_{10}H_{15}NO$ 165.24	91-68-9 78	 276; 170^{15}	4-13-00-00969	H_2O 3; EtOH 3; eth 3; CS_2 3
9298	Phenol, 4,4'-(1,2-diethyl-1,2-ethanediyl)bis-, (*R**,*S**)- Hexestrol	$C_{18}H_{22}O_2$ 270.37	84-16-2 186.5	4621	4-06-00-06761	ace 4; eth 4; EtOH 4
9299	Phenol, 4,4'-(1,2-diethyl-1,2-ethanediyl)bis[2-methyl- Methestrol	$C_{20}H_{26}O_2$ 298.43	130-73-4 145	5888	4-06-00-06786	
9300	Phenol, 4,4'-(1,2-diethyl-1,2-ethenediyl)bis-, (*E*)- Diethylstilbestrol	$C_{18}H_{20}O_2$ 268.36	56-53-1 170.5	3118	4-06-00-06856	eth 4; EtOH 4; chl 4
9301	Phenol, 4,4'-(1,2-diethyl-1,2-ethenediyl)bis-, dipropanoate, (*E*)- Diethylstilbestrol dipropionate	$C_{24}H_{28}O_4$ 380.48	130-80-3 104	3119	4-06-00-06857	bz 4; eth 4; EtOH 4
9302	Phenol, 4,4'-(1,2-diethylidene-1,2-ethanediyl)bis- Dienestrol	$C_{18}H_{18}O_2$ 266.34	84-17-3 227.5	3094 sub 130	4-06-00-06916	ace 4; eth 4; EtOH 4
9303	Phenol, 4,4'-(1,2-diethyl-3-methyl-1,3-propanediyl)bis- Benzestrol	$C_{20}H_{26}O_2$ 298.43	85-95-0 164	1082	3-06-00-05538	ace 4; eth 4; EtOH 4; HOAc 4
9304	Phenol, 2,6-diiodo-4-nitro- Disophenol	$C_6H_3I_2NO_3$ 390.90	305-85-1 157	3365	4-06-00-01368	EtOH 4
9305	Phenol, 2,6-dimethoxy-	$C_8H_{10}O_3$ 154.17	91-10-1 56.5	 261	4-06-00-07329	eth 4; EtOH 4
9306	Phenol, 3,5-dimethoxy-	$C_8H_{10}O_3$ 154.17	500-99-2 37	 199^{35}; 170^{10}	4-06-00-07362	eth 3; bz 3; lig 2
9307	Phenol, 2,3-dimethyl- 2,3-Xylenol	$C_8H_{10}O$ 122.17	526-75-0 72.8	9989 216.9	4-06-00-03096	H_2O 2; EtOH 3; eth 3 1.5420^{20}

No.	Name Synonym	Mol. Form. Mol. Wt.	CAS RN mp/°C	Merck No. bp/°C	Beil. Ref. den/g cm^{-3}	Solubility n_D
9308	Phenol, 2,4-dimethyl- 2,4-Xylenol	$C_8H_{10}O$ 122.17	105-67-9 24.5	9989 210.9	4-06-00-03126 0.9650^{20}	H_2O 2; EtOH 5; eth 5; ctc 3 1.5420^{14}
9309	Phenol, 2,5-dimethyl- 2,5-Xylenol	$C_8H_{10}O$ 122.17	95-87-4 74.8	9989 211.1	4-06-00-03164	H_2O 3; EtOH 3; eth 4; chl 2
9310	Phenol, 2,6-dimethyl- 2,6-Xylenol	$C_8H_{10}O$ 122.17	576-26-1 45.7	9989 201.0	4-06-00-03112	H_2O 3; EtOH 3; eth 3; ctc 3
9311	Phenol, 3,4-dimethyl- 3,4-Xylenol	$C_8H_{10}O$ 122.17	95-65-8 60.8	9989 227	4-06-00-03099 0.9830^{20}	H_2O 2; EtOH 3; eth 5; ctc 3
9312	Phenol, 3,5-dimethyl- 3,5-Xylenol	$C_8H_{10}O$ 122.17	108-68-9 63.6	9989 221.7	4-06-00-03141 0.9680^{20}	H_2O 3; EtOH 3; ctc 3
9313	Phenol, 2,4-dimethyl-, acetate	$C_{10}H_{12}O_2$ 164.20	877-53-2	 226	2-06-00-00459 1.0298^{15}	eth 4; EtOH 4 1.4990^{15}
9314	Phenol, 2,5-dimethyl-, acetate	$C_{10}H_{12}O_2$ 164.20	877-48-5 <-20	 237	2-06-00-00467 1.0624^{15}	eth 4; EtOH 4
9315	Phenol, 3,4-dimethyl-, acetate	$C_{10}H_{12}O_2$ 164.20	22618-23-1 22	 235	4-06-00-03102	bz 4; eth 4; EtOH 4
9316	Phenol, 3,5-dimethyl-, acetate	$C_{10}H_{12}O_2$ 164.20	877-82-7 <-20	 237	3-06-00-01757 1.0624^{15}	eth 4; EtOH 4
9317	Phenol, 2-(dimethylamino)-	$C_8H_{11}NO$ 137.18	3743-22-4 45	 199.5	4-13-00-00815	eth 4; EtOH 4; HOAc 4
9318	Phenol, 3-(dimethylamino)-	$C_8H_{11}NO$ 137.18	99-07-0 86	 266.5	4-13-00-00959	H_2O 1; EtOH 3; eth 3; ace 3 1.5895^{26}
9319	Phenol, 4-(dimethylamino)-	$C_8H_{11}NO$ 137.18	619-60-3 77	 165^{30}	4-13-00-01040	H_2O 2; EtOH 3; eth 3
9320	Phenol, 4-[2-(dimethylamino)ethyl]- Hordenine	$C_{10}H_{15}NO$ 165.24	539-15-1 117.5	4666 173^{11}	4-13-00-01790	eth 4; EtOH 4; chl 4
9321	Phenol, 2-(1,1-dimethylethyl)-	$C_{10}H_{14}O$ 150.22	88-18-6 -6.8	 223	4-06-00-03292 0.9783^{20}	EtOH 3; eth 4; ctc 3; alk 3 1.5160^{20}
9322	Phenol, 3-(1,1-dimethylethyl)-	$C_{10}H_{14}O$ 150.22	585-34-2 42.3	 240	4-06-00-03294	EtOH 3; eth 4; alk 3
9323	Phenol, 4-(1,1-dimethylethyl)- *p-tert*-Butylphenol	$C_{10}H_{14}O$ 150.22	98-54-4 98	1584 237	4-06-00-03296 0.908^{80}	H_2O 3; EtOH 3; eth 3; chl 3 1.4787^{114}
9324	Phenol, 2-(1,1-dimethylethyl)-4,5-dimethyl- 3,4-Xylenol, 6-*tert*-butyl-	$C_{12}H_{18}O$ 178.27	1445-23-4 46	 258.5	3-06-00-02019 0.920^{80}	 1.5222^{20}
9325	Phenol, 2-(1,1-dimethylethyl)-4,6-dimethyl-	$C_{12}H_{18}O$ 178.27	1879-09-0 22.3	 249	4-06-00-03442 0.917^{80}	alk 1 1.5183^{20}
9326	Phenol, 4-(1,1-dimethylethyl)-2,5-dimethyl-	$C_{12}H_{18}O$ 178.27	17696-37-6 71.2	 264	3-06-00-02019 0.939^{80}	alk 3 1.5311^{20}
9327	Phenol, 4-(1,1-dimethylethyl)-2,6-dimethyl-	$C_{12}H_{18}O$ 178.27	879-97-0 82.4	 248	3-06-00-02019 0.916^{80}	alk 3
9328	Phenol, 2-(1,1-dimethylethyl)-4,6-dinitro- 2-*tert*-Butyl-4,6-dinitrophenol	$C_{10}H_{12}N_2O_5$ 240.22	1420-07-1		4-06-00-03293	
9329	Phenol, (1,1-dimethylethyl)-4-methoxy- Butylated hydroxyanisole	$C_{11}H_{16}O_2$ 180.25	25013-16-5 51	1547 268		H_2O 1; peth 3; EtOH 3
9330	Phenol, 2-(1,1-dimethylethyl)-4-methyl-	$C_{11}H_{16}O$ 164.25	2409-55-4 51.5	 237	4-06-00-03397 0.9247^{75}	H_2O 2; ace 3; bz 3; chl 3 1.4969^{75}
9331	Phenol, 2-(1,1-dimethylethyl)-5-methyl-	$C_{11}H_{16}O$ 164.25	88-60-8 46.5	 127^{11}	4-06-00-03400 0.922^{80}	H_2O 1; EtOH 3; eth 3; ace 3 1.5250^{20}
9332	Phenol, 2-(1,1-dimethylethyl)-6-methyl-	$C_{11}H_{16}O$ 164.25	2219-82-1 31	 230	4-06-00-03396 0.9240^{80}	 1.5195^{20}
9333	Phenol, 4-(1,1-dimethylethyl)-2-methyl-	$C_{11}H_{16}O$ 164.25	98-27-1 27.5	 237; 132^{20}	4-06-00-03398 0.965^{20}	H_2O 1; eth 3; ace 3; bz 3 1.5230^{20}
9334	Phenol, 3,6-dimethyl-2-nitro-	$C_8H_9NO_3$ 167.16	71608-10-1 34.5	 236 dec; 150^{15}	4-06-00-03169	EtOH 4; eth 3; ace 3; bz 3
9335	Phenol, 2,4-dimethyl-, phosphate (3:1)	$C_{24}H_{27}O_4P$ 410.45	3862-12-2	 233.5	0-06-00-00488 1.142^{38}	H_2O 1; bz 3; chl 3; hx 3 1.5550^{20}
9336	Phenol, 2,5-dimethyl-, phosphate (3:1)	$C_{24}H_{27}O_4P$ 410.45	19074-59-0 79.8	 $260-5^{8}$	3-06-00-01773 1.197^{25}	H_2O 1; EtOH 2; eth 3; bz 3
9337	Phenol, 2,6-dimethyl-, phosphate (3:1)	$C_{24}H_{27}O_4P$ 410.45	121-06-2 137.8	 $262-4^{6}$	4-06-00-03121	H_2O 1; EtOH 2; bz 3; hx 2
9338	Phenol, 3,4-dimethyl-, phosphate (3:1)	$C_{24}H_{27}O_4P$ 410.45	3862-11-1 72	 $260-3^{7}$	0-06-00-00482	H_2O 1; EtOH 2; bz 3; chl 2
9339	Phenol, 4-(1,1-dimethylpropyl)- *p-tert*-Pentylphenol	$C_{11}H_{16}O$ 164.25	80-46-6 95	7098 262.5	4-06-00-03383	
9340	Phenol, 4-(1,1-dimethylpropyl)-2-methyl-	$C_{12}H_{18}O$ 178.27	71745-63-6 15	 260	3-06-00-02010 0.969^{20}	chl 2 1.5240^{20}
9341	Phenol, 2,3-dinitro- 2,3-Dinitrophenol	$C_6H_4N_2O_5$ 184.11	66-56-8 144.5		4-06-00-01369 1.681^{20}	H_2O 2; EtOH 4; eth 4; bz 3
9342	Phenol, 2,4-dinitro- 2,4-Dinitrophenol	$C_6H_4N_2O_5$ 184.11	51-28-5 115.5	3274 sub	4-06-00-01369 1.683^{24}	H_2O 2; EtOH 3; eth 3; ace 3
9343	Phenol, 2,5-dinitro- 2,5-Dinitrophenol	$C_6H_4N_2O_5$ 184.11	329-71-5 108	3275	4-06-00-01383	bz 4; eth 4
9344	Phenol, 2,6-dinitro- 2,6-Dinitrophenol	$C_6H_4N_2O_5$ 184.11	573-56-8 63.5	3276	4-06-00-01383	H_2O 1; EtOH 4; eth 4; ace 4
9345	Phenol, 3,4-dinitro-	$C_6H_4N_2O_5$ 184.11	577-71-9 134		4-06-00-01384 1.672^{25}	bz 4; eth 4; EtOH 4
9346	Phenol, 3,5-dinitro- 3,5-Dinitrophenol	$C_6H_4N_2O_5$ 184.11	586-11-8 125.1		4-06-00-01385 1.702^{25}	bz 4; eth 4; EtOH 4; chl 4
9347	Phenol, 2,6-dipropyl- 2,6-Dipropylphenol	$C_{12}H_{18}O$ 178.27	6626-32-0 27	 256	3-06-00-02013 1.6450^{0}	eth 4

No.	Name Synonym	Mol. Form. Mol. Wt.	CAS RN mp/°C	Merck No. bp/°C	Beil. Ref. den/g cm^{-3}	Solubility n_D
9348	Phenol, 2-ethenyl-	C_8H_8O 120.15	695-84-1 29.5	 101^{14}	4-06-00-03771 1.0609^{18}	eth 4; EtOH 4 1.5851^{20}
9349	Phenol, 3-ethenyl-	C_8H_8O 120.15	620-18-8	 $114\text{-}6^{16}$	4-06-00-03773 1.0353^{21}	 1.5804^{21}
9350	Phenol, 2-ethoxy-	$C_8H_{10}O_2$ 138.17	94-71-3 29	 217	4-06-00-05565 1.0903^{25}	H_2O 2; EtOH 5; eth 5; ctc 2
9351	Phenol, 3-ethoxy-	$C_8H_{10}O_2$ 138.17	621-34-1	 131^{10}	4-06-00-05664 1.0705^{4}	H_2O 1; EtOH 3; eth 3; bz 3
9352	Phenol, 4-ethoxy-	$C_8H_{10}O_2$ 138.17	622-62-8 66.5	 246.5	4-06-00-05719	H_2O 2; EtOH 4; eth 4; chl 3
9353	Phenol, 2-ethyl- *o*-Ethylphenol	$C_8H_{10}O$ 122.17	90-00-6 18	7302 204.5	4-06-00-03011 1.0146^{25}	ace 4; bz 4; eth 4; EtOH 4 1.5367^{20}
9354	Phenol, 3-ethyl- *m*-Ethylphenol	$C_8H_{10}O$ 122.17	620-17-7 -4	 218.4	4-06-00-03016 1.0283^{20}	H_2O 2; EtOH 4; eth 4; chl 2
9355	Phenol, 4-ethyl- *p*-Ethylphenol	$C_8H_{10}O$ 122.17	123-07-9 45.0	 217.9	4-06-00-03020	H_2O 2; EtOH 4; eth 4; ace 3 1.5239^{25}
9356	Phenol, 3-(ethylamino)-	$C_8H_{11}NO$ 137.18	621-31-8 62	 176^{12}	0-13-00-00408	bz 4; eth 4; EtOH 4
9357	Phenol, 4-ethyl-2-methoxy-	$C_9H_{12}O_2$ 152.19	2785-89-9 -7	 236.5	4-06-00-05927 1.0931^{18}	
9358	Phenol, 4-[1-ethyl-2-(4-methoxyphenyl)-1-butenyl]- Mestilbol	$C_{19}H_{22}O_2$ 282.38	7773-60-6 117.5	5818 $185\text{-}95^{0.3}$	4-06-00-06856	ace 4; eth 4; EtOH 4
9359	Phenol, 4-[1-ethyl-2-(4-methoxyphenyl)-1-butenyl]-, (*E*)-	$C_{19}H_{22}O_2$ 282.38	18839-90-2 117	 $185\text{-}95^{0.3}$	4-06-00-06856	EtOH 4
9360	Phenol, 2-fluoro-	C_6H_5FO 112.10	367-12-4 16.1	 151.5	4-06-00-00770 1.120^{25}	H_2O 3 1.5144^{20}
9361	Phenol, 3-fluoro-	C_6H_5FO 112.10	372-20-3 13.7	 178	4-06-00-00772 1.238^{25}	 1.5140^{20}
9362	Phenol, 4-fluoro-	C_6H_5FO 112.10	371-41-5 48	 185.5	4-06-00-00773 1.1889^{56}	H_2O 2; ace 3; peth 3
9363	Phenol, 4,4'-(3H-2,1-benzoxathiol-3-ylidene)bis-, *S*,*S*-dioxide Phenolsulfonphthalein	$C_{19}H_{14}O_5S$ 354.38	143-74-8 >300	7213	5-19-03-00457	H_2O 2; EtOH 2; eth 1; ace 2
9364	Phenol, 4,4'-(3H-2,1-benzoxathiol-3-ylidene)bis[2-bromo-6-methyl-, *S*,*S*-dioxide Bromcresol purple	$C_{21}H_{16}Br_2O_5S$ 540.23	115-40-2 241.5	1376	5-19-03-00460	
9365	Phenol, 4,4'-(3H-2,1-benzoxathiol-3-ylidene)bis[2-bromo-3-methyl-6-(1-methylethyl)-, *S*,*S*-dioxide Bromthymol blue	$C_{27}H_{28}Br_2O_5S$ 624.39	76-59-5 201	1435	5-19-03-00461	eth 4; EtOH 4
9366	Phenol, 4,4'-(3H-2,1-benzoxathiol-3-ylidene)bis[2,6-dibromo-, *S*,*S*-dioxide Bromphenol blue	$C_{19}H_{10}Br_4O_5S$ 669.97	115-39-9 279 dec	1434	5-19-03-00458	H_2O 2; EtOH 3; bz 3; HOAc 3
9367	Phenol, 4,4'-(3H-2,1-benzoxathiol-3-ylidene)bis[2,6-dibromo-3-methyl-, *S*,*S*-dioxide Bromcresol green	$C_{21}H_{14}Br_4O_5S$ 698.02	76-60-8 218.5	1375	5-19-03-00460	H_2O 2; EtOH 4; eth 4; bz 3
9368	Phenol, 4,4'-(3H-2,1-benzoxathiol-3-ylidene)bis[2-methyl-, *S*,*S*-dioxide Cresol red	$C_{21}H_{18}O_5S$ 382.44	1733-12-6 >300	2583	5-19-03-00460	H_2O 4; EtOH 4
9369	Phenol, 4,4'-(3H-2,1-benzoxathiol-3-ylidene)bis[5-methyl-2-(1-methylethyl)-, *S*,*S*-dioxide Thymol blue	$C_{27}H_{30}O_5S$ 466.60	76-61-9 221-4 dec	9334	5-19-03-00462	H_2O 2; EtOH 3; ace 2; bz 2
9370	Phenol, 4-(3-hydroxy-1-propenyl)-2-methoxy- Coniferyl alcohol	$C_{10}H_{12}O_3$ 180.20	458-35-5 74	2499 $163\text{-}5^{3}$	2-06-00-01093	H_2O 1; EtOH 3; eth 4; alk 3
9371	Phenol, 2-iodo- *o*-Iodophenol	C_6H_5IO 220.01	533-58-4 43	4928 186^{160}; 91^{2}	4-06-00-01070 1.8757^{80}	H_2O 3; EtOH 4; eth 4; CS_2 4
9372	Phenol, 3-iodo-	C_6H_5IO 220.01	626-02-8 118	 186^{100}	4-06-00-01073	H_2O 2; EtOH 3; eth 3
9373	Phenol, 4-iodo- *p*-Iodophenol	C_6H_5IO 220.01	540-38-5 93.5	4929 139^{5} dec	4-06-00-01074 1.8573^{112}	H_2O 2; EtOH 4; eth 4
9374	Phenol, 2-iodo-4-methyl- *p*-Cresol, 2-iodo-	C_7H_7IO 234.04	16188-57-1 35	 117^{12}	4-06-00-02147 1.684^{25}	 1.5331^{25}
9375	Phenol, 2-iodo-6-methyl-	C_7H_7IO 234.04	24885-45-8 15.5	 $70\text{-}5^{2}$	4-06-00-02008 1.661^{25}	 1.610^{25}
9376	Phenol, 2-mercapto-	C_6H_6OS 126.18	1121-24-0 5.5	 217; 89^{8}	4-06-00-05633 1.2371^{0}	bz 4; eth 4; EtOH 4
9377	Phenol, 3-mercapto-	C_6H_6OS 126.18	40248-84-8 16.5	 168^{35}	2-06-00-00827	bz 4; EtOH 4
9378	Phenol, 4-mercapto-	C_6H_6OS 126.18	637-89-8 29.5	 167^{45}; 135^{11}	4-06-00-05790 1.1285^{25}	H_2O 3; EtOH 3; alk 3; con sulf 3 1.5101^{25}
9379	Phenol, 2-methoxy- Guaiacol	$C_7H_8O_2$ 124.14	90-05-1 32	4457 205	4-06-00-05563 1.1287^{21}	H_2O 2; EtOH 3; eth 3; ctc 3 1.5429^{20}
9380	Phenol, 3-methoxy-	$C_7H_8O_2$ 124.14	150-19-6 <-17	 114^{5}	4-06-00-05662 1.131^{25}	H_2O 2; EtOH 5; eth 5; chl 2 1.5510^{20}

No.	Name Synonym	Mol. Form. Mol. Wt.	CAS RN mp/°C	Merck No. bp/°C	Beil. Ref. den/g cm^{-3}	Solubility n_D
9381	Phenol, 4-methoxy-	$C_7H_8O_2$	150-76-5		4-06-00-05717	H_2O 3; EtOH 4; eth 4; bz 3
	4-Methoxyphenol	124.14	57	243		
9382	Phenol, 2-methoxy-, acetate	$C_9H_{10}O_3$	613-70-7		4-06-00-05581	H_2O 1; EtOH 3; eth 3
		166.18	31.5	$123\text{-}4^{13}$	1.1285^{25}	1.5101^{25}
9383	Phenol, 2-methoxy-, benzoate	$C_{14}H_{12}O_3$	531-37-3	4458	4-09-00-00371	eth 4; chl 4
	Guaiacol benzoate	228.25	57.5			
9384	Phenol, 2-methoxy-, carbonate (2:1)	$C_{15}H_{14}O_5$	553-17-3	4459	3-06-00-04233	H_2O 1; EtOH 2; eth 3; chl 4
	Guaiacol carbonate	274.27	89			
9385	Phenol, 2-methoxy-4-methyl-	$C_8H_{10}O_2$	93-51-6	2573	4-06-00-05878	eth 4; EtOH 4
	Creosol	138.17	5.5	221	1.098^{20}	1.5353^{25}
9386	Phenol, 2-methoxy-3-nitro-	$C_7H_7NO_4$	20734-71-8		3-06-00-04263	H_2O 2; EtOH 4; chl 3; peth 3
		169.14	70	sub		
9387	Phenol, 2-methoxy-, phosphate (3:1)	$C_{21}H_{21}O_7P$	563-03-1	4460	3-06-00-04246	ace 4; tol 4; chl 4
	Guaiacol phosphate	416.37	91	$275\text{-}80^{3}$		
9388	Phenol, 2-methoxy-4-(1-propenyl)-	$C_{10}H_{12}O_2$	97-54-1	5054	3-06-00-04993	eth 4; EtOH 4
	Isoeugenol	164.20		266	1.080^{25}	1.5739^{19}
9389	Phenol, 2-methoxy-4-(2-propenyl)-	$C_{10}H_{12}O_2$	97-53-0	3855	4-06-00-06337	H_2O 1; EtOH 5; eth 5; chl 3
	Eugenol	164.20	-7.5	253.2	1.0652^{20}	1.5405^{20}
9390	Phenol, 2-methoxy-5-(2-propenyl)-	$C_{10}H_{12}O_2$	501-19-9		3-06-00-05024	eth 4; EtOH 4
		164.20	8.5	253.5	1.0613^{25}	1.5413^{20}
9391	Phenol, 2-methoxy-6-(1-propenyl)-	$C_{10}H_{12}O_2$	1076-55-7		3-06-00-04991	H_2O 2; EtOH 3; eth 3; ace 3
		164.20	81	267.5		
9392	Phenol, 2-methoxy-6-(2-propenyl)-	$C_{10}H_{12}O_2$	579-60-2		4-06-00-06334	ace 3; bz 3
		164.20		250.5	1.062^{20}	1.5545^{20}
9393	Phenol, 2-methoxy-4-(1-propenyl)-, acetate, (*E*)-	$C_{12}H_{14}O_3$	5912-87-8		4-06-00-06330	eth 4
		206.24	80.5	282.5	1.0251^{100}	1.5052^{100}
9394	Phenol, 2-methoxy-4-(1-propenyl)-, acetate, (*Z*)-	$C_{12}H_{14}O_3$	97412-23-2		3-06-00-05007	H_2O 1; eth 3
	Isoeugenol, acetate (*cis*)	206.24		$160\text{-}2^{13}$	1.0947^{10}	1.5418^{20}
9395	Phenol, 2-methoxy-4-(2-propenyl)-, acetate	$C_{12}H_{14}O_3$	93-28-7		3-06-00-05029	H_2O 1; EtOH 3; ctc 2
		206.24	30.5	281; $127\text{-}8^{6}$	1.0806^{20}	1.5205^{20}
9396	Phenol, 2-methoxy-4-(1-propenyl)-, (*E*)-	$C_{10}H_{12}O_2$	5932-68-3		4-06-00-06324	H_2O 2; EtOH 3; eth 3; chl 3
		164.20	33.5	$141\text{-}2^{13}$	1.0852^{20}	1.5784^{20}
9397	Phenol, 2-methoxy-4-(1-propenyl)-, (*Z*)-	$C_{10}H_{12}O_2$	5912-86-7		4-06-00-06324	H_2O 2; EtOH 3; eth 3
		164.20		133^{13}; $90^{0.5}$	[illegible]	[illegible]
9398	Phenol, 2-methoxy-5-[(1,2,3,4-tetrahydro-6,7-dimethoxy-2-methyl-1-isoquinolinyl)methyl]-, (±)-	$C_{20}H_{25}NO_4$	85-64-3	5248	5-21-06-00048	H_2O 2; EtOH 2; eth 2; bz 3
	Laudanine	343.42	167		1.26^{20}	
9399	Phenol, 2-methoxy-5-[(1,2,3,4-tetrahydro-6,7-dimethoxy-2-methyl-1-isoquinolinyl)methyl]-, (*R*)-	$C_{20}H_{25}NO_4$	301-21-3	5247	5-21-06-00048	H_2O 4; bz 4
	Laudanidine	343.42	184.5			
9400	Phenol, 2-methyl-	C_7H_8O	95-48-7	2580	4-06-00-01940	H_2O 3; EtOH 4; eth 4; ace 5
	o-Cresol	108.14	29.8	191.0	1.135^{25}	1.5361^{20}
9401	Phenol, 3-methyl-	C_7H_8O	108-39-4	2579	4-06-00-02035	H_2O 2; EtOH 5; eth 5; ace 5
	m-Cresol	108.14	11.8	202.2	1.0341^{20}	1.5438^{20}
9402	Phenol, 4-methyl-	C_7H_8O	106-44-5	2581	4-06-00-02093	H_2O 2; EtOH 5; eth 5; ace 5
	p-Cresol	108.14	35.5	201.9	1.0185^{40}	1.5312^{20}
9403	Phenol, 4-[2-(methylamino)propyl]-	$C_{10}H_{15}NO$	370-14-9	7304	3-13-00-01710	eth 4; EtOH 4
	Pholedrine	165.24	161			
9404	Phenol, 4-(3-methylbutyl)-	$C_{11}H_{16}O$	1805-61-4		4-06-00-03378	
		164.25	29	255; 128^{7}	0.9562^{23}	1.505^{27}
9405	Phenol, 2-methyl-4,6-dinitro-	$C_7H_6N_2O_5$	534-52-1	3272	4-06-00-02014	H_2O 2; EtOH 3; eth 3; ace 3
	Dinitrocresol	198.14	86.5			
9406	Phenol, 3,3'-methylenebis-	$C_{13}H_{12}O_2$	10193-50-7		3-06-00-05411	eth 4; EtOH 4; HOAc 4
	Diphenylmethane, 3,3'-dihydroxy	200.24	102.5	235^{3}		
9407	Phenol, 4,4'-methylenebis-	$C_{13}H_{12}O_2$	620-92-8		4-06-00-06664	EtOH 3; eth 3; chl 3; DMSO-d_6 2
		200.24	162.5	sub		
9408	Phenol, 2,2'-methylenebis[4-chloro-	$C_{13}H_{10}Cl_2O_2$	97-23-4	3059	4-06-00-06658	H_2O 1; EtOH 3; ace 3
	Dichlorophene	269.13	177.5			
9409	Phenol, 2,2'-methylenebis[3,4,6-trichloro-	$C_{13}H_6Cl_6O_2$	70-30-4	4602	4-06-00-06659	H_2O 1; EtOH 3; eth 3; ace 3
	Hexachlorophene	406.91	166.5			
9410	Phenol, 2-(1-methylethyl)-	$C_9H_{12}O$	88-69-7		4-06-00-03210	H_2O 2; EtOH 3; eth 3; bz 3
		136.19	15.5	213.5	1.012^{20}	1.5315^{20}
9411	Phenol, 3-(1-methylethyl)-	$C_9H_{12}O$	618-45-1		4-06-00-03214	eth 4
		136.19	26	228		1.5261^{20}
9412	Phenol, 4-(1-methylethyl)-	$C_9H_{12}O$	99-89-8		4-06-00-03215	H_2O 2; EtOH 3; chl 3
		136.19	62.3	230; 110^{10}	0.990^{20}	1.5228^{20}
9413	Phenol, 4,4'-(1-methylethylidene)bis-	$C_{15}H_{16}O_2$	80-05-7	1311	4-06-00-06717	H_2O 1; EtOH 4; eth 4; bz 4
	Bisphenol A	228.29	153	220^{4}; 222^{3}		
9414	Phenol, 2-methyl-4-(1-methylethyl)-	$C_{10}H_{14}O$	1740-97-2		4-06-00-03330	H_2O 3; EtOH 3; bz 3; chl 3
		150.22	8.6	230; 83^{3}	0.9793^{25}	1.5253^{20}
9415	Phenol, 2-methyl-5-(1-methylethyl)-	$C_{10}H_{14}O$	499-75-2	1881	4-06-00-03331	H_2O 2; EtOH 3; eth 3; ace 4
	Carvacrol	150.22	1	237.7	0.9772^{20}	1.5230^{20}
9416	Phenol, 2-methyl-6-(1-methylethyl)-	$C_{10}H_{14}O$	3228-04-4		3-06-00-01882	EtOH 3; bz 3; chl 3; os 3
		150.22	-14.5	225.5	0.9782^{25}	1.5239^{15}
9417	Phenol, 4-methyl-2-(1-methylethyl)-	$C_{10}H_{14}O$	4427-56-9		4-06-00-03329	H_2O 2; EtOH 3; bz 3; chl 3
		150.22	36.5	228.5	0.9910^{20}	1.5275^{20}

No.	Name	Synonym	Mol. Form.	Mol. Wt.	CAS RN	mp/°C	Merck No.	bp/°C	Beil. Ref.	den/g cm^{-3}	Solubility	n_D
9418	Phenol, 5-methyl-2-(1-methylethyl)-	Thymol	$C_{10}H_{14}O$	150.22	89-83-8	51.5	9333	232.5	4-06-00-03334	0.970^{25}	H_2O 1; EtOH 4; eth 4; chl 4	1.5227^{20}
9419	Phenol, 2-methyl-5-(1-methylethyl)-, acetate	Carvacrol, acetate	$C_{12}H_{16}O_2$	192.26	6380-28-5			246.5	2-06-00-00494	0.9896^{25}	eth 4; EtOH 4	1.4913^{28}
9420	Phenol, 5-methyl-2-(1-methylethyl)-, acetate	Thymol, acetate	$C_{12}H_{16}O_2$	192.26	528-79-0			245	4-06-00-03337	1.009^{9}	bz 4; eth 4; EtOH 4; chl 4	
9421	Phenol, 5-methyl-2-(1-methylpropyl)-	6-*sec*-Butyl-*m*-cresol	$C_{11}H_{16}O$	164.25	29472-95-5			$89\text{-}90^{2}$	4-06-00-03394	0.9612^{25}		1.5188^{20}
9422	Phenol, 2-methyl-4-nitro-		$C_7H_7NO_3$	153.14	99-53-6	96		188^{9}; $132^{0.2}$	4-06-00-02011		H_2O 2; EtOH 4; eth 4; bz 4	
9423	Phenol, 2-methyl-6-nitro-		$C_7H_7NO_3$	153.14	13073-29-5	70		255 dec; 102^{9}	4-06-00-02009		H_2O 1; EtOH 4; eth 4; peth 3	
9424	Phenol, 4-methyl-2-nitro-		$C_7H_7NO_3$	153.14	119-33-5	36.5		125^{22}	4-06-00-02149	1.2399^{20}	ace 4; bz 4; eth 4; EtOH 4	1.5744^{40}
9425	Phenol, 5-methyl-2-pentyl-	6-*n*-Amyl-*m*-cresol	$C_{12}H_{18}O$	178.27	1300-94-3	24	643	$137\text{-}9^{15}$	3-06-00-02005		ace 4; eth 4; EtOH 4	
9426	Phenol, 4-methyl-2-(phenylazo)-		$C_{13}H_{12}N_2O$	212.25	952-47-6	108.5		sub	4-16-00-00197		EtOH 3; eth 3; bz 3; chl 3	
9427	Phenol, 2-methyl-6-(2-propenyl)-		$C_{10}H_{12}O$	148.20	3354-58-3			232	4-06-00-03844	0.992^{25}		1.5381^{20}
9428	Phenol, 4-methyl-2-(1-propenyl)-		$C_{10}H_{12}O$	148.20	53889-94-4			$126\text{-}8^{14}$	4-06-00-03844	1.019^{20}		1.5727^{20}
9429	Phenol, 4-methyl-2-(2-propenyl)-		$C_{10}H_{12}O$	148.20	6628-06-4			237	4-06-00-03845	0.96^{25}		1.5385^{20}
9430	Phenol, 2-(1-methylpropyl)-	*o*-*sec*-Butylphenol	$C_{10}H_{14}O$	150.22	89-72-5	16		228; 116^{21}	3-06-00-01852	0.9804^{25}		1.5200^{25}
9431	Phenol, 4-(1-methylpropyl)-		$C_{10}H_{14}O$	150.22	99-71-8	61.5		241	4-06-00-03279	0.986^{20}	H_2O 1; EtOH 3; eth 4; alk 3	1.5182^{21}
9432	Phenol, 4-(2-methylpropyl)-		$C_{10}H_{14}O$	150.22	4167-74-2	51.5		237	4-06-00-03288	0.9778^{20}	ace 4; eth 4; EtOH 4	1.5319^{25}
9433	Phenol, 4,4'-(1-methylpropylidene)bis-	Bisphenol B	$C_{16}H_{18}O_2$	242.32	77-40-7	120.5	1312		4-06-00-06738		ace 4; MeOH 4	
9434	Phenol, 3-methyl-2,4,6-trinitro-	3-Methylpicric acid	$C_7H_5N_3O_7$	243.13	602-99-3	109.5		exp 150	4-06-00-02079		H_2O 2; EtOH 3; eth 3; ace 3	
9435	Phenol, 2-nitro-	*o*-Nitrophenol	$C_6H_5NO_3$	139.11	88-75-5	44.8	6541	216	4-06-00-01246	1.2942^{40}	H_2O 2; EtOH 4; eth 4; ace 4	1.5723^{50}
9436	Phenol, 3-nitro-	*m*-Nitrophenol	$C_6H_5NO_3$	139.11	554-84-7	96.8	6540	194^{70}	4-06-00-01269	1.2797^{100}	H_2O 2; EtOH 4; eth 4; ace 4	
9437	Phenol, 4-nitro-	*p*-Nitrophenol	$C_6H_5NO_3$	139.11	100-02-7	113.8	6542		4-06-00-01279	1.479^{20}	H_2O 2; EtOH 4; eth 4; ace 4	
9438	Phenol, 4-nitroso-	4-Nitrosophenol	$C_6H_5NO_2$	123.11	104-91-6	144 dec	6563		4-07-00-02073		H_2O 2; EtOH 3; eth 3; ace 3	
9439	Phenol, pentabromo-	Pentabromophenol	C_6HBr_5O	488.59	608-71-9	229.5		sub	3-06-00-00766		H_2O 1; EtOH 3; eth 2; bz 3	
9440	Phenol, pentachloro-	Pentachlorophenol	C_6HCl_5O	266.34	87-86-5	174	7059	310 dec	4-06-00-01025	1.978^{22}	H_2O 2; EtOH 4; eth 4; bz 3	
9441	Phenol, 3-pentadecyl-		$C_{21}H_{36}O$	304.52	501-24-6	53.5		230^{8}; $197^{1.5}$	4-06-00-03556		ace 4; bz 4; EtOH 4	
9442	Phenol, pentafluoro-	Pentafluorophenol	C_6HF_5O	184.07	771-61-9	32.8		145.6	4-06-00-00782			1.4263^{20}
9443	Phenol, pentamethyl-	Pentamethylphenol	$C_{11}H_{16}O$	164.25	2819-86-5	128		267	4-06-00-03412		H_2O 1; EtOH 3	
9444	Phenol, 4-pentyl-		$C_{11}H_{16}O$	164.25	14938-35-3	23		250.5	4-06-00-03370	0.960^{20}	eth 4; EtOH 4	1.5272^{25}
9445	Phenol, 2-(phenylamino)-		$C_{12}H_{11}NO$	185.23	644-71-3	69.5		185^{20}	4-13-00-00822		H_2O 2; EtOH 3; eth 3; bz 2	
9446	Phenol, 3-(phenylamino)-		$C_{12}H_{11}NO$	185.23	101-18-8	81.5		340	3-13-00-00944		H_2O 2; EtOH 4; eth 4; ace 4	
9447	Phenol, 4-(phenylamino)-		$C_{12}H_{11}NO$	185.23	122-37-2	73		330	4-13-00-01052		H_2O 2; EtOH 4; eth 4; bz 4	
9448	Phenol, 4-(phenylazo)-		$C_{12}H_{10}N_2O$	198.22	1689-82-3	155		225^{20} dec	2-16-00-00038		H_2O 1; EtOH 4; eth 4; bz 3	
9449	Phenol, 4-(phenylazo)-, acetate (ester)		$C_{14}H_{12}N_2O_2$	240.26	13102-31-3	89		360 dec	1-16-00-00236		EtOH 4; lig 4	
9450	Phenol, 2-[(phenylimino)methyl]-		$C_{13}H_{11}NO$	197.24	779-84-0	49.5			4-12-00-00339	1.087^{25}	H_2O 1; EtOH 3	
9451	Phenol, 2-(phenylmethoxy)-		$C_{13}H_{12}O_2$	200.24	6272-38-4			$173\text{-}4^{13}$	4-06-00-05572	1.154^{22}	eth 4; EtOH 4	1.5906^{18}
9452	Phenol, 3-(phenylmethoxy)-		$C_{13}H_{12}O_2$	200.24	3769-41-3	69.2		200^{5}	4-06-00-05669		H_2O 4	
9453	Phenol, 4-(phenylmethoxy)-	Monobenzone	$C_{13}H_{12}O_2$	200.24	103-16-2	122	6159		4-06-00-05728		H_2O 2; EtOH 4; eth 4; ace 3	
9454	Phenol, 2-(phenylmethyl)-	*o*-Benzylphenol	$C_{13}H_{12}O$	184.24	28994-41-4	21	1158	312	4-06-00-04628		ace 4; bz 4; EtOH 4	1.5994^{20}
9455	Phenol, 4-(phenylmethyl)-	*p*-Benzylphenol	$C_{13}H_{12}O$	184.24	101-53-1	84	1159	322	4-06-00-04640		H_2O 3; EtOH 3; eth 3; bz 3	
9456	Phenol, 2-(1-propenyl)-		$C_9H_{10}O$	134.18	6380-21-8	-6		220	3-06-00-02393	1.0246^{15}	eth 4	1.5181^{20}
9457	Phenol, 2-(2-propenyl)-		$C_9H_{10}O$	134.18	1745-81-9	-6		220	4-06-00-03807	1.0246^{15}	eth 4	1.5181^{20}

No.	Name Synonym	Mol. Form. Mol. Wt.	CAS RN mp/°C	Merck No. bp/°C	Beil. Ref. den/g cm^{-3}	Solubility n_D
9458	Phenol, 4-(2-propenyl)- Chavicol	$C_9H_{10}O$ 134.18	501-92-8 15.8	2039 238	4-06-00-03817 1.0203^{15}	eth 4; EtOH 4; chl 4 1.5441^{18}
9459	Phenol, 2-(1-propenyl)-, (*E*)-	$C_9H_{10}O$ 134.18	23619-59-2 37.5	 230.5	4-06-00-03793 1.0441^{16}	eth 4; lig 4 1.5823^{21}
9460	Phenol, 2-propoxy-	$C_9H_{12}O_2$ 152.19	6280-96-2	 227	4-06-00-05566 1.0523^{25}	EtOH 4 1.5176^{25}
9461	Phenol, 2-propyl-	$C_9H_{12}O$ 136.19	644-35-9 7	 220	4-06-00-03176 1.015^{20}	eth 4; EtOH 4
9462	Phenol, 3-propyl-	$C_9H_{12}O$ 136.19	621-27-2 26	 228	4-06-00-03180 0.987^{20}	H_2O 2; EtOH 3; eth 3 1.5223^{20}
9463	Phenol, 4-propyl-	$C_9H_{12}O$ 136.19	645-56-7 22	 232.6	4-06-00-03181 1.009^{20}	H_2O 2; EtOH 3; ctc 2 1.5379^{25}
9464	Phenol, 4,4'-propylidenebis-	$C_{15}H_{16}O_2$ 228.29	1576-13-2 132	 275^{20}	3-06-00-05457	eth 4; EtOH 4
9465	Phenol, 4,4'-(2-pyridinylmethylene)bis-, diacetate (ester) Bisacodyl	$C_{22}H_{19}NO_4$ 361.40	603-50-9 133.5	1253	5-21-05-00415	
9466	Phenol, 4,4'-sulfonylbis-	$C_{12}H_{10}O_4S$ 250.28	80-09-1 240.5		4-06-00-05809 1.3663^{15}	H_2O 1; EtOH 3; eth 3; bz 2
9467	Phenol, 2,3,4,6-tetrabromo- 2,3,4,6-Tetrabromophenol	$C_6H_2Br_4O$ 409.70	14400-94-3 113.5	 sub	3-06-00-00766	bz 4; EtOH 4
9468	Phenol, 2,3,4,5-tetrabromo-6-methyl- 3,4,5,6-Tetrabromo-*o*-cresol	$C_7H_4Br_4O$ 423.72	576-55-6 208	9120 dec	3-06-00-01272	H_2O 1; EtOH 3; eth 3; bz 3
9469	Phenol, 2,3,4,5-tetrachloro- 2,3,4,5-Tetrachlorophenol	$C_6H_2Cl_4O$ 231.89	4901-51-3 116.5	 sub	4-06-00-01020	EtOH 4
9470	Phenol, 2,3,4,6-tetrachloro- 2,3,4,6-Tetrachlorophenate	$C_6H_2Cl_4O$ 231.89	58-90-2 70	 150^{15}	4-06-00-01021	H_2O 1; EtOH 3; bz 3; chl 3
9471	Phenol, 2,3,5,6-tetrachloro- 2,3,5,6-Tetrachlorophenol	$C_6H_2Cl_4O$ 231.89	935-95-5 115		4-06-00-01025	H_2O 2; bz 4; lig 3
9472	Phenol, 2,3,5,6-tetrachloro-4-methoxy- Drosophilin A	$C_7H_4Cl_4O_2$ 261.92	484-67-3 116	3441	3 06 00 04436	
9473	Phenol, 2,3,4,5-tetramethyl- Prehnitenol	$C_{10}H_{14}O$ 150.22	488-70-0 85.3	 266	4-06-00-03359	H_2O 2; EtOH 4; eth 4; lig 2
9474	Phenol, 2,3,4,6-tetramethyl-	$C_{10}H_{14}O$ 150.22	3238-38-8 80.5	 240	4-06-00-03360	EtOH 3
9475	Phenol, 2,3,5,6-tetramethyl-	$C_{10}H_{14}O$ 150.22	527-35-5 118.5	 247	4-06-00-03363	chl 3; peth 3; HOAc 3
9476	Phenol, 2,3,4,6-tetranitro- 2,3,4,6-Tetranitrophenol	$C_6H_2N_4O_9$ 274.10	641-16-7 140 dec	 exp	2-06-00-00284	H_2O 4; chl 3
9477	Phenol, 2,2'-thiobis[4-chloro- Fenticlor	$C_{12}H_8Cl_2O_2S$ 287.17	97-24-5 174	3947	4-06-00-05645	H_2O 1; EtOH 3; eth 3; gl HOAc 3
9478	Phenol, 2,2'-thiobis[4,6-dichloro- Bithionol	$C_{12}H_6Cl_4O_2S$ 356.06	97-18-7 188	1316	4-06-00-05646 1.73^{25}	ace 4
9479	Phenol, 2,4,6-tribromo- 2,4,6-Tribromophenol	$C_6H_3Br_3O$ 330.80	118-79-6 95.5	9526 286	4 06 00 01067 2.55^{20}	H_2O 2; EtOH 4; eth 3; bz 3
9480	Phenol, 2,4,6-tribromo-3-methyl- 2,4,6-Tribromo-*m*-cresol	$C_7H_5Br_3O$ 344.83	4619-74-3 84	9524	3-06-00-01324	EtOH 3; eth 3; bz 3; chl 2
9481	Phenol, 2,3,4-trichloro- 2,3,4-Trichlorophenol	$C_6H_3Cl_3O$ 197.45	15950-66-0 83.5	 sub	3-06-00-00716	EtOH 3; eth 3; bz 3; alk 3
9482	Phenol, 2,3,5-trichloro- 2,3,5-Trichlorophenol	$C_6H_3Cl_3O$ 197.45	933-78-8 62	 $248\text{-}9^{250}$	3-06-00-00716	eth 4; EtOH 4
9483	Phenol, 2,3,6-trichloro- 2,3,6-Trichlorophenol	$C_6H_3Cl_3O$ 197.45	933-75-5 58		4-06-00-00962	H_2O 2; EtOH 4; eth 4; bz 4
9484	Phenol, 2,4,5-trichloro- 2,4,5-Trichlorophenol	$C_6H_3Cl_3O$ 197.45	95-95-4 69	9555 247	4-06-00-00962	H_2O 2; EtOH 4; eth 4; bz 4
9485	Phenol, 2,4,6-trichloro- 2,4,6-Trichlorophenol	$C_6H_3Cl_3O$ 197.45	88-06-2 69	9556 246	4-06-00-01005 1.4901^{75}	H_2O 2; EtOH 3; eth 3; HOAc 3
9486	Phenol, 3,4,5-trichloro- 3,4,5-Trichlorophenol	$C_6H_3Cl_3O$ 197.45	609-19-8 101	 275	3-06-00-00729	H_2O 2; eth 3; lig 2
9487	Phenol, 2,3,4-trichloro-6-methyl- 4,5,6-Trichloro-*o*-cresol	$C_7H_5Cl_3O$ 211.47	551-78-0 77	9548	3-06-00-01268	
9488	Phenol, 2,3,6-trichloro-4-methyl- 2,3,6-Trichloro-*p*-cresol	$C_7H_5Cl_3O$ 211.47	551-77-9 66.5	9546	0-06-00-00404	EtOH 4
9489	Phenol, 2,4,6-trichloro-3-methyl- 2,4,6-Trichloro-*m*-cresol	$C_7H_5Cl_3O$ 211.47	551-76-8 46	9547 265	4-06-00-02070	EtOH 4; MeOH 4; chl 4
9490	Phenol, 3,4,6-trichloro-2-nitro- 3,4,6-Trichloro-2-nitrophenol	$C_6H_2Cl_3NO_3$ 242.44	82-62-2 92.5	9554	3-06-00-00842	
9491	Phenol, 3-(trifluoromethyl)-	$C_7H_5F_3O$ 162.11	98-17-9 -1.8	 178	4-06-00-02060 1.3418^{25}	
9492	Phenol, 2,4,6-triiodo- 2,4,6-Triiodophenol	$C_6H_3I_3O$ 471.80	609-23-4 159.8	 sub	4-06-00-01085	H_2O 1; EtOH 2; eth 3; ace 3
9493	Phenol, 2,3,4-trimethyl-	$C_9H_{12}O$ 136.19	526-85-2 81	 236	4-06-00-03245	bz 4; eth 4; EtOH 4
9494	Phenol, 2,4,5-trimethyl-	$C_9H_{12}O$ 136.19	496-78-6 72	 232	4-06-00-03247	H_2O 1; EtOH 4; eth 4
9495	Phenol, 2,4,6-trimethyl-	$C_9H_{12}O$ 136.19	527-60-6 73	 220	4-06-00-03253	eth 4; EtOH 4
9496	Phenol, 2,4,5-trimethyl-, acetate 2,4,5-Trimethylphenyl acetate	$C_{11}H_{14}O_2$ 178.23	69305-42-6 34	 245.5	2-06-00-00482	eth 4; EtOH 4

No.	Name Synonym	Mol. Form. Mol. Wt.	CAS RN mp/°C	Merck No. bp/°C	Beil. Ref. den/g cm^{-3}	Solubility n_D
9497	Phenol, 2,4,6-trinitro- Picric acid	$C_6H_3N_3O_7$ 229.11	88-89-1 122.5	7380 exp 300	4-06-00-01388	H_2O 2; EtOH 3; eth 3; ace 4 1.763
9498	Phenol, 2,4,6-tris(1,1-dimethylethyl)-	$C_{18}H_{30}O$ 262.44	732-26-3 131	 278	4-06-00-03539 0.864^{27}	H_2O 1; EtOH 3; ace 3; ctc 3
9499	10H-Phenothiazine Thiodiphenylamine	$C_{12}H_9NS$ 199.28	92-84-2 187.5	7220 371	4-27-00-01214	ace 4; bz 4; eth 4; EtOH 4
9500	10H-Phenothiazine-2-acetic acid, 10-methyl- Metiazinic acid	$C_{15}H_{13}NO_2S$ 271.34	13993-65-2 144	6057		chl 3
9501	10H-Phenothiazine, 2-chloro-10-[3-(4-methyl-1-piperazinyl)propyl]- Phenothiazine, 2-chloro-10(3(4-methyl-1-piperazinyl)propyl)	$C_{20}H_{24}ClN_3S$ 373.95	58-38-8 228	7768	4-27-00-01307	
9502	10H-Phenothiazine, 10-[(dimethylamino)acetyl]- Ahistan	$C_{16}H_{16}N_2OS$ 284.38	518-61-6 144.5	179	4-27-00-01272	
9503	10H-Phenothiazine-10-ethanamine, *N,N*-diethyl- Diethazine	$C_{18}H_{22}N_2S$ 298.45	60-91-3	3098 $167^{0.5}$	4-27-00-01240	H_2O 1; dil HCl 3
9504	10H-Phenothiazine-10-ethanamine, *N,N*,α-trimethyl- Promethazine	$C_{17}H_{20}N_2S$ 284.43	60-87-7 60	7797 $190\text{-}3^{0.5}$	4-27-00-01253	H_2O 1; dil HCl 4
9505	10H-Phenothiazine-10-propanamine, 2-chloro-*N,N*-dimethyl- Chlorpromazine	$C_{17}H_{19}ClN_2S$ 318.87	50-53-3	2186 $200\text{-}5^{0.8}$	4-27-00-01300	H_2O 1; EtOH 4; eth 4; bz 4
9506	Phenothiazin-5-ium, 3,7-bis(dimethylamino)-, chloride Methylene blue	$C_{16}H_{18}ClN_3S$ 319.86	61-73-4	5979	4-27-00-05152	H_2O 3; EtOH 3; eth 1; chl 3
9507	10H-Phenoxazine Phenoxazine	$C_{12}H_9NO$ 183.21	135-67-1 156	7222 dec	4-27-00-01209	bz 4; eth 4; EtOH 4
9508	3H-Phenoxazin-3-one, 7-hydroxy-, 10-oxide Resazurin	$C_{12}H_7NO_4$ 229.19	550-82-3	8144 sub	4-27-00-03594	H_2O 1; EtOH 2; eth 1; alk 3
9509	*DL*-Phenylalanine	$C_9H_{11}NO_2$ 165.19	150-30-1 284-8 dec	7242 sub	4-14-00-01553	H_2O 4; EtOH 1; eth 1; bz 1
9510	*L*-Phenylalanine (*S*)-α-Aminobenzenepropanoic acid	$C_9H_{11}NO_2$ 165.19	63-91-2 283 dec	7242	4-14-00-01552	H_2O 4; EtOH 1; eth 1; bz 1
9511	*L*-Phenylalanine, *N*-*L*-α-aspartyl-, 1-methyl ester Aspartame	$C_{14}H_{18}N_2O_5$ 294.31	22839-47-0 246.5	861		
9512	*L*-Phenylalanine, ethyl ester	$C_{11}H_{15}NO_2$ 193.25	3081-24-1	 148^{13}	4-14-00-01556 1.065^{15}	H_2O 2
9513	Phorate Phosphorodithioic acid, *O,O*-diethyl *S*-[(ethylthio)methyl] ester	$C_7H_{17}O_2PS_3$ 260.38	298-02-2 <-15	7305 $119^{0.8}$	 1.16^{25}	
9514	Phosalone Phosphorodithioic acid, *S*-[(6-chloro-2-oxo-3(2H)-benzoxazolyl)methyl]*O,O*-diethyl ester	$C_{12}H_{15}ClNO_4PS_2$ 367.81	2310-17-0 46	7308		
9515	Phosmet Phosphorodithioic acid, *S*-[(1,3-dihydro-1,3-dioxo-2H-isoindol-2-yl)methyl] *O,O*-dimethyl ester	$C_{11}H_{12}NO_4PS_2$ 317.33	732-11-6 72	7311 dec		
9516	Phosphine, dicyclohexyl- Dicyclohexylphosphine	$C_{12}H_{23}P$ 198.29	829-84-5	 281; 129^8	4-16-00-00947 0.904^{25}	 1.5163^{20}
9517	Phosphine oxide, triethyl-	$C_6H_{15}OP$ 134.16	597-50-2 48	 243	4-04-00-03463	H_2O 4; eth 4; EtOH 4
9518	Phosphine oxide, triphenyl-	$C_{18}H_{15}OP$ 278.29	791-28-6 156.5	 >360	4-16-00-01011 1.2124^{23}	H_2O 2; EtOH 4; eth 2; bz 4
9519	Phosphine, phenyl- Monophenylphosphine	C_6H_7P 110.10	638-21-1	 160.5	4-16-00-00948 1.001^{15}	 1.5796^{20}
9520	Phosphine, tributyl- Tributylphosphine	$C_{12}H_{27}P$ 202.32	998-40-3	 150^{50}	4-04-00-03436 0.812^{25}	 1.4619^{20}
9521	Phosphine, triethyl- Triethyl phosphine	$C_6H_{15}P$ 118.16	554-70-1 -88	9590 129	4-04-00-03431 0.8006^{19}	H_2O 1; EtOH 5; eth 5 1.458^{15}
9522	Phosphine, trimethyl- Trimethylphosphine	C_3H_9P 76.08	594-09-2 -85	 37.5	4-04-00-03429	H_2O 1; eth 3
9523	Phosphine, triphenyl- Triphenylphosphine	$C_{18}H_{15}P$ 262.29	603-35-0 80	9657 188^1	4-16-00-00951 1.0749^{80}	H_2O 1; EtOH 3; eth 4; bz 3 1.6358^{80}
9524	Phosphine, tris(chloromethyl)- Tris(chloromethyl)phosphine	$C_3H_6Cl_3P$ 179.41	13482-62-7	 100^7	3-01-00-02608 1.414^{20}	ace 4; bz 4; eth 4; EtOH 4
9525	Phosphinic acid, diethyl- Diethylphosphinic acid	$C_4H_{11}O_2P$ 122.10	813-76-3 18.5	 320	4-04-00-03474	H_2O 4; EtOH 4; eth 4
9526	Phosphinic acid, diethyl-, anhydride	$C_8H_{20}O_3P_2$ 226.19	7495-97-8	 188^{14}	4-04-00-03477 1.1053^{20}	 1.4647^{20}
9527	Phosphinic acid, diethyl-, ethyl ester Ethyl diethylphosphinate	$C_6H_{15}O_2P$ 150.16	4775-09-1	 95^{14}	4-04-00-03475 0.9908^{20}	eth 4; EtOH 4 1.4337^{20}
9528	Phosphinic acid, dimethyl- Dimethylphosphinic acid	$C_2H_7O_2P$ 94.05	3283-12-3 92	 377	4-04-00-03470	H_2O 4; EtOH 4; eth 4; bz 3
9529	Phosphinic chloride, diethyl-	$C_4H_{10}ClOP$ 140.55	1112-37-4	 104^{15}	4-04-00-03476 1.1394^{20}	 1.4647^{20}
9530	Phosphonic acid, acetyl-, diethyl ester *O,O*-Diethyl acetylphosphonate	$C_6H_{13}O_4P$ 180.14	919-19-7	 $114\text{-}5^{20}$	4-02-00-00440 1.1005^{20}	 1.4200^{26}

No.	Name Synonym	Mol. Form. Mol. Wt.	CAS RN mp/°C	Merck No. bp/°C	Beil. Ref. den/g cm^{-3}	Solubility n_D
9531	Phosphonic acid, bis(1,1-dimethylethyl) ester	$C_8H_{19}O_3P$ 194.21	13086-84-5	 $70\text{-}2^{10}$	4-01-00-01623 0.975^{25}	 1.4168^{25}
9532	Phosphonic acid, dibutyl ester	$C_8H_{19}O_3P$ 194.21	1809-19-4	 131^{19}; $77^{1.5}$	4-01-00-01525 0.9485^{20}	 1.4220^{20}
9533	Phosphonic acid, (1,1-dimethylethyl)-	$C_4H_{11}O_3P$ 138.10	4923-84-6 192	 dec	4-04-00-03557	H_2O 4; EtOH 4
9534	Phosphonic acid, diphenyl ester	$C_{12}H_{11}O_3P$ 234.19	4712-55-4 12	 $218\text{-}9^{26}$	4-06-00-00703 1.223^{25}	 1.5575^{20}
9535	Phosphonic acid, ethenyl-, diethyl ester	$C_6H_{13}O_3P$ 164.14	682-30-4	 110^{2}	4-04-00-03569 1.068^{25}	 1.4290^{20}
9536	Phosphonic acid, ethyl-, diethyl ester Diethyl ethylphosphonate	$C_6H_{15}O_3P$ 166.16	78-38-6	 198; 90^{16}	4-04-00-03524 1.0259^{20}	H_2O 2; EtOH 3; eth 3 1.4163^{20}
9537	Phosphonic acid, ethyl-, dimethyl ester	$C_4H_{11}O_3P$ 138.10	6163-75-3	 82^{18}	4-04-00-03523 1.1029^{30}	H_2O 4; bz 4; EtOH 4 1.4128^{30}
9538	Phosphonic acid, methyl-	CH_5O_3P 96.02	993-13-5 108.5	 dec	4-04-00-03498	H_2O 4; EtOH 4; eth 4; bz 1
9539	Phosphonic acid, methyl-, diethyl ester	$C_5H_{13}O_3P$ 152.13	683-08-9	 194	4-04-00-03500 1.0406^{30}	H_2O 3; EtOH 3; eth 3; bz 1 1.4101^{30}
9540	Phosphonic acid, methyl-, dimethyl ester	$C_3H_9O_3P$ 124.08	756-79-6	 181; 79.5^{20}	4-04-00-03499 1.1507^{20}	H_2O 3; EtOH 3; eth 3 1.4099^{30}
9541	Phosphonic acid, methyl-, diphenyl ester Diphenyl methylphosphonate	$C_{13}H_{13}O_3P$ 248.22	7526-26-3 35	 205^{13}	4-06-00-00703 1.2051^{20}	H_2O 1
9542	Phosphonic acid, (1-methylethyl)- 2-Propanephosphonic acid	$C_3H_9O_3P$ 124.08	4721-37-3 74.5	 dec	4-04-00-03548	H_2O 4; eth 4; EtOH 4
9543	Phosphonic acid, methyl-, monoethyl ester	$C_3H_9O_3P$ 124.08	1832-53-7	 $106\text{-}7^{0.1}$	4-04-00-03499 1.1800^{20}	 1.4258^{20}
9544	Phosphonic acid, (2-methylpropyl)-	$C_4H_{11}O_3P$ 138.10	4721-34-0 119	 dec	4-04-00-03556	H_2O 4; eth 4; EtOH 4; xyl 4
9545	Phosphonic acid, (phenylmethyl)-, diethyl ester	$C_{11}H_{17}O_3P$ 228.23	1080-32-6	 110^{2}	4-16-00-01093	ctc 3 1.4930^{20}
9546	Phosphonic acid, 1,3-propanediylbis-, tetraethyl ester	$C_{11}H_{26}O_6P_2$ 316.27	22401-25-8	 $185\text{-}6^{3}$	4-04-00-03582 1.1227^{20}	 1.4500^{20}
9547	Phosphonic acid, propyl-	$C_3H_9O_3P$ 124.08	4672-38-2 73.8	 dec	4-04-00-03542	H_2O 4; eth 4; EtOH 4
9548	Phosphonic dichloride, phenyl-	$C_6H_5Cl_2OP$ 194.98	824-72-6 1	 258	4-16-00-01074 1.197^{25}	DMSO-d_6 2 1.5581^{25}
9549	Phosphonothioic acid, methyl-, *O,O*-bis(1-methylethyl) ester Phosphonic acid, thiono, methyl, diisopropyl ester	$C_7H_{17}O_2PS$ 196.25	66295-45-2	 42^{1}	4-04-00-03516 0.9885^{25}	 1.4512^{25}
9550	Phosphonothioic acid, methyl-, *O,O*-dimethyl ester	$C_3H_9O_2PS$ 140.14	681-06-1	 43^{4}	4-04-00-03514 1.1375^{25}	 1.4738^{25}
9551	Phosphonothioic acid, phenyl-, *O*-ethyl *O* (4-nitrophenyl) ester *O*-Ethyl *O*-*p*-nitrophenyl benzenethiophosphonate	$C_{14}H_{14}NO_4PS$ 323.31	2104-64-5 36	3576	4-16-00-01089 1.27^{25}	bz 4, eth 4; EtOH 4 1.5978^{30}
9552	Phosphonothioic dichloride, phenyl- Dichlorophenylphosphine sulfide	$C_6H_5Cl_2PS$ 211.05	3497-00-5	 205^{130}	4-16-00-01090 1.376^{13}	
9553	Phosphonous dichloride, (2,5-dimethylphenyl)- Phosphonous dichloride, 2,5-xylyl-	$C_8H_9Cl_2P$ 207.04	57150-66-0 -30	 253.5	3-16-00-00848 1.25^{18}	
9554	Phosphonous dichloride, (4-ethylphenyl)-	$C_8H_9Cl_2P$ 207.04	5274-50-0	 251	4-16-00-00974 1.237^{20}	 1.584^{20}
9555	Phosphonous dichloride, [4-(1-methylethyl)phenyl]- Benzene, 1-chlorophosphino-4-isopropyl	$C_9H_{11}Cl_2P$ 221.07	82906-78-3	 269	3-16-00-00848 1.190^{12}	ace 4 1.5677^{25}
9556	Phosphonous dichloride, phenyl- Dichlorophenylphosphine	$C_6H_5Cl_2P$ 178.98	644-97-3 -51	 225; 142^{57}	4-16-00-00972 1.356^{20}	bz 4 1.6030^{20}
9557	Phosphoramidothioic dichloride, dimethyl- Phosphoric acid, amide,thiono,dichloride,*N,N*-dimethyl	$C_2H_6Cl_2NPS$ 178.02	1498-65-3	 77.5^{10}	4-04-00-00293 1.3731^{20}	 1.5390^{20}
9558	Phosphoric acid, diethyl ester	$C_4H_{11}O_4P$ 154.10	598-02-7	 [illegible] dec	4-01-00-01339 [illegible]	eth 4 [illegible]
9559	Phosphoric acid, dimethyl ester	$C_2H_7O_4P$ 126.05	813-78-5	 174 dec	4-01-00-01259 1.335^{25}	H_2O 4; ace 4; EtOH 4 1.408^{25}
9560	Phosphoric acid, 2-ethylhexyl diphenyl ester	$C_{20}H_{27}O_4P$ 362.41	1241-94-7	 232^{5}	4-06-00-00718 1.090^{25}	 1.510^{25}
9561	Phosphoric acid, monoethyl ester	$C_2H_7O_4P$ 126.05	1623-14-9	 dec	4-01-00-01338 1.430^{25}	H_2O 4; ace 4; eth 4; EtOH 4 1.427
9562	Phosphoric acid tributyl ester Tributyl phosphate	$C_{12}H_{27}O_4P$ 266.32	126-73-8	9531 289	4-01-00-01531 0.9727^{25}	H_2O 3; EtOH 5; eth 3; bz 3 1.4224^{25}
9563	Phosphoric acid, triethyl ester Ethyl phosphate	$C_6H_{15}O_4P$ 182.16	78-40-0 -56.4	9589 215.5	4-01-00-01339 1.0695^{20}	H_2O 3; EtOH 4; eth 3; bz 3 1.4053^{20}
9564	Phosphoric acid, trimethyl ester	$C_3H_9O_4P$ 140.08	512-56-1 -46	 197.2	4-01-00-01259 1.2144^{20}	H_2O 4; EtOH 2; eth 3 1.3967^{20}
9565	Phosphoric acid, tripentyl ester	$C_{15}H_{33}O_4P$ 308.40	2528-38-3	 225^{50}; 167^{5}	4-01-00-01645 0.9608^{20}	H_2O 1; EtOH 3; eth 3; tol 3 1.4319^{20}
9566	Phosphoric acid, triphenyl ester Triphenyl phosphate	$C_{18}H_{15}O_4P$ 326.29	115-86-6 50.5	9656 245^{11}	4-06-00-00720 1.2055^{50}	H_2O 1; EtOH 3; eth 4; bz 4

No.	Name Synonym	Mol. Form. Mol. Wt.	CAS RN mp/°C	Merck No. bp/°C	Beil. Ref. den/g cm^{-3}	Solubility n_D
9567	Phosphoric acid, tripropyl ester Tripropyl phosphate	$C_9H_{21}O_4P$ 224.24	513-08-6	 252	4-01-00-01428 1.0121^{20}	H_2O 2; EtOH 3; eth 3; chl 2 1.4165^{20}
9568	Phosphoric acid, tris(1-methylethyl) ester	$C_9H_{21}O_4P$ 224.24	513-02-0	 219	4-01-00-01478 0.9867^{20}	EtOH 4 1.4057^{20}
9569	Phosphoric acid, tris(2-methylphenyl) ester Tri-*o*-tolyl phosphate	$C_{21}H_{21}O_4P$ 368.37	78-30-8 11	9676 410	4-06-00-01979 1.1955^{20}	H_2O 1; EtOH 4; eth 4; ctc 4 1.5575^{20}
9570	Phosphoric acid, tris(3-methylphenyl) ester	$C_{21}H_{21}O_4P$ 368.37	563-04-2 25.5	 260^{15}	4-06-00-02057 1.150^{25}	H_2O 1; EtOH 2; eth 3; ctc 4 1.5575^{20}
9571	Phosphoric acid, tris(4-methylphenyl) ester	$C_{21}H_{21}O_4P$ 368.37	78-32-0 77.5	 224^{35}	4-06-00-02130 1.247^{25}	EtOH 3; eth 3; bz 3; chl 3
9572	Phosphoric acid, tris(2-methylpropyl) ester Triisobutyl phosphate	$C_{12}H_{27}O_4P$ 266.32	126-71-6	 264	4-01-00-01598 0.9681^{20}	H_2O 4; bz 4; eth 4; EtOH 4 1.4193^{20}
9573	Phosphoric triamide, hexamethyl- Tris(dimethylamino)phosphine oxide	$C_6H_{18}N_3OP$ 179.20	680-31-9	4568 232.5	4-04-00-00284 1.03^{20}	EtOH 3; eth 3 1.4579^{20}
9574	Phosphorochloridic acid, diethyl ester Diethyl chlorophosphate	$C_4H_{10}ClO_3P$ 172.55	814-49-3	 93.5	4-01-00-01345 1.205^{19}	 1.4170^{20}
9575	Phosphorochloridic acid, diphenyl ester	$C_{12}H_{10}ClO_3P$ 268.64	2524-64-3	 $314-6^{272}$	4-06-00-00737 1.296^{25}	tfa 3 1.5500^{20}
9576	Phosphorochloridothioic acid, *O,O*-diethyl ester Diethyl chlorothiophosphate	$C_4H_{10}ClO_2PS$ 188.61	2524-04-1	 45^{3}	4-01-00-01352	ctc 3
9577	Phosphorodichloridic acid, ethyl ester Ethyl dichlorophosphate	$C_2H_5Cl_2O_2P$ 162.94	1498-51-7	 $60-5^{10}$	4-01-00-01345	 1.4338^{20}
9578	Phosphorodichloridothioic acid, *O*-(2,4,5-trichlorophenyl) ester Phenol, 2,4,5-trichloro-, *O*-phosphorodichloridothioate	$C_6H_2Cl_5OPS$ 330.38	16805-78-0	 110^{1}	4-06-00-00996 1.6653^{20}	 1.6084^{20}
9579	Phosphorodithioic acid, *O,O,S*-triethyl ester	$C_6H_{15}O_2PS_2$ 214.29	2524-09-6	 128^{20}	4-01-00-01402 1.1336^{0}	 1.5043
9580	Phosphorofluoridic acid, bis(1-methylethyl) ester Isoflurophate	$C_6H_{14}FO_3P$ 184.15	55-91-4	5060 62^{9}	4-01-00-01480 1.055^{25}	H_2O 2; eth 3; oils 4; lig 2 1.3830^{25}
9581	Phosphorothioic acid, *O*-(2,4-dichlorophenyl) *O,O*-diethyl ester Dichlofenthion	$C_{10}H_{13}Cl_2O_3PS$ 315.16	97-17-6	3030	4-06-00-00939	ctc 3; CS_2 3
9582	Phosphorothioic acid, *O,O*-diethyl *O*-[6-methyl-2-(1-methylethyl)-4-pyrimidinyl] ester Diazinon	$C_{12}H_{21}N_2O_3PS$ 304.35	333-41-5	2978 $85-90^{0.05}$	5-23-11-00187 1.1088^{20}	 1.4922^{20}
9583	Phosphorothioic acid, *O,O*-diethyl *O*-(4-nitrophenyl) ester Parathion	$C_{10}H_{14}NO_5PS$ 291.26	56-38-2 6.1	6983 375	4-06-00-01337 1.2681^{20}	H_2O 1; EtOH 5; eth 3; ace 3 1.5370^{25}
9584	Phosphorothioic acid, *O,O*-dimethyl *O*-(2,4,5-trichlorophenyl) ester Ronnel	$C_8H_8Cl_3O_3PS$ 321.55	299-84-3 41	8239 $151-4^{0.4}$	4-06-00-00994 1.44^{32}	 1.5333^{35}
9585	Phosphorothioic acid, *O,O,O*-triethyl ester	$C_6H_{15}O_3PS$ 198.22	126-68-1	 100^{16}	4-01-00-01348 1.082^{25}	 1.4480^{20}
9586	Phosphorous acid, bis(phenylmethyl) ester Dibenzyl phosphite	$C_{14}H_{15}O_3P$ 262.25	17176-77-1 -5	2996 $162^{0.1}$	4-06-00-02571	 1.5521^{18}
9587	Phosphorous acid, diethyl ester	$C_4H_{11}O_3P$ 138.10	123-22-8	 87^{20}	 1.0720^{20}	EtOH 3; eth 3 1.4101^{20}
9588	Phosphorous acid, dimethyl ester	$C_2H_7O_3P$ 110.05	868-85-9	 170.5	 1.2002^{20}	EtOH 3; ctc 2; py 3 1.4036^{20}
9589	Phosphorous acid, 2-ethylhexyl diphenyl ester	$C_{20}H_{27}O_3P$ 346.41	15647-08-2	 $148-56^{0.15}$	4-06-00-00695 1.054^{20}	 1.5207^{27}
9590	Phosphorous acid, tributyl ester	$C_{12}H_{27}O_3P$ 250.32	102-85-2	 122^{12}	4-01-00-01527 0.9259^{20}	EtOH 3; eth 4; ctc 2 1.4321^{19}
9591	Phosphorous acid, triethyl ester Triethyl phosphite	$C_6H_{15}O_3P$ 166.16	122-52-1	 157.9	4-01-00-01333 0.9629^{20}	H_2O 1; EtOH 4; eth 4 1.4127^{20}
9592	Phosphorous acid, trimethyl ester Trimethyl phosphite	$C_3H_9O_3P$ 124.08	121-45-9	 111.5	4-01-00-01256 1.0518^{20}	EtOH 4; eth 4; ctc 2 1.4095^{20}
9593	Phosphorous acid, triphenyl ester	$C_{18}H_{15}O_3P$ 310.29	101-02-0 25	 360	4-06-00-00695 1.1842^{20}	H_2O 1; EtOH 4; os 4 1.5900^{20}
9594	Phosphorous acid, tripropyl ester Propyl phosphite	$C_9H_{21}O_3P$ 208.24	923-99-9	 206.5	4-01-00-01426 0.9417^{20}	eth 4; EtOH 4 1.4282^{20}
9595	Phosphorous acid, tris(1-methylethyl) ester	$C_9H_{21}O_3P$ 208.24	116-17-6	 $60-1^{10}$	4-01-00-01476 0.9685^{18}	EtOH 3; eth 3; chl 3 1.4085^{25}
9596	Phosphorous acid, tris(2-methylphenyl) ester	$C_{21}H_{21}O_3P$ 352.37	2622-08-4	 238^{11} 198^{2}	4-06-00-01977 1.1423^{20}	eth 3; chl 2 1.5740^{28}
9597	Phosphorous acid, tris(4-methylphenyl) ester	$C_{21}H_{21}O_3P$ 352.37	620-42-8	 $250-5^{10}$	4-06-00-02128 1.1280^{25}	eth 4 1.5703^{28}
9598	*p*-Phthalaldehydic acid, 3-hydroxy- 3-Hydroxybenzoic acid, 4-formyl	$C_8H_6O_4$ 166.13	619-12-5 234	 sub	4-10-00-03623	eth 4; EtOH 4
9599	Phthalazine 2,3-Benzodiazine	$C_8H_6N_2$ 130.15	253-52-1 90.5	7344 316	5-23-07-00149	H_2O 3; EtOH 3; eth 2; bz 3
9600	1,4-Phthalazinedione, 5-amino-2,3-dihydro- Luminol	$C_8H_7N_3O_2$ 177.16	521-31-3 330.5	5470	4-25-00-04192	H_2O 1; EtOH 2; eth 2; alk 4
9601	Picene Benzo[a]chrysene	$C_{22}H_{14}$ 278.35	213-46-7 368	7368 519	4-05-00-02724	H_2O 1; EtOH 2; bz 2; chl 2
9602	Piperazine Diethylenediamine	$C_4H_{10}N_2$ 86.14	110-85-0 106	7431 146	5-23-01-00030	H_2O 4; EtOH 3; eth 1; chl 3 1.446^{113}

No.	Name Synonym	Mol. Form. Mol. Wt.	CAS RN mp/°C	Merck No. bp/°C	Beil. Ref. den/g cm^{-3}	Solubility n_D
9603	1-Piperazinecarboxamide, *N,N*-diethyl-4-methyl- Diethylcarbamazine	$C_{10}H_{21}N_3O$ 199.30	90-89-1 48	3106 109-11^{3}	4-23-00-00225	
9604	1-Piperazinecarboxylic acid, ethyl ester	$C_7H_{14}N_2O_2$ 158.20	120-43-4	 237	5-23-01-00055	H_2O 4; eth 4; EtOH 4 1.4760^{25}
9605	Piperazine, 1-[(4-chlorophenyl)phenylmethyl]-4-[(3-methylphenyl)methyl]- Meclizine	$C_{25}H_{27}ClN_2$ 390.96	569-65-3	5657 230	5-23-01-00235	CS_2 3
9606	1,4-Piperazinedicarboxylic acid, diethyl ester *N,N'*-Piperazinedicarboxylic acid, diethyl ester	$C_{10}H_{18}N_2O_4$ 230.26	5470-28-0 47	 315	4-23-00-00254	eth 4; EtOH 4
9607	Piperazine, dihydrochloride Diethylenediamine dihydrochloride	$C_4H_{12}Cl_2N_2$ 159.06	142-64-3		5-23-01-00030	H_2O 2; EtOH 1
9608	Piperazine, 1,4-dimethyl-	$C_6H_{14}N_2$ 114.19	106-58-1	 131	5-23-01-00128 0.8600^{20}	H_2O 4; EtOH 4; eth 4 1.4474^{20}
9609	Piperazine, 2,5-dimethyl-, *cis*- *cis*-2,5-Dimethylpiperazine	$C_6H_{14}N_2$ 114.19	6284-84-0 114	 162	5-23-03-00304	H_2O 4; EtOH 4; eth 2; bz 2 1.4720^{20}
9610	Piperazine, 2,5-dimethyl-, *trans*- *trans*-2,5-Dimethylpiperazine	$C_6H_{14}N_2$ 114.19	2815-34-1 118.5	 162	5-23-03-00304	H_2O 4; EtOH 4; eth 2; bz 2
9611	Piperazine, 2,6-dimethyl-, *cis*- *cis*-2,6-Dimethylpiperazine	$C_6H_{14}N_2$ 114.19	21655-48-1 114	 160	5-23-03-00326	H_2O 3; EtOH 3; eth 1; bz 2
9612	2,5-Piperazinedione Glycine anhydride	$C_4H_6N_2O_2$ 114.10	106-57-0 312 dec	7434 sub 260	5-24-05-00294	H_2O 2; EtOH 2; HCl 3
9613	Piperazine, 1,4-diphenyl-	$C_{16}H_{18}N_2$ 238.33	613-39-8 165	 300 dec; 232^{12}	5-23-01-00173	H_2O 1; EtOH 2
9614	1,4-Piperazinedipropanamine 1,4-Bis(3-aminopropyl)piperazine	$C_{10}H_{24}N_4$ 200.33	7209-38-3 15	 150-2^{2}	5-23-01-00356 0.973^{25}	 1.5015^{20}
9615	1-Piperazineethanamine 1-(2-Aminoethyl)piperazine	$C_6H_{15}N_3$ 129.21	140-31-8	 220	5-23-01-00257 0.985^{25}	 1.4983^{20}
9616	1-Piperazineethanol	$C_6H_{14}N_2O$ 130.19	103-76-4	 246	5-23-01-00406 1.061^{25}	 1.5065^{20}
9617	1-Piperazineethanol, 4-[3-(2-chloro-10H-phenothiazin-10-yl)propyl]- Perphenazine	$C_{21}H_{26}ClN_3OS$ 403.98	58-39-9 97	7135	4-27-00-01307	
9618	Piperazine, 1-methyl-	$C_5H_{12}N_2$ 100.16	109-01-3	 138	5-23-01-00082	H_2O 4; eth 4; EtOH 4 1.4378^{20}
9619	Piperazine, 2-methyl- 2-Methylpiperazine	$C_5H_{12}N_2$ 100.16	109-07-9 62	 153	5-23-03-00267	H_2O 4; EtOH 3; eth 3; bz 3
9620	Piperazine, 1-phenyl-	$C_{10}H_{14}N_2$ 162.23	92-54-6	 286.5; 161^{15}	5-23-01-00160 1.0621^{20}	H_2O 1; EtOH 5; eth 5; chl 3 1.5875^{20}
9621	Piperazine, 1-(phenylmethyl)-	$C_{11}H_{16}N_2$ 176.26	2759-28-6	 145-7^{12}	5-23-01-00193	H_2O 3; EtOH 3; eth 3; chl 2 1.5430^{28}
9622	Piperazine, 1,2,4-trimethyl- 1,2,4-Trimethylpiperazine	$C_7H_{16}N_2$ 128.22	120-85-4	 149.5	5-23-03-00271	ctc 3 1.4433^{20}
9623	Piperazinium, 4-(2-cyclohexyl-2-hydroxy-2-phenylethyl)-1,1-dimethyl-, methyl sulfate (salt) Hexocyclium methyl sulfate	$C_{21}H_{36}N_2O_5S$ 428.59	115-63-9	4627		
9624	1-Piperidinamine	$C_5H_{12}N_2$ 100.16	2213-43-6	 147	5-20-03-00516 0.928^{25}	 1.4750^{20}
9625	1-Piperidinamine, 2,6-dimethyl-	$C_7H_{16}N_2$ 128.22	39135-39-2	 65-80^{30}	5-20-04-00187 0.865^{25}	 1.4650^{20}
9626	3-Piperidinamine, 1-ethyl-	$C_7H_{16}N_2$ 128.22	6789-94-2	 155; 82^{40}	5-22-08-00026 0.923^{25}	 1.4715^{20}
9627	4-Piperidinamine, 1-methyl-*N*-phenyl-*N*-(phenylmethyl)- Bamipine	$C_{19}H_{24}N_2$ 280.41	4945-47-5 115	966	5-22-08-00036	
9628	4-Piperidinamine, 2,2,6,6-tetramethyl-	$C_9H_{20}N_2$ 156.27	36768-62-4 17	 188.5	5-22-08-00162 0.912^{25}	 1.4706^{20}
9629	Piperidine Azacyclohexane	$C_5H_{11}N$ 85.15	110-89-4 -11.03	7438 106.2	5-20-02-00003 0.8606^{20}	H_2O 5; EtOH 5; eth 3; ace 3 1.4530^{20}
9630	Piperidine, 1-acetyl-	$C_7H_{13}NO$ 127.19	618-42-8 -13.4	 226.5	5-20-02-00440 1.011^{9}	H_2O 4; EtOH 4 1.4790^{25}
9631	Piperidine, 1-acetyl-3-methyl-	$C_8H_{15}NO$ 141.21	4593-16-2 -13.6	 239	5-20-04-00107 0.9684^{25}	H_2O 4 1.4731^{25}
9632	Piperidine, 1-[5-(1,3-benzodioxol-5-yl)-1-oxo-2,4-pentadienyl]-, (*E,E*)- Piperine	$C_{17}H_{19}NO_3$ 285.34	94-62-2 131.5	7442	5-20-03-00469	H_2O 1; EtOH 3; eth 2; bz 3
9633	Piperidine, 1-[5-(1,3-benzodioxol-5-yl)-1-oxo-2,4-pentadienyl]-, (*Z,Z*)- Chavicine	$C_{17}H_{19}NO_3$ 285.34	495-91-0	2038	5-20-03-00469	eth 4; EtOH 4; peth 4
9634	Piperidine, 1-benzoyl-	$C_{12}H_{15}NO$ 189.26	776-75-0 49	 320.5	5-20-02-00450	H_2O 1; EtOH 3; eth 3; ctc 2
9635	Piperidine, 1-butyl-	$C_9H_{19}N$ 141.26	4945-48-6	 176	5-20-02-00029 0.8245^{20}	 1.4467^{20}
9636	Piperidine, 3-butyl- 3-Butylpiperidine	$C_9H_{19}N$ 141.26	13603-21-9	 196.5	5-20-04-00230 0.8620^{18}	eth 4; EtOH 4; chl 4
9637	Piperidine, 4-butyl- 4-*n*-Butylpiperidine	$C_9H_{19}N$ 141.26	24152-39-4	 194	4-20-00-01635 0.879^{23}	 1.4472^{23}
9638	1-Piperidinecarboxaldehyde	$C_6H_{11}NO$ 113.16	2591-86-8 -30.8	 222.5	5-20-02-00432 1.0158^{25}	H_2O 5; EtOH 5; eth 5; bz 5 1.4805^{25}

No.	Name	Synonym	Mol. Form.	Mol. Wt.	CAS RN	mp/°C	Merck No.	bp/°C	Beil. Ref.	den/g cm^{-3}	Solubility	n_D
9639	2-Piperidinecarboxamide, *N*-(2,6-dimethylphenyl)-1-methyl-	Mepivacaine	$C_{15}H_{22}N_2O$	246.35	96-88-8		5748		5-22-01-00223		CS_2 3	
9640	2-Piperidinecarboxylic acid	Pipecolic acid	$C_6H_{11}NO_2$	129.16	535-75-1	264	7425		5-22-01-00220		H_2O 3; EtOH 2	
9641	3-Piperidinecarboxylic acid	Nipecotic acid	$C_6H_{11}NO_2$	129.16	498-95-3	261 dec	6478		5-22-01-00235		H_2O 4	
9642	4-Piperidinecarboxylic acid	Isonipecotic acid	$C_6H_{11}NO_2$	129.16	498-94-2	336	5075		5-22-01-00244			
9643	4-Piperidinecarboxylic acid, ethyl ester		$C_8H_{15}NO_2$	157.21	1126-09-6			$100\text{-}1^{10}$	4-22-00-00128		H_2O 4; bz 4; eth 4; EtOH 4	1.4591^{20}
9644	4-Piperidinecarboxylic acid, methyl ester	Methyl isonipecotate	$C_7H_{13}NO_2$	143.19	2971-79-1			$107\text{-}10^{22}$	4-22-00-00128		H_2O 4; bz 4; eth 4; EtOH 4	1.4635^{25}
9645	4-Piperidinecarboxylic acid, 1-methyl-, methyl ester	Methyl *N*-methylisonipecotinate	$C_8H_{15}NO_2$	157.21	1690-75-1			98^{20}; 84^{10}	5-22-01-00245		H_2O 3; EtOH 3; eth 3; bz 3	1.4539^{24}
9646	2,6-Piperidinediethanol, 1-methyl-α,α'-diphenyl-, [2α(*R**),6α(*S**)]-	Lobelanidine	$C_{22}H_{29}NO_2$	339.48	552-72-7	150	5429		4-21-00-02386		H_2O 1; EtOH 3; eth 2; ace 4	
9647	Piperidine, 2,5-diethyl-		$C_9H_{19}N$	141.26	116836-21-6			190	0-20-00-00128	0.8722^{25}	eth 4; chl 4	
9648	Piperidine, 1,2-dimethyl-, (±)-	Piperidine, *N*,2-dimethyl (*DL*)	$C_7H_{15}N$	113.20	2512-81-4			127.5	5-20-04-00072	0.824^{15}	H_2O 4; eth 4; EtOH 4	1.4395^{20}
9649	Piperidine, 2,5-dimethyl-	2,5-Dimethylpiperidine	$C_7H_{15}N$	113.20	34893-50-0			137	5-20-04-00179	0.8317^{20}	EtOH 4	1.4452^{25}
9650	Piperidine, 2,6-dimethyl-		$C_7H_{15}N$	113.20	504-03-0			127	5-20-04-00180	0.8158^{25}	H_2O 5; EtOH 5; eth 5; ctc 2	1.4377^{20}
9651	Piperidine, 3,3-dimethyl-	3,3-Dimethylpiperidine	$C_7H_{15}N$	113.20	1193-12-0			137	5-20-04-00189		EtOH 4	1.4452^{25}
9652	Piperidine, 3,5-dimethyl-	3,5-Lupetidine	$C_7H_{15}N$	113.20	35794-11-7			144	5-20-04-00191	0.853^{25}		1.4454^{20}
9653	Piperidine, 1-(1,1-dimethylethyl)-	Piperidine, *N-tert*-butyl	$C_9H_{19}N$	141.26	14446-69-6			166	5-20-02-00031	0.8465^{20}	eth 4; EtOH 4	1.4532^{20}
9654	2,4-Piperidinedione, 3,3-diethyl-	Piperidione	$C_9H_{15}NO_2$	169.22	77-03-2	104	7439		5-21-09-00609		H_2O 4; EtOH 4; chl 4; MeOH 4	
9655	2,6-Piperidinedione, 4-[2-(3,5-dimethyl-2-oxocyclohexyl)-2-hydroxyethyl]-,	Cycloheximide	$C_{15}H_{23}NO_4$	281.35	66-81-9	119	2734		5-21-13-00434		EtOH 4	
9656	2,6-Piperidinedione, 4-ethyl-4-methyl-	Bemegride	$C_8H_{13}NO_2$	155.20	64-65-3	126.5	1036	sub 100	5-21-09-00601		chl 3	
9657	Piperidine, 1-[2-[3-(2,2-diphenylethyl)-1,2,4-oxadiazol-5-yl]ethyl]-, monohydrochloride	Prenoxdiazine hydrochloride	$C_{23}H_{28}ClN_3O$	397.95	982-43-4	186.5	7743					
9658	Piperidine, 1-(3,3-diphenylpropyl)-	Fenpiprane	$C_{20}H_{25}N$	279.43	3540-95-2	41.5	3934	$210\text{-}20^{8}$	5-20-02-00091			
9659	Piperidine, 1,1'-dithiobis-		$C_{10}H_{20}N_2S_2$	232.41	10220-20-9	64.2		160.1^{11}	5-20-03-00487		acid 1	
9660	Piperidine, 1-dodecyl-		$C_{17}H_{35}N$	253.47	5917-47-5			161^{5}; $115^{0.6}$	4-20-00-00321	0.8378^{20}		1.4588^{20}
9661	Piperidine, 1,1'-(1,2-ethanediyl)bis-		$C_{12}H_{24}N_2$	196.34	1932-04-3	-0.5		265	5-20-03-00148	0.9160^{25}		1.4853^{25}
9662	1-Piperidineethanol		$C_7H_{15}NO$	129.20	3040-44-6	17.9		202; 90^{12}	5-20-02-00101	0.9703^{25}	H_2O 5; EtOH 4	1.4749^{20}
9663	2-Piperidineethanol	2-(2-Hydroxyethyl)piperidine	$C_7H_{15}NO$	129.20	1484-84-0	69		202; 145^{36}	5-21-01-00121	1.01^{27}	H_2O 4	
9664	4-Piperidineethanol	4-(2-Hydroxyethyl)piperidine	$C_7H_{15}NO$	129.20	622-26-4	132.5		227.5	4-21-00-00093	1.0059^{15}	H_2O 4; eth 4; EtOH 4	1.4907^{20}
9665	1-Piperidineethanol, 2-(diphenylmethyl)-	Diphemethoxidine	$C_{20}H_{25}NO$	295.42	13862-07-2	106.5	3303	$180\text{-}1^{0.1}$	5-20-08-00173			
9666	Piperidine, 3-ethenyl-	Piperidine, 3-vinyl-	$C_7H_{13}N$	111.19	57502-49-5			153.5	4-20-00-01946	0.9274^{25}		1.4731^{25}
9667	Piperidine, 1-ethyl-	*N*-Ethylpiperidine	$C_7H_{15}N$	113.20	766-09-6			130.8	5-20-02-00025	0.8237^{20}		1.4480^{20}
9668	Piperidine, 2-ethyl-, (±)-	Piperidine, 2-ethyl (*DL*)	$C_7H_{15}N$	113.20	78738-37-1			142.5	5-20-04-00169	0.8649^{0}		1.4494^{21}
9669	Piperidine, 3-ethyl-, (±)-	Piperidine, 3-ethyl (*DL*)	$C_7H_{15}N$	113.20	59433-08-8			152.6	5-20-04-00173	0.8565^{23}	ace 4	1.4531^{20}
9670	Piperidine, 4-ethyl-	4-Ethylpiperidine	$C_7H_{15}N$	113.20	3230-23-7			154	5-20-04-00174	0.8759^{25}		1.4503^{25}
9671	Piperidine, 1-(1-ethyl-1-methylpropyl)-	Piperidine, *N*-(3-methyl-3-pentyl)	$C_{11}H_{23}N$	169.31	14045-26-2			214	5-20-02-00033	0.8614^{20}		1.4637^{20}
9672	Piperidine, 1-hexyl-	1-Hexylpiperidine	$C_{11}H_{23}N$	169.31	7335-01-5			219.2	5-20-02-00032	0.8292^{20}		1.4522^{20}
9673	4-Piperidinemethanamine	4-(Aminomethyl)piperidine	$C_6H_{14}N_2$	114.19	7144-05-0	25		200; 31^{10}	5-22-08-00128			1.4900^{20}
9674	2-Piperidinemethanol		$C_6H_{13}NO$	115.18	3433-37-2	69		104^{10}; 80^{1}	5-21-01-00098		chl 2	
9675	3-Piperidinemethanol		$C_6H_{13}NO$	115.18	4606-65-9	61		$106\text{-}7^{3.5}$	5-21-01-00107	1.0263^{20}	chl 2	1.4964^{20}

No.	Name Synonym	Mol. Form. Mol. Wt.	CAS RN mp/°C	Merck No. bp/°C	Beil. Ref. den/g cm^{-3}	Solubility n_D
9676	2-Piperidinemethanol, α,α-diphenyl- Pipradrol	$C_{18}H_{21}NO$ 267.37	467-60-7	7455	5-21-04-00070	
9677	2-Piperidinemethanol, α-ethyl-, (*R**,*S**)-(±)- Conhydrine (*DL*) (lower m.p.)	$C_8H_{17}NO$ 143.23	3238-62-8 121	2495 226	5-21-01-00138	H_2O 2; bz 4; eth 4; EtOH 4
9678	3-Piperidinemethanol, 1-methyl-	$C_7H_{15}NO$ 129.20	7583-53-1	 142.5; 114^{20}	5-21-01-00108 1.013^{25}	 1.4772^{20}
9679	Piperidine, 1-methyl- *N*-Methylpiperidine	$C_6H_{13}N$ 99.18	626-67-5	 107	5-20-02-00021 0.8159^{20}	H_2O 4; EtOH 5; eth 5; ctc 3 1.4355^{20}
9680	Piperidine, 2-methyl-, (±)-	$C_6H_{13}N$ 99.18	3000-79-1 -4.9	 118	5-20-04-00071 0.8436^{24}	H_2O 4; EtOH 3; eth 3; chl 2 1.4459^{20}
9681	Piperidine, 3-methyl-, (±)- Piperidine, 3-methyl (*DL*)	$C_6H_{13}N$ 99.18	53152-98-0 -24	 125.5	5-20-04-00099 0.8446^{26}	H_2O 4; chl 2 1.4470^{20}
9682	Piperidine, 4-methyl-	$C_6H_{13}N$ 99.18	626-58-4	 130	5-20-04-00116 0.8674^{25}	H_2O 4; chl 2 1.4458^{20}
9683	Piperidine, 1,1'-methylenebis-	$C_{11}H_{22}N_2$ 182.31	880-09-1	 $122\text{-}3^{15}$	5-20-02-00293 0.915^{25}	 1.4820^{20}
9684	Piperidine, 1-(1-methylethyl)-	$C_8H_{17}N$ 127.23	766-79-0	 149.5	5-20-02-00028 0.8389^{20}	H_2O 4 1.4491^{20}
9685	Piperidine, 1-(2-methylpropyl)-	$C_9H_{19}N$ 141.26	10315-89-6	 160.5	5-20-02-00030 0.8161^{25}	H_2O 4 1.4382^{25}
9686	Piperidine, 1-methyl-2-propyl-, (*S*)- Methylconiine	$C_9H_{19}N$ 141.26	35305-13-6	5968 174	5-20-04-00206 0.8326^{22}	ace 4; EtOH 4 1.4538^{12}
9687	Piperidine, 1-nitro-	$C_5H_{10}N_2O_2$ 130.15	7119-94-0 -5.5	 245; 121^{20}	5-20-03-00531 1.1519^{26}	 1.4954^{26}
9688	Piperidine, 1-nitroso- *N*-Nitrosopiperidine	$C_5H_{10}N_2O$ 114.15	100-75-4	 219; 109^{20}	5-20-03-00530 1.0631^{18}	H_2O 3; HCl 3 1.4933^{18}
9689	Piperidine, 1-nonyl-	$C_{14}H_{29}N$ 211.39	30538-80-8	 $135\text{-}7^{11}$	4-20-00-00319 0.8313^{25}	 1.4538^{25}
9690	Piperidine, 1-octyl-	$C_{13}H_{27}N$ 197.36	7335-02-6	 137^{13}; 89^{1}	5-20-02-00034 0.8324^{20}	 1.4544^{20}
9691	Piperidine, 1,2,2,6,6-pentamethyl- Pempidine	$C_{10}H_{21}N$ 155.28	79-55-0	7022 147	5-20-04-00238 0.8580^{0}	 1.4550^{21}
9692	Piperidine, 1-pentyl-	$C_{10}H_{21}N$ 155.28	10324-58-0	 198.2	5-20-02-00031 0.8282^{20}	 1.4498^{20}
9693	Piperidine, 1-phenyl-	$C_{11}H_{15}N$ 161.25	4096-20-2 4.7	 258	5-20-02-00057 0.9944^{25}	EtOH 4; eth 4; bz 4; chl 4 1.5598^{25}
9694	Piperidine, 3-phenyl-	$C_{11}H_{15}N$ 161.25	3973-62-4 14.5	 257; 140^{19}	5-20-06-00405 1.0010^{25}	 1.5473^{25}
9695	Piperidine, 4-phenyl- 4-Phenylpiperidine	$C_{11}H_{15}N$ 161.25	771-99-3 60.5	 257	5-20-06-00408 0.9996^{16}	chl 3
9696	Piperidine, 1-(1-phenylcyclohexyl)- Phencyclidine	$C_{17}H_{25}N$ 243.39	77-10-1 46.5	7179 $135\text{-}7^{1.0}$	5-20-02-00078	
9697	Piperidine, 1-(1-phenylethyl)-	$C_{13}H_{19}N$ 189.30	7529-63-7 -13	 272	5-20-02-00064 0.945^{25}	 1.5218^{25}
9698	Piperidine, 1-(phenylmethyl)-	$C_{12}H_{17}N$ 175.27	2905-56-8	 245	5-20-02-00062 0.9625^{16}	 1.5227^{20}
9699	Piperidine, 2-(phenylmethyl)- 2-Benzylpiperidine	$C_{12}H_{17}N$ 175.27	32838-55-4 32	 267.5	5-20-06-00438 0.9660^{25}	eth 4; EtOH 4
9700	Piperidine, 4-(phenylmethyl)- 4-Benzylpiperidine	$C_{12}H_{17}N$ 175.27	31252-42-3 16.8	 270; 150^{17}	5-20-06-00439 0.9970^{20}	H_2O 1; EtOH 3; eth 3 1.5337^{25}
9701	Piperidine, 4,4'-(1,3-propanediyl)bis-	$C_{13}H_{26}N_2$ 210.36	16898-52-5 67.1	 329	5-23-04-00101	H_2O 4
9702	Piperidine, 4,4'-(1,3-propanediyl)bis[1-methyl- Propane, 1,3-bis(1-methyl-4-piperidyl)	$C_{15}H_{30}N_2$ 238.42	64168-11-2 13.7	 215^{50}	4-23-00-00536 0.8962^{25}	 1.4804^{25}
9703	1-Piperidinepropanenitrile	$C_8H_{14}N_2$ 138.21	3088-41-3 -6.8	 145^{50}	5-20-03-00047 0.9403^{25}	 1.4676^{25}
9704	1-Piperidinepropanoic acid, ethyl ester Piperidine, *N*-carboxyethyl, ethyl ester	$C_{10}H_{19}NO_2$ 185.27	19653-33-9	 139^{50}	5-20-03-00042 0.9627^{25}	H_2O 4 1.4525^{25}
9705	1-Piperidinepropanol, α-cyclohexyl-α-phenyl-, hydrochloride Trihexyphenidyl hydrochloride	$C_{20}H_{32}ClNO$ 337.93	52-49-3 258.5	9607		
9706	Piperidine, 2-(1-propenyl)-, (±)-	$C_8H_{15}N$ 125.21	5913-85-9 6	 168	5-20-04-00346 0.8716^{15}	eth 4; EtOH 4
9707	Piperidine, 1-propyl-	$C_8H_{17}N$ [illegible]	5470-02-0	 [illegible]	5-20-02-00027 [illegible]	H_2O 4; eth 4; EtOH 4 [illegible]
9708	Piperidine, 2-propyl-, (*S*)- Coniine	$C_8H_{17}N$ 127.23	458-88-8 -2	2500 166.5	5-20-04-00205 0.8440^{20}	H_2O 2; EtOH 5; eth 4; bz 3 1.4512^{22}
9709	Piperidine, 4-propyl- 4-Propylpiperidine	$C_8H_{17}N$ 127.23	22398-09-0	 173	5-20-04-00208 0.864^{22}	EtOH 3; chl 2 1.4465^{23}
9710	Piperidine, 2,2,4,6-tetramethyl- 2,2,4,6-Tetramethylpiperidine	$C_9H_{19}N$ 141.26	6292-82-6	 156	4-20-00-01638 0.816^{25}	 1.4374^{25}
9711	Piperidine, 2,2,6,6-tetramethyl- Norpempidine	$C_9H_{19}N$ 141.26	768-66-1 28	 156	5-20-04-00237 0.8367^{16}	eth 4 1.4455^{20}
9712	Piperidine, 2,2,4-trimethyl-	$C_8H_{17}N$ 127.23	101257-71-0	 148	4-20-00-01624 0.832^{15}	eth 4; EtOH 4 1.4458^{20}
9713	Piperidine, 2,3,6-trimethyl- 2,3,6-Trimethylpiperidine	$C_8H_{17}N$ 127.23	50402-72-7	 36^{5}	4-20-00-01625 0.8302^{20}	 1.4434^{20}
9714	Piperidine, 2,4,6-trimethyl- 2,4,6-Trimethylpiperidine	$C_8H_{17}N$ 127.23	21974-48-1 <20	 145	5-20-04-00220 0.8315^{19}	H_2O 2 1.4412^{20}
9715	3-Piperidinol, 1-ethyl- 1-Ethyl-3-piperidinol	$C_7H_{15}NO$ 129.20	13444-24-1	3799 $93\text{-}5^{15}$	5-21-01-00039	 1.4777^{14}

No.	Name Synonym	Mol. Form. Mol. Wt.	CAS RN mp/°C	Merck No. bp/°C	Beil. Ref. den/g cm^{-3}	Solubility n_D
9716	3-Piperidinol, 1-methyl-	$C_6H_{13}NO$ 115.18	3554-74-3	$76\text{-}8^{11}$	5-21-01-00029 0.999^{25}	1.4735^{20}
9717	4-Piperidinol, 1-methyl-	$C_6H_{13}NO$ 115.18	106-52-5 29	200	5-21-01-00052	1.4775^{20}
9718	3-Piperidinol, 6-propyl-, (3*S*-*trans*)- Pseudoconhydrine	$C_8H_{17}NO$ 143.23	140-55-6 106	7928 236	5-21-01-00138	H_2O 4; eth 4; EtOH 4
9719	2-Piperidinone	C_5H_9NO 99.13	675-20-7 39.5	256	5-21-06-00396	H_2O 4; EtOH 4; eth 4; dil acid 3
9720	4-Piperidinone, 1-acetyl-	$C_7H_{11}NO_2$ 141.17	32161-06-1	218; $124^{0.2}$	5-21-06-00426 1.146^{25}	1.5026^{20}
9721	2-Piperidinone, 1-methyl-	$C_6H_{11}NO$ 113.16	931-20-4	105^{12}; 94^{9}	5-21-06-00398 1.0263^{25}	1.4820^{20}
9722	2-Piperidinone, 3-methyl-	$C_6H_{11}NO$ 113.16	3768-43-2 55.3	249.5	5-21-06-00479	bz 4; eth 4; EtOH 4
9723	3-Piperidinone, 1-methyl-	$C_6H_{11}NO$ 113.16	5519-50-6	$63\text{-}4^{13}$	5-21-06-00417 0.9655^{25}	1.4559^{25}
9724	4-Piperidinone, 1-methyl-	$C_6H_{11}NO$ 113.16	1445-73-4	$56\text{-}8^{11}$	5-21-06-00419 0.9696^{25}	1.4580^{25}
9725	4-Piperidinone, 1,2,3,6-tetramethyl- 1,2,5,6-Tetramethyl-4-piperidinone	$C_9H_{17}NO$ 155.24	53381-90-1	$86\text{-}8^{7.5}$	4-21-00-03272 0.9499^{20}	1.4680^{20}
9726	4-Piperidinone, 2,2,6,6-tetramethyl-	$C_9H_{17}NO$ 155.24	826-36-8 36	205	5-21-06-00538	H_2O 3; EtOH 3; eth 3; chl 2
9727	Piperonyl butoxide 1,3-Benzodioxole, 5-[[2-(2-butoxyethoxy)ethoxy]methyl]-6-propyl-	$C_{19}H_{30}O_5$ 338.44	51-03-6	7446 180^{1}	1.05^{25}	
9728	Pirimicarb Carbamic acid, dimethyl-, 2-(dimethylamino)-5,6-dimethyl-4-pyrimidinyl ester	$C_{11}H_{18}N_4O_2$ 238.29	23103-98-2 90.5	7468		
9729	Pirimiphos-ethyl Phosphorothioic acid, *O*-[2-(diethylamino)-6-methyl-4-pyrimidinyl] *O*,*O*-diethyl ester	$C_{13}H_{24}N_3O_3PS$ 333.39	23505-41-1	7469 dec >130	1.14^{20}	
9730	Pirimiphos-methyl Phosphorothioic acid, *O*-[2-(diethylamino)-6-methyl-4-pyrimidinyl] *O*,*O*-dimethyl ester	$C_{11}H_{20}N_3O_3PS$ 305.34	29232-93-7 15	dec	1.17^{20}	
9731	Plumbane, tetraethyl Tetraethyl lead	$C_8H_{20}Pb$ 323.45	78-00-2	9136 200 dec	1.653^{20}	H_2O 1; bz 3; EtOH 2 1.5198^{20}
9732	Plumbane, tetramethyl Tetramethyl lead	$C_4H_{12}Pb$ 267.34	75-74-1 -27.5	110	1.995^{20}	
9733	Plumbane, tetraphenyl-	$C_{24}H_{20}Pb$ 515.62	595-89-1 228.3	126^{13}	4-16-00-01615 1.5298^{20}	chl 3
9734	21H,23H-Porphine	$C_{20}H_{14}N_4$ 310.36	101-60-0 360	7574 sub 300	4-26-00-01900 1.336^{25}	H_2O 1; EtOH 2; eth 1; ace 1
9735	21H,23H-Porphine-2,18-dipropanoic acid, 7,12-bis(1-hydroxyethyl)-3,8,13,17-tetramethyl- Hematoporphyrin	$C_{34}H_{38}N_4O_6$ 598.70	14459-29-1 172.5	4555	4-26-00-03157	H_2O 1; EtOH 3; eth 2; chl 2
9736	Pregna-4,6-diene-21-carboxylic acid, 17-hydroxy-3-oxo-, γ-lactone, (17α)- Canrenone	$C_{22}H_{28}O_3$ 340.46	976-71-6 150	1753	5-17-11-00476	
9737	Pregna-1,4-diene-3,20-dione, 9-fluoro-11,17,21-trihydroxy-16-methyl-, (11β,16α)- Dexamethasone	$C_{22}H_{29}FO_5$ 392.47	50-02-2 262	2922		
9738	Pregna-1,4-diene-3,20-dione, 11,17,21-trihydroxy-, (11β)- Prednisolone	$C_{21}H_{28}O_5$ 360.45	50-24-8 235	7719	4-08-00-03467	
9739	Pregna-1,4-diene-3,11,20-trione, 17,21-dihydroxy- Prednisone	$C_{21}H_{26}O_5$ 358.43	53-03-2 233-5 dec	7727	4-08-00-03531	
9740	Pregnane, (5α)- Allopregnane	$C_{21}H_{36}$ 288.52	641-85-0 84.5	257	4-05-00-01215	
9741	Pregnane, (5β)- 17β-Ethyletiocholane	$C_{21}H_{36}$ 288.52	481-26-5 83.5	7731	4-05-00-01215 1.032^{15}	H_2O 1; chl 3; MeOH 3
9742	Pregnane-3,20-diol, (3α,5β,20*S*)- Pregnanediol	$C_{21}H_{36}O_2$ 320.52	80-92-2 243.5	7732	4-06-00-06111 1.15^{25}	EtOH 2; eth 2; ace 3; os 2
9743	Pregnane-3,20-dione, (5β)- 5β-Pregnane-3,20-dione	$C_{21}H_{32}O_2$ 316.48	128-23-4 123	7733	4-07-00-02198	H_2O 1; EtOH 4; eth 3; ace 3
9744	Pregnane-11,20-dione, 3,17,21-trihydroxy-, (3α,5β)- Tetrahydrocortisone	$C_{21}H_{32}O_5$ 364.48	53-05-4 190	9143	4-08-00-03395	
9745	Pregnan-20-one, 3-hydroxy-, (3α,5β)- Pregnan-3α-ol-20-one	$C_{21}H_{34}O_2$ 318.50	128-20-1 149.5	7734		EtOH 4
9746	Pregn-4-en-18-al, 11,21-dihydroxy-3,20-dioxo-, (11β)- Aldosterone	$C_{21}H_{28}O_5$ 360.45	52-39-1 166.5	218	4-08-00-03491	
9747	Pregn-4-ene-3,20-dione, 21-(acetyloxy)-11,17-dihydroxy-, (11β)- Hydrocortisone acetate	$C_{23}H_{23}O_6$ 395.43	50-03-3 223 dec	4711	4-08-00-03424 1.289^{20}	
9748	Pregn-4-ene-3,20-dione, 11,21-dihydroxy-, (11β)- Corticosterone	$C_{21}H_{30}O_4$ 346.47	50-22-6 181	2532 $190^{0.01}$	4-08-00-02907	H_2O 1; EtOH 3; eth 3; ace 3

No.	Name Synonym	Mol. Form. Mol. Wt.	CAS RN mp/°C	Merck No. bp/°C	Beil. Ref. den/g cm^{-3}	Solubility n_D
9749	Pregn-4-ene-3,20-dione, 17,21-dihydroxy- 11-Deoxy-17-hydroxycorticosterone	$C_{21}H_{30}O_4$ 346.47	152-58-9 215	2912	4-08-00-02913	ace 4; EtOH 4; chl 4
9750	Pregn-4-ene-3,20-dione, 17-hydroxy- 17α-Hydroxyprogesterone	$C_{21}H_{30}O_3$ 330.47	68-96-2	4773	4-08-00-02189	chl 2
9751	Pregn-4-ene-3,20-dione, 21-hydroxy- Deoxycorticosterone	$C_{21}H_{30}O_3$ 330.47	64-85-7 141.5	2882	4-08-00-02195	H_2O 2; EtOH 4; eth 2; ace 4
9752	Pregn-4-ene-3,20-dione, 17-hydroxy-6-methyl-, (6α)- Medroxyprogesterone	$C_{22}H_{32}O_3$ 344.49	520-85-4 214.5	5677	4-08-00-02211	chl 4
9753	Pregn-4-ene-3,20-dione, 11,17,21-trihydroxy-, (11β)- Hydrocortisone	$C_{21}H_{30}O_5$ 362.47	50-23-7 220	4710	4-08-00-03422	H_2O 2; EtOH 3; diox 3; HOAc 3
9754	Pregn-4-ene-3,11,20-trione, 17,21-dihydroxy- Cortisone	$C_{21}H_{28}O_5$ 360.45	53-06-5 222	2533	4-08-00-03480	H_2O 2; EtOH 3; eth 2; ace 3
9755	Pregn-4-ene-3,11,20-trione, 21-hydroxy- 11-Dehydrocorticosterone	$C_{21}H_{28}O_4$ 344.45	72-23-1 183.5	2859	4-08-00-02998	H_2O 1; EtOH 3; ace 3; bz 3
9756	Pregn-5-en-20-one, 3-hydroxy-, (3β)- Pregnenolone	$C_{21}H_{32}O_2$ 316.48	145-13-1 192	7739	4-08-00-01019	
9757	Pregn-4-en-3-one, 11,17,20,21-tetrahydroxy-, (11β,20*R*)- 4-Pregnene-11β,17α,20β,21-tetrol-3-one	$C_{21}H_{32}O_5$ 364.48	116-58-5 125 dec	7736	4-08-00-03384	ace 4; EtOH 4
9758	Pregn-4-en-20-yn-3-one, 17-hydroxy-, (17α)- Ethisterone	$C_{21}H_{28}O_2$ 312.45	434-03-7 272	3696	4-08-00-01225	
9759	Procyazine Propanenitrile, 2-[[4-chloro-6-(cyclopropylamino)-1,3,5-triazin-2-yl]amino]-2-methyl-	$C_{10}H_{13}ClN_6$ 252.71	32889-48-8 170			
9760	Procymidone *N*-(3,5-Dichlorophenyl)-1,2-dimethylcyclopropane-1,2-dicarboximide	$C_{13}H_{11}Cl_2NO_2$ 284.14	32809-16-8 166		1.452^{25}	
9761	Prodiamine 1,3-Benzenediamine, 2,4-dinitro-*N*3,*N*3-dipropyl-6-(trifluoromethyl)-	$C_{13}H_{17}F_3N_4O_4$ 350.30	29091-21-2 124	7773	1.47^{25}	
9762	Profenofos Phosphorothioic acid, *O*-(4-bromo-2-chlorophenyl) *O*-ethyl *S*-propyl ester	$C_{11}H_{15}BrClO_3PS$ 373.63	41198-08-7	$110^{0.001}$	1.455^{20}	
9763	Profluralin Benzenamine, *N*-(cyclopropylmethyl)-2,6-dinitro-*N*-propyl-4-(trifluoromethyl)-	$C_{14}H_{16}F_3N_3O_4$ 347.29	26399-36-0 34	7781		
9764	*L*-Proline 2-Pyrrolidinecarboxylic acid	$C_5H_9NO_2$ 115.13	147-85-3 220-2 dec	7790	5-22-01-00031	H_2O 4; EtOH 2; eth 1; ace 2
9765	*DL*-Proline	$C_5H_9NO_2$ 115.13	609-36-9 205 dec		5-22-01-00033	H_2O 4; EtOH 4
9766	*L*-Proline, 1-acetyl-4-hydroxy-, *trans*- Oxaceprol	$C_7H_{11}NO_4$ 173.17	33996-33-7 132	6857	5-22-05-00013	H_2O 4; MeOH 4
9767	*L*-Proline, 4-hydroxy-, *trans*- 4-Hydroxy-*L*-proline	$C_5H_9NO_3$ 131.13	51-35-4 274	4775	5-22-06-00007	H_2O 4; EtOH 2
9768	*L*-Proline, 5-oxo- *L*-Pyroglutamic acid	$C_5H_7NO_3$ 129.12	98-79-3 184.7	8012	5-22-06-00007	DMSO 3
9769	Promecarb Phenol, 3-methyl-5-(1-methylethyl)-, methylcarbamate	$C_{12}H_{17}NO_2$ 207.27	2631-37-0 87	7794 $117^{0.01}$		
9770	Prometryn 1,3,5-Triazine-2,4-diamine, *N,N'*-bis(1-methylethyl)-6-(methylthio)-	$C_{10}H_{19}N_5S$ 241.36	7287-19-6 119	7800	1.157^{20}	
9771	Propachlor Acetamide, 2-chloro-*N*-(1-methylethyl)-*N*-phenyl-	$C_{11}H_{14}ClNO$ 211.69	1918-16-7 77	7805 $110^{0.03}$	1.242^{25}	
9772	1,2-Propadiene Allene	C_3H_4 40.06	463-49-0 -136.2	-34.4	4-01-00-00966 *0.584^{25}	bz 4; peth 4 1.4168
9773	1,2-Propadiene-1,3-dione Carbon suboxide	C_3O_2 68.03	504-64-3 -107	1821 6.8	4-01-00-03764 1.114^{0}	eth 3; bz 3; CS_2 3 1.4538^{0}
9774	Propanal Propionaldehyde	C_3H_6O 58.08	123-38-6 -80	7835 48	4-01-00-03165 0.8657^{25}	H_2O 3; EtOH 5; eth 5 1.3636^{20}
9775	Propanal, 2-bromo-	C_3H_5BrO 136.98	19967-57-0	109.5	4-01-00-03177 1.592^{20}	eth 4 1.4813^{20}
9776	Propanal, 2-bromo-2-methyl-	C_4H_7BrO 151.00	13206-46-7	113	4-01-00-03267 1.383^{25}	1.4518^{25}
9777	Propanal, 2-chloro-	C_3H_5ClO 92.52	683-50-1	86	3-01-00-02691 1.182^{15}	bz 4; eth 4 1.431^{17}
9778	Propanal, 3-chloro-	C_3H_5ClO 92.52	19434-65-2 <20	130; 45^{10}	4-01-00-03174 1.268^{15}	eth 4; EtOH 4 1.475^{25}
9779	Propanal, 2-chloro-2-methyl-	C_4H_7ClO 106.55	917-93-1	90	4-01-00-03267 1.053^{15}	eth 4; EtOH 4 1.4160^{16}
9780	Propanal, 2,3-dibromo-	$C_3H_4Br_2O$ 215.87	5221-17-0	$73\text{-}5^{10}$	4-01-00-03178 2.198^{15}	eth 4 1.5082^{20}
9781	Propanal, 2,3-dichloro-	$C_3H_4Cl_2O$ 126.97	10140-89-3	73^{50}; 48^{14}	4-01-00-03174 1.400^{20}	1.4762^{20}
9782	Propanal, 2,3-dihydroxy-, (±)-	$C_3H_6O_3$ 90.08	56-82-6 145	$140\text{-}50^{0.8}$	3-01-00-03282 1.453^{18}	H_2O 3; EtOH 2; eth 2; bz 1

No.	Name Synonym	Mol. Form. Mol. Wt.	CAS RN mp/°C	Merck No. bp/°C	Beil. Ref. den/g cm^{-3}	Solubility n_D
9783	Propanal, 2,2-dimethyl- Pivaldehyde	$C_5H_{10}O$ 86.13	630-19-3 6	 77.5	4-01-00-03295 0.7923^{17}	EtOH 3; eth 3 1.3791^{20}
9784	Propanal, 2-(hydroxyimino)-, oxime	$C_3H_6N_2O_2$ 102.09	1804-15-5 157	 sub	4-01-00-03633	H_2O 2; EtOH 3; eth 3; bz 3
9785	Propanal, 2-methyl- Isobutanal	C_4H_8O 72.11	78-84-2 -65.9	5038 64.5	4-01-00-03262 0.7891^{20}	H_2O 3; eth 3; ace 3; ctc 2 1.3730^{20}
9786	Propanal, 3-(methylthio)- 3-(Methylthio)propionaldehyde	C_4H_8OS 104.17	3268-49-3	 62^{11}	4-01-00-03974	
9787	Propanal, oxime	C_3H_7NO 73.09	627-39-4 40	 131.5	4-01-00-03170 0.9258^{20}	 1.4287^{20}
9788	Propanal, 2-oxo- Pyruvaldehyde	$C_3H_4O_2$ 72.06	78-98-8	8030 72	4-01-00-03631 1.0455^{20}	EtOH 3; eth 3; bz 3 1.4002^{18}
9789	Propanal, 2-oxo-, 1-oxime Isonitrosoacetone	$C_3H_5NO_2$ 87.08	306-44-5 69	5076 sub	3-01-00-03092 1.0744^{67}	H_2O 3; eth 3; bz 2; ctc 2
9790	Propanal, 2,2,3-trichloro-	$C_3H_3Cl_3O$ 161.41	7789-90-4	 63-5^{45}	3-01-00-02693 1.470^{25}	eth 4 1.473^{25}
9791	Propanamide Propionamide	C_3H_7NO 73.09	79-05-0 81.3	7836 213	4-02-00-00725 0.9262^{110}	H_2O 4; EtOH 4; eth 4; chl 4 1.4180^{110}
9792	Propanamide, 2-bromo-2-methyl- Isobutyramide, 2-bromo	C_4H_8BrNO 166.02	7462-74-0 148	 145^{17}	4-02-00-00863	EtOH 4; chl 4
9793	Propanamide, *N*,*N*-diethyl-	$C_7H_{15}NO$ 129.20	1114-51-8	 191	4-04-00-00353 0.8972^{20}	EtOH 4 1.4425^{20}
9794	Propanamide, *N*,*N*-diethyl-2,2-dimethyl- Pivalamide, *N*,*N*-diethyl	$C_9H_{19}NO$ 157.26	24331-72-4	 206	3-04-00-00211 0.891^{15}	eth 4; EtOH 4
9795	Propanamide, *N*-(2,3-dihydro-1,5-dimethyl-3-oxo-2-phenyl-1H-pyrazol-4-yl)-2-(dimethylamino)- Aminopropylon	$C_{16}H_{22}N_4O_2$ 302.38	3690-04-8 181	484	4-25-00-03595	H_2O 4
9796	Propanamide, *N*-(4-ethoxyphenyl)-2-hydroxy- *p*-Lactophenetide	$C_{11}H_{15}NO_3$ 209.25	539-08-2 118	5220	3-13-00-01135	H_2O 3; EtOH 4; eth 2; bz 2
9797	Propanamide, 2-hydroxy-, (±)-	$C_3H_7NO_2$ 89.09	65144-02-7 75.5		4-03-00-00674 1.1381^{80}	H_2O 4; EtOH 4
9798	Propanamide, 2-methyl-	C_4H_9NO 87.12	563-83-7 128	 218	4-02-00-00852	chl 3
9799	Propanamide, *N*-methyl-	C_4H_9NO 87.12	1187-58-2 -30.9	 148	4-04-00-00183 0.9305^{25}	 1.4345^{25}
9800	Propanamide, *N*-phenyl-	$C_9H_{11}NO$ 149.19	620-71-3 105.5	 222.2	4-12-00-00385 1.175^{25}	H_2O 2; EtOH 4; eth 4
9801	Propanamide, *N*-phenyl-*N*-[1-(2-phenylethyl)-4-piperidinyl]- Fentanyl	$C_{22}H_{28}N_2O$ 336.48	437-38-7 87.5	3944	5-22-08-00049	
9802	Propanamide, *N*-propyl-	$C_6H_{13}NO$ 115.18	3217-86-5 154	 215; 108^9	4-04-00-00477 0.8985^{25}	H_2O 2; eth 2
9803	1-Propanamine Propylamine	C_3H_9N 59.11	107-10-8 -83	7855 47.2	4-04-00-00464 0.7173^{20}	H_2O 3; EtOH 4; eth 4; ace 4 1.3870^{20}
9804	2-Propanamine Isopropylamine	C_3H_9N 59.11	75-31-0 -95.1	5097 31.7	4-04-00-00504 0.6891^{20}	H_2O 5; EtOH 5; eth 5; ace 4 1.3742^{20}
9805	1-Propanamine, *N*,*N*-dichloro- *N*,*N*-Dichloropropylamine	$C_3H_7Cl_2N$ 128.00	69947-01-9	 117	0-04-00-00145 1.1454^{23}	 1.4525^{23}
9806	1-Propanamine, 2,2-dimethyl-	$C_5H_{13}N$ 87.16	5813-64-9	 82	4-04-00-00707 0.7455^{20}	eth 4 1.4023^{20}
9807	1-Propanamine, *N*,*N*-dimethyl- Dimethylpropylamine	$C_5H_{13}N$ 87.16	926-63-6	 66	4-04-00-00467 0.7152^{20}	bz 4; eth 4; EtOH 4 1.3860^{20}
9808	1-Propanamine, *N*,*N*-dipropyl- Tripropylamine	$C_9H_{21}N$ 143.27	102-69-2 -93.5	 156	4-04-00-00470 0.7558^{20}	eth 4; EtOH 4 1.4181^{20}
9809	1-Propanamine, *N*-ethyl-	$C_5H_{13}N$ 87.16	20193-20-8	 81	4-04-00-00468 0.7204^{17}	H_2O 2; ace 4; EtOH 4 1.3858^{25}
9810	2-Propanamine, *N*-ethyl- Ethylisopropylamine	$C_5H_{13}N$ 87.16	19961-27-4	 69.6	4-04-00-00508	 1.3872^{25}
9811	2-Propanamine, *N*-ethyl-*N*-(1-methylethyl)-	$C_8H_{19}N$ 129.25	7087-68-5	 126.5	4-04-00-00511 0.742^{25}	ctc 3 1.4138^{20}
9812	1-Propanamine, *N*-ethyl-*N*-propyl-	$C_8H_{19}N$ 129.25	20634-92-8	 138	2-04-00-00622 0.807^{24}	ace 4; bz 4; eth 4
9813	1-Propanamine, 1,1,2,2,3,3,3-heptafluoro-*N*,*N*-bis(heptafluoropropyl)-	$C_9F_{21}N$ 521.07	338-83-0	 130	4-02-00-00742 1.822^4	 1.279^{25}
9814	1-Propanamine, *N*-hydroxy-*N*-propyl- *N*,*N*-Dipropylhydroxylamine	$C_6H_{15}NO$ 117.19	7446-43-7 29	 154.5	4-04-00-03308	bz 4; eth 4; EtOH 4; chl 4
9815	2-Propanamine, *N*,*N*'-methanetetraylbis-	$C_7H_{14}N_2$ 126.20	693-13-0	 147	4-04-00-00531 0.806^{25}	 1.4320^{20}
9816	1-Propanamine, 3-methoxy-	$C_4H_{11}NO$ 89.14	5332-73-0	 117.5	4-04-00-01623 0.8727^{20}	H_2O 3; ace 3; bz 3; ctc 3 1.4391^{20}
9817	1-Propanamine, 2-methyl- Isobutylamine	$C_4H_{11}N$ 73.14	78-81-9 -86.7	5016 67.7	4-04-00-00625 0.724^{25}	 1.3988^{19}
9818	2-Propanamine, 2-methyl- *tert*-Butylamine	$C_4H_{11}N$ 73.14	75-64-9 -66.9	1545 44.0	4-04-00-00657 0.6958^{20}	H_2O 5; EtOH 5; eth 5; chl 3 1.3784^{20}
9819	2-Propanamine, *N*-methyl- Isopropylmethylamine	$C_4H_{11}N$ 73.14	4747-21-1	 50.4		
9820	1-Propanamine, 2-methyl-*N*,*N*-bis(2-methylpropyl)- Triisobutylamine	$C_{12}H_{27}N$ 185.35	1116-40-1 -21.8	 191.5	4-04-00-00631 0.7684^{20}	eth 4; EtOH 4 1.4252^{17}

No.	Name Synonym	Mol. Form. Mol. Wt.	CAS RN mp/°C	Merck No. bp/°C	Beil. Ref. den/g cm^{-3}	Solubility n_D
9821	2-Propanamine, *N*-(1-methylethyl)- Diisopropylamine	$C_6H_{15}N$ 101.19	108-18-9 -61	3181 83.9	4-04-00-00510 0.7153^{20}	ace 4; bz 4; eth 4; EtOH 4 1.3924^{20}
9822	2-Propanamine, *N*-(1-methylethyl)-*N*-nitroso-	$C_6H_{14}N_2O$ 130.19	601-77-4 48	 194.5	4-04-00-03388 0.9422^{20}	H_2O 2; EtOH 3; eth 3; bz 3
9823	1-Propanamine, 2-methyl-*N*-(2-methylpropyl)- Diisobutylamine	$C_8H_{19}N$ 129.25	110-96-3 -73.5	 139.6	4-04-00-00630	H_2O 2; EtOH 3; eth 3; ace 3 1.4090^{20}
9824	1-Propanamine, *N*-nitro-	$C_3H_8N_2O_2$ 104.11	627-07-6 -21	 $128\text{-}9^{40}$	4-04-00-03405 1.1046^{15}	eth 4; EtOH 4 1.4610^{20}
9825	1-Propanamine, *N*-nitroso-*N*-propyl- Dipropylnitrosamine	$C_6H_{14}N_2O$ 130.19	621-64-7	 206; 113^{40}	4-04-00-03388 0.9163^{20}	H_2O 2; EtOH 5; eth 5 1.4437^{20}
9826	1-Propanamine, *N*-propyl- Dipropylamine	$C_6H_{15}N$ 101.19	142-84-7 -63	3350 109.3	4-04-00-00469 0.7400^{20}	H_2O 3; EtOH 3; eth 5; ace 4 1.4050^{20}
9827	2-Propanamine, *N*-propyl Isopropylpropylamine	$C_6H_{15}N$ 101.19	21968-17-2	 96.9		
9828	1-Propanamine, 3-(triethoxysilyl)-	$C_9H_{23}NO_3Si$ 221.37	919-30-2	 $122\text{-}3^{31}$	4-04-00-04273 0.9506^{20}	 1.4225^{20}
9829	1-Propanamine, *N*,*N*,2-trimethyl-	$C_6H_{15}N$ 101.19	7239-24-9	 80.5	4-04-00-00627 0.7097^{20}	H_2O 4 1.3907^{20}
9830	1-Propanaminium, 2-(acetyloxy)-*N*,*N*,*N*-trimethyl-, chloride Methacholine chloride	$C_8H_{18}ClNO_2$ 195.69	62-51-1 172	5847		H_2O 4; EtOH 4; chl 4
9831	1-Propanaminium, 2 hydroxy-*N*,*N*,*N*-trimethyl-, chloride (2-Hydroxypropyl)trimethyl ammonium chloride	$C_6H_{16}ClNO$ 153.65	2382-43-6 165	 dec	3-04-00-00754	H_2O 4; EtOH 4
9832	2-Propanaminium, *N*-methyl-*N*-(1-methylethyl)-*N*-[2-[(9H-xanthen-9-ylcarbonyl)oxy]ethyl]-, bromide Propantheline bromide	$C_{23}H_{30}BrNO_3$ 448.40	50-34-0 159.5	7816	5-18-06-00590	
9833	1-Propanaminium, *N*,*N*,*N*-tripropyl-, iodide Tetrapropylammonium iodide	$C_{12}H_{28}IN$ 313.27	631-40-3 280 dec		4-04-00-00471 1.3138^{25}	H_2O 4; EtOH 3; eth 2; chl 4
9834	Propane	C_3H_8 44.10	74-98-6 -187.6	7809 -42.1	4-01-00-00176 $*0.493^{25}$	H_2O 3; EtOH 3; eth 4; ace 2
9835	Propane, 2,2-bis(ethylsulfonyl)- Sulfonmethane	$C_7H_{16}O_4S_2$ 228.33	115-24-2 125.8	8939 300 dec	3-01-00-02754	bz 4; EtOH 4; chl 4
9836	Propane, 1-bromo- Propyl bromide	C_3H_7Br 122.99	106-94-5 -110	7857 71.1	4-01-00-00205 1.3537^{20}	H_2O 2; EtOH 3; eth 3; ace 3 1.4343^{20}
9837	Propane, 2-bromo- Isopropyl bromide	C_3H_7Br 122.99	75-26-3 -89	5098 59.5	4-01-00-00208 1.3140^{20}	H_2O 2; EtOH 5; eth 5; ace 3 1.4251^{20}
9838	Propane, 1-bromo-2-chloro- 1-Bromo-2-chloropropane	C_3H_6BrCl 157.44	3017-96-7	 118	4-01-00-00212 1.531^{20}	ace 4; bz 4; eth 4; EtOH 4 1.4745^{20}
9839	Propane, 1-bromo-3-chloro- 1-Bromo-3-chloropropane	C_3H_6BrCl 157.44	109-70-6 -58.9	 143.3	4-01-00-00212 1.5969^{20}	H_2O 1; EtOH 4; eth 4; chl 4 1.4864^{20}
9840	Propane, 2-bromo-1-chloro- 2-Bromo-1-chloropropane	C_3H_6BrCl 157.44	3017-95-6	 118	4-01-00-00212 1.537^{20}	H_2O 1; EtOH 4; eth 4; ace 3 1.4795^{20}
9841	Propane, 2-bromo-2-chloro- 2-Bromo-2-chloropropane	C_3H_6BrCl 157.44	2310-98-7	 95	4-01-00-00213 1.474^{22}	ace 4; bz 4; eth 4; EtOH 4 1.4575^{20}
9842	Propane, 1-bromo-3-chloro-2-methyl-	C_4H_8BrCl 171.46	6974-77-2	 155	4-01-00-00298 1.4839^{20}	 1.4796^{25}
9843	Propane, 1-bromo-2,2-dimethyl-	$C_5H_{11}Br$ 151.05	630-17-1	 106	4-01-00-00337 1.1997^{20}	H_2O 1; EtOH 3; eth 3; ace 3 1.4370^{20}
9844	Propane, 1-bromo-3-fluoro-	C_3H_6BrF 140.98	352-91-0	 101.4	4-01-00-00210 1.542^{25}	bz 4; eth 4; EtOH 4; chl 4 1.4290^{25}
9845	Propane, 1-bromo-2-methyl- Isobutyl bromide	C_4H_9Br 137.02	78-77-3 -119	5019 91.1	4-01-00-00294 1.272^{15}	EtOH 4; eth 4; ace 4; bz 4 1.4348^{20}
9846	Propane, 2-bromo-2-methyl- *tert*-Butyl bromide	C_4H_9Br 137.02	507-19-7 -16.2	1555 73.3	4-01-00-00295 1.4278^{20}	H_2O 1; ctc 2 1.4278^{20}
9847	Propane, 2-bromo-2-nitro-	$C_3H_6BrNO_2$ 167.99	5447-97-2	 153; $73\text{-}5^{50}$	4-01-00-00233 1.6562^{25}	eth 4; EtOH 4
9848	Propane, 1-chloro- Propyl chloride	C_3H_7Cl 78.54	540-54-5 -122.8	7859 46.5	4-01-00-00189 0.8899^{20}	H_2O 2; EtOH 5; eth 5; bz 3 1.3879^{20}
9849	Propane, 2-chloro- Isopropyl chloride	C_3H_7Cl 78.54	75-29-6 -117.2	5099 35.7	4-01-00-00191 0.8617^{20}	H_2O 2; EtOH 5; eth 5; bz 3 1.3777^{20}
9850	Propane, 3-chloro-1,1-[illegible]	[illegible] 166.65	35573-93-4	 [illegible]	3-01-00-02592 [illegible]^{19}	ace 4; bz 4 [illegible]^{20}
9851	Propane, 1-chloro-2,2-difluoro- 2,2-Difluoropropyl chloride	$C_3H_5ClF_2$ 114.52	420-99-5 -56.2	 55	4-01-00-00193 1.2001^{20}	eth 3; bz 3; chl 3 1.3520^{20}
9852	Propane, 3-chloro-1,1-dimethoxy-	$C_5H_{11}ClO_2$ 138.59	35502-06-8	 86^{100}; 51^{19}	3-01-00-02692 1.064^{15}	 1.4163^{20}
9853	Propane, 1-chloro-2,2-dimethyl-	$C_5H_{11}Cl$ 106.60	753-89-9 -20	 84.3	4-01-00-00336 0.8660^{20}	bz 4; eth 4; EtOH 4; chl 4 1.4044^{20}
9854	Propane, 1-(1-chloroethoxy)- 1-Chloroethyl propyl ether	$C_5H_{11}ClO$ 122.59	692-35-3	 114	3-01-00-02655 0.9322^{20}	eth 4 1.4013^{20}
9855	Propane, 2-[(2-chloroethyl)thio]-2-methyl-	$C_6H_{13}ClS$ 152.69	4303-44-0 -49	 $81\text{-}2^{30}$	3-01-00-01592 1.0001^{25}	
9856	Propane, 1-chloro-3-fluoro- 1-Chloro-3-fluoropropane	C_3H_6ClF 96.53	462-38-4	 80	4-01-00-00193	bz 4; eth 4; EtOH 4; chl 4 1.3871^{25}
9857	Propane, 1-chloro-3-iodo-	C_3H_6ClI 204.44	6940-76-7	 171	4-01-00-00226 1.904^{20}	H_2O 1; eth 3; bz 3; ctc 2 1.5472^{20}
9858	Propane, 1-(chloromethoxy)-	C_4H_9ClO 108.57	3587-57-3	 109	3-01-00-02589 0.9884^{20}	eth 4; EtOH 4 1.4125^{20}

No.	Name	Synonym	Mol. Form.	Mol. Wt.	CAS RN	mp/°C	Merck No.	bp/°C	Beil. Ref.	den/g cm^{-3}	Solubility	n_D
9859	Propane, 1-chloro-2-methoxy-		C_4H_9ClO	108.57	5390-72-7			103.5	3-01-00-01470	1.009^{20}		1.4137^{20}
9860	Propane, 1-chloro-3-methoxy-		C_4H_9ClO	108.57	36215-07-3			111	4-01-00-01441	1.0013^{20}		1.4131^{20}
9861	Propane, 2-chloro-1-methoxy-		C_4H_9ClO	108.57	5390-71-6			99	3-01-00-01424	0.9946^{20}		1.4075^{20}
9862	Propane, 1-chloro-2-methyl-	Isobutyl chloride	C_4H_9Cl	92.57	513-36-0	-130.3	5022	68.5	4-01-00-00287	0.8773^{20}	eth 3; ace 3; ctc 2; chl 3	1.3984^{20}
9863	Propane, 2-chloro-2-methyl-	*tert*-Butyl chloride	C_4H_9Cl	92.57	507-20-0	-26	1562	50.9	4-01-00-00288	0.8420^{20}	H_2O 2; EtOH 5; eth 5; bz 3	1.3857^{20}
9864	Propane, 1-chloro-3-(methylthio)-	3-Chloropropyl methyl sulfide	C_4H_9ClS	124.63	13012-59-4			$71\text{-}2^{29}$	4-01-00-01458	1.0844^{20}		1.4833^{20}
9865	Propane, 2-chloro-1-(methylthio)-	2-Chloropropyl methyl sulfide	C_4H_9ClS	124.63	19987-13-4			67^{37}	4-01-00-01457	1.076^{20}		1.4905^{20}
9866	Propane, 1-chloro-1-nitro-	1-Chloro-1-nitropropane	$C_3H_6ClNO_2$	123.54	600-25-9			142	4-01-00-00232	1.207^{20}	H_2O 2; EtOH 3; eth 3; chl 2	1.4251^{20}
9867	Propane, 1-chloro-2-nitro-		$C_3H_6ClNO_2$	123.54	2425-66-3			172.5	4-01-00-00233	1.245^{22}	eth 4; EtOH 4; chl 4	1.4432^{25}
9868	Propane, 1-chloro-3-nitro-	1-Chloro-3-nitropropane	$C_3H_6ClNO_2$	123.54	16694-52-3			197 dec; 115^{40}	3-01-00-00259	1.267^{20}	eth 4; EtOH 4	
9869	Propane, 2-chloro-1-nitro-		$C_3H_6ClNO_2$	123.54	503-76-4			172	4-01-00-00232	1.2361^{15}	eth 4; EtOH 4	1.4447^{20}
9870	Propane, 2-chloro-2-nitro-	2-Chloro-2-nitropropane	$C_3H_6ClNO_2$	123.54	594-71-8			134 dec; 57^{50}	4-01-00-00233	1.230^{19}	H_2O 2; EtOH 3; eth 3; ctc 3	1.4378^{19}
9871	Propane, 3-chloro-1,1,1-trifluoro-		$C_3H_4ClF_3$	132.51	460-35-5	-106.5		45.1	4-01-00-00193	1.3253^{20}		1.3350^{20}
9872	1,2-Propanediamine	Propylenediamine	$C_3H_{10}N_2$	74.13	78-90-0		7865	119.5	2-04-00-00698	0.878^{15}	H_2O 4; eth 1	1.4460^{20}
9873	1,3-Propanediamine	1,3-Diaminopropane	$C_3H_{10}N_2$	74.13	109-76-2			139.8	4-04-00-01258	0.884^{25}	H_2O 3; EtOH 5; eth 5	1.4600^{20}
9874	1,3-Propanediamine, *N*-(3-aminopropyl)-	Bis(3-aminopropyl)amine	$C_6H_{17}N_3$	131.22	56-18-8	-14		151^{50}	4-04-00-01278	0.938^{25}	chl 3	1.4810^{20}
9875	1,3-Propanediamine, *N*-(3-aminopropyl)-*N*-methyl-		$C_7H_{19}N_3$	145.25	105-83-9			232.5; 112^{6}	4-04-00-01279	0.9023^{20}		1.4705^{25}
9876	1,3-Propanediamine, *N*,*N*-diethyl-	3-(Diethylamino)propylamine	$C_7H_{18}N_2$	130.23	104-78-9			168.5	4-04-00-01260	0.822^{20}		1.443^{20}
9877	1,3-Propanediamine, *N*,*N*-diethyl-*N*'-(6-methoxy-8-quinolinyl)-	Plasmocid	$C_{17}H_{25}N_3O$	287.41	551-01-9		7495	$182^{1.0}$	5-22-12-00126	1.0569^{24}		1.5855^{24}
9878	1,2-Propanediamine, 2-methyl-		$C_4H_{12}N_2$	88.15	811-93-8			123	4-04-00-01306	0.841^{25}	ctc 3	1.4410^{20}
9879	1,3-Propanediamine, *N*,*N*,*N*',*N*'-tetramethyl-		$C_7H_{18}N_2$	130.23	110-95-2			144	4-04-00-01259	0.7837^{18}	H_2O 5; EtOH 5; eth 5	
9880	Propane, 1,1-dibromo-	1,1-Dibromopropane	$C_3H_6Br_2$	201.89	598-17-4			133.5	4-01-00-00215	1.982^{20}	eth 4; EtOH 4; chl 4	1.5100^{20}
9881	Propane, 1,2-dibromo-	Propylene dibromide	$C_3H_6Br_2$	201.89	78-75-1	-55.2	7866	141.9	4-01-00-00215	1.9324^{20}	EtOH 3; eth 3; ctc 2; chl 3	1.5201^{20}
9882	Propane, 1,3-dibromo-	1,3-Dibromopropane	$C_3H_6Br_2$	201.89	109-64-8	-34.2	9628	167.3	4-01-00-00216	1.9822^{20}	EtOH 3; eth 3; ctc 2; chl 3	1.5232^{20}
9883	Propane, 2,2-dibromo-	2,2-Dibromopropane	$C_3H_6Br_2$	201.89	594-16-1			115	4-01-00-00217	1.7825^{20}	eth 4; EtOH 4; chl 4	
9884	Propane, 1,3-dibromo-2,2-bis(bromomethyl)-		$C_5H_8Br_4$	387.73	3229-00-3	163		305.5	4-01-00-00337	2.596^{15}	EtOH 3; eth 2; bz 3; chl 2	
9885	Propane, 1,2-dibromo-3-chloro-	Dibromochloropropane	$C_3H_5Br_2Cl$	236.33	96-12-8		3003	196	3-01-00-00250	2.093^{14}		1.553^{14}
9886	Propane, 1,1-dibromo-2,2-dimethyl-	Neopentane, 1,1-dibromo	$C_5H_{10}Br_2$	229.94	2443-91-6			$64\text{-}5^{43}$	0-01-00-00141	1.7880^{0}		1.5085
9887	Propane, 1,3-dibromo-2,2-dimethyl-		$C_5H_{10}Br_2$	229.94	5434-27-5			187 dec; 80^{26}	4-01-00-00337	1.7049^{0}		1.5090
9888	Propane, 1,2-dibromo-3-methoxy-		$C_4H_8Br_2O$	231.91	5836-66-8			185	4-01-00-01447	1.8320^{12}	eth 4	1.5123^{20}
9889	Propane, 1,2-dibromo-2-methyl-		$C_4H_8Br_2$	215.92	594-34-3	10.5		150	4-01-00-00298	1.7827^{20}	EtOH 3; eth 3; chl 3	1.5119^{20}
9890	Propane, 1,3-dibromo-2-methyl-		$C_4H_8Br_2$	215.92	28148-04-1			177.5; 62^{13}	4-01-00-00299	1.8204^{20}		1.5068^{20}
9891	Propane, 1,1-dichloro-	Propylidene chloride	$C_3H_6Cl_2$	112.99	78-99-9		7874	88.1	4-01-00-00195	1.1321^{20}	EtOH 3; eth 3; bz 3; chl 3	1.4289^{20}
9892	Propane, 1,2-dichloro-, (±)-	Propane, 1,2-dichloro (*DL*)	$C_3H_6Cl_2$	112.99	26198-63-0	-100.4	7867	96.4	4-01-00-00195	1.1560^{20}	H_2O 2; EtOH 3; eth 3; bz 3	1.4394^{20}
9893	Propane, 1,3-dichloro-	Trimethylene dichloride	$C_3H_6Cl_2$	112.99	142-28-9	-99.5		120.9	4-01-00-00196	1.1876^{20}	H_2O 2; EtOH 4; eth 4; bz 3	1.4487^{20}
9894	Propane, 2,2-dichloro-		$C_3H_6Cl_2$	112.99	594-20-7	-33.8		69.3	4-01-00-00196	1.1136^{20}	H_2O 1; EtOH 3; eth 5; bz 3	1.4148^{20}
9895	Propane, 1,3-dichloro-2,2-bis(chloromethyl)-		$C_5H_8Cl_4$	209.93	3228-99-7	97		110^{12}	4-01-00-00336		H_2O 1; eth 3; ctc 3; chl 3	

No.	Name Synonym	Mol. Form. Mol. Wt.	CAS RN mp/°C	Merck No. bp/°C	Beil. Ref. den/g cm^{-3}	Solubility n_D
9896	Propane, 1,3-dichloro-2-(ethoxymethoxy)- Formaldehyde, (2,2'-dichloroisopropyl) ethyl acetal	$C_6H_{12}Cl_2O_2$ 187.07	89583-61-9	 $96-8^{16}$	3-01-00-02574 1.181^{17}	eth 4 1.4491^{17}
9897	Propane, 1,2-dichloro-2-fluoro-	$C_3H_5Cl_2F$ 130.98	420-97-3 -91.7	 88.6	4-01-00-00197 1.2624^{20}	H_2O 2; ace 3; bz 3 1.4099^{20}
9898	Propane, 1,2-dichloro-1,1,2,3,3,3-hexafluoro- 1,2-Dichlorohexafluoropropane	$C_3Cl_2F_6$ 220.93	661-97-2	 34.1		
9899	Propane, 1,1-dichloro-2-methyl- Isobutane, 1,1-dichloro	$C_4H_8Cl_2$ 127.01	598-76-5	 105.5	4-01-00-00292 1.011^{12}	bz 4; eth 4; EtOH 4; chl 4 1.4330^{25}
9900	Propane, 1,2-dichloro-2-methyl- Isobutane, 1,2-dichloro	$C_4H_8Cl_2$ 127.01	594-37-6	 106.5	4-01-00-00292 1.093^{20}	H_2O 1; EtOH 5; eth 5; ace 5 1.4370^{20}
9901	Propane, 1,3-dichloro-2-methyl-	$C_4H_8Cl_2$ 127.01	616-19-3	 136.4	3-01-00-00319 1.131^{20}	H_2O 1; EtOH 3; eth 4; bz 4 1.4488^{25}
9902	Propane, 1,1-dichloro-1-nitro- 1,1-Dichloro-1-nitropropane	$C_3H_5Cl_2NO_2$ 157.98	595-44-8	 145	3-01-00-00259 1.312^{20}	ctc 3
9903	Propane, 1,1-diethoxy-	$C_7H_{16}O_2$ 132.20	4744-08-5	 123	4-01-00-03168 0.8232^{20}	H_2O 3; EtOH 4; eth 4; ace 3 1.3924^{19}
9904	Propane, 2,2-diethoxy-	$C_7H_{16}O_2$ 132.20	126-84-1	 114	4-01-00-03200 0.8200^{21}	EtOH 3; eth 4; ace 3; bz 3 1.3891^{20}
9905	Propane, 1,3-difluoro- 1,3-Difluoropropane	$C_3H_6F_2$ 80.08	462-39-5	 41.6	3-01-00-00218 1.0057^{25}	bz 4 1.3190^{26}
9906	Propane, 2,2-difluoro- Dimethyldifluoromethane	$C_3H_6F_2$ 80.08	420-45-1 -104.8	 -0.4	4-01-00-00187 0.9205^{20}	
9907	Propane, 1,3-diiodo- Trimethylene diiodide	$C_3H_6I_2$ 295.89	627-31-6 -20	 227 dec; 110^{19}	4-01-00-00228 2.5754^{20}	H_2O 1; eth 3; ctc 3; chl 3 1.6423^{20}
9908	Propane, 2,2-diiodo- 2,2-Diiodopropane	$C_3H_6I_2$ 295.89	630-13-7	 173; 156^{500}	3-01-00-00255 2.5754^{20}	eth 4; chl 4 1.651^{20}
9909	Propane, 2,2-dimethoxy-	$C_5H_{12}O_2$ 104.15	77-76-9 -47	 83	4-01-00-03199 0.847^{25}	 1.3780^{20}
9910	Propane, 2,2-dimethyl- Neopentane	C_5H_{12} 72.15	463-82-1 -16.6	6372 9.4	4-01-00-00333 $*0.5852^{25}$	H_2O 1; EtOH 3; eth 3; ctc 3 1.3476^{6}
9911	Propanedinitrile Malononitrile	$C_3H_2N_2$ 66.06	109-77-3 32	5592 218.5	4-02-00-01892 1.1910^{20}	H_2O 3; EtOH 4; eth 4; ace 3 1.4146^{34}
9912	Propanedinitrile, dichloro-	$C_3Cl_2N_2$ 134.95	13063-43-9	 97	3-02-00-01638 1.312^{25}	 1.4312^{20}
9913	Propanedinitrile, dimethyl-	$C_5H_6N_2$ 94.12	7321-55-3 31.5	 169.5; 62^{22}	3-02-00-01704	H_2O 3; EtOH 2
9914	Propanedinitrile, (ethoxymethylene)-	$C_6H_6N_2O$ 122.13	123-06-8 66	 160^{12}	4-03-00-01194	EtOH 3; eth 3; chl 2
9915	Propanedinitrile, oxo- Carbonyl cyanide	C_3N_2O 80.05	1115-12-4 -36	 65.5	4-03-00-01806 1.124^{20}	eth 3; ace 3; ctc 3; chl 3 1.3919^{20}
9916	Propanedinitrile, phenyl- Phenylmalononitrile	$C_9H_6N_2$ 142.16	3041-40-5 70.5	 152^{21}	4-09-00-03341	EtOH 4
9917	Propanedinitrile, (phenylmethyl)-	$C_{10}H_8N_2$ 156.19	1867-37-4 91	 174^{23}	4 09 00-03359	bz 4; eth 4; EtOH 4
9918	Propane, 1,1-dinitro- 1,1-Dinitropropane	$C_3H_6N_2O_4$ 134.09	601-76-3 -42	 184	4-01-00-00234 1.2610^{25}	alk 3 1.4339^{20}
9919	Propane, 1,3-dinitro- 1,3-Dinitropropane	$C_3H_6N_2O_4$ 134.09	6125-21-9 -21.4	 103^{1}	4-01-00-00234 1.353^{26}	H_2O 1; eth 3 1.4654^{20}
9920	Propane, 2,2-dinitro- 2,2-Dinitropropane	$C_3H_6N_2O_4$ 134.09	595-49-3 53	 185.5	4-01-00-00234 1.30^{25}	H_2O 2
9921	Propanedioic acid Malonic acid	$C_3H_4O_4$ 104.06	141-82-2 135 dec	5591 sub	4-02-00-01874 1.619^{10}	H_2O 4; EtOH 3; eth 3; bz 1
9922	Propanedioic acid, (acetylamino)-, diethyl ester	$C_9H_{15}NO_5$ 217.22	1068-90-2 96.3	 185^{20}	4-04-00-02993	H_2O 2; EtOH 3; eth 2; tfa 3
9923	Propanedioic acid, acetyl-, diethyl ester	$C_9H_{14}O_5$ 202.21	570-08-1	 232	4-03-00-01819 1.0834^{26}	H_2O 1; ace 3 1.4435^{25}
9924	Propanedioic acid, amino-, diethyl ester Diethyl aminomalonate	$C_7H_{13}NO_4$ 175.18	6829-40-9	 $122-3^{16}$	4-04-00-02991 1.100^{16}	H_2O 4; EtOH 4; eth 4; ace 3 1.4353^{16}
9925	Propanedioic acid, bis(1,1-dimethylethyl) ester Di-*tert*-butyl malonate	$C_{11}H_{20}O_4$ 216.28	541-16-2 -6	3021 113^{31}; 66^{2}	4-02-00-01884 1.4184^{20}	ace 3; chl 3 1.4184^{29}
9926	Propanedioic acid, bis(phenylmethyl)-, diethyl ester Diethyl dibenzylmalonate	[illegible] 340.42	597-55-7 14	 $234-5^{23}$	3-09-00-04552 1.093^{20}	eth 4; EtOH 4
9927	Propanedioic acid, bis(phenylmethyl) ester	$C_{17}H_{16}O_4$ 284.31	15014-25-2	 $188^{0.2}$	4-06-00-02270 1.137^{25}	 1.5447^{20}
9928	Propanedioic acid, bromo-, diethyl ester	$C_7H_{11}BrO_4$ 239.07	685-87-0 -54	 254 dec	4-02-00-01904 1.4022^{25}	H_2O 1; EtOH 5; eth 5; ace 3 1.4521^{20}
9929	Propanedioic acid, butyl- *n*-Butylmalonic acid	$C_7H_{12}O_4$ 160.17	534-59-8 104.5	1574	4-02-00-02011	H_2O 4; EtOH 3; eth 3
9930	Propanedioic acid, butyl-, diethyl ester	$C_{11}H_{20}O_4$ 216.28	133-08-4	 237.5	4-02-00-02011	EtOH 4; eth 4 1.4250^{20}
9931	Propanedioic acid, chloro-, diethyl ester	$C_7H_{11}ClO_4$ 194.61	14064-10-9	 222	4-02-00-01903 1.2040^{20}	H_2O 1; EtOH 5; eth 5; chl 5 1.4327^{20}
9932	Propanedioic acid, (3-chloropropyl)-, diethyl ester	$C_{10}H_{17}ClO_4$ 236.70	18719-43-2	 $147-9^{10}$	4-02-00-01992	H_2O 1; EtOH 3; eth 3; ctc 2 1.4429^{20}

No.	Name Synonym	Mol. Form. Mol. Wt.	CAS RN mp/°C	Merck No. bp/°C	Beil. Ref. den/g cm^{-3}	Solubility n_D
9933	Propanedioic acid, (2-cyanoethyl)-, diethyl ester	$C_{10}H_{15}NO_4$ 213.23	17216-62-5	$115\text{-}6^{2}$	4-02-00-02365 1.078^{25}	1.4368^{20}
9934	Propanedioic acid, cyclohexyl-, diethyl ester	$C_{13}H_{22}O_4$ 242.32	2163-44-2	$163\text{-}5^{20}$	3-09-00-03834 1.0281^{19}	ace 4; bz 4; eth 4; EtOH 4 1.4478^{25}
9935	Propanedioic acid, 2-cyclopenten-1-yl-, diethyl ester Malonic acid, 2-cyclopentenyl, diethyl ester	$C_{12}H_{18}O_4$ 226.27	53608-93-8	141^{10}	3-09-00-03951 1.0507^{20}	H_2O 1; eth 3; ace 3 1.4536^{20}
9936	Propanedioic acid, cyclopentylidene-, diethyl ester Malonic acid, cyclopentylidene, diethyl ester	$C_{12}H_{18}O_4$ 226.27	41589-42-8	140^{10}	3-09-00-03952 1.0616^{20}	ace 4; eth 4 1.4724^{20}
9937	Propanedioic acid, dibutyl-, diethyl ester Diethyl dibutylmalonate	$C_{15}H_{28}O_4$ 272.38	596-75-8	$153\text{-}4^{14}$	4-02-00-02119 0.9457^{20}	H_2O 1; EtOH 3; eth 3; ctc 3 1.4341^{20}
9938	Propanedioic acid, dibutyl ester	$C_{11}H_{20}O_4$ 216.28	1190-39-2 -83	251.5	4-02-00-01884 0.9824^{20}	H_2O 1; EtOH 3; eth 3; ace 3 1.4262^{20}
9939	Propanedioic acid, diethyl- Diethylmalonic acid	$C_7H_{12}O_4$ 160.17	510-20-3 127 dec	3114	4-02-00-02026	H_2O 4; EtOH 4; eth 4; bz 2
9940	Propanedioic acid, diethyl-, diethyl ester Diethyl 2,2-diethylmalonate	$C_{11}H_{20}O_4$ 216.28	77-25-8	3746 230	4-02-00-02026 0.9643^{30}	H_2O 1; EtOH 5; eth 5; ctc 3 1.4240^{20}
9941	Propanedioic acid, diethyl ester Diethyl malonate	$C_7H_{12}O_4$ 160.17	105-53-3 -50	3779 200	4-02-00-01881 1.0551^{20}	H_2O 2; EtOH 5; eth 5; ace 4 1.4139^{20}
9942	Propanedioic acid, dimethyl-, diethyl ester Diethyl dimethylmalonate	$C_9H_{16}O_4$ 188.22	1619-62-1 -30.4	197	4-02-00-01955 0.9964^{20}	H_2O 1; EtOH 5; eth 5; ctc 2 1.4129^{20}
9943	Propanedioic acid, dimethyl ester Methyl malonate	$C_5H_8O_4$ 132.12	108-59-8 -61.9	6009 181.4	4-02-00-01880 1.528^{20}	H_2O 2; EtOH 5; eth 5; ace 4 1.4135^{20}
9944	Propanedioic acid, di-2-propenyl-, diethyl ester	$C_{13}H_{20}O_4$ 240.30	3195-24-2	243.5	3-02-00-02003 0.9943^{20}	EtOH 3; eth 3; ace 3; ctc 2 1.4445^{22}
9945	Propanedioic acid, dipropyl ester Dipropyl malonate	$C_9H_{16}O_4$ 188.22	1117-19-7 -77.1	229	3-02-00-01619 1.0097^{20}	ace 4; bz 4; eth 4; EtOH 4 1.4206^{20}
9946	Propanedioic acid, di-2-propynyl-, diethyl ester Diethyl dipropargylmalonate	$C_{13}H_{16}O_4$ 236.27	2689-88-5 45.5	129^{11}	4-02-00-02340	chl 3
9947	Propanedioic acid, (ethoxymethylene)-, diethyl ester	$C_{10}H_{16}O_5$ 216.23	87-13-8	280 dec; 165^{19}	4-03-00-01192	H_2O 1; EtOH 3; eth 3; chl 2 1.4600^{20}
9948	Propanedioic acid, ethyl-	$C_5H_8O_4$ 132.12	601-75-2 114	$180^{0.05}$	4-02-00-01952	H_2O 4; EtOH 3; eth 3; ace 1
9949	Propanedioic acid, ethyl-, diethyl ester	$C_9H_{16}O_4$ 188.22	133-13-1	208; 98^{12}	4-02-00-01953 1.0047^{20}	H_2O 2; EtOH 4; eth 4; ace 3 1.4166^{20}
9950	Propanedioic acid, ethylidene-, diethyl ester	$C_9H_{14}O_4$ 186.21	1462-12-0	$115\text{-}8^{17}$	4-02-00-02234 1.0194^{17}	eth 4; EtOH 4 1.4308^{17}
9951	Propanedioic acid, ethyl(1-methylethyl)-, diethyl ester Malonic acid, ethylisopropyl-, diethyl ester	$C_{12}H_{22}O_4$ 230.30	2049-66-3	234; 109^{11}	4-02-00-02052	ace 4; eth 4; EtOH 4 1.4280^{25}
9952	Propanedioic acid, ethylphenyl-, diethyl ester	$C_{15}H_{20}O_4$ 264.32	76-67-5	170^{19}	4-09-00-03385 1.071^{20}	H_2O 1; EtOH 3; eth 3; chl 2 1.4896^{25}
9953	Propanedioic acid, (formylamino)-, diethyl ester	$C_8H_{13}NO_5$ 203.19	6326-44-9 48.5	173^{11}	4-04-00-02992	EtOH 4
9954	Propanedioic acid, hexadecyl-, diethyl ester	$C_{23}H_{44}O_4$ 384.60	41433-81-2	238^{14}	4-02-00-02183 0.9118^{20}	EtOH 4 1.4433^{20}
9955	Propanedioic acid, hexyl-, diethyl ester	$C_{13}H_{24}O_4$ 244.33	5398-10-7	269; 143^{15}	3-02-00-01791 0.9577^{21}	ace 4; bz 4; eth 4; EtOH 4 1.4278^{21}
9956	Propanedioic acid, hydroxy- Tartronic acid	$C_3H_4O_5$ 120.06	80-69-3 157	9042 sub	4-03-00-01120	H_2O 3; EtOH 3; eth 2
9957	Propanedioic acid, hydroxy-, diethyl ester	$C_7H_{12}O_5$ 176.17	13937-08-1 -2.5	221	3-03-00-00905 1.1520^{15}	ace 4; bz 4; eth 4; EtOH 4
9958	Propanedioic acid, (hydroxyimino)-, diethyl ester	$C_7H_{11}NO_5$ 189.17	6829-41-0	172^{12}	4-03-00-01805 1.1821^{18}	ace 4; bz 4; eth 4; EtOH 4 1.4544^{18}
9959	Propanedioic acid, (hydroxymethyl)- β-Isomalic acid	$C_4H_6O_5$ 134.09	4360-96-7 142 dec	170 dec	0-03-00-00441	H_2O 4; EtOH 4; eth 4
9960	Propanedioic acid, methyl-	$C_4H_6O_4$ 118.09	516-05-2 135 dec		4-02-00-01932 1.455^{20}	H_2O 4; EtOH 4; eth 4; bz 2
9961	Propanedioic acid, (3-methylbutyl)-, diethyl ester	$C_{12}H_{22}O_4$ 230.30	5398-08-3	241	3-02-00-01777 0.9580^{25}	ace 4; eth 4; EtOH 4 1.4255^{25}
9962	Propanedioic acid, methyl-, diethyl ester	$C_8H_{14}O_4$ 174.20	609-08-5	201	4-02-00-01932 1.0225^{20}	H_2O 2; EtOH 4; eth 4; ace 3 1.4126^{20}
9963	Propanedioic acid, methyl-, dimethyl ester	$C_6H_{10}O_4$ 146.14	609-02-9	174	3-02-00-01681 1.0977^{20}	ace 4; eth 4; EtOH 4; chl 4 1.4128^{20}
9964	Propanedioic acid, (1-methylethyl)-, diethyl ester	$C_{10}H_{18}O_4$ 202.25	759-36-4	215	4-02-00-02001 0.9961^{20}	H_2O 2; EtOH 4; eth 4; ctc 2 1.4188^{21}
9965	Propanedioic acid, (1-methylethylidene)-	$C_6H_8O_4$ 144.13	4441-90-1 170.5	dec	3-02-00-01948	EtOH 4
9966	Propanedioic acid, (1-methylethylidene)-, diethyl ester Diethyl isopropylidenemalonate	$C_{10}H_{16}O_4$ 200.23	6802-75-1	176.5; 116^{14}	4-02-00-02244 1.0282^{18}	ace 4; EtOH 4 1.4486^{17}

No.	Name Synonym	Mol. Form. Mol. Wt.	CAS RN mp/°C	Merck No. bp/°C	Beil. Ref. den/g cm^{-3}	Solubility n_D
9967	Propanedioic acid, (1-methylpropyl)-, diethyl ester	$C_{11}H_{20}O_4$ 216.28	83-27-2		4-02-00-02019 0.988^{15}	H_2O 1; EtOH 4; eth 4; ctc 2 1.4248^{20}
9968	Propanedioic acid, (2-methylpropyl)-	$C_7H_{12}O_4$ 160.17	4361-06-2 112 dec	 dec	4-02-00-02023	eth 4; EtOH 4
9969	Propanedioic acid, (2-methylpropyl)-, diethyl ester Diethyl isobutylmalonate	$C_{11}H_{20}O_4$ 216.28	10203-58-4		3-02-00-01756 0.9804^{20}	H_2O 1; EtOH 4; eth 4; chl 3 1.4236^{20}
9970	Propanedioic acid, 1-naphthalenyl-, diethyl ester	$C_{17}H_{18}O_4$ 286.33	6341-60-2 61.3	 $176^{1.0}$	4-09-00-03536	EtOH 3; bz 3; chl 2
9971	Propanedioic acid, octyl-	$C_{11}H_{20}O_4$ 216.28	760-55-4 116		4-02-00-02114 1.173^{17}	ace 4; EtOH 4
9972	Propanedioic acid, (1-oxobutyl)-, diethyl ester	$C_{11}H_{18}O_5$ 230.26	21633-79-4	 249.5	4-03-00-01833 1.056^{20}	 1.4451^{20}
9973	Propanedioic acid, oxo-, diethyl ester	$C_7H_{10}O_5$ 174.15	609-09-6 -30	 210; 105^{19}	3-03-00-01356 1.1419^{16}	H_2O 4; EtOH 3; eth 3; chl 3 1.4310^{22}
9974	Propanedioic acid, pentyl-, diethyl ester	$C_{12}H_{22}O_4$ 230.30	6065-59-4	 134-6^{14}	4-02-00-02034 0.9652^{20}	H_2O 1; EtOH 4; eth 4; ctc 2 1.4253^{20}
9975	Propanedioic acid, phenyl-, diethyl ester	$C_{13}H_{16}O_4$ 236.27	83-13-6 16.5	 205 dec; 168^{12}	4-09-00-03339 1.0950^{20}	ace 4; EtOH 4 1.4977^{20}
9976	Propanedioic acid, phenyl-, dimethyl ester	$C_{11}H_{12}O_4$ 208.21	37434-59-6 51.6	 147^{13}	4-09-00-03339	eth 4; EtOH 4
9977	Propanedioic acid, (phenylmethyl)-, diethyl ester	$C_{14}H_{18}O_4$ 250.29	607-81-8	 300	4-09-00-03357 1.076^{15}	H_2O 1; chl 2 1.4872^{20}
9978	Propanedioic acid, (phenylmethylene)-, diethyl ester	$C_{14}H_{16}O_4$ 248.28	5292-53-5 32	 216^{30}; 196^{14}	4-09-00-03461 1.1045^{20}	H_2O 1; EtOH 3; eth 3; ace 3 1.5389^{20}
9979	Propanedioic acid, 2-propenyl-	$C_6H_8O_4$ 144.13	2583-25-7 105	 180 dec	3-02-00-01945	H_2O 4; eth 4; EtOH 4
9980	Propanedioic acid, 2-propenyl-, diethyl ester	$C_{10}H_{16}O_4$ 200.23	2049-80-1	 222.5; 93^6	4-02-00-02240 1.0098^{20}	H_2O 1; EtOH 4; eth 4; ctc 3 1.4305^{20}
9981	Propanedioic acid, propyl-, diethyl ester	$C_{10}H_{18}O_4$ 202.25	2163-48-6	 221; 114^{22}	4-02-00-01991 0.9873^{20}	H_2O 2; EtOH 4; eth 4 1.4197^{20}
9982	1,2-Propanediol 1,2-Propylene glycol	$C_3H_8O_2$ 76.10	57-55-6 -60	7868 187.6	3-01-00-02142 1.0361^{20}	H_2O 5; EtOH 5; eth 3; bz 3 1.4324^{20}
9983	1,3-Propanediol Trimethylene glycol	$C_3H_8O_2$ 76.10	504-63-2 -26.7	9629 214.4	4-01-00-02493 1.0538^{20}	H_2O 5; EtOH 5; eth 4; bz 2 1.4398^{20}
9984	1,2-Propanediol, 3-amino-, (±)- 1,2-Propanediol, 3-amino(*DL*)	$C_3H_9NO_2$ 91.11	13552-31-3	 265 dec; 145^9	3-04-00-00839 1.1752^{20}	H_2O 3; EtOH 3; eth 1; bz 1 1.4910^{25}
9985	1,3-Propanediol, 2-amino-2-ethyl- 2-Amino-2-ethyl-1,3-propanediol	$C_5H_{13}NO_2$ 119.16	115-70-8 37.5	451 152-3^{10}	3-04-00-00850 1.099^{20}	H_2O 5 1.490^{20}
9986	1,3-Propanediol, 2-amino-2-(hydroxymethyl)- Tromethamine	$C_4H_{11}NO_3$ 121.14	77-86-1 171.5	9684 219-20^{10}	4-04-00-01903	H_2O 4; MeOH 3
9987	1,3-Propanediol, 2-amino-2-methyl- 2-Amino-2-methyl-1,3-propanediol	$C_4H_{11}NO_2$ 105.14	115-69-5 110	460 151-2^{10}	4-04-00-01881	H_2O 4; EtOH 3
9988	1,3-Propanediol, 2,2-bis[(acetyloxy)methyl]-, diacetate Pentaerythrityl tetraacetate	$C_{13}H_{20}O_8$ 304.30	597-71-7 83.5	7065	4-02-00-00264 1.273^{18}	H_2O 3; EtOH 4; eth 4
9989	1,3-Propanediol, 2,2-bis(hydroxymethyl)- Pentaerythritol	$C_5H_{12}O_4$ 136.15	115-77-5 260	7062 sub	4-01-00-02812	H_2O 3; eth 1; bz 1 1.548
9990	1,3-Propanediol, 2,2-bis[(nitrooxy)methyl]-, dinitrate (ester) Pentaerythrityl tetranitrate	$C_5H_8N_4O_{12}$ 316.14	78-11-5 140.5	7066	4-01-00-02816 1.773^{20}	H_2O 2; EtOH 2; eth 2; ace 4
9991	1,3-Propanediol, 2-bromo-2-nitro- Bronopol	$C_3H_6BrNO_4$ 199.99	52-51-7 131.5	1437	4-01-00-02501	
9992	1,3-Propanediol, 2-butyl-2-ethyl-	$C_9H_{20}O_2$ 160.26	115-84-4 43.8	 262	4-01-00-02611 0.927^{50}	H_2O 2; EtOH 3; ace 2 1.4587^{25}
9993	1,3-Propanediol, 2-butyl-2-methyl-	$C_8H_{18}O_2$ 146.23	3121-83-3 43.8	 262	4-01-00-02599 0.927^{50}	EtOH 4 1.4587^{25}
9994	1,2-Propanediol, 3-chloro- α-Chlorohydrin	$C_3H_7ClO_2$ 110.54	96-24-2	2145 213 dec; 116^{11}	4-01-00-02484 1.325^{18}	H_2O 3; EtOH 3; eth 3 1.4809[illegible]
9995	1,3-Propanediol, 2-chloro- Glycerol β-chlorohydrin	$C_3H_7ClO_2$ 110.54	497-04-1	 146^{18}	4-01-00-02499 1.3219^{20}	H_2O 4; ace 4; EtOH 4 1.4831^{20}
9996	1,2-Propanediol, 3-chloro-, diacetate	$C_7H_{11}ClO_4$ 194.61	869-50-1	 245	3-02-00-00313 1.199^{25}	eth 4; EtOH 4 1.4407^{20}
9997	1,2-Propanediol, 3-chloro-, dinitrate Clonitrate	$C_3H_5ClN_2O_6$ 200.54	2612-33-1	2390 192.5	4-01-00-02487 1.5112^9	ace 4; EtOH 4; chl 4
9998	1,2-Propanediol, 3-chloro-2-methyl- Isobutylene glycol, 3-chloro	$C_4H_9ClO_2$ 124.57	597-33-1	 114-7^{20}	4-01-00-02536 1.2362^{20}	H_2O 4; eth 4; EtOH 4 1.4748^{20}
9999	1,2-Propanediol, 3-(2-chlorophenoxy)-	$C_9H_{11}ClO_3$ 202.64	5112-21-0 71.5	 250^{19}	4-06-00-00790	H_2O 3; eth 4; bz 3
10000	1,2-Propanediol, 3-(4-chlorophenoxy)- Chlorphenesin	$C_9H_{11}ClO_3$ 202.64	104-29-0 78	2178 214-5^{19}	4-06-00-00831	H_2O 1; EtOH 4; eth 4; bz 3

No.	Name	Synonym	Mol. Form.	Mol. Wt.	CAS RN	mp/°C	Merck No.	bp/°C	Beil. Ref.	den/g cm^{-3}	Solubility	n_D
10001	1,2-Propanediol, 3-(4-chlorophenoxy)-, 1-carbamate	Chlorphenesin carbamate	$C_{10}H_{12}ClNO_4$	245.66	886-74-8	90	2179				ace 4; EtOH 4; diox 4	
10002	1,2-Propanediol, diacetate		$C_7H_{12}O_4$	160.17	623-84-7			190.5	4-02-00-00220	1.059^{20}	H_2O 4; EtOH 3; eth 3	1.4173^{20}
10003	1,3-Propanediol, diacetate		$C_7H_{12}O_4$	160.17	628-66-0			209.5	4-02-00-00221	1.070^{14}	H_2O 4; EtOH 3	1.4192
10004	1,3-Propanediol, 2,2-diethyl-		$C_7H_{16}O_2$	132.20	115-76-4	61.5		240.5	4-01-00-02589	1.050^{20}	H_2O 4; EtOH 4; eth 4; chl 3	1.4574^{25}
10005	1,3-Propanediol, 2,2-dimethyl-	Neopentyl glycol	$C_5H_{12}O_2$	104.15	126-30-7	130	6374	208	4-01-00-02551		H_2O 3; EtOH 4; eth 4; bz 3	
10006	1,3-Propanediol, 2-ethyl-2-(hydroxymethyl)-	Trimethylolpropane	$C_6H_{14}O_3$	134.18	77-99-6	58		160^{5}	4-01-00-02786		H_2O 4; EtOH 4	
10007	1,3-Propanediol, 2-ethyl-2-methyl-		$C_6H_{14}O_2$	118.18	77-84-9	40.3		225.8	4-01-00-02574		chl 3	
10008	1,3-Propanediol, 2-ethyl-2-nitro-	2-Nitro-2-ethyl-1,3-propanediol	$C_5H_{11}NO_4$	149.15	597-09-1	57.5		dec	4-01-00-02550		H_2O 4; eth 4; EtOH 4	
10009	1,2-Propanediol, 3-(hexadecyloxy)-		$C_{19}H_{40}O_3$	316.52	53584-29-5	64		$205^{1.0}$	3-01-00-02322		ace 4; chl 4; peth 4	
10010	1,2-Propanediol, 3-(hexadecyloxy)-, (*S*)-	Chimyl alcohol	$C_{19}H_{40}O_3$	316.52	506-03-6	64	2049	$120^{0.005}$	3-01-00-02322		ace 4; peth 4; chl 4	
10011	1,3-Propanediol, 2-(hydroxymethyl)-2-methyl-		$C_5H_{12}O_3$	120.15	77-85-0	204		$135\text{-}7^{15}$	4-01-00-02780		H_2O 5; EtOH 5; eth 1; bz 1	
10012	1,3-Propanediol, 2-(hydroxymethyl)-2-nitro-	Tris(hydroxymethyl)nitromethane	$C_4H_9NO_5$	151.12	126-11-4	165	9667	dec	4-01-00-02777		H_2O 4; eth 4; EtOH 4	
10013	1,2-Propanediol, 3-mercapto-	Thioglycerol	$C_3H_8O_2S$	108.16	96-27-5		9264	$100\text{-}1^{1}$	3-01-00-02339	1.2455^{20}	H_2O 2; EtOH 5; eth 2; ace 4	1.5268^{20}
10014	1,2-Propanediol, 3-methoxy-		$C_4H_{10}O_3$	106.12	623-39-2			220	4-01-00-02755	1.114^{20}	H_2O 4; EtOH 4; eth 3; ace 4	1.442^{25}
10015	1,3-Propanediol, 2-methoxy-	Glycerol, 2-methyl ether	$C_4H_{10}O_3$	106.12	761-06-8			232	3-01-00-02317	1.124^{25}	H_2O 4; ace 4; eth 4; EtOH 4	1.4505^{12}
10016	1,2-Propanediol, 3-(2-methoxyphenoxy)-	Guaifenesin	$C_{10}H_{14}O_4$	198.22	93-14-1	78.5	4465	215^{19}; $127^{0.2}$	4-06-00-05576		H_2O 3; EtOH 4; bz 3; chl 3	
10017	1,2-Propanediol, 2-methyl-		$C_4H_{10}O_2$	90.12	558-43-0			176	4-01-00-02533	1.0024^{20}	H_2O 4; eth 4; EtOH 4	1.4350^{20}
10018	1,3-Propanediol, 2-methyl-2-nitro-		$C_4H_9NO_4$	135.12	77-49-6	150.1		dec	4-01-00-02537		H_2O 4; EtOH 4; DMSO-d_6 2	
10019	1,2-Propanediol, 3-(2-methylphenoxy)-	Mephenesin	$C_{10}H_{14}O_3$	182.22	59-47-2	70 dec	5737		4-06-00-01952		H_2O 2; EtOH 3; eth 2	
10020	1,3-Propanediol, 2-methyl-2-propyl-		$C_7H_{16}O_2$	132.20	78-26-2	62.5		234; 121^{10}	4-01-00-02585		H_2O 3; chl 2; os 3; hx 3	
10021	1,3-Propanediol, 2-methyl-2-propyl-, dicarbamate	Meprobamate	$C_9H_{18}N_2O_4$	218.25	57-53-4	105	5751		4-03-00-00073		bz 4; eth 4; EtOH 4	
10022	1,2-Propanediol, 1-nitrate		$C_3H_7NO_4$	121.09	20266-65-3			$148\text{-}9^{4}$	4-01-00-02476	1.2417^{20}		1.4368^{20}
10023	1,2-Propanediol, 3-(octadecyloxy)-	Batyl alcohol	$C_{21}H_{44}O_3$	344.58	544-62-7	70.5	1021	217^{2}	4-01-00-02758		eth 4	
10024	Propanediol, oxybis-		$C_6H_{14}O_5$	166.17	59113-36-9			267^{15}; 225^{3}	3-01-00-02327	1.2751^{20}	H_2O 4; EtOH 4	1.4890^{20}
10025	1,2-Propanediol, 3,3'-oxybis-		$C_6H_{14}O_5$	166.17	627-82-7			267^{15}; 225^{3}	3-01-00-02327	1.2751^{20}	H_2O 4; EtOH 4	1.4890^{20}
10026	1,2-Propanediol, 3-phenoxy-	Phenylglyceryl ether	$C_9H_{12}O_3$	168.19	538-43-2	67.5	7261	200^{22}	4-06-00-00589	1.225^{20}	H_2O 4; bz 4; eth 4; EtOH 4	
10027	1,3-Propanedione, 1,3-diphenyl-	Dibenzoylmethane	$C_{15}H_{12}O_2$	224.26	120-46-7	70.5	2991		4-07-00-02512		EtOH 3; eth 3; chl 3; dil NaOH 3	
10028	1,2-Propanedione, 1-phenyl-		$C_9H_8O_2$	148.16	579-07-7	<20		222; 102^{12}	4-07-00-02143	1.1006^{20}	H_2O 3; EtOH 3; eth 3	1.537^{10}
10029	Propanedioyl dichloride		$C_3H_2Cl_2O_2$	140.95	1663-67-8			58^{26}	4-02-00-01887	1.4509^{20}	eth 3; AcOEt 3	1.4639^{20}
10030	1,2-Propanedithiol		$C_3H_8S_2$	108.23	814-67-5			152	4-01-00-02492	1.08^{20}	chl 3	1.532^{20}
10031	1,3-Propanedithiol	Trimethylene dimercaptan	$C_3H_8S_2$	108.23	109-80-8	-79	7811	172.9	4-01-00-02503	1.0772^{20}	H_2O 2; EtOH 5; eth 5; bz 5	1.5392^{20}
10032	Propane, 1,1'-[1,2-ethanediylbis(oxy)]bis[2-methyl-	Ethylene glycol diisobutyl ether	$C_{10}H_{22}O_2$	174.28	54922-86-0			171.3	4-01-00-02382	0.8200^{20}		1.4021^{20}
10033	Propane, 1-(ethenyloxy)-		$C_5H_{10}O$	86.13	764-47-6			65	4-01-00-02052	0.7674^{20}		1.3908^{20}
10034	Propane, 2-(ethenyloxy)-		$C_5H_{10}O$	86.13	926-65-8	-140		55.5	4-01-00-02052	0.7534^{20}	ace 4; bz 4; eth 4; EtOH 4	1.3840^{20}
10035	Propane, 1-(ethenyloxy)-2-methyl-	Isobutyl vinyl ether	$C_6H_{12}O$	100.16	109-53-5	-112		83	4-01-00-02054	0.7645^{20}	H_2O 2; EtOH 4; eth 5; ace 4	1.3966^{20}
10036	Propane, 2-(ethenyloxy)-2-methyl-		$C_6H_{12}O$	100.16	926-02-3	-112		75	4-01-00-02055	0.7691^{20}		1.3922^{20}
10037	Propane, 1-ethoxy-	Ethyl propyl ether	$C_5H_{12}O$	88.15	628-32-0	-127.5		63.2	4-01-00-01421	0.7386^{20}	eth 4; EtOH 4; HOAc 4	1.3695^{20}
10038	Propane, 2-ethoxy-	Ethyl isopropyl ether	$C_5H_{12}O$	88.15	625-54-7			54.1	4-01-00-01471	0.720^{25}	H_2O 3; EtOH 5; eth 5; ace 3	1.3698^{25}

No.	Name Synonym	Mol. Form. Mol. Wt.	CAS RN mp/°C	Merck No. bp/°C	Beil. Ref. den/g cm^{-3}	Solubility n_D
10039	Propane, 1-ethoxy-2-methyl-	$C_6H_{14}O$ 102.18	627-02-1	 81	4-01-00-01593 0.751^{20}	H_2O 1; EtOH 5; eth 5; ace 3 1.3739^{25}
10040	Propane, 2-ethoxy-2-methyl- Ethyl *tert*-butyl ether	$C_6H_{14}O$ 102.18	637-92-3 -94	3732 73.1	4-01-00-01615 0.7519^{25}	H_2O 1; EtOH 4; eth 4 1.3794^{20}
10041	Propane, 1,1'-[ethylidenebis(oxy)]bis[2-methyl-	$C_{10}H_{22}O_2$ 174.28	5669-09-0	 171.3	4-01-00-03107 0.8200^{20}	 1.4021^{20}
10042	Propane, 1-(ethylthio)- Ethyl propyl sulfide	$C_5H_{12}S$ 104.22	4110-50-3 -117	 118.6	4-01-00-01451 0.8370^{20}	EtOH 3 1.4462^{20}
10043	Propane, 2-(ethylthio)- Ethyl isopropyl sulfide	$C_5H_{12}S$ 104.22	5145-99-3 -122.2	 107.5	4-01-00-01501 0.8246^{20}	
10044	Propane, 1-(ethylthio)-2-methyl- Ethyl isobutyl sulfide	$C_6H_{14}S$ 118.24	1613-45-2	 134.2	4-01-00-01606 0.8306^{20}	chl 3 1.4450^{20}
10045	Propane, 1-(ethynyloxy)-	C_5H_8O 84.12	42842-08-0	 75	4-01-00-02213 0.8080^{20}	eth 4; EtOH 4 1.3935^{20}
10046	Propane, 1-fluoro- Propyl fluoride	C_3H_7F 62.09	460-13-9 -159	 2.5	4-01-00-00187 0.7956^{20}	H_2O 2; EtOH 4; eth 4 1.3115^{20}
10047	Propane, 2-fluoro- Isopropyl fluoride	C_3H_7F 62.09	420-26-8		4-01-00-00187	
10048	Propane, 1,1,1,2,2,3,3-heptachloro- 1,1,1,2,2,3,3-Heptachloropropane	C_3HCl_7 285.21	594-89-8 29.4	 247.5	4-01-00-00205 1.8048^{34}	ctc 2; chl 3
10049	Propane, 1,1,1,2,3,3,3-heptachloro- 1,1,1,2,3,3,3-Heptachloropropane	C_3HCl_7 285.21	3849-33-0 11	 249	4-01-00-00205 1.7921^{34}	chl 4 1.5427^{21}
10050	Propane, 1,1,1,2,3,3,3-heptafluoro-2-(trifluoromethyl)- Perfluoroisobutane	C_4F_{10} 238.03	354-92-7	 0		
10051	Propane, 1,1,1,2,2,3-hexachloro- 1,1,1,2,2,3-Hexachloropropane	$C_3H_2Cl_6$ 250.77	24425-97-6 26	 216	4-01-00-00204 1.7187^{25}	 1.5282^{25}
10052	Propane, 1,1,1,3,3,3-hexachloro- 1,1,1,3,3,3-Hexachloropropane	$C_3H_2Cl_6$ 250.77	3607-78-1 -27	 206	4-01-00-00205 1.6800^{20}	 1.5179^{20}
10053	Propane, 1,1,2,2,3,3-hexachloro- 1,1,2,2,3,3-Hexachloropropane	$C_3H_2Cl_6$ 250.77	15600-01-8	 218.5	4-01-00-00205 1.7137^{34}	 1.5262^{18}
10054	Propane, 1-iodo- Propyl iodide	C_3H_7I 169.99	107-08-4 -101.3	7875 102.6	4-01-00-00222 1.7489^{20}	H_2O 2; EtOH 5; eth 5; ctc 2 1.5058^{20}
10055	Propane, 2-iodo- Isopropyl iodide	C_3H_7I 169.99	75-30-9 -90	5102 89.5	4-01-00-00223 1.7042^{20}	H_2O 2; EtOH 5; eth 5; bz 5 1.5028^{20}
10056	Propane, 1-iodo-2,2-dimethyl-	$C_5H_{11}I$ 198.05	15501-33-4	 128 dec	4-01-00-00338 1.4940^{20}	H_2O 1; EtOH 3; eth 3 1.4890^{20}
10057	Propane, 1-iodo-2-methyl- Isobutyl iodide	C_4H_9I 184.02	513-38-2	5027 121.1	4-01-00-00299 1.6035^{20}	 1.4959^{20}
10058	Propane, 2-iodo-2-methyl- *tert*-Butyl iodide	C_4H_9I 184.02	558-17-8 -38.2	 100.1	4-01-00-00300 1.571^{25}	EtOH 5; eth 5 1.4918^{20}
10059	Propane, 1-isocyanato- Propyl isocyanate	C_4H_7NO 85.11	110-78-1	 83.5	4-04-00-00491 0.908^{25}	 1.3970^{20}
10060	Propane, 2-isocyanato- Isopropyl isocyanate	C_4H_7NO 85.11	1795-48-8	 74.5	4-04-00-00530 0.866^{25}	 1.3825^{20}
10061	Propane, 1-isocyanato-2-methyl- Isobutyl isocyanate	C_5H_9NO 99.13	1873-29-6	 106	4-04-00-00653	
10062	Propane, 2-isocyanato-2-methyl- *tert*-Butyl isocyanate	C_5H_9NO 99.13	1609-86-5 85.5		4-04-00-00669 0.8670^{7}	 1.4061^{20}
10063	Propane, 2-Isocyano- Isopropyl isocyanide	C_4H_7N 69.11	598-45-8	 87	0-04-00-00154 0.7596^{25}	H_2O 1; EtOH 5; eth 5
10064	Propane, 1-isothiocyanato-	C_4H_7NS 101.17	628-30-8	 153	4-04-00-00491 0.9781^{16}	H_2O 2; EtOH 5; eth 5 1.5085^{16}
10065	Propane, 1-isothiocyanato-2-methyl-	C_5H_9NS 115.20	591-82-2	 160	4-04-00-00653 0.9631^{14}	 1.5005^{14}
10066	Propane, 2-isothiocyanato-2-methyl- *tert*-Butyl isothiocyanate	C_5H_9NS 115.20	590-42-1 10.5	 140	4-04-00-00669 0.9187^{10}	
10067	Propane, 1-isothiocyanato-3-(methylsulfonyl)- Cheirolin	$C_5H_9NO_2S_2$ 179.26	505-34-0 47.5	2040 200^{3}	4-04-00-01664	EtOH 4; chl 4
10068	Propane, 1,1',1'',1'''-[methanetetrayltetrakis(oxy)]tetrakis-	$C_{13}H_{28}O_4$ 248.36	597-72-8	 224.2	4-03-00-00007 0.897^{20}	EtOH 4; eth 4 1.4100^{20}
10069	Propane, 1-methoxy- Methyl propyl ether	$C_4H_{10}O$ 74.12	557-17-5	6031 39.1	4-01-00-01421 0.7356^{13}	H_2O 3; EtOH 5; eth 5; ace 3 1.3579^{25}
10070	Propane, 2-methoxy- Isopropyl methyl ether	$C_4H_{10}O$ 74.12	598-53-8	 30.7	4-01-00-01471 0.7237^{15}	H_2O 3; EtOH 5; eth 5 1.3576^{20}
10071	Propane, 1-methoxy-2-methyl- Isobutyl methyl ether	$C_5H_{12}O$ 88.15	625-44-5	 58.6	4-01-00-01593 0.7311^{20}	eth 4; EtOH 4
10072	Propane, 2-methoxy-2-methyl- *tert*-Butyl methyl ether	$C_5H_{12}O$ 88.15	1634-04-4 -108.6	5954 55.2	4-01-00-01615 0.7405^{20}	H_2O 3; EtOH 4; eth 4 1.3690^{20}
10073	Propane, 2-methyl- Isobutane	C_4H_{10} 58.12	75-28-5 -138.3	 -11.7	4-01-00-00282 $^{*}0.5510^{25}$	H_2O 2; EtOH 3; eth 3; chl 3 1.3518^{-25}
10074	Propane, 1,1'-[methylenebis(oxy)]bis-	$C_7H_{16}O_2$ 132.20	505-84-0 -97.3	 140.5	4-01-00-03029 0.8345^{20}	ace 4; bz 4; eth 4; EtOH 4 1.3939^{19}
10075	Propane, 1,1'-[methylenebis(oxy)]bis[2-methyl-	$C_9H_{20}O_2$ 160.26	2568-91-4	 165.5	4-01-00-03030 0.8268^{20}	 1.4108^{17}
10076	Propane, 1-(1-methylethoxy)-	$C_6H_{14}O$ 102.18	627-08-7	 83	4-01-00-01471 0.7370^{20}	H_2O 2; EtOH 4; eth 3; ace 3 1.376^{21}
10077	Propane, 1-[(1-methylethyl)thio]- Isopropyl propyl sulfide	$C_6H_{14}S$ 118.24	5008-73-1	 132.1	4-01-00-01501 0.8269^{20}	

No.	Name Synonym	Mol. Form. Mol. Wt.	CAS RN mp/°C	Merck No. bp/°C	Beil. Ref. den/g cm^{-3}	Solubility n_D
10078	Propane, 1,1',1''-[methylidynetris(oxy)]tris-	$C_{10}H_{22}O_3$ 190.28	621-76-1	191; 93^{30}	4-02-00-00027 0.8805^{20}	eth 4; EtOH 4 1.4072^{20}
10079	Propane, 2,2',2''-[methylidynetris(oxy)]tris-	$C_{10}H_{22}O_3$ 190.28	4447-60-3	167	4-02-00-00028 0.8621^{20}	eth 4; EtOH 4 1.4000^{20}
10080	Propane, 1,1',1''-[methylidynetris(oxy)]tris[2-methyl- Orthoformic acid, triisobutyl ester	$C_{13}H_{28}O_3$ 232.36	16754-49-7	223	4-02-00-00029 0.8582^{20}	eth 4; EtOH 4 1.4120^{20}
10081	Propane, 2-methyl-2-(1-methylethoxy)-	$C_7H_{16}O$ 116.20	17348-59-3 -88	87.6	4-01-00-01616 0.7365^{25}	chl 3
10082	Propane, 2-methyl-1-(methylthio)-	$C_5H_{12}S$ 104.22	5008-69-5	112.5	4-01-00-01606 0.8335^{20}	EtOH 3; eth 3; ace 3 1.4433^{20}
10083	Propane, 2-methyl-2-(methylthio)- *tert*-Butyl methyl sulfide	$C_5H_{12}S$ 104.22	6163-64-0		4-01-00-01636	
10084	Propane, 2-methyl-1-nitro-	$C_4H_9NO_2$ 103.12	625-74-1	140.5	4-01-00-00301 0.9597^{25}	H_2O 2; EtOH 5; eth 5 1.4066^{25}
10085	Propane, 2-methyl-2-nitro-	$C_4H_9NO_2$ 103.12	594-70-7 26.23	127.16	4-01-00-00301 0.9501^{28}	EtOH 5; eth 5; ace 5; bz 5 1.4015^{20}
10086	Propane, 2-methyl-1-propoxy-	$C_7H_{16}O$ 116.20	15268-49-2	107	4-01-00-01594 0.7549^{20}	eth 4; EtOH 4 1.3852^{25}
10087	Propane, 2-methyl-2-propoxy-	$C_7H_{16}O$ 116.20	29072-93-3	100	3-01-00-01578 0.7472^{25}	1.3830^{25}
10088	Propane, 1-(methylthio)- Methyl propyl sulfide	$C_4H_{10}S$ 90.19	3877-15-4 -113	95.6	4-01-00-01450 0.8424^{20}	H_2O 3; EtOH 3; eth 3; ace 3 1.4442^{20}
10089	Propane, 2-(methylthio)- Isopropyl methyl sulfide	$C_4H_{10}S$ 90.19	1551-21-9 -101.5	84.8	4-01-00-01500 0.8291^{20}	EtOH 3; eth 3; ace 3 1.4932^{20}
10090	Propanenitrile Ethyl cyanide	C_3H_5N 55.08	107-12-0 -92.8	7839 97.1	4-02-00-00728 0.7818^{20}	H_2O 4; EtOH 3; eth 3; ace 3 1.3655^{20}
10091	Propanenitrile, 2-(acetyloxy)-	$C_5H_7NO_2$ 113.12	15657-96-2 -75	172.5	4-03-00-00676 1.0278^{20}	H_2O 4 1.4027^{20}
10092	Propanenitrile, 3-amino- 3-Aminopropionitrile	$C_3H_6N_2$ 70.09	151-18-8	481 185; 88^{20}	4-04-00-02530 0.9584^{20}	1.4396^{20}
10093	Propanenitrile, 2,2'-azobis[2-methyl- 2,2'-Azobis(isobutyronitrile)	$C_8H_{12}N_4$ 164.21	78-67-1	931	4-04-00-03377	H_2O 1; EtOH 2; eth 2
10094	Propanenitrile, 2-bromo-	C_3H_4BrN 133.98	19481-82-4	59^{24}	4-02-00-00764 1.5505^{20}	ace 4; eth 4 1.4585^{20}
10095	Propanenitrile, 3-bromo-	C_3H_4BrN 133.98	2417-90-5	92^{25}; 69^{7}	4-02-00-00766 1.6152^{20}	EtOH 4; eth 4; ctc 2 1.4800^{20}
10096	Propanenitrile, 2-bromo-2-methyl-	C_4H_6BrN 148.00	41658-69-9	139.5	4-02-00-00863 1.4796^{15}	ace 4; eth 4; EtOH 4 1.4379^{15}
10097	Propanenitrile, 3-chloro- β-Chloropropionitrile	C_3H_4ClN 89.52	542-76-7 -51	2160 175.5	4-02-00-00751 1.1573^{20}	ctc 2 1.4360^{20}
10098	Propanenitrile, 3-chloro-2-hydroxy-2-methyl- Lactonitrile, 3-chloro-2-methyl-	C_4H_6ClNO 119.55	33401-05-7	110^{27}	3-03-00-00599 1.2027^{15}	H_2O 4; EtOH 4; ace 4 1.4356^{11}
10099	Propanenitrile, 2,3-dichloro-	$C_3H_3Cl_2N$ 123.97	2601-89-0	$58\text{-}9^{8}$	4-02-00-00757 1.3500^{20}	1.4640^{20}
10100	Propanenitrile, 2,2-dimethyl- *tert*-Butyl cyanide	C_5H_9N 83.13	630-18-2 15	106.1	4-02-00-00913 0.7586^{25}	1.3774^{20}
10101	Propanenitrile, 3-(dimethylamino)- 3-(Dimethylamino)propionitrile	$C_5H_{10}N_2$ 98.15	1738-25-6	173	4-04-00-02533 0.8705^{20}	
10102	Propanenitrile, 2-ethoxy- 2-Ethoxypropionitrile	C_5H_9NO 99.13	14631-45-9	136	4-03-00-00675 0.8743^{20}	eth 4; EtOH 4 1.3890^{22}
10103	Propanenitrile, 3-ethoxy-	C_5H_9NO 99.13	2141-62-0	171	4-03-00-00708 0.9285^{15}	eth 4; EtOH 4 1.4068^{20}
10104	Propanenitrile, 2-(β-*D*-glucopyranosyloxy)-2-methyl- Linamarin	$C_{10}H_{17}NO_6$ 247.25	554-35-8 145	5375	5-17-07-00397	ace 4
10105	Propanenitrile, 2-hydroxy- Acetaldehyde, cyanohydrin	C_3H_5NO 71.08	78-97-7 -40	183	3-03-00-00451 0.9877^{20}	H_2O 5; EtOH 5; eth 3; chl 3 1.4058^{18}
10106	Propanenitrile, 3-hydroxy- Hydracrylonitrile	C_3H_5NO 71.08	109-78-4 -46	3751 221	4-03-00-00708 1.0404^{25}	H_2O 5; EtOH 5; eth 2; chl 3 1.4248^{20}
10107	Propanenitrile, 3-[2-(2-hydroxyethoxy)ethoxy]-	$C_7H_{13}NO_3$ 159.19	10143-54-1	186^{9}	4-03-00-00713 1.089^{32}	1.4452^{22}
10108	Propanenitrile, 2-hydroxy-2-methyl- Acetone cyanohydrin	C_4H_7NO 85.11	75-86-5 -19	59 82^{23}	4-03-00-00785 0.932^{19}	H_2O 4; EtOH 4; eth 4; ace 3 1.3992^{20}
10109	Propanenitrile, 2-methoxy- 2-Methoxypropionitrile	C_4H_7NO 85.11	33695-59-9	119	4-03-00-00675 0.8928^{20}	EtOH 4 1.3818^{20}
10110	Propanenitrile, 3-methoxy-	C_4H_7NO 85.11	110-67-8	163	4-03-00-00708 0.9379^{20}	EtOH 3; eth 3; chl 3 1.4043^{20}
10111	Propanenitrile, 2-methyl- Isobutyronitrile	C_4H_7N 69.11	78-82-0 -71.5	103.9	4-02-00-00853 0.7704^{20}	H_2O 2; EtOH 4; eth 4; ace 3 1.3720^{20}
10112	Propanenitrile, 3-(methylamino)-	$C_4H_8N_2$ 84.12	693-05-0	$101\text{-}4^{49}$	3-04-00-01264 0.8992^{20}	H_2O 3; ace 3; bz 3; chl 3 1.4320^{20}
10113	Propanenitrile, 2-oxo-	C_3H_3NO 69.06	631-57-2	92.3	4-03-00-01515 0.9745^{20}	eth 3; ace 3; CH_3CN 3 1.3764^{20}
10114	Propanenitrile, 3,3'-oxybis-	$C_6H_8N_2O$ 124.14	1656-48-0	$110\text{-}12^{0.5}$	4-03-00-00720 1.043^{25}	1.4405^{20}
10115	Propanenitrile, 3,3,3-trichloro-2-hydroxy- Chlorocyanohydrin	$C_3H_2Cl_3NO$ 174.41	513-96-2 61	2134 217 dec	4-03-00-00680	H_2O 4; eth 4; EtOH 4
10116	Propane, 1-nitro- 1-Nitropropane	$C_3H_7NO_2$ 89.09	108-03-2 -108	6548 131.1	4-01-00-00229 0.9961^{25}	H_2O 2; EtOH 5; eth 5; chl 3 1.4018^{20}

No.	Name Synonym	Mol. Form. Mol. Wt.	CAS RN mp/°C	Merck No. bp/°C	Beil. Ref. den/g cm^{-3}	Solubility n_D
10117	Propane, 2-nitro- 2-Nitropropane	$C_3H_7NO_2$ 89.09	79-46-9 -91.3	6549 120.2	4-01-00-00230 0.9821^{25}	H_2O 2; chl 3 1.3944^{20}
10118	Propane, octafluoro- Perfluoropropane	C_3F_8 188.02	76-19-7 -147.6	 -36.6	4-01-00-00189	
10119	Propane, 1,1'-oxybis- Dipropyl ether	$C_6H_{14}O$ 102.18	111-43-3 -126.1	7870 90.0	4-01-00-01422 0.7466^{20}	eth 4; EtOH 4 1.3809^{20}
10120	Propane, 2,2'-oxybis- Diisopropyl ether	$C_6H_{14}O$ 102.18	108-20-3 -86.8	5100 68.5	4-01-00-01471 0.7241^{20}	H_2O 2; EtOH 5; eth 5; ace 3 1.3679^{20}
10121	Propane, 1,1'-oxybis[2-chloro-	$C_6H_{12}Cl_2O$ 171.07	54460-96-7	 188	2-01-00-00370 1.109^{20}	EtOH 3; eth 3 1.4467^{20}
10122	Propane, 1,1'-oxybis[3-chloro-	$C_6H_{12}Cl_2O$ 171.07	629-36-7	 216; 90.5^{11}	4-01-00-01442 1.138^{20}	EtOH 3; eth 3 1.4158^{20}
10123	Propane, 2,2'-oxybis[1-chloro- 2,2'-Dichlorodiisopropyl ether	$C_6H_{12}Cl_2O$ 171.07	108-60-1	 187	3-01-00-01470 1.103^{20}	H_2O 1; EtOH 5; eth 5; ace 5 1.4505^{20}
10124	Propane, 2,2'-oxybis[2-methyl- Di-*tert*-Butyl ether	$C_8H_{18}O$ 130.23	6163-66-2	3020 107.2	4-01-00-01616 0.7658^{20}	 1.3949^{20}
10125	Propane, 1,1,1,2,3-pentachloro- 1,1,1,2,3-Pentachloropropane	$C_3H_3Cl_5$ 216.32	21700-31-2	 179.5	4-01-00-00203	eth 4; EtOH 4 1.5130^{20}
10126	Propane, 1,1,2,3,3-pentachloro- 1,1,2,3,3-Pentachloropropane	$C_3H_3Cl_5$ 216.32	15104-61-7	 199; 99^{20}	4-01-00-00204 1.6086^{39}	H_2O I; EtOH 4; eth 3; chl 3 1.5131^{17}
10127	Propane, 1,1,1,2,3-pentachloro-2-methyl- Isobutane, 1,1,1,2,3-pentachloro	$C_4H_5Cl_5$ 230.35	4749-31-9 73.5	 215	4-01-00-00294 1.5686^{25}	chl 4 1.5165^{20}
10128	Propane, 1,1,1,2,2-pentafluoro 1,1,1,2,2-Pentafluoropropane	$C_3H_3F_5$ 134.05	1814-88-6	 -17.4		
10129	Propane, 1,1'-sulfinylbis-	$C_6H_{14}OS$ 134.24	4253-91-2 22.5	 80^{2}	4-01-00-01453 0.9654^{20}	eth 4; EtOH 4 1.4663^{20}
10130	2-Propanesulfonic acid Isopropanesulfonic acid	$C_3H_8O_3S$ 124.16	14159-48-9 -37	 $159^{1.4}$	4-04-00-00042 1.187^{25}	H_2O 4 1.4332^{20}
10131	Propane, 1,1'-sulfonylbis-	$C_6H_{14}O_2S$ 150.24	598-03-8 29.5		4-01-00-01453 1.0278^{50}	H_2O 2; EtOH 3; eth 3 1.4456^{30}
10132	Propane, 1,1'-sulfonylbis[2-methyl-	$C_8H_{18}O_2S$ 178.30	10495-45-1 17	 265	4-01-00-01607 1.0056^{18}	
10133	1-Propanesulfonyl chloride	$C_3H_7ClO_2S$ 142.61	10147-36-1	 180 dec; 77^{12}	4-04-00-00039 1.2826^{15}	 1.452^{20}
10134	1-Propanesulfonyl chloride, 2-methyl	$C_4H_9ClO_2S$ 156.63	35432-36-1	 190	4-04-00-00049 1.2014^{20}	 1.4520^{25}
10135	Propane, 1,1,1,2-tetrabromo-	$C_3H_4Br_4$ 359.68	62127-49-5	 122^{15}	3-01-00-00251 2.679^{20}	 1.6187^{20}
10136	Propane, 1,1,2,2-tetrabromo- 1,1,2,2-Tetrabromopropane	$C_3H_4Br_4$ 359.68	34570-59-7	 $105\text{-}7^{9}$	4-01-00-00222 2.687^{20}	 1.6166^{20}
10137	Propane, 1,1,3,3-tetrabromo-	$C_3H_4Br_4$ 359.68	51525-97-4	 $154\text{-}6^{19}$	2-01-00-00077 2.702^{21}	 1.6225^{21}
10138	Propane, 1,2,2,3-tetrabromo- 1,2,2,3-Tetrabromopropane	$C_3H_4Br_4$ 359.68	54268-02-9 10.5	 169^{80}; 115^{9}	3-01-00-00251 2.703^{20}	 1.6200^{20}
10139	1,1,2,3-Propanetetracarboxylic acid, tetraethyl ester Tetraethyl 1,1,2,3-propanetetracarboxylate	$C_{15}H_{24}O_8$ 332.35	635-03-0	 $203\text{-}4^{18}$	3-02-00-02077 1.1184^{20}	EtOH 4 1.4395^{20}
10140	1,1,3,3-Propanetetracarboxylic acid, tetraethyl ester 1,1,3,3-Tetracarbethoxypropane	$C_{15}H_{24}O_8$ 332.35	2121-66-6 -30	 305 dec; 195^{8}	4-02-00-02417 1.116^{20}	EtOH 3 1.4398^{20}
10141	Propane, 1,1,1,2-tetrachloro- 1,1,1,2-Tetrachloropropane	$C_3H_4Cl_4$ 181.88	812-03-3 -64	 152.5	4-01-00-00201 1.473^{20}	H_2O 1; EtOH 4; eth 3; chl 3 1.4867^{20}
10142	Propane, 1,1,1,3-tetrachloro- 1,1,1,3-Tetrachloropropane	$C_3H_4Cl_4$ 181.88	1070-78-6	 157	4-01-00-00201 1.4509^{20}	H_2O 1; EtOH 4; eth 4; bz 4 1.4825^{20}
10143	Propane, 1,1,2,2-tetrachloro- 1,1,2,2-Tetrachloropropane	$C_3H_4Cl_4$ 181.88	13116-60-4	 153	4-01-00-00201 1.47^{12}	H_2O 1; EtOH 5; eth 5; chl 3 1.4850^{25}
10144	Propane, 1,1,2,3-tetrachloro- 1,1,2,3-Tetrachloropropane	$C_3H_4Cl_4$ 181.88	18495-30-2	 179.5	4-01-00-00202 1.513^{17}	H_2O 1; EtOH 3; eth 4; chl 3 1.5037^{17}
10145	Propane, 1,2,2,3-tetrachloro- 1,2,2,3-Tetrachloropropane	$C_3H_4Cl_4$ 181.88	13116-53-5	 165	4-01-00-00202 1.500^{18}	H_2O 1; EtOH 4; eth 4; chl 3 1.4940^{18}
10146	Propane, 1,1,2,3-tetrachloro-2-(chloromethyl)- 1,1,2,3-Tetrachloro-2-(chloromethyl)propane	$C_4H_5Cl_5$ 230.35	66997-45-3	 227; 95^{9}	3-01-00-00321 1.5686^{25}	bz 4; EtOH 4; chl 4 1.5165^{25}
10147	Propane, 1,1,1,2-tetrachloro-2-methyl- Isobutane, 1,1,1,2-tetrachloro	$C_4H_8Cl_4$ 195.90	7086-07-9 178.5	 192^{117}	4-01-00-00293	eth 4; EtOH 4; chl 4
10148	Propane, 1,1,2,3-tetrachloro-2-methyl-	$C_4H_6Cl_4$ 195.90	18963-01-4 -46	 190.5	4-01-00-00294 1.4393^{25}	bz 4; eth 4; chl 4 1.4963^{20}
10149	Propane, 1,2,2,3-tetrachloro-1,1,3,3-tetrafluoro- Propane, 1,2,2,3-tetrachloro-tetrafluoro	$C_3Cl_4F_4$ 253.84	677-68-9 -42.9	 112	4-01-00-00203 1.7199^{20}	 1.3958^{20}
10150	Propane, 1,1,3,3-tetramethoxy-	$C_7H_{16}O_4$ 164.20	102-52-3	 183; 66^{12}	4-01-00-03635 0.997^{25}	 1.4081^{20}
10151	Propane, 1,1'-thiobis- Dipropyl sulfide	$C_6H_{14}S$ 118.24	111-47-7 -102.5	7881 142.9	4-01-00-01452 0.814^{17}	H_2O 1; EtOH 3; eth 3 1.4487^{20}
10152	Propane, 2,2'-thiobis- Diisopropyl sulfide	$C_6H_{14}S$ 118.24	625-80-9 -78.1	 120.1	4-01-00-01502 0.8142^{20}	H_2O 1; EtOH 3; eth 3 1.4438^{20}
10153	Propane, 1,1'-thiobis[3-chloro- Bis(3-chloropropyl) sulfide	$C_6H_{12}Cl_2S$ 187.13	55882-21-8	 162^{43}	3-01-00-01438 1.1774^{25}	eth 4; EtOH 4; tol 4 1.5075^{20}
10154	Propane, 1,1'-thiobis[2-methyl- Diisobutyl sulfide	$C_8H_{18}S$ 146.30	592-65-4 -105.5	5035 171	4-01-00-01607 0.8363^{10}	

No.	Name	Synonym	Mol. Form.	Mol. Wt.	CAS RN	mp/°C	Merck No.	bp/°C	Beil. Ref.	den/g cm^{-3}	Solubility	n_D
10155	Propane, 2,2'-thiobis[2-methyl-	Di-*tert*-butyl sulfide	$C_8H_{18}S$	146.30	107-47-1	-9.0		149.1	4-01-00-01637	0.815^{25}		1.4506^{20}
10156	1-Propanethiol	Propyl mercaptan	C_3H_8S	76.16	107-03-9	-113.3		67.8	4-01-00-01449	0.8411^{20}	H_2O 2; EtOH 3; eth 3; ace 3	1.4380^{20}
10157	2-Propanethiol	Isopropyl mercaptan	C_3H_8S	76.16	75-33-2	-130.5		52.6	4-01-00-01498	0.8143^{20}	H_2O 2; EtOH 5; eth 5; ace 4	1.4255^{20}
10158	1-Propanethiol, 2-methyl-	Isobutyl mercaptan	$C_4H_{10}S$	90.19	513-44-0	<-70	5030	88.5	4-01-00-01605	0.8357^{20}	H_2O 2; EtOH 4; eth 4; ace 4	1.4387^{20}
10159	2-Propanethiol, 2-methyl-	*tert*-Butyl mercaptan	$C_4H_{10}S$	90.19	75-66-1	-0.5	1577	64.3	4-01-00-01634	0.7943^{25}	H_2O 1; ctc 3; hp 3	1.4232^{20}
10160	Propane, 1,1,1-tribromo-		$C_3H_5Br_3$	280.78	62127-61-1			$77\text{-}83^{25}$	4-01-00-00221	2.35^{25}		1.5606^{25}
10161	Propane, 1,1,2-tribromo-		$C_3H_5Br_3$	280.78	14602-62-1			200.5	3-01-00-00251	2.3547^{20}	H_2O 1; EtOH 3; eth 4; chl 3	1.5790^{20}
10162	Propane, 1,2,2-tribromo-		$C_3H_5Br_3$	280.78	14476-30-3			190.5	2-01-00-00077	2.2984^{20}	eth 4; EtOH 4; chl 4	1.5670^{20}
10163	Propane, 1,2,3-tribromo-	1,2,3-Tribromopropane	$C_3H_5Br_3$	280.78	96-11-7	16.9	9527	222.1	4-01-00-00221	2.4208^{20}	H_2O 1; EtOH 4; eth 4; ctc 2	1.5862^{20}
10164	Propane, 1,2,3-tribromo-2-methyl-		$C_4H_7Br_3$	294.81	631-28-7			178^{235}; 108^{18}	4-01-00-00299	2.2105^{14}	H_2O 1; eth 3; chl 3	1.5701^{21}
10165	1,2,3-Propanetricarboxylic acid	Tricarballylic acid	$C_6H_8O_6$	176.13	99-14-9	166	9534		4-02-00-02366		H_2O 4; EtOH 4; eth 2	
10166	1,2,3-Propanetricarboxylic acid, 2-(acetyloxy)-, triethyl ester		$C_{14}H_{22}O_8$	318.32	77-89-4			214^{40}	3-03-00-01105	1.135^{25}		1.4380
10167	1,2,3-Propanetricarboxylic acid, 2-hydroxy-	Citric acid	$C_6H_8O_7$	192.13	77-92-9	153	2328	dec	4-03-00-01272	1.665^{20}	H_2O 4; EtOH 4; eth 3; bz 1	
10168	1,2,3-Propanetricarboxylic acid, 2-hydroxy-, tributyl ester	Butyl citrate	$C_{18}H_{32}O_7$	360.45	77-94-1	-20	1564	233^{22}	4-03-00-01276	1.043^{20}		1.4460^{20}
10169	1,2,3-Propanetricarboxylic acid, 2-hydroxy-, triethyl ester		$C_{12}H_{20}O_7$	276.29	77-93-0			294	4-03-00-01276	1.1369^{20}	H_2O 1; EtOH 3; eth 3; ctc 2	1.4455^{20}
10170	1,2,3-Propanetricarboxylic acid, 2-hydroxy-, trimethyl ester	Trimethyl citrate	$C_9H_{14}O_7$	234.21	1587-20-8	79.3		285; 176^{16}	4-03-00-01276		eth 4; EtOH 4	
10171	Propane, 1,1,1-trichloro-	1,1,1-Trichloropropane	$C_3H_5Cl_3$	147.43	7789-89-1			108	4-01-00-00198	1.287^{23}	H_2O 1; EtOH 3; eth 4; chl 4	
10172	Propane, 1,1,2-trichloro-	1,1,2-Trichloropropane	$C_3H_5Cl_3$	147.43	598-77-6			132.0; 117^{500}	4-01-00-00199	1.372^{15}	H_2O 1; EtOH 3; eth 4; ctc 2	
10173	Propane, 1,1,3-trichloro-	1,1,3-Trichloropropane	$C_3H_5Cl_3$	147.43	20395-25-9	-59		145.5	4-01-00-00199	1.3557^{20}	eth 4; EtOH 4; chl 4	1.4718^{20}
10174	Propane, 1,2,2-trichloro-	1,2,2-Trichloropropane	$C_3H_5Cl_3$	147.43	3175-23-3			124	4-01-00-00199	1.318^{25}	H_2O 1; EtOH 3; eth 3; chl 4	1.4609^{20}
10175	Propane, 1,2,3-trichloro-	1,2,3-Trichloropropane	$C_3H_5Cl_3$	147.43	96-18-4	-14.7		157	4-01-00-00199	1.3889^{20}	H_2O 2; EtOH 3; eth 3; ctc 2	1.4852^{20}
10176	Propane, 1,2,3-trichloro-2-(chloromethyl)-		$C_4H_6Cl_4$	195.90	18963-00-3			211; 87^{9}	3-01-00-00321	1.5036^{25}	chl 4	1.508^{20}
10177	Propane, 1,1,2-trichloro-2-methyl-	Isobutane, 1,1,2-trichloro	$C_4H_7Cl_3$	161.46	29559-52-2	-6		163	4-01-00-00293	1.2588^{20}	eth 4; chl 4	1.4666^{20}
10178	Propane, 1,2,3-trichloro-2-methyl-		$C_4H_7Cl_3$	161.46	1871-58-5			162.5	4-01-00-00293	1.3012^{25}	chl 4	1.4765^{20}
10179	Propane, 1,1,1-trichloro-2,2,3,3,3-pentafluoro-		$C_3Cl_3F_5$	237.38	4259-43-2	-80		74	4-01-00-00200	1.643^{25}		
10180	Propane, 1,2,2-trichloro-1,1,3,3,3-pentafluoro-		$C_3Cl_3F_5$	237.38	1599-41-3	-4.3		72.0	3-01-00-00233	1.6681^{20}		1.3519^{20}
10181	Propane, 1,2,3-trichloro-1,1,2,3,3-pentafluoro-		$C_3Cl_3F_5$	237.38	76-17-5	-72		73.7	4-01-00-00200	1.6631^{20}		1.3512^{20}
10182	Propane, 1,1,1-triethoxy-		$C_9H_{20}O_3$	176.26	115-80-0			171	4-02-00-00707		eth 4; EtOH 4	1.4000^{25}
10183	Propane, 1,1,3-triethoxy-		$C_9H_{20}O_3$	176.26	7789-92-6			185 dec; 78^{14}	4-01-00-03971	0.898^{19}	H_2O 2; EtOH 3; ctc 3	1.4067^{20}
10184	Propane, 1,1,1-trifluoro-3-nitro-	Propane, 1-nitro-3,3,3-trifluoro	$C_3H_4F_3NO_2$	143.07	461-35-8			135	4-01-00-00232	1.4178^{20}	eth 4	1.3549^{20}
10185	Propane, 1,2,3-trimethoxy-	Glycerol trimethyl ether	$C_6H_{14}O_3$	134.18	20637-49-4			148	4-01-00-02755	0.9460^{15}	H_2O 5; eth 3; ace 4; bz 3	1.4055^{15}
10186	1,2,3-Propanetriol	Glycerol	$C_3H_8O_3$	92.09	56-81-5	18.2	4379	290	4-01-00-02751	1.2613^{20}	H_2O 5; EtOH 5; eth 2; bz 1	1.4746^{20}
10187	1,2,3-Propanetriol, 1-acetate, (±)-	Glycerol, 1-monoacetate(*DL*)	$C_5H_{10}O_4$	134.13	93713-40-7			158^{165} 130^{3}	4-02-00-00251	1.2060^{20}	H_2O 3; EtOH 3; eth 2; bz 1	1.4157^{20}
10188	1,2,3-Propanetriol, 1-(4-aminobenzoate)	Glyceryl *p*-aminobenzoate	$C_{10}H_{13}NO_4$	211.22	136-44-7		4382				H_2O 1; EtOH 3	
10189	1,2,3-Propanetriol, diacetate	Diacetin	$C_7H_{12}O_5$	176.17	25395-31-7		2941	259	4-02-00-00252	1.184^{16}	H_2O 4; eth 4; EtOH 4	1.44^{20}
10190	1,2,3-Propanetriol, 1,3-diacetate	1,3-Diacetin	$C_7H_{12}O_5$	176.17	105-70-4		2941	173^{40}	4-02-00-00252	1.179^{15}	H_2O 4; EtOH 4; eth 2; CS_2 1	1.4395^{20}

No.	Name Synonym	Mol. Form. Mol. Wt.	CAS RN mp/°C	Merck No. bp/°C	Beil. Ref. den/g cm^{-3}	Solubility n_D
10191	1,2,3-Propanetriol, 1,3-dinitrate	$C_3H_6N_2O_7$ 182.09	623-87-0 26	148^{15}; $116^{0.6}$	2-01-00-00591 1.523^{20}	H_2O 4; eth 4; EtOH 4 1.4715^{20}
10192	1,2,3-Propanetriol, 1-nitrate	$C_3H_7NO_5$ 137.09	624-43-1 61	 157.5	4-01-00-02761 1.4164^{20}	H_2O 4; EtOH 4; eth 2 1.4698^{20}
10193	1,2,3-Propanetriol, 2-nitrate	$C_3H_7NO_5$ 137.09	620-12-2 54	 157.5	2-01-00-00591 1.40^{22}	H_2O 4; eth 4; EtOH 4
10194	1,2,3-Propanetriol, 1-phenyl- Glycerol, 1-phenyl	$C_9H_{12}O_3$ 168.19	63157-81-3 100.5	 185	4-06-00-07407 1.2213^{13}	 1.5606^{15}
10195	1,2,3-Propanetriol, triacetate Triacetin	$C_9H_{14}O_6$ 218.21	102-76-1 -78	9504 259	4-02-00-00253 1.1583^{20}	H_2O 2; EtOH 5; eth 5; ace 4 1.4301^{20}
10196	1,2,3-Propanetriol, tribenzoate	$C_{24}H_{20}O_6$ 404.42	614-33-5 76		3-09-00-00666 1.228^{12}	H_2O 1; EtOH 3; eth 4; ace 4
10197	1,2,3-Propanetriol, trinitrate Nitroglycerin	$C_3H_5N_3O_9$ 227.09	55-63-0 13.5	6528 $93^{0.31}$; exp 218	4-01-00-02762 1.5931^{20}	H_2O 2; EtOH 3; eth 5; ace 4 1.4786^{12}
10198	1,2,3-Propanetriol, tripropanoate	$C_{12}H_{20}O_6$ 260.29	139-45-7	 $175\text{-}6^{20}$	4-02-00-00717 1.099^{20}	H_2O 1; EtOH 3; eth 4; chl 3 1.4318^{19}
10199	Propanetrione, diphenyl-	$C_{15}H_{10}O_3$ 238.24	643-75-4 70	 289^{175}; 179^{4}	4-07-00-02822	eth 4
10200	Propanil Propanamide, *N*-(3,4-dichlorophenyl)-	$C_9H_9Cl_2NO$ 218.08	709-98-8 92	7814	 1.25^{25}	
10201	Propanoic acid Propionic acid	$C_3H_6O_2$ 74.08	79-09-4 -20.7	7837 141.1	4-02-00-00695 0.9930^{20}	H_2O 5; EtOH 5; eth 3; chl 2 1.3809^{20}
10202	Propanoic acid, 2-(acetyloxy)-, (±)-	$C_5H_8O_4$ 132.12	3853-80-3	 $167\text{-}70^{78}$	4-03-00-00638 1.1758^{20}	bz 4; EtOH 4 1.4240^{20}
10203	Propanoic acid, anhydride Propionic anhydride	$C_6H_{10}O_3$ 130.14	123-62-6 -45	7838 170; 67.5^{18}	4-02-00-00722 1.0110^{20}	eth 5; ctc 2 1.4038^{20}
10204	Propanoic acid, 2-bromo-, (±)-	$C_3H_5BrO_2$ 152.98	10327-08-9 25.7	 203.5	 1.7000^{20}	H_2O 4, EtOH 4; eth 4; chl 2 1.4753^{20}
10205	Propanoic acid, 3-bromo- β-Bromopropionic acid	$C_3H_5BrO_2$ 152.98	590-92-1 62.5	1421 $140\text{-}2^{45}$	4-02-00-00764 1.48^{25}	H_2O 3; EtOH 3; eth 3; bz 3
10206	Propanoic acid, 3-bromo-, butyl ester	$C_7H_{13}BrO_2$ 209.08	6973-79-1	 130^{26}	4-02-00-00765 1.4549^{20}	ace 4; eth 4; EtOH 4 1.3051^{20}
10207	Propanoic acid, 2-bromo-, ethyl ester Ethyl α-bromopropionate	$C_5H_9BrO_2$ 181.03	535-11-5	3731 160 dec; 71^{26}	4-02-00-00762 1.4135^{20}	H_2O 1; EtOH 5; eth 5; chl 3 1.4490^{20}
10208	Propanoic acid, 3-bromo-, ethyl ester	$C_5H_9BrO_2$ 181.03	539-74-2	 179; 65^{15}	4-02-00-00765 1.4123^{18}	EtOH 3; eth 3; ace 3; ctc 2 1.4516^{20}
10209	Propanoic acid, 2-bromo-2-methyl- α-Bromoisobutyric acid	$C_4H_7BrO_2$ 167.00	2052-01-9 48.5	1408 199; 115^{24}	4-02-00-00862 1.4969^{60}	
10210	Propanoic acid, 3-bromo-, 3-methylbutyl ester Isopentyl 3-bromopropionate	$C_8H_{15}BrO_2$ 223.11	100983-12-8	 $110\text{-}1^{11}$	2-02-00-00231 1.2217^{15}	eth 4; EtOH 4 1.4556^{9}
10211	Propanoic acid, 2-bromo-, methyl ester, (±)-	$C_4H_7BrO_2$ 167.00	57885-43-5	 144	4-02-00-00762 1.4966^{25}	eth 4; EtOH 4 1.4451^{22}
10212	Propanoic acid, 3-bromo-, methyl ester	$C_4H_7BrO_2$ 167.00	3395-91-3	 105.5^{60}	4-02-00-00765 1.4897^{15}	EtOH 3; eth 3; ace 3; os 4 1.4542^{20}
10213	Propanoic acid, 2-bromo-2-methyl-, ethyl ester	$C_6H_{11}BrO_2$ 195.06	600-00-0	 163	4-02-00-00862 1.3263^{20}	H_2O 1; EtOH 3; eth 5 1.4446^{20}
10214	Propanoic acid, butyl ester Butyl propionate	$C_7H_{14}O_2$ 130.19	590-01-2 -89	1587 146.8	4-02-00-00708 0.8754^{20}	H_2O 2; EtOH 5; eth 5; ctc 2 1.4014^{20}
10215	Propanoic acid, 2-chloro-	$C_3H_5ClO_2$ 108.52	598-78-7	 185	4-02-00-00745 1.2585^{20}	H_2O 5; EtOH 5; eth 5; ace 3 1.4380^{20}
10216	Propanoic acid, 3-chloro- β-Chloropropionic acid	$C_3H_5ClO_2$ 108.52	107-94-8 41	2159 204 dec	4-02-00-00748	H_2O 3; EtOH 3; eth 5; chl 3
10217	Propanoic acid, 2-chloro-, butyl ester	$C_7H_{13}ClO_2$ 164.63	54819-86-2	 184	4-02-00-00746 1.0253^{20}	eth 4 1.4263^{20}
10218	Propanoic acid, 3-chloro-, butyl ester	$C_7H_{13}ClO_2$ 164.63	27387-79-7	 104^{22}	4-02-00-00749 1.0370^{20}	H_2O 4; eth 4 1.4321^{20}
10219	Propanoic acid, 3-chloro-2,2-dimethyl-	$C_5H_9ClO_2$ 136.58	13511-38-1 41.5	 110^{10}	4-02-00-00914	ctc 4
10220	Propanoic acid, 2-chloro-, ethyl ester Ethyl α-chloropropionate	$C_5H_9ClO_2$ 136.58	535-13-7	3743 117	4-02-00-00746 1.0700^{20}	H_2O 1; EtOH 5; eth 5; ctc 2 1.4170^{20}
10221	Propanoic acid, 3-chloro-, ethyl ester	$C_5H_9ClO_2$ 136.58	623-71-2	 162	4-02-00-00749 1.1086^{20}	H_2O 2; EtOH 5; eth 5; os 3 1.4254^{20}
10222	Propanoic acid, 2-chloro-, methyl ester Methyl 2-chloropropionate	$C_4H_7ClO_2$ 122.55	17639-93-9	 132.5	4-02-00-00746 1.0750^{25}	
10223	Propanoic acid, 2-chloro-, 1-methylethyl ester Isopropyl 2-chloropropionate	$C_6H_{11}ClO_2$ 150.60	40058-87-5	 151.5	4-02-00-00746 1.0315^{20}	H_2O 1; EtOH 3; eth 3 1.4149^{20}
10224	Propanoic acid, 2-chloro-, 2-methylpropyl ester, (*R*)- Propionic acid, 2-chloro, isobutyl ester (*d*)	$C_7H_{13}ClO_2$ 164.63	114489-96-2	 176	0-02-00-00248 1.0312^{20}	 1.4247^{20}
10225	Propanoic acid, 3-chloro-, methyl ester	$C_4H_7ClO_2$ 122.55	6001-87-2	 156	4-02-00-00748 1.1861^{15}	EtOH 3; ctc 3 1.4263^{20}
10226	Propanoic acid, 3-chloro-, 3-methylbutyl ester	$C_8H_{15}ClO_2$ 178.66	62108-70-7	 208; 87^{12}	4-02-00-00749 1.0171^{20}	eth 4; EtOH 4 1.4343^{20}
10227	Propanoic acid, 3-chloro-, 2-methylpropyl ester	$C_7H_{13}ClO_2$ 164.63	62108-68-3	 191.3	4-02-00-00749 1.0323^{20}	eth 4; EtOH 4 1.4295^{20}

No.	Name Synonym	Mol. Form. Mol. Wt.	CAS RN mp/°C	Merck No. bp/°C	Beil. Ref. den/g cm^{-3}	Solubility n_D
10228	Propanoic acid, 2-chloro-2-methyl-	$C_4H_7ClO_2$ 122.55	594-58-1 31	 118^{50}; 93^{20}	4-02-00-00858 1.1790^{20}	EtOH 4 1.450^{20}
10229	Propanoic acid, 2-chloro-2-methyl-, methyl ester	$C_5H_9ClO_2$ 136.58	22421-97-2	 135	4-02-00-00859 1.0883^{15}	eth 4 1.4122^{21}
10230	Propanoic acid, 3-chloro-2-methyl-	$C_4H_7ClO_2$ 122.55	16674-04-7	 $128\text{-}33^{50}$	4-02-00-00860 1.0153^{20}	eth 4; EtOH 4 1.4310^{20}
10231	Propanoic acid, 2-(4-chlorophenoxy)-2-methyl-, ethyl ester Clofibrate	$C_{12}H_{15}ClO_3$ 242.70	637-07-0	2374 $148\text{-}50^{20}$	4-06-00-00851	
10232	Propanoic acid, 3-chloro-, propyl ester	$C_6H_{11}ClO_2$ 150.60	62108-66-1	 180	4-02-00-00749 1.0656^{20}	eth 4; EtOH 4 1.4290^{20}
10233	Propanoic acid, 2-cyano-	$C_4H_5NO_2$ 99.09	632-07-5 35	 143^{11}	3-02-00-01683 1.1400^{20}	
10234	Propanoic acid, 3-cyano-, methyl ester	$C_5H_7NO_2$ 113.12	4107-62-4	 211	3-02-00-01672	chl 3 1.4250^{20}
10235	Propanoic acid, cyclohexyl ester	$C_9H_{16}O_2$ 156.22	6222-35-1	 193; 93^{35}	4-06-00-00037 0.9359^{20}	H_2O 1; EtOH 3; eth 3; ace 3 1.4403^{20}
10236	Propanoic acid, 2-diazo-, ethyl ester Ethyl 2-diazopropionate	$C_5H_8N_2O_2$ 128.13	6111-99-5	 $65\text{-}8^{41}$	2-03-00-00403 1.086^{12}	 1.4472^{18}
10237	Propanoic acid, 2-diazo-, methyl ester Methyl 2-diazopropionate	$C_4H_6N_2O_2$ 114.10	34757-14-7	 $43\text{-}5^{11}$	2-03-00-00403 1.1011^{13}	 1.4487^{20}
10238	Propanoic acid, 2,3-dibromo-	$C_3H_4Br_2O_2$ 231.87	600-05-5 66.5	 160^{20}; 138^{12}	4-02-00-00767	bz 4; eth 4; EtOH 4
10239	Propanoic acid, 2,3-dibromo-, butyl ester Butyl 2,3-dibromopropionate	$C_7H_{12}Br_2O_2$ 287.98	21179-48-6	 237.5	4-02-00-00768 1.6107^{20}	 1.4890^{20}
10240	Propanoic acid, 2,3-dibromo-, ethyl ester	$C_5H_8Br_2O_2$ 259.93	3674-13-3	 214.5	4-02-00-00767 1.7966^{20}	EtOH 3; eth 3 1.5007^{20}
10241	Propanoic acid, 2,3-dibromo-, methyl ester	$C_4H_6Br_2O_2$ 245.90	1729-67-5	 206	4-02-00-00767 1.9333^{20}	EtOH 3 1.5127^{20}
10242	Propanoic acid, 2,3-dibromo-2-methyl-, methyl ester	$C_5H_8Br_2O_2$ 259.93	3673-79-8	 203	4-02-00-00864 1.805^{20}	 1.505^{20}
10243	Propanoic acid, 2,3-dibromo-, propyl ester Propyl 2,3-dibromopropionate	$C_6H_{10}Br_2O_2$ 273.95	79762-76-8	 222.5	3-02-00-00572 1.6799^{20}	 1.4950^{20}
10244	Propanoic acid, 2,2-dichloro- Dalapon	$C_3H_4Cl_2O_2$ 142.97	75-99-0	2806 187.5; 92^{14}	4-02-00-00753 1.389^{12}	H_2O 4; EtOH 4; eth 3; ctc 3
10245	Propanoic acid, 2,3-dichloro-	$C_3H_4Cl_2O_2$ 142.97	565-64-0 50	 210; 130^{26}	4-02-00-00756	H_2O 3; EtOH 3; eth 3; peth 2 1.4650^{20}
10246	Propanoic acid, 2,3-dichloro-, ethyl ester	$C_5H_8Cl_2O_2$ 171.02	6628-21-3	 183.5	4-02-00-00756 1.2401^{20}	eth 4; EtOH 4 1.4482^{20}
10247	Propanoic acid, 3,3-dichloro-2-hydroxy-2-methyl-	$C_4H_6Cl_2O_3$ 173.00	31340-44-0 82.5	 dec	2-03-00-00224	H_2O 3; EtOH 3; eth 3
10248	Propanoic acid, 2,3-dichloro-, methyl ester	$C_4H_6Cl_2O_2$ 157.00	3674-09-7	 92^{50}	4-02-00-00756 1.3282^{20}	ace 4; eth 4; EtOH 4
10249	Propanoic acid, 2,3-dichloro-, 1-methylethyl ester	$C_6H_{10}Cl_2O_2$ 185.05	54774-99-1	 $61\text{-}2^{5}$	4-02-00-00756 1.2010^{25}	 1.4470
10250	Propanoic acid, 2-(2,4-dichlorophenoxy)- Dichlorprop	$C_9H_8Cl_2O_3$ 235.07	120-36-5 117.5	3068	3-06-00-00707	H_2O 2; EtOH 3; eth 3; lig 2
10251	Propanoic acid, 2,2-dichloro-3,3,3-trifluoro-, methyl ester Propionic acid, 2,2-dichloro-3,3,3-trifluoro-, methyl ester	$C_4H_3Cl_2F_3O_2$ 210.97	378-68-7	 $116\text{-}7^{625}$	4-02-00-00758 1.5092^{20}	 1.3806^{20}
10252	Propanoic acid, 2,3-dihydroxy- Glyceric acid	$C_3H_6O_4$ 106.08	473-81-4	4378 dec	4-03-00-01050	H_2O 4; ace 4; EtOH 4
10253	Propanoic acid, 2,3-dihydroxy-, ethyl ester, (±)- Glyceric acid, ethyl ester (*DL*)	$C_5H_{10}O_4$ 134.13	89300-69-6	 235; 120^{14}	4-03-00-01052 1.1897^{15}	H_2O 4; eth 4; EtOH 4
10254	Propanoic acid, 2,3-dihydroxy-, methyl ester, (±)- Methyl *DL*-glycerate	$C_4H_8O_4$ 120.11	15909-76-9	 241.5; 119^{14}	4-03-00-01052 1.2802^{15}	H_2O 4; EtOH 4 1.4502^{20}
10255	Propanoic acid, 2,3-dihydroxy-2-methyl-, ethyl ester Glyceric acid, 2-methyl, ethyl ester	$C_6H_{12}O_4$ 148.16	67535-07-3	 95^{10}	3-03-00-00860 1.114^{25}	 1.4370^{25}
10256	Propanoic acid, 2,3-dihydroxy-2-methyl-, methyl ester Glyceric acid, 2-methyl-, methyl ester	$C_5H_{10}O_4$ 134.13	19860-56-1	 89^{10}	4-03-00-01060 1.185^{25}	 1.4438^{25}
10257	Propanoic acid, 2,3-dihydroxy-, propyl ester	$C_6H_{12}O_4$ 148.16	116435-95-1	 $132\text{-}4^{3}$	3-03-00-00854 1.1537^{20}	 1.4503^{20}
10258	Propanoic acid, 2,2-dimethyl- Pivalic acid	$C_5H_{10}O_2$ 102.13	75-98-9 35	7482 164	4-02-00-00908 0.905^{50}	H_2O 2; EtOH 4; eth 4 1.3931^{30}
10259	Propanoic acid, 2,2-dimethyl-, ethyl ester Ethyl 2,2-dimethylpropionate	$C_7H_{14}O_2$ 130.19	3938-95-2 -89.5	 118.4	4-02-00-00910 0.856^{20}	EtOH 3; eth 3 1.3906^{20}
10260	Propanoic acid, 2,2-dimethyl-, methyl ester Methyl 2,2-dimethylpropionate	$C_6H_{12}O_2$ 116.16	598-98-1	 101.1	4-02-00-00909 0.891^{0}	eth 4; EtOH 4 1.3905^{20}
10261	Propanoic acid, ethenyl ester Vinyl propionate	$C_5H_8O_2$ 100.12	105-38-4	 91.2	4-02-00-00711	

No.	Name	Synonym	Mol. Form.	Mol. Wt.	CAS RN	mp/°C	Merck No.	bp/°C	Beil. Ref.	den/g cm^{-3}	Solubility	n_D
10262	Propanoic acid, 3-ethoxy-, ethyl ester		$C_7H_{14}O_3$	146.19	763-69-9			$48\text{-}9^{5}$	4-03-00-00697	0.92^{25}		1.4065^{20}
10263	Propanoic acid, ethyl ester	Ethyl propionate	$C_5H_{10}O_2$	102.13	105-37-3	-73.9	3801	99.1	4-02-00-00705	0.8917^{20}	H_2O 2; EtOH 5; eth 5; ace 3	1.3839^{20}
10264	Propanoic acid, heptyl ester		$C_{10}H_{20}O_2$	172.27	2216-81-1	-50.9		210	4-02-00-00710	0.8679^{20}	H_2O 1; EtOH 3; eth 3; ace 3	1.4201^{15}
10265	Propanoic acid, hexyl ester		$C_9H_{18}O_2$	158.24	2445-76-3	-57.5		190	4-02-00-00709	0.8698^{20}	H_2O 1; EtOH 3; eth 3; ace 3	1.4162^{15}
10266	Propanoic acid, 2-hydroxy-, (±)-	Lactic acid (*DL*)	$C_3H_6O_2$	90.08	598-82-3	18	5215	122^{15}	2-03-00-00192	1.2060^{21}	H_2O 4; eth 2; EtOH 4	1.4392^{20}
10267	Propanoic acid, 3-hydroxy-	Hydracrylic acid	$C_3H_6O_3$	90.08	503-66-2		4681	dec	4-03-00-00689		H_2O 4; EtOH 3; eth 5	1.4489^{20}
10268	Propanoic acid, 2-hydroxy-, butyl ester, (*R*)-	Lactic acid, butyl ester (*d*)	$C_7H_{14}O_3$	146.19	34451-18-8			77^{10}	3-03-00-00450	0.9744^{27}	eth 4; EtOH 4	
10269	Propanoic acid, 2-hydroxy-, ethyl ester, (±)-	Ethyl lactate	$C_5H_{10}O_3$	118.13	2676-33-7	-26	3773	154.5		1.0328^{20}	H_2O 4; eth 4; EtOH 4	1.4124^{20}
10270	Propanoic acid, 3-hydroxy-, ethyl ester		$C_5H_{10}O_3$	118.13	623-72-3			187.5	2-03-00-00213	1.059^{20}		1.4271^{23}
10271	Propanoic acid, 2-hydroxy-, hexadecyl ester	Cetyl lactate	$C_{19}H_{38}O_3$	314.51	35274-05-6	41	2022	219^{10}; 170^{1}	4-03-00-00660			1.4410^{40}
10272	Propanoic acid, 3-hydroxy-2-(hydroxymethyl)-2-methyl-	Dimethylolpropionic acid	$C_5H_{10}O_4$	134.13	4767-03-7	190	3241		4-03-00-01066			
10273	Propanoic acid, 2-hydroxy-, methyl ester, (±)-	Lactic acid, methyl ester (*DL*)	$C_4H_8O_3$	104.11	2155-30-8			144.8	4-03-00-00640	1.0928^{20}	H_2O 4; eth 4; EtOH 4	1.4141^{20}
10274	Propanoic acid, 2-hydroxy-, 3-methylbutyl ester		$C_8H_{16}O_3$	160.21	19329-89-6			202.4	3-03-00-00483	0.9589^{25}	eth 4; EtOH 4	1.4240^{25}
10275	Propanoic acid, 2-hydroxy-, 1-methylethyl ester	Isopropyl lactate	$C_6H_{12}O_3$	132.16	617-51-6			167	3-03-00-00479	0.9980^{20}	H_2O 4; bz 4; eth 4; EtOH 4	1.4082^{25}
10276	Propanoic acid, 3-hydroxy-, methyl ester		$C_4H_8O_3$	104.11	6149-41-3			179	3-03-00-00525	1.105^{16}	H_2O 5; EtOH 5; eth 5	1.43^{23}
10277	Propanoic acid, 3-hydroxy-, 1-methylethyl ester	Hydracrylic acid, isopropyl ester	$C_6H_{12}O_3$	132.16	84098-45-3			128.6^{82}; 95^{12}	1-03-00-00113	[illegible]		[illegible]
10278	Propanoic acid, 2-hydroxy-2-methyl-		$C_4H_8O_3$	104.11	594-61-6	82.5		212	4-03-00-00782		H_2O 4; EtOH 4; eth 4; bz 2	
10279	Propanoic acid, 2-hydroxy-2-methyl-, ethyl ester		$C_6H_{12}O_3$	132.16	80-55-7			150	4-03-00-00783	0.987^{20}	H_2O 5; EtOH 5; ctc 3	1.4080^{20}
10280	Propanoic acid, 2-hydroxy-2-methyl-, methyl ester		$C_5H_{10}O_3$	118.13	2110-78-3			137	4-03-00-00783		H_2O 4; EtOH 4	1.4056^{20}
10281	Propanoic acid, 2-hydroxy-, 2-propenyl ester		$C_6H_{10}O_3$	130.14	5349-55-3			$56\text{-}60^{8}$	3-03-00-00486	1.0452^{20}	py 3	1.4369^{20}
10282	Propanoic acid, 2-iodo-, (±)-		$C_3H_5IO_2$	199.98	18791-49-6	46		$93\text{-}6^{0.2}$	3-02-00-00573	2.073^{18}	H_2O 2; EtOH 4; eth 4	
10283	Propanoic acid, 3-iodo-, ethyl ester		$C_5H_9IO_2$	228.03	6414-69-3			202	4-02-00-00770	1.6226^{30}		1.4961^{25}
10284	Propanoic acid, 3-iodo-, methyl ester		$C_4H_7IO_2$	214.00	5029-66-3			188	3-02-00-00574	1.8408^{7}	EtOH 4	
10285	Propanoic acid, 2-mercapto-	Thiolactic acid	$C_3H_6O_2S$	106.15	79-42-5	12	9269	$106\text{-}7^{15}$	4-03-00-00682	1.1938^{20}	H_2O 4; eth 4; EtOH 4	1.4810^{20}
10286	Propanoic acid, 3-mercapto-		$C_3H_6O_2S$	106.15	107-96-0	18		111^{15}; 86^{3}	4-03-00-00726	1.218^{21}	H_2O 3; EtOH 3; eth 3; ctc 3	1.494^{20}
10287	Propanoic acid, 3-mercapto-, methyl ester		$C_4H_8O_2S$	120.17	2935-90-2			$54\text{-}5^{14}$	3-03-00-00553	1.085^{25}		1.4640^{20}
10288	Propanoic acid, 2-mercapto-2-methyl-		$C_4H_8O_2S$	120.17	4695-31-2	47		101^{15}	4-03-00-00797		H_2O 4	
10289	Propanoic acid, 3-methoxy-, methyl ester		$C_5H_{10}O_3$	118.13	3852-09-3			142.8	4-03-00-00693	1.0139^{15}		1.4030^{20}
10290	Propanoic acid, 2-methyl-	Isobutyric acid	$C_4H_8O_2$	[illegible]	79-31-2	[illegible]	5039	[illegible]	4-02-00-00843	[illegible]	H_2O 4; EtOH 5; eth 5; ctc 2	[illegible]
10291	Propanoic acid, 2-methyl-, anhydride	Isobutyric anhydride	$C_8H_{14}O_3$	158.20	97-72-3	-53.5		183; 89^{32}	4-02-00-00851	0.9535^{20}	eth 5; chl 3	1.4061^{19}
10292	Propanoic acid, 2-methyl-, 2-chloroethyl ester	2-Chloroethyl isobutyrate	$C_6H_{11}ClO_2$	150.60	33662-96-3			170	4-02-00-00847	1.062^{25}	eth 4; EtOH 4	1.4109^{16}
10293	Propanoic acid, 2-methyl-, cyclohexyl ester	Isobutyric acid, cyclohexyl ester	$C_{10}H_{18}O_2$	170.25	1129-47-1			204	3-06-00-00024	0.9489^{0}	eth 4; EtOH 4	
10294	Propanoic acid, 2-methyl-, 1,1-dimethylethyl ester		$C_8H_{16}O_2$	144.21	16889-72-8			126.7	3-02-00-00648		ace 4; eth 4; EtOH 4	1.3921^{20}
10295	Propanoic acid, 2-methyl-, 3,7-dimethyl-2,6-octadienyl ester, (*E*)-		$C_{14}H_{24}O_2$	224.34	2345-26-8			$135\text{-}7^{13}$	2-02-00-00261	0.8997^{15}		1.4576^{20}
10296	Propanoic acid, methyl ester	Methyl propionate	$C_4H_8O_2$	88.11	554-12-1	-87.5	6030	79.8	4-02-00-00704	0.9150^{20}	H_2O 2; EtOH 5; eth 5; ace 3	1.3775^{20}
10297	Propanoic acid, 2-methyl-, ethenyl ester		$C_6H_{10}O_2$	114.14	2424-98-8			104.5	4-02-00-00848	0.8932^{20}		1.4061^{20}

No.	Name Synonym	Mol. Form. Mol. Wt.	CAS RN mp/°C	Merck No. bp/°C	Beil. Ref. den/g cm^{-3}	Solubility n_D
10298	Propanoic acid, 2-methyl-, ethyl ester Ethyl isobutanoate	$C_6H_{12}O_2$ 116.16	97-62-1 -88.2	3770 110.1	4-02-00-00846 0.868^{20}	H_2O 2; EtOH 5; eth 5; ace 3 1.3869^{18}
10299	Propanoic acid, 1-methylethyl ester Isopropyl propionate	$C_6H_{12}O_2$ 116.16	637-78-5	 109.5	4-02-00-00708 0.8660^{20}	H_2O 2; EtOH 5; eth 5 1.3872^{20}
10300	Propanoic acid, 2-methyl-, 2-furanylmethyl ester	$C_9H_{12}O_3$ 168.19	6270-55-9	 $85\text{-}6^{15}$	2-17-00-00115 1.0313^{20}	eth 4; EtOH 4
10301	Propanoic acid, 2-methyl-, methyl ester Methyl isobutyrate	$C_5H_{10}O_2$ 102.13	547-63-7 -84.7	6003 92.5	4-02-00-00846 0.8906^{20}	H_2O 2; EtOH 5; eth 5; ace 3 1.3840^{20}
10302	Propanoic acid, 2-methyl-, 3-methylbutyl ester	$C_9H_{18}O_2$ 158.24	2050-01-3	 168.5	3-02-00-00649 0.8627^{20}	H_2O 2; EtOH 3; eth 3; ace 3
10303	Propanoic acid, 2-methyl-, 1-methylethyl ester Isopropyl isobutyrate	$C_7H_{14}O_2$ 130.19	617-50-5	 120.7	4-02-00-00847 0.8471^{21}	H_2O 1; EtOH 3; eth 3; ace 3
10304	Propanoic acid, 2-methyl-, 2-methylpropyl ester Isobutyl isobutanoate	$C_8H_{16}O_2$ 144.21	97-85-8 -80.7	5028 148.6	4-02-00-00847 0.8542^{20}	H_2O 2; EtOH 3; eth 5; ace 3 1.3999^{20}
10305	Propanoic acid, 2-methyl-, 2-nitropentyl ester Isobutyric acid, (2-nitropentyl) ester	$C_9H_{17}NO_4$ 203.24	116530-63-3	 249.5; 121^{10}	3-02-00-00648 1.0310^{20}	ace 4; eth 4 1.4315^{20}
10306	Propanoic acid, 2-methyl-, 2-phenylethyl ester	$C_{12}H_{16}O_2$ 192.26	103-48-0	 $122\text{-}4^{15}$	4-06-00-03073 0.9950^{15}	 1.4871^{20}
10307	Propanoic acid, 2-methyl-, phenylmethyl ester	$C_{11}H_{14}O_2$ 178.23	103-28-6	 $114\text{-}5^{20}$	4-06-00-02266 1.0075^{15}	 1.4883^{20}
10308	Propanoic acid, 2-methyl-, propyl ester Propyl isobutyrate	$C_7H_{14}O_2$ 130.19	644-49-5	 135.4	4-02-00-00847 0.8843^{0}	H_2O 2; EtOH 3; eth 4; ace 3 1.3955^{20}
10309	Propanoic acid, 1-methylpropyl ester	$C_7H_{14}O_2$ 130.19	591-34-4	 133	4-02-00-00709 0.8657^{20}	EtOH 3; eth 3 1.3952^{20}
10310	Propanoic acid, 2-methylpropyl ester Isobutyl propionate	$C_7H_{14}O_2$ 130.19	540-42-1 -71.4	5033 137	4-02-00-00709 0.888^{0}	H_2O 2; EtOH 4; eth 4; ace 3 1.3973^{20}
10311	Propanoic acid, 3-(methylthio)-, methyl ester	$C_5H_{10}O_2S$ 134.20	13532-18-8	 $74\text{-}5^{13}$	4-03-00-00737 1.077^{25}	 1.4650^{20}
10312	Propanoic acid, 2-nitro-, ethyl ester	$C_5H_9NO_4$ 147.13	2531-80-8	 190.5	4-02-00-00771	bz 4; eth 4; EtOH 4 1.4210^{20}
10313	Propanoic acid, octyl ester	$C_{11}H_{22}O_2$ 186.29	142-60-9 -42.6	 228	4-02-00-00710 0.8663^{20}	H_2O 1; EtOH 3; eth 3; bz 3 1.4221^{15}
10314	Propanoic acid, 2-oxo- Pyruvic acid	$C_3H_4O_3$ 88.06	127-17-3 13.8	8032 165 dec; 54^{10}	4-03-00-01505 1.2272^{20}	H_2O 5; EtOH 5; eth 5; ace 3 1.4280^{20}
10315	Propanoic acid, 2-oxo-, ethyl ester	$C_5H_8O_3$ 116.12	617-35-6 -50	 155	4-03-00-01513 1.0596^{15}	H_2O 2; EtOH 5; eth 5; ace 3 1.4052^{20}
10316	Propanoic acid, 2-oxo-, methyl ester	$C_4H_6O_3$ 102.09	600-22-6	 135.5	4-01-00-04146 1.154^{0}	H_2O 2; EtOH 5; eth 5; ace 3 1.4046^{25}
10317	Propanoic acid, pentafluoro-, methyl ester Pentafluoropropionic acid, methyl ester	$C_4H_3F_5O_2$ 178.06	378-75-6	 59.5	4-02-00-00740 1.390^{25}	 1.2869^{25}
10318	Propanoic acid, pentyl ester	$C_8H_{16}O_2$ 144.21	624-54-4 -73.1	 168.6	4-02-00-00709 0.8761^{25}	H_2O 1; EtOH 5; eth 5; bz 3 1.4096^{15}
10319	Propanoic acid, 2-phenoxy-, (±)-	$C_9H_{10}O_3$ 166.18	1912-21-6 115.5	 265.5	4-06-00-00642 1.1865^{20}	eth 4; EtOH 4 1.5184^{20}
10320	Propanoic acid, 3-phenoxy-	$C_9H_{10}O_3$ 166.18	7170-38-9 97.5	 240; 188^{26}	4-06-00-00643	H_2O 3; chl 2; lig 3
10321	Propanoic acid, 2-phenoxy-, ethyl ester	$C_{11}H_{14}O_3$ 194.23	42412-84-0	 243.5	4-06-00-00643 1.360^{17}	H_2O 1; EtOH 3; eth 3 1.360^{17}
10322	Propanoic acid, 3-phenoxy-, ethyl ester	$C_{11}H_{14}O_3$ 194.23	22409-91-2 24	 142^{11}	4-06-00-00644 1.0745^{20}	 1.5007^{18}
10323	Propanoic acid, phenyl ester	$C_9H_{10}O_2$ 150.18	637-27-4 20	 211	4-06-00-00615 1.0436^{25}	H_2O 1; EtOH 4; eth 4; bz 3 1.4980^{20}
10324	Propanoic acid, 2-phenylethyl ester 2-Phenethyl propionate	$C_{11}H_{14}O_2$ 178.23	122-70-3		3-06-00-01709	
10325	Propanoic acid, 2-propenyl ester	$C_6H_{10}O_2$ 114.14	2408-20-0	 124	4-02-00-00711 0.9140^{20}	EtOH 3; eth 3; ace 3 1.4105^{20}
10326	Propanoic acid, propyl ester Propyl propionate	$C_6H_{12}O_2$ 116.16	106-36-5 -75.9	7880 122.5	4-02-00-00707 0.8809^{20}	H_2O 2; EtOH 5; eth 5; ace 3 1.3935^{20}
10327	Propanoic acid, sodium salt	$C_3H_5NaO_2$ 96.06	137-40-6	8623	4-02-00-00695	H_2O 2
10328	Propanoic acid, 2,3,3,3-tetrafluoro-, ethyl ester Ethyl 2,3,3,3-tetrafluoropropionate	$C_5H_6F_4O_2$ 174.10	399-92-8	 108.5	4-02-00-00737 1.289^{20}	 1.3260^{20}
10329	Propanoic acid, 3,3'-thiobis- 3,3'-Thiodipropionic acid	$C_6H_{10}O_4S$ 178.21	111-17-1 129	9261	4-03-00-00735	H_2O 4; EtOH 4
10330	Propanoic acid, 3,3'-thiobis-, diethyl ester	$C_{10}H_{18}O_4S$ 234.32	673-79-0	 121^{2}	4-03-00-00741 1.095^{25}	 1.4655^{20}
10331	Propanoic acid, 3,3'-thiobis-, dimethyl ester	$C_8H_{14}O_4S$ 206.26	4131-74-2	 148^{18}	4-03-00-00738 1.198^{25}	 1.4740^{20}
10332	Propanoic acid, 2,2,3-trichloro-	$C_3H_3Cl_3O_2$ 177.41	3278-46-4 65.5	 140^{40}	4-02-00-00759	H_2O 3; EtOH 3; bz 3; CS_2 3
10333	Propanoic acid, 3,3,3-trichloro-2-hydroxy- Lactic acid, 3,3,3-trichloro-	$C_3H_3Cl_3O_3$ 193.41	599-01-9 125	 140^{45}	3-03-00-00501	H_2O 4; eth 4; EtOH 4; chl 4
10334	1-Propanol Propyl alcohol	C_3H_8O 60.10	71-23-8 -126.1	7854 97.2	4-01-00-01413 0.8035^{20}	H_2O 5; EtOH 5; eth 5; ace 3 1.3850^{20}
10335	2-Propanol Isopropyl alcohol	C_3H_8O 60.10	67-63-0 -89.5	5096 82.3	4-01-00-01461 0.7855^{20}	H_2O 5; EtOH 5; eth 5; ace 3 1.3776^{20}

No.	Name Synonym	Mol. Form. Mol. Wt.	CAS RN mp/°C	Merck No. bp/°C	Beil. Ref. den/g cm^{-3}	Solubility n_D
10336	1-Propanol, 2-amino-, (±)-	C_3H_9NO 75.11	6168-72-5	 174.5	3-04-00-00737	H_2O 4; EtOH 4; eth 4; chl 2 1.4502^{20}
10337	1-Propanol, 3-amino- Propanolamine	C_3H_9NO 75.11	156-87-6 11	 187.5	4-04-00-01623 0.9824^{26}	H_2O 3; EtOH 3; eth 3 1.4617^{20}
10338	2-Propanol, 1-amino-, (±)- *DL*-1-Amino-2-propanol	C_3H_9NO 75.11	1674-56-2 1.7	 159.4	4-04-00-01665 0.9611^{20}	H_2O 5; EtOH 5; eth 5; ace 5 1.4479^{20}
10339	2-Propanol, 1-amino-3-(diethylamino)- 3-(Diethylamino)-2-hydroxypropylamine	$C_7H_{18}N_2O$ 146.23	6322-01-6	 223	3-04-00-00767 0.931^{20}	EtOH 3; eth 3; chl 2 1.465^{20}
10340	2-Propanol, 1-[(2-aminoethyl)amino]- *N*-(2-Hydroxypropyl)ethylenediamine	$C_5H_{14}N_2O$ 118.18	123-84-2	450 94^{3}	4-04-00-01684 0.9837^{25}	 1.4738^{20}
10341	1-Propanol, 2-amino-2-methyl- 2-Aminoisobutanol	$C_4H_{11}NO$ 89.14	124-68-5 25.5	461 165.5	4-04-00-01740 0.934^{20}	H_2O 5; ctc 3 1.449^{20}
10342	1-Propanol, 2-[(2-amino-2-methylpropyl)amino]-2-methyl- 1-Propanol, 2-[2-(2-amino-2-methylpropyl)amino]-2-methyl-	$C_8H_{20}N_2O$ 160.26	72622-74-3 46	 231	3-04-00-00788 0.9360^{20}	chl 2 1.4635^{25}
10343	2-Propanol, 1,3-bis(dimethylamino)-	$C_7H_{18}N_2O$ 146.23	5966-51-8	 181.5	4-04-00-01695 0.8788^{20}	H_2O 4 1.4418^{20}
10344	1-Propanol, 3-bromo-	C_3H_7BrO 138.99	627-18-9	 105^{185}; 62^{5}	4-01-00-01446 1.5374^{20}	H_2O 3; EtOH 5; eth 5 1.4834^{25}
10345	2-Propanol, 1-bromo-	C_3H_7BrO 138.99	19686-73-8	 146.5	3-01-00-01474 1.5585^{30}	H_2O 3; EtOH 4; eth 4 1.4801^{20}
10346	2-Propanol, 1-butoxy-	$C_7H_{16}O_2$ 132.20	5131-66-8	 171.5; 71^{20}	4-01-00-02471 0.882^{20}	EtOH 3; eth 3; bz 3; ctc 3 1.4168^{20}
10347	1-Propanol, 2-chloro- Propylene chlorohydrin	C_3H_7ClO 94.54	78-89-7	7863 133.5	4-01-00-01440 1.103^{20}	H_2O 4; eth 4; EtOH 4 1.4390^{20}
10348	1-Propanol, 3-chloro- 3-Chloro-1-propanol	C_3H_7ClO 94.54	627-30-5	 165	4-01-00-01441 1.1309^{20}	H_2O 4; EtOH 3; eth 3; ctc 2 1.4459^{20}
10349	2-Propanol, 1-chloro- *sec*-Propylene chlorohydrin	C_3H_7ClO 94.54	127-00-4	7864 127	4-01-00-01490 1.113^{20}	H_2O 5; EtOH 5; eth 5; ctc 2 1.4392^{20}
10350	1-Propanol, 3-chloro-, acetate	$C_5H_9ClO_2$ 136.58	628-09-1	 164; 62^{10}	4-02-00-00140 1.250^{19}	H_2O 1; EtOH 3; eth 3; ace 3 1.431^{20}
10351	2-Propanol, 1-chloro-, acetate	$C_5H_9ClO_2$ 136.58	623-60-9	 149.5	4-02-00-00142 1.0788^{20}	eth 4; EtOH 4 1.4223^{20}
10352	1-Propanol, 2-chloro-2-methyl- Isobutyl alcohol, 2-chloro	C_4H_9ClO 108.57	558-38-3	 132 dec; 60^{50}	4-01-00-01603 1.0472^{20}	 1.4388^{20}
10353	1-Propanol, 3-chloro-2-methyl- Isobutyl alcohol, 3-chloro	C_4H_9ClO 108.57	10317-10-9	 $76\text{-}8^{11}$	3-01-00-01564 1.083^{25}	eth 4; EtOH 4 1.4460^{25}
10354	2-Propanol, 1-chloro-2-methyl- Isobutylene chlorohydrin	C_4H_9ClO 108.57	558-42-9 -20	 128.5	4-01-00-01628 1.0628^{20}	H_2O 4; EtOH 4 1.4380^{24}
10355	2-Propanol, 1-chloro-3-(1-methylethoxy)- 2-Propanol, 1-chloro-3-isopropoxy-	$C_6H_{13}ClO_2$ 152.62	4288-84-0	 87^{20}	4-01-00-02485 1.0530^{25}	EtOH 3; eth 3 1.4370^{25}
10356	2-Propanol, 1-chloro-3-propoxy-	$C_6H_{13}ClO_2$ 152.62	6943-58-4	 $92\text{-}5^{16}$	3-01-00-02153 1.0526^{25}	ace 4; eth 4 1.4378^{25}
10357	2-Propanol, 1,3-dibromo-	$C_3H_6Br_2O$ 217.89	96-21-9	 219 dec; 105^{16}	4-01-00-01496 2.1201^{25}	ace 4; eth 4; EtOH 4 1.5495^{25}
10358	1-Propanol, 2,3-dibromo-, phosphate (3:1) Tris(2,3-dibromopropyl) phosphate	$C_9H_{15}Br_6O_4P$ 697.61	126-72-7	9665		chl 3
10359	1-Propanol, 2,3-dichloro-	$C_3H_6Cl_2O$ 128.99	616-23-9	 184	4-01-00-01442 1.3607^{20}	H_2O 2; EtOH 5; eth 5; ace 5 1.4819^{20}
10360	1-Propanol, 3,3-dichloro-	$C_3H_6Cl_2O$ 128.99	83682-72-8	 $82\text{-}3^{20}$	3-01-00-01427 1.316^{25}	 1.467^{20}
10361	2-Propanol, 1,1-dichloro-	$C_3H_6Cl_2O$ 128.99	53894-19-2	 147	2-01-00-00383 1.3334^{22}	ace 4; eth 4; EtOH 4
10362	2-Propanol, 1,3-dichloro- 1,3-Dichloro-2-propanol	$C_3H_6Cl_2O$ 128.99	96-23-1	3063 176	4-01-00-01491 1.3506^{17}	H_2O 4; EtOH 4; eth 5; ace 3 1.4837^{20}
10363	2-Propanol, 1,3-dichloro-, acetate	$C_5H_8Cl_2O_2$ 171.02	3674-10-0	 205	3-02-00-00233 1.281^{20}	eth 4; EtOH 4; chl 4 1.4542^{20}
10364	2-Propanol, 1,1-dichloro-2-methyl- 2,2-Dichloro-1,1-dimethylethanol	$C_4H_8Cl_2O$ 143.01	4773-53-9 8	 152.5	4-01-00-01629 1.2363^{19}	bz 4; EtOH 4 1.4598^{19}
10365	2-Propanol, 1,3-dichloro-2-methyl- β-Methylglycerol α,γ-dichlorohydrin	$C_4H_8Cl_2O$ 143.01	597-32-0	 174.5	4-01-00-01629 1.2745^{20}	H_2O 3; EtOH 4; ace 3 1.4744^{21}
10366	1-Propanol, 2-(diethylamino)- 2-Diethylamino-1-propanol	$C_7H_{17}NO$ 131.22	611-12-1	 168; 78^{12}	4-04-00-01618 0.8649^{27}	EtOH 3; eth 3; ace 3; bz 3 1.4332^{20}
10367	1-Propanol, 3-(diethylamino)-	$C_7H_{17}NO$ 131.22	622-93-5	 189.5	4-04-00-01633 0.8600^{20}	EtOH 3; eth 3; ace 3; bz 3 1.4439^{20}
10368	2-Propanol, 1-(diethylamino)-, (±)- 2-Propanol, 1-diethylamino (*DL*)	$C_7H_{17}NO$ 131.22	78738-36-0 13.5	 159; 63^{22}	3-04-00-00759 0.8511^{20}	H_2O 4; ace 4; EtOH 4 1.4255^{20}
10369	2-Propanol, 1,3-difluoro- 1,3-Difluoro-2-propanol	$C_3H_6F_2O$ 96.08	453-13-4	 $54\text{-}5^{34}$	4-01-00-01489 1.24^{25}	 1.3725^{20}
10370	1-Propanol, 2,3-dimercapto- Dimercaprol	$C_3H_8OS_2$ 124.23	59-52-9	3196 $82\text{-}4^{0.8}$	4-01-00-02770 1.2463^{20}	EtOH 3; eth 3; chl 2; oils 3 1.5749^{20}
10371	2-Propanol, 1,3-dimercapto-	$C_3H_8OS_2$ 124.23	584-04-3	 94^{12}	4-01-00-02773 1.2386^{20}	 1.5700^{20}
10372	2-Propanol, 1,3-dimethoxy- Glycerol 1,3-dimethyl ether	$C_5H_{12}O_3$ 120.15	623-69-8	 169	3-01-00-02318 1.0085^{20}	H_2O 5; EtOH 4; eth 3 1.4192^{20}
10373	1-Propanol, 2,2-dimethyl- Neopentyl alcohol	$C_5H_{12}O$ 88.15	75-84-3 52.5	6373 113.5	4-01-00-01690 0.812^{20}	H_2O 2; EtOH 4; eth 4; ctc 3

No.	Name Synonym	Mol. Form. Mol. Wt.	CAS RN mp/°C	Merck No. bp/°C	Beil. Ref. den/g cm^{-3}	Solubility n_D
10374	1-Propanol, 2-(dimethylamino)- 2-Dimethylamino-1-propanol	$C_5H_{13}NO$ 103.16	15521-18-3	 150.3	4-04-00-01617 0.8820^{26}	H_2O 3
10375	1-Propanol, 3-(dimethylamino)-	$C_5H_{13}NO$ 103.16	3179-63-3	 163.5	4-04-00-01628 0.872^{25}	ctc 3 1.4360^{20}
10376	2-Propanol, 1-(dimethylamino)-	$C_5H_{13}NO$ 103.16	108-16-7	 124.5	3-04-00-00756 0.837^{25}	ctc 3 1.4193^{20}
10377	2-Propanol, 1,3-diphenoxy-	$C_{15}H_{16}O_3$ 244.29	622-04-8 81.5	 224^{17}; 175^{2}	4-06-00-00590 1.179^{24}	bz 4; eth 4; EtOH 4; chl 4
10378	2-Propanol, 1,3-diphenoxy-, acetate Glycerol, 1,3-diphenyl ether-2-acetate	$C_{17}H_{18}O_4$ 286.33	71159-31-4 33	 190^{160}	3-06-00-00583	bz 4; eth 4; EtOH 4; chl 4
10379	2-Propanol, 1,1',1'',1'''-(1,2-ethanediyldinitrilo)tetrakis- ENTPROL	$C_{14}H_{32}N_2O_4$ 292.42	102-60-3	3551	4-04-00-01685 1.030^{25}	chl 2 1.478^{25}
10380	1-Propanol, 2-ethoxy- Propylene glycol β-monoethyl ether	$C_5H_{12}O_2$ 104.15	19089-47-5	 140.5	3-01-00-02147 0.9044^{20}	H_2O 3; EtOH 3; eth 3 1.4122^{20}
10381	2-Propanol, 1-ethoxy- 1-Ethoxy-2-propanol	$C_5H_{12}O_2$ 104.15	1569-02-4	 131	4-01-00-02471 0.9028^{20}	H_2O 3; EtOH 3; eth 3 1.4075^{20}
10382	1-Propanol, 3-fluoro- 3-Hydroxypropyl fluoride	C_3H_7FO 78.09	462-43-1	 127.8	4-01-00-01437 1.0390^{25}	H_2O 4; eth 4; EtOH 4 1.3771^{25}
10383	2-Propanol, 1-[(2-hydroxyethyl)amino]-	$C_5H_{13}NO_2$ 119.16	6579-55-1	 $120\text{-}3^{1}$	4-04-00-01678 1.0455^{20}	H_2O 3; EtOH 3; ace 3 1.4695^{20}
10384	2-Propanol, 1,1'-iminobis- Diisopropanolamine	$C_6H_{15}NO_2$ 133.19	110-97-4 44.5	 250; 151^{23}	3-04-00-00761 0.989^{20}	H_2O 3; EtOH 3; eth 2
10385	1-Propanol, 3-iodo-	C_3H_7IO 185.99	627-32-7	 115^{38}; 88^{4}	4-01-00-01448 1.9976^{20}	 1.5585^{20}
10386	1-Propanol, 2-methoxy- 2-Methoxy-1-propanol	$C_4H_{10}O_2$ 90.12	1589-47-5	 130	4-01-00-02471 0.938^{20}	 1.4070^{20}
10387	2-Propanol, 1-methoxy- 1-Methoxy-2-propanol	$C_4H_{10}O_2$ 90.12	107-98-2	 119	3-01-00-02146 0.9620^{20}	 1.4034^{20}
10388	1-Propanol, 2-methyl- Isobutyl alcohol	$C_4H_{10}O$ 74.12	78-83-1 -108	5015 107.8	4-01-00-01588 0.8018^{20}	H_2O 2; EtOH 3; eth 3; ace 3 1.3955^{20}
10389	2-Propanol, 2-methyl- *tert*-Butyl alcohol	$C_4H_{10}O$ 74.12	75-65-0 25.4	1542 82.4	4-01-00-01609 0.7887^{20}	H_2O 5; EtOH 5; eth 5; chl 3 1.3878^{20}
10390	2-Propanol, 1-(1-methylethoxy)- 1-Isopropoxy-2-propanol	$C_6H_{14}O_2$ 118.18	3944-36-3	 137.5	4-01-00-02471 0.879^{20}	 1.4070^{20}
10391	1-Propanol, 2-methyl-2-nitro-	$C_4H_9NO_3$ 119.12	76-39-1 89.5	 94^{10}	4-01-00-01604	H_2O 2; EtOH 4; eth 4; chl 3
10392	1-Propanol, 3-[methyl(phenylmethyl)amino]-	$C_{11}H_{17}NO$ 179.26	5814-42-6	 $132\text{-}5^{4}$	4-12-00-02208 1.001^{25}	ctc 2; CS_2 3 1.5230^{20}
10393	2-Propanol, 1,1',1''-nitrilotris-	$C_9H_{21}NO_3$ 191.27	122-20-3 45	 $170\text{-}80^{10}$	4-04-00-01680 1.0^{20}	H_2O 3; EtOH 3; chl 2
10394	1-Propanol, 2-nitro-	$C_3H_7NO_3$ 105.09	2902-96-7	 100^{12}	4-01-00-01448 1.1841^{25}	H_2O 3; EtOH 3; eth 3; chl 2 1.4379^{20}
10395	1-Propanol, 3-nitro-	$C_3H_7NO_3$ 105.09	25182-84-7	 $138\text{-}40^{32}$	3-01-00-01429 1.173^{13}	H_2O 4; eth 4; EtOH 4
10396	2-Propanol, 1-nitro- 1-Nitro-2-propanol	$C_3H_7NO_3$ 105.09	3156-73-8	 112^{13}; 68^{1}	4-01-00-01496 1.1906^{20}	H_2O 3; EtOH 3; eth 3; bz 3 1.4383^{20}
10397	Propanol, oxybis- Dipropylene glycol	$C_6H_{14}O_3$ 134.18	25265-71-8	 230.5	4-01-00-02473 1.0206^{20}	H_2O 5; EtOH 3
10398	1-Propanol, 2-phenoxy-	$C_9H_{12}O_2$ 152.19	4169-04-4	 244	4-06-00-00582 0.9801^{25}	EtOH 3; eth 3 1.4760^{25}
10399	2-Propanol, 1-phenoxy-	$C_9H_{12}O_2$ 152.19	770-35-4	 134.5^{20}	4-06-00-00582 1.0622^{20}	 1.5232^{20}
10400	1-Propanol, 3-(phenylmethoxy)-	$C_{10}H_{14}O_2$ 166.22	4799-68-2	 172^{43}	4-06-00-02243 1.0474^{20}	EtOH 3; eth 3
10401	2-Propanol, 1-propoxy- 1,2-Propylene glycol 1-propyl ether	$C_6H_{14}O_2$ 118.18	1569-01-3	 150	3-01-00-02147 0.8886^{20}	 1.4130^{20}
10402	2-Propanol, 1,1,3,3-tetrachloro- 1,1,3,3-Tetrachloro-2-propanol	$C_3H_4Cl_4O$ 197.87	18992-39-7	 $80\text{-}90^{14}$	3-01-00-01474 1.612^{20}	eth 3 1.5133^{20}
10403	2-Propanol, 1,1,1,3-tetrachloro-	$C_3H_4Cl_4O$ 197.87	88947-16-4	 $95\text{-}6^{17}$	3-01-00-01474 1.610^{20}	eth 4 1.5145^{20}
10404	1-Propanol, 2,2,3,3-tetrafluoro- 2,2,3,3-Tetrafluoro-1-propanol	$C_3H_4F_4O$ 132.06	76-37-9 -15	9139 109.5	4-01-00-01438 1.4853^{20}	EtOH 3; ace 3; chl 3 1.3197^{20}
10405	2-Propanol, 1,1,1-tribromo-2-methyl- Tribromo-*tert*-butyl alcohol	$C_4H_7Br_3O$ 310.81	76-08-4 169	9523 sub	3-01-00-01588	H_2O 2; EtOH 3; eth 3; chl 2
10406	2-Propanol, 1,1,1-trichloro- 1,1,1-Trichloro-2-propanol	$C_3H_5Cl_3O$ 163.43	76-00-6 50.5	9557 163; $53\text{-}5^{12}$	3-01-00-01474	ace 4; bz 4; eth 4; EtOH 4
10407	2-Propanol, 1,1,1-trichloro-2-methyl- Chlorobutanol	$C_4H_7Cl_3O$ 177.46	57-15-8 97	2129 167	4-01-00-01629	H_2O 1; EtOH 3; eth 3; ace 3
10408	2-Propanol, 1,1,1-trichloro-3-nitro- 2-Propanol, 1-nitro-3,3,3-trichloro	$C_3H_4Cl_3NO_3$ 208.43	760-40-7 45.5	 $105^{3.5}$	4-01-00-01497	eth 4; EtOH 4
10409	2-Propanol, 1,1,1-trifluoro-, (±)- 2-Propanol, 1,1,1-trifluoro(*DL*)	$C_3H_5F_3O$ 114.07	17556-48-8 -52	 78	4-01-00-01489 1.2632^{25}	EtOH 4; eth 4; ace 3; bz 3 1.3130^{25}
10410	1-Propanol, 3-(trimethylsilyl)-	$C_6H_{16}OSi$ 132.28	2917-47-7	 141; 82^{24}	4-04-00-03953 0.822^{25}	 1.4298^{20}
10411	2-Propanone Acetone	C_3H_6O 58.08	67-64-1 -94.8	58 56.0	4-01-00-03180 0.7899^{20}	H_2O 5; EtOH 5; eth 5; ace 5 1.3588^{20}
10412	2-Propanone, 1-(acetyloxy)-	$C_5H_8O_3$ 116.12	592-20-1	 171; 63^{11}	4-02-00-00297 1.0757^{20}	H_2O 4; eth 4; EtOH 4 1.4141^{20}
10413	1-Propanone, 1-(4-aminophenyl)- *p*-Aminopropiophenone	$C_9H_{11}NO$ 149.19	70-69-9 140	482	4-14-00-00139	DMSO 3

No.	Name / Synonym	Mol. Form. / Mol. Wt.	CAS RN / mp/°C	Merck No. / bp/°C	Beil. Ref. / den/g cm^{-3}	Solubility / n_D
10414	1-Propanone, 2-amino-1-phenyl-	$C_9H_{11}NO$ 149.19	5265-18-9 46.5	$93^{0.8}$	3-14-00-00147	ace 4; eth 4; EtOH 4
10415	2-Propanone, 1,3-bis(1-methyl-2-pyrrolidinyl)-, (*R**,*S**)- Cuscohygrine	$C_{13}H_{24}N_2O$ 224.35	454-14-8	2685 169^{23}; 122^{2}	5-24-01-00458 0.9733^{20}	H_2O 4; bz 4; eth 4; EtOH 4 1.4832^{20}
10416	2-Propanone, 1-bromo- Bromoacetone	C_3H_5BrO 136.98	598-31-2 -36.5	1389 138; 31.5^{8}	4-01-00-03223 1.634^{23}	H_2O 2; EtOH 3; eth 3; ace 3 1.4697^{15}
10417	2-Propanone, 1-bromo-3-chloro- 1-Bromo-3-chloroacetone	C_3H_4BrClO 171.42	53535-68-5 34.5	179	4-01-00-03224	eth 4; EtOH 4
10418	1-Propanone, 1-(4-bromophenyl)-	C_9H_9BrO 213.07	10342-83-3 48	169^{15}; 137^{2}	4-07-00-00684	H_2O 1; EtOH 3; eth 3; ace 3
10419	1-Propanone, 2-bromo-1-phenyl-	C_9H_9BrO 213.07	2114-00-3	247.5	4-07-00-00684 1.4298^{20}	H_2O 1; EtOH 3; eth 3; ace 3 1.5720^{20}
10420	2-Propanone, 1-chloro- Chloroacetone	C_3H_5ClO 92.52	78-95-5 -44.5	2113 119	4-01-00-03215 1.15^{20}	H_2O 3; EtOH 3; eth 3; chl 3
10421	2-Propanone, 1-chloro-1,1,3,3,3-pentafluoro- Chloropentafluoroacetone	C_3ClF_5O 182.48	79-53-8 -133	8		
10422	1-Propanone, 1-(4-chlorophenyl)-	C_9H_9ClO 168.62	6285-05-8 37.3	135^{31}; 114^{2}	4-07-00-00683	H_2O 1; EtOH 3; chl 2; CS_2 3
10423	1-Propanone, 1-cyclohexyl- Cyclohexyl ethyl ketone	$C_9H_{16}O$ 140.23	1123-86-0	196	4-07-00-00064 0.9105^{20}	1.4530^{20}
10424	2-Propanone, 1,3-dibromo-	$C_3H_4Br_2O$ 215.87	816-39-7 26	97^{22}	4-01-00-03225 2.1670^{18}	eth 4; CS_2 4
10425	2-Propanone, 1,1-dichloro- 1,1-Dichloroacetone	$C_3H_4Cl_2O$ 126.97	513-88-2	3038 120	4-01-00-03218 1.304^{18}	H_2O 2; EtOH 3; eth 5
10426	2-Propanone, 1,3-dichloro- 1,3-Dichloroacetone	$C_3H_4Cl_2O$ 126.97	534-07-6 45	3039 173.4	4-01-00-03219 1.3826^{46}	H_2O 3; EtOH 3; eth 3 1.4716^{40}
10427	2-Propanone, 1-(diethylamino)-	$C_7H_{15}NO$ 129.20	1620-14-0	157 dec	4-04-00-01925 0.8620^{20}	H_2O 5; EtOH 5; eth 5; ctc 3 1.4249^{20}
10428	1-Propanone, 2-(diethylamino)-1-phenyl- Diethylpropion	$C_{13}H_{19}NO$ 205.30	90-84-6	3116	4-14-00-00144	
10429	2-Propanone, 1,3-dihydroxy- Dihydroxyacetone	$C_3H_6O_3$ 90.08	96-26-4 90	3166	4-01-00-04119	H_2O 3; EtOH 3; eth 3; ace 3
10430	1-Propanone, 1-(2,4-dihydroxyphenyl)-	$C_9H_{10}O_3$ 166.18	5792-36-9 99	176^{6}	4-08-00-01837	bz 4; eth 4; EtOH 4
10431	1-Propanone, 2,2-dimethyl-1-phenyl-	$C_{11}H_{14}O$ 162.23	938-16-9	220	4-07-00-00710 0.963^{26}	[illegible] 3 1.5086^{19}
10432	2-Propanone, 1,1-diphenyl-	$C_{15}H_{14}O$ 210.28	781-35-1 46	307; 174^{10}	4-07-00-01426	EtOH 3; eth 3; bz 3; chl 3 1.5361^{16}
10433	2-Propanone, 1,3-diphenyl-	$C_{15}H_{14}O$ 210.28	102-04-5 35	331	4-07-00-01420 1.195^{0}	H_2O 1; EtOH 3; eth 3; peth 3
10434	1-Propanone, 3-(3-ethenyl-4-piperidinyl)-1-(6-methoxy-4-quinolinyl)-, (3*R*-*cis*)- Viquidil	$C_{20}H_{24}N_2O_2$ 324.42	84-55-9 60	9908	4-25-00-00181	eth 4; EtOH 4; chl 4
10435	1-Propanone, 3-(3-ethenyl-4-piperidinyl)-1-(4-quinolinyl)-, (3*R*-*cis*)- Cinchotoxine	$C_{19}H_{22}N_2O$ 294.40	69-24-9 59	2291	4-24-00-00651	H_2O 1; EtOH 4; eth 4; ace 4
10436	1-Propanone, 3-ethoxy-1-phenyl-	$C_{11}H_{14}O_2$ 178.23	34008-71-4 11.5	135^{18}	4-08-00-00453 1.0386^{20}	1.5190^{20}
10437	2-Propanone, 1-fluoro-	C_3H_5FO 76.07	430-51-3	77	4-01-00-03213 1.0288^{20}	1.3700^{20}
10438	1-Propanone, 1-(2-furanyl)-	$C_7H_8O_2$ 124.14	3194-15-8 28	88^{14}	5-17-09-00415 1.0626^{28}	eth 3; ctc 2 1.4922^{25}
10439	2-Propanone, 1-(2-furanyl)-	$C_7H_8O_2$ 124.14	6975-60-6 29	179.5	5-17-09-00418 1.104^{20}	1.5035^{20}
10440	2-Propanone, 1,1,1,3,3,3-hexachloro- Hexachloroacetone	C_3Cl_6O 264.75	116-16-5 -2	203	4-01-00-03223 1.7434^{12}	H_2O 2; ace 3 1.5112^{20}
10441	2-Propanone, 1,1,1,3,3,3-hexafluoro- Perfluoroacetone	C_3F_6O 166.02	684-16-2 -125	-27.4	4-01-00-03215	
10442	2-Propanone, 1-(1H-indol-3-yl)- 3-Indolylacetone	$C_{11}H_{11}NO$ 173.21	1201-26-9 116	4873	5-21-08-00327	
10443	2-Propanone, 1-hydroxy-	$C_3H_6O_2$ 74.08	116-09-6 -17	145.5	4-01-00-03977 1.0805^{20}	H_2O 4; EtOH 4; eth 4 1.4295^{20}
10444	1-Propanone, 1-(2-hydroxy-3-methylphenyl)- 2-Methyl-6-propionylphenol	$C_{10}H_{12}O_2$ 164.20	3338-15-6 22.5	124^{17}	4-08-00-00517 1.0000^{20}	ctc 3
10445	1-Propanone, 1-(2-hydroxy-4-methylphenyl)- 2'-Hydroxy-4'-methylpropiophenone	$C_{10}H_{12}O_2$ 164.20	2886-52-4 44	115^{10}	4-08-00-00519	chl 3
10446	1-Propanone, 1-(2-hydroxy-5-methylphenyl)-	$C_{10}H_{12}O_2$ 164.20	938-45-4 2	$129\text{-}30^{16.5}$	4-08-00-00517 1.0841^{14}	chl 3 1.549^{13}
10447	1-Propanone, 1-(2-hydroxy-5-methylphenyl)-2-methyl-	$C_{11}H_{14}O_2$ 178.23	64207-03-0	251; 125^{11}	4-08-00-00545 1.0460^{16}	1.538^{16}
10448	1-Propanone, 1-(4-hydroxy-3-methylphenyl)-2-methyl- *o*-Cresol, 4-isobutyryl	$C_{11}H_{14}O_2$ 178.23	73206-57-2 122	182^{12}	2-08-00-00130	eth 4; EtOH 4
10449	1-Propanone, 1-(2-hydroxyphenyl)-	$C_9H_{10}O_2$ 150.18	610-99-1	150^{80}; 115^{15}	4-08-00-00436	H_2O 2; EtOH 3; eth 3; ctc 3 1.5501^{20}
10450	1-Propanone, 1-(4-hydroxyphenyl)- Paroxypropione	$C_9H_{10}O_2$ 150.18	70-70-2 149	6992	4-08-00-00441	H_2O 2; EtOH 3; eth 3; ace 2

No.	Name Synonym	Mol. Form. Mol. Wt.	CAS RN mp/°C	Merck No. bp/°C	Beil. Ref. den/g cm^{-3}	Solubility n_D
10451	1-Propanone, 2-hydroxy-1-phenyl-	$C_9H_{10}O_2$ 150.18	5650-40-8	 251	4-08-00-00450 1.1085^{18}	 1.536^{23}
10452	1-Propanone, 3-hydroxy-1-phenyl-	$C_9H_{10}O_2$ 150.18	5650-41-9	 $98^{0.2}$	4-08-00-00453 1.1170^{20}	 1.5450^{20}
10453	1-Propanone, 3-(4-hydroxyphenyl)-1-(2,4,6-trihydroxyphenyl)- Phloretin	$C_{15}H_{14}O_5$ 274.27	60-82-2 263-4 dec	7299	4-08-00-03518	H_2O 2; EtOH 5; eth 1; ace 3
10454	2-Propanone, 1-iodo-	C_3H_5IO 183.98	3019-04-3	 62^{12}	4-01-00-03226 2.17^{15}	EtOH 3
10455	2-Propanone, 1-methoxy-	$C_4H_8O_2$ 88.11	5878-19-3	 116	4-01-00-03977 0.957^{25}	 1.3970^{20}
10456	1-Propanone, 1-(3-methoxyphenyl)- 3-Methoxypropiophenone	$C_{10}H_{12}O_2$ 164.20	37951-49-8	 259	4-08-00-00441 1.0812^{25}	eth 4; EtOH 4 1.5230^{25}
10457	1-Propanone, 2-methoxy-1-phenyl- Propiophenone, 2-methoxy-	$C_{10}H_{12}O_2$ 164.20	6493-83-0	 $79\text{-}80^{1}$	4-08-00-00451 1.0704^{20}	 1.5225^{20}
10458	1-Propanone, 3-methoxy-1-phenyl-	$C_{10}H_{12}O_2$ 164.20	55563-72-9	 $125\text{-}6^{16}$	4-08-00-00453 1.0602^{20}	 1.5250^{20}
10459	2-Propanone, 1-(methoxyphenyl)- 2-Propanone, (methoxyphenyl)-	$C_{10}H_{12}O_2$ 164.20	31116-84-4 <-15	 267-9 142^{14}	2-08-00-00104 1.0670^{18}	H_2O 2; EtOH 3; eth 3 1.5253^{20}
10460	2-Propanone, 1-(3-methoxyphenyl)-	$C_{10}H_{12}O_2$ 164.20	3027-13-2	 259	4-08-00-00459 1.0812^{6}	EtOH 3; eth 3 1.5230^{25}
10461	2-Propanone, 1-(4-methoxyphenyl)-	$C_{10}H_{12}O_2$ 164.20	122-84-9 <-15	 268	4-08-00-00460 1.0694^{17}	eth 4; EtOH 4 1.5253^{20}
10462	1-Propanone, 2-methyl-1,2-di-3-pyridinyl- Metyrapone	$C_{14}H_{14}N_2O$ 226.28	54-36-4 50.5	6083	5-24-04-00048	
10463	2-Propanone, (1-methylethylidene)hydrazone	$C_6H_{12}N_2$ 112.17	627-70-3 -12.5	 133	4-01-00-03207 0.8390^{20}	H_2O 5; EtOH 5; eth 5; ace 3 1.4535^{20}
10464	1-Propanone, 2-methyl-1-(1-naphthalenyl)- Isopropyl 1-naphthyl ketone	$C_{14}H_{14}O$ 198.26	61838-78-6	 307	3-07-00-01985 1.0761^{25}	bz 4; eth 4; EtOH 4 1.5948^{20}
10465	1-Propanone, 2-methyl-1-(2-naphthalenyl)- 2-Isobutyronaphthone	$C_{14}H_{14}O$ 198.26	107574-57-2	 313	3-07-00-01985 1.0617^{25}	bz 4; eth 4; EtOH 4
10466	1-Propanone, 1-(2-methylphenyl)-	$C_{10}H_{12}O$ 148.20	2040-14-4 -27.6	 219.5	4-07-00-00720 1.0119^{0}	
10467	1-Propanone, 1-(3-methylphenyl)-	$C_{10}H_{12}O$ 148.20	51772-30-6	 235; 132^{31}	4-07-00-00721 1.0059^{0}	H_2O 1; EtOH 3; eth 3; ace 3
10468	1-Propanone, 1-(4-methylphenyl)-	$C_{10}H_{12}O$ 148.20	5337-93-9 7.2	 236	4-07-00-00722 0.9926^{20}	H_2O 1; EtOH 3; eth 3; ace 3 1.5278^{20}
10469	1-Propanone, 2-methyl-1-phenyl-	$C_{10}H_{12}O$ 148.20	611-70-1 -1.3	 220	4-07-00-00716 0.9863^{11}	eth 4; EtOH 4 1.5172^{20}
10470	2-Propanone, 1-(1-methyl-2-piperidinyl)-, (±)-	$C_9H_{17}NO$ 155.24	18747-42-7	 $96\text{-}8^{13}$	5-21-06-00523 0.9478^{20}	H_2O 4; lig 4 1.4674^{20}
10471	2-Propanone, 1-(1-methyl-2-pyrrolidinyl)-, (*R*)- Hygrine	$C_8H_{15}NO$ 141.21	496-49-1	4788 76.5^{11}	5-21-06-00511	EtOH 4; chl 4 1.4555^{20}
10472	1-Propanone, 2-methyl-1-(2-thienyl)- Isopropyl 2-thienyl ketone	$C_8H_{10}OS$ 154.23	36448-60-9	 229	5-17-09-00439 1.0894^{20}	 1.5405^{20}
10473	1-Propanone, 1-(1-naphthalenyl)-	$C_{13}H_{12}O$ 184.24	2876-63-3	 306	4-07-00-01298 1.0971^{20}	 1.6108^{20}
10474	1-Propanone, 1-(2-naphthalenyl)-	$C_{13}H_{12}O$ 184.24	6315-96-4 60	 313	4-07-00-01299	EtOH 4; chl 4
10475	2-Propanone, 1-nitro-	$C_3H_5NO_3$ 103.08	10230-68-9 50.3	 103^{24}	4-01-00-03226	bz 4; eth 4; EtOH 4
10476	2-Propanone, oxime Acetoxime	C_3H_7NO 73.09	127-06-0 61	68 136; 61^{20}	4-01-00-03202 0.9113^{62}	H_2O 3; EtOH 3; eth 3; chl 3 1.4156^{20}
10477	2-Propanone, 1,1,1,3,3-pentabromo- Pentabromoacetone	C_3HBr_5O 452.56	79-49-2 79.5	7056 sub	4-01-00-03226	H_2O 1; EtOH 4; eth 4; ace 4
10478	2-Propanone, 1,1,1,3,3-pentachloro-	C_3HCl_5O 230.30	1768-31-6 2.1	 192^{253}; 98^{40}	4-01-00-03222 1.69^{15}	
10479	2-Propanone, 1-phenoxy-	$C_9H_{10}O_2$ 150.18	621-87-4	 229.5	4-06-00-00604 1.0903^{20}	eth 3; ace 3 1.5228^{20}
10480	1-Propanone, 1-phenyl- Propiophenone	$C_9H_{10}O$ 134.18	93-55-0 18.6	7842 217.5	4-07-00-00680 1.0096^{20}	H_2O 1; EtOH 3; eth 3; chl 3 1.5269^{20}
10481	2-Propanone, 1-phenyl- Phenylacetone	$C_9H_{10}O$ 134.18	103-79-7 -15	7240 216.5	4-07-00-00687 1.0157^{20}	H_2O 1; EtOH 4; eth 4; bz 5 1.5168^{20}
10482	2-Propanone, phenylhydrazone Acetone, phenylhydrazone	$C_9H_{12}N_2$ 148.21	103-02-6 42	 163^{50}	4-15-00-00060	EtOH 3; eth 3; dil acid 3
10483	1-Propanone, 1-phenyl-, oxime	$C_9H_{11}NO$ 149.19	2157-50-8 54.8	 245; 165^{38}	4-07-00-00682 1.0639^{18}	H_2O 1; EtOH 3; eth 3; chl 2
10484	1-Propanone, 1-(2-piperidinyl)- Piperidine, 2(1-propionyl)	$C_8H_{15}NO$ 141.21	97073-23-9	 94^{18}	5-21-06-00522 0.9380^{20}	 1.4621^{20}
10485	2-Propanone, 1-(2-piperidinyl)-, (*R*)- (-)-Pelletierine	$C_8H_{15}NO$ 141.21	2858-66-4	7015 195; 125^{100}	5-21-06-00522 0.988^{20}	eth 4; EtOH 4; chl 4 1.4683^{20}
10486	2-Propanone, 1,1,1,3-tetrachloro-	$C_3H_2Cl_4O$ 195.86	16995-35-0 65	 183	4-01-00-03222 1.624^{25}	H_2O 2; eth 4; ace 4 1.497^{18}
10487	2-Propanone, 1,1,3,3-tetrachloro-	$C_3H_2Cl_4O$ 195.86	632-21-3	 183	4-01-00-03222	H_2O 4; EtOH 4; eth 4; ace 4 1.4944^{20}
10488	2-Propanone, 1,1,1-trichloro-	$C_3H_3Cl_3O$ 161.41	918-00-3	 149; 28^{10}	4-01-00-03221 1.435^{20}	H_2O 1; EtOH 4; eth 4 1.4635^{17}
10489	1-Propanone, 1-(2,4,6-trihydroxyphenyl)- Flopropione	$C_9H_{10}O_4$ 182.18	2295-58-1 175.5	4039	3-08-00-03413	eth 4; EtOH 4

No.	Name	Synonym	Mol. Form.	Mol. Wt.	CAS RN	mp/°C	Merck No.	bp/°C	Beil. Ref.	den/g cm^{-3}	Solubility	n_D
10490	Propanoyl bromide	Propionic acid bromide	C_3H_5BrO	136.98	598-22-1			102	4-02-00-00724	1.5210^{16}	eth 4	1.4578^{16}
10491	Propanoyl bromide, 2-bromo-2-methyl-		$C_4H_6Br_2O$	229.90	20769-85-1			163	3-02-00-00661	1.4067^{14}	ace 4; CS_2 4	
10492	Propanoyl bromide, 2-methyl-		C_4H_7BrO	151.00	2736-37-0			117	4-02-00-00852	1.4067^{15}		1.4552^{25}
10493	Propanoyl chloride	Propionyl chloride	C_3H_5ClO	92.52	79-03-8	-94	7840	80	4-02-00-00724	1.0646^{20}	eth 3	1.4032^{20}
10494	Propanoyl chloride, 2-(acetyloxy)-, (±)-	Lactyl chloride, acetate (*DL*)	$C_5H_7ClO_3$	150.56	55057-45-9			150 dec; 56^{11}	4-03-00-00674	1.1920^{17}		1.4241^{17}
10495	Propanoyl chloride, 2-bromo-		C_3H_4BrClO	171.42	7148-74-5			132	4-02-00-00764	1.697^{11}	eth 3; ctc 2; chl 3	1.4780^{20}
10496	Propanoyl chloride, 2-chloro-, (*S*)-	Propanoic acid, chloride, 2-chloro(*d*)	$C_3H_4Cl_2O$	126.97	70110-24-6			110	4-02-00-00747	1.2394^{7}		1.4400^{20}
10497	Propanoyl chloride, 3-chloro-		$C_3H_4Cl_2O$	126.97	625-36-5			144	4-02-00-00750	1.3307^{13}	H_2O 2; EtOH 4; eth 4; chl 4	1.4549^{20}
10498	Propanoyl chloride, 2-chloro-2-methyl-		$C_4H_6Cl_2O$	141.00	13222-26-9			126.5	4-02-00-00859		eth 4	1.4369^{20}
10499	Propanoyl chloride, 3-chloro-2-methyl-	Isobutyryl chloride, 3-chloro	$C_4H_6Cl_2O$	141.00	7623-10-1	-90		92	4-02-00-00860	1.0174^{20}	eth 4	1.4079^{20}
10500	Propanoyl chloride, 2,2-dichloro-		$C_3H_3Cl_3O$	161.41	26073-26-7			118; 70^{89}	4-02-00-00755	1.4062^{20}		1.4524^{20}
10501	Propanoyl chloride, 2,3-dichloro-		$C_3H_3Cl_3O$	161.41	7623-13-4			$52\text{-}4^{16}$	4-02-00-00757	1.4757^{20}		1.4764^{20}
10502	Propanoyl chloride, 3,3-dichloro-	3,3-Dichloropropionyl chloride	$C_3H_3Cl_3O$	161.41	17880-86-3			$43\text{-}4^{10}$	3-02-00-00562	1.4557^{20}	eth 4; Diox 4	1.4738^{20}
10503	Propanoyl chloride, 2,2-dimethyl-	Pivalic acid chloride	C_5H_9ClO	120.58	3282-30-2			107	4-02-00-00912	1.003^{20}	eth 4	1.4139^{20}
10504	Propanoyl chloride, 2-methyl-	Isobutyric acid chloride	C_4H_7ClO	106.55	79-30-1	-90		92	4-02-00-00852		eth 3	1.4079^{20}
10505	Propanoyl chloride, 2-phenoxy-		$C_9H_9ClO_2$	184.62	122-35-0			147; 116^{10}	4-06-00-00643	1.1865^{20}	eth 3	1.5178^{20}
10506	Propanoyl fluoride	Propionyl fluoride	C_3H_5FO	76.07	430-71-7			44	4-02-00-00724	0.972^{15}		1.329^{13}
10507	Propargite	Sulfurous acid, 2-[4-(1,1-dimethylethyl)phenoxy]cyclohexyl 2-propynyl ester	$C_{19}H_{26}O_4S$	350.48	2312-35-8		7818			1.10^{25}		
10508	Propazine	1,3,5-Triazine-2,4-diamine, 6-chloro-*N*,*N'*-bis(1-methylethyl)-	$C_9H_{16}ClN_5$	229.71	139-40-2	213	7822			1.162^{20}		
10509	2-Propenal	Acrolein	C_3H_4O	56.06	107-02-8	-87.7	122	52.6	4-01-00-03435	0.840^{20}	H_2O 4; EtOH 3; eth 3; ace 3	1.4017^{20}
10510	2-Propenal, 3-bromo-3-phenyl-		C_9H_7BrO	211.06	14804-59-2			$144\text{-}6^{12}$	3-07-00-01383	1.492^{20}	eth 4	
10511	2-Propenal, 2-chloro-		C_3H_3ClO	90.51	683-51-2			40^{30}	4-01-00-03440	1.199^{20}	eth 4; EtOH 4	1.463^{20}
10512	2-Propenal, 3-(2-furanyl)-		$C_7H_6O_2$	122.12	623-30-3	54		135^{14}	5-17-09-00570		H_2O 1; EtOH 5; eth 3; chl 2	
10513	2-Propenal, 3-(2-furanyl)-, (*E*)-		$C_7H_6O_2$	122.12	39511-08-5	50		143^{37}; 100^{12}	5-17-09-00570		bz 4; eth 4; EtOH 4	1.5286^{20}
10514	2-Propenal, 3-(4-hydroxy-3-methoxyphenyl)-		$C_{10}H_{10}O_3$	178.19	458-36-6	84			4-08-00-01979	1.1562^{102}	bz 4; eth 4; EtOH 4	
10515	2-Propenal, 2-methyl-	Methacrolein	C_4H_6O	70.09	78-85-3			68.4	4-01-00-03455	0.840^{25}	H_2O 5; EtOH 5; eth 5	1.4144^{20}
10516	2-Propenal, 2-methyl-3-phenyl-		$C_{10}H_{10}O$	146.19	101-39-3			150^{100}	2-07-00-00291	1.0407^{17}		1.6057^{17}
10517	2-Propenal, 3-(4-methylphenyl)-		$C_{10}H_{10}O$	146.19	1504-75-2	41.5		154^{25}	4-07-00-01013	0.9670^{23}	EtOH 4	
10518	2-Propenal, 3-phenyl-, (*E*)-	*trans*-Cinnamaldehyde	C_9H_8O	132.16	14371-10-9	-7.5		253 dec; 127^{16}	4-07-00-00984	1.0497^{20}	H_2O 2; EtOH 3; eth 3; chl 3	1.6195^{20}
10519	2-Propenamide	Acrylamide	C_3H_5NO	71.08	79-06-1	84.5	123	192.6	4-02-00-01471		H_2O 4; EtOH 3; eth 3; ace 3	
10520	2-Propenamide, *N*-(3,4-dichlorophenyl)-2-methyl-	Dicryl	$C_{10}H_9Cl_2NO$	230.09	2164-09-2	128	3078				ace 4; EtOH 4	
10521	2-Propenamide, *N*-(1,1-dimethyl-3-oxobutyl)-	Diacetone acrylamide	$C_9H_{15}NO_2$	169.22	2873-97-4		2943				chl 3	
10522	2-Propen-1-amine	Allylamine	C_3H_7N	57.10	107-11-9	-88.2	285	53.3	4-04-00-01057	0.758^{20}	H_2O 5; EtOH 5; eth 5; chl 3	1.4205^{20}
10523	2-Propen-1-amine, *N*,*N*-di-2-propenyl-	Triallylamine	$C_9H_{15}N$	137.22	102-70-5	94		155.5	4-04-00-01061	0.809^{20}	EtOH 3; eth 3; ace 3; bz 3	1.4502^{20}
10524	2-Propen-1-amine, *N*-methyl-		C_4H_9N	71.12	627-37-2			65	4-04-00-01058		H_2O 4; ace 4; eth 4; EtOH 4	1.4065^{20}
10525	2-Propen-1-amine, *N*-(1-methylethyl)-		$C_6H_{13}N$	99.18	35000-22-7			96.5	4-04-00-01059	0.7400^{30}		1.4140^{25}
10526	2-Propen-1-amine, *N*-2-propenyl-	Diallylamine	$C_6H_{11}N$	97.16	124-02-7		2951	111	4-04-00-01060		EtOH 3; eth 3	1.4387^{20}

No.	Name Synonym	Mol. Form. Mol. Wt.	CAS RN mp/°C	Merck No. bp/°C	Beil. Ref. den/g cm^{-3}	Solubility n_D
10527	2-Propen-1-amine, *N*-propyl-	$C_6H_{13}N$ 99.18	5666-21-7	 112	0-04-00-00207 0.7708^{18}	H_2O 4 1.4140^{25}
10528	1-Propene Propylene	C_3H_6 42.08	115-07-1 -185.2	7862 -47.6	4-01-00-00725 $*0.505^{25}$	H_2O 4; EtOH 4; HOAc 4 1.3567^{-70}
10529	1-Propene, 1-bromo-, (*Z*)-	C_3H_5Br 120.98	590-13-6 -113	 57.8	4-01-00-00754 1.4291^{20}	H_2O 1; eth 3; ace 3; chl 3 1.4560^{20}
10530	1-Propene, 2-bromo-	C_3H_5Br 120.98	557-93-7 -126	 48.4	4-01-00-00754 1.3965^{16}	H_2O 1; eth 3; ace 3; chl 3 1.4467^{16}
10531	1-Propene, 3-bromo- Allyl bromide	C_3H_5Br 120.98	106-95-6 -119	286 70.1	4-01-00-00754 1.398^{20}	H_2O 1; EtOH 5; eth 5; ctc 3 1.4697^{20}
10532	1-Propene, 3-bromo-3,3-difluoro- Propylene, 3-bromo-3,3-difluoro	$C_3H_3BrF_2$ 156.96	420-90-6	 42	4-01-00-00756 1.543^{25}	 1.3773^{25}
10533	1-Propene, 1-chloro-, (*E*)- *trans*-1-Chloropropene	C_3H_5Cl 76.53	16136-85-9 -99	 37.4	4-01-00-00737 0.9349^{20}	H_2O 1; eth 3; ace 3; bz 3 1.4054^{20}
10534	1-Propene, 1-chloro-, (*Z*)- *cis*-1-Chloropropene	C_3H_5Cl 76.53	16136-84-8 -134.8	 32.8	4-01-00-00737 0.9347^{20}	H_2O 1; eth 3; ace 3; bz 3 1.4055^{20}
10535	1-Propene, 2-chloro- Isopropenyl chloride	C_3H_5Cl 76.53	557-98-2 -137.4	 22.6	4-01-00-00737 0.9017^{20}	H_2O 1; eth 3; ace 3; bz 3 1.3973^{20}
10536	1-Propene, 3-chloro- Allyl chloride	C_3H_5Cl 76.53	107-05-1 -134.5	287 45.1	4-01-00-00738 0.9376^{20}	H_2O 1; EtOH 5; eth 5; ace 5 1.4157^{20}
10537	1-Propene, 3-chloro-2-(chloromethyl)-	$C_4H_6Cl_2$ 125.00	1871-57-4 -14	 138	4-01-00-00805 1.1782^{20}	EtOH 4; chl 4 1.4753
10538	1-Propene, 2-chloro-3-isothiocyanato-	C_4H_4ClNS 133.60	14214-31-4	 182	0-04-00-00219 1.27^{12}	EtOH 2
10539	1-Propene, 1-chloro-2-methyl- 1-Chloro-2-methylpropene	C_4H_7Cl 90.55	513-37-1	2147 68	4-01-00-00803 0.9186^{20}	chl 3 1.4221^{20}
10540	1-Propene, 3-chloro-2-methyl- 3-Chloro-2-methylpropene	C_4H_7Cl 90.55	563-47-3	2148 71.5	4-01-00-00803 0.9165^{20}	EtOH 5; eth 5; ace 3; chl 4 1.4291^{20}
10541	1-Propene, 2-chloro-1,1,3,3,3-pentafluoro-	C_3ClF_5 166.48	2804-50-4 -130	 6.8	4-01-00-00741 1.5920^{20}	
10542	1-Propene, 1,1-dibromo-	$C_3H_4Br_2$ 199.87	13195-80-7	 125	4-01-00-00759 1.9767^{20}	H_2O 2; bz 3; ctc 3; chl 3 1.5260^{20}
10543	1-Propene, 1,3-dibromo-, (*E*)- Propylene, 1,3-dibromo (*trans*)	$C_3H_4Br_2$ 199.87	32121-07-6	 66^{25}	4-01-00-00760 1.9791^{20}	 1.5495^{20}
10544	1-Propene, 1,3-dibromo-, (*Z*)- Propylene, 1,3-dibromo (*cis*)	$C_3H_4Br_2$ 199.87	32121-06-5	 60^{25}	4-01-00-00760 2.0599^{20}	 1.5550^{20}
10545	1-Propene, 2,3-dibromo- 2,3-Dibromopropene	$C_3H_4Br_2$ 199.87	513-31-5	3009 141; 37.7^{11}	4-01-00-00760 2.0345^{25}	H_2O 1; eth 3; ace 3; chl 3 1.5416^{25}
10546	1-Propene, 1,3-dibromo-2-methyl- Isobutylene, 1,3-dibromo	$C_4H_6Br_2$ 213.90	35911-17-2	 102^{100}	3-01-00-00769 1.8939^{4}	 1.5396^{15}
10547	1-Propene, 1,1-dichloro-	$C_3H_4Cl_2$ 110.97	563-58-6	 76.5	4-01-00-00742 1.1864^{25}	H_2O 1; eth 3; ace 3; chl 3 1.4430^{25}
10548	1-Propene, 1,2-dichloro-, (*E*)-	$C_3H_4Cl_2$ 110.97	7069-38-7	 77	4-01-00-00742 1.1818^{20}	H_2O 1; EtOH 4; ctc 4; MeOH 4 1.4471^{20}
10549	1-Propene, 1,2-dichloro-, (*Z*)-	$C_3H_4Cl_2$ 110.97	6923-20-2	 93	4-01-00-00742	H_2O 1; ace 3; bz 3; chl 3 1.4549^{20}
10550	1-Propene, 1,3-dichloro-, (*E*)- *trans*-1,3-Dichloropropylene	$C_3H_4Cl_2$ 110.97	10061-02-6	3064 112	4-01-00-00744 1.217^{20}	H_2O 1; eth 3; bz 3; chl 3 1.4730^{20}
10551	1-Propene, 1,3-dichloro-, (*Z*)- *cis*-1,3-Dichloropropylene	$C_3H_4Cl_2$ 110.97	10061-01-5	3064 104.3	4-01-00-00743 1.224^{20}	H_2O 1; eth 3; bz 3; chl 3 1.4682^{20}
10552	1-Propene, 2,3-dichloro- 2,3-Dichloropropene	$C_3H_4Cl_2$ 110.97	78-88-6 10	 94	4-01-00-00744 1.211^{20}	H_2O 1; EtOH 5; eth 3; bz 3 1.4603^{20}
10553	1-Propene, 3,3-dichloro-	$C_3H_4Cl_2$ 110.97	563-57-5	 84.4	4-01-00-00745 1.175^{20}	bz 4; eth 4; EtOH 4 1.4510^{20}
10554	1-Propene, 1,1-dichloro-2-fluoro- Propylene, 1,1-dichloro-2-fluoro	$C_3H_3Cl_2F$ 128.96	430-95-5	 78	4-01-00-00745 1.3026^{25}	 1.4196^{25}
10555	1-Propene, 1,1-dichloro-2-methyl-	$C_4H_6Cl_2$ 125.00	6065-93-6 -43	 108.5	4-01-00-00804 1.1447^{20}	bz 4; eth 4; chl 4 1.4580^{20}
10556	1-Propene, 3,3-dichloro-2-methyl-	$C_4H_6Cl_2$ 125.00	22227-75-4	 108	4-01-00-00805 1.136^{24}	bz 4; eth 4; chl 4 1.4523^{24}
10557	1-Propene, 1,2-dichloro-1,3,3,3-tetrafluoro-	$C_3Cl_2F_4$ 182.93	431-53-8 -137	 47.3	4-01-00-00747 1.5468^{25}	 1.3511^{20}
10558	Propene, 1,2-dichloro-3,3,3-trifluoro- Propylene, 1,2-dichloro-3,3,3-trifluoro	$C_3HCl_2F_3$ 164.94	431-27-6 -109.2	 53.7	4-01-00-00746 1.4653^{20}	 1.3670^{20}
10559	1-Propene, 1,1-diethoxy-	$C_7H_{14}O_2$ 130.19	21504-43-8	 133.5	4-01-00-03434 0.8628^{25}	 1.4083^{25}
10560	1-Propene, 3,3-diethoxy- Acrolein, diethyl acetal	$C_7H_{14}O_2$ 130.19	3054-95-3	 123.5	4-01-00-03437 0.8543^{15}	H_2O 2; EtOH 5; eth 5 1.4000^{20}
10561	1-Propene, 1,1-difluoro 1,1-Difluoro-1-propene	$C_3H_4F_2$ 78.06	430-63-7			
10562	1-Propene, 3,3-dimethoxy-	$C_5H_{10}O_2$ 102.13	6044-68-4	 88	4-01-00-03437 0.862^{25}	 1.3954^{20}
10563	2-Propene-1,1-diol, diacetate	$C_7H_{10}O_4$ 158.15	869-29-4 -37.6	 180	4-02-00-00291 1.0760^{20}	ace 4; bz 4; eth 4; EtOH 4 1.4193^{20}
10564	2-Propene-1,1-diol, 2-methyl-, diacetate Methacrolein diacetate	$C_8H_{12}O_4$ 172.18	10476-95-6	 191	4-02-00-00292	 1.4241^{20}
10565	1-Propene, 3-(ethenyloxy)-	C_5H_8O 84.12	3917-15-5	 66	4-01-00-02085 0.8050^{20}	H_2O 1; eth 3; ace 3; chl 3 1.4062^{20}

No.	Name Synonym	Mol. Form. Mol. Wt.	CAS RN mp/°C	Merck No. bp/°C	Beil. Ref. den/g cm^{-3}	Solubility n_D
10566	1-Propene, 3-ethoxy- Allyl ethyl ether	$C_5H_{10}O$ 86.13	557-31-3	291 67.6	4-01-00-02083 0.7651^{20}	H_2O 1; EtOH 5; eth 5; ace 3 1.3881^{20}
10567	1-Propene, 1-fluoro- (*Z*) *cis*-1-Fluoropropene	C_3H_5F 60.07	19184-10-2			
10568	1-Propene, 1-fluoro- (*E*) *trans*-1-Fluoropropene	C_3H_5F 60.07	20327-65-5			
10569	1-Propene, 2-fluoro- 2-Fluoropropene	C_3H_5F 60.07	1184-60-7	 -24	4-01-00-00733	
10570	1-Propene, 3-fluoro- 3-Fluoropropene	C_3H_5F 60.07	818-92-8	 -3	4-01-00-00733	H_2O 2; EtOH 4; eth 4; chl 3
10571	1-Propene, 1,1,2,3,3,3-hexachloro- Hexachloropropylene	C_3Cl_6 248.75	1888-71-7 -72.9	 209.5	4-01-00-00753 1.7632^{20}	H_2O 1; ctc 3; chl 3 1.5091^{20}
10572	1-Propene, 1,1,2,3,3,3-hexafluoro- Perfluoropropene	C_3F_6 150.02	116-15-4 -156.5	 -29.6	4-01-00-00735	 1.583^{-40}
10573	1-Propene, 3-iodo- Allyl iodide	C_3H_5I 167.98	556-56-9 -99.3	292 103	4-01-00-00761 1.848^{12}	H_2O 1; EtOH 3; eth 3; chl 3 1.5540^{21}
10574	1-Propene, 3-isocyano- Allyl isocyanide	C_4H_5N 67.09	2835-21-4	 98	0-04-00-00208 0.794^{17}	eth 4; EtOH 4
10575	1-Propene, 3-isothiocyanato- Allyl isothiocyanate	C_4H_5NS 99.16	57-06-7 -80	293 152	4-04-00-01081 1.0126^{20}	bz 4; eth 4; EtOH 4 1.5306^{20}
10576	1-Propene, 3-methoxy-	C_4H_8O 72.11	627-40-7	 44	4-01-00-02083 0.77^{11}	H_2O 1; EtOH 5; eth 5; ace 3 1.3778^{20}
10577	1-Propene, 3-methoxy-2-methyl-	$C_5H_{10}O$ 86.13	22418-49-1	 68	3-01-00-01904 0.7698^{20}	 1.3964^{20}
10578	1-Propene, 2-methyl- Isobutene	C_4H_8 56.11	115-11-7 -140.4	5024 -6.9	4-01-00-00796 $*0.589^{25}$	H_2O 1; EtOH 4; eth 4; bz 3 1.3926^{-25}
10579	1-Propene, 3-(1-methylethoxy)-	$C_6H_{12}O$ 100.16	6140-80-3	 83.5	4-01-00-02083 0.7764^{20}	ace 4; eth 4; EtOH 4 1.3946^{20}
10580	1-Propene, 2-methyl-, tetramer Propene, 2-methyl(tetramer)	$C_{16}H_{32}$ 224.43	15220-85-6 -98	 244; 109^{15}	3-01-00-00763 0.7944^{20}	 1.4482^{20}
10581	1-Propene, 3-(methylthio)-	C_4H_8S 88.17	10152-76-8	 92	4-01-00-02095 0.8767^{20}	 1.4714^{20}
10582	1-Propene, 2-methyl-, trimer Isobutylene (trimer)	$C_{12}H_{24}$ 168.32	7756-94-7 -76	 180	3-01-00-00763 0.7590^{20}	 1.4314^{20}
10583	2-Propenenitrile Acrylonitrile	C_3H_3N 53.06	107-13-1 -83.5	125 77.3	4-02-00-01473 0.8060^{20}	ace 4; bz 4; eth 4; EtOH 4 1.3911^{20}
10584	2-Propenenitrile, 2-chloro-	C_3H_2ClN 87.51	920-37-6 -65	 88.5	4-02-00-01484 1.096^{25}	 1.4290^{20}
10585	2-Propenenitrile, 3-(2-furanyl)- 2-Furanacrylonitrile	C_7H_5NO 119.12	7187-01-1 38	4208 96	4-18-00-04152	tol 4 1.5824^{25}
10586	2-Propenenitrile, 2-methyl- Methylacrylonitrile	C_4H_5N 67.09	126-98-7 -35.8	5850 90.3	4-02-00-01539 0.8001^{20}	H_2O 1; EtOH 5; eth 5; ace 5 1.4003^{20}
10587	2-Propenenitrile, 3-phenyl-, (*E*)-	C_9H_7N 129.16	1885-38-7 22	 263.8	4-09-00-02027 1.0304^{20}	H_2O 1; EtOH 3; ace 3; ctc 3 1.6013^{20}
10588	2-Propenenitrile, 3-phenyl-, (*Z*)-	C_9H_7N 129.16	24840-05-9 -4.4	 249; 139^{30}	4-09-00-02027 1.0289^{20}	H_2O 1; EtOH 3; bz 4 1.5843^{20}
10589	1-Propene, 1-nitro-	$C_3H_5NO_2$ 87.08	3156-70-5	 60^{34}; 37^{10}	3-01-00-00715 1.0661^{20}	eth 3; ace 3; chl 3 1.4527^{20}
10590	1-Propene, 2-nitro-	$C_3H_5NO_2$ 87.08	4749-28-4	 57^{100}; 32^{30}	4-01-00-00764 1.0643^{20}	eth 3; ace 3; chl 3 1.4358^{20}
10591	1-Propene, 3-nitro-	$C_3H_5NO_2$ 87.08	625-46-7	 127.5	4-01-00-00764 1.051^{22}	eth 4; EtOH 4
10592	1-Propene, 3,3'-oxybis- Allyl ether	$C_6H_{10}O$ 98.14	557-40-4 -6	290 94	4-01-00-02086 0.8260^{20}	H_2O 1; EtOH 5; eth 5; ace 4 1.4163^{20}
10593	1-Propene, 1,1,2,3,3-pentachloro-	C_3HCl_5 214.30	1600-37-9	 185	4-01-00-00753 1.6317^{34}	eth 4 1.5313^{20}
10594	1-Propene, 3-propoxy-	$C_6H_{12}O$ 100.16	1471-03-0	 91	4-01-00-02083 0.7764^{20}	ace 4; eth 4; EtOH 4 1.3919^{20}
10595	2-Propene-1-sulfinothioic acid, *S*-2-propenyl ester Allicin	$C_6H_{10}OS_2$ 162.28	539-86-6	249 dec	4-04-00-00007 1.112^{20}	H_2O 4 1.561^{20}
10596	1-Propene, 3,3'-sulfinylbis-	$C_6H_{10}OS$ 130.21	14180-63-3 23.5	 114^{12}; $89^{2.6}$	4-01-00-02097 1.0261^{20}	EtOH 3; eth 3; ace 3 1.5115^{20}
10597	1-Propene, 3,3'-sulfonylbis-	$C_6H_{10}O_2S$ 146.21	16841-48-8	 109^{3}	4-01-00-02098 1.1213^{20}	 1.4893^{20}
10598	1-Propene, 1,2,3,3-tetrachloro-	$C_3H_2Cl_4$ 179.86	20589-85-9	 162	3-01-00-00708 1.537^{19}	bz 4; chl 4 1.5121^{19}
10599	1-Propene, 3,3'-thiobis- Diallyl sulfide	$C_6H_{10}S$ 114.21	592-88-1 -85	295 138.6	4-01-00-02097 0.8877^{27}	eth 4; EtOH 4 1.4870^{25}
10600	2-Propene-1-thiol	C_3H_6S 74.15	870-23-5	 65	4-01-00-02095 0.925^{23}	H_2O 1; EtOH 5; eth 5; chl 3 1.4832^{20}
10601	1-Propene-1,2,3-tricarboxylic acid, triethyl ester, (*E*)- Aconitic acid, triethyl ester (*trans*)	$C_{12}H_{18}O_6$ 258.27	68077-28-1	 275 dec; 159^{9}	4-02-00-02405 1.1064^{20}	eth 4; EtOH 4
10602	1-Propene, 1,1,2-trichloro-	$C_3H_3Cl_3$ 145.41	21400-25-9	 118	4-01-00-00747 1.382^{20}	H_2O 1; EtOH 3; eth 3; bz 3 1.4827^{20}
10603	1-Propene, 1,2,3-trichloro-	$C_3H_3Cl_3$ 145.41	96-19-5	 142	4-01-00-00748 1.412^{20}	H_2O 1; EtOH 4; eth 4; bz 3 1.5030^{20}
10604	1-Propene, 3,3,3-trichloro-	$C_3H_3Cl_3$ 145.41	2233-00-3 -30	 114.5	4-01-00-00749 1.367^{20}	H_2O 1; EtOH 3; eth 3; bz 3 1.4827^{20}

No.	Name	Synonym	Mol. Form.	Mol. Wt.	CAS RN	mp/°C	Merck No.	bp/°C	Beil. Ref.	den/g cm^{-3}	Solubility	n_D
10605	1-Propene, 1,1,3-trichloro-2-methyl-		$C_4H_5Cl_3$	159.44	31702-33-7			156	4-01-00-00805	1.346^{20}	ace 4; bz 4; chl 4	1.4990^{20}
10606	1-Propene, 3,3,3-trichloro-2-methyl-		$C_4H_5Cl_3$	159.44	4749-27-3			133	4-01-00-00805	1.293^{20}	H_2O 1; ace 3; bz 3; chl 4	1.4770^{20}
10607	1-Propene, 1,1,2-trichloro-3,3,3-trifluoro-		$C_3Cl_3F_3$	199.39	431-52-7	-114.6		88.1	4-01-00-00751	1.617^{20}		
10608	1-Propene, 3,3,3-trifluoro-	3,3,3-Trifluoropropene	$C_3H_3F_3$	96.05	677-21-4			-17	4-01-00-00734	0		
10609	2-Propenoic acid	Acrylic acid	$C_3H_4O_2$	72.06	79-10-7	12.3	124	141	4-02-00-01455	1.0511^{20}	H_2O 5; EtOH 5; eth 5; ace 3	1.4224^{20}
10610	2-Propenoic acid, 2-bromo-		$C_3H_3BrO_2$	150.96	10443-65-9	71.5		dec	4-02-00-01486		H_2O 4; EtOH 4	
10611	2-Propenoic acid, 2-bromoethyl ester		$C_5H_7BrO_2$	179.01	4823-47-6			156.5; 53^{5}	4-02-00-01462	1.4581^{25}		1.4660^{25}
10612	2-Propenoic acid, 3-bromo-2-methyl-	Methacrylic acid, 3-bromo	$C_4H_5BrO_2$	164.99	89123-63-7	64		229	0-02-00-00424		H_2O 4	
10613	2-Propenoic acid, 2-bromo-3-phenyl-, ethyl ester		$C_{11}H_{11}BrO_2$	255.11	57296-00-1			295	4-09-00-02042	1.3885^{25}		1.5845^{25}
10614	2-Propenoic acid, 3-bromo-3-phenyl-, ethyl ester, (*Z*)-	Cinnamic acid, β-bromo-, ethyl ester	$C_{11}H_{11}BrO_2$	255.11	1504-70-7			167^{12}	3-09-00-02733	1.4044^{10}		1.5841^{10}
10615	2-Propenoic acid, 2-bromo-3-phenyl-, methyl ester	Methyl α-bromocinnamate	$C_{10}H_9BrO_2$	241.08	24127-62-6	23		$120\text{-}1^{0.6}$	4-09-00-02042	1.4975^{20}		
10616	2-Propenoic acid, butyl ester	Butyl acrylate	$C_7H_{12}O_2$	128.17	141-32-2	-64.6	1539	145	4-02-00-01463	0.8898^{20}	H_2O 1; EtOH 3; eth 3; ace 3	1.4185^{20}
10617	2-Propenoic acid, 3-chloro-, (*E*)-		$C_3H_3ClO_2$	106.51	2345-61-1	86		94^{18}	4-02-00-01481		EtOH 3; eth 3; chl 2	
10618	2-Propenoic acid, 3-chloro-, (*Z*)-		$C_3H_3ClO_2$	106.51	1609-93-4	63.5		106^{17}	4-02-00-01481		EtOH 3; eth 3	
10619	2-Propenoic acid, chloro-, ethyl ester	Acrylic acid, chloro-, ethyl ester	$C_5H_7ClO_2$	134.56	30600-19-2			$51\text{-}3^{18}$	3-02-00-01245	1.1404^{20}	EtOH 4; eth 4	1.4384^{20}
10620	2-Propenoic acid, 2-chloroethyl ester		$C_5H_7ClO_2$	134.56	2206-89-5			$51\text{-}3^{18}$	3-02-00-01225	1.1604^{20}	eth 4; EtOH 4	1.4384^{20}
10621	2-Propenoic acid, 2-chloro-, methyl ester	Methyl 2-chloroacrylate	$C_4H_5ClO_2$	120.54	80-63-7			52^{51}	4-02-00-01482	1.189^{20}	eth 4	1.4420^{20}
10622	2-Propenoic acid, 2-chloro-3-phenyl-, methyl ester, (*E*)-	Cinnamic acid, α-chloro-, methyl ester, (*E*)-	$C_{10}H_9ClO_2$	196.63	14737-94-1	33		$108\text{-}9^{0.5}$	4-09-00-02035		ace 4	
10623	2-Propenoic acid, 3-chloro-3-phenyl-, methyl ester	Cinnamic acid, β-chloro-, methyl ester	$C_{10}H_9ClO_2$	196.63	87541-87-5	29		$113\text{-}4^{0.5}$		1.2248^{21}	ace 4	1.5781^{21}
10624	2-Propenoic acid, 2-cyano-, methyl ester	Mecrylate	$C_5H_5NO_2$	111.10	137-05-3		5667	$47\text{-}8^{2}$	4-02-00-02226	1.1012^{20}		1.4430
10625	2-Propenoic acid, 2-cyano-3-phenyl-, ethyl ester		$C_{12}H_{11}NO_2$	201.22	2025-40-3	51		188^{15}	2-09-00-00640	1.1076^{25}	ace 4; chl 4	1.5033
10626	2-Propenoic acid, cyclohexyl ester		$C_9H_{14}O_2$	154.21	3066-71-5			183; 88^{20}	4-06-00-00038	1.0275^{20}	H_2O 1; EtOH 5; eth 5; chl 3	1.4673^{20}
10627	2-Propenoic acid, 2,3-dibromo-, (*Z*)-		$C_3H_2Br_2O_2$	229.86	24557-10-6	85.5		148^{6}	3-02-00-01249		H_2O 2; EtOH 3; eth 3; bz 2	
10628	2-Propenoic acid, 3,3-dibromo-	Acrylic acid, 3,3-dibromo-	$C_3H_2Br_2O_2$	229.86	1578-21-8	86.3		245	4-02-00-01487		eth 4; EtOH 4	
10629	2-Propenoic acid, 2,3-dibromo-3-phenyl-, (*Z*)-		$C_9H_6Br_2O_2$	305.95	708-82-7	100		$124^{0.5}$	4-09-00-02043		H_2O 1; EtOH 3; eth 3; chl 3	
10630	2-Propenoic acid, 3,3-dichloro-	Acrylic acid, 3,3-dichloro-	$C_3H_2Cl_2O_2$	140.95	1561-20-2	77.5		sub	4-02-00-01484		eth 4; chl 4	
10631	2-Propenoic acid, 2-(diethylamino)ethyl ester		$C_9H_{17}NO_2$	171.24	2426-54-2	<-60		81^{10}	4-04-00-01477	0.937^{20}		1.4376^{25}
10632	2-Propenoic acid, 3-(3,4-dihydroxyphenyl)-	Caffeic acid	$C_9H_8O_4$	180.16	331-39-5	225 dec	1634		2-10-00-00294		EtOH 4	
10633	2-Propenoic acid, 2-(dimethylamino)ethyl ester		$C_7H_{13}NO_2$	143.19	2439-35-2	<-60		95^{50}	4-04-00-01431	0.938^{20}		
10634	2-Propenoic acid, 2-ethoxyethyl ester		$C_7H_{12}O_3$	144.17	106-74-1	-47		174	4-02-00-01469	0.983^{20}		1.4274^{20}
10635	2-Propenoic acid, ethyl ester	Ethyl acrylate	$C_5H_8O_2$	100.12	140-88-5	-71.2	3715	99.4	4-02-00-01460	0.9234^{20}	H_2O 2; EtOH 5; eth 5; chl 3	1.4068^{20}
10636	2-Propenoic acid, 2-ethylhexyl ester	2-Ethylhexyl acrylate	$C_{11}H_{20}O_2$	184.28	103-11-7	-90		$123\text{-}7^{60}$	4-02-00-01467	0.880^{25}		1.4332^{25}
10637	2-Propenoic acid, 2-fluoro-3-phenyl-	Cinnamic acid, α-fluoro-	$C_9H_7FO_2$	166.15	350-90-3	157.6		290	1-09-00-00237		EtOH 3; eth 3	
10638	2-Propenoic acid, 3-(2-furanyl)-	2-Furanacrylic acid	$C_7H_6O_3$	138.12	539-47-9	141	4207	286	5-18-06-00306		eth 4; EtOH 4	
10639	2-Propenoic acid, 3-(2-furanyl)-, ethyl ester		$C_9H_{10}O_3$	166.18	623-20-1	24.5		232	5-18-06-00307	1.0891^{25}	H_2O 1; EtOH 5; eth 5; chl 2	1.5286^{20}
10640	2-Propenoic acid, 3-(2-furanyl)-, methyl ester		$C_8H_8O_3$	152.15	623-18-7	36		227; 117^{23}	4-18-00-04144		bz 4; eth 4; EtOH 4	1.4447^{20}

No.	Name Synonym	Mol. Form. Mol. Wt.	CAS RN mp/°C	Merck No. bp/°C	Beil. Ref. den/g cm^{-3}	Solubility n_D
10641	2-Propenoic acid, 3-(2-furanyl)-, pentyl ester Acrylic acid, 3-(2-furyl), pentyl ester	$C_{12}H_{13}O_3$ 208.26	2438-19-9	 $116\text{-}8^{2}$	4-18-00-04146 1.0322^{20}	eth 4; EtOH 4 1.5289^{14}
10642	2-Propenoic acid, 3-(2-furanyl)-, propyl ester Acrylic acid, 3-(2-furyl), propyl ester	$C_{10}H_{12}O_3$ 180.20	623-22-3	 $91\text{-}4^{3}$	4-18-00-04146 1.0744^{20}	bz 4; eth 4; EtOH 4 1.5392^{24}
10643	2-Propenoic acid, hexyl ester	$C_9H_{16}O_2$ 156.22	2499-95-8 -45	 40^{1}	4-02-00-01466 0.878^{20}	
10644	2-Propenoic acid, 3-(4-hydroxyphenyl)- *p*-Coumaric acid	$C_9H_8O_3$ 164.16	7400-08-0 211.5	2561	4-10-00-01005	eth 4; EtOH 4
10645	2-Propenoic acid, 2-isocyano-3-phenyl-, ethyl ester Malonomononitrile, benzylidene,monoethyl ester	$C_{12}H_{11}NO_2$ 201.22	52744-85-1 51	 188^{15}; $118^{0.4}$	 1.0762^{25}	H_2O 1; EtOH 2; eth 3; bz 4 1.5033
10646	2-Propenoic acid, isodecyl ester	$C_{13}H_{24}O_2$ 212.33	1330-61-6 -100	 158^{50}	 0.885^{20}	 1.4416^{20}
10647	2-Propenoic acid, 3-(2-methoxyphenyl)-, methyl ester	$C_{11}H_{12}O_3$ 192.21	15854-58-7	 304; $161\text{-}3^{3}$	3-10-00-00835 1.1366^{17}	 1.5854^{16}
10648	2-Propenoic acid, 2-methyl- Methacrylic acid	$C_4H_6O_2$ 86.09	79-41-4 16	5849 162.5	4-02-00-01518 1.0153^{20}	H_2O 3; EtOH 5; eth 5; chl 3 1.4314^{20}
10649	2-Propenoic acid, 2-methyl-, anhydride Methacrylic acid anhydride	$C_8H_{10}O_3$ 154.17	760-93-0	 89^{5}	4-02-00-01537	EtOH 5; eth 5 1.4540^{20}
10650	2-Propenoic acid, 2-methyl-, butyl ester Butyl methacrylate	$C_8H_{14}O_2$ 142.20	97-88-1	 160	4-02-00-01525 0.8936^{20}	eth 4; EtOH 4 1.4240^{20}
10651	2-Propenoic acid, methylbutyl ester Propenoic acid, 2-methyl,butyl ester	$C_8H_{14}O_2$ 142.20	90638-48-5	 160; 71^{37}	3-02-00-01286 0.8936^{20}	H_2O 1; EtOH 5; eth 5; ctc 2 1.4240^{20}
10652	2-Propenoic acid, 2-methylbutyl ester Acrylic acid, (2-methylbutyl) ester	$C_8H_{14}O_2$ 142.20	44914-03-6	 160; 45^{10}	3-02-00-01228 0.8936^{20}	eth 4; EtOH 4 1.4240^{20}
10653	2-Propenoic acid, 2-methyl-, cyclohexyl ester	$C_{10}H_{16}O_2$ 168.24	101-43-9	 210	4-06-00-00039 0.9626^{20}	 1.4578^{20}
10654	2-Propenoic acid, 2-methyl-, 2-(dimethylamino)ethyl ester β-Dimethylaminoethyl methacrylate	$C_8H_{15}NO_2$ 157.21	2867-47-2	 $62\text{-}5^{6}$	4-04-00-01432	
10655	2-Propenoic acid, methyl ester Methyl acrylate	$C_4H_6O_2$ 86.09	96-33-3 <-75	5935 80.7	4-02-00-01457 0.9535^{20}	H_2O 2; EtOH 3; eth 3; ace 3 1.4040^{20}
10656	2-Propenoic acid, 2-methyl-, 1,2-ethanediyl ester	$C_{10}H_{14}O_4$ 198.22	97-90-5 -40	 260	4-02-00-01532 1.053^{20}	bz 4; EtOH 4; lig 4 1.4532^{25}
10657	2-Propenoic acid, 2-methyl-, 1,2-ethanediylbis(oxy-2,1-ethanediyl) ester	$C_{14}H_{22}O_6$ 286.33	109-16-0		4-02-00-01531 1.072^{25}	ace 4; eth 4; EtOH 4; peth 4 1.4595^{25}
10658	2-Propenoic acid, 2-methyl-, ethyl ester Ethyl methacrylate	$C_6H_{10}O_2$ 114.14	97-63-2	 117	4-02-00-01523 0.9135^{20}	H_2O 2; EtOH 5; eth 5; chl 2 1.4147^{20}
10659	2-Propenoic acid, 2-methyl-, 2-ethylhexyl ester 2-Ethylhexyl methacrylate	$C_{12}H_{22}O_2$ 198.31	688-84-6	 110^{14}	4-02-00-01528 0.880^{25}	 1.436^{25}
10660	2-Propenoic acid, 2-methyl-, hexyl ester	$C_{10}H_{18}O_2$ 170.25	142-09-6	 162; 86^{17}	4-02-00-01527 0.880^{25}	ace 4; bz 4; eth 4; EtOH 4 1.429^{25}
10661	2-Propenoic acid, 2-methyl-, 2-hydroxyethyl ester	$C_6H_{10}O_3$ 130.14	868-77-9	 $67^{3.5}$	4-02-00-01530 1.034^{25}	 1.4515^{20}
10662	2-Propenoic acid, 2-methyl-, 2-hydroxypropyl ester	$C_7H_{12}O_3$ 144.17	923-26-2	 90^{9}	4-02-00-01532 1.066^{25}	 1.4458^{20}
10663	2-Propenoic acid, 2-methyl-, methyl ester Methyl methacrylate	$C_5H_8O_2$ 100.12	80-62-6 -48	5849 100.5	4-02-00-01519 0.9440^{20}	H_2O 2; EtOH 5; eth 5; ace 5 1.4142^{20}
10664	2-Propenoic acid, 2-methyl-, 1-methylethyl ester	$C_7H_{12}O_2$ 128.17	4655-34-9	 125	4-02-00-01525 0.8847^{20}	ace 4; bz 4; eth 4; EtOH 4 1.4122^{20}
10665	2-Propenoic acid, 2-methyl-, 1-methylpropyl ester	$C_8H_{14}O_2$ 142.20	2998-18-7	 147; 35^{4}	4-02-00-01526 0.886^{20}	 1.4179^{20}
10666	2-Propenoic acid, 2-methyl-, 2-methylpropyl ester Isobutyl methacrylate	$C_8H_{14}O_2$ 142.20	97-86-9	 155	4-02-00-01526 0.8858^{20}	H_2O 1; EtOH 5; eth 5 1.4199^{20}
10667	2-Propenoic acid, 2-methyi-, octadecyl ester	$C_{22}H_{42}O_2$ 338.57	32360-05-7	 195^{6}	4-02-00-01529 0.880^{25}	 1.429^{25}
10668	2-Propenoic acid, 2-methyl-, oxiranylmethyl ester	$C_7H_{10}O_3$ 142.15	106-91-2	 75^{10}	5-17-03-00035 1.07^{25}	bz 4; eth 4; EtOH 4 1.448^{25}
10669	2-Propenoic acid, 2-methyl-, oxydi-2,1-ethanediyl ester	$C_{12}H_{18}O_5$ 242.27	2358-84-1	 >200	4-02-00-01531 1.056^{20}	 1.4571^{25}
10670	2-Propenoic acid, 2-methyl-3-phenyl-, (*Z*)-	$C_{10}H_{10}O_2$ 162.19	15250-29-0 94	 288	4-09-00-02063	H_2O 2; EtOH 4; eth 4; bz 4
10671	2-Propenoic acid, 2-methyl-, 2-propenyl ester	$C_7H_{10}O_2$ 126.16	96-05-9	 55^{30}; 60^{43}	4-02-00-01529 0.927^{20}	 1.4360^{20}
10672	2-Propenoic acid, 2-methyl-, propyl ester Propyl methacrylate	$C_7H_{12}O_2$ 128.17	2210-28-8	 141	4-02-00-01524 0.9022^{20}	H_2O 1; EtOH 5; eth 5 1.4190^{20}
10673	2-Propenoic acid, 2-methylpropyl ester Isobutyl acrylate	$C_7H_{12}O_2$ 128.17	106-63-8 -61	 132	4-02-00-01465 0.8896^{20}	H_2O 2; EtOH 3; eth 3; MeOH 3 1.4150^{20}

No.	Name Synonym	Mol. Form. Mol. Wt.	CAS RN mp/°C	Merck No. bp/°C	Beil. Ref. den/g cm^{-3}	Solubility n_D
10674	2-Propenoic acid, 2-methyl-, (tetrahydro-2-furanyl)methyl ester	$C_9H_{14}O_3$ 170.21	2455-24-5	 265; 81^{4}	4-17-00-01105 1.040^{25}	 1.4554^{25}
10675	2-Propenoic acid, 3-(1-naphthalenyl)-, (*E*)- Acrylic acid, 3-(1-naphtyl) (*trans*)	$C_{13}H_{10}O_2$ 198.22	2006-14-6 211.5	 sub	4-09-00-02486	H_2O 2; EtOH 3; eth 3; chl 3
10676	2-Propenoic acid, 3-(2-nitrophenyl)-, (*E*)-	$C_9H_7NO_4$ 193.16	1013-96-3 244.3	 sub	4-09-00-02043	H_2O 1; EtOH 3; bz 1
10677	2-Propenoic acid, 3-(2-nitrophenyl)-, methyl ester, (*E*)- Cinnamic acid, 2-nitro, methyl ester (*trans*)	$C_{10}H_9NO_4$ 207.19	39228-29-0 73	 187^{15}	4-09-00-02044	EtOH 4
10678	2-Propenoic acid, 3-(3-nitrophenyl)-, methyl ester, (*E*)- Cinnamic acid, 3-nitro, methyl ester (*trans*)	$C_{10}H_9NO_4$ 207.19	659-04-1 125.3	 dec	3-09-00-02742	bz 4; eth 4; chl 4
10679	2-Propenoic acid, 3-(4-nitrophenyl)-, methyl ester, (*E*)-	$C_{10}H_9NO_4$ 207.19	637-57-0 162	 283.5	4-09-00-02046	EtOH 4
10680	2-Propenoic acid, oxiranylmethyl ester	$C_6H_8O_3$ 128.13	106-90-1	 53^{10}	4-17-00-01005 1.1^{20}	bz 4 1.4490^{20}
10681	2-Propenoic acid, 2-phenoxyethyl ester	$C_{11}H_{12}O_3$ 192.21	48145-04-6	 110^{2}	3-06-00-00572 1.090^{25}	ace 4; eth 4; chl 4
10682	2-Propenoic acid, 3-phenyl-, (*E*)- *trans*-Cinnamic acid	$C_9H_8O_2$ 148.16	140-10-3 133	2300 300	4-09-00-02002 1.2475^{4}	H_2O 1; EtOH 4; eth 3; ace 3
10683	2-Propenoic acid, 3-phenyl-, (*Z*)-	$C_9H_8O_2$ 148.16	102-94-3 42		4-09-00-02001	EtOH 4; HOAc 4; lig 4
10684	2-Propenoic acid, 3-phenyl-, anhydride Cinnamic anhydride	$C_{18}H_{14}O_3$ 278.31	538-56-7 136	2300		bz 4
10685	2-Propenoic acid, 3-phenyl-, ethyl ester, (*E*)-	$C_{11}H_{12}O_2$ 176.22	4192-77-2 10	2300 271.5	4-09-00-02006 1.0491^{20}	H_2O 1; EtOH 4; eth 4; ace 4 1.5598^{20}
10686	2-Propenoic acid, 3-phenyl-, methyl ester, (*E*)-	$C_{10}H_{10}O_2$ 162.19	1754-62-7 36.5	2300 261.9	4-09-00-02005 1.042^{36}	H_2O 1; EtOH 4; eth 4; ace 3 1.5766^{22}
10687	2-Propenoic acid, 3-phenyl-, 1-methylethyl ester, (*E*)- Cinnamic acid, isopropyl ester (*trans*)	$C_{12}H_{14}O_2$ 190.24	60512-85-8	 269	4-09-00-02008 1.0320^{20}	H_2O 1; EtOH 3; eth 3; ace 3 1.5455^{20}
10688	2-Propenoic acid, phenylmethyl ester	$C_{10}H_{10}O_2$ 162.19	2495-35-4	 228	3-06-00-01481 1.0573^{20}	H_2O 1; EtOH 3; eth 3; ace 3 1.5143^{20}
10689	2-Propenoic acid, 3-phenyl-, phenyl ester, (*E*)- Cinnamic acid, phenyl ester (*trans*)	$C_{15}H_{12}O_2$ 224.26	25695-77-6 74.3	 205^{15}	4-09-00-02011 1.0650^{100}	bz 4; eth 4
10690	2-Propenoic acid, 3-phenyl-, phenylmethyl ester, (*E*)- Cinnamic acid, benzyl ester (*trans*)	$C_{16}H_{14}O_2$ 238.29	78277-23-3 39	 350 dec; 244^{5}	4-09-00-02012 1.109^{15}	H_2O 1; EtOH 3; eth 3; bz 2
10691	2-Propenoic acid, 3-phenyl-, 3-phenyl-2-propenyl ester Cinnamyl cinnamate	$C_{18}H_{16}O_2$ 264.32	122-69-0 44	2307	4-09-00-02013 1.1565^{4}	H_2O 1; EtOH 3; eth 4; chl 3
10692	2-Propenoic acid, 3-phenyl-, 2-propenyl ester Cinnamic acid, allyl ester	$C_{12}H_{12}O_2$ 188.23	1866-31-5	 268 dec; 163^{17}	4-09-00-02010 1.048^{23}	H_2O 1; EtOH 4; eth 5; ctc 2 1.530^{20}
10693	2-Propenoic acid, 3-phenyl-, propyl ester, (*E*)- Cinnamic acid, propyl ester (*trans*)	$C_{12}H_{14}O_2$ 190.24	74513-58-9	 285	4-09-00-02007 1.0433^{0}	H_2O 1
10694	2-Propenoic acid, 2-propenyl ester	$C_6H_8O_2$ 112.13	999-55-3	 121	4-02-00-01468 0.9441^{20}	H_2O 2; EtOH 3; eth 3; acid 3 1.4320^{20}
10695	2-Propenoic acid, (tetrahydro-2-furanyl)methyl ester	$C_8H_{12}O_3$ 156.18	2399-48-6 <-60	 96^{6}	4-17-00-01104 1.061^{20}	
10696	2-Propenoic acid, 2,3,3-trichloro-	$C_3HCl_3O_2$ 175.40	2257-35-4 76	 222	4-02-00-01486	H_2O 3; EtOH 3; eth 3; ctc 3 1.5271^{18}
10697	2-Propen-1-ol Allyl alcohol	C_3H_6O 58.08	107-18-6 -129	284 97.0	4-01-00-02079 0.8540^{20}	H_2O 5; EtOH 5; eth 5; chl 3 1.4135^{20}
10698	1-Propen-2-ol, acetate Isopropenyl acetate	$C_5H_8O_2$ 100.12	108-22-5 -92.9	5091 94	4-02-00-00179 0.9090^{20}	H_2O 2; EtOH 3; eth 4; ace 3 1.4033^{20}
10699	2-Propen-1-ol, 2-bromo-	C_3H_5BrO 136.98	598-19-6	 154; 62^{11}	4-01-00-02094 1.621^{18}	eth 4; chl 4 1.500^{18}
10700	2-Propen-1-ol, 2-chloro-	C_3H_5ClO 92.52	5976-47-6	 135	4-01-00-02091 1.1618^{20}	EtOH 3; ctc 2 1.4588^{20}
10701	2-Propen-1-ol, 3-chloro- Allyl alcohol, 3-chloro	C_3H_5ClO 92.52	29560-84-7	 147	4-01-00-02090 1.1769^{20}	 1.4738^{20}
10702	2-Propen-1-ol, 2-methyl- Methallyl alcohol	C_4H_8O 72.11	513-42-8	 114.5	4-01-00-02114 0.8515^{20}	H_2O 4; EtOH 5; eth 5 1.4255^{20}
10703	2-Propen-1-ol, 2-methyl-, carbonate (2:1) Bis-(2-methallyl)carbonate	$C_9H_{14}O_3$ 170.21	64057-79-0 201.3	 66^{3}	4-03-00-00011 0.943^{25}	 1.4371^{20}
10704	2-Propen-1-ol, 3-phenyl-, (*E*)-	$C_9H_{10}O$ 134.18	4407-36-7 34	 257.5	4-06-00-03799 1.0440^{20}	H_2O 2; EtOH 4; eth 4; chl 2 1.5819^{20}
10705	2-Propen-1-ol, 3-phenyl-, (*Z*)-	$C_9H_{10}O$ 134.18	4510-34-3 34	 257.5	3-06-00-02401 1.0440^{20}	eth 4; EtOH 4 1.5819^{20}
10706	2-Propen-1-ol, 3-phenyl-, acetate, (*E*)-	$C_{11}H_{12}O_2$ 176.22	21040-45-9	 145^{15}; 114^{1}	4-06-00-03801 1.0567^{20}	H_2O 1; EtOH 3; eth 3; ace 3 1.5425^{20}
10707	2-Propen-1-ol, 3-phenyl-, formate	$C_{10}H_{10}O_2$ 162.19	104-65-4 0	 252	0-06-00-00571 1.086^{25}	
10708	2-Propen-1-one, 1,3-diphenyl- Chalcone	$C_{15}H_{12}O$ 208.26	94-41-7 59	2028 346 dec	4-07-00-01658 1.0712^{62}	bz 4; EtOH 4; chl 4

No.	Name Synonym	Mol. Form. Mol. Wt.	CAS RN mp/°C	Merck No. bp/°C	Beil. Ref. den/g cm^{-3}	Solubility n_D
10709	2-Propen-1-one, 1,3-diphenyl-, (*E*)-	$C_{15}H_{12}O$ 208.26	614-47-1 59	 346 dec	4-07-00-01658 1.0712^{62}	H_2O 1; EtOH 2; eth 3; bz 3
10710	2-Propen-1-one, 3-(2-furanyl)-1-phenyl-	$C_{13}H_{10}O_2$ 198.22	717-21-5 47	 317	5-17-10-00371 1.1140^{20}	EtOH 3; eth 3
10711	2-Propen-1-one, 3-hydroxy-1,3-diphenyl-	$C_{15}H_{12}O_2$ 224.26	1704-15-0 79.8	 220^{18}; $140^{1.0}$	4-07-00-02512	eth 4; EtOH 4; chl 4
10712	2-Propen-1-one, 3-(3-methoxyphenyl)-1-phenyl-	$C_{16}H_{14}O_2$ 238.29	5470-91-7 65	 247^{12}	1-08-00-00579	ace 4; bz 4; eth 4; EtOH 4
10713	2-Propen-1-one, 3-(4-methoxyphenyl)-1-phenyl-	$C_{16}H_{14}O_2$ 238.29	959-33-1 79	 187^{19}	2-08-00-00218	H_2O 1; EtOH 4; eth 3; ctc 3
10714	2-Propenoyl chloride Acrylic acid chloride	C_3H_3ClO 90.51	814-68-6	 75.5	4-02-00-01471 1.1136^{20}	chl 4 1.4343^{20}
10715	2-Propenoyl chloride, 2-methyl- Methacrylic acid chloride	C_4H_5ClO 104.54	920-46-7 -60	 96	4-02-00-01537 1.0871^{20}	eth 3; ace 3; chl 3 1.4435^{20}
10716	2-Propenoyl chloride, 3-phenyl- Cinnamoyl chloride	C_9H_7ClO 166.61	102-92-1 36	2303 257	2-09-00-00395 1.1617^{45}	lig 4 1.614^{42}
10717	2-Propenoyl chloride, 3-phenyl-, (*E*)-	C_9H_7ClO 166.61	17082-09-6 37.5	 257.5	2-09-00-00390 1.1617^{45}	H_2O 1; EtOH 3; ctc 3; lig 3 1.614^{42}
10718	2-Propenoyl chloride, 2,3,3-trichloro-	C_3Cl_4O 193.84	815-58-7	 158	4-02-00-01486	bz 4 1.5271^{18}
10719	Propetamphos 2-Butenoic acid, 3-[[(ethylamino)methoxyphosphinothioyl]oxy]-, 1-methylethyl ester, (*E*)-	$C_{10}H_{20}NO_4PS$ 281.31	31218-83-4	7827 $88^{0.005}$	 1.1294^{20}	
10720	Propiconazole 1-[[2-(2,4-Dichlorophenyl)-4-propyl-1,3-dioxolan-2-yl]methyl]-1H-1,2,4-triazole	$C_{15}H_{17}Cl_2N_3O_2$ 342.22	60207-90-1	 $180^{0.1}$	 1.27^{20}	
10721	Propionic acid, 2,2-dichloro-3,3,3-trifluoro-, propyl ester Propanoic acid, 2,2-dichloro-3,3,3-trifluoro, propyl ester	$C_6H_7Cl_2F_3O_2$ 239.02	357-49-3	 144.5^{625}	4-02-00-00758 1.3531^{20}	 1.3888^{20}
10722	Propoxur Phenol, 2-(1-methylethoxy)-, methylcarbamate	$C_{11}H_{15}NO_3$ 209.25	114-26-1 87	7849 dec	 1.12^{20}	
10723	2-Propynal Propargyl aldehyde	C_3H_2O 54.05	624-67-9	 60	4-01-00-03537 0.9152^{20}	H_2O 5; EtOH 3; eth 3; ace 3 1.4033^{25}
10724	2-Propynal, 3-phenyl-	C_9H_6O 130.15	2579-22-8	 127^{28}; 104^{11}	4-07-00-01173 1.0680^{12}	 1.6079^{12}
10725	2-Propyn-1-amine	C_3H_5N 55.08	2450-71-7	 83	4-04-00-01135 0.803^{25}	 1.4480^{20}
10726	2-Propyn-1-amine, *N*-methyl-	C_4H_7N 69.11	35161-71-8	 83	0-04-00-00228 0.819^{25}	 1.4332^{20}
10727	1-Propyne Methylacetylene	C_3H_4 40.06	74-99-7 -102.7	 -23.2	4-01-00-00958 $^{*}0.607^{25}$	H_2O 2; EtOH 4; bz 3; chl 3 1.3863^{-40}
10728	1-Propyne, 3-bromo- Propargyl bromide	C_3H_3Br 118.96	106-96-7	 89	4-01-00-00964 1.579^{19}	EtOH 3; eth 3; bz 3; ctc 3 1.4922^{20}
10729	1-Propyne, 3-chloro- Propargyl chloride	C_3H_3Cl 74.51	624-65-7	7820 58	4-01-00-00963 1.030^{25}	H_2O 1; EtOH 5; eth 5; bz 3 1.4349^{20}
10730	1-Propyne, 1,3-dibromo- Propyne, 1,3-dibromo-	$C_3H_2Br_2$ 197.86	627-16-7	 $73\text{-}4^{30}$	4-01-00-00965 2.1894^{20}	eth 4; chl 4 1.5690^{20}
10731	1-Propyne, 3,3-diethoxy-	$C_7H_{12}O_2$ 128.17	10160-87-9	 139	4-01-00-03538 0.8942^{22}	ace 4; eth 4; EtOH 4; chl 4 1.4140^{20}
10732	1-Propyne, 1-ethoxy-	C_5H_8O 84.12	14273-06-4	 84	3-01-00-01969 0.8276^{20}	eth 4; EtOH 4 1.4039^{20}
10733	1-Propyne, 3-ethoxy-	C_5H_8O 84.12	628-33-1	 82	4-01-00-02215 0.8326^{20}	eth 4; EtOH 4 1.4039^{20}
10734	1-Propyne, 1-iodo-	C_3H_3I 165.96	624-66-8	 110	4-01-00-00965 2.08^{22}	eth 4; EtOH 4
10735	1-Propyne, 3-iodo-	C_3H_3I 165.96	659-86-9	 115	4-01-00-00965 2.0177^{25}	eth 3
10736	1-Propyne, 3-methoxy-	C_4H_6O 70.09	627-41-8	 63	4-01-00-02215 0.83^{12}	eth 4; EtOH 4 1.5035^{20}
10737	2-Propynenitrile Cyanoacetylene	C_3HN 51.05	1070-71-9 5	 42.5	4-02-00-01689 0.8167^{17}	H_2O 2; EtOH 3 1.3868^{25}
10738	2-Propynoic acid Propiolic acid	$C_3H_2O_2$ 70.05	471-25-0 9	7833 144 dec; 72^{50}	4-02-00-01687 1.1380^{20}	H_2O 4; eth 4; EtOH 4; chl 4 1.4306^{20}
10739	2-Propynoic acid, ethyl ester	$C_5H_6O_2$ 98.10	623-47-2	 120	4-02-00-01688 0.9560^{25}	H_2O 1; EtOH 4; eth 4; chl 4 1.4105^{20}
10740	2-Propynoic acid, 3-(2-nitrophenyl)- *o*-Nitrophenylpropiolic acid	$C_9H_5NO_4$ 191.14	530-85-8 157 dec	6546 exp	4-09-00-02330	eth 4; EtOH 4
10741	2-Propynoic acid, 3-phenyl-, ethyl ester	$C_{11}H_{10}O_2$ 174.20	2216-94-6	 265; $128^{1.6}$	4-09-00-02328 1.055^{25}	eth 3 1.5520^{20}
10742	2-Propynoic acid, 3-phenyl-, methyl ester	$C_{10}H_8O_2$ 160.17	4891-38-7 26	 158^{48}; 132^{16}	4-09-00-02328 1.0830^{25}	 1.5618^{25}

No.	Name	Synonym	Mol. Form.	Mol. Wt.	CAS RN	mp/°C	Merck No.	bp/°C	Beil. Ref.	den/g cm^{-3}	Solubility	n_D
10743	2-Propyn-1-ol	Propargyl alcohol	C_3H_4O	56.06	107-19-7	-51.8	7819	113.6	4-01-00-02214	0.9478^{20}	H_2O 3; EtOH 5; eth 5; chl 3	1.4322^{20}
10744	2-Propyn-1-ol, acetate	Propargyl acetate	$C_5H_6O_2$	98.10	627-09-8			121.5	4-02-00-00197	0.9982^{20}	H_2O 2; EtOH 3; eth 3	1.4187^{20}
10745	2-Propyn-1-ol, 3-phenyl-		C_9H_8O	132.16	1504-58-1			$137\text{-}8^{15}$	4-06-00-04065	1.07^{18}	eth 3; ace 3; bz 3	1.5873^{28}
10746	Propyzamide	Benzamide, 3,5-dichloro-*N*-(1,1-dimethyl-2-propynyl)-	$C_{12}H_{11}Cl_2NO$	256.13	23950-58-5	155	7886					
10747	Pteridine	Pyrazino[2,3-d]pyrimidine	$C_6H_4N_4$	132.12	91-18-9	139.5	7947	sub 125	4-26-00-01770		H_2O 4; EtOH 3; eth 2; bz 2	
10748	2,4(1H,3H)-Pteridinedione	Lumazine	$C_6H_4N_4O_2$	164.12	487-21-8	348.5	5467		4-26-00-02489		HOAc 4	
10749	4,6-Pteridinedione, 2-amino-1,5-dihydro-	Xanthopterin	$C_6H_5N_5O_2$	179.14	119-44-8	>410 dec	9973	$98\text{-}100^{18}$	4-26-00-04000	1.559^{25}	H_2O 1; EtOH 2; eth 2; acid 4	
10750	2,4,7-Pteridinetriamine, 6-phenyl-	Triamterene	$C_{12}H_{11}N_7$	253.27	396-01-0		9515		4-26-00-03837			
10751	1H-Purin-6-amine	Adenine	$C_5H_5N_5$	135.13	73-24-5	360 dec	141	sub 220	4-26-00-03561		H_2O 3; EtOH 2; eth 1; chl 1	
10752	1H-Purine	7H-Imidazo[4,5-d]pyrimidine	$C_5H_4N_4$	120.11	120-73-0	216.5	7959		4-26-00-01736		H_2O 4; EtOH 4; eth 2; ace 3	
10753	1H-Purine, 6-chloro-	6-Chloropurine	$C_5H_3ClN_4$	154.56	87-42-3	175-7 dec	2161		4-26-00-01742			
10754	1H-Purine-2,6-diamine	2,6-Diaminopurine	$C_5H_6N_6$	150.14	1904-98-9	302	2963		4-26-00-03785			
10755	1H-Purine-2,6-dione, 3,7-dihydro-	Xanthine	$C_5H_4N_4O_2$	152.11	69-89-6	dec	9968	sub	4-26-00-02327			
10756	1H-Purine-2,6-dione, 3,7-dihydro-1,3-dimethyl-	Theophylline	$C_7H_8N_4O_2$	180.17	58-55-9	273	9212		4-26-00-02331		H_2O 4; EtOH 2; eth 2; chl 2	
10757	1H-Purine-2,6-dione, 3,7-dihydro-3,7-dimethyl-	Theobromine	$C_7H_8N_4O_2$	180.17	83-67-0	357	9209	sub 290	4-26-00-02336		H_2O 2; EtOH 2; eth 1; bz 1	
10758	1H-Purine-2,6-dione, 3,7-dihydro-7-(2-hydroxyethyl)-1,3-dimethyl-	Etofylline	$C_9H_{12}N_4O_3$	224.22	519-37-9	158	3835		4-26-00-02357		H_2O 4; EtOH 3; eth 2; bz 2	
10759	1H-Purine-2,6-dione, 3,7-dihydro-8-methoxy-1,3,7-trimethyl-		$C_9H_{12}N_4O_3$	224.22	569-34-6	176			4-26-00-02727	1.399^{25}	H_2O 3; EtOH 4; bz 3	
10760	1H-Purine-2,6-dione, 3,7-dihydro-1,3,7-trimethyl-	Caffeine	$C_8H_{10}N_4O_2$	194.19	58-08-2	238	1635	sub 90	4-26-00-02338	1.23^{19}	H_2O 2; EtOH 2; eth 1; ace 2	
10761	1H-Purine-2,6-dione, 7-(2,3-dihydroxypropyl)-3,7-dihydro-1,3-dimethyl-	Dyphylline	$C_{10}H_{14}N_4O_4$	254.25	479-18-5	161.5	3459		4-26-00-02370			
10762	6H-Purine-6-thione, 2-amino-1,7-dihydro-	Thioguanine	$C_5H_5N_5S$	167.19	154-42-7	>360	9266		4-26-00-03926			
10763	6H-Purine-6-thione, 1,7-dihydro-	6-Mercaptopurine	$C_5H_4N_4S$	152.18	50-44-2	313 dec	5762				H_2O 1; alk 3	
10764	1H-Purine-2,6,8(3H)-trione, 7,9-dihydro-	Uric acid	$C_5H_4N_4O_3$	168.11	69-93-2	dec	9790	dec	4-26-00-02619	1.89^{25}	H_2O 1; EtOH 1; eth 1; alk 4	
10765	1H-Purine-2,6,8(3H)-trione, 4,9-dihydro-3-methyl-	3-Methyluric acid	$C_6H_6N_4O_3$	182.14	605-99-2	>350			4-26-00-02621	1.6104^{25}	H_2O 2; EtOH 2; alk 3	1.6334^{25}
10766	1H-Purine-2,6,8(3H)-trione, 7,9-dihydro-7-methyl-	7-Methyluric acid	$C_6H_6N_4O_3$	182.14	612-37-3	370-80 dec			4-26-00-02622	1.706^{25}	H_2O 2; NaOH 3	
10767	2H-Purin-2-one, 6-amino-1,3-dihydro-	Isoguanine	$C_5H_5N_5O$	151.13	3373-53-3	>360						
10768	6H-Purin-6-one, 2-amino-1,7-dihydro-	Guanine	$C_5H_5N_5O$	151.13	73-40-5	360 dec	4477	sub	4-26-00-03890		H_2O 1; EtOH 2; eth 2; alk 3	
10769	6H-Purin-6-one, 2-amino-1,7-dihydro-7-methyl	7-Methylguanine	$C_6H_7N_5O$	165.15	578-76-7	370						
10770	6H-Purin-6-one, 1,7-dihydro-	Hypoxanthine	$C_5H_4N_4O$	136.11	68-94-0	150 dec	4805		4-26-00-02081		H_2O 2; alk 3; dil acid 3	
10771	Pyran		C_5H_6O	82.10	33941-07-0			80	0-17-00-00036		EtOH 3; eth 3; bz 3	1.4559^{20}
10772	2H-Pyran-5-carboxylic acid, 3,4-dihydro-3-(hydroxymethyl)-2-methyl-4-[(9-methyl-9H-pyrido[3,4-b]indol-1-yl)methyl]-, methyl ester, [2S-(2α,3β,4β)]-	Alstonidine	$C_{22}H_{24}N_2O_4$	380.44	25394-75-6	189	313		4-27-00-07962		ace 4; EtOH 4	
10773	2H-Pyran-5-carboxylic acid, 2-oxo-	Coumalic acid	$C_6H_4O_4$	140.10	500-05-0	205-10 dec	2558	218^{120}	5-18-08-00120		H_2O 2; EtOH 3; eth 2; ace 2	
10774	2H-Pyran-4-carboxylic acid, tetrahydro-	Pyran-4-carboxylic acid, tetrahydro-	$C_6H_{10}O_3$	130.14	5337-03-1	88		144^{11}	5-18-06-00023		H_2O 3; peth 2	
10775	4H-Pyran-2,6-dicarboxylic acid, 3-hydroxy-4-oxo-	Meconic acid	$C_7H_4O_7$	200.10	497-59-6	120 dec	5664	120 dec	5-18-09-00319		H_2O 2; EtOH 3; eth 2; ace 2	
10776	4H-Pyran-2,6-dicarboxylic acid, 4-oxo-	Chelidonic acid	$C_7H_4O_6$	184.11	99-32-1	262	2042		5-18-08-00646		H_2O 3; EtOH 2	

No.	Name Synonym	Mol. Form. Mol. Wt.	CAS RN mp/°C	Merck No. bp/°C	Beil. Ref. den/g cm^{-3}	Solubility n_D
10777	2H-Pyran, 3,4-dihydro- 3,4-Dihydro-2H-pyran	C_5H_8O 84.12	110-87-2	 86	5-17-01-00181 0.921^{19}	H_2O 3; EtOH 3; chl 2 1.4402^{19}
10778	2H-Pyran, 3,6-dihydro-4-methyl-	$C_6H_{10}O$ 98.14	16302-35-5	 117.5	5-17-01-00194 0.912^{25}	 1.4495^{20}
10779	2H-Pyran-2,4(3H)-dione, 3-acetyl-6-methyl- Dehydroacetic acid	$C_8H_8O_4$ 168.15	520-45-6 109	2855 270	4-17-00-06699	H_2O 4; EtOH 2; eth 4; chl 2
10780	2H-Pyran-2,6(3H)-dione, dihydro-	$C_5H_6O_3$ 114.10	108-55-4 56.3	 158^{15}	5-17-11-00009 1.4110^{20}	
10781	2H-Pyran-2,6(3H)-dione, dihydro-3-phenyl-	$C_{11}H_{10}O_3$ 190.20	2959-96-8 95	 224^{13}	5-17-11-00279	EtOH 4
10782	2H-Pyran-2,6(3H)-dione, dihydro-4-phenyl-	$C_{11}H_{10}O_3$ 190.20	4160-80-9 105.6	 217^{15}	5-17-11-00279	bz 4; eth 4; chl 4
10783	2H-Pyran, 2-ethoxy-3,4-dihydro-	$C_7H_{12}O_2$ 128.17	103-75-3	 42^{16}	5-17-03-00196 0.957^{25}	 1.4394^{20}
10784	2H-Pyran-2-methanol, tetrahydro-	$C_6H_{12}O_2$ 116.16	100-72-1	 185	5-17-03-00136 1.027^{25}	 1.458^{20}
10785	4H,5H-Pyrano[4,3-b]pyran-4,5-dione, 2,3-dihydro-3-hydroxy-2-methyl-7-(1-propenyl)-, [2*S*-[2α,3β,7(*E*)]]- Radicinin	$C_{12}H_{12}O_5$ 236.22	10088-95-6 221.5	8117	4-19-00-02746	chl 2
10786	Pyrano[3,2-c][2]benzopyran-6(2H)-one, 3,4,4a,10b-tetrahydro-3,4,8,10-tetrahydroxy-2-(hydroxymethyl)-9-methoxy-[2R-(2α,3β,4α,4aα,10bβ)]- Bergenin	$C_{14}H_{16}O_9$ 328.28	477-90-7 238	1174	5-19-07-00132	H_2O 4; EtOH 4
10787	2H-Pyran-4-ol, tetrahydro-	$C_5H_{10}O_2$ 102.13	2081-44-9	 87^{15}	5-17-03-00111 1.071^{25}	 1.4600^{20}
10788	2H-Pyran-2-one	$C_5H_4O_2$ 96.09	504-31-4 8.5	 207.5	5-17-09-00288 1.200^{20}	H_2O 5; ace 4 1.5270^{25}
10789	4H-Pyran-4-one	$C_5H_4O_2$ 96.09	108-97-4 32.5	 212.5	5-17-09-00290 1.190^{25}	H_2O 4; EtOH 3; eth 4; bz 3 1.5238
10790	2H-Pyran-2-one, 6-[2-(1,3-benzodioxol-5-yl)ethenyl]-5,6-dihydro-4-methoxy-, [*R*-(*E*)]- Methysticin	$C_{15}H_{14}O_5$ 274.27	495-85-2 137	6056	5-19-10-00456	
10791	2H-Pyran-2-one, 5,6-dihydro-6-methyl-, (*S*)- Parasorbic acid	$C_6H_8O_2$ 112.13	10048-32-5	6981 100^{15}	5-17-09-00130 1.079^{19}	H_2O 4; eth 4; EtOH 4 1.4730^{20}
10792	2H-Pyran-2-one, 4,6-dimethyl-	$C_7H_8O_2$ 124.14	675-09-2 51.5	 245	5-17-09-00409	H_2O 4; eth 4; EtOH 4
10793	4H-Pyran-4-one, 2,6-dimethyl-	$C_7H_8O_2$ 124.14	1004-36-0 132	 251; 140^{25}	5-17-09-00410 0.9953^{137}	H_2O 3; EtOH 3; eth 3; ace 3
10794	2H-Pyran-2-one, 3-hydroxy-	$C_5H_4O_3$ 112.08	496-64-0 93.8	 112^{20}; 102^{15}	5-18-01-00095	H_2O 3; EtOH 3; eth 3; bz 3
10795	4H-Pyran-4-one, 5-hydroxy-2-(hydroxymethyl)- Kojic acid	$C_6H_6O_4$ 142.11	501-30-4 153.5	5197	5-18-02-00516	H_2O 2; EtOH 3; eth 3; ace 3
10796	4H-Pyran-4-one, 3-hydroxy-2-methyl- Maltol	$C_6H_6O_3$ 126.11	118-71-8 161.5	5594 sub 93	5-18-01-00114	H_2O 2; eth 2; bz 2; chl 4
10797	4H-Pyran-4-one, 2-methoxy-3,5-dimethyl-6-[tetrahydro-4-[2-methyl-3-(4-nitrophenyl)-2-propenylidene]-2-furanyl] Aureothin	$C_{22}H_{23}NO_6$ 397.43	2825-00-5 158	897	5-19-06-00099	ace 4; EtOH 4; chl 4
10798	2H-Pyran-2-one, tetrahydro-	$C_5H_8O_2$ 100.12	542-28-9 -12.5	 219	5-17-09-00017 1.0794^{20}	H_2O 3; EtOH 5; eth 5; ctc 2 1.4503^{20}
10799	4H-Pyran-4-one, tetrahydro-	$C_5H_8O_2$ 100.12	29943-42-8	 166.5	5-17-09-00021 1.084^{25}	 1.4520^{20}
10800	2H-Pyran, tetrahydro- Oxane	$C_5H_{10}O$ 86.13	142-68-7 -45	9149 88	5-17-01-00064 0.8814^{20}	EtOH 3; eth 3; bz 3; ctc 3 1.4200^{20}
10801	Pyrazine 1,4-Diazine	$C_4H_4N_2$ 80.09	290-37-9 55	7971 115	5-23-05-00351 1.0311^{61}	H_2O 3; EtOH 3; eth 3; ace 3 1.4953^{61}
10802	Pyrazinecarboxamide Pyrazinamide	$C_5H_5N_3O$ 123.11	98-96-4 192	7970 sub	4-25-00-00772	H_2O 3; EtOH 3
10803	Pyrazinecarboxylic acid Pyrazinoic acid	$C_5H_4N_2O_2$ 124.10	98-97-5 225 dec	7973 sub	4-25-00-00771	
10804	2,3-Pyrazinedicarboxylic acid 2,3-Dicarboxypyrazine	$C_6H_4N_2O_4$ 168.11	89-01-0 193 dec	7972	4-25-00-01064	H_2O 4; EtOH 2; eth 2; ace 3
10805	Pyrazine, 2,3-dimethyl- 2,3-Dimethylpyrazine	$C_6H_8N_2$ 108.14	5910-89-4	 156	5-23-05-00402 1.0281^{0}	H_2O 3; EtOH 3; eth 3
10806	Pyrazine, 2,5-dimethyl- 2,5-Dimethylpyrazine	$C_6H_8N_2$ 108.14	123-32-0 15	 155	5-23-05-00403 0.9887^{20}	H_2O 5; EtOH 5; eth 5; ace 3 1.4980^{20}
10807	Pyrazine, 2,6-dimethyl- 2,6-Dimethylpyrazine	$C_6H_8N_2$ 108.14	108-50-9 47.5	 155.6	5-23-05-00406 0.9647^{50}	H_2O 3; EtOH 3; eth 3; ctc 2
10808	Pyrazineethanol	$C_6H_8N_2O$ 124.14	6705-31-3	 129^{10}	5-23-11-00173 1.163^{20}	 1.5378^{20}
10809	Pyrazine, 3-ethyl-2,5-dimethyl-	$C_8H_{12}N_2$ 136.20	13360-65-1	 180.5	5-23-05-00440 0.9657^{24}	H_2O 2; EtOH 2; eth 2 1.5014^{24}
10810	Pyrazine, methyl- 2-Methylpyrazine	$C_5H_6N_2$ 94.12	109-08-0 -29	 137	5-23-05-00386 1.03^{20}	H_2O 5; EtOH 5; eth 5; ace 3 1.5042^{20}
10811	2(1H)-Pyrazinone, 1-hydroxy-6-(1-methylpropyl)-3-(2-methylpropyl)-, (+)- Aspergillic acid	$C_{12}H_{20}N_2O_2$ 224.30	490-02-8 98	863	4-24-00-00235	bz 4; eth 4; EtOH 4

No.	Name Synonym	Mol. Form. Mol. Wt.	CAS RN mp/°C	Merck No. bp/°C	Beil. Ref. den/g cm^{-3}	Solubility n_D
10812	1H-Pyrazol-4-amine 4-Aminopyrazole	$C_3H_5N_3$ 83.09	28466-26-4 81	 sub	4-25-00-02029	H_2O 4; EtOH 4; chl 4
10813	1H-Pyrazol-4-amine, 3-methyl-1-phenyl- Pyrazole, 4-amino-3-methyl-1-phenyl-	$C_{10}H_{11}N_3$ 173.22	103095-51-8 88.8	 314.5	4-25-00-02033	bz 4; eth 4; EtOH 4; chl 4
10814	1H-Pyrazol-5-amine, 3-methyl-1-phenyl-	$C_{10}H_{11}N_3$ 173.22	1131-18-6 116	 333	4-25-00-02034	H_2O 3; EtOH 3; bz 2; chl 3
10815	1H-Pyrazole 1,2-Diazole	$C_3H_4N_2$ 68.08	288-13-1 68	7974 187	5-23-04-00122	H_2O 3; EtOH 3; eth 3; bz 3 1.4203
10816	1H-Pyrazole, 4-bromo-1,3-dimethyl- 4-Bromo-1,3-dimethylpyrazole	$C_5H_7BrN_2$ 175.03	5775-82-6	 $76\text{-}7^{10}$	5-23-05-00022 1.4975^{15}	ace 4; eth 4; EtOH 4 1.5214^{15}
10817	1H-Pyrazole, 4-bromo-3-methyl- 4-Bromo-3-methylpyrazole	$C_4H_5BrN_2$ 161.00	13808-64-5 76.5		5-23-05-00022 1.5638^{100}	ace 4; eth 4; EtOH 4 1.5182^{100}
10818	1H-Pyrazole, 3-chloro-1,5-dimethyl- 3-Chloro-1,5-dimethylpyrazole	$C_5H_7ClN_2$ 130.58	51500-32-4 47.5	 211	2-23-00-00049 1.0823^{100}	ace 4; bz 4; eth 4; EtOH 4 1.4648^{100}
10819	1H-Pyrazole, 4-chloro-3,5-dimethyl-	$C_5H_7ClN_2$ 130.58	15953-73-8 117.5	 221	5-23-05-00151	ace 4; bz 4; eth 4; EtOH 4
10820	1H-Pyrazole, 5-chloro-1,3-dimethyl- 5-Chloro-1,3-dimethylpyrazole	$C_5H_7ClN_2$ 130.58	54454-10-3	 157.5	5-23-05-00019 1.1367^{18}	H_2O 4 1.4877^{18}
10821	1H-Pyrazole, 3-chloro-5-methyl-	$C_4H_5ClN_2$ 116.55	15953-45-4 118.5	 258; 138^{15}	5-23-05-00019	H_2O 4
10822	1H-Pyrazole, 4,5-dihydro- 2-Pyrazoline	$C_3H_6N_2$ 70.09	109-98-8	7975 144	5-23-03-00361 1.0200^{17}	H_2O 4; eth 4; EtOH 4 1.4796^{17}
10823	1H-Pyrazole, 4,5-dihydro-1-phenyl- 1-Phenyl-2-pyrazoline	$C_9H_{10}N_2$ 146.19	936-53-8 52	 273; 151^{17}	5-23-03-00363 1.0689^{58}	H_2O 1; EtOH 3; eth 3; bz 3 1.6015^{58}
10824	1H-Pyrazole, 4,5-dihydro-3-phenyl- 2-Pyrazoline, 3-phenyl	$C_9H_{10}N_2$ 146.19	936-48-1 45.8	 164^{17}	5-23-06-00396 1.0892^{54}	
10825	1H-Pyrazole, 1,3-dimethyl-	$C_5H_8N_2$ 96.13	694-48-4	 137	5-23-05-00012 0.9561^{17}	H_2O 4 1.4734^{15}
10826	1H-Pyrazole, 1,5-dimethyl-	$C_5H_8N_2$ 96.13	694-31-5	 149	5-23-05-00006 0.9813^{17}	H_2O 4; EtOH 4; eth 4 1.4782^{16}
10827	1H-Pyrazole, 3,4-dimethyl-	$C_5H_8N_2$ 96.13	2820-37-3 58	 110^{11}	5-23-05-00108 0.9325^{99}	EtOH 3; eth 3; ace 3; bz 3
10828	1H-Pyrazole, 3,5-dimethyl-	$C_5H_8N_2$ 96.13	67-51-6 107.5	 218	5-23-05-00110 0.8839^{16}	H_2O 3; EtOH 4; eth 4; ace 3
10829	1H-Pyrazole, 3,5-dimethyl-1-phenyl-	$C_{11}H_{12}N_2$ 172.23	1131-16-4	 273; $145^{12.5}$	5-23-05-00127 1.0566^{20}	eth 4; EtOH 4; chl 4 1.5738^{19}
10830	1H-Pyrazole, 1,3-diphenyl- 1,3-Diphenylpyrazole	$C_{15}H_{12}N_2$ 220.27	4492-01-7 85.8	 341^{270}	5-23-07-00178 1.0794^{101}	
10831	1H-Pyrazole, 1,5-diphenyl- 1,5-Diphenylpyrazole	$C_{15}H_{12}N_2$ 220.27	6831-89-6 55.5	 338; 167^{12}	5-23-07-00176 1.0696^{100}	bz 4; eth 4; EtOH 4
10832	1H-Pyrazole, 1-ethyl-	$C_5H_8N_2$ 96.13	2817-71-2	 139	5-23-04-00135 0.9537^{20}	ace 4; bz 4; eth 4; EtOH 4 1.4700^{20}
10833	1H-Pyrazole, 1-ethyl-3-methyl- 1-Ethyl-3-methylpyrazole	$C_6H_{10}N_2$ 110.16	30433-57-9 115	 152	5-23-05-00013 0.936^{20}	
10834	1H-Pyrazole, 1-methyl- 1-Methylpyrazole	$C_4H_6N_2$ 82.11	930-36-9	 127	5-23-04-00134 0.9929^{13}	 1.4787^{13}
10835	1H-Pyrazole, 3-methyl-	$C_4H_6N_2$ 82.11	1453-58-3 36.5	 204; 108^{25}	5-23-05-00004 1.0203^{16}	H_2O 5; EtOH 5; eth 5 1.4915^{20}
10836	1H-Pyrazole, 4-methyl- 4-Methylpyrazole	$C_4H_6N_2$ 82.11	7554-65-6	10146 206; 95^{13}	5-23-05-00031 1.015^{20}	
10837	1H-Pyrazole, 1-methyl-3-phenyl-	$C_{10}H_{10}N_2$ 158.20	3463-26-1 56	 280.5	5-23-07-00178 1.0232^{99}	
10838	1H-Pyrazole, 3-methyl-1-phenyl-	$C_{10}H_{10}N_2$ 158.20	1128-54-7 37	 255	5-23-05-00013 1.076^{20}	
10839	1H-Pyrazole, 1,3,4-trimethyl- 1,3,4-Trimethylpyrazole	$C_6H_{10}N_2$ 110.16	15802-99-0	 160	5-23-05-00109 0.9567^{17}	 1.4866^{17}
10840	1H-Pyrazole, 1,3,5-trimethyl- 1,3,5-Trimethylpyrazole	$C_6H_{10}N_2$ 110.16	1072-91-9	 170	5-23-05-00125 0.9130^{57}	 1.4589^{57}
10841	1H-Pyrazole, 1,4,5-trimethyl- 1,4,5-Trimethylpyrazole	$C_6H_{10}N_2$ 110.16	15802-97-8	 176.5	5-23-05-00108 0.9685^{17}	 1.4849^{17}
10842	1H-Pyrazole, 3,4,5-trimethyl-	$C_6H_{10}N_2$ 110.16	5519-42-6 138.5	 233	5-23-05-00198	bz 4; eth 4; EtOH 4
10843	3,5-Pyrazolidinedione, 1,4-diphenyl- Phenopyrazone	$C_{15}H_{12}N_2O_2$ 252.27	3426-01-5 233.5	7217	5-24-08-00003	
10844	3-Pyrazolidinone, 1-phenyl- 1-Phenyl-3-pyrazolidinone	$C_9H_{10}N_2O$ 162.19	92-43-3 126	7281	5-24-01-00011	eth 1; lig 1
10845	3H-Pyrazol-3-one, 4-amino-1,2-dihydro-1,5- dimethyl-2-phenyl- Ampyrone	$C_{11}H_{13}N_3O$ 203.24	83-07-8 109	625	4-25-00-03554	H_2O 3; EtOH 3; eth 2; bz 3
10846	3H-Pyrazol-3-one, 4-bromo-1,2-dihydro-1,5- dimethyl-2-phenyl- 4-Bromoantipyrine	$C_{11}H_{11}BrN_2O$ 267.13	5426-65-3 122	 300^9	5-24-01-00343 1.5900^{20}	H_2O 3; EtOH 4; eth 3; chl 4
10847	3H-Pyrazol-3-one, 1,2-dihydro-1,5-dimethyl-2- phenyl- Antipyrine	$C_{11}H_{12}N_2O$ 188.23	60-80-0 114	748 319	5-24-01-00330	H_2O 4; EtOH 4
10848	3H-Pyrazol-3-one, 2,4-dihydro-5-methyl-4- nitro-2-(4-nitrophenyl)- Picrolonic acid	$C_{10}H_8N_4O_5$ 264.20	550-74-3 116	7383 dec		H_2O 2; EtOH 3; eth 3; MeOH 3

No.	Name Synonym	Mol. Form. Mol. Wt.	CAS RN mp/°C	Merck No. bp/°C	Beil. Ref. den/g cm^{-3}	Solubility n_D
10849	3H-Pyrazol-3-one, 1,2-dihydro-5-methyl-2-phenyl-	$C_{10}H_{10}N_2O$ 174.20	19735-89-8 128	 287^{105}; 191^{17}	5-24-01-00317 1.2600^{20}	H_2O 3; EtOH 3; bz 2; peth 1 1.637
10850	3H-Pyrazol-3-one, 1,2-dihydro-2-(1-methyl-4-piperidinyl)-5-phenyl-4-(phenylmethyl)- Benzpiperylon	$C_{22}H_{25}N_3O$ 347.46	53-89-4 181-3 dec	1131	5-24-04-00398	
10851	3H-Pyrazol-3-one, 4-(dimethylamino)-1,2-dihydro-1,5-dimethyl-2-phenyl- Aminopyrine	$C_{13}H_{17}N_3O$ 231.30	58-15-1 134.5	488	4-24-00-00096	H_2O 4; bz 4; EtOH 4
10852	4H-Pyrazolo[3,4-d]pyrimidin-4-one, 1,5-dihydro- Allopurinol	$C_5H_4N_4O$ 136.11	315-30-0 350	278	4-26-00-02076	
10853	Pyrene Benzo[def]phenanthrene	$C_{16}H_{10}$ 202.26	129-00-0 151.2	7977 404	4-05-00-02467 1.271^{23}	H_2O 1; EtOH 3; eth 3; bz 3
10854	1-Pyrenecarboxylic acid 3-Pyrenecarboxylic acid	$C_{17}H_{10}O_2$ 246.27	19694-02-1 274	 sub	3-09-00-03575	eth 4; chl 4
10855	Pyridazine 1,2-Diazabenzene	$C_4H_4N_2$ 80.09	289-80-5 -8	7982 208	5-23-05-00321 1.1035^{23}	H_2O 5; EtOH 5; eth 4; ace 4 1.5218^{20}
10856	Pyridazine, 3-methyl- 3-Methyl-1,2-diazine	$C_5H_6N_2$ 94.12	1632-76-4 184	 214	5-23-05-00371 1.0450^{26}	 1.5145^{20}
10857	2-Pyridinamine 2 Aminopyridine	$C_5H_6N_2$ 94.12	504-29-0 57.5	486 105^{20}	5-22-08-00280	EtOH 3; eth 3; ace 3; bz 3
10858	3-Pyridinamine β-Aminopyridine	$C_5H_6N_2$ 94.12	462-08-8 64.5	487 252	5-22-09-00003	H_2O 3; EtOH 3; eth 3; lig 2
10859	4-Pyridinamine 4-Aminopyridine	$C_5H_6N_2$ 94.12	504-24-5 158.5	 273	5-22-09-00106	H_2O 3; EtOH 4; eth 3; bz 3
10860	3-Pyridinamine, 6-butoxy-	$C_9H_{14}N_2O$ 166.22	539-23-1	 $148\text{-}50^{12}$	4-22-00-05574 1.037^{25}	EtOH 3; eth 3; chl 2; dil acid 3 1.5373^{20}
10861	2-Pyridinamine, 5-chloro-	$C_5H_5ClN_2$ 128.56	1072-98-6 137	 127^{11}	5-22-08-00451	H_2O 3; EtOH 3; peth 1; lig 1
10862	3-Pyridinamine, 2,6-dichloro- Pyridine, 3-amino-2,6-dichloro	$C_5H_4Cl_2N_2$ 163.01	62476-56-6 119	 $110^{0.3}$	4-22-00-04093	eth 4; EtOH 4
10863	2-Pyridinamine, *N,N*-dimethyl-	$C_7H_{10}N_2$ 122.17	5683-33-0 [illegible]	 100	5-22-08-00287 1.0110^{14}	EtOH 3; eth 3; bz 3 [illegible]
10864	4-Pyridinamine, 2,6-dimethyl-	$C_7H_{10}N_2$ 122.17	3512-80-9 192.5	 246	5-22-09-00417	H_2O 3; EtOH 3; eth 2; ace 3
10865	3-Pyridinamine, 6-methoxy-	$C_6H_8N_2O$ 124.14	6628-77-9 30	 125^{10}; 87^{1}	5-22-11-00408	 1.5745^{20}
10866	2-Pyridinamine, 3-methyl-	$C_6H_8N_2$ 108.14	1603-40-3 33.5	 222; 95^{8}	5-22-09-00272	H_2O 4; EtOH 3; eth 3; ace 3
10867	2-Pyridinamine, 4-methyl- 2-Amino-4-picoline	$C_6H_8N_2$ 108.14	695-34-1 100	478 $115\text{-}7^{11}$	5-22-09-00325	H_2O 4; EtOH 3; eth 3; ace 3
10868	2-Pyridinamine, 6-methyl-	$C_6H_8N_2$ 108.14	1824-81-3 41	 208.5	5-22-09-00210	H_2O 4; EtOH 3; eth 3; ace 3
10869	2-Pyridinamine, *N*-methyl-	$C_6H_8N_2$ 108.14	4597-87-9 15	 200.5	5-22-08-00286 1.048^{29}	H_2O 3; EtOH 4; eth 4; bz 3
10870	3-Pyridinamine, 4-methyl-	$C_6H_8N_2$ 108.14	3430-27-1 106	 255	5-22-09-00344	H_2O 2; EtOH 2; eth 2; ace 2
10871	2-Pyridinamine, *N*-2-pyridinyl-	$C_{10}H_9N_3$ 171.20	1202-34-2 90.5	 307.5	5-22-08-00415	H_2O 2; EtOH 4; eth 4; ace 4
10872	Pyridine Azine	C_5H_5N 79.10	110-86-1 -41.6	7983 115.2	5-20-05-00160 0.9819^{20}	H_2O 5; EtOH 5; eth 5; ace 5 1.5095^{20}
10873	1(4H)-Pyridineacetic acid, 3,5-diiodo-4-oxo-, propyl ester Propyliodone	$C_{10}H_{11}I_2NO_3$ 447.01	587-61-1 186	7876	5-21-07-00164	
10874	Pyridine, 2,6-bis(1,1-dimethylethyl)- 2,6-Di-*tert*-butylpyridine	$C_{13}H_{21}N$ 191.32	585-48-8	3022 $120\text{-}1^{20}$	5-20-06-00181	
10875	Pyridine, 2-bromo-	C_5H_4BrN 158.00	109-04-6	 193; 75^{13}	5-20-05-00422 1.657^{15}	H_2O 2; EtOH 3; eth 3; ctc 3 1.5734^{20}
10876	Pyridine, 3-bromo- 3-Bromopyridine	C_5H_4BrN 158.00	626-55-1 -27.3	 173; 69^{18}	5-20-05-00426 1.645^{0}	H_2O 3; EtOH 4; eth 4 1.5694^{20}
10877	Pyridine, 4-bromo- 4-Bromopyridine	C_5H_4BrN 158.00	1120-87-2 0.5	 $28\text{-}30^{0.1}$	5-20-05-00431 1.6450^{0}	ace 3; bz 3 1.5694^{20}
10878	Pyridine, 2-butyl- 2-Butylpyridine	$C_9H_{13}N$ 135.21	5058-19-5 97	 192	5-20-06-00118 0.9135^{15}	
10879	Pyridine, 4-butyl-	$C_9H_{13}N$ 135.21	5335-75-1	 209; 98^{20}	5-20-06-00120 0.9157^{20}	 1.4946^{20}
10880	Pyridine, 4-(1-butylpentyl)-	$C_{14}H_{23}N$ 205.34	2961-47-9	 181^{50}	5-20-06-00189 0.8878^{25}	 1.4846^{25}
10881	2-Pyridinecarbonitrile	$C_6H_4N_2$ 104.11	100-70-9 29	 224.5	5-22-02-00019 1.0810^{25}	H_2O 3; EtOH 4; eth 4; bz 4 1.5242^{25}
10882	3-Pyridinecarbonitrile	$C_6H_4N_2$ 104.11	100-54-9 51	 206.9; 170^{300}	5-22-02-00115 1.1590^{25}	H_2O 4; EtOH 4; eth 4; bz 4
10883	3-Pyridinecarbonitrile, 1,2-dihydro-4-methoxy-1-methyl-2-oxo- Ricinine	$C_8H_8N_2O_2$ 164.16	524-40-3 201.5	8212 sub 170	5-22-07-00318	H_2O 3; EtOH 2; bz 2; chl 3

No.	Name Synonym	Mol. Form. Mol. Wt.	CAS RN mp/°C	Merck No. bp/°C	Beil. Ref. den/g cm^{-3}	Solubility n_D
10884	3-Pyridinecarbonitrile, 1,2-dihydro-1-methyl-2-oxo-	$C_7H_6N_2O$ 134.14	767-88-4 140	 243^{28}	5-22-06-00129	EtOH 4
10885	4-Pyridinecarbothioamide, 2-ethyl- Ethionamide	$C_8H_{10}N_2S$ 166.25	536-33-4 163	3692	5-22-02-00360	
10886	4-Pyridinecarbothioamide, 2-propyl- Protionamide	$C_9H_{12}N_2S$ 180.27	14222-60-7 136.7	7905	5-22-02-00376	
10887	2-Pyridinecarboxaldehyde	C_6H_5NO 107.11	1121-60-4	 180; 62^{13}	5-21-07-00293 1.1255^{18}	H_2O 3; EtOH 3; eth 3; ctc 2 1.5389^{18}
10888	3-Pyridinecarboxaldehyde	C_6H_5NO 107.11	500-22-1	 89.5^{14}	5-21-07-00334 1.1394^{25}	H_2O 3; EtOH 3; eth 2; ace 3
10889	4-Pyridinecarboxaldehyde	C_6H_5NO 107.11	872-85-5	 $77\text{-}8^{12}$	5-21-07-00351	H_2O 3; eth 3; ctc 3 1.5423^{20}
10890	3-Pyridinecarboxamide Niacinamide	$C_6H_6N_2O$ 122.13	98-92-0 130	6398 $157^{0.0005}$	5-22-02-00080 1.400^{25}	H_2O 4; EtOH 4; chl 2; glycerol 4 1.466
10891	3-Pyridinecarboxamide, *N,N*-diethyl- Nikethamide	$C_{10}H_{14}N_2O$ 178.23	59-26-7 25	6459 280 dec; 175^{25}	5-22-02-00084 1.060^{25}	DMSO-d_6 2 1.525^{20}
10892	4-Pyridinecarboxamide, *N,N*-diethyl- Isonicotinic acid diethylamide	$C_{10}H_{14}N_2O$ 178.23	530-40-5	5074 $119\text{-}20^{1}$	5-22-02-00196	H_2O 4; ace 4; eth 4; EtOH 4 1.525^{20}
10893	2-Pyridinecarboxylic acid Picolinic acid	$C_6H_5NO_2$ 123.11	98-98-6 136.5	7375 sub	5-22-02-00003	H_2O 2; EtOH 3; eth 1; bz 2
10894	3-Pyridinecarboxylic acid Nicotinic acid	$C_6H_5NO_2$ 123.11	59-67-6 236.6	6435 sub	5-22-02-00057 1.473^{25}	H_2O 2; EtOH 2; eth 2
10895	4-Pyridinecarboxylic acid Isonicotinic acid	$C_6H_5NO_2$ 123.11	55-22-1 315	5073 sub 260	5-22-02-00188	H_2O 2; EtOH 2; eth 2; bz 2
10896	2-Pyridinecarboxylic acid, 4-amino-3,5,6-trichloro- Picloram	$C_6H_3Cl_3N_2O_2$ 241.46	1918-02-1 218.5	7370	5-22-13-00585	
10897	3-Pyridinecarboxylic acid, 6-amino- 6-Aminonicotinic acid	$C_6H_6N_2O_2$ 138.13	3167-49-5 312	468	4-22-00-06726	
10898	2-Pyridinecarboxylic acid, 5-butyl- Fusaric acid	$C_{10}H_{13}NO_2$ 179.22	536-69-6 97	4227	5-22-02-00384	
10899	3-Pyridinecarboxylic acid, 1,6-dihydro-6-oxo-	$C_6H_5NO_3$ 139.11	5006-66-6 310 dec	 sub	5-22-06-00119	H_2O 2; EtOH 1; eth 1; bz 1
10900	4-Pyridinecarboxylic acid, 1,2-dihydro-6-hydroxy-2-oxo- Citrazinic acid	$C_6H_5NO_4$ 155.11	99-11-6 >300 dec	2327	5-22-07-00024	H_2O 3; alk 3; HCl 2
10901	2-Pyridinecarboxylic acid, ethyl ester	$C_8H_9NO_2$ 151.17	2524-52-9 1	 243	5-22-02-00004 1.1194^{20}	H_2O 4; eth 4; EtOH 4 1.5104^{20}
10902	3-Pyridinecarboxylic acid, ethyl ester	$C_8H_9NO_2$ 151.17	614-18-6 8.5	 224	5-22-02-00060 1.1070^{20}	H_2O 4; EtOH 4; eth 4; bz 4 1.5024^{20}
10903	4-Pyridinecarboxylic acid, ethyl ester	$C_8H_9NO_2$ 151.17	1570-45-2 23	 219.5	5-22-02-00190 1.0091^{15}	H_2O 2; EtOH 3; eth 4; bz 3 1.5017^{20}
10904	4-Pyridinecarboxylic acid, [1-(2-furanyl)ethylidene]hydrazide Furonazide	$C_{12}H_{11}N_3O_2$ 229.24	3460-67-1 202.3	4220	5-22-02-00264	
10905	4-Pyridinecarboxylic acid, hydrazide Isoniazid	$C_6H_7N_3O$ 137.14	54-85-3 171.4	5071	5-22-02-00219	H_2O 4; EtOH 4
10906	3-Pyridinecarboxylic acid, methyl ester Methyl nicotinate	$C_7H_7NO_2$ 137.14	93-60-7 42.5	6014 204	5-22-02-00059	H_2O 3; EtOH 3; bz 3
10907	3-Pyridinecarboxylic acid, 1-methylethyl ester	$C_9H_{11}NO_2$ 165.19	553-60-6	 126^{30}; 92.5^{5}	5-22-02-00060 1.0624^{20}	 1.4926^{20}
10908	4-Pyridinecarboxylic acid, methyl ester	$C_7H_7NO_2$ 137.14	2459-09-8 16.1	 208	5-22-02-00189 1.1599^{20}	H_2O 2; EtOH 3; eth 3; bz 3 1.5135^{20}
10909	3-Pyridinecarboxylic acid, 1-oxide Oxiniacic acid	$C_6H_5NO_3$ 139.11	2398-81-4 254-5 dec	6895	5-22-02-00140	H_2O 4; MeOH 4
10910	4-Pyridinecarboxylic acid, 2-[3-oxo-3-[(phenylmethyl)amino]propyl]hydrazide Nialamide	$C_{16}H_{18}N_4O_2$ 298.34	51-12-7 151.6	6399	5-22-02-00251	
10911	3-Pyridinecarboxylic acid, phenylmethyl ester Nicotinic acid benzyl ester	$C_{13}H_{11}NO_2$ 213.24	94-44-0	6436 170^{3}	5-22-02-00062	
10912	3-Pyridinecarboxylic acid, 1,2,5,6-tetrahydro- Guvacine	$C_6H_9NO_2$ 127.14	498-96-4 295 dec	4495	5-22-01-00322	H_2O 4
10913	3-Pyridinecarboxylic acid, 1,2,5,6-tetrahydro-1-methyl- Arecaidine	$C_7H_{11}NO_2$ 141.17	499-04-7 232 dec	801	5-22-01-00322	H_2O 4; EtOH 1; eth 1; bz 1
10914	3-Pyridinecarboxylic acid, 1,2,5,6-tetrahydro-1-methyl-, methyl ester Arecoline	$C_8H_{13}NO_2$ 155.20	63-75-2	802 209	5-22-01-00322 1.0485^{20}	H_2O 5; EtOH 5; eth 5; chl 3 1.486^{20}
10915	Pyridine, 2-chloro- 2-Chloropyridine	C_5H_4ClN 113.55	109-09-1	 170	5-20-05-00402 1.205^{15}	H_2O 2; EtOH 3; eth 3 1.5320^{20}
10916	Pyridine, 3-chloro- 3-Chloropyridine	C_5H_4ClN 113.55	626-60-8	 148; 86^{100}	5-20-05-00406	H_2O 2 1.5304^{20}
10917	Pyridine, 4-chloro- 4-Chloropyridine	C_5H_4ClN 113.55	626-61-9 -43.5	 147.5	5-20-05-00410 1.2000^{25}	H_2O 3; EtOH 5
10918	Pyridine, 2-chloro-6-methoxy-	C_6H_6ClNO 143.57	17228-64-7	 185.5	5-21-02-00018 1.207^{25}	 1.5263^{20}
10919	Pyridine, 2-chloro-4-methyl-	C_6H_6ClN 127.57	3678-62-4 115	 192	5-20-05-00579 1.1459^{20}	 1.5293^{8}

PHYSICAL CONSTANTS OF ORGANIC COMPOUNDS (continued)

No.	Name Synonym	Mol. Form. Mol. Wt.	CAS RN mp/°C	Merck No. bp/°C	Beil. Ref. den/g cm^{-3}	Solubility n_D
10920	Pyridine, 2-chloro-6-methyl-	C_6H_6ClN 127.57	18368-63-3	$183.5; 64^{10}$	5-20-05-00498 1.167^{25}	1.5270^{20}
10921	Pyridine, 2-[(4-chlorophenyl)methyl]-	$C_{12}H_{10}ClN$ 203.67	4350-41-8	$181\text{-}3^{20}$	5-20-07-00558 1.39^{25}	1.5868^{20}
10922	Pyridine, 4-(3-cyclohexen-1-yl)- 4-(Cyclohex-3-enyl)pyridine	$C_{11}H_{13}N$ 159.23	70644-46-1 22.1	226	4-20-00-03239 1.0222^{25}	1.5466^{25}
10923	2,3-Pyridinediamine	$C_5H_7N_3$ 109.13	452-58-4 120.8	$148\text{-}50^{5}$	5-22-11-00241	H_2O 3; EtOH 3; bz 3
10924	2,5-Pyridinediamine 2,5-Diaminopyridine	$C_5H_7N_3$ 109.13	4318-76-7 110.3	182^{12}	5-22-11-00252	H_2O 4; EtOH 4
10925	2,6-Pyridinediamine	$C_5H_7N_3$ 109.13	141-86-6 121.5	$285; 148^{5}$	5-22-11-00255	H_2O 2; ace 2
10926	Pyridine, 3,5-dibromo- 3,5-Dibromopyridine	$C_5H_3Br_2N$ 236.89	625-92-3 112	222	5-20-05-00436	H_2O 2; EtOH 3; eth 3
10927	2,3-Pyridinedicarboxylic acid Quinolinic acid	$C_7H_5NO_4$ 167.12	89-00-9 228.5	8102	5-22-04-00115	H_2O 2; EtOH 1; eth 1; bz 1
10928	2,4-Pyridinedicarboxylic acid	$C_7H_5NO_4$ 167.12	499-80-9 249		5-22-04-00119 0.942^{25}	H_2O 2; EtOH 3; eth 1; bz 1
10929	2,5-Pyridinedicarboxylic acid Isocinchomeronic acid	$C_7H_5NO_4$ 167.12	100-26-5 254	5043	5-22-04-00124	H_2O 3; EtOH 2; eth 1; bz 1
10930	3,4-Pyridinedicarboxylic acid Cinchomeronic acid	$C_7H_5NO_4$ 167.12	490-11-9 256	2285 sub	5-22-04-00135	H_2O 2; EtOH 2; eth 1; bz 2
10931	3,5-Pyridinedicarboxylic acid	$C_7H_5NO_4$ 167.12	499-81-0 324	sub	5-22-04-00139	H_2O 1; eth 2; HCl 3; HOAc 2
10932	3,5-Pyridinedicarboxylic acid, 2,6-dimethyl-, diethyl ester	$C_{13}H_{17}NO_4$ 251.28	1149-24-2 71	$301; 208^{40}$	5-22-04-00162	H_2O 1; EtOH 3; eth 3; bz 3
10933	Pyridine, 2,4-diethyl- 2,4-Diethylpyridine	$C_9H_{13}N$ 135.21	626-21-1 112	187.5	4-20-00-02825 0.9338^{25}	
10934	3,4-Pyridinedimethanol, 5-hydroxy-6-methyl-, hydrochloride Pyridoxine hydrochloride	$C_8H_{12}ClNO_3$ 205.64	58-56-0 207	7995 sub	5-21-05-00492	H_2O 4
10935	Pyridine, 2,6-dimethoxy- 2,6-Dimethoxypyridine	$C_7H_9NO_2$ 139.15	6231-18-1	179	5-21-04-00416 1.053^{25}	1.5029^{20}
10936	Pyridine, 2,3-dimethyl- 2,3-Lutidine	C_7H_9N 107.16	583-61-9	161.2	5-20-06-00015 0.9319^{25}	H_2O 3; EtOH 3; eth 3 1.5057^{20}
10937	Pyridine, 2,4-dimethyl- 2,4-Lutidine	C_7H_9N 107.16	108-47-4 -64	158.5	5-20-06-00019 0.9309^{20}	H_2O 4; EtOH 4; eth 4; ace 3 1.5010^{20}
10938	Pyridine, 2,5-dimethyl- 2,5-Lutidine	C_7H_9N 107.16	589-93-5 -16	157.0	5-20-06-00027 0.9297^{20}	H_2O 2; EtOH 4; eth 5; ace 3 1.5006^{20}
10939	Pyridine, 2,6-dimethyl- 2,6-Lutidine	C_7H_9N 107.16	108-48-5 -6.1	5485 144.1	5-20-06-00032 0.9226^{20}	H_2O 5; EtOH 2; eth 3; ace 3 1.4953^{20}
10940	Pyridine, 3,4-dimethyl- 3,4-Lutidine	C_7H_9N 107.16	583-58-4 -11	179.1	5-20-06-00054 0.9281^{20}	H_2O 2; EtOH 3; eth 3; ace 3 1.5096^{20}
10941	Pyridine, 3,5-dimethyl- 3,5-Lutidine	C_7H_9N 107.16	591-22-0 -6.6	172	5-20-06-00060 0.9419^{20}	H_2O 3; EtOH 3; eth 3; ace 3 1.5061^{20}
10942	Pyridine, 2-(1,1-dimethylethyl)- 2-*tert*-Butylpyridine	$C_9H_{13}N$ 135.21	5944-41-2 -33	170	5-20-06-00121	1.4891^{20}
10943	Pyridine, 3-(1,1-dimethylethyl)- 3-*tert*-Butylpyridine	$C_9H_{13}N$ 135.21	38031-78-6 -44	194.8	5-20-06-00122	1.4965^{20}
10944	Pyridine, 4-(1,1-dimethylethyl)-	$C_9H_{13}N$ 135.21	3978-81-2 -41	196.5	5-20-06-00123 0.915^{25}	ctc 3; CS_2 3 1.4958^{20}
10945	Pyridine, 2,6-dimethyl-, 1-oxide	C_7H_9NO 123.15	1073-23-0	$115\text{-}9^{18}$	5-20-06-00040 1.073^{25}	1.5706^{20}
10946	2,4(1H,3H)-Pyridinedione, 3,3-diethyl- Pyrithyldione	$C_9H_{13}NO_2$ 167.21	77-04-3 90.7	8005	5-21-10-00086	
10947	Pyridine, 2,6-diphenyl- 2,6-Diphenylpyridine	$C_{17}H_{13}N$ 231.30	3558-69-8 82	397	5-20-08-00459	eth 4; EtOH 4
10948	2-Pyridineethanamine	$C_7H_{10}N_2$ 122.17	2706-56-1	131.5^{50}	5-22-09-00368 1.0220^{25}	1.5335^{25}
10949	4-Pyridineethanamine	$C_7H_{10}N_2$ 122.17	13258-63-4	121^{10}	4-22-00-04201 1.0302^{25}	H_2O 4 1.5381^{25}
10950	2-Pyridineethanol	C_7H_9NO 123.15	103-74-2 -7.8	190^{200}; 170^{100}	5-21-02-00206 1.091^{25}	H_2O 4; EtOH 4; eth 2; chl 4 1.5366^{20}
10951	2-Pyridineethanol, α-methyl- 2-(2-Hydroxypropyl)pyridine	$C_8H_{11}NO$ 137.18	5307-19-7 33	124^{20}; 110^{10}	5-21-02-00234	H_2O 4; EtOH 4; chl 4
10952	Pyridine, 2-ethenyl-	C_7H_7N 105.14	100-69-6	159.5	5-20-06-00211 0.9983^{20}	H_2O 2; EtOH 4; eth 4; ace 4 1.5495^{20}
10953	Pyridine, 3-ethenyl-	C_7H_7N 105.14	1121-55-7	162	5-20-06-00212	H_2O 2; EtOH 3; eth 3 1.5530^{20}
10954	Pyridine, 4-ethenyl-	C_7H_7N 105.14	100-43-6	121^{150}; 65^{15}	5-20-06-00213 0.9800^{20}	H_2O 3; EtOH 3; eth 2; chl 3 1.5449^{20}
10955	Pyridine, 5-ethenyl-2-methyl- 5-Vinyl-2-picoline	C_8H_9N 119.17	140-76-1		5-20-06-00218	
10956	Pyridine, 2-ethyl- 2-Ethylpyridine	C_7H_9N 107.16	100-71-0 -63.1	148.6	5-20-06-00003 0.9502^{25}	H_2O 3; EtOH 5; eth 4; ace 4 1.4964^{20}
10957	Pyridine, 3-ethyl- 3-Ethylpyridine	C_7H_9N 107.16	536-78-7 -76.9	3802 165	5-20-06-00007 0.9539^{25}	H_2O 3; EtOH 3; eth 3; ace 4 1.5021^{20}

No.	Name Synonym	Mol. Form. Mol. Wt.	CAS RN mp/°C	Merck No. bp/°C	Beil. Ref. den/g cm^{-3}	Solubility n_D
10958	Pyridine, 4-ethyl- 4-Ethylpyridine	C_7H_9N 107.16	536-75-4 -90.5	3803 168.3	5-20-06-00010 0.9417^{20}	H_2O 3; EtOH 3; eth 3; ace 4 1.5009^{20}
10959	Pyridine, 2-ethyl-3,5-dimethyl-	$C_9H_{13}N$ 135.21	1123-96-2	 198; 86^{15}	5-20-06-00129 0.9338^{25}	H_2O 3; EtOH 4; eth 4; ace 4
10960	Pyridine, 3-ethyl-2,6-dimethyl-	$C_9H_{13}N$ 135.21	23580-52-1	 $83\text{-}5^{23}$	5-20-06-00129 0.9120^{18}	H_2O 2
10961	Pyridine, 4-ethyl-2,6-dimethyl-	$C_9H_{13}N$ 135.21	36917-36-9 97	 187.5	5-20-06-00130 0.9089^{25}	 1.4964^{25}
10962	Pyridine, 2-ethyl-4-methyl-	$C_8H_{11}N$ 121.18	2150-18-7	 175	5-20-06-00086 0.9238^{20}	ace 4; eth 4; EtOH 4
10963	Pyridine, 2-ethyl-6-methyl-	$C_8H_{11}N$ 121.18	1122-69-6	 160.5	5-20-06-00089 0.9207^{25}	H_2O 2; EtOH 3; eth 3; ace 4 1.4920^{25}
10964	Pyridine, 3-ethyl-4-methyl- 3-Ethyl-4-picoline	$C_8H_{11}N$ 121.18	529-21-5	3796 198	5-20-06-00090 0.9286^{17}	H_2O 2; EtOH 3; eth 3; ace 4
10965	Pyridine, 4-ethyl-2-methyl- 4-Ethyl-2-picoline	$C_8H_{11}N$ 121.18	536-88-9	3797 179	5-20-06-00085 0.9130^{25}	ace 4; bz 4; eth 4; EtOH 4
10966	Pyridine, 5-ethyl-2-methyl- 5-Ethyl-2-picoline	$C_8H_{11}N$ 121.18	104-90-5	3798 178.3	5-20-06-00086 0.9202^{20}	H_2O 2; EtOH 3; eth 3; ace 4 1.4971^{20}
10967	Pyridine, 2-(1-ethylpropyl)-	$C_{10}H_{15}N$ 149.24	7399-50-0	 195.4	5-20-06-00145 0.8981^{20}	 1.4850^{25}
10968	Pyridine, 4-(1-ethylpropyl)- 4-(3-Pentyl)pyridine	$C_{10}H_{15}N$ 149.24	35182-51-5 125.5	 217; 80^{12}	5-20-06-00145 0.9085^{25}	 1.40905^{25}
10969	Pyridine, hydrochloride Pyridinium chloride	C_5H_6ClN 115.56	628-13-7 146	 222	5-20-05-00160	H_2O 4; EtOH 4; chl 4
10970	Pyridine, 2-iodo- 2-Iodopyridine	C_5H_4IN 205.00	5029-67-4	 93^{13}	5-20-05-00438 1.9735^{20}	EtOH 3; eth 3; ace 3; bz 3 1.6366^{20}
10971	2-Pyridinemethanamine	$C_6H_8N_2$ 108.14	3731-51-9	 203; 91^{17}	5-22-09-00232 1.0525^{25}	H_2O 4 1.5431^{25}
10972	3-Pyridinemethanamine	$C_6H_8N_2$ 108.14	3731-52-0 -21.1	 226	5-22-09-00301 1.064^{20}	H_2O 4; eth 4; EtOH 4 1.552^{20}
10973	4-Pyridinemethanamine	$C_6H_8N_2$ 108.14	3731-53-1 -7.6	 230; 103^{11}	5-22-09-00345 1.072^{20}	H_2O 4 1.5495^{25}
10974	2-Pyridinemethanamine, *N*-(2-pyridinylmethyl)-	$C_{12}H_{13}N_3$ 199.26	1539-42-0	 200^{10}	5-22-09-00267 1.1074^{25}	 1.5757^{25}
10975	2-Pyridinemethanol	C_6H_7NO 109.13	586-98-1	 112^{16}, 102.5^{8}	5-21-02-00150 1.1317^{20}	H_2O 5; EtOH 4; eth 4; ace 4 1.5444^{20}
10976	3-Pyridinemethanol Nicotinyl alcohol	C_6H_7NO 109.13	100-55-0 -6.5	6438 266	5-21-02-00172 1.131^{20}	H_2O 4; eth 4 1.5455^{20}
10977	4-Pyridinemethanol 4-Picolyl alcohol	C_6H_7NO 109.13	586-95-8 53	 $140\text{-}2^{12}$	5-21-02-00191	chl 3
10978	3-Pyridinemethanol, 4-(aminomethyl)-5-hydroxy-6-methyl-, dihydrochloride Pyridoxamine dihydrochloride	$C_8H_{14}Cl_2N_2O_2$ 241.12	524-36-7 226-7 dec	7993	5-22-12-00324	H_2O 4; EtOH 2
10979	Pyridine, 2-(2-methoxyethyl)- Metyridine	$C_8H_{11}NO$ 137.18	114-91-0	6084 203; 96^{17}	5-21-02-00206 0.988^{20}	H_2O 4; EtOH 4 1.4975^{20}
10980	Pyridine, 2-methyl- 2-Picoline	C_6H_7N 93.13	109-06-8 -66.7	7372 129.3	5-20-05-00464 0.9443^{20}	H_2O 4; EtOH 5; eth 5; ace 4 1.4957^{20}
10981	Pyridine, 3-methyl- 3-Picoline	C_6H_7N 93.13	108-99-6 -18.1	7373 144.1	5-20-05-00506 0.9566^{20}	H_2O 5; EtOH 5; eth 5; ace 4 1.5040^{20}
10982	Pyridine, 4-methyl- 4-Picoline	C_6H_7N 93.13	108-89-4 3.66	7374 145.3	5-20-05-00543 0.9548^{20}	H_2O 5; EtOH 5; eth 5; ace 3 1.5037^{20}
10983	Pyridine, 2-[2-(1-methylethoxy)ethyl]- Pyridine, 2-(2-isopropoxyethyl)	$C_{10}H_{15}NO$ 165.24	70715-19-4	 133^{50}	 0.9502^{25}	H_2O 4 1.4820^{25}
10984	Pyridine, 2-(1-methylethyl)-	$C_8H_{11}N$ 121.18	644-98-4 -141	 159.8	5-20-06-00080 0.9342^{25}	H_2O 2; EtOH 5; eth 5; ace 4 1.4915^{20}
10985	Pyridine, 3-(1-methylethyl)- 3-Isopropylpyridine	$C_8H_{11}N$ 121.18	6304-18-3 -45.8	 177.5	5-20-06-00081 0.9217^{16}	
10986	Pyridine, 4-(1-methylethyl)-	$C_8H_{11}N$ 121.18	696-30-0 -54.9	 178	5-20-06-00082 0.9382^{25}	H_2O 2; EtOH 5; eth 5; ace 4 1.4962^{20}
10987	Pyridine, 3-(1-methyl-1H-pyrrol-2-yl)-	$C_{10}H_{10}N_2$ 158.20	487-19-4	 281; 150^{16}	5-23-07-00227 1.241^{20}	EtOH 4 1.6057^{20}
10988	Pyridine, 3-methyl-, 1-oxide	C_6H_7NO 109.13	1003-73-2 39	 148^{15}	5-20-05-00517	chl 3
10989	Pyridine, 2-(4-methylphenyl)-	$C_{12}H_{11}N$ 169.23	4467-06-5	 $170\text{-}80^{20}$	5-20-07-00569 0.999^{25}	 1.6125^{20}
10990	Pyridine, 2-methyl-5-phenyl-	$C_{12}H_{11}N$ 169.23	3256-88-0	 189^{50}	5-20-07-00573 1.0590^{25}	 1.6055^{25}
10991	Pyridine, 3-methyl-2-phenyl- 3-Picoline, 2-phenyl-	$C_{12}H_{11}N$ 169.23	10273-90-2	 $148\text{-}9^{16}$	5-20-07-00571 1.065^{25}	 1.6026^{20}
10992	Pyridine, 3-(1-methyl-2-pyrrolidinyl)-, (*S*)- Nicotine	$C_{10}H_{14}N_2$ 162.23	54-11-5 -79	6434 247; 125^{18}	5-23-06-00064 1.0097^{20}	H_2O 5; EtOH 4; eth 4; chl 4 1.5282^{20}
10993	Pyridine, 4-nitro-, 1-oxide 4-Nitropyridine *N*-oxide	$C_5H_4N_2O_3$ 140.10	1124-33-0 160.5		5-20-05-00448	
10994	Pyridine, 1-oxide Pyridine *N*-oxide	C_5H_5NO 95.10	694-59-7 65.5	7984 $146\text{-}7^{13}$	5-20-05-00217	
10995	Pyridine, pentachloro-	C_5Cl_5N 251.33	2176-62-7 125.5	 280	5-20-05-00422	bz 4; EtOH 4; lig 4
10996	Pyridine, 2-(3-pentenyl)-	$C_{10}H_{13}N$ 147.22	2057-43-4	 216; 93^{12}	5-20-06-00355 0.9234^{25}	 1.5076^{25}

No.	Name Synonym	Mol. Form. Mol. Wt.	CAS RN mp/°C	Merck No. bp/°C	Beil. Ref. den/g cm^{-3}	Solubility n_D
10997	Pyridine, 2-phenyl- 2-Phenylpyridine	$C_{11}H_9N$ 155.20	1008-89-5	 271	5-20-07-00545 1.0833^{25}	H_2O 2; EtOH 5; eth 5 1.6210^{20}
10998	Pyridine, 3-phenyl- 3-Phenylpyridine	$C_{11}H_9N$ 155.20	1008-88-4 164	 272	5-20-07-00548	H_2O 2; EtOH 3; eth 3 1.6123^{25}
10999	Pyridine, 4-phenyl- 4-Phenylpyridine	$C_{11}H_9N$ 155.20	939-23-1 77.5	 281	5-20-07-00549	H_2O 3; EtOH 3; eth 3
11000	Pyridine, 2-(2-phenylethenyl)-, (*E*)-	$C_{13}H_{11}N$ 181.24	538-49-8 91.5	 325; 194^{14}	5-20-08-00057 1.0147^{100}	H_2O 1; EtOH 3; eth 3; bz 3
11001	Pyridine, 4-(2-phenylethenyl)-, (*E*)-	$C_{13}H_{11}N$ 181.24	5097-93-8 131	 208^{33}	5-20-08-00063	H_2O 1; EtOH 3; eth 3; chl 3
11002	Pyridine, 2-(2-phenylethenyl)-, (*Z*)-	$C_{13}H_{11}N$ 181.24	1519-59-1 -50	 141^{10}	4-20-00-03874	EtOH 3; eth 3; bz 3
11003	Pyridine, 2-(2-phenylethyl)-	$C_{13}H_{13}N$ 183.25	2116-62-3 -3	 289	5-20-07-00588 1.0465^{0}	
11004	Pyridine, 2-(phenylmethyl)- 2-Benzylpyridine	$C_{12}H_{11}N$ 169.23	101-82-6 12.5	 277; 149^{16}	5-20-07-00556 1.067^{0}	H_2O 1; EtOH 3; eth 3; chl 3 1.5785^{20}
11005	Pyridine, 3-(phenylmethyl)- 3-Benzylpyridine	$C_{12}H_{11}N$ 169.23	620-95-1 34	 287	5-20-07-00560 1.061^{25}	eth 4; EtOH 4
11006	Pyridine, 4-(phenylmethyl)-	$C_{12}H_{11}N$ 169.23	2116-65-6 12.4	 288; 180^{31}	5-20-07-00561 1.0612^{20}	H_2O 1; EtOH 3; eth 4; ctc 3 1.5818^{20}
11007	Pyridine, 4-(3-phenylpropyl)-	$C_{14}H_{15}N$ 197.28	2057-49-0	 322; 150^{5}	5-20-07-00607 1.024^{25}	bz 4; eth 4; py 4; EtOH 4 1.5616^{25}
11008	Pyridine, 3-(2-piperidinyl)-, (*S*)- Anabasine	$C_{10}H_{14}N_2$ 162.23	494-52-0 9	655 276	5-23-06-00096 1.0436^{20}	H_2O 5; EtOH 3; eth 3; bz 3 1.5430^{20}
11009	Pyridine, 4-(4-piperidinyl)-	$C_{10}H_{14}N_2$ 162.23	581-45-3 79	 292	5-23-06-00115	H_2O 2; EtOH 4; eth 3; bz 3
11010	2-Pyridinepropanamine, *N*,*N*-dimethyl-γ-phenyl- Pheniramine	$C_{16}H_{20}N_2$ 240.35	86-21-5	7198 181^{13}; $135^{0.5}$	5-22-10-00487 1.0081^{25}	bz 4; eth 4; EtOH 4; chl 4 1.5519^{25}
11011	1(4H)-Pyridinepropanoic acid, α-amino-3-hydroxy-4-oxo-, (*S*)- Mimosine	$C_8H_{10}N_2O_4$ 198.18	500-44-7 228 dec	6118	5-21-12-00107	H_2O 2; EtOH 1; eth 1; ace 1
11012	2-Pyridinepropanol 2-(3-Hydroxypropyl)pyridine	$C_8H_{11}NO$ 137.18	2859-68-9 34	 260.2; 116^{4}	5-21-02-00236 1.060^{25}	H_2O 4 1.5298^{20}
11013	3-Pyridinepropanol	$C_8H_{11}NO$ 137.18	2859-67-8	 284; 130^{3}	5-21-02-00237 1.063^{25}	H_2O 4 1.5313^{20}
11014	Pyridine, 2-(2-propenyl)-	C_8H_9N 119.17	2835-33-8	 190; 63^{15}	5-20-06-00216 0.959^{20}	eth 4; EtOH 4
11015	Pyridine, 2-propyl-	$C_8H_{11}N$ 121.18	622-39-9 2	 167	5-20-06-00076 0.9119^{20}	H_2O 2; EtOH 5; eth 5; ace 4 1.4925^{20}
11016	Pyridine, 4-propyl-	$C_8H_{11}N$ 121.18	1122-81-2	 185	5-20-06-00078 0.9381^{15}	eth 4; EtOH 4 1.4966^{20}
11017	Pyridine, 3-(2-pyrrolidinyl)-, (*S*)- Nornicotine	$C_9H_{12}N_2$ 148.21	494-97-3	6631 270	4-23-00-00998 1.0737^{19}	H_2O 4; ace 4; eth 4; EtOH 4 1.5378^{18}
11018	3-Pyridinesulfonic acid 3-Pyridylsulfonic acid	$C_5H_5NO_3S$ 159.17	636-73-7 357 dec		5-22-07-00552 1.713^{25}	H_2O 4; EtOH 2; eth 1
11019	Pyridine, 2,3,5,6-tetrachloro- 2,3,5,6-Tetrachloropyridine	C_5HCl_4N 216.88	2402-79-1 90.5	 250.5	5-20-05-00421	eth 4; EtOH 4; peth 4
11020	Pyridine, 1,2,3,6-tetrahydro- Δ 3-Piperidine	C_5H_9N 83.13	694-05-3 -48	 108	5-20-04-00282 0.911^{25}	chl 3 1.4800^{20}
11021	Pyridine, 2,3,4,5-tetrahydro-6-propyl- γ-Coniceine	$C_8H_{15}N$ 125.21	1604-01-9	2497 174	5-20-04-00345 0.8753^{15}	 1.4661^{16}
11022	Pyridine, 2,3,4,6-tetramethyl- 2,3,4,6-Tetramethylpyridine	$C_9H_{13}N$ 135.21	20820-82-0	 203	5-20-06-00132 0.9322^{25}	eth 4; EtOH 4 1.5087^{25}
11023	Pyridine, 2-(trichloromethyl)- 2-(Trichloromethyl)pyridine	$C_6H_4Cl_3N$ 196.46	4377-37-1 -10	 $125\text{-}6^{25}$	5-20-05-00499 1.4526^{25}	 1.5596^{25}
11024	Pyridine, 2,3,4-trimethyl- 2,3,4-Collidine	$C_8H_{11}N$ 121.18	2233-29-6	 192.5	5-20-06-00091 0.9127^{15}	H_2O 3; EtOH 3; eth 3 1.5150^{20}
11025	Pyridine, 2,3,5-trimethyl- 2,3,5-Collidine	$C_8H_{11}N$ 121.18	695-98-7	 184	5-20-06-00092 0.9352^{19}	H_2O 2; EtOH 3; eth 3; ace 3 1.5057^{25}
11026	Pyridine, 2,3,6-trimethyl- 2,3,6-Collidine	$C_8H_{11}N$ 121.18	1462-84-6	 171.6	5-20-06-00092 0.9220^{25}	H_2O 3; EtOH 3; eth 3; ace 3 1.5053^{20}
11027	Pyridine, 2,4,5-trimethyl- 2,4,5-Collidine	$C_8H_{11}N$ 121.18	1122-39-0	 188	5-20-06-00093 0.9330^{25}	ace 4; bz 4; eth 4; EtOH 4 1.5054^{25}
11028	Pyridine, 2,4,6-trimethyl- 2,4,6-Collidine	$C_8H_{11}N$ 121.18	108-75-8 -46	9632 170.6	5-20-06-00093 0.9166^{22}	H_2O 3; EtOH 3; eth 3; ace 3 1.4959^{25}
11029	Pyridinium, 3-carboxy-1-methyl-, hydroxide, inner salt Trigonelline	$C_7H_7NO_2$ 137.14	535-83-1	9606	5-22-02-00143	H_2O 4
11030	Pyridinium, 1-hexadecyl-, chloride Hexadecyl pyridinium chloride	$C_{21}H_{38}ClN$ 339.99	123-03-5 80	2024	5-20-05-00233	H_2O 4; chl 4
11031	2-Pyridinol	C_5H_5NO 95.10	72762-00-6 107.8		5-21-07-00106 1.3910^{20}	H_2O 4; bz 4; EtOH 4
11032	4-Pyridinol	C_5H_5NO 95.10	626-64-2 149.8	 >350; 257^{10}	5-21-07-00152	H_2O 3; EtOH 3; eth 1; bz 1
11033	2(1H)-Pyridinone	C_5H_5NO 95.10	142-08-5 107.8	 280	5-21-07-00106 1.3910^{20}	H_2O 3; EtOH 3; eth 2; bz 3
11034	2(1H)-Pyridinone, hydrazone	$C_5H_7N_3$ 109.13	4930-98-7 46.6	 185^{140}; 90^{1}	5-22-14-00486	chl 3

No.	Name Synonym	Mol. Form. Mol. Wt.	CAS RN mp/°C	Merck No. bp/°C	Beil. Ref. den/g cm^{-3}	Solubility n_D
11035	2(1H)-Pyridinone, 1-methyl-	C_6H_7NO 109.13	694-85-9 31	 250	5-21-07-00109 1.1120^{20}	H_2O 5; peth 2; lig 2
11036	2(1H)-Pyridinone, 3-methyl-	C_6H_7NO 109.13	1003-56-1 141.5	 289	5-21-07-00199	H_2O 4; EtOH 4; chl 4
11037	2(1H)-Pyridinone, 4-methyl-	C_6H_7NO 109.13	13466-41-6 130	 308	5-21-07-00201	H_2O 3; EtOH 3; eth 2; bz 2
11038	3H-Pyrido[3,4-b]indole, 4,9-dihydro-7-methoxy-1-methyl- Harmaline	$C_{13}H_{14}N_2O_{13}$ 406.26	304-21-2 250 dec	4528	5-23-12-00148	H_2O 2; EtOH 2; eth 2; chl 3
11039	9H-Pyrido[3,4-b]indole, 7-methoxy-1-methyl- Harmine	$C_{13}H_{12}N_2O$ 212.25	442-51-3 273	4531 sub	5-23-12-00237	H_2O 2; EtOH 2; eth 2; chl 2
11040	9H-Pyrido[3,4-b]indole, 1-methyl- Harman	$C_{12}H_{10}N_2$ 182.22	486-84-0 236.5	4530	5-23-08-00261	
11041	9H-Pyrido[3,4-b]indole, 1-[(2-methylphenyl)methyl]-	$C_{19}H_{16}N_2$ 272.35	525-15-5 217.8	 $150^{0.01}$	5-23-09-00536	EtOH 3; eth 3; bz 3; chl 3
11042	2-Pyrimidinamine	$C_4H_5N_3$ 95.10	109-12-6 127.5	 sub	4-25-00-02071	H_2O 3; chl 2
11043	4-Pyrimidinamine, 2-chloro-*N*,*N*,6-trimethyl- Crimidine	$C_7H_{10}ClN_3$ 171.63	535-89-7 87	2590 $140\text{-}7^4$	4-25-00-02184	EtOH 4
11044	2-Pyrimidinamine, 4,5-dimethyl- 2-Amino-4,5-dimethylpyrimidine	$C_6H_9N_3$ 123.16	1193-74-4 215.8	 sub	4-25-00-02203	ace 4; bz 4; EtOH 4; chl 4
11045	4-Pyrimidinamine, 2,6-dimethyl- Kyanmethin	$C_6H_9N_3$ 123.16	461-98-3 183	5205 sub	4-25-00-02192	H_2O 2; EtOH 2; bz 2; chl 2
11046	2-Pyrimidinamine, 4-methyl-	$C_5H_7N_3$ 109.13	108-52-1 160.3	 sub	4-25-00-02152	H_2O 3; EtOH 3; chl 2
11047	2-Pyrimidinamine, 5-methyl-	$C_5H_7N_3$ 109.13	50840-23-8 193.5	 sub	5-25-10-00191	EtOH 4
11048	4-Pyrimidinamine, 6-methyl-	$C_5H_7N_3$ 109.13	3435-28-7 197	 sub	4-25-00-02182	H_2O 3; EtOH 3
11049	Pyrimidine 1,3-Diazine	$C_4H_4N_2$ 80.09	289-95-2 22	7998 123.8	5-23-05-00334	H_2O 5; EtOH 3 1.4998^{20}
11050	4-Pyrimidinecarboxylic acid, 1,2,3,6-tetrahydro-2,6-dioxo- Orotic acid	$C_5H_4N_2O_4$ 156.10	65-86-1 345.5	6828	4-25-00-01759	H_2O 2; os 1
11051	2,4-Pyrimidinediamine, 5-(4-chlorophenyl)-6-ethyl- Pyrimethamine	$C_{12}H_{13}ClN_4$ 248.71	58-14-0 233.5	7997	4-25-00-03014	
11052	Pyrimidine, 2,4-dichloro-5-methyl- 2,4-Dichloro-5-methylpyrimidine	$C_5H_4Cl_2N_2$ 163.01	1780-31-0 26	 235	5-23-05-00385	H_2O 2; EtOH 4; eth 4; bz 4
11053	Pyrimidine, 2,4-dichloro-6-methyl- 2,6-Dichloro-4-methylpyrimidine	$C_5H_4Cl_2N_2$ 163.01	5424-21-5 46.5	 219	5-23-05-00381	bz 4; eth 4; EtOH 4; chl 4
11054	Pyrimidine, 2,4-dichloro-5-nitro- 2,4-Dichloro-5-nitropyrimidine	$C_4HCl_2N_3O_2$ 193.98	49845-33-2 29.3	 154^{58}, 135^{17}	5-23-05-00350	EtOH 2; eth 2
11055	Pyrimidine, 2,4-dimethyl- 2,4-Dimethylpyrimidine	$C_6H_8N_2$ 108.14	14331-54-5 -4	 150.5	5-23-05-00397 1.168^{14}	 1.4880^{25}
11056	Pyrimidine, 2,5-dimethyl- 2,5-Dimethylpyrimidine	$C_6H_8N_2$ 108.14	22868-76-4 19	 59^{13}	5-23-05-00399	H_2O 4
11057	Pyrimidine, 4,5-dimethyl- 4,5-Dimethylpyrimidine	$C_6H_8N_2$ 108.14	694-81-5 3	 177	5-23-05-00399	H_2O 4
11058	Pyrimidine, 4,6-dimethyl- 4,6-Dimethylpyrimidine	$C_6H_8N_2$ 108.14	1558-17-4 25	 159	5-23-05-00400	H_2O 4 1.4880^{20}
11059	2,4(1H,3H)-Pyrimidinedione Uracil	$C_4H_4N_2O_2$ 112.09	66-22-8 338	9761	5-24-06-00043	H_2O 2; EtOH 4; eth 4
11060	2,4(1H,3H)-Pyrimidinedione, 6-amino-3-ethyl-1-(2-propenyl)- Aminometradine	$C_9H_{13}N_3O_2$ 195.22	642-44-4 143	463	4-25-00-04113	
11061	2,4(1H,3H)-Pyrimidinedione, 5-[bis(2-chloroethyl)amino]- Uracil mustard	$C_8H_{11}Cl_2N_3O_2$ 252.10	66-75-1 206 dec	9762	4-25-00-04127	H_2O 2
11062	2,4(1H,3H)-Pyrimidinedione, 5-bromo- 5-Bromouracil	$C_4H_3BrN_2O_2$ 190.98	51-20-7 310	1430	5-24-06-00322	
11063	4,6(1H,5H)-Pyrimidinedione, dihydro-5-(2-methylpropyl)-5-(2-propenyl)-2-thioxo-, monosodium salt Barbituric acid, 5-allyl-5-isobutyl-2-thio, sodium salt	$C_{11}H_{15}N_2NaO_2S$ 262.31	510-90-7	1518		H_2O 4; EtOH 2; eth 1; bz 1
11064	4,6(1H,5H)-Pyrimidinedione, 5-ethyldihydro-5-(2-methyl-2-propenyl)-2-thioxo- Methallatal	$C_{10}H_{14}N_2O_2S$ 226.30	115-56-0 160.5	5855	5-24-09-00224	
11065	4,6(1H,5H)-Pyrimidinedione, 5-ethyldihydro-5-(1-methylpropyl)-2-thioxo- Barbituric acid, 5-*sec*-butyl-5-ethyl-2-thio-	$C_{10}H_{16}N_2O_2S$ 228.32	2095-57-0 169	9250	5-24-09-00164	
11066	4,6(1H,5H)-Pyrimidinedione, 5-ethyldihydro-5-phenyl- Primidone	$C_{12}H_{14}N_2O_2$ 218.26	125-33-7 281.5	7753	5-24-08-00102	
11067	2,4(1H,3H)-Pyrimidinedione, 5-fluoro- 5-Fluorouracil	$C_4H_3FN_2O_2$ 130.08	51-21-8 283	4109 sub 190	5-24-06-00271	
11068	2,4(1H,3H)-Pyrimidinedione, 5-methyl- Thymine	$C_5H_6N_2O_2$ 126.11	65-71-4 316	9332	5-24-07-00055	H_2O 2; EtOH 2; eth 2

No.	Name Synonym	Mol. Form. Mol. Wt.	CAS RN mp/°C	Merck No. bp/°C	Beil. Ref. den/g cm^{-3}	Solubility n_D
11069	2,4(1H,3H)-Pyrimidinedione, 6-methyl- 6-Methyluracil	$C_5H_6N_2O_2$ 126.11	626-48-2 270-80 dec	6051	5-24-07-00006	H_2O 3; EtOH 3; eth 2; tfa 2
11070	Pyrimidine, 2-methyl- 2-Methyl-1,3-diazine	$C_5H_6N_2$ 94.12	5053-43-0 -4	 138	5-23-05-00376	H_2O 5
11071	Pyrimidine, 4-methyl- 4-Methyl-1,3-diazine	$C_5H_6N_2$ 94.12	3438-46-8 32	 142	5-23-05-00379 1.030^{16}	H_2O 5 1.500^{20}
11072	Pyrimidine, 5-methyl- 5-Methyl-1,3-diazine	$C_5H_6N_2$ 94.12	2036-41-1 30.5	 153	5-23-05-00383	H_2O 4
11073	2,4,5,6(1H,3H)-Pyrimidinetetrone Alloxan	$C_4H_2N_2O_4$ 142.07	50-71-5 256 dec	281 sub	5-24-09-00382	H_2O 4; EtOH 3; ace 3; bz 3
11074	2,4,5,6(1H,3H)-Pyrimidinetetrone, 5-oxime Violuric acid	$C_4H_3N_3O_4$ 157.09	87-39-8 203-4 dec	9904	5-24-09-00386	H_2O 2; EtOH 3
11075	Pyrimidine, 2,4,6-trichloro- 2,4,6-Trichloropyrimidine	$C_4HCl_3N_2$ 183.42	3764-01-0 22.5	 212.5	5-23-05-00347	 1.5700^{20}
11076	2,4,6(1H,3H,5H)-Pyrimidinetrione Barbituric acid	$C_4H_4N_2O_3$ 128.09	67-52-7 248	973 260 dec	5-24-09-00078	H_2O 3; EtOH 2; eth 3
11077	2,4,6(1H,3H,5H)-Pyrimidinetrione, 5-amino- Uramil	$C_4H_5N_3O_3$ 143.10	118-78-5 >400	9763	4-25-00-04228	H_2O 3; eth 1; bz 1; chl 3
11078	2,4,6(1H,3H,5H)-Pyrimidinetrione, 5-(2-bromo-2-propenyl)-5-(1-methylethyl)- Propallylonal	$C_{10}H_{13}BrN_2O_3$ 289.13	545-93-7 181	7807	4-24-00-01994	H_2O 2; EtOH 4; eth 2; ace 4
11079	2,4,6(1H,3H,5H)-Pyrimidinetrione, 5-(2-bromo-2-propenyl)-1-methyl-5-(1-methylethyl)- Narcobarbital	$C_{11}H_{15}BrN_2O_3$ 303.16	125-55-3 115	6342	4-24-00-01995	H_2O 2; EtOH 3; py 3
11080	2,4,6(1H,3H,5H)-Pyrimidinetrione, 5-(2-bromo-2-propenyl)-5-(1-methylpropyl)- Butallylonal	$C_{11}H_{15}BrN_2O_3$ 303.16	1142-70-7 131.5	1503	5-24-09-00231	eth 4; EtOH 4
11081	2,4,6(1H,3H,5H)-Pyrimidinetrione, 5-butyl-1-cyclohexyl- Bucolome	$C_{14}H_{22}N_2O_3$ 266.34	841-73-6 84	1452 $185\text{-}7^{0.8}$	5-24-09-00132	
11082	2,4,6(1H,3H,5H)-Pyrimidinetrione, 5-butyl-5-ethyl- Butethal	$C_{10}H_{16}N_2O_3$ 212.25	77-28-1 128.5	1515	5-24-09-00161	
11083	2,4,6(1H,3H,5H)-Pyrimidinetrione, 5-(1-cyclohexen-1-yl)-1,5-dimethyl- Hexobarbital	$C_{12}H_{16}N_2O_3$ 236.27	56-29-1 146.5	4625	5-24-09-00253	
11084	2,4,6(1H,3H,5H)-Pyrimidinetrione, 5-(1-cyclohexen-1-yl)-5-ethyl- Cyclobarbital	$C_{12}H_{16}N_2O_3$ 236.27	52-31-3 173	2717	5-24-09-00258	H_2O 1; EtOH 4; eth 3; dil alk 3
11085	2,4,6(1H,3H,5H)-Pyrimidinetrione, 5-(2-cyclopenten-1-yl)-5-(2-propenyl)- Cyclopentobarbital	$C_{12}H_{14}N_2O_3$ 234.25	76-68-6 139.5	2751	5-24-09-00269	H_2O 2; EtOH 4
11086	2,4,6(1H,3H,5H)-Pyrimidinetrione, 5,5-dibromo-	$C_4H_2Br_2N_2O_3$ 285.88	511-67-1 235	 dec	4-24-00-01881	eth 4; EtOH 4
11087	2,4,6(1H,3H,5H)-Pyrimidinetrione, 5,5-diethyl- Barbital	$C_8H_{12}N_2O_3$ 184.19	57-44-3 190	972	5-24-09-00137 1.220^{25}	H_2O 2; EtOH 3; eth 3; ace 3
11088	2,4,6(1H,3H,5H)-Pyrimidinetrione, 5,5-diethyl-1-methyl- Metharbital	$C_9H_{14}N_2O_3$ 198.22	50-11-3 150.5	5873	5-24-09-00144	H_2O 3; chl 2
11089	2,4,6(1H,3H,5H)-Pyrimidinetrione, 5,5-diethyl-1-phenyl- Phenetharbital	$C_{14}H_{16}N_2O_3$ 260.29	357-67-5 178	7183	4-24-00-01908	EtOH 4
11090	2,4,6(1H,3H,5H)-Pyrimidinetrione, 5-(2,2-dimethylpropyl)-5-(2-propenyl)- Nealbarbital	$C_{12}H_{18}N_2O_3$ 238.29	561-83-1 156	6350	5-24-09-00238	ace 4; eth 4; EtOH 4
11091	2,4,6(1H,3H,5H)-Pyrimidinetrione, 5,5-di-2-propenyl- Allobarbital	$C_{10}H_{12}N_2O_3$ 208.22	52-43-7 172	252	5-24-09-00245	H_2O 2; EtOH 3; eth 3; bz 3
11092	2,4,6(1H,3H,5H)-Pyrimidinetrione, 5-ethyl-5-(3-methylbutyl)- Amobarbital	$C_{11}H_{18}N_2O_3$ 226.28	57-43-2 157	601	5-24-09-00173	bz 4; EtOH 4; chl 4
11093	2,4,6(1H,3H,5H)-Pyrimidinetrione, 5-ethyl-5-(1-methylethyl)- Probarbital	$C_9H_{14}N_2O_3$ 198.22	76-76-6 200	7759	5-24-09-00159	H_2O 2; EtOH 1; eth 3
11094	2,4,6(1H,3H,5H)-Pyrimidinetrione, 5-ethyl-1-methyl-5-phenyl- Mephobarbital	$C_{13}H_{14}N_2O_3$ 246.27	115-38-8 176	5742	5-24-09-00294	H_2O 2; EtOH 4; eth 2; chl 3
11095	2,4,6(1H,3H,5H)-Pyrimidinetrione, 5-ethyl-5-phenyl- Phenobarbital	$C_{12}H_{12}N_2O_3$ 232.24	50-06-6 174	7201	5-24-09-00286	H_2O 1; EtOH 3; eth 3; bz 1
11096	2,4,6(1H,3H,5H)-Pyrimidinetrione, 5-ethyl-5-(1-piperidinyl)- 5-Ethyl-5-(1-piperidyl)barbituric acid	$C_{11}H_{17}N_3O_3$ 239.27	509-87-5 215	3800	4-25-00-04239	ace 4; eth 4; EtOH 4
11097	2,4,6(1H,3H,5H)-Pyrimidinetrione, 5-[(hexahydro-2,4,6-trioxo-5-pyrimidinyl)imino]-, monoammonium salt Murexide	$C_8H_{10}N_6O_7$ 302.20	3051-09-0	6215	4-25-00-04236	H_2O 2; EtOH 1; eth 1; alk 3
11098	2,4,6(1H,3H,5H)-Pyrimidinetrione, 5-(1-methylethyl)-5-(2-propenyl)- Aprobarbital	$C_{10}H_{14}N_2O_3$ 210.23	77-02-1 141	782	5-24-09-00224	ace 4; eth 4; EtOH 4; chl 4

No.	Name Synonym	Mol. Form. Mol. Wt.	CAS RN mp/°C	Merck No. bp/°C	Beil. Ref. den/g cm^{-3}	Solubility n_D
11099	2,4,6(1H,3H,5H)-Pyrimidinetrione, 1-methyl-5-(1-methylethyl)-5-(2-propenyl)- Enallylpropymal	$C_{11}H_{16}N_2O_3$ 224.26	1861-21-8 56.5	3523 176-8^{12}	5-24-09-00225	bz 4; eth 4; EtOH 4; chl 4
11100	2,4,6(1H,3H,5H)-Pyrimidinetrione, 5-methyl-5-phenyl- Phenylmethylbarbituric acid	$C_{11}H_{10}N_2O_3$ 218.21	76-94-8 220	7275	5-24-09-00283	H_2O 1; EtOH 3; eth 3; alk 3
11101	2,4,6(1H,3H,5H)-Pyrimidinetrione, 5-(2-methylpropyl)-5-(2-propenyl)- Butalbital	$C_{11}H_{16}N_2O_3$ 224.26	77-26-9 138.5	1502	5-24-09-00231	H_2O 2; EtOH 3; eth 3; ace 3
11102	2,4,6(1H,3H,5H)-Pyrimidinetrione, 5-nitro- 5-Nitrobarbituric acid	$C_4H_3N_3O_5$ 173.09	480-68-2 180.5	6507	5-24-09-00105	H_2O 3; EtOH 3; eth 1
11103	2,4,6(1H,3H,5H)-Pyrimidinetrione, 5-phenyl-5-(2-propenyl)- Phenallymal	$C_{13}H_{12}N_2O_3$ 244.25	115-43-5 156.5	7162	5-24-09-00334	H_2O 2; EtOH 4; eth 4; bz 2
11104	2(1H)-Pyrimidinone, 4-amino- Cytosine	$C_4H_5N_3O$ 111.10	71-30-7 320-5 dec	2801	4-25-00-03654	H_2O 3; EtOH 2; eth 1; chl 2
11105	2(1H)-Pyrimidinone, 4-amino-5-(hydroxymethyl)- 5-Hydroxymethylcytosine	$C_5H_7N_3O_2$ 141.13	1123-95-1 >300 dec			
11106	2(1H)-Pyrimidinone, 4-amino-5-methyl- 5-Methylcytosine	$C_5H_7N_3O$ 125.13	554-01-8 270 dec	5970	4-25-00-03727	H_2O 3; EtOH 2; eth 1; acid 3
11107	4(1H)-Pyrimidinone, 2,6-diamino-5-(β-*D*-glucopyranosyloxy)- Vicine	$C_{10}H_{16}N_4O_7$ 304.26	152-93-2 239-42 dec	9880	4-25-00-04285	H_2O 2; EtOH 2; acid 4; alk 4
11108	4(1H)-Pyrimidinone, 2,3-dihydro-6-methyl-2-thioxo- Methylthiouracil	$C_5H_6N_2OS$ 142.18	56-04-2 330 dec	6048 sub	5-24-07-00049	H_2O 1; EtOH 2; eth 2; bz 2
11109	4(1H)-Pyrimidinone, 2,3-dihydro-6-propyl-2-thioxo- Propylthiouracil	$C_7H_{10}N_2OS$ 170.24	51-52-5 219	7882	5-24-07-00257	H_2O 2; EtOH 2; eth 1; bz 1
11110	4(1H)-Pyrimidinone, 2,3-dihydro-2-thioxo- 2-Thiouracil	$C_4H_4N_2OS$ 128.15	141-90-2 >340 dec	9298	5-24-06-00361	H_2O 2; EtOH 2; anh HF 3
11111	5H-Pyrimido[4,5-d]thiazolo[3,2-a]pyrimidine-8-ethanol, 2,7-dimethyl- Thiochrome	$C_{12}H_{14}N_4OS$ 262.34	92-35-3 228.8	 sub	4-27-00-09599	H_2O 3; EtOH 2; eth 2; ace 2
11112	1H-Pyrrole Azole	C_4H_5N 67.09	109-97-7 -23.4	8025 129.7	5-20-05-00003 0.9698^{20}	H_2O 2; EtOH 3; eth 3; ace 3 1.5085^{20}
11113	1H-Pyrrole, 1-butyl-	$C_8H_{13}N$ 123.20	589-33-3	 170.5	5-20-05-00010 0.8747^{20}	 1.4727^{20}
11114	1H-Pyrrole-2-carboxaldehyde	C_5H_5NO 95.10	1003-29-8 46.5	 218	5-21-07-00169	chl 2; os 4; lig 2 1.5939^{16}
11115	1H-Pyrrole-3-carboxylic acid, 2,4-dimethyl-, ethyl ester	$C_9H_{13}NO_2$ 167.21	2199-51-1 78.5	 291	5-22-01-00427	eth 4; EtOH 4
11116	1H-Pyrrole-3-carboxylic acid, 2,5-dimethyl-, ethyl ester	$C_9H_{13}NO_2$ 167.21	2199-52-2 117.5	 291; 130^{15}	5-22-01-00432	EtOH 4
11117	1H-Pyrrole, 3,4-diethyl-2-methyl- 3,4-Diethyl-2-methylpyrrole	$C_9H_{15}N$ 137.22	34874-30-1	 202.5	4-20-00-02171 0.9100^{17}	 1.4988^{17}
11118	1H-Pyrrole, 2,5-dihydro- 3-Pyrroline	C_4H_7N 69.11	109-96-6	8028 90.5	5-20-04-00274 0.9097^{20}	H_2O 4; ace 4; eth 4; EtOH 4 1.4664^{20}
11119	1H-Pyrrole, 2,4-dimethyl- 2,4-Dimethylpyrrole	C_6H_9N 95.14	625-82-1	 168	5-20-05-00063 0.9236^{20}	H_2O 2; EtOH 4; eth 4; bz 4 1.5048^{20}
11120	1H-Pyrrole, 2,5-dimethyl-	C_6H_9N 95.14	625-84-3	 171; 51^{8}	5-20-05-00064 0.9353^{20}	H_2O 1; EtOH 4; eth 4 1.5036^{20}
11121	1H-Pyrrole-2,5-dione	$C_4H_3NO_2$ 97.07	541-59-3 94	 sub	5-21-10-00003 1.2493^{106}	H_2O 3; EtOH 3; eth 3
11122	1H-Pyrrole-2,5-dione, 1-ethyl- *N*-Ethylmaleimide	$C_6H_7NO_2$ 125.13	128-53-0 45.5	3778	5-21-10-00006	H_2O 2; EtOH 4; eth 4; chl 3
11123	1H-Pyrrole-2,5-dione, 1-phenyl- *N*-Phenylmaleimide	$C_{10}H_7NO_2$ 173.17	941-69-5 90.5	7270 162^{12}	5-21-10-00009	bz 4; eth 4; EtOH 4
11124	1H-Pyrrole, 1-ethyl-	C_6H_9N 95.14	617-92-5	 129.5	5-20-05-00010 0.9009^{20}	EtOH 4 1.4841^{20}
11125	1H-Pyrrole, 2-ethyl-	C_6H_9N 95.14	1551-06-0	 164	5-20-05-00061 0.9042^{20}	H_2O 1; EtOH 3; chl 3 1.4942^{20}
11126	1H-Pyrrole, 3-ethyl-2,4-dimethyl-	$C_8H_{13}N$ 123.20	517-22-6 0	 199; 96^{16}	5-20-05-00099 0.913^{20}	H_2O 2; EtOH 3; eth 3; bz 3 1.4961^{20}
11127	1H-Pyrrole, 4-ethyl-2,3-dimethyl-	$C_8H_{13}N$ 123.20	491-18-9 16.5	 200; 113^{16}	4-20-00-02152 0.915^{20}	H_2O 4
11128	1H-Pyrrole, 3-ethyl-4-methyl-	$C_7H_{11}N$ 109.17	488-92-6 3	 70^{11}	5-20-05-00083 0.9059^{20}	eth 4; EtOH 4 1.4913^{20}
11129	Pyrrole, 2-isopropyl-	$C_7H_{11}N$ 109.17	7696-51-7	 172	5-20-05-00081 0.908^{25}	eth 4; EtOH 4 1.491^{25}
11130	1H-Pyrrole, 1-methyl-	C_5H_7N 81.12	96-54-8	 115	5-20-05-00008 0.9145^{15}	H_2O 1; EtOH 5; eth 5 1.4875^{20}
11131	1H-Pyrrole, 2-methyl-	C_5H_7N 81.12	636-41-9 -35.6	 147	5-20-05-00053 0.9446^{15}	H_2O 1; EtOH 5; eth 5 1.5035^{16}
11132	1H-Pyrrole, 3-methyl- 3-Methylpyrrole	C_5H_7N 81.12	616-43-3 -48.4	 143; 45^{11}	5-20-05-00056	EtOH 5; eth 5 1.4970^{20}
11133	1H-Pyrrole, 2,2'-methylenebis-	$C_9H_{10}N_2$ 146.19	21211-65-4 73	 164^{12}	5-23-06-00468	bz 4; eth 4; EtOH 4

No.	Name Synonym	Mol. Form. Mol. Wt.	CAS RN mp/°C	Merck No. bp/°C	Beil. Ref. den/g cm^{-3}	Solubility n_D
11134	1H-Pyrrole, 1-phenyl-	$C_{10}H_9N$ 143.19	635-90-5 62	 234	5-20-05-00012	H_2O 1; EtOH 3; eth 3; ace 3
11135	1H-Pyrrole, 2-phenyl- 2-Phenylpyrrole	$C_{10}H_9N$ 143.19	3042-22-6 129	 272	5-20-07-00369	H_2O 1; EtOH 4; eth 4; bz 4
11136	1H-Pyrrole, 1-(phenylmethyl)-	$C_{11}H_{11}N$ 157.22	2051-97-0 15	 247	5-20-05-00013 1.0183^{20}	H_2O 1; EtOH 4; eth 4 1.5655^{24}
11137	1H-Pyrrole-1-propanenitrile	$C_7H_8N_2$ 120.15	43036-06-2	 $132\text{-}3^{10}$	5-20-05-00023 1.048^{25}	 1.5103^{20}
11138	1H-Pyrrole, 1-propyl-	$C_7H_{11}N$ 109.17	5145-64-2	 145.5	5-20-05-00010 0.8833^{20}	eth 4; EtOH 4
11139	1H-Pyrrole, 3-propyl- 3-Propylpyrrole	$C_7H_{11}N$ 109.17	1551-09-3	 90^{30}; 49^2	5-20-05-00081	eth 3; ace 3; chl 2 1.4900^{25}
11140	1H-Pyrrole, 2,3,4,5-tetraiodo- Iodopyrrole	C_4HI_4N 570.68	87-58-1 150 dec	4934	5-20-05-00045	ace 4; eth 4; chl 4
11141	1H-Pyrrole, 2,3,4,5-tetramethyl-	$C_8H_{13}N$ 123.20	1003-90-3 111	 130^7	5-20-05-00099	EtOH 5; eth 5; ace 5; bz 4
11142	1H-Pyrrole, 1,2,5-trimethyl- 1,2,5-Trimethylpyrrole	$C_7H_{11}N$ 109.17	930-87-0	 171	5-20-05-00065 0.807^{25}	 1.4969^{20}
11143	Pyrrolidine Azacyclopentane	C_4H_9N 71.12	123-75-1 -57.8	8026 86.5	5-20-01-00162 0.8586^{20}	H_2O 5; EtOH 3; eth 3; bz 2 1.4431^{20}
11144	3-Pyrrolidineacetic acid, 2-carboxy-4-(1-methylethenyl)-, [2*S*-(2α,3β,4β)]- Kainic acid	$C_{10}H_{15}NO_4$ 213.23	487-79-6	5157	4-22-00-01523	
11145	1-Pyrrolidineacetic acid, 2,5-dioxo-, ethyl ester Aminoacetic acid, *N*-succinyl, ethyl ester	$C_8H_{11}NO_4$ 185.18	14181-05-6 67	 198^{32}	5-21-09-00474	H_2O 4; EtOH 4
11146	Pyrrolidine, 1-butyl-	$C_8H_{17}N$ 127.23	767-10-2	 154.5	5-20-01-00170 0.816^{25}	EtOH 4 1.4373^{25}
11147	Pyrrolidine, 2-butyl- 2-Butylpyrrolidine	$C_8H_{17}N$ 127.23	3446-98-8	 174; 67^{18}	5-20-04-00222 0.8277^{20}	H_2O 4; EtOH 4 1.4490^{20}
11148	1,2-Pyrrolidinedicarboxylic acid, 1-(phenylmethyl) ester, (*S*)-	$C_{13}H_{15}NO_4$ 249.27	1148-11-4 77		5-22-01-00084	chl 2 1.5310^{20}
11149	Pyrrolidine, 2,2-dimethyl- 2,2-Dimethylpyrrolidine	$C_6H_{13}N$ 99.18	35018-15-6	 106	4-20-00-01529 0.8211^{25}	 1.4304^{25}
11150	Pyrrolidine, 2,4-dimethyl-	$C_6H_{13}N$ 99.18	13603-04-8	 116	4-20-00-01537 0.8297^{16}	H_2O 4; eth 4; EtOH 4 1.4325^{16}
11151	Pyrrolidine, 2,5-dimethyl-, *cis*- *cis*-2,5-Dimethylpyrrolidine	$C_6H_{13}N$ 99.18	39713-71-8	 106.5	5-20-04-00146 0.8205^{20}	H_2O 4; eth 4; EtOH 4 1.4299^{20}
11152	2,5-Pyrrolidinedione Succinimide	$C_4H_5NO_2$ 99.09	123-56-8 126.5	8842 287 dec	5-21-09-00438 1.418^{25}	H_2O 3; EtOH 2; eth 2; ace 2
11153	2,5-Pyrrolidinedione, 1-bromo- *N*-Bromosuccinimide	$C_4H_4BrNO_2$ 177.99	128-08-5 174	1428	5-21-09-00543 2.098^{25}	H_2O 2; eth 2; ace 4
11154	2,5-Pyrrolidinedione, 1-chloro- *N*-Chlorosuccinimide	$C_4H_4ClNO_2$ 133.53	128-09-6 150	2165	5-21-09-00543 1.65^{25}	H_2O 2; EtOH 2; ace 3; bz 2
11155	2,5-Pyrrolidinedione, 1-(chloromethyl)-	$C_5H_6ClNO_2$ 147.56	54553-14-9 58	 158^{12}	5-21-09-00460	ace 4; bz 4
11156	2,5-Pyrrolidinedione, 1,3-dimethyl-3-phenyl- Methsuximide	$C_{12}H_{13}NO_2$ 203.24	77-41-8 52.5	5928 $121\text{-}2^{0.1}$	5-21-11-00209	
11157	2,5-Pyrrolidinedione, 1-ethyl-	$C_6H_9NO_2$ 127.14	2314-78-5 26	 236	5-21-09-00442	eth 4; EtOH 4
11158	2,5-Pyrrolidinedione, 3-ethyl-3-methyl- Ethosuximide	$C_7H_{11}NO_2$ 141.17	77-67-8 64.5	3704	5-21-09-00595	H_2O 4
11159	2,5-Pyrrolidinedione, 1-iodo- *N*-Iodosuccinimide	$C_4H_4INO_2$ 224.99	516-12-1 200.5	4936	5-21-09-00544 2.245^{25}	H_2O 4; EtOH 3; eth 2; ace 3
11160	2,5-Pyrrolidinedione, 1-methyl-	$C_5H_7NO_2$ 113.12	1121-07-9 71	 234	5-21-09-00441	H_2O 3; EtOH 3; eth 4
11161	2,5-Pyrrolidinedione, 1-methyl-3-phenyl- Phensuximide	$C_{11}H_{11}NO_2$ 189.21	86-34-0 72	7231	5-21-11-00188	EtOH 4; MeOH 4
11162	2,5-Pyrrolidinedione, 1-phenyl- Succinanil	$C_{10}H_9NO_2$ 175.19	83-25-0 156	8838 400	5-21-09-00446 1.356^{25}	H_2O 1; EtOH 3; eth 3
11163	2,5-Pyrrolidinedione, 1-(phenylmethyl)- *N*-Benzylsuccinimide	$C_{11}H_{11}NO_2$ 189.21	2142-06-5 103	 393	5-21-09-00449	bz 4; eth 4; EtOH 4; chl 4
11164	1-Pyrrolidineethanamine	$C_6H_{14}N_2$ 114.19	7154-73-6	 $66\text{-}70^{23}$	5-20-01-00405 0.901^{25}	 1.4687^{20}
11165	1-Pyrrolidineethanol	$C_6H_{13}NO$ 115.18	2955-88-6	 $79\text{-}81^{13}$	4-20-00-00092 0.985^{25}	 1.4713^{20}
11166	Pyrrolidine, 1-ethyl-	$C_6H_{13}N$ 99.18	7335-06-0	 104	5-20-01-00168 0.8156^{20}	 1.4113^{15}
11167	2-Pyrrolidinemethanamine, 1-ethyl-	$C_7H_{16}N_2$ 128.22	26116-12-1	 $58\text{-}60^{16}$	5-22-08-00096 0.887^{25}	 1.4665^{20}
11168	Pyrrolidine, 1-methyl- *N*-Methylpyrrolidine	$C_5H_{11}N$ 85.15	120-94-5	 81	5-20-01-00166 0.8188^{20}	H_2O 4; eth 4 1.4247^{20}
11169	Pyrrolidine, 1-[2-(3-methylbutoxy)-2-phenylethyl]- Amixetrine	$C_{17}H_{27}NO$ 261.41	24622-72-8	507 121^2		 1.4978^{22}
11170	Pyrrolidine, 1-nitroso- *N*-Nitrosopyrrolidine	$C_4H_8N_2O$ 100.12	930-55-2	6564 214	5-20-01-00521 1.085^{25}	 1.4880^{25}
11171	Pyrrolidine, 1-phenyl-	$C_{10}H_{13}N$ 147.22	4096-21-3	 $119\text{-}20^{12}$	5-20-01-00182 1.0260^{25}	eth 3 1.5813^{20}

No.	Name Synonym	Mol. Form. Mol. Wt.	CAS RN mp/°C	Merck No. bp/°C	Beil. Ref. den/g cm^{-3}	Solubility n_D
11172	Pyrrolidine, 1-(phenylmethyl)-	$C_{11}H_{15}N$ 161.25	29897-82-3	 $45^{0.7}$	5-20-01-00184 0.965^{25}	 1.5270^{20}
11173	Pyrrolidine, 1-propyl-	$C_7H_{15}N$ 113.20	7335-07-1 105	 132.5	5-20-01-00169 0.8171^{20}	 1.4389^{20}
11174	Pyrrolidine, 3-propyl-	$C_7H_{15}N$ 113.20	116632-47-4	 159	4-20-00-01598 0.8450^{20}	 1.4469^{20}
11175	Pyrrolidine, 2,2,4-trimethyl- 2,2,4-Trimethylpyrrolidine	$C_7H_{15}N$ 113.20	35018-28-1	 119	5-20-04-00198 0.8063^{25}	 1.4259^{25}
11176	Pyrrolidine, 2,2,5-trimethyl-	$C_7H_{15}N$ 113.20	6496-48-6	 112	4-20-00-01603 0.7980^{25}	 1.4223^{25}
11177	Pyrrolidinium, 2-carboxy-1,1-dimethyl-, hydroxide, inner salt, (*S*)- Stachydrine	$C_7H_{13}NO_2$ 143.19	471-87-4 235	8729		H_2O 4; EtOH 4; EtOH 4
11178	Pyrrolidinium, 2-carboxy-4-hydroxy-1,1-dimethyl-, hydroxide, inner salt, (2*S*-*trans*)- Betonicine	$C_7H_{13}NO_3$ 159.19	515-25-3 252 dec	1209		EtOH 4
11179	Pyrrolidinium, 3-[(cyclopentylhydroxyphenylacetyl)oxy]-1,1-dimethyl-, bromide Glycopyrrolate	$C_{19}H_{28}BrNO_3$ 398.34	596-51-0 192.5	4397	5-21-01-00015	
11180	2-Pyrrolidinone 2-Pyrrolidone	C_4H_7NO 85.11	616-45-5 25	8027 251; 133^{12}	5-21-06-00317 1.120^{20}	H_2O 4; EtOH 4; eth 4; bz 4 1.4806^{30}
11181	2-Pyrrolidinone, 1,5-dimethyl-	$C_6H_{11}NO$ 113.16	5075-92-3	 217; 87.5^{10}	5-21-06-00431	H_2O 4; eth 4 1.4650^{20}
11182	2-Pyrrolidinone, 3,3-dimethyl- 3,3-Dimethyl-2-pyrrolidone	$C_6H_{11}NO$ 113.16	4831-43-0 66	 237	0-21-00-00242	H_2O 4; ace 4; eth 4; EtOH 4
11183	2-Pyrrolidinone, 1-ethenyl-	C_6H_9NO 111.14	88-12-0 13.5	 193^{400}; 93^{11}	5-21-06-00330 1.04^{20}	
11184	2-Pyrrolidinone, 1-(2-hydroxyethyl)-	$C_6H_{11}NO_2$ 129.16	3445-11-2 20	 295	5-21-06-00341 1.1435^{20}	
11185	2-Pyrrolidinone, 1-methyl- *N*-Methyl-2-pyrrolidone	C_5H_9NO 99.13	872-50-4 -24	 202	5-21-06-00321 1.0230^{25}	H_2O 4; eth 3; ace 3; chl 3 1.4684^{20}
11186	2-Pyrrolidinone, 5-methyl-	C_5H_9NO 99.13	108-27-0 43	 248	5-21-06-00431 1.0458^{20}	
11187	3-Pyrrolidinone, 1-methyl-	C_5H_9NO 99.13	68165-06-0	 $46\text{-}7^{18}$	5-21-06-00389 0.9646^{25}	 1.4431^{25}
11188	2-Pyrrolidinone, 1-[4-(1-pyrrolidinyl)-2-butynyl]- Oxotremorine	$C_{12}H_{18}N_2O$ 206.29	70-22-4	6904 $124^{0.1}$	5-21-06-00370 0.991^{25}	 1.5160^{20}
11189	2-Pyrroline, 2-methyl-	C_5H_9N 83.13	100791-95-5	 96	0-20-00-00135 0.8995^{22}	H_2O 4; eth 4; EtOH 4
11190	Pyrrolo[1,2-a]pyrimidine, 2,3,4,6,7,8-hexahydro-	$C_7H_{12}N_2$ 124.19	3001-72-7	 $95\text{-}8^{7}$	5-23-05-00239 1.005^{25}	 1.5190^{20}
11191	Pyrrolo[2,3-b]indol-5-ol, 1,2,3,3a,8,8a-hexahydro-1,3a,8-trimethyl-, methylcarbamate (ester), (3a*S*-*cis*)- Physostigmine	$C_{15}H_{21}N_3O_2$ 275.35	57-47-6 105.5	7357	5-23-11-00401	H_2O 2; EtOH 3; eth 3; bz 3
11192	Pyrrolo[2,1-b]quinazolin-3-ol, 1,2,3,9-tetrahydro-, (*R*)- Vasicine	$C_{11}H_{12}N_2O$ 188.23	6159-55-3 211.5	9842	5-23-12-00040	H_2O 2; EtOH 3; eth 2; ace 3
11193	1,1':2',1'':2'',1'''-Quaterphenyl	$C_{24}H_{18}$ 306.41	641-96-3 118.5	 420; $193^{1.0}$	4-05-00-02729	H_2O 1; EtOH 2; eth 3; ace 3
11194	1,1':4',1'':4'',1'''-Quaterphenyl	$C_{24}H_{18}$ 306.41	135-70-6 320	 428^{18}	4-05-00-02730	H_2O 1; EtOH 1; eth 1; bz 3
11195	Quinazoline 1,3-Benzodiazine	$C_8H_6N_2$ 130.15	253-82-7 48	8062 241	5-23-07-00129	H_2O 4; EtOH 3; eth 3; ace 3
11196	Quinazoline, 3,4-dihydro-3-phenyl- Orexine	$C_{14}H_{12}N_2$ 208.26	612-97-5 96.5	 dec	4-23-00-01108 1.290^{4}	bz 4; eth 4; EtOH 4; chl 4
11197	6-Quinazolinesulfonamide, 7-chloro-2-ethyl-1,2,3,4-tetrahydro-4-oxo- Quinethazone	$C_{10}H_{12}ClN_3O_3S$ 289.74	73-49-4	8067		tfa 3
11198	4(1H)-Quinazolinone, 2,3-dihydro-1-(4-morpholinylacetyl)-3-phenyl- Moquizone	$C_{20}H_{21}N_3O_3$ 351.41	19395-58-5 136	6174		chl 3
11199	4(3H)-Quinazolinone, 2-methyl-3-(2-methylphenyl)- Methaqualone	$C_{16}H_{14}N_2O$ 250.30	72-44-6 120	5872	5-24-03-00132	eth 4; EtOH 4; chl 4
11200	4(1H)-Quinazolinone, 1-methyl-2-(phenylmethyl)- Glycosine	$C_{16}H_{14}N_2O$ 250.30	6873-15-0 161.5	4398	5-24-04-00363	
11201	Quinclorac 3,7-Dichloroquinoline-8-carboxylic acid	$C_{10}H_5Cl_2NO_2$ 242.06	84087-01-4 274		 1.75	
11202	2-Quinolinamine 2-Aminoquinoline	$C_9H_8N_2$ 144.18	580-22-3 131.5	 sub	5-22-10-00220	H_2O 4; EtOH 3; eth 3; ace 3
11203	4-Quinolinamine	$C_9H_8N_2$ 144.18	578-68-7 154.8	 180^{12}	5-22-10-00241	H_2O 3; EtOH 4; eth 4; bz 3
11204	5-Quinolinamine	$C_9H_8N_2$ 144.18	611-34-7 110	 310; 184^{10}	5-22-10-00297	H_2O 2; EtOH 4; eth 4; bz 3
11205	6-Quinolinamine 6-Aminoquinoline	$C_9H_8N_2$ 144.18	580-15-4 114	 187^{12}	5-22-10-00303	H_2O 2; EtOH 3; eth 2; NH_3 3

No.	Name Synonym	Mol. Form. Mol. Wt.	CAS RN mp/°C	Merck No. bp/°C	Beil. Ref. den/g cm^{-3}	Solubility n_D
11206	8-Quinolinamine	$C_9H_8N_2$ 144.18	578-66-5 70	 157^{19}	5-22-10-00316	H_2O 4; EtOH 4
11207	4-Quinolinamine, 3-bromo- 4-Amino-3-bromoquinoline	$C_9H_7BrN_2$ 223.07	36825-36-2 203	 sub	4-22-00-04662	eth 4; EtOH 4
11208	2-Quinolinamine, 4-methyl- Lepidine, 2-amino-	$C_{10}H_{10}N_2$ 158.20	27063-27-0 133	 320	5-22-10-00375	eth 4; EtOH 4; chl 4
11209	3-Quinolinamine, 2-methyl-	$C_{10}H_{10}N_2$ 158.20	21352-22-7 160	 278; 198^{16}	5-22-10-00347	bz 4; eth 4; EtOH 4
11210	4-Quinolinamine, 2-methyl-	$C_{10}H_{10}N_2$ 158.20	6628-04-2 169	 333	5-22-10-00347	H_2O 2; EtOH 4; eth 4; ace 4
11211	7-Quinolinamine, 2-methyl- 7-Amino-2-methylquinoline	$C_{10}H_{10}N_2$ 158.20	64334-96-9 148	 304	5-22-10-00364	ace 4; bz 4; eth 4; EtOH 4
11212	8-Quinolinamine, 6-methyl- 8-Amino-6-methylquinoline	$C_{10}H_{10}N_2$ 158.20	68420-93-9 73	 sub	4-22-00-04822	ace 4; bz 4; eth 4; EtOH 4
11213	6-Quinolinamine, *N*,*N*,2-trimethyl-	$C_{12}H_{14}N_2$ 186.26	92-99-9 101	 319	2-22-00-00361	ctc 3; CS_2 3
11214	Quinoline 1-Azanaphthalene	C_9H_7N 129.16	91-22-5 -14.78	8097 237.1	5-20-07-00276 1.0977^{15}	H_2O 1; EtOH 5; eth 5; ace 5 1.6268^{20}
11215	Quinoline, 6-amino-3-bromo-	$C_9H_7BrN_2$ 223.07	7101-96-4 106	 sub	5-22-10-00312	EtOH 4
11216	Quinoline, 2-[2-(1,3-benzodioxol-5-yl)ethyl]-4-methoxy- Cusparine	$C_{19}H_{17}NO_3$ 307.35	529-92-0 92(form a); 116(form b)	2687	2-27-00-00545	ace 4; bz 4; eth 4; EtOH 4
11217	Quinoline, 3-bromo- 3-Bromoquinoline	C_9H_6BrN 208.06	5332-24-1 13.3	 275	5-20-07-00318	chl 3; HOAc 4 1.6641^{20}
11218	Quinoline, 5-bromo- 5-Bromoquinoline	C_9H_6BrN 208.06	4964-71-0 52	 280; 106^{12}	5-20-07-00319	acid 3
11219	Quinoline, 6-bromo- 6-Bromoquinoline	C_9H_6BrN 208.06	5332-25-2 24	 281	5-20-07-00319	EtOH 3; eth 3; acid 3
11220	Quinoline, 7-bromo- 7-Bromoquinoline	C_9H_6BrN 208.06	4965-36-0 34	 290	4-20-00-03390	EtOH 3; eth 3; acid 4
11221	Quinoline, 6-bromo-5-nitro- 6-Bromo-5-nitroquinoline	$C_9H_5BrN_2O_2$ 253.05	98203-04-4 130	 sub	0-20-00-00376	eth 4; EtOH 4
11222	Quinoline, 2-butyl- 2-Butylquinoline	$C_{13}H_{15}N$ 185.27	7661-39-4	 153^{14}	5-20-07-00482 1.003^{20}	 1.5699^{20}
11223	2-Quinolinecarbonitrile 2-Cyanoquinoline	$C_{10}H_6N_2$ 154.17	1436-43-7 95	 160^{23}	5-22-03-00190	H_2O 3; EtOH 4; eth 4; bz 4
11224	4-Quinolinecarbonitrile	$C_{10}H_6N_2$ 154.17	2973-27-5 103.5	 242.5	5-22-03-00206	ace 4; bz 4; eth 4; EtOH 4
11225	4-Quinolinecarboxaldehyde Cinchoninaldehyde	$C_{10}H_7NO$ 157.17	4363-93-3 51	 122^4	5-21-08-00443	eth 4; tol 4
11226	4-Quinolinecarboxamide, 2-butoxy-*N*-[2-(diethylamino)ethyl]-, monohydrochloride Dibucaine hydrochloride	$C_{20}H_{30}ClN_3O_2$ 379.93	61-12-1 94 dec	3016	5-22-05-00278	chl 3
11227	2-Quinolinecarboxylic acid Quinaldic acid	$C_{10}H_7NO_2$ 173.17	93-10-7 156	8054	5-22-03-00183	H_2O 3; bz 4
11228	7-Quinolinecarboxylic acid	$C_{10}H_7NO_2$ 173.17	1078-30-4 249.5	 sub	5-22-03-00217	EtOH 4
11229	8-Quinolinecarboxylic acid 8-Carboxyquinoline	$C_{10}H_7NO_2$ 173.17	86-59-9 187	8099 sub	5-22-03-00217	EtOH 4
11230	2-Quinolinecarboxylic acid, 4,8-dihydroxy- Xanthurenic acid	$C_{10}H_7NO_4$ 205.17	59-00-7 289	9977	4-22-00-02513	H_2O 1; EtOH 3; eth 2; bz 2
11231	1(2H)-Quinolinecarboxylic acid, 2-ethoxy-, ethyl ester EEDQ	$C_{14}H_{17}NO_3$ 247.29	16357-59-8 56.5	3488 $125\text{-}8^{0.1}$	5-21-03-00028	chl 3
11232	2-Quinolinecarboxylic acid, 4-hydroxy- Kynurenic acid	$C_{10}H_7NO_3$ 189.17	492-27-3 282.5	5206	5-22-06-00280	H_2O 2; EtOH 3; eth 1; os 2
11233	4-Quinolinecarboxylic acid, 3-hydroxy-2-phenyl- Oxycinchophen	$C_{16}H_{11}NO_3$ 265.27	485-89-2 206-7 dec	6910	4-22-00-02383	bz 4; EtOH 4; HOAc 4
11234	4-Quinolinecarboxylic acid, 6-methoxy- Quininic acid	$C_{11}H_9NO_3$ 203.20	86-68-0 285 dec	8092 sub	4-22-00-02297	H_2O 2; EtOH 3; eth 2; bz 2
11235	4-Quinolinecarboxylic acid, 2-phenyl Cinchophen	$C_{16}H_{11}NO_2$ 249.27	132-60-5 214.5	2290	5-22-03-00484	H_2O 1; EtOH 3; eth 3; ace 2
11236	4-Quinolinecarboxylic acid, 2-phenyl-, 2-propenyl ester	$C_{19}H_{15}NO_2$ 289.33	524-34-5 31.5	 262^{15}	4-22-00-01361	ace 4; eth 4; EtOH 4
11237	Quinoline, 2-chloro- 2-Chloroquinoline	C_9H_6ClN 163.61	612-62-4 38	 266; 153^{22}	5-20-07-00312 1.2464^{25}	H_2O 1; EtOH 4; eth 4; bz 3 1.6342^{25}
11238	Quinoline, 4-chloro- 4-Chloroquinoline	C_9H_6ClN 163.61	611-35-8 34.5	 262; 130^{15}	5-20-07-00314 1.251^{25}	H_2O 2; EtOH 4; eth 4; dil HCl 3
11239	Quinoline, 5-chloro- 5-Chloroquinoline	C_9H_6ClN 163.61	635-27-8 45	 256	5-20-07-00314	EtOH 3
11240	Quinoline, 6-chloro- 6-Chloroquinoline	C_9H_6ClN 163.61	612-57-7 43.8	 263	5-20-07-00315	 1.6110^{56}
11241	Quinoline, 7-chloro- 7-Chloroquinoline	C_9H_6ClN 163.61	612-61-3 31.5	 267.5	5-20-07-00315 1.2158^{58}	H_2O 2; EtOH 4; eth 4; ace 4 1.6108^{58}
11242	Quinoline, 8-chloro- 8-Chloroquinoline	C_9H_6ClN 163.61	611-33-6 -20	 288.5	5-20-07-00315 1.2834^{14}	H_2O 3; EtOH 4; eth 4; ace 4 1.6408^{14}

No.	Name Synonym	Mol. Form. Mol. Wt.	CAS RN mp/°C	Merck No. bp/°C	Beil. Ref. den/g cm^{-3}	Solubility n_D
11243	Quinoline, 2-chloro-4-methyl-	$C_{10}H_8ClN$ 177.63	634-47-9 59	296	5-20-07-00396	H_2O 1; EtOH 4; eth 4; chl 4
11244	Quinoline, 2-chloro-8-methyl- 2-Chloro-8-methylquinoline	$C_{10}H_8ClN$ 177.63	4225-85-8 61	287	5-20-07-00407	bz 4; eth 4; EtOH 4; chl 4
11245	Quinoline, 4-chloro-2-methyl-	$C_{10}H_8ClN$ 177.63	4295-06-1 42.5	269.5	5-20-07-00383 0.881^{25}	H_2O 2; EtOH 3; eth 3; bz 3 1.6224^{20}
11246	Quinoline, decahydro-, *cis*- *cis*-Decahydroquinoline	$C_9H_{17}N$ 139.24	10343-99-4 -40	206; 90^{20}	5-20-04-00405 0.9426^{20}	H_2O 2; EtOH 3; eth 3 1.4926^{20}
11247	Quinoline, decahydro-, *trans*-(±)- Quinoline, decahydro (*trans*, *DL*)	$C_9H_{17}N$ 139.24	105728-23-2 48	204	5-20-04-00405 0.9610^{22}	ace 4; bz 4; eth 4; EtOH 4 1.4692^{56}
11248	Quinoline, 2,4-dichloro- 2,4-Dichloroquinoline	$C_9H_5Cl_2N$ 198.05	703-61-7 67.5	281	4-20-00-03382	bz 4; eth 4; EtOH 4; chl 4
11249	Quinoline, 2,7-dichloro- 2,7-Dichloroquinoline	$C_9H_5Cl_2N$ 198.05	613-77-4 120	100^{2}	4-20-00-03383	eth 4; EtOH 4
11250	Quinoline, 4,5-dichloro-	$C_9H_5Cl_2N$ 198.05	21617-18-5 118	$134^{5.5}$	4-20-00-03383	eth 4; EtOH 4
11251	Quinoline, 4,7-dichloro- 4,7-Dichloroquinoline	$C_9H_5Cl_2N$ 198.05	86-98-6 93	148^{10}	5-20-07-00316	chl 2
11252	Quinoline, 5,8-dichloro-	$C_9H_5Cl_2N$ 198.05	703-32-2 97.5	sub	5-20-07-00317	eth 4; EtOH 4
11253	Quinoline, 2-[2-(3,4-dimethoxyphenyl)ethyl]-4-methoxy- Galipine	$C_{20}H_{21}NO_3$ 323.39	525-68-8 115.5	4247	4-21-00-02656	ace 4; bz 4; eth 4; EtOH 4
11254	Quinoline, 2,3-dimethyl- 2,3-Dimethylquinoline	$C_{11}H_{11}N$ 157.22	1721-89-7 68.5	263	5-20-07-00431 1.1013^{25}	eth 4; EtOH 4; lig 4
11255	Quinoline, 2,4-dimethyl- 4-Methylquinaldine	$C_{11}H_{11}N$ 157.22	1198-37-4	265	5-20-07-00431 1.0611^{15}	H_2O 2; EtOH 4; eth 4; chl 2 1.6075^{20}
11256	Quinoline, 2,5-dimethyl-	$C_{11}H_{11}N$ 157.22	26190-82-9 61	264.5	4-20-00-03524	bz 4; eth 4; EtOH 4
11257	Quinoline, 2,6-dimethyl-	$C_{11}H_{11}N$ 157.22	877-43-0 60	266.5	5-20-07-00433	H_2O 2; EtOH 2; eth 2; bz 4
11258	Quinoline, 2,7-dimethyl- *m*-Toluquinaldine	$C_{11}H_{11}N$ 157.22	93-37-8 61	264.5	5-20-07-00435	H_2O 2; EtOH 3; eth 3; chl 3
11259	Quinoline, 2,8-dimethyl-	$C_{11}H_{11}N$ 157.22	1463-17-8 27	255.3	5-20-07-00436 1.0394^{20}	H_2O 2; EtOH 4; eth 4; ctc 3 1.6022^{20}
11260	Quinoline, 3,4-dimethyl- 3,4-Dimethylquinoline	$C_{11}H_{11}N$ 157.22	2436-92-2 73.5	291; 114^{4}	5-20-07-00437	eth 4; EtOH 4
11261	Quinoline, 3,8-dimethyl- 3,8-Dimethylquinoline	$C_{11}H_{11}N$ 157.22	20668-29-5	270	5-20-07-00437 1.049^{20}	1.6063^{20}
11262	Quinoline, 4,6-dimethyl- 6-Methyllepidine	$C_{11}H_{11}N$ 157.22	826-77-7 22	273.5; 140^{12}	5-20-07-00437 1.0577^{20}	eth 4; EtOH 4 1.6100^{20}
11263	Quinoline, 5,8-dimethyl- 5,8-Dimethylquinoline	$C_{11}H_{11}N$ 157.22	2623-50-9 4.5	266	5-20-07-00438 1.070^{21}	eth 4; EtOH 4
11264	Quinoline, 6,8-dimethyl- 6,8-Dimethylquinoline	$C_{11}H_{11}N$ 157.22	2436-93-3	268.5	5-20-07-00439 1.0665^{4}	eth 4; EtOH 4
11265	Quinoline, 6-ethoxy-1,2-dihydro-2,2,4-trimethyl- Ethoxyquin	$C_{14}H_{19}NO$ 217.31	91-53-2	3710 $123\text{-}5^{2}$	5-21-03-00095 1.026^{25}	1.569^{25}
11266	Quinoline, 2-ethyl- 2-Ethylquinoline	$C_{11}H_{11}N$ 157.22	1613-34-9	245.5; 134^{16}	5-20-07-00428 1.050^{17}	eth 4; EtOH 4; chl 4 1.5979^{23}
11267	Quinoline, 3-ethyl- 3-Ethylquinoline	$C_{11}H_{11}N$ 157.22	1873-54-7	266	5-20-07-00428 1.0508^{20}	1.6160^{20}
11268	Quinoline, 6-ethyl- 6-Ethylquinoline	$C_{11}H_{11}N$ 157.22	19655-60-8	$135\text{-}6^{14}$	5-20-07-00429 1.043^{25}	1.6009^{25}
11269	Quinoline, 2-ethyl-3-methyl- 2-Ethyl-3-methylquinoline	$C_{12}H_{13}N$ 171.24	27356-52-1 57	269; 140^{12}	5-20-07-00456	bz 4; eth 4; EtOH 4
11270	Quinoline, 2-ethyl-4-methyl- Lepidine, 2-ethyl-	$C_{12}H_{13}N$ 171.24	33357-44-7	272.5	5-20-07-00456 1.025^{20}	1.5941^{20}
11271	Quinoline, 3-ethyl-6-methyl-	$C_{12}H_{13}N$ 171.24	105688-74-2	284.5	4-20-00-03555 1.0280^{20}	1.5955^{20}
11272	Quinoline, 4-iodo- 4-Iodoquinoline	C_9H_6IN 255.06	16560-43-3 98.5	sub	5-20-07-00322	eth 4; EtOH 4
11273	Quinoline, 6-iodo-	C_9H_6IN 255.06	13327-31-6 91	$120^{1.0}$	5-20-07-00322	eth 4; EtOH 4
11274	Quinoline, 4-methoxy- 4-Methoxyquinoline	$C_{10}H_9NO$ 159.19	607-31-8 41	245	5-21-03-00229	H_2O 1; EtOH 3; eth 3; bz 3
11275	Quinoline, 6-methoxy- 6-Methoxyquinoline	$C_{10}H_9NO$ 159.19	5263-87-6 26.5	306; 153^{12}	5-21-03-00244 1.152^{20}	EtOH 3; eth 3; chl 3; dil HCl 3
11276	Quinoline, 8-methoxy- 8-Methoxyquinoline	$C_{10}H_9NO$ 159.19	938-33-0 49.5	283; 164^{14}	5-21-03-00254 1.034^{29}	EtOH 3; eth 3; bz 3; peth 3
11277	Quinoline, 2-methyl- Quinaldine	$C_{10}H_9N$ 143.19	91-63-4 -1.5	8055 246.5	5-20-07-00375 1.06^{25}	H_2O 2; EtOH 3; eth 3; ace 3 1.6116^{20}
11278	Quinoline, 3-methyl- 3-Methylquinoline	$C_{10}H_9N$ 143.19	612-58-8 16.5	259.8	5-20-07-00388 1.0673^{20}	ace 4; eth 4; EtOH 4 1.6171^{20}
11279	Quinoline, 4-methyl- Lepidine	$C_{10}H_9N$ 143.19	491-35-0 9.5	5324 262	5-20-07-00389 1.083^{20}	H_2O 2; EtOH 3; eth 3; ace 3 1.6200^{20}
11280	Quinoline, 5-methyl- 5-Methylquinoline	$C_{10}H_9N$ 143.19	7661-55-4 19	262.7	5-20-07-00399 1.0832^{20}	H_2O 2; EtOH 5; eth 5; ace 3 1.6219^{20}
11281	Quinoline, 6-methyl- 6-Methylquinoline	$C_{10}H_9N$ 143.19	91-62-3 -22	258.6	5-20-07-00400 1.0654^{20}	H_2O 2; EtOH 3; eth 3; ace 3 1.6157^{20}

No.	Name Synonym	Mol. Form. Mol. Wt.	CAS RN mp/°C	Merck No. bp/°C	Beil. Ref. den/g cm^{-3}	Solubility n_D
11282	Quinoline, 7-methyl- *m*-Toluquinoline	$C_{10}H_9N$ 143.19	612-60-2 39	 257.6	5-20-07-00402 1.0609^{20}	H_2O 2; EtOH 3; eth 3; ace 3 1.6150^{20}
11283	Quinoline, 8-methyl- 8-Methylquinoline	$C_{10}H_9N$ 143.19	611-32-5 -80	 247.5	5-20-07-00405 1.0719^{20}	H_2O 2; EtOH 5; eth 5; ace 3 1.6164^{20}
11284	Quinoline, 5-nitro- 5-Nitroquinoline	$C_9H_6N_2O_2$ 174.16	607-34-1 74	 sub	5-20-07-00325	H_2O 2; EtOH 3; bz 3; chl 2
11285	Quinoline, 6-nitro- 6-Nitroquinoline	$C_9H_6N_2O_2$ 174.16	613-50-3 153.5	 $170^{0.2}$	5-20-07-00326	H_2O 3; EtOH 3; eth 2; bz 4
11286	Quinoline, 7-nitro- 7-Nitroquinoline	$C_9H_6N_2O_2$ 174.16	613-51-4 132.5	 sub	5-20-07-00327	H_2O 3; EtOH 3; eth 3; chl 4
11287	Quinoline, 4-nitro-, 1-oxide 4-Nitroquinoline oxide	$C_9H_6N_2O_3$ 190.16	56-57-5 154		5-20-07-00324	
11288	Quinoline, 2-phenyl- 2-Phenylquinoline	$C_{15}H_{11}N$ 205.26	612-96-4 86	 363; 194^{6}	5-20-08-00355	H_2O 2; EtOH 4; eth 4; ace 4
11289	Quinoline, 3-phenyl- 3-Phenylquinoline	$C_{15}H_{11}N$ 205.26	1666-96-2 52	 205^{12}	5-20-08-00359	H_2O 2; EtOH 3; eth 3; ace 4
11290	Quinoline, 4-phenyl- 4-Phenylquinoline	$C_{15}H_{11}N$ 205.26	605-03-8 61.5	 260^{77}; $156^{0.3}$	5-20-08-00360	H_2O 1; EtOH 4; eth 4; ace 4
11291	Quinoline, 6-phenyl- 6-Phenylquincline	$C_{15}H_{11}N$ 205.26	612-95-3 110	 260^{77}	5-20-08-00361 1.1945^{20}	H_2O 2; EtOH 4; eth 2; ace 4
11292	Quinoline, 2-propyl- 2-Propylquinoline	$C_{12}H_{13}N$ 171.24	1613-32-7	 $130\text{-}1^{10}$	5-20-07-00453 1.038^{17}	 1.5886^{23}
11293	5-Quinolinesulfonic acid, 8-ethoxy- Actinoquinol	$C_{11}H_{11}NO_4S$ 253.28	15301-40-3 286 dec	134	5-22-07-00581	alk 3
11294	5-Quinolinesulfonic acid, 8-hydroxy- 8-Hydroxy-5-quinolinesulfonic acid	$C_9H_7NO_4S$ 225.22	84-88-8 322.5	4780	5-22-07-00581	H_2O 2
11295	5-Quinolinesulfonic acid, 8-hydroxy-7-iodo- Ferron	$C_9H_6INO_4S$ 351.12	547-91-1 260 dec	4757	5-22-07-00582	H_2O 2; EtOH 2; eth 1; bz 1
11296	Quinoline, 1,2,3,4-tetrahydro-	$C_9H_{11}N$ 133.19	635-46-1 20	 251	5-20-06-00293 1.0588^{20}	H_2O 3; EtOH 5; eth 5; chl 3 1.6062^{19}
11297	Quinoline, 5,6,7,8-tetrahydro- 2,3-Cyclohexenopyridine	$C_9H_{11}N$ 133.19	10500-57-9	 222	5-20-06-00292 1.0304^{13}	H_2O 2; EtOH 3; eth 3; ace 3 1.5435^{20}
11298	Quinoline, 1,2,3,4-tetrahydro-6-methoxy-	$C_{10}H_{13}NO$ 163.22	120-15-0 42.5	 284; 128^{1}	5-21-02-00314	chl 3 1.5718^{20}
11299	Quinoline, 1,2,3,4-tetrahydro-1-methyl-	$C_{10}H_{13}N$ 147.22	491-34-9	 246	5-20-06-00294 1.022^{20}	EtOH 4 1.5802^{23}
11300	Quinoline, 2,3,6-trimethyl-	$C_{12}H_{13}N$ 171.24	2437-73-2 86.5	 285	4-20-00-03557	eth 4; EtOH 4
11301	Quinoline, 2,3,8-trimethyl-	$C_{12}H_{13}N$ 171.24	4945-28-2	 283	4-20-00-03557 1.0069^{20}	H_2O 1; EtOH 3; eth 3; bz 2
11302	Quinoline, 2,4,6-trimethyl- 4,6-Dimethylquinaldine	$C_{12}H_{13}N$ 171.24	2243-89-2 68	 280	5-20-07-00458	ace 4; bz 4; eth 4; EtOH 4
11303	Quinoline, 2,4,7-trimethyl- 2,4,7-Trimethylquinoline	$C_{12}H_{13}N$ 171.24	71633-43-7 63.5	 280.5	5-20-07-00458 1.0337^{20}	H_2O 2; EtOH 4; eth 4; ace 4 1.5973^{24}
11304	Quinoline, 2,5,7-trimethyl-	$C_{12}H_{13}N$ 171.24	102871-67-0	 287; $107^{1.2}$	4-20-00-03559	H_2O 1; EtOH 4; eth 4 1.5980^{20}
11305	5-Quinolinol 5-Hydroxyquinoline	C_9H_7NO 145.16	578-67-6 226 dec	 sub	5-21-03-00240	H_2O 3; EtOH 2; eth 2; bz 3
11306	6-Quinolinol	C_9H_7NO 145.16	580-16-5 195	 360	5-21-03-00244	H_2O 1; EtOH 2; eth 2; bz 1
11307	7-Quinolinol 7-Hydroxyquinoline	C_9H_7NO 145.16	580-20-1 239	 sub	5-21-03-00249	EtOH 4
11308	8-Quinolinol 8-Hydroxyquinoline	C_9H_7NO 145.16	148-24-3 75.5	4778 267	5-21-03-00252 1.034^{20}	H_2O 1; EtOH 4; eth 1; ace 3
11309	8-Quinolinol, benzoate (ester) Benzoxiquine	$C_{16}H_{11}NO_2$ 249.27	86-75-9	1122	5-21-03-00259	chl 2
11310	8-Quinolinol, 5-bromo-	C_9H_6BrNO 224.06	1198-14-7 124	 sub	5-21-03-00288	H_2O 4; EtOH 4; bz 4; chl 4
11311	8-Quinolinol, 5-chloro- Cloxyquin	C_9H_6ClNO 179.61	130-16-5 130	2416	5-21-03-00279	
11312	8-Quinolinol, 5-chloro-7-iodo- Iodochlorhydroxyquin	C_9H_5ClINO 305.50	130-26-7 178.5	4924	5-21-03-00294	EtOH 2; HOAc 3
11313	8-Quinolinol, 5,7-dibromo- Broxyquinoline	$C_9H_5Br_2NO$ 302.95	521-74-4 196	1441 sub	5-21-03-00290	H_2O 1; EtOH 3; eth 2; ace 3
11314	8-Quinolinol, 5,7-dichloro- Chloroxine	$C_9H_5Cl_2NO$ 214.05	773-76-2 179.5	2175	5-21-03-00286	EtOH 2; ace 2; bz 3; chl 2
11315	8-Quinolinol, 5,7-diiodo- Iodoquinol	$C_9H_5I_2NO$ 396.95	83-73-8 210	4935	5-21-03-00296	H_2O 2; EtOH 4; eth 2; bz 2
11316	4-Quinolinol, 2,8-dimethyl-	$C_{11}H_{11}NO$ 173.21	15644-80-1 262.3	 sub	5-21-08-00321	EtOH 4
11317	6-Quinolinol, 2,4-dimethyl- 2,4-Dimethyl-6-hydroxyquinoline	$C_{11}H_{11}NO$ 173.21	64165-34-0 214	 360 dec	2-21-00-00068	ace 4; EtOH 4
11318	8-Quinolinol, 2,4-dimethyl- Quinoline, 2,4-dimethyl-8-hydroxy	$C_{11}H_{11}NO$ 173.21	115310-98-0 65	 281	4-21-00-01292	ace 4; bz 4; eth 4; EtOH 4
11319	6-Quinolinol, 2-methyl-	$C_{10}H_9NO$ 159.19	613-21-8 213	 304.5	5-21-03-00339 1.1665^{25}	eth 4; EtOH 4
11320	7-Quinolinol, 6-methyl- Quinoline, 7-hydroxy-6-methyl	$C_{10}H_9NO$ 159.19	84583-53-9 244	 sub	0-21-00-00111	bz 4; eth 4; EtOH 4
11321	8-Quinolinol, 2-methyl-	$C_{10}H_9NO$ 159.19	826-81-3 73.8	 267	5-21-03-00341	H_2O 1; EtOH 3; eth 3; bz 3

PHYSICAL CONSTANTS OF ORGANIC COMPOUNDS (continued)

No.	Name Synonym	Mol. Form. Mol. Wt.	CAS RN mp/°C	Merck No. bp/°C	Beil. Ref. den/g cm^{-3}	Solubility n_D
11322	8-Quinolinol, 6-methyl-	$C_{10}H_9NO$ 159.19	20984-33-2 95.5	 sub	4-21-00-01275	EtOH 4
11323	8-Quinolinol, 5-nitro- Nitroxoline	$C_9H_6N_2O_3$ 190.16	4008-48-4 180	6577	5-21-03-00297	
11324	8-Quinolinol, sulfate (2:1) (salt) 8-Hydroxyquinoline sulfate	$C_9H_8NO_3S$ 210.23	134-31-6 177.5	4779	5-21-03-00252	H_2O 4; EtOH 3; eth 1
11325	2(1H)-Quinolinone Carbostyril	C_9H_7NO 145.16	59-31-4 199.5	1831 sub	5-21-08-00217	H_2O 2; EtOH 4; eth 4
11326	2(1H)-Quinolinone, 4-bromo-	C_9H_6BrNO 224.06	938-39-6 266.5	 sub	5-21-08-00229	EtOH 4
11327	2(1H)-Quinolinone, 3,4-dihydro- Hydrocarbostyril	C_9H_9NO 147.18	553-03-7 163.5	4702 201^{45}	5-21-08-00057	eth 4; EtOH 4
11328	2(1H)-Quinolinone, 1-methyl-	$C_{10}H_9NO$ 159.19	606-43-9 74	 325	5-21-08-00218	H_2O 2; EtOH 3; eth 3; ace 3
11329	2(1H)-Quinolinone, 6-methyl-	$C_{10}H_9NO$ 159.19	4053-34-3 237	 240^{12}	5-21-08-00289	ace 4; bz 4; eth 4; EtOH 4
11330	4(1H)-Quinolinone, 1-methyl- Echinopsine	$C_{10}H_9NO$ 159.19	83-54-5	3472	5-21-08-00210	H_2O 3; EtOH 4; eth 2; bz 4
11331	2H-Quinolizine-1-methanol, octahydro-, (1*R*-*trans*)- Lupinine	$C_{10}H_{19}NO$ 169.27	486-70-4 70	5479 270	5-21-01-00338	H_2O 3; EtOH 3; eth 3; bz 3
11332	6-Quinoxalinamine 6-Aminoquinoxaline	$C_8H_7N_3$ 145.16	6298-37-9 159	 sub	4-25-00-02611	H_2O 4; EtOH 4; eth 3; bz 2
11333	Quinoxaline 1,4-Benzodiazine	$C_8H_6N_2$ 130.15	91-19-0 28	8107 229.5	5-23-07-00135 1.1334^{48}	H_2O 3; EtOH 5; eth 5; ace 5 1.6231^{48}
11334	Quinoxaline, 6-chloro- 6-Chloroquinoxaline	$C_8H_5ClN_2$ 164.59	5448-43-1 64	 117^{10}	5-23-07-00144	H_2O 2
11335	Quinoxaline, 2,6-dimethyl-	$C_{10}H_{10}N_2$ 158.20	60814-29-1 54	 268	0-23-00-00192	H_2O 4; ace 4; bz 4; EtOH 4
11336	Quinoxaline, 6-methoxy- 6-Methoxyquinoxaline	$C_9H_8N_2O$ 160.18	6639-82-3 58.9	 128^7	5-23-11-00461	ace 4; eth 4
11337	Quinoxaline, 2-methyl- 2-Methylquinoxaline	$C_9H_8N_2$ 144.18	7251-61-8 180.5	 244	5-23-07-00219	H_2O 5; EtOH 4; eth 5; ace 5
11338	Quinoxaline, 6-methyl- 6-Methylquinoxaline	$C_9H_8N_2$ 144.18	6344-72-5 218.5	 249; 141.5^{29}	5-23-07-00223 1.1164^{20}	H_2O 5; EtOH 5; eth 5; ace 5 1.6211^{18}
11339	Quinoxaline, 1,2,3,4-tetrahydro- 1,2,3,4-Tetrahydroquinoxaline	$C_8H_{10}N_2$ 134.18	3476-89-9 99	 289	5-23-06-00024	H_2O 3; EtOH 4; eth 4; bz 4
11340	(-)-*cis*-Resmethrin Cyclopropanecarboxylic acid, 2,2-dimethyl-3-(2-methyl-1-propenyl)-, [5-(phenylmethyl)- 3-furanyl] methyl ester	$C_{22}H_{26}O_3$ 338.45	10453-86-8 75			
11341	Retinol Vitamin A	$C_{20}H_{30}O$ 286.46	68-26-8 63.5	9918 $137\text{-}8^{10\text{-}6}$	4-06-00-04133	H_2O 1; EtOH 3; eth 3; ace 3
11342	Rheadan, 8-methoxy-16-methyl-2,3:10,11-bis[methylenebis(oxy)]-, (8β)- Rheadine	$C_{21}H_{21}NO_6$ 383.40	2718-25-4 257	8174 sub	4-27-00-06897	
11343	Ribitol Adonitol	$C_5H_{12}O_5$ 152.15	488-81-3 104	157	4-01-00-02832	H_2O 3; EtOH 3; eth 1; lig 1
11344	Riboflavin Riboflavine	$C_{17}H_{20}N_4O_6$ 376.37	83-88-5 280 dec	8201	4-26-00-02542	H_2O 1; EtOH 2; eth 1; ace 1
11345	*D*-Ribose	$C_5H_{10}O_5$ 150.13	50-69-1 95	8205	4-01-00-04211	H_2O 3; EtOH 2
11346	Sarpagan-17-al Vellosimine	$C_{19}H_{20}N_2O$ 292.38	6874-98-2 305.5	9846 sub 180		
11347	Sarpagan-10,17-diol Sarpagine	$C_{19}H_{22}N_2O_2$ 310.40	482-68-8 320	8336	5-23-13-00401	H_2O 1; EtOH 3
11348	9,10-Secocholesta-5,7,10(19)-trien-3-ol, (3β,5Z,7E)- Vitamin D3	$C_{27}H_{44}O$ 384.65	67-97-0 84.5	9929	4-06-00-04149	H_2O 1; os 3
11349	9,10-Secoergosta-5(10),6,8,22-tetraen-3-ol, (3β,6E,22E)- Tachysterol	$C_{28}H_{44}O$ 396.66	115-61-7	9002	4-06-00-04404	H_2O 1; EtOH 3; eth 3; ace 3
11350	9,10-Secoergosta-5,7,10(19),22-tetraen-3-ol, (3β,5Z,7E,22E)- Vitamin D2	$C_{28}H_{44}O$ 396.66	50-14-6 116.5	9928 sub	4-06-00-04404	H_2O 1; EtOH 3; eth 3; ace 3
11351	16,19-Secostrychnidine-10,16-dione, 4-hydroxy-19-methyl- Vomicine	$C_{22}H_{24}N_2O_4$ 380.44	125-15-5 282	9946	4-27-00-07873	EtOH 2; eth 2; ace 2; chl 4
11352	9,10-Secostrychnidin-10-oic acid, 2,3-dihydro-4-nitro-2,3-dioxo- Cacotheline	$C_{21}H_{21}N_3O_7$ 427.41	561-20-6 >300	1604	4-27-00-08014	H_2O 2
11353	Selenanthrene Selanthrene	$C_{12}H_8Se_2$ 310.12	262-30-6 181	 223^{11}	5-19-02-00057	EtOH 2; eth 2; CS_2 3
11354	Selenourea Carbamimidoselenoic acid	CH_4N_2Se 123.02	630-10-4	 200 dec	4-03-00-00435	H_2O 4
11355	*D*-Serine	$C_3H_7NO_3$ 105.09	312-84-5 229 dec	 dec	4-04-00-03118	H_2O 4; EtOH 1; eth 1; bz 1
11356	*DL*-Serine	$C_3H_7NO_3$ 105.09	302-84-1 246 dec		4-04-00-03119 1.603^{22}	H_2O 3; EtOH 1; eth 1; bz 1
11357	*L*-Serine (*S*)-2-Amino-3-hydroxypropanoic acid	$C_3H_7NO_3$ 105.09	56-45-1 228 dec	8411 sub 150	4-04-00-03118 1.6^{22}	H_2O 4; EtOH 1; eth 1; bz 1

No.	Name Synonym	Mol. Form. Mol. Wt.	CAS RN mp/°C	Merck No. bp/°C	Beil. Ref. den/g cm^{-3}	Solubility n_D
11358	Sesone Ethanol, 2-(2,4-dichlorophenoxy)-, hydrogen sulfate, sodium salt	$C_8H_8Cl_2NaO_5S$ 310.11	136-78-7 245 dec	3372		
11359	Sethoxydim (±)-2-[1-(Ethoxyimino)butyl]-5-[2-(ethylthio)propyl]-3-hydroxy-2-cyclohexen-1-one	$C_{17}H_{29}NO_3S$ 327.49	74051-80-2	8424 $>90^{0.00003}$	1.043^{25}	
11360	Silacyclohexane, 1,1-diphenyl- 1,1-Diphenylsilacyclohexane	$C_{17}H_{20}Si$ 252.43	18002-79-4	$193-8^{5}$	4-27-00-09833 1.0319^{25}	1.5779^{20}
11361	Silanamine, pentamethyl-	$C_5H_{15}NSi$ 117.27	2083-91-2	86	4-04-00-04010 0.7400^{20}	1.4379^{24}
11362	Silanamine, 1,1,1-trimethyl-*N*-phenyl-*N*-(trimethylsilyl)-	$C_{12}H_{23}NSi_2$ 237.49	4147-89-1 16.5	$98-9^{11}$	0.8951^{23}	1.4855^{20}
11363	Silanamine, 1,1,1-trimethyl-*N*-(trimethylsilyl)- Hexamethyldisilazane	$C_6H_{19}NSi_2$ 161.39	999-97-3	10158 125	4-04-00-04014 0.7741^{25}	1.4090^{20}
11364	Silane, [1,1'-biphenyl]-4-yltrichloro-	$C_{12}H_9Cl_3Si$ 287.65	18030-61-0	$209-11^{30}$	4-16-00-01572 1.3190^{20}	ctc 3; CS_2 3
11365	Silane, (bromomethyl)chlorodimethyl- (Bromomethyl)dimethylchlorosilane	$C_3H_8BrClSi$ 187.54	16532-02-8	131	4-04-00-04024 1.375^{25}	1.4630^{25}
11366	Silane, (bromomethyl)trimethyl- Bromomethyltrimethylsilane	$C_4H_{11}BrSi$ 167.12	18243-41-9	116.5	4-04-00-03878 1.170^{25}	1.4460^{20}
11367	Silane, (4-bromophenoxy)trimethyl-	$C_9H_{13}BrOSi$ 245.19	17878-44-3	126^{25}	4-06-00-01056 1.2619^{20}	1.5145^{20}
11368	Silane, bromotriethyl-	$C_6H_{15}BrSi$ 195.17	1112-48-7	$66-7^{24}$	4-04-00-04051 1.1403^{20}	1.4561^{20}
11369	Silane, butylchlorodimethyl-	$C_6H_{15}ClSi$ 150.72	1000-50-6	139	4-04-00-04068 0.8751^{20}	1.5145^{20}
11370	Silane, butyltrichloro- Trichlorobutylsilane	$C_4H_9Cl_3Si$ 191.56	7521-80-4	148.5	4-04-00-04246 1.1606^{20}	eth 3; bz 3; tol 3; AcOEt 3 1.4363^{20}
11371	Silane, butyltrimethyl- Trimethylbutylsilane	$C_7H_{18}Si$ 130.31	1000-49-3	115	4-04-00-03907 0.7353^{0}	H_2O 1
11372	Silane, chloro(chloromethyl)dimethyl-	$C_3H_8Cl_2Si$ 143.09	1719-57-9	115.5	4-04-00-04024 1.0865^{20}	1.4360^{20}
11373	Silane, chloro(dichloromethyl)dimethyl- (Dichloromethyl)dimethylchlorosilane	$C_3H_7Cl_3Si$ 177.53	18171-59-0 -48	149	4-04-00-04031 1.2369^{20}	1.461^{20}
11374	Silane, chlorodimethylphenyl-	$C_8H_{11}ClSi$ 170.71	768-33-2	$80-4^{16}$	4-16-00-01475 1.032^{20}	1.5082^{20}
11375	Silane, chlorodimethyl-2-propenyl-	$C_5H_{11}ClSi$ 134.68	4028-23-3	111	4-04-00-04080 0.8964^{20}	1.4195^{20}
11376	Silane, chloroethenyldimethyl-	C_4H_9ClSi 120.65	1719-58-0	83.5	4-04-00-04080 0.8744^{20}	1.4141^{20}
11377	Silane, chloroethyldimethyl-	$C_4H_{11}ClSi$ 122.67	6917-76-6	89.5	4-04-00-04035 0.8675^{20}	1.4105^{20}
11378	Silane, chloromethyl Chloromethylsilane	CH_5ClSi 80.59	993-00-0 -135	7; -45^{83}		
11379	Silane, (chloromethyl)dimethylphenyl-	$C_9H_{13}ClSi$ 184.74	1833-51-8	225	4-16-00-01376 1.0240^{25}	ctc 3; CS_2 3
11380	Silane, chloromethyldiphenyl-	$C_{13}H_{13}ClSi$ 232.78	144-79-6	295	4-16-00-01479 1.1277^{20}	1.5742^{20}
11381	Silane, chloromethylphenyl-	C_7H_9ClSi 156.69	1631-82-9	113^{100}	4-16-00-01473 1.0540^{20}	1.5171^{20}
11382	Silane, (chloromethyl)trimethyl-	$C_4H_{11}ClSi$ 122.67	2344-80-1	98.5	4-04-00-03877 0.879^{25}	1.4175^{20}
11383	Silane, (4-chlorophenoxy)trimethyl-	$C_9H_{13}ClOSi$ 200.74	17005-59-3	214; 101^{14}	4-06-00-00878 1.0320^{20}	1.4930^{20}
11384	Silane, chlorophenyl- Phenylchlorosilane	C_6H_7ClSi 142.66	4206-75-1	162.5	4-16-00-01472 1.0683^{20}	1.5340^{20}
11385	Silane, (3-chlorophenyl)trimethyl-	$C_9H_{13}ClSi$ 184.74	4405-42-9	206.5	4-16-00-01386 1.0071^{20}	1.5108^{20}
11386	Silane, (3-chloropropyl)trimethyl-	$C_6H_{15}ClSi$ 150.72	2344-83-4	151	4-04-00-03905 0.8789^{20}	1.4319^{20}
11387	Silane, chlorotriethoxy-	$C_6H_{15}ClO_3Si$ 198.72	4667-99-6 -51	156	4-01-00-01363 1.030^{20}	EtOH 4 1.3999^{20}
11388	Silane, chlorotriethyl-	$C_6H_{15}ClSi$ 150.72	994-30-9	144.5	4-04-00-04051 0.8967^{20}	1.4314^{20}
11389	Silane, chlorotrimethyl- Trimethylchlorosilane	C_3H_9ClSi 108.64	75-77-4 -40	60	4-04-00-04007 0.856^{25}	1.3870^{20}
11390	Silane, dichloro(chloromethyl)methyl- Dichloro(chloromethyl)methylsilane	$C_2H_5Cl_3Si$ 163.51	1558-33-4	121.5	4-04-00-04135 1.2858^{20}	1.4500^{20}
11391	Silane, dichloro(dichloromethyl)methyl- Dichloromethyl methyl dichlorosilane	$C_2H_4Cl_4Si$ 197.95	1558-31-2	149	4-04-00-04144 1.4116^{20}	1.4700^{20}
11392	Silane, dichlorodiethoxy-	$C_4H_{10}Cl_2O_2Si$ 189.11	4667-38-3 -130	135.9	4-01-00-01364 1.1290^{20}	EtOH 4
11393	Silane, dichlorodiethyl- Dichlorodiethylsilane	$C_4H_{10}Cl_2Si$ 157.11	1719-53-5 -96.5	129 dec	4-04-00-04158 1.0504^{20}	1.4309^{20}
11394	Silane, dichlorodimethyl- Dichlorodimethylsilane	$C_2H_6Cl_2Si$ 129.06	75-78-5 -16	70.3	4-04-00-04111 1.064^{25}	H_2O 6; EtOH 6 1.4038^{20}
11395	Silane, dichlorodiphenyl- Dichlorodiphenylsilane	$C_{12}H_{10}Cl_2Si$ 253.20	80-10-4	305	4-16-00-01526 1.204^{25}	EtOH 3; eth 3; ace 3; bz 3 1.5800^{20}

No.	Name Synonym	Mol. Form. Mol. Wt.	CAS RN mp/°C	Merck No. bp/°C	Beil. Ref. den/g cm^{-3}	Solubility n_D
11396	Silane, dichloroethenylmethyl-	$C_3H_6Cl_2Si$ 141.07	124-70-9	 92.5	4-04-00-04184 1.0868^{20}	H_2O 6 1.4270^{20}
11397	Silane, dichloroethylmethyl- Methylethyldichlorosilane	$C_3H_8Cl_2Si$ 143.09	4525-44-4	 101	4-04-00-04152 1.0630^{20}	 1.4197^{20}
11398	Silane, dichloromethyl- Dichloromethylsilane	CH_4Cl_2Si 115.03	75-54-7 -93	 41	4-04-00-04096 1.105^{25}	
11399	Silane, dichloromethyl(1-methylethyl)-	$C_4H_{10}Cl_2Si$ 157.11	18236-89-0	 121.5	4-04-00-04171 1.0385^{20}	 1.4270^{20}
11400	Silane, dichloromethyl(4-methylphenyl)-	$C_8H_{10}Cl_2Si$ 205.16	18236-57-2 57		 1.1619^{20}	 1.5170^{20}
11401	Silane, dichloromethylphenyl- Dichloromethylphenylsilane	$C_7H_8Cl_2Si$ 191.13	149-74-6	 206.5	4-16-00-01517 1.1866^{20}	 1.5180^{20}
11402	Silane, dichloromethyl-2-propenyl-	$C_4H_8Cl_2Si$ 155.10	1873-92-3	 119.5	4-04-00-04188 1.0758^{20}	 1.4419^{20}
11403	Silane, dichlorophenyl- Phenyldichlorosilane	$C_6H_6Cl_2Si$ 177.10	1631-84-1	 181	4-16-00-01514 1.221^{25}	H_2O 6
11404	Silane, diethenyldiphenyl-	$C_{16}H_{16}Si$ 236.39	17937-68-7	 $130-1^{0.05}$	4-16-00-01368 1.0092^{25}	 1.5350^{25}
11405	Silane, diethoxydimethyl- Dimethyldiethoxysilane	$C_6H_{16}O_2Si$ 148.28	78-62-6 -87	 114	4-04-00-04101 0.865^{25}	ctc 3 1.3811^{20}
11406	Silane, diethoxydiphenyl- Diphenyldiethoxysilane	$C_{16}H_{20}O_2Si$ 272.42	2553-19-7	 167^{15}	4-16-00-01524 1.0329^{20}	 1.5269^{20}
11407	Silane, diethoxymethylphenyl-	$C_{11}H_{18}O_2Si$ 210.35	775-56-4	 218	4-16-00-01515 0.9627^{20}	 1.4690^{20}
11408	Silane, diethoxymethyl-2-propenyl- Silane, allyldiethoxymethyl-	$C_8H_{18}O_2Si$ 174.32	18388-45-9	 155	4-04-00-04187 0.8572^{25}	 1.4104^{20}
11409	Silane, diethyl- 3-Silapentane	$C_4H_{12}Si$ 88.22	542-91-6 -134.3	3117 57	4-04-00-03894 0.6843^{20}	H_2O 1 1.3921^{20}
11410	Silane, diethyldifluoro- Diethyldifluorosilane	$C_4H_{10}F_2Si$ 124.21	358-06-5 -78.7	 61	4-04-00-04157 0.9348^{20}	 1.3385^{20}
11411	Silane, difluorodiphenyl-	$C_{12}H_{10}F_2Si$ 220.29	312-40-3	 157^{50}	4-16-00-01525 1.145^{17}	 1.5221^{25}
11412	Silane, dimethoxydimethyl-	$C_4H_{12}O_2Si$ 120.22	1112-39-6	 82	4-04-00-04100 0.8646^{20}	H_2O 6 1.3708^{20}
11413	Silane, dimethoxydiphenyl-	$C_{14}H_{16}O_2Si$ 244.37	6843-66-9	 161^{15}	4-16-00-01524 1.0771^{20}	 1.5447^{20}
11414	Silane, dimethyl- 2-Silapropane	C_2H_8Si 60.17	1111-74-6 -150	 -20	4-04-00-03874 0.68^{-80}	
11415	Silane, dimethyldiphenoxy- Diphenoxydimethylsilane	$C_{14}H_{16}O_2Si$ 244.37	3440-02-6 -23	 $130-2^{5}$	4-06-00-00764 1.0599^{25}	 1.5330^{20}
11416	Silane, dimethyldiphenyl- Diphenyl(dimethyl)silane	$C_{14}H_{16}Si$ 212.37	778-24-5	 277; 173^{45}	4-16-00-01366 0.9867^{20}	 1.5644^{20}
11417	Silane, dimethyldi-2-propenyl-	$C_8H_{16}Si$ 140.30	1113-12-8	 68^{50}	4-04-00-03929 0.7679^{20}	 1.4420^{20}
11418	Silane, dimethylphenyl-	$C_8H_{12}Si$ 136.27	766-77-8	 156.5	4-16-00-01360 0.8891^{20}	H_2O 1 1.4995^{20}
11419	Silanediol, dimethyl-, diacetate	$C_6H_{12}O_4Si$ 176.24	2182-66-3 -12.5	 165	4-04-00-04108 1.0540^{20}	 1.4030^{20}
11420	Silane, diphenyl- Diphenylsilane	$C_{12}H_{12}Si$ 184.31	775-12-2	 $95-7^{13}$	4-16-00-01366 0.9969^{20}	ctc 3; CS_2 3 1.5800^{20}
11421	Silane, 1,2-ethenediylbis[trimethyl-, (*E*)-	$C_8H_{20}Si_2$ 172.42	18178-59-1 -18.5	 145.5	4-04-00-03948 0.7589^{20}	H_2O 1 1.4310^{20}
11422	Silane, ethenyldiethoxymethyl-	$C_7H_{16}O_2Si$ 160.29	5507-44-8	 133	4-04-00-04183 0.8620^{20}	 1.4001^{20}
11423	Silane, ethenylethoxydimethyl-	$C_6H_{14}OSi$ 130.26	5356-83-2	 99	4-04-00-04079 0.790^{20}	 1.3983^{20}
11424	Silane, ethenyltriethoxy-	$C_8H_{18}O_3Si$ 190.31	78-08-0	 148^{535}, 62^{20}	4-04-00-04256 0.9027^{20}	chl 3 1.3960^{25}
11425	Silane, ethenyltrimethyl-	$C_5H_{12}Si$ 100.24	754-05-2	 55	4-04-00-03922 0.6910^{20}	H_2O 1 1.3914^{20}
11426	Silane, ethenyltris(1-methylethoxy)-	$C_{11}H_{24}O_3Si$ 232.40	18023-33-1	 179.5; 77^{20}	4-04-00-04257 0.8627^{25}	ctc 3 1.3981^{20}
11427	Silane, ethenyltris(2-propenyloxy)- Tris(allyloxy)vinylsilane	$C_{11}H_{18}O_3Si$ 226.35	17988-31-7	 $90.3^{7.5}$	4-04-00-04257 0.9435^{25}	 1.4380^{20}
11428	Silane, ethoxytriethyl-	$C_8H_{20}OSi$ 160.33	597-67-1	 154.5	4-04-00-04042 0.8160^{20}	eth 4; EtOH 4 1.4140^{20}
11429	Silane, ethoxytrifluoro- Ethoxytrifluorosilicon	$C_2H_5F_3OSi$ 130.14	460-55-9 -122	 -7	3-01-00-01336 1.3200^{-63}	
11430	Silane, ethoxytrimethyl-	$C_5H_{14}OSi$ 118.25	1825-62-3	 76	4-04-00-03994 0.7573^{20}	H_2O 1; EtOH 3; eth 3; ace 3 1.3741^{20}
11431	Silane, ethoxytriphenyl-	$C_{20}H_{20}OSi$ 304.46	1516-80-9 65	 344	4-16-00-01481	chl 3
11432	Silane, ethyltrifluoro- Trifluoroethylsilane	$C_2H_5F_3Si$ 114.14	353-89-9 -105	 -4.4	4-04-00-04226 1.227^{-76}	
11433	Silane, ethyltrimethoxy- Ethyltrimethoxysilane	$C_5H_{14}O_3Si$ 150.25	5314-55-6	 124.3	4-04-00-04223 0.9488^{20}	EtOH 4 1.3838^{20}
11434	Silane, 1,2-ethynediylbis[trimethyl-	$C_8H_{18}Si_2$ 170.40	14630-40-1 26	 134	4-04-00-03950 0.770^{20}	 1.413^{20}
11435	Silane, ethynyl Ethynylsilane	C_2H_4Si 56.14	1066-27-9	 -22.5		

No.	Name Synonym	Mol. Form. Mol. Wt.	CAS RN mp/°C	Merck No. bp/°C	Beil. Ref. den/g cm^{-3}	Solubility n_D
11436	Silane, methoxy Methyl silyl ether	CH_6OSi 62.14	2171-96-2			
11437	Silane, methyl- Methylsilane	CH_6Si 46.14	992-94-9 -156.5	 -57.5	4-04-00-03873	
11438	Silane, methyldiphenyl-	$C_{13}H_{14}Si$ 198.34	776-76-1	 93.5^{1}	4-16-00-01366 0.9973^{20}	ctc 3 1.5694^{20}
11439	Silane, methylenebis-	CH_8Si_2 76.25	1759-88-2	 15	4-01-00-03072 0.6979^{4}	 1.4115^{4}
11440	Silane, methylenebis[trichloro-	$CH_2Cl_6Si_2$ 282.91	4142-85-2	 183; 64^{10}	4-01-00-03077 1.5567^{20}	eth 4; EtOH 4 1.4740^{20}
11441	Silane, (2-methylphenoxy)triphenyl-	$C_{25}H_{22}OSi$ 366.53	18858-65-6	 192	4-16-00-01482 0.9287^{20}	 1.4830^{20}
11442	Silane, methylphenyl-	$C_7H_{10}Si$ 122.24	766-08-5	 140	4-16-00-01360 0.8895^{20}	 1.5058^{20}
11443	Silane, methyltriphenoxy- Methyltriphenoxysilane	$C_{19}H_{18}O_3Si$ 322.44	3439-97-2	 269^{100}; 210^{12}	4-06-00-00765 1.135^{20}	 1.5599^{20}
11444	Silane, methyltriphenyl- Triphenylmethylsilane	$C_{19}H_{18}Si$ 274.44	791-29-7 69.5	 198^{9}	4-16-00-01370 1.0880^{20}	ctc 3; CS_2 3
11445	Silane, methyltri-*p*-tolyl- Silane, methyl tri-(4-tolyl)	$C_{22}H_{24}Si$ 316.52	18752-92-6 92	 198^{4}	4-16-00-01399	chl 2
11446	Silane, phenyl-	C_6H_8Si 108.22	694-53-1	 119	4-16-00-01360 0.8681^{20}	H_2O 1 1.5125^{20}
11447	Silane, [1,3-phenylenebis(oxy)]bis[trimethyl-	$C_{12}H_{22}O_2Si_2$ 254.48	4520-29-0	 240	4-06-00-05683 0.950^{20}	 1.4748^{20}
11448	Silane, phenyltripropyl- Phenyltripropylsilane	$C_{15}H_{26}Si$ 234.46	78938-11-1	 $146\text{-}7^{18}$	4-16-00-01362 0.8799^{20}	 1.4950^{20}
11449	Silane, tetraethenyl- Tetravinylsilane	$C_8H_{12}Si$ 136.27	1112-55-6	 130.2	4-04-00-03925 0.7999^{20}	 1.4625^{20}
11450	Silane, tetraethyl- Tetraethylsilane	$C_8H_{20}Si$ 144.33	631-36-7	 154.7	4-04-00-03895 0.7658^{20}	H_2O 1 1.4268^{20}
11451	Silane, tetramethyl- Tetramethylsilane	$C_4H_{12}Si$ 88.22	75-76-3 -99.0	 26.6	4-04-00-03875 0.648^{19}	H_2O 1; EtOH 4; eth 4; sulf 1 1.3587^{20}
11452	Silane, tetraphenyl- Tetraphenylsilane	$C_{24}H_{20}Si$ 336.51	1048-08-4 236.5	 228^{3}	4-16-00-01372 1.078^{20}	ctc 3; CS_2 3
11453	Silane, tribromomethyl-	CH_3Br_3Si 282.83	4095-09-4 -28.4	 132	4-04-00-04213 2.2130^{25}	 1.5152^{25}
11454	Silane, tributyl- Tributylsilane	$C_{12}H_{28}Si$ 200.44	998-41-4	 221	4-04-00-03908 0.7794^{20}	 1.4380^{20}
11455	Silane, tributylphenyl- Silane, phenyl-tributyl	$C_{18}H_{32}Si$ 276.54	18510-29-7	 $140\text{-}2^{4}$	4-16-00-01362 0.8753^{20}	 1.4915^{20}
11456	Silane, trichloro(chloromethyl)- (Chloromethyl)trichlorosilane	CH_2Cl_4Si 183.92	1558-25-4	 118	4-01-00-03074 1.4650^{20}	 1.4555^{20}
11457	Silane, trichloro(4 chlorophenyl)-	$C_6H_4Cl_4Si$ 245.99	825-94-5	 $115\text{-}7^{20}$	4-16-00-01563 1.4316^{20}	 1.5418^{20}
11458	Silane, trichloro(3-chloropropyl)-	$C_3H_6Cl_4Si$ 211.98	2550-06-3	 181.5	4-04-00-04243 1.3590^{20}	 1.4668^{20}
11459	Silane, trichloro(dichloromethyl)- (Dichloromethyl)trichlorosilane	$CHCl_5Si$ 218.37	1558-24-3	 145	4-02-00-00088 1.5518^{20}	 1.4714^{20}
11460	Silane, trichlorododecyl- Dodecyltrichlorosilane	$C_{12}H_{25}Cl_3Si$ 303.77	4484-72-4	 155^{10}	4-04-00-04255	 1.4581^{20}
11461	Silane, trichloroethenyl- Vinyltrichlorosilane	$C_2H_3Cl_3Si$ 161.49	75-94-5 -95	 91.5	4-04-00-04258 1.2426^{20}	chl 4 1.4295^{20}
11462	Silane, trichloroethoxy-	$C_2H_5Cl_3OSi$ 179.50	1825-82-7 -135	 101.9	4-01-00-01364 1.2274^{20}	EtOH 4 1.4045^{20}
11463	Silane, trichloroethyl- Ethyltrichlorosilane	$C_2H_5Cl_3Si$ 163.51	115-21-9 -105.6	 100.5	4-04-00-04227 1.2373^{20}	ctc 3 1.4256^{20}
11464	Silane, trichlorohexyl- Hexyltrichlorosilane	$C_6H_{13}Cl_3Si$ 219.61	928-65-4	 190	4-04-00-04251 1.1100^{20}	H_2O 6
11465	Silane, trichloromethyl- Methyltrichlorosilane	CH_3Cl_3Si 149.48	75-79-6 -90	 65.6	4-04-00-04212 1.273^{20}	H_2O 6; EtOH 6 1.4106^{20}
11466	Silane, trichloro(1-methylethyl)-	$C_3H_7Cl_3Si$ 177.53	4170-46-1 -87.7	 121	4-04-00-04244 1.1934^{20}	H_2O 6 1.4319^{20}
11467	Silane, trichloro(2-methylphenyl)-	$C_7H_7Cl_3Si$ 225.58	13835-81-9 -25.5	 226.7	4-16-00-01565 1.306^{25}	 1.5336^{25}
11468	Silane, trichloro(3-methylphenyl)-	$C_7H_7Cl_3Si$ 225.58	13688-75-0 -26	 226.5	4-16-00-01565 1.306^{25}	 1.5336^{25}
11469	Silane, trichloro(2-methylpropyl)-	$C_4H_9Cl_3Si$ 191.56	18169-57-8	 143.3	4-04-00-04248 1.154^{20}	H_2O 6
11470	Silane, trichlorooctadecyl- Octadecyltrichlorosilane	$C_{18}H_{37}Cl_3Si$ 387.94	112-04-9	 223^{10}	4-04-00-04256 0.984^{25}	 1.4602^{20}
11471	Silane, trichlorooctyl- Octyltrichlorosilane	$C_8H_{17}Cl_3Si$ 247.67	5283-66-9	 232	4-04-00-04253	H_2O 6; EtOH 6; ctc 3 1.4480^{20}
11472	Silane, trichloropentyl- Amyltrichlorosilane	$C_5H_{11}Cl_3Si$ 205.59	107-72-2	 172; 60.5^{15}	4-04-00-04249 1.1330^{20}	 1.4503^{20}
11473	Silane, trichlorophenyl- Phenyltrichlorosilane	$C_6H_5Cl_3Si$ 211.55	98-13-5	 201	4-16-00-01560 1.321^{20}	ctc 3; chl 3; CS_2 3 1.5230^{20}
11474	Silane, trichloro(2-phenylethyl)-	$C_8H_9Cl_3Si$ 239.60	940-41-0	 242; $98\text{-}9^{5}$	4-16-00-01568 1.2397^{20}	 1.5185^{20}
11475	Silane, trichloro-2-propenyl- Allyltrichlorosilane	$C_3H_5Cl_3Si$ 175.52	107-37-9 35	 117.5	4-04-00-04261 1.2011^{20}	 1.4460^{20}

No.	Name Synonym	Mol. Form. Mol. Wt.	CAS RN mp/°C	Merck No. bp/°C	Beil. Ref. den/g cm^{-3}	Solubility n_D
11476	Silane, trichloropropyl- Propyltrichlorosilane	$C_3H_7Cl_3Si$ 177.53	141-57-1	123.5	4-04-00-04239 1.195^{20}	1.4310^{20}
11477	Silane, triethoxy- Triethoxysilane	$C_6H_{16}O_3Si$ 164.28	998-30-1	133.5	4-01-00-01359 0.8745^{20}	
11478	Silane, triethoxyethyl-	$C_8H_{20}O_3Si$ 192.33	78-07-9	158.5	4-04-00-04223 0.8963^{20}	H_2O 1; EtOH 5; eth 5; chl 3 1.3955^{20}
11479	Silane, triethoxymethyl-	$C_7H_{18}O_3Si$ 178.30	2031-67-6	142	4-04-00-04204 0.8948^{25}	1.3832^{20}
11480	Silane, triethoxypentyl-	$C_{11}H_{26}O_3Si$ 234.41	2761-24-2	95^{13}	4-04-00-04248 0.8862^{20}	1.4059^{20}
11481	Silane, triethoxyphenyl-	$C_{12}H_{20}O_3Si$ 240.37	780-69-8	$112\text{-}3^{10}$	4-16-00-01556 0.996^{25}	1.4604^{20}
11482	Silane, triethoxy-2-propenyl-	$C_9H_{20}O_3Si$ 204.34	2550-04-1	100^{50}	4-04-00-04261 0.9030^{20}	1.4072^{20}
11483	Silane, triethyl- Triethylsilane	$C_6H_{16}Si$ 116.28	617-86-7	109	4-04-00-03895 0.7302^{20}	H_2O 1; sulf 1 1.447^{20}
11484	Silane, triethylfluoro-	$C_6H_{15}FSi$ 134.27	358-43-0	110	4-04-00-04051 0.8354^{25}	peth 4 1.3900^{25}
11485	Silane, triethylphenyl-	$C_{12}H_{20}Si$ 192.38	2987-77-1	236	4-16-00-01362 0.8915^{20}	1.4999^{20}
11486	Silane, trifluorophenyl-	$C_6H_5F_3Si$ 162.19	368-47-8 -18	101.5	4-16-00-01559 1.2169^{20}	bz 4; EtOH 4 1.4110^{20}
11487	Silane, trimethoxymethyl-	$C_4H_{12}O_3Si$ 136.22	1185-55-3	102.5	4-04-00-04203 0.9548^{20}	chl 3 1.3696^{20}
11488	Silane, trimethoxyphenyl-	$C_9H_{14}O_3Si$ 198.29	2996-92-1	130^{45}	4-16-00-01556 1.064^{20}	ctc 3; CS_2 3 1.4734^{20}
11489	Silane, trimethyl- Trimethylsilane	$C_3H_{10}Si$ 74.20	993-07-7 -135.9	6.7		
11490	Silane, trimethyl(4-methylphenyl)-	$C_{10}H_{16}Si$ 164.32	3728-43-6	192; 73^{10}	4-16-00-01396 0.860^{25}	1.4900^{20}
11491	Silane, trimethyl(2-methylpropyl)-	$C_7H_{18}Si$ 130.31	1118-09-8	108.5	1-04-00-00580 0.7330^{0}	
11492	Silane, trimethylphenoxy-	$C_9H_{14}OSi$ 166.30	1529-17-5 -55	119	4-06-00-00762 0.8681^{20}	1.5125^{20}
11493	Silane, trimethylphenyl-	$C_9H_{14}Si$ 150.30	768-32-1	169.5	4-16-00-01361 0.8722^{20}	ctc 3; CS_2 3 1.4907^{20}
11494	Silane, trimethyl(phenylmethyl)-	$C_{10}H_{16}Si$ 164.32	770-09-2	190.5	4-16-00-01401 0.8933^{20}	1.4941^{20}
11495	Silane, trimethyl-2-propenyl-	$C_6H_{14}Si$ 114.26	762-72-1	86	4-04-00-03927 0.7158^{25}	H_2O 1 1.4074^{20}
11496	Silane, trimethylpropyl- Trimethylpropylsilane	$C_6H_{16}Si$ 116.28	3510-70-1	89	4-04-00-03901 0.7196^{0}	1.3929^{20}
11497	Silane, trimethyl[4-[(trimethylsilyl)oxy]phenyl]- Silane, (oxy-*p*-phenylene)bis[trimethyl-	$C_{12}H_{22}OSi_2$ 238.48	18036-81-2	132^{25}	4-16-00-01441 0.900^{25}	1.4794^{25}
11498	Silanetriol, ethenyl-, triacetate	$C_8H_{12}O_6Si$ 232.27	4130-08-9	115^{10}	4-04-00-04258 1.167^{20}	1.4226^{20}
11499	Silanetriol, methyl-, triacetate	$C_7H_{12}O_6Si$ 220.25	4253-34-3 40.5	$110\text{-}12^{17}$	4-04-00-04208 1.1750^{20}	1.4083^{20}
11500	Silane, tripropyl- Tripropylsilane	$C_9H_{22}Si$ 158.36	998-29-8	172	4-04-00-03902 0.7723^{0}	H_2O 1 1.4280^{20}
11501	Silanol, ethyldimethyl- Silanol, dimethyl ethyl	$C_4H_{12}OSi$ 104.22	5906-73-0	120; 58^{50}	4-04-00-04034 0.8332^{20}	1.4070^{20}
11502	Silanol, methyldiphenyl-	$C_{13}H_{14}OSi$ 214.34	778-25-6 167	184^{24}; 148^{3}	4-16-00-01478 1.0840^{25}	ctc 3; CS_2 3
11503	Silanol, triethyl-	$C_6H_{16}OSi$ 132.28	597-52-4	154	4-04-00-04040 0.8647^{20}	H_2O 1; EtOH 5; eth 5 1.4329^{20}
11504	Silanol, triphenyl-	$C_{18}H_{16}OSi$ 276.41	791-31-1 154.8		4-16-00-01480 1.1777^{20}	ctc 3; CS_2 3
11505	Silicic acid (H4SiO4), tetrabutyl ester	$C_{16}H_{36}O_4Si$ 320.54	4766-57-8	120^{3}	4-01-00-01541 0.8990^{20}	1.4128^{20}
11506	Silicic acid (H4SiO4), tetraethyl ester Ethyl silicate	$C_8H_{20}O_4Si$ 208.33	78-10-4 -82.5	3805 168.8	4-01-00-01360 0.9320^{20}	H_2O 6 1.3928^{20}
11507	Silicic acid (H4SiO4), tetrakis(2-ethylbutyl) ester	$C_{24}H_{52}O_4Si$ 432.76	78-13-7		4-01-00-01725 0.8920^{20}	H_2O 1; EtOH 2; eth 3; bz 3 1.4307^{20}
11508	Silicic acid (H4SiO4), tetramethyl ester Methyl silicate	$C_4H_{12}O_4Si$ 152.22	681-84-5 -2	121	4-01-00-01266 1.0232^{20}	EtOH 4 1.3683^{20}
11509	Silicic acid (H4SiO4), tetraphenyl ester	$C_{24}H_{20}O_4Si$ 400.51	1174-72-7 49	417; 236^{1}	4-06-00-00766 1.1412^{60}	
11510	Silicic acid (H4SiO4), tetrapropyl ester	$C_{12}H_{28}O_4Si$ 264.44	682-01-9	226	4-01-00-01435 0.9158^{20}	ctc 3; CS_2 3 1.4012^{20}
11511	Silicic acid (H4SiO4), triethyl phenyl ester	$C_{12}H_{20}O_4Si$ 256.37	18023-36-4	$120\text{-}2^{10}$	4-06-00-00765 1.0283^{20}	1.4525^{20}
11512	Silvex Propanoic acid, 2-(2,4,5-trichlorophenoxy)-	$C_9H_7Cl_3O_3$ 269.51	93-72-1 181.6	8483		
11513	Simazine 1,3,5-Triazine-2,4-diamine, 6-chloro-*N,N'*-diethyl-	$C_7H_{12}ClN_5$ 201.66	122-34-9 226	8485	1.302^{20}	
11514	Sodium dimethyldithiocarbamate Carbamodithioic acid, dimethyl-, sodium salt	$C_3H_7NNaS_2$ 144.22	128-04-1			

No.	Name Synonym	Mol. Form. Mol. Wt.	CAS RN mp/°C	Merck No. bp/°C	Beil. Ref. den/g cm^{-3}	Solubility n_D
11515	Solanid-5-ene-3,12-diol, (3β,12α)- Rubijervine	$C_{27}H_{43}NO_2$ 413.64	79-58-3 242	8264	5-21-05-00208	bz 4; EtOH 4; chl 4
11516	Solanid-5-ene-3,18-diol, (3β)- Isorubijervine	$C_{27}H_{43}NO_2$ 413.64	468-45-1 242.5	5111	5-21-05-00208	bz 4; chl 4
11517	Sonar 2-Naphthalenol, 1,1'-methylenebis-, monosodium salt	$C_{21}H_{16}NaO_2$ 323.35	38827-66-6 132			
11518	*DL*-Sorbose	$C_6H_{12}O_6$ 180.16	65732-90-3 162.5	8681	4-01-00-04410 1.638^{17}	H_2O 4; MeOH 4
11519	*L*-Sorbose *L*-Sorbinose	$C_6H_{12}O_6$ 180.16	87-79-6 165	8681	4-01-00-04412 1.612^{17}	H_2O 3; EtOH 2; eth 2; MeOH 2
11520	Spiro[benzofuran-2(3H),1'-[2]cyclohexene]-3,4'-dione, 7-chloro-2',4,6-trimethoxy- Griseofulvin, (+)-	$C_{17}H_{17}ClO_6$ 352.77	126-07-8 220	4453	5-18-05-00150	H_2O 1; EtOH 2; eth 2; ace 2
11521	Spiro[4.5]dec-6-en-8-one, 6,10-dimethyl-2-(1-methylethylidene)-, (5*R*-*cis*)-	$C_{15}H_{22}O$ 218.34	18444-79-6 44.5	 $141\text{-}2^{2}$	4-07-00-00822 1.0001^{20}	ace 3 1.5309^{20}
11522	Spiro[isobenzofuran-1(3H),9'-[9H]xanthen]-3-one, 3',6'-dihydroxy- Fluorescein	$C_{20}H_{12}O_5$ 332.31	2321-07-5 315 dec	4085	5-19-06-00456	H_2O 2; EtOH 2; eth 2; ace 4
11523	Spiro[isobenzofuran-1(3H),9'-[9H]xanthen]-3-one, 3',6'-dihydroxy-4',5'-diiodo- Fluorescein, 4',5'-diiodo	$C_{20}H_{10}I_2O_5$ 584.11	38577-97-8	3174	5-19-06-00466	
11524	Spiro[isobenzofuran-1(3H),9'-[9H]xanthen]-3-one, 3',6'-dihydroxy-2',4',5',7'-tetraiodo-, disodium salt Erythrosin	$C_{20}H_8I_4O_5$ 835.90	16423-68-0	3640	4-19-00-02923	eth 4; EtOH 4
11525	Spiro[isobenzofuran-1(3H),9'-[9H]xanthen]-3-one, 2',4',5',7'-tetrabromo-3',6'-dihydroxy-, disodium salt Eosin	$C_{20}H_6Br_4Na_2O_5$ 691.86	17372-87-1 295.5	3555		EtOH 4
11526	Spiro[isobenzofuran-1(3H),9'-[9H]xanthen]-3-one, 3',4',5',6'-tetrahydroxy- Gallein	$C_{20}H_{12}O_5$ 332.31	2103-64-2 >300	4250	5-19-07-00083	ace 4; EtOH 4
11527	Spiropentane Spiro[2.2]pentane	C_5H_8 68.12	157-40-4 -134.6	 39	4-05-00-00218 0.7266^{20}	 1.4120^{20}
11528	Spirosolan-3-ol, (3β,5α,22β,25*S*)- Tomatidine	$C_{27}H_{45}NO_2$ 415.66	77-59-8 210.5	9470		EtOH 3; eth 3
11529	Spirosol-5-en-3-ol, (3β,22α,25*R*)- Solasodine	$C_{27}H_{43}NO_2$ 413.64	126-17-0 202	8665	4-27-00-02000	EtOH 3; eth 2; ace 3; bz 3
11530	Spirostan-2,3-diol, (2α,3β,5α,25*R*)- Gitogenin	$C_{27}H_{44}O_4$ 432.64	511-96-6 271.5	4325	5-19-03-00338	H_2O 1; EtOH 3; eth 2; chl 3
11531	Spirostan-3-ol, (3β,5β,25*S*)- Sarsasapogenin	$C_{27}H_{44}O_3$ 416.64	126-19-2 200.5	8338	5-19-02-00672	EtOH 3; ace 3; bz 3; chl 3
11532	Spirostan-3-ol, (3β,5α,25*R*)- Tigogenin	$C_{27}H_{44}O_3$ 416.64	77-60-1 205.5	9367		EtOH 3; eth 3; ace 3; ctc 3
11533	Spirostan-3-ol, (3β,5β,25*R*)- Smilagenin	$C_{27}H_{44}O_3$ 416.64	126-18-1 185	8508	5-19-02-00673	ace 4; bz 4; EtOH 4
11534	Spirostan-12-one, 3-hydroxy-, (3β,5α,25*R*)- Hecogenin	$C_{27}H_{42}O_4$ 430.63	467-55-0 266.5	4537	5-19-05-00592	ace 4; eth 4; EtOH 4
11535	Spirostan-2,3,15-triol, (2α,3β,5α,15β,25*R*)- Digitogenin	$C_{27}H_{44}O_5$ 448.64	511-34-2 281.5	3143	5-19-03-00603	chl 4
11536	Spirost-5-en-3-ol, (3β,25*R*)- Diosgenin	$C_{27}H_{42}O_3$ 414.63	512-04-9 205.5	3289	5-19-03-00030	EtOH 4
11537	Spiro[5.5]undecane Spir[5.]undecane	$C_{11}H_{20}$ 152.28	180-43-8	 207	4-05-00-00328 0.8783^{20}	 1.4731
11538	Stannane, (acetyloxy)tributyl- Acetoxytributylstannane	$C_{14}H_{30}O_2Sn$ 349.10	56-36-0 84.7			
11539	Stannane, (acetyloxy)triphenyl- Triphenyltin acetate	$C_{20}H_{18}O_2Sn$ 409.07	900-95-8 121.5			
11540	Stannane, chlorotrimethyl-	C_3H_9ClSn 199.27	1066-45-1 38.5	 148		H_2O 3; chl 3; os 3
11541	Stannane, chlorotriphenyl- Triphenyltin chloride	$C_{18}H_{15}ClSn$ 385.48	639-58-7 103.5			chl 3
11542	Stannane, chlorotripropyl-	$C_9H_{21}ClSn$ 283.43	2279-76-7 -23.5	 123^{13}	 1.2678^{28}	ctc 3; os 3 1.49102^{28}
11543	Stannane, tetraethenyl-	$C_8H_{12}Sn$ 226.89	1112-56-7	 $55\text{-}7^{17}$	4-04-00-04316 1.2651^{25}	ctc 3 1.4993^{25}
11544	Stannane, tetramethyl- Tetramethylstannane	$C_4H_{12}Sn$ 178.85	594-27-4 -54.8	 78	4-04-00-04307 1.314^{25}	H_2O 1; ctc 3; CS_2 3; os 3 1.4386
11545	Stannane, tetraphenyl-	$C_{24}H_{20}Sn$ 427.13	595-90-4 228	 420	4-16-00-01592	chl 2
11546	Stannane, tetrapropyl-	$C_{12}H_{28}Sn$ 291.06	2176-98-9 -109.1	 228	4-04-00-04310 1.1065^{20}	 1.4745^{20}
11547	Stibine, dichlorophenyl-	$C_6H_5Cl_2Sb$ 269.77	5035-52-9 69.5	 $110\text{-}5^{10}$	4-16-00-01205	ace 4; EtOH 4; HOAc 4
11548	Stibine oxide, dihydroxyphenyl- Benzenestibonic acid	$C_6H_7O_3Sb$ 248.88	535-46-6 139	1077	3-16-00-01178	

No.	Name Synonym	Mol. Form. Mol. Wt.	CAS RN mp/°C	Merck No. bp/°C	Beil. Ref. den/g cm^{-3}	Solubility n_D
11549	Stibine, triethyl- Triethylstibine	$C_6H_{15}Sb$ 208.94	617-85-6 -98	 161.4	4-04-00-03685 1.3224^{15}	H_2O 1; EtOH 3; eth 3
11550	Stibine, trimethyl- Trimethylstibine	C_3H_9Sb 166.86	594-10-5 -62	 80.6	4-04-00-03684 1.523^{15}	H_2O 1; EtOH 3; eth 3; CS_2 3 1.42^{15}
11551	Stibine, triphenyl-	$C_{18}H_{15}Sb$ 353.07	603-36-1 53.5	 >360	4-16-00-01198 1.4343^{25}	H_2O 1; EtOH 3; eth 4; ace 4 1.6948^{42}
11552	Stibine, tris(3-methylphenyl)-	$C_{21}H_{21}Sb$ 395.15	35569-54-1 72		4-16-00-01199 1.3957^{16}	bz 4; eth 4; EtOH 4; chl 4
11553	Stibine, tris(4-methylphenyl)-	$C_{21}H_{21}Sb$ 395.15	5395-43-7 127.5	 218^{3}	4-16-00-01199 1.3595^{16}	bz 4; eth 4; chl 4
11554	Stigmasta-5,7-dien-3-ol, (3β)- 7-Dehydrositosterol	$C_{29}H_{48}O$ 412.70	521-04-0 144.5	2863	3-06-00-02855	bz 4; eth 4; EtOH 4
11555	Stigmasta-5,22-dien-3-ol, (3β,22E)- Stigmasterol	$C_{29}H_{48}O$ 412.70	83-48-7 170	8771	4-06-00-04170	bz 4; eth 4; EtOH 4
11556	Stigmasta-7,24(28)-dien-3-ol, 4-methyl-, (3β,4α,5α,24Z)- α1-Sitosterol	$C_{30}H_{50}O$ 426.73	474-40-8 166	8499	3-06-00-02876	EtOH 4; chl 4
11557	Stigmastan-3-ol, (3β,5α)- Stigmastanol	$C_{29}H_{52}O$ 416.73	83-45-4 144	8770	3-06-00-02172	
11558	Stigmast-5-en-3-ol, (3β)- β-Sitosterol	$C_{29}H_{50}O$ 414.72	83-46-5 140	8500	4-06-00-04037	EtOH 3; eth 3; HOAc 3
11559	Stigmast-5-en-3-ol, (3β,24*S*)- γ-Sitosterol	$C_{29}H_{50}O$ 414.72	83-47-6	8501		
11560	Stilbene, α-ethyl- 1-Butene, 1,2-diphenyl	$C_{16}H_{16}$ 208.30	22692-70-2 57	 296.5	4-05-00-02206 1.0124^{18}	 1.593^{18}
11561	*D*-Streptamine, *O*-2-deoxy-2-(methylamino)-α-*L*-glucopyranosyl-(12)-*O*-5-deoxy-3-*C*-formyl-α-L-lyxofuranosyl-(1-4)-*N,N*' 55-bis(aminoiminomethyl)-, sulfate (2:3)(salt) Streptomycin, sulfate	$C_{42}H_{84}N_{14}O_{36}S_3$ 1457.40	3810-74-0		5-18-11-00082	
11562	Streptomycin *N*-Methyl-*L*-glucosamidinostreptosidostreptidine	$C_{21}H_{39}N_7O_{12}$ 581.58	57-92-1	8786		
11563	Strychnidin-10-one Strychnine	$C_{21}H_{22}N_2O_2$ 334.42	57-24-9 287	8822 270^{5}	4-27-00-07537 1.36^{20}	H_2O 2; EtOH 2; eth 1; ace 2
11564	Strychnidin-10-one, 2,3-dimethoxy- Brucine	$C_{23}H_{26}N_2O_4$ 394.47	357-57-3 178	1443	4-27-00-07875	H_2O 2; EtOH 4; eth 2; bz 2
11565	Strychnidin-10-one, mononitrate Strychnine nitrate	$C_{21}H_{23}N_3O_5$ 397.43	66-32-0 295	8823	2-27-00-00723 1.627^{25}	H_2O 4; MeOH 4
11566	Strychnidin-10-one, sulfate (2:1) Strychnine, sulfate	$C_{42}H_{46}N_4O_8S$ 766.92	60-41-3 200 dec	8826		H_2O 3; EtOH 3; eth 1; chl 2
11567	Succinimide, *N*-(ethoxymethyl)-	$C_7H_{11}NO_3$ 157.17	98431-97-1 31.5	 262	2-21-00-00304	H_2O 4; EtOH 4
11568	Sulfamic acid, cyclohexyl- Cyclamic acid	$C_6H_{13}NO_3S$ 179.24	100-88-9 169.5	2707	4-12-00-00102	alk 4
11569	Sulfamide, tetramethyl-	$C_4H_{12}N_2O_2S$ 152.22	3768-63-6 73	 225	4-04-00-00270	EtOH 4
11570	Sulfide, *m*-tolyl *p*-tolyl 3,4'-Ditolyl sulfide	$C_{14}H_{14}S$ 214.33	107770-92-3 28	 179^{11}	4-06-00-02173	eth 4; EtOH 4
11571	Sulfoxylic acid, diethyl ester	$C_4H_{10}O_2S$ 122.19	10297-38-8	 118	3-01-00-01314 0.9940^{20}	H_2O 3; EtOH 4; eth 4; ace 4 1.4234^{20}
11572	Sulfur cyanide	C_2N_2S 84.10	627-52-1 63.5	 sub 30	4-03-00-00339	eth 4; EtOH 4
11573	Sulfuric acid, diethyl ester Diethyl sulfate	$C_4H_{10}O_4S$ 154.19	64-67-5 -24	3120 208	4-01-00-01326 1.172^{25}	H_2O 1; EtOH 5; eth 5 1.3989^{20}
11574	Sulfuric acid, dimethyl ester Dimethyl sulfate	$C_2H_6O_4S$ 126.13	77-78-1 -27	3244 188 dec; 76^{15}	4-01-00-01251 1.3322^{20}	H_2O 3; EtOH 5; eth 3; bz 3 1.3874^{20}
11575	Sulfuric acid, dipropyl ester	$C_6H_{14}O_4S$ 182.24	598-05-0	 121^{30}	4-01-00-01423 1.1064^{20}	peth 4 1.4135^{20}
11576	Sulfuric acid, monoethyl ester Ethyl sulfate	$C_2H_6O_4S$ 126.13	540-82-9	3808 280 dec	4-01-00-01324 1.3657^{20}	H_2O 4 1.4105^{20}
11577	Sulfuric acid, monomethyl ester Methyl sulfate	CH_4O_4S 112.11	75-93-4 <-30	6041 135 dec	4-01-00-01250	H_2O 4; eth 4; EtOH 4
11578	Sulfurous acid, bis(2-methylpropyl) ester	$C_8H_{18}O_3S$ 194.30	18748-27-1	 210; 92-4^{13}	4-01-00-01595 0.9862^{20}	 1.4268^{20}
11579	Sulfurous acid, 2-chloroethyl 2-[4-(1,1-dimethylethyl)phenoxy]-1-methylethyl ester Aramite	$C_{15}H_{23}ClO_4S$ 334.86	140-57-8 -37.3	794 195^{2}	4-06-00-03301 1.143^{20}	ace 4; bz 4; eth 4; EtOH 4 1.5100^{20}
11580	Sulfurous acid, dibutyl ester Butyl sulfite	$C_8H_{18}O_3S$ 194.30	626-85-7	 230	4-01-00-01522 0.9957^{20}	EtOH 3; eth 3 1.4310^{20}
11581	Sulfurous acid, diethyl ester Ethyl sulfite	$C_4H_{10}O_3S$ 138.19	623-81-4	 158; 51^{13}	4-01-00-01324 1.1^{20}	EtOH 3; eth 3 1.4310^{20}
11582	Sulfurous acid, dimethyl ester	$C_2H_6O_3S$ 110.13	616-42-2	 126	4-01-00-01250 1.2129^{20}	H_2O 3; EtOH 3; eth 3 1.4083^{20}
11583	Sulfurous diamide, tetramethyl-	$C_4H_{12}N_2OS$ 136.22	3768-60-3 31	 209; 74^{13}	4-04-00-00269	eth 4
11584	Sulfur, pentafluoro(trifluoromethyl)-, (OC-6-21)-	CF_8S 196.06	373-80-8 -87	 -20	4-03-00-00035	H_2O 1; EtOH 1; CS_2 3

No.	Name Synonym	Mol. Form. Mol. Wt.	CAS RN mp/°C	Merck No. bp/°C	Beil. Ref. den/g cm^{-3}	Solubility n_D
11585	Sulfuryl chloride isocyanate	$CClNO_3S$ 141.53	1189-71-5 -44	 107	4-03-00-00086 1.626^{25}	 1.4467^{20}
11586	Sutan Carbamothioic acid, bis(2-methylpropyl)-, *S*-ethyl ester	$C_{11}H_{23}NOS$ 217.38	2008-41-5	1546 138^{21}	 0.9402^{25}	
11587	*D*-Tagatose Tagatose, *D*-	$C_6H_{12}O_6$ 180.16	87-81-0 134.5	9005	4-01-00-04414	H_2O 4
11588	Tartar emetic Antimonate(2-), bis[μ-[2,3-dihydroxybutanedioato(4-)-O1,O2:O3,O4]]di-, dipotassium, trihydrate, stereoisomer	$C_8H_{10}K_2O_{15}Sb_2$ 667.87	28300-74-5	732	 2.6	
11589	Tebuconazole (*RS*)-1-*p*-Chlorophenyl-4,4-dimethyl-3-(1H-1,2,4-triazol-1-ylmethyl)pentan-3-ol	$C_{16}H_{23}ClN_3O$ 308.83	107534-96-3 102.4			
11590	Tebuthiuron Urea, *N*-[5-(1,1-dimethylethyl)-1,3,4-thiadiazol-2-yl]-*N,N'*-dimethyl-	$C_9H_{16}N_4OS$ 228.32	34014-18-1 163 dec	9053		
11591	Terbacil 2,4(1H,3H)-Pyrimidinedione, 5-chloro-3-(1,1-dimethylethyl)-6-methyl-	$C_9H_{13}ClN_2O_2$ 216.67	5902-51-2 176	9085 sub 175	 1.34^{25}	
11592	Terbufos Phosphorodithioic acid, *S*-[[(1,1-dimethylethyl)thio]methyl] *O,O*-diethyl ester	$C_9H_{21}O_2PS_3$ 288.44	13071-79-9 -29.2	9088 $69^{0.01}$	 1.105^{24}	
11593	Terbuthylazine 1,3,5-Triazine-2,4-diamine, 6-chloro-*N*-(1,1-dimethylethyl)-*N'*-ethyl-	$C_9H_{16}ClN_5$ 229.71	5915-41-3 178		 1.188^{20}	
11594	Terbutryn 1,3,5-Triazine-2,4-diamine, *N*-(1,1-dimethylethyl)-*N'*-ethyl-6-(methylthio)-	$C_{10}H_{19}N_5S$ 241.36	886-50-0 104	 $157^{0.06}$	 1.115^{20}	
11595	1,1':2',1''-Terphenyl *o*-Terphenyl	$C_{18}H_{14}$ 230.31	84-15-1 56.2	 332	4-05-00-02478	H_2O 1; ace 3; bz 3; chl 3
11596	1,1':3',1''-Terphenyl *m*-Terphenyl	$C_{18}H_{14}$ 230.31	92-06-8 87	 363	4-05-00-02480	H_2O 1; EtOH 3; eth 3; bz 3
11597	1,1':4',1''-Terphenyl *p*-Terphenyl	$C_{18}H_{14}$ 230.31	92-94-4 210.1	 376	4-05-00-02483	H_2O 1; EtOH 2; eth 3; bz 3
11598	1,1':3',1''-Terphenyl, 5'-phenyl-	$C_{24}H_{18}$ 306.41	612-71-5 176	 462	4-05-00-02732 1.199^{30}	H_2O 1; EtOH 3; eth 3; bz 4
11599	Terrazole 1,2,4-Thiadiazole, 5-ethoxy-3-(trichloromethyl)-	$C_5H_5Cl_3N_2OS$ 247.53	2593-15-9 19.9	 95^{1}	 1.503^{25}	
11600	2,2':5',2''-Terthiophene α-Terthienyl	$C_{12}H_8S_3$ 248.39	1081-34-1	9106	5-19-09-00226	
11601	1,3,5,7-Tetraazatricyclo[3.3.1.13,7]decane Methenamine	$C_6H_{12}N_4$ 140.19	100-97-0 >250	5879 sub	4-26-00-01680 1.331^{-5}	H_2O 4; EtOH 3; eth 2; ace 3
11602	2,6,10,14,18,22-Tetracosahexaene, 2,6,10,15,19,23-hexamethyl-, (all-E)- Squalene	$C_{30}H_{50}$ 410.73	111-02-4 <-20	8727 280^{17}; 241^{4}	3-01-00-01071 0.8584^{20}	H_2O 1; EtOH 2; eth 3; ace 3 1.4990^{20}
11603	Tetracosane	$C_{24}H_{50}$ 338.66	646-31-1 54	 391.3	4-01-00-00578 0.7991^{20}	H_2O 1; EtOH 2; eth 4 1.4283^{70}
11604	Tetracosane, 11-decyl-	$C_{34}H_{70}$ 478.93	55429-84-0 10.8	 301^{10}	4-01-00-00598 0.8161^{20}	 1.4556^{20}
11605	Tetracosane, 3-ethyl-	$C_{26}H_{54}$ 366.71	55282-17-2 30.1	 255.5^{10}	4-01-00-00584 0.7949^{40}	 1.4436^{40}
11606	Tetracosane, 2,6,10,15,19,23-hexamethyl- Squalane	$C_{30}H_{62}$ 422.82	111-01-3 -38	8726 350	4-01-00-00593 0.8115^{15}	H_2O 1; EtOH 2; eth 3; ace 2 1.4530^{15}
11607	Tetracosanoic acid Lignoceric acid	$C_{24}H_{48}O_2$ 368.64	557-59-5 84.2	5364 272^{10}	4-02-00-01301 0.8207^{100}	bz 4; eth 4 1.4287^{100}
11608	15-Tetracosenoic acid	$C_{24}H_{46}O_2$ 366.63	14490-79-0 61		4-02-00-01681	EtOH 3; chl 3; ace 3
11609	5,9-Tetradecadien-7-yne, 6,9-dimethyl-	$C_{16}H_{26}$ 218.38	101427-54-7	 $95\text{-}8^{0.3}$	4-01-00-01133 0.8241^{20}	bz 4; eth 4 1.4866^{20}
11610	6,8-Tetradecadiyne	$C_{14}H_{22}$ 190.33	16387-72-7 2	 $118\text{-}9^{4}$	3-01-00-01067 0.8699^{16}	eth 4
11611	Tetradecanamide	$C_{14}H_{29}NO$ 227.39	638-58-4 104	 217^{12}	4-02-00-01138	EtOH 4
11612	1-Tetradecanamine	$C_{14}H_{31}N$ 213.41	2016-42-4 83.1	 291.2	4-04-00-00812 0.8079^{20}	H_2O 1; EtOH 4; eth 4; ace 3 1.4463^{20}
11613	Tetradecane	$C_{14}H_{30}$ 198.39	629-59-4 5.8	 253.5	4-01-00-00520 0.7628^{20}	H_2O 1; EtOH 4; eth 4; ctc 3 1.4290^{20}
11614	Tetradecane, 1-bromo-	$C_{14}H_{29}Br$ 277.29	112-71-0 5.6	 307	4-01-00-00523 1.0170^{20}	ace 4; bz 4; EtOH 4 1.4603^{20}
11615	Tetradecane, 1-chloro-	$C_{14}H_{29}Cl$ 232.84	2425-54-9 4.9	 292	3-01-00-00550 0.8665^{20}	H_2O 1; EtOH 3; ace 4; bz 4 1.4473^{20}
11616	Tetradecane, 1,14-dibromo- Tetradecamethylene dibromide	$C_{14}H_{28}Br_2$ 356.18	37688-96-3 50.4	 190^{8}	4-01-00-00523	eth 4; EtOH 4; chl 4
11617	Tetradecane, 1,1-dimethoxy- Myristaldehyde, dimethyl acetal	$C_{16}H_{34}O_2$ 258.44	14620-53-2	 $134\text{-}6^{4}$	4-01-00-03389	eth 4; EtOH 4 1.4342^{25}
11618	1,14-Tetradecanediol	$C_{14}H_{30}O_2$ 230.39	19812-64-7 85.8	 200^{9}	4-01-00-02631	eth 4; EtOH 4

No.	Name	Synonym	Mol. Form.	Mol. Wt.	CAS RN	mp/°C	Merck No.	bp/°C	Beil. Ref.	den/g cm^{-3}	Solubility	n_D
11619	Tetradecanenitrile	Myristonitrile	$C_{14}H_{27}N$	209.38	629-63-0	19		226^{100}; 119^{1}	4-02-00-01139	0.8281^{19}	H_2O 1; EtOH 5; eth 5; ace 5	1.4392^{23}
11620	1-Tetradecanesulfonic acid	Tetradecylsulfonic acid	$C_{14}H_{30}O_3S$	278.46	7314-37-6	65			4-04-00-00066	0.9996^{25}	H_2O 4	
11621	1-Tetradecanethiol		$C_{14}H_{30}S$	230.46	2079-95-0			$176\text{-}80^{22}$	4-01-00-01867	0.8469^{20}	H_2O 1; EtOH 3; eth 3; ctc 3	1.4597^{20}
11622	Tetradecanoic acid	Myristic acid	$C_{14}H_{28}O_2$	228.38	544-63-8	53.9	6246	250^{100}	4-02-00-01126	0.8622^{54}	H_2O 1; EtOH 3; eth 2; ace 3	1.4723^{70}
11623	Tetradecanoic acid, anhydride		$C_{28}H_{54}O_3$	438.73	626-29-9	53.4			4-02-00-01138	0.8502^{70}	eth 4; EtOH 4	1.4335^{70}
11624	Tetradecanoic acid, ethyl ester		$C_{16}H_{32}O_2$	256.43	124-06-1	12.3		295	4-02-00-01131	0.8573^{25}	H_2O 1; EtOH 3; eth 2; ctc 3	1.4362^{20}
11625	Tetradecanoic acid, methyl ester		$C_{15}H_{30}O_2$	242.40	124-10-7	19		295; 155^{7}	4-02-00-01131	0.8671^{20}	H_2O 1; EtOH 5; eth 5; ace 5	1.425^{45}
11626	Tetradecanoic acid, 1-methylethyl ester	Isopropyl myristate	$C_{17}H_{34}O_2$	270.46	110-27-0		5103	193^{20}; 140^{2}	4-02-00-01132	0.8532^{20}	H_2O 1; EtOH 3; eth 3; ace 4	1.4325^{25}
11627	Tetradecanoic acid, phenylmethyl ester		$C_{21}H_{34}O_2$	318.50	31161-71-4	20.5		229.3^{11}	2-06-00-00417	0.9293^{25}	bz 4; eth 4; EtOH 4; chl 4	
11628	Tetradecanoic acid, 1,2,3-propanetriyl ester	Trimyristin	$C_{45}H_{86}O_6$	723.17	555-45-3	56.5	9638	311	4-02-00-01135	0.8848^{60}	H_2O 1; EtOH 2; eth 3; ace 3	1.4428^{60}
11629	Tetradecanoic acid, propyl ester		$C_{17}H_{34}O_2$	270.46	14303-70-9			147^{2}	4-02-00-01132	0.8592^{20}	ace 4; bz 4; eth 4; EtOH 4	1.4356^{25}
11630	1-Tetradecanol	Tetradecyl alcohol	$C_{14}H_{30}O$	214.39	112-72-1	39.5	6248	289	4-01-00-01864	0.8236^{38}	H_2O 1; EtOH 4; eth 4; ace 4	
11631	2-Tetradecanol		$C_{14}H_{30}O$	214.39	4706-81-4	34		284	4-01-00-01867	0.8315^{20}		1.4444^{20}
11632	3-Tetradecanol		$C_{14}H_{30}O$	214.39	1653-32-3	31.5		173^{25}; 146^{10}	4-01-00-01868	0.8098^{53}	bz 4; eth 4; EtOH 4	1.4340^{45}
11633	2-Tetradecanone	Dodecylmethylketone	$C_{14}H_{28}O$	212.38	2345-27-9	33.5		205^{100}; 134^{13}	4-01-00-03389		H_2O 1; EtOH 3; ace 3; os 3	
11634	3-Tetradecanone		$C_{14}H_{28}O$	212.38	629-23-2	34		152^{16}	4-01-00-03389		H_2O 1; EtOH 3; ace 3; os 3	
11635	Tetradecanoyl chloride	Myristoyl chloride	$C_{14}H_{27}ClO$	246.82	112-64-1	-1		171^{16}	4-02-00-01138	0.9078^{25}	eth 3	
11636	1-Tetradecene		$C_{14}H_{28}$	196.38	1120-36-1	-12		233	4-01-00-00924	0.7745^{25}	H_2O 1; EtOH 4; eth 4; bz 3	1.4351^{20}
11637	4-Tetradecenoic acid	Tuduic acid	$C_{14}H_{26}O_2$	226.36	544-65-0	18.5		$185\text{-}8^{13}$	4-02-00-01626	0.9024^{20}	bz 4; peth 4	1.4559^{20}
11638	5-Tetradecenoic acid	Physoteric acid	$C_{14}H_{26}O_2$	226.36	544-66-1	20		$190\text{-}5^{15}$	3-02-00-01373	0.9046^{20}		1.4552^{20}
11639	9-Tetradecenoic acid		$C_{14}H_{26}O_2$	226.36	13147-06-3	-4		$144^{0.6}$	4-02-00-01626	0.9018^{20}		1.4519^{20}
11640	2-Tetradecyne		$C_{14}H_{26}$	194.36	638-60-8	6.5		252.5	0-01-00-00262	0.8000^{20}	eth 4; EtOH 4	
11641	7-Tetradecyne		$C_{14}H_{26}$	194.36	35216-11-6			144^{30}	3-01-00-01027	0.7991^{20}	eth 4; EtOH 4	1.4330^{25}
11642	2,5,8,11-Tetraoxadodecane	Triglyme	$C_8H_{18}O_4$	178.23	112-49-2	-45	9604	216	4-01-00-02401	0.986^{20}	H_2O 4; bz 4	1.4224^{20}
11643	2,4,8,10-Tetraoxaspiro[5.5]undecane		$C_7H_{12}O_4$	160.17	126-54-5	48.3		147^{53}; 68^{1}	5-19-11-00342		H_2O 4; ace 4; eth 4; EtOH 4	
11644	Tetraphenylene	Tetrabenzocyclooctatetraene	$C_{24}H_{16}$	304.39	212-74-8	233		sub 200	4-05-00-02773		EtOH 3; eth 2; AcOEt 3; $PhNO_2$ 3	
11645	Tetraphosphoric acid, hexaethyl ester	Ethyl tetraphosphate	$C_{12}H_{30}O_{13}P_4$	506.26	757-58-4	-40		150 dec		1.2917^{27}	ace 4; bz 4; EtOH 4	1.4273^{27}
11646	Tetrasiloxane, decamethyl-	Decamethyltetrasiloxane	$C_{10}H_{30}O_3Si_4$	310.69	141-62-8	-76	2843	194	4-04-00-04119	0.8536^{25}	H_2O 1; EtOH 2; bz 3; peth 3	1.3895^{20}
11647	Tetrasiloxane, 1,1,1,3,5,7,7,7-octamethyl-	1,1,1,3,5,7,7,7-Octamethyltetrasiloxane	$C_8H_{26}O_3Si_4$	282.63	16066-09-4			170	4-04-00-04098	0.8559^{20}		1.3854^{20}
11648	Tetrasul	*p*-Chlorophenyl 2,4,5-trichlorophenyl sulfide	$C_{12}H_6Cl_4S$	324.06	2227-13-6							
11649	Tetrasulfide, bis(1,1-dimethylethyl)	Di-*tert*-Butyl tetrasulfide	$C_8H_{18}S_4$	242.49	5943-35-1	2.3		$70^{0.2}$	4-01-00-01638	1.0690^{20}		1.5660^{20}
11650	Tetratriacontane		$C_{34}H_{70}$	478.93	14167-59-0	72.6		285.4^{3}	4-01-00-00597	0.7728^{90}		1.4296^{90}
11651	1,2,4,5-Tetrazine	*sym*-Tetrazine	$C_2H_2N_4$	82.06	290-96-0	99		sub	4-26-00-01710		H_2O 3; EtOH 3; eth 3; sulf 3	
11652	1H-Tetrazole		CH_2N_4	70.05	288-94-8	157		sub	4-26-00-01652	1.4060^{20}	H_2O 2	
11653	2H-Tetrazolium, 2,3,5-triphenyl-, chloride	Triphenyltetrazolium chloride	$C_{19}H_{15}ClN_4$	334.81	298-96-4	243 dec	9658		4-26-00-01774		H_2O 3; EtOH 3; eth 1; ace 3	
11654	5H-Tetrazolo[1,5-a]azepine, 6,7,8,9-tetrahydro-	Pentylenetetrazole	$C_6H_{10}N_4$	138.17	54-95-5	59.5	7097	194^{12}	4-26-00-01712		H_2O 4; EtOH 4; eth 3; ace 4	
11655	4-Thia-1-azabicyclo[3.2.0]heptane-2-carboxylic acid, 3,3-dimethyl-7-oxo-6-[(phenylacetyl)amino]-	Penicillin G, sodium salt	$C_{16}H_{17}N_2NaO_4S$	356.38	69-57-8		1157		4-27-00-05861			

No.	Name Synonym	Mol. Form. Mol. Wt.	CAS RN mp/°C	Merck No. bp/°C	Beil. Ref. den/g cm^{-3}	Solubility n_D
11656	Thiabendazole 1H-Benzimidazole, 2-(4-thiazolyl)-	$C_{10}H_7N_3S$ 201.25	148-79-8	9217 sub 305		
11657	1,2,4-Thiadiazole 1-Thia-2,4-diazacyclopentadiene	$C_2H_2N_2S$ 86.12	288-92-6 -33	 121	4-27-00-07093 1.3298^{20}	
11658	1,2,5-Thiadiazole Piazthiole	$C_2H_2N_2S$ 86.12	288-39-1 -50.1	 94	 1.268^{25}	 1.5150^{25}
11659	1,3,4-Thiadiazole, 2,5-dimethyl- 2,5-Dimethyl-1,3,4-thiadiazole	$C_4H_6N_2S$ 114.17	27464-82-0 65	 202.5	4-27-00-07097	H_2O 2; EtOH 2; eth 2
11660	Thianthrene	$C_{12}H_8S_2$ 216.33	92-85-3 159.3	 365	5-19-02-00049 1.4420^{20}	H_2O 1; EtOH 2; eth 3; bz 3
11661	Thianthrene, 2,7-dimethyl- Mesulphen	$C_{14}H_{12}S_2$ 244.38	135-58-0 123	5821 184^{3}	5-19-02-00072	ace 4; eth 4; peth 4; chl 4
11662	2-Thiazolamine 2-Aminothiazole	$C_3H_4N_2S$ 100.14	96-50-4 93	494 140^{11}	4-27-00-04574	H_2O 2; EtOH 2; eth 2; chl 2
11663	2-Thiazolamine, 5-[(4-aminophenyl)sulfonyl]- Thiazolsulfone	$C_9H_9N_3O_2S_2$ 255.32	473-30-3 219-21 dec	9237	4-27-00-05401	ace 4; eth 4; EtOH 4; diox 4
11664	2-Thiazolamine, 4,5-dihydro-	$C_3H_6N_2S$ 102.16	1779-81-3 85.3	 dec	4-27-00-04555	H_2O 4; EtOH 4; bz 4; chl 4
11665	2-Thiazolamine, 4-methyl- 2-Amino-4-methylthiazole	$C_4H_6N_2S$ 114.17	1603-91-4 45.5	462 125^{20}; $70^{0.4}$	4-27-00-04687	H_2O 4; EtOH 4; eth 4
11666	2-Thiazolamine, 5-nitro- 2-Amino-5-nitrothiazole	$C_3H_3N_3O_2S$ 145.14	121-66-4 202 dec	469	4-27-00-04675	
11667	Thiazole	C_3H_3NS 85.13	288-47-1	9235 118	4-27-00-00960 1.1998^{17}	H_2O 2; EtOH 3; eth 3; ace 3 1.5969^{20}
11668	Thiazole, 2-bromo- 2-Bromo-1,3-thiazole	C_3H_2BrNS 164.03	3034-53-5	 171	4-27-00-00962 1.82^{25}	 1.5927^{20}
11669	5-Thiazolecarboxylic acid, 4-methyl- 4-Methyl-5-thiazolecarboxylic acid	$C_5H_5NO_2S$ 143.17	20485-41-0 280 dec	 sub 250	4-27-00-04008	H_2O 2; EtOH 3; eth 3; bz 2
11670	Thiazole, 5-(2-chloroethyl)-4-methyl- Clomethiazole	C_6H_8ClNS 161.65	533-45-9	2382 92^{7}	4-27-00-00990 1.233^{25}	
11671	2,4-Thiazolediamine, 5-phenyl- Amiphenazole	$C_9H_9N_3S$ 191.26	490-55-1 163-4 dec	498	4-27-00-05139	
11672	Thiazole, 4,5-dihydro-2-methyl-	C_4H_7NS 101.17	2346-00-1 -101	 145	4-27-00-00921 1.067^{25}	 1.5200^{20}
11673	Thiazole, 4,5-dihydro-2,4,4-trimethyl-	$C_6H_{11}NS$ 129.23	4145-94-2	 147	4-27-00-00931 0.969^{25}	 1.4825^{25}
11674	Thiazole, 2,4-dimethyl- 2,4-Dimethylthiazole	C_5H_7NS 113.18	541-58-2	3249 146; 71^{50}	4-27-00-00981 1.0562^{15}	H_2O 2; EtOH 3; eth 3; chl 3 1.5091^{20}
11675	Thiazole, 4,5-dimethyl- 4,5-Dimethylthiazole	C_5H_7NS 113.18	3581-91-7 83.5	 158	4-27-00-00986 1.0699^{20}	eth 4; EtOH 4
11676	Thiazole, 2,4-diphenyl- 2,4-Diphenylthiazole	$C_{15}H_{11}NS$ 237.33	1826-14-8 92.5	 >360	4-27-00-01435 1.1554^{98}	eth 4; EtOH 4
11677	5-Thiazoleethanol, 4-methyl- 4-Methyl-5-thiazoleethanol	C_6H_9NOS 143.21	137-00-8	6046 135^{7}	4-27-00-01754 1.196^{24}	H_2O 4; EtOH 3; eth 3; bz 3
11678	Thiazole, 2-methyl- 2-Methylthiazole	C_4H_5NS 99.16	3581-87-1	 128.5	4-27-00-00967	H_2O 5; EtOH 3; ace 3 1.510
11679	Thiazole, 4-methyl- 4-Methylthiazole	C_4H_5NS 99.16	693-95-8	 133.3	4-27-00-00969 1.112^{25}	H_2O 3; EtOH 3; eth 3
11680	2(3H)-Thiazolethione, 4-methyl-	$C_4H_5NS_2$ 131.22	5685-06-3 89.3	 188^{3}	4-27-00-02584	EtOH 4
11681	Thiazolidine	C_3H_7NS 89.16	504-78-9	 164.5	4-27-00-00009 1.131^{25}	H_2O 5; EtOH 3; eth 4; ace 4 1.551^{20}
11682	4-Thiazolidinecarboxylic acid Timonacic	$C_4H_7NO_2S$ 133.17	444-27-9 196.5	9375		H_2O 4
11683	2,4-Thiazolidinedione	$C_3H_3NO_2S$ 117.13	2295-31-0 128	 179^{19}	4-27-00-03144	eth 4
11684	4-Thiazolidinone, 5-[[4-(dimethylamino)phenyl]methylene]-2-thioxo- *p*-(Dimethylamino)benzal-5-rhodanine	$C_{12}H_{12}N_2OS_2$ 264.37	536-17-4 270 dec	3220	4-27-00-05697	H_2O 1; EtOH 2; eth 4; ace 3
11685	4-Thiazolidinone, 2-thioxo- Rhodanine	$C_3H_3NOS_2$ 133.19	141-84-4 170	8184	4-27-00-03188 0.868^{25}	H_2O 2; EtOH 4; eth 4
11686	Thiazolium, 3-[(4-amino-2-methyl-5-pyrimidinyl)methyl]-5-(2-hydroxyethyl)-4-methyl- chloride, monohydrochlorid Thiamine hydrochloride	$C_{12}H_{18}Cl_2N_4OS$ 337.27	67-03-8 248 dec	9222	4-27-00-01764	H_2O 4; EtOH 2; eth 1; ace 2
11687	Thidiazuron Urea, *N*-phenyl-*N*'-1,2,3-thiadiazol-5-yl-	$C_9H_8N_4OS$ 220.25	51707-55-2 211 dec			
11688	Thieno[2,3-b]thiophene	$C_6H_4S_2$ 140.23	250-84-0 6.5	 225	5-19-01-00431	eth 4
11689	1H-Thieno[3,4-d]imidazole-4-pentanoic acid, hexahydro-2-oxo-, [3a*S*-(3aα,4β,6aα)]- Biotin	$C_{10}H_{16}N_2O_3S$ 244.31	58-85-5 232 dec	1244	4-27-00-07979	H_2O 3; EtOH 3; eth 2; chl 2
11690	Thiepane	$C_6H_{12}S$ 116.23	4753-80-4 0.5	 173.5	5-17-01-00090 0.9883^{20}	H_2O 1; eth 3; ace 3; chl 3 1.5044^{18}
11691	Thietane Trimethylene sulfide	C_3H_6S 74.15	287-27-4 -73.2	 95	5-17-01-00014 1.0200^{20}	H_2O 1; EtOH 4; ace 3; bz 4 1.5102^{20}
11692	Thietane, 1,1-dioxide Trimethylene sulfone	$C_3H_6O_2S$ 106.15	5687-92-3 75.5	 91.2^{14}	5-17-01-00015	H_2O 3; EtOH 3; eth 2; peth 2 1.5156^{20}
11693	Thietane, 2-methyl-	C_4H_8S 88.17	17837-41-1	 107	5-17-01-00055 0.9571^{20}	H_2O 1; EtOH 3; eth 3; ace 3 1.4852^{20}

No.	Name Synonym	Mol. Form. Mol. Wt.	CAS RN mp/°C	Merck No. bp/°C	Beil. Ref. den/g cm^{-3}	Solubility n_D
11694	Thietane, 1-oxide 1,3-Propanesulfone	$C_3H_6O_3S$ 122.14	13153-11-2 31.5	 180^{30}	5-17-01-00015 1.392^{25}	
11695	Thiirane	C_2H_4S 60.12	420-12-2	 57 dec	5-17-01-00009 1.0368^{0}	EtOH 2; eth 2; ace 3; chl 3 1.4935^{20}
11696	Thiirane, methyl-	C_3H_6S 74.15	1072-43-1 -91	 72.5	5-17-01-00025 0.941^{20}	chl 3 1.472^{20}
11697	Thiobencarb Carbamothioic acid, diethyl-, *S*-[(4-chlorophenyl)methyl] ester	$C_{12}H_{16}ClNOS$ 257.78	28249-77-6 3.3	 $127^{0.008}$	 1.16^{20}	
11698	Thioboric acid, trimethyl ester	$C_3H_9BS_3$ 152.11	997-49-9 5	 218.2	4-01-00-01287 1.126^{20}	ace 4; eth 4 1.5788^{20}
11699	Thiocyanic acid, butyl ester Butyl thiocyanate	C_5H_9NS 115.20	628-83-1	 186	4-03-00-00329 0.9563^{15}	H_2O 1; EtOH 3; eth 3 1.4360^{20}
11700	Thiocyanic acid, chloromethyl ester	C_2H_2ClNS 107.56	3268-79-9	 185	2-03-00-00124 1.37^{15}	os 5
11701	Thiocyanic acid, 1,1-dimethylethyl ester *tert*-Butyl thiocyanate	C_5H_9NS 115.20	37985-18-5 10.5	 140 dec; 40^{10}	4-03-00-00330 0.9187^{10}	
11702	Thiocyanic acid, 1,2-ethanediyl ester	$C_4H_4N_2S_2$ 144.22	629-17-4 90	 dec	4-03-00-00333 1.4200^{0}	H_2O 2; EtOH 3; eth 3; ace 4
11703	Thiocyanic acid, ethyl ester Ethyl thiocyanate	C_3H_5NS 87.15	542-90-5 -85.5	 146	4-03-00-00328 1.007^{23}	H_2O 1; EtOH 5; eth 5; chl 3 1.4684^{15}
11704	Thiocyanic acid, heptyl ester Heptyl thiocyanate	$C_8H_{15}NS$ 157.28	5416-94-4	 235	4-03-00-00331 0.92^{20}	eth 4; EtOH 4
11705	Thiocyanic acid, methyl ester Methyl thiocyanate	C_2H_3NS 73.12	556-64-9 -5	6047 132.9	4-03-00-00327 1.0678^{25}	H_2O 2; EtOH 5; eth 5; ctc 3 1.4669^{25}
11706	Thiocyanic acid, 1-methylethyl ester Isopropyl thiocyanate	C_4H_7NS 101.17	625-59-2	 153	4-03-00-00329 0.9784^{20}	H_2O 1; EtOH 5; eth 5
11707	Thiocyanic acid, 2-methylpropyl ester Isobutyl thiocyanate	C_5H_9NS 115.20	591-84-4 -59	5036 175.4	4-03-00-00330	eth 4; EtOH 4
11708	Thiocyanic acid, octyl ester	$C_9H_{17}NS$ 171.31	19942-78-0 105	 $141\text{-}2^{19}$	4-03-00-00331 0.9149^{25}	H_2O 1; EtOH 4; eth 4; bz 3 1.4649^{20}
11709	Thiocyanic acid, phenyl ester Phenyl thiocyanate	C_7H_5NS 135.19	5285-87-0	 232.5	4-06-00-01536 1.153^{18}	H_2O 1; EtOH 3; eth 3
11710	Thiocyanic acid, phenylmethyl ester	C_8H_7NS 149.22	3012-37-1 43	 232	4-06-00-02680	H_2O 1; EtOH 3; eth 3; chl 3
11711	Thiocyanic acid, potassium salt KSCN	CKNS 97.18	333-20-0	7687	4-03-00-00299	
11712	Thiocyanic acid, 2-propenyl ester	C_4H_5NS 99.16	764-49-8	 161	4-03-00-00332 1.056^{15}	eth 4; EtOH 4
11713	Thiodicarb Dimethyl *N,N'*-[thiobis[[(methylimino)carbonyl]oxyl]]bis[ethanimidothioate]	$C_{10}H_{18}N_4O_4S_3$ 354.48	59669-26-0 173	9258	 1.4^{20}	
11714	Thiodicarbonic diamide, tetraethyl- Sulfiram	$C_{10}H_{20}N_2S_3$ 264.48	95-05-6	8928 $225\text{-}40^{3}$	4-04-00-00397 1.12^{20}	chl 3
11715	Thiodicarbonic diamide, tetramethyl-	$C_6H_{12}N_2S_3$ 208.37	97-74-5 109.5		4-04-00-00238 1.37^{25}	H_2O 1; EtOH 3; eth 2; ace 3
11716	Thiodiphosphoric acid, tetraethyl ester Sulfotep	$C_8H_{20}O_5P_2S_2$ 322.32	3689-24-5	8945 $136\text{-}9^{2}$	4-01-00-01351 1.196^{25}	H_2O 1; EtOH 3 1.4753^{25}
11717	Thiofanox 2-Butanone, 3,3-dimethyl-1-(methylthio)-, *O*-[(methylamino)carbonyl]oxime	$C_9H_{18}N_2O_2S$ 218.32	39196-18-4 57			
11718	Thioimidodicarbonic diamide 2,4-Dithiobiuret	$C_2H_5N_3S_2$ 135.21	541-53-7 181 dec	3378	4-03-00-00356	ace 4
11719	Thiomorpholine Thiamorpholine	C_4H_9NS 103.19	123-90-0	9229 175; 110^{100}	4-27-00-00636 1.0882^{20}	H_2O 4; ace 4; eth 4; EtOH 4 1.5386^{20}
11720	Thionazin Phosphorothioic acid, *O,O*-diethyl *O*-pyrazinyl ester	$C_8H_{13}N_2O_3PS$ 248.24	297-97-2 -1.7	9275 80		
11721	Thioperoxydicarbonic diamide ([(H2N)C(*S*)]2S2), tetraethyl- Disulfiram	$C_{10}H_{20}N_2S_4$ 296.55	97-77-8 71.5	3370 117^{17}	4-04-00-00398	H_2O 1; EtOH 3; eth 2; chl 4
11722	Thioperoxydicarbonic diamide ([(H2N)C(*S*)]2S2), tetramethyl- Thiram	$C_6H_{12}N_2S_4$ 240.44	137-26-8 155.6	9304 129^{20}	4-04-00-00242	chl 4
11723	Thiophanate-methyl Carbamic acid, [1,2-phenylenebis(iminocarbonothioyl)]bis-, dimethyl ester	$C_{12}H_{14}N_4O_4S_2$ 342.40	23564-05-8 172 dec			
11724	Thiophene Thiofuran	C_4H_4S 84.14	110-02-1 -39.4	9283 84.0	5-17-01-00297 1.0649^{20}	EtOH 5; eth 5; ace 5; bz 5 1.5289^{20}
11725	2-Thiopheneacetonitrile	C_6H_5NS 123.18	20893-30-5	 $115\text{-}20^{22}$	5-18-06-00208 1.157^{25}	 1.5425^{20}
11726	Thiophene, 2-bromo- 2-Thienyl bromide	C_4H_3BrS 163.04	1003-09-4	 150	5-17-01-00305 1.684^{20}	H_2O 1; eth 4; ace 4; ctc 3 1.5868^{20}
11727	Thiophene, 3-bromo-	C_4H_3BrS 163.04	872-31-1	 159.5	5-17-01-00306 1.735^{20}	H_2O 1; ace 3; bz 3; chl 2 1.5919^{20}
11728	Thiophene, 2-bromo-5-chloro- 2-Bromo-5-chlorothiophene	C_4H_2BrClS 197.48	2873-18-9 -20	 70^{18}	5-17-01-00307 1.803^{25}	eth 4 1.5925^{25}

PHYSICAL CONSTANTS OF ORGANIC COMPOUNDS (continued)

No.	Name Synonym	Mol. Form. Mol. Wt.	CAS RN mp/°C	Merck No. bp/°C	Beil. Ref. den/g cm^{-3}	Solubility n_D
11729	Thiophene, 2-bromo-3-methyl- 2-Bromo-3-methylthiophene	C_5H_5BrS 177.06	14282-76-9	 176; $27^{1.8}$	5-17-01-00333 1.5844^{18}	bz 4; eth 4 1.5714^{20}
11730	Thiophene, 2-bromo-5-methyl-	C_5H_5BrS 177.06	765-58-2	 178; $29^{1.8}$	5-17-01-00327 1.5529^{20}	eth 3; bz 3 1.5673^{20}
11731	Thiophene, 2-butyl- 2-Butylthiophene	$C_8H_{12}S$ 140.25	1455-20-5	 181.5	5-17-01-00369 0.9537^{20}	 1.5090^{20}
11732	2-Thiophenecarbonitrile 2-Cyanothiophene	C_5H_3NS 109.15	1003-31-2	 192	5-18-06-00171 1.172^{25}	chl 3 1.5629^{20}
11733	2-Thiophenecarboxaldehyde	C_5H_4OS 112.15	98-03-3	 197; 85^{16}	5-17-09-00349 1.2127^{21}	H_2O 1; EtOH 4; eth 3; chl 2 1.5920^{20}
11734	3-Thiophenecarboxaldehyde 3-Formylthiophene	C_5H_4OS 112.15	498-62-4	 86.7^{20}	5-17-09-00373	H_2O 1; EtOH 4; eth 3 1.5855^{20}
11735	2-Thiophenecarboxaldehyde, 5-chloro-	C_5H_4ClOS 147.60	7283-96-7	 77.5^{5}	5-17-09-00360	chl 2 1.6036^{25}
11736	2-Thiophenecarboxaldehyde, 5-methyl-	C_6H_6OS 126.18	13679-70-4	 114^{25}	5-17-09-00406	chl 3 1.5825^{20}
11737	2-Thiophenecarboxylic acid 2-Carboxythiophene	$C_5H_4O_2S$ 128.15	527-72-0 129.5	9284 260 dec	5-18-06-00158	H_2O 4; EtOH 4; eth 4; chl 3
11738	3-Thiophenecarboxylic acid 3-Thenoic acid	$C_5H_4O_2S$ 128.15	88-13-1 138	9206	5-18-06-00199	H_2O 3
11739	2-Thiophenecarboxylic acid, ethyl ester	$C_7H_8O_2S$ 156.21	2810-04-0	 218	5-18-06-00159 1.1623^{16}	EtOH 3; ace 3; ctc 2; os 3 1.5248^{20}
11740	Thiophene, 2-chloro- 2-Thienyl chloride	C_4H_3ClS 118.59	96-43-5 -71.9	 128.3	5-17-01-00303 1.2863^{20}	H_2O 1; EtOH 5; eth 5; chl 2 1.5487^{20}
11741	Thiophene, 2-chloro-5-methyl- 2-Chloro-5-methylthiophene	C_5H_5ClS 132.61	17249-82-0	 155; 55^{19}	5-17-01-00326 1.2147^{25}	ace 4; bz 4; eth 4; EtOH 4 1.5372^{20}
11742	Thiophene, 2,3-dibromo- 2,3-Dibromothiophene	$C_4H_2Br_2S$ 241.93	3140-93-0 -17.5	 218.5; 89^{13}	5-17-01-00308	 1.6304^{22}
11743	Thiophene, 2,4-dibromo- 2,4-Dibromothiophene	$C_4H_2Br_2S$ 241.93	3140-92-9 -26.3	 210	5-17-01-00308 2.1488^{20}	
11744	Thiophene, 2,5-dibromo- 2,5-Dibromothiophene	$C_4H_2Br_2S$ 241.93	3141-27-3 -6	 210.3	5-17-01-00308 2.142^{23}	H_2O 1; EtOH 4; eth 4; ctc 3 1.6288^{20}
11745	2,5-Thiophenedicarboxylic acid 2,5-Dicarboxythiophene	$C_6H_4O_4S$ 172.16	4282-31-9 359	 sub 150	5-18-07-00049	H_2O 2; EtOH 3; eth 3
11746	Thiophene, 2,3-dichloro- 2,3-Dichlorothiophene	$C_4H_2Cl_2S$ 153.03	17249-79-5 -26.2	 173.5	5-17-01-00304 1.4605^{20}	H_2O 1; EtOH 5; eth 3; ace 3 1.5651^{20}
11747	Thiophene, 2,4-dichloro- 2,4-Dichlorothiophene	$C_4H_2Cl_2S$ 153.03	17249-75-1 -34	 167.6	5-17-01-00304 1.4553^{20}	H_2O 1; EtOH 4; eth 3; ace 3 1.5660^{20}
11748	Thiophene, 2,5-dichloro- 5-Chloro-2-thienyl chloride	$C_4H_2Cl_2S$ 153.03	3172-52-9 -40.5	 162	5-17-01-00304 1.4422^{20}	H_2O 1; EtOH 5; eth 5; ctc 3 1.5626^{20}
11749	Thiophene, 3,4-dichloro- 3,4-Dichlorothiophene	$C_4H_2Cl_2S$ 153.03	17249-76-2 1	 185	5-17-01-00305 1.4867^{20}	H_2O 1; EtOH 4; eth 4; ace 3 1.5821^{19}
11750	Thiophene, 2,3-dihydro- 2,3-Dihydrothiophene	C_4H_6S 86.16	1120-59-8	 112.1		
11751	Thiophene, 2,5-dihydro- 2,5-Dihydrothiophene	C_4H_6S 86.16	1708-32-3	 122.4		
11752	Thiophene, 2,5-dihydro-, 1,1-dioxide 3-Sulfolene	$C_4H_6O_2S$ 118.16	77-79 2 64.5	8935	5-17-01-00177	chl 3
11753	Thiophene, 2,5-diiodo- 2,5-Diiodothiophene	$C_4H_2I_2S$ 335.93	625-88-7 41.5	 139^{15}	5-17-01-00313	H_2O 1; EtOH 4
11754	Thiophene, 2,3-dimethyl- 2,3-Dimethylthiophene	C_6H_8S 112.20	632-16-6 -49	 141.6	5-17-01-00342 1.0021^{20}	H_2O 1; EtOH 4; eth 4; bz 3 1.5192^{20}
11755	Thiophene, 2,4-dimethyl- 2,4-Dimethylthiophene	C_6H_8S 112.20	638-00-6	 140.7	5-17-01-00344 0.9938^{20}	H_2O 1; EtOH 3; eth 3; bz 3 1.5104^{20}
11756	Thiophene, 2,5-dimethyl- 2,5-Dimethylthiophene	C_6H_8S 112.20	638-02-8 -62.6	 136.5	5-17-01-00345 0.9850^{20}	H_2O 1; EtOH 3; eth 3; bz 3 1.5129^{20}
11757	Thiophene, 3,4-dimethyl- 3,4-Dimethylthiophene	C_6H_8S 112.20	632-15-5	 145	5-17-01-00347 0.993^{25}	H_2O 1; EtOH 3; eth 4 1.5206^{20}
11758	Thiophene, 2,5-dinitro- 2,5-Dinitrothiophene	$C_4H_2N_2O_4S$ 174.14	59434-05-8 81	 290	5-17-01-00315	eth 4; EtOH 4
11759	Thiophene, 2-ethyl- 2-Ethylthiophene	C_6H_8S 112.20	872-55-9	 134	5-17-01-00339 0.9930^{20}	H_2O 1; EtOH 4; eth 4 1.5122^{20}
11760	Thiophene, 3-ethyl- 3-Ethylthiophene	C_6H_8S 112.20	1795-01-3 -89.1	 136	5-17-01-00341 0.9980^{20}	H_2O 1; EtOH 3; eth 4 1.5146^{20}
11761	Thiophene, 2-ethyl-3-methyl- 2-Ethyl-3-methylthiophene	$C_7H_{10}S$ 126.22	31805-48-8	 161	5-17-01-00357 0.9815^{20}	 1.5105^{20}
11762	Thiophene, 2-ethyl-5-methyl- 2-Ethyl-5-methylthiophene	$C_7H_{10}S$ 126.22	40323-88-4 -68.4	 160	5-17-01-00357 0.9663^{20}	 1.5073^{20}
11763	Thiophene, 3-ethyl-4-methyl-	$C_7H_{10}S$ 126.22	66577-03-5	 165.5	4-17-00-00301 0.9976^{20}	 1.5186^{20}
11764	Thiophene, 4-ethyl-2-methyl- 5-Methyl-3-ethylthiophene	$C_7H_{10}S$ 126.22	13678-54-1 -60	 163	4-17-00-00299 0.9742^{20}	 1.5098^{20}
11765	Thiophene, 2-ethyltetrahydro-	$C_6H_{12}S$ 116.23	1551-32-2	 157	5-17-01-00096 0.944^{20}	ace 4; bz 4; eth 4; EtOH 4 1.490^{20}
11766	Thiophene, 3-ethyltetrahydro- 3-Ethyltetrahydrothiophene	$C_6H_{12}S$ 116.23	62184-67-2	 165	5-17-01-00096 0.950^{20}	 1.491^{20}
11767	Thiophene, 2-iodo-	C_4H_3IS 210.04	3437-95-4 -40	 181	5-17-01-00309 2.0595^{25}	EtOH 4; eth 4; chl 2 1.6465^{25}
11768	2-Thiophenemethanol	C_5H_6OS 114.17	636-72-6	 207; 86^{10}	5-17-03-00358 1.2053^{16}	EtOH 3; ace 3 1.5280^{20}

No.	Name Synonym	Mol. Form. Mol. Wt.	CAS RN mp/°C	Merck No. bp/°C	Beil. Ref. den/g cm^{-3}	Solubility n_D
11769	Thiophene, 2-methyl- 2-Methylthiophene	C_5H_6S 98.17	554-14-3 -63.4	 112.6	5-17-01-00324 1.0193^{20}	H_2O 1; EtOH 5; eth 5; ace 5 1.5203^{20}
11770	Thiophene, 3-methyl- 3-Thiotolene	C_5H_6S 98.17	616-44-4 -69	 115.5	5-17-01-00331 1.0218^{20}	H_2O 1; EtOH 5; eth 5; ace 5 1.5204^{20}
11771	Thiophene, 2-(3-methylbutyl)- 2-Isopentylthiophene	$C_9H_{14}S$ 154.28	26963-33-7	 $74\text{-}5^{11}$	4-17-00-00316 0.9481^{12}	 1.5014^{12}
11772	Thiophene, 2-(1-methylethyl)- 2-Isopropylthiophene	$C_7H_{10}S$ 126.22	4095-22-1	 153	5-17-01-00356 0.9678^{20}	 1.5038^{20}
11773	Thiophene, 3-(1-methylethyl)- 3-Isopropylthiophene	$C_7H_{10}S$ 126.22	29488-27-5	 157	5-17-01-00356 0.9733^{20}	 1.5052^{20}
11774	Thiophene, 2-methyl-5-phenyl- 2-Methyl-5-phenylthiophene	$C_{11}H_{10}S$ 174.27	5069-26-1 51	 271	5-17-02-00153	EtOH 4; eth 4; lig 4
11775	Thiophene, 2-nitro- 2-Nitrothiophene	$C_4H_3NO_2S$ 129.14	609-40-5 46.5	 224.5	5-17-01-00313 1.3644^{43}	H_2O 1; EtOH 4; alk 3; peth 2
11776	Thiophene, 3-nitro- 3-Nitrothiophene	$C_4H_3NO_2S$ 129.14	822-84-4 78.5	 225	5-17-01-00313	H_2O 1; EtOH 4; ctc 2; peth 3
11777	Thiophene-2-ol	C_4H_4OS 100.14	17236-58-7 9	 218	 1.255^{20}	H_2O 3; ace 3; chl 3 1.5644^{20}
11778	Thiophene, 2-propyl- 2-Propylthiophene	$C_7H_{10}S$ 126.22	1551-27-5	 158	5-17-01-00355 0.9683^{20}	H_2O 1; EtOH 3; eth 3; bz 3 1.5076^{20}
11779	Thiophene, 3-propyl- 3-Propylthiophene	$C_7H_{10}S$ 126.22	1518-75-8	 163.2	5-17-01-00355 0.9738^{20}	H_2O 1; EtOH 3; eth 3; bz 3 1.5057^{20}
11780	2-Thiophenesulfonyl chloride	$C_4H_3ClO_2S_2$ 182.65	16629-19-9 28	 100^6	5-18-09-00426	eth 3
11781	Thiophene, tetrabromo- 2,3,4,5-Tetrabromothiophene	C_4Br_4S 399.73	3958-03-0 117.5	 326	5-17-01-00309	H_2O 1; EtOH 3; eth 4
11782	Thiophene, tetrachloro-	C_4Cl_4S 221.92	6012-97-1 30.5	 233.4	5-17-01-00305 1.7036^{30}	H_2O 1; EtOH 4; eth 5 1.5915^{30}
11783	Thiophene, tetrahydro- Thiacyclopentane	C_4H_8S 88.17	110-01-0 -96.1	 121.0	5-17-01-00036 0.9987^{20}	H_2O 1; EtOH 5; eth 5; ace 5 1.4871^{18}
11784	Thiophene, tetrahydro-2,4-dimethyl-, 1,1-dioxide 2,4-Dimethylsulfolane	$C_6H_{12}O_2S$ 148.23	1003-78-7 -3	3245 281	5-17-01-00097 1.1362^{20}	lig 4 1.4732^{20}
11785	Thiophene, tetrahydro-2,5-dimethyl-, *cis*-	$C_6H_{12}S$ 116.23	5161-13-7 -89	 142.3	5-17-01-00098 0.9222^{20}	ace 4; bz 4; eth 4; EtOH 4 1.4799^{20}
11786	Thiophene, tetrahydro-, 1,1-dioxide Sulfolane	$C_4H_8O_2S$ 120.17	126-33-0 27.6	8934 287.3	5-17-01-00039 1.2723^{18}	chl 3 1.4833^{18}
11787	Thiophene, tetrahydro-2-methyl-	$C_5H_{10}S$ 102.20	1795-09-1 -100.7	 134	5-17-01-00080 0.9564^{18}	H_2O 1; EtOH 4; eth 4; ace 4 1.4886^{15}
11788	Thiophene, tetrahydro-3-methyl-	$C_5H_{10}S$ 102.20	4740-00-5	 139	5-17-01-00082 0.9596^{18}	H_2O 1; EtOH 4; eth 4; ace 4 1.4886^{18}
11789	Thiophene, tetrahydro-3-methyl-, 1,1-dioxide 3-Methylsulfolane	$C_5H_{10}O_2S$ 134.20	872-93-5 1	 276	5-17-01-00082 1.188^{25}	 1.4772^{20}
11790	Thiophene, 2,3,5-tribromo- 2,3,5-Tribromothiophene	C_4HBr_3S 320.83	3141-24-0 29	 260	5-17-01-00309	chl 3
11791	Thiophene, 2,3,5-trichloro- 2,3,5-Trichlorothiophene	C_4HCl_3S 187.48	17249-77-3 -16.1	 198.7	5-17-01-00305 1.5856^{20}	H_2O 1; EtOH 5; eth 5 1.5791^{20}
11792	Thiophene, 2,3,4-trimethyl- 2,3,4-Trimethylthiophene	$C_7H_{10}S$ 126.22	1795-04-6 -26.2	 162; 75^{30}	5-17-01-00359 0.995^{20}	 1.5208^{20}
11793	Thiophene, 2,3,5-trimethyl- 2,3,5-Trimethylthiophene	$C_7H_{10}S$ 126.22	1795-05-7	 164.5	5-17-01-00359 0.9753^{20}	H_2O 1; EtOH 4; eth 4; ace 4 1.5112^{20}
11794	2(3H)-Thiophenone, dihydro-	C_4H_6OS 102.16	1003-10-7	 $39\text{-}40^1$	5-17-09-00012 1.18^{25}	 1.5230^{20}
11795	2H-Thiopyran, tetrahydro- Thiacyclohexane	$C_5H_{10}S$ 102.20	1613-51-0 19	 141.8	5-17-01-00068 0.9861^{20}	H_2O 1; EtOH 3; eth 3; ace 3 1.5067^{20}
11796	2H-Thiopyran, tetrahydro-2-methyl-	$C_6H_{12}S$ 116.23	5161-16-0 -58	 153	5-17-01-00093 0.9428^{20}	 1.4905^{20}
11797	2H-Thiopyran, tetrahydro-3-methyl-	$C_6H_{12}S$ 116.23	5258-50-4 -60	 158	5-17-01-00094 0.9473^{20}	 1.4922^{20}
11798	2H-Thiopyran, tetrahydro-4-methyl-	$C_6H_{12}S$ 116.23	5161-17-1 -28	 158.6	5-17-01-00094 0.9471^{20}	 1.4923^{20}
11799	Thiourea Thiocarbamide	CH_4N_2S 76.12	62-56-6 182	9299	4-03-00-00342 1.405^{25}	H_2O 3; EtOH 3; eth 1
11800	Thiourea, *N,N'*-bis[4-(3-methylbutoxy) phenyl]- Tiocarlide	$C_{23}H_{32}N_2O_2S$ 400.59	910-86-1 146	9382	4-13-00-01187	
11801	Thiourea, (2-chlorophenyl)- 1-(2-Chlorophenyl)thiourea	$C_7H_7ClN_2S$ 186.66	5344-82-1 146		3-12-00-01296	bz 4; EtOH 4
11802	Thiourea, *N*-(3,5-dichlorophenyl)-*N'*-(4-fluorophenyl)- Loflucarban	$C_{13}H_9Cl_2FN_2S$ 315.20	790-69-2 163.5	5440	4-12-00-01275	
11803	Thiourea, *N,N'*-diethyl-	$C_5H_{12}N_2S$ 132.23	105-55-5 78	 dec	4-04-00-00375	H_2O 3; EtOH 3; eth 4; ctc 2
11804	Thiourea, *N,N'*-dimethyl- *N,N'*-Dimethylthiourea	$C_3H_8N_2S$ 104.18	534-13-4 62	3250	4-04-00-00217	H_2O 4; EtOH 4; eth 2; ace 4
11805	Thiourea, *N,N'*-diphenyl- *sym*-Diphenylthiourea	$C_{13}H_{12}N_2S$ 228.32	102-08-9 154.5	3337	4-12-00-00810 1.32^{25}	H_2O 2; EtOH 4; eth 4; chl 4
11806	Thiourea, (2-methylphenyl)- *o*-Tolylthiourea	$C_8H_{10}N_2S$ 166.25	614-78-8 162		4-12-00-01766	H_2O 4; EtOH 4; eth 2

No.	Name Synonym	Mol. Form. Mol. Wt.	CAS RN mp/°C	Merck No. bp/°C	Beil. Ref. den/g cm^{-3}	Solubility n_D
11807	Thiourea, 1-naphthalenyl- ANTU	$C_{11}H_{10}N_2S$ 202.28	86-88-4 198	755	4-12-00-03086	H_2O 1; EtOH 2; eth 2; ace 2
11808	Thiourea, phenyl- Phenylthiourea	$C_7H_8N_2S$ 152.22	103-85-5 154	7286	4-12-00-00804	H_2O 2; EtOH 3; NaOH 3
11809	Thiourea, 2-propenyl- Thiosinamine	$C_4H_8N_2S$ 116.19	109-57-9 78	9293	4-04-00-01072 1.217^{20}	H_2O 3; EtOH 3; eth 2; bz 1 1.5936^{78}
11810	Thiourea, tetramethyl-	$C_5H_{12}N_2S$ 132.23	2782-91-4 79.3	 245	4-04-00-00232	H_2O 3; EtOH 3; eth 2; chl 3
11811	9H-Thioxanthene Dibenzothiapyran	$C_{13}H_{10}S$ 198.29	261-31-4 128.5	9300 341	5-17-02-00254	chl 3
11812	9H-Thioxanthen-9-one Thioxanthone	$C_{13}H_8OS$ 212.27	492-22-8 209	9301 373	5-17-10-00437	H_2O 1; EtOH 2; bz 3; chl 3
11813	*L*-Threonine [*R*-(*R**,*S**)]-2-Amino-3-hydroxybutanoic acid	$C_4H_9NO_3$ 119.12	72-19-5 256 dec	9316	4-04-00-03171	H_2O 3; EtOH 1; eth 1; chl 1
11814	Thujane, (1*S*,4*R*,5*S*)-(+)- Sabinane (*d*)	$C_{10}H_{18}$ 138.25	7712-66-5	 157	4-05-00-00318 0.8139^{20}	 1.4376^{20}
11815	4(10)-Thujene, (+)- 4(10)-Thujene (*d*)	$C_{10}H_{16}$ 136.24	2514-91-2	 164	4-05-00-00451 0.842^{20}	ace 4; bz 4; eth 4; EtOH 4 1.4678^{20}
11816	Thymidine Thymine 2-desoxyriboside	$C_{10}H_{14}N_2O_5$ 242.23	50-89-5 186.5	9331	5-24-07-00109	H_2O 3; EtOH 3; ace 3; chl 2
11817	Toluene, 3-chloro-2-iodo-	C_7H_6ClI 252.48	5100-98-1 27.5	 123^{14}	3-05-00-00726	 1.608^{23}
11818	*o*-Toluenesulfonic acid, thio-, *S*-phenyl ester Benzenesulfonic acid, thiolo,2-methyl,phenyl ester	$C_{13}H_{12}O_2S_2$ 264.37	96097-69-7 59.5	 241.6	4-11-00-00237	H_2O 2; EtOH 4; eth 3; chl 2 1.53412^{20}
11819	Triacontane	$C_{30}H_{62}$ 422.82	638-68-6 65.8	 449.7	4-01-00-00592 0.8097^{20}	H_2O 1; EtOH 2; eth 3; bz 4 1.4352^{70}
11820	Triacontanoic acid	$C_{30}H_{60}O_2$ 452.81	506-50-3 93.6		4-02-00-01321	bz 4; CS_2 4; chl 4 1.4323^{100}
11821	1-Triacontanol Myricyl alcohol	$C_{30}H_{62}O$ 438.82	593-50-0 88	9506	4-01-00-01918 0.777^{95}	bz 4; eth 4; EtOH 4
11822	Triadimenol Mercury, chloro(2-methoxyethyl)-	C_3H_7ClHgO 295.13	123-88-6			
11823	Triallate Carbamothioic acid, bis(1-methylethyl)-, *S*-(2,3,3-trichloro-2-propenyl) ester	$C_{10}H_{16}Cl_3NOS$ 304.67	2303-17-5 29	9510 $117^{0.0003}$	 1.270^{25}	
11824	Triasulfuron 2-(2-Chloroethoxy)-*N*-[[(4-methoxy-6-methyl-1,3,5-triazi*N*-2-yl)amino]carbonyl]benzene sulfonamide	$C_{14}H_{16}ClN_5O_5S$ 401.83	82097-50-5 186			
11825	3,5,7-Triaza-1-azoniatricyclo[3.3.1.13,7]decane, 1-(2-propenyl)-, iodide Hexamethylenetetramine, allyl iodide	$C_9H_{17}IN_4$ 308.17	36895-62-2	5880		
11826	1-Triazene, 1,3-diphenyl- Diazoaminobenzene	$C_{12}H_{11}N_3$ 197.24	136-35-6 98	2981	4-16-00-00904	H_2O 1; EtOH 4; eth 4; bz 4
11827	1,3,5-Triazine	$C_3H_3N_3$ 81.08	290-87-9 86	 114	4-26-00-00063 1.38^{25}	EtOH 3; eth 3
11828	1,3,5-Triazine-2,4-diamine, 6-chloro-*N*,*N*,*N*',*N*'-tetraethyl-	$C_{11}H_{20}ClN_5$ 257.77	580-48-3 27	 $154\text{-}6^{9}$	4-26-00-01208 1.0956^{20}	bz 4; chl 4; EtOH 4; lig 4 1.5320^{20}
11829	1,3,5-Triazine-2,4-diamine, 6-phenyl- Benzoguanamine	$C_9H_9N_5$ 187.20	91-76-9 226.5	1099	4-26-00-01244	EtOH 3; eth 3; tfa 2
11830	1,3,5-Triazine-2,4-diamine, *N*-phenyl- Amanozine	$C_9H_9N_5$ 187.20	537-17-7 235.5	379	4-26-00-01195	
11831	1,2,4-Triazine-3,5(2H,4H)-dione, 6-methyl- 6-Azathymine	$C_4H_5N_3O_2$ 127.10	932-53-6 211	919	4-26-00-00556	H_2O 3; EtOH 3; ace 3
11832	1,2,4-Triazine-3,5(2H,4H)-dione, 2-β-*D*-ribofuranosyl- 6-Azauridine	$C_8H_{11}N_3O_6$ 245.19	54-25-1 158	920	4-26-00-00554	H_2O 3
11833	1,3,5-Triazine, hexahydro-1,3,5-trinitro- Cyclonite	$C_3H_6N_6O_6$ 222.12	121-82-4 205.5	2742	4-26-00-00022 1.82^{20}	H_2O 1; EtOH 1; eth 2; ace 3
11834	1,3,5-Triazine, hexahydro-1,3,5-triphenyl-	$C_{21}H_{21}N_3$ 315.42	91-78-1 111	 100, 00^{20}	4-26-00-00007	H_2O 1; EtOH 2; eth 3; ace 3
11835	1,3,5-Triazine-2,4,6-triamine Melamine	$C_3H_6N_6$ 126.12	108-78-1 345 dec	5691 sub	4-26-00-01253 1.573^{16}	H_2O 2; EtOH 2; eth 1 1.872^{20}
11836	1,3,5-Triazine-2,4,6-tricarbonitrile	C_6N_6 156.11	7615-57-8 119	 262; $119^{1.0}$	1-26-00-00091	bz 4
11837	1,3,5-Triazine, 2,4,6-trichloro- Cyanuric acid trichloride	$C_3Cl_3N_3$ 184.41	108-77-0 154	 192	4-26-00-00066	EtOH 4
11838	1,3,5-Triazine-2,4,6(1H,3H,5H)-trione, 1,3-dichloro- Dichlorocyanuric acid	$C_3HCl_2N_3O_3$ 197.96	2782-57-2			
11839	1,3,5-Triazine-2,4,6(1H,3H,5H)-trione, dihydrate Cyanuric acid, dihydrate	$C_3H_7N_3O_5$ 165.11	6202-04-6 >360 dec	 dec	4-26-00-00632 2.500^{20}	H_2O 3; EtOH 2; eth 1; ace 1 2.500^{20}
11840	1,3,5-Triazine-2,4,6(1H,3H,5H)-trione, 1,3,5-trichloro- Symclosene	$C_3Cl_3N_3O_3$ 232.41	87-90-1 246.7 dec	8993	4-26-00-00642	

No.	Name Synonym	Mol. Form. Mol. Wt.	CAS RN mp/°C	Merck No. bp/°C	Beil. Ref. den/g cm^{-3}	Solubility n_D
11841	1,3,5-Triazine-2,4,6(1H,3H,5H)-trione, 1,3,5-trimethyl-	$C_6H_9N_3O_3$ 171.16	827-16-7 176.5	 274	4-26-00-00635	H_2O 1; EtOH 3
11842	1,3,5-Triazine-2,4,6(1H,3H,5H)-trione, 1,3,5-tri-2-propenyl-	$C_{12}H_{15}N_3O_3$ 249.27	1025-15-6 20.5	 149^{4}; $105^{0.5}$	4-26-00-00637 1.1590^{20}	
11843	1,3,5-Triazine, 2,4,6-tri-2-pyridinyl- 2,4,6-Tripyridyl-s-triazine	$C_{18}H_{12}N_6$ 312.33	3682-35-7 210	9664	4-26-00-04192	
11844	1,3,5-Triazine, 2,4,6-tris(1-aziridinyl)- Triethylenemelamine	$C_9H_{12}N_6$ 204.23	51-18-3	9586	4-26-00-01268	
11845	1H-1,2,4-Triazol-3-amine Amitrole	$C_2H_4N_4$ 84.08	61-82-5 159	506	4-26-00-01073	H_2O 4; EtOH 4; eth 1; ace 1
11846	1H-1,2,3-Triazole	$C_2H_3N_3$ 69.07	288-36-8 23	 204	4-26-00-00029 1.1861^{25}	H_2O 3; eth 3; ace 3; os 3 1.4854^{25}
11847	1H-1,2,4-Triazole, 1-butyl 1-Butyl-*s*-triazole	$C_6H_{11}N_3$ 125.17	6086-22-2			
11848	4H-1,2,4-Triazole, 4-cyclohexyl-3-ethyl- Hexazole	$C_{10}H_{17}N_3$ 179.27	4671-03-8 89.5	4618 227^{10}	4-26-00-00049	H_2O 4; bz 4; chl 4
11849	1H-1,2,3-Triazole, 1,5-dimethyl- 1,5-Dimethyl-1,2,3-triazole	$C_4H_7N_3$ 97.12	15922-53-9 -4	 255	1-26-00-00005	H_2O 4
11850	1,2,4-Triazolidine-3,5-dione Urazole	$C_2H_3N_3O_2$ 101.06	3232-84-6 249-50 dec	9780	4-26-00-00538	
11851	7H-1,2,3-Triazolo[4,5-d]pyrimidin-7-one, 5-amino-1,4-dihydro- 8-Azaguanine	$C_4H_4N_6O$ 152.12	134-58-7 300	912	4-26-00-04171	
11852	*S,S,S*-Tributyl phosphorotrothioate *S,S,S*-Tributyl trithiophosphate	$C_{12}H_{27}OPS_3$ 314.52	78-48-8 <-25	 $150^{0.3}$	 1.057^{20}	
11853	Trichlorfon Phosphonic acid, (2,2,2-trichloro-1-hydroxyethyl)-, dimethyl ester	$C_4H_8Cl_3O_4P$ 257.44	52-68-6 77	9536 $100^{0.1}$	 1.73^{20}	
11854	Triclopyr Acetic acid, [(3,5,6-trichloro-2-pyridinyl)oxy]-	$C_7H_4Cl_3NO_3$ 256.47	55335-06-3 149	9572 dec 290		
11855	Tricosane	$C_{23}H_{48}$ 324.63	638-67-5 47.6	 380	4-01-00-00576 0.7785^{48}	H_2O 1; EtOH 2; eth 3; ctc 3 1.4468^{20}
11856	Tricosane, 2-methyl- 2-Methyltricosane	$C_{24}H_{50}$ 338.66	1928-30-9 37.6	 205^{3}	3-01-00-00576 0.7539^{90}	 1.4201^{90}
11857	12-Tricosanone Diundecyl ketone	$C_{23}H_{46}O$ 338.62	540-09-0 70.2		4-01-00-03408 0.8086^{69}	bz 4; eth 4; chl 4 1.4283^{80}
11858	Tricyclazole 1,2,4-Triazolo[3,4-b]benzothiazole, 5-methyl-	$C_9H_7N_3S$ 189.24	41814-78-2 187			
11859	Tricyclo[3.3.1.13,7]decan-1-amine Amantadine	$C_{10}H_{17}N$ 151.25	768-94-5 180	380		H_2O 2
11860	Tricyclo[3.3.1.13,7]decane Adamantane	$C_{10}H_{16}$ 136.24	281-23-2 268	140 sub	4-05-00-00469 1.07^{25}	bz 3; ctc 3 1.568
11861	Tricyclo[4.4.0.02,7]dec-3-ene, 1,3-dimethyl-8-(1-methylethyl)-, stereoisomer Copaene	$C_{15}H_{24}$ 204.36	3856-25-5	2510 248.5	4-05-00-01189 0.8996^{20}	H_2O 1; eth 3; ace 3; HOAc 3 1.4894^{20}
11862	Tricyclo[2.2.1.02,6]heptane-1-carboxylic acid, 7,7-dimethyl- Tricyclenic acid	$C_{10}H_{14}O_2$ 166.22	512-60-7 151.3	 263	4-09-00-00246	bz 4; EtOH 4; peth 4
11863	Tricyclo[2.2.1.02,6]heptane-1-carboxylic acid, 7,7-dimethyl-, ethyl ester Tricyclenic acid, ethyl ester	$C_{12}H_{18}O_2$ 194.27	56694-98-5	 $100\text{-}1^{10}$	2-09-00-00064 1.0143^{20}	 1.4230^{20}
11864	Tricyclo[2.2.1.02,6]heptane-1-carboxylic acid, 7,7-dimethyl-, methyl ester Tricyclenic acid, methyl ester	$C_{11}H_{16}O_2$ 180.25	56694-97-4 45.5	 99^{14}	2-09-00-00064 1.0255^{42}	 1.4695^{42}
11865	Tricyclo[2.2.1.02,6]heptane, 3,3-dimethyl-	C_9H_{14} 122.21	473-02-9 42.5	 138	4-05-00-00430 0.8710^{40}	bz 4; eth 4; EtOH 4 1.4514^{40}
11866	Tricyclo[2.2.1.02,6]heptane, 1,3,3-trimethyl- Cyclofenchene	$C_{10}H_{16}$ 136.24	488-97-1	 145	3-05-00-00392 0.859^{20}	 1.4503^{22}
11867	Tricyclo[2.2.1.02,6]heptane, 1,7,7-trimethyl-	$C_{10}H_{16}$ 136.24	508-32-7 67.5	 152.5	4-05-00-00468 0.8668^{80}	 1.4296^{80}
11868	1,12-Tridecadiyne	$C_{13}H_{20}$ 176.30	38628-39-6 -3	 115.5^{12}	2-01-00-00248 0.8262^{21}	 1.454^{20}
11869	Tridecanal	$C_{13}H_{26}O$ 198.35	10486-19-8 14	 156^{13}	4-01-00-03386 0.8356^{18}	H_2O 1; EtOH 3; os 4 1.4384^{18}
11870	1-Tridecanamine Tridecane, 1-amino	$C_{13}H_{29}N$ 199.38	2869-34-3 27.4	 275.8	4-04-00-00810 0.8049^{20}	H_2O 2; EtOH 3; eth 3 1.4443^{20}
11871	Tridecane	$C_{13}H_{28}$ 184.37	629-50-5 -5.3	 235.4	4-01-00-00512 0.7564^{20}	H_2O 1; EtOH 4; eth 4; ctc 3 1.4256^{20}
11872	Tridecane, 1-bromo-	$C_{13}H_{27}Br$ 263.26	765-09-3 6.2	 296	4-01-00-00514 1.0177^{20}	H_2O 1; chl 4 1.4593^{20}
11873	Tridecane, 1-bromo-12-methyl- Tridecane, 2-methyl-13-bromo	$C_{14}H_{29}Br$ 277.29	111772-90-8	 $120\text{-}2^{3}$	3-01-00-00551 1.0241^{20}	 1.4598^{20}
11874	Tridecane, 1,13-dibromo- Tridecamethylene dibromide	$C_{13}H_{26}Br_2$ 342.16	31772-05-1 9	 $188\text{-}92^{12}$	3-01-00-00548 1.276^{15}	eth 4; chl 4 1.4880^{27}
11875	1,12-Tridecanediol	$C_{13}H_{28}O_2$ 216.36	99706-68-0 60.5	 188^{8}	2-01-00-00563	EtOH 4
11876	1,13-Tridecanediol	$C_{13}H_{28}O_2$ 216.36	13362-52-2 76.5	 195^{10}	4-01-00-02630	EtOH 4

No.	Name Synonym	Mol. Form. Mol. Wt.	CAS RN mp/°C	Merck No. bp/°C	Beil. Ref. den/g cm^{-3}	Solubility n_D
11877	Tridecane, 2,2,4,10,12,12-hexamethyl-7-(3,5,5-trimethylhexyl)-	$C_{28}H_{58}$ 394.77	3035-75-4	 $166^{0.5}$	4-01-00-00590 0.8137^{20}	 1.4558^{20}
11878	Tridecane, 7-hexyl-	$C_{19}H_{40}$ 268.53	7225-66-3 -28.3	 170.5^{10}	4-01-00-00562 0.7877^{20}	 1.4409^{20}
11879	Tridecane, 7-methyl- 7-Methyltridecane	$C_{14}H_{30}$ 198.39	26730-14-3 -37.2	 115.5^{10}	4-01-00-00525 0.7634^{20}	 1.4291^{20}
11880	Tridecanenitrile	$C_{13}H_{25}N$ 195.35	629-60-7 9.7	 293	3-02-00-00906 0.8257^{20}	EtOH 4; eth 4 1.4378^{20}
11881	Tridecanoic acid	$C_{13}H_{26}O_2$ 214.35	638-53-9 44.5	 236^{100}, 140^{1}	4-02-00-01117 0.8458^{80}	H_2O 1; EtOH 4; eth 4; ace 3 1.4286^{60}
11882	Tridecanoic acid, methyl ester Methyl tridecanoate	$C_{14}H_{28}O_2$ 228.38	1731-88-0 6.5	 92^{1}	4-02-00-01118	EtOH 5; ctc 3 1.4405^{20}
11883	1-Tridecanol Tridecyl alcohol	$C_{13}H_{28}O$ 200.36	112-70-9 32.5	 152^{14}	4-01-00-01860 0.8223^{31}	H_2O 1; EtOH 3; eth 3
11884	2-Tridecanol 2-Hydroxytridecane	$C_{13}H_{28}O$ 200.36	1653-31-2 23	 161^{30}, 145^{12}	4-01-00-01861	bz 4; eth 4; EtOH 4 1.4188^{70}
11885	3-Tridecanol	$C_{13}H_{28}O$ 200.36	10289-68-6 32	 139^{12}	4-01-00-01861 0.8139^{46}	
11886	1-Tridecanol, 12-methyl- 12-Methyl-1-tridecanol	$C_{14}H_{30}O$ 214.39	21987-21-3 10.5	 $145\text{-}50^{6}$	4-01-00-01869 0.8414^{20}	 1.4437^{20}
11887	3-Tridecanol, 2-methyl-	$C_{14}H_{30}O$ 214.39	98930-89-3 20	 274	1-01-00-00219 0.8390^{20}	 1.4460^{20}
11888	4-Tridecanol, 2-methyl Acetic acid, trifluoro,(1,2-dimethyl-1-ethylbutyl) ester	$C_{14}H_{30}O$ 214.39	36691-77-7	 $147\text{-}8^{14}$	4-01-00-01869	chl 2 1.4404^{25}
11889	2-Tridecanone Methyl undecyl ketone	$C_{13}H_{26}O$ 198.35	593-08-8 30.5	 263	4-01-00-03386 0.8217^{30}	H_2O 1; EtOH 4; eth 4; ace 4 1.4318^{20}
11890	3-Tridecanone	$C_{13}H_{26}O$ 198.35	1534-26-5 31	 140^{17}	4-01-00-03387	H_2O 1; ace 3
11891	7-Tridecanone Dihexyl ketone	$C_{13}H_{26}O$ 198.35	462-18-0 33	 261	4-01-00-03315 0.825^{30}	EtOH 3; eth 4; chl 3; lig 3
11892	3-Tridecanone, 2-methyl- Decyl isopropyl ketone	$C_{14}H_{28}O$ 212.38	40239-35-8 47	 267; 142^{17}	4-01-00-03390 0.8314^{20}	
11893	1-Tridecene	$C_{13}H_{26}$ 182.35	2437-56-1 -13	 232.8	4-01-00-00921 0.7658^{20}	H_2O 1; EtOH 4; eth 4; bz 3 1.4340^{20}
11894	2-Tridecenoic acid	$C_{13}H_{24}O_2$ 212.33	6969-16-0	 $167\text{-}71^{2}$	4-02-00-01622 0.8995^{30}	 1.4612^{20}
11895	12-Tridecenoic acid (10-Undecenyl)acetic acid	$C_{13}H_{24}O_2$ 212.33	6006-06-0 38.5	 192^{20}	2-02-00-00422	bz 4; eth 4; EtOH 4; peth 4
11896	12-Tridecenoic acid, methyl ester	$C_{14}H_{26}O_2$ 226.36	29780-00-5	 143^{8}	4-02-00-01622 0.8803^{20}	 1.4438^{20}
11897	1-Tridecyne	$C_{13}H_{24}$ 180.33	26186-02-7	 94.5^{25}	4-01-00-01069 0.7729^{20}	bz 4; eth 4 1.4309^{20}
11898	Tridemorph Morpholine, 2,6-dimethyl-4-tridecyl-	$C_{19}H_{39}NO$ 297.52	24602-86-6	9576 $141^{1.3}$	 0.86	
11899	Tridiphane (±)-2-(3,5-Dichlorophenyl)-2-(2,2,2-trichloroethyl)oxirane	$C_{10}H_7Cl_5O$ 320.43	58138-08-2 42.8			
11900	Triflumizole (*E*)-1-[1-[[4-Chloro-2-(trifluoromethyl)phenyl]imino]-2-propoxyethyl]-1H-imidazole	$C_{15}H_{15}ClF_3N_3O$ 345.75	68694-11-1 63.5			
11901	Trifluralin Benzenamine, 2,6-dinitro-*N*,*N*-dipropyl-4-(trifluoromethyl)-	$C_{13}H_{16}F_3N_3O_4$ 335.28	1582-09-8 49	9598 $140^{4.2}$		
11902	Triforine Formamide, *N*,*N* '-[1,4-piperazinediylbis(2,2,2-trichloroethylidene)]bis-	$C_{10}H_{14}Cl_6N_4O_2$ 434.96	26644-46-2 155 dec	9602		
11903	2,5,7-Trioxabicyclo[2.2.1]heptane, 1,3,3,4,6,6-hexamethyl- Bicylo[2.2.1]heptane, 1,3,3,4,6,6-hexamethyl-2,5,7-trioxa	$C_{10}H_{18}O_3$ 186.25	7045-89-8	 166	5-19-09-00139 0.9702^{17}	H_2O 1
11904	2H,4aH-1,4,5-Trioxadicyclopent[a,hi]indene-7-carboxylic acid, 3-ethylidene-3,3a,7a,9b-tetrahydro-2-oxo-, methyl ester, [3aS-(3E,3aα,4aβ,7aβ,9aR*,9bβ)]- Plumericin	$C_{15}H_{14}O_6$ 290.27	77-16-7	7512	5-19-11-00097	chl 3
11905	1,3,5-Trioxane Formaldehyde, trimer	$C_3H_6O_3$ 90.08	110-88-3 60.2	9646 114.5	5-19-09-00103 1.17^{65}	H_2O 4; EtOH 3; eth 3; bz 3
11906	1,3,5-Trioxane, 2,4,6-triethenyl- Metacrolein	$C_9H_{12}O_3$ 168.19	588-01-2 50	 170	5-19-09-00165	eth 4; EtOH 4
11907	1,3,5-Trioxane-2,4,6-triimine Cyamelide	$C_3H_3N_3O_3$ 129.08	462-02-2 dec	2689 dec	3-03-00-00030 1.127^{15}	eth 4; EtOH 4
11908	1,3,5-Trioxane, 2,4,6-trimethyl- Paraldehyde	$C_6H_{12}O_3$ 132.16	123-63-7 12.6	6975 124.3	5-19-09-00112 0.9943^{20}	H_2O 2; EtOH 5; eth 5; chl 5 1.4049^{20}

PHYSICAL CONSTANTS OF ORGANIC COMPOUNDS (continued)

No.	Name	Synonym	Mol. Form.	Mol. Wt.	CAS RN	mp/°C	Merck No.	bp/°C	Beil. Ref.	den/g cm^{-3}	Solubility	n_D
11909	2,6,7-Trioxa-1-phosphabicyclo[2.2.2]octane, 4-ethyl-	Trimethylolpropane phosphite	$C_6H_{11}O_3P$	162.13	824-11-3	53.7					chl 3	
11910	Triphenylene	9,10-Benzophenanthrene	$C_{18}H_{12}$	228.29	217-59-4	199	9654	425	4-05-00-02556		H_2O 1; EtOH 3; bz 4; chl 4	
11911	Triphenyltin hydroxide	Stannane, hydroxytriphenyl-	$C_{18}H_{16}OSn$	367.03	76-87-9	119	9659			1.54^{20}		
11912	Tripropylamine, eicosafluoro-2-(trifluoromethyl)-	Dipropylisobutylamine, perfluoro	$C_{10}F_{23}N$	571.08	307-11-9			147	4-02-00-00858	1.84^{25}	eth 4; EtOH 4	1.283^{25}
11913	Trisiloxane, 1,5-dichloro-1,1,3,3,5,5-hexamethyl-		$C_6H_{18}Cl_2O_2Si_3$	277.37	3582-71-6	-53		184	4-04-00-04123	1.018^{20}	H_2O 6	
11914	Trisiloxane, octamethyl-	Octamethyltrisiloxane	$C_8H_{24}O_2Si_3$	236.53	107-51-7	-80	6669	153; $50\text{-}2^{17}$	4-04-00-04115	0.8200^{20}	EtOH 2; bz 3; peth 3	1.3840^{20}
11915	Trisiloxane, 1,1,3,5,5-pentamethyl-1,3,5-triphenyl-	1,1,3,5,5-Pentamethyl-1,3,5-triphenyltrisiloxane	$C_{23}H_{30}O_2Si_3$	422.75	80-14-8			$169^{0.7}$	3-16-00-01210	1.0227^{20}	bz 4; lig 4	1.5280^{20}
11916	Trisulfide, diethyl		$C_4H_{10}S_3$	154.32	3600-24-6			$96\text{-}7^{26}$	4-01-00-01398	1.114^{20}		1.5689^{13}
11917	Trisulfide, di-2-propenyl		$C_6H_{10}S_3$	178.34	2050-87-5			$112\text{-}22^{16}$	0-01-00-00441	1.0845^{15}	eth 4	
11918	Tritetracontane		$C_{43}H_{88}$	605.17	7098-21-7	85.5		332^{3}	3-01-00-00593	0.7812^{90}		1.4340^{90}
11919	1,3,5-Trithiane		$C_3H_6S_3$	138.28	291-21-4	220		sub	5-19-09-00105	1.6374^{24}	H_2O 2; EtOH 2; eth 2; bz 3	
11920	1,3,5-Trithiane, 2,4,6-trimethyl-	Thioacetaldehyde trimer	$C_6H_{12}S_3$	180.36	2765-04-0		9245		4-19-00-04718			
11921	1,3,5-Trithiane, 2,4,6-trimethyl-, (2α,4α,6β)-		$C_6H_{12}S_3$	180.36	23769-39-3	101		246.5	2-19-00-00399		H_2O 1; EtOH 3; eth 3; ace 3	
11922	*L*-Tryptophan	*l*-α-Aminoindole-3-propionic acid	$C_{11}H_{12}N_2O_2$	204.23	73-22-3	290-2 dec	9707		5-22-14-00014		H_2O 2; EtOH 3; eth 1; chl 1	
11923	*L*-Tryptophan, *N*-methyl-	*L*-Abrine	$C_{12}H_{14}N_2O_2$	218.26	526-31-8	295 dec	5		5-22-14-00040		H_2O 2; MeOH 2; eth 1; alk 3	
11924	Tubocuraran-7',12'-diol, 6,6'-dimethoxy-2,2'-dimethyl-, (1'α)-	Bebeerine	$C_{36}H_{38}N_2O_6$	594.71	477-60-1	221	1024		2-27-00-00896		EtOH 3; eth 3; ace 4; chl 4	
11925	Tubocuraran-7',12'-diol, 6,6'-dimethoxy-2,2'-dimethyl-, (1β)-	Curine	$C_{36}H_{38}N_2O_6$	594.71	436-05-5	221	2682		4-27-00-08725		ace 4; bz 4; py 4	
11926	*L*-Tyrosine	*L*-Phenylalanine, 4-hydroxy-	$C_9H_{11}NO_3$	181.19	60-18-4	343 dec	9747	sub	4-14-00-02264		H_2O 2; EtOH 1; eth 1; HOAc 2	
11927	*L*-Tyrosine, 3,5-dibromo-	Tyrosine, 3,5-dibromo(*L*)	$C_9H_9Br_2NO_3$	338.98	300-38-9	245	3014		4-14-00-02365		H_2O 2; EtOH 2; eth 1; alk 3	
11928	Tyrosine, 3,5-diiodo-	3,5-Diiodotyrosine	$C_9H_9I_2NO_3$	432.98	66-02-4	213 dec	3177		2-14-00-00384		H_2O 2; EtOH 1; eth 1; bz 1	
11929	*L*-Tyrosine, 3-hydroxy-	Levodopa	$C_9H_{11}NO_4$	197.19	59-92-7	285 dec	5344		4-14-00-02492		H_2O 3; EtOH 1; eth 1; ace 1	
11930	*L*-Tyrosine, *O*-(4-hydroxy-3,5-diiodophenyl)-3,5-diiodo-	Thyroxine	$C_{15}H_{11}I_4NO_4$	776.87	51-48-9	235.5	9348		4-14-00-02373		H_2O 2; EtOH 1; bz 1	
11931	*L*-Tyrosine, 3-hydroxy-α-methyl-	Methyldopa	$C_{10}H_{13}NO_4$	211.22	555-30-6	300 dec	5974					
11932	*L*-Tyrosine, 3-iodo-	3-Iodotyrosine	$C_9H_{10}INO_3$	307.09	70-78-0	204-6 dec	4937		4-14-00-02369			
11933	*L*-Tyrosine, *N*-methyl-	Surinamine	$C_{10}H_{13}NO_3$	195.22	537-49-5	293	8988		3-14-00-01513			
11934	1,10-Undecadiyne		$C_{11}H_{16}$	148.25	4117-15-1	-17		83^{12}	2-01-00-00248	0.8182^{21}	ace 4; bz 4	1.453^{21}
11935	Undecanal		$C_{11}H_{22}O$	170.30	112-44-7	-4		117^{18}	4-01-00-03374	0.8251^{23}	H_2O 1; EtOH 3; eth 3	1.4520^{20}
11936	Undecanal, 2-methyl-		$C_{12}H_{24}O$	184.32	110-41-8			114^{10}	4-01-00-03383	0.830^{15}	H_2O 2; EtOH 3; eth 3	1.4321^{20}
11937	1-Undecanamine		$C_{11}H_{25}N$	171.33	7307-55-3	17		242	4-04-00-00792	0.7979^{20}	H_2O 3; EtOH 3; eth 1; ctc 2	1.4398^{20}
11938	Undecane	Hendecane	$C_{11}H_{24}$	156.31	1120-21-4	-25.6		195.9	4-01-00-00487	0.7402^{20}	H_2O 1; EtOH 5; eth 5	1.4398^{20}
11939	Undecane, 1-bromo-11-fluoro-	1-Bromo-11-fluoroundecane	$C_{11}H_{22}BrF$	253.20	463-33-2			$95^{0.6}$	4-01-00-00490		eth 4; EtOH 4	1.4518^{25}
11940	Undecane, 1,11-dibromo-	1,11-Dibromoundecane	$C_{11}H_{22}Br_2$	314.10	16696-65-4	166		$170\text{-}5^{15}$	3-01-00-00536	1.332^{15}		
11941	Undecanenitrile	Decyl cyanide	$C_{11}H_{21}N$	167.29	2244-07-7			253	3-02-00-00860	0.8254^{30}	H_2O 1; EtOH 3; eth 3; ctc 3	1.4293^{30}
11942	1-Undecanethiol	Undecyl mercaptan	$C_{11}H_{24}S$	188.38	5332-52-5	-3		257.4	4-01-00-01838	0.8448^{20}		1.4585^{20}
11943	Undecanoic acid		$C_{11}H_{22}O_2$	186.29	112-37-8	28.6		280	4-02-00-01068	0.8907^{20}	H_2O 1; EtOH 4; eth 3; ace 4	1.4294^{45}

No.	Name Synonym	Mol. Form. Mol. Wt.	CAS RN mp/°C	Merck No. bp/°C	Beil. Ref. den/g cm^{-3}	Solubility n_D
11944	Undecanoic acid, 11-bromo- 11-Bromoundecanoic acid	$C_{11}H_{21}BrO_2$ 265.19	2834-05-1 57	188^{18}	4-02-00-01073	ace 4; bz 4; eth 4; EtOH 4
11945	Undecanoic acid, 2-bromo- 2-Bromoundecanoic acid	$C_{11}H_{21}BrO_2$ 265.19	2623-84-9 10	180^{14}	4-02-00-01073 1.1586^{24}	
11946	Undecanoic acid, ethyl ester Ethyl undecylate	$C_{13}H_{26}O_2$ 214.35	627-90-7 -15	131^{14}	3-02-00-00858 0.8633^{20}	H_2O 1; EtOH 3; eth 3; ace 3 1.4285^{20}
11947	Undecanoic acid, 11-fluoro-	$C_{11}H_{21}FO_2$ 204.28	463-17-2 36	$113^{0.25}$	4-02-00-01071	eth 4; EtOH 4; lig 4
11948	Undecanoic acid, 11-hydroxy- 11-Hydroxyundecanoic acid	$C_{11}H_{22}O_3$ 202.29	3669-80-5 65.5	$148\text{-}50^{4}$	4-03-00-00903	eth 4; EtOH 4
11949	Undecanoic acid, 11-hydroxy-, methyl ester Methyl 11-hydroxyhendecanoate	$C_{12}H_{24}O_3$ 216.32	24724-07-0 27.5	168^{8}	2-03-00-00243 0.9542^{20}	bz 4; eth 4; EtOH 4
11950	1-Undecanol	$C_{11}H_{24}O$ 172.31	112-42-5 19	243	4-01-00-01835 0.8298^{20}	H_2O 1; EtOH 3; eth 4 1.4392^{20}
11951	2-Undecanol, (±)-	$C_{11}H_{24}O$ 172.31	113666-64-1 12	228	4-01-00-01838 0.8268^{19}	H_2O 1; EtOH 3; eth 3 1.4369^{20}
11952	3-Undecanol, (*R*)-	$C_{11}H_{24}O$ 172.31	107494-37-1 17	229	2-01-00-00462 0.8295^{20}	bz 4; eth 4; EtOH 4 1.4367^{20}
11953	5-Undecanol 1-Butylheptyl alcohol	$C_{11}H_{24}O$ 172.31	37493-70-2 -3.5	229	4-01-00-01839 0.8292^{20}	eth 4; EtOH 4 1.4354^{24}
11954	6-Undecanol	$C_{11}H_{24}O$ 172.31	23708-56-7 25	228	4-01-00-01839 0.8334^{20}	H_2O 1; EtOH 3; ace 3 1.4374^{20}
11955	1-Undecanol, 2-methyl- 2-Methyl-1-undecanol	$C_{12}H_{26}O$ 186.34	10522-26-6	$129\text{-}31^{12}$	4-01-00-01855 0.8300^{15}	eth 4; EtOH 4 1.4382^{20}
11956	3-Undecanol, 2-methyl- 3-Hydroxy-2-methylundecane	$C_{12}H_{26}O$ 186.34	60671-36-5	135^{26}	1-01-00-00217 0.8327^{20}	1.4405^{20}
11957	5-Undecanol, 2-methyl-	$C_{12}H_{26}O$ 186.34	33978-71-1	$130\text{-}3^{20}$	4-01-00-01855 0.8235^{20}	eth 4; EtOH 4 1.4346^{20}
11958	2-Undecanone Methyl nonyl ketone	$C_{11}H_{22}O$ 170.30	112-12-9 15	231.5	4-01-00-03374 0.8250^{20}	H_2O 1; EtOH 3; eth 3; ace 3 1.4291^{20}
11959	3-Undecanone Ethyl octyl ketone	$C_{11}H_{22}O$ 170.30	2216-87-7 12	227	4-01-00-03375 0.8272^{20}	H_2O 1; EtOH 3; eth 3; ctc 2 1.4296^{20}
11960	4-Undecanone Heptyl propyl ketone	$C_{11}H_{22}O$ 170.30	14476-37-0 4.5	106^{13}	4-01-00-03375 0.8274^{25}	eth 4; EtOH 4 1.4248^{24}
11961	5-Undecanone Butyl hexyl ketone	$C_{11}H_{22}O$ 170.30	33083-83-9 2	227	4-01-00-03375 0.8278^{19}	H_2O 1; EtOH 3; eth 3; ctc 3 1.4275^{18}
11962	6-Undecanone Diamyl ketone	$C_{11}H_{22}O$ 170.30	927-49-1 14.5	228	4-01-00-03376 0.8308^{20}	H_2O 1; EtOH 4; eth 4 1.4270^{20}
11963	4-Undecanone, 7-ethyl-2-methyl-	$C_{14}H_{28}O$ 212.38	6976-00-7	252.5	3-01-00-02920 0.8362^{20}	H_2O 1; EtOH 4; eth 4; ace 4 1.4370^{20}
11964	Undecasiloxane, tetracosamethyl- Tetracosamethylhendecasiloxane	$C_{24}H_{72}O_{10}Si_{11}$ 829.77	107-53-9	9128 322.8; 202^{47}	3-04-00-01881 0.9247^{25}	bz 4 1.3994^{20}
11965	3,5,9-Undecatrien-2-one, 6,10-dimethyl- Pseudoionone	$C_{13}H_{20}O$ 192.30	141-10-6	7931 143 5^{12}	3-01-00-03067 0.8984^{20}	EtOH 3; eth 3; chl 3; MeOH 3 1.5335^{20}
11966	1-Undecene	$C_{11}H_{22}$ 154.30	821-95-4 -49.2	192.7	4-01-00-00910 0.7503^{20}	H_2O 1; eth 3; chl 3; lig 3 1.4261^{20}
11967	3-Undecene	$C_{11}H_{22}$ 154.30	60669-40-1 -62	193.5	0.7516^{20}	1.4290^{20}
11968	2-Undecene, (*E*)-	$C_{11}H_{22}$ 154.30	693-61-8 -48.3	192.5	4-01-00-00911 0.7528^{20}	1.4292^{20}
11969	2-Undecene, (*Z*)-	$C_{11}H_{22}$ 154.30	821-96-5 -66.5	196.1	0.7576^{20}	
11970	4-Undecene, (*E*)-	$C_{11}H_{22}$ 154.30	693-62-9 -63.7	193	0.7508^{20}	1.4285^{20}
11971	4-Undecene, (*Z*)-	$C_{11}H_{22}$ 154.30	821-98-7 -97	192.6	0.7541^{20}	1.4302^{20}
11972	5-Undecene, (*E*)-	$C_{11}H_{22}$ 154.30	764-97-6 -61.1	192	4-01-00-00911 0.7497^{20}	eth 4; chl 4; lig 4 1.4285^{20}
11973	5-Undecene, (*Z*)-	$C_{11}H_{22}$ 154.30	764-96-5 -106.5	192.3	4-01-00-00911 0.7537^{20}	1.4302^{20}
11974	10-Undecenoic acid Undecylenic acid	$C_{11}H_{20}O_2$ 184.28	112-38-9 24.5	9760 275	4-02-00-01612 0.9072^{24}	H_2O 1; EtOH 3; eth 3; ctc 2 1.4486^{24}
11975	9-Undecenoic acid, (*E*)- *trans*-9-Undecenoic acid	$C_{11}H_{20}O_2$ 184.28	37973-84-5 19	273	4-02-00-01616 0.9118^{25}	eth 4 1.4519^{20}
11976	9-Undecenoic acid, (*Z*)-	$C_{11}H_{20}O_2$ 184.28	116836-30-7 2.5	156^{4}; $119^{0.8}$	4-02-00-01616 0.9150^{20}	1.4530^{20}
11977	10-Undecenoic acid, ethyl ester	$C_{13}H_{24}O_2$ 212.33	692-86-4 -38	264.5	4-02-00-01614 0.8827^{15}	H_2O 1; EtOH 3; eth 3; ctc 2 1.4449^{25}
11978	10-Undecenoic acid, methyl ester	$C_{12}H_{22}O_2$ 198.31	111-81-9 -27.5	248	4-02-00-01613 0.889^{15}	H_2O 1; EtOH 3; eth 3; ctc 2 1.4393^{20}
11979	10-Undecen-1-ol	$C_{11}H_{22}O$ 170.30	112-43-6 -2	250	4-01-00-02194 0.8495^{15}	H_2O 1; EtOH 3; eth 3; ctc 2 1.4500^{20}
11980	2-Undecen-4-ol	$C_{11}H_{22}O$ 170.30	22381-86-8	$114\text{-}5^{13}$	4-01-00-02194 0.843^{22}	1.4485^{22}
11981	9-Undecen-1-ol	$C_{11}H_{22}O$ 170.30	112-46-9 6	248.5	3-01-00-01957 0.8507^{15}	1.4535^{19}
11982	10-Undecenoyl chloride	$C_{11}H_{19}ClO$ 202.72	38460-95-6	$120\text{-}2^{10}$	4-02-00-01615 0.944^{20}	1.454^{20}
11983	1-Undecen-3-yne	$C_{11}H_{18}$ 150.26	74744-28-8	74^{9}	3-01-00-01054 0.7962^{20}	1.4606^{20}

No.	Name	Synonym	Mol. Form.	Mol. Wt.	CAS RN	mp/°C	Merck No.	bp/°C	Beil. Ref.	den/g cm^{-3}	Solubility	n_D
11984	1-Undecyne		$C_{11}H_{20}$	152.28	2243-98-3	-25		196	4-01-00-01064	0.7728^{20}	ace 4; bz 4; eth 4; EtOH 4	1.4306^{20}
11985	2-Undecyne		$C_{11}H_{20}$	152.28	60212-29-5	-30.1		204.2	0-01-00-00261	0.7827^{20}		1.4391^{20}
11986	4-Undecyne		$C_{11}H_{20}$	152.28	60212-31-9	-74.7		198.5	3-01-00-01022	0.7752^{20}		1.4369
11987	5-Undecyne		$C_{11}H_{20}$	152.28	2294-72-6	-74.1		198	4-01-00-01064	0.7753^{20}		1.4369^{20}
11988	6-Undecynoic acid		$C_{11}H_{18}O_2$	182.26	55182-83-7			$124\text{-}5^{0.17}$	4-02-00-01739	0.9537^{25}		1.4566^{25}
11989	9-Undecynoic acid		$C_{11}H_{18}O_2$	182.26	22202-65-9	61		170^{15}, $132^{0.1}$	3-02-00-01471		H_2O 4; eth 4; EtOH 4	
11990	9-Undecynoic acid, methyl ester	Methyl 9-undecynoate	$C_{12}H_{20}O_2$	196.29	18937-76-3			$93\text{-}4^{1}$	3-02-00-01472	0.9245^{17}		1.4732^{19}
11991	10-Undecynoic acid, methyl ester	Methyl 10-undecynoate	$C_{12}H_{20}O_2$	196.29	2777-66-4			121^{15}	4-02-00-01738	0.9177^{20}		1.4421^{21}
11992	Urea	Carbamide	CH_4N_2O	60.06	57-13-6	132.7	9781	dec	4-03-00-00094	1.3230^{20}	H_2O 4; EtOH 4; eth 1; bz 1	1.484
11993	Urea, (aminoiminomethyl)-		$C_2H_6N_4O$	102.10	141-83-3	105		160 dec	4-03-00-00155		H_2O 3; EtOH 2; eth 1; bz 1	
11994	Urea, *N*,*N*'-bis(2-chloroethyl)-*N*-nitroso-	Carmustine	$C_5H_9Cl_2N_3O_2$	214.05	154-93-8	31	1852				H_2O 4; EtOH 4	
11995	Urea, *N*,*N*'-bis(hydroxymethyl)-	Oxymethurea	$C_3H_8N_2O_3$	120.11	140-95-4	126	6921	149^{25}	4-03-00-00107		H_2O 3; EtOH 3; eth 1; MeOH 3	
11996	Urea, *N*,*N*-bis(2-methylpropyl)-	1,1-Diisobutylurea	$C_9H_{20}N_2O$	172.27	77464-06-3	73		180^{25}	0-04-00-00170		eth 4; EtOH 4	
11997	Urea, *N*,*N*'-bis(4-nitrophenyl)-	4,4'-Dinitrocarbanilide	$C_{13}H_{10}N_4O_5$	302.25	587-90-6	312 dec	3271		4-12-00-01646			
11998	Urea, (4-bromophenyl)-		$C_7H_7BrN_2O$	215.05	1967-25-5	226		260	4-12-00-01511		H_2O 2; EtOH 4; eth 4; bz 4	
11999	Urea, *N*'-(4-chlorophenyl)-*N*,*N*-dimethyl-	Monuron	$C_9H_{11}ClN_2O$	198.65	150-68-5	170.5	6169		4-12-00-01191		H_2O 1; EtOH 2; ace 2	
12000	Urea, *N*'-(3,4-dichlorophenyl)-*N*,*N*-dimethyl-	Diuron	$C_9H_{10}Cl_2N_2O$	233.10	330-54-1	158	3388		4-12-00-01263			
12001	Urea, *N*,*N*'-diethyl-		$C_5H_{12}N_2O$	116.16	623-76-7	112.5		263	4-04-00-00370	1.0415^{25}	H_2O 4; EtOH 4; eth 4	1.4616^{40}
12002	Urea, *N*,*N*-diethyl-		$C_5H_{12}N_2O$	116.16	634-95-7	75		$94\text{-}6^{0.02}$	4-04-00-00380		H_2O 4; EtOH 4; eth 3; bz 4	
12003	Urea, *N*,*N*'-diethyl-*N*,*N*'-diphenyl-	*N*,*N*'-Diethylcarbanilide	$C_{17}H_{20}N_2O$	268.36	85-98-3	79	3107		4-12-00-00844		H_2O 1; EtOH 4; chl 3	
12004	Urea, *N*,*N*'-dimethyl-		$C_3H_8N_2O$	88.11	96-31-1	108		269	4-04-00-00207	1.142^{25}	H_2O 4; EtOH 4; eth 1; chl 2	
12005	Urea, *N*,*N*-dimethyl-		$C_3H_8N_2O$	88.11	598-94-7	183.5			4-04-00-00224	1.2555^{25}	H_2O 3; EtOH 2; eth 1; tfa 2	
12006	Urea, *N*,*N*'-dimethyl-*N*,*N*'-diphenyl-		$C_{15}H_{16}N_2O$	240.30	611-92-7	122		350	4-12-00-00839		H_2O 4; EtOH 4; eth 2; ace 4	
12007	Urea, *N*,*N*-dimethyl-*N*'-[3-(trifluoromethyl)phenyl]-	Fluometuron	$C_{10}H_{11}F_3N_2O$	232.21	2164-17-2	164	4080				ace 4; EtOH 4	
12008	Urea, *N*,*N*'-di-1-naphthalenyl-		$C_{21}H_{16}N_2O$	312.37	607-56-7	296		sub	4-12-00-03078		py 4	
12009	Urea, (2,5-dioxo-4-imidazolidinyl)-	Allantoin	$C_4H_6N_4O_3$	158.12	97-59-6	239	246		4-25-00-04071		H_2O 2; EtOH 3; eth 1; MeOH 1	
12010	Urea, *N*,*N*'-diphenyl-	Carbanilide	$C_{13}H_{12}N_2O$	212.25	102-07-8	236	1787	262	4-12-00-00741	1.239^{25}	H_2O 2; EtOH 2; eth 3; ace 2	
12011	Urea, *N*,*N*-diphenyl-	1,1-Diphenylurea	$C_{13}H_{12}N_2O$	212.25	603-54-3	189		dec	4-12-00-00852	1.276^{25}	H_2O 2; EtOH 3; eth 3; chl 3	
12012	Urea, *N*,*N*'-dipropyl-		$C_7H_{16}N_2O$	144.22	623-95-0	105		255	4-04-00-00482		eth 4; EtOH 4	
12013	Urea, (4-ethoxyphenyl)-	Dulcin	$C_9H_{12}N_2O_2$	180.21	150-69-6	173.5	3448	dec	4-13-00-01154		H_2O 2; EtOH 3; AcOEt 4	
12014	Urea, ethyl-		$C_3H_8N_2O$	88.11	625-52-5	92.5		dec	4-04-00-00369	1.2130^{18}	H_2O 4; EtOH 4; eth 3; bz 4	
12015	Urea, *N*-ethyl-*N*-nitroso-	*N*-Ethyl-*N*-nitrosourea	$C_3H_7N_3O_2$	117.11	759-73-9	98-100 dec	3788		3-04-00-00233		chl 3	
12016	Urea, hydroxy-	Hydroxyurea	$CH_4N_2O_2$	76.06	127-07-1	141	4785	dec	4-03-00-00170		H_2O 4	
12017	Urea, methyl-		$C_2H_6N_2O$	74.08	598-50-5	103		dec	4-04-00-00205	1.2040^{0}	H_2O 4; EtOH 4; eth 1; bz 1	
12018	Urea, *N*-methyl-*N*,*N*'-diphenyl-		$C_{14}H_{14}N_2O$	226.28	612-01-1	106		203	4-12-00-00839		H_2O 2; EtOH 2; eth 4; bz 3	
12019	Urea, *N*-methyl-*N*-nitroso-	Methylnitrosourea	$C_2H_5N_3O_2$	103.08	684-93-5	123-4 dec			4-04-00-03385		H_2O 2; EtOH 2; eth 2	
12020	Urea, monohydrochloride	Urea hydrochloride	CH_5ClN_2O	96.52	506-89-8	145 dec	9782		4-03-00-00094		H_2O 3	
12021	Urea, mononitrate	Urea nitrate	$CH_5N_3O_4$	123.07	124-47-0	152 dec	9784		4-03-00-00094	1.690^{20}	EtOH 4	

No.	Name Synonym	Mol. Form. Mol. Wt.	CAS RN mp/°C	Merck No. bp/°C	Beil. Ref. den/g cm^{-3}	Solubility n_D
12022	Urea, 1-naphthalenyl- 1-Naphthylurea	$C_{11}H_{10}N_2O$ 186.21	6950-84-1 219.5		4-12-00-03076	H_2O 1; EtOH 4; eth 3
12023	Urea, 2-naphthalenyl- 2-Naphthylurea	$C_{11}H_{10}N_2O$ 186.21	13114-62-0 219		4-12-00-03149	EtOH 4
12024	Urea, nitro- Nitrourea	$CH_3N_3O_3$ 105.05	556-89-8 158-9 dec	6573	4-03-00-00248	ace 4; EtOH 4
12025	Urea, (4-nitrophenyl)- 4-Nitrophenylurea	$C_7H_7N_3O_3$ 181.15	556-10-5 238	6547	4-12-00-01645	H_2O 4; EtOH 4
12026	Urea, phenyl- Phenylurea	$C_7H_8N_2O$ 136.15	64-10-8 147	7290 238	4-12-00-00734 1.302^{25}	H_2O 2; EtOH 3; eth 2; AcOEt 3
12027	Urea, (phenylmethyl)- Benzylurea	$C_8H_{10}N_2O$ 150.18	538-32-9 148	1165 200 dec	4-12-00-02253	ace 4; EtOH 4
12028	Urea, 2-propenyl- Allylurea	$C_4H_8N_2O$ 100.12	557-11-9 85	296	4-04-00-01070	H_2O 5; EtOH 5; eth 2; chl 2
12029	Urea, tetraethyl- Urea, 1,1,3,3-tetraethyl-	$C_9H_{20}N_2O$ 172.27	1187-03-7	 209	4-04-00-00380 0.919^{20}	H_2O 1; alk 1; acid 1 1.4474^{20}
12030	Urea, tetramethyl- Tetramethylurea	$C_5H_{12}N_2O$ 116.16	632-22-4 -1.2	9160 176.5	4-04-00-00225 0.9687^{20}	EtOH 2; eth 2; ctc 2 1.4496^{23}
12031	Urea, tetraphenyl- Tetraphenylurea	$C_{25}H_{20}N_2O$ 364.45	632-89-3 183		4-12-00-00853 1.222^{25}	H_2O 1; EtOH 2
12032	Urea, trimethyl- Trimethylurea	$C_4H_{10}N_2O$ 102.14	632-14-4 75.5	 232.5	4-04-00-00224 1.1900^{20}	H_2O 3, EtOH 3, eth 2, bz 2
12033	Uridine 1-β-*D*-Ribofuranosyluracil	$C_9H_{12}N_2O_6$ 244.20	58-96-8 165	9792	5-24-06-00132	H_2O 3; EtOH 3; py 3
12034	Uridine, 2'-deoxy-5-fluoro- Floxuridine	$C_9H_{11}FN_2O_5$ 246.20	50-91-9	4045	5-24-06-00291	
12035	5'-Uridylic acid Uridine 5'-phosphoric acid	$C_9H_{13}N_2O_9P$ 324.18	58-97-9 202 dec	9796	5-24-06-00173	H_2O 4; MeOH 3
12036	Urs-12-ene-27,28-dioic acid, 3-hydroxy-, (3β)- Quinovic acid	$C_{30}H_{46}O_5$ 486.69	465-74-7 298 dec	8104	4-10-00-02197	
12037	Urs-12-en-28-oic acid, 3-hydroxy-, (3β)- Ursolic acid	$C_{30}H_{48}O_3$ 456.71	77-52-1 284	9802	4-10-00-01160	ace 4; eth 4; chl 4
12038	Urs-12-en-3-ol, (3β)- α-Amyrin	$C_{30}H_{50}O$ 426.73	638-95-9 186	653 $243^{0.5}$	4-06-00-04191	EtOH 3; eth 3; bz 3; chl 3
12039	Valeric acid, 3-chloro-, ethyl ester	$C_7H_{13}ClO_2$ 164.63	6513-13-9	 189	3-02-00-00678 1.0330^{20}	eth 4; EtOH 4 1.4278^{20}
12040	*DL*-Valine	$C_5H_{11}NO_2$ 117.15	516-06-3 298 dec	9818 sub	3-04-00-01370 1.316^{25}	H_2O 3; EtOH 2; eth 2; bz 2
12041	*L*-Valine 2-Aminoisovaleric acid	$C_5H_{11}NO_2$ 117.15	72-18-4 315	9818 sub	4-04-00-02659 1.23^{25}	H_2O 4
12042	*D*-Valine, 3-[(2-amino-2-carboxyethyl)dithio]-, (*R*)- Penicillamine cysteine disulfide	$C_8H_{16}N_2O_4S_2$ 268.36	18840-45-4 195	7030	4-04-00-03231	
12043	*D*-Valine, 3-mercapto- Penicillamine	$C_5H_{11}NO_2S$ 149.21	52-67-5 198.5	7029	4-04-00-03228	
12044	Veratraman-3,23-diol, 14,15,16,17-tetradehydro-, (3β,23β)- Veratramine	$C_{27}H_{39}NO_2$ 409.61	60-70-8 206	9853	5-21-05-00297	EtOH 3; bz 3; chl 3; dil acid 3
12045	Veratraman-11-one, 17,23-epoxy-3-(β-*D*-glucopyranosyloxy)-, (3β,23β)- Pseudojervine	$C_{33}H_{49}NO_8$ 587.75	36069-05-3 304-5 dec	7932	4-27-00-03592	H_2O 1; EtOH 3; eth 1; bz 1
12046	Veratraman-11-one, 17,23-epoxy-3-hydroxy-, (3β,23β)- Jervine	$C_{27}H_{39}NO_3$ 425.61	469-59-0 243 dec	5145	4-27-00-03590	H_2O 1; EtOH 3; eth 2; ace 3
12047	Vernolate Carbamothioic acid, dipropyl-, *S*-propyl ester	$C_{10}H_{21}NOS$ 203.35	1929-77-7	9866 150^{30}	 0.952^{20}	
12048	Vinclozolin 2,4-Oxazolidinedione, 3-(3,5-dichlorophenyl)-5-ethenyl-5-methyl-	$C_{12}H_9Cl_2NO_3$ 286.11	50471-44-8 108	9890 $131^{0.05}$	 1.51	
12049	9H-Xanthene 10H-9-Oxaanthracene	$C_{13}H_{10}O$ 182.22	92-83-1 100.5	 311	5-17-02-00252	H_2O 1; EtOH 2; eth 3; bz 3
12050	9H-Xanthen-9-one Xanthone	$C_{13}H_8O_2$ 196.21	90-47-1 174	9971 351; 146^3	5-17-10-00430	H_2O 1; EtOH 3; eth 3; bz 3
12051	9H-Xanthen-9-one, 1,3-dihydroxy-	$C_{13}H_8O_4$ 228.20	3875-68-1 259	 sub	5-18-04-00004	EtOH 4
12052	9H-Xanthen-9-one, 2,7-dihydroxy- Xanthone, 2,7-dihydroxy	$C_{13}H_8O_4$ 228.20	64632-72-0 330	 sub	1-18-00-00357	eth 4; EtOH 4
12053	9H-Xanthen-9-one, 3,6-dihydroxy-	$C_{13}H_8O_4$ 228.20	1214-24-0 350 dec	 sub	5-18-04-00010	EtOH 4; HOAc 4
12054	9H-Xanthen-9-one, 1,7-dihydroxy-3-methoxy- Gentisin	$C_{14}H_{10}O_5$ 258.23	437-50-3 266.5	4291	5-18-04-00497	H_2O 1; EtOH 4; ace 1; py 3
12055	*peri*-Xanthenoxanthene Dinaphthalene dioxide	$C_{20}H_{10}O_2$ 282.30	191-28-6 243.5	 400^{20-25}	5-19-02-00242	bz 3; chl 3
12056	Xanthylium, 9-(2-carboxyphenyl)-3,6-bis(diethylamino)-, chloride Rhodamine, B	$C_{28}H_{32}ClN_2O_3$ 480.03	81-88-9 165	8182	5-19-08-00669	H_2O 3; EtOH 3; eth 3; bz 3
12057	*o*-Xylene, 4-propyl-	$C_{11}H_{16}$ 148.25	3982-66-9 -20	 208.9	3-05-00-01005 0.8715^{20}	ace 4; bz 4; eth 4; EtOH 4 1.5000^{20}

No.	Name Synonym	Mol. Form. Mol. Wt.	CAS RN mp/°C	Merck No. bp/°C	Beil. Ref. den/g cm^{-3}	Solubility n_D
12058	Xylitol Xylite	$C_5H_{12}O_5$ 152.15	87-99-0 93.5	9992 216	4-01-00-02832	H_2O 4; py 4; EtOH 4
12059	*D*-Xylose	$C_5H_{10}O_5$ 150.13	58-86-6 90.5	9995	4-01-00-04223 1.525^{20}	H_2O 4; EtOH 3; eth 2
12060	Yohimban-16-carboxylic acid, 11,17-dimethoxy-18-[[1-oxo-3-(3,4,5-trimethoxyphenyl)-2-propenyl]oxy]-, methyl ester, (3β,16β,17α,18β,20α)- Rescinnamine	$C_{35}H_{42}N_2O_9$ 634.73	24815-24-5 238.5	8146	4-25-00-01323	H_2O 1; EtOH 2; ace 3; chl 3
12061	Yohimban-16-carboxylic acid, 11,17-dimethoxy-18-[(3,4,5-trimethoxybenzoyl)oxy]-, methyl ester, Reserpine	$C_{33}H_{40}N_2O_9$ 608.69	50-55-5 264.5	8149	4-25-00-01319	H_2O 2; EtOH 3; eth 2; ace 2
12062	Yohimban-16-carboxylic acid, 17-hydroxy-, methyl ester, (16α,17α)- Yohimbine	$C_{21}H_{26}N_2O_3$ 354.45	146-48-5 241	10011 sub 160	4-25-00-01237	H_2O 2; EtOH 3; eth 3; bz 2
12063	Yohimban-16-carboxylic acid, 18-hydroxy-11,17-dimethoxy-, (3β,16β,17α,18β,20 Reserpic acid	$C_{22}H_{28}N_2O_5$ 400.47	83-60-3 242	8147	4-25-00-01305	
12064	Yohimban-16-carboxylic acid, 17-hydroxy-18-[(3,4,5-trimethoxybenzoyl)oxy]-, methyl ester, (3β,16β Raunescine	$C_{31}H_{36}N_2O_8$ 564.64	117-73-7 165	8130	4-25-00-01281	H_2O 1; EtOH 3; chl 3; HOAc 3
12065	Yohimban-16-carboxylic acid, 17-methoxy-18-[(3,4,5-trimethoxybenzoyl)oxy]-, methyl ester, (3β,16β Deserpidine	$C_{32}H_{38}N_2O_8$ 578.66	131-01-1 230.5	2901	4-25-00-01282	H_2O 1; EtOH 3; chl 3
12066	Zinc, dimethyl Dimethyl zinc	C_2H_6Zn 95.46	544-97-8 -40	3252 46	1.386^{10}	eth 3; peth 5
12067	Zinc, [[1,2-ethanediylbis[carbamodithioato]] (2-)]- Zineb	$C_4H_6N_2S_4Zn$ 275.76	12122-67-7 157 dec	10071		
12068	Ziram Zinc, bis(dimethylcarbamodithioato-*S*,*S*')-, (T-4)-	$C_6H_{12}N_2S_4Zn$ 305.83	137-30-4			

STRUCTURAL FORMULAS OF ORGANIC COMPOUNDS

In numeric order as they occur in the Table of Physical Constants of Organic Compounds

1

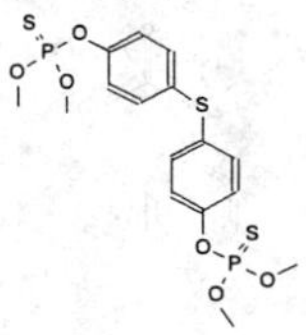

3

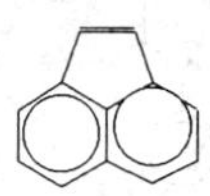

4

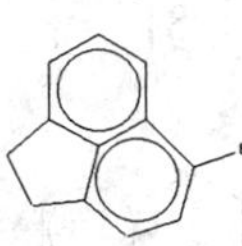

5

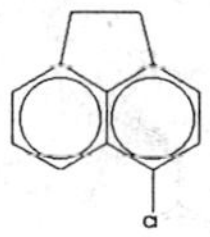

6

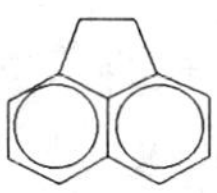

8

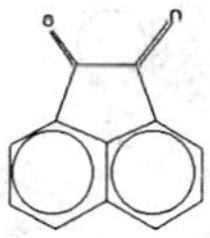

9

10

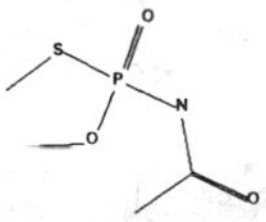

11

12

14

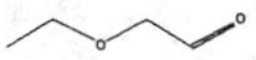

15

16

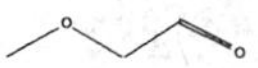

17

18

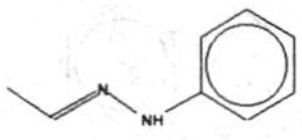

19

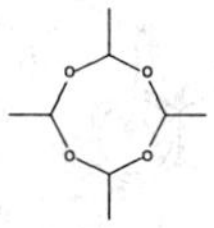

20

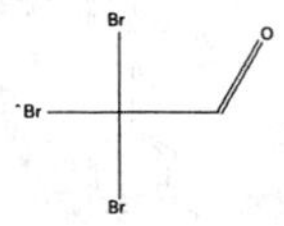

21

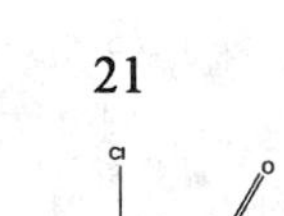

22

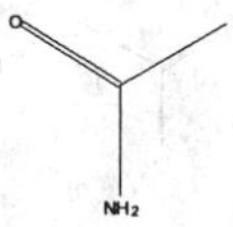

23

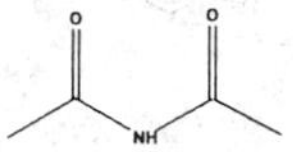

24

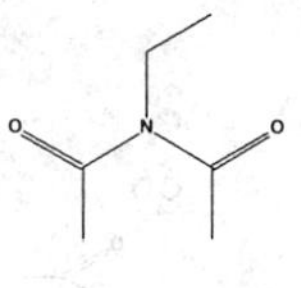

25

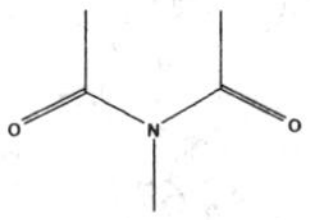

26

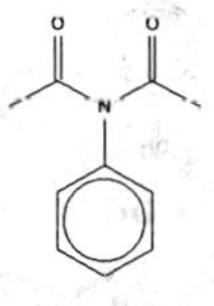

27

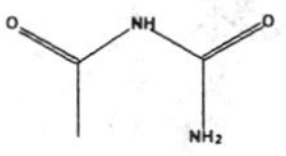

28

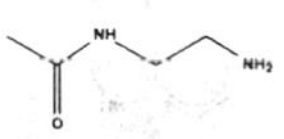

29

30

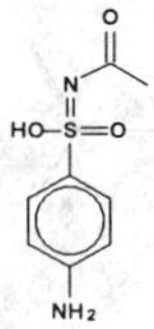

31

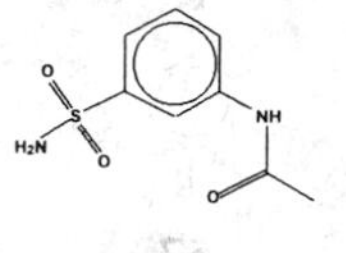

32

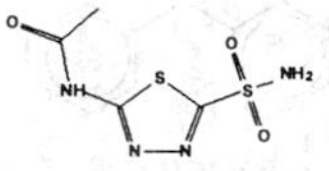

33

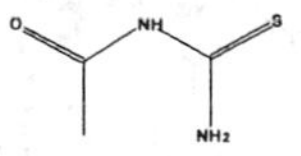

34

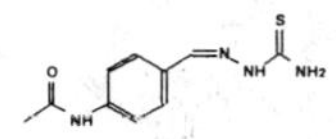

35

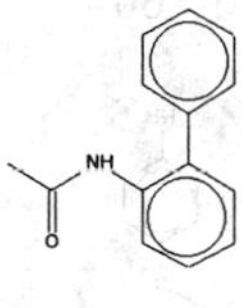

37

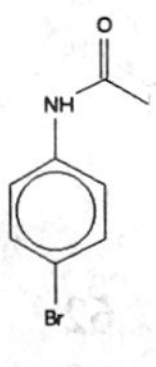

38

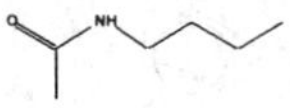

39

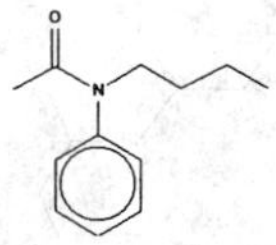

40

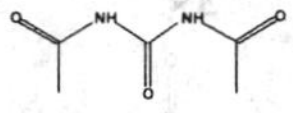

41

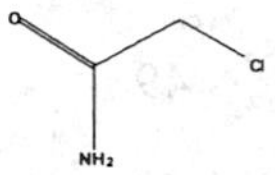

42

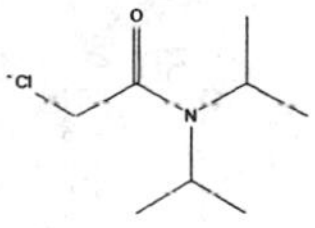

43

44

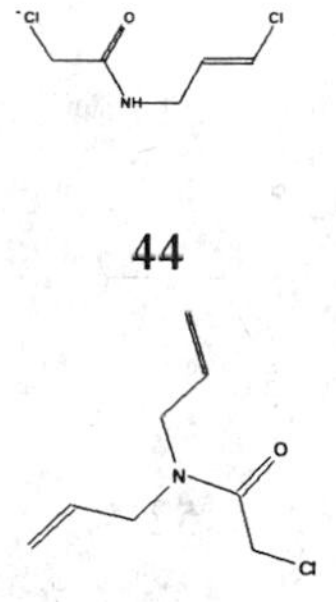

45

46

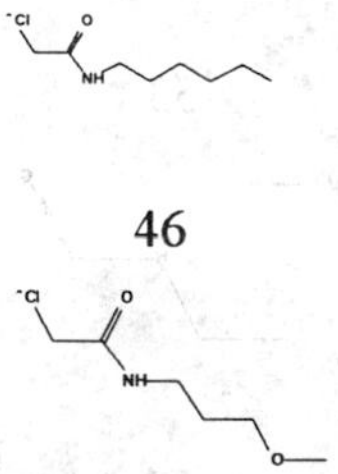

STRUCTURAL FORMULAS OF ORGANIC COMPOUNDS (continued)

In numeric order as they occur in the Table of Physical Constants of Organic Compounds

47

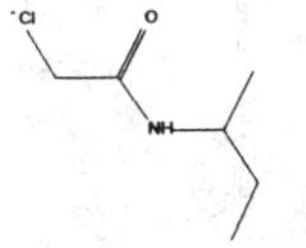

48

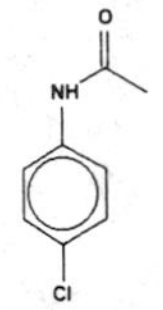

49

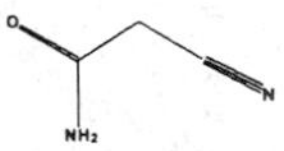

50

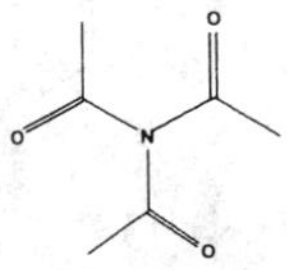

51

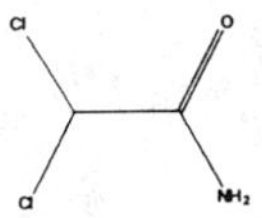

52

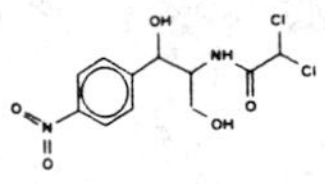

53

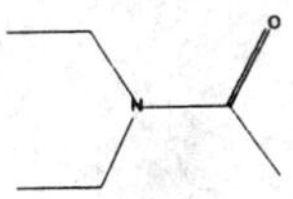

54

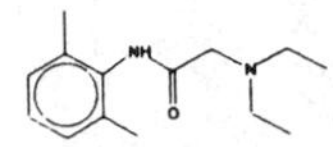

55

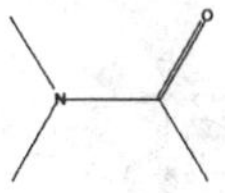

56

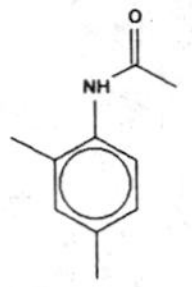

57

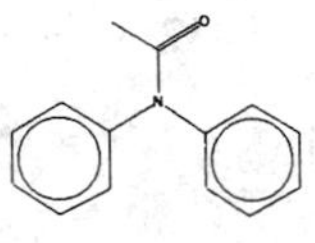

58

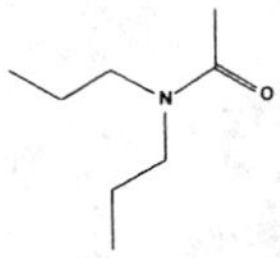

59

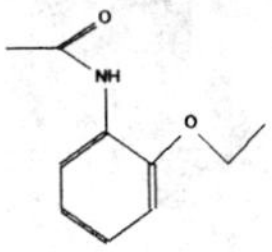

60

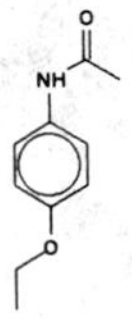

61

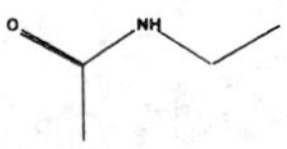

62

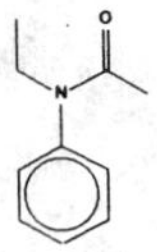

63

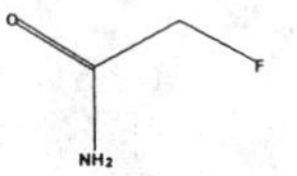

64

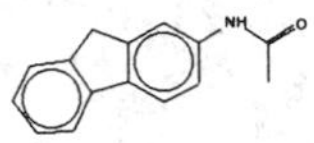

65

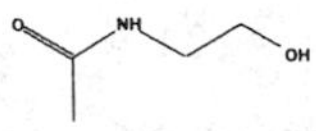

66

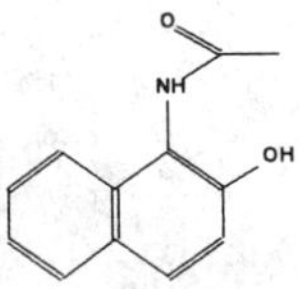

67

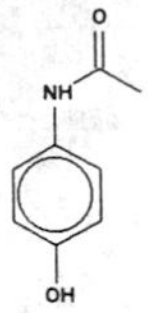

68

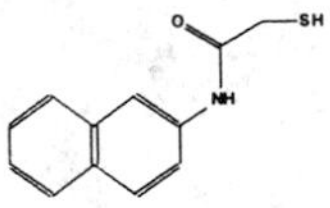

69

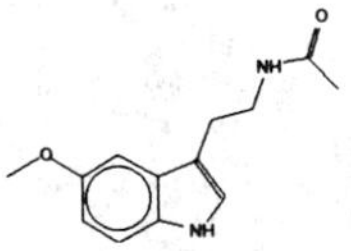

70

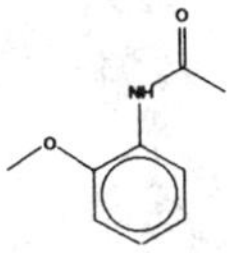

71

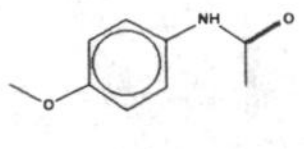

72

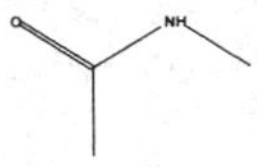

73

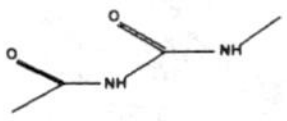

74

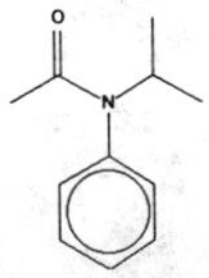

75

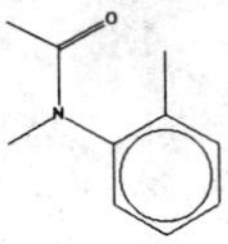

76

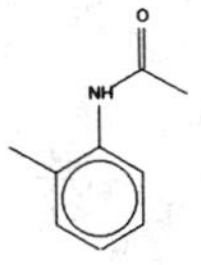

77

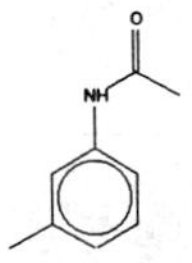

78

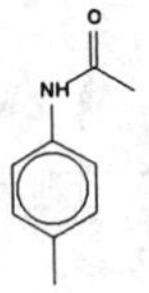

79

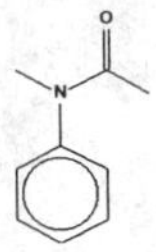

80

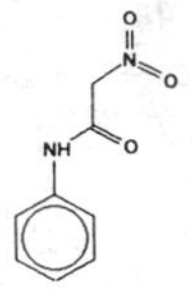

81

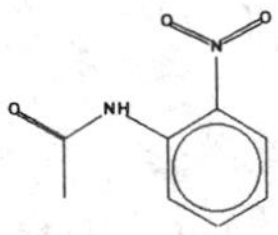

82

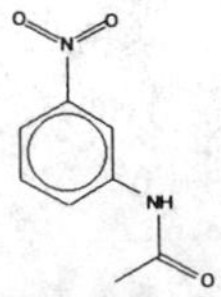

83

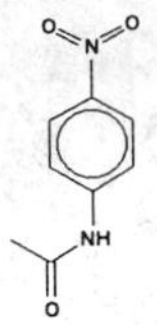

84

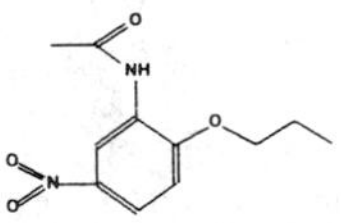

85

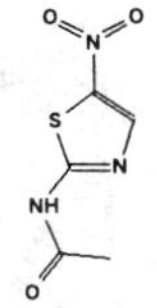

STRUCTURAL FORMULAS OF ORGANIC COMPOUNDS (continued)

In numeric order as they occur in the Table of Physical Constants of Organic Compounds

86

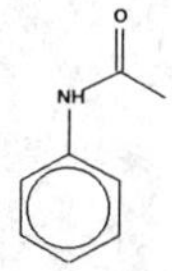

87

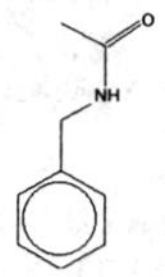

88

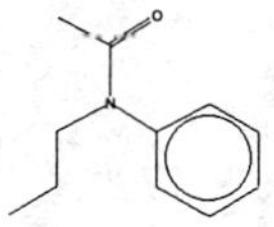

89

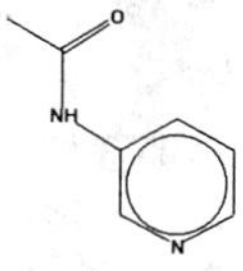

90

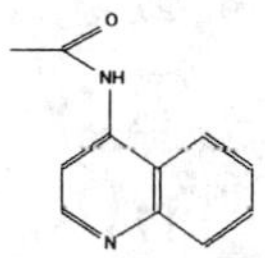

91

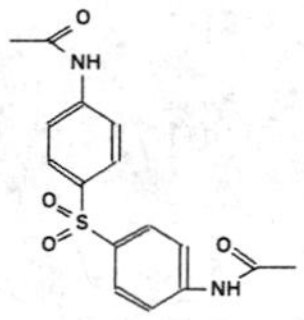

92

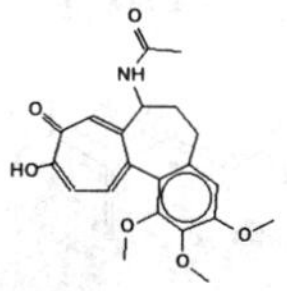

93

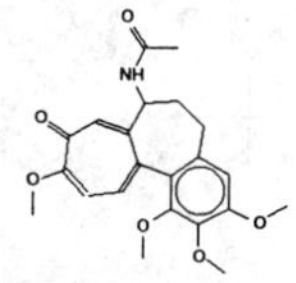

95

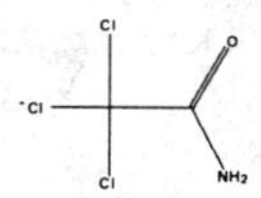

96

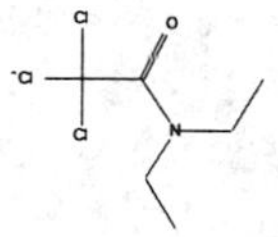

97

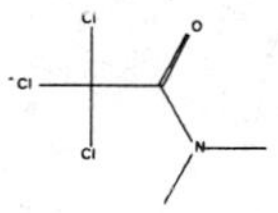

98

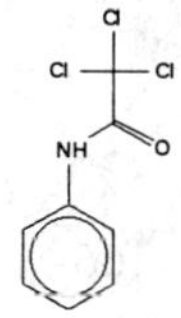

99

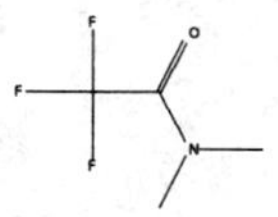

100

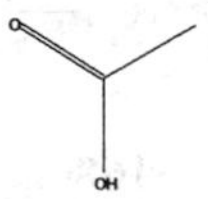

101

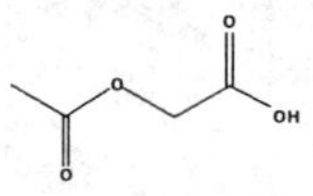

102

103

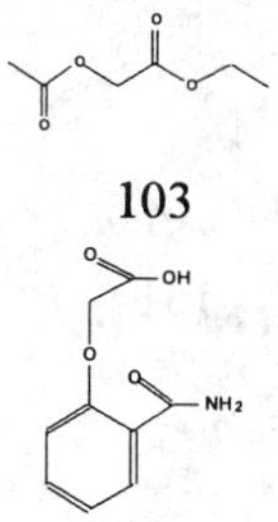

104

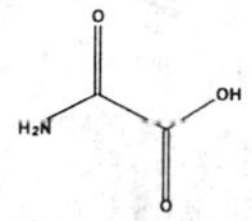

105

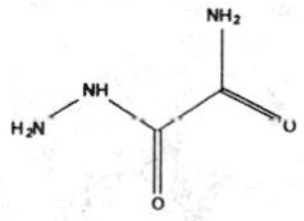

107

108

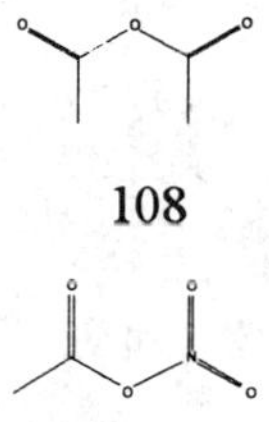

109

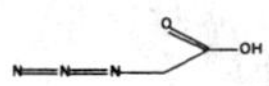

111

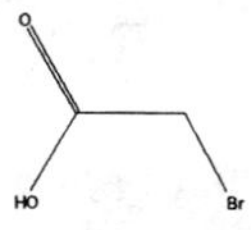

112

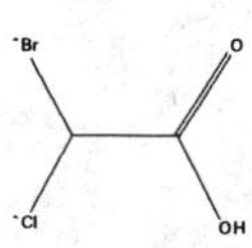

113

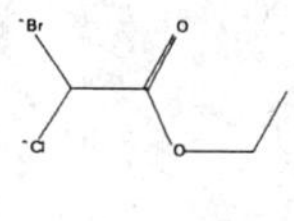

114

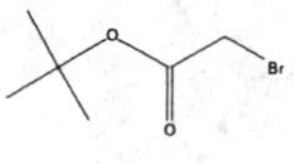

115

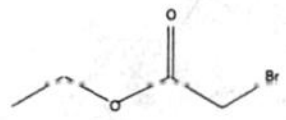

116

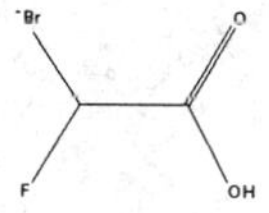

117

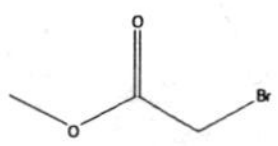

118

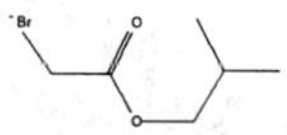

119

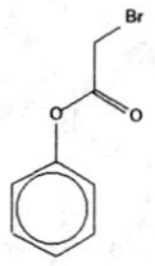

120

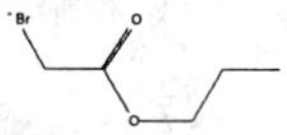

121

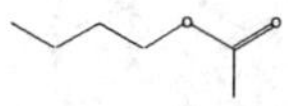

123

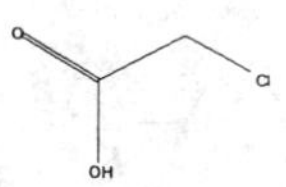

124

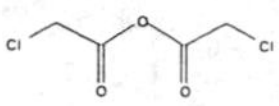

125

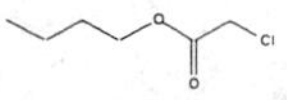

126

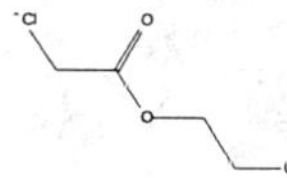

127

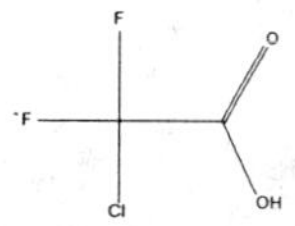

128

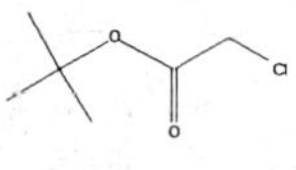

129

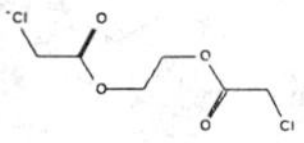

130

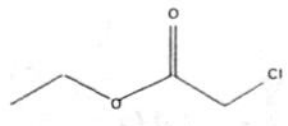

131

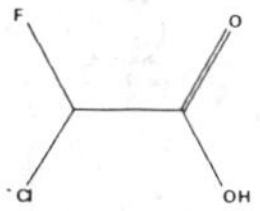

STRUCTURAL FORMULAS OF ORGANIC COMPOUNDS (continued)

In numeric order as they occur in the Table of Physical Constants of Organic Compounds

132 141 150 161 169

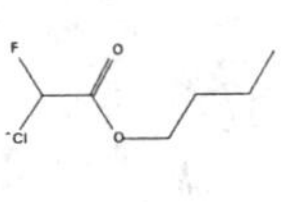

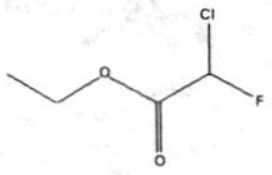

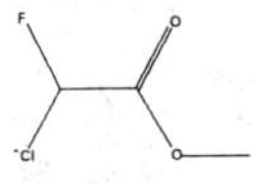

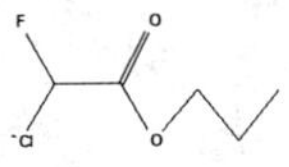

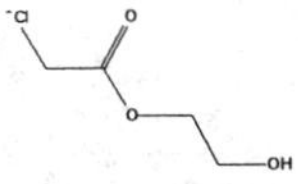

133 142 151 162 170

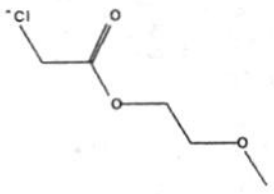

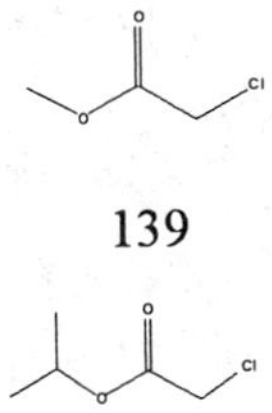

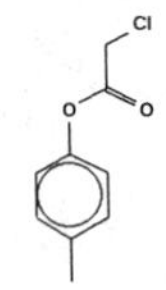

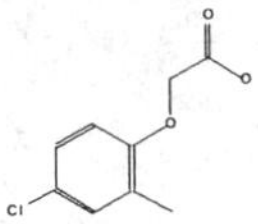

134 143 152 163 171

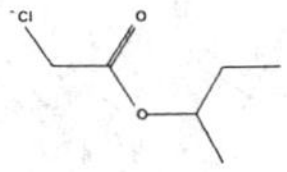

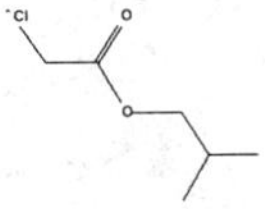

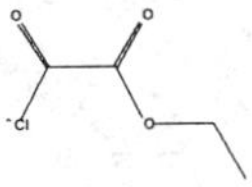

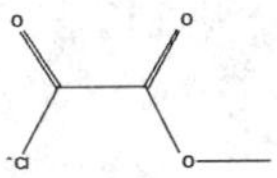

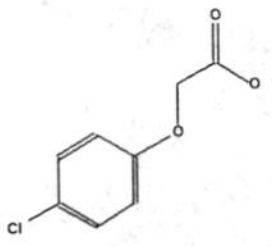

135 144 153 164 172

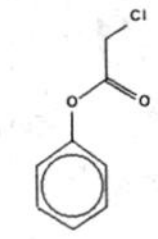

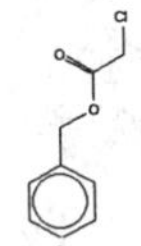

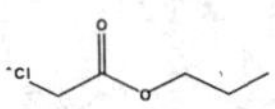

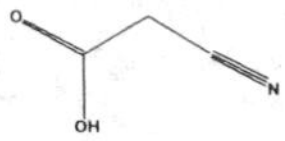

136 145 154 165 173

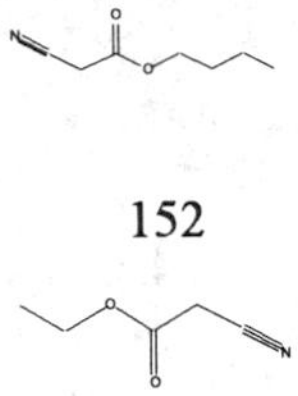

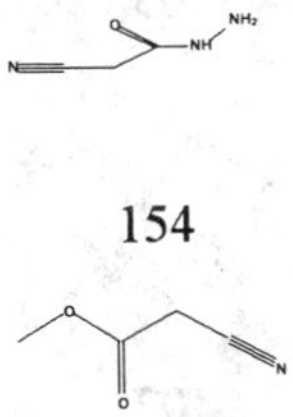

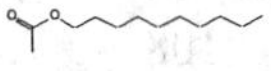

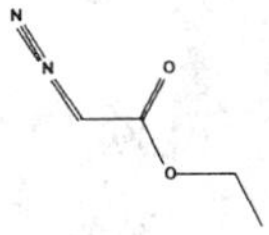

137 146 157 166 174

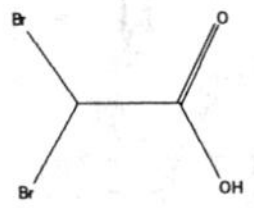

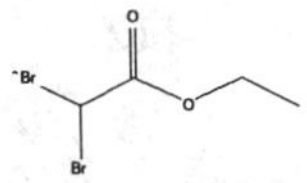

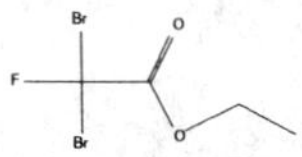

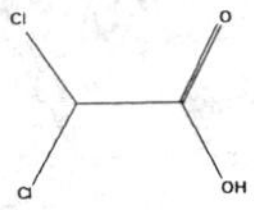

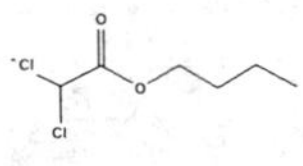

138 147 158 167 175

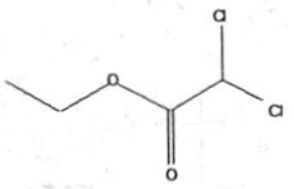

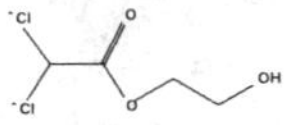

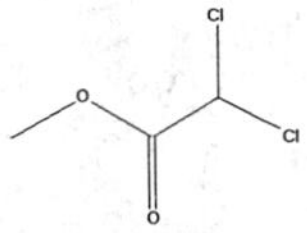

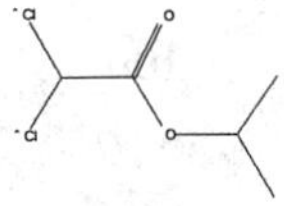

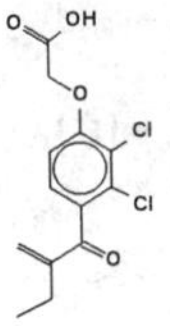

139 148 159 168 176

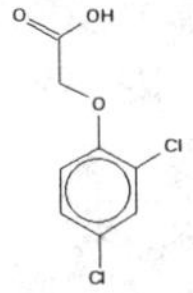

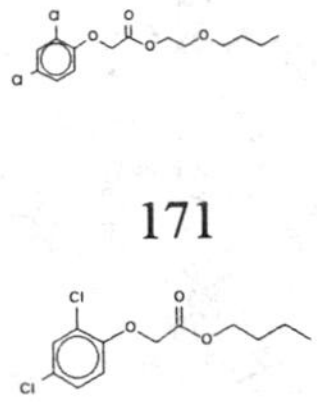

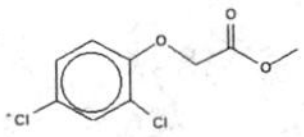

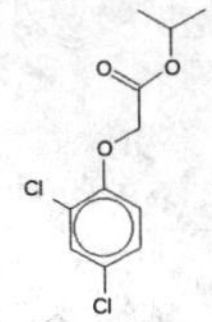

140 149 160 177

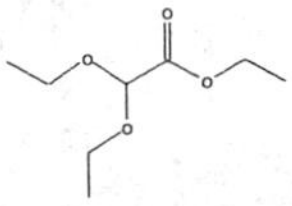

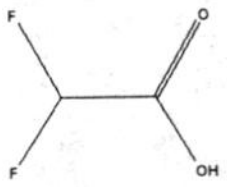

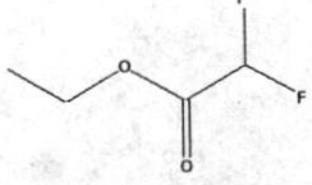

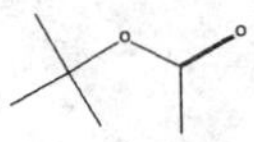

STRUCTURAL FORMULAS OF ORGANIC COMPOUNDS (continued)

In numeric order as they occur in the Table of Physical Constants of Organic Compounds

178

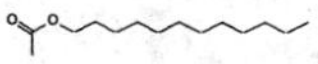

179

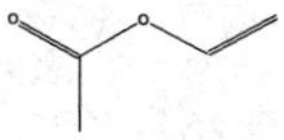

180

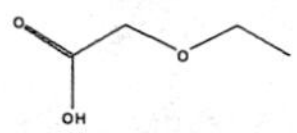

181

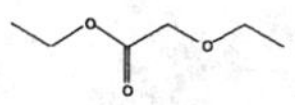

182

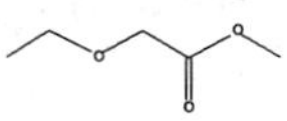

183

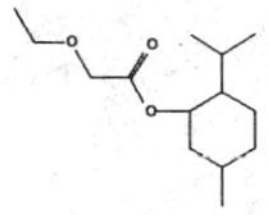

184

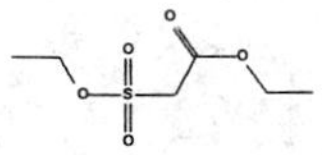

185

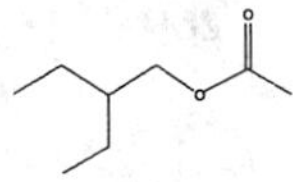

186

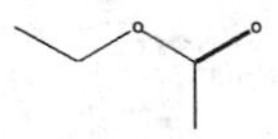

187

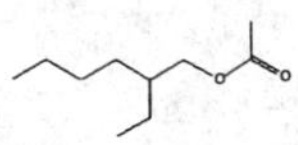

188

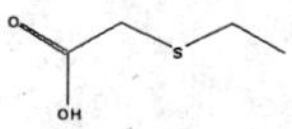

189

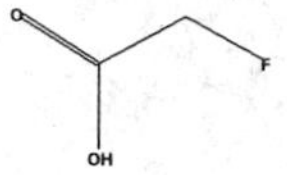

190

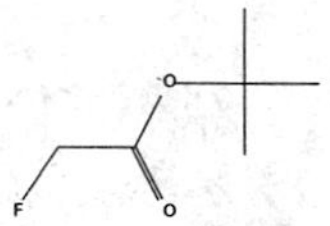

191

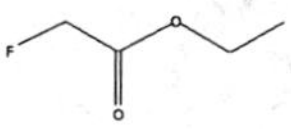

192

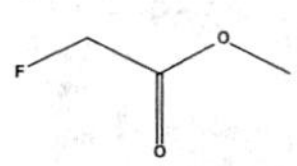

193

194

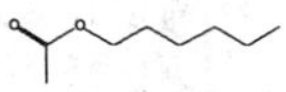

195

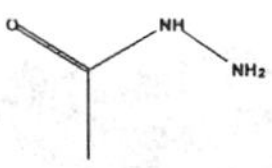

196

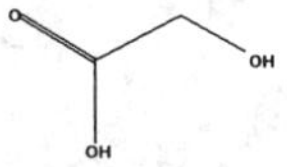

197

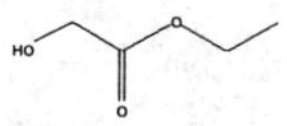

198

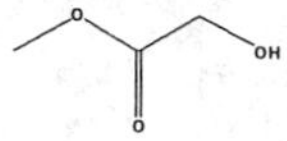

199

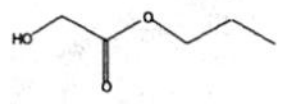

200

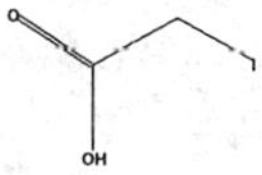

201

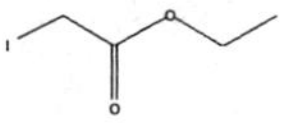

202

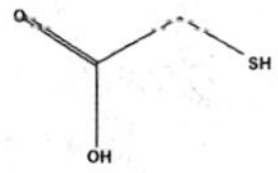

203

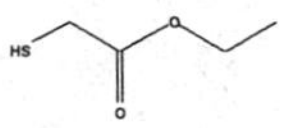

204

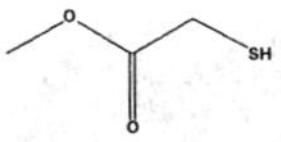

205

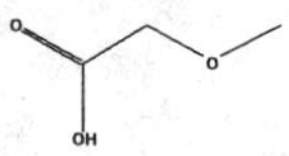

206

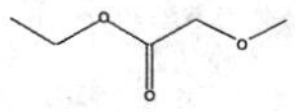

207

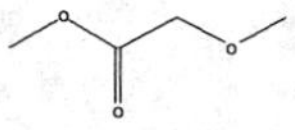

208

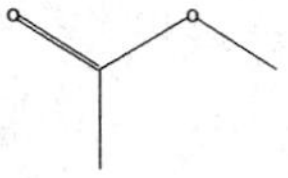

209

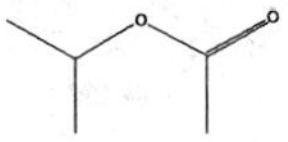

211

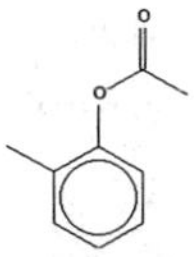

212

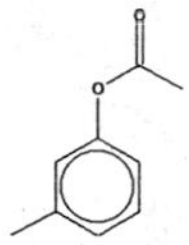

213

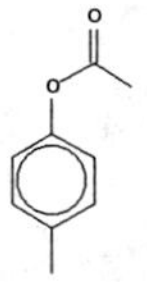

214

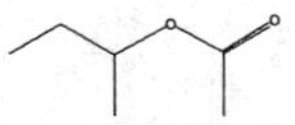

215

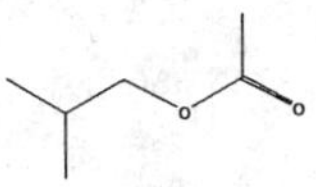

216

217

218

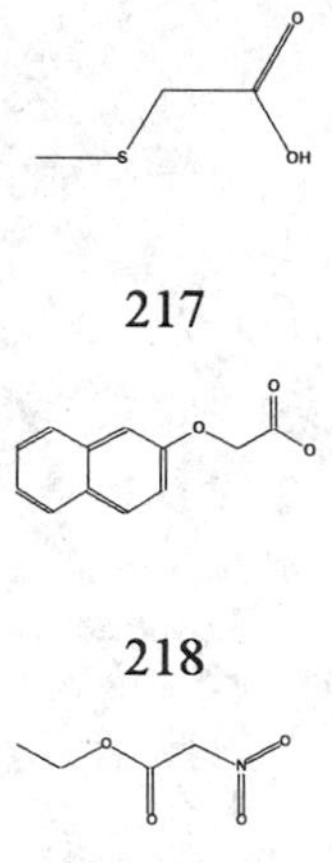

219

220

221

222

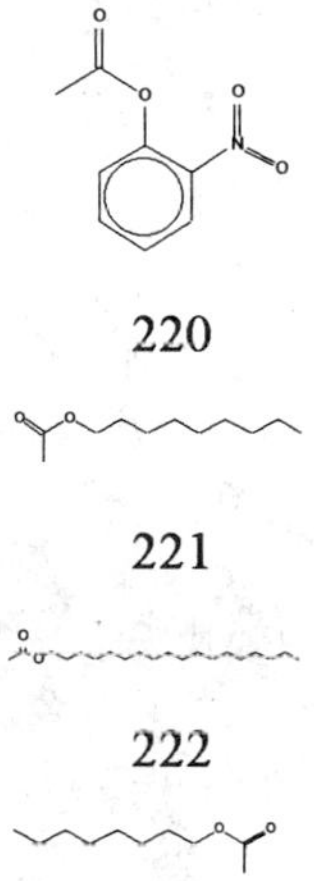

223

224

225

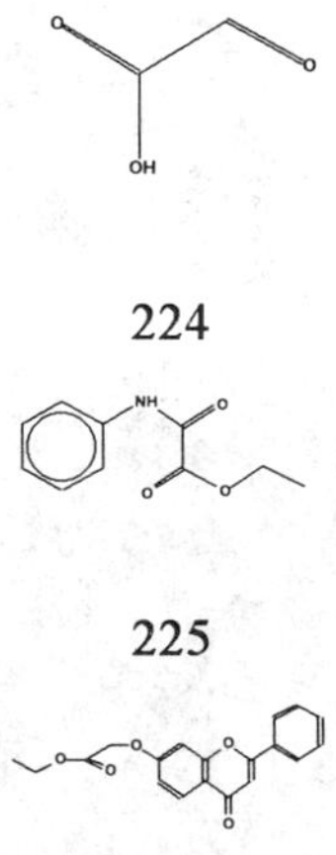

STRUCTURAL FORMULAS OF ORGANIC COMPOUNDS (continued)

In numeric order as they occur in the Table of Physical Constants of Organic Compounds

226

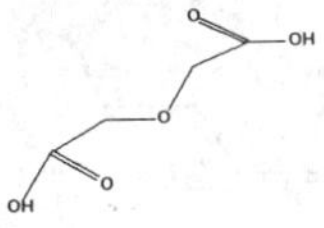

227

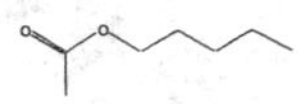

228

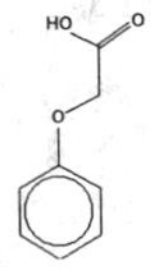

229

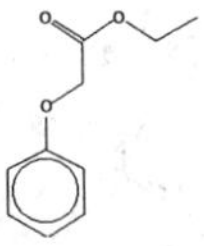

230

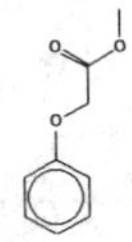

231

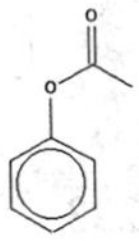

232

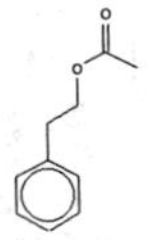

233

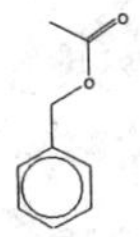

235

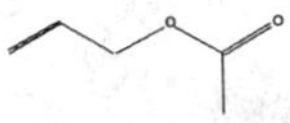

236

238

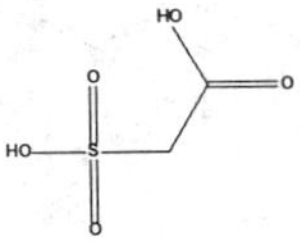

239

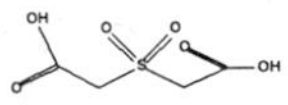

240

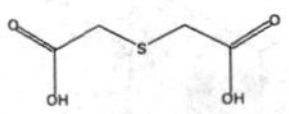

241

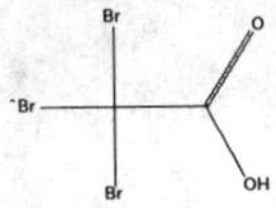

242

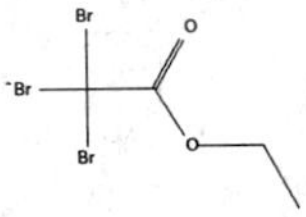

243

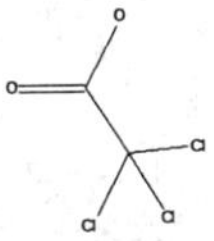

244

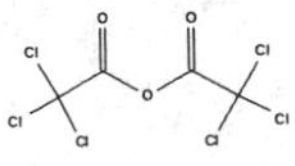

245

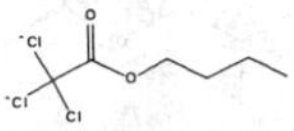

246

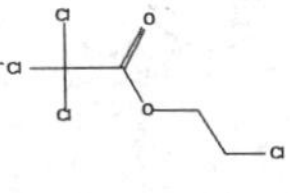

247

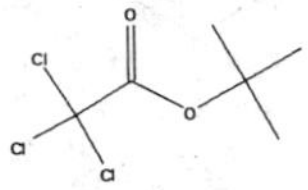

248

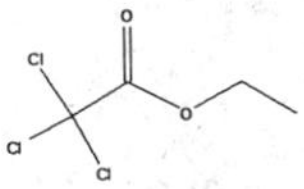

249

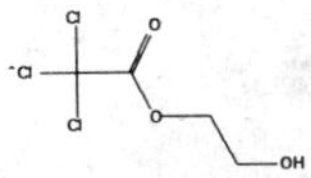

250

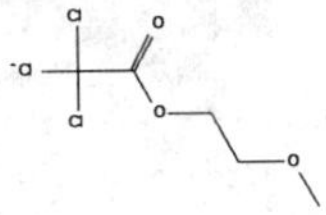

252

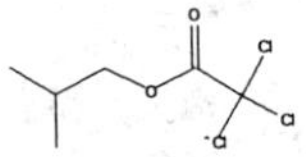

253

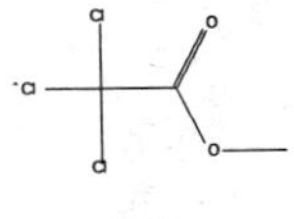

254

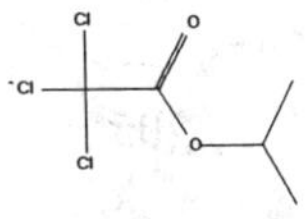

255

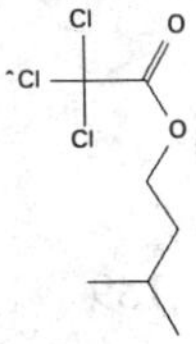

256

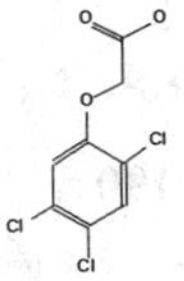

257

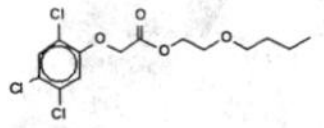

258

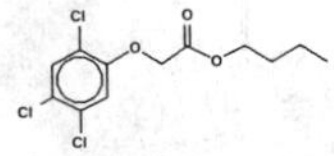

259

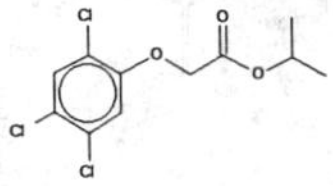

260

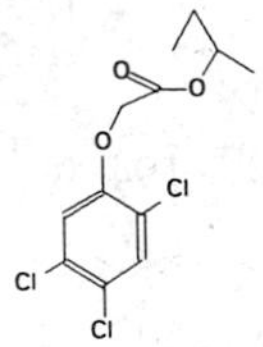

261

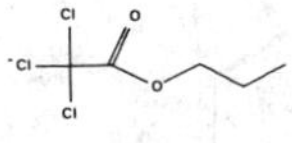

262

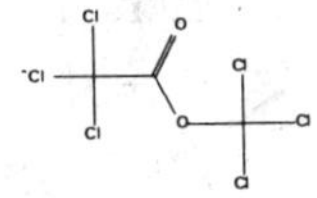

263

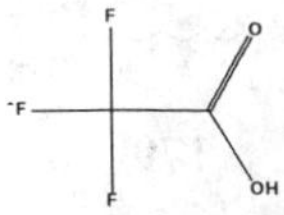

264

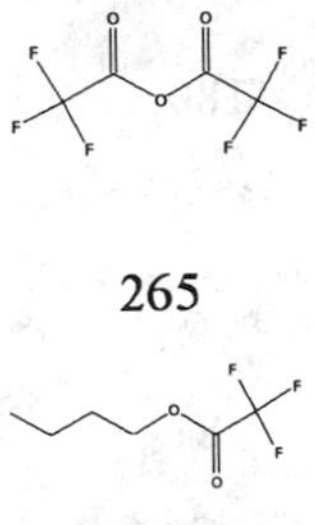

265

266

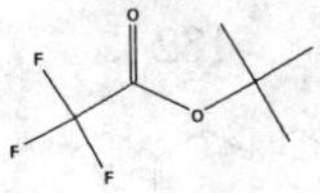

267

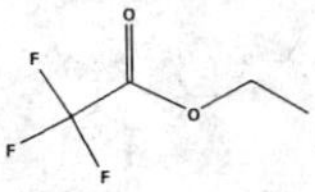

268

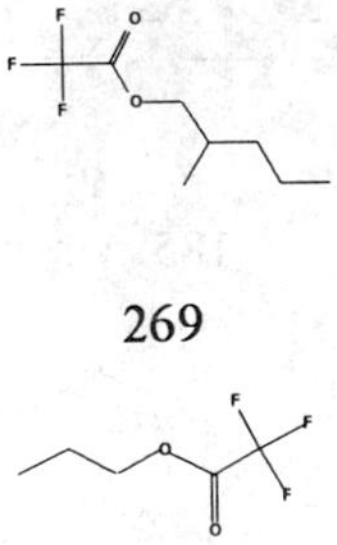

269

270

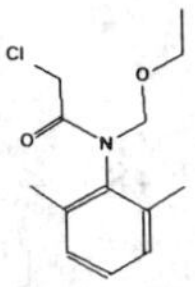

271

STRUCTURAL FORMULAS OF ORGANIC COMPOUNDS (continued)

In numeric order as they occur in the Table of Physical Constants of Organic Compounds

272

273

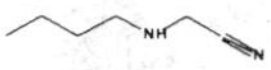

274

277

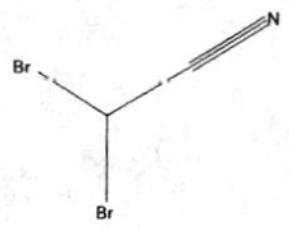

278

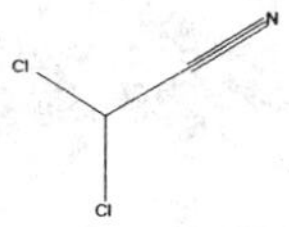

279

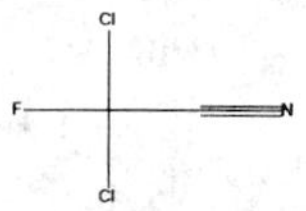

280

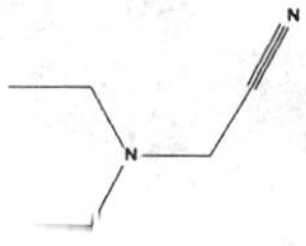

281

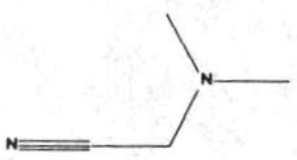

282

283

284

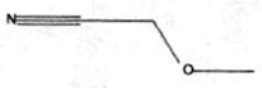

285

286

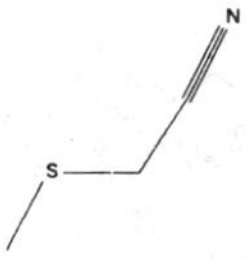

287

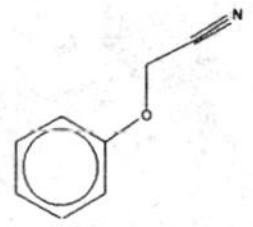

288

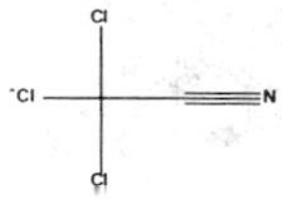

289

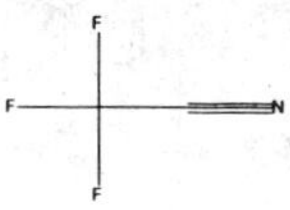

290

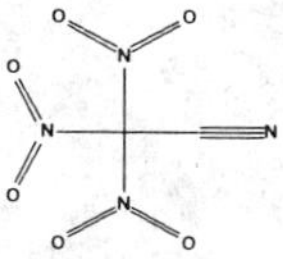

291

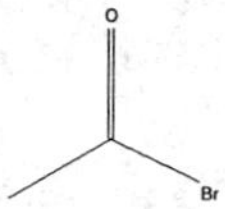

292

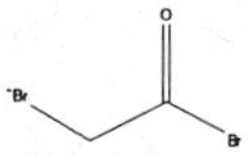

293

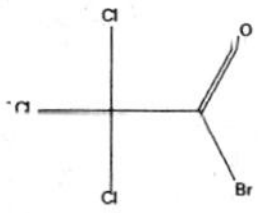

294

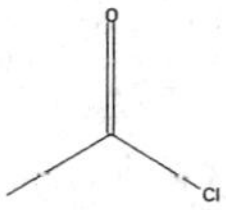

295

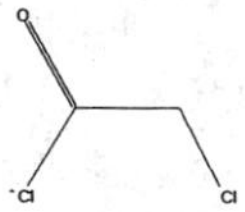

296

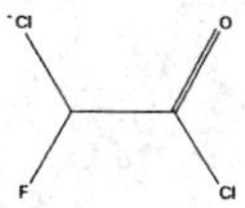

297

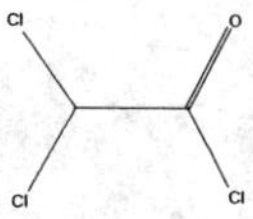

298

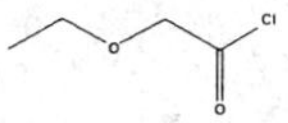

299

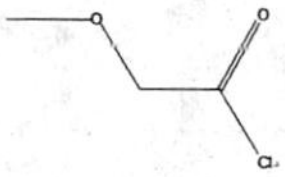

300

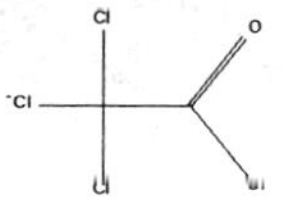

301

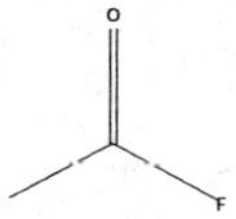

302

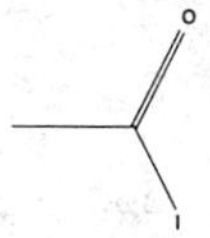

303

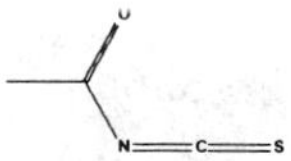

304

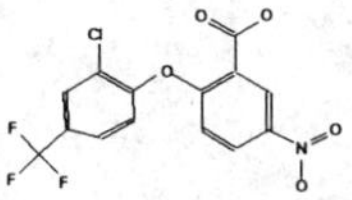

306

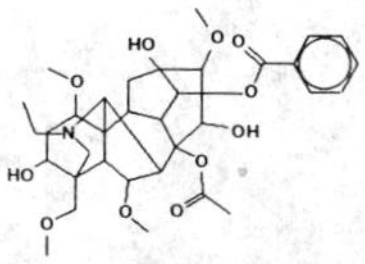

312

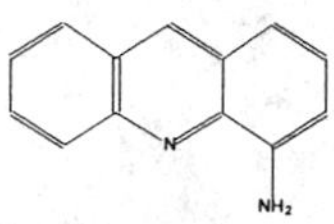

313

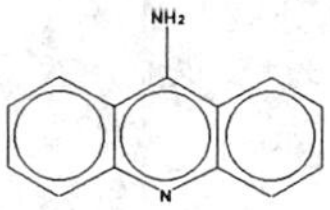

314

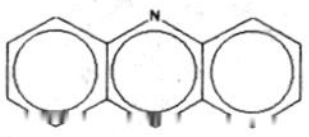

315

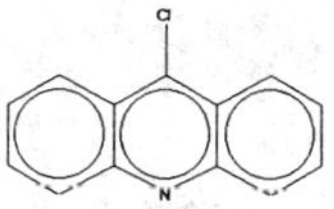

316

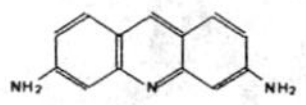

317

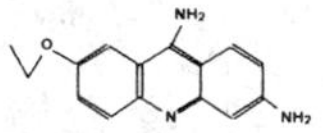

318

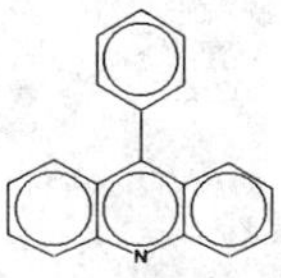

STRUCTURAL FORMULAS OF ORGANIC COMPOUNDS (continued)

In numeric order as they occur in the Table of Physical Constants of Organic Compounds

319 327 337 345 353

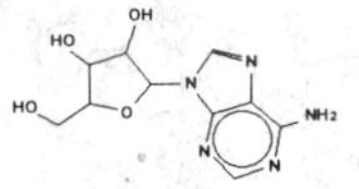

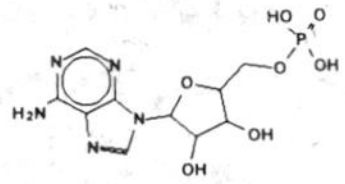

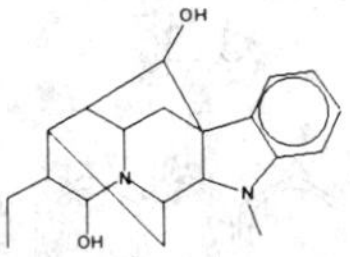

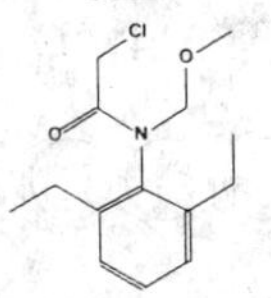

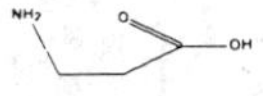

320 329 338 346 354

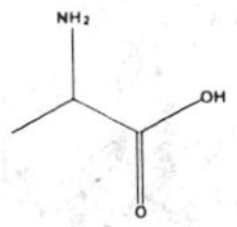

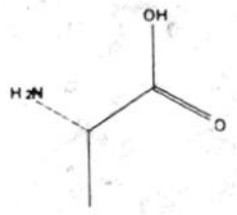

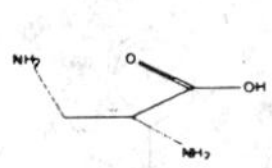

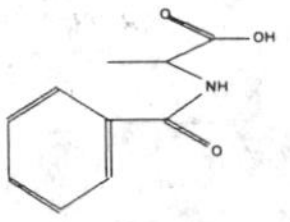

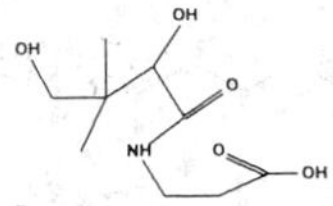

321 331 339 347 355

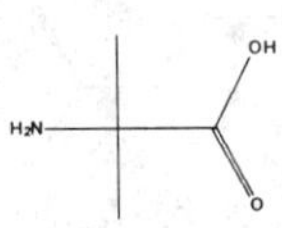

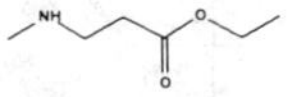

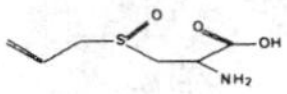

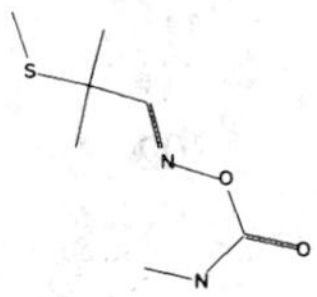

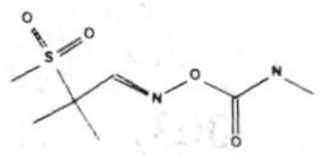

322 332 340 348 356

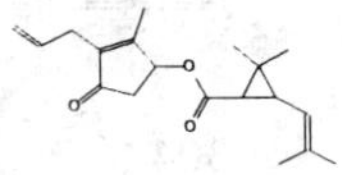

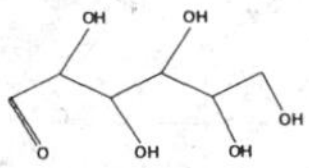

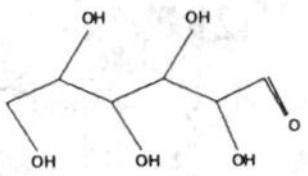

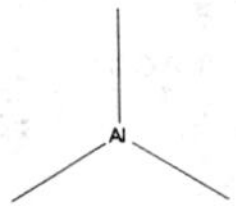

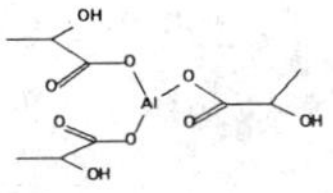

323 333 341 349 357

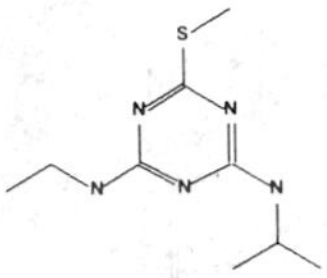

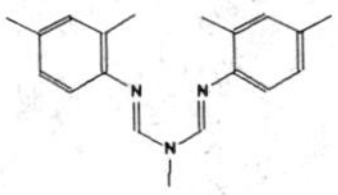

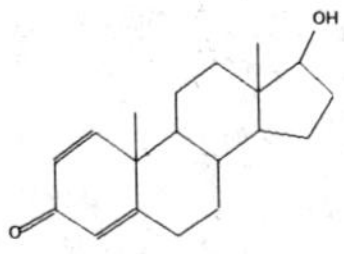

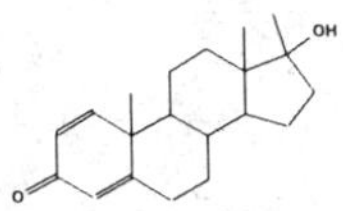

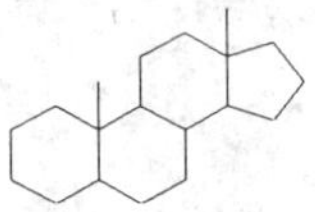

324 334 350 358

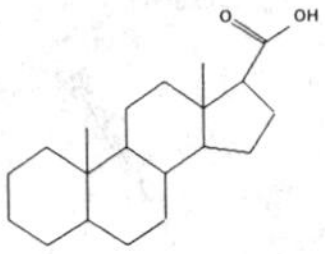

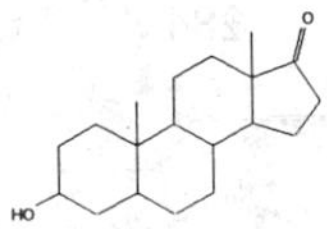

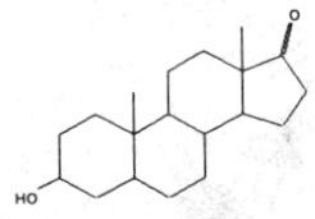

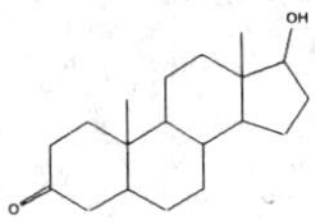

325 342 351

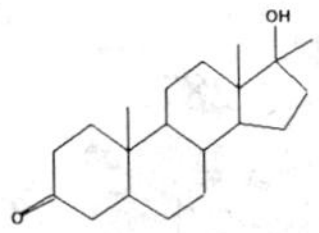

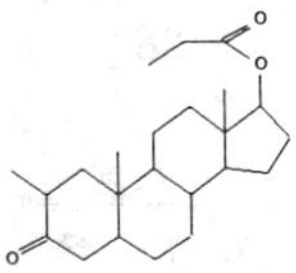

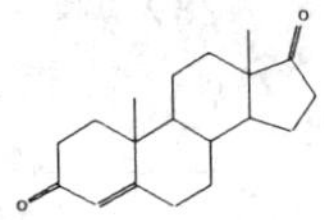

335 359

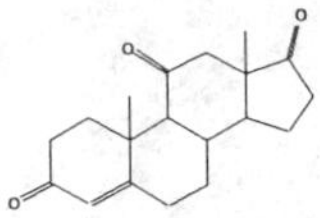

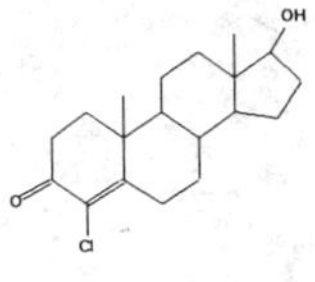

326 343 352

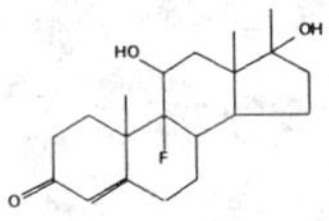

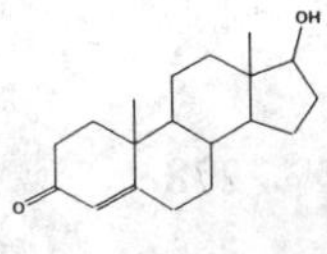

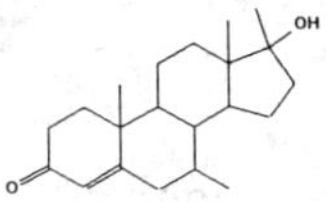

336 360

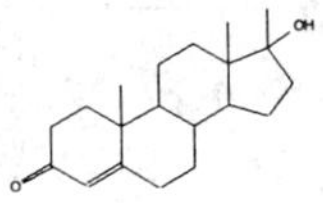

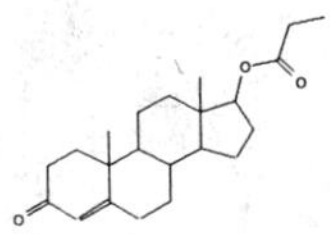

344

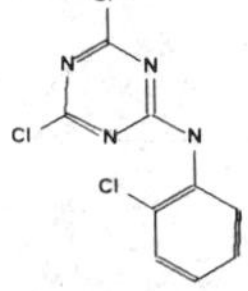

STRUCTURAL FORMULAS OF ORGANIC COMPOUNDS (continued)

In numeric order as they occur in the Table of Physical Constants of Organic Compounds

362

364

365

366

367

368

369

370

371

372

373

374

375

376

377

379

380

381

382

383

384

385

386

387

388

389

390

391

392

393

394

395

396

397

398

399

400

401

402

403

404

405

STRUCTURAL FORMULAS OF ORGANIC COMPOUNDS (continued)

In numeric order as they occur in the Table of Physical Constants of Organic Compounds

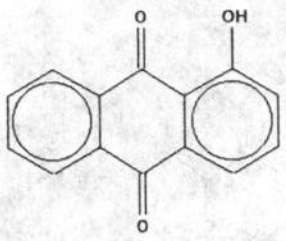

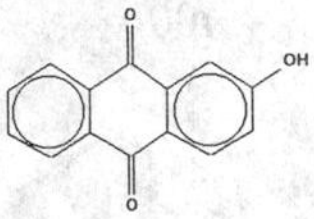

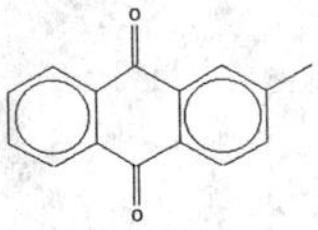

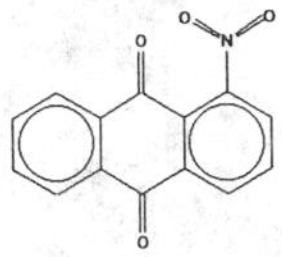

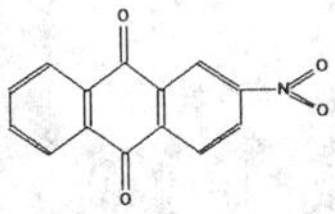

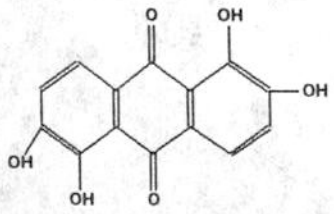

406

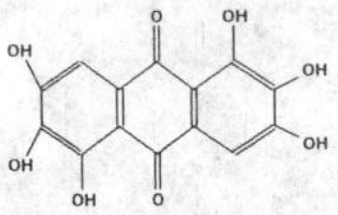

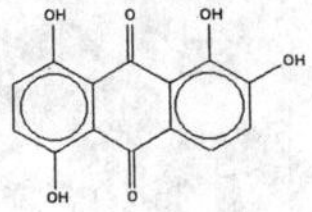

414

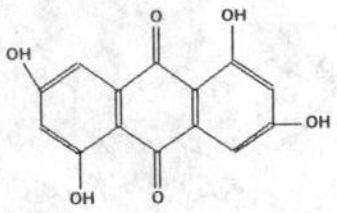

423

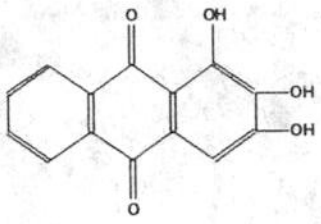

432

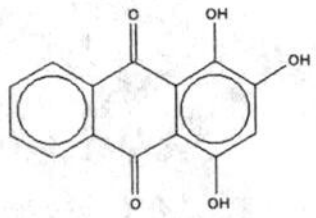

441

424

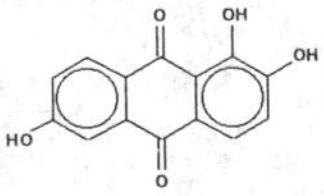

407

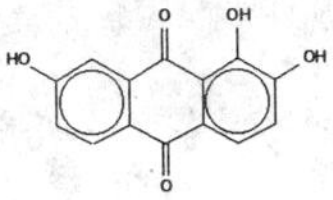

415

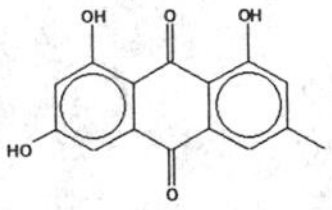

433

442

425

408

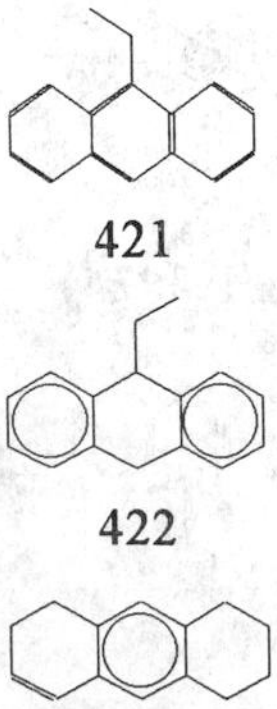

416

426

434

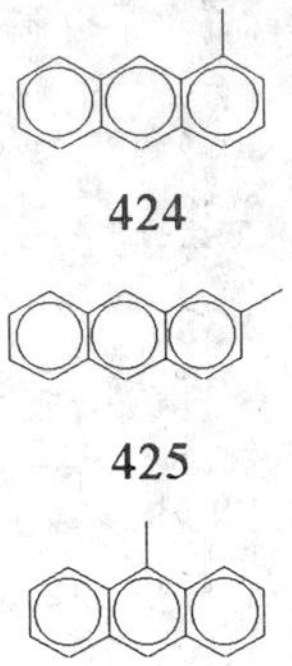

443

409

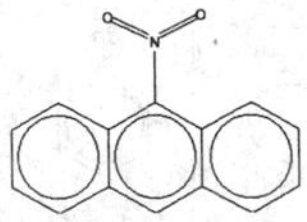

417

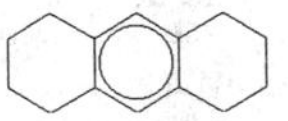

427

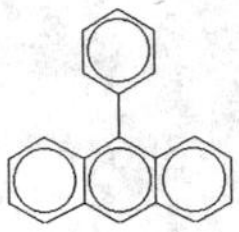

435

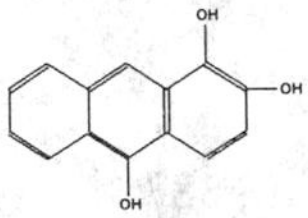

444

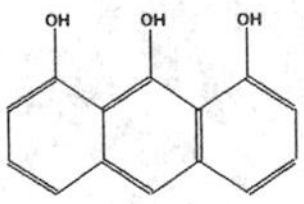

428

410

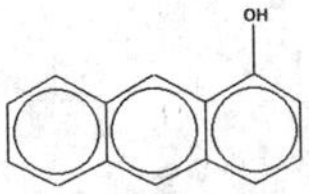

418

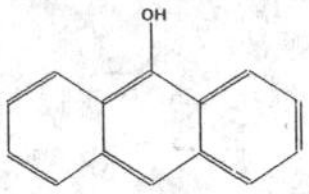

436

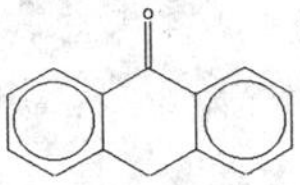

445

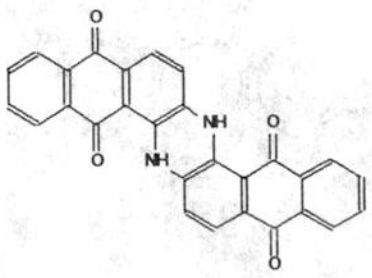

429

411

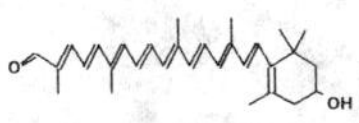

419

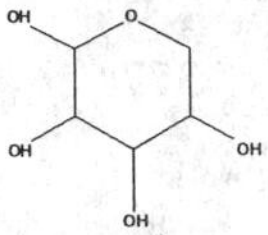

437

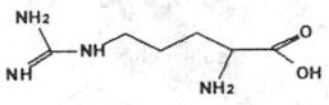

446

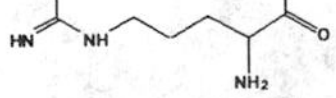

412

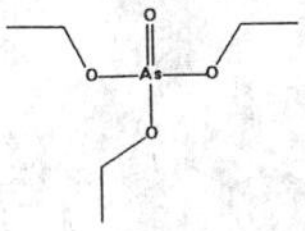

430

438

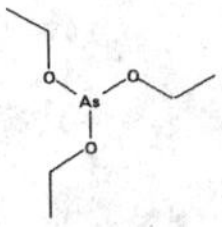

420

447

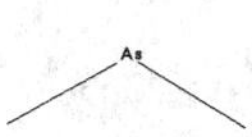

439

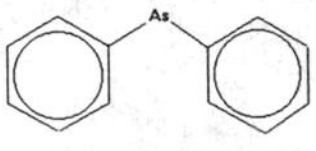

421

431

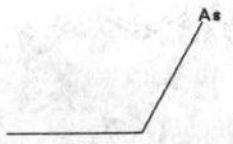

413

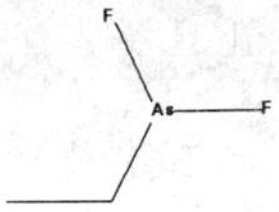

448

422

440

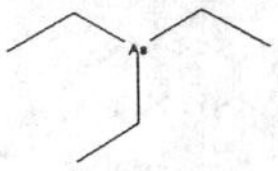

STRUCTURAL FORMULAS OF ORGANIC COMPOUNDS (continued)

In numeric order as they occur in the Table of Physical Constants of Organic Compounds

449

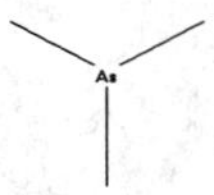

450

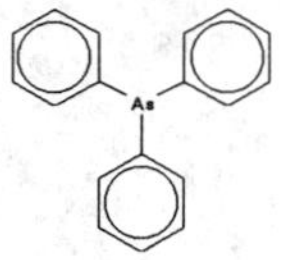

451

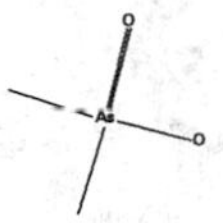

452

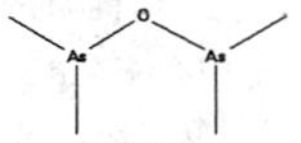

453

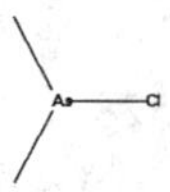

454

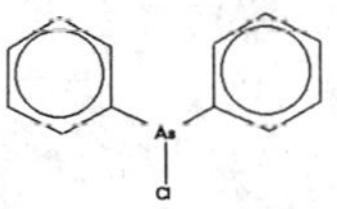

455

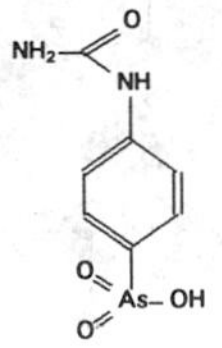

456

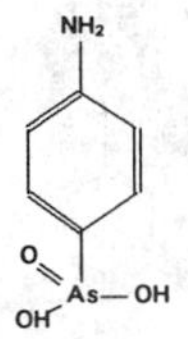

457

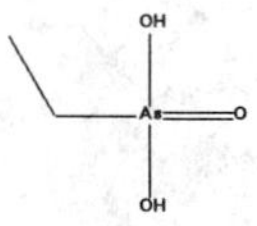

458

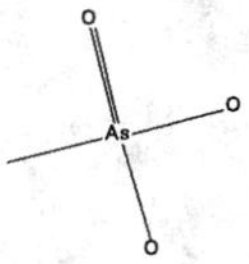

459

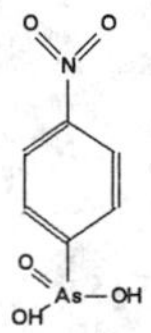

460

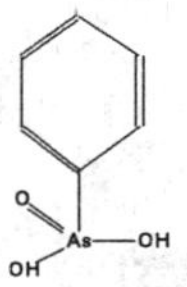

461

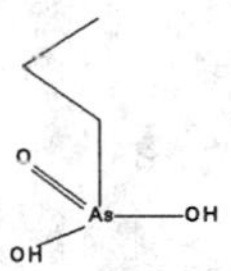

462

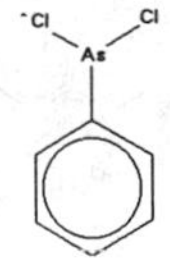

463

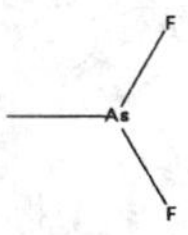

464

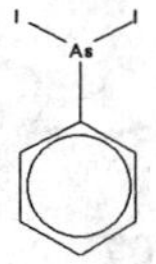

465

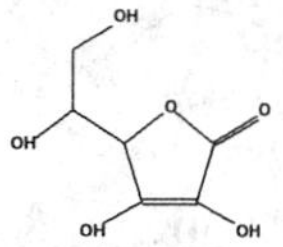

466

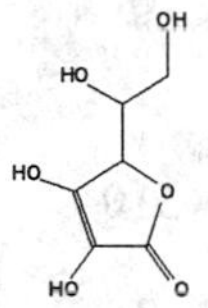

467

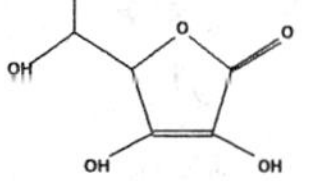

468

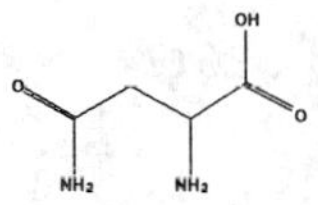

470

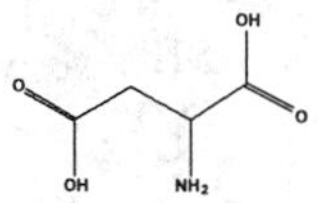

471

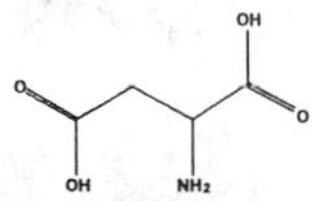

472

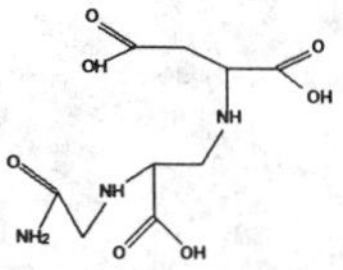

473

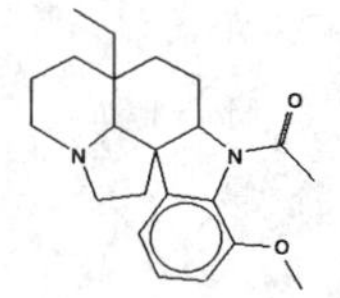

474

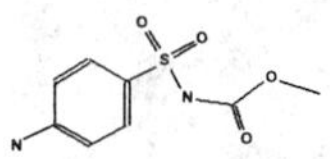

475

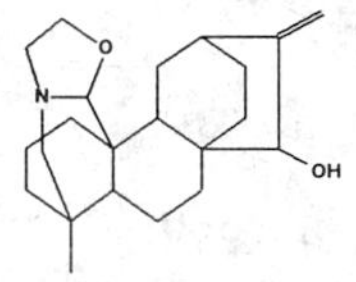

476

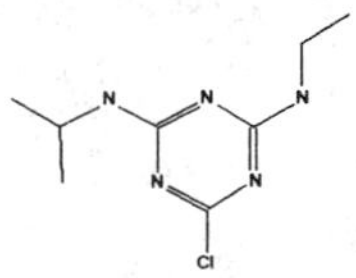

477

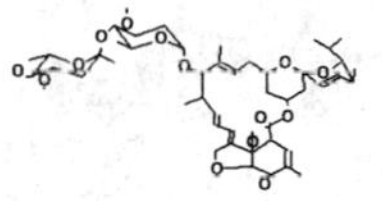

478

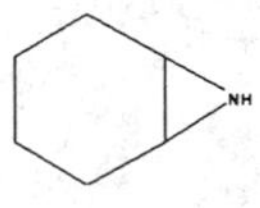

479

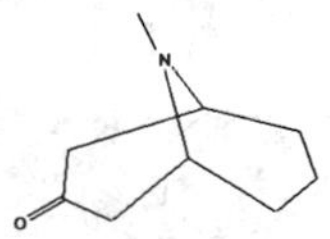

480

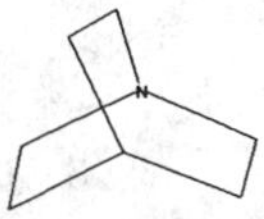

481

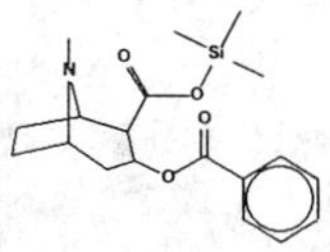

482

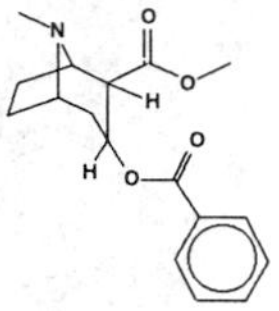

483

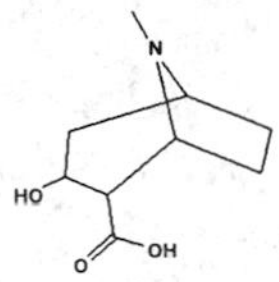

484

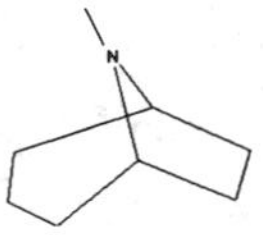

485

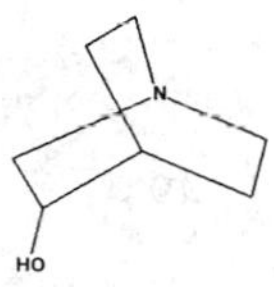

486

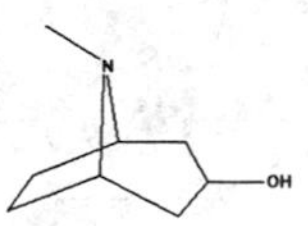

487

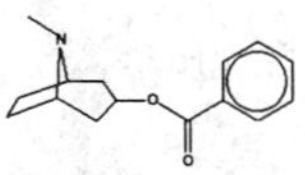

488

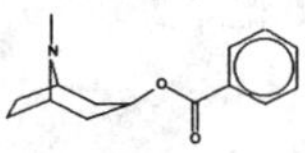

STRUCTURAL FORMULAS OF ORGANIC COMPOUNDS (continued)

In numeric order as they occur in the Table of Physical Constants of Organic Compounds

489

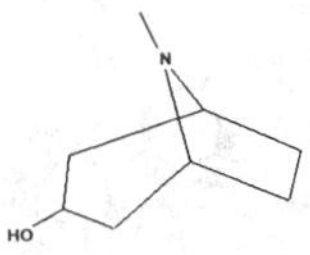

490

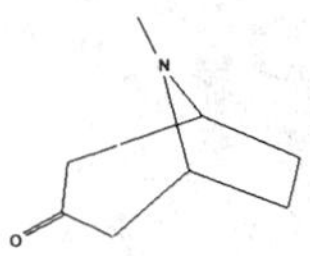

491

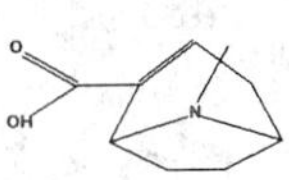

492

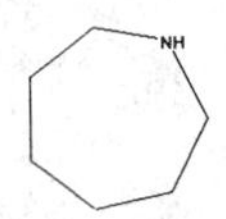

493

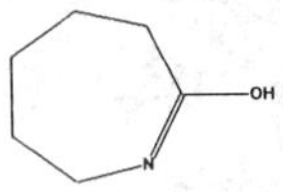

494

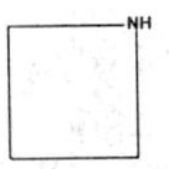

495

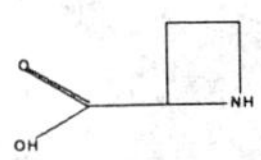

496

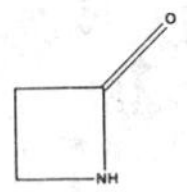

497

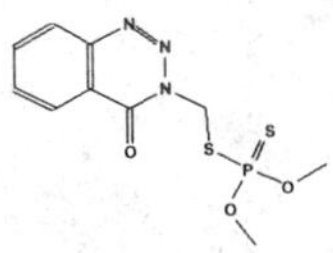

498

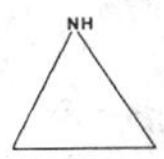

499

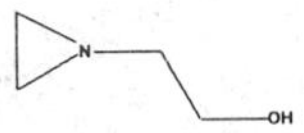

500

501

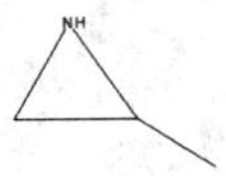

502

503

504

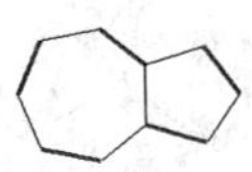

505

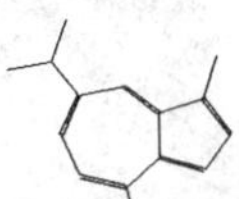

506

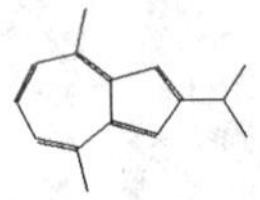

507

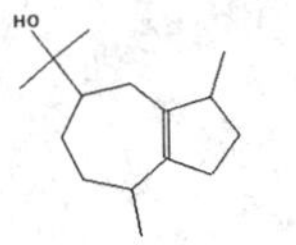

509

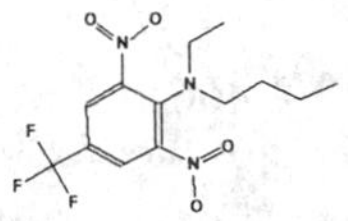

510

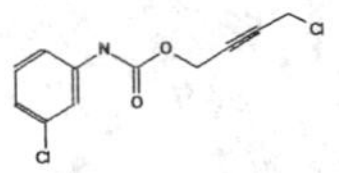

511

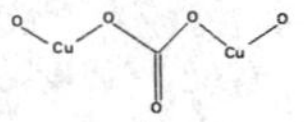

512

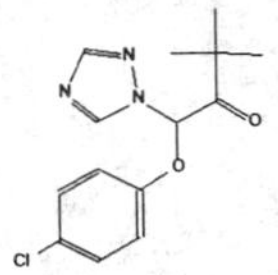

513

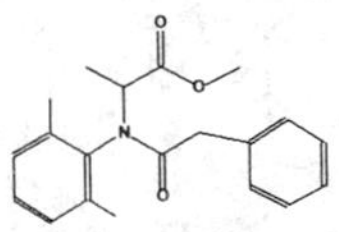

514

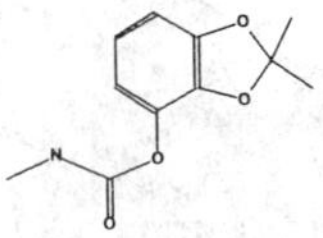

515

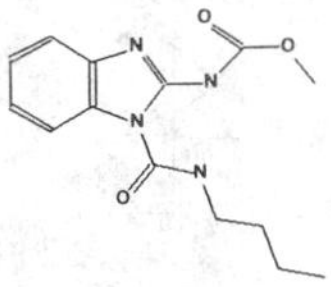

516

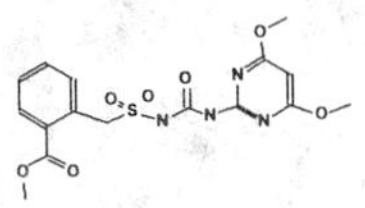

517

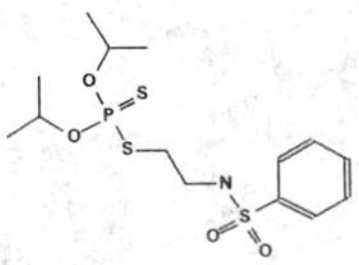

518

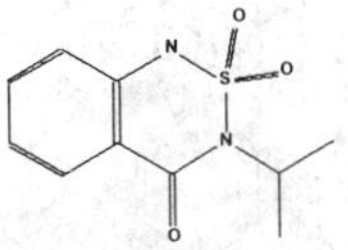

519

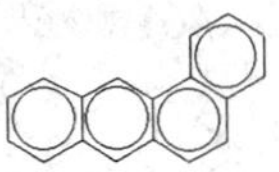

520

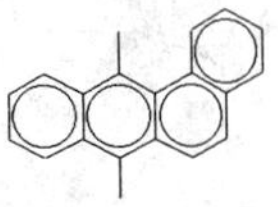

521

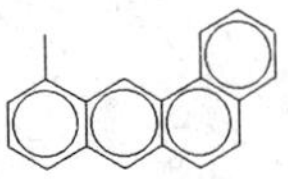

522

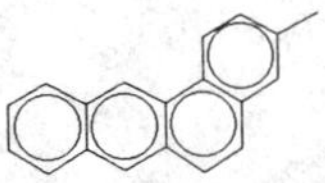

523

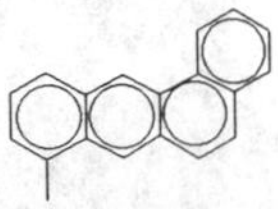

524

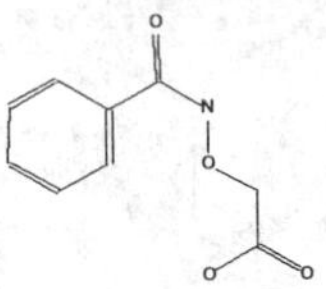

525

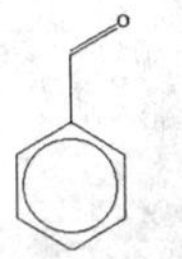

526

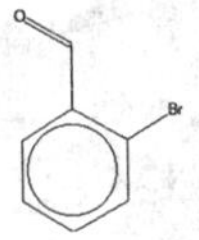

527

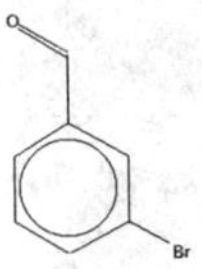

528

529

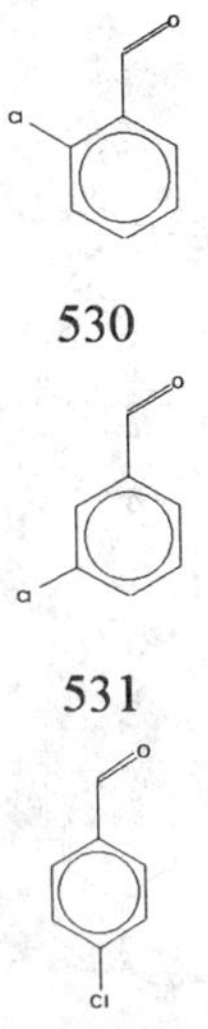

530

531

STRUCTURAL FORMULAS OF ORGANIC COMPOUNDS (continued)

In numeric order as they occur in the Table of Physical Constants of Organic Compounds

532 Cl OH

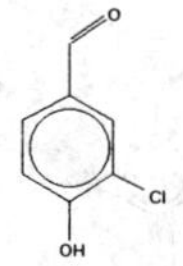

533 HO Cl

535 Cl Cl

536 Cl Cl O

537 O Cl Cl

538

539 N O

540 OH OH

541 HO OH

542 OH HO O

543 O O O

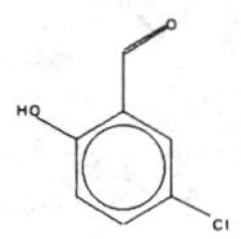

544 O O

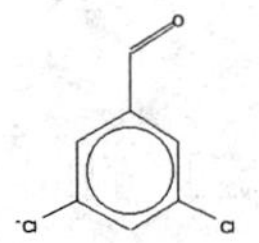

545 O O O

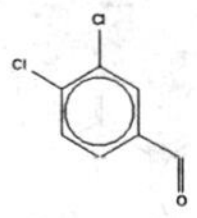

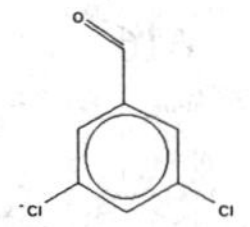

546

547

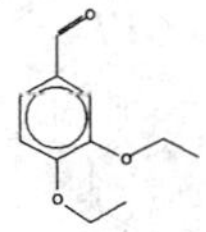

548

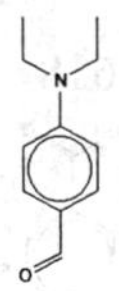

549

550 N O

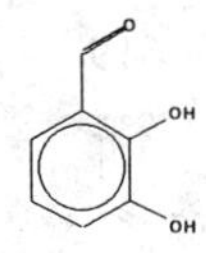

551 O N O O N O

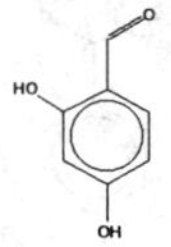

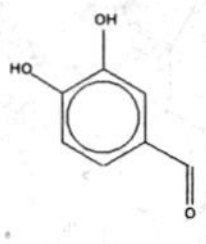

552 O

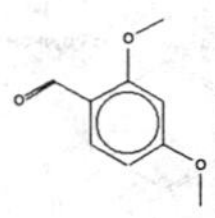

553 O O

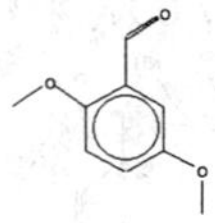

554 O O

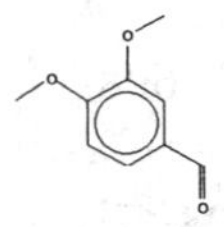

555 OH O

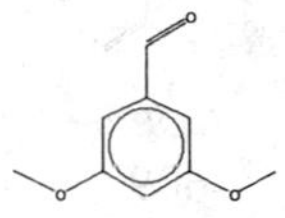

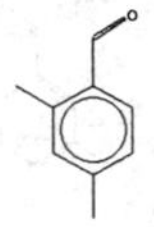

556 OH O

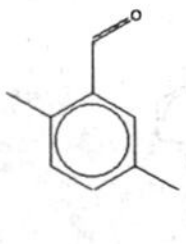

557 O O O

558

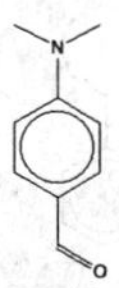

559 O F

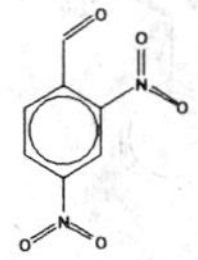

560 O F

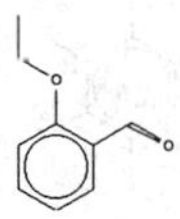

561 F

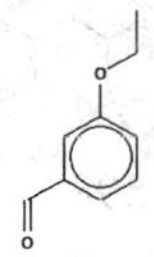

562 OH OH O OH OH O

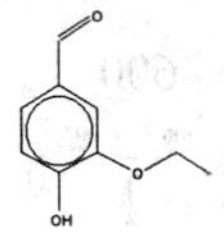

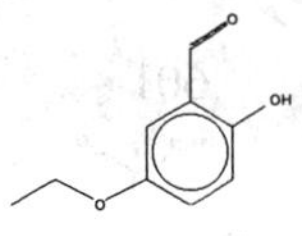

563 NH2 N

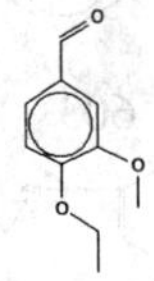

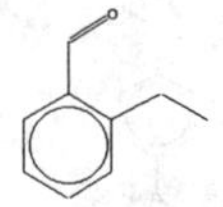

564 OH

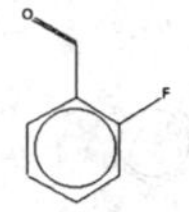

565 HO

566 OH

567 O OH O O

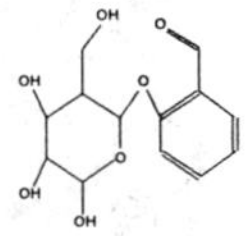

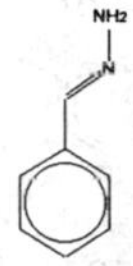

568 O HO O O

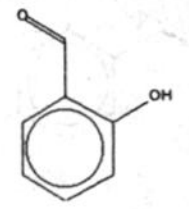

569 OH O O O

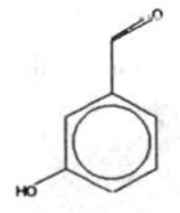

570 OH O O

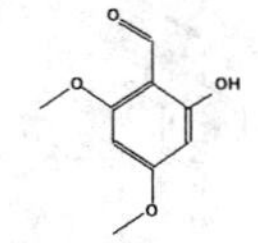

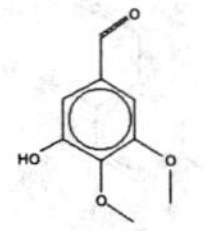

571 OH O

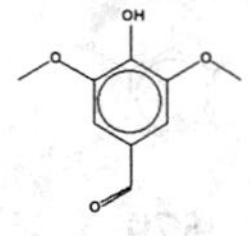

572 HO O O

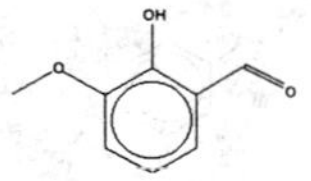

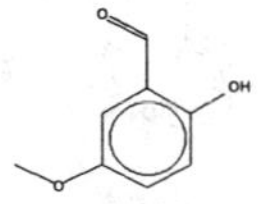

573 O OH

574 HO O O

575 OH

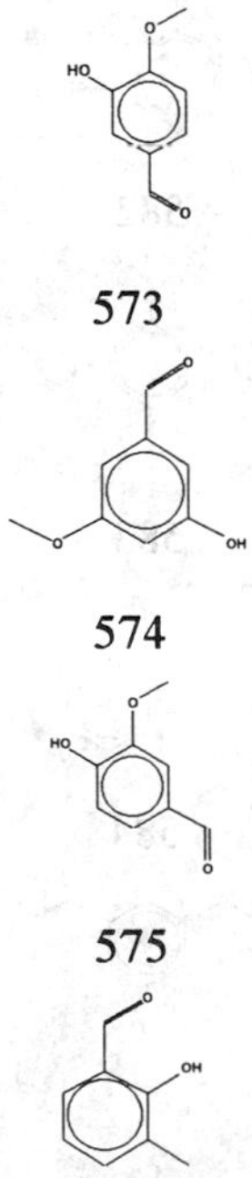

STRUCTURAL FORMULAS OF ORGANIC COMPOUNDS (continued)

In numeric order as they occur in the Table of Physical Constants of Organic Compounds

576

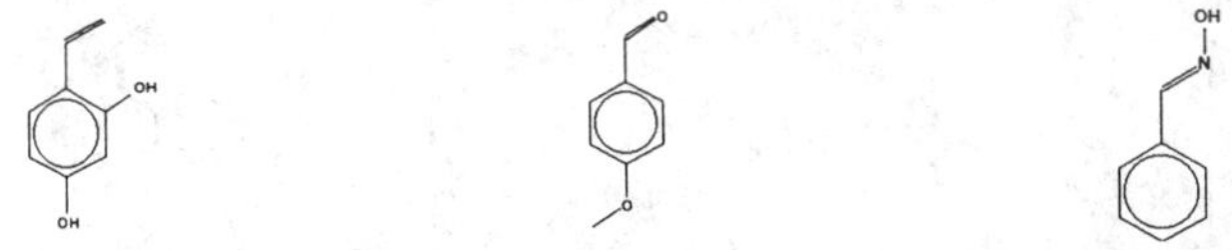

577

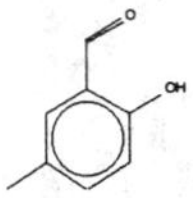

578

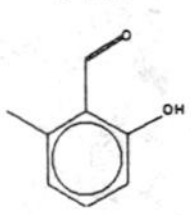

579

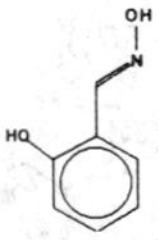

580

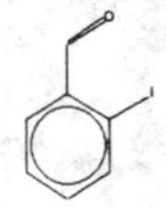

581

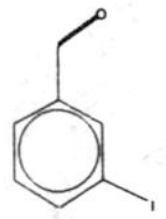

582

583

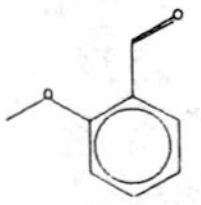

584

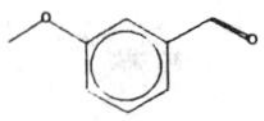

585

586

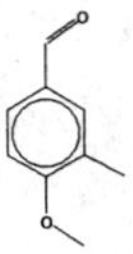

587

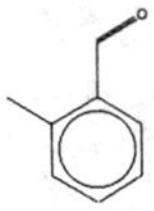

588

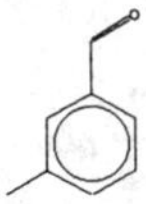

589

590

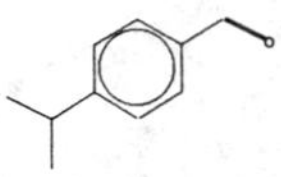

591

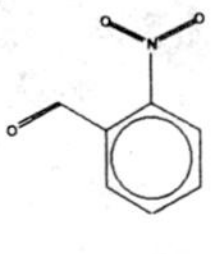

592

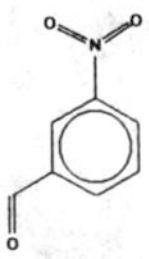

593

594

595

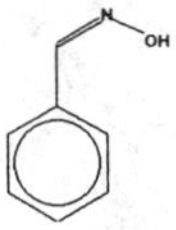

596

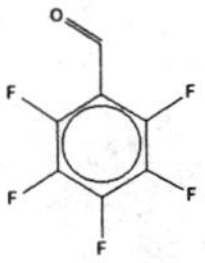

597

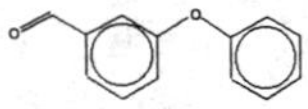

598

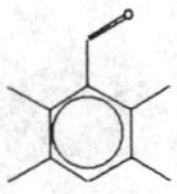

599

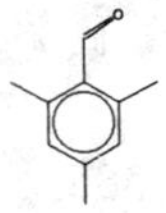

600

601

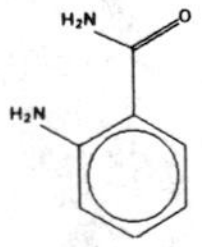

603

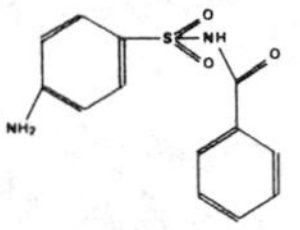

604

605

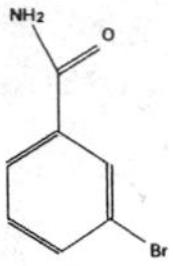

606

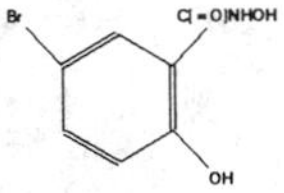

607

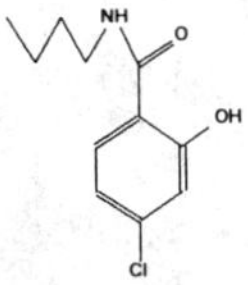

608

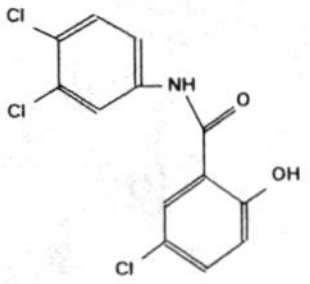

609

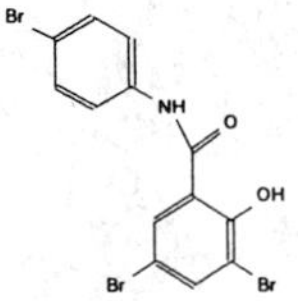

610

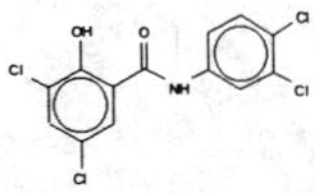

611

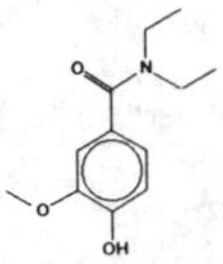

612

613

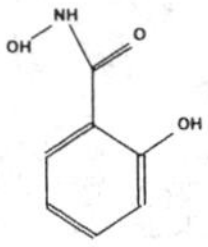

614

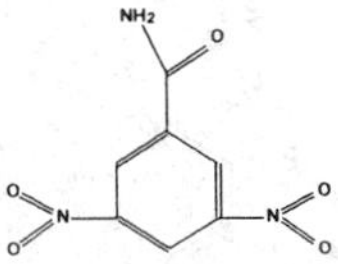

615

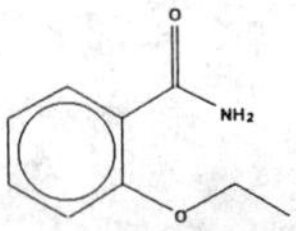

616

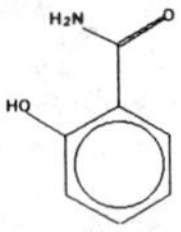

617

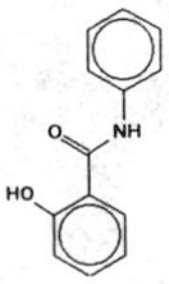

618

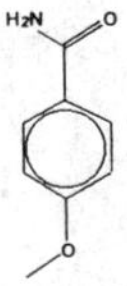

619

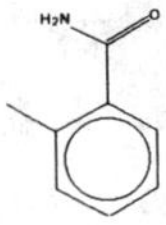

STRUCTURAL FORMULAS OF ORGANIC COMPOUNDS (continued)

In numeric order as they occur in the Table of Physical Constants of Organic Compounds

620

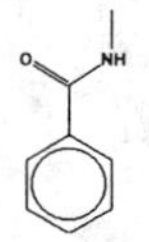

NH

621

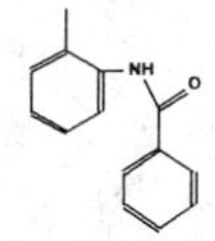

NH

622

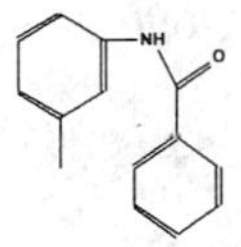

NH

623

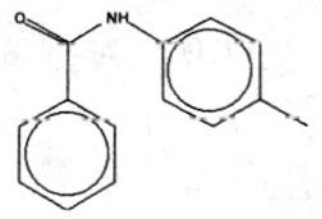

NH

624

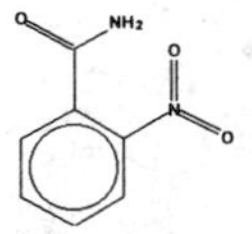

NH_2

625

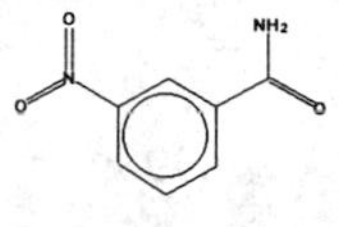

NH_2

626

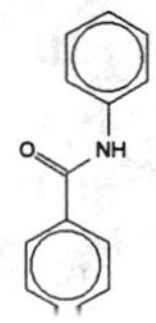

NH

627

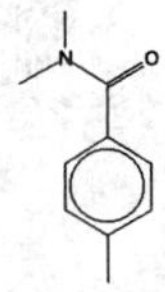

N

628

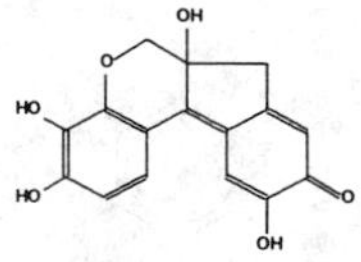

OH HO HO OH

629

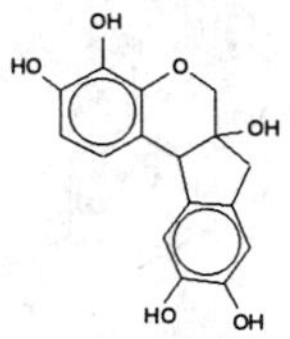

OH HO OH HO OH

630

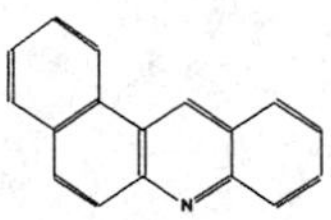

N

631

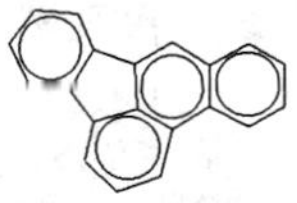

632

633

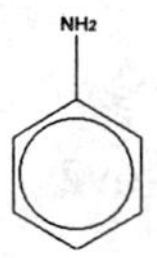

NH_2

635

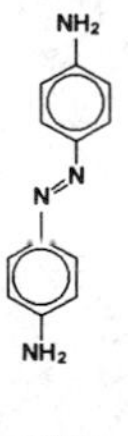

NH_2 N N NH_2

636

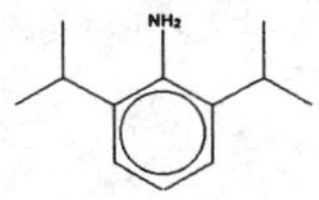

NH_2

637

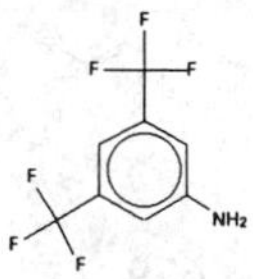

F F F F F F NH_2

638

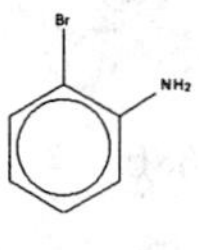

Br NH_2

639

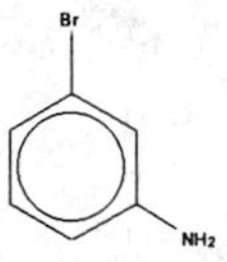

Br NH_2

640

Br NH_2

641

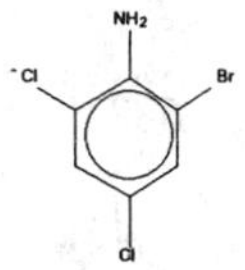

NH_2 Cl Br Cl

642

N Br

643

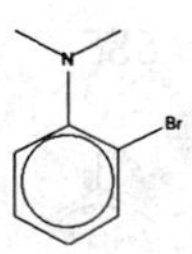

N Br

644

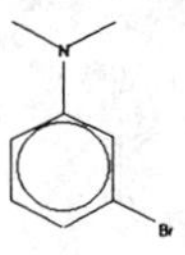

N Br

645

N Br

646

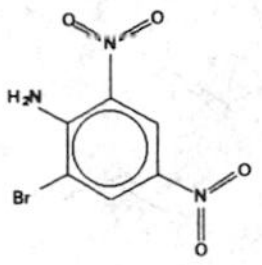

H_2N Br N

647

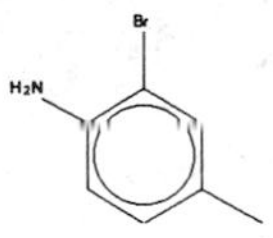

Br H_2N

648

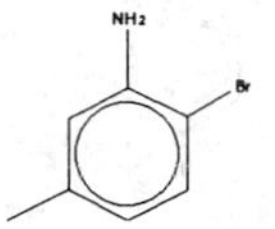

NH_2 Br

649

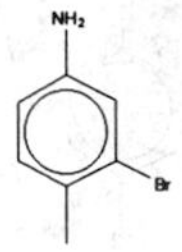

NH_2 Br

650

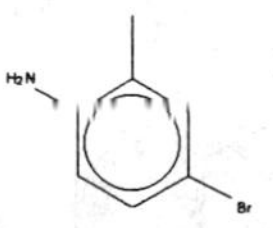

H_2N Br

651

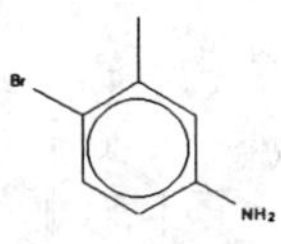

Br NH_2

652

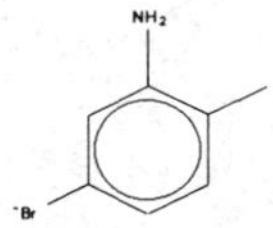

NH_2 Br

653

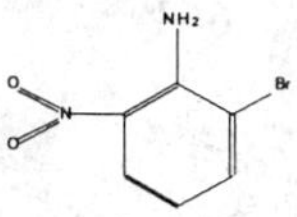

NH_2 N Br

654

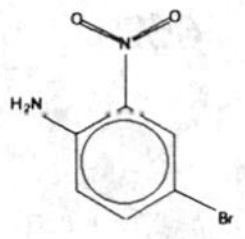

N H_2N Br

655

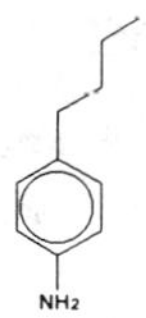

NH_2

656

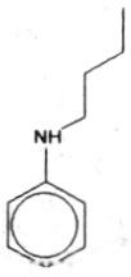

NH

657

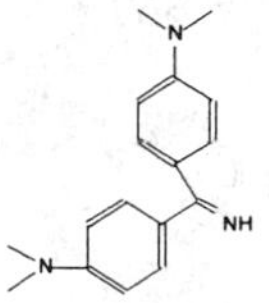

N NH N

658

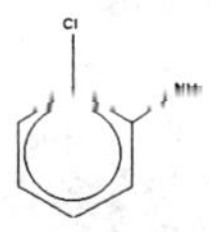

Cl

659

Cl NH_2

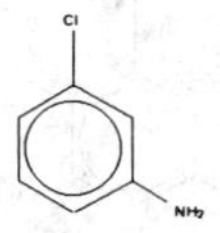

660

670

678

686

694

679

661

671

695

687

680

662

672

696

688

681

663

697

673

689

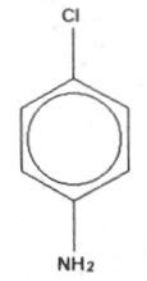

682

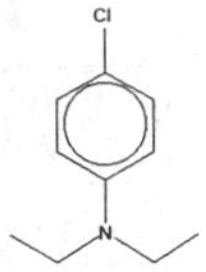

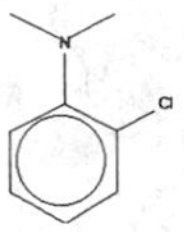

666

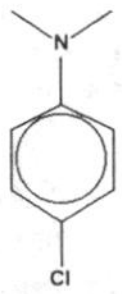

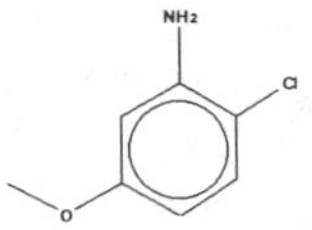

674

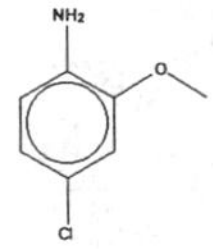

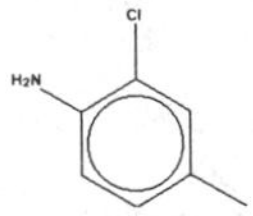

698

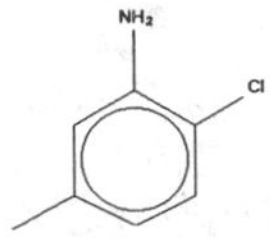

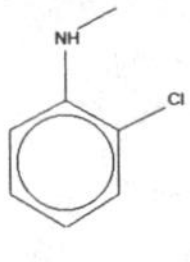

690

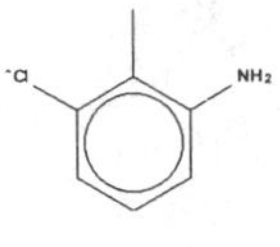

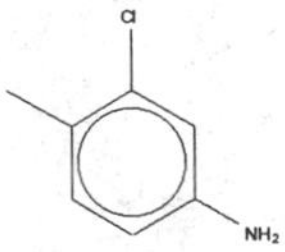

683

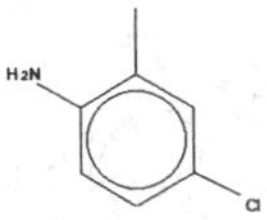

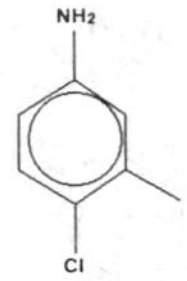

667

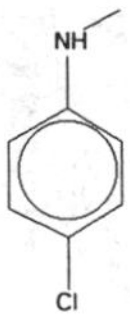

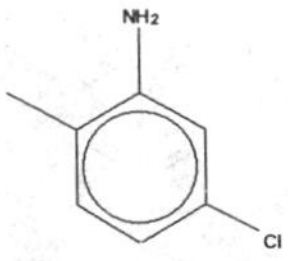

675

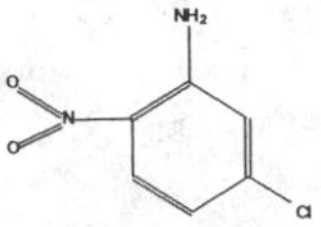

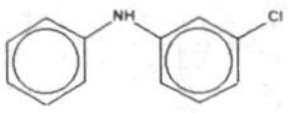

699

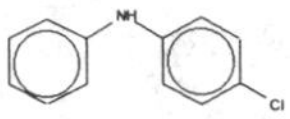

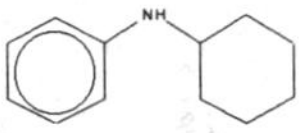

692

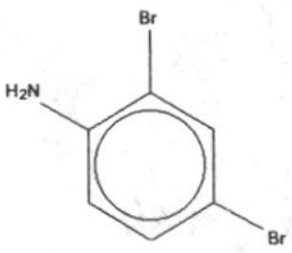

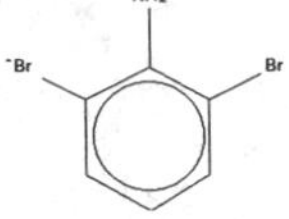

684

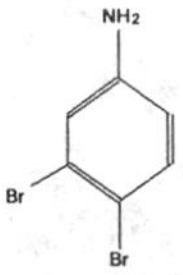

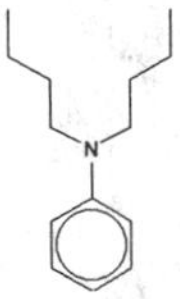

668

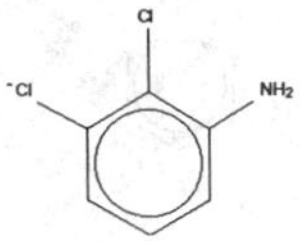

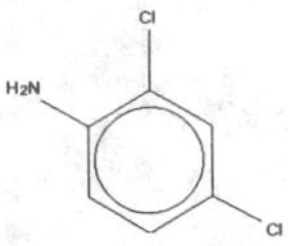

676

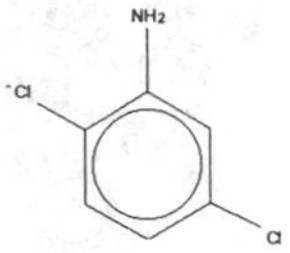

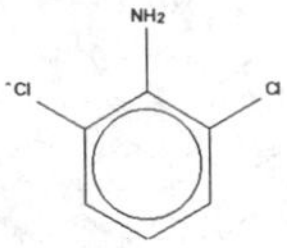

700

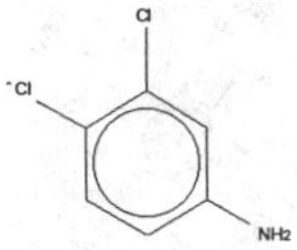

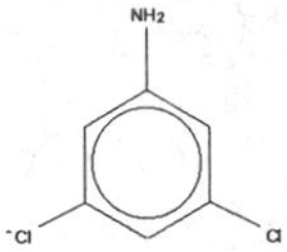

693

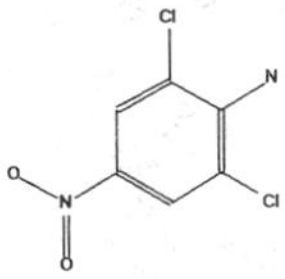

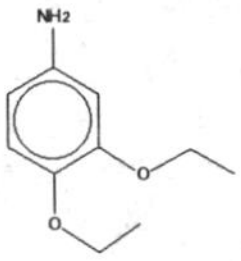

685

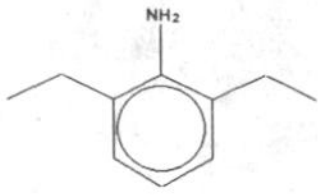

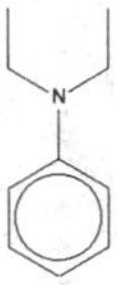

669

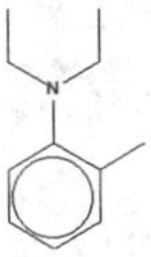

677

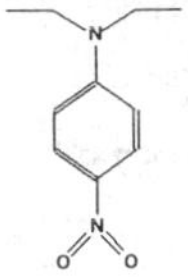

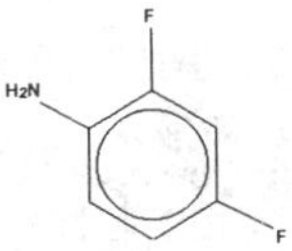

STRUCTURAL FORMULAS OF ORGANIC COMPOUNDS (continued)

In numeric order as they occur in the Table of Physical Constants of Organic Compounds

701
NH2 I I

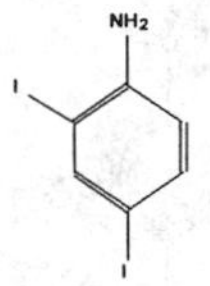

702
H2N O O

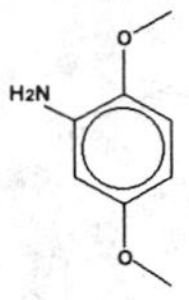

703
NH2 O O

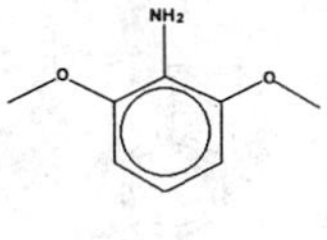

704
NH2

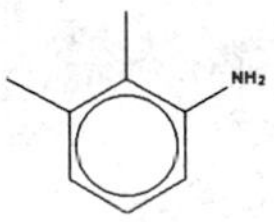

705
NH2

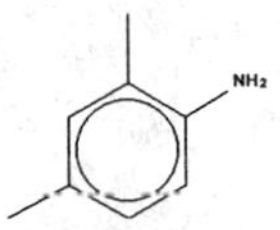

706
NH2

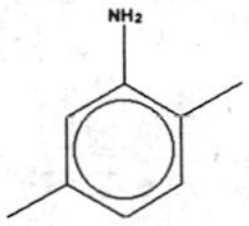

707
NH2

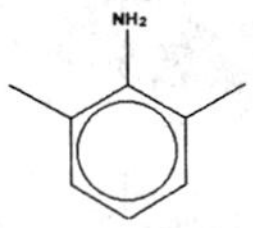

708
NH2

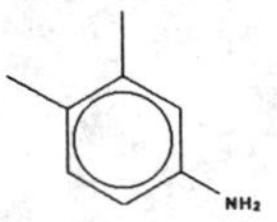

709
NH2

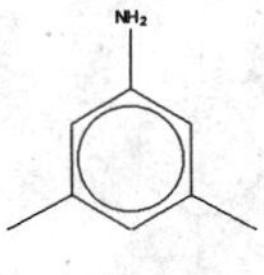

710
NH

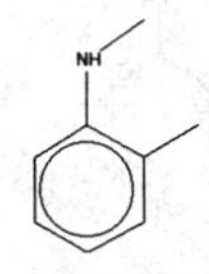

711
NH

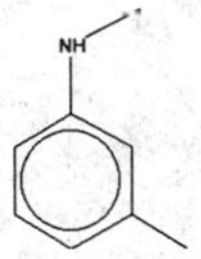

712
NH

713
N

714
NH2

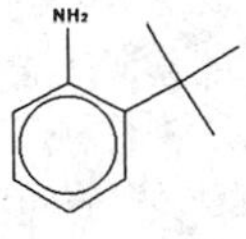

715
NH2

716
NH

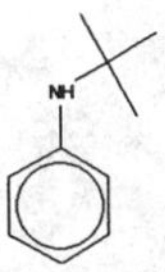

718
N O N O

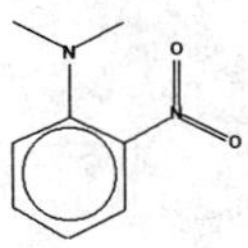

719
N O N O

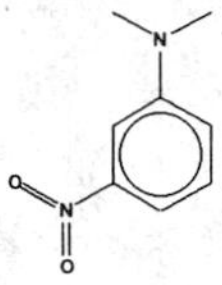

720
N N O

721
N N N

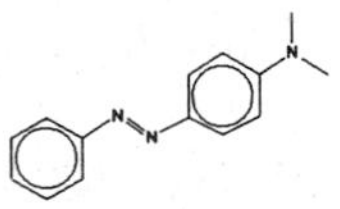

722
N

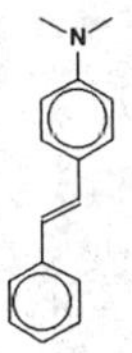

723
N N N F F F

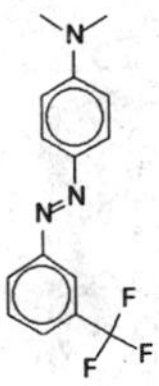

724
NH2 O N O O N O

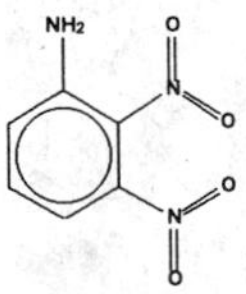

725
O N O H2N O N O

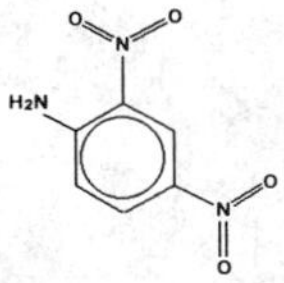

726
O NH2 O N N O O

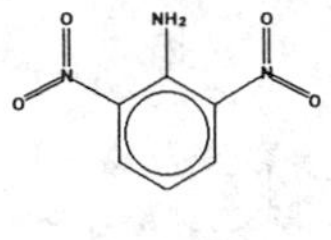

727
O N O O N O NH2

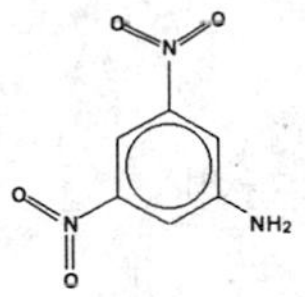

728
N

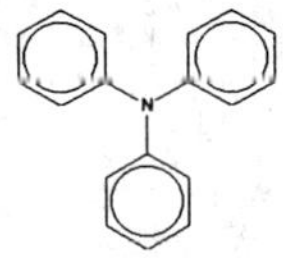

729
NH2

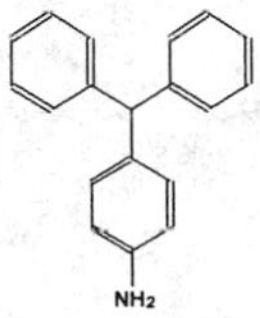

730
N

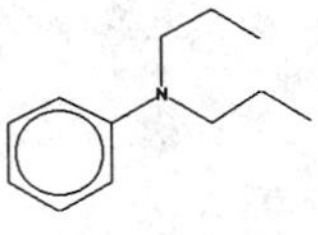

731
NH2 NH2

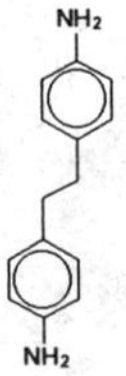

732
NH2 NH2

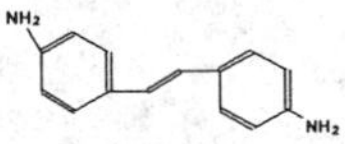

733
NH2

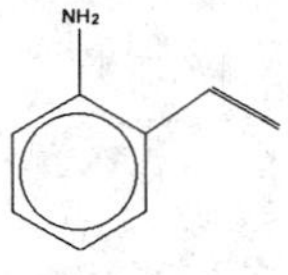

734
NH2

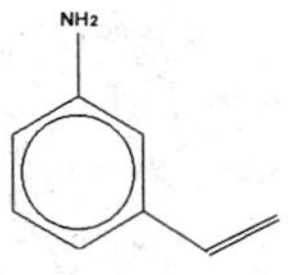

735
NH2

736
O NH2

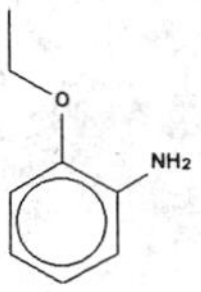

737
O NH2

738
N O

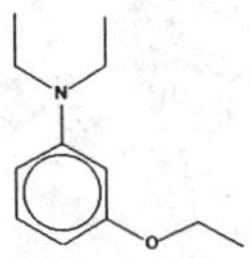

STRUCTURAL FORMULAS OF ORGANIC COMPOUNDS (continued)

In numeric order as they occur in the Table of Physical Constants of Organic Compounds

740 NH2

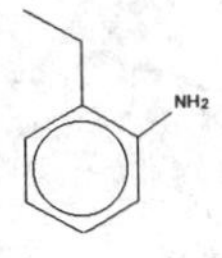

741 NH2

742 NH2

743 NH

744 NH

745 NH2

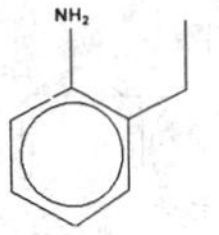

746 NH

747 NH

748 NH

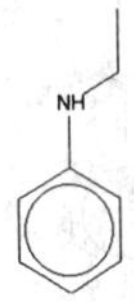

749 N

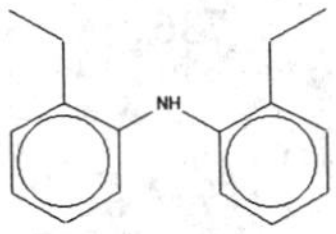

750 O N N

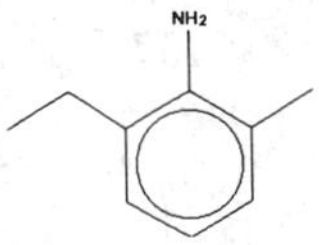

751 N

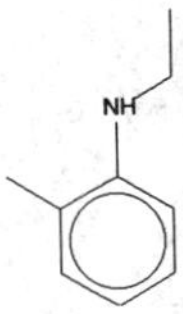

752 F NH2

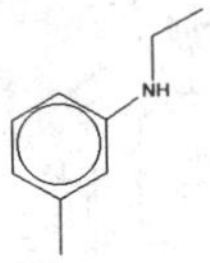

753 F NH2

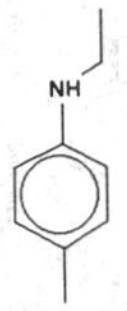

754 F NH2

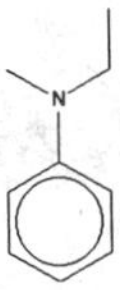

755 H2N F

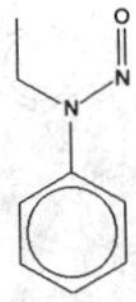

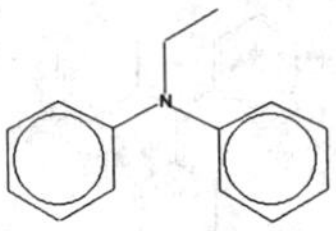

756 NH2

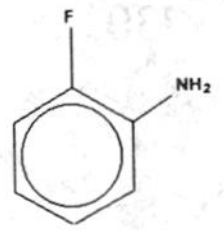

757 OH NH

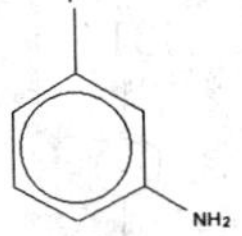

758 OH NH

759 O N HO N

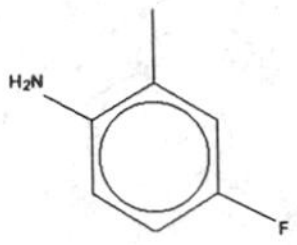

760 H2N I

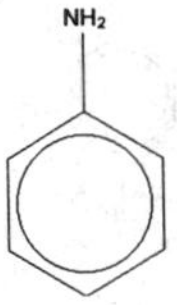

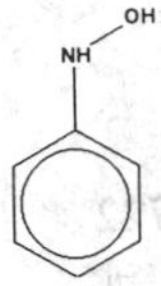

761 I NH2

762 NH2 I

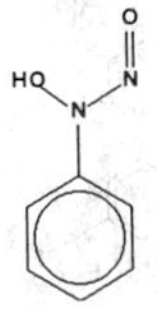

763 NH2 I

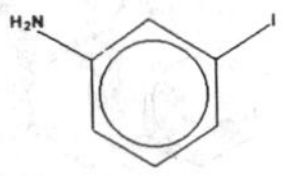

764 N C N

765 N C N

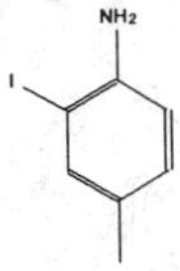

766 O NH2

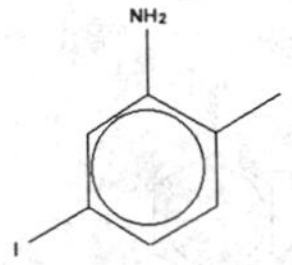

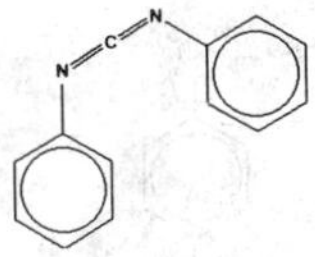

767 O NH2

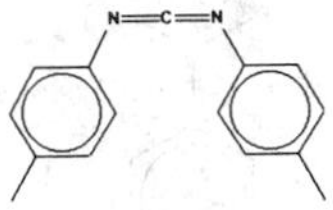

768 O NH2

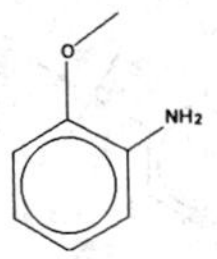

769 N O

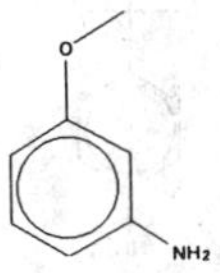

770 O H2N

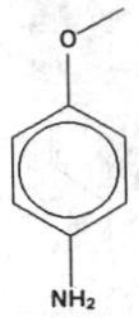

771 NH2 O

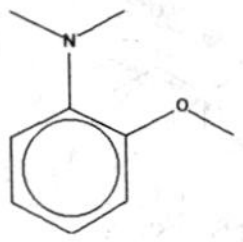

772 H2N O

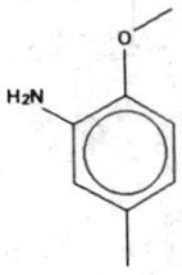

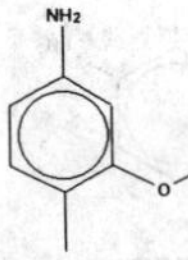

773 NH2 O

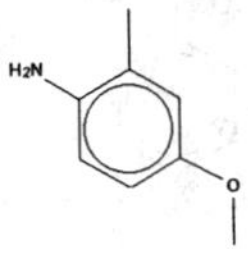

775 O N O H2N O

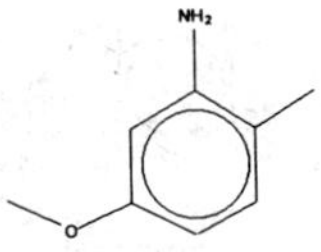

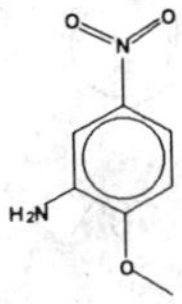

STRUCTURAL FORMULAS OF ORGANIC COMPOUNDS (continued)

In numeric order as they occur in the Table of Physical Constants of Organic Compounds

776

777

778

779

780

781

782

783

784

785

786

787

788

789

790

791

793

794

795

796

797

798

799

800

801

802

803

804

805

806

807

808

809

810

811

812

813

STRUCTURAL FORMULAS OF ORGANIC COMPOUNDS (continued)

In numeric order as they occur in the Table of Physical Constants of Organic Compounds

814
NH2

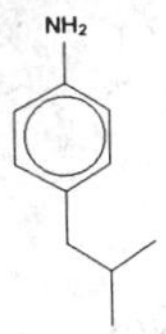

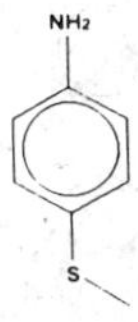

822
NH2

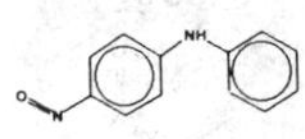

830

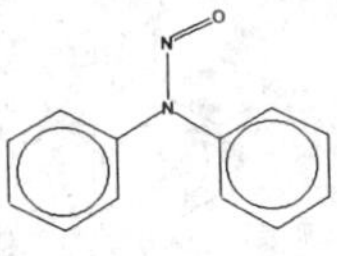

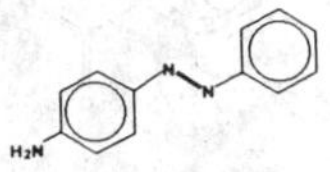

838
H2N

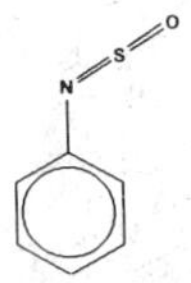

845
H2N NH2

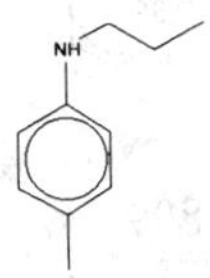

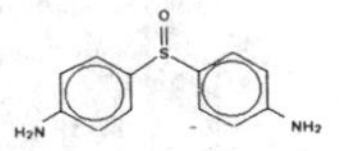

846
NH2
O=S=O
NH2

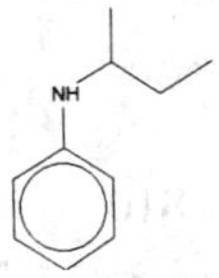

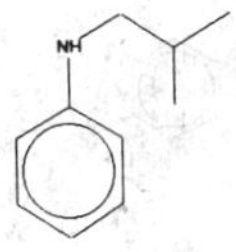

815
NH

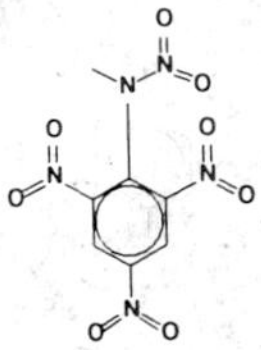

823
NH2

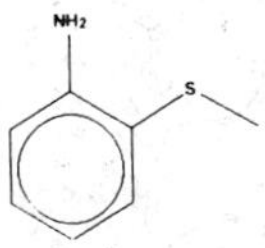

831
NH2

839
NH2

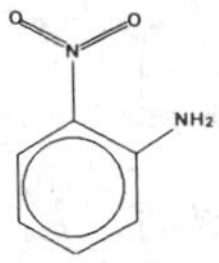

816
NH

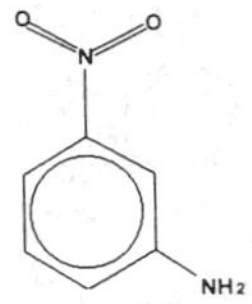

824
NH2

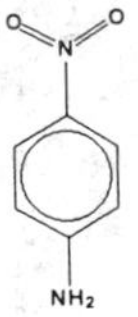

832
H2N NH2

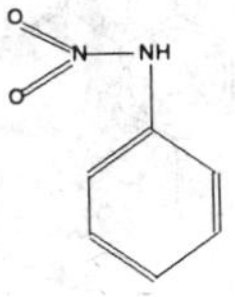

840
N

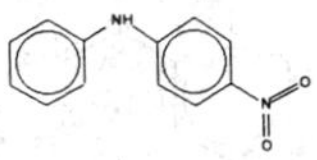

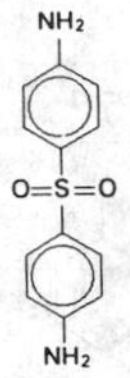

847
NH2

817
NH

825
NH

833
NH2

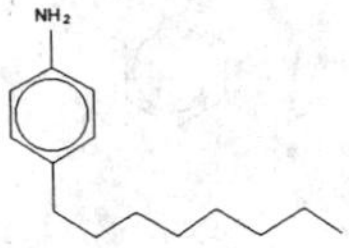

841
NH

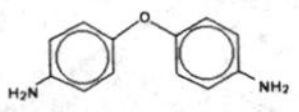

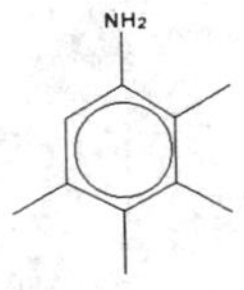

848
NH2

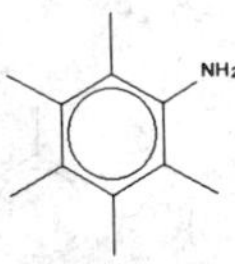

818

826
NH

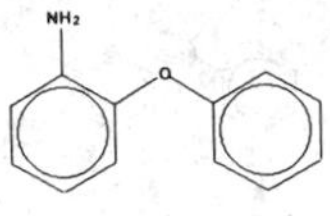

834
NH2

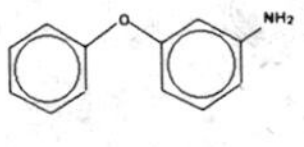

842
H2N

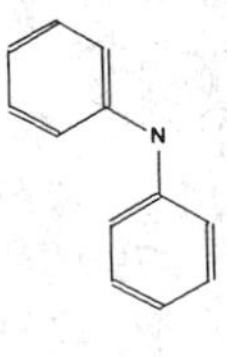

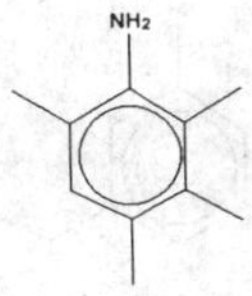

849
N

819
NH2
S

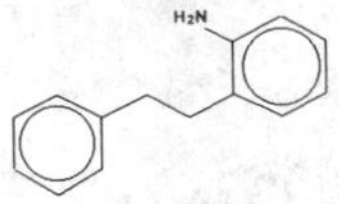

828
NH2
N

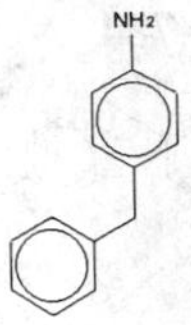

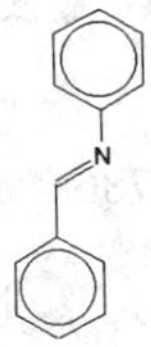

835
NH2

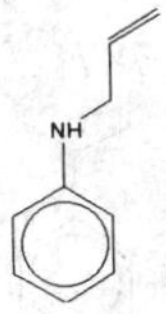

843
NH

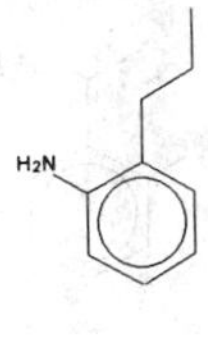

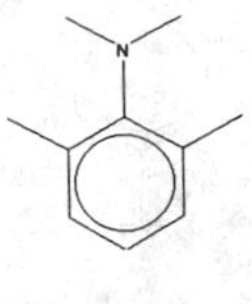

850
S
H2N NH2

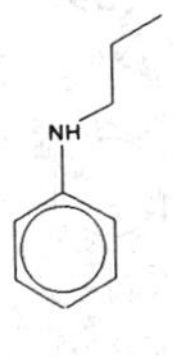

820
NH2
S

829
NH

836
N

844
N=S=O

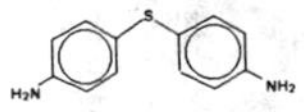

851
NH2
Br Br
Br

837
H2N

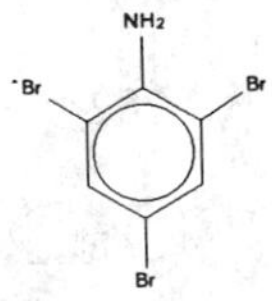

STRUCTURAL FORMULAS OF ORGANIC COMPOUNDS (continued)

In numeric order as they occur in the Table of Physical Constants of Organic Compounds

852

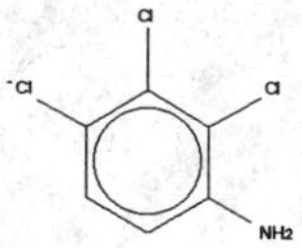

853

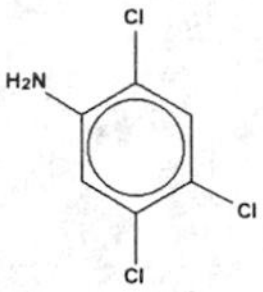

854

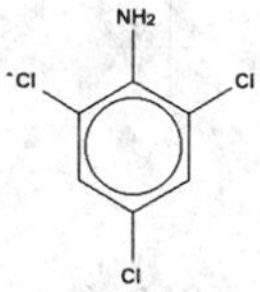

855

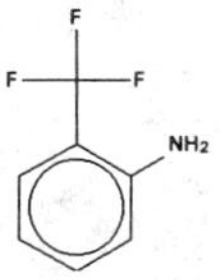

856

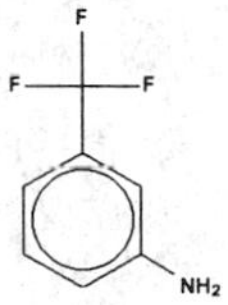

857

858

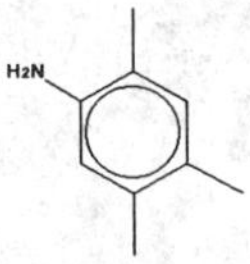

859

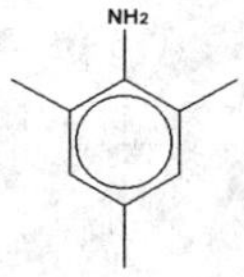

860

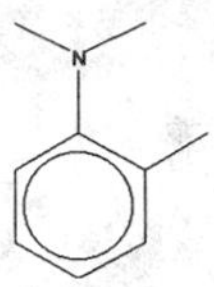

861

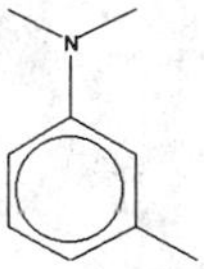

862

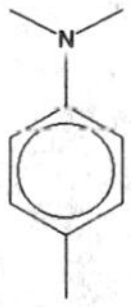

863

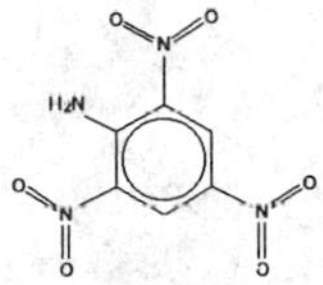

864

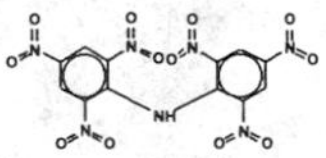

867

868

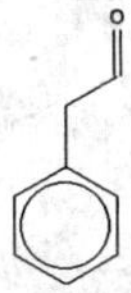

869

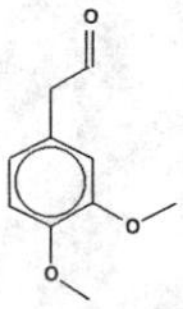

870

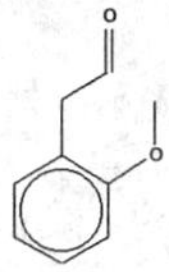

871

872

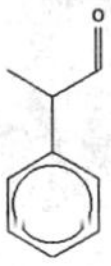

873

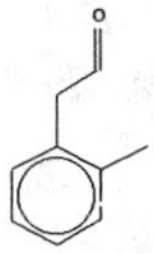

874

875

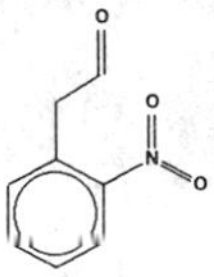

877

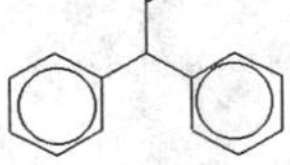

878

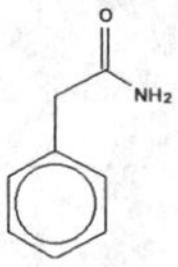

879

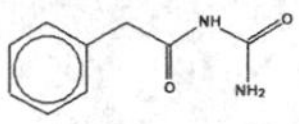

880

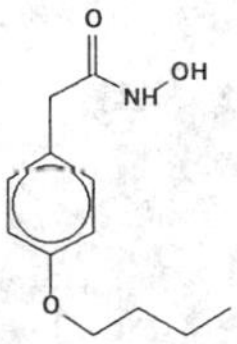

881

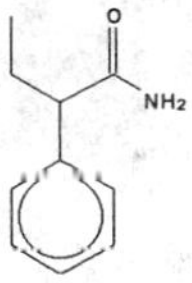

882

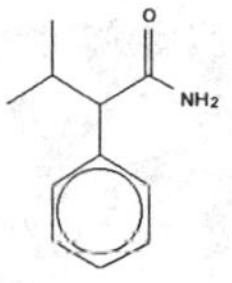

883

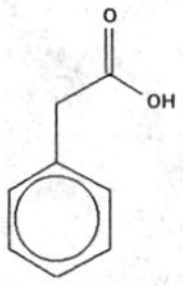

884

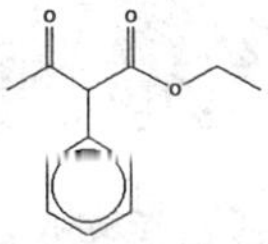

886

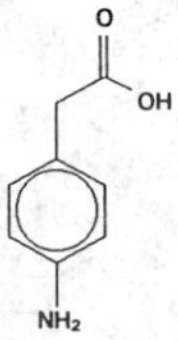

887

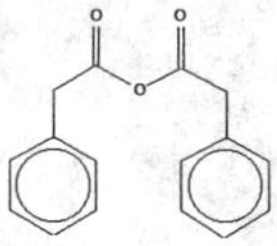

888

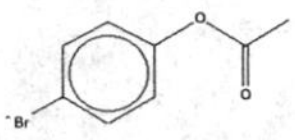

890

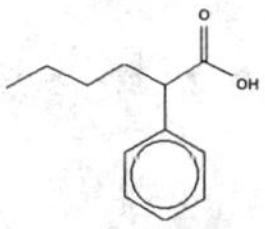

891

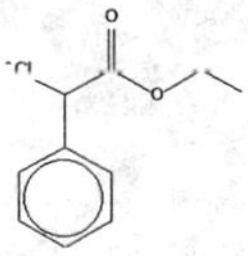

892

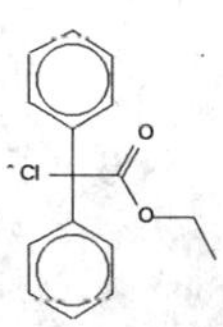

893

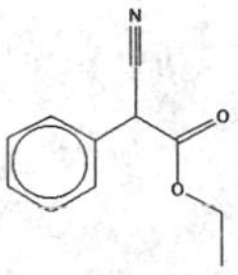

894

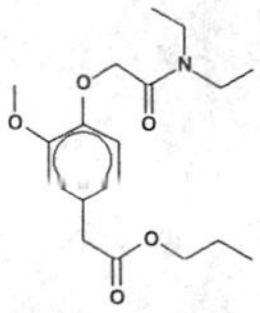

895

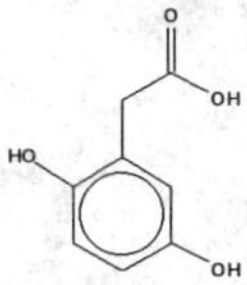

STRUCTURAL FORMULAS OF ORGANIC COMPOUNDS (continued)

In numeric order as they occur in the Table of Physical Constants of Organic Compounds

896

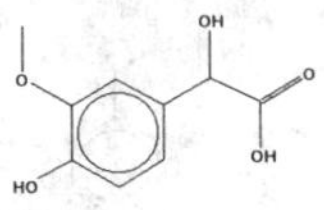

897

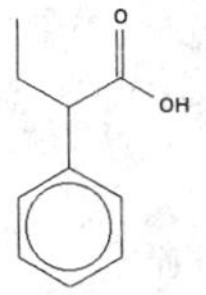

898

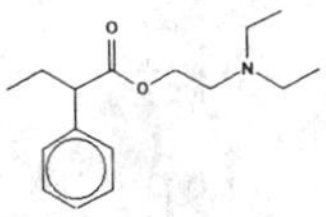

899

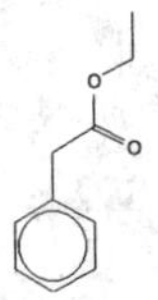

900

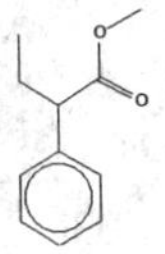

901

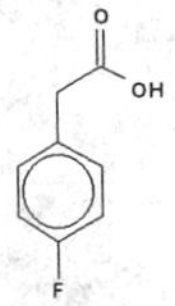

902

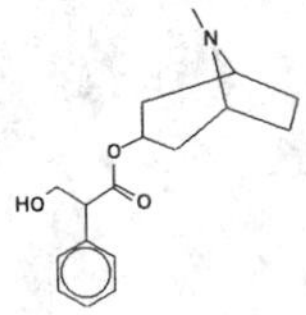

903

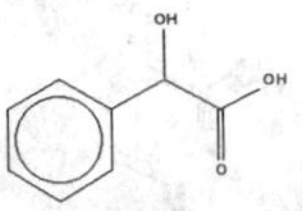

904

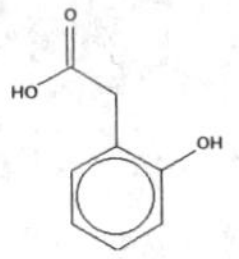

905

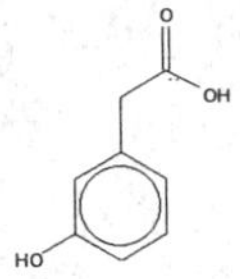

906

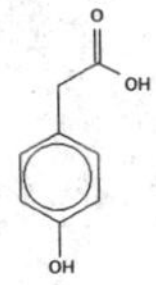

907

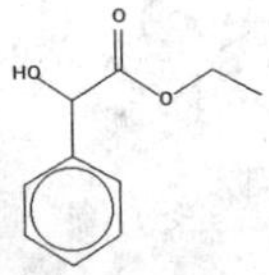

908

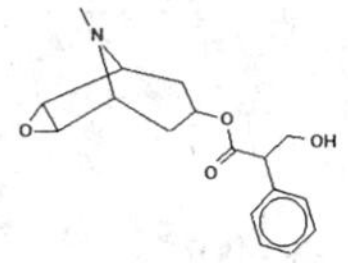

909

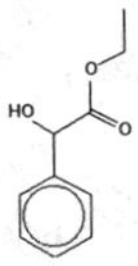

910

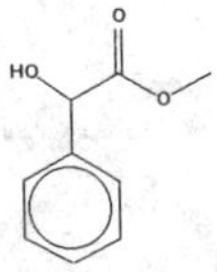

911

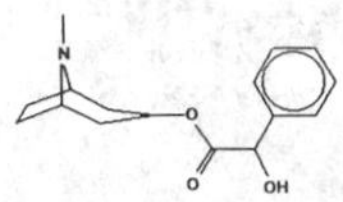

912

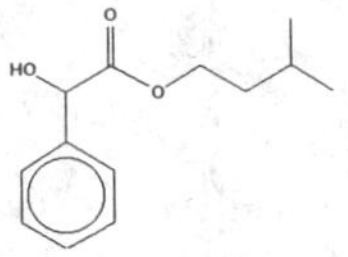

913

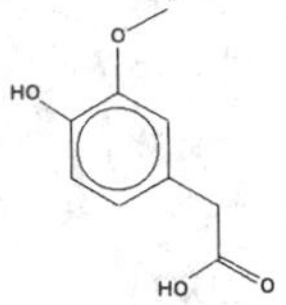

914

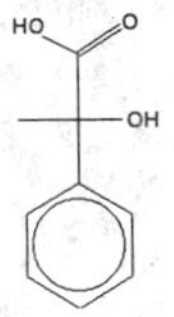

915

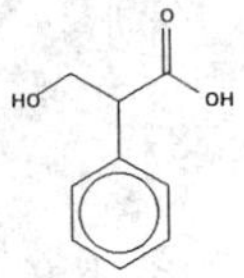

916

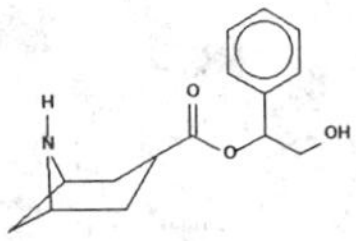

917

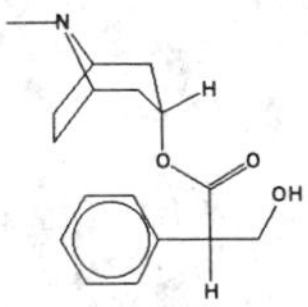

918

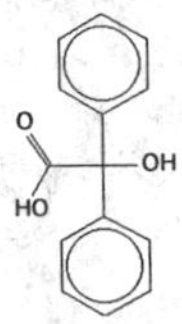

919

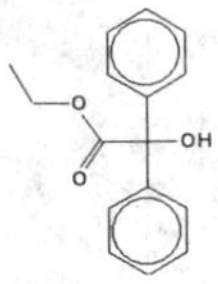

920

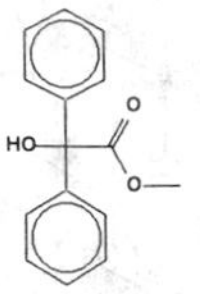

921

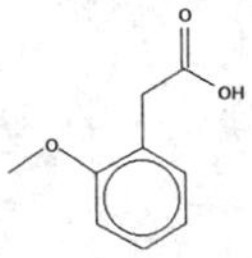

923

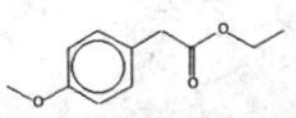

924

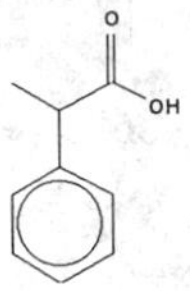

925

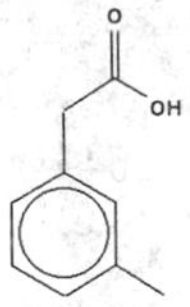

926

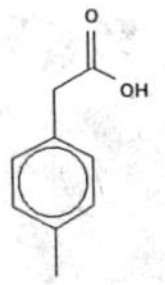

927

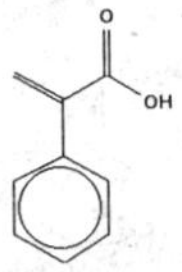

928

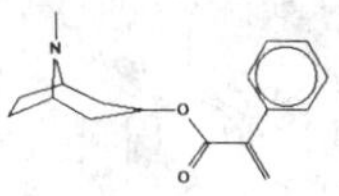

930

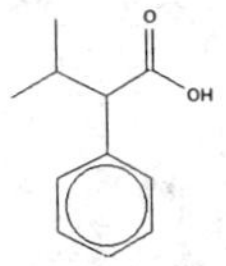

931

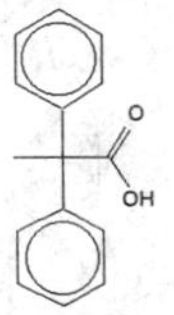

932

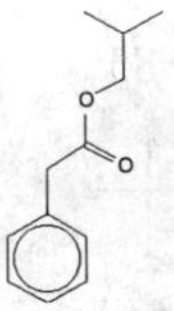

933

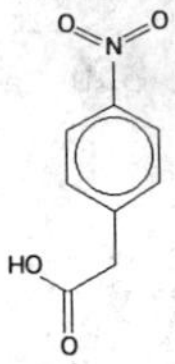

934

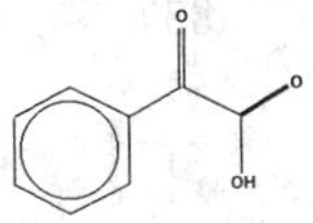

935

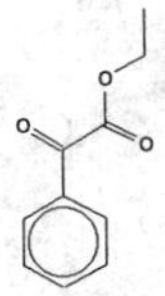

STRUCTURAL FORMULAS OF ORGANIC COMPOUNDS (continued)

In numeric order as they occur in the Table of Physical Constants of Organic Compounds

937

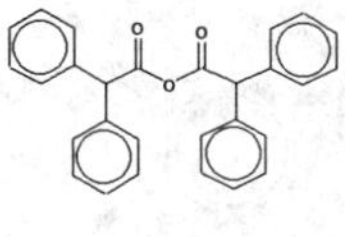

946

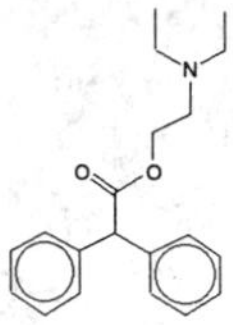

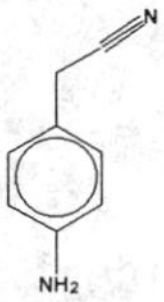

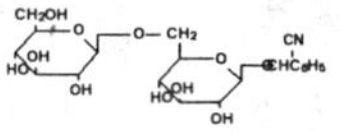

956

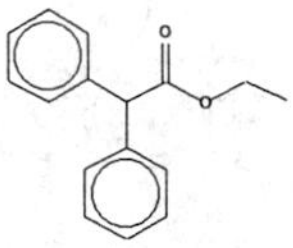

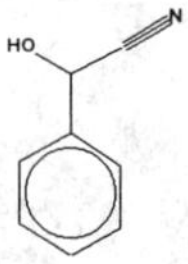

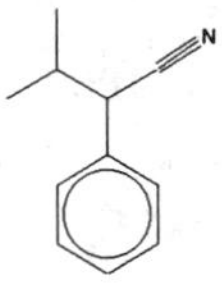

964

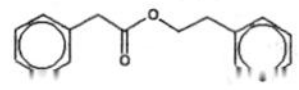

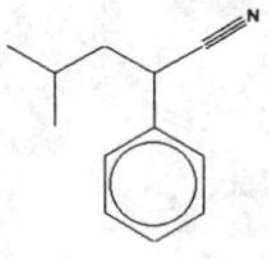

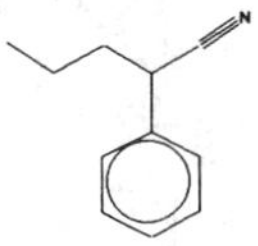

971

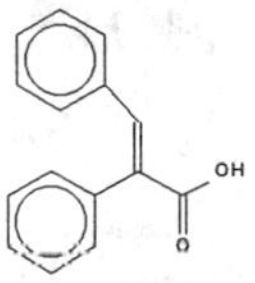

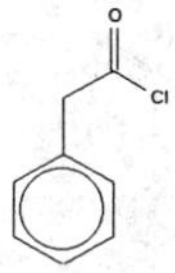

938

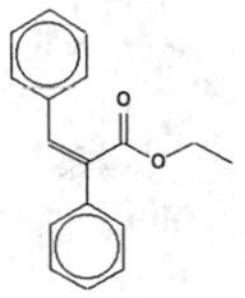

947

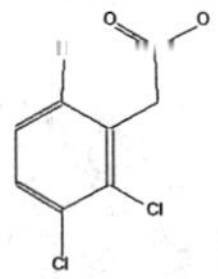

957

965

972

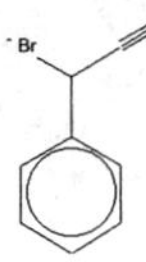

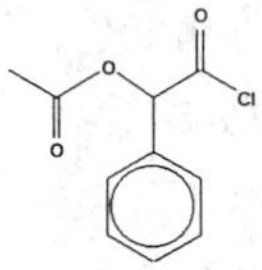

948

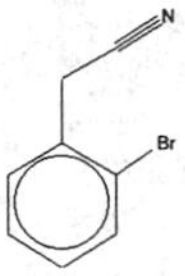

939

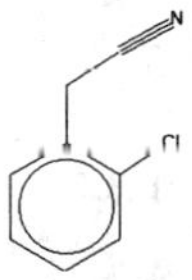

959

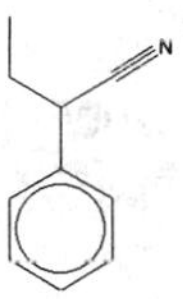

966

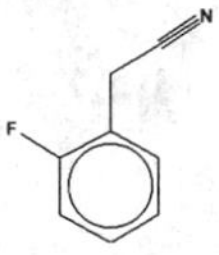

973

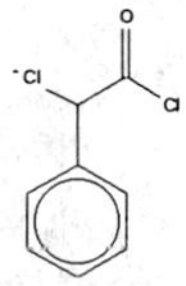

949

940

960

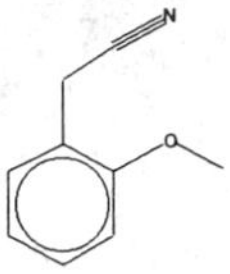

967

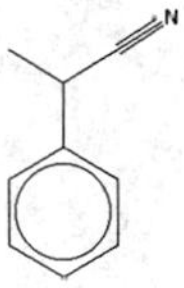

974

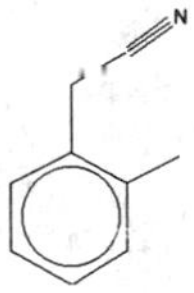

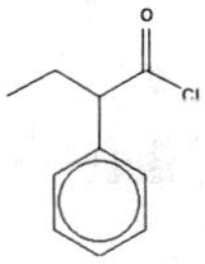

941

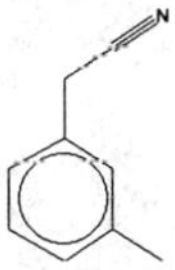

951

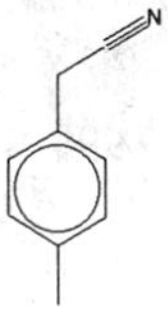

961

975

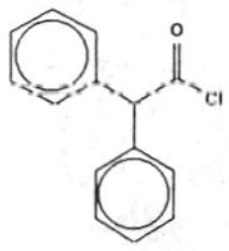

968

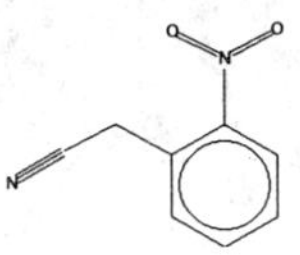

942

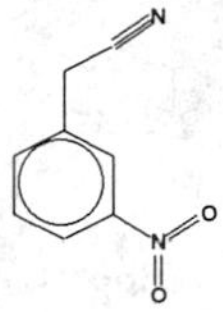

976

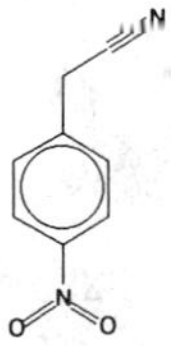

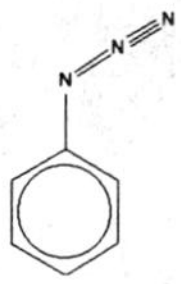

952

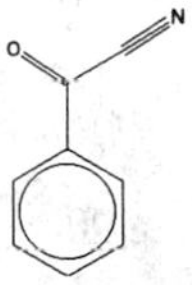

962

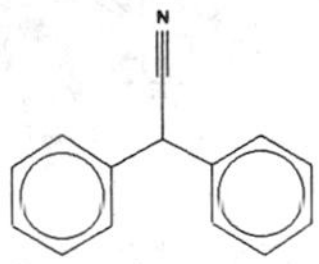

969

944

953

977

963

945

970

954

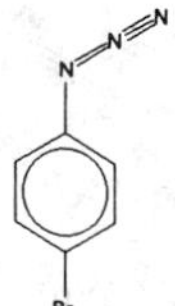

STRUCTURAL FORMULAS OF ORGANIC COMPOUNDS (continued)

In numeric order as they occur in the Table of Physical Constants of Organic Compounds

978

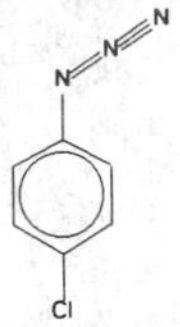

979

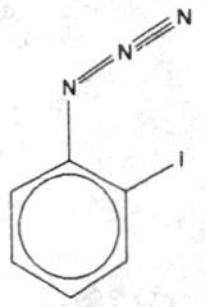

980

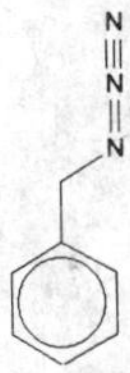

981

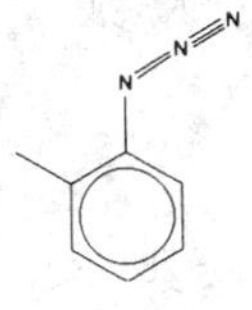

982

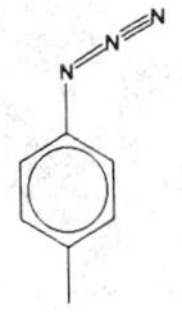

983

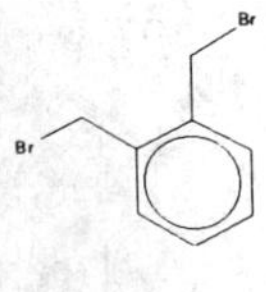

984

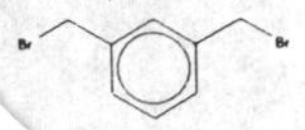

985

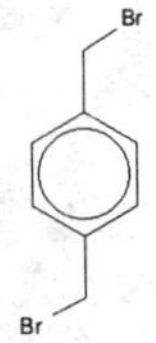

986

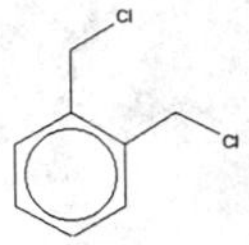

987

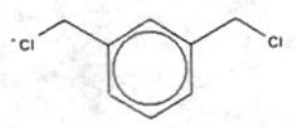

988

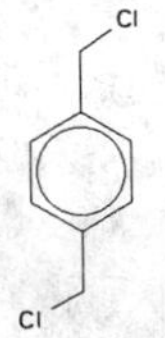

989

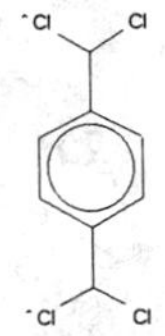

990

991

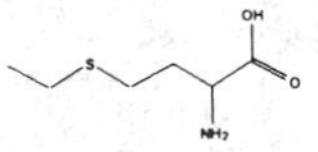

992

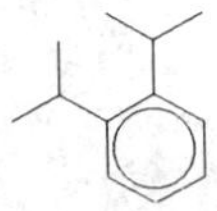

993

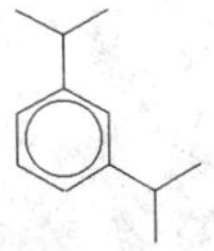

994

995

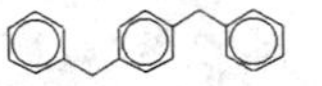

996

997

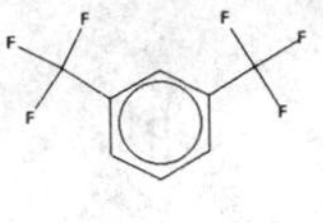

998

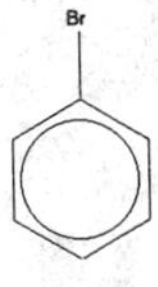

999

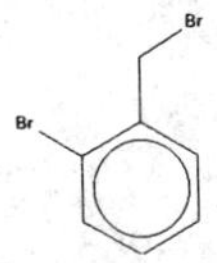

1000

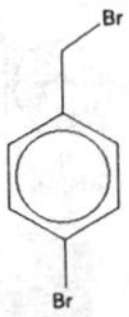

1001

1002

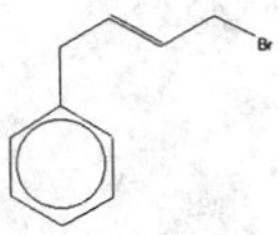

1003

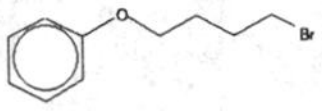

1004

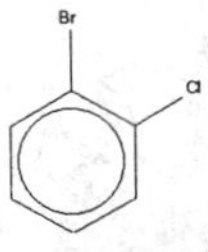

1005

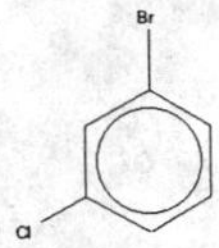

1006

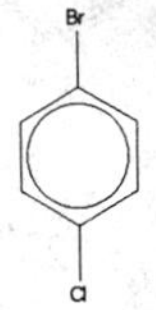

1007

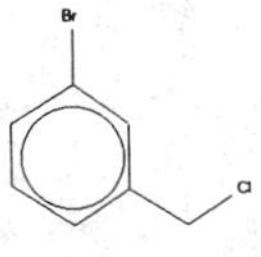

1008

1010

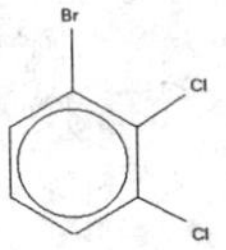

1011

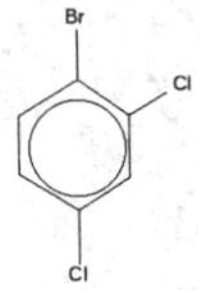

1012

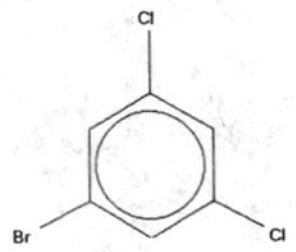

1013

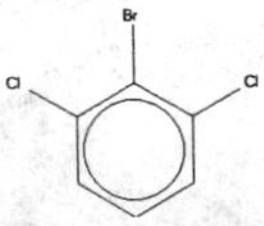

1014

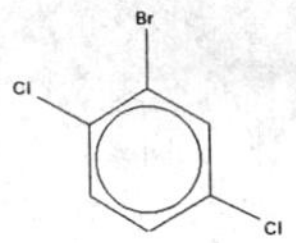

1015

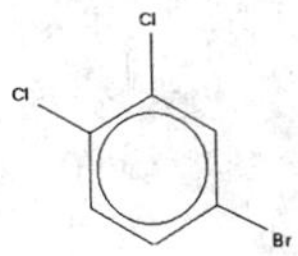

1016

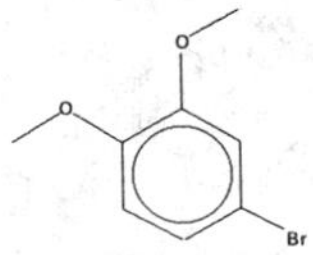

STRUCTURAL FORMULAS OF ORGANIC COMPOUNDS (continued)

In numeric order as they occur in the Table of Physical Constants of Organic Compounds

1017

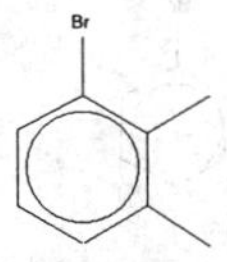

1018

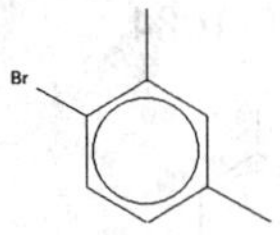

1019

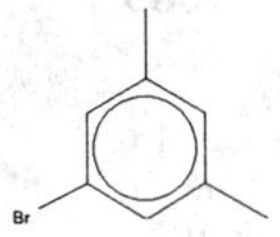

1020

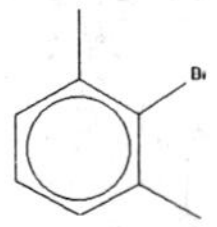

1021

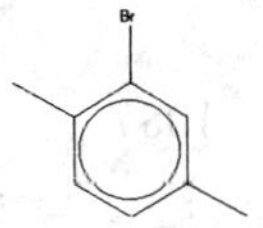

1022

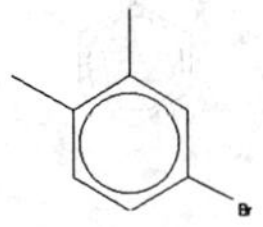

1023

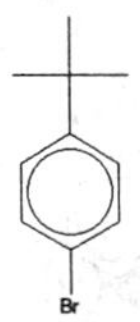

1024

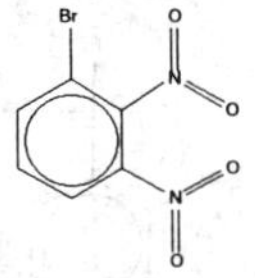

1025

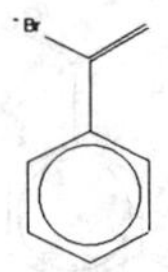

1026

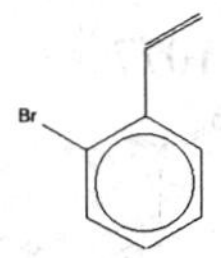

1027

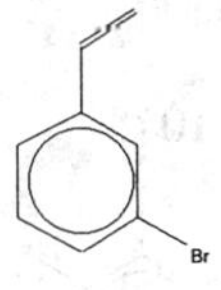

1028

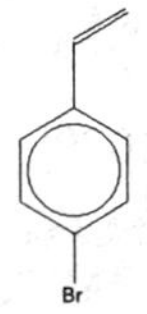

1029

1030

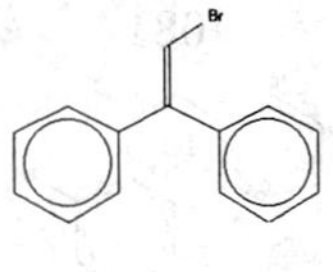

1031

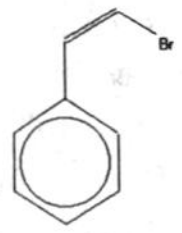

1032

1033

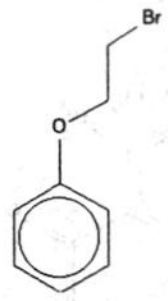

1034

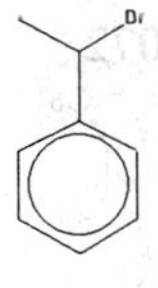

1035

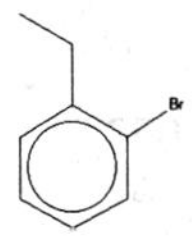

1036

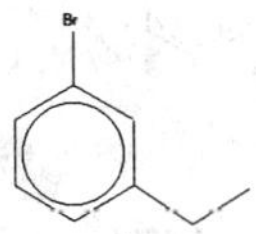

1037

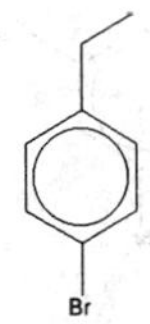

1038

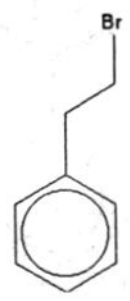

1039

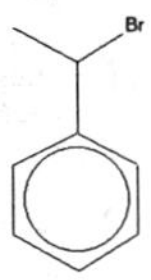

1040

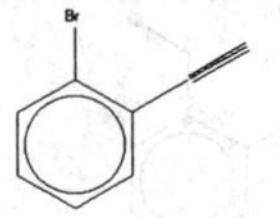

1041

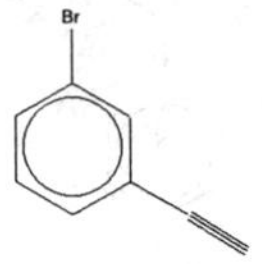

1042

1043

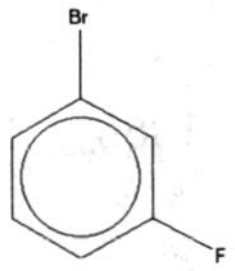

1044

1045

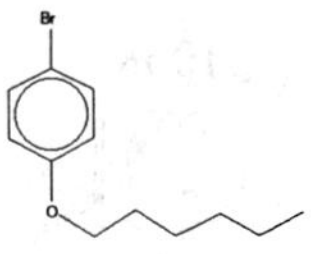

1046

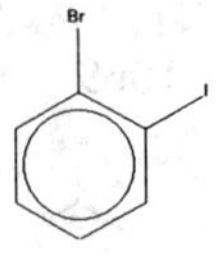

1047

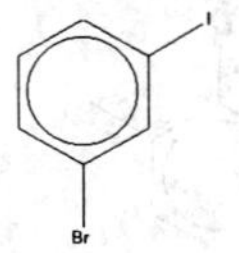

1048

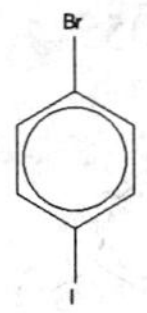

1049

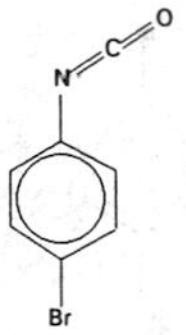

1050

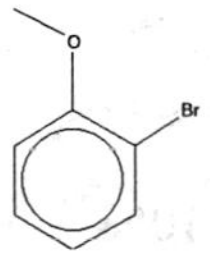

1051

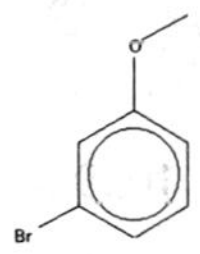

1052

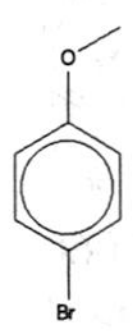

1053

In numeric order as they occur in the Table of Physical Constants of Organic Compounds

1054 Br

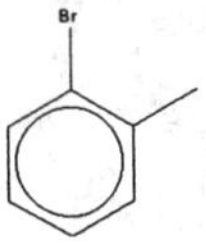

1055 Br

1056 Br

1057 Br

1058 Br

1059 Br

1060 Br

1061 Br

1062 Br

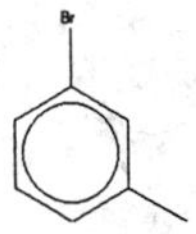

1063 Br

1064 Br

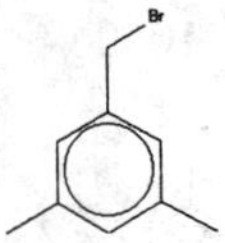

1065 Br

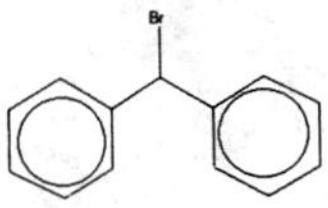

1066 Br N O O

1067 O N O Br

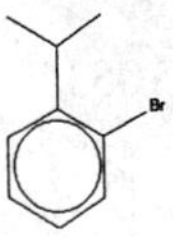

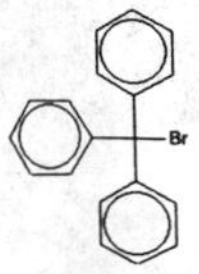

1068 Br O N O

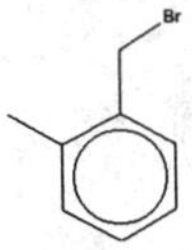

1069 Br O N O

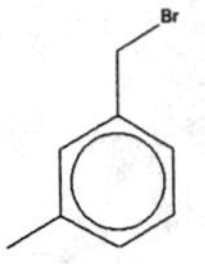

1070 Br N O O

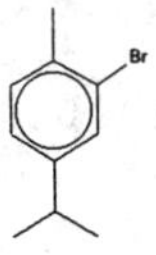

1071 Br N O O

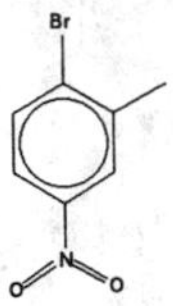

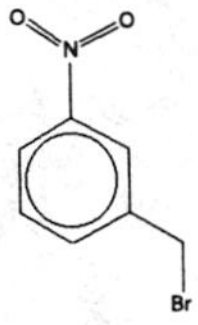

1072 O N O Br

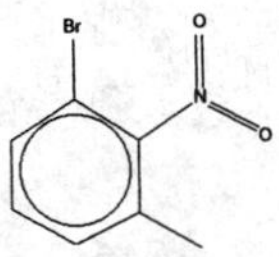

1073 O N O Br

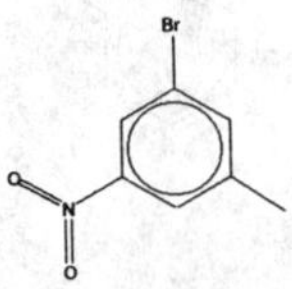

1074 O N O Br

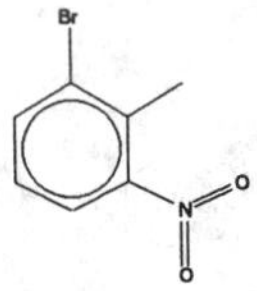

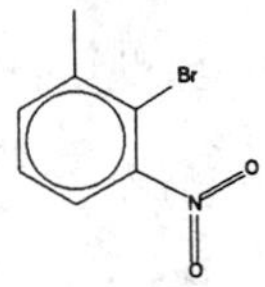

1075 O N O Br

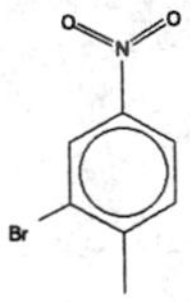

1076 O N O Br

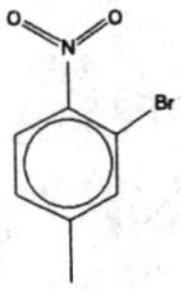

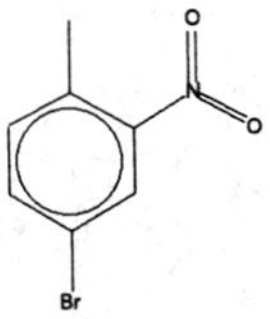

1077 O N O Br

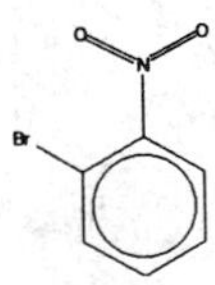

1078 F F F F F Br

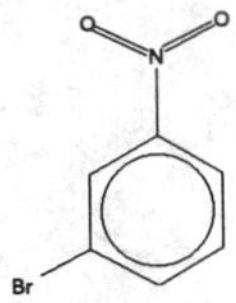

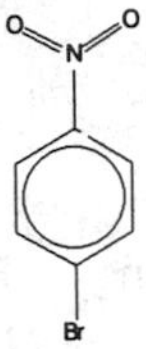

1079 O Br

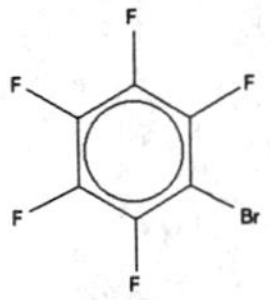

1080 Br

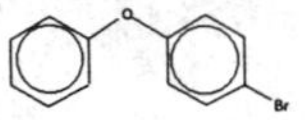

1081 Br

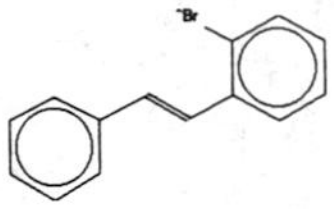

1082 Br

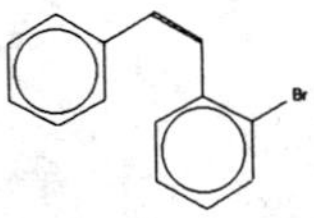

1083 Br

1084 Br

1085 Br O

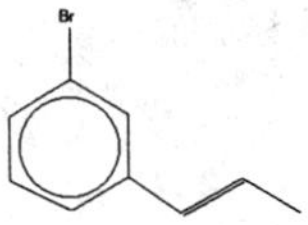

1086 Br

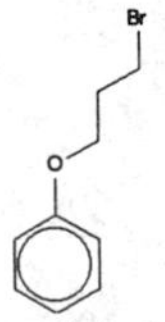

1087 Br

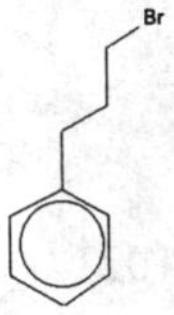

1088 Br

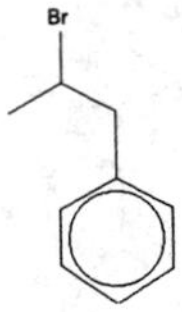

1089 F F F Br

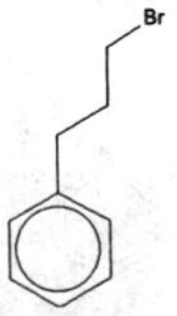

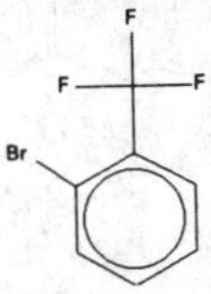

STRUCTURAL FORMULAS OF ORGANIC COMPOUNDS (continued)

In numeric order as they occur in the Table of Physical Constants of Organic Compounds

1090

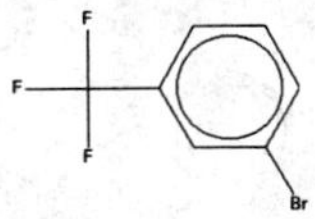

F F F Br

1091

F F F Br

1092

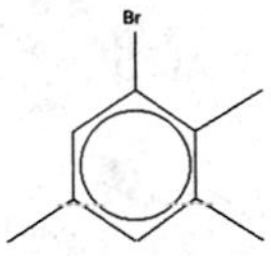

Br

1093

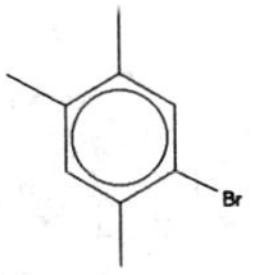

Br

1094

Br

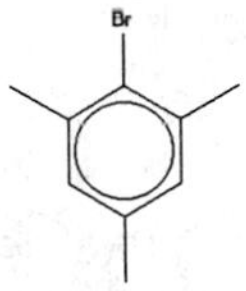

1095

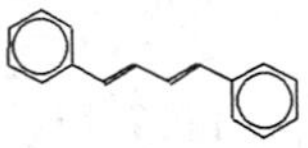

1096

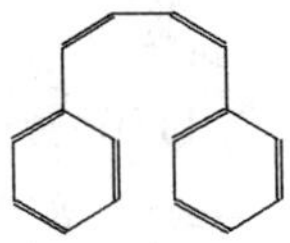

1097

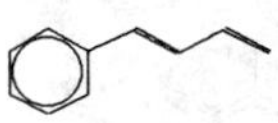

1098

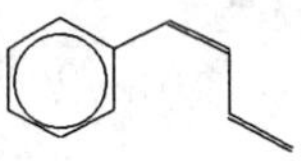

1099

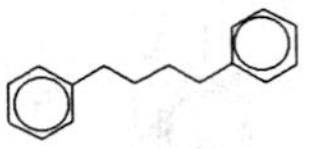

1100

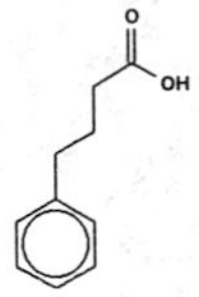

O OH

1101

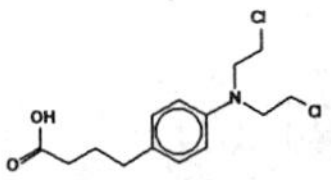

OH O N Cl Cl

1102

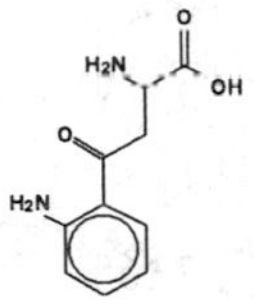

H₂N O OH O H₂N

1103

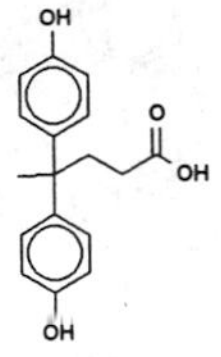

OH O OH OH

1104

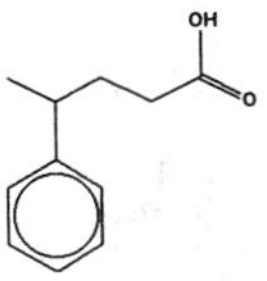

OH O

1105

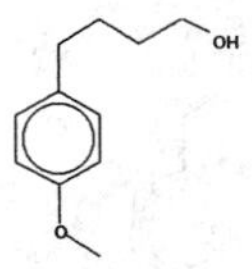

OH O

1106

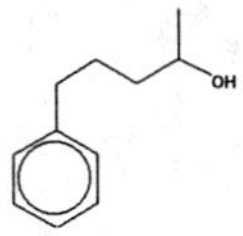

OH

1108

OH

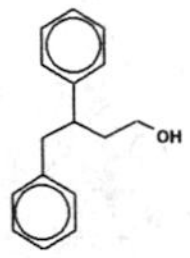

1109

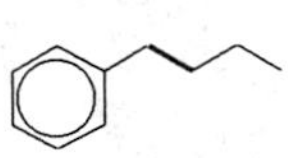

1110

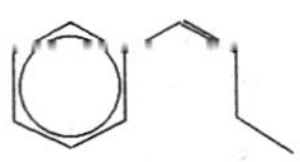

1111

1112

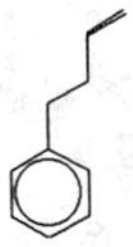

1113

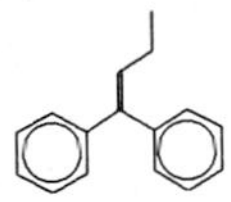

1114

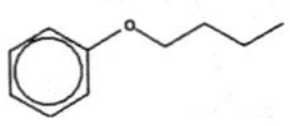

O

1115

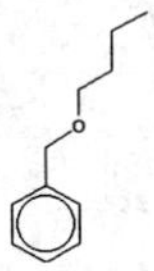

O

1116

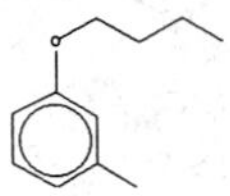

O

1117

O

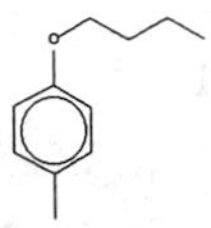

1118

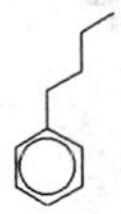

1119

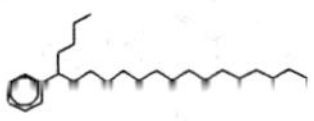

1120

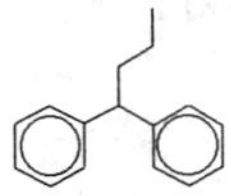

1121

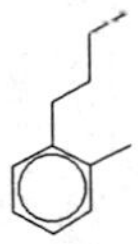

1122

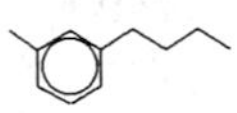

1123

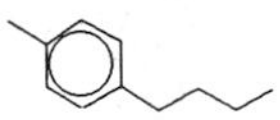

1124

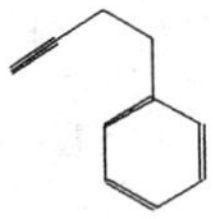

1125

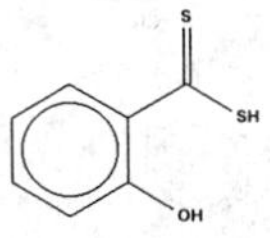

S SH OH

1126

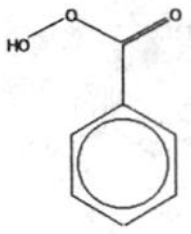

HO O O

1127

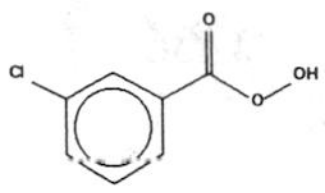

Cl O O OH

1128

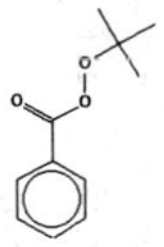

O O O

1129

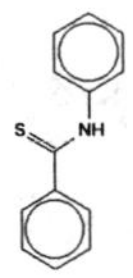

S NH

1130

HO S

1131

Cl

1132

Cl Cl

1133

Cl Cl

In numeric order as they occur in the Table of Physical Constants of Organic Compounds

1134

1135

1136

1137

1138

1139

1140

1141

1142

1143

1144

1145

1146

1147

1148

1149

1150

1151

1153

1154

1155

1156

1157

1158

1159

1160

1161

1162

1163

1164

1166

1167

1168

1169

1170

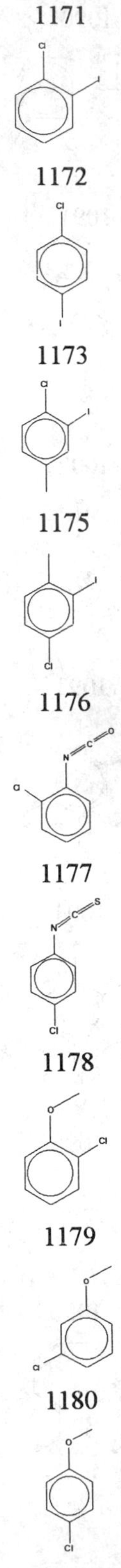

STRUCTURAL FORMULAS OF ORGANIC COMPOUNDS (continued)

In numeric order as they occur in the Table of Physical Constants of Organic Compounds

1181

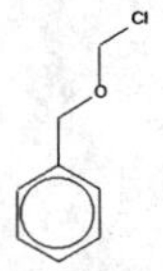

1182

1183

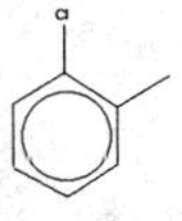

1184

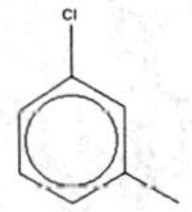

1185

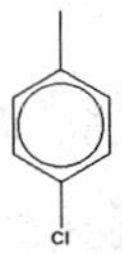

1186

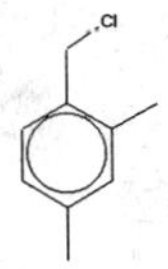

1187

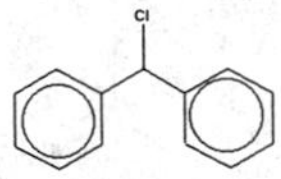

1189

1190

1191

1192

1193

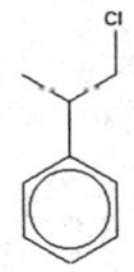

1194

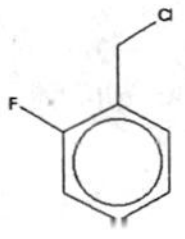

1195

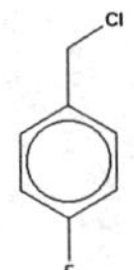

1196

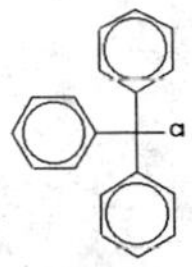

1197

1198

1199

1200

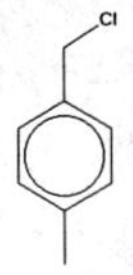

1201

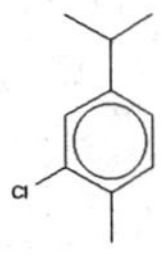

1202

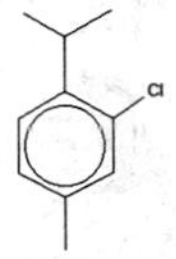

1203

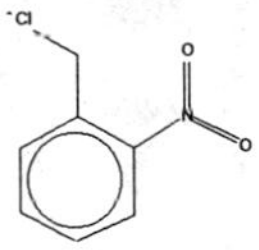

1204

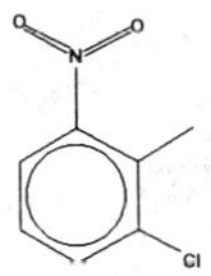

1205

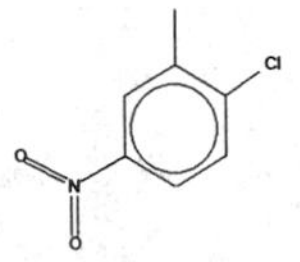

1206

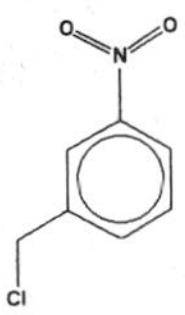

1207

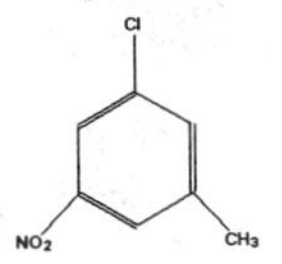

1208

1209

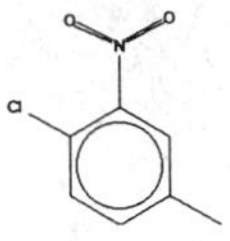

1210

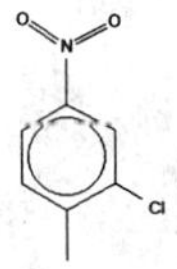

1211

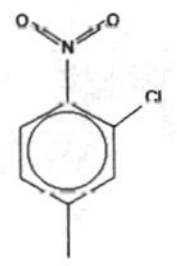

1212

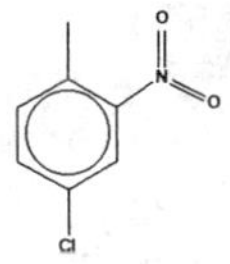

1213

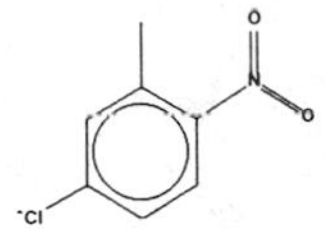

1214

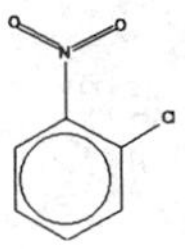

1213

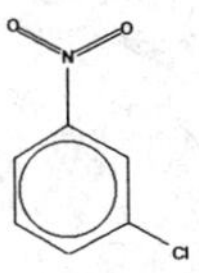

1216

1217

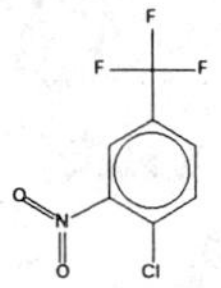

1218

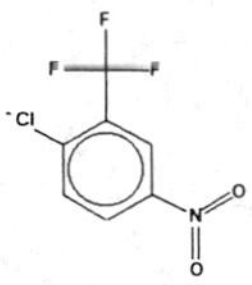

1219

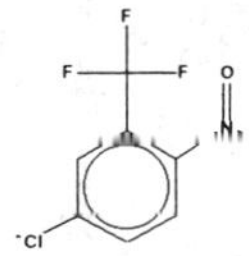

1220

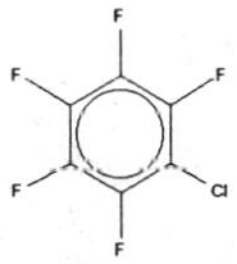

1221

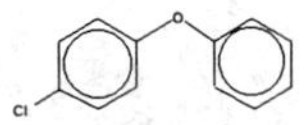

1222

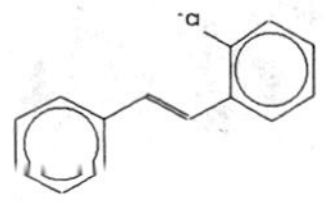

1223

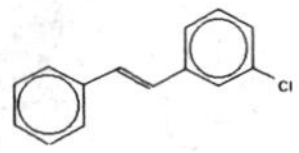

STRUCTURAL FORMULAS OF ORGANIC COMPOUNDS (continued)

In numeric order as they occur in the Table of Physical Constants of Organic Compounds

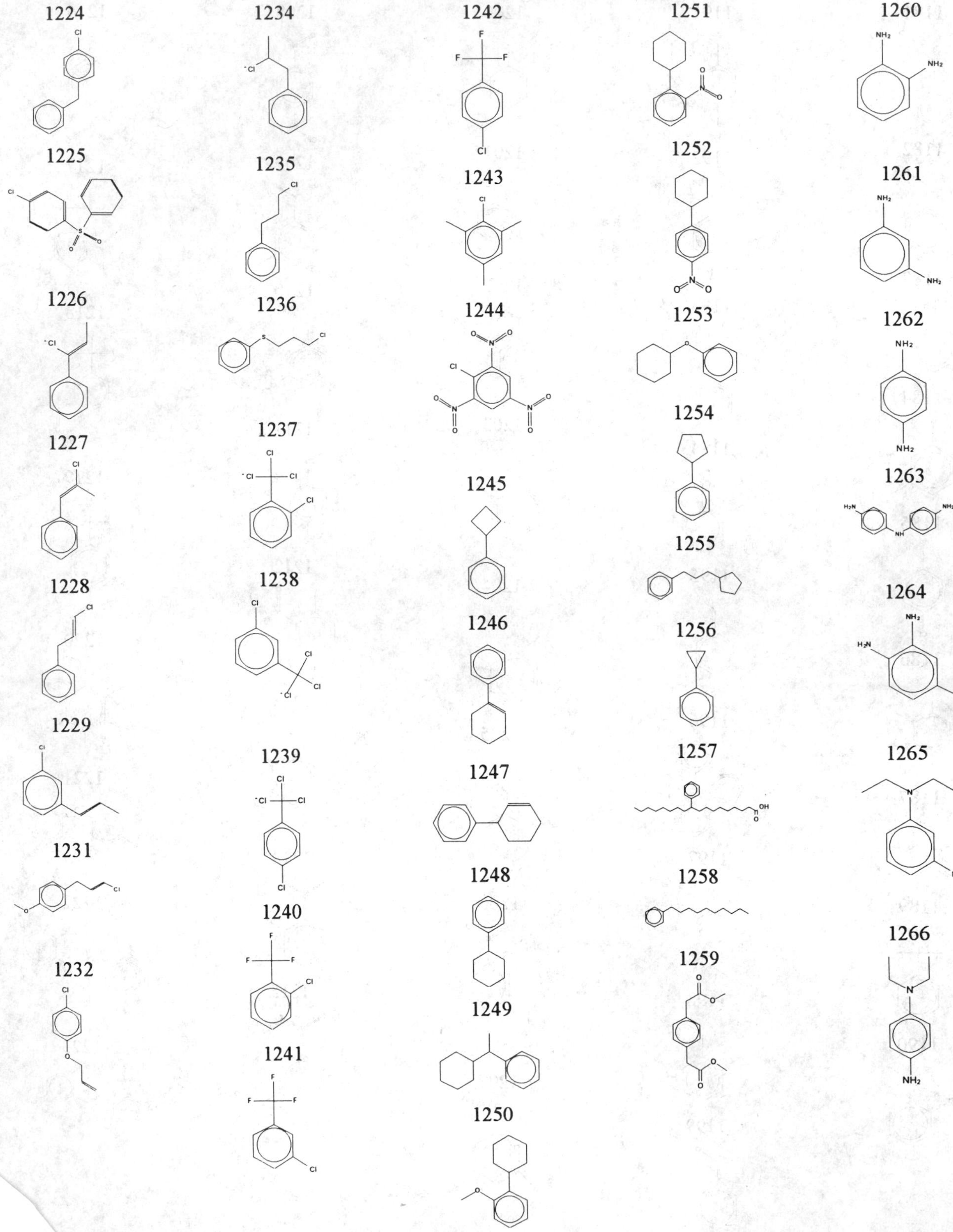

STRUCTURAL FORMULAS OF ORGANIC COMPOUNDS (continued)

In numeric order as they occur in the Table of Physical Constants of Organic Compounds

1267

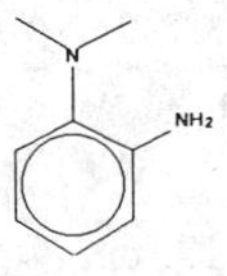

N

NH2

1268

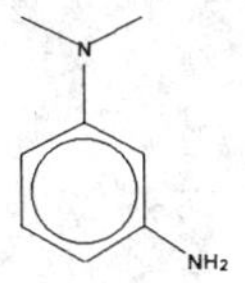

N

NH2

1270

N

NH2

1271

NH

NH

1272

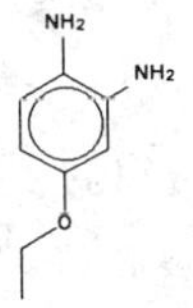

NH2

NH2

O

1273

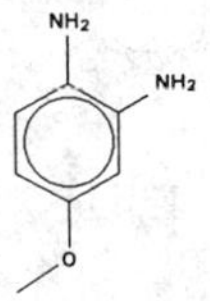

NH2

NH2

O

1274

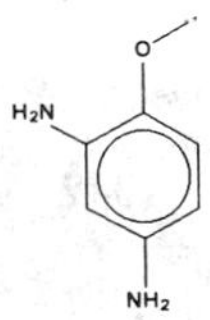

O

H2N

NH2

1275

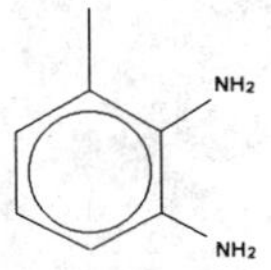

NH2

NH2

1276

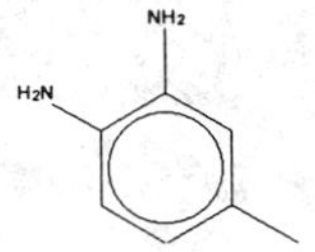

NH2

H2N

1277

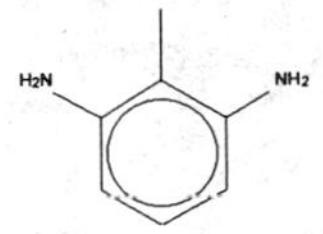

H2N

NH2

1278

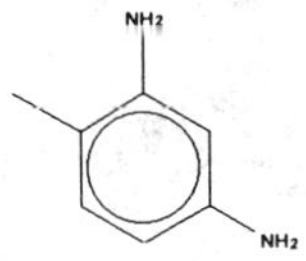

NH2

NH2

1279

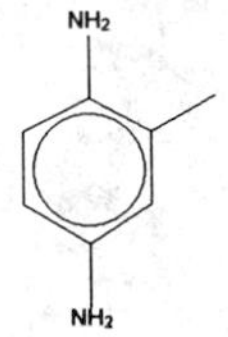

NH2

NH2

1280

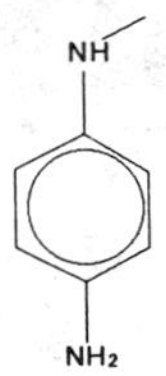

NH

NH2

1281

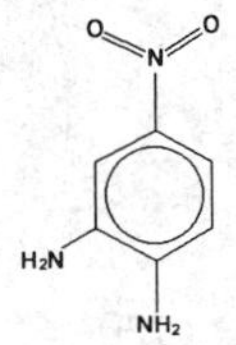

O

N

O

H2N

NH2

1282

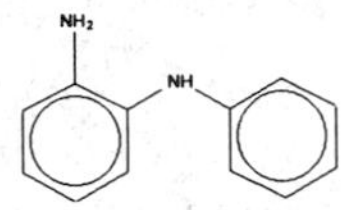

NH2

NH

1283

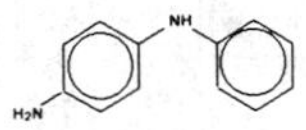

NH

H2N

1284

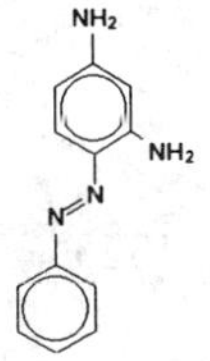

NH2

NH2

N

N

1285

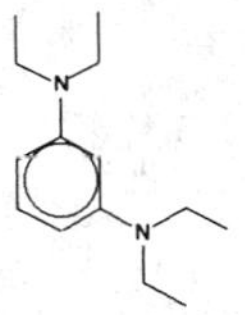

N

N

1286

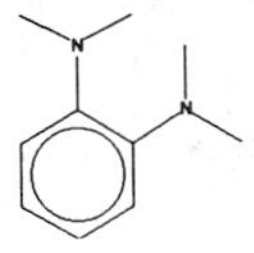

N

N

1287

N

N

1288

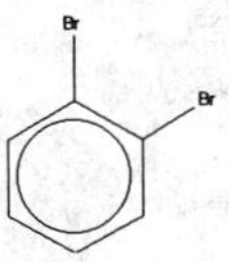

Br

Br

1289

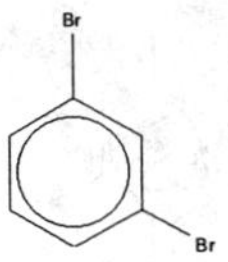

Br

Br

1290

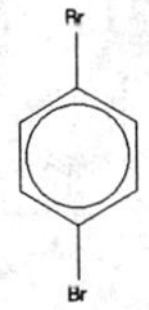

Br

Br

1291

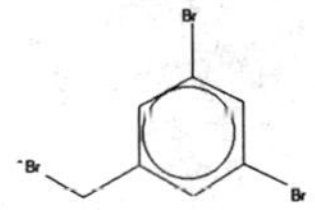

Br

Br

Br

1292

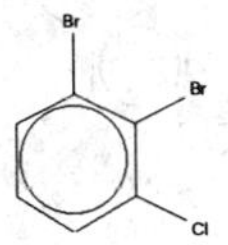

Br

Br

Cl

1293

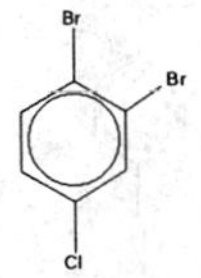

Br

Br

Cl

1294

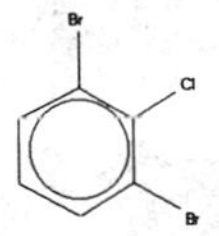

Br

Cl

Br

1295

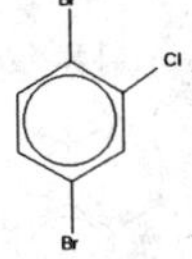

Br

Cl

Br

1297

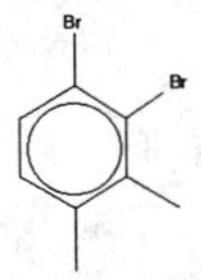

Br

Br

1298

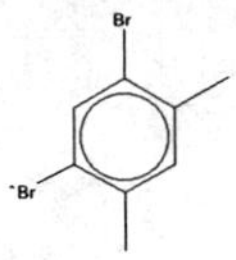

Br

Br

1299

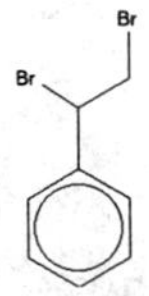

Br

Br

1300

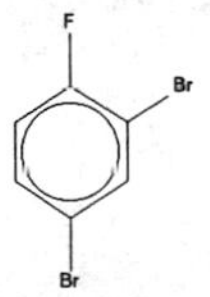

F

Br

Br

1301

Br

Br

1302

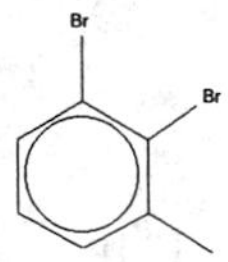

Br

Br

1303

Br

Br

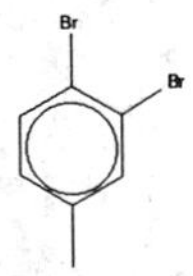

STRUCTURAL FORMULAS OF ORGANIC COMPOUNDS (continued)

In numeric order as they occur in the Table of Physical Constants of Organic Compounds

1304
Br Br

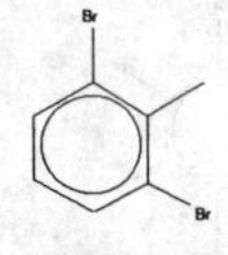

1305
Br Br

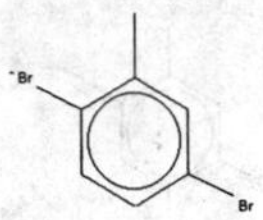

1306
Br Br

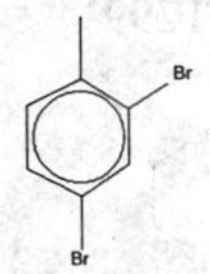

1307
Br Br N O O

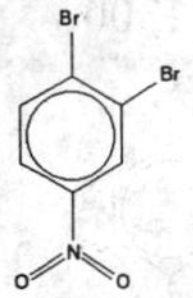

1308
NO_2 Br Br

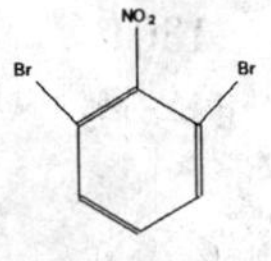

1309
Br Br NO_2

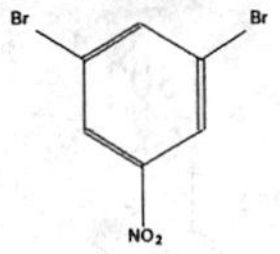

1310
O N O Br Br

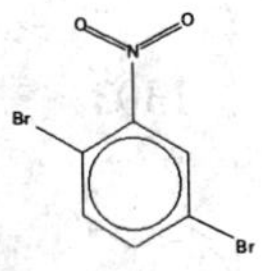

1311
NO_2 Br Br

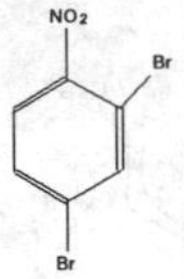

1312
F F F F Br Br

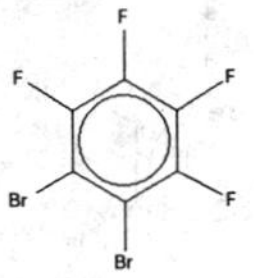

1313
Br Br

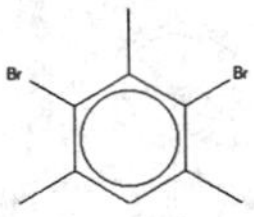

1314
O O

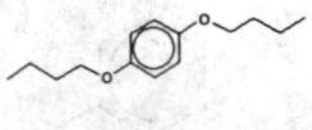

1315
N N

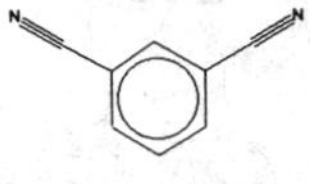

1316
N N

1317
N Cl Cl Cl Cl N

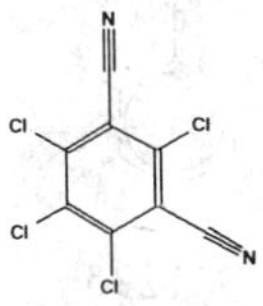

1318
Cl O Cl O

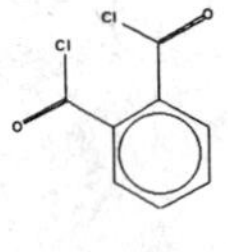

1319
Cl O Cl O

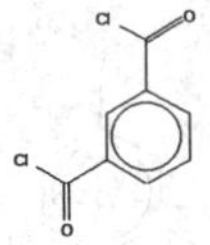

1320
Cl O O Cl

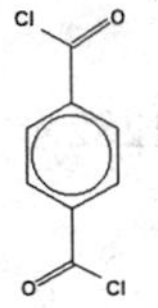

1321
O F O F

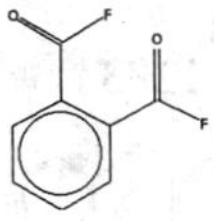

1322
O O

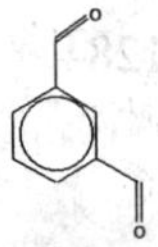

1323
O O

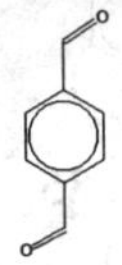

1324
H_2N O NH_2 O

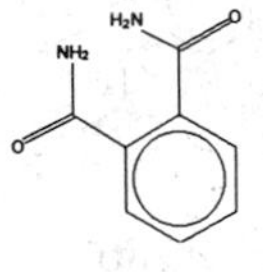

1325
N O O N

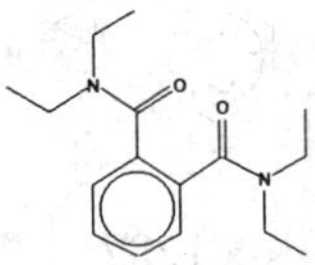

1327
O OH O OH

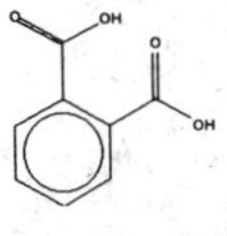

1328
O OH OH O

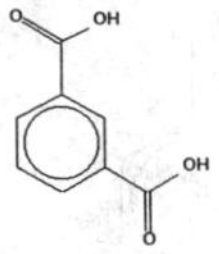

1329
HO O O OH

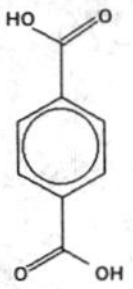

1330
HO O NH_2 O OH

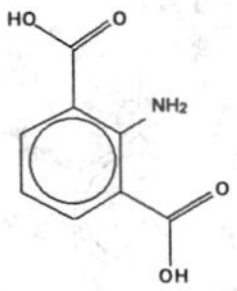

1331
HO O OH H_2N O

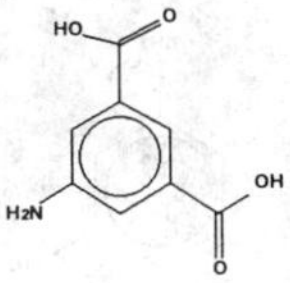

1332
HO O O OH O

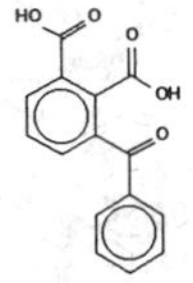

1333
O O O O O O

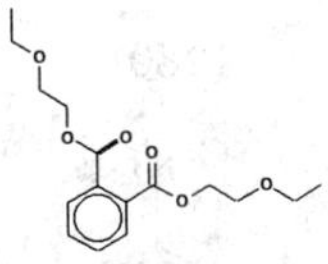

1334
O O O O

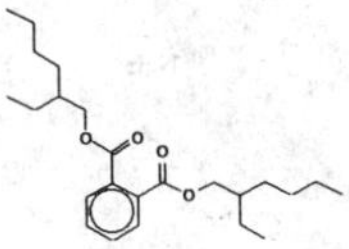

1335
O O O O

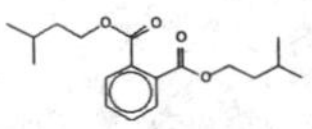

1336
O O O O

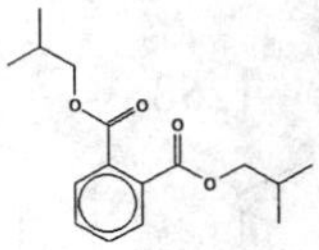

1337
O O O O

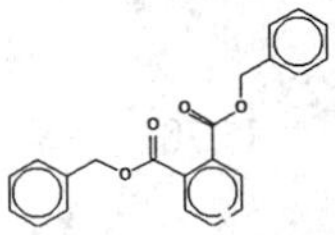

1338
O O O O

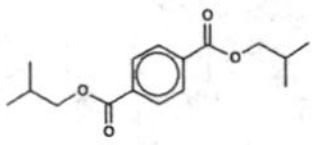

1339
O O O O

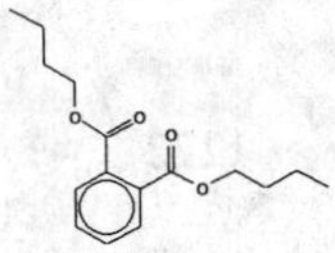

1340
HO O O OH Cl Cl

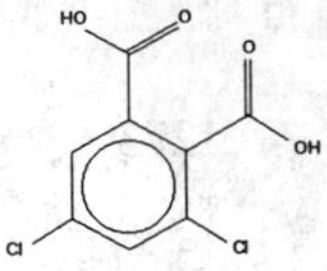

1341
O O O O

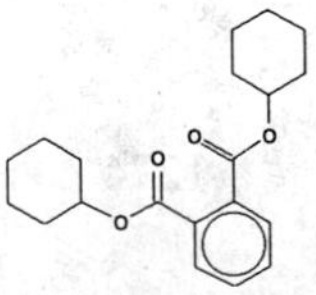

1342
O O O O

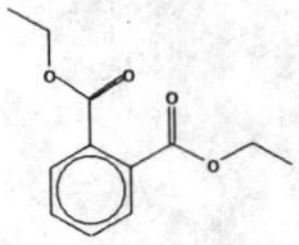

STRUCTURAL FORMULAS OF ORGANIC COMPOUNDS (continued)

In numeric order as they occur in the Table of Physical Constants of Organic Compounds

1343

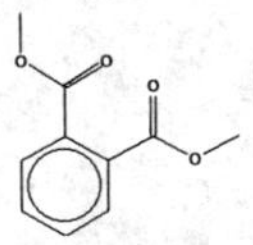

1344

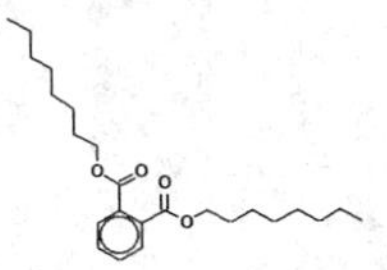

1345

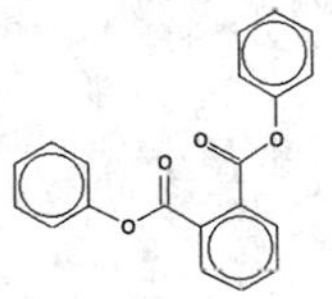

1346

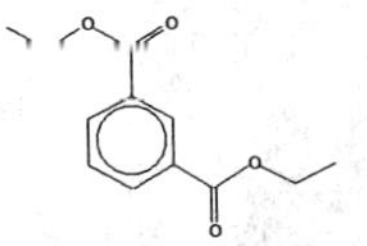

1347

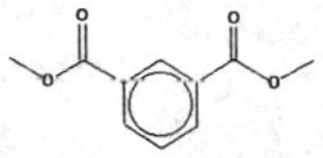

1348

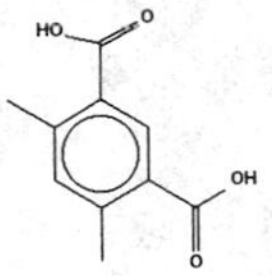

1349

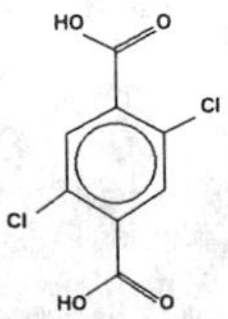

1350

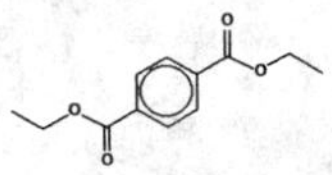

1351

1352

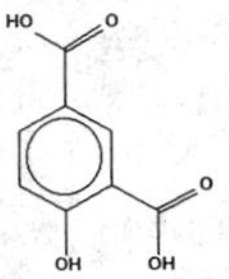

1353

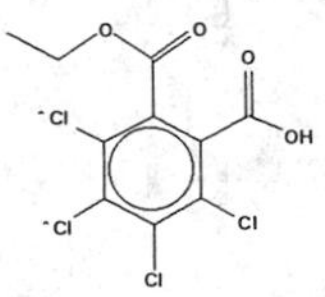

1354

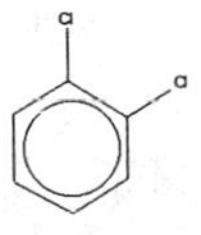

1355

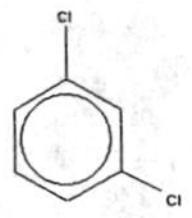

1356

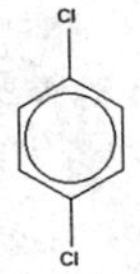

1357

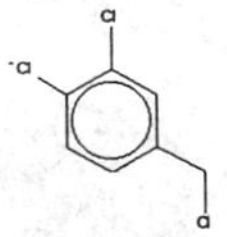

1358

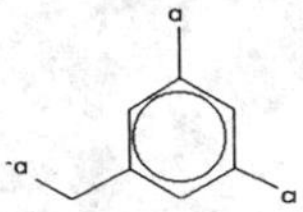

1359

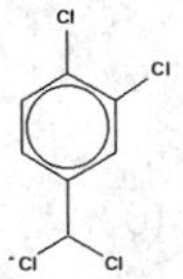

1361

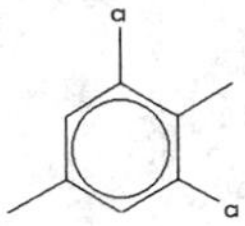

1362

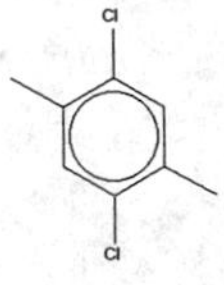

1363

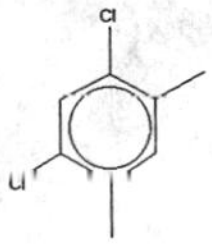

1364

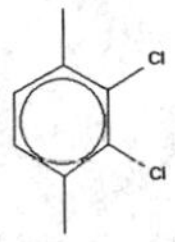

1365

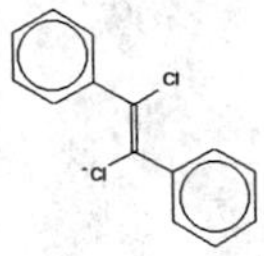

1366

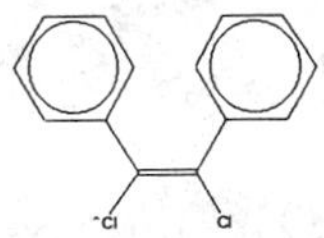

1367

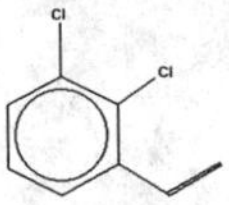

1368

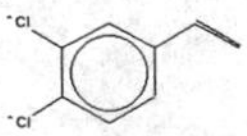

1369

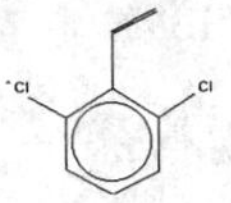

1370

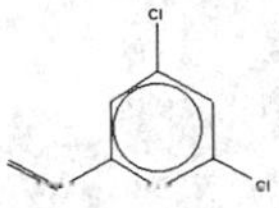

1371

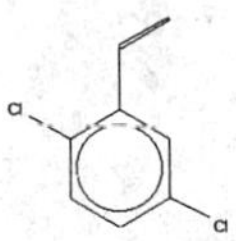

1372

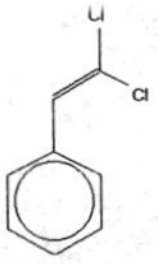

1373

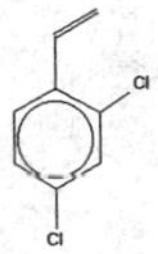

1374

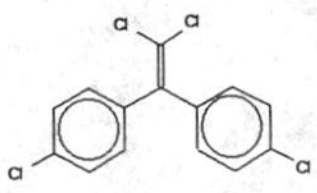

1375

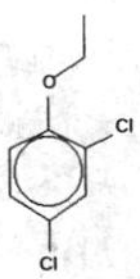

1376

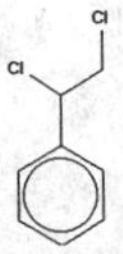

1377

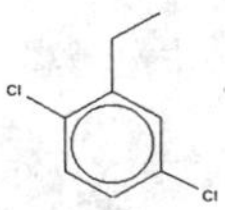

1378

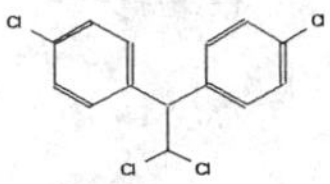

1379

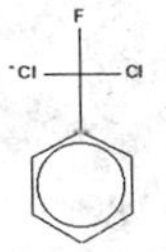

1380

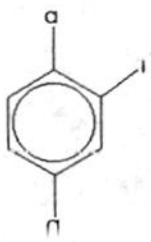

1381

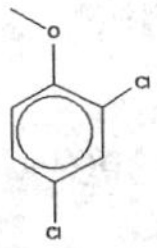

1382

1383

1384

STRUCTURAL FORMULAS OF ORGANIC COMPOUNDS (continued)

In numeric order as they occur in the Table of Physical Constants of Organic Compounds

1385

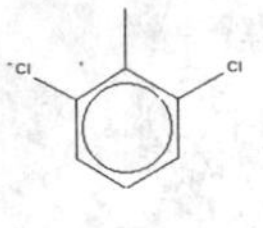

Cl Cl

1386

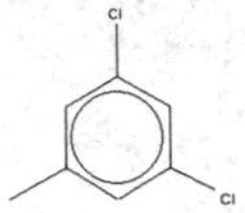

Cl Cl

1387

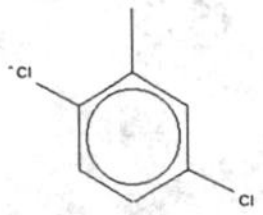

Cl Cl

1388

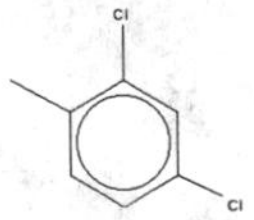

Cl Cl

1389

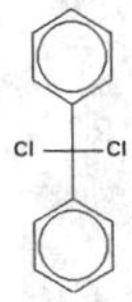

Cl Cl

1390

Cl Cl

1391

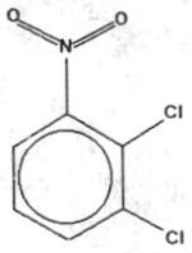

O N O Cl Cl

1392

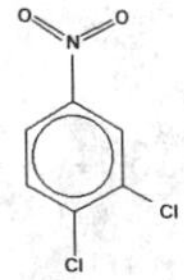

O N O Cl Cl

1393

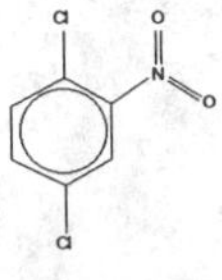

Cl O N O Cl

1394

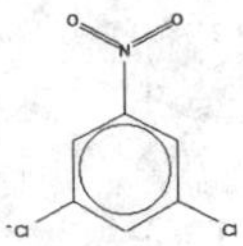

O N O Cl Cl

1395

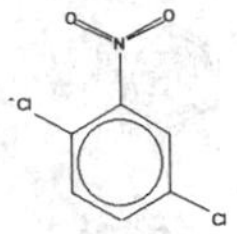

O N O Cl Cl

1396

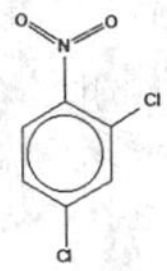

O N O Cl Cl

1397

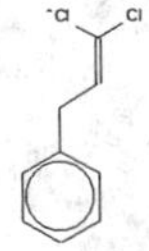

Cl Cl

1398

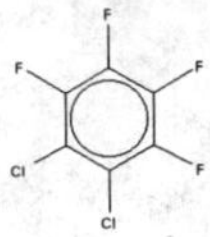

F F F Cl F Cl

1399

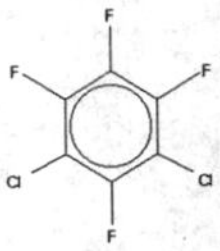

F F F Cl Cl F

1400

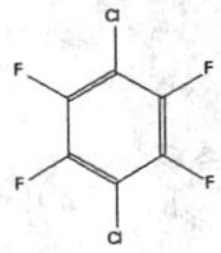

Cl F F F F Cl

1401

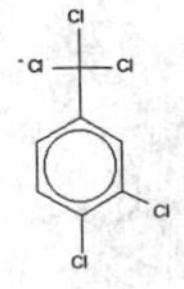

Cl Cl Cl Cl Cl

1402

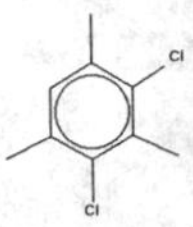

Cl Cl

1403

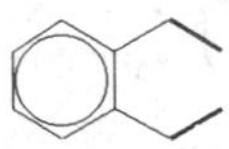

1404

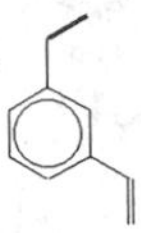

1405

1406

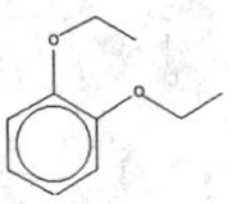

O O

1407

O O

1408

O O

1409

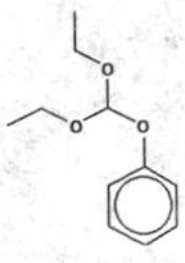

O O O

1410

O O

1411

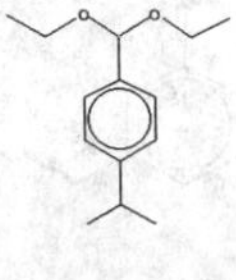

O O

1412

O O

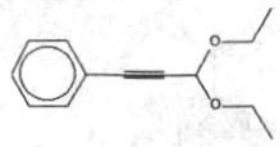

1413

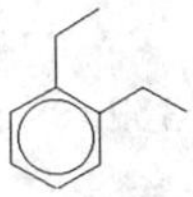

1414

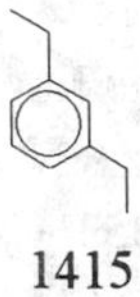

1415

1416

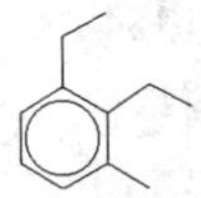

1417

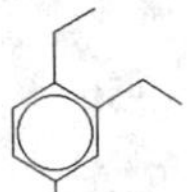

1418

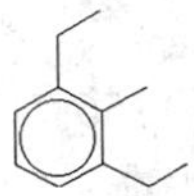

1419

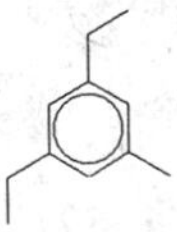

1420

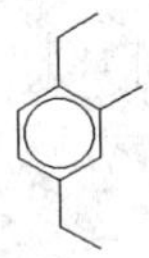

1421

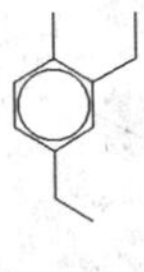

1422

F F

1423

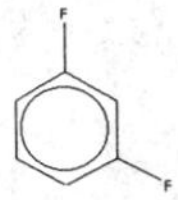

F F

1424

F F

1425

F F

1426

O N O F F

1427

I

1428

I I

1429

I I

1430

O C N N C O

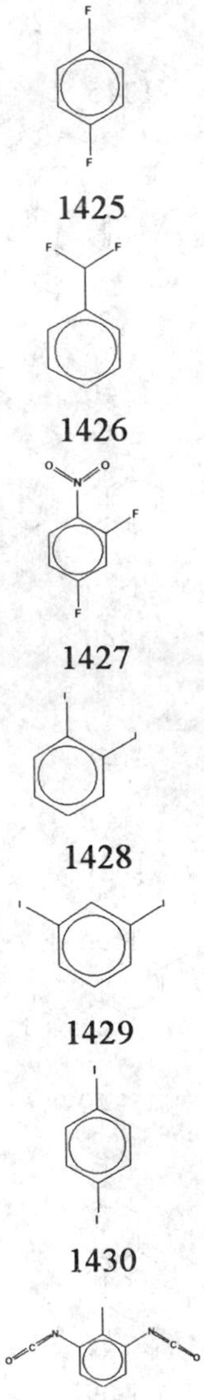

In numeric order as they occur in the Table of Physical Constants of Organic Compounds

1431

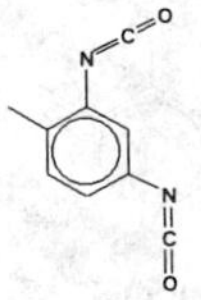

N=C=O
N=C=O

1432

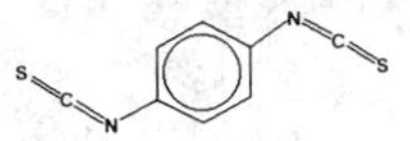

N=C=S
S=C=N

1433

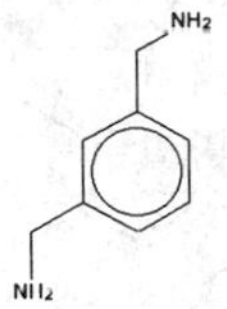

NH2
NH2

1434

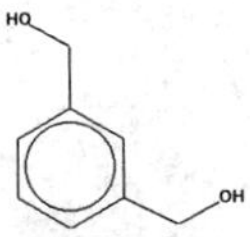

HO
OH

1435

OH
HO

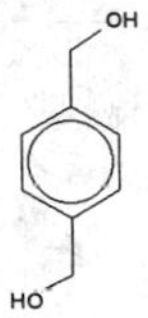

1436

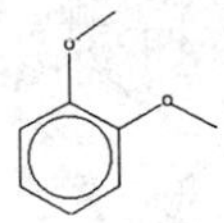

1437

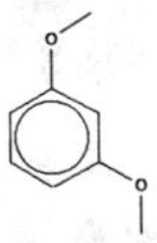

1438

1439

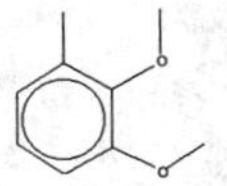

1440

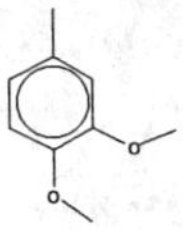

1441

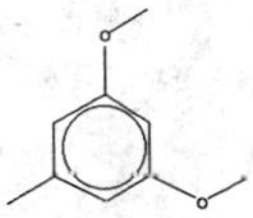

1442

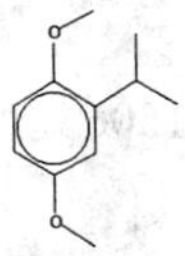

1443

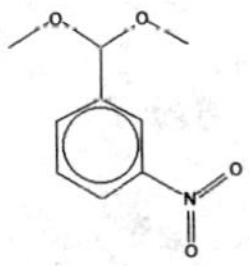

N=O
O

1444

N=O
O

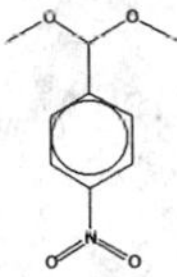

1445

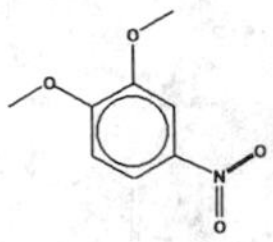

1447

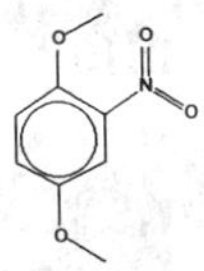

1448

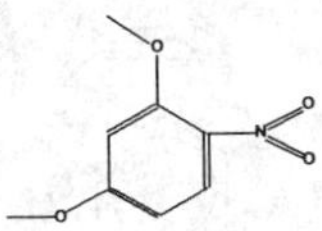

1449

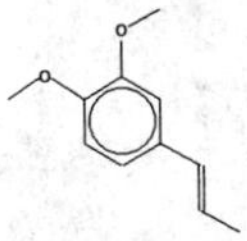

1450

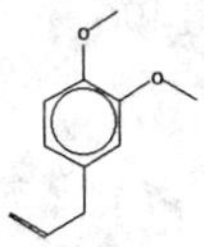

1451

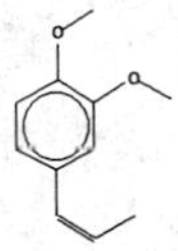

1452

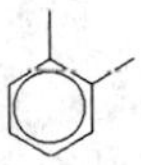

1453

1454

1455

1456

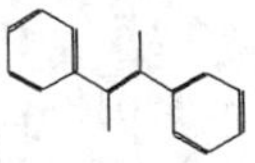

1457

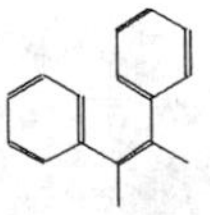

1458

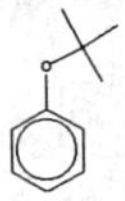

1459

1460

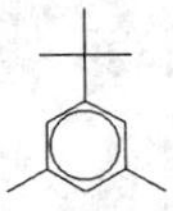

1461

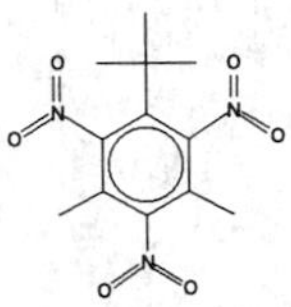

1462

1463

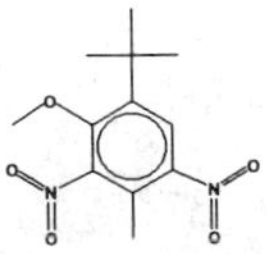

1464

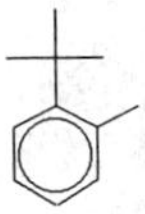

1465

1466

1467

1468

1469

1470

1471

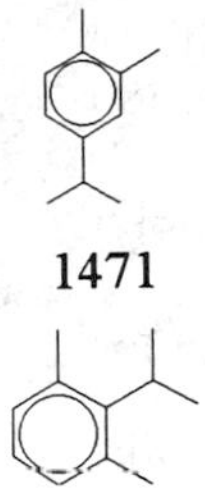

1472

1473

1474

1475

1476

1477

STRUCTURAL FORMULAS OF ORGANIC COMPOUNDS (continued)

In numeric order as they occur in the Table of Physical Constants of Organic Compounds

1478

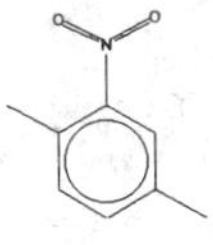

1479

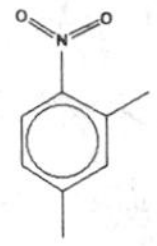

1480

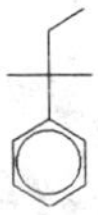

1481

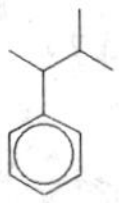

1482

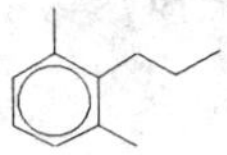

1483

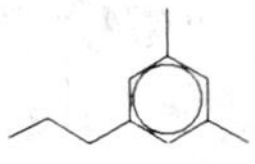

1484

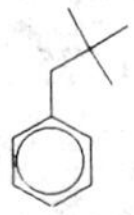

1486

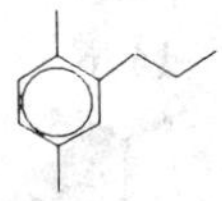

1487

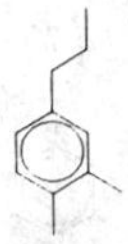

1488

1489

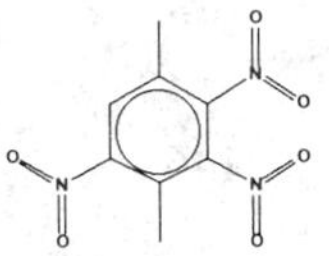

1490

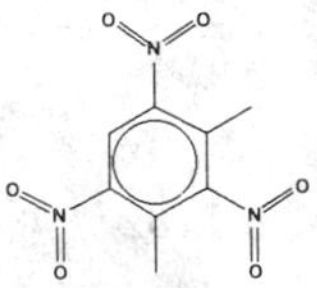

1491

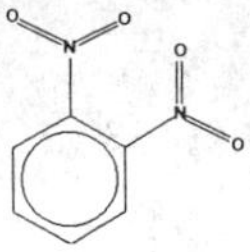

1492

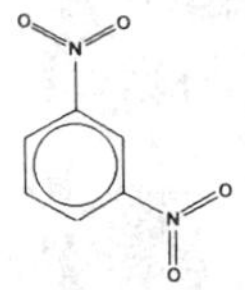

1493

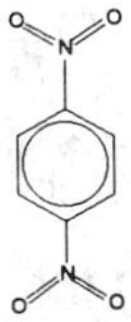

1494

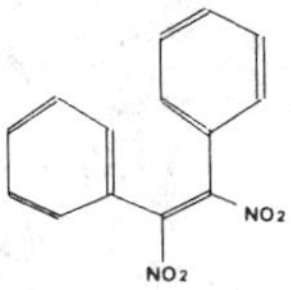

1495

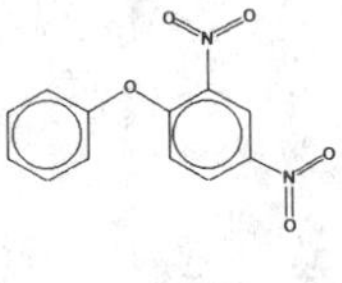

1496

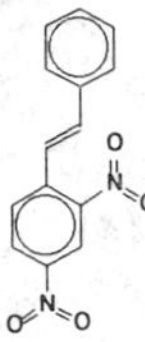

1497

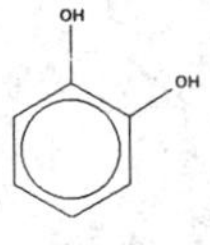

1498

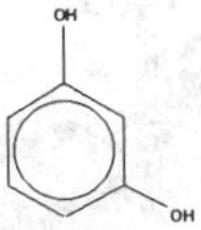

1499

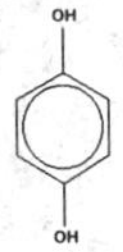

1500

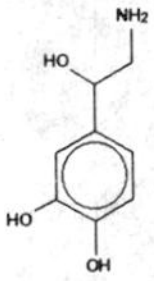

1501

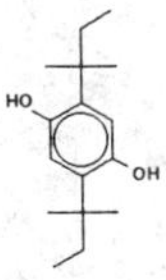

1502

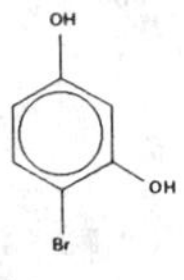

1503

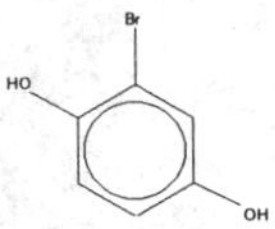

1504

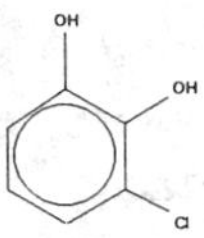

1505

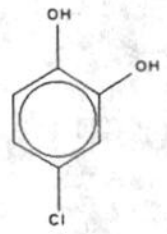

1506

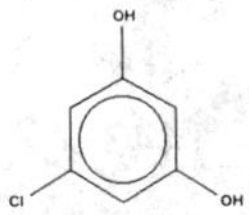

1507

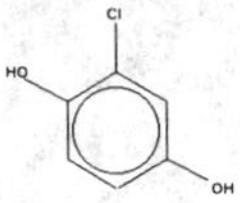

1508

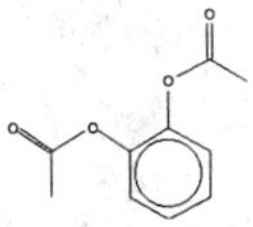

1509

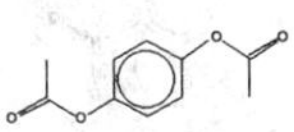

1510

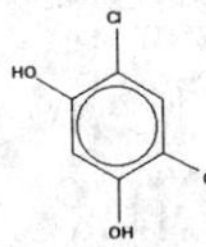

1511

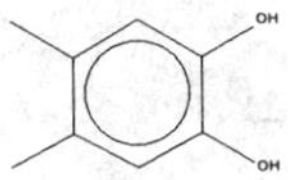

1512

1513

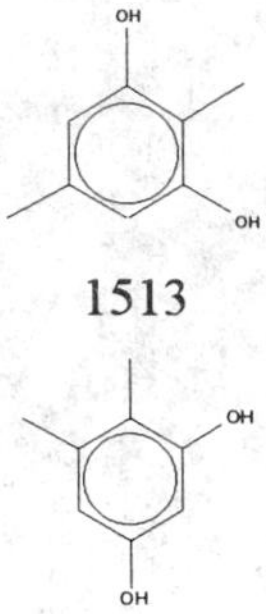

1514

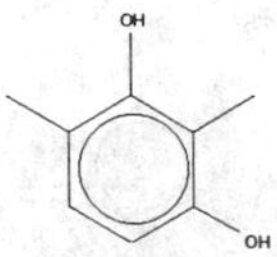

1515

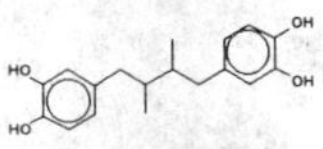

1516

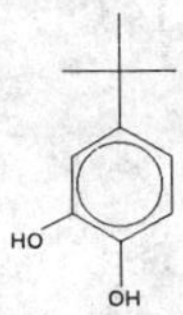

1517

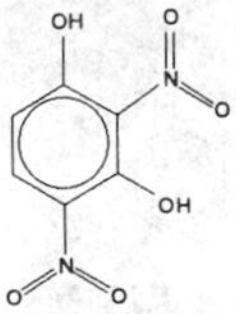

1518

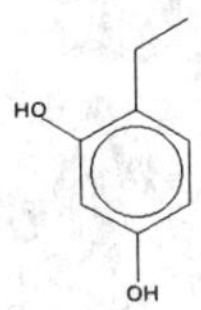

1519

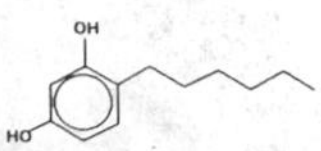

In numeric order as they occur in the Table of Physical Constants of Organic Compounds

1520

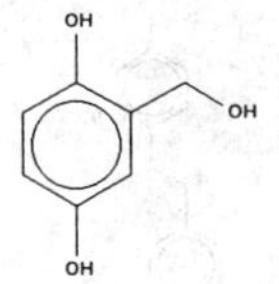

1521

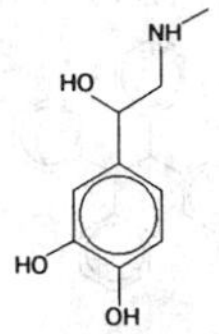

1522

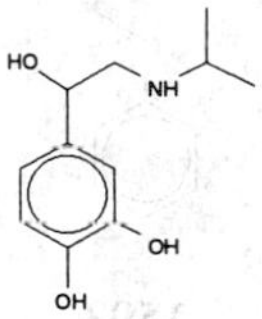

1523

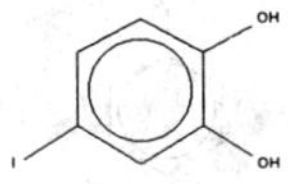

1524

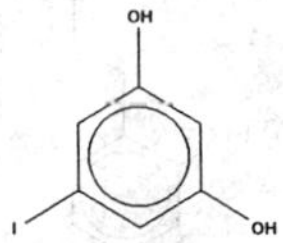

1525

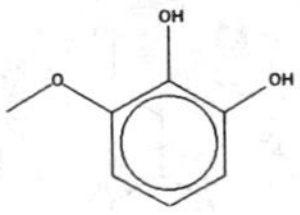

1526

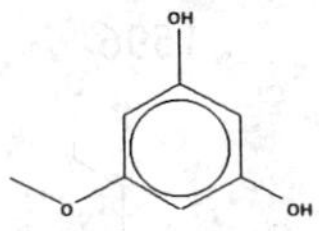

1527

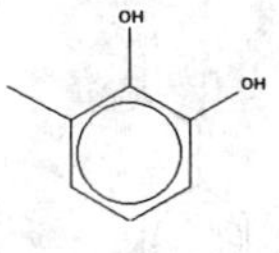

1528

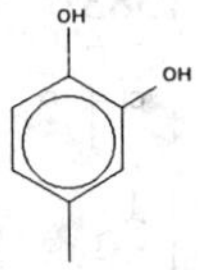

1529

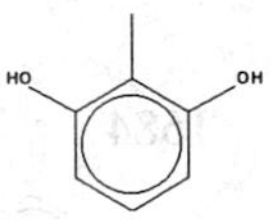

1530

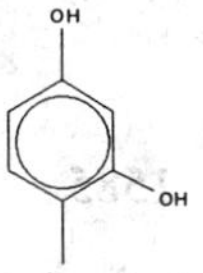

1531

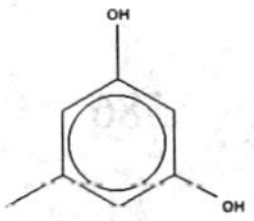

1532

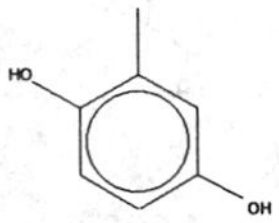

1533

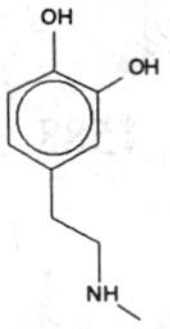

1534

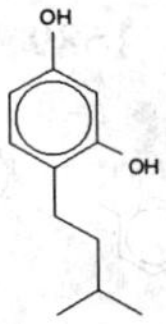

1535

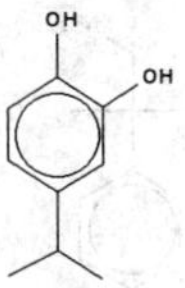

1536

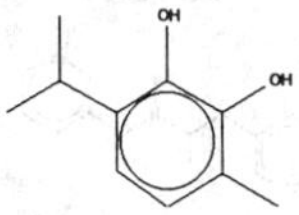

1537

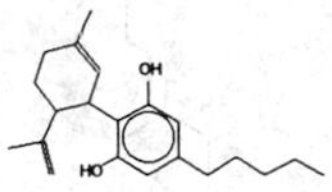

1538

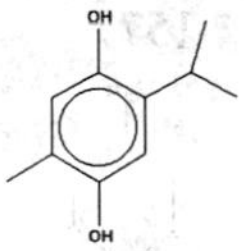

1539

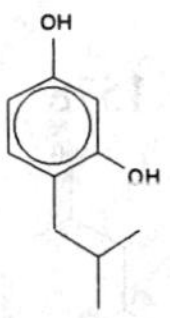

1540

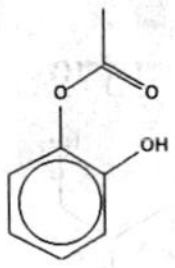

1541

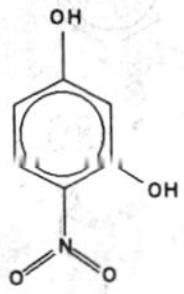

1542

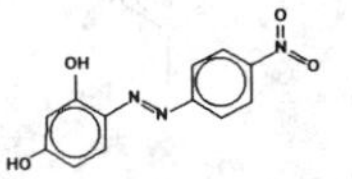

1543

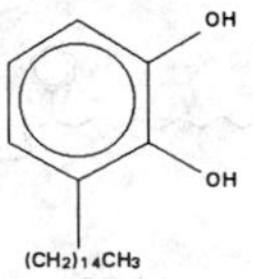

$(CH_2)_{14}CH_3$

1544

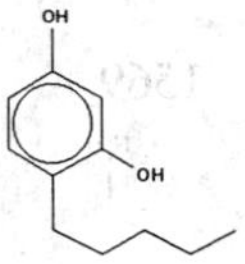

1545

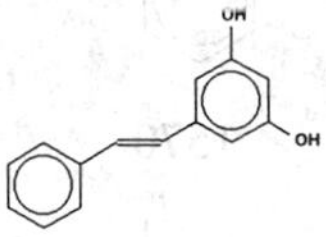

1546

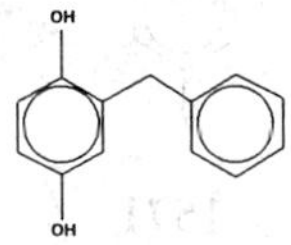

1547

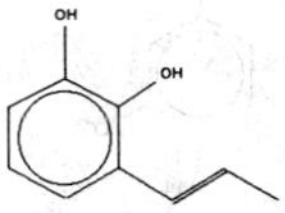

1548

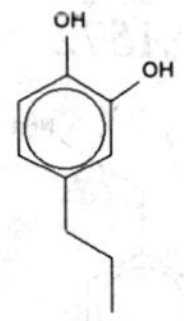

1549

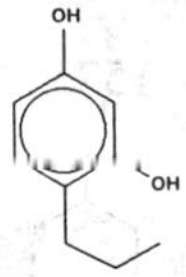

1550

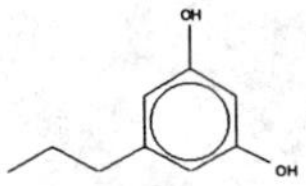

1551

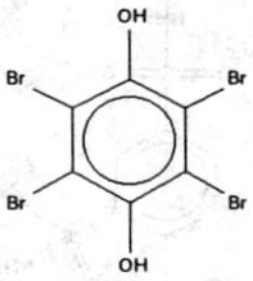

1552

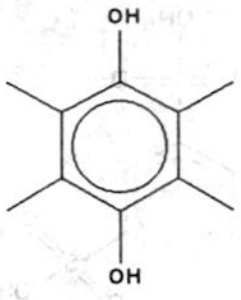

1553

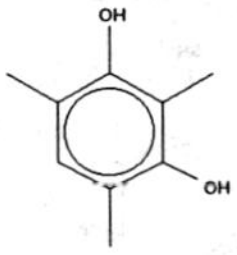

1554

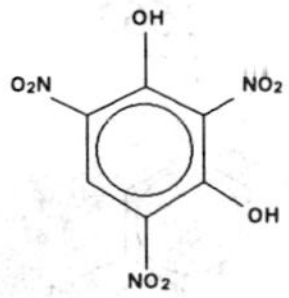

1555

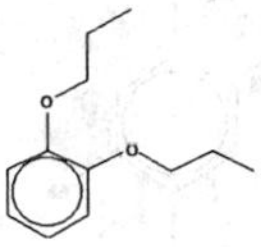

1556

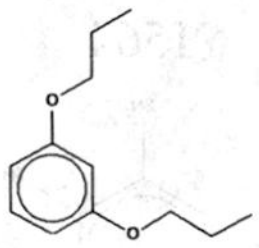

1557

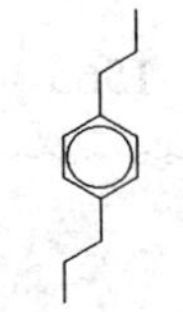

In numeric order as they occur in the Table of Physical Constants of Organic Compounds

1558

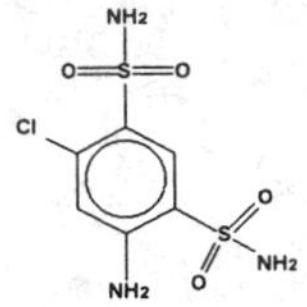

1560

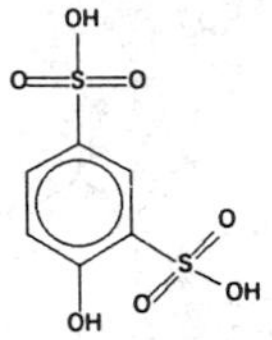

1561

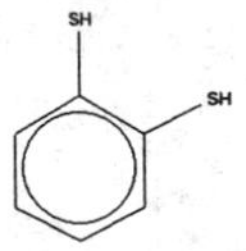

1562

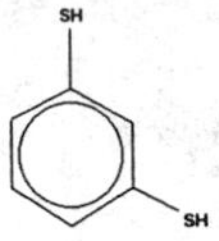

1563

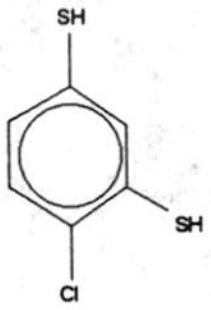

1564

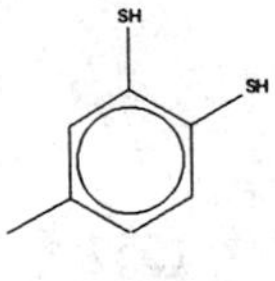

1565

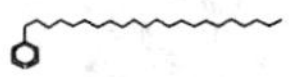

1566

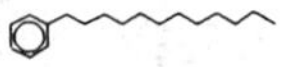

1567

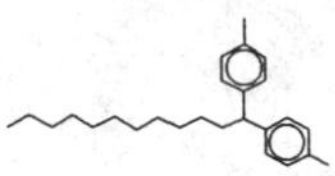

1568

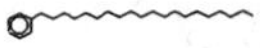

1569

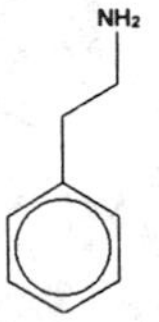

1570

1571

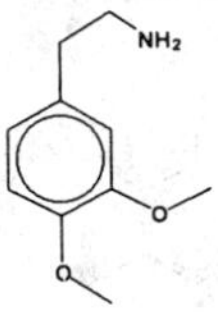

1572

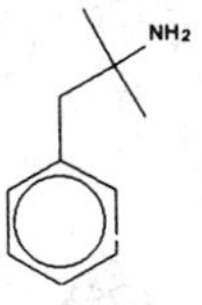

1573

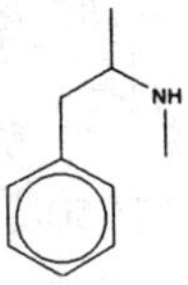

1574

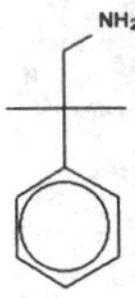

1575

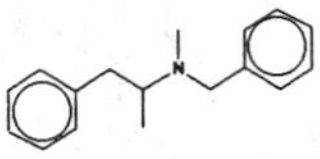

1576

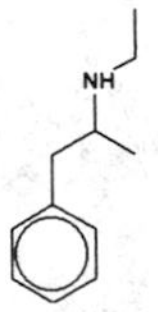

1577

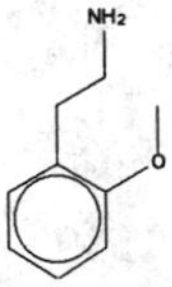

1578

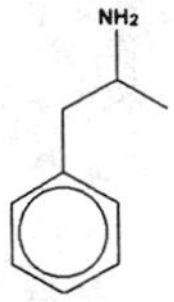

1579

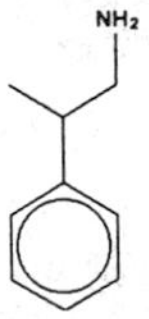

1580

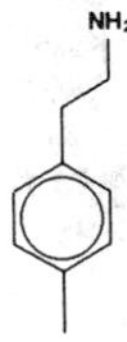

1581

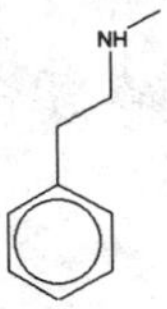

1583

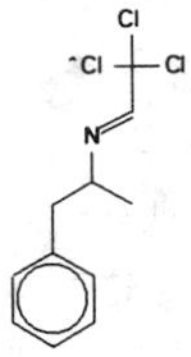

1584

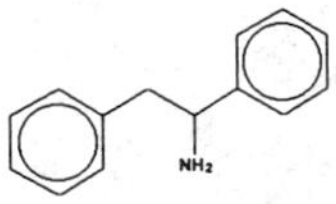

1585

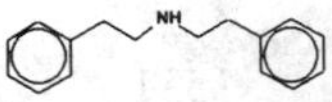

1586

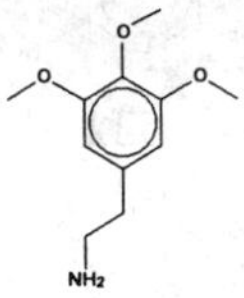

1587

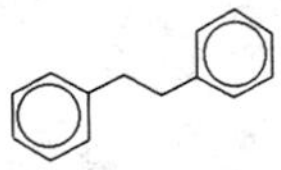

1588

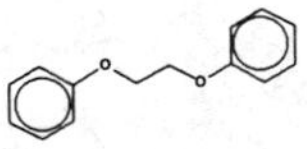

1589

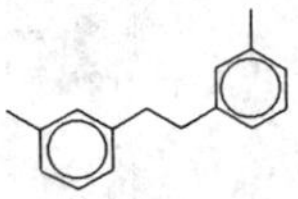

1590

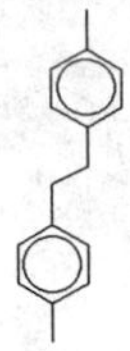

1591

1592

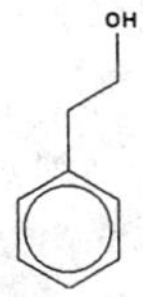

1593

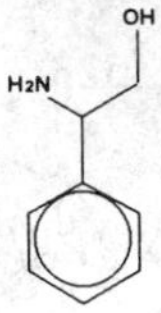

1594

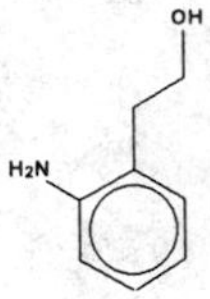

1595

1596

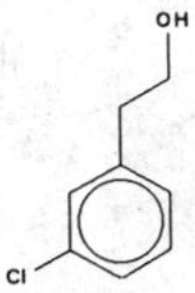

In numeric order as they occur in the Table of Physical Constants of Organic Compounds

1597

OH

Cl

1598

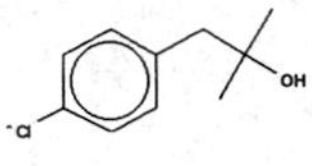

Cl

OH

1599

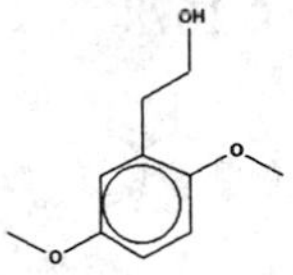

OH

O

O

1600

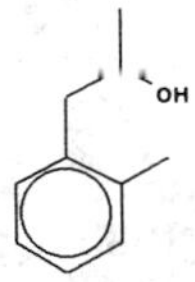

OH

1601

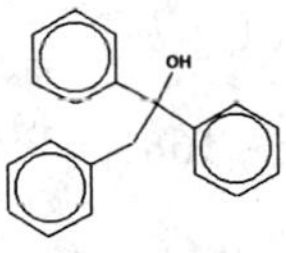

OH

1602

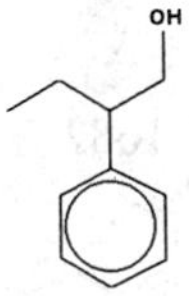

OH

1603

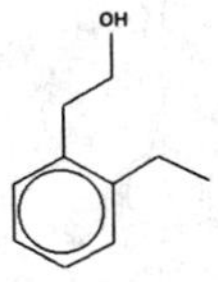

OH

1604

OH

1606

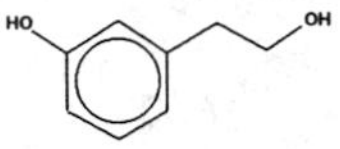

HO

OH

1607

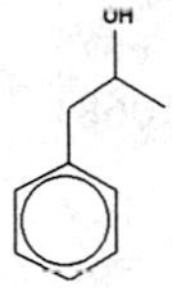

OH

1608

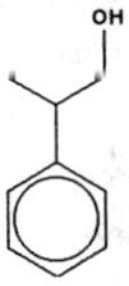

OH

1609

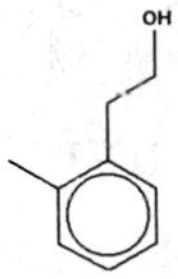

OH

1610

OH

1612

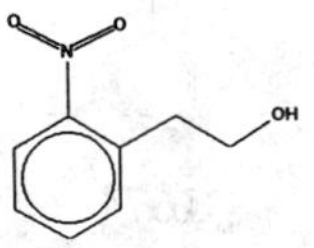

O

N

O

OH

1613

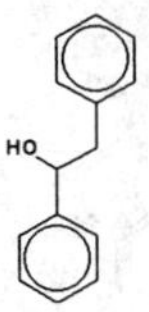

HO

1614

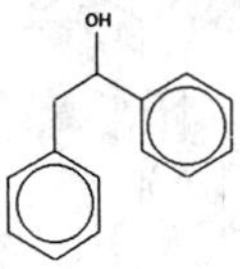

OH

1615

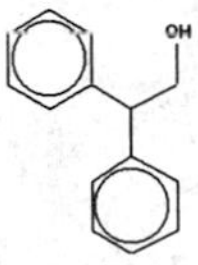

OH

1616

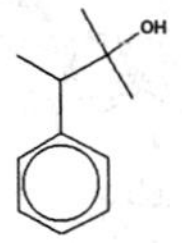

OH

1617

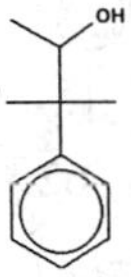

OH

1618

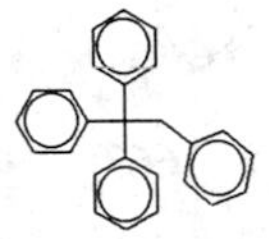

1619

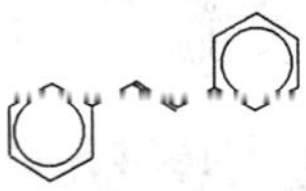

1620

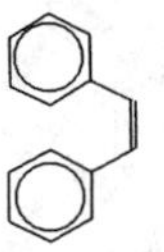

1621

O

O

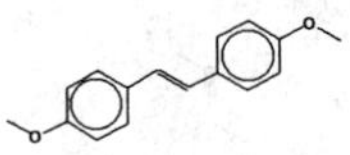

1622

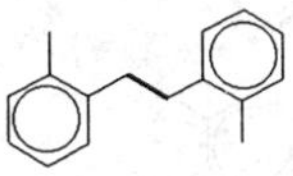

1623

1624

1626

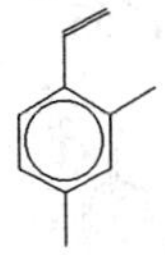

1627

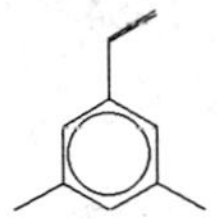

1628

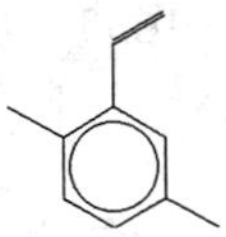

1629

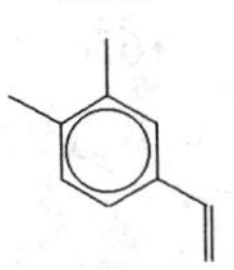

1630

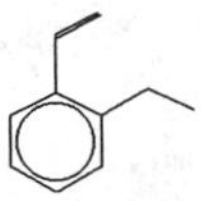

1631

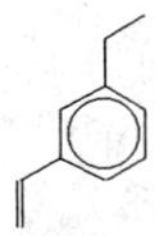

1632

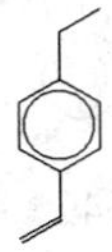

1633

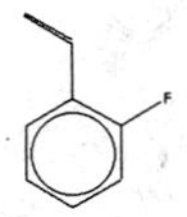

F

1634

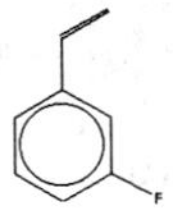

F

1635

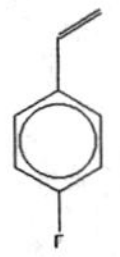

F

1636

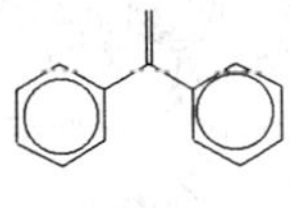

1637

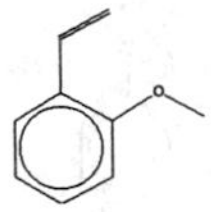

O

1638

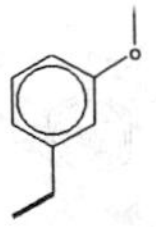

O

1639

O

STRUCTURAL FORMULAS OF ORGANIC COMPOUNDS (continued)

In numeric order as they occur in the Table of Physical Constants of Organic Compounds

1640

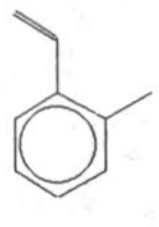

1641

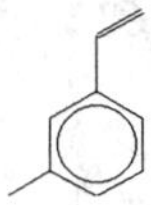

1642

1643

1644

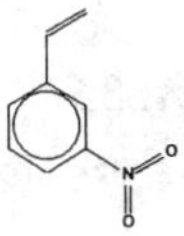

1645

1646

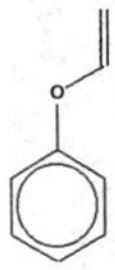

1647

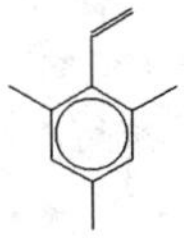

1648

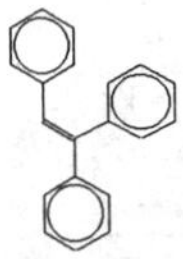

1649

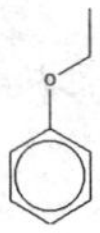

1650

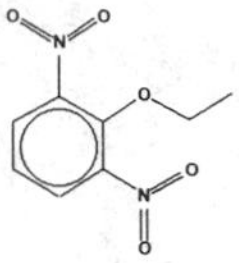

1651

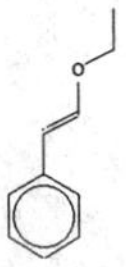

1652

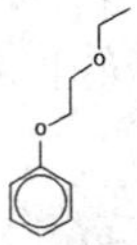

1653

1654

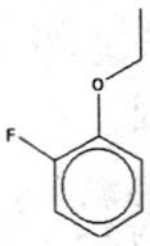

1655

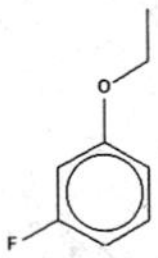

1656

1657

1658

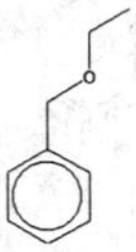

1659

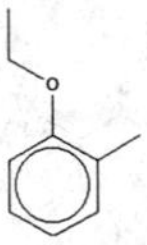

1660

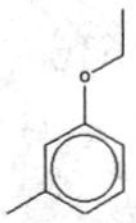

1661

1662

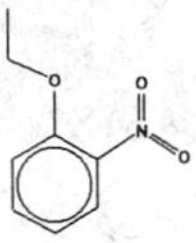

1663

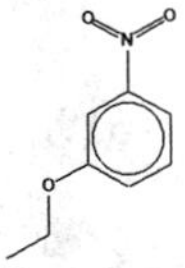

1664

1665

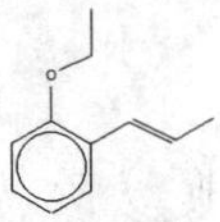

1666

1667

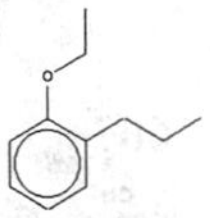

1668

1669

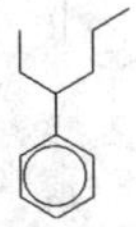

1670

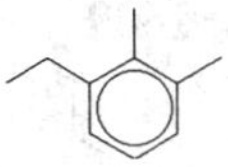

1671

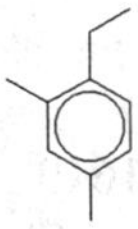

1672

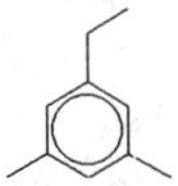

1673

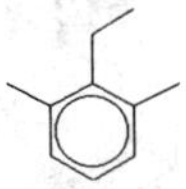

1674

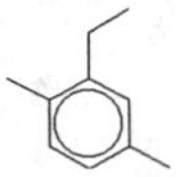

1675

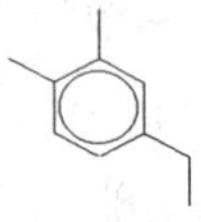

1676

1677

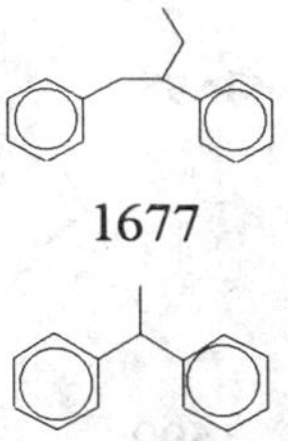

1678

1679

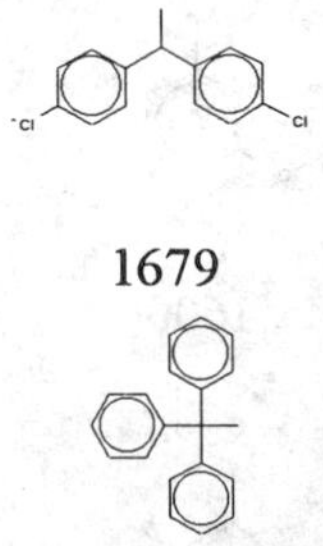

1680

1681

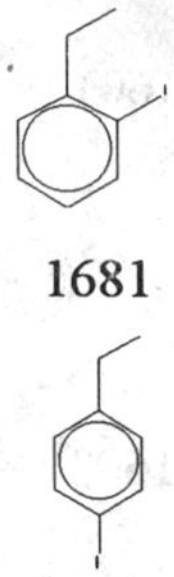

1682

1683

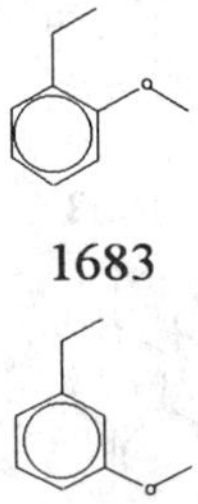

STRUCTURAL FORMULAS OF ORGANIC COMPOUNDS (continued)

In numeric order as they occur in the Table of Physical Constants of Organic Compounds

1684

1685

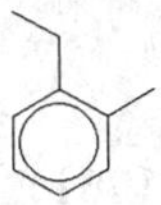

1686

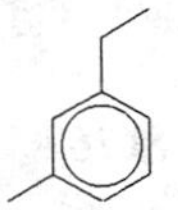

1687

1689

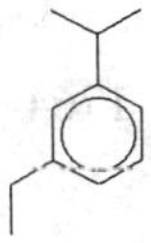

1690

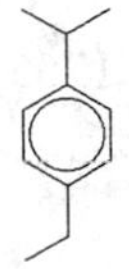

1691

1692

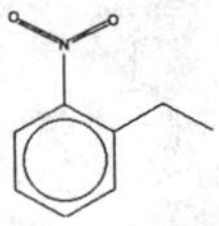

1693

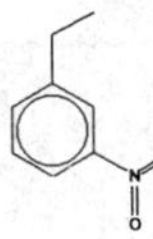

1694

1695

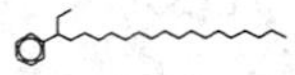

1696

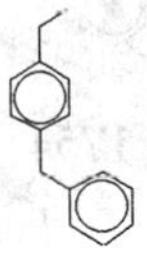

1697

1698

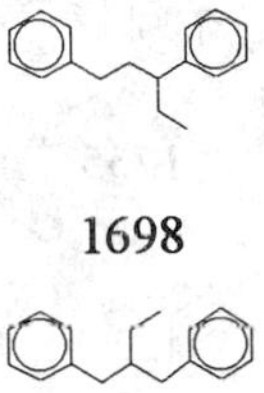

1699

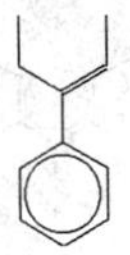

1700

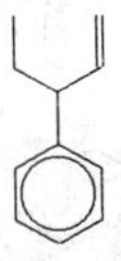

1701

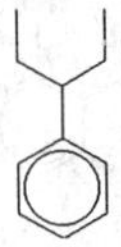

1702

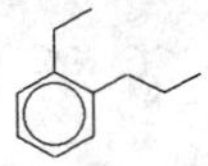

1703

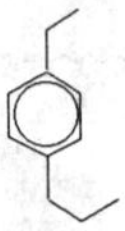

1704

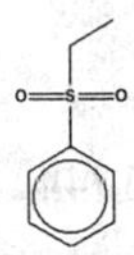

1705

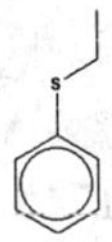

1706

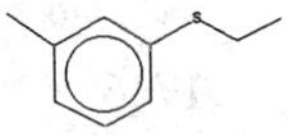

1707

1708

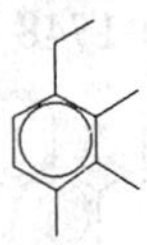

1709

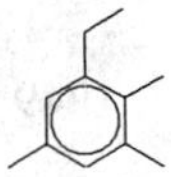

1710

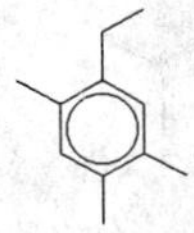

1711

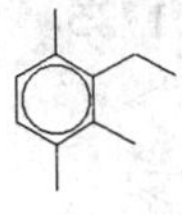

1712

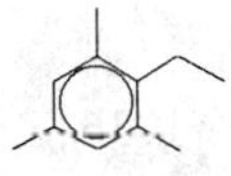

1713

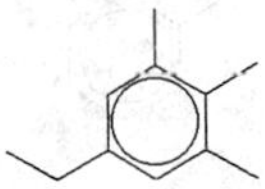

1714

1715

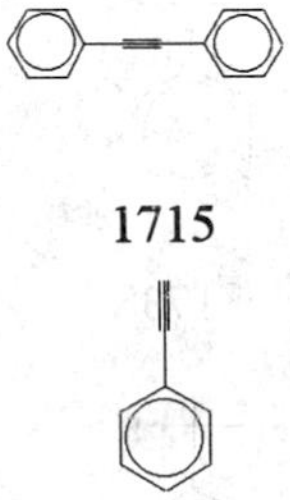

1716

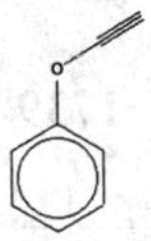

1717

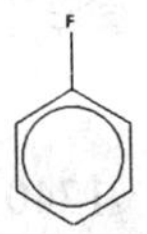

1718

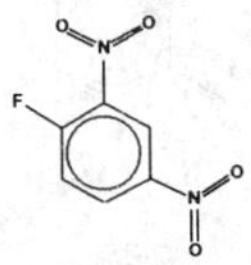

1719

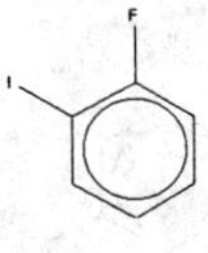

1720

1721

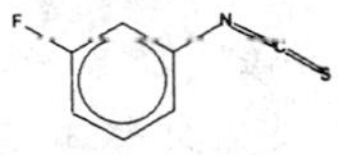

1722

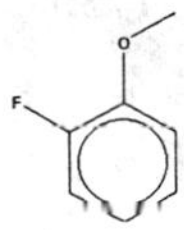

1723

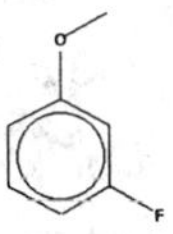

1724

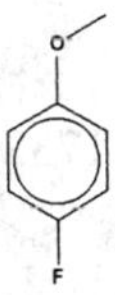

1725

1726

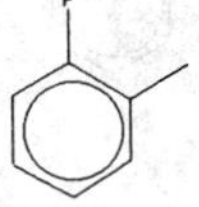

1727

F

1728

F

1729

F N O O

1730

F N O O

1731

O N O F

1732

O N O F

1733

O N O F

1734

O N O F

1735

O N O F

1736

Cl Cl Cl F

1737

F F F F

1738

F F F F

1739

F F F F

1740

F

1741

1742

1743

1744

1745

1746

1747

O

1748

HO O O OH HO O O OH O OH OH O OH

1749

Cl Cl Cl Cl Cl Cl

1750

1751

1752

O

1753

1754

1755

1756

1757

F F F F F F

1758

1759

OH O

1760

1761

1762

1763

1764

1765

O

1766

I

1767

I O O

1768

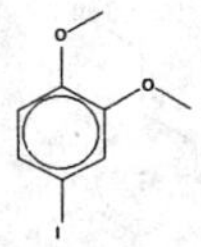

1769

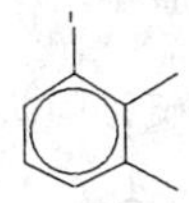

1770

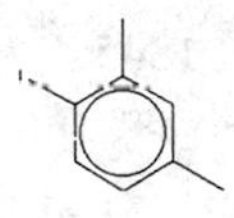

1771

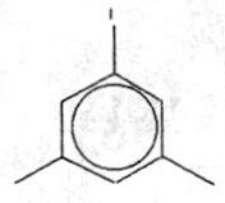

1772

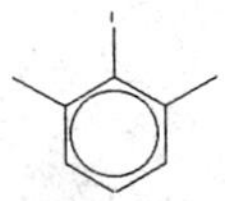

1773

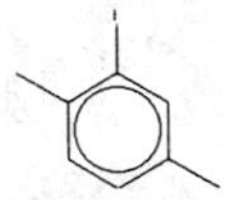

1774

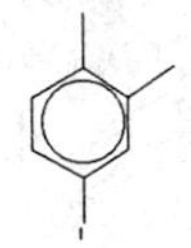

1775

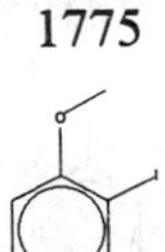

1776

1777

1778

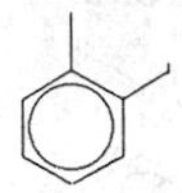

1779

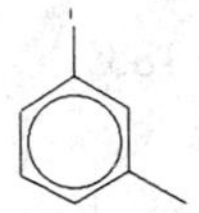

1780

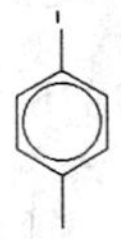

1781

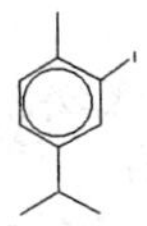

1782

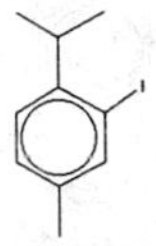

1783

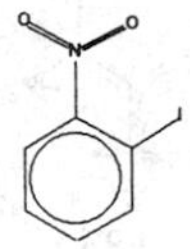

1784

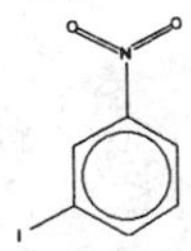

1785

1786

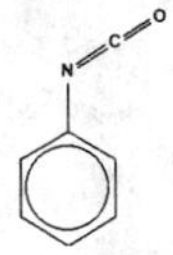

1787

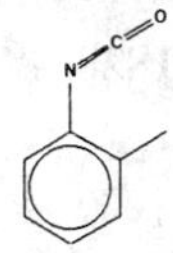

1788

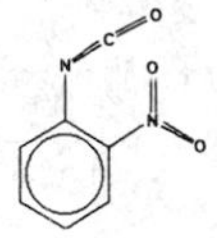

1789

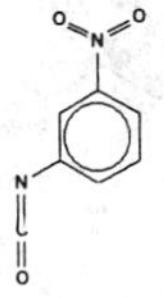

1790

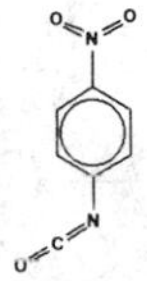

1791

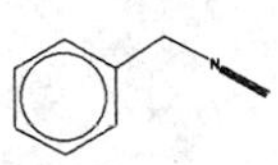

1792

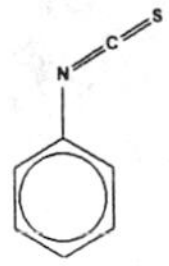

1793

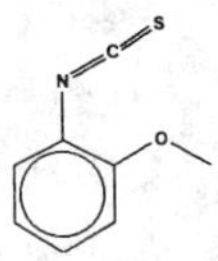

1794

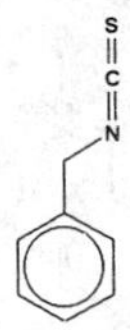

1795

1796

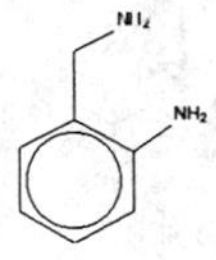

1797

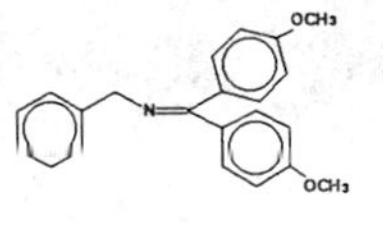

1798

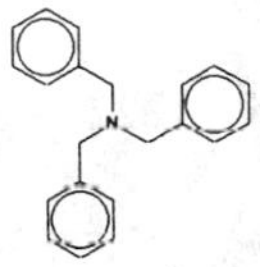

1799

1800

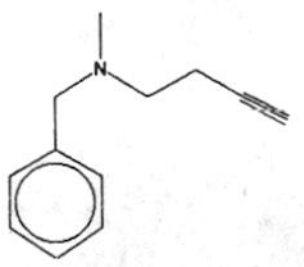

1801

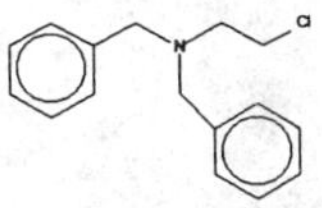

1802

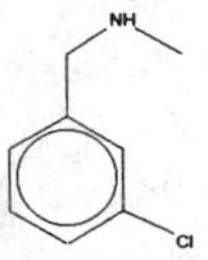

1803

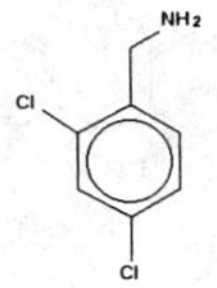

1804

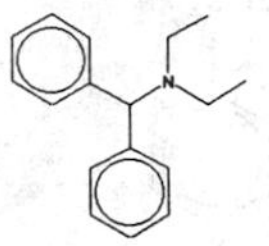

1805

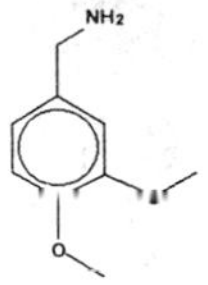

1806

1807

1808

1809

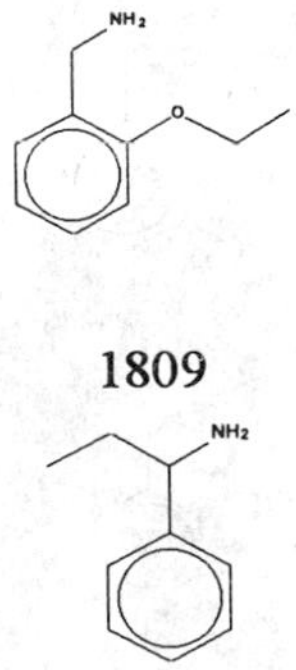

STRUCTURAL FORMULAS OF ORGANIC COMPOUNDS (continued)

In numeric order as they occur in the Table of Physical Constants of Organic Compounds

1810

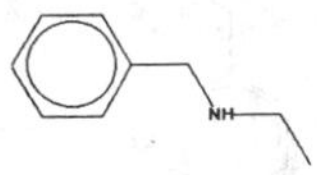

NH

1811

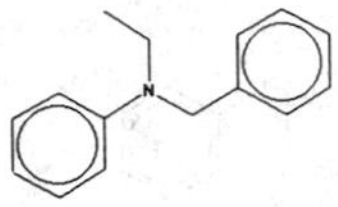

N

1812

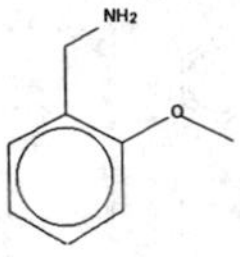

NH_2

O

1813

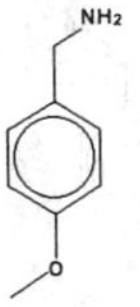

NH_2

O

1814

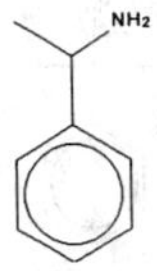

NH_2

1815

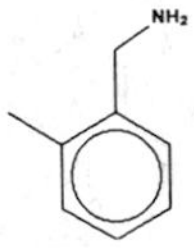

NH_2

1816

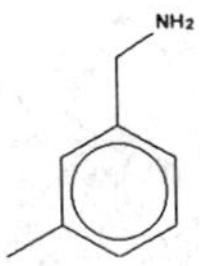

NH_2

1817

NH_2

1818

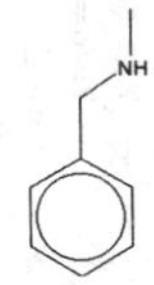

NH

1819

NH_2

1820

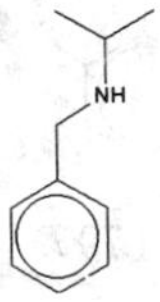

NH

1821

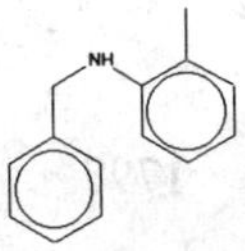

NH

1822

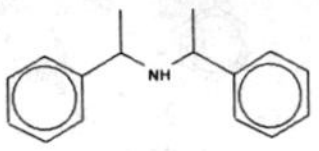

NH

1823

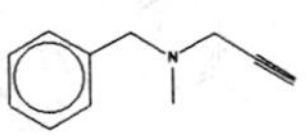

N

1826

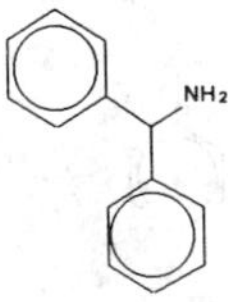

NH_2

1827

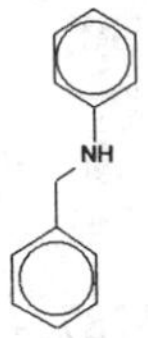

NH

1828

NH

1829

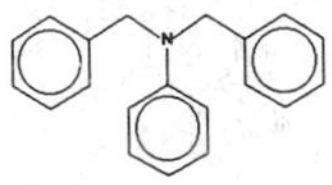

N

1830

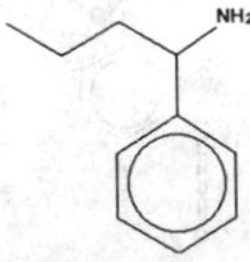

NH_2

1831

1832

SH

1833

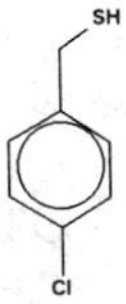

SH

Cl

1834

SH

1835

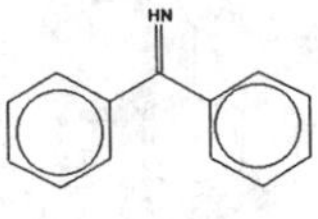

HN

1836

OH

1837

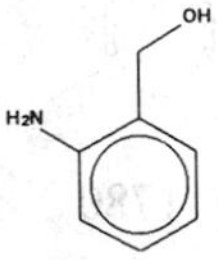

OH

H_2N

1838

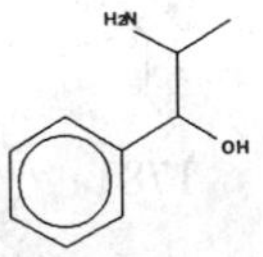

H_2N

OH

1839

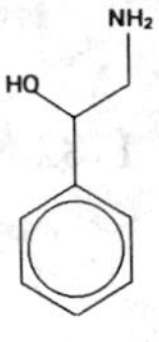

NH_2

HO

1840

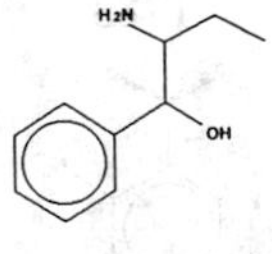

H_2N

OH

1841

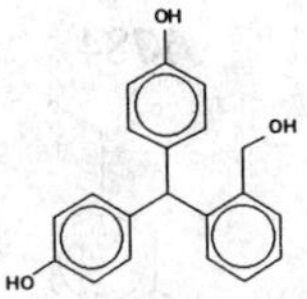

OH

OH

HO

1842

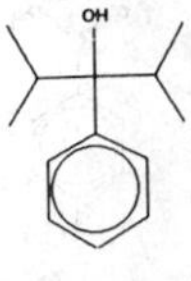

OH

1843

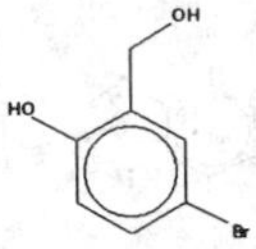

OH

HO

Br

1844

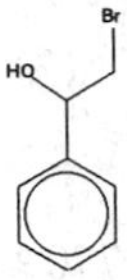

Br

HO

1845

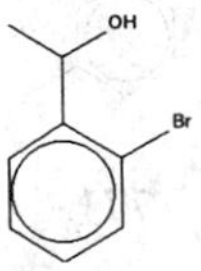

OH

Br

1846

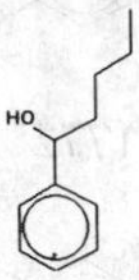

HO

1847

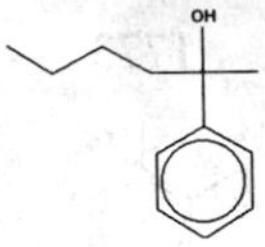

OH

1848

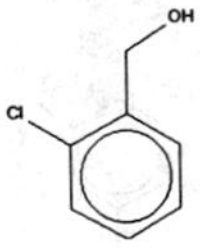

OH

Cl

1849

OH

Cl

STRUCTURAL FORMULAS OF ORGANIC COMPOUNDS (continued)

In numeric order as they occur in the Table of Physical Constants of Organic Compounds

1850

1851

1852

1853

1854

1855

1856

1857

1858

1859

1860

1861

1862

1863

1864

1865

1866

1867

1868

1869

1870

1871

1872

1873

1874

1875

1876

1877

1878

1879

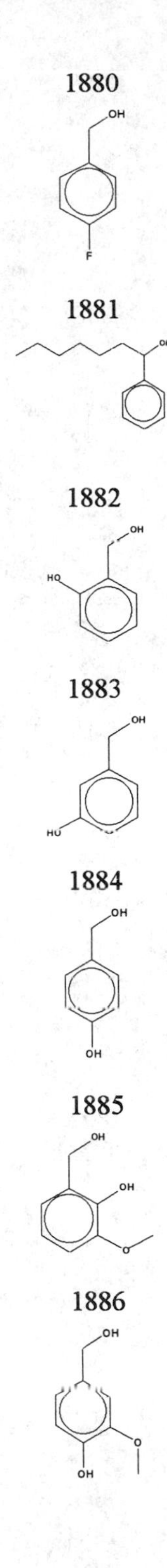

STRUCTURAL FORMULAS OF ORGANIC COMPOUNDS (continued)

In numeric order as they occur in the Table of Physical Constants of Organic Compounds

1887

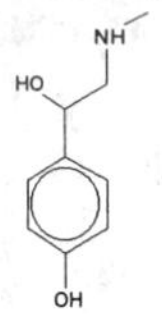

1888

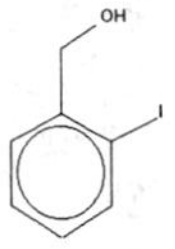

1889

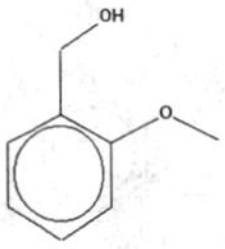

1890

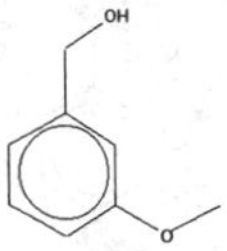

1891

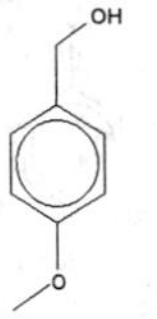

1892

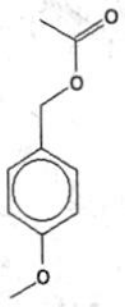

1893

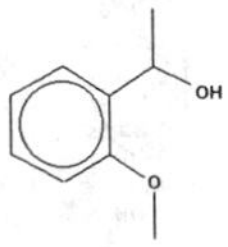

1894

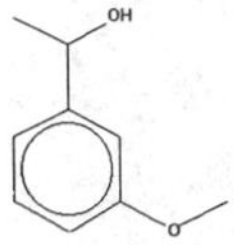

1895

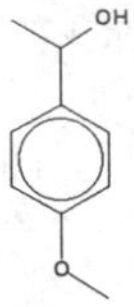

1896

1897

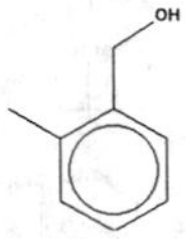

1898

1899

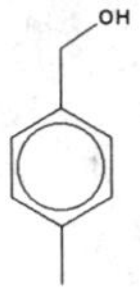

1900

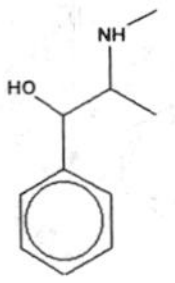

1901

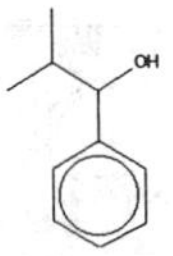

1902

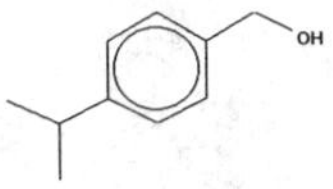

1903

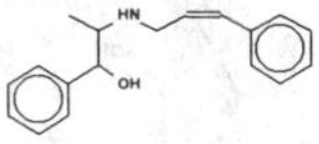

1904

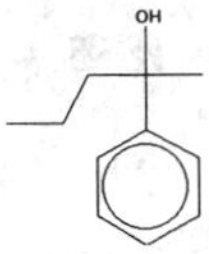

1905

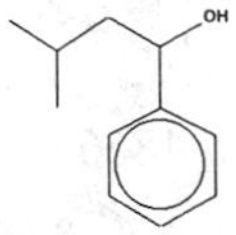

1906

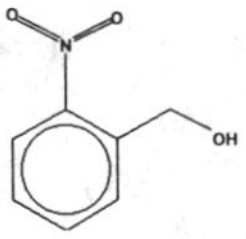

1907

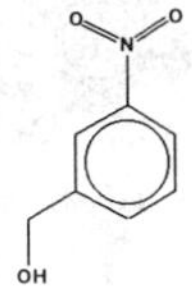

1908

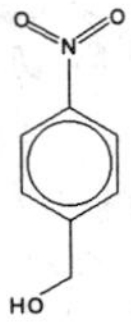

1909

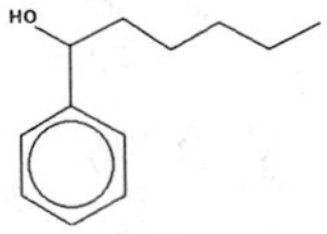

1910

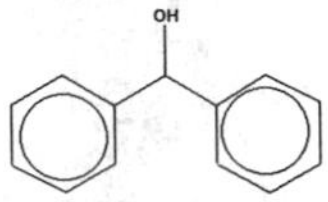

1911

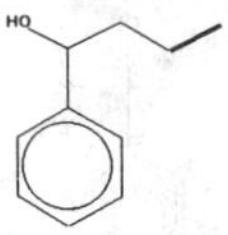

1912

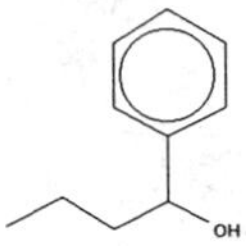

1913

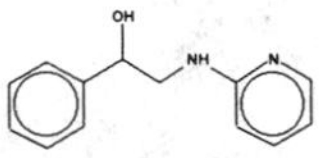

1914

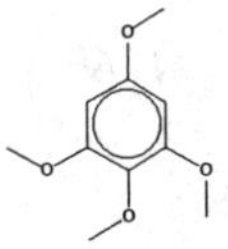

1915

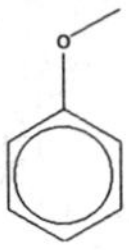

1916

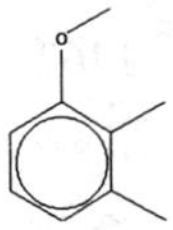

1917

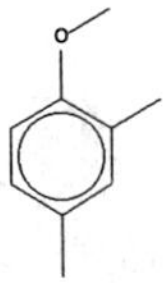

1918

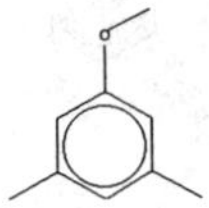

1919

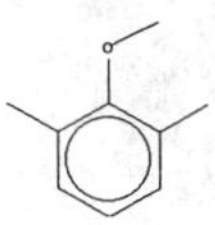

1920

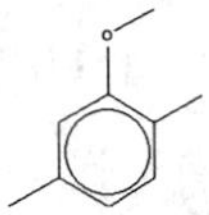

1921

1923

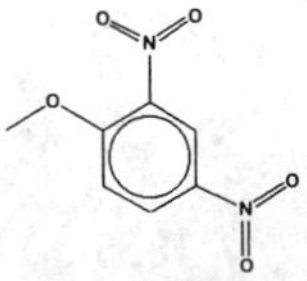

1924

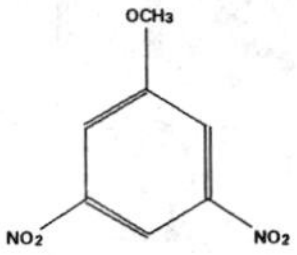

1925

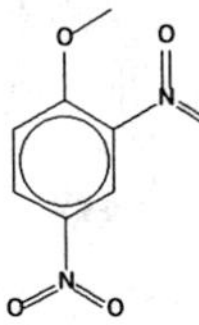

1926

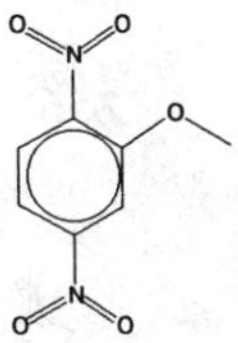

STRUCTURAL FORMULAS OF ORGANIC COMPOUNDS (continued)

In numeric order as they occur in the Table of Physical Constants of Organic Compounds

1927

NO2 NO2 OCH3

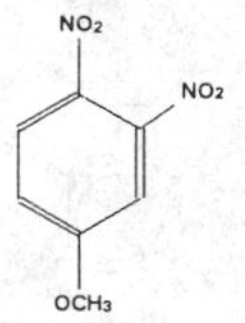

1928

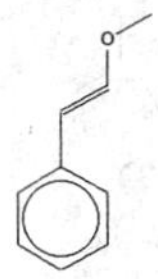

1929

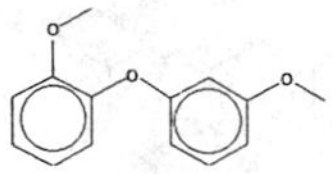

1930

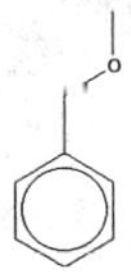

1931

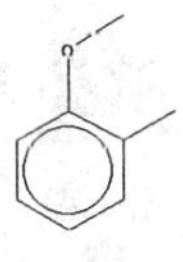

1932

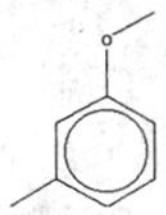

1933

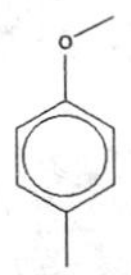

1934

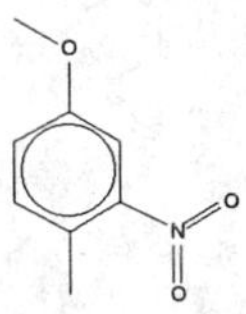

1935

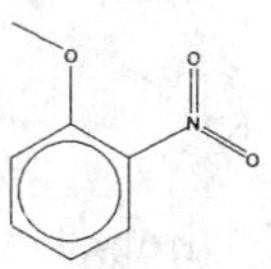

1936

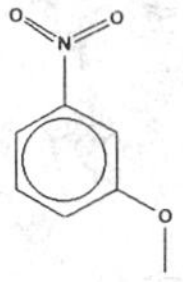

1937

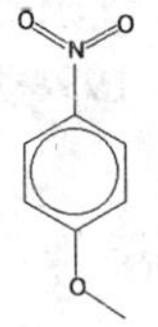

1938

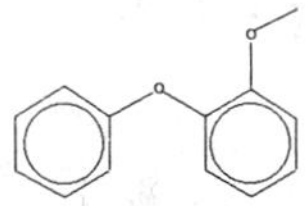

1939

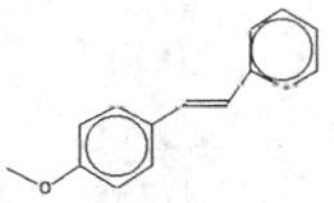

1940

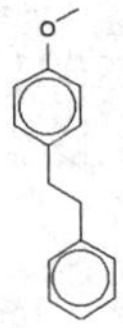

1941

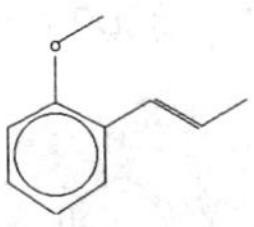

1942

1943

1944

1945

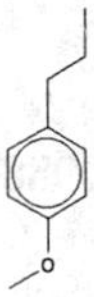

1946

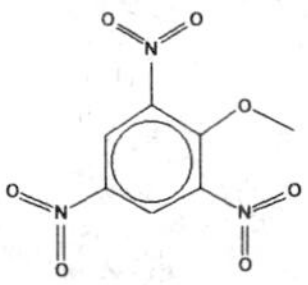

1947

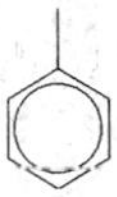

1948

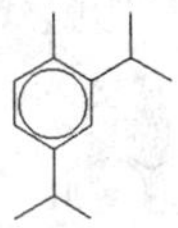

1949

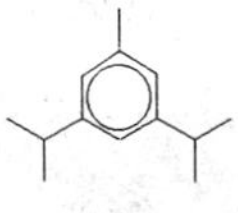

1950

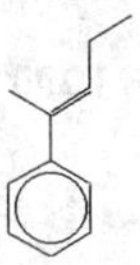

1951

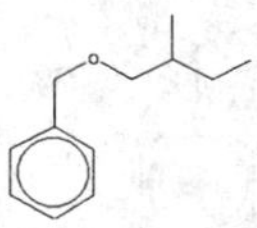

1952

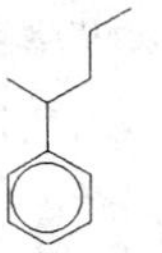

1953

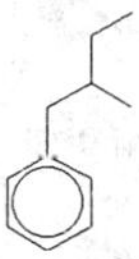

1954

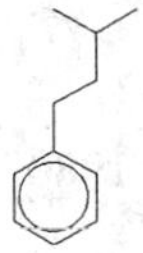

1955

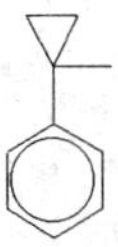

1956

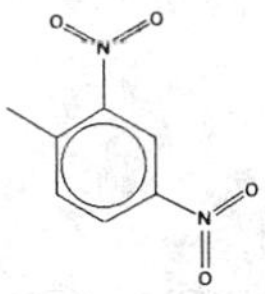

1957

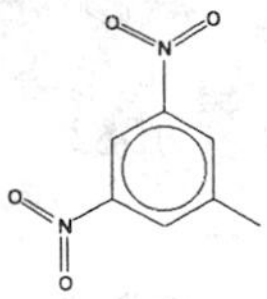

1958

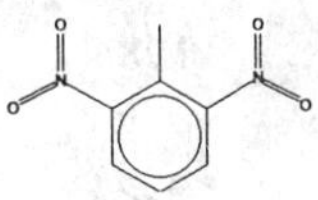

1959

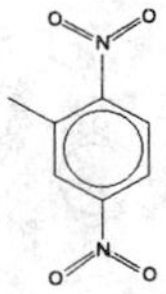

1960

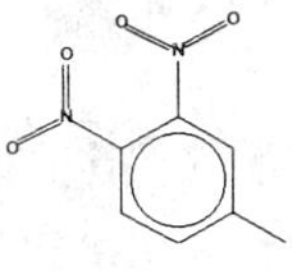

1961

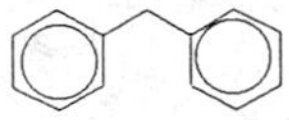

1962

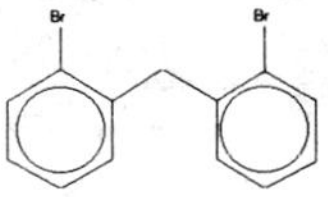

Br Br

1963

Cl Cl

1964

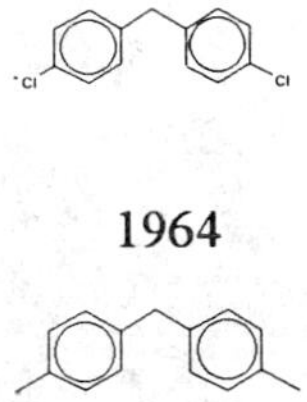

1965

Cl Cl

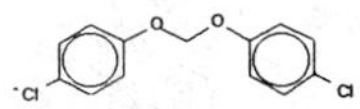

1966

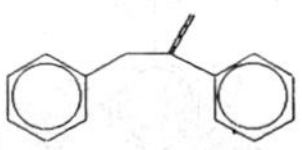

1967

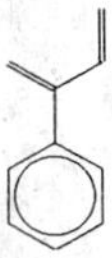

STRUCTURAL FORMULAS OF ORGANIC COMPOUNDS (continued)

In numeric order as they occur in the Table of Physical Constants of Organic Compounds

1968

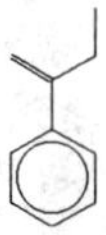

1969

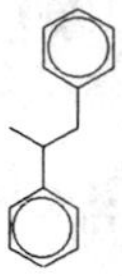

1970

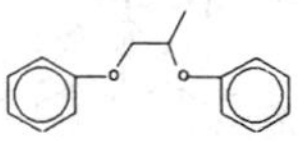

O O

1971

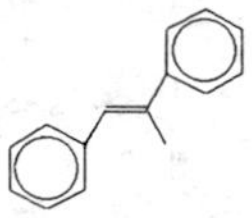

1972

1973

1974

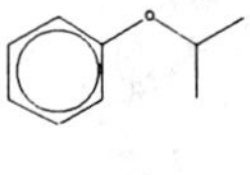

O

1975

1976

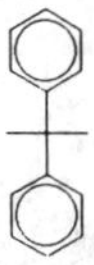

1977

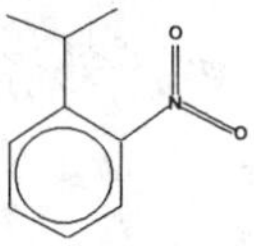

O N O

1978

O N O

1979

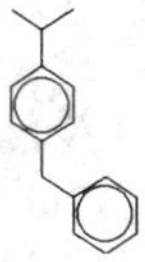

1980

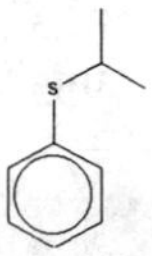

S

1981

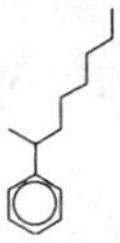

1982

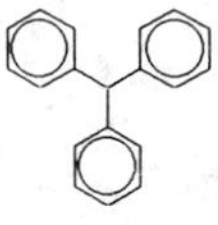

1983

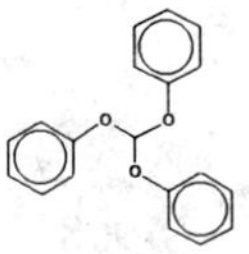

O O O

1984

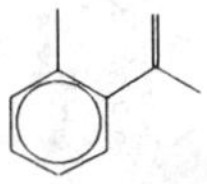

1985

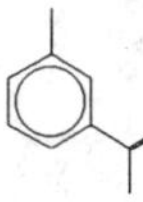

1986

1987

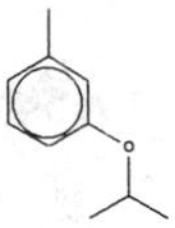

O

1988

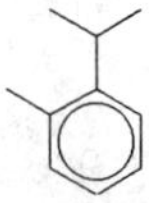

1989

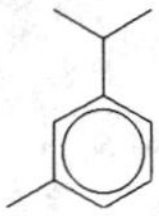

1990

1991

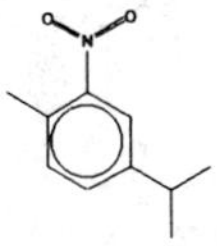

O N O

1992

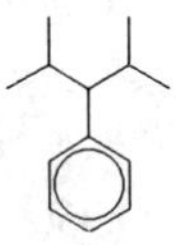

1993

S

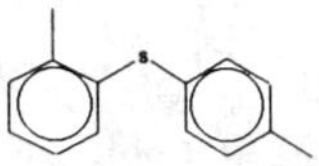

1995

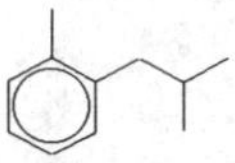

1996

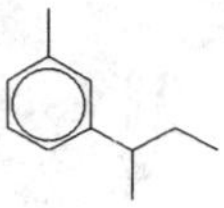

1997

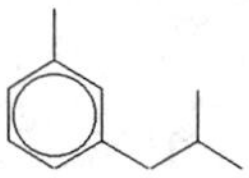

1998

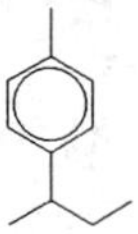

1999

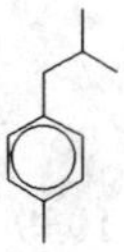

2000

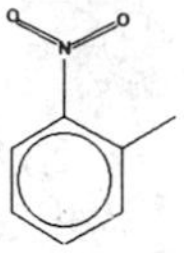

O N O

2001

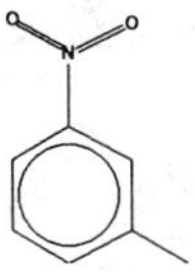

O N O

2002

O N O

2003

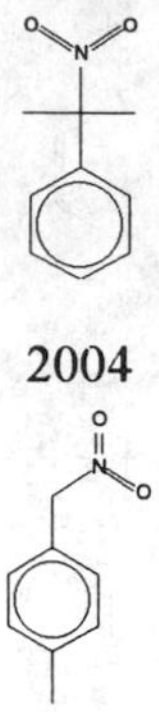

O N O

2004

O N O

2005

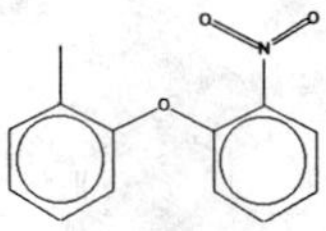

O N O
O

2006

O
N O O

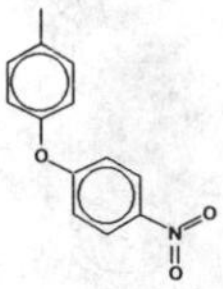

2007

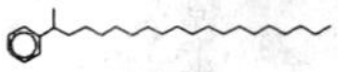

2009

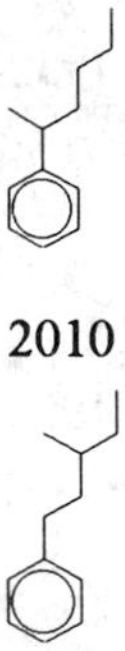

2010

2011

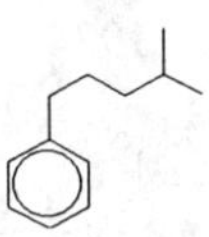

STRUCTURAL FORMULAS OF ORGANIC COMPOUNDS (continued)

In numeric order as they occur in the Table of Physical Constants of Organic Compounds

2012

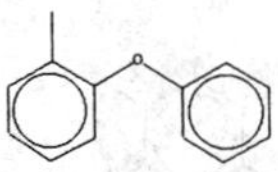

2013

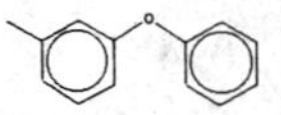

2014

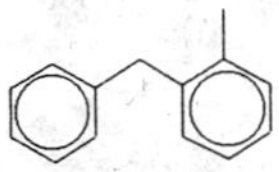

2015

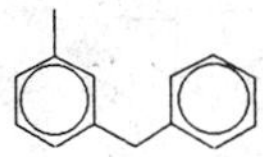

2016

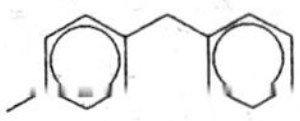

2017

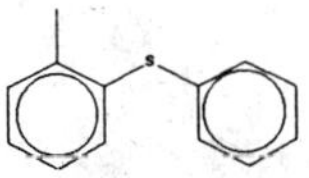

2018

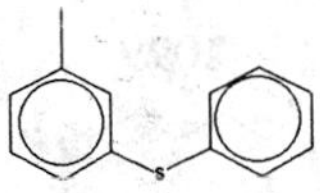

2019

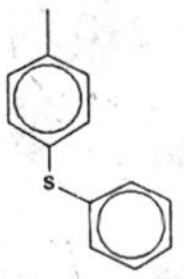

2021

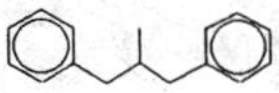

2022

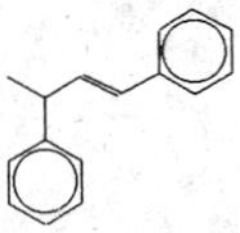

2023

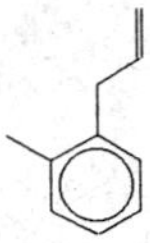

2024

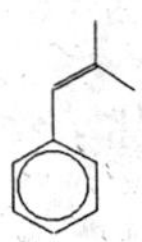

2025

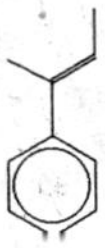

2026

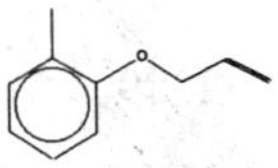

2027

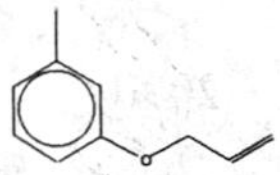

2028

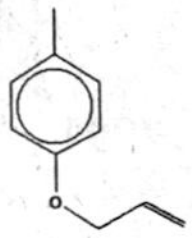

2029

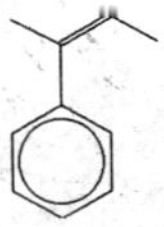

2030

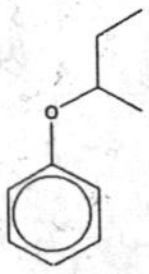

2031

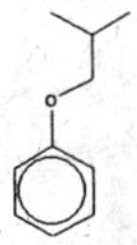

2033

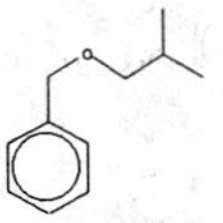

2034

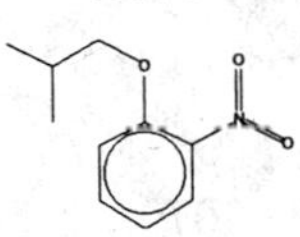

2035

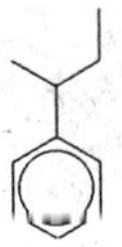

2036

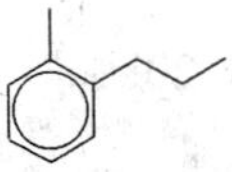

2037

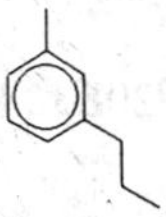

2038

2039

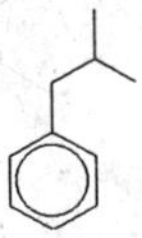

2040

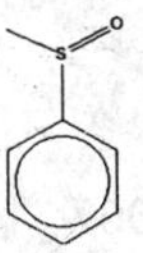

2041

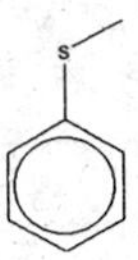

2042

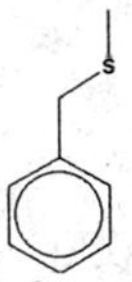

2044

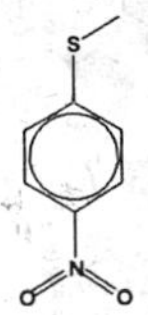

2045

2047

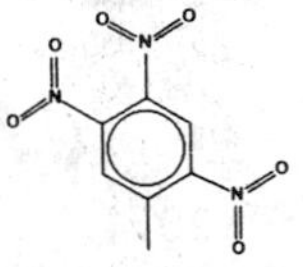

2048

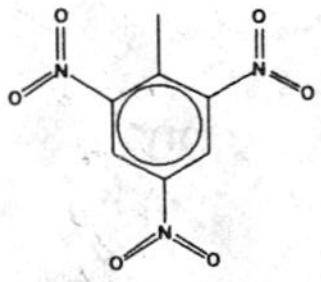

2049

2050

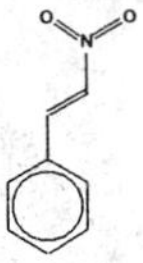

2051

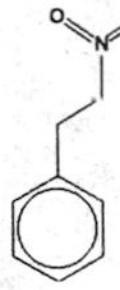

2052

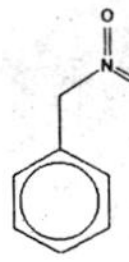

2053

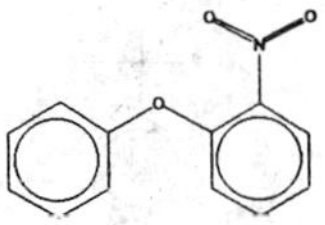

2054

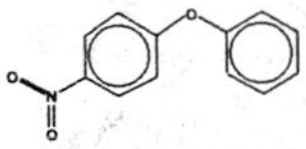

2055

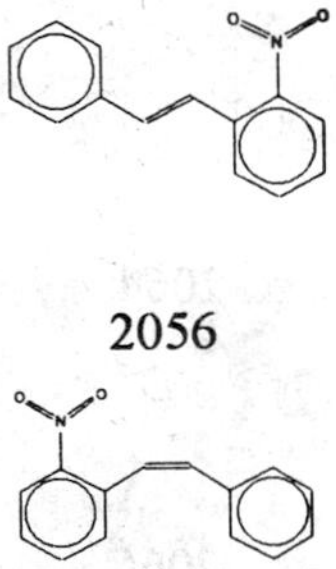

2056

STRUCTURAL FORMULAS OF ORGANIC COMPOUNDS (continued)

In numeric order as they occur in the Table of Physical Constants of Organic Compounds

2057

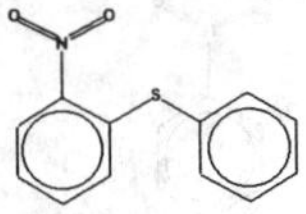

2058

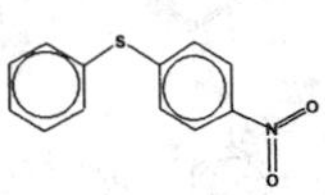

2059

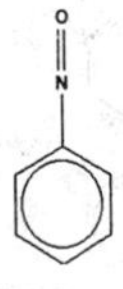

2060

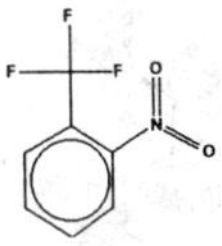

2061

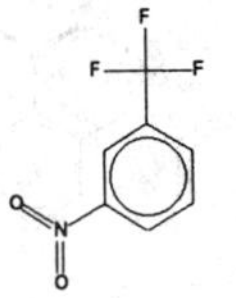

2062

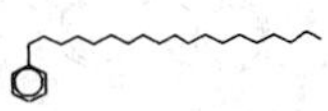

2063

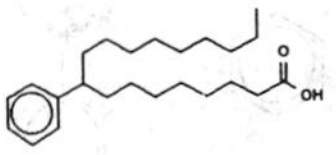

2064

2065

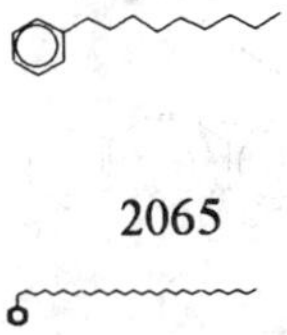

2066

2067

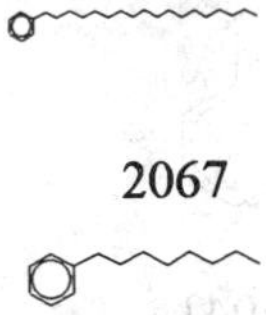

2068

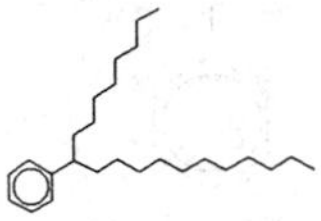

2069

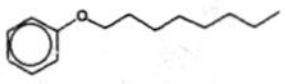

2070

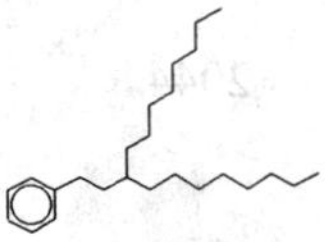

2071

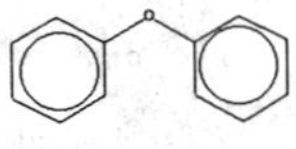

2072

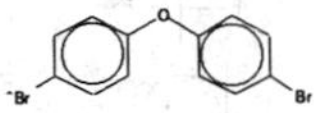

2073

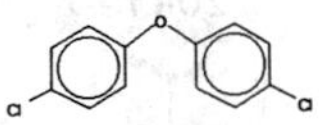

2074

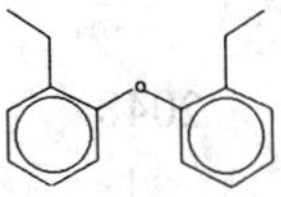

2075

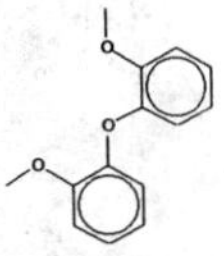

2076

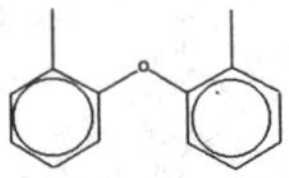

2077

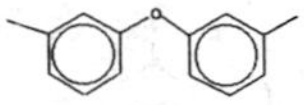

2078

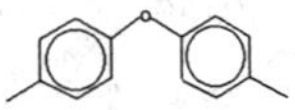

2079

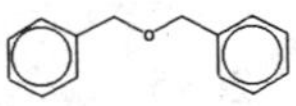

2080

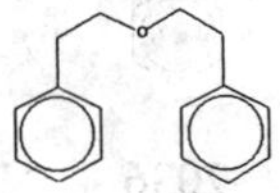

2081

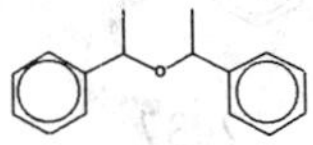

2083

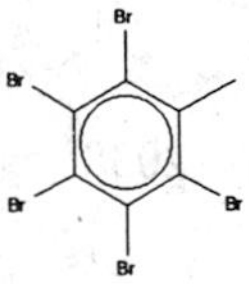

2084

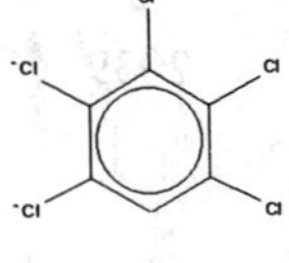

2085

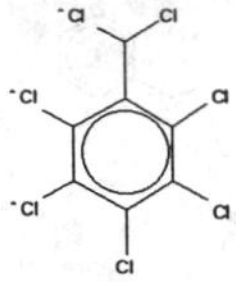

2086

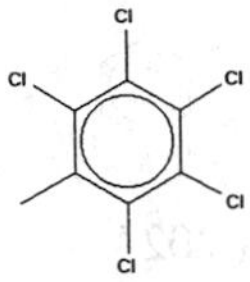

2087

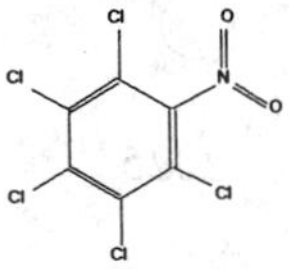

2088

2089

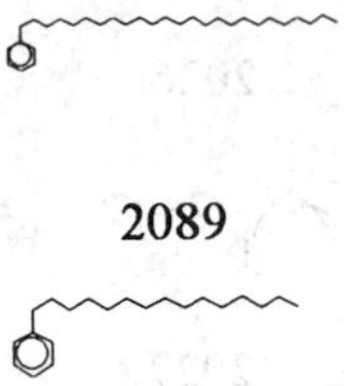

2090

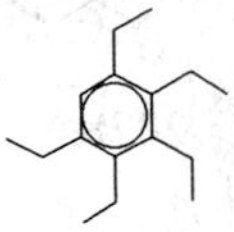

2091

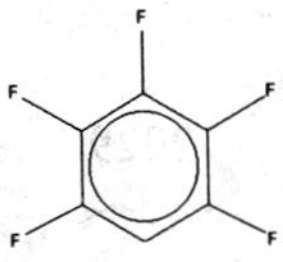

2092

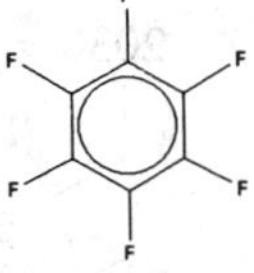

2093

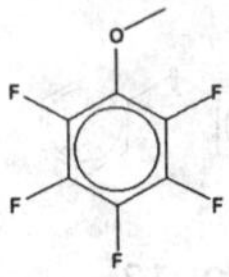

2094

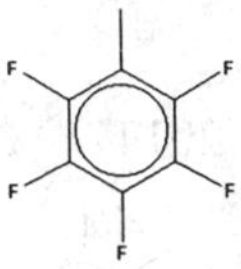

2095

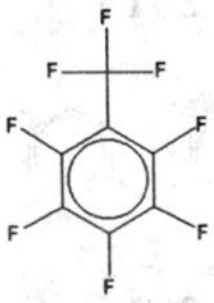

2097

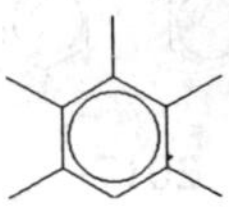

2098

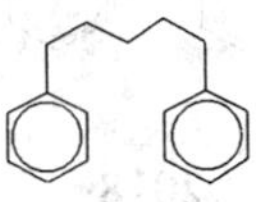

2099

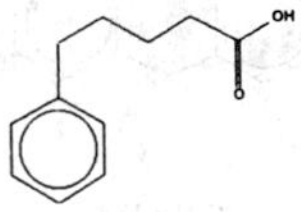

2100

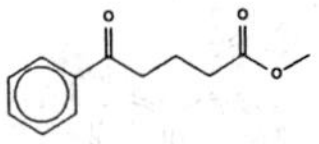

2101

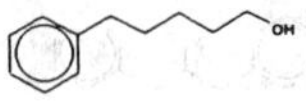

2102

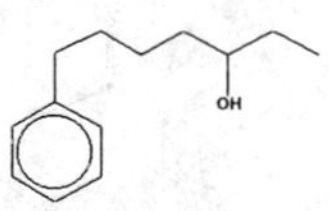

OH

2103

OH

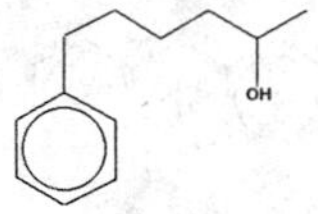

2104

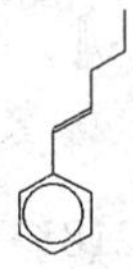

2105

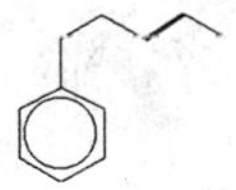

2106

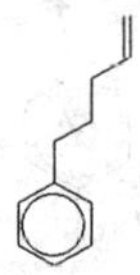

2107

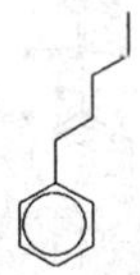

2108

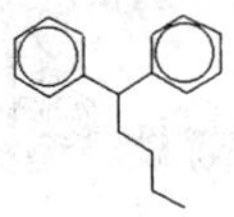

2110

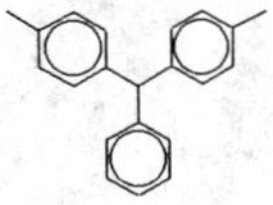

2111

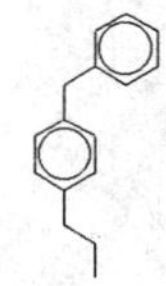

2113

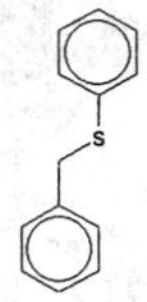

S

2114

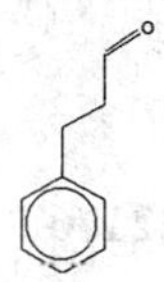

O

2116

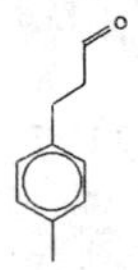

O

2117

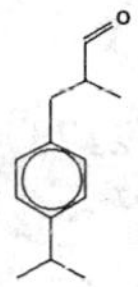

O

2118

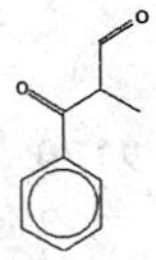

O

O

2119

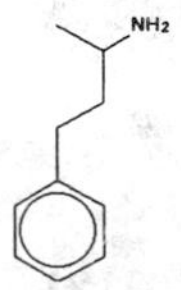

NH_2

2121

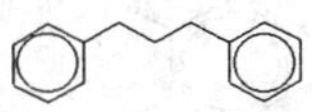

2122

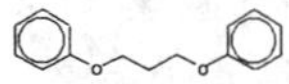

O O

2123

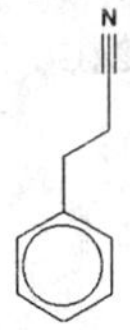

N

2124

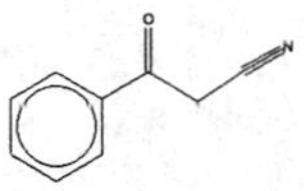

O

N

2125

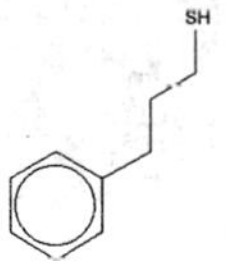

SH

2126

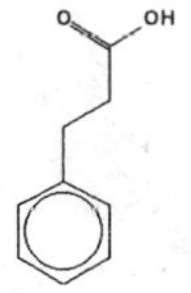

O OH

2127

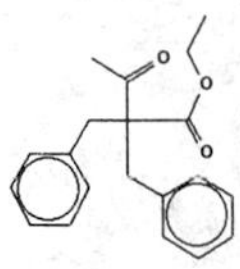

O

O

O

2129

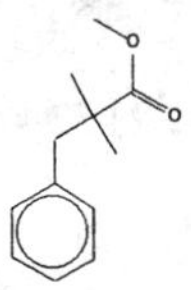

O

O

2130

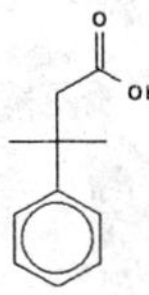

O

OH

2131

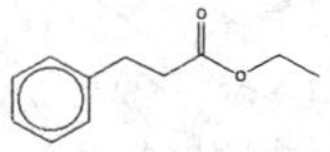

O

O

2132

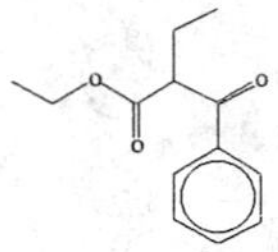

O

O O

2133

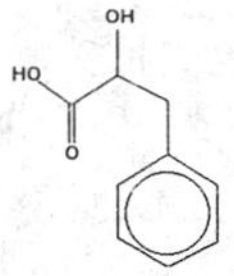

OH

HO

O

2134

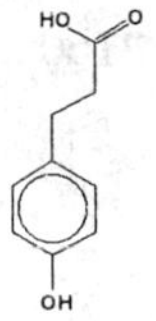

HO O

OH

2137

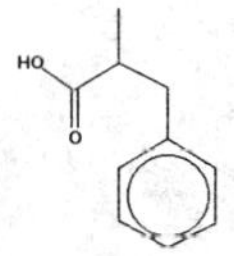

HO

O

2138

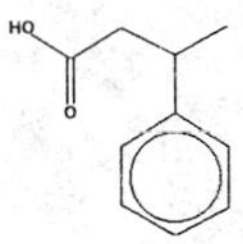

HO

O

2139

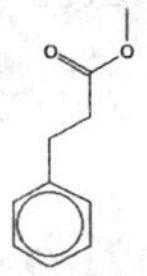

O

O

2140

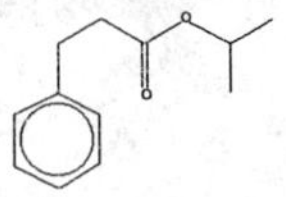

O

O

2141

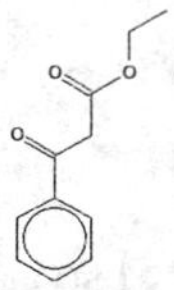

O

O

O

2142

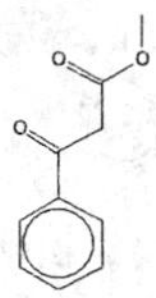

O

O

O

2143

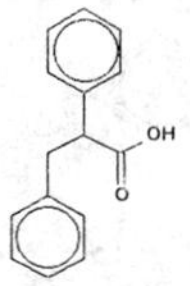

OH

O

2144

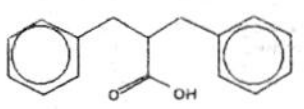

O OH

2145

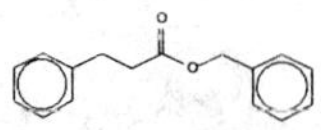

O

O

2146

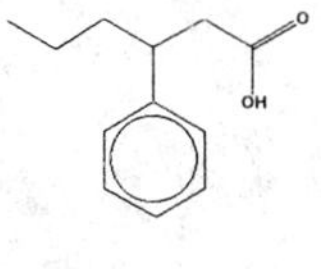

O

OH

2147

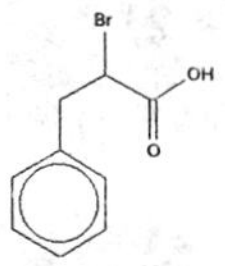

Br

OH

O

2148

OH

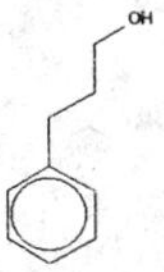

STRUCTURAL FORMULAS OF ORGANIC COMPOUNDS (continued)

In numeric order as they occur in the Table of Physical Constants of Organic Compounds

2149

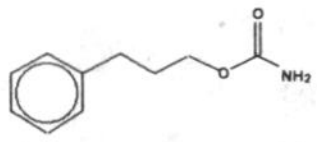

2150

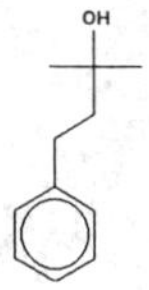

2151

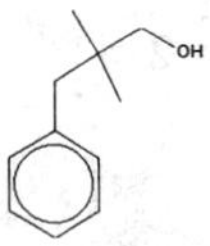

2152

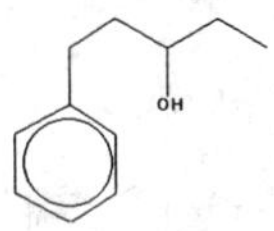

2153

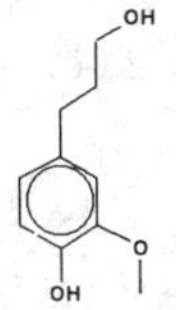

2154

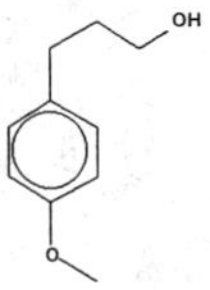

2155

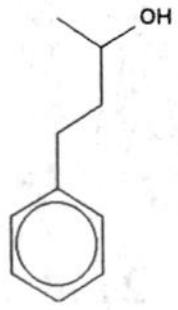

2156

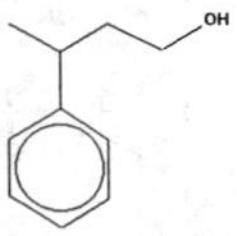

2157

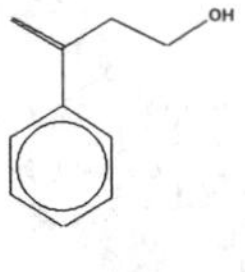

2158

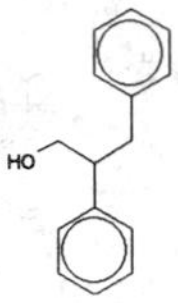

2159

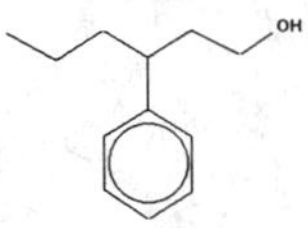

2160

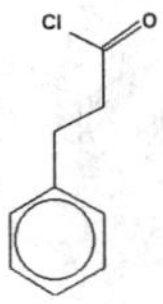

2161

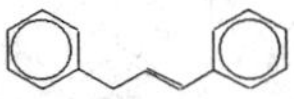

2162

2163

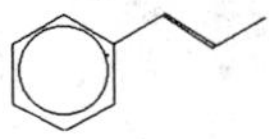

2164

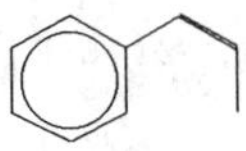

2165

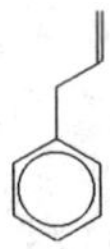

2166

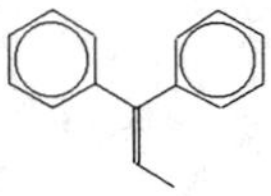

2167

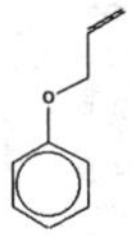

2168

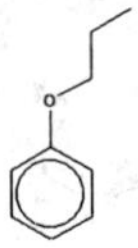

2169

2170

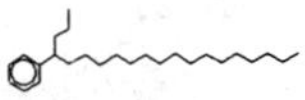

2171

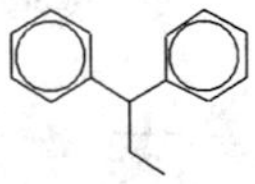

2172

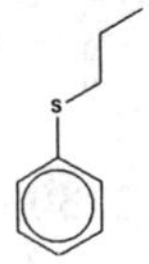

2173

2174

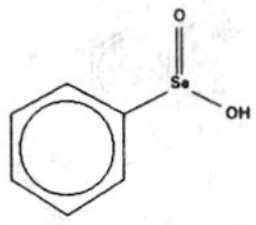

2175

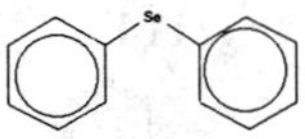

2176

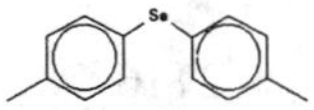

2177

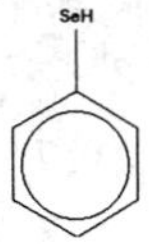

2178

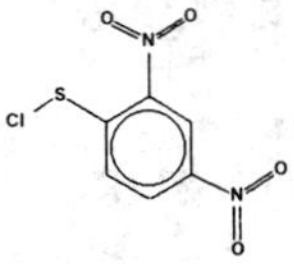

2179

2181

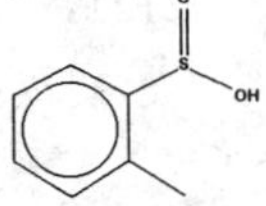

2182

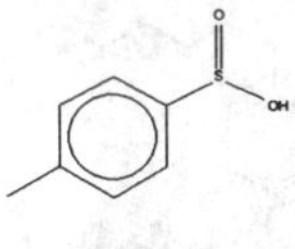

2183

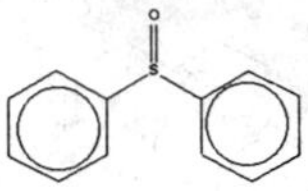

2184

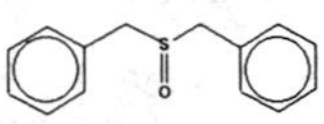

2185

2186

2187

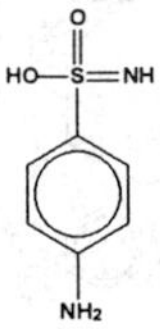

2188

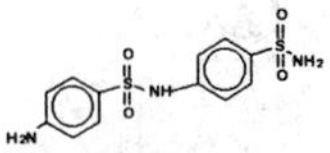

2189

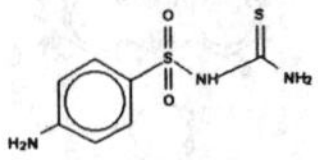

STRUCTURAL FORMULAS OF ORGANIC COMPOUNDS (continued)

In numeric order as they occur in the Table of Physical Constants of Organic Compounds

2190

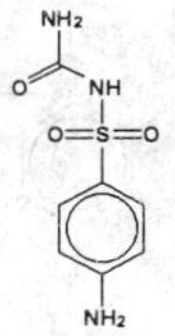

2191

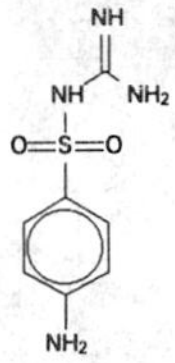

2192

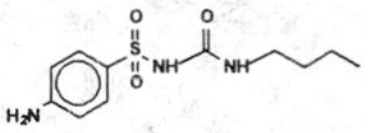

2193

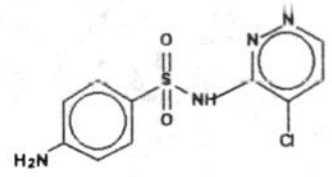

2194

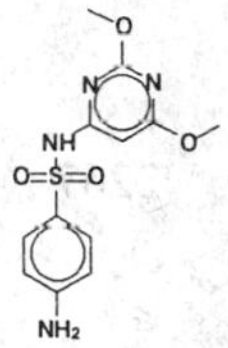

2195

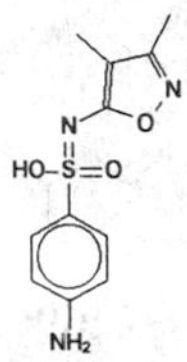

2196

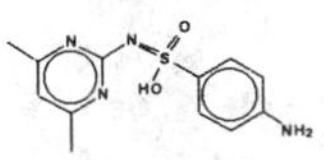

2197

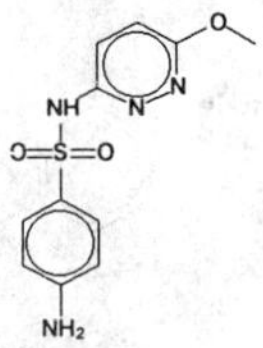

2198

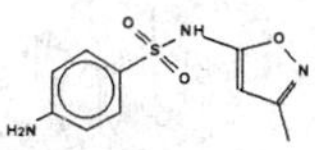

2199

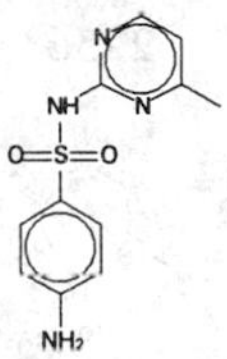

2200

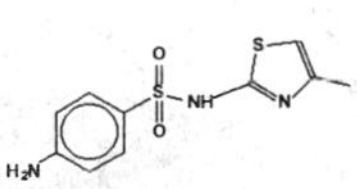

2202

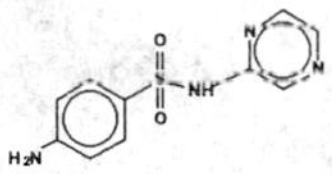

2203

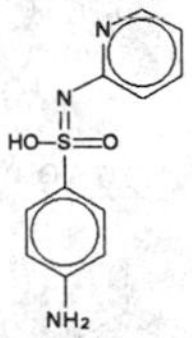

2204

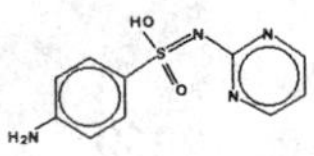

2205

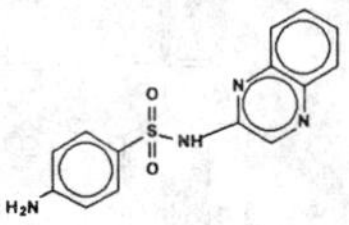

2206

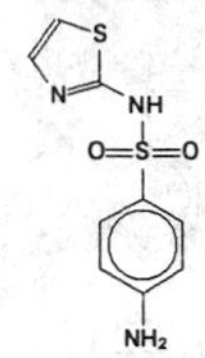

2207

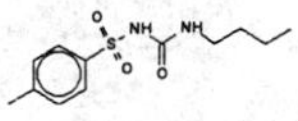

2208

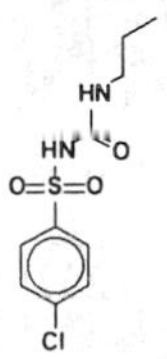

2210

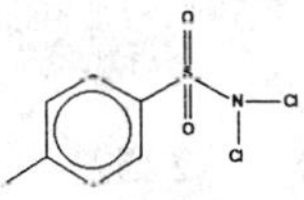

2211

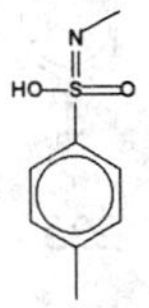

2212

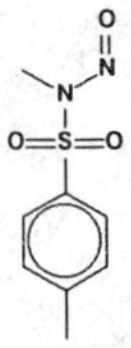

2213

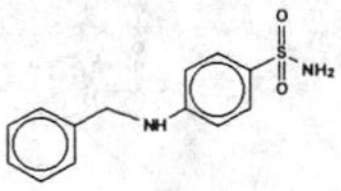

2214

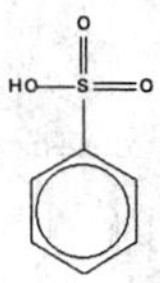

2215

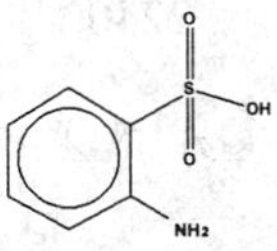

2216

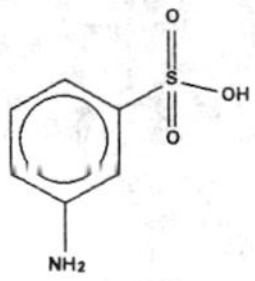

2217

2220

2221

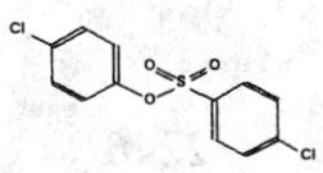

2223

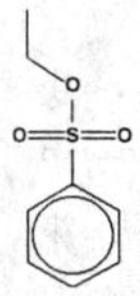

2224

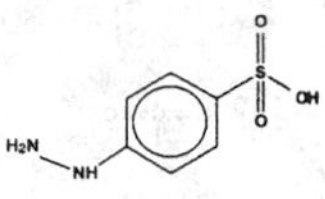

2225

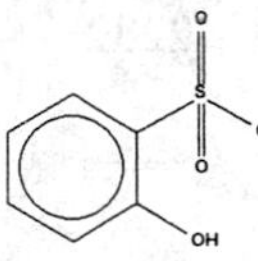

2226

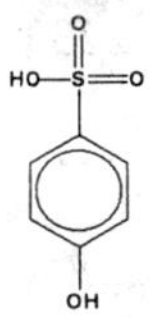

2227

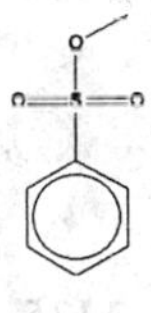

2228

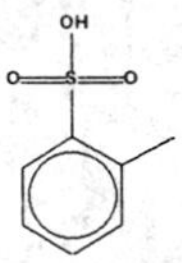

2229

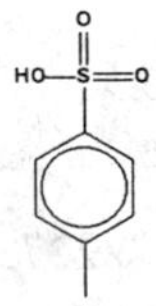

STRUCTURAL FORMULAS OF ORGANIC COMPOUNDS (continued)

In numeric order as they occur in the Table of Physical Constants of Organic Compounds

2230

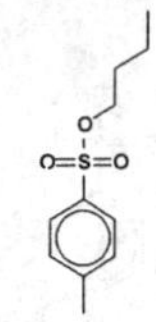

2231

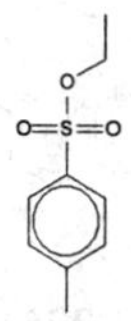

2232

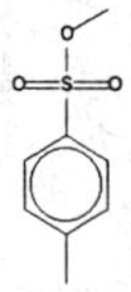

2233

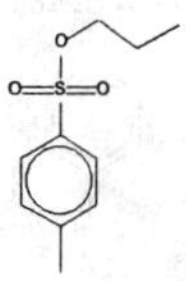

2234

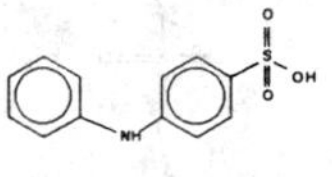

2235

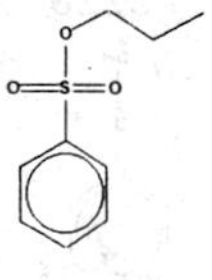

2236

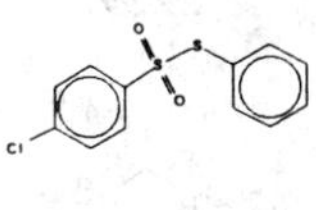

2237

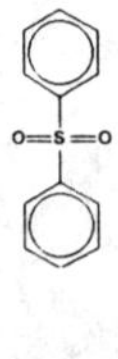

2238

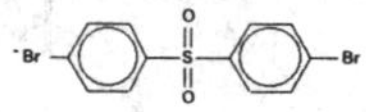

2239

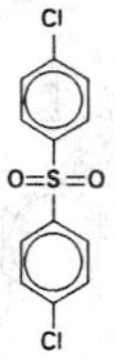

2240

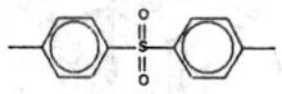

2242

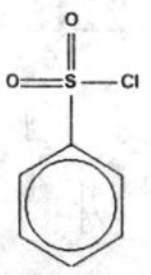

2243

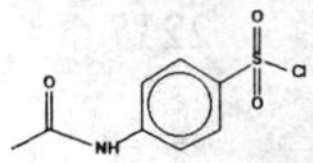

2244

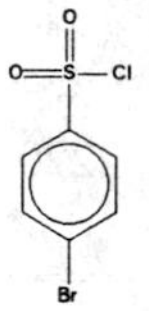

2245

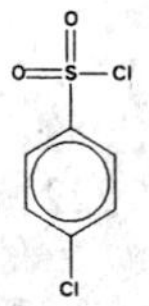

2246

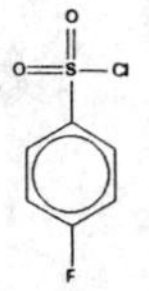

2247

2248

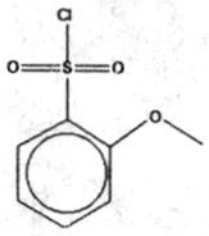

2249

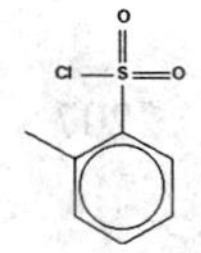

2250

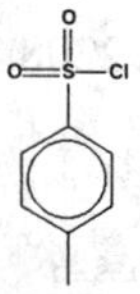

2251

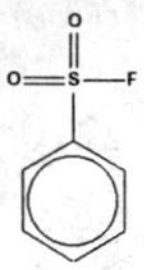

2252

2253

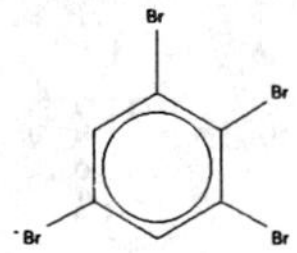

2254

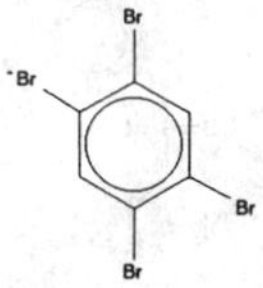

2255

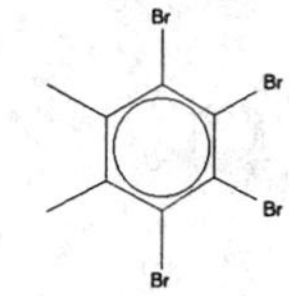

2256

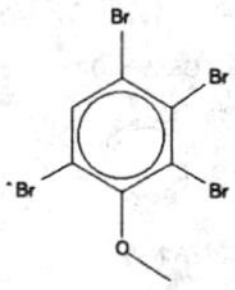

2257

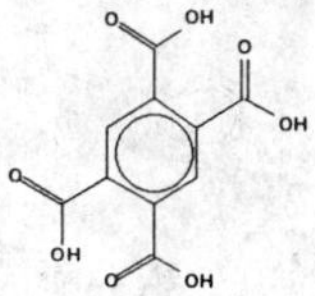

2258

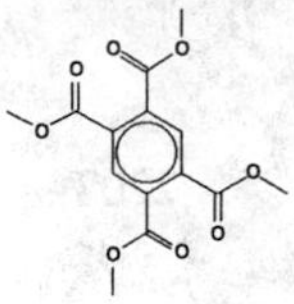

2259

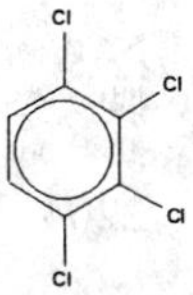

2260

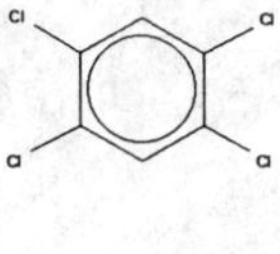

2261

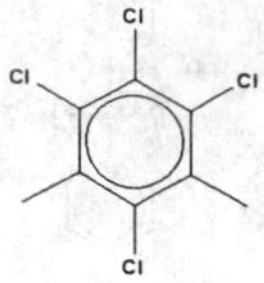

2262

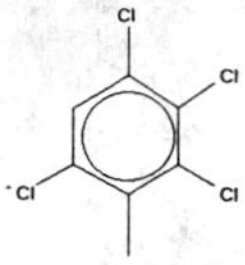

2264

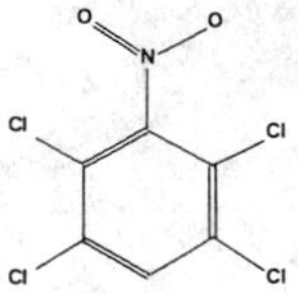

2265

2266

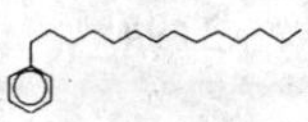

2267

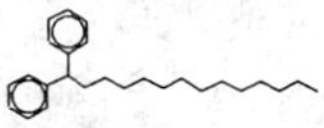

2268

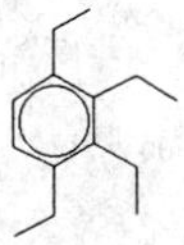

2269

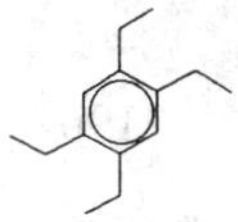

In numeric order as they occur in the Table of Physical Constants of Organic Compounds

2270

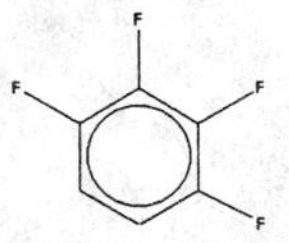

F F F F

2271

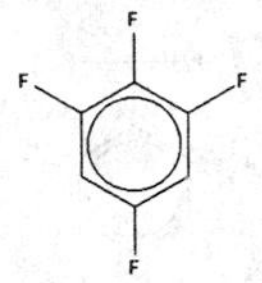

F F F F

2272

F F F F

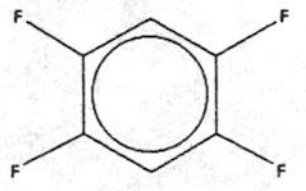

2276

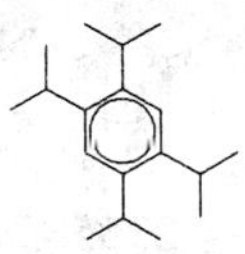

2277

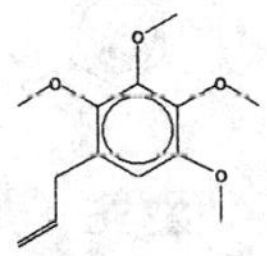

2278

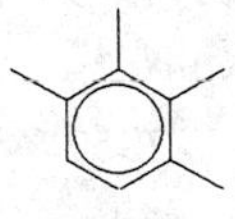

2279

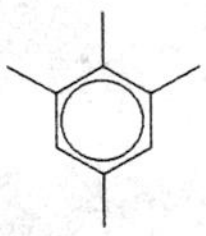

2280

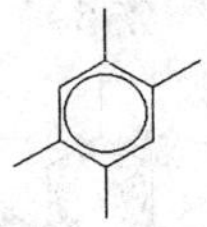

2281

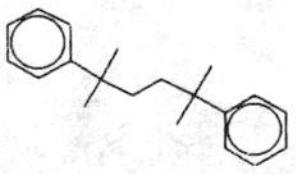

2283

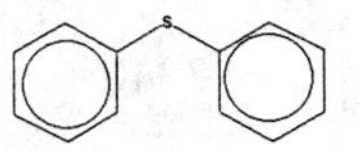

S

2284

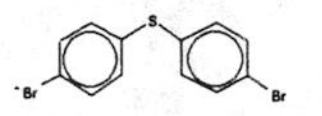

S Br Br

2285

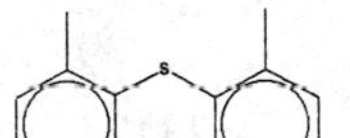

S

2286

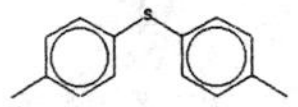

S

2287

S

2289

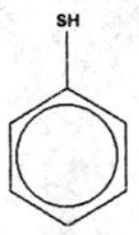

SH

2290

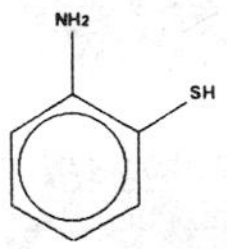

NH_2 SH

2291

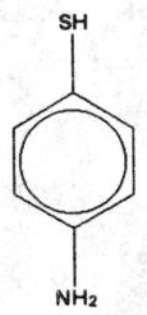

SH NH_2

2292

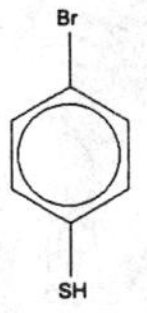

Br SH

2293

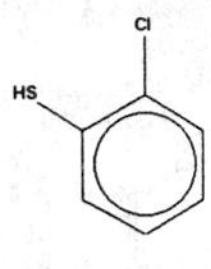

Cl HS

2294

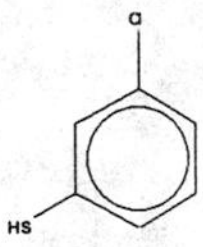

Cl HS

2295

Cl SH

2296

SH O

2297

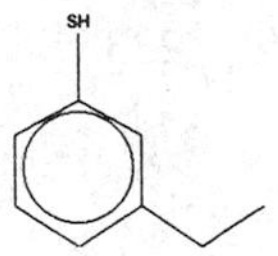

SH

2299

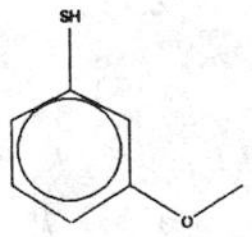

SH O

2300

SH O

2301

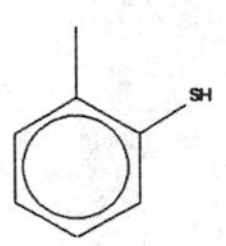

SH

2302

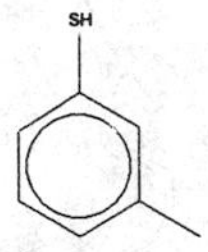

SH

2303

SH

2304

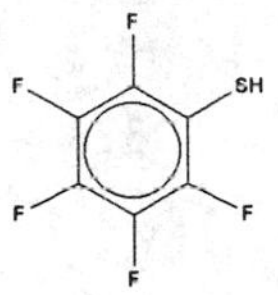

F F SH F F F

2305

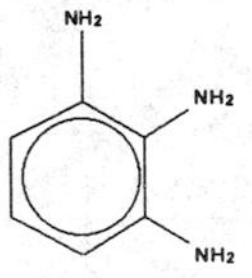

NH_2 NH_2 NH_2

2306

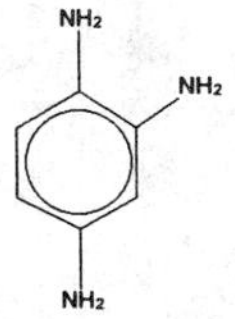

NH_2 NH_2 NH_2

2308

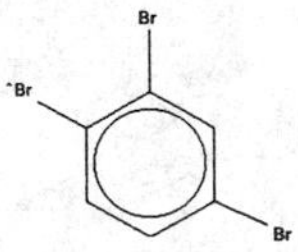

Br Br Br

2309

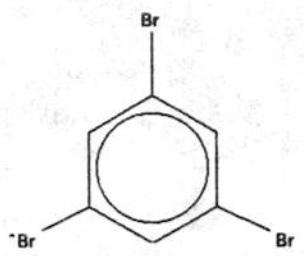

Br Br Br

2310

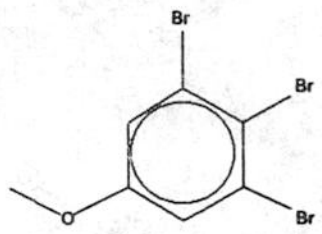

Br Br Br O

2311

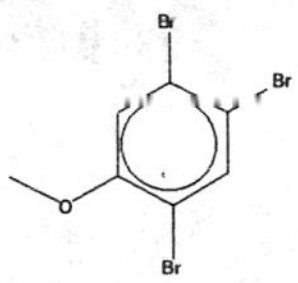

Br Br O Br

2312

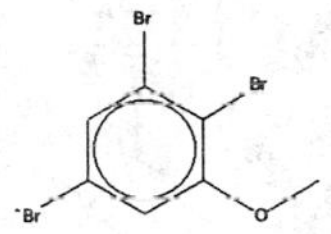

Br Br Br O

2313

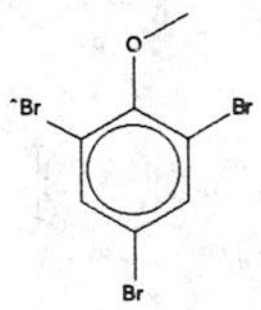

O Br Br Br

2314

Br Br Br

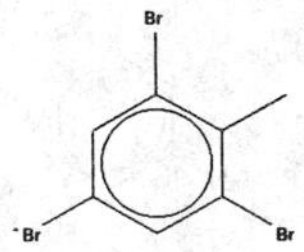

STRUCTURAL FORMULAS OF ORGANIC COMPOUNDS (continued)

In numeric order as they occur in the Table of Physical Constants of Organic Compounds

2316

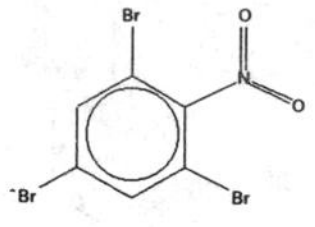

2317

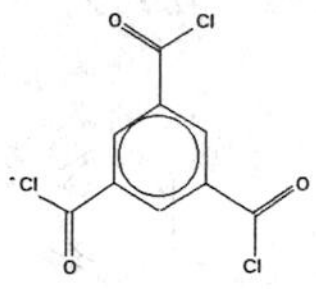

2318

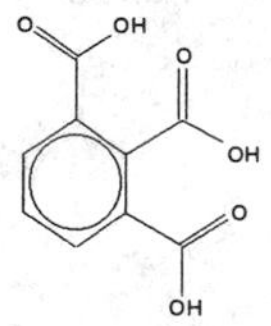

2319

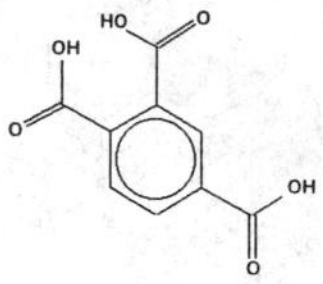

2321

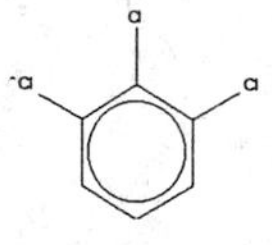

2322

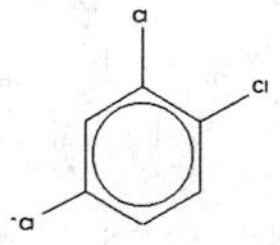

2323

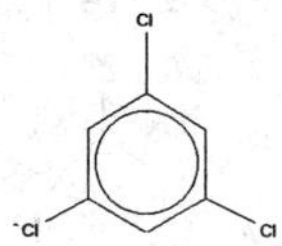

2324

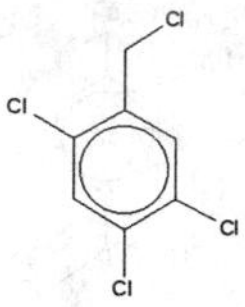

2325

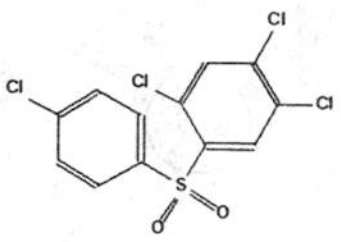

2326

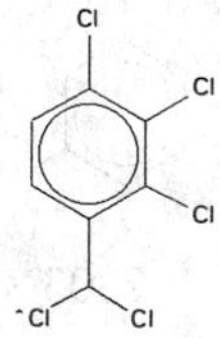

2327

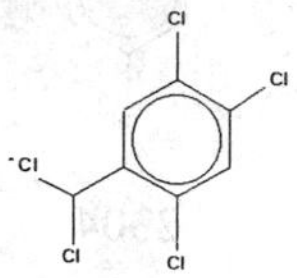

2328

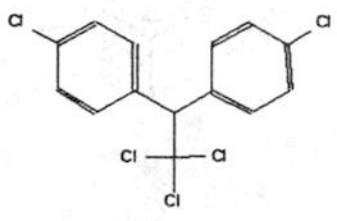

2329

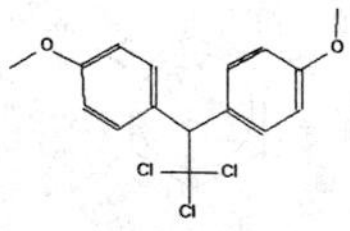

2330

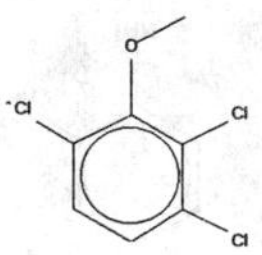

2331

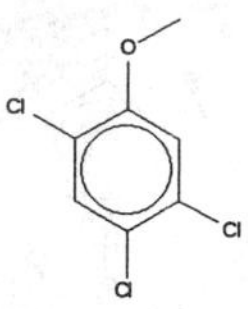

2332

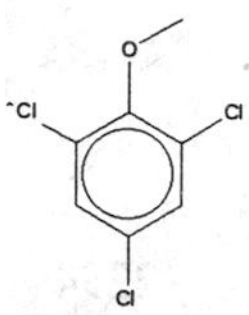

2333

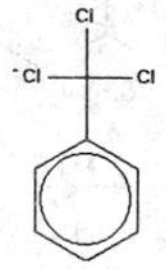

2334

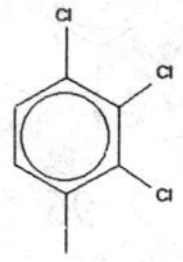

2335

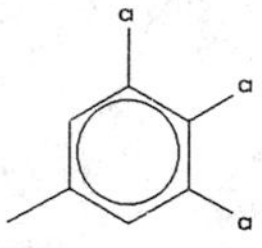

2336

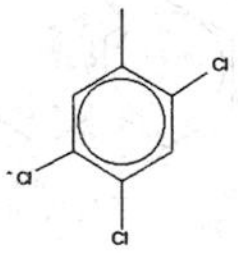

2337

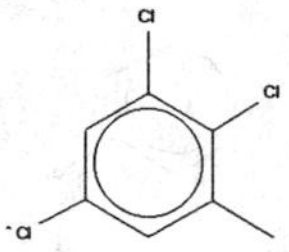

2339

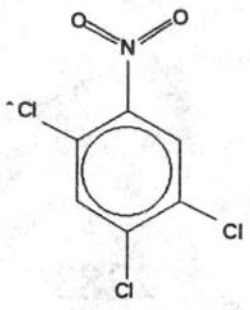

2340

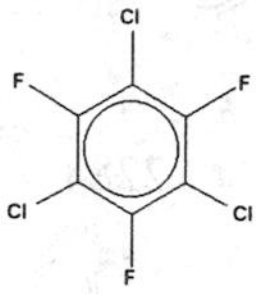

2341

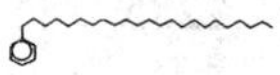

2342

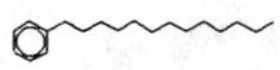

2343

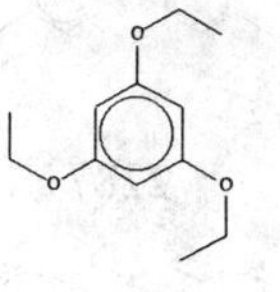

2344

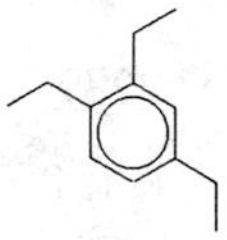

2345

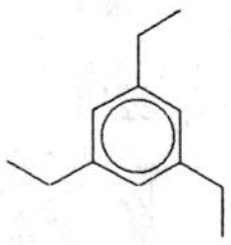

2346

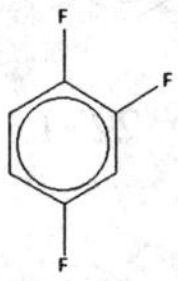

2347

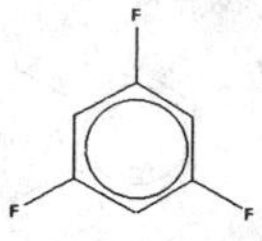

2348

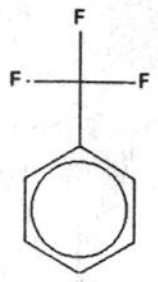

2352

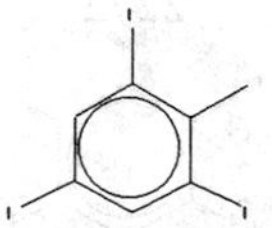

2354

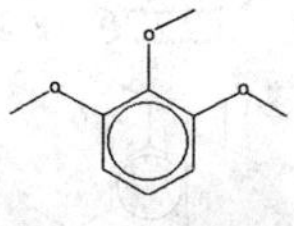

2355

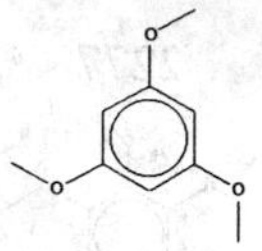

2356

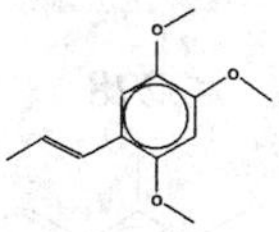

2357

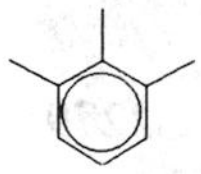

2358

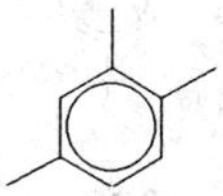

2359

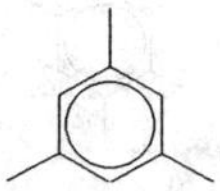

STRUCTURAL FORMULAS OF ORGANIC COMPOUNDS (continued)

In numeric order as they occur in the Table of Physical Constants of Organic Compounds

2360

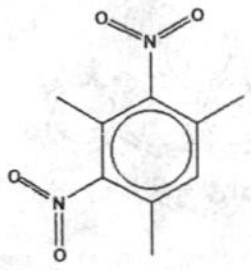

2361

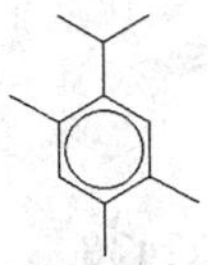

2362

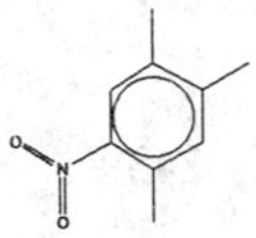

2363

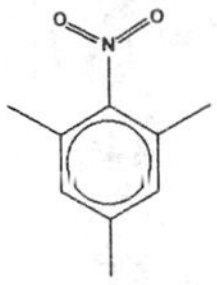

2364

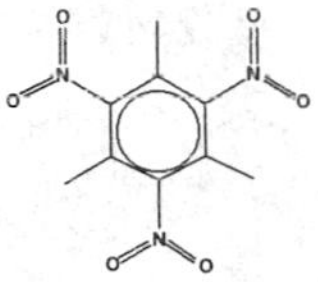

2365

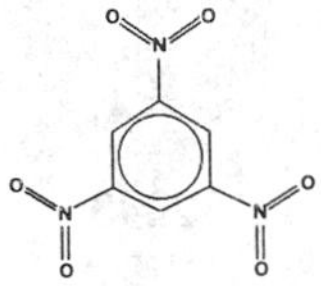

2366

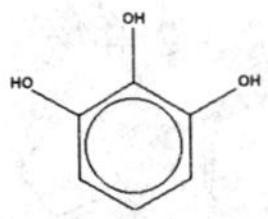

2367

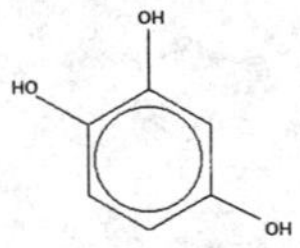

2368

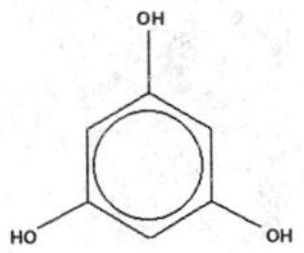

2369

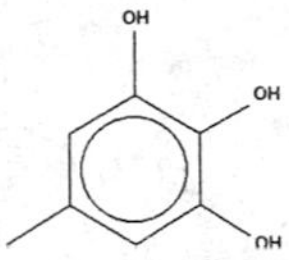

2370

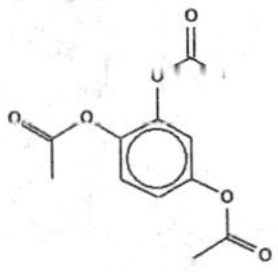

2371

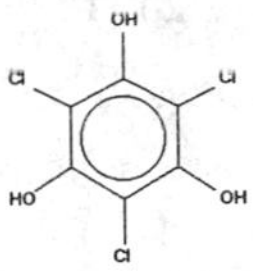

2372

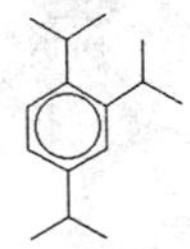

2373

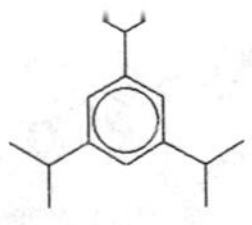

2374

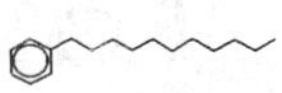

2375

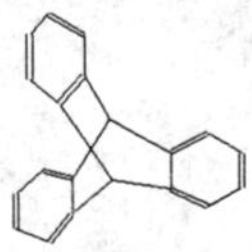

2376

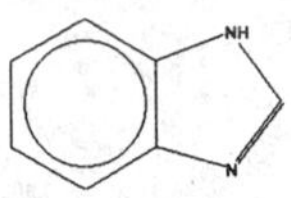

2377

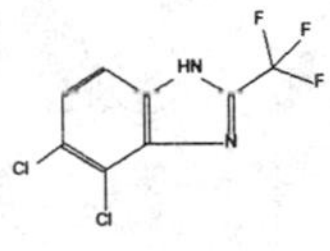

2378

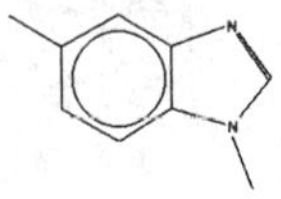

2379

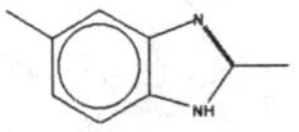

2380

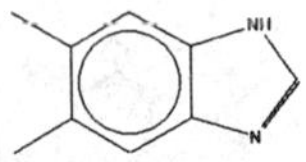

2381

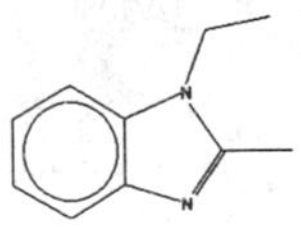

2382

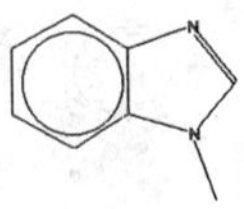

2383

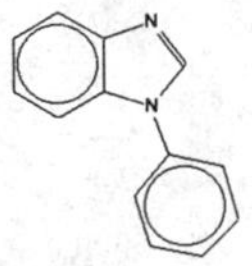

2384

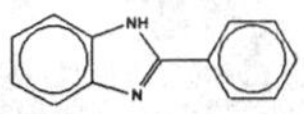

2385

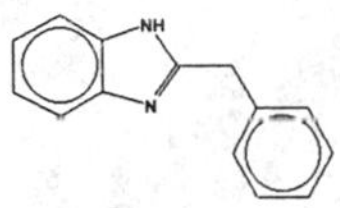

2386

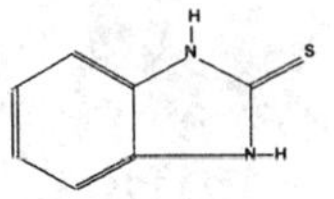

2387

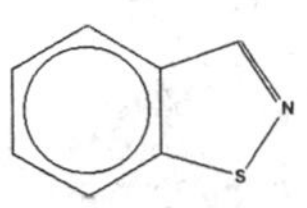

2388

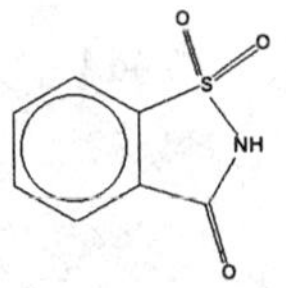

2389

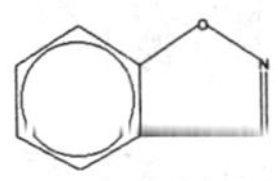

2390

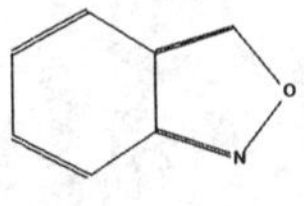

2391

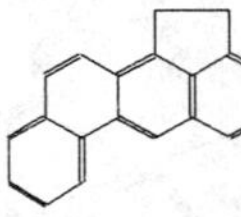

2392

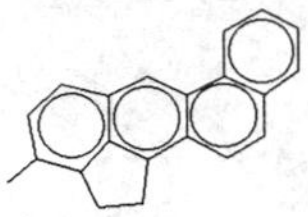

2393

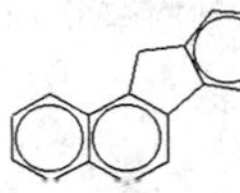

2394

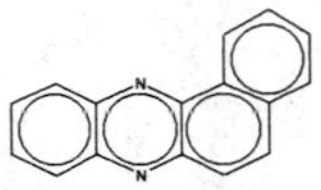

2395

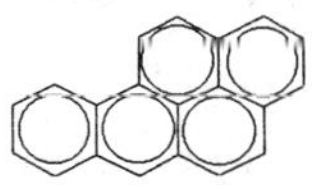

2396

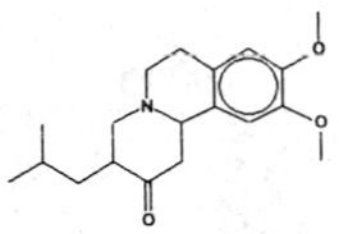

2397

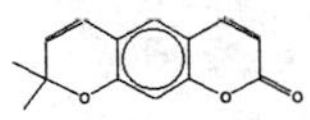

2398

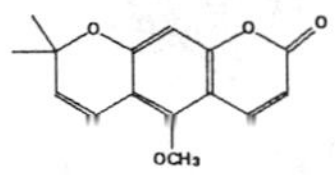

2399

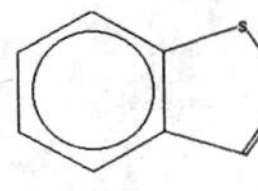

STRUCTURAL FORMULAS OF ORGANIC COMPOUNDS (continued)

In numeric order as they occur in the Table of Physical Constants of Organic Compounds

2400

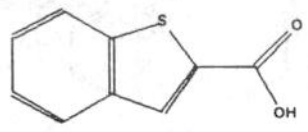

2401

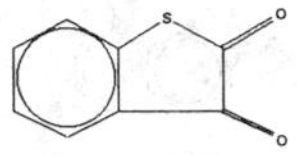

2402

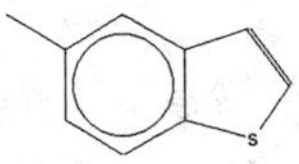

2403

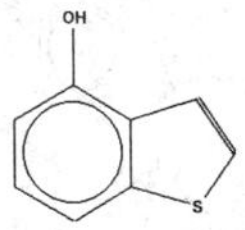

2404

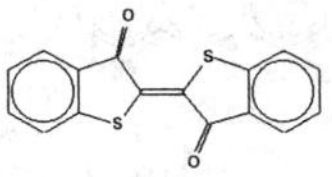

2405

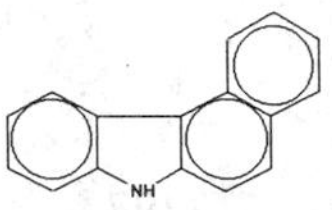

2406

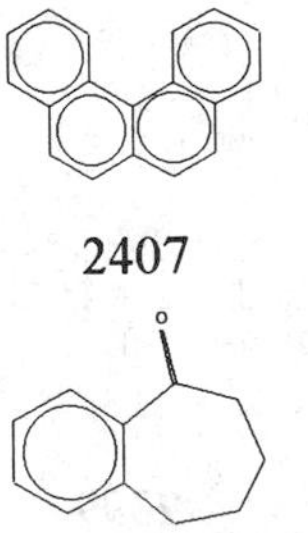

2407

2408

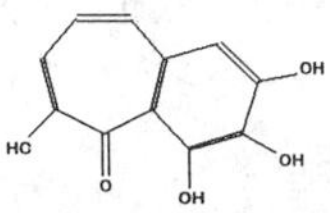

2409

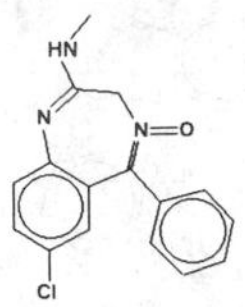

2410

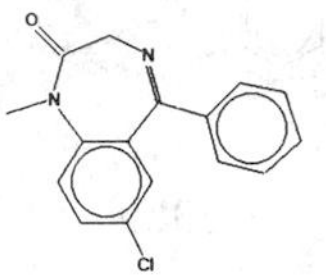

2411

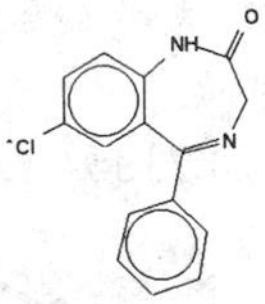

2412

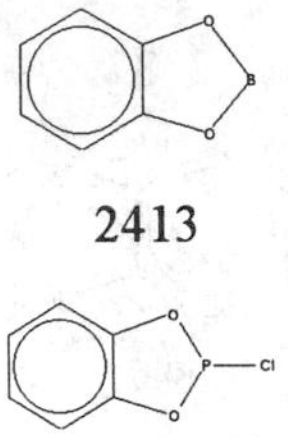

2413

2414

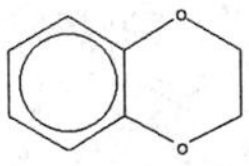

2415

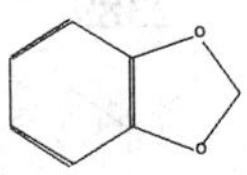

2416

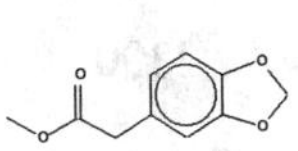

2417

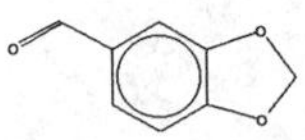

2418

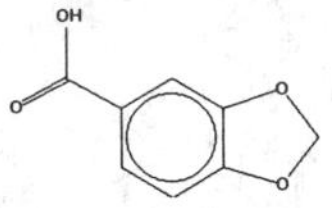

2419

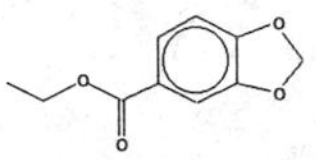

2420

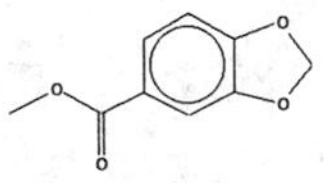

2421

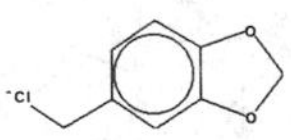

2422

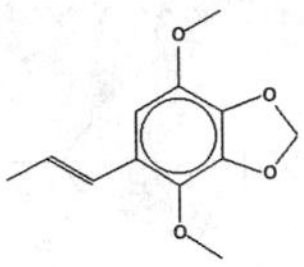

2423

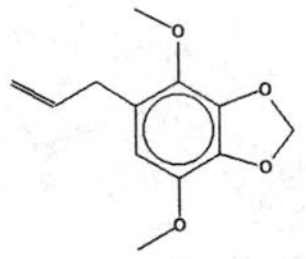

2424

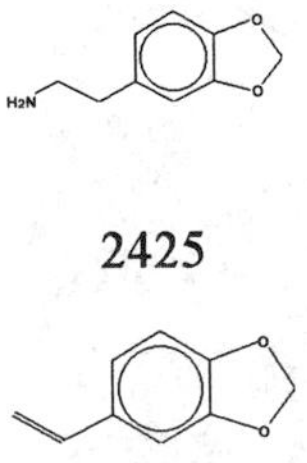

2425

2426

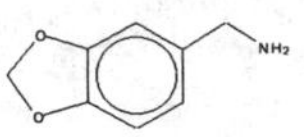

2427

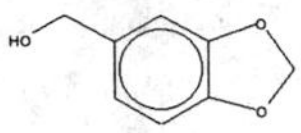

2428

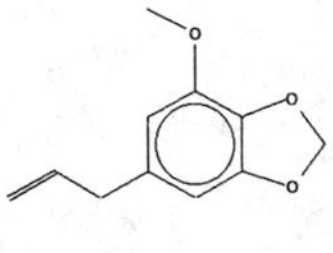

2429

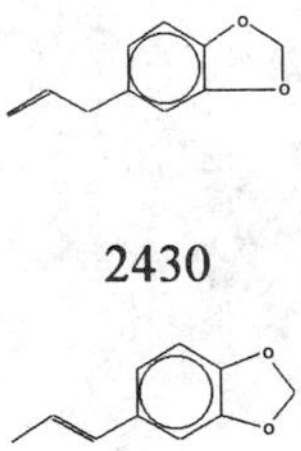

2430

2431

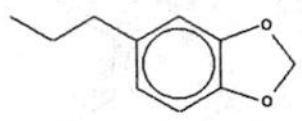

2432

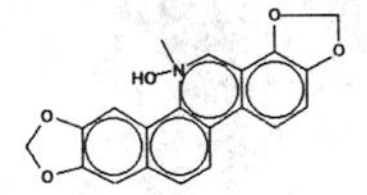

2434

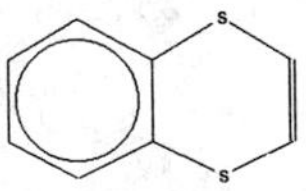

2435

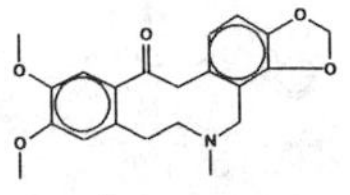

2436

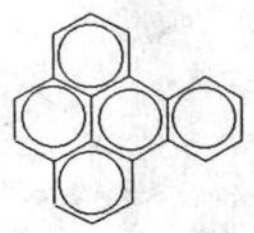

2437

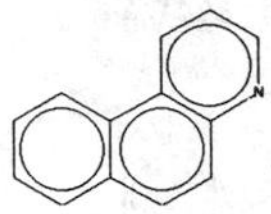

2438

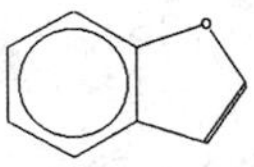

2439

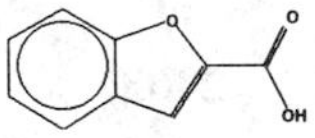

2440

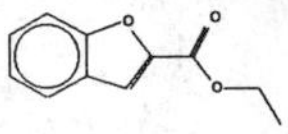

2441

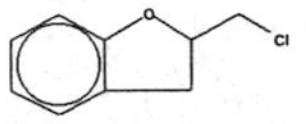

2442

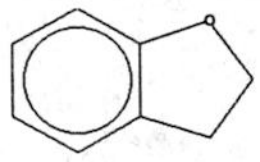

2443

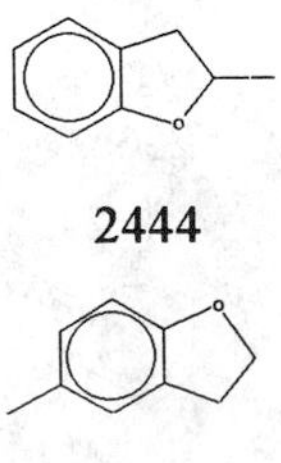

2444

2445

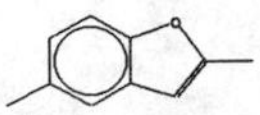

STRUCTURAL FORMULAS OF ORGANIC COMPOUNDS (continued)

In numeric order as they occur in the Table of Physical Constants of Organic Compounds

2446

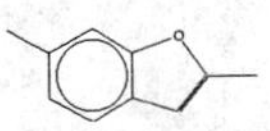

2447

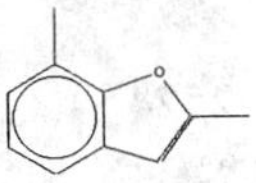

2448

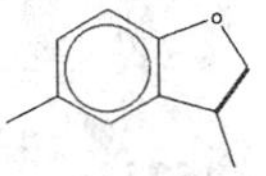

2449

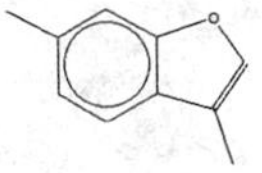

2451

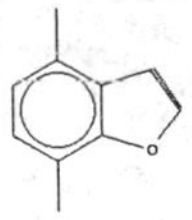

2452

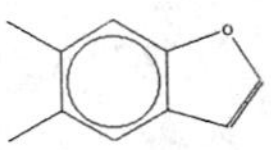

2453

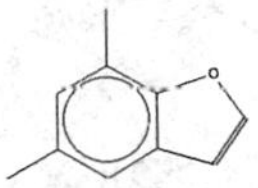

2454

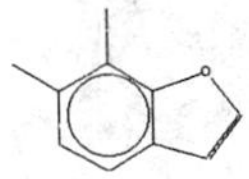

2455

2456

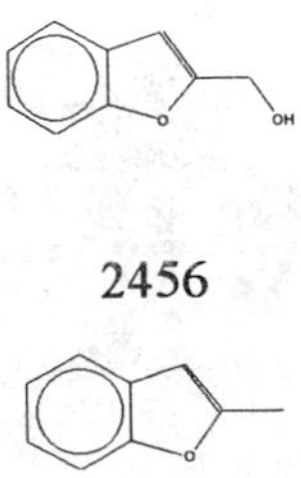

2457

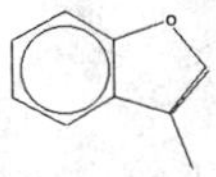

2458

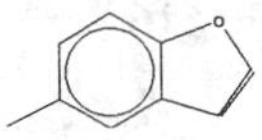

2459

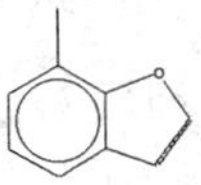

2460

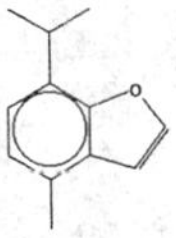

2461

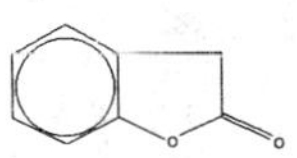

2462

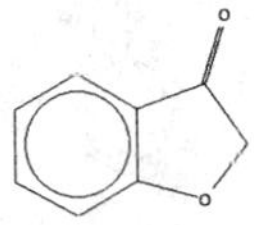

2463

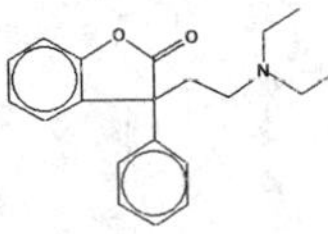

2464

2465

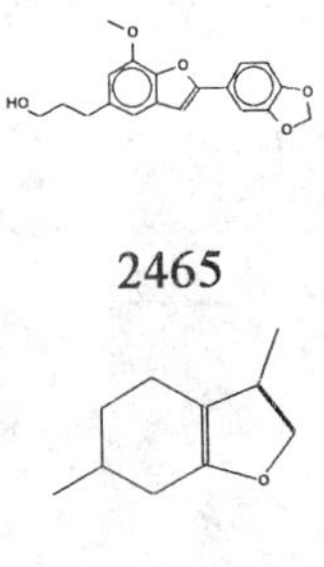

2466

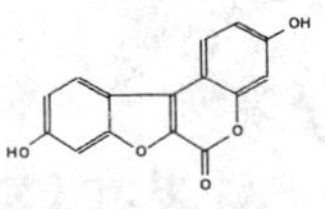

2468

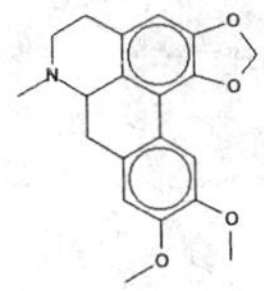

2469

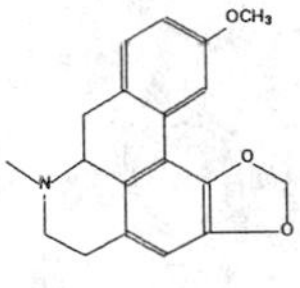

2470

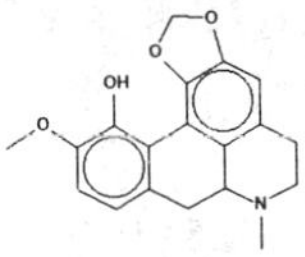

2471

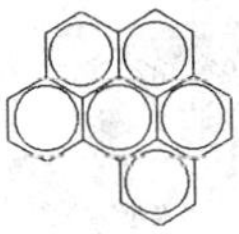

2472

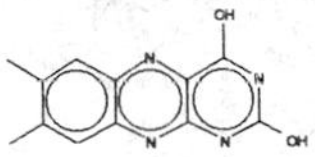

2474

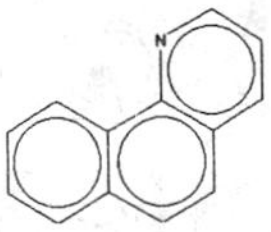

2475

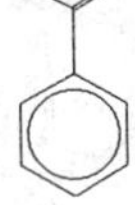

2476

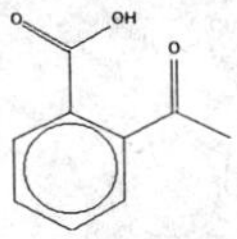

2477

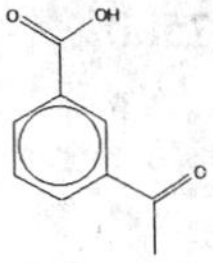

2478

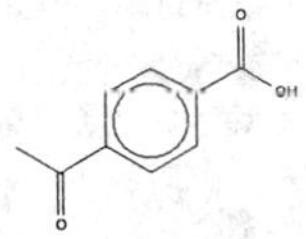

2479

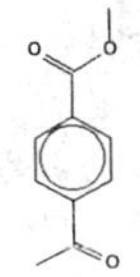

2480

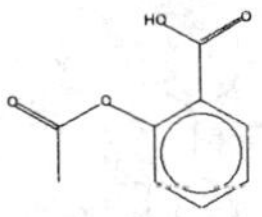

2481

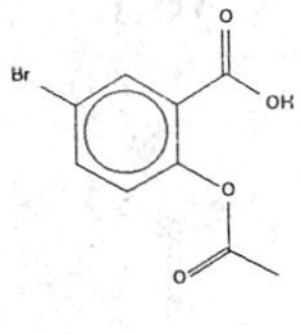

2482

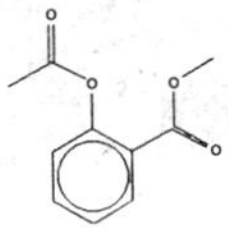

2483

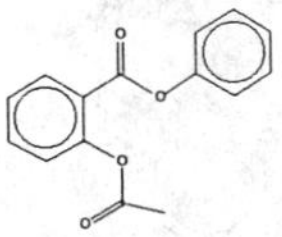

2484

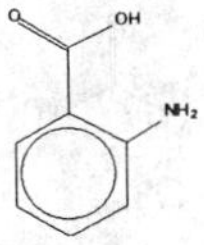

2485

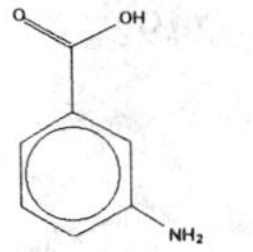

2486

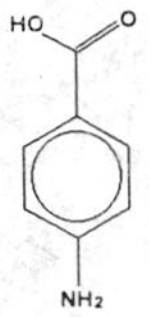

2487

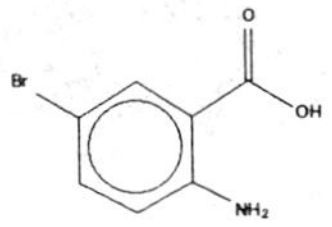

2488

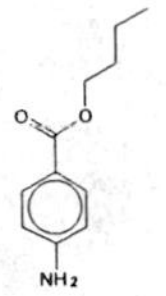

2489

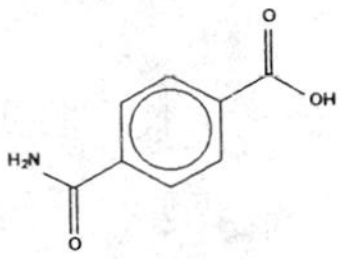

2490

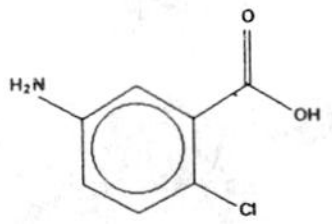

STRUCTURAL FORMULAS OF ORGANIC COMPOUNDS (continued)

In numeric order as they occur in the Table of Physical Constants of Organic Compounds

2491

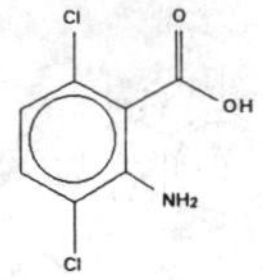

Cl, O, OH, NH2, Cl

2492

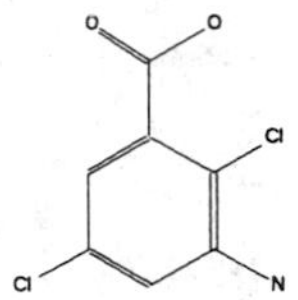

O, O, Cl, Cl, N

2493

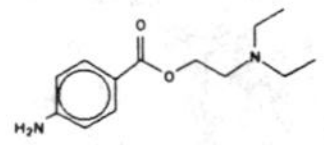

O, O, N, H2N

2494

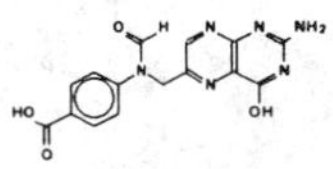

O, H, N, N, N, NH2, N, N, HO, O, OH

2495

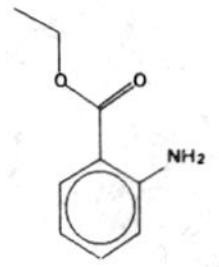

O, O, NH2

2496

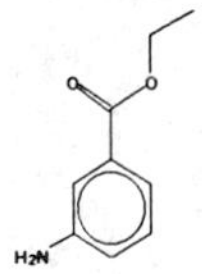

O, O, H2N

2497

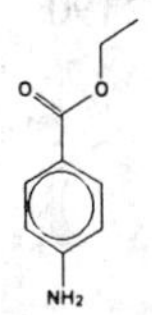

O, O, NH2

2498

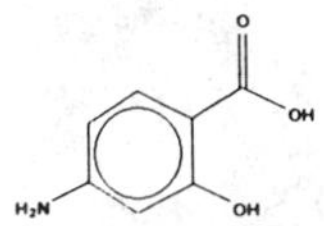

O, OH, H2N, OH

2499

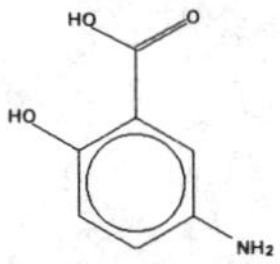

HO, O, HO, NH2

2500

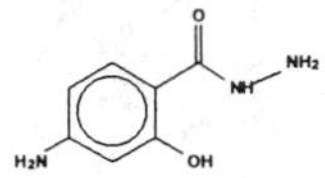

O, NH, NH2, H2N, OH

2501

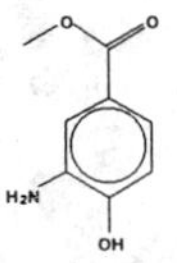

O, O, H2N, OH

2502

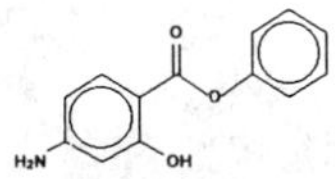

O, O, H2N, OH

2503

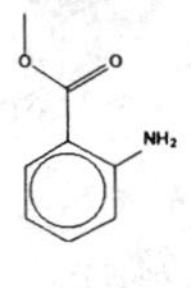

O, O, NH2

2504

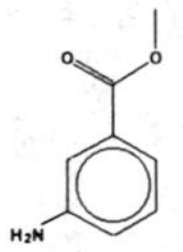

O, O, H2N

2505

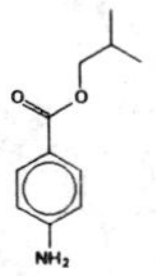

O, O, NH2

2506

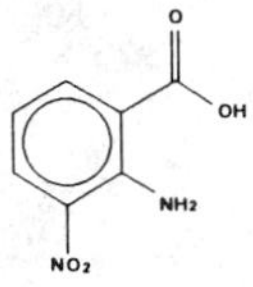

O, OH, NH2, NO2

2507

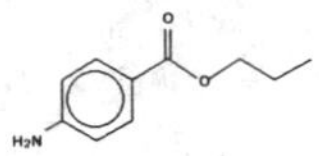

O, O, H2N

2508

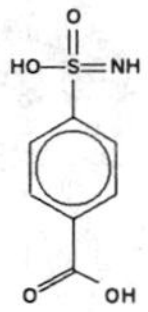

O, HO—S=NH, O, OH

2509

HO, O

2510

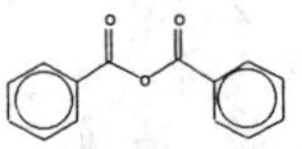

O, O, O

2511

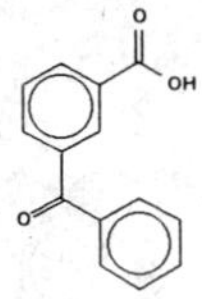

O, OH, O

2512

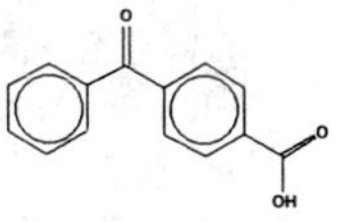

O, O, OH

2513

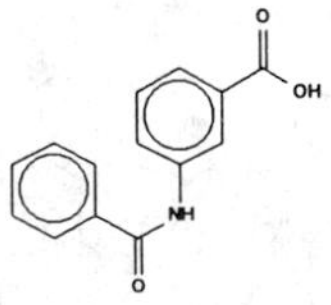

O, OH, NH, O

2514

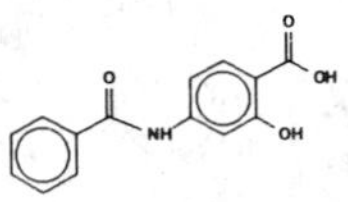

O, OH, NH, OH

2515

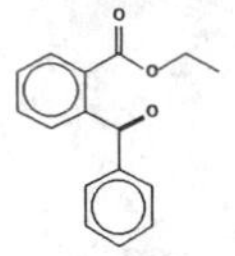

O, O, O

2516

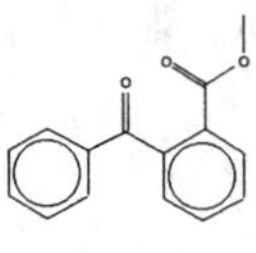

O, O, O

2517

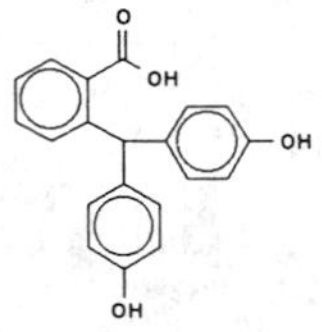

O, OH, OH, OH

2518

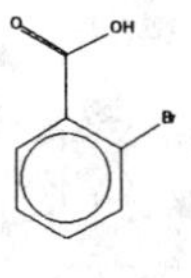

O, OH, Br

2519

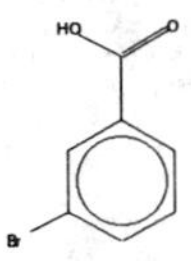

HO, O, Br

2520

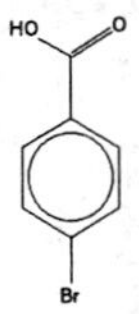

HO, O, Br

2521

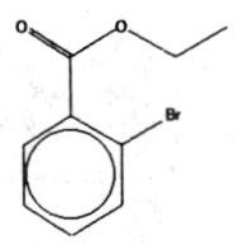

O, O, Br

2522

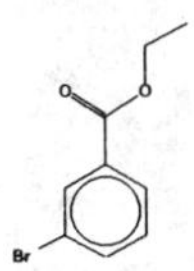

O, O, Br

2523

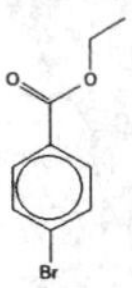

O, O, Br

2524

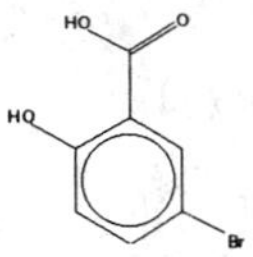

HO, O, HO, Br

2525

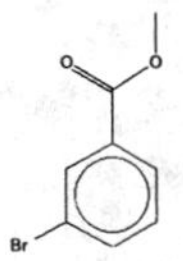

O, O, Br

2526

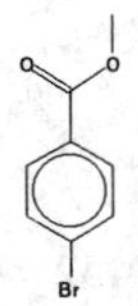

O, O, Br

2527

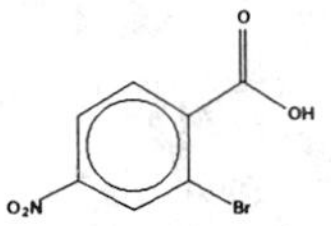

O, OH, O2N, Br

2528

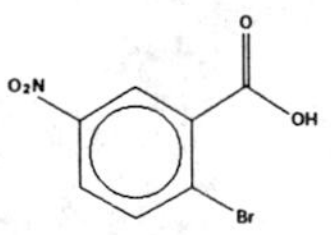

O, O2N, OH, Br

2529

O, OH, Br, NO2

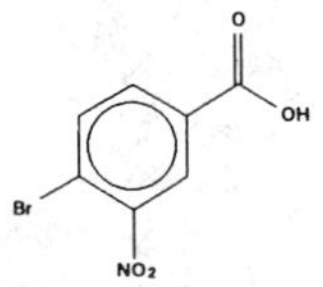

STRUCTURAL FORMULAS OF ORGANIC COMPOUNDS (continued)

In numeric order as they occur in the Table of Physical Constants of Organic Compounds

2530

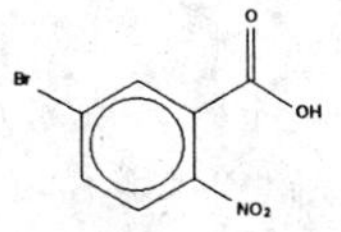

2532

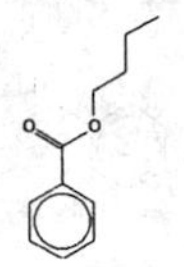

2533

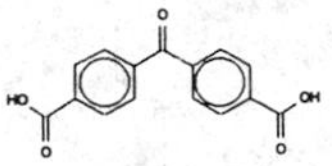

2534

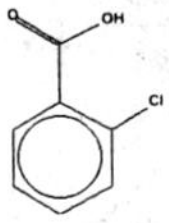

2535

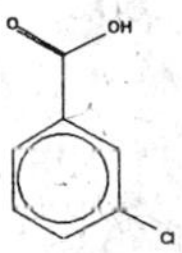

2536

2537

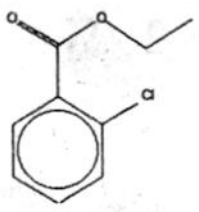

2538

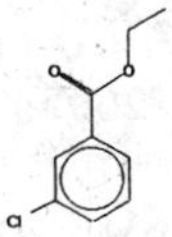

2539

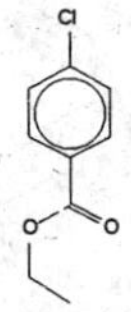

2540

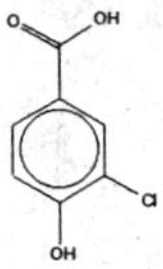

2541

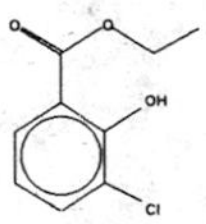

2542

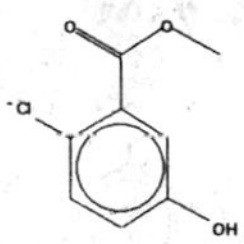

2543

2544

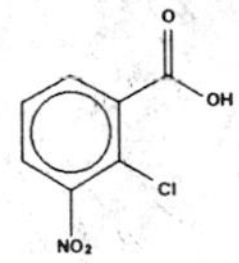

2545

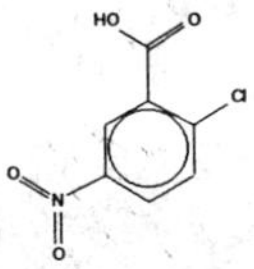

2546

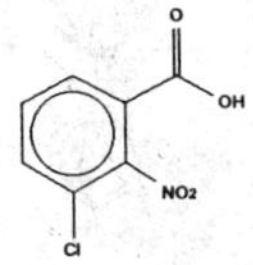

2547

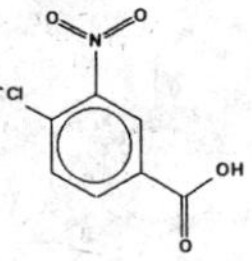

2548

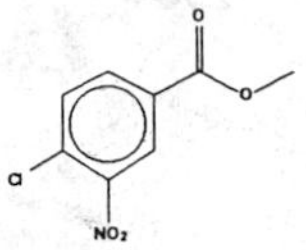

2549

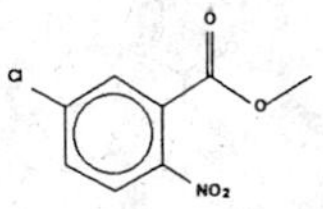

2550

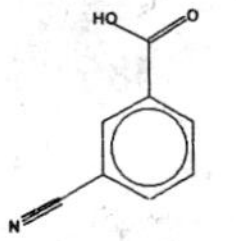

2551

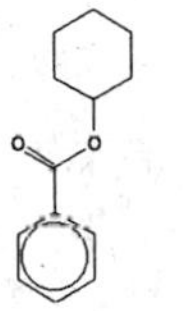

2552

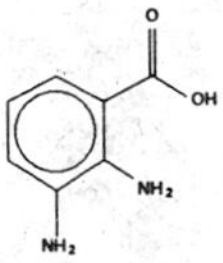

2553

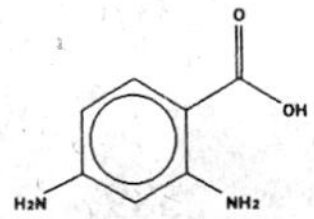

2554

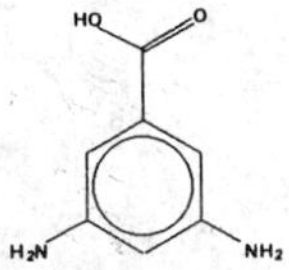

2555

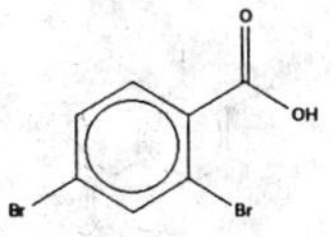

2556

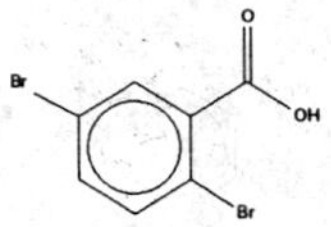

2557

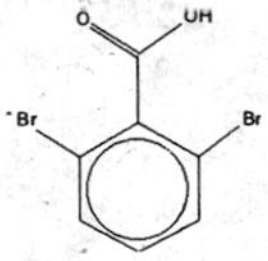

2558

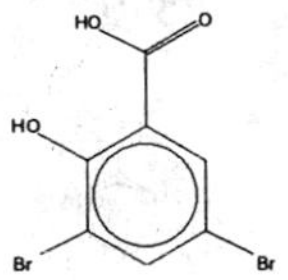

2559

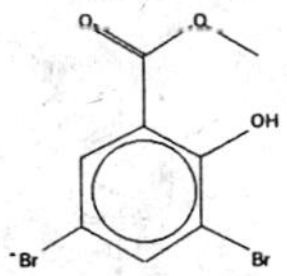

2560

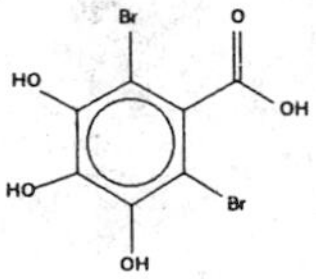

2561

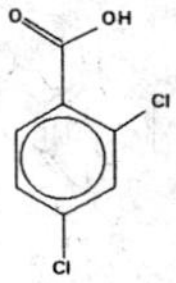

2562

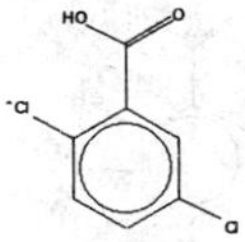

2563

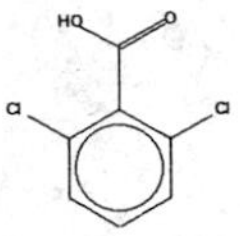

2564

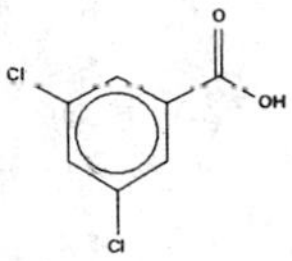

2565

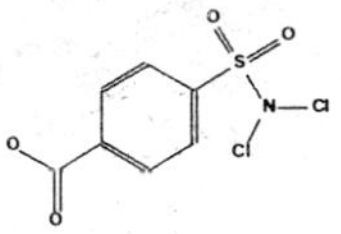

2566

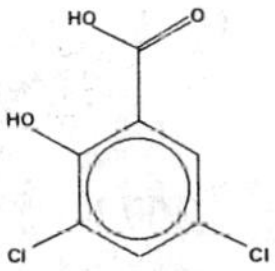

2567

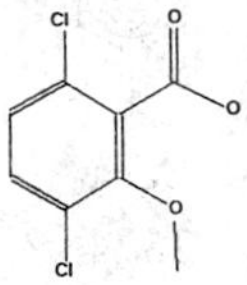

2568

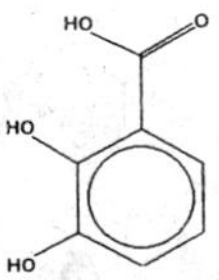

STRUCTURAL FORMULAS OF ORGANIC COMPOUNDS (continued)

In numeric order as they occur in the Table of Physical Constants of Organic Compounds

2569

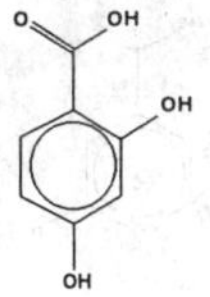

2570

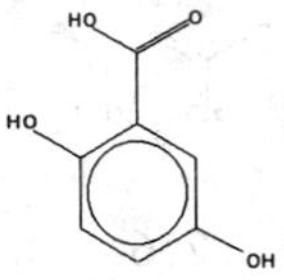

2571

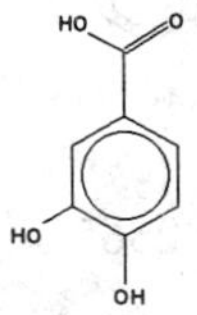

2572

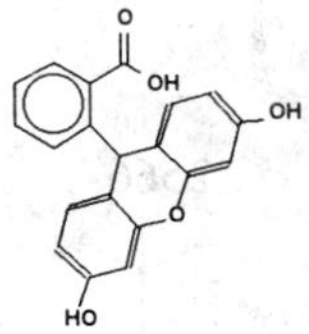

2573

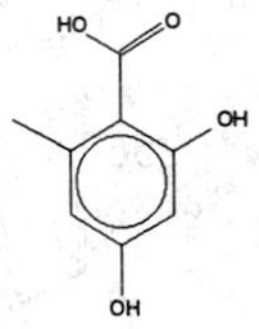

2574

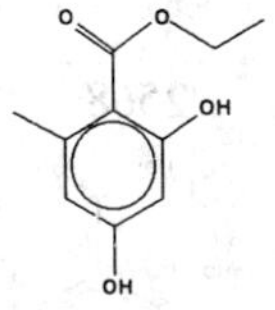

2575

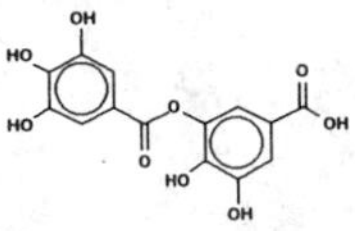

2576

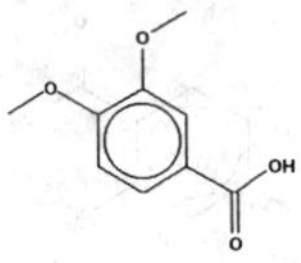

2577

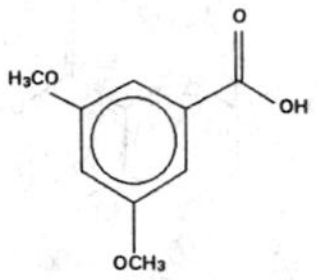

2578

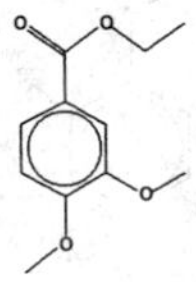

2579

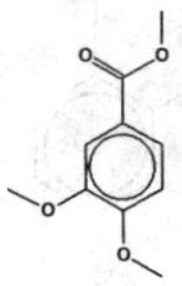

2580

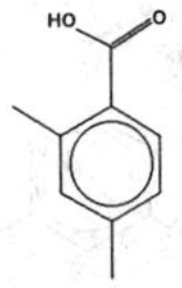

2581

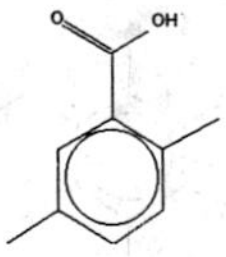

2582

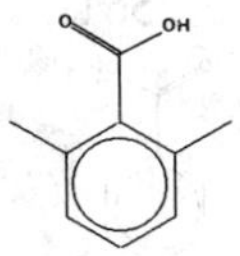

2583

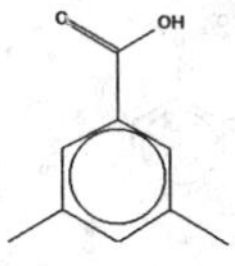

2584

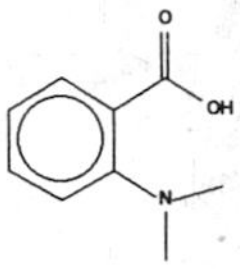

2585

2586

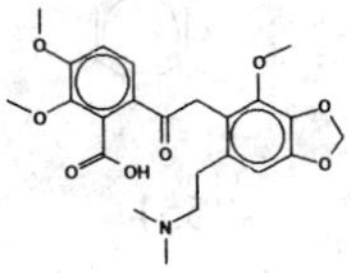

2587

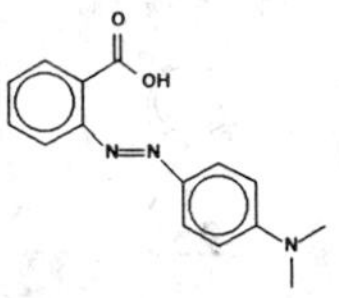

2588

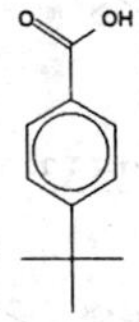

2589

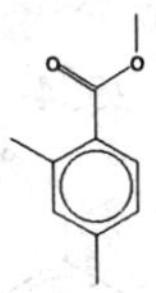

2590

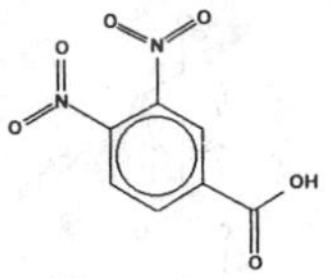

2591

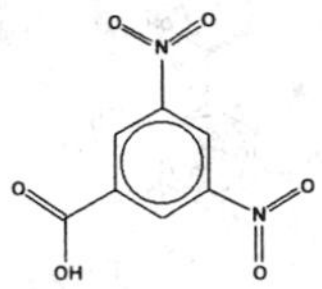

2592

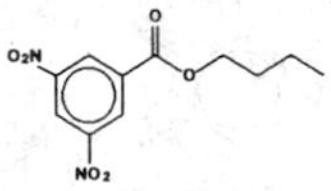

2593

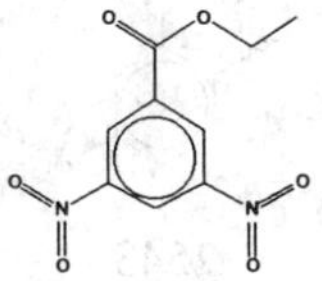

2594

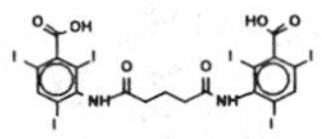

2595

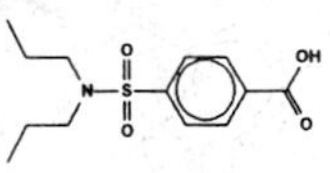

2596

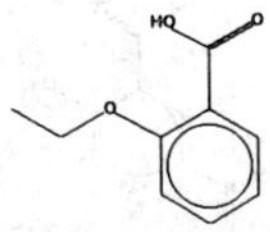

2597

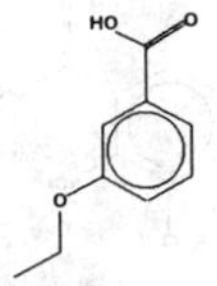

2598

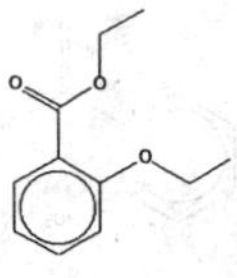

2599

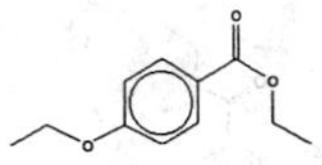

2600

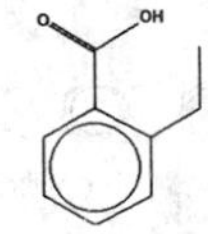

2601

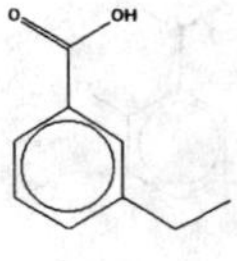

2602

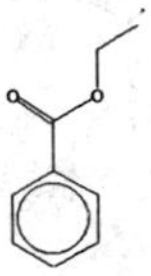

2603

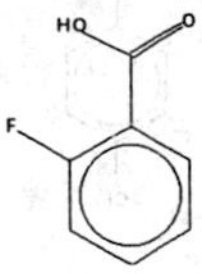

2604

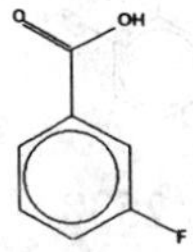

2605

STRUCTURAL FORMULAS OF ORGANIC COMPOUNDS (continued)

In numeric order as they occur in the Table of Physical Constants of Organic Compounds

2606

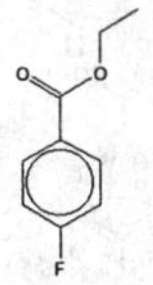

2607

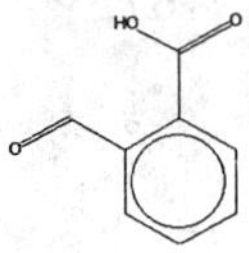

2608

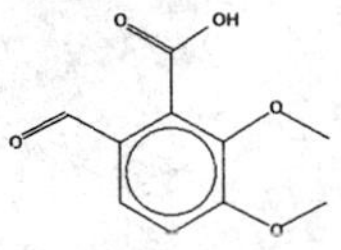

2609

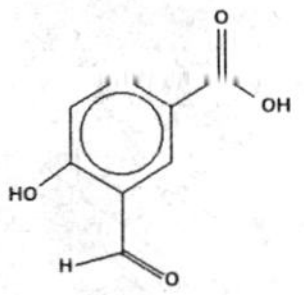

2610

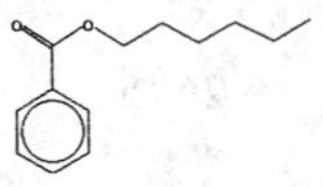

2611

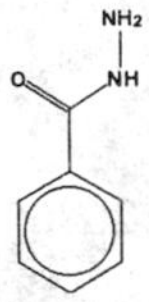

2612

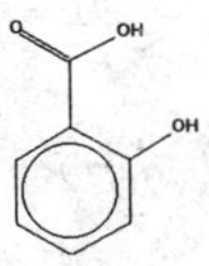

2613

2614

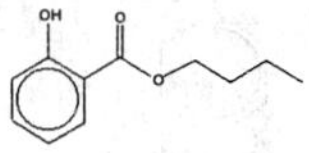

2615

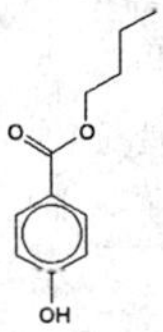

2616

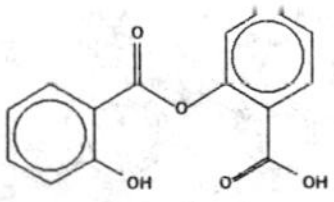

2617

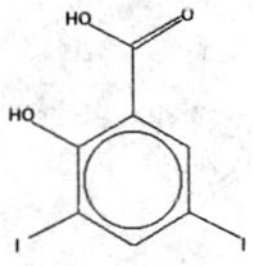

2618

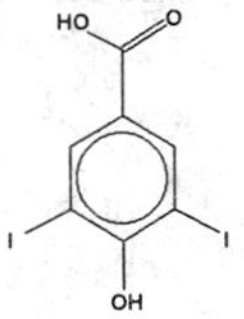

2619

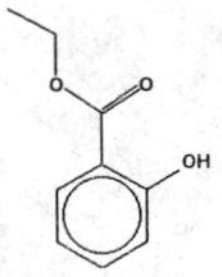

2620

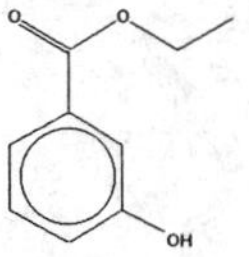

2621

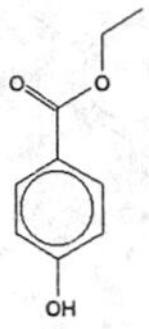

2622

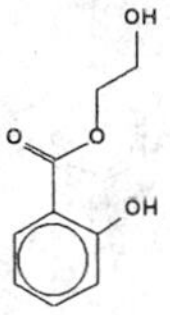

2624

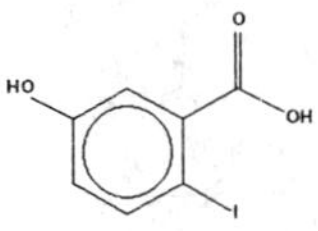

2625

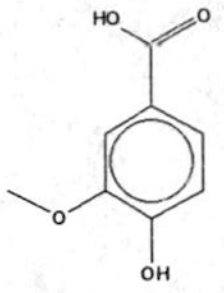

2626

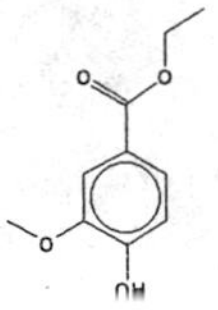

2627

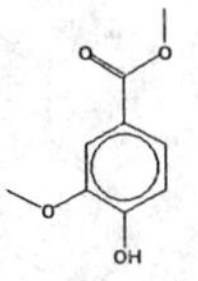

2628

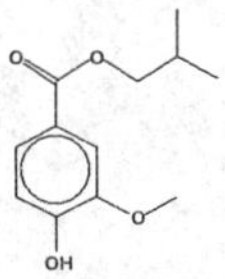

2629

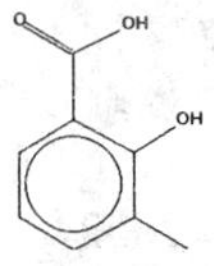

2630

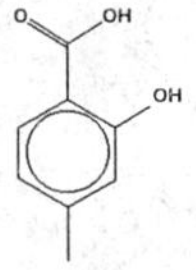

2631

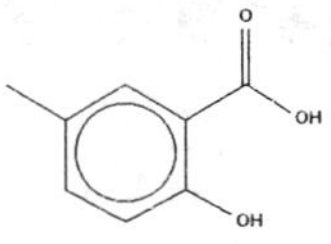

2632

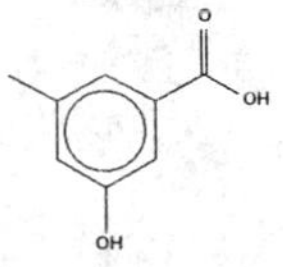

2633

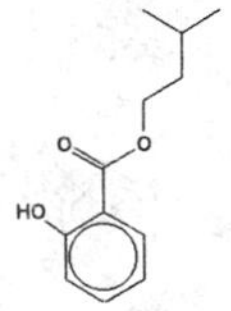

2634

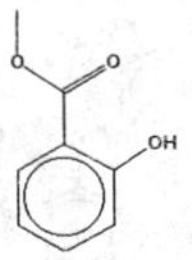

2635

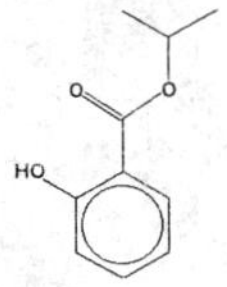

2636

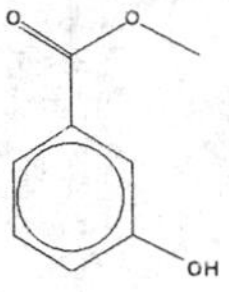

2637

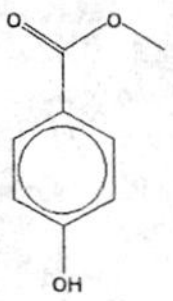

2638

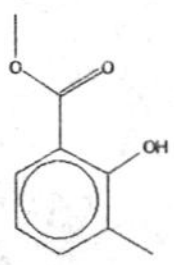

2639

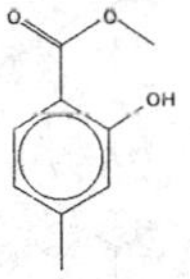

2640

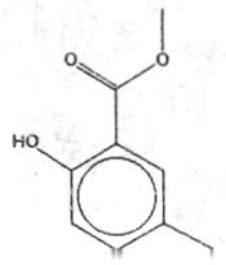

2641

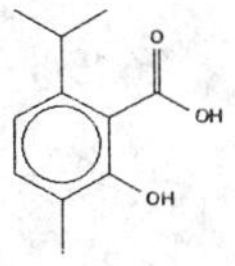

STRUCTURAL FORMULAS OF ORGANIC COMPOUNDS (continued)

In numeric order as they occur in the Table of Physical Constants of Organic Compounds

2642

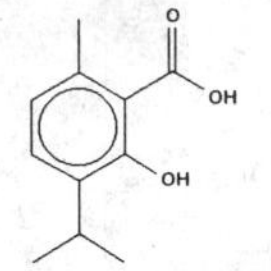

2643

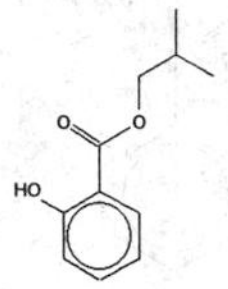

2644

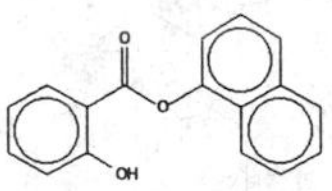

2645

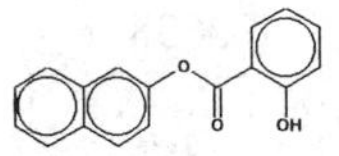

2646

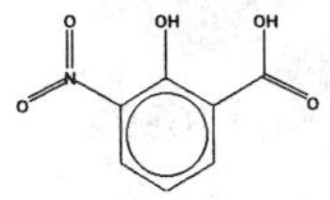

2647

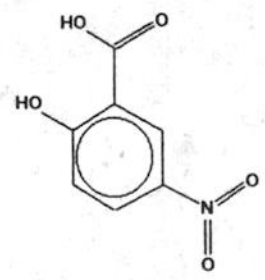

2648

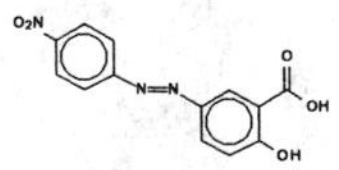

2649

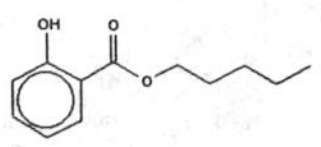

2650

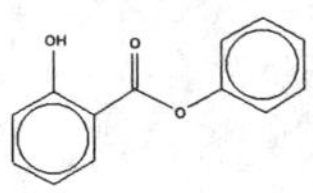

2651

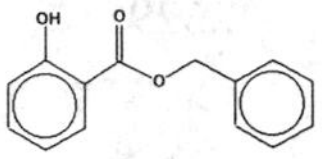

2652

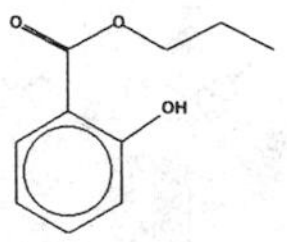

2653

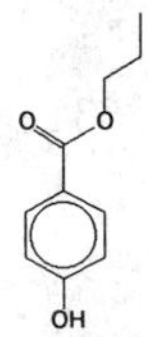

2654

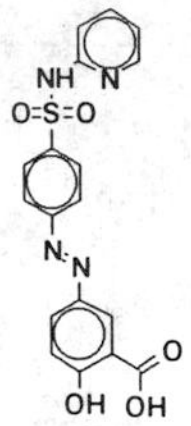

2655

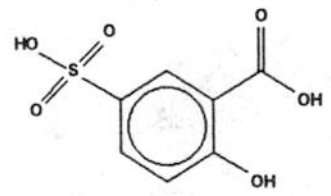

2656

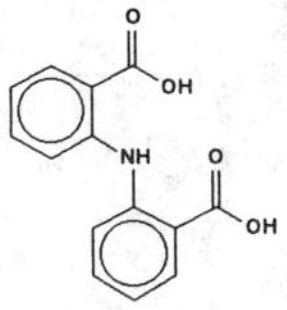

2657

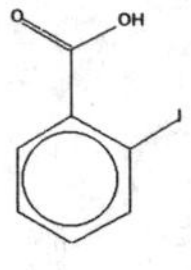

2658

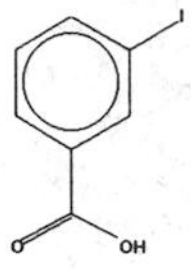

2659

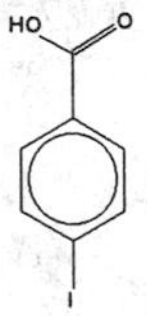

2660

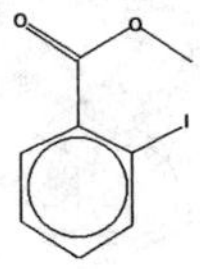

2661

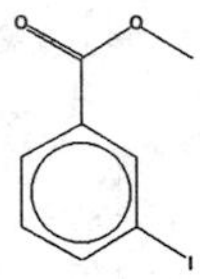

2662

2663

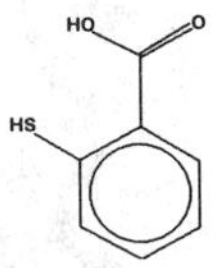

2664

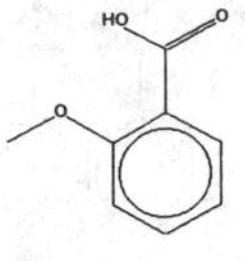

2665

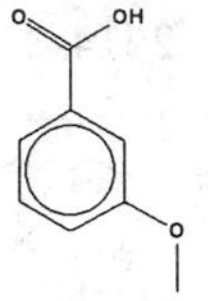

2666

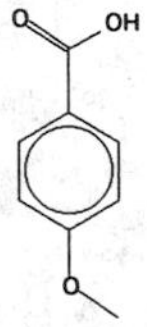

2667

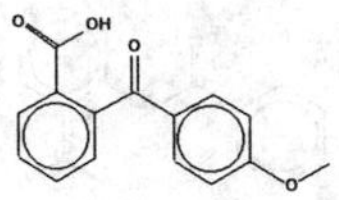

2668

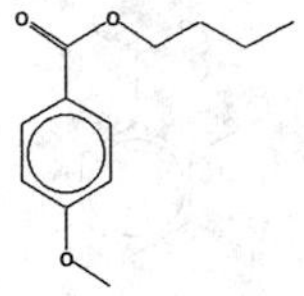

2669

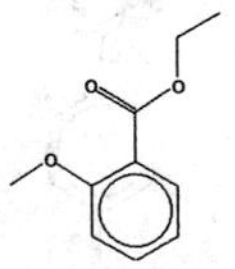

2670

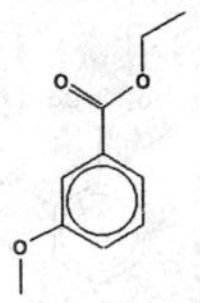

2671

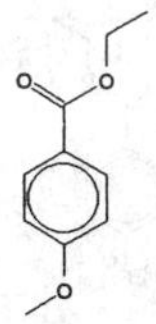

2672

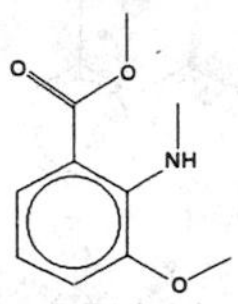

2673

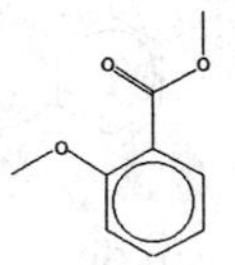

2674

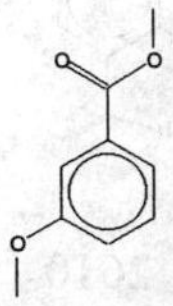

2675

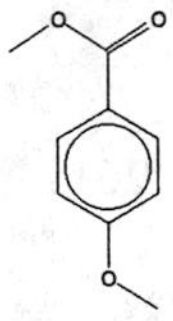

2676

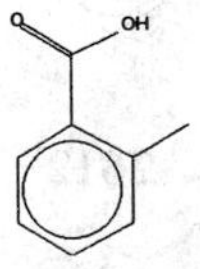

2677

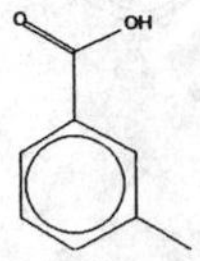

STRUCTURAL FORMULAS OF ORGANIC COMPOUNDS (continued)

In numeric order as they occur in the Table of Physical Constants of Organic Compounds

2678

2679

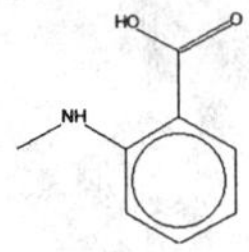

2680

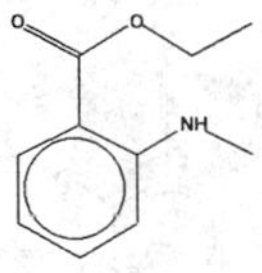

2681

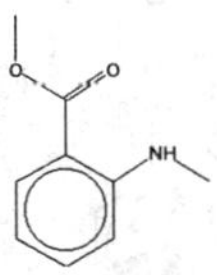

2682

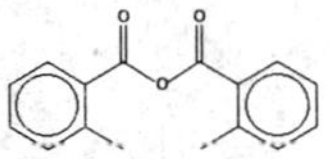

2683

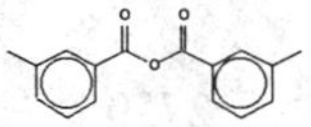

2684

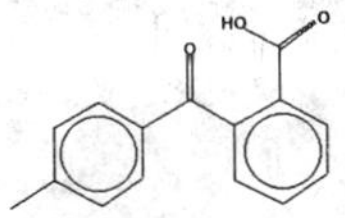

2685

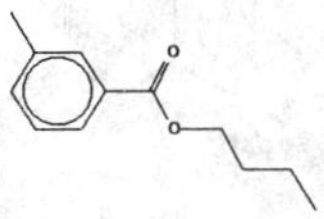

2686

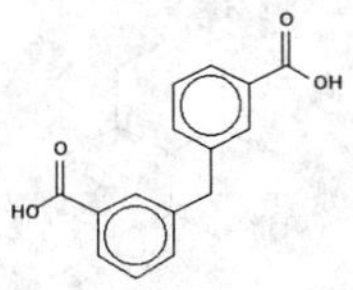

2687

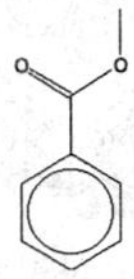

2688

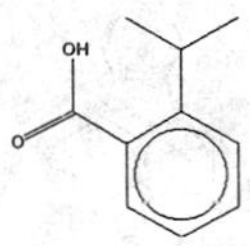

2689

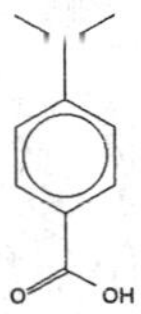

2690

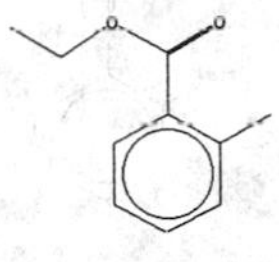

2691

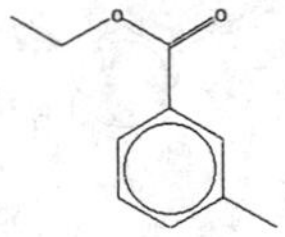

2692

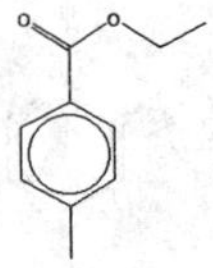

2693

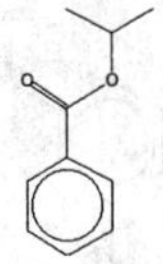

2694

2695

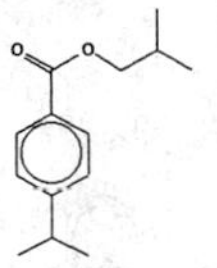

2696

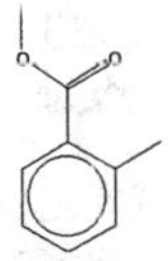

2697

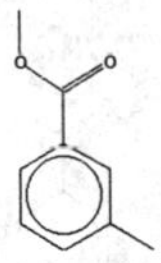

2698

2699

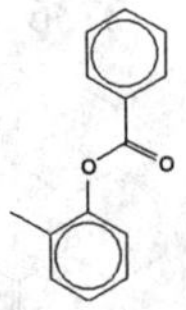

2700

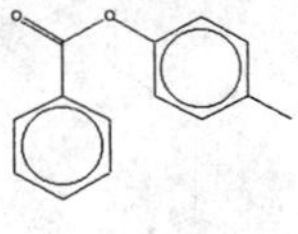

2701

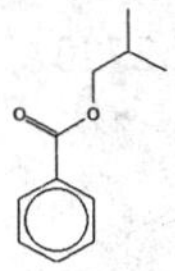

2702

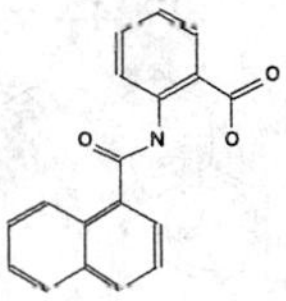

2703

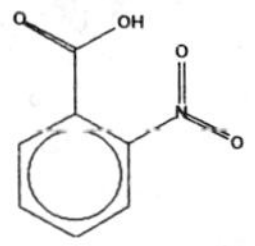

2704

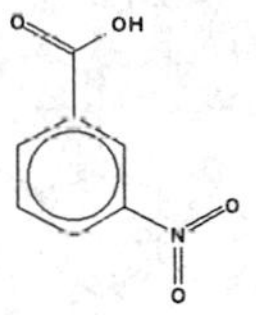

2705

2706

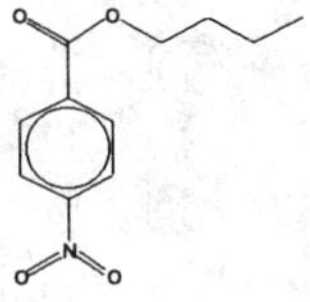

2707

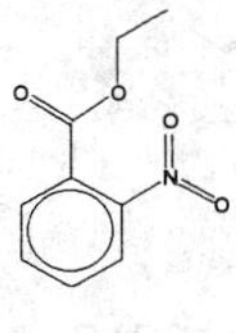

2708

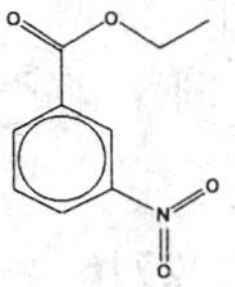

2709

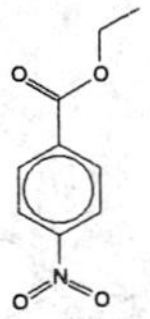

2710

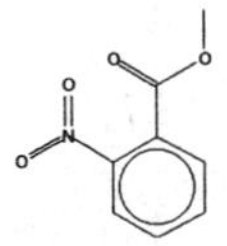

2711

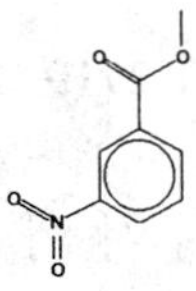

2712

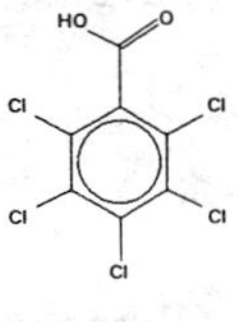

2713

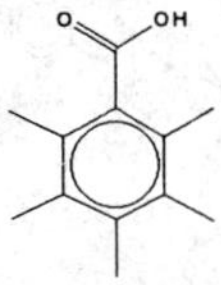

STRUCTURAL FORMULAS OF ORGANIC COMPOUNDS (continued)

In numeric order as they occur in the Table of Physical Constants of Organic Compounds

2714

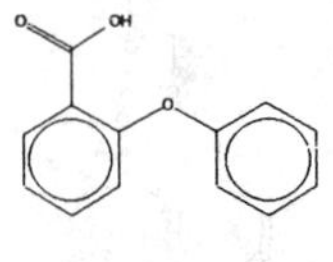

2715

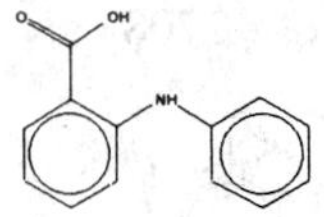

2716

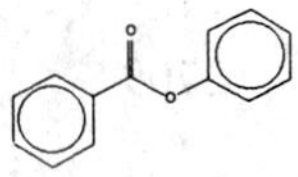

2717

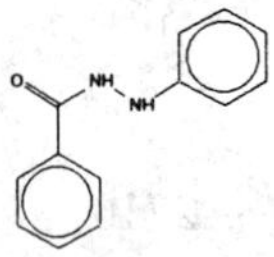

2718

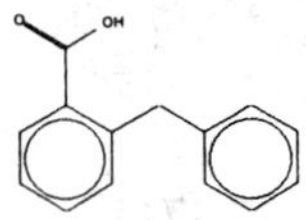

2719

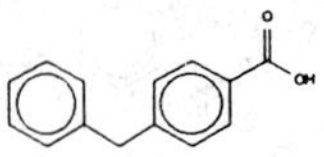

2720

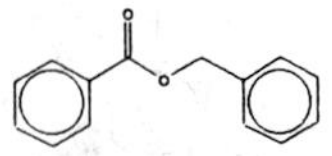

2721

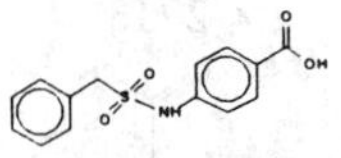

2722

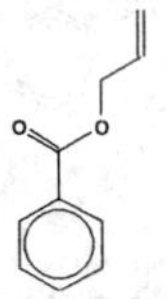

2723

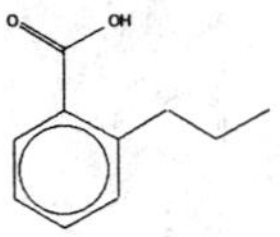

2724

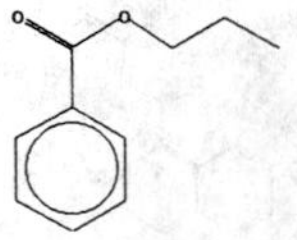

2726

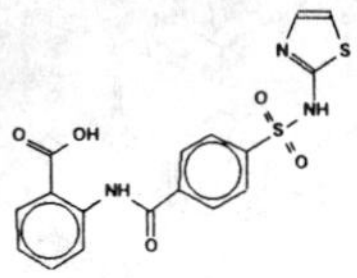

2727

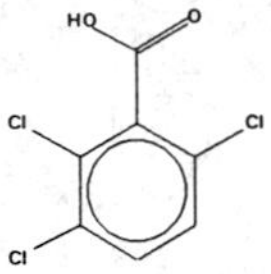

2728

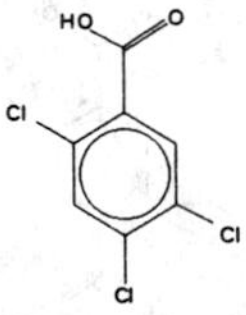

2729

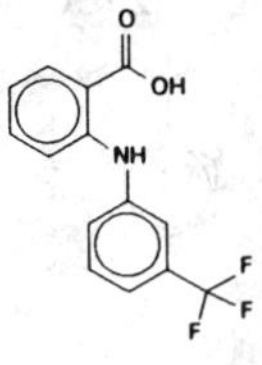

2730

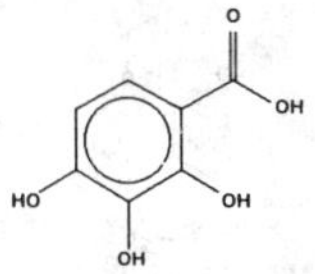

2731

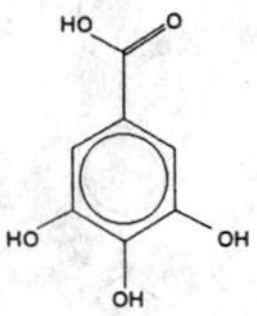

2732

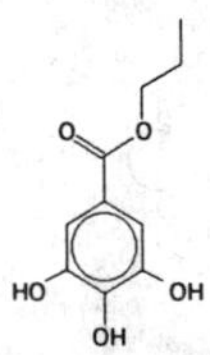

2733

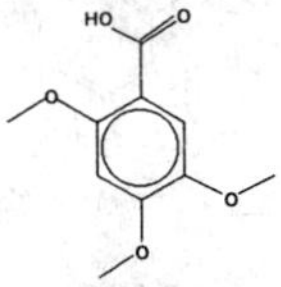

2734

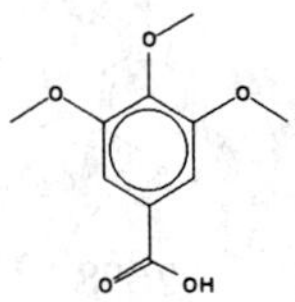

2735

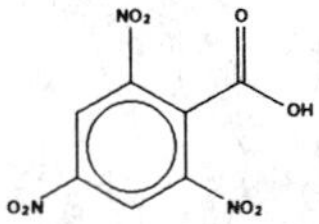

2736

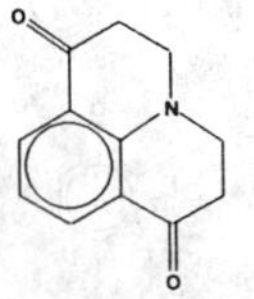

2737

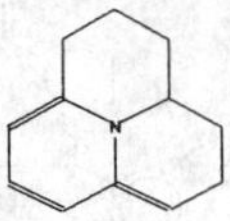

2738

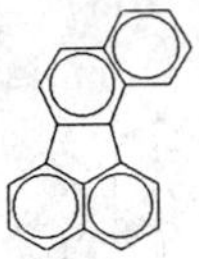

2739

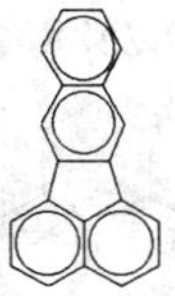

2740

2741

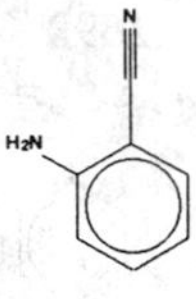

2742

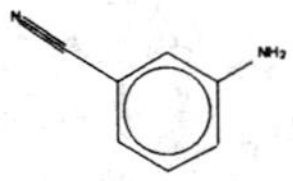

2744

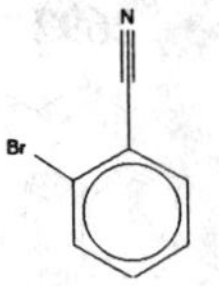

2745

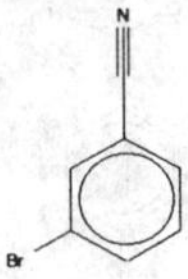

2746

2747

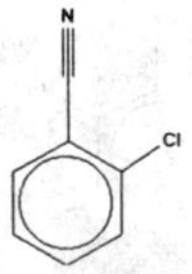

2748

2749

2750

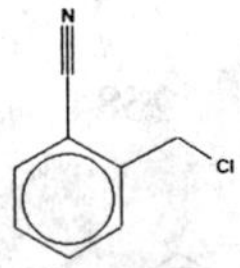

2751

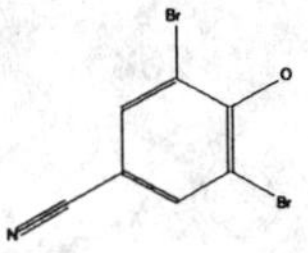

STRUCTURAL FORMULAS OF ORGANIC COMPOUNDS (continued)

In numeric order as they occur in the Table of Physical Constants of Organic Compounds

2752

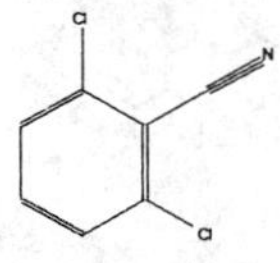

2753

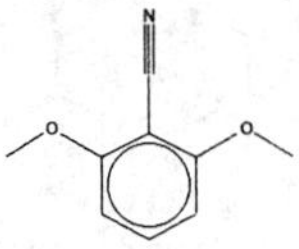

2754

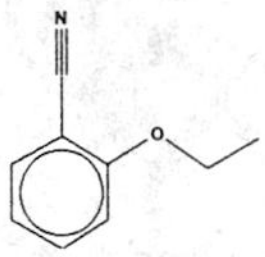

2755

2756

2757

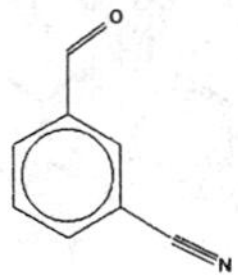

2758

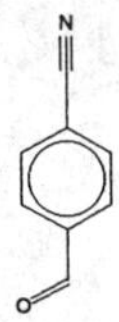

2759

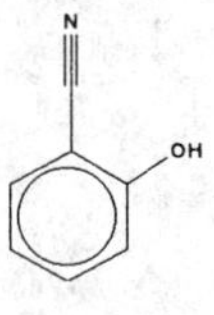

2760

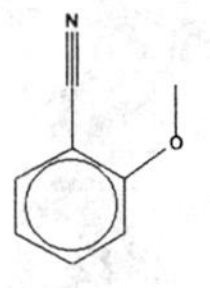

2761

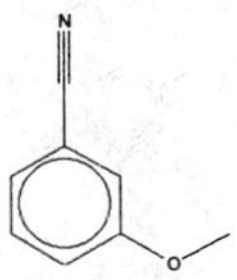

2762

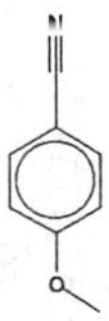

2763

2764

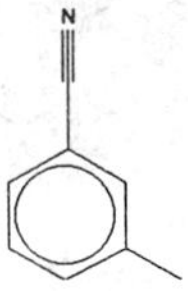

2765

2766

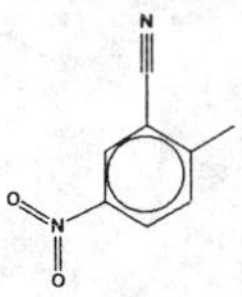

2767

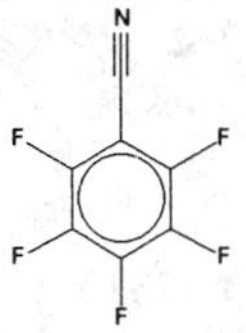

2768

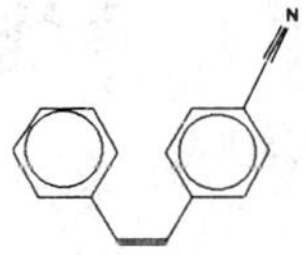

2769

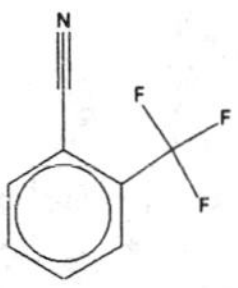

2770

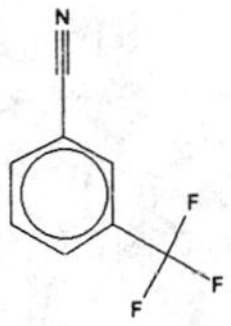

2771

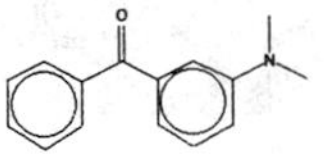

2772

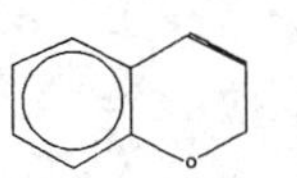

2773

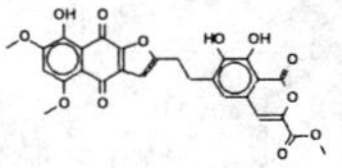

2774

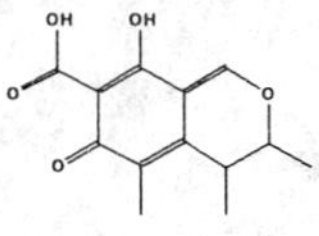

2775

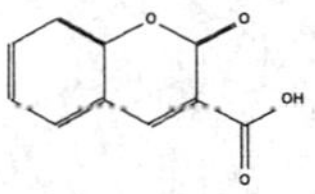

2776

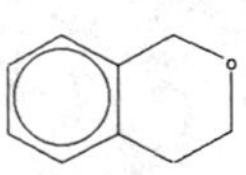

2777

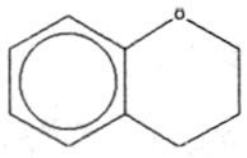

2778

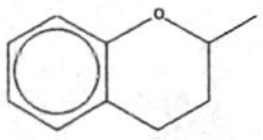

2780

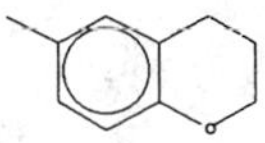

2781

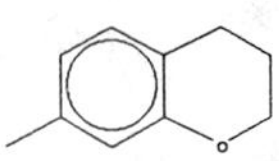

2782

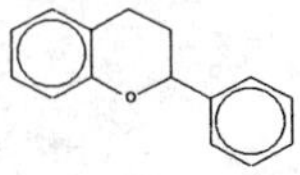

2783

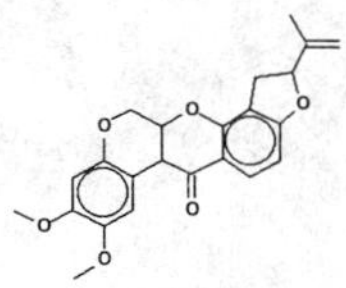

2784

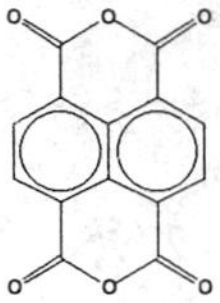

2785

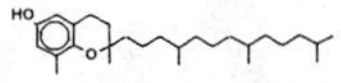

2787

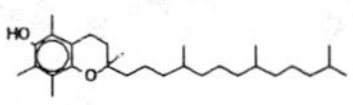

2788

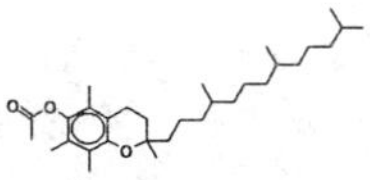

2789

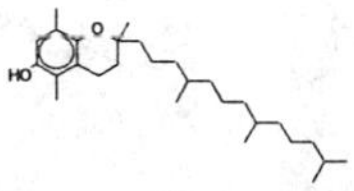

2790

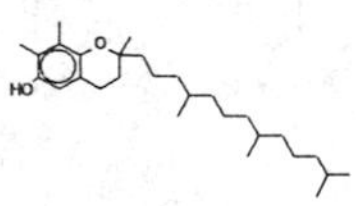

2791

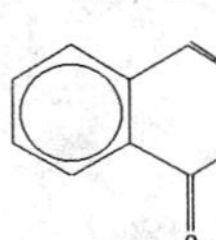

STRUCTURAL FORMULAS OF ORGANIC COMPOUNDS (continued)

In numeric order as they occur in the Table of Physical Constants of Organic Compounds

2792

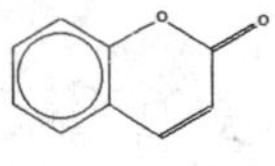

2793

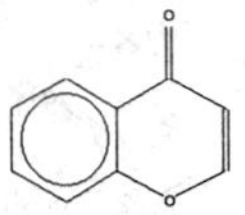

2795

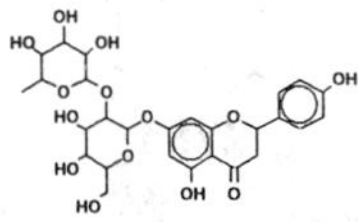

2797

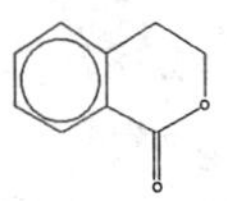

2798

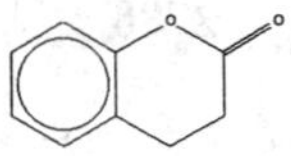

2799

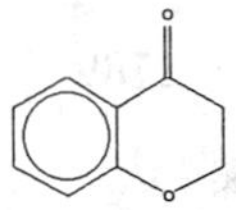

2800

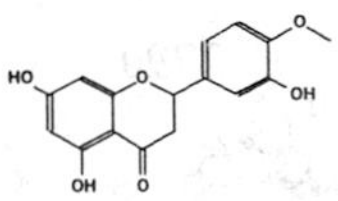

2801

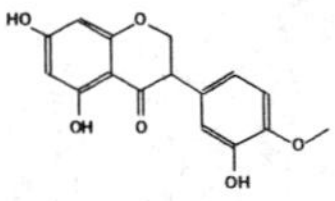

2802

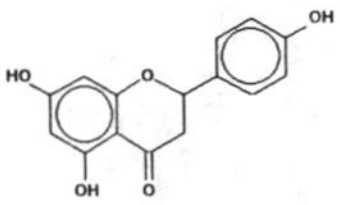

2803

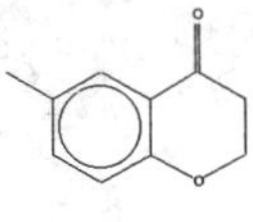

2804

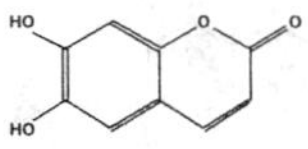

2805

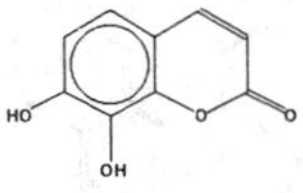

2806

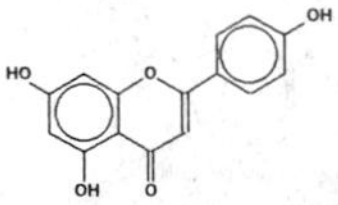

2807

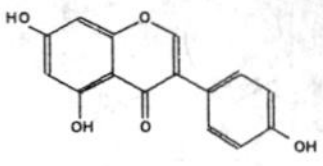

2808

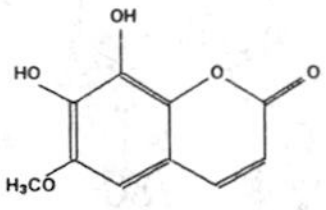

2809

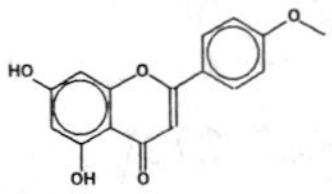

2810

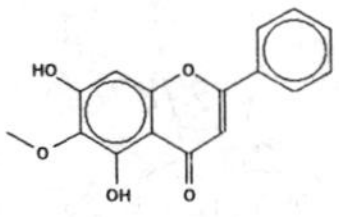

2811

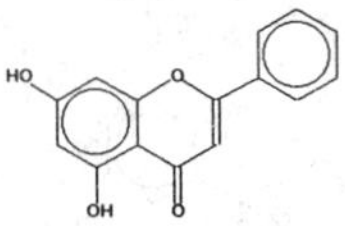

2812

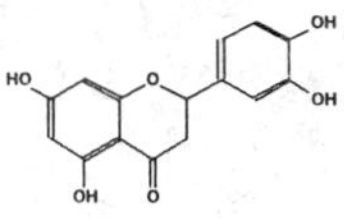

2813

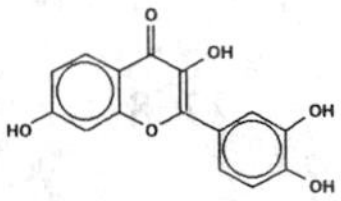

2814

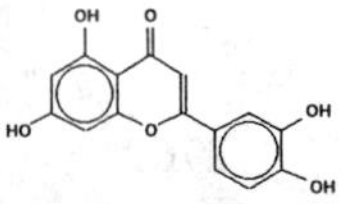

2815

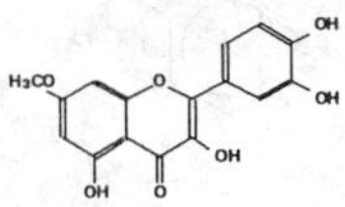

2816

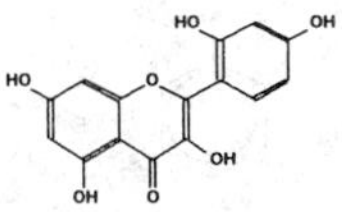

2817

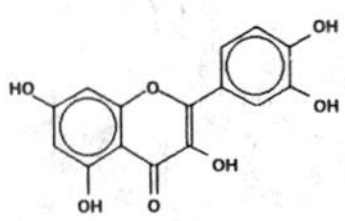

2818

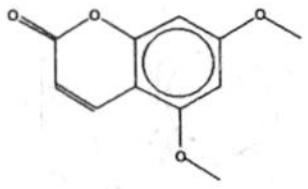

2819

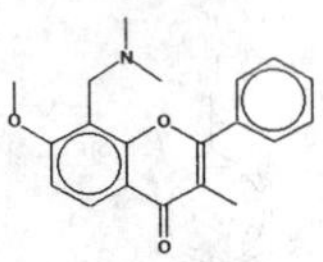

2820

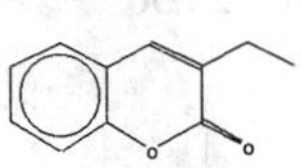

2822

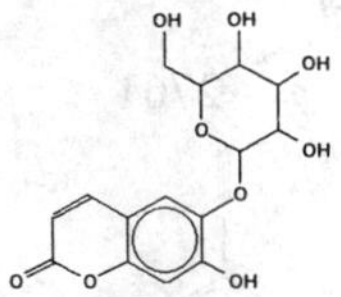

2825

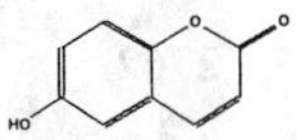

2826

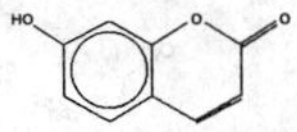

2827

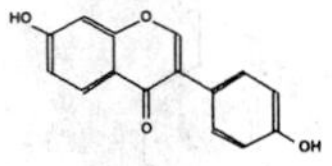

2828

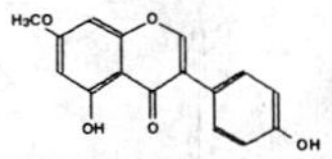

2829

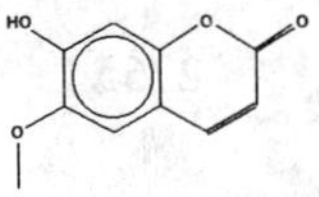

2830

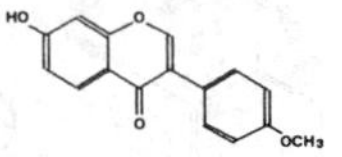

2831

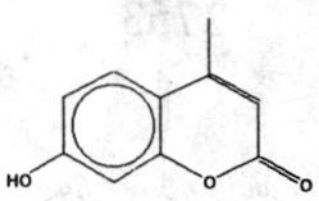

2832

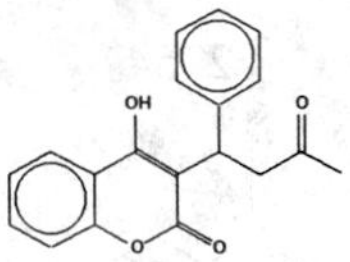

2833

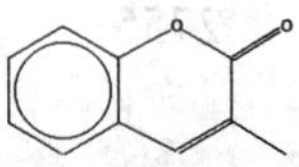

2834

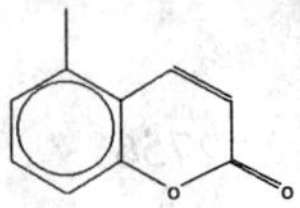

2835

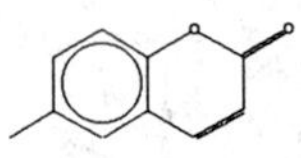

2836

2837

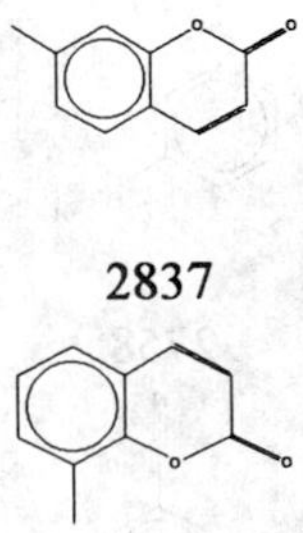

2838

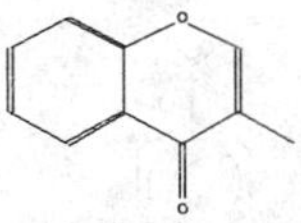

2839

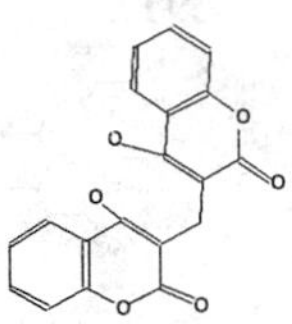

2840

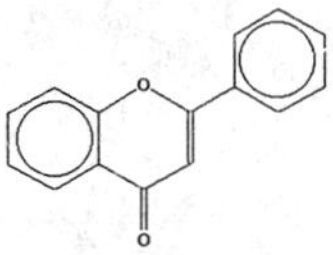

2841

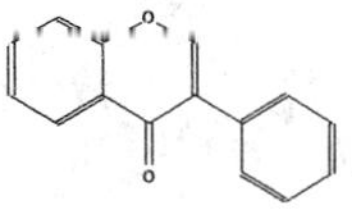

2842

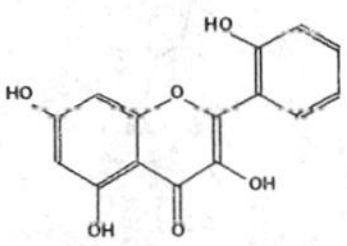

2843

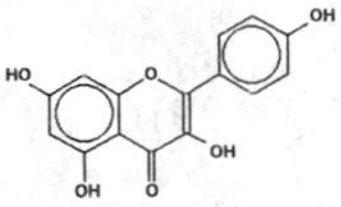

2844

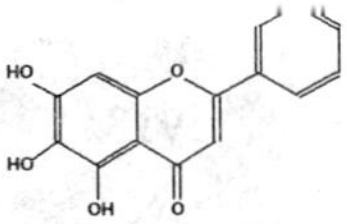

2845

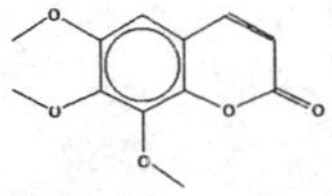

2847

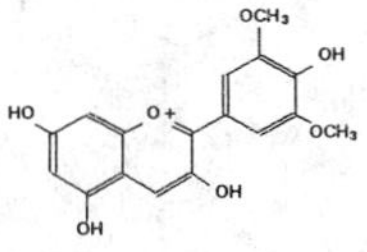

2849

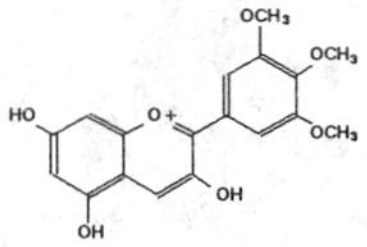

2850

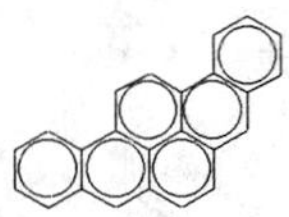

2851

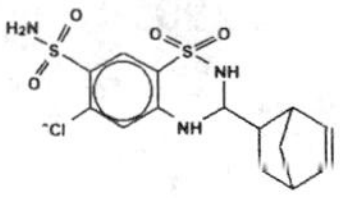

2852

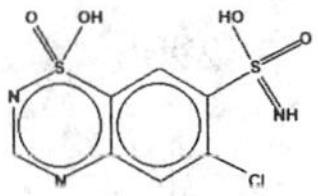

2853

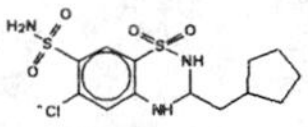

2854

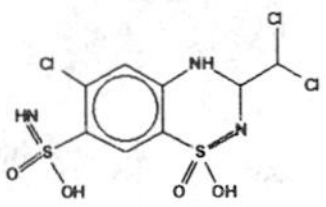

2855

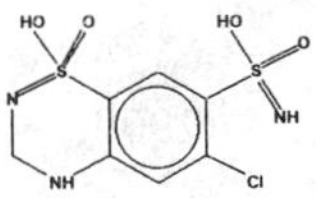

2856

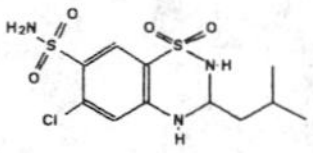

2857

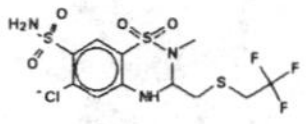

2858

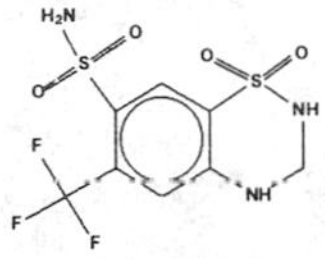

2859

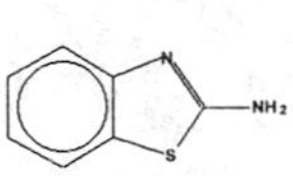

2860

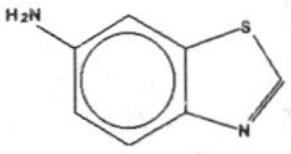

2861

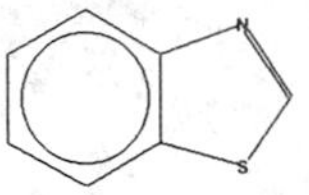

2862

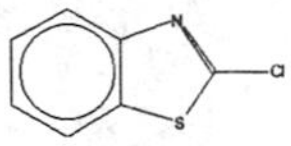

2863

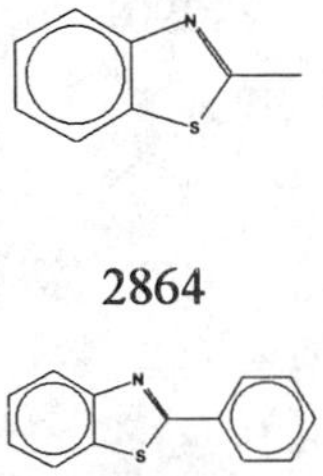

2864

2865

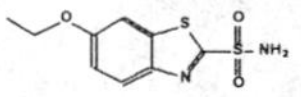

2866

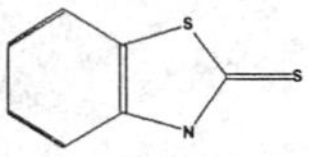

2867

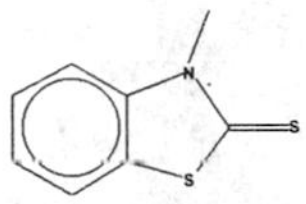

2868

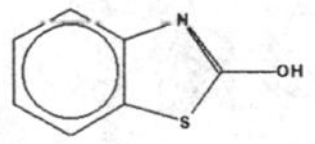

2869

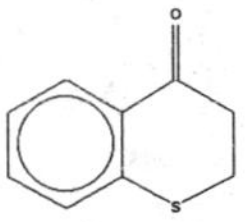

2870

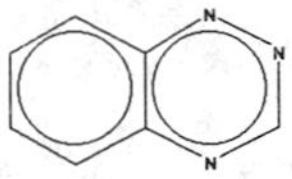

2871

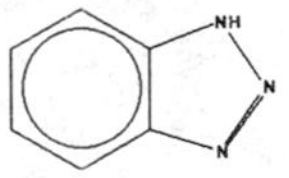

2872

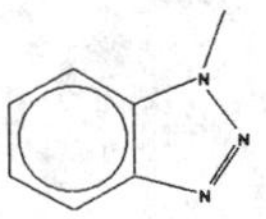

2873

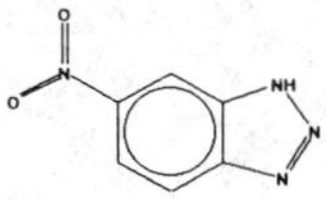

2874

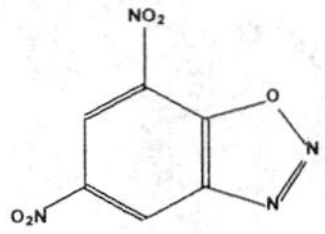

2875

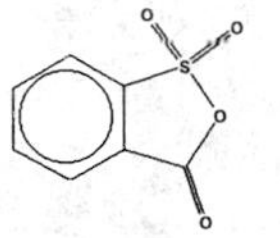

2876

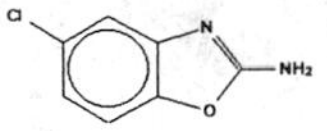

2877

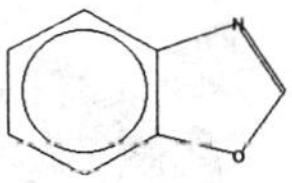

2878

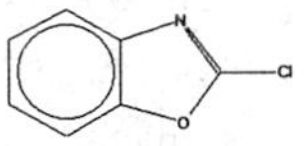

2879

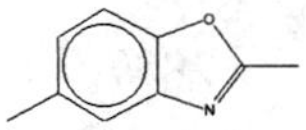

2880

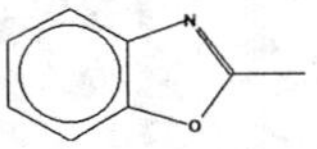

STRUCTURAL FORMULAS OF ORGANIC COMPOUNDS (continued)

In numeric order as they occur in the Table of Physical Constants of Organic Compounds

2881

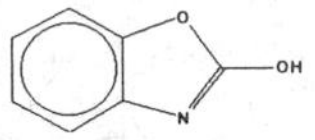

2882

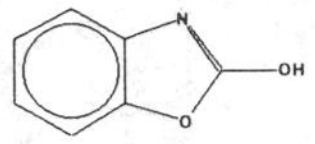

2883

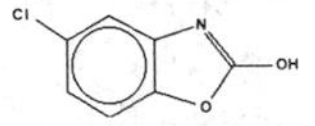

2884

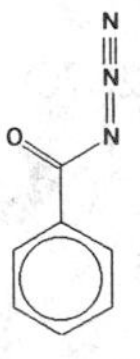

2885

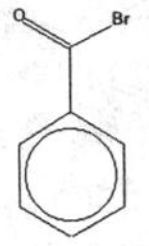

2886

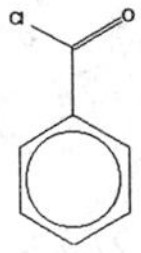

2887

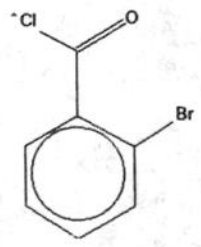

2888

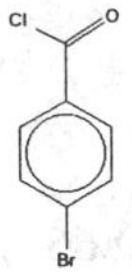

2889

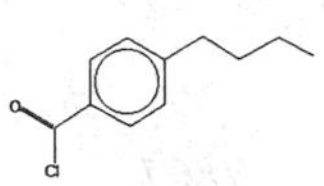

2890

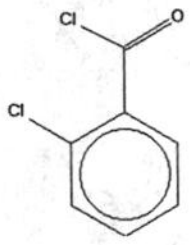

2891

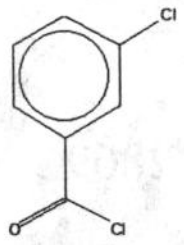

2892

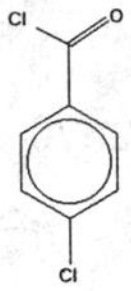

2893

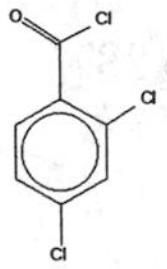

2894

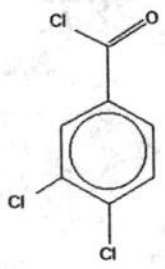

2895

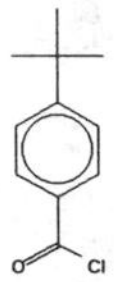

2896

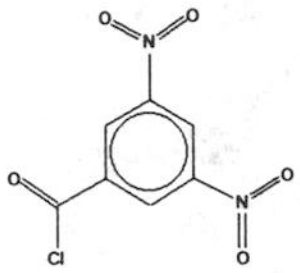

2897

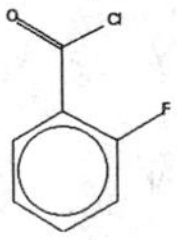

2898

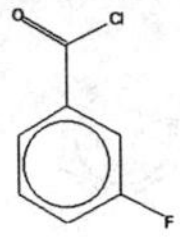

2899

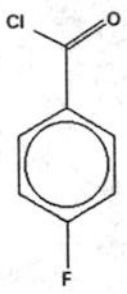

2900

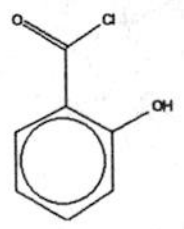

2901

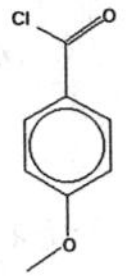

2902

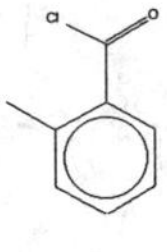

2903

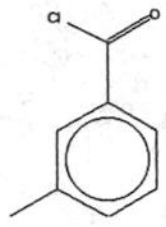

2904

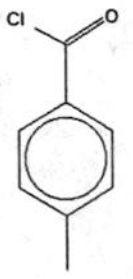

2905

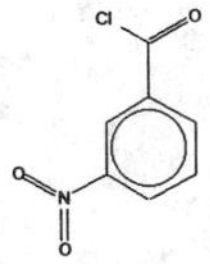

2906

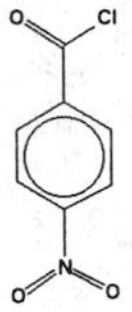

2907

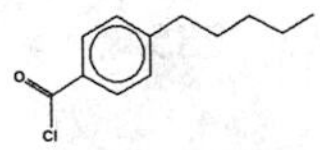

2908

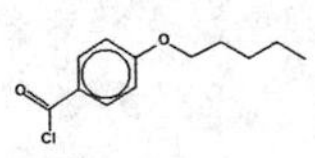

2909

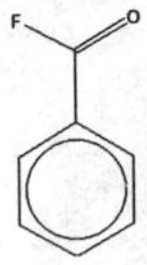

2910

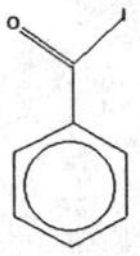

2911

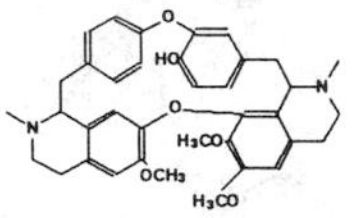

2912

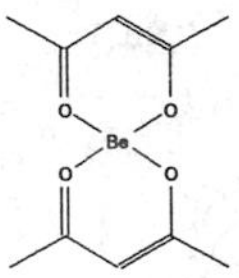

2913

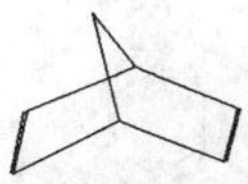

2914

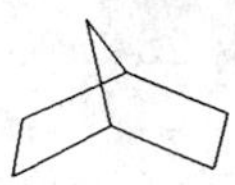

2915

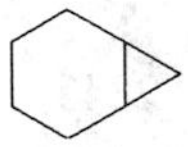

2916

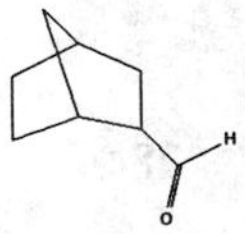

2918

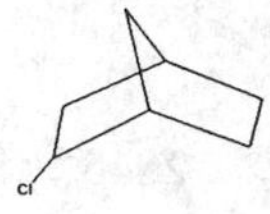

In numeric order as they occur in the Table of Physical Constants of Organic Compounds

2919

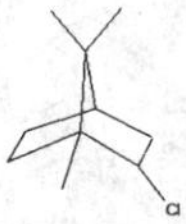

2920

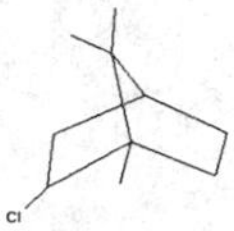

2921

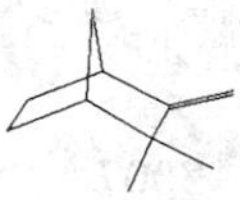

2922

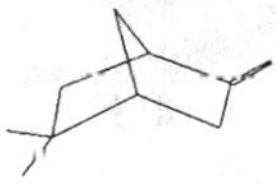

2923

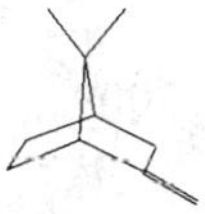

2925

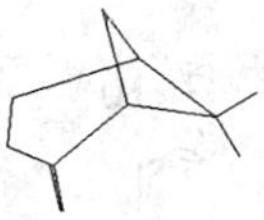

2926

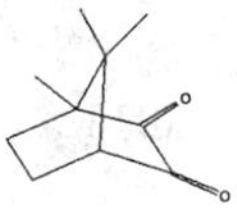

2927

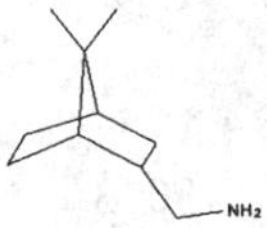

2928

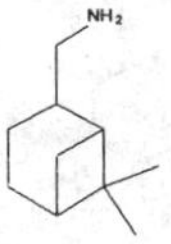

2929

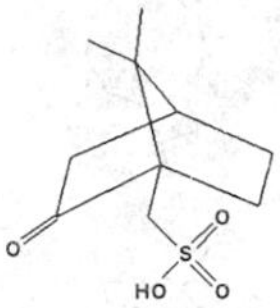

2930

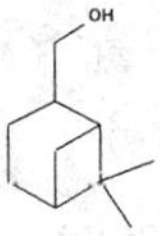

2932

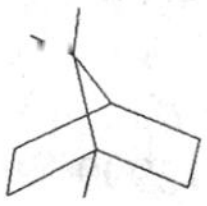

2933

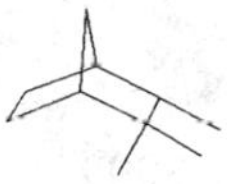

2934

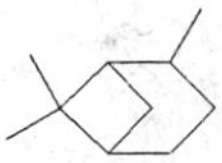

2935

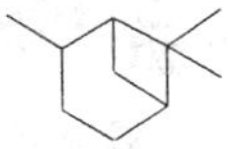

2936

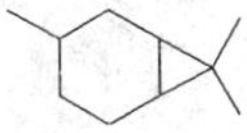

2937

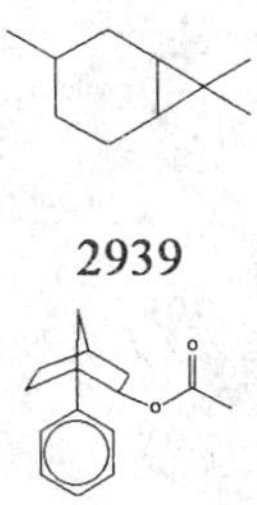

2939

2941

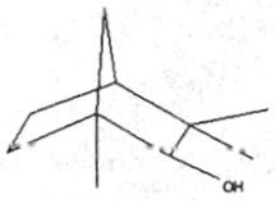

2944

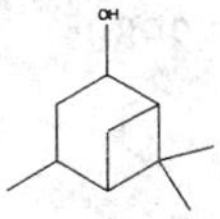

2945

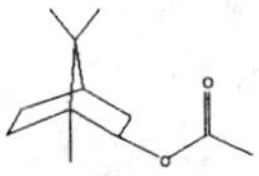

2946

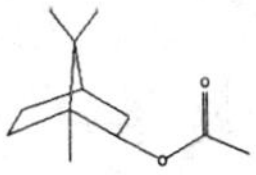

2947

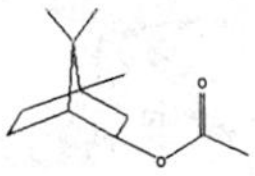

2948

2950

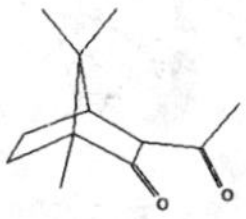

2951

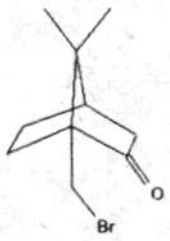

2952

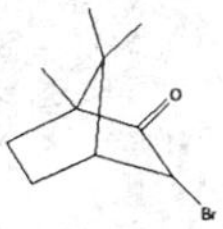

2953

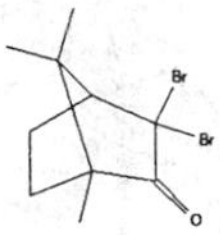

2954

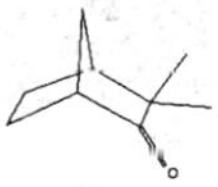

2955

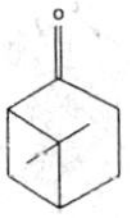

2957

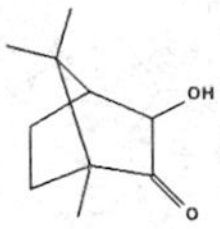

2958

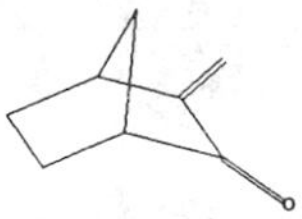

2959

2960

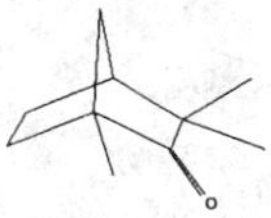

2961

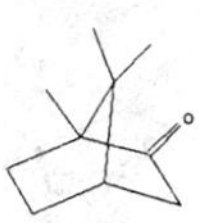

2962

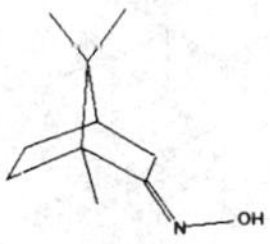

2963

2964

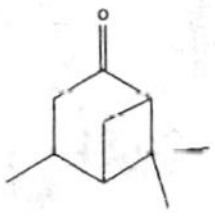

2966

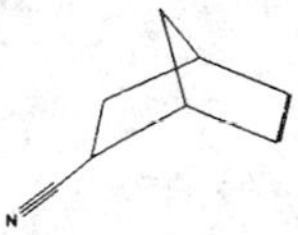

2967

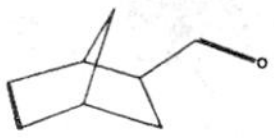

2968

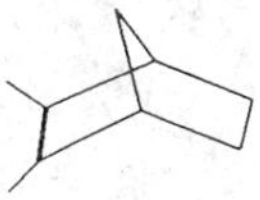

STRUCTURAL FORMULAS OF ORGANIC COMPOUNDS (continued)

In numeric order as they occur in the Table of Physical Constants of Organic Compounds

2969

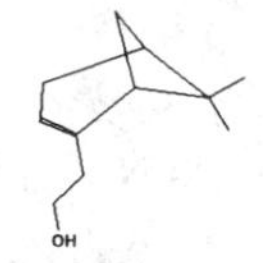

OH

2970

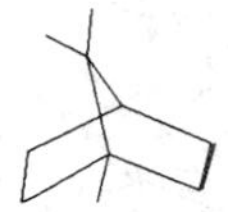

2971

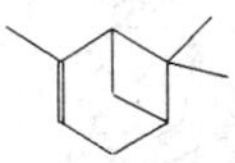

2972

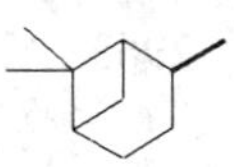

2973

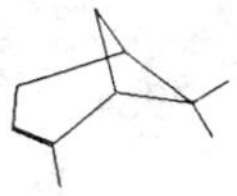

2974

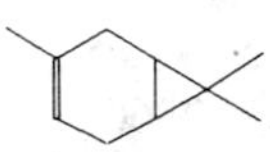

2975

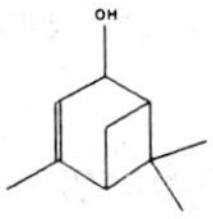

OH

2976

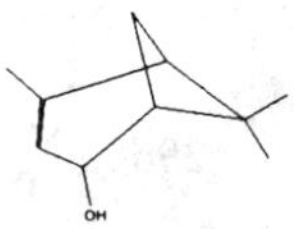

OH

2977

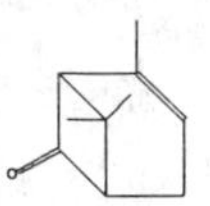

O

2978

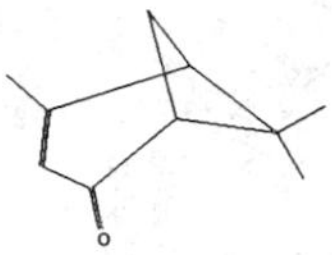

O

2980

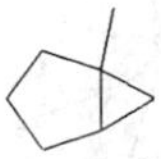

2981

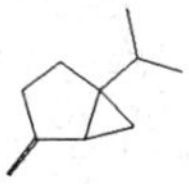

2982

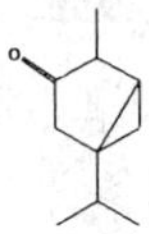

O

2983

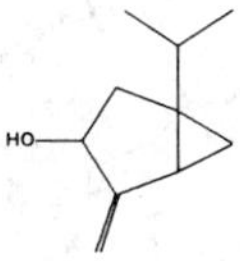

HO

2984

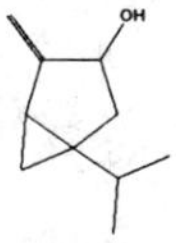

OH

2985

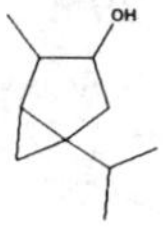

OH

2986

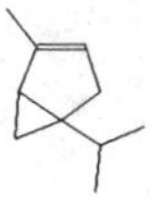

2987

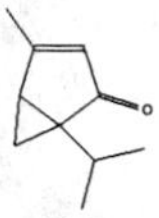

O

2988

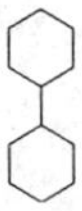

2989

O

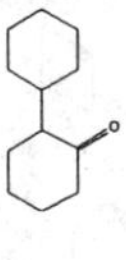

2990

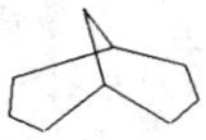

2991

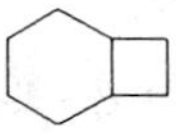

2992

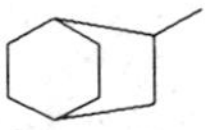

2993

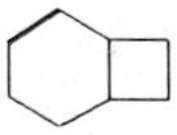

2994

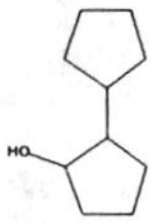

HO

2995

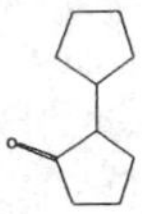

O

2996

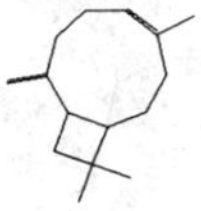

2997

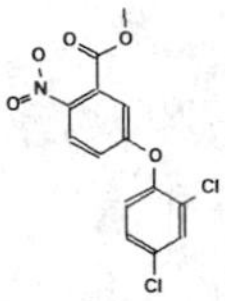

O O N O O O Cl Cl

2998

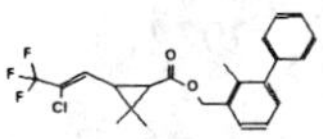

F F F Cl O O

2999

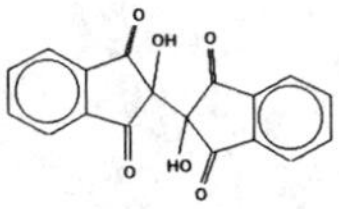

O OH O O HO O

3000

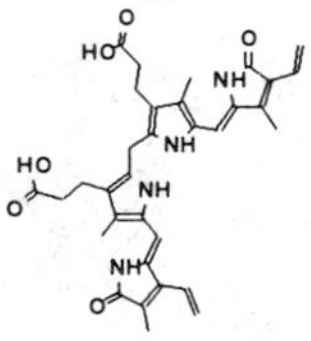

HO O O NH NH HO O NH NH O

3001

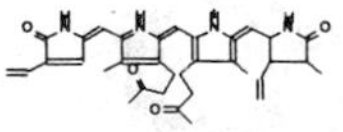

3002

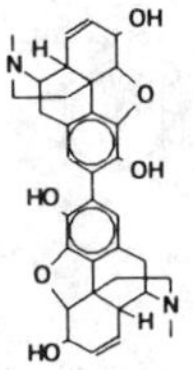

N H OH O OH HO O H N HO

3003

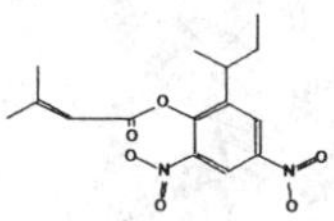

O O O N O N O O

3004

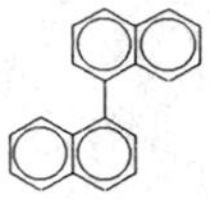

3005

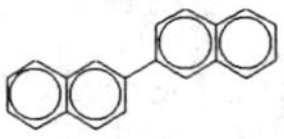

3007

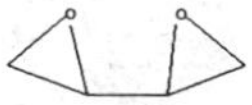

O O

3008

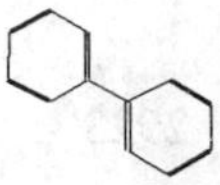

3009

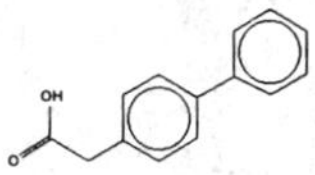

OH O

3010

NH_2

3011

NH_2

3013

H_2N

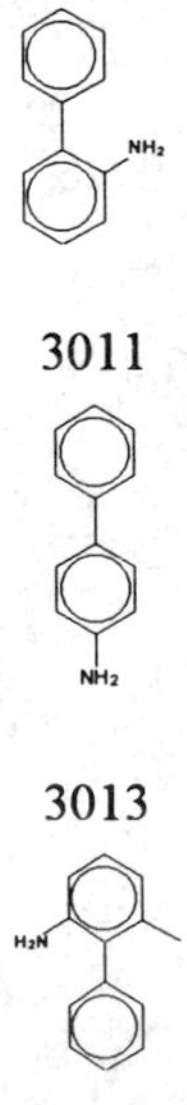

STRUCTURAL FORMULAS OF ORGANIC COMPOUNDS (continued)

In numeric order as they occur in the Table of Physical Constants of Organic Compounds

3014

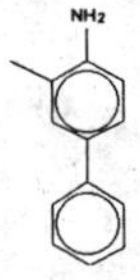

3015

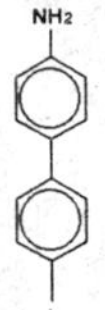

3016

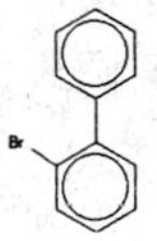

3017

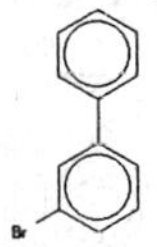

3018

3019

3020

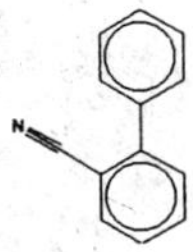

3021

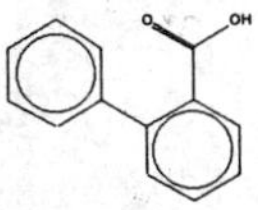

3022

3023

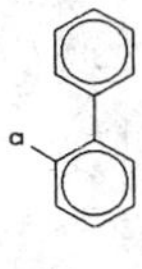

3024

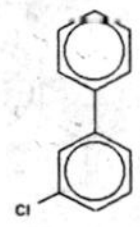

3025

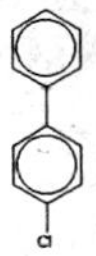

3026

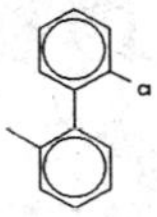

3027

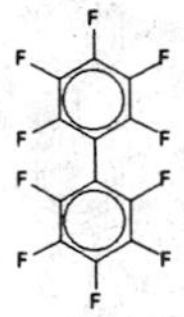

3028

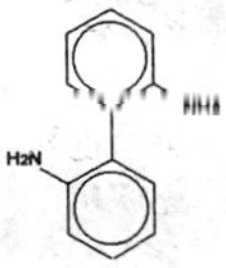

3029

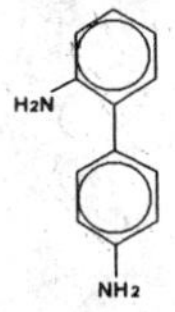

3030

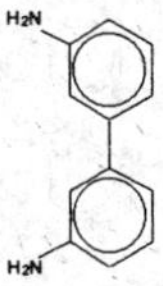

3031

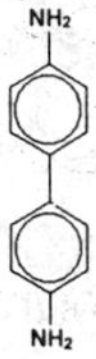

3032

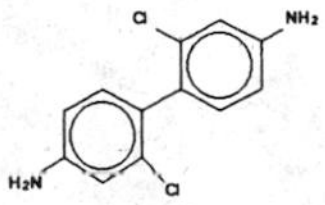

3033

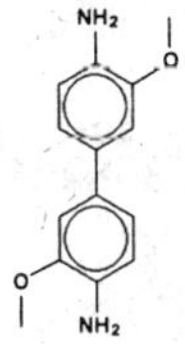

3034

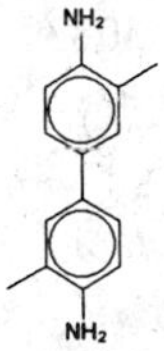

3035

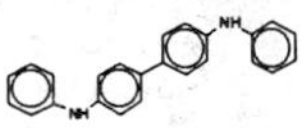

3037

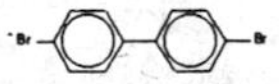

3038

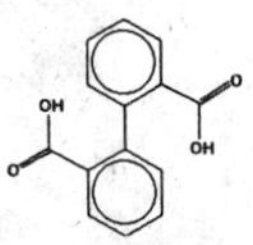

3039

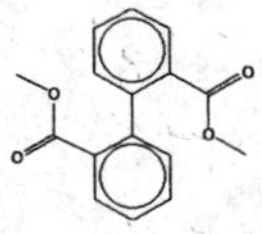

3040

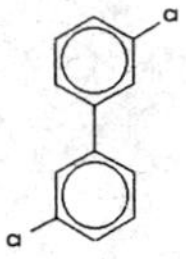

3041

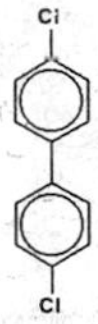

3042

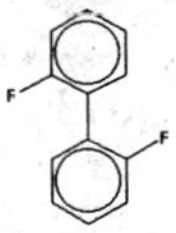

3043

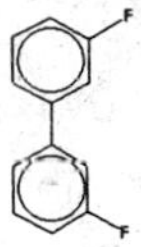

3044

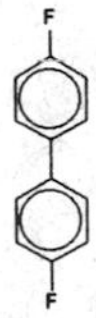

3045

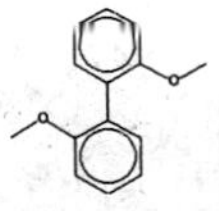

3046

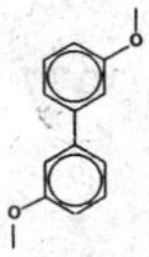

3050

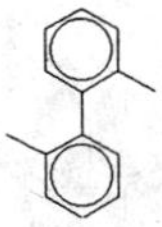

3051

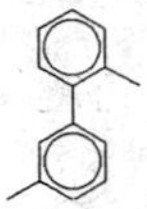

3052

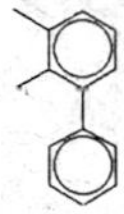

3053

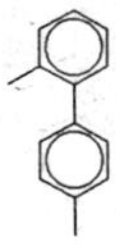

3054

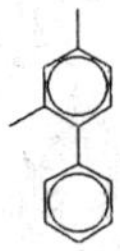

3055

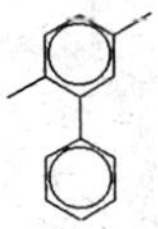

3056

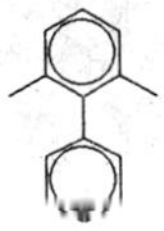

3057

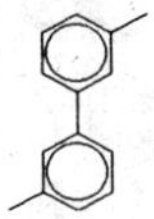

STRUCTURAL FORMULAS OF ORGANIC COMPOUNDS (continued)

In numeric order as they occur in the Table of Physical Constants of Organic Compounds

3058

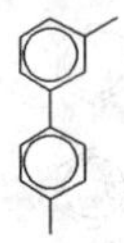

3059

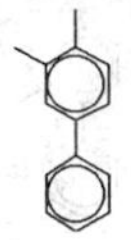

3060

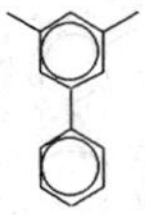

3061

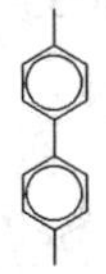

3062

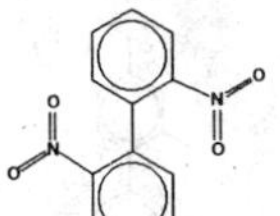

3063

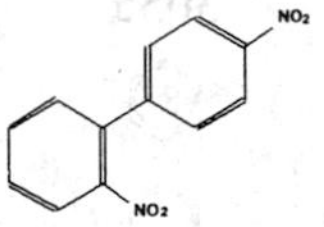

NO₂

NO₂

3064

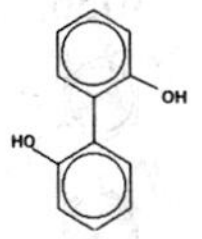

OH

HO

3065

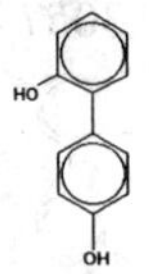

HO

OH

3066

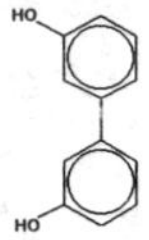

HO

HO

3070

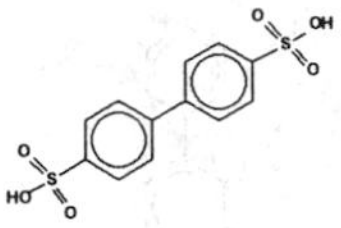

3071

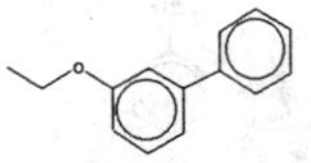

3072

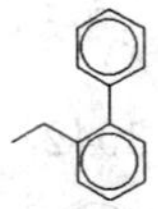

3073

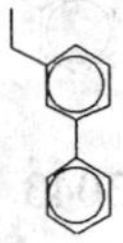

3074

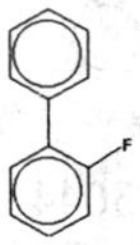

F

3075

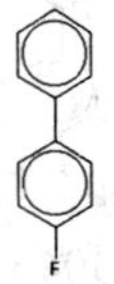

F

3076

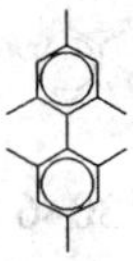

3077

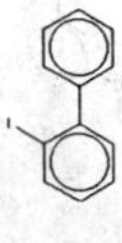

I

3078

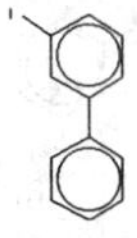

I

3079

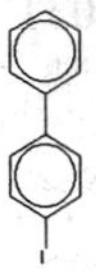

I

3080

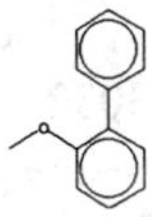

O

3081

O

3082

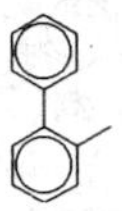

3083

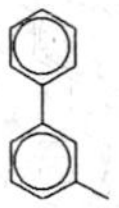

3084

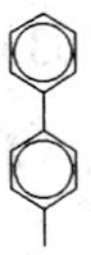

3085

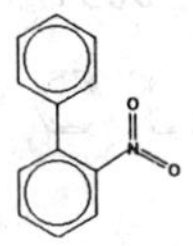

3086

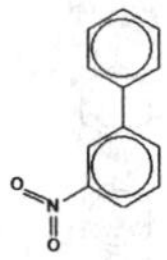

3087

3089

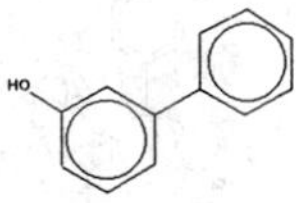

HO

3090

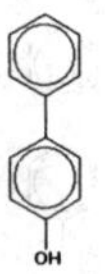

OH

3091

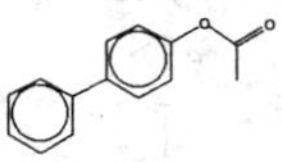

O

3092

HO

Cl

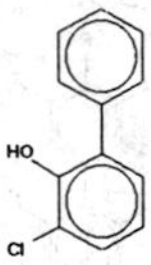

3093

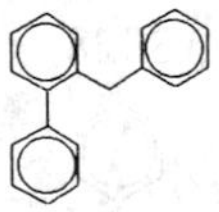

3094

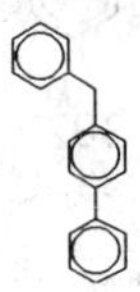

3095

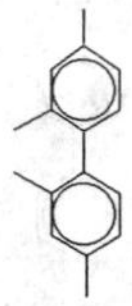

3096

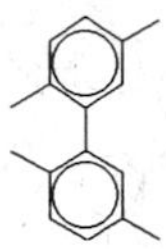

3098

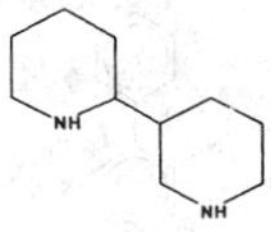

NH

NH

3099

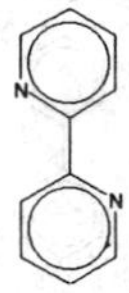

N

N

3100

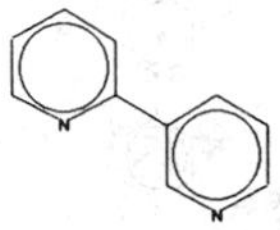

N

N

3101

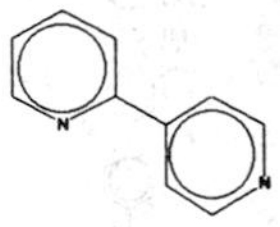

N

N

3102

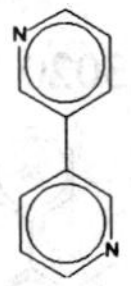

N

N

STRUCTURAL FORMULAS OF ORGANIC COMPOUNDS (continued)

In numeric order as they occur in the Table of Physical Constants of Organic Compounds

3103

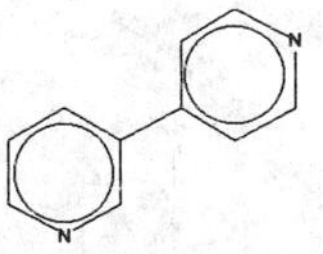

3104

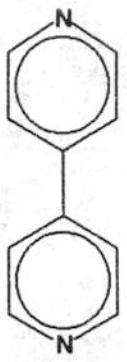

3105

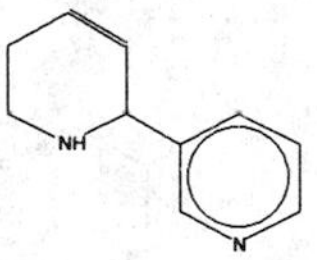

3108

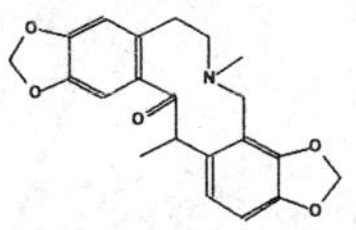

3109

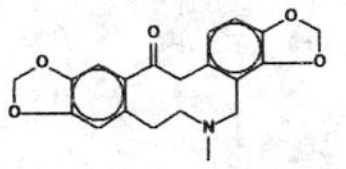

3111

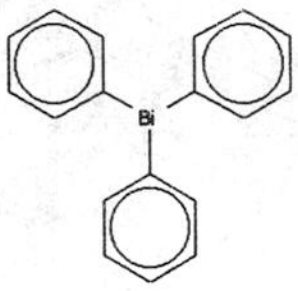

3112

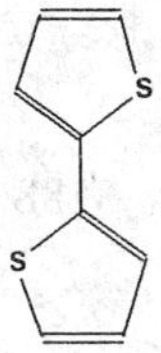

3113

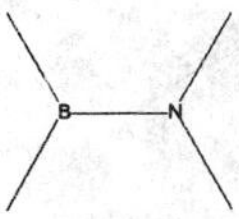

3114

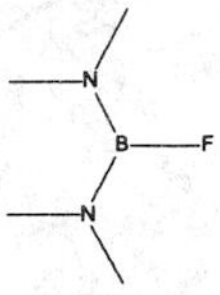

3115

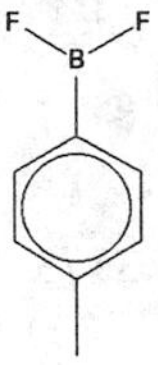

3116

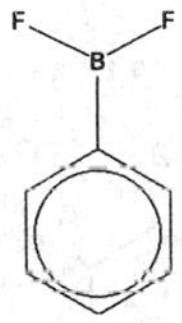

3117

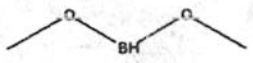

3118

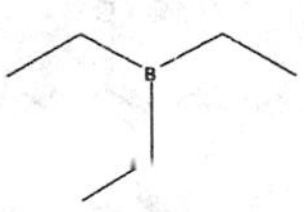

3119

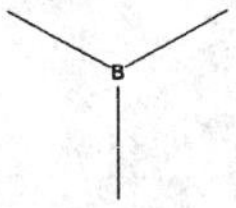

3120

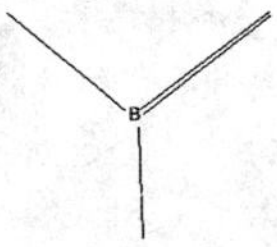

3121

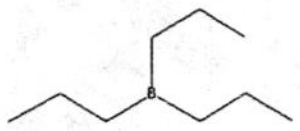

3122

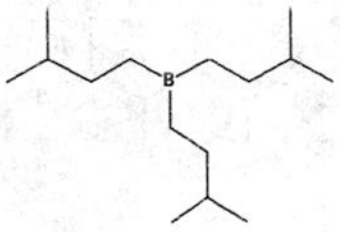

3123

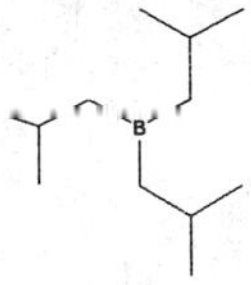

3124

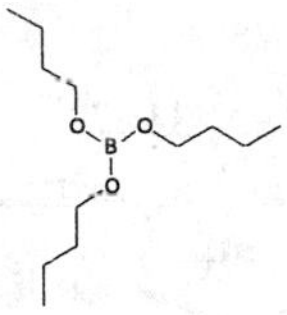

3125

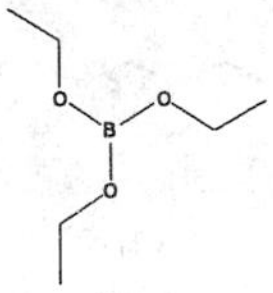

3126

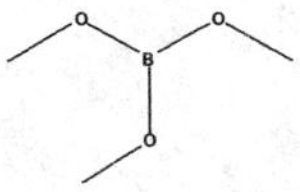

3127

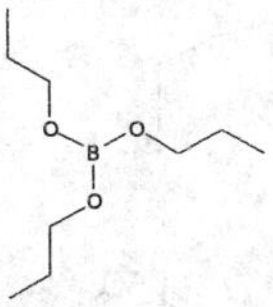

3128

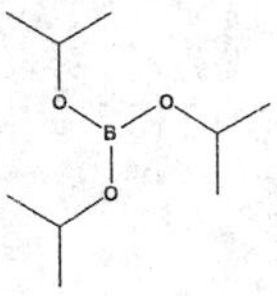

3129

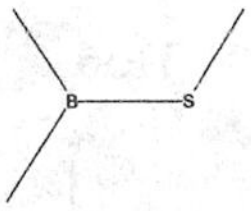

3130

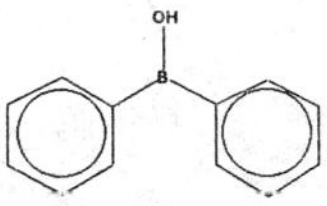

3131

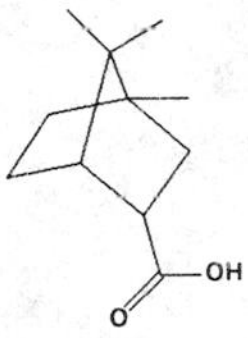

3132

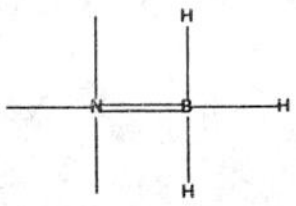

3134

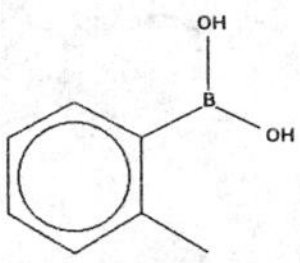

3135

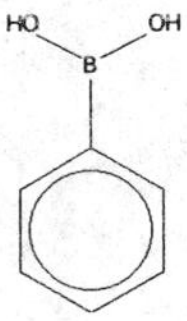

3136

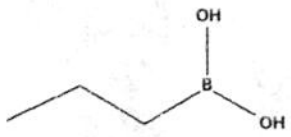

3137

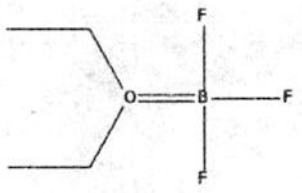

3138

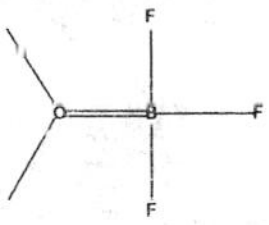

3139

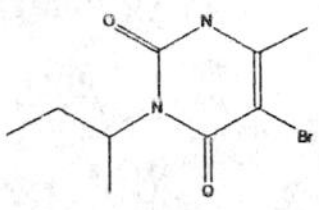

3140

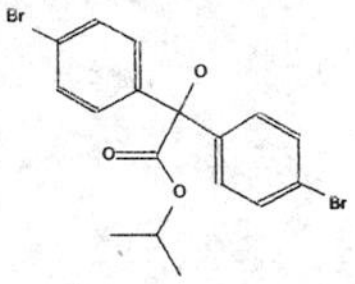

3141

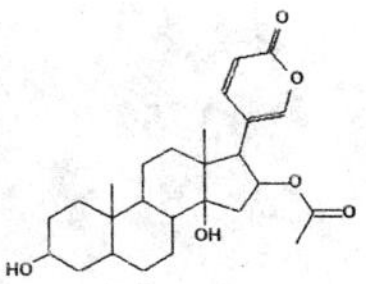

STRUCTURAL FORMULAS OF ORGANIC COMPOUNDS (continued)

In numeric order as they occur in the Table of Physical Constants of Organic Compounds

3143

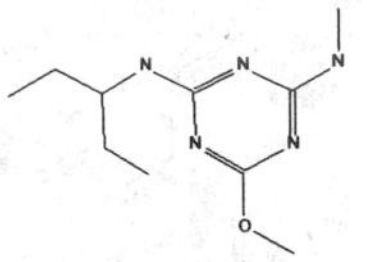

3144

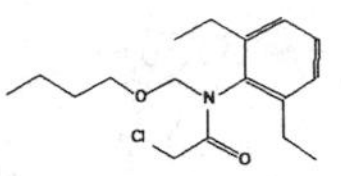

3145

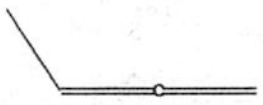

3146

3147

3148

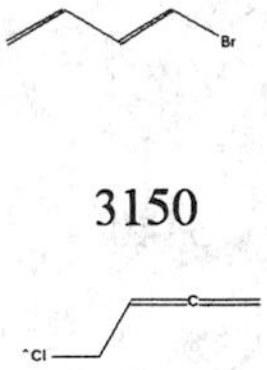

3150

3151

3152

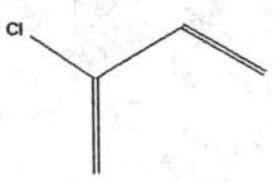

3153

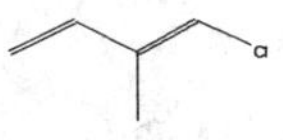

3154

3155

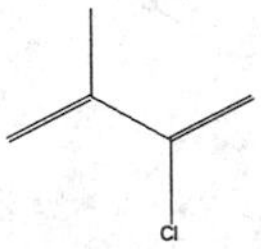

3156

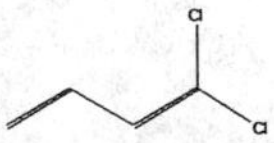

3157

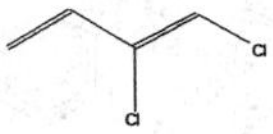

3158

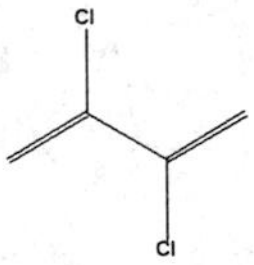

3159

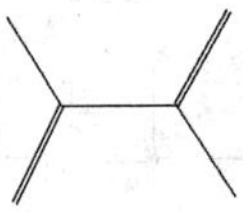

3160

3161

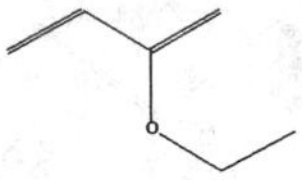

3162

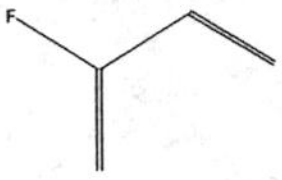

3163

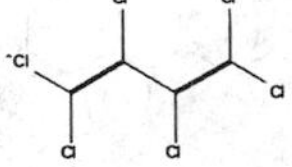

3164

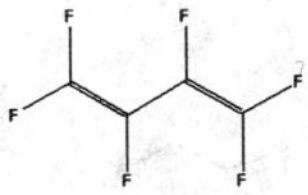

3166

3167

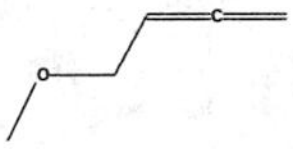

3168

3169

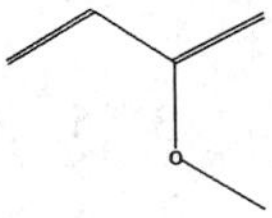

3170

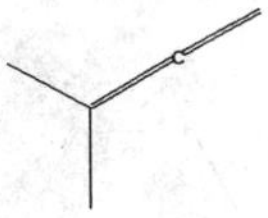

3171

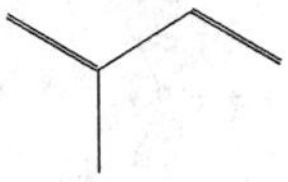

3172

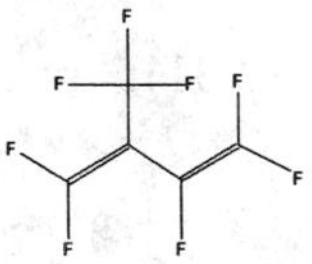

3173

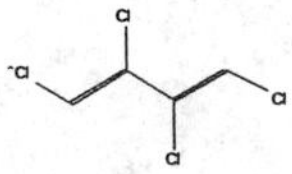

3174

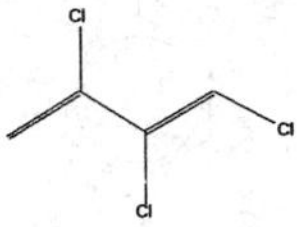

3175

3176

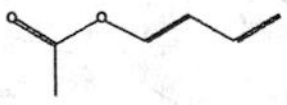

3177

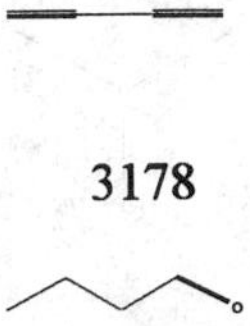

3178

3179

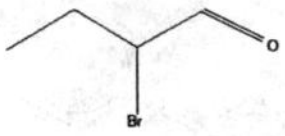

3180

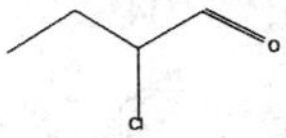

3181

3182

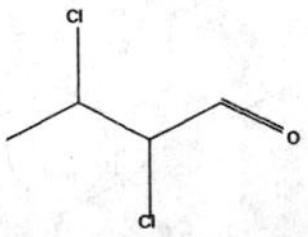

3183

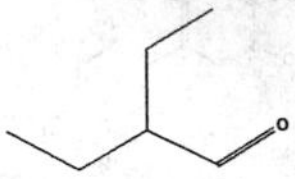

3184

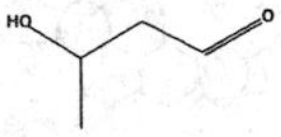

3185

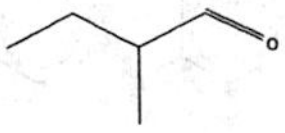

3186

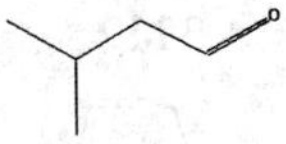

3188

STRUCTURAL FORMULAS OF ORGANIC COMPOUNDS (continued)

In numeric order as they occur in the Table of Physical Constants of Organic Compounds

3189

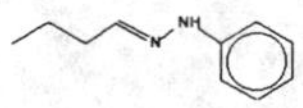

3190

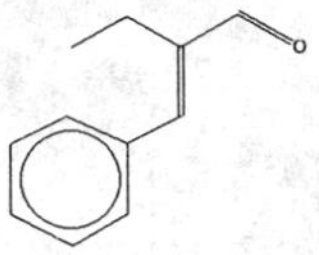

3191

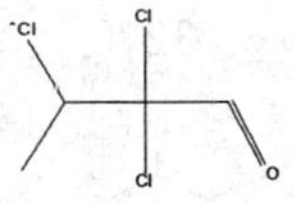

3192

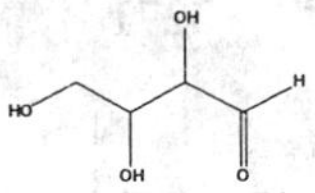

3194

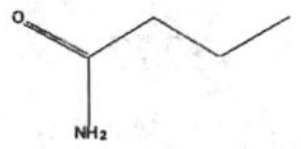

3195

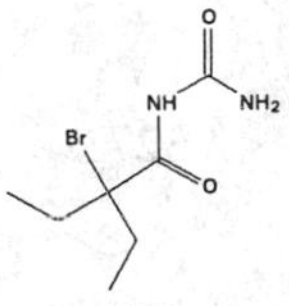

3196

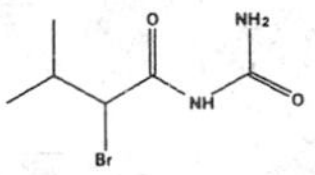

3197

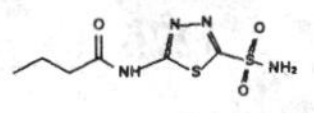

3198

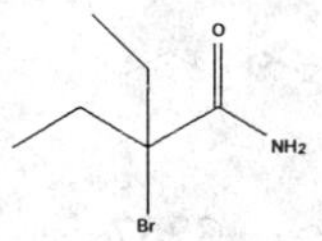

3199

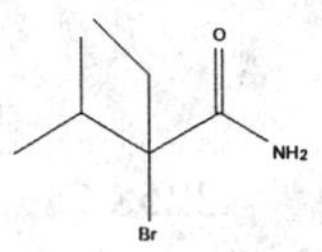

3200

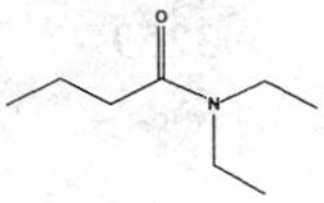

3201

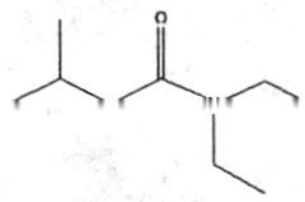

3202

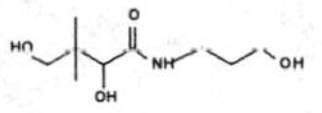

3203

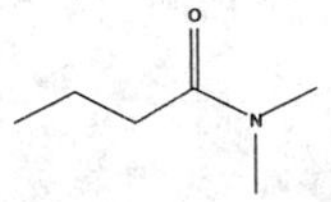

3205

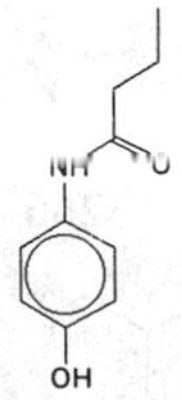

3206

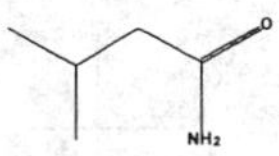

3207

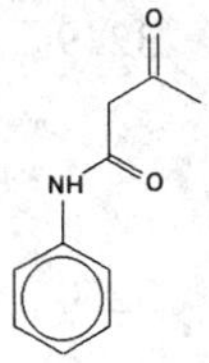

3208

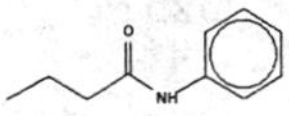

3209

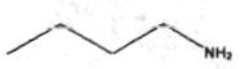

3210

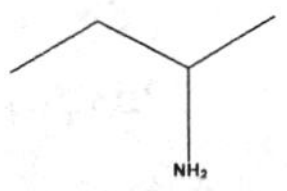

3211

3212

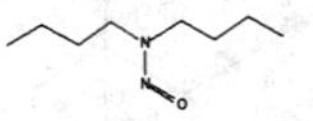

3213

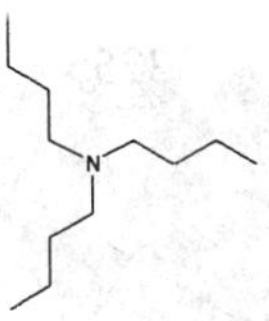

3214

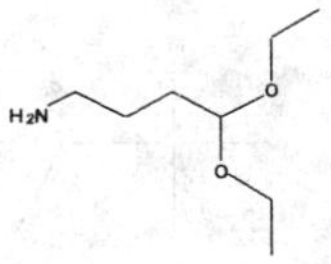

3215

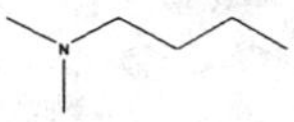

3216

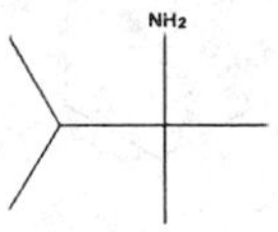

3217

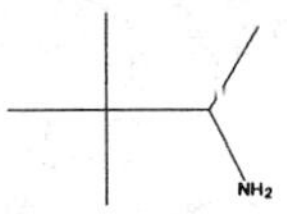

3218

3220

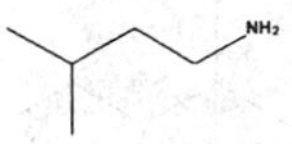

3221

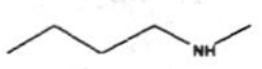

3222

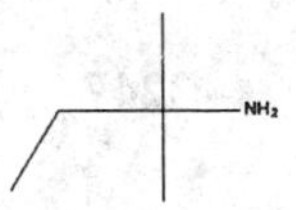

3223

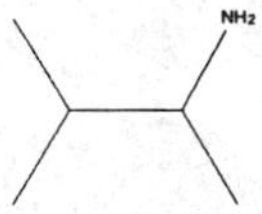

3224

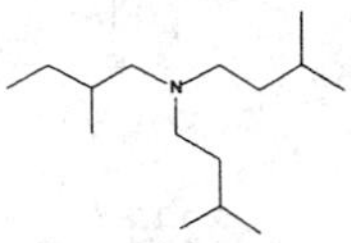

3225

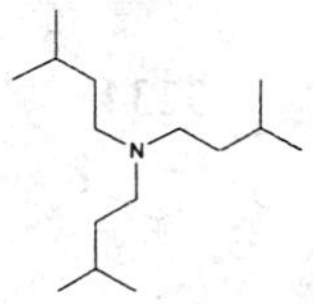

3226

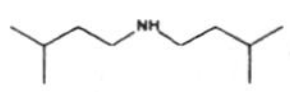

3227

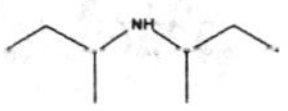

3228

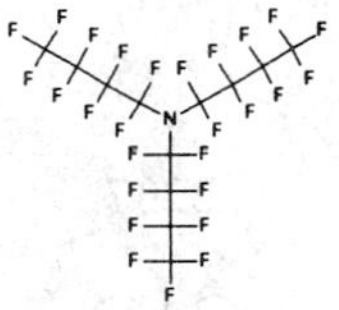

3229

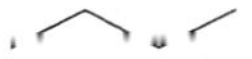

3230

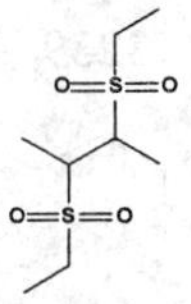

STRUCTURAL FORMULAS OF ORGANIC COMPOUNDS (continued)

In numeric order as they occur in the Table of Physical Constants of Organic Compounds

3231

3232

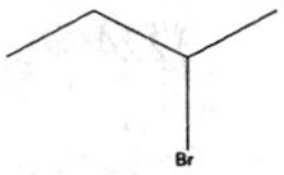

3233

3234

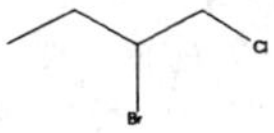

3235

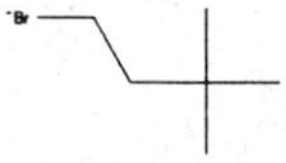

3236

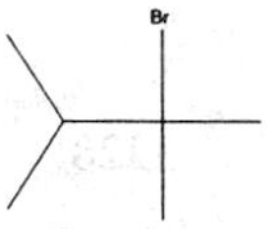

3237

3238

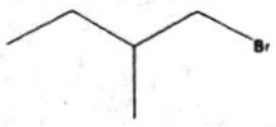

3239

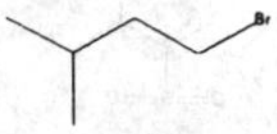

3240

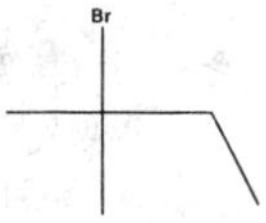

3241

3242

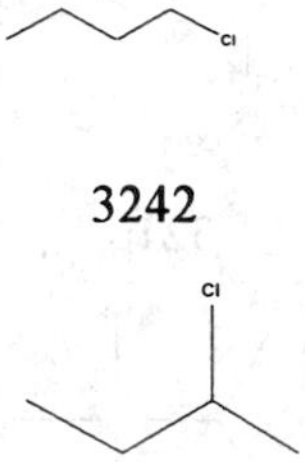

3243

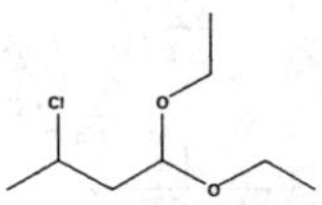

3244

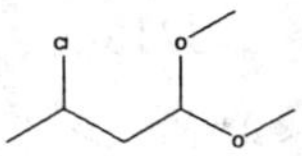

3245

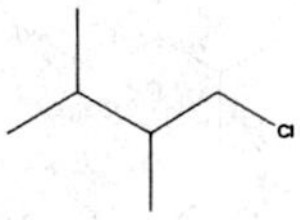

3246

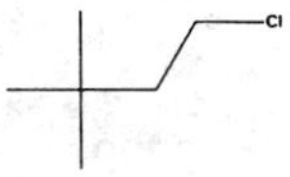

3247

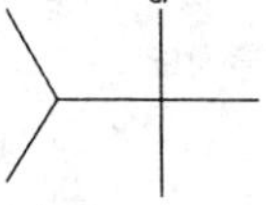

3248

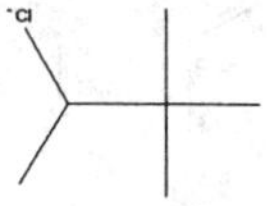

3249

3250

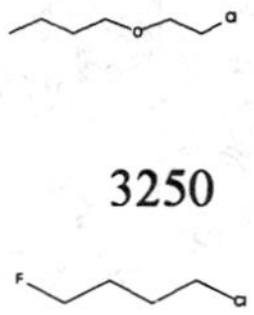

3251

3252

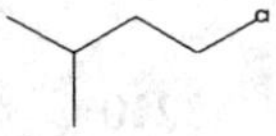

3253

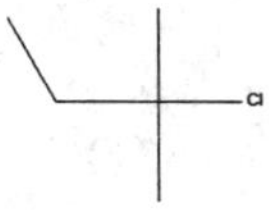

3254

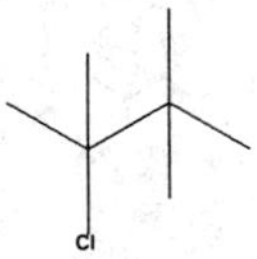

3255

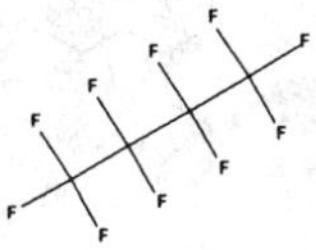

3256

3257

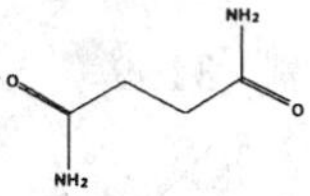

3258

3259

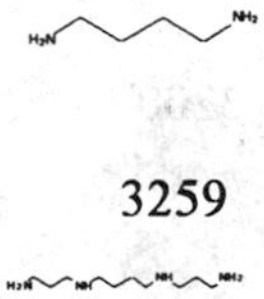

3262

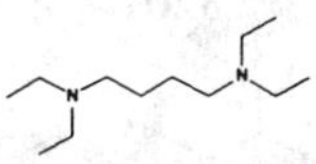

3263

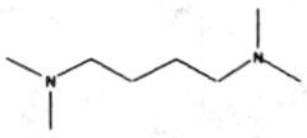

3264

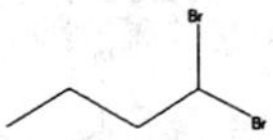

3265

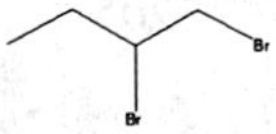

3266

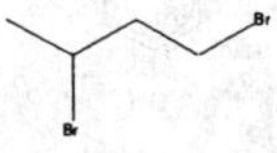

3267

3268

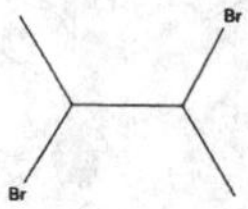

3269

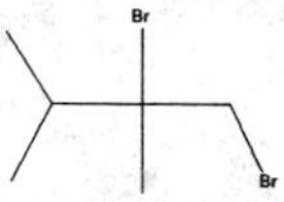

3270

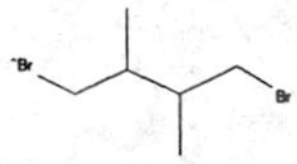

3271

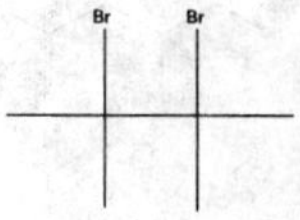

3272

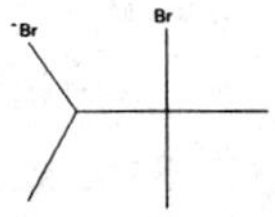

3273

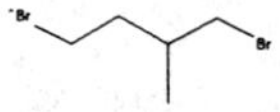

3274

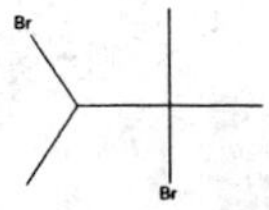

3275

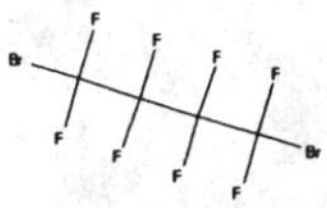

STRUCTURAL FORMULAS OF ORGANIC COMPOUNDS (continued)

In numeric order as they occur in the Table of Physical Constants of Organic Compounds

3276

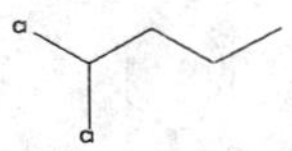

3277

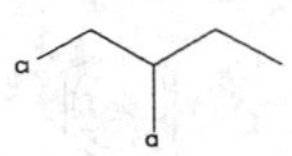

3278

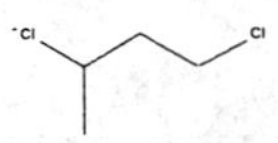

3279

3280

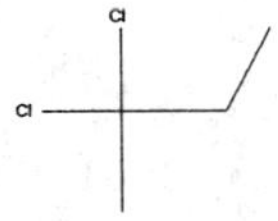

3281

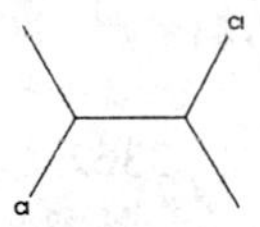

3282

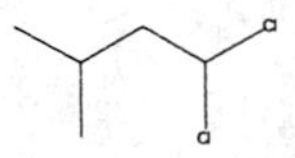

3283

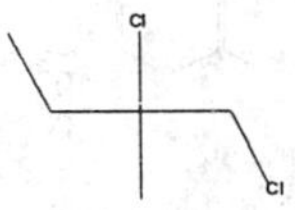

3284

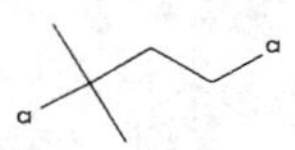

3285

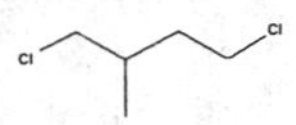

3286

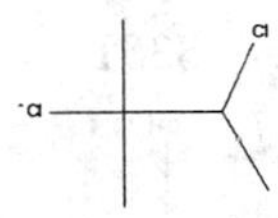

3287

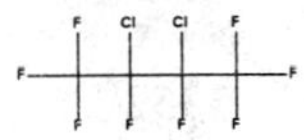

3288

3289

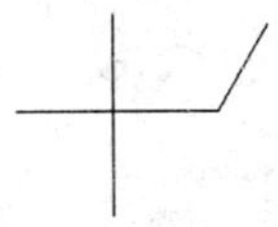

3290

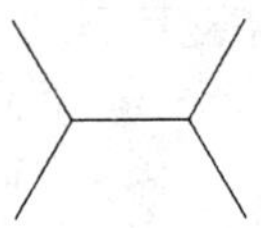

3291

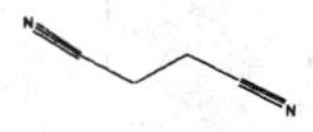

3292

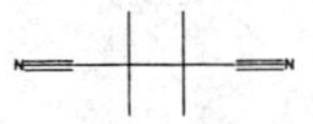

3293

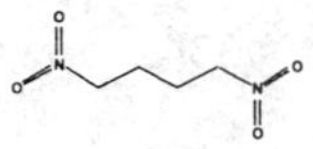

3294

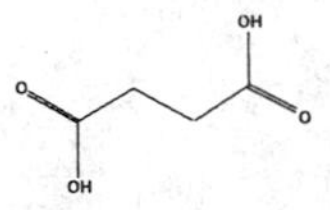

3295

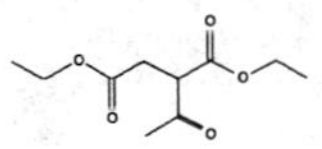

3296

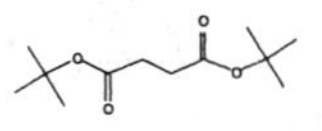

3297

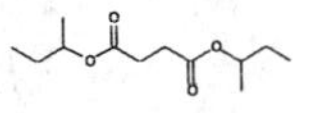

3298

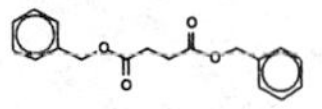

3299

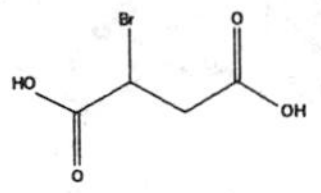

3300

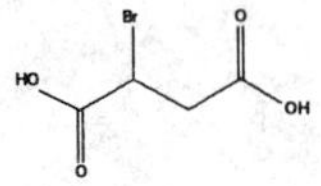

3301

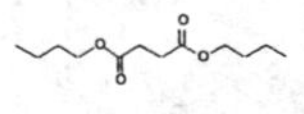

3302

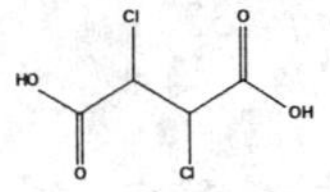

3303

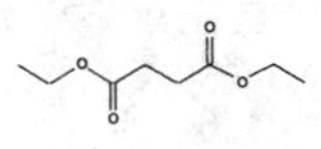

3304

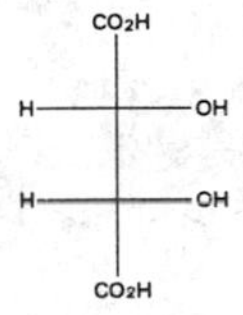

3305

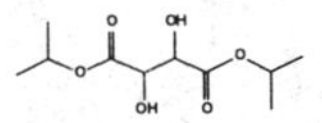

3307

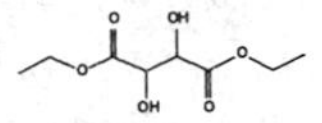

3308

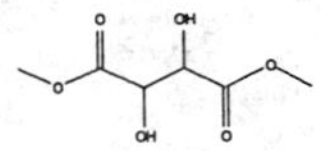

3309

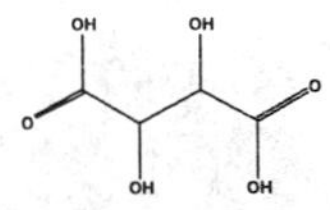

3310

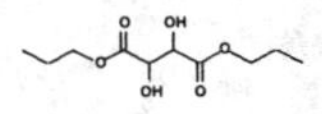

3311

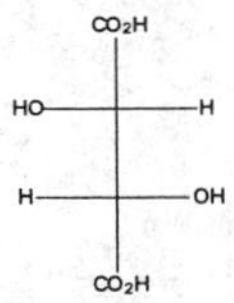

3312

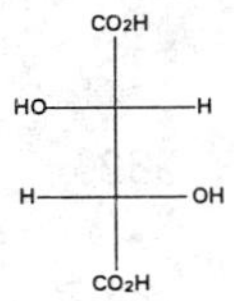

3313

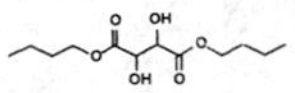

3316

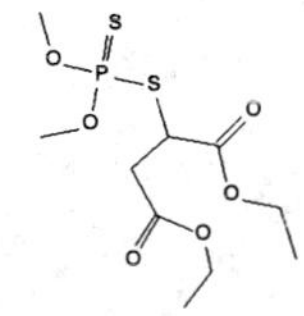

3317

3318

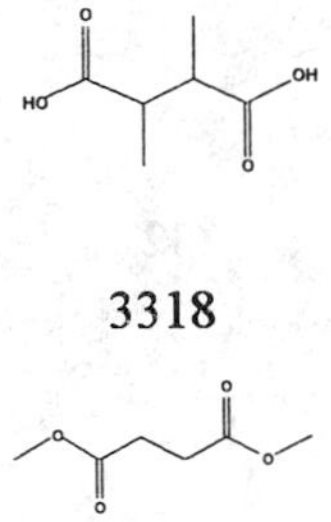

3319

3320

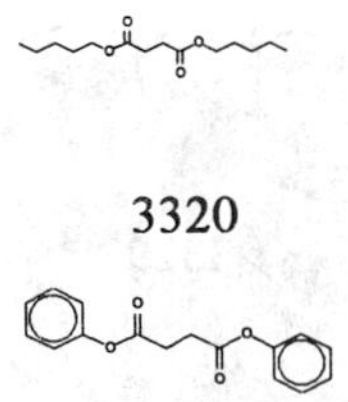

STRUCTURAL FORMULAS OF ORGANIC COMPOUNDS (continued)

In numeric order as they occur in the Table of Physical Constants of Organic Compounds

3321 3322 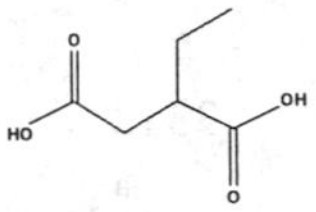 3323 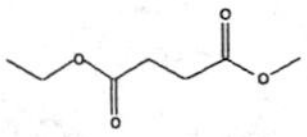 3324 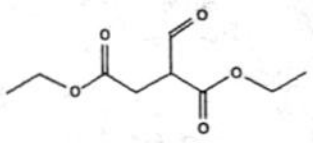 3325 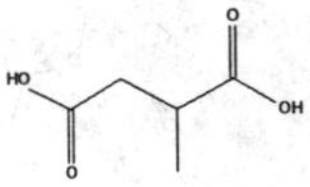 3326 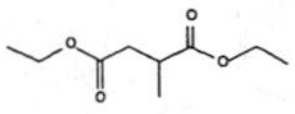 3327 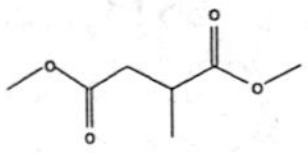 3328 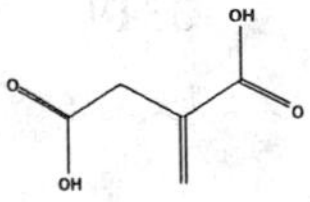 3329

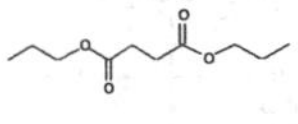

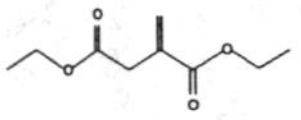

3330 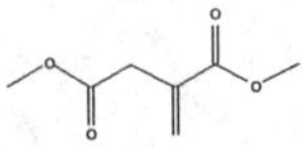 3331 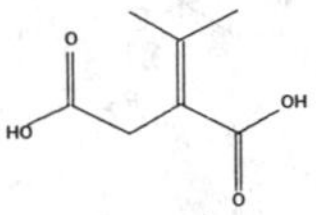 3332 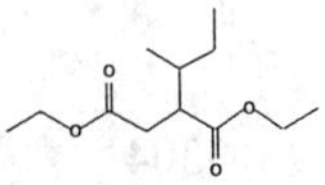 3333 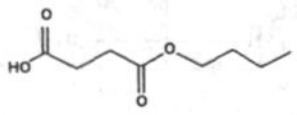 3334 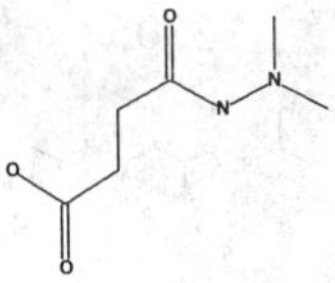 3336 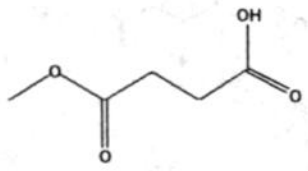 3339 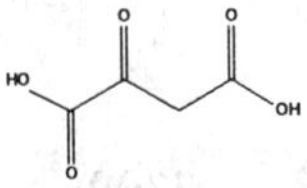 3340

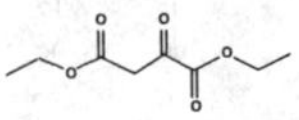

3341 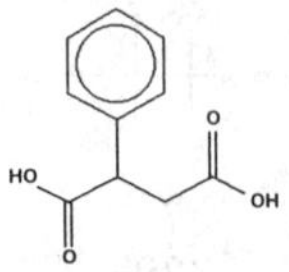 3342 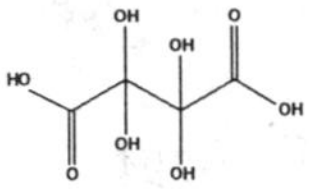 3343 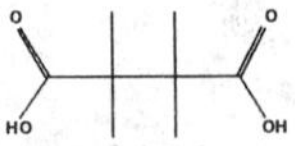 3344 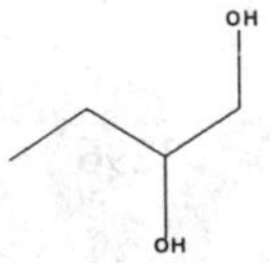 3345 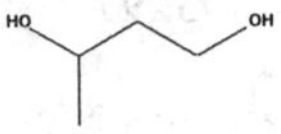 3346  3347 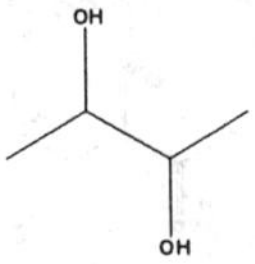 3348

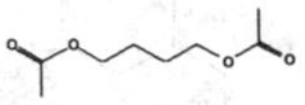

3349 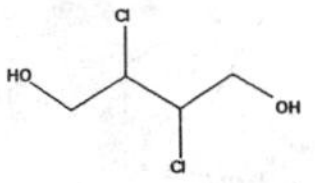 3350 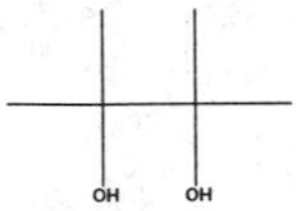 3351 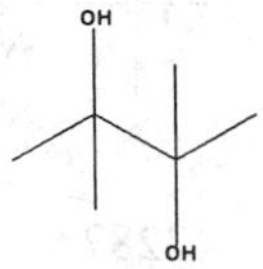 3352 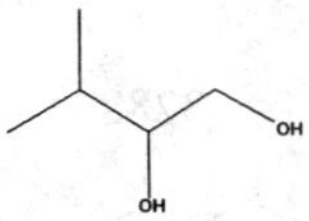 3353 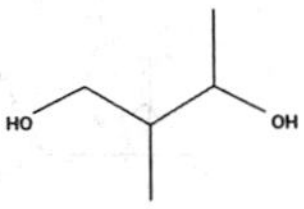 3354 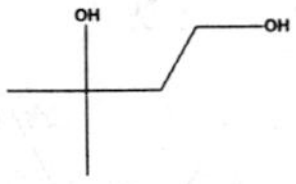 3355

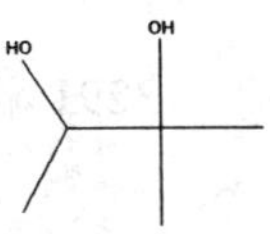

3356 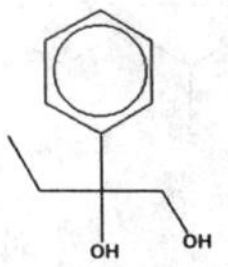 3357 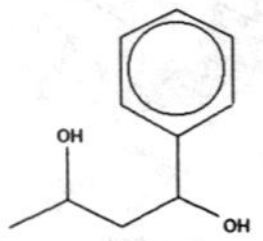 3359 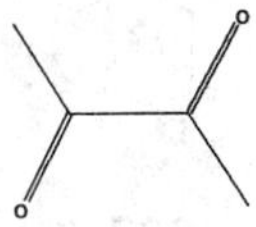 3360 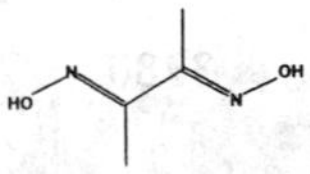 3361 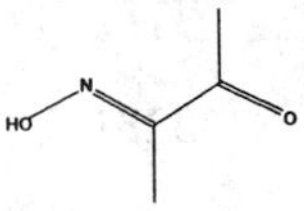 3362 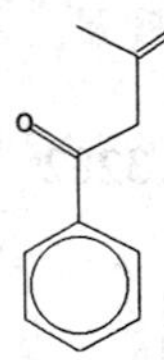 3363

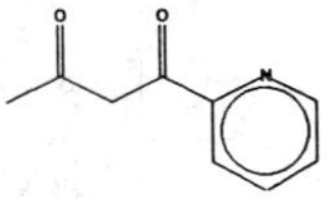

STRUCTURAL FORMULAS OF ORGANIC COMPOUNDS (continued)

In numeric order as they occur in the Table of Physical Constants of Organic Compounds

3364

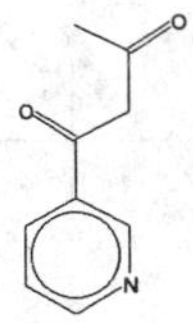

3365

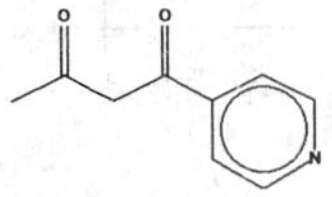

3366

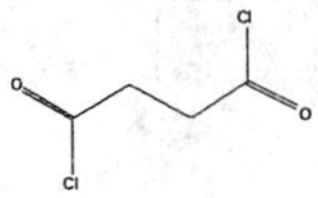

3367

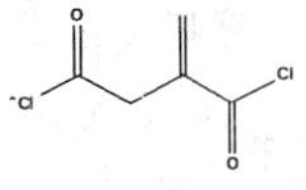

3368

3369

3370

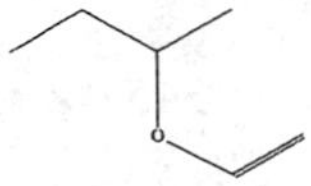

3371

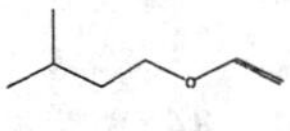

3372

3373

3374

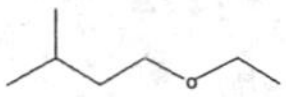

3375

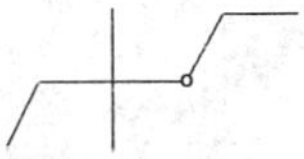

3376

3377

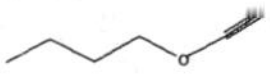

3378

3379

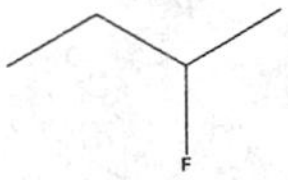

3381

3382

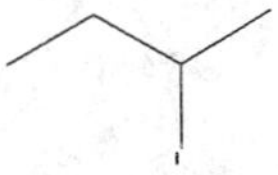

3383

3384

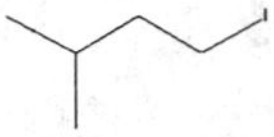

3385

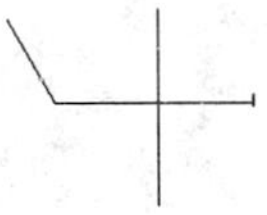

3387

3388

3390

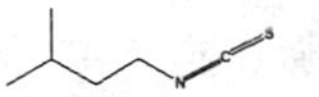

3391

3392

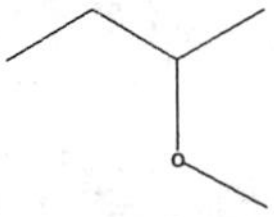

3394

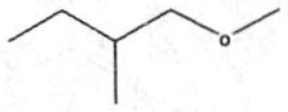

3395

3396

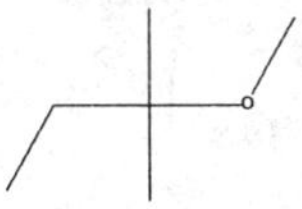

3397

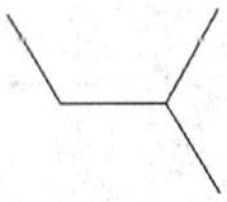

3398

3399

3400

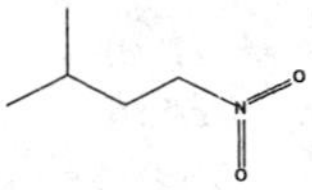

3401

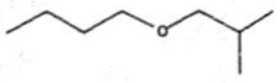

3402

3403

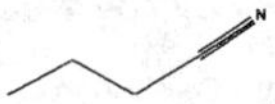

3404

3405

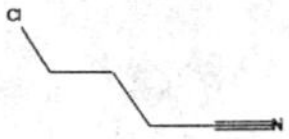

3406

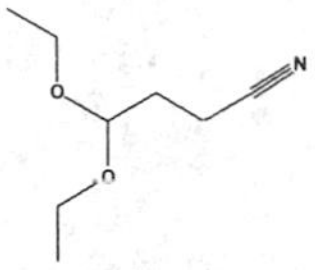

3407

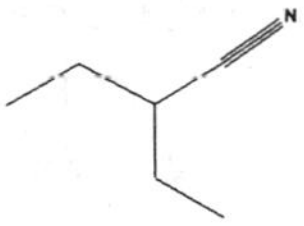

3408

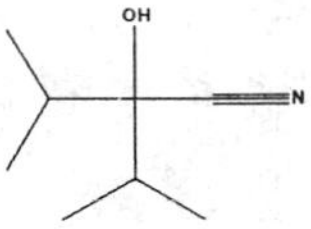

3409

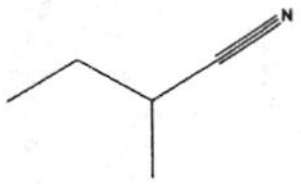

3410

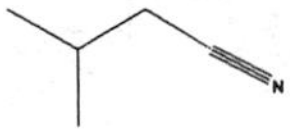

3411

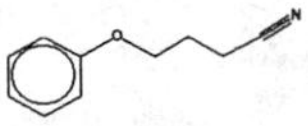

STRUCTURAL FORMULAS OF ORGANIC COMPOUNDS (continued)

In numeric order as they occur in the Table of Physical Constants of Organic Compounds

3412

3414

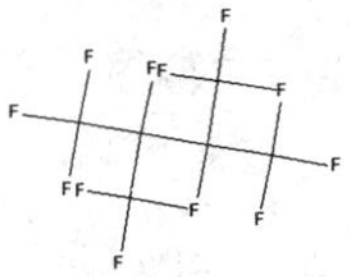

3415

3416

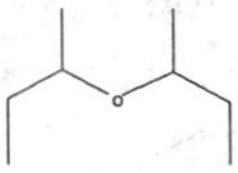

3417

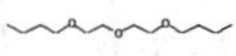

3418

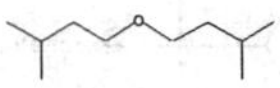

3419

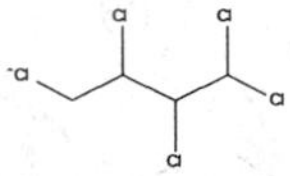

3420

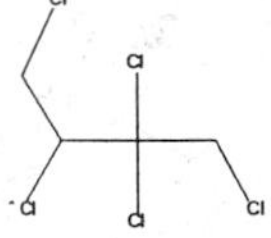

3421

3422

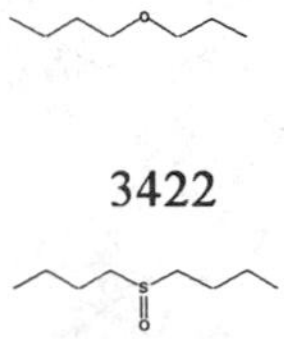

3423

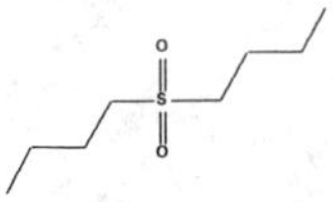

3426

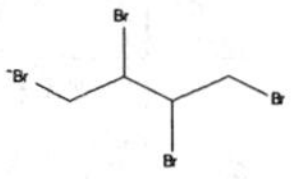

3427

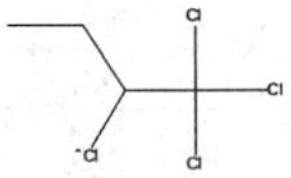

3428

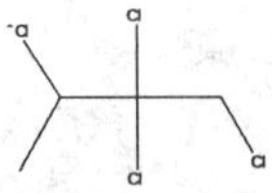

3429

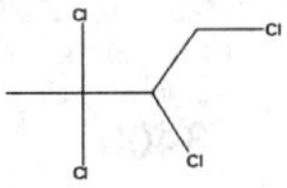

3430

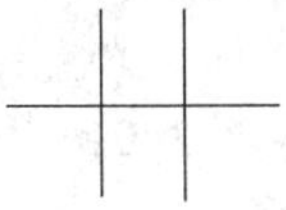

3431

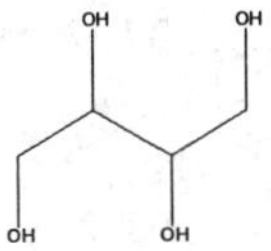

3433

3434

3435

3436

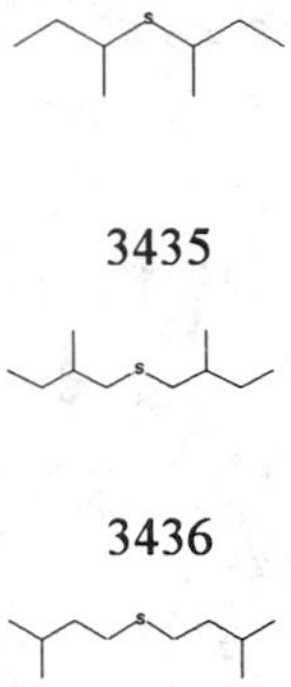

3437

3438

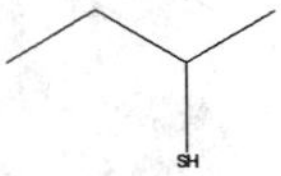

3439

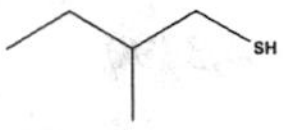

3440

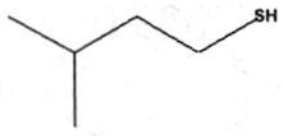

3441

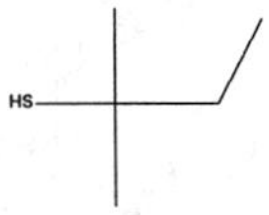

3442

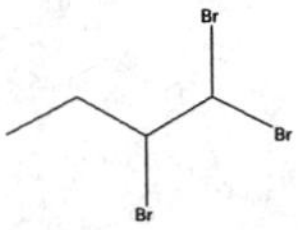

3443

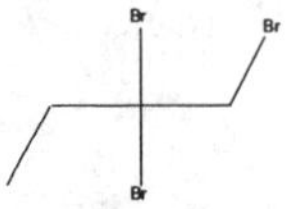

3444

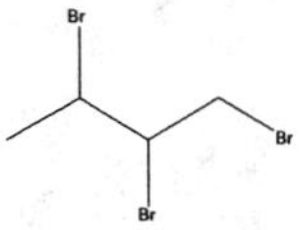

3445

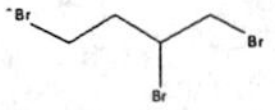

3446

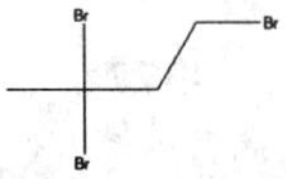

3447

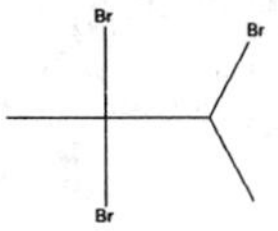

3448

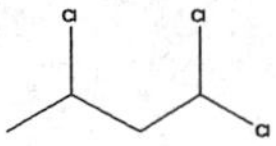

3449

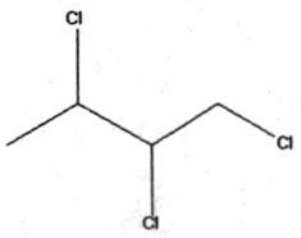

3450

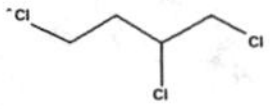

3451

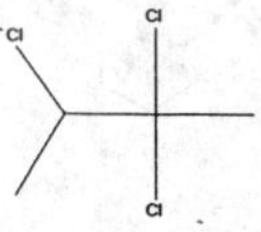

3452

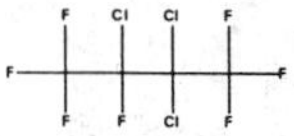

3453

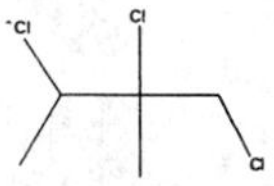

3454

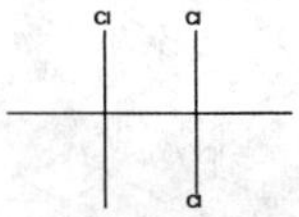

3455

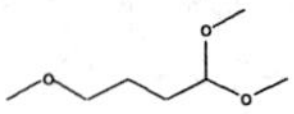

3456

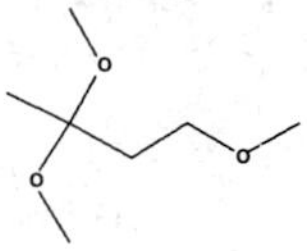

3457

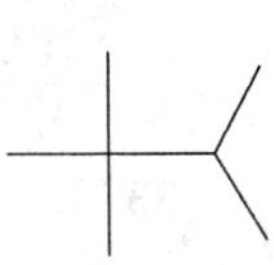

3458

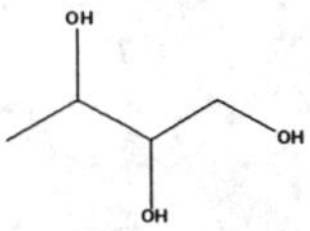

STRUCTURAL FORMULAS OF ORGANIC COMPOUNDS (continued)

In numeric order as they occur in the Table of Physical Constants of Organic Compounds

3459

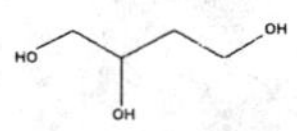

3460

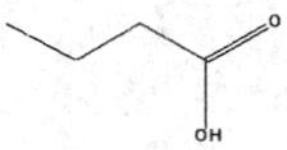

3461

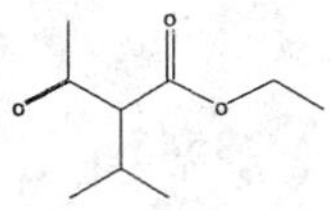

3462

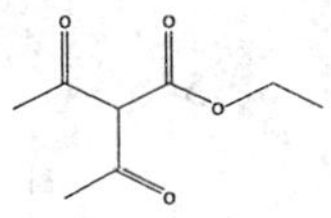

3464

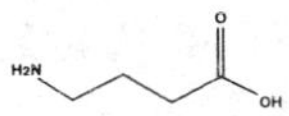

3465

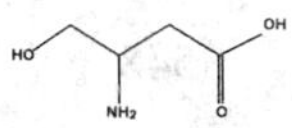

3467

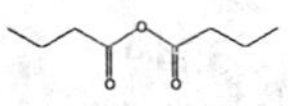

3468

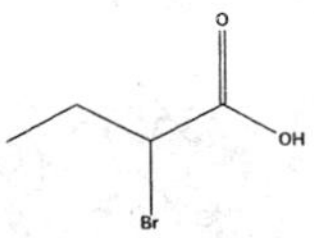

3469

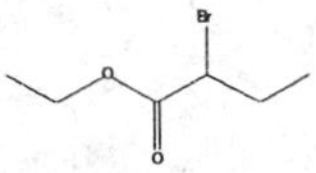

3470

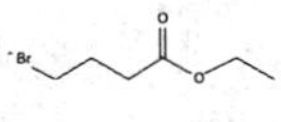

3471

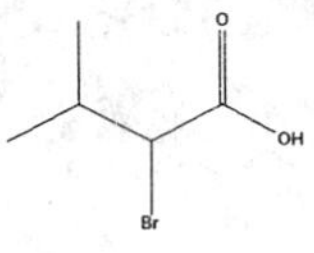

3472

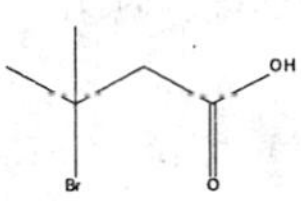

3473

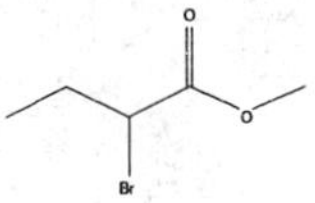

3474

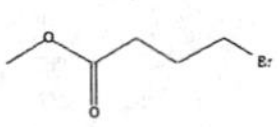

3475

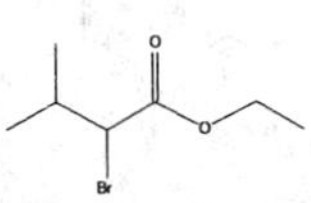

3476

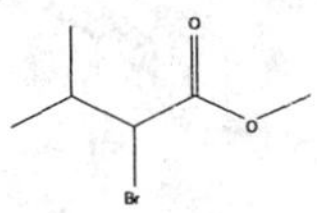

3477

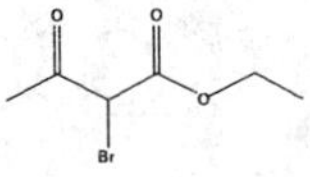

3478

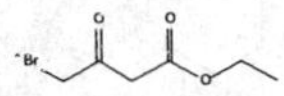

3479

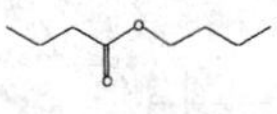

3480

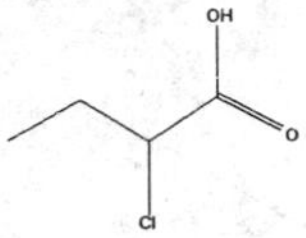

3481

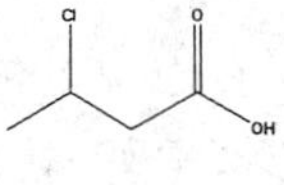

3482

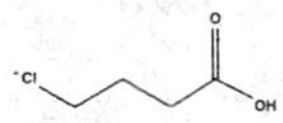

3483

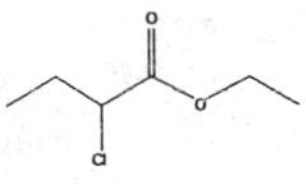

3484

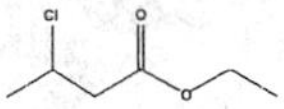

3485

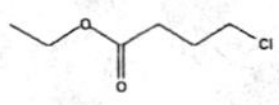

3486

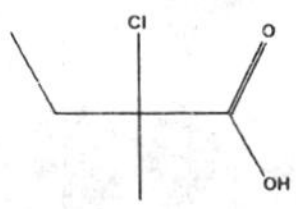

3487

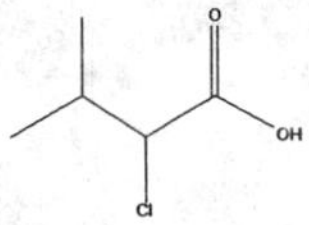

3488

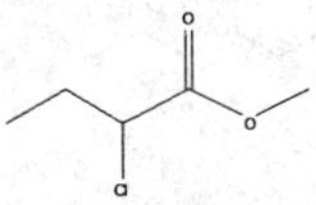

3489

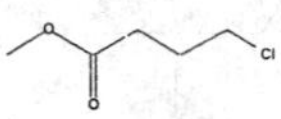

3490

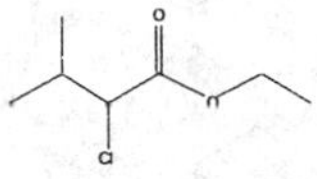

3492

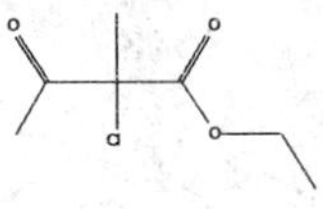

3493

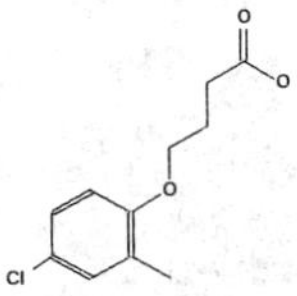

3494

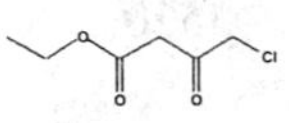

3495

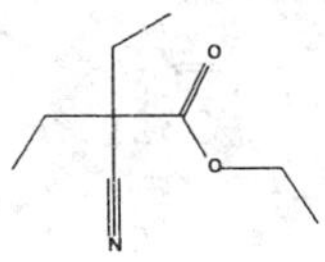

3497

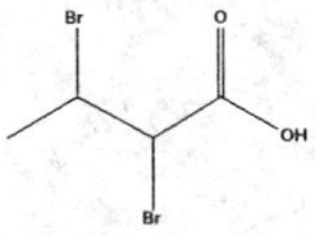

3498

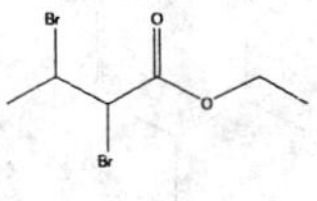

3499

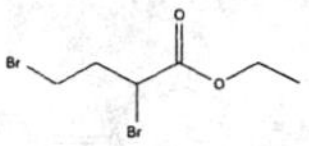

3500

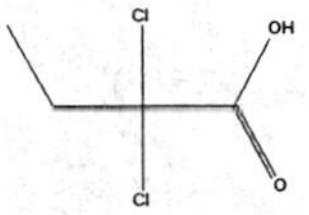

3502

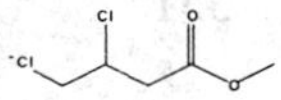

3503

3504

3505

3506

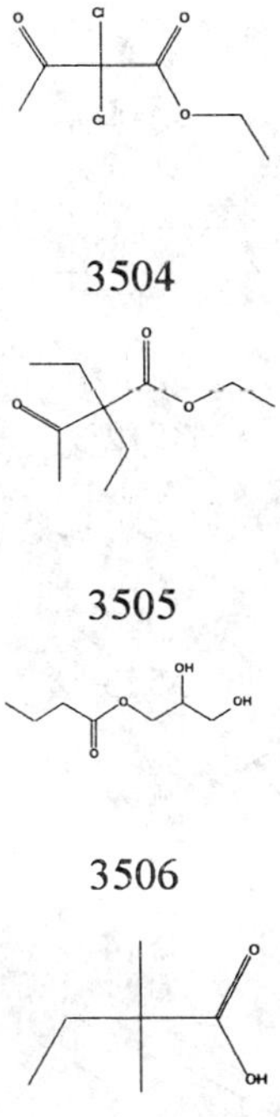

In numeric order as they occur in the Table of Physical Constants of Organic Compounds

3507

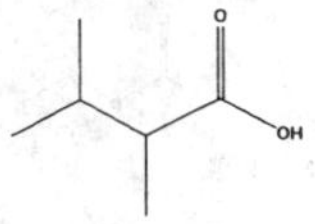

OH, O

3508

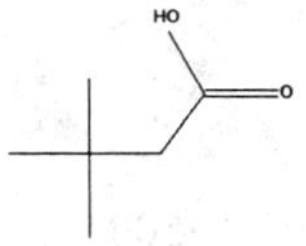

HO, O

3509

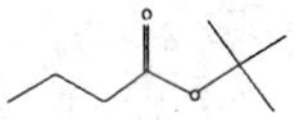

O, O

3510

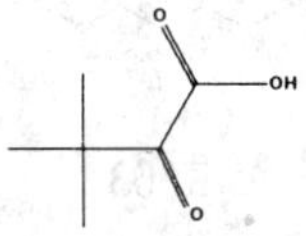

O, OH, O

3511

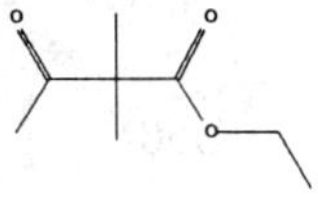

O, O, O

3512

O, O, O, O

3513

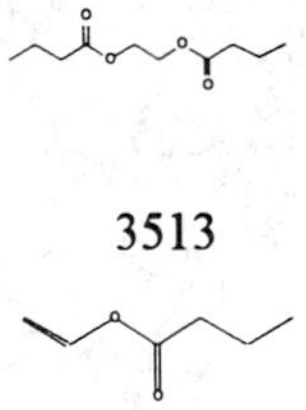

O, O

3514

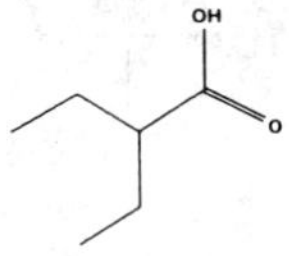

OH, O

3515

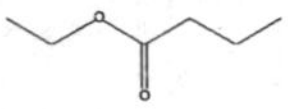

O, O

3516

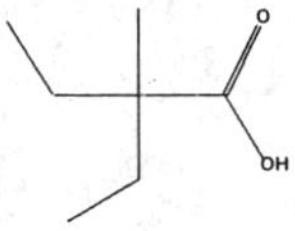

O, OH

3517

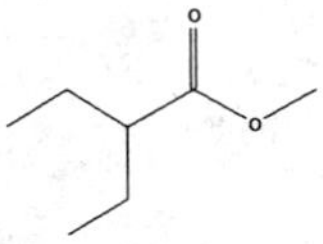

O, O

3518

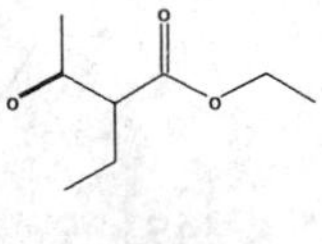

O, O, O

3519

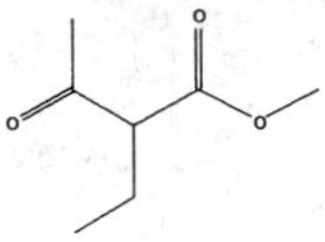

O, O, O

3520

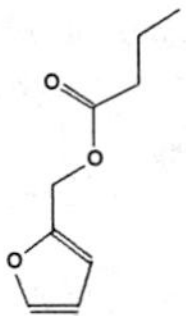

O, O, O

3521

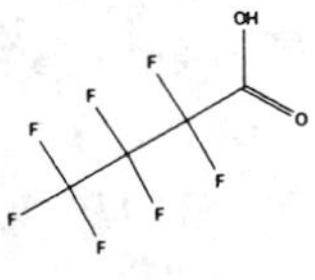

OH, F, F, F, F, F, F, F, O

3522

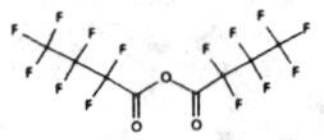

F, F, F, F, F, F, F, O, O, O, F, F, F, F, F, F, F

3523

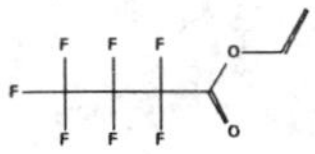

F, F, F, F, F, F, F, O, O

3524

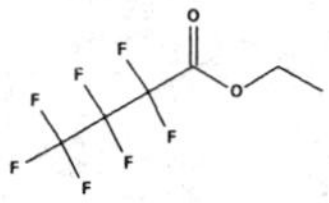

O, F, F, F, F, F, F, F, O

3525

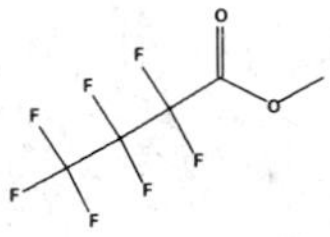

O, F, F, F, F, F, F, F, O

3526

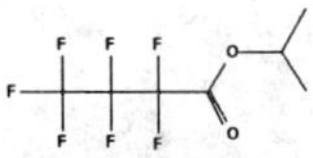

F, F, F, F, F, F, F, O, O

3527

O, O

3528

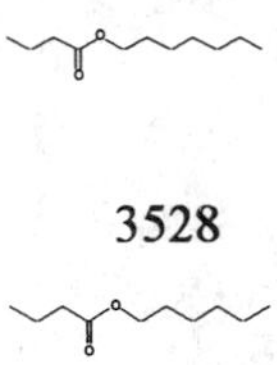

O, O

3529

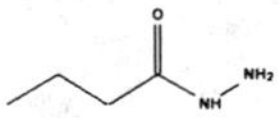

O, NH, NH$_2$

3530

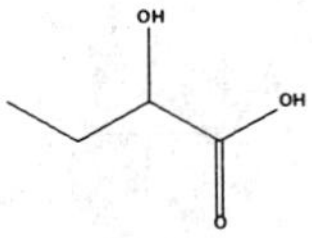

OH, OH, O

3531

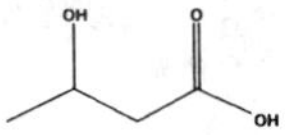

OH, O, OH

3532

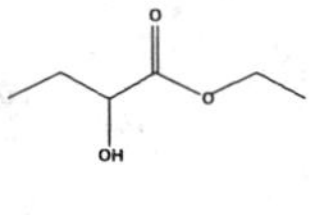

O, O, OH

3533

OH, O, O

3534

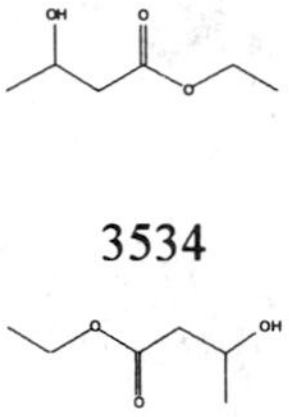

O, O, OH

3535

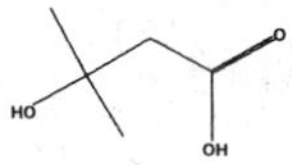

HO, O, OH

3537

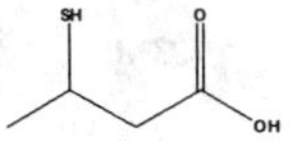

SH, O, OH

3538

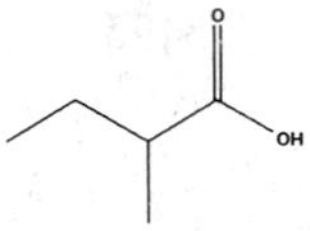

O, OH

3539

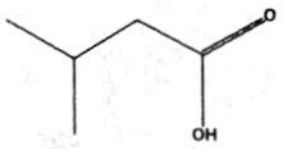

O, OH

3540

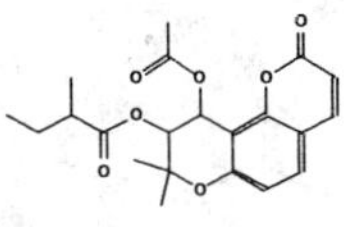

O, O, O, O, O, O

3541

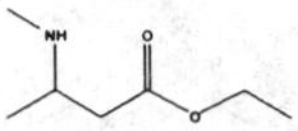

NH, O, O

3542

O, O, O

3543

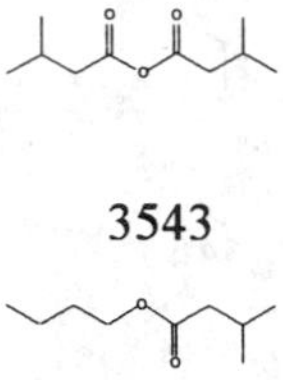

O, O

3544

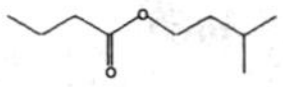

O, O

3545

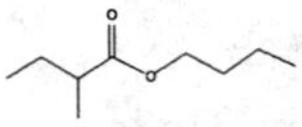

O, O

3546

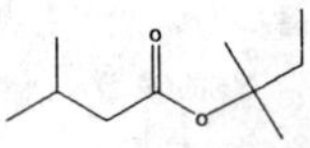

O, O

3547

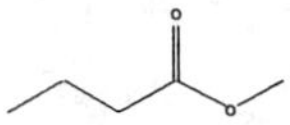

O, O

3548

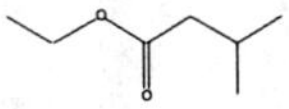

O, O

3549

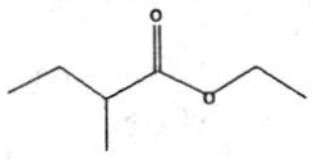

O, O

3551

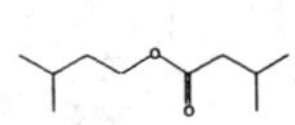

O, O

3552

O, O

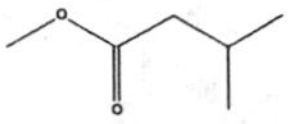

STRUCTURAL FORMULAS OF ORGANIC COMPOUNDS (continued)

In numeric order as they occur in the Table of Physical Constants of Organic Compounds

3553

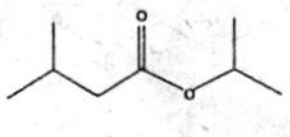

3555

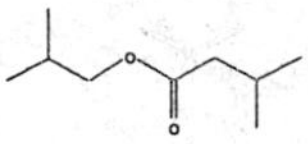

3556

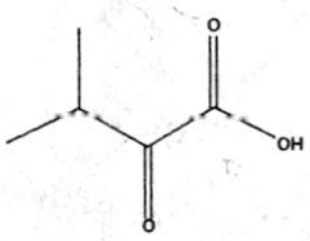

OH

3557

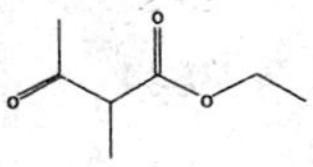

3558

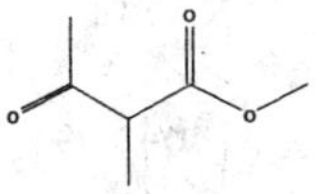

3560

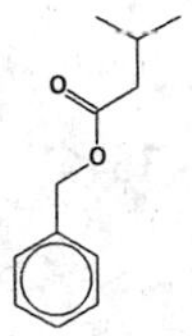

3561

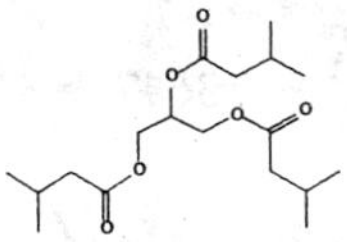

3562

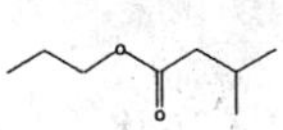

3563

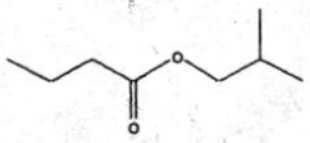

3565

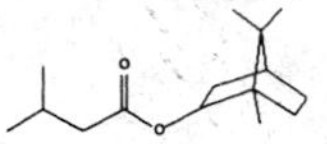

3566

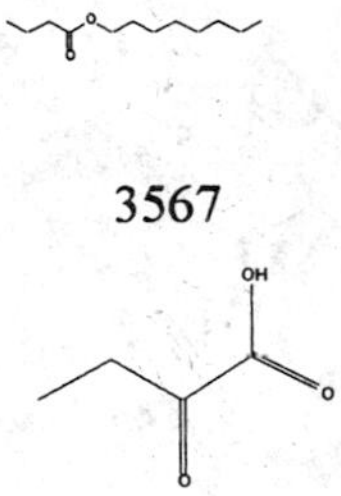

3567

OH

3568

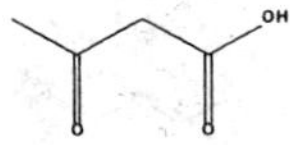

OH

3569

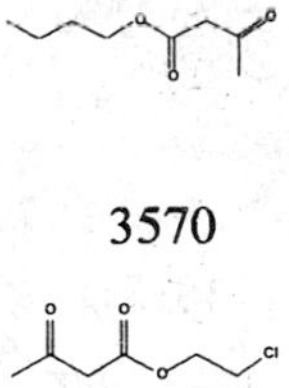

3570

Cl

3571

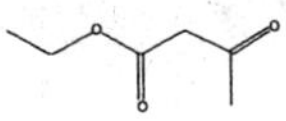

3572

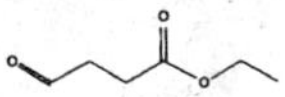

3573

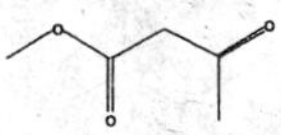

3574

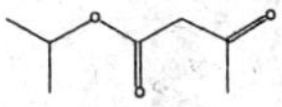

3575

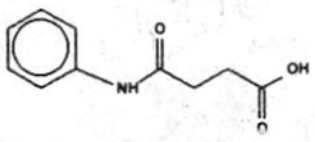

NH

OH

3576

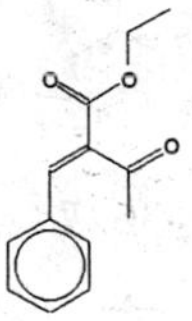

3577

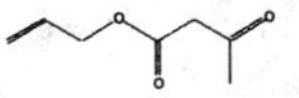

3578

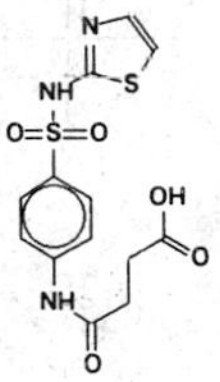

N

S

NH

O=S=O

OH

NH

3579

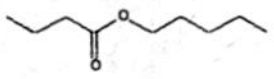

3580

HO

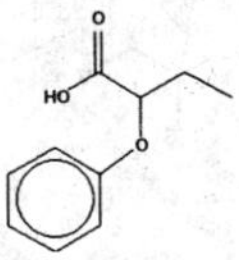

3581

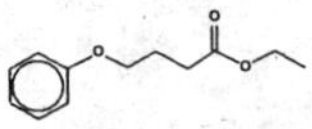

3582

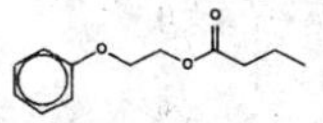

3584

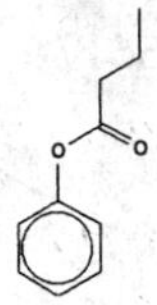

3585

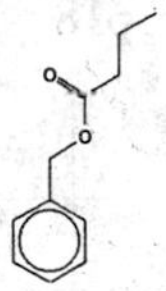

3586

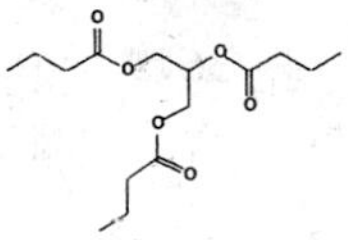

3587

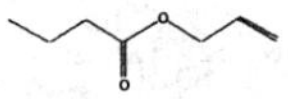

3588

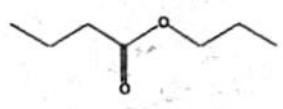

3590

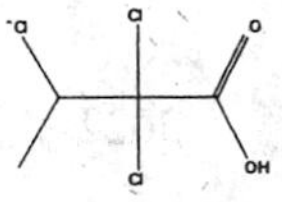

Cl

Cl

Cl

OH

3592

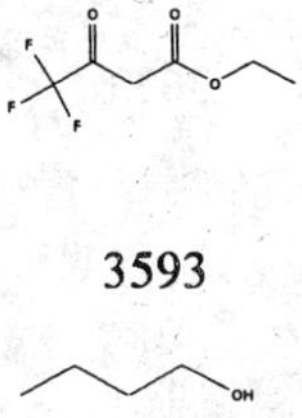

F

F

F

3593

OH

3594

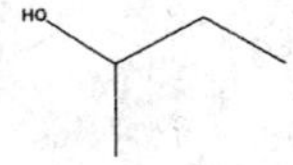

HO

3595

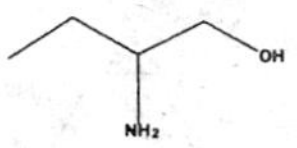

OH

NH_2

3596

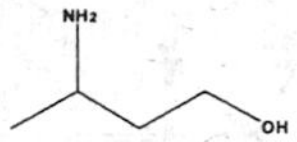

NH_2

OH

3597

H_2N

OH

3598

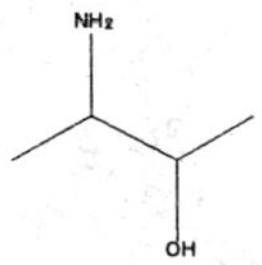

NH_2

OH

3599

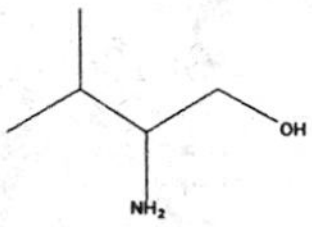

OH

NH_2

3601

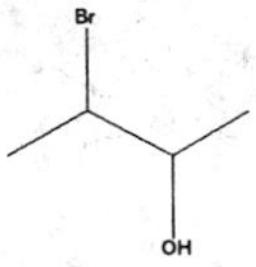

Br

OH

3604

Cl

OH

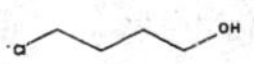

3605

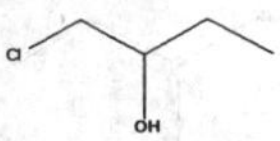

Cl OH

3606

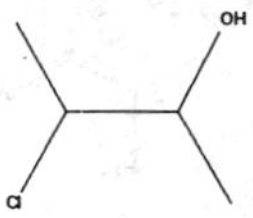

OH Cl

3608

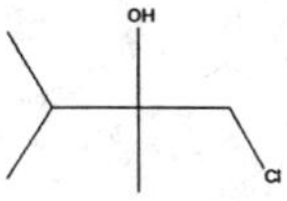

OH Cl

3609

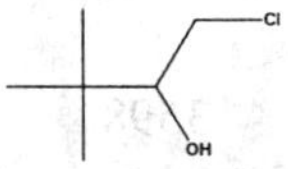

Cl OH

3612

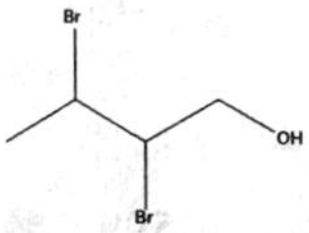

Br OH Br

3613

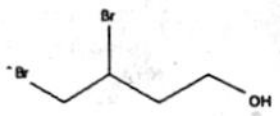

Br Br OH

3614

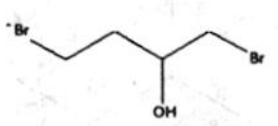

Br Br OH

3615

OH

3616

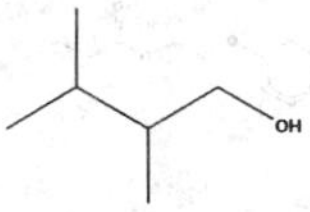

OH

3617

OH

3618

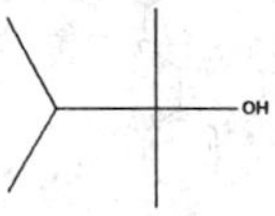

OH

3619

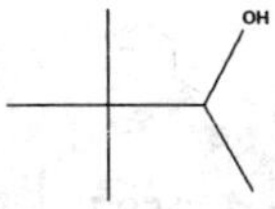

OH

3620

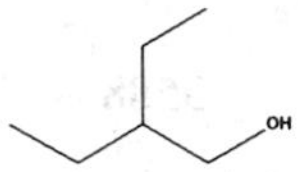

OH

3621

F OH

3622

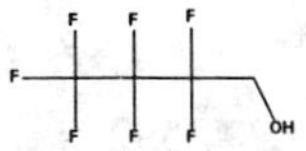

F F F F F F F OH

3623

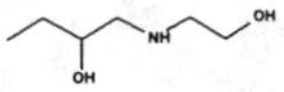

NH OH OH

3624

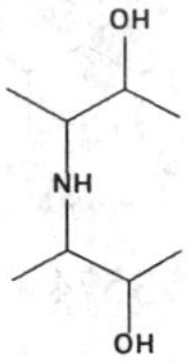

OH NH OH

3625

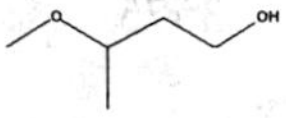

O OH

3626

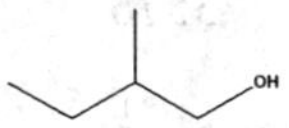

OH

3627

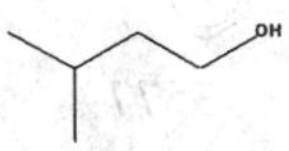

OH

3628

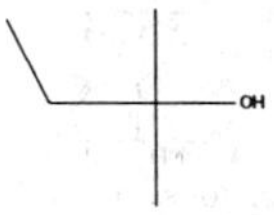

OH

3629

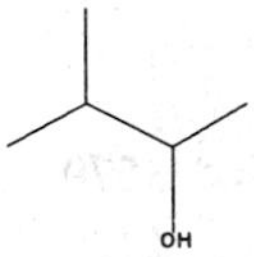

OH

3630

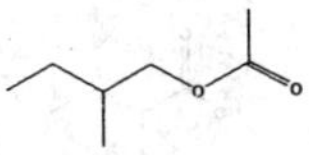

O O

3631

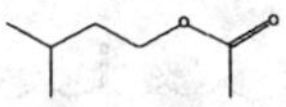

O O

3632

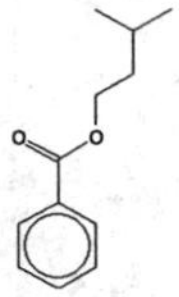

O O

3633

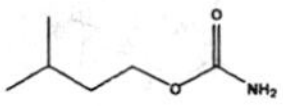

O O NH₂

3634

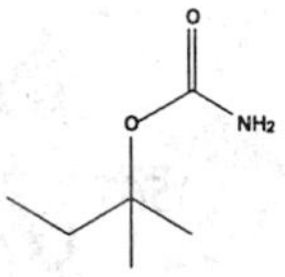

O O NH₂

3635

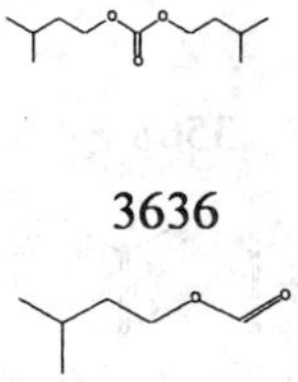

O O O

3636

O O

3637

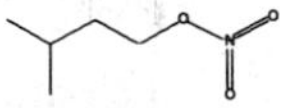

O N O O

3638

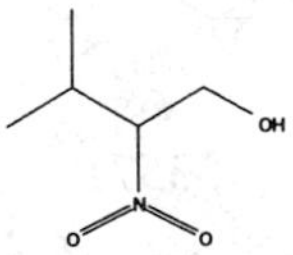

OH O N O

3639

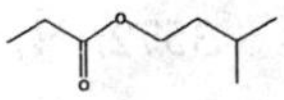

O O

3640

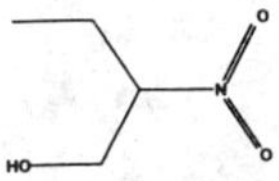

O N O HO

3641

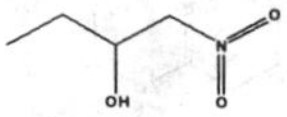

N O O OH

3642

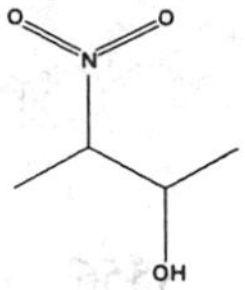

O N O OH

3643

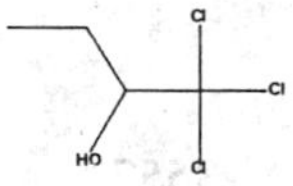

Cl Cl Cl HO

3644

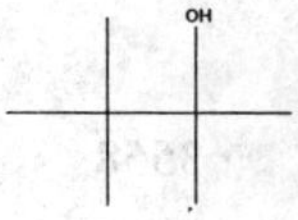

OH

3645

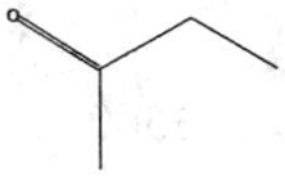

O

3646

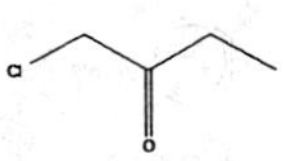

Cl O

3647

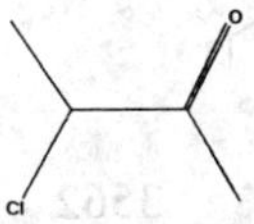

O Cl

3648

O Cl

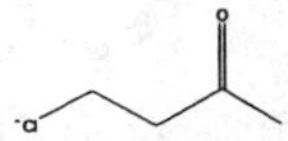

STRUCTURAL FORMULAS OF ORGANIC COMPOUNDS (continued)

In numeric order as they occur in the Table of Physical Constants of Organic Compounds

3649 Cl O

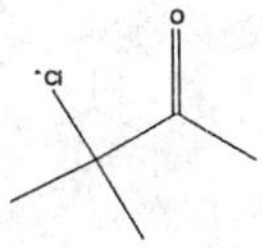

3650 Cl O

3651 Cl HO N O F

3653 O Br Br

3654 O Cl Cl

3655 O Cl Cl

3656 N O

3657 O

3658 O

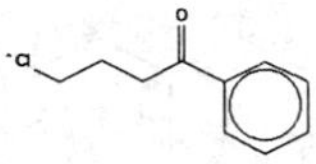

3659 O O

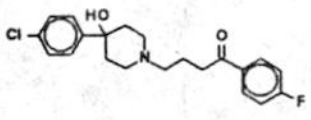

3660 OH O

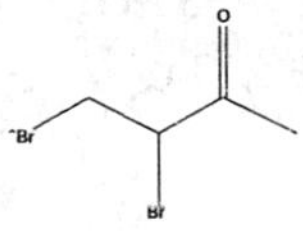

3661 O HO

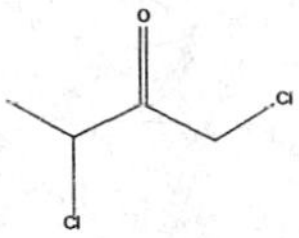

3662 O OH

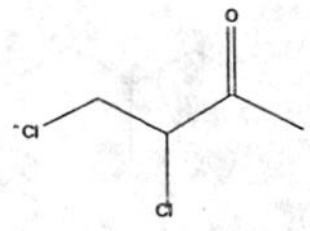

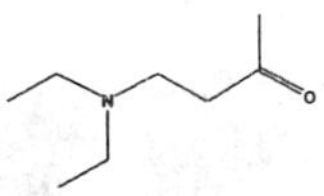

3663 O HO O

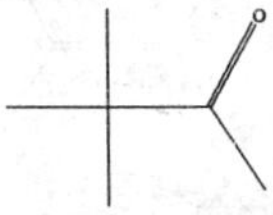

3664 HO O

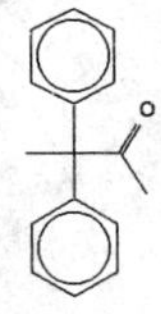

3665 OH O

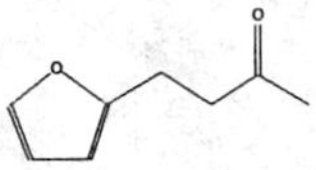

3666 O OH

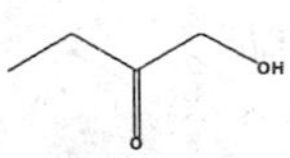

3667 O OH

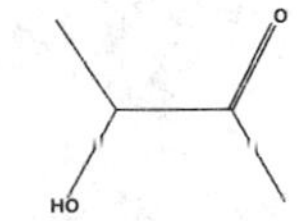

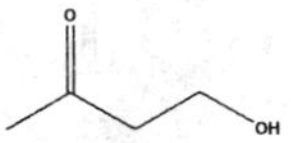

3668 O

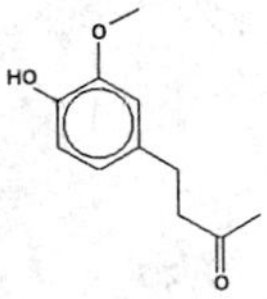

3669 O

3670 O

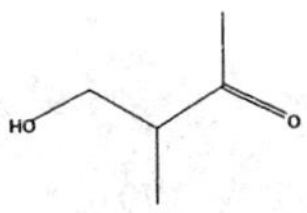

3671 O

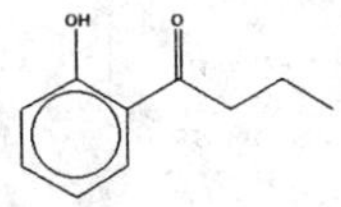

3673 N N

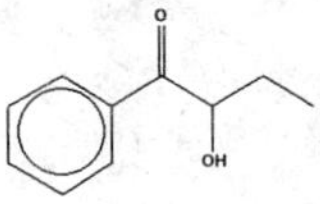

3675 HO N

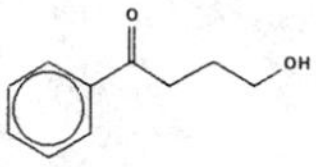

3676 O

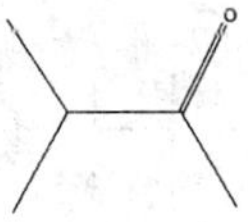

3677 O

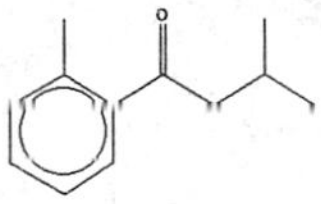

3678 O

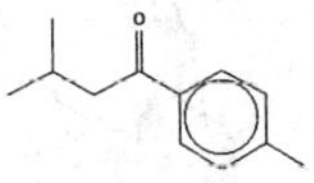

3679 O HO OH OH

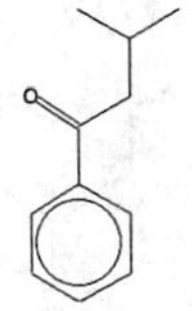

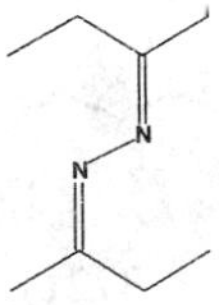

3680 O Br

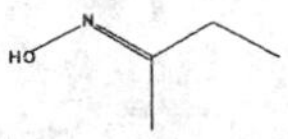

3681 O Cl

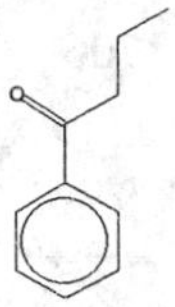

3682 O Cl Br

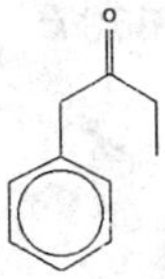

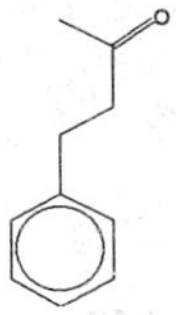

3683 O Cl Cl

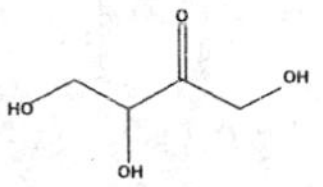

3684 Cl O Cl

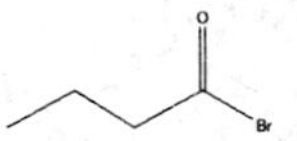

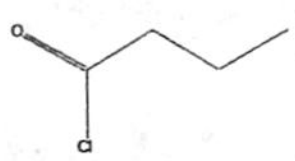

3685 Cl O Cl

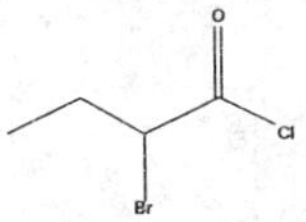

3688 O Cl

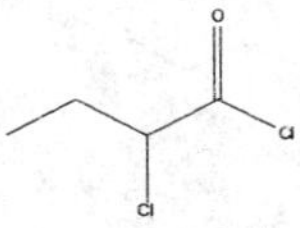

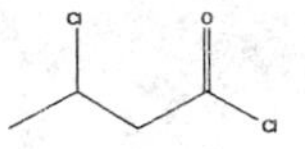

3690 O Cl

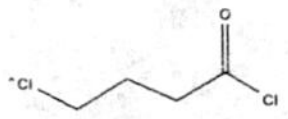

3691 O Cl

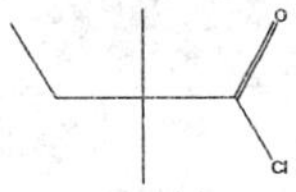

3692 F F F F F F F O Cl

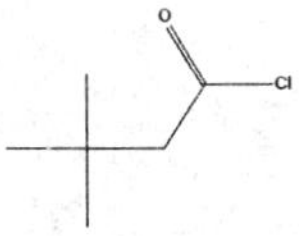

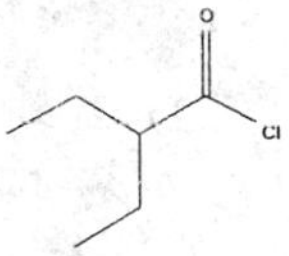

3693 O Cl

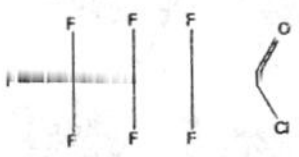

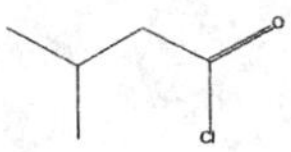

STRUCTURAL FORMULAS OF ORGANIC COMPOUNDS (continued)

In numeric order as they occur in the Table of Physical Constants of Organic Compounds

3695

3696

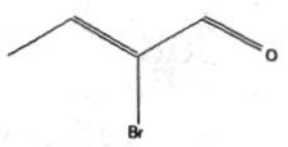

Br

3697

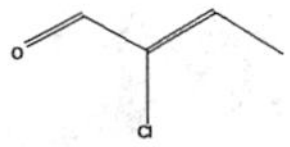

Cl

3698

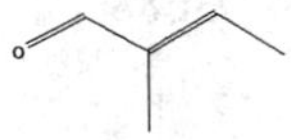

3699

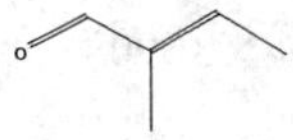

3700

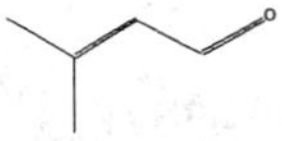

3702

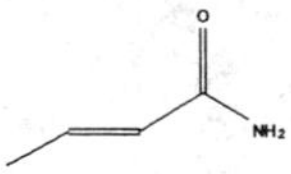

NH2

3704

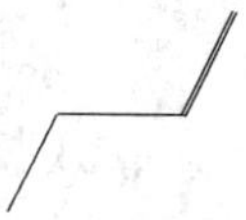

3705

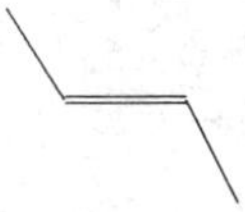

3706

3707

Br

3708

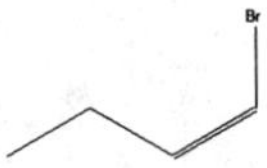

Br

3709

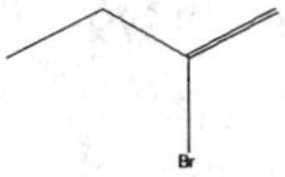

Br

3710

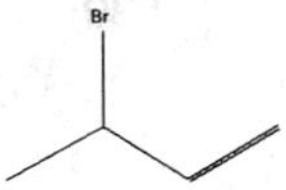

Br

3711

Br

3712

Br

3713

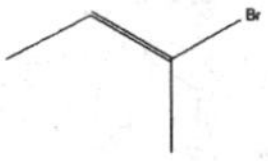

Br

3714

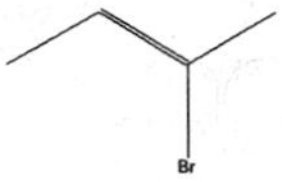

Br

3715

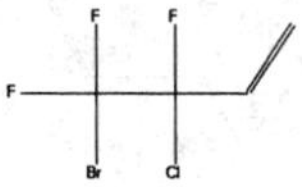

F F F Br Cl

3716

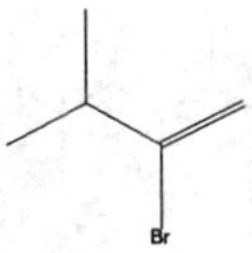

Br

3717

Br

3718

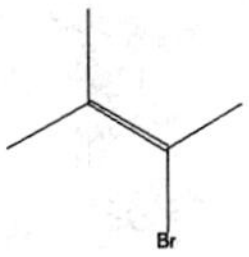

Br

3720

Cl

3721

Cl

3722

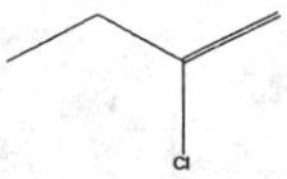

Cl

3723

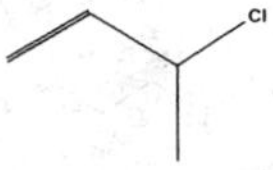

Cl

3724

Cl

3725

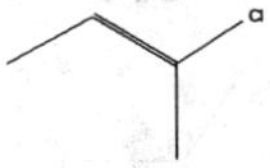

Cl

3726

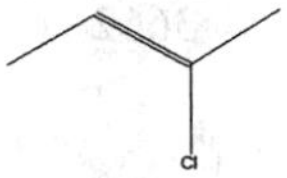

Cl

3727

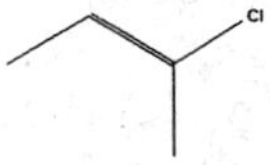

Cl

3728

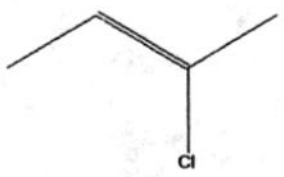

Cl

3729

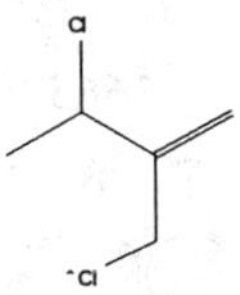

Cl Cl

3730

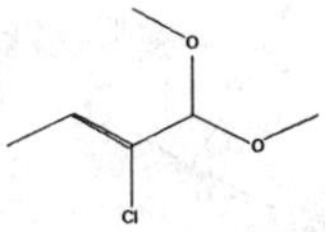

Cl

3731

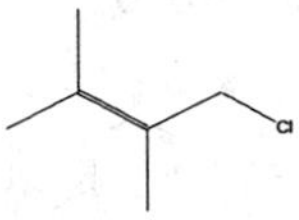

Cl

3732

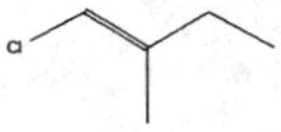

Cl

3733

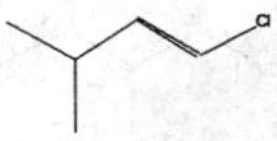

Cl

3734

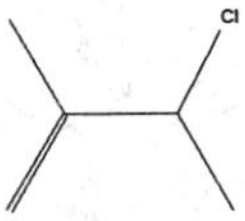

Cl

3735

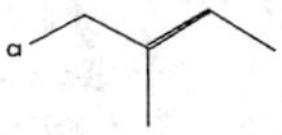

Cl

3736

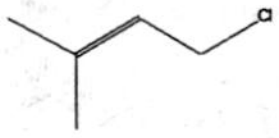

Cl

3737

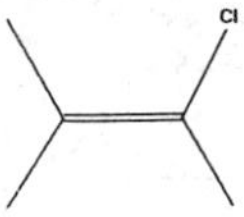

Cl

STRUCTURAL FORMULAS OF ORGANIC COMPOUNDS (continued)

In numeric order as they occur in the Table of Physical Constants of Organic Compounds

3738

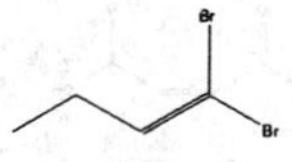

Br Br

3739

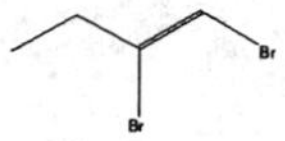

Br Br

3740

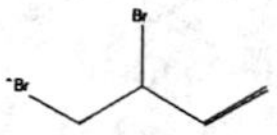

Br Br

3741

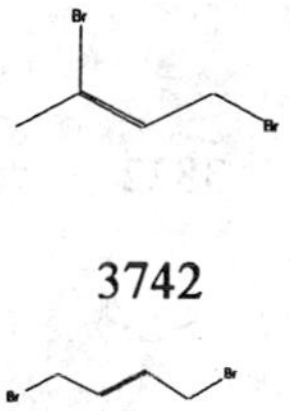

Br Br

3742

Br Br

3743

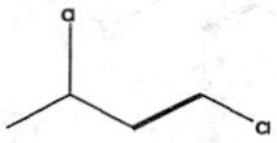

Cl Cl

3744

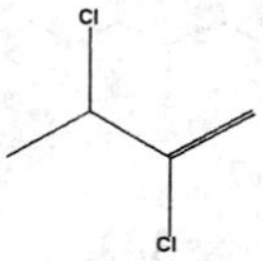

Cl Cl

3745

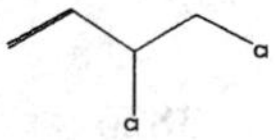

Cl Cl

3746

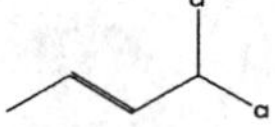

Cl Cl

3747

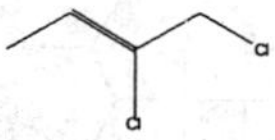

Cl Cl

3748

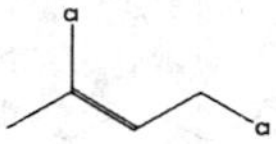

Cl Cl

3749

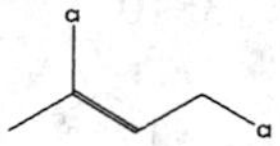

Cl Cl

3750

Cl Cl

3751

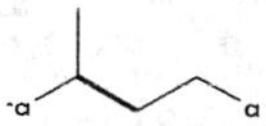

Cl Cl

3752

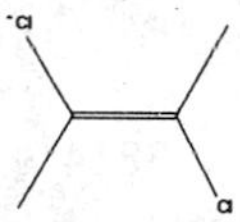

Cl Cl

3753

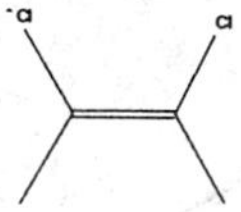

Cl Cl

3754

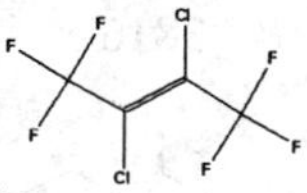

F F F Cl F F F Cl

3755

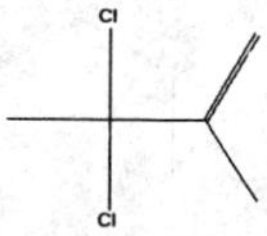

Cl Cl

3756

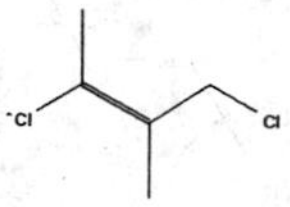

Cl Cl

3757

Cl Cl

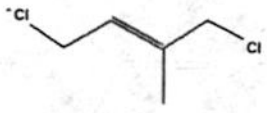

3758

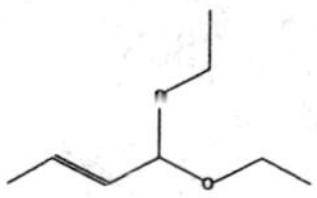

3759

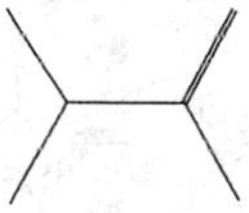

3760

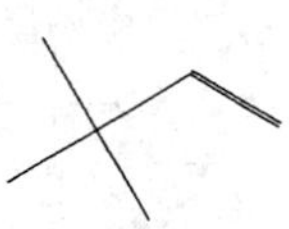

3761

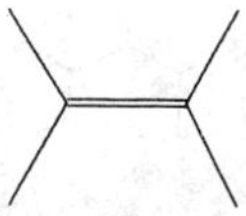

3762

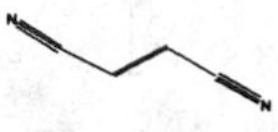

N N

3763

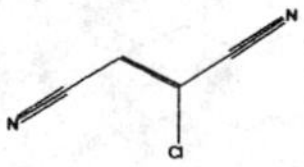

N N Cl

3764

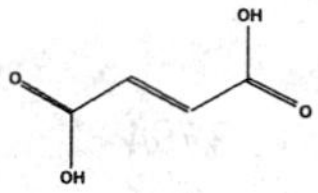

OH O O OH

3765

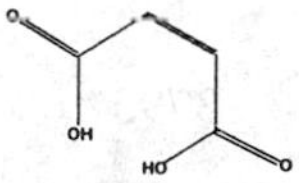

O OH HO O

3766

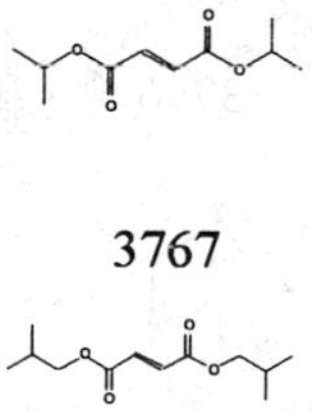

O O O O

3767

O O O O

3769

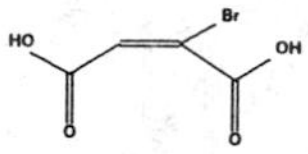

HO Br OH O O

3770

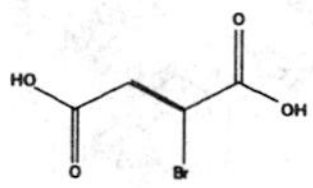

HO O O OH Br

3771

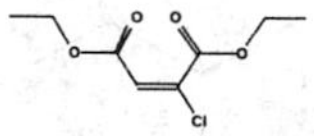

O O O O Cl

3772

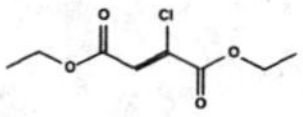

O Cl O O O

3773

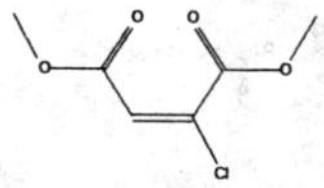

O O O O Cl

3774

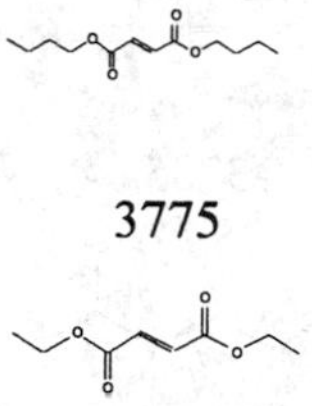

O O O O

3775

O O O O

3776

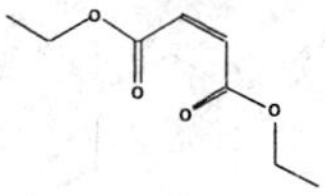

O O O O

3777

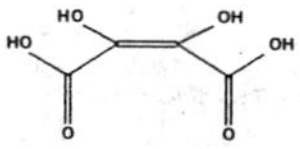

HO OH HO OH O O

3778

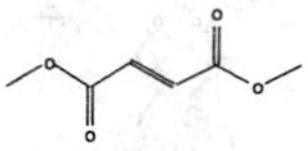

O O O O

3779

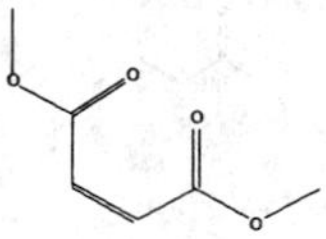

O O O O

3781

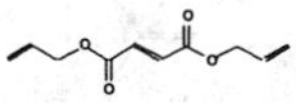

O O O O

3782

O O O O

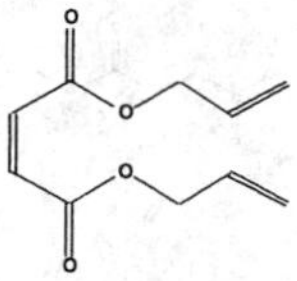

STRUCTURAL FORMULAS OF ORGANIC COMPOUNDS (continued)

In numeric order as they occur in the Table of Physical Constants of Organic Compounds

3783

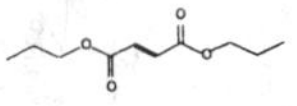

3784

3785

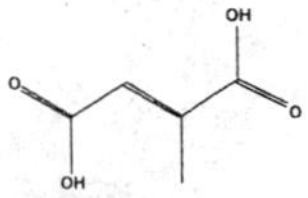

3786

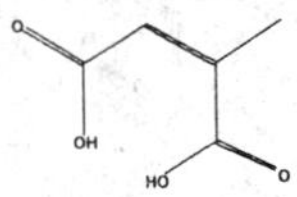

3787

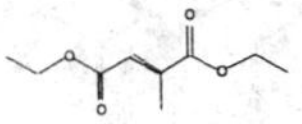

3788

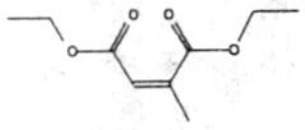

3789

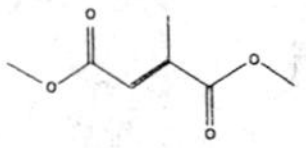

3790

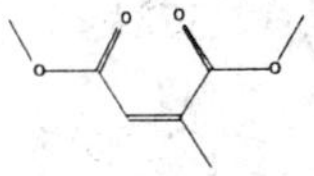

3791

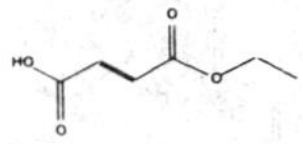

3792

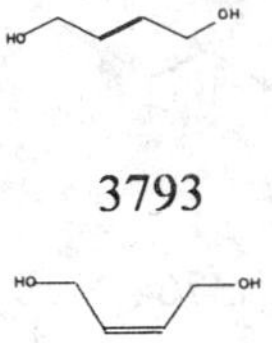

3793

3794

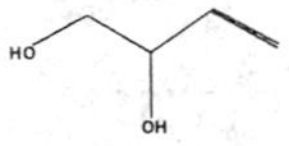

3795

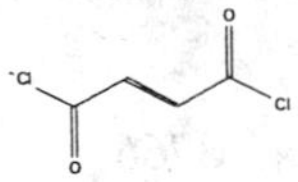

3796

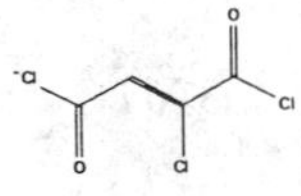

3797

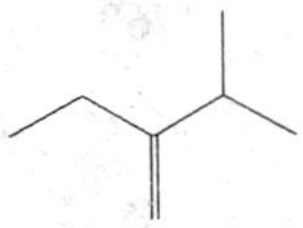

3798

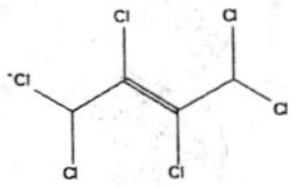

3799

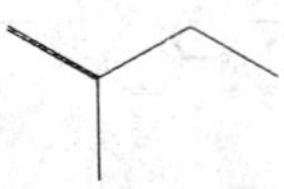

3800

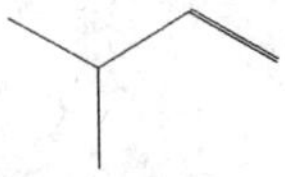

3801

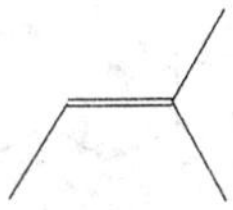

3802

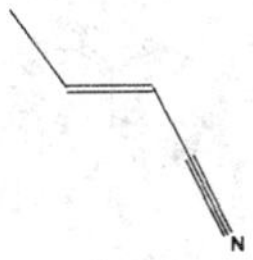

3803

3804

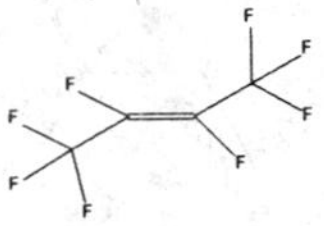

3805

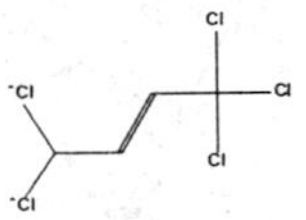

3806

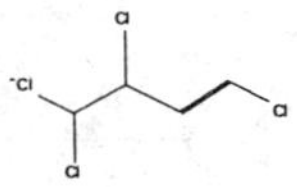

3807

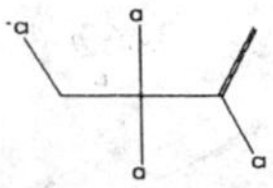

3808

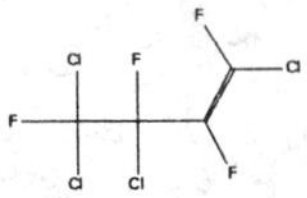

3809

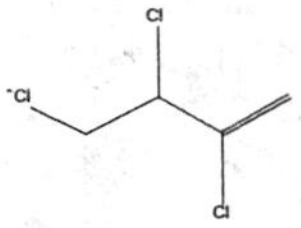

3810

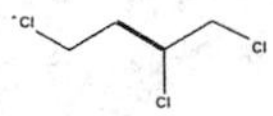

3812

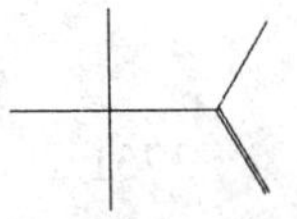

3813

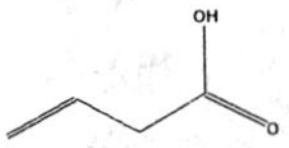

3814

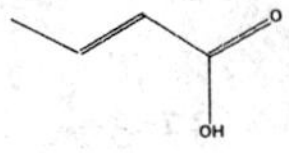

3815

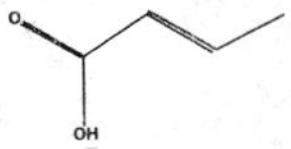

3816

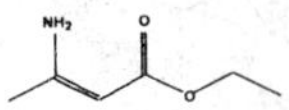

3817

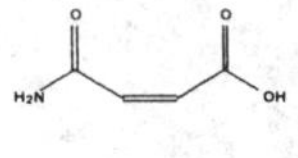

3819

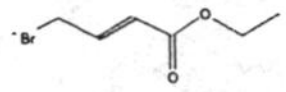

3820

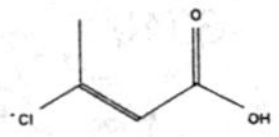

3821

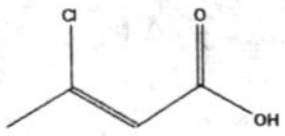

3822

3823

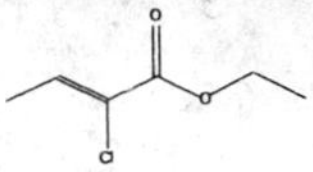

3824

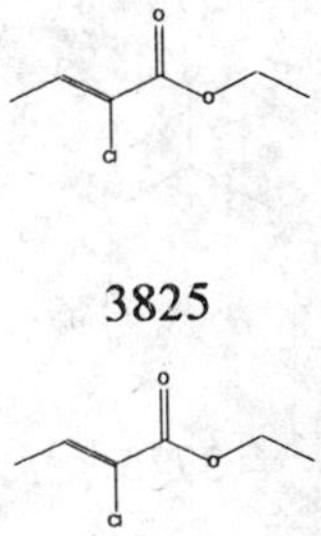

3825

3826

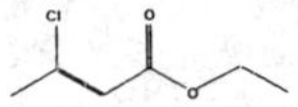

STRUCTURAL FORMULAS OF ORGANIC COMPOUNDS (continued)

In numeric order as they occur in the Table of Physical Constants of Organic Compounds

3827

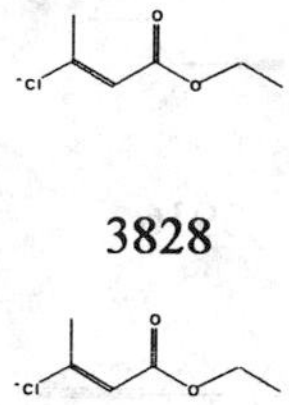

3828

3829

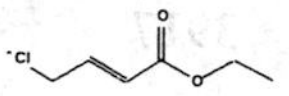

3830

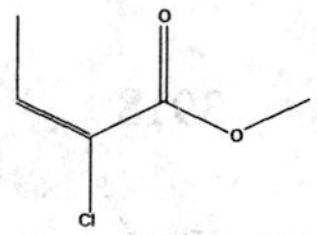

3831

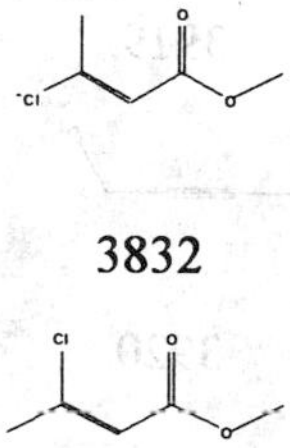

3832

3833

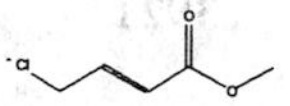

3834

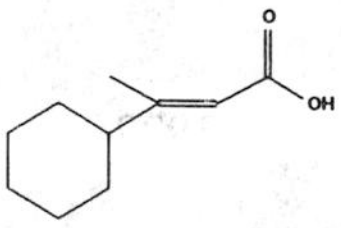

3835

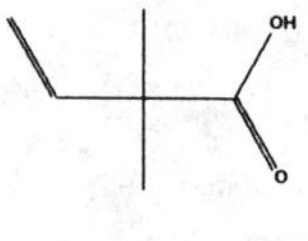

3836

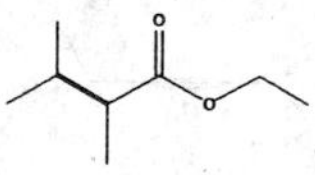

3837

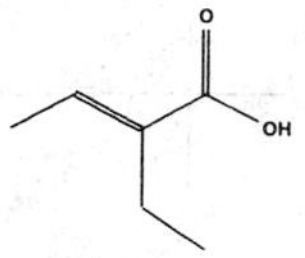

3838

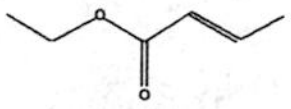

3839

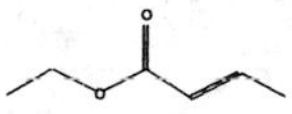

3840

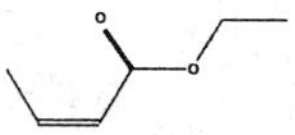

3841

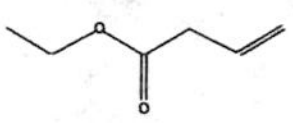

3842

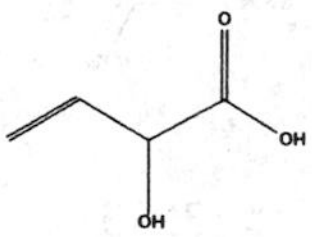

3843

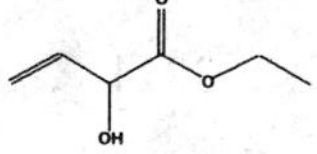

3844

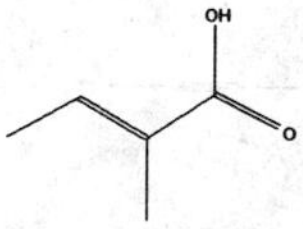

3845

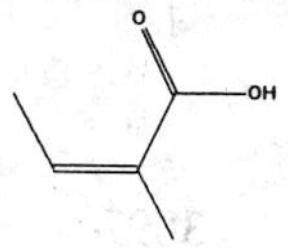

3846

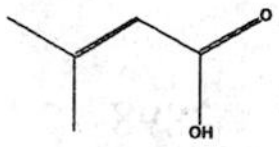

3847

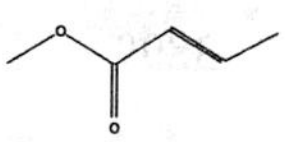

3848

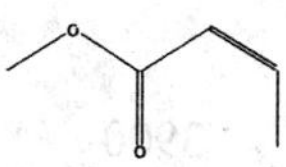

3849

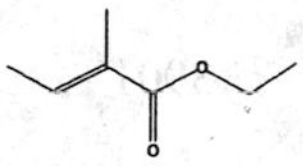

3850

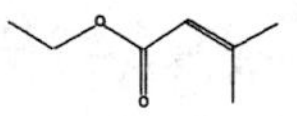

3851

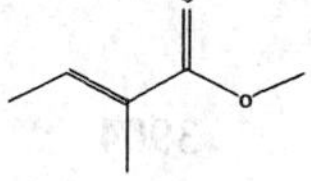

3852

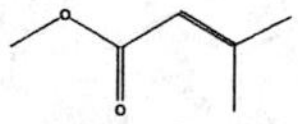

3856

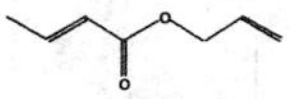

3857

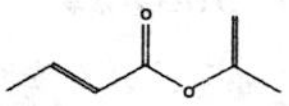

3858

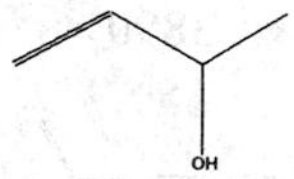

3859

3860

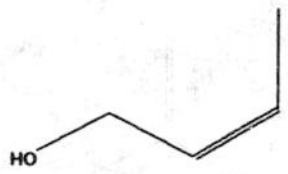

3861

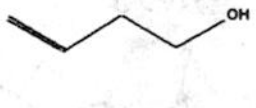

3862

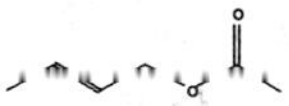

3864

3865

3866

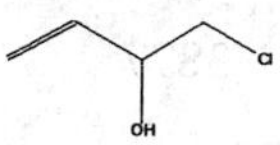

3868

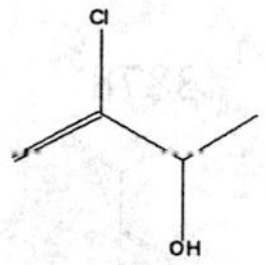

3869

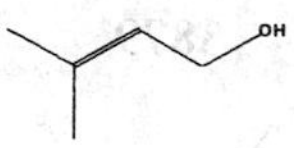

3870

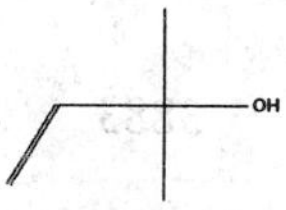

3871

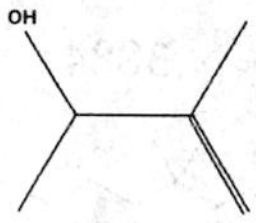

3873

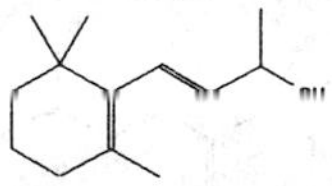

3874

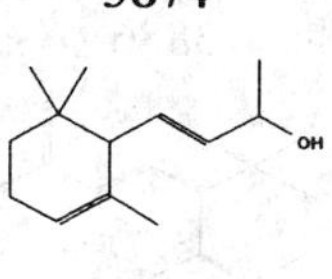

STRUCTURAL FORMULAS OF ORGANIC COMPOUNDS (continued)

In numeric order as they occur in the Table of Physical Constants of Organic Compounds

3875

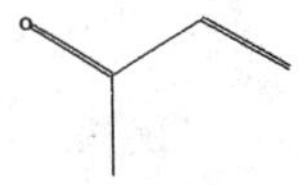

3877

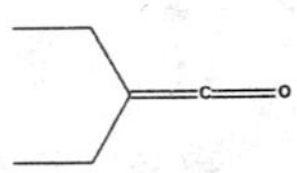

3878

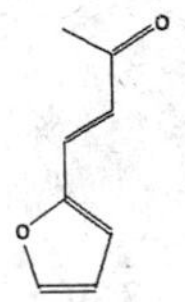

3879

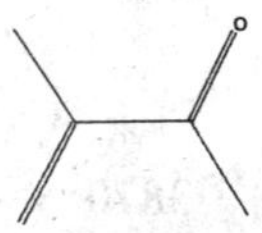

3883

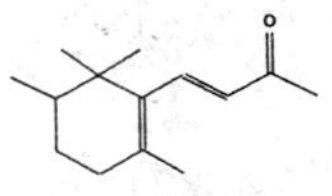

3884

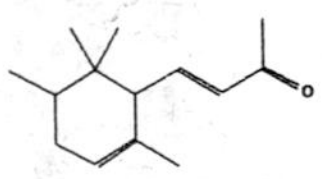

3885

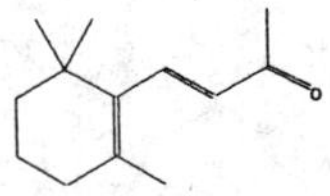

3886

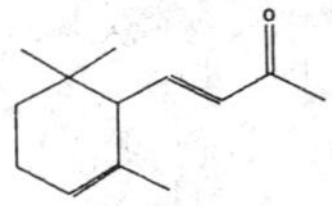

3887

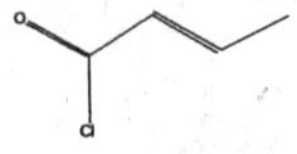

3888

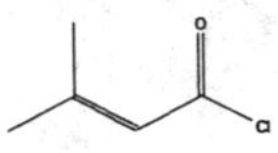

3889

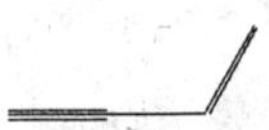

3890

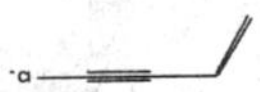

3891

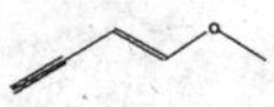

3892

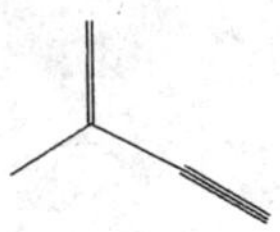

3893

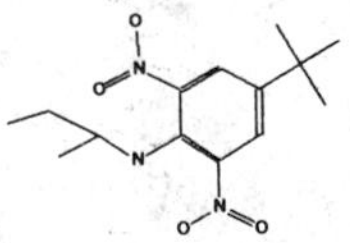

3894

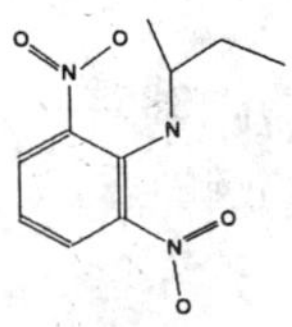

3895

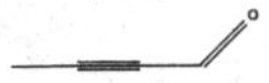

3896

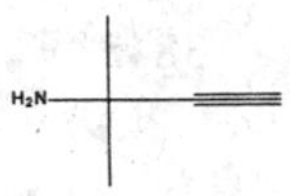

3897

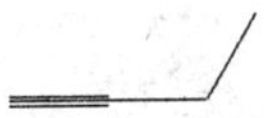

3898

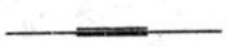

3899

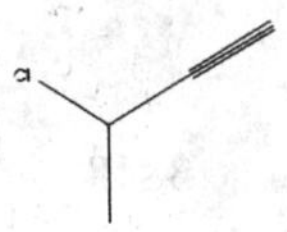

3900

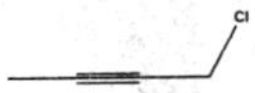

3901

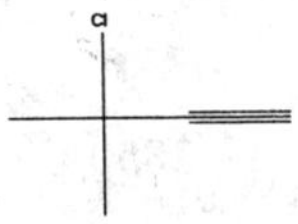

3902

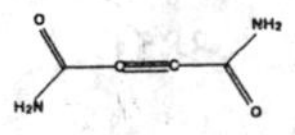

3904

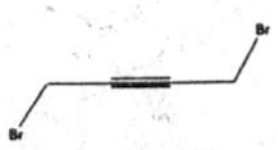

3905

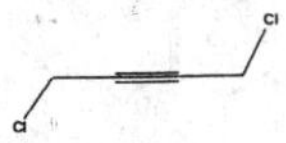

3907

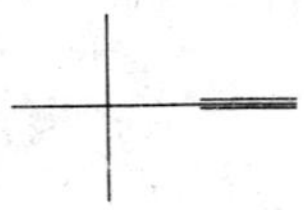

3908

3909

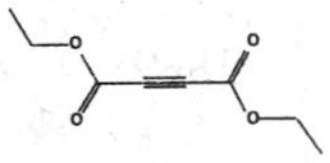

3910

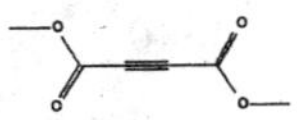

3911

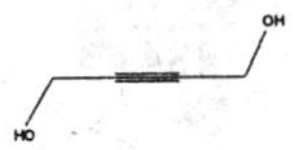

3912

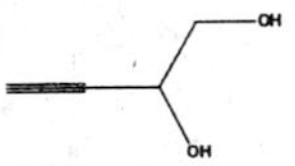

3913

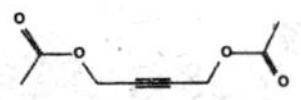

3914

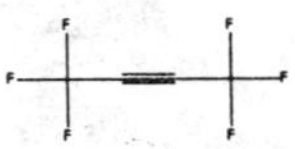

3915

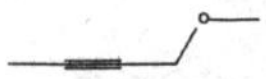

3916

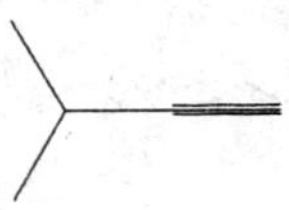

3917

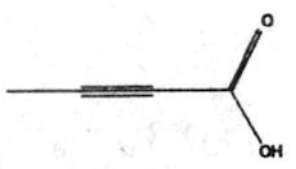

3918

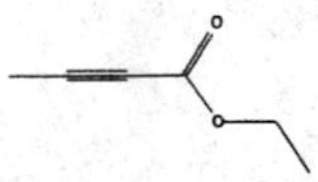

3919

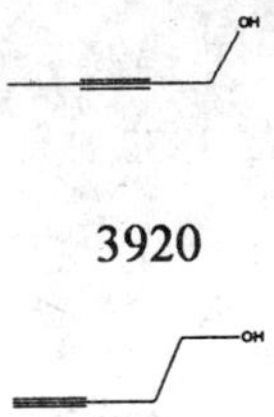

3920

3921

3922

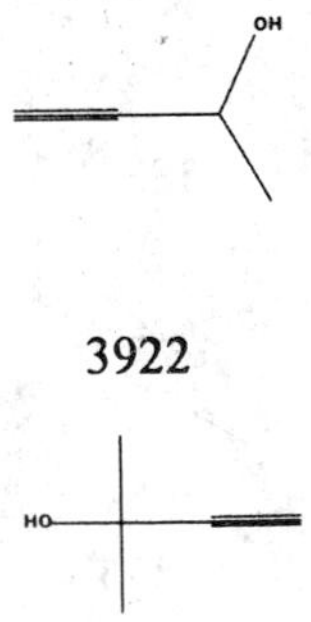

STRUCTURAL FORMULAS OF ORGANIC COMPOUNDS (continued)

In numeric order as they occur in the Table of Physical Constants of Organic Compounds

3923

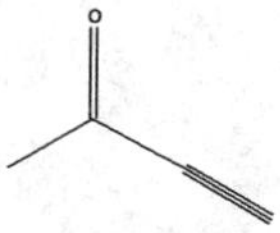

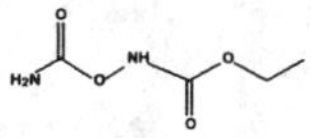

3933

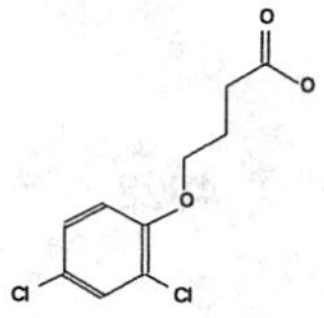

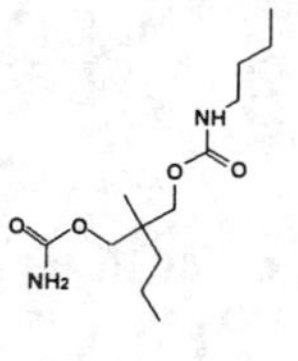

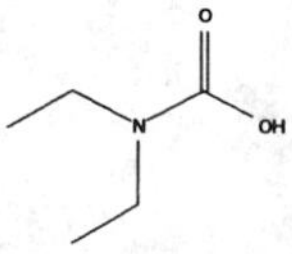

3942

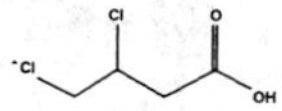

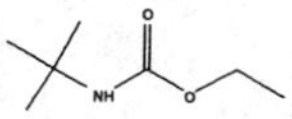

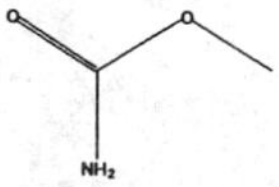

3954

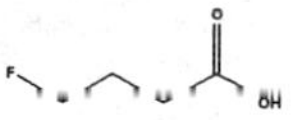

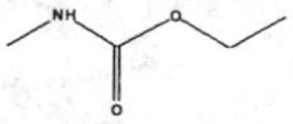

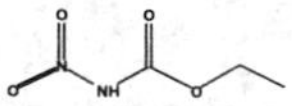

3964

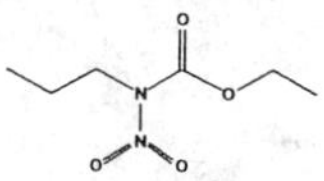

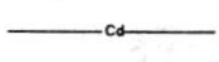

3925

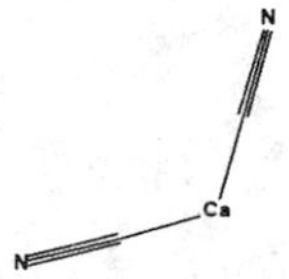

3934

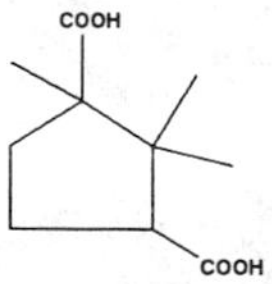

3943

3955

3966

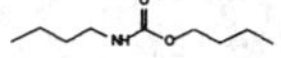

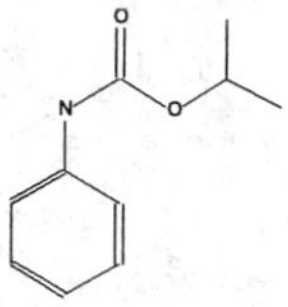

3926

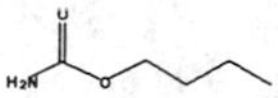

3935

3944

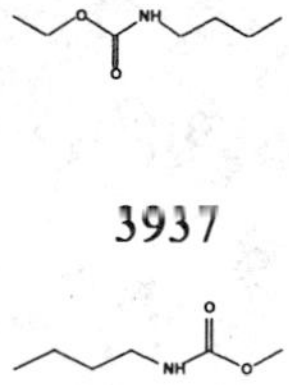

3956

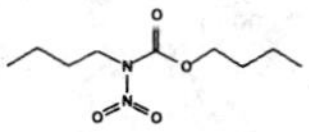

3969

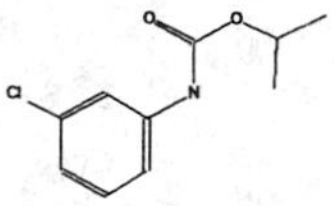

3927

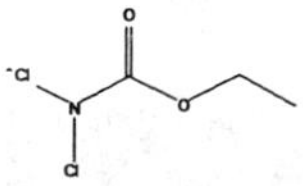

3936

3947

3957

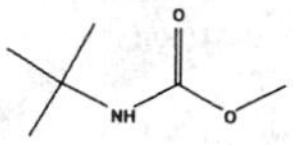

3970

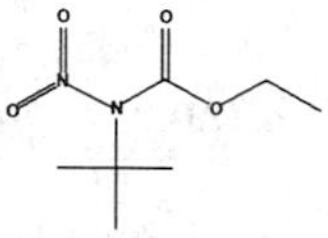

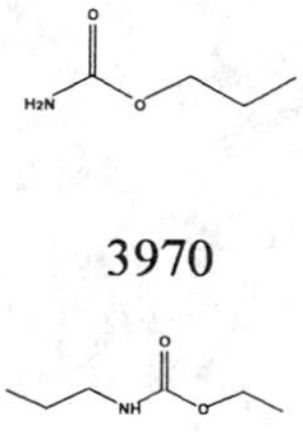

3929

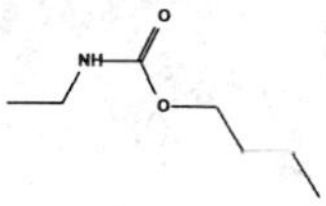

3937

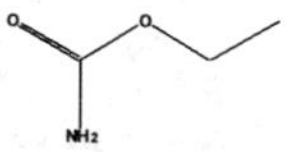

3948

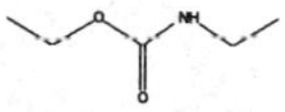

3959

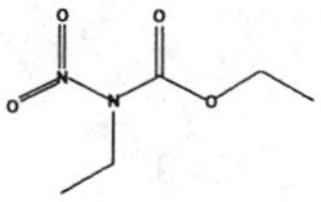

3971

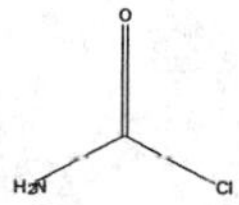

3930

3938

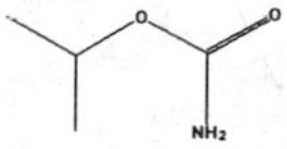

3949

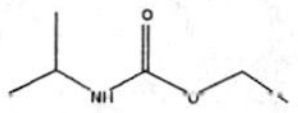

3960

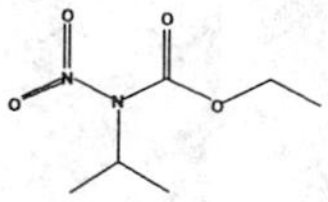

3972

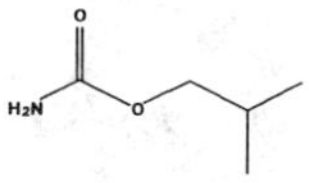

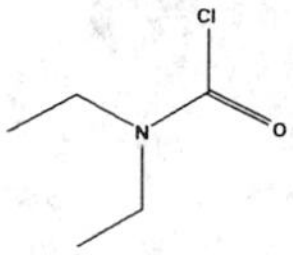

3931

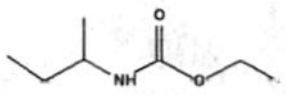

3939

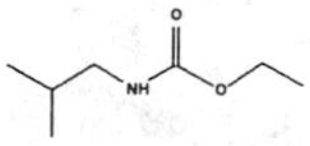

3950

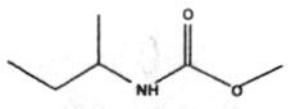

3961

3973

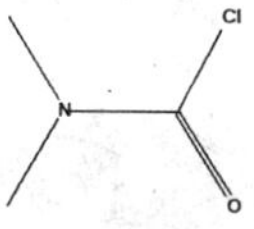

3932

3940

3953

3962

3976

3941

3963

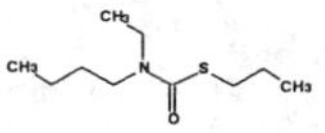

STRUCTURAL FORMULAS OF ORGANIC COMPOUNDS (continued)

In numeric order as they occur in the Table of Physical Constants of Organic Compounds

3978

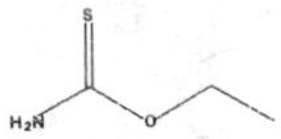

3979

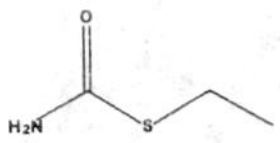

3980

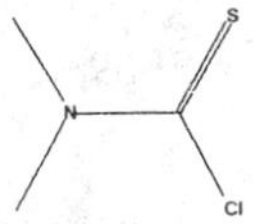

3982

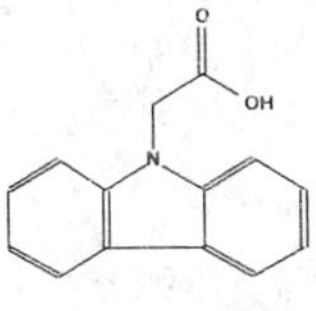

3989

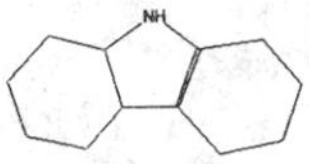

3990

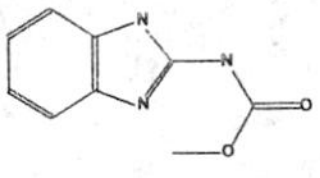

3991

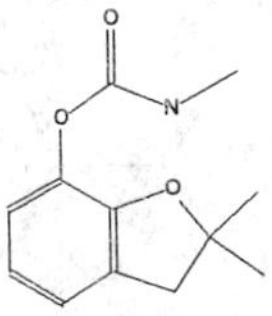

3992

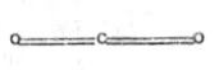

3993

3994

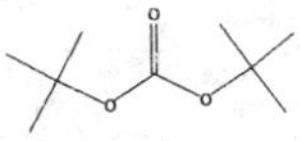

3995

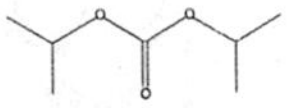

3997

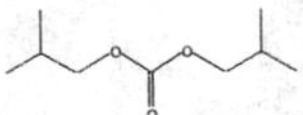

3998

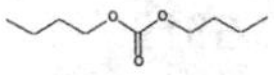

3999

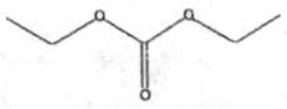

4000

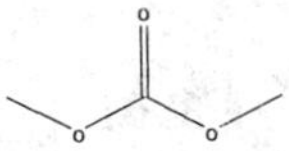

4002

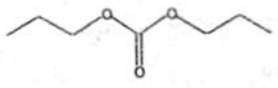

4003

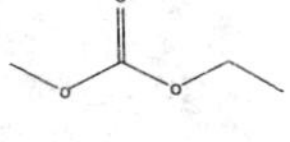

4005

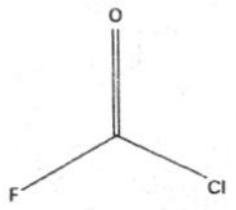

4006

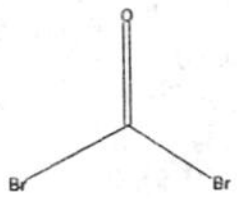

4007

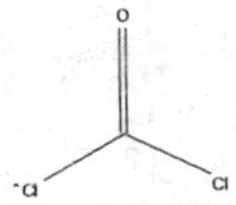

4008

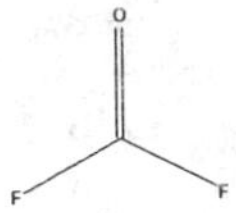

4009

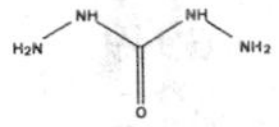

4012

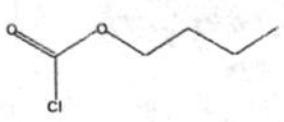

4013

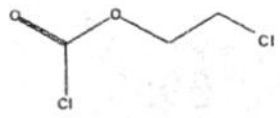

4014

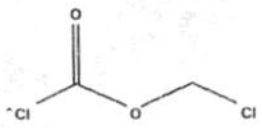

4015

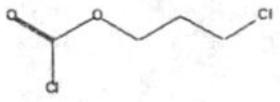

4016

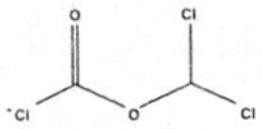

4017

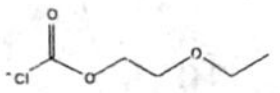

4018

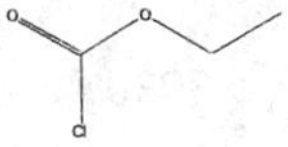

4019

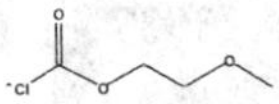

4020

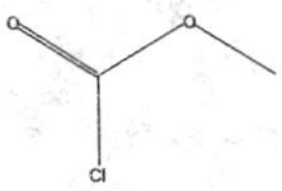

4021

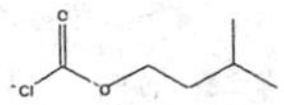

4022

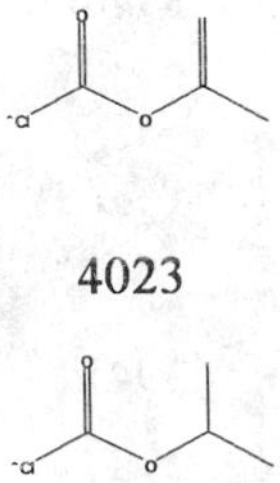

4023

4024

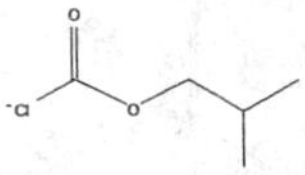

4025

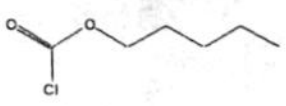

4028

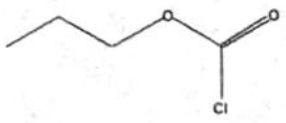

4029

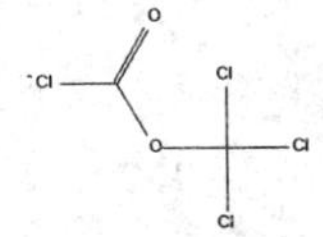

4030

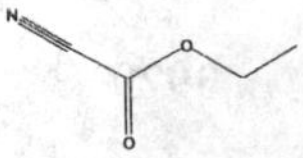

4031

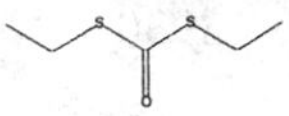

4032

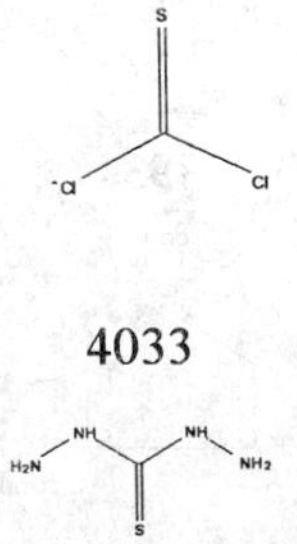

4033

STRUCTURAL FORMULAS OF ORGANIC COMPOUNDS (continued)

In numeric order as they occur in the Table of Physical Constants of Organic Compounds

4034

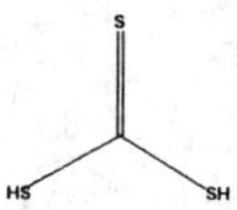

4035

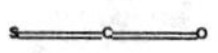

4036

4037

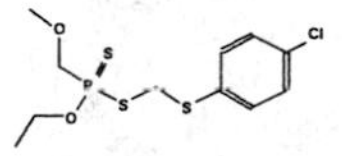

4038

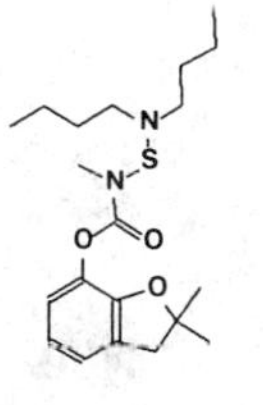

4039

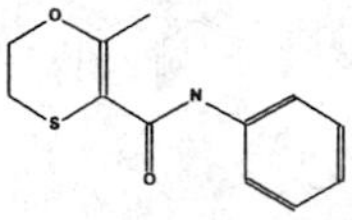

4040

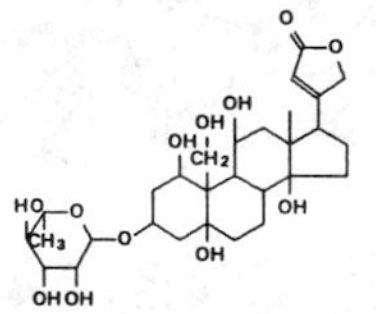

4043

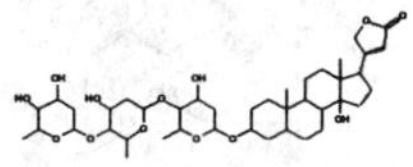

4044

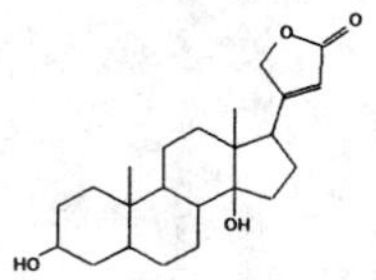

4046

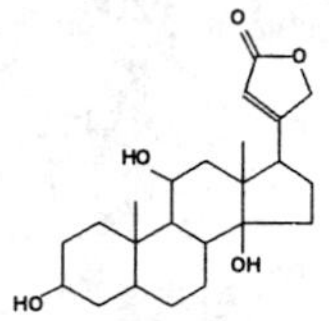

4047

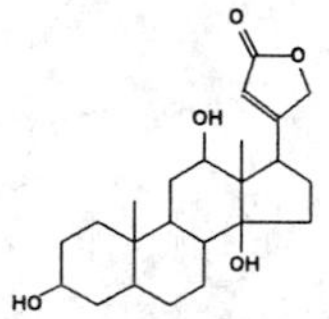

4048

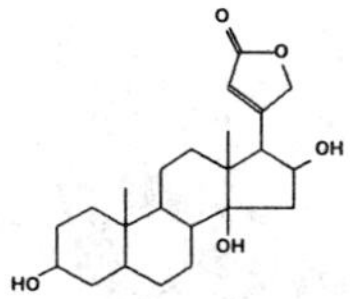

4049

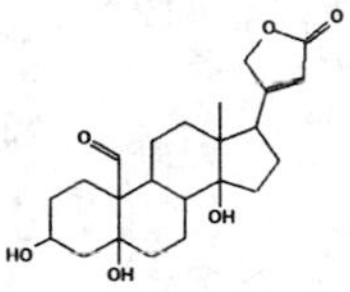

4050

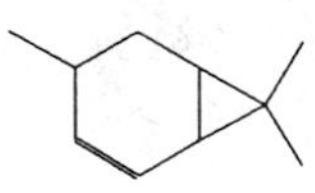

4065

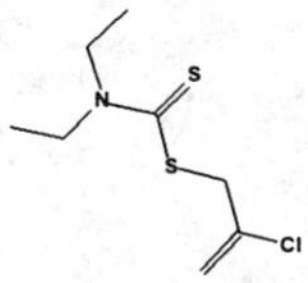

4066

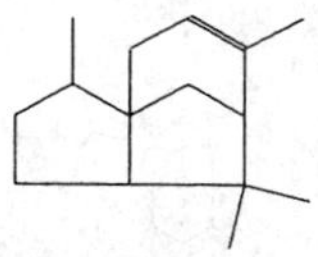

4067

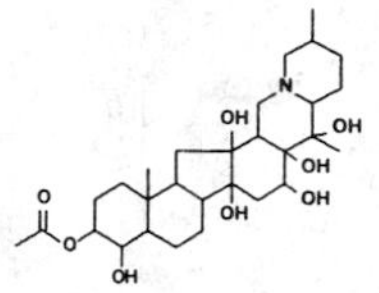

4068

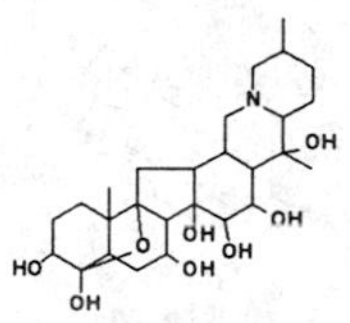

4069

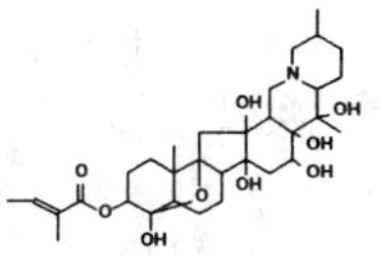

4071

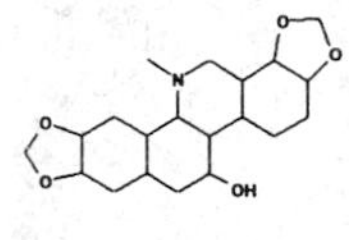

4072

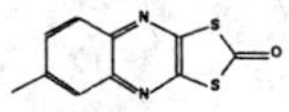

4075

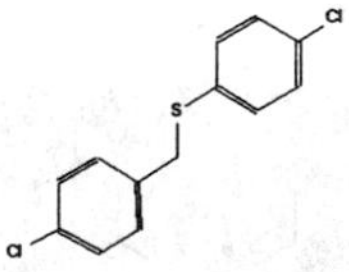

4076

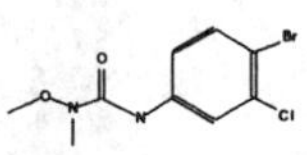

4077

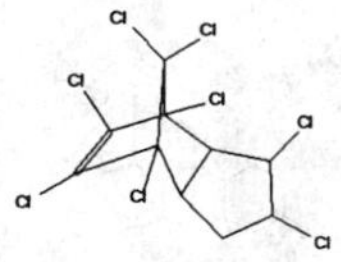

4078

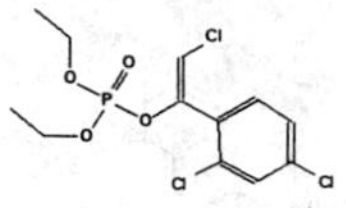

4079

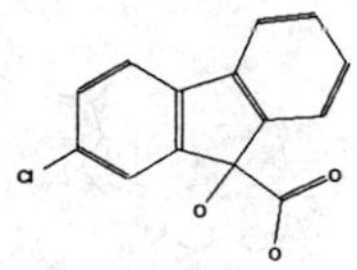

4080

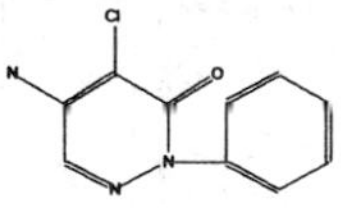

4081

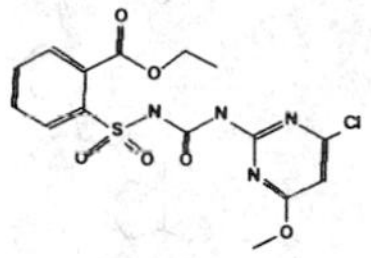

4082

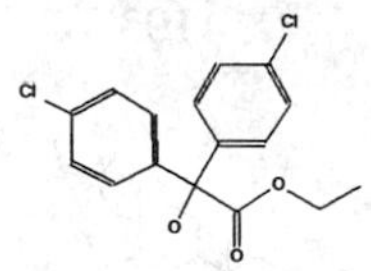

4083

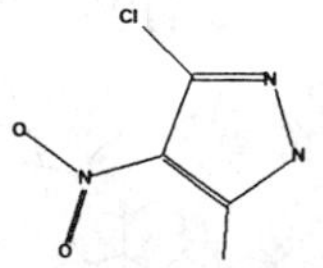

4084

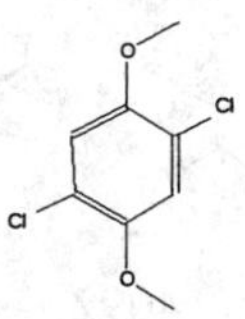

4085

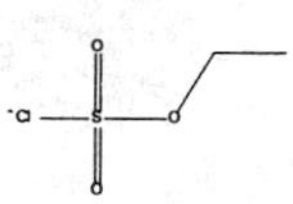

4086

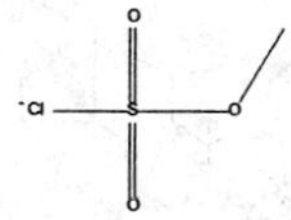

4087

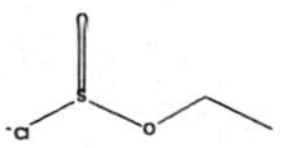

4088

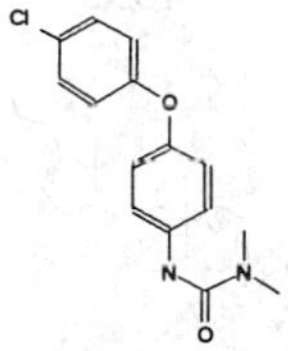

4089

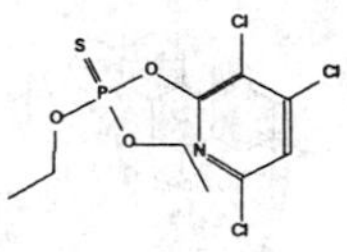

4090

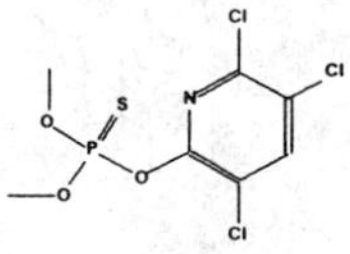

STRUCTURAL FORMULAS OF ORGANIC COMPOUNDS (continued)

In numeric order as they occur in the Table of Physical Constants of Organic Compounds

4091

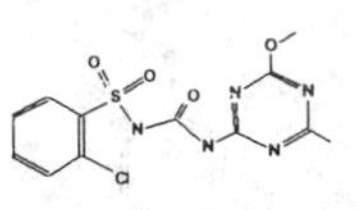

4092

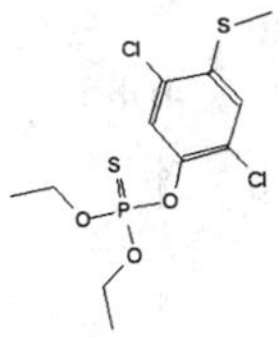

4093

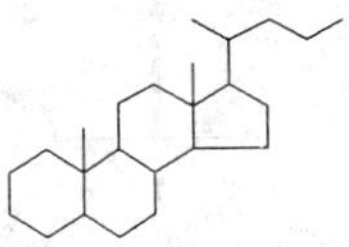

4094

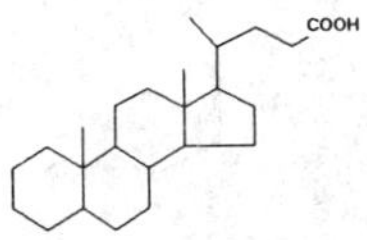

4095

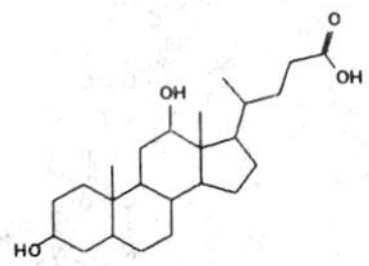

4096

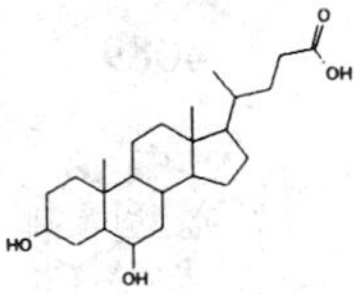

4097

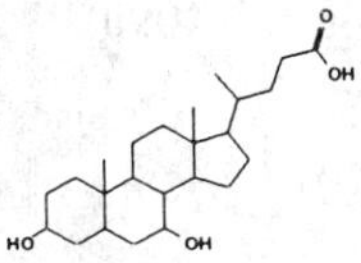

4098

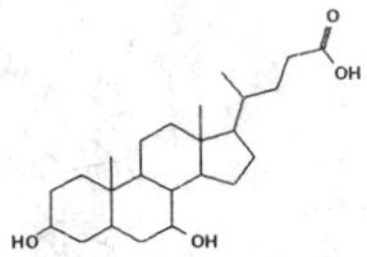

4099

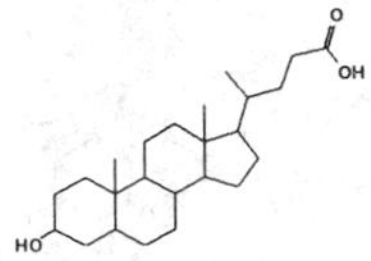

4100

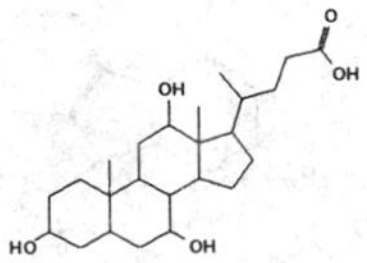

4101

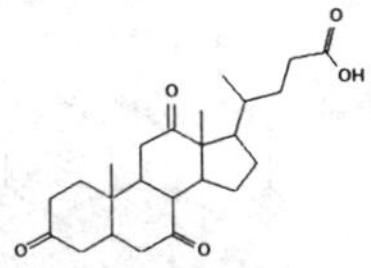

4102

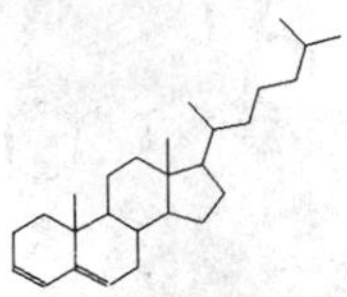

4103

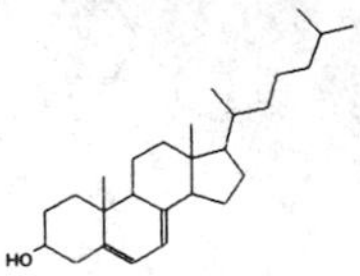

4106

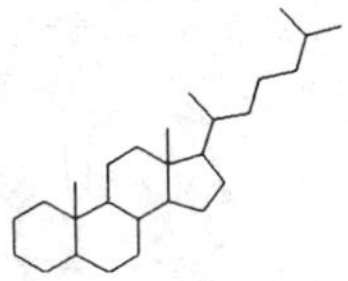

4107

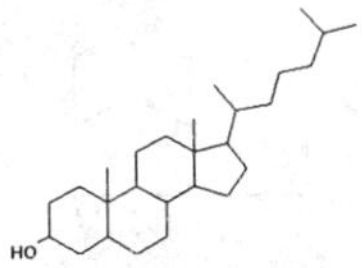

4108

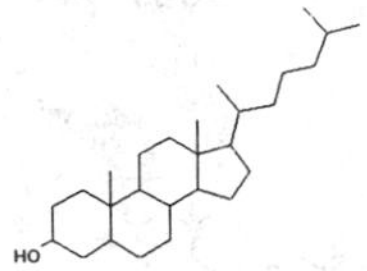

4110

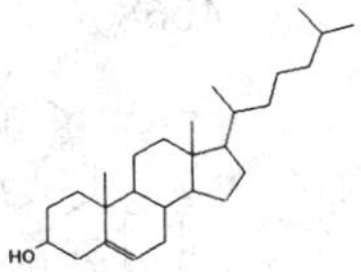

4111

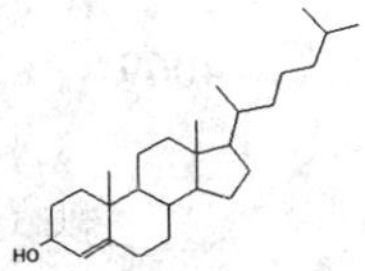

4113

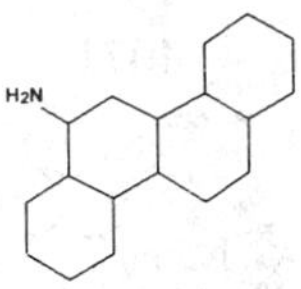

4116

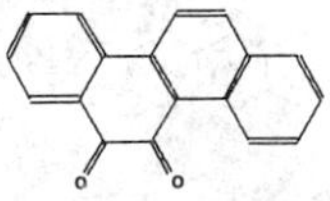

4118

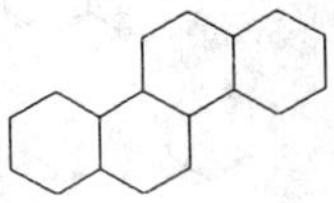

4119

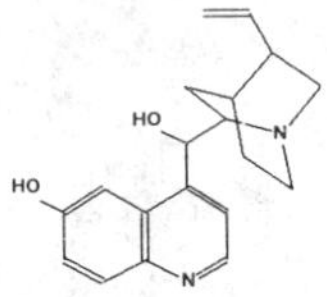

4120

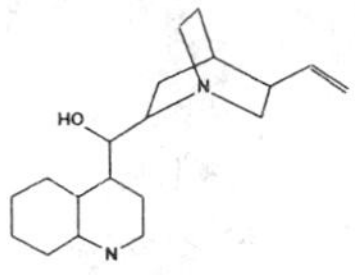

4121

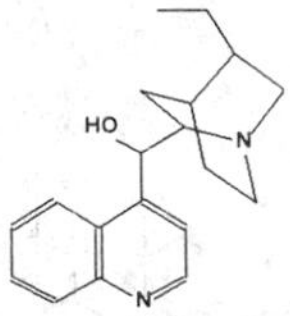

4122

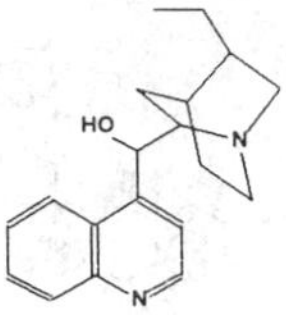

4123

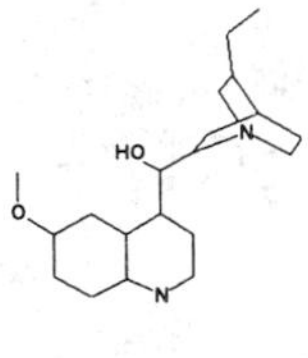

4124

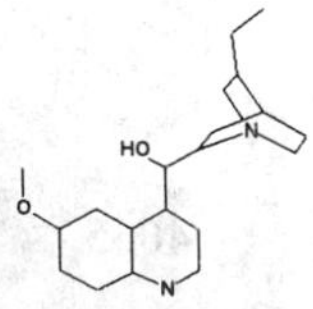

4125

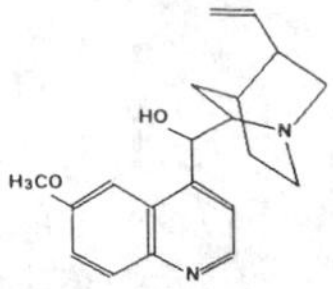

4126

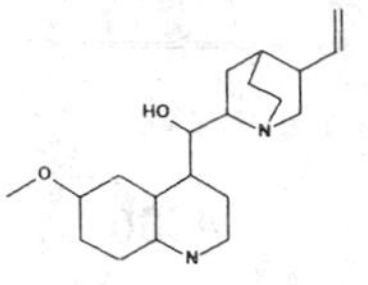

4127

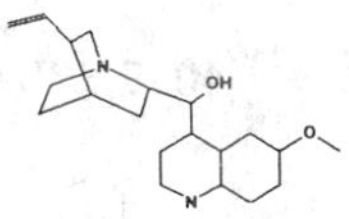

4128

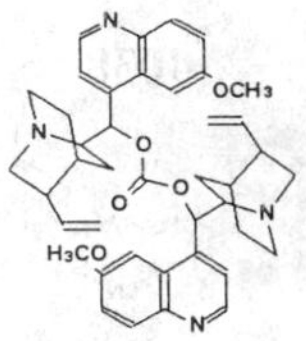

4130

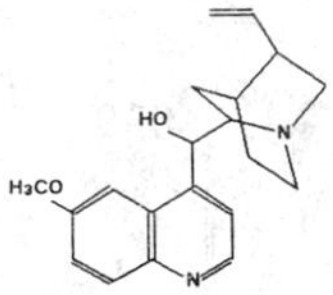

4133

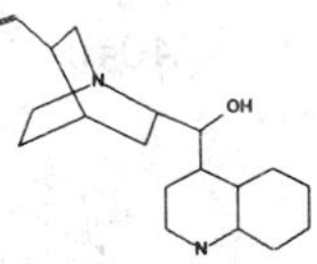

4134

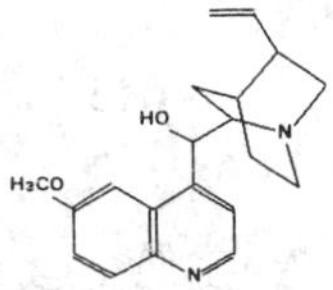

STRUCTURAL FORMULAS OF ORGANIC COMPOUNDS (continued)

In numeric order as they occur in the Table of Physical Constants of Organic Compounds

4137

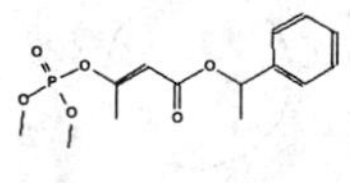

4138

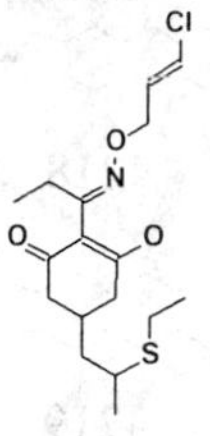

4139

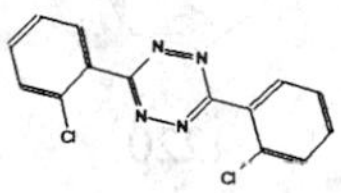

4140

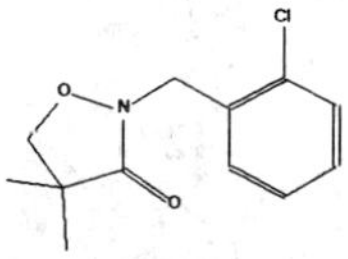

4141

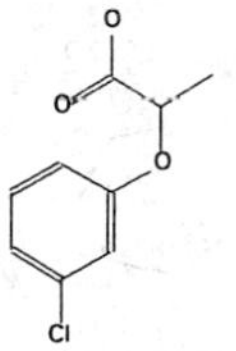

4142

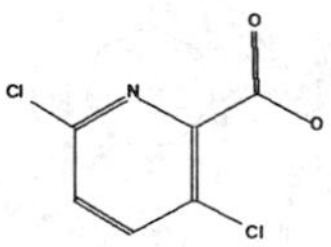

4143

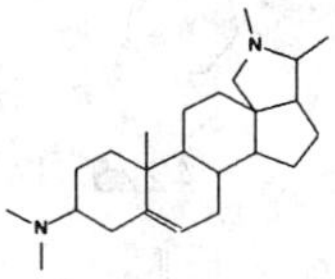

4144

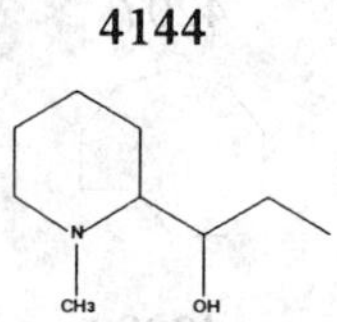

4146

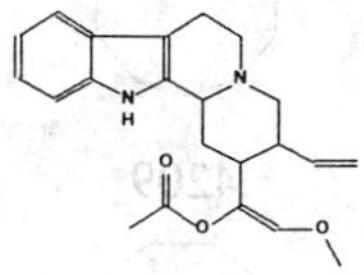

4147

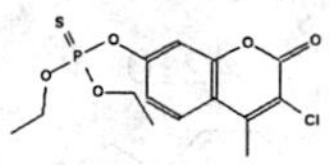

4150

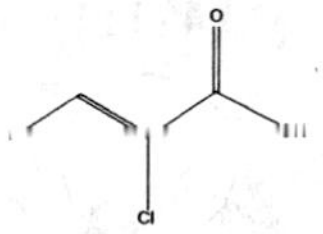

4151

4152

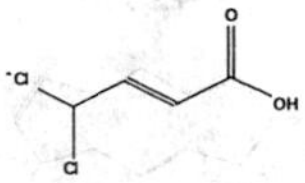

4153

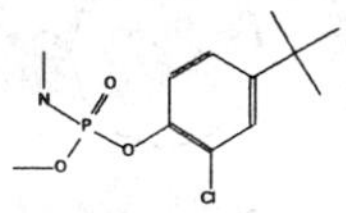

4154

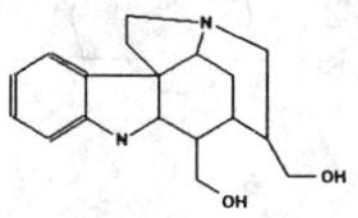

4155

4156

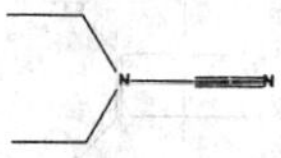

4157

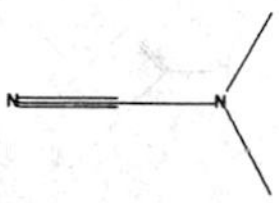

4159

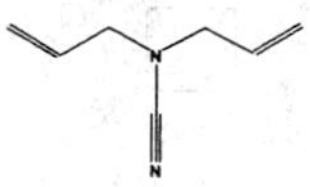

4161

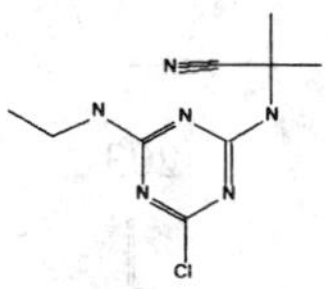

4162

4163

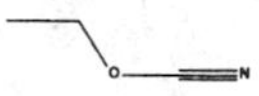

4164

4165

4166

4167

4168

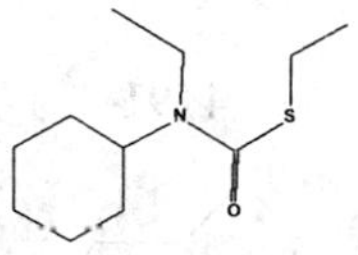

4169

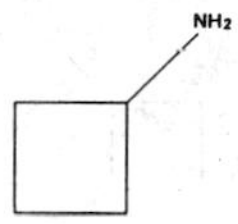

4170

4171

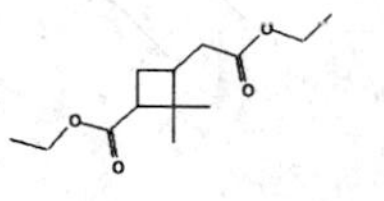

4172

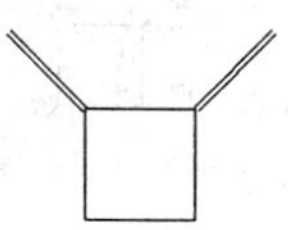

4173

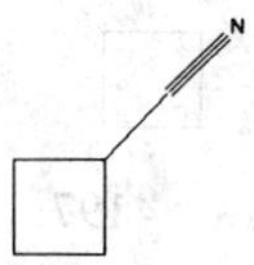

4174

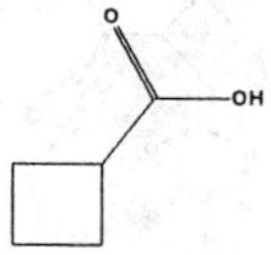

4175

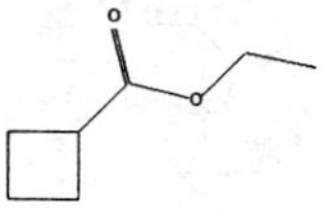

4176

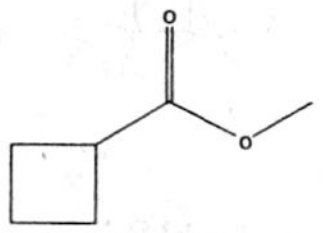

4177

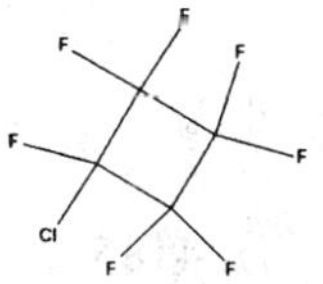

4178

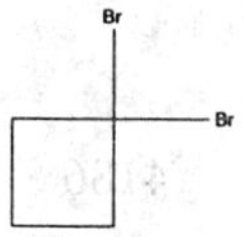

4179

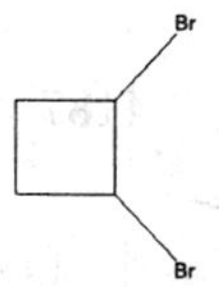

4180

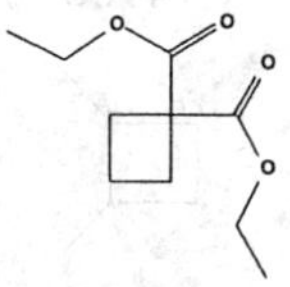

STRUCTURAL FORMULAS OF ORGANIC COMPOUNDS (continued)

In numeric order as they occur in the Table of Physical Constants of Organic Compounds

4181

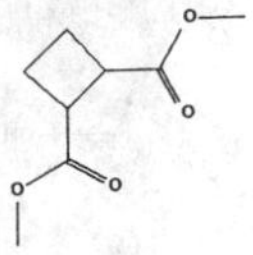

4182

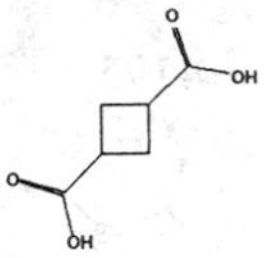

4183

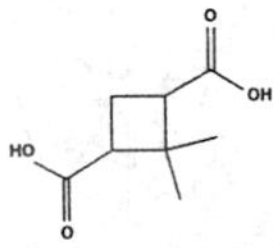

4184

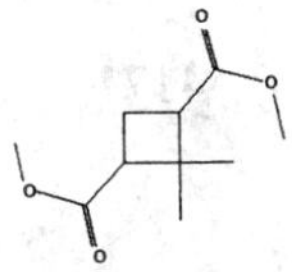

4185

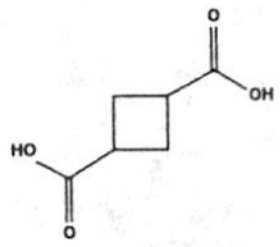

4186

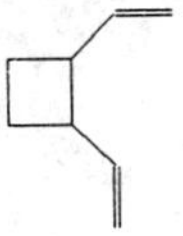

4187

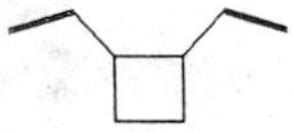

4188

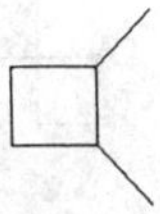

4189

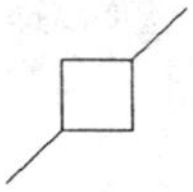

4190

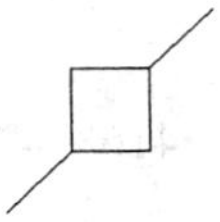

4191

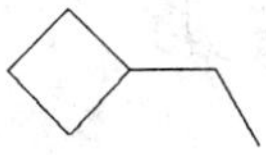

4192

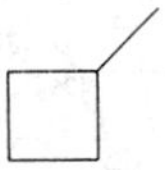

4193

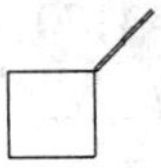

4194

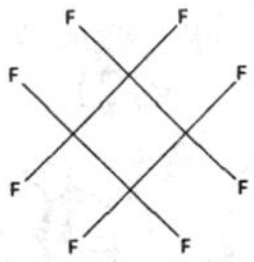

4195

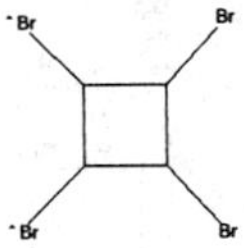

4196

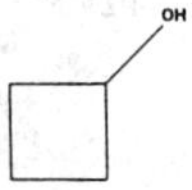

4197

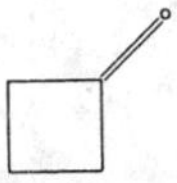

4198

4199

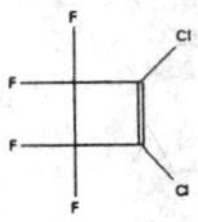

4200

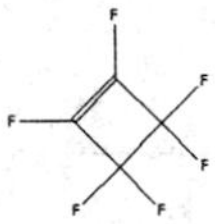

4201

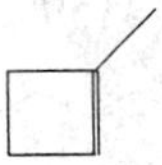

4202

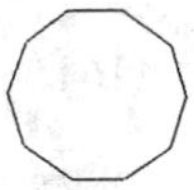

4203

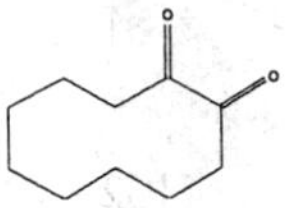

4204

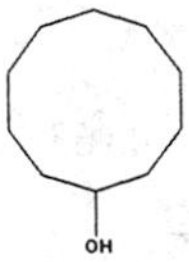

4205

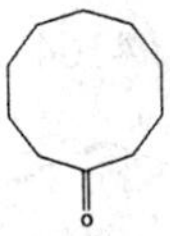

4206

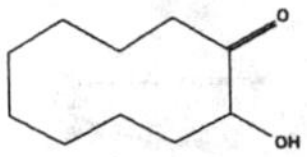

4207

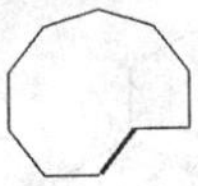

4208

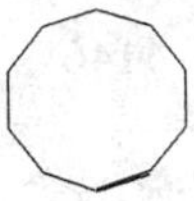

4209

4210

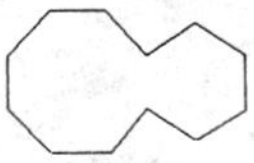

4211

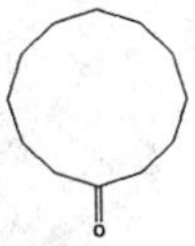

4212

4213

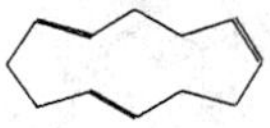

4214

4215

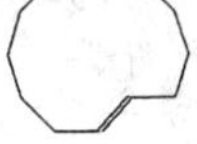

4216

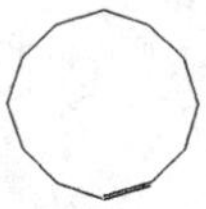

4217

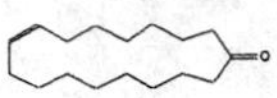

4218

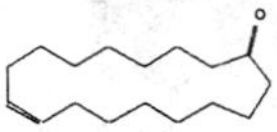

4219

4220

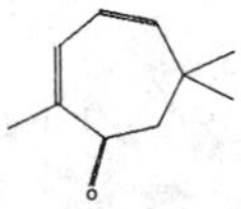

4221

4222

4223

4224

4225

In numeric order as they occur in the Table of Physical Constants of Organic Compounds

4226

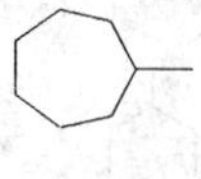

4227

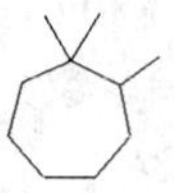

4228

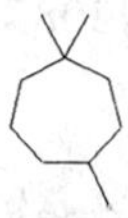

4229

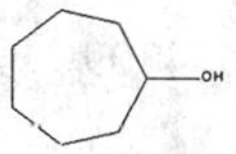

OH

4230

OH

4231

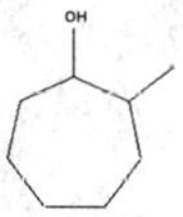

OH

4233

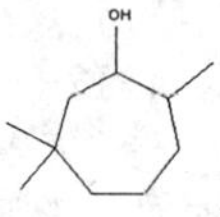

OH

4234

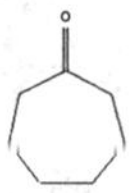

O

4235

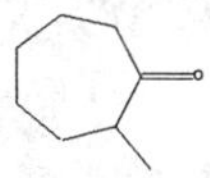

O

4236

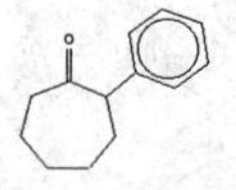

O

4237

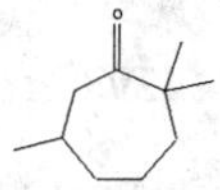

O

4238

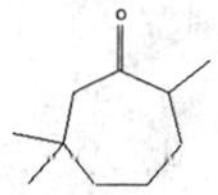

O

4239

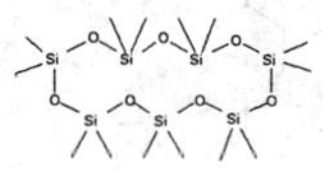

Si O Si O Si O Si O Si O Si O

4240

4241

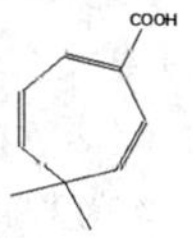

COOH

4242

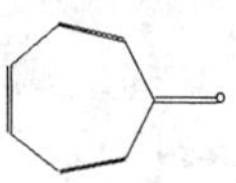

O

4243

O OH

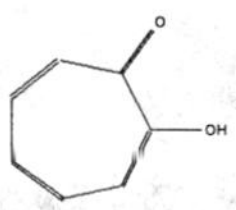

4244

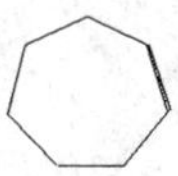

4245

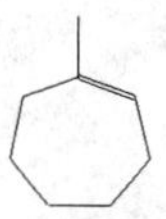

4246

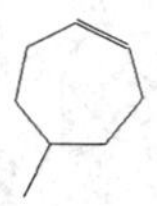

4247

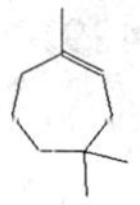

4248

4249

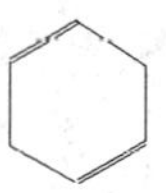

4250

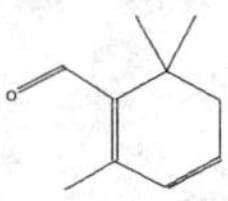

O

4251

N N O O Cl Cl

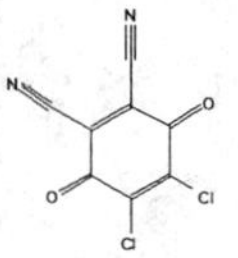

4252

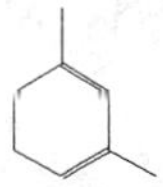

4253

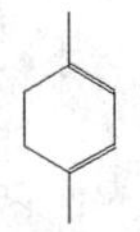

4254

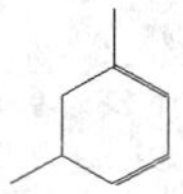

4255

4256

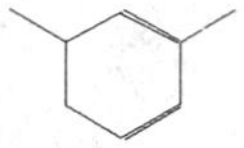

4257

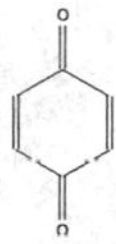

O O

4258

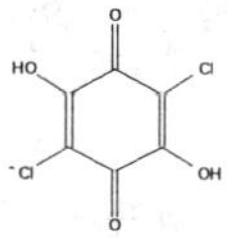

O HO Cl Cl OH O

4259

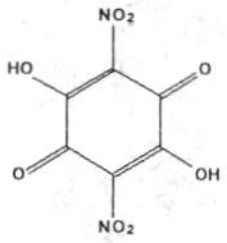

NO_2 HO O O OH NO_2

4260

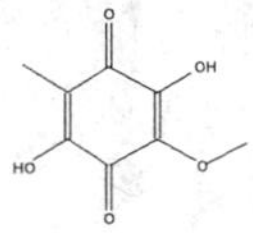

O OH HO O

4261

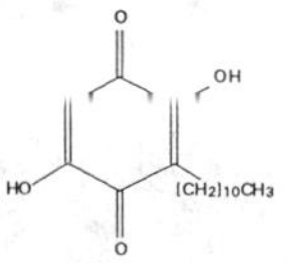

O OH HO $(CH_2)_{10}CH_3$ O

4262

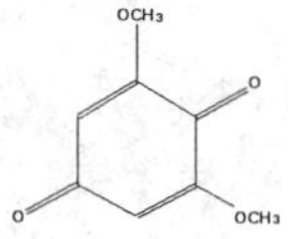

OCH_3 O O OCH_3

4263

O O

4264

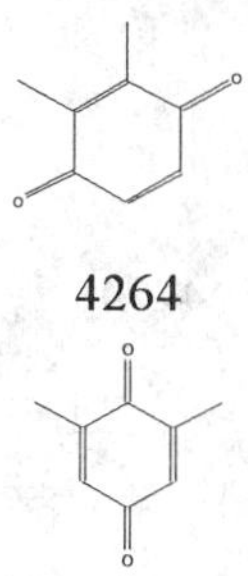

O O

4265

O O OCH_3 OH

4266

O O

4267

O O

4268

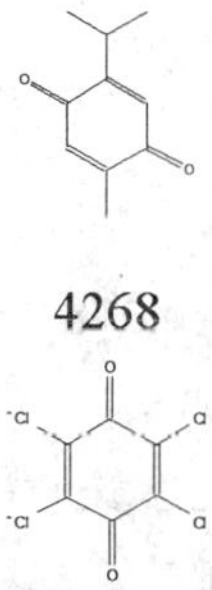

O Cl Cl Cl Cl O

4269

OH HO O O OH OH

4270

O O

4271

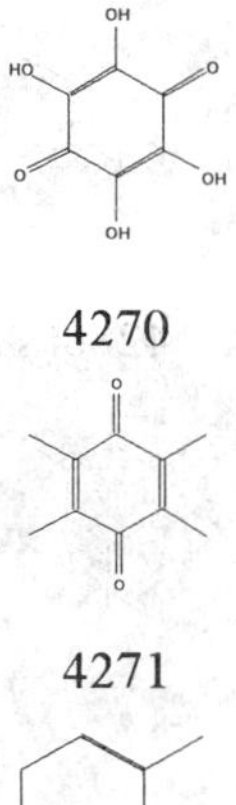

STRUCTURAL FORMULAS OF ORGANIC COMPOUNDS (continued)

In numeric order as they occur in the Table of Physical Constants of Organic Compounds

4273

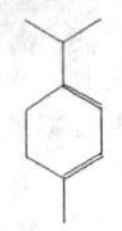

4274

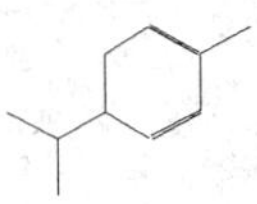

4275

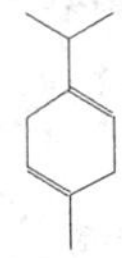

4276

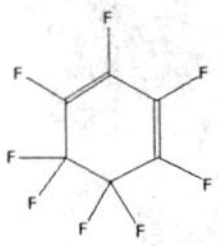

4277

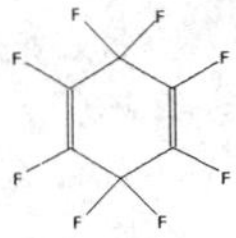

4278

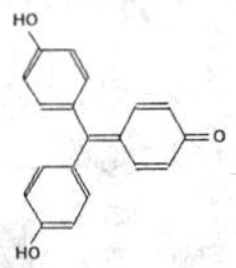

4279

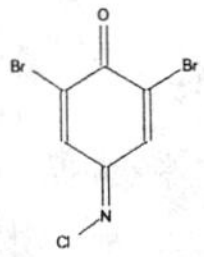

4280

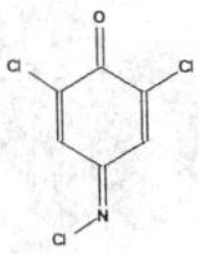

4281

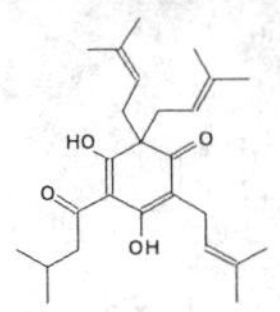

4282

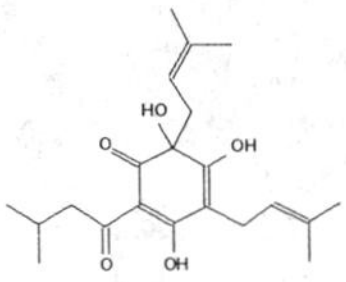

4283

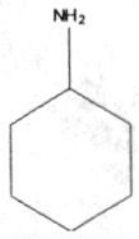

4284

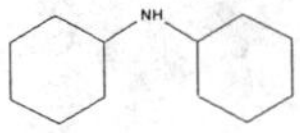

4285

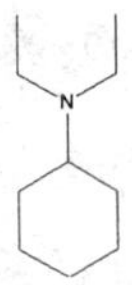

4286

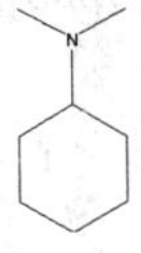

4287

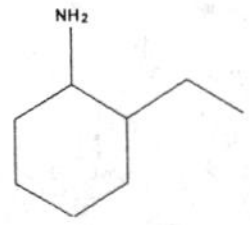

4288

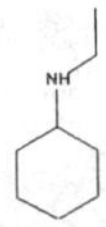

4289

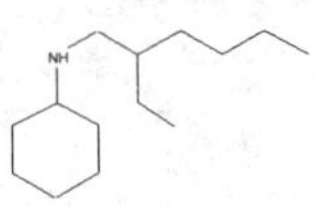

4290

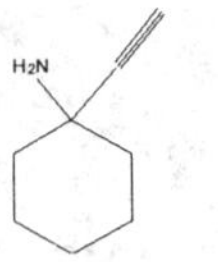

4291

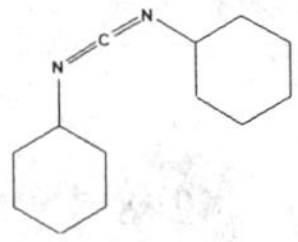

4292

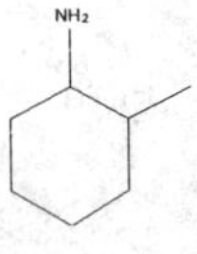

4293

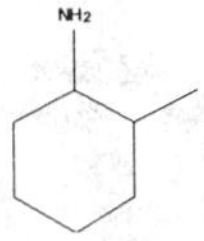

4294

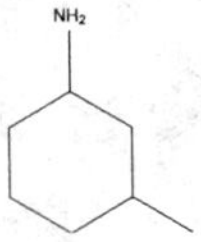

4295

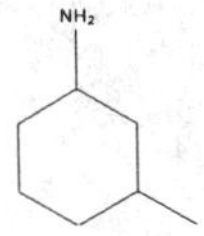

4296

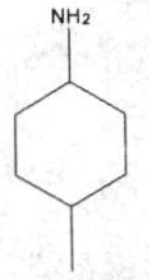

4297

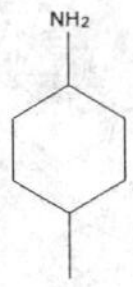

4298

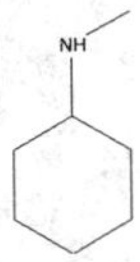

4299

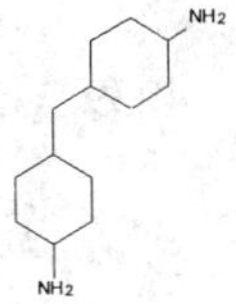

4300

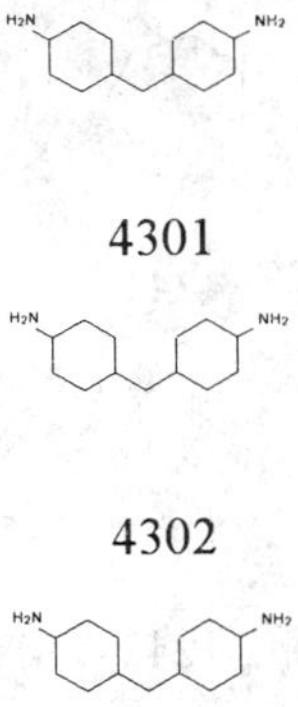

4301

4302

4303

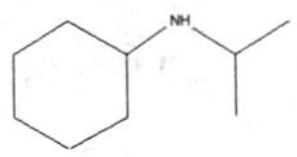

4304

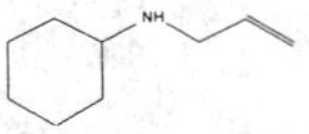

4305

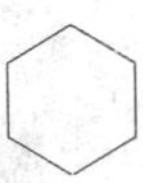

4306

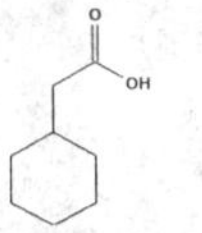

4307

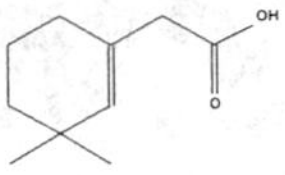

4308

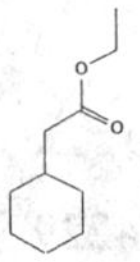

4309

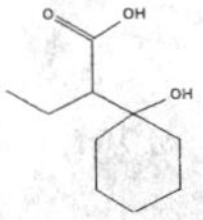

4310

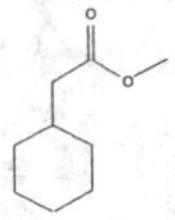

4311

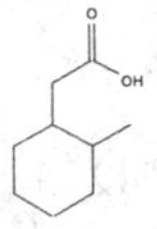

4312

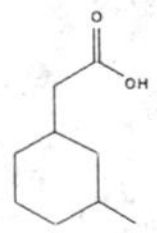

4313

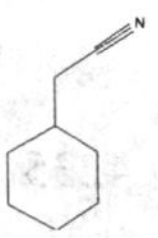

4314

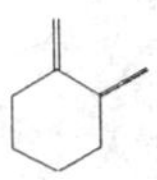

STRUCTURAL FORMULAS OF ORGANIC COMPOUNDS (continued)

In numeric order as they occur in the Table of Physical Constants of Organic Compounds

4315

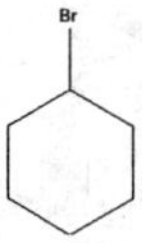

Br

4316

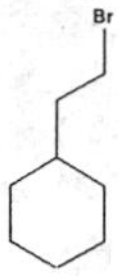

Br

4317

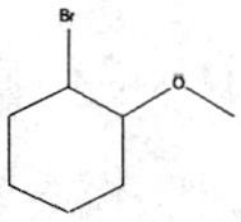

Br

4318

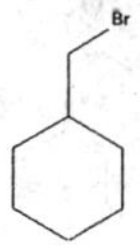

Br

4319

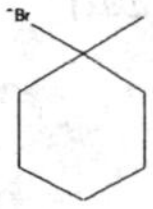

Br

4320

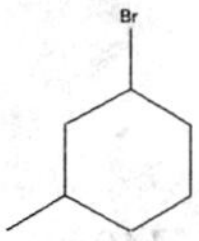

Br

4322

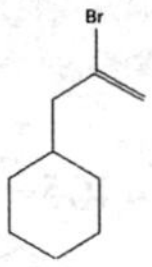

Br

4323

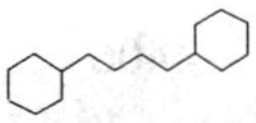

4324

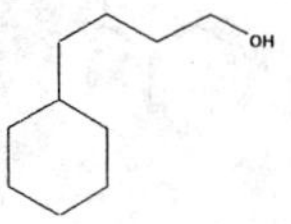

OH

4325

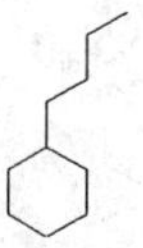

4326

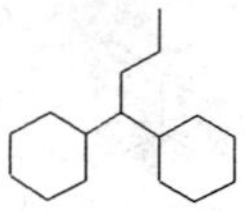

4327

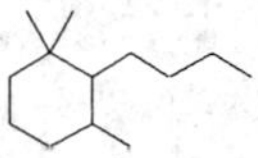

4328

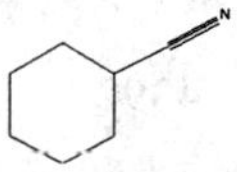

N

4329

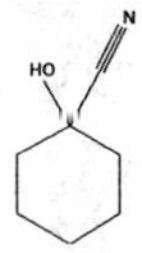

HO

N

4330

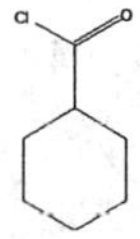

Cl

O

4331

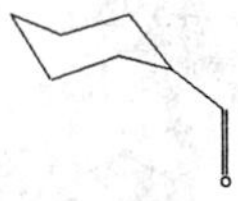

4332

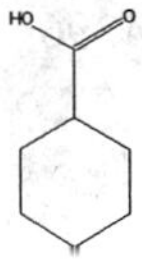

HO

O

4333

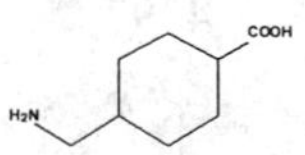

COOH

H_2N

4334

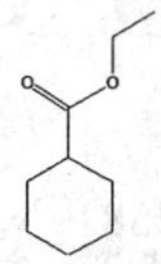

O

O

4335

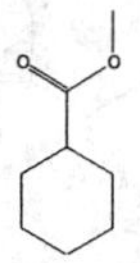

O

O

4336

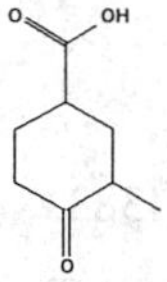

O

OH

O

4337

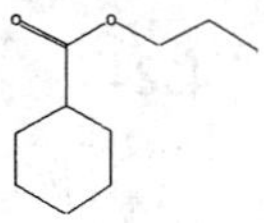

O

O

4339

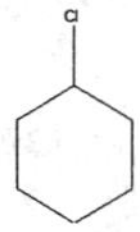

Cl

4340

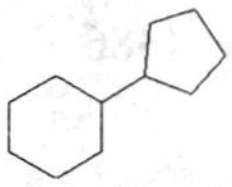

4341

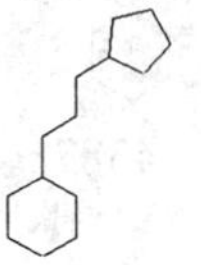

4342

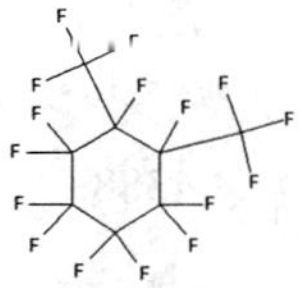

F F F F F F F F F F F F F F F F

4343

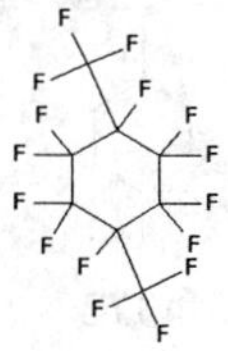

F F F F F F F F F F F F F F F F

4344

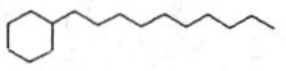

4345

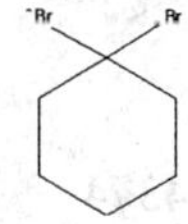

Br Br

4346

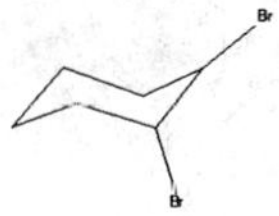

Br

Br

4347

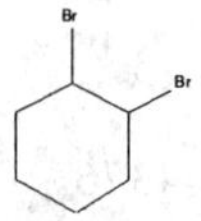

Br

Br

4348

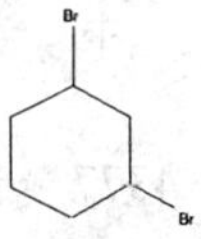

Br

Br

4349

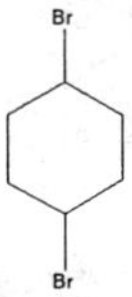

Br

Br

4350

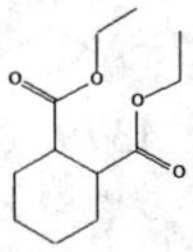

O O O O

4351

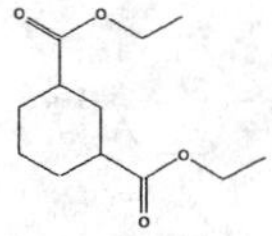

O O O O

4352

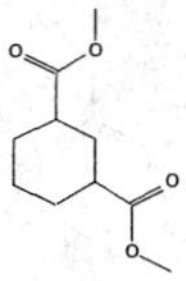

O O O O

4353

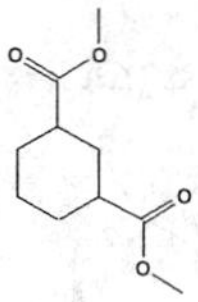

O O O O

4354

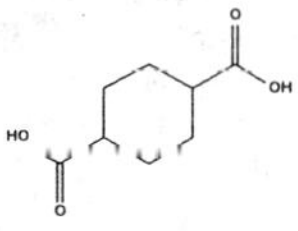

O OH HO O

4356

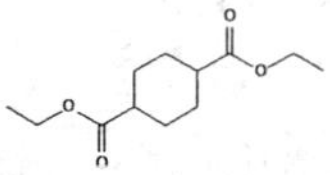

O O O O

4357

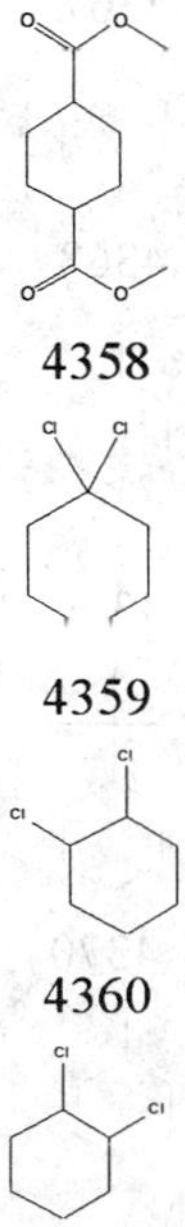

O O O O

4358

Cl Cl

4359

Cl Cl

4360

Cl Cl

In numeric order as they occur in the Table of Physical Constants of Organic Compounds

4361 Cl Cl

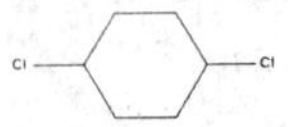

4371

4380

4389

4398

4362 NH2 NH2

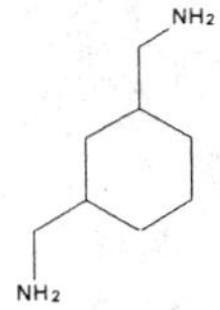

4372 OH OH

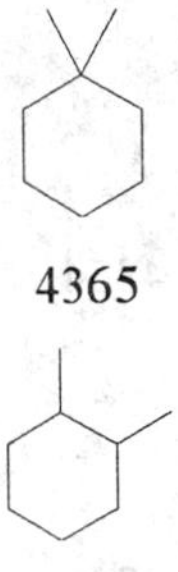

4381 F F F F F F F F F F F F

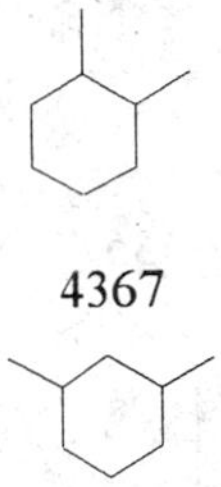

4390

4399

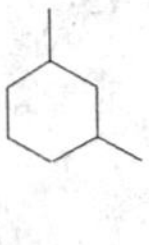

4364

4373 OH OH

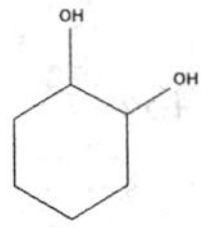

4382

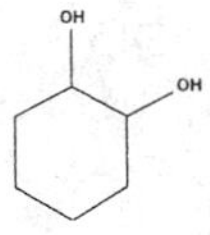

4391

4400

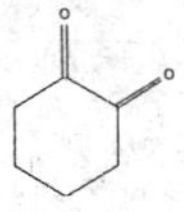

4383

4374 OH OH

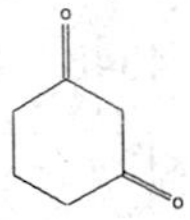

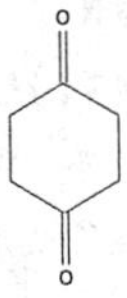

4392

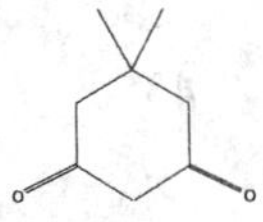

4401

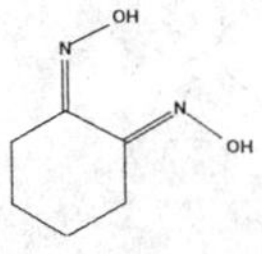

4365

4384

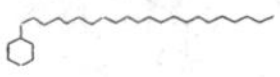

4393

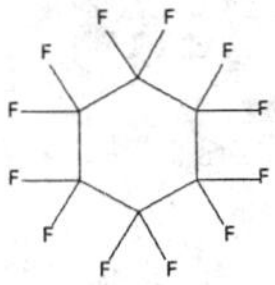

4375 O O

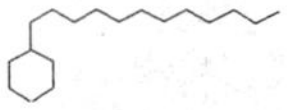

4385 NH

4402

4366

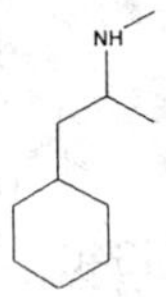

4394

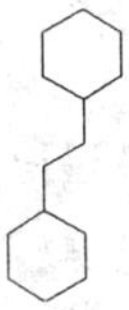

4403 F

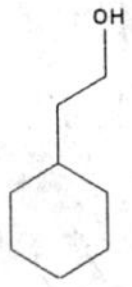

4376 O O

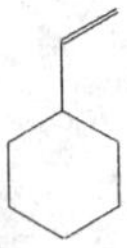

4386

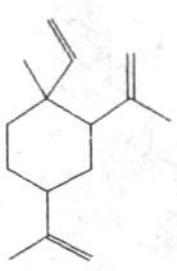

4367

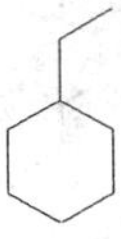

4395

4404

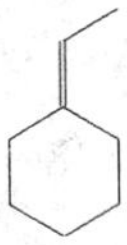

4368

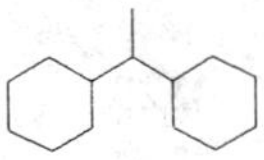

4377 O O

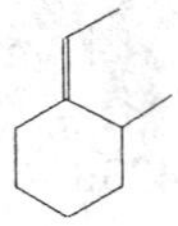

4387 OH

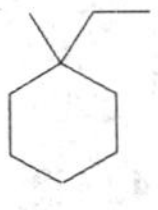

4396

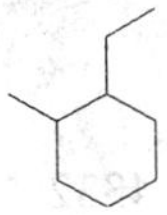

4405

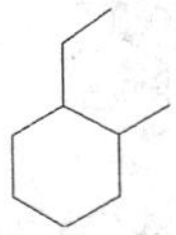

4369

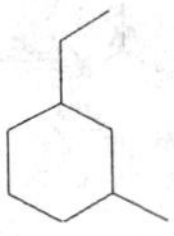

4378 O O

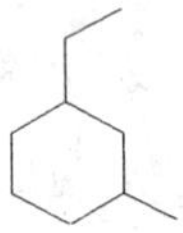

4406 Cl Cl Cl Cl Cl Cl

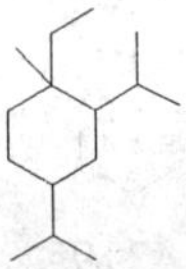

4397

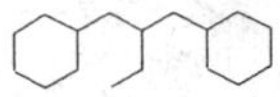

4388

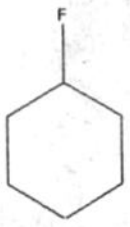

4370

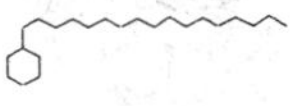

4379 N OH N OH

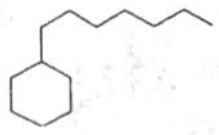

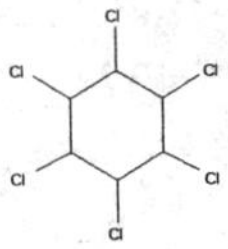

STRUCTURAL FORMULAS OF ORGANIC COMPOUNDS (continued)

In numeric order as they occur in the Table of Physical Constants of Organic Compounds

4407

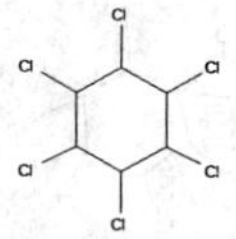

4408

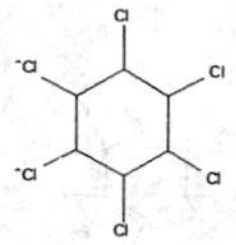

4409

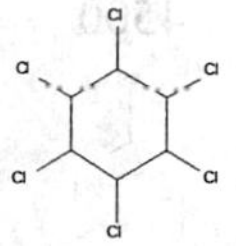

4410

4411

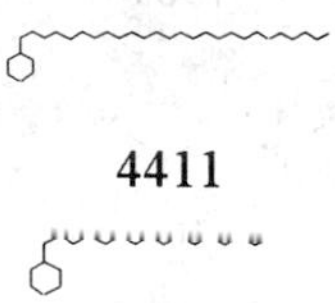

4413

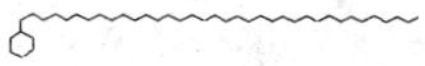

4414

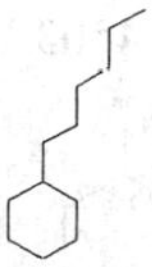

4415

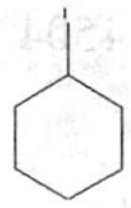

4416

4417

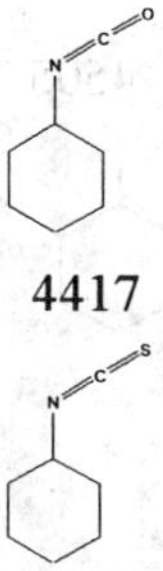

4418

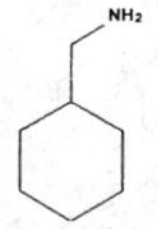

4419

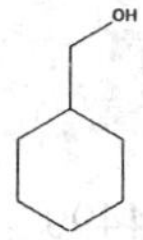

4420

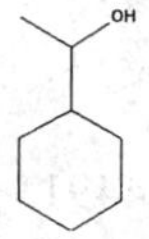

4421

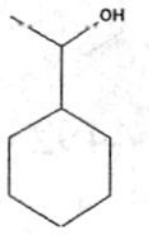

4422

4423

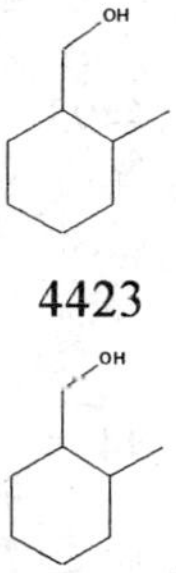

4424

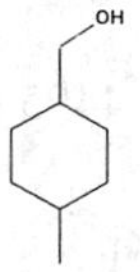

4425

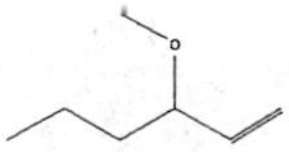

4426

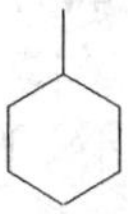

4427

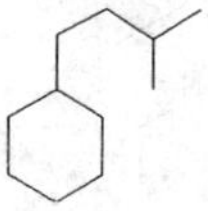

4428

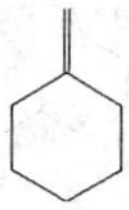

4429

4430

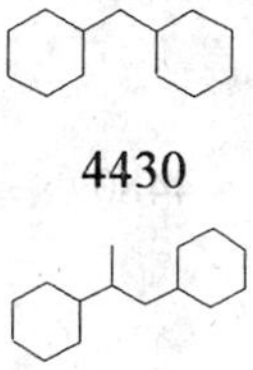

4431

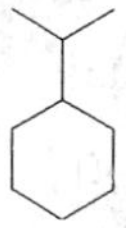

4432

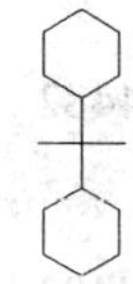

4433

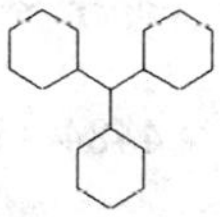

4434

4435

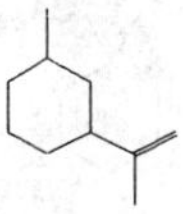

4436

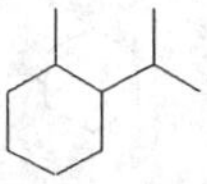

4437

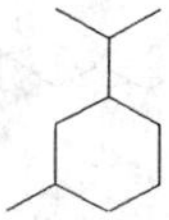

4438

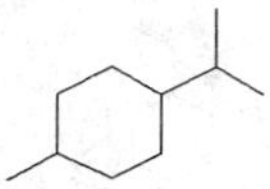

4439

4440

4441

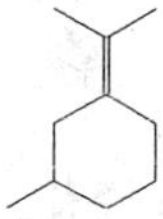

4442

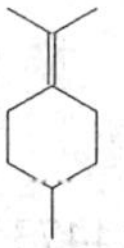

4443

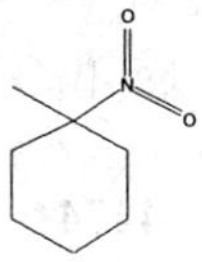

4444

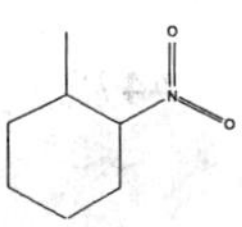

4445

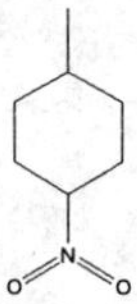

4446

4447

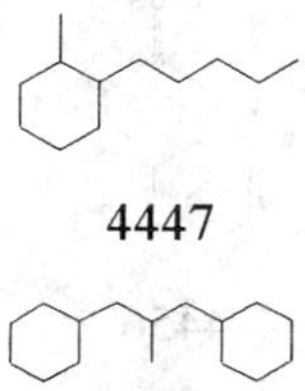

4448

4449

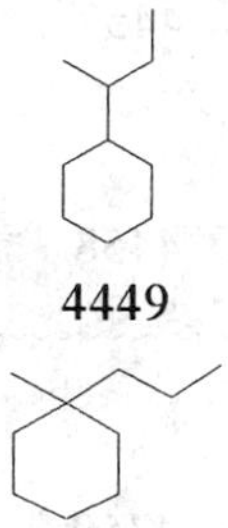

4450

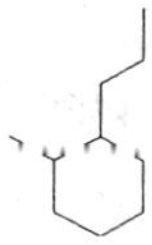

4451

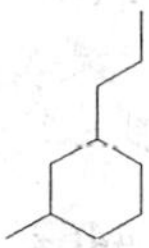

4452

4453

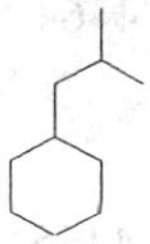

4454

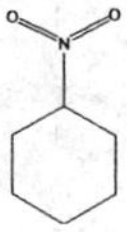

STRUCTURAL FORMULAS OF ORGANIC COMPOUNDS (continued)

In numeric order as they occur in the Table of Physical Constants of Organic Compounds

4455

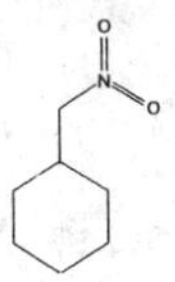

4456

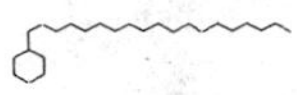

4457

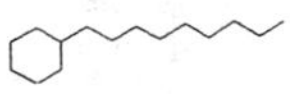

4458

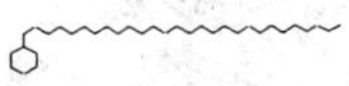

4459

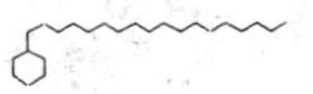

4460

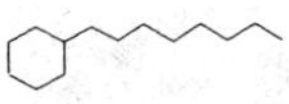

4461

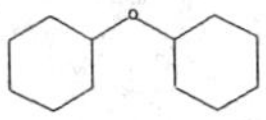

4462

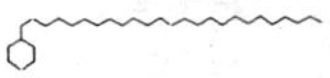

4463

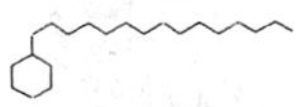

4464

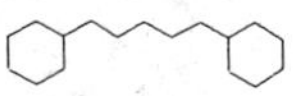

4465

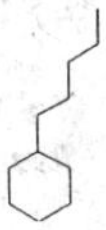

4466

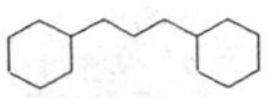

4467

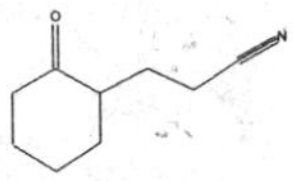

4468

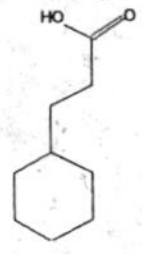

4469

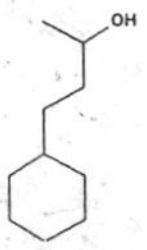

4470

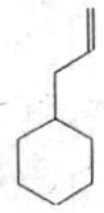

4471

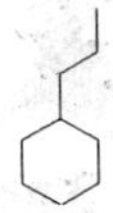

4472

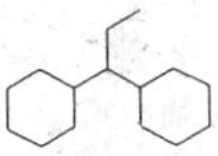

4473

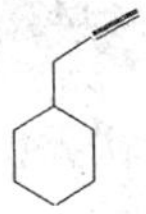

4474

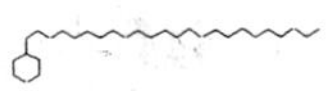

4475

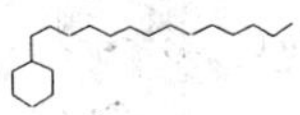

4476

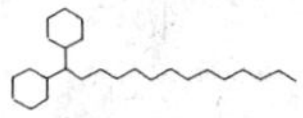

4477

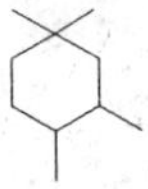

4478

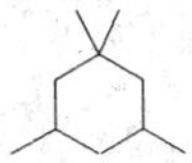

4479

4480

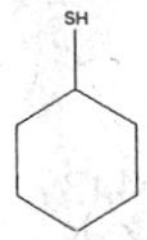

4481

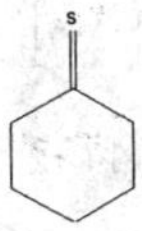

4482

4483

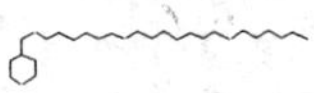

4484

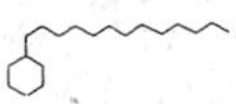

4485

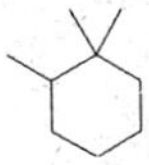

4486

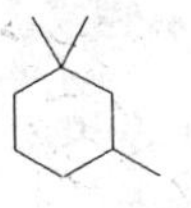

4487

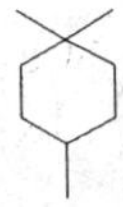

4488

4489

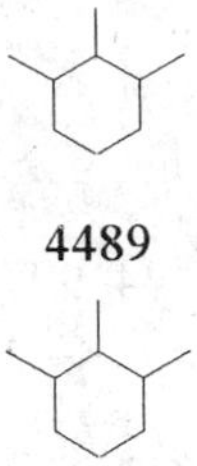

4490

4491

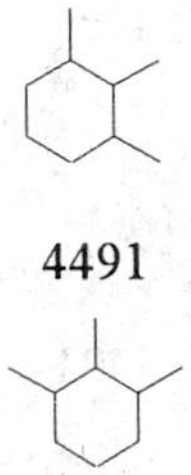

4492

4493

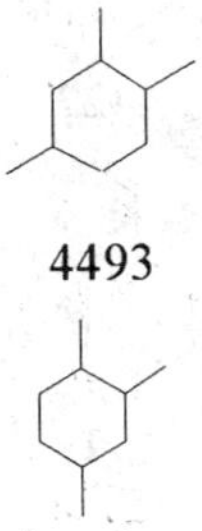

4494

4495

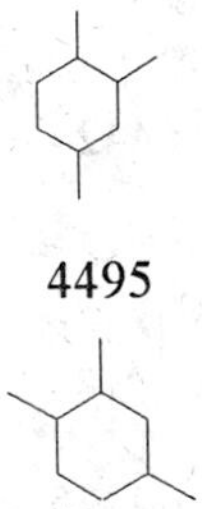

4496

4497

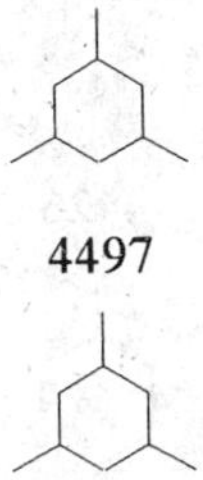

4498

4499

4500

4501

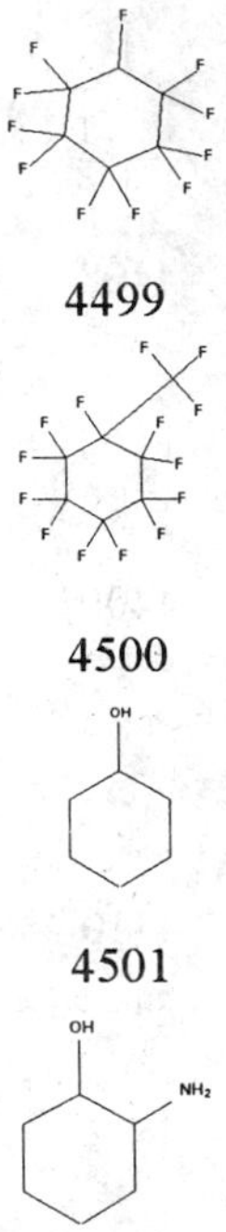

4502

4503

4504

4505

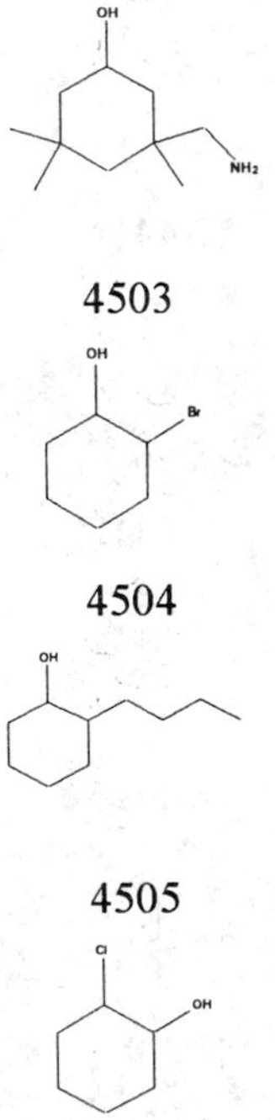

STRUCTURAL FORMULAS OF ORGANIC COMPOUNDS (continued)

In numeric order as they occur in the Table of Physical Constants of Organic Compounds

4507
Cl
OH

4508
OH

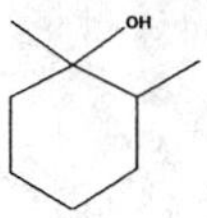

4509
OH

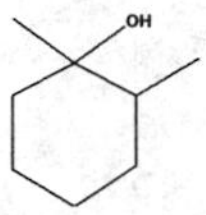

4510
OH

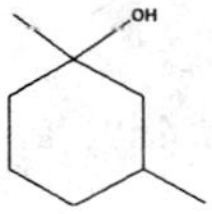

4511
OH

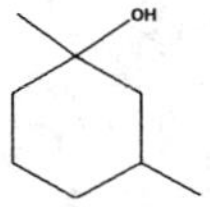

4513
HO

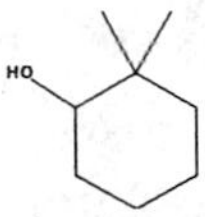

4514
OH

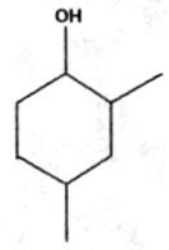

4515
OH

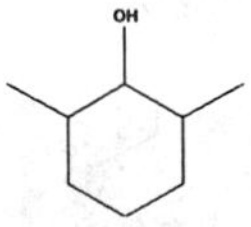

4516
OH

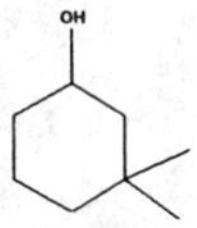

4517
OH

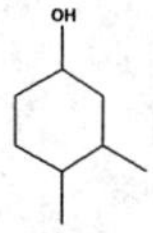

4518
OH

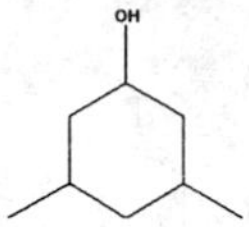

4519
OH

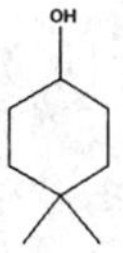

4521
HO

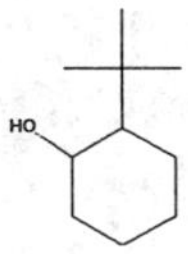

4522
HO

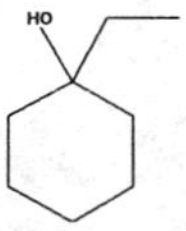

4523
OH

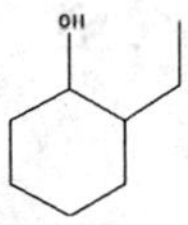

4525
OH

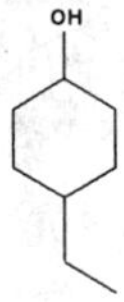

4526
OH

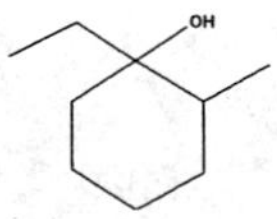

4527
OH

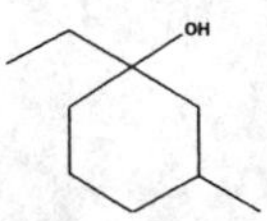

4528
HO
OH

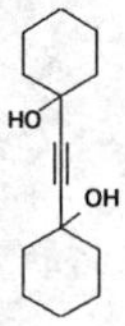

4529
HO

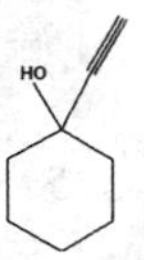

4530
OH

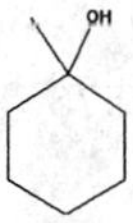

4531
OH

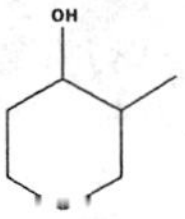

4532
OH

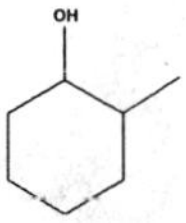

4533
OH

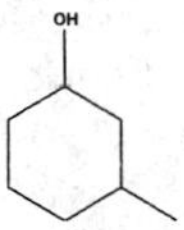

4534
OH

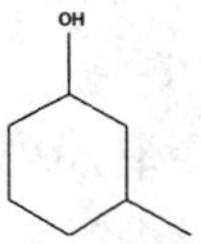

4535
OH

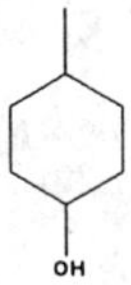

4536
OH

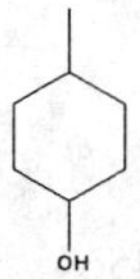

4537
OH

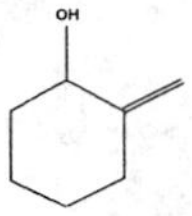

4538
OH

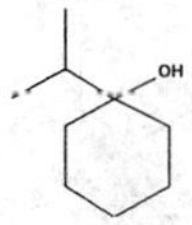

4539
OH

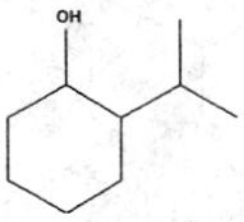

4540
OH

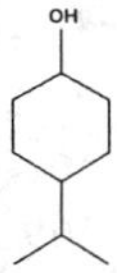

4541
OH

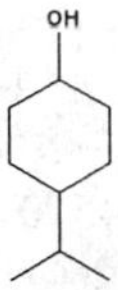

4542
OH

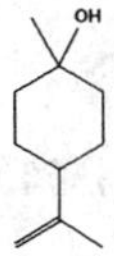

4543
OH

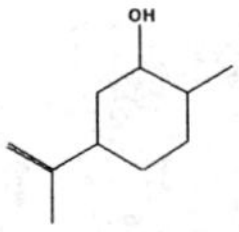

4544
OH

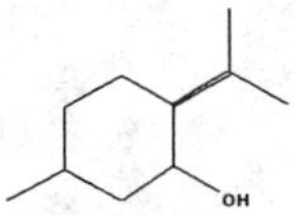

4545
O
O

4546
OH

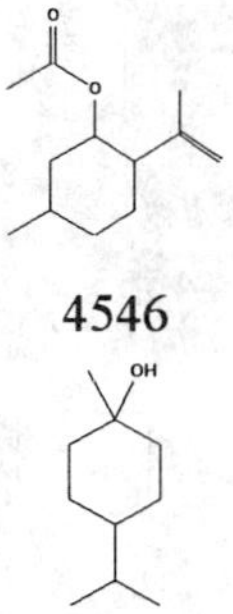

4547
OH

4548
OH

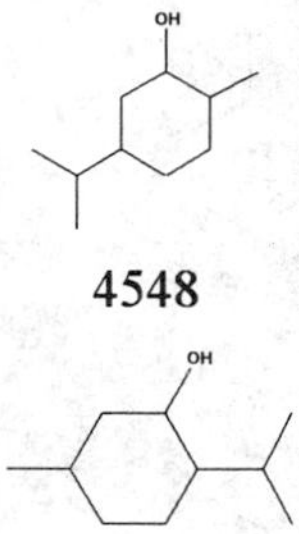

4549
O
O

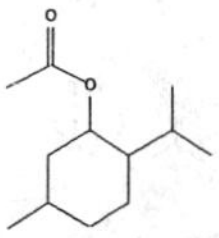

4550
OH

4551
OH

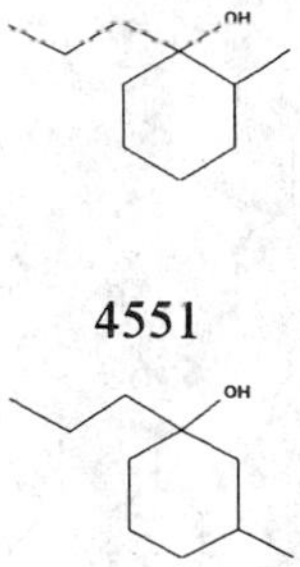

4552
OH

4553
OH

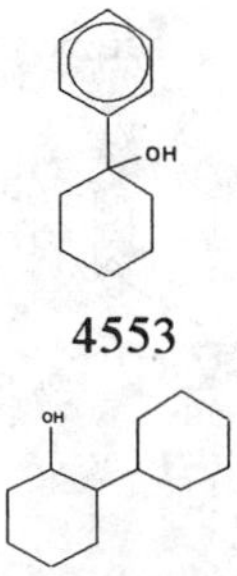

STRUCTURAL FORMULAS OF ORGANIC COMPOUNDS (continued)

In numeric order as they occur in the Table of Physical Constants of Organic Compounds

4554

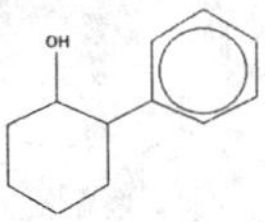

4555

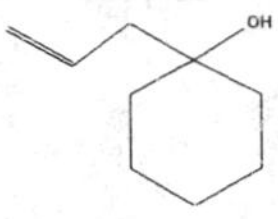

4556

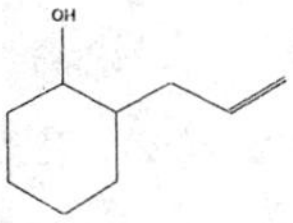

4557

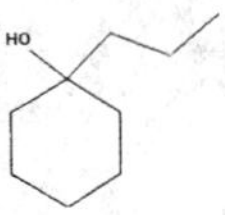

4558

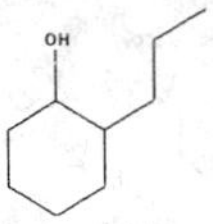

4559

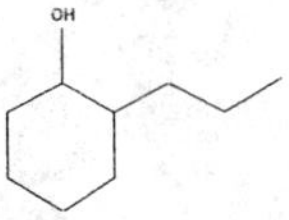

4560

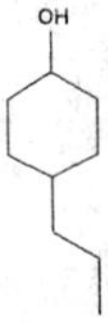

4562

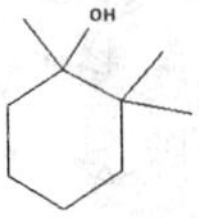

4563

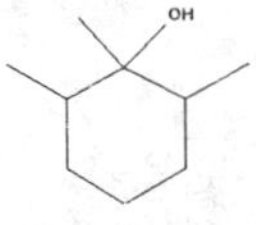

4564

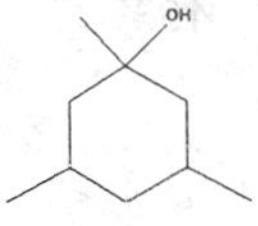

4565

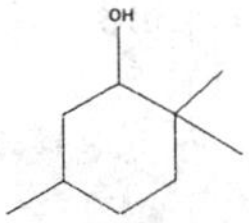

4566

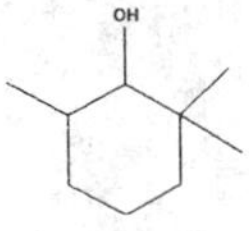

4568

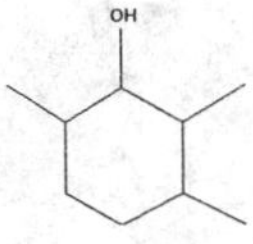

4569

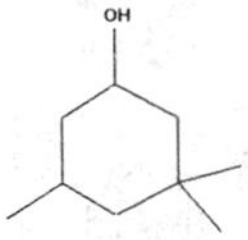

4570

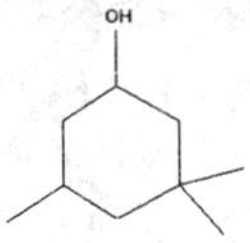

4571

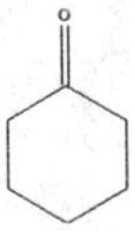

4572

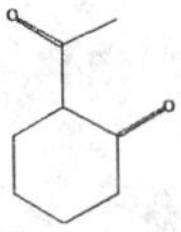

4573

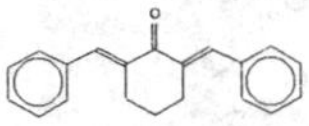

4574

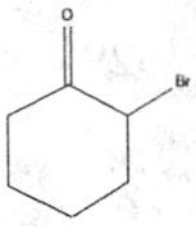

4575

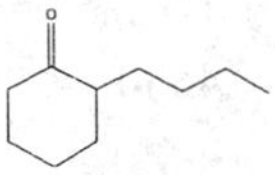

4576

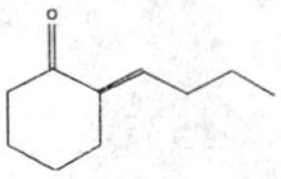

4577

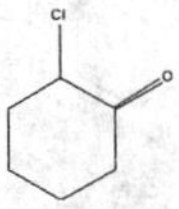

4578

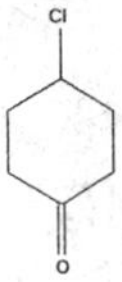

4579

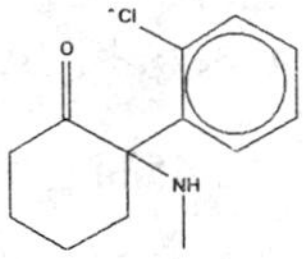

4580

4581

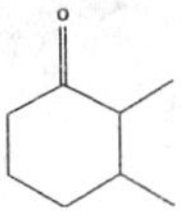

4582

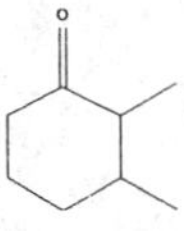

4583

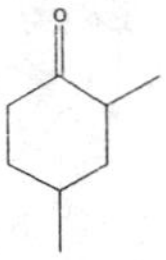

4585

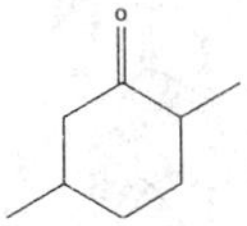

4586

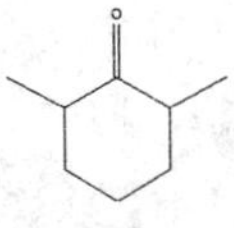

4587

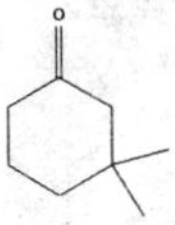

4588

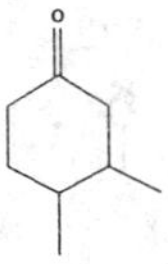

4589

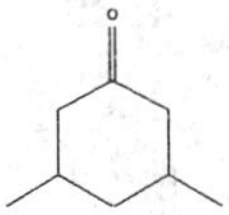

4590

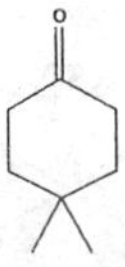

4591

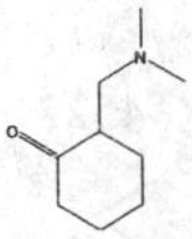

4592

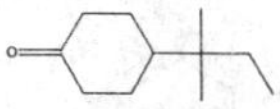

4593

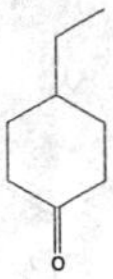

4594

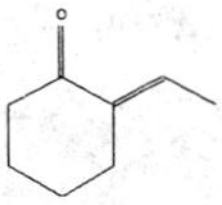

4595

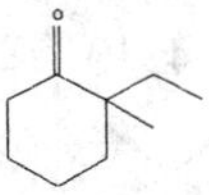

4596

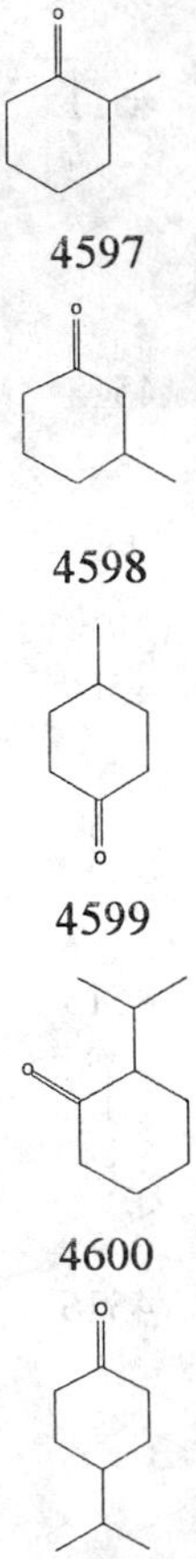

4597

4598

4599

4600

STRUCTURAL FORMULAS OF ORGANIC COMPOUNDS (continued)

In numeric order as they occur in the Table of Physical Constants of Organic Compounds

4601

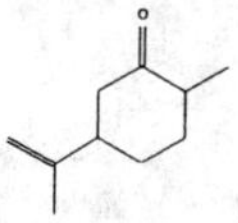

4602

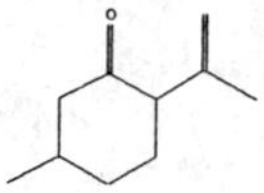

4603

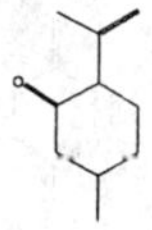

4604

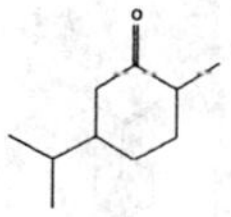

4605

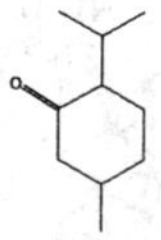

4606

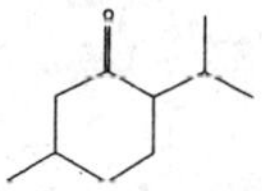

4607

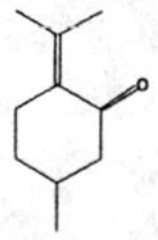

4608

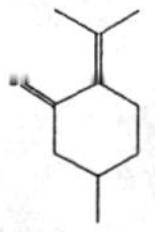

4609

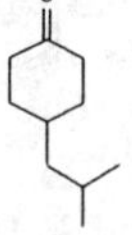

4610

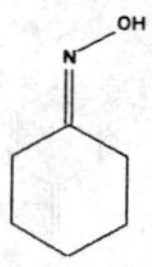

4611

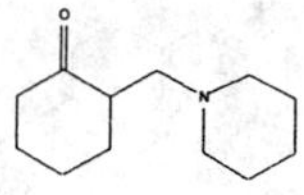

4612

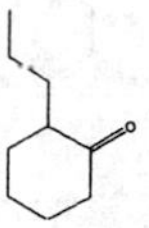

4613

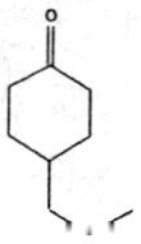

4614

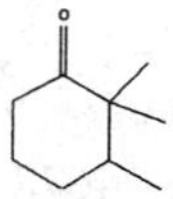

4615

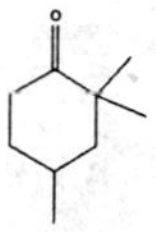

4616

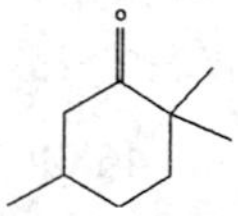

4617

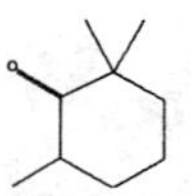

4618

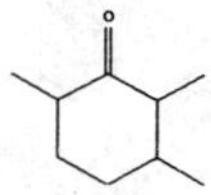

4619

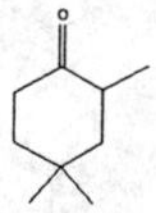

4620

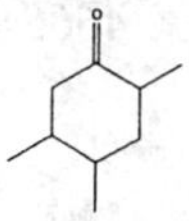

4621

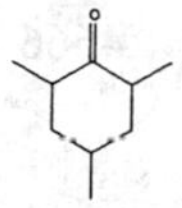

4622

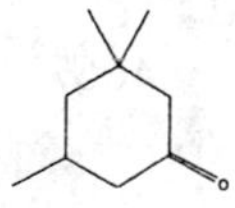

4623

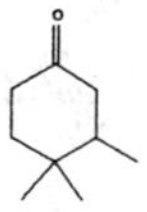

4624

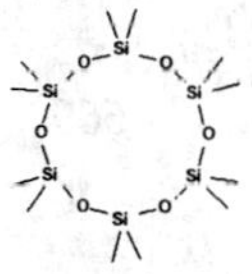

4625

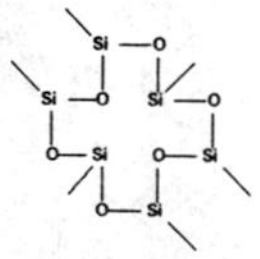

4626

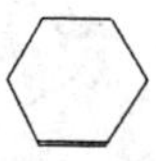

4627

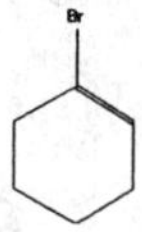

4628

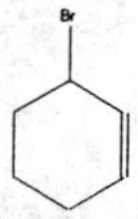

4629

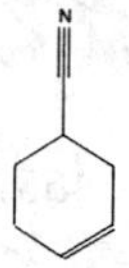

4630

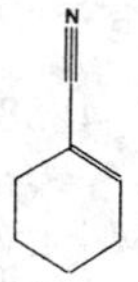

4631

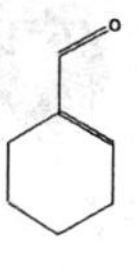

4632

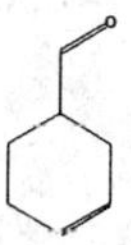

4633

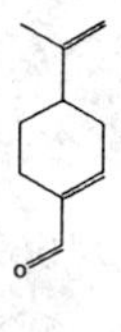

4634

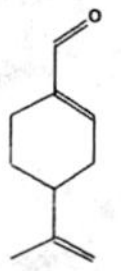

4635

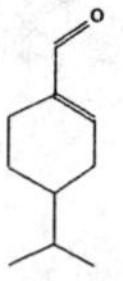

4636

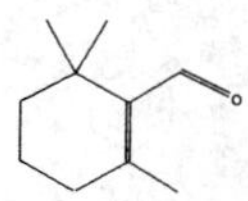

4637

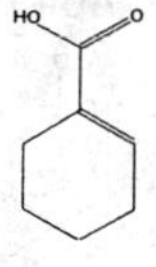

4638

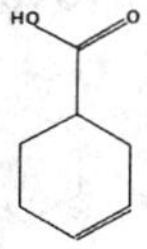

4639

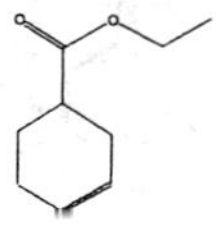

4640

4642

4643

4644

4645

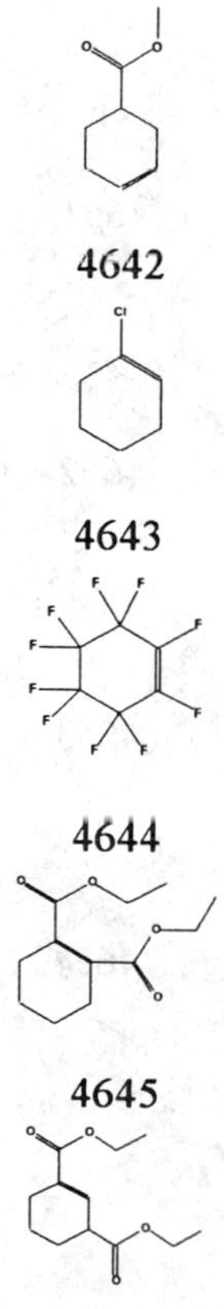

4646

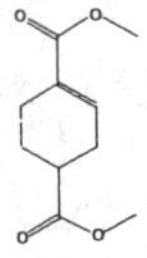

4647

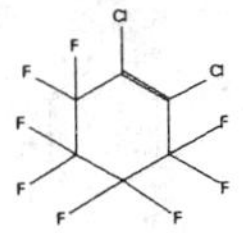

4648

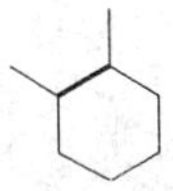

4649

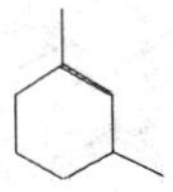

4650

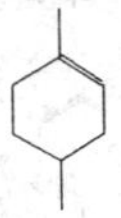

4651

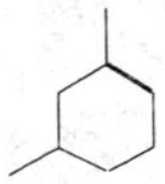

4652

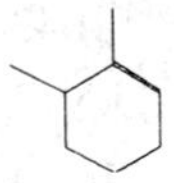

4653

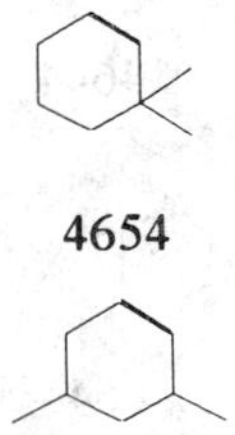

4654

4655

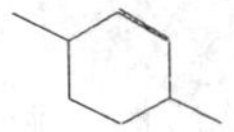

4656

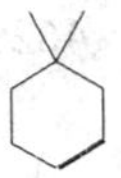

4657

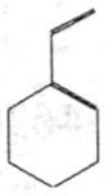

4658

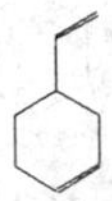

4659

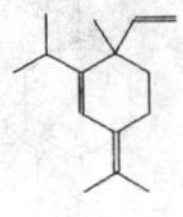

4660

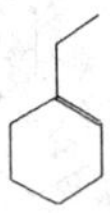

4661

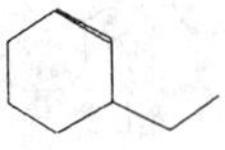

4662

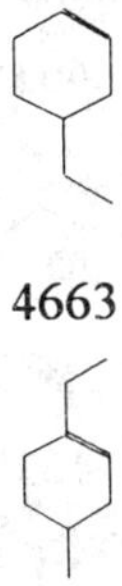

4663

4664

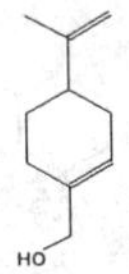

4665

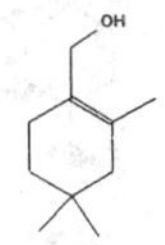

4666

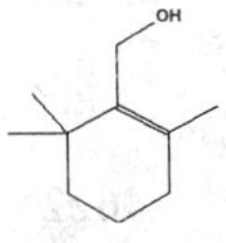

4667

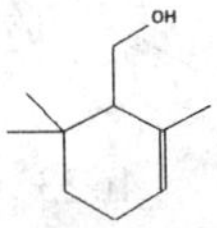

4668

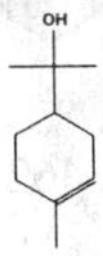

4669

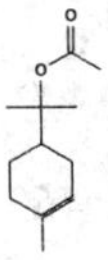

4670

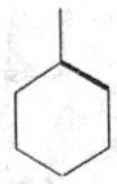

4671

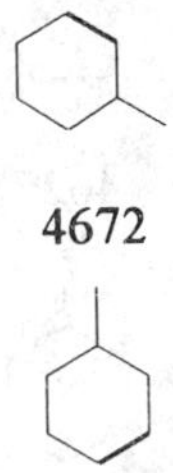

4672

4673

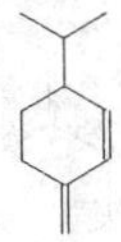

4674

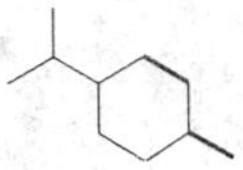

4675

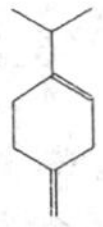

4676

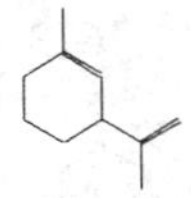

4677

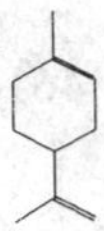

4678

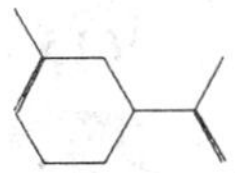

4679

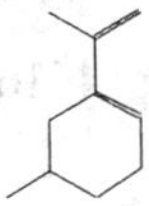

4680

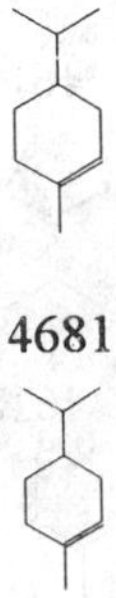

4681

4682

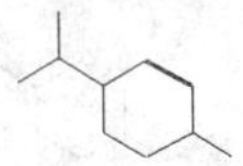

4683

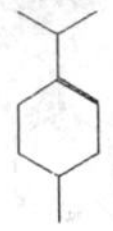

4684

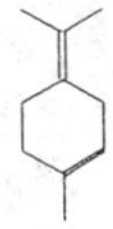

4685

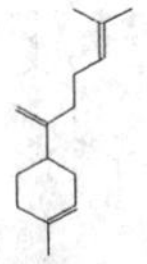

4686

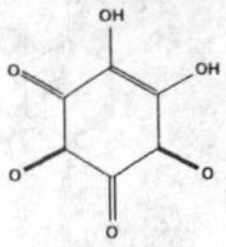

4687

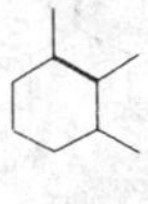

4688

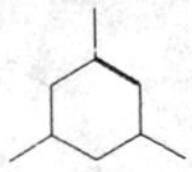

4689

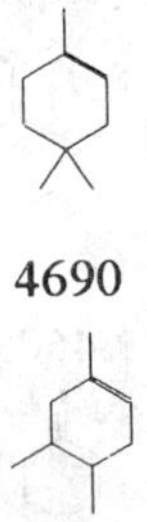

4690

STRUCTURAL FORMULAS OF ORGANIC COMPOUNDS (continued)

In numeric order as they occur in the Table of Physical Constants of Organic Compounds

4691

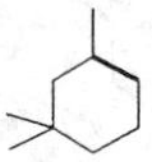

4692

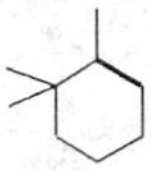

4694

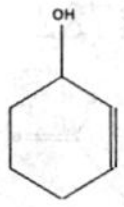

OH

4695

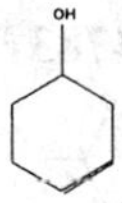

OH

4696

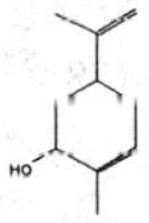

HO

4697

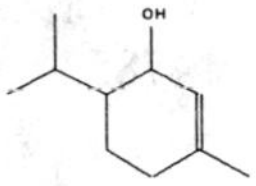

OH

4698

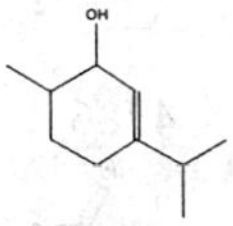

OH

4699

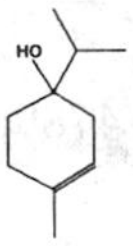

HO

4700

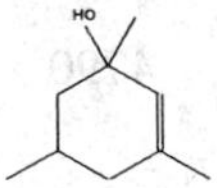

HO

4701

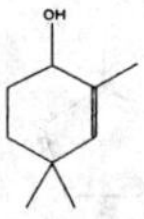

OH

4702

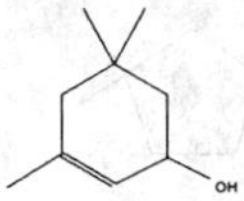

OH

4703

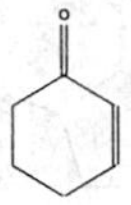

O

4704

O

4705

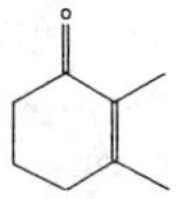

O

4706

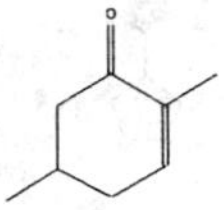

O

4707

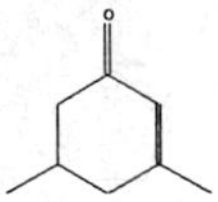

O

4708

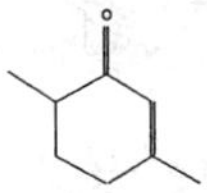

O

4709

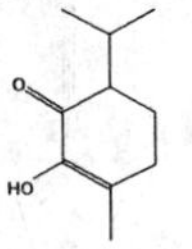

O

HO

4710

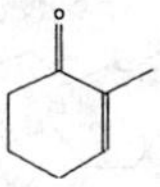

O

4711

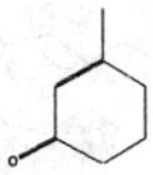

O

4712

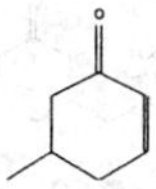

O

4713

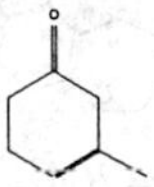

O

4714

O

4715

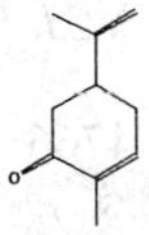

O

4716

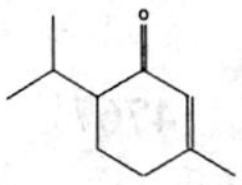

O

4718

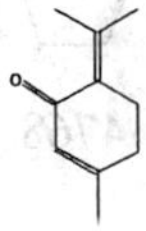

O

4719

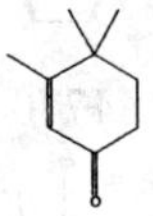

O

4720

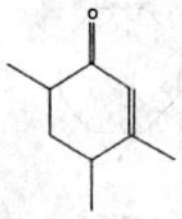

O

4721

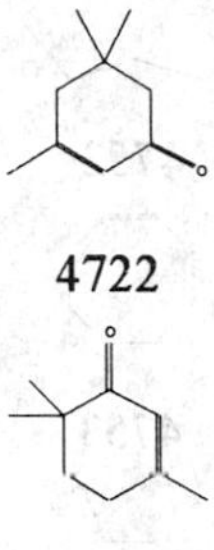

O

4722

O

4723

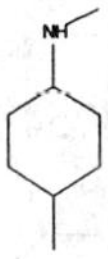

NH

4724

4725

O

4726

Si O Si O Si O Si O Si O Si O Si O Si

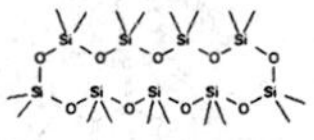

4727

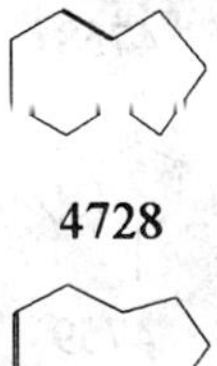

4728

4729

4731

4732

4733

4734

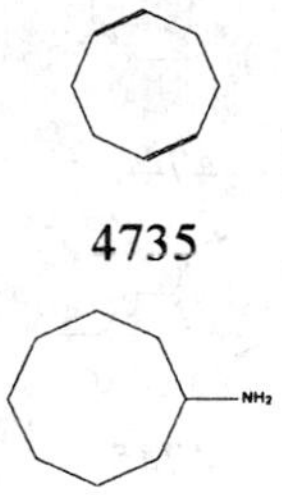

4735

NH_2

4736

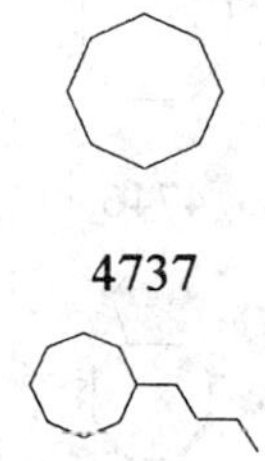

4737

4738

O

4739

OH

4740

O

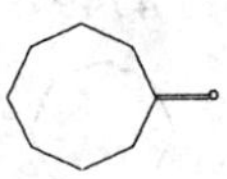

4741

4742

4743

4744

4745

4746

4747

4748

4749

4750

4751

4752

4753

4754

4755

4756

4757

4758

4759

4760

4761

4762

4763

4764

4765

4766

4767

4768

4770

4771

4772

4773

4774

4775

4776

4777

4778

4779

4780

4781

4782

4783

4784

4785

4786

4789

4790

STRUCTURAL FORMULAS OF ORGANIC COMPOUNDS (continued)

In numeric order as they occur in the Table of Physical Constants of Organic Compounds

4791

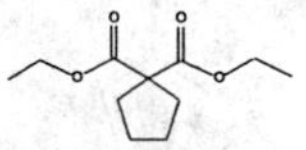

4792

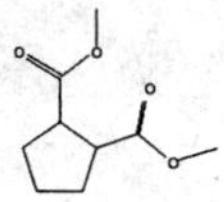

4793

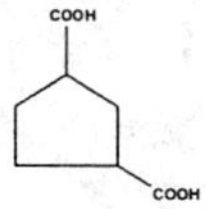

COOH

COOH

4794

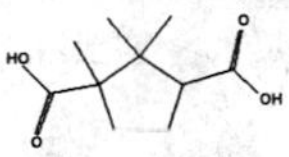

HO

O

O

OH

4796

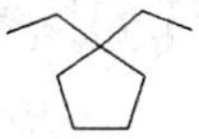

4797

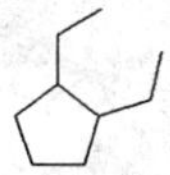

4798

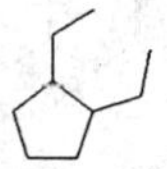

4799

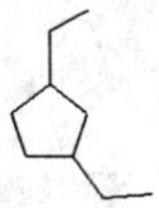

4800

4801

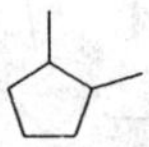

4802

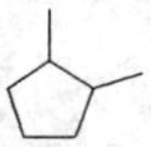

4803

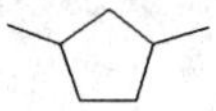

4804

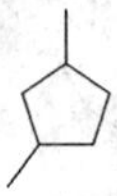

4805

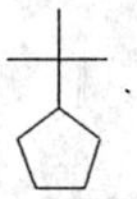

4806

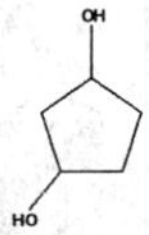

OH

HO

4807

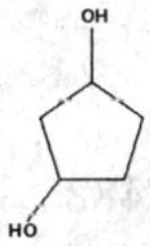

OH

HO

4808

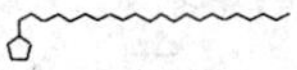

4809

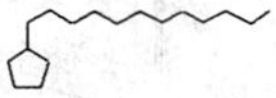

4810

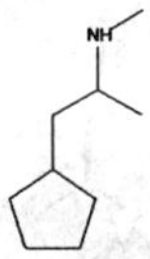

NH

4811

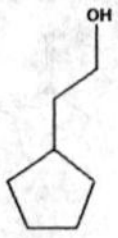

OH

4812

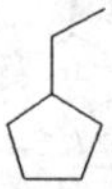

4813

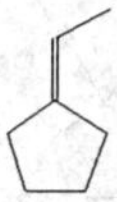

4814

4815

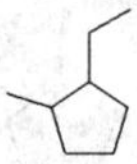

4816

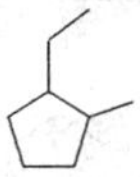

4817

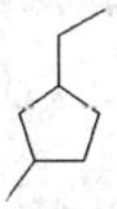

4818

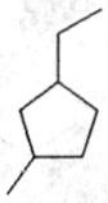

4819

4820

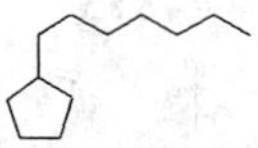

4821

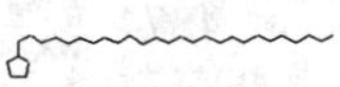

4822

4823

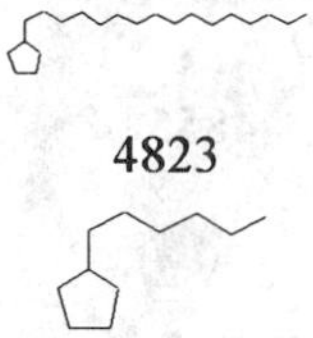

4824

4825

OH

4826

4827

4828

4829

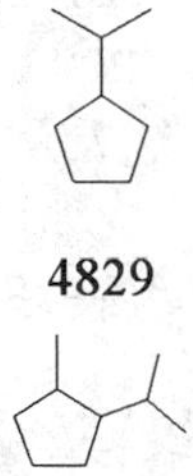

4830

O

N

O

4831

O

N

O

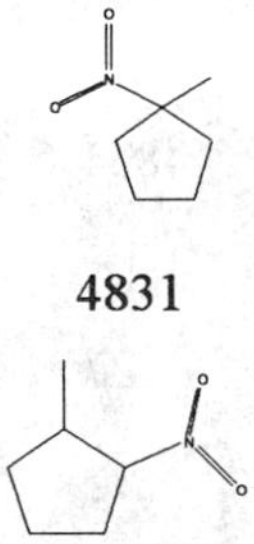

4832

4833

4834

4835

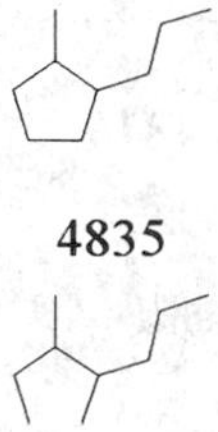

4836

4837

4838

4839

O

N

O

4840

O

N

O

4841

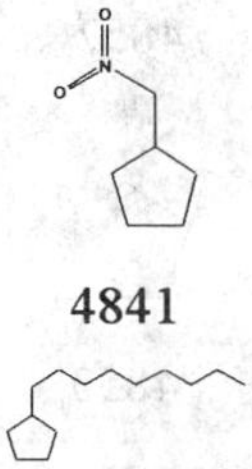

STRUCTURAL FORMULAS OF ORGANIC COMPOUNDS (continued)

In numeric order as they occur in the Table of Physical Constants of Organic Compounds

4842

4843

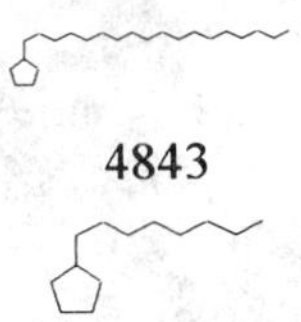

4844

4845

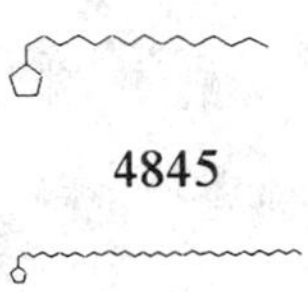

4846

4847

HO O

4848

4849

4850

4851

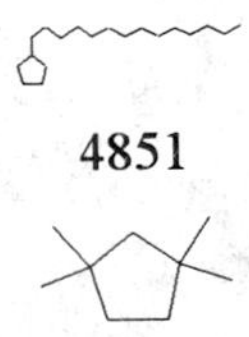

4852

SH

4853

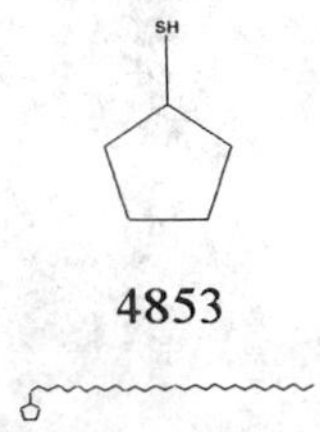

4854

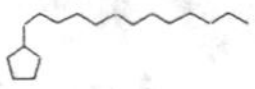

4855

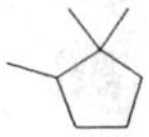

4856

4857

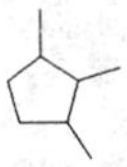

4858

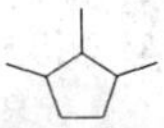

4859

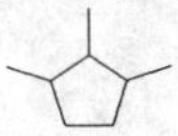

4860

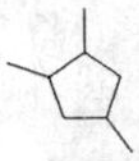

4861

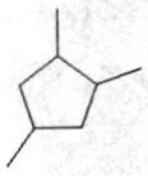

4862

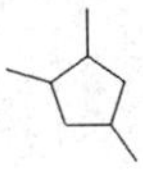

4863

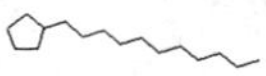

4864

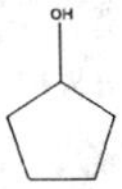

OH

4865

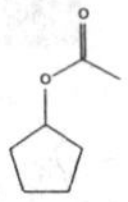

O O

4866

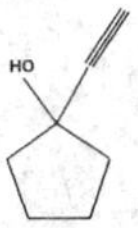

HO

4867

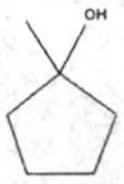

OH

4868

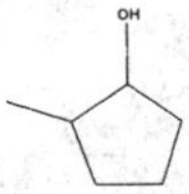

OH

4869

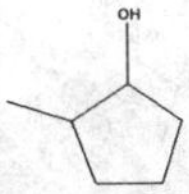

OH

4870

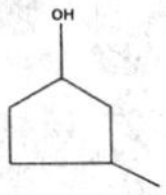

OH

4871

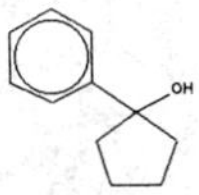

OH

4872

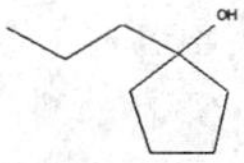

OH

4873

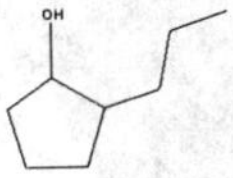

OH

4874

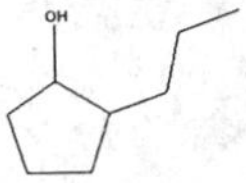

OH

4875

OH

4876

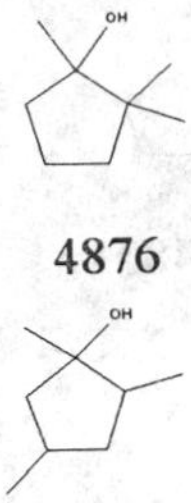

OH

4877

OH

4878

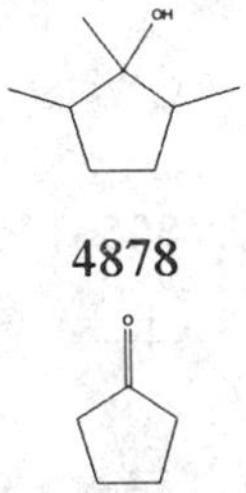

O

4879

O O

4880

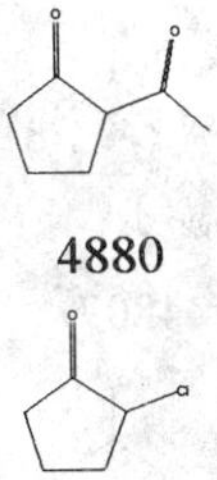

O Cl

4881

O

4882

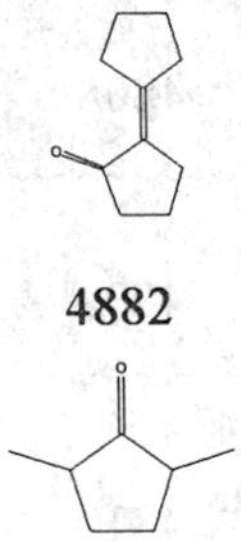

O

4883

O

4884

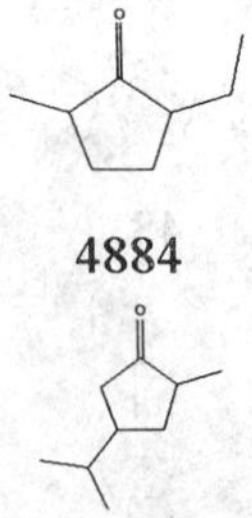

O

4885

O

4886

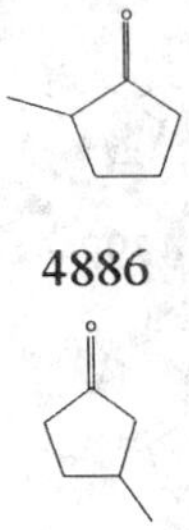

O

4887

O

4888

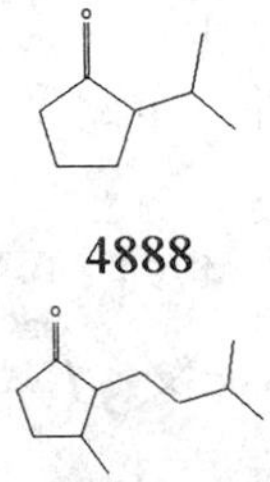

O

4889

O

4890

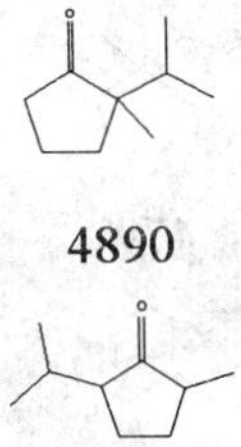

O

4891

O

4892

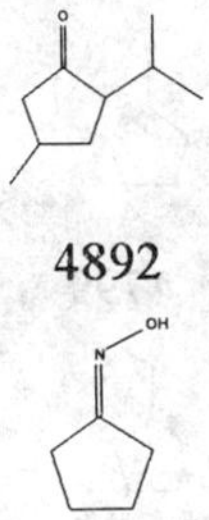

N OH

4893

O

4894

O

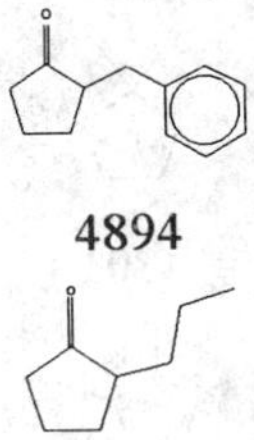

STRUCTURAL FORMULAS OF ORGANIC COMPOUNDS (continued)

In numeric order as they occur in the Table of Physical Constants of Organic Compounds

4895

4896

4897

4898

4899

4900

4901

4902

4903

4904

4905

4906

4907

4908

4909

4910

4911

4912

4913

4914

4915

4916

4917

4918

4919

4920

4921

4922

4923

4924

4925

4926

4927

4928

4929

4930

4931

4932

4933

4934

4935

4936

4937

4938

4939

4940

4941

STRUCTURAL FORMULAS OF ORGANIC COMPOUNDS (continued)

In numeric order as they occur in the Table of Physical Constants of Organic Compounds

4942

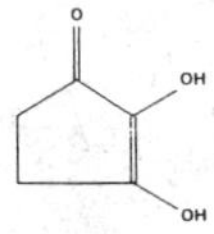

4943

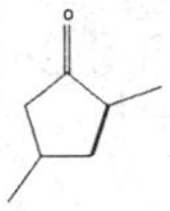

4944

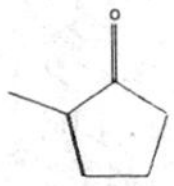

4945

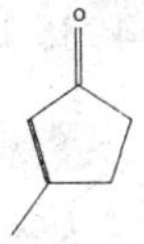

4946

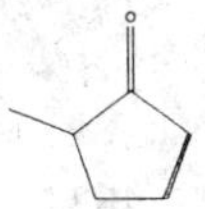

4947

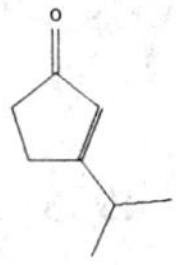

4948

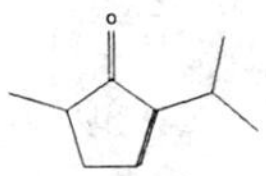

4949

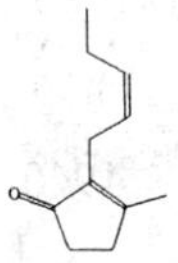

4950

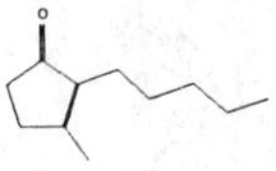

4951

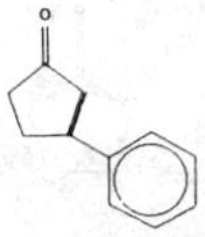

4952

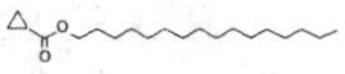

4953

4954

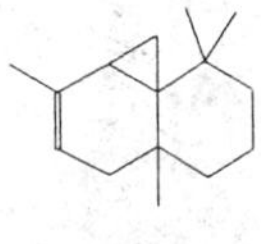

4955

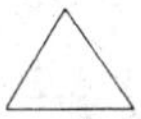

4956

4957

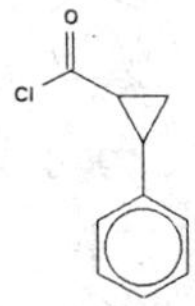

4958

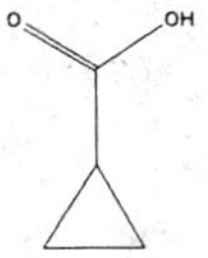

4959

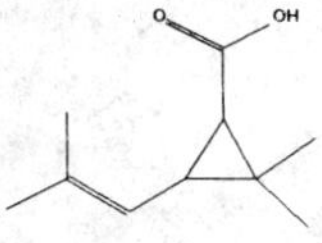

4960

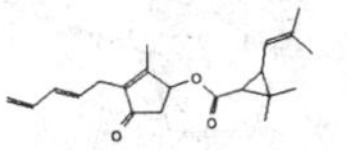

4961

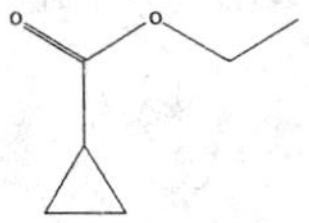

4962

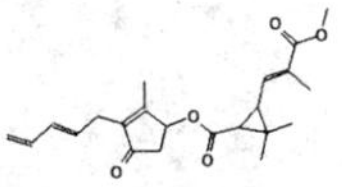

4963

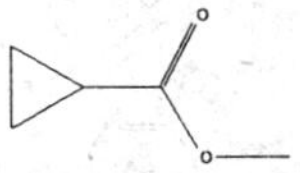

4964

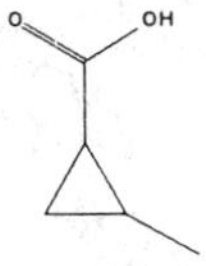

4966

4967

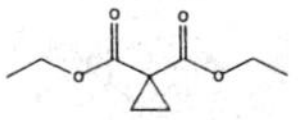

4968

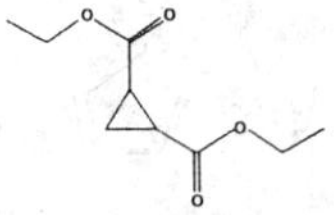

4969

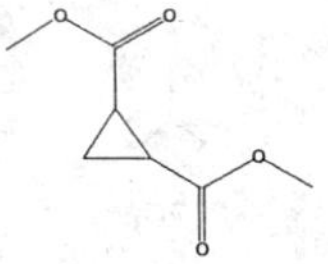

4970

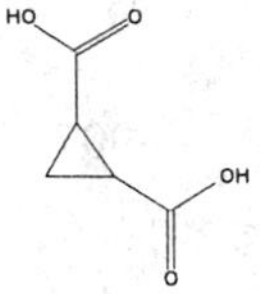

4971

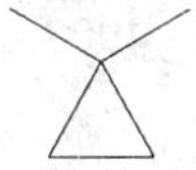

4972

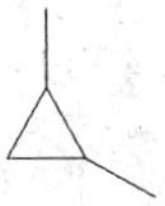

4973

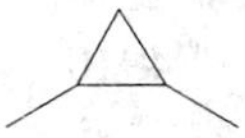

4974

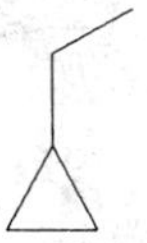

4975

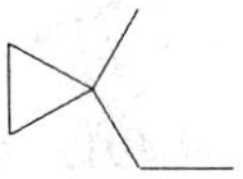

4976

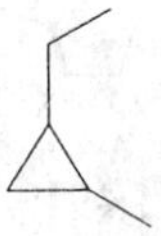

4977

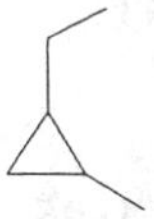

4978

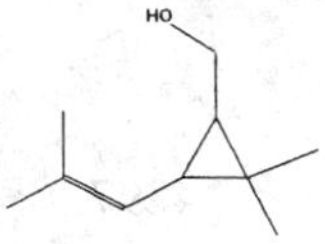

4979

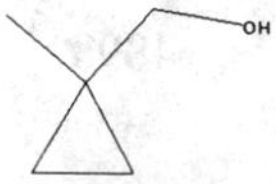

4980

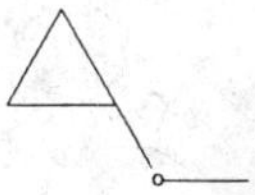

4981

4982

4983

4984

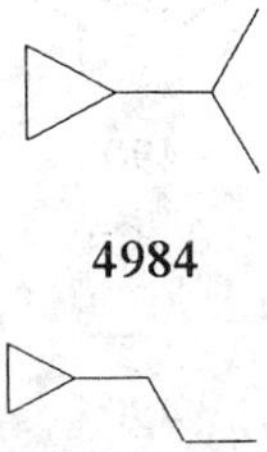

STRUCTURAL FORMULAS OF ORGANIC COMPOUNDS (continued)

In numeric order as they occur in the Table of Physical Constants of Organic Compounds

4985

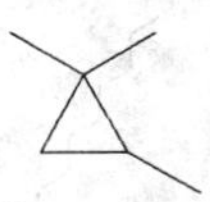

4986

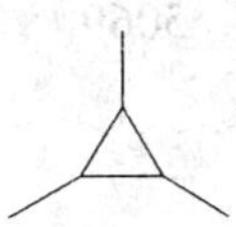

4987

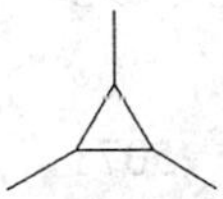

4988

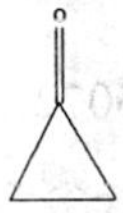

4989

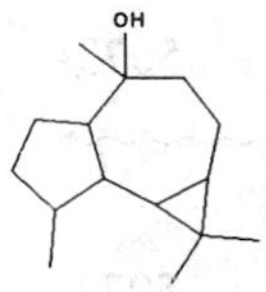

4990

4991

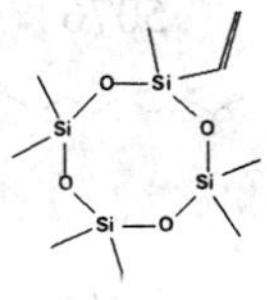

4992

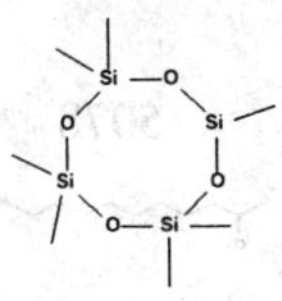

4993

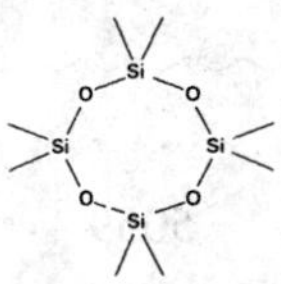

4994

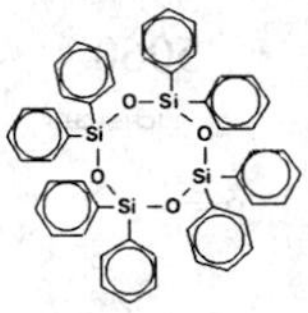

4995

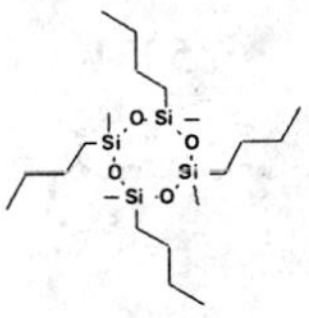

4996

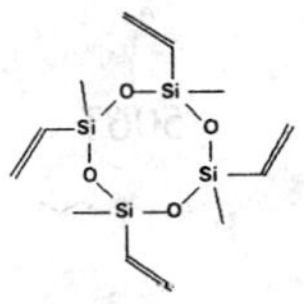

4997

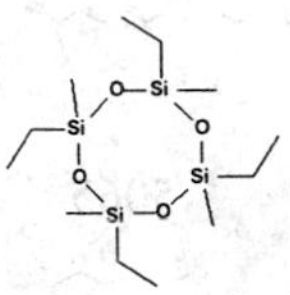

4998

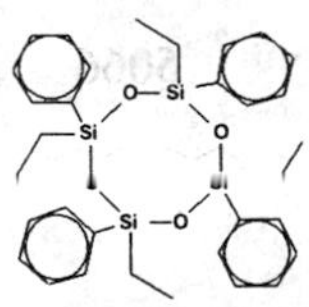

4999

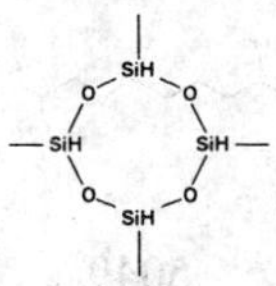

5000

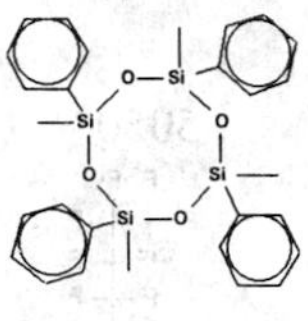

5001

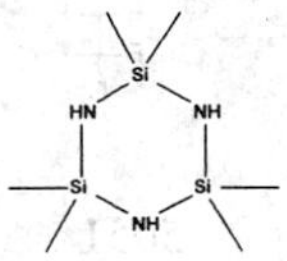

5002

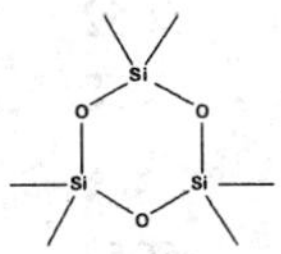

5003

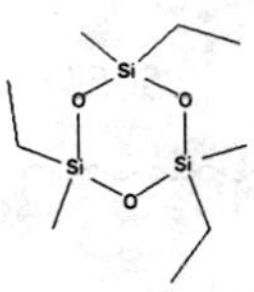

5004

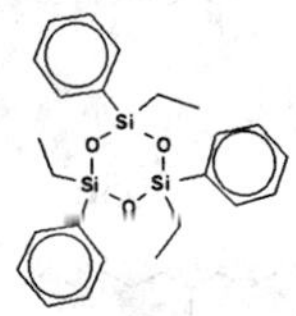

5005

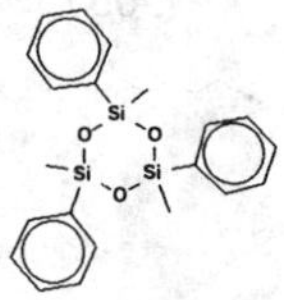

5006

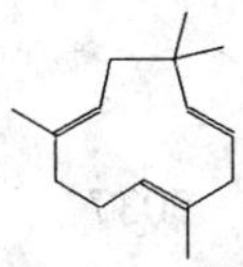

5008

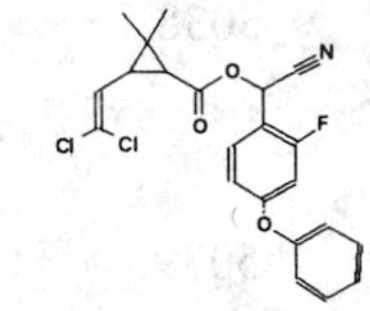

5009

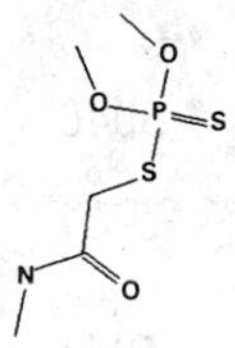

5010

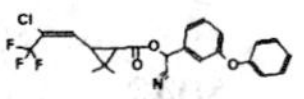

5011

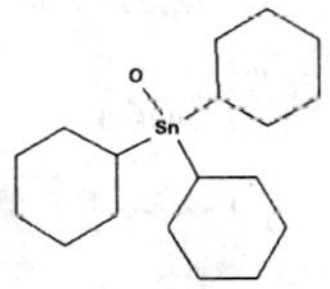

5012

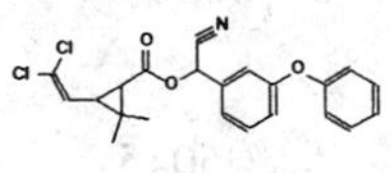

5013

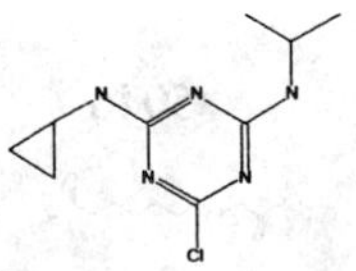

5014

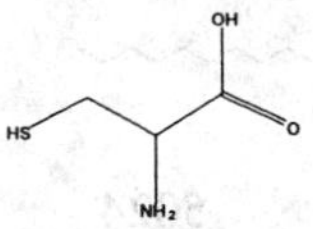

5015

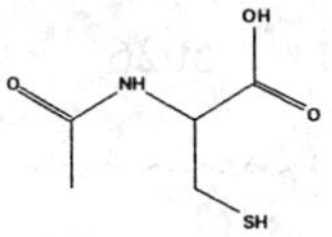

5016

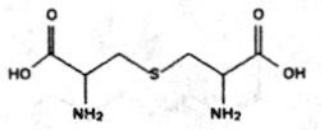

5017

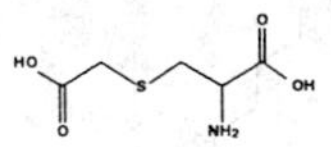

5018

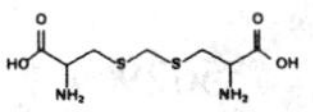

5020

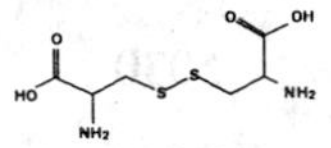

5021

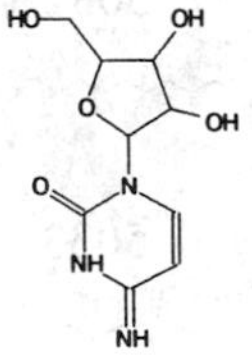

5022

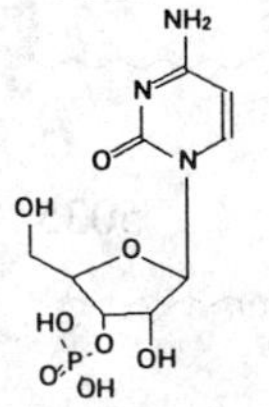

STRUCTURAL FORMULAS OF ORGANIC COMPOUNDS (continued)

In numeric order as they occur in the Table of Physical Constants of Organic Compounds

5023

5024

5026

5027

5028

5029

5030

5032

5033

5034

5035

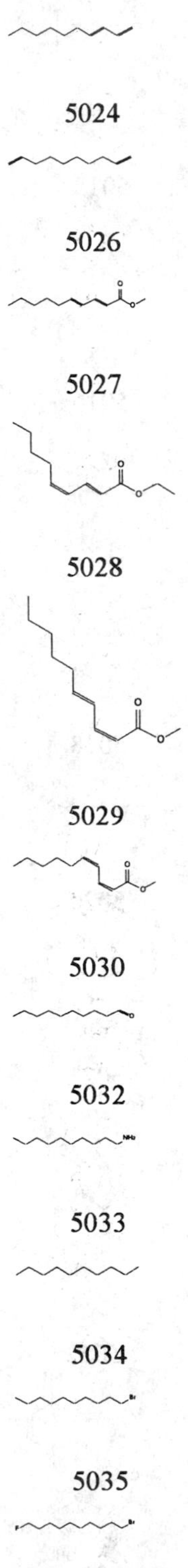

5036

5037

5038

5039

5040

5041

5042

5043

5044

5045

5046

5047

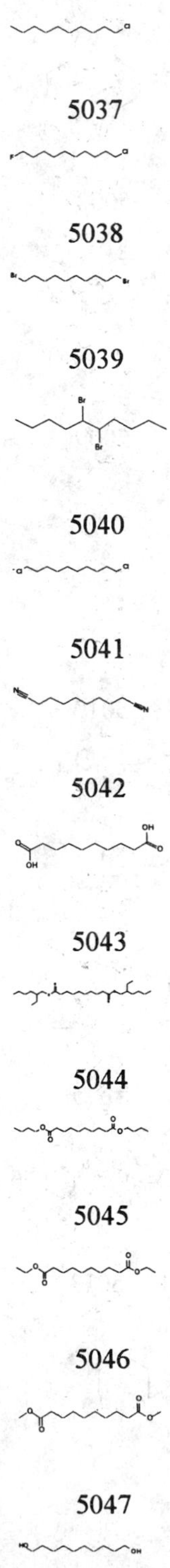

5048

5049

5050

5051

5052

5053

5054

5055

5056

5057

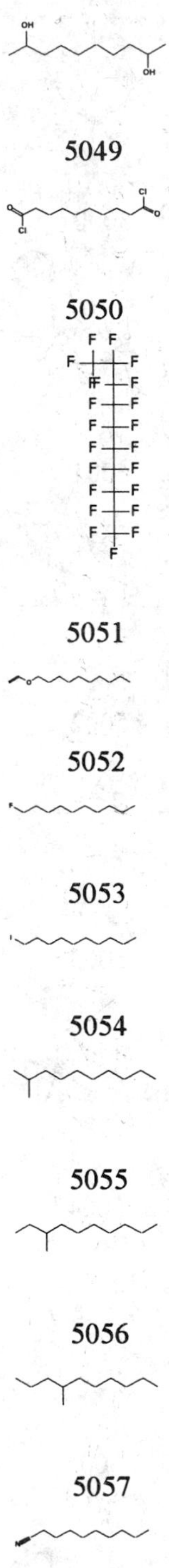

5058

5059

5060

5061

5062

5063

5064

5065

5066

5067

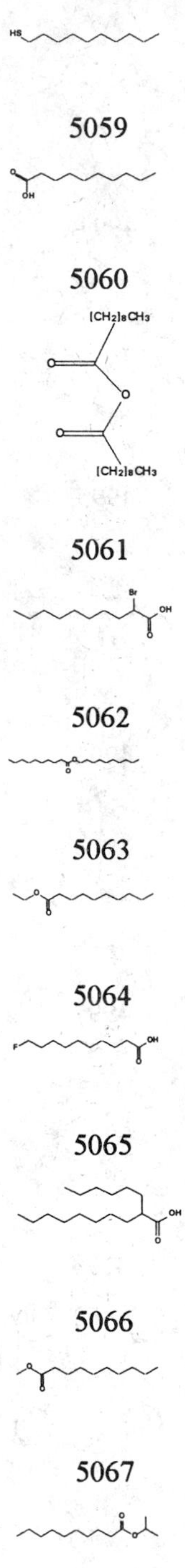

5068

5069

5070

5071

5072

5073

5074

5075

5076

5077

5078

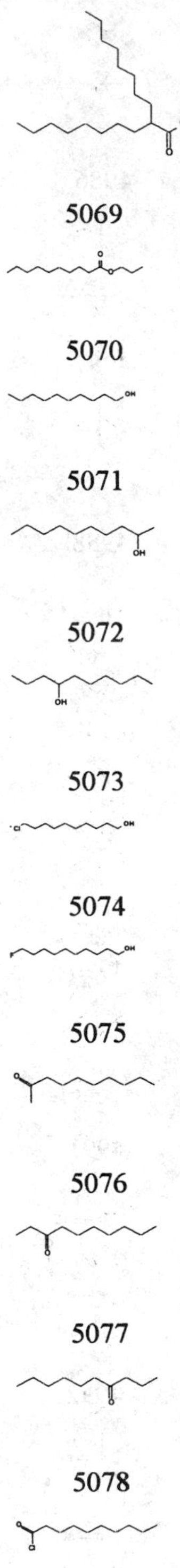

STRUCTURAL FORMULAS OF ORGANIC COMPOUNDS (continued)

In numeric order as they occur in the Table of Physical Constants of Organic Compounds

5080

5081

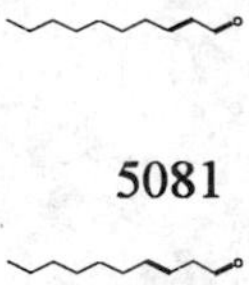

5082

5083

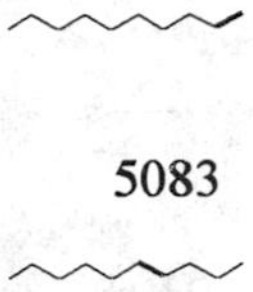

5084

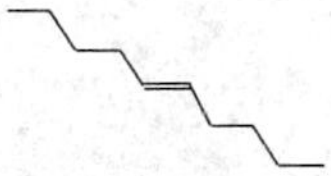

5085

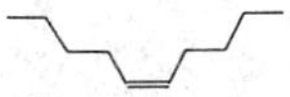

5086

Br

5087

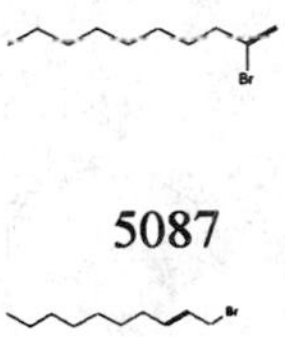

Br

5088

OH

5089

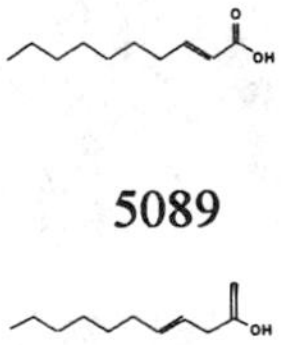

OH

5090

OH

5091

OH

5092

HO

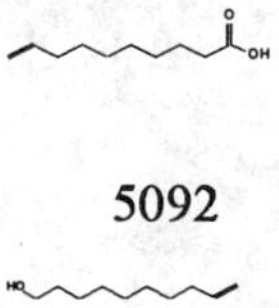

5093

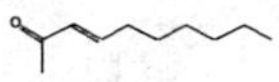

5094

5095

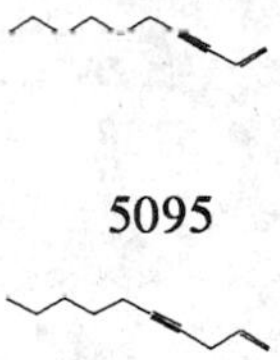

5097

5098

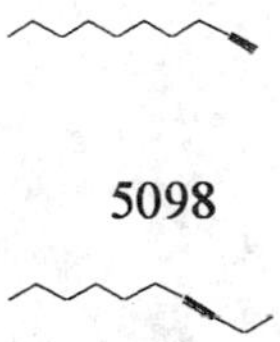

5099

5100

5101

5102

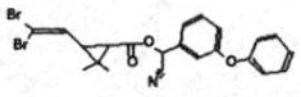

Br Br

5103

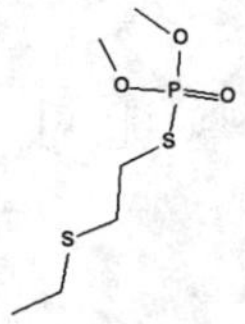

O O P O S S

5104

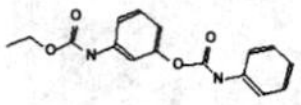

5105

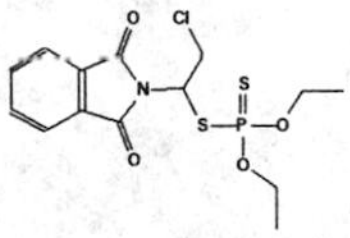

O Cl N O S P S O O

5106

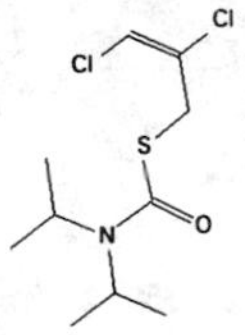

Cl Cl S N O

5109

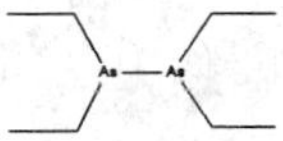

As As

5110

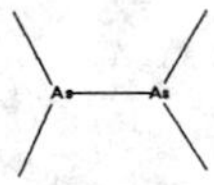

As As

5111

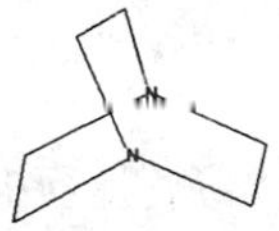

N

5112

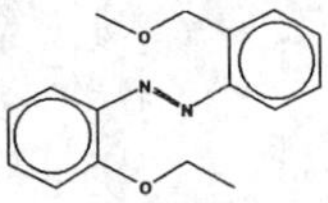

N N O O

5113

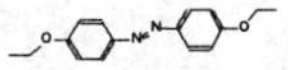

O N N O

5114

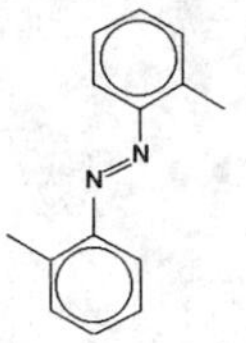

N N

5115

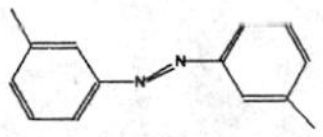

N N

5116

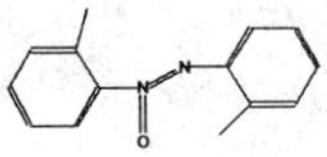

N N O

5118

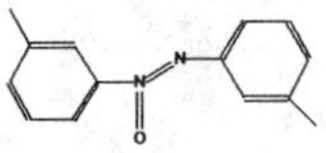

N N O

5120

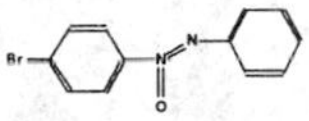

Br N N O

5121

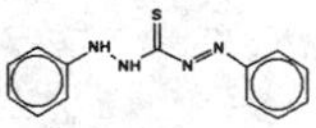

NH NH S N N

5122

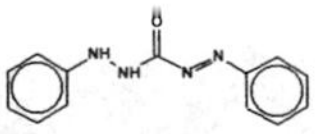

NH NH O N N

5123

N N

5124

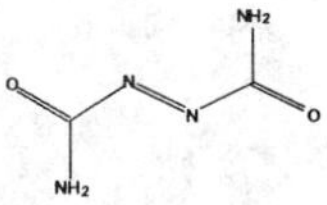

NH_2 O N N O NH_2

5125

N N

5126

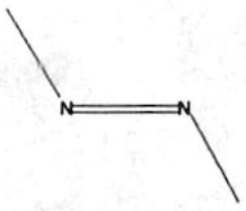

N N

5127

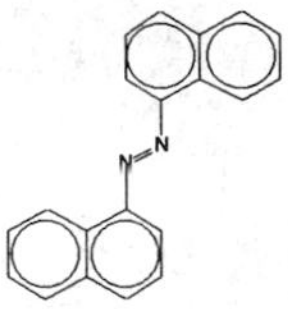

N N

5128

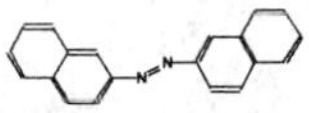

N N

5129

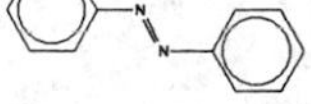

N N

5130

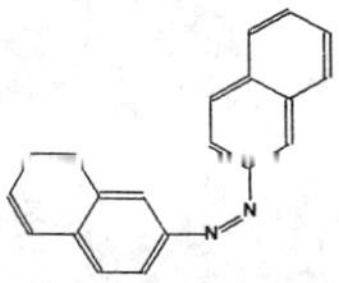

N N

5131

N N O

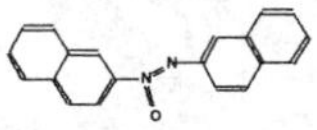

STRUCTURAL FORMULAS OF ORGANIC COMPOUNDS (continued)

In numeric order as they occur in the Table of Physical Constants of Organic Compounds

5132

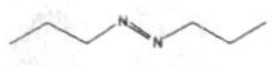

5133

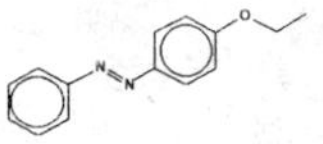

5134

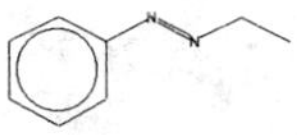

5135

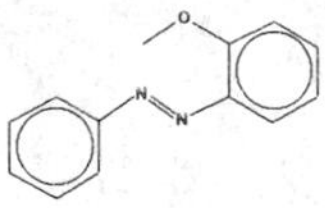

5136

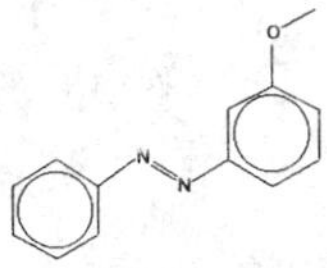

5137

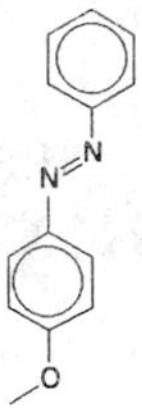

5140

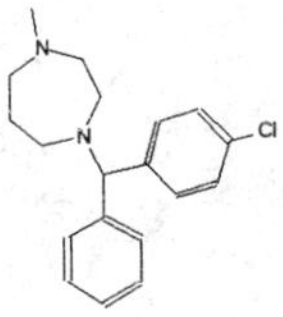

5141

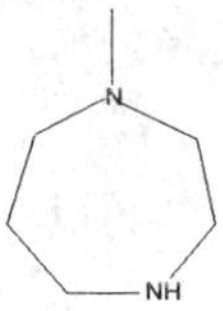

5142

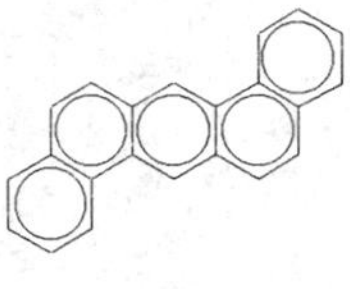

5143

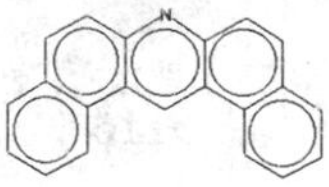

5144

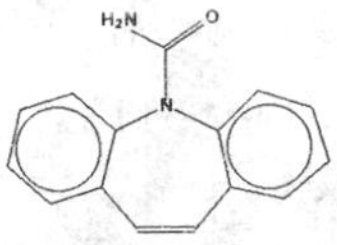

5145

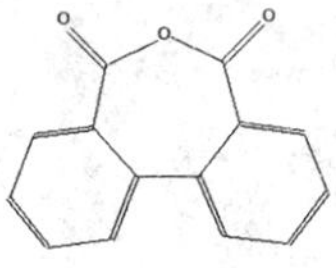

5146

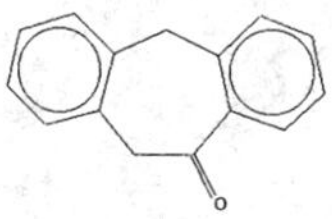

5147

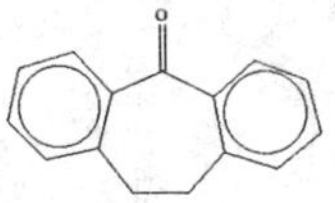

5148

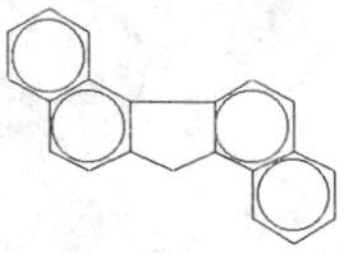

5149

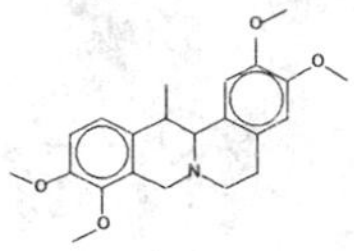

5150

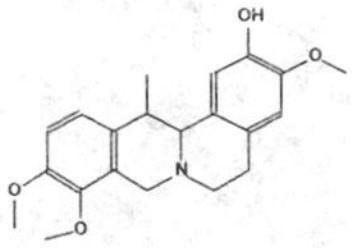

5151

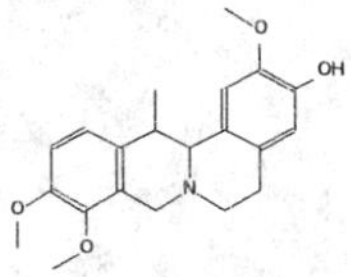

5152

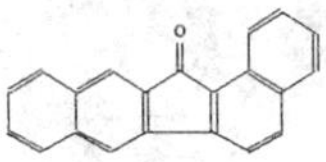

5153

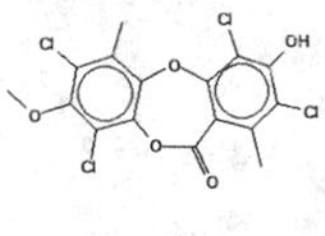

5154

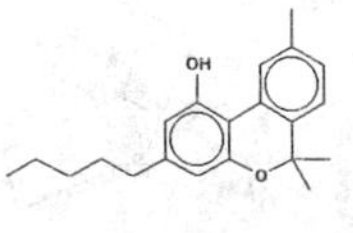

5155

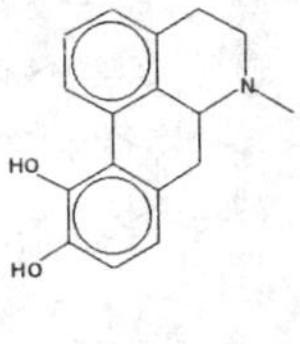

5156

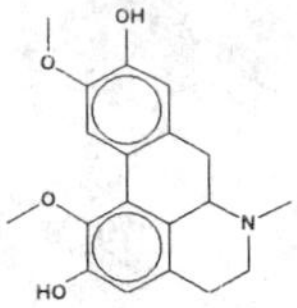

5157

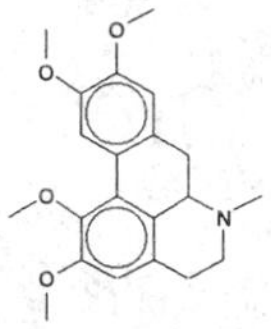

5158

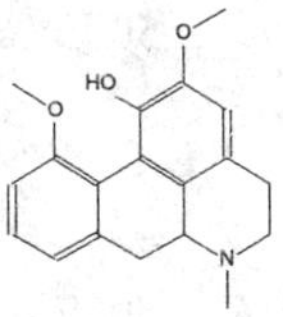

5159

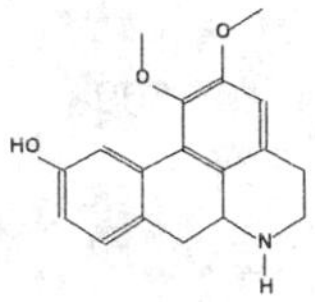

5160

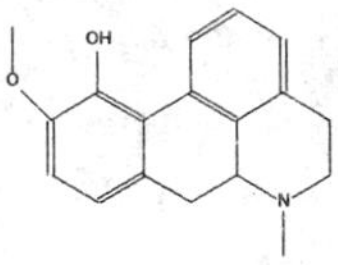

5161

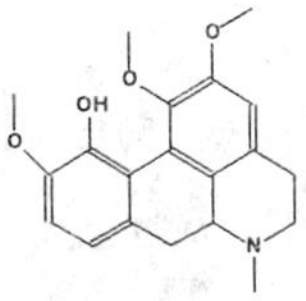

5162

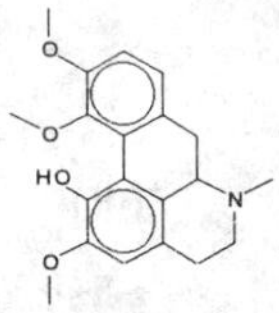

5163

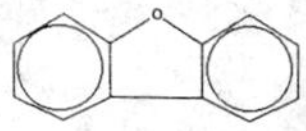

5164

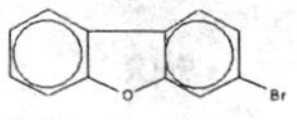

5165

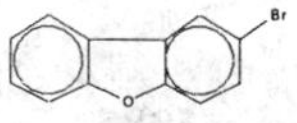

5166

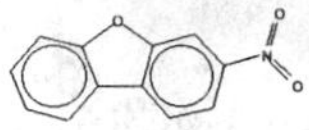

5167

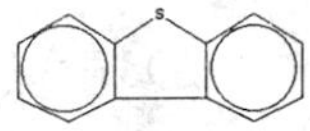

5168

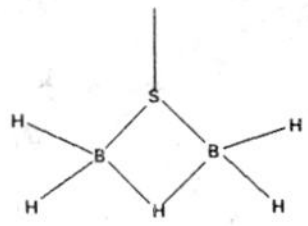

5169

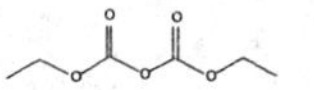

In numeric order as they occur in the Table of Physical Constants of Organic Compounds

5170

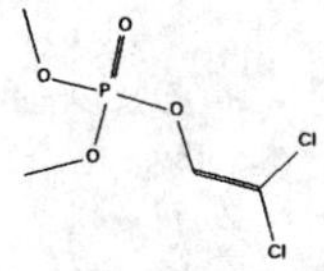

5171

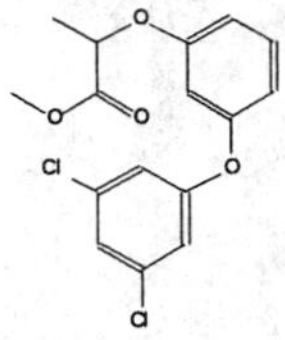

5172

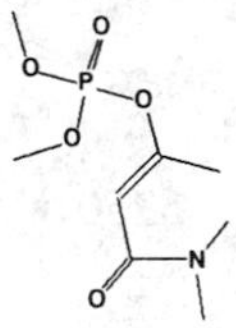

5173

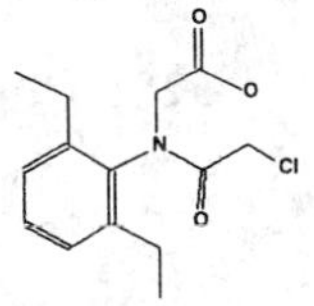

5174

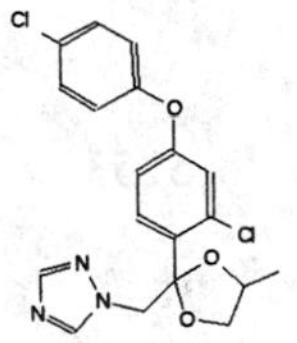

5175

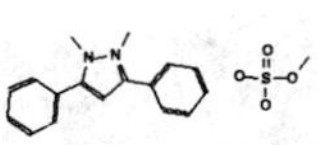

5176

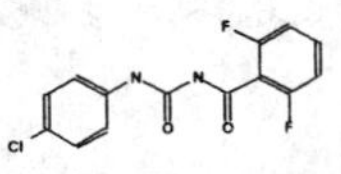

5177

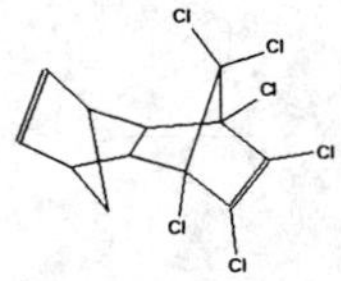

5178

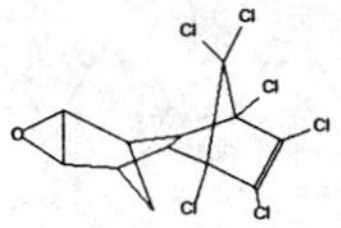

5179

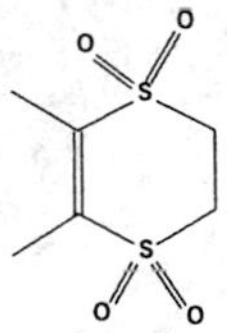

5180

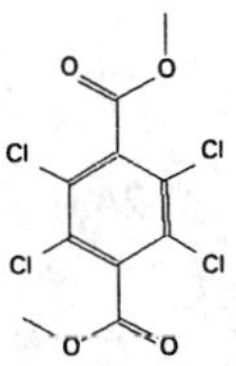

5181

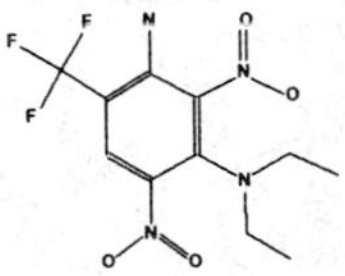

5182

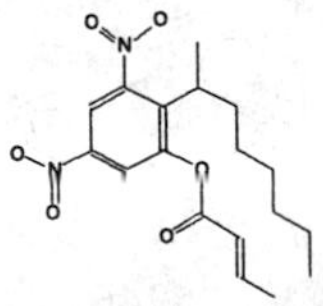

5183

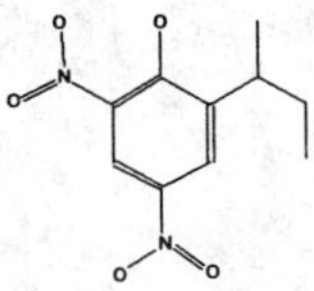

5184

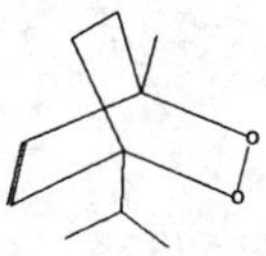

5186

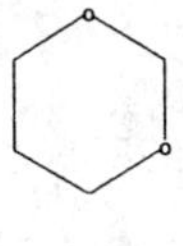

5187

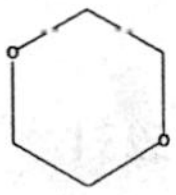

5188

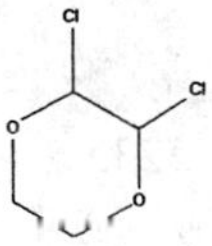

5189

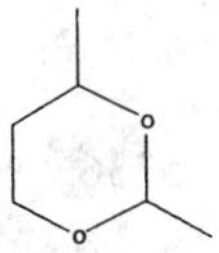

5190

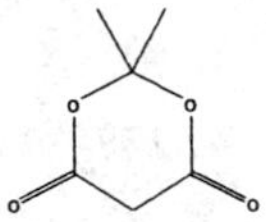

5192

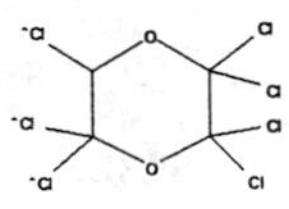

5193

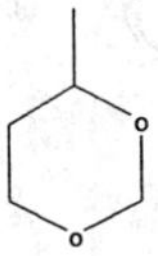

5194

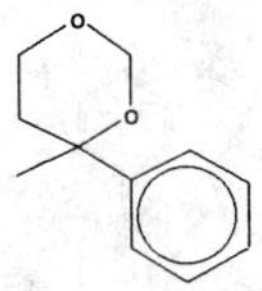

5195

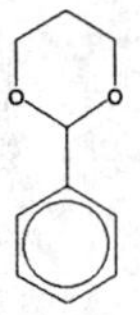

5196

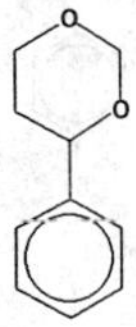

5197

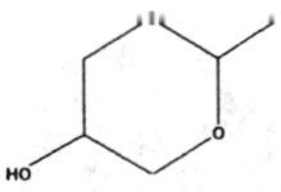

5200

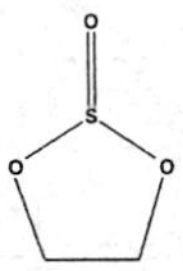

5201

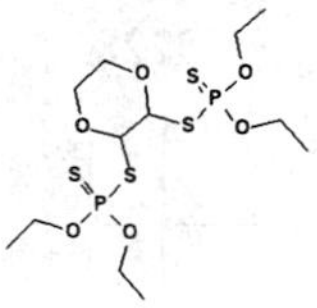

5202

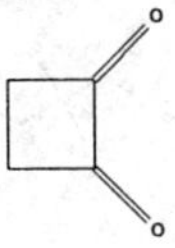

5203

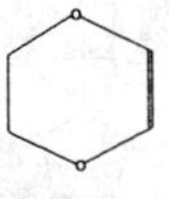

5204

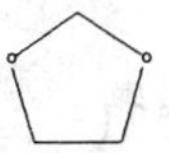

5205

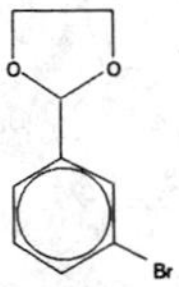

5206

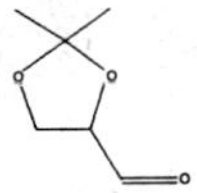

5207

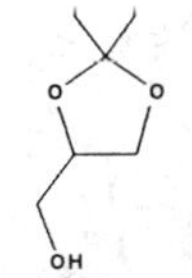

5208

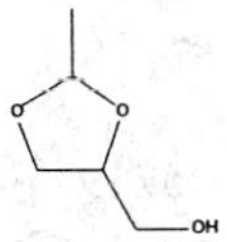

5209

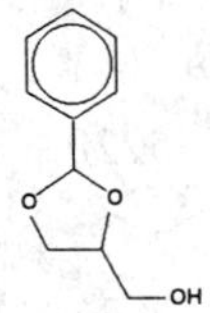

5210

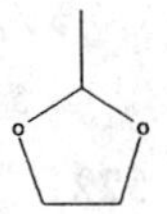

5211

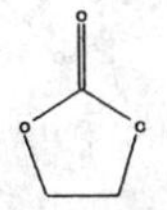

STRUCTURAL FORMULAS OF ORGANIC COMPOUNDS (continued)

In numeric order as they occur in the Table of Physical Constants of Organic Compounds

5212

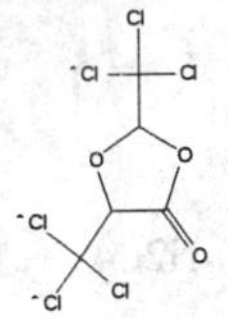

5213

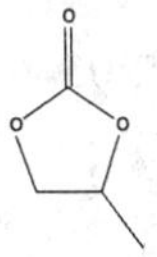

5214

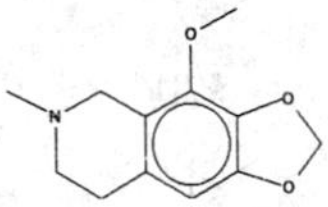

5215

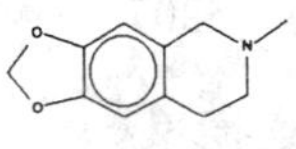

5217

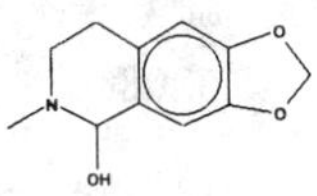

5219

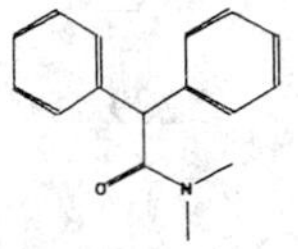

5220

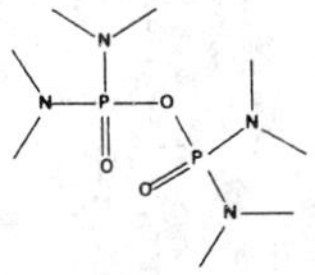

5221

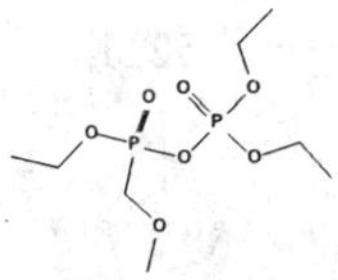

5222

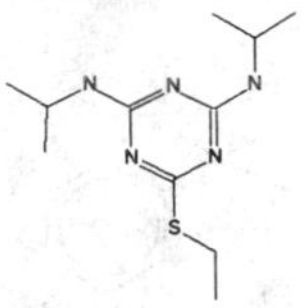

5224

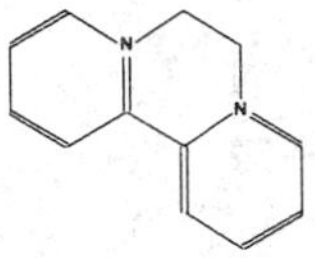

5225

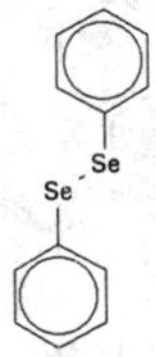

5226

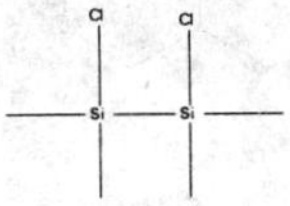

5227

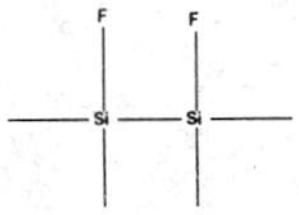

5228

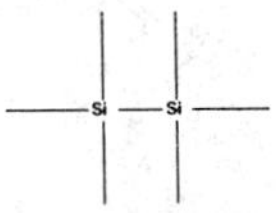

5229

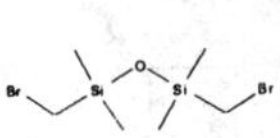

5230

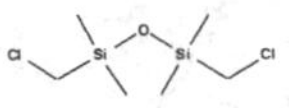

5232

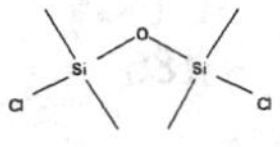

5233

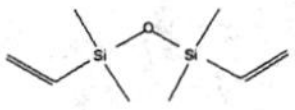

5234

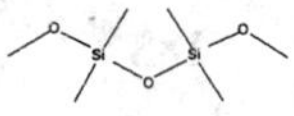

5235

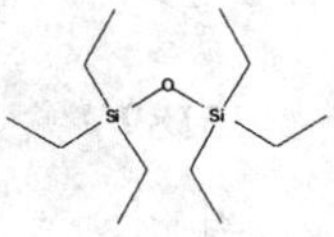

5236

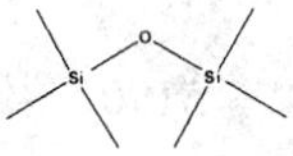

5237

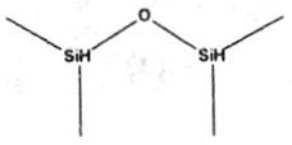

5238

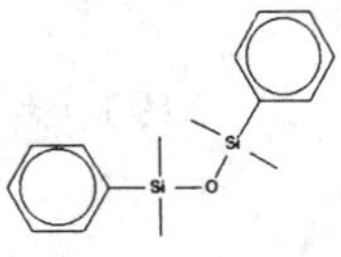

5239

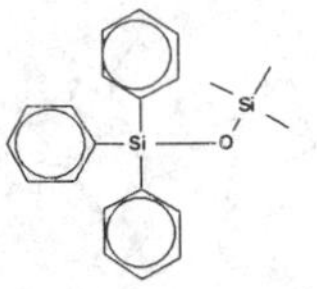

5240

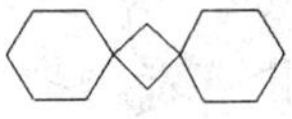

5241

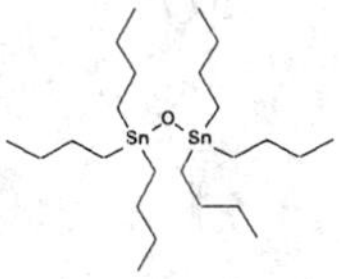

5242

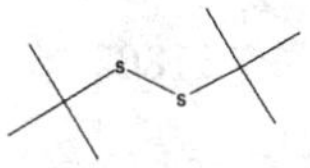

5243

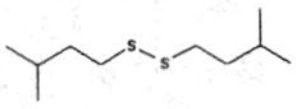

5244

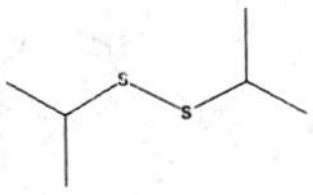

5245

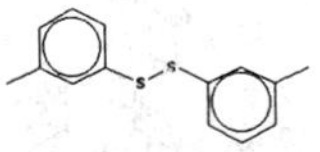

5246

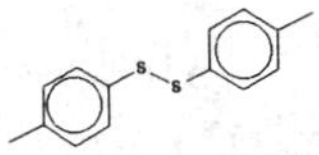

5247

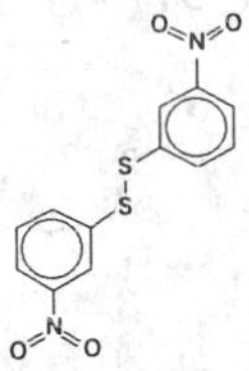

5248

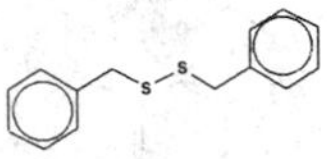

5249

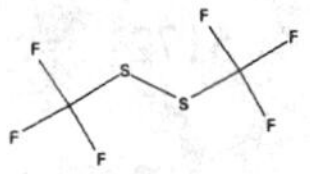

5250

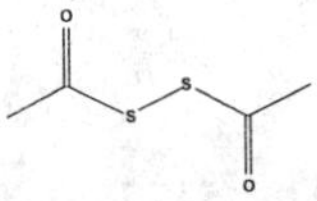

5251

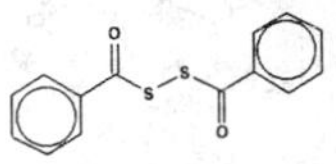

5252

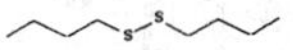

5253

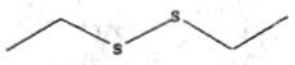

5254

5257

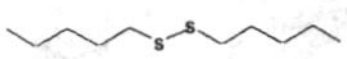

STRUCTURAL FORMULAS OF ORGANIC COMPOUNDS (continued)

In numeric order as they occur in the Table of Physical Constants of Organic Compounds

5258

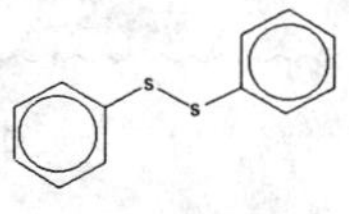

S S

5259

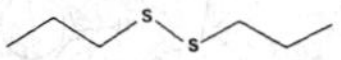

S S

5260

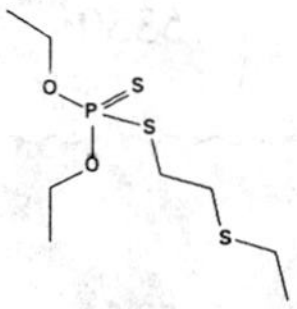

O P S S O S

5261

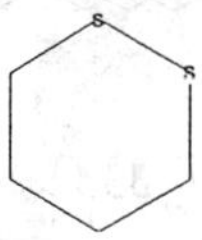

S S

5262

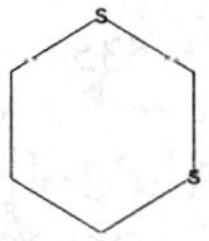

S S

5263

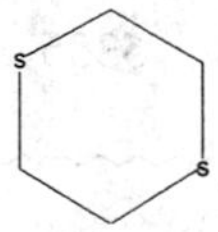

S S

5264

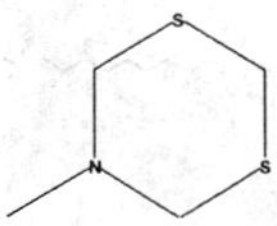

S N S

5265

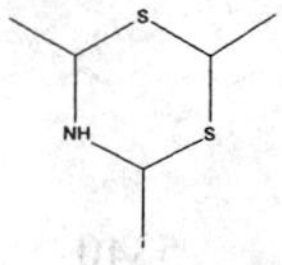

S NH S

5266

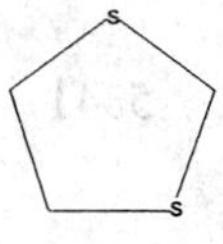

S S

5267

F F F N F F S O S O

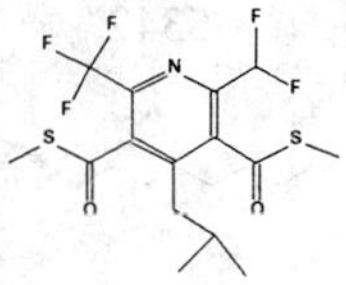

5268

5269

5270

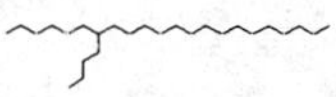

5271

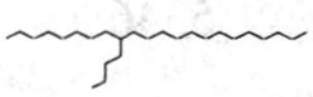

5272

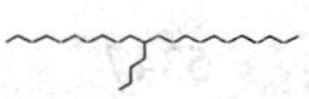

5274

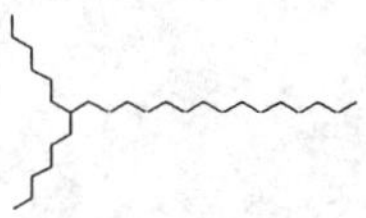

5275

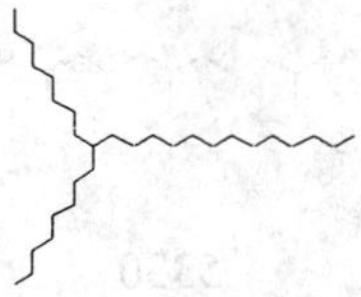

5276

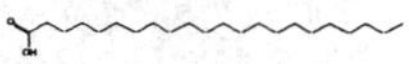

O OH

5277

O

5278

O

5279

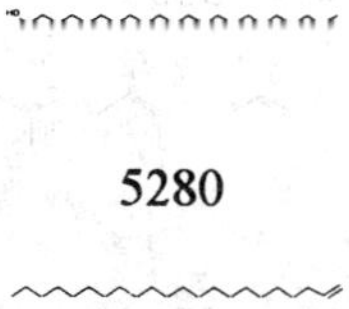

HO

5280

5281

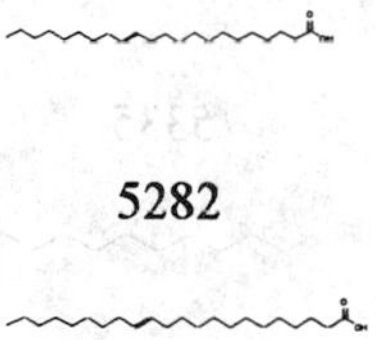

O OH

5282

O OH

5283

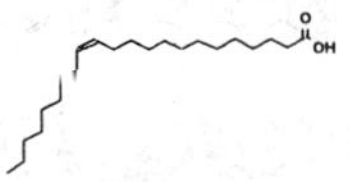

O OH

5284

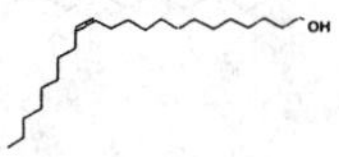

OH

5285

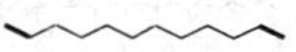

5286

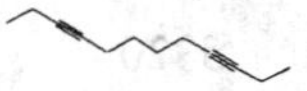

5287

5288

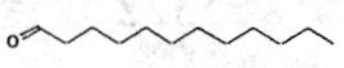

O

5289

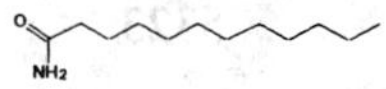

O NH2

5290

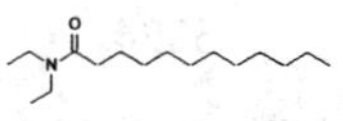

O N

5291

H2N

5292

5293

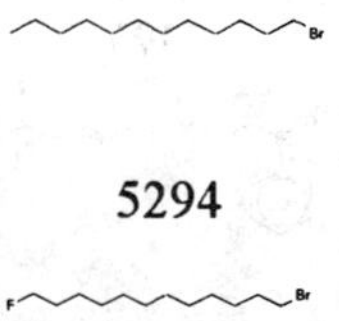

Br

5294

F Br

5295

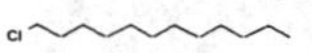

Cl

5296

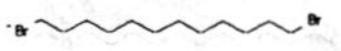

Br Br

5297

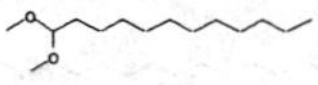

O O

5298

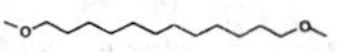

O O

5299

OH O O OH

5300

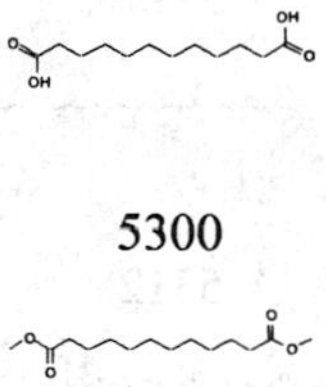

O O O O

5301

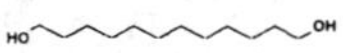

HO OH

5302

I

5303

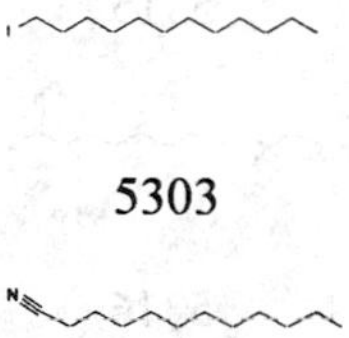

N

5304

HS

5305

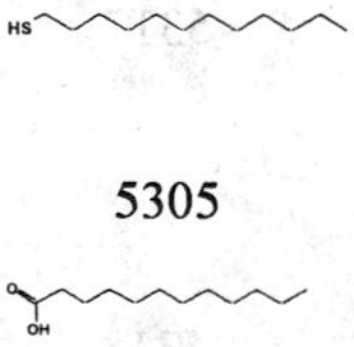

O OH

5307

O OH Br

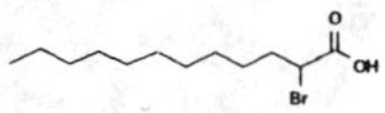

STRUCTURAL FORMULAS OF ORGANIC COMPOUNDS (continued)

In numeric order as they occur in the Table of Physical Constants of Organic Compounds

5308

5309

5310

5311

5312

5314

5315

5316

5317

5318

5319

5320

5321

5322

5323

5324

5325

5326

5327

5328

5329

5330

5331

5332

5333

5334

5335

5337

5338

5339

5340

5341

5342

5343

5344

5345

5346

5347

5348

5349

5350

5351

5352

5354

5355

5356

5357

STRUCTURAL FORMULAS OF ORGANIC COMPOUNDS (continued)

In numeric order as they occur in the Table of Physical Constants of Organic Compounds

5358

HO

5359

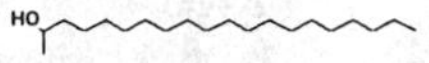

HO

5361

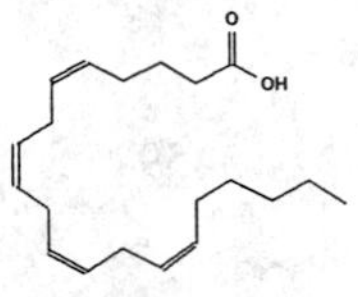

O

OH

5362

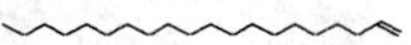

5363

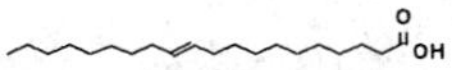

O

OH

5364

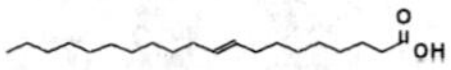

O

OH

5365

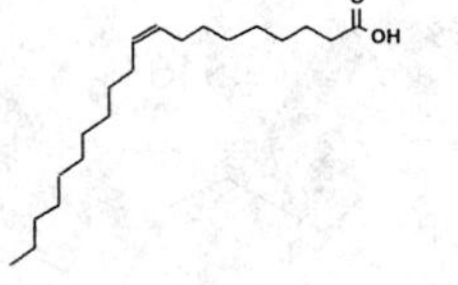

O

OH

5366

5367

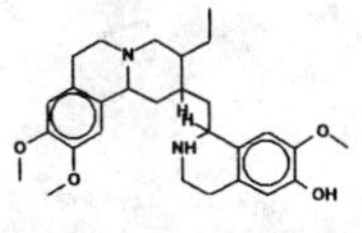

O

O

NH

H

O

OH

5368

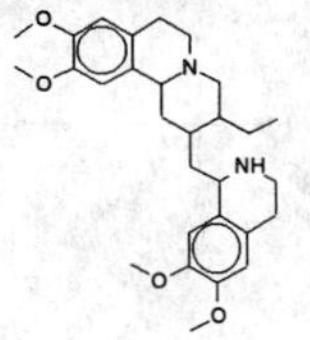

O

O

N

NH

O

O

5369

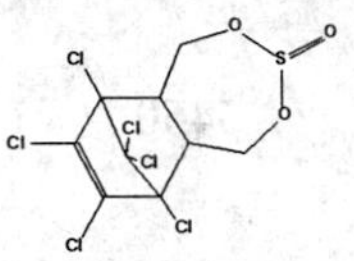

Cl

Cl

Cl

Cl

Cl

Cl

O

S

O

O

5370

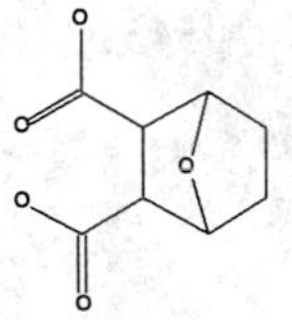

O

O

O

O

O

5371

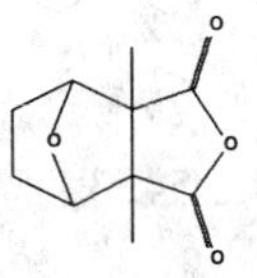

O

O

O

O

5372

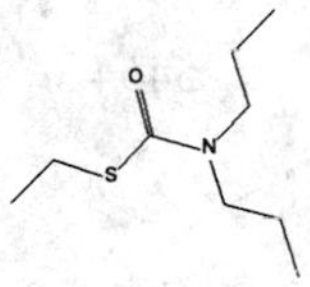

O

S

N

5373

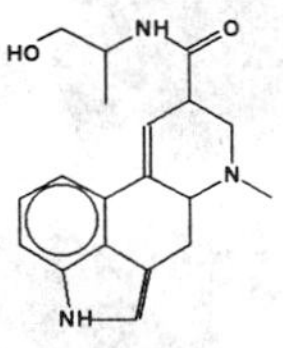

HO

NH

O

N

NH

5374

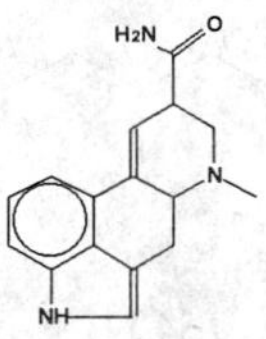

H_2N

O

N

NH

5376

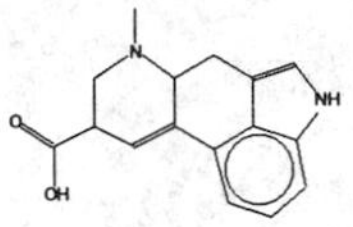

N

NH

O

OH

5379

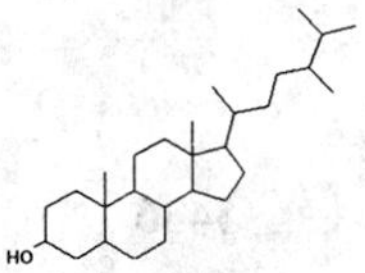

HO

5383

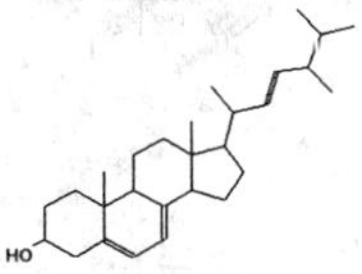

HO

5384

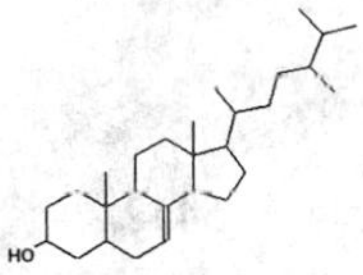

HO

5385

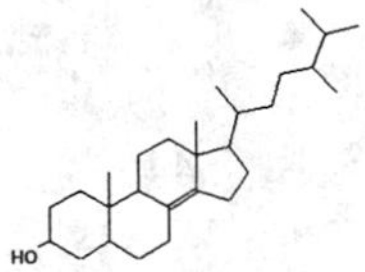

HO

5386

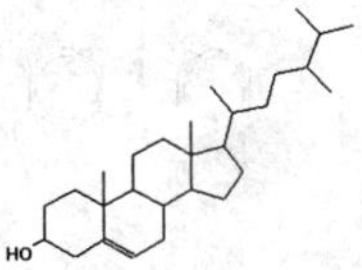

HO

5387

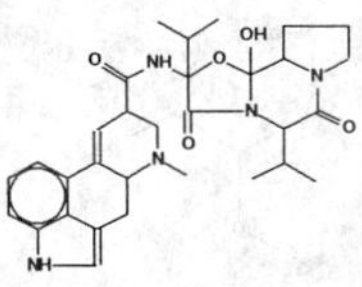

O

NH

OH

O

N

N

O

O

N

NH

5391

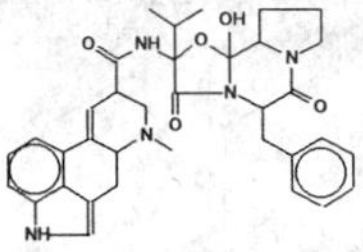

O

NH

OH

O

N

N

O

O

N

NH

5393

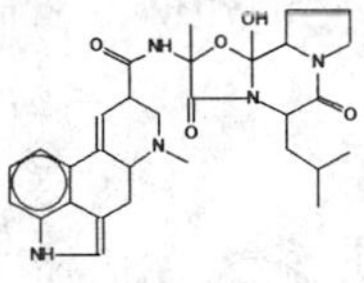

O

NH

OH

O

N

N

O

O

N

NH

5394

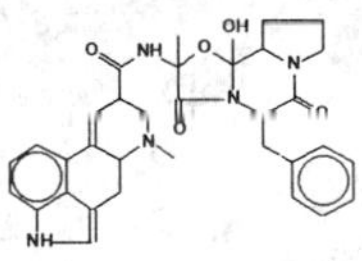

O

NH

OH

O

N

N

O

O

N

NH

5396

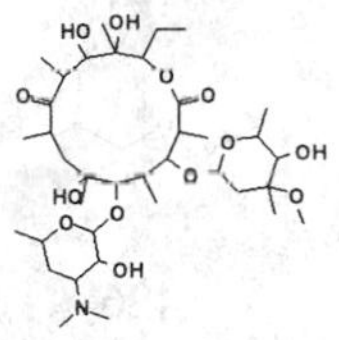

HO

OH

O

O

O

OH

O

O

HO

O

O

OH

N

5397

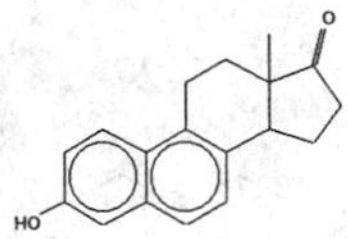

O

HO

5398

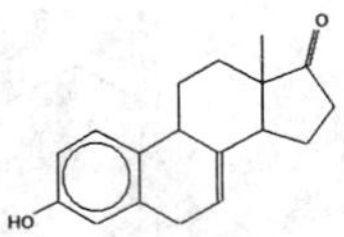

O

HO

5399

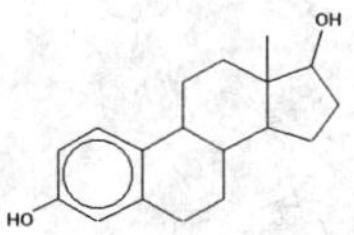

OH

HO

5400

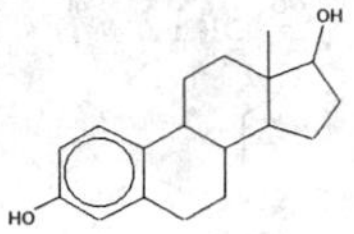

OH

HO

5403

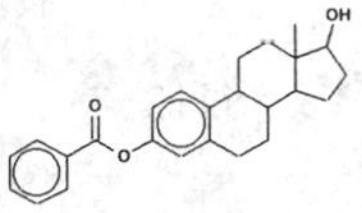

OH

O

O

5404

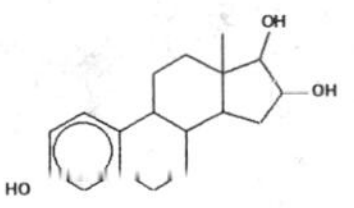

OH

OH

HO

5405

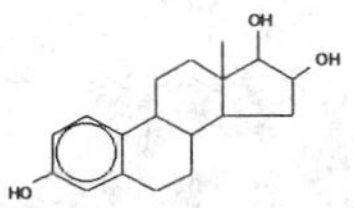

OH

OH

HO

5406

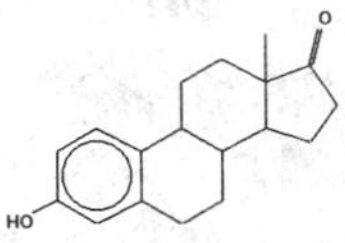

O

HO

5409

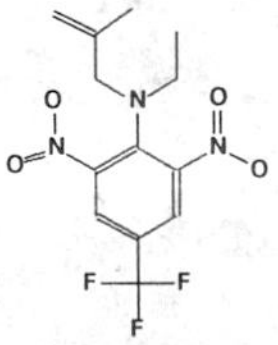

O

N

O

N

N

O

O

F

F

F

5410

NH_2

STRUCTURAL FORMULAS OF ORGANIC COMPOUNDS (continued)

In numeric order as they occur in the Table of Physical Constants of Organic Compounds

5411

5412

5413

5414

5415

5416

5417

5419

5420

5421

5422

5423

5424

5426

5427

5429

5430

5431

5432

5436

5441

5442

5445

5448

5449

5450

5451

5452

5453

5454

5455

5456

5457

5458

5459

5460

5461

5462

5463

5464

5465

STRUCTURAL FORMULAS OF ORGANIC COMPOUNDS (continued)

In numeric order as they occur in the Table of Physical Constants of Organic Compounds

5466

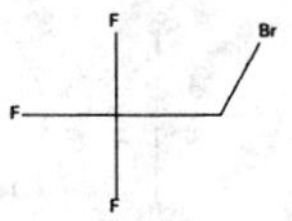

5467

5468

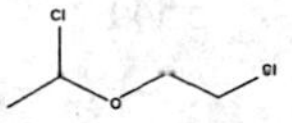

5469

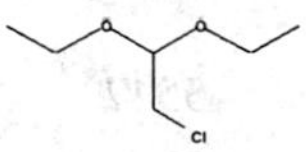

5470

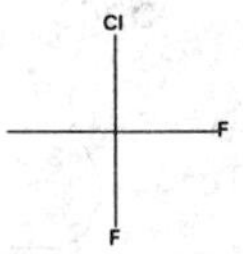

5471

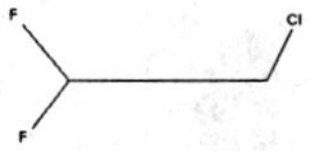

5472

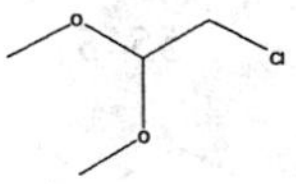

5473

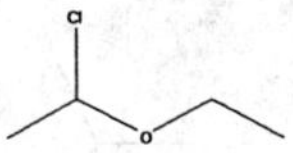

5474

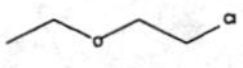

5475

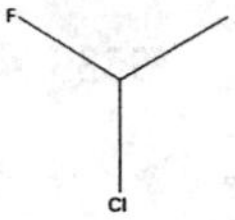

5476

5477

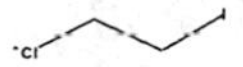

5478

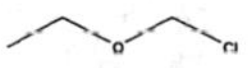

5479

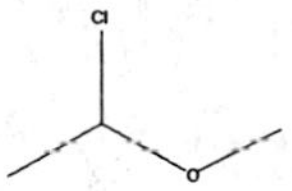

5480

5481

5482

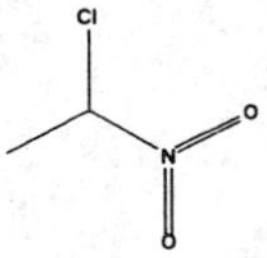

5483

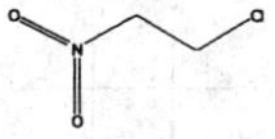

5484

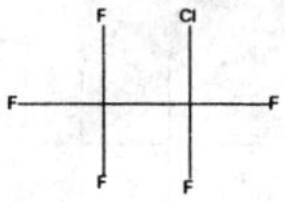

5485

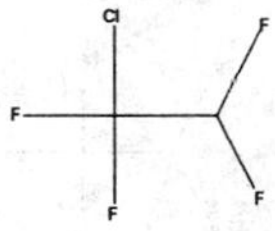

5486

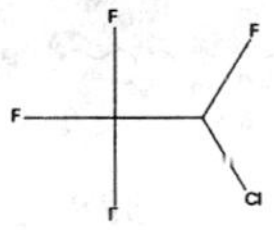

5487

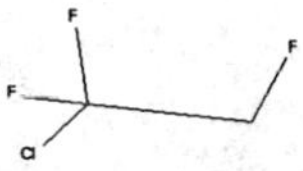

5488

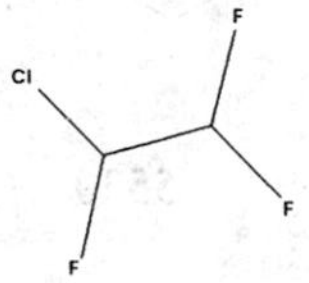

5489

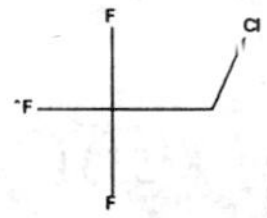

5490

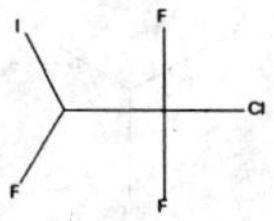

5491

5492

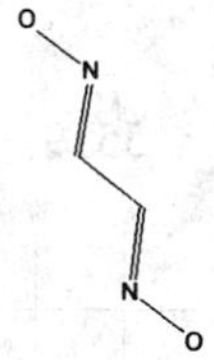

5493

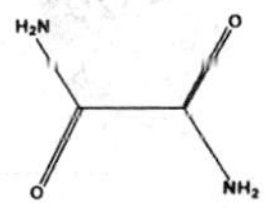

5495

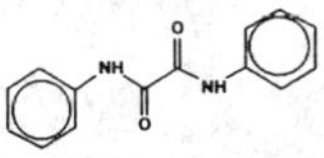

5496

5497

5498

5499

5500

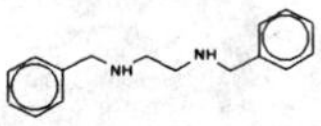

5501

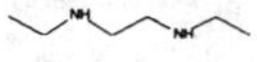

5502

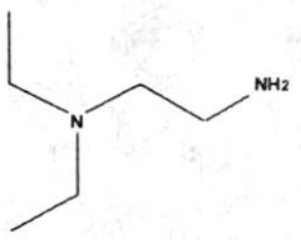

5503

5505

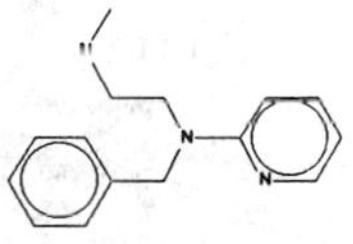

5506

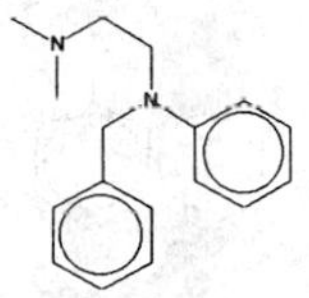

5507

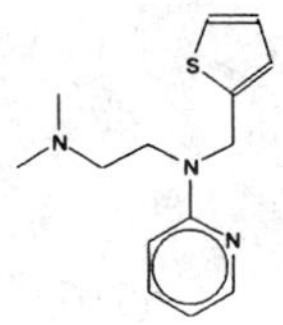

5508

STRUCTURAL FORMULAS OF ORGANIC COMPOUNDS (continued)

In numeric order as they occur in the Table of Physical Constants of Organic Compounds

5509

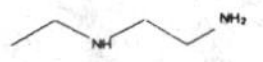

5510

5511

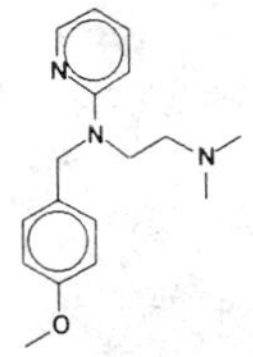

5512

5513

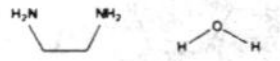

5514

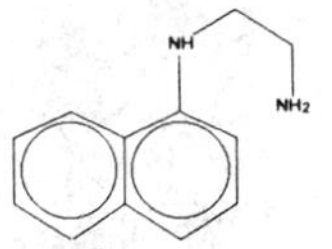

5515

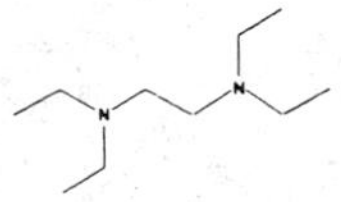

5516

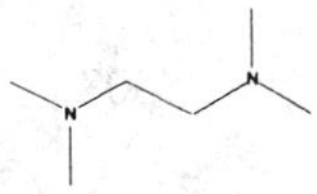

5517

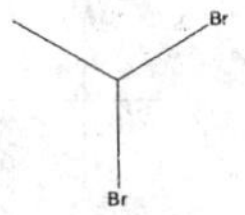

5518

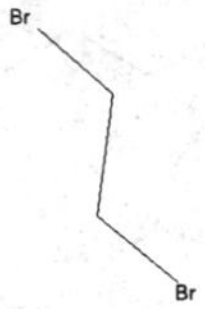

5519

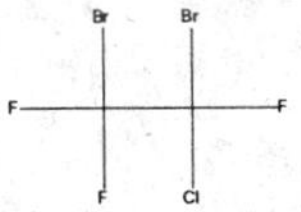

5520

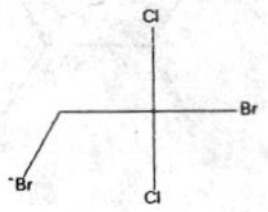

5521

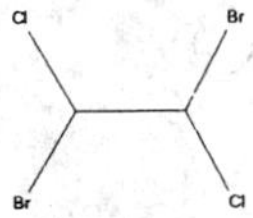

5522

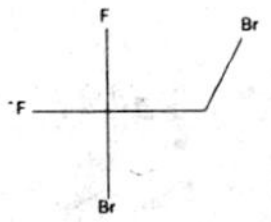

5523

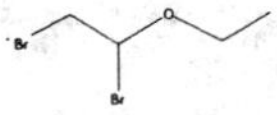

5524

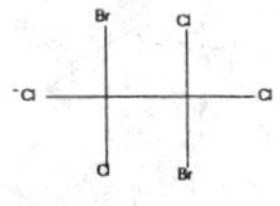

5525

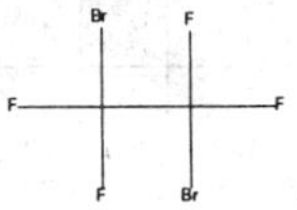

5526

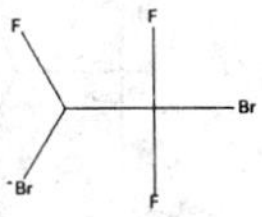

5527

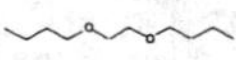

5528

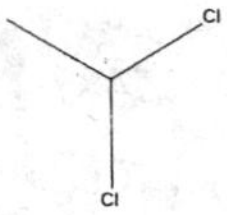

5529

5530

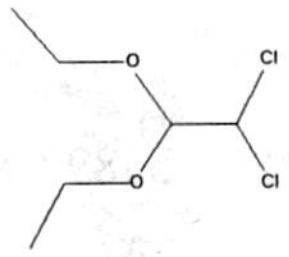

5531

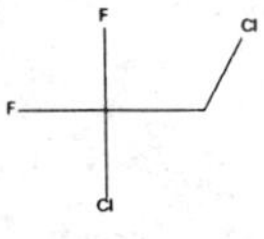

5532

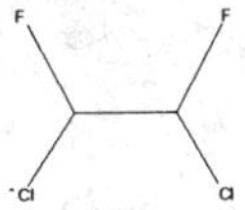

5533

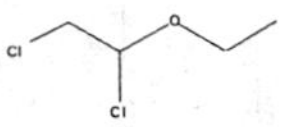

5534

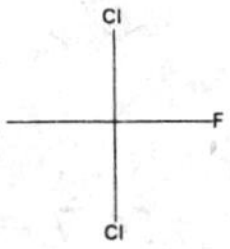

5535

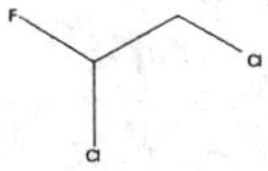

5536

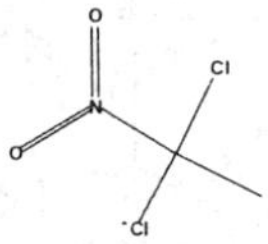

5537

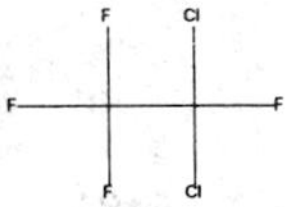

5538

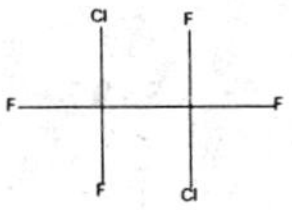

5539

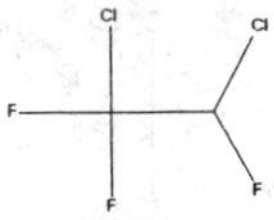

5540

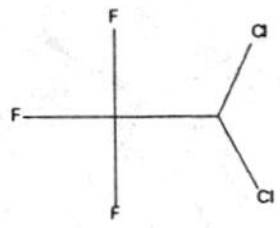

5541

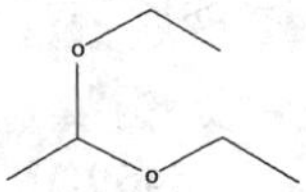

5542

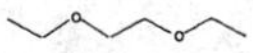

5543

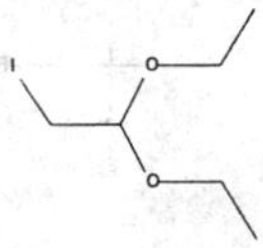

5544

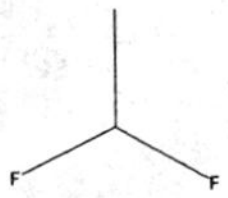

5545

5546

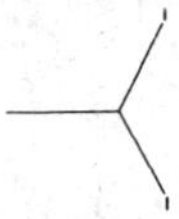

STRUCTURAL FORMULAS OF ORGANIC COMPOUNDS (continued)

In numeric order as they occur in the Table of Physical Constants of Organic Compounds

5547

5548

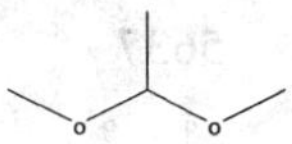

5549

5550

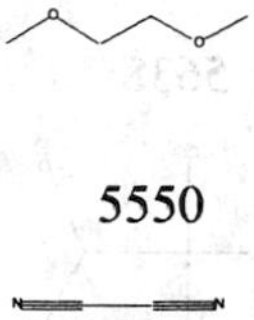

5551

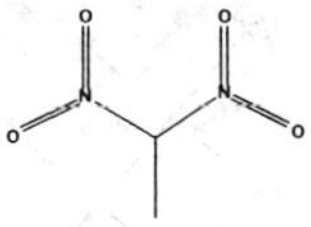

5552

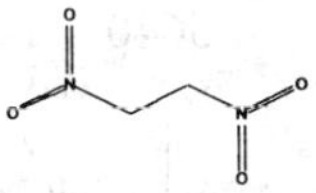

5553

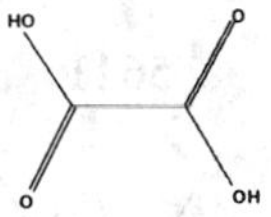

5554

5555

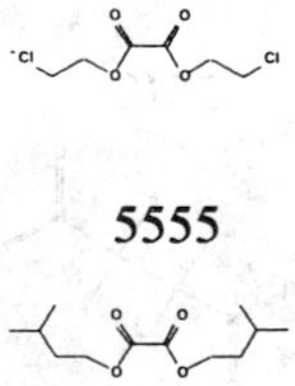

5556

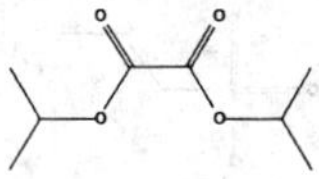

5557

5558

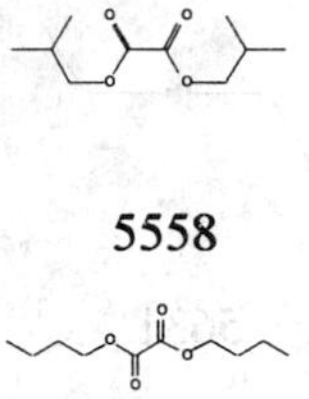

5559

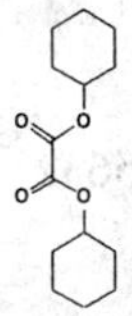

5560

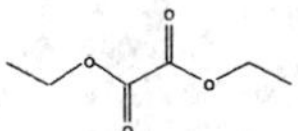

5562

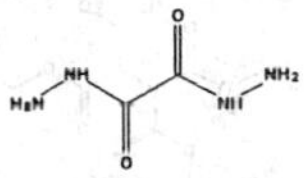

5563

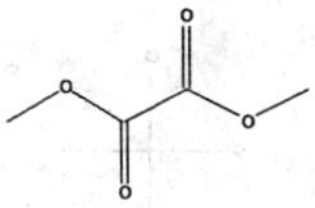

5564

5565

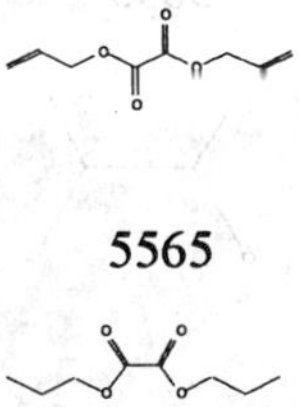

5567

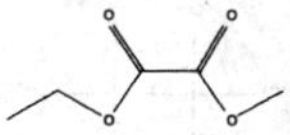

5569

5570

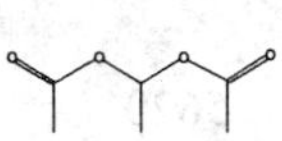

5571

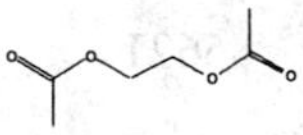

5572

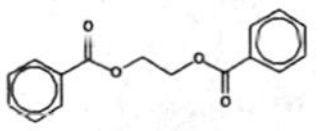

5573

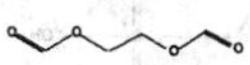

5574

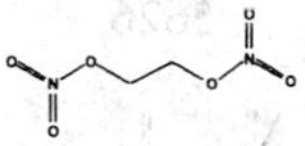

5576

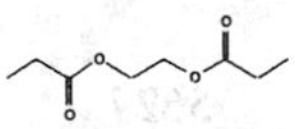

5577

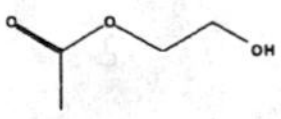

5578

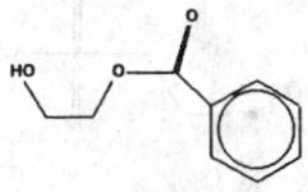

5579

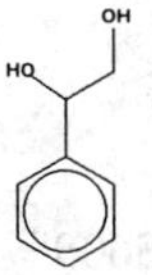

5580

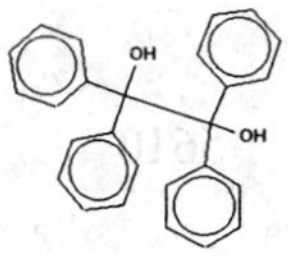

5582

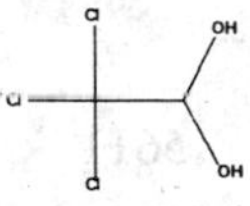

5583

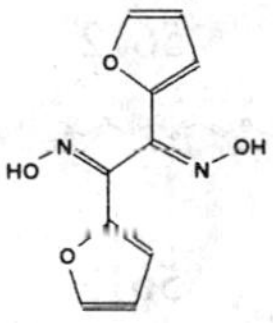

5584

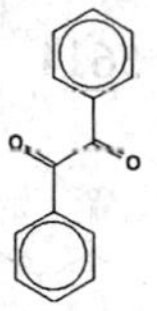

5585

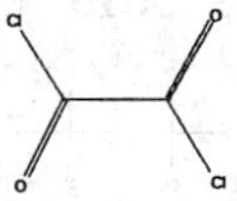

5586

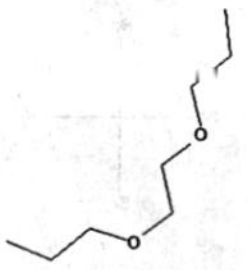

5588

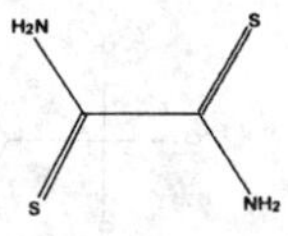

5589

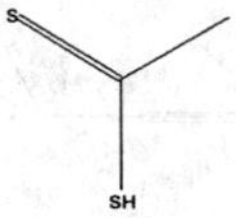

5590

5591

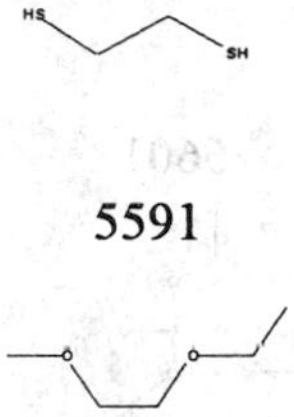

5592

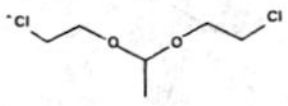

5593

5595

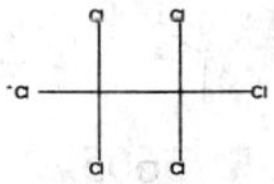

5596

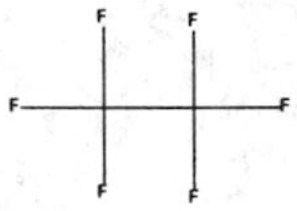

5597

5598

STRUCTURAL FORMULAS OF ORGANIC COMPOUNDS (continued)

In numeric order as they occur in the Table of Physical Constants of Organic Compounds

5599

5600

5601

5602

5603

5604

5605

5606

5607

5608

5609

5610

5611

5612

5613

5614

5615

5616

5617

5618

5619

5620

5621

5622

5626

5627

5628

5629

5630

5631

5632

5633

5634

5635

5636

5637

5638

5639

5640

5641

5642

STRUCTURAL FORMULAS OF ORGANIC COMPOUNDS (continued)

In numeric order as they occur in the Table of Physical Constants of Organic Compounds

5643

5644

5645

5646

5647

5648

5650

5651

5652

5653

5654

5655

5656

5657

5658

5659

5660

5661

5662

5663

5664

5665

5666

5667

5668

5669

5672

5673

5674

5675

5676

5677

5678

5679

5680

5681

5682

5683

5684

5685

In numeric order as they occur in the Table of Physical Constants of Organic Compounds

5686

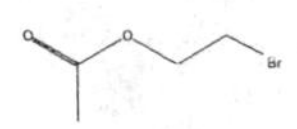

5687

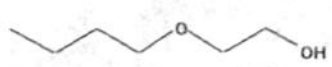

5688

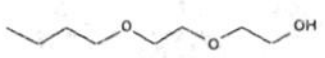

5689

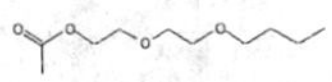

5690

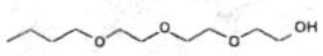

5691

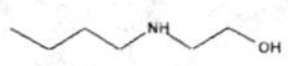

5692

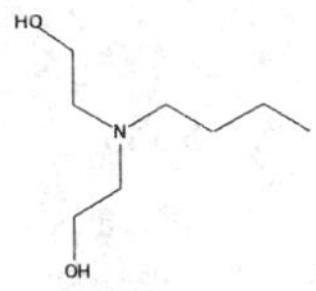

5693

5694

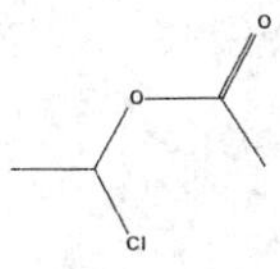

5695

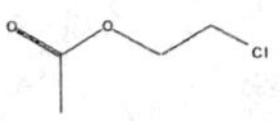

5696

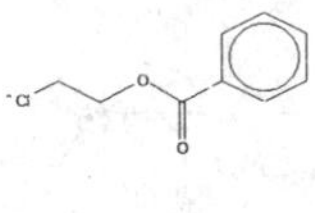

5697

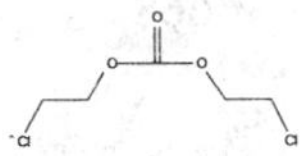

5698

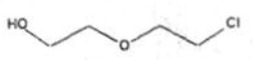

5699

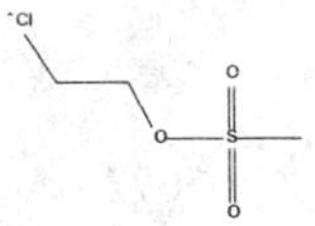

5700

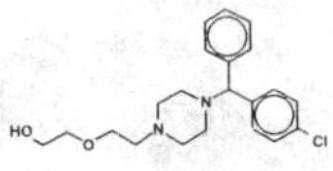

5701

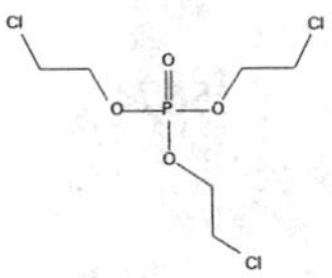

5702

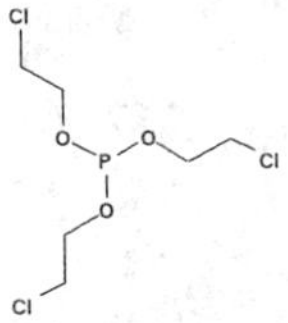

5703

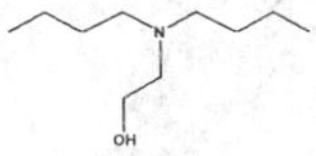

5704

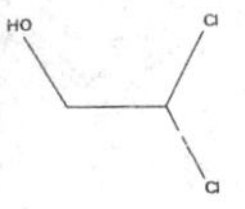

5705

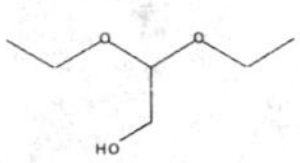

5706

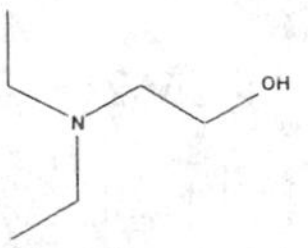

5707

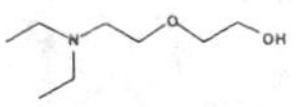

5708

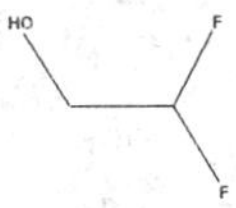

5709

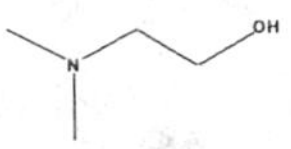

5710

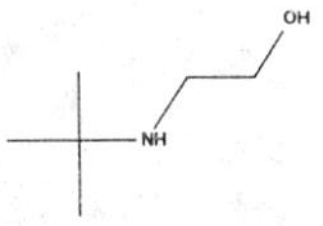

5711

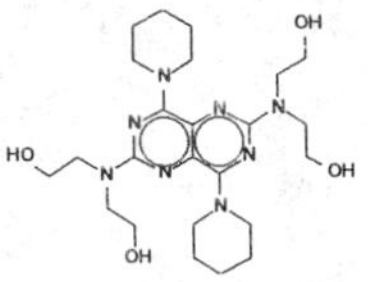

5712

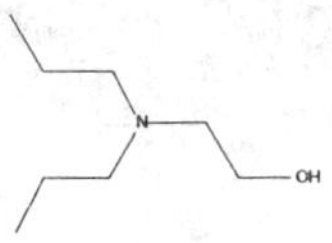

5713

5714

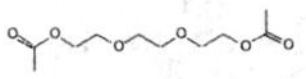

5715

5716

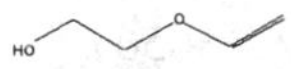

5717

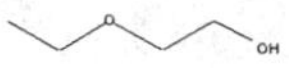

5718

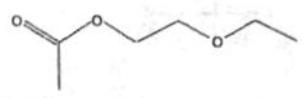

5719

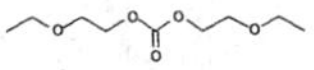

5720

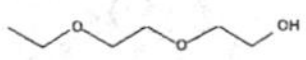

5721

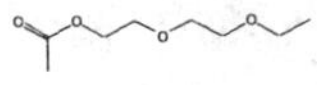

5722

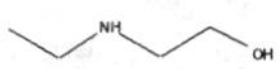

5723

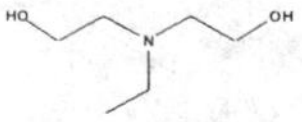

5724

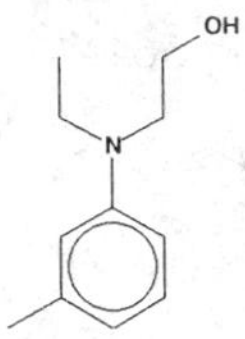

5726

5727

5728

5729

5730

5731

5732

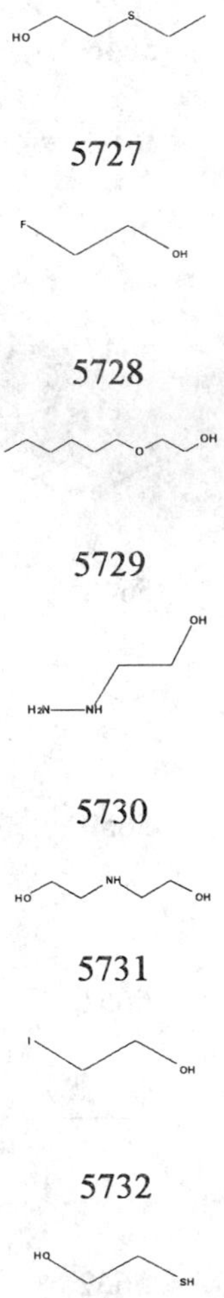

5733

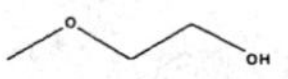

5734

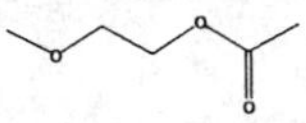

5735

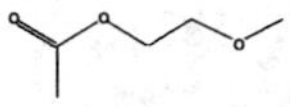

5736

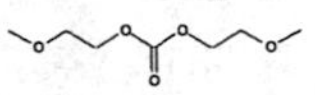

5737

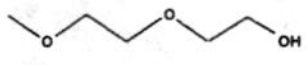

5738

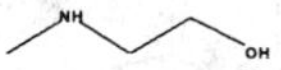

5739

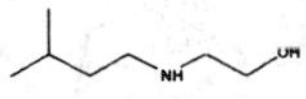

5740

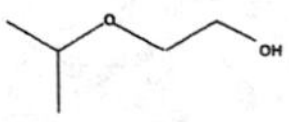

5741

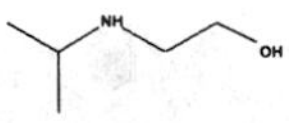

5742

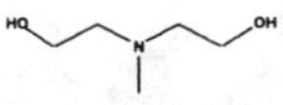

5743

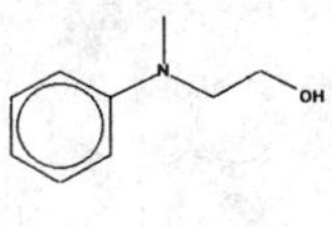

5744

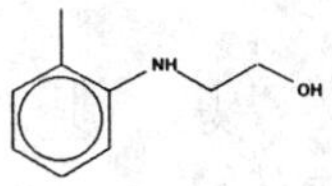

5745

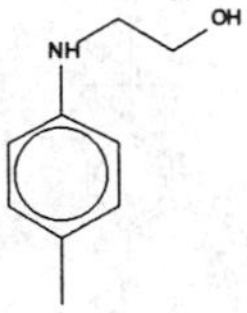

5746

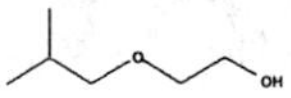

5747

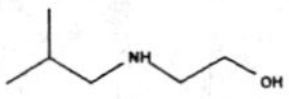

5748

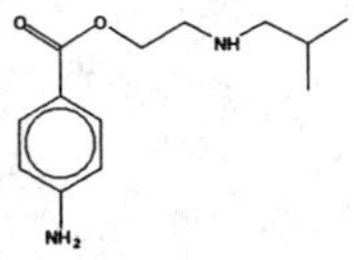

5749

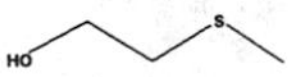

5750

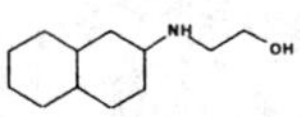

5751

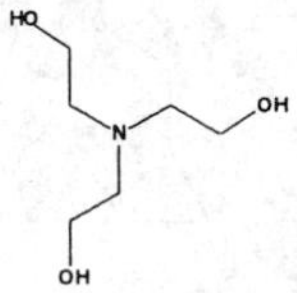

5752

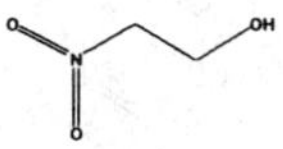

5753

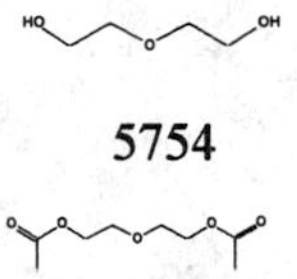

5754

5755

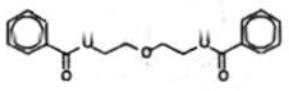

5756

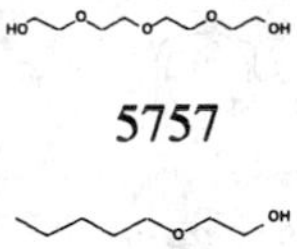

5757

5758

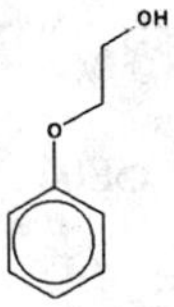

5759

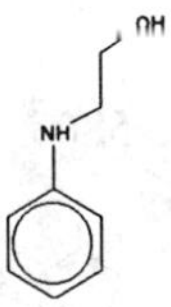

5760

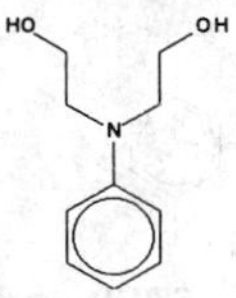

5761

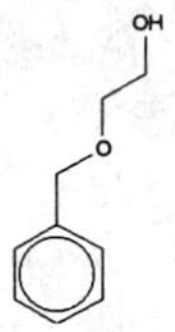

5762

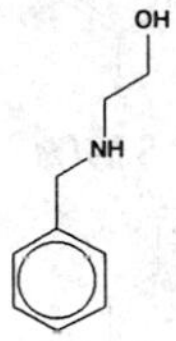

5763

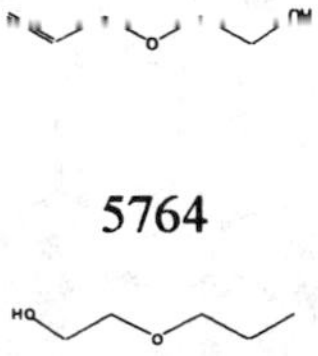

5764

5765

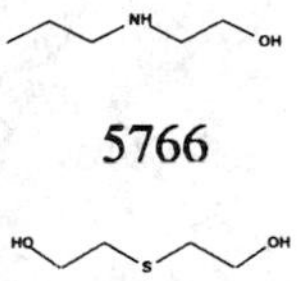

5766

5767

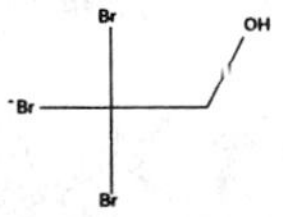

5768

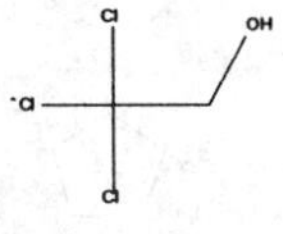

5769

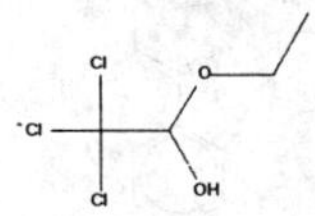

5770

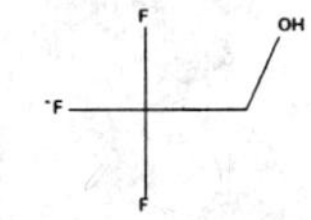

5771

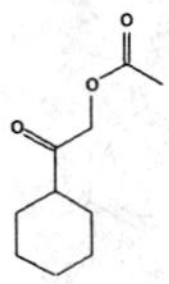

5772

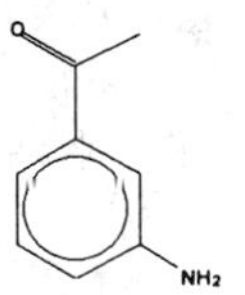

5773

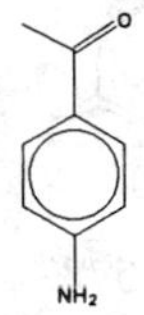

5774

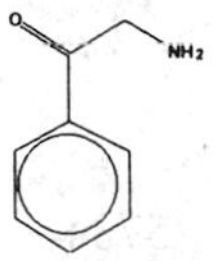

5775

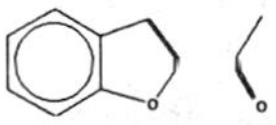

5776

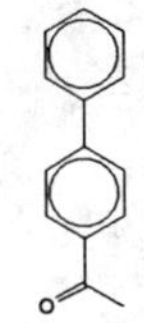

STRUCTURAL FORMULAS OF ORGANIC COMPOUNDS (continued)

In numeric order as they occur in the Table of Physical Constants of Organic Compounds

5777

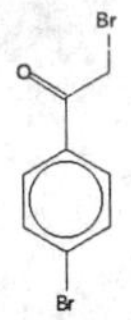

5779

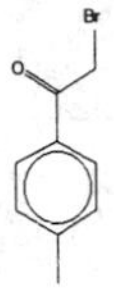

5780

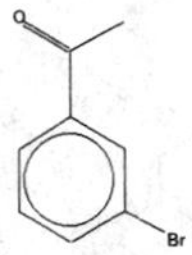

5781

5782

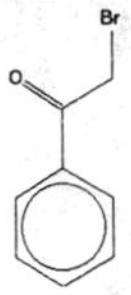

5783

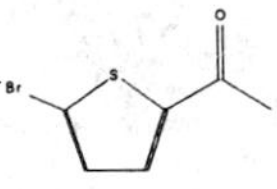

5784

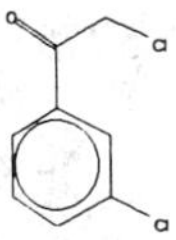

5786

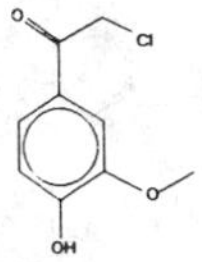

5787

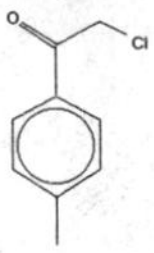

5788

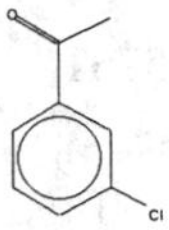

5789

5790

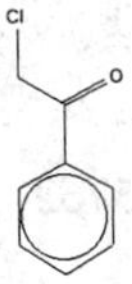

5791

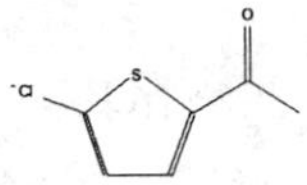

5792

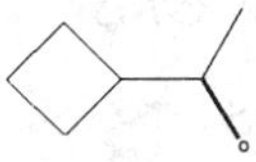

5793

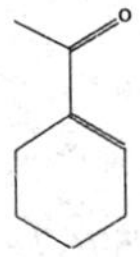

5794

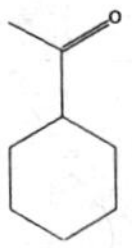

5795

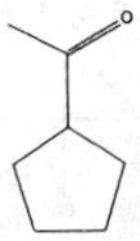

5796

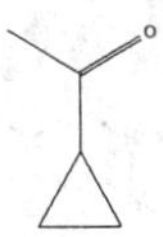

5797

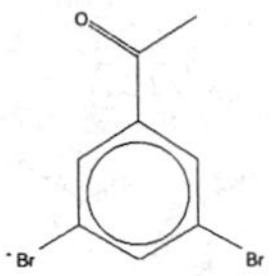

5798

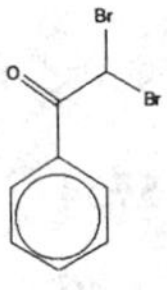

5800

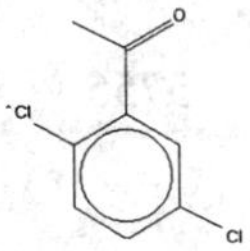

5801

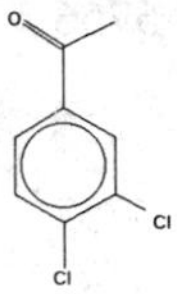

5802

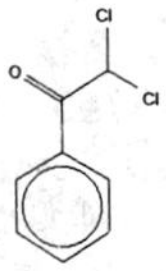

5803

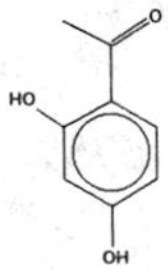

5805

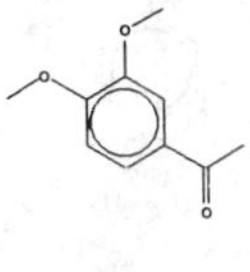

5806

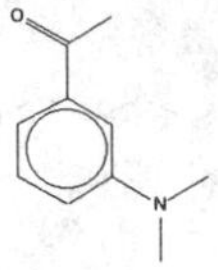

5807

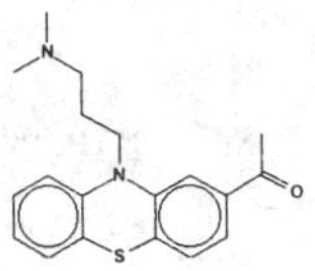

5808

5809

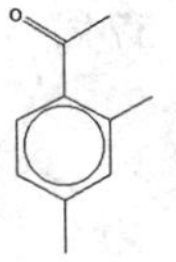

5810

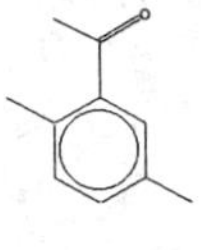

5811

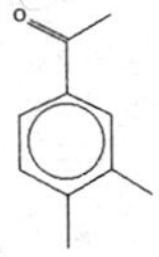

5812

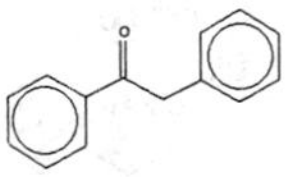

5813

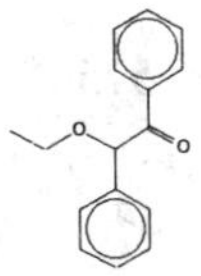

5814

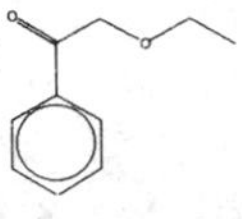

5815

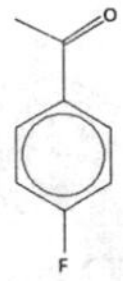

5816

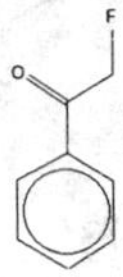

5817

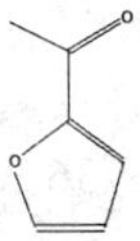

5818

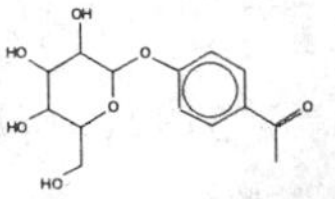

5819

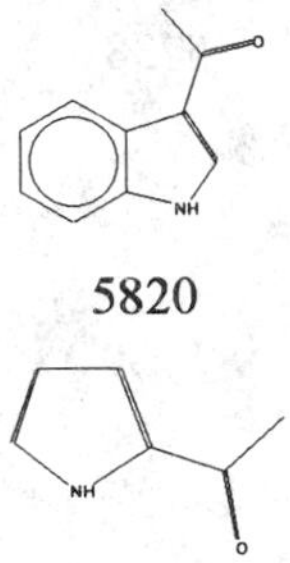

5820

5821

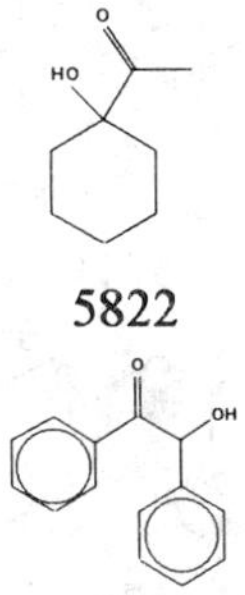

5822

In numeric order as they occur in the Table of Physical Constants of Organic Compounds

5823

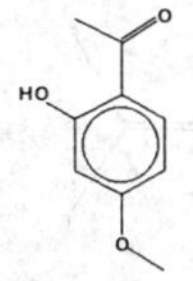

HO

O

5824

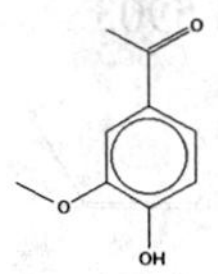

O

OH

5826

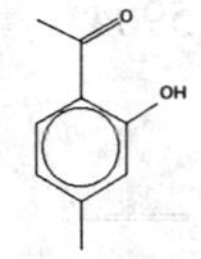

OH

5827

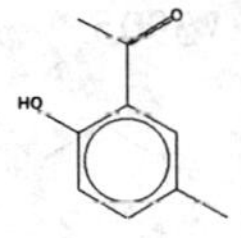

HO

5828

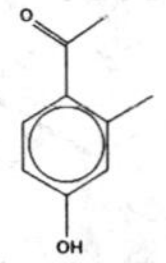

O

OH

5829

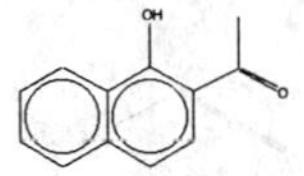

OH

5830

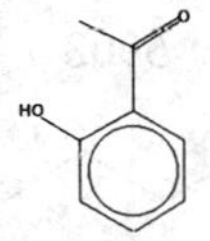

HO

5831

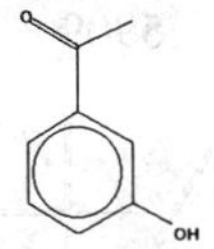

O

OH

5832

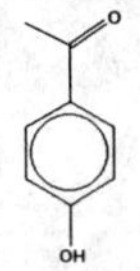

OH

5833

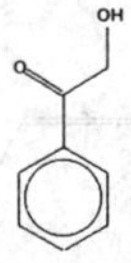

OH

5834

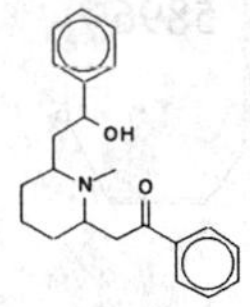

OH

N

O

5835

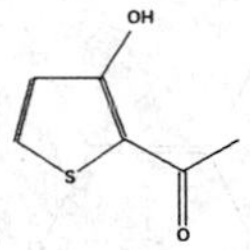

OH

S

O

5836

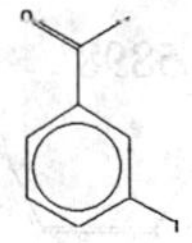

5837

5838

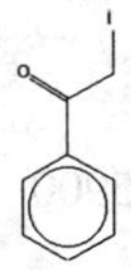

5839

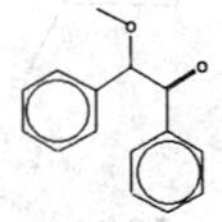

O

5840

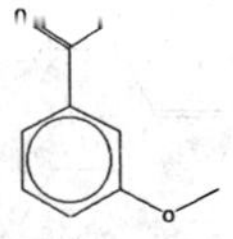

O

5841

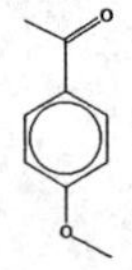

O

5842

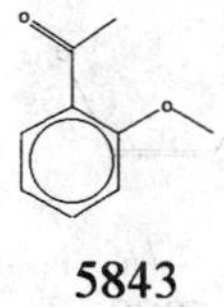

O

5843

O

5844

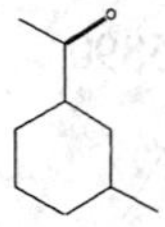

5845

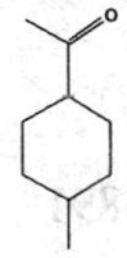

O

5846

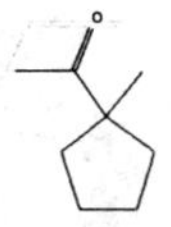

O

5847

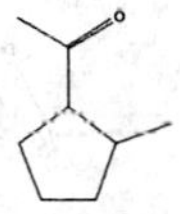

5848

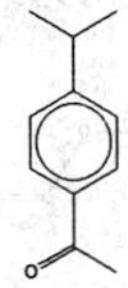

O

5849

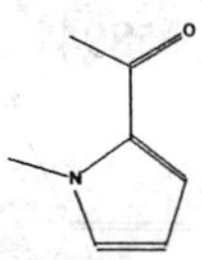

N

5850

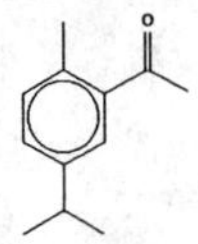

O

5851

5852

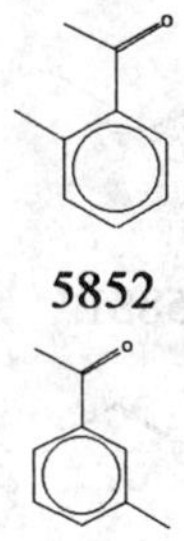

5853

5855

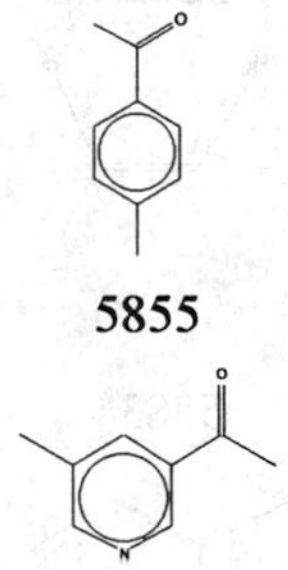

O

N

5856

S

O

5857

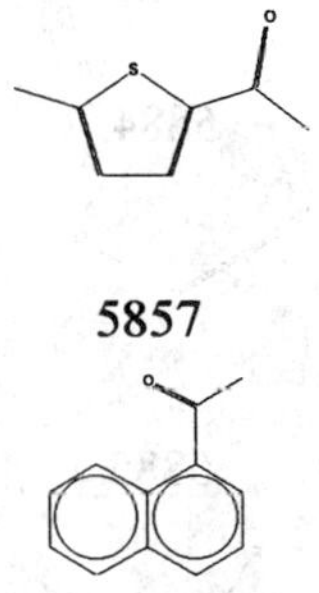

5858

O

5859

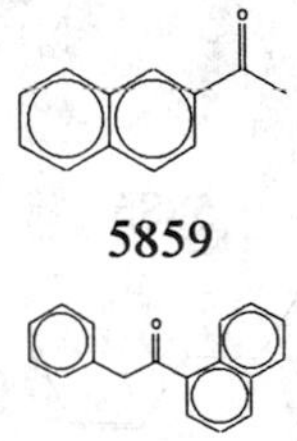

O

5860

N

O

5861

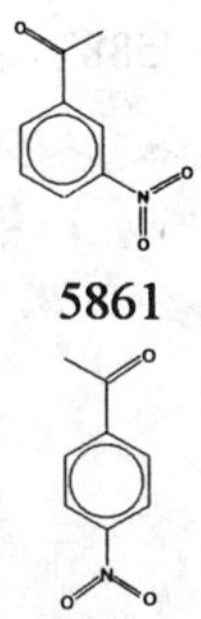

O

N

O

5862

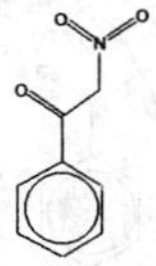

O

N

O

5863

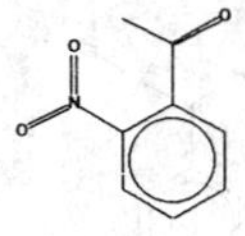

O

N

O

5864

5865

5866

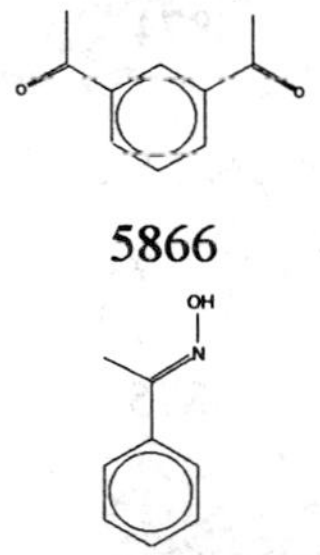

OH

N

5867

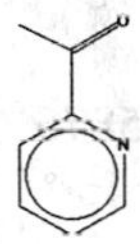

N

5868

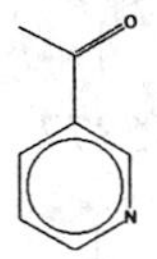

N

5869

N

5870

S

O

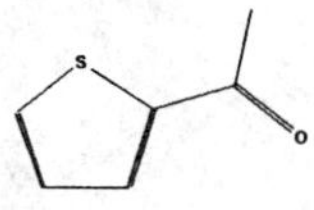

STRUCTURAL FORMULAS OF ORGANIC COMPOUNDS (continued)

In numeric order as they occur in the Table of Physical Constants of Organic Compounds

5871

5872

5873

5874

5875

5876

5877

5879

5880

5881

5882

5883

5884

5885

5886

5887

5888

5889

5890

5891

5892

5893

5894

5895

5896

5897

5898

5899

5900

5901

5902

5903

5904

5905

5906

5907

5908

5909

STRUCTURAL FORMULAS OF ORGANIC COMPOUNDS (continued)

In numeric order as they occur in the Table of Physical Constants of Organic Compounds

5910

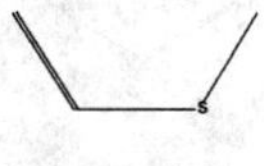

5911

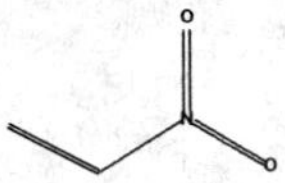

5912

5913

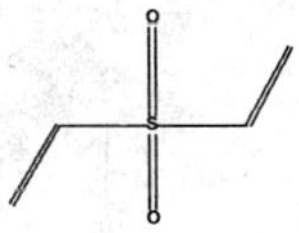

5914

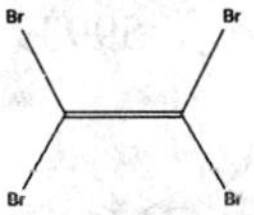

5915

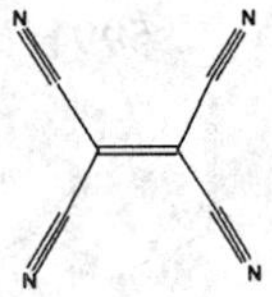

5916

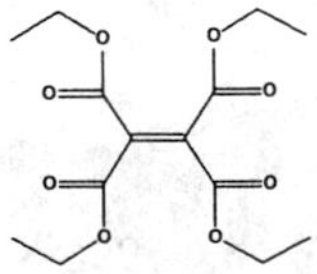

5917

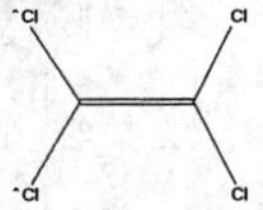

5918

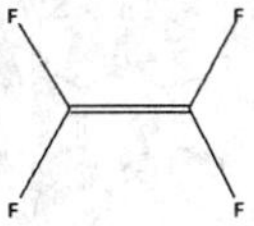

5919

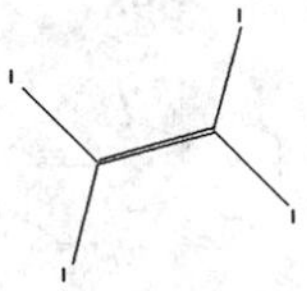

5920

5921

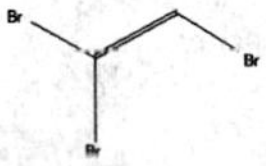

5922

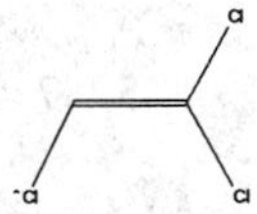

5923

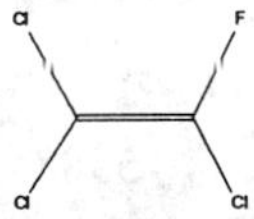

5924

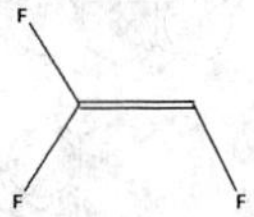

5925

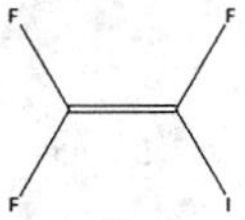

5926

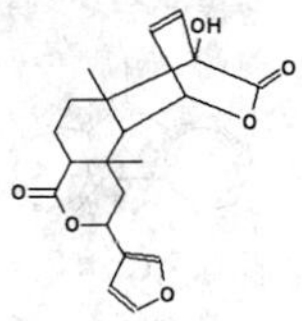

5927

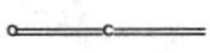

5929

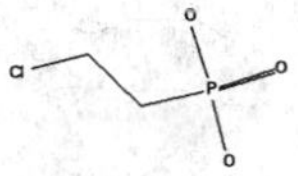

5930

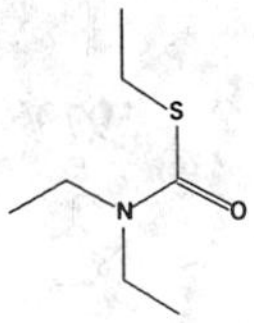

5931

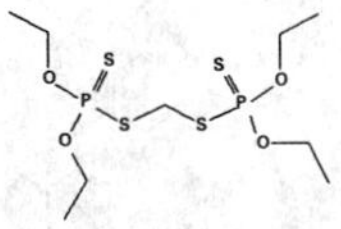

5932

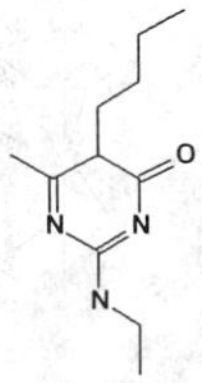

5933

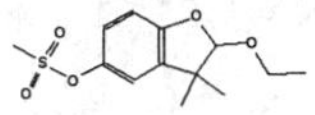

5934

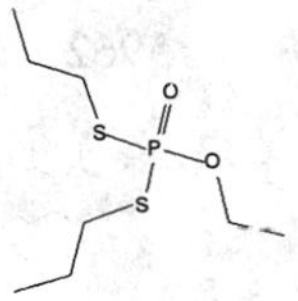

5935

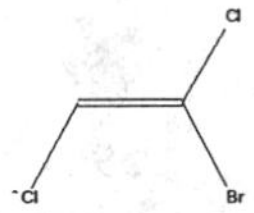

5936

5937

5938

5939

5940

5941

5942

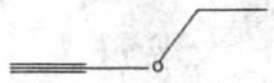

5943

5944

5945

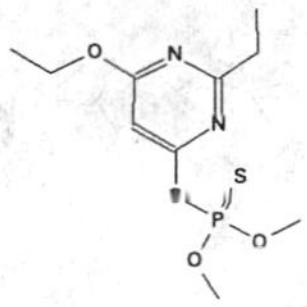

5946

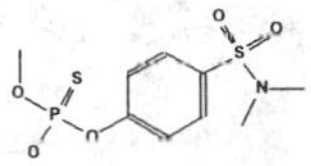

5947

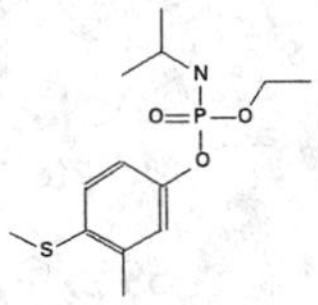

5948

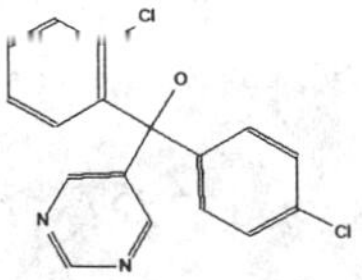

In numeric order as they occur in the Table of Physical Constants of Organic Compounds

5950

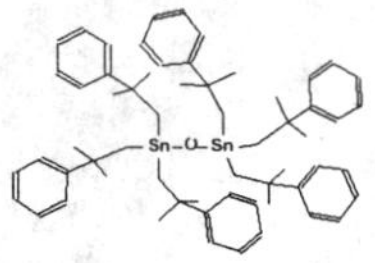

5951

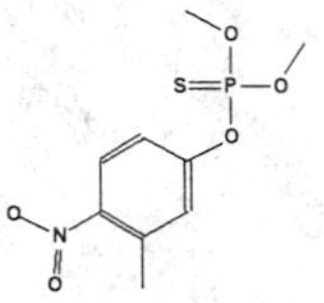

5952

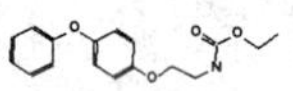

5953

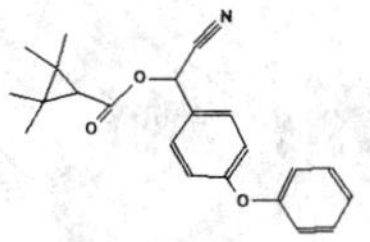

5954

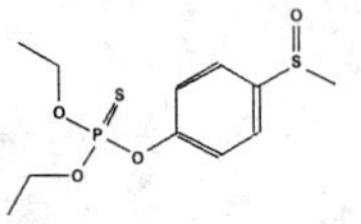

5955

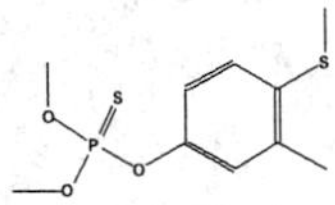

5956

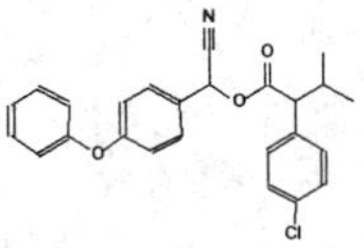

5957

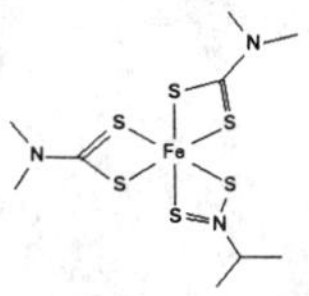

5960

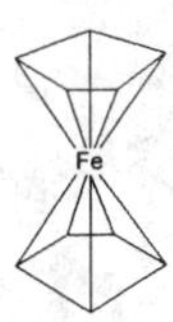

5961

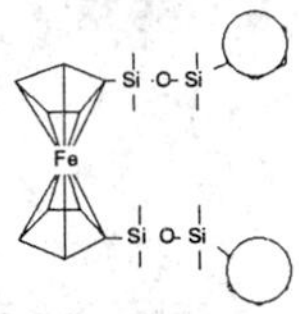

5962

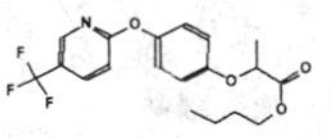

5963

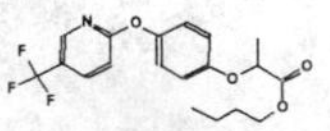

5964

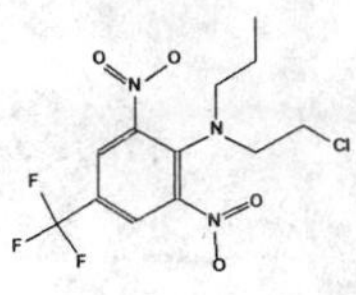

5965

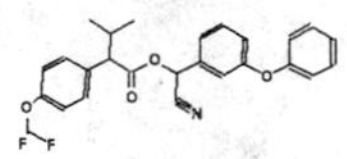

5967

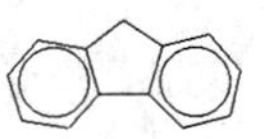

5968

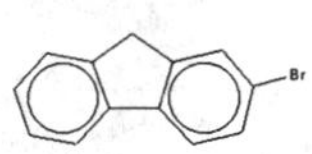

5970

5971

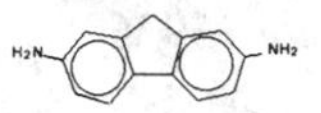

5973

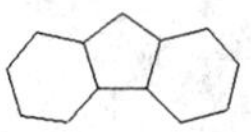

5974

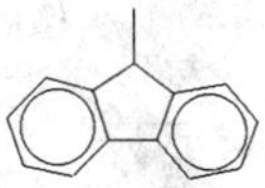

5975

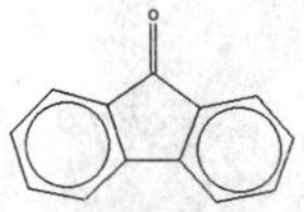

5977

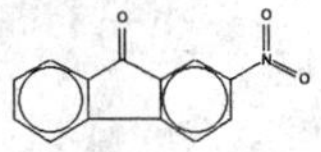

5978

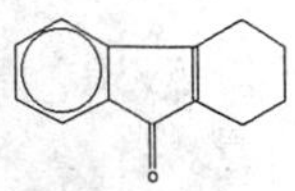

5979

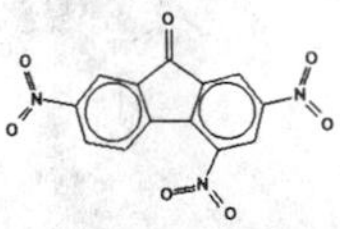

5980

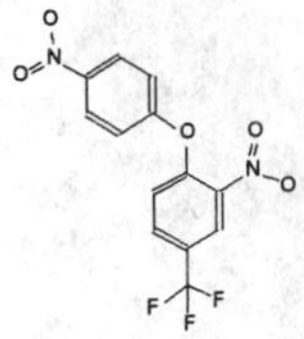

5981

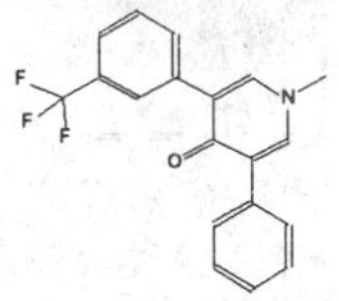

5982

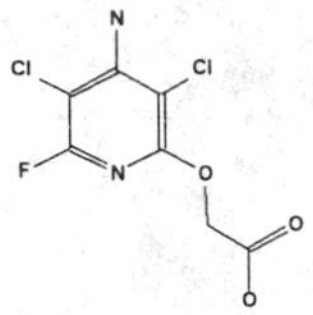

5984

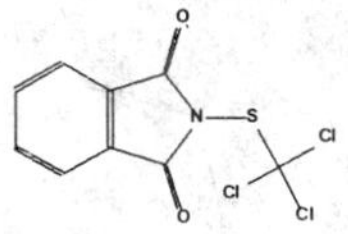

5985

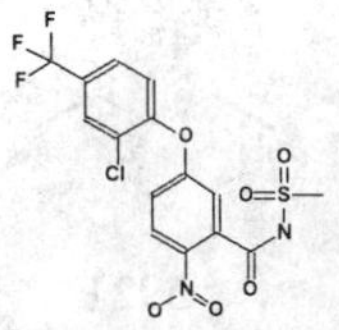

5986

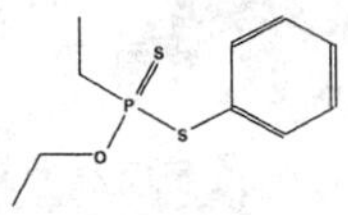

5987

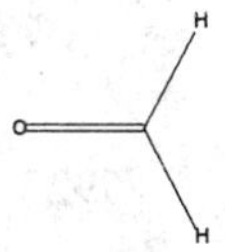

5988

5989

5990

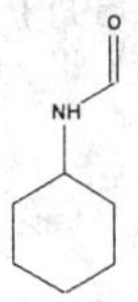

5991

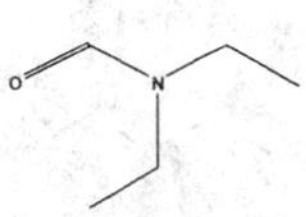

5992

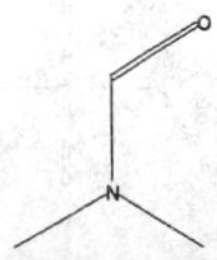

5993

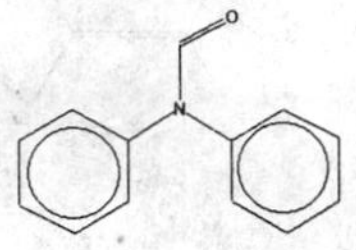

5994

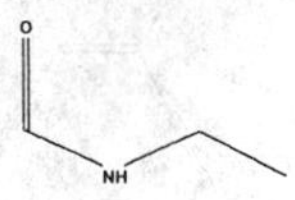

5995

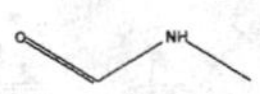

5996

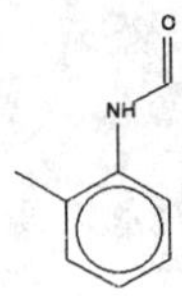

In numeric order as they occur in the Table of Physical Constants of Organic Compounds

5997

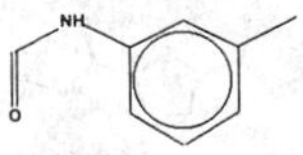

5999

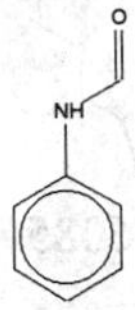

6000

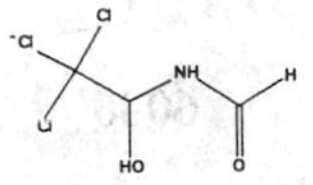

6001

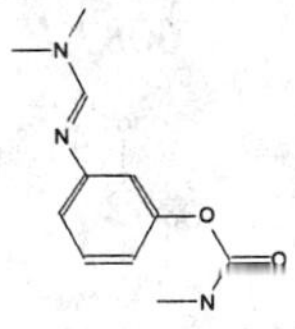

6002

6004

6005

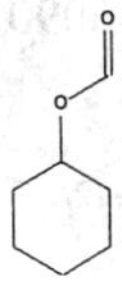

6006

6007

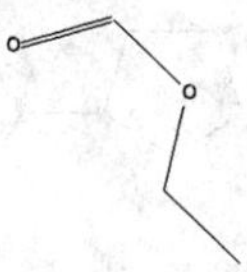

6008

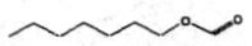

6009

6010

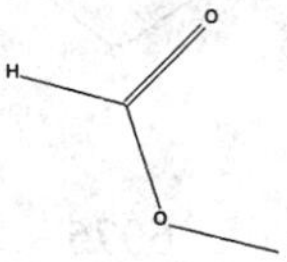

6011

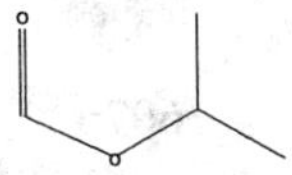

6012

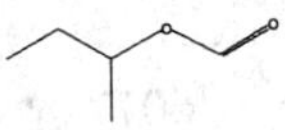

6013

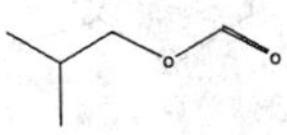

6014

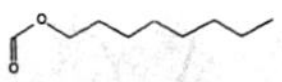

6015

6016

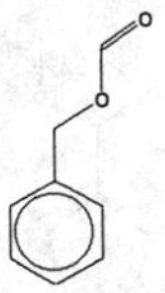

6017

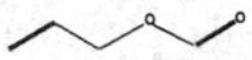

6018

6019

6020

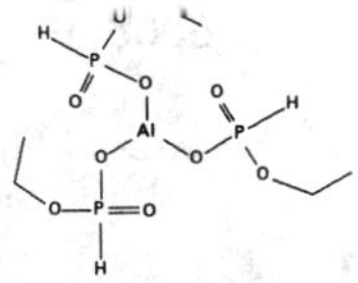

6021

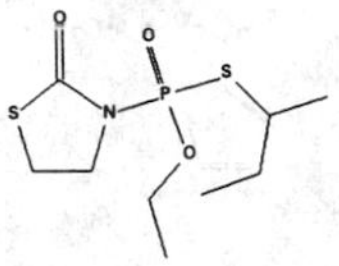

6024

6025

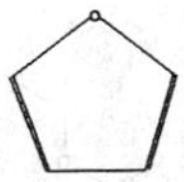

6026

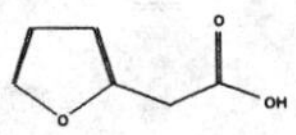

6027

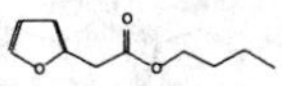

6028

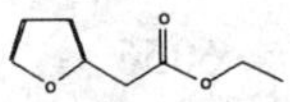

6029

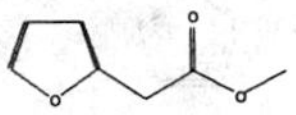

6031

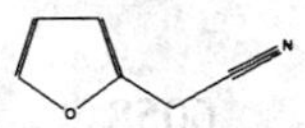

6032

6033

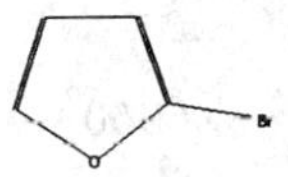

6034

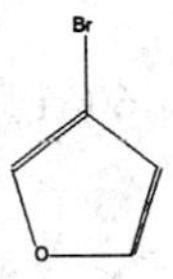

6035

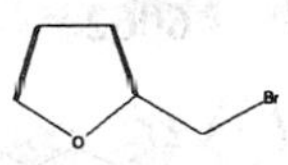

6036

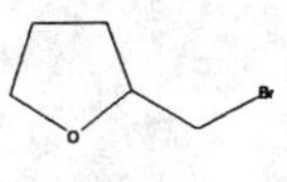

6037

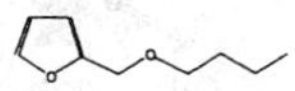

6039

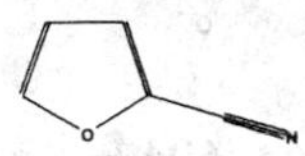

6040

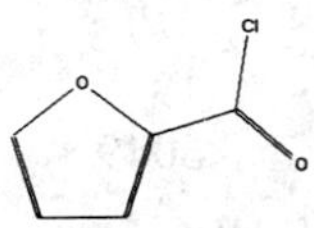

6041

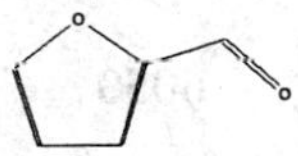

6042

6043

6044

6045

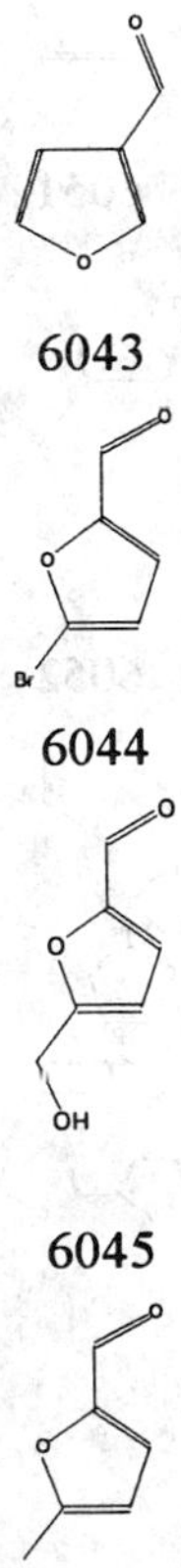

STRUCTURAL FORMULAS OF ORGANIC COMPOUNDS (continued)

In numeric order as they occur in the Table of Physical Constants of Organic Compounds

6046

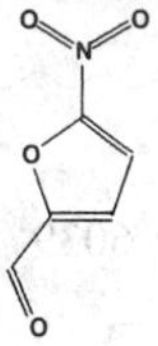

6047

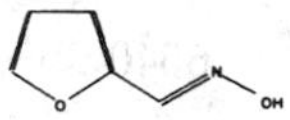

6049

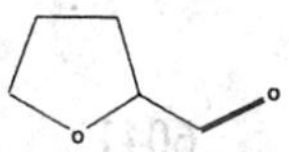

6050

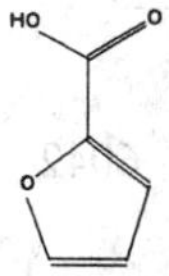

6051

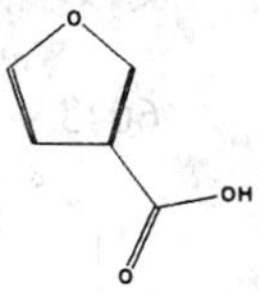

6052

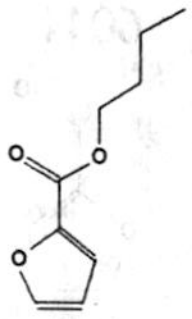

6053

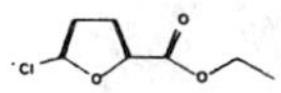

6055

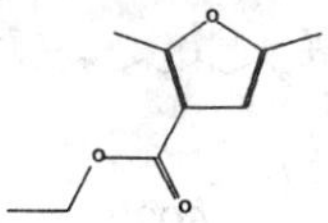

6056

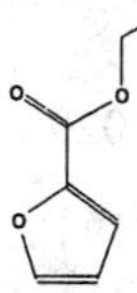

6057

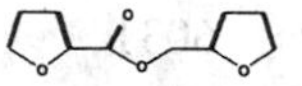

6058

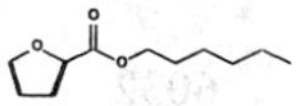

6059

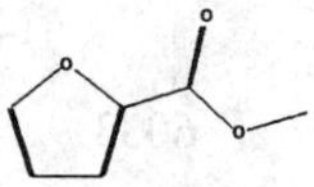

6060

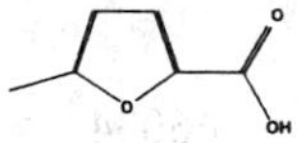

6061

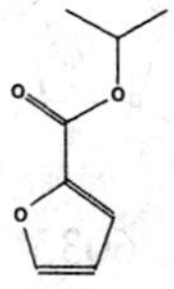

6062

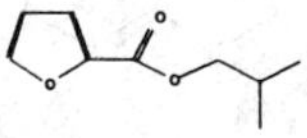

6063

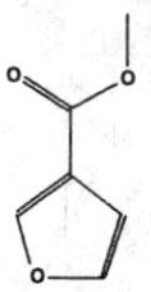

6064

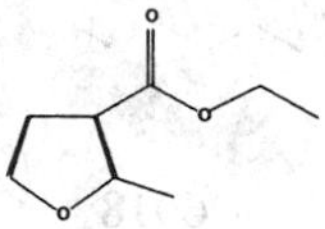

6066

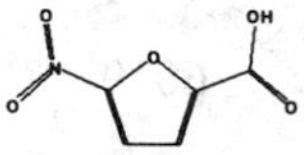

6067

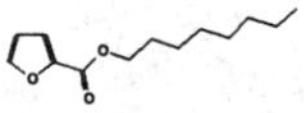

6068

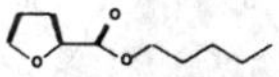

6069

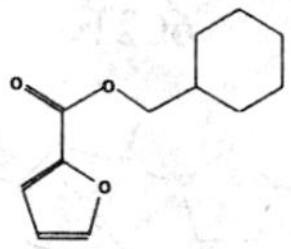

6070

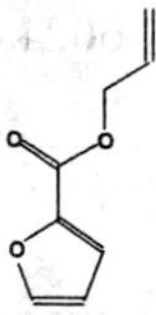

6071

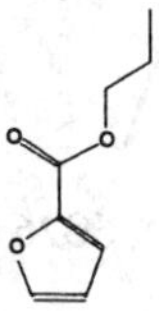

6072

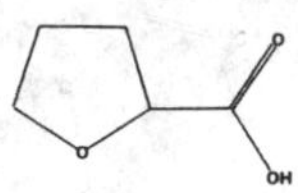

6073

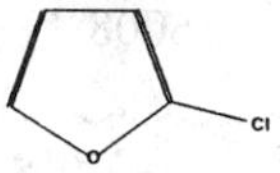

6074

6075

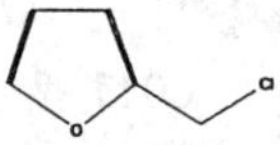

6076

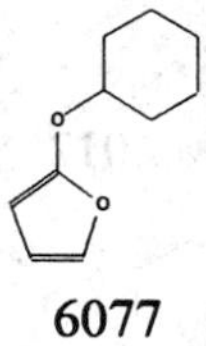

6077

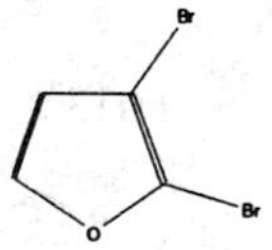

6081

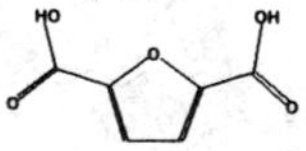

6082

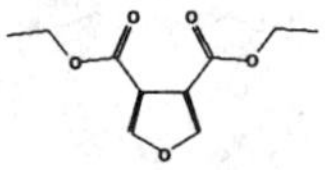

6083

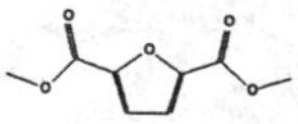

6084

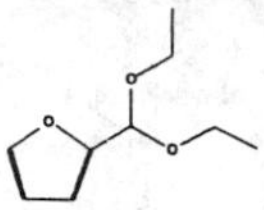

6085

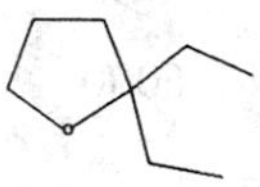

6086

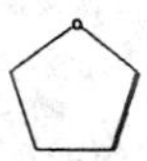

6087

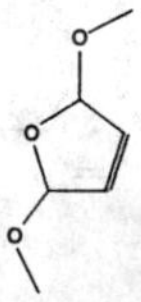

6088

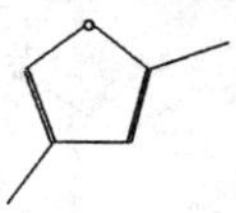

6089

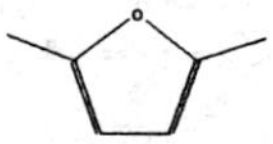

6090

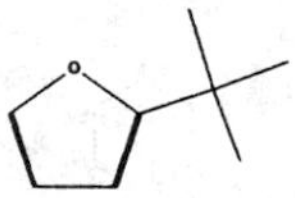

6091

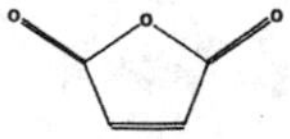

STRUCTURAL FORMULAS OF ORGANIC COMPOUNDS (continued)

In numeric order as they occur in the Table of Physical Constants of Organic Compounds

6092

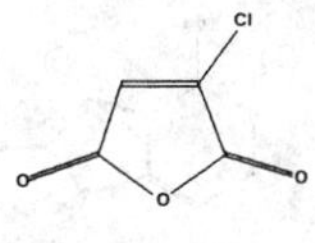

Cl O O O

6093

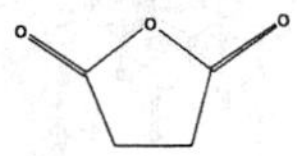

O O O

6094

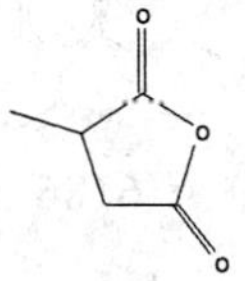

O O O

6095

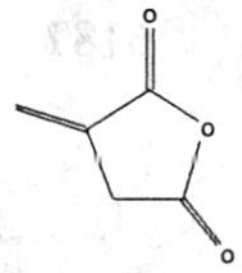

O O O

6096

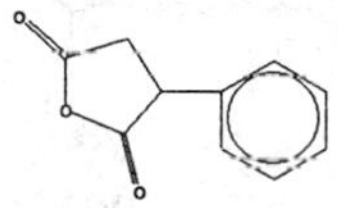

O O O

6097

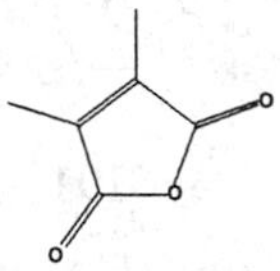

O O O

6098

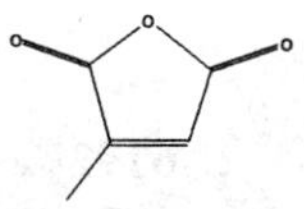

O O O

6099

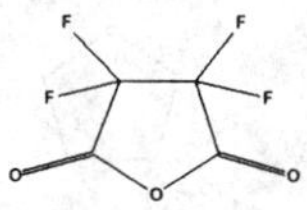

F F F F O O O

6101

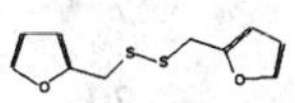

S S O O

6102

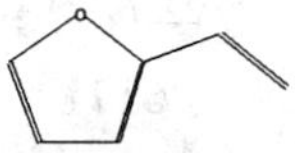

O

6103

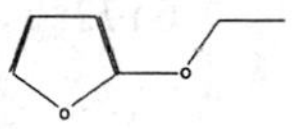

O O

6104

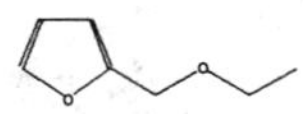

O O

6105

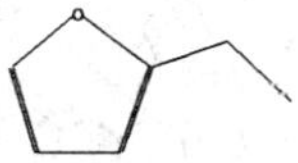

O

6106

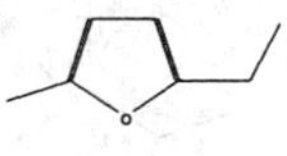

O

6107

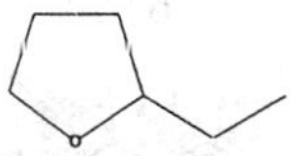

O

6108

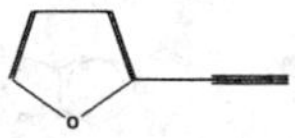

O

6109

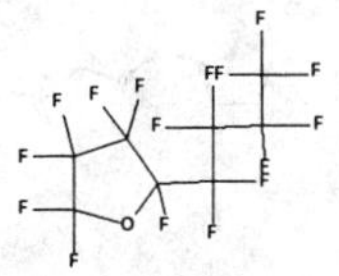

F F F F FF F F F F F F F F F F F O F F

6110

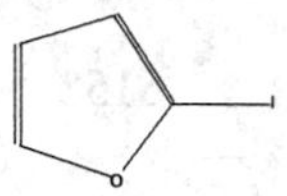

O I

6111

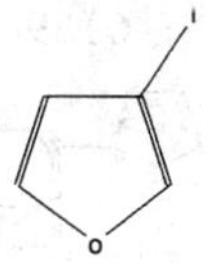

I O

6112

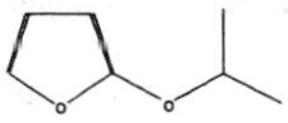

O O

6113

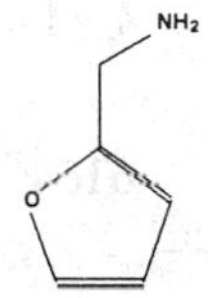

NH2 O

6114

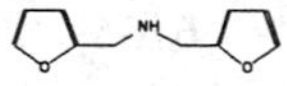

NH O O

6115

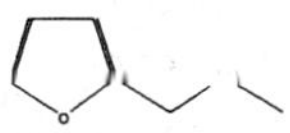

O

6116

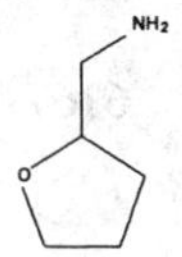

NH2 O

6117

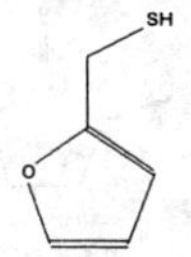

SH O

6118

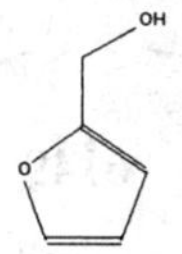

OH O

6119

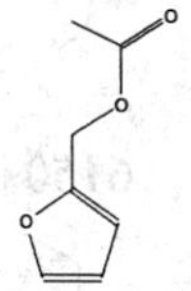

O O O

6120

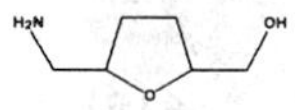

H2N O OH

6121

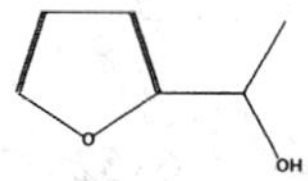

O OH

6122

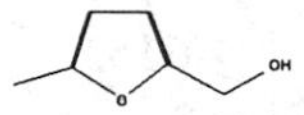

O OH

6123

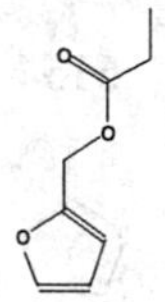

O O O

6124

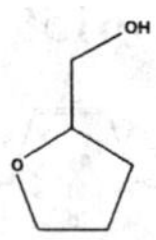

OH O

6125

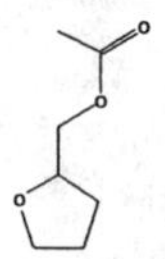

O O O

6126

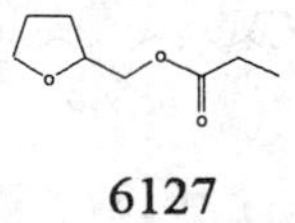

O O O

6127

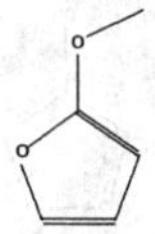

O O

6128

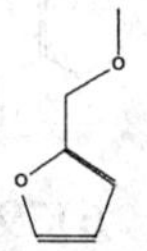

O O

6129

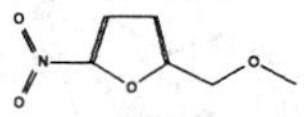

O N O O O

6130

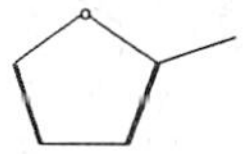

O

6131

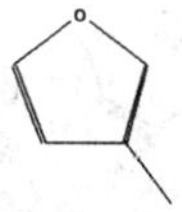

O

6132

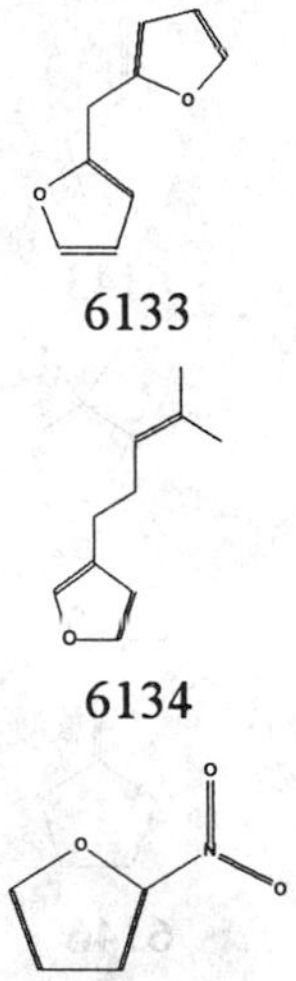

O O

6133

O

6134

O N O O

STRUCTURAL FORMULAS OF ORGANIC COMPOUNDS (continued)

In numeric order as they occur in the Table of Physical Constants of Organic Compounds

6135

6137

6138

6139

6140

6141

6142

6143

6145

6146

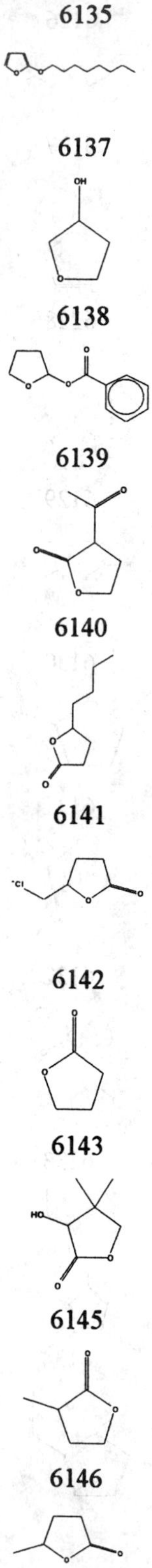

6147

6148

6149

6150

6151

6152

6153

6154

6155

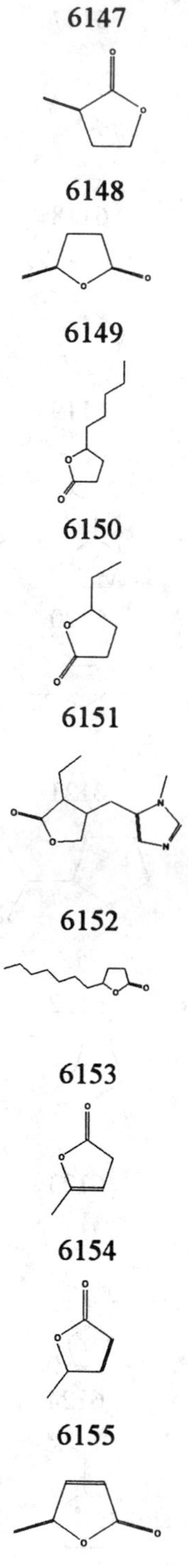

6156

6157

6158

6159

6160

6163

6164

6165

6166

6167

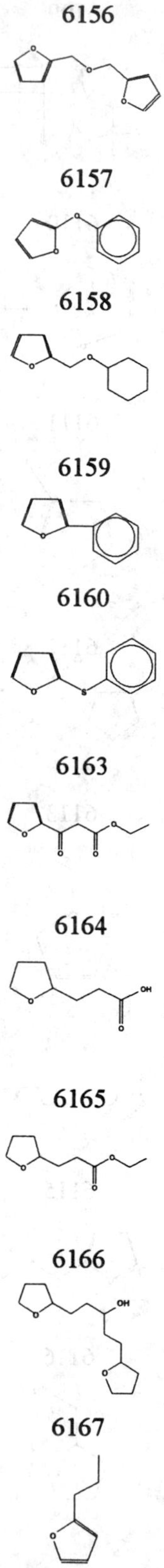

6168

6169

6170

6171

6172

6174

6175

6176

6180

6181

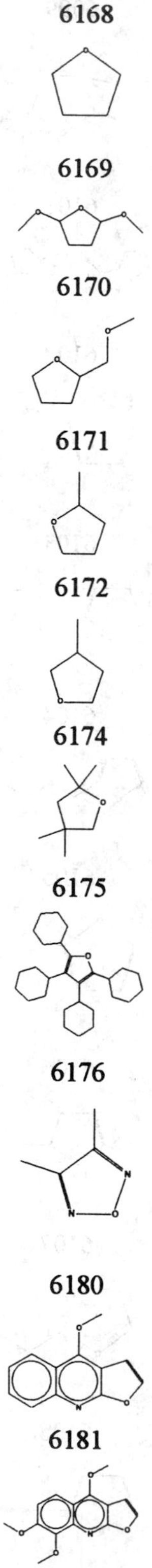

6183

6184

6185

6186

6187

6188

6190

6191

6192

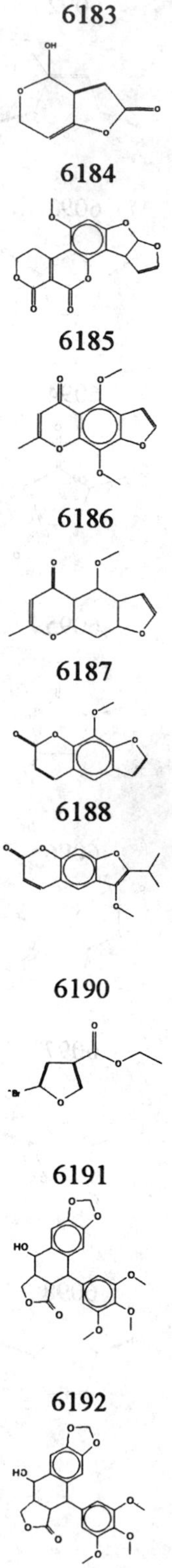

In numeric order as they occur in the Table of Physical Constants of Organic Compounds

6194

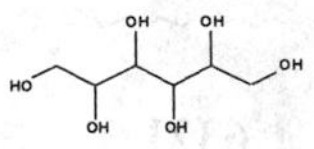

6197

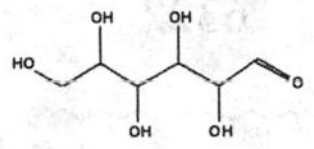

6199

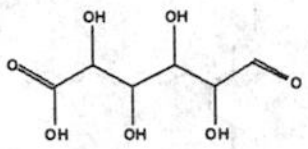

6200

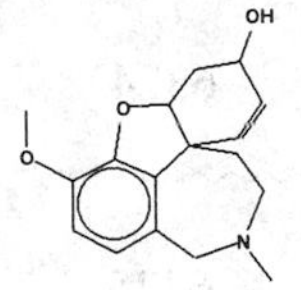

6201

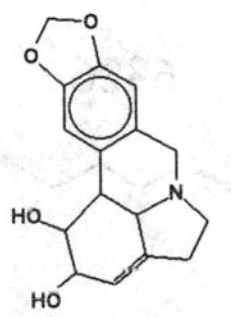

6202

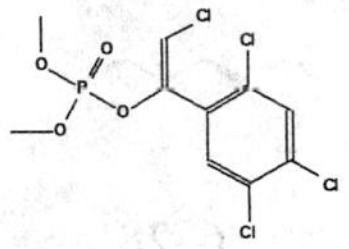

6203

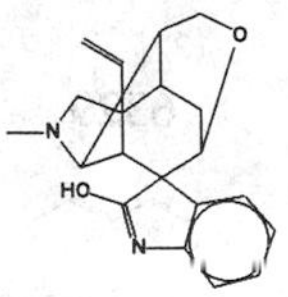

6204

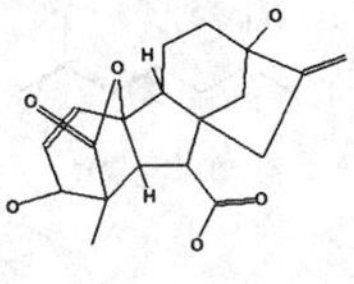

6206

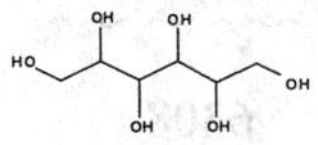

6208

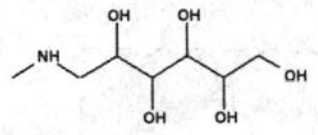

6212

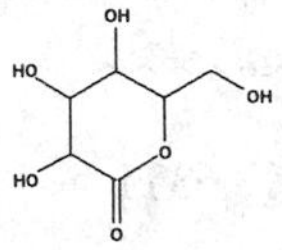

6214

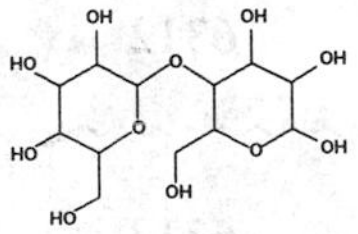

6216

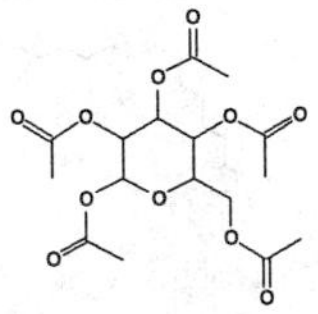

6217

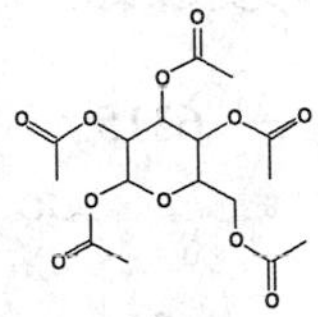

6218

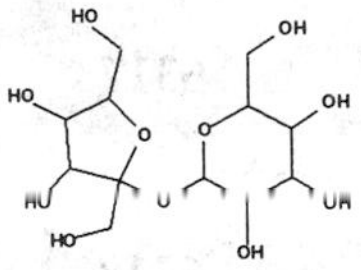

6220

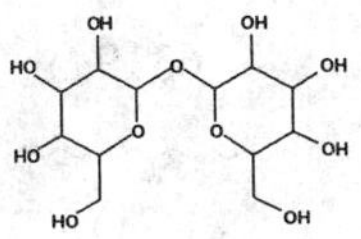

6221

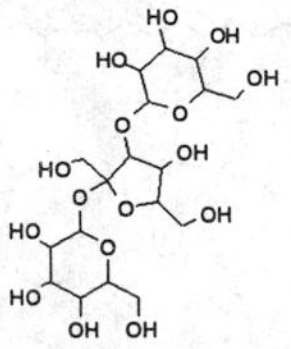

6222

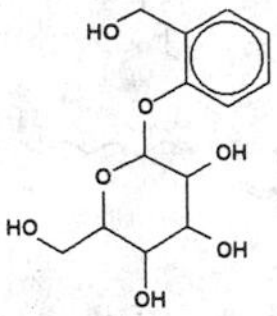

6227

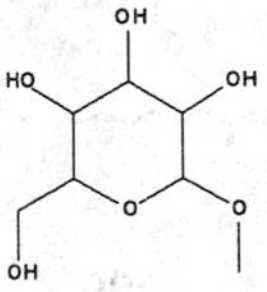

6228

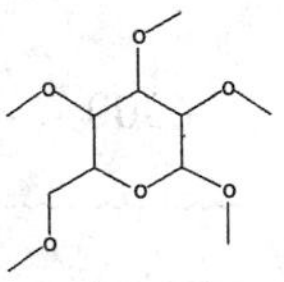

6229

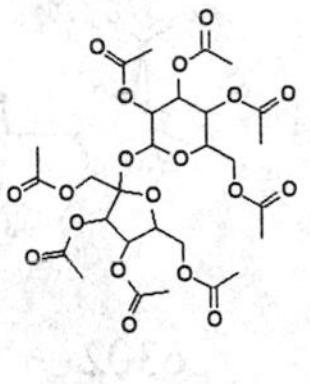

6238

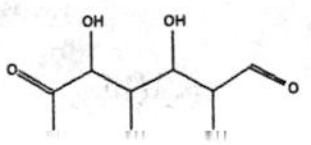

6239

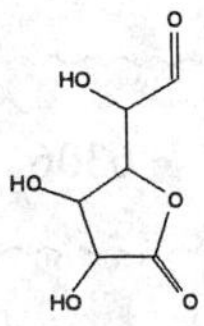

6240

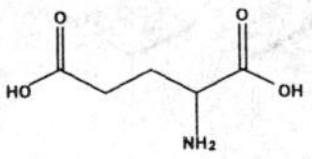

6241

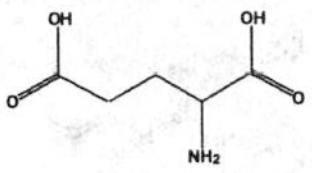

6243

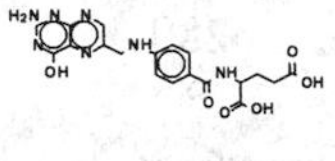

6249

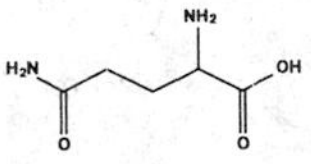

6250

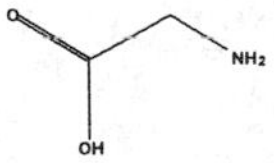

6251

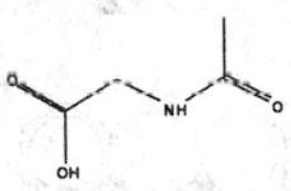

6252

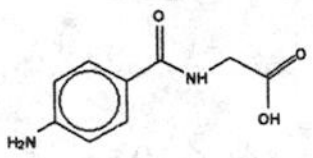

6254

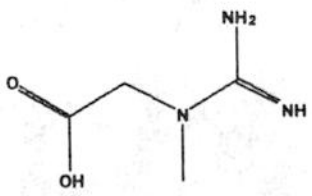

6255

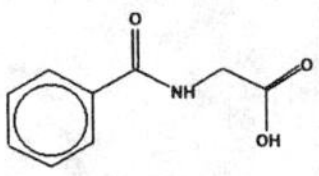

6260

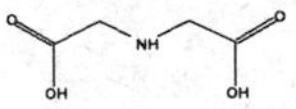

6262

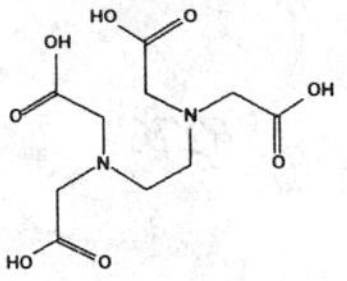

6263

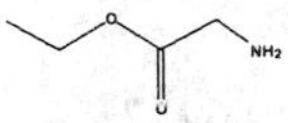

6265

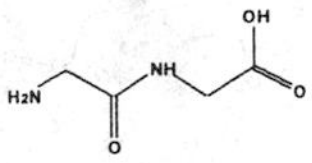

6266

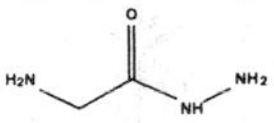

6267

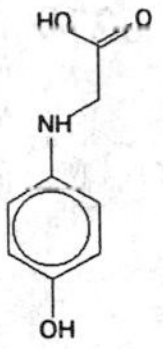

6268

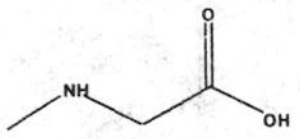

6272

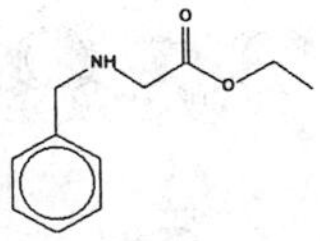

STRUCTURAL FORMULAS OF ORGANIC COMPOUNDS (continued)

In numeric order as they occur in the Table of Physical Constants of Organic Compounds

6275

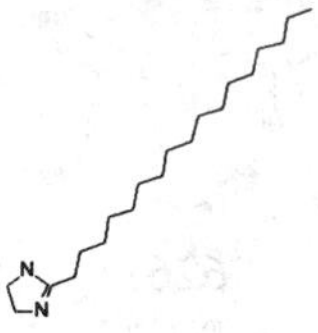

N
N

6276

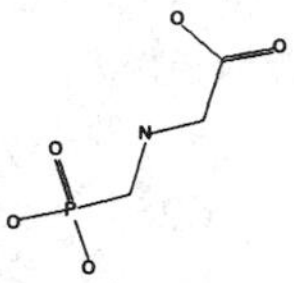

O
O
N
O
P
O
O

6277

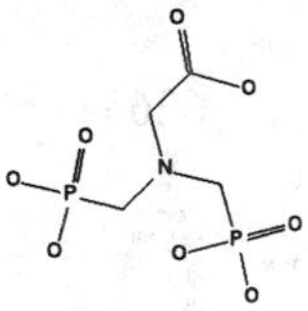

O
O
O
N
O
P
O
O
P
O
O

6278

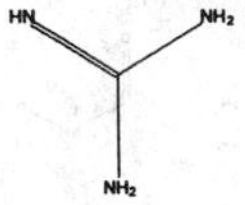

HN
NH₂
NH₂

6279

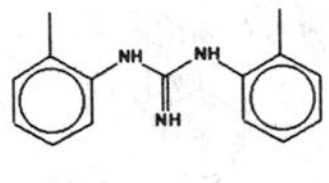

NH
NH
NH

6280

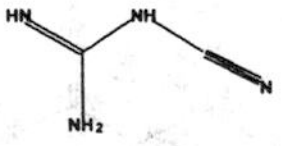

HN
NH
NH₂
N

6282

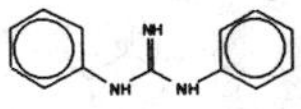

NH
NH
NH

6284

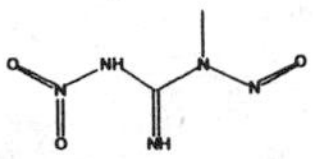

O
NH
N
O
N
O
NH

6287

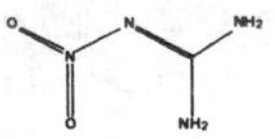

O
N
N
NH₂
O
NH₂

6288

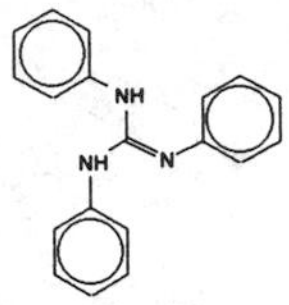

NH
N
NH

6289

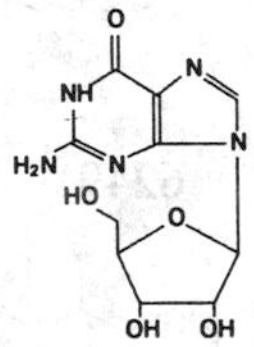

O
NH
N
N
H₂N
N
N
HO
O
OH
OH

6291

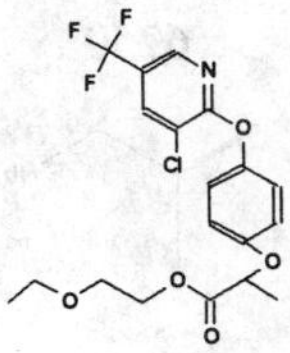

F
F
F
N
O
Cl
O
O
O
O
O

6292

O

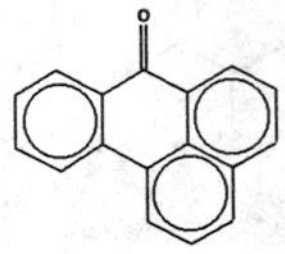

6294

6295

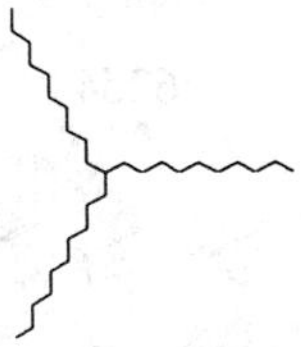

6296

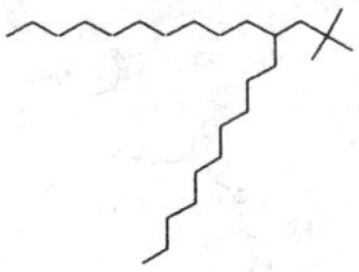

6297

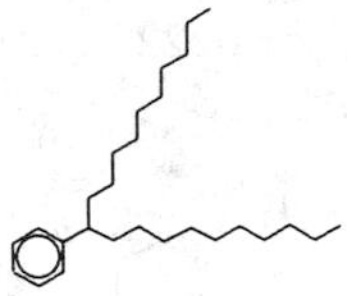

6298

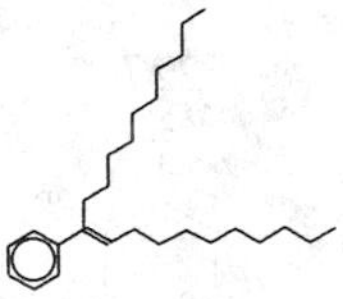

6299

6300

S

6301

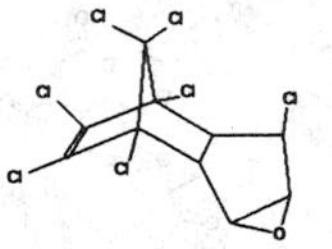

Cl
Cl
Cl
Cl
Cl
Cl
Cl
Cl
O

6302

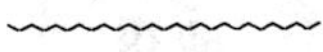

6304

O

6305

NH₂

6306

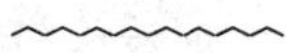

6307

Br

6308

Br
Br

6309

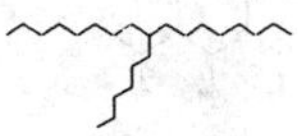

6310

6311

6312

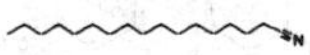

N

6313

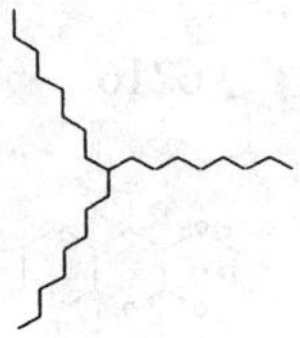

6314

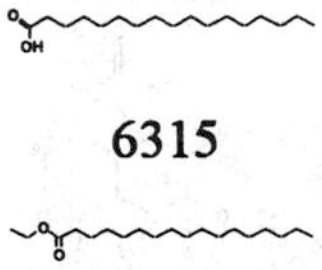

O
OH

6315

O
O

6316

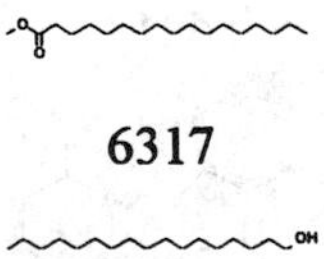

O
O

6317

OH

6318

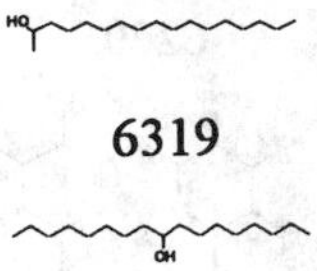

HO

6319

OH

6320

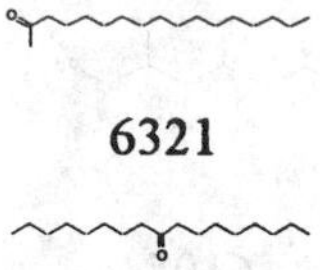

O

6321

O

6322

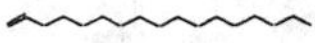

6323

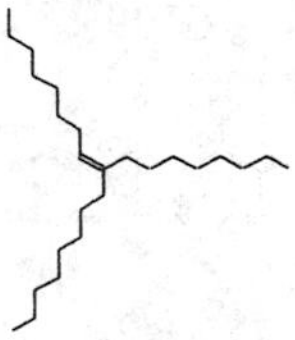

6324

OH
O

6325

O

6326

C

6327

6328

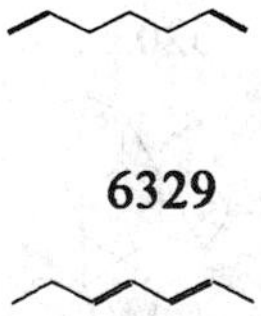

6329

6330

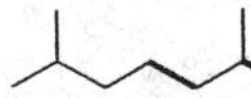

STRUCTURAL FORMULAS OF ORGANIC COMPOUNDS (continued)

In numeric order as they occur in the Table of Physical Constants of Organic Compounds

6331

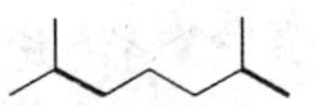

6332

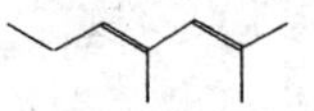

6333

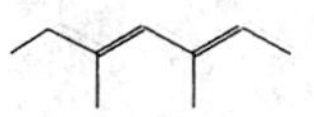

6334

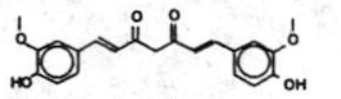

O O HO OH

6335

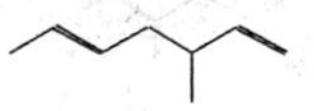

6336

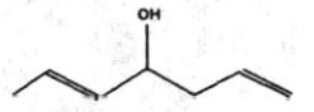

OH

6337

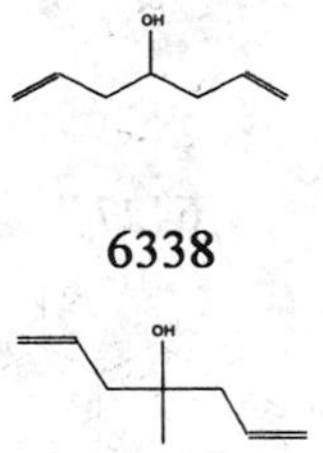

OH

6338

OH

6339

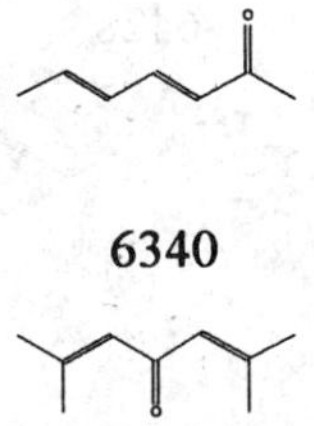

O

6340

O

6341

6342

6343

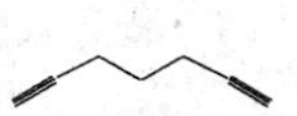

6344

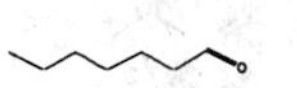

O

6345

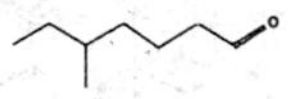

O

6346

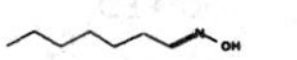

OH

6347

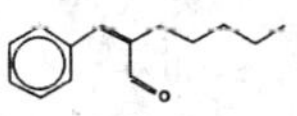

O

6348

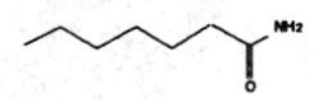

NH_2 O

6349

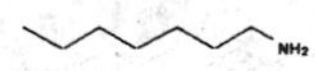

NH_2

6350

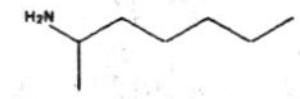

H_2N

6351

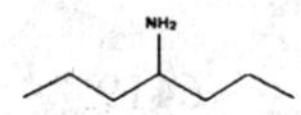

NH_2

6352

NH

6353

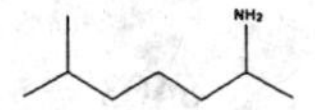

NH_2

6354

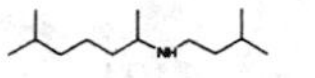

NH

6355

6356

Br

6357

Br

6358

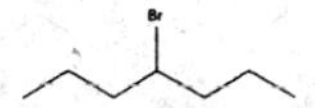

Br

6359

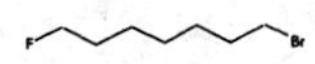

F Br

6360

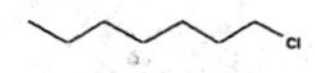

Cl

6361

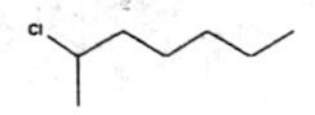

Cl

6362

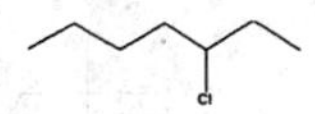

Cl

6363

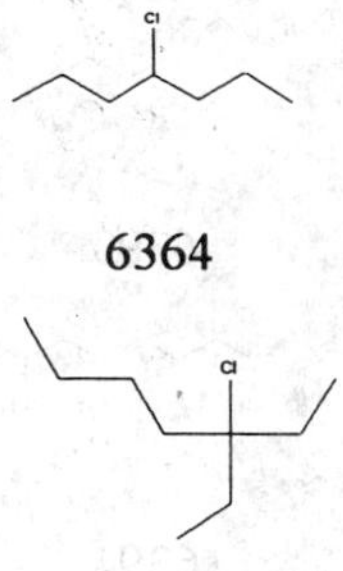

Cl

6364

Cl

6365

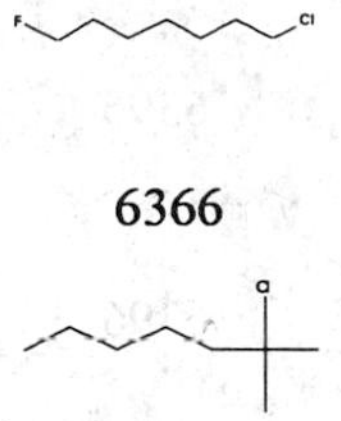

F Cl

6366

Cl

6367

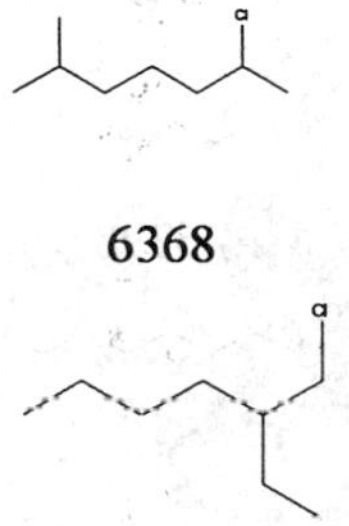

Cl

6368

Cl

6369

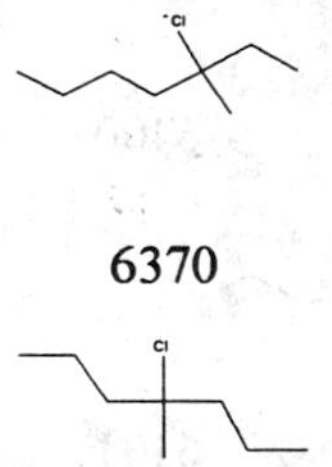

Cl

6370

Cl

6371

H_2N NH_2

6372

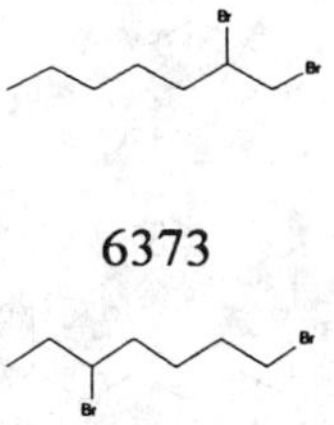

Br Br

6373

Br Br

6374

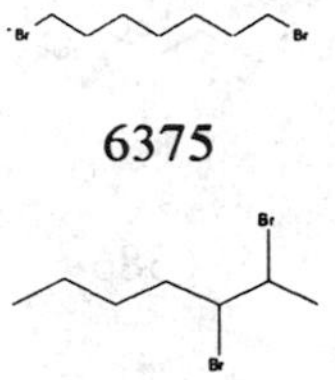

Br Br

6375

Br Br

6376

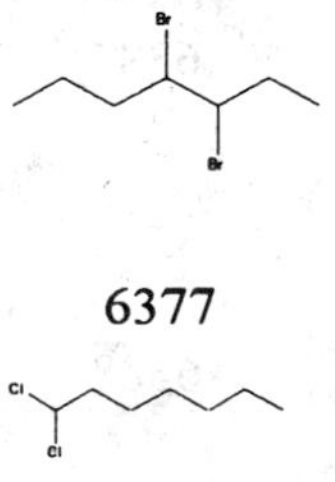

Br Br

6377

Cl Cl

6378

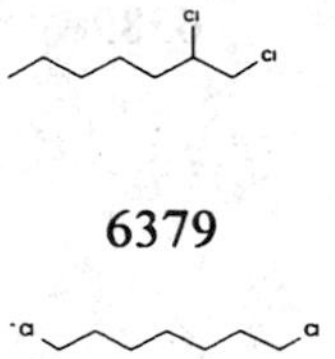

Cl Cl

6379

Cl Cl

6380

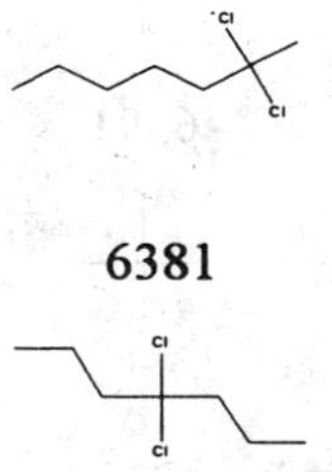

Cl Cl

6381

Cl Cl

STRUCTURAL FORMULAS OF ORGANIC COMPOUNDS (continued)

In numeric order as they occur in the Table of Physical Constants of Organic Compounds

6383

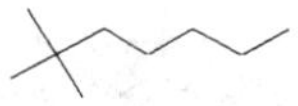

6384

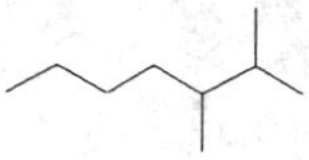

6385

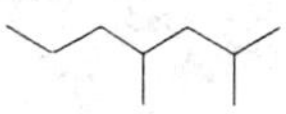

6386

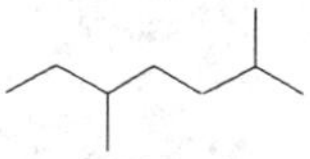

6387

6388

6389

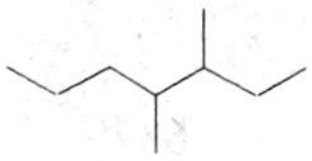

6390

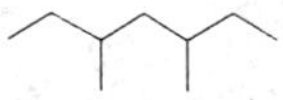

6391

6392

N N

6393

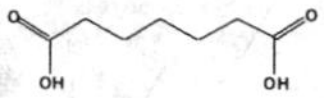

O O OH OH

6394

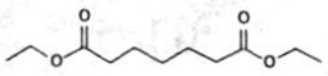

6395

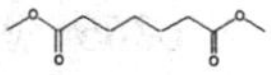

6396

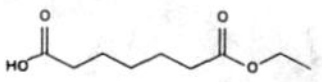

HO

6397

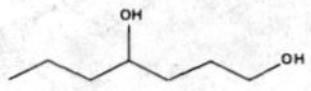

OH OH

6398

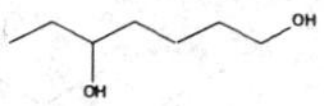

OH OH

6399

HO OH

6400

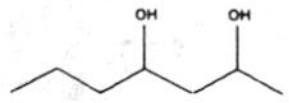

OH OH

6401

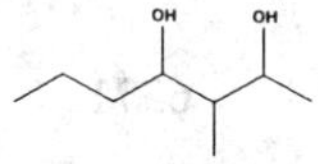

OH OH

6402

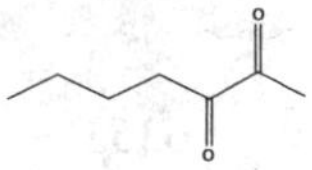

O O

6403

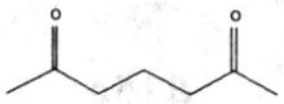

O O

6404

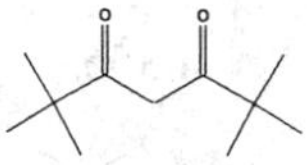

O O

6405

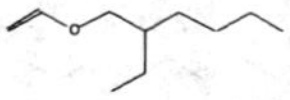

O

6406

O

6407

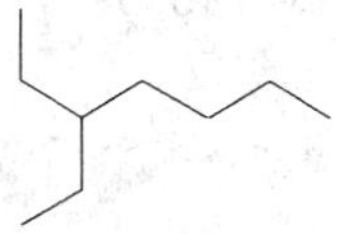

6408

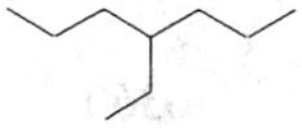

6409

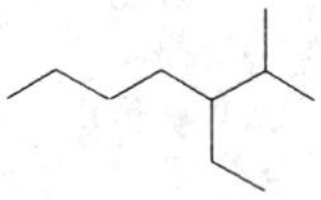

6410

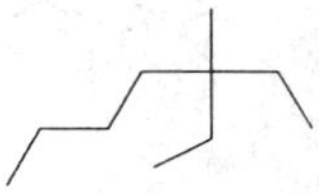

6411

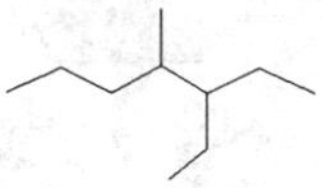

6412

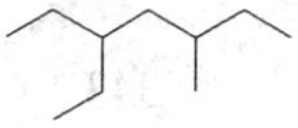

6413

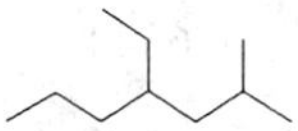

6414

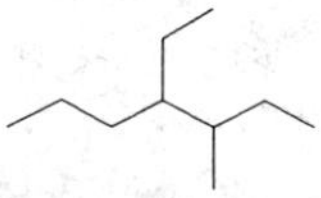

6415

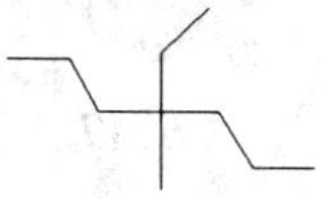

6416

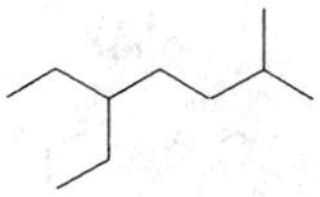

6417

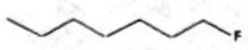

F

6418

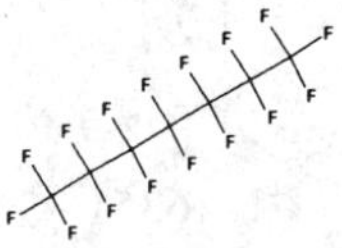

F F F F F F F F F F F F F F F F F

6419

I

6420

I

6421

O

6422

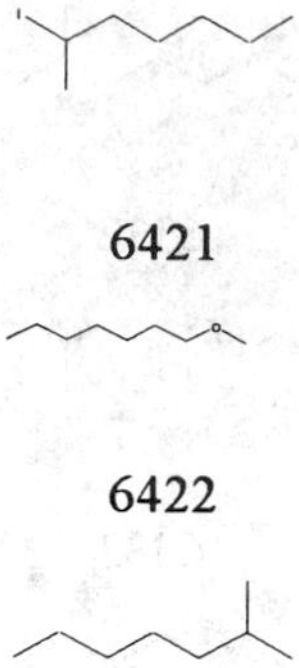

6423

6424

6425

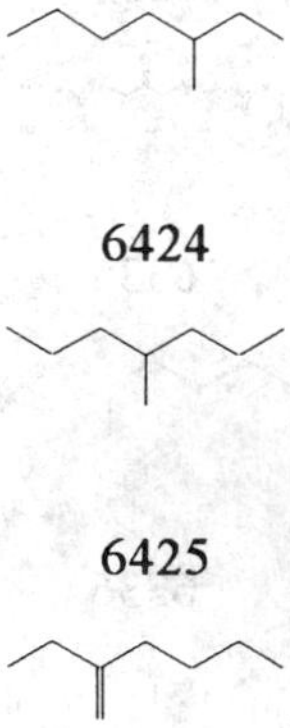

6426

6427

6428

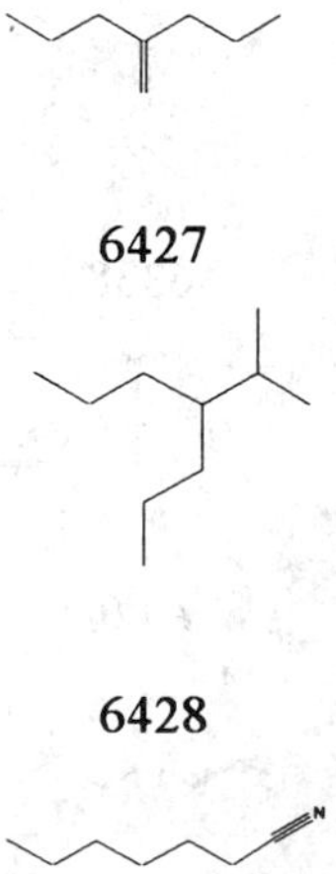

N

STRUCTURAL FORMULAS OF ORGANIC COMPOUNDS (continued)

In numeric order as they occur in the Table of Physical Constants of Organic Compounds

6429

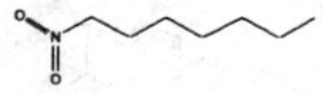

6430

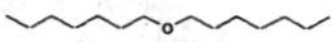

6431

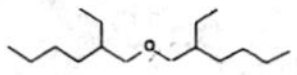

6432

6433

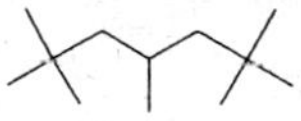

6434

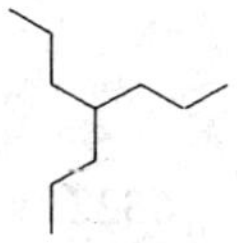

6435

6436

6437

6438

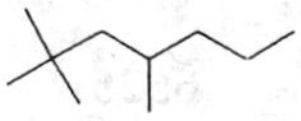

6439

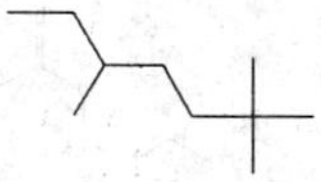

6440

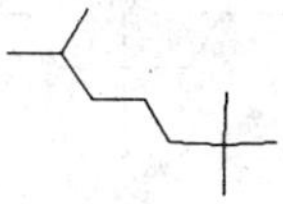

6441

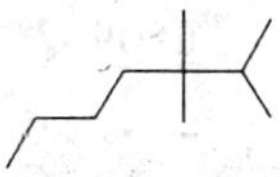

6442

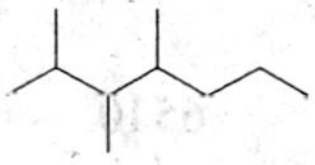

6443

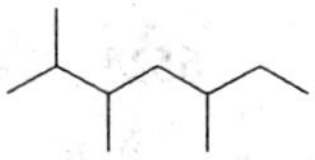

6444

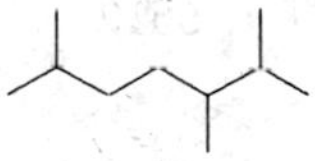

6445

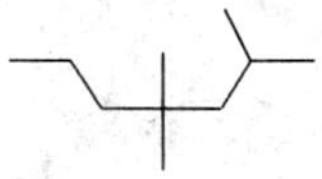

6446

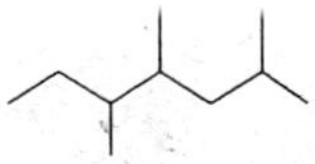

6447

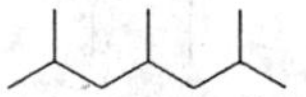

6448

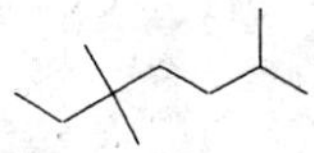

6449

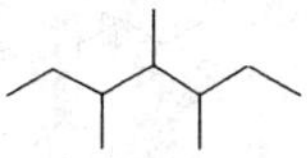

6450

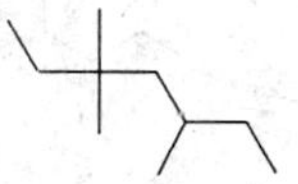

6451

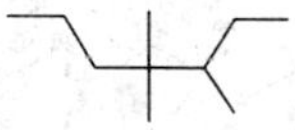

6452

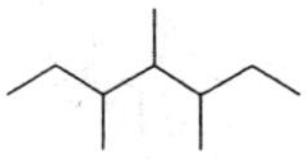

6453

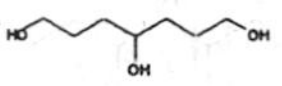

6454

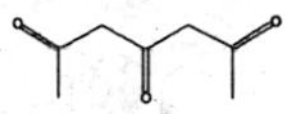

6455

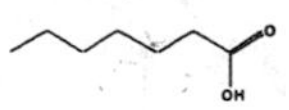

6456

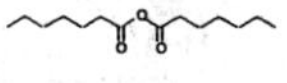

6457

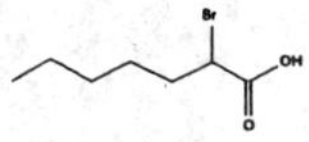

6459

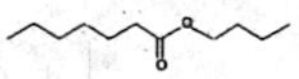

6461

6462

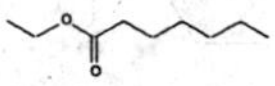

6463

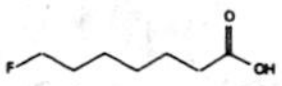

6464

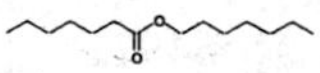

6465

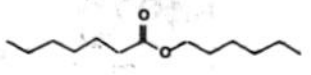

6466

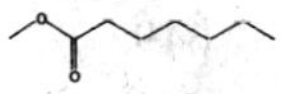

6467

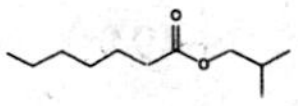

6468

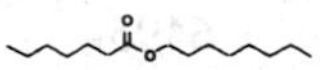

6469

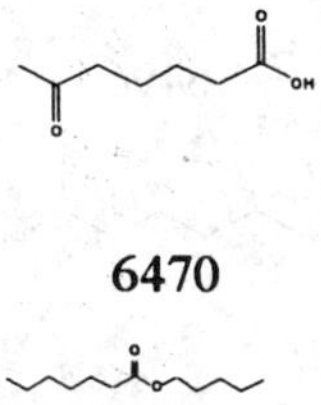

6470

6471

6472

6473

6474

6475

6476

6477

6478

6479

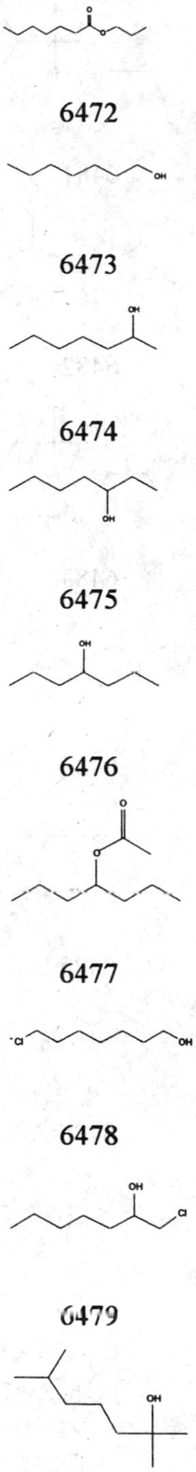

STRUCTURAL FORMULAS OF ORGANIC COMPOUNDS (continued)

In numeric order as they occur in the Table of Physical Constants of Organic Compounds

6480 OH

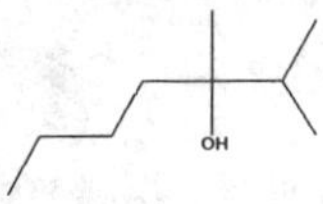

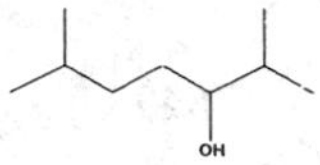

6481 OH

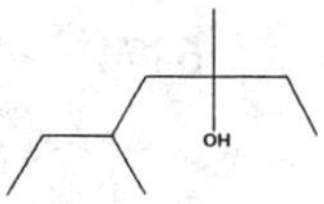

6482 OH

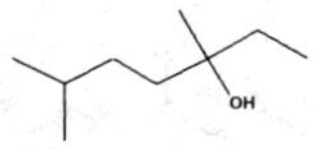

6483 OH

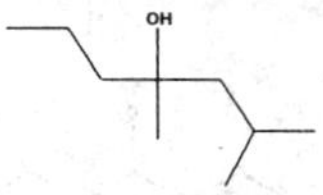

6484 OH

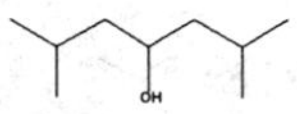

6485 OH

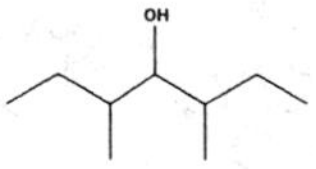

6486 OH

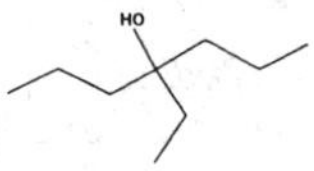

6487 HO

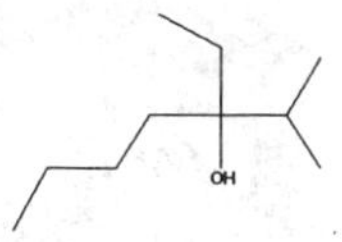

6488 OH

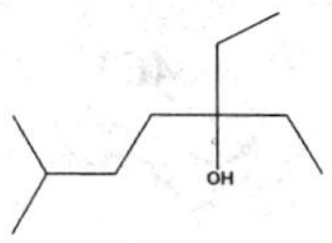

6489 OH

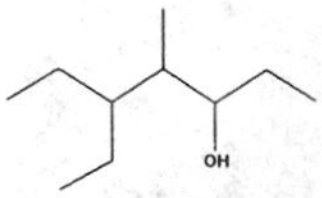

6490 OH

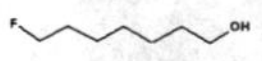

6491 F OH

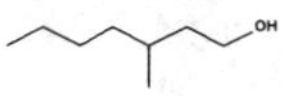

6492 OH

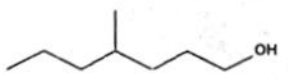

6493 OH

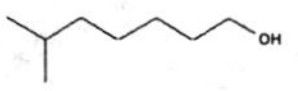

6494 OH

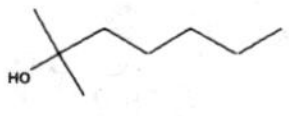

6495 HO

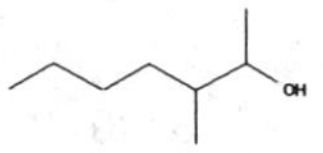

6496 OH

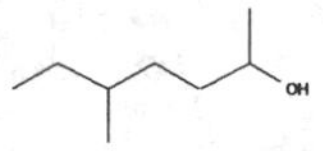

6497 OH

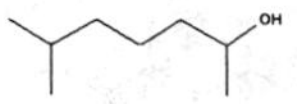

6498 OH

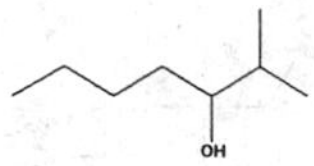

6499 OH

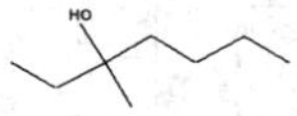

6500 HO

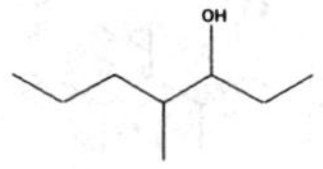

6501 OH

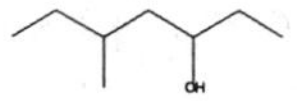

6502 OH

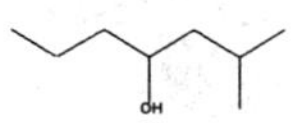

6503 OH

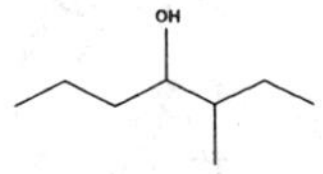

6504 OH

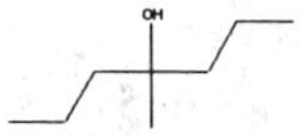

6505 OH

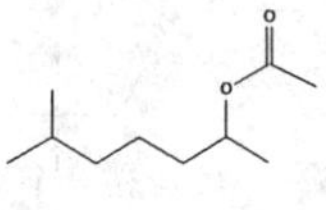

6506 O O

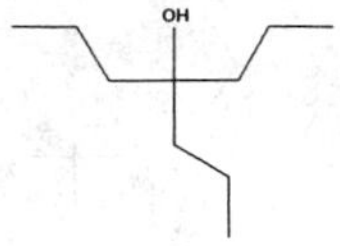

6508 OH

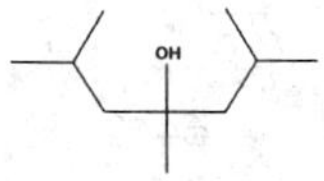

6509 OH

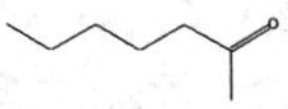

6510 O

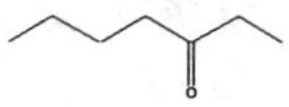

6511 O

6512 O

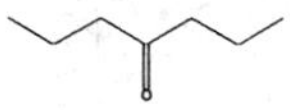

6513 O Cl

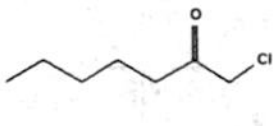

6514 O

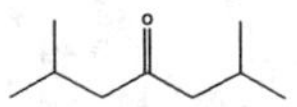

6515 O

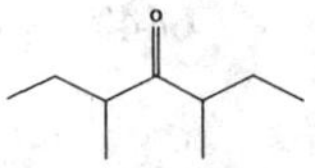

6517 O OH

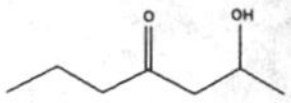

6518 OH O

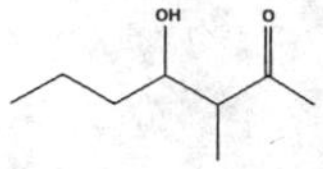

6519 OH O

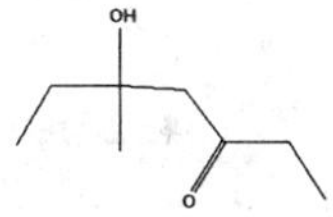

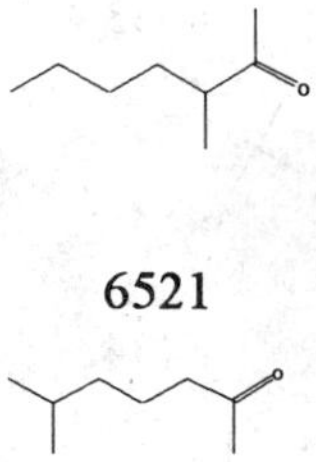

6520 O

6521 O

6522 O

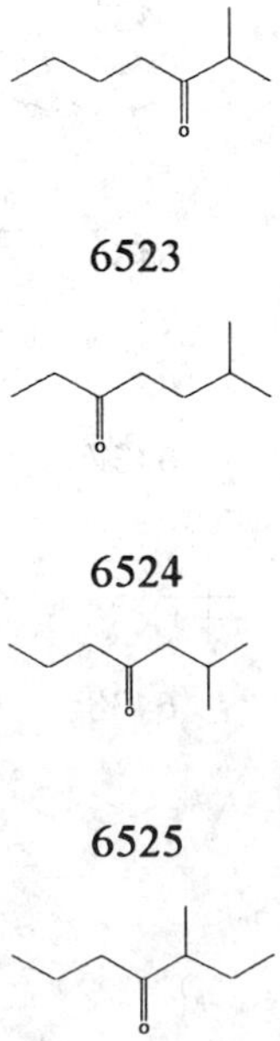

6523 O

6524 O

6525 O

STRUCTURAL FORMULAS OF ORGANIC COMPOUNDS (continued)

In numeric order as they occur in the Table of Physical Constants of Organic Compounds

6526

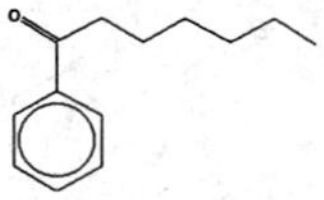

O

6527

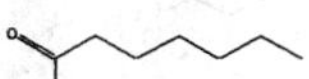

O
Cl

6528

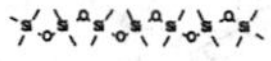

6529

6530

O

6531

O

6532

NH

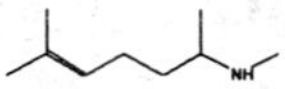

6533

6534

6535

6536

6537

6538

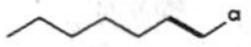

Cl

6539

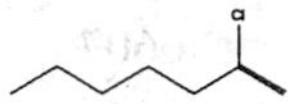

Cl

6540

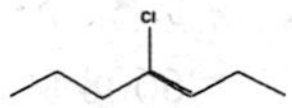

Cl

6541

Cl

6542

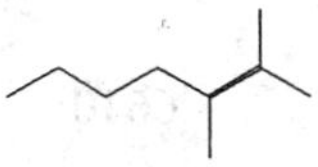

6543

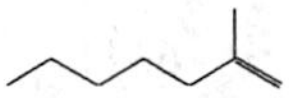

6544

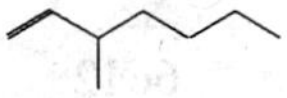

6545

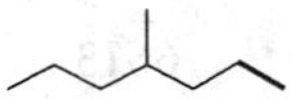

6546

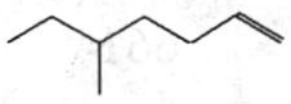

6547

6548

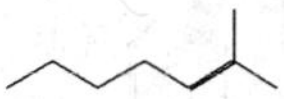

6549

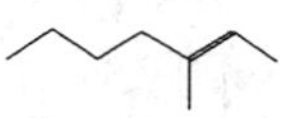

6550

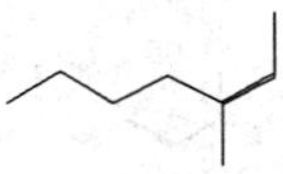

6551

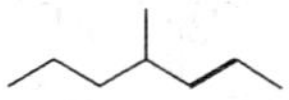

6552

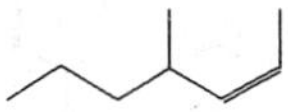

6553

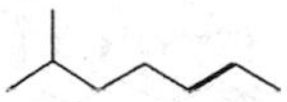

6554

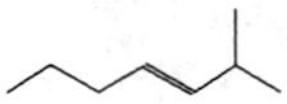

6555

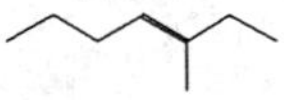

6556

6557

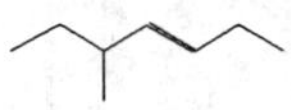

6558

6559

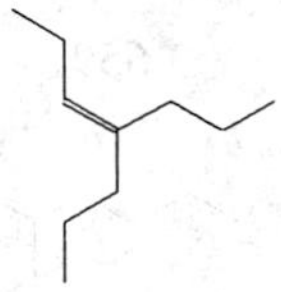

6560

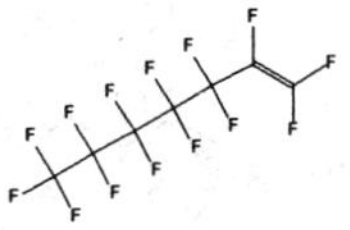

F F F F F F F F F F F F F

6561

OH
O

6562

OH
O

6563

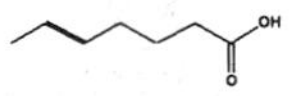

OH
O

6564

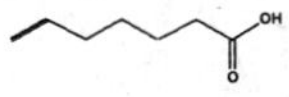

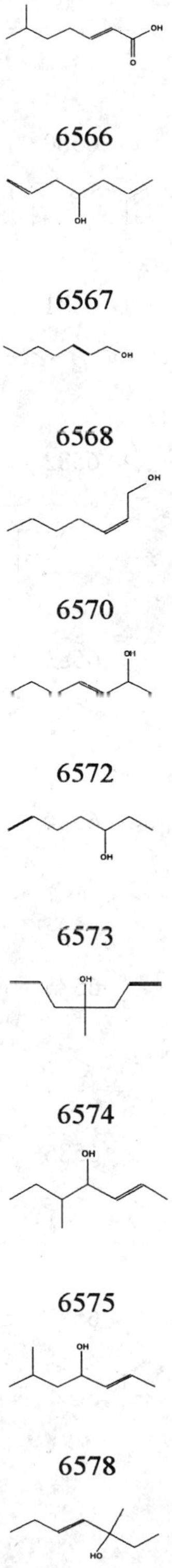

OH
O

6565

OH
O

6566

OH

6567

OH

6568

OH

6570

OH

6572

OH

6573

OH

6574

OH

6575

OH

6578

HO

STRUCTURAL FORMULAS OF ORGANIC COMPOUNDS (continued)

In numeric order as they occur in the Table of Physical Constants of Organic Compounds

6579

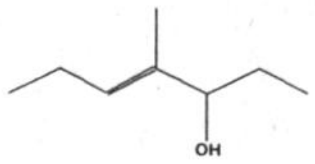

OH

6580

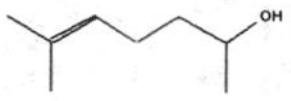

OH

6581

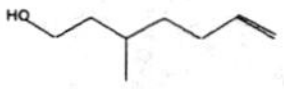

HO

6582

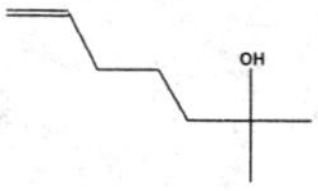

OH

6583

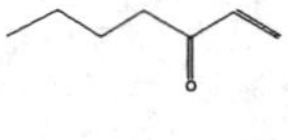

O

6584

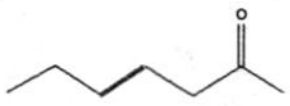

O

6585

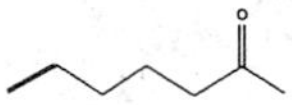

O

6586

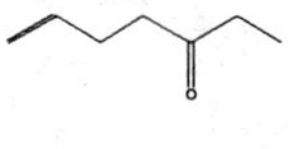

O

6587

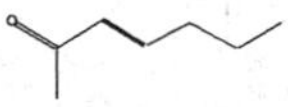

O

6588

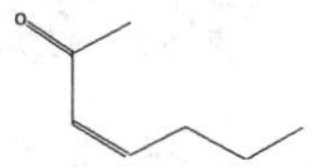

O

6589

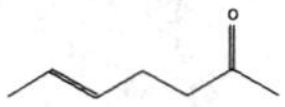

O

6590

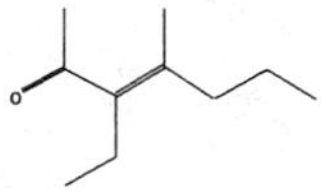

O

6591

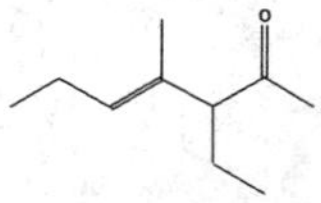

O

6592

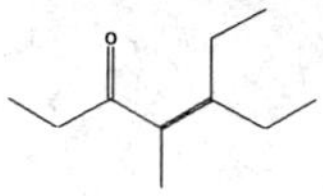

O

6593

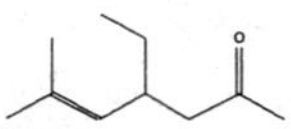

O

6594

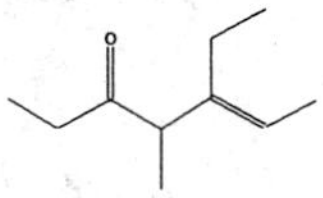

O

6595

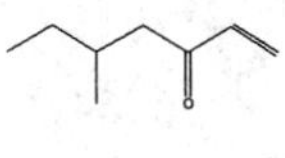

O

6596

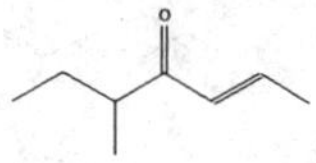

O

6597

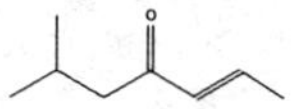

O

6598

O

6599

O

6600

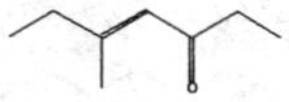

O

6601

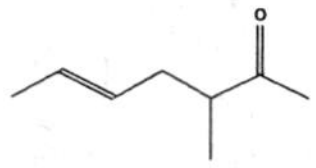

O

6602

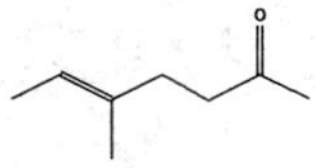

O

6603

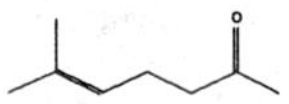

O

6604

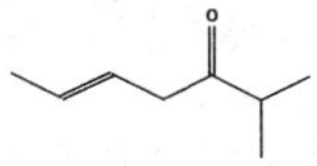

O

6605

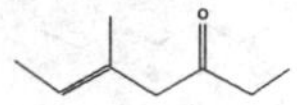

O

6606

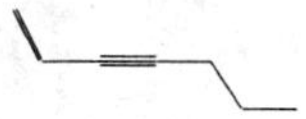

6607

6608

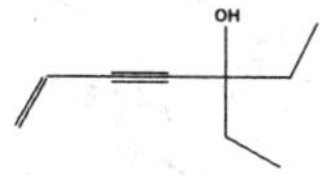

OH

6609

6610

6611

6612

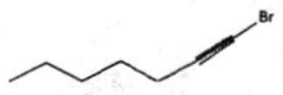

Br

6613

Br

6614

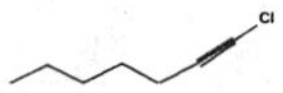

Cl

6615

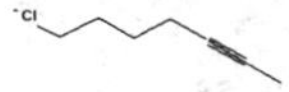

Cl

6616

Cl

6617

6618

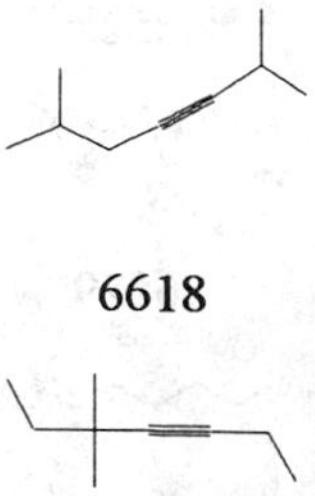

6619

6620

6621

6622

6623

6625

STRUCTURAL FORMULAS OF ORGANIC COMPOUNDS (continued)

In numeric order as they occur in the Table of Physical Constants of Organic Compounds

6626

6627

6628

6629

6630

6631

6632

6633

6634

6635

6636

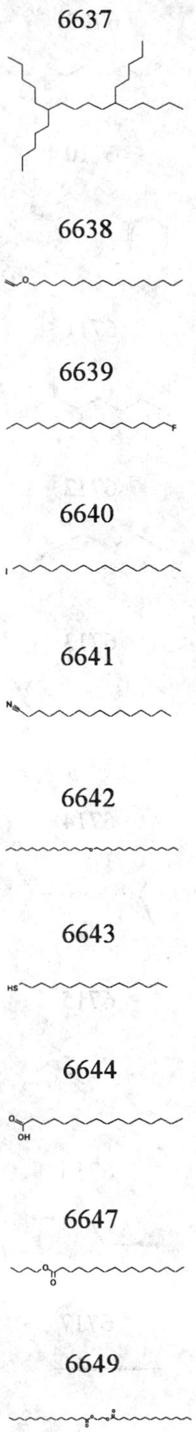

6637

6638

6639

6640

6641

6642

6643

6644

6647

6649

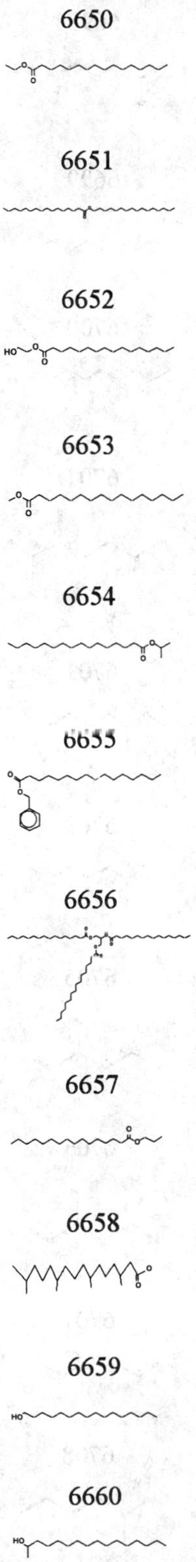

6650

6651

6652

6653

6654

6655

6656

6657

6658

6659

6660

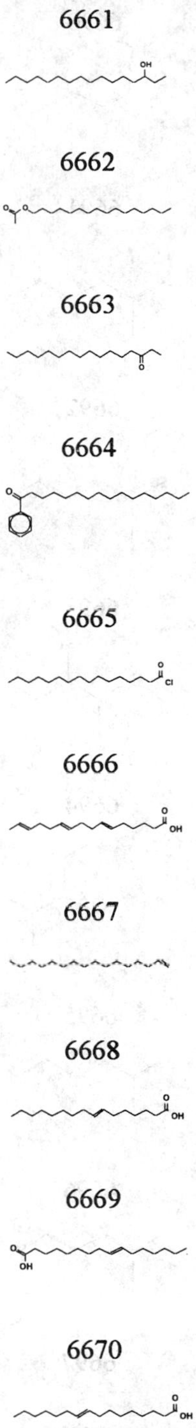

6661

6662

6663

6664

6665

6666

6667

6668

6669

6670

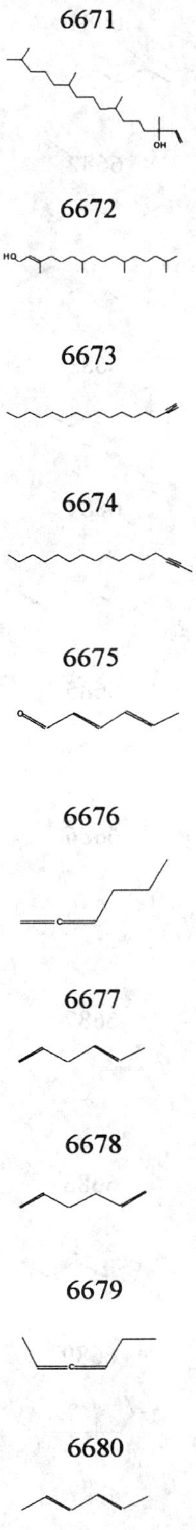

6671

6672

6673

6674

6675

6676

6677

6678

6679

6680

STRUCTURAL FORMULAS OF ORGANIC COMPOUNDS (continued)

In numeric order as they occur in the Table of Physical Constants of Organic Compounds

6681

6682

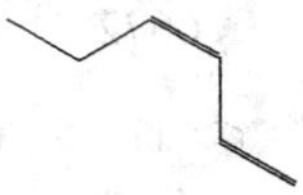

6683

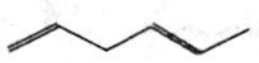

6684

6685

6686

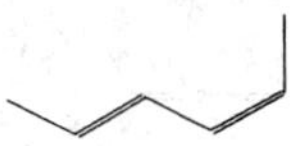

6687

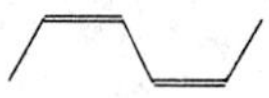

6688

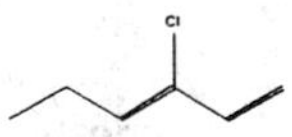

Cl

6689

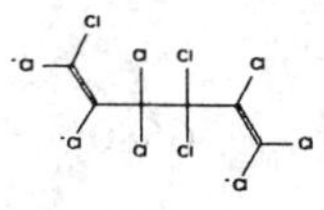

Cl Cl Cl Cl Cl Cl Cl Cl Cl Cl

6690

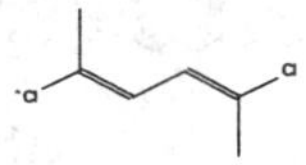

Cl Cl

6691

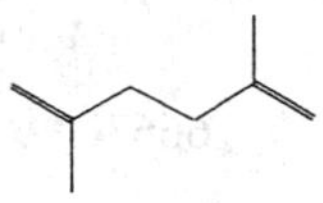

6692

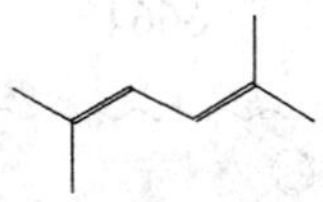

6693

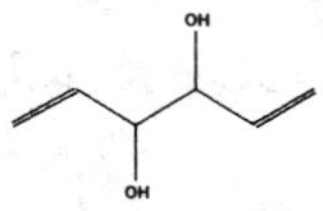

OH OH

6694

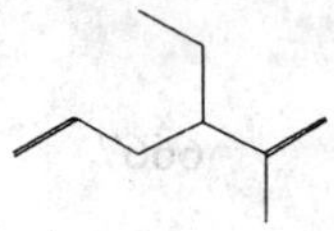

6695

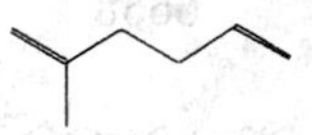

6696

6697

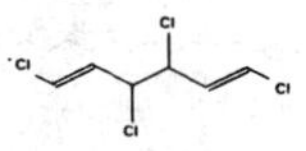

Cl Cl Cl Cl

6698

O OH

6699

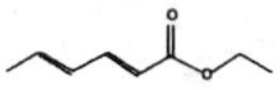

O O

6700

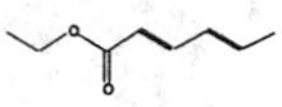

O O

6701

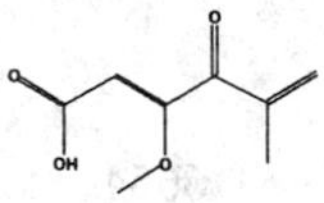

O O OH O

6702

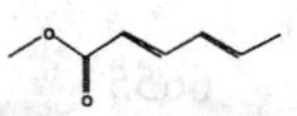

O O

6703

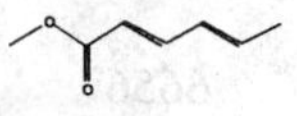

O O

6705

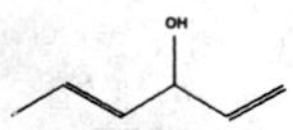

OH

6706

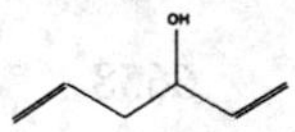

OH

6707

HO

6708

HO

6709

HO

6710

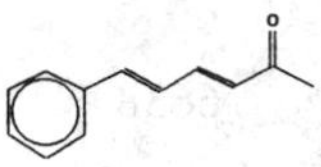

O

6711

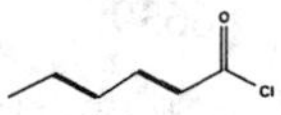

O Cl

6712

6713

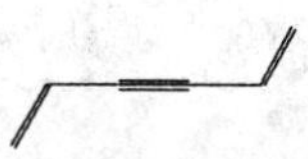

6714

6715

6716

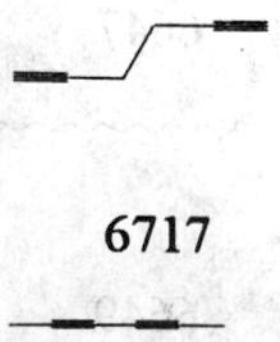

6717

6718

O

6719

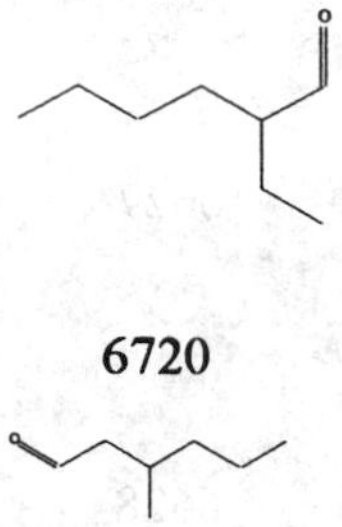

O

6720

O

6721

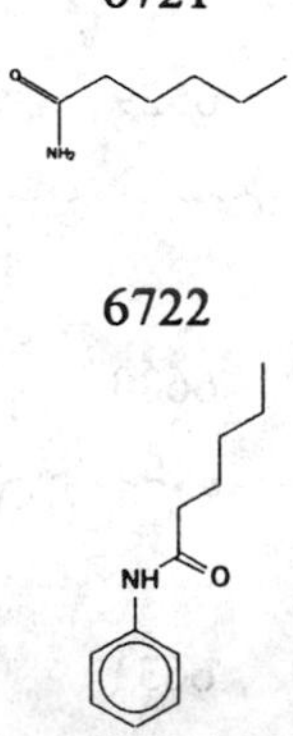

O NH_2

6722

NH O

6723

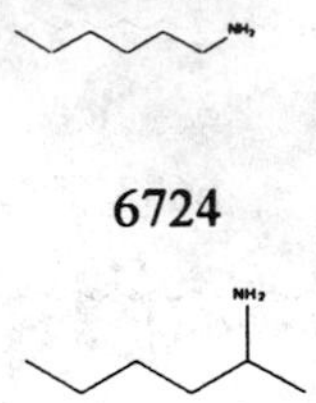

NH_2

6724

NH_2

6725

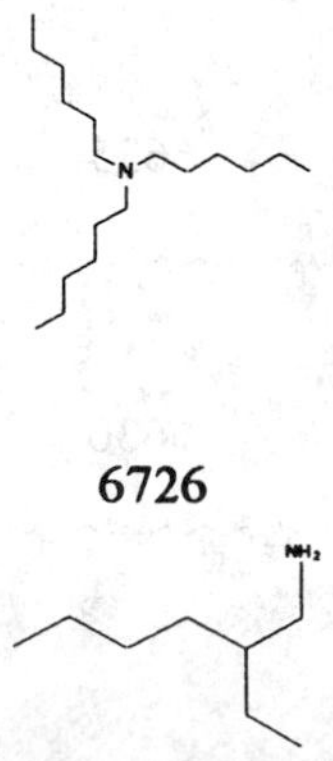

N

6726

NH_2

6727 NH

6728 NH2

6729 NH2

6730 NH2

6731

6732 Br

6733 Br

6734 Br

6735 Br

6736 O Br

6737 Cl

6738 Cl

6739 Cl

6740 Cl

6741 Cl

6742 F Cl

6743 Cl

6744 Cl

6745 Cl

6746 Cl

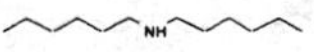

6747 Cl

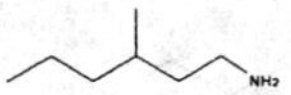

6748 O O

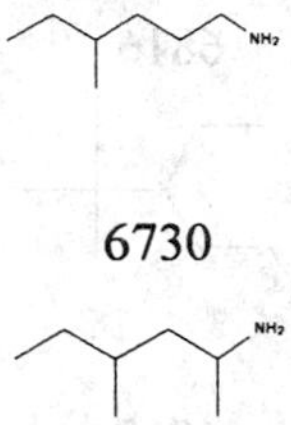

6749 H2N NH2

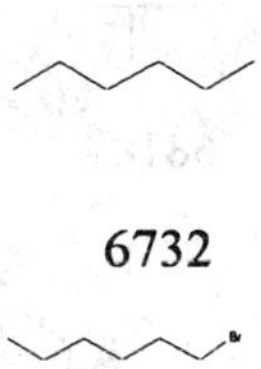

6750 NH NH

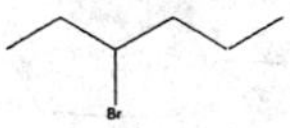

6751 H2N NH2

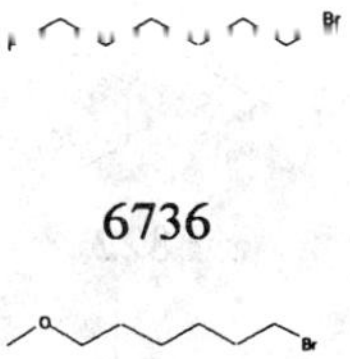

6752 N N

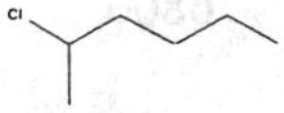

6754 Br Br

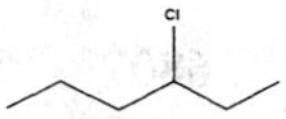

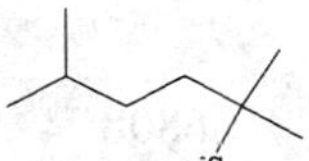

6755 Br Br

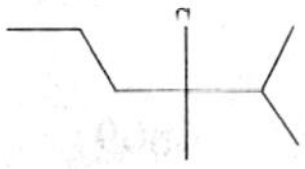

6756 Br Br

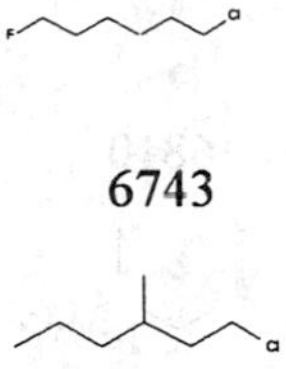

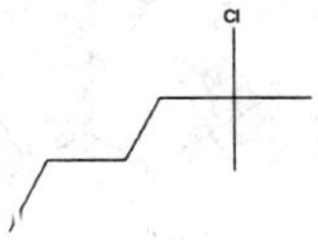

6757 Br Br

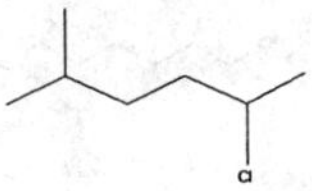

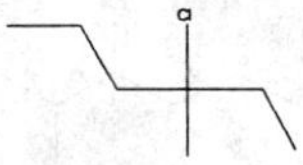

6758 Br Br

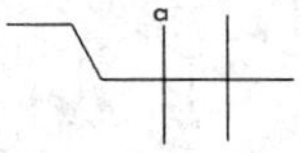

6759 Br Br

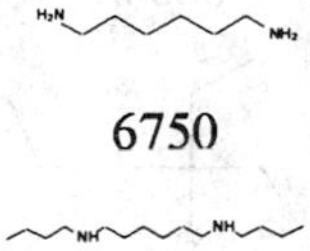

6760 Br Br

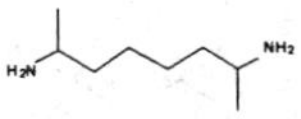

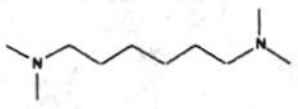

6761 Cl Cl

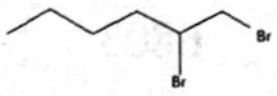

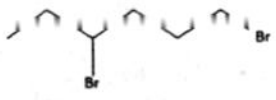

6762 I Cl

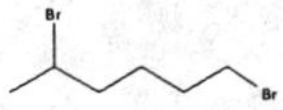

6763 Cl Cl

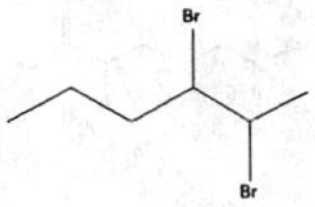

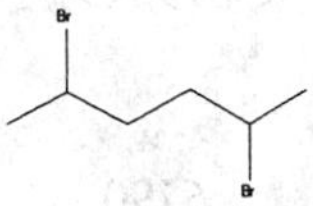

6764 Cl Cl

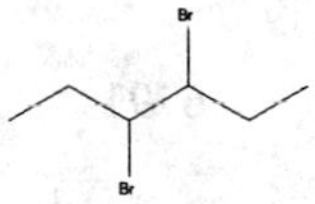

6766 Cl Cl

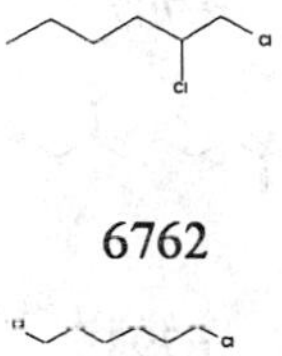

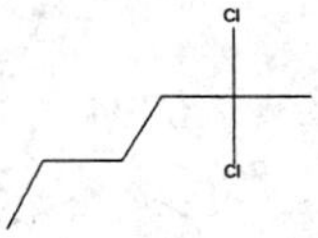

6767

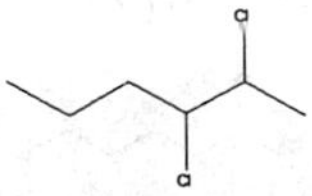

6768

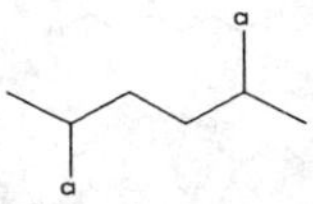

6769 I I

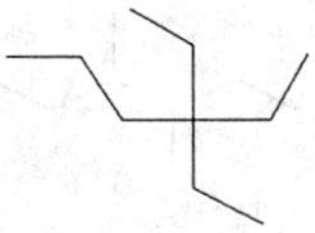

6770 O=C=N N=C=O

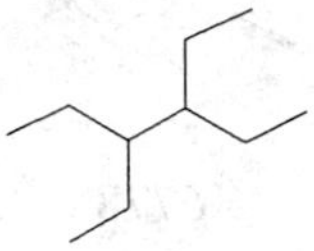

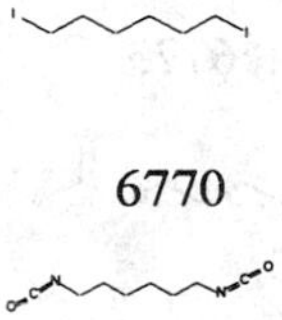

6771

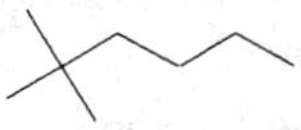

6772

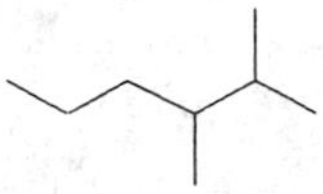

6773

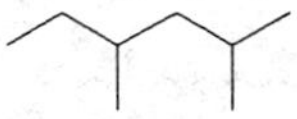

6774

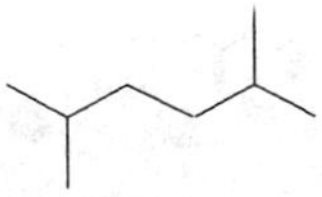

6775

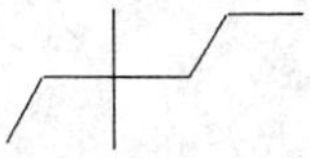

STRUCTURAL FORMULAS OF ORGANIC COMPOUNDS (continued)

In numeric order as they occur in the Table of Physical Constants of Organic Compounds

6776

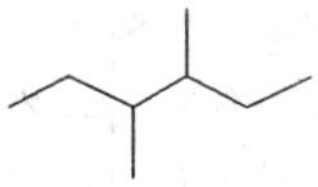

6777

6778

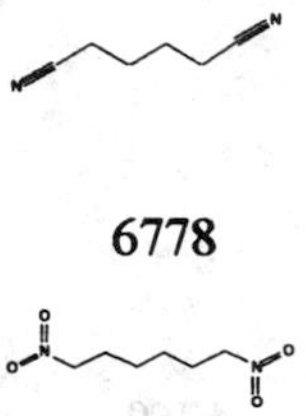

6779

6781

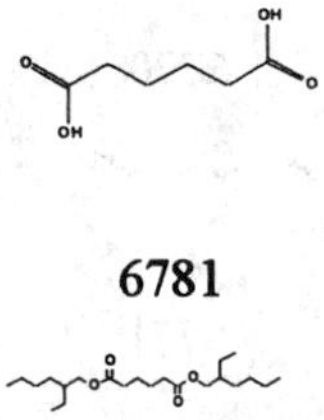

6782

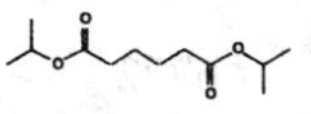

6783

6784

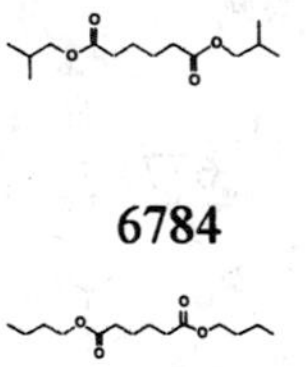

6785

6786

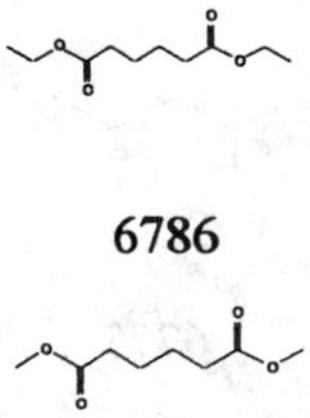

6787

6788

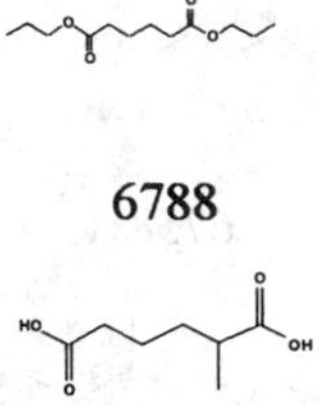

6789

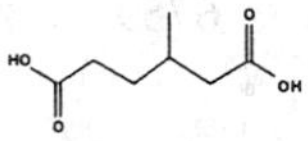

6790

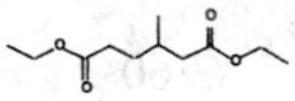

6791

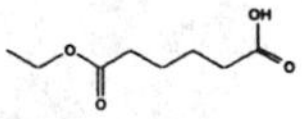

6792

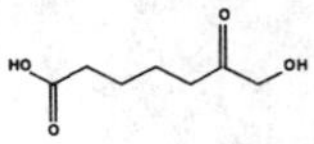

6793

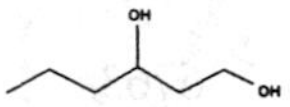

6794

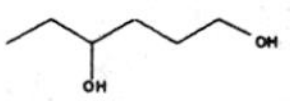

6795

6796

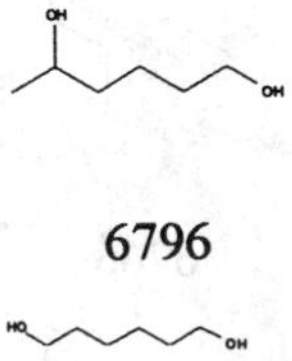

6797

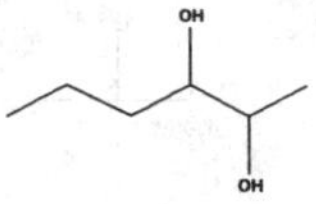

6798

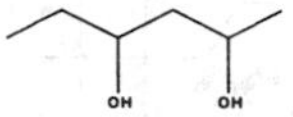

6799

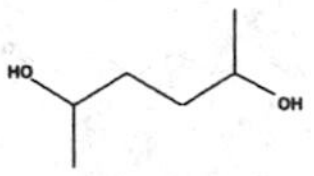

6800

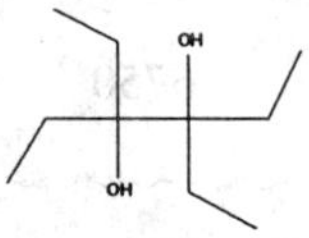

6801

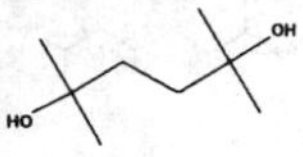

6802

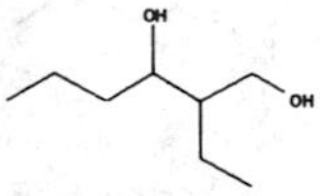

6803

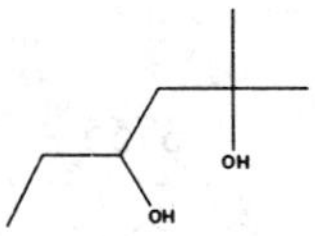

6804

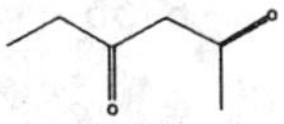

6805

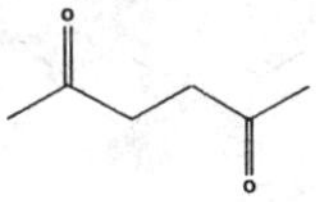

6806

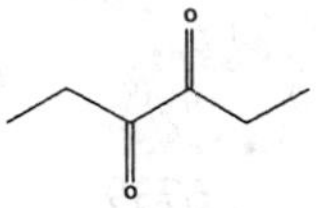

6807

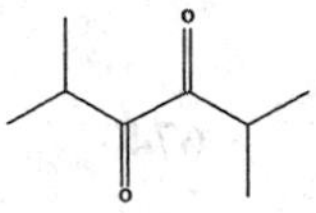

6808

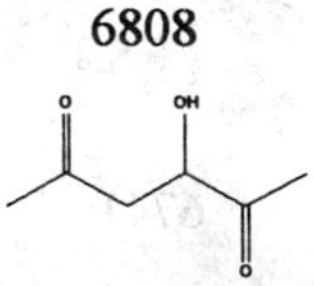

6809

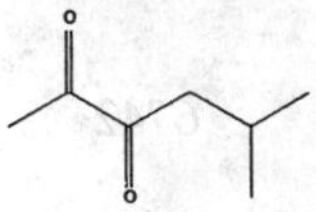

6810

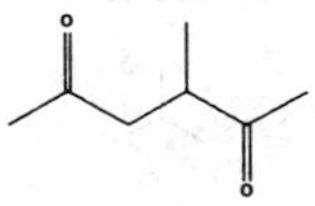

6811

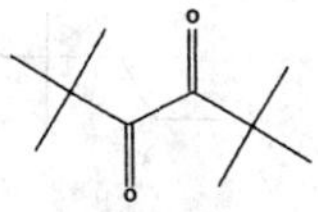

6812

6813

6814

6815

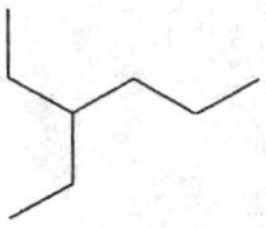

6816

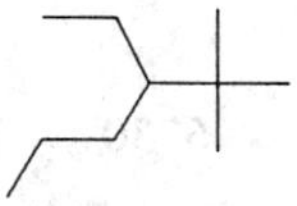

6817

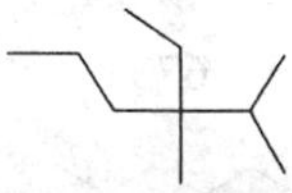

6818

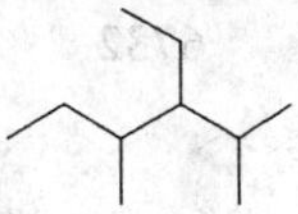

6819

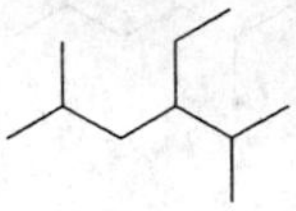

6820

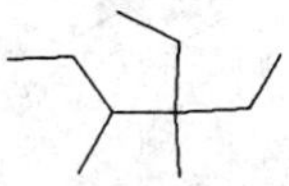

6821

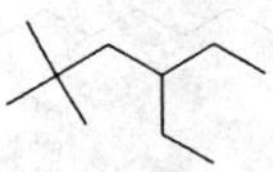

6822

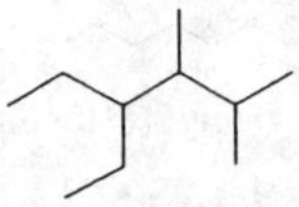

STRUCTURAL FORMULAS OF ORGANIC COMPOUNDS (continued)

In numeric order as they occur in the Table of Physical Constants of Organic Compounds

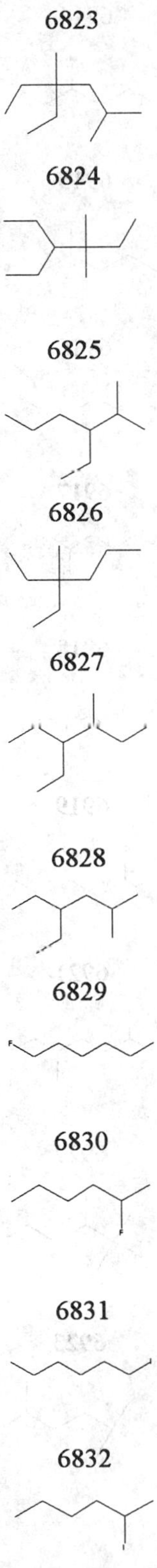

6823

6824

6825

6826

6827

6828

6829

F

6830

F

6831

I

6832

I

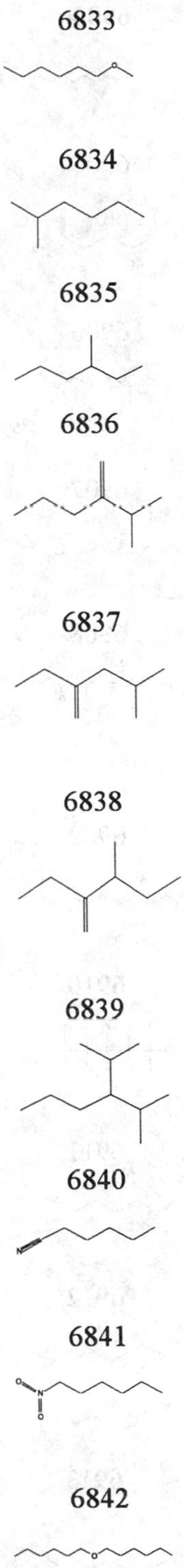

6833

O

6834

6835

6836

6837

6838

6839

6840

N

6841

O N O

6842

O

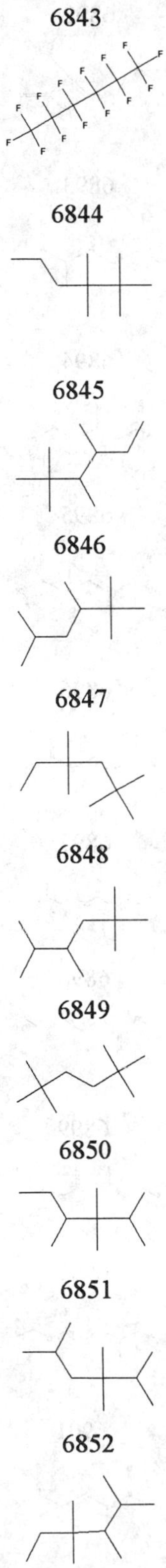

6843

F F F F F F F F F F F F F F

6844

6845

6846

6847

6848

6849

6850

6851

6852

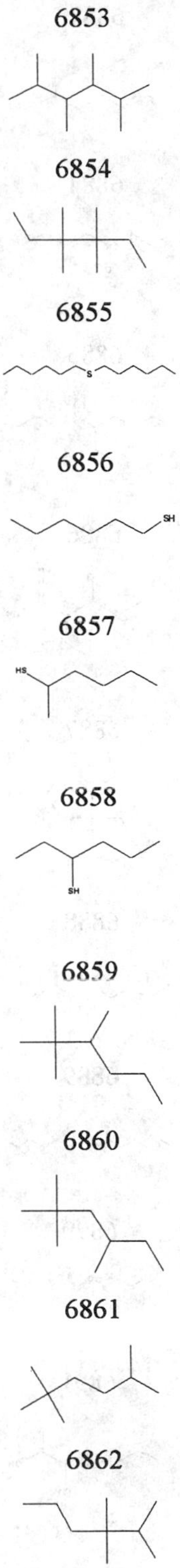

6853

6854

6855

S

6856

SH

6857

HS

6858

SH

6859

6860

6861

6862

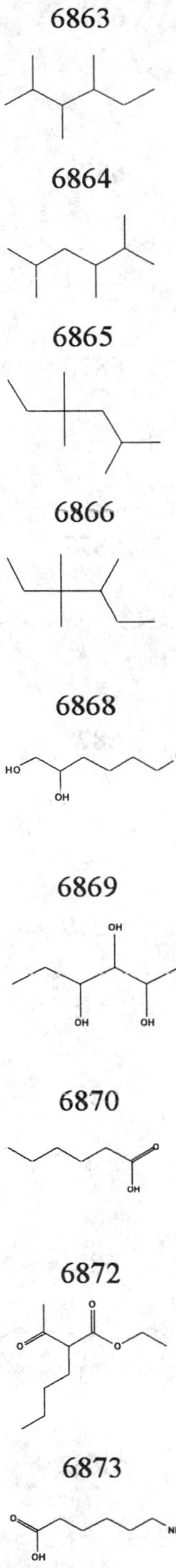

6863

OH

6864

6865

6866

6868

HO OH OH

6869

OH OH OH

6870

O OH

6872

O O O

6873

O OH NH$_2$

STRUCTURAL FORMULAS OF ORGANIC COMPOUNDS (continued)

In numeric order as they occur in the Table of Physical Constants of Organic Compounds

6874

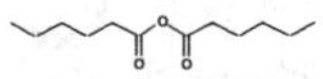

6875

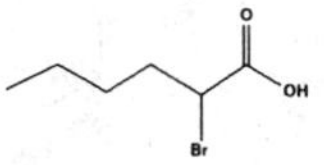

Br OH O

6876

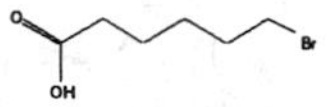

O OH Br

6877

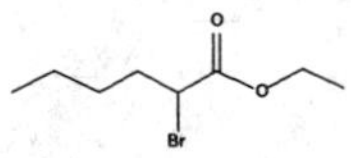

O Br

6878

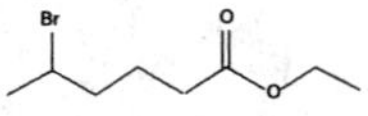

Br O

6879

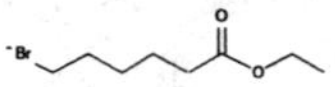

Br O

6880

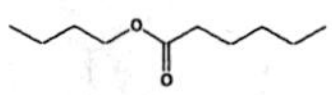

O O

6881

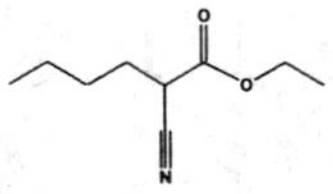

O N

6882

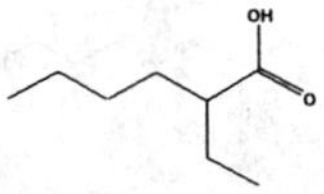

OH O

6883

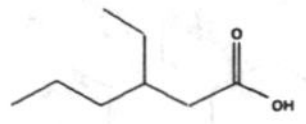

O OH

6884

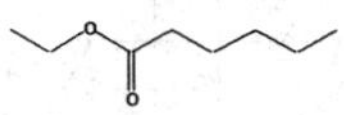

O O

6885

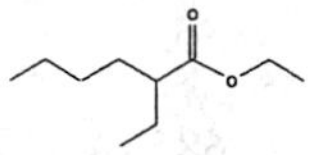

O O

6886

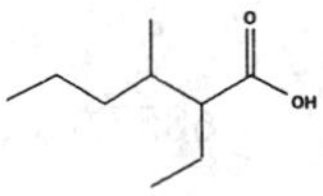

O OH

6887

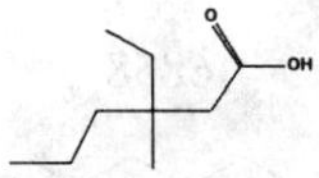

O OH

6888

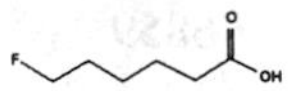

F O OH

6889

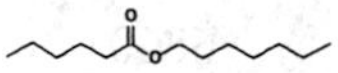

O O

6890

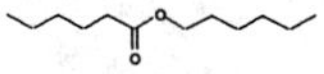

O O

6891

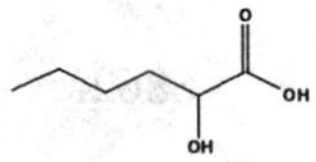

O OH OH

6892

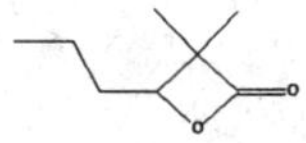

O O

6893

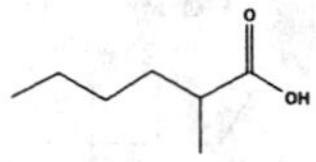

O OH

6894

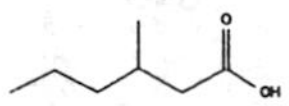

O OH

6895

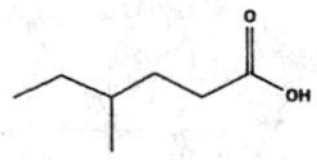

O OH

6896

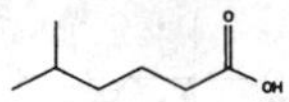

O OH

6897

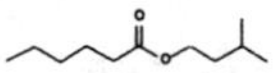

O O

6898

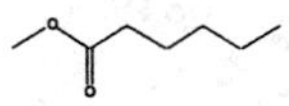

O O

6899

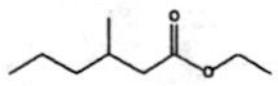

O O

6900

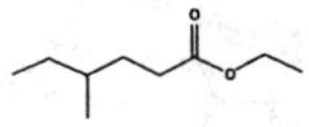

O O

6901

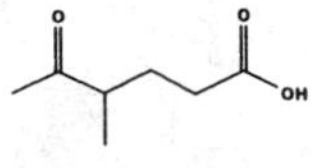

O O OH

6904

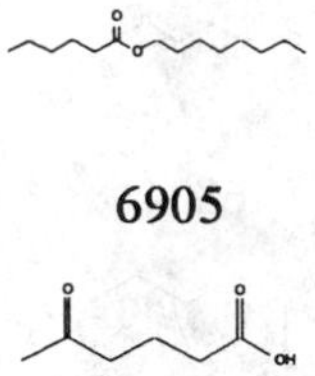

O O

6905

O O OH

6906

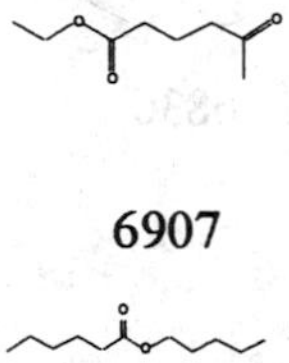

O O O

6907

O O

6908

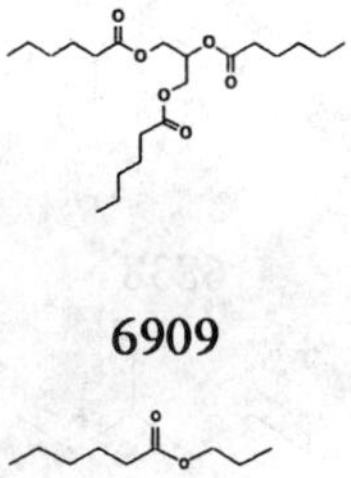

O O O O O O

6909

O O

6910

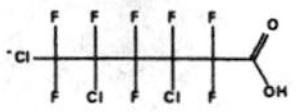

Cl F F F F F Cl Cl F F F F O OH

6911

OH

6912

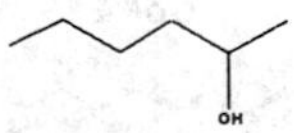

OH

6913

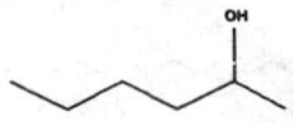

OH

6914

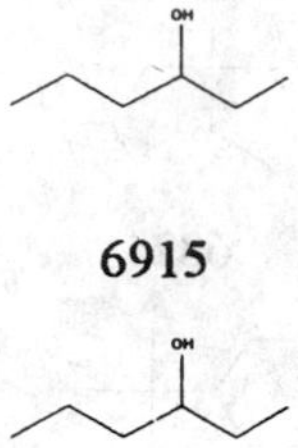

OH

6915

OH

6916

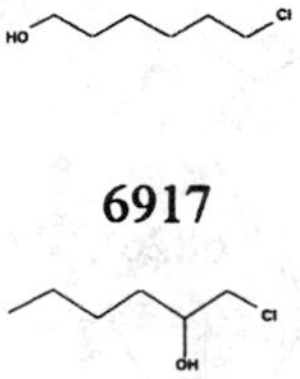

HO Cl

6917

Cl OH

6918

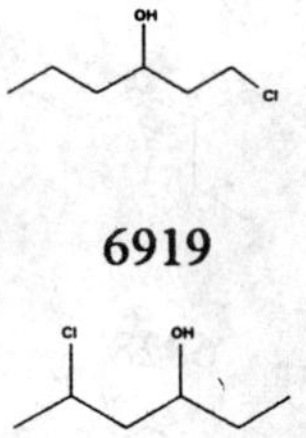

OH Cl

6919

Cl OH

6921

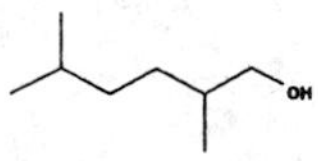

OH

6922

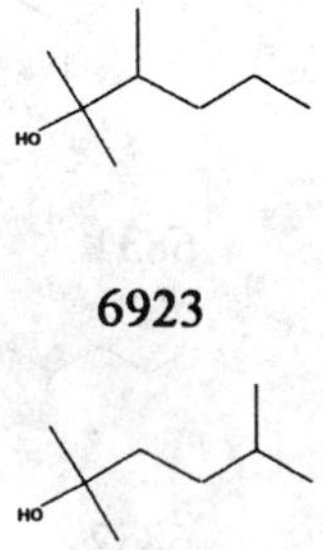

HO

6923

HO

STRUCTURAL FORMULAS OF ORGANIC COMPOUNDS (continued)

In numeric order as they occur in the Table of Physical Constants of Organic Compounds

6924 OH

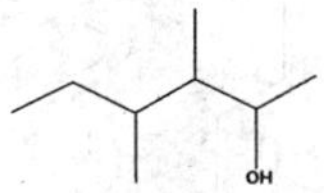

6925 OH

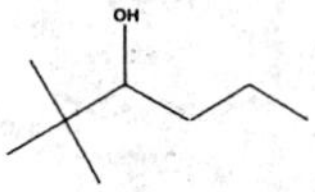

6926 HO

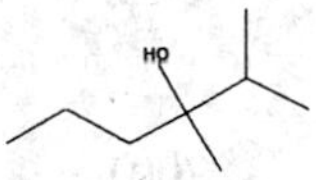

6927 OH

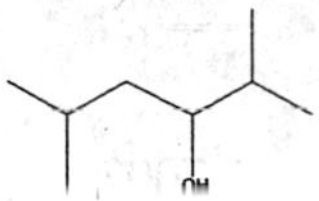

6928 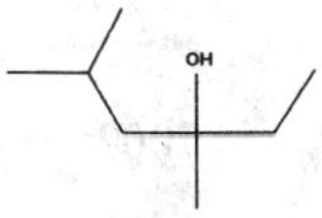 OH

6929 OH

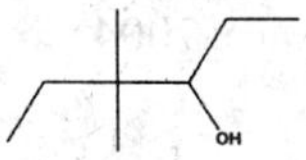

6930 OH

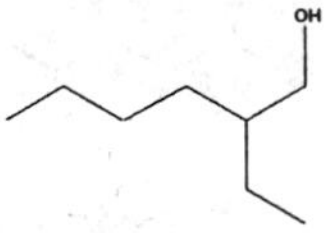

6931 OH

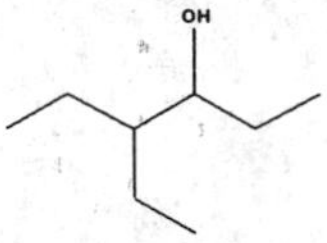

6932 OH

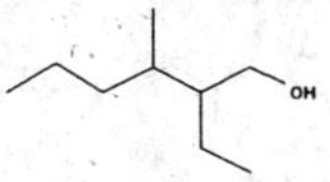

6933 OH

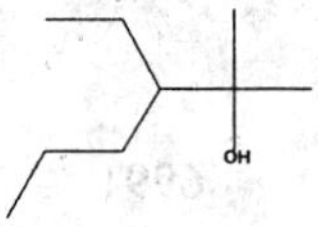

6934 OH

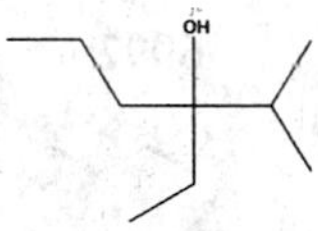

6935 OH

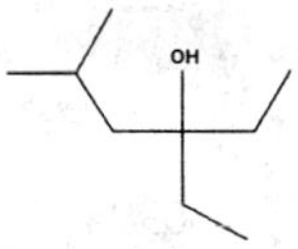

6936 F OH

6937 OH

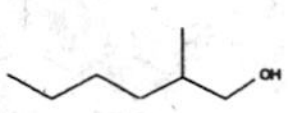

6938 OH

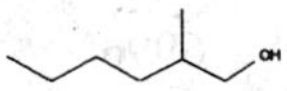

6939 OH

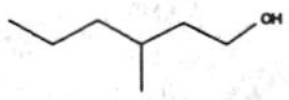

6940 OH

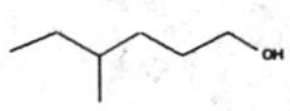

6941 OH

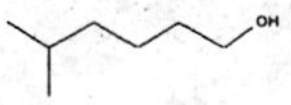

6942 HO

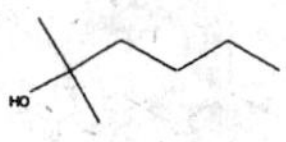

6943 OH

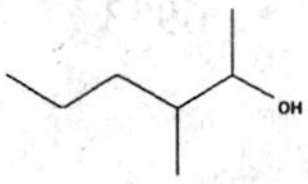

6944 OH

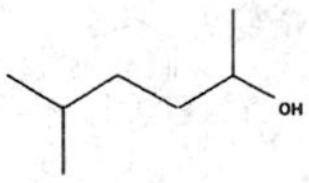

6945 OH

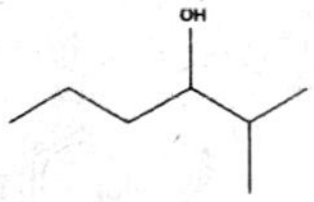

6946 OH

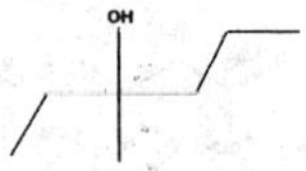

6947 OH

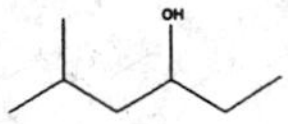

6948 OH

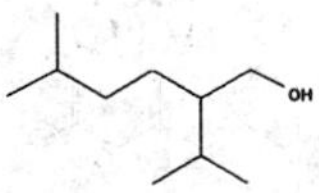

6949 OH

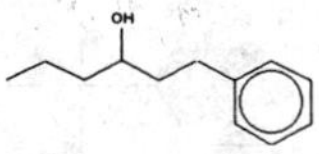

6950 OH

6951 OH

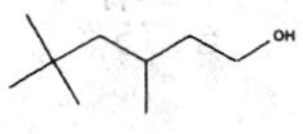

6952 OH

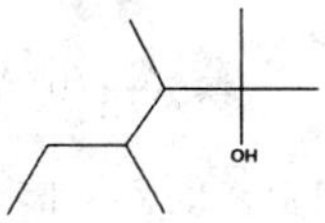

6953 OH

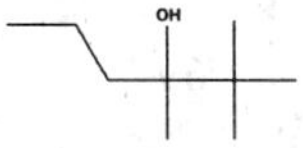

6954 OH

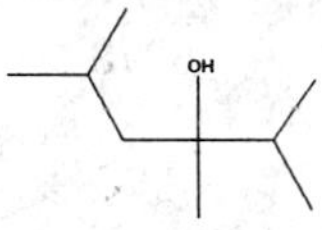

6955 OH

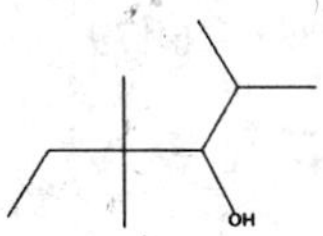

6956 HO

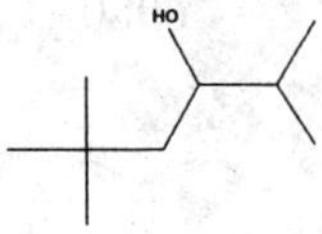

6957 OH

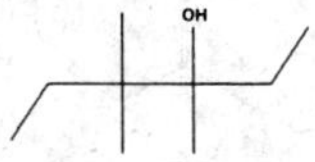

6958 OH

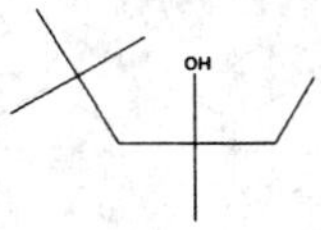

6959 O

6960 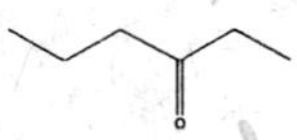 O

6961 O Br

6962 O OH OH

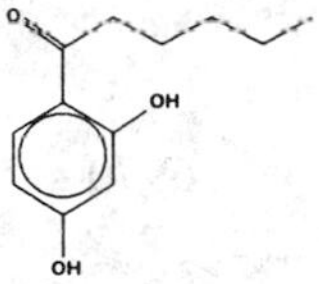

6963 O

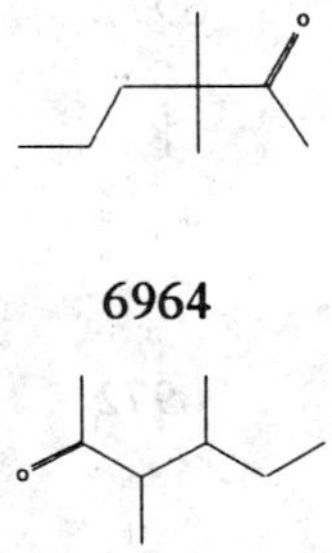

6964 O

STRUCTURAL FORMULAS OF ORGANIC COMPOUNDS (continued)

In numeric order as they occur in the Table of Physical Constants of Organic Compounds

6965

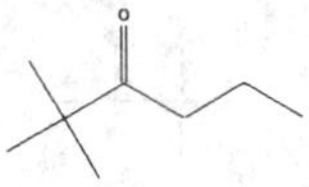

6966

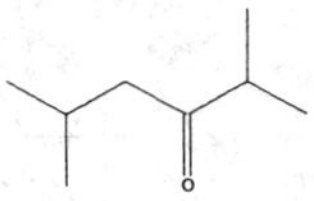

6967

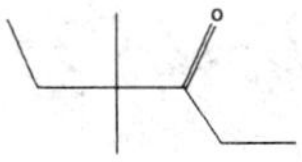

6968

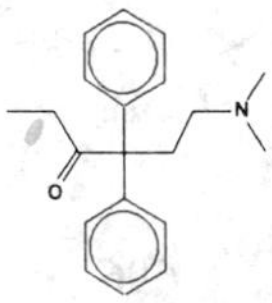

6969

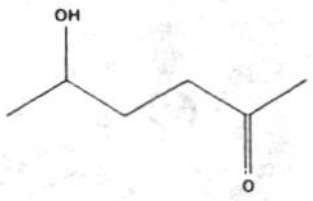

6970

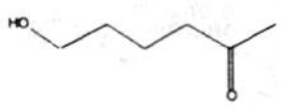

6971

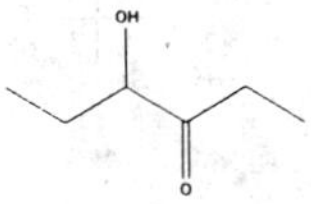

6972

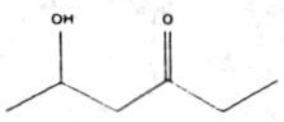

6973

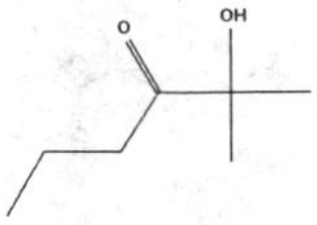

6974

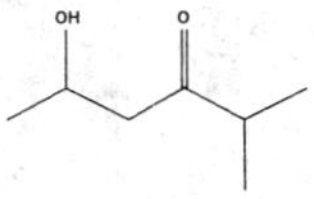

6975

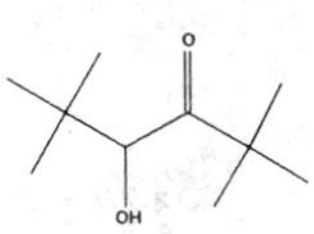

6976

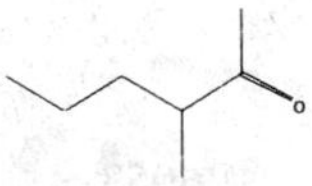

6977

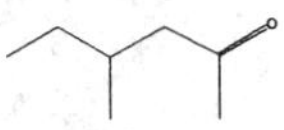

6978

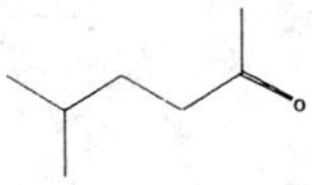

6979

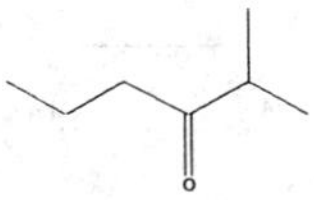

6980

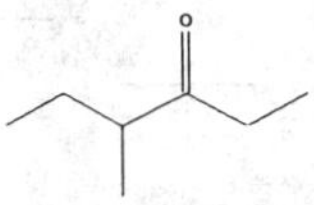

6981

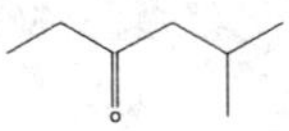

6982

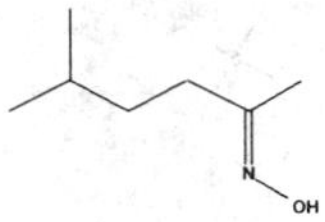

6983

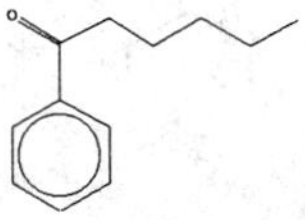

6984

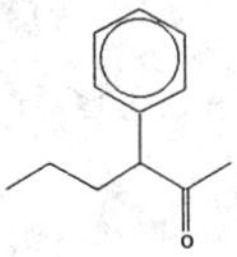

6985

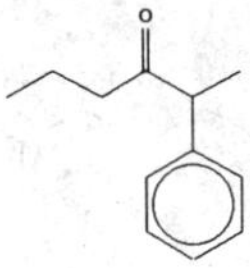

6986

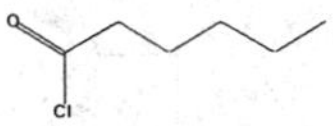

6987

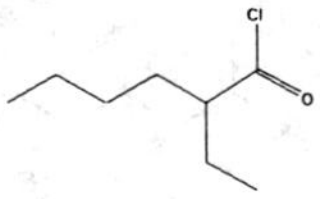

6988

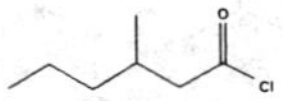

6989

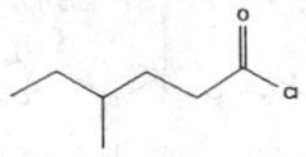

6990

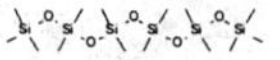

6991

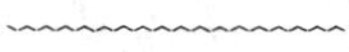

6992

6993

6994

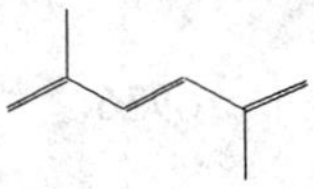

6995

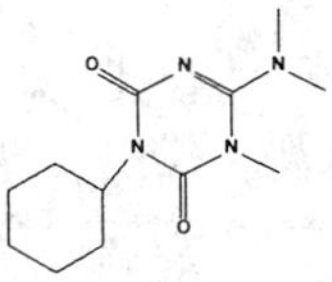

6996

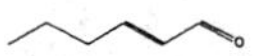

6997

6998

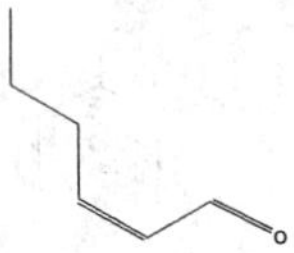

6999

7000

7001

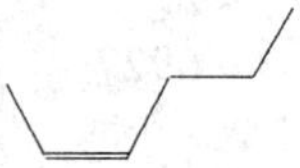

7002

7003

7004

7005

7007

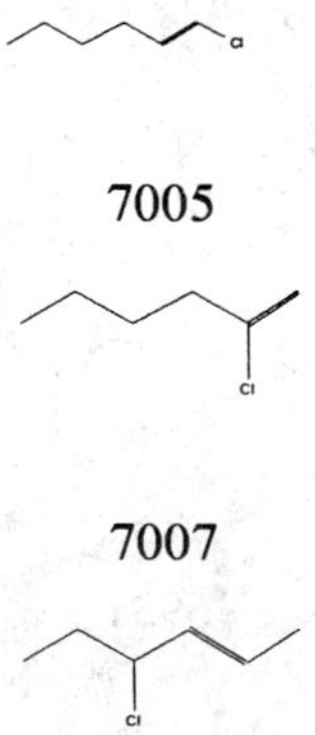

STRUCTURAL FORMULAS OF ORGANIC COMPOUNDS (continued)

In numeric order as they occur in the Table of Physical Constants of Organic Compounds

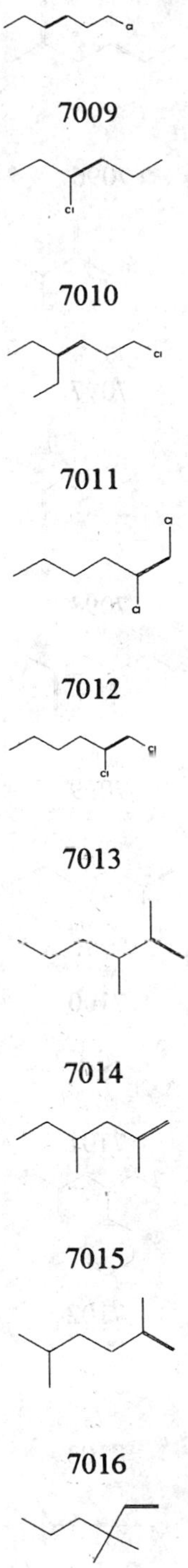

7008

Cl

7009

Cl

7010

Cl

7011

Cl

Cl

7012

Cl

Cl

7013

7014

7015

7016

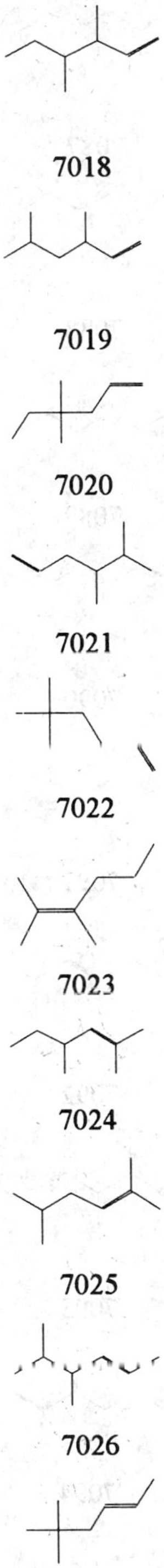

7017

7018

7019

7020

7021

7022

7023

7024

7025

7026

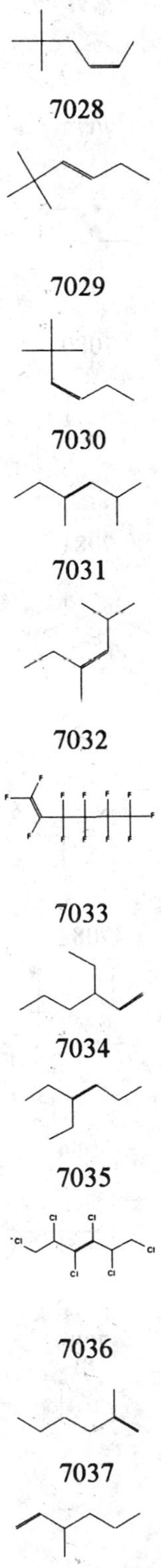

7027

7028

7029

7030

7031

7032

F F F F F F F F F F F F F

7033

7034

7035

Cl Cl Cl Cl Cl Cl

7036

7037

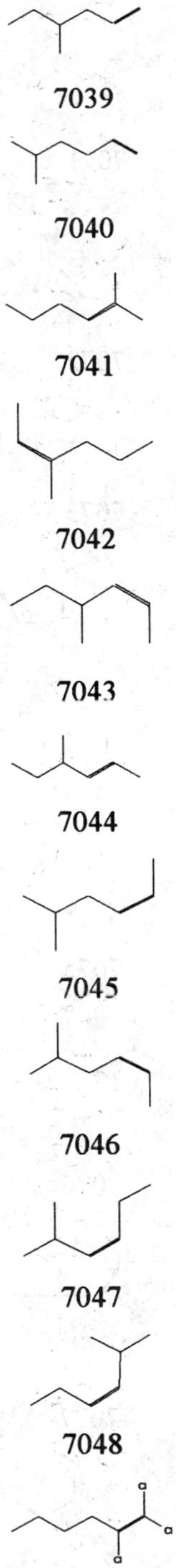

7038

7039

7040

7041

7042

7043

7044

7045

7046

7047

7048

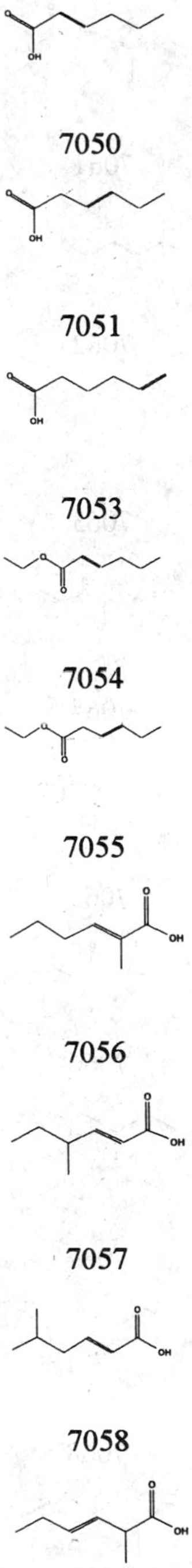

Cl Cl Cl

7049

O OH

7050

O OH

7051

O OH

7053

O O

7054

O O

7055

O OH

7056

O OH

7057

O OH

7058

O OH

STRUCTURAL FORMULAS OF ORGANIC COMPOUNDS (continued)

In numeric order as they occur in the Table of Physical Constants of Organic Compounds

7060 O OH

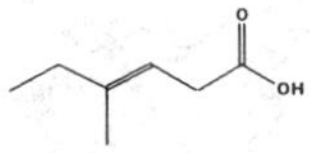

7061 O OH

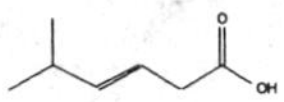

7062 O O

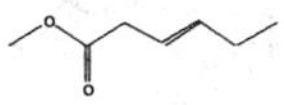

7063 O O

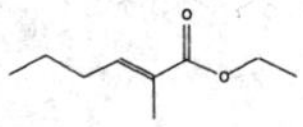

7064 O O

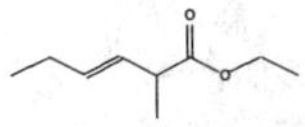

7065 O O

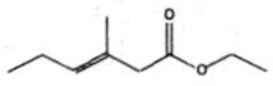

7066 OH

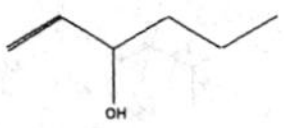

7067 OH

7068 HO

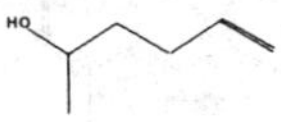

7069 OH

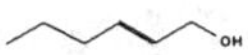

7070 HO

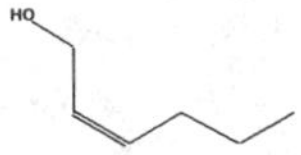

7071 HO

7072 HO

7073 OH

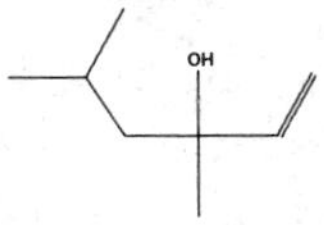

7074 HO

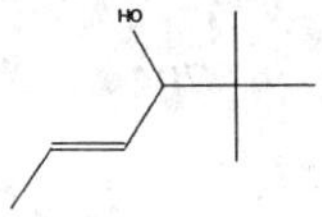

7075 OH

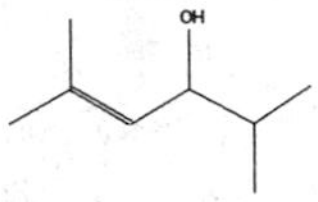

7076 OH

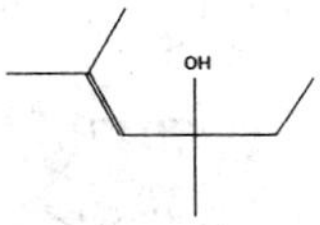

7077 OH

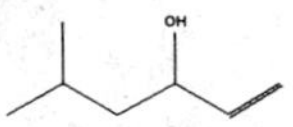

7078 OH

7079 OH

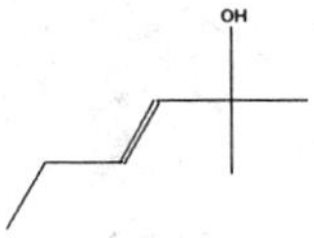

7080 OH

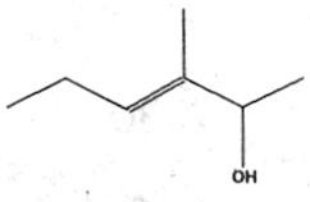

7081 OH

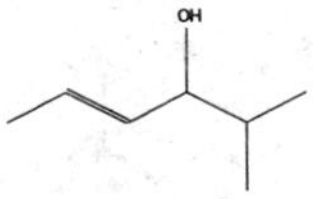

7082 OH

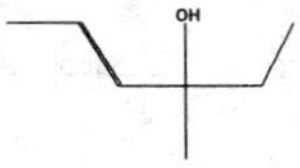

7083 OH

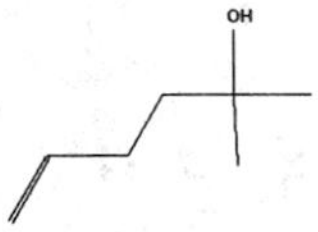

7084 OH

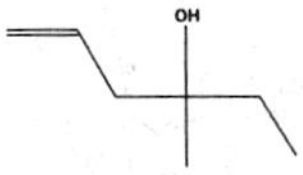

7085 OH

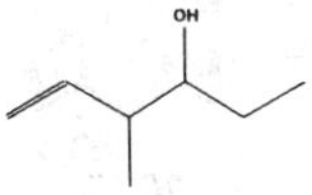

7086 O

7087 O

7088 O

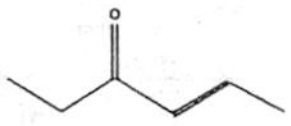

7089 O

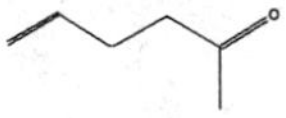

7090 O

7091 O

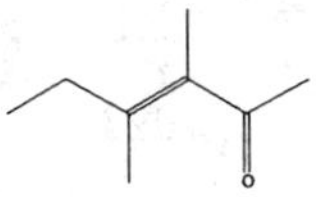

7092 O

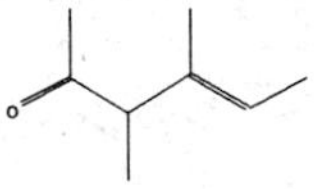

7093 O

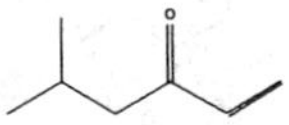

7094 O

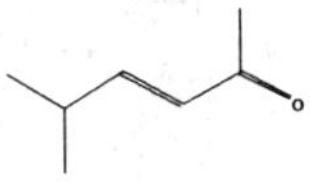

7095 O

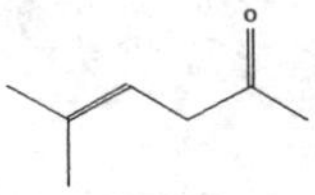

7096 O

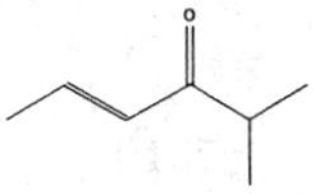

7097 O

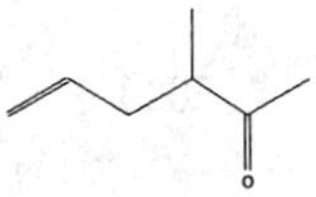

7098 O

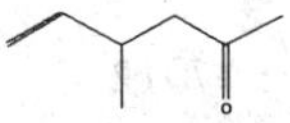

7099 O

7100 O

7101 O

7102 O

7103

7104

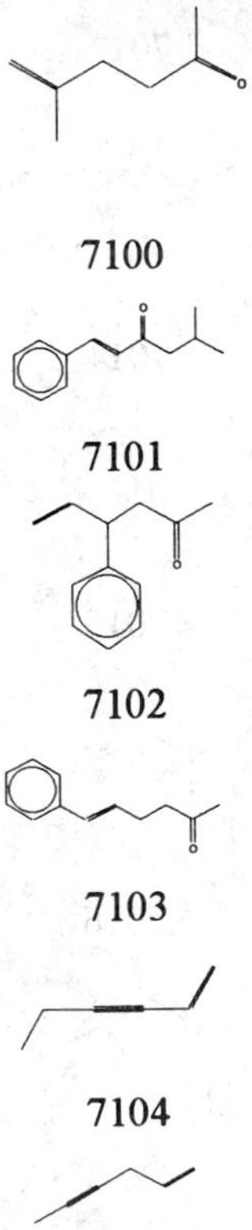

STRUCTURAL FORMULAS OF ORGANIC COMPOUNDS (continued)

In numeric order as they occur in the Table of Physical Constants of Organic Compounds

7105

7106

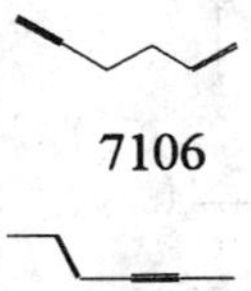

7107

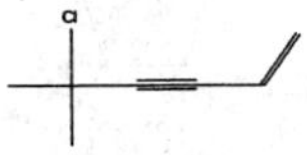

7108

7109

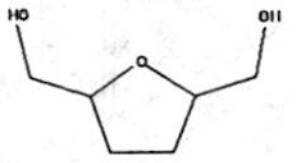

7113

7114

7115

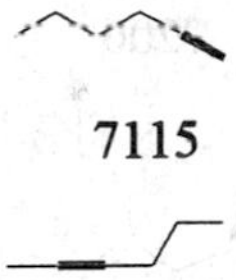

7116

7117

7118

7119

7120

7121

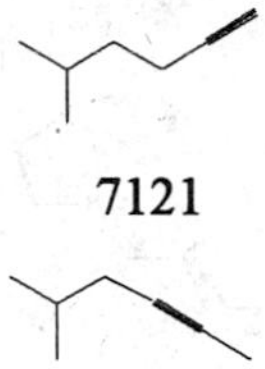

7122

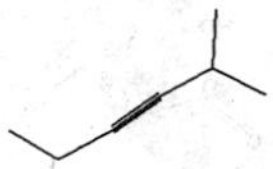

7123

7124

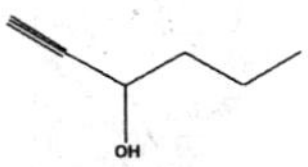

7125

7126

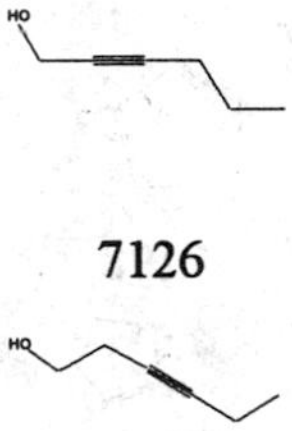

7127

7128

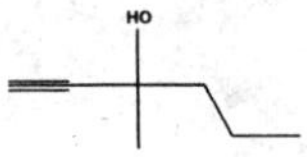

7129

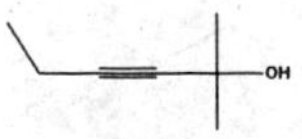

7130

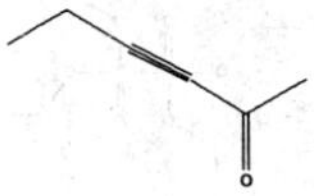

7131

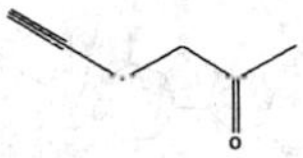

7132

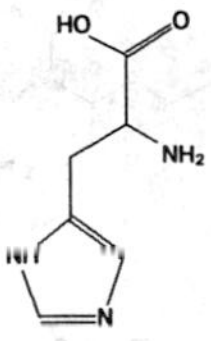

7136

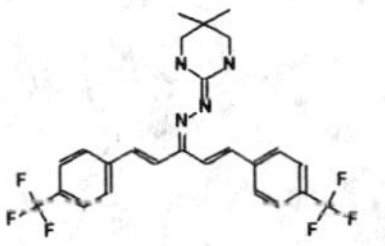

7137

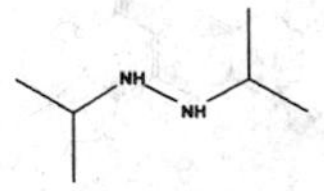

7138

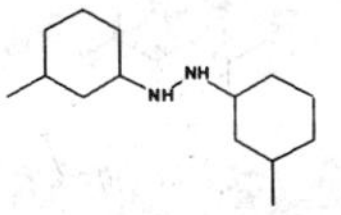

7140

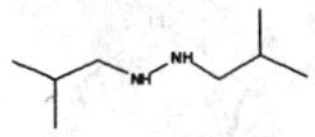

7142

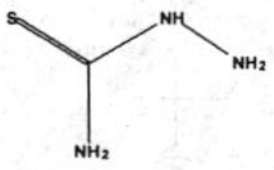

7144

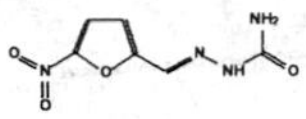

7145

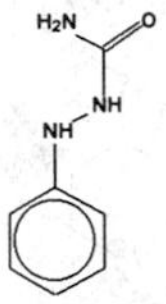

7146

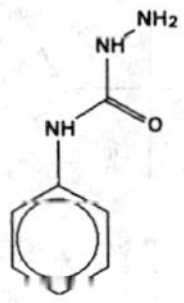

7147

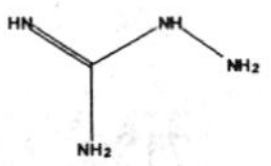

7148

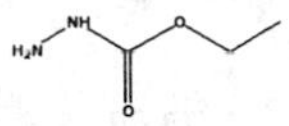

7149

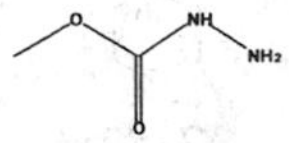

7150

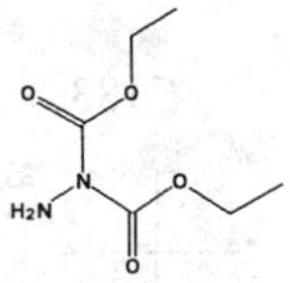

7151

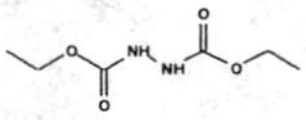

7152

7153

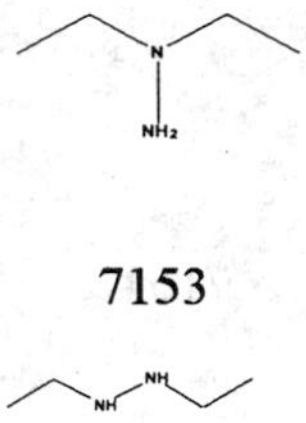

7154

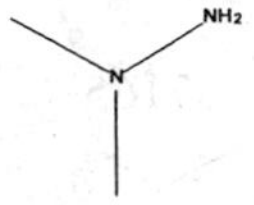

7155

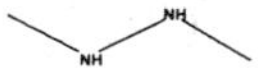

7156

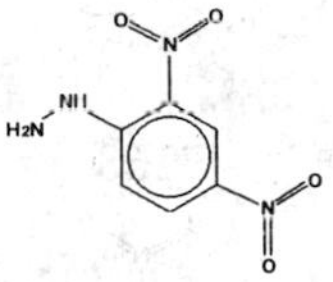

7158

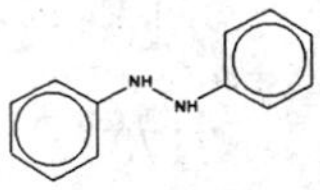

7159

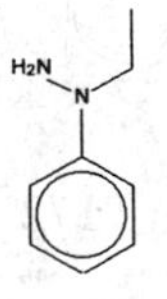

7160

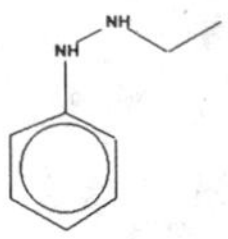

NH NH

7161

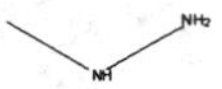

NH_2 NH

7164

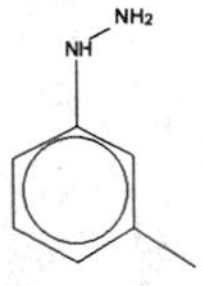

NH_2 NH

7165

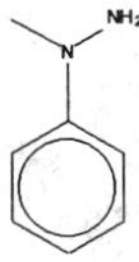

NH_2 N

7167

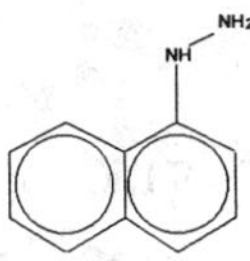

NH_2 NH

7168

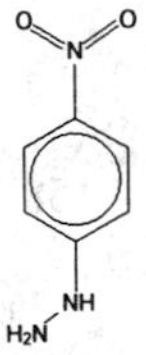

O N O

NH H_2N

7170

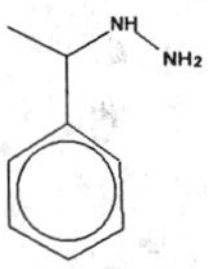

NH NH_2

7172

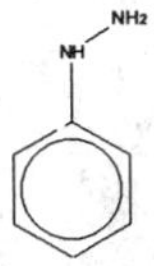

NH_2 NH

7175

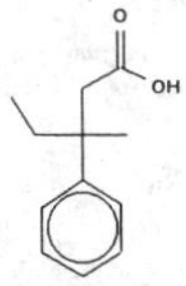

O OH

7176

N

7177

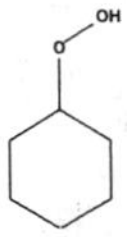

OH O

7178

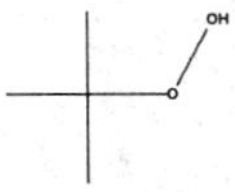

OH O

7179

OH O

7180

OH O

7181

OH O

7182

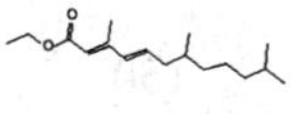

O O

7183

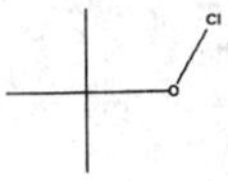

Cl O

7184

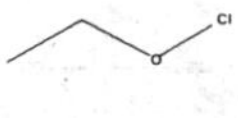

Cl O

7185

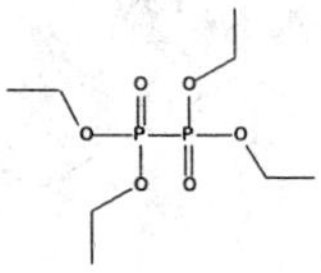

O O O P P O O O O O

7186

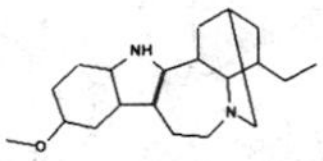

NH N O

7187

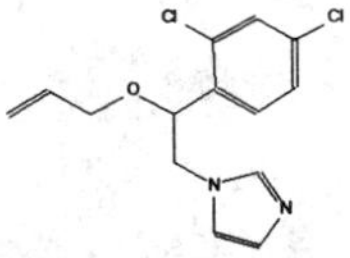

Cl Cl O N N

7188

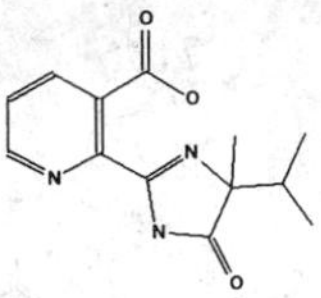

O O N N N O

7189

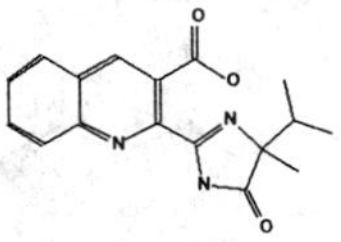

O O N N N O

7190

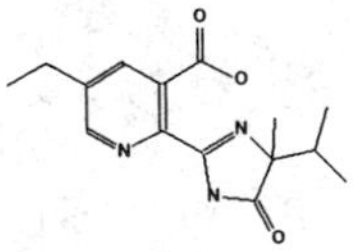

O O N N N O

7192

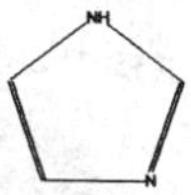

NH N

7193

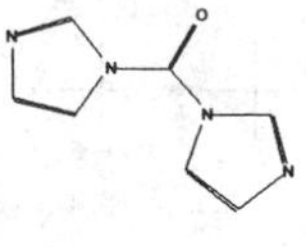

N N O N N

7194

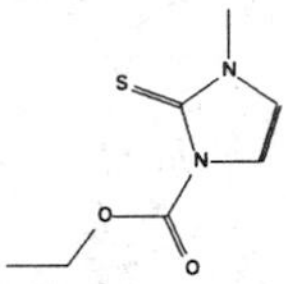

S N N O O

7195

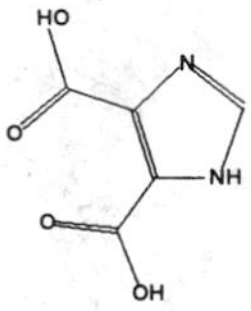

HO O N NH O OH

7196

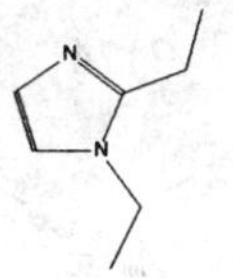

N N

7197

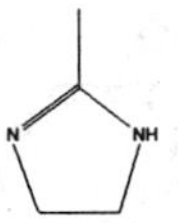

N NH

7198

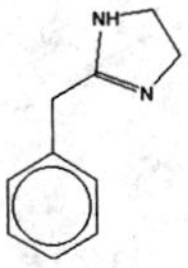

NH N

7199

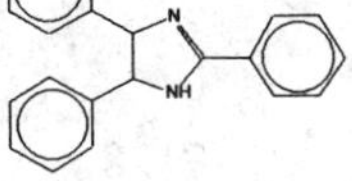

N NH

7200

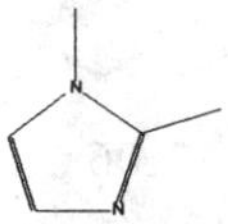

N N

7201

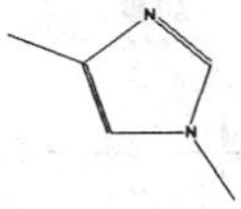

N N

7202

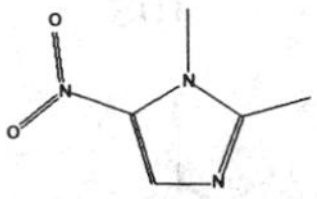

O N O N N

7203

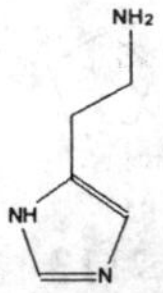

NH_2 NH N

7205

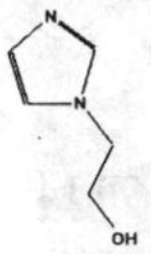

N N OH

7206

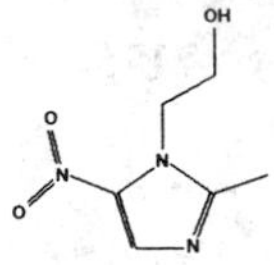

OH O N O N N

7207

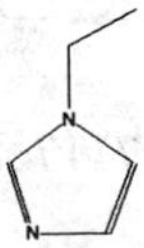

N N

7208

N N

7209

NH N

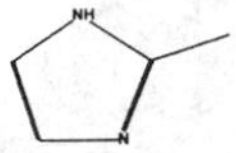

7210

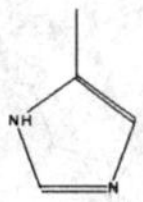

NH N

7211

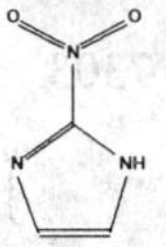

O N O N NH

7212

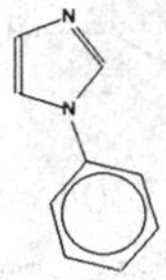

N N

7213

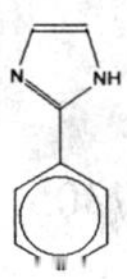

N NH

7214

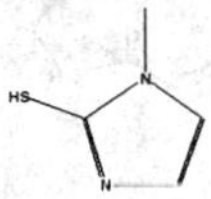

HS N N

7215

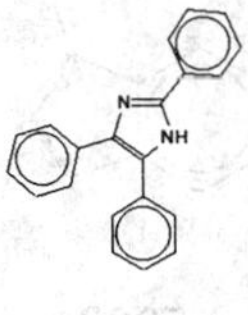

N NH

7216

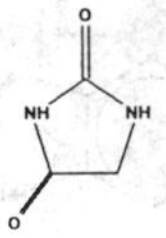

O NH NH O

7217

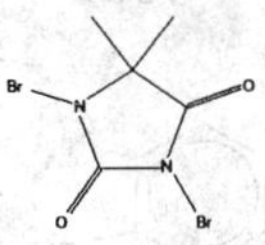

Br N O O N Br

7218

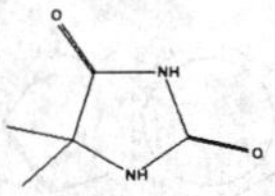

O NH NH O

7219

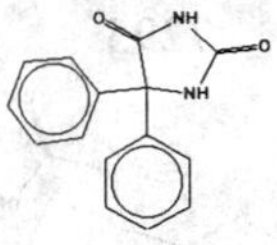

NH O O NH

7222

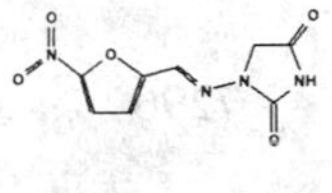

O N O O N N O NH O

7224

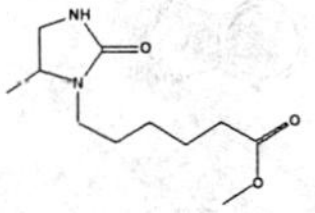

NH O N O O

7225

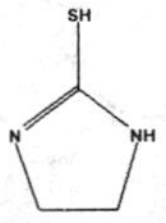

SH N NH

7226

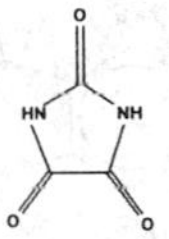

O HN NH O O

7227

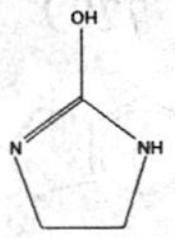

OH N NH

7228

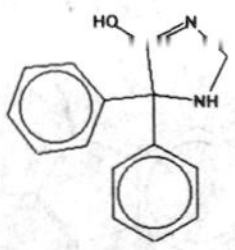

HO N NH

7230

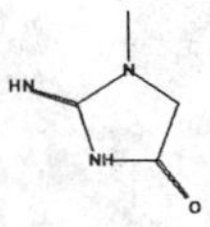

HN N NH O

7231

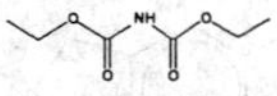

O NH O O O

7232

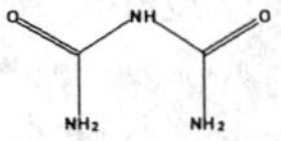

O NH O NH2 NH2

7233

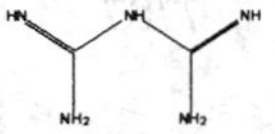

HN NH NH NH2 NH2

7235

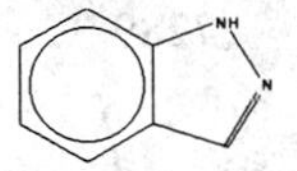

NH N

7236

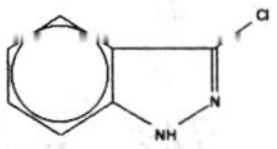

Cl N NH

7237

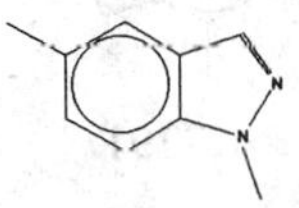

N N

7238

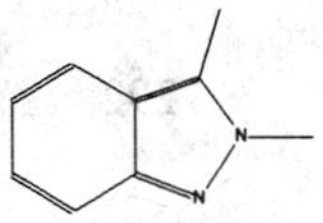

N N

7239

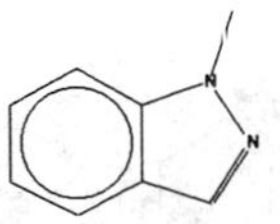

N N

7240

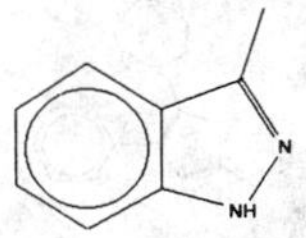

N NH

7241

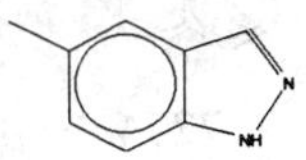

N NH

7242

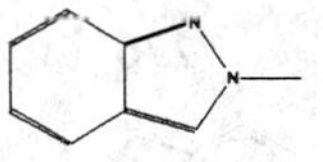

N

7243

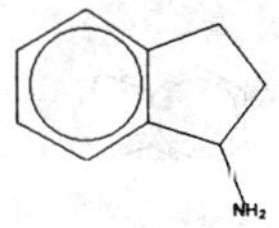

NH2

7244

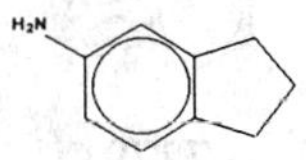

H2N

7245

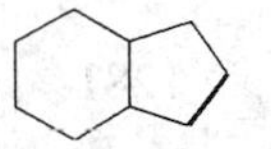

7246

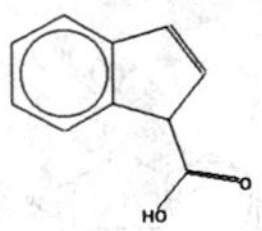

HO O

7247

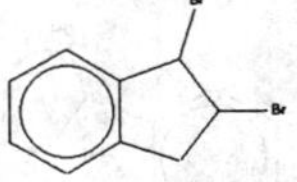

Br Br

7249

Cl Cl

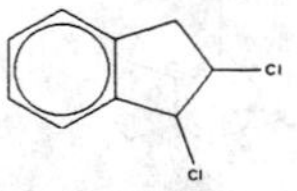

7250

7251

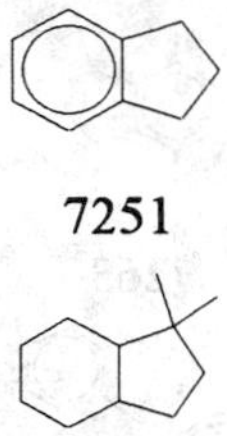

7252

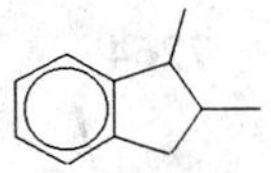

7253

7254

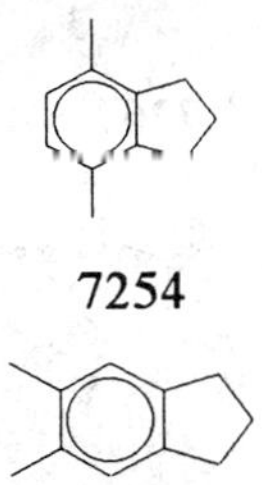

7255

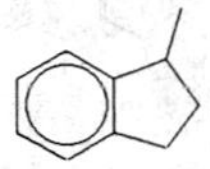

7257

7258

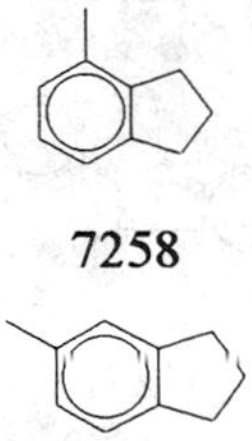

7259

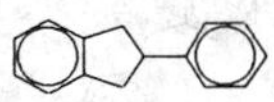

STRUCTURAL FORMULAS OF ORGANIC COMPOUNDS (continued)

In numeric order as they occur in the Table of Physical Constants of Organic Compounds

7260

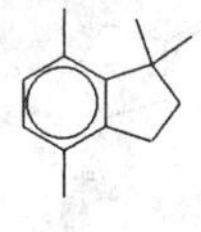

7262

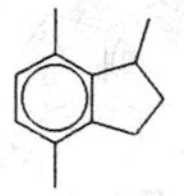

7263

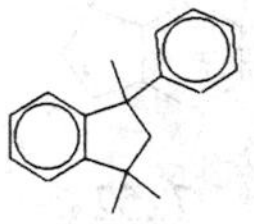

7264

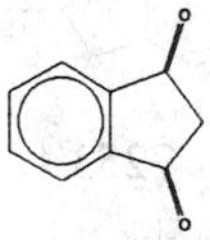

O

O

7267

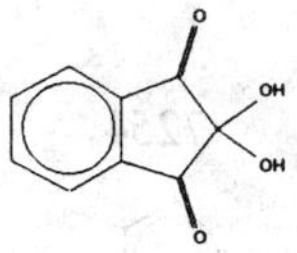

O

OH

OH

O

7268

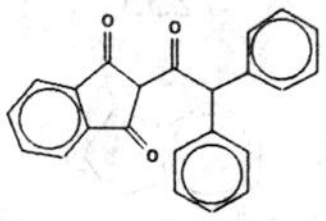

O O

O

7269

O

O

O

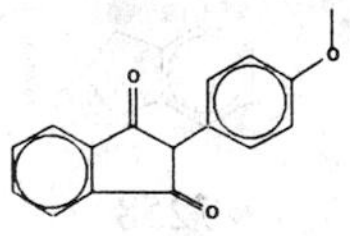

7271

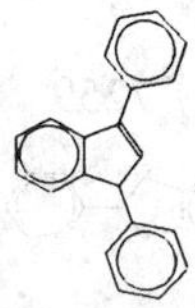

7272

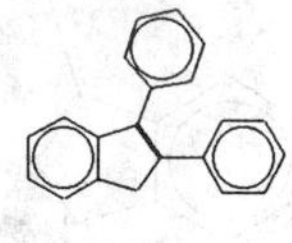

7273

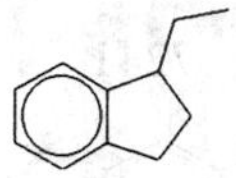

7274

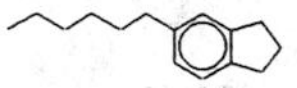

7276

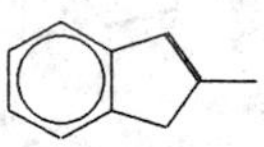

7277

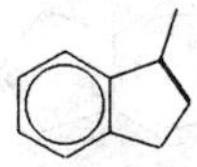

7278

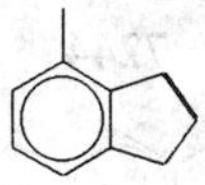

7279

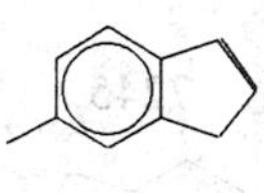

7281

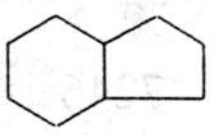

7282

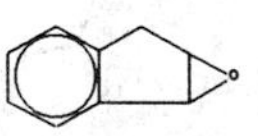

O

7284

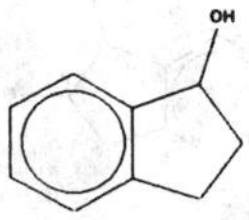

OH

7285

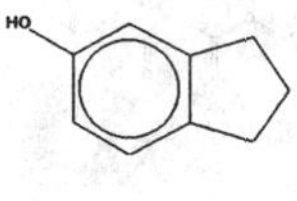

HO

7286

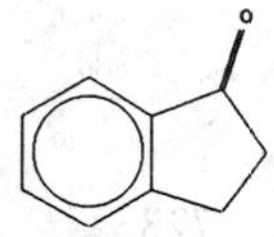

O

7287

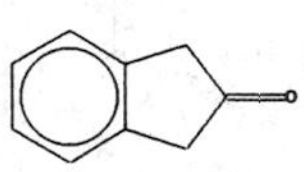

O

7289

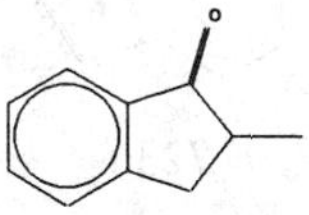

O

7290

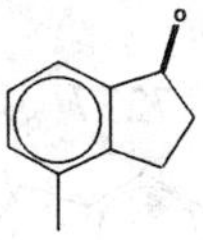

O

7291

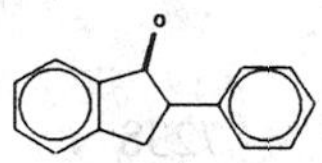

O

7292

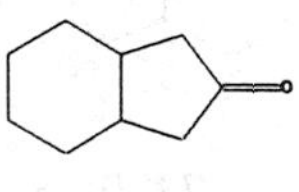

O

7293

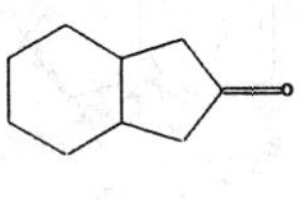

O

7294

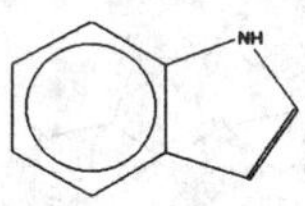

NH

7295

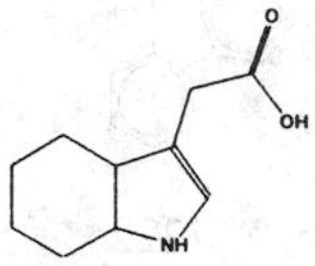

O

OH

NH

7296

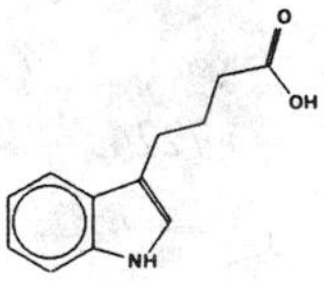

O

OH

NH

7297

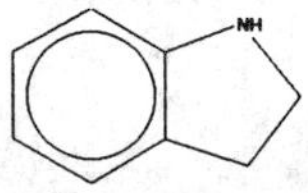

NH

7298

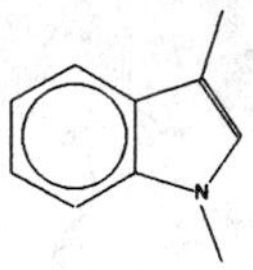

N

7299

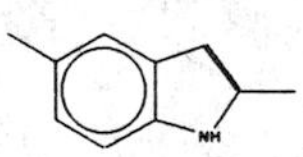

NH

7300

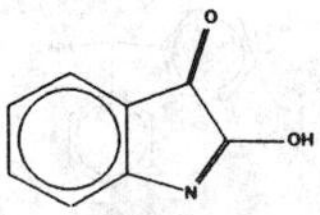

O

OH

N

7302

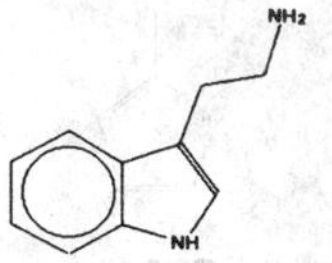

NH_2

NH

7303

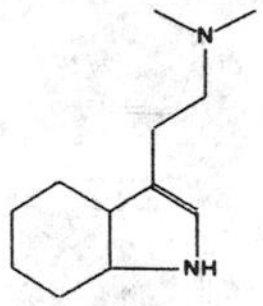

N

NH

7304

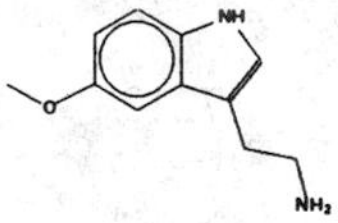

NH

O

NH_2

7305

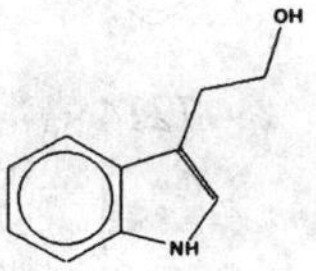

OH

NH

7307

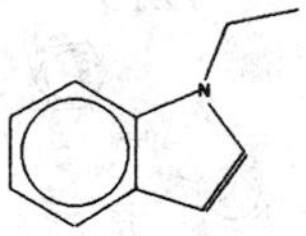

N

7308

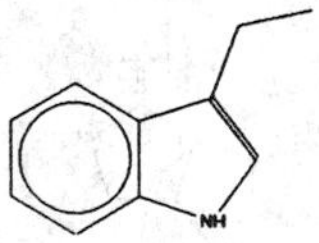

NH

7309

NH

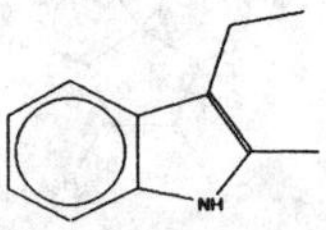

STRUCTURAL FORMULAS OF ORGANIC COMPOUNDS (continued)

In numeric order as they occur in the Table of Physical Constants of Organic Compounds

7310

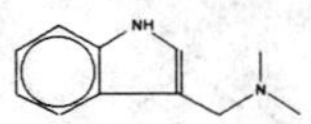

7312

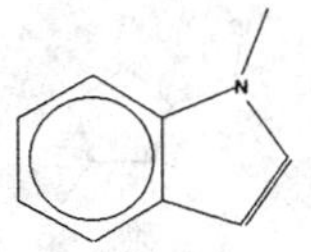

7313

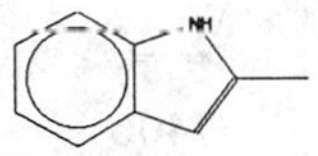

7314

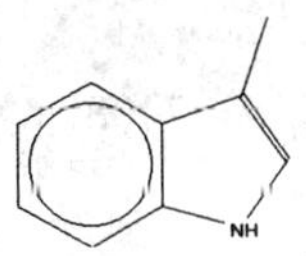

7315

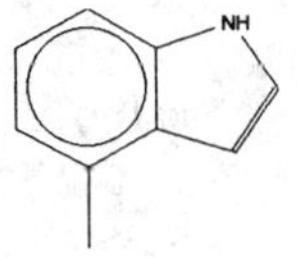

7316

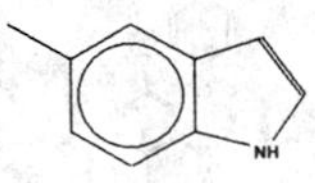

7317

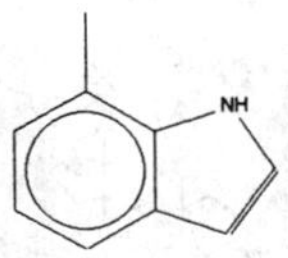

7319

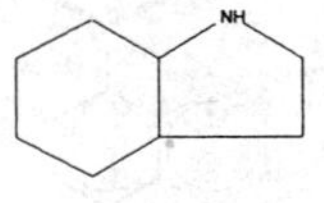

7320

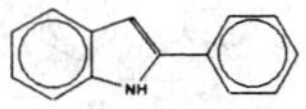

7321

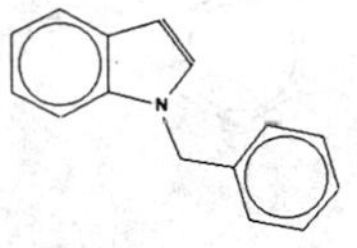

7322

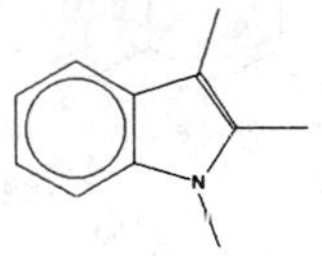

7323

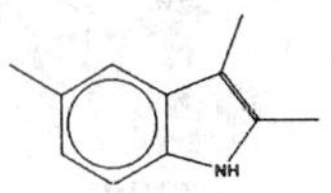

7324

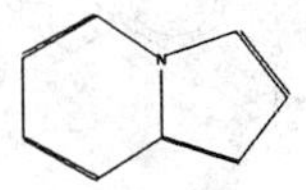

7325

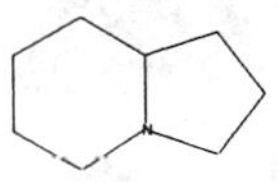

7326

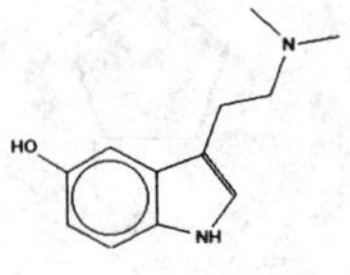

7327

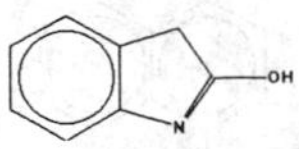

7328

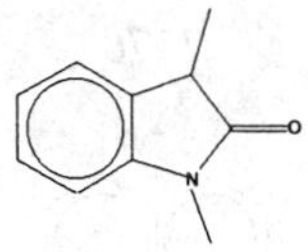

7329

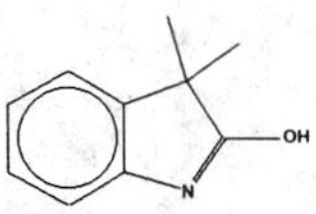

7330

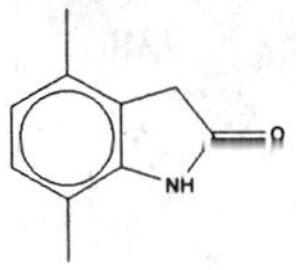

7331

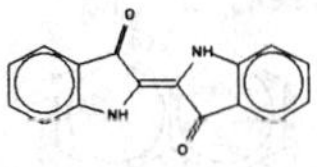

7335

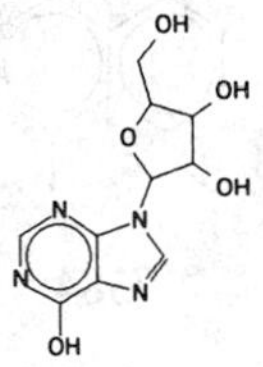

7336

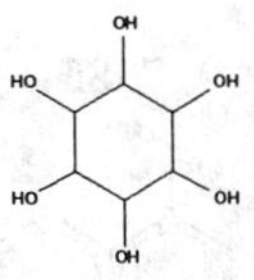

7337

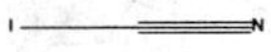

7338

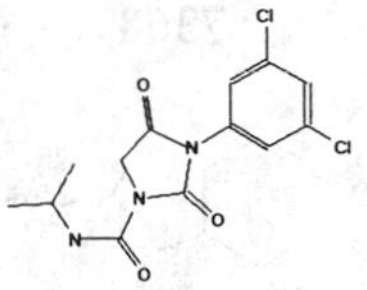

7339

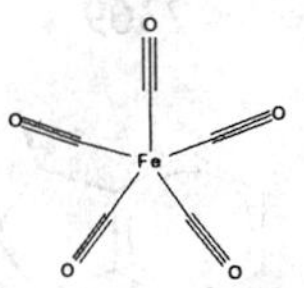

7340

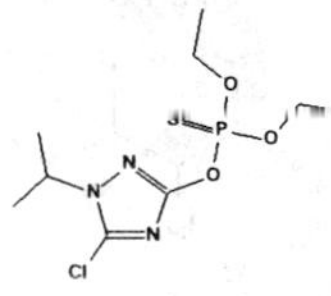

7342

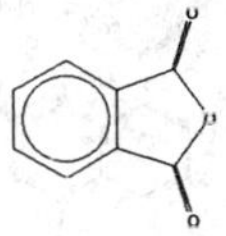

7343

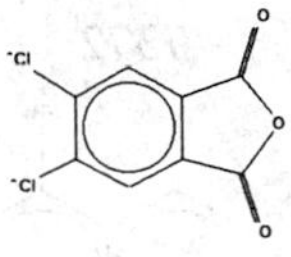

7344

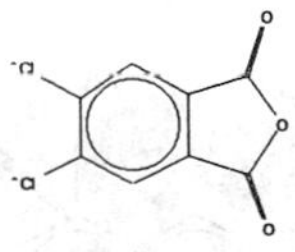

7345

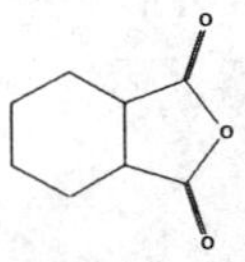

7346

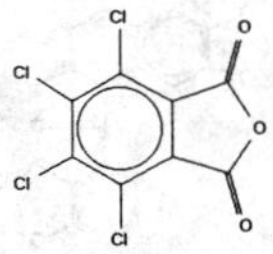

7347

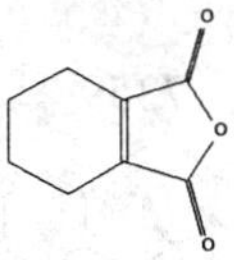

7350

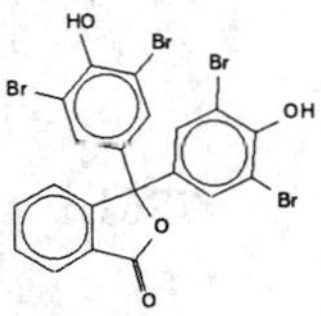

7351

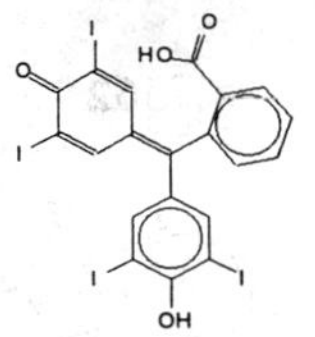

7354

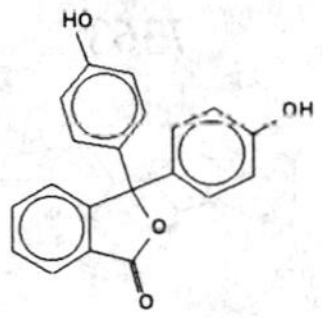

7355

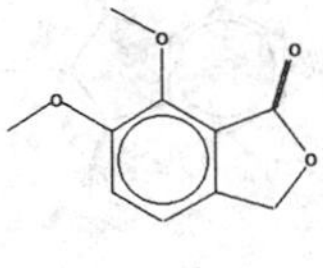

7356

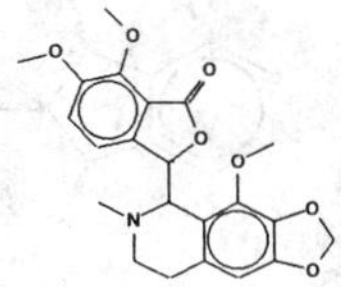

7357

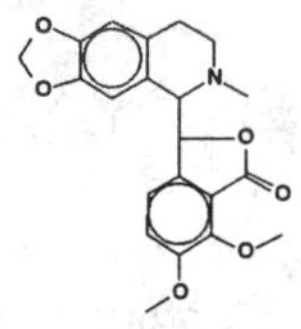

7359

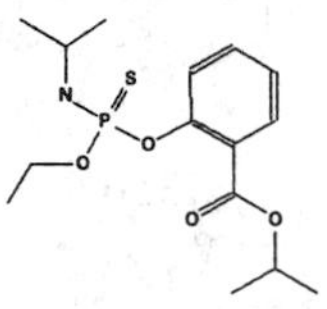

7361

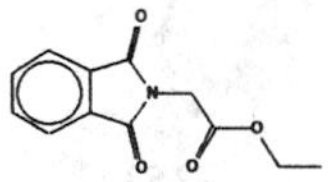

7362

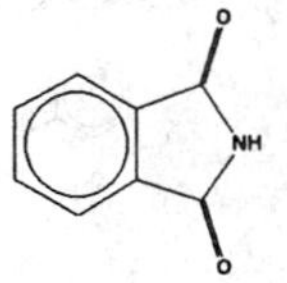

7363

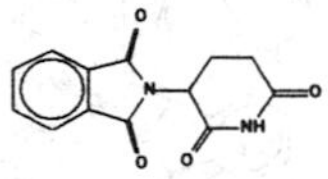

7364

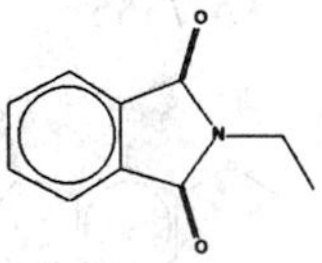

7365

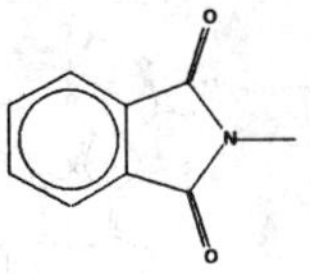

7367

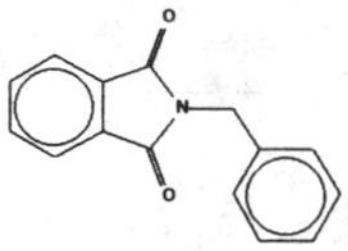

7368

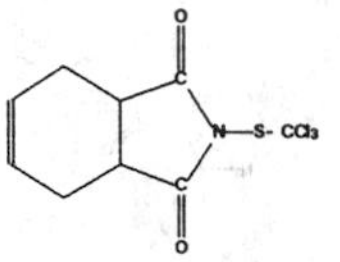

7369

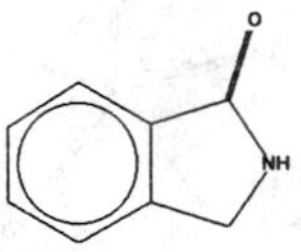

7370

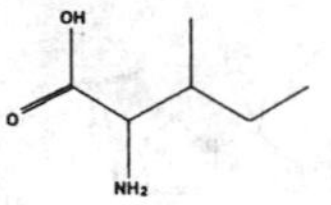

7371

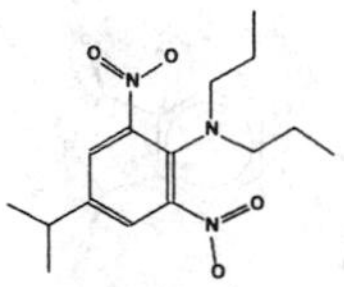

7372

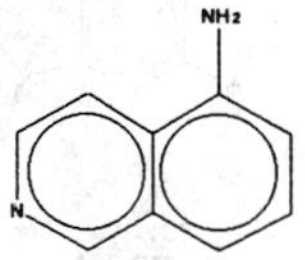

7374

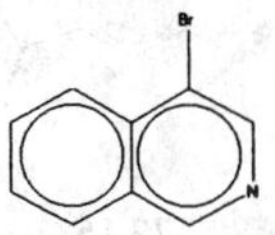

7375

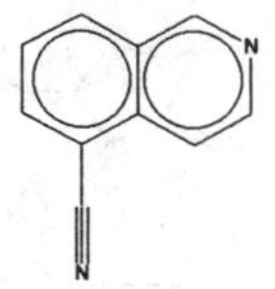

7376

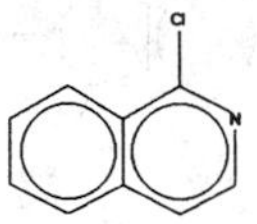

7378

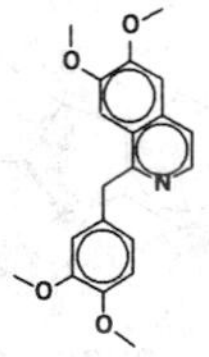

7380

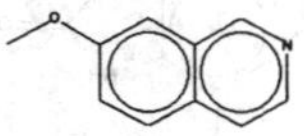

7381

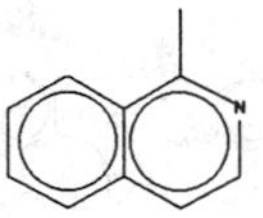

7383

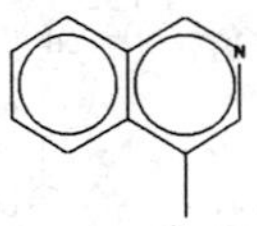

7384

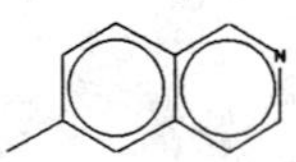

7385

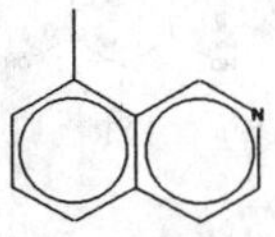

7386

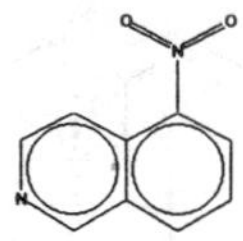

7387

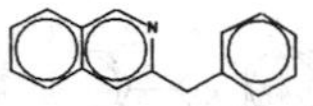

7388

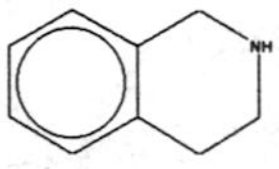

7389

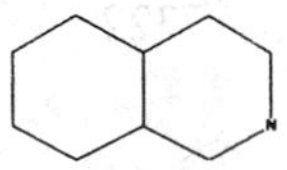

7390

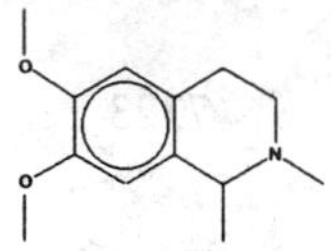

7398

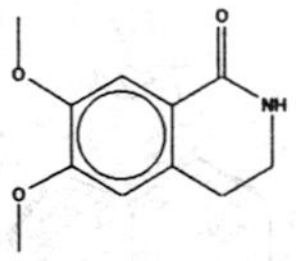

7399

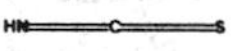

7402

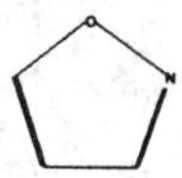

7403

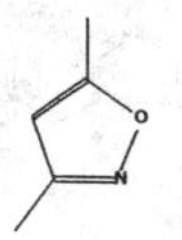

7404

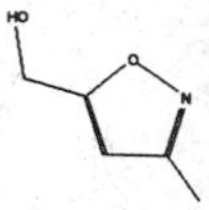

7405

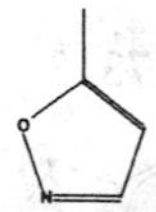

7406

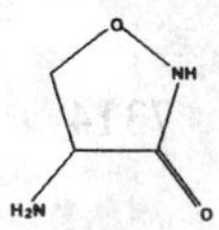

7408

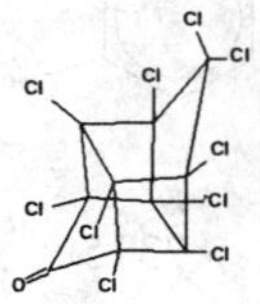

7409

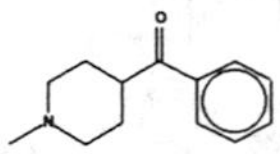

7410

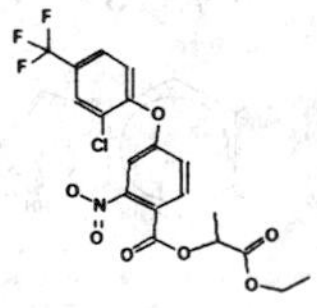

7411

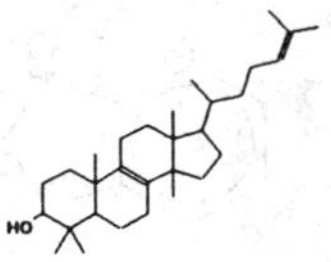

STRUCTURAL FORMULAS OF ORGANIC COMPOUNDS (continued)

In numeric order as they occur in the Table of Physical Constants of Organic Compounds

7412

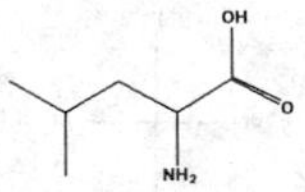

7413

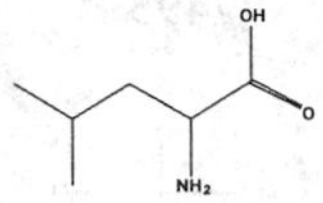

7415

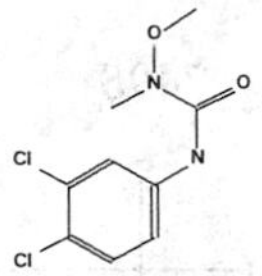

7416

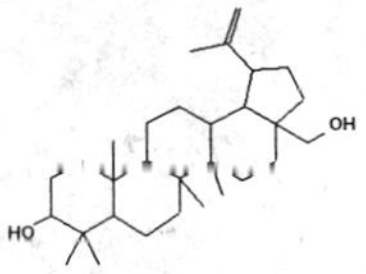

7417

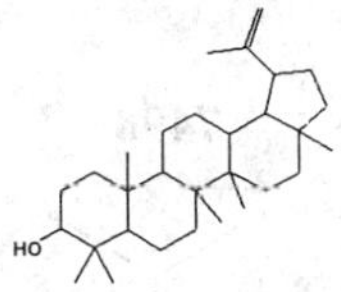

7418

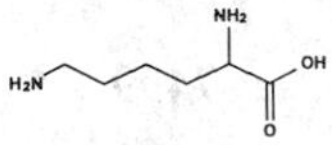

7419

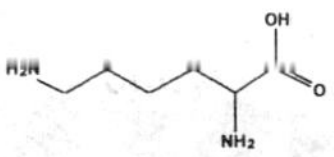

7420

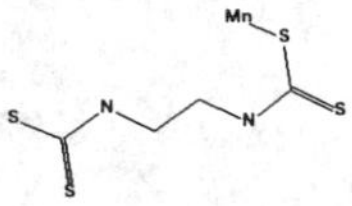

7421

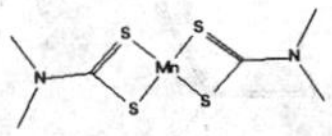

7422

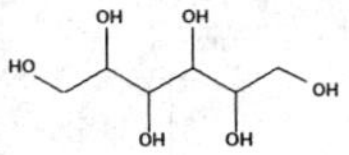

7423

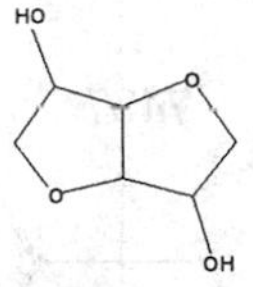

7425

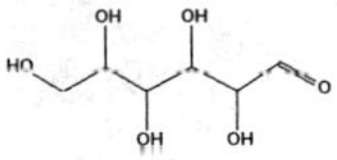

7426

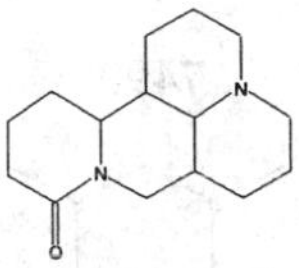

7427

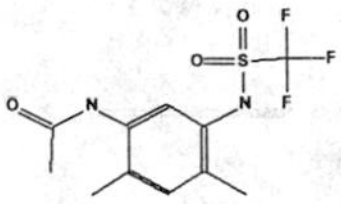

7430

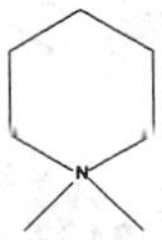

7431

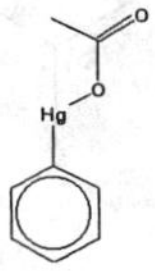

7432

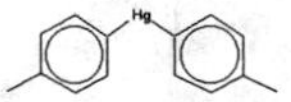

7433

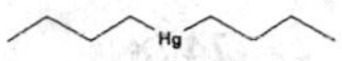

7434

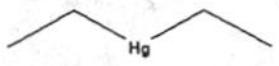

7435

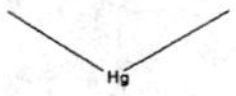

7436

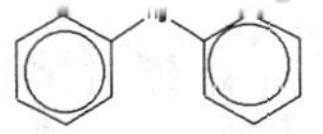

7438

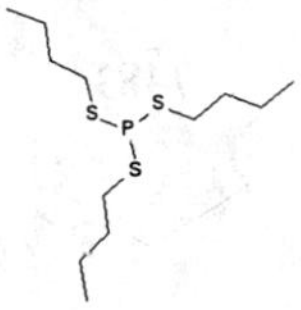

7439

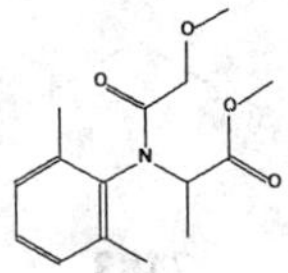

7441

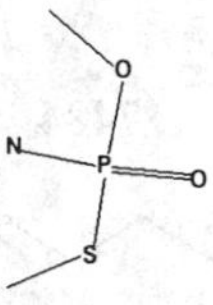

7442

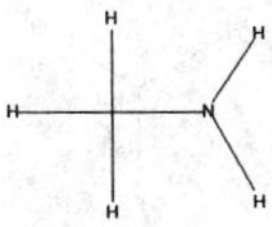

7443

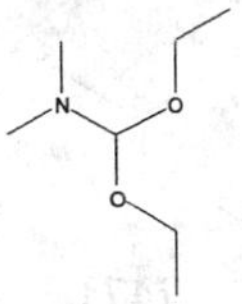

7444

7446

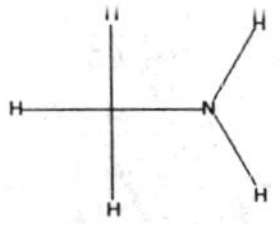

7447

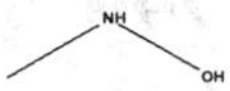

7448

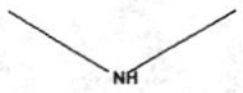

7449

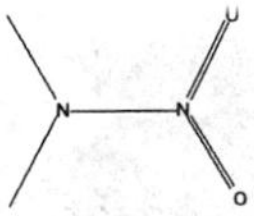

7450

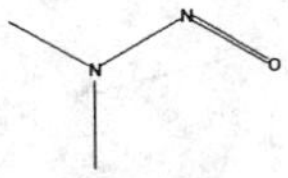

7451

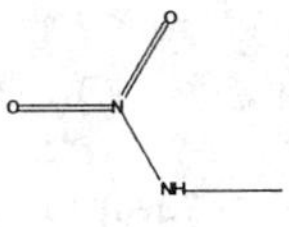

7452

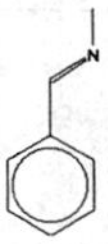

7454

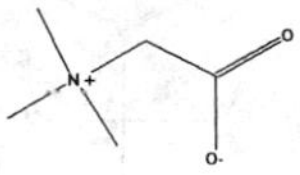

7455

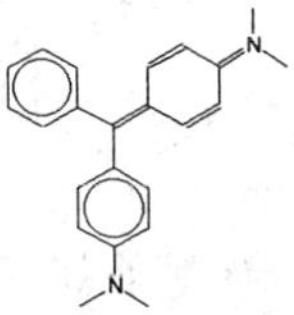

7459

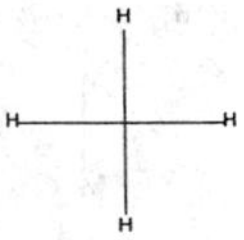

7460

7461

In numeric order as they occur in the Table of Physical Constants of Organic Compounds

7462

Br

7463

Cl Br

7464

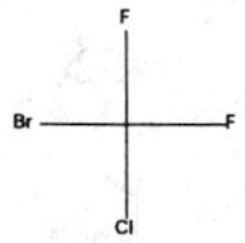

F Br F Cl

7465

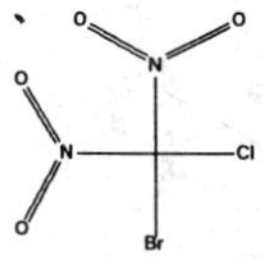

O O N O N Cl O Br

7466

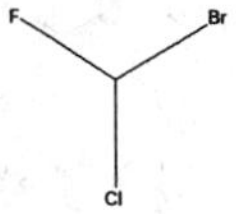

F Br Cl

7467

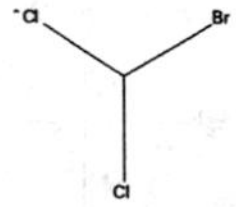

Cl Br Cl

7468

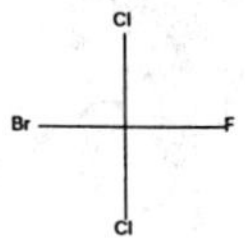

Cl Br F Cl

7469

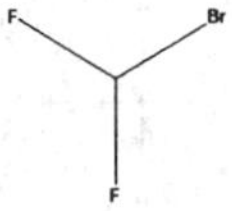

F Br F

7470

F Br

7471

Br I

7472

O Br

7473

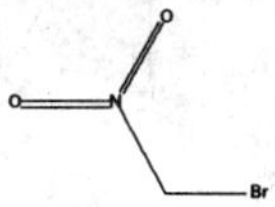

O O N Br

7474

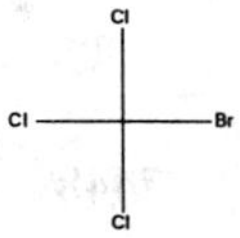

Cl Cl Br Cl

7475

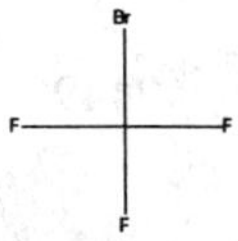

Br F F F

7476

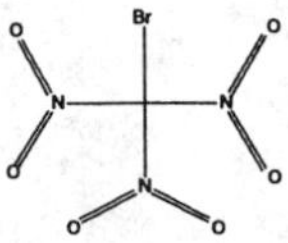

Br O O N N O O N O O

7477

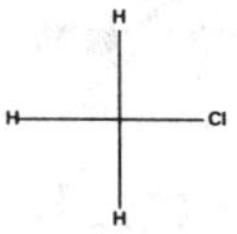

H H Cl H

7478

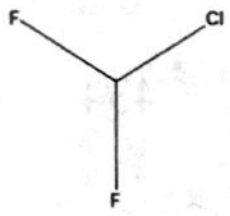

F Cl F

7479

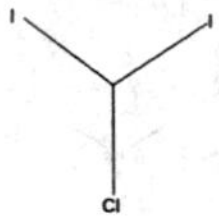

I I Cl

7480

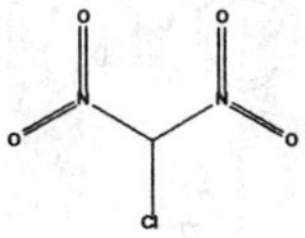

O O N N O O Cl

7481

F Cl

7482

Cl I

7483

O Cl

7484

S Cl

7485

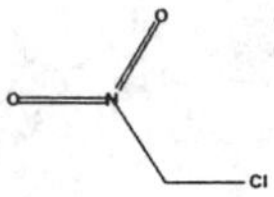

O O N Cl

7486

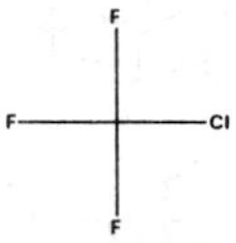

F F Cl F

7487

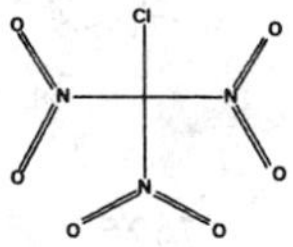

Cl O O N N O O N O O

7489

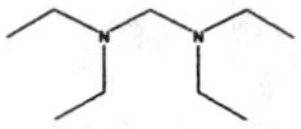

N N

7490

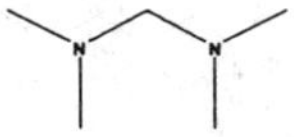

N N

7491

N N

7492

Br Br

7493

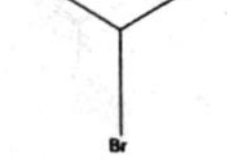

Cl Br Br

7494

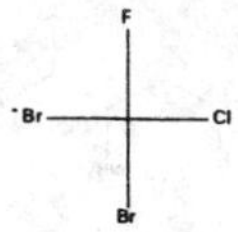

F Br Cl Br

7495

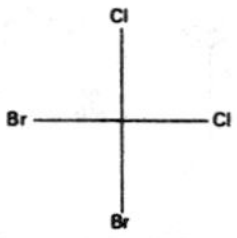

Cl Br Cl Br

7496

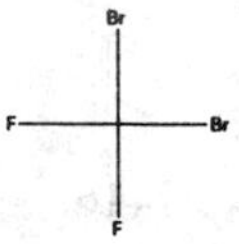

Br F Br F

7497

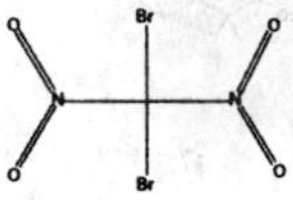

Br O O N N O O Br

7498

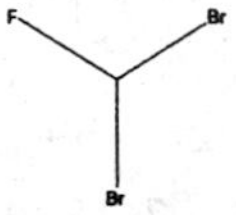

F Br Br

7499

Cl Cl

7500

Cl F Cl F

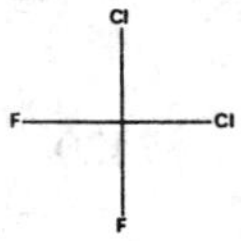

STRUCTURAL FORMULAS OF ORGANIC COMPOUNDS (continued)

In numeric order as they occur in the Table of Physical Constants of Organic Compounds

7501

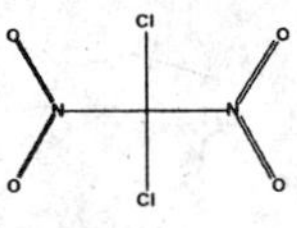

7502

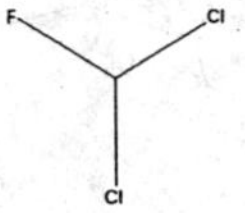

7503

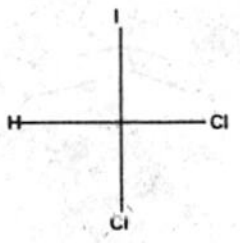

7504

7505

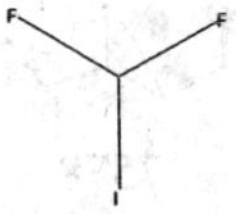

7506

7507

7508

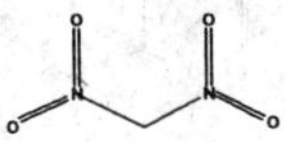

7509

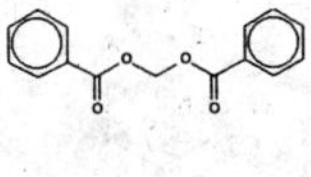

7510

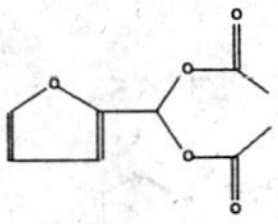

7512

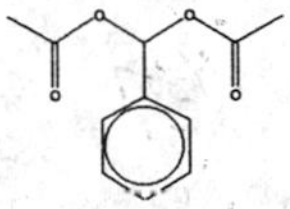

7513

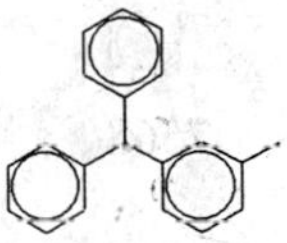

7515

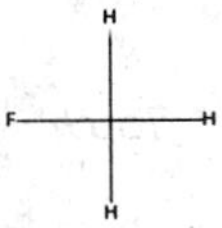

7516

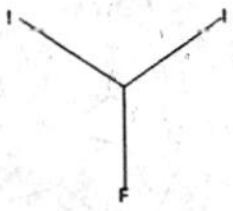

7517

7518

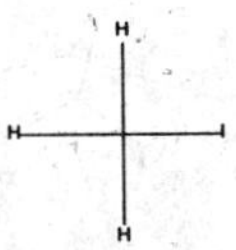

7519

7520

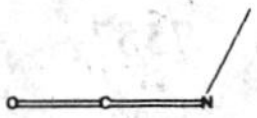

7521

7522

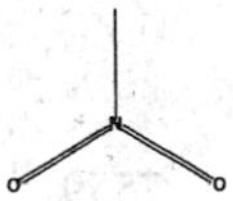

7523

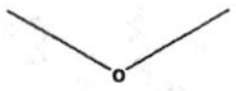

7524

7525

7526

7527

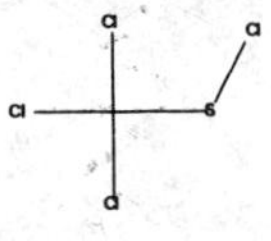

7528

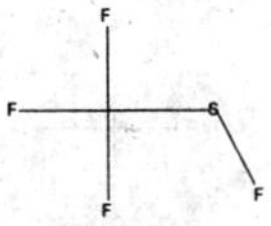

7529

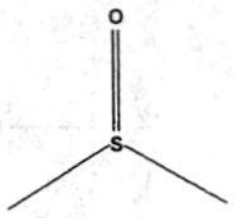

7530

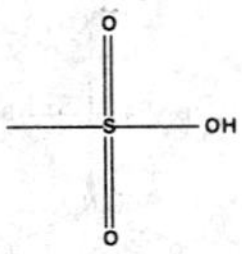

7531

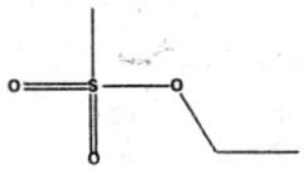

7532

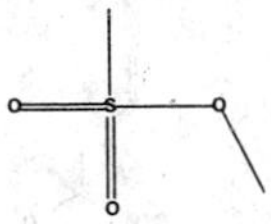

7533

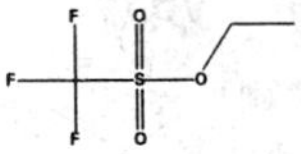

7534

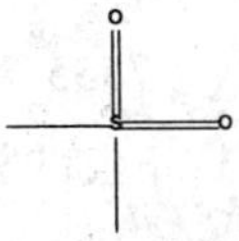

7535

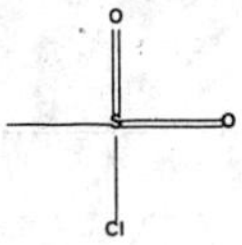

7536

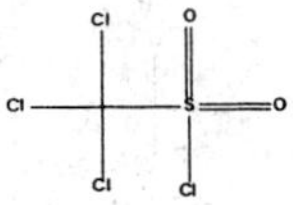

7537

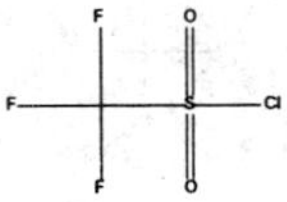

7538

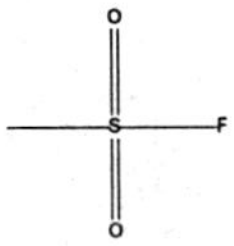

7539

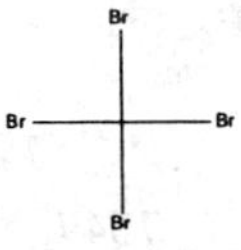

7540

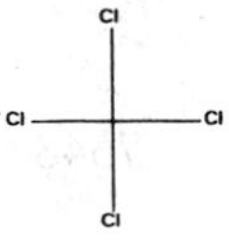

7541

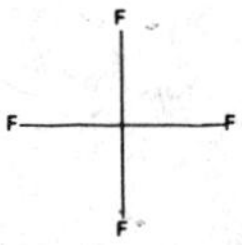

STRUCTURAL FORMULAS OF ORGANIC COMPOUNDS (continued)

In numeric order as they occur in the Table of Physical Constants of Organic Compounds

7542

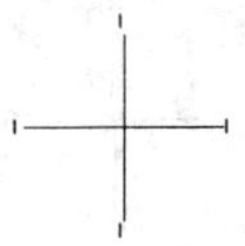

7543

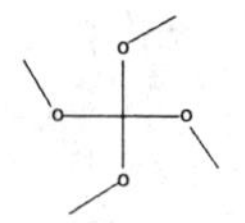

7544

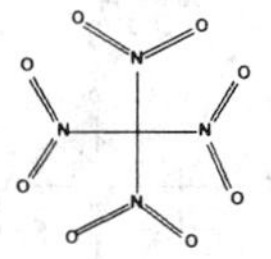

7545

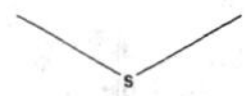

7546

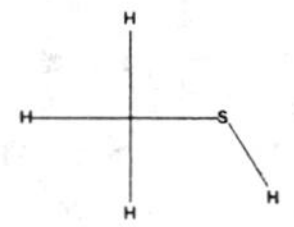

7547

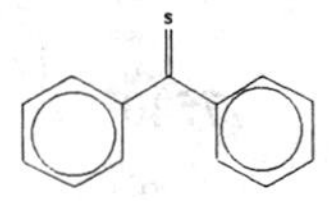

7548

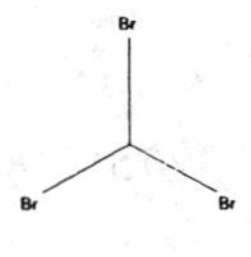

7549

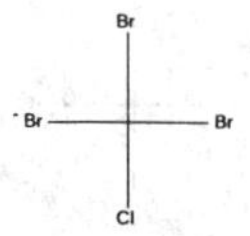

7550

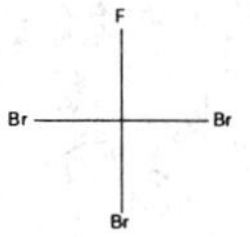

7551

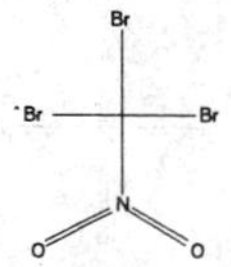

7552

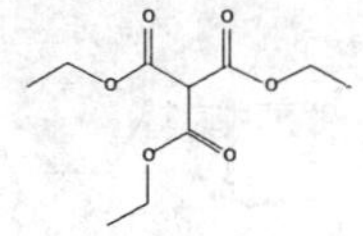

7553

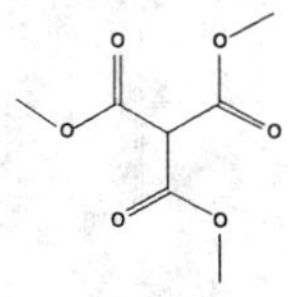

7554

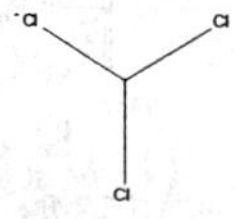

7555

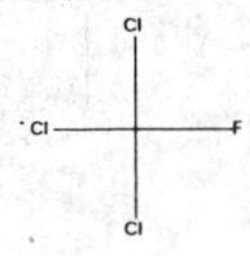

7556

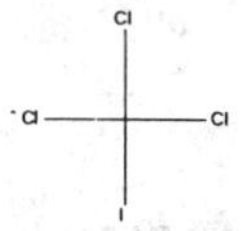

7557

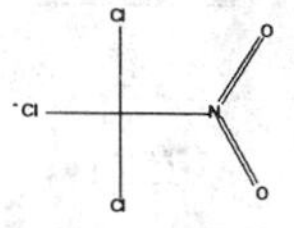

7558

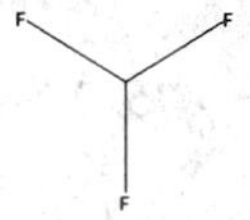

7559

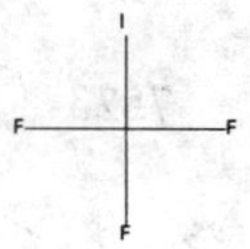

7560

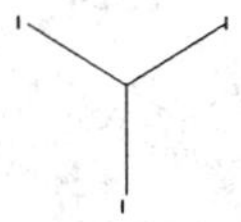

7561

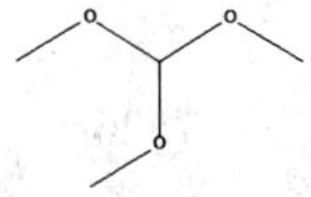

7562

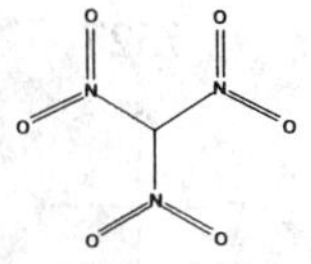

7564

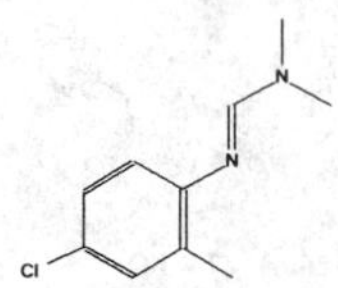

7565

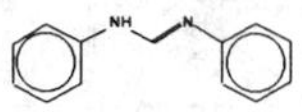

7567

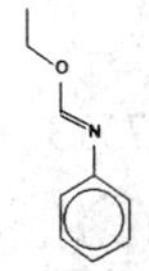

7568

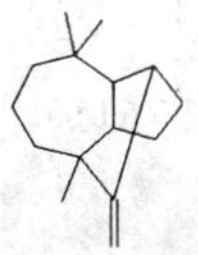

7569

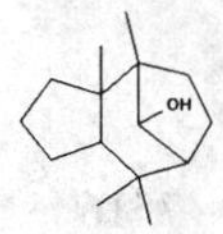

7570

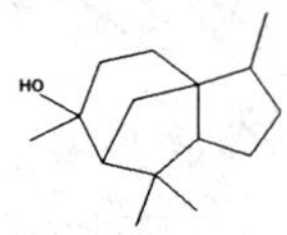

7571

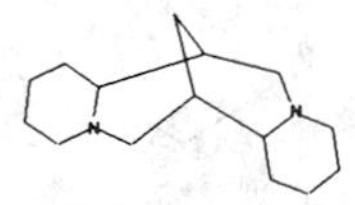

7572

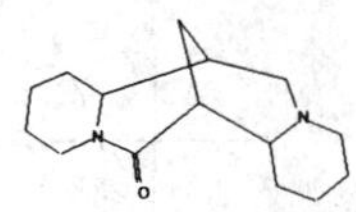

7573

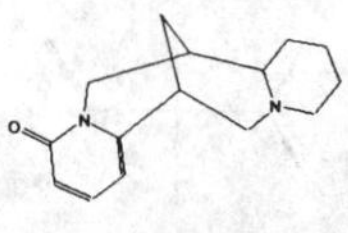

7574

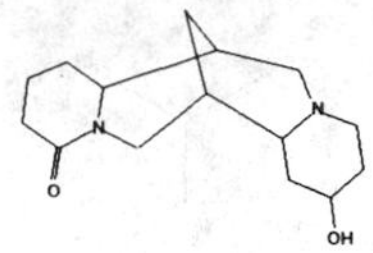

7575

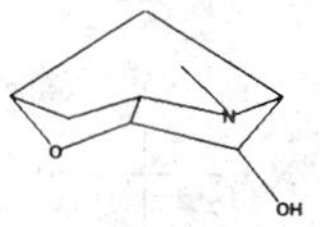

7576

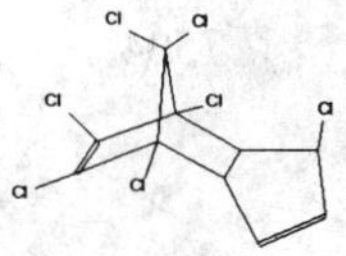

7577

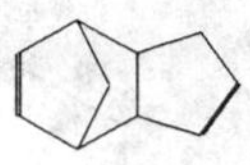

7578

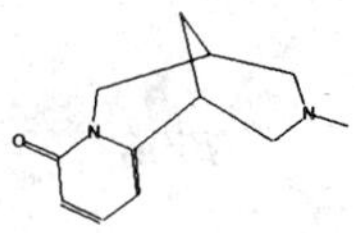

7579

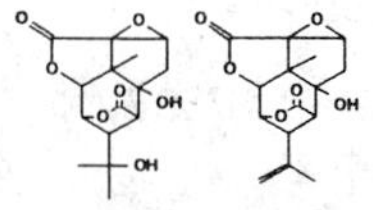

7580

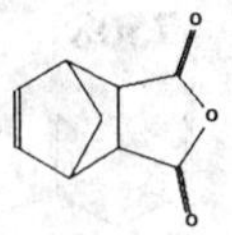

STRUCTURAL FORMULAS OF ORGANIC COMPOUNDS (continued)

In numeric order as they occur in the Table of Physical Constants of Organic Compounds

7581

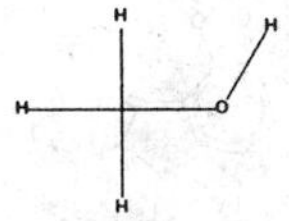

7582

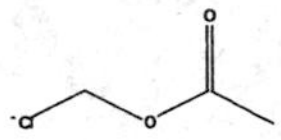

7585

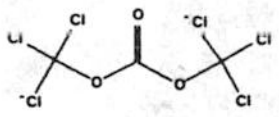

7586

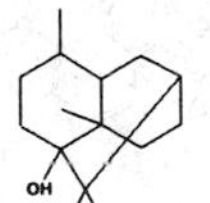

7587

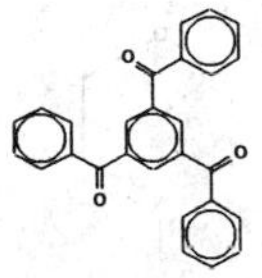

7588

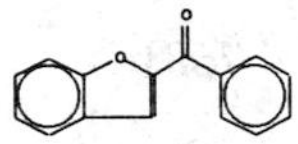

7590

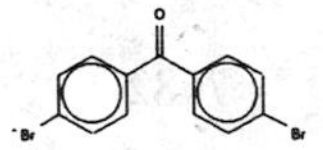

7591

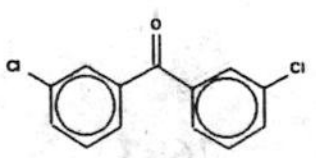

7592

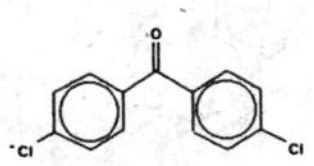

7593

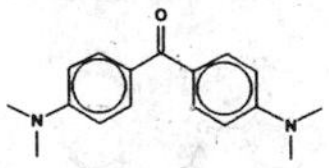

7594

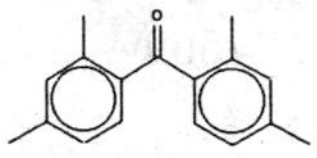

7595

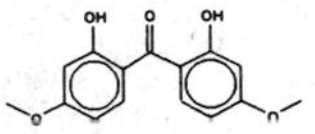

7597

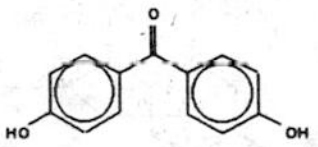

7598

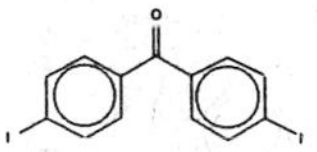

7599

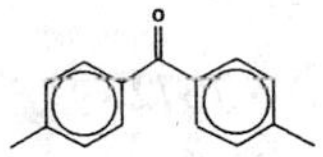

7600

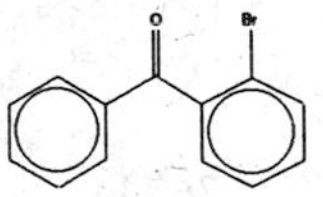

7601

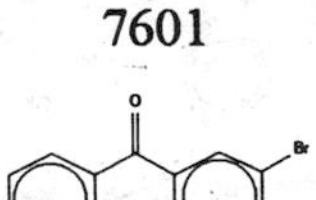

7602

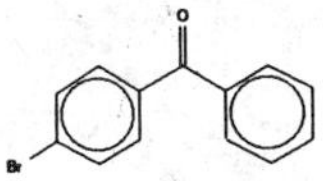

7604

7605

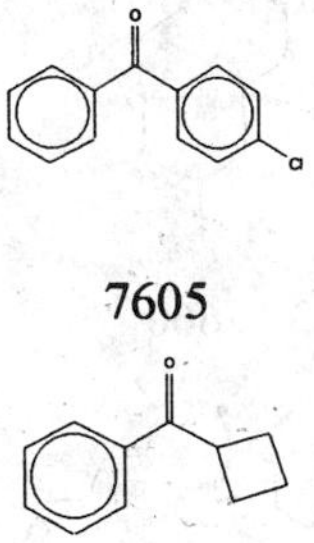

7606

7607

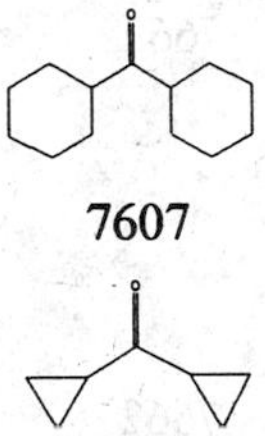

7609

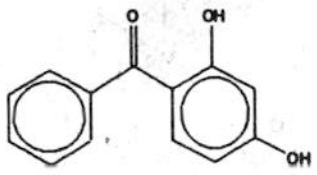

7612

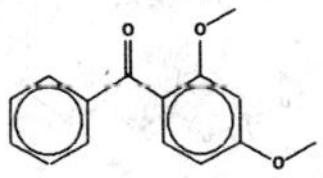

7614

7615

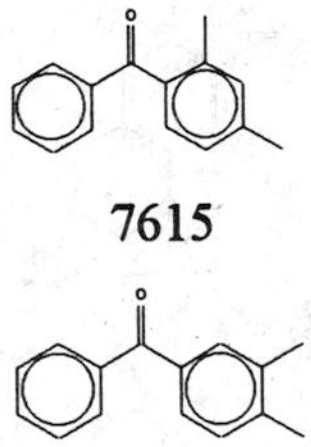

7616

7617

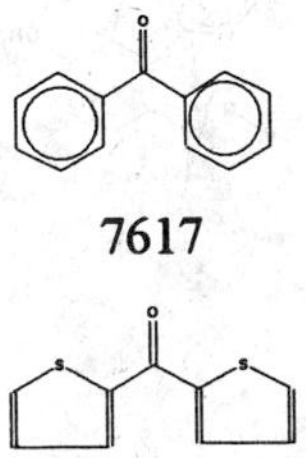

7618

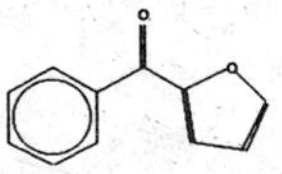

7619

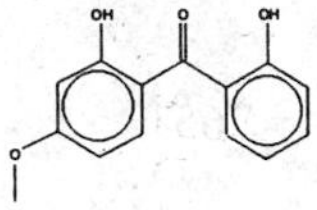

7621

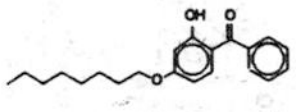

7622

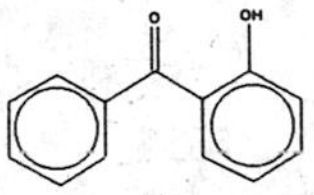

7623

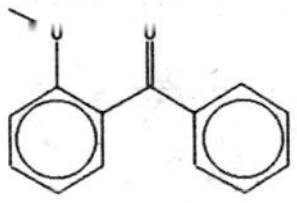

7624

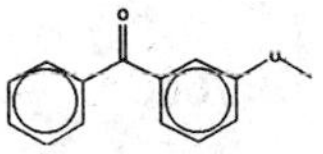

7625

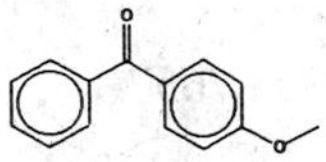

7626

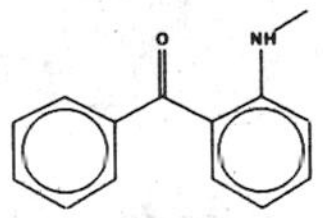

7629

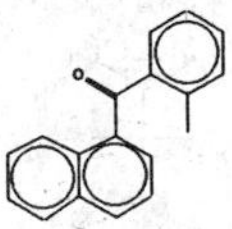

7631

7632

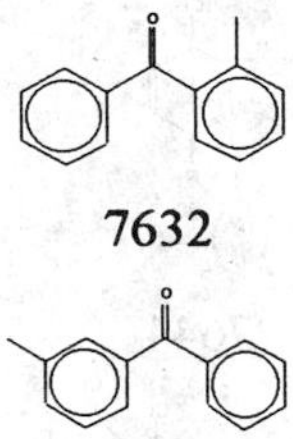

7633

7635

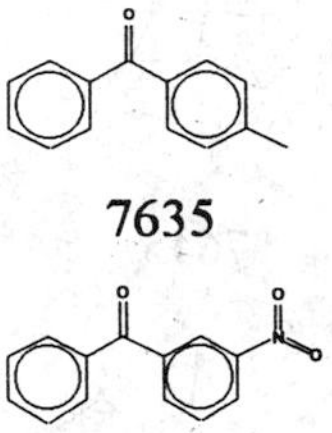

7636

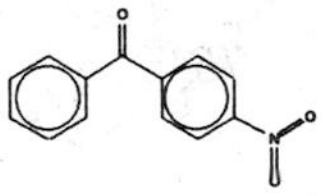

7639

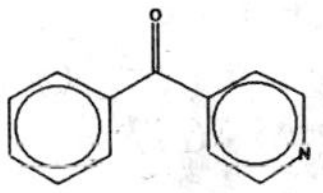

7640

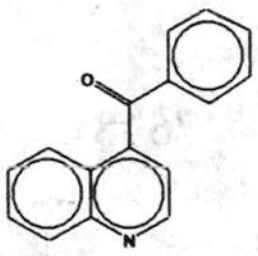

7641

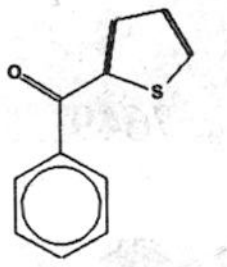

7642

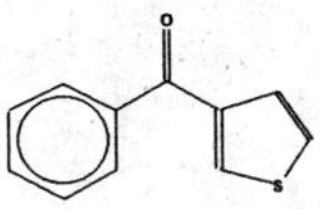

STRUCTURAL FORMULAS OF ORGANIC COMPOUNDS (continued)

In numeric order as they occur in the Table of Physical Constants of Organic Compounds

7643

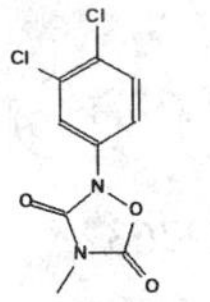

7644

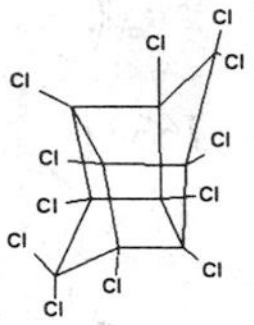

7645

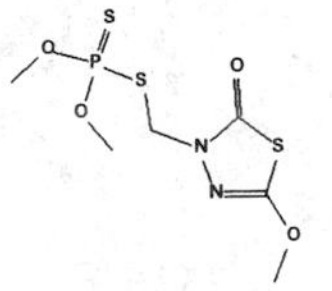

7646

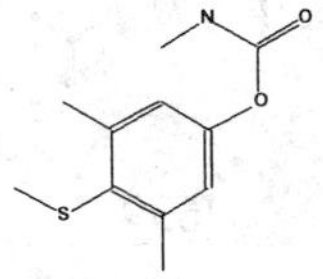

7647

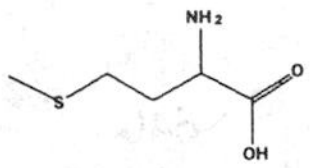

7648

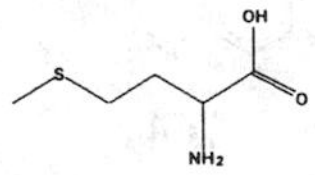

7649

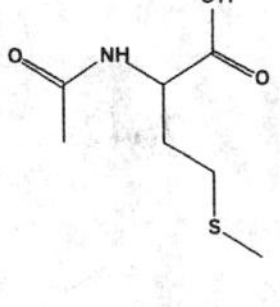

7650

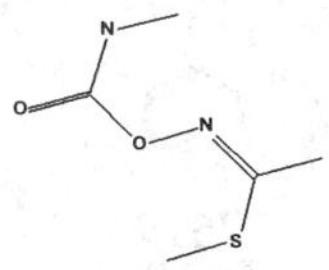

7651

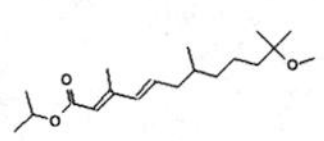

7653

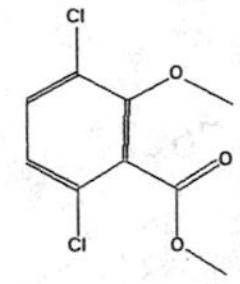

7654

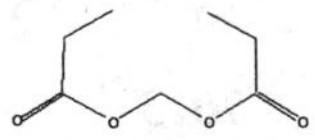

7655

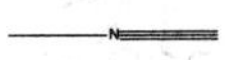

7656

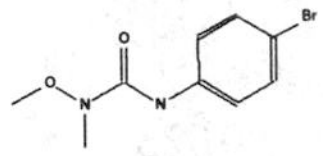

7657

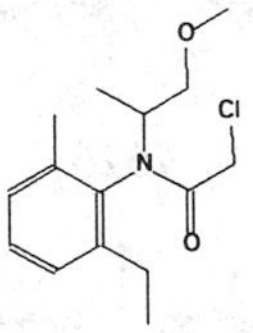

7658

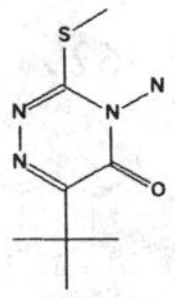

7659

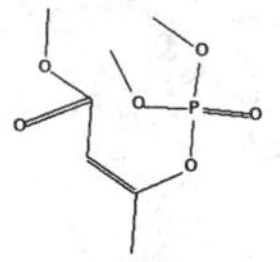

7660

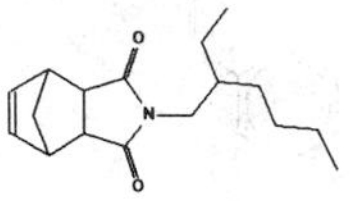

7661

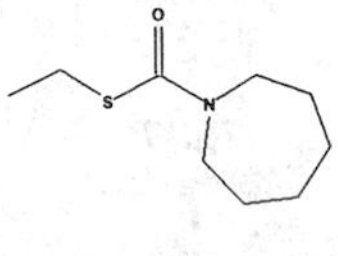

7662

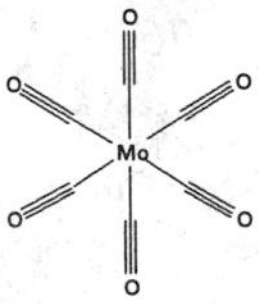

7663

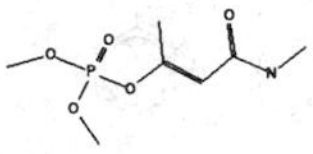

7665

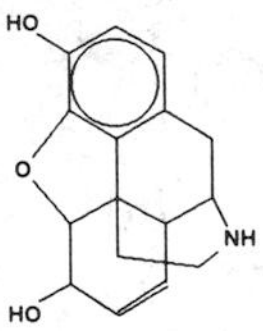

7666

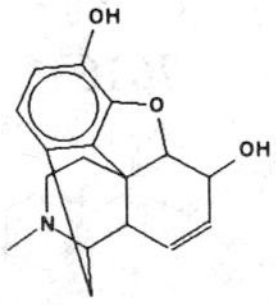

7667

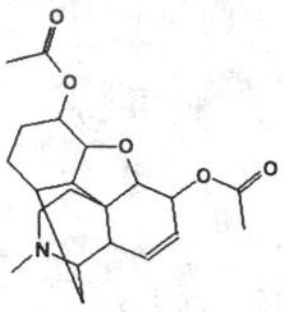

7669

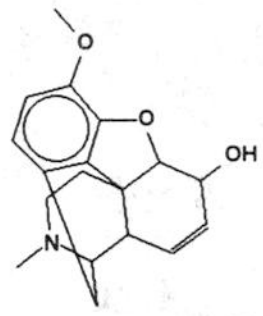

7670

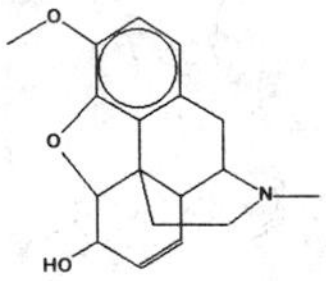

7671

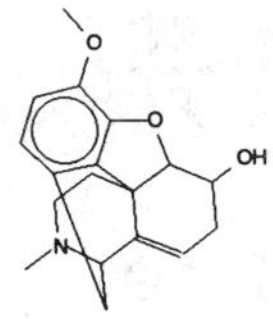

7672

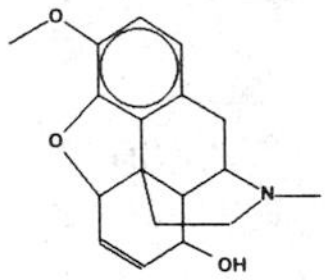

7674

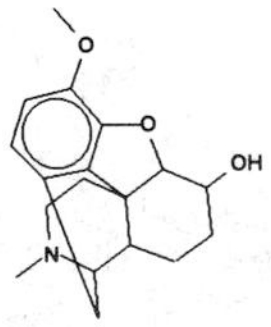

7675

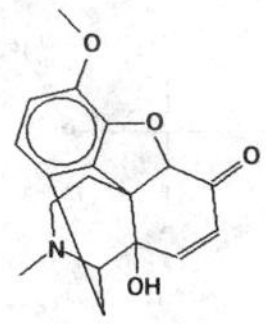

7678

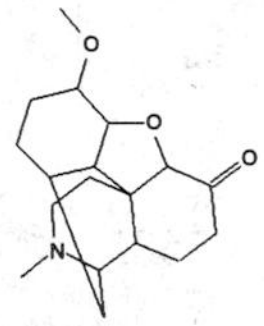

7679

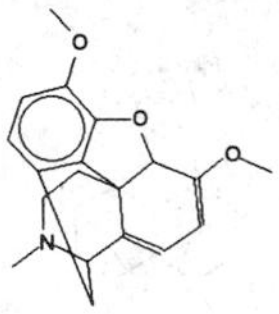

7680

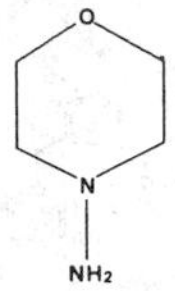

7681

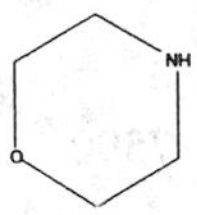

7682

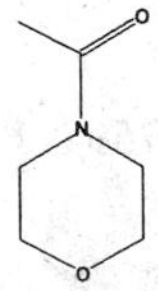

7683

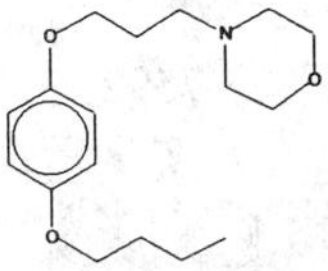

STRUCTURAL FORMULAS OF ORGANIC COMPOUNDS (continued)

In numeric order as they occur in the Table of Physical Constants of Organic Compounds

7684

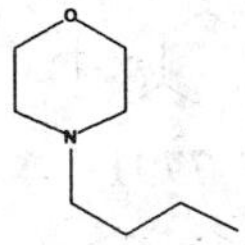

7685

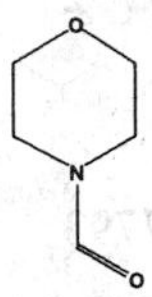

7686

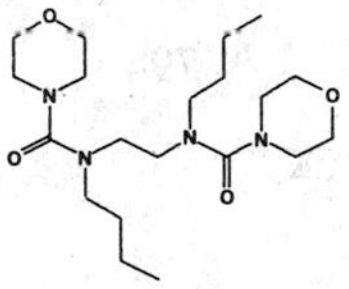

7687

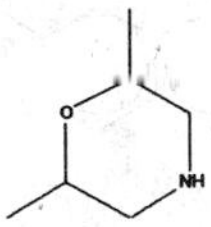

7689

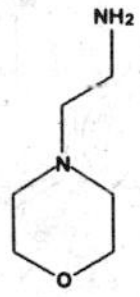

7690

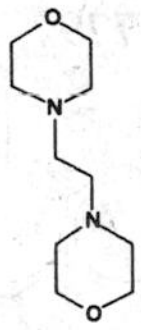

7691

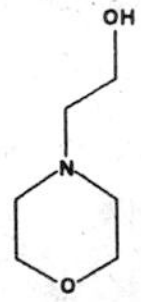

7692

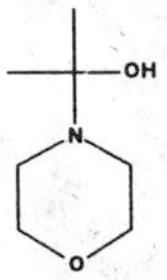

7693

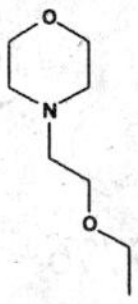

7694

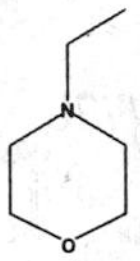

7696

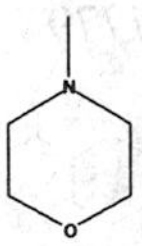

7697

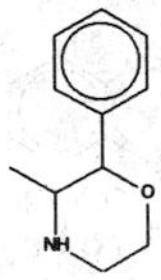

7698

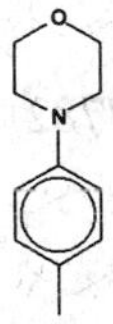

7699

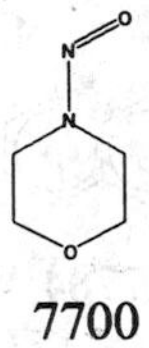

7700

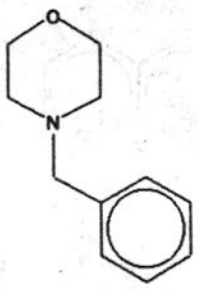

7701

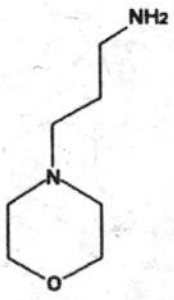

7702

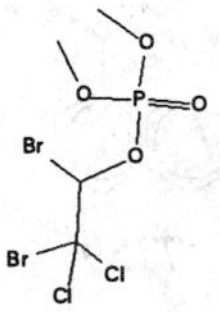

7703

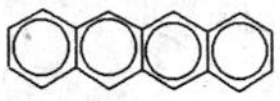

7704

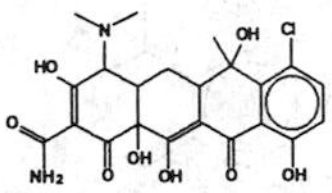

7705

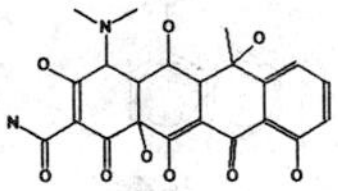

7706

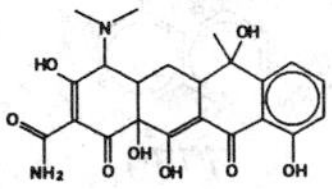

7708

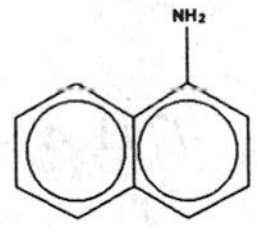

7709

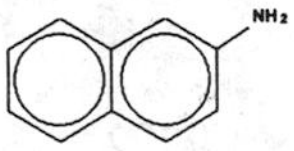

7710

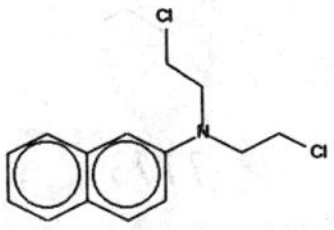

7712

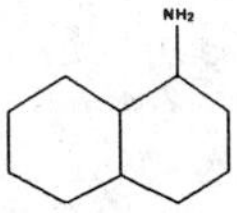

7713

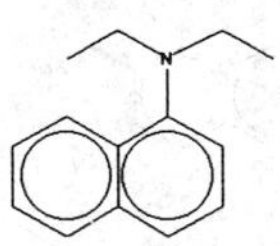

7714

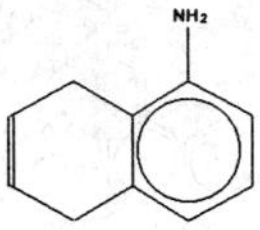

7715

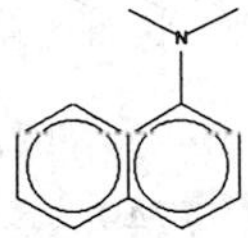

7716

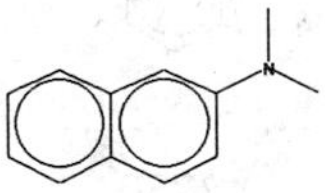

7717

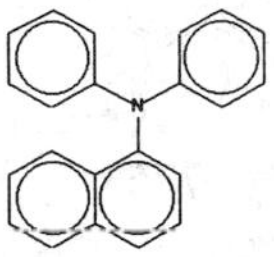

7718

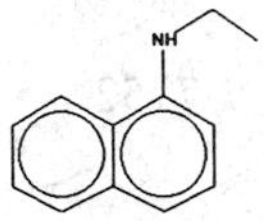

7720

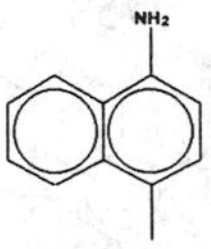

7721

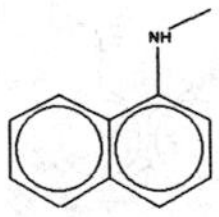

7722

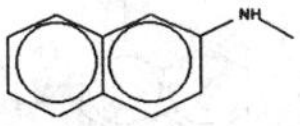

7723

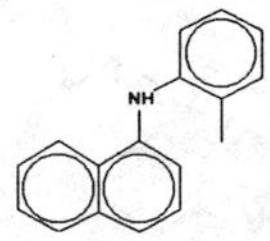

7724

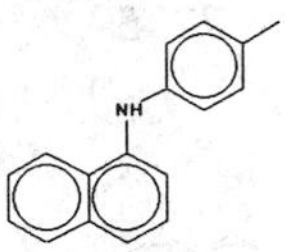

7725

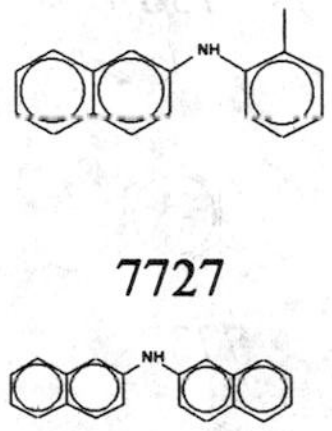

7727

7728

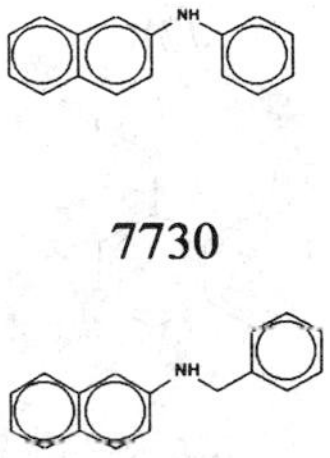

7730

7731

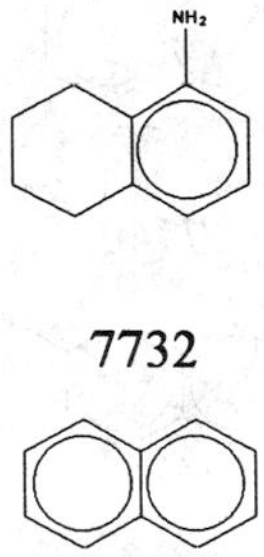

7732

7733

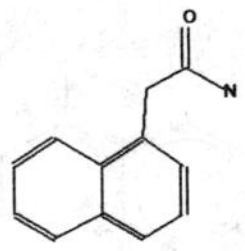

STRUCTURAL FORMULAS OF ORGANIC COMPOUNDS (continued)

In numeric order as they occur in the Table of Physical Constants of Organic Compounds

7734

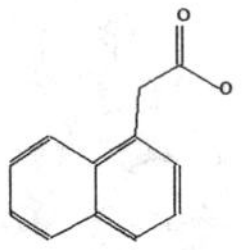

7735

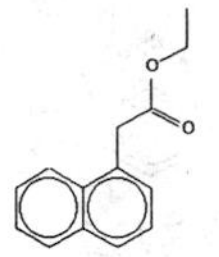

7736

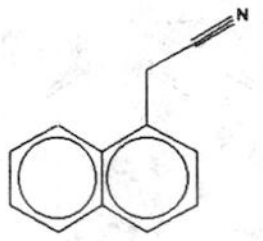

7737

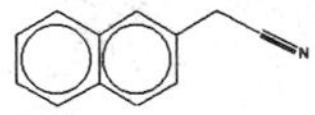

7738

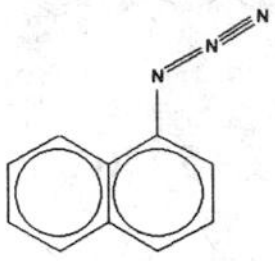

7739

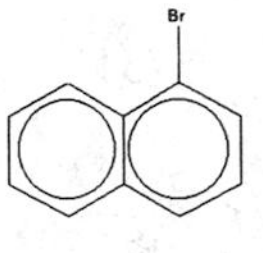

7740

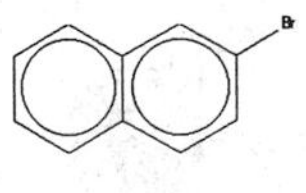

7742

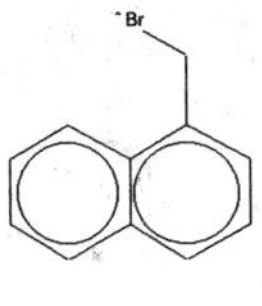

7743

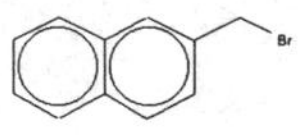

7746

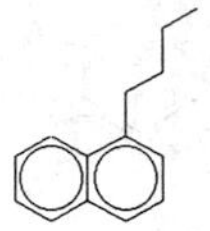

7747

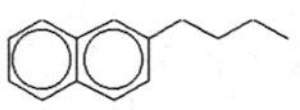

7749

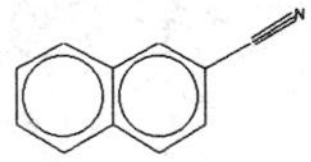

7750

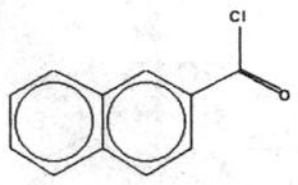

7751

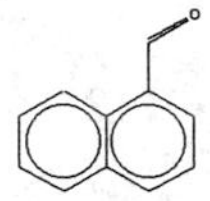

7752

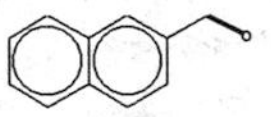

7753

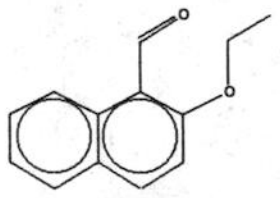

7754

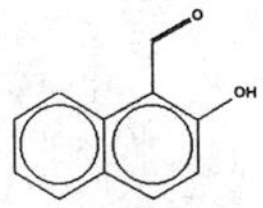

7756

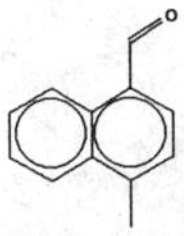

7758

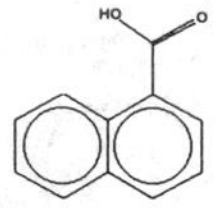

7759

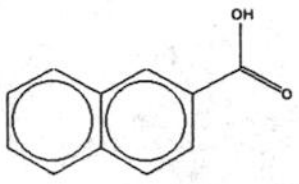

7764

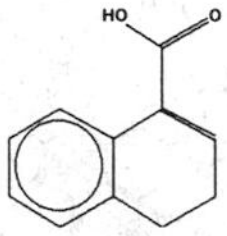

7765

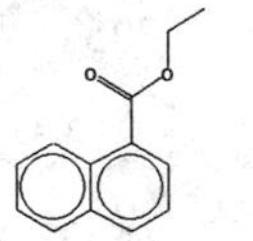

7766

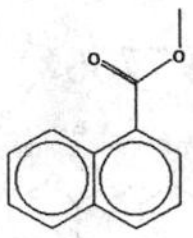

7770

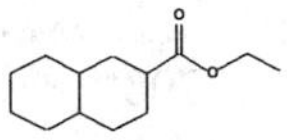

7771

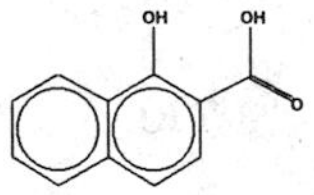

7772

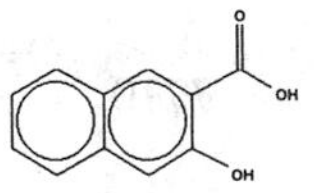

7773

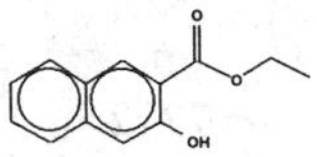

7775

7776

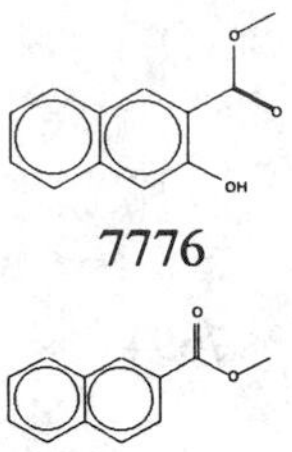

7778

7779

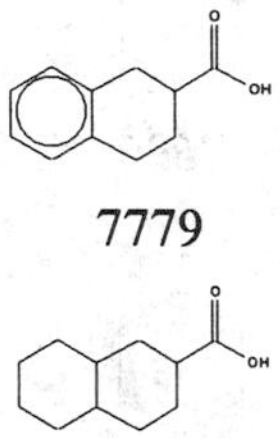

7780

7781

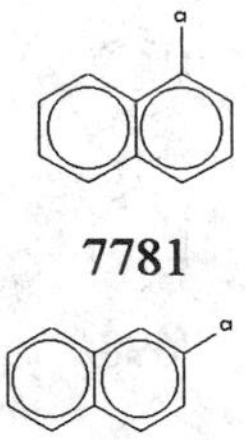

7782

7783

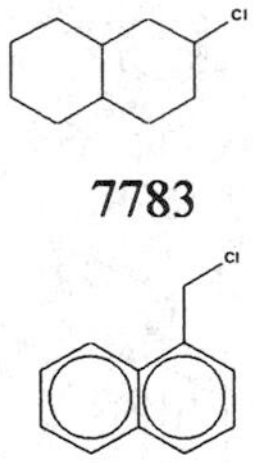

7784

7785

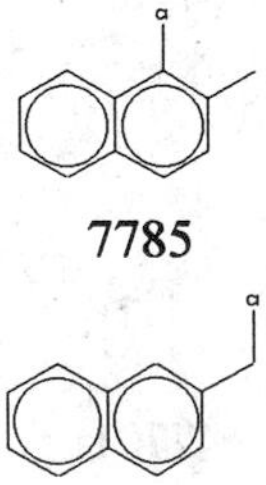

7789

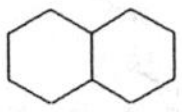

7790

7791

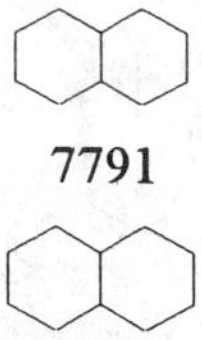

7792

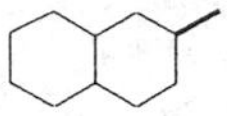

7793

7794

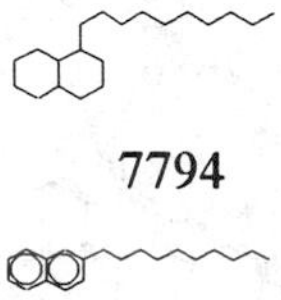

7795

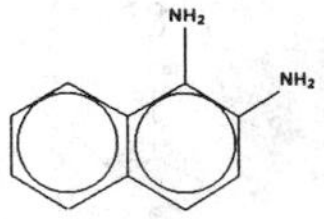

7796

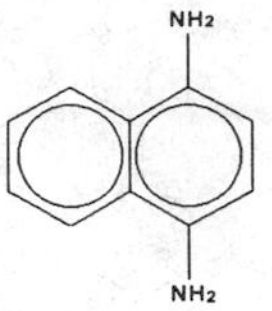

7797

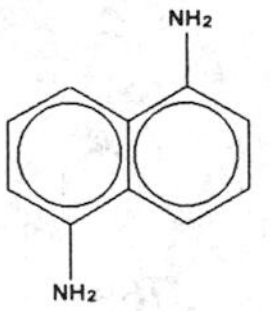

7798

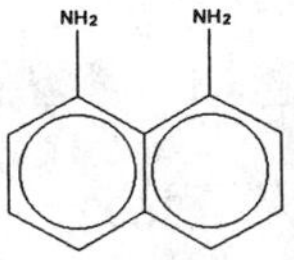

In numeric order as they occur in the Table of Physical Constants of Organic Compounds

7799 NH2 NH2

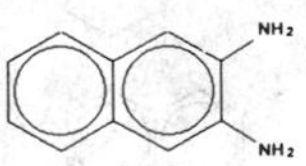

7801 Br Br

7802 Br Br

7803 Br Br

7804 Cl O Cl O

7806 O O O O

7807 Cl Cl

7808 Cl Cl

7809 Cl Cl

7810 Cl Cl

7811 Cl Cl

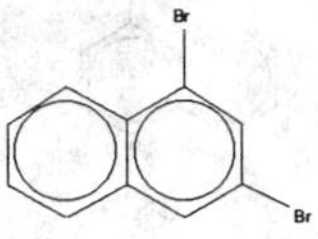

7812 Cl Cl

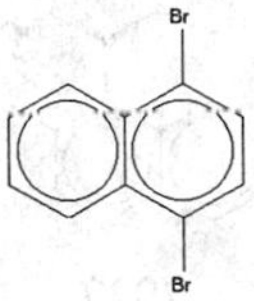

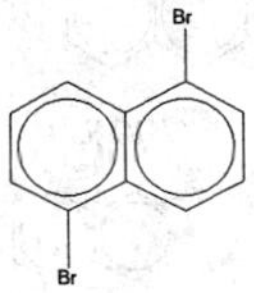

7813 Cl Cl

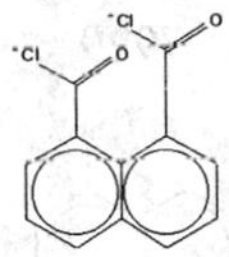

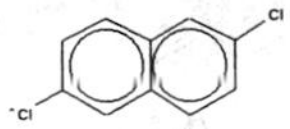

7814

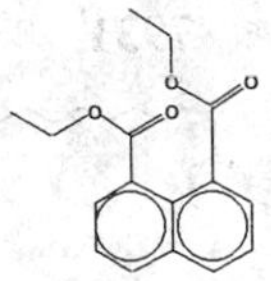

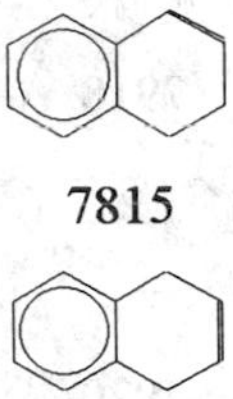

7815

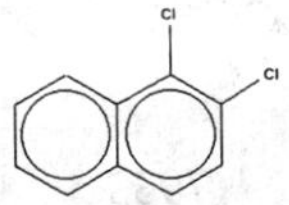

7816

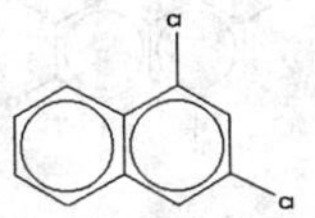

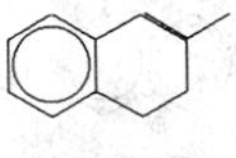

7817

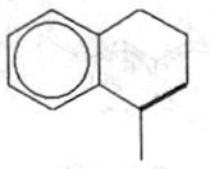

7818

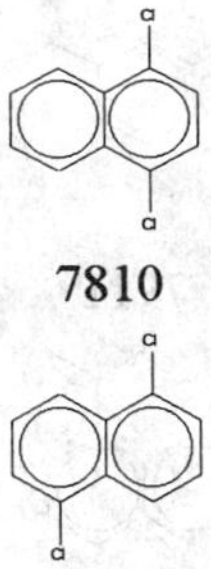

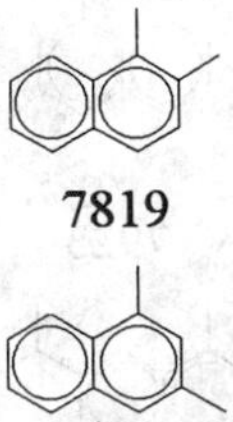

7819

7820

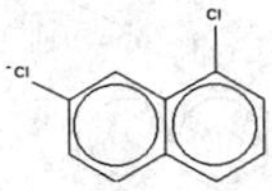

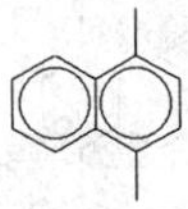

7821

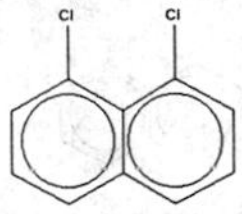

7822

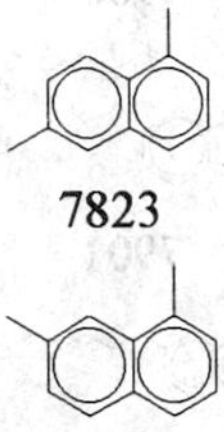

7823

7824

7825

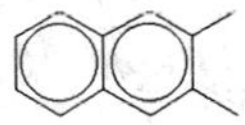

7826

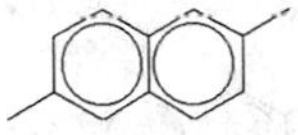

7827

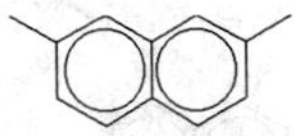

7828

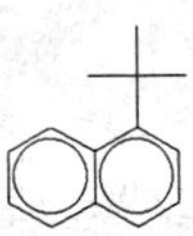

7829

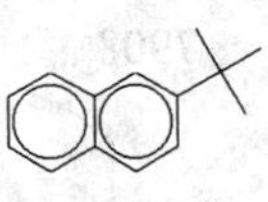

7830

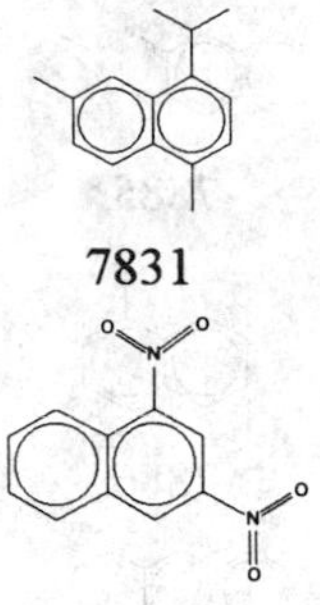

7831 O N O N O O

7832 O N O N O O

7833 O N O O N O

7834 OH OH

7835 OH OH

7836 OH HO

7838 OH HO

7839 HO OH

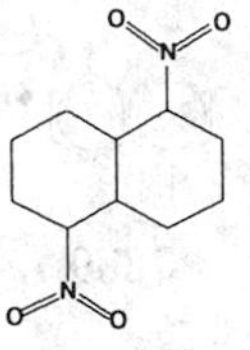

7842 O O

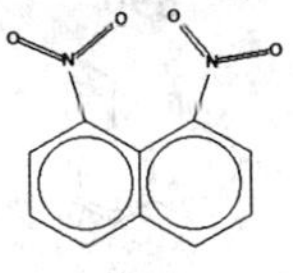

7843 O O

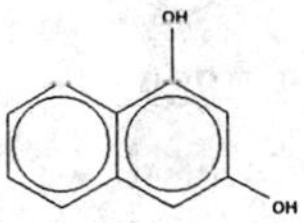

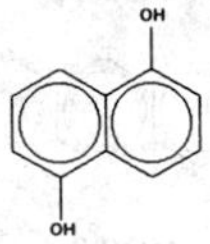

7846 O Br O

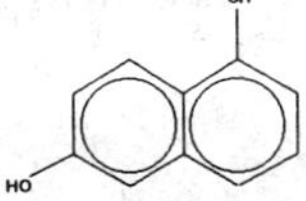

7848 O OH Cl O

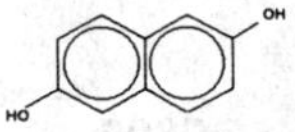

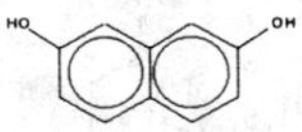

7850 O Cl Cl O

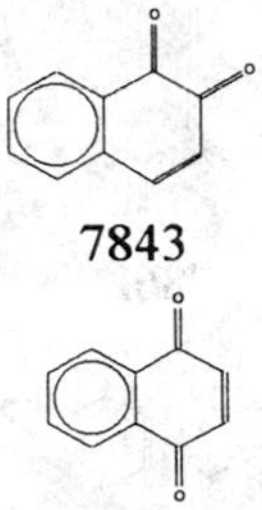

7852 O OH OH O

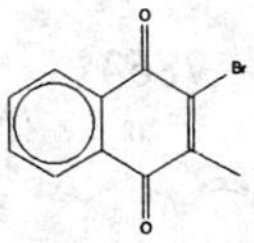

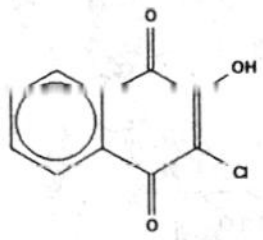

7853 OH O OH O

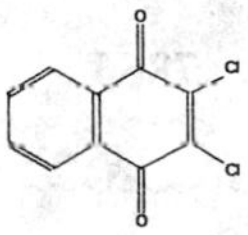

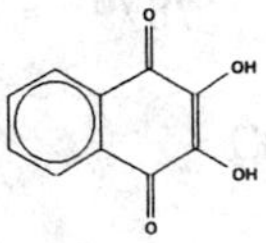

7859 O OH O

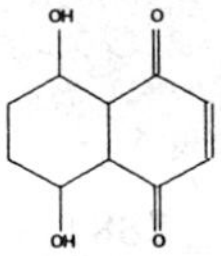

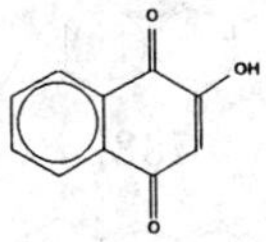

7860 OH O O

7861 O OH O

7862 O OH O

7863 O OH O

7864 O O

7865 O O

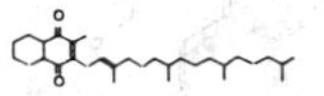

7879

7880

7881

7882

7884 O

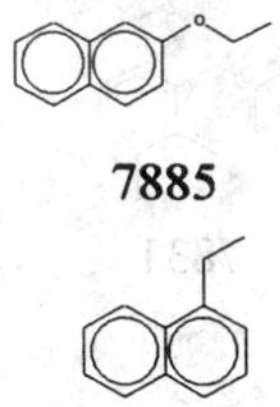

7885

7886

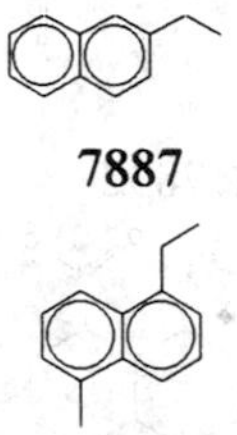

7887

7889

7890

7891

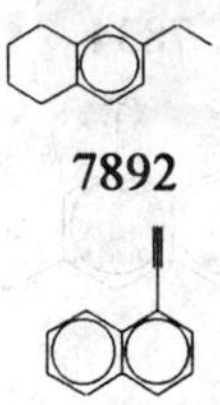

7892

7894 F

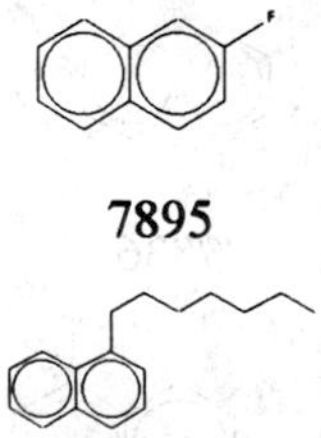

7895

7896

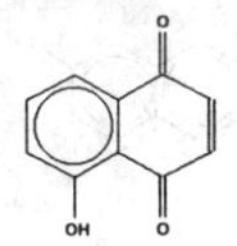

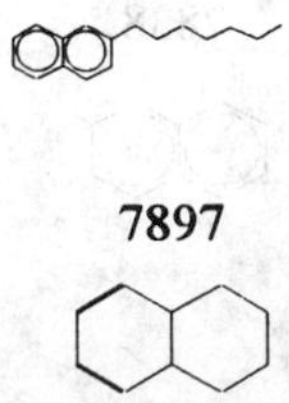

7897

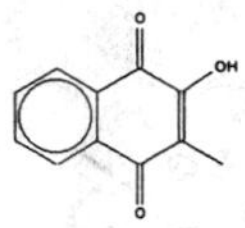

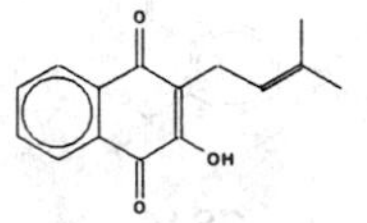

7898

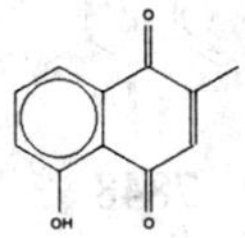

7899

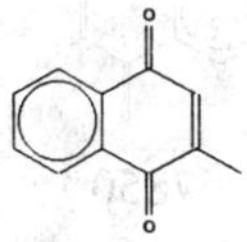

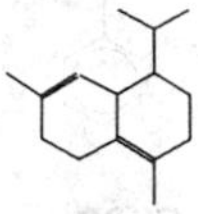

7900

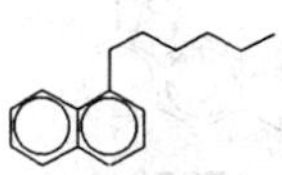

7901

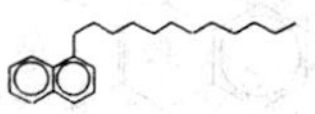

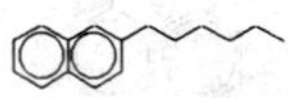

7902 I

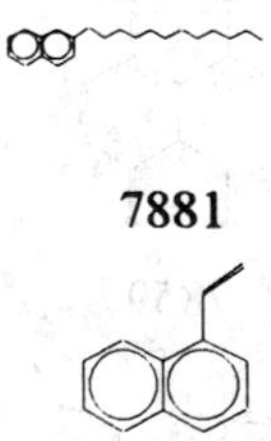

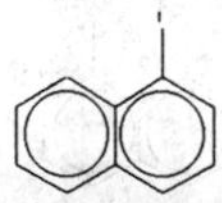

7903 I

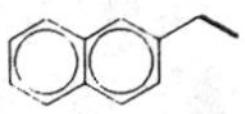

7904 N C O

7906 N C S

7907 NH₂

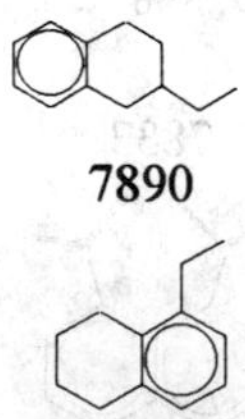

7908 OH

7909 OH

7911 HO

7913 O

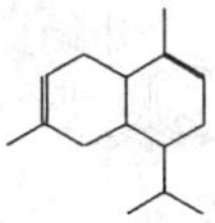

7915

7916

7917 O

7918 O

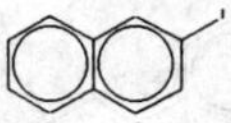

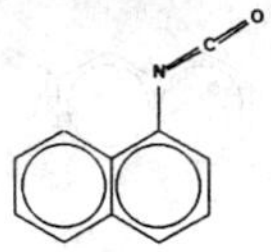

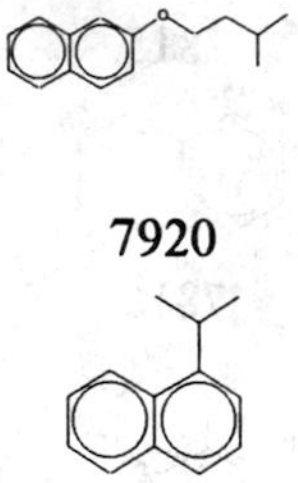

7920

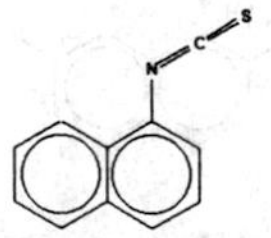

7921

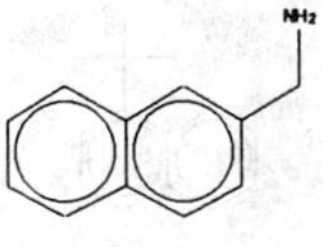

7922

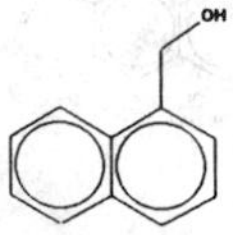

7923

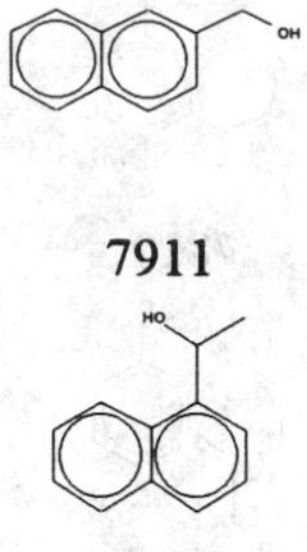

7924

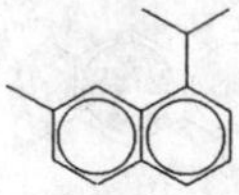

7925 O N O

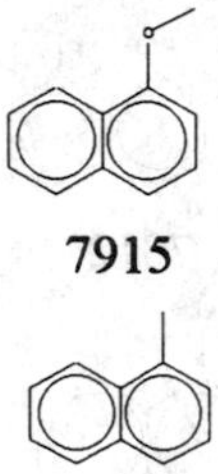

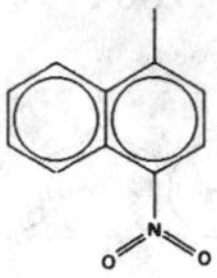

7926 O N O

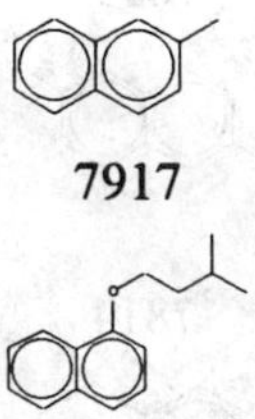

7927

7928 O

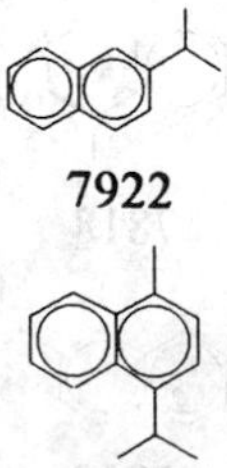

7929 O N O

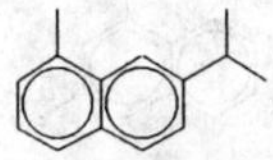

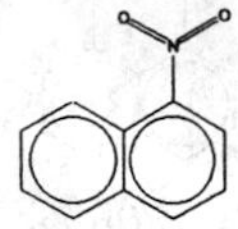

7931

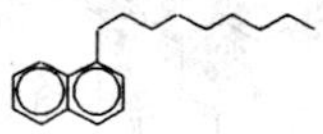

7932

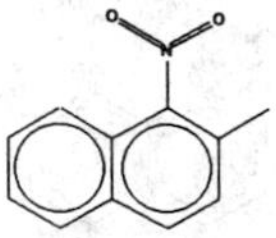

7933 Cl Cl Cl Cl Cl Cl Cl Cl

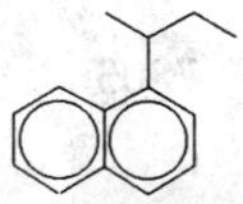

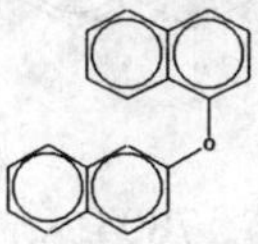

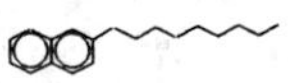

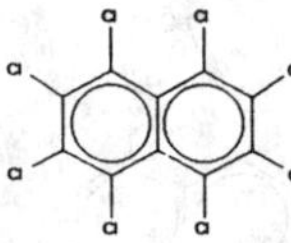

STRUCTURAL FORMULAS OF ORGANIC COMPOUNDS (continued)

In numeric order as they occur in the Table of Physical Constants of Organic Compounds

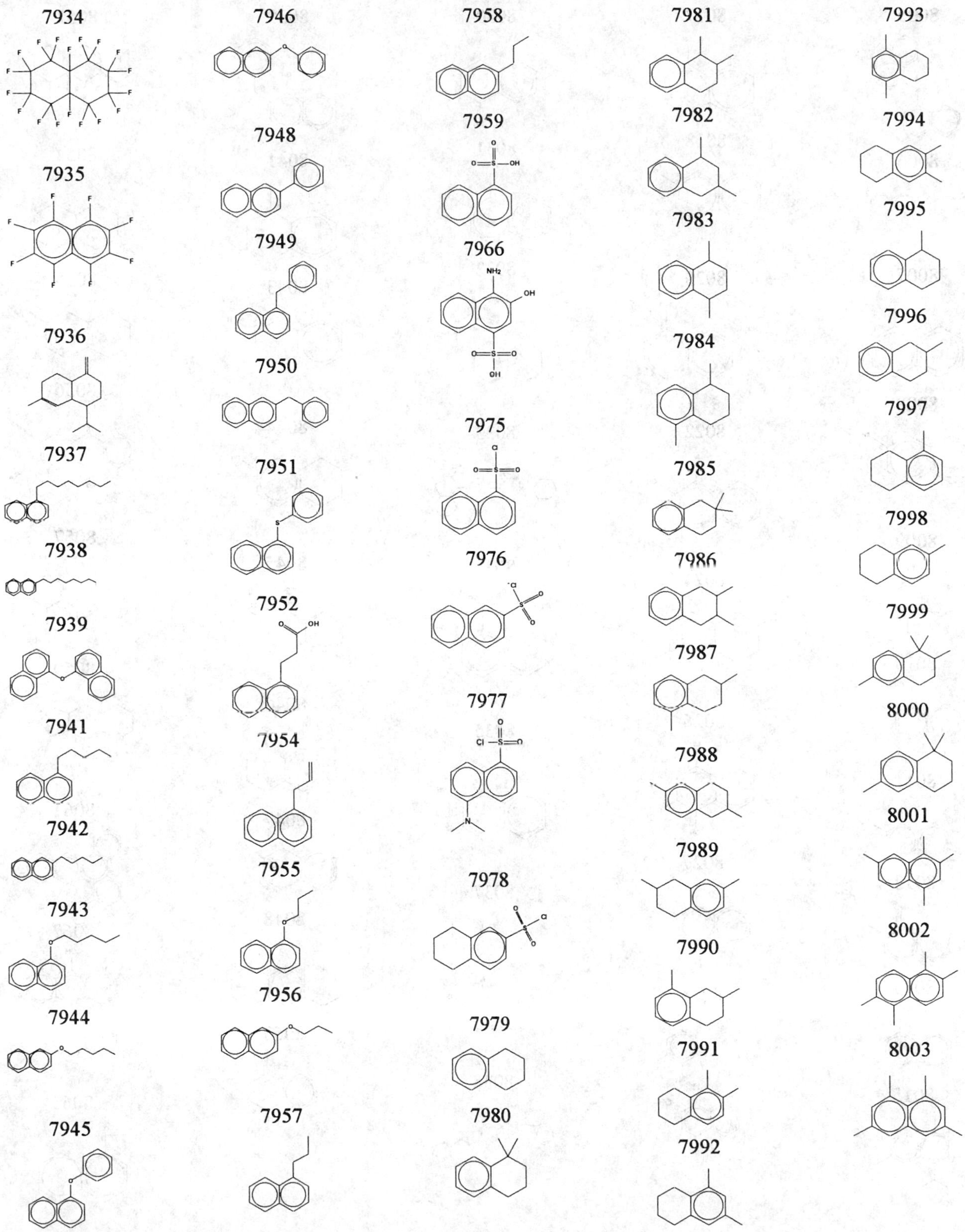

7934

7935

7936

7937

7938

7939

7941

7942

7943

7944

7945

7946

7948

7949

7950

7951

7952

7954

7955

7956

7957

7958

7959

7966

7975

7976

7977

7978

7979

7980

7981

7982

7983

7984

7985

7986

7987

7988

7989

7990

7991

7992

7993

7994

7995

7996

7997

7998

7999

8000

8001

8002

8003

8005

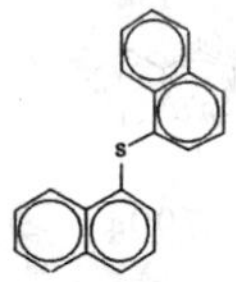

8006

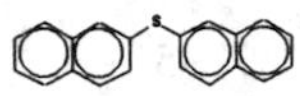

8007

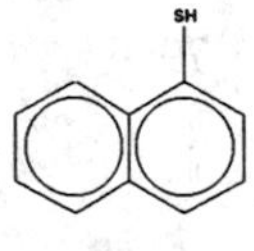

8008

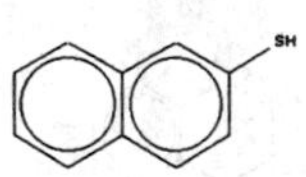

8009

8010

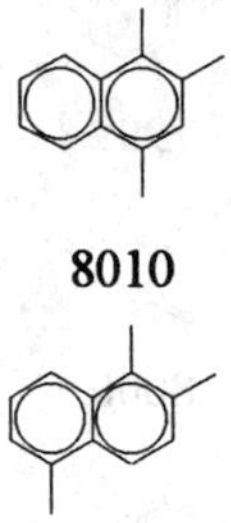

8011

8012

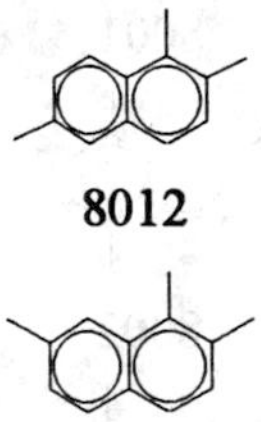

8013

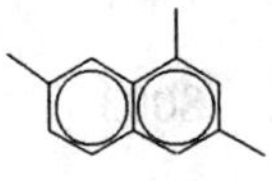

8015

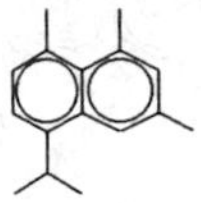

8017

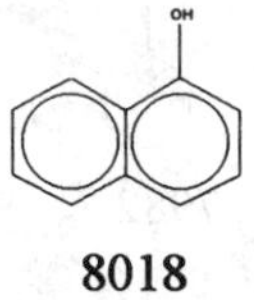

8018

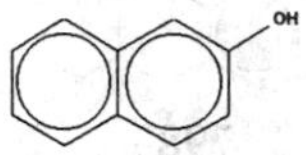

8020

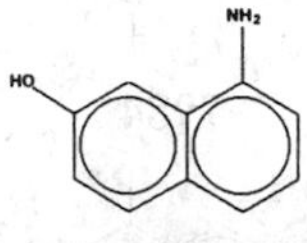

8022

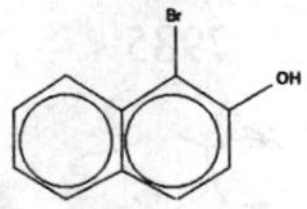

8024

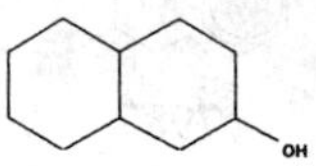

8026

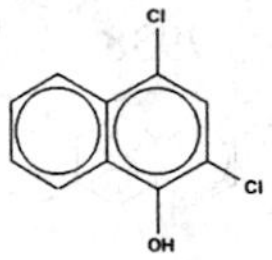

8027

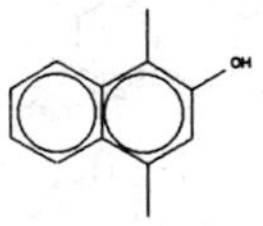

8028

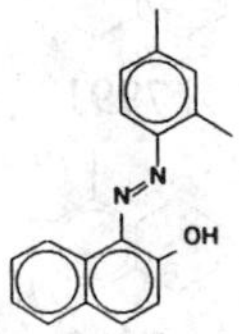

8030

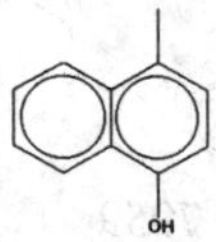

8031

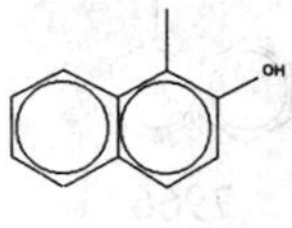

8032

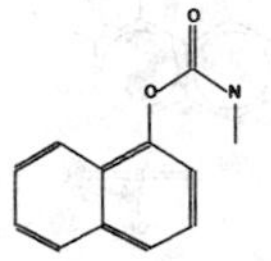

8033

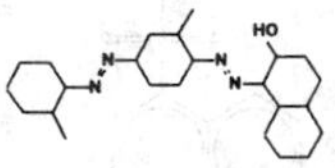

8034

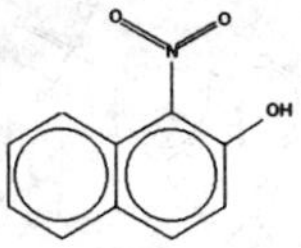

8035

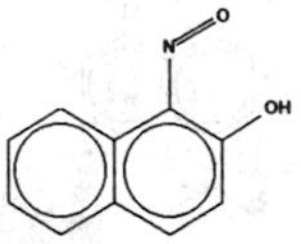

8037

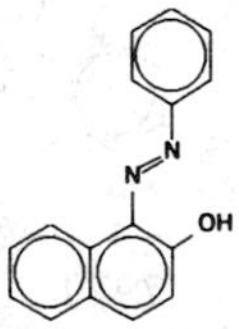

8038

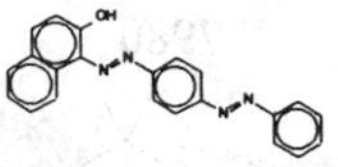

8039

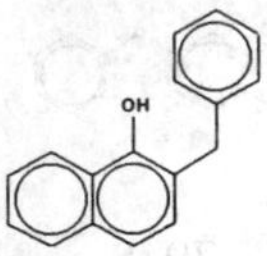

8041

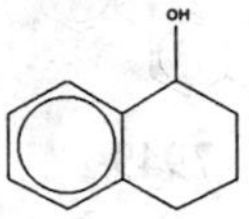

8043

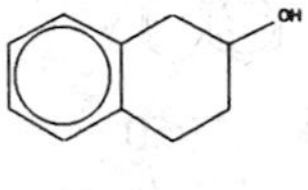

8044

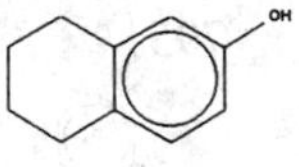

8045

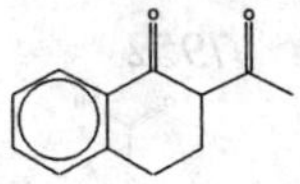

8046

8047

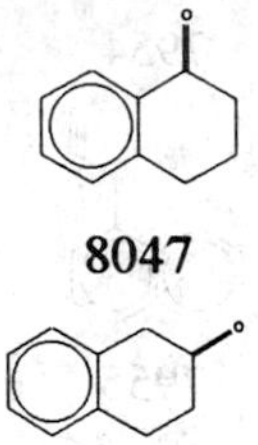

8048

8050

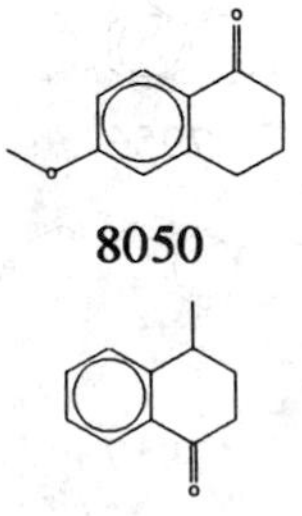

8052

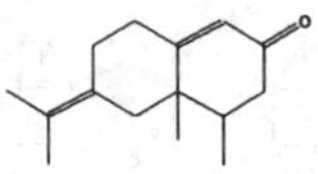

8054

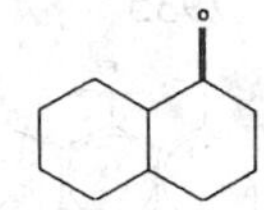

8055

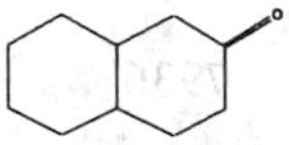

8056

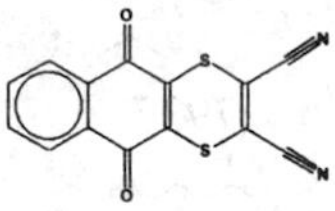

8057

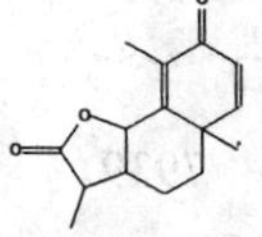

8059

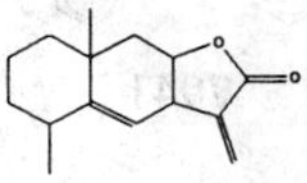

8061

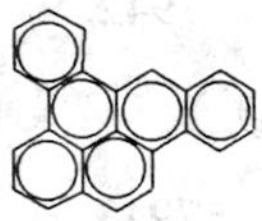

8062

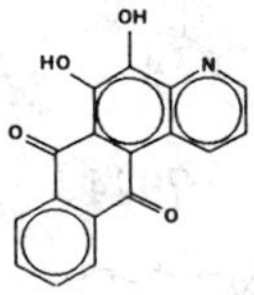

8063

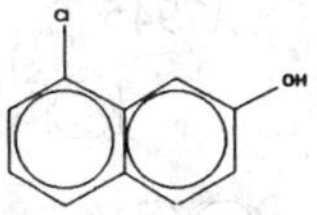

STRUCTURAL FORMULAS OF ORGANIC COMPOUNDS (continued)

In numeric order as they occur in the Table of Physical Constants of Organic Compounds

8064

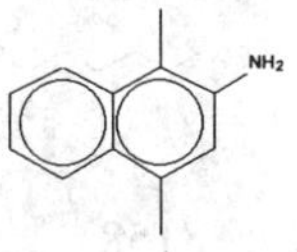

8065

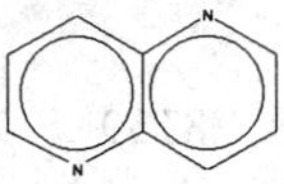

8066

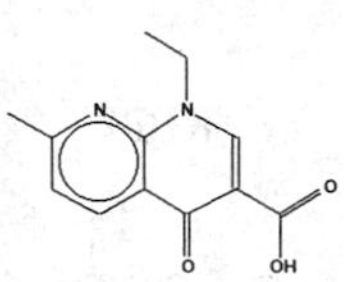

8067

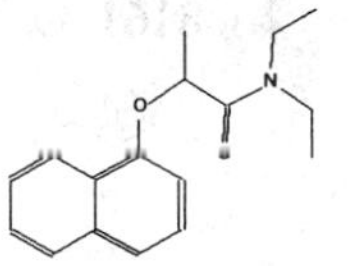

8068

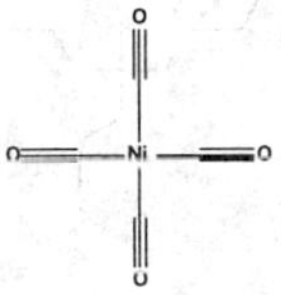

8069

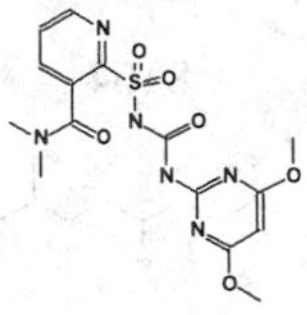

8070

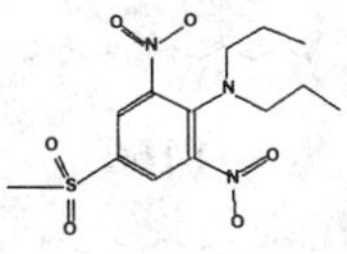

8071

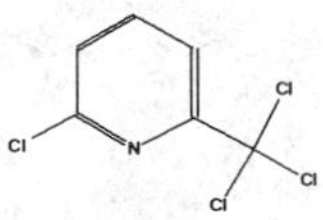

8072

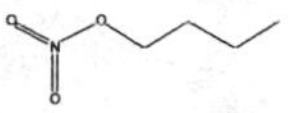

8073

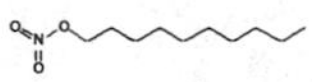

8074

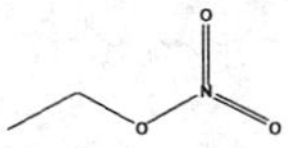

8075

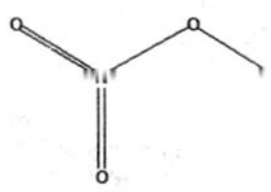

8076

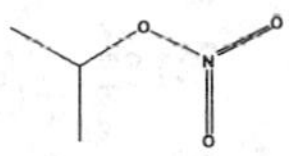

8077

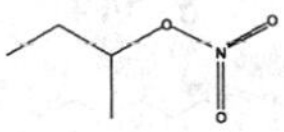

8078

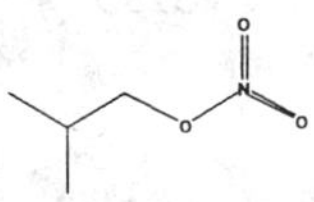

8079

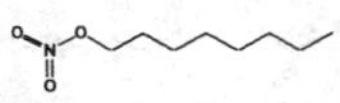

8080

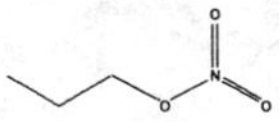

8081

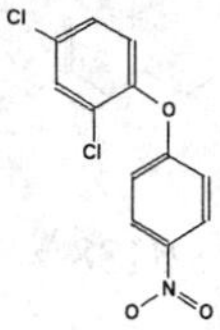

8082

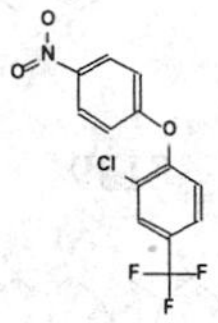

8083

8084

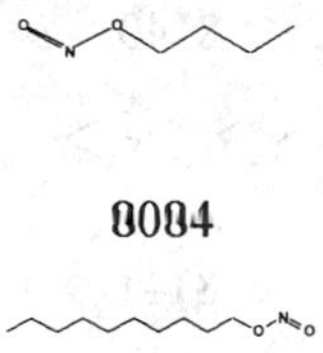

8085

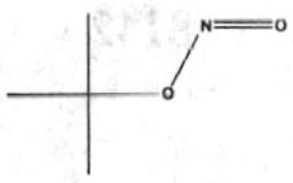

8086

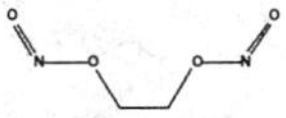

8087

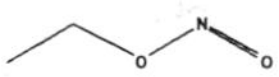

8088

8089

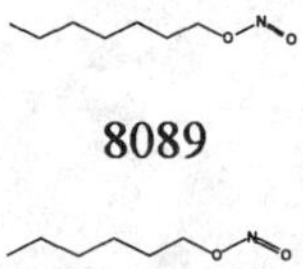

8090

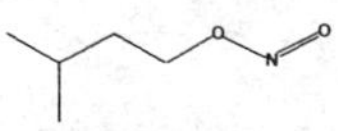

8091

8092

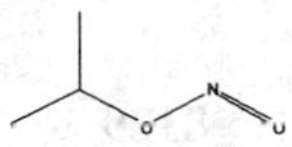

8093

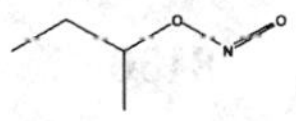

8094

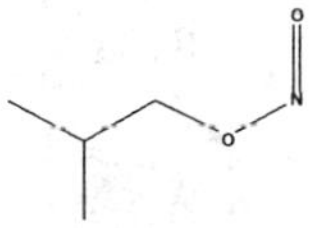

8095

8096

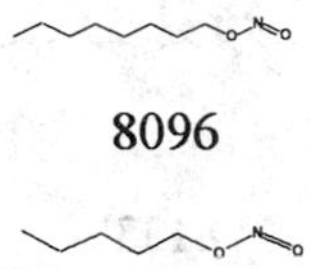

8097

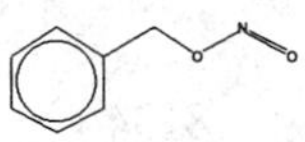

8098

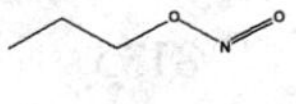

8099

8101

8103

8104

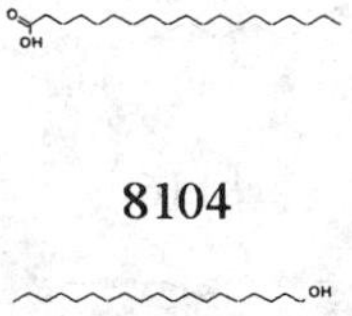

8105

8106

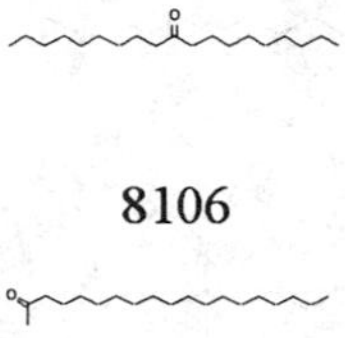

8107

8108

8109

8110

8111

8112

8113

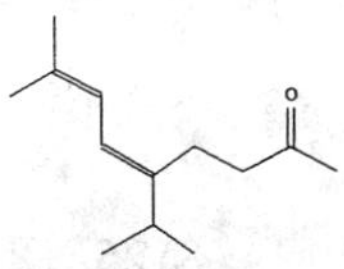

8114

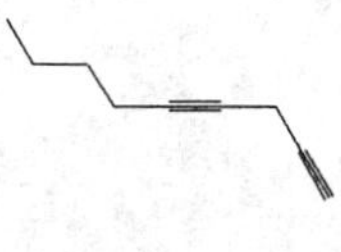

STRUCTURAL FORMULAS OF ORGANIC COMPOUNDS (continued)

In numeric order as they occur in the Table of Physical Constants of Organic Compounds

8115

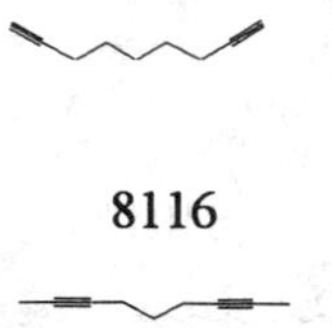

8116

8117

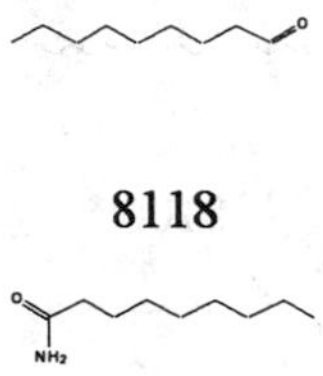

8118

NH2

8119

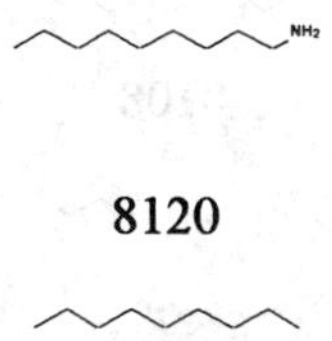

NH2

8120

8121

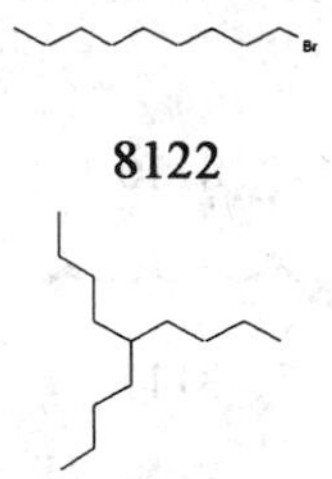

Br

8122

8123

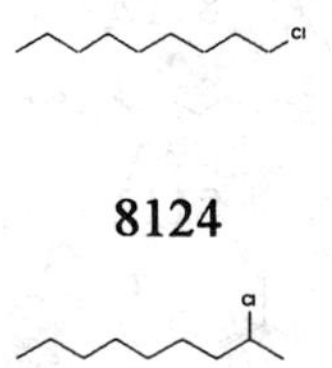

Cl

8124

Cl

8125

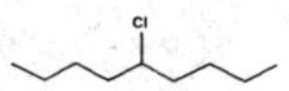

Cl

8126

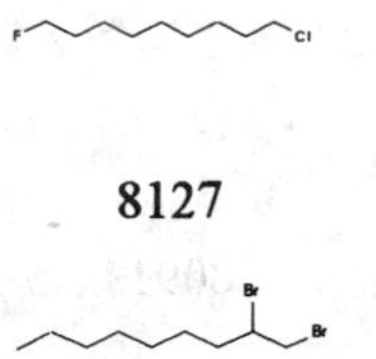

F Cl

8127

Br Br

8128

Cl Cl

8129

N N

8130

O O
OH OH

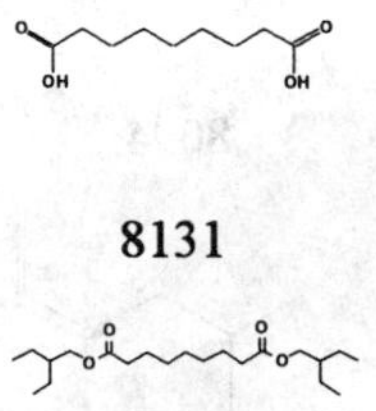

8131

8132

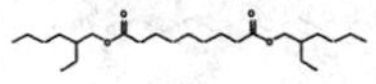

8133

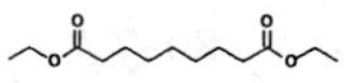

8134

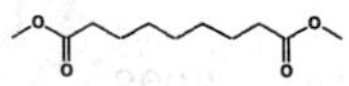

8135

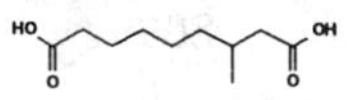

HO OH

8136

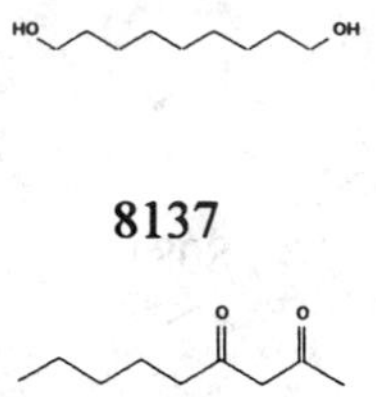

HO OH

8137

O O

8138

O O
Cl Cl

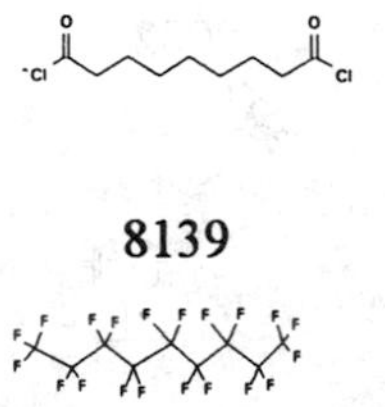

8139

8140

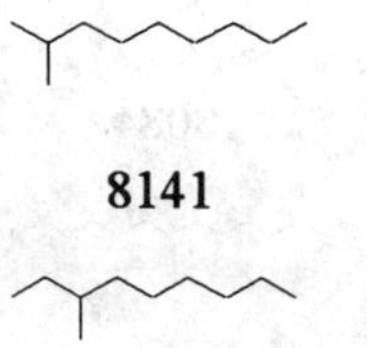

8141

8142

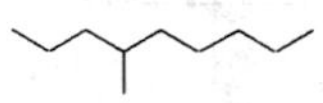

8143

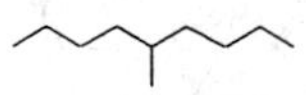

8144

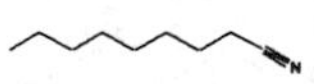

N

8145

SH

8146

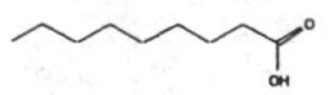

O
OH

8147

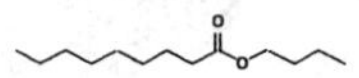

8148

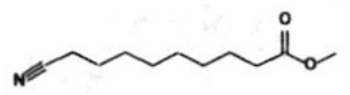

N O

8149

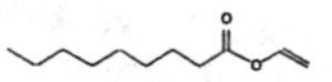

8150

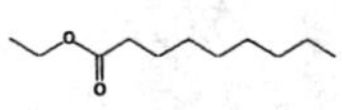

8151

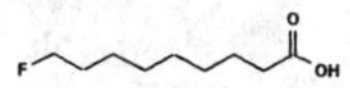

O
F OH

8152

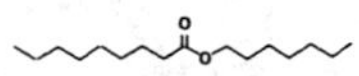

8154

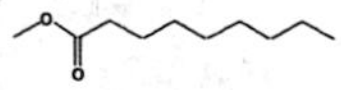

8155

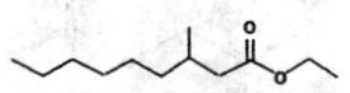

8156

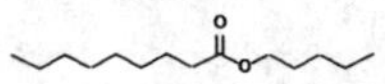

O

8157

OH

8158

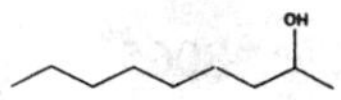

OH

8159

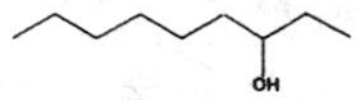

OH

8160

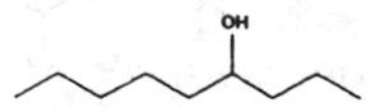

OH

8161

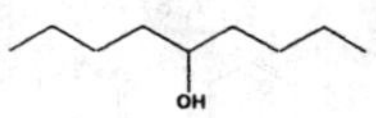

OH

8162

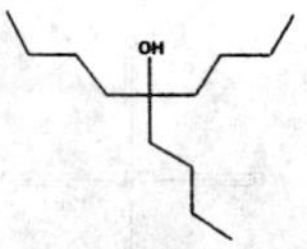

OH

8163

Cl OH

8164

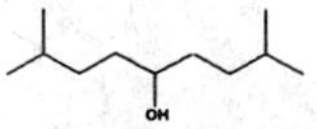

OH

8165

F OH

8166

OH

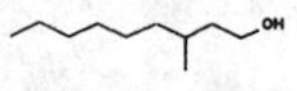

STRUCTURAL FORMULAS OF ORGANIC COMPOUNDS (continued)

In numeric order as they occur in the Table of Physical Constants of Organic Compounds

8167 OH

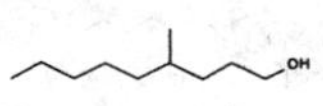

8168 O

8169 OH

8170 HO

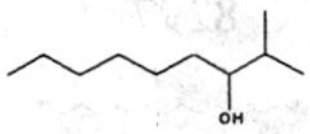

8171 HO

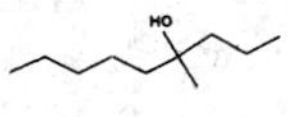

8172

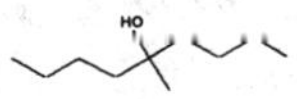

8173 O

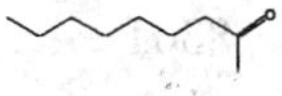

8174 O

8175 O

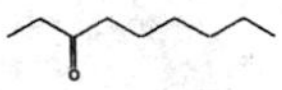

8176 O

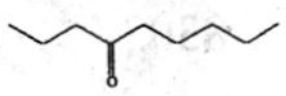

8177 O Cl

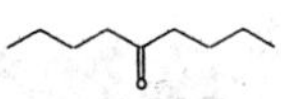

8178 Si O Si O Si O Si O Si O Si O Si O Si O Si

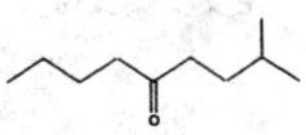

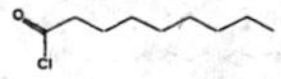

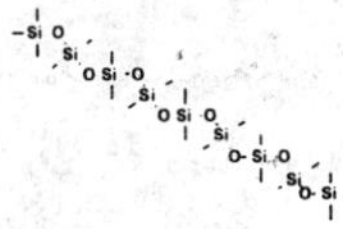

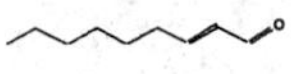

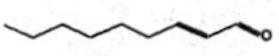

8179 O

8180 O

8181 O OH NH O

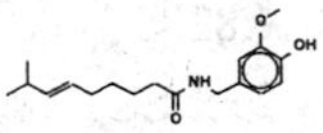

8182

8183

8184

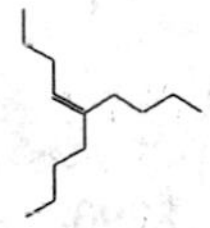

8185

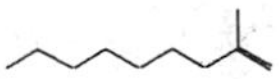

8186

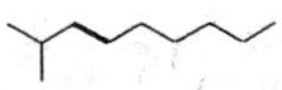

8187 OH O

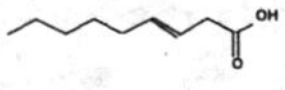

8188 O OH

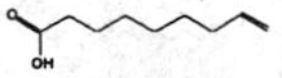

8189 OH

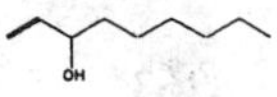

8190 OH

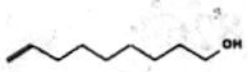

8191 O

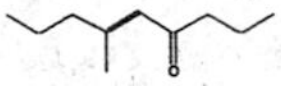

8192 O

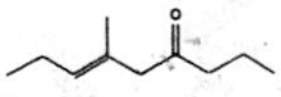

8193

8194

8195

8196

8197

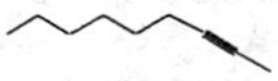

8198

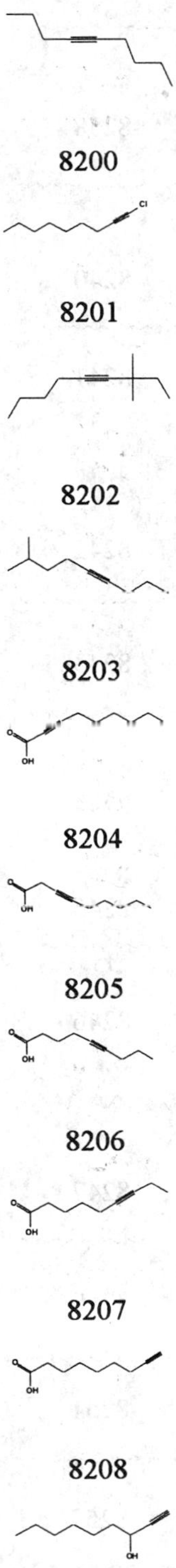

8199

8200 Cl

8201

8202

8203 O OH

8204 O OH

8205 O OH

8206 O OH

8207 O OH

8208 OH

8209 OH

8212 O N N

8213 F F F N N N O Cl

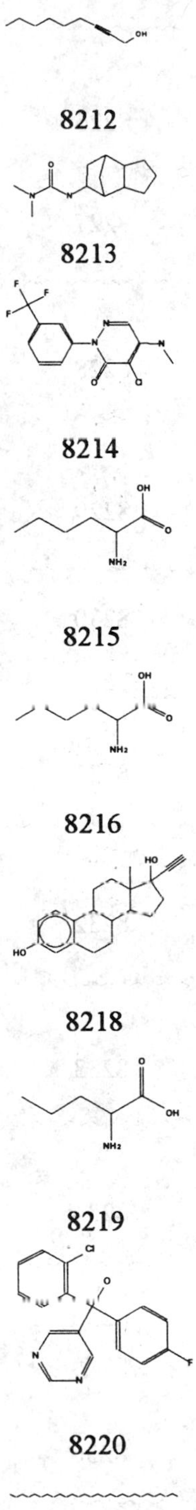

8214 OH O NH2

8215 OH O NH2

8216 HO HO

8218 O OH NH2

8219 Cl O N N F

8220

8223

8226

8227

8228

8229

8230

8231

8232

8233

8234

8235

8236

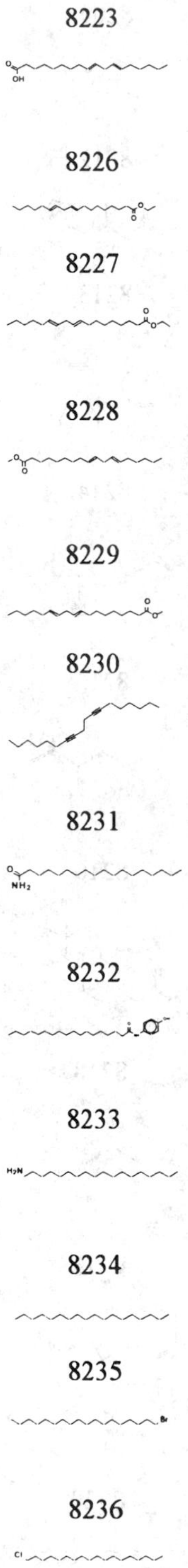

8237

8238

8240

8241

8242

8243

8244

8245

8246

8247

8250

8251

8257

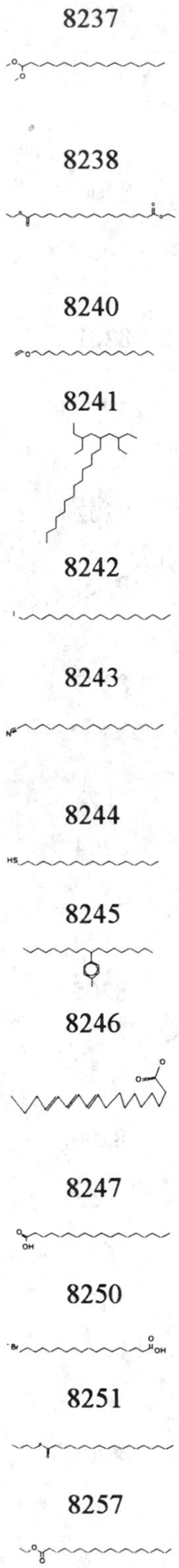

8260

8262

8263

8264

8265

8266

8267

8268

8269

8271

8273

8278

8279

8280

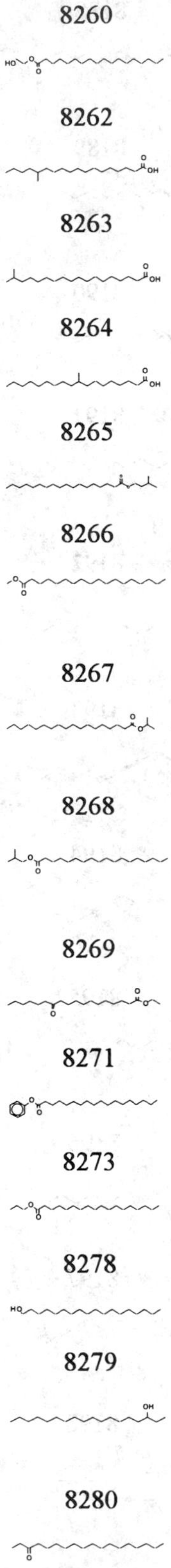

8281

8282

8283

8284

8285

8287

8288

8289

8290

8291

8292

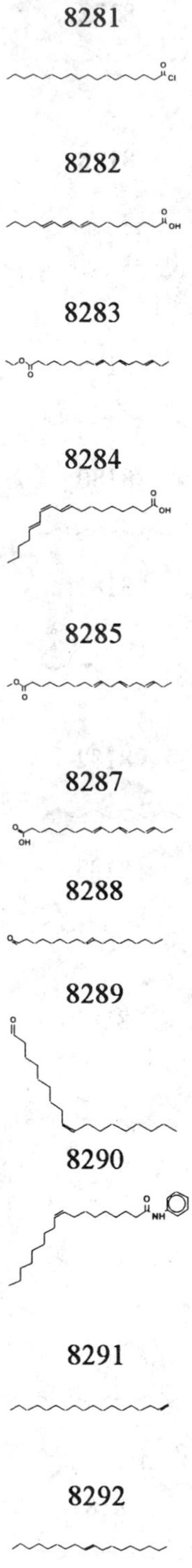

8293

8294

8297

8298

8299

8301

8302

8303

8304

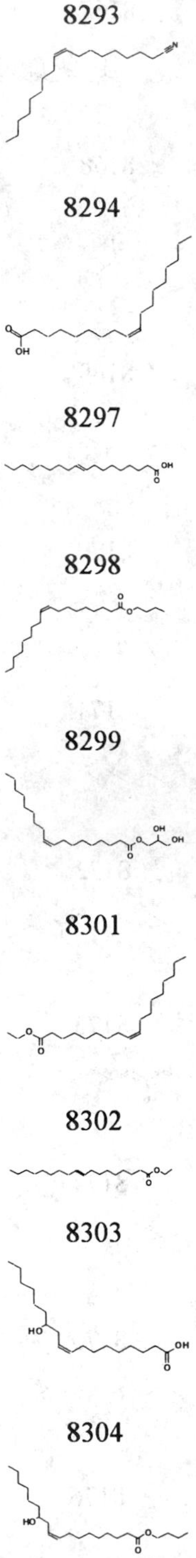

STRUCTURAL FORMULAS OF ORGANIC COMPOUNDS (continued)

In numeric order as they occur in the Table of Physical Constants of Organic Compounds

8305

HO O O

8306

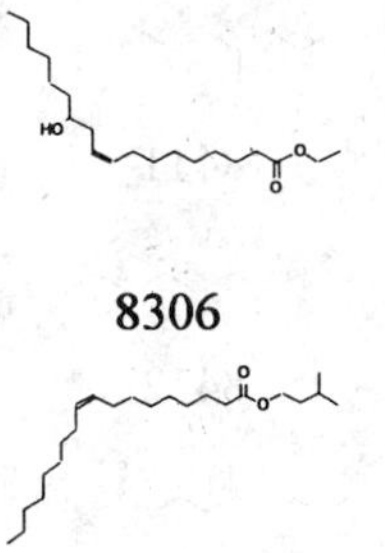

O O

8307

O O

8308

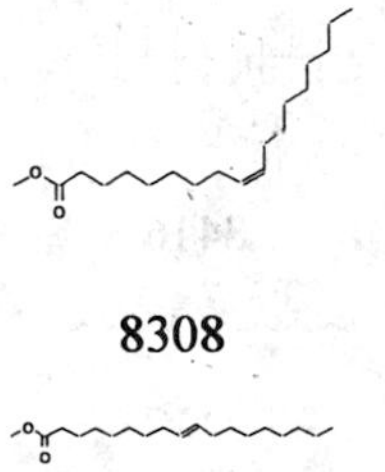

O O

8309

O O

8310

O O O O O O

8311

OH

8312

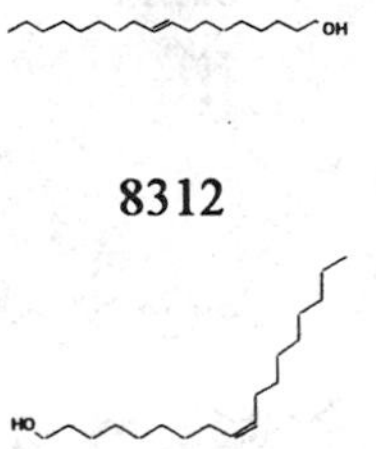

HO

8313

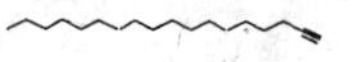

8314

8315

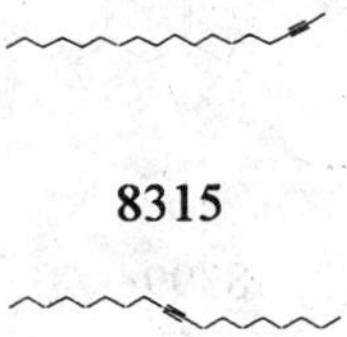

8316

O OH

8317

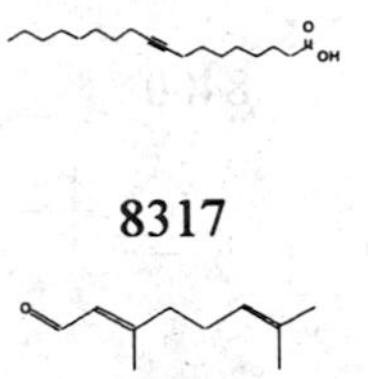

O

8318

O

8319

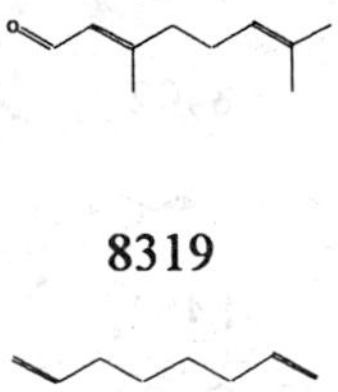

8320

8321

Br

8322

8324

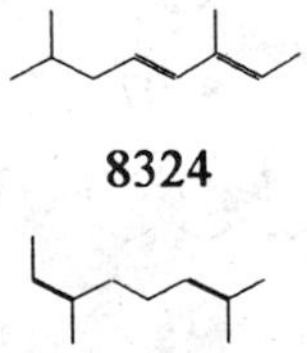

8325

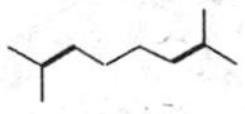

8327

8328

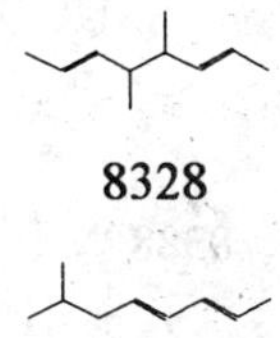

8329

8330

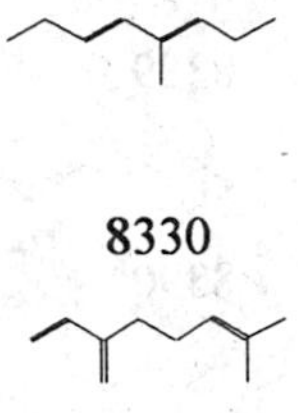

8331

O OH

8332

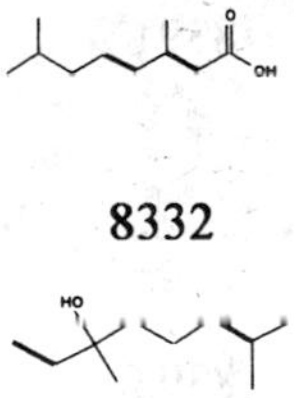

HO

8333

OH

8334

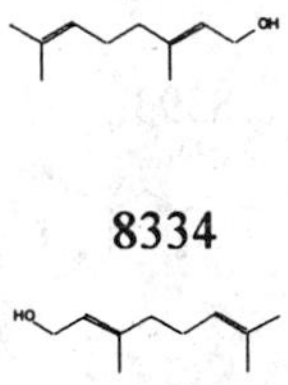

HO

8335

O O

8336

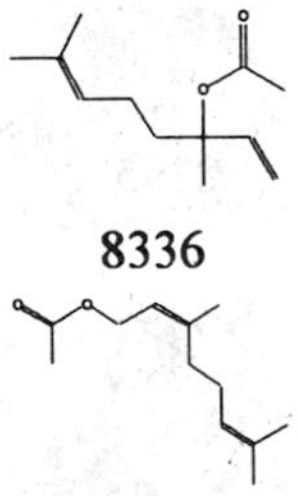

O O

8337

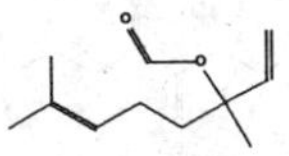

O O

8338

O O

8339

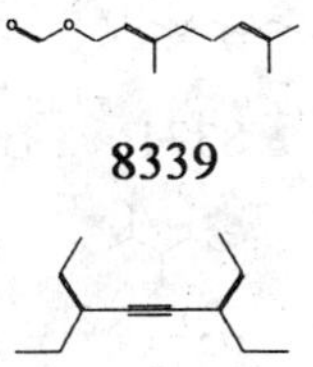

8340

8341

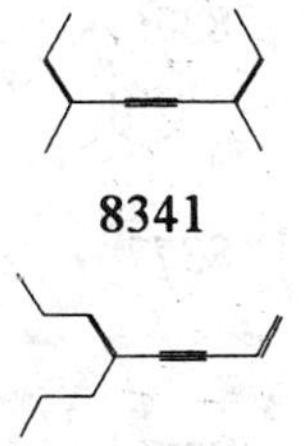

8342

8343

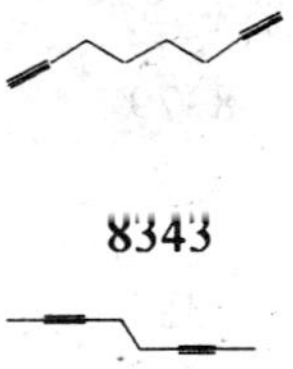

8344

8345

8346

O

8347

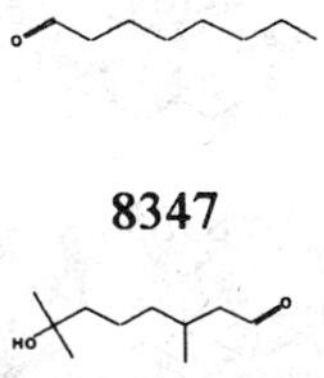

HO O

8348

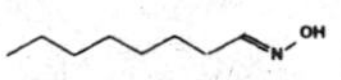

N OH

8349

O NH₂

8350

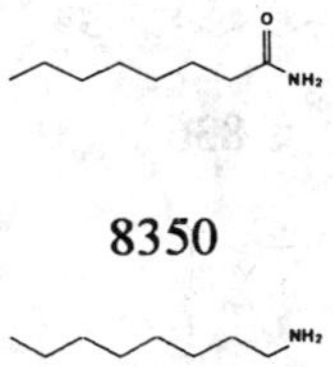

NH₂

8351

NH₂

8352

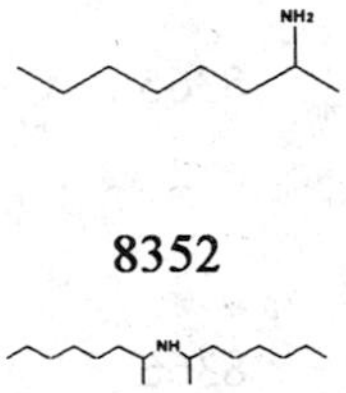

NH

8353

N

8354

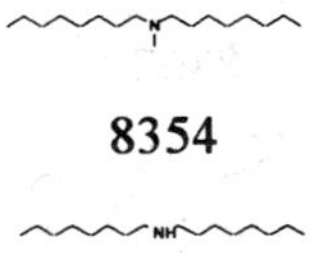

NH

8355

8356

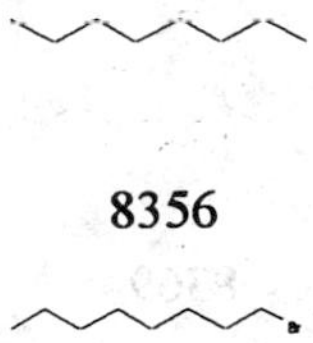

Br

8357

Br

8358

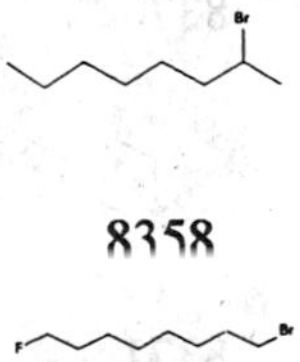

F Br

8359

Cl

8360

Cl

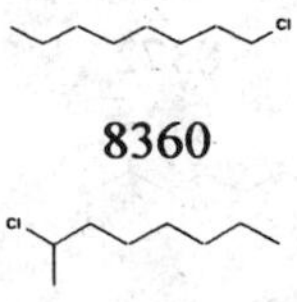

STRUCTURAL FORMULAS OF ORGANIC COMPOUNDS (continued)

In numeric order as they occur in the Table of Physical Constants of Organic Compounds

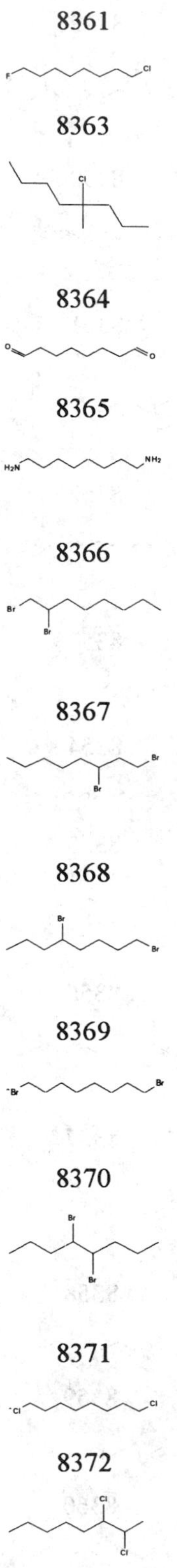

8361

8363

8364

8365

8366

8367

8368

8369

8370

8371

8372

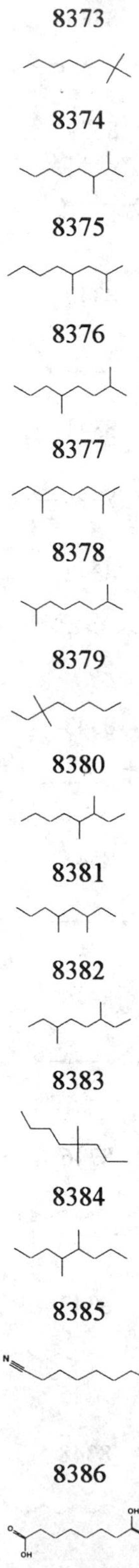

8373

8374

8375

8376

8377

8378

8379

8380

8381

8382

8383

8384

8385

8386

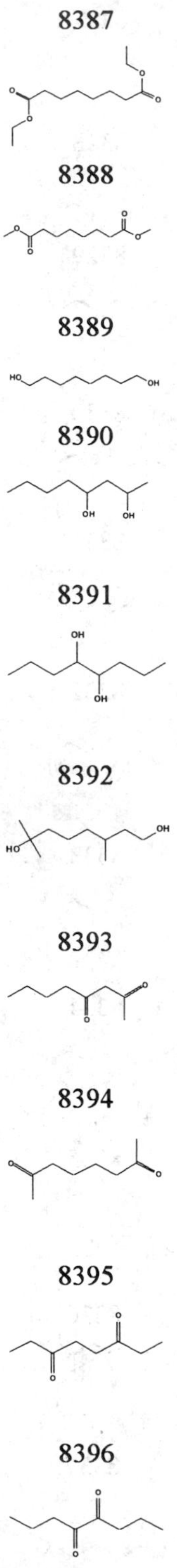

8387

8388

8389

8390

8391

8392

8393

8394

8395

8396

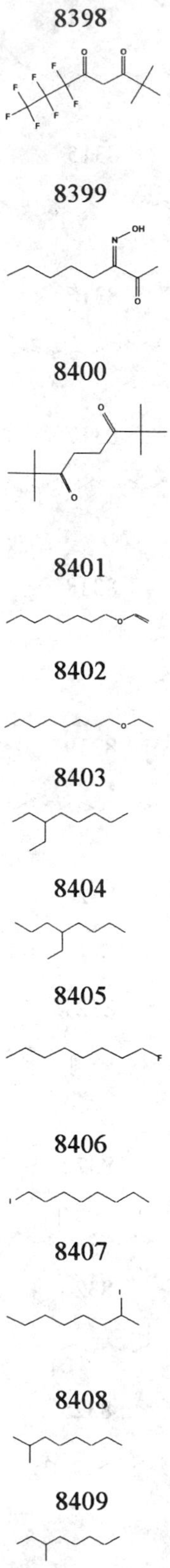

8398

8399

8400

8401

8402

8403

8404

8405

8406

8407

8408

8409

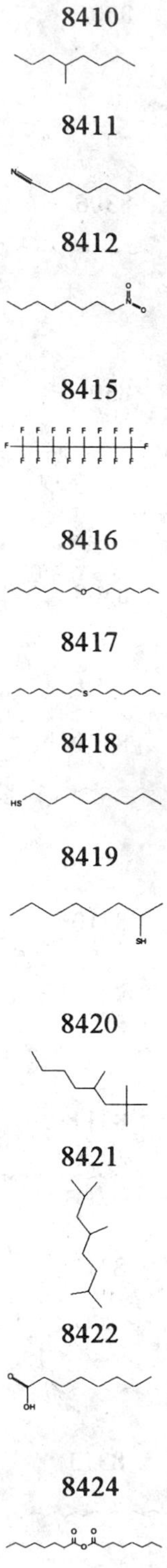

8410

8411

8412

8415

8416

8417

8418

8419

8420

8421

8422

8424

STRUCTURAL FORMULAS OF ORGANIC COMPOUNDS (continued)

In numeric order as they occur in the Table of Physical Constants of Organic Compounds

8425

Br OH O

8426

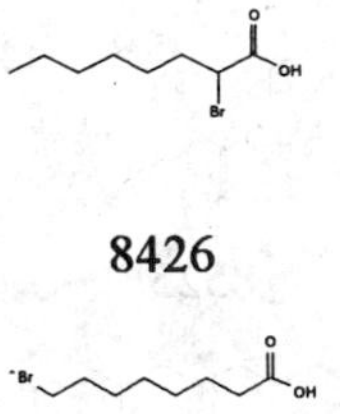

Br OH O

8427

O O

8428

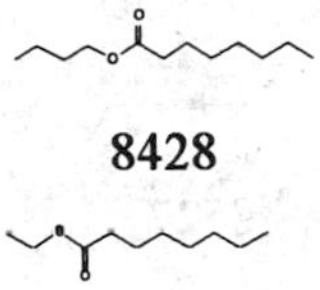

S O

8429

F OH O

8430

O O

8431

O O

8432

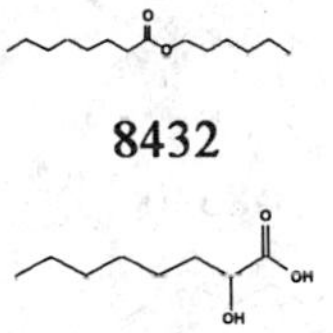

OH OH O

8433

OH O

8434

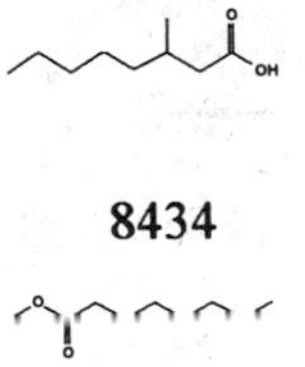

O O

8435

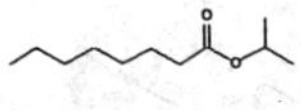

O O

8436

O O O

8437

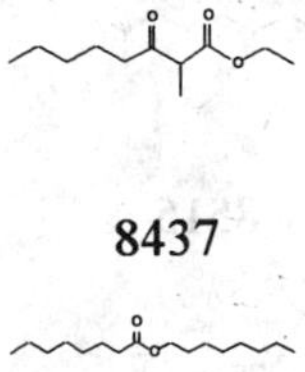

O O

8438

F F F F F F F F F F F F F F F F F O O

8439

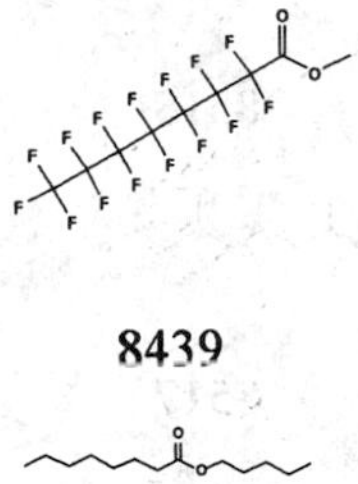

O O

8440

O O O O O O

8441

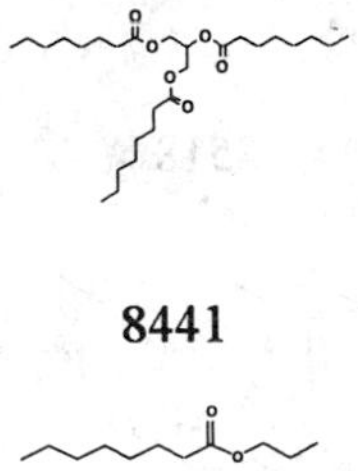

O O

8442

Cl F Cl F Cl F Cl F F F F F F F OH O

8443

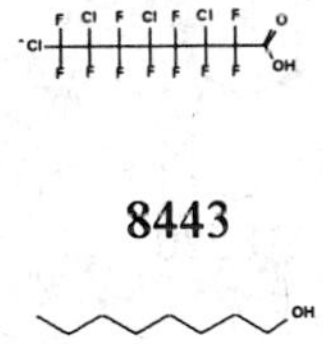

OH

8444

OH

8445

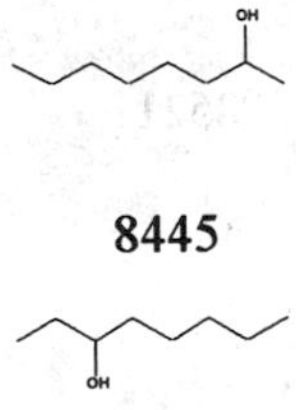

OH

8446

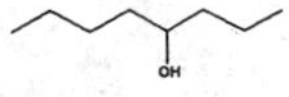

OH

8447

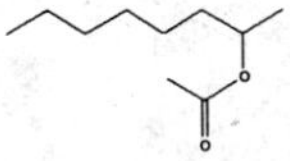

O O

8449

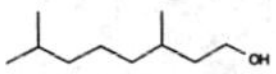

OH

8450

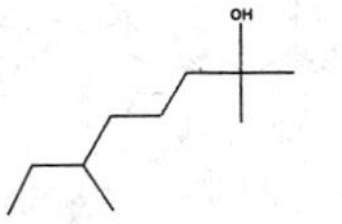

OH

8451

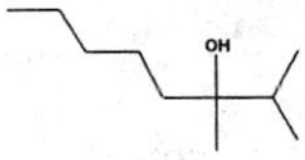

OH

8452

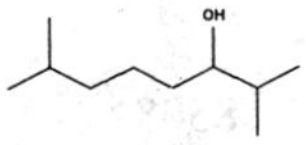

OH

8453

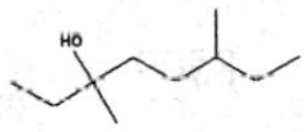

HO

8454

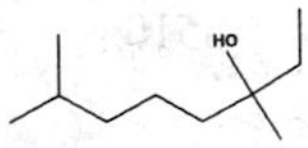

HO

8455

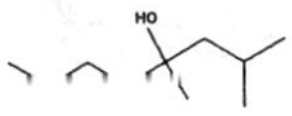

HO

8456

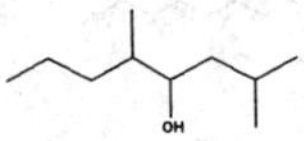

OH

8457

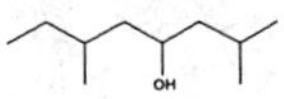

OH

8458

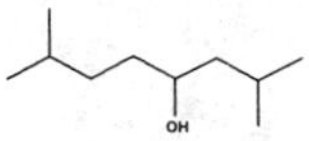

OH

8459

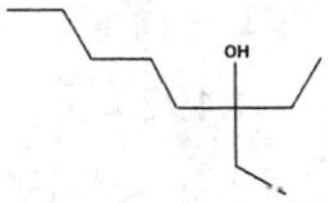

OH

8460

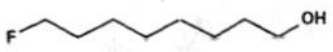

F OH

8461

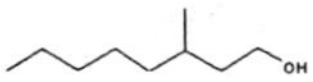

OH

8462

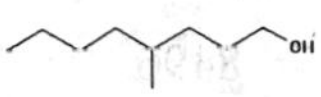

OH

8464

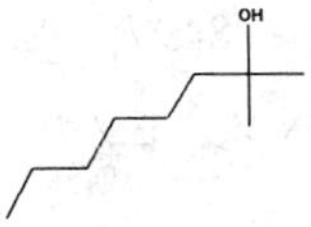

OH

8465

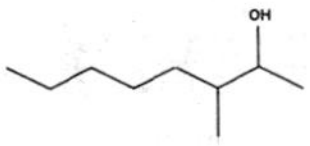

OH

8466

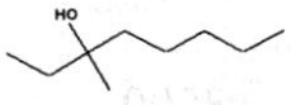

HO

8468

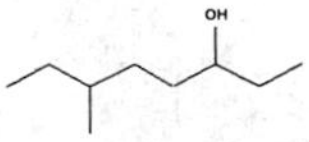

OH

8469

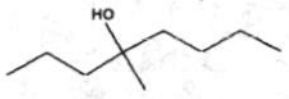

HO

8470

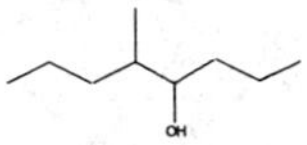

OH

8471

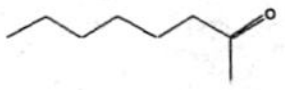

O

8472

O

8473

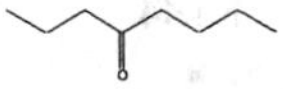

O

8474

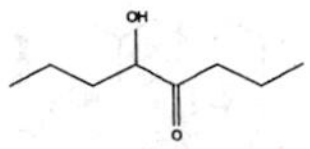

OH O

8475

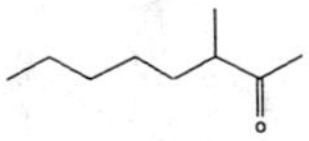

O

8476

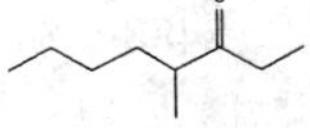

O

8477

O

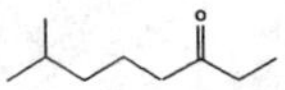

STRUCTURAL FORMULAS OF ORGANIC COMPOUNDS (continued)

In numeric order as they occur in the Table of Physical Constants of Organic Compounds

8478

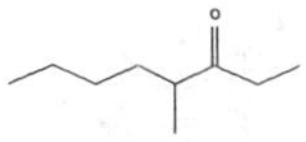

8479

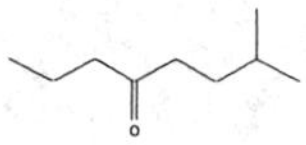

8480

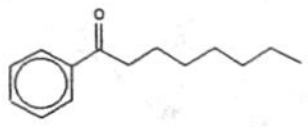

8481

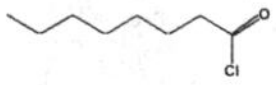

8482

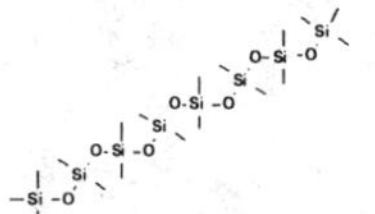

8484

8485

8486

8487

8488

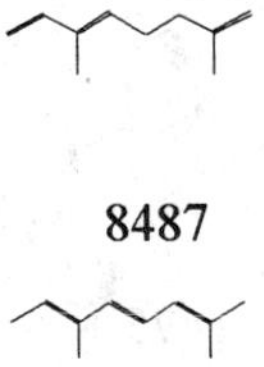

8489

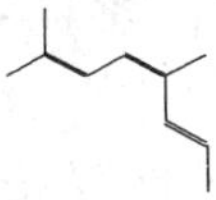

8490

8491

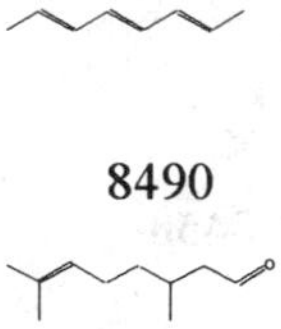

8492

8493

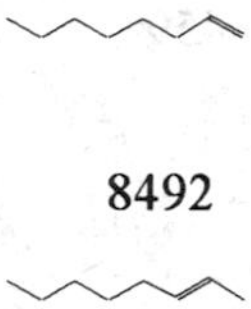

8494

8495

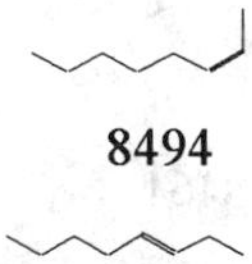

8496

8497

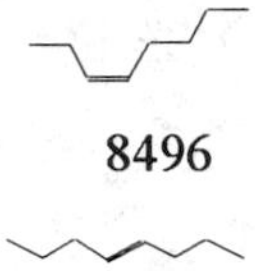

8498

8499

8500

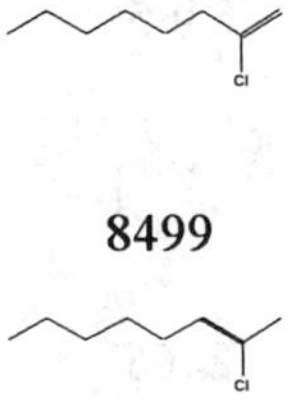

8502

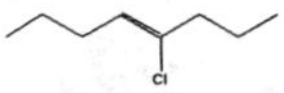

8503

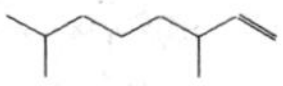

8504

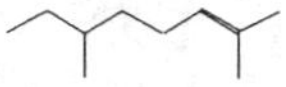

8505

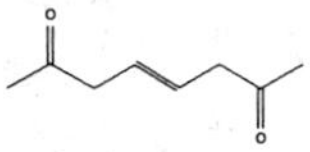

8507

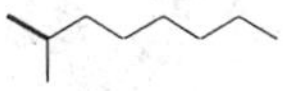

8508

8509

8510

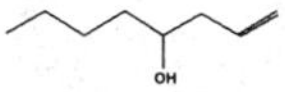

8511

8512

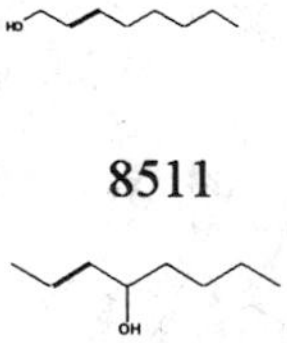

8513

8514

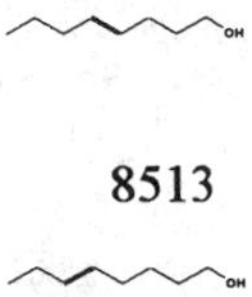

8515

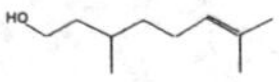

8516

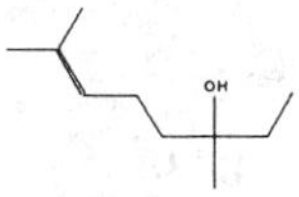

8517

8518

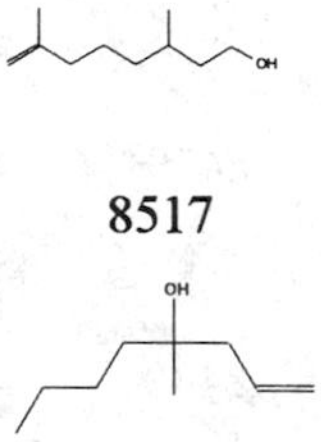

8519

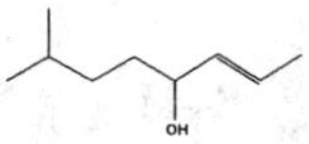

8520

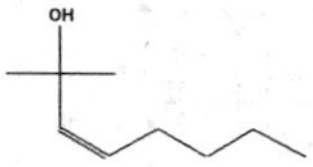

8521

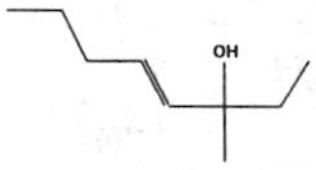

8523

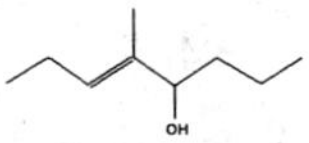

8524

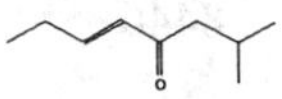

8525

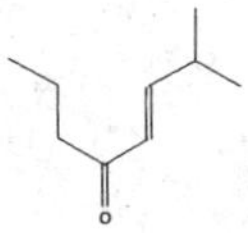

8526

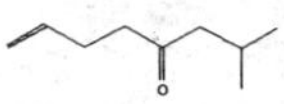

8527

8528

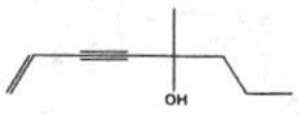

8529

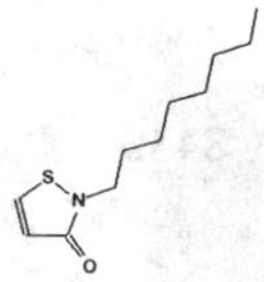

8530

8531

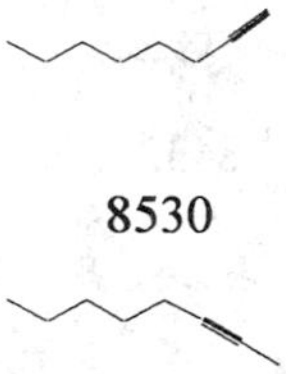

8532

8533

In numeric order as they occur in the Table of Physical Constants of Organic Compounds

8534

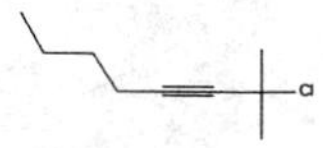

8535

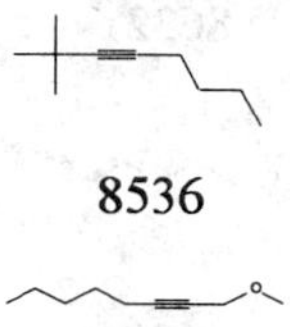

8536

8537

8538

8539

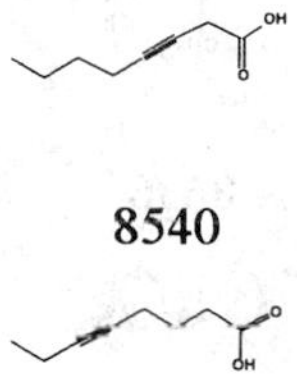

8540

8541

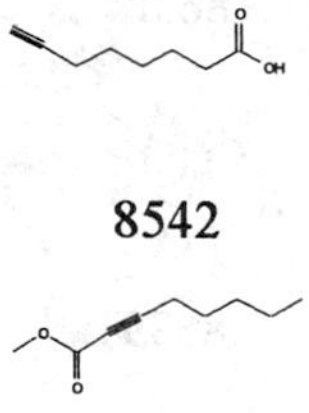

8542

8543

8544

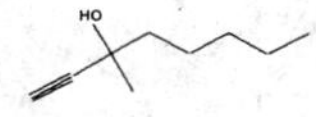

8545

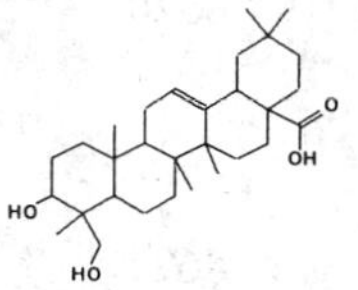

8547

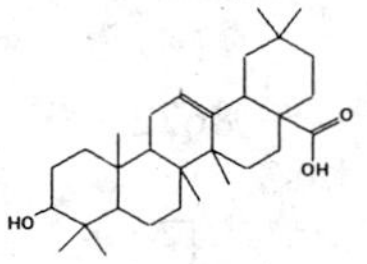

8548

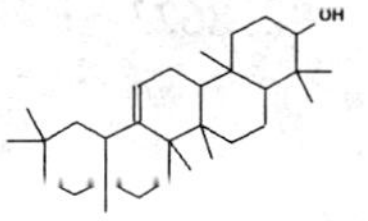

8552

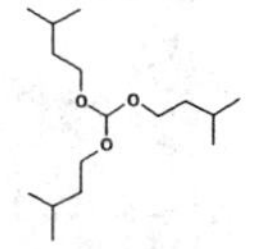

8553

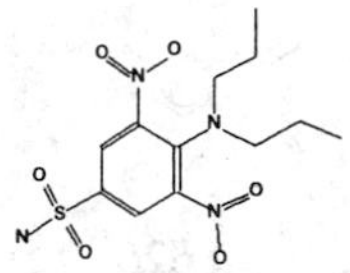

8554

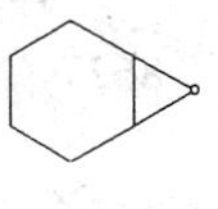

8555

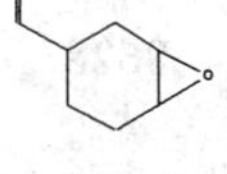

8556

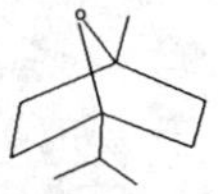

8557

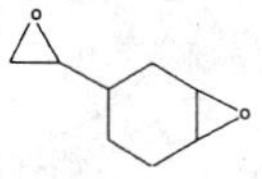

8558

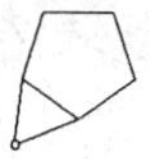

8559

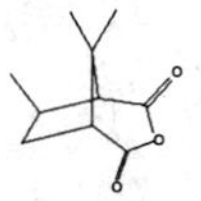

8560

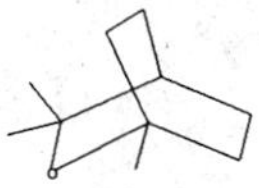

8562

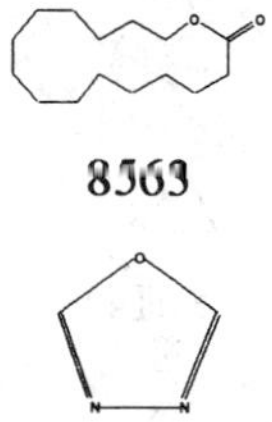

8563

8564

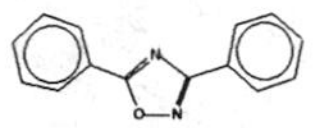

8565

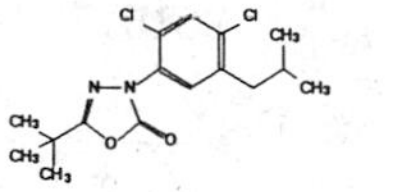

8566

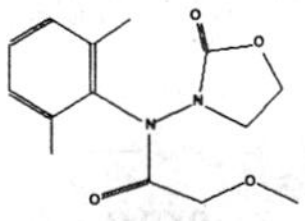

8568

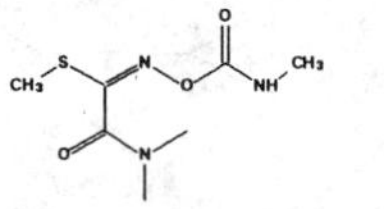

8569

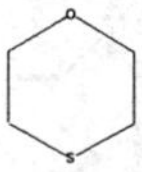

8570

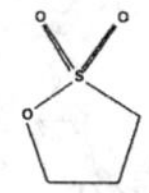

8571

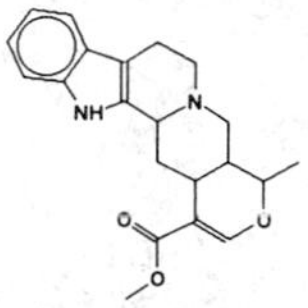

8574

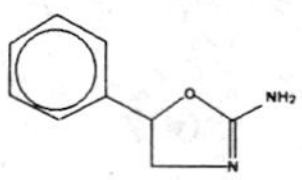

8575

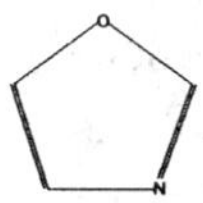

8576

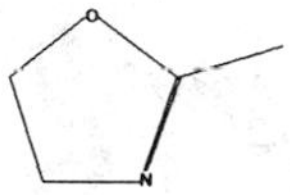

8577

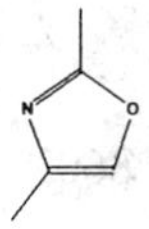

8578

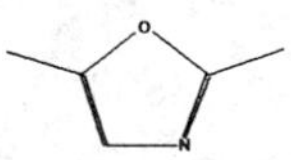

8579

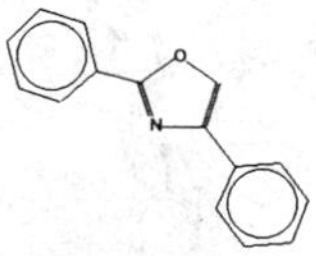

8580

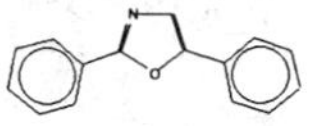

8581

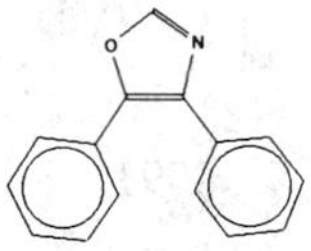

8582

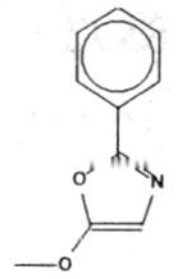

8583

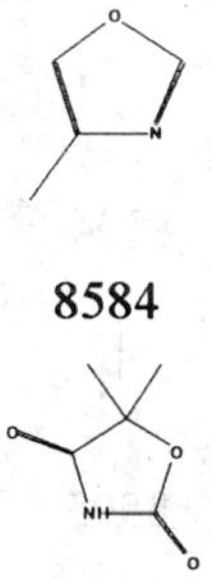

8584

8585

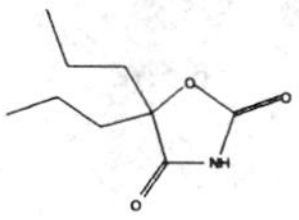

8586

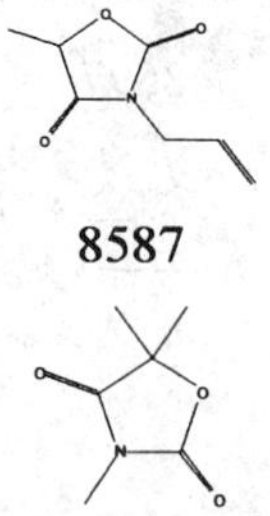

8587

STRUCTURAL FORMULAS OF ORGANIC COMPOUNDS (continued)

In numeric order as they occur in the Table of Physical Constants of Organic Compounds

8588

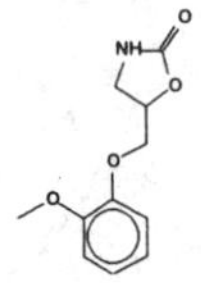

8589

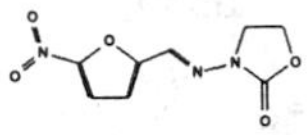

8590

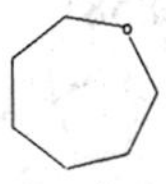

8591

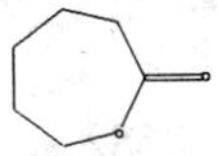

8592

8593

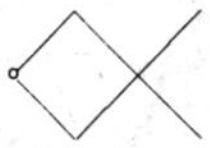

8594

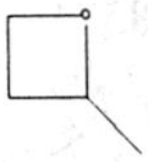

8595

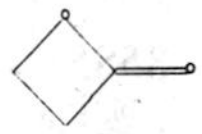

8596

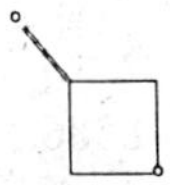

8597

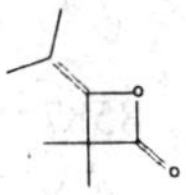

8598

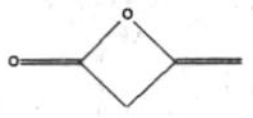

8599

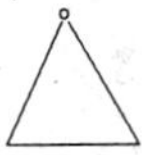

8600

8601

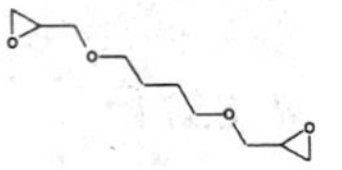

8602

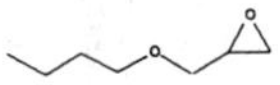

8603

8604

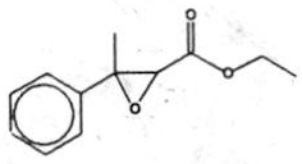

8605

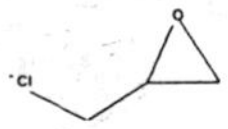

8607

8608

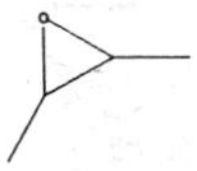

8609

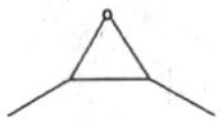

8610

8611

8612

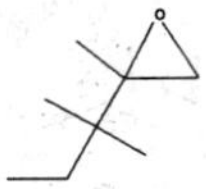

8613

8614

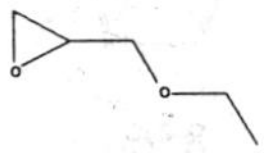

8615

8616

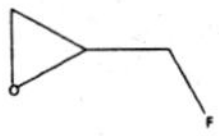

8617

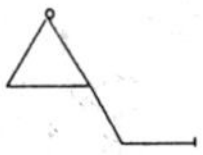

8618

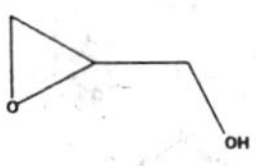

8619

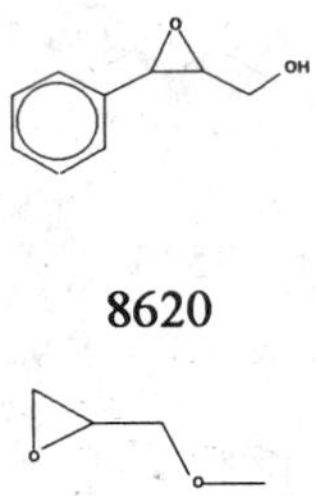

8620

8621

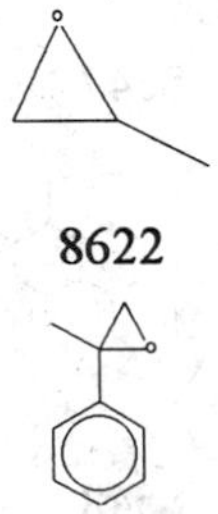

8622

8623

8624

8625

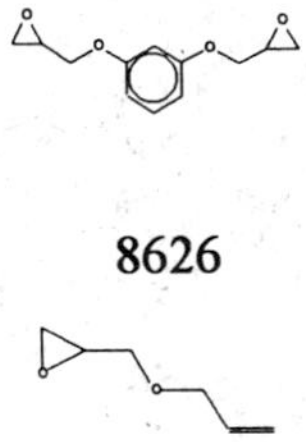

8626

8627

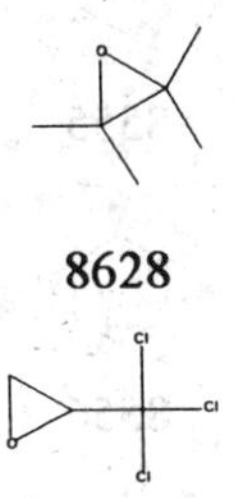

8628

8629

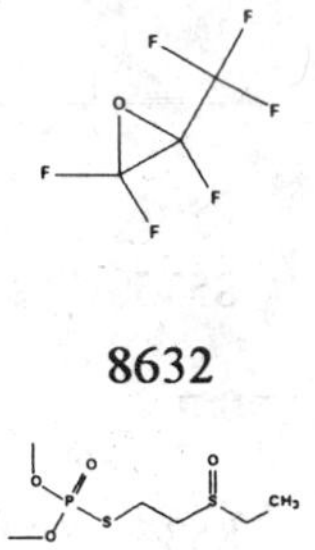

8632

8633

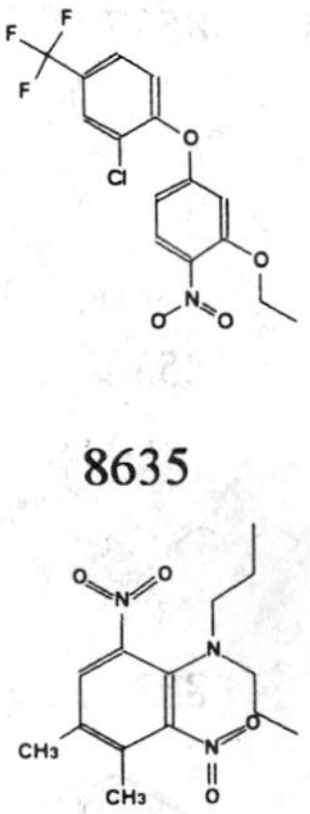

8635

8636

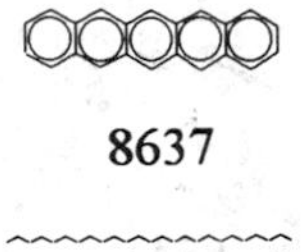

8637

8638

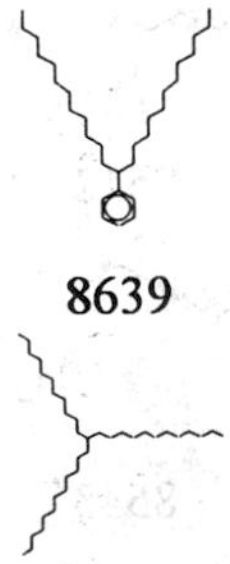

8639

8640

8641

STRUCTURAL FORMULAS OF ORGANIC COMPOUNDS (continued)

In numeric order as they occur in the Table of Physical Constants of Organic Compounds

8642

8643

Br

8644

8645

OH

8646

8647

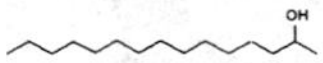

OH

8648

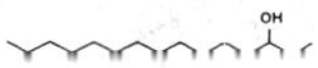

OH

8649

OH

8650

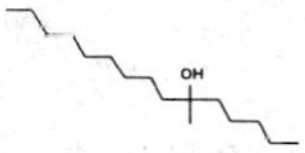

OH

8651

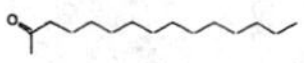

8652

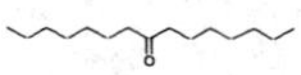

O

8653

8654

8655

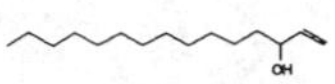

OH

8656

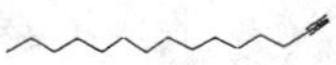

8657

O

8658

C

8659

8660

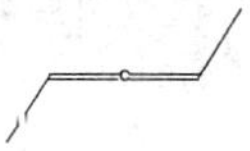

C

8661

8662

8663

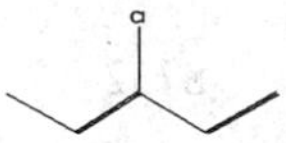

Cl

8664

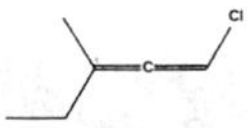

C Cl

8665

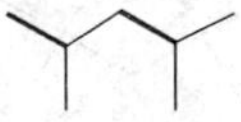

8666

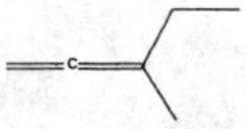

C

8667

O

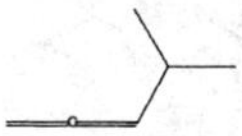

8668

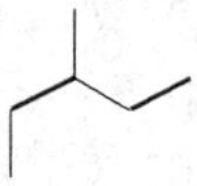

8669

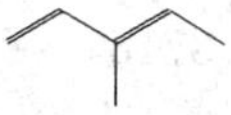

8670

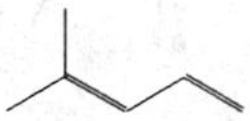

8671

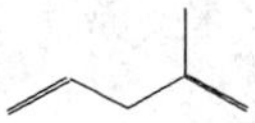

8672

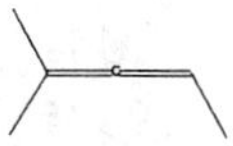

C

8673

N

8674

N

8675

N

8676

OH

O

8679

OH O OH

O

8680

O

O

8681

OH

8682

O

O O

8683

O

8684

8685

O

O

8686

8687

8688

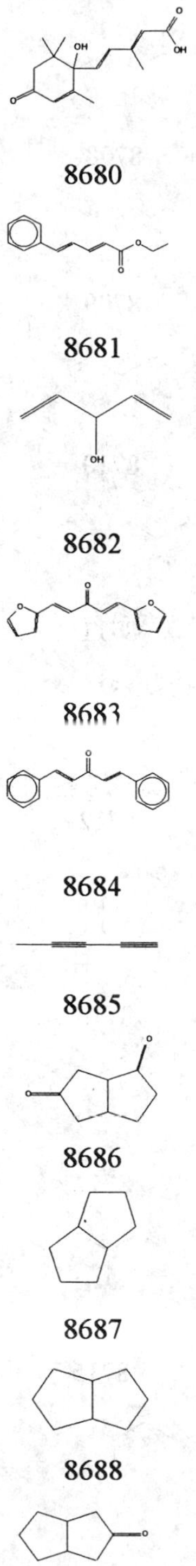

O

8689

O

8690

O

8691

O

OH

8692

O

HO

8693

O

8694

O

8695

O

O

8696

O

O

8697

O

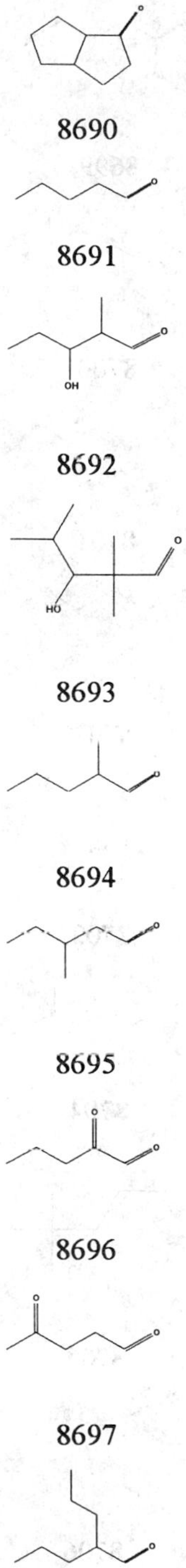

STRUCTURAL FORMULAS OF ORGANIC COMPOUNDS (continued)

In numeric order as they occur in the Table of Physical Constants of Organic Compounds

8698

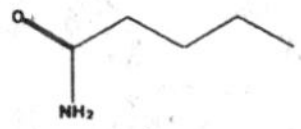

8699

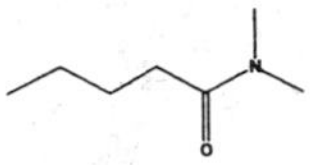

8700

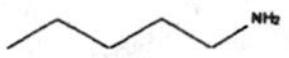

8701

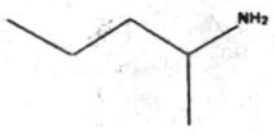

8702

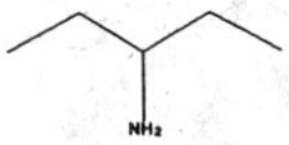

8703

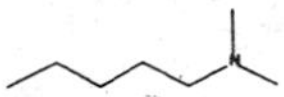

8704

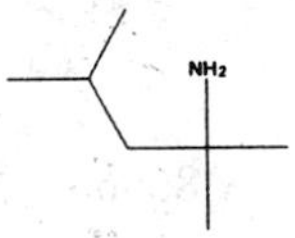

8705

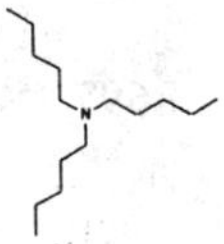

8706

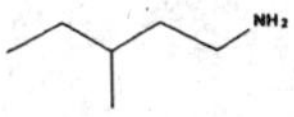

8707

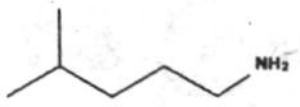

8708

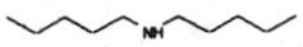

8709

8710

8711

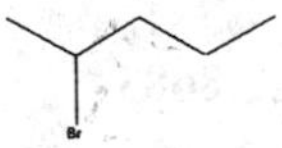

8712

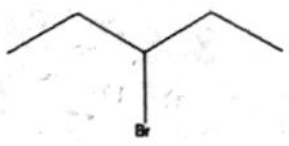

8713

8714

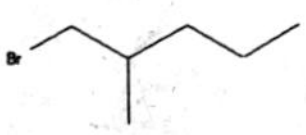

8715

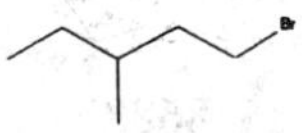

8716

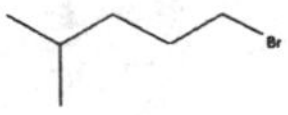

8717

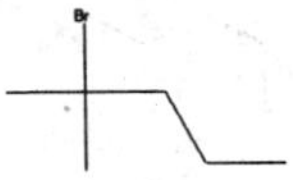

8718

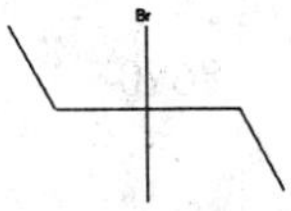

8719

8720

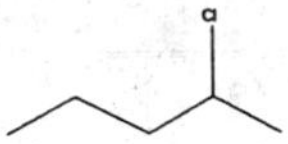

8721

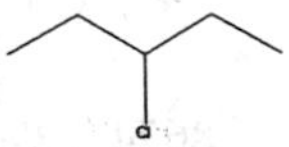

8722

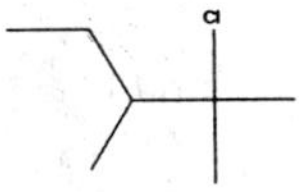

8723

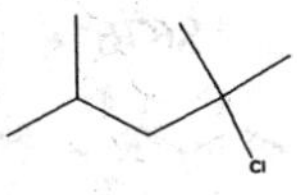

8726

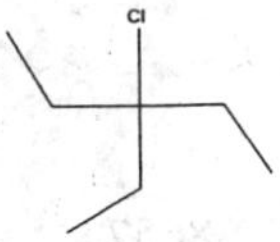

8727

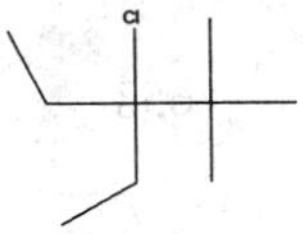

8728

8729

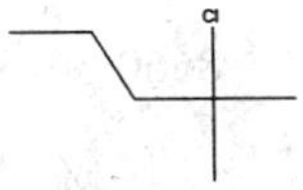

8730

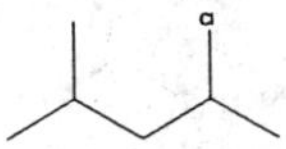

8731

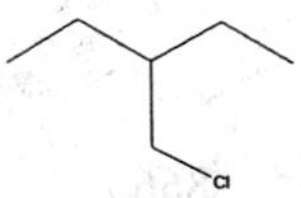

8732

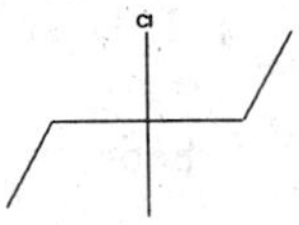

8733

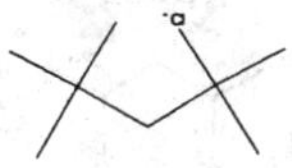

8734

8735

8736

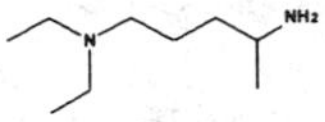

8737

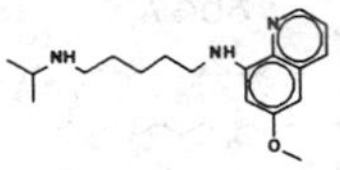

8739

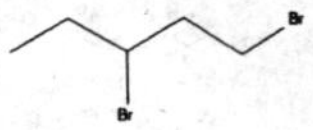

8740

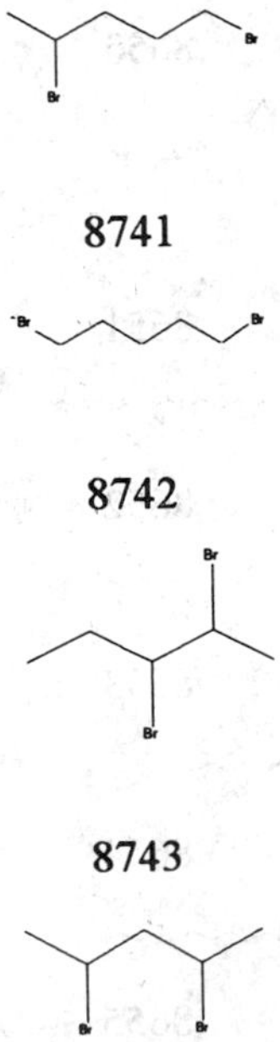

8741

8742

8743

STRUCTURAL FORMULAS OF ORGANIC COMPOUNDS (continued)

In numeric order as they occur in the Table of Physical Constants of Organic Compounds

8744

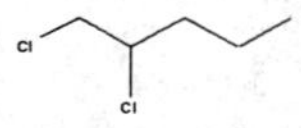

Cl, Cl

8745

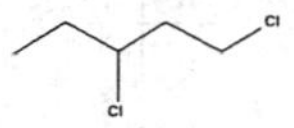

Cl, Cl

8746

Cl, Cl

8747

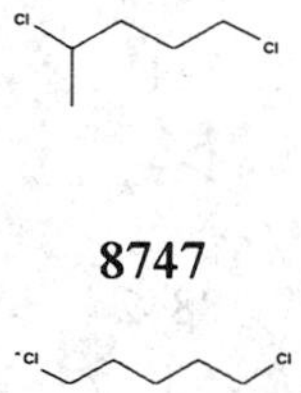

Cl, Cl

8748

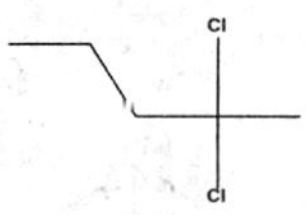

Cl, Cl

8749

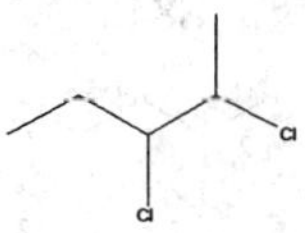

Cl, Cl

8750

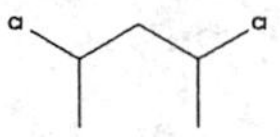

Cl, Cl

8751

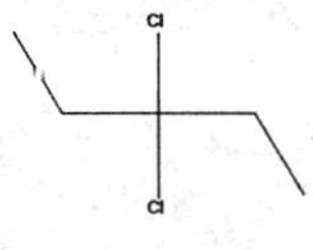

Cl, Cl

8752

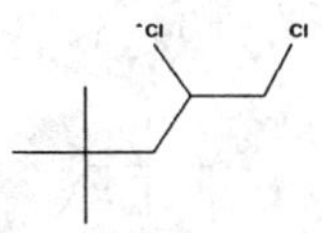

Cl, Cl

8753

Cl, Cl

8754

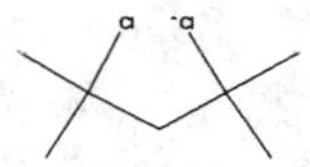

Cl, Cl

8755

O, O

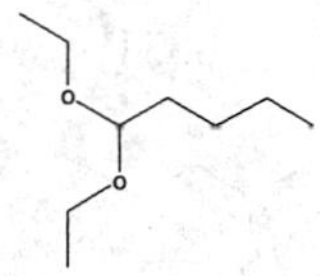

8756

8757

8758

8760

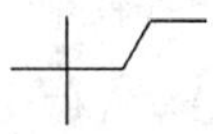

8761

8762

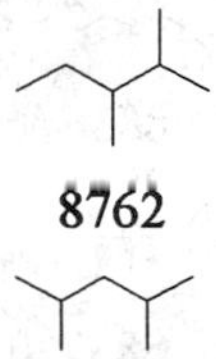

8763

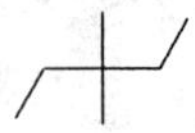

8764

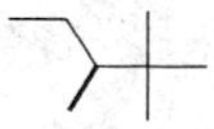

8765

8766

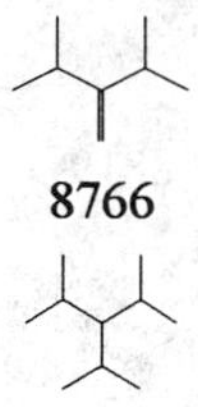

8767

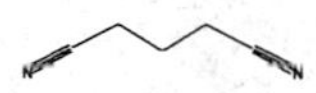

N, N

8768

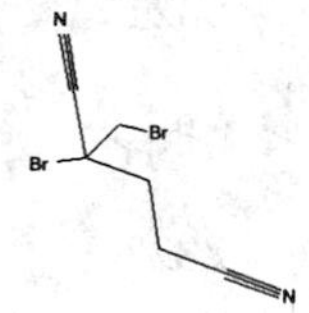

N, Br, Br, N

8769

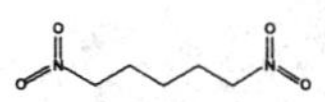

O, N, O, N, O

8770

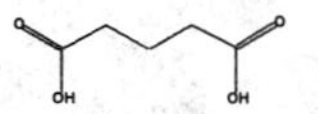

O, OH, O, OH

8771

O, O, O, O, O

8772

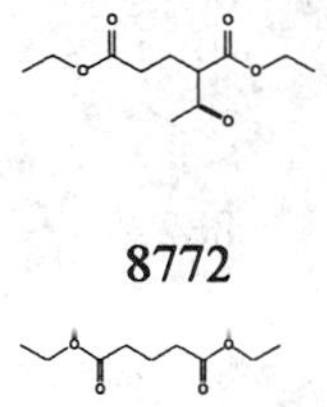

O, O, O, O

8773

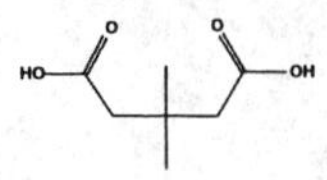

O, O, HO, OH

8774

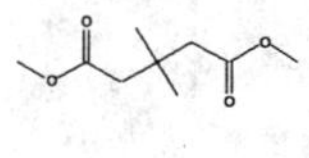

O, O, O, O

8775

O, O, O, O

8776

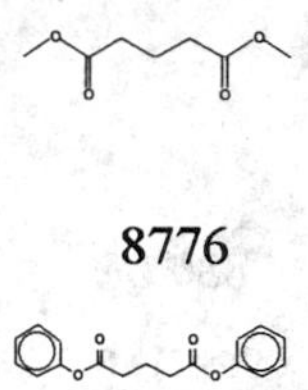

O, O, O, O

8777

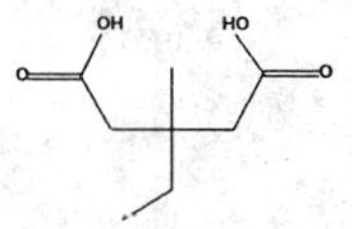

OH, HO, O, O

8778

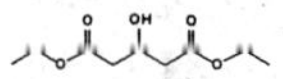

O, OH, O, O, O

8780

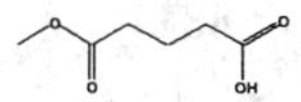

O, O, O, OH

8781

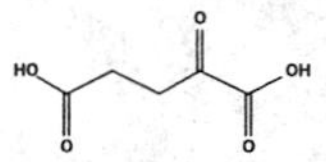

O, HO, OH, O, O

8783

O, O, O, O, O

8784

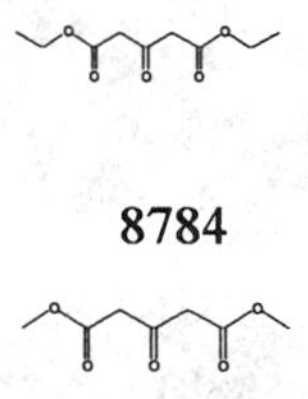

O, O, O, O, O

8785

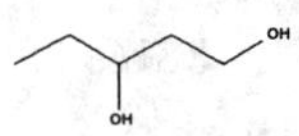

OH, OH

8786

OH, OH

8787

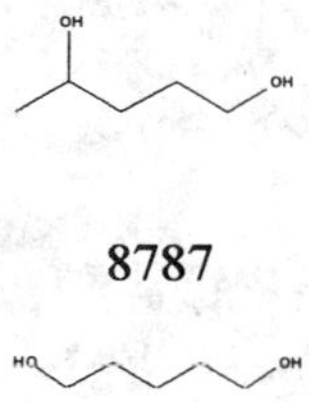

HO, OH

8788

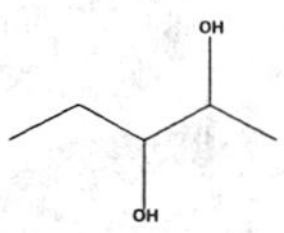

OH, OH

8789

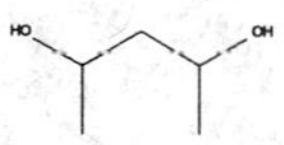

HO, OH

8790

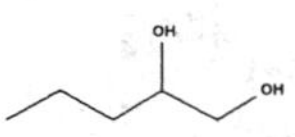

OH, OH

8791

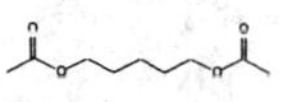

O, O, O, O

8792

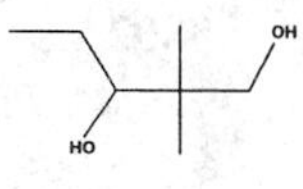

OH, HO

8793

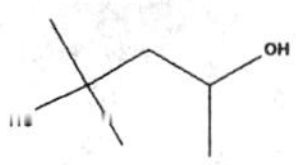

OH

8794

HO, OH

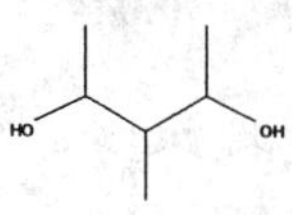

STRUCTURAL FORMULAS OF ORGANIC COMPOUNDS (continued)

In numeric order as they occur in the Table of Physical Constants of Organic Compounds

8795

OH HO

8796

HO HO

8797

O O

8798

O O

8799

O O

8801

O O

8802

F F F O O F F F

8803

O N HO

8804

O O

8805

O O

8806

O O Cl Cl

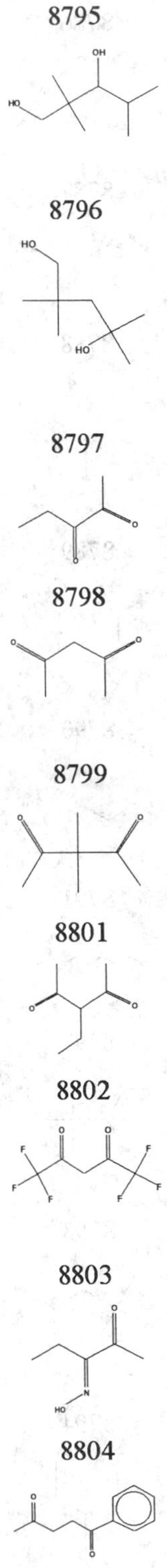

8807

F F F F F F F F F F F F F F

8808

O

8809

8810

8811

8812

8813

8814

8815

8816

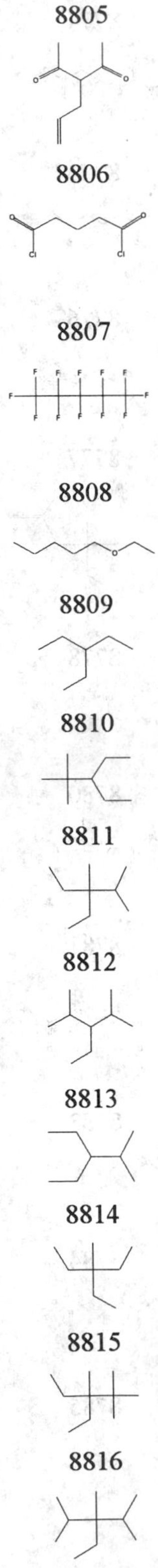

8817

F

8818

I

8819

I

8820

I

8821

N

8822

O

8823

8824

8825

8826

S

8827

N

8828

Br N

8829

OH N

8830

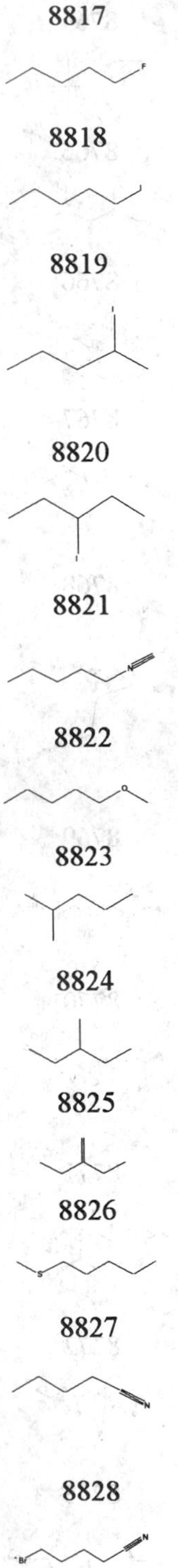

NH N

8831

N

8832

O N O

8833

O N O

8834

N O O

8835

O

8836

8837

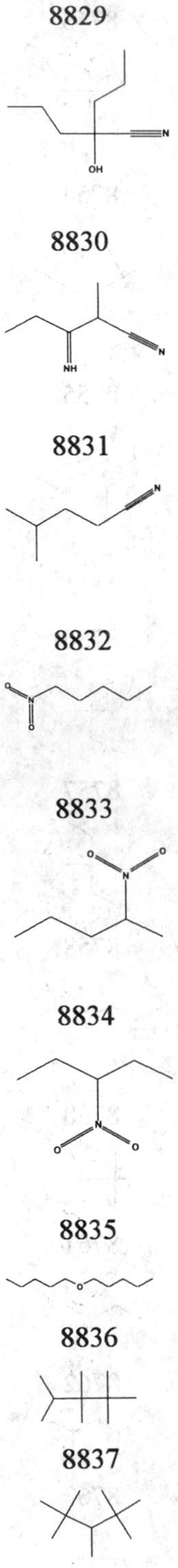

8838

Cl Cl Cl Cl

8839

8840

8841

8842

8843

S

8844

SH

8845

HS

8846

SH

8847

8848

8849

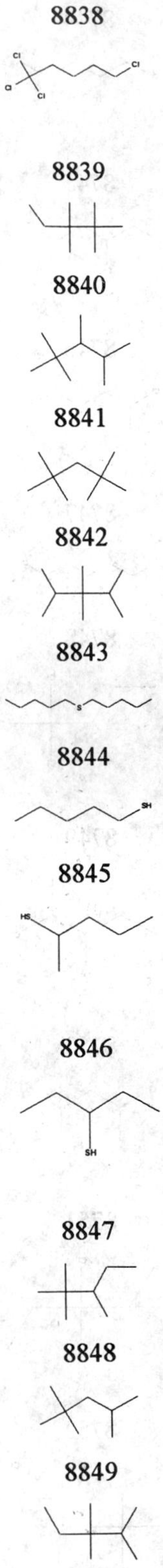

8850

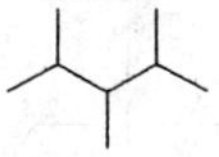

8851

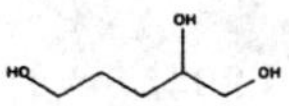

8852

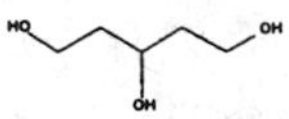

8853

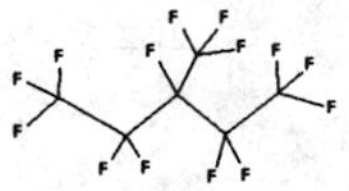

8854

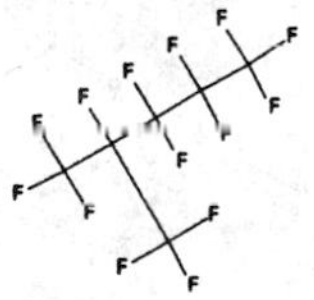

8855

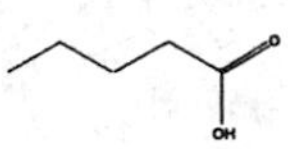

8856

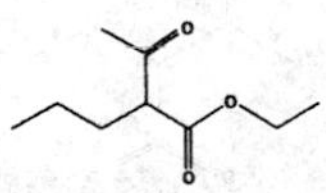

8857

8858

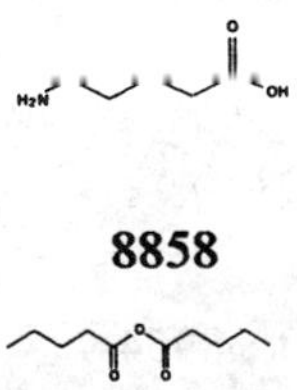

8859

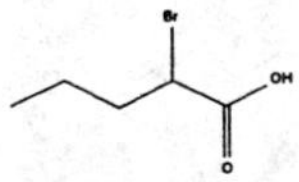

8860

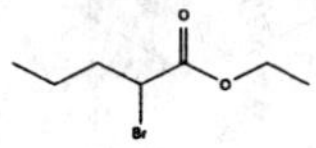

8861

8862

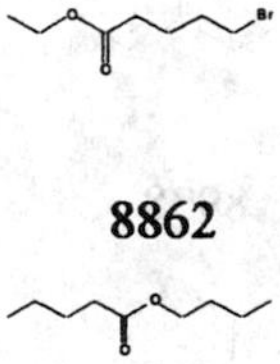

8865

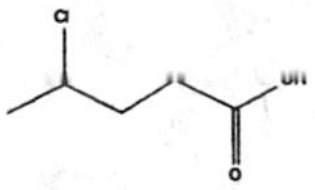

8866

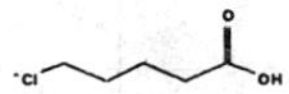

8867

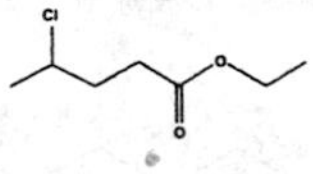

8868

8869

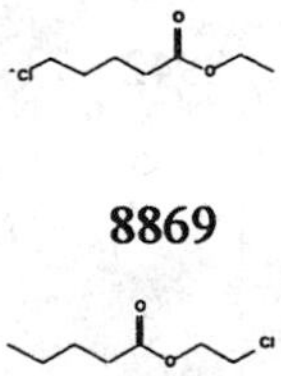

8870

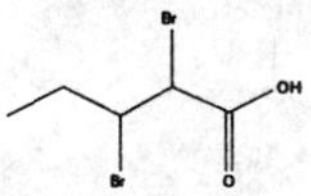

8871

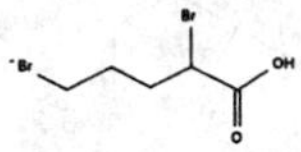

8872

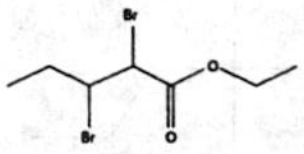

8873

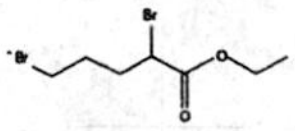

8874

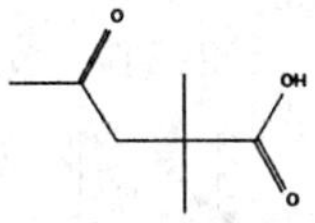

8875

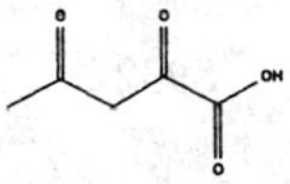

8876

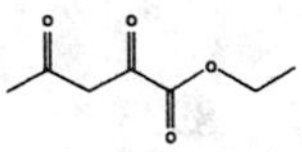

8878

8879

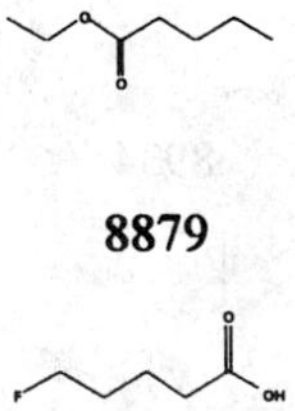

8881

8882

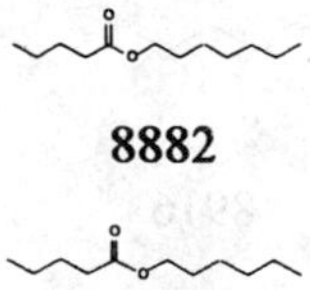

8884

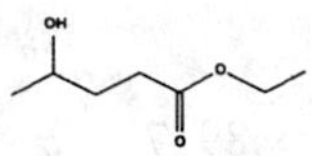

8885

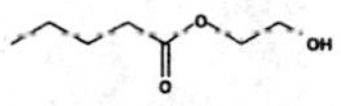

8886

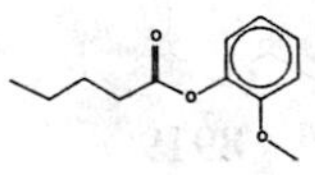

8887

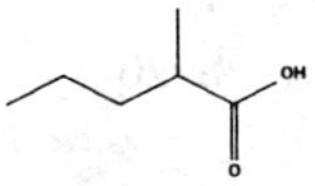

8888

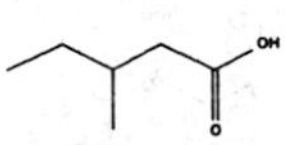

8889

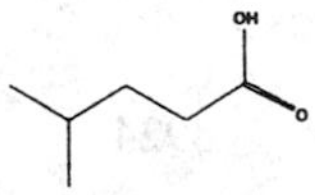

8891

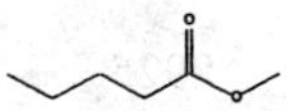

8892

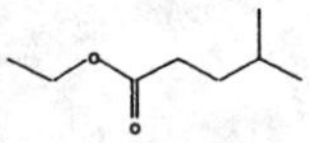

8893

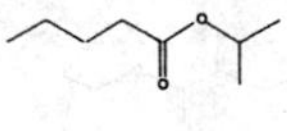

8894

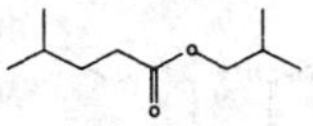

8895

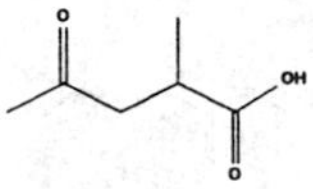

8896

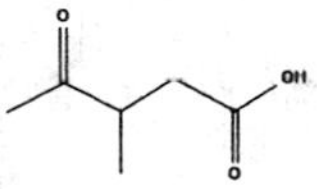

8897

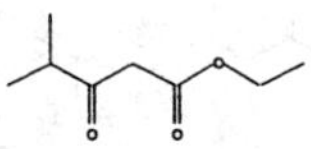

8898

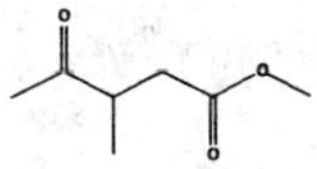

8900

8903

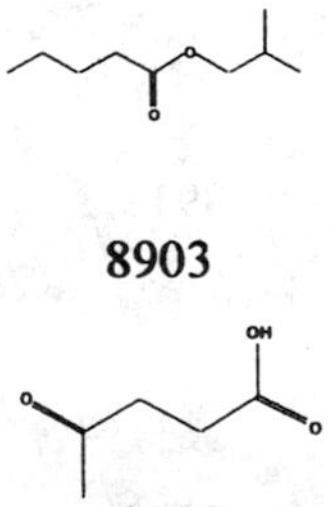

8904

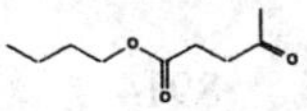

STRUCTURAL FORMULAS OF ORGANIC COMPOUNDS (continued)

In numeric order as they occur in the Table of Physical Constants of Organic Compounds

8905

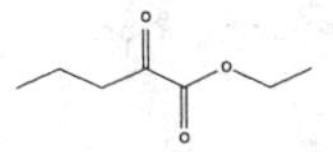

8906

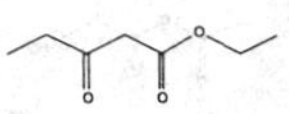

8907

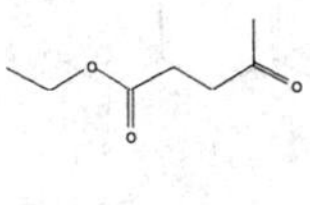

8908

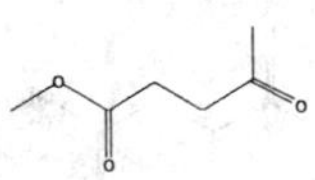

8909

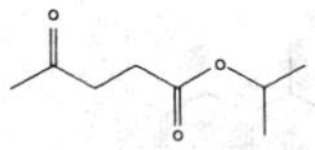

8910

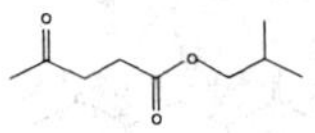

8911

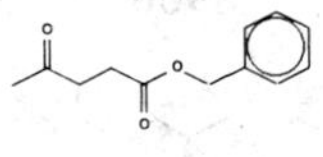

8912

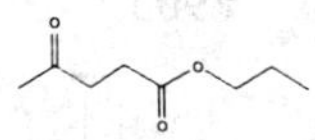

8913

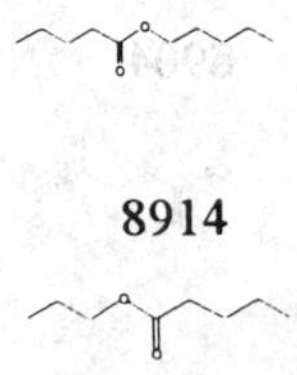

8914

8915

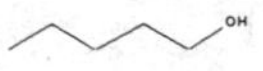

8916

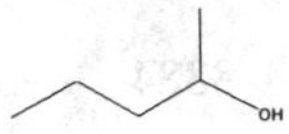

8917

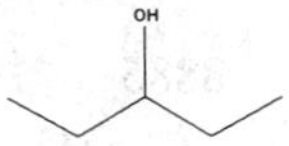

8918

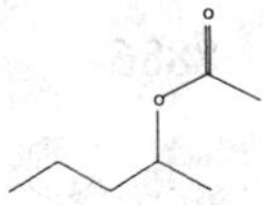

8919

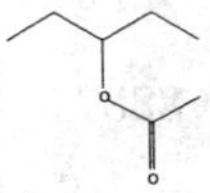

8920

8922

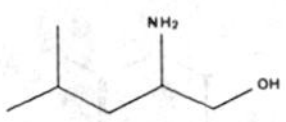

8923

8924

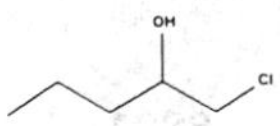

8925

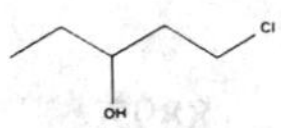

8927

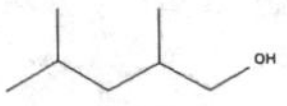

8928

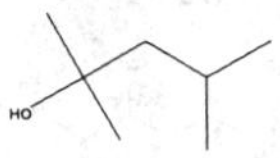

8929

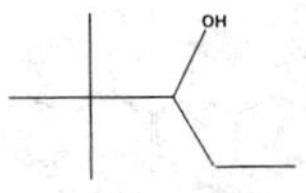

8930

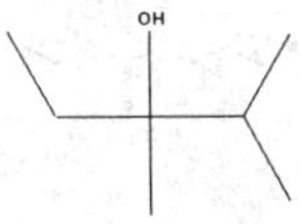

8931

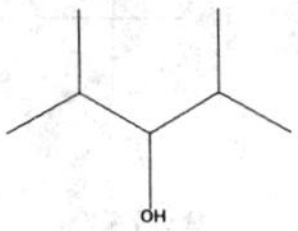

8932

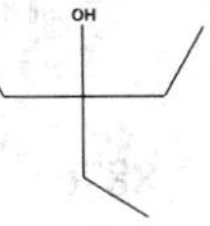

8933

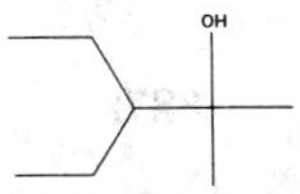

8934

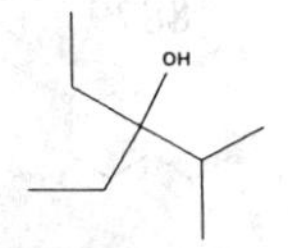

8935

8936

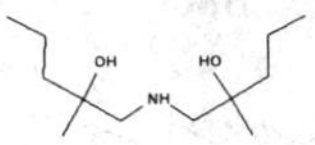

8937

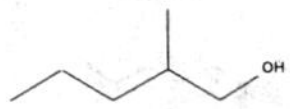

8938

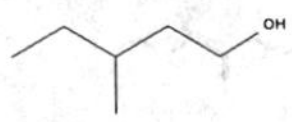

8939

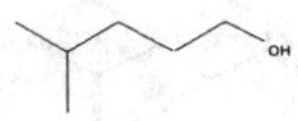

8940

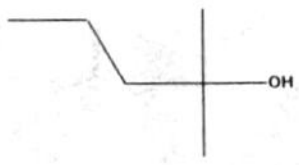

8941

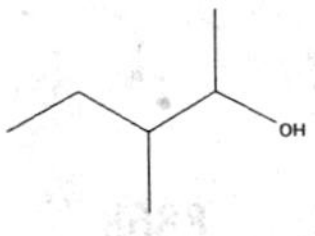

8942

8943

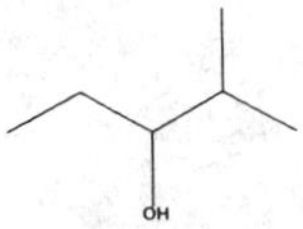

8944

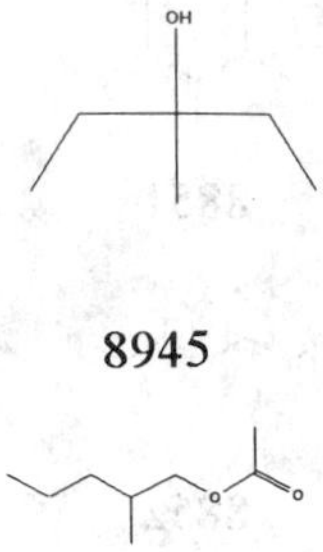

8945

8946

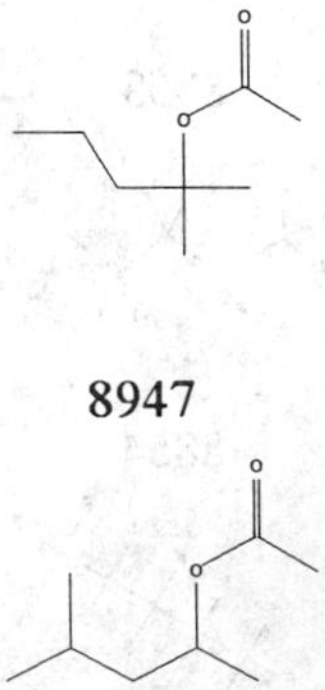

8947

8950

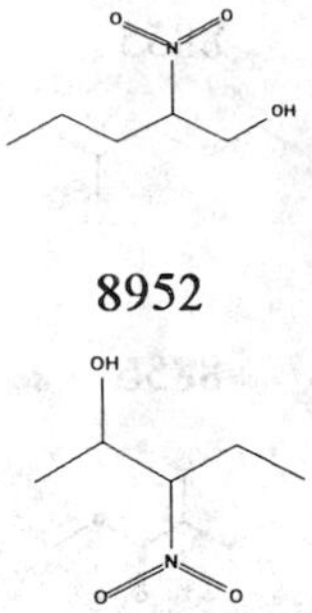

8952

8953

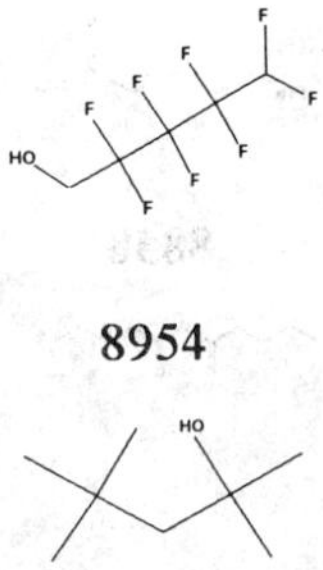

8954

STRUCTURAL FORMULAS OF ORGANIC COMPOUNDS (continued)

In numeric order as they occur in the Table of Physical Constants of Organic Compounds

8955

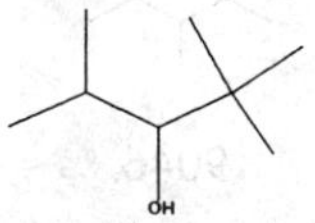

OH

8956

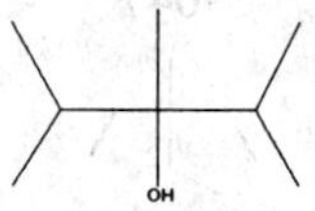

OH

8957

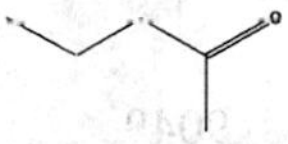

8958

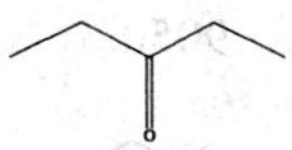

8959

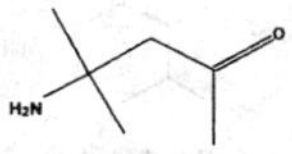

H_2N

8961

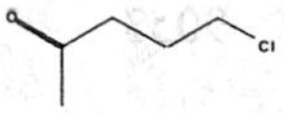

Cl

8962

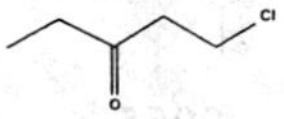

Cl

8963

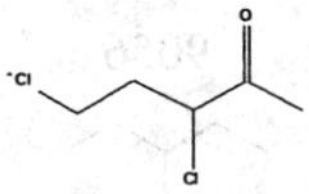

Cl

Cl

8964

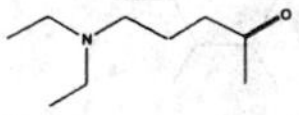

N

8965

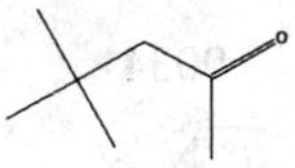

8966

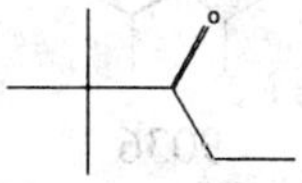

8967

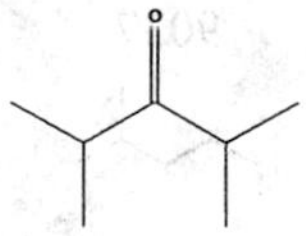

8968

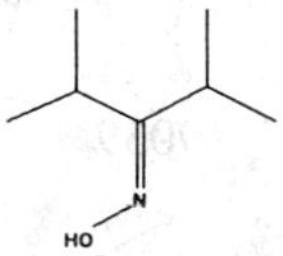

N

HO

8969

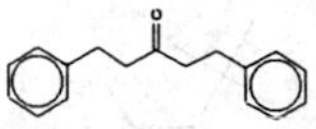

8970

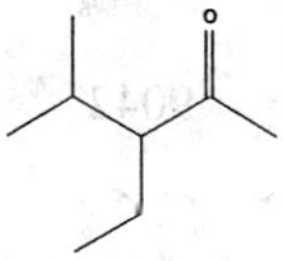

8971

HO

8972

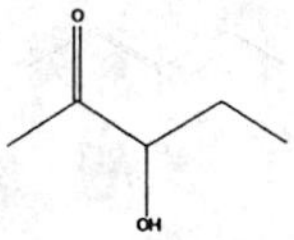

OH

8973

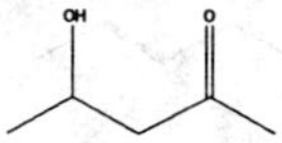

OH

8974

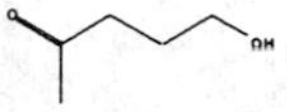

OH

8975

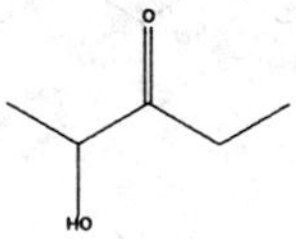

HO

8976

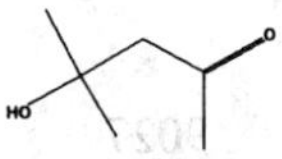

HO

8977

NH

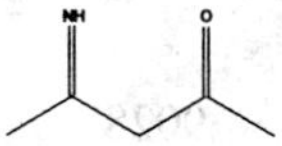

8978

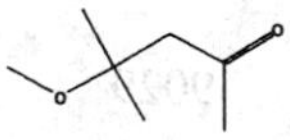

8979

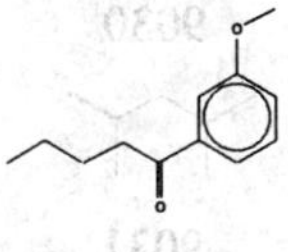

8980

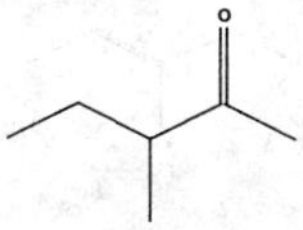

8981

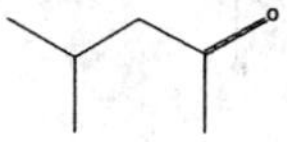

8982

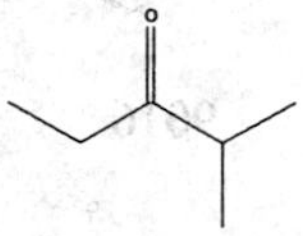

8983

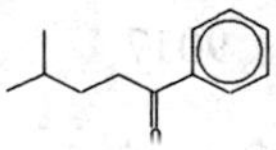

8984

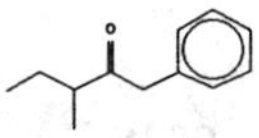

8985

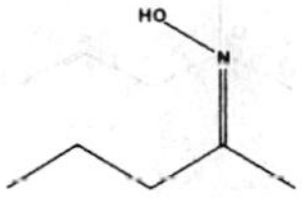

HO

N

8986

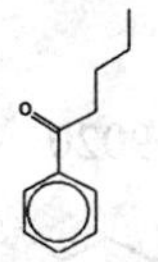

8987

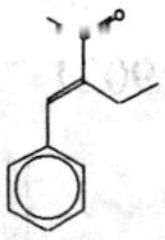

8988

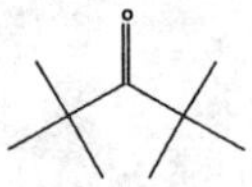

8989

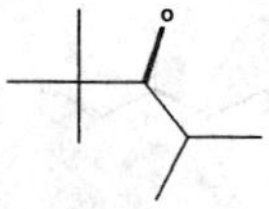

8990

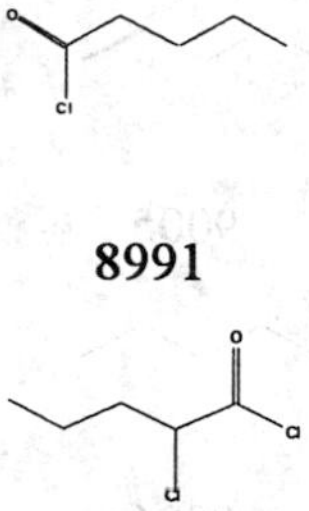

Cl

8991

Cl

Cl

8992

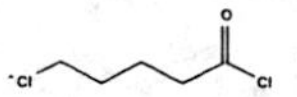

Cl

Cl

8995

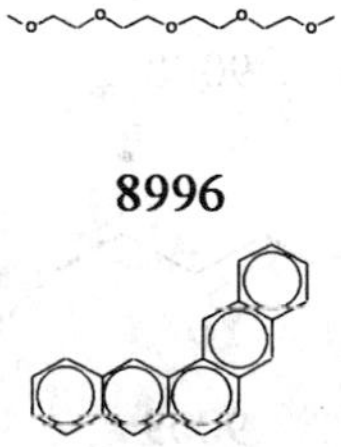

8996

8997

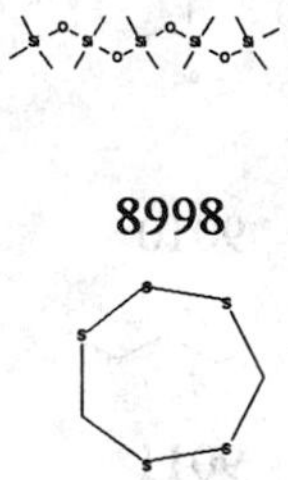

Si

8998

S

8999

9000

9001

STRUCTURAL FORMULAS OF ORGANIC COMPOUNDS (continued)

In numeric order as they occur in the Table of Physical Constants of Organic Compounds

9002

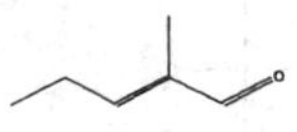

9004

9005

9006

9007

Br

9008

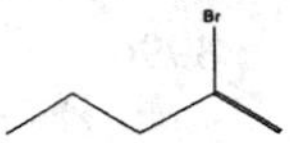

Br

9009

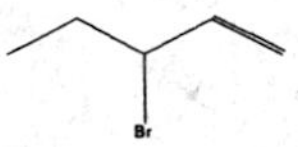

Br

9010

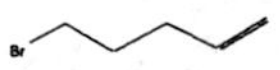

Br

9011

Br

9012

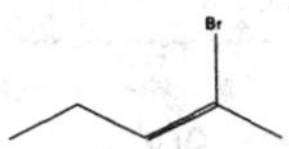

Br

9013

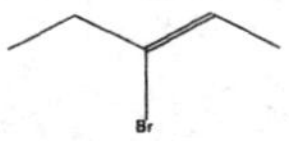

Br

9014

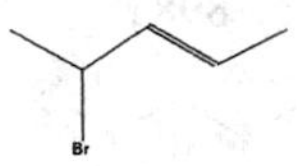

Br

9015

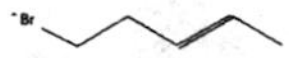

Br

9016

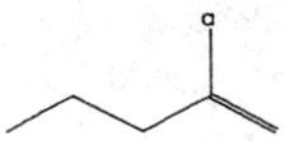

Cl

9017

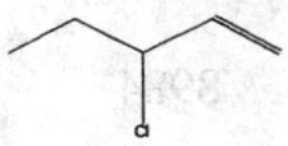

Cl

9018

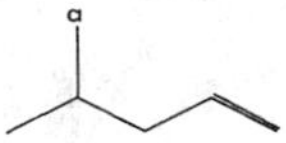

Cl

9019

Cl

9020

Cl

9021

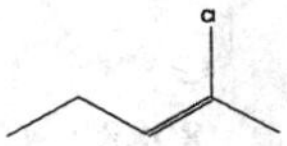

Cl

9022

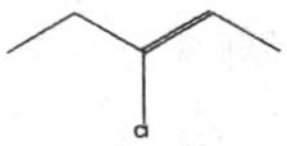

Cl

9023

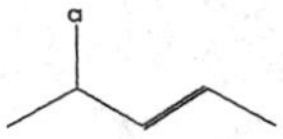

Cl

9024

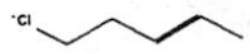

Cl

9025

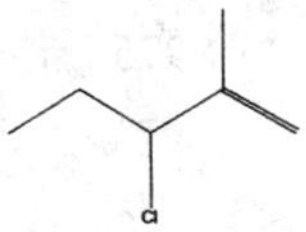

Cl

9026

Cl

9027

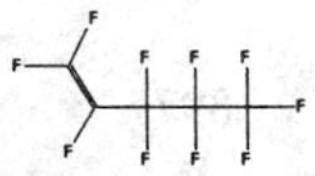

F F F F F F F F F F

9028

Cl Cl

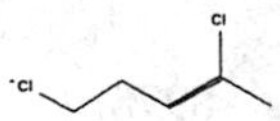

9029

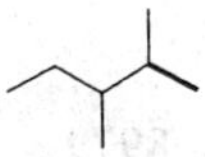

9030

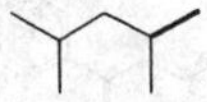

9031

9032

9033

9034

9035

9036

9037

9038

9039

9040

9041

O O O O

9042

O O O O

9043

9044

9045

9046

9047

9048

9049

9050

9051

9052

9053

9054

9055

9056

9057

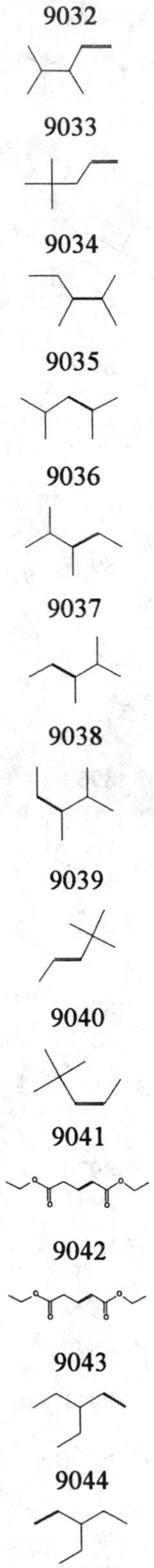

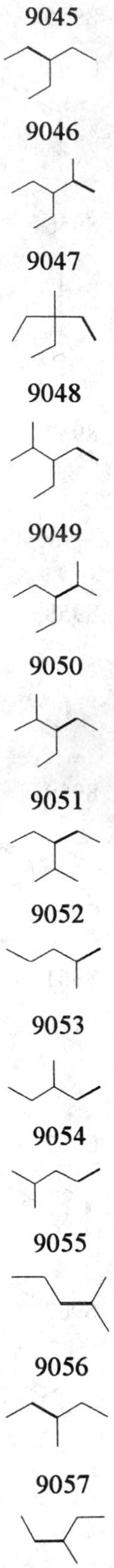

In numeric order as they occur in the Table of Physical Constants of Organic Compounds

9058

9059

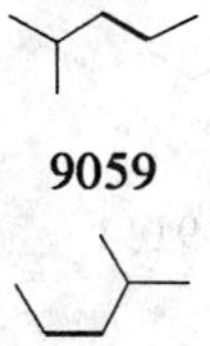

9060

9061

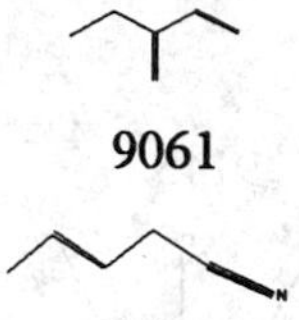

N

9062

N

9065

9066

9067

9068

9069

9070

9071

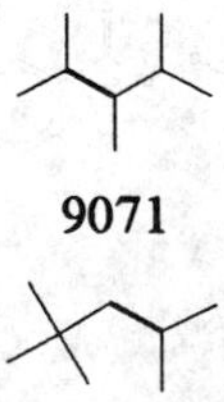

9072

9073

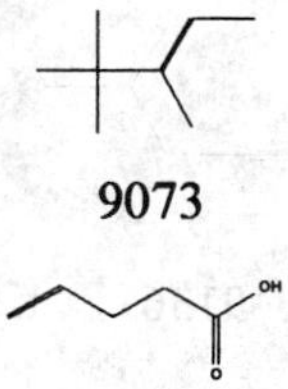

OH O

9074

O O O

9075

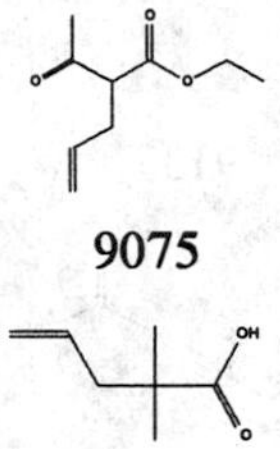

OH O

9076

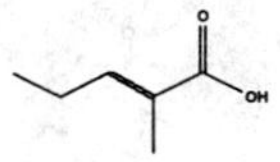

O OH

9077

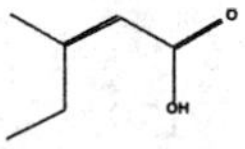

O OH

9078

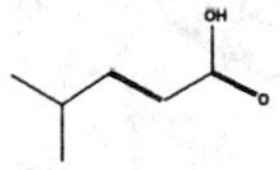

OH O

9079

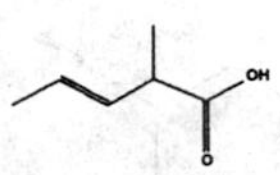

OH O

9080

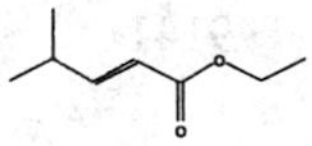

O O

9081

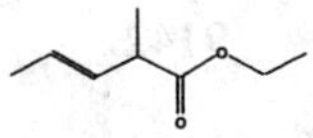

O O

9082

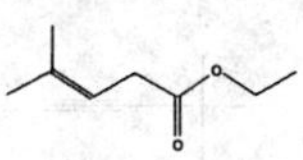

O O

9083

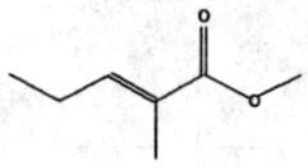

O O

9084

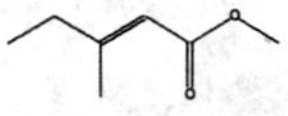

O O

9085

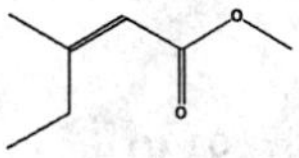

O O

9086

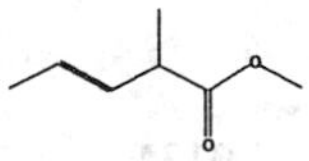

O O

9087

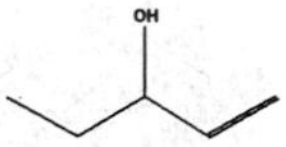

OH

9088

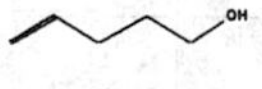

OH

9089

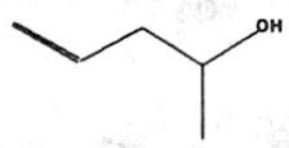

OH

9090

OH

9091

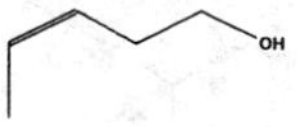

OH

9092

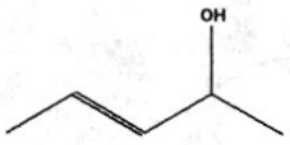

OH

9093

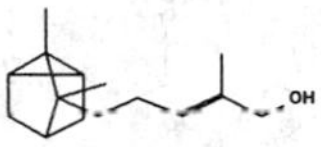

OH

9094

HO

9095

OH

9096

OH

9097

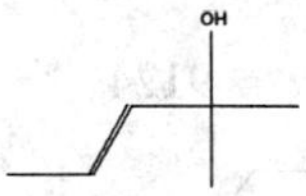

OH

9098

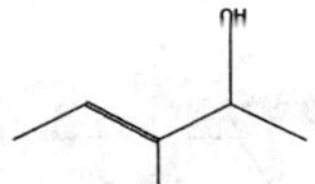

OH

9099

OH

9100

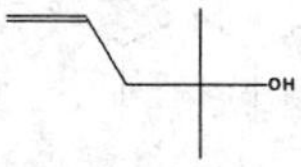

OH

9101

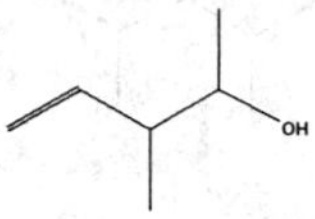

OH

9102

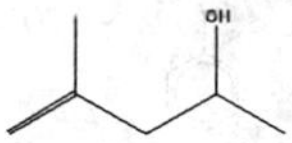

OH

9103

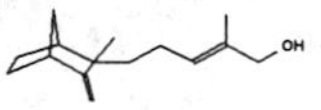

OH

9104

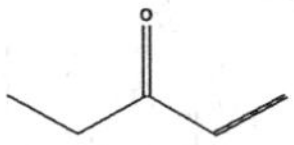

O

9105

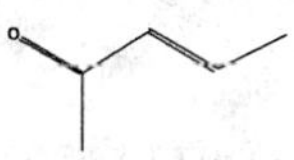

O

9106

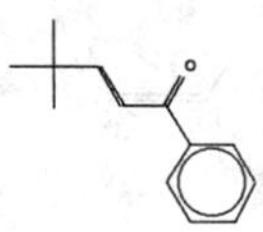

O

9107

O

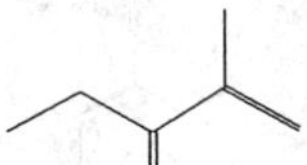

In numeric order as they occur in the Table of Physical Constants of Organic Compounds

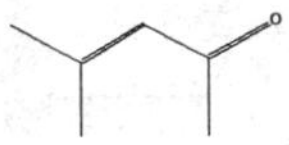

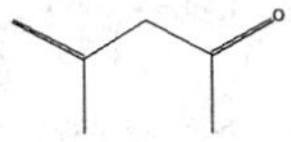

9108 O

9109 O

9110 O

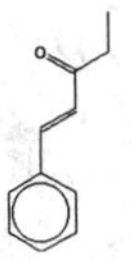

9111

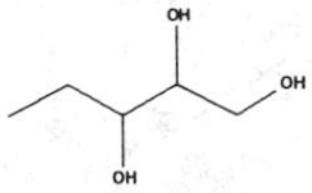

9112

9113

9114

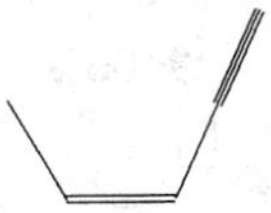

9115

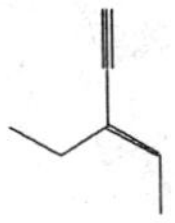

9116

9117

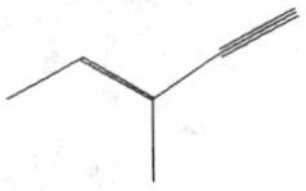

9118 OH

9119 OH OH OH

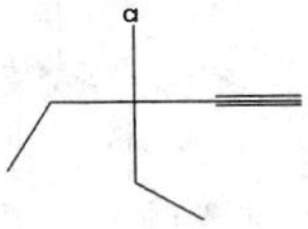

9121 NH_2

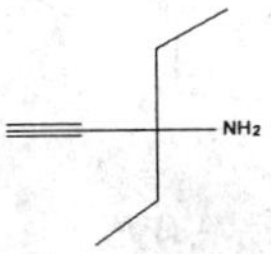

9122

9123

9124 Br

9125 Cl

9126 Cl

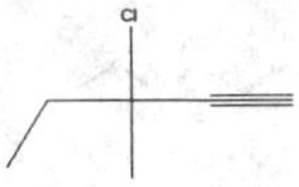

9128

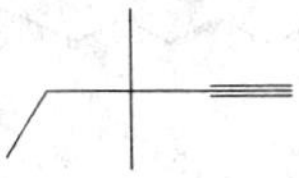

9129

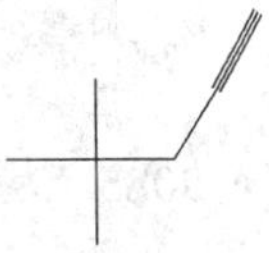

9130

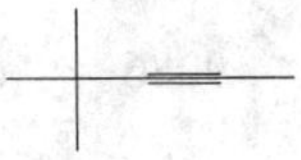

9131

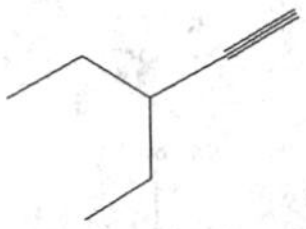

9132

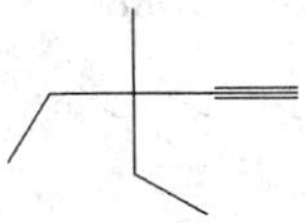

9133 I

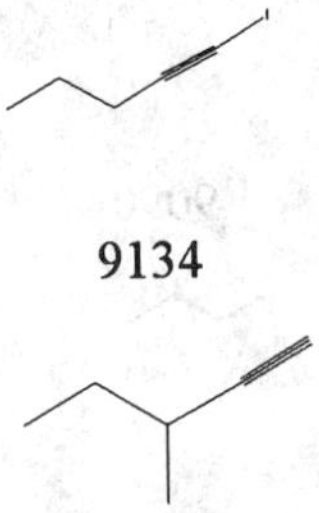

9134

9135

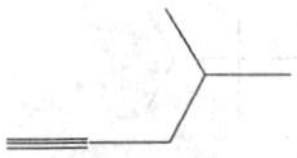

9136

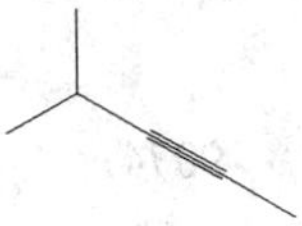

9137 O OH

9138 O OH

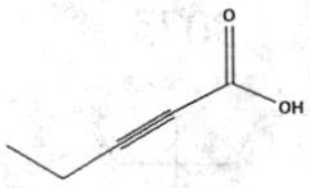

9139 O O

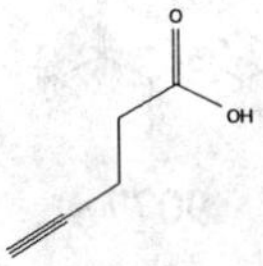

9140 OH

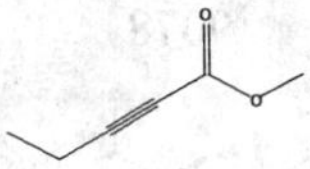

9141 HO

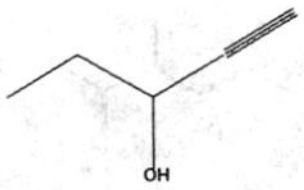

9142 HO

9143 OH

9144 OH

9145 OH

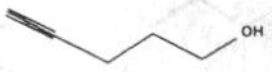

9146 OH

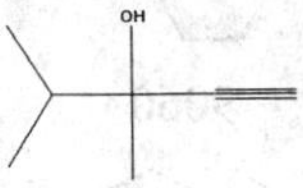

9147 O Cl

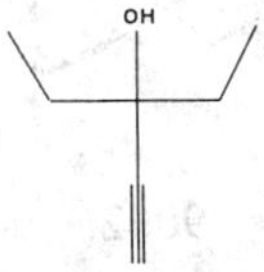

9149 O S O NH S O O F F F CH_3

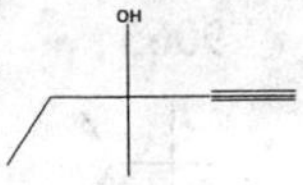

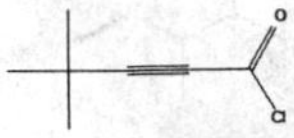

9150 Cl Cl O O O

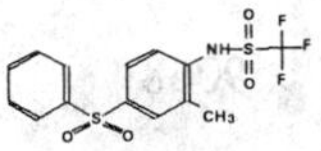

9151 O O O O

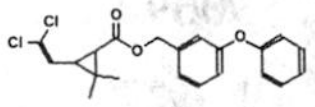

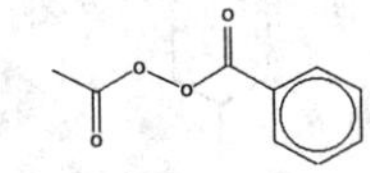

STRUCTURAL FORMULAS OF ORGANIC COMPOUNDS (continued)

In numeric order as they occur in the Table of Physical Constants of Organic Compounds

9153

9154

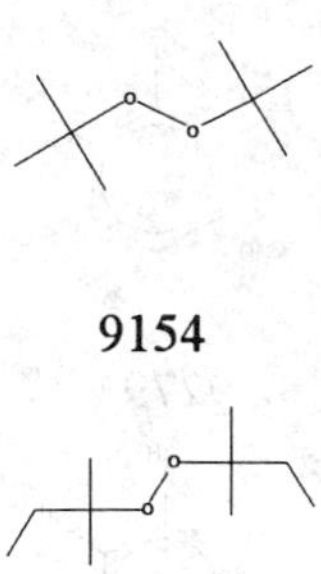

9156

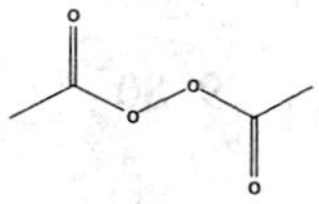

9158

9159

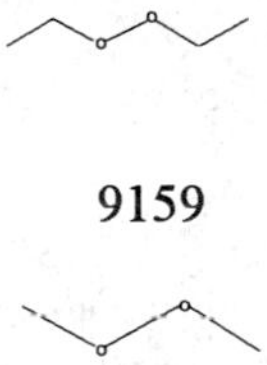

9160

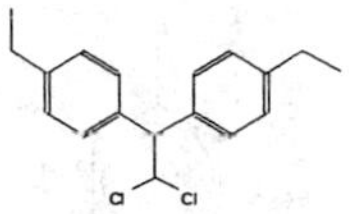

Cl Cl

9161

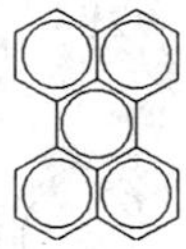

9162

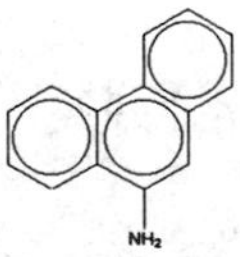

NH₂

9164

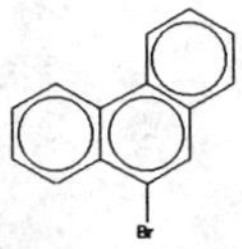

Br

9167

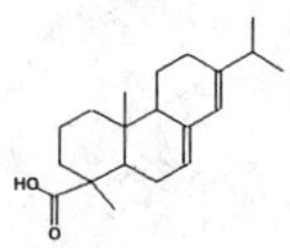

HO O

9168

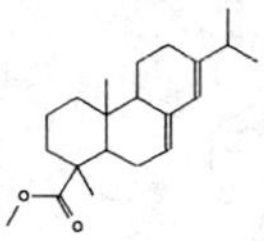

O O

9170

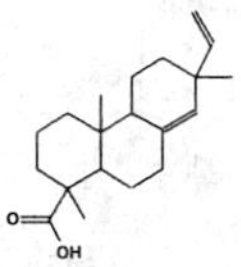

O OH

9171

9172

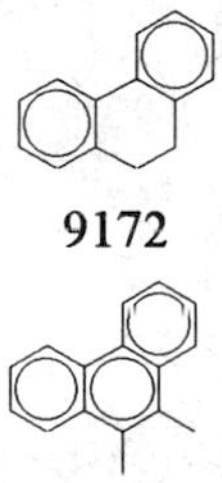

9173

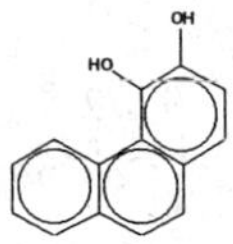

HO OH

9174

O O

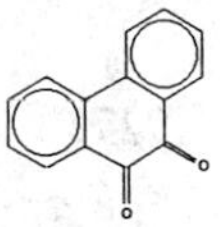

9179

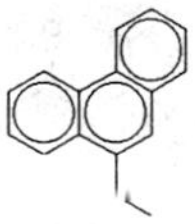

9180

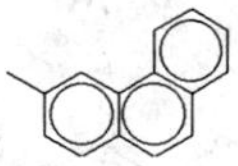

9181

9182

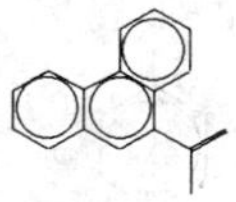

9183

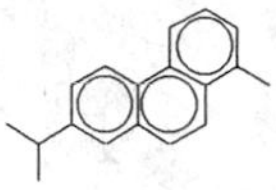

9184

9185

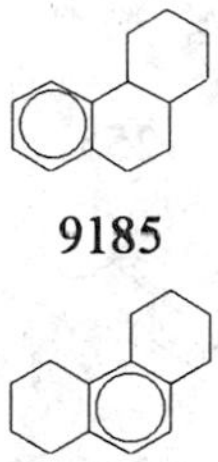

9186

9187

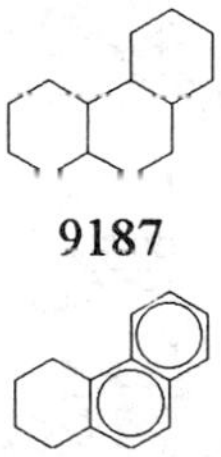

9188

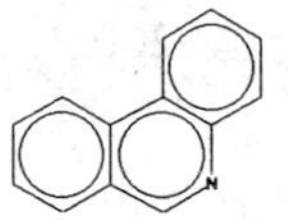

N

9189

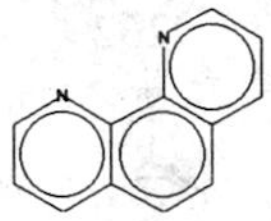

N N

9190

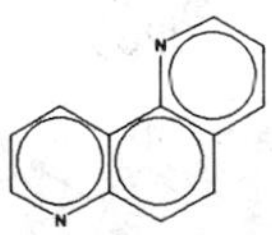

N N

9191

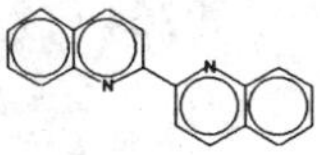

N N

9192

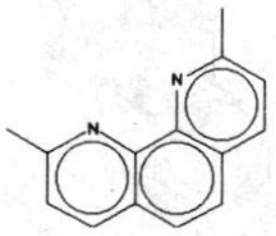

N N

9193

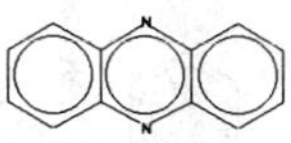

N N

9195

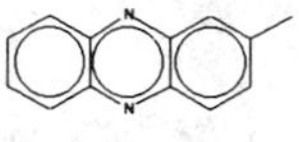

N N

9198

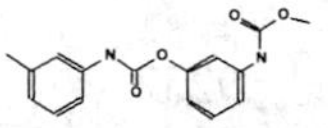

N O O N O O

9199

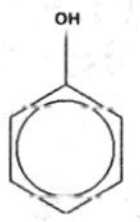

OH

9201

NH₂ OH

9202

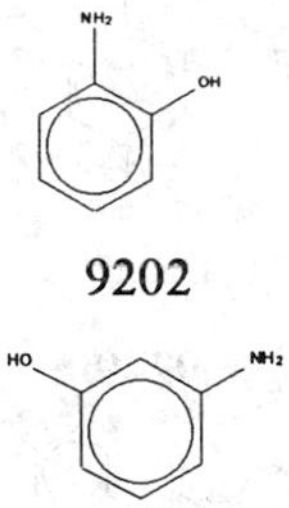

HO NH₂

9204

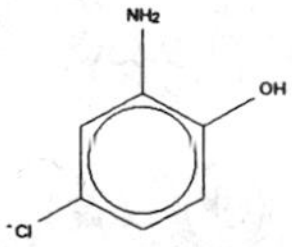

NH₂ OH Cl

9205

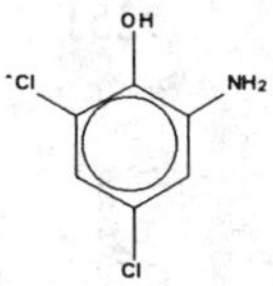

OH Cl NH₂ Cl

9206

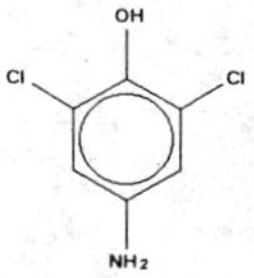

OH Cl Cl NH₂

9208

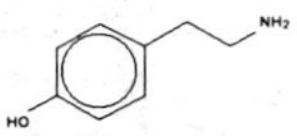

NH₂ HO

9210

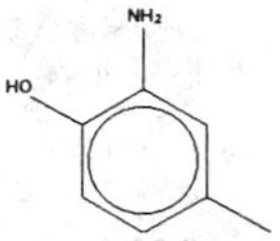

NH₂ HO

9211

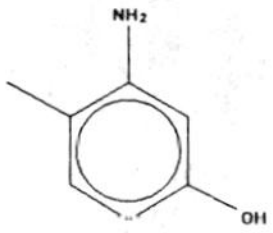

NH₂ OH

9212

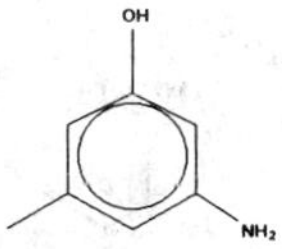

OH NH₂

9213

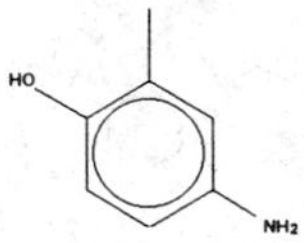

HO NH₂

9216

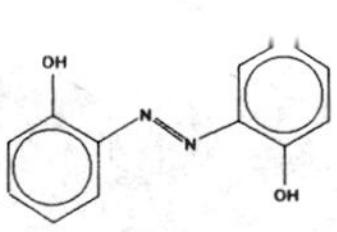

OH N N OH

9217

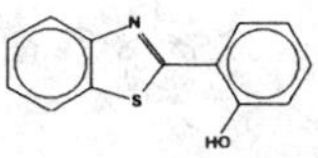

N S HO

STRUCTURAL FORMULAS OF ORGANIC COMPOUNDS (continued)

In numeric order as they occur in the Table of Physical Constants of Organic Compounds

9218

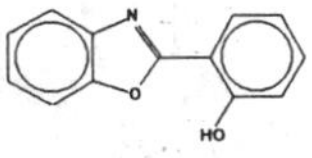

9219

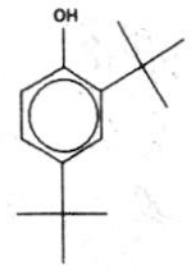

9220

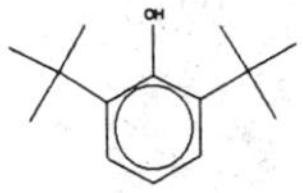

9221

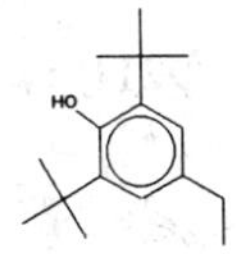

9223

9224

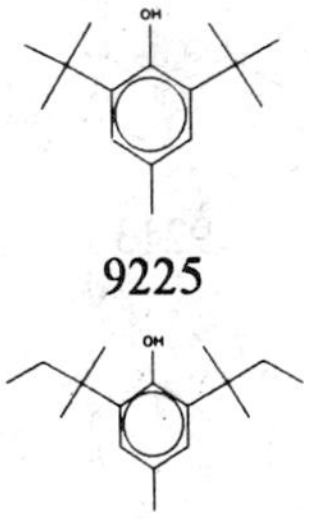

9225

9226

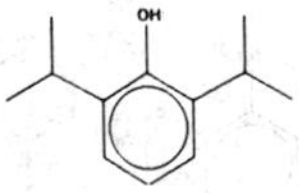

9227

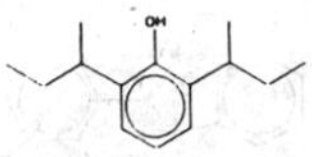

9228

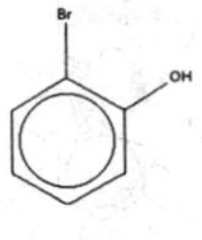

9230

9231

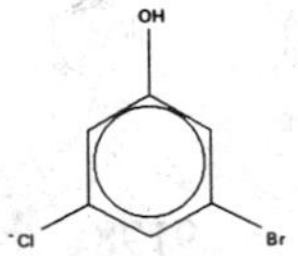

9232

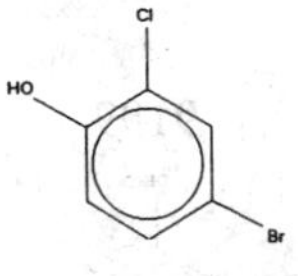

9233

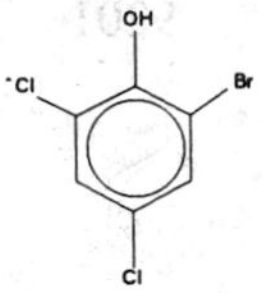

9249

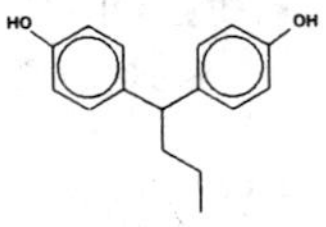

9250

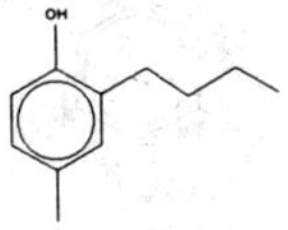

9251

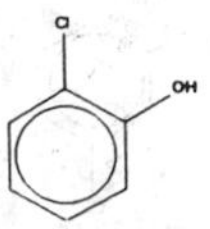

9253

9254

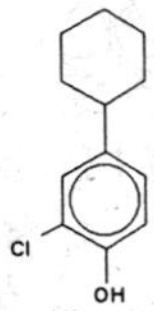

9255

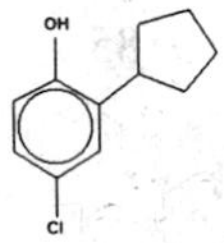

9256

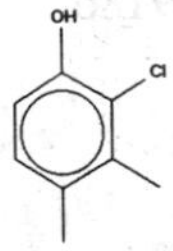

9257

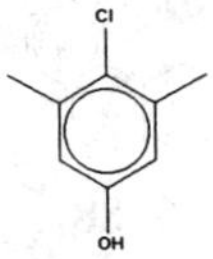

9260

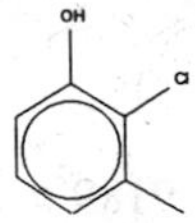

9262

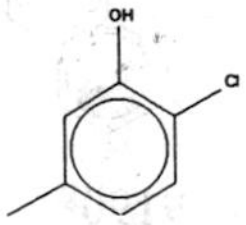

9263

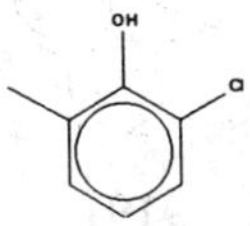

9264

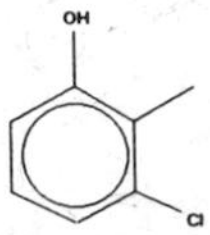

9265

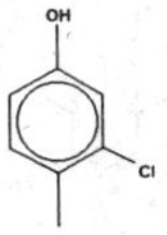

9266

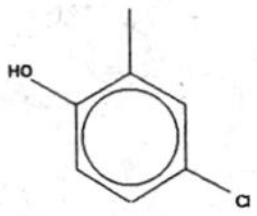

9267

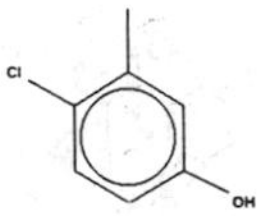

9270

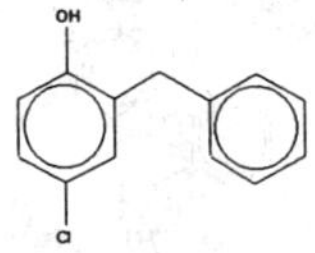

9271

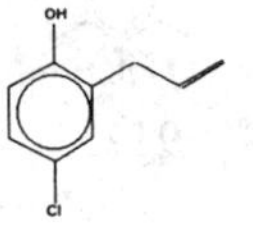

9272

9273

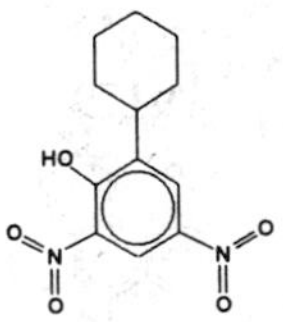

9276

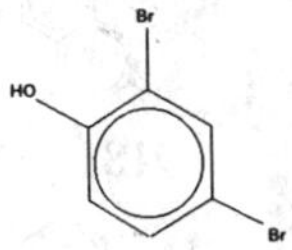

9277

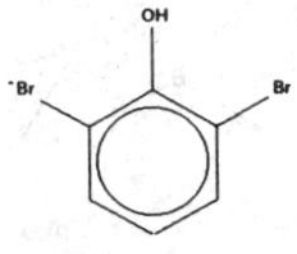

9279

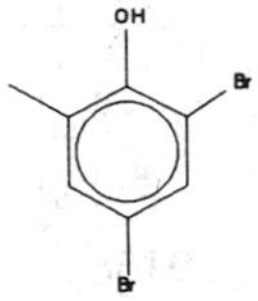

9280

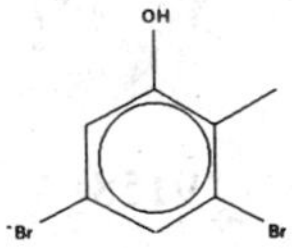

9282

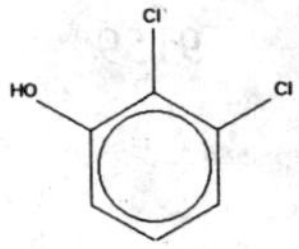

9283

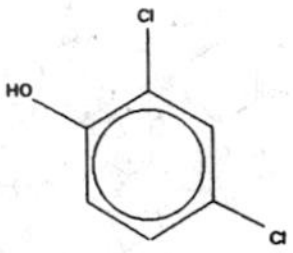

9284

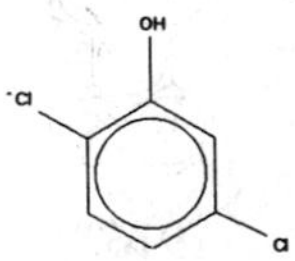

9285

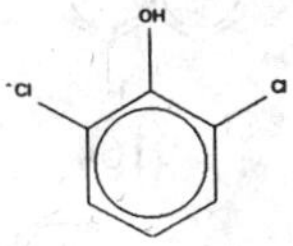

STRUCTURAL FORMULAS OF ORGANIC COMPOUNDS (continued)

In numeric order as they occur in the Table of Physical Constants of Organic Compounds

9287

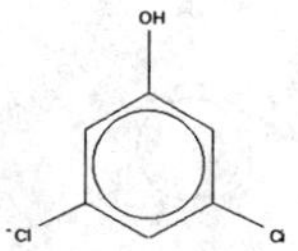

9288

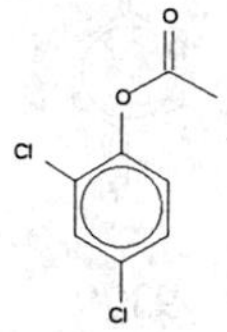

9289

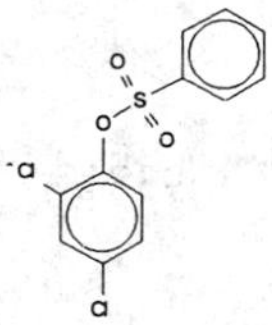

9290

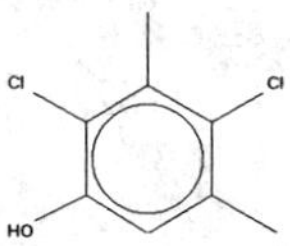

9291

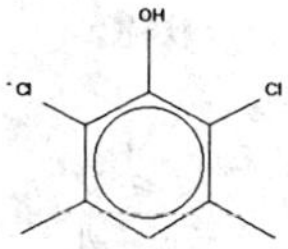

9292

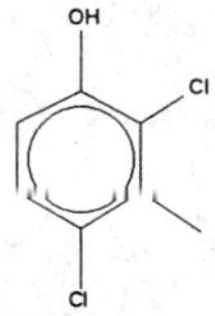

9294

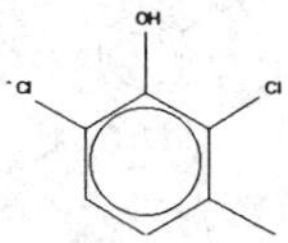

9295

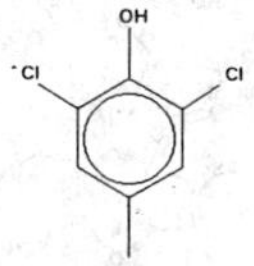

9296

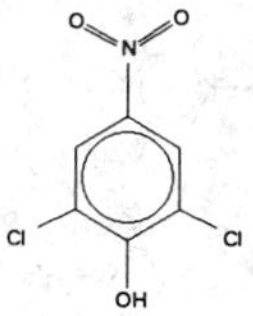

9297

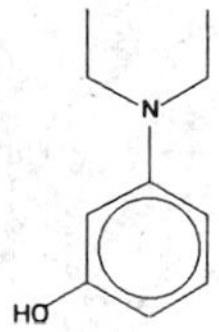

9298

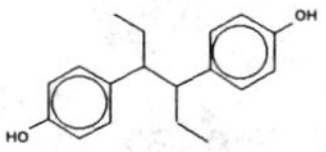

9300

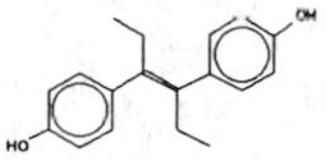

9305

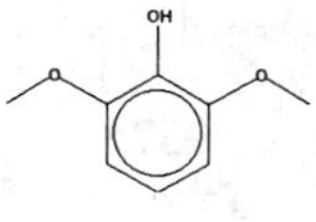

9306

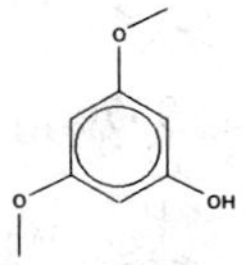

9307

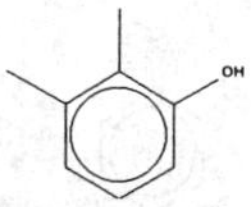

9308

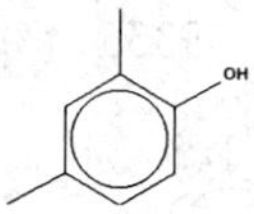

9309

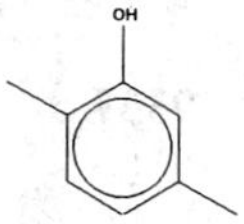

9310

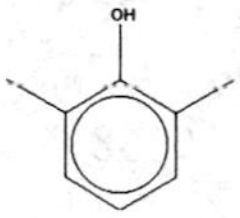

9312

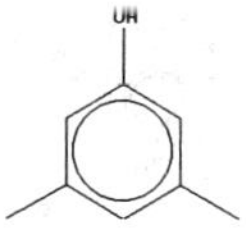

9313

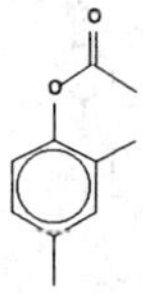

9314

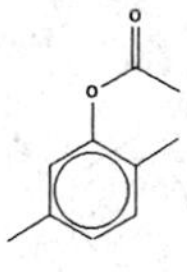

9315

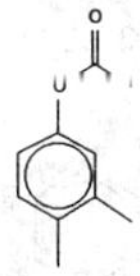

9316

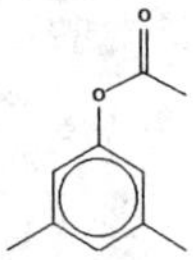

9317

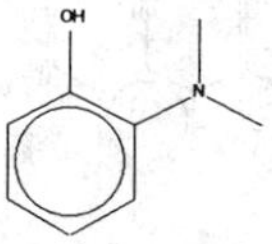

9319

9320

9321

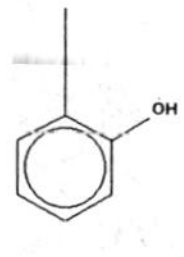

9322

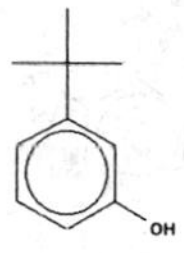

9323

9324

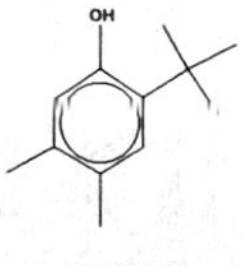

9326

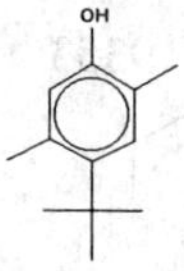

9327

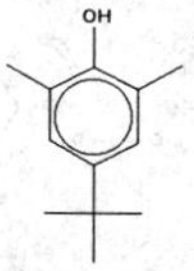

9328

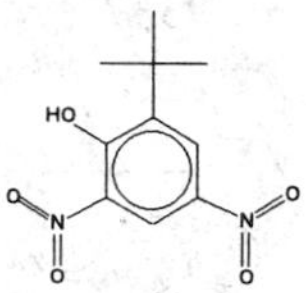

9330

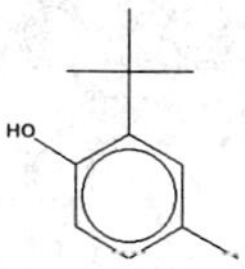

9331

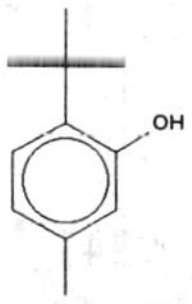

9333

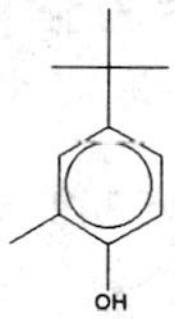

9334

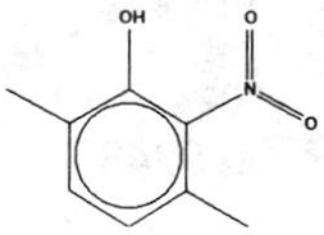

9335

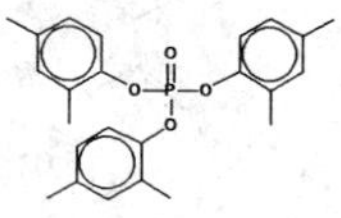

9336

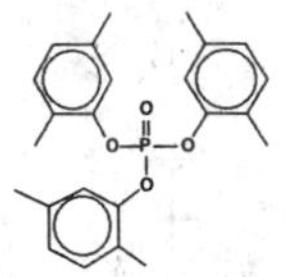

9337

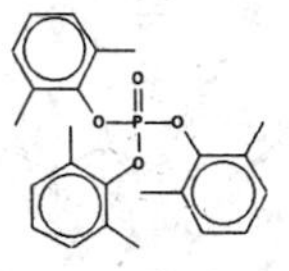

9338

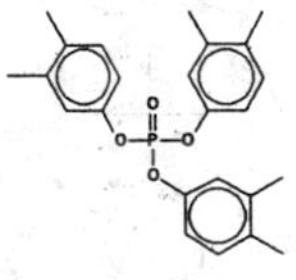

9340

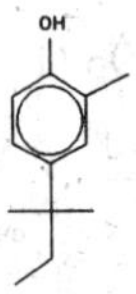

9341

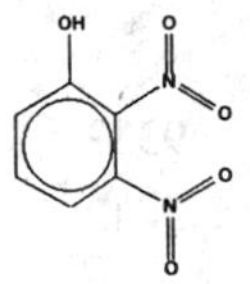

9342

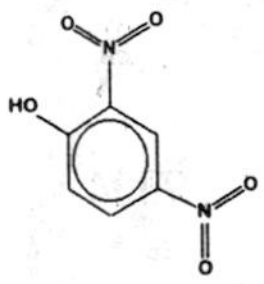

9343

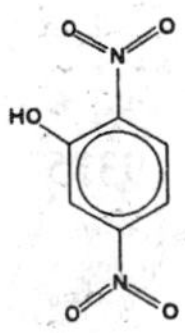

9344

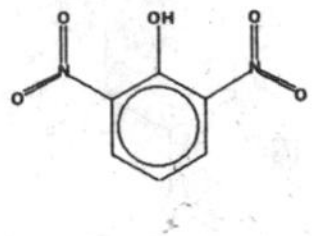

9345

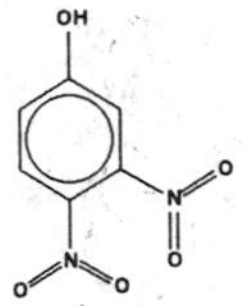

9348

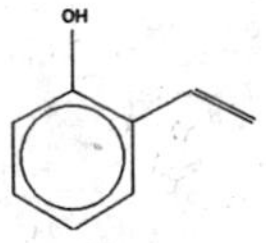

9349

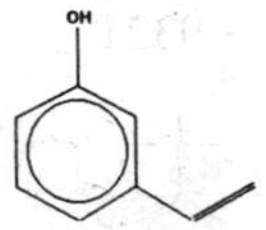

9350

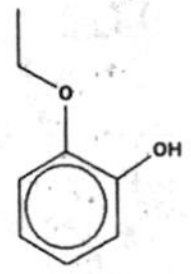

9351

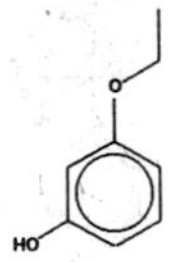

9352

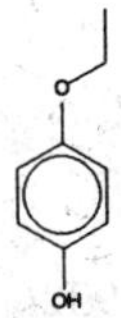

9353

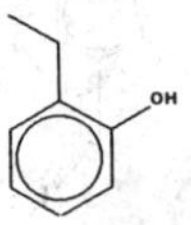

9355

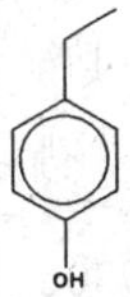

9356

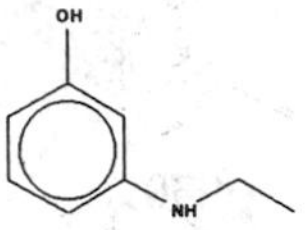

9357

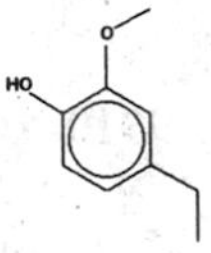

9358

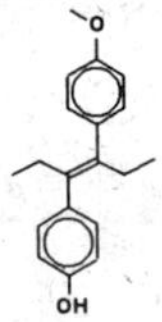

9359

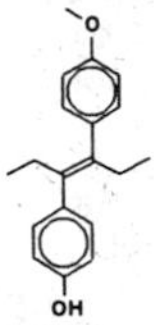

9360

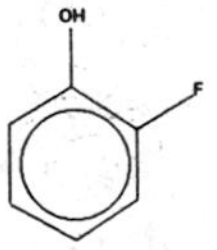

9362

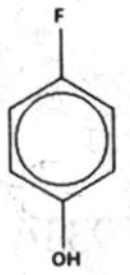

9364

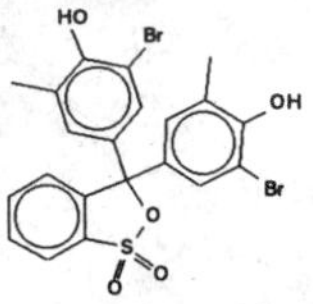

9365

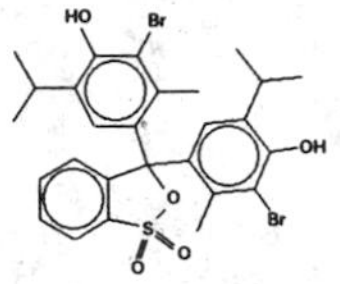

9370

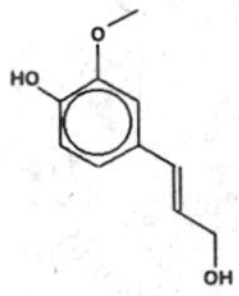

9371

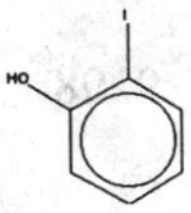

9372

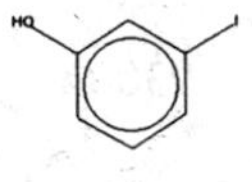

9374

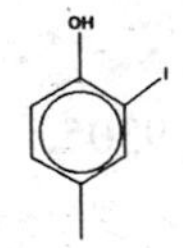

9375

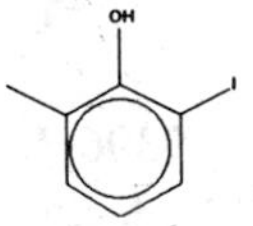

9376

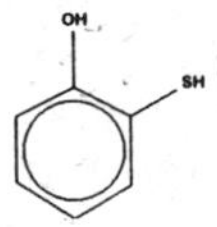

9377

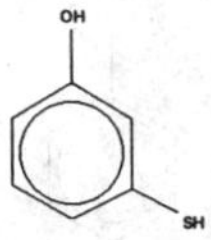

9378

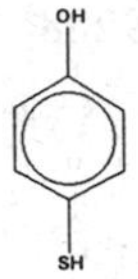

9379

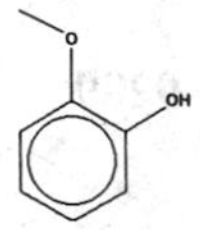

9381

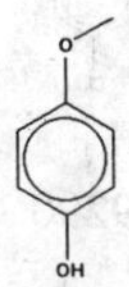

9382

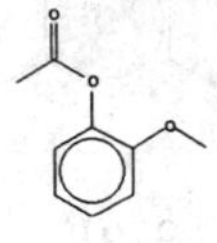

9385

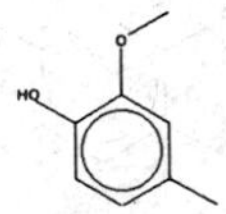

9387

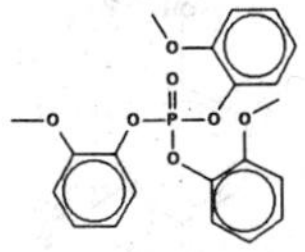

9388

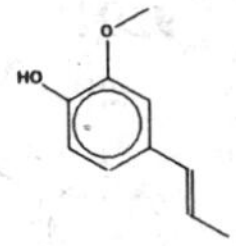

STRUCTURAL FORMULAS OF ORGANIC COMPOUNDS (continued)

In numeric order as they occur in the Table of Physical Constants of Organic Compounds

9389

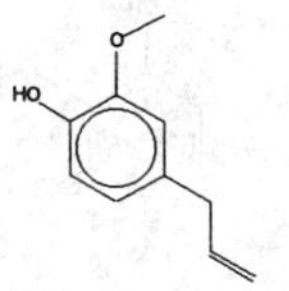

HO O OH

9391

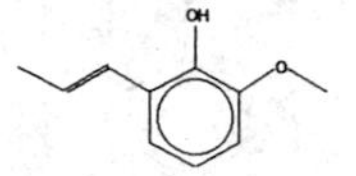

OH O

9392

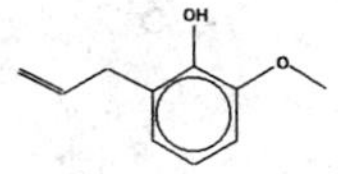

OH O

9394

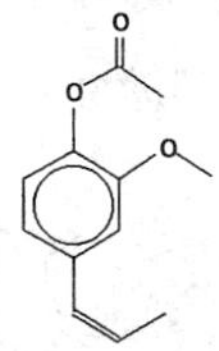

O O O

9395

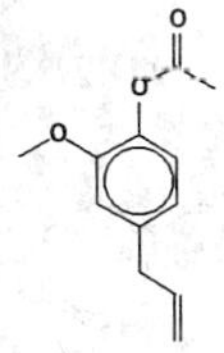

O O O

9396

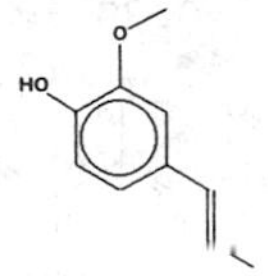

HO O

9397

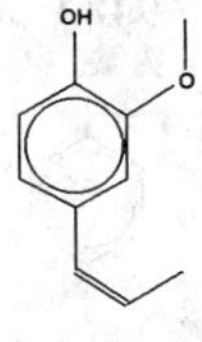

OH O

9401

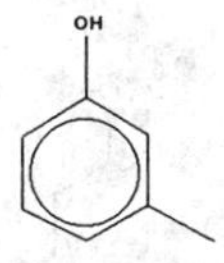

OH

9402

OH

9403

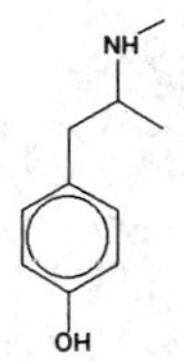

NH OH

9404

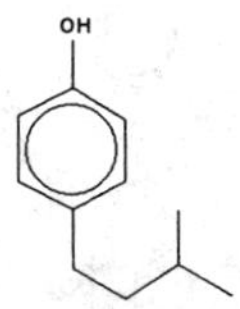

OH

9405

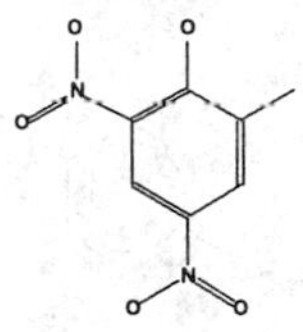

O O N O O N O O

9407

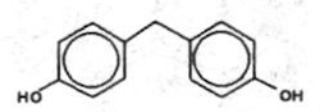

HO OH

9409

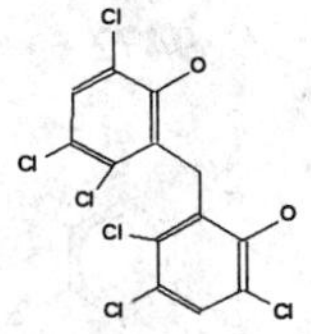

Cl O Cl Cl Cl O Cl Cl

9410

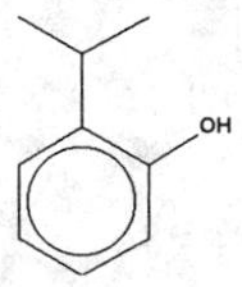

OH

9411

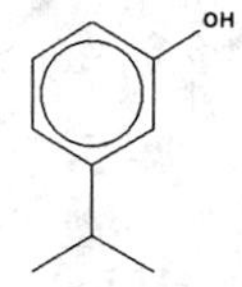

OH

9412

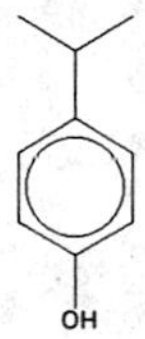

OH

9413

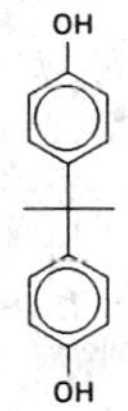

OH OH

9414

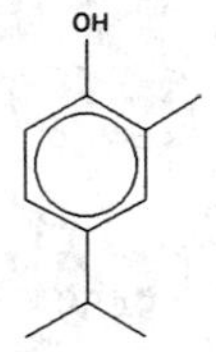

OH

9416

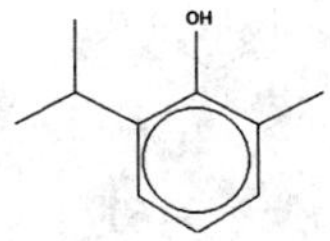

OH

9417

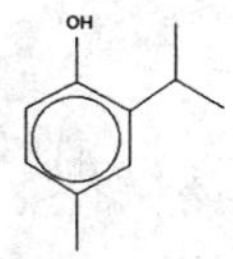

OH

9418

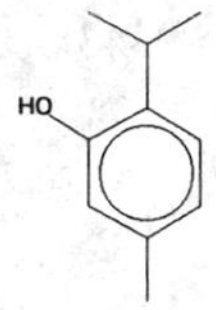

HO

9419

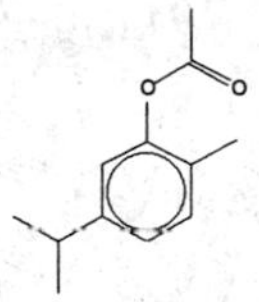

O O

9420

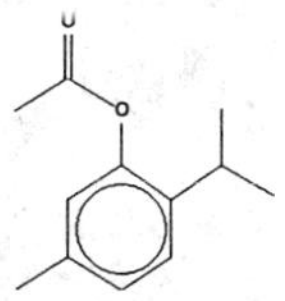

O O

9421

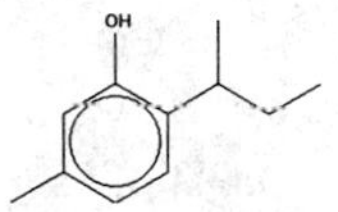

OH

9423

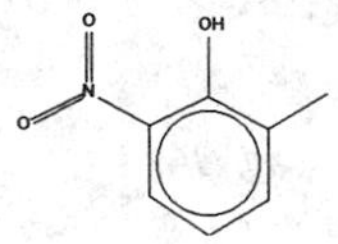

O N O OH

9424

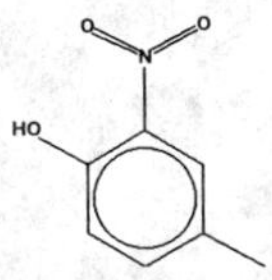

O N O HO

9425

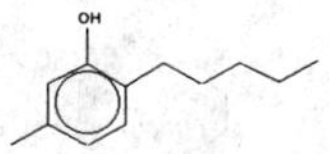

OH

9426

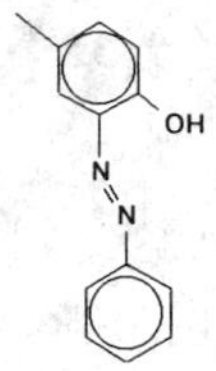

OH N N

9427

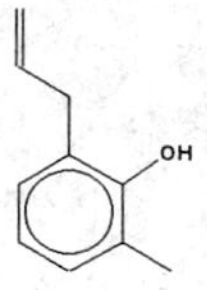

OH

9428

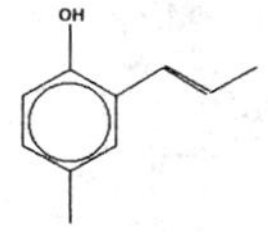

OH

9430

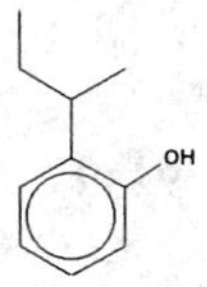

OH

9431

OH

9432

OH

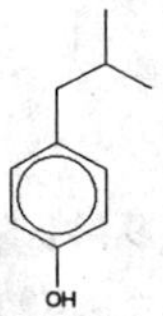

9434

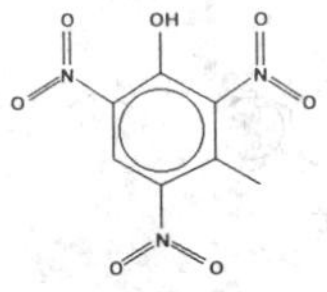

9435

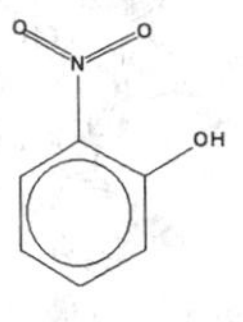

9436

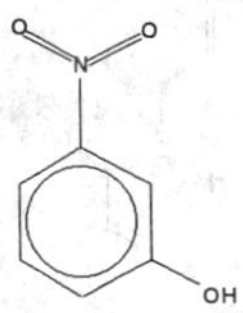

9438

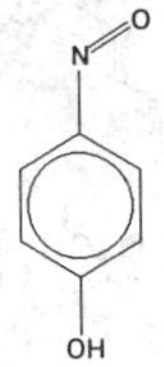

9439

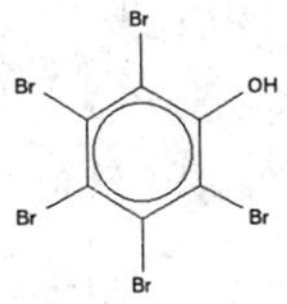

9440

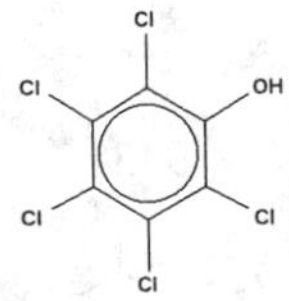

9441

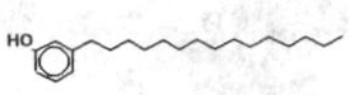

9442

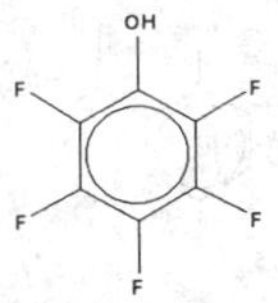

9443

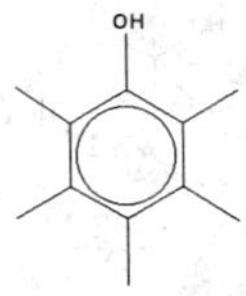

9445

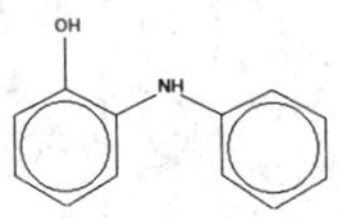

9446

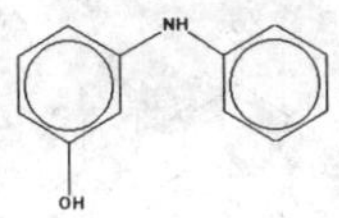

9447

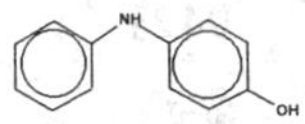

9448

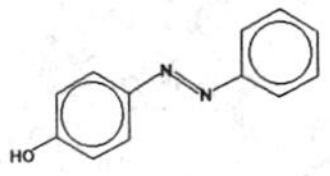

9451

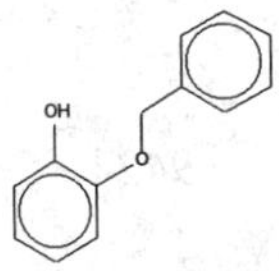

9452

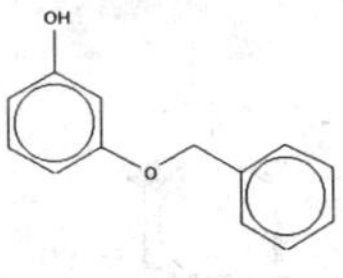

9454

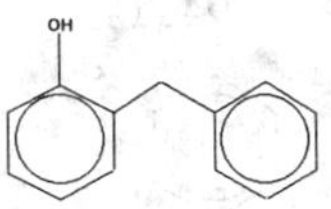

9455

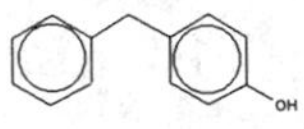

9456

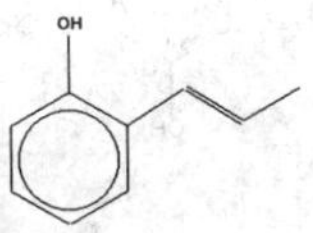

9457

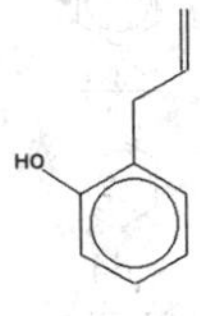

9458

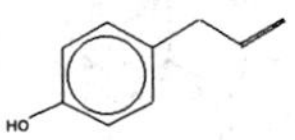

9459

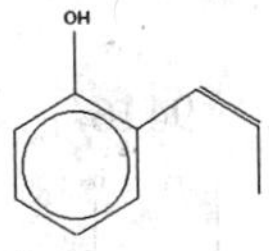

9461

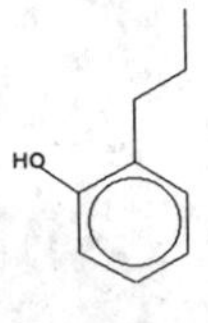

9462

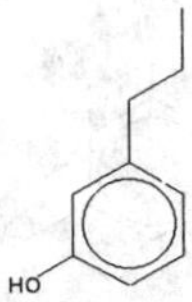

9463

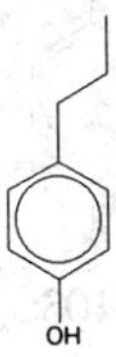

9465

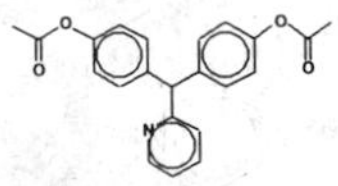

9466

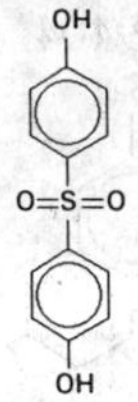

9469

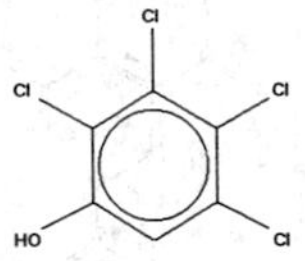

9471

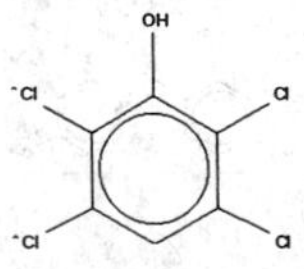

9472

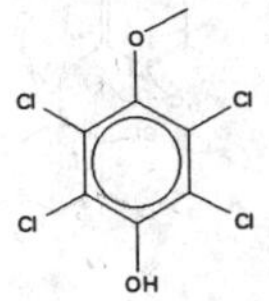

9473

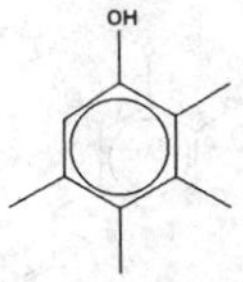

9474

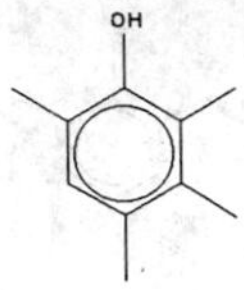

9475

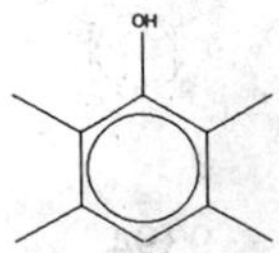

9479

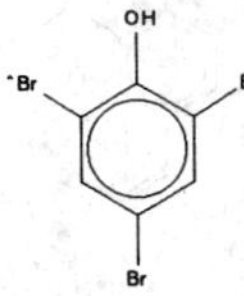

9482

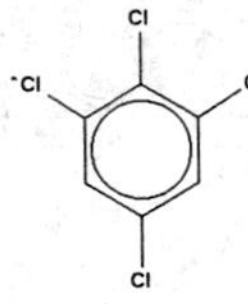

9483

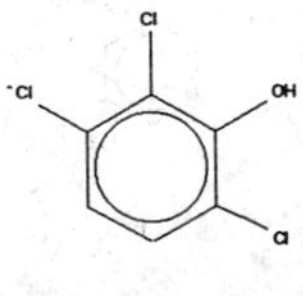

9484

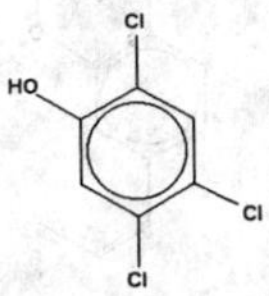

STRUCTURAL FORMULAS OF ORGANIC COMPOUNDS (continued)

In numeric order as they occur in the Table of Physical Constants of Organic Compounds

9485

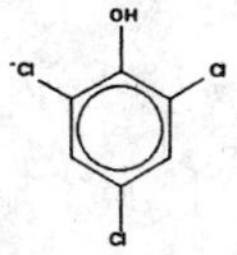

9486

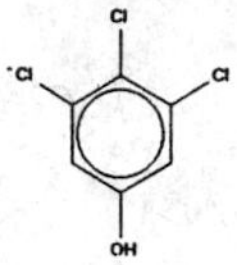

9489

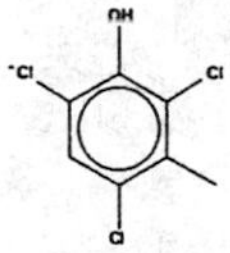

9492

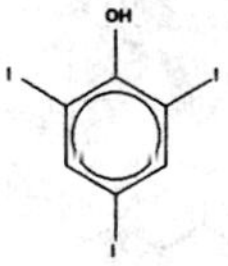

9493

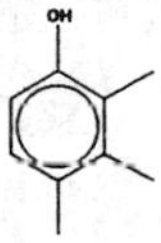

9494

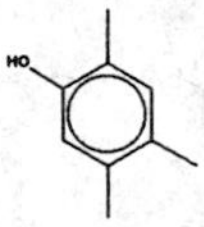

9495

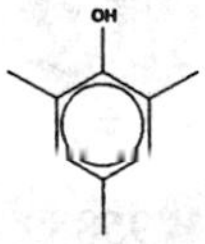

9496

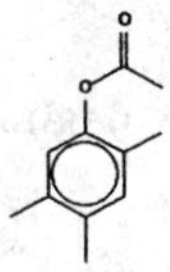

9497

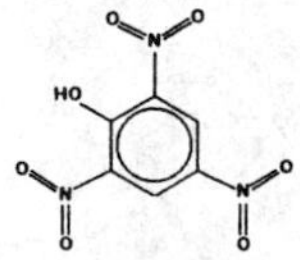

9499

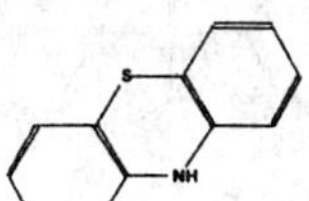

9501

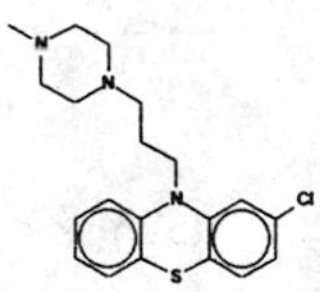

9503

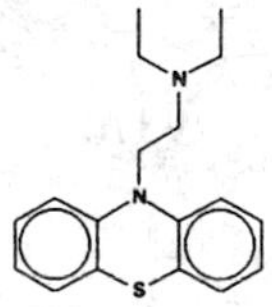

9504

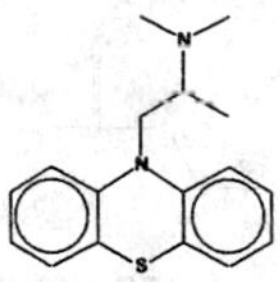

9505

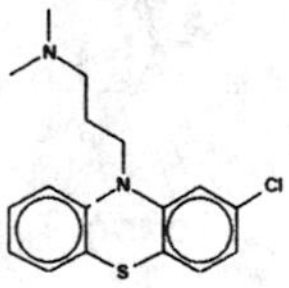

9507

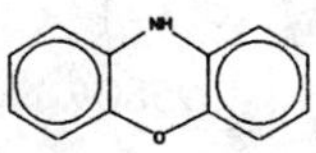

9510

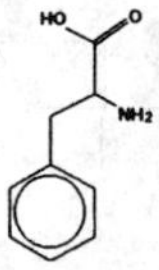

9512

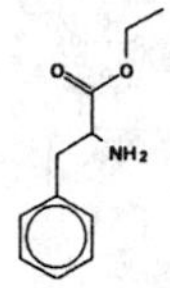

9513

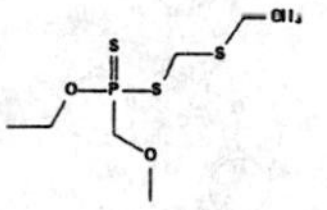

9514

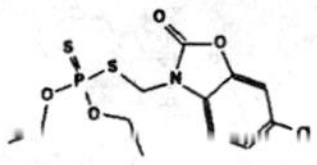

9515

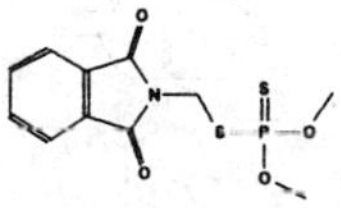

9517

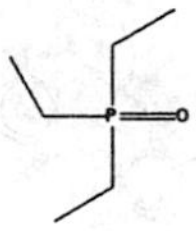

9518

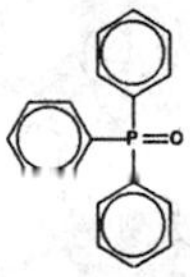

9519

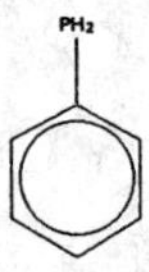

9520

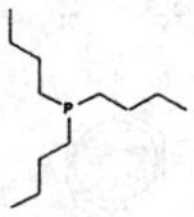

9521

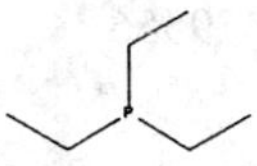

9522

9523

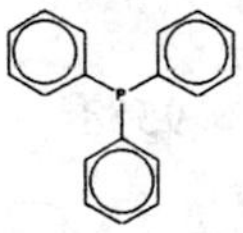

9524

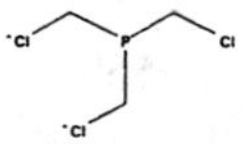

9525

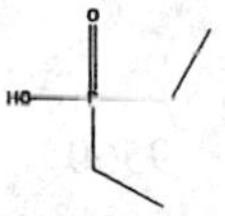

9527

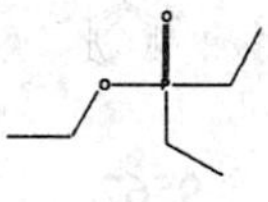

9528

9529

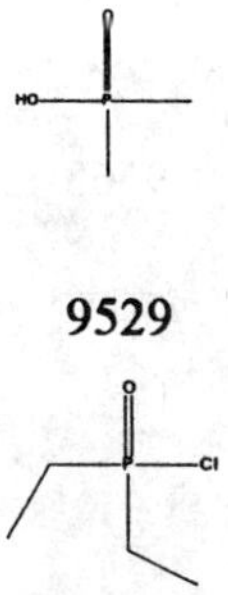

9530

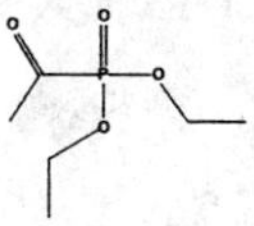

9532

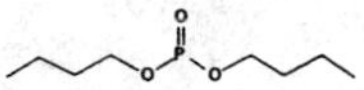

9535

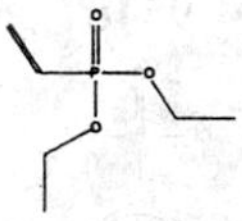

9536

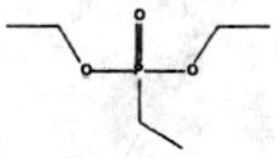

9537

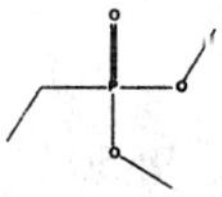

9539

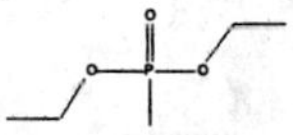

9540

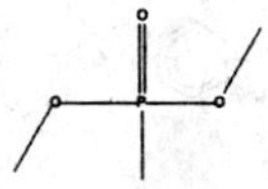

9541

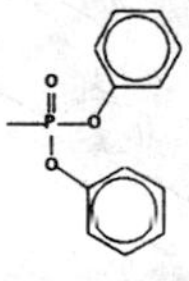

9543

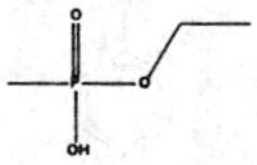

STRUCTURAL FORMULAS OF ORGANIC COMPOUNDS (continued)

In numeric order as they occur in the Table of Physical Constants of Organic Compounds

9545

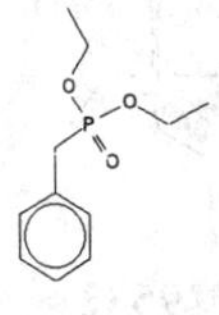

9546

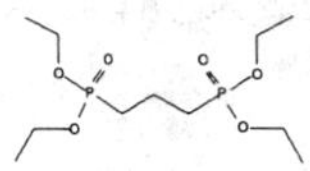

9548

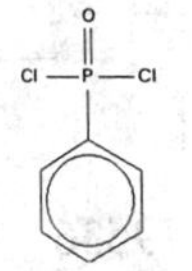

9549

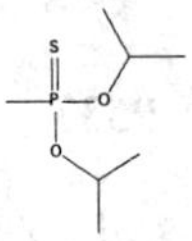

9550

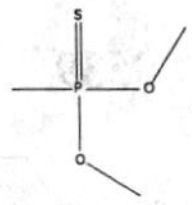

9551

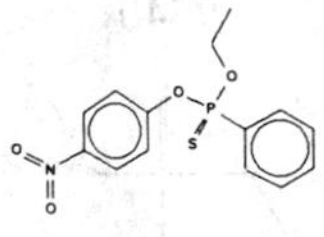

9553

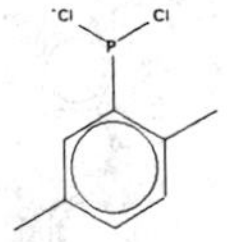

9554

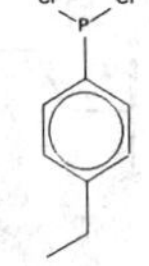

9555

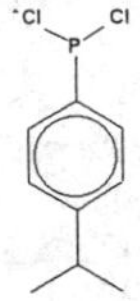

9556

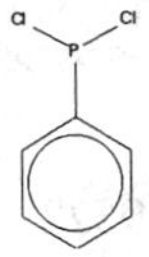

9557

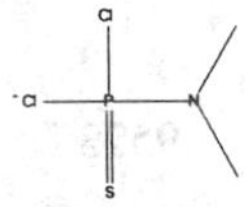

9558

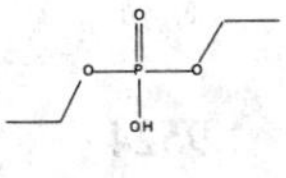

9559

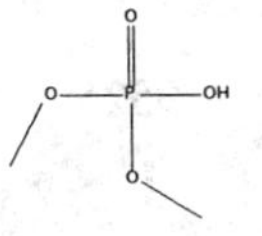

9560

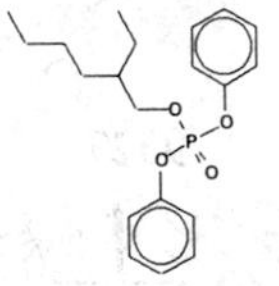

9562

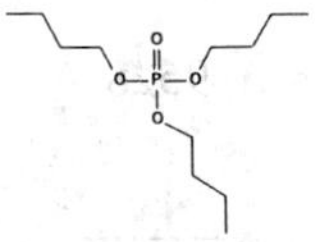

9563

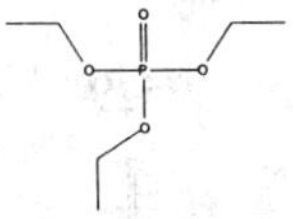

9564

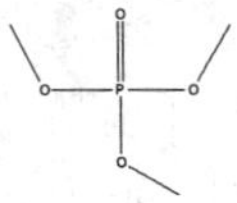

9565

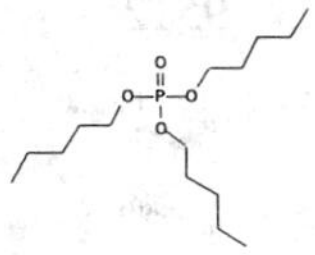

9566

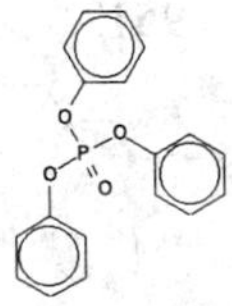

9567

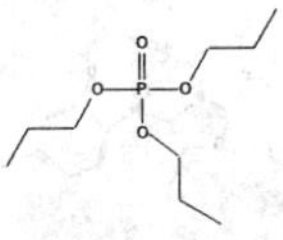

9568

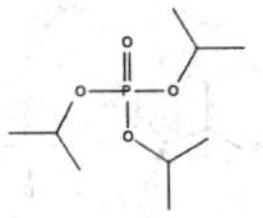

9569

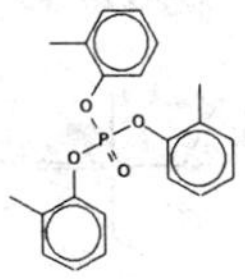

9570

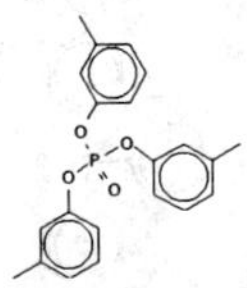

9571

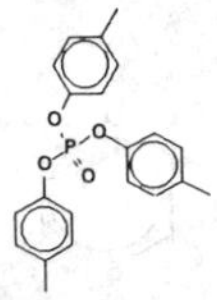

9572

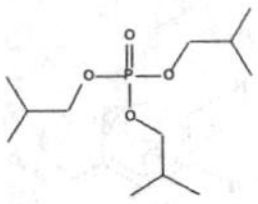

9573

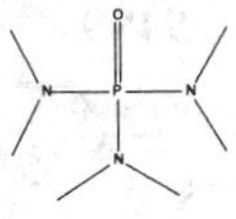

9574

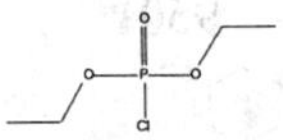

9575

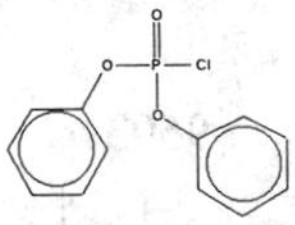

9576

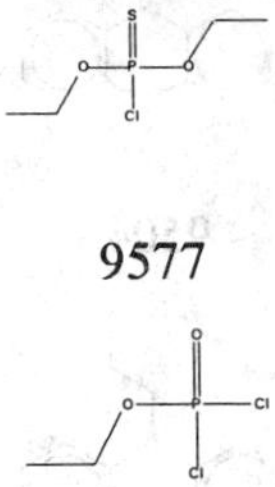

9577

9578

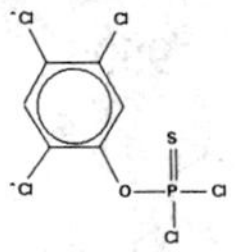

9579

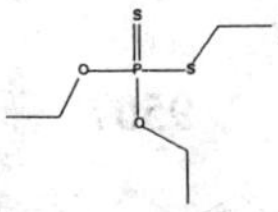

9580

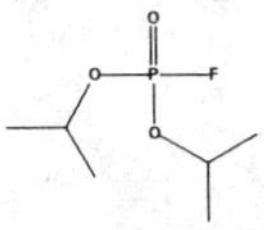

9581

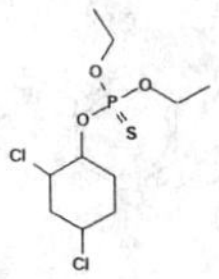

9582

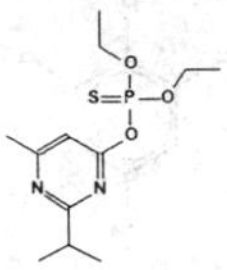

9583

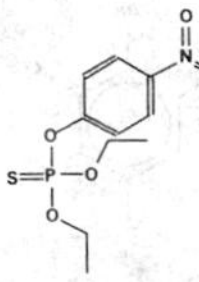

9584

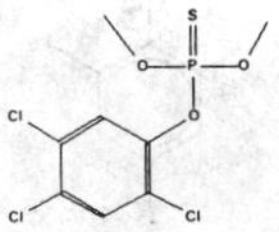

9585

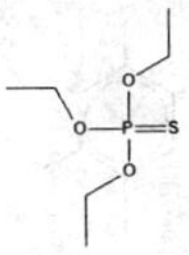

9586

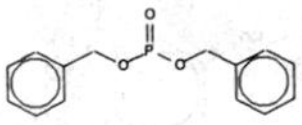

9587

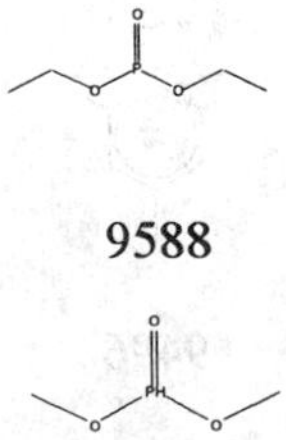

9588

9589

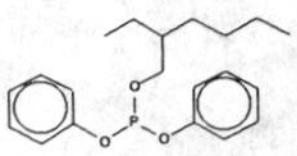

STRUCTURAL FORMULAS OF ORGANIC COMPOUNDS (continued)

In numeric order as they occur in the Table of Physical Constants of Organic Compounds

9590

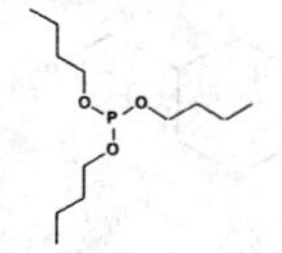

9591

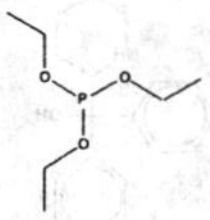

9592

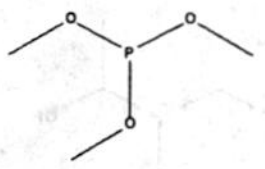

9593

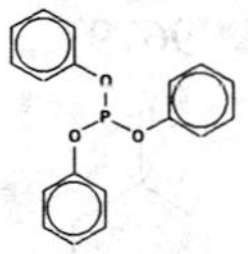

9594

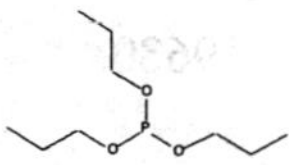

9595

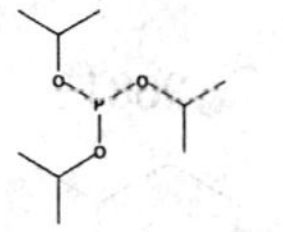

9596

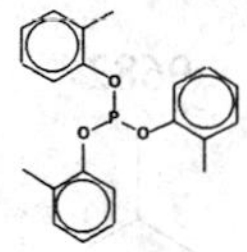

9597

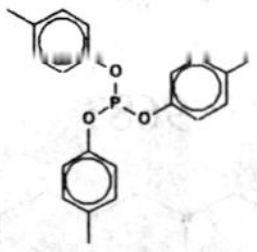

9599

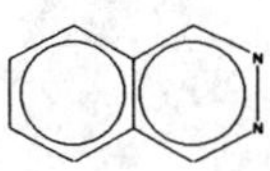

9600

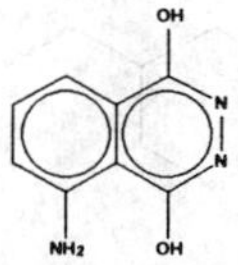

9601

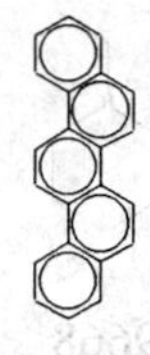

9602

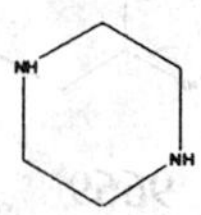

9603

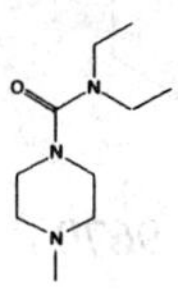

9604

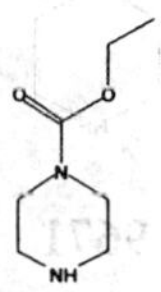

9605

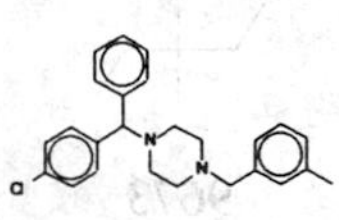

9606

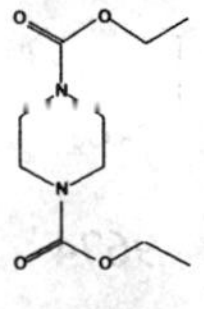

9608

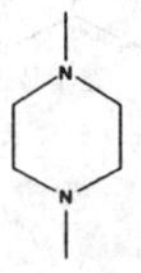

9609

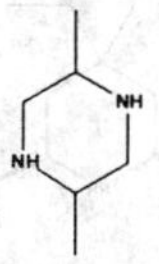

9610

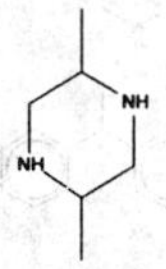

9611

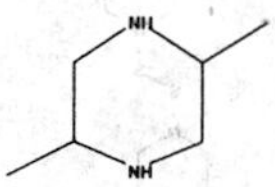

9612

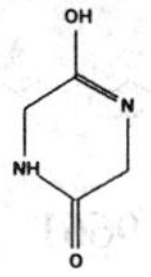

9613

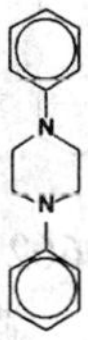

9614

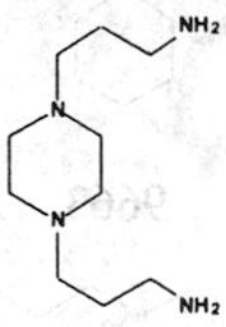

9615

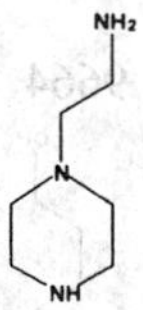

9616

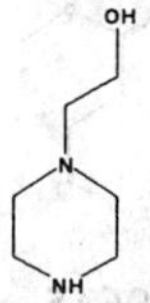

9617

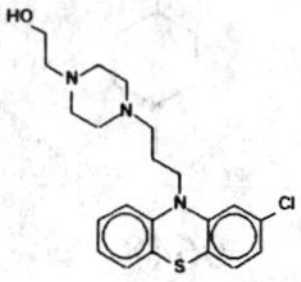

9618

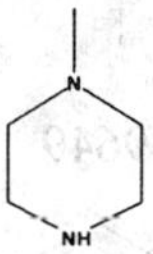

9619

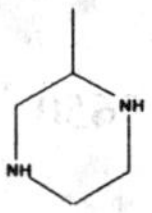

9620

9621

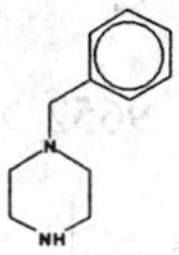

9622

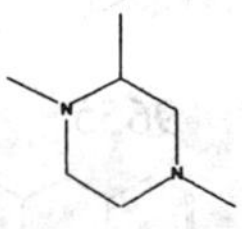

9624

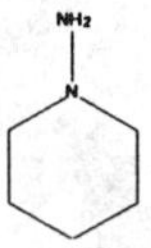

9625

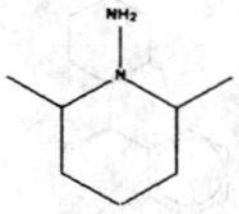

9626

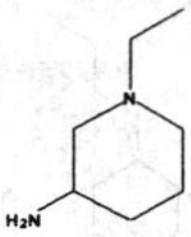

9627

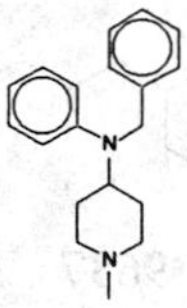

9628

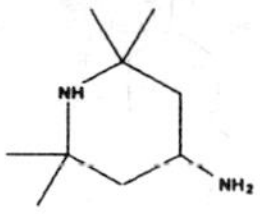

9629

9630

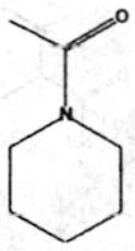

9631

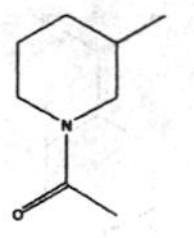

9632

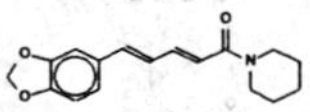

STRUCTURAL FORMULAS OF ORGANIC COMPOUNDS (continued)

In numeric order as they occur in the Table of Physical Constants of Organic Compounds

9634

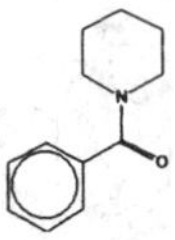

9635

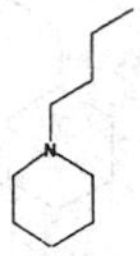

9636

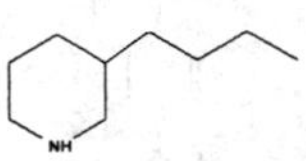

9637

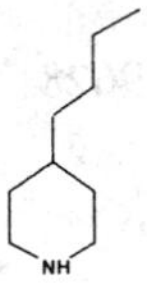

9638

9639

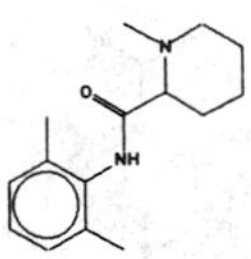

9642

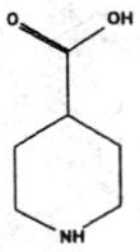

9643

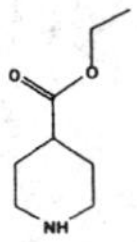

9644

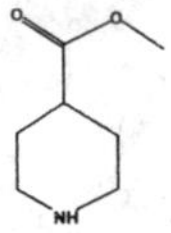

9645

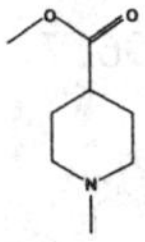

9648

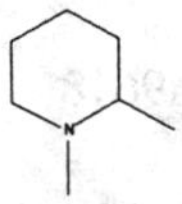

9649

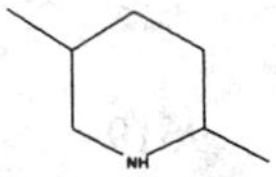

9650

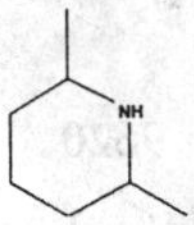

9651

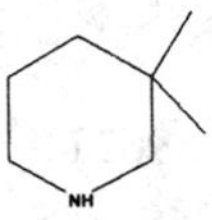

9652

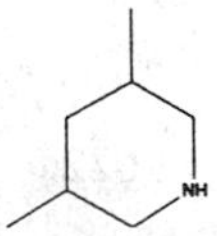

9655

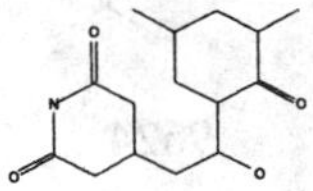

9656

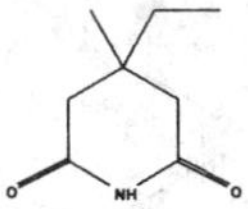

9658

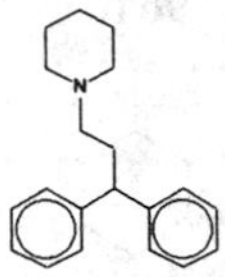

9659

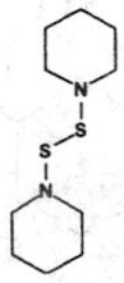

9660

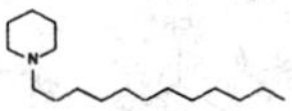

9661

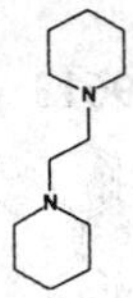

9662

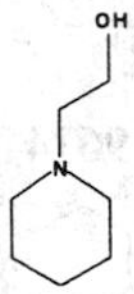

9663

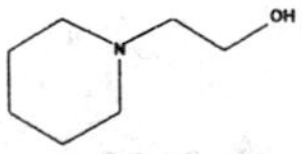

9664

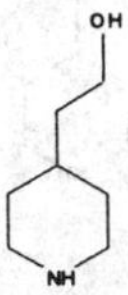

9666

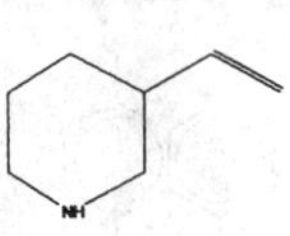

9667

9668

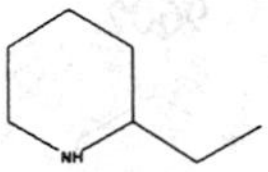

9669

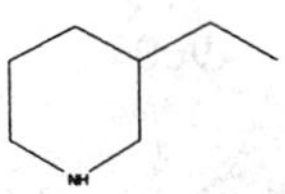

9670

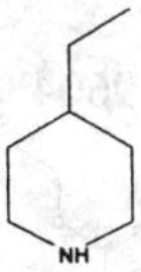

9671

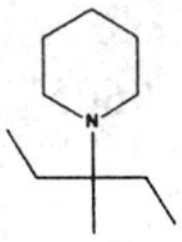

9673

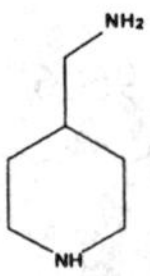

9674

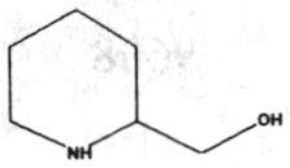

9675

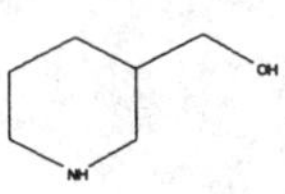

9676

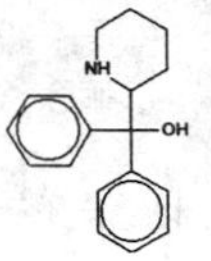

9678

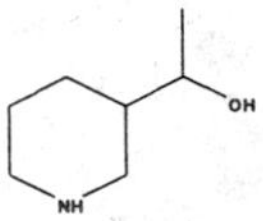

9679

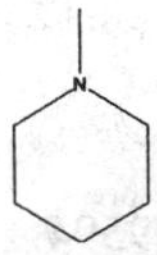

9680

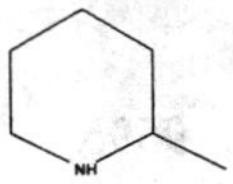

9681

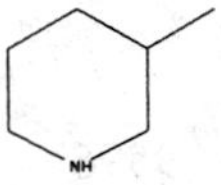

9682

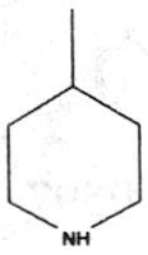

9683

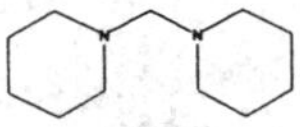

In numeric order as they occur in the Table of Physical Constants of Organic Compounds

9684

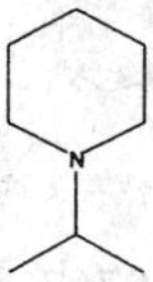

N

9685

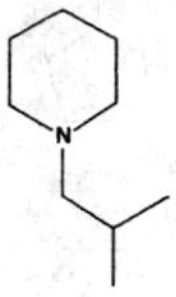

N

9686

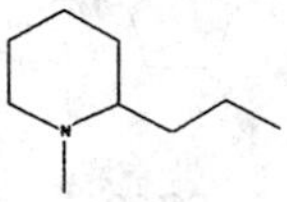

N

9687

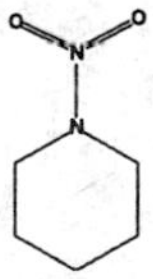

O N O

N

9688

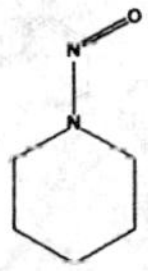

N O

N

9689

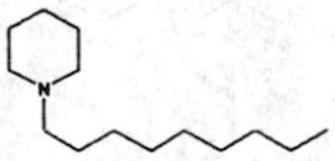

N

9690

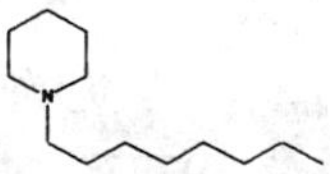

N

9691

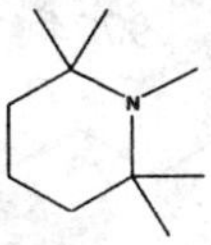

N

9692

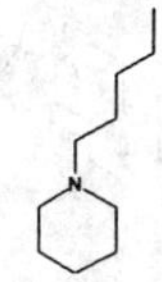

N

9693

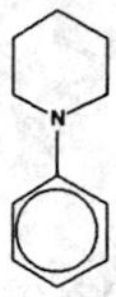

N

9695

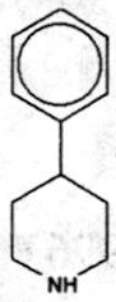

NH

9696

N

9697

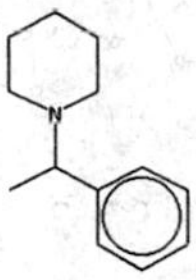

N

9698

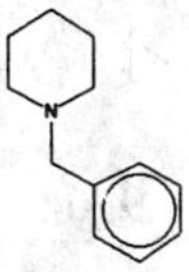

N

9699

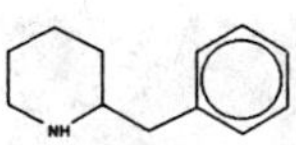

NH

9700

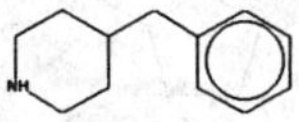

NH

9701

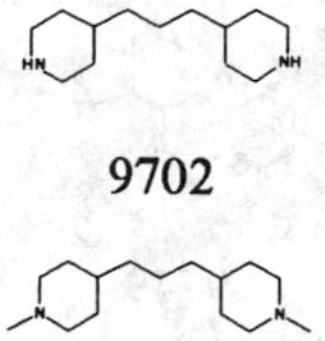

HN NH

9702

N N

9703

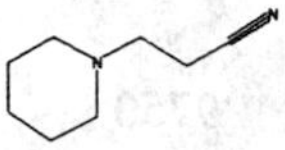

N N

9704

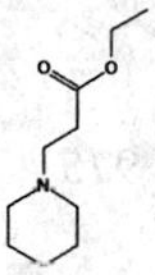

O O

N

9705

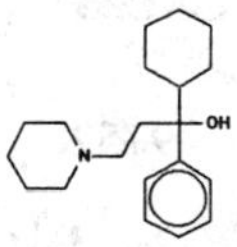

OH

N

9706

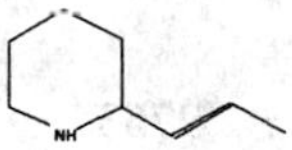

NH

9707

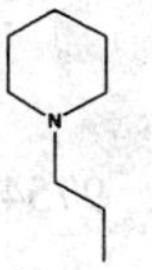

N

9708

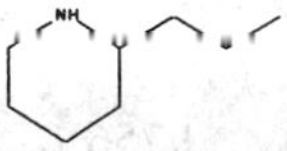

NH

9709

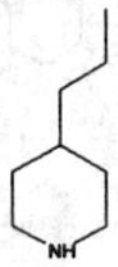

NH

9710

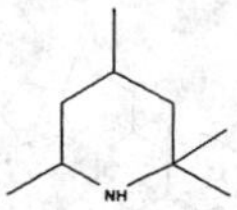

NH

9711

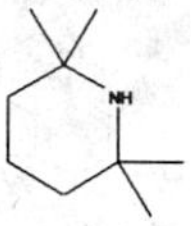

NH

9712

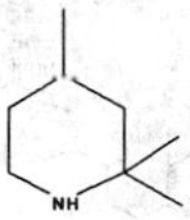

NH

9713

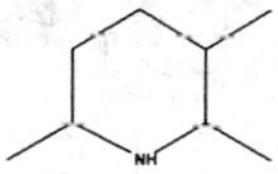

NH

9714

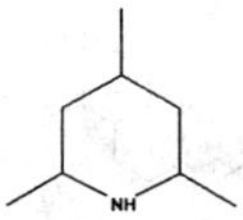

NH

9715

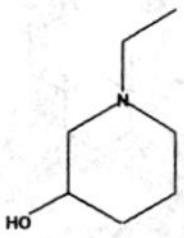

N

HO

9716

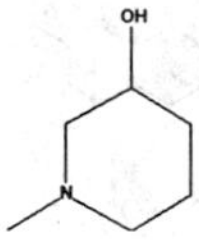

OH

N

9717

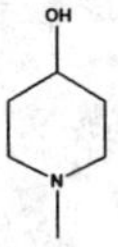

OH

N

9718

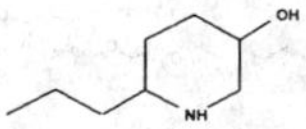

OH

NH

9719

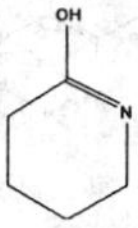

OH

N

9720

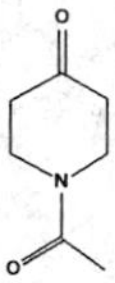

O

N

O

9721

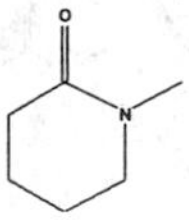

O

N

9722

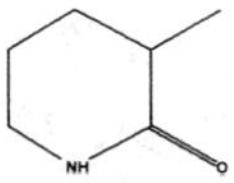

NH O

9723

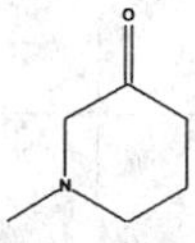

O

N

9724

O

N

9726

NH

O

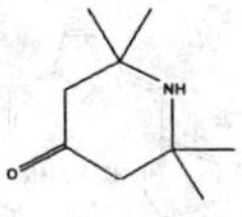

STRUCTURAL FORMULAS OF ORGANIC COMPOUNDS (continued)

In numeric order as they occur in the Table of Physical Constants of Organic Compounds

9727

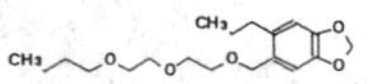

9728

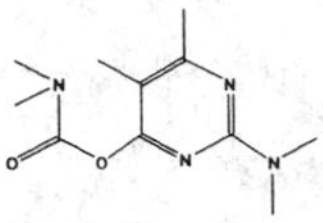

9729

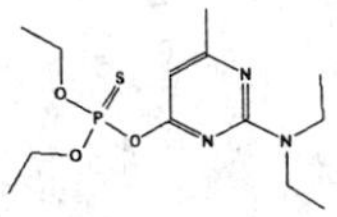

9730

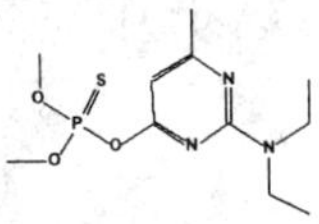

9731

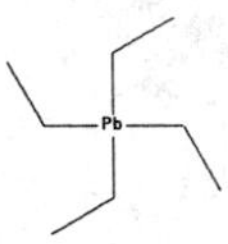

9732

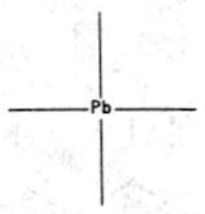

9733

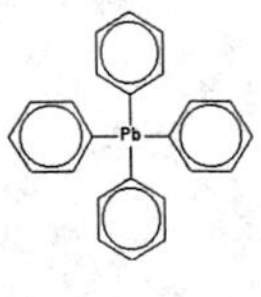

9734

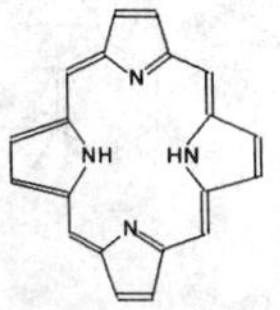

9736

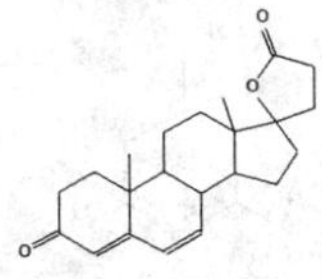

9738

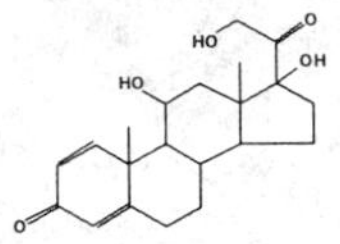

9739

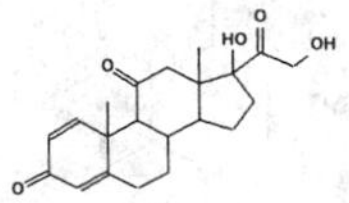

9740

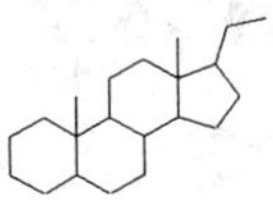

9741

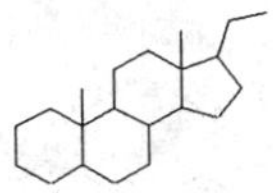

9743

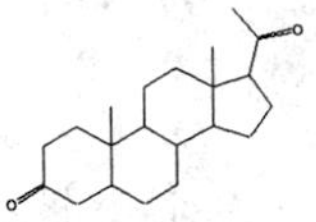

9745

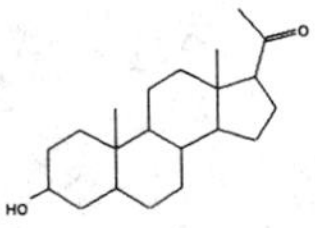

9746

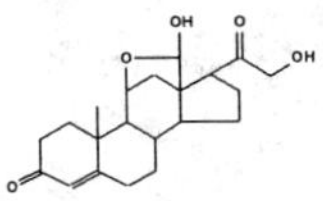

9748

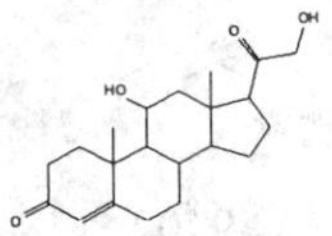

9749

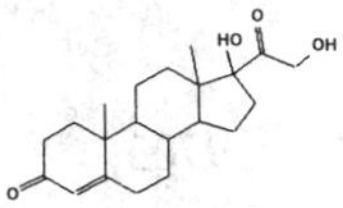

9750

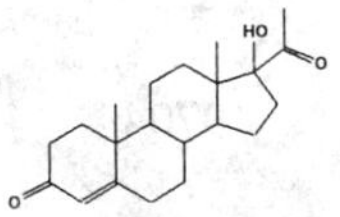

9751

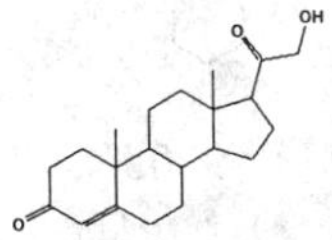

9752

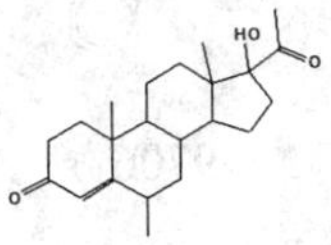

9753

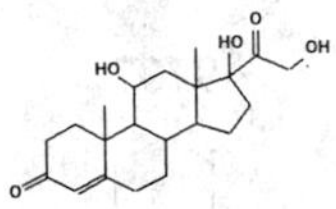

9754

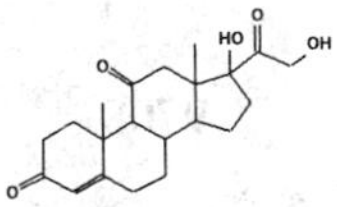

9756

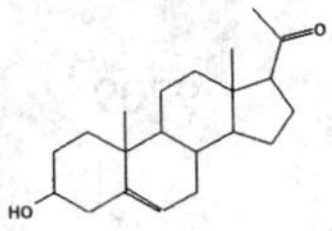

9758

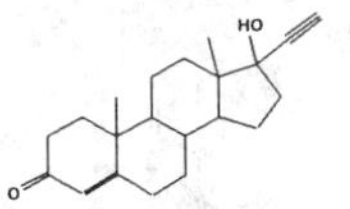

9759

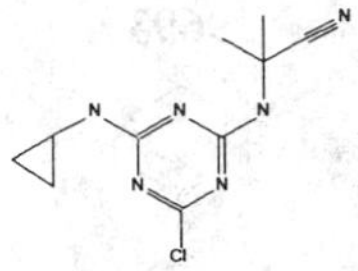

9760

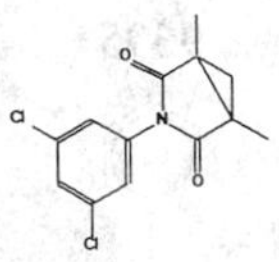

9761

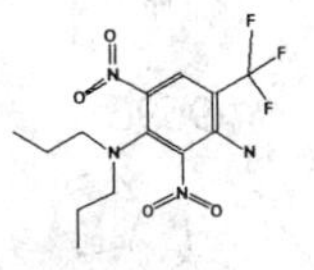

9762

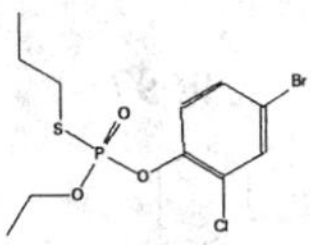

9763

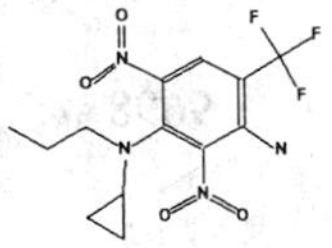

9764

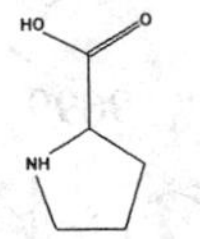

9765

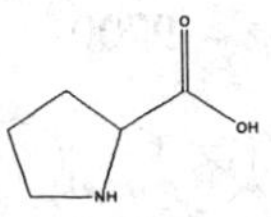

9766

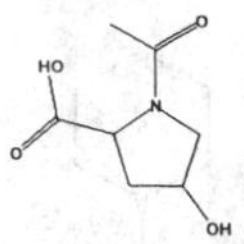

9767

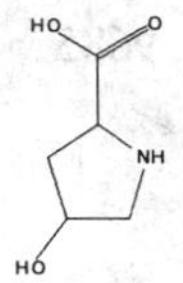

9768

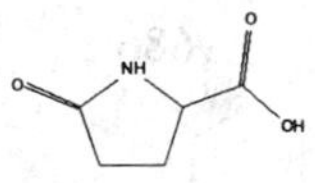

9769

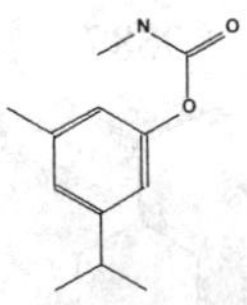

9770

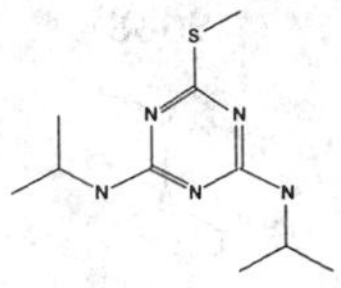

9771

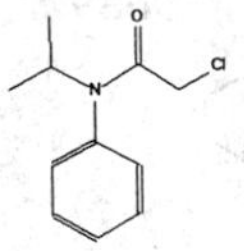

9772

9773

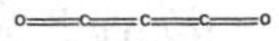

9774

STRUCTURAL FORMULAS OF ORGANIC COMPOUNDS (continued)

In numeric order as they occur in the Table of Physical Constants of Organic Compounds

9775

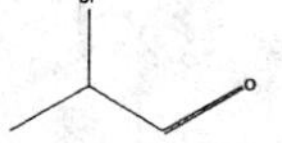

9776

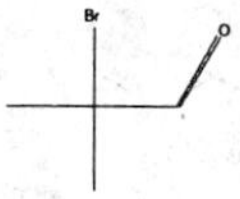

9777

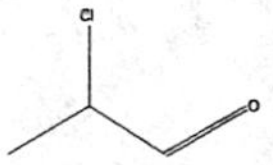

9778

9779

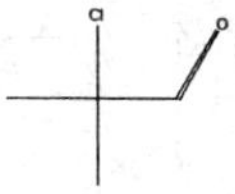

9780

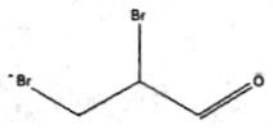

9781

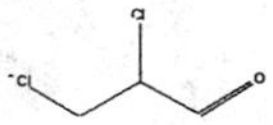

9782

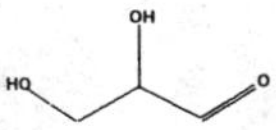

9783

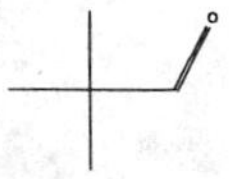

9784

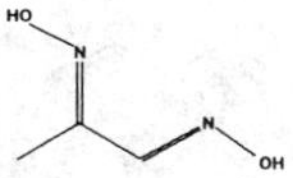

9785

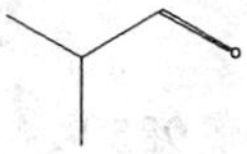

9786

9787

9788

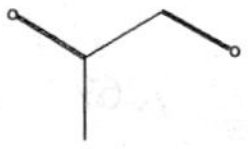

9790

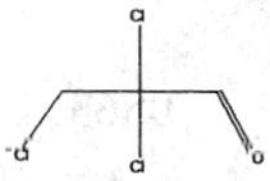

9791

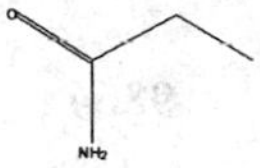

9792

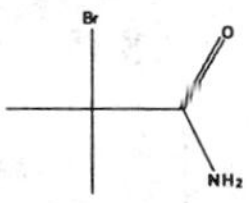

9793

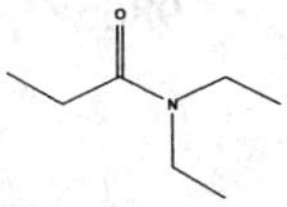

9798

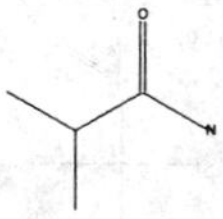

9799

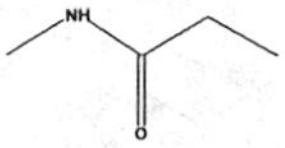

9800

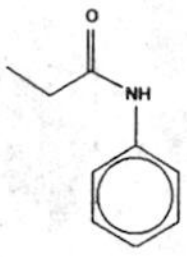

9801

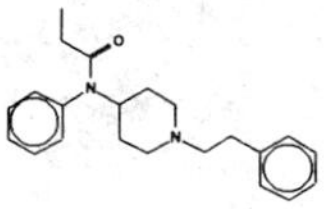

9803

9804

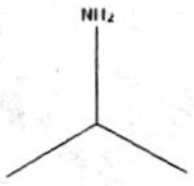

9805

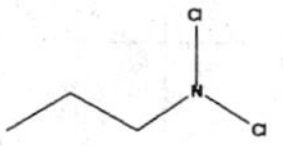

9806

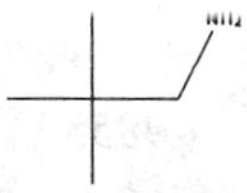

9807

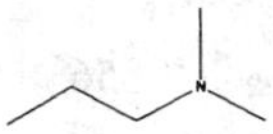

9808

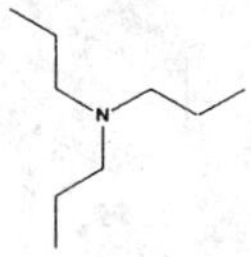

9809

9810

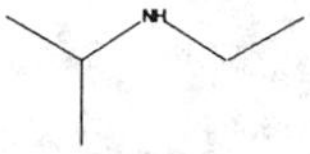

9811

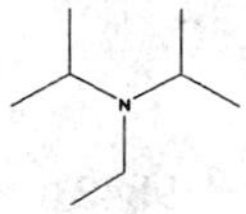

9812

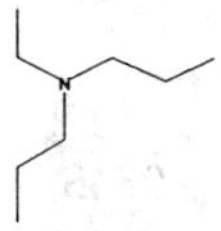

9813

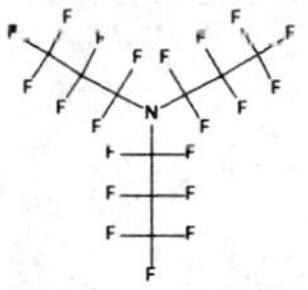

9814

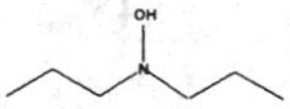

9815

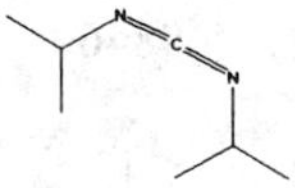

9816

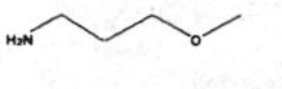

9817

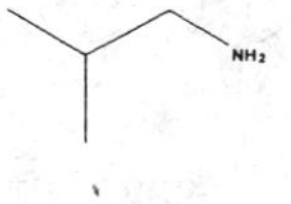

9818

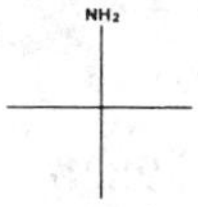

9819

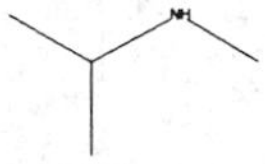

9820

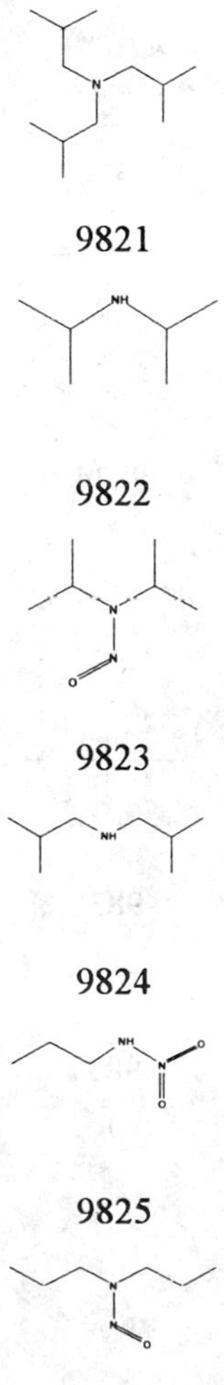

9821

9822

9823

9824

9825

In numeric order as they occur in the Table of Physical Constants of Organic Compounds

9826

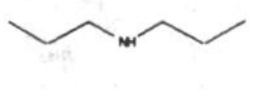

NH

9827

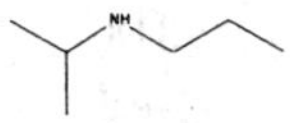

NH

9828

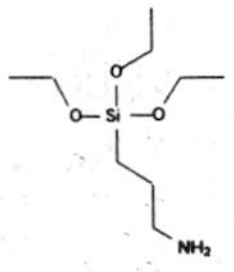

O Si O O NH2

9829

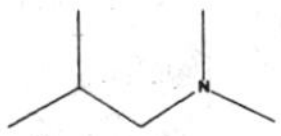

N

9832

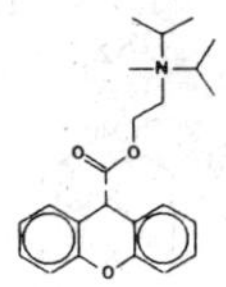

N O O O

9834

9835

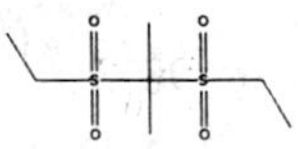

O S O O S O

9836

Br

9837

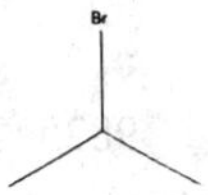

Br

9838

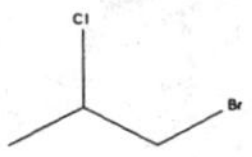

Cl Br

9839

Cl Br

9840

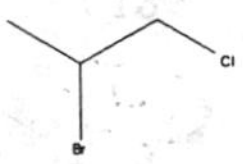

Cl Br

9841

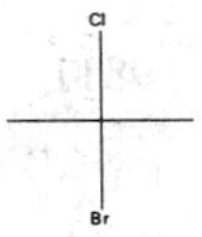

Cl Br

9842

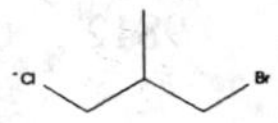

Cl Br

9843

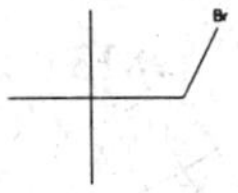

Br

9844

F Br

9845

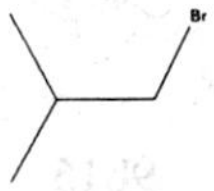

Br

9846

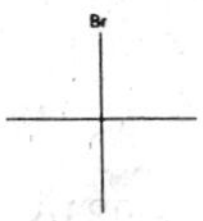

Br

9847

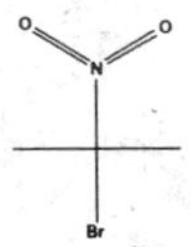

O N O Br

9848

Cl

9849

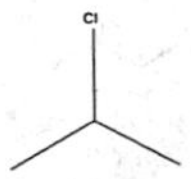

Cl

9850

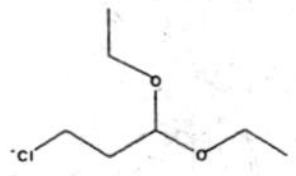

Cl O O

9851

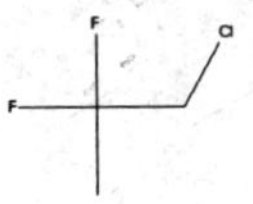

F F Cl

9852

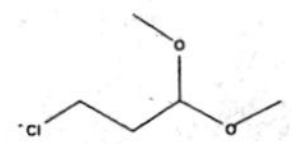

O Cl O

9853

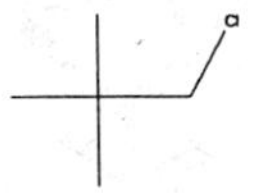

Cl

9854

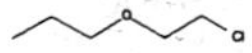

O Cl

9855

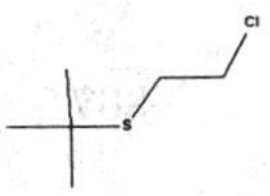

Cl S

9856

F Cl

9857

I Cl

9858

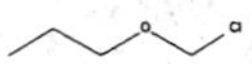

O Cl

9859

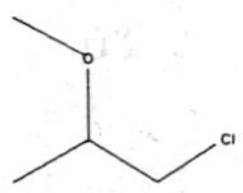

O Cl

9860

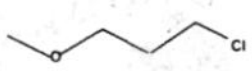

O Cl

9861

Cl O

9862

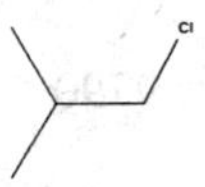

Cl

9863

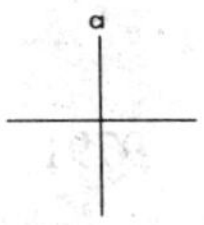

Cl

9864

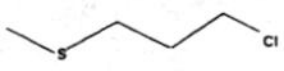

S Cl

9865

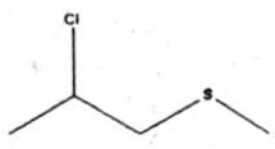

Cl S

9866

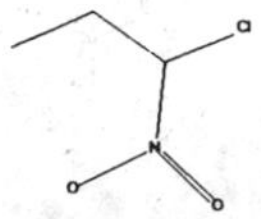

Cl N O O

9867

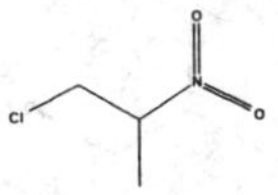

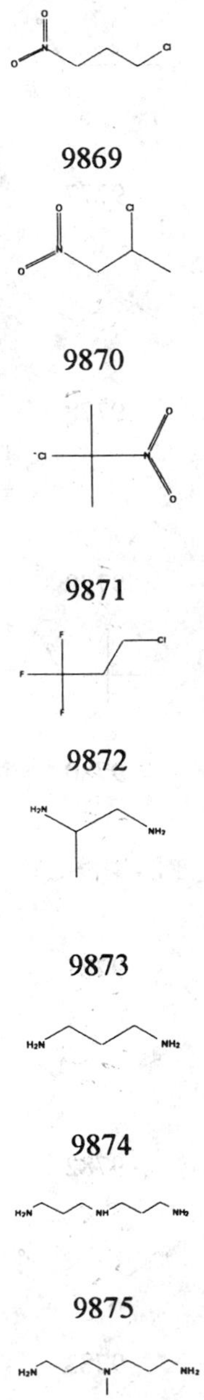

O N O Cl

9868

O N O Cl

9869

O N O Cl

9870

O Cl N O

9871

F F F Cl

9872

H2N NH2

9873

H2N NH2

9874

H2N NH NH2

9875

H2N N NH2

9876

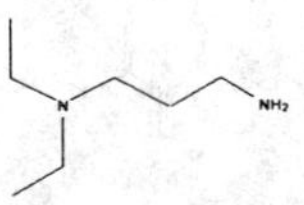

9877

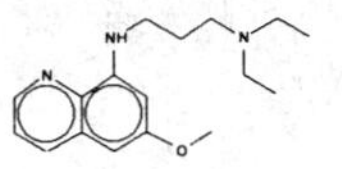

9878

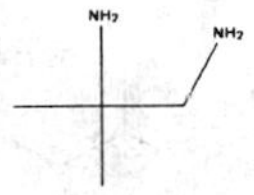

9879

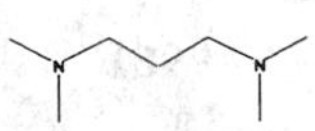

9880

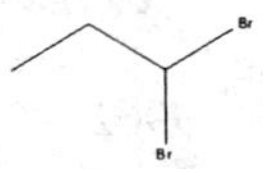

9881

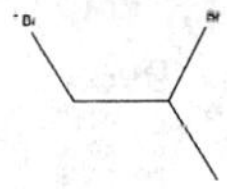

9882

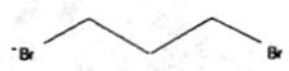

9883

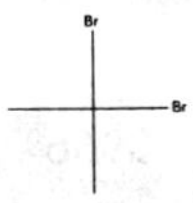

9884

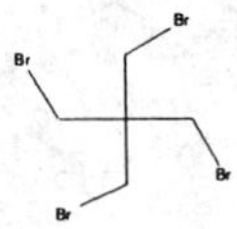

9885

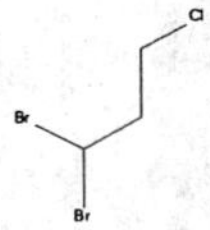

9886

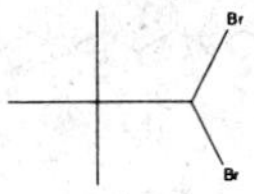

9887

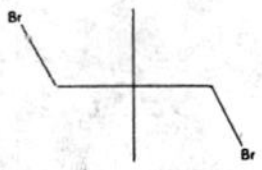

9888

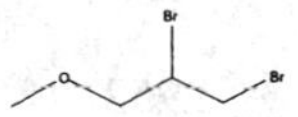

9889

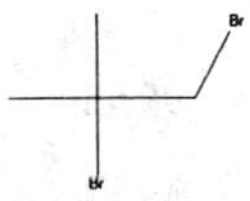

9890

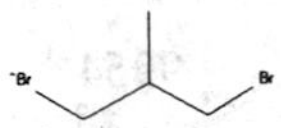

9891

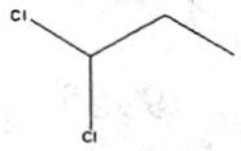

9892

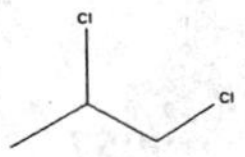

9893

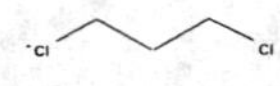

9894

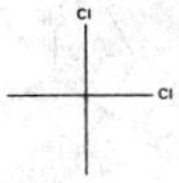

9895

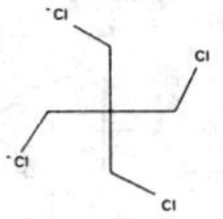

9896

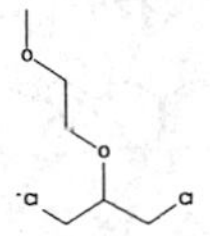

9897

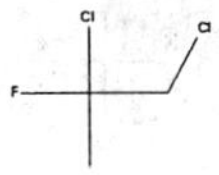

9898

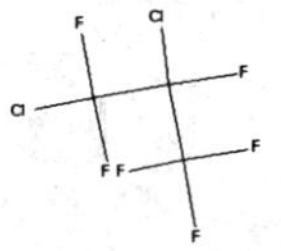

9899

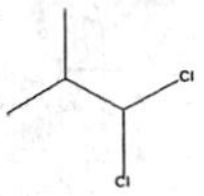

9900

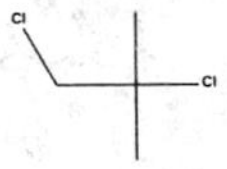

9901

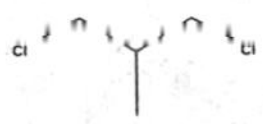

9902

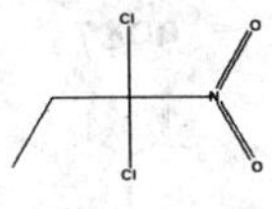

9903

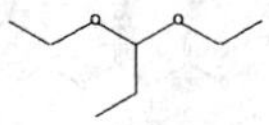

9904

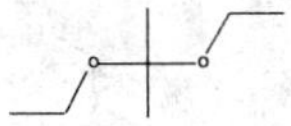

9905

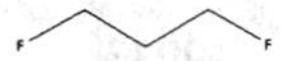

9906

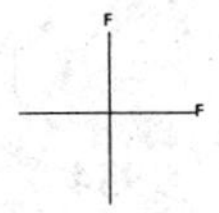

9907

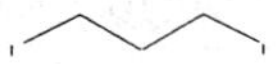

9908

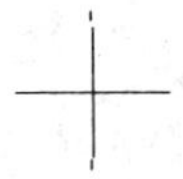

9909

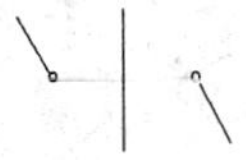

9910

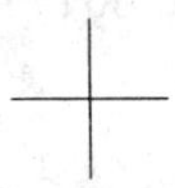

9911

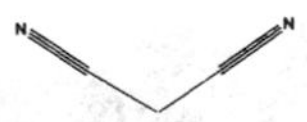

9912

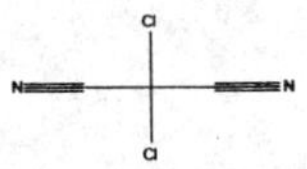

9913

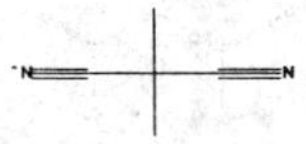

9914

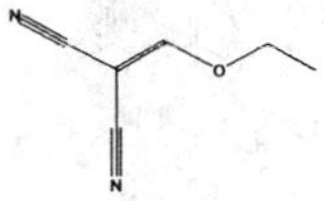

9915

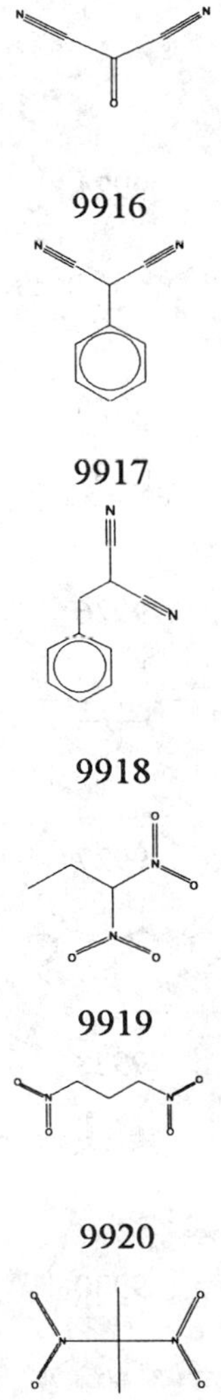

9916

9917

9918

9919

9920

STRUCTURAL FORMULAS OF ORGANIC COMPOUNDS (continued)

In numeric order as they occur in the Table of Physical Constants of Organic Compounds

9921

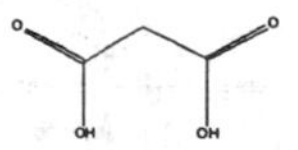

9922

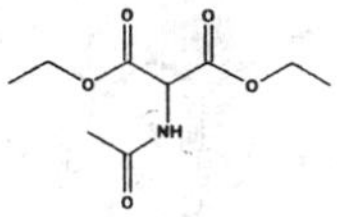

9923

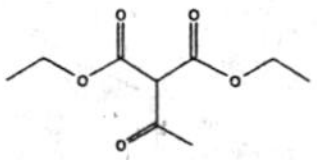

9924

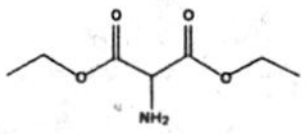

9925

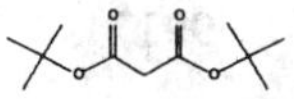

9926

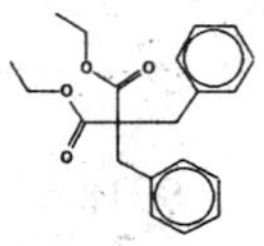

9927

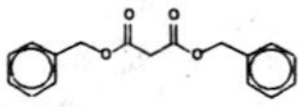

9928

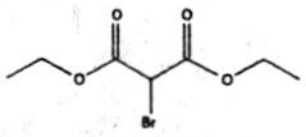

9930

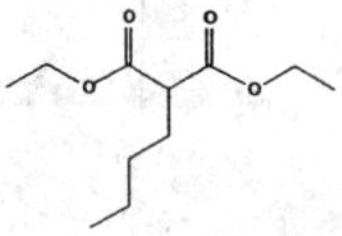

9931

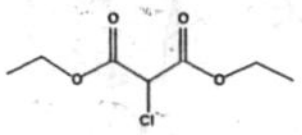

9932

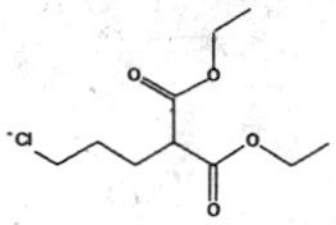

9933

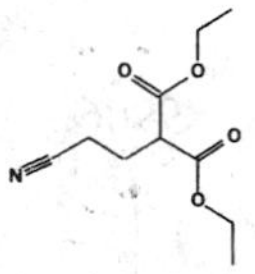

9934

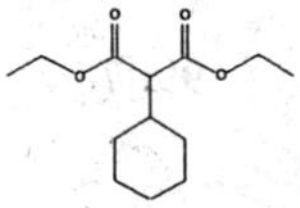

9935

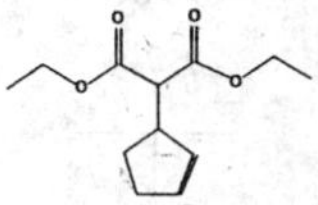

9936

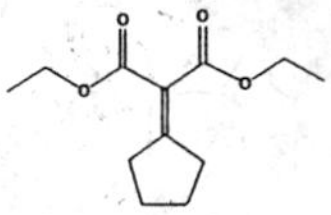

9937

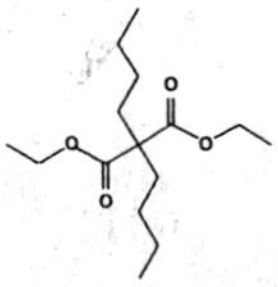

9938

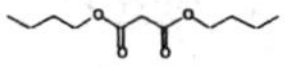

9939

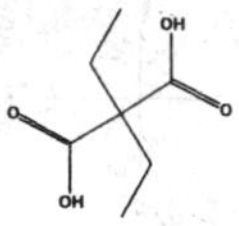

9940

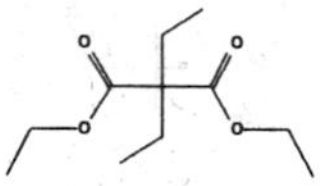

9941

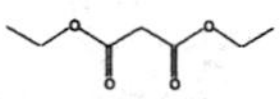

9942

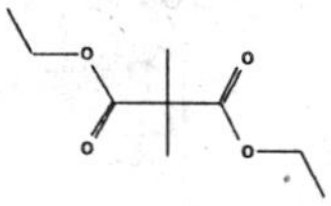

9943

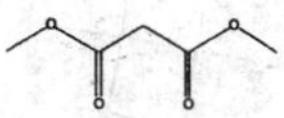

9944

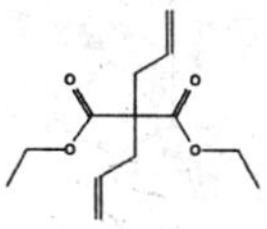

9945

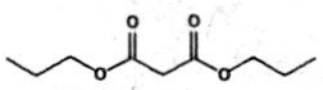

9946

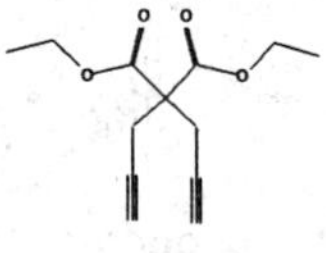

9947

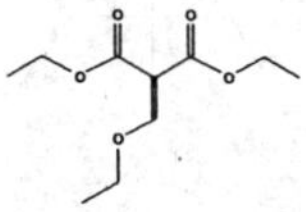

9948

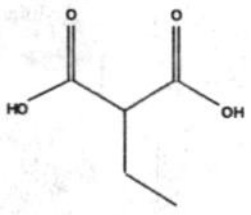

9949

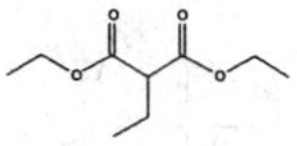

9950

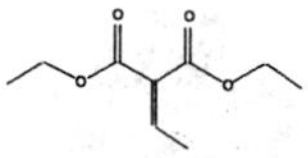

9951

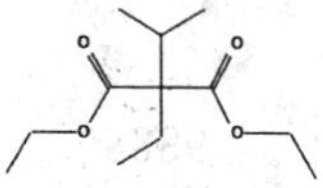

9952

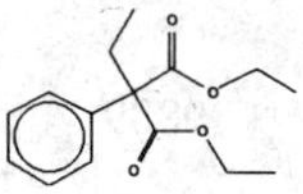

9953

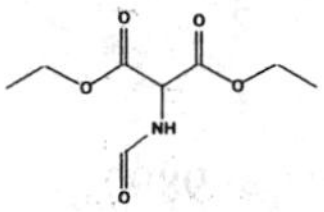

9954

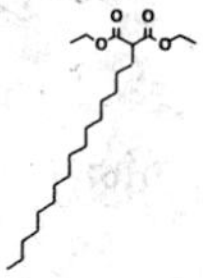

9955

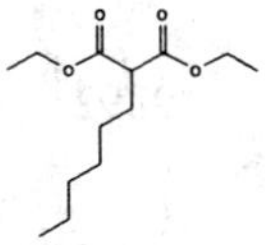

9957

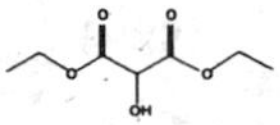

9958

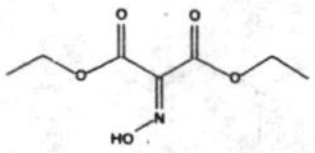

9959

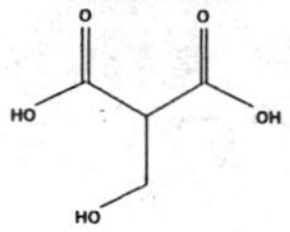

9960

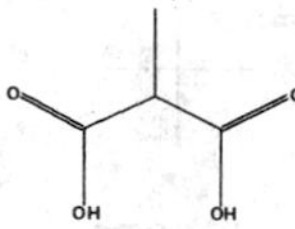

9961

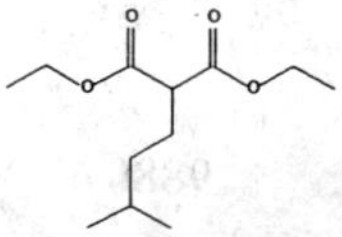

9962

9963

9964

9966

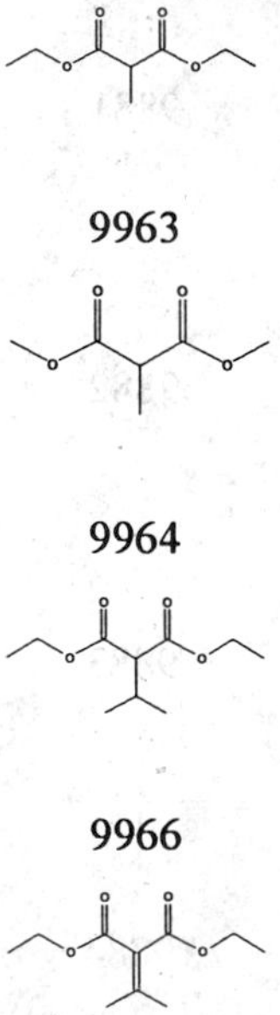

STRUCTURAL FORMULAS OF ORGANIC COMPOUNDS (continued)

In numeric order as they occur in the Table of Physical Constants of Organic Compounds

9967

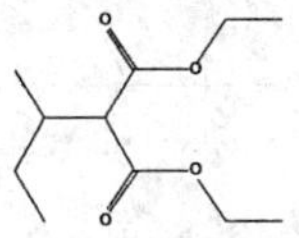

9969

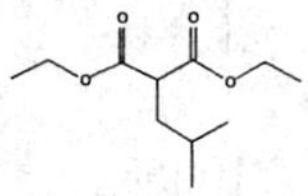

9970

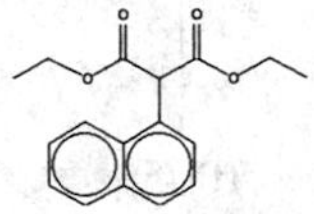

9972

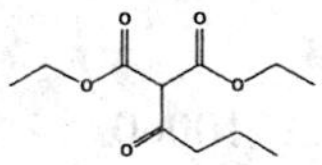

9973

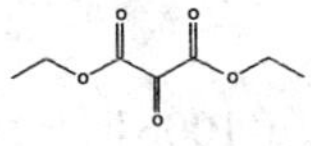

9974

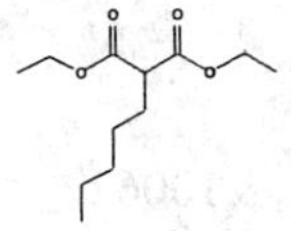

9975

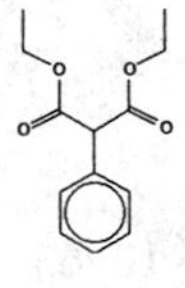

9976

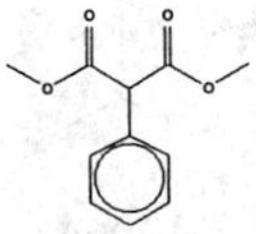

9977

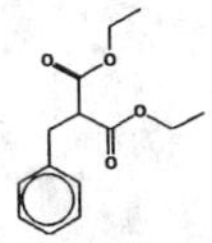

9978

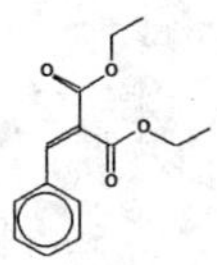

9980

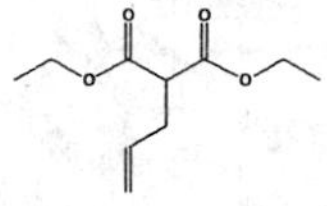

9981

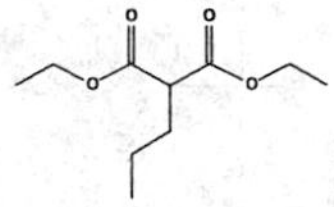

9982

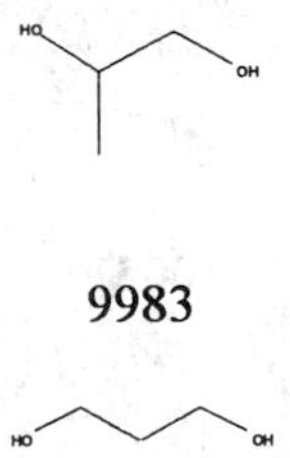

9983

9985

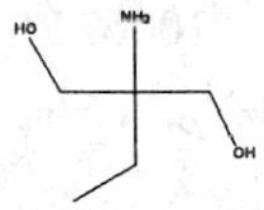

9986

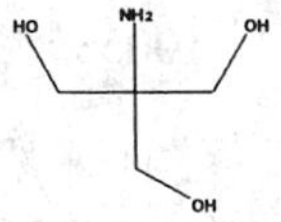

9987

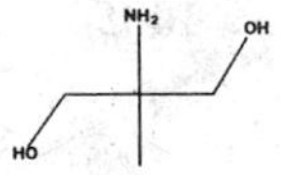

9989

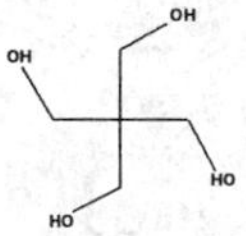

9990

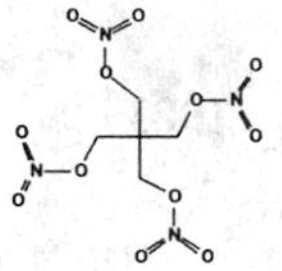

9991

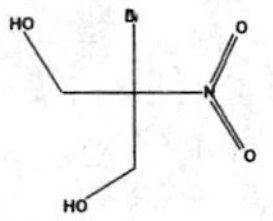

9992

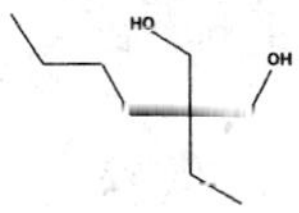

9994

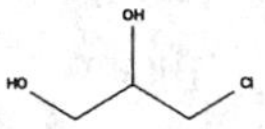

9995

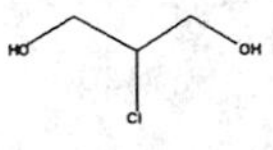

9997

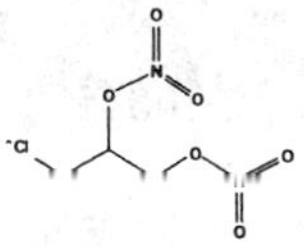

9998

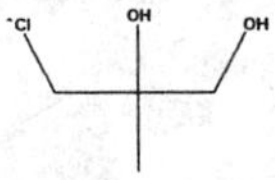

9999

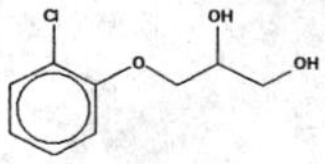

10000

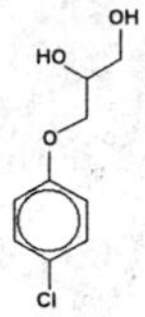

10001

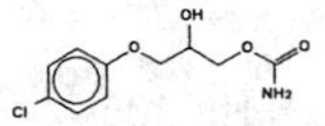

10002

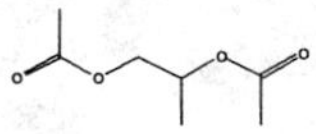

10003

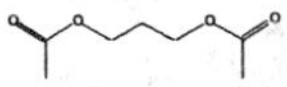

10004

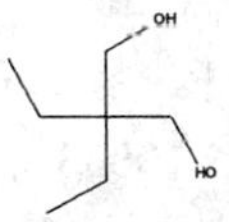

10005

10006

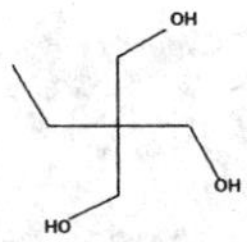

10007

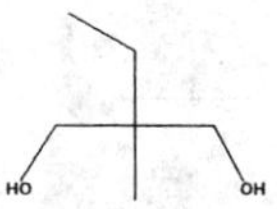

10008

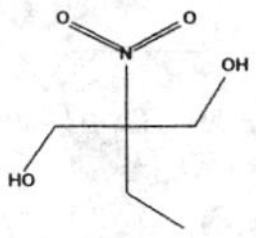

10009

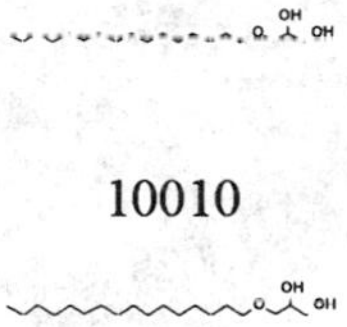

10010

10011

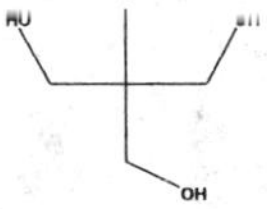

10012

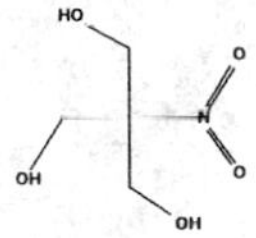

10013

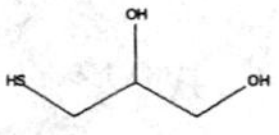

10014

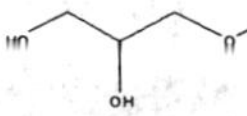

10016

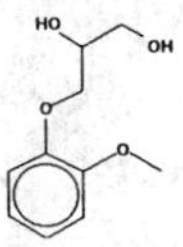

STRUCTURAL FORMULAS OF ORGANIC COMPOUNDS (continued)

In numeric order as they occur in the Table of Physical Constants of Organic Compounds

10017

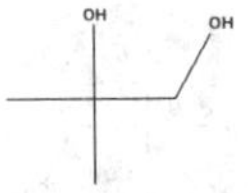

10018

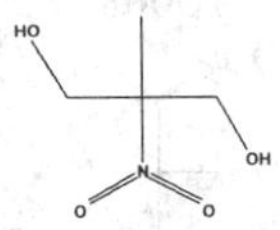

10019

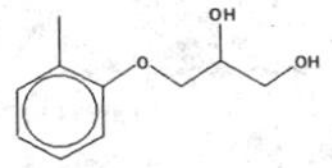

10020

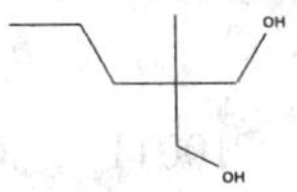

10021

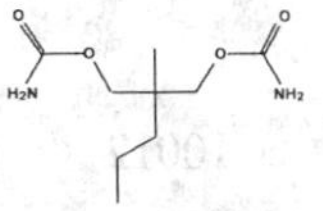

10022

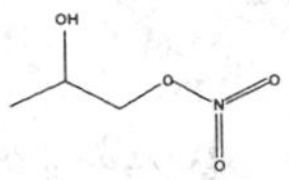

10023

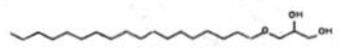

10025

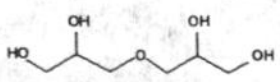

10026

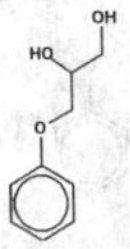

10027

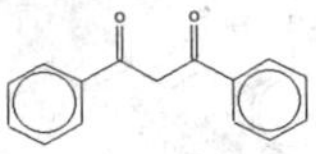

10028

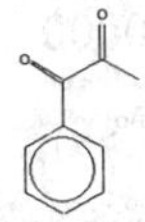

10029

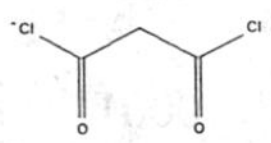

10030

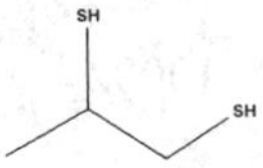

10031

10032

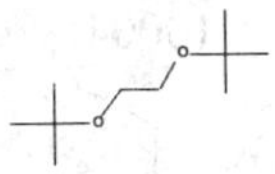

10033

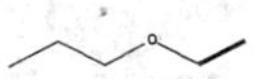

10034

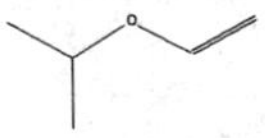

10035

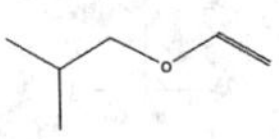

10036

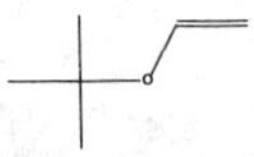

10037

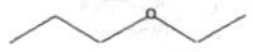

10038

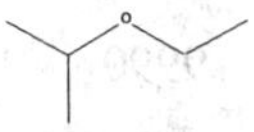

10039

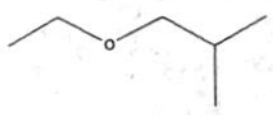

10040

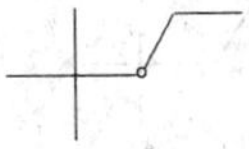

10041

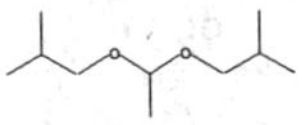

10042

10043

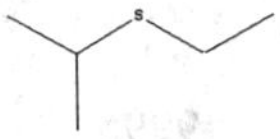

10044

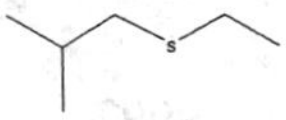

10045

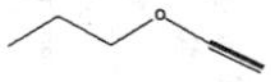

10046

10047

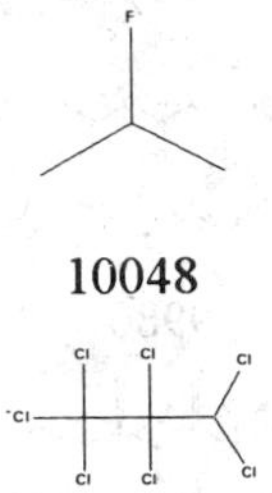

10048

10049

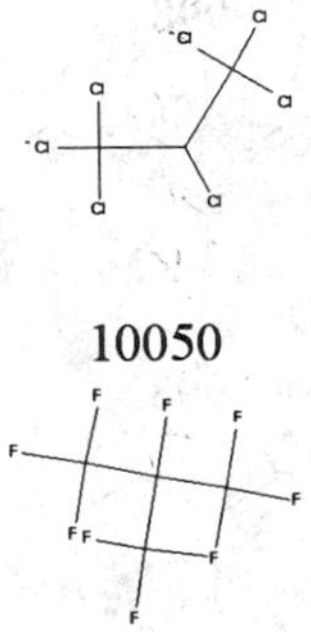

10050

10051

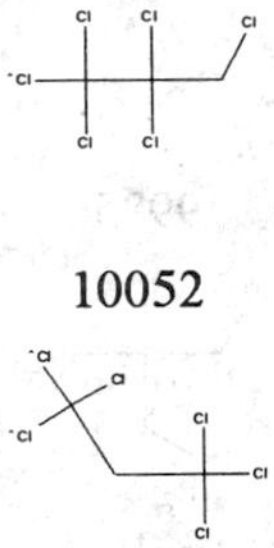

10052

10053

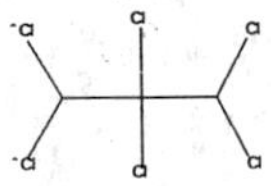

10054

10055

10056

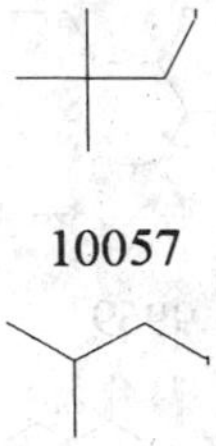

10057

10058

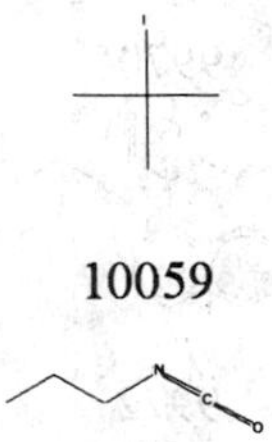

10059

10060

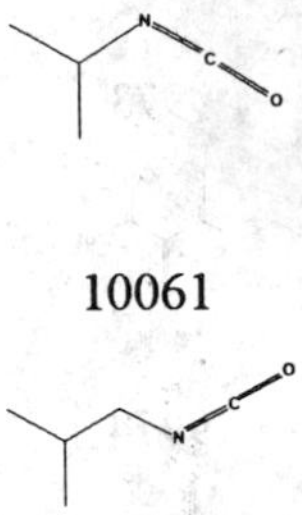

10061

10062

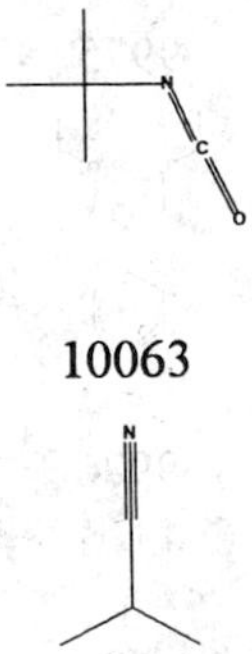

10063

10064

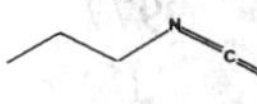

STRUCTURAL FORMULAS OF ORGANIC COMPOUNDS (continued)

In numeric order as they occur in the Table of Physical Constants of Organic Compounds

10065

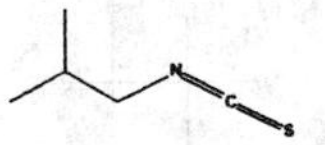

10066

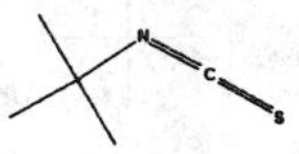

10067

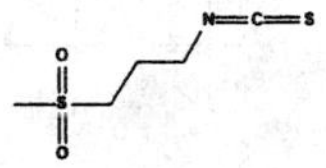

10068

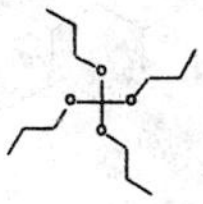

10069

10070

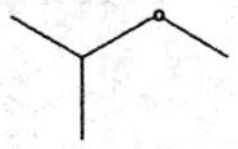

10071

10072

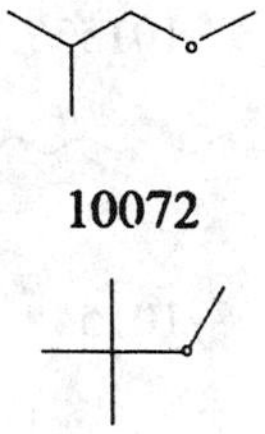

10073

10074

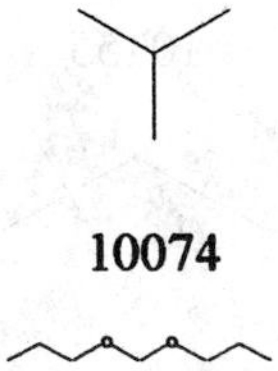

10075

10076

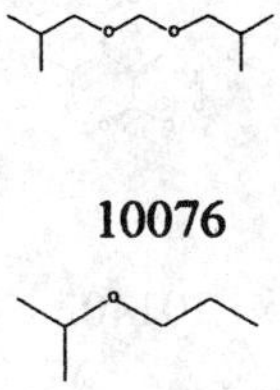

10077

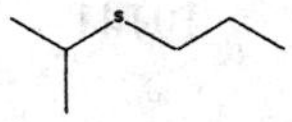

10078

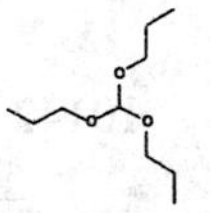

10079

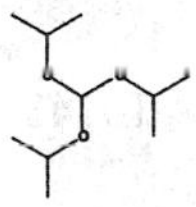

10080

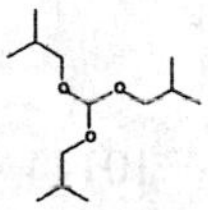

10081

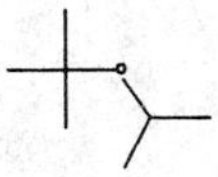

10082

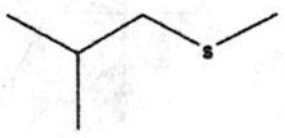

10083

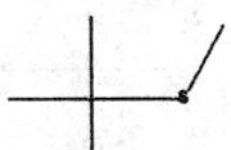

10084

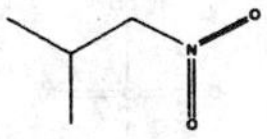

10085

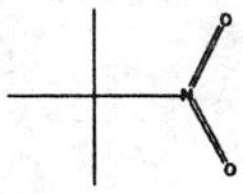

10086

10087

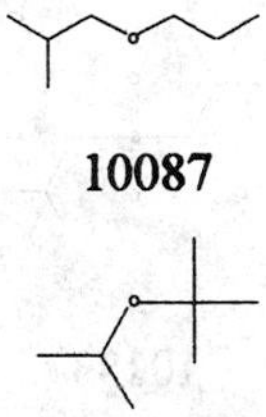

10088

10089

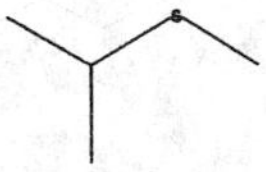

10090

10091

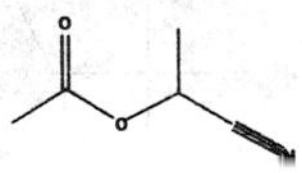

10092

10093

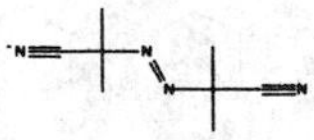

10094

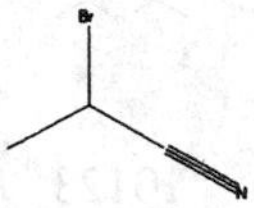

10095

10096

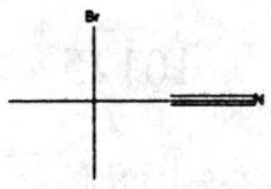

10097

10098

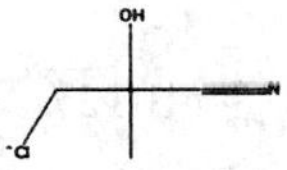

10099

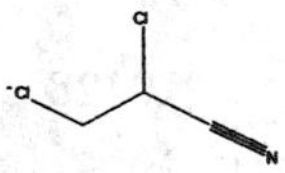

10100

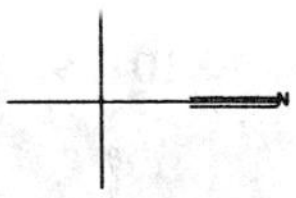

10101

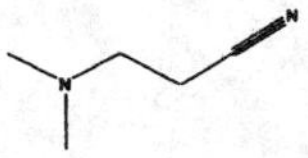

10102

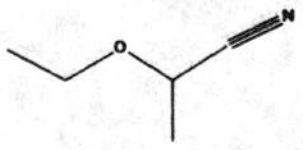

10103

10105

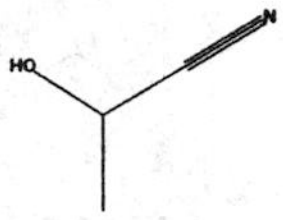

10106

10107

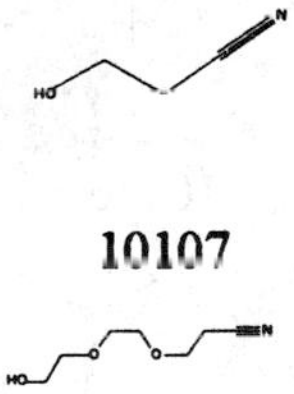

10108

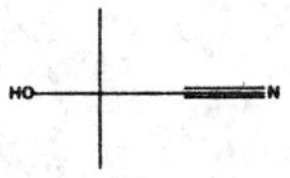

10109

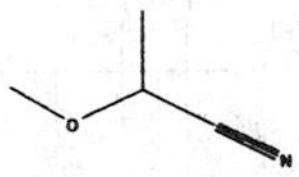

10110

10111

STRUCTURAL FORMULAS OF ORGANIC COMPOUNDS (continued)

In numeric order as they occur in the Table of Physical Constants of Organic Compounds

10112

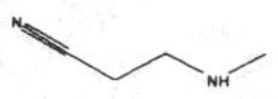

10113

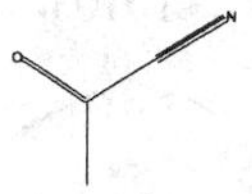

10114

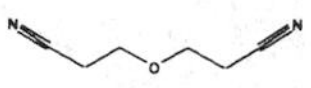

10115

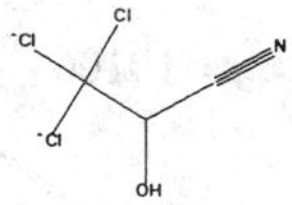

10116

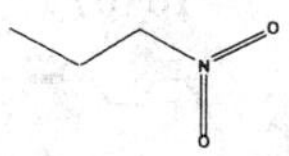

10117

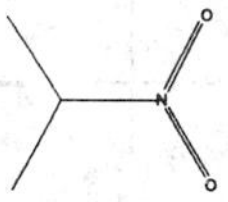

10118

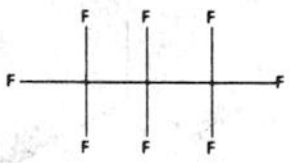

10119

10120

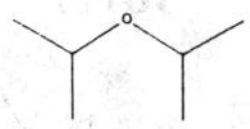

10121

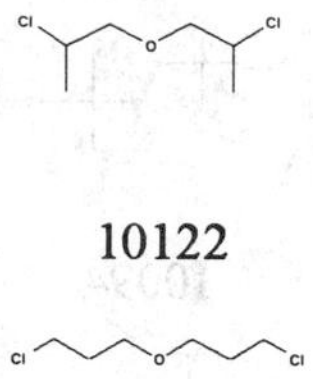

10122

10123

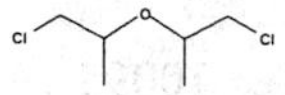

10124

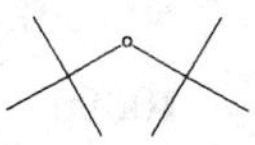

10125

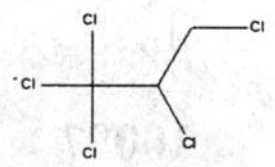

10126

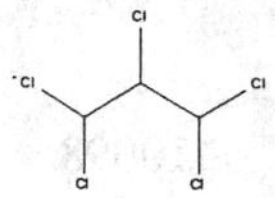

10127

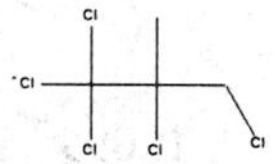

10128

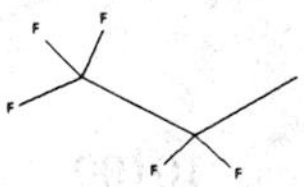

10129

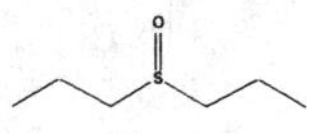

10130

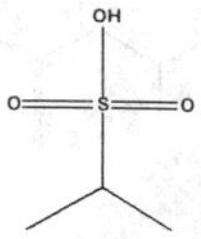

10131

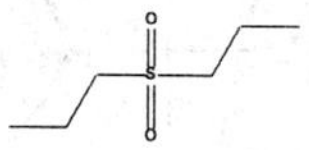

10132

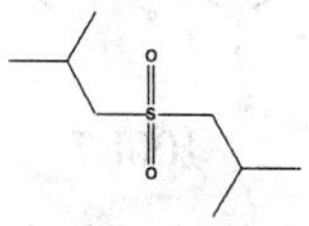

10133

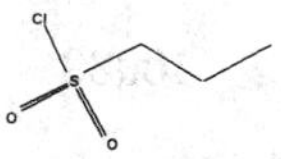

10134

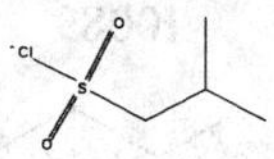

10135

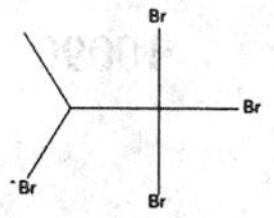

10136

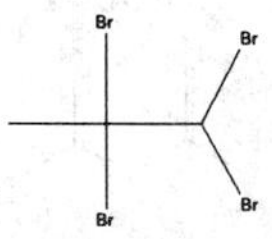

10138

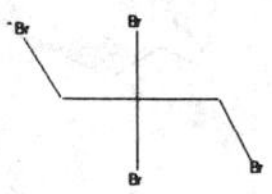

10139

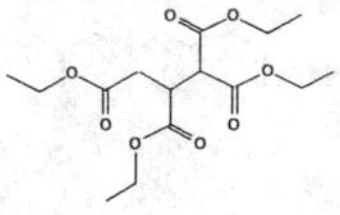

10140

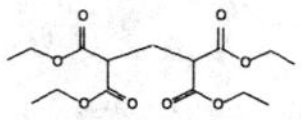

10141

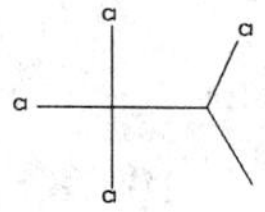

10142

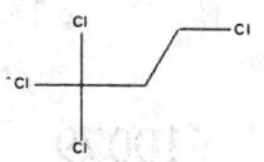

10143

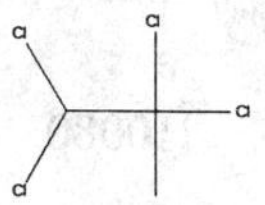

10144

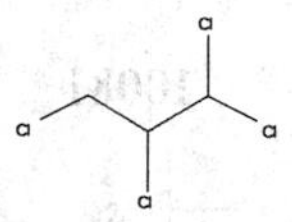

10145

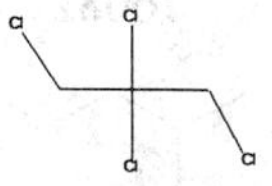

10146

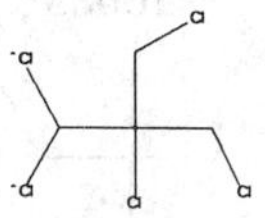

10147

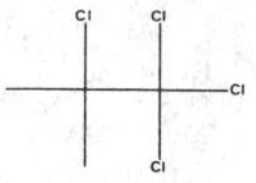

10148

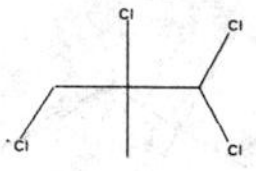

10149

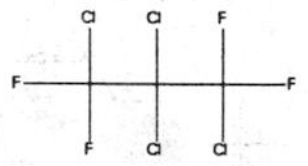

10150

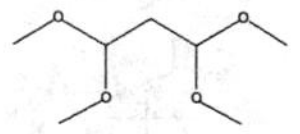

10151

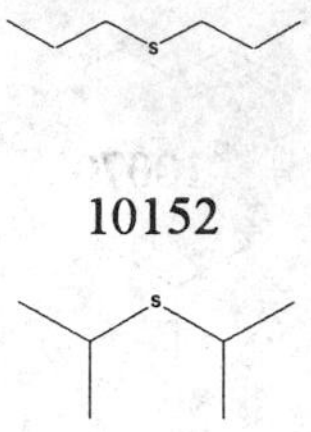

10152

10153

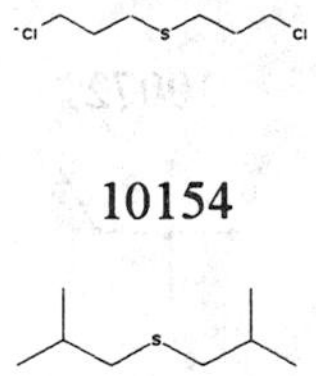

10154

10155

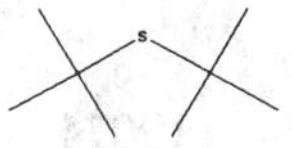

STRUCTURAL FORMULAS OF ORGANIC COMPOUNDS (continued)

In numeric order as they occur in the Table of Physical Constants of Organic Compounds

10156

10157

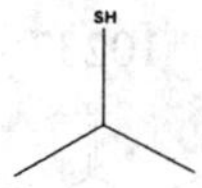

10158

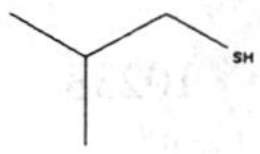

10159

10160

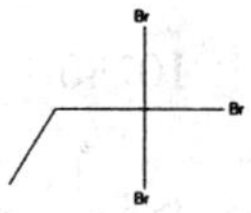

10161

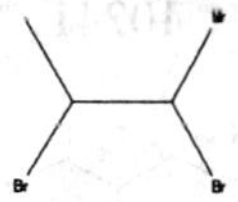

10162

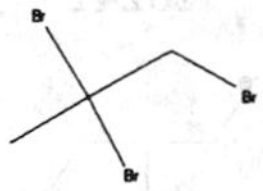

10163

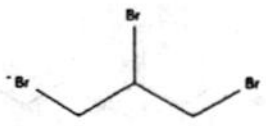

10164

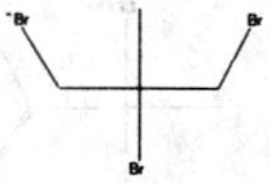

10166

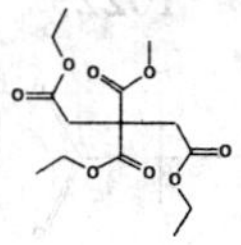

10167

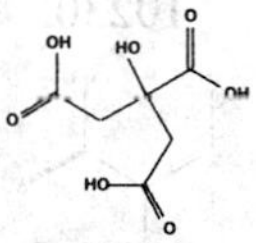

10168

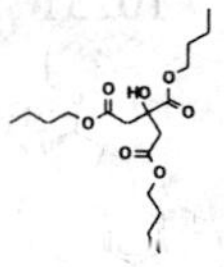

10169

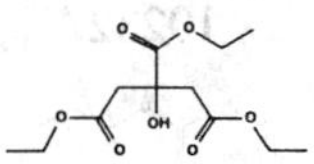

10170

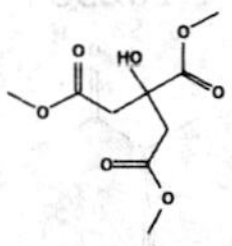

10171

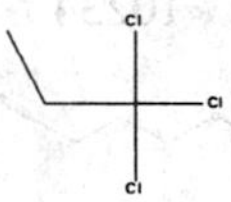

10172

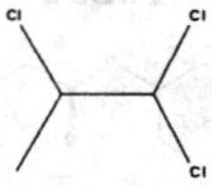

10173

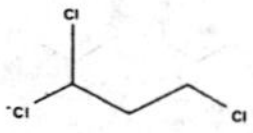

10174

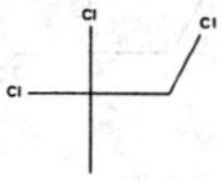

10175

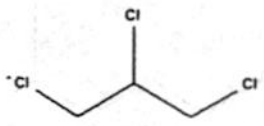

10176

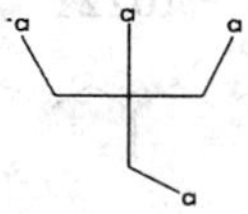

10177

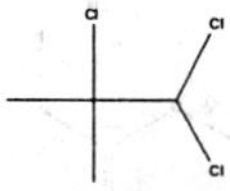

10178

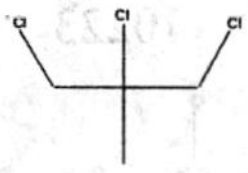

10179

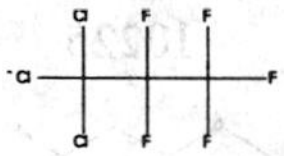

10180

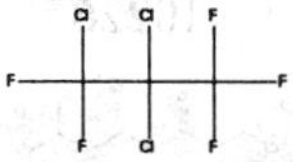

10181

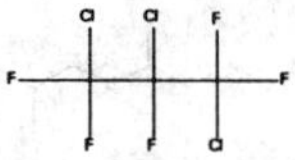

10182

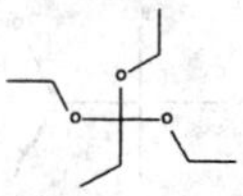

10183

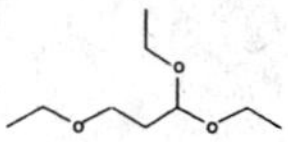

10184

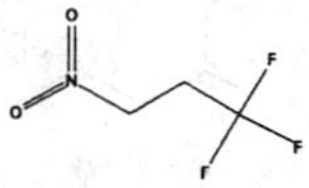

10185

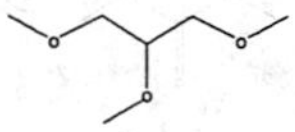

10186

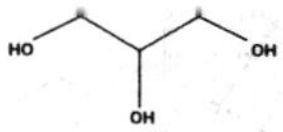

10187

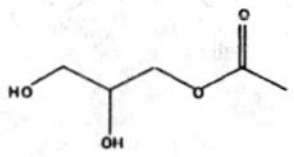

10188

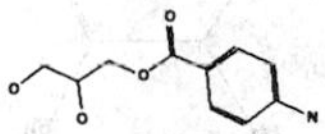

10189

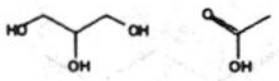

10190

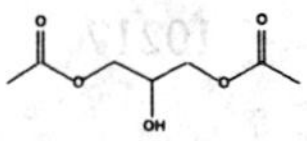

10191

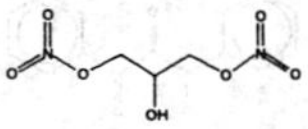

10192

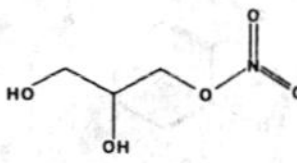

10193

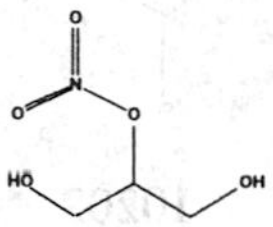

10194

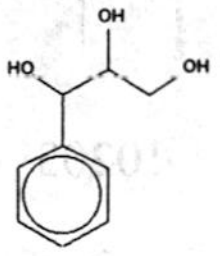

10195

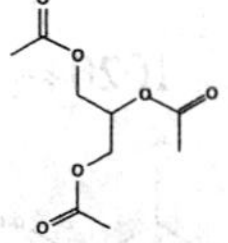

10196

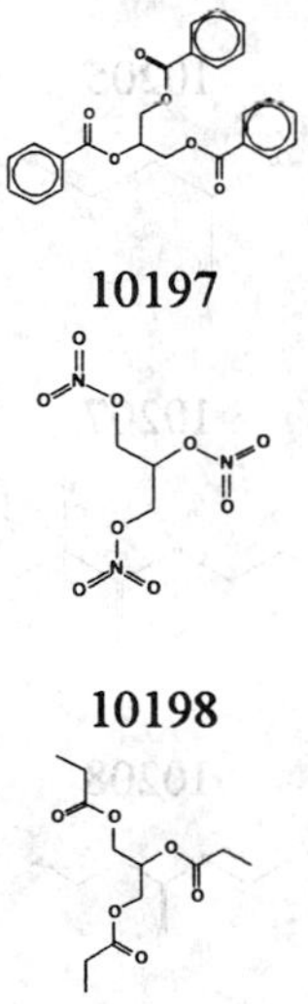

10197

10198

STRUCTURAL FORMULAS OF ORGANIC COMPOUNDS (continued)

In numeric order as they occur in the Table of Physical Constants of Organic Compounds

10199

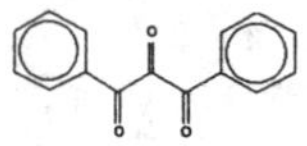

10200

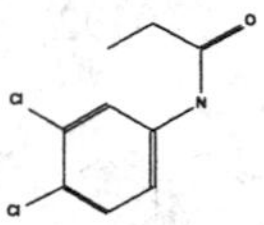

10201

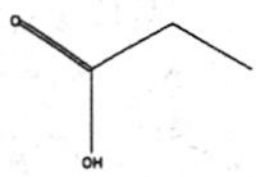

10202

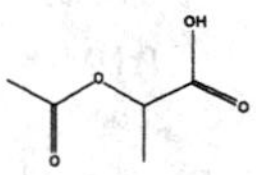

10203

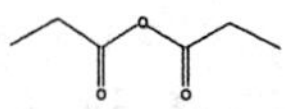

10204

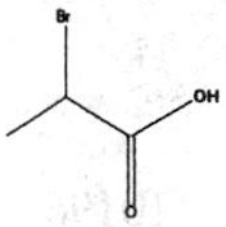

10205

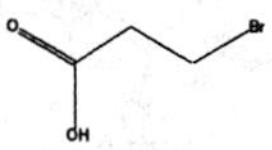

10207

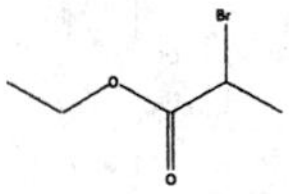

10208

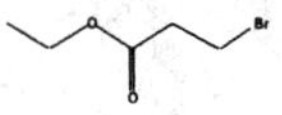

10209

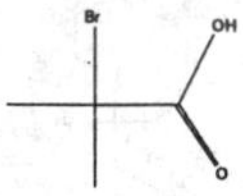

10210

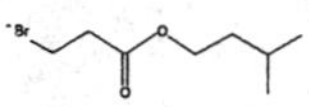

10211

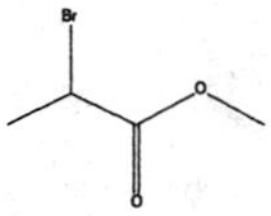

10212

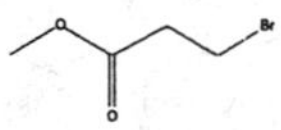

10213

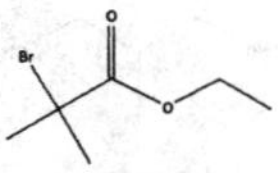

10214

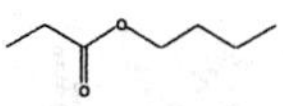

10215

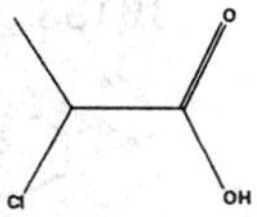

10216

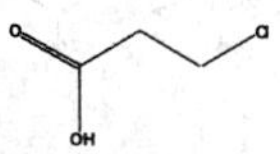

10217

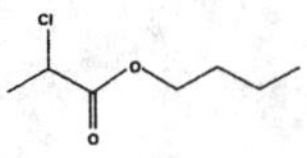

10218

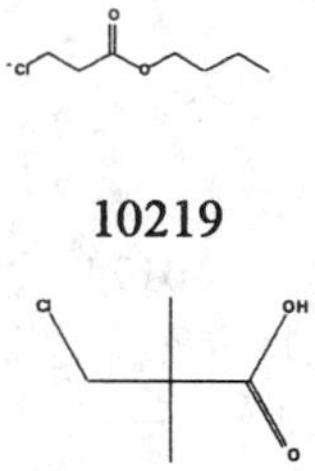

10219

10220

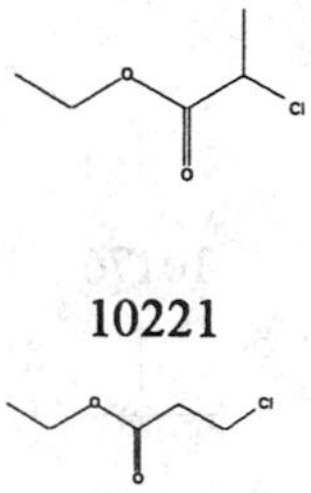

10221

10222

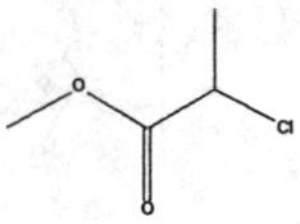

10223

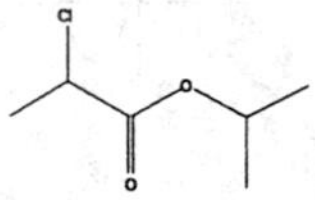

10225

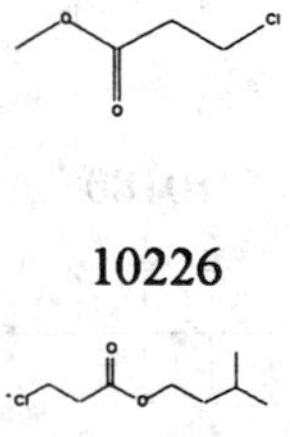

10226

10227

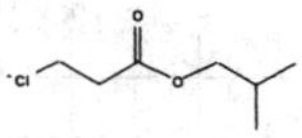

10228

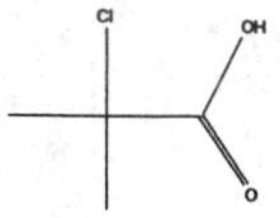

10229

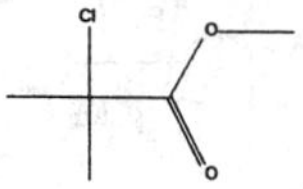

10230

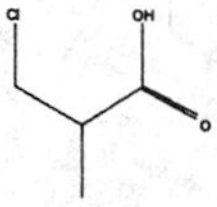

10231

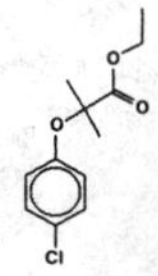

10232

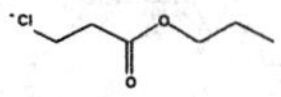

10233

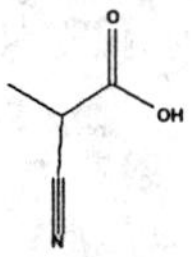

10234

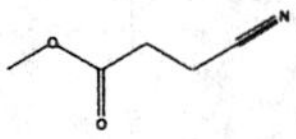

10235

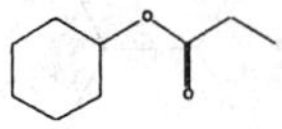

10236

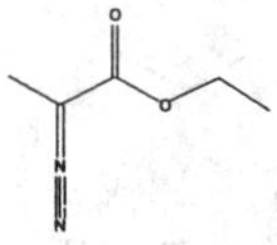

10237

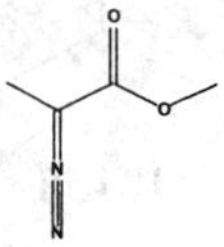

10238

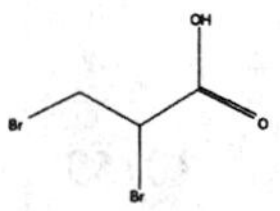

10239

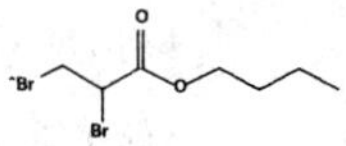

10240

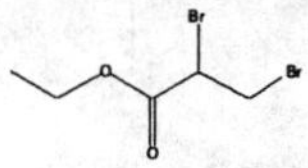

10241

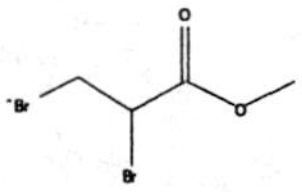

10242

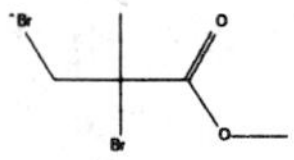

10243

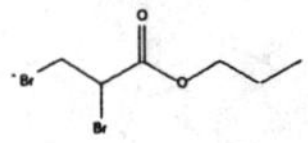

STRUCTURAL FORMULAS OF ORGANIC COMPOUNDS (continued)

In numeric order as they occur in the Table of Physical Constants of Organic Compounds

10244

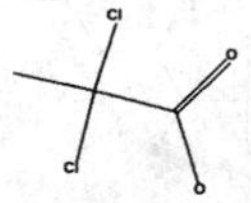

Cl, Cl, O, O

10245

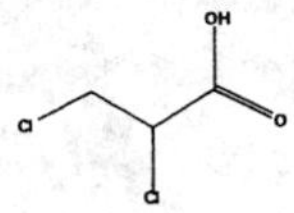

OH, Cl, O, Cl

10246

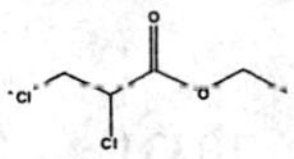

O, Cl, O, Cl

10248

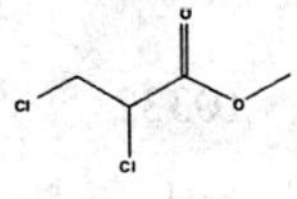

O, Cl, O, Cl

10249

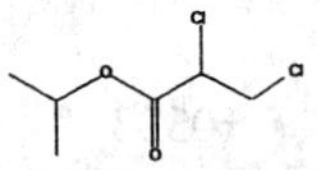

Cl, O, Cl, O

10250

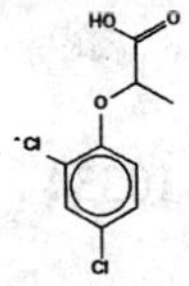

HO, O, O, Cl, Cl

10251

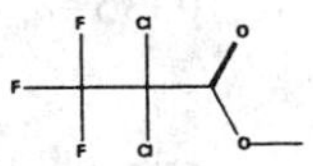

F, Cl, O, F, F, Cl, O

10253

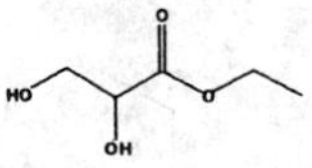

O, HO, O, OH

10254

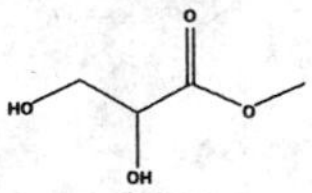

O, HO, O, OH

10255

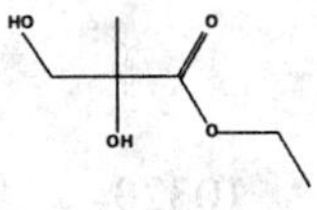

HO, O, OH, O

10256

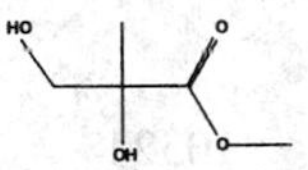

HO, O, OH, O

10258

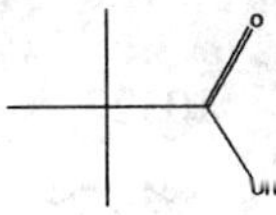

O, OH

10259

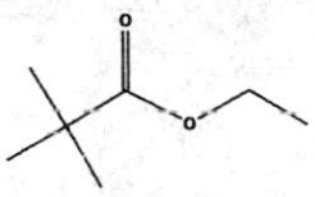

O, O

10260

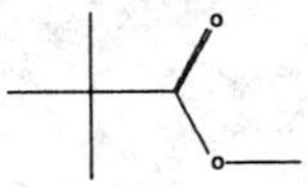

O, O

10261

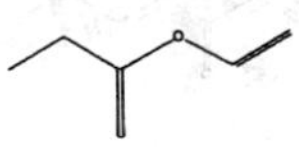

O, O

10262

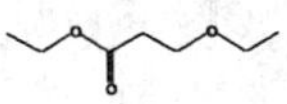

O, O, O

10263

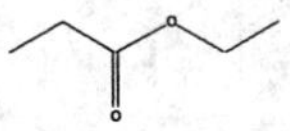

O, O

10264

O, O

10265

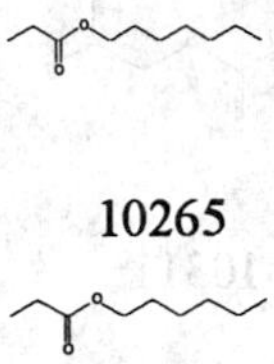

O, O

10268

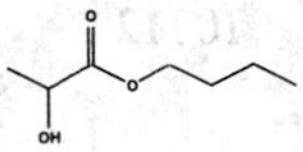

O, O, OH

10269

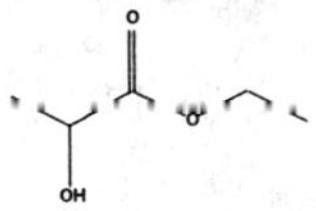

O, O, OH

10270

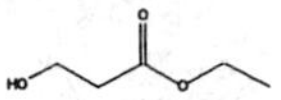

O, HO, O

10271

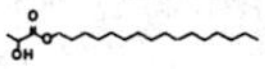

O, OH

10273

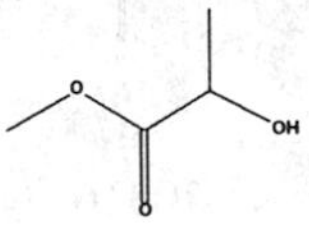

O, OH, O

10274

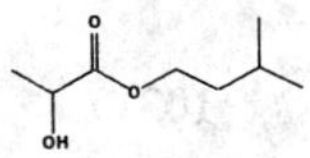

O, O, OH

10275

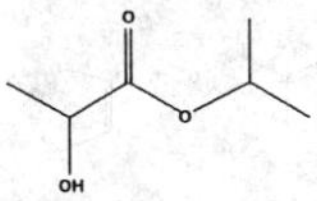

O, O, OH

10276

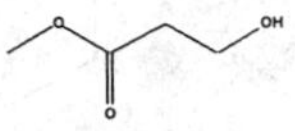

O, OH, O

10277

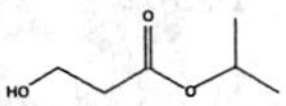

O, HO, O

10278

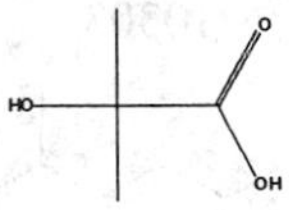

O, HO, OH

10279

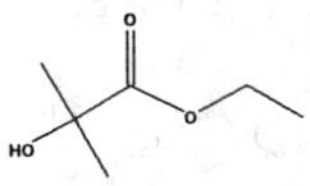

O, O, HO

10280

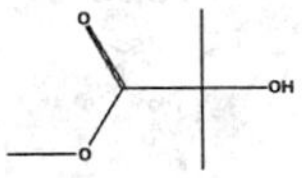

O, OH, O

10281

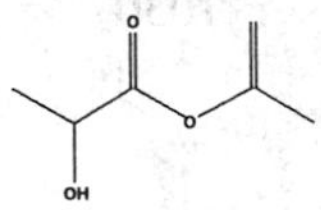

O, O, OH

10282

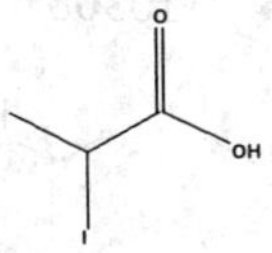

O, OH, I

10283

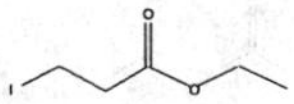

O, I, O

10285

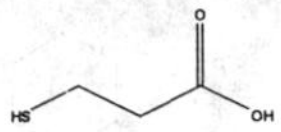

O, HS, OH

10286

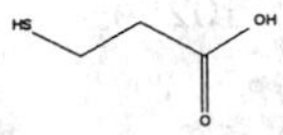

HS, OH, O

10287

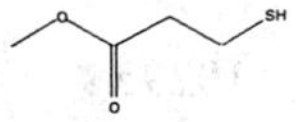

O, SH, O

10288

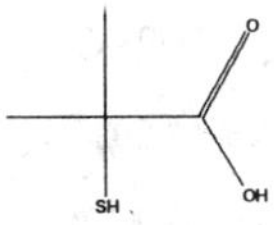

O, SH, OH

10289

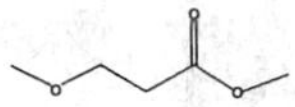

O, O, O

10290

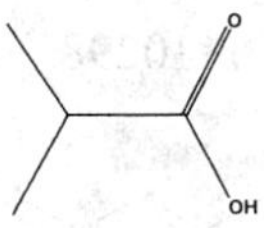

O, OH

10291

O, O, O

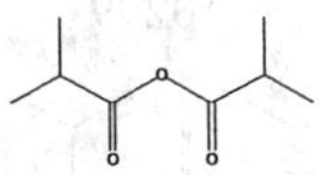

In numeric order as they occur in the Table of Physical Constants of Organic Compounds

10292

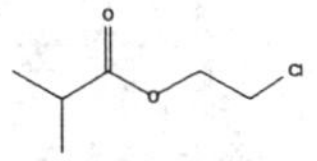

10293

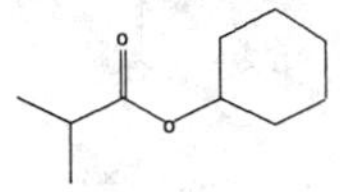

10294

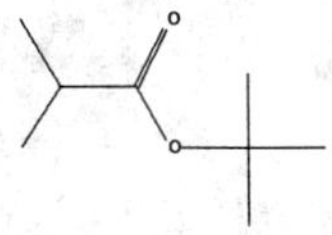

10295

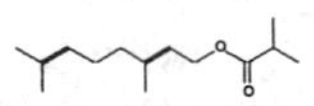

10296

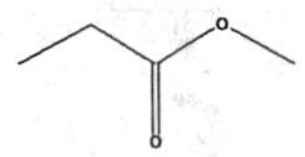

10297

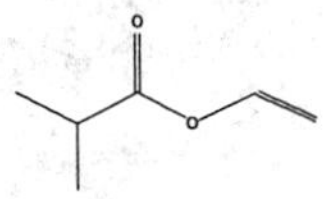

10298

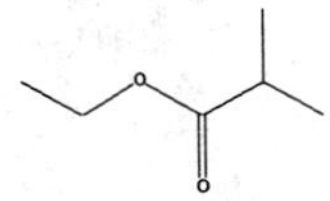

10299

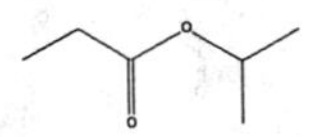

10300

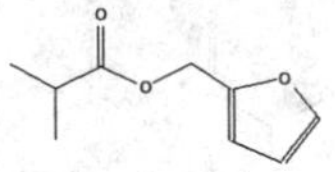

10301

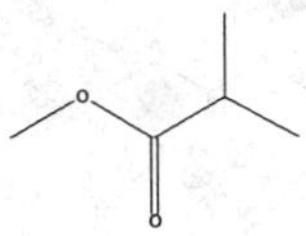

10302

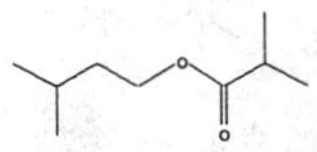

10303

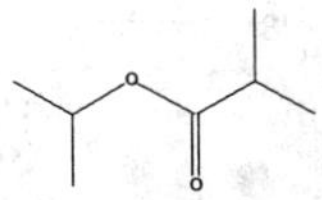

10304

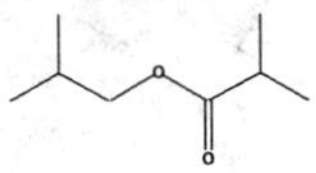

10306

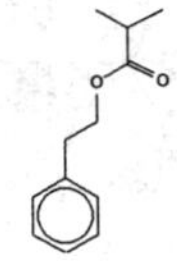

10307

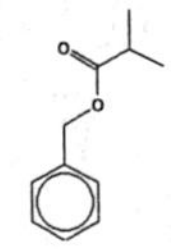

10308

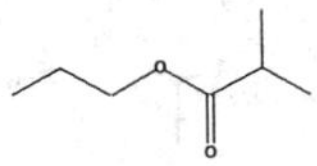

10309

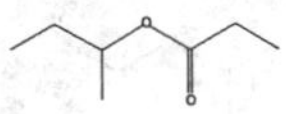

10310

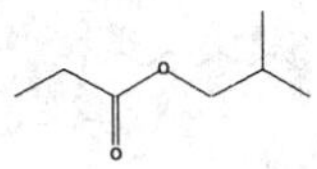

10311

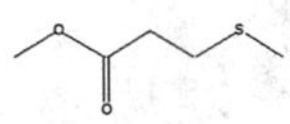

10312

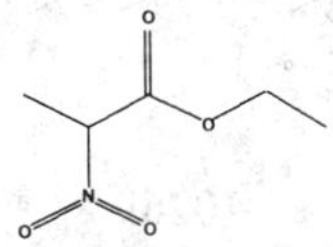

10313

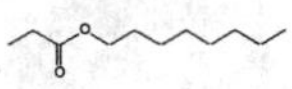

10314

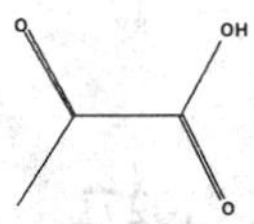

10315

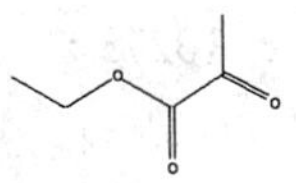

10316

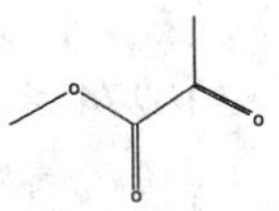

10317

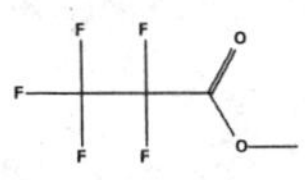

10318

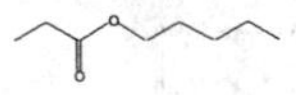

10319

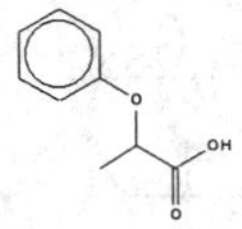

10320

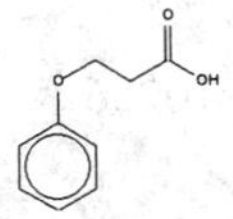

10321

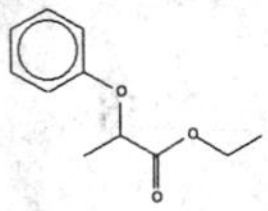

10322

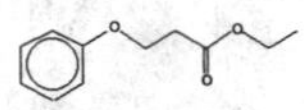

10323

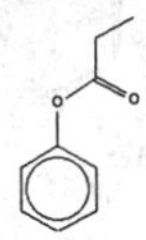

10324

10325

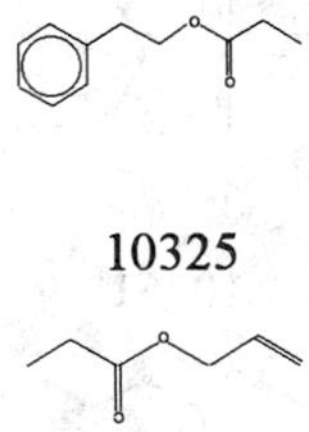

10326

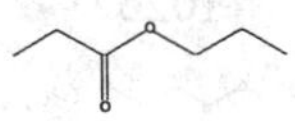

10328

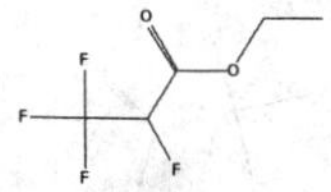

10329

10330

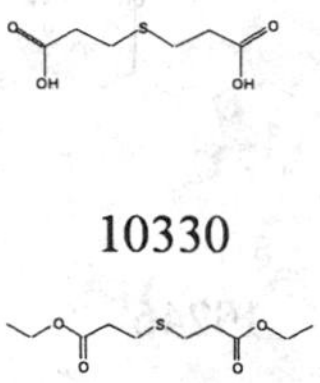

10331

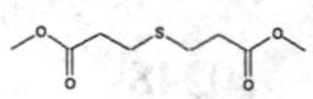

10332

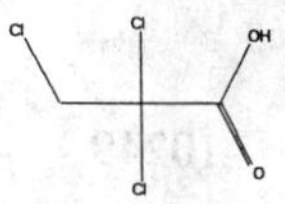

10333

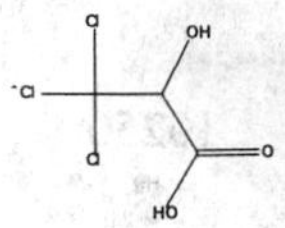

10334

10335

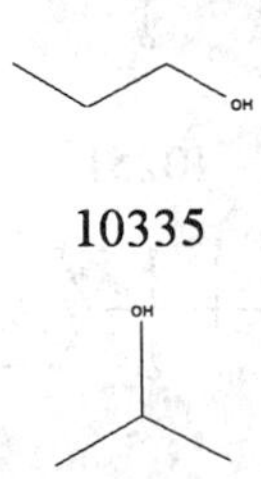

10336

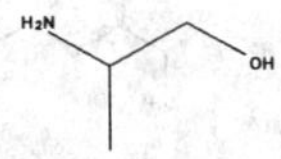

STRUCTURAL FORMULAS OF ORGANIC COMPOUNDS (continued)

In numeric order as they occur in the Table of Physical Constants of Organic Compounds

10337

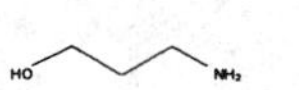

10338

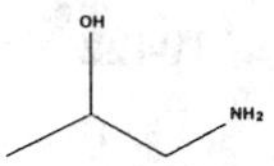

10339

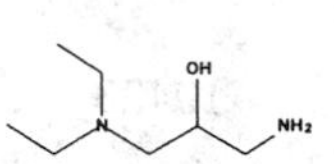

10340

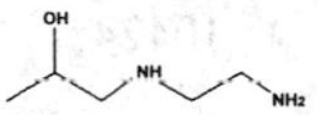

10341

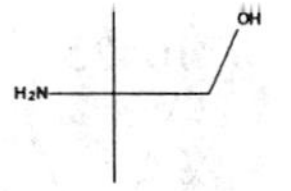

10342

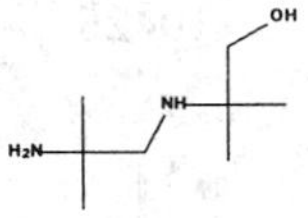

10343

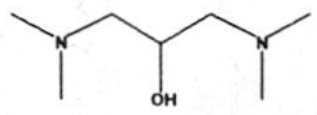

10344

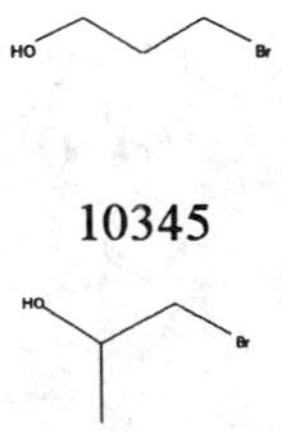

10345

10346

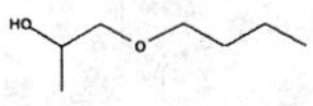

10347

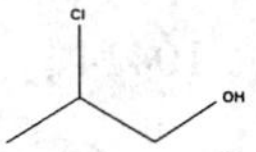

10348

10349

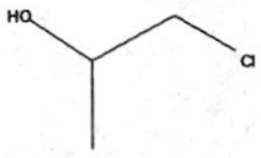

10350

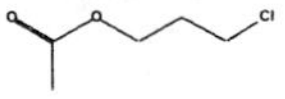

10351

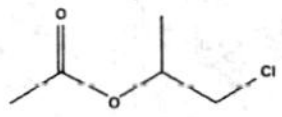

10352

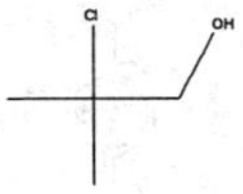

10353

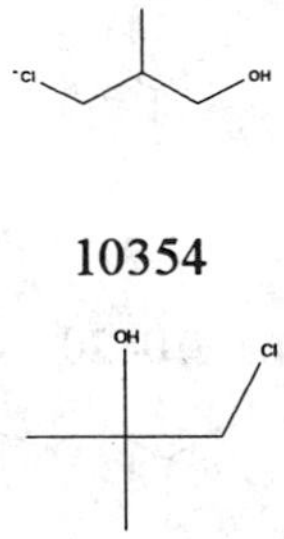

10354

10355

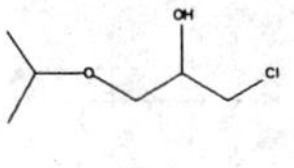

10356

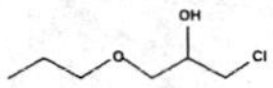

10357

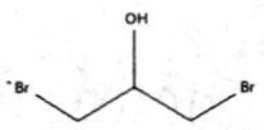

10358

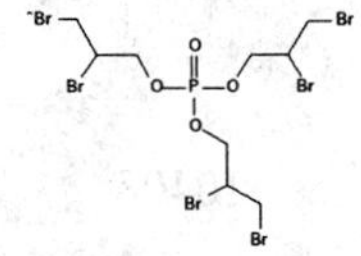

10359

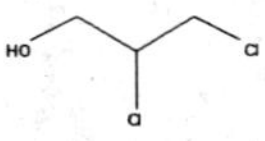

10360

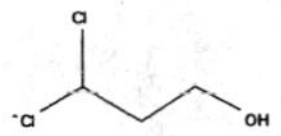

10361

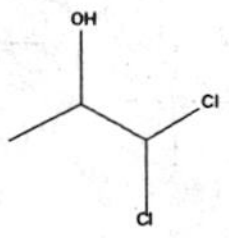

10362

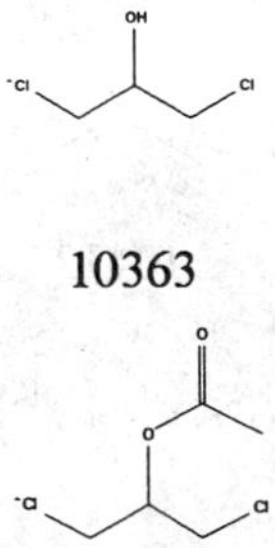

10363

10364

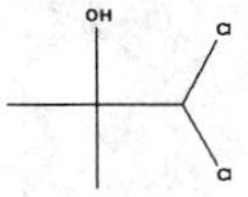

10365

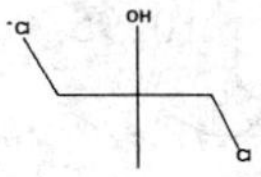

10366

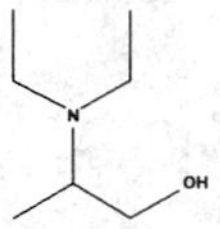

10367

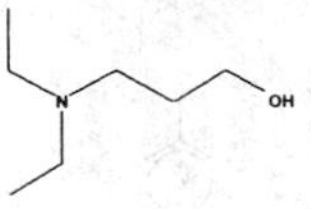

10368

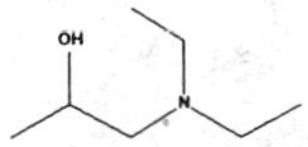

10369

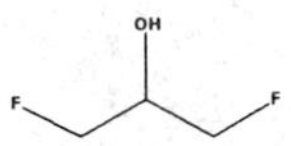

10370

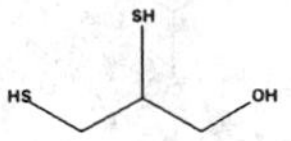

10371

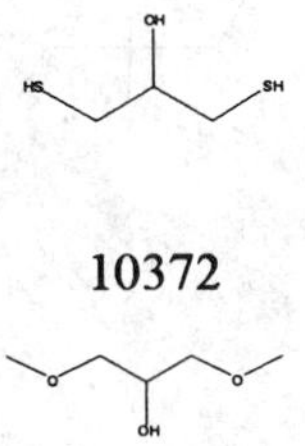

10372

10373

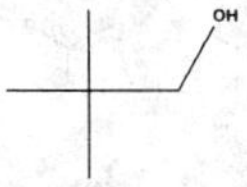

10374

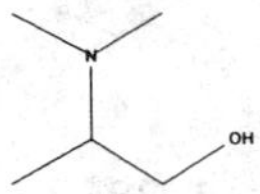

10375

10376

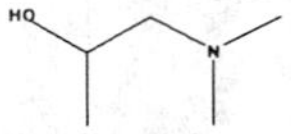

10378

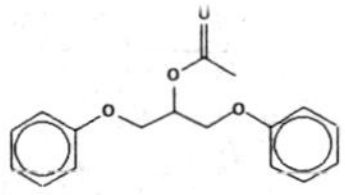

10380

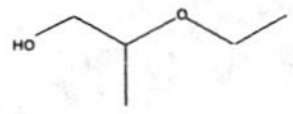

10381

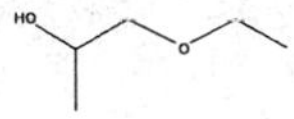

10382

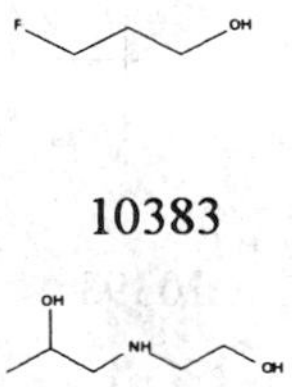

10383

STRUCTURAL FORMULAS OF ORGANIC COMPOUNDS (continued)

In numeric order as they occur in the Table of Physical Constants of Organic Compounds

10384

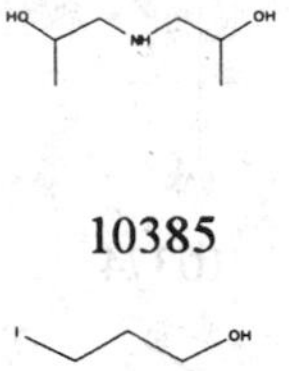

10385

10386

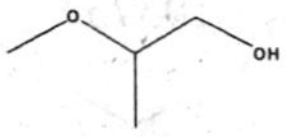

10387

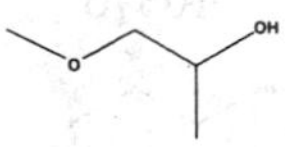

10388

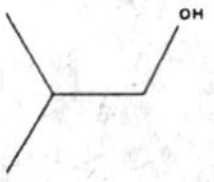

10389

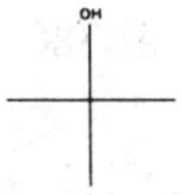

10390

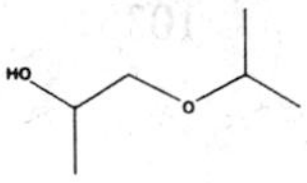

10391

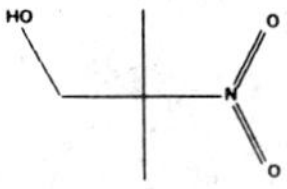

10393

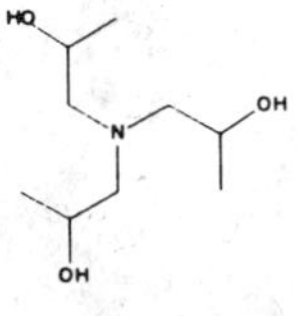

10394

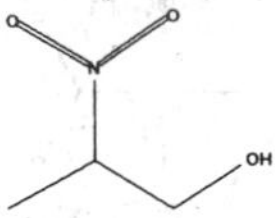

10395

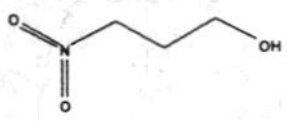

10396

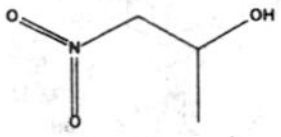

10398

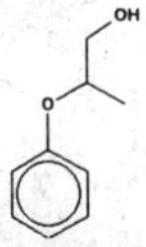

10399

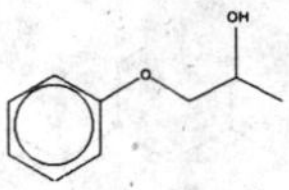

10401

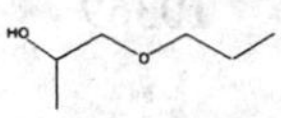

10402

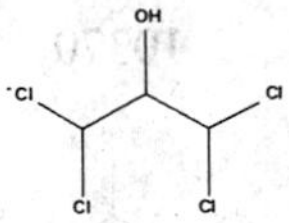

10404

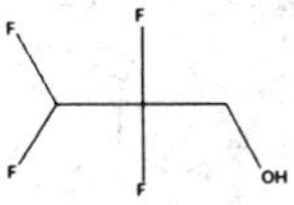

10405

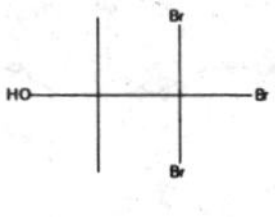

10406

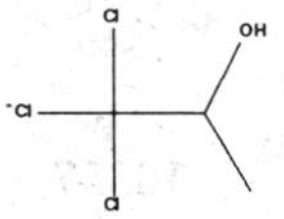

10407

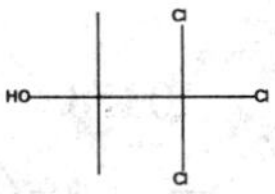

10408

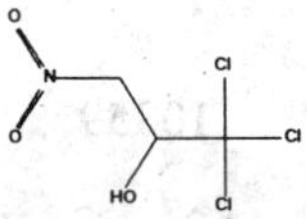

10409

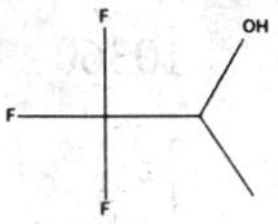

10410

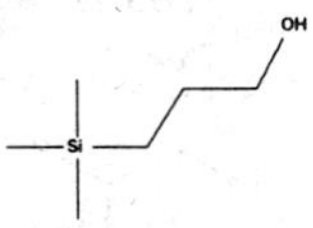

10411

10412

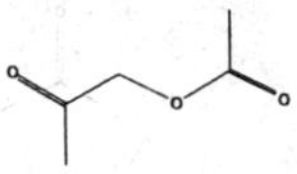

10413

10414

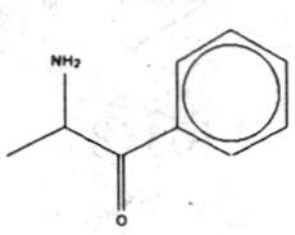

10415

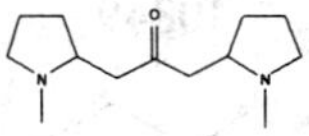

10416

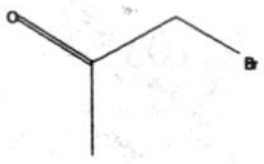

10417

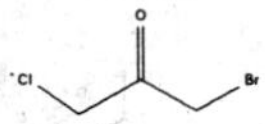

10418

10419

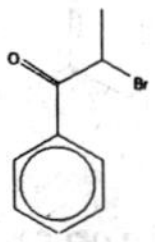

10420

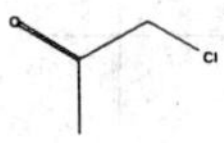

10421

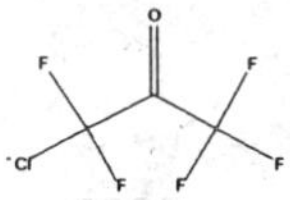

10422

10423

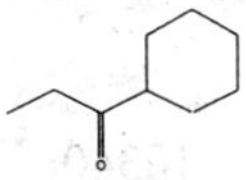

10424

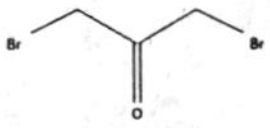

10425

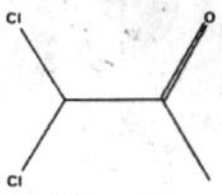

10426

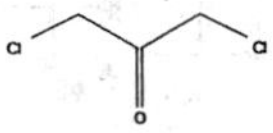

10427

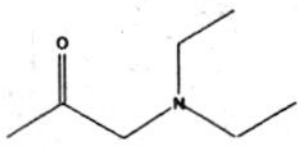

10428

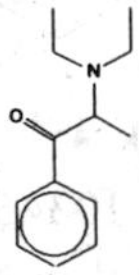

STRUCTURAL FORMULAS OF ORGANIC COMPOUNDS (continued)

In numeric order as they occur in the Table of Physical Constants of Organic Compounds

10429

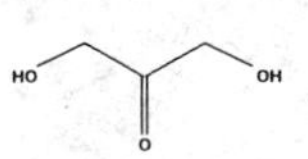

HO OH O

10430

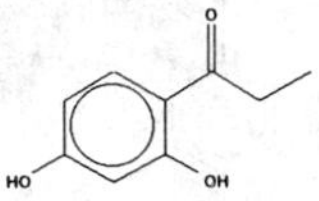

O HO OH

10431

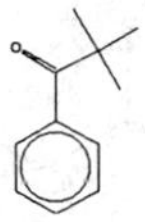

O

10432

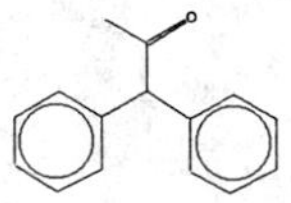

O

10433

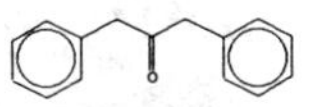

O

10436

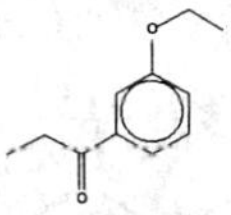

O O

10437

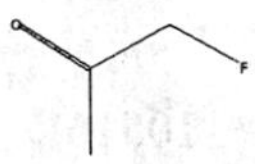

O F

10438

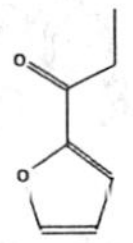

O O

10439

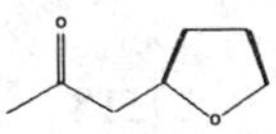

O O

10440

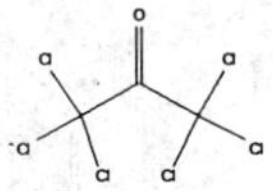

O Cl Cl Cl Cl Cl Cl

10441

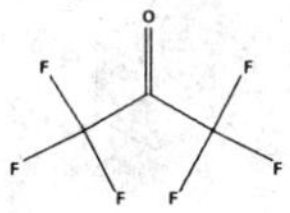

O F F F F F F

10442

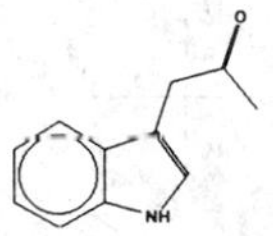

O NH

10443

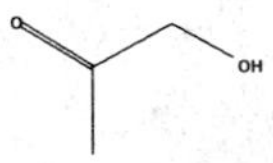

O OH

10444

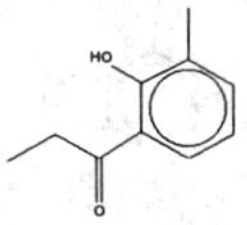

HO O

10445

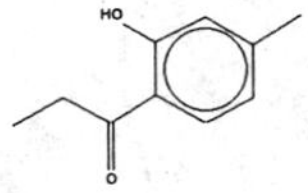

HO O

10446

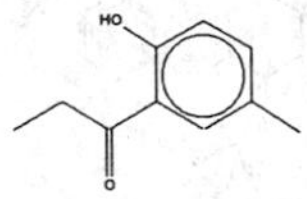

HO O

10447

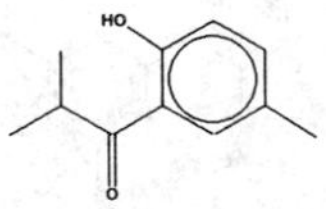

HO O

10448

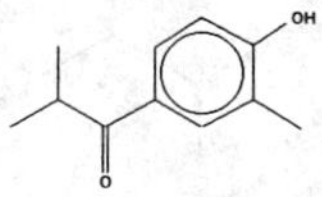

OH O

10449

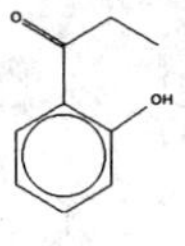

O OH

10450

O OH

10451

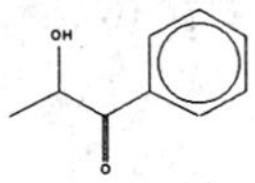

OH O

10452

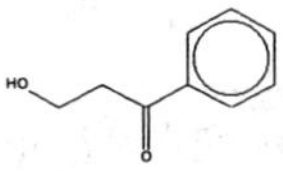

HO O

10454

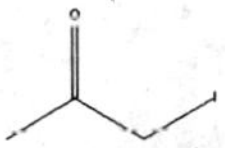

O

10455

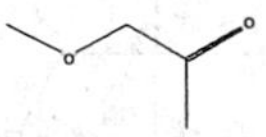

O O

10457

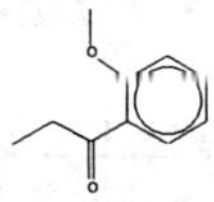

O O

10458

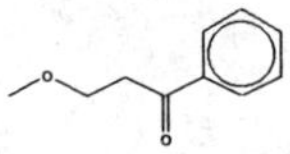

O O

10460

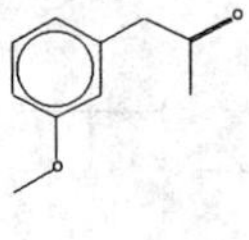

O O

10461

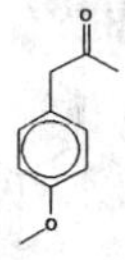

O O

10462

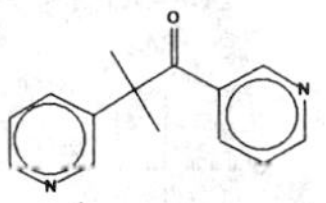

O N N

10463

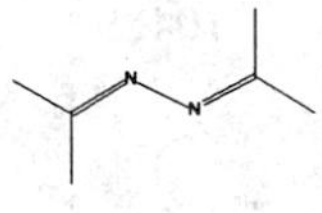

N N

10464

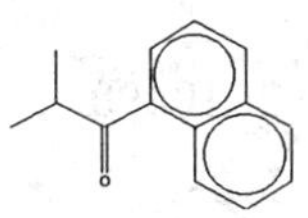

O

10465

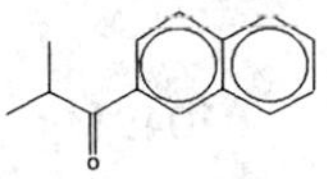

O

10466

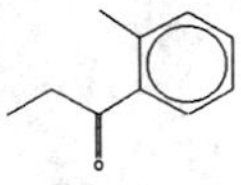

O

10467

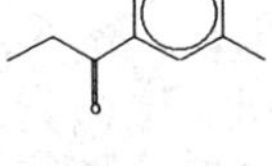

O

10468

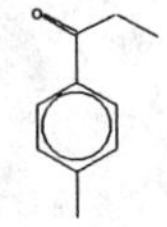

O

10469

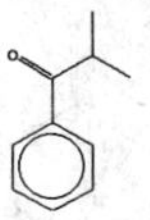

O

10470

O N

10471

O N

10472

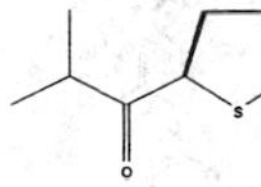

S O

10473

O

10474

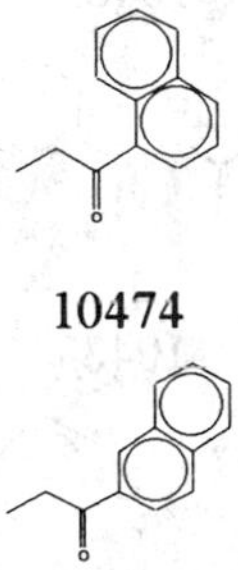

O

10475

O N O O

10476

N OH

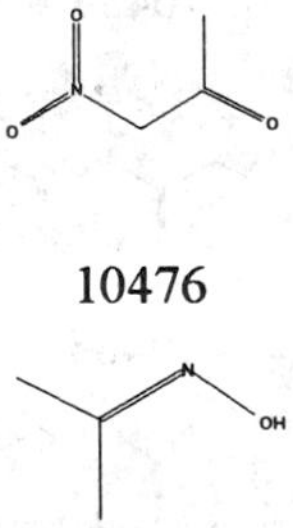

STRUCTURAL FORMULAS OF ORGANIC COMPOUNDS (continued)

In numeric order as they occur in the Table of Physical Constants of Organic Compounds

10478

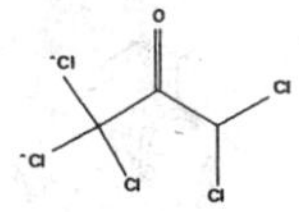

10479

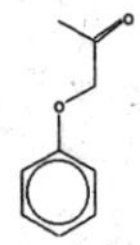

10480

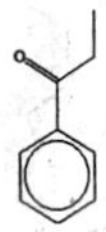

10481

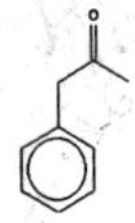

10482

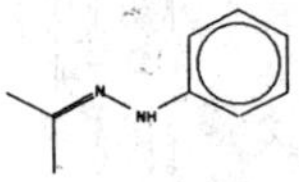

10483

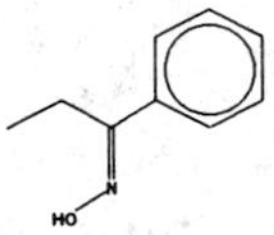

10484

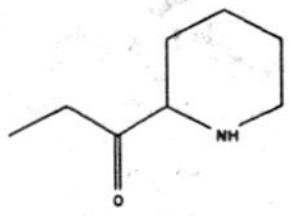

10485

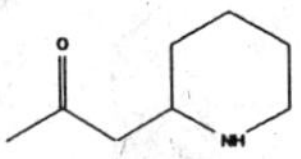

10486

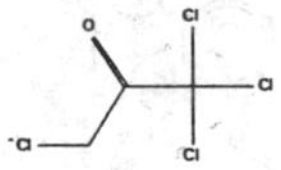

10487

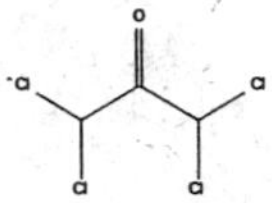

10488

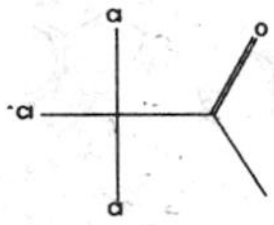

10489

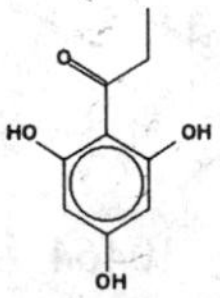

10490

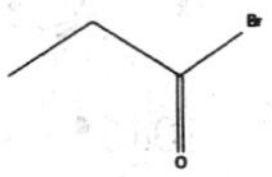

10491

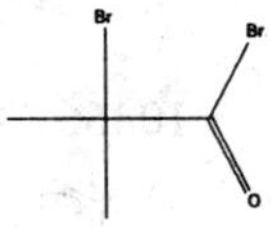

10492

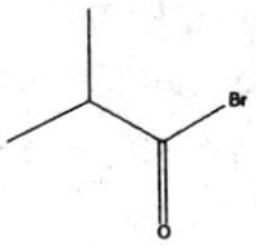

10493

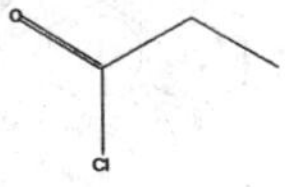

10494

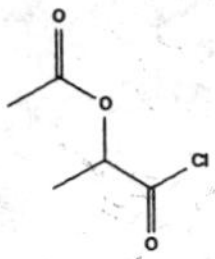

10495

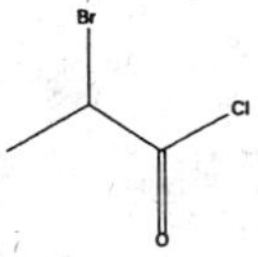

10496

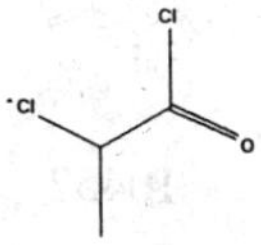

10497

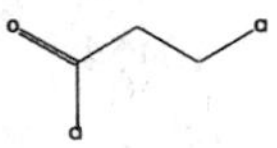

10498

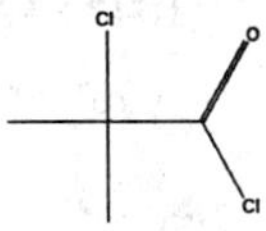

10500

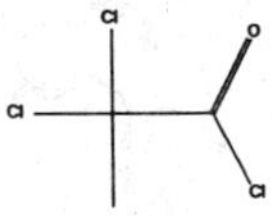

10501

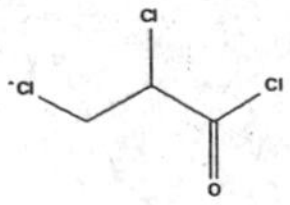

10502

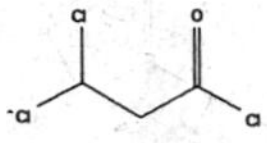

10503

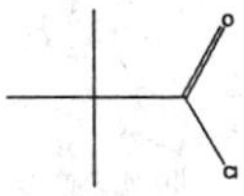

10504

10505

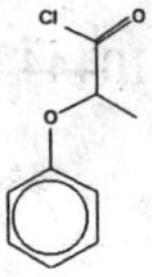

10506

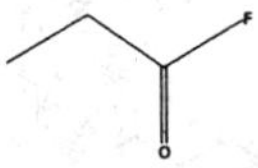

10507

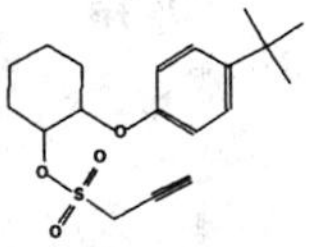

10508

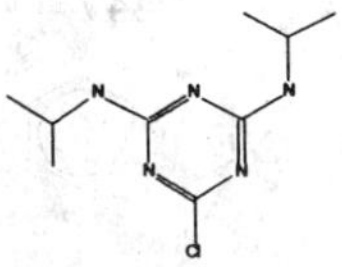

10509

10510

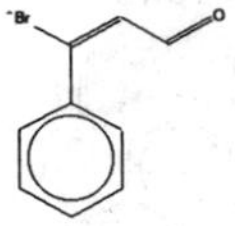

10511

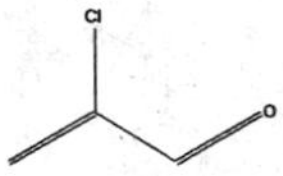

10512

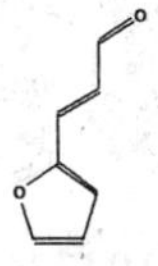

10514

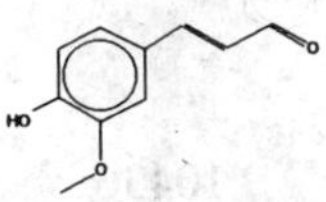

10515

10516

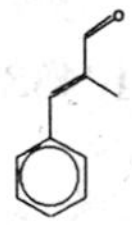

10517

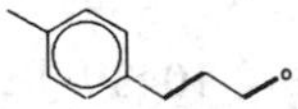

10518

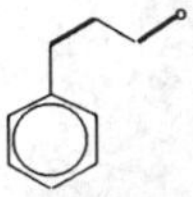

STRUCTURAL FORMULAS OF ORGANIC COMPOUNDS (continued)

In numeric order as they occur in the Table of Physical Constants of Organic Compounds

10519

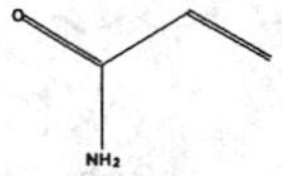

NH_2

10520

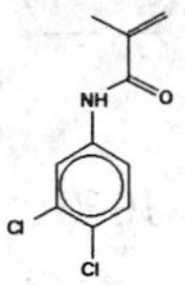

NH O Cl Cl

10521

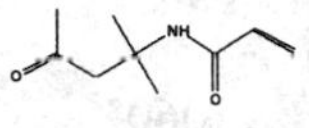

O NH O

10522

NH_2

10523

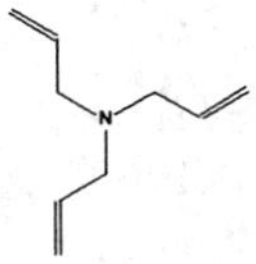

N

10524

NH

10525

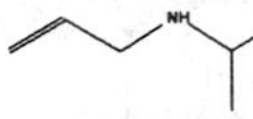

NH

10526

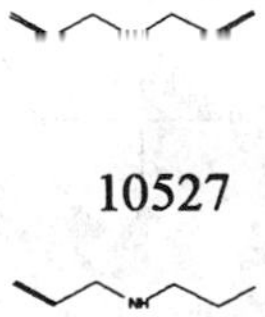

10527

NH

10528

10529

Br

10530

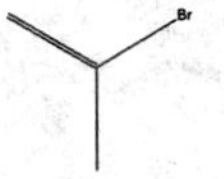

Br

10531

Br

10532

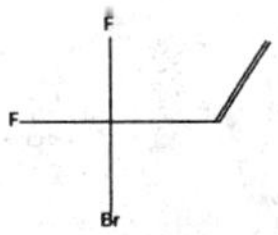

F F Br

10533

Cl

10534

Cl

10535

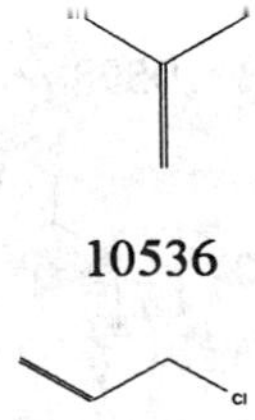

10536

Cl

10537

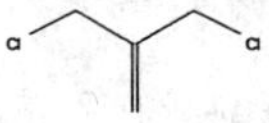

Cl Cl

10538

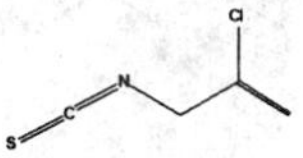

S C N Cl

10539

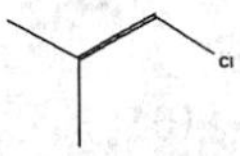

Cl

10540

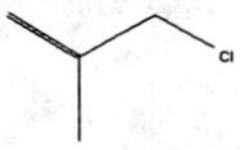

Cl

10541

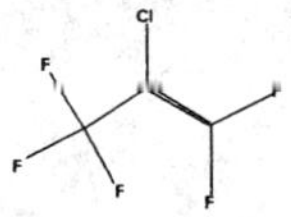

Cl F F F F F

10542

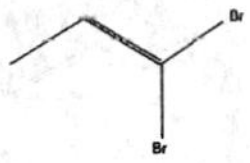

Br Br

10543

Br Br

10544

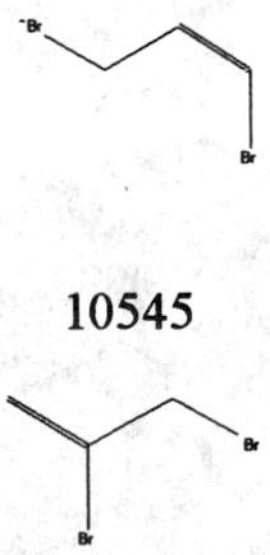

Br Br

10545

Br Br

10546

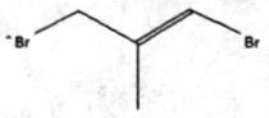

Br Br

10547

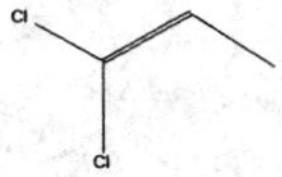

Cl Cl

10548

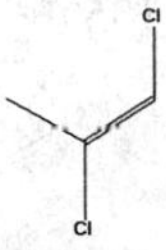

Cl Cl

10549

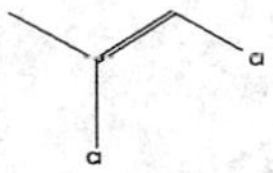

Cl Cl

10550

Cl Cl

10551

Cl Cl

10552

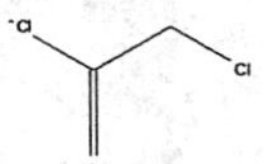

Cl Cl

10553

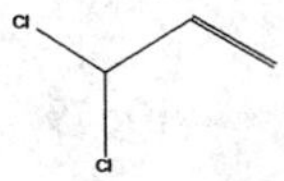

Cl Cl

10554

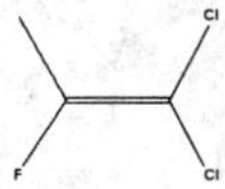

Cl F Cl

10555

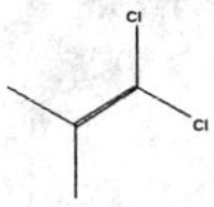

Cl Cl

10556

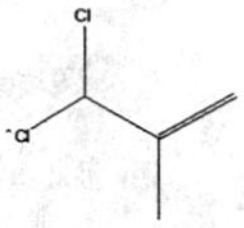

Cl Cl

10557

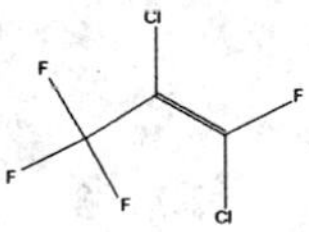

Cl F F F F Cl

10558

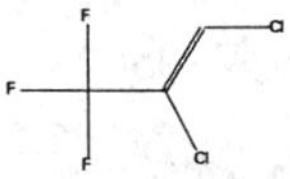

F F F Cl Cl

10559

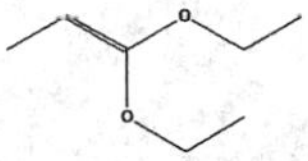

O O

10560

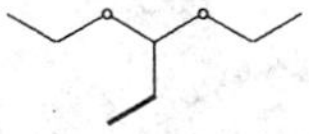

O O

10561

F F

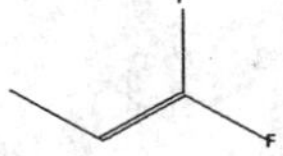

STRUCTURAL FORMULAS OF ORGANIC COMPOUNDS (continued)

In numeric order as they occur in the Table of Physical Constants of Organic Compounds

10562

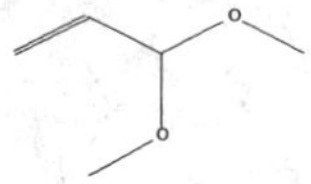

10563

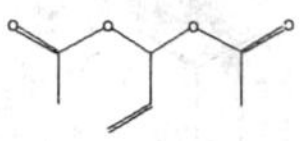

10564

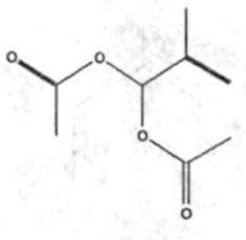

10565

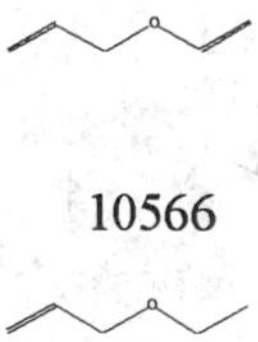

10566

10567

10568

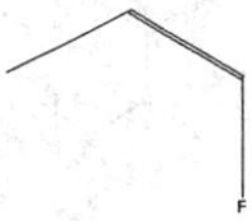

10569

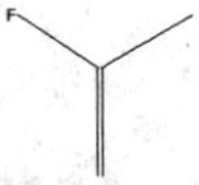

10570

10571

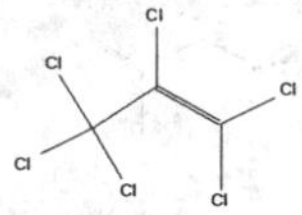

10572

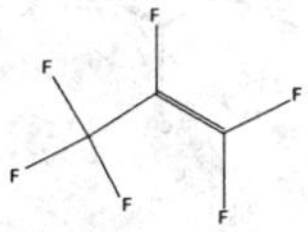

10573

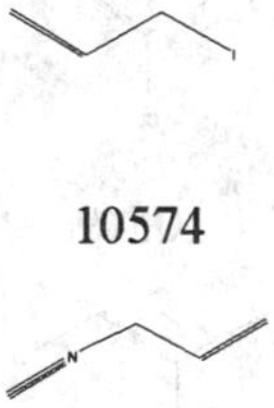

10574

10575

10576

10577

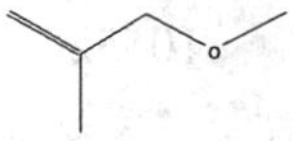

10578

10579

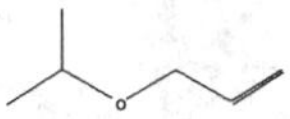

10580

10581

10582

10583

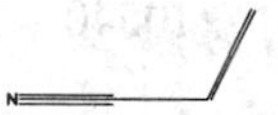

10584

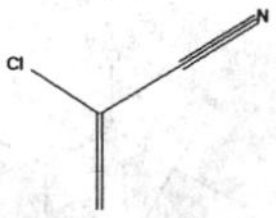

10585

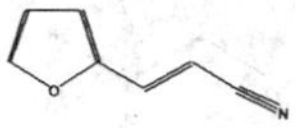

10586

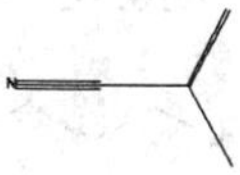

10587

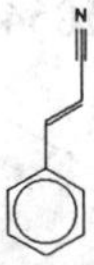

10588

10589

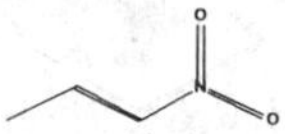

10590

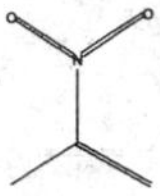

10591

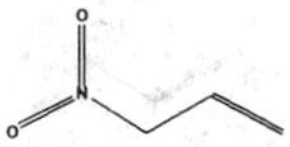

10592

10593

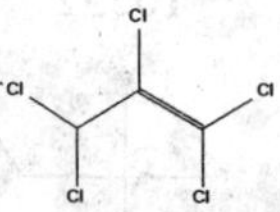

10594

10596

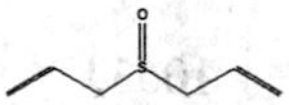

10597

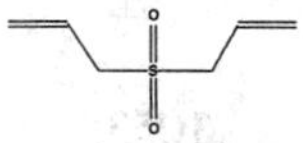

10598

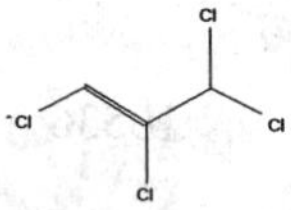

10599

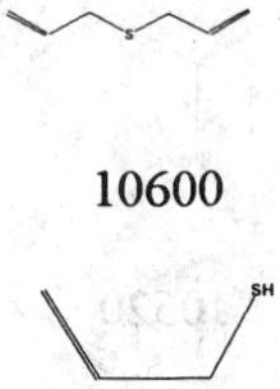

10600

10601

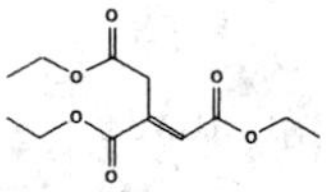

10602

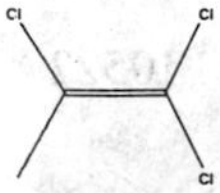

10603

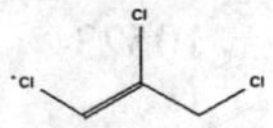

10604

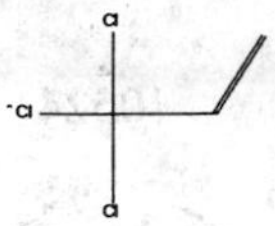

10605

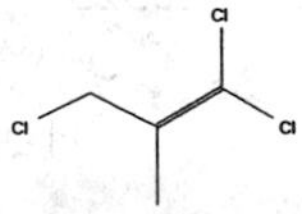

10606

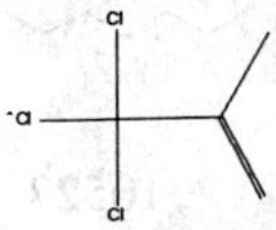

STRUCTURAL FORMULAS OF ORGANIC COMPOUNDS (continued)

In numeric order as they occur in the Table of Physical Constants of Organic Compounds

10607

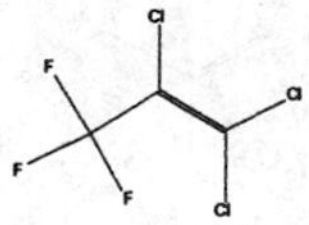

10608

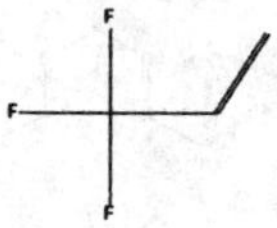

10609

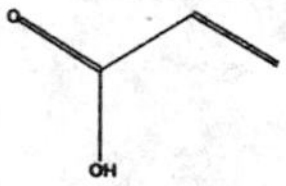

10611

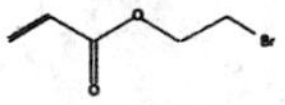

10613

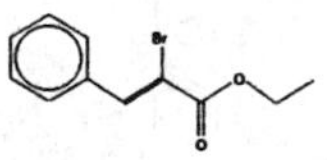

10614

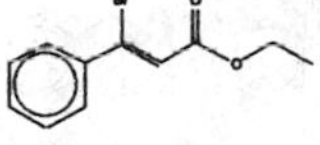

10616

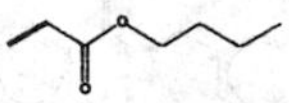

10617

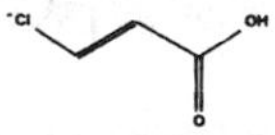

10618

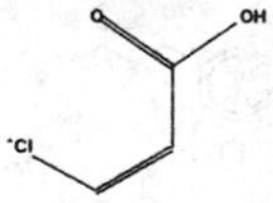

10620

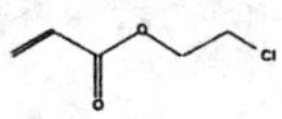

10621

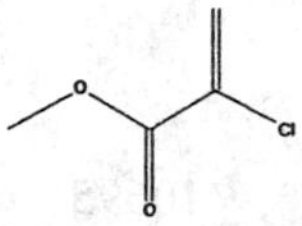

10622

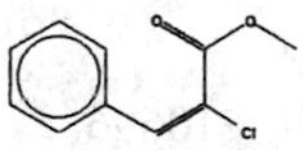

10624

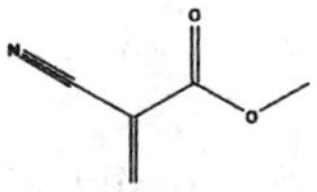

10625

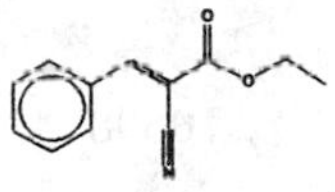

10626

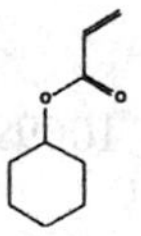

10627

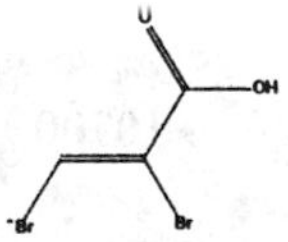

10628

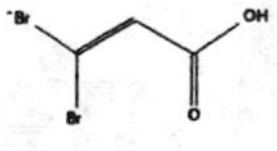

10629

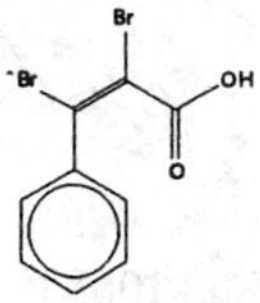

10631

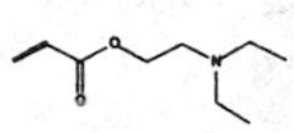

10632

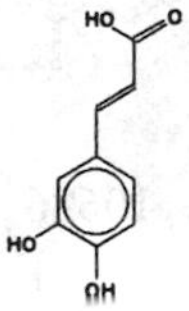

10633

10634

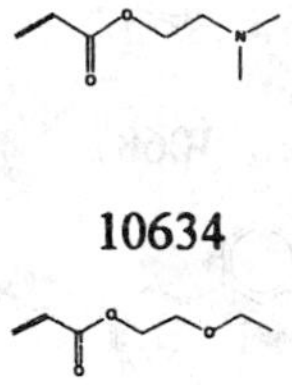

10635

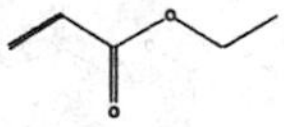

10636

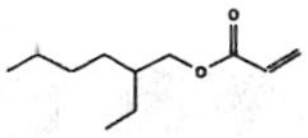

10637

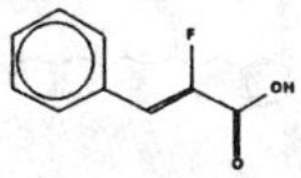

10638

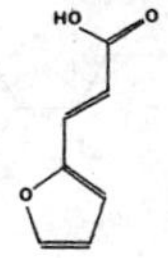

10639

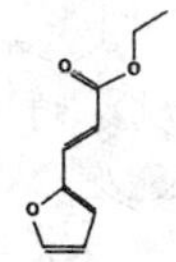

10640

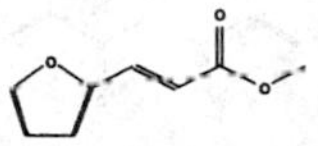

10643

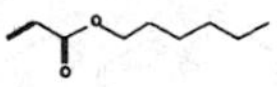

10644

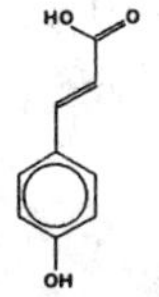

10645

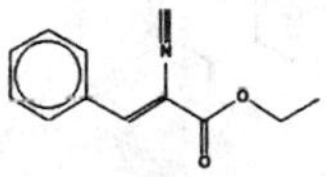

10647

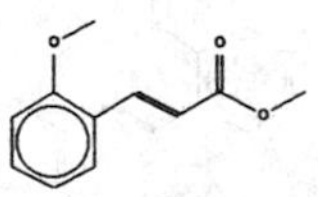

10648

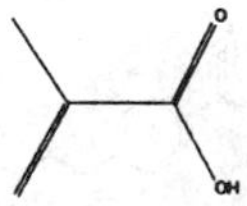

10649

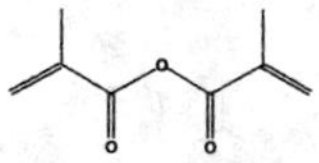

10650

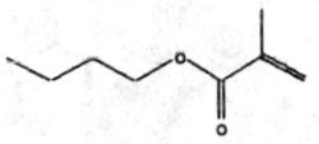

10652

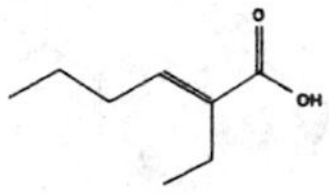

10653

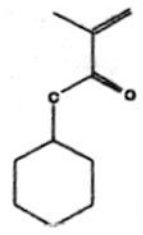

10654

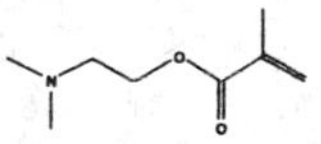

10655

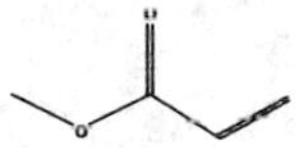

10656

10657

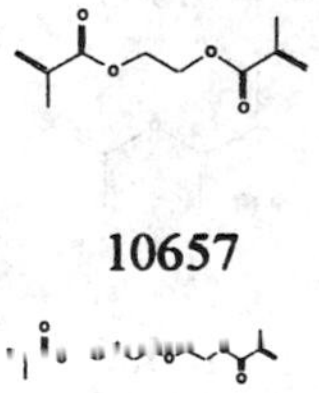

10658

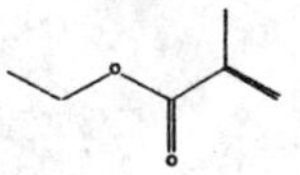

STRUCTURAL FORMULAS OF ORGANIC COMPOUNDS (continued)

In numeric order as they occur in the Table of Physical Constants of Organic Compounds

10659

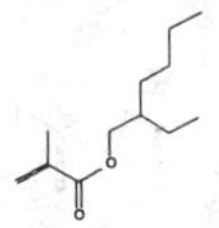

10660

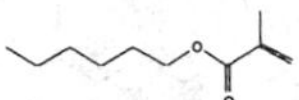

10661

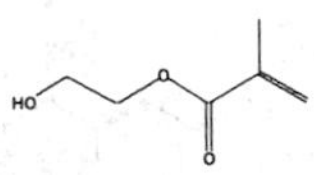

10662

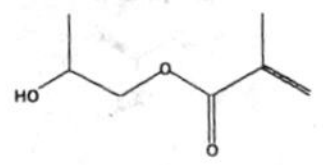

10663

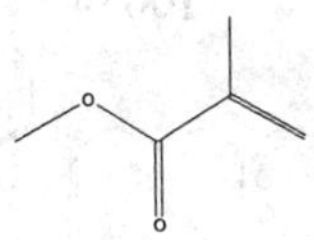

10664

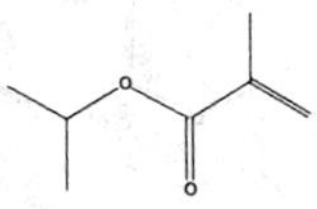

10665

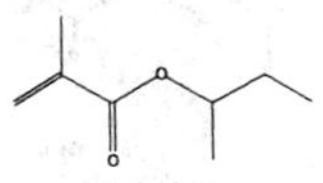

10666

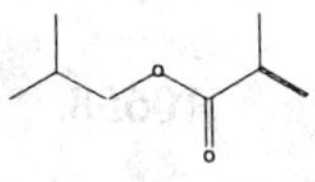

10668

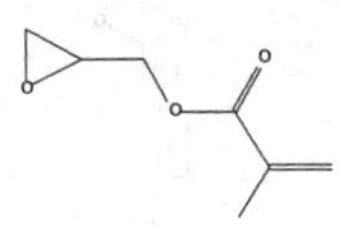

10670

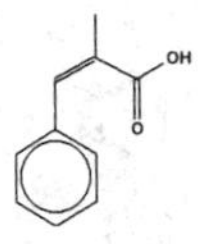

10671

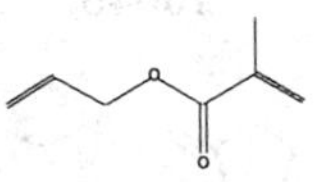

10672

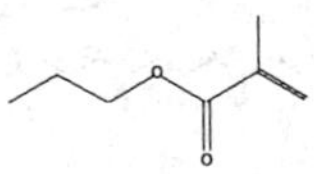

10673

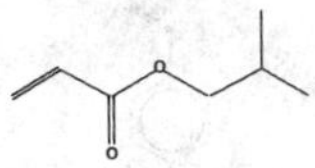

10674

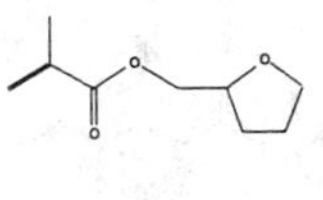

10677

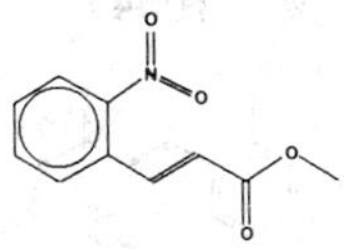

10679

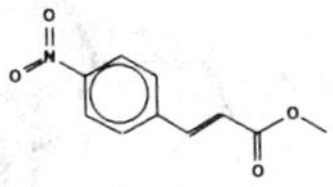

10680

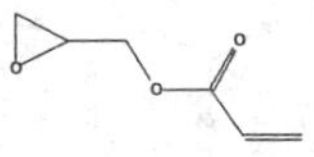

10681

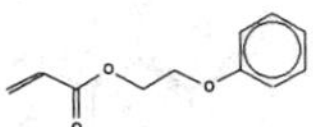

10683

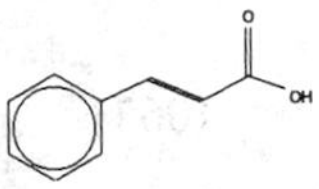

10685

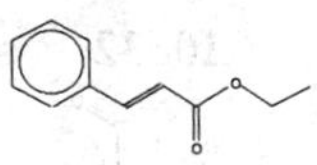

10686

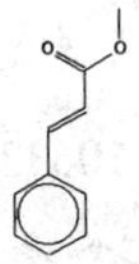

10687

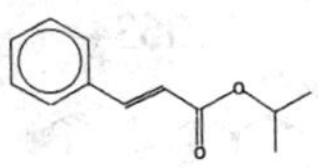

10688

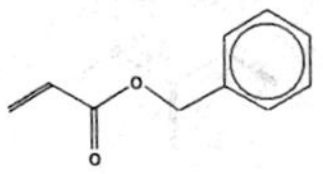

10689

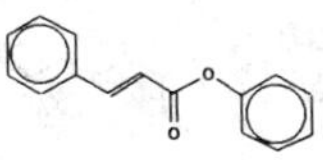

10691

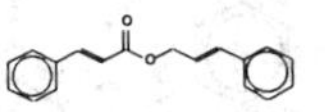

10692

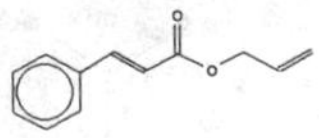

10693

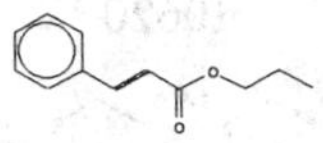

10694

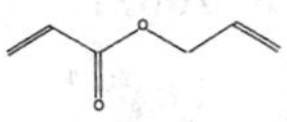

10695

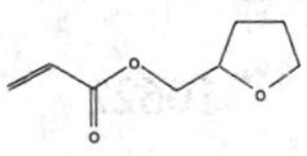

10696

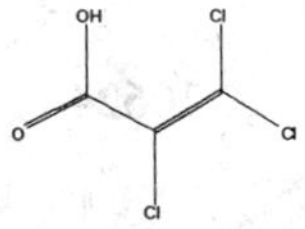

10697

10698

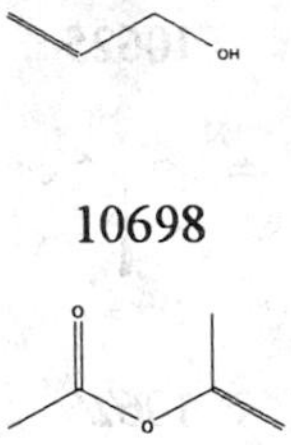

10699

10700

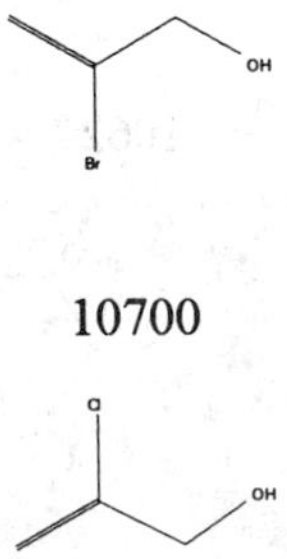

10702

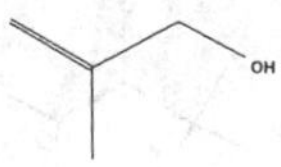

10703

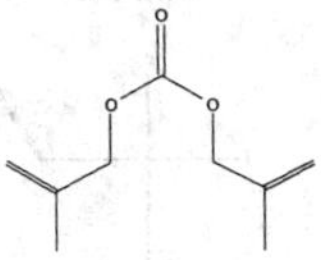

10704

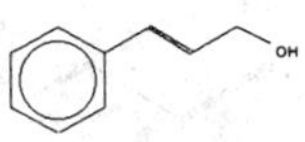

10706

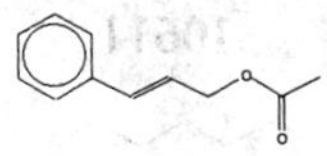

10707

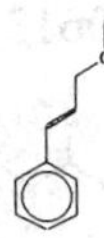

10708

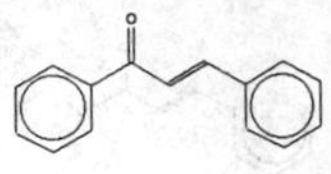

10709

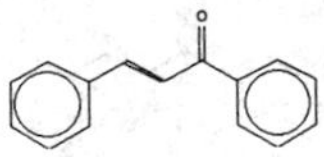

10710

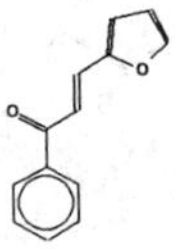

STRUCTURAL FORMULAS OF ORGANIC COMPOUNDS (continued)

In numeric order as they occur in the Table of Physical Constants of Organic Compounds

10711

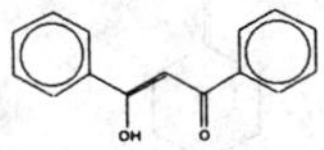

10713

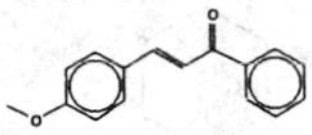

10714

10715

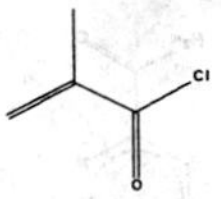

10716

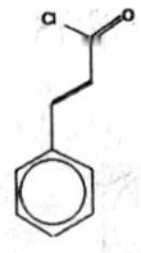

10717

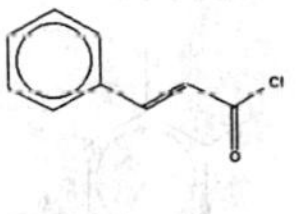

10718

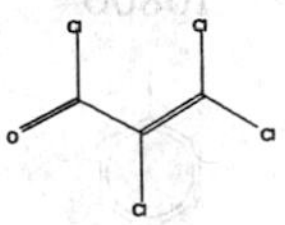

10719

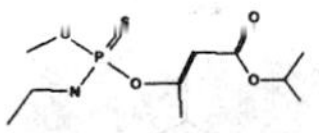

10720

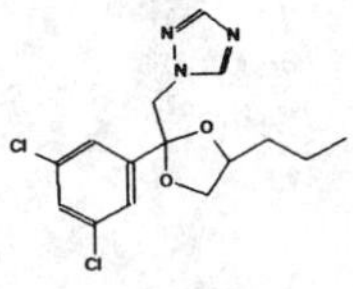

10721

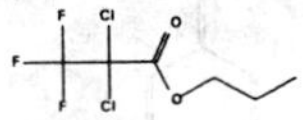

10722

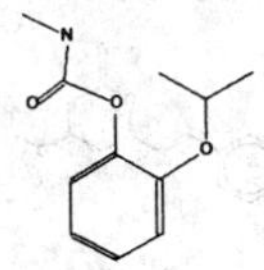

10723

10724

10725

10726

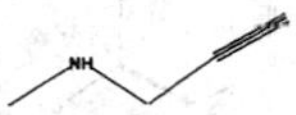

10727

10728

10729

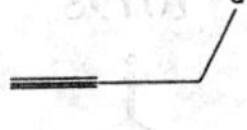

10730

10731

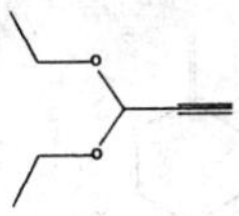

10732

10733

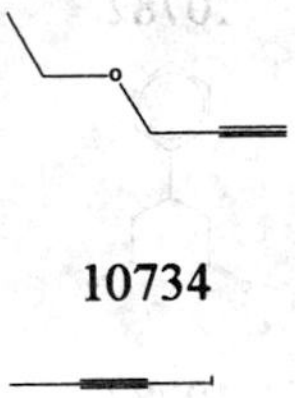

10734

10735

10736

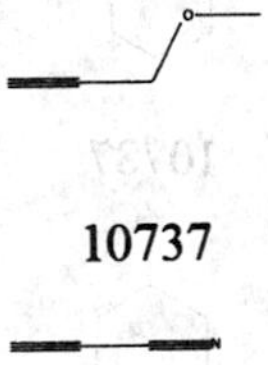

10737

10738

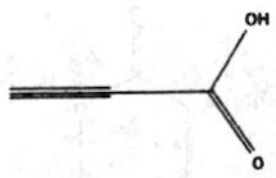

10739

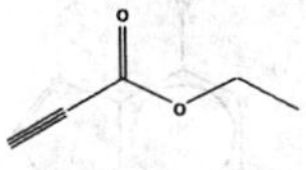

10743

10744

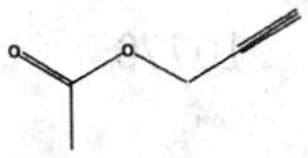

10745

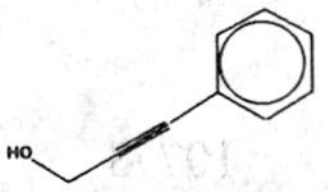

10746

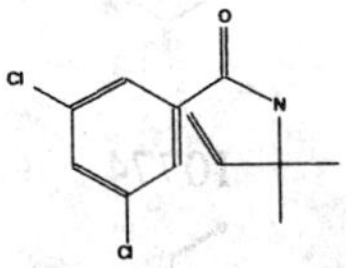

10747

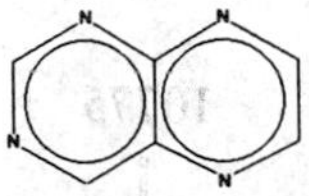

10748

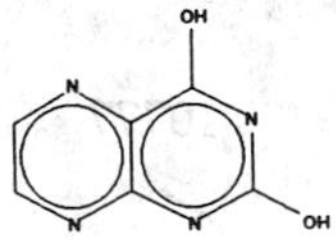

10749

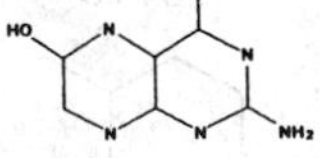

10750

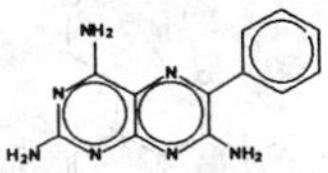

10751

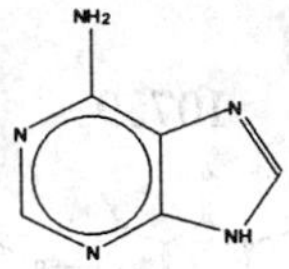

10752

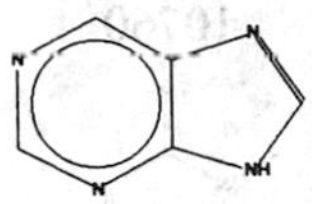

10753

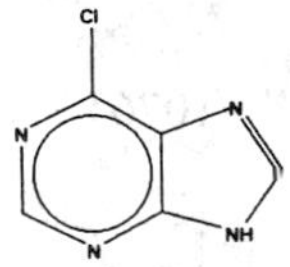

10755

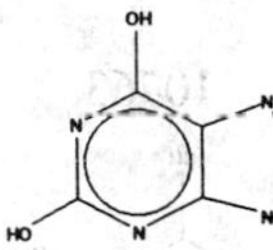

10756

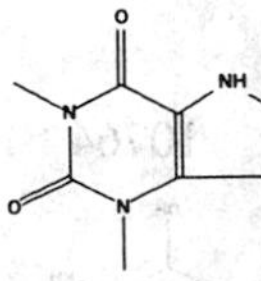

10757

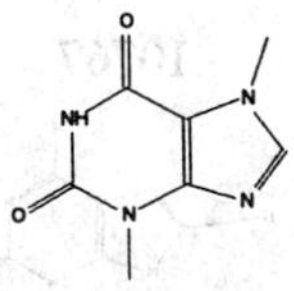

10758

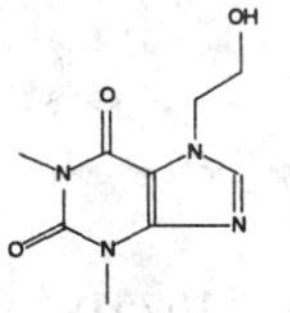

10759

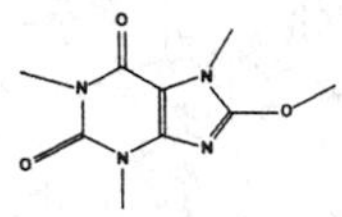

10760

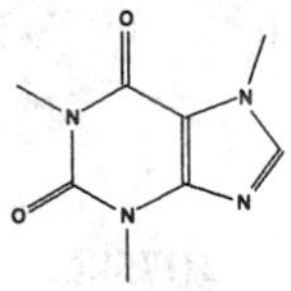

10761

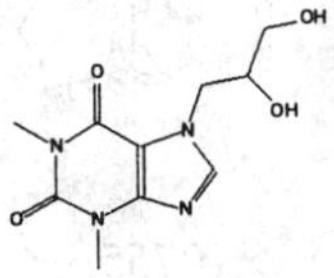

10763

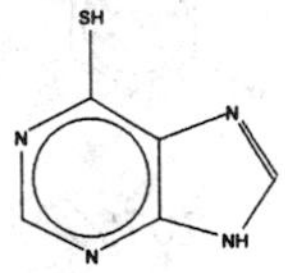

10764

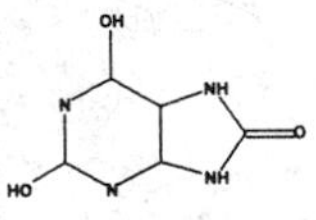

10767

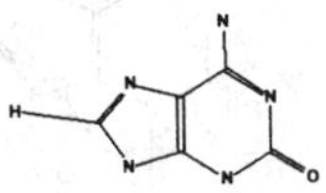

10768

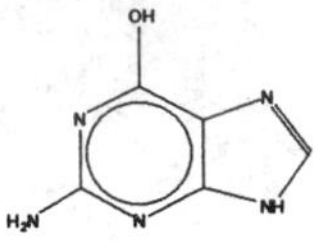

10769

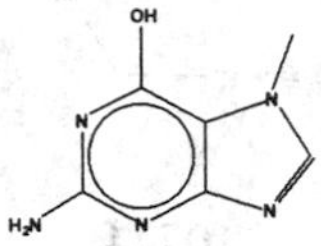

10770

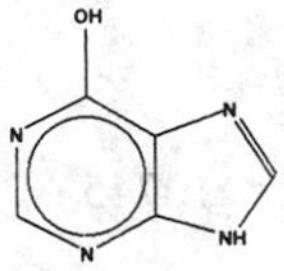

10773

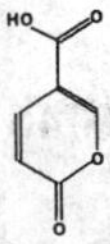

10774

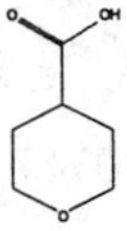

10775

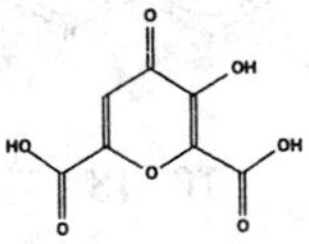

10777

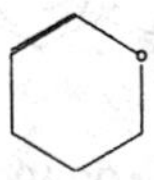

10778

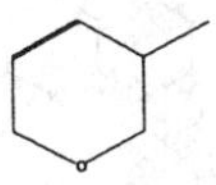

10779

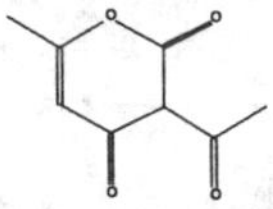

10780

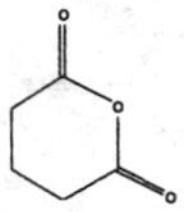

10781

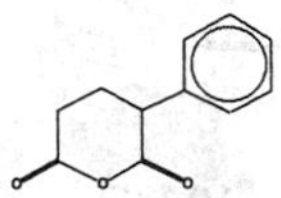

10782

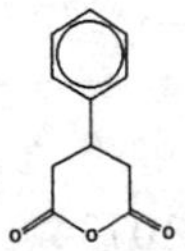

10783

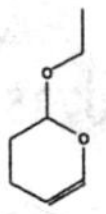

10784

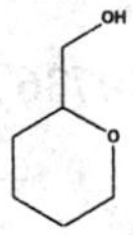

10787

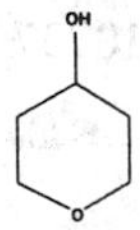

10788

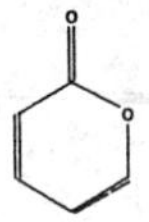

10789

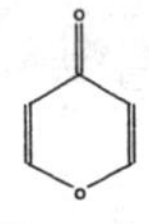

10790

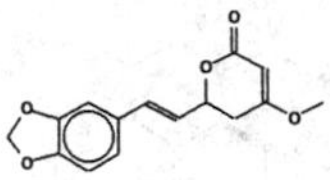

10791

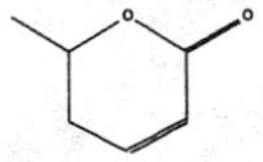

10792

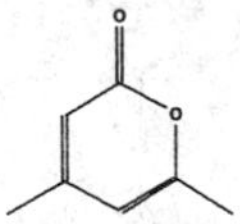

10793

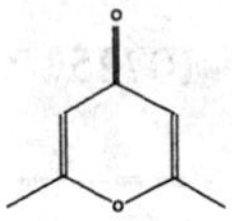

10794

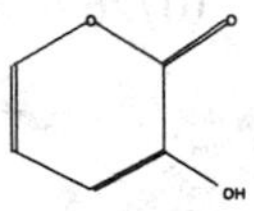

10795

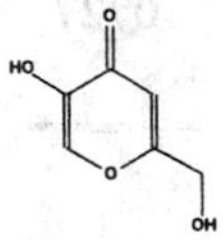

10796

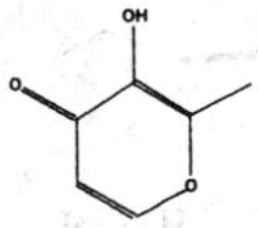

10798

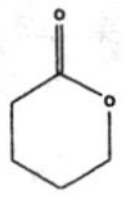

10799

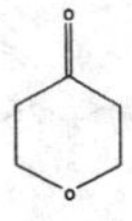

10800

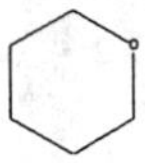

10801

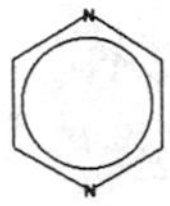

10802

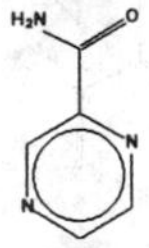

10803

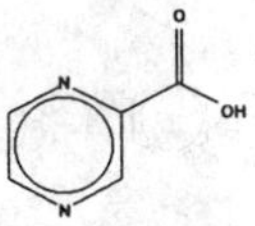

10805

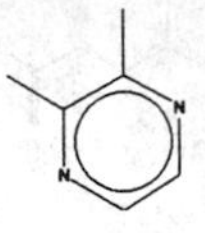

10806

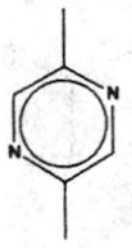

10807

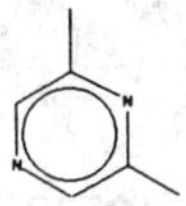

10809

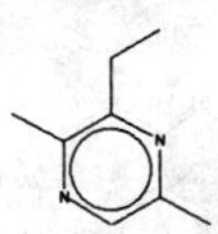

10810

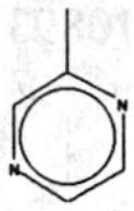

10811

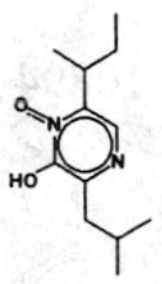

10813

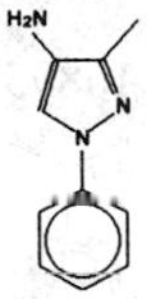

10814

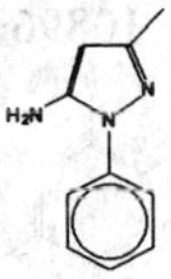

10815

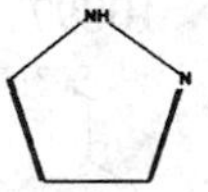

10816

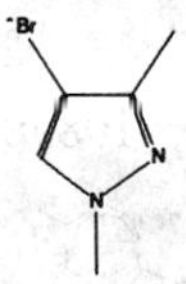

10818

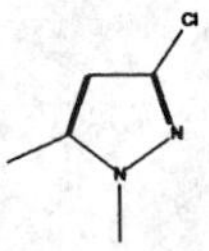

10819

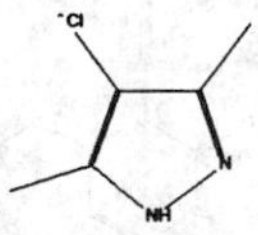

10820

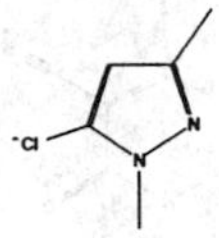

10821

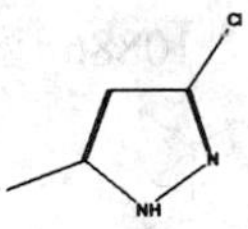

10822

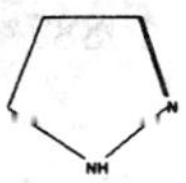

10823

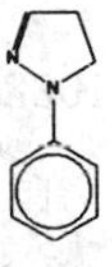

10824

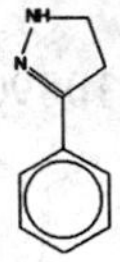

10825

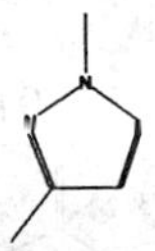

10826

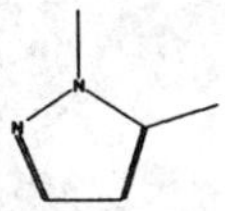

10827

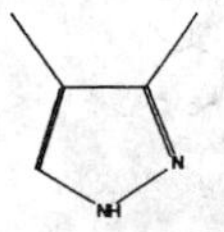

10828

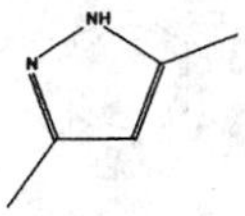

10829

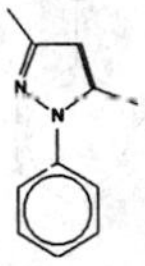

10830

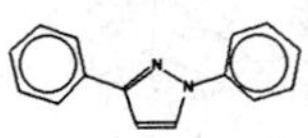

10831

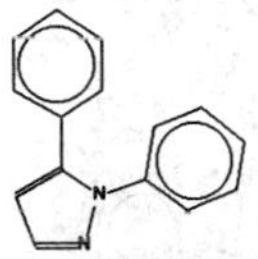

10832

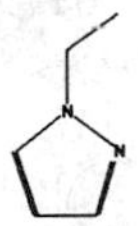

10833

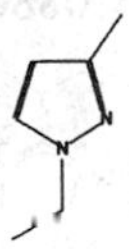

10834

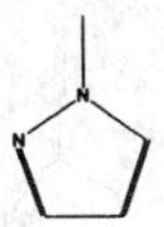

10835

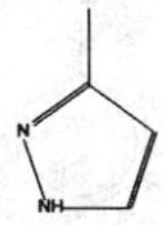

10836

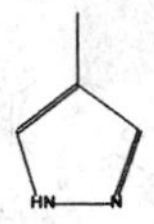

10837

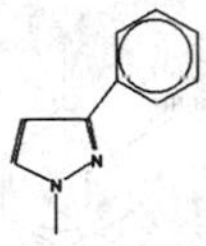

10838

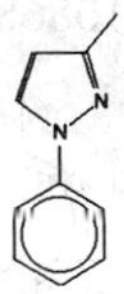

10839

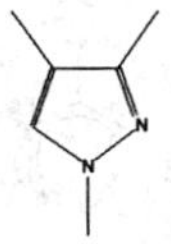

10840

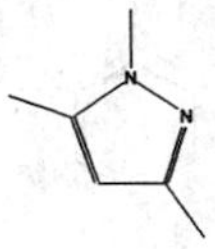

10841

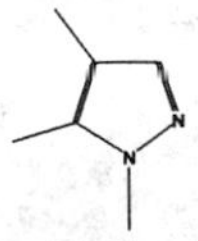

10842

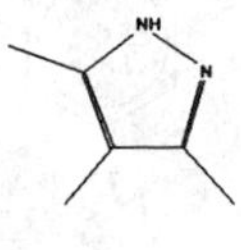

10844

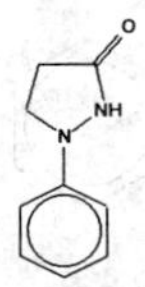

10845

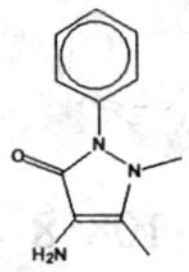

10846

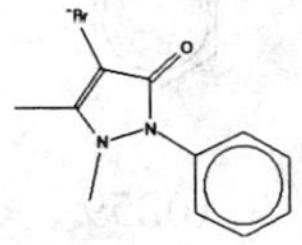

10847

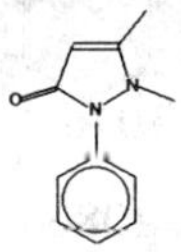

10849

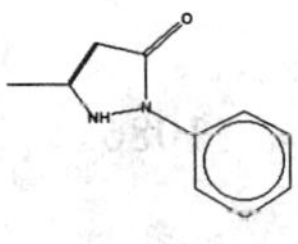

10851

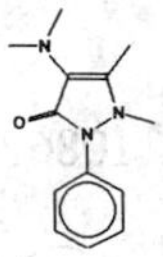

10852

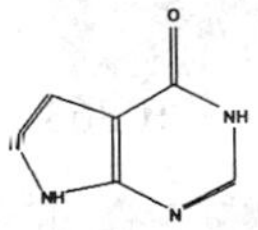

10853

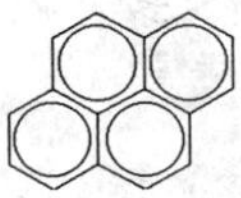

In numeric order as they occur in the Table of Physical Constants of Organic Compounds

10855

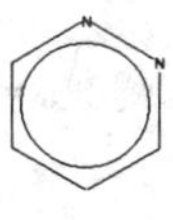

10856

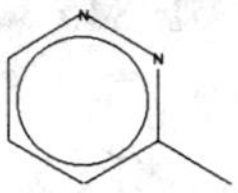

10858

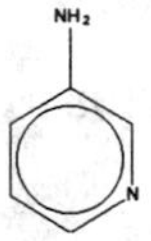

10859

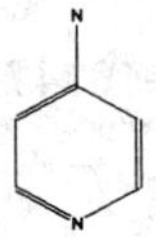

10860

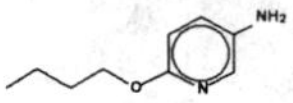

10861

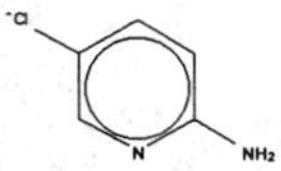

10862

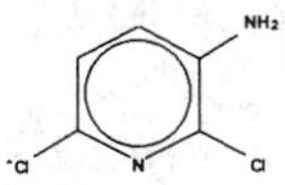

10863

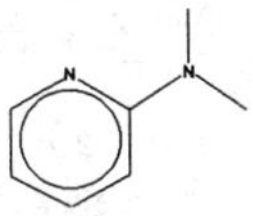

10864

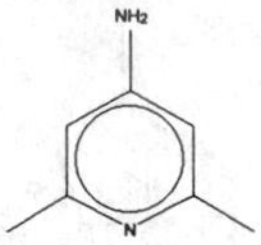

10865

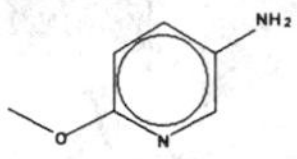

10866

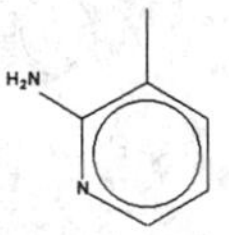

10868

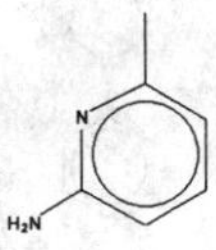

10869

10870

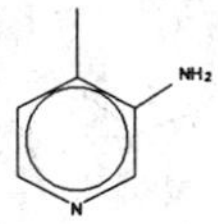

10871

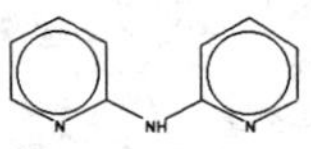

10872

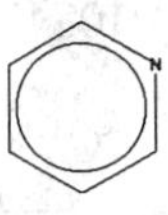

10873

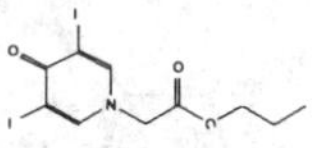

10875

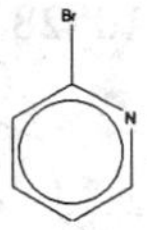

10876

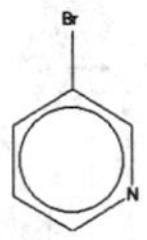

10877

10878

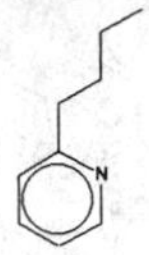

10879

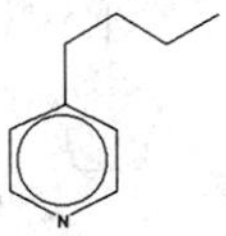

10880

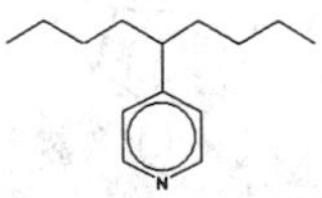

10882

10883

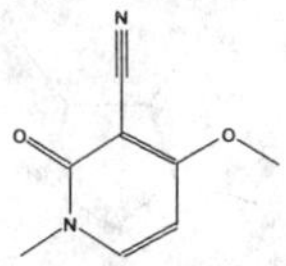

10884

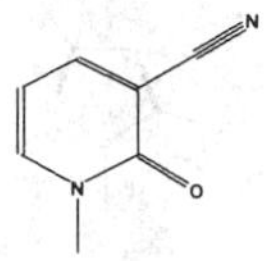

10886

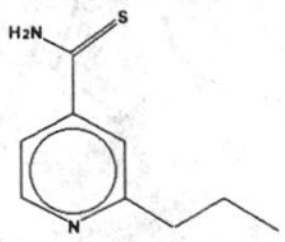

10887

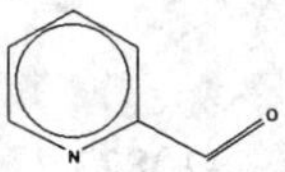

10888

10890

10891

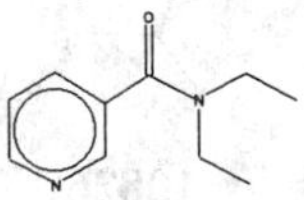

10892

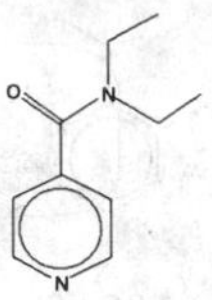

10893

10894

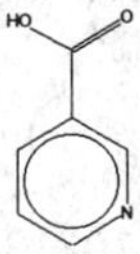

10895

10896

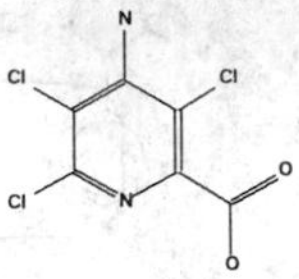

10897

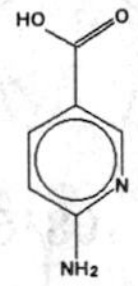

10898

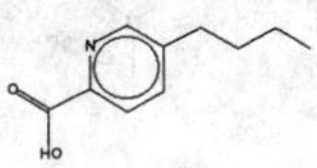

In numeric order as they occur in the Table of Physical Constants of Organic Compounds

10899

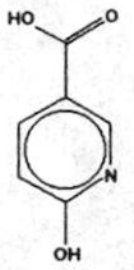

HO O N OH

10901

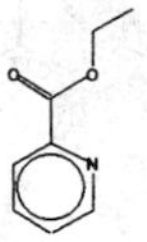

O O N

10903

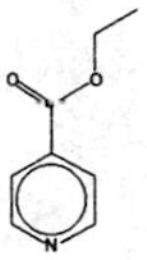

O O N

10905

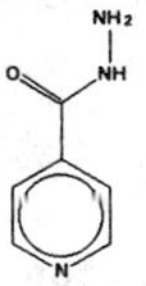

NH2 O NH N

10906

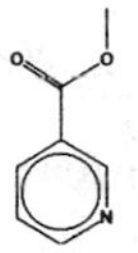

O O N

10907

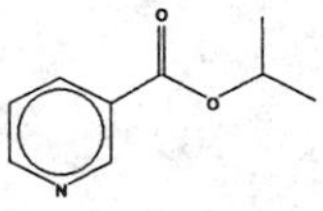

O O N

10908

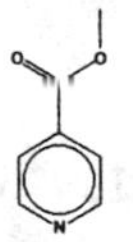

O O N

10909

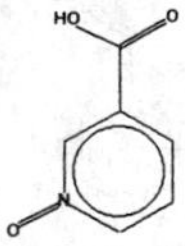

HO O N O

10911

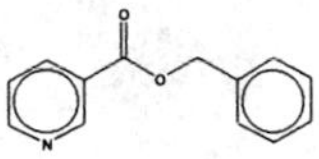

O O N

10914

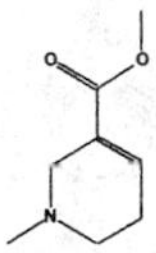

O O N

10915

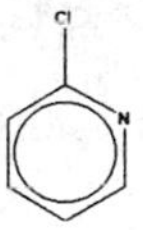

Cl N

10916

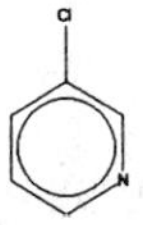

Cl N

10917

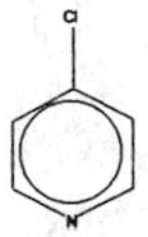

Cl N

10918

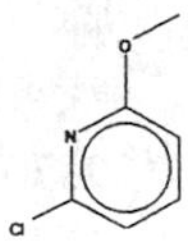

O N Cl

10920

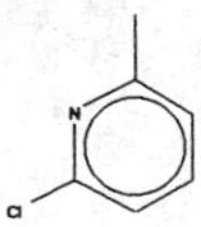

N Cl

10921

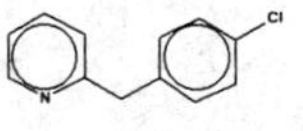

N Cl

10922

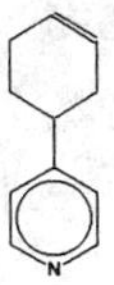

N

10923

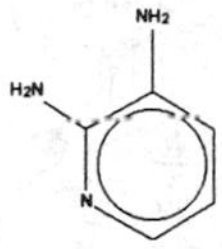

NH2 H2N N

10924

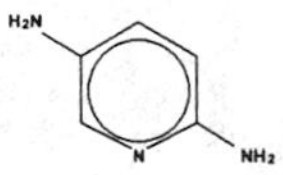

H2N N NH2

10925

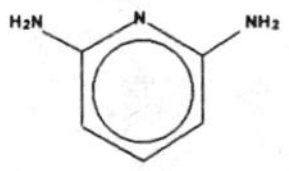

H2N N NH2

10926

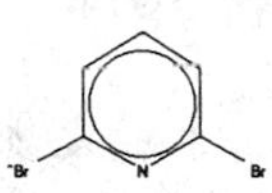

Br N Br

10927

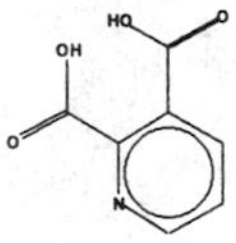

HO O OH O N

10929

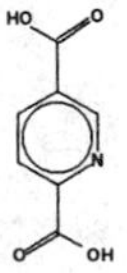

HO O N O OH

10932

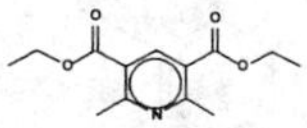

O O O O N

10933

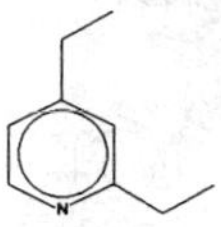

N

10935

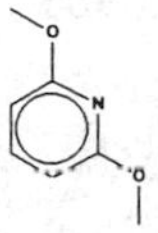

O N O

10936

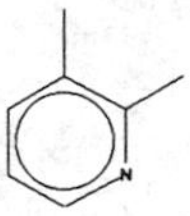

N

10937

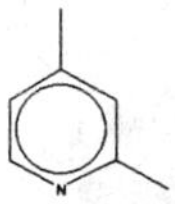

N

10938

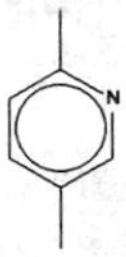

N

10939

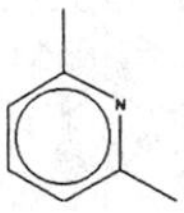

N

10940

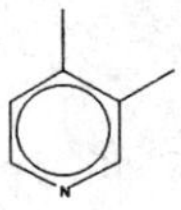

N

10941

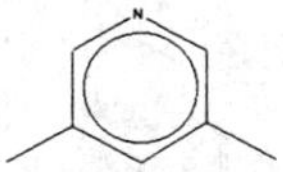

N

10942

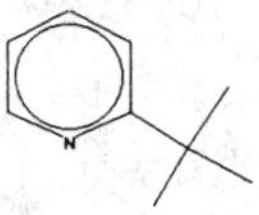

N

10943

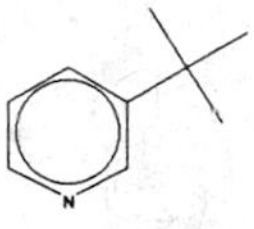

N

10944

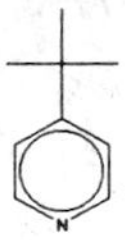

N

10945

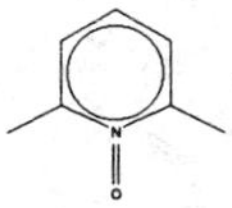

N O

10946

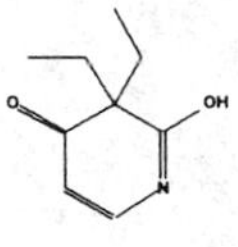

O OH N

10947

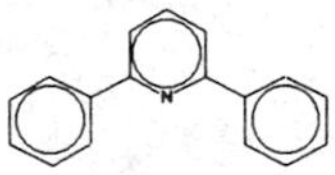

N

10948

NH2 N

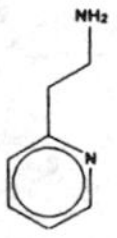

10949

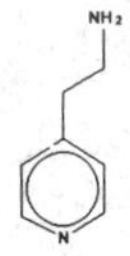

NH_2

N

10951

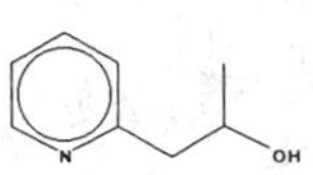

N OH

10952

N

10953

N

10954

N

10955

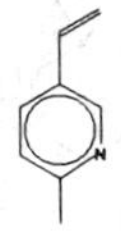

N

10956

N

10958

N

10959

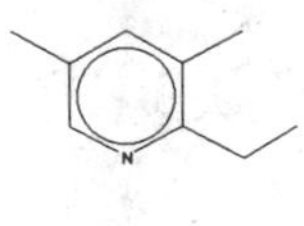

N

10960

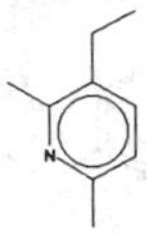

N

10961

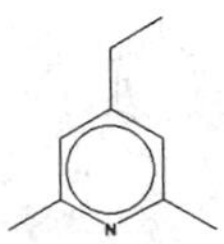

N

10962

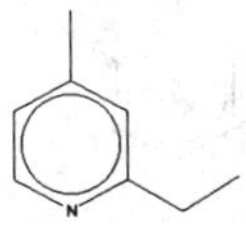

N

10963

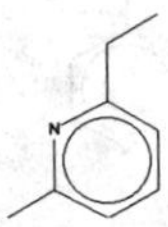

N

10965

N

10966

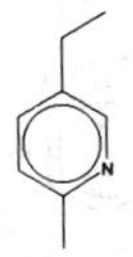

N

10967

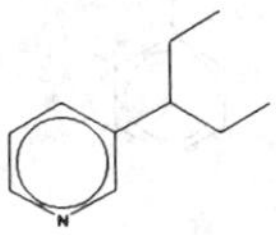

N

10968

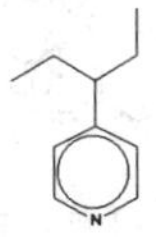

N

10969

N

10970

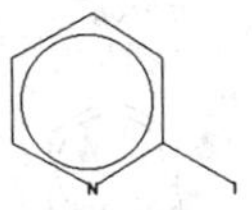

N

10972

NH_2

N

10973

NH_2

N

10974

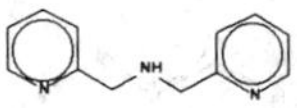

N NH N

10975

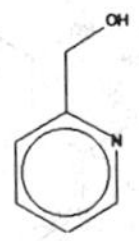

OH

N

10976

OH

N

10977

OH

N

10980

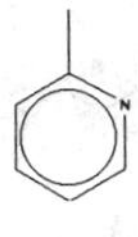

N

10981

N

10982

N

10983

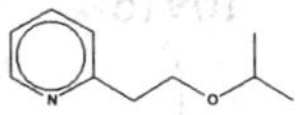

N O

10984

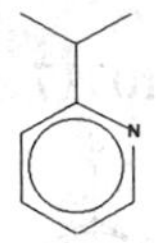

N

10985

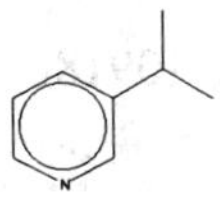

N

10987

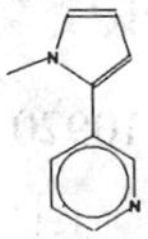

N

N

10988

O

N

10989

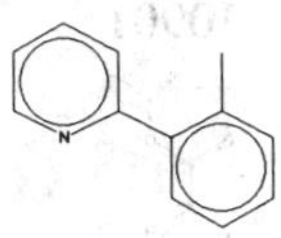

N

10990

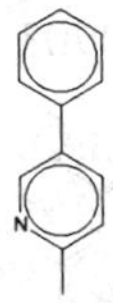

N

10991

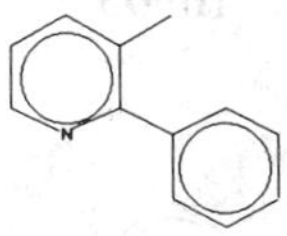

N

10992

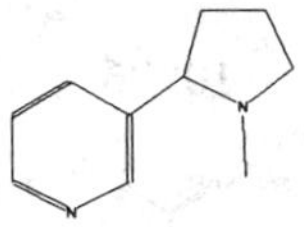

N

N

10994

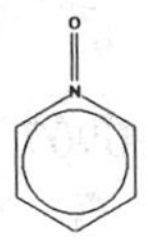

O

N

10995

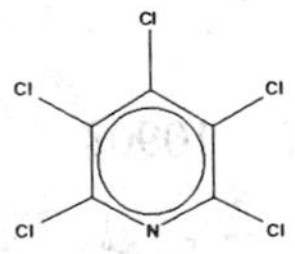

Cl

Cl Cl

Cl N Cl

10996

N

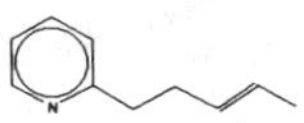

STRUCTURAL FORMULAS OF ORGANIC COMPOUNDS (continued)

In numeric order as they occur in the Table of Physical Constants of Organic Compounds

10997

10998

10999

11001

11002

11003

11004

11005

11006

11007

11008

11009

11010

11012

11013

11014

11016

11017

11018

11019

11020

11021

11023

11024

11025

11026

11027

11028

11029

11031

11032

11033

11034

11035

11036

11038

11039

11040

11041

11042

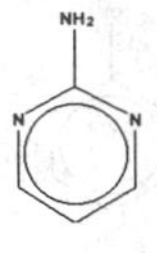

11043

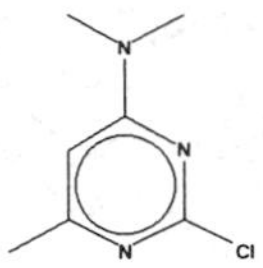

11045

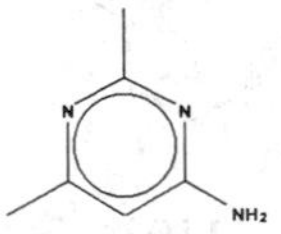

11049

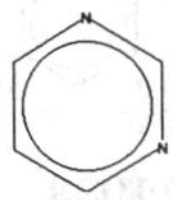

11050

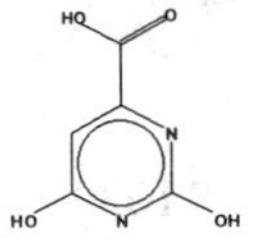

11051

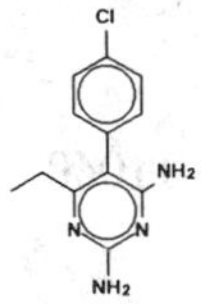

11052

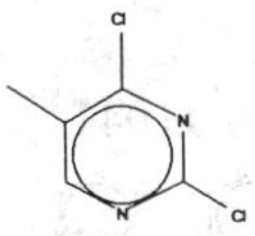

11054

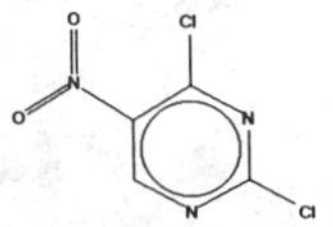

11055

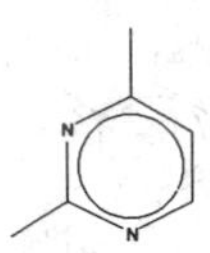

11056

11057

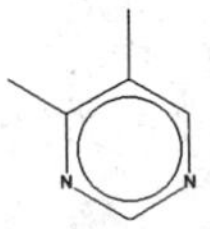

11058

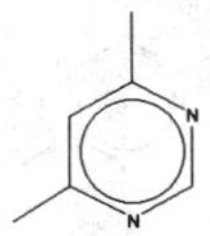

11059

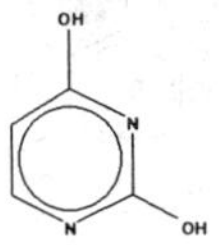

11062

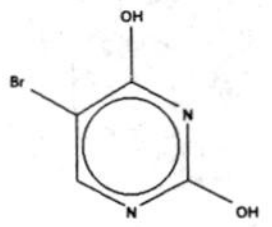

11064

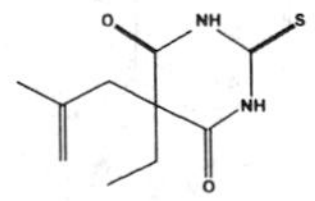

11066

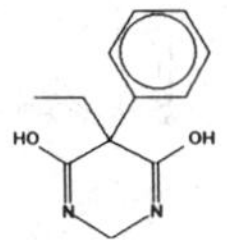

11067

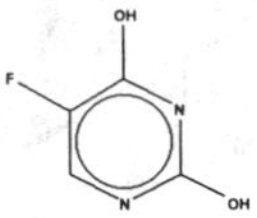

11068

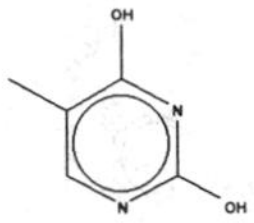

11069

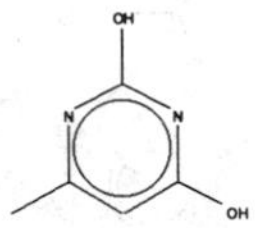

11071

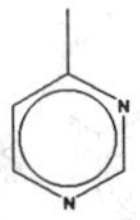

11072

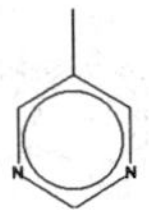

11075

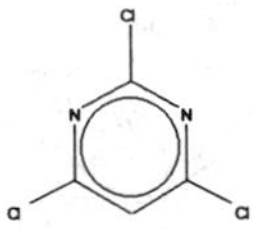

11076

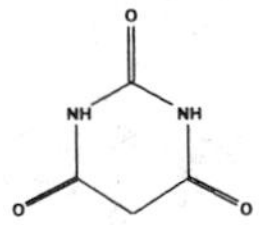

11078

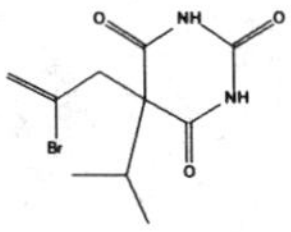

11080

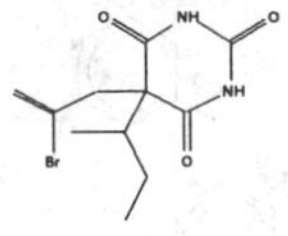

11081

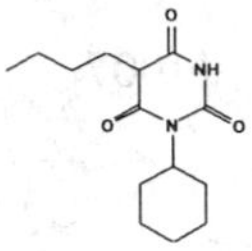

11082

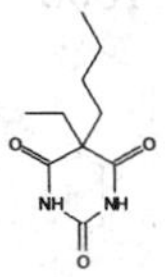

11083

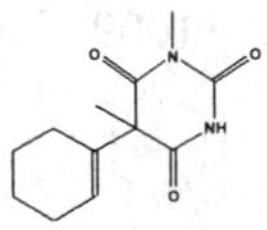

11084

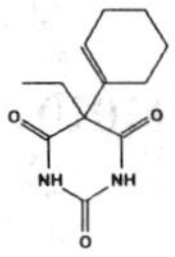

11085

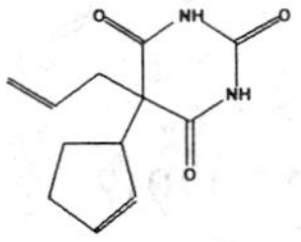

11087

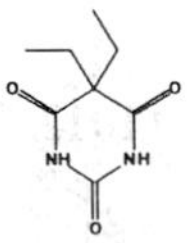

11088

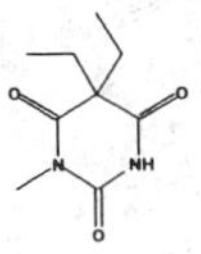

11090

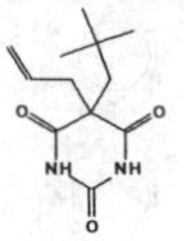

11091

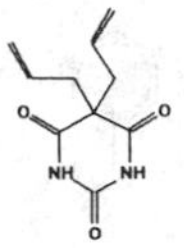

11092

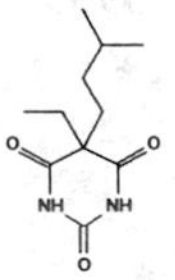

11093

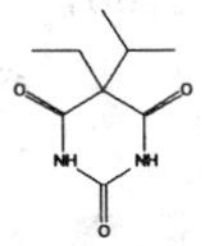

11094

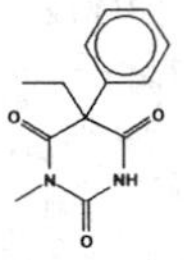

11095

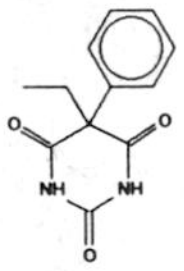

11098

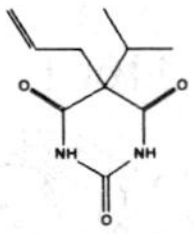

11099

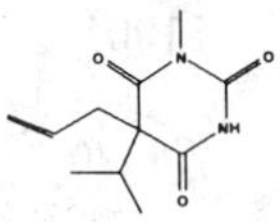

STRUCTURAL FORMULAS OF ORGANIC COMPOUNDS (continued)

In numeric order as they occur in the Table of Physical Constants of Organic Compounds

11100

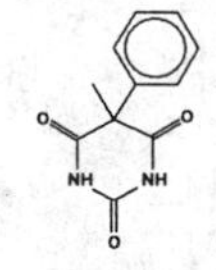

11101

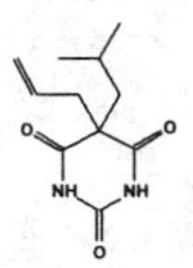

11103

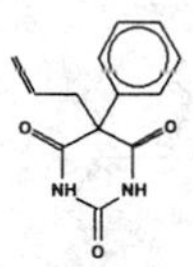

11104

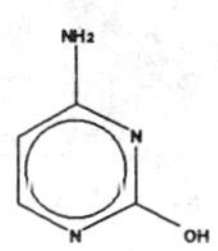

11105

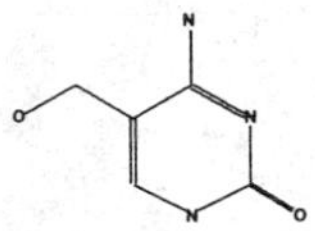

11106

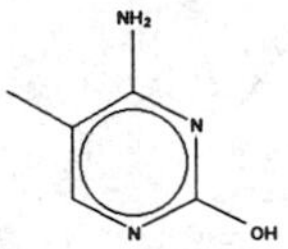

11109

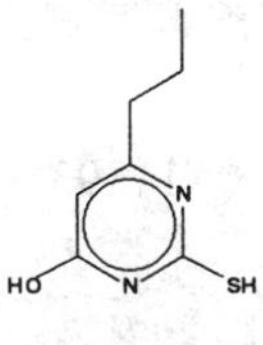

11110

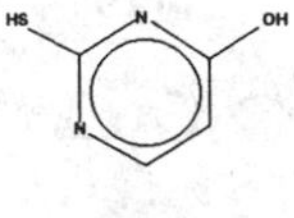

11112

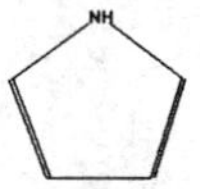

11113

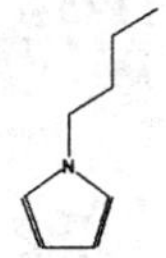

11114

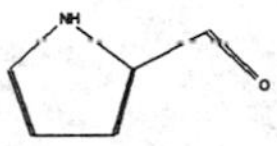

11115

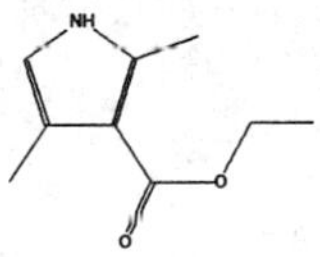

11116

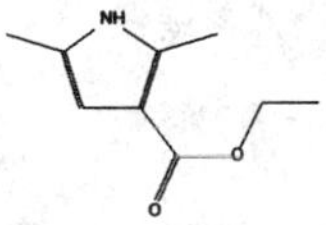

11117

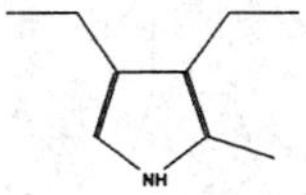

11118

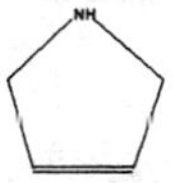

11119

11120

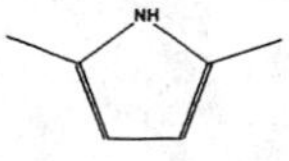

11121

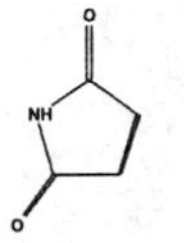

11122

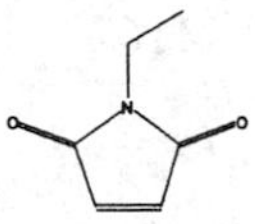

11124

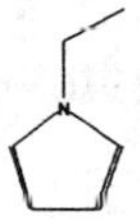

11125

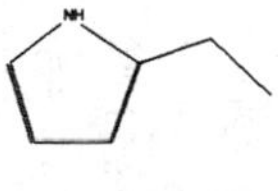

11126

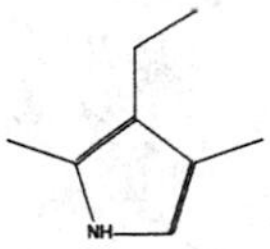

11127

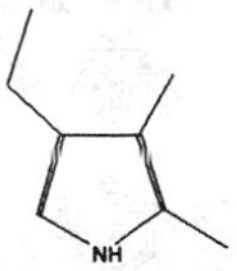

11128

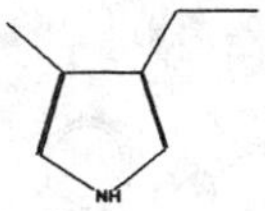

11129

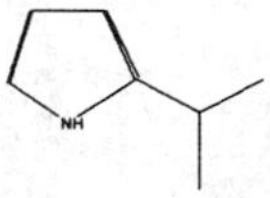

11130

11131

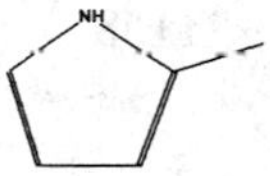

11132

11133

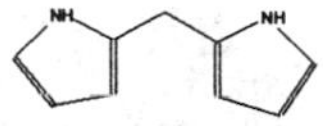

11134

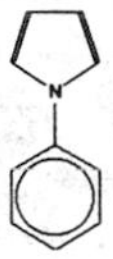

11135

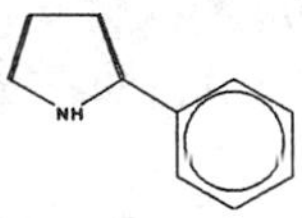

11136

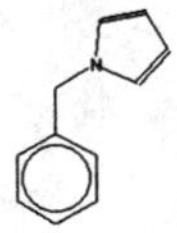

11137

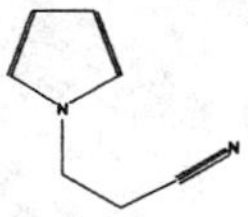

11138

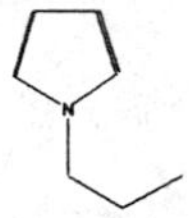

11139

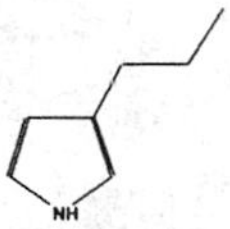

11141

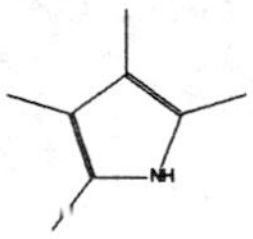

11142

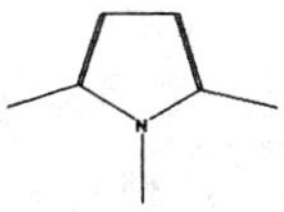

11143

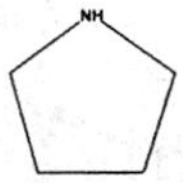

11146

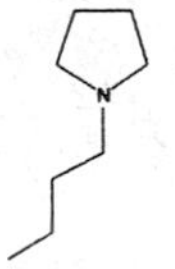

11147

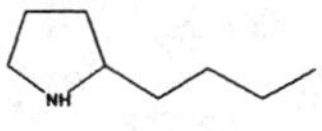

STRUCTURAL FORMULAS OF ORGANIC COMPOUNDS (continued)

In numeric order as they occur in the Table of Physical Constants of Organic Compounds

11148

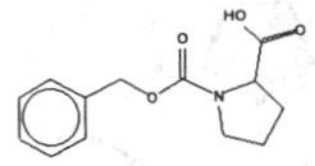

11149

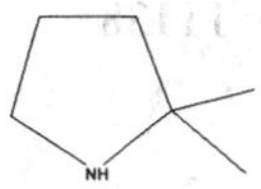

11150

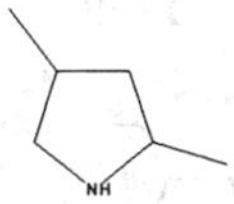

11151

11152

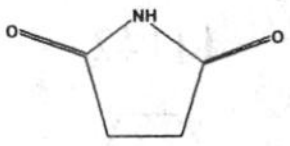

11153

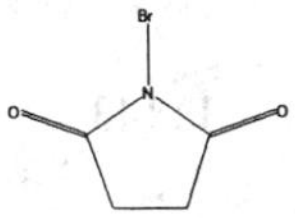

11154

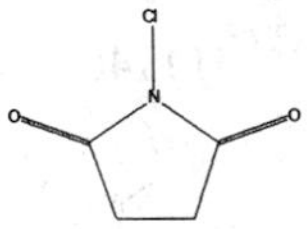

11155

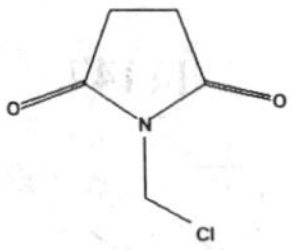

11156

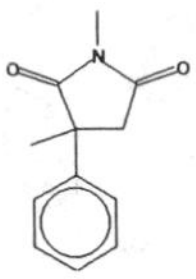

11157

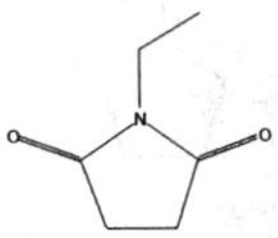

11158

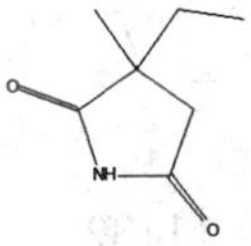

11160

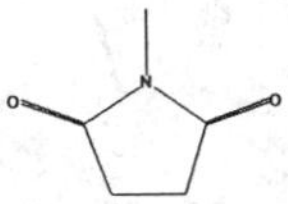

11161

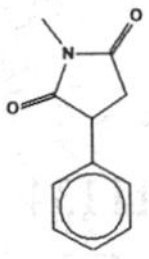

11163

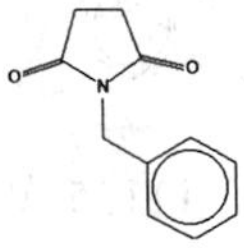

11164

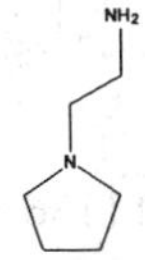

11165

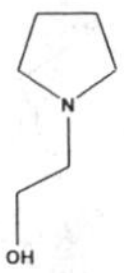

11166

11167

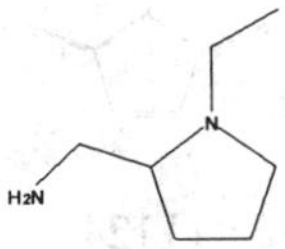

11168

11169

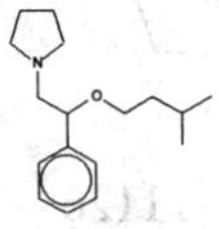

11170

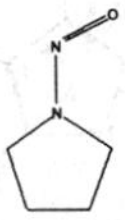

11171

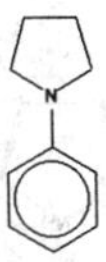

11172

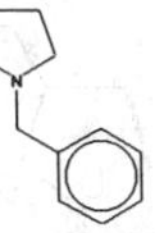

11173

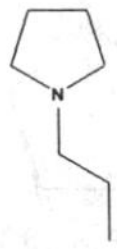

11175

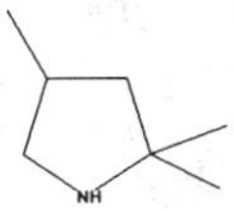

11176

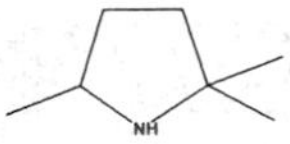

11180

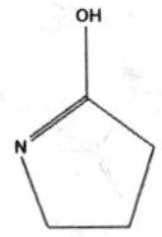

11181

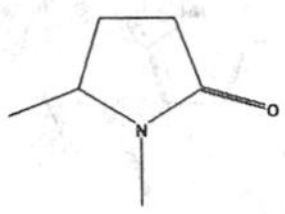

11182

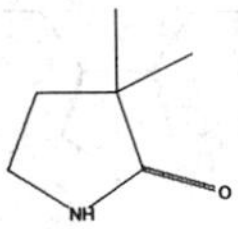

11183

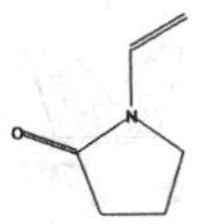

11184

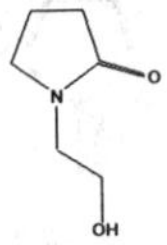

11185

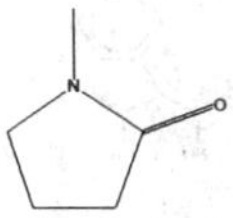

11186

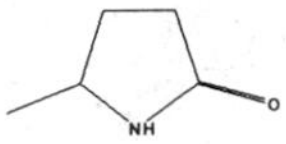

11187

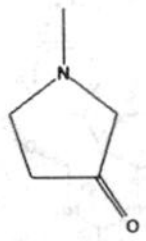

11188

11189

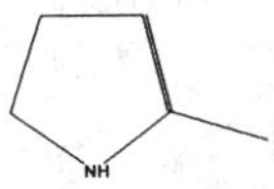

11190

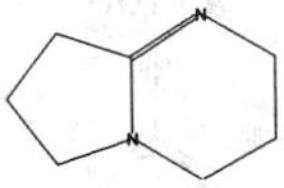

11191

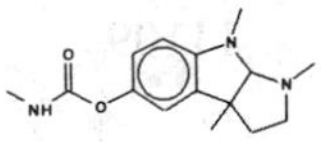

11193

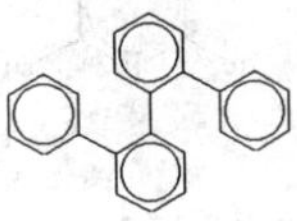

11194

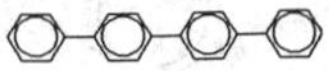

STRUCTURAL FORMULAS OF ORGANIC COMPOUNDS (continued)

In numeric order as they occur in the Table of Physical Constants of Organic Compounds

11195

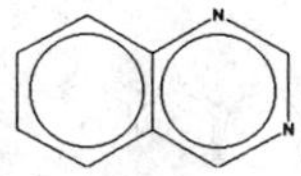

11199

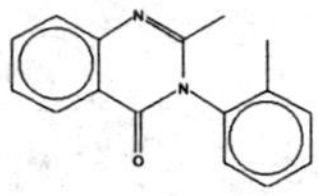

11201

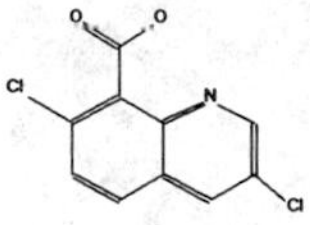

11202

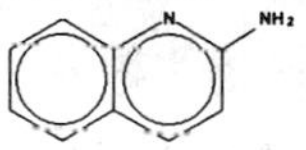

11203

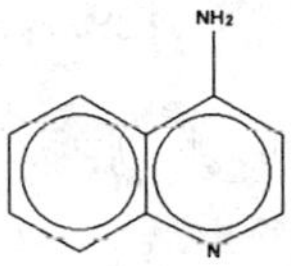

11204

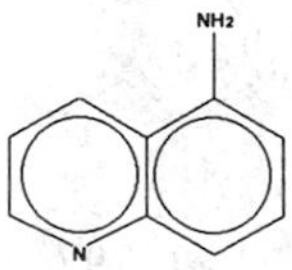

11206

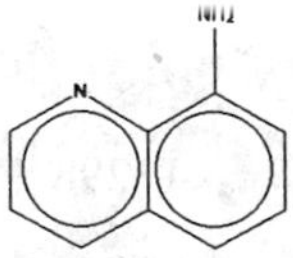

11208

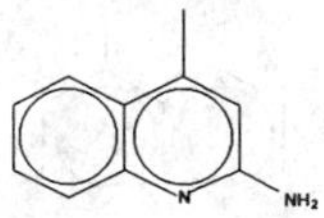

11209

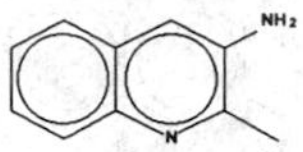

11210

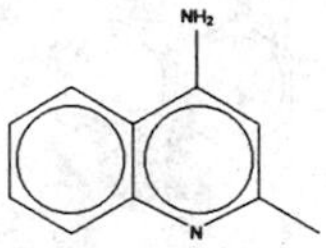

11211

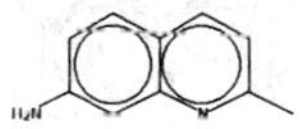

11213

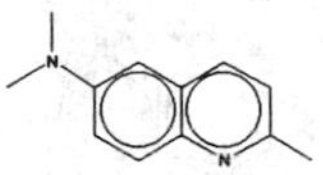

11214

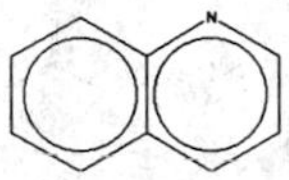

11217

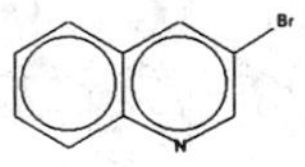

11218

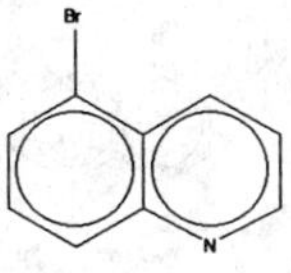

11219

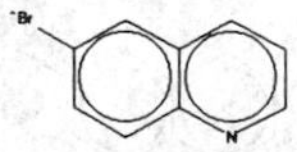

11220

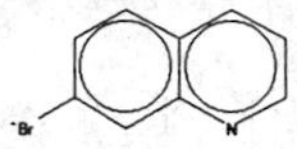

11222

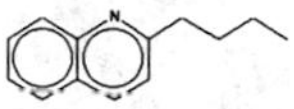

11223

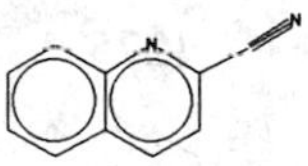

11224

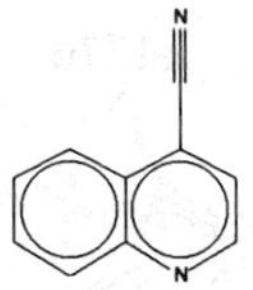

11225

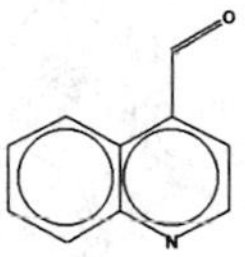

11227

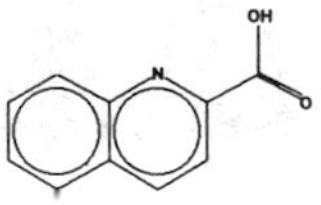

11230

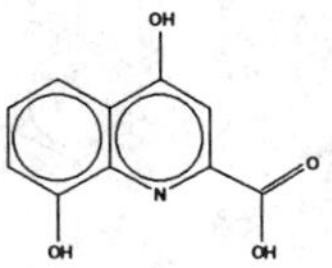

11231

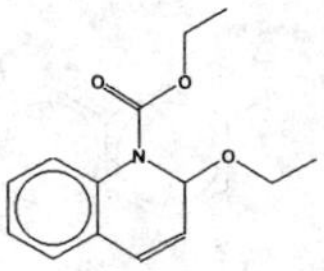

11232

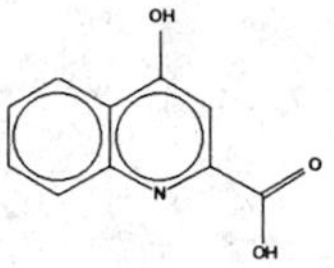

11236

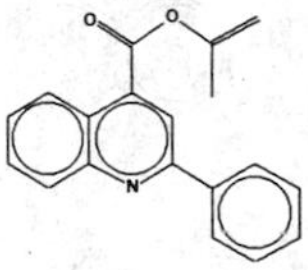

11237

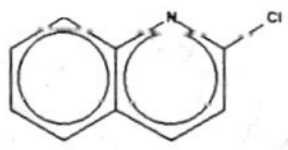

11238

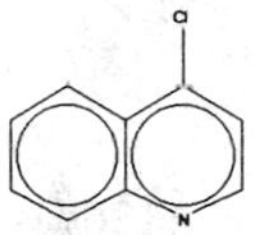

11239

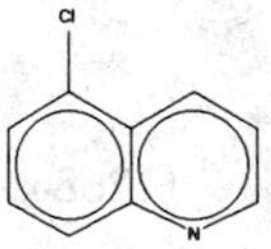

11240

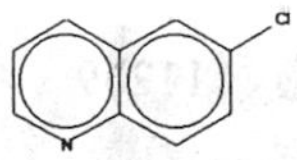

11241

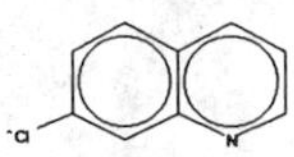

11242

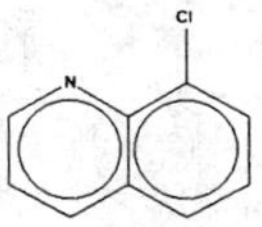

11243

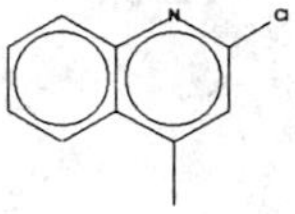

11244

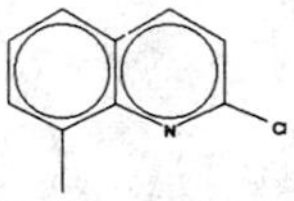

11245

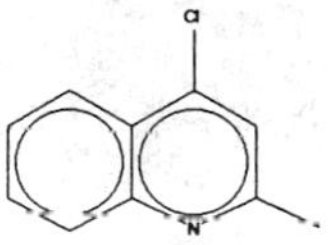

11246

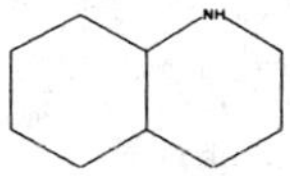

11247

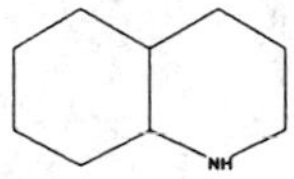

11248

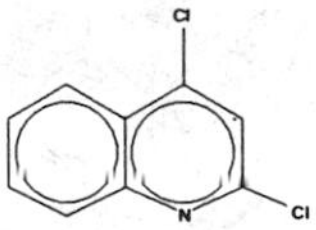

11250

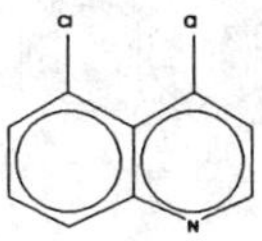

STRUCTURAL FORMULAS OF ORGANIC COMPOUNDS (continued)

In numeric order as they occur in the Table of Physical Constants of Organic Compounds

11251

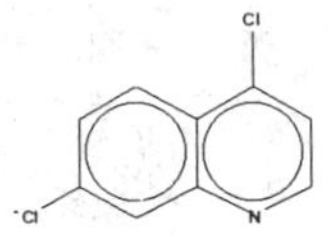

11254

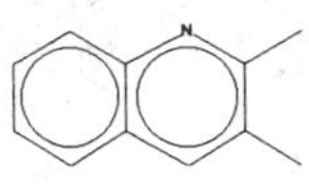

11255

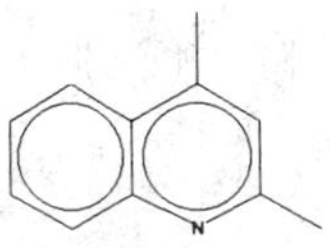

11257

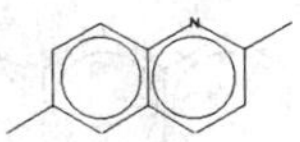

11258

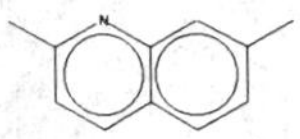

11259

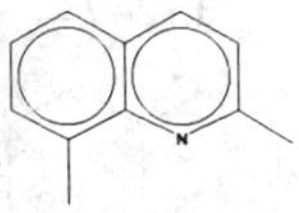

11260

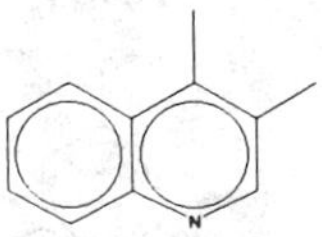

11261

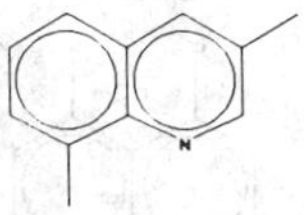

11262

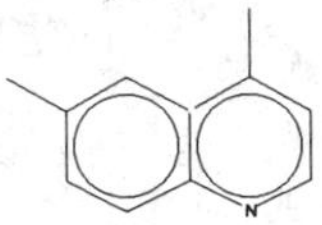

11263

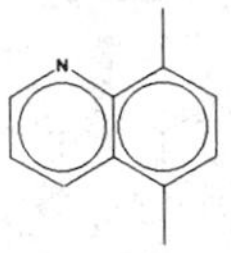

11264

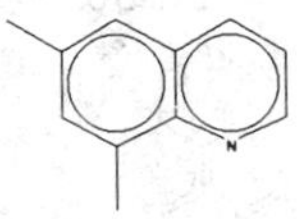

11265

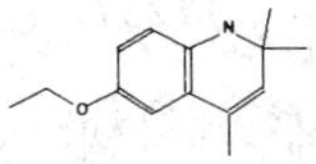

11266

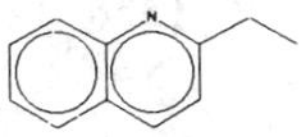

11267

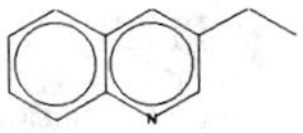

11268

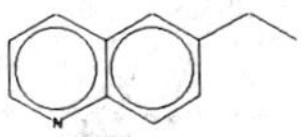

11269

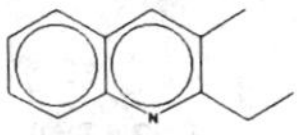

11270

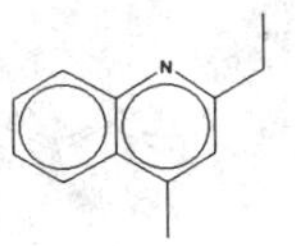

11271

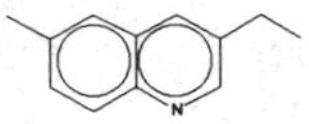

11274

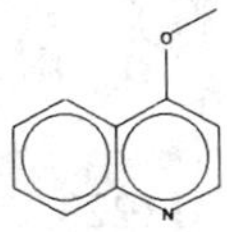

11275

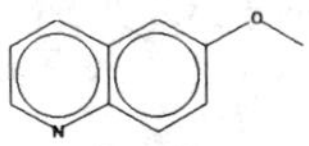

11276

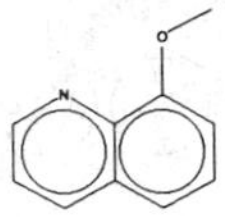

11277

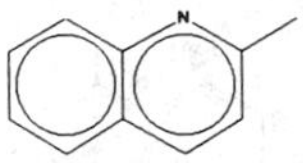

11278

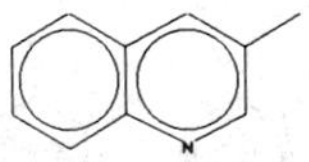

11279

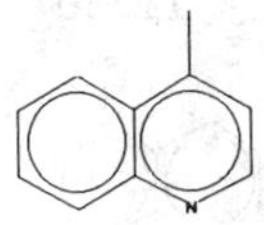

11280

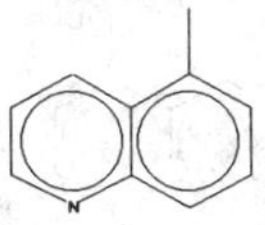

11281

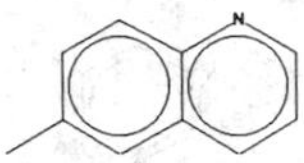

11282

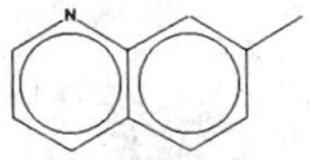

11283

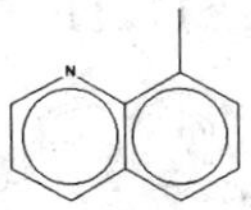

11284

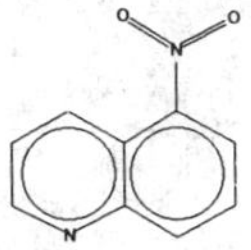

11285

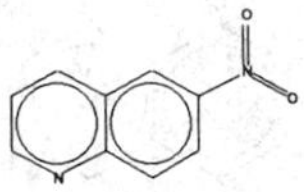

11286

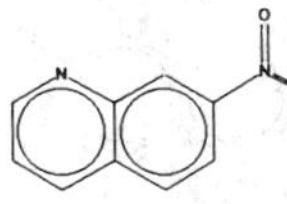

11287

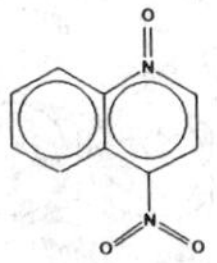

11288

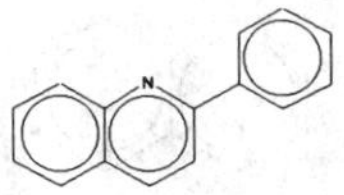

11289

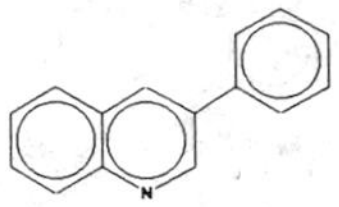

11290

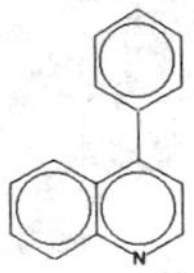

11291

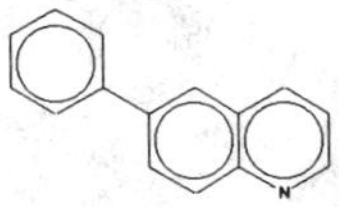

11292

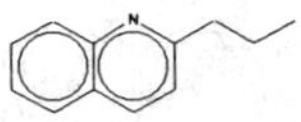

11296

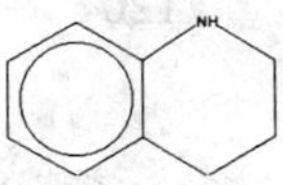

11297

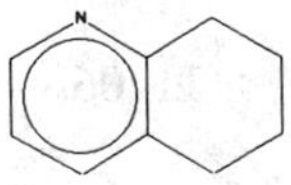

11298

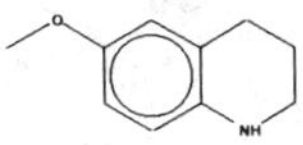

STRUCTURAL FORMULAS OF ORGANIC COMPOUNDS (continued)

In numeric order as they occur in the Table of Physical Constants of Organic Compounds

11299

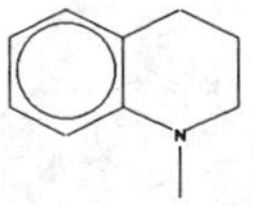

11300

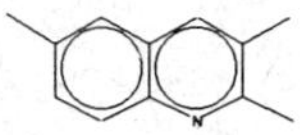

11301

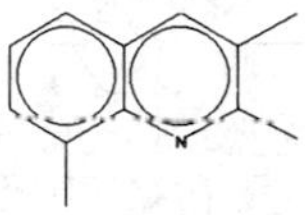

11302

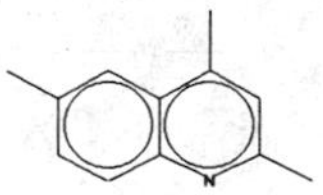

11303

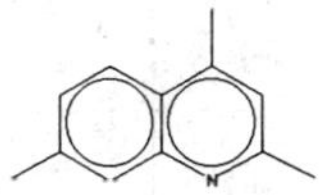

11304

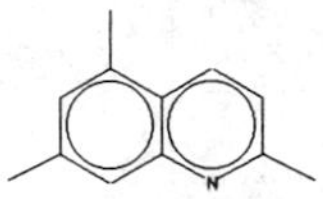

11305

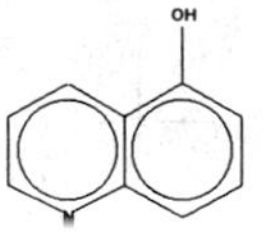

11306

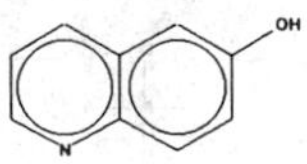

11307

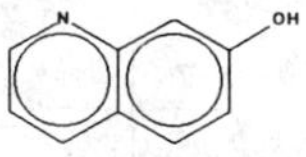

11308

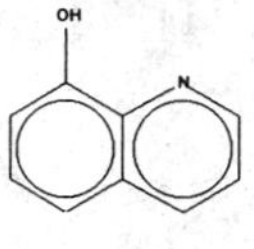

11311

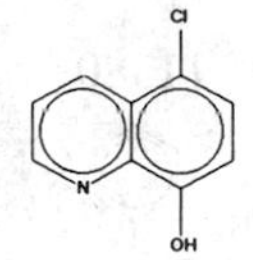

11312

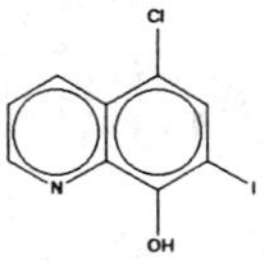

11313

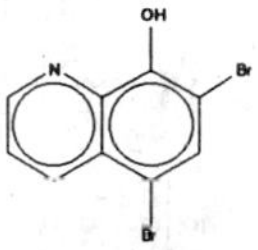

11314

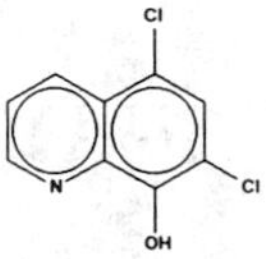

11315

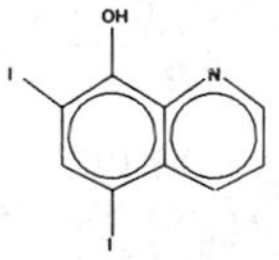

11317

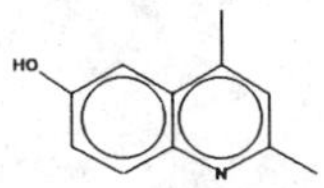

11318

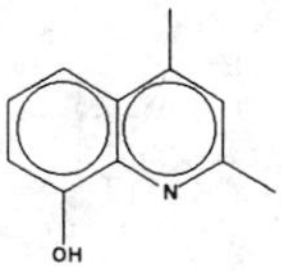

11319

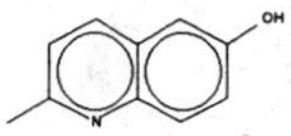

11320

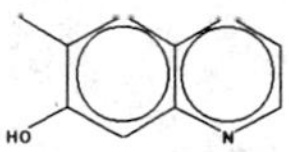

11321

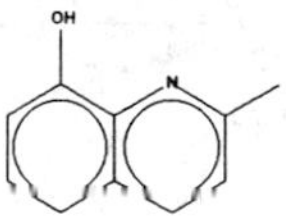

11325

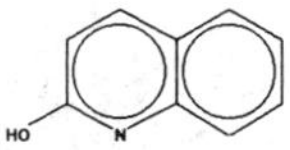

11327

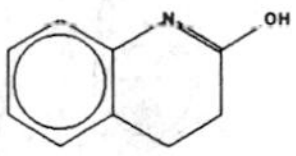

11328

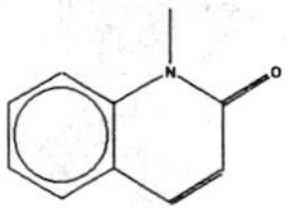

11329

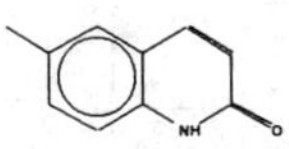

11330

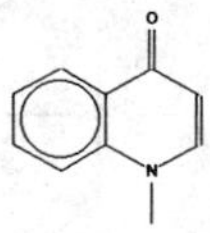

11331

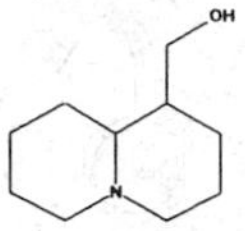

11333

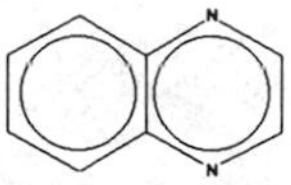

11334

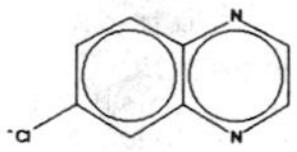

11335

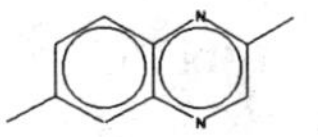

11337

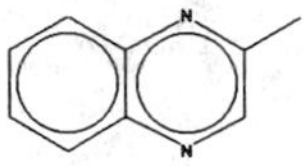

11338

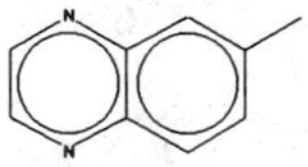

11339

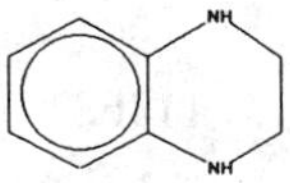

11340

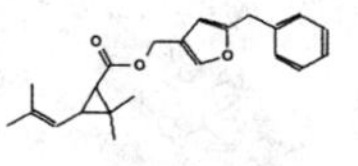

11341

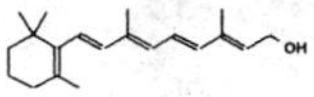

11342

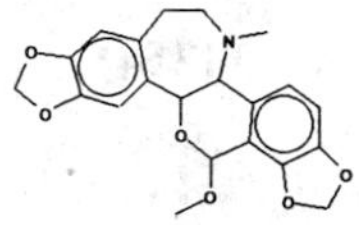

11343

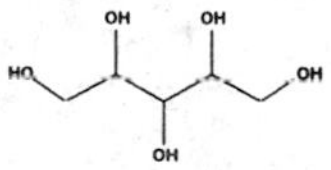

11344

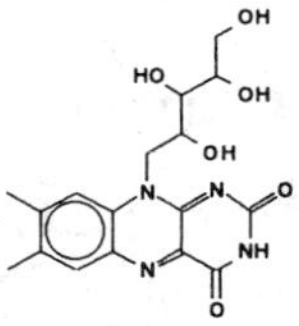

11345

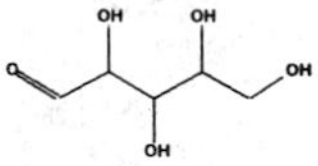

11348

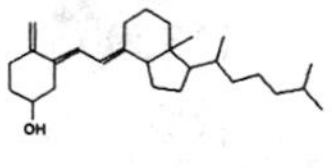

11350

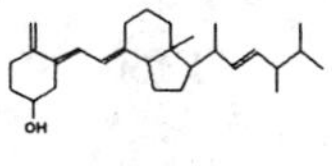

In numeric order as they occur in the Table of Physical Constants of Organic Compounds

11351

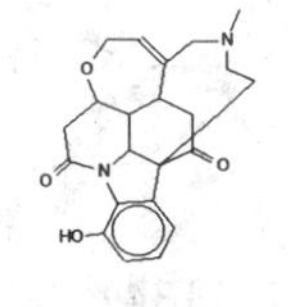

11353

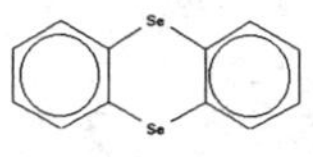

11354

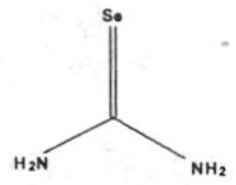

11356

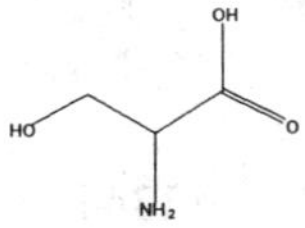

11357

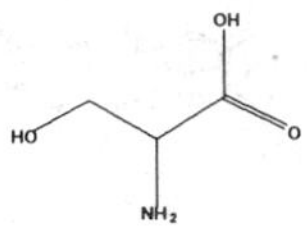

11358

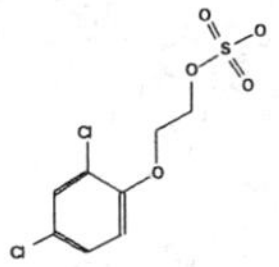

11359

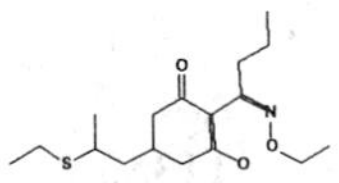

11360

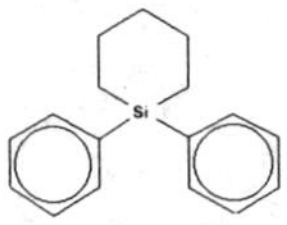

11361

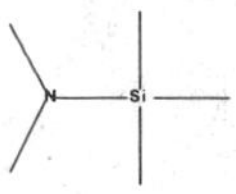

11362

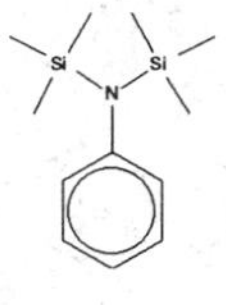

11363

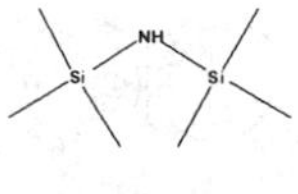

11364

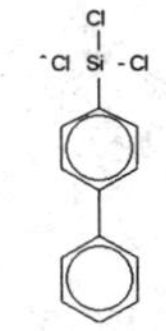

11365

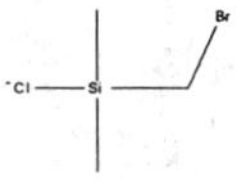

11366

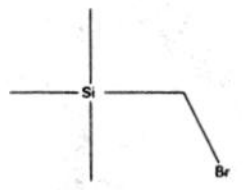

11367

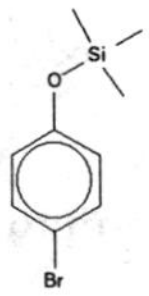

11368

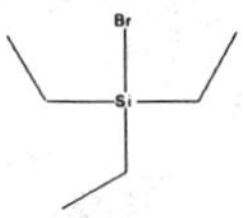

11369

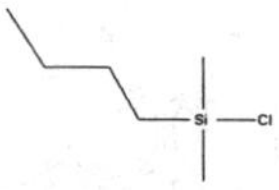

11370

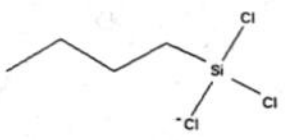

11371

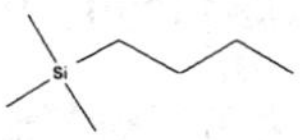

11372

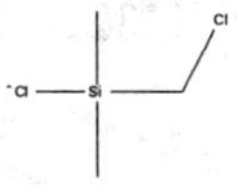

11373

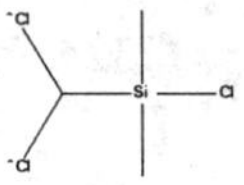

11374

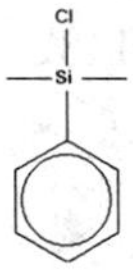

11375

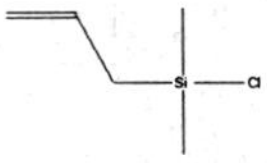

11376

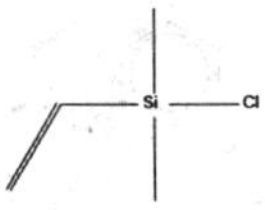

11377

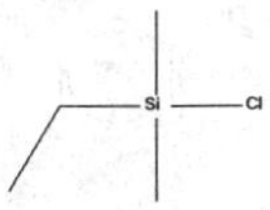

11378

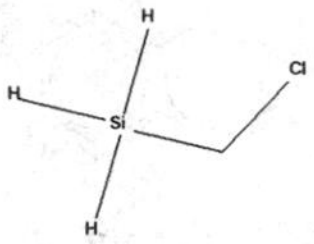

11380

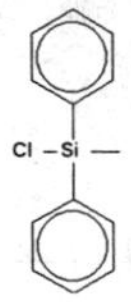

11381

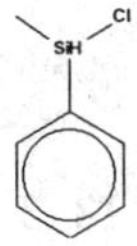

11382

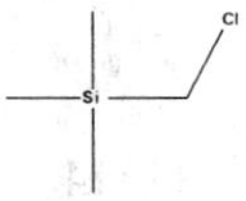

11383

11384

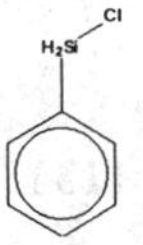

11385

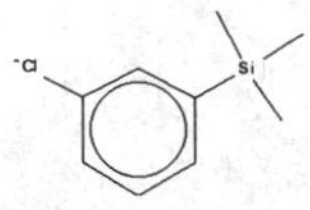

11386

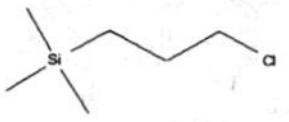

11387

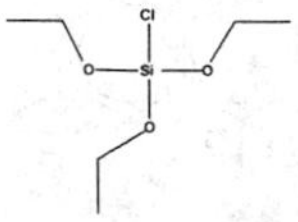

11388

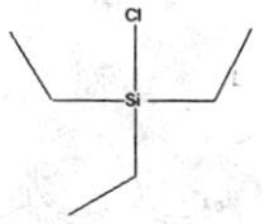

11389

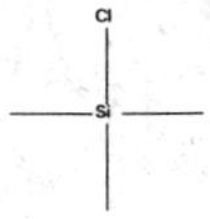

11390

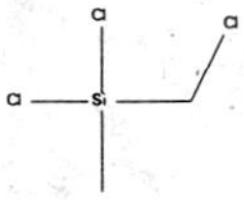

11391

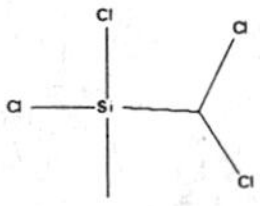

11392

11393

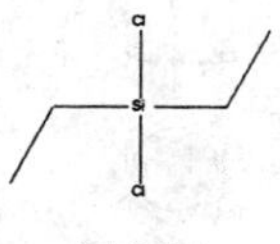

11394

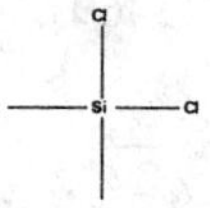

11395

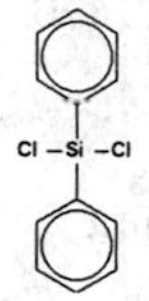

11396

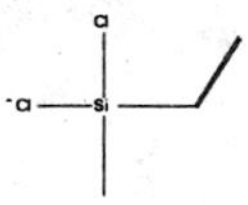

11397

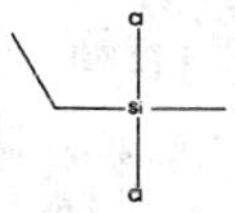

11398

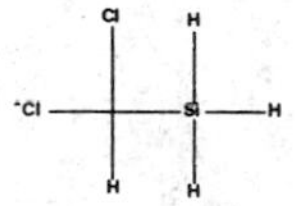

11399

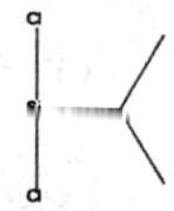

11400

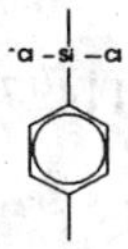

11401

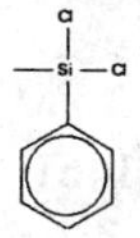

11402

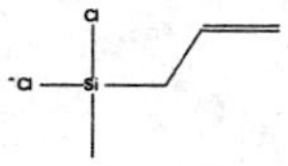

11403

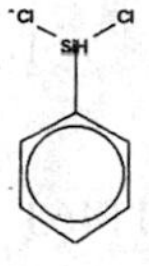

11404

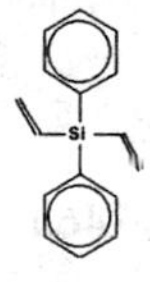

11405

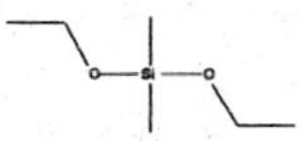

11406

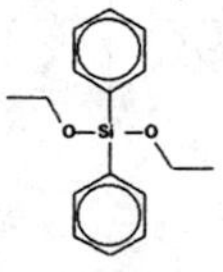

11407

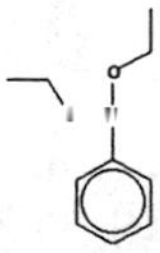

11408

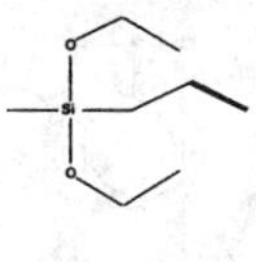

11409

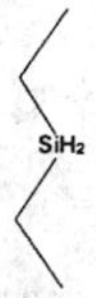

11410

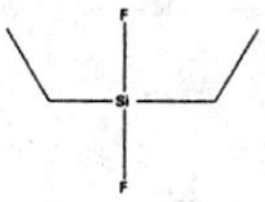

11411

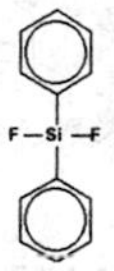

11412

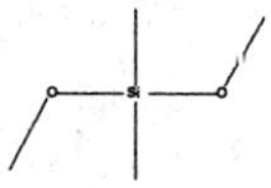

11413

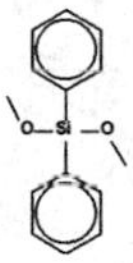

11414

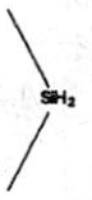

11415

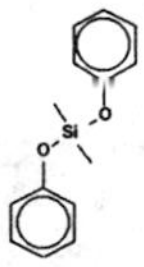

11417

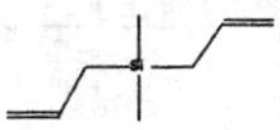

11418

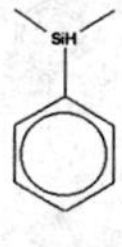

11419

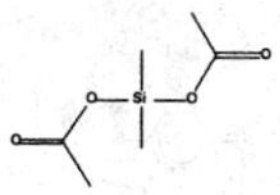

11420

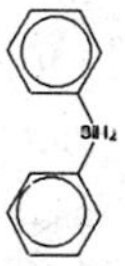

11421

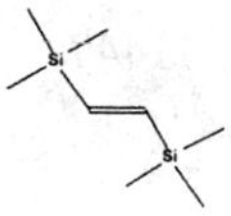

11422

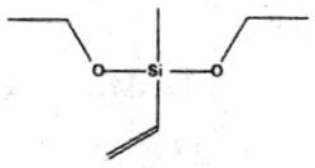

11423

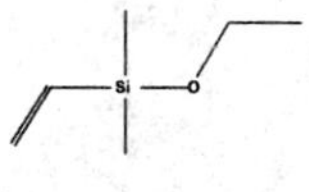

11424

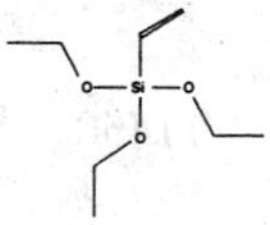

11425

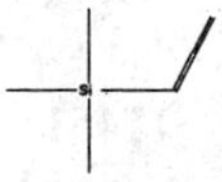

11426

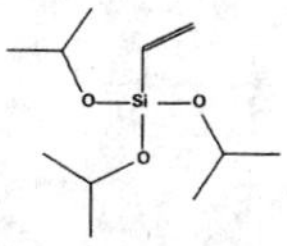

11427

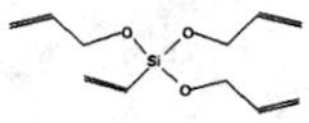

11428

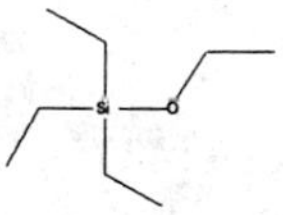

11429

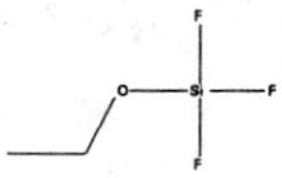

11430

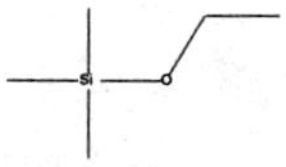

11431

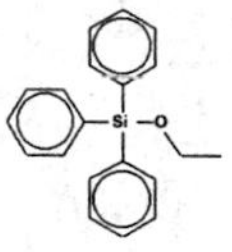

11432

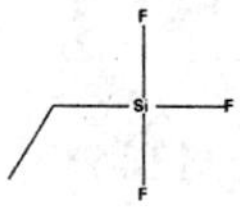

11433

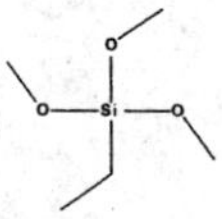

STRUCTURAL FORMULAS OF ORGANIC COMPOUNDS (continued)

In numeric order as they occur in the Table of Physical Constants of Organic Compounds

11434

11435

11436

11437

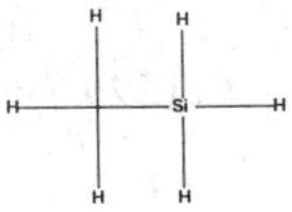

11438

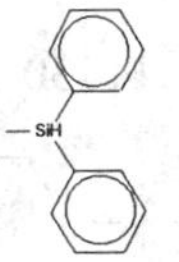

11439

11440

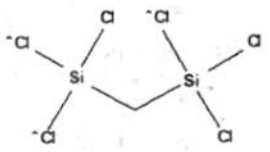

11443

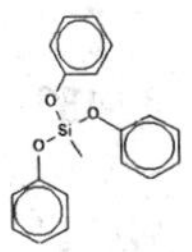

11444

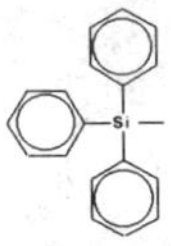

11445

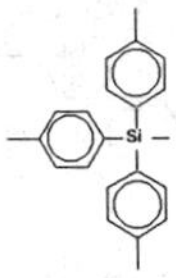

11446

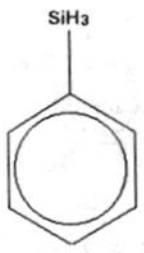

11447

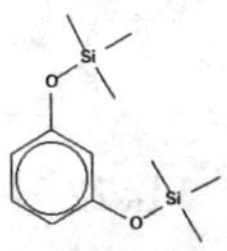

11448

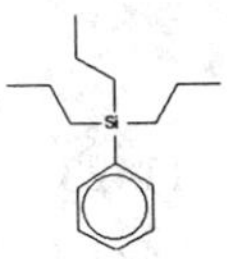

11449

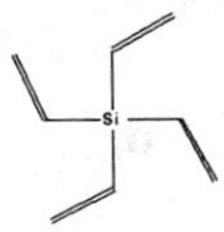

11450

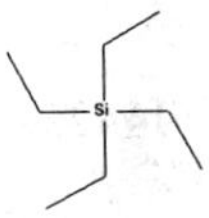

11451

11452

11453

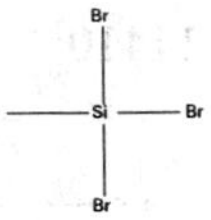

11454

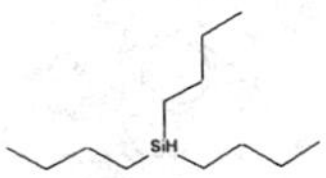

11455

11456

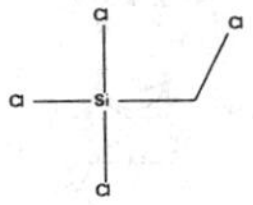

11457

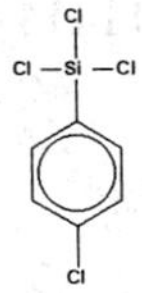

11458

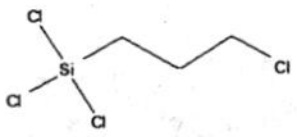

11459

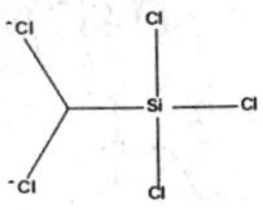

11460

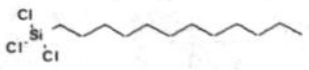

11461

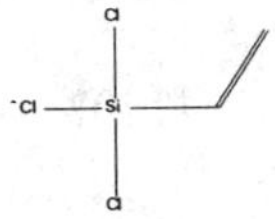

11462

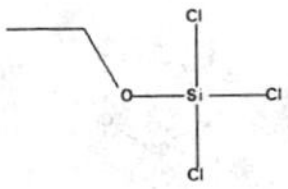

11463

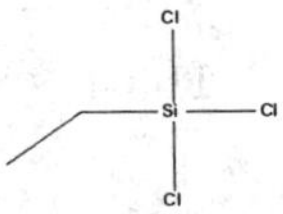

11464

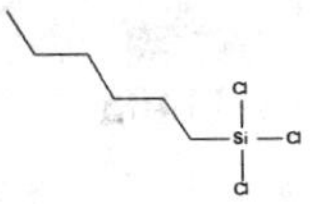

11465

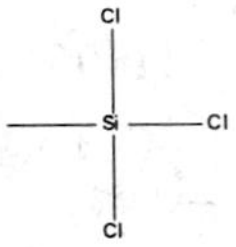

11466

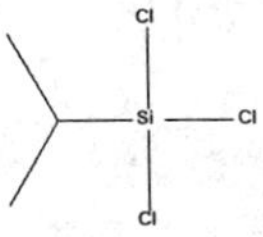

11467

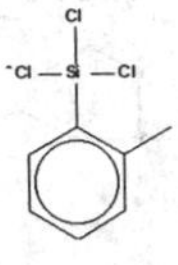

11468

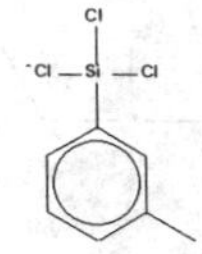

11469

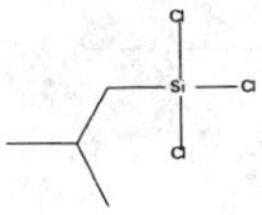

11470

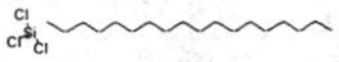

11471

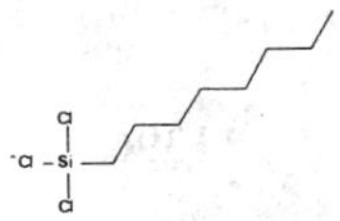

11472

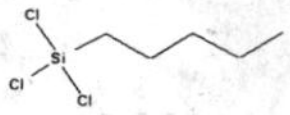

11474

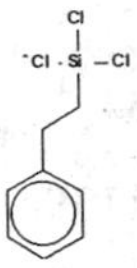

11475

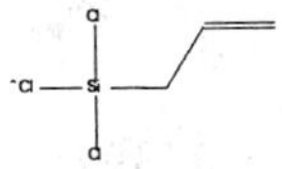

11476

11477

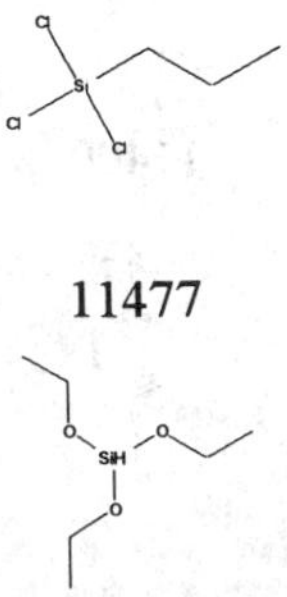

STRUCTURAL FORMULAS OF ORGANIC COMPOUNDS (continued)

In numeric order as they occur in the Table of Physical Constants of Organic Compounds

11478

11479

11480

11481

11482

11483

11484

11485

11486

11487

11488

11489

11490

11491

11492

11493

11494

11495

11496

11497

11498

11499

11500

11501

11502

11503

11504

11505

11506

11507

11508

11509

11510

11511

11512

11513

11514

11517

11519

11520

11521

11527

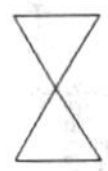

11528

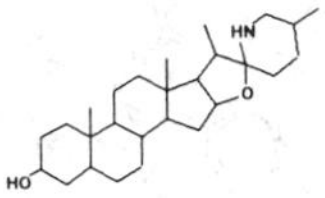

11529

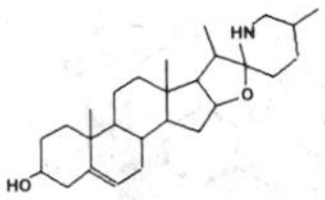

11531

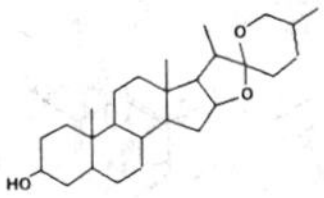

11532

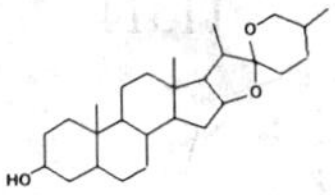

11536

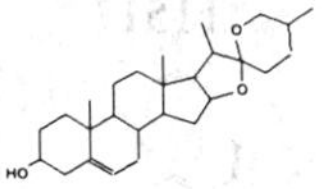

11537

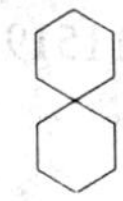

11539

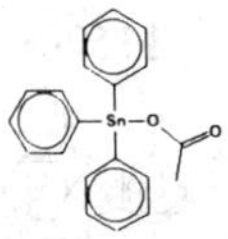

11540

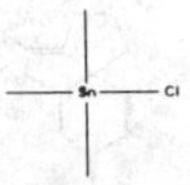

11541

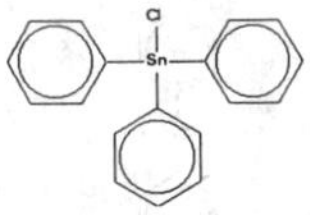

11542

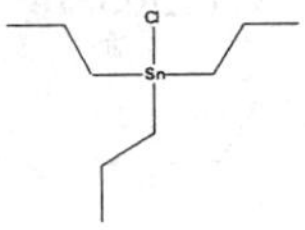

11543

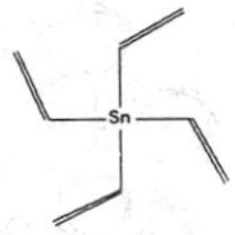

11544

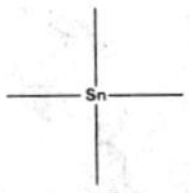

11545

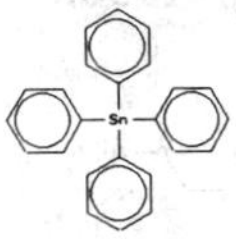

11546

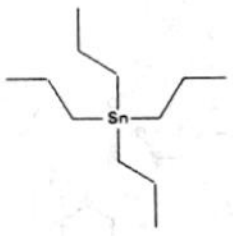

11549

11550

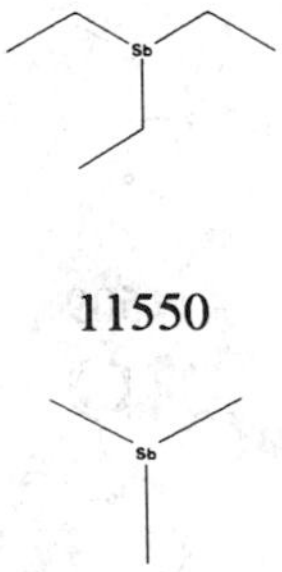

11551

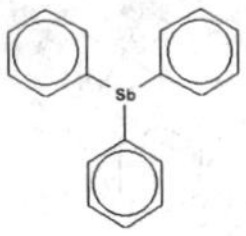

11555

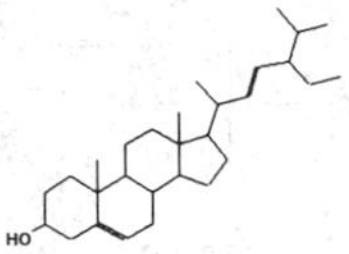

11556

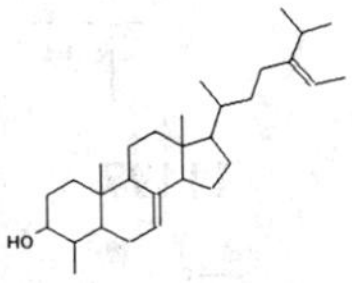

11558

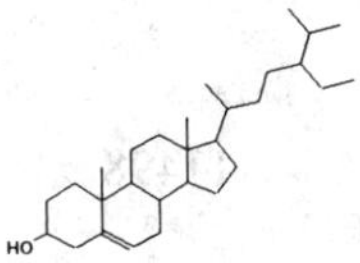

11559

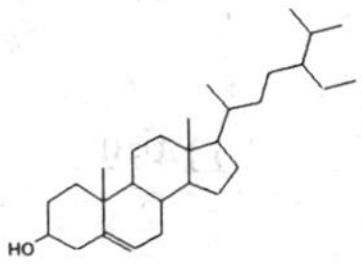

11560

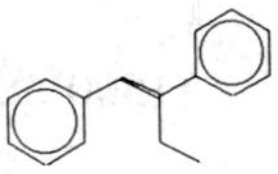

11562

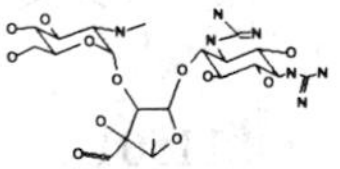

11563

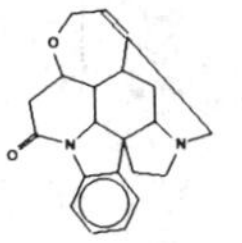

11564

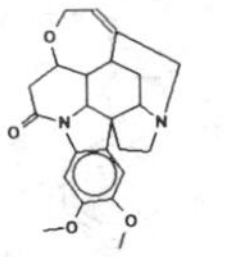

11567

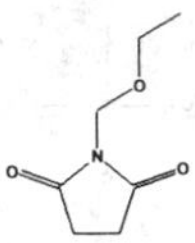

11568

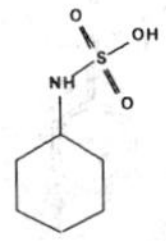

11569

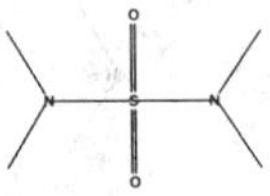

11570

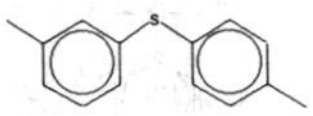

11571

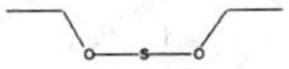

11572

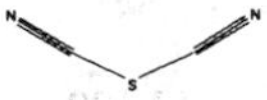

11573

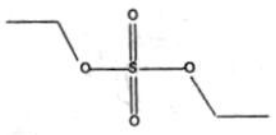

11574

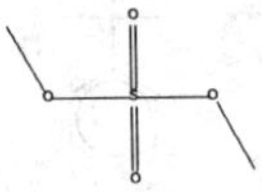

11575

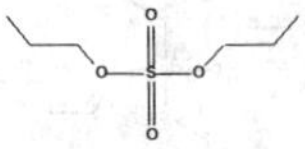

11576

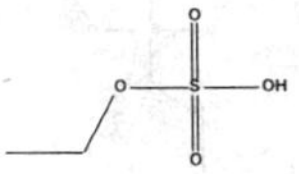

11577

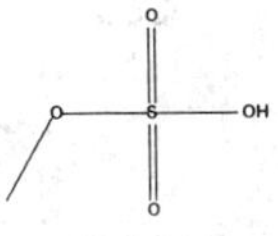

11578

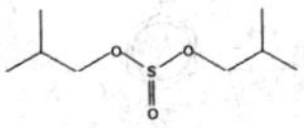

11579

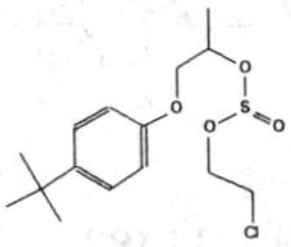

11580

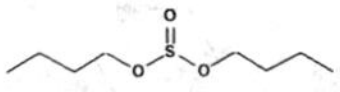

11581

11582

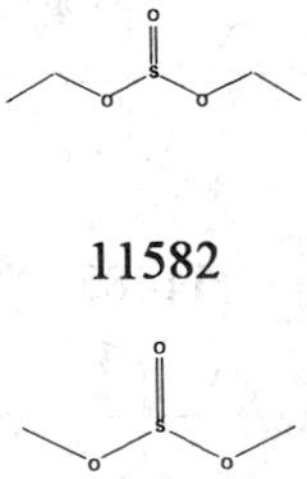

STRUCTURAL FORMULAS OF ORGANIC COMPOUNDS (continued)

In numeric order as they occur in the Table of Physical Constants of Organic Compounds

11583

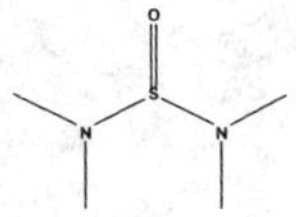

11584

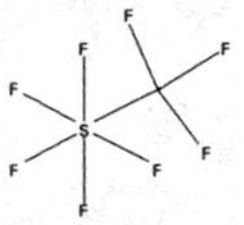

11585

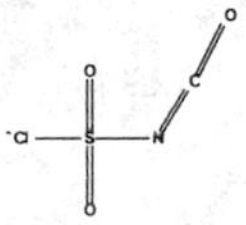

11586

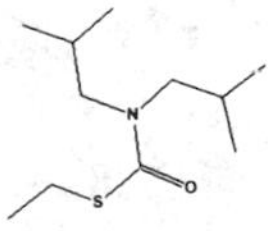

11588

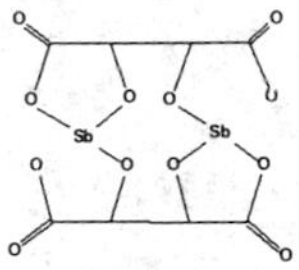

11590

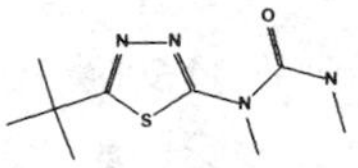

11591

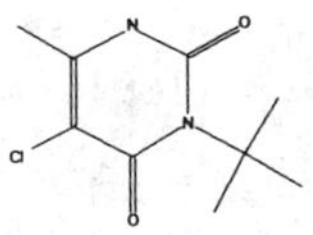

11592

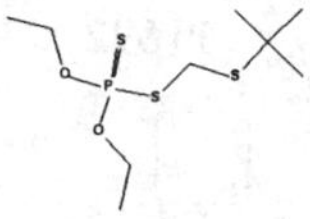

11593

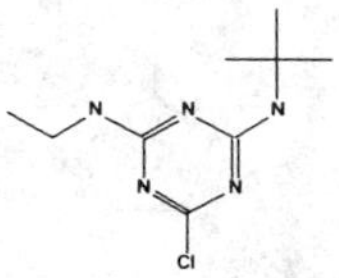

11594

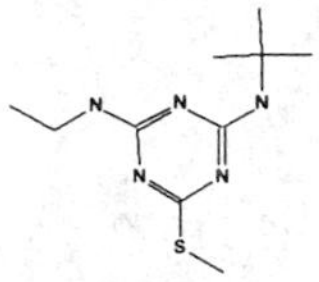

11595

11596

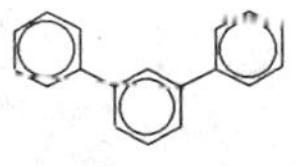

11598

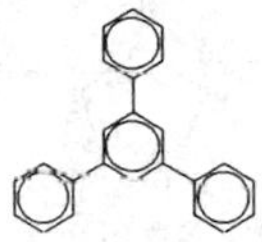

11599

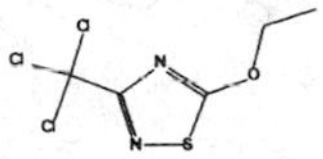

11600

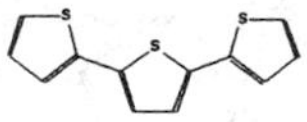

11601

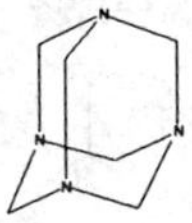

11602

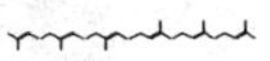

11603

11604

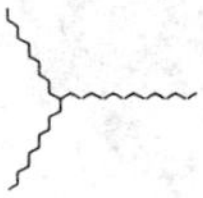

11605

11606

11607

11609

11610

11611

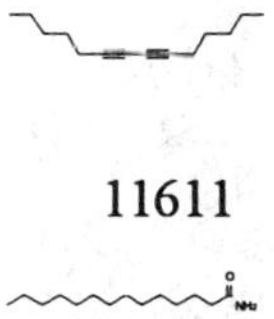

11612

11613

11614

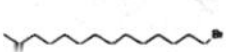

11615

11616

11617

11618

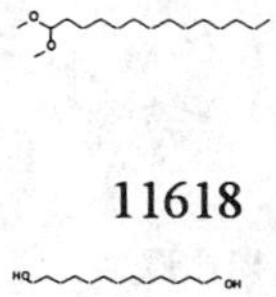

11619

11621

11622

11624

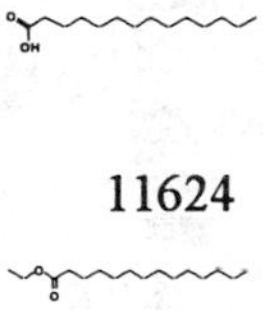

11625

11626

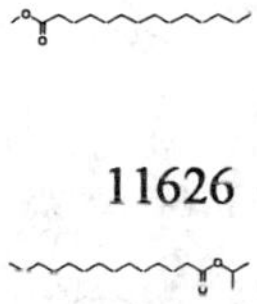

11627

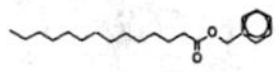

11628

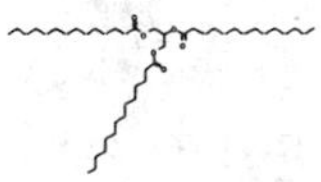

11629

11630

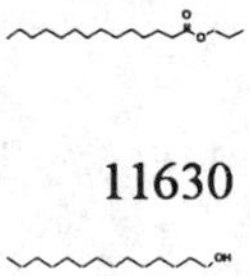

11631

11632

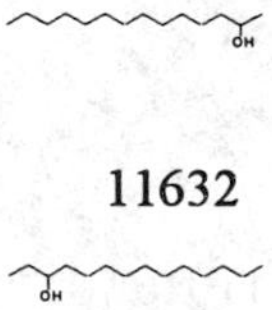

11633

11634

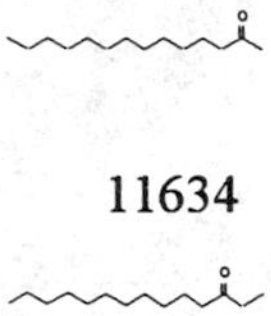

11635

11636

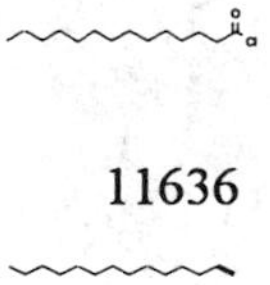

11637

11638

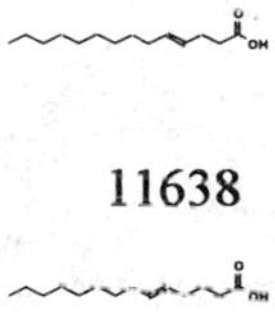

11640

11641

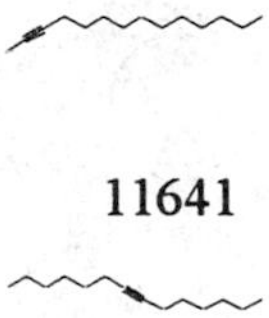

11642

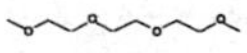

11643

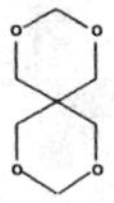

STRUCTURAL FORMULAS OF ORGANIC COMPOUNDS (continued)

In numeric order as they occur in the Table of Physical Constants of Organic Compounds

11644

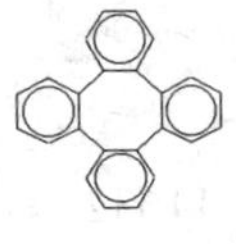

11646

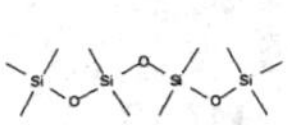

11647

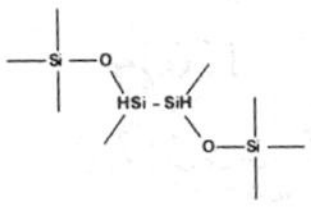

11648

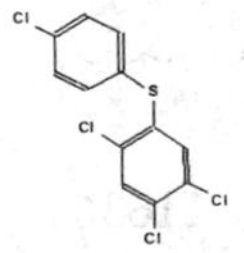

11649

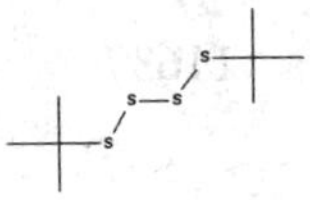

11650

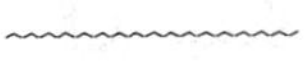

11651

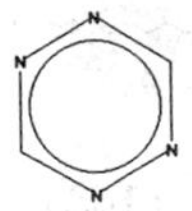

11652

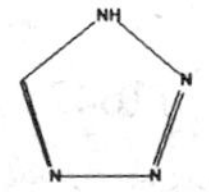

11654

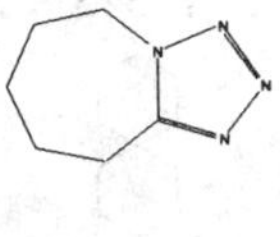

11656

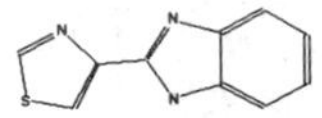

11657

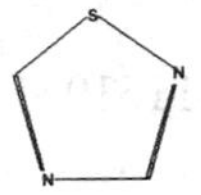

11658

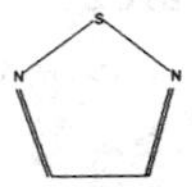

11659

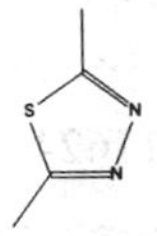

11660

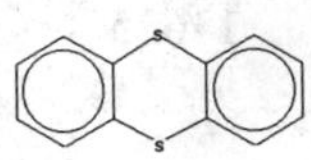

11661

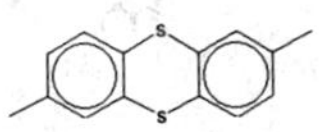

11662

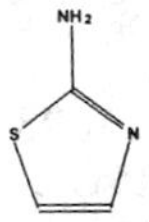

11664

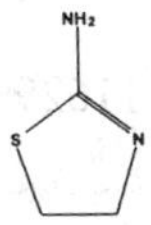

11665

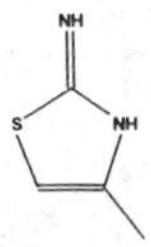

11666

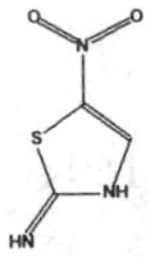

11667

11668

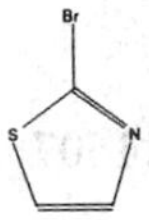

11670

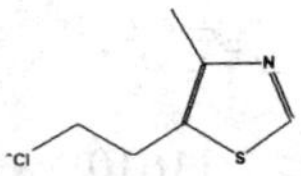

11671

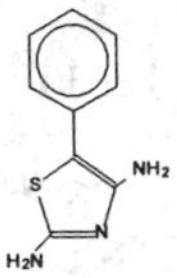

11672

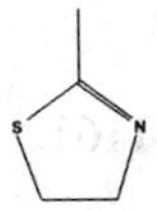

11673

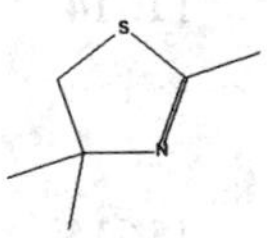

11674

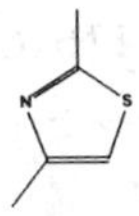

11675

11676

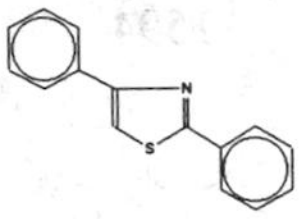

11677

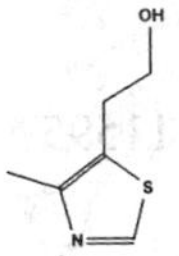

11678

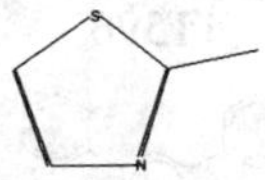

11679

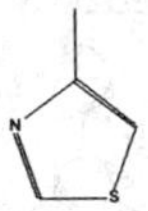

11680

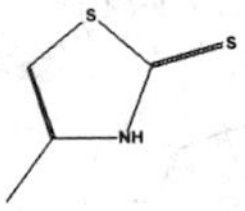

11681

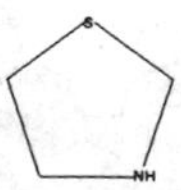

11682

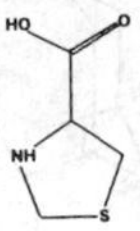

11683

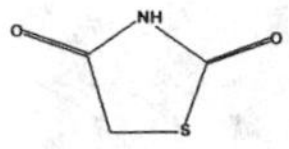

11684

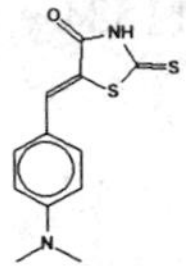

11685

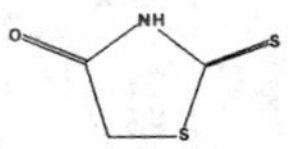

11687

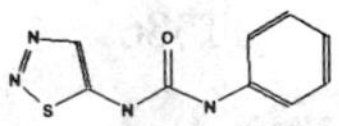

11688

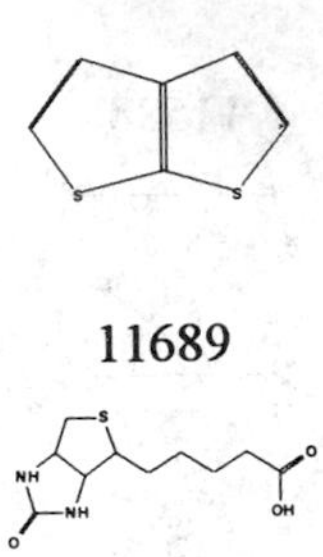

11689

11690

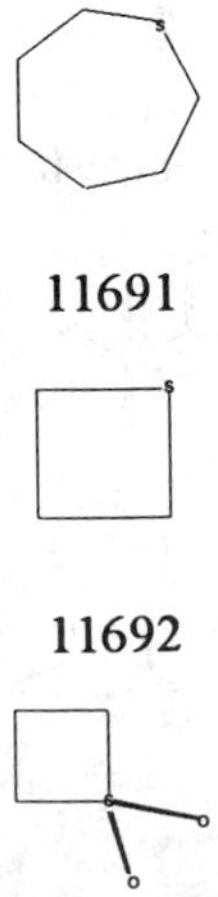

11691

11692

STRUCTURAL FORMULAS OF ORGANIC COMPOUNDS (continued)

In numeric order as they occur in the Table of Physical Constants of Organic Compounds

11693

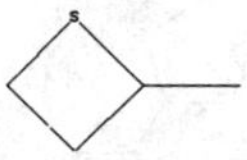

11694

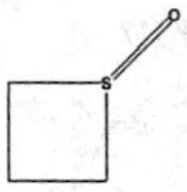

11695

11696

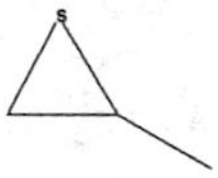

11697

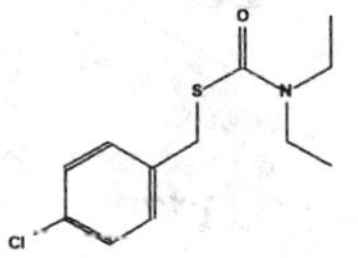

11698

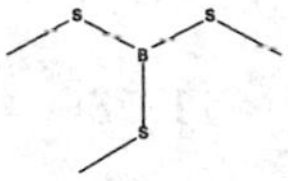

11699

11700

11701

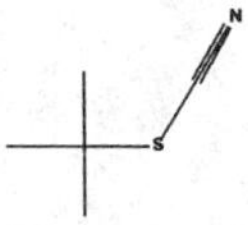

11703

11704

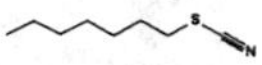

11705

11706

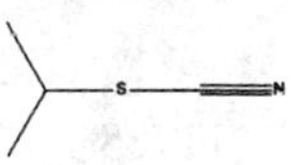

11707

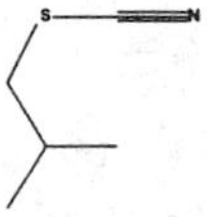

11708

11709

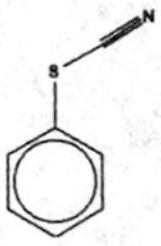

11710

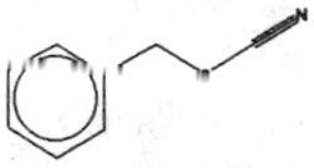

11712

11713

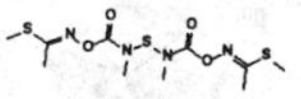

11714

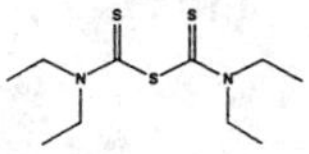

11715

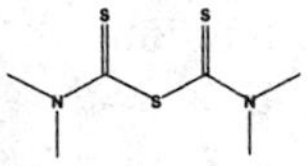

11716

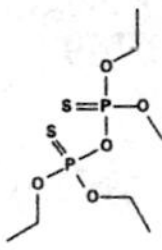

11717

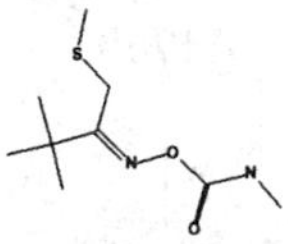

11719

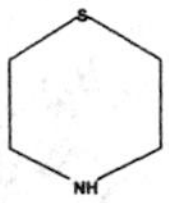

11720

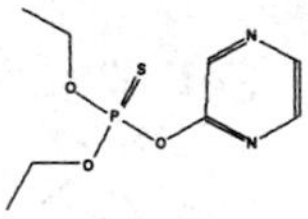

11721

11722

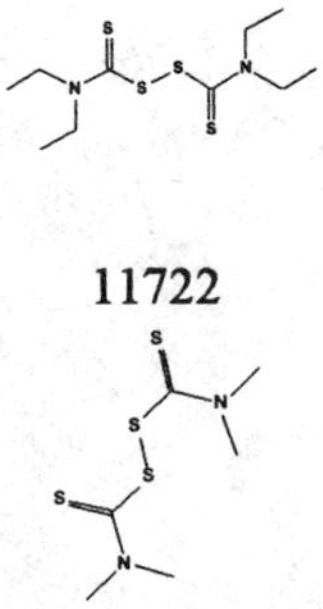

11723

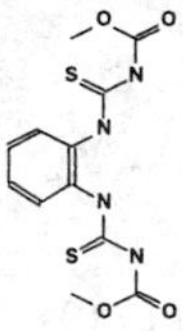

11724

11725

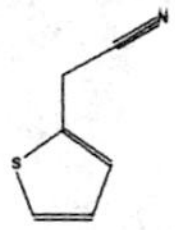

11726

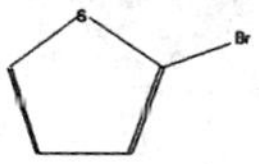

11727

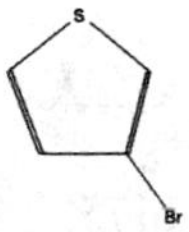

11728

11729

11730

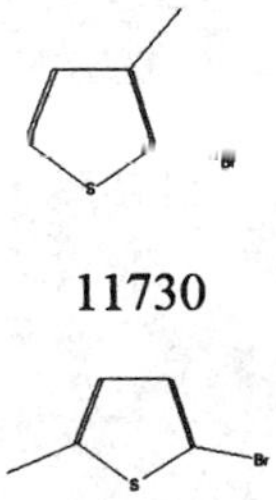

11731

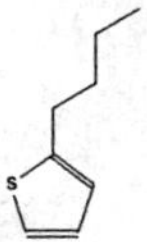

11732

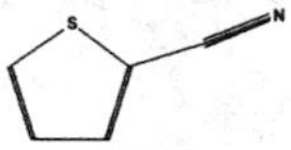

11733

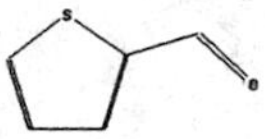

11734

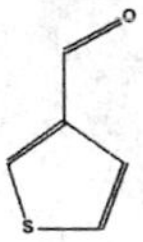

11735

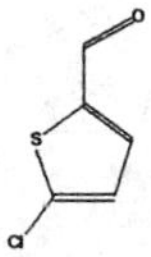

11736

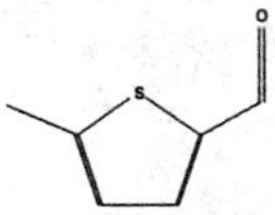

11737

11738

STRUCTURAL FORMULAS OF ORGANIC COMPOUNDS (continued)

In numeric order as they occur in the Table of Physical Constants of Organic Compounds

11739

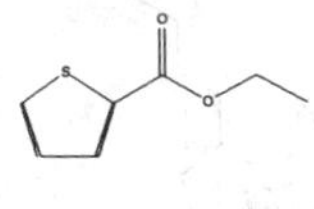

11740

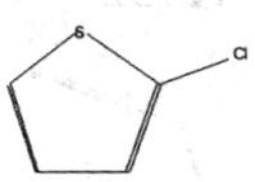

11741

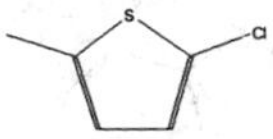

11742

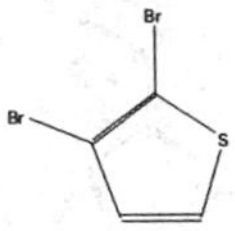

11743

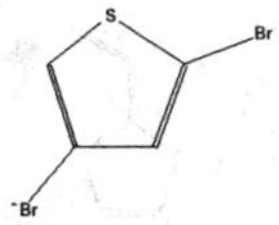

11744

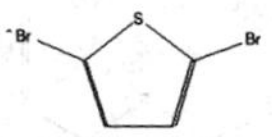

11746

11747

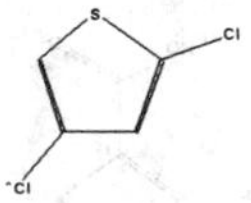

11748

11749

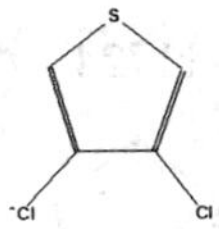

11750

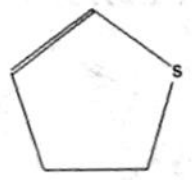

11751

11752

11753

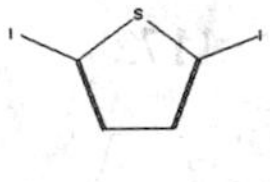

11754

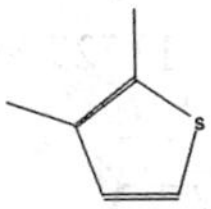

11755

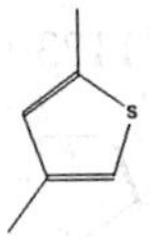

11756

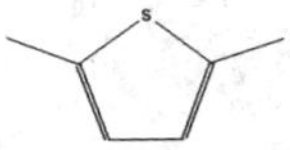

11757

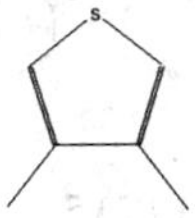

11758

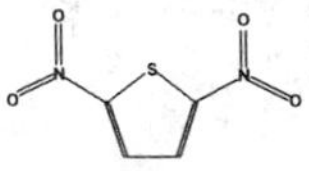

11759

11760

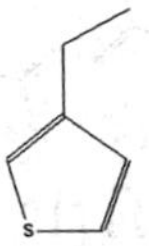

11761

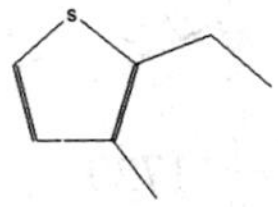

11762

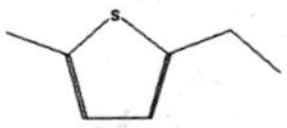

11763

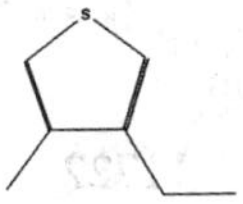

11764

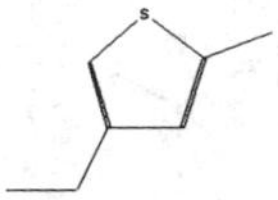

11765

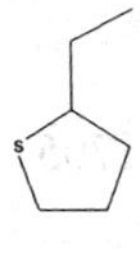

11767

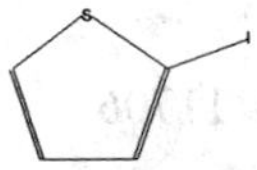

11768

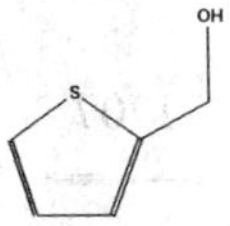

11769

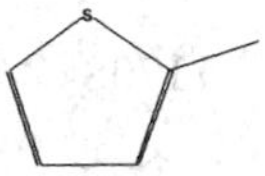

11770

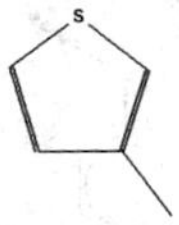

11771

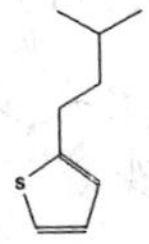

11772

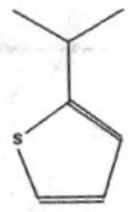

11773

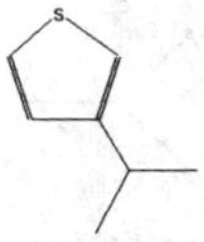

11774

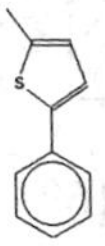

11775

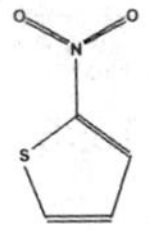

11776

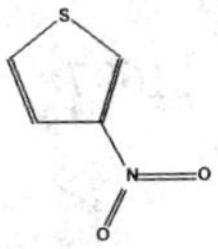

11777

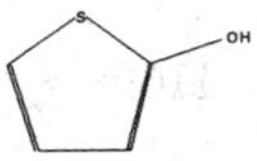

11778

11779

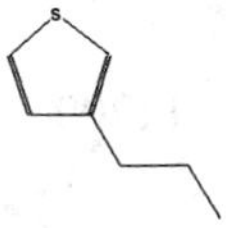

11780

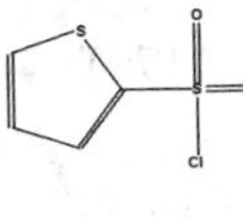

STRUCTURAL FORMULAS OF ORGANIC COMPOUNDS (continued)

In numeric order as they occur in the Table of Physical Constants of Organic Compounds

11781

11782

11783

11784

11785

11786

11787

11788

11789

11790

11791

11792

11793

11794

11795

11796

11797

11798

11799

11800

11801

11803

11804

11805

11806

11807

11808

11809

11810

11812

11813

11816

11817

11819

11822

11823

11824

11826

11827

11828

11829

STRUCTURAL FORMULAS OF ORGANIC COMPOUNDS (continued)

In numeric order as they occur in the Table of Physical Constants of Organic Compounds

11831

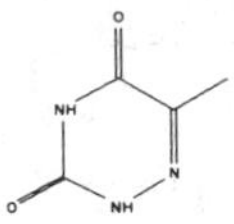

11833

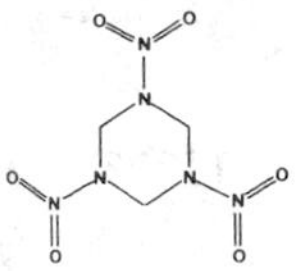

11834

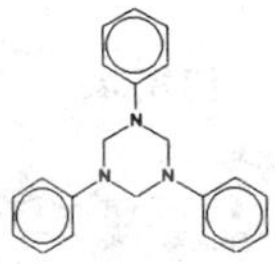

11835

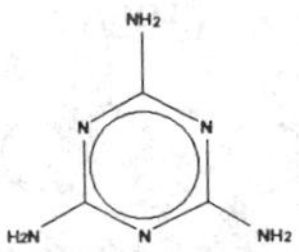

11836

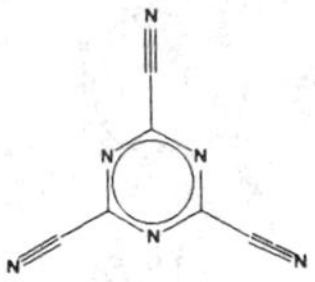

11837

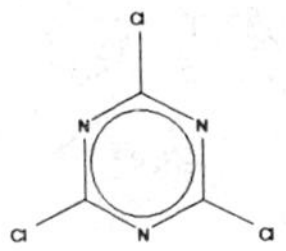

11840

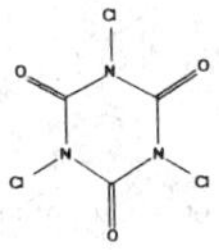

11841

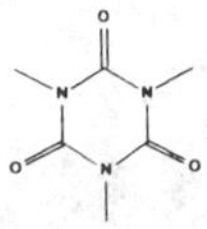

11842

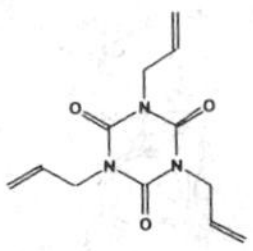

11845

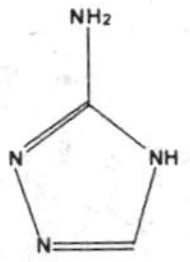

11846

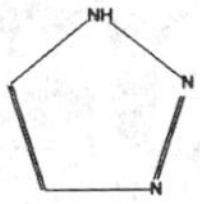

11847

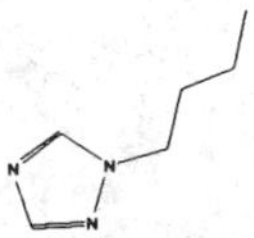

11848

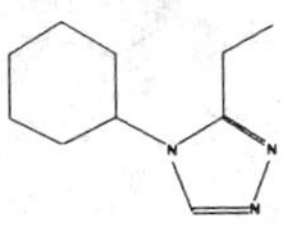

11849

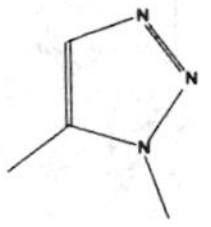

11852

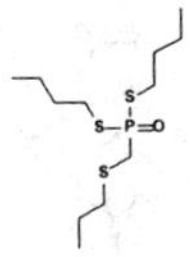

11853

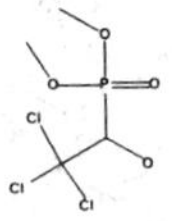

11854

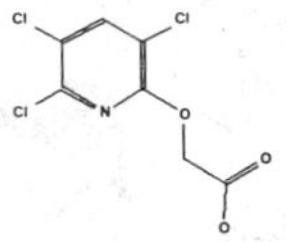

11855

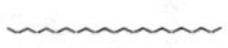

11856

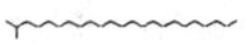

11857

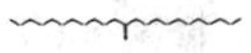

11858

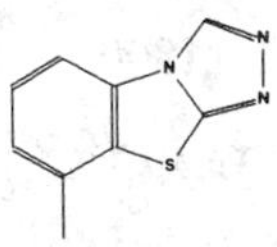

11859

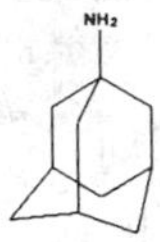

11860

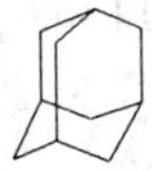

11861

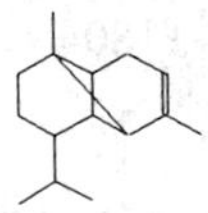

11862

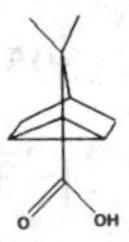

11863

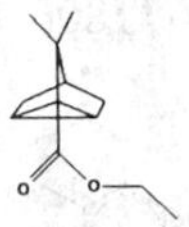

11865

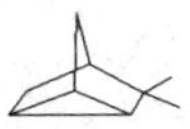

11866

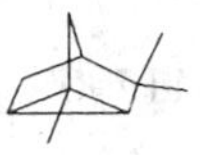

11867

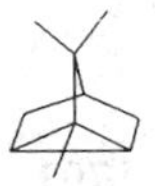

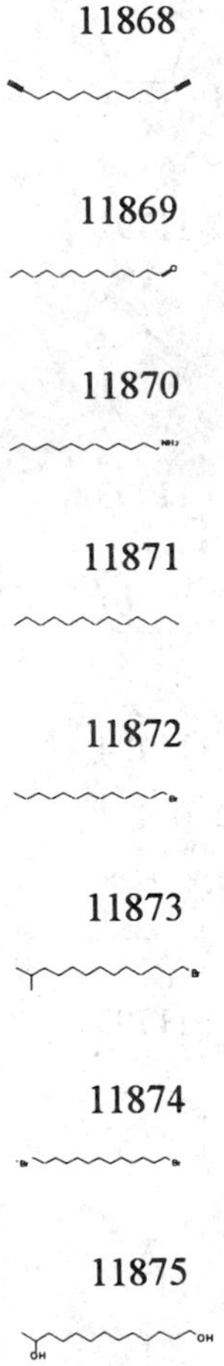

11868
11869
11870
11871
11872
11873
11874
11875

11877

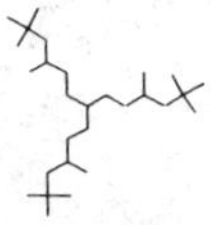

11878

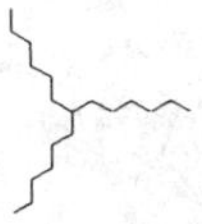

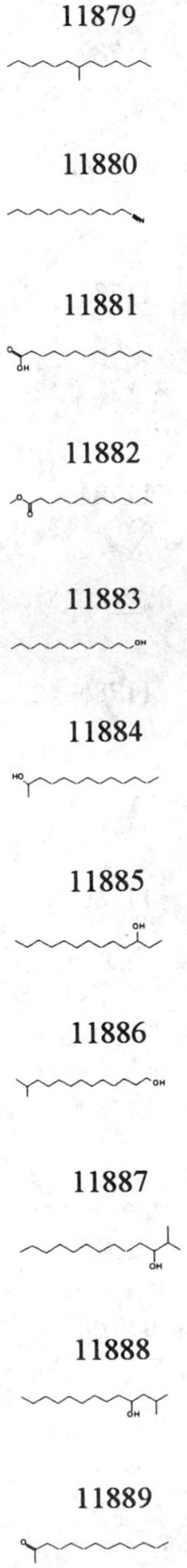

11879
11880
11881
11882
11883
11884
11885
11886
11887
11888
11889

STRUCTURAL FORMULAS OF ORGANIC COMPOUNDS (continued)

In numeric order as they occur in the Table of Physical Constants of Organic Compounds

11890

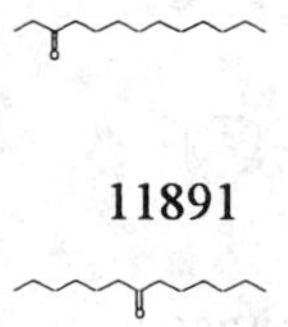

11891

11892

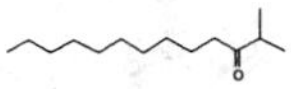

11893

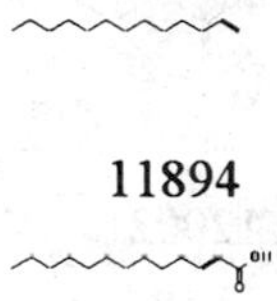

11894

11895

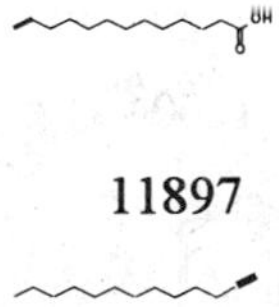

11897

11898

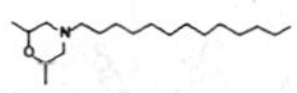

11899

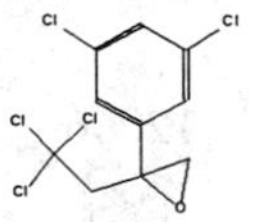

11900

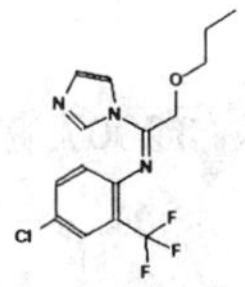

11901

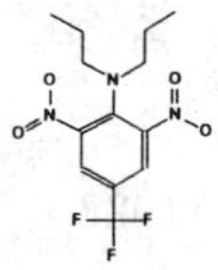

11902

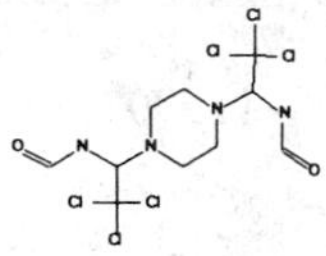

11903

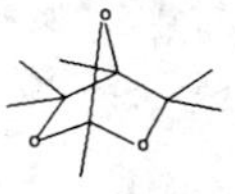

11904

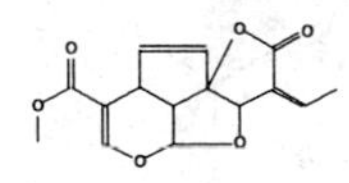

11905

11906

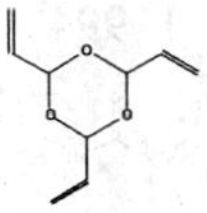

11908

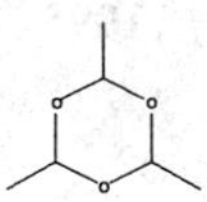

11909

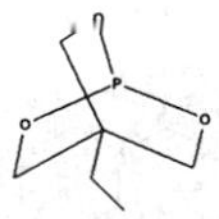

11910

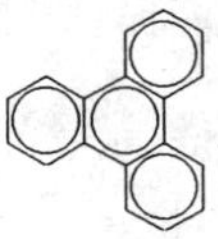

11911

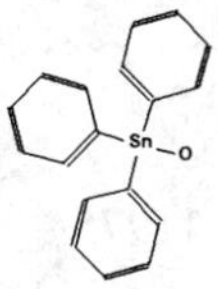

11912

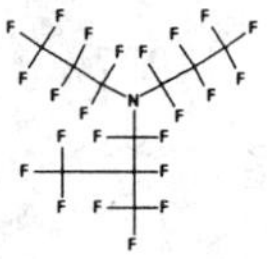

11913

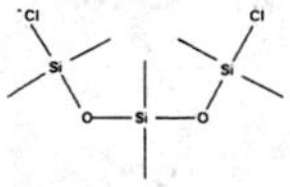

11914

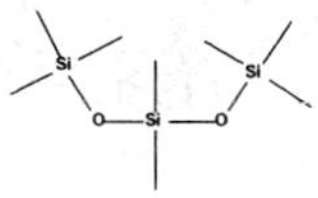

11915

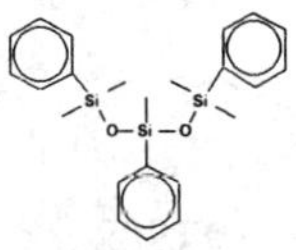

11916

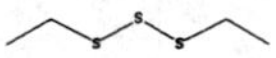

11917

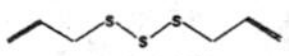

11919

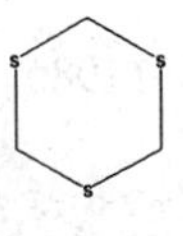

11920

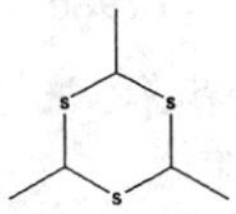

11921

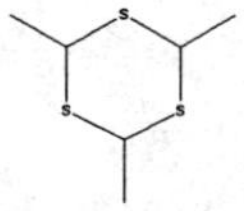

11922

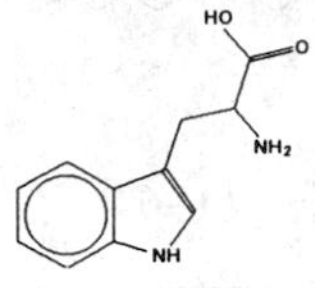

11926

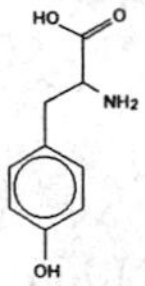

11929

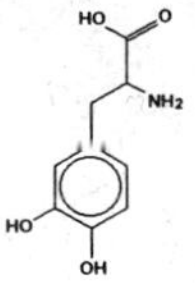

11931

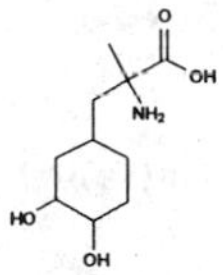

11934

11935

11936

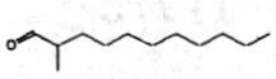

11937

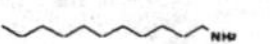

11938

11939

11940

11941

11942

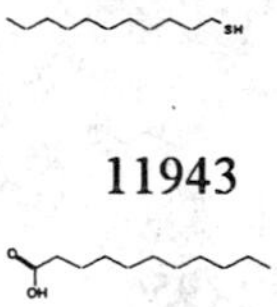

11943

11944

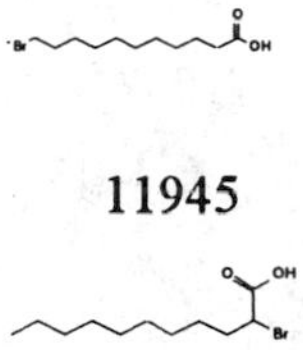

11945

11946

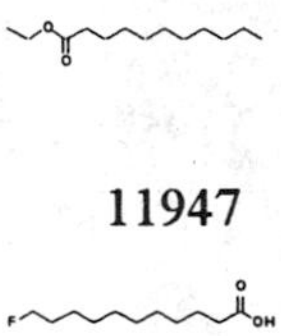

11947

11948

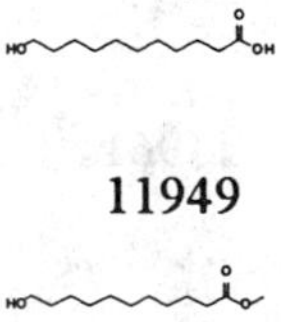

11949

STRUCTURAL FORMULAS OF ORGANIC COMPOUNDS (continued)

In numeric order as they occur in the Table of Physical Constants of Organic Compounds

11950

11951

11953

11954

11955

11956

11957

11958

11959

11960

11961

11962

11963

11964

11965

11966

11967

11968

11969

11970

11971

11972

11973

11974

11975

11977

11978

11979

11980

11981

11982

11983

11984

11985

11986

11987

11988

11990

11991

11992

11993

11994

11995

11996

11999

12000

12001

12002

12004

12005

12006

12007

STRUCTURAL FORMULAS OF ORGANIC COMPOUNDS (continued)

In numeric order as they occur in the Table of Physical Constants of Organic Compounds

12009

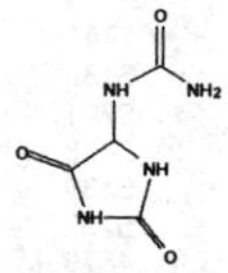

12010

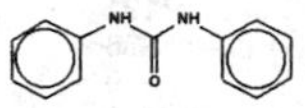

12012

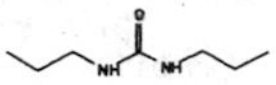

12013

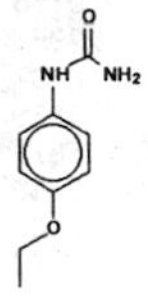

12014

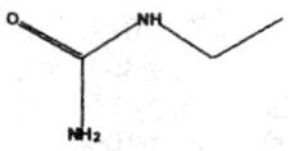

12015

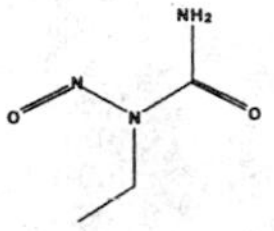

12016

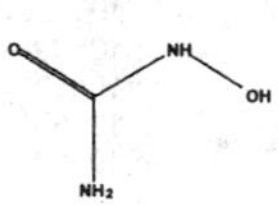

12017

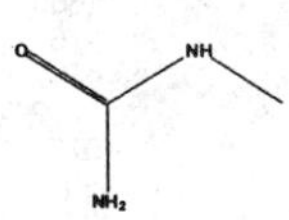

12018

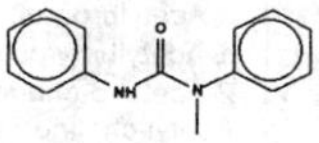

12019

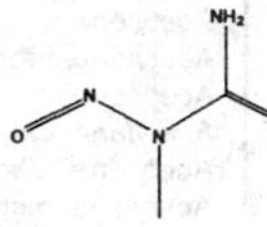

12026

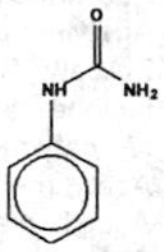

12027

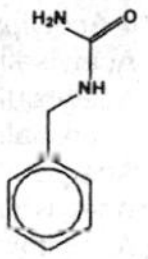

12028

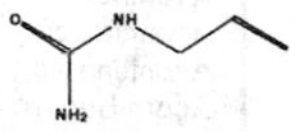

12029

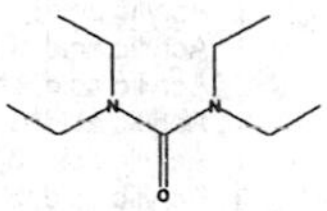

12030

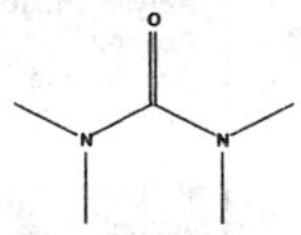

12031

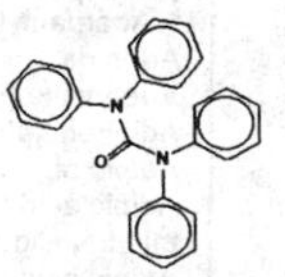

12032

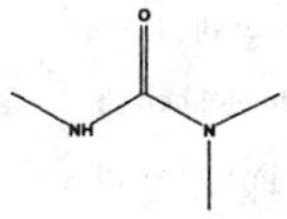

12033

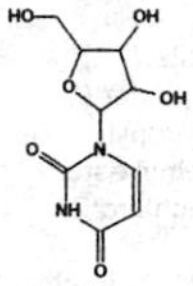

12034

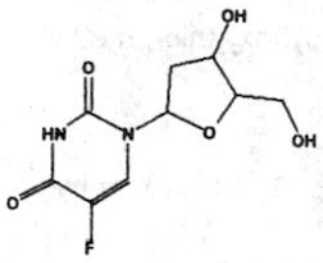

12037

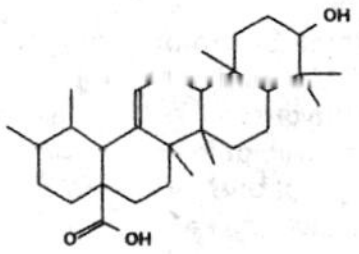

12038

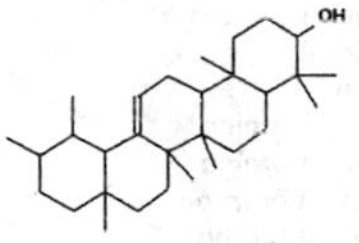

12039

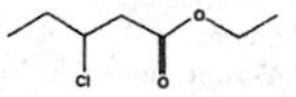

12040

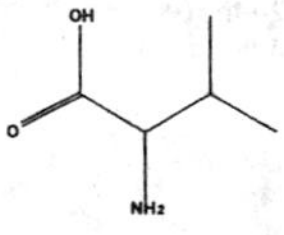

12041

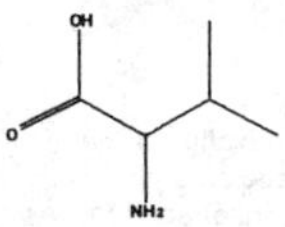

12043

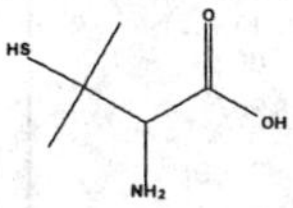

12044

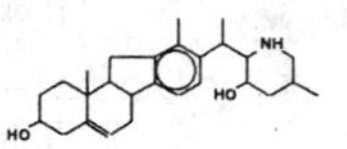

12047

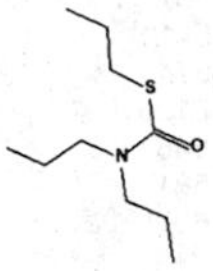

12048

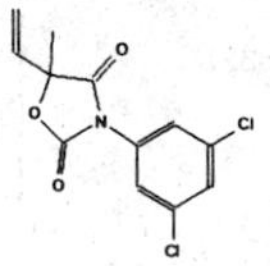

12049

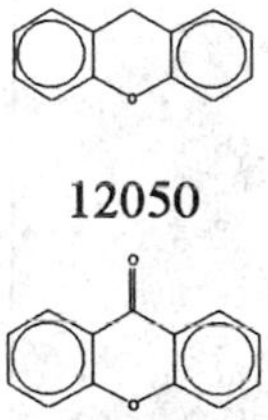

12050

12055

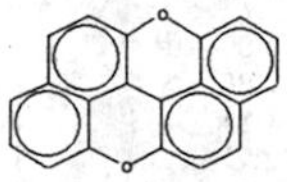

12056

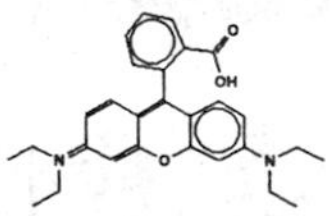

12057

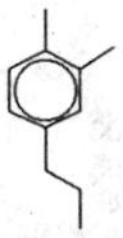

12058

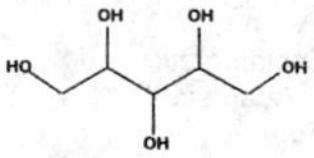

12059

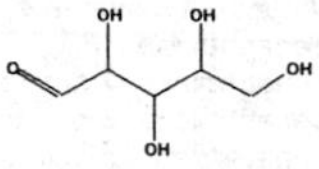

12061

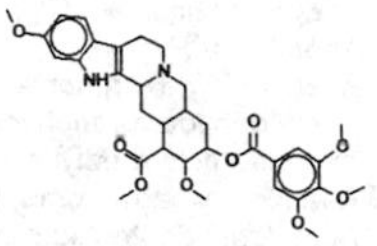

12062

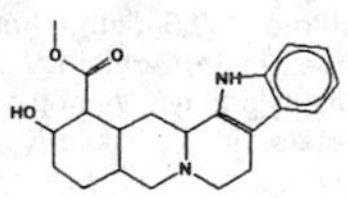

12065

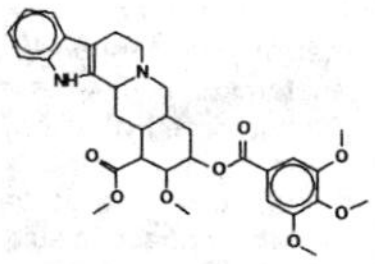

12066

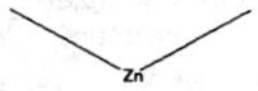

12067

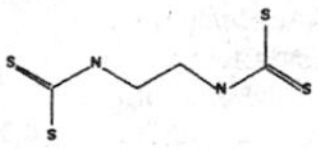

12068

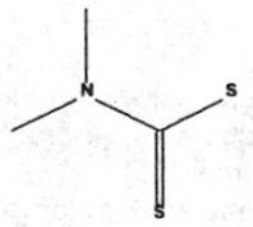

SYNONYM INDEX OF ORGANIC COMPOUNDS

SYNONYM INDEX OF ORGANIC COMPOUNDS (continued)

β-Aminopyridine : 10858
4-Aminopyridine : 10859
Aminopyrine : 10851
2-Aminoquinoline : 11202
4-Aminoquinoline : 11203
5-Aminoquinoline : 11204
6-Aminoquinoline : 11205
8-Aminoquinoline : 11206
6-Aminoquinoxaline : 11332
Aminorex : 8574
4-Amino-1-β-*D*-ribofuranosyl-2(1H)-pyrimidinone : 5021
p-Aminosalicylic acid : 2498
5-Aminosalicylic acid : 2499
p-Aminosalicylic acid hydrazide : 2500
α-Aminosuccinamic acid : 468
L-Aminosuccinic acid : 471
2-Amino-1,3-thiazole : 11662
2-Aminothiazole : 11662
Aminoxafen : 8574
Aminozide : 3334
Amiphenazole : 11671
Amitrole : 11845
Amixetrine : 11169
Amizepin : 5144
Ammonium, acetate : 106
Ammonium, (3-carbamoyl-3,3-diphenylpropyl) diisopropyl methyl,iodide : 2120
Ammonium, diethyl(2-hydroxyethyl)methyl-, bromide, xanthene-9-carboxylate : 5445
Ammonium, dimethyl ethyl 3-hydroxyphenyl,chloride : 865
Ammonium formate : 6003
Ammonium, pentamethylenebis trimethyl, bromide : 8738
Ammonium, phenyl trimethyl,iodide : 866
Ammonium, tetraethyl,bromide : 5447
Ammonium, tetramethyl,bromide : 7456
Ammonium, tetramethyl,chloride : 7457
Ammonium, tetramethyl,iodide : 7458
Ammonium, tetrapropyl,iodide : 9833
Amobarbital : 11092
Amolanone : 2463
Amphecloral : 1583
Amphetamine : 1578
Ampyrone : 10845
Amudane : 11520
Amygdalin : 954
Amyl acetate : 227
sec-Amyl acetate (*R*) : 8918
Amylacetylene : 6609
tert-Amyl alcohol : 3628
Amyl alcohol : 8915
sec-Amyl alcohol : 8916
iso-Amylamine : 3223
Amylamine : 8700
Amylbenzene : 2107
d-Amyl bromide : 3238
tert-Amyl bromide : 3240
Amyl bromide : 8710
Amyl butyrate : 3579
Amyl caproate : 6907
tert-Amyl carbamate : 3634
Amyl chloride : 8719
6-*n*-Amyl-*m*-cresol : 9425
Amylene : 3801
α-Amylene : 9004
Amylene dichloride : 3286
2,4-Amylene glycol : 8789
Amyl ether : 8835
Amyl ethyl ketone : 8472
Amylethylmethylcarbinol : 8466
Amyl formate : 6015
Amyl heptanoate : 6470
Amyl hexanoate : 6907
Amyl hexyl ketone : 5326
Amyl iodide : 8818
Amylisopropylmethylcarbinol : 8451
tert-Amyl isovalerate : 3546
Amyl ketone : 11962
Amyl mercaptan : 8844
sec-Amylmercaptan : 8845
Amyl nitrite : 8090
: 8096
Amylol : 8915
4-Amylresorcinol : 1544
Amyl salicylate : 2649
Amyl stearate : 8270
Amyl sulfide : 8843
Amyltrichlorosilane : 11472
Amyl valerate : 8913
β-Amyrin : 8548
α-Amyrin : 12038
Anabasine : 11008
Anagyrine : 7573
Analeptin : 1887
Anastil : 9379
Anatabine : 3105
Anchovyxanthin, all-*trans*- : 4057
Androfluorone : 355
4-Androstene-3,17-dione : 352
Androsterone : 348
Anemonin : 5185
Anesthesin : 2497
Anethole : 1942
Angelic acid : 3845
Anhalamine : 7393
Anhalonidine : 7395
Anhydroecgonine : 491
Anhydron : 2851
Aniline : 633
Aniline, 2-bromo : 638
Aniline, 3-bromo : 639
Aniline, 4-bromo : 640
Aniline, 2-bromo-6-nitro- : 653
Aniline, 4-butyl : 655
Aniline, *N*-butyl- : 656
Aniline, 2-*tert*-butyl : 714
Aniline, 4-*tert*-butyl : 715
Aniline, *N*-*tert*-butyl- : 716
Aniline-4-*sec*-butyl : 813
Aniline, *N*-*sec*-butyl- : 816
Aniline-3-carboxylic acid : 2485
Aniline-4-carboxylic acid : 2486
Aniline, 2-chloro : 658
Aniline, 3-chloro : 659
Aniline, 4-chloro : 660
Aniline, 2-chloro, hydrochloride : 665
Aniline, 2,4-dibromo- : 681
Aniline, 2,6-dibromo- : 682
Aniline, 3,4-dibromo- : 683
Aniline, 2,3-dichloro- : 685
Aniline, 2,4-dichloro- : 686
Aniline, 2,5-dichloro- : 687
Aniline, 2,6-dichloro- : 688
Aniline, 3,4-dichloro- : 689
Aniline, 3,5-dichloro- : 690
Aniline, 2,6-dichloro-4-ethoxy : 691
Aniline, 3,4-diethoxy- : 693
Aniline, *N*,*N*-diethyl-4-nitroso : 699
Aniline, 2,4-difluoro- : 700
Aniline, 2,4-diiodo- : 701
Aniline, 2,6-diisopropyl- : 636
Aniline, 2,6-dimethoxy- : 703
Aniline, *N*,*N*-dimethyl-*p*-[(α,α,α-trifluoro-*m*-tolyl)azo]- : 723
Aniline, 2,3-dinitro- : 724
Aniline, 2,6-dinitro- : 726
Aniline, 3,5-dinitro- : 727
Aniline, hydrochloride : 756
Aniline, 2-methylthio : 819
Aniline nitrate : 821
Aniline, 5-nitro-2-propoxy : 827
p-Anilinesulfonamide : 2187
2-Anilinesulfonic acid : 2215
3-Anilinesulfonic acid : 2216
Aniline-4-sulfonic acid : 2217
Aniline, 2,3,4,5-tetramethyl : 847
Aniline, 2,3,4-trichloro- : 852
p-Anilinobenzenesulfonic acid : 2234
o-Anilinobenzoic acid : 2715
Anilotic acid : 2647
p-Anisaldehyde : 585
Anise alcohol : 1891
p-Anisic acid : 2666
m-Anisic acid, 2-(methylamino)-, methyl ester : 2672
o-Anisidine : 766
m-Anisidine : 767
p-Anisidine : 768
m-Anisidine, 4-nitro- : 776
Anisindione : 7269
Anisole : 1915
Anisole, 3,5-dinitro- : 1924
Anisole, 2,5-dinitro- : 1926
Anisole, 2,3-dinitro : 1922
Anisole, 4-nitro : 1937
Anisole, 2,3,4,6-tetrabromo : 2256
Anisole, 3,4,5-tribromo : 2310
Anisole, 2,4,5-tribromo : 2311
Anisole, 2,3,5-tribromo : 2312
o-(*p*-Anisoyl)benzoic acid : 2667
p-Anisoyl chloride : 2901
p-Anisyl alcohol : 1891
o-Anisyl benzoate : 9383
1-(*p*-Anisyl)ethanol : 1895
[6]Annulene : 867
[8]Annulene : 4742
Anthorine : 475
Anthracene, 9,10-dibenzyl- : 367
Anthracene, 9,3-dimethyl : 378
Anthragallol : 415
Anthralin : 430
Anthranil : 2390
o-Anthranilic acid : 2484
Anthranilic acid, *N*-methyl-, ethyl ester : 2680
Anthranilic acid, 3-nitro- : 2506
Anthranilic acid, *N*-(α,α,α-trifluoro-*m*-tolyl)- : 2729
Anthranol : 432
Anthraquinone : 379
Anthraquinone, 2-amino-1-hydroxy : 383
Anthraquinone, 2,7-dibromo- : 390
9,10-Anthraquinone, 1,7-dihydroxy : 394
Anthrarobin : 429
Anthrarufin : 393
1-Anthroic acid : 368
2-Anthroic acid : 369
9-Anthroic acid : 370
Anthrone : 433
Anthropodeoxycholic acid : 4097
Anticanitic vitamin : 2486
Antifebrin : 86
Antimonate(2-), bis[μ-[2,3-dihydroxybutanedioato(4-)-O1,O2:O3,O4]]di-, dipotassium, trihydrate, stereoisomer : 11588
Antipyrine : 10847
Antipyrine, 4-bromo- : 10846
ANTU : 11807
Apacil : 2498
Apacizin : 2500
Aphylline : 7572
Apigenin : 2806
Apigenin 4'-methyl ether : 2809
Apiole : 2423
Apoatropine : 928
Apocodeine : 5160
Apocynin : 5824
Apomorphine : 5155
6aα-Aporphine-2,9-diol, 1,10-dimethoxy- : 5156
6aα-Aporphine, 1,2,9,10-tetramethoxy- : 5157
6aα-Aporphin-11-ol, 10-methoxy-1,2-(methylenedioxy)- : 2470
6aα-Aporphin-11-ol, 1,2,10-trimethoxy- : 5161
Aprobarbital : 11098
Arabinose (*D*)(β) : 436
Arachic alcohol : 5358
Arachidic acid : 5355
Arachidonic acid : 5361
Aramite : 11579
Arborine : 11200
Arbutin : 6224
Arecaidine : 10913
Arecoline : 10914
Argobyl : 10489
Armstrong's acid : 7867
Arsanilic acid : 456
Arsine, difluoromethyl- : 463
Arsine, oxybis[dimethyl- : 452
Ascaridole : 5184
Ascensil : 10867
Asculetine : 2804
Asmacoril : 2840
Aspartame : 9511

SYNONYM INDEX OF ORGANIC COMPOUNDS (continued)

SYNONYM INDEX OF ORGANIC COMPOUNDS (continued)

Ether, benzyl isopentyl : 1951
Ether, benzyl methyl : 1930
Ether, benzyl (3-methylbutyl) : 1951
Ether, bis(2-bromoethyl) : 5609
Ether, bis(bromomethyl) : 7524
Ether, bis(*p*-bromophenyl) : 2072
Ether, bis(2-butoxyethyl) : 3417
Ether, bis(1-chloroethyl) : 5610
Ether, bis(chloroethyl) : 5611
Ether, bis(2-chloroethyl) : 5611
Ether, bis(chloromethyl) : 7525
Ether, bis(2-chloro-1-methylethyl) : 10123
Ether, bis(*p*-chlorophenyl) : 2073
Ether, bis(2-chloropropyl) : 10121
Ether, bis(3-chloropropyl) : 10122
Ether, bis(2-ethoxyethyl) : 5612
Ether, bis[2-(2-methoxyethoxy)ethyl] : 8995
Ether, bis(2-methoxyethyl) : 5613
Ether, bis(*o*-methoxyphenyl) : 2075
Ether, 2-bromethyl phenyl : 1033
Ether, 4-bromobutyl phenyl : 1003
Ether, 2-bromocyclohexyl methyl : 4317
Ether, 2-bromoethyl ethyl : 5461
Ether, 2-bromoethyl methyl : 5465
Ether, 6-bromohexyl methyl : 6736
Ether, bromomethyl ethyl : 5464
Ether, bromomethyl methyl : 7472
Ether, 4-bromophenyl hexyl : 1045
Ether, *p*-bromophenyl hexyl : 1045
Ether, *p*-bromophenyl phenyl : 1079
Ether, 3-bromopropyl phenyl : 1085
Ether, 1,3-butadienyl ethyl : 3160
Ether, 2,3-butadienyl methyl : 3167
Ether, 1,3-butadienyl methyl : 3168
Ether, 1-buten-3-ynyl methyl : 3891
Ether, butyl 2-chloroethyl : 3249
Ether, butyl ethyl : 3372
Ether, *sec*-butyl ethyl : 3373
Ether, *tert*-butyl ethyl : 10040
Ether, butyl ethynyl : 3377
Ether, butyl isobutyl : 3401
Ether, butyl isopropyl : 3399
Ether, *tert*-butyl isopropyl : 10081
Ether, butyl methyl : 3391
Ether, *tert*-butyl methyl : 10072
Ether, butyl phenyl : 1114
Ether, *tert*-butyl phenyl : 1458
Ether, *sec*-butyl phenyl : 2030
Ether, butyl propyl : 3421
Ether, *tert*-butyl propyl : 10087
Ether, butyl *m*-tolyl : 1116
Ether, butyl *p*-tolyl : 1117
Ether, butyl vinyl : 3369
Ether, *sec*-butyl vinyl : 3370
Ether, *tert*-butyl vinyl : 10036
Ether, 2-butynyl methyl : 3915
Ether chloratus : 5467
Ether, 1-chloroethyl 2-chloroethyl : 5468
Ether, 1-chloroethyl ethyl : 5473
Ether, 2-chloroethyl ethyl : 5474
Ether, 1-chloroethyl methyl : 5479
Ether, 2-chloroethyl methyl : 5480
Ether, 1-chloroethyl propyl : 9854
Ether, 2-chloroethyl vinyl : 5887
Ether, chloromethyl ethyl : 5478
Ether, 2-chloro-1-methylethyl methyl : 9859
Ether, chloromethyl methyl : 7483
Ether, chloromethyl propyl : 9858
Ether, *p*-chlorophenyl phenyl : 1221
Ether, 3-chloropropyl methyl : 9860
Ether, 2-chloropropyl methyl : 9861
Ether, 2-chloro-α,α,α-trifluoro-*p*-tolyl 3-ethoxy-4-nitrophenyl : 8633
Ether, 2-chloro-α,α,α-trifluoro-*p*-tolyl *p*-nitrophenyl : 8082
Ether, 2-chlorovinyl ethyl : 5886
Ether, cinnamyl ethyl : 1666
Ether cyanatus : 10090
Ether, cyclohexyl methyl : 4425
Ether, cyclohexyl phenyl : 1253
Ether, cyclopropyl methyl : 4980
Ether, decyl vinyl : 5051
Ether, diallyl : 10592
Ether, dibenzyl : 2079
Ether, 1,2-dibromoethyl ethyl : 5523
Ether, 2,3-dibromopropyl methyl : 9888
Ether, dibutyl : 3415
Ether, di-*tert*-butyl : 10124
Ether, 1,2-dichloroethyl ethyl : 5533
Ether, 2,4-dichlorophenyl *p*-nitrophenyl : 8081
Ether, 1,2-dichlorovinyl ethyl : 5899
Ether, dicyclohexyl : 4461
Ether, diethyl : 5608
Ether, diethyl,2-amino : 5423
Ether, diethyl,2-bromo : 5461
Ether, diethyl,1-chloro : 5473
Ether, diethyl,2-chloro : 5474
Ether, diethyl,2,2'-dibromo : 5609
Ether, diethyl,1,2,-dichloro : 5533
Ether, diethyl,2,2,'-dichloro : 5611
Ether, diethyl,2,2'-dicyano : 10114
Ether, diethyl,2,2'-diphenyl : 2074
Ether, diethyl,1,1'-diphenyl(*DL*) : 2081
Ether, difurfuryl : 6156
Ether, dihexdecyl : 6642
Ether, dihexyl : 6842
Ether, dihexyl,2,2'-diethyl : 6431
Ether, diisopropyl : 10120
Ether, diisopropyl,β,β'-dichloro : 10123
Ether, dimethyl : 7523
Ether, dimethyl,chloro : 7483
Ether, 2,2'-dinaphthyl : 7940
Ether, 2,4-dinitrophenyl phenyl : 1495
Ether, 2,4'-dinitro-4-trifluoromethyldiphenyl : 5980
Ether, dioctyl : 8416
Ether, dipentyl : 8835
Ether, diphenyl : 2071
Ether, diphenyl,4-bromo : 1079
Ether, diphenyl,4-chloro : 1221
Ether, diphenyl,4,4'-diamino : 832
Ether, diphenyl,4,4'-dibromo : 2072
Ether, diphenyl,4,4'-dichloro : 2073
Ether, diphenyl,2,2'-dimethoxy : 2075
Ether, diphenyl,2,2'-dimethyl : 2076
Ether, diphenyl,3,3'-dimethyl : 2077
Ether, diphenyl,4,4'-dimethyl : 2078
Ether, diphenyl,2-methoxy : 1938
Ether, diphenyl,3-methyl : 2013
Ether, diphenyl,2-nitro : 2053
Ether, diphenyl,4-nitro : 2054
Ether, dipropyl : 10119
Ether, dipropyl,2,2'-dichloro : 10121
Ether, dipropyl,3,3'-dichloro : 10122
Ether, divinyl : 5912
Ether, 2,3-epoxypropyl ethyl : 8614
Ether, 2,3-epoxypropyl phenyl : 8623
Ether, ethyl ethynyl : 5942
Ether, ethyl furfuryl : 6104
Ether, ethyl heptyl : 6406
Ether, ethyl hexyl : 6814
Ether, 2-ethylhexyl vinyl : 6405
Ether, ethyl isobutyl : 10039
Ether, ethyl isopentyl : 3374
Ether, ethyl isopropyl : 10038
Ether, ethyl methyl : 5602
Ether, ethyl 1-methyleneallyl : 3161
Ether, ethyl octyl : 8402
Ether, ethyl *tert*-pentyl : 3375
Ether, ethyl pentyl : 8808
Ether, ethyl propyl : 10037
Ether, ethyl 1-propynyl : 10732
Ether, ethyl 2-propynyl : 10733
Ether, ethyl styryl : 1651
Ether, ethyl vinyl : 5905
Ether, ethynyl methyl : 5944
Ether, ethynyl phenyl : 1716
Ether, ethynyl propyl : 10045
Ether, furfuryl methyl : 6128
Ether, 2-furyl phenyl : 6157
Ether, heptyl methyl : 6421
Ether, heptyl phenyl : 1747
Ether, hexadecyl phenyl : 1752
Ether, hexadecyl vinyl : 6638
Ether, hexyl phenyl : 1765
Ether hydrochloric : 5467
Ether, iodomethyl methyl : 7519
Ether, isobutyl methyl : 10071
Ether, isobutyl phenyl : 2031
Ether, isobutyl propyl : 10086
Ether, isobutyl vinyl : 10035
Ether, isopentyl methyl : 3395
Ether, isopentyl vinyl : 3371
Ether, isopropyl methyl : 10070
Ether, isopropyl phenyl : 1974
Ether, isopropyl propyl : 10076
Ether, isopropyl *m*-tolyl : 1987
Ether, isopropyl vinyl : 10034
Ether, *m*-methoxyphenyl *o*-methoxyphenyl : 1929
Ether, methyl 2-methylallyl : 10577
Ether, methyl 2-methylbutyl : 3394
Ether, methyl 1-methyleneallyl : 3169
Ether, methyl 2-octynyl : 8536
Ether, methyl *tert*-pentyl : 3396
Ether, methyl pentyl : 8822
Ether, methyl propyl : 10069
Ether, methyl 2-propynyl : 10736
Ether, methyl styryl : 1928
Ether, methyl vinyl : 5908
Ether muriatic : 5467
Ether, 1-naphthyl 2-naphthyl : 7928
Ether, 1-naphthyl phenyl : 7945
Ether, 2-naphthyl phenyl : 7946
Ether, *o*-nitrophenyl phenyl : 2053
Ether, *p*-nitrophenyl phenyl : 2054
Ether, *p*-nitrophenyl *p*-tolyl : 2006
Ether, *p*-nitrophenyl α,α,α-trifluoro-2-nitro-*p*-tolyl : 5980
Ether, octadecyl vinyl : 8240
Ether, octyl phenyl : 2069
Ether, octyl vinyl : 8401
Ether, phenyl *m*-tolyl : 2013
Ether, phenyl propyl : 2168
Ether, phenyl *o*-tolyl : 2012
Ether, phenyl vinyl : 1646
Ether, propyl vinyl : 10033
Ethide : 5536
17α-Ethinyl-17β-estradiol : 8216
Ethionamide : 10885
Ethionine : 991
Ethisterone : 9758
Ethohexadiol : 6802
Ethosuximide : 11158
α-Ethoxyacetaldehyde : 14
2-Ethoxyacetic acid : 180
Ethoxyacetic acid methyl ester : 182
2-Ethoxyacetophenone : 5814
o-Ethoxybenzonitrile : 2754
6-Ethoxy-2-benzothiazolesulfonamide : 2865
2-Ethoxybenzylamine : 1808
3-Ethoxybiphenyl : 3071
N-(Ethoxycarbonyl)aniline : 3965
Ethoxycarbonyl chloride : 4018
Ethoxydiethylphosphine oxide : 9527
2-Ethoxy-2,3-dihydro-3,3-dimethyl-5-benzofuranyl methanesulfonate : 5933
2-Ethoxy-1,3-dinitrobenzene : 1650
2-Ethoxyethyl acetate : 5718
(±)-2-Ethoxy-1-ethyl-2-oxoethyl 5-[2-chloro-4-(trifluoromethyl)phenoxy]-2-nitrobenzoate : 7410
5-Ethoxy-2-hydroxybenzaldehyde : 556
(±)-2-[1-(Ethoxyimino)butyl]-5-[2-(ethylthio)propyl]-3-hydroxy-2-cyclohexen-1-one : 11359
1-Ethoxy-2-methoxyethane : 5591
o-Ethoxy-β-methylstyrene : 1665
2-Ethoxynaphthalene : 7884
2-Ethoxy-1-propanol : 10380
1-Ethoxy-2-propanol : 10381
2-Ethoxypropionitrile : 10102
Ethoxyquin : 11265
8-Ethoxy-5-quinolinesulfonic acid : 11293
Ethoxytrifluorosilicon : 11429
Ethoxzolamide : 2865
N-Ethylacetamide : 61
Ethyl acetate : 186
Ethylacetic acid : 3460
Ethyl acetoacetate : 3571
Ethyl acetylacetate : 3571
Ethylacetylene : 3897
Ethyl acrylate : 10635
Ethyl adipate : 6785
Ethyl alcohol : 5672
Ethylallene : 8658
3-Ethyl-1-allyl-6-aminouracil : 11060
1-Ethylallyl bromide : 9009
1-Ethylallyl chloride : 9017

SYNONYM INDEX OF ORGANIC COMPOUNDS (continued)

4-Hydroxy-1-pentanol : 8786
1-Hydroxy-2-pentanone : 8971
2-Hydroxy-3-pentanone : 8975
3-Hydroxy-1-pentyne : 9140
1-Hydroxypent-2-yn-4-ene : 9118
3-Hydroxy-9,10-phenanthraquinone : 9176
2-(β-Hydroxyphenethyl-amino)pyridine : 1913
Hydroxyphenylamine : 9200
o-Hydroxyphenylamine : 9201
m-Hydroxyphenylamine : 9202
N-(4-Hydroxyphenyl)glycine : 6267
1-Hydroxy-3-phenylpropane : 2148
1-(*m*-Hydroxyphenyl)-1-propanol : 1872
DL-2-Hydroxy-3-phenylpropionic acid : 2133
4-Hydroxyphenylpropionic acid : 2134
3-Hydroxy-3-phenylprop-1-yne : 1876
4-Hydroxyphenylsulfonic acid : 2226
17α-Hydroxyprogesterone : 9750
4-Hydroxy-*L*-proline : 9767
3-Hydroxy-1-propene : 10697
2-Hydroxypropionitrile : 10105
(3-Hydroxypropyl)benzene : 2148
1-Hydroxy-1-propylcyclohexane : 4557
3-Hydroxypropylene oxide : 8618
N-(2-Hydroxypropyl)ethylenediamine : 10340
3-Hydroxypropyl fluoride : 10382
2-(2-Hydroxypropyl)pyridine : 10951
2-(3-Hydroxypropyl)pyridine : 11012
(2-Hydroxypropyl)trimethyl ammonium chloride : 9831
4-Hydroxypyrazolo[3,4-d]pyrimidine : 10852
5-Hydroxyquinoline : 11305
7-Hydroxyquinoline : 11307
8-Hydroxyquinoline : 11308
8-Hydroxyquinoline sulfate : 11324
8-Hydroxy-5-quinolinesulfonic acid : 11294
p-Hydroxysalicylic acid : 2569
5-Hydroxysalicylic acid : 2570
2α-Hydroxytigogenin : 11530
2-Hydroxy-*m*-toluic acid : 2629
2-Hydroxy-*p*-toluic acid : 2630
6-Hydroxy-*m*-toluic acid : 2631
2-Hydroxytridecane : 11884
11-Hydroxyundecanoic acid : 11948
4-Hydroxyundecanoic acid lactone : 6152
Hydroxyurea : 12016
Hydroxyzine : 5700
Hygrine : 10471
Hymecromone : 2831
Hyodeoxycholic acid : 4096
Hyoscyamine : 917
Hypothiazide : 2855
Hypoxanthine : 10770
Hypoxanthine riboside : 7335
Ibogaine : 7186
Ibrotamide : 3199
Imecromone : 2831
1H-Imidazole, 1-[2-(2,4-dichlorophenyl)-2-(2-propenyloxy)ethyl]- : 7187
1H-Imidazole, 2-heptadecyl-4,5-dihydro-, monoacetate : 6275
2-Imidazolidine, 2-benzyl : 7198
1-Imidazolidinecarboxamide, 3-(3,5-dichlorophenyl)-*N*-(1-methylethyl)-2,4-dioxo- : 7338
7H-Imidazo[4,5-d]pyrimidine : 10752
2-Iminobenzothiazoline : 2859
Iminodiacetic acid : 6260
2,2′-Iminodibenzoic acid : 2656
Iminourea : 6278
Imperatorin : 6189
Indaconitine : 307
Indan : 7250
Indan, 1-amino (*DL*) : 7243
Indan, 1,2-dibromo- : 7247
Indan, 1,2-dichloro : 7249
Indan, 1,1-diemthyl : 7251
Indan, 1,2-dimethyl- : 7252
Indan, 4,7-dimethyl- : 7253
Indan, 5,6-dimethyl- : 7254
Indan, 1-ethyl- : 7273
Indan, 5-hexyl- : 7274
Indan, 1-methyl- : 7255
Indan, 2-methyl- : 7256
Indan, 4-methyl- : 7257
Indan, 5-methyl- : 7258
2-Indanone : 7287
1-Indanone, 4-methyl- : 7290
1-Indanone, 2-phenyl- : 7291
Indan, 2-phenyl- : 7259
Indan, 1-phenyl-1,3,3-trimethyl : 7263
Indan, 1,1,4,7-tetramethyl- : 7260
Indanthrene : 434
Indan, 1,1,5-trimethyl- : 7261
Indan, 1,4,7-trimethyl : 7262
1-Indenecarboxylic acid : 7246
Indene, 1,3-diphenyl- : 7271
Indene, 2,3-diphenyl- : 7272
Indene, 4-methyl- : 7278
Indene, 6-methyl- : 7279
Indene, 7-methyl- : 7280
Indigo : 7331
Indigo Yellow : 2843
Indoleacetic acid : 7295
Indolebutyric acid : 7296
Indole, 7-methyl- : 7317
Indole, 1,2,3-trimethyl- : 7322
Indole, 2,3,5-trimethyl- : 7323
2-Indolinone, 1-ethyl-3-methyl- : 7332
1H-Indolo[3,2,1-de]pyrido[3,2,1-ij][1,5]naphthyridine, eburnamenine-14-carboxylic acid deriv. : 5349
Indolo[4,3-fg]quinoline, ergoline-8-carboxamide deriv. : 5373
3-Indolylacetone : 10442
Indonaphthene : 7245
Indoxazene : 2389
Inosine, 2-amino- : 6289
Iodipamide : 2594
5-Iodoacenaphthene : 7
Iodoacetic acid : 200
Iodoacetonitrile : 283
Iodoalphionic acid : 2135
p-Iodoaniline : 761
o-Iodoanisole : 1775
Iodobenzene : 1766
p-Iodobenzenesulfonyl chloride : 2247
o-Iodobenzoic acid : 2657
4-Iodo-1,2-butadiene : 3166
4-Iodocatechol : 1523
Iodochlorhydroxyquin : 11312
Iodocyanide (ICN) : 7337
Iodocyclohexane : 4415
Iodocyclopentane : 4824
2-Iodo-*p*-cymene : 1781
1-Iododecane : 5053
Iododifluoromethane : 7505
1-Iodo-3,5-dimethylbenzene : 1771
4-Iododiphenyl : 3079
1-Iodododecane : 5302
Iodoethane : 5597
2-Iodoethanol : 5731
Iodoethene : 5907
Iodoethylene : 5907
Iodoform : 7560
2-Iodofuran : 6110
3-Iodofuran : 6111
1-Iodoheptane : 6419
2-Iodoheptane : 6420
1-Iodo-1-heptyne : 6620
1-Iodohexadecane : 6640
1-Iodohexane : 6831
Iodomethane : 7518
2-Iodo-4-methylaniline : 762
4-Iodo-2-methylbutane : 3384
(Iodomethyl) methyl ether : 7519
1-Iodonaphthalene : 7902
1-Iodooctadecane : 8242
1-Iodooctane : 8406
1-Iodopentane : 8818
3-Iodopentane : 8820
1-Iodo-1-pentyne : 9133
o-Iodophenol : 9371
p-Iodophenol : 9373
1-Iodopropane : 10054
2-Iodopropane : 10055
3-Iodopropene : 10573
3-Iodopropylene : 10573
1-Iodopropyne : 10734
3-Iodopropyne : 10735
2-Iodopyridine : 10970
Iodopyrrole : 11140
Iodoquinol : 11315
4-Iodoquinoline : 11272
Iodostarin : 8300
N-Iodosuccinimide : 11159
Iodotrifluoroethylene : 5925
Iodotrifluoromethane : 7559
3-Iodotyrosine : 11932
3-Iodoveratrole : 1767
4-Iodo-*o*-xylene : 1774
β-Ionol : 3873
α-Ionol : 3874
Irene : 7999
Iron dicyclopentadienyl : 5960
β-Irone : 3883
α-Irone : 3884
Iron pentacarbonyl : 7339
Iron, tris(dimethylcarbamodithioato-*S,S′*)-, (OC-6-11)- : 5957
Isatin : 7300
Isoamylaldehyde : 3186
Isoamylamine : 3220
Isoamyl benzoate : 3632
Isoamyl bromide : 3239
Isoamyl chloride : 3252
Isoamyl cyanide : 8831
Isoamyl ether : 3418
Isoamyl iodide : 3384
Isoamyl isovalerate : 3551
Isoamyl 3-methylbutyrate : 3551
Isoamyl nitrate : 3637
Isoamyl nitrite : 8090
Isoamyl salicylate : 2633
Isoamyl sulfide : 3436
Isobutamben : 2505
Isobutanal : 9785
Isobutane : 10073
Isobutane, 1,1-dichloro : 9899
Isobutane, 1,2-dichloro : 9900
Isobutane, 1,1,1,2,3-pentachloro : 10127
Isobutane, 1,1,1,2-tetrachloro : 10147
Isobutane, 1,1,2-trichloro : 10177
Isobutene : 10578
2-Isobutoxyethanol : 5746
1-Isobutoxy-2-nitrobenzene : 2034
Isobutryic anhydride : 10291
Isobutyl acetate : 215
Isobutyl acrylate : 10673
Isobutyl alcohol : 10388
Isobutyl alcohol, 2-chloro : 10352
Isobutyl alcohol, 3-chloro : 10353
Isobutylamine : 9817
Isobutyl *p*-aminobenzoate : 2505
Isobutylbenzene : 2039
Isobutyl bromide : 9845
Isobutyl butanoate : 3563
Isobutyl carbamate : 3959
Isobutyl chloride : 9862
Isobutyl chlorocarbonate : 4024
Isobutyl chloroformate : 4024
Isobutyl cyanide : 3410
Isobutylcyclopentane : 4838
Isobutyldimethylcarbinol : 8928
Isobutylene chlorohydrin : 10354
Isobutylene, 1,3-dibromo : 10546
Isobutylene glycol, 3-chloro : 9998
Isobutylene (trimer) : 10582
Isobutylethene : 9054
Isobutyl formate : 6013
Isobutylhydrochlorothiazide : 2856
Isobutyl iodide : 10057
Isobutyl isobutanoate : 10304
Isobutyl isocyanate : 10061
Isobutyl isovalerate : 3555
Isobutyl ketone : 6514
Isobutyl mercaptan : 10158
Isobutyl methacrylate : 10666
Isobutyl methyl ether : 10071
Isobutyl methyl ketone : 8981
Isobutyl nitrate : 8078
Isobutyl nitrite : 8094
Isobutyloxycarbonyl chloride : 4024
Isobutyl propionate : 10310
Isobutyl propyl ketone : 6524
4-Isobutylresorcinol : 1539
Isobutyl salicylate : 2643
Isobutyl stearate : 8268

SYNONYM INDEX OF ORGANIC COMPOUNDS (continued)

SYNONYM INDEX OF ORGANIC COMPOUNDS (continued)

2,4,5-T Butoxyethyl ester : 257
2,4,5-T Butyl ester : 258
2,4,5-T *sec*-Butyl ester : 260
2,4,5-T Isopropyl ester : 259
Tabutrex : 3301
Tachysterol : 11349
Tagatose, *D*- : 11587
L-Tartaric acid : 3309
(+)-Tartaric acid : 3309
DL-Tartaric acid : 3311
(±)-Tartaric acid : 3311
(*R*,*S*)-Tartaric acid : 3304
meso-Tartaric acid : 3304
(-)-*D*-Tartaric acid : 3312
D-Tartaric acid : 3312
(*S*,*S*)-Tartaric acid : 3312
Tartaric acid, diethyl ester (*DL*) : 3307
Tartaric acid, diisopropyl ester (*DL*) : 3305
Tartaric acid, dipropyl ester, (+)- : 3310
Tartaric acid, disodium salt : 3314
Tartaric acid, ethyl ester : 3315
Tartronic acid : 9956
Taurine : 5623
Taurocholic acid : 5625
Tebanyl : 2502
Tegosept P : 2653
Tephrosin : 3110
Terephthalic acid : 1329
Terephthalic acid, 2,5-dichloro- : 1349
Terephthalic acid, diethyl ester : 1350
Terpenolic acid : 6030
Terpenylic acid : 6030
o-Terphenyl : 11595
m-Terphenyl : 11596
p-Terphenyl : 11597
β-Terpineol : 4542
Terpinolene : 4684
α-Terthienyl : 11600
Testosterone : 356
Testosterone-17-propionate : 359
1,4,7,10-Tetraazadecane : 5499
1,5,10,14-Tetraazatetradecane : 3259
Tetrabenzocyclooctatetraene : 11644
2,3,4,6-Tetrabromoanisole : 2256
1,2,3,5-Tetrabromobenzene : 2253
2,3,5,6-Tetrabromobenzene : 2254
1,2,4,5-Tetrabromobenzene : 2254
3,4,5,6-Tetrabromo-*o*-cresol : 9468
1,2,3,4-Tetrabromocyclobutane : 4195
1,2,2,2-Tetrabromoethane : 5631
Tetrabromoethylene : 5914
2,3,4,6-Tetrabromophenol : 9467
3',3",5',5"-Tetrabromophenolphthalein : 7350
1,1,2,2-Tetrabromopropane : 10136
1,2,2,3-Tetrabromopropane : 10138
2,3,4,5-Tetrabromothiophene : 11781
Tetracaine hydrochloride : 2531
1,1,3,3-Tetracarbethoxypropane : 10140
1,2,4,5-Tetracarboxybenzene : 2257
Tetracene : 7703
1,2,3,4-Tetrachlorobenzene : 2259
1,2,4,5-Tetrachlorobenzene : 2260
s-Tetrachlorobenzene : 2260
Tetrachloro-*p*-benzoquinone : 4268
1,1,1,2-Tetrachlorobutane : 3427
1,2,2,3-Tetrachlorobutane : 3428
1,2,3,3-Tetrachlorobutane : 3429
1,3,4,4-Tetrachloro-1-butene : 3806
2,3,3,4-Tetrachloro-1-butene : 3807
1,1,2,3-Tetrachloro-2-(chloromethyl)propane : 10146
Tetrachloro-1,2-difluoroethane : 5637
1,1,1,2-Tetrachloroethane : 5634
Tetrachloroethylene : 5917
1,1,1,2-Tetrachloro-2-fluoroethane : 5639
1,1,1,2-Tetrachloro-2-methylpropane : 10147
1,1,1,5-Tetrachloropentane : 8838
2,3,4,6-Tetrachlorophenate : 9470
2,3,4,5-Tetrachlorophenol : 9469
2,3,5,6-Tetrachlorophenol : 9471
Tetrachlorophthalic anhydride : 7346
1,1,1,2-Tetrachloropropane : 10141
1,1,1,3-Tetrachloropropane : 10142
1,1,2,2-Tetrachloropropane : 10143
1,1,2,3-Tetrachloropropane : 10144
1,2,2,3-Tetrachloropropane : 10145
1,1,3,3-Tetrachloro-2-propanol : 10402
2,3,5,6-Tetrachloropyridine : 11019
Tetrachloro-*p*-quinone : 4268
3,3',4',5-Tetrachlorosalicylanilide : 610
2,3,4,6-Tetrachlorotoluene : 2262
2,3,5,6-Tetrachlorotoluene : 2263
Tetracosamethylhendecasiloxane : 11964
Tetracosamethylundecasiloxane : 11964
Tetracosylbenzene : 2265
Tetracosylcyclohexane : 4474
Tetracyanoethylene : 5915
Tetracycline : 7706
Tetradecamethylcycloheptasiloxane : 4239
Tetradecamethylene dibromide : 11616
Tetradecamethylhexasiloxane : 6990
Tetradecane, 1-cyclopentyl- : 4850
Tetradecane, 1,1-dicyclohexyl : 4476
Tetradecane, 1,1-diphenyl- : 2267
Tetradecanoic acid, chloride : 11635
Tetradecyl alcohol : 11630
Tetradecylsulfonic acid : 11620
Tetradifon : 2325
Tetraethylammonium bromide : 5447
1,2,3,4-Tetraethylbenzene : 2268
1,2,4,5-Tetraethylbenzene : 2269
N,*N*,*N*',*N*'-Tetraethyl-1,4-butanediamine : 3262
Tetraethyldiarsenic : 5109
Tetraethyl diphosphate : 5221
Tetraethylene glycol : 5756
Tetraethylene glycol dimethyl ether : 8995
Tetraethylenepentamine : 5498
Tetraethyl lead : 9731
Tetraethylmethane : 8756
N,*N*,*N*',*N*'-Tetraethyl-*m*-phenylenediamine : 1285
N,*N*,*N*',*N*'-Tetraethylphthalamide : 1325
Tetraethyl 1,1,2,3-propanetetracarboxylate : 10139
Tetraethyl 1,1,3,3-propanetetracarboxylate : 10140
Tetraethyl pyrophosphate : 5221
Tetraethylsilane : 11450
Tetraethyl silicate : 11506
2,3,4,5-Tetrafluorobenzene : 2270
1,2,3,4-Tetrafluorobenzene : 2270
1,2,3,5-Tetrafluorobenzene : 2271
2,3,5,6-Tetrafluorobenzene : 2272
1,2,4,5-Tetrafluorobenzene : 2272
1,1,1,2-Tetrafluoroethane : 5640
1,1,2,2-Tetrafluoroethane : 5641
Tetrafluoroethylene : 5918
2,2,3,3-Tetrafluoro-1-propanol : 10404
Tetrahelicene : 2406
(1-4)-Tetrahydroacenaphthene : 9
Tetrahydrobenzene : 4626
2,3,6,7-Tetrahydro-1H,5H-benzo[ij]quinolizine : 2737
Tetrahydrocortisone : 9744
1,2,3,4-Tetrahydro-1,2-dimethylnaphthalene : 7981
Tetrahydro-5,5-dimethyl-2-(1H)-pyrimidinone[3[4-(trifluoromethyl-styryl)-cinnamylidene- hydrazone : 7136
Tetrahydroeucarvone : 4238
Tetrahydrofuran : 6168
Tetrahydro-2,5-furandimethanol : 7109
Tetrahydrofurfuryl alcohol : 6124
Tetrahydrofurfurylamine : 6116
Tetrahydro-2H-pyran : 10800
5,6,7,8-Tetrahydroisoquinoline : 7389
Tetrahydro-3-methylfuran : 6172
Tetrahydromyrcenol : 8450
5,6,7,8-Tetrahydronaphthalene-2-sulfonyl chloride : 7978
1,2,3,4-Tetrahydro-2-naphthoic acid : 7778
Tetrahydro-1,4-oxazine : 7681
1,2,3,4-Tetrahydrophenanthrene : 9187
3,4,5,6-Tetrahydrophthalic anhydride : 7347
1,2,3,6-Tetrahydropyridine : 11020
5,6,7,8-Tetrahydroquinoline : 11297
Tetrahydroquinone : 4377
1,2,3,4-Tetrahydroquinoxaline : 11339
D-Tetrahydroxyadipic acid : 6205
Tetrahydroxy-1,4-benzoquinone : 4269
3,3',5,5'-Tetrahydroxybiphenyl : 3097
3,4',5,7-Tetrahydroxy-3',5'-dimethoxyflavylium chloride : 2847
3,3',4',7-Tetrahydroxyflavone : 2813
3',4',5,7-Tetrahydroxyflavone : 2814
2',3,5,7-Tetrahydroxyflavone : 2842
3,4',5,7-Tetrahydroxyflavone : 2843
3,5,3',4'-Tetrahydroxy-7-methoxyflavone : 2815
N,*N*,*N*',*N*'-Tetra(2-hydroxypropyl)ethylenediamine : 10379
1,2,4,5-Tetraiodobenzene : 2275
1,2,3,4-Tetraiodobenzene : 2273
1,2,3,5-Tetraiodobenzene : 2274
Tetraiodoethylene : 5919
Tetraiodomethane : 7542
1,2,4,5-Tetraisopropylbenzene : 2276
Tetrakisdimethylaminophosphonous anhydride : 5220
Tetrakis(2-hydroxypropyl)-ethylenediamine : 10379
Tetralin : 7979
Tetralin, 1,1-dimethyl : 7980
Tetralin, 1,2-dimethyl : 7981
Tetralin, 1,3-dimethyl : 7982
Tetralin, 1,4-dimethyl : 7983
Tetralin, 2,2-dimethyl : 7985
Tetralin, 2,3-dimethyl : 7986
Tetralin, 2,5-dimethyl : 7987
Tetralin, 2,7-dimethyl : 7989
Tetralin, 2,8-dimethyl : 7990
Tetralin, 5,6-dimethyl : 7991
Tetralin, 5,7-dimethyl : 7992
Tetralin, 1-ethyl : 7888
Tetralin, 2-ethyl : 7889
Tetralin, 6-ethyl : 7891
6-Tetralinsulfonyl chloride : 7978
2-Tetralol : 8043
1-Tetralone, 7-ethyl : 8051
Tetramethylammonium bromide : 7456
Tetramethylammonium chloride : 7457
Tetramethylammonium iodide : 7458
2,3,5,6-Tetramethylbenzaldehyde : 598
2,2',4,4'-Tetramethylbenzophenone : 7594
Tetramethyl-*p*-benzoquinone : 4270
2,2,3,3-Tetramethylbutane : 3430
1,1,4,4-Tetramethyl-1,4-butanediol : 6801
1,1,3,4-Tetramethylcyclohexane : 4477
cis-1,1,3,5-Tetramethylcyclohexane : 4478
4-(2,5,6,6-Tetramethyl-2-cyclohexen-1-yl)-3-butene-2-one : 3884
1,1,3,3-Tetramethylcyclopentane : 4851
N,*N*,*N*',*N*'-Tetramethyl-1,4-diaminobutane : 3263
N,*N*,*N*',*N*'-Tetramethyldiaminoethane : 5516
Tetramethylene : 4170
2,3-Tetramethyleneaziridine : 478
1,4-Tetramethylenediamine : 3258
Tetramethylenedithiol : 3368
Tetramethylene glycol : 3346
Tetramethylethylene glycol : 3350
2,2,6,6-Tetramethyl-3,5-heptanedione : 6404
2,2,3,3-Tetramethylhexane : 6844
2,2,3,4-Tetramethylhexane : 6845
2,2,3,5-Tetramethylhexane : 6846
2,2,4,4-Tetramethylhexane : 6847
2,2,4,5-Tetramethylhexane : 6848
2,2,5,5-Tetramethylhexane : 6849
2,3,3,4-Tetramethylhexane : 6850
2,3,3,5-Tetramethylhexane : 6851
2,3,4,4-Tetramethylhexane : 6852
2,3,4,5-Tetramethylhexane : 6853
3,3,4,4-Tetramethylhexane : 6854
2,2,5,5-Tetramethyl-3,4-hexanedione : 6811
1,1,4,7-Tetramethylindan : 7260
Tetramethyl lead : 9732
Tetramethylmethanediamine : 7490
1,3,6,8-Tetramethylnaphthalene : 8003
2,2,7,7-Tetramethyl-3,6-octanedione : 8400
2,2,3,3-Tetramethylpentane : 8839
2,2,3,4-Tetramethylpentane : 8840
2,2,4,4-Tetramethylpentane : 8841
2,3,3,4-Tetramethylpentane : 8842
2,3,4,5-Tetramethylphenol : 9473
Tetramethyl-*p*-phenylenediamine : 1287

SYNONYM INDEX OF ORGANIC COMPOUNDS (continued)

MOLECULAR FORMULA INDEX OF ORGANIC COMPOUNDS

5632 - Ethane, 1,1,2,2-tetrabromo-
$C_2H_2ClFO_2$
131 - Acetic acid, chlorofluoro-
$C_2H_2ClF_3$
5487 - Ethane, 1-chloro-1,1,2-trifluoro
5488 - Ethane, 1-chloro-1,2,2-trifluoro
5489 - Ethane, 2-chloro-1,1,1-trifluoro-
C_2H_2ClI
5888 - Ethene, 1-chloro-2-iodo-
C_2H_2ClN
274 - Acetonitrile, chloro-
C_2H_2ClNS
11700 - Thiocyanic acid, chloromethyl ester
$C_2H_2Cl_2$
5894 - Ethene, 1,1-dichloro-
5895 - Ethene, 1,2-dichloro-, (*E*)-
5896 - Ethene, 1,2-dichloro-, (*Z*)-
$C_2H_2Cl_2F_2$
5531 - Ethane, 1,2-dichloro-1,1-difluoro-
5532 - Ethane, 1,2-dichloro-1,2-difluoro-
$C_2H_2Cl_2O$
295 - Acetyl chloride, chloro-
$C_2H_2Cl_2O_2$
162 - Acetic acid, dichloro-
4014 - Carbonochloridic acid, chloromethyl ester
$C_2H_2Cl_3F$
5661 - Ethane, 1,1,2-trichloro-2-fluoro-
$C_2H_2Cl_3NO$
95 - Acetamide, 2,2,2 trichloro-
$C_2H_2Cl_4$
5634 - Ethane, 1,1,1,2-tetrachloro-
5635 - Ethane, 1,1,2,2-tetrachloro-
$C_2H_2F_2$
5902 - Ethene, 1,1-difluoro-
5903 - Ethene, 1,2-difluoro-, (*Z*)
$C_2H_2F_2O_2$
175 - Acetic acid, difluoro-
$C_2H_2F_3I$
5667 - Ethane, 1,1,1-trifluoro-2-iodo-
$C_2H_2F_3NO_2$
5668 - Ethane, 1,1,1-trifluoro-2-nitro-
$C_2H_2F_4$
5640 - Ethane, 1,1,1,2-tetrafluoro-
5641 - Ethane, 1,1,2,2-tetrafluoro-
C_2H_2IN
283 - Acetonitrile, iodo-
$C_2H_2I_2$
5904 - Ethene, 1,2-diiodo-, (*Z*)-
$C_2H_2N_2O$
8563 - 1,3,4-Oxadiazole
$C_2H_2N_2S$
11657 - 1,2,4-Thiadiazole
11658 - 1,2,5-Thiadiazole
$C_2H_2N_4$
11651 - 1,2,4,5-Tetrazine
C_2H_2O
5927 - Ethenone
$C_2H_2O_2$
5491 - Ethanedial
$C_2H_2O_3$
223 - Acetic acid, oxo-
$C_2H_2O_4$
5553 - Ethanedioic acid
C_2H_3Br
5880 - Ethene, bromo-
C_2H_3BrO
291 - Acetyl bromide
$C_2H_3BrO_2$
111 - Acetic acid, bromo-
$C_2H_3Br_3$
5653 - Ethane, 1,1,2-tribromo-
$C_2H_3Br_3O$
5767 - Ethanol, 2,2,2-tribromo-
$C_2H_3Br_3O_2$
5581 - 1,1-Ethanediol, 2,2,2-tribromo-
C_2H_3Cl
5884 - Ethene, chloro-
$C_2H_3ClF_2$
5470 - Ethane, 1-chloro-1,1-difluoro
5471 - Ethane, 2-chloro-1,1-difluoro
C_2H_3ClO
12 - Acetaldehyde, chloro-
294 - Acetyl chloride
$C_2H_3ClO_2$
123 - Acetic acid, chloro-
4020 - Carbonochloridic acid, methyl ester
$C_2H_3Cl_2F$
5534 - Ethane, 1,1-dichloro-1-fluoro-
5535 - Ethane, 1,2-dichloro-1-fluoro-
$C_2H_3Cl_2NO$
51 - Acetamide, 2,2-dichloro-
$C_2H_3Cl_2NO_2$
5536 - Ethane, 1,1-dichloro-1-nitro-
$C_2H_3Cl_3$
5655 - Ethane, 1,1,1-trichloro-
5656 - Ethane, 1,1,2-trichloro-
$C_2H_3Cl_3O$
5768 - Ethanol, 2,2,2-trichloro-
$C_2H_3Cl_3O_2$
5582 - 1,1-Ethanediol, 2,2,2-trichloro-
$C_2H_3Cl_3Si$
11461 - Silane, trichloroethenyl-
C_2H_3F
5906 - Ethene, fluoro-
C_2H_3FO
301 - Acetyl fluoride
$C_2H_3FO_2$
189 - Acetic acid, fluoro-
$C_2H_3F_3$
5665 - Ethane, 1,1,1-trifluoro-
5666 - Ethane, 1,1,2-trifluoro-
$C_2H_3F_3O$
5770 - Ethanol, 2,2,2-trifluoro-
C_2H_3I
5907 - Ethene, iodo-
C_2H_3IO
302 - Acetyl iodide
$C_2H_3IO_2$
200 - Acetic acid, iodo-
$C_2H_3KO_2$
234 - Acetic acid, potassium salt
C_2H_3N
271 - Acetonitrile
7655 - Methylisonitrile
C_2H_3NO
282 - Acetonitrile, hydroxy-
7520 - Methane, isocyanato-
$C_2H_3NO_2$
5911 - Ethene, nitro-
$C_2H_3NO_3$
104 - Acetic acid, aminooxo-
$C_2H_3NO_4$
108 - Acetic acid, anhydride with nitric acid
C_2H_3NS
7521 - Methane, isothiocyanato-
11705 - Thiocyanic acid, methyl ester
$C_2H_3N_3$
11846 - 1H-1,2,3-Triazole
$C_2H_3N_3O_2$
109 - Acetic acid, azido-
11850 - 1,2,4-Triazolidine-3,5-dione
$C_2H_3NaO_2$
237 - Acetic acid, sodium salt
C_2H_4
5879 - Ethene
C_2H_4BrCl
5455 - Ethane, 1-bromo-1-chloro-
5456 - Ethane, 1-bromo-2-chloro-
$C_2H_4BrClO_2S$
5628 - Ethanesulfonyl chloride, 2-bromo-
C_2H_4BrF
5462 - Ethane, 1-bromo-2-fluoro-
C_2H_4BrI
5463 - Ethane, 1-bromo-2-iodo-
C_2H_4BrNO
36 - Acetamide, *N*-bromo-
$C_2H_4Br_2$
5517 - Ethane, 1,1-dibromo-
5518 - Ethane, 1,2-dibromo-
$C_2H_4Br_2O$
7524 - Methane, oxybis[bromo-
C_2H_4ClF
5475 - Ethane, 1-chloro-1-fluoro-
5476 - Ethane, 1-chloro-2-fluoro-
C_2H_4ClI
5477 - Ethane, 1-chloro-2-iodo-
C_2H_4ClNO
41 - Acetamide, 2-chloro-
$C_2H_4ClNO_2$
5482 - Ethane, 1-chloro-1-nitro-
5483 - Ethane, 1-chloro-2-nitro-
$C_2H_4Cl_2$
5528 - Ethane, 1,1-dichloro-
5529 - Ethane, 1,2-dichloro-
$C_2H_4Cl_2O$
5704 - Ethanol, 2,2-dichloro-
7525 - Methane, oxybis[chloro-
$C_2H_4Cl_2O_2S$
5629 - Ethanesulfonyl chloride, 2-chloro-
$C_2H_4Cl_3NO$
5680 - Ethanol, 1-amino-2,2,2-trichloro-
$C_2H_4Cl_4Si$
11391 - Silane, dichloro(dichloromethyl)methyl-
C_2H_4FNO
63 - Acetamide, 2-fluoro
$C_2H_4F_2$
5544 - Ethane, 1,1-difluoro-
5545 - Ethane, 1,2-difluoro-
$C_2H_4F_2O$
5708 - Ethanol, 2,2-difluoro-
$C_2H_4I_2$
5546 - Ethane, 1,1-diiodo-
5547 - Ethane, 1,2-diiodo-
$C_2H_4N_2$
272 - Acetonitrile, amino-
$C_2H_4N_2O_2$
5492 - Ethanedial, dioxime
5493 - Ethanediamide
$C_2H_4N_2O_4$
5551 - Ethane, 1,1-dinitro-
5552 - Ethane, 1,2-dinitro-
8086 - Nitrous acid, 1,2-ethanediyl ester
$C_2H_4N_2O_6$
5574 - 1,2-Ethanediol, dinitrate
$C_2H_4N_2S_2$
5588 - Ethanedithioamide
$C_2H_4N_4$
6280 - Guanidine, cyano-
11845 - 1H-1,2,4-Triazol-3-amine
$C_2H_4N_4O_2$
5124 - Diazenedicarboxamide
C_2H_4O
8599 - Oxirane
11 - Acetaldehyde
$(C_2H_4O)_x$
7440 - Metaldehyde
C_2H_4OS
5646 - Ethanethioic acid
$C_2H_4O_2$
15 - Acetaldehyde, hydroxy-
100 - Acetic acid
6010 - Formic acid, methyl ester
$C_2H_4O_2S$
202 - Acetic acid, mercapto-
$C_2H_4O_3$
196 - Acetic acid, hydroxy-
5619 - Ethaneperoxoic acid
$C_2H_4O_3S$
5200 - 1,3,2-Dioxathiolane, 2-oxide
$C_2H_4O_4S$
5199 - 1,3,2-Dioxathiolane, 2,2-dioxide

$C_2H_4O_5S$
238 - Acetic acid, sulfo-
C_2H_4S
11695 - Thiirane
$C_2H_4S_2$
5589 - Ethane(dithioic) acid
$C_2H_4S_5$
8998 - 1,2,3,5,6-Pentathiepane
C_2H_4Si
11435 - Silane, ethynyl
$C_2H_5AsF_2$
445 - Arsine, ethyldifluoro-
C_2H_5Br
5454 - Ethane, bromo-
C_2H_5BrO
5685 - Ethanol, 2-bromo-
7472 - Methane, bromomethoxy-
C_2H_5Cl
5467 - Ethane, chloro-
C_2H_5ClO
5693 - Ethanol, 2-chloro-
7184 - Hypochlorous acid, ethyl ester
7483 - Methane, chloromethoxy-
$C_2H_5ClO_2S$
4087 - Chlorosulfurous acid, ethyl ester
5627 - Ethanesulfonyl chloride
$C_2H_5ClO_3S$
4085 - Chlorosulfuric acid, ethyl ester
C_2H_5ClS
5651 - Ethanethiol, 2-chloro-
7484 - Methane, chloro(methylthio)-
$C_2H_5Cl_2O_2P$
9577 - Phosphorodichloridic acid, ethyl ester
$C_2H_5Cl_3OSi$
11462 - Silane, trichloroethoxy-
$C_2H_5Cl_3Si$
11390 - Silane, dichloro(chloromethyl)methyl-
11463 - Silane, trichloroethyl-
C_2H_5F
5593 - Ethane, fluoro-
C_2H_5FO
5727 - Ethanol, 2-fluoro-
$C_2H_5F_3OSi$
11429 - Silane, ethoxytrifluoro-
$C_2H_5F_3Si$
11432 - Silane, ethyltrifluoro-
C_2H_5I
5597 - Ethane, iodo-
C_2H_5IO
5731 - Ethanol, 2-iodo-
7519 - Methane, iodomethoxy-
C_2H_5N
498 - Aziridine
C_2H_5NO
17 - Acetaldehyde, oxime
22 - Acetamide
5995 - Formamide, *N*-methyl-
$C_2H_5NO_2$
3953 - Carbamic acid, methyl ester
5607 - Ethane, nitro-
6250 - Glycine
8087 - Nitrous acid, ethyl ester
$C_2H_5NO_3$
5752 - Ethanol, 2-nitro-
8074 - Nitric acid, ethyl ester
C_2H_5NS
5643 - Ethanethioamide
$C_2H_5N_3O$
5681 - Ethanol, 2-azido-
$C_2H_5N_3O_2$
105 - Acetic acid, aminooxo-, hydrazide
7232 - Imidodicarbonic diamide
12019 - Urea, *N*-methyl-*N*-nitroso-
$C_2H_5N_3S_2$
11718 - Thioimidodicarbonic diamide
$C_2H_5N_5O_3$
6284 - Guanidine, *N*-methyl-*N*′-nitro-*N*-nitroso-

C_2H_6
5448 - Ethane
C_2H_6AsCl
453 - Arsinous chloride, dimethyl-
$C_2H_6BF_3O$
3138 - Boron, trifluoro[oxybis[methane]]-, (T-4)-
C_2H_6Cd
3929 - Cadmium, dimethyl-
$C_2H_6ClO_3P$
5929 - Ethephon
$C_2H_6Cl_2NPS$
9557 - Phosphoramidothioic dichloride, dimethyl-
$C_2H_6Cl_2Si$
11394 - Silane, dichlorodimethyl-
C_2H_6Hg
7435 - Mercury, dimethyl-
$C_2H_6N_2$
5125 - Diazene, dimethyl-
5126 - Diazene, dimethyl-, (*E*)-
$C_2H_6N_2O$
195 - Acetic acid, hydrazide
7450 - Methanamine, *N*-methyl-*N*-nitroso-
12017 - Urea, methyl-
$C_2H_6N_2O_2$
7149 - Hydrazinecarboxylic acid, methyl ester
7449 - Methanamine, *N*-methyl-*N*-nitro-
$C_2H_6N_4O$
11993 - Urea, (aminoiminomethyl)-
$C_2H_6N_4O_2$
5562 - Ethanedioic acid, dihydrazide
C_2H_6O
5672 - Ethanol
7523 - Methane, oxybis-
C_2H_6OS
5732 - Ethanol, 2-mercapto-
7529 - Methane, sulfinylbis-
$C_2H_6O_2$
5569 - 1,2-Ethanediol
7179 - Hydroperoxide, ethyl
9159 - Peroxide, dimethyl
$C_2H_6O_2S$
7534 - Methane, sulfonylbis-
$C_2H_6O_3S$
5622 - Ethanesulfonic acid
7532 - Methanesulfonic acid, methyl ester
11582 - Sulfurous acid, dimethyl ester
$C_2H_6O_4S$
11574 - Sulfuric acid, dimethyl ester
11576 - Sulfuric acid, monoethyl ester
$C_2H_6O_6$
5561 - Ethanedioic acid, dihydrate
$C_2H_6O_6S_2$
5587 - 1,2-Ethanedisulfonic acid
C_2H_6S
5648 - Ethanethiol
7545 - Methane, thiobis-
$C_2H_6S_2$
5254 - Disulfide, dimethyl
5590 - 1,2-Ethanedithiol
C_2H_6Se
7526 - Methane, selenobis-
C_2H_6Zn
12066 - Zinc, dimethyl
C_2H_7As
442 - Arsine, dimethyl-
444 - Arsine, ethyl-
$C_2H_7AsO_2$
451 - Arsinic acid, dimethyl-
$C_2H_7AsO_3$
457 - Arsonic acid, ethyl-
$C_2H_7BO_2$
3117 - Borane, dimethoxy-
3133 - Boronic acid, ethyl-
$C_2H_7ClN_2$
5671 - Ethanimidamide, monohydrochloride
C_2H_7N
5410 - Ethanamine
7448 - Methanamine, *N*-methyl-
C_2H_7NO
5434 - Ethanamine, *N*-hydroxy-
5674 - Ethanol, 1-amino-
5675 - Ethanol, 2-amino-
$C_2H_7NO_2$
106 - Acetic acid, ammonium salt
$C_2H_7NO_3S$
5623 - Ethanesulfonic acid, 2-amino-
C_2H_7NS
5649 - Ethanethiol, 2-amino-
$C_2H_7N_3O$
6266 - Glycine, hydrazide
$C_2H_7N_5$
7233 - Imidodicarbonimidic diamide
$C_2H_7O_2P$
9528 - Phosphinic acid, dimethyl-
$C_2H_7O_3P$
9588 - Phosphorous acid, dimethyl ester
$C_2H_7O_4P$
9559 - Phosphoric acid, dimethyl ester
9561 - Phosphoric acid, monoethyl ester
C_2H_8ClN
5433 - Ethanamine, hydrochloride
$C_2H_8NO_2PS$
7441 - Methamidophos
$C_2H_8N_2$
5496 - 1,2-Ethanediamine
7154 - Hydrazine, 1,1-dimethyl-
7155 - Hydrazine, 1,2-dimethyl-
$C_2H_8N_2O$
5729 - Ethanol, 2-hydrazino-
C_2H_8Si
11414 - Silane, dimethyl-
$C_2H_{10}N_2O$
5513 - 1,2-Ethanediamine, monohydrate
C_2I_2
5941 - Ethyne, diiodo-
C_2I_4
5919 - Ethene, tetraiodo-
C_2N_2
5550 - Ethanedinitrile
C_2N_2S
11572 - Sulfur cyanide
$C_2N_4O_6$
290 - Acetonitrile, trinitro-
$C_2Na_2O_4$
5566 - Ethanedioic acid, disodium salt
C_2O_4Pb
5568 - Ethanedioic acid, lead salt
C_3ClF_5
10541 - 1-Propene, 2-chloro-1,1,3,3,3-pentafluoro-
C_3ClF_5O
10421 - 2-Propanone, 1-chloro-1,1,3,3,3-pentafluoro-
$C_3Cl_2F_4$
10557 - 1-Propene, 1,2-dichloro-1,3,3,3-tetrafluoro-
$C_3Cl_2F_6$
9898 - Propane, 1,2-dichloro-1,1,2,3,3,3-hexafluoro-
$C_3Cl_2N_2$
9912 - Propanedinitrile, dichloro-
$C_3Cl_3F_3$
10607 - 1-Propene, 1,1,2-trichloro-3,3,3-trifluoro-
$C_3Cl_3F_5$
10179 - Propane, 1,1,1-trichloro-2,2,3,3,3-pentafluoro-
10180 - Propane, 1,2,2-trichloro-1,1,3,3,3-pentafluoro-
10181 - Propane, 1,2,3-trichloro-1,1,2,3,3-pentafluoro-
$C_3Cl_3N_3$
11837 - 1,3,5-Triazine, 2,4,6-trichloro-
$C_3Cl_3N_3O_3$
11840 - 1,3,5-Triazine-2,4,6(1H,3H,5H)-trione, 1,3,5-trichloro-
$C_3Cl_4F_4$
10149 - Propane, 1,2,2,3-tetrachloro-1,1,3,3-tetrafluoro-

$C_3H_6O_3$
10267 - Propanoic acid, 3-hydroxy-
10429 - 2-Propanone, 1,3-dihydroxy-
11905 - 1,3,5-Trioxane
$C_3H_6O_3S$
8570 - 1,2-Oxathiolane, 2,2-dioxide
11694 - Thietane, 1-oxide
$C_3H_6O_4$
10252 - Propanoic acid, 2,3-dihydroxy-
C_3H_6S
5910 - Ethene, (methylthio)-
10600 - 2-Propene-1-thiol
11691 - Thietane
11696 - Thiirane, methyl-
$C_3H_6S_2$
5266 - 1,3-Dithiolane
$C_3H_6S_3$
11919 - 1,3,5-Trithiane
C_3H_7Br
9836 - Propane, 1-bromo-
9837 - Propane, 2-bromo-
C_3H_7BrO
5464 - Ethane, (bromomethoxy)-
5465 - Ethane, 1-bromo-2-methoxy-
10344 - 1-Propanol, 3-bromo
10345 - 2-Propanol, 1-bromo-
C_3H_7Cl
9848 - Propane, 1-chloro-
9849 - Propane, 2-chloro-
C_3H_7ClHgO
11822 - Triadimonol
C_3H_7ClO
5478 - Ethane, (chloromethoxy)-
5479 - Ethane, 1-chloro-1-methoxy-
5480 - Ethane, 1-chloro-2-methoxy-
10347 - 1-Propanol, 2-chloro-
10348 - 1-Propanol, 3-chloro-
10349 - 2-Propanol, 1-chloro-
$C_3H_7ClO_2$
9994 - 1,2-Propanediol, 3-chloro-
9995 - 1,3-Propanediol, 2-chloro-
$C_3H_7ClO_2S$
10133 - 1-Propanesulfonyl chloride
$C_3H_7ClO_3S$
5699 - Ethanol, 2-chloro-, methanesulfonate
C_3H_7ClS
5481 - Ethane, 1-chloro-2-(methylthio)-
$C_3H_7Cl_2N$
9805 - 1-Propanamine, *N*,*N*-dichloro-
$C_3H_7Cl_3Si$
11373 - Silane, chloro(dichloromethyl)dimethyl-
11466 - Silane, trichloro(1-methylethyl)-
11476 - Silane, trichloropropyl-
C_3H_7F
10046 - Propane, 1-fluoro-
10047 - Propane, 2-fluoro-
C_3H_7FO
10382 - 1-Propanol, 3-fluoro-
C_3H_7I
10054 - Propane, 1-iodo-
10055 - Propane, 2-iodo-
C_3H_7IO
10285 - 1-Propanol, 3-iodo-
C_3H_7N
494 - Azetidine
500 - Aziridine, 1-methyl-
501 - Aziridine, 2-methyl-
4953 - Cyclopropanamine
10522 - 2-Propen-1-amine
$C_3H_7NNaS_2$
11514 - Sodium dimethyldithiocarbamate
C_3H_7NO
72 - Acetamide, *N*-methyl-
5992 - Formamide, *N*,*N*-dimethyl-
5994 - Formamide, *N*-ethyl-
9787 - Propanal, oxime
9791 - Propanamide
10476 - 2-Propanone, oxime
C_3H_7NOS
3978 - Carbamothioic acid, *O*-ethyl ester
3979 - Carbamothioic acid, *S*-ethyl ester
$C_3H_7NO_2$
323 - β-Alanine
324 - *DL*-Alanine
325 - *L*-Alanine
3948 - Carbamic acid, ethyl ester
6268 - Glycine, *N*-methyl-
8092 - Nitrous acid, 1-methylethyl ester
8098 - Nitrous acid, propyl ester
9797 - Propanamide, 2-hydroxy-, (±)-
10116 - Propane, 1-nitro-
10117 - Propane, 2-nitro-
$C_3H_7NO_2S$
5014 - *L*-Cysteine
$C_3H_7NO_3$
8076 - Nitric acid, 1-methylethyl ester
8080 - Nitric acid, propyl ester
10394 - 1-Propanol, 2-nitro-
10395 - 1-Propanol, 3-nitro-
10396 - 2-Propanol, 1-nitro-
11355 - *D*-Serine
11356 - *DL*-Serine
11357 - *L*-Serine
$C_3H_7NO_4$
10022 - 1,2-Propanediol, 1-nitrate
$C_3H_7NO_5$
10192 - 1,2,3-Propanetriol, 1-nitrate
10193 - 1,2,3-Propanetriol, 2-nitrate
C_3H_7NS
11681 - Thiazolidine
$C_3H_7N_3O_2$
6253 - Glycine, *N*-(aminoiminomethyl)-
12015 - Urea, *N*-ethyl-*N*-nitroso-
$C_3H_7N_3O_5$
11839 - 1,3,5-Triazine-2,4,6(1H,3H,5H)-trione, dihydrate
C_3H_8
9834 - Propane
$C_3H_8BrClSi$
11365 - Silane, (bromomethyl)chlorodimethyl-
$C_3H_8Cl_2Si$
11372 - Silane, chloro(chloromethyl)dimethyl-
11397 - Silane, dichloroethylmethyl-
$C_3H_8NO_5P$
6276 - Glyphosate
$C_3H_8N_2O$
12004 - Urea, *N*,*N*′-dimethyl-
12005 - Urea, *N*,*N*-dimethyl-
12014 - Urea, ethyl-
$C_3H_8N_2O_2$
326 - Alanine, 3-amino-
7148 - Hydrazinecarboxylic acid, ethyl ester
9824 - 1-Propanamine, *N*-nitro-
$C_3H_8N_2O_3$
11995 - Urea, *N*,*N*′-bis(hydroxymethyl)-
$C_3H_8N_2S$
11804 - Thiourea, *N*,*N*′-dimethyl-
C_3H_8O
5602 - Ethane, methoxy-
10334 - 1-Propanol
10335 - 2-Propanol
C_3H_8OS
5749 - Ethanol, 2-(methylthio)-
$C_3H_8OS_2$
10370 - 1-Propanol, 2,3-dimercapto-
10371 - 2-Propanol, 1,3-dimercapto-
$C_3H_8O_2$
5733 - Ethanol, 2-methoxy-
7507 - Methane, dimethoxy-
9982 - 1,2-Propanediol
9983 - 1,3-Propanediol
$C_3H_8O_2S$
10013 - 1,2-Propanediol, 3-mercapto-
$C_3H_8O_3$
10186 - 1,2,3-Propanetriol
$C_3H_8O_3S$
7531 - Methanesulfonic acid, ethyl ester
10130 - 2-Propanesulfonic acid
C_3H_8S
5606 - Ethane, (methylthio)-
10156 - 1-Propanethiol
10157 - 2-Propanethiol
$C_3H_8S_2$
7461 - Methane, bis(methylthio)-
10030 - 1,2-Propanedithiol
10031 - 1,3-Propanedithiol
C_3H_9Al
339 - Aluminum, trimethyl
C_3H_9As
449 - Arsine, trimethyl-
$C_3H_9AsO_3$
461 - Arsonic acid, propyl-
C_3H_9B
3119 - Borane, trimethyl-
$C_3H_9BO_2$
3136 - Boronic acid, propyl-
$C_3H_9BO_3$
3126 - Boric acid, trimethyl ester
C_3H_9BS
3129 - Borinic acid, dimethylthio-, methyl ester
$C_3H_9BS_3$
11698 - Thioboric acid, trimethyl ester
C_3H_9ClSi
11389 - Silane, chlorotrimethyl-
C_3H_9ClSn
11540 - Stannane, chlorotrimethyl-
C_3H_9N
7444 - Methanamine, *N*,*N*-dimethyl-
9803 - 1-Propanamine
9804 - 2-Propanamine
C_3H_9NO
5738 - Ethanol, 2-(methylamino)-
10336 - 1-Propanol, 2-amino-, (±)-
10337 - 1-Propanol, 3-amino-
10338 - 2-Propanol, 1-amino-, (±)-
$C_3H_9NO_2$
9984 - 1,2-Propanediol, 3-amino-, (±)-
$C_3H_9NO_3S$
5624 - Ethanesulfonic acid, 2-(methylamino)-
$C_3H_9O_2PS$
9550 - Phosphonothioic acid, methyl-, *O*,*O*-dimethyl ester
$C_3H_9O_3P$
9540 - Phosphonic acid, methyl-, dimethyl ester
9542 - Phosphonic acid, (1-methylethyl)-
9543 - Phosphonic acid, methyl-, monoethyl ester
9547 - Phosphonic acid, propyl-
9592 - Phosphorous acid, trimethyl ester
$C_3H_9O_4P$
9564 - Phosphoric acid, trimethyl ester
C_3H_9P
9522 - Phosphine, trimethyl-
C_3H_9Sb
11550 - Stibine, trimethyl-
$C_3H_{10}ClN$
5435 - Ethanamine, *N*-methyl-, hydrochloride
7445 - Methanamine, *N*,*N*-dimethyl-, hydrochloride
$C_3H_{10}N_2$
5512 - 1,2-Ethanediamine, *N*-methyl-
9872 - 1,2-Propanediamine
9873 - 1,3-Propanediamine
$C_3H_{10}Si$
11489 - Silane, trimethyl-
$C_3H_{12}BN$
3120 - Borane, trimethyl-, monoammoniate
3132 - Boron, (*N*,*N*-dimethylmethanamine)trihydro-, (T-4)-
C_3N_2O
9915 - Propanedinitrile, oxo-

3920 - 3-Butyn-1-ol
3921 - 3-Butyn-2-ol
4197 - Cyclobutanone
5912 - Ethene, 1,1′-oxybis-
5942 - Ethyne, ethoxy-
6086 - Furan, 2,3-dihydro-
8613 - Oxirane, ethenyl-
10515 - 2-Propenal, 2-methyl-
10736 - 1-Propyne, 3-methoxy-

C_4H_6OS
11794 - 2(3H)-Thiophenone, dihydro-

$C_4H_6O_2$
179 - Acetic acid ethenyl ester
3007 - 2,2′-Bioxirane, (*R**,*S**)-
3256 - Butanedial
3359 - 2,3-Butanedione
3813 - 3-Butenoic acid
3814 - 2-Butenoic acid, (*E*)-
3815 - 2-Butenoic acid, (*Z*)-
3911 - 2-Butyne-1,4-diol
3912 - 3-Butyne-1,2-diol
4958 - Cyclopropanecarboxylic acid
5203 - 1,4-Dioxin, 2,3-dihydro-
6017 - Formic acid, 2-propenyl ester
6142 - 2(3H)-Furanone, dihydro-
10648 - 2-Propenoic acid, 2-methyl-
10655 - 2-Propenoic acid, methyl ester

$C_4H_6O_2S$
5913 - Ethene, 1,1′-sulfonylbis-
11752 - Thiophene, 2,5-dihydro-, 1,1-dioxide

$C_4H_6O_2S_2$
5250 - Disulfide, diacetyl

$C_4H_6O_3$
107 - Acetic acid, anhydride
3567 - Butanoic acid, 2-oxo-
3568 - Butanoic acid, 3-oxo-
3842 - 3-Butenoic acid, 2-hydroxy-
5213 - 1,3-Dioxolan-2-one, 4-methyl-
10316 - Propanoic acid, 2-oxo-, methyl ester

$C_4H_6O_4$
101 - Acetic acid, (acetyloxy)-
3294 - Butanedioic acid
5563 - Ethanedioic acid, dimethyl ester
5573 - 1,2-Ethanediol, diformate
9156 - Peroxide, diacetyl
9960 - Propanedioic acid, methyl-

$C_4H_6O_4S$
240 - Acetic acid, 2,2′-thiobis-

$C_4H_6O_5$
226 - Acetic acid, 2,2′-oxybis-
9959 - Propanedioic acid, (hydroxymethyl)-

$C_4H_6O_6$
3304 - Butanedioic acid, 2,3-dihydroxy-, (*R**,*S**)-
3309 - Butanedioic acid, 2,3-dihydroxy- [*R*-(*R**,*R**)]-
3311 - Butanedioic acid, 2,3-dihydroxy-, (*R**,*R**)-(±)-
3312 - Butanedioic acid, 2,3-dihydroxy-, [*S*-(*R**,*R**)]-

$C_4H_6O_6S$
239 - Acetic acid, 2,2′-sulfonylbis-

$C_4H_6O_8$
3342 - Butanedioic acid, tetrahydroxy-

C_4H_6S
5920 - Ethene, 1,1′-thiobis-
11750 - Thiophene, 2,3-dihydro-
11751 - Thiophene, 2,5-dihydro-

C_4H_7Br
3707 - 1-Butene, 1-bromo-, (*E*)-
3708 - 1-Butene, 1-bromo-, (*Z*)-
3709 - 1-Butene, 2-bromo-
3710 - 1-Butene, 3-bromo-
3711 - 1-Butene, 4-bromo-
3712 - 2-Butene, 1-bromo-
3713 - 2-Butene, 2-bromo-, (*E*)-
3714 - 2-Butene, 2-bromo-, (*Z*)-

C_4H_7BrO
3179 - Butanal, 2-bromo-
3680 - Butanoyl bromide
9776 - Propanal, 2-bromo-2-methyl-
10492 - Propanoyl bromide, 2-methyl-

$C_4H_7BrO_2$
115 - Acetic acid, bromo-, ethyl ester
3468 - Butanoic acid, 2-bromo-, (±)-
5686 - Ethanol, 2-bromo-, acetate
10209 - Propanoic acid, 2-bromo-2-methyl-
10211 - Propanoic acid, 2-bromo-, methyl ester, (±)-
10212 - Propanoic acid, 3-bromo-, methyl ester

$C_4H_7Br_2Cl_2O_4P$
7702 - Naled

$C_4H_7Br_3$
3442 - Butane, 1,1,2-tribromo-
3443 - Butane, 1,2,2-tribromo-
3444 - Butane, 1,2,3-tribromo-
3445 - Butane, 1,2,4-tribromo-
3446 - Butane, 1,3,3-tribromo-
3447 - Butane, 2,2,3-tribromo-
10164 - Propane, 1,2,3-tribromo-2-methyl-

$C_4H_7Br_3O$
10405 - 2-Propanol, 1,1,1-tribromo-2-methyl-

C_4H_7Cl
3720 - 1-Butene, 1-chloro-, (*E*)-
3721 - 1-Butene, 1-chloro-, (*Z*)-
3722 - 1-Butene, 2-chloro-
3723 - 1-Butene, 3-chloro-
3724 - 1-Butene, 4-chloro-
3725 - 2-Butene, 1-chloro-, (*E*)-
3726 - 2-Butene, 1-chloro-, (*Z*)-
3727 - 2-Butene, 2-chloro-, (*E*)-
3728 - 2-Butene, 2-chloro-, (*Z*)-
4966 - Cyclopropane, (chloromethyl)-
10539 - 1-Propene, 1-chloro-2-methyl-
10540 - 1-Propene, 3-chloro-2-methyl-

C_4H_7ClO
3180 - Butanal, 2-chloro-
3181 - Butanal, 4-chloro-
3646 - 2-Butanone, 1-chloro-
3647 - 2-Butanone, 3-chloro-
3648 - 2-Butanone, 4-chloro-
3681 - Butanoyl chloride
3863 - 2-Buten-1-ol, 2-chloro-
3864 - 2-Buten-1-ol, 3-chloro-
3865 - 2-Buten-1-ol, 4-chloro-
3866 - 3-Buten-2-ol, 1-chloro-
3867 - 3-Buten-1-ol, 2-chloro-
3868 - 3-Buten-2-ol, 3-chloro-
5886 - Ethene, 1-chloro-2-ethoxy-
5887 - Ethene, (2-chloroethoxy)-
8606 - Oxirane, 2-(chloromethyl)-2-methyl-
9779 - Propanal, 2-chloro-2-methyl-
10504 - Propanoyl chloride, 2-methyl-

$C_4H_7ClO_2$
130 - Acetic acid, chloro-, ethyl ester
298 - Acetyl chloride, ethoxy-
3480 - Butanoic acid, 2-chloro-
3481 - Butanoic acid, 3-chloro-, (±)-
3482 - Butanoic acid, 4-chloro-
4023 - Carbonochloridic acid, 1-methylethyl ester
4028 - Carbonochloridic acid, propyl ester
5694 - Ethanol, 1-chloro-, acetate
5695 - Ethanol, 2-chloro-, acetate
10222 - Propanoic acid, 2-chloro-, methyl ester
10225 - Propanoic acid, 3-chloro-, methyl ester
10228 - Propanoic acid, 2-chloro-2-methyl-
10230 - Propanoic acid, 3-chloro-2-methyl-

$C_4H_7ClO_3$
136 - Acetic acid, chloro-, 2-hydroxyethyl ester
4019 - Carbonochloridic acid, 2-methoxyethyl ester

$C_4H_7Cl_2O_4P$
5170 - Dichlorvos

$C_4H_7Cl_3$
3448 - Butane, 1,1,3-trichloro-
3449 - Butane, 1,2,3-trichloro-
3450 - Butane, 1,2,4-trichloro-
3451 - Butane, 2,2,3-trichloro-
10177 - Propane, 1,1,2-trichloro-2-methyl-
10178 - Propane, 1,2,3-trichloro-2-methyl-

$C_4H_7Cl_3O$
3643 - 2-Butanol, 1,1,1-trichloro-
10407 - 2-Propanol, 1,1,1-trichloro-2-methyl-

$C_4H_7Cl_3O_2$
5769 - Ethanol, 2,2,2-trichloro-1-ethoxy-

$C_4H_7FO_2$
191 - Acetic acid, fluoro-, ethyl ester
3927 - Butyric acid, 4-fluoro-

$C_4H_7IO_2$
201 - Acetic acid, iodo-, ethyl ester
10284 - Propanoic acid, 3-iodo-, methyl ester

C_4H_7N
3403 - Butanenitrile
10063 - Propane, 2-isocyano-
10111 - Propanenitrile, 2-methyl-
10726 - 2-Propyn-1-amine, *N*-methyl-
11118 - 1H-Pyrrole, 2,5-dihydro-

C_4H_7NO
3702 - 2-Butenamide, (*E*)-
8576 - Oxazole, 4,5-dihydro-2-methyl-
10059 - Propane, 1-isocyanato-
10060 - Propane, 2-isocyanato-
10108 - Propanenitrile, 2-hydroxy-2-methyl-
10109 - Propanenitrile, 2-methoxy-
10110 - Propanenitrile, 3-methoxy-
11180 - 2-Pyrrolidinone

$C_4H_7NO_2$
23 - Acetamide, *N*-acetyl-
495 - 2-Azetidinecarboxylic acid
3361 - 2,3-Butanedione, monooxime

$C_4H_7NO_2S$
11682 - 4-Thiazolidinecarboxylic acid

$C_4H_7NO_3$
6251 - Glycine, *N*-acetyl-

$C_4H_7NO_4$
218 - Acetic acid, nitro-, ethyl ester
470 - *DL*-Aspartic acid
471 - *L*-Aspartic acid
3466 - Butanoic acid, 4-amino-2-hydroxy-4-oxo-
6260 - Glycine, *N*-(carboxymethyl)-

C_4H_7NS
10064 - Propane, 1-isothiocyanato-
11672 - Thiazole, 4,5-dihydro-2-methyl-
11706 - Thiocyanic acid, 1-methylethyl ester

$C_4H_7N_3$
11849 - 1H-1,2,3-Triazole, 1,5-dimethyl-

$C_4H_7N_3O$
7230 - 4H-Imidazol-4-one, 2-amino-1,5-dihydro-1-methyl-

C_4H_8
3704 - 1-Butene
3705 - 2-Butene, (*E*)-
3706 - 2-Butene, (*Z*)-
4170 - Cyclobutane
4981 - Cyclopropane, methyl-
10578 - 1-Propene, 2-methyl-

C_4H_8BrCl
3233 - Butane, 1-bromo-4-chloro-
3234 - Butane, 2-bromo-1-chloro-
9842 - Propane, 1-bromo-3-chloro-2-methyl-

C_4H_8BrF
3237 - Butane, 1-bromo-4-fluoro-

C_4H_8BrNO
9792 - Propanamide, 2-bromo-2-methyl-

$C_4H_8Br_2$
3264 - Butane, 1,1-dibromo-
3265 - Butane, 1,2-dibromo-
3266 - Butane, 1,3-dibromo-
3267 - Butane, 1,4-dibromo-
3268 - Butane, 2,3-dibromo-
9889 - Propane, 1,2-dibromo-2-methyl-
9890 - Propane, 1,3-dibromo-2-methyl-

$C_4H_8Br_2O$
3612 - 1-Butanol, 2,3-dibromo-
3613 - 1-Butanol, 3,4-dibromo-
3614 - 2-Butanol, 1,4-dibromo-
5523 - Ethane, 1,2-dibromo-1-ethoxy-

C_5F_{10}
9027 - 1-Pentene, 1,1,2,3,3,4,4,5,5,5-decafluoro-
C_5F_{12}
8807 - Pentane, dodecafluoro-
C_5FeO_5
7339 - Iron carbonyl
C_5HCl_4N
11019 - Pyridine, 2,3,5,6-tetrachloro-
$C_5H_2Cl_6O_3$
5212 - 1,3-Dioxolan-4-one, 2,5-bis(trichloromethyl)-
$C_5H_2F_6O_2$
8802 - 2,4-Pentanedione, 1,1,1,5,5,5-hexafluoro-
$C_5H_2O_5$
4939 - 4-Cyclopentene-1,2,3-trione, 4,5-dihydroxy-
$C_5H_3BrO_2$
6043 - 2-Furancarboxaldehyde, 5-bromo-
$C_5H_3Br_2N$
10926 - Pyridine, 3,5-dibromo-
$C_5H_3ClN_4$
10753 - 1H-Purine, 6-chloro-
$C_5H_3ClO_2$
6040 - 2-Furancarbonyl chloride
$C_5H_3F_7O_2$
3525 - Butanoic acid, heptafluoro-, methyl ester
C_5H_3NO
6039 - 2-Furancarbonitrile
$C_5H_3NO_4$
6046 - 2-Furancarboxaldehyde, 5-nitro-
$C_5H_3NO_5$
6066 - 2-Furancarboxylic acid, 5-nitro-
C_5H_3NS
11732 - 2-Thiophenecarbonitrile
C_5H_4
8684 - 1,3-Pentadiyne
C_5H_4BrN
10875 - Pyridine, 2-bromo-
10876 - Pyridine, 3-bromo-
10877 - Pyridine, 4-bromo-
C_5H_4ClN
10915 - Pyridine, 2-chloro-
10916 - Pyridine, 3-chloro-
10917 - Pyridine, 4-chloro-
C_5H_4ClOS
11735 - 2-Thiophenecarboxaldehyde, 5-chloro-
$C_5H_4Cl_2N_2$
10862 - 3-Pyridinamine, 2,6-dichloro-
11052 - Pyrimidine, 2,4-dichloro-5-methyl-
11053 - Pyrimidine, 2,4-dichloro-6-methyl-
$C_5H_4Cl_2O_2$
3367 - Butanedioyl dichloride, methylene-
$C_5H_4F_8O$
8953 - 1-Pentanol, 2,2,3,3,4,4,5,5-octafluoro-
C_5H_4IN
10970 - Pyridine, 2-iodo-
$C_5H_4N_2O_2$
10803 - Pyrazinecarboxylic acid
$C_5H_4N_2O_3$
10993 - Pyridine, 4-nitro-, 1-oxide
$C_5H_4N_2O_4$
7195 - 1H-Imidazole-4,5-dicarboxylic acid
11050 - 4-Pyrimidinecarboxylic acid, 1,2,3,6-tetrahydro-2,6-dioxo-
$C_5H_4N_4$
10752 - 1H-Purine
$C_5H_4N_4O$
10770 - 6H-Purin-6-one, 1,7-dihydro-
10852 - 4H-Pyrazolo[3,4-d]pyrimidin-4-one, 1,5-dihydro-
$C_5H_4N_4O_2$
10755 - 1H-Purine-2,6-dione, 3,7-dihydro-
$C_5H_4N_4O_3$
10764 - 1H-Purine-2,6,8(3H)-trione, 7,9-dihydro-
$C_5H_4N_4S$
10763 - 6H-Purine-6-thione, 1,7-dihydro-
C_5H_4OS
11733 - 2-Thiophenecarboxaldehyde
11734 - 3-Thiophenecarboxaldehyde
$C_5H_4O_2$
6041 - 2-Furancarboxaldehyde
6042 - 3-Furancarboxaldehyde
6155 - 2(5H)-Furanone, 5-methylene-
10788 - 2H-Pyran-2-one
10789 - 4H-Pyran-4-one
$C_5H_4O_2S$
11737 - 2-Thiophenecarboxylic acid
11738 - 3-Thiophenecarboxylic acid
$C_5H_4O_3$
6050 - 2-Furancarboxylic acid
6051 - 3-Furancarboxylic acid
6095 - 2,5-Furandione, dihydro-3-methylene-
6098 - 2,5-Furandione, 3-methyl-
10794 - 2H-Pyran-2-one, 3-hydroxy-
$C_5H_4O_4$
6054 - 3-Furancarboxylic acid, 4,5-dihydro-5-oxo-
C_5H_5BrO
6035 - Furan, 2-(bromomethyl)-
C_5H_5BrS
11729 - Thiophene, 2-bromo-3-methyl-
11730 - Thiophene, 2-bromo-5-methyl-
$C_5H_5ClN_2$
10861 - 2-Pyridinamine, 5-chloro-
C_5H_5ClO
6075 - Furan, 2-(chloromethyl)-
C_5H_5ClS
11741 - Thiophene, 2-chloro-5-methyl-
$C_5H_5Cl_3N_2OS$
11599 - Terrazole
C_5H_5N
8673 - 2,4-Pentadienenitrile
8674 - 2,4-Pentadienenitrile, (*E*)-
8675 - 2,4-Pentadienenitrile, (*Z*)-
10872 - Pyridine
C_5H_5NO
10994 - Pyridine, 1-oxide
11031 - 2-Pyridinol
11032 - 4-Pyridinol
11033 - 2(1H)-Pyridinone
11114 - 1H-Pyrrole-2-carboxaldehyde
$C_5H_5NO_2$
6047 - 2-Furancarboxaldehyde, oxime, (*E*)-
6048 - 2-Furancarboxaldehyde, oxime, (*Z*)-
10624 - 2-Propenoic acid, 2-cyano-, methyl ester
$C_5H_5NO_2S$
11669 - 5-Thiazolecarboxylic acid, 4-methyl-
$C_5H_5NO_3S$
11018 - 3-Pyridinesulfonic acid
$C_5H_5N_3O$
10802 - Pyrazinecarboxamide
$C_5H_5N_3O_3S$
85 - Acetamide, *N*-(5-nitro-2-thiazolyl)-
$C_5H_5N_5$
10751 - 1H-Purin-6-amine
$C_5H_5N_5O$
10767 - 2H-Purin-2-one, 6-amino-1,3-dihydro-
10768 - 6H-Purin-6-one, 2-amino-1,7-dihydro-
$C_5H_5N_5S$
10762 - 6H-Purine-6-thione, 2-amino-1,7-dihydro-
C_5H_6
3892 - 1-Buten-3-yne, 2-methyl-
4764 - 1,3-Cyclopentadiene
9111 - 1-Penten-3-yne
9112 - 1-Penten-4-yne
9113 - 3-Penten-1-yne, (*E*)-
9114 - 3-Penten-1-yne, (*Z*)-
$C_5H_6Br_2N_2O_2$
7217 - 2,4-Imidazolidinedione, 1,3-dibromo-5,5-dimethyl-
C_5H_6ClN
10969 - Pyridine, hydrochloride
$C_5H_6ClNO_2$
11155 - 2,5-Pyrrolidinedione, 1-(chloromethyl)-
$C_5H_6Cl_2O_2$
8806 - Pentanedioyl dichloride
$C_5H_6F_4O_2$
10328 - Propanoic acid, 2,3,3,3-tetrafluoro-, ethyl ester
$C_5H_6N_2$
8767 - Pentanedinitrile
9913 - Propanedinitrile, dimethyl-
10810 - Pyrazine, methyl-
10856 - Pyridazine, 3-methyl-
10857 - 2-Pyridinamine
10858 - 3-Pyridinamine
10859 - 4-Pyridinamine
11070 - Pyrimidine, 2-methyl-
11071 - Pyrimidine, 4-methyl-
11072 - Pyrimidine, 5-methyl-
$C_5H_6N_2OS$
11108 - 4(1H)-Pyrimidinone, 2,3-dihydro-6-methyl-2-thioxo-
$C_5H_6N_2O_2$
11068 - 2,4(1H,3H)-Pyrimidinedione, 5-methyl-
11069 - 2,4(1H,3H)-Pyrimidinedione, 6-methyl-
$C_5H_6N_6$
10754 - 1H-Purine-2,6-diamine
C_5H_6O
3891 - 1-Buten-3-yne, 1-methoxy-
4941 - 2-Cyclopenten-1-one
6130 - Furan, 2-methyl
6131 - Furan, 3-methyl-
9118 - 4-Penten-2-yn-1-ol
10771 - Pyran
C_5H_6OS
6117 - 2-Furanmethanethiol
11768 - 2-Thiophenemethanol
$C_5H_6O_2$
6118 - 2-Furanmethanol
6127 - Furan, 2-methoxy-
6147 - 2(3H)-Furanone, dihydro-3-methylene-
6148 - 2(3H)-Furanone, dihydro-5-methylene-
6150 - 2(3H)-Furanone, 5-methyl-
6154 - 2(5H)-Furanone, 5-methyl-
8676 - 2,4-Pentadienoic acid
9137 - 2-Pentynoic acid
9138 - 4-Pentynoic acid
10739 - 2-Propynoic acid, ethyl ester
10744 - 2-Propyn-1-ol, acetate
$C_5H_6O_3$
4942 - 2-Cyclopenten-1-one, 2,3-dihydroxy-
6094 - 2,5-Furandione, dihydro-3-methyl-
10780 - 2H-Pyran-2,6(3H)-dione, dihydro-
$C_5H_6O_4$
3328 - Butanedioic acid, methylene-
3785 - 2-Butenedioic acid, 2-methyl-, (*E*)-
3786 - 2-Butenedioic acid, 2-methyl-, (*Z*)-
4970 - 1,2-Cyclopropanedicarboxylic acid, *trans*-(±)-
8875 - Pentanoic acid, 2,4-dioxo-
$C_5H_6O_5$
8781 - Pentanedioic acid, 2-oxo-
8782 - Pentanedioic acid, 3-oxo-
C_5H_6S
11769 - Thiophene, 2-methyl-
11770 - Thiophene, 3-methyl-
C_5H_7Br
9124 - 1-Pentyne, 1-bromo-
$C_5H_7BrN_2$
10816 - 1H-Pyrazole, 4-bromo-1,3-dimethyl-
$C_5H_7BrO_2$
10611 - 2-Propenoic acid, 2-bromoethyl ester
C_5H_7Cl
3153 - 1,3-Butadiene, 1-chloro-2-methyl-
3154 - 1,3-Butadiene, 1-chloro-3-methyl-
3155 - 1,3-Butadiene, 2-chloro-3-methyl-
3901 - 1-Butyne, 3-chloro-3-methyl-
4914 - Cyclopentene, 3-chloro-
8663 - 1,3-Pentadiene, 3-chloro-
$C_5H_7ClN_2$
10818 - 1H-Pyrazole, 3-chloro-1,5-dimethyl-
10819 - 1H-Pyrazole, 4-chloro-3,5-dimethyl-
10820 - 1H-Pyrazole, 5-chloro-1,3-dimethyl-

MOLECULAR FORMULA INDEX OF ORGANIC COMPOUNDS (continued)

5820 - Ethanone, 1-(1H-pyrrol-2-yl)-
9200 - Phenol, amino-
9201 - Phenol, 2-amino-
9202 - Phenol, 3-amino-
9203 - Phenol, 4-amino-
10975 - 2-Pyridinemethanol
10976 - 3-Pyridinemethanol
10977 - 4-Pyridinemethanol
10988 - Pyridine, 3-methyl-, 1-oxide
11035 - 2(1H)-Pyridinone, 1-methyl-
11036 - 2(1H)-Pyridinone, 3-methyl-
11037 - 2(1H)-Pyridinone, 4-methyl-

$C_6H_7NO_2$
11122 - 1H-Pyrrole-2,5-dione, 1-ethyl-

$C_6H_7NO_3S$
2215 - Benzenesulfonic acid, 2-amino-
2216 - Benzenesulfonic acid, 3-amino-
2217 - Benzenesulfonic acid, 4-amino-

$C_6H_7NO_4$
6129 - Furan, 2-(methoxymethyl)-5-nitro-

C_6H_7NS
2290 - Benzenethiol, 2-amino-
2291 - Benzenethiol, 4-amino-

$C_6H_7N_3O$
10905 - 4-Pyridinecarboxylic acid, hydrazide

$C_6H_7N_3O_2$
1281 - 1,2-Benzenediamine, 4-nitro-
7168 - Hydrazine, (4-nitrophenyl)-

$C_6H_7N_5O$
10769 - 6H-Purin-6-one, 2-amino-1,7-dihydro-7-methyl

$C_6H_7O_3Sb$
11548 - Stibine oxide, dihydroxyphenyl-

C_6H_7P
9519 - Phosphine, phenyl-

C_6H_8
4172 - Cyclobutane, 1,2-bis(methylene)-
4248 - 1,3-Cyclohexadiene
4249 - 1,4-Cyclohexadiene
6992 - 1,3,5-Hexatriene, (*E*)-
6993 - 1,3,5-Hexatriene, (*Z*)-
7103 - 1-Hexen-3-yne
7104 - 1-Hexen-4-yne
7105 - 1-Hexen-5-yne
7106 - 2-Hexen-4-yne
9116 - 1-Penten-3-yne, 2-methyl-
9117 - 3-Penten-1-yne, 3-methyl-

$C_6H_8AsNO_3$
456 - Arsonic acid, (4-aminophenyl)-

C_6H_8ClN
756 - Benzenamine, hydrochloride

C_6H_8ClNS
11670 - Thiazole, 5-(2-chloroethyl)-4-methyl-

$C_6H_8ClN_3O_4S_2$
1558 - 1,3-Benzenedisulfonamide, 4-amino-6-chloro-

$C_6H_8Cl_2$
6690 - 2,4-Hexadiene, 1,3-dichloro-

$C_6H_8Cl_2O_3$
3503 - Butanoic acid, 2,2-dichloro-3-oxo-, ethyl ester

$C_6H_8Cl_2O_4$
129 - Acetic acid, chloro-, 1,2-ethanediyl ester
5554 - Ethanedioic acid, bis(2-chloroethyl) ester

$C_6H_8N_2$
1260 - 1,2-Benzenediamine
1261 - 1,3-Benzenediamine
1262 - 1,4-Benzenediamine
6777 - Hexanedinitrile
7169 - Hydrazine, phenyl-
10805 - Pyrazine, 2,3-dimethyl-
10806 - Pyrazine, 2,5-dimethyl-
10807 - Pyrazine, 2,6-dimethyl-
10866 - 2-Pyridinamine, 3-methyl-
10867 - 2-Pyridinamine, 4-methyl-
10868 - 2-Pyridinamine, 6-methyl-
10869 - 2-Pyridinamine, *N*-methyl-
10870 - 3-Pyridinamine, 4-methyl-
10971 - 2-Pyridinemethanamine
10972 - 3-Pyridinemethanamine
10973 - 4-Pyridinemethanamine
11055 - Pyrimidine, 2,4-dimethyl-
11056 - Pyrimidine, 2,5-dimethyl-
11057 - Pyrimidine, 4,5-dimethyl-
11058 - Pyrimidine, 4,6-dimethyl-

$C_6H_8N_2O$
9274 - Phenol, 2,4-diamino-
10114 - Propanenitrile, 3,3′-oxybis-
10808 - Pyrazineethanol
10865 - 3-Pyridinamine, 6-methoxy-

$C_6H_8N_2O_2S$
2187 - Benzenesulfonamide, 4-amino-

$C_6H_8N_2O_3$
821 - Benzenamine, nitrate

$C_6H_8N_2O_3S$
2224 - Benzenesulfonic acid, 4-hydrazino-

$C_6H_8N_6O_{18}$
7424 - *D*-Mannitol, hexanitrate

C_6H_8O
4703 - 2-Cyclohexen-1-one
4704 - 3-Cyclohexen-1-one
4908 - 1-Cyclopentene-1-carboxaldehyde
4944 - 2-Cyclopenten-1-one, 2-methyl-
4945 - 2-Cyclopenten-1-one, 3-methyl-
4946 - 2-Cyclopenten-1-one, 5-methyl-
6088 - Furan, 2,4-dimethyl-
6089 - Furan, 2,5-dimethyl-
6105 - Furan, 2-ethyl-
6675 - 2,4-Hexadienal, (*E,E*)-
7113 - 3-Hexynal
7130 - 3-Hexyn-2-one
7131 - 5-Hexyn-2-one

$C_6H_8O_2$
3176 - 1,3-Butadien-1-ol, acetate
3918 - 2-Butynoic acid, ethyl ester
4375 - 1,2-Cyclohexanedione
4376 - 1,3-Cyclohexanedione
4377 - 1,4-Cyclohexanedione
4909 - 1-Cyclopentene-1-carboxylic acid
6103 - Furan, 2-ethoxy-
6121 - 2-Furanmethanol, α-methyl-
6122 - 2-Furanmethanol, 5-methyl-
6128 - Furan, 2-(methoxymethyl)-
6698 - 2,4-Hexadienoic acid
7123 - 2-Hexynoic acid
9139 - 2-Pentynoic acid, methyl ester
10694 - 2-Propenoic acid, 2-propenyl ester
10791 - 2H-Pyran-2-one, 5,6-dihydro-6-methyl-, (*S*)-

$C_6H_8O_3$
6139 - 2(3H)-Furanone, 3-acetyldihydro-
10680 - 2-Propenoic acid, oxiranylmethyl ester

$C_6H_8O_4$
3778 - 2-Butenedioic acid (*E*)-, dimethyl ester
3779 - 2-Butenedioic acid (*Z*)-, dimethyl ester
3791 - 2-Butenedioic acid (*E*)-, monoethyl ester
4182 - 1,3-Cyclobutanedicarboxylic acid, *cis*-
4185 - 1,3-Cyclobutanedicarboxylic acid, *trans*-
5190 - 1,3-Dioxane-4,6-dione, 2,2-dimethyl-
9965 - Propanedioic acid, (1-methylethylidene)-
9979 - Propanedioic acid, 2-propenyl-

$C_6H_8O_5$
467 - *L*-Ascorbic acid, 6-deoxy-

$C_6H_8O_6$
465 - *DL*-Ascorbic acid
466 - *L*-Ascorbic acid
6239 - *D*-Glucuronic acid, γ-lactone
10165 - 1,2,3-Propanetricarboxylic acid

$C_6H_8O_7$
10167 - 1,2,3-Propanetricarboxylic acid, 2-hydroxy-

C_6H_8S
11754 - Thiophene, 2,3-dimethyl-
11755 - Thiophene, 2,4-dimethyl-
11756 - Thiophene, 2,5-dimethyl-
11757 - Thiophene, 3,4-dimethyl-
11759 - Thiophene, 2-ethyl-
11760 - Thiophene, 3-ethyl-

C_6H_8Si
11446 - Silane, phenyl-

C_6H_9Br
4627 - Cyclohexene, 1-bromo-
4628 - Cyclohexene, 3-bromo-

C_6H_9BrO
4574 - Cyclohexanone, 2-bromo-

$C_6H_9BrO_2$
3819 - 2-Butenoic acid, 4-bromo-, ethyl ester, (*E*)-

$C_6H_9BrO_3$
3477 - Butanoic acid, 2-bromo-3-oxo-, ethyl ester
3478 - Butanoic acid, 4-bromo-3-oxo-, ethyl ester

C_6H_9Cl
4642 - Cyclohexene, 1-chloro-
6688 - 1,3-Hexadiene, 3-chloro-
8664 - 1,2-Pentadiene, 1-chloro-3-methyl-
9126 - 1-Pentyne, 3-chloro-3-methyl-
9127 - 2-Pentyne, 4-chloro-4-methyl-

$C_6H_9ClN_2$
7172 - Hydrazine, phenyl-, monohydrochloride

C_6H_9ClO
4577 - Cyclohexanone, 2-chloro-
4578 - Cyclohexanone, 4-chloro-

$C_6H_9ClO_2$
3823 - 2-Butenoic acid, 2-chloro-, ethyl ester
3824 - 2-Butenoic acid, 2-chloro-, ethyl ester, (*E*)-
3825 - 2-Butenoic acid, 2-chloro-, ethyl ester, (*Z*)-
3826 - 2-Butenoic acid, 3-chloro-, ethyl ester
3827 - 2-Butenoic acid, 3-chloro-, ethyl ester, (*E*)-
3828 - 2-Butenoic acid, 3-chloro-, ethyl ester, (*Z*)-
3829 - 2-Butenoic acid, 4-chloro-, ethyl ester

$C_6H_9ClO_3$
3494 - Butanoic acid, 4-chloro-3-oxo-, ethyl ester
3570 - Butanoic acid, 3-oxo-, 2-chloroethyl ester

$C_6H_9Cl_3$
7048 - 1-Hexene, 1,1,2-trichloro-

$C_6H_9Cl_3O_2$
245 - Acetic acid, trichloro-, butyl ester
247 - Acetic acid, trichloro-, 1,1-dimethylethyl ester
251 - Acetic acid, trichloro-, 1-methylpropyl ester
252 - Acetic acid, trichloro-, 2-methylpropyl ester
3591 - Butanoic acid, 2,2,3-trichloro-, ethyl ester

$C_6H_9Cl_3O_3$
13 - Acetaldehyde, chloro-, trimer

$C_6H_9F_3O_2$
265 - Acetic acid, trifluoro-, butyl ester
266 - Acetic acid, trifluoro-, 1,1-dimethylethyl ester

C_6H_9N
4779 - Cyclopentanecarbonitrile
11119 - 1H-Pyrrole, 2,4-dimethyl-
11120 - 1H-Pyrrole, 2,5-dimethyl-
11124 - 1H-Pyrrole, 1-ethyl-
11125 - 1H-Pyrrole, 2-ethyl-

C_6H_9NO
6115 - 2-Furanmethanamine, *N*-methyl-
11183 - 2-Pyrrolidinone, 1-ethenyl-

C_6H_9NOS
11677 - 5-Thiazoleethanol, 4-methyl-

$C_6H_9NO_2$
10912 - 3-Pyridinecarboxylic acid, 1,2,5,6-tetrahydro-
11157 - 2,5-Pyrrolidinedione, 1-ethyl-

$C_6H_9NO_3$
50 - Acetamide, *N,N*-diacetyl-
8587 - 2,4-Oxazolidinedione, 3,5,5-trimethyl-

$C_6H_9NO_6$
6256 - Glycine, *N,N*-bis(carboxymethyl)-

$C_6H_9N_3$
2305 - 1,2,3-Benzenetriamine
2306 - 1,2,4-Benzenetriamine
11044 - 2-Pyrimidinamine, 4,5-dimethyl-
11045 - 4-Pyrimidinamine, 2,6-dimethyl-

$C_6H_9N_3O_2$
759 - Benzenamine, *N*-hydroxy-*N*-nitroso-, ammonium salt
7132 - *L*-Histidine

$C_6H_9N_3O_3$
7206 - 1H-Imidazole-1-ethanol, 2-methyl-5-nitro-

6718 - Hexanal
6959 - 2-Hexanone
6960 - 3-Hexanone
7066 - 1-Hexen-3-ol
7067 - 4-Hexen-2-ol
7068 - 5-Hexen-2-ol
7069 - 2-Hexen-1-ol, (*E*)-
7070 - 2-Hexen-1-ol, (*Z*)-
7071 - 3-Hexen-1-ol, (*Z*)-
7072 - 4-Hexen-1-ol, (*E*)-
8590 - Oxepane
8627 - Oxirane, tetramethyl-
8693 - Pentanal, 2-methyl-
8694 - Pentanal, 3-methyl-
8980 - 2-Pentanone, 3-methyl-, (±)-
8981 - 2-Pentanone, 4-methyl-
8982 - 3-Pentanone, 2-methyl-
9094 - 2-Penten-1-ol, 2-methyl-
9095 - 2-Penten-1-ol, 4-methyl-
9096 - 3-Penten-1-ol, 4-methyl-
9097 - 3-Penten-2-ol, 2-methyl-
9098 - 3-Penten-2-ol, 3-methyl-
9099 - 3-Penten-2-ol, 4-methyl-
9100 - 4-Penten-2-ol, 2-methyl-
9101 - 4-Penten-2-ol, 3-methyl-
9102 - 4-Penten-2-ol, 4-methyl-
10035 - Propane, 1-(ethenyloxy)-2-methyl-
10036 - Propane, 2-(ethenyloxy)-2-methyl-
10579 - 1-Propene, 3-(1-methylethoxy)-
10594 - 1-Propene, 3-propoxy-

$C_6H_{12}O_2$
121 - Acetic acid, butyl ester
177 - Acetic acid, 1,1-dimethylethyl ester
214 - Acetic acid, 1-methylpropyl ester
215 - Acetic acid, 2-methylpropyl ester
3506 - Butanoic acid, 2,2-dimethyl-
3507 - Butanoic acid, 2,3-dimethyl-
3508 - Butanoic acid, 3,3-dimethyl-
3514 - Butanoic acid, 2-ethyl-
3515 - Butanoic acid, ethyl ester
3552 - Butanoic acid, 2-methyl-, methyl ester
3636 - 1-Butanol, 3-methyl-, formate
4372 - 1,2-Cyclohexanediol, *cis*-
4373 - 1,2-Cyclohexanediol, *trans*-(±)-
4374 - 1,4-Cyclohexanediol, *trans*-
5189 - 1,3-Dioxane, 2,4-dimethyl-
5901 - Ethene, 1,1-diethoxy-
6015 - Formic acid, pentyl ester
6170 - Furan, tetrahydro-2-(methoxymethyl)-
6870 - Hexanoic acid
6969 - 2-Hexanone, 5-hydroxy-
6970 - 2-Hexanone, 6-hydroxy-
6971 - 3-Hexanone, 4-hydroxy-
6972 - 3-Hexanone, 5-hydroxy-
7177 - Hydroperoxide, cyclohexyl
8691 - Pentanal, 3-hydroxy-2-methyl-
8887 - Pentanoic acid, 2-methyl-, (±)-
8888 - Pentanoic acid, 3-methyl-, (±)-
8889 - Pentanoic acid, 4-methyl-
8891 - Pentanoic acid, methyl ester
8976 - 2-Pentanone, 4-hydroxy-4-methyl-
10260 - Propanoic acid, 2,2-dimethyl-, methyl ester
10298 - Propanoic acid, 2-methyl-, ethyl ester
10299 - Propanoic acid, 1-methylethyl ester
10326 - Propanoic acid, propyl ester
10784 - 2H-Pyran-2-methanol, tetrahydro-

$C_6H_{12}O_2S$
11784 - Thiophene, tetrahydro-2,4-dimethyl-, 1,1-dioxide

$C_6H_{12}O_3$
181 - Acetic acid, ethoxy-, ethyl ester
3532 - Butanoic acid, 2-hydroxy-, ethyl ester, (±)-
3533 - Butanoic acid, 3-hydroxy-, ethyl ester, (±)-
3534 - Butanoic acid, 3-hydroxy-, ethyl ester
5207 - 1,3-Dioxolane-4-methanol, 2,2-dimethyl-
5718 - Ethanol, 2-ethoxy-, acetate
6169 - Furan, tetrahydro-2,5-dimethoxy-
6891 - Hexanoic acid, 2-hydroxy-, (±)-
7109 - Hexitol, 2,5-anhydro-3,4-dideoxy-
10275 - Propanoic acid, 2-hydroxy-, 1-methylethyl ester
10277 - Propanoic acid, 3-hydroxy-, 1-methylethyl ester
10279 - Propanoic acid, 2-hydroxy-2-methyl-, ethyl ester
11908 - 1,3,5-Trioxane, 2,4,6-trimethyl-

$C_6H_{12}O_4$
5673 - Ethanol, 2-[2-(acetyloxy)ethoxy]-
7111 - *D-ribo*-Hexose, 2,6-dideoxy-
10255 - Propanoic acid, 2,3-dihydroxy-2-methyl-, ethyl ester
10257 - Propanoic acid, 2,3-dihydroxy-, propyl ester

$C_6H_{12}O_4Si$
11419 - Silanediol, dimethyl-, diacetate

$C_6H_{12}O_5$
4074 - *D*-chiro-Inositol, 2-deoxy-
6235 - *D*-Glucose, 6-deoxy-
7110 - *D-arabino*-Hexose, 2-deoxy-

$C_6H_{12}O_5S$
184 - Acetic acid, (ethoxysulfonyl)-, ethyl ester

$C_6H_{12}O_6$
337 - *D*-Allose
338 - *D*-Altrose
4073 - *DL*-chiro-Inositol
6023 - β-*D*-Fructose
6197 - *D*-Galactose
6232 - β-*D*-Glucose
6290 - *D*-Gulose
7336 - *scyllo*-Inositol
7425 - *D*-Mannose
11518 - *DL*-Sorbose
11519 - *L*-Sorbose
11587 - *D*-Tagatose

$C_6H_{12}O_7$
6209 - *D*-Gluconic acid

$C_6H_{12}S$
4480 - Cyclohexanethiol
11690 - Thiepane
11765 - Thiophene, 2-ethyltetrahydro-
11766 - Thiophene, 3-ethyltetrahydro-
11785 - Thiophene, tetrahydro-2,5-dimethyl-, *cis*-
11796 - 2H-Thiopyran, tetrahydro-2-methyl-
11797 - 2H-Thiopyran, tetrahydro-3-methyl-
11798 - 2H-Thiopyran, tetrahydro-4-methyl-

$C_6H_{12}S_3$
11920 - 1,3,5-Trithiane, 2,4,6-trimethyl-
11921 - 1,3,5-Trithiane, 2,4,6-trimethyl-, (2α,4α,6β)-

$C_6H_{13}Br$
3235 - Butane, 1-bromo-3,3-dimethyl-
3236 - Butane, 2-bromo-2,3-dimethyl-
6732 - Hexane, 1-bromo-
6733 - Hexane, 2-bromo-
6734 - Hexane, 3-bromo-
8714 - Pentane, 1-bromo-2-methyl-
8715 - Pentane, 1-bromo-3-methyl-
8716 - Pentane, 1-bromo-4-methyl-
8717 - Pentane, 2-bromo-2-methyl-
8718 - Pentane, 3-bromo-3-methyl-

$C_6H_{13}BrO_2$
5459 - Ethane, 2-bromo-1,1-diethoxy-

$C_6H_{13}Cl$
3245 - Butane, 1-chloro-2,3-dimethyl-
3246 - Butane, 1-chloro-3,3-dimethyl-
3247 - Butane, 2-chloro-2,3-dimethyl-
3248 - Butane, 3-chloro-2,2-dimethyl-
6737 - Hexane, 1-chloro-
6738 - Hexane, 2-chloro-
6739 - Hexane, 3-chloro-
8729 - Pentane, 2-chloro-2-methyl-
8730 - Pentane, 2-chloro-4-methyl-
8731 - Pentane, 3-(chloromethyl)-
8732 - Pentane, 3-chloro-3-methyl-

$C_6H_{13}ClO$
3249 - Butane, 1-(2-chloroethoxy)-
3608 - 2-Butanol, 1-chloro-2,3-dimethyl-
3609 - 2-Butanol, 1-chloro-3,3-dimethyl-
6916 - 1-Hexanol, 6-chloro-
6917 - 2-Hexanol, 1-chloro-
6918 - 3-Hexanol, 1-chloro-
6919 - 3-Hexanol, 5-chloro-

$C_6H_{13}ClO_2$
3244 - Butane, 3-chloro-1,1-dimethoxy-
5469 - Ethane, 2-chloro-1,1-diethoxy-
10355 - 2-Propanol, 1-chloro-3-(1-methylethoxy)-
10356 - 2-Propanol, 1-chloro-3-propoxy-

$C_6H_{13}ClS$
9855 - Propane, 2-[(2-chloroethyl)thio]-2-methyl-

$C_6H_{13}Cl_3Si$
11464 - Silane, trichlorohexyl-

$C_6H_{13}F$
6829 - Hexane, 1-fluoro-
6830 - Hexane, 2-fluoro-

$C_6H_{13}FO$
6936 - 1-Hexanol, 6-fluoro-

$C_6H_{13}I$
6831 - Hexane, 1-iodo-
6832 - Hexane, 2-iodo-, (*R*)-

$C_6H_{13}IO_2$
5543 - Ethane, 1,1-diethoxy-2-iodo-

$C_6H_{13}N$
492 - 1H-Azepine, hexahydro
4283 - Cyclohexanamine
4772 - Cyclopentanamine, 1-methyl-
4773 - Cyclopentanamine, 2-methyl-
9679 - Piperidine, 1-methyl-
9680 - Piperidine, 2-methyl-, (±)-
9681 - Piperidine, 3-methyl-, (±)-
9682 - Piperidine, 4-methyl-
10525 - 2-Propen-1-amine, *N*-(1-methylethyl)-
10527 - 2-Propen-1-amine, *N*-propyl-
11149 - Pyrrolidine, 2,2-dimethyl-
11150 - Pyrrolidine, 2,4-dimethyl-
11151 - Pyrrolidine, 2,5-dimethyl-, *cis*-
11166 - Pyrrolidine, 1-ethyl-

$C_6H_{13}NO$
38 - Acetamide, *N*-butyl-
53 - Acetamide, *N,N*-diethyl-
3203 - Butanamide, *N,N*-dimethyl-
4501 - Cyclohexanol, 2-amino-, *trans*-(±)-
6721 - Hexanamide
7687 - Morpholine, 2,6-dimethyl-
7694 - Morpholine, 4-ethyl-
8959 - 2-Pentanone, 4-amino-4-methyl-
9674 - 2-Piperidinemethanol
9675 - 3-Piperidinemethanol
9716 - 3-Piperidinol, 1-methyl-
9717 - 4-Piperidinol, 1-methyl-
9802 - Propanamide, *N*-propyl-
11165 - 1-Pyrrolidineethanol

$C_6H_{13}NO_2$
332 - β-Alanine, *N*-methyl-, ethyl ester
3633 - 1-Butanol, 3-methyl-, carbamate
3634 - 2-Butanol, 2-methyl-, carbamate
3937 - Carbamic acid, butyl-, methyl ester
3943 - Carbamic acid, (1,1-dimethylethyl)-, methyl ester
3956 - Carbamic acid, (1-methylethyl)-, ethyl ester
3962 - Carbamic acid, (1-methylpropyl)-, methyl ester
3970 - Carbamic acid, propyl-, ethyl ester
6120 - 2-Furanmethanol, 5-(aminomethyl)tetrahydro-
6841 - Hexane, 1-nitro-
6873 - Hexanoic acid, 6-amino-
7370 - *L*-Isoleucine
7412 - *DL*-Leucine
7413 - *L*-Leucine
7691 - 4-Morpholineethanol
8089 - Nitrous acid, hexyl ester
8214 - Norleucine
8215 - *DL*-Norleucine

$C_6H_{13}NO_3S$
11568 - Sulfamic acid, cyclohexyl-

$C_6H_{13}NO_4$
6259 - Glycine, *N,N*-bis(2-hydroxyethyl)-
$C_6H_{13}NO_5$
6233 - *D*-Glucose, 2-amino-2-deoxy-
$C_6H_{13}NS_2$
5265 - 4H-1,3,5-Dithiazine, dihydro-2,4,6-trimethyl-, (2α,4α,6α)-
$C_6H_{13}N_3$
6283 - Guanidine, (3-methyl-2-butenyl)-
$C_6H_{13}N_3O_3$
8551 - *L*-Ornithine, *N*5-(aminocarbonyl)-
$C_6H_{13}O_3P$
9535 - Phosphonic acid, ethenyl-, diethyl ester
$C_6H_{13}O_4P$
9530 - Phosphonic acid, acetyl-, diethyl ester
C_6H_{14}
3289 - Butane, 2,2-dimethyl-
3290 - Butane, 2,3-dimethyl-
6731 - Hexane
8823 - Pentane, 2-methyl-
8824 - Pentane, 3-methyl-
$C_6H_{14}Cl_4OSi_2$
5231 - Disiloxane, 1,3-bis(dichloromethyl)-1,1,3,3-tetramethyl-
$C_6H_{14}FO_3P$
9580 - Phosphorofluoridic acid, bis(1-methylethyl) ester
$C_6H_{14}N_2$
5132 - Diazene, dipropyl
5141 - 1H-1,4-Diazepine, hexahydro-1-methyl-
9608 - Piperazine, 1,4-dimethyl-
9609 - Piperazine, 2,5-dimethyl-, *cis-*
9610 - Piperazine, 2,5-dimethyl-, *trans-*
9611 - Piperazine, 2,6-dimethyl-, *cis-*
9673 - 4-Piperidinemethanamine
11164 - 1-Pyrrolidineethanamine
$C_6H_{14}N_2O$
7689 - 4-Morpholineethanamine
9616 - 1-Piperazineethanol
9822 - 2-Propanamine, *N*-(1-methylethyl)-*N*-nitroso-
9825 - 1-Propanamine, *N*-nitroso-*N*-propyl-
$C_6H_{14}N_2O_2$
7418 - *DL*-Lysine
7419 - *L*-Lysine
$C_6H_{14}N_4O_2$
437 - *DL*-Arginine
438 - *L*-Arginine
$C_6H_{14}O$
3372 - Butane, 1-ethoxy-
3373 - Butane, 2-ethoxy-
3394 - Butane, 1-methoxy-2-methyl-
3395 - Butane, 1-methoxy-3-methyl-
3396 - Butane, 2-methoxy-2-methyl-
3615 - 1-Butanol, 2,2-dimethyl-
3616 - 1-Butanol, 2,3-dimethyl-, (±)-
3617 - 1-Butanol, 3,3-dimethyl-
3618 - 2-Butanol, 2,3-dimethyl-
3619 - 2-Butanol, 3,3-dimethyl-, (±)-
3620 - 1-Butanol, 2-ethyl-
6911 - 1-Hexanol
6912 - 2-Hexanol, (±)-
6913 - 2-Hexanol, (*R*)-
6914 - 3-Hexanol, (±)-
6915 - 3-Hexanol, (*R*)-
8822 - Pentane, 1-methoxy-
8937 - 1-Pentanol, 2-methyl-
8938 - 1-Pentanol, 3-methyl-, (±)-
8939 - 1-Pentanol, 4-methyl-
8940 - 2-Pentanol, 2-methyl-
8941 - 2-Pentanol, 3-methyl-
8942 - 2-Pentanol, 4-methyl-
8943 - 3-Pentanol, 2-methyl-
8944 - 3-Pentanol, 3-methyl-
10039 - Propane, 1-ethoxy-2-methyl-
10040 - Propane, 2-ethoxy-2-methyl-
10076 - Propane, 1-(1-methylethoxy)-
10119 - Propane, 1,1′-oxybis-
10120 - Propane, 2,2′-oxybis-
$C_6H_{14}OS$
10129 - Propane, 1,1′-sulfinylbis-
$C_6H_{14}OSi$
11423 - Silane, ethenylethoxydimethyl-
$C_6H_{14}O_2$
3350 - 2,3-Butanediol, 2,3-dimethyl-
5541 - Ethane, 1,1-diethoxy-
5542 - Ethane, 1,2-diethoxy-
5687 - Ethanol, 2-butoxy-
5746 - Ethanol, 2-(2-methylpropoxy)-
6793 - 1,3-Hexanediol
6794 - 1,4-Hexanediol
6795 - 1,5-Hexanediol
6796 - 1,6-Hexanediol
6797 - 2,3-Hexanediol
6798 - 2,4-Hexanediol
6799 - 2,5-Hexanediol
8793 - 2,4-Pentanediol, 2-methyl-
8794 - 2,4-Pentanediol, 3-methyl-
10007 - 1,3-Propanediol, 2-ethyl-2-methyl-
10390 - 2-Propanol, 1-(1-methylethoxy)-
10401 - 2-Propanol, 1-propoxy-
$C_6H_{14}O_2S$
10131 - Propane, 1,1′-sulfonylbis-
$C_6H_{14}O_3$
5613 - Ethane, 1,1′-oxybis[2-methoxy-
5614 - Ethane, 1,1′-[oxybis(methyleneoxy)]bis-
5705 - Ethanol, 2,2-diethoxy-
5720 - Ethanol, 2-(2-ethoxyethoxy)-
6867 - 1,2,5-Hexanetriol
6868 - 1,2,6-Hexanetriol
6869 - 2,3,4-Hexanetriol
10006 - 1,3-Propanediol, 2-ethyl-2-(hydroxymethyl)-
10185 - Propane, 1,2,3-trimethoxy-
10397 - Propanol, oxybis-
$C_6H_{14}O_4$
5713 - Ethanol, 2,2′-[1,2-ethanediylbis(oxy)]bis-
$C_6H_{14}O_4S$
11575 - Sulfuric acid, dipropyl ester
$C_6H_{14}O_5$
10024 - Propanediol, oxybis-
10025 - 1,2-Propanediol, 3,3′-oxybis-
$C_6H_{14}O_6$
6194 - Galactitol
6206 - *D*-Glucitol
7422 - *D*-Mannitol
$C_6H_{14}S$
3376 - Butane, 1-(ethylthio)-
6856 - 1-Hexanethiol
6857 - 2-Hexanethiol
6858 - 3-Hexanethiol
8826 - Pentane, 1-(methylthio)-
10044 - Propane, 1-(ethylthio)-2-methyl-
10077 - Propane, 1-[(1-methylethyl)thio]-
10151 - Propane, 1,1′-thiobis-
10152 - Propane, 2,2′-thiobis-
$C_6H_{14}S_2$
5244 - Disulfide, bis(1-methylethyl)
5259 - Disulfide, dipropyl
5450 - Ethane, 1,1-bis(ethylthio)-
5451 - Ethane, 1,2-bis(ethylthio)-
6812 - 1,6-Hexanedithiol
$C_6H_{14}Si$
11495 - Silane, trimethyl-2-propenyl-
$C_6H_{15}As$
448 - Arsine, triethyl-
$C_6H_{15}AsO_3$
440 - Arsenous acid, triethyl ester
$C_6H_{15}AsO_4$
439 - Arsenic acid, triethyl ester
$C_6H_{15}B$
3118 - Borane, triethyl
$C_6H_{15}BO_3$
3125 - Boric acid, triethyl ester
$C_6H_{15}BrSi$
11368 - Silane, bromotriethyl-
$C_6H_{15}ClN_2O_2$
5440 - Ethanaminium, 2-[(aminocarbonyl)oxy]-*N,N,N*-trimethyl-, chloride
$C_6H_{15}ClO_3Si$
11387 - Silane, chlorotriethoxy-
$C_6H_{15}ClSi$
11369 - Silane, butylchlorodimethyl-
11386 - Silane, (3-chloropropyl)trimethyl-
11388 - Silane, chlorotriethyl-
$C_6H_{15}FSi$
11484 - Silane, triethylfluoro-
$C_6H_{15}N$
3215 - 1-Butanamine, *N,N*-dimethyl-
3216 - 2-Butanamine, 2,3-dimethyl-
3217 - 2-Butanamine, 3,3-dimethyl-
3218 - 1-Butanamine, *N*-ethyl-
3219 - 2-Butanamine, *N*-ethyl-, (±)-
5417 - Ethanamine, *N,N*-diethyl-
6723 - 1-Hexanamine
6724 - 2-Hexanamine, (±)-
8706 - 1-Pentanamine, 3-methyl-
8707 - 1-Pentanamine, 4-methyl-
9821 - 2-Propanamine, *N*-(1-methylethyl)-
9826 - 1-Propanamine, *N*-propyl-
9827 - 2-Propanamine, *N*-propyl
9829 - 1-Propanamine, *N,N*,2-trimethyl-
$C_6H_{15}NO$
5424 - Ethanamine, 2-ethoxy-*N,N*-dimethyl-
5691 - Ethanol, 2-(butylamino)-
5706 - Ethanol, 2-(diethylamino)-
5710 - Ethanol, 2-[(1,1-dimethylethyl)amino]-
5747 - Ethanol, 2-[(2-methylpropyl)amino]-
8922 - 1-Pentanol, 2-amino-4-methyl-, (±)-
9814 - 1-Propanamine, *N*-hydroxy-*N*-propyl-
$C_6H_{15}NO_2$
3623 - 2-Butanol, 1-[(2-hydroxyethyl)amino]-
5414 - Ethanamine, 2,2-diethoxy-
5723 - Ethanol, 2,2′-(ethylimino)bis-
10384 - 2-Propanol, 1,1′-iminobis-
$C_6H_{15}NO_3$
5751 - Ethanol, 2,2′,2″-nitrilotris-
$C_6H_{15}NO_5$
6207 - *D*-Glucitol, 1-amino-1-deoxy-
$C_6H_{15}NS$
5650 - Ethanethiol, 2-(butylamino)-
$C_6H_{15}N_2$
5510 - 1,2-Ethanediamine, *N*′-ethyl-*N,N*-dimethyl-
$C_6H_{15}N_3$
9615 - 1-Piperazineethanamine
$C_6H_{15}OP$
9517 - Phosphine oxide, triethyl-
$C_6H_{15}O_2P$
9527 - Phosphinic acid, diethyl-, ethyl ester
$C_6H_{15}O_2PS_2$
9579 - Phosphorodithioic acid, *O,O,S*-triethyl ester
$C_6H_{15}O_3P$
9536 - Phosphonic acid, ethyl-, diethyl ester
9591 - Phosphorous acid, triethyl ester
$C_6H_{15}O_3PS$
9585 - Phosphorothioic acid, *O,O,O*-triethyl ester
$C_6H_{15}O_3PS_2$
5103 - Demeton S methyl
$C_6H_{15}O_4P$
9563 - Phosphoric acid, triethyl ester
$C_6H_{15}O_4PS_2$
8632 - Oxydemeton-methyl
$C_6H_{15}P$
9521 - Phosphine, triethyl-
$C_6H_{15}Sb$
11549 - Stibine, triethyl-
$C_6H_{16}Br_2OSi_2$
5229 - Disiloxane, 1,3-bis(bromomethyl)-1,1,3,3-tetramethyl-
$C_6H_{16}ClN$
5418 - Ethanamine, *N,N*-diethyl-, hydrochloride

$C_6H_{16}ClNO$
9831 - 1-Propanaminium, 2-hydroxy-*N,N,N*-trimethyl-, chloride
$C_6H_{16}Cl_2OSi_2$
5230 - Disiloxane, 1,3-bis(chloromethyl)-1,1,3,3-tetramethyl-
$C_6H_{16}N_2$
5501 - 1,2-Ethanediamine, *N,N'*-diethyl-
5502 - 1,2-Ethanediamine, *N,N*-diethyl-
5516 - 1,2-Ethanediamine, *N,N,N',N'*-tetramethyl-
6749 - 1,6-Hexanediamine
7137 - Hydrazine, 1,2-bis(1-methylethyl)-
$C_6H_{16}N_2O$
5677 - Ethanol, 2-[(2-amino-2-methylpropyl)amino]-
$C_6H_{16}N_2O_2$
5715 - Ethanol, 2,2'-(1,2-ethanediyldiimino)bis-
$C_6H_{16}OSi$
10410 - 1-Propanol, 3-(trimethylsilyl)-
11503 - Silanol, triethyl-
$C_6H_{16}O_2Si$
11405 - Silane, diethoxydimethyl-
$C_6H_{16}O_3Si$
11477 - Silane, triethoxy-
$C_6H_{16}Si$
11483 - Silane, triethyl-
11496 - Silane, trimethylpropyl-
$C_6H_{17}N_3$
9874 - 1,3-Propanediamine, *N*-(3-aminopropyl)-
$C_6H_{18}AlO_9P_3$
6020 - Fosetyl-Al
$C_6H_{18}Cl_2O_2Si_3$
11913 - Trisiloxane, 1,5-dichloro-1,1,3,3,5,5-hexamethyl-
$C_6H_{18}N_3OP$
9573 - Phosphoric triamide, hexamethyl-
$C_6H_{18}N_4$
5499 - 1,2-Ethanediamine, *N,N'*-bis(2-aminoethyl)-
$C_6H_{18}OSi_2$
5236 - Disiloxane, hexamethyl-
$C_6H_{18}O_3Si_2$
5234 - Disiloxane, 1,3-dimethoxy-1,1,3,3-tetramethyl-
$C_6H_{18}O_3Si_3$
5002 - Cyclotrisiloxane, hexamethyl-
$C_6H_{18}Si_2$
5228 - Disilane, hexamethyl-
$C_6H_{19}NSi_2$
11363 - Silanamine, 1,1,1-trimethyl-*N*-(trimethylsilyl)-
$C_6H_{21}N_3Si_3$
5001 - Cyclotrisilazane, 2,2,4,4,6,6-hexamethyl-
$C_6H_{24}O_6Si_6$
4625 - Cyclohexasiloxane, 2,4,6,8,10,12-hexamethyl-
$C_6H_{26}O_8$
3351 - 2,3-Butanediol, 2,3-dimethyl-, hexahydrate
C_6MoO_6
7662 - Molybdenum carbonyl
C_6N_4
5915 - Ethenetetracarbonitrile
C_6N_6
11836 - 1,3,5-Triazine-2,4,6-tricarbonitrile
C_7F_5N
2767 - Benzonitrile, pentafluoro-
C_7F_8
2095 - Benzene, pentafluoro(trifluoromethyl)-
C_7F_{14}
4499 - Cyclohexane, undecafluoro(trifluoromethyl)-
6560 - 1-Heptene, 1,1,2,3,3,4,4,5,5,6,6,7,7,7-tetradecafluoro-
C_7F_{16}
6418 - Heptane, hexadecafluoro-
$C_7HCl_5O_2$
2712 - Benzoic acid, pentachloro-
C_7HCl_7
2085 - Benzene, pentachloro(dichloromethyl)-
C_7HF_5O
596 - Benzaldehyde, pentafluoro-
C_7HF_{15}
6432 - Heptane, 1,1,1,2,2,3,3,4,4,5,5,6,6,7,7-pentadecafluoro-
$C_7H_3Br_2NO$
2751 - Benzonitrile, 3,5-dibromo-4-hydroxy-
$C_7H_3Br_5$
2083 - Benzene, pentabromomethyl-
$C_7H_3ClF_3NO_2$
1217 - Benzene, 1-chloro-2-nitro-4-(trifluoromethyl)-
1218 - Benzene, 1-chloro-4-nitro-2-(trifluoromethyl)-
1219 - Benzene, 4-chloro-1-nitro-2-(trifluoromethyl)-
$C_7H_3ClN_2O_5$
2896 - Benzoyl chloride, 3,5-dinitro-
$C_7H_3Cl_2N$
2752 - Benzonitrile, 2,6-dichloro-
$C_7H_3Cl_3O$
2893 - Benzoyl chloride, 2,4-dichloro-
2894 - Benzoyl chloride, 3,4-dichloro-
$C_7H_3Cl_3O_2$
2727 - Benzoic acid, 2,3,6-trichloro-
2728 - Benzoic acid, 2,4,5-trichloro-
$C_7H_3Cl_5$
1401 - Benzene, 1,2-dichloro-4-(trichloromethyl)-
2086 - Benzene, pentachloromethyl-
2326 - Benzene, 1,2,3-trichloro-4-(dichloromethyl)-
2327 - Benzene, 1,2,4-trichloro-5-(dichloromethyl)-
$C_7H_3F_5$
2094 - Benzene, pentafluoromethyl-
$C_7H_3F_5O$
2093 - Benzene, pentafluoromethoxy-
$C_7H_3N_3O_8$
2735 - Benzoic acid, 2,4,6-trinitro-
C_7H_4BrClO
2887 - Benzoyl chloride, 2-bromo-
2888 - Benzoyl chloride, 4-bromo-
$C_7H_4BrF_3$
1089 - Benzene, 1-bromo-2-(trifluoromethyl)-
1090 - Benzene, 1-bromo-3-(trifluoromethyl)-
1091 - Benzene, 1-bromo-4-(trifluoromethyl)-
C_7H_4BrN
2744 - Benzonitrile, 2-bromo-
2745 - Benzonitrile, 3-bromo-
2746 - Benzonitrile, 4-bromo-
C_7H_4BrNO
1049 - Benzene, 1-bromo-4-isocyanato-
$C_7H_4BrNO_4$
2527 - Benzoic acid, 2-bromo-4-nitro-
2528 - Benzoic acid, 2-bromo-5-nitro-
2529 - Benzoic acid, 4-bromo-3-nitro-
2530 - Benzoic acid, 5-bromo-2-nitro-
$C_7H_4Br_2O_2$
534 - Benzaldehyde, 3,5-dibromo-2-hydroxy-
2555 - Benzoic acid, 2,4-dibromo-
2556 - Benzoic acid, 2,5-dibromo-
2557 - Benzoic acid, 2,6-dibromo-
$C_7H_4Br_2O_3$
2558 - Benzoic acid, 3,5-dibromo-2-hydroxy-
$C_7H_4Br_2O_5$
2560 - Benzoic acid, 2,6-dibromo-3,4,5-trihydroxy-
$C_7H_4Br_4O$
2256 - Benzene, 1,2,3,5-tetrabromo-4-methoxy-
9468 - Phenol, 2,3,4,5-tetrabromo-6-methyl-
C_7H_4ClFO
2897 - Benzoyl chloride, 2-fluoro-
2898 - Benzoyl chloride, 3-fluoro-
2899 - Benzoyl chloride, 4-fluoro-
$C_7H_4ClF_3$
1240 - Benzene, 1-chloro-2-(trifluoromethyl)-
1241 - Benzene, 1-chloro-3-(trifluoromethyl)-
1242 - Benzene, 1-chloro-4-(trifluoromethyl)-
C_7H_4ClN
2747 - Benzonitrile, 2-chloro-
2748 - Benzonitrile, 3-chloro-
2749 - Benzonitrile, 4-chloro-
C_7H_4ClNO
1176 - Benzene, 1-chloro-2-isocyanato-
2878 - Benzoxazole, 2-chloro-
$C_7H_4ClNO_2$
2883 - 2(3H)-Benzoxazolone, 5-chloro-
$C_7H_4ClNO_3$
2905 - Benzoyl chloride, 3-nitro-
2906 - Benzoyl chloride, 4-nitro-
$C_7H_4ClNO_4$
2544 - Benzoic acid, 2-chloro-3-nitro-
2545 - Benzoic acid, 2-chloro-5-nitro-
2546 - Benzoic acid, 3-chloro-2-nitro-
2547 - Benzoic acid, 4-chloro-3-nitro-
C_7H_4ClNS
1177 - Benzene, 1-chloro-4-isothiocyanato-
2862 - Benzothiazole, 2-chloro-
$C_7H_4Cl_2O$
535 - Benzaldehyde, 2,5-dichloro-
536 - Benzaldehyde, 3,4-dichloro-
537 - Benzaldehyde, 3,5-dichloro-
2890 - Benzoyl chloride, 2-chloro-
2891 - Benzoyl chloride, 3-chloro-
2892 - Benzoyl chloride, 4-chloro-
$C_7H_4Cl_2O_2$
2561 - Benzoic acid, 2,4-dichloro-
2562 - Benzoic acid, 2,5-dichloro-
2563 - Benzoic acid, 2,6-dichloro-
2564 - Benzoic acid, 3,5-dichloro-
$C_7H_4Cl_2O_3$
2566 - Benzoic acid, 3,5-dichloro-2-hydroxy
$C_7H_4Cl_3F$
1736 - Benzene, 1-fluoro-2-(trichloromethyl)-
$C_7H_4Cl_3NO_3$
11854 - Triclopyr
$C_7H_4Cl_4$
1237 - Benzene, 1-chloro-2-(trichloromethyl)-
1238 - Benzene, 1-chloro-3-(trichloromethyl)-
1239 - Benzene, 1-chloro-4-(trichloromethyl)-
1359 - Benzene, 1,2-dichloro-4-(dichloromethyl)-
2262 - Benzene, 1,2,3,5-tetrachloro-4-methyl-
2263 - Benzene, 1,2,4,5-tetrachloro-3-methyl-
2324 - Benzene, 1,2,4-trichloro-5-(chloromethyl)-
$C_7H_4Cl_4O_2$
9472 - Phenol, 2,3,5,6-tetrachloro-4-methoxy-
C_7H_4FN
2756 - Benzonitrile, 4-fluoro
C_7H_4FNS
1721 - Benzene, 1-fluoro-3-isothiocyanato-
$C_7H_4F_3NO_2$
2060 - Benzene, 1-nitro-2-(trifluoromethyl)-
2061 - Benzene, 1-nitro-3-(trifluoromethyl)-
$C_7H_4F_4$
1737 - Benzene, 1-fluoro-2-(trifluoromethyl)-
1738 - Benzene, 1-fluoro-3-(trifluoromethyl)-
1739 - Benzene, 1-fluoro-4-(trifluoromethyl)-
$C_7H_4I_2O_3$
2617 - Benzoic acid, 2-hydroxy-3,5-diiodo-
2618 - Benzoic acid, 4-hydroxy-3,5-diiodo-
$C_7H_4N_2O_3$
1788 - Benzene, 1-isocyanato-2-nitro-
1789 - Benzene, 1-isocyanato-3-nitro-
1790 - Benzene, 1-isocyanato-4-nitro-
$C_7H_4N_2O_5$
551 - Benzaldehyde, 2,4-dinitro-
$C_7H_4N_2O_6$
2590 - Benzoic acid, 3,4-dinitro-
2591 - Benzoic acid, 3,5-dinitro-
$C_7H_4O_4S$
[illegible] - 3H-2,1-Benzoxathiol-3-one, 1,1-dioxide
$C_7H_4O_6$
10776 - 4H-Pyran-2,6-dicarboxylic acid, 4-oxo-
$C_7H_4O_7$
10775 - 4H-Pyran-2,6-dicarboxylic acid, 3-hydroxy-4-oxo-
C_7H_5BrO
526 - Benzaldehyde, 2-bromo-
527 - Benzaldehyde, 3-bromo-
528 - Benzaldehyde, 4-bromo-
2885 - Benzoyl bromide

1731 - Benzene, 4-fluoro-1-methyl-2-nitro-
1732 - Benzene, 4-fluoro-2-methyl-1-nitro-

$C_7H_6F_2$
1425 - Benzene, (difluoromethyl)-

$C_7H_6F_3N$
855 - Benzenamine, 2-(trifluoromethyl)-
856 - Benzenamine, 3-(trifluoromethyl)-
857 - Benzenamine, 4-(trifluoromethyl)-

$C_7H_6N_2$
2376 - 1H-Benzimidazole
2741 - Benzonitrile, 2-amino-
2742 - Benzonitrile, 3-amino-
7235 - 1H-Indazole

$C_7H_6N_2O$
10884 - 3-Pyridinecarbonitrile, 1,2-dihydro-1-methyl-2-oxo-

$C_7H_6N_2O_3$
624 - Benzamide, 2-nitro-
625 - Benzamide, 3-nitro-

$C_7H_6N_2O_4$
1956 - Benzene, 1-methyl-2,4-dinitro-
1957 - Benzene, 1-methyl-3,5-dinitro-
1958 - Benzene, 2-methyl-1,3-dinitro-
1959 - Benzene, 2-methyl-1,4-dinitro-
1960 - Benzene, 4-methyl-1,2-dinitro-
2506 - Benzoic acid, 2-amino-3-nitro-

$C_7H_6N_2O_5$
1922 - Benzene, 1-methoxy-2,3-dinitro-
1923 - Benzene, 1-methoxy-2,4 dinitro-
1924 - Benzene, 1-methoxy-3,5-dinitro-
1925 - Benzene, 2-methoxy-1,3-dinitro-
1926 - Benzene, 2-methoxy-1,4-dinitro-
1927 - Benzene, 4-methoxy-1,2-dinitro-
9405 - Phenol, 2-methyl-4,6-dinitro-

$C_7H_6N_2S$
2386 - 2H-Benzimidazole-2-thione, 1,3-dihydro-
2859 - 2-Benzothiazolamine
2860 - 6-Benzothiazolamine

$C_7H_6N_4O$
7193 - 1H-Imidazole, 1,1′-carbonylbis-

C_7H_6O
525 - Benzaldehyde
4242 - 2,4,6-Cycloheptatrien-1-one

C_7H_6OS
1130 - Benzenecarbothioic acid

$C_7H_6OS_2$
1125 - Benzenecarbodithioic acid, 2-hydroxy-

$C_7H_6O_2$
564 - Benzaldehyde, 2-hydroxy-
565 - Benzaldehyde, 3-hydroxy-
566 - Benzaldehyde, 4-hydroxy-
2415 - 1,3-Benzodioxole
2475 - Benzoic acid
4243 - 2,4,6-Cycloheptatrien-1-one, 2-hydroxy-
4266 - 2,5-Cyclohexadiene-1,4-dione, 2-methyl-
10512 - 2-Propenal, 3-(2-furanyl)-
10513 - 2-Propenal, 3-(2-furanyl)-, (*E*)-

$C_7H_6O_2S$
2663 - Benzoic acid, 2-mercapto-

$C_7H_6O_3$
540 - Benzaldehyde, 2,3-dihydroxy-
541 - Benzaldehyde, 2,4-dihydroxy-
542 - Benzaldehyde, 3,4-dihydroxy-
1126 - Benzenecarboperoxoic acid
2612 - Benzoic acid, 2-hydroxy-
2613 - Benzoic acid, 4-hydroxy-
10638 - 2-Propenoic acid, 3-(2-furanyl)-

$C_7H_6O_4$
2568 - Benzoic acid, 2,3-dihydroxy-
2569 - Benzoic acid, 2,4-dihydroxy-
2570 - Benzoic acid, 2,5-dihydroxy-
2571 - Benzoic acid, 3,4-dihydroxy-
6183 - 4H-Furo[3,2-c]pyran-2(6H)-one, 4-hydroxy-

$C_7H_6O_5$
2730 - Benzoic acid, 2,3,4-trihydroxy-
2731 - Benzoic acid, 3,4,5-trihydroxy-

$C_7H_6O_6S$
2655 - Benzoic acid, 2-hydroxy-5-sulfo-

$C_7H_7BF_2$
3115 - Borane, difluoro(4-methylphenyl)-

C_7H_7Br
1053 - Benzene, (bromomethyl)-
1054 - Benzene, 1-bromo-2-methyl-
1055 - Benzene, 1-bromo-3-methyl-
1056 - Benzene, 1-bromo-4-methyl-

$C_7H_7BrN_2O$
11998 - Urea, (4-bromophenyl)-

C_7H_7BrO
1050 - Benzene, 1-bromo-2-methoxy-
1051 - Benzene, 1-bromo-3-methoxy-
1052 - Benzene, 1-bromo-4-methoxy-
9236 - Phenol, 2-bromo-4-methyl-
9237 - Phenol, 2-bromo-5-methyl-
9238 - Phenol, 3-bromo-2-methyl-
9239 - Phenol, 3-bromo-4-methyl-
9240 - Phenol, 3-bromo-5-methyl-
9241 - Phenol, 4-bromo-2-methyl-
9242 - Phenol, 4-bromo-3-methyl-

$C_7H_7BrO_2$
1843 - Benzenemethanol, 5-bromo-2-hydroxy-

$C_7H_7BrO_3$
6190 - 3-Furoic acid, 5-bromo-, ethyl ester

C_7H_7Cl
1182 - Benzene, (chloromethyl)-
1183 - Benzene, 1-chloro-2-methyl-
1184 - Benzene, 1-chloro-3-methyl-
1185 - Benzene, 1-chloro-4-methyl-

$C_7H_7ClN_2S$
11801 - Thiourea, (2-chlorophenyl)-

C_7H_7ClO
1178 - Benzene, 1-chloro-2-methoxy-
1179 - Benzene, 1-chloro-3-methoxy-
1180 - Benzene, 1-chloro-4-methoxy-
1848 - Benzenemethanol, 2-chloro-
1849 - Benzenemethanol, 4-chloro-
9260 - Phenol, 2-chloro-3-methyl
9261 - Phenol, 2-chloro-4-methyl-
9262 - Phenol, 2-chloro-5-methyl-
9263 - Phenol, 2-chloro-6-methyl-
9264 - Phenol, 3-chloro-2-methyl-
9265 - Phenol, 3-chloro-4-methyl-
9266 - Phenol, 4-chloro-2-methyl-
9267 - Phenol, 4-chloro-3-methyl-

C_7H_7ClOS
2186 - Benzenesulfinyl chloride, 4-methyl-

$C_7H_7ClO_2S$
2249 - Benzenesulfonyl chloride, 2-methyl-
2250 - Benzenesulfonyl chloride, 4-methyl-

$C_7H_7ClO_3$
6053 - 2-Furancarboxylic acid, 5-chloro-, ethyl ester

$C_7H_7ClO_3S$
2248 - Benzenesulfonyl chloride, 2-methoxy-

C_7H_7ClS
1833 - Benzenemethanethiol, 4-chloro-

$C_7H_7Cl_2N$
1803 - Benzenemethanamine, 2,4-dichloro-

$C_7H_7Cl_2NO_2S$
2210 - Benzenesulfonamide, *N*,*N*-dichloro-4-methyl-

$C_7H_7Cl_3NO_3PS$
4090 - Chlorpyrifos-methyl

$C_7H_7Cl_3Si$
11467 - Silane, trichloro(2-methylphenyl)-
11468 - Silane, trichloro(3-methylphenyl)-

C_7H_7F
1725 - Benzene, (fluoromethyl)-
1726 - Benzene, 1-fluoro-2-methyl-
1727 - Benzene, 1-fluoro-3-methyl-
1728 - Benzene, 1-fluoro-4-methyl-

C_7H_7FO
1722 - Benzene, 1-fluoro-2-methoxy-
1723 - Benzene, 1-fluoro-3-methoxy-
1724 - Benzene, 1-fluoro-4-methoxy-
1879 - Benzenemethanol, 3-fluoro-
1880 - Benzenemethanol, 4-fluoro-

$C_7H_7F_7O_2$
3526 - Butanoic acid, heptafluoro-, 1-methylethyl ester

C_7H_7I
1777 - Benzene, (iodomethyl)-
1778 - Benzene, 1-iodo-2-methyl-
1779 - Benzene, 1-iodo-3-methyl-
1780 - Benzene, 1-iodo-4-methyl-

C_7H_7IO
1775 - Benzene, 1-iodo-2-methoxy-
1776 - Benzene, 1-iodo-4-methoxy-
1888 - Benzenemethanol, 2-iodo-
9374 - Phenol, 2-iodo-4-methyl-
9375 - Phenol, 2-iodo-6-methyl-

C_7H_7N
10952 - Pyridine, 2-ethenyl-
10953 - Pyridine, 3-ethenyl-
10954 - Pyridine, 4-ethenyl-

C_7H_7NO
594 - Benzaldehyde, oxime, (*E*)-
595 - Benzaldehyde, oxime, (*Z*)-
600 - Benzamide
5867 - Ethanone, 1-(2-pyridinyl)-
5868 - Ethanone, 1-(3-pyridinyl)-
5869 - Ethanone, 1-(4-pyridinyl)-
5999 - Formamide, *N*-phenyl-

$C_7H_7NO_2$
579 - Benzaldehyde, 2-hydroxy-, oxime
616 - Benzamide, 2-hydroxy-
2000 - Benzene, 1-methyl-2-nitro-
2001 - Benzene, 1-methyl-3-nitro-
2002 - Benzene, 1-methyl-4-nitro-
2052 - Benzene, (nitromethyl)-
2484 - Benzoic acid, 2-amino-
2485 - Benzoic acid, 3-amino-
2486 - Benzoic acid, 4-amino-
8097 - Nitrous acid, phenylmethyl ester
10906 - 3-Pyridinecarboxylic acid, methyl ester
10908 - 4-Pyridinecarboxylic acid, methyl ester
11029 - Pyridinium, 3-carboxy-1-methyl-, hydroxide, inner salt

$C_7H_7NO_2S$
2043 - Benzene, 1-(methylthio)-2-nitro-
2044 - Benzene, 1-(methylthio)-4-nitro-

$C_7H_7NO_3$
613 - Benzamide, *N*,2-dihydroxy-
1906 - Benzenemethanol, 2-nitro-
1907 - Benzenemethanol, 3-nitro-
1908 - Benzenemethanol, 4-nitro-
1935 - Benzene, 1-methoxy-2-nitro-
1936 - Benzene, 1-methoxy-3-nitro-
1937 - Benzene, 1-methoxy-4-nitro-
2498 - Benzoic acid, 4-amino-2-hydroxy-
2499 - Benzoic acid, 5-amino-2-hydroxy-
9422 - Phenol, 2-methyl-4-nitro-
9423 - Phenol, 2-methyl-6-nitro-
9424 - Phenol, 4-methyl-2-nitro-

$C_7H_7NO_4$
9386 - Phenol, 2-methoxy-3-nitro-

$C_7H_7NO_4S$
2508 - Benzoic acid, 4-(aminosulfonyl)-

$C_7H_7N_3$
980 - Benzene, (azidomethyl)-
981 - Benzene, 1-azido-2-methyl-
982 - Benzene, 1-azido-4-methyl-
2872 - 1H-Benzotriazole, 1-methyl-

$C_7H_7N_3O_3$
12025 - Urea, (4-nitrophenyl)-

C_7H_8
1947 - Benzene, methyl-
2913 - Bicyclo[2.2.1]hepta-2,5-diene
4240 - 1,3,5-Cycloheptatriene
4765 - 1,3-Cyclopentadiene, 5-ethylidene-
6341 - 1,6-Heptadien-3-yne
6342 - 1,5-Heptadiyne
6343 - 1,6-Heptadiyne

C_7H_8BrN
647 - Benzenamine, 2-bromo-4-methyl-
648 - Benzenamine, 2-bromo-5-methyl-
649 - Benzenamine, 3-bromo-4-methyl-
650 - Benzenamine, 4-bromo-2-methyl-
651 - Benzenamine, 4-bromo-3-methyl-
652 - Benzenamine, 5-bromo-2-methyl-
1799 - Benzenemethanamine, 4-bromo-
C_7H_8ClN
668 - Benzenamine, 2-chloro-4-methyl-
669 - Benzenamine, 2-chloro-5-methyl-
670 - Benzenamine, 2-chloro-*N*-methyl-
671 - Benzenamine, 3-chloro-2-methyl-
672 - Benzenamine, 3-chloro-4-methyl-
673 - Benzenamine, 4-chloro-2-methyl-
674 - Benzenamine, 4-chloro-3-methyl-
675 - Benzenamine, 4-chloro-*N*-methyl-
676 - Benzenamine, 5-chloro-2-methyl-
C_7H_8ClNO
666 - Benzenamine, 2-chloro-5-methoxy-
667 - Benzenamine, 4-chloro-2-methoxy-
$C_7H_8ClN_3O_4S_2$
2855 - 2H-1,2,4-Benzothiadiazine-7-sulfonamide, 6-chloro-3,4-dihydro-, 1,1-dioxide
$C_7H_8Cl_2Si$
11401 - Silane, dichloromethylphenyl-
C_7H_8FN
755 - Benzenamine, 4-fluoro-2-methyl-
C_7H_8IN
762 - Benzenamine, 2-iodo-4-methyl-
763 - Benzenamine, 5-iodo-2-methyl-
$C_7H_8N_2$
563 - Benzaldehyde, hydrazone
11137 - 1H-Pyrrole-1-propanenitrile
$C_7H_8N_2O$
89 - Acetamide, *N*-3-pyridinyl-
601 - Benzamide, 2-amino-
806 - Benzenamine, *N*-methyl-*N*-nitroso-
2611 - Benzoic acid, hydrazide
12026 - Urea, phenyl-
$C_7H_8N_2O_2$
799 - Benzenamine, 2-methyl-3-nitro-
800 - Benzenamine, 2-methyl-4-nitro-
801 - Benzenamine, 2-methyl-5-nitro-
802 - Benzenamine, 2-methyl-6-nitro-
803 - Benzenamine, 4-methyl-2-nitro-
804 - Benzenamine, *N*-methyl-2-nitro-
805 - Benzenamine, *N*-methyl-4-nitro-
2552 - Benzoic acid, 2,3-diamino-
2553 - Benzoic acid, 2,4-diamino-
2554 - Benzoic acid, 3,5-diamino-
$C_7H_8N_2O_3$
775 - Benzenamine, 2-methoxy-5-nitro-
776 - Benzenamine, 3-methoxy-4-nitro-
777 - Benzenamine, 5-methoxy-2-nitro-
$C_7H_8N_2S$
11808 - Thiourea, phenyl-
$C_7H_8N_4O_2$
10756 - 1H-Purine-2,6-dione, 3,7-dihydro-1,3-dimethyl-
10757 - 1H-Purine-2,6-dione, 3,7-dihydro-3,7-dimethyl-
C_7H_8O
1836 - Benzenemethanol
1915 - Benzene, methoxy-
9400 - Phenol, 2-methyl-
9401 - Phenol, 3-methyl-
9402 - Phenol, 4-methyl-
C_7H_8OS
2040 - Benzene, (methylsulfinyl)-
2299 - Benzenethiol, 3-methoxy-
2300 - Benzenethiol, 4-methoxy-
5856 - Ethanone, 1-(5-methyl-2-thienyl)-
$C_7H_8O_2$
1527 - 1,2-Benzenediol, 3-methyl-
1528 - 1,2-Benzenediol, 4-methyl-
1529 - 1,3-Benzenediol, 2-methyl-
1530 - 1,3-Benzenediol, 4-methyl-
1531 - 1,3-Benzenediol, 5-methyl-
1532 - 1,4-Benzenediol, 2-methyl-
1882 - Benzenemethanol, 2-hydroxy-
1883 - Benzenemethanol, 3-hydroxy-
1884 - Benzenemethanol, 4-hydroxy-
9379 - Phenol, 2-methoxy-
9380 - Phenol, 3-methoxy-
9381 - Phenol, 4-methoxy-
10438 - 1-Propanone, 1-(2-furanyl)-
10439 - 2-Propanone, 1-(2-furanyl)-
10792 - 2H-Pyran-2-one, 4,6-dimethyl-
10793 - 4H-Pyran-4-one, 2,6-dimethyl-
$C_7H_8O_2S$
2181 - Benzenesulfinic acid, 2-methyl-
2182 - Benzenesulfinic acid, 4-methyl-
11739 - 2-Thiophenecarboxylic acid, ethyl ester
$C_7H_8O_3$
1520 - 1,4-Benzenediol, 2-(hydroxymethyl)-
1525 - 1,2-Benzenediol, 3-methoxy-
1526 - 1,3-Benzenediol, 5-methoxy-
2369 - 1,2,3-Benzenetriol, 5-methyl-
6029 - 2-Furanacetic acid, methyl ester
6056 - 2-Furancarboxylic acid, ethyl ester
6119 - 2-Furanmethanol, acetate
6161 - 2-Furanpropanoic acid
$C_7H_8O_3S$
2227 - Benzenesulfonic acid, methyl ester
2228 - Benzenesulfonic acid, 2-methyl-
2229 - Benzenesulfonic acid, 4-methyl-
C_7H_8S
1832 - Benzenemethanethiol
2041 - Benzene, (methylthio)-
2301 - Benzenethiol, 2-methyl-
2302 - Benzenethiol, 3-methyl-
2303 - Benzenethiol, 4-methyl-
$C_7H_8S_2$
1564 - 1,2-Benzenedithiol, 4-methyl-
$C_7H_9AsN_2O_4$
455 - Arsonic acid, [4-[(aminocarbonyl)amino]phenyl]-
$C_7H_9BO_2$
3134 - Boronic acid, (2-methylphenyl)-
C_7H_9Cl
7107 - 1-Hexen-3-yne, 5-chloro-5-methyl-
C_7H_9ClO
9147 - 2-Pentynoyl chloride, 4,4-dimethyl-
C_7H_9ClSi
11381 - Silane, chloromethylphenyl-
C_7H_9N
778 - Benzenamine, 2-methyl-
779 - Benzenamine, 3-methyl-
780 - Benzenamine, 4-methyl-
781 - Benzenamine, *N*-methyl-
1795 - Benzenemethanamine
4629 - 3-Cyclohexene-1-carbonitrile
4630 - 1-Cyclohexenecarbonitrile
4905 - 1-Cyclopentene-1-acetonitrile
10936 - Pyridine, 2,3-dimethyl-
10937 - Pyridine, 2,4-dimethyl-
10938 - Pyridine, 2,5-dimethyl-
10939 - Pyridine, 2,6-dimethyl-
10940 - Pyridine, 3,4-dimethyl-
10941 - Pyridine, 3,5-dimethyl-
10956 - Pyridine, 2-ethyl-
10957 - Pyridine, 3-ethyl-
10958 - Pyridine, 4-ethyl-
C_7H_9NO
758 - Benzenamine, *N*-hydroxy-4-methyl-
766 - Benzenamine, 2-methoxy-
767 - Benzenamine, 3-methoxy-
768 - Benzenamine, 4-methoxy-
1837 - Benzenemethanol, 2-amino-
5849 - Ethanone, 1-(1-methyl-1H-pyrrol-2-yl)-
9209 - Phenol, 2-amino-3-methyl-
9210 - Phenol, 2-amino-4-methyl-
9211 - Phenol, 3-amino-4-methyl-
9212 - Phenol, 3-amino-5-methyl-
9213 - Phenol, 4-amino-2-methyl-
10945 - Pyridine, 2,6-dimethyl-, 1-oxide
10950 - 2-Pyridineethanol
$C_7H_9NO_2$
2509 - Benzoic acid, ammonium salt
10935 - Pyridine, 2,6-dimethoxy-
$C_7H_9NO_3$
8586 - 2,4-Oxazolidinedione, 5-methyl-3-(2-propenyl)-
C_7H_9NS
819 - Benzenamine, 2-(methylthio)-
820 - Benzenamine, 4-(methylthio)-
$C_7H_9N_3O$
7145 - Hydrazinecarboxamide, 2-phenyl-
7146 - Hydrazinecarboxamide, *N*-phenyl-
$C_7H_9N_3O_2$
2500 - Benzoic acid, 4-amino-2-hydroxy-, hydrazide
$C_7H_9N_3O_2S_2$
2189 - Benzenesulfonamide, 4-amino-*N*-(aminothioxomethyl)-
$C_7H_9N_3O_3S$
2190 - Benzenesulfonamide, 4-amino-*N*-(aminocarbonyl)-
C_7H_{10}
4219 - 1,3-Cycloheptadiene
4271 - 1,3-Cyclohexadiene, 2-methyl-
4272 - 1,3-Cyclohexadiene, 5-methyl-, (±)-
6529 - 1,3,5-Heptatriene
6606 - 1-Hepten-3-yne
9115 - 3-Penten-1-yne, 3-ethyl-
$C_7H_{10}ClN$
791 - Benzenamine, 2-methyl-, hydrochloride
792 - Benzenamine, *N*-methyl-, hydrochloride
$C_7H_{10}ClN_3$
11043 - 4-Pyrimidinamine, 2-chloro-*N*,*N*,6-trimethyl-
$C_7H_{10}N_2$
1275 - 1,2-Benzenediamine, 3-methyl-
1276 - 1,2-Benzenediamine, 4-methyl-
1277 - 1,3-Benzenediamine, 2-methyl-
1278 - 1,3-Benzenediamine, 4-methyl-
1279 - 1,4-Benzenediamine, 2-methyl-
1280 - 1,4-Benzenediamine, *N*-methyl-
1796 - Benzenemethanamine, 2-amino-
4159 - Cyanamide, di-2-propenyl-
6392 - Heptanedinitrile
7164 - Hydrazine, (3-methylphenyl)-
7165 - Hydrazine, 1-methyl-1-phenyl-
7166 - Hydrazine, 1-methyl-2-phenyl-
10863 - 2-Pyridinamine, *N*,*N*-dimethyl-
10864 - 4-Pyridinamine, 2,6-dimethyl-
10948 - 2-Pyridineethanamine
10949 - 4-Pyridineethanamine
$C_7H_{10}N_2O$
1273 - 1,2-Benzenediamine, 4-methoxy-
1274 - 1,3-Benzenediamine, 4-methoxy-
$C_7H_{10}N_2OS$
11109 - 4(1H)-Pyrimidinone, 2,3-dihydro-6-propyl-2-thioxo-
$C_7H_{10}N_2O_2S$
7194 - 1H-Imidazole-1-carboxylic acid, 2,3-dihydro-3-methyl-2-thioxo-, ethyl ester
$C_7H_{10}N_4O_2S$
2191 - Benzenesulfonamide, 4-amino-*N*-(aminoiminomethyl)-
$C_7H_{10}O$
4631 - 1-Cyclohexene-1-carboxaldehyde
4632 - 3-Cyclohexene-1-carboxaldehyde
4710 - 2-Cyclohexen-1-one, 2-methyl-
4711 - 2-Cyclohexen-1-one, 3-methyl-
4712 - 2-Cyclohexen-1-one, 5-methyl-
4713 - 3-Cyclohexen-1-one, 3-methyl-
4714 - 3-Cyclohexen-1-one, 4-methyl-
4866 - Cyclopentanol, 1-ethynyl-
4943 - 2-Cyclopenten-1-one, 2,4-dimethyl-
6106 - Furan, 2-ethyl-5-methyl-
6167 - Furan, 2-propyl-

6325 - 2,4-Heptadienal, (*E,E*)-
6339 - 3,5-Heptadien-2-one
7607 - Methanone, dicyclopropyl-

$C_7H_{10}O_2$
3856 - 2-Butenoic acid, 2-propenyl ester
3857 - 2-Butenoic acid, 2-propenyl ester, (*E*)-
4225 - 1,2-Cycloheptanedione
4637 - 1-Cyclohexene-1-carboxylic acid
4638 - 3-Cyclohexene-1-carboxylic acid
4879 - Cyclopentanone, 2-acetyl-
4904 - 3-Cyclopentene-1-acetic acid
6104 - Furan, 2-(ethoxymethyl)-
6112 - Furan, 2-isopropoxy-
6702 - 2,4-Hexadienoic acid, methyl ester
6703 - 2,4-Hexadienoic acid, methyl ester, (*E,E*)-
10671 - 2-Propenoic acid, 2-methyl-, 2-propenyl ester

$C_7H_{10}O_3$
3577 - Butanoic acid, 3-oxo-, 2-propenyl ester
6454 - 2,4,6-Heptanetrione
10668 - 2-Propenoic acid, 2-methyl-, oxiranylmethyl ester

$C_7H_{10}O_4$
3330 - Butanedioic acid, methylene-, dimethyl ester
3331 - Butanedioic acid, (1-methylethylidene)-
3789 - 2-Butenedioic acid, 2-methyl-, dimethyl ester, (*E*)-
3790 - 2-Butenedioic acid, 2-methyl-, dimethyl ester, (*Z*)-
4793 - 1,3-Cyclopentanedicarboxylic acid, *cis*-
4969 - 1,2-Cyclopropanedicarboxylic acid, dimethyl ester
8876 - Pentanoic acid, 2,4-dioxo-, ethyl ester
10563 - 2-Propene-1,1-diol, diacetate

$C_7H_{10}O_5$
8784 - Pentanedioic acid, 3-oxo-, dimethyl ester
9973 - Propanedioic acid, oxo-, diethyl ester

$C_7H_{10}O_6$
7553 - Methanetricarboxylic acid, trimethyl ester

$C_7H_{10}S$
11761 - Thiophene, 2-ethyl-3-methyl-
11762 - Thiophene, 2-ethyl-5-methyl-
11763 - Thiophene, 3-ethyl-4-methyl-
11764 - Thiophene, 4-ethyl-2-methyl-
11772 - Thiophene, 2-(1-methylethyl)-
11773 - Thiophene, 3-(1-methylethyl)-
11778 - Thiophene, 2-propyl-
11779 - Thiophene, 3-propyl-
11792 - Thiophene, 2,3,4-trimethyl-
11793 - Thiophene, 2,3,5-trimethyl-

$C_7H_{10}Si$
11442 - Silane, methylphenyl-

$C_7H_{11}Br$
6612 - 1-Heptyne, 1-bromo-
6613 - 2-Heptyne, 1-bromo-

$C_7H_{11}BrO_4$
9928 - Propanedioic acid, bromo-, diethyl ester

$C_7H_{11}Cl$
2918 - Bicyclo[2.2.1]heptane, 2-chloro-
6614 - 1-Heptyne, 1-chloro-
6615 - 2-Heptyne, 7-chloro-
6616 - 3-Heptyne, 7-chloro-
9125 - 1-Pentyne, 3-chloro-3-ethyl-

$C_7H_{11}ClO$
4330 - Cyclohexanecarbonyl chloride

$C_7H_{11}ClO_3$
3492 - Butanoic acid, 2-chloro-2-methyl-3-oxo-, ethyl ester

$C_7H_{11}ClO_4$
9931 - Propanedioic acid, chloro-, diethyl ester
9996 - 1,2-Propanediol, 3-chloro-, diacetate

$C_7H_{11}Cl_3O_2$
255 - Acetic acid, trichloro-, 3-methylbutyl ester

$C_7H_{11}I$
6620 - 1-Heptyne, 1-iodo-

$C_7H_{11}N$
4328 - Cyclohexanecarbonitrile
11128 - 1H-Pyrrole, 3-ethyl-4-methyl-
11129 - Pyrrole, 2-isopropyl-
11138 - 1H-Pyrrole, 1-propyl-
11139 - 1H-Pyrrole, 3-propyl-
11142 - 1H-Pyrrole, 1,2,5-trimethyl-

$C_7H_{11}NO$
4329 - Cyclohexanecarbonitrile, 1-hydroxy-
4416 - Cyclohexane, isocyanato-

$C_7H_{11}NO_2$
151 - Acetic acid, cyano-, butyl ester
9720 - 4-Piperidinone, 1-acetyl-
10913 - 3-Pyridinecarboxylic acid, 1,2,5,6-tetrahydro-1-methyl-
11158 - 2,5-Pyrrolidinedione, 3-ethyl-3-methyl-

$C_7H_{11}NO_3$
11567 - Succinimide, *N*-(ethoxymethyl)-

$C_7H_{11}NO_4$
9766 - *L*-Proline, 1-acetyl-4-hydroxy-, *trans*-

$C_7H_{11}NO_5$
9958 - Propanedioic acid, (hydroxyimino)-, diethyl ester

$C_7H_{11}NS$
4417 - Cyclohexane, isothiocyanato-

C_7H_{12}
2914 - Bicyclo[2.2.1]heptane
2915 - Bicyclo[4.1.0]heptane
2980 - Bicyclo[3,1,0]hexane, 1-methyl-
4244 - Cycloheptene
4428 - Cyclohexane, methylene-
4670 - Cyclohexene, 1-methyl-
4671 - Cyclohexene, 3-methyl-, (±)-
4672 - Cyclohexene, 4-methyl-
4813 - Cyclopentane, ethylidene-
4918 - Cyclopentene, 1,2-dimethyl-
4919 - Cyclopentene, 1,3-dimethyl-
4920 - Cyclopentene, 1,4-dimethyl-
4921 - Cyclopentene, 1,5-dimethyl-
4922 - Cyclopentene, 4,4-dimethyl-
4924 - Cyclopentene, 1-ethyl-
4925 - Cyclopentene, 3-ethyl-
4926 - Cyclopentene, 4-ethyl-
6326 - 1,2-Heptadiene
6327 - 1,4-Heptadiene
6328 - 1,5-Heptadiene
6329 - 2,4-Heptadiene
6609 - 1-Heptyne
6610 - 2-Heptyne
6611 - 3-Heptyne
6695 - 1,5-Hexadiene, 2-methyl-
6696 - 2,4-Hexadiene, 2-methyl-
7120 - 1-Hexyne, 5-methyl-
7121 - 2-Hexyne, 5-methyl-
7122 - 3-Hexyne, 2-methyl-
8665 - 1,3-Pentadiene, 2,4-dimethyl-
9128 - 1-Pentyne, 3,3-dimethyl-
9129 - 1-Pentyne, 4,4-dimethyl-
9130 - 2-Pentyne, 4,4-dimethyl-
9131 - 1-Pentyne, 3-ethyl-

$C_7H_{12}Br_2O_2$
8872 - Pentanoic acid, 2,3-dibromo-, ethyl ester
8873 - Pentanoic acid, 2,5-dibromo-, ethyl ester
10239 - Propanoic acid, 2,3-dibromo-, butyl ester

$C_7H_{12}ClN_5$
11513 - Simazine

$C_7H_{12}N_2$
7196 - 1H-Imidazole, 1,2-diethyl-
11190 - Pyrrolo[1,2-a]pyrimidine, 2,3,4,6,7,8-hexahydro-

$C_7H_{12}O$
4234 - Cycloheptanone
4331 - Cyclohexanecarboxaldehyde
4537 - Cyclohexanol, 2-methylene-
4596 - Cyclohexanone, 2-methyl-, (±)-
4597 - Cyclohexanone, 3-methyl-, (±)-
4598 - Cyclohexanone, 4-methyl-
4882 - Cyclopentanone, 2,5-dimethyl-
5795 - Ethanone, 1-cyclopentyl-
6336 - 1,5-Heptadien-4-ol
6337 - 1,6-Heptadien-4-ol
6530 - 2-Heptenal
6531 - 3-Heptenal
6583 - 1-Hepten-3-one
6584 - 4-Hepten-2-one
6585 - 6-Hepten-2-one
6586 - 6-Hepten-3-one
6587 - 3-Hepten-2-one, (*E*)-
6588 - 3-Hepten-2-one, (*Z*)-
6589 - 5-Hepten-2-one, (*E*)-
7093 - 1-Hexen-3-one, 5-methyl-
7094 - 3-Hexen-2-one, 5-methyl-
7095 - 4-Hexen-2-one, 5-methyl-
7096 - 4-Hexen-3-one, 2-methyl-
7097 - 5-Hexen-2-one, 3-methyl-
7098 - 5-Hexen-2-one, 4-methyl-
7099 - 5-Hexen-2-one, 5-methyl-
7128 - 1-Hexyn-3-ol, 3-methyl-
7129 - 3-Hexyn-2-ol, 2-methyl-
9144 - 1-Pentyn-3-ol, 3,4-dimethyl-
9145 - 1-Pentyn-3-ol, 3-ethyl-

$C_7H_{12}O_2$
3587 - Butanoic acid, 2-propenyl ester
3849 - 2-Butenoic acid, 2-methyl-, ethyl ester, (*E*)-
3850 - 2-Butenoic acid, 3-methyl-, ethyl ester
4175 - Cyclobutanecarboxylic acid, ethyl ester
4332 - Cyclohexanecarboxylic acid
4775 - Cyclopentaneacetic acid
4783 - Cyclopentanecarboxylic acid, 1-methyl-
4865 - Cyclopentanol, acetate
6005 - Formic acid, cyclohexyl ester
6402 - 2,3-Heptanedione
6403 - 2,6-Heptanedione
6561 - 2-Heptenoic acid
6562 - 4-Heptenoic acid
6563 - 5-Heptenoic acid
6564 - 6-Heptenoic acid
6809 - 2,3-Hexanedione, 5-methyl-
6810 - 2,5-Hexanedione, 3-methyl-
7055 - 2-Hexenoic acid, 2-methyl-, (*E*)-
7056 - 2-Hexenoic acid, 4-methyl-
7057 - 2-Hexenoic acid, 5-methyl-
7058 - 3-Hexenoic acid, 2-methyl-
7059 - 3-Hexenoic acid, 3-methyl-
7060 - 3-Hexenoic acid, 4-methyl-
7061 - 4-Hexenoic acid, 5-methyl-
7062 - 3-Hexenoic acid, methyl ester
8799 - 2,4-Pentanedione, 3,3-dimethyl-
8801 - 2,4-Pentanedione, 3-ethyl-
9075 - 4-Pentenoic acid, 2,2-dimethyl-
9083 - 2-Pentenoic acid, 2-methyl-, methyl ester
9084 - 2-Pentenoic acid, 3-methyl-, methyl ester, (*E*)-
9085 - 2-Pentenoic acid, 3-methyl-, methyl ester, (*Z*)-
9086 - 3-Pentenoic acid, 3-methyl-, methyl ester
10616 - 2-Propenoic acid, butyl ester
10664 - 2-Propenoic acid, 2-methyl-, 1-methylethyl ester
10672 - 2-Propenoic acid, 2-methyl-, propyl ester
10673 - 2-Propenoic acid, 2-methylpropyl ester
10731 - 1-Propyne, 3,3-diethoxy-
10783 - 2H-Pyran, 2-ethoxy-3,4-dihydro-

$C_7H_{12}O_3$
3519 - Butanoic acid, 2-ethyl-3-oxo-, methyl ester
3557 - Butanoic acid, 2-methyl-3-oxo-, ethyl ester
3574 - Butanoic acid, 3-oxo-, 1-methylethyl ester
6123 - 2-Furanmethanol, tetrahydro-, acetate
6164 - 2-Furanpropanoic acid, tetrahydro-
6469 - Heptanoic acid, 6-oxo-
6901 - Hexanoic acid, 4-methyl-5-oxo-
8874 - Pentanoic acid, 2,2-dimethyl-4-oxo-
8898 - Pentanoic acid, 3-methyl-4-oxo-, methyl ester
8905 - Pentanoic acid, 2-oxo-, ethyl ester
8906 - Pentanoic acid, 3-oxo-, ethyl ester
8907 - Pentanoic acid, 4-oxo-, ethyl ester
10634 - 2-Propenoic acid, 2-ethoxyethyl ester
10662 - 2-Propenoic acid, 2-methyl-, 2-hydroxypropyl ester

3550 - Butanoic acid, 2-methyl-, ethyl ester, (*S*)-
3588 - Butanoic acid, propyl ester
3630 - 1-Butanol, 2-methyl-, acetate
3631 - 1-Butanol, 3-methyl-, acetate
6009 - Formic acid, hexyl ester
6455 - Heptanoic acid
6517 - 4-Heptanone, 2-hydroxy-
6893 - Hexanoic acid, 2-methyl-, (±)-
6894 - Hexanoic acid, 3-methyl-
6895 - Hexanoic acid, 4-methyl-, (±)-
6896 - Hexanoic acid, 5-methyl-
6898 - Hexanoic acid, methyl ester
6973 - 3-Hexanone, 2-hydroxy-2-methyl-
6974 - 3-Hexanone, 5-hydroxy-2-methyl-
8602 - Oxirane, (butoxymethyl)-
8610 - Oxirane, [(1,1-dimethylethoxy)methyl]-
8877 - Pentanoic acid, 2-ethyl-, (±)-
8878 - Pentanoic acid, ethyl ester
8918 - 2-Pentanol, acetate, (*R*)-
8919 - 3-Pentanol, acetate
8978 - 2-Pentanone, 4-methoxy-4-methyl-
10214 - Propanoic acid, butyl ester
10259 - Propanoic acid, 2,2-dimethyl-, ethyl ester
10303 - Propanoic acid, 2-methyl-, 1-methylethyl ester
10308 - Propanoic acid, 2-methyl-, propyl ester
10309 - Propanoic acid, 1-methylpropyl ester
10310 - Propanoic acid, 2-methylpropyl ester
10559 - 1-Propene, 1,1-diethoxy-
10560 - 1-Propene, 3,3-diethoxy-

$C_7H_{14}O_3$
3995 - Carbonic acid, bis(1-methylethyl) ester
4002 - Carbonic acid, dipropyl ester
8884 - Pentanoic acid, 4-hydroxy-, ethyl ester
8885 - Pentanoic acid, 2-hydroxyethyl ester
10262 - Propanoic acid, 3-ethoxy-, ethyl ester
10268 - Propanoic acid, 2-hydroxy-, butyl ester, (*R*)-

$C_7H_{14}O_4$
3505 - Butanoic acid, 2,3-dihydroxypropyl ester
7112 - *ribo*-Hexose, 2,6-dideoxy-3-*O*-methyl-

$C_7H_{14}O_5$
5736 - Ethanol, 2-methoxy-, carbonate (2:1)
6198 - Galactose, 6-deoxy-3-*O*-methyl-

$C_7H_{14}O_6$
6227 - α-*D*-Glucopyranoside, methyl

$C_7H_{15}Br$
6356 - Heptane, 1-bromo-
6357 - Heptane, 2-bromo-
6358 - Heptane, 4-bromo-

$C_7H_{15}BrO$
6736 - Hexane, 1-bromo-6-methoxy-

$C_7H_{15}Cl$
3254 - Butane, 2-chloro-2,3,3-trimethyl-
6360 - Heptane, 1-chloro-
6361 - Heptane, 2-chloro-
6362 - Heptane, 3-chloro-
6363 - Heptane, 4-chloro-
6743 - Hexane, 1-chloro-3-methyl-
6744 - Hexane, 2-chloro-2-methyl-
6745 - Hexane, 2-chloro-5-methyl-
6746 - Hexane, 3-chloro-3-methyl-
8722 - Pentane, 2-chloro-2,3-dimethyl-
8723 - Pentane, 2-chloro-2,4-dimethyl-
8724 - Pentane, 3-chloro-2,3-dimethyl-
8725 - Pentane, 4-chloro-2,2-dimethyl-
8726 - Pentane, 3-chloro-3-ethyl-

$C_7H_{15}ClO$
6477 - 1-Heptanol, 7-chloro-
6478 - 2-Heptanol, 1-chloro-

$C_7H_{15}ClO_2$
9850 - Propane, 3-chloro-1,1-diethoxy-

$C_7H_{15}F$
6417 - Heptane, 1-fluoro-

$C_7H_{15}FO$
6491 - 1-Heptanol, 7-fluoro-

$C_7H_{15}I$
6419 - Heptane, 1-iodo-
6420 - Heptane, 2-iodo-

$C_7H_{15}N$
502 - Azocine, octahydro-
4292 - Cyclohexanamine, 2-methyl-, *cis*-
4293 - Cyclohexanamine, 2-methyl-, *trans*-
4294 - Cyclohexanamine, 3-methyl-, *cis*-
4295 - Cyclohexanamine, 3-methyl-, *trans*-
4296 - Cyclohexanamine, 4-methyl-, *cis*-
4297 - Cyclohexanamine, 4-methyl-, *trans*-
4298 - Cyclohexanamine, *N*-methyl-
4418 - Cyclohexanemethanamine
9648 - Piperidine, 1,2-dimethyl-, (±)-
9649 - Piperidine, 2,5-dimethyl-
9650 - Piperidine, 2,6-dimethyl-
9651 - Piperidine, 3,3-dimethyl-
9652 - Piperidine, 3,5-dimethyl-
9667 - Piperidine, 1-ethyl-
9668 - Piperidine, 2-ethyl-, (±)-
9669 - Piperidine, 3-ethyl-, (±)-
9670 - Piperidine, 4-ethyl-
11173 - Pyrrolidine, 1-propyl-
11174 - Pyrrolidine, 3-propyl-
11175 - Pyrrolidine, 2,2,4-trimethyl-
11176 - Pyrrolidine, 2,2,5-trimethyl-

$C_7H_{15}NO$
6346 - Heptanal, oxime
6348 - Heptanamide
6982 - 2-Hexanone, 5-methyl-, oxime
8699 - Pentanamide, *N,N*-dimethyl-
8968 - 3-Pentanone, 2,4-dimethyl-, oxime
9662 - 1-Piperidineethanol
9663 - 2-Piperidineethanol
9664 - 4-Piperidineethanol
9678 - 3-Piperidinemethanol, 1-methyl-
9715 - 3-Piperidinol, 1-ethyl-
9793 - Propanamide, *N,N*-diethyl-
10427 - 2-Propanone, 1-(diethylamino)-

$C_7H_{15}NOS$
5930 - Ethiolate

$C_7H_{15}NO_2$
3541 - Butanoic acid, 3-(methylamino)-, ethyl ester
3936 - Carbamic acid, butyl-, ethyl ester
3942 - Carbamic acid, (1,1-dimethylethyl)-, ethyl ester
3947 - Carbamic acid, ethyl-, butyl ester
3960 - Carbamic acid, (1-methylpropyl)-, ethyl ester
3961 - Carbamic acid, (2-methylpropyl)-, ethyl ester
6429 - Heptane, 1-nitro-
7692 - 4-Morpholineethanol, α-methyl-
8088 - Nitrous acid, heptyl ester

$C_7H_{15}NO_5$
6215 - α-*L*-Glucopyranose, 2-(methylamino)-2-deoxy-

C_7H_{16}
3457 - Butane, 2,2,3-trimethyl-
6355 - Heptane
6834 - Hexane, 2-methyl-
6835 - Hexane, 3-methyl-, (*S*)-
6941 - 1-Hexanol, 5-methyl-
8760 - Pentane, 2,2-dimethyl-
8761 - Pentane, 2,3-dimethyl-
8762 - Pentane, 2,4-dimethyl-
8763 - Pentane, 3,3-dimethyl-
8809 - Pentane, 3-ethyl-

$C_7H_{16}BrNO_2$
5438 - Ethanaminium, 2-(acetyloxy)-*N,N,N*-trimethyl-, bromide

$C_7H_{16}NCl$
7430 - Mepiquat chloride

$C_7H_{16}ClNO_2$
5439 - Ethanaminium, 2-(acetyloxy)-*N,N,N*-trimethyl-, chloride

$C_7H_{16}N_2$
9622 - Piperazine, 1,2,4-trimethyl-
9625 - 1-Piperidinamine, 2,6-dimethyl-
9626 - 3-Piperidinamine, 1-ethyl-
11167 - 2-Pyrrolidinemethanamine, 1-ethyl-

$C_7H_{16}N_2O$
7701 - 4-Morpholinepropanamine
12012 - Urea, *N,N′*-dipropyl-

$C_7H_{16}O$
3374 - Butane, 1-ethoxy-3-methyl-
3375 - Butane, 2-ethoxy-2-methyl-
3399 - Butane, 1-(1-methylethoxy)-
3421 - Butane, 1-propoxy-
3644 - 2-Butanol, 2,3,3-trimethyl-
6472 - 1-Heptanol
6473 - 2-Heptanol, (±)-
6474 - 3-Heptanol, (*S*)-
6475 - 4-Heptanol
6833 - Hexane, 1-methoxy-
6937 - 1-Hexanol, 2-methyl-, (±)-
6938 - 1-Hexanol, 2-methyl-, (*R*)-
6939 - 1-Hexanol, 3-methyl-, (±)-
6940 - 1-Hexanol, 4-methyl-, (±)-
6942 - 2-Hexanol, 2-methyl-
6943 - 2-Hexanol, 3-methyl-
6944 - 2-Hexanol, 5-methyl-
6945 - 3-Hexanol, 2-methyl-, (±)-
6946 - 3-Hexanol, 3-methyl-
6947 - 3-Hexanol, 5-methyl-, (±)-
8808 - Pentane, 1-ethoxy-
8927 - 1-Pentanol, 2,4-dimethyl-, (±)-
8928 - 2-Pentanol, 2,4-dimethyl-
8929 - 3-Pentanol, 2,2-dimethyl-
8930 - 3-Pentanol, 2,3-dimethyl-
8931 - 3-Pentanol, 2,4-dimethyl-
8932 - 3-Pentanol, 3-ethyl-
10081 - Propane, 2-methyl-2-(1-methylethoxy)-
10086 - Propane, 2-methyl-1-propoxy-
10087 - Propane, 2-methyl-2-propoxy-

$C_7H_{16}O_2$
5757 - Ethanol, 2-(pentyloxy)-
6397 - 1,4-Heptanediol
6398 - 1,5-Heptanediol
6399 - 1,7-Heptanediol
6400 - 2,4-Heptanediol
6803 - 2,4-Hexanediol, 2-methyl-
8792 - 1,3-Pentanediol, 2,2-dimethyl-
9903 - Propane, 1,1-diethoxy-
9904 - Propane, 2,2-diethoxy-
10004 - 1,3-Propanediol, 2,2-diethyl-
10020 - 1,3-Propanediol, 2-methyl-2-propyl-
10074 - Propane, 1,1′-[methylenebis(oxy)]bis-
10346 - 2-Propanol, 1-butoxy-

$C_7H_{16}O_2Si$
11422 - Silane, ethenyldiethoxymethyl-

$C_7H_{16}O_3$
3455 - Butane, 1,1,3-trimethoxy-
3456 - Butane, 1,3,3-trimethoxy-
5604 - Ethane, 1,1′,1″-[methylidynetris(oxy)]tris-
6453 - 1,4,7-Heptanetriol

$C_7H_{16}O_4$
10150 - Propane, 1,1,3,3-tetramethoxy-

$C_7H_{16}O_4S_2$
9835 - Propane, 2,2-bis(ethylsulfonyl)-

$C_7H_{16}S$
6436 - 1-Heptanethiol

$C_7H_{16}S_3$
5605 - Ethane, 1,1′,1″-[methylidynetris(thio)]tris-

$C_7H_{17}N$
6349 - 1-Heptanamine
6350 - 2-Heptanamine
6351 - 4-Heptanamine
6728 - 1-Hexanamine, 3-methyl-
6729 - 1-Hexanamine, 4-methyl-
6730 - 2-Hexanamine, 4-methyl-
8703 - 1-Pentanamine, *N,N*-dimethyl-
8704 - 2-Pentanamine, 2,4-dimethyl-

$C_7H_{17}NO$
5739 - Ethanol, 2-[(3-methylbutyl)amino]-
10366 - 1-Propanol, 2-(diethylamino)-
10367 - 1-Propanol, 3-(diethylamino)-

10368 - 2-Propanol, 1-(diethylamino)-, (±)-
$C_7H_{17}NO_2$
7443 - Methanamine, 1,1-diethoxy-*N,N*-dimethyl-
$C_7H_{17}NO_5$
6208 - *D*-Glucitol, 1-deoxy-1-(methylamino)-
$C_7H_{17}O_2PS$
9549 - Phosphonothioic acid, methyl-, *O,O*-bis(1-methylethyl) ester
$C_7H_{17}O_2PS_3$
9513 - Phorate
$C_7H_{18}N_2$
6371 - 1,7-Heptanediamine
9876 - 1,3-Propanediamine, *N,N*-diethyl-
9879 - 1,3-Propanediamine, *N,N,N',N'*-tetramethyl-
$C_7H_{18}N_2O$
10339 - 2-Propanol, 1-amino-3-(diethylamino)-
10343 - 2-Propanol, 1,3-bis(dimethylamino)-
$C_7H_{18}N_2O_2$
5679 - Ethanol, 2,2′-[(3-aminopropyl)imino]bis-
$C_7H_{18}O_3Si$
11479 - Silane, triethoxymethyl-
$C_7H_{18}Si$
11371 - Silane, butyltrimethyl-
11491 - Silane, trimethyl(2-methylpropyl)-
$C_7H_{19}N_3$
9875 - 1,3-Propanediamine, *N*-(3-aminopropyl)-*N*-methyl-
$C_7H_{22}O_4Si_4$
4992 - Cyclotetrasiloxane, heptamethyl-
$C_8Cl_2N_2O_2$
4251 - 1,4-Cyclohexadiene-1,2-dicarbonitrile, 4,5-dichloro-3,6-dioxo-
$C_8Cl_4N_2$
1317 - 1,3-Benzenedicarbonitrile, 2,4,5,6-tetrachloro-
$C_8Cl_4O_3$
7346 - 1,3-Isobenzofurandione, 4,5,6,7-tetrachloro-
$C_8F_{14}O_3$
3522 - Butanoic acid, heptafluoro-, anhydride
C_8F_{16}
4342 - Cyclohexane, 1,1,2,2,3,3,4,4,5,6-decafluoro-5,6-bis(trifluoromethyl)-
4343 - Cyclohexane, 1,1,2,2,3,4,4,5,5,6-decafluoro-3,6-bis(trifluoromethyl)-
$C_8F_{16}O$
6109 - Furan, 2,2,3,3,4,4,5-heptafluorotetrahydro-5-(nonafluorobutyl)-
C_8F_{18}
8415 - Octane, octadecafluoro-
$C_8HCl_4F_{11}O_2$
8442 - Octanoic acid, 3,5,7,8-tetrachloro-2,2,3,4,4,5,6,6,7,8,8-undecafluoro-
$C_8H_2Cl_2O_3$
7343 - 1,3-Isobenzofurandione, 4,5-dichloro-
7344 - 1,3-Isobenzofurandione, 5,6-dichloro-
$C_8H_3Cl_2F_3N_2$
2377 - 1H-Benzimidazole, 4,5-dichloro-2-(trifluoromethyl)-
$C_8H_4Cl_2O_2$
1318 - 1,2-Benzenedicarbonyl dichloride
1319 - 1,3-Benzenedicarbonyl dichloride
1320 - 1,4-Benzenedicarbonyl dichloride
$C_8H_4Cl_2O_4$
1340 - 1,2-Benzenedicarboxylic acid, 3,5-dichloro-
1349 - 1,4-Benzenedicarboxylic acid, 2,5-dichloro-
$C_8H_4Cl_6$
996 - Benzene, 1,4-bis(trichloromethyl)-
$C_8H_4F_2O_2$
1321 - 1,2-Benzenedicarbonyl difluoride
$C_8H_4F_3N$
2769 - Benzonitrile, 2-(trifluoromethyl)-
2770 - Benzonitrile, 3-(trifluoromethyl)-
$C_8H_4F_6$
997 - Benzene, 1,3-bis(trifluoromethyl)-
$C_8H_4N_2$
1315 - 1,3-Benzenedicarbonitrile
1316 - 1,4-Benzenedicarbonitrile
$C_8H_4N_2S_2$
1432 - Benzene, 1,4-diisothiocyanato-
$C_8H_4O_2S$
2401 - Benzo[b]thiophene-2,3-dione
$C_8H_4O_3$
7342 - 1,3-Isobenzofurandione
C_8H_5Br
1040 - Benzene, 1-bromo-2-ethynyl-
1041 - Benzene, 1-bromo-3-ethynyl-
1042 - Benzene, 1-bromo-4-ethynyl-
$C_8H_5ClN_2$
11334 - Quinoxaline, 6-chloro-
$C_8H_5Cl_3O$
5871 - Ethanone, 2,2,2-trichloro-1-phenyl-
$C_8H_5Cl_3O_2$
944 - Benzeneacetic acid, 2,3,6-trichloro-
$C_8H_5Cl_3O_3$
256 - Acetic acid, (2,4,5-trichlorophenoxy)-
$C_8H_5F_3O$
5872 - Ethanone, 2,2,2-trifluoro-1-phenyl-
$C_8H_5F_6N$
637 - Benzenamine, 3,5-bis(trifluoromethyl)-
C_8H_5NO
968 - Benzeneacetonitrile, α-oxo-
2757 - Benzonitrile, 3-formyl-
2758 - Benzonitrile, 4-formyl-
$C_8H_5NO_2$
2550 - Benzoic acid, 3-cyano-
7300 - 1H-Indole-2,3-dione
7362 - 1H-Isoindole-1,3(2H)-dione
C_8H_6
1715 - Benzene, ethynyl-
C_8H_6BrClO
5778 - Ethanone, 2-bromo-1-(4-chlorophenyl)-
C_8H_6BrN
947 - Benzeneacetonitrile, α-bromo-
948 - Benzeneacetonitrile, 2-bromo-
$C_8H_6Br_2O$
5777 - Ethanone, 2-bromo-1-(4-bromophenyl)-
5797 - Ethanone, 1-(3,5-dibromophenyl)-
5798 - Ethanone, 2,2-dibromo-1-phenyl-
$C_8H_6Br_2O_3$
2559 - Benzoic acid, 3,5-dibromo-2-hydroxy-, methyl ester
$C_8H_6Br_4$
2255 - Benzene, 1,2,3,4-tetrabromo-5,6-dimethyl-
C_8H_6ClN
949 - Benzeneacetonitrile, 2-chloro-
950 - Benzeneacetonitrile, 3-chloro-
2750 - Benzonitrile, 2-(chloromethyl)-
$C_8H_6ClNO_4$
2548 - Benzoic acid, 4-chloro-3-nitro-, methyl ester
2549 - Benzoic acid, 5-chloro-2-nitro-, methyl ester
$C_8H_6Cl_2$
1367 - Benzene, 1,2-dichloro-3-ethenyl-
1368 - Benzene, 1,2-dichloro-4-ethenyl-
1369 - Benzene, 1,3-dichloro-2-ethenyl-
1370 - Benzene, 1,3-dichloro-5-ethenyl-
1371 - Benzene, 1,4-dichloro-2-ethenyl-
1372 - Benzene, (2,2-dichloroethenyl)-
1373 - Benzene, 2,4-dichloro-1-ethenyl-
$C_8H_6Cl_2O$
973 - Benzeneacetyl chloride, α-chloro-
5784 - Ethanone, 2-chloro-1-(4-chlorophenyl)-
5799 - Ethanone, 1-(2,4-dichlorophenyl)-
5800 - Ethanone, 1-(2,5-dichlorophenyl)-
5801 - Ethanone, 1-(3,4-dichlorophenyl)-
5802 - Ethanone, 2,2-dichloro-1-phenyl-
$C_8H_6Cl_2O_2$
9288 - Phenol, 2,4-dichloro-, acetate
$C_8H_6Cl_2O_3$
169 - Acetic acid, (2,4-dichlorophenoxy)-
2567 - Benzoic acid, 3,6-dichloro-2-methoxy-
$C_8H_6Cl_3NO$
98 - Acetamide, 2,2,2-trichloro-*N*-phenyl-
$C_8H_6Cl_4$
989 - Benzene, 1,4-bis(dichloromethyl)-
2261 - Benzene, 1,2,3,5-tetrachloro-4,6-dimethyl-
C_8H_6FN
952 - Benzeneacetonitrile, 2-fluoro-
953 - Benzeneacetonitrile, 4-fluoro-
$C_8H_6N_2$
4136 - Cinnoline
8065 - 1,5-Naphthyridine
9599 - Phthalazine
11195 - Quinazoline
11333 - Quinoxaline
$C_8H_6N_2O_2$
965 - Benzeneacetonitrile, 2-nitro-
966 - Benzeneacetonitrile, 3-nitro-
967 - Benzeneacetonitrile, 4-nitro-
2766 - Benzonitrile, 2-methyl-5-nitro-
$C_8H_6N_4O_5$
7222 - 2,4-Imidazolidinedione, 1-[[(5-nitro-2-furanyl)methylene]amino]-
$C_8H_6N_4O_8$
3106 -[5,5′Bipyrimidine]2,2′,4,4′,6,6′(1H,1′H,3H,3′H,5H,5′H)-hexone-5,5′-dihydroxy
C_8H_6O
1716 - Benzene, (ethynyloxy)-
2438 - Benzofuran
C_8H_6OS
2403 - Benzo[b]thiophene-4-ol
$C_8H_6O_2$
1322 - 1,3-Benzenedicarboxaldehyde
1323 - 1,4-Benzenedicarboxaldehyde
2461 - 2(3H)-Benzofuranone
2462 - 3(2H)-Benzofuranone
7349 - 1(3H)-Isobenzofuranone
$C_8H_6O_3$
934 - Benzeneacetic acid, α-oxo-
2417 - 1,3-Benzodioxole-5-carboxaldehyde
2607 - Benzoic acid, 2-formyl-
$C_8H_6O_4$
1327 - 1,2-Benzenedicarboxylic acid
1328 - 1,3-Benzenedicarboxylic acid
1329 - 1,4-Benzenedicarboxylic acid
2418 - 1,3-Benzodioxole-5-carboxylic acid
2609 - Benzoic acid, 3-formyl-4-hydroxy-
9598 - *p*-Phthalaldehydic acid, 3-hydroxy-
$C_8H_6O_5$
1352 - 1,3-Benzenedicarboxylic acid, 4-hydroxy-
C_8H_6S
2399 - Benzo[b]thiophene
$C_8H_6S_2$
2434 - 1,4-Benzodithiin
3112 - 2,2′-Bithiophene
C_8H_7Br
1025 - Benzene, (1-bromoethenyl)-
1026 - Benzene, 1-bromo-2-ethenyl-
1027 - Benzene, 1-bromo-3-ethenyl-
1028 - Benzene, 1-bromo-4-ethenyl-
1029 - Benzene, (2-bromoethenyl)-, (*E*)-
1031 - Benzene, (2-bromoethenyl)-, (*Z*)-
4743 - 1,3,5,7-Cyclooctatetraene, 1-bromo-
C_8H_7BrO
5780 - Ethanone, 1-(3-bromophenyl)-
5781 - Ethanone, 1-(4-bromophenyl)-
5782 - Ethanone, 2-bromo-1-phenyl-
$C_8H_7BrO_2$
119 - Acetic acid, bromo-, phenyl ester
888 - Benzeneacetic acid, 4-bromo-
2525 - Benzoic acid, 3-bromo-, methyl ester
2526 - Benzoic acid, 4-bromo-, methyl ester
$C_8H_7BrO_3$
889 - Benzeneacetic acid, 4-bromo-α-hydroxy-, (±)-
C_8H_7Cl
1148 - Benzene, (1-chloroethenyl)-
1149 - Benzene, 1-chloro-2-ethenyl-
1150 - Benzene, 1-chloro-3-ethenyl-
1151 - Benzene, 1-chloro-4-ethenyl-
1153 - Benzene, (2-chloroethenyl)-, (*E*)-
1155 - Benzene, (2-chloroethenyl)-, (*Z*)-
4745 - 1,3,5,7-Cyclooctatetraene, 1-chloro-

C_8H_7ClO
971 - Benzeneacetyl chloride
2902 - Benzoyl chloride, 2-methyl-
2903 - Benzoyl chloride, 3-methyl-
2904 - Benzoyl chloride, 4-methyl-
5788 - Ethanone, 1-(3-chlorophenyl)-
5789 - Ethanone, 1-(4-chlorophenyl)-
5790 - Ethanone, 2-chloro-1-phenyl-
$C_8H_7ClO_2$
147 - Acetic acid, chloro-, phenyl ester
2421 - 1,3-Benzodioxole, 5-(chloromethyl)-
2543 - Benzoic acid, 4-chloro-, methyl ester
2901 - Benzoyl chloride, 4-methoxy-
4027 - Carbonochloridic acid, phenylmethyl ester
$C_8H_7ClO_3$
146 - Acetic acid, (4-chlorophenoxy)-
2542 - Benzoic acid, 5-chloro-2-hydroxy-, methyl ester
C_8H_7F
1633 - Benzene, 1-ethenyl-2-fluoro-
1634 - Benzene, 1-ethenyl-3-fluoro-
1635 - Benzene, 1-ethenyl-4-fluoro-
C_8H_7FO
5815 - Ethanone, 1-(4-fluorophenyl)-
5816 - Ethanone, 2-fluoro-1-phenyl-
$C_8H_7FO_2$
901 - Benzeneacetic acid, 4-fluoro-
$C_8H_7F_3$
2045 - Benzene, 1-methyl-4-(trifluoromethyl)-
C_8H_7IO
5836 - Ethanone, 1-(3-iodophenyl)-
5837 - Ethanone, 1-(4-iodophenyl)-
5838 - Ethanone, 2-iodo-1-phenyl-
$C_8H_7IO_2$
2660 - Benzoic acid, 2-iodo-, methyl ester
2661 - Benzoic acid, 3-iodo-, methyl ester
2662 - Benzoic acid, 4-iodo-, methyl ester
C_8H_7N
945 - Benzeneacetonitrile
1791 - Benzene, (isocyanomethyl)-
2763 - Benzonitrile, 2-methyl-
2764 - Benzonitrile, 3-methyl-
2765 - Benzonitrile, 4-methyl-
7294 - 1H-Indole
7324 - Indolizine
C_8H_7NO
287 - Acetonitrile, phenoxy-
956 - Benzeneacetonitrile, α-hydroxy-, (±)-
1787 - Benzene, 1-isocyanato-2-methyl-
2760 - Benzonitrile, 2-methoxy-
2761 - Benzonitrile, 3-methoxy-
2762 - Benzonitrile, 4-methoxy-
2880 - Benzoxazole, 2-methyl-
7327 - 2H-Indol-2-one, 1,3-dihydro-
7369 - 1H-Isoindol-1-one, 2,3-dihydro-
C_8H_7NOS
1793 - Benzene, 1-isothiocyanato-2-methoxy-
$C_8H_7NO_2$
876 - Benzeneacetaldehyde, α-oxo-, aldoxime
1644 - Benzene, 1-ethenyl-3-nitro-
1645 - Benzene, 1-ethenyl-4-nitro-
2050 - Benzene, (2-nitroethenyl)-, (*E*)-
8414 - Octane, 3-nitro-
$C_8H_7NO_3$
875 - Benzeneacetaldehyde, 2-nitro-
2489 - Benzoic acid, 4-(aminocarbonyl)-
5860 - Ethanone, 1-(3-nitrophenyl)-
5861 - Ethanone, 1-(4-nitrophenyl)-
5862 - Ethanone, 2-nitro-1-phenyl-
$C_8H_7NO_4$
219 - Acetic acid, 2-nitrophenyl ester
933 - Benzeneacetic acid, 4-nitro-
1330 - 1,3-Benzenedicarboxylic acid, 2-amino-
1331 - 1,3-Benzenedicarboxylic acid, 5-amino-
2710 - Benzoic acid, 2-nitro-, methyl ester
2711 - Benzoic acid, 3-nitro-, methyl ester
C_8H_7NS
1794 - Benzene, (isothiocyanatomethyl)-
2863 - Benzothiazole, 2-methyl-
11710 - Thiocyanic acid, phenylmethyl ester
$C_8H_7NS_2$
2867 - 2(3H)-Benzothiazolethione, 3-methyl-
$C_8H_7N_3$
11332 - 6-Quinoxalinamine
$C_8H_7N_3O$
5863 - Ethanone, 1-(2-nitrophenyl)-
$C_8H_7N_3O_2$
9600 - 1,4-Phthalazinedione, 5-amino-2,3-dihydro-
$C_8H_7N_3O_5$
8589 - 2-Oxazolidinone, 3-[[(5-nitro-2-furanyl)methylene]amino]-
$C_8H_7N_3O_6$
1489 - Benzene, 1,4-dimethyl-2,3,5-trinitro-
1490 - Benzene, 2,4-dimethyl-1,3,5-trinitro-
C_8H_8
1624 - Benzene, ethenyl-
4742 - 1,3,5,7-Cyclooctatetraene
4769 - 1,3-Cyclopentadiene, 5-(2-propenylidene)-, (*E*)-
C_8H_8BrNO
37 - Acetamide, *N*-(4-bromophenyl)-
$C_8H_8Br_2$
983 - Benzene, 1,2-bis(bromomethyl)-
984 - Benzene, 1,3-bis(bromomethyl)-
985 - Benzene, 1,4-bis(bromomethyl)-
1297 - Benzene, 1,2-dibromo-3,4-dimethyl-
1298 - Benzene, 1,5-dibromo-2,4-dimethyl-
1299 - Benzene, (1,2-dibromoethyl)-
C_8H_8ClN
7652 - Methylamine, *N*-(*o*-chlorobenzylidene)-
C_8H_8ClNO
48 - Acetamide, *N*-(4-chlorophenyl)-
3974 - Carbamic chloride, methylphenyl-
$C_8H_8ClNO_3S$
2243 - Benzenesulfonyl chloride, 4-(acetylamino)-
$C_8H_8Cl_2$
986 - Benzene, 1,2-bis(chloromethyl)-
987 - Benzene, 1,3-bis(chloromethyl)-
988 - Benzene, 1,4-bis(chloromethyl)-
1360 - Benzene, 1,2-dichloro-3,4-dimethyl-
1361 - Benzene, 1,3-dichloro-2,5-dimethyl-
1362 - Benzene, 1,4-dichloro-2,5-dimethyl-
1363 - Benzene, 1,5-dichloro-2,4-dimethyl-
1364 - Benzene, 2,3-dichloro-1,4-dimethyl-
1376 - Benzene, (1,2-dichloroethyl)-
1377 - Benzene, 1,4-dichloro-2-ethyl-
1390 - Benzene, 1-(dichloromethyl)-4-methyl-
$C_8H_8Cl_2NaO_5S$
11358 - Sesone
$C_8H_8Cl_2O$
1375 - Benzene, 2,4-dichloro-1-ethoxy-
1854 - Benzenemethanol, 2,4-dichloro-α-methyl-
9290 - Phenol, 2,4-dichloro-3,5-dimethyl-
9291 - Phenol, 2,6-dichloro-3,5-dimethyl-
$C_8H_8Cl_2O_2$
4084 - Chloroneb
$C_8H_8Cl_3N_3O_4S_2$
2854 - 2H-1,2,4-Benzothiadiazine-7-sulfonamide, 6-chloro-3-(dichloromethyl)-3,4-dihydro-, 1,1-dioxide
$C_8H_8Cl_3O_3PS$
9584 - Phosphorothioic acid, *O*,*O*-dimethyl *O*-(2,4,5-trichlorophenyl) ester
$C_8H_8F_3N_3O_4S_2$
2858 - 2H-1,2,4-Benzothiadiazine-7-sulfonamide, 3,4-dihydro-6-(trifluoromethyl)-, 1,1-dioxide
$C_8H_8HgO_2$
7431 - Mercury, (acetato-*O*)phenyl-
$C_8H_8N_2$
946 - Benzeneacetonitrile, 4-amino-
2382 - 1H-Benzimidazole, 1-methyl-
2743 - Benzonitrile, 5-amino-2-methyl-
4160 - Cyanamide, methylphenyl-
7239 - 1H-Indazole, 1-methyl-
7240 - 1H-Indazole, 3-methyl-
7241 - 1H-Indazole, 5-methyl-
7242 - 2H-Indazole, 2-methyl-
$C_8H_8N_2O_2$
1324 - 1,2-Benzenedicarboxamide
10883 - 3-Pyridinecarbonitrile, 1,2-dihydro-4-methoxy-1-methyl-2-oxo-
$C_8H_8N_2O_3$
80 - Acetamide, 2-nitro-*N*-phenyl-
81 - Acetamide, *N*-(2-nitrophenyl)-
82 - Acetamide, *N*-(3-nitrophenyl)-
83 - Acetamide, *N*-(4-nitrophenyl)-
$C_8H_8N_2O_5$
1650 - Benzene, 2-ethoxy-1,3-dinitro-
C_8H_8O
587 - Benzaldehyde, 2-methyl-
588 - Benzaldehyde, 3-methyl-
589 - Benzaldehyde, 4-methyl-
868 - Benzeneacetaldehyde
1646 - Benzene, (ethenyloxy)-
2442 - Benzofuran, 2,3-dihydro-
5864 - Ethanone, 1-phenyl-
8624 - Oxirane, phenyl-
9348 - Phenol, 2-ethenyl-
9349 - Phenol, 3-ethenyl-
$C_8H_8O_2$
231 - Acetic acid, phenyl ester
575 - Benzaldehyde, 2-hydroxy-3-methyl-
576 - Benzaldehyde, 2-hydroxy-4-methyl-
577 - Benzaldehyde, 2-hydroxy-5-methyl-
578 - Benzaldehyde, 2-hydroxy-6-methyl-
583 - Benzaldehyde, 2-methoxy-
584 - Benzaldehyde, 3-methoxy-
585 - Benzaldehyde, 4-methoxy-
883 - Benzeneacetic acid
2414 - 1,4-Benzodioxin, 2,3-dihydro-
2676 - Benzoic acid, 2-methyl-
2677 - Benzoic acid, 3-methyl-
2678 - Benzoic acid, 4-methyl-
2687 - Benzoic acid, methyl ester
3878 - 3-Buten-2-one, 4-(2-furanyl)-
4263 - 2,5-Cyclohexadiene-1,4-dione, 2,3-dimethyl-
4264 - 2,5-Cyclohexadiene-1,4-dione, 2,6-dimethyl-
5830 - Ethanone, 1-(2-hydroxyphenyl)-
5831 - Ethanone, 1-(3-hydroxyphenyl)-
5832 - Ethanone, 1-(4-hydroxyphenyl)-
5833 - Ethanone, 2-hydroxy-1-phenyl-
6016 - Formic acid, phenylmethyl ester
$C_8H_8O_3$
228 - Acetic acid, phenoxy-
570 - Benzaldehyde, 2-hydroxy-3-methoxy-
571 - Benzaldehyde, 2-hydroxy-5-methoxy-
572 - Benzaldehyde, 3-hydroxy-4-methoxy-
573 - Benzaldehyde, 3-hydroxy-5-methoxy-
574 - Benzaldehyde, 4-hydroxy-3-methoxy-
903 - Benzeneacetic acid, α-hydroxy-, (*S*)-
904 - Benzeneacetic acid, 2-hydroxy-
905 - Benzeneacetic acid, 3-hydroxy-
906 - Benzeneacetic acid, 4-hydroxy-
1540 - 1,2-Benzenediol, monoacetate
2427 - 1,3-Benzodioxole-5-methanol
2629 - Benzoic acid, 2-hydroxy-3-methyl-
2630 - Benzoic acid, 2-hydroxy-4-methyl-
2631 - Benzoic acid, 2-hydroxy-5-methyl-
2632 - Benzoic acid, 3-hydroxy-5-methyl-
2634 - Benzoic acid, 2-hydroxy-, methyl ester
2636 - Benzoic acid, 3-hydroxy-, methyl ester
2637 - Benzoic acid, 4-hydroxy-, methyl ester
2664 - Benzoic acid, 2-methoxy-
2665 - Benzoic acid, 3-methoxy-
2666 - Benzoic acid, 4-methoxy-
5803 - Ethanone, 1-(2,4-dihydroxyphenyl)-
6070 - 2-Furancarboxylic acid, 2-propenyl ester
7347 - 1,3-Isobenzofurandione, 4,5,6,7-tetrahydro-
10640 - 2-Propenoic acid, 3-(2-furanyl)-, methyl ester

7076 - 4-Hexen-3-ol, 3,5-dimethyl-
8346 - Octanal
8471 - 2-Octanone
8472 - 3-Octanone
8473 - 4-Octanone
8510 - 2-Octen-1-ol
8511 - 2-Octen-4-ol
8512 - 4-Octen-1-ol
8513 - 5-Octen-1-ol
8612 - Oxirane, 2-(2,2-dimethylpropyl)-2-methyl-
8697 - Pentanal, 2-propyl-
8970 - 2-Pentanone, 3-ethyl-4-methyl-
8989 - 3-Pentanone, 2,2,4-trimethyl-

$C_8H_{16}O_2$
185 - Acetic acid, 2-ethylbutyl ester
194 - Acetic acid, hexyl ester
3479 - Butanoic acid, butyl ester
3509 - Butanoic acid, 1,1-dimethylethyl ester
3553 - Butanoic acid, 3-methyl-, 1-methylethyl ester
3562 - Butanoic acid, 3-methyl-, propyl ester
3563 - Butanoic acid, 2-methylpropyl ester
3564 - Butanoic acid, 1-methylpropyl ester, (*S*)-
3639 - 1-Butanol, 3-methyl-, propanoate
3758 - 2-Butene, 1,1-diethoxy-
4363 - 1,3-Cyclohexanedimethanol
6008 - Formic acid, heptyl ester
6466 - Heptanoic acid, methyl ester
6518 - 2-Heptanone, 4-hydroxy-3-methyl-
6519 - 3-Heptanone, 5-hydroxy-5-methyl-
6882 - Hexanoic acid, 2-ethyl-
6883 - Hexanoic acid, 3-ethyl-
6884 - Hexanoic acid, ethyl ester
8422 - Octanoic acid
8474 - 4-Octanone, 5-hydroxy-
8692 - Pentanal, 3-hydroxy-2,2,4-trimethyl-
8892 - Pentanoic acid, 4-methyl-, ethyl ester
8893 - Pentanoic acid, 1-methylethyl ester
8914 - Pentanoic acid, propyl ester
8945 - 1-Pentanol, 2-methyl-, acetate
8946 - 2-Pentanol, 2-methyl-, acetate
8947 - 2-Pentanol, 4-methyl-, acetate
8948 - 3-Pentanol, 2-methyl-, acetate
8949 - 3-Pentanol, 3-methyl-, acetate
10294 - Propanoic acid, 2-methyl-, 1,1-dimethylethyl ester
10304 - Propanoic acid, 2-methyl-, 2-methylpropyl ester
10318 - Propanoic acid, pentyl ester

$C_8H_{16}O_3$
3536 - Butanoic acid, 2-hydroxy-, 2-methylpropyl ester
8432 - Octanoic acid, 2-hydroxy-
10274 - Propanoic acid, 2-hydroxy-, 3-methylbutyl ester

$C_8H_{16}O_4$
174 - Acetic acid, diethoxy-, ethyl ester
5721 - Ethanol, 2-(2-ethoxyethoxy)-, acetate

$C_8H_{16}Si$
11417 - Silane, dimethyldi-2-propenyl-

$C_8H_{17}Br$
8356 - Octane, 1-bromo-
8357 - Octane, 2-bromo-, (±)-

$C_8H_{17}Cl$
6366 - Heptane, 2-chloro-2-methyl-
6367 - Heptane, 2-chloro-6-methyl-
6368 - Heptane, 3-(chloromethyl)-
6369 - Heptane, 3-chloro-3-methyl-
6370 - Heptane, 4-chloro-4-methyl-
6740 - Hexane, 2-chloro-2,5-dimethyl-
6741 - Hexane, 3-chloro-2,3-dimethyl-
8359 - Octane, 1-chloro-
8360 - Octane, 2-chloro-
8733 - Pentane, 2-chloro-2,4,4-trimethyl-

$C_8H_{17}ClO$
8448 - 1-Octanol, 8-chloro-

$C_8H_{17}ClO_2$
3243 - Butane, 3-chloro-1,1-diethoxy-

$C_8H_{17}Cl_3Si$
11471 - Silane, trichlorooctyl-

$C_8H_{17}F$
8405 - Octane, 1-fluoro-

$C_8H_{17}FO$
8460 - 1-Octanol, 8-fluoro-

$C_8H_{17}I$
8406 - Octane, 1-iodo-
8407 - Octane, 2-iodo-, (±)-

$C_8H_{17}N$
503 - 1*H*-Azonine, octahydro-
4286 - Cyclohexanamine, *N,N*-dimethyl-
4287 - Cyclohexanamine, 2-ethyl-
4288 - Cyclohexanamine, *N*-ethyl-
4723 - Cyclohexylamine, *N*,4-dimethyl-
4735 - Cyclooctanamine
9684 - Piperidine, 1-(1-methylethyl)-
9707 - Piperidine, 1-propyl-
9708 - Piperidine, 2-propyl-, (*S*)-
9709 - Piperidine, 4-propyl-
9712 - Piperidine, 2,2,4-trimethyl-
9713 - Piperidine, 2,3,6-trimethyl-
9714 - Piperidine, 2,4,6-trimethyl-
11146 - Pyrrolidine, 1-butyl-
11147 - Pyrrolidine, 2-butyl-

$C_8H_{17}NO$
58 - Acetamide, *N,N*-dipropyl-
3200 - Butanamide, *N,N*-diethyl-
3656 - 2-Butanone, 4-(diethylamino)-
4520 - Cyclohexanol, 3-(dimethylamino)-
7684 - Morpholine, 4-butyl-
8348 - Octanal, oxime
8349 - Octanamide
9677 - 2-Piperidinemethanol, α-ethyl-, (*R**,*S**)-(±)-
9718 - 3-Piperidinol, 6-propyl-, (3*S-trans*)-

$C_8H_{17}NO_2$
3952 - Carbamic acid, (3-methylbutyl)-, ethyl ester
7693 - Morpholine, 4-(2-ethoxyethyl)-
8095 - Nitrous acid, octyl ester
8412 - Octane, 1-nitro-
8413 - Octane, 2-nitro-, (+)-
8423 - Octanoic acid, 2-amino-, (±)-

$C_8H_{17}NO_3$
8079 - Nitric acid, octyl ester

C_8H_{18}
3430 - Butane, 2,2,3,3-tetramethyl-
6422 - Heptane, 2-methyl-
6423 - Heptane, 3-methyl-, (*S*)-
6424 - Heptane, 4-methyl-
6771 - Hexane, 2,2-dimethyl-
6772 - Hexane, 2,3-dimethyl-
6773 - Hexane, 2,4-dimethyl-
6774 - Hexane, 2,5-dimethyl-
6775 - Hexane, 3,3-dimethyl-
6776 - Hexane, 3,4-dimethyl-
6815 - Hexane, 3-ethyl-
8355 - Octane
8813 - Pentane, 3-ethyl-2-methyl-
8814 - Pentane, 3-ethyl-3-methyl-
8847 - Pentane, 2,2,3-trimethyl-
8848 - Pentane, 2,2,4-trimethyl-
8849 - Pentane, 2,3,3-trimethyl-
8850 - Pentane, 2,3,4-trimethyl-

$C_8H_{18}ClNO_2$
9830 - 1-Propanaminium, 2-(acetyloxy)-*N,N,N*-trimethyl-, chloride

$C_8H_{18}Hg$
7433 - Mercury, dibutyl-

$C_8H_{18}N_2$
4362 - 1,3-Cyclohexanedimethanamine
5123 - Diazene, dibutyl

$C_8H_{18}N_2O$
3212 - 1-Butanamine, *N*-butyl-*N*-nitroso-

$C_8H_{18}O$
3401 - Butane, 1-(2-methylpropoxy)-
3415 - Butane, 1,1′-oxybis-
3416 - Butane, 2,2′-oxybis-, (±)-
6421 - Heptane, 1-methoxy-
6492 - 1-Heptanol, 3-methyl-
6493 - 1-Heptanol, 4-methyl-
6494 - 1-Heptanol, 6-methyl-
6495 - 2-Heptanol, 2-methyl-
6496 - 2-Heptanol, 3-methyl-
6497 - 2-Heptanol, 5-methyl-
6498 - 2-Heptanol, 6-methyl-
6499 - 3-Heptanol, 2-methyl-, (±)-
6500 - 3-Heptanol, 3-methyl-
6501 - 3-Heptanol, 4-methyl-
6502 - 3-Heptanol, 5-methyl-
6503 - 4-Heptanol, 2-methyl-
6504 - 4-Heptanol, 3-methyl-, (*R**,*S**)-(±)-
6505 - 4-Heptanol, 4-methyl-
6814 - Hexane, 1-ethoxy-
6921 - 1-Hexanol, 2,5-dimethyl-
6922 - 2-Hexanol, 2,3-dimethyl-
6923 - 2-Hexanol, 2,5-dimethyl-
6924 - 2-Hexanol, 3,4-dimethyl-
6925 - 3-Hexanol, 2,2-dimethyl-
6926 - 3-Hexanol, 2,3-dimethyl-
6927 - 3-Hexanol, 2,5-dimethyl-
6928 - 3-Hexanol, 3,5-dimethyl-, (±)-
6929 - 3-Hexanol, 4,4-dimethyl-
6930 - 1-Hexanol, 2-ethyl-
6931 - 3-Hexanol, 3-ethyl-
8443 - 1-Octanol
8444 - 2-Octanol, (±)-
8445 - 3-Octanol
8446 - 4-Octanol, (±)-
8933 - 2-Pentanol, 3-ethyl-2-methyl-
8934 - 3-Pentanol, 3-ethyl-2-methyl-
8954 - 2-Pentanol, 2,4,4-trimethyl-
8955 - 3-Pentanol, 2,2,4-trimethyl-
8956 - 3-Pentanol, 2,3,4-trimethyl-
10124 - Propane, 2,2′-oxybis[2-methyl-

$C_8H_{18}OS$
3422 - Butane, 1,1′-sulfinylbis-

$C_8H_{18}OSi_2$
5233 - Disiloxane, 1,3-diethenyl-1,1,3,3-tetramethyl-

$C_8H_{18}O_2$
5586 - Ethane, 1,1-diproxy-
5728 - Ethanol, 2-(hexyloxy)-
6401 - 2,4-Heptanediol, 3-methyl-
6801 - 2,5-Hexanediol, 2,5-dimethyl-
6802 - 1,3-Hexanediol, 2-ethyl-
8389 - 1,8-Octanediol
8390 - 2,4-Octanediol
8391 - 4,5-Octanediol, (±)-
8795 - 1,3-Pentanediol, 2,2,4-trimethyl-
8796 - 1,4-Pentanediol, 2,2,4-trimethyl-
9153 - Peroxide, bis(1,1-dimethylethyl)
9993 - 1,3-Propanediol, 2-butyl-2-methyl-

$C_8H_{18}O_2S$
3423 - Butane, 1,1′-sulfonylbis-
10132 - Propane, 1,1′-sulfonylbis[2-methyl-

$C_8H_{18}O_2Si$
11408 - Silane, diethoxymethyl-2-propenyl-

$C_8H_{18}O_3$
5612 - Ethane, 1,1′-oxybis[2-ethoxy-
5664 - Ethane, 1,1,1-triethoxy-
5688 - Ethanol, 2-(2-butoxyethoxy)-
6920 - 1-Hexanol, 3,5-dimethoxy-

$C_8H_{18}O_3S$
11578 - Sulfurous acid, bis(2-methylpropyl) ester
11580 - Sulfurous acid, dibutyl ester

$C_8H_{18}O_3Si$
11424 - Silane, ethenyltriethoxy-

$C_8H_{18}O_4$
11642 - 2,5,8,11-Tetraoxadodecane

$C_8H_{18}O_4S_2$
3230 - Butane, 2,2-bis(ethylsulfonyl)-

$C_8H_{18}O_5$
5756 - Ethanol, 2,2′-[oxybis(2,1-ethanediyloxy)]bis-

$C_8H_{18}S$
3433 - Butane, 1,1′-thiobis-

3434 - Butane, 2,2′-thiobis-
8418 - 1-Octanethiol
8419 - 2-Octanethiol, (±)-
10154 - Propane, 1,1′-thiobis[2-methyl-
10155 - Propane, 2,2′-thiobis[2-methyl-

$C_8H_{18}S_2$
5242 - Disulfide, bis(1,1-dimethylethyl)
5252 - Disulfide, dibutyl

$C_8H_{18}S_4$
11649 - Tetrasulfide, bis(1,1-dimethylethyl)

$C_8H_{18}Si_2$
11434 - Silane, 1,2-ethynediylbis[trimethyl-

$C_8H_{19}N$
3211 - 1-Butanamine, *N*-butyl-
3227 - 2-Butanamine, *N*-(1-methylpropyl)-
6353 - 2-Heptanamine, 6-methyl-, (±)-
6726 - 1-Hexanamine, 2-ethyl-
8350 - 1-Octanamine
8351 - 2-Octanamine, (±)-
9811 - 2-Propanamine, *N*-ethyl-*N*-(1-methylethyl)-
9812 - 1-Propanamine, *N*-ethyl-*N*-propyl-
9823 - 1-Propanamine, 2-methyl-*N*-(2-methylpropyl)-

$C_8H_{19}NO$
5683 - Ethanol, 2-[bis(1-methylethyl)amino]-
5712 - Ethanol, 2-(dipropylamino)-

$C_8H_{19}NO_2$
3214 - 1-Butanamine, 4,4-diethoxy-
3624 - 2-Butanol, 3,3′-iminobis-
5416 - Ethanamine, 2,2-diethoxy-*N*,*N*-dimethyl-
5425 - Ethanamine, 2-ethoxy-*N*-(2-ethoxyethyl)-
5692 - Ethanol, 2,2′-(butylimino)bis-
5707 - Ethanol, 2-[2-(diethylamino)ethoxy]-

$C_8H_{19}O_2PS_2$
5934 - Ethoprop

$C_8H_{19}O_2PS_3$
5260 - Disulfoton

$C_8H_{19}O_3P$
9531 - Phosphonic acid, bis(1,1-dimethylethyl) ester
9532 - Phosphonic acid, dibutyl ester

$C_8H_{20}As_2$
5103 - Diarsine, tetraethyl-

$C_8H_{20}BrN$
5447 - Ethanaminium, *N*,*N*,*N*-triethyl-, bromide

$C_8H_{20}N_2$
3263 - 1,4-Butanediamine, *N*,*N*,*N*′,*N*′-tetramethyl-
6751 - 2,5-Hexanediamine, 2,5-dimethyl-
7140 - Hydrazine, 1,2-bis(2-methylpropyl)-
8365 - 1,8-Octanediamine

$C_8H_{20}N_2O$
10342 - 1-Propanol, 2-[(2-amino-2-methylpropyl)amino]-2-methyl-

$C_8H_{20}OSi$
11428 - Silane, ethoxytriethyl-

$C_8H_{20}O_3P_2$
9526 - Phosphinic acid, diethyl-, anhydride

$C_8H_{20}O_3Si$
11478 - Silane, triethoxyethyl-

$C_8H_{20}O_4Si$
11506 - Silicic acid (H4SiO4), tetraethyl ester

$C_8H_{20}O_5P_2S_2$
11716 - Thiodiphosphoric acid ([(HO)2P(*S*)]2O), tetraethyl ester

$C_8H_{20}O_6P_2$
7185 - Hypophosphoric acid, tetraethyl ester

$C_8H_{20}O_7P_2$
5221 - Diphosphoric acid, tetraethyl ester

$C_8H_{20}Pb$
9731 - Plumbane, tetraethyl

$C_8H_{20}Si$
11450 - Silane, tetraethyl-

$C_8H_{20}Si_2$
11421 - Silane, 1,2-ethenediylbis[trimethyl-, (*E*)-

$C_8H_{23}N_5$
5498 - 1,2-Ethanediamine, *N*-(2-aminoethyl)-*N*′-[2-[(2-aminoethyl)amino]ethyl]-

$C_8H_{24}N_4O_3P_2$
5220 - Diphosphoramide, octamethyl-

$C_8H_{24}O_2Si_3$
11914 - Trisiloxane, octamethyl-

$C_8H_{24}O_4Si_4$
4993 - Cyclotetrasiloxane, octamethyl-

$C_8H_{26}O_3Si_4$
11647 - Tetrasiloxane, 1,1,1,3,5,7,7,7-octamethyl-

$C_8I_4O_3$
7348 - 1,3-Isobenzofurandione, 4,5,6,7-tetraiodo-

C_9F_{20}
8139 - Nonane, icosofluoro

$C_9F_{21}N$
9813 - 1-Propanamine, 1,1,2,2,3,3,3-heptafluoro-*N*,*N*-bis(heptafluoropropyl)-

$C_9H_3Cl_3O_3$
2317 - 1,3,5-Benzenetricarbonyl trichloride

$C_9H_3F_{15}O_2$
8438 - Octanoic acid, pentadecafluoro-, methyl ester

$C_9H_4Cl_3NO_2S$
5984 - Folpet

$C_9H_4O_5$
7341 - 5-Isobenzofurancarboxylic acid, 1,3-dihydro-1,3-dioxo-

$C_9H_5BrN_2O_2$
11221 - Quinoline, 6-bromo-5-nitro-

$C_9H_5Br_2NO$
11313 - 8-Quinolinol, 5,7-dibromo-

C_9H_5ClINO
11312 - 8-Quinolinol, 5-chloro-7-iodo-

$C_9H_5ClO_6$
2320 - 1,3,5-Benzenetricarboxylic acid, 2-chloro-

$C_9H_5Cl_2N$
11248 - Quinoline, 2,4-dichloro-
11249 - Quinoline, 2,7-dichloro-
11250 - Quinoline, 4,5-dichloro-
11251 - Quinoline, 4,7-dichloro-
11252 - Quinoline, 5,8-dichloro-

$C_9H_5Cl_2NO$
11314 - 8-Quinolinol, 5,7-dichloro-

$C_9H_5Cl_3N_4$
760 - Anilazine

$C_9H_5I_2NO$
11315 - 8-Quinolinol, 5,7-diiodo-

$C_9H_5NO_4$
10740 - 2-Propynoic acid, 3-(2-nitrophenyl)-

C_9H_6BrN
7374 - Isoquinoline, 4-bromo-
11217 - Quinoline, 3-bromo-
11218 - Quinoline, 5-bromo-
11219 - Quinoline, 6-bromo
11220 - Quinoline, 7-bromo-

C_9H_6BrNO
11310 - 8-Quinolinol, 5-bromo-
11326 - 2(1H)-Quinolinone, 4-bromo-

$C_9H_6Br_2O_2$
10629 - 2-Propenoic acid, 2,3-dibromo-3-phenyl-, (*Z*)-

C_9H_6ClN
7376 - Isoquinoline, 1-chloro-
11237 - Quinoline, 2-chloro-
11238 - Quinoline, 4-chloro-
11239 - Quinoline, 5-chloro-
11240 - Quinoline, 6-chloro-
11241 - Quinoline, 7-chloro-
11242 - Quinoline, 8-chloro-

C_9H_6ClNO
11311 - 8-Quinolinol, 5-chloro-

$C_9H_6Cl_2N_2O_3$
7643 - Methazole

$C_9H_6Cl_6O_3S$
5369 - Endosulfan

C_9H_6IN
11272 - Quinoline, 4-iodo-
11273 - Quinoline, 6-iodo-

$C_9H_6INO_4S$
11295 - 5-Quinolinesulfonic acid, 8-hydroxy-7-iodo-

$C_9H_6N_2$
9916 - Propanedinitrile, phenyl-

$C_9H_6N_2O_2$
1430 - Benzene, 1,3-diisocyanato-2-methyl-
1431 - Benzene, 2,4-diisocyanato-1-methyl-
7386 - Isoquinoline, 5-nitro-
11284 - Quinoline, 5-nitro-
11285 - Quinoline, 6-nitro-
11286 - Quinoline, 7-nitro-

$C_9H_6N_2O_3$
11287 - Quinoline, 4-nitro-, 1-oxide
11323 - 8-Quinolinol, 5-nitro-

C_9H_6O
10724 - 2-Propynal, 3-phenyl-

$C_9H_6OS_2$
7617 - Methanone, di-2-thienyl-

$C_9H_6O_2$
2791 - 1H-2-Benzopyran-1-one
2792 - 2H-1-Benzopyran-2-one
2793 - 4H-1-Benzopyran-4-one
7264 - 1H-Indene-1,3(2H)-dione

$C_9H_6O_2S$
2400 - Benzo[b]thiophene-2-carboxylic acid

$C_9H_6O_3$
2439 - 2-Benzofurancarboxylic acid
2775 - 2H-1-Benzopyran-3-carboxylic acid, 2-oxo-
2825 - 2H-1-Benzopyran-2-one, 6-hydroxy-
2826 - 2H-1-Benzopyran-2-one, 7-hydroxy-

$C_9H_6O_4$
2804 - 2H-1-Benzopyran-2-one, 6,7-dihydroxy-
2805 - 2H-1-Benzopyran-2-one, 7,8-dihydroxy-
7267 - 1H-Indene-1,3(2H)-dione, 2,2-dihydroxy-

$C_9H_6O_6$
2318 - 1,2,3-Benzenetricarboxylic acid
2319 - 1,2,4-Benzenetricarboxylic acid

$C_9H_7BrN_2$
11207 - 4-Quinolinamine, 3-bromo-
11215 - Quinoline, 6-amino-3-bromo-

C_9H_7BrO
10510 - 2-Propenal, 3-bromo-3-phenyl-

$C_9H_7BrO_4$
2481 - Benzoic acid, 2-(acetyloxy)-5-bromo-

C_9H_7ClO
10716 - 2-Propenoyl chloride, 3-phenyl-
10717 - 2-Propenoyl chloride, 3-phenyl-, (*E*)-

$C_9H_7Cl_3O_3$
11512 - Silvex

$C_9H_7FO_2$
10637 - 2-Propenoic acid, 2-fluoro-3-phenyl-

C_9H_7N
7373 - Isoquinoline
10587 - 2-Propenenitrile, 3-phenyl-, (*E*)-
10588 - 2-Propenenitrile, 3-phenyl-, (*Z*)-
11214 - Quinoline

C_9H_7NO
2124 - Benzenepropanenitrile, β-oxo-
11305 - 5-Quinolinol
11306 - 6-Quinolinol
11307 - 7-Quinolinol
11308 - 8-Quinolinol
11325 - 2(1H)-Quinolinone

$C_9H_7NO_2$
7365 - 1H-Isoindole-1,3(2H)-dione, 2-methyl-

$C_9H_7NO_4$
10676 - 2-Propenoic acid, 3-(2-nitrophenyl)-, (*E*)

$C_9H_7NO_4S$
11294 - 5-Quinolinesulfonic acid, 8-hydroxy-

$C_9H_7N_3S$
11858 - Tricyclazole

C_9H_8
2173 - Benzene, 1-propynyl-
7245 - 1H-Indene

$C_9H_8Br_2$
7247 - 1H-Indene, 1,2-dibromo-2,3-dihydro-

$C_9H_8Cl_2$
1397 - Benzene, (3,3-dichloro-2-propenyl)-
7249 - 1H-Indene, 1,2-dichloro-2,3-dihydro-

MOLECULAR FORMULA INDEX OF ORGANIC COMPOUNDS (continued)

5852 - Ethanone, 1-(3-methylphenyl)-
5853 - Ethanone, 1-(4-methylphenyl)-
7284 - 1H-Inden-1-ol, 2,3-dihydro-
7285 - 1H-Inden-5-ol, 2,3-dihydro-
8622 - Oxirane, 2-methyl-2-phenyl-
9456 - Phenol, 2-(1-propenyl)-
9457 - Phenol, 2-(2-propenyl)-
9458 - Phenol, 4-(2-propenyl)-
9459 - Phenol, 2-(1-propenyl)-, (*E*)-
10480 - 1-Propanone, 1-phenyl-
10481 - 2-Propanone, 1-phenyl-
10704 - 2-Propen-1-ol, 3-phenyl-, (*E*)-
10705 - 2-Propen-1-ol, 3-phenyl-, (*Z*)-

$C_9H_{10}O_2$

211 - Acetic acid, 2-methylphenyl ester
212 - Acetic acid, 3-methylphenyl ester
213 - Acetic acid, 4-methylphenyl ester
233 - Acetic acid, phenylmethyl ester
552 - Benzaldehyde, 2-ethoxy-
553 - Benzaldehyde, 3-ethoxy-
554 - Benzaldehyde, 4-ethoxy-
586 - Benzaldehyde, 4-methoxy-3-methyl-
870 - Benzeneacetaldehyde, 2-methoxy-
871 - Benzeneacetaldehyde, 4-methoxy-
924 - Benzeneacetic acid, α-methyl-, (±)-
925 - Benzeneacetic acid, 3-methyl-
926 - Benzeneacetic acid, 4-methyl-
929 - Benzeneacetic acid, methyl ester
1547 - 1,2-Benzenediol, 3-(2-propenyl)-
2126 - Benzenepropanoic acid
2580 - Benzoic acid, 2,4-dimethyl-
2581 - Benzoic acid, 2,5-dimethyl-
2582 - Benzoic acid, 2,6-dimethyl-
2583 - Benzoic acid, 3,5-dimethyl-
2600 - Benzoic acid, 2-ethyl-
2601 - Benzoic acid, 3-ethyl-
2602 - Benzoic acid, ethyl ester
2696 - Benzoic acid, 2-methyl-, methyl ester
2697 - Benzoic acid, 3-methyl-, methyl ester
2698 - Benzoic acid, 4-methyl-, methyl ester
5826 - Ethanone, 1-(2-hydroxy-4-methylphenyl)-
5827 - Ethanone, 1-(2-hydroxy-5-methylphenyl)-
5828 - Ethanone, 1-(4-hydroxy-2-methylphenyl)-
5840 - Ethanone, 1-(3-methoxyphenyl)-
5841 - Ethanone, 1-(4-methoxyphenyl)-
5842 - Ethanone, 2-methoxy-1-phenyl-
8619 - Oxiranemethanol, 3-phenyl-
8623 - Oxirane, (phenoxymethyl)-
10323 - Propanoic acid, phenyl ester
10449 - 1-Propanone, 1-(2-hydroxyphenyl)-
10450 - 1-Propanone, 1-(4-hydroxyphenyl)-
10451 - 1-Propanone, 2-hydroxy-1-phenyl-
10452 - 1-Propanone, 3-hydroxy-1-phenyl-
10479 - 2-Propanone, 1-phenoxy-

$C_9H_{10}O_3$

230 - Acetic acid, phenoxy-, methyl ester
543 - Benzaldehyde, 2,4-dimethoxy-
544 - Benzaldehyde, 2,5-dimethoxy-
545 - Benzaldehyde, 3,4-dimethoxy-
546 - Benzaldehyde, 3,5-dimethoxy-
555 - Benzaldehyde, 3-ethoxy-4-hydroxy-
556 - Benzaldehyde, 5-ethoxy-2-hydroxy-
910 - Benzeneacetic acid, α-hydroxy-, methyl ester, (±)-
914 - Benzeneacetic acid, α-hydroxy-α-methyl-
915 - Benzeneacetic acid, α-(hydroxymethyl)-
921 - Benzeneacetic acid, 2-methoxy-
922 - Benzeneacetic acid, 4-methoxy-
2133 - Benzenepropanoic acid, α-hydroxy-, (±)-
2134 - Benzenepropanoic acid, 4-hydroxy-
2596 - Benzoic acid, 2-ethoxy-
2597 - Benzoic acid, 3-ethoxy-
2619 - Benzoic acid, 2-hydroxy-, ethyl ester
2620 - Benzoic acid, 3-hydroxy-, ethyl ester
2621 - Benzoic acid, 4-hydroxy-, ethyl ester
2638 - Benzoic acid, 2-hydroxy-3-methyl-, methyl ester
2639 - Benzoic acid, 2-hydroxy-4-methyl-, methyl ester
2640 - Benzoic acid, 2-hydroxy-5-methyl-, methyl ester
2673 - Benzoic acid, 2-methoxy-, methyl ester
2674 - Benzoic acid, 3-methoxy-, methyl ester
2675 - Benzoic acid, 4-methoxy-, methyl ester
5578 - 1,2-Ethanediol, monobenzoate
5823 - Ethanone, 1-(2-hydroxy-4-methoxyphenyl)-
5824 - Ethanone, 1-(4-hydroxy-3-methoxyphenyl)-
9382 - Phenol, 2-methoxy-, acetate
10319 - Propanoic acid, 2-phenoxy-, (±)-
10320 - Propanoic acid, 3-phenoxy-
10430 - 1-Propanone, 1-(2,4-dihydroxyphenyl)-
10639 - 2-Propenoic acid, 3-(2-furanyl)-, ethyl ester

$C_9H_{10}O_4$

567 - Benzaldehyde, 2-hydroxy-4,6-dimethoxy-
568 - Benzaldehyde, 3-hydroxy-4,5-dimethoxy-
569 - Benzaldehyde, 4-hydroxy-3,5-dimethoxy-
913 - Benzeneacetic acid, 4-hydroxy-3-methoxy-
2576 - Benzoic acid, 3,4-dimethoxy-
2577 - Benzoic acid, 3,5-dimethoxy-
2622 - Benzoic acid, 2-hydroxy-, 2-hydroxyethyl ester
2627 - Benzoic acid, 4-hydroxy-3-methoxy-, methyl ester
6163 - 2-Furanpropanoic acid, β-oxo-, ethyl ester
10489 - 1-Propanone, 1-(2,4,6-trihydroxyphenyl)-

$C_9H_{10}O_5$

896 - Benzeneacetic acid, α,4-dihydroxy-3-methoxy-
7510 - Methanediol, 2-furanyl-, diacetate

$C_9H_{11}Br$

1057 - Benzene, 1-(bromomethyl)-3,5-dimethyl-
1059 - Benzene, 1-bromo-4-(1-methylethyl)-
1060 - Benzene, (2-bromo-1-methylethyl)-
1086 - Benzene, (1-bromopropyl)-, (±)-
1087 - Benzene, (2-bromopropyl)-
1088 - Benzene, (3-bromopropyl)-
1092 - Benzene, 1-bromo-2,3,5-trimethyl-
1093 - Benzene, 1-bromo-2,4,5-trimethyl-
1094 - Benzene, 2-bromo-1,3,5-trimethyl-

$C_9H_{11}BrN_2O_2$

7656 - Metobromuron

$C_9H_{11}BrO$

1085 - Benzene, (3-bromopropoxy)-

$C_9H_{11}Cl$

1186 - Benzene, 1-(chloromethyl)-2,4-dimethyl-
1189 - Benzene, (1-chloro-1-methylethyl)-
1190 - Benzene, 1-chloro-2-(1-methylethyl)-
1191 - Benzene, 1-(chloromethyl)-4-ethyl-
1192 - Benzene, 1-chloro-4-(1-methylethyl)-
1193 - Benzene, (2-chloro-1-methylethyl)-
1234 - Benzene, (2-chloropropyl)-, (*R*)-
1235 - Benzene, (3-chloropropyl)-
1243 - Benzene, 2-chloro-1,3,5-trimethyl-

$C_9H_{11}ClN_2O$

11999 - Urea, *N'*-(4-chlorophenyl)-*N*,*N*-dimethyl-

$C_9H_{11}ClO$

1233 - Benzene, (3-chloropropoxy)-

$C_9H_{11}ClO_3$

9999 - 1,2-Propanediol, 3-(2-chlorophenoxy)-
10000 - 1,2-Propanediol, 3-(4-chlorophenoxy)-

$C_9H_{11}ClS$

1236 - Benzene, [(3-chloropropyl)thio]-

$C_9H_{11}Cl_2P$

9555 - Phosphonous dichloride, [4-(1-methylethyl)phenyl]-

$C_9H_{11}Cl_3NO_3PS$

4089 - Chlorpyrifos

$C_9H_{11}F$

1740 - Benzene, 2-fluoro-1,3,5-trimethyl-

$C_9H_{11}FN_2O_5$

12034 - Uridine, 2′-deoxy-5-fluoro-

$C_9H_{11}N$

787 - Benzenamine, 2-(1-methylethenyl)-
841 - Benzenamine, *N*-2-propenyl-
5437 - Ethanamine, *N*-(phenylmethylene)-
7243 - 1H-Inden-1-amine, 2,3-dihydro-, (±)-
7244 - 1H-Inden-5-amine, 2,3-dihydro-
7388 - Isoquinoline, 1,2,3,4-tetrahydro-
7389 - Isoquinoline, 5,6,7,8-tetrahydro-
11296 - Quinoline, 1,2,3,4-tetrahydro-
11297 - Quinoline, 5,6,7,8-tetrahydro-

$C_9H_{11}NO$

76 - Acetamide, *N*-(2-methylphenyl)-
77 - Acetamide, *N*-(3-methylphenyl)-
78 - Acetamide, *N*-(4-methylphenyl)-
79 - Acetamide, *N*-methyl-*N*-phenyl-
87 - Acetamide, *N*-(phenylmethyl)-
550 - Benzaldehyde, 4-(dimethylamino)-
7567 - Methanimidic acid, *N*-phenyl-, ethyl ester
9800 - Propanamide, *N*-phenyl-
10413 - 1-Propanone, 1-(4-aminophenyl)-
10414 - 1-Propanone, 2-amino-1-phenyl-
10483 - 1-Propanone, 1-phenyl-, oxime

$C_9H_{11}NO_2$

70 - Acetamide, *N*-(2-methoxyphenyl)-
71 - Acetamide, *N*-(4-methoxyphenyl)-
615 - Benzamide, 2-ethoxy-
1977 - Benzene, 1-(1-methylethyl)-2-nitro-
1978 - Benzene, 1-(1-methylethyl)-4-nitro-
2003 - Benzene, (1-methyl-1-nitroethyl)-
2362 - Benzene, 1,2,4-trimethyl-5-nitro-
2363 - Benzene, 1,3,5-trimethyl-2-nitro-
2424 - 1,3-Benzodioxole-5-ethanamine
2495 - Benzoic acid, 2-amino-, ethyl ester
2496 - Benzoic acid, 3-amino-, ethyl ester
2497 - Benzoic acid, 4-amino-, ethyl ester
2584 - Benzoic acid, 2-(dimethylamino)-
2585 - Benzoic acid, 4-(dimethylamino)-
2681 - Benzoic acid, 2-(methylamino)-, methyl ester
3958 - Carbamic acid, methylphenyl-, methyl ester
3965 - Carbamic acid, phenyl-, ethyl ester
9509 - *DL*-Phenylalanine
9510 - *L*-Phenylalanine
10907 - 3-Pyridinecarboxylic acid, 1-methylethyl ester

$C_9H_{11}NO_3$

5804 - Ethanone, 1-(3,4-dihydroxyphenyl)-2-(methylamino)-
11926 - *L*-Tyrosine

$C_9H_{11}NO_4$

1443 - Benzene, 1-(dimethoxymethyl)-3-nitro-
1444 - Benzene, 1-(dimethoxymethyl)-4-nitro-
11929 - *L*-Tyrosine, 3-hydroxy-

C_9H_{12}

1685 - Benzene, 1-ethyl-2-methyl-
1686 - Benzene, 1-ethyl-3-methyl-
1687 - Benzene, 1-ethyl-4-methyl-
1975 - Benzene, (1-methylethyl)-
2169 - Benzene, propyl-
2357 - Benzene, 1,2,3-trimethyl-
2358 - Benzene, 1,2,4-trimethyl-
2359 - Benzene, 1,3,5-trimethyl-
8114 - 1,4-Nonadiyne
8115 - 1,8-Nonadiyne
8116 - 2,7-Nonadiyne

$C_9H_{12}NO_5PS$

5951 - Fenitrothion

$C_9H_{12}N_2$

10482 - 2-Propanone, phenylhydrazone
11017 - Pyridine, 3-(2-pyrrolidinyl)-, (*S*)-

$C_9H_{12}N_2O_2$

12013 - Urea, (4-ethoxyphenyl)-

$C_9H_{12}N_2O_3$

827 - Benzenamine, 5-nitro-2-propoxy-

$C_9H_{12}N_2O_6$

12033 - Uridine

$C_9H_{12}N_2S$

10886 - 4-Pyridinecarbothioamide, 2-propyl-

$C_9H_{12}N_4O_3$

10758 - 1H-Purine-2,6-dione, 3,7-dihydro-7-(2-hydroxyethyl)-1,3-dimethyl-
10759 - 1H-Purine-2,6-dione, 3,7-dihydro-8-methoxy-1,3,7-trimethyl-

$C_9H_{14}O_3$
6084 - Furan, 2-(diethoxymethyl)-
9074 - 4-Pentenoic acid, 2-acetyl-, ethyl ester
10674 - 2-Propenoic acid, 2-methyl-, (tetrahydro-2-furanyl)methyl ester
10703 - 2-Propen-1-ol, 2-methyl-, carbonate (2:1)
$C_9H_{14}O_3Si$
11488 - Silane, trimethoxyphenyl-
$C_9H_{14}O_4$
3329 - Butanedioic acid, methylene-, diethyl ester
3787 - 2-Butenedioic acid, 2-methyl-, diethyl ester, (*E*)-
3788 - 2-Butenedioic acid, 2-methyl-, diethyl ester, (*Z*)-
4792 - 1,2-Cyclopentanedicarboxylic acid, dimethyl ester, *trans*-(±)-
4967 - 1,1-Cyclopropanedicarboxylic acid, diethyl ester
4968 - 1,2-Cyclopropanedicarboxylic acid, diethyl ester, *cis*-
9041 - 2-Pentenedioic acid, diethyl ester
9042 - 2-Pentenedioic acid, diethyl ester, (*E*)-
9950 - Propanedioic acid, ethylidene-, diethyl ester
$C_9H_{14}O_5$
3324 - Butanedioic acid, formyl-, diethyl ester
8783 - Pentanedioic acid, 3-oxo-, diethyl ester
9923 - Propanedioic acid, acetyl-, diethyl ester
$C_9H_{14}O_6$
10195 - 1,2,3-Propanetriol, triacetate
$C_9H_{14}O_7$
10170 - 1,2,3-Propanetricarboxylic acid, 2-hydroxy-, trimethyl ester
$C_9H_{14}S$
11771 - Thiophene, 2-(3-methylbutyl)-
$C_9H_{14}Si$
11493 - Silane, trimethylphenyl-
$C_9H_{15}AlO_9$
340 - Aluminum, tris(2-hydroxypropanoato-*O*1,*O*2)-
$C_9H_{15}Br$
4322 - Cyclohexane, (2-bromo-2-propenyl)-
$C_9H_{15}Br_6O_4P$
10358 - 1-Propanol, 2,3-dibromo-, phosphate (3:1)
$C_9H_{15}Cl$
8200 - 1-Nonyne, 1-chloro-
8534 - 3-Octyne, 2-chloro-2-methyl-
$C_9H_{15}N$
10523 - 2-Propen-1-amine, *N*,*N*-di-2-propenyl-
11117 - 1H-Pyrrole, 3,4-diethyl-2-methyl-
$C_9H_{15}NO$
479 - 9-Azabicyclo[3.3.1]nonan-3-one, 9-methyl-
$C_9H_{15}NO_2$
3495 - Butanoic acid, 2-cyano-2-ethyl-, ethyl ester
6881 - Hexanoic acid, 2-cyano-, ethyl ester
9654 - 2,4-Piperidinedione, 3,3-diethyl-
10521 - 2-Propenamide, *N*-(1,1-dimethyl-3-oxobutyl)-
$C_9H_{15}NO_3$
483 - 8-Azabicyclo[3.2.1]octane-2-carboxylic acid, 3-hydroxy-8-methyl-, [1*R*-(*exo*,*exo*)]-
8585 - 2,4-Oxazolidinedione, 5,5-dipropyl-
$C_9H_{15}NO_5$
9922 - Propanedioic acid, (acetylamino)-, diethyl ester
$C_9H_{15}N_3O_2S$
7204 - 1H-Imidazole-4-ethanaminium, α-carboxy-2,3-dihydro-*N*,*N*,*N*-trimethyl-2-thioxo-, hydroxide, in
$C_9H_{15}N_3O_7$
472 - *L*-Aspartic acid, *N*-[2-[(2-amino-2-oxoethyl)amino]-2-carboxyethyl]-
C_9H_{16}
2990 - Bicyclo[3.3.1]nonane
2992 - Bicyclo[2.2.2]octane, 2-methyl-
4393 - Cyclohexane, 1-ethylidene-2-methyl-
4470 - Cyclohexane, 2-propenyl-
4663 - Cyclohexene, 1-ethyl-4-methyl-
4687 - Cyclohexene, 1,2,3-trimethyl-
4688 - Cyclohexene, 1,3,5-trimethyl-
4689 - Cyclohexene, 1,4,4-trimethyl-
4690 - Cyclohexene, 1,4,5-trimethyl-
4691 - Cyclohexene, 1,5,5-trimethyl-
4692 - Cyclohexene, 1,6,6-trimethyl-
4693 - Cyclohexene, 1,5,6-trimethyl-
4727 - Cyclononene, (*E*)-
4728 - Cyclononene, (*Z*)-
4906 - Cyclopentene, 1-butyl-
4927 - Cyclopentene, 4-isopropyl-1-methyl-
4931 - Cyclopentene, 3-methyl-1-(1-methylethyl)-
6330 - 1,3-Heptadiene, 2,6-dimethyl-
6331 - 1,5-Heptadiene, 2,6-dimethyl-
6332 - 2,4-Heptadiene, 2,4-dimethyl-
6333 - 2,4-Heptadiene, 3,5-dimethyl-
6617 - 3-Heptyne, 2,6-dimethyl-
6618 - 3-Heptyne, 5,5-dimethyl-
6694 - 1,5-Hexadiene, 3-ethyl-2-methyl-
7281 - 1H-Indene, octahydro-
8110 - 1,8-Nonadiene
8111 - 2,7-Nonadiene
8196 - 1-Nonyne
8197 - 2-Nonyne
8198 - 3-Nonyne
8199 - 4-Nonyne
8328 - 2,4-Octadiene, 7-methyl-
8329 - 3,5-Octadiene, 4-methyl-
8537 - 3-Octyne, 7-methyl-
$C_9H_{16}ClN_5$
10508 - Propazine
11593 - Terbuthylazine
$C_9H_{16}N_4OS$
11590 - Tebuthiuron
$C_9H_{16}O$
2938 - Bicyclo[2.2.1]heptan-2-ol, 1,7-dimethyl-, (*exo*,*syn*)-
4555 - Cyclohexanol, 1-(2-propenyl)-
4556 - Cyclohexanol, 2-(2-propenyl)-, *trans*-
4595 - Cyclohexanone, 2-ethyl-2-methyl-
4599 - Cyclohexanone, 2-(1-methylethyl)-
4600 - Cyclohexanone, 4-(1-methylethyl)-
4612 - Cyclohexanone, 2-propyl-
4613 - Cyclohexanone, 4-propyl-
4614 - Cyclohexanone, 2,2,3-trimethyl-
4615 - Cyclohexanone, 2,2,4-trimethyl-
4616 - Cyclohexanone, 2,2,5-trimethyl-
4617 - Cyclohexanone, 2,2,6-trimethyl-
4618 - Cyclohexanone, 2,3,6-trimethyl-
4619 - Cyclohexanone, 2,4,4-trimethyl-
4620 - Cyclohexanone, 2,4,5-trimethyl-
4621 - Cyclohexanone, 2,4,6-trimethyl-
4622 - Cyclohexanone, 3,3,5-trimethyl-
4623 - Cyclohexanone, 3,4,4-trimethyl-
4700 - 2-Cyclohexen-1-ol, 1,3,5-trimethyl-
4701 - 2-Cyclohexen-1-ol, 2,4,4-trimethyl-
4702 - 2-Cyclohexen-1-ol, 3,5,5-trimethyl-
4725 - Cyclononanone
4738 - Cyclooctanecarboxaldehyde
4884 - Cyclopentanone, 4-isopropyl-2-methyl-
4889 - Cyclopentanone, 2-methyl-2-(1-methylethyl)-
4890 - Cyclopentanone, 2-methyl-5-(1-methylethyl)-
4891 - Cyclopentanone, 4-methyl-2-(1-methylethyl)-
5843 - Ethanone, 1-(1-methylcyclohexyl)-
5844 - Ethanone, 1-(3-methylcyclohexyl)-
5845 - Ethanone, 1-(4-methylcyclohexyl)-
8112 - 2,6-Nonadien-1-ol
8179 - 2-Nonenal
8180 - 2-Nonenal, (*E*)-
8208 - 1-Nonyn-3-ol
8209 - 2-Nonyn-1-ol
8522 - 1-Octen-3-one, 5-methyl-
8523 - 5-Octen-4-one, 2-methyl-
8524 - 5-Octen-4-one, 7-methyl-
8525 - 7-Octen-4-one, 2-methyl-
8536 - 2-Octyne, 1-methoxy-
8544 - 1-Octyn-3-ol, 3-methyl-
10423 - 1-Propanone, 1-cyclohexyl-
$C_9H_{16}O_2$
4310 - Cyclohexaneacetic acid, methyl ester
4311 - Cyclohexaneacetic acid, 2-methyl-
4312 - Cyclohexaneacetic acid, 3-methyl-
4334 - Cyclohexanecarboxylic acid, ethyl ester
4468 - Cyclohexanepropanoic acid
6149 - 2(3H)-Furanone, dihydro-5-pentyl-
7063 - 2-Hexenoic acid, 2-methyl-, ethyl ester
7064 - 3-Hexenoic acid, 2-methyl-, ethyl ester
7065 - 3-Hexenoic acid, 3-methyl-, ethyl ester
8137 - 2,4-Nonanedione
8187 - 3-Nonenoic acid
8188 - 8-Nonenoic acid
10235 - Propanoic acid, cyclohexyl ester
10643 - 2-Propenoic acid, hexyl ester
$C_9H_{16}O_3$
3461 - Butanoic acid, 2-acetyl-3-methyl-, ethyl ester
6165 - 2-Furanpropanoic acid, tetrahydro-, ethyl ester
6460 - Heptanoic acid, 2,2-dimethyl-6-oxo-
6461 - Heptanoic acid, 4,4-dimethyl-6-oxo-
6902 - Hexanoic acid, 4-methyl-5-oxo-, ethyl ester
8856 - Pentanoic acid, 2-acetyl-, ethyl ester
8904 - Pentanoic acid, 4-oxo-, butyl ester
8910 - Pentanoic acid, 4-oxo-, 2-methylpropyl ester
$C_9H_{16}O_4$
3326 - Butanedioic acid, methyl-, diethyl ester
3337 - Butanedioic acid, monopentyl ester
6395 - Heptanedioic acid, dimethyl ester
6396 - Heptanedioic acid, monoethyl ester
8130 - Nonanedioic acid
8772 - Pentanedioic acid, diethyl ester
8774 - Pentanedioic acid, 3,3-dimethyl-, dimethyl ester
8791 - 1,5-Pentanediol, diacetate
9942 - Propanedioic acid, dimethyl-, diethyl ester
9945 - Propanedioic acid, dipropyl ester
9949 - Propanedioic acid, ethyl-, diethyl ester
$C_9H_{16}O_5$
8778 - Pentanedioic acid, 3-hydroxy-, diethyl ester
$C_9H_{17}ClN_3O_3PS$
7340 - Isazophos
$C_9H_{17}ClO$
8177 - Nonanoyl chloride
$C_9H_{17}FO_2$
8151 - Nonanoic acid, 9-fluoro-
$C_9H_{17}IN_4$
11825 - 3,5,7-Triaza-1-azoniatricyclo[3.3.1.13,7]decane, 1-(2-propenyl)-, iodide
$C_9H_{17}N$
4304 - Cyclohexanamine, *N*-2-propenyl-
8144 - Nonanenitrile
11246 - Quinoline, decahydro-, *cis*-
11247 - Quinoline, decahydro-, *trans*-(±)-
$C_9H_{17}NO$
4591 - Cyclohexanone, 2-[(dimethylamino)methyl]-
9003 - 4-Pentenamide, 2,2-diethyl-
9725 - 4-Piperidinone, 1,2,3,6-tetramethyl-
9726 - 4-Piperidinone, 2,2,6,6-tetramethyl-
10470 - 2-Propanone, 1-(1-methyl-2-piperidinyl)-, (±)-
$C_9H_{17}NOS$
7661 - Molinate
$C_9H_{17}NO_2$
10631 - 2-Propenoic acid, 2-(diethylamino)ethyl ester
$C_9H_{17}NO_4$
10305 - Propanoic acid, 2-methyl-, 2-nitropentyl ester
$C_9H_{17}NO_5$
329 - β-Alanine, *N*-(2,4-dihydroxy-3,3-dimethyl-1-oxobutyl)-, (*R*)-
$C_9H_{17}NS$
11708 - Thiocyanic acid, octyl ester
$C_9H_{17}N_5S$
341 - Ametryn
C_9H_{18}
4394 - Cyclohexane, 1-ethyl-1-methyl-
4395 - Cyclohexane, 1-ethyl-2-methyl-, *cis*-
4396 - Cyclohexane, 1-ethyl-2-methyl-, *trans*-

4397 - Cyclohexane, 1-ethyl-3-methyl-, *cis-*
4398 - Cyclohexane, 1-ethyl-3-methyl-, *trans-*
4399 - Cyclohexane, 1-ethyl-4-methyl-, *cis-*
4400 - Cyclohexane, 1-ethyl-4-methyl-, *trans-*
4431 - Cyclohexane, (1-methylethyl)-
4471 - Cyclohexane, propyl-
4485 - Cyclohexane, 1,1,2-trimethyl-
4486 - Cyclohexane, 1,1,3-trimethyl-
4487 - Cyclohexane, 1,1,4-trimethyl-
4488 - Cyclohexane, 1,2,3-trimethyl-
4489 - Cyclohexane, 1,2,3-trimethyl-, (1α,2β,3α)-
4490 - Cyclohexane, 1,2,3-trimethyl-, (1α,2α,3α)-
4491 - Cyclohexane, 1,2,3-trimethyl-, (1α,2α,3β)-
4492 - Cyclohexane, 1,2,4-trimethyl-, (1α,2α,4α)-
4493 - Cyclohexane, 1,2,4-trimethyl-, (1α,2β,4α)-
4494 - Cyclohexane, 1,2,4-trimethyl-, (1α,2α,4β)-
4495 - Cyclohexane, 1,2,4-trimethyl-, (1α,2β,4β)-
4496 - Cyclohexane, 1,3,5-trimethyl-, (1α,3α,5α)-
4497 - Cyclohexane, 1,3,5-trimethyl-, (1α,3α,5β)-
4724 - Cyclononane
4778 - Cyclopentane, butyl-
4796 - Cyclopentane, 1,1-diethyl-
4797 - Cyclopentane, 1,2-diethyl-, *cis-*
4798 - Cyclopentane, 1,2-diethyl-, *trans-*
4799 - Cyclopentane, 1,3-diethyl-
4805 - Cyclopentane, (1,1-dimethylethyl)-
4829 - Cyclopentane, 1-methyl-2-(1-methylethyl)-
4832 - Cyclopentane, (1-methylpropyl)-
4833 - Cyclopentane, 1-methyl-1-propyl-
4834 - Cyclopentane, 1-methyl-2-propyl-, *cis-*
4835 - Cyclopentane, 1-methyl-2-propyl-, *trans-*
4836 - Cyclopentane, 1-methyl-3-propyl-, *cis-*
4837 - Cyclopentane, 1-methyl-3-propyl-, *trans-*
4838 - Cyclopentane, (2-methylpropyl)-
4851 - Cyclopentane, 1,1,3,3-tetramethyl-
6542 - 2-Heptene, 2,3-dimethyl-
8182 - 1-Nonene
8183 - 3-Nonene, (*E*)-
8505 - 1-Octene, 2-methyl-
8506 - 2-Octene, 3-methyl-
8507 - 3-Octene, 7-methyl-
8508 - 4-Octene, 2-methyl-

$C_9H_{18}Br_2$
8127 - Nonane, 1,2-dibromo-

$C_9H_{18}ClF$
8126 - Nonane, 1-chloro-9-fluoro-

$C_9H_{18}Cl_2$
6382 - Heptane, 2,6-dichloro-2,6-dimethyl-
8128 - Nonane, 1,9-dichloro-

$C_9H_{18}FeN_3S_6$
5957 - Ferbam

$C_9H_{18}NO_3PS_2$
6021 - Fosthiazate

$C_9H_{18}N_2O_2S$
11717 - Thiofanox

$C_9H_{18}N_2O_4$
3938 - Carbamic acid, *N*-butyl-*N*-nitro,butyl ester
10021 - 1,3-Propanediol, 2-methyl-2-propyl-, dicarbamate

$C_9H_{18}N_6O_6$
7584 - Methanol, (1,3,5-triazine-2,4,6-triyltrinitrilo)hexakis-

$C_9H_{18}O$
4526 - Cyclohexanol, 1-ethyl-2-methyl-
4527 - Cyclohexanol, 1-ethyl-3-methyl-
4538 - Cyclohexanol, 1-(1-methylethyl)-
4539 - Cyclohexanol, 2-(1-methylethyl)-, *cis-*
4540 - Cyclohexanol, 4-(1-methylethyl)-, *cis-*
4541 - Cyclohexanol, 4-(1-methylethyl)-, *trans-*
4557 - Cyclohexanol, 1-propyl-
4558 - Cyclohexanol, 2-propyl-, *cis-*
4559 - Cyclohexanol, 2-propyl-, *trans-*
4560 - Cyclohexanol, 4-propyl-, *trans-*
4561 - Cyclohexanol, 4-propyl-, *cis-*
4562 - Cyclohexanol, 1,2,2-trimethyl-
4563 - Cyclohexanol, 1,2,6-trimethyl-
4564 - Cyclohexanol, 1,3,5-trimethyl-
4565 - Cyclohexanol, 2,2,5-trimethyl-
4566 - Cyclohexanol, 2,2,6-trimethyl-
4567 - Cyclohexanol, 2,3,3-trimethyl-
4568 - Cyclohexanol, 2,3,6-trimethyl-
4569 - Cyclohexanol, 3,3,5-trimethyl-, *cis-*
4570 - Cyclohexanol, 3,3,5-trimethyl-, *trans-*
6514 - 4-Heptanone, 2,6-dimethyl-
6515 - 4-Heptanone, 3,5-dimethyl-
8117 - Nonanal
8172 - 2-Nonanone
8173 - 3-Nonanone
8174 - 4-Nonanone
8175 - 5-Nonanone
8189 - 1-Nonen-3-ol
8190 - 8-Nonen-1-ol
8475 - 2-Octanone, 3-methyl-
8476 - 3-Octanone, 4-methyl-
8477 - 3-Octanone, 7-methyl-
8478 - 4-Octanone, 3-methyl-
8479 - 4-Octanone, 7-methyl-
8517 - 1-Octen-4-ol, 4-methyl-
8518 - 2-Octen-4-ol, 7-methyl-
8519 - 3-Octen-2-ol, 2-methyl-, (*Z*)-
8520 - 4-Octen-3-ol, 3-methyl-
8521 - 5-Octen-4-ol, 5-methyl-
8988 - 3-Pentanone, 2,2,4,4-tetramethyl-

$C_9H_{18}O_2$
193 - Acetic acid, heptyl ester
3543 - Butanoic acid, 3-methyl-, butyl ester
3544 - Butanoic acid, 3-methylbutyl ester
3545 - Butanoic acid, 2-methyl-, butyl ester
3555 - Butanoic acid, 3-methyl-, 2-methylpropyl ester
3579 - Butanoic acid, pentyl ester
6014 - Formic acid, octyl ester
6462 - Heptanoic acid, ethyl ester
6476 - 4-Heptanol, acetate
6886 - Hexanoic acid, 2-ethyl-3-methyl-
6887 - Hexanoic acid, 3-ethyl-3-methyl-
6899 - Hexanoic acid, 3-methyl-, ethyl ester
6900 - Hexanoic acid, 4-methyl-, ethyl ester
6909 - Hexanoic acid, propyl ester
8146 - Nonanoic acid
8433 - Octanoic acid, 3-methyl-
8434 - Octanoic acid, methyl ester
8862 - Pentanoic acid, butyl ester
8899 - Pentanoic acid, 1-methylpropyl ester, (+)-
8900 - Pentanoic acid, 2-methylpropyl ester
10265 - Propanoic acid, hexyl ester
10302 - Propanoic acid, 2-methyl-, 3-methylbutyl ester

$C_9H_{18}O_3$
3994 - Carbonic acid, bis(1,1-dimethylethyl) ester
3997 - Carbonic acid, bis(2-methylpropyl) ester
3998 - Carbonic acid, dibutyl ester

$C_9H_{18}O_5$
5719 - Ethanol, 2-ethoxy-, carbonate (2:1)

$C_9H_{19}Br$
8121 - Nonane, 1-bromo-

$C_9H_{19}Cl$
6364 - Heptane, 3-chloro-3-ethyl-
6747 - Hexane, 3-chloro-2,2,3-trimethyl-
8123 - Nonane, 1-chloro-
8124 - Nonane, 2-chloro-
8125 - Nonane, 5-chloro-
8362 - Octane, 3-chloro-3-methyl-
8363 - Octane, 4-chloro-4-methyl-
8727 - Pentane, 3-chloro-3-ethyl-2,2-dimethyl-

$C_9H_{19}ClO$
8163 - 1-Nonanol, 9-chloro-

$C_9H_{19}FO$
8165 - 1-Nonanol, 9-fluoro-

$C_9H_{19}N$
4303 - Cyclohexanamine, *N*-(1-methylethyl)-
4810 - Cyclopentaneethanamine, *N*,α-dimethyl-
6532 - 5-Hepten-2-amine, *N*,6-dimethyl-
9635 - Piperidine, 1-butyl-
9636 - Piperidine, 3-butyl-
9637 - Piperidine, 4-butyl-
9647 - Piperidine, 2,5-diethyl-
9653 - Piperidine, 1-(1,1-dimethylethyl)-
9685 - Piperidine, 1-(2-methylpropyl)-
9686 - Piperidine, 1-methyl-2-propyl-, (*S*)-
9710 - Piperidine, 2,2,4,6-tetramethyl-
9711 - Piperidine, 2,2,6,6-tetramethyl-

$C_9H_{19}NO$
3201 - Butanamide, *N*,*N*-diethyl-3-methyl-
4144 - Conhydrine, *N*-methyl-
8118 - Nonanamide
8964 - 2-Pentanone, 5-(diethylamino)-
9794 - Propanamide, *N*,*N*-diethyl-2,2-dimethyl-

$C_9H_{19}NOS$
5372 - EPTC

$C_9H_{19}NO_2$
328 - β-Alanine, *N*,*N*-diethyl-, ethyl ester
3934 - Carbamic acid, butyl-, butyl ester

$C_9H_{19}NO_4$
3202 - Butanamide, 2,4-dihydroxy-*N*-(3-hydroxypropyl)-3,3-dimethyl-, (*R*)-

C_9H_{20}
6383 - Heptane, 2,2-dimethyl-
6384 - Heptane, 2,3-dimethyl-
6385 - Heptane, 2,4-dimethyl-
6386 - Heptane, 2,5-dimethyl-
6387 - Heptane, 2,6-dimethyl-
6388 - Heptane, 3,3-dimethyl-
6389 - Heptane, 3,4-dimethyl-
6390 - Heptane, 3,5-dimethyl-
6391 - Heptane, 4,4-dimethyl-
6407 - Heptane, 3-ethyl-
6408 - Heptane, 4-ethyl-
6825 - Hexane, 3-ethyl-2-methyl-
6826 - Hexane, 3-ethyl-3-methyl-
6827 - Hexane, 3-ethyl-4-methyl-
6828 - Hexane, 4-ethyl-2-methyl-
6859 - Hexane, 2,2,3-trimethyl-
6860 - Hexane, 2,2,4-trimethyl-
6861 - Hexane, 2,2,5-trimethyl-
6862 - Hexane, 2,3,3-trimethyl-
6863 - Hexane, 2,3,4-trimethyl-
6864 - Hexane, 2,3,5-trimethyl-
6865 - Hexane, 2,4,4-trimethyl-
6866 - Hexane, 3,3,4-trimethyl-
8120 - Nonane
8408 - Octane, 2-methyl-
8409 - Octane, 3-methyl-
8410 - Octane, 4-methyl-
8756 - Pentane, 3,3-diethyl-
8810 - Pentane, 3-ethyl-2,2-dimethyl-
8811 - Pentane, 3-ethyl-2,3-dimethyl-
8812 - Pentane, 3-ethyl-2,4-dimethyl-
8839 - Pentane, 2,2,3,3-tetramethyl-
8840 - Pentane, 2,2,3,4-tetramethyl-
8841 - Pentane, 2,2,4,4-tetramethyl-
8842 - Pentane, 2,3,3,4-tetramethyl-

$C_9H_{20}N_2$
9628 - 4-Piperidinamine, 2,2,6,6-tetramethyl-

$C_9H_{20}N_2O$
11996 - Urea, *N*,*N*-bis(2-methylpropyl)-
12029 - Urea, tetraethyl-

$C_9H_{20}O$
6406 - Heptane, 1-ethoxy-
6479 - 2-Heptanol, 2,6-dimethyl-
6480 - 3-Heptanol, 2,3-dimethyl-
6481 - 3-Heptanol, 2,6-dimethyl-
6482 - 3-Heptanol, 3,5-dimethyl-
6483 - 3-Heptanol, 3,6-dimethyl-
6484 - 4-Heptanol, 2,4-dimethyl-
6485 - 4-Heptanol, 2,6-dimethyl-
6486 - 4-Heptanol, 3,5-dimethyl-
6487 - 4-Heptanol, 4-ethyl-
6932 - 1-Hexanol, 2-ethyl-3-methyl-
6933 - 2-Hexanol, 3-ethyl-2-methyl-
6934 - 3-Hexanol, 3-ethyl-2-methyl-
6935 - 3-Hexanol, 3-ethyl-5-methyl-

6951 - 1-Hexanol, 3,5,5-trimethyl-
6952 - 2-Hexanol, 2,3,4-trimethyl-
6953 - 3-Hexanol, 2,2,3-trimethyl-
6954 - 3-Hexanol, 2,3,5-trimethyl-
6955 - 3-Hexanol, 2,4,4-trimethyl-
6956 - 3-Hexanol, 2,5,5-trimethyl-
6957 - 3-Hexanol, 3,4,4-trimethyl-
6958 - 3-Hexanol, 3,5,5-trimethyl-
8157 - 1-Nonanol
8158 - 2-Nonanol, (±)-
8159 - 3-Nonanol, (±)-
8160 - 4-Nonanol, (*S*)-
8161 - 5-Nonanol
8461 - 1-Octanol, 3-methyl-
8462 - 1-Octanol, 4-methyl-
8463 - 1-Octanol, 7-methyl-
8464 - 2-Octanol, 2-methyl-
8465 - 2-Octanol, 3-methyl-
8466 - 3-Octanol, 3-methyl-
8467 - 3-Octanol, 4-methyl-
8468 - 3-Octanol, 6-methyl-
8469 - 4-Octanol, 4-methyl-
8470 - 4-Octanol, 5-methyl-

$C_9H_{20}O_2$
3398 - Butane, 1,1′-[methylenebis(oxy)]bis-
8136 - 1,9-Nonanediol
8755 - Pentane, 1,1-diethoxy-
9992 - 1,3 Propanediol, 2-butyl-2-ethyl-
10075 - Propane, 1,1′-[methylenebis(oxy)]bis[2-methyl-

$C_9H_{20}O_3$
10182 - Propane, 1,1,1-triethoxy-
10183 - Propane, 1,1,3-triethoxy-

$C_9H_{20}O_3Si$
11482 - Silane, triethoxy-2-propenyl-

$C_9H_{20}O_4$
5601 - Ethane, 1,1′,1″1‴-[methanetetrayl tetrakis(oxy)]tetrakis-

$C_9H_{20}S$
8146 - 1-Nonanethiol

$C_9H_{21}B$
3121 - Borane, tripropyl-

$C_9H_{21}BO_3$
3127 - Boric acid, tripropyl ester
3128 - Boric acid, tris(1-methylethyl) ester

$C_9H_{21}ClSn$
11542 - Stannane, chlorotripropyl-

$C_9H_{21}N$
8119 - 1-Nonanamine
9808 - 1-Propanamine, *N,N*-dipropyl

$C_9H_{21}NO_3$
10393 - 2-Propanol, 1,1′,1″-nitrilotris-

$C_9H_{21}NO_4$
5678 - Ethanol, 2-[2-[2-(3-aminopropoxy) ethoxy]ethoxy]-

$C_9H_{21}O_2PS_3$
11592 - Terbufos

$C_9H_{21}O_3P$
9594 - Phosphorous acid, tripropyl ester
9595 - Phosphorous acid, tris(1-methylethyl) ester

$C_9H_{21}O_4P$
9567 - Phosphoric acid, tripropyl ester
9568 - Phosphoric acid, tris(1-methylethyl) ester

$C_9H_{22}N_2$
7489 - Methanediamine, *N,N,N′,N′*-tetraethyl-
8736 - 1,4-Pentanediamine, *N1,N1*-diethyl

$C_9H_{22}O_4P_2S_4$
5931 - Ethion

$C_9H_{22}Si$
11500 - Silane, tripropyl-

$C_9H_{23}NO_3Si$
9828 - 1-Propanamine, 3-(triethoxysilyl)-

$C_9H_{24}O_3Si_3$
5003 - Cyclotrisiloxane, 2,4,6-triethyl-2,4,6-trimethyl-

$C_9H_{24}O_4Si_4$
4991 - Cyclotetrasiloxane, ethenylheptamethyl-

$C_{10}Cl_8$
7933 - Naphthalene, octachloro-

$C_{10}Cl_{10}O$
7408 - Kepone

$C_{10}Cl_{12}$
7644 - 1,3,4-Metheno-1H-cyclobuta[cd]pentalene, 1,1a,2,2,3,3a,4,5,5,5a,5b, 6-dodecachlorooctahydro-

$C_{10}F_8$
7935 - Naphthalene, octafluoro-

$C_{10}F_{18}$
7934 - Naphthalene, octadecafluorodecahydro-

$C_{10}F_{22}$
5050 - Decane, docosafluoro-

$C_{10}F_{23}N$
11912 - Tripropylamine, eicosafluoro-2-(trifluoromethyl)-

$C_{10}H_4Cl_2O_2$
7849 - 1,2-Naphthalenedione, 3,4-dichloro-
7850 - 1,4-Naphthalenedione, 2,3-dichloro-
7851 - 1,4-Naphthalenedione, 5,6-dichloro-

$C_{10}H_4N_4O_8$
8004 - Naphthalene, 1,3,6,8-tetranitro-

$C_{10}H_5BrO_2$
7844 - 1,2-Naphthalenedione, 3-bromo-

$C_{10}H_5BrO_3$
7845 - 1,4-Naphthalenedione, 2-bromo-3-hydroxy-

$C_{10}H_5ClO_2$
7847 - 1,4-Naphthalenedione, 5-chloro-

$C_{10}H_5ClO_3$
7848 - 1,4-Naphthalenedione, 2-chloro-3-hydroxy-

$C_{10}H_5Cl_2NO_2$
11201 - Quinclorac

$C_{10}H_5Cl_7$
7576 - 4,7-Methano-1H-indene, 1,4,5,6,7,8,8-heptachloro-3a,4,7,7a-tetrahydro-

$C_{10}H_5Cl_7O$
6301 - Heptachlor epoxide

$C_{10}H_5N_3O_6$
8016 - Naphthalene, 1,3,5-trinitro-

$C_{10}H_5NaO_5S$
7970 - 1-Naphthalenesulfonic acid, 3,4-dihydro-3,4-dioxo-, sodium salt

$C_{10}H_6BrCl$
7741 - Naphthalene, 1-bromo-4-chloro-

$C_{10}H_6BrNO_2$
7744 - Naphthalene, 1-bromo-3-nitro-
7745 - Naphthalene, 2-bromo-1-nitro-

$C_{10}H_6Br_2$
7801 - Naphthalene, 1,3-dibromo-
7802 - Naphthalene, 1,4-dibromo-
7803 - Naphthalene, 1,5-dibromo-

$C_{10}H_6ClNO_2$
7786 - Naphthalene, 1-chloro-5-nitro-
7787 - Naphthalene, 1-chloro-8-nitro-
7788 - Naphthalene, 2-chloro-1-nitro-

$C_{10}H_6Cl_2$
7807 - Naphthalene, 1,2-dichloro-
7808 - Naphthalene, 1,3-dichloro-
7809 - Naphthalene, 1,4-dichloro-
7810 - Naphthalene, 1,5-dichloro-
7811 - Naphthalene, 1,7-dichloro-
7812 - Naphthalene, 1,8-dichloro-
7813 - Naphthalene, 2,6-dichloro-

$C_{10}H_6Cl_2O$
8025 - 1-Naphthalenol, 2,3-dichloro-
8026 - 1-Naphthalenol, 2,4-dichloro-

$C_{10}H_6Cl_4O_4$
1353 - 1,2-Benzenedicarboxylic acid, 3,4,5,6-tetrachloro-, monoethyl ester
5180 - Dimethyl tetrachloroterephthalate

$C_{10}H_6Cl_8$
4077 - Chlordane

$C_{10}H_6N_2$
7375 - 5-Isoquinolinecarbonitrile
11223 - 2-Quinolinecarbonitrile
11224 - 4-Quinolinecarbonitrile

$C_{10}H_6N_2OS_2$
4072 - Chinomethionat

$C_{10}H_6N_2O_2$
5223 - 5H,10H-Dipyrrolo[1,2-a:1′,2′-d]pyrazine-5,10-dione

$C_{10}H_6N_2O_4$
7831 - Naphthalene, 1,3-dinitro-
7832 - Naphthalene, 1,5-dinitro-
7833 - Naphthalene, 1,8-dinitro-

$C_{10}H_6O_2$
7842 - 1,2-Naphthalenedione
7843 - 1,4-Naphthalenedione

$C_{10}H_6O_3$
7859 - 1,4-Naphthalenedione, 2-hydroxy-
7860 - 1,4-Naphthalenedione, 5-hydroxy-

$C_{10}H_6O_3S$
7840 - 1,8-Naphthalenediol, cyclic sulfite

$C_{10}H_6O_4$
7852 - 1,4-Naphthalenedione, 2,3-dihydroxy-
7853 - 1,4-Naphthalenedione, 5,8-dihydroxy-

$C_{10}H_6O_8$
2257 - 1,2,4,5-Benzenetetracarboxylic acid

$C_{10}H_7Br$
7739 - Naphthalene, 1-bromo-
7740 - Naphthalene, 2-bromo-

$C_{10}H_7BrO$
8022 - 2-Naphthalenol, 1-bromo-

$C_{10}H_7Cl$
7780 - Naphthalene, 1-chloro-
7781 - Naphthalene, 2-chloro-

$C_{10}H_7ClO$
8023 - 2-Naphthalenol, 6-chloro-
8063 - 2-Naphthol, 8-chloro-

$C_{10}H_7ClO_2S$
7975 - 1-Naphthalenesulfonyl chloride
7976 - 2-Naphthalenesulfonyl chloride

$C_{10}H_7Cl_5O$
11899 - Tridiphane

$C_{10}H_7F$
7893 - Naphthalene, 1-fluoro-
7894 - Naphthalene, 2-fluoro-

$C_{10}H_7I$
7902 - Naphthalene, 1-iodo-
7903 - Naphthalene, 2-iodo-

$C_{10}H_7NO$
11225 - 4-Quinolinecarboxaldehyde

$C_{10}H_7NO_2$
7929 - Naphthalene, 1-nitro-
7930 - Naphthalene, 2-nitro-
8035 - 2-Naphthalenol, 1-nitroso-
11123 - 1H-Pyrrole-2,5-dione, 1-phenyl-
11227 - 2-Quinolinecarboxylic acid
11228 - 7-Quinolinecarboxylic acid
11229 - 8-Quinolinecarboxylic acid

$C_{10}H_7NO_3$
8034 - 2-Naphthalenol, 1-nitro-
11232 - 2-Quinolinecarboxylic acid, 4-hydroxy-

$C_{10}H_7NO_4$
11230 - 2-Quinolinecarboxylic acid, 4,8-dihydroxy-

$C_{10}H_7N_3$
7738 - Naphthalene, 1-azido-

$C_{10}H_7N_3S$
11656 - Thiabendazole

$C_{10}H_8$
504 - Azulene
7732 - Naphthalene

$C_{10}H_8BrN$
7711 - 1-Naphthalenamine, 5-bromo-

$C_{10}H_8ClN$
11243 - Quinoline, 2-chloro-4-methyl-
11244 - Quinoline, 2-chloro-8-methyl-
11245 - Quinoline, 4-chloro-2-methyl-

$C_{10}H_8ClN_3O$
4080 - Chloridazon

$C_{10}H_8N_2$
3099 - 2,2′-Bipyridine
3100 - 2,3′-Bipyridine

9322 - Phenol, 3-(1,1-dimethylethyl)-
9323 - Phenol, 4-(1,1-dimethylethyl)-
9414 - Phenol, 2-methyl-4-(1-methylethyl)-
9415 - Phenol, 2-methyl-5-(1-methylethyl)-
9416 - Phenol, 2-methyl-6-(1-methylethyl)-
9417 - Phenol, 4-methyl-2-(1-methylethyl)-
9418 - Phenol, 5-methyl-2-(1-methylethyl)-
9430 - Phenol, 2-(1-methylpropyl)-
9431 - Phenol, 4-(1-methylpropyl)-
9432 - Phenol, 4-(2-methylpropyl)-
9473 - Phenol, 2,3,4,5-tetramethyl-
9474 - Phenol, 2,3,4,6-tetramethyl-
9475 - Phenol, 2,3,5,6-tetramethyl-

$C_{10}H_{14}O_2$
1406 - Benzene, 1,2-diethoxy-
1407 - Benzene, 1,3-diethoxy-
1408 - Benzene, 1,4-diethoxy-
1516 - 1,2-Benzenediol, 4-(1,1-dimethylethyl)-
1536 - 1,2-Benzenediol, 3-methyl-6-(1-methylethyl)-
1538 - 1,4-Benzenediol, 2-methyl-5-(1-methylethyl)-
1539 - 1,3-Benzenediol, 4-(2-methylpropyl)-
1552 - 1,4-Benzenediol, 2,3,5,6-tetramethyl-
1652 - Benzene, (2-ethoxyethoxy)-
1873 - Benzenemethanol, α-ethyl-4-methoxy-
2154 - Benzenepropanol, 4-methoxy-
2926 - Bicyclo[2.2.1]heptane-2,3-dione, 1,7,7-trimethyl-, (1*R*)-
3356 - 1,2-Butanediol, 2-phenyl-
3357 - 1,3-Butanediol, 1-phenyl-
3358 - 1,3-Butanediol, 3-phenyl-
4758 - Cyclopenta[c]pyran-1(4aH)-one, 5,6,7,7a-tetrahydro-4,7-dimethyl-
6076 - Furan, 2-(cyclohexyloxy)-
9245 - Phenol, 2-butoxy-
10400 - 1-Propanol, 3-(phenylmethoxy)-
11862 - Tricyclo[2.2.1.02,6]heptane-1-carboxylic acid, 7,7-dimethyl-

$C_{10}H_{14}O_3$
1599 - Benzeneethanol, 2,5-dimethoxy-
2153 - Benzenepropanol, 4-hydroxy-3-methoxy-
6027 - 2-Furanacetic acid, butyl ester
6068 - 2-Furancarboxylic acid, pentyl ester
8559 - 3-Oxabicyclo[3.2.1]octane-2,4-dione, 1,8,8-trimethyl-, (±)-
8880 - Pentanoic acid, 2-furanylmethyl ester
10019 - 1,2-Propanediol, 3-(2-methylphenoxy)-

$C_{10}H_{14}O_3S$
2219 - Benzenesulfonic acid, *sec*-butyl ester
2233 - Benzenesulfonic acid, 4-methyl-, propyl ester

$C_{10}H_{14}O_4$
1914 - Benzenemethanol, 3,4,5-trimethoxy-
4646 - 1-Cyclohexene-1,4-dicarboxylic acid, dimethyl ester
10016 - 1,2-Propanediol, 3-(2-methoxyphenoxy)-
10656 - 2-Propenoic acid, 2-methyl-, 1,2-ethanediyl ester

$C_{10}H_{15}BrO$
2951 - Bicyclo[2.2.1]heptan-2-one, 1-(bromomethyl)-7,7-dimethyl-, (1*S*)-
2952 - Bicyclo[2.2.1]heptan-2-one, 3-bromo-1,7,7-trimethyl-, (1*R-endo*)-

$C_{10}H_{15}N$
655 - Benzenamine, 4-butyl-
656 - Benzenamine, *N*-butyl-
694 - Benzenamine, 2,6-diethyl-
695 - Benzenamine, *N*,*N*-diethyl-
714 - Benzenamine, 2-(1,1-dimethylethyl)-
715 - Benzenamine, 4-(1,1-dimethylethyl)-
716 - Benzenamine, *N*-(1,1-dimethylethyl)-
793 - Benzenamine, 2-methyl-5-(1-methylethyl)-
794 - Benzenamine, 4-methyl-*N*-(1-methylethyl)-
812 - Benzenamine, 2-(1-methylpropyl)-
813 - Benzenamine, 4-(1-methylpropyl)-
814 - Benzenamine, 4-(2-methylpropyl)-
815 - Benzenamine, 4-methyl-*N*-propyl-
816 - Benzenamine, *N*-(1-methylpropyl)-
817 - Benzenamine, *N*-(2-methylpropyl)-
847 - Benzenamine, 2,3,4,5-tetramethyl-
848 - Benzenamine, 2,3,4,6-tetramethyl-
849 - Benzenamine, *N*,*N*,2,6-tetramethyl-
1572 - Benzeneethanamine, α,α-dimethyl-
1573 - Benzeneethanamine, *N*,α-dimethyl-, (*S*)-
1574 - Benzeneethanamine, *N*,β-dimethyl-
1819 - Benzenemethanamine, 4-(1-methylethyl)-
1820 - Benzenemethanamine, *N*-(1-methylethyl)-
1830 - Benzenemethanamine, α-propyl-
2119 - Benzenepropanamine, α-methyl-
10967 - Pyridine, 2-(1-ethylpropyl)-
10968 - Pyridine, 4-(1-ethylpropyl)-

$C_{10}H_{15}NO$
1840 - Benzenemethanol, α-(1-aminopropyl)-
1900 - Benzenemethanol, α-[1-(methylamino)ethyl]-, (*R**,*S**)-(±)-
9297 - Phenol, 3-(diethylamino)-
9320 - Phenol, 4-[2-(dimethylamino)ethyl]-
9403 - Phenol, 4-[2-(methylamino)propyl]-
10983 - Pyridine, 2-[2-(1-methylethoxy)ethyl]-

$C_{10}H_{15}NO_2$
693 - Benzenamine, 3,4-diethoxy-
1571 - Benzeneethanamine, 3,4-dimethoxy-
5760 - Ethanol, 2,2′-(phenylimino)bis-

$C_{10}H_{15}NO_4$
9933 - Propanedioic acid, (2-cyanoethyl)-, diethyl ester
11144 - 3-Pyrrolidineacetic acid, 2-carboxy-4-(1-methylethenyl)-, [2*S*-(2α,3β,4β)]-

$C_{10}H_{15}N_3$
6281 - Guanidine, *N*,*N*′-dimethyl-*N*″-(phenylmethyl)-

$C_{10}H_{15}OPS_2$
5986 - Fonofos

$C_{10}H_{15}O_3PS_2$
5955 - Fenthion

$C_{10}H_{16}$
2921 - Bicyclo[2.2.1]heptane, 2,2-dimethyl-3-methylene-, (±)-
2922 - Bicyclo[2.2.1]heptane, 2,2-dimethyl-5-methylene-
2923 - Bicyclo[2.2.1]heptane, 7,7-dimethyl-2-methylene-, (±)-
2924 - Bicyclo[3.1.1]heptane, 6,6-dimethyl-2-methylene-, (1*R*)-
2925 - Bicyclo[3.1.1]heptane, 6,6-dimethyl-2-methylene-
2970 - Bicyclo[2.2.1]hept-2-ene, 1,7,7-trimethyl-
2971 - Bicyclo[3.1.1]hept-2-ene, 2,6,6-trimethyl-, (1*S*)-
2972 - Bicyclo[3.1.1]hept-2-ene, 2,6,6-trimethyl-, (±)-
2973 - Bicyclo[3.1.1]hept-2-ene, 2,6,6-trimethyl
2974 - Bicyclo[4.1.0]hept-3-ene, 3,7,7-trimethyl-, (1*S*)-
2981 - Bicyclo[3.1.0]hexane, 4-methylene-1-(1-methylethyl)-, (±)-
2986 - Bicyclo[3.1.0]hex-2-ene, 2-methyl-5-(1-methylethyl)-
4050 - 4-Carene, (1*S*,3*R*,6*R*)-(-)-
4209 - Cyclodecyne
4273 - 1,3-Cyclohexadiene, 1-methyl-4-(1-methylethyl)-
4274 - 1,3-Cyclohexadiene, 2-methyl-5-(1-methylethyl)-, (±)-
4275 - 1,4-Cyclohexadiene, 1-methyl-4-(1-methylethyl)-
4673 - Cyclohexene, 3-methylene-6-(1-methylethyl)-
4674 - Cyclohexene, 3-methylene-6-(1-methylethyl)-, (+)-
4675 - Cyclohexene, 4-methylene-1-(1-methylethyl)-
4676 - Cyclohexene, 1-methyl-3-(1-methylethenyl)-
4677 - Cyclohexene, 1-methyl-4-(1-methylethenyl)-
4678 - Cyclohexene, 1-methyl-5-(1-methylethenyl)-
4679 - Cyclohexene, 5-methyl-1-(1-methylethenyl)-, (±)-
4684 - Cyclohexene, 1-methyl-4-(1-methylethylidene)-
5094 - 1-Decen-3-yne
5095 - 1-Decen-4-yne
5096 - 2-Decen-4-yne
8330 - 1,6-Octadiene, 7-methyl-3-methylene-
8485 - 1,3,6-Octatriene, 3,7-dimethyl-
8486 - 1,3,7-Octatriene, 3,7-dimethyl-
8487 - 2,4,6-Octatriene, 2,6-dimethyl-, (*E*,*E*)-
8488 - 2,4,6-Octatriene, 2,6-dimethyl-, (*E*,*Z*)-
11815 - 4(10)-Thujene, (+)-
11860 - Tricyclo[3.3.1.13,7]decane
11866 - Tricyclo[2.2.1.02,6]heptane, 1,3,3-trimethyl-
11867 - Tricyclo[2.2.1.02,6]heptane, 1,7,7-trimethyl-

$C_{10}H_{16}ClNO$
865 - Benzenaminium, *N*-ethyl-3-hydroxy-*N*,*N*-dimethyl-, chloride

$C_{10}H_{16}Cl_2O_2$
5049 - Decanedioyl dichloride

$C_{10}H_{16}Cl_3NOS$
11823 - Triallate

$C_{10}H_{16}NO_5PS_2$
5946 - Famphur

$C_{10}H_{16}N_2$
1265 - 1,3-Benzenediamine, *N*,*N*-diethyl-
1266 - 1,4-Benzenediamine, *N*,*N*-diethyl-
1286 - 1,2-Benzenediamine, *N*,*N*,*N*′,*N*′-tetramethyl-
1287 - 1,4-Benzenediamine, *N*,*N*,*N*′,*N*′-tetramethyl-
5041 - Decanedinitrile

$C_{10}H_{16}N_2OS$
7229 - 4-Imidazolidinone, 5-(2-methylpropyl)-3-(2-propenyl)-2-thioxo-

$C_{10}H_{16}N_2O_2S$
11065 - 4,6(1H,5H)-Pyrimidinedione, 5-ethyldihydro-5-(1-methylpropyl)-2-thioxo-

$C_{10}H_{16}N_2O_3$
11082 - 2,4,6(1H,3H,5H)-Pyrimidinetrione, 5-butyl-5-ethyl-

$C_{10}H_{16}N_2O_3S$
11689 - 1H-Thieno[3,4-d]imidazole-4-pentanoic acid, hexahydro-2-oxo-, [3a*S*-(3aα,4β,6aα)]-

$C_{10}H_{16}N_2O_8$
6262 - Glycine, *N*,*N*′-1,2-ethanediylbis[*N*-(carboxymethyl)-

$C_{10}H_{16}N_4O_7$
11107 - 4(1H)-Pyrimidinone, 2,6-diamino-5-(β-*D*-glucopyranosyloxy)-

$C_{10}H_{16}O$
2960 - Bicyclo[2.2.1]heptan-2-one, 1,3,3-trimethyl-, (±)-
2961 - Bicyclo[2.2.1]heptan-2-one, 1,7,7-trimethyl-, (1*R*)
2963 - Bicyclo[2.2.1]heptan-2-one, 4,7,7-trimethyl-, (1*S*)-
2964 - Bicyclo[3.1.1]heptan-2-one, 4,6,6-trimethyl-, [1*R*-(1α,4α,5α)]-
2965 - Bicyclo[3.1.1]heptan-3-one, 2,6,6-trimethyl-, [1*S*-(1α,2β,5α)]-
2975 - Bicyclo[3.1.1]hept-3-en-2-ol, 4,6,6-trimethyl-, (1α,2β,5α)-
2976 - Bicyclo[3.1.1]hept-3-en-2-ol, 4,6,6-trimethyl-, (1α,2α,5α)-
2983 - Bicyclo[3.1.0]hexan-3-ol, 4-methylene-1-(1-methylethyl)-
2984 - Bicyclo[3.1.0]hexan-3-ol, 4-methylene-1-(1-methylethyl)-, [1*S*-(1α,3β,5α)]-
2995 - [1,1′-Bicyclopentyl]-2-one
4576 - Cyclohexanone, 2-butylidene-
4601 - Cyclohexanone, 2-methyl-5-(1-methylethenyl)-, (2*S*-*trans*)-
4602 - Cyclohexanone, 5-methyl-2-(1-methylethenyl)-
4603 - Cyclohexanone, 5-methyl-2-(1-methylethenyl)-, *trans*-
4607 - Cyclohexanone, 5-methyl-2-(1-methylethylidene)-
4608 - Cyclohexanone, 5-methyl-2-(1-methylethylidene)-, (*R*)-
4635 - 1-Cyclohexene-1-carboxaldehyde, 4-(1-methylethyl)-

4636 - 1-Cyclohexene-1-carboxaldehyde, 2,6,6-trimethyl-
4664 - 1-Cyclohexene-1-methanol, 4-(1-methylethenyl)-
4696 - 2-Cyclohexen-1-ol, 2-methyl-5-(1-methylethenyl)-
4716 - 2-Cyclohexen-1-one, 3-methyl-6-(1-methylethyl)-, (±)-
4717 - 2-Cyclohexen-1-one, 6-methyl-3-(1-methylethyl)-, (±)-
8053 - 1(2H)-Naphthalenone, octahydro-, *cis*-
8054 - 1(2H)-Naphthalenone, octahydro-, *trans*-
8055 - 2(1H)-Naphthalenone, octahydro-
8317 - 2,6-Octadienal, 3,7-dimethyl-, (*E*)-
8318 - 2,6-Octadienal, 3,7-dimethyl-, (*Z*)-
8561 - 6-Oxabicyclo[3.2.1]oct-3-ene, 4,7,7-trimethyl-, (±)-

$C_{10}H_{16}O_2$
2957 - Bicyclo[2.2.1]heptan-2-one, 3-hydroxy-1,7,7-trimethyl-
3834 - 2-Butenoic acid, 3-cyclohexyl-
4203 - 1,2-Cyclodecanedione
4307 - Δ1,α-Cyclohexaneacetic acid, 3,3-dimethyl-
4709 - 2-Cyclohexen-1-one, 2-hydroxy-3-methyl-6-(1-methylethyl)-
4911 - 1-Cyclopentene-1-carboxylic acid, 2,3,3-trimethyl-, methyl ester
4959 - Cyclopropanecarboxylic acid, 2,2-dimethyl-3-(2-methyl-1-propenyl)-, (1*R*-*cis*)-
5025 - 2,4-Decadienoic acid, (*E*,*E*)-
5184 - 2,3-Dioxabicyclo[2.2.2]oct-5-ene, 1-methyl-4-(1-methylethyl)-
8331 - 2,4-Octadienoic acid, 3,7-dimethyl-
10653 - 2-Propenoic acid, 2-methyl-, cyclohexyl ester

$C_{10}H_{16}O_4$
3766 - 2-Butenedioic acid (*E*)-, bis(1-methylethyl) ester
3783 - 2-Butenedioic acid (*E*)-, dipropyl ester
3784 - 2-Butenedioic acid (*Z*)-, dipropyl ester
3931 - Camphoric acid, *trans*-(±)-
4180 - 1,1-Cyclobutanedicarboxylic acid, diethyl ester
4184 - 1,3-Cyclobutanedicarboxylic acid, 2,2-dimethyl-, dimethyl ester
4352 - 1,3-Cyclohexanedicarboxylic acid, dimethyl ester, *cis*-
4353 - 1,3-Cyclohexanedicarboxylic acid, dimethyl ester, *trans*-
4357 - 1,4-Cyclohexanedicarboxylic acid, dimethyl ester, *cis*-
4794 - 1,3-Cyclopentanedicarboxylic acid, 1,2,2-trimethyl-, (1*R*-*cis*)-
9966 - Propanedioic acid, (1-methylethylidene)-, diethyl ester
9980 - Propanedioic acid, 2-propenyl-, diethyl ester

$C_{10}H_{16}O_4S$
2929 - Bicyclo[2.2.1]heptane-1-methanesulfonic acid, 7,7-dimethyl-2-oxo-, (1*S*)-

$C_{10}H_{16}O_5$
3295 - Butanedioic acid, acetyl-, diethyl ester
9947 - Propanedioic acid, (ethoxymethylene)-, diethyl ester

$C_{10}H_{16}O_6$
7552 - Methanetricarboxylic acid, triethyl ester

$C_{10}H_{16}Si$
11490 - Silane, trimethyl(4-methylphenyl)-
11494 - Silane, trimethyl(phenylmethyl)-

$C_{10}H_{17}Br$
8321 - 2,6-Octadiene, 1-bromo-3,7-dimethyl-, (*E*)-

$C_{10}H_{17}Cl$
2919 - Bicyclo[2.2.1]heptane, 2-chloro-1,7,7-trimethyl-
2920 - Bicyclo[2.2.1]heptane, 2-chloro-1,7,7-trimethyl-, *endo*-
7782 - Naphthalene, 2-chlorodecahydro-

$C_{10}H_{17}ClO_4$
9932 - Propanedioic acid, (3-chloropropyl)-, diethyl ester

$C_{10}H_{17}Cl_2NOS$
5106 - Diallate

$C_{10}H_{17}N$
11859 - Tricyclo[3.3.1.13,7]decan-1-amine

$C_{10}H_{17}NO$
2962 - Bicyclo[2.2.1]heptan-2-one, 1,7,7-trimethyl-, oxime, (1*S*)-

$C_{10}H_{17}NO_6$
6258 - Glycine, *N*,*N*-bis(ethoxycarbonyl)-, ethyl ester
10104 - Propanenitrile, 2-(β-*D*-glucopyranosyloxy)-2-methyl-

$C_{10}H_{17}N_2O_4PS$
5945 - Etrimfos

$C_{10}H_{17}N_3$
11848 - 4H-1,2,4-Triazole, 4-cyclohexyl-3-ethyl-

$C_{10}H_{17}N_3O_6S$
6264 - Glycine, *N*-(*N*-*L*-γ-glutamyl-*L*-cysteinyl)-

$C_{10}H_{18}$
2931 - Bicyclo[2.2.1]heptane, 1,3,3-trimethyl-, (+)-
2932 - Bicyclo[2.2.1]heptane, 1,7,7-trimethyl-
2933 - Bicyclo[2.2.1]heptane, 2,2,3-trimethyl-
2934 - Bicyclo[3.1.1]heptane, 2,6,6-trimethyl-, (1α,2α,5α)-(±)-
2935 - Bicyclo[3.1.1]heptane, 2,6,6-trimethyl-, [1*R*-(1α,2β,5α)]-
2936 - Bicyclo[4.1.0]heptane, 3,7,7-trimethyl-, [1*S*-(1α,3β,6α)]-
2937 - Bicyclo[4.1.0]heptane, 3,7,7-trimethyl-, [1*S*-(1α,3α,6α)]-
2982 - Bicyclo[3.1.0]hexane, 4-methyl-1-(1-methylethyl)-
4207 - Cyclodecene, (*E*)-
4208 - Cyclodecene, (*Z*)-
4247 - Cycloheptene, 1,4,4-trimethyl-
4435 - Cyclohexane, 1-methyl-3-(1-methylethenyl)-
4441 - Cyclohexane, 1-methyl-3-(1-methylethylidene)-
4442 - Cyclohexane, 1-methyl-4-(1-methylethylidene)-
4680 - Cyclohexene, 1-methyl-4-(1-methylethyl)-
4681 - Cyclohexene, 1-methyl-4-(1-methylethyl)-, (*R*)-
4682 - Cyclohexene, 3-methyl-6-(1-methylethyl)-, (3*R*-*trans*)-
4683 - Cyclohexene, 4-methyl-1-(1-methylethyl)-, (*R*)-
4923 - Cyclopentene, 1,5-dimethyl-4-(1-methylethyl)-
4934 - Cyclopentene, 1-pentyl-
5023 - 1,3-Decadiene
5024 - 1,9-Decadiene
5097 - 1-Decyne
5098 - 3-Decyne
5099 - 4-Decyne
5100 - 5-Decyne
6619 - 3-Heptyne, 5-ethyl-5-methyl-
7429 - *p*-Menth-3-ene, (*S*)-(-)-
7789 - Naphthalene, decahydro-
7790 - Naphthalene, decahydro-, *cis*-
7791 - Naphthalene, decahydro-, *trans*-
8202 - 4-Nonyne, 8-methyl-
8322 - 2,4-Octadiene, 3,7-dimethyl-
8323 - 2,5-Octadiene, 2,6-dimethyl-
8324 - 2,6-Octadiene, 2,6-dimethyl-, (*Z*)-
8325 - 2,6-Octadiene, 2,7-dimethyl-
8326 - 2,6-Octadiene, 3,6-dimethyl-
8327 - 2,6-Octadiene, 4,5-dimethyl-
8535 - 3-Octyne, 2,2-dimethyl-
11814 - Thujane, (1*S*,4*R*,5*S*)-(+)-

$C_{10}H_{18}N_2O_3$
7223 - 4-Imidazolidinehexanoic acid, 5-methyl-2-oxo-, (4*R*-*cis*)-

$C_{10}H_{18}N_2O_4$
9606 - 1,4-Piperazinedicarboxylic acid, diethyl ester

$C_{10}H_{18}N_4O_4S_3$
11713 - Thiodicarb

$C_{10}H_{18}O$
2930 - Bicyclo[3.1.1]heptane-2-methanol, 6,6-dimethyl-, [1*S*-(1α,2β,5α)]-
2940 - Bicyclo[2.2.1]heptan-2-ol, 1,3,3-trimethyl-, *endo*-(±)-
2941 - Bicyclo[2.2.1]heptan-2-ol, 1,3,3-trimethyl-, (1*S*-*endo*)-
2942 - Bicyclo[2.2.1]heptan-2-ol, 1,7,7-trimethyl-, *endo*-(±)-
2943 - Bicyclo[2.2.1]heptan-2-ol, 1,7,7-trimethyl-, *exo*-(±)-
2944 - Bicyclo[3.1.1]heptan-2-ol, 4,6,6-trimethyl-, [1*R*-(1α,2α,4α,5α)]-
2985 - Bicyclo[3.1.0]hexan-3-ol, 4-methyl-1-(1-methylethyl)-
2994 - [1,1′-Bicyclopentyl]-2-ol
3652 - 1-Butanone, 1-cyclohexyl-
4205 - Cyclodecanone
4237 - Cycloheptanone, 2,2,6-trimethyl-
4238 - Cycloheptanone, 2,6,6-trimethyl-
4542 - Cyclohexanol, 1-methyl-4-(1-methylethenyl)-
4543 - Cyclohexanol, 2-methyl-5-(1-methylethenyl)-, [1*S*-(1α,2β,5α)]-
4544 - Cyclohexanol, 5-methyl-2-(1-methylethenyl)-, [1*R*-(1α,2β,5α)]-
4575 - Cyclohexanone, 2-butyl-
4604 - Cyclohexanone, 2-methyl-5-(1-methylethyl)-, (2*S*-*trans*)-
4605 - Cyclohexanone, 5-methyl-2-(1-methylethyl)-, (2*R*-*cis*)-
4606 - Cyclohexanone, 5-methyl-2-(1-methylethyl)-, (2*S*-*trans*)-
4609 - Cyclohexanone, 4-(2-methylpropyl)-
4665 - 1-Cyclohexene-1-methanol, 2,4,4-trimethyl-
4666 - 1-Cyclohexene-1-methanol, 2,6,6-trimethyl-
4667 - 2-Cyclohexene-1-methanol, 2,6,6-trimethyl-
4668 - 3-Cyclohexene-1-methanol, α,α,4-trimethyl-, (±)-
4697 - 2-Cyclohexen-1-ol, 3-methyl-6-(1-methylethyl)-
4698 - 2-Cyclohexen-1-ol, 6-methyl-3-(1-methylethyl)-
4699 - 3-Cyclohexen-1-ol, 4-methyl-1-(1-methylethyl)-
4978 - Cyclopropanemethanol, 2,2-dimethyl-3-(2-methyl-1-propenyl)-
5080 - 2-Decenal
5081 - 3-Decenal
5093 - 3-Decen-2-one
6590 - 3-Hepten-2-one, 3-ethyl-4-methyl-
6591 - 4-Hepten-2-one, 3-ethyl-4-methyl-
6592 - 4-Hepten-3-one, 5-ethyl-4-methyl-
6593 - 5-Hepten-2-one, 4-ethyl-6-methyl-
6594 - 5-Hepten-3-one, 5-ethyl-4-methyl-
8024 - 2-Naphthalenol, decahydro-
8191 - 5-Nonen-4-one, 6-methyl-
8192 - 6-Nonen-4-one, 6-methyl-
8332 - 1,6-Octadien-3-ol, 3,7-dimethyl-, (±)-
8333 - 2,6-Octadien-1-ol, 3,7-dimethyl-, (*E*)-
8334 - 2,6-Octadien-1-ol, 3,7-dimethyl-, (*Z*)-
8490 - 6-Octenal, 3,7-dimethyl-
8556 - 7-Oxabicyclo[2.2.1]heptane, 1-methyl-4-(1-methylethyl)-
8560 - 2-Oxabicyclo[2.2.2]octane, 1,3,3-trimethyl-

$C_{10}H_{18}O_2$
3496 - Butanoic acid, cyclohexyl ester
4206 - Cyclodecanone, 2-hydroxy-
4224 - Cycloheptanecarboxylic acid, ethyl ester
4308 - Cyclohexaneacetic acid, ethyl ester
4337 - Cyclohexanecarboxylic acid, propyl ester
4784 - Cyclopentanecarboxylic acid, 2-methyl-5-(1-methylethyl)-
4786 - Cyclopentanecarboxylic acid, 1,2,2,3-tetramethyl-
5088 - 2-Decenoic acid
5089 - 3-Decenoic acid

5090 - 4-Decenoic acid
5091 - 9-Decenoic acid
6811 - 3,4-Hexanedione, 2,2,5,5-tetramethyl-
10293 - Propanoic acid, 2-methyl-, cyclohexyl ester
10660 - 2-Propenoic acid, 2-methyl-, hexyl ester

$C_{10}H_{18}O_3$
3504 - Butanoic acid, 2,2-diethyl-3-oxo-, ethyl ester
3542 - Butanoic acid, 3-methyl-, anhydride
4309 - Cyclohexaneacetic acid, α-ethyl-1-hydroxy-
6872 - Hexanoic acid, 2-acetyl-, ethyl ester
8858 - Pentanoic acid, anhydride
11903 - 2,5,7-Trioxabicyclo[2.2.1]heptane, 1,3,3,4,6,6-hexamethyl-

$C_{10}H_{18}O_4$
3321 - Butanedioic acid, dipropyl ester
3512 - Butanoic acid, 1,2-ethanediyl ester
5042 - Decanedioic acid
5557 - Ethanedioic acid, bis(2-methylpropyl) ester
5558 - Ethanedioic acid, dibutyl ester
6785 - Hexanedioic acid, diethyl ester
8135 - Nonanedioic acid, 3-methyl-
8388 - Octanedioic acid, dimethyl ester
8601 - Oxirane, 2,2′-[1,4-butanediylbis (oxymethylene)]bis-
9964 - Propanedioic acid, (1-methylethyl)-, diethyl ester
9981 - Propanedioic acid, propyl-, diethyl ester

$C_{10}H_{18}O_4S$
10330 - Propanoic acid, 3,3′-thiobis-, diethyl ester

$C_{10}H_{18}O_6$
3305 - Butanedioic acid, 2,3-dihydroxy-, bis(1-methylethyl) ester, (*R**,*R**)-(±)-
3310 - Butanedioic acid, 2,3-dihydroxy- [*R*-(*R**,*R**)]-, dipropyl ester
5714 - Ethanol, 2,2′-[1,2-ethanediylbis(oxy)]bis-, diacetate

$C_{10}H_{19}Br$
4321 - Cyclohexane, 1-bromo-4-methyl-1-(1-methylethyl)-
5086 - 1-Decene, 2-bromo-
5087 - 2-Decene, 1-bromo-

$C_{10}H_{19}BrO_2$
5061 - Decanoic acid, 2-bromo-

$C_{10}H_{19}ClO$
5078 - Decanoyl chloride

$C_{10}H_{19}FO_2$
5064 - Decanoic acid, 10-fluoro-

$C_{10}H_{19}N$
2927 - Bicyclo[2.2.1]heptane-1-methanamine, 7,7-dimethyl-
2928 - Bicyclo[3.1.1]heptane-2-methanamine, 6,6-dimethyl-, [1*S*-(1α,2β,5α)]-
5057 - Decanenitrile
7712 - 1-Naphthalenamine, decahydro-

$C_{10}H_{19}NO$
11331 - 2H-Quinolizine-1-methanol, octahydro-, (1*R*-*trans*)-

$C_{10}H_{19}NO_2$
9704 - 1-Piperidinepropanoic acid, ethyl ester

$C_{10}H_{19}N_5O$
3143 - *sec*-Bumeton

$C_{10}H_{19}N_5S$
9770 - Prometryn
11594 - Terbutryn

$C_{10}H_{19}O_6PS_2$
3316 - Butanedioic acid, [(dimethoxyphosphinothioyl)thio]-, diethyl ester

$C_{10}H_{20}$
4202 - Cyclodecane
4227 - Cycloheptane, 1,1,2-trimethyl-
4228 - Cycloheptane, 1,1,4-trimethyl-
4325 - Cyclohexane, butyl-
4369 - Cyclohexane, (1,1-dimethylethyl)-
4436 - Cyclohexane, 1-methyl-2-(1-methylethyl)-
4437 - Cyclohexane, 1-methyl-3-(1-methylethyl)-
4438 - Cyclohexane, 1-methyl-4-(1-methylethyl)-
4439 - Cyclohexane, 1-methyl-4-(1-methylethyl)-, *cis*-
4440 - Cyclohexane, 1-methyl-4-(1-methylethyl)-, *trans*-
4448 - Cyclohexane, (1-methylpropyl)-
4449 - Cyclohexane, 1-methyl-1-propyl-
4450 - Cyclohexane, 1-methyl-2-propyl-
4451 - Cyclohexane, 1-methyl-3-propyl-
4452 - Cyclohexane, 1-methyl-4-propyl-
4453 - Cyclohexane, (2-methylpropyl)-
4477 - Cyclohexane, 1,1,3,4-tetramethyl-
4478 - Cyclohexane, 1,1,3,5-tetramethyl-, *cis*-
4846 - Cyclopentane, pentyl-
5082 - 1-Decene
5083 - 4-Decene
5084 - 5-Decene, (*E*)-
5085 - 5-Decene, (*Z*)-
6559 - 3-Heptene, 4-propyl-
8185 - 1-Nonene, 2-methyl-
8186 - 3-Nonene, 2-methyl-
8502 - 1-Octene, 3,7-dimethyl-
8503 - 2-Octene, 2,6-dimethyl-

$C_{10}H_{20}BrF$
5035 - Decane, 1-bromo-10-fluoro-

$C_{10}H_{20}Br_2$
5038 - Decane, 1,10-dibromo-
5039 - Decane, 5,6-dibromo-

$C_{10}H_{20}ClF$
5037 - Decane, 1-chloro-10-fluoro-

$C_{10}H_{20}Cl_2$
5040 - Decane, 1,10-dichloro-

$C_{10}H_{20}NO_4PS$
10719 - Propetamphos

$C_{10}H_{20}N_2$
3098 - 2,3′-Bipiperidine

$C_{10}H_{20}N_2O_2$
7690 - Morpholine, 4,4′-(1,2-ethanediyl)bis-

$C_{10}H_{20}N_2S_2$
9659 - Piperidine, 1,1′-dithiobis-

$C_{10}H_{20}N_2S_3$
11714 - Thiodicarbonic diamide ([(H2N)C(*S*)]2*S*), tetraethyl-

$C_{10}H_{20}N_2S_4$
11721 - Thioperoxydicarbonic diamide ([(H2N)C(*S*)]2S2), tetraethyl-

$C_{10}H_{20}O$
4204 - Cyclodecanol
4233 - Cycloheptanol, 2,6,6-trimethyl-
4324 - Cyclohexanebutanol
4469 - Cyclohexanepropanol, α-methyl-
4504 - Cyclohexanol, 2-butyl-, *trans*-
4521 - Cyclohexanol, 2-(1,1-dimethylethyl)-
4546 - Cyclohexanol, 1-methyl-4-(1-methylethyl)-
4547 - Cyclohexanol, 2-methyl-5-(1-methylethyl)-, [1*R*-(1α,2α,5β)]-
4548 - Cyclohexanol, 5-methyl-2-(1-methylethyl)-, (1α,2β,5α)-(±)-
4550 - Cyclohexanol, 2-methyl-1-propyl-
4551 - Cyclohexanol, 3-methyl-1-propyl-
5030 - Decanal
5075 - 2-Decanone
5076 - 3-Decanone
5077 - 4-Decanone
5092 - 9-Decen-1-ol
6405 - Heptane, 3-[(ethenyloxy)methyl]-
7428 - *m*-Menthan-6-ol
8176 - 5-Nonanone, 2-methyl-
8401 - Octane, 1-(ethenyloxy)-
8514 - 6-Octen-1-ol, 3,7-dimethyl-, (±)-
8515 - 6-Octen-3-ol, 3,7-dimethyl-
8516 - 7-Octen-1-ol, 3,7-dimethyl-, (*S*)-

$C_{10}H_{20}O_2$
187 - Acetic acid, 2-ethylhexyl ester
222 - Acetic acid, octyl ester
3528 - Butanoic acid, hexyl ester
3546 - Butanoic acid, 3-methyl-, 1,1-dimethylpropyl ester
3551 - Butanoic acid, 3-methyl-, 3-methylbutyl ester
3559 - Butanoic acid, 1-methylpentyl ester, (*S*)-
5059 - Decanoic acid
6471 - Heptanoic acid, propyl ester
6506 - 2-Heptanol, 6-methyl-, acetate
6507 - 3-Heptanol, 6-methyl-, acetate
6880 - Hexanoic acid, butyl ester
6885 - Hexanoic acid, 2-ethyl-, ethyl ester
6975 - 3-Hexanone, 4-hydroxy-2,2,5,5-tetramethyl-
8153 - Nonanoic acid, 3-methyl-
8154 - Nonanoic acid, methyl ester
8347 - Octanal, 7-hydroxy-3,7-dimethyl-
8428 - Octanoic acid, ethyl ester
8447 - 2-Octanol, acetate, (±)-
8894 - Pentanoic acid, 4-methyl-, 2-methylpropyl ester
8913 - Pentanoic acid, pentyl ester
10264 - Propanoic acid, heptyl ester

$C_{10}H_{20}O_3$
3811 - 2-Butene, 1,1,3-triethoxy-

$C_{10}H_{20}O_4$
5689 - Ethanol, 2-(2-butoxyethoxy)-, acetate

$C_{10}H_{21}Br$
5034 - Decane, 1-bromo-

$C_{10}H_{21}Cl$
5036 - Decane, 1-chloro-

$C_{10}H_{21}ClO$
5073 - 1-Decanol, 10-chloro-

$C_{10}H_{21}F$
5052 - Decane, 1-fluoro-

$C_{10}H_{21}FO$
5074 - 1-Decanol, 10-fluoro-

$C_{10}H_{21}I$
5053 - Decane, 1-iodo-

$C_{10}H_{21}N$
4285 - Cyclohexanamine, *N*,*N*-diethyl-
4385 - Cyclohexaneethanamine, *N*,α-dimethyl-
9691 - Piperidine, 1,2,2,6,6-pentamethyl-
9692 - Piperidine, 1-pentyl-

$C_{10}H_{21}NO$
4502 - Cyclohexanol, 3-(aminomethyl)-3,5,5-trimethyl-
5031 - Decanamide

$C_{10}H_{21}NOS$
3976 - Carbamothioic acid, butylethyl-, *S*-propyl ester
12047 - Vernolate

$C_{10}H_{21}NO_2$
8084 - Nitrous acid, decyl ester

$C_{10}H_{21}NO_3$
8073 - Nitric acid, decyl ester

$C_{10}H_{21}N_3O$
9603 - 1-Piperazinecarboxamide, *N*,*N*-diethyl-4-methyl-

$C_{10}H_{22}$
5033 - Decane
6409 - Heptane, 3-ethyl-2-methyl-
6410 - Heptane, 3-ethyl-3-methyl-
6411 - Heptane, 3-ethyl-4-methyl-
6412 - Heptane, 3-ethyl-5-methyl-
6413 - Heptane, 4-ethyl-2-methyl-
6414 - Heptane, 4-ethyl-3-methyl-
6415 - Heptane, 4-ethyl-4-methyl-
6416 - Heptane, 5-ethyl-2-methyl-
6427 - Heptane, 4-(1-methylethyl)-
6434 - Heptane, 4-propyl-
6437 - Heptane, 2,2,3-trimethyl-
6438 - Heptane, 2,2,4-trimethyl-
6439 - Heptane, 2,2,5-trimethyl-
6440 - Heptane, 2,2,6-trimethyl-
6441 - Heptane, 2,3,3-trimethyl-
6442 - Heptane, 2,3,4-trimethyl-
6443 - Heptane, 2,3,5-trimethyl-
6444 - Heptane, 2,3,6-trimethyl-
6445 - Heptane, 2,4,4-trimethyl-
6446 - Heptane, 2,4,5-trimethyl-
6447 - Heptane, 2,4,6-trimethyl-
6448 - Heptane, 2,5,5-trimethyl-
6449 - Heptane, 3,3,4-trimethyl-

6450 - Heptane, 3,3,5-trimethyl-
6451 - Heptane, 3,4,4-trimethyl-
6452 - Heptane, 3,4,5-trimethyl-
6767 - Hexane, 3,3-diethyl-
6768 - Hexane, 3,4-diethyl-
6816 - Hexane, 3-ethyl-2,2-dimethyl-
6817 - Hexane, 3-ethyl-2,3-dimethyl-
6818 - Hexane, 3-ethyl-2,4-dimethyl-
6819 - Hexane, 3-ethyl-2,5-dimethyl-
6820 - Hexane, 3-ethyl-3,4-dimethyl-
6821 - Hexane, 4-ethyl-2,2-dimethyl-
6822 - Hexane, 4-ethyl-2,3-dimethyl-
6823 - Hexane, 4-ethyl-2,4-dimethyl-
6824 - Hexane, 4-ethyl-3,3-dimethyl-
6839 - Hexane, 2-methyl-3-(1-methylethyl)-
6844 - Hexane, 2,2,3,3-tetramethyl-
6845 - Hexane, 2,2,3,4-tetramethyl-
6846 - Hexane, 2,2,3,5-tetramethyl-
6847 - Hexane, 2,2,4,4-tetramethyl-
6848 - Hexane, 2,2,4,5-tetramethyl-
6849 - Hexane, 2,2,5,5-tetramethyl-
6850 - Hexane, 2,3,3,4-tetramethyl-
6851 - Hexane, 2,3,3,5-tetramethyl-
6852 - Hexane, 2,3,4,4-tetramethyl-
6853 - Hexane, 2,3,4,5-tetramethyl-
6854 - Hexane, 3,3,4,4-tetramethyl-
8140 - Nonane, 2-methyl-
8141 - Nonane, 3-methyl-
8142 - Nonane, 4-methyl-
8143 - Nonane, 5-methyl-
8373 - Octane, 2,2-dimethyl-
8374 - Octane, 2,3-dimethyl-
8375 - Octane, 2,4-dimethyl-
8376 - Octane, 2,5-dimethyl-
8377 - Octane, 2,6-dimethyl-
8378 - Octane, 2,7-dimethyl-
8379 - Octane, 3,3-dimethyl-
8380 - Octane, 3,4-dimethyl-
8381 - Octane, 3,5-dimethyl-
8382 - Octane, 3,6-dimethyl-
8383 - Octane, 4,4-dimethyl-
8384 - Octane, 4,5-dimethyl-
8403 - Octane, 3-ethyl-
8404 - Octane, 4-ethyl-
8757 - Pentane, 3,3-diethyl-2-methyl-
8766 - Pentane, 2,4-dimethyl-3-(1-methylethyl)-
8815 - Pentane, 3-ethyl-2,2,3-trimethyl-
8816 - Pentane, 3-ethyl-2,3,4-trimethyl-
8836 - Pentane, 2,2,3,3,4-pentamethyl-
8837 - Pentane, 2,2,3,4,4-pentamethyl-

$C_{10}H_{22}O$
3418 - Butane, 1,1′-oxybis[3-methyl-
5070 - 1-Decanol
5071 - 2-Decanol, (±)-
5072 - 4-Decanol
6488 - 3-Heptanol, 3-ethyl-2-methyl-
6489 - 3-Heptanol, 3-ethyl-6-methyl-
6490 - 3-Heptanol, 5-ethyl-4-methyl-
6508 - 4-Heptanol, 4-propyl-
6509 - 4-Heptanol, 2,4,6-trimethyl-
6948 - 1-Hexanol, 5-methyl-2-(1-methylethyl)-
6950 - 3-Hexanol, 2,2,5,5-tetramethyl-
8166 - 1-Nonanol, 3-methyl-
8167 - 1-Nonanol, 4-methyl-
8168 - 1-Nonanol, 8-methyl
8169 - 3-Nonanol, 2-methyl-
8170 - 4-Nonanol, 4-methyl-
8171 - 5-Nonanol, 5-methyl-
8402 - Octane, 1-ethoxy-
8449 - 1-Octanol, 3,7-dimethyl-, (±)-
8450 - 2-Octanol, 2,6-dimethyl-
8451 - 3-Octanol, 2,3-dimethyl-
8452 - 3-Octanol, 2,7-dimethyl-
8453 - 3-Octanol, 3,6-dimethyl-
8454 - 3-Octanol, 3,6-dimethyl-(+-)-
8455 - 4-Octanol, 2,4-dimethyl-
8456 - 4-Octanol, 2,5-dimethyl-
8457 - 4-Octanol, 2,6-dimethyl-
8458 - 4-Octanol, 2,7-dimethyl-
8459 - 3-Octanol, 3-ethyl-
8835 - Pentane, 1,1′-oxybis-

$C_{10}H_{22}O_2$
5047 - 1,10-Decanediol
5048 - 2,9-Decanediol
5527 - Ethane, 1,2-dibutoxy
6800 - 3,4-Hexanediol, 3,4-diethyl-
8392 - 1,7-Octanediol, 3,7-dimethyl-
9154 - Peroxide, bis(1,1-dimethylpropyl)
10032 - Propane, 1,1′-[1,2-ethanediylbis(oxy)]bis[2-methyl-
10041 - Propane, 1,1′-[ethylidenebis(oxy)]bis[2-methyl-

$C_{10}H_{22}O_3$
10078 - Propane, 1,1′,1″-[methylidynetris(oxy)]tris-
10079 - Propane, 2,2′,2″-[methylidynetris(oxy)]tris-

$C_{10}H_{22}O_4$
5690 - Ethanol, 2-[2-(2-butoxyethoxy)ethoxy]-

$C_{10}H_{22}O_5$
8995 - 2,5,8,11,14-Pentaoxapentadecane

$C_{10}H_{22}S$
3435 - Butane, 1,1′-thiobis[2-methyl-
3436 - Butane, 1,1′-thiobis[3-methyl-
5058 - 1-Decanethiol
8843 - Pentane, 1,1′-thiobis-

$C_{10}H_{22}S_2$
5243 - Disulfide, bis(3-methylbutyl)
5257 - Disulfide, dipentyl

$C_{10}H_{23}N$
3226 - 1-Butanamine, 3-methyl-*N*-(3-methylbutyl)-
5032 - 1-Decanamine
8708 - 1-Pentanamine, *N*-pentyl-

$C_{10}H_{23}NO$
5684 - Ethanol, 2-[bis(2-methylpropyl)amino]-
5703 - Ethanol, 2-(dibutylamino)-

$C_{10}H_{23}NO_2$
5415 - Ethanamine, 2,2-diethoxy-*N,N*-diethyl-

$C_{10}H_{24}N_2$
5515 - 1,2-Ethanediamine, *N,N,N′,N′*-tetraethyl-
6752 - 1,6-Hexanediamine, *N,N,N′,N′*-tetramethyl-

$C_{10}H_{24}N_4$
9614 - 1,4-Piperazinedipropanamine

$C_{10}H_{26}N_4$
3259 - 1,4-Butanediamine, *N,N′*-bis(3-aminopropyl)-

$C_{10}H_{30}O_3Si_4$
11646 - Tetrasiloxane, decamethyl-

$C_{10}H_{30}O_5Si_5$
4901 - Cyclopentasiloxane, decamethyl-

$C_{11}H_7BrO_2$
7761 - 1-Naphthalenecarboxylic acid, 5-bromo-
7769 - 2-Naphthalenecarboxylic acid, 5-bromo-
7846 - 1,4-Naphthalenedione, 2-bromo-3-methyl-

$C_{11}H_7ClO$
7750 - 2-Naphthalenecarbonyl chloride

$C_{11}H_7ClO_2$
7762 - 1-Naphthalenecarboxylic acid, 5-chloro-
7763 - 1-Naphthalenecarboxylic acid, 8-chloro-

$C_{11}H_7N$
7748 - 1-Naphthalenecarbonitrile
7749 - 2-Naphthalenecarbonitrile

$C_{11}H_7NO$
7904 - Naphthalene, 1-isocyanato-

$C_{11}H_7NO_4$
7767 - 1-Naphthalenecarboxylic acid, 5-nitro-

$C_{11}H_7NS$
7906 - Naphthalene, 1-isothiocyanato-

$C_{11}H_8O$
7751 - 1-Naphthalenecarboxaldehyde
7752 - 2-Naphthalenecarboxaldehyde

$C_{11}H_8OS$
7641 - Methanone, phenyl-2-thienyl-
7642 - Methanone, phenyl-3-thienyl-

$C_{11}H_8O_2$
7618 - Methanone, 2-furanylphenyl-
7754 - 1-Naphthalenecarboxaldehyde, 2-hydroxy-
7758 - 1-Naphthalenecarboxylic acid
7759 - 2-Naphthalenecarboxylic acid
7864 - 1,4-Naphthalenedione, 2-methyl-

$C_{11}H_8O_3$
7771 - 2-Naphthalenecarboxylic acid, 1-hydroxy-
7772 - 2-Naphthalenecarboxylic acid, 3-hydroxy-
7861 - 1,4-Naphthalenedione, 2-hydroxy-3-methyl-
7863 - 1,4-Naphthalenedione, 5-hydroxy-2-methyl-

$C_{11}H_8O_4$
7856 - 1,4-Naphthalenedione, 3,5-dihydroxy-2-methyl-

$C_{11}H_8O_5$
2408 - 5H-Benzocyclohepten-5-one, 2,3,4,6-tetrahydroxy-

$C_{11}H_9Br$
7742 - Naphthalene, 1-(bromomethyl)-
7743 - Naphthalene, 2-(bromomethyl)-

$C_{11}H_9Cl$
7783 - Naphthalene, 1-(chloromethyl)-
7784 - Naphthalene, 1-chloro-2-methyl-
7785 - Naphthalene, 2-(chloromethyl)-

$C_{11}H_9Cl_2NO_2$
510 - Barban

$C_{11}H_9N$
10997 - Pyridine, 2-phenyl-
10998 - Pyridine, 3-phenyl-
10999 - Pyridine, 4-phenyl-

$C_{11}H_9NO$
7637 - Methanone, phenyl-1H-pyrrol-2-yl-
7757 - 1-Naphthalenecarboxamide

$C_{11}H_9NO_2$
7760 - 1-Naphthalenecarboxylic acid, 5-amino-
7768 - 2-Naphthalenecarboxylic acid, 3-amino-
7925 - Naphthalene, 1-methyl-4-nitro-
7926 - Naphthalene, 2-methyl-1-nitro-

$C_{11}H_9NO_3$
11234 - 4-Quinolinecarboxylic acid, 6-methoxy-

$C_{11}H_{10}$
7915 - Naphthalene, 1-methyl-
7916 - Naphthalene, 2-methyl-

$C_{11}H_{10}NO_3$
7695 - Morpholine, 4-(2-hydroxybenzoyl)-

$C_{11}H_{10}N_2O$
90 - Acetamide, *N*-4-quinolinyl-
12022 - Urea, 1-naphthalenyl-
12023 - Urea, 2-naphthalenyl-

$C_{11}H_{10}N_2O_3$
11100 - 2,4,6(1H,3H,5H)-Pyrimidinetrione, 5-methyl-5-phenyl-

$C_{11}H_{10}N_2S$
11807 - Thiourea, 1-naphthalenyl-

$C_{11}H_{10}O$
4951 - 2-Cyclopenten-1-one, 3-phenyl-
7908 - 1-Naphthalenemethanol
7909 - 2-Naphthalenemethanol
7913 - Naphthalene, 1-methoxy-
7914 - Naphthalene, 2-methoxy-
8030 - 1-Naphthalenol, 4-methyl-
8031 - 2-Naphthalenol, 1-methyl-
8657 - 2,4-Pentadienal, 5-phenyl-

$C_{11}H_{10}O_2$
2820 - 2H-1-Benzopyran-2-one, 3-ethyl-
6158 - Furan, 2-(phenoxymethyl)-
7764 - 1-Naphthalenecarboxylic acid, 3,4-dihydro-
10741 - 2-Propynoic acid, 3-phenyl-, ethyl ester

$C_{11}H_{10}O_3$
2440 - 2-Benzofurancarboxylic acid, ethyl ester
10781 - 2H-Pyran-2,6(3H)-dione, dihydro-3-phenyl-
10782 - 2H-Pyran-2,6(3H)-dione, dihydro-4-phenyl-

$C_{11}H_{10}O_4$
2818 - 2H-1-Benzopyran-2-one, 5,7-dimethoxy-

$C_{11}H_{10}S$
11774 - Thiophene, 2-methyl-5-phenyl-

$C_{11}H_{11}BrN_2O$
10846 - 3H-Pyrazol-3-one, 4-bromo-1,2-dihydro-1,5-dimethyl-2-phenyl-

$C_{11}H_{11}BrO_2$
10613 - 2-Propenoic acid, 2-bromo-3-phenyl-, ethyl ester
10614 - 2-Propenoic acid, 3-bromo-3-phenyl-, ethyl ester, (*Z*)-
$C_{11}H_{11}Cl_3O_3$
259 - Acetic acid, (2,4,5-trichlorophenoxy)-, 1-methylethyl ester
$C_{11}H_{11}N$
7720 - 1-Naphthalenamine, 4-methyl-
7721 - 1-Naphthalenamine, *N*-methyl-
7722 - 2-Naphthalenamine, *N*-methyl-
7907 - 2-Naphthalenemethanamine
11136 - 1H-Pyrrole, 1-(phenylmethyl)-
11254 - Quinoline, 2,3-dimethyl-
11255 - Quinoline, 2,4-dimethyl-
11256 - Quinoline, 2,5-dimethyl-
11257 - Quinoline, 2,6-dimethyl-
11258 - Quinoline, 2,7-dimethyl-
11259 - Quinoline, 2,8-dimethyl-
11260 - Quinoline, 3,4-dimethyl-
11261 - Quinoline, 3,8-dimethyl-
11262 - Quinoline, 4,6-dimethyl-
11263 - Quinoline, 5,8-dimethyl-
11264 - Quinoline, 6,8-dimethyl-
11266 - Quinoline, 2-ethyl-
11267 - Quinoline, 3-ethyl-
11268 - Quinoline, 6-ethyl-
$C_{11}H_{11}NO$
10442 - 2-Propanone, 1-(1H-indol-3-yl)-
11316 - 4-Quinolinol, 2,8-dimethyl-
11317 - 6-Quinolinol, 2,4-dimethyl-
11318 - 8-Quinolinol, 2,4-dimethyl-
$C_{11}H_{11}NO_2$
893 - Benzeneacetic acid, α-cyano-, ethyl ester
11161 - 2,5-Pyrrolidinedione, 1-methyl-3-phenyl-
11163 - 2,5-Pyrrolidinedione, 1-(phenylmethyl)-
$C_{11}H_{11}NO_4S$
7407 - Isoxazolium, 2-ethyl-5-(3-sulfophenyl)-, hydroxide, inner salt
11293 - 5-Quinolinesulfonic acid, 8-ethoxy-
$C_{11}H_{11}NO_6$
7511 - Methanediol, (3-nitrophenyl)-, diacetate (ester)
$C_{11}H_{11}N_3O_2S$
2203 - Benzenesulfonamide, 4-amino-*N*-2-pyridinyl-
$C_{11}H_{12}$
7816 - Naphthalene, 1,2-dihydro-3-methyl-
7817 - Naphthalene, 1,2-dihydro-4-methyl-
$C_{11}H_{12}Cl_2N_2O_5$
52 - Acetamide, 2,2-dichloro-*N*-[2-hydroxy-1-(hydroxymethyl)-2-(4-nitrophenyl)ethyl]-, [*R*-(*R**,*R**)]-
$C_{11}H_{12}Cl_2O_3$
173 - Acetic acid, (2,4-dichlorophenoxy)-, 1-methylethyl ester
$C_{11}H_{12}Cl_3N$
1583 - Benzeneethanamine, α-methyl-*N*-(2,2,2-trichloroethylidene)-
$C_{11}H_{12}NO_4PS_2$
9515 - Phosmet
$C_{11}H_{12}N_2$
7800 - 1,4-Naphthalenediamine, 2-methyl-
10829 - 1H-Pyrazole, 3,5-dimethyl-1-phenyl-
$C_{11}H_{12}N_2O$
10847 - 3H-Pyrazol-3-one, 1,2-dihydro-1,5-dimethyl-2-phenyl-
11192 - Pyrrolo[2,1-b]quinazolin-3-ol, 1,2,3,9-tetrahydro-, (*R*)-
$C_{11}H_{12}N_2O_2$
11922 - *L*-Tryptophan
$C_{11}H_{12}N_2O_6$
2592 - Benzoic acid, 3,5-dinitro-, butyl ester
$C_{11}H_{12}N_4O_2S$
2199 - Benzenesulfonamide, 4-amino-*N*-(4-methyl-2-pyrimidinyl)-
$C_{11}H_{12}N_4O_3S$
2197 - Benzenesulfonamide, 4-amino-*N*-(6-methoxy-3-pyridazinyl)-
$C_{11}H_{12}O$
2407 - 5H-Benzocyclohepten-5-one, 6,7,8,9-tetrahydro-
3190 - Butanal, 2-(phenylmethylene)-
7288 - 1H-Inden-1-one, 2,3-dihydro-3,3-dimethyl-
7605 - Methanone, cyclobutylphenyl-
8049 - 1(2H)-Naphthalenone, 3,4-dihydro-2-methyl-
8050 - 1(2H)-Naphthalenone, 3,4-dihydro-4-methyl-
9110 - 1-Penten-3-one, 1-phenyl-
$C_{11}H_{12}O_2$
7778 - 2-Naphthalenecarboxylic acid, 1,2,3,4-tetrahydro-
7779 - 2-Naphthalenecarboxylic acid, 5,6,7,8-tetrahydro-
8048 - 1(2H)-Naphthalenone, 3,4-dihydro-6-methoxy-
8804 - 1,4-Pentanedione, 1-phenyl-
10685 - 2-Propenoic acid, 3-phenyl-, ethyl ester, (*E*)-
10706 - 2-Propen-1-ol, 3-phenyl-, acetate, (*E*)-
$C_{11}H_{12}O_3$
2141 - Benzenepropanoic acid, β-oxo-, ethyl ester
2428 - 1,3-Benzodioxole, 4-methoxy-6-(2-propenyl)-
6138 - 2-Furanol, tetrahydro-, benzoate
10647 - 2-Propenoic acid, 3-(2-methoxyphenyl)-, methyl ester
10601 - 2-Propenoic acid, 2-phenoxyethyl ester
$C_{11}H_{12}O_4$
7512 - Methanediol, phenyl-, diacetate
9976 - Propanedioic acid, phenyl-, dimethyl ester
$C_{11}H_{13}ClF_3N_3O_4S_3$
2857 - 2H-1,2,4-Benzothiadiazine-7-sulfonamide, 6-chloro-3,4-dihydro-2-methyl-3-[[(2,2,2-trifluoroethyl)thio]methyl]-
$C_{11}H_{13}ClO$
2889 - Benzoyl chloride, 4-butyl-
2895 - Benzoyl chloride, 4-(1,1-dimethylethyl)-
9255 - Phenol, 4-chloro-2-cyclopentyl-
$C_{11}H_{13}ClO_3$
3493 - Butanoic acid, 4-(4-chloro-2-methylphenoxy)-
$C_{11}H_{13}F_3N_2O_3S$
7427 - Mefluidide
$C_{11}H_{13}F_3N_4O_4$
5181 - Dinitramine
$C_{11}H_{13}N$
963 - Benzeneacetonitrile, α-(1-methylethyl)-
970 - Benzeneacetonitrile, α-propyl-
1823 - Benzenemethanamine, *N*-methyl-*N*-2-propynyl-
7309 - 1H-Indole, 3-ethyl-2-methyl-
7322 - 1H-Indole, 1,2,3-trimethyl-
7323 - 1H-Indole, 2,3,5-trimethyl-
10922 - Pyridine, 4-(3-cyclohexen-1-yl)-
$C_{11}H_{13}NO$
3703 - 3-Butenamide, 3-methyl-4-phenyl-
7332 - 2H-Indol-2-one, 1-ethyl-1,3-dihydro-3-methyl-
$C_{11}H_{13}NO_2$
5215 - 1,3-Dioxolo[4,5-g]isoquinoline, 5,6,7,8-tetrahydro-6-methyl-
$C_{11}H_{13}NO_3$
5217 - 1,3-Dioxolo[4,5-g]isoquinolin-5-ol, 5,6,7,8-tetrahydro-6-methyl-
7398 - 1(2H)-Isoquinolinone, 3,4-dihydro-6,7-dimethoxy-
$C_{11}H_{13}NO_4$
514 - Bendiocarb
2706 - Benzoic acid, 4-nitro-, butyl ester
8588 - 2-Oxazolidinone, 5-[(2-methoxyphenoxy)methyl]-
$C_{11}H_{13}N_3O$
10845 - 3H-Pyrazol-3-one, 4-amino-1,2-dihydro-1,5-dimethyl-2-phenyl-
$C_{11}H_{13}N_3O_3S$
2195 - Benzenesulfonamide, 4-amino-*N*-(3,4-dimethyl-5-isoxazolyl)-
$C_{11}H_{14}$
1254 - Benzene, cyclopentyl-
1643 - Benzene, 1-ethenyl-4-(1-methylethyl)-
1647 - Benzene, 2-ethenyl-1,3,5-trimethyl-
1699 - Benzene, (1-ethyl-1-propenyl)-
1700 - Benzene, (1-ethyl-2-propenyl)-
1950 - Benzene, (1-methyl-1-butenyl)-
2104 - Benzene, 1-pentenyl-
2105 - Benzene, 3-pentenyl-
2106 - Benzene, 4-pentenyl-
4748 - 1,3,5,7-Cyclooctatetraene, propyl-
7251 - 1H-Indene, 2,3-dihydro-1,1-dimethyl-
7252 - 1H-Indene, 2,3-dihydro-1,2-dimethyl-
7253 - 1H-Indene, 2,3-dihydro-4,7-dimethyl-
7254 - 1H-Indene, 2,3-dihydro-5,6-dimethyl-
7273 - 1H-Indene, 1-ethyl-2,3-dihydro-
7995 - Naphthalene, 1,2,3,4-tetrahydro-1-methyl-
7996 - Naphthalene, 1,2,3,4-tetrahydro-2-methyl-
7997 - Naphthalene, 1,2,3,4-tetrahydro-5-methyl-
7998 - Naphthalene, 1,2,3,4-tetrahydro-6-methyl-
$C_{11}H_{14}ClNO$
9771 - Propachlor
$C_{11}H_{14}ClNO_2$
607 - Benzamide, *N*-butyl-4-chloro-2-hydroxy-
$C_{11}H_{14}N_2$
7310 - 1H-Indole-3-methanamine, *N*,*N*-dimethyl-
$C_{11}H_{14}N_2O$
7304 - 1H-Indole-3-ethanamine, 5-methoxy-
$C_{11}H_{14}N_2O_4$
84 - Acetamide, *N*-(5-nitro-2-propoxyphenyl)-
$C_{11}H_{14}O$
598 - Benzaldehyde, 2,3,5,6-tetramethyl-
1665 - Benzene, 1-ethoxy-2-(1-propenyl)-
1666 - Benzene, (3-ethoxy-1-propenyl)-
3671 - 1-Butanone, 3-methyl-1-phenyl-
3672 - 1-Butanone, 1-(4-methylphenyl)-
4871 - Cyclopentanol, 1-phenyl-
5848 - Ethanone, 1-[4-(1-methylethyl)phenyl]-
5875 - Ethanone, 1-(2,4,5-trimethylphenyl)-
5876 - Ethanone, 1-(2,4,6-trimethylphenyl)-
5877 - Ethanone, 1-(3,4,5-trimethylphenyl)-
8986 - 1-Pentanone, 1-phenyl-
10431 - 1-Propanone, 2,2-dimethyl-1-phenyl-
$C_{11}H_{14}O_2$
900 - Benzeneacetic acid, α-ethyl-, methyl ester
930 - Benzeneacetic acid, α-(1-methylethyl)-
943 - Benzeneacetic acid, α-propyl-, (±)-
1104 - Benzenebutanoic acid, γ-methyl-
1449 - Benzene, 1,2-dimethoxy-4-(1-propenyl)-
1450 - Benzene, 1,2-dimethoxy-4-(2-propenyl)-
1451 - Benzene, 1,2-dimethoxy-4-(1-propenyl)-, (*Z*)-
2099 - Benzenepentanoic acid
2130 - Benzenepropanoic acid, β,β-dimethyl-
2131 - Benzenepropanoic acid, ethyl ester
2532 - Benzoic acid, butyl ester
2588 - Benzoic acid, 4-(1,1-dimethylethyl)-
2694 - Benzoic acid, 4-(1-methylethyl)-, methyl ester
2701 - Benzoic acid, 2-methylpropyl ester
3585 - Butanoic acid, phenylmethyl ester
5194 - 1,3-Dioxane, 4-methyl-4-phenyl-
9496 - Phenol, 2,4,5-trimethyl-, acetate
10307 - Propanoic acid, 2-methyl-, phenylmethyl ester
10324 - Propanoic acid, 2-phenylethyl ester
10436 - 1-Propanone, 3-ethoxy-1-phenyl-
10447 - 1-Propanone, 1-(2-hydroxy-5-methylphenyl)-2-methyl-
10448 - 1-Propanone, 1-(4-hydroxy-3-methylphenyl)-2-methyl-
$C_{11}H_{14}O_3$
538 - Benzaldehyde, 3,4-diethoxy-
923 - Benzeneacetic acid, 4-methoxy-, ethyl ester
1128 - Benzenecarboperoxoic acid, 1,1-dimethylethyl ester
2598 - Benzoic acid, 2-ethoxy-, ethyl ester
2599 - Benzoic acid, 4-ethoxy-, ethyl ester
2614 - Benzoic acid, 2-hydroxy-, butyl ester

2615 - Benzoic acid, 4-hydroxy-, butyl ester
2641 - Benzoic acid, 2-hydroxy-3-methyl-6-(1-methylethyl)-
2642 - Benzoic acid, 2-hydroxy-6-methyl-3-(1-methylethyl)-
2643 - Benzoic acid, 2-hydroxy-, 2-methylpropyl ester
3663 - 2-Butanone, 4-(4-hydroxy-3-methoxyphenyl)-
10321 - Propanoic acid, 2-phenoxy-, ethyl ester
10322 - Propanoic acid, 3-phenoxy-, ethyl ester

$C_{11}H_{14}O_4$
2578 - Benzoic acid, 3,4-dimethoxy-, ethyl ester
5874 - Ethanone, 1-(2,3,4-trimethoxyphenyl)-

$C_{11}H_{15}BrClO_3PS$
9762 - Profenofos

$C_{11}H_{15}BrN_2O_3$
11079 - 2,4,6(1H,3H,5H)-Pyrimidinetrione, 5-(2-bromo-2-propenyl)-1-methyl-5-(1-methylethyl)-
11080 - 2,4,6(1H,3H,5H)-Pyrimidinetrione, 5-(2-bromo-2-propenyl)-5-(1-methylpropyl)-

$C_{11}H_{15}Cl_2O_3PS_2$
4092 - Chlorthiophos

$C_{11}H_{15}N$
9693 - Piperidine, 1-phenyl-
9694 - Piperidine, 3-phenyl-
9695 - Piperidine, 4-phenyl-
11172 - Pyrrolidine, 1-(phenylmethyl)-

$C_{11}H_{15}NO$
74 - Acetamide, *N*-(1-methylethyl)-*N*-phenyl-
88 - Acetamide, *N*-phenyl-*N*-propyl-
539 - Benzaldehyde, 4-(diethylamino)-
882 - Benzeneacetamide, α-(1-methylethyl)-
7697 - Morpholine, 3-methyl-2-phenyl-
7698 - Morpholine, 4-(4-methylphenyl)-
7700 - Morpholine, 4-(phenylmethyl)-
8960 - 1-Pentanone, 1-(4-aminophenyl)-

$C_{11}H_{15}NO_2$
2488 - Benzoic acid, 4-amino-, butyl ester
2505 - Benzoic acid, 4-amino-, 2-methylpropyl ester
3968 - Carbamic acid, phenyl-, 2-methylpropyl ester
6272 - Glycine, *N*-(phenylmethyl)-, ethyl ester
7397 - 6-Isoquinolinol, 1,2,3,4-tetrahydro-7-methoxy-1-methyl-, (*S*)-
9512 - *L*-Phenylalanine, ethyl ester

$C_{11}H_{15}NO_2S$
7646 - Methiocarb

$C_{11}H_{15}NO_3$
7393 - 8-Isoquinolinol, 1,2,3,4-tetrahydro-6,7-dimethoxy-
9796 - Propanamide, *N*-(4-ethoxyphenyl)-2-hydroxy-
10722 - Propoxur

$C_{11}H_{15}N_2NaO_2S$
11063 - 4,6(1H,5H)-Pyrimidinedione, dihydro-5-(2-methylpropyl)-5-(2-propenyl)-2-thioxo-, monosodium salt

$C_{11}H_{16}$
1121 - Benzene, 1-butyl-2-methyl-
1122 - Benzene, 1-butyl-3-methyl-
1123 - Benzene, 1-butyl-4-methyl-
1416 - Benzene, 1,2-diethyl-3-methyl-
1417 - Benzene, 1,2-diethyl-4-methyl-
1418 - Benzene, 1,3-diethyl-2-methyl-
1419 - Benzene, 1,3-diethyl-5-methyl-
1420 - Benzene, 1,4-diethyl-2-methyl-
1421 - Benzene, 2,4-diethyl-1-methyl-
1464 - Benzene, 1-(1,1-dimethylethyl)-2-methyl-
1465 - Benzene, 1-(1,1-dimethylethyl)-3-methyl-
1466 - Benzene, 1-(1,1-dimethylethyl)-4-methyl-
1469 - Benzene, 1,2-dimethyl-3-(1-methylethyl)-
1470 - Benzene, 1,2-dimethyl-4-(1-methylethyl)-
1471 - Benzene, 1,3-dimethyl-2-(1-methylethyl)-
1472 - Benzene, 2,4-dimethyl-1-(1-methylethyl)-
1480 - Benzene, (1,1-dimethylpropyl)-
1481 - Benzene, (1,2-dimethylpropyl)-
1482 - Benzene, 1,3-dimethyl-2-propyl-
1483 - Benzene, 1,3-dimethyl-5-propyl-
1484 - Benzene, (2,2-dimethylpropyl)-
1485 - Benzene, 1,2-dimethyl-3-propyl-
1486 - Benzene, 1,4-dimethyl-2-propyl-
1487 - Benzene, 2,4-dimethyl-1-propyl-
1688 - Benzene, 1-ethyl-2-(1-methylethyl)-
1689 - Benzene, 1-ethyl-3-(1-methylethyl)-
1690 - Benzene, 1-ethyl-4-(1-methylethyl)-
1701 - Benzene, (1-ethylpropyl)-
1702 - Benzene, 1-ethyl-2-propyl-
1703 - Benzene, 1-ethyl-4-propyl-
1708 - Benzene, 1-ethyl-2,3,4-trimethyl-
1709 - Benzene, 1-ethyl-2,3,5-trimethyl-
1710 - Benzene, 1-ethyl-2,4,5-trimethyl-
1711 - Benzene, 2-ethyl-1,3,4-trimethyl-
1712 - Benzene, 2-ethyl-1,3,5-trimethyl-
1713 - Benzene, 5-ethyl-1,2,3-trimethyl-
1952 - Benzene, (1-methylbutyl)-
1953 - Benzene, (2-methylbutyl)-
1954 - Benzene, (3-methylbutyl)-
1994 - Benzene, 1-methyl-2-(1-methylpropyl)-
1995 - Benzene, 1-methyl-2-(2-methylpropyl)-
1996 - Benzene, 1-methyl-3-(1-methylpropyl)-
1997 - Benzene, 1-methyl-3-(2-methylpropyl)-
1998 - Benzene, 1-methyl-4-(1-methylpropyl)-
1999 - Benzene, 1-methyl-4-(2-methylpropyl)-
2097 - Benzene, pentamethyl-
2107 - Benzene, pentyl-
8341 - 1,5-Octadien-3-yne, 5-propyl-
11934 - 1,10-Undecadiyne
12057 - *o*-Xylene, 4-propyl-

$C_{11}H_{16}ClN_3O_2$
6001 - Formetanate hydrochloride

$C_{11}H_{16}ClN_3O_4S_2$
2856 - 2H-1,2,4-Benzothiadiazine-7-sulfonamide, 6-chloro-3,4-dihydro-3-(2-methylpropyl)-, 1,1-dioxide

$C_{11}H_{16}ClO_2PS_3$
4037 - Carbophenothion

$C_{11}H_{16}N_2$
9621 - Piperazine, 1-(phenylmethyl)-

$C_{11}H_{16}N_2O_2$
6151 - 2(3H)-Furanone, 3-ethyldihydro-4-[(1-methyl-1H-imidazol-5-yl)methyl]-, (3*S*-*cis*)-

$C_{11}H_{16}N_2O_3$
11099 - 2,4,6(1H,3H,5H)-Pyrimidinetrione, 1-methyl-5-(1-methylethyl)-5-(2-propenyl)-
11101 - 2,4,6(1H,3H,5H)-Pyrimidinetrione, 5-(2-methylpropyl)-5-(2-propenyl)-

$C_{11}H_{16}O$
363 - Anisole, 3-isopropyl-2-methyl-
1106 - Benzenebutanol, α-methyl-
1107 - Benzenebutanol, β-methyl-, (±)-
1115 - Benzene, (butoxymethyl)-
1116 - Benzene, 1-butoxy-3-methyl-
1117 - Benzene, 1-butoxy-4-methyl-
1462 - Benzene, 1-(1,1-dimethylethyl)-4-methoxy-
1611 - Benzeneethanol, β-(1-methylethyl)-
1616 - Benzeneethanol, α,α,β-trimethyl-
1617 - Benzeneethanol, α,β,β-trimethyl-
1667 - Benzene, 1-ethoxy-2-propyl-
1846 - Benzenemethanol, α-butyl-
1855 - Benzenemethanol, α,α-diethyl-
1864 - Benzenemethanol, α-(1,1-dimethylethyl)-
1865 - Benzenemethanol, 4-(1,1-dimethylethyl)-
1904 - Benzenemethanol, α-methyl-α-propyl-
1905 - Benzenemethanol, α-(2-methylpropyl)-
2033 - Benzene, [(2-methylpropoxy)methyl]-
2101 - Benzenepentanol
2109 - Benzene, (pentyloxy)-
2150 - Benzenepropanol, α,α-dimethyl-
2151 - Benzenepropanol, β,β-dimethyl-
2152 - Benzenepropanol, α-ethyl-, (*S*)-
4949 - 2-Cyclopenten-1-one, 3-methyl-2-(2-pentenyl)-, (*Z*)-
9250 - Phenol, 2-butyl-4-methyl-
9330 - Phenol, 2-(1,1-dimethylethyl)-4-methyl-
9331 - Phenol, 2-(1,1-dimethylethyl)-5-methyl-
9332 - Phenol, 2-(1,1-dimethylethyl)-6-methyl-
9333 - Phenol, 4-(1,1-dimethylethyl)-2-methyl-
9339 - Phenol, 4-(1,1-dimethylpropyl)-
9404 - Phenol, 4-(3-methylbutyl)-
9421 - Phenol, 5-methyl-2-(1-methylpropyl)-
9443 - Phenol, pentamethyl-
9444 - Phenol, 4-pentyl-

$C_{11}H_{16}O_2$
1105 - Benzenebutanol, 4-methoxy-
1410 - Benzene, 1,4-diethoxy-2-methyl-
1442 - Benzene, 1,4-dimethoxy-2-(1-methylethyl)-
1534 - 1,3-Benzenediol, 4-(3-methylbutyl)-
1544 - 1,3-Benzenediol, 4-pentyl-
9329 - Phenol, (1,1-dimethylethyl)-4-methoxy-
11864 - Tricyclo[2.2.1.02,6]heptane-1-carboxylic acid, 7,7-dimethyl-, methyl ester

$C_{11}H_{16}O_3$
1409 - Benzene, (diethoxymethoxy)-
2917 - Bicyclo[2.2.1]heptane-2-carboxylic acid, 4,7,7-trimethyl-3-oxo-
4641 - 2-Cyclohexene-1-carboxylic acid, 2,6-dimethyl-4-oxo-, ethyl ester
6058 - 2-Furancarboxylic acid, hexyl ester

$C_{11}H_{16}O_3S$
2230 - Benzenesulfonic acid, 4-methyl-, butyl ester

$C_{11}H_{16}O_4$
4915 - 1-Cyclopentene-1,2-dicarboxylic acid, diethyl ester
4916 - 1-Cyclopentene-1,3-dicarboxylic acid, diethyl ester

$C_{11}H_{17}Cl_3N_2O_2S$
7220 - 2,4-Imidazolidinedione, 5-(1-ethylpentyl)-3-[(trichloromethyl)thio]-

$C_{11}H_{17}N$
362 - Aniline, *p*-isopentyl-
696 - Benzenamine, *N*,*N*-diethyl-2-methyl-
697 - Benzenamine, *N*,*N*-diethyl-4-methyl-
783 - Benzenamine, *N*-(3-methylbutyl)-
833 - Benzenamine, 2,3,4,5,6-pentamethyl-
1576 - Benzeneethanamine, *N*-ethyl-α-methyl-

$C_{11}H_{17}NO$
1863 - Benzenemethanol, α-[1-(dimethylamino)ethyl]-, [*R*-(*R**,*S**)]-
1871 - Benzenemethanol, α-[1-(ethylamino)ethyl]-, (*R**,*S**)-
5724 - Ethanol, 2-[ethyl(3-methylphenyl)amino]-
10392 - 1-Propanol, 3-[methyl(phenylmethyl)amino]-

$C_{11}H_{17}NO_3$
1522 - 1,2-Benzenediol, 4-[1-hydroxy-2-[(1-methylethyl)amino]ethyl]-
1586 - Benzeneethanamine, 3,4,5-trimethoxy-

$C_{11}H_{17}N_3O_3$
11096 - 2,4,6(1H,3H,5H)-Pyrimidinetrione, 5-ethyl-5-(1-piperidinyl)-

$C_{11}H_{17}N_3O_3S$
2192 - Benzenesulfonamide, 4-amino-*N*-[(butylamino)carbonyl]-

$C_{11}H_{17}O_3P$
9545 - Phosphonic acid, (phenylmethyl)-, diethyl ester

$C_{11}H_{17}O_4PS_2$
5954 - Fensulfothion

$C_{11}H_{18}$
7792 - Naphthalene, decahydro-2-methylene-, *trans*-
11983 - 1-Undecen-3-yne

$C_{11}H_{18}N_2$
5504 - 1,2-Ethanediamine, *N*,*N*-dimethyl-*N*′-(phenylmethyl)-
7162 - Hydrazine, 1-(3-methylbutyl)-1-phenyl-

$C_{11}H_{18}N_2O_3$
11092 - 2,4,6(1H,3H,5H)-Pyrimidinetrione, 5-ethyl-5-(3-methylbutyl)-

$C_{11}H_{18}N_4O_2$
9728 - Pirimicarb

$C_{11}H_{18}O$
2959 - Bicyclo[2.2.1]heptan-2-one, 1,4,7,7-tetramethyl-
2969 - Bicyclo[3.1.1]hept-2-ene-2-ethanol, 6,6-dimethyl-

4950 - 2-Cyclopenten-1-one, 3-methyl-2-pentyl-

$C_{11}H_{18}O_2$
2949 - Bicyclo[2.2.1]heptan-2-ol, 1,7,7-trimethyl-, formate, (1*R*-*endo*)-
3131 - 3-Bornanecarboxylic acid
4965 - Cyclopropanecarboxylic acid, 2,2-dimethyl-3-(2-methyl-1-propenyl)-, methyl ester, *trans*-(±)-
5026 - 2,4-Decadienoic acid, methyl ester, (*E,E*)-
5027 - 2,4-Decadienoic acid, methyl ester, (*E,Z*)-
5028 - 2,4-Decadienoic acid, methyl ester, (*Z,E*)-
5029 - 2,4-Decadienoic acid, methyl ester, (*Z,Z*)-
8337 - 1,6-Octadien-3-ol, 3,7-dimethyl-, formate
8338 - 2,6-Octadien-1-ol, 3,7-dimethyl-, formate, (*E*)-
11988 - 6-Undecynoic acid
11989 - 9-Undecynoic acid

$C_{11}H_{18}O_2Si$
11407 - Silane, diethoxymethylphenyl-

$C_{11}H_{18}O_3Si$
11427 - Silane, ethenyltris(2-propenyloxy)-

$C_{11}H_{18}O_4$
4791 - 1,1-Cyclopentanedicarboxylic acid, diethyl ester

$C_{11}H_{18}O_5$
8771 - Pentanedioic acid, 2-acetyl-, diethyl ester
9972 - Propanedioic acid, (1-oxobutyl)-, diethyl ester

$C_{11}H_{18}O_6$
5654 - 1,1,2-Ethanetricarboxylic acid, triethyl ester

$C_{11}H_{19}ClO$
11982 - 10-Undecenoyl chloride

$C_{11}H_{19}NOS$
8528 - Octhilinone

$C_{11}H_{19}NO_2$
8148 - Nonanoic acid, 9-cyano-, methyl ester

$C_{11}H_{19}N_3O$
5932 - Ethirimol

$C_{11}H_{20}$
4340 - Cyclohexane, cyclopentyl-
8201 - 4-Nonyne, 3,3-dimethyl-
11537 - Spiro[5.5]undecane
11984 - 1-Undecyne
11985 - 2-Undecyne
11986 - 4-Undecyne
11987 - 5-Undecyne

$C_{11}H_{20}ClN_5$
11828 - 1,3,5-Triazine-2,4-diamine, 6-chloro-*N,N,N′,N′*-tetraethyl-

$C_{11}H_{20}N_2O_3$
7224 - 4-Imidazolidinehexanoic acid, 5-methyl-2-oxo-, methyl ester, (4*R*-*cis*)-

$C_{11}H_{20}N_3O_3PS$
9730 - Pirimiphos-methyl

$C_{11}H_{20}O$
4592 - Cyclohexanone, 4-(1,1-dimethylpropyl)-
4888 - Cyclopentanone, 3-methyl-2-(3-methylbutyl)-

$C_{11}H_{20}O_2$
6152 - 2(3H)-Furanone, 5-heptyldihydro-
6404 - 3,5-Heptanedione, 2,2,6,6-tetramethyl-
8149 - Nonanoic acid, ethenyl ester
10636 - 2-Propenoic acid, 2-ethylhexyl ester
11974 - 10-Undecenoic acid
11975 - 9-Undecenoic acid, (*E*)-
11976 - 9-Undecenoic acid, (*Z*)-

$C_{11}H_{20}O_3$
8436 - Octanoic acid, 2-methyl-3-oxo-, ethyl ester

$C_{11}H_{20}O_4$
6394 - Heptanedioic acid, diethyl ester
6700 - Hexanedioic acid, 3-methyl-, diethyl ester
8134 - Nonanedioic acid, dimethyl ester
9925 - Propanedioic acid, bis(1,1-dimethylethyl) ester
9930 - Propanedioic acid, butyl-, diethyl ester
9938 - Propanedioic acid, dibutyl ester
9940 - Propanedioic acid, diethyl-, diethyl ester
9967 - Propanedioic acid, (1-methylpropyl)-, diethyl ester
9969 - Propanedioic acid, (2-methylpropyl)-, diethyl ester
9971 - Propanedioic acid, octyl-

$C_{11}H_{20}O_{10}$
6234 - *D*-Glucose, 6-*O*-α-*L*-arabinopyranosyl-
6237 - *D*-Glucose, 6-*O*-β-*D*-xylopyranosyl-

$C_{11}H_{21}BrO_2$
11944 - Undecanoic acid, 11-bromo-
11945 - Undecanoic acid, 2-bromo-

$C_{11}H_{21}FO_2$
11947 - Undecanoic acid, 11-fluoro-

$C_{11}H_{21}N$
11941 - Undecanenitrile

$C_{11}H_{21}NOS$
4168 - Cycloate

$C_{11}H_{21}NO_4$
330 - β-Alanine, *N*-(3-ethoxy-3-oxopropyl)-*N*-methyl-, ethyl ester

$C_{11}H_{21}N_5S$
5222 - Dipropetryn

$C_{11}H_{22}$
4427 - Cyclohexane, (3-methylbutyl)-
4465 - Cyclohexane, pentyl-
4823 - Cyclopentane, hexyl-
11966 - 1-Undecene
11967 - 3-Undecene
11968 - 2-Undecene, (*E*)-
11969 - 2-Undecene, (*Z*)-
11970 - 4-Undecene, (*E*)-
11971 - 4-Undecene, (*Z*)-
11972 - 5-Undecene, (*E*)-
11973 - 5-Undecene, (*Z*)-

$C_{11}H_{22}BrF$
11939 - Undecane, 1-bromo-11-fluoro-

$C_{11}H_{22}Br_2$
11940 - Undecane, 1,11-dibromo-

$C_{11}H_{22}N_2$
9683 - Piperidine, 1,1′-methylenebis-

$C_{11}H_{22}O$
11935 - Undecanal
11958 - 2-Undecanone
11959 - 3-Undecanone
11960 - 4-Undecanone
11961 - 5-Undecanone
11962 - 6-Undecanone
11979 - 10-Undecen-1-ol
11980 - 2-Undecen-4-ol
11981 - 9-Undecen-1-ol

$C_{11}H_{22}O_2$
220 - Acetic acid, nonyl ester
3527 - Butanoic acid, heptyl ester
5066 - Decanoic acid, methyl ester
6459 - Heptanoic acid, butyl ester
6467 - Heptanoic acid, 2-methylpropyl ester
6897 - Hexanoic acid, 3-methylbutyl ester
6907 - Hexanoic acid, pentyl ester
8150 - Nonanoic acid, ethyl ester
8435 - Octanoic acid, 1-methylethyl ester
8441 - Octanoic acid, propyl ester
8882 - Pentanoic acid, hexyl ester
10313 - Propanoic acid, octyl ester
11943 - Undecanoic acid

$C_{11}H_{22}O_3$
3635 - 1-Butanol, 3-methyl-, carbonate (2:1)
11948 - Undecanoic acid, 11-hydroxy-

$C_{11}H_{22}O_6$
6228 - α-*D*-Glucopyranoside, methyl 2,3,4,6-tetra-*O*-methyl-

$C_{11}H_{23}N$
9671 - Piperidine, 1-(1-ethyl-1-methylpropyl)-
9672 - Piperidine, 1-hexyl-

$C_{11}H_{23}NOS$
11586 - Sutan

$C_{11}H_{24}$
5054 - Decane, 2-methyl-
5055 - Decane, 3-methyl-
5056 - Decane, 4-methyl-
8420 - Octane, 2,2,4-trimethyl-
8421 - Octane, 2,4,7-trimethyl
11938 - Undecane

$C_{11}H_{24}O$
8164 - 5-Nonanol, 2,8-dimethyl-
11950 - 1-Undecanol
11951 - 2-Undecanol, (±)-
11952 - 3-Undecanol, (*R*)-
11953 - 5-Undecanol
11954 - 6-Undecanol

$C_{11}H_{24}O_3Si$
11426 - Silane, ethenyltris(1-methylethoxy)-

$C_{11}H_{24}S$
11942 - 1-Undecanethiol

$C_{11}H_{25}N$
11937 - 1-Undecanamine

$C_{11}H_{26}O_3Si$
11480 - Silane, triethoxypentyl-

$C_{11}H_{26}O_6P_2$
9546 - Phosphonic acid, 1,3-propanediylbis-, tetraethyl ester

$C_{11}H_{28}Br_2N_2$
8738 - 1,5-Pentanediaminium, *N,N,N,N′,N′,N′*-hexamethyl-, dibromide

$C_{12}F_{10}$
3027 - 1,1′-Biphenyl, 2,2′,3,3′,4,4′,5,5′,6,6′-decafluoro-

$C_{12}F_{27}N$
3228 - 1-Butanamine, 1,1,2,2,3,3,4,4,4-nonafluoro-*N,N*-bis(nonafluorobutyl)-

$C_{12}H_4N_6O_{12}S$
2288 - Benzene, 1,1′-thiobis[2,4,6-trinitro-

$C_{12}H_5N_7O_{12}$
864 - Benzenamine, 2,4,6-trinitro-*N*-(2,4,6-trinitrophenyl)-

$C_{12}H_6Cl_2O_2$
7804 - 1,8-Naphthalenedicarbonyl dichloride

$C_{12}H_6Cl_4O_2S$
2325 - Benzene, 1,2,4-trichloro-5-[(4-chlorophenyl)sulfonyl]-
9478 - Phenol, 2,2′-thiobis[4,6-dichloro-

$C_{12}H_6Cl_4S$
11648 - Tetrasul

$C_{12}H_6O_2$
8 - 1,2-Acenaphthylenedione

$C_{12}H_6O_{12}$
1748 - Benzenehexacarboxylic acid

$C_{12}H_7BrO$
5164 - Dibenzofuran, 2-bromo-
5165 - Dibenzofuran, 3-bromo-

$C_{12}H_7Cl_2NO_3$
8081 - Nitrofen

$C_{12}H_7NO_3$
5166 - Dibenzofuran, 3-nitro-

$C_{12}H_7NO_4$
9508 - 3H-Phenoxazin-3-one, 7-hydroxy-, 10-oxide

$C_{12}H_8$
3 - Acenaphthylene
7892 - Naphthalene, 1-ethynyl-

$C_{12}H_8Br_2$
3037 - 1,1′-Biphenyl, 4,4′-dibromo-

$C_{12}H_8Br_2O$
2072 - Benzene, 1,1′-oxybis[4-bromo-

$C_{12}H_8Br_2O_2S$
2238 - Benzene, 1,1′-sulfonylbis[4-bromo-

$C_{12}H_8Br_2S$
2284 - Benzene, 1,1′-thiobis[4-bromo-

$C_{12}H_8Cl_2$
3040 - 1,1′-Biphenyl, 3,3′-dichloro-
3041 - 1,1′-Biphenyl, 4,4′-dichloro-

$C_{12}H_8Cl_2O$
2073 - Benzene, 1,1′-oxybis[4-chloro-

$C_{12}H_8Cl_2O_2S$
2239 - Benzene, 1,1′-sulfonylbis[4-chloro-
9477 - Phenol, 2,2′-thiobis[4-chloro-

$C_{12}H_8Cl_2O_3S$
2221 - Benzenesulfonic acid, 4-chloro-, 4-chlorophenyl ester
9289 - Phenol, 2,4-dichloro-, benzenesulfonate

$C_{12}H_8Cl_6$
5177 - 1,4 5,8-Dimethanonaphthalene,1,2,3,4,10,10-hexachloro-1,4,4a,5,8,8a-hexahydro-,
$C_{12}H_8Cl_6O$
5178 - 2,7 3,6-Dimethanonaphth[2,3-b]oxirene, 3,4,5,6,9,9-hexachloro-1a,2,2a,3,6,6a,7,7a-octahydro-,(1aα,2β,2aα,3β,6β,6aα,7β,7aα)-
$C_{12}H_8F_2$
3042 - 1,1′-Biphenyl, 2,2′-difluoro-
3043 - 1,1′-Biphenyl, 3,3′-difluoro-
3044 - 1,1′-Biphenyl, 4,4′-difluoro-
$C_{12}H_8N_2$
9189 - 1,10-Phenanthroline
9190 - 1,7-Phenanthroline
9191 - 4,7-Phenanthroline
9193 - Phenazine
$C_{12}H_8N_2O$
9197 - 1-Phenazinol
$C_{12}H_8N_2O_4$
3062 - 1,1′-Biphenyl, 2,2′-dinitro-
3063 - 1,1′-Biphenyl, 2,4′-dinitro-
$C_{12}H_8N_2O_4S_2$
5247 - Disulfide, bis(3-nitrophenyl)
$C_{12}H_8N_2O_5$
1495 - Benzene, 2,4-dinitro-1-phenoxy-
$C_{12}H_8O$
5163 - Dibenzofuran
$C_{12}H_8O_3$
8060 - Naphtho[1,2-c]furan-1,3-dione, 4,5-dihydro-
$C_{12}H_8O_4$
6187 - 7H-Furo[3,2-g][1]benzopyran-7-one, 9-methoxy-
7805 - 1,8-Naphthalenedicarboxylic acid
$C_{12}H_8S$
5167 - Dibenzothiophene
$C_{12}H_8S_2$
11660 - Thianthrene
$C_{12}H_8S_3$
11600 - 2,2′:5′,2″-Terthiophene
$C_{12}H_8Se_2$
11353 - Selenanthrene
$C_{12}H_9Br$
4 - Acenaphthylene, 5-bromo-1,2-dihydro-
3016 - 1,1′-Biphenyl, 2-bromo-
3017 - 1,1′-Biphenyl, 3-bromo-
3018 - 1,1′-Biphenyl, 4-bromo-
$C_{12}H_9BrN_2O$
5120 - Diazene, (4-bromophenyl)phenyl-, 1-oxide
$C_{12}H_9BrO$
1079 - Benzene, 1-bromo-4-phenoxy-
$C_{12}H_9Cl$
5 - Acenaphthylene, 5-chloro-1,2-dihydro-
3023 - 1,1′-Biphenyl, 2-chloro-
3024 - 1,1′-Biphenyl, 3-chloro-
3025 - 1,1′-Biphenyl, 4-chloro-
$C_{12}H_9ClF_3N_3O$
8213 - Norflurazon
$C_{12}H_9ClO$
1221 - Benzene, 1-chloro-4-phenoxy-
3092 - [1,1′-Biphenyl]-2-ol, 3-chloro-
$C_{12}H_9ClO_2S$
1225 - Benzene, 1-chloro-4-(phenylsulfonyl)-
$C_{12}H_9ClO_2S_2$
2236 - Benzenesulfonothioic acid, 4-chloro-, *S*-phenyl ester
$C_{12}H_9Cl_2NO_3$
12048 - Vinclozolin
$C_{12}H_9Cl_3Si$
11364 - Silane, [1,1′-biphenyl]-4-yltrichloro-
$C_{12}H_9F$
3074 - 1,1′-Biphenyl, 2-fluoro-
3075 - 1,1′-Biphenyl, 4-fluoro-
$C_{12}H_9I$
7 - Acenaphthylene, 1,2-dihydro-5-iodo-
3077 - 1,1′-Biphenyl, 2-iodo-
3078 - 1,1′-Biphenyl, 3-iodo-
3079 - 1,1′-Biphenyl, 4-iodo-
$C_{12}H_9N$
3981 - 9H-Carbazole
7736 - 1-Naphthaleneacetonitrile
7737 - 2-Naphthaleneacetonitrile
$C_{12}H_9NO$
7638 - Methanone, phenyl-2-pyridinyl-
7639 - Methanone, phenyl-4-pyridinyl-
9507 - 10H-Phenoxazine
$C_{12}H_9NO_2$
3085 - 1,1′-Biphenyl, 2-nitro-
3086 - 1,1′-Biphenyl, 3-nitro-
3087 - 1,1′-Biphenyl, 4-nitro-
6180 - Furo[2,3-b]quinoline, 4-methoxy-
$C_{12}H_9NO_2S$
2057 - Benzene, 1-nitro-2-(phenylthio)-
2058 - Benzene, 1-nitro-4-(phenylthio)-
$C_{12}H_9NO_3$
2053 - Benzene, 1-nitro-2-phenoxy-
2054 - Benzene, 1-nitro-4-phenoxy-
$C_{12}H_9NS$
9499 - 10H-Phenothiazine
$C_{12}H_9N_3O_4$
1542 - 1,3-Benzenediol, 4-[(4-nitrophenyl)azo]-
$C_{12}H_{10}$
6 - Acenaphthylene, 1,2-dihydro-
3008 - 1,1′-Biphenyl
7881 - Naphthalene, 1-ethenyl-
7882 - Naphthalene, 2-ethenyl-
$C_{12}H_{10}AsCl$
454 - Arsinous chloride, diphenyl-
$C_{12}H_{10}ClN$
678 - Benzenamine, 3-chloro-*N*-phenyl-
679 - Benzenamine, 4-chloro-*N*-phenyl-
3012 - [1,1′-Biphenyl]-4-amine, 4′-chloro-
10921 - Pyridine, 2-[(4-chlorophenyl)methyl]-
$C_{12}H_{10}ClO_3P$
9575 - Phosphorochloridic acid, diphenyl ester
$C_{12}H_{10}Cl_2N_2$
3032 - [1,1′-Biphenyl]-4,4′-diamine, 2,2′-dichloro-
$C_{12}H_{10}Cl_2Si$
11395 - Silane, dichlorodiphenyl-
$C_{12}H_{10}F_2Si$
11411 - Silane, difluorodiphenyl-
$C_{12}H_{10}Hg$
7436 - Mercury, diphenyl-
$C_{12}H_{10}N_2$
5129 - Diazene, diphenyl-, (*E*)-
5130 - Diazene, diphenyl-, (*Z*)-
11040 - 9H-Pyrido[3,4-b]indole, 1-methyl-
$C_{12}H_{10}N_2O$
829 - Benzenamine, 4-nitroso-*N*-phenyl-
830 - Benzenamine, *N*-nitroso-*N*-phenyl-
5131 - Diazene, diphenyl-, 1-oxide, (*Z*)-
9448 - Phenol, 4-(phenylazo)-
$C_{12}H_{10}N_2O_2$
826 - Benzenamine, 4-nitro-*N*-phenyl-
9216 - Phenol, 2,2′-azobis-
$C_{12}H_{10}N_4$
9194 - 2,3-Phenazinediamine
$C_{12}H_{10}N_4O_2$
2472 - Benzo[g]pteridine-2,4(1H,3H)-dione, 7,8-dimethyl-
$C_{12}H_{10}O$
2071 - Benzene, 1,1′-oxybis-
3088 - [1,1′-Biphenyl]-2-ol
3089 - [1,1′-Biphenyl]-3-ol
3090 - [1,1′-Biphenyl]-4-ol
5857 - Ethanone, 1-(1-naphthalenyl)-
5858 - Ethanone, 1-(2-naphthalenyl)-
7756 - 1-Naphthalenecarboxaldehyde, 4-methyl-
$C_{12}H_{10}OS$
2183 - Benzene, 1,1′-sulfinylbis-
$C_{12}H_{10}O_2$
3064 - [1,1′-Biphenyl]-2,2′-diol
3065 - [1,1′-Biphenyl]-2,4′-diol
3066 - [1,1′-Biphenyl]-3,3′-diol
5829 - Ethanone, 1-(1-hydroxy-2-naphthalenyl)-
7734 - 1-Naphthaleneacetic acid
7755 - 1-Naphthalenecarboxaldehyde, 4-methoxy-
7766 - 1-Naphthalenecarboxylic acid, methyl ester
7776 - 2-Naphthalenecarboxylic acid, methyl ester
$C_{12}H_{10}O_2S$
2237 - Benzene, 1,1′-sulfonylbis-
$C_{12}H_{10}O_3$
217 - Acetic acid, (2-naphthalenyloxy)-
6069 - 2-Furancarboxylic acid, phenylmethyl ester
7775 - 2-Naphthalenecarboxylic acid, 3-hydroxy-, methyl ester
$C_{12}H_{10}O_4$
3097 - [1,1′-Biphenyl]-3,3′,5,5′-tetrol
8677 - 2,4-Pentadienoic acid, 5-(1,3-benzodioxol-5-yl)-
8678 - 2,4-Pentadienoic acid, 5-(1,3-benzodioxol-5-yl)-, (*E,E*)-
$C_{12}H_{10}O_4S$
9466 - Phenol, 4,4′-sulfonylbis-
$C_{12}H_{10}O_6S_2$
3070 - [1,1′-Biphenyl]-4,4′-disulfonic acid
$C_{12}H_{10}O_7$
7857 - 1,4-Naphthalenedione, 2-ethyl-3,5,6,7,8-pentahydroxy-
7858 - 1,4-Naphthalenedione, 6-ethyl-2,3,5,7,8-pentahydroxy-
$C_{12}H_{10}S$
2283 - Benzene, 1,1′-thiobis-
$C_{12}H_{10}S_2$
5258 - Disulfide, diphenyl
$C_{12}H_{10}Se$
2175 - Benzene, 1,1′-selenobis-
$C_{12}H_{10}Se_2$
5225 - Diselenide, diphenyl
$C_{12}H_{11}As$
443 - Arsine, diphenyl-
$C_{12}H_{11}BO$
3130 - Borinic acid, diphenyl-
$C_{12}H_{11}Cl_2NO$
10746 - Propyzamide
$C_{12}H_{11}N$
2 - 1-Acenaphthylenamine, 1,2-dihydro-
836 - Benzenamine, *N*-phenyl-
3010 - [1,1′-Biphenyl]-2-amine
3011 - [1,1′-Biphenyl]-4-amine
10989 - Pyridine, 2-(4-methylphenyl)-
10990 - Pyridine, 2-methyl-5-phenyl-
10991 - Pyridine, 3-methyl-2-phenyl-
11004 - Pyridine, 2-(phenylmethyl)-
11005 - Pyridine, 3-(phenylmethyl)-
11006 - Pyridine, 4-(phenylmethyl)-
$C_{12}H_{11}NO$
834 - Benzenamine, 2-phenoxy-
835 - Benzenamine, 3-phenoxy-
7733 - 1-Naphthaleneacetamide
9445 - Phenol, 2-(phenylamino)-
9446 - Phenol, 3-(phenylamino)-
9447 - Phenol, 4-(phenylamino)-
$C_{12}H_{11}NOS$
68 - Acetamide, 2-mercapto-*N*-2-naphthalenyl-
$C_{12}H_{11}NO_2$
66 - Acetamide, *N*-(2-hydroxy-1-naphthalenyl)-
2736 - 1H,5H-Benzo[ij]quinolizine-1,6(7H)-dione, 2,3-dihydro-
8032 - 1-Naphthalenol, methylcarbamate
10625 - 2-Propenoic acid, 2-cyano-3-phenyl-, ethyl ester
10645 - 2-Propenoic acid, 2-isocyano-3-phenyl-, ethyl ester
$C_{12}H_{11}NO_3S$
2234 - Benzenesulfonic acid, 4-(phenylamino)-
$C_{12}H_{11}NO_4$
5218 - 1,3-Dioxolo[4,5-h]quinolin-8(9H)-one, 6-methoxy-9-methyl-
7361 - 2H-Isoindole-2-acetic acid, 1,3-dihydro-1,3-dioxo-, ethyl ester

$C_{12}H_{11}N_3$
837 - Benzenamine, 4-(phenylazo)-
11826 - 1-Triazene, 1,3-diphenyl-
$C_{12}H_{11}N_3O_2$
10904 - 4-Pyridinecarboxylic acid, [1-(2-furanyl)ethylidene]hydrazide
$C_{12}H_{11}N_7$
10750 - 2,4,7-Pteridinetriamine, 6-phenyl-
$C_{12}H_{11}O_3P$
9534 - Phosphonic acid, diphenyl ester
$C_{12}H_{12}$
7818 - Naphthalene, 1,2-dimethyl-
7819 - Naphthalene, 1,3-dimethyl-
7820 - Naphthalene, 1,4-dimethyl-
7821 - Naphthalene, 1,5-dimethyl-
7822 - Naphthalene, 1,6-dimethyl-
7823 - Naphthalene, 1,7-dimethyl-
7824 - Naphthalene, 1,8-dimethyl-
7825 - Naphthalene, 2,3-dimethyl-
7826 - Naphthalene, 2,6-dimethyl-
7827 - Naphthalene, 2,7-dimethyl-
7885 - Naphthalene, 1-ethyl-
7886 - Naphthalene, 2-ethyl-
$C_{12}H_{12}N_2Br$
5224 - Diquat dibromide
$C_{12}H_{12}ClNO_2S$
7977 - 1-Naphthalenesulfonyl chloride, 5-(dimethylamino)-
$C_{12}H_{12}ClN_5O_4S$
4091 - Chlorsulfuron
$C_{12}H_{12}N_2$
1282 - 1,2-Benzenediamine, *N*-phenyl-
1283 - 1,4-Benzenediamine, *N*-phenyl-
3028 - [1,1′-Biphenyl]-2,2′-diamine
3029 - [1,1′-Biphenyl]-2,4′-diamine
3030 - [1,1′-Biphenyl]-3,3′-diamine
3031 - [1,1′-Biphenyl]-4,4′-diamine
7157 - Hydrazine, 1,1-diphenyl-
7158 - Hydrazine, 1,2-diphenyl-
$C_{12}H_{12}N_2O$
832 - Benzenamine, 4,4′-oxybis-
$C_{12}H_{12}N_2OS$
845 - Benzenamine, 4,4′-sulfinylbis-
$C_{12}H_{12}N_2OS_2$
11684 - 4-Thiazolidinone, 5-[[4-(dimethylamino)phenyl]methylene]-2-thioxo-
$C_{12}H_{12}N_2O_2S$
846 - Benzenamine, 4,4′-sulfonylbis-
$C_{12}H_{12}N_2O_3$
8066 - 1,8-Naphthyridine-3-carboxylic acid, 1-ethyl-1,4-dihydro-7-methyl-4-oxo-
11095 - 2,4,6(1H,3H,5H)-Pyrimidinetrione, 5 ethyl-5-phenyl-
$C_{12}H_{12}N_2S$
850 - Benzenamine, 4,4′-thiobis-
$C_{12}H_{12}N_4$
635 - Benzenamine, 4,4′-azobis-
$C_{12}H_{12}O$
6710 - 3,5-Hexadien-2-one, 6-phenyl-
7883 - Naphthalene, 1-ethoxy-
7884 - Naphthalene, 2-ethoxy-
7911 - 1-Naphthalenemethanol, α-methyl-, (±)-
8027 - 2-Naphthalenol, 1,4-dimethyl-
$C_{12}H_{12}O_2$
8045 - 1(2H)-Naphthalenone, 2-acetyl-3,4-dihydro-
1000[illegible] - 2-Propenoic acid, 3-phenyl-, 2-propenyl ester
$C_{12}H_{12}O_5$
2845 - 2H-1-Benzopyran-2-one, 6,7,8-trimethoxy-
10785 - 4H,5H-Pyrano[4,3-b]pyran-4,5-dione, 2,3-dihydro-3-hydroxy-2-methyl-7-(1-propenyl)-, [2*S*-[2α,3β,7(*E*)]]-
$C_{12}H_{12}O_6$
2370 - 1,2,4-Benzenetriol, triacetate
$C_{12}H_{12}Si$
11420 - Silane, diphenyl-
$C_{12}H_{13}ClF_3N_3O_4$
5964 - Fluchloralin
$C_{12}H_{13}ClN_4$
1284 - 1,3-Benzenediamine, 4-(phenylazo)-, monohydrochloride
11051 - 2,4-Pyrimidinediamine, 5-(4-chlorophenyl)-6-ethyl-
$C_{12}H_{13}Cl_3O_3$
258 - Acetic acid, (2,4,5-trichlorophenoxy)-, butyl ester
260 - Acetic acid, (2,4,5-trichlorophenoxy)-, 1-methylpropyl ester
$C_{12}H_{13}N$
3989 - 1H-Carbazole, 2,3,4,9-tetrahydro-
7715 - 1-Naphthalenamine, *N,N*-dimethyl-
7716 - 2-Naphthalenamine, *N,N*-dimethyl-
7718 - 1-Naphthalenamine, *N*-ethyl-
7719 - 2-Naphthalenamine, *N*-ethyl-
8064 - 2-Naphthylamine, 1,4-dimethyl-
11269 - Quinoline, 2-ethyl-3-methyl-
11270 - Quinoline, 2-ethyl-4-methyl-
11271 - Quinoline, 3-ethyl-6-methyl-
11292 - Quinoline, 2-propyl-
11300 - Quinoline, 2,3,6-trimethyl-
11301 - Quinoline, 2,3,8-trimethyl-
11302 - Quinoline, 2,4,6-trimethyl-
11303 - Quinoline, 2,4,7-trimethyl-
11304 - Quinoline, 2,5,7-trimethyl-
$C_{12}H_{13}NO$
5750 - Ethanol, 2-(2-naphthalenylamino)-
$C_{12}H_{13}NO_2$
7296 - 1H-Indole-3-butanoic acid
11156 - 2,5-Pyrrolidinedione, 1,3-dimethyl-3-phenyl-
$C_{12}H_{13}NO_2S$
4039 - Carboxin
$C_{12}H_{13}N_3$
1263 - 1,4-Benzenediamine, *N*-(4-aminophenyl)-
10974 - 2-Pyridinemethanamine, *N*-(2-pyridinylmethyl)-
$C_{12}H_{13}N_3O_4S_2$
2188 - Benzenesulfonamide, 4-amino-*N*-[4-(aminosulfonyl)phenyl]-
$C_{12}H_{14}$
9 - Acenaphthylene, 1,2,2a,3,4,5-hexahydro-
1246 - Benzene, 1-cyclohexen-1-yl-
1247 - Benzene, 2-cyclohexen-1-yl-
1625 - Benzene, (1-ethenyl-3-butenyl)-
1753 - Benzene, 1,3-hexadienyl-
1754 - Benzene, 1,5-hexadienyl-
1755 - Benzene, 3,5-hexadienyl-
$C_{12}H_{14}ClNO_2$
4140 - Clomazone
$C_{12}H_{14}ClN_5O_2S$
2209 - Benzenesulfonamide, 4-[(2,4-diaminophenyl)azo]-
$C_{12}H_{14}Cl_2O_3$
171 - Acetic acid, (2,4-dichlorophenoxy)-, butyl ester
$C_{12}H_{14}Cl_3O_4P$
4078 - Chlorfenvinphos
$C_{12}H_{14}N_2$
5514 - 1,2-Ethanediamine, *N*-1-naphthalenyl-
11213 - 6-Quinolinamine, *N,N*,2-trimethyl-
$C_{12}H_{14}N_2O_2$
11066 - 4,6(1H,5H)-Pyrimidinedione, 5-ethyldihydro-5-phenyl-
110[illegible] - *L*-Tryptophan, *N*-methyl-
$C_{12}H_{14}N_2O_3$
11085 - 2,4,6(1H,3H,5H)-Pyrimidinetrione, 5-(2-cyclopenten-1-yl)-5-(2-propenyl)-
$C_{12}H_{14}N_2O_5$
6242 - *L*-Glutamic acid, *N*-(4-aminobenzoyl)-
9273 - Phenol, 2-cyclohexyl-4,6-dinitro-
$C_{12}H_{14}N_4OS$
11111 - 5H-Pyrimido[4,5-d]thiazolo[3,2-a]pyrimidine-8-ethanol, 2,7-dimethyl-
$C_{12}H_{14}N_4O_2S$
2196 - Benzenesulfonamide, 4-amino-*N*-(4,6-dimethyl-2-pyrimidinyl)-
$C_{12}H_{14}N_4O_4S$
2194 - Benzenesulfonamide, 4-amino-*N*-(2,6-dimethoxy-4-pyrimidinyl)-
$C_{12}H_{14}N_4O_4S_2$
11723 - Thiophanate-methyl
$C_{12}H_{14}O$
2460 - Benzofuran, 4-methyl-7-(1-methylethyl)-
4893 - Cyclopentanone, 2-(phenylmethyl)-
7101 - 5-Hexen-2-one, 4-phenyl-
7102 - 5-Hexen-2-one, 6-phenyl-
8051 - 1(2H)-Naphthalenone, 7-ethyl-3,4-dihydro-
8987 - 2-Pentanone, 3-(phenylmethylene)-
$C_{12}H_{14}O_2$
4135 - Cinnamic acid, β-propyl-
6038 - Furan, 2,2′-*sec*-butylidenedi-
10687 - 2-Propenoic acid, 3-phenyl-, 1-methylethyl ester, (*E*)-
10693 - 2-Propenoic acid, 3-phenyl-, propyl ester, (*E*)-
$C_{12}H_{14}O_3$
884 - Benzeneacetic acid, α-acetyl-, ethyl ester
2100 - Benzenepentanoic acid, δ-oxo-, methyl ester
8604 - Oxiranecarboxylic acid, 3-methyl-3-phenyl-, ethyl ester
8911 - Pentanoic acid, 4-oxo-, phenylmethyl ester
9393 - Phenol, 2-methoxy-4-(1-propenyl)-, acetate, (*E*)-
9394 - Phenol, 2-methoxy-4-(1-propenyl)-, acetate, (*Z*)-
9395 - Phenol, 2-methoxy-4-(2-propenyl)-, acetate
$C_{12}H_{14}O_4$
1259 - 1,4-Benzenediacetic acid, dimethyl ester
1342 - 1,2-Benzenedicarboxylic acid, diethyl ester
1346 - 1,3-Benzenedicarboxylic acid, diethyl ester
1350 - 1,4-Benzenedicarboxylic acid, diethyl ester
2422 - 1,3-Benzodioxole, 4,7-dimethoxy-5-(1-propenyl)-, (*E*)-
2423 - 1,3-Benzodioxole, 4,7-dimethoxy-5-(2-propenyl)-
8625 - Oxirane, 2,2′-[1,3-phenylenebis(oxymethylene)]bis-
$C_{12}H_{15}Br$
1009 - Benzene, 1-bromo-4-cyclohexyl-
$C_{12}H_{15}ClNO_4PS_2$
9514 - Phosalone
$C_{12}H_{15}ClO$
2907 - Benzoyl chloride, 4-pentyl-
9254 - Phenol, 2-chloro-4-cyclohexyl-
$C_{12}H_{15}ClO_2$
2908 - Benzoyl chloride, 4-(pentyloxy)-
$C_{12}H_{15}ClO_3$
10231 - Propanoic acid, 2-(4-chlorophenoxy)-2-methyl-, ethyl ester
$C_{12}H_{15}N$
964 - Benzeneacetonitrile, α-(2-methylpropyl)-
1800 - Benzenemethanamine, *N*-3-butynyl-*N*-methyl-
2737 - 1H,5H-Benzo[ij]quinolizine, 2,3,6,7-tetrahydro-
$C_{12}H_{15}NO$
9634 - Piperidine, 1-benzoyl-
$C_{12}H_{15}NO_2$
1251 - Benzene, 1-cyclohexyl-2-nitro-
1252 - Benzene, 1-cyclohexyl-4-nitro-
$C_{12}H_{15}NO_3$
3991 - Carbofuran
5214 - 1,3-Dioxolo[4,5-g]isoquinoline, 5,6,7,8-tetrahydro-4-methoxy-6-methyl-
$C_{12}H_{15}NO_4$
5216 - 1,3-Dioxolo[4,5-g]isoquinolin-5-ol, 5,6,7,8-tetrahydro-4-methoxy-6-methyl-
$C_{12}H_{15}N_3O_3$
11842 - 1,3,5-Triazine-2,4,6(1H,3H,5H)-trione, 1,3,5-tri-2-propenyl-

$C_{12}H_{15}N_3O_6$
1461 - Benzene, 1-(1,1-dimethylethyl)-3,5-dimethyl-2,4,6-trinitro-
$C_{12}H_{16}$
1248 - Benzene, cyclohexyl-
1760 - Benzene, 2-hexenyl-
1761 - Benzene, 3-hexenyl-
1762 - Benzene, 5-hexenyl-
1973 - Benzene, 1-(1-methylethenyl)-4-(1-methylethyl)-
2008 - Benzene, (1-methyl-1-pentenyl)-
4744 - 1,3,5,7-Cyclooctatetraene, 1-butyl-
7261 - 1H-Indene, 2,3-dihydro-1,1,5-trimethyl-
7262 - 1H-Indene, 2,3-dihydro-1,4,7-trimethyl-
7888 - Naphthalene, 1-ethyl-1,2,3,4-tetrahydro-
7889 - Naphthalene, 2-ethyl-1,2,3,4-tetrahydro-
7890 - Naphthalene, 5-ethyl-1,2,3,4-tetrahydro-
7891 - Naphthalene, 6-ethyl-1,2,3,4-tetrahydro-
7980 - Naphthalene, 1,2,3,4-tetrahydro-1,1-dimethyl-
7981 - Naphthalene, 1,2,3,4-tetrahydro-1,2-dimethyl-
7982 - Naphthalene, 1,2,3,4-tetrahydro-1,3-dimethyl-
7983 - Naphthalene, 1,2,3,4-tetrahydro-1,4-dimethyl-
7984 - Naphthalene, 1,2,3,4-tetrahydro-1,5-dimethyl-
7985 - Naphthalene, 1,2,3,4-tetrahydro-2,2-dimethyl-
7986 - Naphthalene, 1,2,3,4-tetrahydro-2,3-dimethyl-
7987 - Naphthalene, 1,2,3,4-tetrahydro-2,5-dimethyl-
7988 - Naphthalene, 1,2,3,4-tetrahydro-2,6-dimethyl-
7989 - Naphthalene, 1,2,3,4-tetrahydro-2,7-dimethyl-
7990 - Naphthalene, 1,2,3,4-tetrahydro-2,8-dimethyl-
7991 - Naphthalene, 1,2,3,4-tetrahydro-5,6-dimethyl-
7992 - Naphthalene, 1,2,3,4-tetrahydro-5,7-dimethyl-
7993 - Naphthalene, 1,2,3,4-tetrahydro-5,8-dimethyl-
7994 - Naphthalene, 1,2,3,4-tetrahydro-6,7-dimethyl-
$C_{12}H_{16}ClNOS$
11697 - Thiobencarb
$C_{12}H_{16}N_2$
7303 - 1H-Indole-3-ethanamine, *N,N*-dimethyl-
$C_{12}H_{16}N_2O$
7326 - 1H-Indol-5-ol, 3-[2-(dimethylamino)ethyl]-
7578 - 1,5-Methano-8H-pyrido[1,2-a][1,5]diazocin-8-one, 1,2,3,4,5,6-hexahydro-3-methyl-, (1*R*)-
$C_{12}H_{16}N_2O_3$
11083 - 2,4,6(1H,3H,5H)-Pyrimidinetrione, 5-(1-cyclohexen-1-yl)-1,5-dimethyl-
11084 - 2,4,6(1H,3H,5H)-Pyrimidinetrione, 5-(1-cyclohexen-1-yl)-5-ethyl-
$C_{12}H_{16}N_2O_5$
1463 - Benzene, 1-(1,1-dimethylethyl)-2-methoxy-4-methyl-3,5-dinitro-
$C_{12}H_{16}O$
1253 - Benzene, (cyclohexyloxy)-
3669 - 1-Butanone, 3-methyl-1-(2-methylphenyl)-
3670 - 1-Butanone, 3-methyl-1-(4-methylphenyl)-
4552 - Cyclohexanol, 1-phenyl-
4553 - Cyclohexanol, 2-phenyl-, *cis*-(±)-
4554 - Cyclohexanol, 2-phenyl-, *trans*-(±)-
5808 - Ethanone, 1-[4-(1,1-dimethylethyl)phenyl]-
5850 - Ethanone, 1-[2-methyl-5-(1-methylethyl)phenyl]-
6983 - 1-Hexanone, 1-phenyl-
6984 - 2-Hexanone, 3-phenyl-
6985 - 3-Hexanone, 2-phenyl-
8983 - 1-Pentanone, 4-methyl-1-phenyl-
8984 - 2-Pentanone, 3-methyl-1-phenyl-
9272 - Phenol, 4-cyclohexyl-
$C_{12}H_{16}O_2$
890 - Benzeneacetic acid, α-butyl-
932 - Benzeneacetic acid, 2-methylpropyl ester
1759 - Benzenehexanoic acid
2129 - Benzenepropanoic acid, α,α-dimethyl-, methyl ester
2140 - Benzenepropanoic acid, 1-methylethyl ester
2146 - Benzenepropanoic acid, β-propyl-
2685 - Benzoic acid, 3-methyl-, butyl ester
2713 - Benzoic acid, pentamethyl-
3560 - Butanoic acid, 3-methyl-, phenylmethyl ester
3632 - 1-Butanol, 3-methyl-, benzoate
7175 - Hydrocinnamic acid, β-ethyl-β-methyl-
8979 - 1-Pentanone, 1-(3-methoxyphenyl)-
9419 - Phenol, 2-methyl-5-(1-methylethyl)-, acetate
9420 - Phenol, 5-methyl-2-(1-methylethyl)-, acetate
10306 - Propanoic acid, 2-methyl-, 2-phenylethyl ester
$C_{12}H_{16}O_3$
2356 - Benzene, 1,2,4-trimethoxy-5-(1-propenyl)-
2633 - Benzoic acid, 2-hydroxy-, 3-methylbutyl ester
2649 - Benzoic acid, 2-hydroxy-, pentyl ester
2668 - Benzoic acid, 4-methoxy-, butyl ester
3581 - Butanoic acid, 4-phenoxy-, ethyl ester
3582 - Butanoic acid, 2-phenoxyethyl ester
6962 - 1-Hexanone, 1-(2,4-dihydroxyphenyl)-
8886 - Pentanoic acid, 2-methoxyphenyl ester
10641 - 2-Propenoic acid, 3-(2-furanyl)-, pentyl ester
$C_{12}H_{16}O_4$
2628 - Benzoic acid, 4-hydroxy-3-methoxy-, 2-methylpropyl ester
$C_{12}H_{16}O_7$
6224 - β-*D*-Glucopyranoside, 4-hydroxyphenyl
$C_{12}H_{17}BrO$
1045 - Benzene, 1-bromo-4-(hexyloxy)-
$C_{12}H_{17}N$
680 - Benzenamine, *N*-cyclohexyl-
9698 - Piperidine, 1-(phenylmethyl)-
9699 - Piperidine, 2-(phenylmethyl)-
9700 - Piperidine, 4-(phenylmethyl)-
$C_{12}H_{17}NO$
39 - Acetamide, *N*-butyl-*N*-phenyl-
612 - Benzamide, *N,N*-diethyl-3-methyl-
6722 - Hexanamide, *N*-phenyl-
$C_{12}H_{17}NO_2$
9769 - Promecarb
$C_{12}H_{17}NO_3$
611 - Benzamide, *N,N*-diethyl-4-hydroxy-3-methoxy-
880 - Benzeneacetamide, 4-butoxy-*N*-hydroxy-
7395 - 8-Isoquinolinol, 1,2,3,4-tetrahydro-6,7-dimethoxy-1-methyl-, (*S*)-
$C_{12}H_{17}N_3O_4$
6247 - *L*-Glutamic acid, 5-[2-[4-(hydroxymethyl)phenyl]hydrazide]
$C_{12}H_{18}$
992 - Benzene, 1,2-bis(1-methylethyl)-
993 - Benzene, 1,3-bis(1-methylethyl)-
994 - Benzene, 1,4-bis(1-methylethyl)-
1455 - Benzene, (1,1-dimethylbutyl)-
1460 - Benzene, 1-(1,1-dimethylethyl)-3,5-dimethyl-
1473 - Benzene, 2,4-dimethyl-1-(1-methylpropyl)-
1557 - Benzene, 1,4-dipropyl-
1669 - Benzene, (1-ethylbutyl)-
1691 - Benzene, (1-ethyl-1-methylpropyl)-
1758 - Benzene, hexamethyl-
1763 - Benzene, hexyl-
2009 - Benzene, (1-methylpentyl)-
2010 - Benzene, (3-methylpentyl)-
2011 - Benzene, (4-methylpentyl)-
2344 - Benzene, 1,2,4-triethyl-
2345 - Benzene, 1,3,5-triethyl-
2361 - Benzene, 1,2,4-trimethyl-5-(1-methylethyl)-
4212 - 1,5,9-Cyclododecatriene
4213 - 1,5,9-Cyclododecatriene, (*Z,E,E*)-
4214 - 1,5,9-Cyclododecatriene, (*Z,Z,Z*)-
5286 - 3,9-Dodecadiyne
5287 - 5,7-Dodecadiyne
8339 - 2,6-Octadien-4-yne, 3,6-diethyl-
$C_{12}H_{18}Cl_2N_4OS$
11686 - Thiazolium, 3-[(4-amino-2-methyl-5-pyrimidinyl)methyl]-5-(2-hydroxyethyl)-4-methyl- chloride, monohydrochlorid
$C_{12}H_{18}N_2O$
11188 - 2-Pyrrolidinone, 1-[4-(1-pyrrolidinyl)-2-butynyl]-
$C_{12}H_{18}N_2O_3$
11090 - 2,4,6(1H,3H,5H)-Pyrimidinetrione, 5-(2,2-dimethylpropyl)-5-(2-propenyl)-
$C_{12}H_{18}N_2O_3S$
2207 - Benzenesulfonamide, *N*-[(butylamino) carbonyl]-4-methyl-
$C_{12}H_{18}N_4O$
7171 - Hydrazine, phenyl-, hemihydrate
$C_{12}H_{18}N_4O_6S$
8553 - Oryzalin
$C_{12}H_{18}O$
1488 - Benzene, 1-(1,1-dimethylpropyl)-4-methoxy-
1765 - Benzene, (hexyloxy)-
1847 - Benzenemethanol, α-butyl-α-methyl-
1875 - Benzenemethanol, α-ethyl-α-propyl-
1909 - Benzenemethanol, α-pentyl-
1951 - Benzene, [(3-methylbutoxy)methyl]-
2103 - Benzenepentanol, α-methyl-
2159 - Benzenepropanol, γ-propyl-
6949 - 3-Hexanol, 1-phenyl-
9226 - Phenol, 2,6-bis(1-methylethyl)-
9324 - Phenol, 2-(1,1-dimethylethyl)-4,5-dimethyl-
9325 - Phenol, 2-(1,1-dimethylethyl)-4,6-dimethyl-
9326 - Phenol, 4-(1,1-dimethylethyl)-2,5-dimethyl-
9327 - Phenol, 4-(1,1-dimethylethyl)-2,6-dimethyl-
9340 - Phenol, 4-(1,1-dimethylpropyl)-2-methyl-
9347 - Phenol, 2,6-dipropyl-
9425 - Phenol, 5-methyl-2-pentyl-
$C_{12}H_{18}O_2$
1519 - 1,3-Benzenediol, 4-hexyl-
1555 - Benzene, 1,2-dipropoxy-
1556 - Benzene, 1,3-dipropoxy-
2950 - Bicyclo[2.2.1]heptan-2-one, 3-(acetyl-17*O*)-1,7,7-trimethyl-
11863 - Tricyclo[2.2.1.02,6]heptane-1-carboxylic acid, 7,7-dimethyl-, ethyl ester
$C_{12}H_{18}O_3$
2343 - Benzene, 1,3,5-triethoxy-
$C_{12}H_{18}O_4$
4644 - 1-Cyclohexene-1,2-dicarboxylic acid, diethyl ester
4645 - 1-Cyclohexene-1,3-dicarboxylic acid, diethyl ester
9935 - Propanedioic acid, 2-cyclopenten-1-yl-, diethyl ester
9936 - Propanedioic acid, cyclopentylidene-, diethyl ester
$C_{12}H_{18}O_5$
10669 - 2-Propenoic acid, 2-methyl-, oxydi-2,1-ethanediyl ester
$C_{12}H_{18}O_6$
10601 - 1-Propene-1,2,3-tricarboxylic acid, triethyl ester, (*E*)-
$C_{12}H_{19}ClNO_3P$
4153 - Crufomate
$C_{12}H_{19}N$
636 - Benzenamine, 2,6-bis(1-methylethyl)-
730 - Benzenamine, *N,N*-dipropyl-
$C_{12}H_{19}NO$
738 - Benzenamine, 3-ethoxy-*N,N*-diethyl-
$C_{12}H_{20}$
5342 - 1-Dodecen-3-yne
$C_{12}H_{20}As_2Cl_2O_5$
9275 - Phenol, 4,4′-(1,2-diarsenediyl)bis[2-amino-, dihydrochloride
$C_{12}H_{20}N_2O_2$
10811 - 2(1H)-Pyrazinone, 1-hydroxy-6-(1-methylpropyl)-3-(2-methylpropyl)-, (+)-
$C_{12}H_{20}N_4O_2$
6995 - Hexazinone
$C_{12}H_{20}O$
2989 - [1,1′-Bicyclohexyl]-2-one
6032 - Furan, 2,5-bis(1,1-dimethylethyl)-
$C_{12}H_{20}O_2$
2945 - Bicyclo[2.2.1]heptan-2-ol, 1,7,7-trimethyl-, acetate, *endo*-(±)-
2946 - Bicyclo[2.2.1]heptan-2-ol, 1,7,7-trimethyl-, acetate, *exo*-(±)-
2947 - Bicyclo[2.2.1]heptan-2-ol, 4,7,7-trimethyl-, acetate, (1*R*-*endo*)-

4545 - Cyclohexanol, 5-methyl-2-(1-methylethenyl)-, acetate, [1*R*-(1α,2β,5α)]-
4669 - 3-Cyclohexene-1-methanol, α,α,4-trimethyl-, acetate
6135 - Furan, 2-(octyloxy)-
8335 - 1,6-Octadien-3-ol, 3,7-dimethyl-, acetate, (*R*)-
8336 - 2,6-Octadien-1-ol, 3,7-dimethyl-, acetate, (*Z*)-
11990 - 9-Undecynoic acid, methyl ester
11991 - 10-Undecynoic acid, methyl ester

$C_{12}H_{20}O_3Si$
11481 - Silane, triethoxyphenyl-

$C_{12}H_{20}O_4$
3767 - 2-Butenedioic acid (*E*)-, bis(2-methylpropyl) ester
3774 - 2-Butenedioic acid (*E*)-, dibutyl ester
4350 - 1,2-Cyclohexanedicarboxylic acid, diethyl ester, *trans*-(±)-
4351 - 1,3-Cyclohexanedicarboxylic acid, diethyl ester, *cis*-
4355 - 1,4-Cyclohexanedicarboxylic acid, diethyl ester, *cis*-
4356 - 1,4-Cyclohexanedicarboxylic acid, diethyl ester, *trans*-
4795 - 1,3-Cyclopentanedicarboxylic acid, 1,2,2-trimethyl-, dimethyl ester, (1*R*-*cis*)-
5336 - 2-Dodecenedioic acid, (*E*)-

$C_{12}H_{20}O_4Si$
11511 - Silicic acid (H_4SiO_4), triethyl phenyl ester

$C_{12}H_{20}O_6$
10198 - 1,2,3-Propanetriol, tripropanoate

$C_{12}H_{20}O_7$
10169 - 1,2,3-Propanetricarboxylic acid, 2-hydroxy-, triethyl ester

$C_{12}H_{20}Si$
11485 - Silane, triethylphenyl-

$C_{12}H_{21}NO$
4611 - Cyclohexanone, 2-(1-piperidinylmethyl)-

$C_{12}H_{21}N_2O_3PS$
9582 - Phosphorothioic acid, *O,O*-diethyl *O*-[6-methyl-2-(1-methylethyl)-4-pyrimidinyl] ester

$C_{12}H_{22}$
2988 - 1,1′-Bicyclohexyl
4215 - Cyclododecene, (*E*)-
4216 - Cyclododecene, (*Z*)-
5101 - 4-Decyne, 3,3-dimethyl-
5285 - 1,11-Dodecadiene
5343 - 1-Dodecyne
5344 - 2-Dodecyne
5345 - 3-Dodecyne
5346 - 6-Dodecyne

$C_{12}H_{22}O$
4211 - Cyclododecanone
4461 - Cyclohexane, 1,1′-oxybis-

$C_{12}H_{22}OSi_2$
11497 - Silane, trimethyl[4-[(trimethylsilyl) oxy]phenyl]-

$C_{12}H_{22}O_2$
2948 - Bicyclo[2.2.1]heptan-2-ol, 1,7,7-trimethyl-, acetate, *endo*-
4412 - Cyclohexanehexanoic acid
4549 - Cyclohexanol, 5-methyl-2-(1-methylethyl)-, acetate, [1*R*-(1α,2β,5α)]-
4787 - Cyclopentanecarboxylic acid, 3-isopropyl-1-methyl-
4788 - Cyclopentanecarboxylic acid, 2-methyl-5-(1-methylethyl)-, ethyl ester
5337 - 2-Dodecenoic acid
5338 - 4-Dodecenoic acid
5339 - 5-Dodecenoic acid
5340 - 11-Dodecenoic acid
8400 - 3,6-Octanedione, 2,2,7,7-tetramethyl-
10659 - 2-Propenoic acid, 2-methyl-, 2-ethylhexyl ester
11978 - 10-Undecenoic acid, methyl ester

$C_{12}H_{22}O_2Si_2$
11447 - Silane, [1,3-phenylenebis(oxy)]bis[trimethyl-

$C_{12}H_{22}O_3$
210 - Acetic acid, [[5-methyl-2-(1-methylethyl)cyclohexyl]oxy]-, [1*R*-(1α,2β,5α)]-
6874 - Hexanoic acid, anhydride

$C_{12}H_{22}O_4$
3296 - Butanedioic acid, bis(1,1-dimethylethyl) ester
3297 - Butanedioic acid, bis(1-methylpropyl) ester
3301 - Butanedioic acid, dibutyl ester
3332 - Butanedioic acid, (1-methylpropyl)-, diethyl ester
5046 - Decanedioic acid, dimethyl ester
5299 - Dodecanedioic acid
5555 - Ethanedioic acid, bis(3-methylbutyl) ester
6782 - Hexanedioic acid, bis(1-methylethyl) ester
6787 - Hexanedioic acid, dipropyl ester
8387 - Octanedioic acid, diethyl ester
9951 - Propanedioic acid, ethyl(1-methylethyl)-, diethyl ester
9961 - Propanedioic acid, (3-methylbutyl)-, diethyl ester
9974 - Propanedioic acid, pentyl-, diethyl ester

$C_{12}H_{22}O_6$
3313 - Butanedioic acid, 2,3-dihydroxy- [*R*-(*R**,*R**)]-, dibutyl ester

$C_{12}H_{22}O_{10}$
6236 - *D*-Glucose, 6-*O*-(6-deoxy-α-*L*-mannopyranosyl)-

$C_{12}H_{22}O_{11}$
6022 - *D*-Fructopyranose, 3-*O*-α-*D*-glucopyranosyl-
6214 - β-*D*-Glucopyranose, 4-*O*-β-*D*-galactopyranosyl-
6218 - α-*D*-Glucopyranoside, β-*D*-fructofuranosyl
6220 - α-*D*-Glucopyranoside, α-*D*-glucopyranosyl

$C_{12}H_{22}O_{12}$
6210 - *D*-Gluconic acid, 4-*O*-β-*D*-galactopyranosyl-

$C_{12}H_{23}BrO_2$
5307 - Dodecanoic acid, 2-bromo-

$C_{12}H_{23}ClO$
5328 - Dodecanoyl chloride

$C_{12}H_{23}N$
4284 - Cyclohexanamine, *N*-cyclohexyl-
5303 - Dodecanenitrile

$C_{12}H_{23}NSi_2$
11362 - Silanamine, 1,1,1-trimethyl-*N*-phenyl-*N*-(trimethylsilyl)-

$C_{12}H_{23}P$
9516 - Phosphine, dicyclohexyl-

$C_{12}H_{24}$
4210 - Cyclododecane
4414 - Cyclohexane, hexyl-
4446 - Cyclohexane, 1-methyl-2-pentyl-
4737 - Cyclooctane, butyl-
4820 - Cyclopentane, heptyl-
5335 - 1-Dodecene
10582 - 1-Propene, 2-methyl-, trimer

$C_{12}H_{24}BrF$
5294 - Dodecane, 1-bromo-12-fluoro-

$C_{12}H_{24}Br_2$
5296 - Dodecane, 1,12-dibromo-

$C_{12}H_{24}N_2$
9661 - Piperidine, 1,1′-(1,2-ethanediyl)bis-

$C_{12}H_{24}O$
5051 - Decane, 1-(ethenyloxy)-
5288 - Dodecanal
5325 - 2-Dodecanone
5326 - 6-Dodecanone
11936 - Undecanal, 2-methyl-

$C_{12}H_{24}O_2$
157 - Acetic acid, decyl ester
3566 - Butanoic acid, octyl ester
5063 - Decanoic acid, ethyl ester
5305 - Dodecanoic acid
6470 - Heptanoic acid, pentyl ester
6890 - Hexanoic acid, hexyl ester
8155 - Nonanoic acid, 3-methyl-, ethyl ester
8427 - Octanoic acid, butyl ester
8881 - Pentanoic acid, heptyl ester

$C_{12}H_{24}O_3$
11949 - Undecanoic acid, 11-hydroxy-, methyl ester

$C_{12}H_{24}O_4Si_4$
4996 - Cyclotetrasiloxane, 2,4,6,8-tetraethenyl-2,4,6,8-tetramethyl-

$C_{12}H_{25}Br$
5293 - Dodecane, 1-bromo-

$C_{12}H_{25}Cl$
5295 - Dodecane, 1-chloro-

$C_{12}H_{25}Cl_3Si$
11460 - Silane, trichlorododecyl-

$C_{12}H_{25}I$
5302 - Dodecane, 1-iodo-

$C_{12}H_{25}NO$
5289 - Dodecanamide

$C_{12}H_{26}$
5292 - Dodecane
6433 - Heptane, 2,2,4,6,6-pentamethyl-

$C_{12}H_{26}O$
5321 - 1-Dodecanol
5322 - 2-Dodecanol
5323 - 3-Dodecanol
5324 - 6-Dodecanol
6842 - Hexane, 1,1′-oxybis-
11955 - 1-Undecanol, 2-methyl-
11956 - 3-Undecanol, 2-methyl-
11957 - 5-Undecanol, 2-methyl-

$C_{12}H_{26}O_2$
5301 - 1,12-Dodecanediol

$C_{12}H_{26}O_3$
3417 - Butane, 1,1′-[oxybis(2,1-ethanediyloxy)]bis-

$C_{12}H_{26}O_6P_2S_4$
5201 - Dioxathion

$C_{12}H_{26}S$
5304 - 1-Dodecanethiol
6855 - Hexane, 1,1′-thiobis-

$C_{12}H_{27}B$
3123 - Borane, tris(2-methylpropyl)-

$C_{12}H_{27}BO_3$
3124 - Boric acid, tributyl ester

$C_{12}H_{27}N$
3213 - 1-Butanamine, *N,N*-dibutyl-
5291 - 1-Dodecanamine
6727 - 1-Hexanamine, *N*-hexyl-
9820 - 1-Propanamine, 2-methyl-*N,N*-bis(2-methylpropyl)-

$C_{12}H_{27}NO$
5682 - Ethanol, 2-[bis(3-methylbutyl)amino]-

$C_{12}H_{27}NO_2$
8936 - 2-Pentanol, 1,1′-iminobis[2-methyl-

$C_{12}H_{27}OPS_3$
11852 - *S,S,S*-Tributyl phosphorotrothioate

$C_{12}H_{27}O_3P$
9590 - Phosphorous acid, tributyl ester

$C_{12}H_{27}O_4P$
9562 - Phosphoric acid tributyl ester
9572 - Phosphoric acid, tris(2-methylpropyl) ester

$C_{12}H_{27}P$
9520 - Phosphine, tributyl-

$C_{12}H_{27}PS_3$
7438 - Merphos

$C_{12}H_{28}IN$
9833 - 1-Propanaminium, *N,N,N*-tripropyl-, iodide

$C_{12}H_{28}N_2$
3262 - 1,4-Butanediamine, *N,N,N′,N′*-tetraethyl-

$C_{12}H_{28}O_4Si$
11510 - Silicic acid (H4SiO4), tetrapropyl ester

$C_{12}H_{28}Si$
11454 - Silane, tributyl-

$C_{12}H_{28}Sn$
11546 - Stannane, tetrapropyl-

$C_{12}H_{30}OSi_2$
5235 - Disiloxane, hexaethyl-

$C_{12}H_{30}O_{13}P_4$
11645 - Tetraphosphoric acid, hexaethyl ester

$C_{12}H_{32}O_4Si_4$
4997 - Cyclotetrasiloxane, 2,4,6,8-tetraethyl-2,4,6,8-tetramethyl-
$C_{12}H_{36}O_4Si_5$
8997 - Pentasiloxane, dodecamethyl-
$C_{12}H_{36}O_6Si_6$
4624 - Cyclohexasiloxane, dodecamethyl-
$C_{13}H_5N_3O_7$
5979 - 9H-Fluoren-9-one, 2,4,7-trinitro-
$C_{13}H_6Cl_6O_2$
9409 - Phenol, 2,2′-methylenebis[3,4,6-trichloro-
$C_{13}H_7ClF_3NO_3$
8082 - Nitrofluorfen
$C_{13}H_7ClO$
5976 - 9H-Fluoren-9-one, 2-chloro-
$C_{13}H_7Cl_4NO_2$
610 - Benzamide, 3,5-dichloro-*N*-(3,4-dichlorophenyl)-2-hydroxy-
$C_{13}H_7F_3N_2O_5$
5980 - Fluorodifen
$C_{13}H_7NO_3$
5977 - 9H-Fluoren-9-one, 2-nitro-
$C_{13}H_8Br_2O$
7590 - Methanone, bis(4-bromophenyl)-
$C_{13}H_8Br_3NO_2$
609 - Benzamide, 3,5-dibromo-*N*-(4-bromophenyl)-2-hydroxy-
$C_{13}H_8ClN$
315 - Acridine, 9-chloro-
$C_{13}H_8Cl_2$
5972 - 9H-Fluorene, 2,7-dichloro-
$C_{13}H_8Cl_2O$
7591 - Methanone, bis(3-chlorophenyl)-
7592 - Methanone, bis(4-chlorophenyl)-
7603 - Methanone, (2-chlorophenyl)(4-chlorophenyl)-
$C_{13}H_8Cl_3NO_2$
608 - Benzamide, 5-chloro-*N*-(3,4-dichlorophenyl)-2-hydroxy-
$C_{13}H_8I_2O$
7598 - Methanone, bis(4-iodophenyl)-
$C_{13}H_8O$
5975 - 9H-Fluoren-9-one
$C_{13}H_8OS$
11812 - 9H-Thioxanthen-9-one
$C_{13}H_8O_2$
12050 - 9H-Xanthen-9-one
$C_{13}H_8O_4$
12051 - 9H-Xanthen-9-one, 1,3-dihydroxy-
12052 - 9H-Xanthen-9-one, 2,7-dihydroxy-
12053 - 9H-Xanthen-9-one, 3,6-dihydroxy-
$C_{13}H_9Br$
5968 - 9H-Fluorene, 2-bromo-
$C_{13}H_9BrO$
7600 - Methanone, (2-bromophenyl)phenyl-
7601 - Methanone, (3-bromophenyl)phenyl-
7602 - Methanone, (4-bromophenyl)phenyl-
$C_{13}H_9ClO$
7604 - Methanone, (4-chlorophenyl)phenyl-
$C_{13}H_9Cl_2FN_2S$
11802 - Thiourea, *N*-(3,5-dichlorophenyl)-*N*′-(4-fluorophenyl)-
$C_{13}H_9N$
314 - Acridine
2437 - Benzo[f]quinoline
2473 - Benzo[g]quinoline
2474 - Benzo[h]quinoline
3020 - [1,1′-Biphenyl]-2-carbonitrile
9188 - Phenanthridine
$C_{13}H_9NOS$
9217 - Phenol, 2-(2-benzothiazolyl)-
$C_{13}H_9NO_2$
9218 - Phenol, 2-(2-benzoxazolyl)-
$C_{13}H_9NO_3$
7635 - Methanone, (3-nitrophenyl)phenyl-
7636 - Methanone, (4-nitrophenyl)phenyl-
$C_{13}H_9NS$
2864 - Benzothiazole, 2-phenyl-
$C_{13}H_9N_3O_5$
2648 - Benzoic acid, 2-hydroxy-5-[(4-nitrophenyl)azo]-
$C_{13}H_{10}$
5967 - 9H-Fluorene
$C_{13}H_{10}Br_2$
1962 - Benzene, 1,1′-methylenebis[2-bromo-
$C_{13}H_{10}Cl_2$
1389 - Benzene, 1,1′-(dichloromethylene)bis-
1963 - Benzene, 1,1′-methylenebis[4-chloro-
$C_{13}H_{10}Cl_2O_2$
1965 - Benzene, 1,1′-[methylenebis(oxy)]bis[4-chloro-
9408 - Phenol, 2,2′-methylenebis[4-chloro-
$C_{13}H_{10}Cl_2S$
4075 - Chlorbenside
$C_{13}H_{10}N_2$
312 - 4-Acridinamine
313 - 9-Acridinamine
764 - Benzenamine, *N,N*′-methanetetraylbis-
2383 - 1H-Benzimidazole, 1-phenyl-
2384 - 1H-Benzimidazole, 2-phenyl-
4158 - Cyanamide, diphenyl-
9195 - Phenazine, 2-methyl-
$C_{13}H_{10}N_2O$
9196 - Phenazinium, 1-hydroxy-5-methyl-, hydroxide, inner salt
$C_{13}H_{10}N_2O_4$
7363 - 1H-Isoindole-1,3(2H)-dione, 2-(2,6-dioxo-3-piperidinyl)-
$C_{13}H_{10}N_4O_5$
11997 - Urea, *N,N*′-bis(4-nitrophenyl)-
$C_{13}H_{10}O$
7616 - Methanone, diphenyl-
12049 - 9H-Xanthene
$C_{13}H_{10}O_2$
597 - Benzaldehyde, 3-phenoxy-
2716 - Benzoic acid, phenyl ester
3021 - [1,1′-Biphenyl]-2-carboxylic acid
3022 - [1,1′-Biphenyl]-4-carboxylic acid
7622 - Methanone, (2-hydroxyphenyl)phenyl-
10675 - 2-Propenoic acid, 3-(1-naphthalenyl)-, (*E*)-
10710 - 2-Propen-1-one, 3-(2-furanyl)-1-phenyl-
$C_{13}H_{10}O_3$
2650 - Benzoic acid, 2-hydroxy-, phenyl ester
2714 - Benzoic acid, 2-phenoxy-
4001 - Carbonic acid, diphenyl ester
7596 - Methanone, bis(2-hydroxyphenyl)-
7597 - Methanone, bis(4-hydroxyphenyl)-
7609 - Methanone, (2,4-dihydroxyphenyl)phenyl-
8682 - 1,4-Pentadien-3-one, 1,5-di-2-furanyl-
$C_{13}H_{10}O_4$
6186 - 5H-Furo[3,2-g][1]benzopyran-5-one, 4-methoxy-7-methyl-
$C_{13}H_{10}O_6$
7610 - Methanone, (3,4-dihydroxyphenyl)(2,4,6-trihydroxyphenyl)-
$C_{13}H_{10}S$
7547 - Methanethione, diphenyl-
11811 - 9H-Thioxanthene
$C_{13}H_{11}Br$
1058 - Benzene, 1,1′-(bromomethylene)bis-
$C_{13}H_{11}Cl$
1187 - Benzene, 1,1′-(chloromethylene)bis-
1224 - Benzene, 1-chloro-4-(phenylmethyl)-
3026 - Biphenyl, 2-chloro-2′-methyl-
$C_{13}H_{11}ClO$
9270 - Phenol, 4-chloro-2-(phenylmethyl)-
$C_{13}H_{11}Cl_2NO_2$
9760 - Procymidone
$C_{13}H_{11}N$
840 - Benzenamine, *N*-(phenylmethylene)-
1835 - Benzenemethanimine, α-phenyl-
3986 - 9H-Carbazole, 3-methyl-
3987 - 9H-Carbazole, 9-methyl-
11000 - Pyridine, 2-(2-phenylethenyl)-, (*E*)-
11001 - Pyridine, 4-(2-phenylethenyl)-, (*E*)-
11002 - Pyridine, 2-(2-phenylethenyl)-, (*Z*)-
$C_{13}H_{11}NO$
626 - Benzamide, *N*-phenyl-
5993 - Formamide, *N,N*-diphenyl-
9450 - Phenol, 2-[(phenylimino)methyl]-
$C_{13}H_{11}NO_2$
617 - Benzamide, 2-hydroxy-*N*-phenyl-
2715 - Benzoic acid, 2-(phenylamino)-
10911 - 3-Pyridinecarboxylic acid, phenylmethyl ester
$C_{13}H_{11}NO_3$
2005 - Benzene, 1-methyl-2-(2-nitrophenoxy)-
2006 - Benzene, 1-methyl-4-(4-nitrophenoxy)-
2502 - Benzoic acid, 4-amino-2-hydroxy-, phenyl ester
6179 - Furo[2,3-b]quinoline, 4,8-dimethoxy-
$C_{13}H_{11}NS$
1129 - Benzenecarbothioamide, *N*-phenyl-
$C_{13}H_{11}N_3$
316 - 3,6-Acridinediamine
$C_{13}H_{12}$
632 - 1H-Benz[e]indene, 2,3-dihydro-
1961 - Benzene, 1,1′-methylenebis-
3082 - 1,1′-Biphenyl, 2-methyl-
3083 - 1,1′-Biphenyl, 3-methyl-
3084 - 1,1′-Biphenyl, 4-methyl-
7954 - Naphthalene, 1-(2-propenyl)-
$C_{13}H_{12}Cl_2N_2$
785 - Benzenamine, 4,4′-methylenebis[2-chloro-
$C_{13}H_{12}Cl_2O_4$
168 - Acetic acid, [2,3-dichloro-4-(2-methylene-1-oxobutyl)phenoxy]-
$C_{13}H_{12}N_2$
5138 - Diazene, (3-methylphenyl)phenyl-
5139 - Diazene, (4-methylphenyl)phenyl-
5971 - 9H-Fluorene-2,7-diamine
7565 - Methanimidamide, *N,N*′-diphenyl-
$C_{13}H_{12}N_2O$
2717 - Benzoic acid, 2-phenylhydrazide
5135 - Diazene, (2-methoxyphenyl)phenyl-
5136 - Diazene, (3-methoxyphenyl)phenyl-
5137 - Diazene, (4-methoxyphenyl)phenyl-
7589 - Methanone, bis(3-aminophenyl)-
9426 - Phenol, 4-methyl-2-(phenylazo)-
11039 - 9H-Pyrido[3,4-b]indole, 7-methoxy-1-methyl-
12010 - Urea, *N,N*′-diphenyl-
12011 - Urea, *N,N*-diphenyl-
$C_{13}H_{12}N_2O_3$
11103 - 2,4,6(1H,3H,5H)-Pyrimidinetrione, 5-phenyl-5-(2-propenyl)-
$C_{13}H_{12}N_2O_3S$
603 - Benzamide, *N*-[(4-aminophenyl)sulfonyl]-
$C_{13}H_{12}N_2S$
11805 - Thiourea, *N,N*′-diphenyl-
$C_{13}H_{12}N_4O$
5122 - Diazenecarboxylic acid, phenyl-, 2-phenylhydrazide
$C_{13}H_{12}N_4S$
5121 - Diazenecarbothioic acid, phenyl-, 2-phenylhydrazide
$C_{13}H_{12}O$
1910 - Benzenemethanol, α-phenyl-
2012 - Benzene, 1-methyl-2-phenoxy-
2013 - Benzene, 1-methyl-3-phenoxy-
3080 - 1,1′-Biphenyl, 2-methoxy-
3081 - 1,1′-Biphenyl, 4-methoxy-
5978 - 9H-Fluoren-9-one, 1,2,3,4-tetrahydro-
9454 - Phenol, 2-(phenylmethyl)-
9455 - Phenol, 4-(phenylmethyl)-
10473 - 1-Propanone, 1-(1-naphthalenyl)-
10474 - 1-Propanone, 1-(2-naphthalenyl)-
$C_{13}H_{12}O_2$
1546 - 1,4-Benzenediol, 2-(phenylmethyl)-
1938 - Benzene, 1-methoxy-2-phenoxy-
7753 - 1-Naphthalenecarboxaldehyde, 2-ethoxy-
7765 - 1-Naphthalenecarboxylic acid, ethyl ester
7770 - 2-Naphthalenecarboxylic acid, ethyl ester
7952 - 1-Naphthalenepropanoic acid

9406 - Phenol, 3,3′-methylenebis-
9407 - Phenol, 4,4′-methylenebis-
9451 - Phenol, 2-(phenylmethoxy)-
9452 - Phenol, 3-(phenylmethoxy)-
9453 - Phenol, 4-(phenylmethoxy)-

$C_{13}H_{12}O_2S$
2112 - Benzene, [(phenylmethyl)sulfonyl]-

$C_{13}H_{12}O_2S_2$
11818 - *o*-Toluenesulfonic acid, thio-, *S*-phenyl ester

$C_{13}H_{12}O_3$
5825 - Ethanone, 1-[6-hydroxy-2-(1-methylethenyl)-5-benzofuranyl]-
7773 - 2-Naphthalenecarboxylic acid, 3-hydroxy-, ethyl ester
7953 - 2-Naphthalenepropanoic acid, 6-hydroxy-

$C_{13}H_{12}S$
2017 - Benzene, 1-methyl-2-(phenylthio)-
2018 - Benzene, 1-methyl-3-(phenylthio)-
2019 - Benzene, 1-methyl-4-(phenylthio)-
2113 - Benzene, [(phenylmethyl)thio]-

$C_{13}H_{13}ClSi$
11380 - Silane, chloromethyldiphenyl-

$C_{13}H_{13}Cl_2N_3O_3$
7338 - Iprodione

$C_{13}H_{13}N$
807 - Benzenamine, *N*-methyl-*N*-phenyl-
808 - Benzenamine, 3-methyl-*N*-phenyl-
839 - Benzenamine, 4-(phenylmethyl)-
1826 - Benzenemethanamine, α-phenyl-
1827 - Benzenemethanamine, *N*-phenyl-
3013 - [1,1′-Biphenyl]-2-amine, 6-methyl-
3014 - [1,1′-Biphenyl]-4-amine, 3-methyl-
3015 - [1,1′-Biphenyl]-4-amine, 4′-methyl-
11003 - Pyridine, 2-(2-phenylethyl)-

$C_{13}H_{13}N_3$
6282 - Guanidine, *N,N′*-diphenyl-

$C_{13}H_{13}N_3O_5S_2$
3578 - Butanoic acid, 4-oxo-4-[[4-[(2-thiazolylamino)sulfonyl]phenyl]amino]-

$C_{13}H_{13}O_3P$
9541 - Phosphonic acid, methyl-, diphenyl ester

$C_{13}H_{14}$
7887 - Naphthalene, 1-ethyl-5-methyl-
7920 - Naphthalene, 1-(1-methylethyl)-
7921 - Naphthalene, 2-(1-methylethyl)-
7957 - Naphthalene, 1-propyl-
7958 - Naphthalene, 2-propyl-
8009 - Naphthalene, 1,2,4-trimethyl-
8010 - Naphthalene, 1,2,5-trimethyl-
8011 - Naphthalene, 1,2,6-trimethyl-
8012 - Naphthalene, 1,2,7-trimethyl-
8013 - Naphthalene, 1,3,7-trimethyl-

$C_{13}H_{14}F_3N_3O_4$
5409 - Ethalfluralin

$C_{13}H_{14}N_2$
784 - Benzenamine, 4,4′-methylenebis-
7174 - Hydrazobenzene, 3-methyl-

$C_{13}H_{14}N_2O$
1913 - Benzenemethanol, α-[(2-pyridinylamino)methyl]-

$C_{13}H_{14}N_2O_2S$
2213 - Benzenesulfonamide, 4-[(phenylmethyl)amino]-

$C_{13}H_{14}N_2O_3$
11094 - 2,4,6(1H,3H,5H)-Pyrimidinetrione, 5-ethyl-1-methyl-5-phenyl-

$C_{13}H_{14}N_2O_{13}$
11038 - 3H-Pyrido[3,4-b]indole, 4,9-dihydro-7-methoxy-1-methyl-

$C_{13}H_{14}N_4O$
4010 - Carbonic dihydrazide, 2,2′-diphenyl-

$C_{13}H_{14}O$
7955 - Naphthalene, 1-propoxy-
7956 - Naphthalene, 2-propoxy-

$C_{13}H_{14}OSi$
11502 - Silanol, methyldiphenyl-

$C_{13}H_{14}O_2$
8680 - 2,4-Pentadienoic acid, 5-phenyl-, ethyl ester

$C_{13}H_{14}O_3$
3576 - Butanoic acid, 3-oxo-2-(phenylmethylene)-, ethyl ester

$C_{13}H_{14}O_5$
2774 - 3H-2-Benzopyran-7-carboxylic acid, 4,6-dihydro-8-hydroxy-3,4,5-trimethyl-6-oxo-, (3*R*-*trans*)-

$C_{13}H_{14}Si$
11438 - Silane, methyldiphenyl-

$C_{13}H_{15}N$
11222 - Quinoline, 2-butyl-

$C_{13}H_{15}NO_4$
11148 - 1,2-Pyrrolidinedicarboxylic acid, 1-(phenylmethyl) ester, (*S*)-

$C_{13}H_{15}N_3O_2$
3945 - Carbamic acid, dimethyl-, 3-methyl-1-phenyl-1H-pyrazol-5-yl ester

$C_{13}H_{15}N_3O_3$
7188 - Imazapyr

$C_{13}H_{16}ClNO$
4579 - Cyclohexanone, 2-(2-chlorophenyl)-2-(methylamino)-, (±)-

$C_{13}H_{16}F_3N_3O_4$
509 - Balan
11901 - Trifluralin

$C_{13}H_{16}N_2O_2$
69 - Acetamide, *N*-[2-(5-methoxy-1H-indol-3-yl)ethyl]-

$C_{13}H_{16}O$
4236 - Cycloheptanone, 2-phenyl-
7100 - 1-Hexen-3-one, 5-methyl-1-phenyl-
9106 - 1-Penten-3-one, 4,4-dimethyl-1-phenyl-

$C_{13}H_{16}O_2$
1412 - Benzene, (3,3-diethoxy-1-propynyl)-
2551 - Benzoic acid, cyclohexyl ester

$C_{13}H_{16}O_3$
2132 - Benzenepropanoic acid, α-ethyl-β-oxo-, ethyl ester

$C_{13}H_{16}O_4$
9946 - Propanedioic acid, di-2-propynyl-, diethyl ester
9975 - Propanedioic acid, phenyl-, diethyl ester

$C_{13}H_{16}O_7$
562 - Benzaldehyde, 2-(β-*D*-glucopyranosyloxy)-

$C_{13}H_{17}F_3N_4O_4$
9761 - Prodiamine

$C_{13}H_{17}NO$
7409 - Ketone, 1-methyl-4-piperidyl phenyl

$C_{13}H_{17}NO_4$
10932 - 3,5-Pyridinedicarboxylic acid, 2,6-dimethyl-, diethyl ester

$C_{13}H_{17}N_3O$
10851 - 3H-Pyrazol-3-one, 4-(dimethylamino)-1,2-dihydro-1,5-dimethyl-2-phenyl-

$C_{13}H_{18}$
7260 - 1H-Indene, 2,3-dihydro-1,1,4,7-tetramethyl-
8000 - Naphthalene, 1,2,3,4-tetrahydro-1,1,6-trimethyl-

$C_{13}H_{18}ClN_3O_4S_2$
2853 - 2H-1,2,4-Benzothiadiazine-7-sulfonamide, 6-chloro-3-(cyclopentylmethyl)-3,4-dihydro-, 1,1-dioxide

$C_{13}H_{18}N_2O_2$
4900 - 1H-Cyclopentapyrimidine-2,4(3H,5H)-dione, 3-cyclohexyl-6,7-dihydro-

$C_{13}H_{18}O$
1250 - Benzene, 1-cyclohexyl-2-methoxy-
2117 - Benzenepropanal, α-methyl-4-(1-methylethyl)-
6526 - 1-Heptanone, 1-phenyl-

$C_{13}H_{18}O_2$
2610 - Benzoic acid, hexyl ester
8611 - Oxirane, [[4-(1,1-dimethylethyl)phenoxy]methyl]-

$C_{13}H_{18}O_3$
912 - Benzeneacetic acid, α-hydroxy-, 3-methylbutyl ester

$C_{13}H_{18}O_4$
2277 - Benzene, 1,2,3,4-tetramethoxy-5-(2-propenyl)-

$C_{13}H_{18}O_5S$
5933 - Ethofumesate

$C_{13}H_{18}O_7$
6222 - β-*D*-Glucopyranoside, 2-(hydroxymethyl)phenyl

$C_{13}H_{19}N$
9697 - Piperidine, 1-(1-phenylethyl)-

$C_{13}H_{19}NO$
10428 - 1-Propanone, 2-(diethylamino)-1-phenyl-

$C_{13}H_{19}NO_2$
7390 - Isoquinoline, 1,2,3,4-tetrahydro-6,7-dimethoxy-1,2-dimethyl-, (±)-
7391 - Isoquinoline, 1,2,3,4-tetrahydro-6,7-dimethoxy-1,2-dimethyl-

$C_{13}H_{19}NO_3$
7394 - 8-Isoquinolinol, 1,2,3,4-tetrahydro-6,7-dimethoxy-1,2-dimethyl-, (*S*)-

$C_{13}H_{19}NO_4S$
2595 - Benzoic acid, 4-[(dipropylamino)sulfonyl]-

$C_{13}H_{19}N_3O_4$
8635 - Pendimethalin

$C_{13}H_{19}N_3O_6S$
8070 - Nitralin

$C_{13}H_{20}$
1745 - Benzene, heptyl-
1948 - Benzene, 1-methyl-2,4-bis(1-methylethyl)-
1949 - Benzene, 1-methyl-3,5-bis(1-methylethyl)-
1992 - Benzene, [2-methyl-1-(1-methylethyl)propyl]-
11868 - 1,12-Tridecadiyne

$C_{13}H_{20}N_2O_2$
2493 - Benzoic acid, 4-amino-, 2-(diethylamino)ethyl ester
5748 - Ethanol, 2-[(2-methylpropyl)amino]-, 4-aminobenzoate (ester)

$C_{13}H_{20}O$
1747 - Benzene, (heptyloxy)-
1842 - Benzenemethanol, α,α-bis(1-methylethyl)-
1867 - Benzenemethanol, α,α-dipropyl-
1881 - Benzenemethanol, α-hexyl-
2102 - Benzenepentanol, α-ethyl-
3885 - 3-Buten-2-one, 4-(2,6,6-trimethyl-1-cyclohexen-1-yl)-, (*E*)-
3886 - 3-Buten-2-one, 4-(2,6,6-trimethyl-2-cyclohexen-1-yl)-, (*E*)-(±)-
4148 - *o*-Cresol, 4-(3-methylpentyl)-
11965 - 3,5,9-Undecatrien-2-one, 6,10-dimethyl-

$C_{13}H_{20}O_3$
6067 - 2-Furancarboxylic acid, octyl ester

$C_{13}H_{20}O_4$
9944 - Propanedioic acid, di-2-propenyl-, diethyl ester

$C_{13}H_{20}O_8$
9988 - 1,3-Propanediol, 2,2-bis[(acetyloxy)methyl]-, diacetate

$C_{13}H_{21}N$
10874 - Pyridine, 2,6-bis(1,1-dimethylethyl)-

$C_{13}H_{22}$
5973 - 1H-Fluorene, dodecahydro-

$C_{13}H_{22}ClN_3O$
602 - Benzamide, 4-amino-*N*-[2-(diethylamino)ethyl]-, monohydrochloride

$C_{13}H_{22}NO_3PS$
5947 - Fenamiphos

$C_{13}H_{22}N_2$
4291 - Cyclohexanamine, *N,N′*-methanetetraylbis-

$C_{13}H_{22}N_2O$
8212 - Norea

$C_{13}H_{22}O$
3873 - 3-Buten-2-ol, 4-(2,6,6-trimethyl-1-cyclohexen-1-yl)-
3874 - 3-Buten-2-ol, 4-(2,6,6-trimethyl-2-cyclohexen-1-yl)-
7606 - Methanone, dicyclohexyl-

8113 - 6,8-Nonadien-2-one, 8-methyl-5-(1-methylethyl)-, [*S*-(*E*)]-
$C_{13}H_{22}O_4$
4171 - Cyclobutaneacetic acid, 3-(ethoxycarbonyl)-2,2-dimethyl-, ethyl ester
9934 - Propanedioic acid, cyclohexyl-, diethyl ester
$C_{13}H_{24}$
4429 - Cyclohexane, 1,1′-methylenebis-
11897 - 1-Tridecyne
$C_{13}H_{24}N_2O$
10415 - 2-Propanone, 1,3-bis(1-methyl-2-pyrrolidinyl)-, (*R**,*S**)-
$C_{13}H_{24}N_3O_3PS$
9729 - Pirimiphos-ethyl
$C_{13}H_{24}O_2$
5341 - 11-Dodecenoic acid, methyl ester
10646 - 2-Propenoic acid, isodecyl ester
11894 - 2-Tridecenoic acid
11895 - 12-Tridecenoic acid
11977 - 10-Undecenoic acid, ethyl ester
$C_{13}H_{24}O_3$
6166 - 2-Furanpropanol, tetrahydro-α-[2-(tetrahydro-2-furanyl)ethyl]-
$C_{13}H_{24}O_4$
8133 - Nonanedioic acid, diethyl ester
9955 - Propanedioic acid, hexyl-, diethyl ester
$C_{13}H_{25}BrO_2$
5309 - Dodecanoic acid, 2-bromo-, methyl ester
$C_{13}H_{25}N$
11880 - Tridecanenitrile
$C_{13}H_{26}$
4327 - Cyclohexane, 2-butyl-1,1,3-trimethyl-
4405 - Cyclohexane, heptyl-
4843 - Cyclopentane, octyl-
8184 - 4-Nonene, 5-butyl-
11893 - 1-Tridecene
$C_{13}H_{26}Br_2$
11874 - Tridecane, 1,13-dibromo-
$C_{13}H_{26}N_2$
4299 - Cyclohexanamine, 4,4′-methylenebis-
4300 - Cyclohexanamine, 4,4′-methylenebis-, [cis(*cis*)]-
4301 - Cyclohexanamine, 4,4′-methylenebis-, [*trans*(*cis*)]-
4302 - Cyclohexanamine, 4,4′-methylenebis-, [trans(*trans*)]-
9701 - Piperidine, 4,4′-(1,3-propanediyl)bis-
$C_{13}H_{26}N_2O_4$
3933 - Carbamic acid, butyl-,2-[[(aminocarbonyl)oxy]methyl]-2-methylpentyl ester
$C_{13}H_{26}O$
11869 - Tridecanal
11889 - 2-Tridecanone
11890 - 3-Tridecanone
11891 - 7-Tridecanone
$C_{13}H_{26}O_2$
5067 - Decanoic acid, 1-methylethyl ester
5069 - Decanoic acid, propyl ester
5314 - Dodecanoic acid, 2-methyl-
5315 - Dodecanoic acid, methyl ester
6465 - Heptanoic acid, hexyl ester
6889 - Hexanoic acid, heptyl ester
8147 - Nonanoic acid, butyl ester
8439 - Octanoic acid, pentyl ester
8901 - Pentanoic acid, octyl ester
11881 - Tridecanoic acid
11946 - Undecanoic acid, ethyl ester
$C_{13}H_{27}Br$
11872 - Tridecane, 1-bromo-
$C_{13}H_{27}N$
9690 - Piperidine, 1-octyl-
$C_{13}H_{28}$
8122 - Nonane, 5-butyl-
11871 - Tridecane
$C_{13}H_{28}O$
8162 - 5-Nonanol, 5-butyl-
11883 - 1-Tridecanol
11884 - 2-Tridecanol
11885 - 3-Tridecanol
$C_{13}H_{28}O_2$
11875 - 1,12-Tridecanediol
11876 - 1,13-Tridecanediol
$C_{13}H_{28}O_3$
10080 - Propane, 1,1′,1″-[methylidynetris(oxy)]tris[2-methyl-
$C_{13}H_{28}O_4$
10068 - Propane, 1,1′,1″,1‴-[methanetetrayltetrakis(oxy)]tetrakis-
$C_{13}H_{29}N$
6354 - 2-Heptanamine, 6-methyl-*N*-(3-methylbutyl)-
11870 - 1-Tridecanamine
$C_{14}H_4N_2O_2S_2$
8056 - Naphtho[2,3-b]-1,4-dithiin-2,3-dicarbonitrile, 5,10-dihydro-5,10-dioxo-
$C_{14}H_4N_4O_{12}$
401 - 9,10-Anthracenedione, 1,8-dihydroxy-2,4,5,7-tetranitro-
$C_{14}H_4O_6$
2784 - [2]Benzopyrano[6,5,4-def][2]benzopyran-1,3,6,8-tetrone
$C_{14}H_6Br_2O_2$
390 - 9,10-Anthracenedione, 2,7-dibromo-
$C_{14}H_6Cl_4O_4$
9152 - Peroxide, bis(2,4-dichlorobenzoyl)
$C_{14}H_6N_2O_6$
405 - 9,10-Anthracenedione, 1,5-dinitro-
$C_{14}H_7BrO_2$
385 - 9,10-Anthracenedione, 1-bromo-
386 - 9,10-Anthracenedione, 2-bromo-
$C_{14}H_7ClF_3NO_5$
304 - Acifluorfen
$C_{14}H_7ClO_2$
387 - 9,10-Anthracenedione, 1-chloro-
388 - 9,10-Anthracenedione, 2-chloro-
$C_{14}H_7NO_4$
410 - 9,10-Anthracenedione, 1-nitro-
411 - 9,10-Anthracenedione, 2-nitro-
$C_{14}H_7NO_6$
399 - 9,10-Anthracenedione, 1,2-dihydroxy-3-nitro-
400 - 9,10-Anthracenedione, 1,2-dihydroxy-4-nitro-
$C_{14}H_8Br_2$
375 - Anthracene, 9,10-dibromo-
$C_{14}H_8Cl_2N_4$
4139 - Clofentezine
$C_{14}H_8Cl_4$
1374 - Benzene, 1,1′-(dichloroethenylidene)bis[4-chloro-
$C_{14}H_8O_2$
379 - 9,10-Anthracenedione
9174 - 9,10-Phenanthrenedione
$C_{14}H_8O_3$
407 - 9,10-Anthracenedione, 1-hydroxy-
408 - 9,10-Anthracenedione, 2-hydroxy-
5145 - Dibenz[c,e]oxepin-5,7-dione
5969 - 9H-Fluorene-2-carboxylic acid, 9-oxo-
9175 - 9,10-Phenanthrenedione, 2-hydroxy-
9176 - 9,10-Phenanthrenedione, 3-hydroxy-
$C_{14}H_8O_4$
391 - 9,10-Anthracenedione, 1,2-dihydroxy-
392 - 9,10-Anthracenedione, 1,4-dihydroxy-
393 - 9,10-Anthracenedione, 1,5-dihydroxy-
394 - 9,10-Anthracenedione, 1,7-dihydroxy-
395 - 9,10-Anthracenedione, 1,8-dihydroxy-
396 - 9,10-Anthracenedione, 2,7-dihydroxy-
$C_{14}H_8O_5$
415 - 9,10-Anthracenedione, 1,2,3-trihydroxy-
416 - 9,10-Anthracenedione, 1,2,4-trihydroxy-
417 - 9,10-Anthracenedione, 1,2,6-trihydroxy-
418 - 9,10-Anthracenedione, 1,2,7-trihydroxy-
$C_{14}H_8O_6$
412 - 9,10-Anthracenedione, 1,2,5,6-tetrahydroxy-
413 - 9,10-Anthracenedione, 1,2,5,8-tetrahydroxy-
414 - 9,10-Anthracenedione, 1,3,5,7-tetrahydroxy-
$C_{14}H_8O_8$
406 - 9,10-Anthracenedione, 1,2,3,5,6,7-hexahydroxy-
$C_{14}H_9Br$
9164 - Phenanthrene, 9-bromo-
$C_{14}H_9Cl$
374 - Anthracene, 1-chloro-
$C_{14}H_9ClF_2N_2O_2$
5176 - Diflubenzuron
$C_{14}H_9ClO_3$
4079 - Chlorflurecol
$C_{14}H_9Cl_2NO_5$
2997 - Bifenox
$C_{14}H_9Cl_5$
2328 - Benzene, 1,1′-(2,2,2-trichloroethylidene)bis[4-chloro-
$C_{14}H_9Cl_5O$
1851 - Benzenemethanol, 4-chloro-α-(4-chlorophenyl)-α-(trichloromethyl)-
$C_{14}H_9NO_2$
380 - 9,10-Anthracenedione, 1-amino-
381 - 9,10-Anthracenedione, 2-amino-
426 - Anthracene, 9-nitro-
7366 - 1H-Isoindole-1,3(2H)-dione, 2-phenyl-
$C_{14}H_9NO_3$
383 - 9,10-Anthracenedione, 2-amino-1-hydroxy-
$C_{14}H_9NO_4$
382 - 9,10-Anthracenedione, 3-amino-1,2-dihydroxy-
$C_{14}H_{10}$
366 - Anthracene
1714 - Benzene, 1,1′-(1,2-ethynediyl)bis-
9163 - Phenanthrene
$C_{14}H_{10}Cl_2$
1365 - Benzene, 1,1′-(1,2-dichloro-1,2-ethenediyl)bis-, (*E*)-
1366 - Benzene, 1,1′-(1,2-dichloro-1,2-ethenediyl)bis-, (*Z*)-
$C_{14}H_{10}Cl_4$
1378 - Benzene, 1,1′-(2,2-dichloroethylidene)bis[4-chloro-
$C_{14}H_{10}F_3NO_2$
2729 - Benzoic acid, 2-[[3-(trifluoromethyl)phenyl]amino]-
$C_{14}H_{10}N_2O$
8564 - 1,2,4-Oxadiazole, 3,5-diphenyl-
$C_{14}H_{10}N_2O_2$
389 - 9,10-Anthracenedione, 1,5-diamino-
$C_{14}H_{10}N_2O_4$
1494 - Benzene, 1,1′-(1,2-dinitro-1,2-ethenediyl)bis-, (*Z*)-
1496 - Benzene, 2,4-dinitro-1-(2-phenylethenyl)-, (*E*)-
$C_{14}H_{10}O$
431 - 1-Anthracenol
432 - 9-Anthracenol
433 - 9(10H)-Anthracenone
5928 - Ethenone, diphenyl-
$C_{14}H_{10}O_2$
5584 - Ethanedione, diphenyl-
9173 - 3,4-Phenanthrenediol
$C_{14}H_{10}O_2S_2$
5251 - Disulfide, dibenzoyl
$C_{14}H_{10}O_3$
429 - 1,2,10-Anthracenetriol
430 - 1,8,9-Anthracenetriol
2510 - Benzoic acid, anhydride
2511 - Benzoic acid, 3-benzoyl-
2512 - Benzoic acid, 4-benzoyl-
$C_{14}H_{10}O_4$
3038 - [1,1′-Biphenyl]-2,2′-dicarboxylic acid
9157 - Peroxide, dibenzoyl
$C_{14}H_{10}O_5$
2616 - Benzoic acid, 2-hydroxy-, 2-carboxyphenyl ester
12054 - 9H-Xanthen-9-one, 1,7-dihydroxy-3-methoxy-

$C_{14}H_{10}O_6$
2979 - [Bi-1,4-cyclohexadien-1-yl]-3,3′,6,6′-tetrone, 2,2′-dihydroxy-4,4′-dimethyl-
$C_{14}H_{10}O_9$
2575 - Benzoic acid, 3,4-dihydroxy-5-[(3,4,5-trihydroxybenzoyl)oxy]-
$C_{14}H_{11}Br$
1030 - Benzene, 1,1′-(bromoethenylidene)bis-
1080 - Benzene, 1-bromo-2-(2-phenylethenyl)-, (*E*)-
1081 - Benzene, 1-bromo-2-(2-phenylethenyl)-, (*Z*)-
$C_{14}H_{11}Cl$
1146 - Benzene, 1,1′-(1-chloro-1,2-ethenediyl)bis-, (*E*)-
1147 - Benzene, 1,1′-(1-chloro-1,2-ethenediyl)bis-, (*Z*)-
1222 - Benzene, 1-chloro-2-(2-phenylethenyl)-, (*E*)-
1223 - Benzene, 1-chloro-3-(2-phenylethenyl)-, (*E*)-
$C_{14}H_{11}ClO$
975 - Benzeneacetyl chloride, α-phenyl-
5785 - Ethanone, 2-chloro-1,2-diphenyl-
$C_{14}H_{11}N$
365 - 2-Anthracenamine
969 - Benzeneacetonitrile, α-phenyl-
7320 - 1H-Indole, 2-phenyl-
9162 - 9-Phenanthrenamine
$C_{14}H_{11}NO$
3983 - 9H-Carbazole, 9-acetyl-
$C_{14}H_{11}NO_2$
2055 - Benzene, 1-nitro-2-(2-phenylethenyl)-, (*E*)-
2056 - Benzene, 1-nitro-2-(2-phenylethenyl)-, (*Z*)-
3982 - 9H-Carbazole-9-acetic acid
$C_{14}H_{11}NO_3$
2513 - Benzoic acid, 3-(benzoylamino)-
7627 - Methanone, (4-methyl-2-nitrophenyl)phenyl-
7630 - Methanone, (4-methylphenyl)(4-nitrophenyl)-
$C_{14}H_{11}NO_4$
2514 - Benzoic acid, 4-(benzoylamino)-2-hydroxy-
2656 - Benzoic acid, 2,2′-iminobis-
$C_{14}H_{12}$
376 - Anthracene, 9,10-dihydro-
1619 - Benzene, 1,1′-(1,2-ethenediyl)bis-, (*E*)-
1620 - Benzene, 1,1′-(1,2-ethenediyl)bis-, (*Z*)-
1636 - Benzene, 1,1′-ethenylidenebis-
5974 - 9H-Fluorene, 9-methyl-
9171 - Phenanthrene, 9,10-dihydro-
$C_{14}H_{12}Cl_2$
1678 - Benzene, 1,1′-ethylidenebis[4-chloro-
$C_{14}H_{12}Cl_2O$
1850 - Benzenemethanol, 4-chloro-α-(4-chlorophenyl)-α-methyl-
$C_{14}H_{12}F_3NO_4S_2$
9149 - Perfluidone
$C_{14}H_{12}N_2$
2385 - 1H-Benzimidazole, 2-(phenylmethyl)-
9192 - 1,10-Phenanthroline, 2,9-dimethyl-
11196 - Quinazoline, 3,4-dihydro-3-phenyl-
$C_{14}H_{12}N_2O_2$
5495 - Ethanediamide, *N,N*′-diphenyl-
9449 - Phenol, 4-(phenylazo)-, acetate (ester)
$C_{14}H_{12}N_2S$
782 - Benzenamine, 4-(6-methyl-2-benzothiazolyl)-
$C_{14}H_{12}N_4O_2S$
2205 - Benzenesulfonamide, 4-amino-*N*-2-quinoxalinyl-
$C_{14}H_{12}O$
877 - Benzeneacetaldehyde, α-phenyl-
5776 - Ethanone, 1-[1,1′-biphenyl]-4-yl-
5812 - Ethanone, 1,2-diphenyl-
7631 - Methanone, (2-methylphenyl)phenyl-
7632 - Methanone, (3-methylphenyl)phenyl-
7633 - Methanone, (4-methylphenyl)phenyl-
$C_{14}H_{12}O_2$
936 - Benzeneacetic acid, α-phenyl-
1545 - 1,3-Benzenediol, 5-(2-phenylethenyl)-, (*E*)-
2699 - Benzoic acid, 2-methylphenyl ester
2700 - Benzoic acid, 4-methylphenyl ester
2718 - Benzoic acid, 2-(phenylmethyl)-
2719 - Benzoic acid, 4-(phenylmethyl)-
2720 - Benzoic acid, phenylmethyl ester
3009 - [1,1′-Biphenyl]-4-acetic acid
3091 - [1,1′-Biphenyl]-4-ol, acetate
5822 - Ethanone, 2-hydroxy-1,2-diphenyl-, (±)-
7623 - Methanone, (2-methoxyphenyl)phenyl-
7624 - Methanone, (3-methoxyphenyl)phenyl-
7625 - Methanone, (4-methoxyphenyl)phenyl-
$C_{14}H_{12}O_3$
918 - Benzeneacetic acid, α-hydroxy-α-phenyl-
2397 - 2H,8H-Benzo[1,2-b:5,4-b′]dipyran-2-one, 8,8-dimethyl-
2651 - Benzoic acid, 2-hydroxy-, phenylmethyl ester
7620 - Methanone, (2-hydroxy-4-methoxyphenyl)phenyl-
9383 - Phenol, 2-methoxy-, benzoate
$C_{14}H_{12}O_4$
7608 - Methanone, (2,6-dihydroxy-4-methoxyphenyl)phenyl-
7619 - Methanone, (2-hydroxy-4-methoxyphenyl)(2-hydroxyphenyl)-
$C_{14}H_{12}O_5$
6185 - 5H-Furo[3,2-g][1]benzopyran-5-one, 4,9-dimethoxy-7-methyl-
$C_{14}H_{12}S_2$
11661 - Thianthrene, 2,7-dimethyl-
$C_{14}H_{13}N$
809 - Benzenamine, 2-methyl-*N*-(phenylmethylene)-
810 - Benzenamine, 3-methyl-*N*-(phenylmethylene)-
811 - Benzenamine, 4-methyl-*N*-(phenylmethylene)-
3985 - 9H-Carbazole, 9-ethyl-
$C_{14}H_{13}NO$
35 - Acetamide, *N*-[1,1′-biphenyl]-2-yl-
57 - Acetamide, *N,N*-diphenyl-
621 - Benzamide, *N*-(2-methylphenyl)-
622 - Benzamide, *N*-(3-methylphenyl)-
623 - Benzamide, *N*-(4-methylphenyl)-
7626 - Methanone, [2-(methylamino)phenyl]phenyl-
$C_{14}H_{13}NO_4$
6181 - Furo[2,3-b]quinoline, 4,7,8-trimethoxy-
$C_{14}H_{13}NO_4S$
2721 - Benzoic acid, 4-[[(phenylmethyl)sulfonyl]amino]-
$C_{14}H_{14}$
1587 - Benzene, 1,1′-(1,2-ethanediyl)bis-
1677 - Benzene, 1,1′-ethylidenebis-
2014 - Benzene, 1-methyl-2-(phenylmethyl)-
2015 - Benzene, 1-methyl-3-(phenylmethyl)-
2016 - Benzene, 1-methyl-4-(phenylmethyl)-
3050 - 1,1′-Biphenyl, 2,2′-dimethyl-
3051 - 1,1′-Biphenyl, 2,3′-dimethyl-
3052 - 1,1′-Biphenyl, 2,3-dimethyl-
3053 - 1,1′-Biphenyl, 2,4′-dimethyl-
3054 - 1,1′-Biphenyl, 2,4-dimethyl-
3055 - 1,1′-Biphenyl, 2,5-dimethyl-
3056 - 1,1′-Biphenyl, 2,6-dimethyl-
3057 - 1,1′-Biphenyl, 3,3′-dimethyl-
3058 - 1,1′-Biphenyl, 3,4′-dimethyl-
3059 - 1,1′-Biphenyl, 3,4-dimethyl-
3060 - 1,1′-Biphenyl, 3,5-dimethyl-
3061 - 1,1′-Biphenyl, 4,4′-dimethyl-
3072 - 1,1′-Biphenyl, 2-ethyl-
3073 - 1,1′-Biphenyl, 3-ethyl-
9187 - Phenanthrene, 1,2,3,4-tetrahydro-
$C_{14}H_{14}Cl_2N_2O$
7187 - Imazalil
$C_{14}H_{14}Hg$
7432 - Mercury, bis(4-methylphenyl)-
$C_{14}H_{14}NO_4PS$
9551 - Phosphonothioic acid, phenyl-, *O*-ethyl *O*-(4-nitrophenyl) ester
$C_{14}H_{14}N_2$
732 - Benzenamine, 4,4′-(1,2-ethenediyl)bis-, (*E*)-
5114 - Diazene, bis(2-methylphenyl)-
5115 - Diazene, bis(3-methylphenyl)-, (*E*)-
$C_{14}H_{14}N_2O$
5116 - Diazene, bis(2-methylphenyl)-, 1-oxide, (*Z*)-
5117 - Diazene, bis(2-methylphenyl)-, 1-oxide, (*E*)-
5118 - Diazene, bis(3-methylphenyl)-, 1-oxide, (*Z*)-
5119 - Diazene, bis(3-methylphenyl)-, 1-oxide, (*E*)-
5133 - Diazene, (4-ethoxyphenyl)phenyl-
10462 - 1-Propanone, 2-methyl-1,2-di-3-pyridinyl-
12018 - Urea, *N*-methyl-*N,N*′-diphenyl-
$C_{14}H_{14}N_3NaO_3S$
2222 - Benzenesulfonic acid, 4-[[4-(dimethylamino)phenyl]azo]-, sodium salt
$C_{14}H_{14}O$
1613 - Benzeneethanol, α-phenyl-
1614 - Benzeneethanol, α-phenyl-, (*S*)-
1615 - Benzeneethanol, β-phenyl-
2076 - Benzene, 1,1′-oxybis[2-methyl-
2077 - Benzene, 1,1′-oxybis[3-methyl-
2078 - Benzene, 1,1′-oxybis[4-methyl-
2079 - Benzene, 1,1′-[oxybis(methylene)]bis-
3071 - 1,1′-Biphenyl, 3-ethoxy-
3674 - 1-Butanone, 1-(1-naphthalenyl)-
10464 - 1-Propanone, 2-methyl-1-(1-naphthalenyl)-
10465 - 1-Propanone, 2-methyl-1-(2-naphthalenyl)-
$C_{14}H_{14}OS$
2184 - Benzene, 1,1′-[sulfinylbis(methylene)]bis-
$C_{14}H_{14}O_2$
1588 - Benzene, 1,1′-[1,2-ethanediylbis(oxy)]bis-
3045 - 1,1′-Biphenyl, 2,2′-dimethoxy-
3046 - 1,1′-Biphenyl, 3,3′-dimethoxy-
3047 - 1,1′-Biphenyl, 4,4′-dimethoxy-
3067 - [1,1′-Biphenyl]-2,2′-diol, 3,3′-dimethyl-
3068 - [1,1′-Biphenyl]-2,2′-diol, 5,5′-dimethyl-
5575 - 1,2-Ethanediol, 1,2-diphenyl-, (*R**,*R**)-(±)-
7735 - 1-Naphthaleneacetic acid, ethyl ester
$C_{14}H_{14}O_2S$
2240 - Benzene, 1,1′-sulfonylbis[4-methyl-
2241 - Benzene, 1,1′-[sulfonylbis(methylene)]bis-
$C_{14}H_{14}O_3$
1920 - Benzene, 1-methoxy-2-(3-methoxyphenoxy)-
2075 - Benzene, 1,1′-oxybis[2-methoxy-
2700 - 2H-1-Benzopyran-7-ol, 3,4-dihydro-3-(4-hydroxyphenyl)-, (*S*)-
$C_{14}H_{14}O_8$
2258 - 1,2,4,5-Benzenetetracarboxylic acid, tetramethyl ester
$C_{14}H_{14}S$
1993 - Benzene, 1-methyl-2-[(4-methylphenyl)thio]-
2285 - Benzene, 1,1′-thiobis[2-methyl-
2286 - Benzene, 1,1′-thiobis[4-methyl-
2287 - Benzene, 1,1′-[thiobis(methylene)]bis-
11570 - Sulfide, *m*-tolyl *p* tolyl
$C_{14}H_{14}S_2$
5245 - Disulfide, bis(3-methylphenyl)
5246 - Disulfide, bis(4-methylphenyl)
5248 - Disulfide, bis(phenylmethyl)
$C_{14}H_{14}Se$
2176 - Benzene, 1,1′-selenobis[4-methyl-
$C_{14}H_{15}Cl_2N$
7710 - 2-Naphthalenamine, *N,N*-bis(2-chloroethyl)-
$C_{14}H_{15}N$
751 - Benzenamine, *N*-ethyl-*N*-phenyl-
795 - Benzenamine, 2-methyl-*N*-(2-methylphenyl)-
796 - Benzenamine, 3-methyl-*N*-(3-methylphenyl)-
797 - Benzenamine, 4-methyl-*N*-(4-methylphenyl)-
838 - Benzenamine, 2-(2-phenylethyl)-
1584 - Benzeneethanamine, α-phenyl-
1821 - Benzenemethanamine, *N*-(2-methylphenyl)-
1824 - Benzenemethanamine, *N*-(3-methylphenyl)-
1825 - Benzenemethanamine, *N*-(4-methylphenyl)-
1828 - Benzenemethanamine, *N*-(phenylmethyl)-
11007 - Pyridine, 4-(3-phenylpropyl)-
$C_{14}H_{15}N_3$
721 - Benzenamine, *N,N*-dimethyl-4-(phenylazo)-
798 - Benzenamine, 2-methyl-4-[(2-methylphenyl)azo]-
$C_{14}H_{15}O_3P$
9586 - Phosphorous acid, bis(phenylmethyl) ester
$C_{14}H_{16}$
422 - Anthracene, 1,2,3,4,5,6-hexahydro-

7746 - Naphthalene, 1-butyl-
7828 - Naphthalene, 1-(1,1-dimethylethyl)-
7829 - Naphthalene, 2-(1,1-dimethylethyl)-
7922 - Naphthalene, 1-methyl-4-(1-methylethyl)-
7923 - Naphthalene, 1-methyl-7-(1-methylethyl)-
7924 - Naphthalene, 7-methyl-1-(1-methylethyl)-
7927 - Naphthalene, 1-(1-methylpropyl)-
8001 - Naphthalene, 1,2,4,7-tetramethyl-
8002 - Naphthalene, 1,2,5,6-tetramethyl-
8003 - Naphthalene, 1,3,6,8-tetramethyl-

$C_{14}H_{16}ClN_3O_2$
512 - Bayleton

$C_{14}H_{16}ClN_3O_4S_2$
2851 - 2H-1,2,4-Benzothiadiazine-7-sulfonamide, 3-bicyclo[2.2.1]hept-5-en-2-yl-6-chloro-3,4-dihydro-, 1,1-dioxide

$C_{14}H_{16}ClN_5O_5S$
11824 - Triasulfuron

$C_{14}H_{16}ClO_5PS$
4147 - Coumaphos

$C_{14}H_{16}F_3N_3O_4$
9763 - Profluralin

$C_{14}H_{16}N_2$
731 - Benzenamine, 4,4′-(1,2-ethanediyl)bis-
3034 - [1,1′-Biphenyl]-4,4′-diamine, 3,3′-dimethyl-
5508 - 1,2-Ethanediamine, *N,N′*-diphenyl-
7138 - Hydrazine, 1,2-bis(3-methylphenyl)-
7139 - Hydrazine, 1,2-bis(4-methylphenyl)-

$C_{14}H_{16}N_2O_2$
3033 - [1,1′-Biphenyl]-4,4′-diamine, 3,3′-dimethoxy-

$C_{14}H_{16}N_2O_3$
11089 - 2,4,6(1H,3H,5H)-Pyrimidinetrione, 5,5-diethyl-1-phenyl-

$C_{14}H_{16}O_2Si$
11413 - Silane, dimethoxydiphenyl-
11415 - Silane, dimethyldiphenoxy-

$C_{14}H_{16}O_4$
9978 - Propanedioic acid, (phenylmethylene)-, diethyl ester

$C_{14}H_{16}O_9$
10786 - Pyrano[3,2-c][2]benzopyran-6(2H)-one, 3,4,4a,10b-tetrahydro-3,4,8,10-tetrahydroxy-2-(hydroxymethyl)-9-methoxy-[2R-(2α,3β,4α,4aα,10bβ)]-

$C_{14}H_{16}Si$
11416 - Silane, dimethyldiphenyl-

$C_{14}H_{17}ClNO_4PS_2$
5105 - Dialifor

$C_{14}H_{17}Cl_3O_4$
257 - Acetic acid, (2,4,5-trichlorophenoxy)-, 2-butoxyethyl ester

$C_{14}H_{17}N$
7713 - 1-Naphthalenamine, *N,N*-diethyl-

$C_{14}H_{17}NO_3$
11231 - 1(2H)-Quinolinecarboxylic acid, 2-ethoxy-, ethyl ester

$C_{14}H_{17}NO_6$
955 - Benzeneacetonitrile, α-(β-*D*-glucopyranosyloxy)-

$C_{14}H_{18}$
427 - Anthracene, 1,2,3,4,5,6,7,8-octahydro-
7747 - Naphthalene, 2-butyl-
9184 - Phenanthrene, 1,2,3,4,4a,9,10,10a-octahydro-
9185 - Phenanthrene, 1,2,3,4,5,6,7,8-octahydro-

$C_{14}H_{18}Cl_2O_4$
170 - Acetic acid, (2,4-dichlorophenoxy)-, 2-butoxyethyl ester

$C_{14}H_{18}N_2O_4$
8566 - Oxadixyl

$C_{14}H_{18}N_2O_5$
9511 - *L*-Phenylalanine, *N*-*L*-α-aspartyl-, 1-methyl ester

$C_{14}H_{18}N_4O_3$
515 - Benomyl

$C_{14}H_{18}O$
6347 - Heptanal, 2-(phenylmethylene)-

$C_{14}H_{18}O_4$
9977 - Propanedioic acid, (phenylmethyl)-, diethyl ester

$C_{14}H_{18}O_7$
5818 - Ethanone, 1-[4-(β-*D*-glucopyranosyloxy) phenyl]-

$C_{14}H_{19}Cl_2NO_2$
1101 - Benzenebutanoic acid, 4-[bis (2-chloroethyl)amino]-

$C_{14}H_{19}NO$
11265 - Quinoline, 6-ethoxy-1,2-dihydro-2,2,4-trimethyl-

$C_{14}H_{19}N_3S$
5507 - 1,2-Ethanediamine, *N,N*-dimethyl-*N′*-2-pyridinyl-*N′*-(2-thienylmethyl)-

$C_{14}H_{19}O_6P$
4137 - Ciodrin

$C_{14}H_{20}$
1249 - Benzene, (1-cyclohexylethyl)-
1255 - Benzene, (3-cyclopentylpropyl)-
7999 - Naphthalene, 1,2,3,4-tetrahydro-1,1,2,6-tetramethyl-

$C_{14}H_{20}ClNO_2$
270 - Acetochlor
322 - Alachlor

$C_{14}H_{20}O$
8480 - 1-Octanone, 1-phenyl-

$C_{14}H_{20}O_2$
2695 - Benzoic acid, 4-(1-methylethyl)-, 2-methylpropyl ester

$C_{14}H_{20}O_8$
5916 - Ethenetetracarboxylic acid, tetraethyl ester

$C_{14}H_{21}N_3O_4$
3893 - Butralin

$C_{14}H_{22}$
990 - Benzene, 1,4-bis(1,1-dimethylethyl)-
1981 - Benzene, (1-methylheptyl)-
2067 - Benzene, octyl-
2268 - Benzene, 1,2,3,4-tetraethyl-
2269 - Benzene, 1,2,4,5-tetraethyl-
9178 - Phenanthrene, dodecahydro-
11610 - 6,8-Tetradecadiyne

$C_{14}H_{22}N_2O$
54 - Acetamide, 2-(diethylamino)-*N*-(2,6-dimethylphenyl)-

$C_{14}H_{22}N_2O_3$
11081 - 2,4,6(1H,3H,5H)-Pyrimidinetrione, 5-butyl-1-cyclohexyl-

$C_{14}H_{22}O$
2069 - Benzene, (octyloxy)-
3883 - 3-Buten-2-one, 4-(2,5,6,6-tetramethyl-1-cyclohexen-1-yl)-
3884 - 3-Buten-2-one, 4-(2,5,6,6-tetramethyl-2-cyclohexen-1-yl)-
9219 - Phenol, 2,4-bis(1,1-dimethylethyl)-
9220 - Phenol, 2,6-bis(1,1-dimethylethyl)-
9227 - Phenol, 2,6-bis(1-methylpropyl)-

$C_{14}H_{22}O_2$
1314 - Benzene, 1,4-dibutoxy-
1411 - Benzene, 1-(diethoxymethyl)-4-(1-methylethyl)-
4528 - Cyclohexanol, 1,1′-(1,2-ethynediyl)bis-

$C_{14}H_{22}O_4$
5559 - Ethanedioic acid, dicyclohexyl ester

$C_{14}H_{22}O_6$
10657 - 2-Propenoic acid, 2-methyl-, 1,2-ethanediylbis(oxy-2,1-ethanediyl) ester

$C_{14}H_{22}O_8$
5633 - 1,1,2,2-Ethanetetracarboxylic acid, tetraethyl ester
10166 - 1,2,3-Propanetricarboxylic acid, 2-(acetyloxy)-, triethyl ester

$C_{14}H_{23}N$
684 - Benzenamine, *N,N*-dibutyl-
831 - Benzenamine, 4-octyl-
10880 - Pyridine, 4-(1-butylpentyl)-

$C_{14}H_{23}N_3O_{10}$
6257 - Glycine, *N,N*-bis[2-[bis(carboxymethyl) amino]ethyl]-

$C_{14}H_{24}$
5240 - Dispiro[5.1.5.1]tetradecane
9186 - Phenanthrene, tetradecahydro-

$C_{14}H_{244}NO_4PS_3$
517 - Bensulide

$C_{14}H_{24}N_2$
1285 - 1,3-Benzenediamine, *N,N,N′,N′*-tetraethyl-

$C_{14}H_{24}O_2$
10295 - Propanoic acid, 2-methyl-, 3,7-dimethyl-2,6-octadienyl ester, (*E*)-

$C_{14}H_{26}$
4341 - Cyclohexane, (3-cyclopentylpropyl)-
4386 - Cyclohexane, 1,1′-(1,2-ethanediyl)bis-
4392 - Cyclohexane, 1,1′-ethylidenebis-
11640 - 2-Tetradecyne
11641 - 7-Tetradecyne

$C_{14}H_{26}O_2$
11637 - 4-Tetradecenoic acid
11638 - 5-Tetradecenoic acid
11639 - 9-Tetradecenoic acid
11896 - 12-Tridecenoic acid, methyl ester

$C_{14}H_{26}O_3$
183 - Acetic acid, ethoxy-, 5-methyl-2-(1-methylethyl)cyclohexyl ester, (1α,2β,5α)-
6456 - Heptanoic acid, anhydride

$C_{14}H_{26}O_4$
3319 - Butanedioic acid, dipentyl ester
5045 - Decanedioic acid, diethyl ester
5300 - Dodecanedioic acid, dimethyl ester
6783 - Hexanedioic acid, bis(2-methylpropyl) ester
6784 - Hexanedioic acid, dibutyl ester

$C_{14}H_{27}BrO_2$
5308 - Dodecanoic acid, 2-bromoethyl ester

$C_{14}H_{27}ClO$
11635 - Tetradecanoyl chloride

$C_{14}H_{27}N$
11619 - Tetradecanenitrile

$C_{14}H_{28}$
4460 - Cyclohexane, octyl-
4841 - Cyclopentane, nonyl-
11636 - 1-Tetradecene

$C_{14}H_{28}Br_2$
11616 - Tetradecane, 1,14-dibromo-

$C_{14}H_{28}O$
11633 - 2-Tetradecanone
11634 - 3-Tetradecanone
11892 - 3-Tridecanone, 2-methyl-
11963 - 4-Undecanone, 7-ethyl-2-methyl-

$C_{14}H_{28}O_2$
178 - Acetic acid, dodecyl ester
5312 - Dodecanoic acid, ethyl ester
6464 - Heptanoic acid, heptyl ester
6904 - Hexanoic acid, octyl ester
8156 - Nonanoic acid, pentyl ester
8431 - Octanoic acid, hexyl ester
11622 - Tetradecanoic acid
11882 - Tridecanoic acid, methyl ester

$C_{14}H_{29}Br$
11614 - Tetradecane, 1-bromo-
11873 - Tridecane, 1-bromo-12-methyl-

$C_{14}H_{29}Cl$
11615 - Tetradecane, 1-chloro-

$C_{14}H_{29}N$
4289 - Cyclohexanamine, *N*-(2-ethylhexyl)-
9689 - Piperidine, 1-nonyl-

$C_{14}H_{29}NO$
11611 - Tetradecanamide

$C_{14}H_{30}$
11613 - Tetradecane
11879 - Tridecane, 7-methyl-

$C_{14}H_{30}O$
6430 - Heptane, 1,1′-oxybis-
11630 - 1-Tetradecanol
11631 - 2-Tetradecanol

11632 - 3-Tetradecanol
11886 - 1-Tridecanol, 12-methyl-
11887 - 3-Tridecanol, 2-methyl-
11888 - 4-Tridecanol, 2-methyl

$C_{14}H_{30}O_2$
5297 - Dodecane, 1,1-dimethoxy-
5298 - Dodecane, 1,12-dimethoxy-
11618 - 1,14-Tetradecanediol

$C_{14}H_{30}O_2Sn$
11538 - Stannane, (acetyloxy)tributyl-

$C_{14}H_{30}O_3S$
11620 - 1-Tetradecanesulfonic acid

$C_{14}H_{30}S$
6435 - Heptane, 1,1′-thiobis-
11621 - 1-Tetradecanethiol

$C_{14}H_{31}N$
6352 - 1-Heptanamine, *N*-heptyl-
11612 - 1-Tetradecanamine

$C_{14}H_{32}N_2$
6750 - 1,6-Hexanediamine, *N,N′*-dibutyl-

$C_{14}H_{32}N_2O_4$
10379 - 2-Propanol, 1,1′,1″,1‴-(1,2-ethanediyldinitrilo)tetrakis-

$C_{14}H_{42}O_5Si_6$
6990 - Hexasiloxane, tetradecamethyl-

$C_{14}H_{42}O_7Si_7$
4239 - Cycloheptasiloxane, tetradecamethyl-

$C_{15}H_8O_4$
372 - 2-Anthracenecarboxylic acid, 9,10-dihydro-9,10-dioxo-

$C_{15}H_8O_5$
2466 - 6H-Benzofuro[3,2-c][1]benzopyran-6-one, 3,9-dihydroxy-

$C_{15}H_8O_6$
371 - 2-Anthracenecarboxylic acid, 9,10-dihydro-4,5-dihydroxy-9,10-dioxo-

$C_{15}H_9BrO_2$
7265 - 1H-Indene-1,3(2H)-dione, 2-(4-bromophenyl)-

$C_{15}H_9ClO_2$
7266 - 1H-Indene-1,3(2H)-dione, 2-(4-chlorophenyl)

$C_{15}H_{10}ClF_3N_2O_6S$
5985 - Fomesafen

$C_{15}H_{10}O_2$
368 - 1-Anthracenecarboxylic acid
369 - 2-Anthracenecarboxylic acid
370 - 9-Anthracenecarboxylic acid
409 - 9,10-Anthracenedione, 2-methyl-
2840 - 4H-1-Benzopyran-4-one, 2-phenyl-
2841 - 4H-1-Benzopyran-4-one, 3-phenyl-
7270 - 1H-Indene-1,3(2H)-dione, 2-phenyl-
7588 - Methanone, 2-benzofuranylphenyl-
9165 - 9-Phenanthrenecarboxylic acid

$C_{15}H_{10}O_3$
10199 - Propanetrione, diphenyl-

$C_{15}H_{10}O_4$
398 - 9,10-Anthracenedione, 1,8-dihydroxy-3-methyl-
2811 - 4H-1-Benzopyran-4-one, 5,7-dihydroxy-2-phenyl-
2827 - 4H-1-Benzopyran-4-one, 7-hydroxy-3-(4-hydroxyphenyl)-

$C_{15}H_{10}O_5$
397 - 9,10-Anthracenedione, 1,8-dihydroxy-3-(hydroxymethyl)-
419 - 9,10-Anthracenedione, 1,3,8-trihydroxy-6-methyl-
1332 - 1,2-Benzenedicarboxylic acid, 3-benzoyl-
2500 - Benzoic acid, 4,4′-carbonylbis-
2806 - 4H-1-Benzopyran-4-one, 5,7-dihydroxy-2-(4-hydroxyphenyl)-
2807 - 4H-1-Benzopyran-4-one, 5,7-dihydroxy-3-(4-hydroxyphenyl)-
2844 - 4H-1-Benzopyran-4-one, 5,6,7-trihydroxy-2-phenyl-

$C_{15}H_{10}O_6$
2813 - 4H-1-Benzopyran-4-one, 2-(3,4-dihydroxyphenyl)-3,7-dihydroxy-
2814 - 4H-1-Benzopyran-4-one, 2-(3,4-dihydroxyphenyl)-5,7-dihydroxy-
2842 - 4H-1-Benzopyran-4-one, 3,5,7-trihydroxy-2-(2-hydroxyphenyl)-
2843 - 4H-1-Benzopyran-4-one, 3,5,7-trihydroxy-2-(4-hydroxyphenyl)-

$C_{15}H_{10}O_7$
2816 - 4H-1-Benzopyran-4-one, 2-(2,4-dihydroxyphenyl)-3,5,7-trihydroxy-
2817 - 4H-1-Benzopyran-4-one, 2-(3,4-dihydroxyphenyl)-3,5,7-trihydroxy-

$C_{15}H_{11}ClF_3NO_4$
8633 - Oxyfluorfen

$C_{15}H_{11}ClN_2O$
2411 - 2H-1,4-Benzodiazepin-2-one, 7-chloro-1,3-dihydro-5-phenyl-

$C_{15}H_{11}ClO_5$
2848 - 1-Benzopyrylium, 3,5,7-trihydroxy-2-(4-hydroxyphenyl)-, chloride

$C_{15}H_{11}ClO_7$
2849 - 1-Benzopyrylium, 3,5,7-trihydroxy-2-(3,4,5-trihydroxyphenyl)-, chloride

$C_{15}H_{11}I_3O_4$
2136 - Benzenepropanoic acid, 4-(4-hydroxy-3-iodophenoxy)-3,5-diiodo-

$C_{15}H_{11}I_4NO_4$
11930 - *L*-Tyrosine, *O*-(4-hydroxy-3,5-diiodophenyl)-3,5-diiodo-

$C_{15}H_{11}N$
2768 - Benzonitrile, 4-(2-phenylethenyl)-, (*Z*)-
11288 - Quinoline, 2-phenyl-
11289 - Quinoline, 3-phenyl-
11290 - Quinoline, 4-phenyl-
11291 - Quinoline, 6-phenyl-

$C_{15}H_{11}NO$
8579 - Oxazole, 2,4-diphenyl-
8580 - Oxazole, 2,5-diphenyl-
8581 - Oxazole, 4,5-diphenyl-

$C_{15}H_{11}NO_2$
384 - 9,10-Anthracenedione, 1-amino-2-methyl-
7367 - 1H-Isoindole-1,3(2H)-dione, 2-(phenylmethyl)-

$C_{15}H_{11}NS$
11676 - Thiazole, 2,4-diphenyl-

$C_{15}H_{12}$
423 - Anthracene, 1-methyl-
424 - Anthracene, 2-methyl-
425 - Anthracene, 9-methyl-
9180 - Phenanthrene, 3-methyl-
9181 - Phenanthrene, 4-methyl-

$C_{15}H_{12}I_2O_3$
2135 - Benzenepropanoic acid, 4-hydroxy-3,5-diiodo-α-phenyl-

$C_{15}H_{12}N_2$
10830 - 1H-Pyrazole, 1,3-diphenyl-
10831 - 1H-Pyrazole, 1,5-diphenyl-

$C_{15}H_{12}N_2O$
5144 - 5H-Dibenz[b,f]azepine-5-carboxamide

$C_{15}H_{12}N_2O_2$
7219 - 2,4-Imidazolidinedione, 5,5-diphenyl-
10843 - 3,5-Pyrazolidinedione, 1,4-diphenyl-

$C_{15}H_{12}N_2O_3$
7488 - Methanediamine, 1-(2-furanyl)-*N,N′*-bis(2-furanylmethylene)-

$C_{15}H_{12}N_6O_4$
2494 - Benzoic acid, 4-[[(2-amino-1,4-dihydro-4-oxo-6-pteridinyl)methyl]formylamino]-

$C_{15}H_{12}O$
5146 - 10H-Dibenzo[a,d]cyclohepten-10-one, 5,11-dihydro-
5147 - 5H-Dibenzo[a,d]cyclohepten-5-one, 10,11-dihydro-
7291 - 1H-Inden-1-one, 2,3-dihydro-2-phenyl-
10708 - 2-Propen-1-one, 1,3-diphenyl-
10709 - 2-Propen-1-one, 1,3-diphenyl-, (*E*)-

$C_{15}H_{12}O_2$
941 - Benzeneacetic acid, α-(phenylmethylene)-
10027 - 1,3-Propanedione, 1,3-diphenyl-
10689 - 2-Propenoic acid, 3-phenyl-, phenyl ester, (*E*)-
10711 - 2-Propen-1-one, 3-hydroxy-1,3-diphenyl-

$C_{15}H_{12}O_3$
2516 - Benzoic acid, 2-benzoyl-, methyl ester
2684 - Benzoic acid, 2-(4-methylbenzoyl)-

$C_{15}H_{12}O_4$
2483 - Benzoic acid, 2-(acetyloxy)-, phenyl ester
2667 - Benzoic acid, 2-(4-methoxybenzoyl)-
7509 - Methanediol, dibenzoate

$C_{15}H_{12}O_5$
2802 - 4H-1-Benzopyran-4-one, 2,3-dihydro-5,7-dihydroxy-2-(4-hydroxyphenyl)-, (*S*)-

$C_{15}H_{12}O_6$
2686 - Benzoic acid, 3,3′-methylenebis[6-hydroxy-
2812 - 4H-1-Benzopyran-4-one, 2-(3,4-dihydroxyphenyl)-2,3-dihydro-5,7-dihydroxy-, (*S*)-

$C_{15}H_{13}N$
7318 - 1H-Indole, 3-methyl-2-phenyl-
7321 - 1H-Indole, 1-(phenylmethyl)-

$C_{15}H_{13}NO$
64 - Acetamide, *N*-9H-fluoren-2-yl-

$C_{15}H_{13}NO_2S$
9500 - 10H-Phenothiazine-2-acetic acid, 10-methyl-

$C_{15}H_{14}$
1966 - Benzene, 1,1′-(1-methylene-1,2-ethanediyl)bis-
1971 - Benzene, 1,1′-(1-methyl-1,2-ethenediyl)bis-
2161 - Benzene, 1,1′-(1-propene-1,3-diyl)bis-
2166 - Benzene, 1,1′-(1-propenylidene)bis-
7259 - 1H-Indene, 2,3-dihydro-2-phenyl-

$C_{15}H_{14}F_3N_3$
723 - Benzenamine, *N,N*-dimethyl-4-[[3-(trifluoromethyl)phenyl]azo]-

$C_{15}H_{14}N_2$
765 - Benzenamine, *N,N′*-methanetetraylbis[4-methyl-

$C_{15}H_{14}N_2O$
7228 - 4-Imidazolidinone, 5,5-diphenyl-

$C_{15}H_{14}N_4O_2S$
2201 - Benzenesulfonamide, 4-amino-*N*-(1-phenyl-1H-pyrazol-5-yl)-

$C_{15}H_{14}O$
1939 - Benzene, 1-methoxy-4-(2-phenylethenyl)-, (*E*)-
2782 - 2H-1-Benzopyran, 3,4-dihydro-2-phenyl-
7599 - Methanone, bis(4-methylphenyl)-
7614 - Methanone, (2,4-dimethylphenyl)phenyl-
7615 - Methanone, (3,4-dimethylphenyl)phenyl-
7628 - Methanone, (2-methylphenyl)(4-methylphenyl)-
10432 - 2-Propanone, 1,1-diphenyl-
10433 - 2-Propanone, 1,3-diphenyl-

$C_{15}H_{14}O_2$
931 - Benzeneacetic acid, α-methyl-α-phenyl-
2143 - Benzenepropanoic acid, α-phenyl-, (±)-
5839 - Ethanone, 2-methoxy-1,2-diphenyl-

$C_{15}H_{14}O_3$
920 - Benzeneacetic acid, α-hydroxy-α-phenyl-, methyl ester
3996 - Carbonic acid, bis(2-methylphenyl) ester
7612 - Methanone, (2,4-dimethoxyphenyl)phenyl-
7862 - 1,4-Naphthalenedione, 2-hydroxy-3-(3-methyl-2-butenyl)-

$C_{15}H_{14}O_4$
2398 - 2H,8H-Benzo[1,2-b:5,4-b′]dipyran-2-one, 5-methoxy-8,8-dimethyl-
6188 - 7H-Furo[3,2-g][1]benzopyran-7-one, 3-methoxy-2-(1-methylethyl)-
7841 - 1,4-Naphthalenediol, 2-methyl-, diacetate

$C_{15}H_{14}O_5$
7595 - Methanone, bis(2-hydroxy-4-methoxyphenyl)-
9384 - Phenol, 2-methoxy-, carbonate (2:1)
10453 - 1-Propanone, 3-(4-hydroxyphenyl)-1-(2,4,6-trihydroxyphenyl)-

10790 - 2H-Pyran-2-one, 6-[2-(1,3-benzodioxol-5-yl)ethenyl]-5,6-dihydro-4-methoxy-, [*R*-(*E*)]-

$C_{15}H_{14}O_6$
2846 - 2H-1-Benzopyran-3,5,7-triol, 2-(3,4-dihydroxyphenyl)-3,4-dihydro-, (2*S*-*cis*)-
7248 - 2H-Indene-2,2-dicarboxylic acid, 1,3-dihydro-1,3-dioxo-, diethyl ester
7855 - 1,4-Naphthalenedione, 5,8-dihydroxy-6-methoxy-2-methyl-3-(2-oxopropyl)-
11904 - 2H,4aH-1,4,5-Trioxadicyclopent[a,hi]indene-7-carboxylic acid, 3-ethylidene-3,3a,7a,9b-tetrahydro-2-oxo-, methyl ester, [3aS-(3E,3aα,4aβ,7aβ,9aR*,9bβ)]-

$C_{15}H_{15}ClF_3N_3O$
11900 - Triflumizole

$C_{15}H_{15}ClN_2O_2$
4088 - Chloroxuron

$C_{15}H_{15}ClN_4O_6S$
4081 - Chlorimuron

$C_{15}H_{15}NO$
2771 - Benzophenone, 3-(dimethylamino)-
7613 - Methanone, [4-(dimethylamino)phenyl]phenyl-

$C_{15}H_{15}NO_2$
3946 - Carbamic acid, diphenyl-, ethyl ester

$C_{15}H_{15}N_3O$
317 - 3,9-Acridinediamine, 7-ethoxy-

$C_{15}H_{15}N_3O_2$
2587 - Benzoic acid, 2-[[4-(dimethylamino) phenyl]azo]-

$C_{15}H_{16}$
1696 - Benzene, 1-ethyl-4-(phenylmethyl)-
1964 - Benzene, 1,1′-methylenebis[4-methyl-
1969 - Benzene, 1,1′-(1-methyl-1,2-ethanediyl)bis-
1976 - Benzene, 1,1′-(1-methylethylidene)bis-
2121 - Benzene, 1,1′-(1,3-propanediyl)bis-
2171 - Benzene, 1,1′-propylidenebis-

$C_{15}H_{16}F_5NO_2S_2$
5267 - Dithiopyr

$C_{15}H_{16}N_2O$
12006 - Urea, *N,N′*-dimethyl-*N,N′*-diphenyl-

$C_{15}H_{16}O$
1940 - Benzene, 1-methoxy-4-(1-phenylethyl)-
2158 - Benzenepropanol, β-phenyl-

$C_{15}H_{16}O_2$
1970 - Benzene, 1,1′-[(1-methyl-1,2-ethanediyl)bis(oxy)]bis-
2122 - Benzene, 1,1′-[1,3-propanediylbis(oxy)]bis-
9413 - Phenol, 4,4′-(1-methylethylidene)bis-
9464 - Phenol, 4,4′-propylidenebis-

$C_{15}H_{16}O_3$
10377 - 2-Propanol, 1,3-diphenoxy-

$C_{15}H_{16}O_8$
2821 - 2H-1-Benzopyran-2-one, 7-(β-*D*-glucopyranosyloxy)-

$C_{15}H_{16}O_9$
2822 - 2H-1-Benzopyran-2-one, 6-(β-*D*-glucopyranosyloxy)-7-hydroxy-

$C_{15}H_{17}Cl_2N_3O_2$
10720 - Propiconazole

$C_{15}H_{17}N$
1811 - Benzenemethanamine, *N*-ethyl-*N*-phenyl-

$C_{15}H_{17}N_3$
6279 - Guanidine, *N,N′*-bis(2-methylphenyl)-

$C_{15}H_{18}$
505 - Azulene, 1,4-dimethyl-7-(1-methylethyl)-
506 - Azulene, 4,8-dimethyl-2-(1-methylethyl)-
7830 - Naphthalene, 1,6-dimethyl-4-(1-methylethyl)-
7941 - Naphthalene, 1-pentyl-
7942 - Naphthalene, 2-pentyl-

$C_{15}H_{18}Cl_2N_2O_3$
8565 - Oxadiazon

$C_{15}H_{18}N_2O_6$
3003 - Binapacryl

$C_{15}H_{18}N_6O_6S$
8069 - Nicosulfuron

$C_{15}H_{18}O$
7917 - Naphthalene, 1-(3-methylbutoxy)-
7918 - Naphthalene, 2-(3-methylbutoxy)-
7943 - Naphthalene, 1-(pentyloxy)-
7944 - Naphthalene, 2-(pentyloxy)-

$C_{15}H_{18}O_2$
2939 - Bicyclo[2.2.1]heptan-2-ol, 1-phenyl-, acetate

$C_{15}H_{18}O_3$
8057 - Naphtho[1,2-b]furan-2,8(3H,4H)-dione, 3a,5,5a,9b-tetrahydro-3,5a,9-trimethyl-,

$C_{15}H_{18}O_6$
8058 - Naphtho[1,2-b]furan-2,8(3H,4H)-dione, 3a,5,5a,9b-tetrahydro-4,5-dihydroxy-9-methoxy-3,5a-dimethyl-

$C_{15}H_{19}NO_2$
487 - 8-Azabicyclo[3.2.1]octan-3-ol, 8-methyl-, benzoate (ester), *endo*-
488 - 8-Azabicyclo[3.2.1]octan-3-ol, 8-methyl-, benzoate (ester), *exo*-

$C_{15}H_{19}N_3O_3$
7190 - Imazethapyr

$C_{15}H_{20}N_2O$
7573 - 7,14-Methano-4H,6H-dipyrido[1,2-a1′,2′-e][1,5]diazocin-4-one, 7,7a,8,9,10,11,13,14-octahydro-,

$C_{15}H_{20}O_2$
6293 - Helenine
8059 - Naphtho[2,3-b]furan-2(3H)-one, 3a,5,6,7,8,8a,9,9a-octahydro-5,8a-dimethyl-3-methylene-,

$C_{15}H_{20}O_4$
8679 - 2,4-Pentadienoic acid, 5-(1-hydroxy-2,6,6-trimethyl-4-oxo-2-cyclohexen-1-yl)-3-methyl-, [*S*-(*Z,E*)]-
9952 - Propanedioic acid, ethylphenyl-, diethyl ester

$C_{15}H_{21}NO_4$
7439 - Metalaxyl

$C_{15}H_{21}N_3O_2$
11191 - Pyrrolo[2,3-b]indol-5-ol, 1,2,3,3a,8,8a-hexahydro-1,3a,8-trimethyl-, methylcarbamate (ester), (3a*S*-*cis*)-

$C_{15}H_{22}$
1468 - Benzene, 1-(1,5-dimethyl-4-hexenyl)-4-methyl-
7274 - 1H-Indene, 5-hexyl-2,3-dihydro-

$C_{15}H_{22}ClNO_2$
7657 - Metolachlor

$C_{15}H_{22}N_2O$
9639 - 2-Piperidinecarboxamide, *N*-(2,6-dimethylphenyl)-1-methyl-

$C_{15}H_{22}O$
8052 - 2(3H)-Naphthalenone, 4,4a,5,6,7,8-hexahydro-4,4a-dimethyl-6-(1-methylethylidene)-, (4*R*-*cis*)-
11521 - Spiro[4.5]dec-6-en-8-one, 6,10-dimethyl-2-(1-methylethylidene)-, (5*R*-*cis*)-

$C_{15}H_{23}ClO_4S$
11579 - Sulfurous acid, 2-chloroethyl 2-[4-(1,1-dimethylethyl)phenoxy]-1-methylethyl ester

$C_{15}H_{23}NO_4$
9655 - 2,6-Piperidinedione, 4-[2-(3,5-dimethyl-2-oxocyclohexyl)-2-hydroxyethyl]-,

$C_{15}H_{23}N_3O_4$
7371 - Isopropalin

$C_{15}H_{24}$
2064 - Benzene, nonyl-
2372 - Benzene, 1,2,4-tris(1-methylethyl)-
2373 - Benzene, 1,3,5-tris(1-methylethyl)-
2996 - Bicyclo[7.2.0]undec-4-ene, 4,11,11-trimethyl-8-methylene-, [1*R*-(1*R**,4E,9*S**)]-
4066 - Cedrene
4389 - Cyclohexane, 1-ethenyl-1-methyl-2,4-bis(1-methylethenyl)-, (1α,2β,4β)-
4659 - Cyclohexene, 6-ethenyl-6-methyl-1-(1-methylethyl)-3-(1-methylethylidene)-, (*S*)-
4685 - Cyclohexene, 1-methyl-4-(5-methyl-1-methylene-4-hexenyl)-, (*S*)-
4954 - Cyclopropa[d]naphthalene, 1,1a,4,4a,5,6,7,8-octahydro-2,4a,8,8-tetramethyl-, [1a*S*-(1aα,4aβ,8a*R**)]-
5006 - 1,4,8-Cycloundecatriene, 2,6,6,9-tetramethyl-, (*E,E,E*)-
5329 - 1,3,6,10-Dodecatetraene, 3,7,11-trimethyl-, (*E,E*)-
5331 - 1,6,10-Dodecatriene, 7,11-dimethyl-3-methylene-, (*E*)-
7568 - 1,4-Methanoazulene, decahydro-4,8,8-trimethyl-9-methylene-, [1*S*-(1α,3aβ,4α,8aβ)]-
7898 - Naphthalene, 1,2,4a,5,8,8a-hexahydro-4,7-dimethyl-1-(1-methylethyl)-, [1*S*-(1α,4aβ,8aα)]-
7899 - Naphthalene, 1,2,3,5,6,8a-hexahydro-4,7-dimethyl-1-(1-methylethyl)-, (1*S*-*cis*)-
7936 - Naphthalene, 1,2,3,4,4a,5,6,8a-octahydro-7-methyl-4-methylene-1-(1-methylethyl)-, (1α,4aβ,8aα)-
11861 - Tricyclo[4.4.0.02,7]dec-3-ene, 1,3-dimethyl-8-(1-methylethyl)-, stereoisomer

$C_{15}H_{24}NO_4PS$
7359 - Isofenphos

$C_{15}H_{24}N_2O$
7426 - Matridin-15-one
7572 - 7,14-Methano-2H,6H-dipyrido[1,2-a 1′,2′-e][1,5]diazocin-6-one, dodecahydro-, [7*R*-(7α,7aβ,14α,14aalp

$C_{15}H_{24}N_2O_2$
7574 - 7,14-Methano-4H,6H-dipyrido[1,2-a 1′,2′-e][1,5]diazocin-4-one, dodecahydro-9-hydroxy-,

$C_{15}H_{24}O$
5330 - 2,6,10-Dodecatrienal, 3,7,11-trimethyl-
9093 - 2-Penten-1-ol, 5-(2,3-dimethyltricyclo [2.2.1.02,6]hept-3-yl)-2-methyl-, stereoisomer
9103 - 2-Penten-1-ol, 2-methyl-5-(2-methyl-3-methylenebicyclo[2.2.1]hept-2-yl)-, [1*S*-[1α,2α(*Z*),4α]
9222 - Phenol, 2,4-bis(1,1-dimethylethyl)-5-methyl-
9223 - Phenol, 2,4-bis(1,1-dimethylethyl)-6-methyl-
9224 - Phenol, 2,6-bis(1,1-dimethylethyl)-4-methyl-

$C_{15}H_{24}O_8$
10139 - 1,1,2,3-Propanetetracarboxylic acid, tetraethyl ester
10140 - 1,1,3,3-Propanetetracarboxylic acid, tetraethyl ester

$C_{15}H_{25}ClN_2O_2$
2531 - Benzoic acid, 4-(butylamino)-, 2-(dimethylamino)ethyl ester, monohydrochloride

$C_{15}H_{26}N_2$
7571 - 7,14-Methano-2H,6H-dipyrido[1,2-a 1′,2′-e][1,5]diazocine, dodecahydro-,

$C_{15}H_{26}O$
507 - 5-Azulenemethanol, 1,2,3,4,5,6,7,8-octahydro-α,α,3,8-tetramethyl-,
4989 - 1H-Cycloprop[e]azulen-4-ol, decahydro-1,1,4,7tetramethyl,(1a*R*,4*R*,4a*S*,7*R*,7a*S*,7b*S*)-
5332 - 1,6,10-Dodecatrien-3-ol, 3,7,11-trimethyl-, [*S*-(*Z*)]-
5333 - 2,6,10-Dodecatrien-1-ol, 3,7,11-trimethyl-, (*E,E*)-
5334 - 2,6,10-Dodecatrien-1-ol, 3,7,11-trimethyl-, (*Z,E*)-
7569 - 1,4-Methanoazulen-9-ol, decahydro-1,5,5,8a-tetramethyl-, [1*R*-(1α,3aβ,4α,8aβ,9*S**)]
7570 - 1H-3a,7-Methanoazulen-6-ol, octahydro-3,6,8,8-tetramethyl-,
7586 - 1,6-Methanonaphthalen-1(2H)-ol, octahydro-4,8a,9,9-tetramethyl-,

$C_{15}H_{26}O_2$
3565 - Butanoic acid, 3-methyl-, 1,7,7-trimethylbicyclo[2.2.1]hept-2-yl ester, (1*R*-*endo*)-

$C_{15}H_{26}O_6$
3586 - Butanoic acid, 1,2,3-propanetriyl ester

$C_{15}H_{26}Si$
- 11448 - Silane, phenyltripropyl-

$C_{15}H_{28}$
- 4430 - Cyclohexane, 1,1′-(1-methyl-1,2-ethanediyl)bis-
- 4432 - Cyclohexane, 1,1′-(1-methylethylidene)bis-
- 4466 - Cyclohexane, 1,1′-(1,3-propanediyl)bis-
- 4472 - Cyclohexane, 1,1′-propylidenebis-
- 8656 - 1-Pentadecyne

$C_{15}H_{28}NNaO_3$
- 6269 - Glycine, *N*-methyl-*N*-(1-oxododecyl)-, sodium salt

$C_{15}H_{28}O$
- 4761 - Cyclopentadecanone

$C_{15}H_{28}O_2$
- 3554 - Butanoic acid, 3-methyl-, 5-methyl-2-(1-methylethyl)cyclohexyl ester
- 8562 - Oxacyclohexadecan-2-one

$C_{15}H_{28}O_4$
- 9937 - Propanedioic acid, dibutyl-, diethyl ester

$C_{15}H_{30}$
- 4401 - Cyclohexane, 1-ethyl-1-methyl-2,4-bis(1-methylethyl)-, [1*R*-(1α,2β,4β)]-
- 4457 - Cyclohexane, nonyl-
- 4759 - Cyclopentadecane
- 4790 - Cyclopentane, decyl-
- 8653 - 1-Pentadecene
- 8654 - 7-Pentadecene

$C_{15}H_{30}N_2$
- 9702 - Piperidine, 4,4′-(1,3-propanediyl)bis[1-methyl-

$C_{15}H_{30}O$
- 4760 - Cyclopentadecanol
- 8640 - Pentadecanal
- 8651 - 2-Pentadecanone
- 8652 - 8-Pentadecanone
- 8655 - 1-Pentadecen-3-ol

$C_{15}H_{30}O_2$
- 5316 - Dodecanoic acid, 1-methylethyl ester
- 5320 - Dodecanoic acid, propyl ester
- 6468 - Heptanoic acid, octyl ester
- 8430 - Octanoic acid, heptyl ester
- 8645 - Pentadecanoic acid
- 11625 - Tetradecanoic acid, methyl ester

$C_{15}H_{30}O_4$
- 5310 - Dodecanoic acid, 2,3-dihydroxypropyl ester, (±)-

$C_{15}H_{31}Br$
- 8643 - Pentadecane, 1-bromo-

$C_{15}H_{32}$
- 8642 - Pentadecane

$C_{15}H_{32}O$
- 8647 - 2-Pentadecanol
- 8648 - 3-Pentadecanol

$C_{15}H_{33}B$
- 3122 - Borane, tris(3-methylbutyl)-

$C_{15}H_{33}N$
- 3224 - 1-Butanamine, 2-methyl-*N*,*N*-bis(2-methylbutyl)-
- 3225 - 1-Butanamine, 3-methyl-*N*,*N*-bis(3-methylbutyl)-
- 8641 - 1-Pentadecanamine
- 8705 - 1-Pentanamine, *N*,*N*-dipentyl-

$C_{13}H_{29}N_3 \cdot C_2H_4O_2$
- 5347 - Dodine

$C_{15}H_{33}O_4P$
- 9565 - Phosphoric acid, tripentyl ester

$C_{16}H_8O_2S_2$
- 2404 - Benzo[b]thiophen-3(2H)-one, 2-(3-oxobenzo[b]thien-2(3H)-ylidene)-

$C_{16}H_9N_3Na_2O_{10}S_2$
- 7878 - 2,7-Naphthalenedisulfonic acid, 4,5-dihydroxy-3-[(4-nitrophenyl)azo]-, disodium salt

$C_{16}H_{10}$
- 5966 - Fluoranthene
- 10853 - Pyrene

$C_{16}H_{10}Cl_4O_5$
- 5153 - 11H-Dibenzo[b,e][1,4]dioxepin-11-one, 2,4,7,9-tetrachloro-3-hydroxy-8-methoxy-1,6-dimethyl-

$C_{16}H_{10}N_2$
- 2394 - Benzo[a]phenazine

$C_{16}H_{10}N_2O_2$
- 7331 - 3H-Indol-3-one, 2-(1,3-dihydro-3-oxo-2H-indol-2-ylidene)-1,2-dihydro-

$C_{16}H_{11}N$
- 2405 - 7H-Benzo[c]carbazole

$C_{16}H_{11}NO$
- 7640 - Methanone, phenyl-4-quinolinyl-

$C_{16}H_{11}NO_2$
- 7866 - 1,4-Naphthalenedione, 2-(phenylamino)-
- 11235 - 4-Quinolinecarboxylic acid, 2-phenyl-
- 11309 - 8-Quinolinol, benzoate (ester)

$C_{16}H_{11}NO_3$
- 11233 - 4-Quinolinecarboxylic acid, 3-hydroxy-2-phenyl-

$C_{16}H_{12}$
- 7947 - Naphthalene, 1-phenyl-
- 7948 - Naphthalene, 2-phenyl-

$C_{16}H_{12}N_2O$
- 8036 - 1-Naphthalenol, 2-(phenylazo)-
- 8037 - 2-Naphthalenol, 1-(phenylazo)-

$C_{16}H_{12}O$
- 6100 - Furan, 2,5-diphenyl-
- 7945 - Naphthalene, 1-phenoxy-
- 7946 - Naphthalene, 2-phenoxy-

$C_{16}H_{12}O_2$
- 402 - 9,10-Anthracenedione, 1,4-dimethyl-
- 403 - 9,10-Anthracenedione, 2,3-dimethyl-
- 404 - 9,10-Anthracenedione, 2,6-dimethyl-

$C_{16}H_{12}O_3$
- 7269 - 1H-Indene-1,3(2H)-dione, 2-(4-methoxyphenyl)-

$C_{16}H_{12}O_4$
- 2830 - 4H-1-Benzopyran-4-one, 7-hydroxy-3-(4-methoxyphenyl)-
- 3780 - 2-Butenedioic acid (*Z*)-, diphenyl ester

$C_{16}H_{12}O_5$
- 2809 - 4H-1-Benzopyran-4-one, 5,7-dihydroxy-2-(4-methoxyphenyl)-
- 2810 - 4H-1-Benzopyran-4-one, 5,7-dihydroxy-6-methoxy-2-phenyl-
- 2828 - 4H-1-Benzopyran-4-one, 5-hydroxy-3-(4-hydroxyphenyl)-7-methoxy-

$C_{16}H_{12}O_6$
- 628 - Benz[b]indeno[1,2-d]pyran-9(6H)-one, 6a,7-dihydro-3,4,6a,10-tetrahydroxy-

$C_{16}H_{12}O_7$
- 2815 - 4H-1-Benzopyran-4-one, 2-(3,4-dihydroxyphenyl)-3,5-dihydroxy-7-methoxy-

$C_{16}H_{12}S$
- 7951 - Naphthalene, 1-(phenylthio)-

$C_{16}H_{13}ClN_2O$
- 2410 - 2H-1,4-Benzodiazepin-2-one, 7-chloro-1,3-dihydro-1-methyl-5-phenyl-

$C_{16}H_{13}N$
- 7387 - Isoquinoline, 3-(phenylmethyl)-
- 7728 - 2-Naphthalenamine, *N*-phenyl-

$C_{16}H_{13}N_3$
- 7729 - 2-Naphthalenamine, 1-(phenylazo)-

$C_{16}H_{14}$
- 377 - Anthracene, 1,3-dimethyl-
- 378 - Anthracene, 2,10-dimethyl-
- 420 - Anthracene, 9-ethyl-
- 1095 - Benzene, 1,1′-(1,3-butadiene-1,4-diyl)bis-, (*E*,*E*)-
- 1096 - Benzene, 1,1′-(1,3-butadiene-1,4-diyl)bis-, (*Z*,*Z*)-
- 9172 - Phenanthrene, 9,10-dimethyl-
- 9179 - Phenanthrene, 9-ethyl-

$C_{16}H_{14}ClN_3O$
- 2409 - 3H-1,4-Benzodiazepin-2-amine, 7-chloro-*N*-methyl-5-phenyl-, 4-oxide

$C_{16}H_{14}Cl_2O_3$
- 4082 - Chlorobenzilate

$C_{16}H_{14}Cl_2O_4$
- 5171 - Diclofop-methyl

$C_{16}H_{14}N_2O$
- 11199 - 4(3H)-Quinazolinone, 2-methyl-3-(2-methylphenyl)-
- 11200 - 4(1H)-Quinazolinone, 1-methyl-2-(phenylmethyl)-

$C_{16}H_{14}O$
- 3876 - 2-Buten-1-one, 1,3-diphenyl-

$C_{16}H_{14}O_2$
- 10690 - 2-Propenoic acid, 3-phenyl-, phenylmethyl ester, (*E*)-
- 10712 - 2-Propen-1-one, 3-(3-methoxyphenyl)-1-phenyl-
- 10713 - 2-Propen-1-one, 3-(4-methoxyphenyl)-1-phenyl-

$C_{16}H_{14}O_3$
- 887 - Benzeneacetic acid, anhydride
- 2515 - Benzoic acid, 2-benzoyl-, ethyl ester
- 2682 - Benzoic acid, 2-methyl-, anhydride
- 2683 - Benzoic acid, 3-methyl-, anhydride

$C_{16}H_{14}O_4$
- 3039 - [1,1′-Biphenyl]-2,2′-dicarboxylic acid, dimethyl ester
- 3320 - Butanedioic acid, diphenyl ester
- 5572 - 1,2-Ethanediol, dibenzoate
- 6189 - 7H-Furo[3,2-g][1]benzopyran-7-one, 9-[(3-methyl-2-butenyl)oxy]-

$C_{16}H_{14}O_6$
- 629 - Benz[b]indeno[1,2-d]pyran-3,4,6a,9,10(6H)-pentol, 7,11b-dihydro-, *cis*-(+)-
- 2800 - 4H-1-Benzopyran-4-one, 2,3-dihydro-5,7-dihydroxy-2-(3-hydroxy-4-methoxyphenyl)-(*S*)-
- 2801 - 4H-1-Benzopyran-4-one, 2,3-dihydro-5,7-dihydroxy-3-(3-hydroxy-4-methoxyphenyl)-

$C_{16}H_{15}ClO_2$
- 892 - Benzeneacetic acid, α-chloro-α-phenyl-, ethyl ester

$C_{16}H_{15}Cl_3O_2$
- 2329 - Benzene, 1,1′-(2,2,2-trichloroethylidene)bis[4-methoxy-

$C_{16}H_{16}$
- 421 - Anthracene, 9-ethyl-9,10-dihydro-
- 1113 - Benzene, 1,1′-(1-butenylidene)bis-
- 1456 - Benzene, 1,1′-(1,2-dimethyl-1,2-ethenediyl)bis-, (*E*)-
- 1457 - Benzene, 1,1′-(1,2-dimethyl-1,2-ethenediyl)bis-, (*Z*)-
- 1622 - Benzene, 1,1′-(1,2-ethenediyl)bis[2-methyl-
- 2022 - Benzene, 1,1′-(3-methyl-1-propene-1,3-diyl)bis-
- 11560 - Stilbene, α-ethyl-

$C_{16}H_{16}N_2OS$
- 9502 - 10H-Phenothiazine, 10-[(dimethylamino)acetyl]-

$C_{16}H_{16}N_2O_2$
- 5375 - Ergoline-8-carboxylic acid, 9,10-didehydro-6-methyl-, (8α)-
- 5376 - Ergoline-8-carboxylic acid, 9,10-didehydro-6-methyl-, (8β)-

$C_{16}H_{16}N_2O_4$
- 5104 - Desmedipham
- 9198 - Phenmedipham

$C_{16}H_{16}N_2O_4S$
- 91 - Acetamide, *N*,*N*′-(sulfonyldi-4,1-phenylene)bis-

$C_{16}H_{16}O$
- 3658 - 2-Butanone, 3,3-diphenyl-

$C_{16}H_{16}O_2$
- 939 - Benzeneacetic acid, α-phenyl-, ethyl ester
- 940 - Benzeneacetic acid, 2-phenylethyl ester
- 1621 - Benzene, 1,1′-(1,2-ethenediyl)bis[4-methoxy-

2144 - Benzenepropanoic acid, α-(phenylmethyl)-
2145 - Benzenepropanoic acid, phenylmethyl ester
3019 - [1,1′-Biphenyl]-4-butanoic acid
5813 - Ethanone, 2-ethoxy-1,2-diphenyl-

$C_{16}H_{16}O_3$
919 - Benzeneacetic acid, α-hydroxy-α-phenyl-, ethyl ester
3583 - Butanoic acid, 2-phenoxy-, phenyl ester

$C_{16}H_{16}O_4$
7806 - 1,8-Naphthalenedicarboxylic acid, diethyl ester

$C_{16}H_{16}O_5$
7854 - 1,4-Naphthalenedione, 5,8-dihydroxy-6-(1-hydroxy-4-methyl-3-pentenyl)-, (*S*)-

$C_{16}H_{16}Si$
11404 - Silane, diethenyldiphenyl-

$C_{16}H_{17}N$
722 - Benzenamine, *N,N*-dimethyl-4-(2-phenylethenyl)-, (*E*)-
3984 - 9H-Carbazole, 9-butyl-

$C_{16}H_{17}NO$
5219 - Diphenamid

$C_{16}H_{17}NO_3$
7665 - Morphinan-3,6-diol, 7,8-didehydro-4,5-epoxy-, (5α,6α)-

$C_{16}H_{17}NO_4$
6201 - Galanthan-1,2-diol, 3,12-didehydro-9,10-[methylenebis(oxy)]-, (1α,2β)-

$C_{16}H_{17}N_2NaO_4S$
11655 - 4-Thia-1-azabicyclo[3.2.0]heptane-2-carboxylic acid, 3,3-dimethyl-7-oxo-6-[(phenylacetyl)amino]-

$C_{16}H_{17}N_3O$
5374 - Ergoline-8-carboxamide, 9,10-didehydro-6-methyl-, (8β)-

$C_{16}H_{18}$
1099 - Benzene, 1,1′-(1,4-butanediyl)bis-
1120 - Benzene, 1,1′-butylidenebis-
1589 - Benzene, 1,1′-(1,2-ethanediyl)bis[3-methyl-
1590 - Benzene, 1,1′-(1,2-ethanediyl)bis[4-methyl-
1676 - Benzene, 1,1′-(1-ethyl-1,2-ethanediyl)bis-
1979 - Benzene, 1-(1-methylethyl)-4-(phenylmethyl)-
2020 - Benzene, 1,1′-(1-methyl-1,3-propanediyl)bis-, (±)-
2021 - Benzene, 1,1′-(2-methyl-1,3-propanediyl)bis-
2111 - Benzene, 1-(phenylmethyl)-4-propyl-
3095 - 1,1′-Biphenyl, 2,2′,4,4′-tetramethyl-
3096 - 1,1′-Biphenyl, 2,2′,5,5′-tetramethyl-

$C_{16}H_{18}ClN_3S$
9506 - Phenothiazin-5-ium, 3,7-bis(dimethylamino)-, chloride

$C_{16}H_{18}N_2$
9613 - Piperazine, 1,4-diphenyl-

$C_{16}H_{18}N_2O_2$
5112 - Diazene, bis(2-ethoxyphenyl)-
5113 - Diazene, bis(4-ethoxyphenyl)-

$C_{16}H_{18}N_2O_3$
6144 - 2(3H)-Furanone, dihydro-3-(hydroxyphenylmethyl)-4-[(1-methyl-1H-imidazol-5-yl)methyl]-,

$C_{16}H_{18}N_4O_2$
10910 - 4-Pyridinecarboxylic acid, 2-[3-oxo-3-[(phenylmethyl)amino]propyl]hydrazide

$C_{16}H_{18}N_4O_7S$
516 - Bensulfuron-methyl

$C_{16}H_{18}O$
1108 - Benzenebutanol, β-phenyl-, (*R*)-
2074 - Benzene, 1,1′-oxybis[2-ethyl-
2080 - Benzene, 1,1′-(oxydi-2,1-ethanediyl)bis-
2081 - Benzene, 1,1′-(oxydiethylidene)bis-, (*R**,*R**)-(±)-

$C_{16}H_{18}O_2$
3048 - 1,1′-Biphenyl, 2,2′-dimethoxy-5,5′-dimethyl-
3049 - 1,1′-Biphenyl, 2,5′-dimethoxy-2′,5-dimethyl-
3069 - [1,1′-Biphenyl]-4,4′-diol, 3,3′,5,5′-tetramethyl-
9249 - Phenol, 4,4′-butylidenebis-
9433 - Phenol, 4,4′-(1-methylpropylidene)bis-

$C_{16}H_{18}O_{10}$
2823 - 2H-1-Benzopyran-2-one, 8-(β-*D*-glucopyranosyloxy)-7-hydroxy-6-methoxy-

$C_{16}H_{19}Cl_2N$
1801 - Benzenemethanamine, *N*-(2-chloroethyl)-*N*-(phenylmethyl)-, hydrochloride

$C_{16}H_{19}N$
744 - Benzenamine, 2-ethyl-*N*-(2-ethylphenyl)-
1585 - Benzeneethanamine, *N*-(2-phenylethyl)-
1822 - Benzenemethanamine, α-methyl-*N*-(1-phenylethyl)-

$C_{16}H_{19}NO_3$
6182 - Furo[2,3-b]quinolin-4(2H)-one, 3,9-dihydro-8-methoxy-9-methyl-2-(1-methylethyl)-, (*R*)-
8567 - 16(15H)-Oxaerythrinan-15-one, 1,2,6,7-tetradehydro-14,17-dihydro-3-methoxy-, (3β)-

$C_{16}H_{19}NO_4$
481 - 8-Azabicyclo[3.2.1]octane-2-carboxylic acid, 3-(benzoyloxy)-8-methyl-, [1*R*-(*exo,exo*)]-

$C_{16}H_{20}$
7900 - Naphthalene, 1-hexyl-
7901 - Naphthalene, 2-hexyl-
7905 - Naphthalene, 1-isopropyl-2,4,7-trimethyl-
8014 - Naphthalene, 1,2,5-trimethyl-8-(1-methylethyl)-
8015 - Naphthalene, 1,3,8-trimethyl-5-(1-methylethyl)-

$C_{16}H_{20}N_2$
3036 - [1,1′-Biphenyl]-4,4′-diamine, tetramethyl-
5500 - 1,2-Ethanediamine, *N,N*′-bis(phenylmethyl)-
11010 - 2-Pyridinepropanamine, *N,N*-dimethyl-γ-phenyl-

$C_{16}H_{20}N_4$
361 - Aniline, 3,3′-azobis[*N,N*-dimethyl-

$C_{16}H_{20}O_2Si$
11406 - Silane, diethoxydiphenyl-

$C_{16}H_{20}O_6P_2S_3$
1 - Abate

$C_{16}H_{21}NO_3$
911 - Benzeneacetic acid, α-hydroxy-, 8-methyl-8-azabicyclo[3.2.1]oct-3-yl ester, *endo*-(±)-
916 - Benzeneacetic acid, α-(hydroxymethyl)-, 8-azabicyclo[3.2.1]oct-3-yl ester, [3(*S*)-*endo*]-

$C_{16}H_{21}N_3$
5505 - 1,2-Ethanediamine, *N,N*-dimethyl-*N*′-(phenylmethyl)-*N*′-2-pyridinyl-

$C_{16}H_{22}ClNO_3$
5173 - Diethatyl-ethyl

$C_{16}H_{22}N_4O_2$
9795 - Propanamide, *N*-(2,3-dihydro-1,5-dimethyl-3-oxo-2-phenyl-1H-pyrazol-4-yl)-2-(dimethylamino)-

$C_{16}H_{22}OSi_2$
5238 - Disiloxane, 1,1,3,3-tetramethyl-1,3-diphenyl-

$C_{16}H_{22}O_4$
1336 - 1,2-Benzenedicarboxylic acid, bis(2-methylpropyl) ester
1338 - 1,4-Benzenedicarboxylic acid, bis(2-methylpropyl) ester
1339 - 1,2-Benzenedicarboxylic acid, dibutyl ester

$C_{16}H_{22}O_6$
1333 - 1,2-Benzenedicarboxylic acid, bis(2-ethoxyethyl) ester

$C_{16}H_{22}O_8$
6226 - β-*D*-Glucopyranoside, 4-(3-hydroxy-1-propenyl)-2-methoxyphenyl

$C_{16}H_{22}O_{11}$
6216 - α-*D*-Glucopyranose, pentaacetate
6217 - β-*D*-Glucopyranose, pentaacetate

$C_{16}H_{23}ClN_3O$
11589 - Tebuconazole

$C_{16}H_{24}N_2O_2$
1325 - 1,2-Benzenedicarboxamide, *N,N,N*′,*N*′-tetraethyl-
1326 - 1,3-Benzenedicarboxamide, *N,N,N*′,*N*′-tetraethyl-

$C_{16}H_{25}NO_2$
898 - Benzeneacetic acid, α-ethyl-, 2-(diethylamino)ethyl ester

$C_{16}H_{26}$
1258 - Benzene, decyl-
2090 - Benzene, pentaethyl-
6627 - 6,10-Hexadecadiyne
11609 - 5,9-Tetradecadien-7-yne, 6,9-dimethyl-

$C_{16}H_{26}O$
9221 - Phenol, 2,6-bis(1,1-dimethylethyl)-4-ethyl-

$C_{16}H_{26}O_2$
1501 - 1,4-Benzenediol, 2,5-bis(1,1-dimethylpropyl)-
6666 - 6,10,14-Hexadecatrienoic acid

$C_{16}H_{28}O_2$
4940 - 2-Cyclopentene-1-undecanoic acid, (*R*)-

$C_{16}H_{30}$
4323 - Cyclohexane, 1,1′-(1,4-butanediyl)bis-
4326 - Cyclohexane, 1,1′-butylidenebis-
4447 - Cyclohexane, 1,1′-(2-methyl-1,3-propanediyl)bis-
6673 - 1-Hexadecyne
6674 - 2-Hexadecyne

$C_{16}H_{30}O$
4762 - Cyclopentadecanone, 2-methyl-
4763 - Cyclopentadecanone, 3-methyl-

$C_{16}H_{30}O_2$
6668 - 7-Hexadecenoic acid
6669 - 9-Hexadecenoic acid
6670 - 9-Hexadecenoic acid, (*Z*)-

$C_{16}H_{30}O_3$
8424 - Octanoic acid, anhydride

$C_{16}H_{31}ClO$
6665 - Hexadecanoyl chloride

$C_{16}H_{31}N$
6641 - Hexadecanenitrile

$C_{16}H_{32}$
4344 - Cyclohexane, decyl-
4863 - Cyclopentane, undecyl-
6667 - 1-Hexadecene
10580 - 1-Propene, 2-methyl-, tetramer

$C_{16}H_{32}Br_2$
6635 - Hexadecane, 1,16-dibromo-

$C_{16}H_{32}O$
6628 - Hexadecanal
6663 - 3-Hexadecanone

$C_{16}H_{32}O_2$
5065 - Decanoic acid, 2-hexyl-
6644 - Hexadecanoic acid
8152 - Nonanoic acid, heptyl ester
8437 - Octanoic acid, octyl ester
8646 - Pentadecanoic acid, methyl ester
11624 - Tetradecanoic acid, ethyl ester

$C_{16}H_{32}O_4$
5313 - Dodecanoic acid, 2-(2-hydroxyethoxy)ethyl ester

$C_{16}H_{33}Br$
6633 - Hexadecane, 1-bromo-

$C_{16}H_{33}Cl$
6634 - Hexadecane, 1-chloro-

$C_{16}H_{33}F$
6639 - Hexadecane, 1-fluoro-

$C_{16}H_{33}I$
6640 - Hexadecane, 1-iodo-

$C_{16}H_{33}NO$
5290 - Dodecanamide, *N,N*-diethyl-
6629 - Hexadecanamide

$C_{16}H_{34}$
6632 - Hexadecane

$C_{16}H_{34}O$
6431 - Heptane, 3,3′-[oxybis(methylene)]bis-
6659 - 1-Hexadecanol
6660 - 2-Hexadecanol
6661 - 3-Hexadecanol
8416 - Octane, 1,1′-oxybis-
8649 - 1-Pentadecanol, 2-methyl-
8650 - 6-Pentadecanol, 6-methyl-

MOLECULAR FORMULA INDEX OF ORGANIC COMPOUNDS (continued)

$C_{17}H_{34}O_2$
6314 - Heptadecanoic acid
6653 - Hexadecanoic acid, methyl ester
11626 - Tetradecanoic acid, 1-methylethyl ester
11629 - Tetradecanoic acid, propyl ester

$C_{17}H_{35}Br$
6307 - Heptadecane, 1-bromo-

$C_{17}H_{35}N$
9660 - Piperidine, 1-dodecyl-

$C_{17}H_{36}$
6306 - Heptadecane

$C_{17}H_{36}O$
6317 - 1-Heptadecanol
6318 - 2-Heptadecanol
6319 - 9-Heptadecanol

$C_{17}H_{37}N$
6305 - 1-Heptadecanamine
8353 - 1-Octanamine, *N*-methyl-*N*-octyl-

$C_{18}H_{10}O_2$
4116 - 5,6-Chrysenedione
7707 - 1,2-Naphthacenedione

$C_{18}H_{10}O_6$
2999 - [2,2′-Bi-1H-indene]-1,1′,3,3′(2H,2′H)-tetrone, 2,2′-dihydroxy-

$C_{18}H_{12}$
519 - Benz[a]anthracene
2406 - Benzo[c]phenanthrene
4114 - Chrysene
7703 - Naphthacene
11910 - Triphenylene

$C_{18}H_{12}N_2$
3107 - 2,3′-Biquinoline

$C_{18}H_{12}N_6$
11843 - 1,3,5-Triazine, 2,4,6-tri-2-pyridinyl-

$C_{18}H_{13}N$
4113 - 6-Chrysenamine

$C_{18}H_{13}NO_3$
2702 - Benzoic acid, 2-[(1-naphthalenylamino)carbonyl]-

$C_{18}H_{13}N_3O$
7334 - Indolo[2′,3′:3,4]pyrido[2,1-b]quinazolin-5(7H)-one, 8,13-dihydro-

$C_{18}H_{14}$
11595 - 1,1′:2′,1″-Terphenyl
11596 - 1,1′:3′,1″-Terphenyl
11597 - 1,1′:4′,1″-Terphenyl

$C_{18}H_{14}N_4O_5S$
2654 - Benzoic acid, 2-hydroxy-5-[[4-[(2-pyridinylamino)sulfonyl]phenyl]azo]-

$C_{18}H_{14}O$
5859 - Ethanone, 1-(1-naphthalenyl)-2-phenyl-
7629 - Methanone, (2-methylphenyl)-1-naphthalenyl-

$C_{18}H_{14}O_3$
10684 - 2-Propenoic acid, 3-phenyl-, anhydride

$C_{18}H_{15}As$
450 - Arsine, triphenyl-

$C_{18}H_{15}Bi$
3111 - Bismuthine, triphenyl-

$C_{18}H_{15}ClSn$
11541 - Stannane, chlorotriphenyl-

$C_{18}H_{15}N$
728 - Benzenamine, *N,N*-diphenyl-

$C_{18}H_{15}OP$
9518 - Phosphine oxide, triphenyl-

$C_{18}H_{15}O_3P$
9593 - Phosphorous acid, triphenyl ester

$C_{18}H_{15}O_4P$
9566 - Phosphoric acid, triphenyl ester

$C_{18}H_{15}P$
9523 - Phosphine, triphenyl-

$C_{18}H_{15}Sb$
11551 - Stibine, triphenyl-

$C_{18}H_{16}N_2$
1271 - 1,4-Benzenediamine, *N,N′*-diphenyl-
7173 - Hydrazine, triphenyl-

$C_{18}H_{16}N_2O$
8028 - 2-Naphthalenol, 1-[(2,4-dimethylphenyl)azo]-
8029 - 2-Naphthalenol, 1-[(2,5-dimethylphenyl)azo]-

$C_{18}H_{16}OSi$
11504 - Silanol, triphenyl-

$C_{18}H_{16}OSn$
11911 - Triphenyltin hydroxide

$C_{18}H_{16}O_2$
9177 - 9,10-Phenanthrenedione, 1-methyl-7-(1-methylethyl)-
10691 - 2-Propenoic acid, 3-phenyl-, 3-phenyl-2-propenyl ester

$C_{18}H_{16}O_4$
3768 - 2-Butenedioic acid (*E*)-, bis(phenylmethyl) ester

$C_{18}H_{18}$
9183 - Phenanthrene, 1-methyl-7-(1-methylethyl)-

$C_{18}H_{18}O_2$
5397 - Estra-1,3,5,7,9-pentaen-17-one, 3-hydroxy-
9302 - Phenol, 4,4′-(1,2-diethylidene-1,2-ethanediyl)bis-

$C_{18}H_{18}O_4$
3298 - Butanedioic acid, bis(phenylmethyl) ester

$C_{18}H_{18}O_5$
5755 - Ethanol, 2,2′-oxybis-, dibenzoate

$C_{18}H_{18}O_6$
3306 - Butanedioic acid, 2,3-dihydroxy- [*R*-(*R**,*R**)]-, bis(phenylmethyl) ester

$C_{18}H_{19}NO_2$
5160 - 4H-Dibenzo[de,g]quinolin-11-ol, 5,6,6a,7-tetrahydro-10-methoxy-6-methyl-, (*R*)-

$C_{18}H_{19}NO_3$
5159 - 4H-Dibenzo[de,g]quinolin-10-ol, 5,6,6a,7-tetrahydro-1,2-dimethoxy-, (*R*)-

$C_{18}H_{19}NO_4$
7675 - Morphinan-6-one, 7,8-didehydro-4,5-epoxy-14-hydroxy-3-methoxy-17-methyl-, (5α)-

$C_{18}H_{20}$
7263 - 1H-Indene, 2,3-dihydro-1,1,3-trimethyl-3-phenyl-

$C_{18}H_{20}Cl_2$
9160 - Perthane

$C_{17}H_{17}N_2{\cdot}CH_3O_4S$
5175 - Difenzoquat methyl sulfate

$C_{18}H_{20}O_2$
5398 - Estra-1,3,5(10),7-tetraen-17-one, 3-hydroxy-
9300 - Phenol, 4,4′-(1,2-diethyl-1,2-ethenediyl)bis-, (*E*)-

$C_{18}H_{21}NO$
9676 - 2-Piperidinemethanol, α,α-diphenyl-

$C_{18}H_{21}NO_3$
7669 - Morphinan-6-ol, 7,8-didehydro-4,5-epoxy-3-methoxy-17-methyl-, (5α,6α)-
7670 - Morphinan-6-ol, 7,8-didehydro-4,5-epoxy-3-methoxy-17-methyl-, (5α,6β)-
7671 - Morphinan-6-ol, 8,14-didehydro-4,5-epoxy-3-methoxy-17-methyl-, (5α,6α)-
7672 - Morphinan-8-ol, 6,7-didehydro-4,5-epoxy-3-methoxy-17-methyl-, (5α,8β)-
7677 - Morphinan-6-one, 7,8-didehydro-4-hydroxy-3-methoxy-17-methyl-
7678 - Morphinan-6-one, 4,5-epoxy-3-methoxy-17-methyl-, (5α)-

$C_{18}H_{22}$
3076 - 1,1′-Biphenyl, 2,2′,4,4′,6,6′-hexamethyl-

$C_{18}H_{22}N_2S$
9503 - 10H-Phenothiazine-10-ethanamine, *N,N*-diethyl-

$C_{18}H_{22}O_2$
5406 - Estra-1,3,5(10)-trien-17-one, 3-hydroxy-
5407 - Estra-1,3,5(10)-trien-17-one, 3-hydroxy-, (8α)-
9298 - Phenol, 4,4′-(1,2-diethyl-1,2-ethanediyl)bis-, (*R**,*S**)-

$C_{18}H_{22}O_4$
1515 - 1,2-Benzenediol, 4,4′-(2,3-dimethyl-1,4-butanediyl)bis-

$C_{18}H_{23}ClN_2O_2$
5670 - Ethanimidamide, *N,N′*-bis(4-ethoxyphenyl)-, monohydrochloride

$C_{18}H_{23}NO_3$
7674 - Morphinan-6-ol, 4,5-epoxy-3-methoxy-17-methyl-, (5α,6α)-

$C_{18}H_{24}$
7937 - Naphthalene, 1-octyl-
7938 - Naphthalene, 2-octyl-

$C_{18}H_{24}NO_7P$
7673 - Morphinan-6-ol, 7,8-didehydro-4,5-epoxy-3-methoxy-17-methyl-, (5α,6α)-, phosphate (1 1) (salt)

$C_{18}H_{24}N_2O_6$
5182 - Dinocap

$C_{18}H_{24}O_2$
5399 - Estra-1,3,5(10)-triene-3,17-diol (17β)-
5400 - Estra-1,3,5(10)-triene-3,17-diol, (17α)-
5401 - Estra-1,3,5(10)-triene-3,17-diol, (8α,17β)-
5402 - Estra-5,7,9-triene-3,17-diol, (3β,17β)-

$C_{18}H_{24}O_3$
5404 - Estra-1,3,5(10)-triene-3,16,17-triol, (16α,17β)-
5405 - Estra-1,3,5(10)-triene-3,16,17-triol, (16β,17β)-

$C_{18}H_{26}O_4$
1335 - 1,2-Benzenedicarboxylic acid, bis(3-methylbutyl) ester

$C_{18}H_{27}NO_3$
8181 - 6-Nonenamide, *N*-[(4-hydroxy-3-methoxyphenyl)methyl]-8-methyl-, (*E*)-

$C_{18}H_{27}NO_5$
894 - Benzeneacetic acid, 4-[2-(diethylamino)-2-oxoethoxy]-3-methoxy-, propyl ester

$C_{18}H_{27}N_3O$
8737 - 1,5-Pentanediamine, *N*-(6-methoxy-8-quinolinyl)-*N′*-(1-methylethyl)-

$C_{18}H_{28}N_2O_4S$
1582 - Benzeneethanamine, α-methyl-, (*S*)-, sulfate (2:1)

$C_{18}H_{28}O$
5327 - 1-Dodecanone, 1-phenyl-

$C_{18}H_{28}O_2$
5317 - Dodecanoic acid, phenyl ester

$C_{18}H_{30}$
1566 - Benzene, dodecyl-
1756 - Benzene, hexaethyl-
2276 - Benzene, 1,2,4,5-tetrakis(1-methylethyl)-
4118 - Chrysene, octadecahydro-
8230 - 7,11-Octadecadiyne

$C_{18}H_{30}O$
9498 - Phenol, 2,4,6-tris(1,1-dimethylethyl)-

$C_{18}H_{30}O_2$
8246 - 9,11,13-Octadecanetrienoic acid (*Z,Z,Z*)
8282 - 9,11,13-Octadecatrienoic acid, (*E,E,E*)-
8284 - 9,11,13-Octadecatrienoic acid, (*E,Z,E*)-
8286 - 9,11,13-Octadecatrienoic acid, (*Z,Z,E*)-
8287 - 9,12,15-Octadecatrienoic acid, (*Z,Z,Z*)-

$C_{18}H_{30}O_4$
4933 - 1-Cyclopentene-1-pentanoic acid, δ-hydroxy-3,5-bis(1-methylpropyl)-β-oxo-

$C_{18}H_{32}$
4731 - 1,3-Cyclooctadecadiene

$C_{18}H_{32}I_2$
8225 - 9,12-Octadecadienoic acid

$C_{18}H_{32}I_2O_2$
8300 - 6-Octadecenoic acid, 6,7-diiodo-

$C_{18}H_{32}N_2O_5S$
8926 - 1-Pentanol, 2-(diethylamino)-4-methyl-, 4-aminobenzoate (ester), monomethanesulfonate (salt)

$C_{18}H_{32}O_2$
4936 - 2-Cyclopentene-1-tridecanoic acid, (*S*)-
8223 - 9,12-Octadecadienoic acid (*Z,Z*)-
8224 - 10,12-Octadecadienoic acid, (*E,E*)-
8316 - 9-Octadecynoic acid

$C_{18}H_{32}O_5$
9120 - Pentonic acid, 5-C-[3,5-bis(1-methylpropyl)-1-cyclopenten-1-yl]-4-deoxy-

$C_{18}H_{32}O_6$
3561 - Butanoic acid, 3-methyl-, 1,2,3-propanetriyl ester

$C_{18}H_{32}O_7$
10168 - 1,2,3-Propanetricarboxylic acid, 2-hydroxy-, tributyl ester

$C_{18}H_{32}O_{16}$
6219 - α-*D*-Glucopyranoside, β-*D*-fructofuranosyl *O*-α-*D*-galactopyranosyl-(1
6221 - α-*D*-Glucopyranoside, *O*-α-*D*-glucopyranosyl-(13)-β-*D*-fructofuranosyl

$C_{18}H_{32}Si$
11455 - Silane, tributylphenyl-

$C_{18}H_{33}N$
8293 - 9-Octadecenenitrile, (*Z*)-

$C_{18}H_{34}$
8313 - 1-Octadecyne
8314 - 2-Octadecyne
8315 - 9-Octadecyne

$C_{18}H_{34}O$
8288 - 9-Octadecenal
8289 - 9-Octadecenal, (*Z*)-

$C_{18}H_{34}OSn$
5011 - Cyhexatin

$C_{18}H_{34}O_2$
8294 - 9-Octadecenoic acid (*Z*)-
8295 - 11-Octadecenoic acid, (*E*)-
8296 - 2-Octadecenoic acid, (*E*)-
8297 - 9-Octadecenoic acid, (*E*)-

$C_{18}H_{34}O_3$
8303 - 9-Octadecenoic acid, 12-hydroxy-, [*R*-(*Z*)]

$C_{18}H_{34}O_4$
5044 - Decanedioic acid, dibutyl ester
6780 - Hexanedioic acid, bis(2-ethylbutyl) ester

$C_{18}H_{34}O_6$
8239 - Octadecanedioic acid, 9,10-dihydroxy-, (*R**,*R**)-(±)-

$C_{18}H_{35}BrO_2$
8250 - Octadecanoic acid, 18-bromo-

$C_{18}H_{35}ClO$
8281 - Octadecanoyl chloride

$C_{18}H_{35}N$
8243 - Octadecanenitrile

$C_{18}H_{36}$
4382 - Cyclohexane, dodecyl-
4854 - Cyclopentane, tridecyl-
8291 - 1-Octadecene
8292 - 9-Octadecene

$C_{18}H_{36}O$
6638 - Hexadecane, 1-(ethenyloxy)-
8280 - 3-Octadecanone
8311 - 9-Octadecen-1-ol, (*E*)-
8312 - 9-Octadecen-1-ol, (*Z*)-

$C_{18}H_{36}O_2$
5068 - Decanoic acid, 2-octyl-
6316 - Heptadecanoic acid, methyl ester
6650 - Hexadecanoic acid, ethyl ester
6662 - 1-Hexadecanol, acetate
8247 - Octadecanoic acid

$C_{18}H_{36}O_3$
6652 - Hexadecanoic acid, 2-hydroxyethyl ester

$C_{18}H_{36}O_4$
8254 - Octadecanoic acid, 9,10-dihydroxy-

$C_{18}H_{36}O_5$
8276 - Octadecanoic acid, 9,10,18-trihydroxy-, (*R**,*R**)-

$C_{18}H_{37}Br$
8235 - Octadecane, 1-bromo-

$C_{18}H_{37}Cl$
[illegible] - Octadecane, 1-chloro-

$C_{18}H_{37}Cl_3Si$
11470 - Silane, trichlorooctadecyl-

$C_{18}H_{37}I$
8242 - Octadecane, 1-iodo-

$C_{18}H_{37}NO$
8231 - Octadecanamide

$C_{18}H_{38}$
6310 - Heptadecane, 2-methyl-
6311 - Heptadecane, 9-methyl-
8234 - Octadecane

$C_{18}H_{38}O$
8278 - 1-Octadecanol
8279 - 3-Octadecanol, (±)-

$C_{18}H_{38}O_2$
6636 - Hexadecane, 1,1-dimethoxy-

$C_{18}H_{38}S$
8244 - 1-Octadecanethiol

$C_{18}H_{39}N$
6725 - 1-Hexanamine, *N*,*N*-dihexyl-
8233 - 1-Octadecanamine

$C_{18}H_{54}O_7Si_8$
8482 - Octasiloxane, octadecamethyl-

$C_{18}H_{54}O_9Si_9$
4726 - Cyclononasiloxane, octadecamethyl-

$C_{19}H_{10}Br_4O_5S$
9366 - Phenol, 4,4′-(3H-2,1-benzoxathiol-3-ylidene)bis[2,6-dibromo-, *S*,*S*-dioxide

$C_{19}H_{12}O_6$
2839 - 2H-1-Benzopyran-2-one, 3,3′-methylenebis[4-hydroxy-

$C_{19}H_{13}N$
318 - Acridine, 9-phenyl-

$C_{19}H_{14}$
521 - Benz[a]anthracene, 11-methyl-
522 - Benz[a]anthracene, 3-methyl-
523 - Benz[a]anthracene, 8-methyl-
4117 - Chrysene, 1-methyl-

$C_{19}H_{14}F_3NO$
5981 - Fluridone

$C_{19}H_{14}O_3$
4278 - 2,5-Cyclohexadien-1-one, 4-[bis(4-hydroxyphenyl)methylene]-

$C_{19}H_{14}O_5S$
9363 - Phenol, 4,4′-(3H-2,1-benzoxathiol-3-ylidene)bi*S*-, *S*,*S*-dioxide

$C_{19}H_{15}Br$
1061 - Benzene, 1,1′,1″-(bromomethylidyne)tris-

$C_{19}H_{15}Cl$
1196 - Benzene, 1,1′,1″-(chloromethylidyne)tris-

$C_{19}H_{15}ClF_3NO_7$
7410 - Lactofen

$C_{19}H_{15}ClN_4$
11653 - 2H-Tetrazolium, 2,3,5-triphenyl-, chloride

$C_{19}H_{15}N$
3988 - 9H-Carbazole, 9-(phenylmethyl)-

$C_{19}H_{15}NO_2$
11236 - 4-Quinolinecarboxylic acid, 2-phenyl-, 2-propenyl ester

$C_{19}H_{16}$
1982 - Benzene, 1,1′,1″-methylidynetris-
3093 - 1,1′-Biphenyl, 2-(phenylmethyl)-
3094 - 1,1′-Biphenyl, 4-(phenylmethyl)-

$C_{19}H_{16}N_2$
11041 - 9H-Pyrido[3,4-b]indole, 1-[(2-methylphenyl)methyl]-

$C_{19}H_{16}O$
1866 - Benzenemethanol, α,α-diphenyl-

$C_{19}H_{16}O_3$
1983 - Benzene, 1,1′,1″-[methylidynetris(oxy)]tris-

$C_{19}H_{16}O_4$
2832 - 2H-1-Benzopyran-2-one, 4-hydroxy-3-(3-oxo-1-phenylbutyl)-

$C_{19}H_{16}O_5$
225 - Acetic acid, [(4-oxo-2-phenyl-4H-1-benzopyran-7-yl)oxy]-, ethyl ester

$C_{19}H_{17}Cl_2N_3O_3$
5174 - Difenoconazole

$C_{19}H_{17}N$
729 - Benzenamine, 4-(diphenylmethyl)-

$C_{19}H_{17}NO_3$
11216 - Quinoline, 2-[2-(1,3-benzodioxol-5-yl)ethyl]-4-methoxy-

$C_{19}H_{17}N_3$
6288 - Guanidine, *N*,*N*′,*N*″-triphenyl-

$C_{19}H_{17}N_3O$
7333 - Indolo[2′,3′:3,4]pyrido[2,1-b]quinazolin-5(7H)-one, 8,13,13b,14-tetrahydro-14-methyl-, (*S*)-

$C_{19}H_{18}O_3Si$
11443 - Silane, methyltriphenoxy-

$C_{19}H_{18}O_5$
2464 - 5-Benzofuranpropanol, 2-(1,3-benzodioxol-5-yl)-7-methoxy-

$C_{19}H_{18}Si$
11444 - Silane, methyltriphenyl-

$C_{19}H_{19}NO_3$
2469 - 5H-Benzo[g]-1,3-benzodioxolo[6,5,4-de]quinoline, 6,7,7a,8-tetrahydro-11-methoxy-7-methyl-, (*R*)-

$C_{19}H_{19}NO_4$
2470 - 5H-Benzo[g]-1,3-benzodioxolo[6,5,4-de]quinolin-12-ol, 6,7,7a,8-tetrahydro-11-methoxy-7-methyl-, (*S*)-

$C_{19}H_{19}N_7O_6$
6243 - *L*-Glutamic acid, *N*-[4-[[(2-amino-1,4-dihydro-4-oxo-6-pteridinyl)methyl]amino]benzoyl]-

$C_{19}H_{20}F_3NO_4$
5962 - Fluazipop-butyl
5963 - Fluazipop-P-butyl

$C_{19}H_{20}N_2O$
11346 - Sarpagan-17-al

$C_{19}H_{21}NO_3$
5158 - 4H-Dibenzo[de,g]quinolin-1-ol, 5,6,6a,7-tetrahydro-2,11-dimethoxy-6-methyl-, (*S*)-
7679 - Morphinan, 6,7,8,14-tetradehydro-4,5-epoxy-3,6-dimethoxy-17-methyl-, (5α)-

$C_{19}H_{21}NO_4$
5156 - 4H-Dibenzo[de,g]quinoline-2,9-diol, 5,6,6a,7-tetrahydro-1,10-dimethoxy-6-methyl-, (*S*)-

$C_{19}H_{22}$
1744 - Benzene, 1,1′-(1-heptenylidene)bis-

$C_{19}H_{22}N_2O$
4120 - Cinchonan-9-ol, (9*S*)-
4133 - Cinchonan-9-ol, (8α,9*R*)-
10435 - 1-Propanone, 3-(3-ethenyl-4-piperidinyl)-1-(4-quinolinyl)-, (3*R*-*cis*)-

$C_{19}H_{22}N_2OS$
5807 - Ethanone, 1-[10-[3-(dimethylamino)propyl]-10H-phenothiazin-2-yl]-

$C_{19}H_{22}N_2O_2$
4119 - Cinchonan-6′,9-diol, (8α,9*R*)-
11347 - Sarpagan-10,17-diol

$C_{19}H_{22}O_2$
9358 - Phenol, 4-[1-ethyl-2-(4-methoxyphenyl)-1-butenyl]-
9359 - Phenol, 4-[1-ethyl-2-(4-methoxyphenyl)-1-butenyl]-, (*E*)-

$C_{19}H_{22}O_6$
6204 - Gibb-3-ene-1,10-dicarboxylic acid, 2,4a,7-trihydroxy-1-methyl-8-methylene-, 1,4a-lactone,

$C_{19}H_{23}ClN_2$
5140 - 1H-1,4-Diazepine, 1-[(4chlorophenyl) phenylmethyl]hexahydro-4-methyl-

$C_{19}H_{23}NO$
1903 - Benzenemethanol, α-[1-[methyl(3-phenyl-2-propenyl)amino]ethyl]-

$C_{19}H_{23}NO_3$
7664 - Morphinan, 6,7-didehydro-4,5-epoxy-3,6-dimethoxy-17-methyl-, (5α)-

$C_{19}H_{23}NO_4$
7676 - Morphinan-6-one, 7,8-didehydro-4-hydroxy-3,7-dimethoxy-17-methyl-, (9α,13α,14α)-

$C_{19}H_{23}N_3$
342 - Amitraz

$C_{19}H_{23}N_3O_2$
5373 - Ergoline-8-carboxamide, 9,10-didehydro-*N*-(2-hydroxy-1-methylethyl)-6-methyl-, [8α(*S*)]-

$C_{19}H_{24}$
1746 - Benzene, 1,1′-heptylidenebis-

$C_{19}H_{24}N_2$
9627 - 4-Piperidinamine, 1-methyl-*N*-phenyl-*N*-(phenylmethyl)-
$C_{19}H_{24}N_2O$
4121 - Cinchonan-9-ol, 10,11-dihydro-, (8α,9*R*)-
4122 - Cinchonan-9-ol, 10,11-dihydro-, (9*S*)-
7306 - 1H-Indole-3-ethanol, 2-(5-ethenyl-1-azabicyclo[2.2.2]oct-2-yl)-
$C_{19}H_{24}N_2OS$
3977 - Carbamothioic acid, diphenyl-, *S*-[2-(diethylamino)ethyl] ester
$C_{19}H_{24}N_2O_2$
6177 - 3aH-Furo[2,3-b]indol-3a-ol, 8a-(5-ethenyl-1-azabicyclo[2.2.2]oct-2-yl)-2,3,8,8a-tetrahydro-
6178 - 3aH-Furo[2,3-b]indol-3a-ol, 8a-(5-ethenyl-1-azabicyclo[2.2.2]oct-2-yl)-2,3,8,8a-tetrahydro-
$C_{19}H_{24}O_3$
353 - Androst-4-ene-3,11,17-trione
$C_{19}H_{26}$
7931 - Naphthalene, 1-nonyl-
7932 - Naphthalene, 2-nonyl-
$C_{19}H_{26}N_2O$
4154 - Curan-17-ol, (16α)-
$C_{19}H_{26}O_2$
343 - Androsta-1,4-dien-3-one, 17-hydroxy-, (17β)-
352 - Androst-4-ene-3,17-dione
$C_{19}H_{26}O_3$
336 - Allethrin
$C_{19}H_{26}O_4S$
10507 - Propargite
$C_{19}H_{27}ClO_2$
354 - Androst-4-en-3-one, 4-chloro-17-hydroxy-, (17β)-
$C_{19}H_{27}NO_3$
2396 - 2H-Benzo[a]quinolizin-2-one, 1,3,4,6,7,11b-hexahydro-9,10-dimethoxy-3-(2-methylpropyl)-
$C_{19}H_{28}BrNO_3$
11179 - Pyrrolidinium, 3-[(cyclopentylhydroxyphenylacetyl)oxy]-1,1-dimethyl-, bromide
$C_{19}H_{28}O_2$
356 - Androst-4-en-3-one, 17-hydroxy-, (17β)-
$C_{19}H_{30}O_2$
347 - Androstan-17-one, 3-hydroxy-, (3β,5α)-
348 - Androstan-17-one, 3-hydroxy-, (3α,5α)-
349 - Androstan-3-one, 17-hydroxy-, (5α,17β)-
5318 - Dodecanoic acid, phenylmethyl ester
$C_{19}H_{30}O_5$
9727 - Piperonyl butoxide
$C_{19}H_{32}$
345 - Androstane
$C_{20}H_{27}O_4P$
9560 - Phosphoric acid, 2-ethylhexyl diphenyl ester
$C_{20}H_{28}$
7793 - Naphthalene, 1-decyl-
7794 - Naphthalene, 2-decyl-
$C_{20}H_{28}O_2$
344 - Androsta-1,4-dien-3-one, 17-hydroxy-17-methyl-, (17β)-
$C_{20}H_{28}O_3S$
7968 - 1-Naphthalenesulfonic acid, 3,6-bis(1,1-dimethylethyl)-, ethyl ester
$C_{20}H_{29}FO_3$
355 - Androst-4-en-3-one, 9-fluoro-11,17-dihydroxy-17-methyl-, (11β,17β)-
$C_{20}H_{30}ClN_3O_2$
11226 - 4-Quinolinecarboxamide, 2-butoxy-*N*-[2-(diethylamino)ethyl]-, monohydrochloride
$C_{20}H_{30}O$
11341 - Retinol
$C_{20}H_{30}O_2$
358 - Androst-4-en-3-one, 17-hydroxy-17-methyl-, (17β)-
9166 - 1-Phenanthrenecarboxylic acid, 1,2,3,4,4a,5,6,9,10,10a-decahydro-1,4a-dimethyl-7-(1-methylethyl)-,
9167 - 1-Phenanthrenecarboxylic acid, 1,2,3,4,4a,4b,5,6,10,10a-decahydro-1,4a-dimethyl-7-(1-methylethyl)-,
9169 - 1-Phenanthrenecarboxylic acid, 1,2,3,4,4a,4b,5,9,10,10a-decahydro-1,4a-dimethyl-7-(1-methylethyl)-,
9170 - 1-Phenanthrenecarboxylic acid, 7-ethenyl-1,2,3,4,4a,4b,5,6,7,9,10,10a-dodecahydro-1,4a,7-trimethyl-, [1R-(1α,4aβ,4bα,7β,10aα)]-
$C_{20}H_{32}ClNO$
9705 - 1-Piperidinepropanol, α-cyclohexyl-α-phenyl-, hydrochloride
$C_{20}H_{32}N_2O_3S$
4038 - Carbosulfan
$C_{20}H_{32}O_2$
346 - Androstane-17-carboxylic acid, (5β,17β)-
350 - Androstan-3-one, 17-hydroxy-17-methyl-, (5α,17β)-
5360 - 5,8,11,14-Eicosatetraenoic acid
5361 - 5,8,11,14-Eicosatetraenoic acid, (all-Z)-
$C_{20}H_{34}$
2266 - Benzene, tetradecyl-
$C_{20}H_{34}O_2$
8283 - 9,12,15-Octadecatrienoic acid, ethyl ester, (*Z*,*Z*,*Z*)-
$C_{20}H_{36}O_2$
8226 - 9,12-Octadecadienoic acid (*Z*,*Z*)-, ethyl ester
8227 - 9,12-Octadecadienoic acid, ethyl ester
$C_{20}H_{38}$
5366 - 1-Eicosyne
$C_{20}H_{38}N_4O_4$
7686 - 4-Morpholinecarboxamide, *N*,*N*′-1,2-ethanediylbis[*N*-butyl-
$C_{20}H_{38}O_2$
4952 - Cycloprate
5363 - 11-Eicosenoic acid
5364 - 9-Eicosenoic acid
5365 - 9-Eicosenoic acid, (*Z*)-
8301 - 9-Octadecenoic acid (*Z*)-, ethyl ester
8302 - 9-Octadecenoic acid, ethyl ester, (*E*)-
$C_{20}H_{38}O_3$
5060 - Decanoic acid, anhydride
8269 - Octadecanoic acid, 12-oxo-, ethyl ester
8305 - 9-Octadecenoic acid, 12-hydroxy-, ethyl ester, [*R*-(*Z*)]-
$C_{20}H_{38}O_4$
5352 - Eicosanedioic acid
$C_{20}H_{40}$
4475 - Cyclohexane, tetradecyl-
4844 - Cyclopentane, pentadecyl-
5362 - 1-Eicosene
$C_{20}H_{40}O$
6671 - 1-Hexadecen-3-ol, 3,7,11,15-tetramethyl-
6672 - 2-Hexadecen-1-ol, 3,7,11,15-tetramethyl-, [*R*-[*R**,*R**-(*E*)]]-
8240 - Octadecane, 1-(ethenyloxy)-
$C_{20}H_{40}O_2$
221 - Acetic acid, octadecyl ester
5062 - Decanoic acid, decyl ester
5355 - Eicosanoic acid
6647 - Hexadecanoic acid, butyl ester
6658 - Hexadecanoic acid, 3,7,11,15-tetramethyl-
8257 - Octadecanoic acid, ethyl ester
$C_{20}H_{40}O_3$
8260 - Octadecanoic acid, 2-hydroxyethyl ester
$C_{20}H_{42}$
5350 - Eicosane
$C_{20}H_{42}O$
5358 - 1-Eicosanol
5359 - 2-Eicosanol
$C_{20}H_{42}O_2$
8237 - Octadecane, 1,1-dimethoxy-
$C_{20}H_{48}O_4Si_4$
4995 - Cyclotetrasiloxane, 2,4,6,8-tetrabutyl-2,4,6,8-tetramethyl-
$C_{20}H_{60}O_8Si_9$
8178 - Nonasiloxane, eicosamethyl-
$C_{21}H_{12}O$
5152 - 13H-Dibenzo[a,h]fluoren-13-one
$C_{21}H_{13}N$
5143 - Dibenz[a,j]acridine
$C_{21}H_{14}$
5148 - 13H-Dibenzo[a,g]fluorene
$C_{21}H_{14}Br_4O_5S$
9367 - Phenol, 4,4′-(3H-2,1-benzoxathiol-3-ylidene)bis[2,6-dibromo-3-methyl-, *S*,*S*-dioxide
$C_{21}H_{14}O$
7634 - Methanone, 1-naphthalenyl-2-naphthalenyl-
$C_{21}H_{16}$
2392 - Benz[j]aceanthrylene, 1,2-dihydro-3-methyl-
7271 - 1H-Indene, 1,3-diphenyl-
7272 - 1H-Indene, 2,3-diphenyl-
7919 - Naphthalene, 1,1′-methylenebis-
$C_{21}H_{16}Br_2O_5S$
9364 - Phenol, 4,4′-(3H-2,1-benzoxathiol-3-ylidene)bis[2-bromo-6-methyl-, *S*,*S*-dioxide
$C_{21}H_{16}N_2$
7215 - 1H-Imidazole, 2,4,5-triphenyl-
$C_{21}H_{16}N_2O$
12008 - Urea, *N*,*N*′-di-1-naphthalenyl-
$C_{21}H_{16}NaO_2$
11517 - Sonar
$C_{21}H_{18}N_2$
7199 - 1H-Imidazole, 4,5-dihydro-2,4,5-triphenyl-, *cis*-
$C_{21}H_{18}O_5S$
9368 - Phenol, 4,4′-(3H-2,1-benzoxathiol-3-ylidene)bis[2-methyl-, *S*,*S*-dioxide
$C_{21}H_{19}NO_5$
2433 - [1,3]Benzodioxolo[5,6-c]phenanthridinium, 1,2-dimethoxy-12-methyl-
$C_{21}H_{20}$
2110 - Benzene, 1,1′-(phenylmethylene)bis[4-methyl-
$C_{21}H_{20}Cl_2O_3$
9150 - Permethrin
$C_{21}H_{20}N_2O_3$
8572 - Oxayohimbanium, 3,4,5,6,16,17 hexadehydro-16-(methoxycarbonyl)-19-methyl-, (19α)-
$C_{21}H_{20}O_6$
6334 - 1,6-Heptadiene-3,5-dione, 1,7-bis(4-hydroxy-3-methoxyphenyl)-, (*E*,*E*)-
$C_{21}H_{20}O_{10}$
2824 - 4H-1-Benzopyran-4-one, 3-[4-(β-*D*-glucopyranosyloxy)phenyl]-5,7-dihydroxy-
$C_{21}H_{20}O_{11}$
2796 - 4H-1-Benzopyran-4-one, 3-[(6-deoxy-α-*L*-mannopyranosyl)oxy]-2-(3,4-dihydroxyphenyl)-5,7-dihydroxy-
$C_{21}H_{21}N$
1798 - Benzenemethanamine, *N*,*N*-bis(phenylmethyl)-
$C_{21}H_{21}NO_5$
3108 - Bis[1,3]benzodioxolo[4,5-c:5′,6′-g]azecin-13(5H)-one, 4,6,7,14-tetrahydro-5,14-dimethyl-, (*R*)-
$C_{21}H_{21}NO_6$
7357 - 1(3H)-Isobenzofuranone, 6,7-dimethoxy-3-(5,6,7,8-tetrahydro-6-methyl-1,3-dioxolo[4,5-g]isoquinolin-5-yl)-, [*S*-
11342 - Rheadan, 8-methoxy-16-methyl-2,3 10,11-bis[methylenebis(oxy)]-, (8β)-
$C_{21}H_{21}N_3$
11834 - 1,3,5-Triazine, hexahydro-1,3,5-triphenyl-
$C_{21}H_{21}N_3O_7$
11352 - 9,10-Secostrychnidin-10-oic acid, 2,3-dihydro-4-nitro-2,3-dioxo-
$C_{21}H_{21}O_3P$
9596 - Phosphorous acid, tris(2-methylphenyl) ester
9597 - Phosphorous acid, tris(4-methylphenyl) ester
$C_{21}H_{21}O_4P$
9569 - Phosphoric acid, tris(2-methylphenyl) ester
9570 - Phosphoric acid, tris(3-methylphenyl) ester
9571 - Phosphoric acid, tris(4-methylphenyl) ester

MOLECULAR FORMULA INDEX OF ORGANIC COMPOUNDS (continued)

$C_{21}H_{21}O_7P$
9387 - Phenol, 2-methoxy-, phosphate (3:1)
$C_{21}H_{21}Sb$
11552 - Stibine, tris(3-methylphenyl)-
11553 - Stibine, tris(4-methylphenyl)-
$C_{21}H_{22}N_2O_2$
11563 - Strychnidin-10-one
$C_{21}H_{23}ClFNO_2$
3651 - 1-Butanone, 4-[4-(4-chlorophenyl)-4-hydroxy-1-piperidinyl]-1-(4-fluorophenyl)-
$C_{21}H_{23}NO_5$
2435 - Benzo[e]-1,3-dioxolo[4,5-l][2]benzazecin-12(5H)-one, 4,6,7,13-tetrahydro-9,10-dimethoxy-5-methyl-
7667 - Morphinan-3,6-diol, 7,8-didehydro-4,5-epoxy-17-methyl- (5α,6α)-, diacetate (ester)
$C_{21}H_{23}NO_6$
92 - Acetamide, *N*-(5,6,7,9-tetrahydro-10-hydroxy-1,2,3-trimethoxy-9-oxobenzo[a]heptalen-7-yl)-, (*S*)-
$C_{21}H_{23}N_3O_5$
11565 - Strychnidin-10-one, mononitrate
$C_{21}H_{24}N_2O_3$
8571 - Oxayohimban-16-carboxylic acid, 16,17-didehydro-19-methyl-, methyl ester, (19α)-
$C_{21}H_{24}OSi_2$
5239 - Disiloxane, 1,1,1-trimethyl-3,3,3-triphenyl-
$C_{21}H_{24}O_3Si_3$
5005 - Cyclotrisiloxane, 2,4,6-trimethyl-2,4,6-triphenyl-
$C_{21}H_{24}O_4$
9926 - Propanedioic acid, bis(phenylmethyl)-, diethyl ester
$C_{21}H_{24}O_7$
3540 - Butanoic acid, 2-methyl-, 10-(acetyloxy)-9,10-dihydro-8,8-dimethyl-2-oxo-2H,8H-benzo[1,2-b 3,4-b′]dipyran-9-yl
$C_{21}H_{25}NO_4$
5150 - 6H-Dibenzo[a,g]quinolizin-2-ol, 5,8,13,13a-tetrahydro-3,9,10-trimethoxy-13-methyl-, (13S-*trans*)-
5151 - 6H-Dibenzo[a,g]quinolizin-3-ol, 5,8,13,13a-tetrahydro-2,9,10-trimethoxy-13-methyl-, (13*S*-*trans*)-
5157 - 4H-Dibenzo[de,g]quinoline, 5,6,6a,7-tetrahydro-1,2,9,10-tetramethoxy-6-methyl-, (*S*)-
$C_{21}H_{26}BrNO_3$
5445 - Ethanaminium, *N*,*N*-diethyl-*N*-methyl-2-[(9H-xanthen-9-ylcarbonyl)oxy]-, bromide
$C_{21}H_{26}ClN_3OS$
9617 - 1-Piperazineethanol, 4-[3-(2-chloro-10H-phenothiazin-10-yl)propyl]-
$C_{21}H_{26}N_2O_3$
5349 - Eburnamenine-14-carboxylic acid, 14,15-dihydro-14-hydroxy-, methyl ester, (3α,14β,16α)-
12062 - Yohimban-16-carboxylic acid, 17-hydroxy-, methyl ester, (16α,17α)-
$C_{21}H_{26}N_2O_4$
4129 - Cinchonan-9-ol, 6′-methoxy-, (8α,9*R*)-, monoformate (salt)
$C_{21}H_{26}O_2$
5154 - 6H-Dibenzo[b,d]pyran-1-ol, 6,6,9-trimethyl-3-pentyl-
$C_{21}H_{26}O_3$
7621 - Methanone, [2-hydroxy-4-(octyloxy)phenyl]phenyl-
$C_{21}H_{26}O_5$
9739 - Pregna-1,4-diene-3,11,20-trione, 17,21-dihydroxy-
$C_{21}H_{27}ClN_2O_2$
5700 - Ethanol, 2-[2-[4-[(4-chlorophenyl)phenylmethyl]-1-piperazinyl]ethoxy]-
$C_{21}H_{27}NO_4$
7379 - Isoquinoline, 1-[(3,4-dimethoxyphenyl)methyl]-1,2,3,4-tetrahydro-6,7-dimethoxy-2-methyl-, (*S*)-
$C_{21}H_{28}ClNO$
6516 - 3-Heptanone, 6-(dimethylamino)-4,4-diphenyl-, hydrochloride
$C_{21}H_{28}O_2$
9758 - Pregn-4-en-20-yn-3-one, 17-hydroxy-, (17α)-
$C_{21}H_{28}O_3$
4960 - Cyclopropanecarboxylic acid, 2,2-dimethyl-3-(2-methyl-1-propenyl)-, 2-methyl-4-oxo-3-(2,4-pentadienyl)-2-cyclopenten-1-yl ester, [1R-[1α[S*(Z)],3β]]-
$C_{21}H_{28}O_4$
9755 - Pregn-4-ene-3,11,20-trione, 21-hydroxy-
$C_{21}H_{28}O_5$
9738 - Pregna-1,4-diene-3,20-dione, 11,17,21-trihydroxy-, (11β)-
9746 - Pregn-4-en-18-al, 11,21-dihydroxy-3,20-dioxo-, (11β)-
9754 - Pregn-4-ene-3,11,20-trione, 17,21-dihydroxy-
$C_{21}H_{30}O_2$
1537 - 1,3-Benzenediol, 2-[3-methyl-6-(1-methylethenyl)-2-cyclohexen-1-yl]-5-pentyl-, (1*R*-*trans*)-
$C_{21}H_{30}O_3$
9750 - Pregn-4-ene-3,20-dione, 17-hydroxy-
9751 - Pregn-4-ene-3,20-dione, 21-hydroxy-
$C_{21}H_{30}O_4$
9748 - Pregn-4-ene-3,20-dione, 11,21-dihydroxy-, (11β)-
9749 - Pregn-4-ene-3,20-dione, 17,21-dihydroxy-
$C_{21}H_{30}O_5$
4282 - 2,4-Cyclohexadien-1-one, 3,5,6-trihydroxy-4,6-bis(3-methyl-2-butenyl)-2-(3-methyl-1-oxobutyl)-, (*R*)-
9753 - Pregn-4-ene-3,20-dione, 11,17,21-trihydroxy-, (11β)-
$C_{21}H_{32}O_2$
9168 - 1-Phenanthrenecarboxylic acid, 1,2,3,4,4a,4b,5,6,10,10a-decahydro-1,4a-dimethyl-7-(1-methylethyl)-, methyl ester, [1R-(1α,4aβ,4bα,10aα)]-
9743 - Pregnane-3,20-dione, (5β)-
9756 - Pregn-5-en-20-one, 3-hydroxy-, (3β)-
$C_{21}H_{32}O_5$
9744 - Pregnane-11,20-dione, 3,17,21-trihydroxy-, (3α,5β)-
9757 - Pregn-4-en-3 one, 11,17,20,21-tetrahydroxy-, (11β,20*R*)-
$C_{21}H_{34}BrNO_3$
5443 - Ethanaminium, 2-[(cyclohexylhydroxyphenylacetyl)oxy]-*N*,*N*-diethyl-*N*-methyl-, bromide
$C_{21}H_{34}O_2$
9745 - Pregnan-20-one, 3-hydroxy-, (3α,5β)-
11627 - Tetradecanoic acid, phenylmethyl ester
$C_{21}H_{36}$
2089 - Benzene, pentadecyl-
9740 - Pregnane, (5α)-
9741 - Pregnane, (5β)-
$C_{21}H_{36}N_2O_5S$
9623 - Piperazinium, 4-(2-cyclohexyl-2-hydroxy-2-phenylethyl)-1,1-dimethyl-, methyl sulfate (salt)
$C_{21}H_{36}O$
9441 - Phenol, 3-pentadecyl-
$C_{21}H_{36}O_2$
1543 - 1,2-Benzenediol, 3-pentadecyl-
9742 - Pregnane-3,20-diol, (3α,5β,20*S*)-
$C_{21}H_{38}ClN$
11030 - Pyridinium, 1-hexadecyl-, chloride
$C_{21}H_{38}O_6$
6908 - Hexanoic acid, 1,2,3-propanetriyl ester
$C_{21}H_{39}N_7O_{12}$
11562 - Streptomycin
$C_{21}H_{40}O_4$
8131 - Nonanedioic acid, bis(2-ethylbutyl) ester
8299 - 9-Octadecenoic acid (*Z*)-, 2,3-dihydroxypropyl ester
$C_{21}H_{41}O_2$
5357 - Eicosanoic acid, methyl ester
$C_{21}H_{42}$
4463 - Cyclohexane, pentadecyl-
4822 - Cyclopentane, hexadecyl-
$C_{21}H_{42}O_2$
8267 - Octadecanoic acid, 1-methylethyl ester
8273 - Octadecanoic acid, propyl ester
$C_{21}H_{42}O_4$
8255 - Octadecanoic acid, 2,3-dihydroxypropyl ester, (±)-
$C_{21}H_{44}$
6294 - Heneicosane
$C_{21}H_{44}O_3$
10023 - 1,2-Propanediol, 3-(octadecyloxy)-
$C_{22}H_{12}$
2471 - Benzo[ghi]perylene
$C_{22}H_{14}$
5142 - Dibenz[a,h]anthracene
8636 - Pentacene
8996 - Pentaphene
9601 - Picene
$C_{22}H_{16}N_4O$
8038 - 2-Naphthalenol, 1-[[4-(phenylazo)phenyl]azo]-
$C_{22}H_{17}N$
7717 - 1-Naphthalenamine, *N*,*N*-diphenyl-
$C_{22}H_{18}Cl_2FNO_3$
5008 - Cyfluthrin
$C_{22}H_{18}O_4$
1337 - 1,2-Benzenedicarboxylic acid, bis(phenylmethyl) ester
7353 - 1(3H)-Isobenzofuranone, 3,3-bis(4-hydroxy-3-methylphenyl)-
$C_{22}H_{19}Br_2NO_3$
5102 - Deltamethrin
$C_{22}H_{19}Cl_2NO_3$
5012 - Cypermethrin
$C_{22}H_{19}NO_4$
9465 - Phenol, 4,4′-(2-pyridinylmethylene)bis-, diacetate (ester)
$C_{22}H_{20}O_{13}$
373 - 2-Anthracenecarboxylic acid, 7-β-*D*-glucopyranosyl-9,10-dihydro-3,5,6,8-tetrahydroxy-1-methyl-9,10-dioxo-
$C_{22}H_{21}NO_2$
1797 - Benzenemethanamine, *N*-[bis(4-methoxyphenyl)methylene]-
$C_{22}H_{22}O_8$
6191 - Furo[3′,4′:6,7]naphtho[2,3-d]-1,3-dioxol-6(5aH)-one, 5,8,8a,9-tetrahydro-9-hydroxy-5-(3,4,5-trimethoxyphenyl)-
6192 - Furo[3′,4′:6,7]naphtho[2,3-d]-1,3-dioxol-6(5aH)-one, 5,8,8a,9-tetrahydro-9-hydroxy-5-(3,4,5-trimetho
$C_{22}H_{23}ClN_2O_8$
7704 - 2-Naphthacenecarboxamide, 7-chloro-4-(dimethylamino)-1,4,4a,5,5a,6,11,12a-octahydro-3,6,10,12,12a-pentahydroxy-6-methyl-1,11-dioxo, [4S(4α,4aα,5aα,6β,12aα)]-
$C_{22}H_{23}NO_3$
5953 - Fenpropathrin
$C_{22}H_{23}NO_6$
10797 - 4H-Pyran-4-one, 2-methoxy-3,5-dimethyl-6-[tetrahydro-4-[2-methyl-3-(4-nitrophenyl)-2-propenylidene]-2-furanyl]
$C_{22}H_{23}NO_7$
7356 - 1(3H)-Isobenzofuranone, 6,7-dimethoxy-3-(5,6,7,8-tetrahydro-4-methoxy-6-methyl-1,3-dioxolo[4,5-g]isoquinolin-5
$C_{22}H_{24}N_2O_4$
3204 - Butanamide, *N*,*N*′-(3,3′-dimethyl[1,1′-biphenyl]-4,4′-diyl)bis[3-oxo-

10772 - 2H-Pyran-5-carboxylic acid, 3,4-dihydro-3-(hydroxymethyl)-2-methyl-4-[(9-methyl-9H-pyrido[3,4-b]indol-1-yl)methyl]-, methyl ester, [2S-(2α,3β,4β)]-
11351 - 16,19-Secostrychnidine-10,16-dione, 4-hydroxy-19-methyl-

$C_{22}H_{24}N_2O_8$
7706 - 2-Naphthacenecarboxamide, 4-(dimethylamino)-1,4,4a,5,5a,6,11,12a-octahydro-3,6,10,12,12a-pentahydroxy-6-methyl-1,11-dioxo-, [4S-(4α,4aα,5aα,6β,12aα)]-

$C_{22}H_{24}N_2O_9$
7705 - 2-Naphthacenecarboxamide, 4-(dimethylamino)-1,4,4a,5,5a,6,11,12a-octahydro-3,5,6,10,12,12a-hexahydroxy-6-methyl-1,11-dioxo-,[4S-(4α,4aα,5aα,6β,12aα)]-

$C_{22}H_{24}Si$
11445 - Silane, methyltri-*p*-tolyl-

$C_{22}H_{25}NO_2$
5854 - Ethanone, 2,2′-(1-methyl-2,6-piperidinediyl)bis[1-phenyl-, *cis*-

$C_{22}H_{25}NO_6$
93 - Acetamide, *N*-(5,6,7,9-tetrahydro-1,2,3,10-tetramethoxy-9-oxobenzo[a]heptalen-7-yl)-, (*S*)-

$C_{22}H_{25}N_3O$
10850 - 3H-Pyrazol-3-one, 1,2-dihydro-2-(1-methyl-4-piperidinyl)-5-phenyl-4-(phenylmethyl)-

$C_{22}H_{26}N_2O_3$
4146 - Corynan-16-carboxylic acid, 16,17,18,19-tetradehydro-17-methoxy-, methyl ester, (16E)-

$C_{22}H_{26}O_3$
11340 - (-)-*cis*-Resmethrin

$C_{22}H_{27}NO_2$
5834 - Ethanone, 2-[6-(2-hydroxy-2-phenylethyl)-1-methyl-2-piperidinyl]-1-phenyl-, [2*R*-[2α,6α(*S**)]]-

$C_{22}H_{27}NO_4$
5149 - 6H-Dibenzo[a,g]quinolizine, 5,8,13,13a-tetrahydro-2,3,9,10-tetramethoxy-13-methyl-, (13*S*-*trans*)-

$C_{22}H_{28}N_2O$
9801 - Propanamide, *N*-phenyl-*N*-[1-(2-phenylethyl)-4-piperidinyl]-

$C_{22}H_{28}N_2O_5$
12063 - Yohimban-16-carboxylic acid, 18-hydroxy-11,17-dimethoxy-, (3β,16β,17α,18β,20

$C_{22}H_{28}O_3$
9736 - Pregna-4,6-diene-21-carboxylic acid, 17-hydroxy-3-oxo-, γ-lactone, (17α)-

$C_{22}H_{28}O_5$
4962 - Cyclopropanecarboxylic acid, 3-(3-methoxy-2-methyl-3-oxo-1-propenyl)-2,2-dimethyl-, 2-methyl-4-oxo-3-(2,4-pentadienyl)-2-cyclopenten-1-yl ester [1R-[1α[S*(Z)],3β(E)]]-

$C_{22}H_{29}FO_5$
9737 - Pregna-1,4-diene-3,20-dione, 9-fluoro-11,17,21-trihydroxy-16-methyl-, (11β,16α)-

$C_{22}H_{29}NO_2$
9646 - 2,6-Piperidinediethanol, 1-methyl-α,α′-diphenyl-, [2α(*R**),6α(*S**)]-

$C_{22}H_{30}N_2O_2$
473 - Aspidospermidine, 1-acetyl-17-methoxy-

$C_{22}H_{30}N_2O_5$
4730 - 2,4(1H)-Cyclo-3,4-secoakuammilanium, 3,17-dihydroxy-16-(methoxycarbonyl)-4-methyl-, (3β,16*R*)-

$C_{22}H_{32}$
7879 - Naphthalene, 1-dodecyl-
7880 - Naphthalene, 2-dodecyl-

$C_{22}H_{32}O_3$
359 - Androst-4-en-3-one, 17-(1-oxopropoxy)-, (17β)-
9752 - Pregn-4-ene-3,20-dione, 17-hydroxy-6-methyl-, (6α)-

$C_{22}H_{33}NO_2$
475 - Atisine

$C_{22}H_{35}NO_3$
5007 - 7,20-Cycloveatchane-1,12,15-triol, 21-ethyl-4-methyl-16-methylene-, (1α,12α,15β)-

$C_{22}H_{36}O$
6664 - 1-Hexadecanone, 1-phenyl-

$C_{22}H_{37}NO$
6630 - Hexadecanamide, *N*-phenyl-

$C_{22}H_{38}$
1751 - Benzene, hexadecyl-

$C_{22}H_{38}O$
1752 - Benzene, (hexadecyloxy)-

$C_{22}H_{40}O_7$
8102 - 1,2,3-Nonadecanetricarboxylic acid, 2-hydroxy-

$C_{22}H_{42}O_2$
5281 - 13-Docosenoic acid
5282 - 13-Docosenoic acid, (*E*)-
5283 - 13-Docosenoic acid, (*Z*)-
8298 - 9-Octadecenoic acid (*Z*)-, butyl ester
10667 - 2-Propenoic acid, 2-methyl-, octadecyl ester

$C_{22}H_{42}O_3$
8304 - 9-Octadecenoic acid, 12-hydroxy-, butyl ester, [*R*-(*Z*)]-

$C_{22}H_{42}O_4$
6781 - Hexanedioic acid, bis(2-ethylhexyl) ester
8238 - Octadecanedioic acid, diethyl ester

$C_{22}H_{44}$
4411 - Cyclohexane, hexadecyl-
5280 - 1-Docosene

$C_{20}H_{40}N_2{\bullet}C_2H_4O_2$
6275 - Glyodin

$C_{22}H_{44}O$
5284 - 13-Docosen-1-ol, (*Z*)-

$C_{22}H_{44}O_2$
5276 - Docosanoic acid
5356 - Eicosanoic acid, ethyl ester
8251 - Octadecanoic acid, butyl ester
8268 - Octadecanoic acid, 2-methylpropyl ester

$C_{22}H_{46}$
5268 - Docosane

$C_{22}H_{46}O$
5279 - 1-Docosanol

$C_{22}H_{66}O_9Si_{10}$
5079 - Decasiloxane, docosamethyl-

$C_{23}H_{16}O_3$
7268 - 1H-Indene-1,3(2H)-dione, 2-(diphenylacetyl)-

$C_{23}H_{16}O_6$
7777 - 2-Naphthalenecarboxylic acid, 4,4′-methylenebis[3-hydroxy-

$C_{23}H_{18}O$
7910 - 1-Naphthalenemethanol, α,α-diphenyl-

$C_{23}H_{19}ClF_3NO_3$
5010 - Cyhalothrin

$C_{23}H_{21}ClO_3$
1154 - Benzene, 1,1′,1″-(1-chloro-1-ethenyl-2-ylidene)tris[4-methoxy-

$C_{23}H_{22}ClF_3O_2$
2998 - Bifenthrin

$C_{23}H_{22}O_6$
2783 - [1]Benzopyrano[3,4-b]furo[2,3-h][1]benzopyran-6(6aH)-one, 1,2,12,12a-tetrahydro-8,9-dimethoxy-2-(1-methylethen

$C_{23}H_{22}O_7$
3110 - 3H-Bis[1]benzopyrano[3,4-b:6′,5′-e]pyran-7(7aH)-one, 13,13a-dihydro-7a-hydroxy-9,10-dimethoxy-3,3-dimethyl

$C_{23}H_{23}O_6$
9747 - Pregn-4-ene-3,20-dione, 21-(acetyloxy)-11,17-dihydroxy-, (11β)-

$C_{23}H_{25}ClN_2$
7455 - Methanaminium, *N*-[4-[[4-(dimethylamino)phenyl]phenylmethylene]-2,5-cyclohexadien-1-ylidene]-*N*-methyl-, chlorid

$C_{23}H_{26}N_2O_4$
11564 - Strychnidin-10-one, 2,3-dimethoxy-

$C_{23}H_{27}NO_8$
2586 - Benzoic acid, 6-[[6-[2-(dimethylamino)ethyl]-4-methoxy-1,3-benzodioxol-5-yl]acetyl]-2,3-dimethoxy-

$C_{23}H_{28}ClN_3O$
9657 - Piperidine, 1-[2-[3-(2,2-diphenylethyl)-1,2,4-oxadiazol-5-yl]ethyl]-, monohydrochloride

$C_{23}H_{30}BrNO_3$
9832 - 2-Propanaminium, *N*-methyl-*N*-(1-methylethyl)-*N*-[2-[(9H-xanthen-9-ylcarbonyl)oxy]ethyl]-, bromide

$C_{23}H_{30}O_2Si_3$
11915 - Trisiloxane, 1,1,3,5,5-pentamethyl-1,3,5-triphenyl-

$C_{23}H_{32}N_2O_2S$
11800 - Thiourea, *N,N*′-bis[4-(3-methylbutoxy)phenyl]-

$C_{23}H_{32}O_6$
4049 - Card-20(22)-enolide, 3,5,14-trihydroxy-19-oxo-, (3β,5β)-

$C_{23}H_{33}IN_2O$
2120 - Benzenepropanaminium, γ-(aminocarbonyl)-*N*-methyl-*N,N*-bis(1-methylethyl)-γ-phenyl-, iod

$C_{23}H_{34}O_4$
4044 - Card-20(22)-enolide, 3,14-dihydroxy-, (3β,5β)-

$C_{23}H_{34}O_5$
4046 - Card-20(22)-enolide, 3,11,14-trihydroxy-, (3β,5β,11α)-
4047 - Card-20(22)-enolide, 3,12,14-trihydroxy-, (3β,5β,12β)-
4048 - Card-20(22)-enolide, 3,14,16-trihydroxy-, (3β,5β,16β)-

$C_{23}H_{36}O_3$
351 - Androstan-3-one, 2-methyl-17-(1-oxopropoxy)-, (2α,5α,17β)-

$C_{23}H_{38}O_2$
6655 - Hexadecanoic acid, phenylmethyl ester
8210 - 24-Norcholan-23-oic acid, (5β)-

$C_{23}H_{40}$
1743 - Benzene, heptadecyl-

$C_{23}H_{44}O_2$
8306 - 9-Octadecenoic acid (*Z*)-, 3-methylbutyl ester

$C_{23}H_{44}O_3$
8275 - Octadecanoic acid, (tetrahydro-2-furanyl)methyl ester

$C_{23}H_{44}O_4$
9954 - Propanedioic acid, hexadecyl-, diethyl ester

$C_{23}H_{46}$
4404 - Cyclohexane, heptadecyl-
4842 - Cyclopentane, octadecyl-

$C_{23}H_{46}O$
11857 - 12-Tricosanone

$C_{23}H_{46}O_2$
5278 - Docosanoic acid, methyl ester
8265 - Octadecanoic acid, 3-methylbutyl ester
8270 - Octadecanoic acid, pentyl ester

$C_{23}H_{48}$
6309 - Heptadecane, 9-hexyl-
11855 - Tricosane

$C_{24}H_{12}$
4145 - Coronene

$C_{24}H_{14}$
2850 - Benzo[rst]pentaphene
8061 - Naphtho[1,2,3,4-def]chrysene

$C_{24}H_{16}$
11644 - Tetraphenylene

$C_{24}H_{18}$
11193 - 1,1′:2′,1″:2″,1‴-Quaterphenyl
11194 - 1,1′:4′,1″:4″,1‴-Quaterphenyl
11598 - 1,1′:3′,1″-Terphenyl, 5′-phenyl-

$C_{24}H_{20}N_2$
3035 - [1,1′-Biphenyl]-4,4′-diamine, *N,N*′-diphenyl-

$C_{24}H_{20}N_4O$
8033 - 2-Naphthalenol, 1-[[2-methyl-4-[(2-methylphenyl)azo]phenyl]azo]-

$C_{24}H_{20}O_4Si$
11509 - Silicic acid (H4SiO4), tetraphenyl ester
$C_{24}H_{20}O_6$
10196 - 1,2,3-Propanetriol, tribenzoate
$C_{24}H_{20}Pb$
9733 - Plumbane, tetraphenyl-
$C_{24}H_{20}Si$
11452 - Silane, tetraphenyl-
$C_{24}H_{20}Sn$
11545 - Stannane, tetraphenyl-
$C_{24}H_{27}O_4P$
9335 - Phenol, 2,4-dimethyl-, phosphate (3:1)
9336 - Phenol, 2,5-dimethyl-, phosphate (3:1)
9337 - Phenol, 2,6-dimethyl-, phosphate (3:1)
9338 - Phenol, 3,4-dimethyl-, phosphate (3:1)
$C_{24}H_{28}O_4$
9301 - Phenol, 4,4′-(1,2-diethyl-1,2-ethenediyl)bis-, dipropanoate, (*E*)-
$C_{24}H_{29}NO_4$
7377 - Isoquinoline, 1-[(3,4-diethoxyphenyl)methyl]-6,7-diethoxy-
$C_{24}H_{30}O_3Si_3$
5004 - Cyclotrisiloxane, 2,4,6-triethyl-2,4,6-triphenyl-
$C_{24}H_{34}O_5$
4101 - Cholan-24-oic acid, 3,7,12-trioxo-, (5β)-
$C_{24}H_{38}O_4$
1334 - 1,2-Benzenedicarboxylic acid, bis(2-ethylhexyl) ester
1344 - 1,2-Benzenedicarboxylic acid, dioctyl ester
$C_{24}H_{39}NO$
8290 - 9-Octadecenamide, *N*-phenyl-, (*Z*)-
$C_{24}H_{40}N_2$
4143 - Con-5-enin-3-amine, *N,N*-dimethyl-, (3β)-
$C_{24}H_{40}N_8O_4$
5711 - Ethanol, 2,2′,2″,2‴-[(4,8-di-1-piperidinylpyrimido[5,4-d]pyrimidine-
$C_{24}H_{40}O_2$
1257 - Benzenedecanoic acid, ι-octyl-
2063 - Benzenenonanoic acid, θ-nonyl-
4094 - Cholan-24-oic acid
8271 - Octadecanoic acid, phenyl ester
$C_{24}H_{40}O_3$
4099 - Cholan-24-oic acid, 3-hydroxy-, (3α,5β)-
$C_{24}H_{40}O_4$
4095 - Cholan-24-oic acid, 3,12-dihydroxy-, (3α,5β,12α)-
4096 - Cholan-24-oic acid, 3,6-dihydroxy-, (3α,5β,6α)-
4097 - Cholan-24-oic acid, 3,7-dihydroxy-, (3α,5β,7α)-
4098 - Cholan-24-oic acid, 3,7-dihydroxy-, (3α,5β,7β)-
$C_{24}H_{40}O_5$
4100 - Cholan-24-oic acid, 3,7,12-trihydroxy-, (3α,5β,7α,12α)-
$C_{24}H_{41}NO_2$
8232 - Octadecanamide, *N*-(4-hydroxyphenyl)-
$C_{24}H_{42}$
2066 - Benzene, octadecyl-
4093 - Cholane
$C_{24}H_{46}O_2$
8253 - Octadecanoic acid, cyclohexyl ester
11608 - 15-Tetracosenoic acid
$C_{24}H_{46}O_3$
5306 - Dodecanoic acid, anhydride
$C_{24}H_{46}O_4$
5353 - Eicosanedioic acid, diethyl ester
9155 - Peroxide, bis(1-oxododecyl)
$C_{24}H_{48}$
4459 - Cyclohexane, octadecyl-
$C_{24}H_{48}O_2$
5277 - Docosanoic acid, ethyl ester
11607 - Tetracosanoic acid
$C_{24}H_{50}$
11603 - Tetracosane
11856 - Tricosane, 2-methyl-
$C_{24}H_{52}O_4Si$
11507 - Silicic acid (H4SiO4), tetrakis(2-ethylbutyl) ester
$C_{24}H_{54}OSn_2$
5241 - Distannoxane, hexabutyl-
$C_{24}H_{72}O_{10}Si_{11}$
11964 - Undecasiloxane, tetracosamethyl-
$C_{25}H_{20}$
1831 - Benzene, 1,1′,1″,1‴-methanetetrayltetrakis-
$C_{25}H_{20}N_2O$
12031 - Urea, tetraphenyl-
$C_{25}H_{22}ClNO_3$
5956 - Fenvalerate
$C_{25}H_{22}OSi$
11441 - Silane, (2-methylphenoxy)triphenyl-
$C_{25}H_{24}F_6N_4$
7136 - Hydramethylnon
$C_{25}H_{27}ClN_2$
9605 - Piperazine, 1-[(4-chlorophenyl)phenylmethyl]-4-[(3-methylphenyl)methyl]-
$C_{25}H_{28}O_3$
5403 - Estra-1,3,5(10)-triene-3,17-diol (17β)-, 3-benzoate
$C_{25}H_{30}ClN_3$
7453 - Methanaminium, *N*-[4-[bis[4-(dimethylamino)phenyl]methylene]-2,5-cyclohexadien-1-ylidene]-*N*-methyl-, chloride
$C_{25}H_{30}O_4$
5108 - 6,6′-Diapo-ψ,ψ-carotenedioic acid, monomethyl ester, 9-*cis*-
$C_{25}H_{40}O_2$
8309 - 9-Octadecenoic acid (*Z*)-, phenylmethyl ester
$C_{25}H_{41}NO_9$
305 - Aconitane-3,8,13,14,15-pentol, 20-ethyl-1,6,16-trimethoxy-4-(methoxymethyl)-,
$C_{25}H_{44}$
2062 - Benzene, nonadecyl-
2070 - Benzene, (3-octylundecyl)-
8245 - Octadecane, 9-*p*-tolyl-
$C_{25}H_{48}O_4$
8132 - Nonanedioic acid, bis(2-ethylhexyl) ester
$C_{25}H_{50}$
4456 - Cyclohexane, nonadecyl-
5351 - Eicosane, 1-cyclopentyl-
6323 - 8-Heptadecene, 9-octyl-
$C_{25}H_{52}$
6313 - Heptadecane, 9-octyl-
8637 - Pentacosane
$C_{26}H_{16}$
6621 - Hexacene
$C_{26}H_{20}$
1623 - Benzene, 1,1′,1″,1‴-(1,2-ethenediylidene)tetrakis-
$C_{26}H_{22}$
1591 - Benzene, 1,1′,1″,1‴-(1,2-ethanediylidene)tetrakis-
1618 - Benzene, 1,1′,1″,1‴-(1-ethanyl-2-ylidyne)tetrakis-
$C_{26}H_{22}ClF_3N_2O_3$
5983 - Fluvalinate
$C_{26}H_{22}O_2$
5580 - 1,2-Ethanediol, 1,1,2,2-tetraphenyl-
$C_{26}H_{23}F_2NO_4$
5965 - Flucythrinate
$C_{26}H_{36}O_6$
3141 - Bufa-20,22-dienolide, 16-(acetyloxy)-3,14-dihydroxy-, (3β,5β,16β)-
$C_{26}H_{38}$
1567 - Benzene, 1,1′-dodecylidenebis[4-methyl-
2267 - Benzene, 1,1′-tetradecylidenebis-
$C_{26}H_{38}O_4$
4281 - 2,4-Cyclohexadien-1-one, 3,5-dihydroxy-2,6,6-tris(3-methyl-2-butenyl)-4-(3-methyl-1-oxobutyl)-
$C_{26}H_{43}NO_6$
6273 - Glycine, *N*-[(3α,5β,7α,12α)-3,7,12-trihydroxy-24-oxocholan-24-yl]-
$C_{26}H_{45}NO_7S$
5625 - Ethanesulfonic acid, 2-[[(3α,5β,7α,12α)-3,7,12-trihydroxy-24-oxocholan-24-yl]amino]-
$C_{26}H_{46}$
1119 - Benzene, (1-butylhexadecyl)-
1568 - Benzene, eicosyl-
1695 - Benzene, (1-ethyloctadecyl)-
2007 - Benzene, (1-methylnonadecyl)-
2068 - Benzene, (1-octyldodecyl)-
2170 - Benzene, (1-propylheptadecyl)-
$C_{26}H_{50}$
4476 - Cyclohexane, 1,1′-tetradecylidenebis-
$C_{26}H_{50}O_4$
5043 - Decanedioic acid, bis(2-ethylhexyl) ester
5311 - Dodecanoic acid, 1,2-ethanediyl ester
$C_{26}H_{52}$
4384 - Cyclohexane, eicosyl-
4819 - Cyclopentane, heneicosyl-
$C_{26}H_{52}O_2$
6624 - Hexacosanoic acid
$C_{26}H_{52}O_6$
8259 - Octadecanoic acid, 2-[2-[2-(2-hydroxyethoxy)ethoxy]ethoxy]ethyl ester
$C_{26}H_{54}$
5269 - Docosane, 5-butyl-
5270 - Docosane, 7-butyl-
5271 - Docosane, 9-butyl-
5272 - Docosane, 11-butyl-
6296 - Heneicosane, 11-(2,2-dimethylpropyl)-
6622 - Hexacosane
6637 - Hexadecane, 6,11-dipentyl-
8241 - Octadecane, 3-ethyl-5-(2-ethylbutyl)-
11605 - Tetracosane, 3-ethyl-
$C_{26}H_{54}O$
6626 - 1-Hexacosanol
$C_{27}H_{18}O_3$
7587 - Methanone, 1,3,5-benzenetriyltris[phenyl-
$C_{27}H_{20}O_{12}$
2773 - 1H-2-Benzopyran-3-carboxylic acid, 6-[2-(4,9-dihydro-8-hydroxy-5,7-dimethoxy-4,9-dioxonaphtho[2,3-b]furan-2-yl
$C_{27}H_{28}Br_2O_5S$
9365 - Phenol, 4,4′-(3H-2,1-benzoxathiol-3-ylidene)bis[2-bromo-3-methyl-6-(1-methylethyl)-, *S,S*-dioxide
$C_{27}H_{30}O_5S$
9369 - Phenol, 4,4′-(3H-2,1-benzoxathiol-3-ylidene)bis[5-methyl-2-(1-methylethyl)-, *S,S*-dioxide
$C_{27}H_{32}N_2O_6$
4131 - Cinchonan-9-ol, 6′-methoxy-, (8α,9*R*)-, mono(2-hydroxybenzoate) (salt)
$C_{27}H_{32}O_{14}$
2795 - 4H-1-Benzopyran-4-one, 7-[[2-*O*-(6-deoxy-α-*L*-mannopyranosyl)-β-*D*-glucopyranosyl]oxy]-2,3-dihydro-5-hy
$C_{27}H_{34}N_2O_4S$
5444 - Ethanaminium, *N*-[4-[[4-(diethylamino)phenyl]phenylmethylene]-2,5-cyclohexadien-1-ylidene]-*N*-ethyl-, sulfate (1
$C_{27}H_{34}O_3$
5408 - Estr-4-en-3-one, 17-(1-oxo-3-phenylpropoxy)-, (17β)-
$C_{27}H_{39}NO_2$
12044 - Veratraman-3,23-diol, 14,15,16,17-tetradehydro-, (3β,23β)-
$C_{27}H_{39}NO_3$
12046 - Veratraman-11-one, 17,23-epoxy-3-hydroxy-, (3β,23β)-
$C_{27}H_{40}O_3$
7774 - 2-Naphthalenecarboxylic acid, 3-hydroxy-, hexadecyl ester

$C_{27}H_{42}Cl_2N_2O_6$
6648 - Hexadecanoic acid, 2-[(dichloroacetyl)amino]-3-hydroxy-3-(4-nitrophenyl)propyl ester, [*R*-(*R**,*R**)]-

$C_{27}H_{42}O_3$
11536 - Spirost-5-en-3-ol, (3β,25*R*)-

$C_{27}H_{42}O_4$
11534 - Spirostan-12-one, 3-hydroxy-, (3β,5α,25*R*)-

$C_{27}H_{43}NO_2$
11515 - Solanid-5-ene-3,12-diol, (3β,12α)-
11516 - Solanid-5-ene-3,18-diol, (3β)-
11529 - Spirosol-5-en-3-ol, (3β,22α,25*R*)-

$C_{27}H_{43}NO_8$
4068 - Cevane-3,4,7,14,15,16,20-heptol, 4,9-epoxy-, (3β,4α,7α,15α,16β)-

$C_{27}H_{43}NO_9$
4070 - Cevane-3,4,6,7,14,15,16,20-octol, 4,9-epoxy-, (3β,4α,6α,7α,15α,16β

$C_{27}H_{44}$
4102 - Cholesta-3,5-diene

$C_{27}H_{44}O$
4103 - Cholesta-5,7-dien-3-ol, (3β)-
4104 - Cholesta-8,24-dien-3-ol, (3β,5α)-
11348 - 9,10-Secocholesta-5,7,10(19)-trien-3-ol, (3β,5Z,7E)-

$C_{27}H_{44}O_3$
11531 - Spirostan-3-ol, (3β,5β,25*S*)-
11532 - Spirostan-3-ol, (3β,5α,25*R*)-
11533 - Spirostan-3-ol, (3β,5β,25*R*)-

$C_{27}H_{44}O_4$
11530 - Spirostan-2,3-diol, (2α,3β,5α,25*R*)-

$C_{27}H_{44}O_5$
11535 - Spirostan-2,3,15-triol, (2α,3β,5α,15β,25*R*)-

$C_{27}H_{45}NO_2$
11528 - Spirosolan-3-ol, (3β,5α,22β,25*S*)-

$C_{27}H_{46}$
6298 - 10-Heneicosene, 11-phenyl-

$C_{27}H_{46}O$
4110 - Cholest-5-en-3-ol (3β)-
4111 - Cholest-4-en-3-ol, (3β)-
4112 - Cholest-5-en-3-ol, (3α)-

$C_{27}H_{46}O_2$
2785 - 2H-1-Benzopyran-6-ol, 3,4-dihydro-2,8-dimethyl-2-(4,8,12-trimethyltridecyl)-, [2*R*-[2*R**(4*R**,8*R**)]]-

$C_{27}H_{48}$
1741 - Benzene, heneicosyl-
4105 - Cholestane, (5α)-
4106 - Cholestane, (5β)-
6297 - Heneicosane, 11-phenyl-

$C_{27}H_{48}O$
4107 - Cholestanol
4108 - Cholestan-3-ol, (3α,5α)-
4109 - Cholestan-3-ol, (3β,5β)-

$C_{27}H_{50}O_6$
8440 - Octanoic acid, 1,2,3-propanetriyl ester

$C_{27}H_{54}$
4808 - Cyclopentane, docosyl-

$C_{27}H_{54}O$
6303 - 14-Heptacosanone

$C_{27}H_{54}O_2$
6625 - Hexacosanoic acid, methyl ester

$C_{27}H_{56}$
6302 - Heptacosane

$C_{28}H_{14}N_2O_4$
434 - 5,9,14,18-Anthrazinetetrone, 6,15-dihydro-

$C_{28}H_{20}O$
6175 - Furan, tetraphenyl-

$C_{28}H_{22}$
367 - Anthracene, 9,10-bis(phenylmethyl)-

$C_{28}H_{22}O_3$
937 - Benzeneacetic acid, α-phenyl-, anhydride

$C_{28}H_{24}N_2O_7$
8549 - Orcein

$C_{28}H_{30}O_4$
7352 - 1(3H)-Isobenzofuranone, 3,3-bis[4-hydroxy-2-methyl-5-(1-methylethyl)phenyl]-

$C_{28}H_{32}ClN_2O_3$
12056 - Xanthylium, 9-(2-carboxyphenyl)-3,6-bis(diethylamino)-, chloride

$C_{28}H_{32}O_4Si_4$
5000 - Cyclotetrasiloxane, 2,4,6,8-tetramethyl-2,4,6,8-tetraphenyl-

$C_{28}H_{34}O_{15}$
2794 - 4H-1-Benzopyran-4-one, 7-[[6-*O*-(6-deoxy-α-*L*-mannopyranosyl)-β-*D*-glucopyranosyl]oxy]-2,3-dih

$C_{28}H_{38}N_2O_4$
5367 - Emetan-6′-ol, 7′,10,11-trimethoxy-

$C_{28}H_{38}O_{19}$
6229 - α-*D*-Glucopyranoside, 1,3,4,6-tetra-*O*-acetyl-β-*D*-fructofuranosyl, tetraacetate

$C_{28}H_{42}O$
5380 - Ergosta-5,7,9(11),22-tetraen-3-ol, (3β,22E)-

$C_{28}H_{44}O$
5381 - Ergosta-5,7,22-trien-3-ol, (3β,10α,22E)-
5382 - Ergosta-5,7,22-trien-3-ol, (3β,9β,10α,22E)-
5383 - Ergosta-5,7,22-trien-3-ol, (3β,22E)-
11349 - 9,10-Secoergosta-5(10),6,8,22-tetraen-3-ol, (3β,6E,22E)-
11350 - 9,10-Secoergosta-5,7,10(19),22-tetraen-3-ol, (3β,5Z,7E,22E)-

$C_{28}H_{48}O$
357 - Androst-4-en-3-one, 17-hydroxy-7,17-dimethyl-, (7β,17β)-
5384 - Ergost-7-en-3-ol, (3β,5α)-
5385 - Ergost-8(14)-en-3-ol, (3β,5α)-
5386 - Ergost-5-en-3-ol, (3β,24*R*)-

$C_{28}H_{48}O_2$
2789 - 2H-1-Benzopyran-6-ol, 3,4-dihydro-2,5,8-trimethyl-2-(4,8,12-trimethyltridecyl)-
2790 - 2H-1-Benzopyran-6-ol, 3,4-dihydro-2,7,8-trimethyl-2-(4,8,12-trimethyltridecyl)-

$C_{28}H_{50}$
1565 - Benzene, docosyl-
5377 - Ergostane, (5α)-
5378 - Ergostane, (5β)-

$C_{28}H_{50}N_2O_4$
4064 - Carpaine

$C_{28}H_{50}O$
5379 - Ergostan-3-ol, (3β,5α)-

$C_{28}H_{54}O_3$
11623 - Tetradecanoic acid, anhydride

$C_{28}H_{56}$
4380 - Cyclohexane, docosyl-

$C_{28}H_{56}O_2$
8221 - Octacosanoic acid

$C_{28}H_{58}$
5274 - Docosane, 7-hexyl-
5354 - Eicosane, 9-octyl-
8220 - Octacosane
11877 - Tridecane, 2,2,4,10,12,12-hexamethyl-7-(3,5,5-trimethylhexyl)-

$C_{28}H_{58}O$
8222 - 1-Octacosanol

$C_{29}H_{40}N_2O_4$
5368 - Emetan, 6′,7′,10,11-tetramethoxy-

$C_{29}H_{44}O_{12}$
4040 - Card-20(22)-enolide, 3-[(6-deoxy-α-*L*-mannopyranosyl)oxy]-1,5,11,14,19-pentahydroxy-, (1β,3β

$C_{29}H_{47}NO_8$
4067 - Cevane-3,4,12,14,16,17,20-heptol, 3-acetate, (3β,4α,16β)-

$C_{29}H_{48}O$
11554 - Stigmasta-5,7-dien-3-ol, (3β)-
11555 - Stigmasta-5,22-dien-3-ol, (3β,22E)-

$C_{29}H_{50}O$
11558 - Stigmast-5-en-3-ol, (3β)-
11559 - Stigmast-5-en-3-ol, (3β,24*S*)-

$C_{29}H_{50}O_2$
2787 - 2H-1-Benzopyran-6-ol, 3,4-dihydro-2,5,7,8-tetramethyl-2-(4,8,12-trimethyltridecyl)-, [2*R*-[2*R**(4*R**,8*R**)]]-

$C_{29}H_{52}$
2341 - Benzene, tricosyl-

$C_{29}H_{52}O$
11557 - Stigmastan-3-ol, (3β,5α)-

$C_{29}H_{58}$
4483 - Cyclohexane, tricosyl-

$C_{29}H_{60}$
8099 - Nonacosane

$C_{29}H_{60}O$
8100 - 1-Nonacosanol

$C_{30}H_{34}O_{13}$
7579 - 3,6-Methano-8H-1,5,7-trioxacyclopenta [ij]cycloprop[a]azulene-4,8(3H)-dione, hexahydro-2a-hydroxy

$C_{30}H_{37}N_5O_5$
5393 - Ergotaman-3′,6′,18-trione, 12′-hydroxy-2′-methyl-5′-(2-methylpropyl)-,

$C_{30}H_{40}O_2$
435 - 8′-Apo-β,ψ-carotenal, 3-hydroxy-, (3*R*)-

$C_{30}H_{42}FeO_2Si_4$
5961 - Ferrocene, 1,1′-bis(1,1,3,3-tetramethyl-3-phenyldisiloxanyl)-

$C_{30}H_{46}O_5$
8546 - Olean-12-en-28-oic acid, 3,16-dihydroxy-23-oxo-, (3β,4α,16α)-
12036 - Urs-12-ene-27,28-dioic acid, 3-hydroxy-, (3β)-

$C_{30}H_{48}O_3$
8547 - Olean-12-en-28-oic acid, 3-hydroxy-, (3β)-
12037 - Urs-12-en-28-oic acid, 3-hydroxy-, (3β)-

$C_{30}H_{48}O_4$
8545 - Olean-12-en-28-oic acid, 3,23-dihydroxy-, (3β,4α)-

$C_{30}H_{50}$
11602 - 2,6,10,14,18,22-Tetracosahexaene, 2,6,10,15,19,23-hexamethyl-, (all-E)-

$C_{30}H_{50}O$
7411 - Lanosta-8,24-dien-3-ol, (3β)-
7417 - Lup-20(29)-en-3-ol, (3β)-
8548 - Olean-12-en-3-ol, (3β)-
11556 - Stigmasta-7,24(28)-dien-3-ol, 4-methyl-, (3β,4α,5α,24Z)-
12038 - Urs-12-en-3-ol, (3β)-

$C_{30}H_{50}O_2$
7416 - Lup-20(29)-ene-3,28-diol, (3β)-

$C_{30}H_{54}$
2265 - Benzene, tetracosyl-

$C_{30}H_{60}$
4474 - Cyclohexane, tetracosyl-

$C_{30}H_{60}I_3N_3O_3$
5441 - Ethanaminium, 2,2′,2″-[1,2,3-benzenetriyltris(oxy)]tris[*N*,*N*,*N*-triethyl-, tr

$C_{30}H_{60}O_2$
11820 - Triacontanoic acid

$C_{30}H_{62}$
5275 - Docosane, 9-octyl-
11606 - Tetracosane, 2,6,10,15,19,23-hexamethyl-
11819 - Triacontane

$C_{30}H_{62}O$
11821 - 1-Triacontanol

$C_{31}H_{36}N_2O_8$
12064 - Yohimban-16-carboxylic acid, 17-hydroxy-18-[(3,4,5-trimethoxybenzoyl)oxy]-, methyl ester, (3β,16β

$C_{31}H_{39}N_5O_5$
5387 - Ergotaman-3′,6′,18-trione, 12′-hydroxy-2′,5′-bis(1-methylethyl)-,
5388 - Ergotaman-3′,6′,18-trione, 12′-hydroxy-2′,5′-bis(1-methylethyl)-,

$C_{31}H_{46}O_2$
7865 - 1,4-Naphthalenedione, 2-methyl-3-(3,7,11,15-tetramethyl-2-hexadecenyl)-, [*R*-[*R**,*R**-(*E*)]]-

$C_{31}H_{52}O_3$
2788 - 2H-1-Benzopyran-6-ol, 3,4-dihydro-2,5,7,8-tetramethyl-2-(4,8,12-trimethyltridecyl)-, acetate, [2*R*-[2*R**(4*R**,8*R**

$C_{31}H_{56}$
2088 - Benzene, pentacosyl-
8638 - Pentacosane, 13-phenyl-

$C_{31}H_{62}$
4462 - Cyclohexane, pentacosyl-
4821 - Cyclopentane, hexacosyl-

$C_{31}H_{62}O$
6300 - 16-Hentriacontanone

$C_{31}H_{64}$
6295 - Heneicosane, 11-decyl-
6299 - Hentriacontane

$C_{32}H_{22}N_6Na_2O_6S_2$
7969 - 1-Naphthalenesulfonic acid, 3,3′-[[1,1′-biphenyl]-4,4′-diylbis(azo)]bis[4-amino-, disodium salt

$C_{32}H_{38}N_2O_8$
12065 - Yohimban-16-carboxylic acid, 17-methoxy-18-[(3,4,5-trimethoxybenzoyl)oxy]-, methyl ester, (3β,16β

$C_{32}H_{40}O_4Si_4$
4998 - Cyclotetrasiloxane, 2,4,6,8-tetraethyl-2,4,6,8-tetraphenyl-

$C_{32}H_{41}N_5O_5$
5389 - Ergotaman-3′,6′,18-trione, 12′-hydroxy-2′-(1-methylethyl)-5′-(2-methylpropyl)-

$C_{32}H_{44}N_2O_8$
309 - Aconitane-4,8,9-triol, 20-ethyl-1,14,16-trimethoxy-, 4-[2-(acetylamino)benzoate], (1α,14α,16β)-

$C_{32}H_{44}O_{12}$
3142 - Bufa-4,20,22-trienolide, 6-(acetyloxy)-3-(β-*D*-glucopyranosyloxy)-8,14-dihydroxy-, (3β,6β)-

$C_{32}H_{49}NO_9$
4069 - Cevane-3,4,12,14,16,17,20-heptol, 4,9-epoxy-, 3-(2-methyl-2-butenoate), [3β(*Z*),4α,16β]-

$C_{32}H_{58}$
1750 - Benzene, hexacosyl-

$C_{32}H_{62}O_3$
6646 - Hexadecanoic acid, anhydride

$C_{32}H_{64}$
4410 - Cyclohexane, hexacosyl-

$C_{32}H_{64}O_2$
6651 - Hexadecanoic acid, hexadecyl ester

$C_{32}H_{66}$
5273 - Docosane, decyl-
5348 - Dotriacontane

$C_{32}H_{66}O$
6642 - Hexadecane, 1,1′-oxybis-

$C_{33}H_{34}N_4O_6$
3001 - 21H-Biline-8,12-dipropanoic acid, 3,18-diethenyl-1,19,22,24-tetrahydro-2,7,13,17-tetramethyl-1,19-dioxo-

$C_{33}H_{35}N_5O_5$
5395 - Ergotaman-3′,6′,18-trione, 12′-hydroxy-2′-methyl-5′-(phenylmethyl)-,

$C_{33}H_{36}N_4O_6$
3000 - 21H-Biline-8,12-dipropanoic acid, 2,17-diethenyl-1,10,19,22,23,24-hexahydro-3,7,13,18-tetramethyl-1,19-dioxo-

$C_{33}H_{40}N_2O_9$
12061 - Yohimban-16-carboxylic acid, 11,17-dimethoxy-18-[(3,4,5-trimethoxybenzoyl)oxy]-, methyl ester,

$C_{33}H_{41}N_5O_5$
5390 - Ergotaman-3′,6′,18-trione, 12′-hydroxy-2′-(1-methylethyl)-5′-(2-methylpropyl)-, (5′α,8α)-

$C_{33}H_{44}O_8$
8211 - 29-Nordammara-1,17(20),24-trien-21-oic acid, 6,16-bis(acetyloxy)-3,7-dioxo-,

$C_{33}H_{45}NO_9$
311 - Aconitane-8,13,14-triol, 1,6,16-trimethoxy-4-(methoxymethyl)-20-methyl-, 8-acetate 14-benzoate,

$C_{33}H_{49}NO_8$
12045 - Veratraman-11-one, 17,23-epoxy-3-(β-*D*-glucopyranosyloxy)-, (3β,23β)-

$C_{33}H_{55}N_5O_5$
5394 - Ergotaman-3′,6′,18-trione, 12′-hydroxy-2′-methyl-5′-(phenylmethyl)-, (5′α)-

$C_{33}H_{60}$
1742 - Benzene, heptacosyl-

$C_{34}H_{24}N_6Na_4O_{14}S_4$
7877 - 1,3-Naphthalenedisulfonic acid, 6,6′-[(3,3′-dimethyl[1,1′-biphenyl]-4,4′-diyl)bis(azo)]bis[4-amino-5-hydroxy-,

$C_{34}H_{32}ClFeN_4O_4$
5958 - Ferrate(2-), chloro[7,12-diethenyl-3,8,13,17-tetramethyl-21H,23H-porphine-2,18-dipropanoato(4-)-*N*21,*N*22,*N*23,

$C_{34}H_{33}FeN_4O_5$
5959 - Ferrate(2-), [7,12-diethenyl-3,8,13,17-tetramethyl-21H,23H-porphine-2,18-dipropanoato(4-)-*N*21,*N*22,*N*23,*N*24

$C_{34}H_{36}N_2O_6$
3002 - [2,2′-Bimorphinan]-3,3′,6,6′-tetrol, 7,7′,8,8′-tetradehydro-4,5:4′,5′-diepoxy-17,17′dimethyl-

$C_{34}H_{38}N_4O_6$
9735 - 21H,23H-Porphine-2,18-dipropanoic acid, 7,12-bis(1-hydroxyethyl)-3,8,13,17-tetramethyl-

$C_{34}H_{47}NO_{10}$
307 - Aconitane-3,8,13,14-tetrol, 20-ethyl-1,6,16-trimethoxy-4-(methoxymethyl)-, 8-acetate 14-benzoate, (1α,3alp

$C_{34}H_{47}NO_{11}$
306 - Aconitane-3,8,13,14,15-pentol, 20-ethyl-1,6,16-trimethoxy-4-(methoxymethyl)-, 8-acetate 14-benzoate,

$C_{34}H_{62}$
2065 - Benzene, octacosyl-

$C_{34}H_{66}O_4$
6649 - Hexadecanoic acid, 1,2-ethanediyl ester

$C_{34}H_{68}$
4458 - Cyclohexane, octacosyl-

$C_{34}H_{68}O_2$
8258 - Octadecanoic acid, hexadecyl ester

$C_{34}H_{70}$
11604 - Tetracosane, 11-decyl-
11650 - Tetratriacontane

$C_{35}H_{39}N_5O_5$
5391 - Ergotaman-3′,6′,18-trione, 12′-hydroxy-2′-(1-methylethyl)-5′-(phenylmethyl)-
5392 - Ergotaman-3′,6′,18-trione, 12′-hydroxy-2′-(1-methylethyl)-5′-(phenylmethyl)-

$C_{35}H_{42}N_2O_9$
12060 - Yohimban-16-carboxylic acid, 11,17-dimethoxy-18-[[1-oxo-3-(3,4,5-trimethoxyphenyl)-2-propenyl]oxy]-, methyl ester, (3β,16β,17α,18β,20α)-

$C_{35}H_{54}O_{14}$
4045 - Card-20(22)-enolide, 3-[(6-*O*-β-*D*-glucopyranosyl-β-D-glucopyranosyl)oxy]-14-hydroxy-,

$C_{35}H_{70}$
4853 - Cyclopentane, triacontyl-

$C_{35}H_{70}O$
9000 - 18-Pentatriacontanone

$C_{35}H_{72}$
8999 - Pentatriacontane

$C_{36}H_{38}N_2O_6$
11924 - Tubocuraran-7′,12′-diol, 6,6′-dimethoxy-2,2′-dimethyl-, (1′α)-
11925 - Tubocuraran-7′,12′-diol, 6,6′-dimethoxy-2,2′-dimethyl-, (1β)-

$C_{36}H_{42}Br_2N_2$
6753 - 1,6-Hexanediaminium, *N,N*′-di-9H-fluoren-9-yl-*N,N,N*′,*N*′-tetramethyl-dibromide

$C_{36}H_{51}NO_{11}$
310 - Aconitane-8,13,14-triol, 20-ethyl-1,6,16-trimethoxy-4-(methoxymethyl)-, 8-acetate 14-(3,4-dimethoxybenzoate),

$C_{36}H_{51}NO_{12}$
308 - Aconitane-3,8,13,14-tetrol, 20-ethyl-1,6,16-trimethoxy-4-(methoxymethyl)-, 8-acetate 14-(3,4-dimethoxybenzoate),(1α,3α,6α,14α,16β)-

$C_{36}H_{70}CaO_4$
8252 - Octadecanoic acid, calcium salt

$C_{36}H_{70}O_3$
8249 - Octadecanoic acid, anhydride

$C_{36}H_{70}O_4Pb$
8261 - Octadecanoic acid, lead salt

$C_{36}H_{70}O_4Zn$
8277 - Octadecanoic acid, zinc salt

$C_{36}H_{72}$
4482 - Cyclohexane, triacontyl-

$C_{36}H_{74}$
6991 - Hexatriacontane
8639 - Pentacosane, 13-undecyl-

$C_{37}H_{38}N_2O_6$
8630 - Oxyacanthan, 6′,12′-dimethoxy-2,2′-dimethyl-6,7-[methylenebis(oxy)]-

$C_{37}H_{40}N_2O_6$
2911 - Berbaman-12-ol, 6,6′,7-trimethoxy-2,2′-dimethyl-
8631 - Oxyacanthan-12′-ol, 6,6′,7-trimethoxy-2,2′-dimethyl-

$C_{37}H_{67}NO_{13}$
5396 - Erythromycin

$C_{38}H_{74}O_4$
8256 - Octadecanoic acid, 1,2-ethanediyl ester

$C_{38}H_{76}$
4383 - Cyclohexane, dotriacontyl-

$C_{38}H_{78}$
6623 - Hexacosane, 13-dodecyl-

$C_{39}H_{74}O_6$
5319 Dodecanoic acid, 1,2,3-propanetriyl ester

$C_{40}H_{36}$
4054 - β,ε Carotene, (6′*R*)-

$C_{40}H_{50}N_4O_8S$
4132 Cinchonan-9-ol, 6′-methoxy-, (8α,9*R*)-, sulfate (2:1) (salt)

$C_{40}H_{52}O_2$
4059 - β,ψ-Caroten-16′-oic acid, 3′,4′-didehydro-

$C_{40}H_{56}$
4051 - β,β-Carotene
4052 - β,ψ-Carotene
4053 - ψ,ψ-Carotene

$C_{40}H_{56}O$
4060 - ψ,ψ-Caroten-16-ol
4061 - β,β-Caroten-3-ol, (3*R*)-
4062 - β,ψ-Caroten-3-ol, (3*R*)-

$C_{40}H_{56}O_2$
4057 - β,β-Carotene-3,3′-diol, (3*R*,3′*R*)-
4058 - β,ε-Carotene-3,3′-diol, (3*R*,3′*R*,6′*R*)-

$C_{40}H_{56}O_3$
4063 - β,κ-Caroten-6′-one, 3,3′-dihydroxy-, (3*R*,3′*S*,5′*R*)-

$C_{40}H_{56}O_4$
4056 - β,β-Carotene-3,3′-diol, 5,6:5′,6′-diepoxy-5,5′,6,6′-tetrahydro-

$C_{40}H_{60}O_6$
4055 - β,β-Carotene, 3′-(acetyloxy)-6′,7′-didehydro-5,6-epoxy-5,5′,6,6′,7,8-hexahydro-3,5′-dihydroxy-8-oxo-,

$C_{40}H_{80}$
4479 - Cyclohexane, tetratriacontyl-
4845 - Cyclopentane, pentatriacontyl-

$C_{41}H_{46}N_4O_5$
4128 - Cinchonan-9-ol, 6′-methoxy-, carbonate (2:1) (este*R*), (8α,9*R*)-

$C_{41}H_{64}O_{13}$
4043 - Card-20(22)-enolide, 3-[(*O*-2,6-dideoxy-β-*D*-ribo-hexopyranosyl-(1-4)-*O*-2,6-dideoxy-β-*D*-ribo-hexopyranosyl-(1-4)-2,6-dideoxy-β-*D*, 14-dihydroxy

CAS NUMBER INDEX OF ORGANIC COMPOUNDS

CAS No.	No.
50-00-0	5987
50-01-1	6285
50-02-2	9737
50-03-3	9747
50-06-6	11095
50-10-2	5443
50-11-3	11088
50-14-6	11350
50-22-6	9748
50-23-7	9753
50-24-8	9738
50-27-1	5404
50-28-2	5399
50-29-3	2328
50-30-6	2563
50-31-7	2727
50-32-8	2395
50-34-0	9832
50-35-1	7363
50-36-2	482
50-42-0	938
50-44-2	10763
50-50-0	5403
50-53-3	9505
50-55-5	12061
50-69-1	11345
50-70-4	6206
50-71-5	11073
50-78-2	2480
50-79-3	2562
50-81-7	400
50-82-8	2728
50-84-0	2561
50-85-1	2630
50-89-5	11816
50-91-9	12034
51-03-6	9727
51-12-7	10910
51-17-2	2376
51-18-3	11844
51-20-7	11062
51-21-8	11067
51-26-3	2136
51-28-5	9342
51-34-3	908
51-35-4	9767
51-36-5	2564
51-41-2	1500
51-43-4	1521
51-45-6	7203
51-48-9	11930
51-52-5	11109
51-55-8	902
51-63-8	1582
51-66-1	71
51-67-2	9208
51-75-2	5413
51-79-6	3948
51-80-9	7490
51-83-2	5440
52-28-8	7673
52-31-3	11084
52-39-1	9746
52-43-7	11091
52-49-3	9705
52-51-7	9991
52-52-8	4782
52-67-5	12043
52-68-6	11853
52-85-7	5946
52-86-8	3651
52-90-4	5014
53-03-2	9739
53-05-4	9744
53-06-5	9754
53-16-7	5406
53-41-8	348
53-46-3	5445
53-70-3	5142
53-89-4	10850
53-96-3	64
54-04-6	1586
54-06-8	7301
54-11-5	10992
54-25-1	11832
54-36-4	10462
54-85-3	10905
54-95-5	11654
55-10-7	896
55-18-5	5432
55-21-0	600
55-22-1	10895
55-38-9	5955
55-43-6	1801
55-63-0	10197
55-73-2	6281
55-91-4	9580
56-03-1	7233
56-04-2	11108
56-12-2	3464
56-18-8	9874
56-23-5	7540
56-25-7	5371
56-29-1	11083
56-33-7	5238
56-35-9	5241
56-36-0	11538
56-38-2	9583
56-40-6	6250
56-41-7	325
56-45-1	11357
56-49-5	2392
56-53-1	9300
56-54-2	4126
56-55-3	519
56-57-5	11287
56-72-4	4147
56-75-7	52
56-81-5	10186
56-82-6	9782
56-84-8	471
56-85-9	6249
56-86-0	6241
56-87-1	7419
56-89-3	5020
57-00-1	6254
57-06-7	10575
57-08-9	6871
57-10-3	6644
57-11-4	8247
57-13-6	11002
57-14-7	7154
57-15-8	10407
57-24-9	11563
57-27-2	7666
57-41-0	7219
57-43-2	11092
57-44-3	11087
57-47-6	11191
57-50-1	6218
57-53-4	10021
57-55-6	9982
57-57-8	8595
57-62-5	7704
57-63-6	8216
57-66-9	2595
57-67-0	2191
57-68-1	2196
57-71-6	3361
57-74-9	4077
57-85-2	359
57-87-4	5383
57-88-5	4110
57-91-0	5400
57-92-1	11562
57-97-6	520
58-00-4	5155
58-05-9	6244
58-08-2	10760
58-14-0	11051
58-15-1	10851
58-18-4	358
58-22-0	356
58-25-3	2409
58-27-5	7864
58-32-2	5711
58-38-8	9501
58-39-9	9617
58-46-8	2396
58-54-8	168
58-55-9	10756
58-56-0	10934
58-61-7	319
58-63-9	7335
58-72-0	1648
58-73-1	5422
58-74-2	7378
58-85-5	11689
58-86-6	12059
58-89-9	4409
58-90-2	9470
58-93-5	2855
58-94-6	2852
58-95-7	2788
58-96-8	12033
58-97-9	12035
59-00-7	11230
59-02-9	2787
59-23-4	6197
59-26-7	10891
59-30-3	6243
59-31-4	11325
59-40-5	2205
59-46-1	2493
59-47-2	10019
59-48-3	7327
59-49-4	2882
59-50-7	9267
59-51-8	7647
59-52-9	10370
59-66-5	32
59-67-6	10894
59-87-0	7144
59-88-1	7172
59-89-2	7699
59-92-7	11929
59-98-3	7198
60-00-4	6262
60-01-5	3586
60-09-3	837
60-10-6	5121
60-11-7	721
60-12-8	1592
60-18-4	11926
60-23-1	5649
60-24-2	5732
60-27-5	7230
60-29-7	5608
60-31-1	5439
60-32-2	6873
60-33-3	8223
60-34-4	7161
60-35-5	22
60-41-3	11566
60-51-5	5009
60-54-8	7706
60-56-0	7214
60-57-1	5178
60-70-8	12044
60-80-0	10847
60-82-2	10453
60-87-7	9504
60-91-3	9503
61-00-7	5807
61-12-1	11226
61-19-8	320
61-50-7	7303
61-54-1	7302
61-73-4	9506
61-78-9	6252
61-80-3	2876
61-82-5	11845
61-90-5	7413
62-23-7	2705
62-38-4	7431
62-44-2	60
62-50-0	7531
62-51-1	9830
62-53-3	633
62-54-4	122
62-55-5	5643
62-56-6	11799
62-57-7	331
62-59-9	4069
62-73-7	5170
62-75-9	7450
62-76-0	5566
62-90-8	5408
63-05-8	352
63-25-2	8032
63-68-3	7648
63-74-1	2187
63-75-2	10914
63-91-2	9510
63-98-9	879
64-04-0	1569
64-10-8	12026
64-17-5	5672
64-18-6	6002
64-19-7	100
64-20-0	7456
64-65-3	9656
64-67-5	11573
64-69-7	200
64-77-7	2207
64-85-7	9751
64-86-8	93
65-29-2	5441
65-45-2	616
65-46-3	5021
65-49-6	2498
65-64-5	7170
65-71-4	11068
65-82-7	7649
65-85-0	2475
65-86-1	11050
66-02-4	11928
66-22-8	11059
66-23-9	5438
66-25-1	6718
66-27-3	7532
66-28-4	4049
66-32-0	11565
66-56-8	9341
66-71-7	9189
66-75-1	11061
66-76-2	2839
66-77-3	7751
66-81-9	9655
66-99-9	7752
67-03-8	11686
67-20-9	7222
67-43-6	6257
67-45-8	0009
67-47-0	6044
67-48-1	5446
67-51-6	10828
67-52-7	11076
67-56-1	7581
67-63-0	10335
67-64-1	10411
67-66-3	7554
67-68-5	7529
67-71-0	7534
67-72-1	5595
67-97-0	11348
68-11-1	202
68-12-2	5992
68-26-8	11341
68-35-9	2204
68-36-0	996
68-41-7	7406
68-88-2	5700
68-94-0	10770
68-96-2	9750
69-24-9	10435
69-57-8	11655
69-65-8	7422
69-72-7	2612
69-89-6	10755
69-91-0	885
69-93-2	10764
70-07-5	8588
70-11-1	5782
70-18-8	6264
70-22-4	11188
70-25-7	6284
70-26-8	8550
70-29-1	5621
70-30-4	9409
70-34-8	1718
70-47-3	468
70-54-2	7418
70-69-9	10413
70-70-2	10450
70-78-0	11932
71-00-1	7132
71-23-8	10334
71-30-7	11104
71-36-3	3593
71-41-0	8915
71-43-2	867
71-44-3	3259
71-55-6	5655
71-63-6	4043
71-81-8	2120
71-91-0	5447
72-14-0	2206
72-18-4	12041
72-19-5	11813
72-23-1	9755
72-43-5	2329
72-44-6	11199
72-48-0	391
72-54-8	1378
72-55-9	1374
72-56-0	9160
72-63-9	344
73-22-3	11922
73-24-5	10751
73-31-4	69
73-32-5	7370
73-40-5	10768
73-49-4	11197
74-11-3	2536
74-31-7	1271
74-30-5	1542
74-79-3	438
74-82-8	7459
74-83-9	7462
74-84-0	5448
74-85-1	5879
74-86-2	5936
74-87-3	7477
74-88-4	7518
74-89-5	7442
74-90-8	7176
74-93-1	7546
74-95-3	7492
74-96-4	5454
74-97-5	7463
74-98-6	9834
74-99-7	10727
75-00-3	5467
75-01-4	5884
75-02-5	5906
75-03-6	5597
75-04-7	5410
75-05-8	271
75-07-0	11
75-08-1	5648
75-09-2	7499
75-10-5	7504
75-11-6	7506
75-12-7	5989
75-15-0	3993
75-17-2	5988
75-18-3	7545
75-19-4	4955
75-21-8	8599
75-22-9	3132
75-24-1	339
75-25-2	7548
75-26-3	9837
75-27-4	7467
75-28-5	10073
75-29-6	9849
75-30-9	10055
75-31-0	9804
75-33-2	10157
75-34-3	5528
75-35-4	5894
75-36-5	294
75-37-6	5544
75-38-7	5902
75-39-8	5674
75-43-4	7502
75-44-5	4007
75-45-6	7478
75-46-7	7558
75-47-8	7560
75-50-3	7444
75-52-5	7522
75-54-7	11398
75-55-8	501

CAS No.	No.
75-56-9	8621
75-57-0	7457
75-58-1	7458
75-60-5	451
75-61-6	7496
75-62-7	7474
75-63-8	7475
75-64-9	9818
75-65-0	10389
75-66-1	10159
75-68-3	5470
75-69-4	7555
75-71-8	7500
75-72-9	7486
75-73-0	7541
75-74-1	9732
75-75-2	7530
75-76-3	11451
75-77-4	11389
75-78-5	11394
75-79-6	11465
75-80-9	5767
75-81-0	5520
75-82-1	5522
75-83-2	3289
75-84-3	10373
75-85-4	3628
75-86-5	10108
75-87-6	21
75-88-7	5489
75-89-8	5770
75-91-2	7178
75-93-4	11577
75-94-5	11461
75-95-6	5616
75-96-7	241
75-97-8	3657
75-98-9	10258
75-99-0	10244
76-00-6	10406
76-01-7	5617
76-02-8	300
76-03-9	243
76-04-0	127
76-05-1	263
76-06-2	7557
76-08-4	10405
76-09-5	3350
76-11-9	5636
76-12-0	5637
76-13-1	5663
76-14-2	5538
76-15-3	5484
76-16-4	5596
76-17-5	10181
76-19-7	10118
76-20-0	3230
76-24-4	3106
76-28-8	4046
76-30-2	3342
76-36-8	3191
76-37-9	10404
76-39-1	10391
76-43-7	355
76-44-8	7576
76-45-9	4070
76-49-3	2948
76-57-3	7669
76-59-5	9365
76-60-8	9367
76-61-9	9369
76-62-0	7350
76-65-3	2463
76-67-5	9952
76-68-6	11085
76-76-6	11093
76-80-2	3110
76-83-5	1196
76-84-6	1866
76-87-9	11911
76-89-1	920
76-93-7	918
76-94-8	11100
77-02-1	11098
77-03-2	9654
77-04-3	10946
77-06-5	6204
77-09-8	7354
77-10-1	9696
77-16-7	11904
77-25-8	9940
77-26-9	11101
77-28-1	11082
77-40-7	9433
77-41-8	11156
77-42-9	9103
77-46-3	91
77-47-4	4766
77-48-5	7217
77-49-6	10018
77-52-1	12037
77-53-2	7570
77-59-8	11528
77-60-1	11532
77-63-4	5000
77-65-6	3195
77-67-8	11158
77-71-4	7218
77-74-7	8944
77-75-8	9146
77-76-9	9909
77-77-0	5913
77-78-1	11574
77-79-2	11752
77-83-8	8604
77-84-9	10007
77-85-0	10011
77-86-1	9986
77-89-4	10166
77-92-9	10167
77-93-0	10169
77-94-1	10168
77-95-2	4338
77-99-6	10006
78-00-2	9731
78-07-9	11478
78-08-0	11424
78-09-1	5601
78-10-4	11506
78-11-5	9990
78-13-7	11507
78-26-2	10020
78-27-3	4529
78-30-8	9569
78-32-0	9571
78-34-2	5201
78-38-6	9536
78-39-7	5664
78-40-0	9563
78-48-8	11852
78-54-6	4528
78-59-1	4721
78-62-6	11405
78-67-1	10093
78-74-0	5653
78-75-1	9881
78-77-3	9845
78-78-4	3397
78-79-5	3171
78-80-8	3892
78-81-9	9817
78-82-0	10111
78-83-1	10388
78-84-2	9785
78-85-3	10515
78-88-6	10552
78-89-7	10347
78-90-0	9872
78-92-2	3594
78-93-3	3645
78-94-4	3875
78-95-5	10420
78-97-7	10105
78-98-8	9788
78-99-9	9891
79-00-5	5656
79-01-6	5922
79-03-8	10493
79-04-9	295
79-05-0	9791
79-06-1	10519
79-07-2	41
79-08-3	111
79-09-4	10201
79-10-7	10609
79-11-8	123
79-14-1	196
79-15-2	36
79-16-3	72
79-17-4	7147
79-19-6	7142
79-20-9	208
79-21-0	5619
79-22-1	4020
79-24-3	5607
79-27-6	5632
79-28-7	5914
79-29-8	3290
79-30-1	10504
79-31-2	10290
79-34-5	5635
79-35-6	5897
79-36-7	297
79-37-8	5585
79-38-9	5889
79-40-3	5588
79-41-4	10648
79-42-5	10285
79-43-6	162
79-44-7	3973
79-46-9	10117
79-49-2	10477
79-53-8	10421
79-54-9	9169
79-55-0	9691
79-57-2	7705
79-58-3	11515
79-63-0	7411
79-69-6	3884
79-70-9	3883
79-74-3	1501
79-77-6	3885
79-83-4	329
80-00-2	1225
80-05-7	9413
80-06-8	1850
80-07-9	2239
80-08-0	846
80-09-1	9466
80-10-4	11395
80-11-5	2212
80-13-7	2565
80-14-8	11915
80-15-9	7181
80-18-2	2227
80-26-2	4669
80-32-0	2193
80-33-1	2221
80-35-3	2197
80-40-0	2231
80-42-2	2235
80-46-6	9339
80-48-8	2232
80-55-7	10279
80-56-8	2973
80-59-1	3844
80-62-6	10663
80-63-7	10621
80-69-3	9956
80-72-8	4942
80-92-2	9742
80-97-7	4107
81-04-9	7867
81-07-2	2388
81-08-3	2875
81-13-0	3202
81-15-2	1461
81-16-3	7961
81-20-9	1476
81-23-2	4101
81-24-3	5625
81-25-4	4100
81-30-1	2784
81-38-9	2469
81-54-9	416
81-61-8	413
81-64-1	392
81-77-6	434
81-81-2	2832
81-88-9	12056
81-90-3	2517
81-92-5	1841
82-02-0	6185
82-05-3	6292
82-12-2	406
82-28-0	384
82-29-1	417
82-34-8	410
82-35-9	405
82-40-6	6182
82-44-0	387
82-45-1	380
82-54-2	5216
82-57-5	6186
82-58-6	5376
82-62-2	9490
82-66-6	7268
82-68-8	2087
82-71-3	1554
82-75-7	7965
82-86-0	8
83-07-8	10845
83-12-5	7270
83-13-6	9975
83-14-7	7394
83-25-0	11162
83-27-2	9967
83-32-9	6
83-33-0	7286
83-34-1	7314
83-40-9	2629
83-41-0	1474
83-42-1	1204
83-44-3	4095
83-45-4	11557
83-46-5	11558
83-47-6	11559
83-48-7	11555
83-49-8	4096
83-53-4	7802
83-54-5	11330
83-56-7	7835
83-60-3	12063
83-66-9	1463
83-67-0	10757
83-68-1	7800
83-72-7	7859
83-73-8	11315
83-74-9	7186
83-79-4	2783
83-81-8	1325
83-88-5	11344
83-95-4	6181
84-11-7	9174
84-15-1	11595
84-16-2	9298
84-17-3	9302
84-26-4	7334
84-31-1	4134
84-52-6	5022
84-54-8	409
84-55-9	10434
84-58-2	4251
84-61-7	1341
84-62-8	1345
84-65-1	379
84-66-2	1342
84-68-4	3032
84-69-5	1336
84-74-2	1339
84-79-7	7862
84-80-0	7865
84-86-6	7962
84-87-7	7971
84-88-8	11294
84-89-9	7963
84-95-7	7713
84-99-1	2398
85-00-7	5224
85-01-8	9163
85-02-9	2437
85-23-4	4260
85-29-0	7603
85-34-7	944
85-38-1	2646
85-41-6	7362
85-42-7	7345
85-44-9	7342
85-46-1	7975
85-47-2	7959
85-55-2	2684
85-64-3	9398
85-66-5	9196
85-73-4	2726
85-74-5	7873
85-75-6	7872
85-82-5	8029
85-83-6	8033
85-84-7	7729
85-86-9	8038
85-90-5	2838
85-91-6	2681
85-95-0	9303
85-97-2	3092
85-98-3	12003
86-00-0	3085
86-21-5	11010
86-26-0	3080
86-28-2	3985
86-29-3	969
86-30-6	830
86-34-0	11161
86-48-6	7771
86-50-0	497
86-52-2	7783
86-53-3	7748
86-55-5	7758
86-56-6	7715
86-57-7	7929
86-59-9	11229
86-60-2	7964
86-65-7	7870
86-68-0	11234
86-73-7	5967
86-74-8	3981
86-75-9	11309
86-76-0	5164
86-78-2	8737
86-80-6	5411
86-84-0	7904
86-86-2	7733
86-87-3	7734
86-88-4	11807
86-89-5	7941
86-98-6	11251
87-00-3	911
87-10-5	609
87-13-8	9947
87-17-2	617
87-19-4	2643
87-20-7	2633
87-24-1	2690
87-25-2	2495
87-28-5	2622
87-31-0	2874
87-39-8	11074
87-40-1	2332
87-41-2	7349
87-42-3	10753
87-44-5	2996
87-47-8	3945
87-51-4	7295
87-52-5	7310
87-58-1	11140
87-59-2	704
87-60-5	671
87-61-6	2321
87-62-7	707
87-64-9	9263
87-65-0	9285
87-66-1	2366
87-68-3	3163
87-69-4	3309
87-73-0	6205
87-79-6	11519
87-81-0	11587
87-83-2	2083
87-85-4	1758
87-86-5	9440
87-88-7	4258
87-90-1	11840
87-92-3	3313
87-99-0	12058
88-04-0	9257
88-05-1	859

CAS No.	No.
88-06-2	9485
88-09-5	3514
88-10-8	3972
88-12-0	11183
88-13-1	11738
88-14-2	6050
88-15-3	5870
88-16-4	1240
88-17-5	855
88-18-6	9321
88-20-0	2228
88-21-1	2215
88-60-8	9331
88-65-3	2518
88-67-5	2657
88-68-6	601
88-69-7	9410
88-72-2	2000
88-73-3	1214
88-74-4	822
88-75-5	9435
88-85-7	5183
88-87-9	9258
88-88-0	1244
88-89-1	9497
88-95-9	1318
88-96-0	1324
88-99-3	1327
89-00-9	10927
89-01-0	10804
89-05-4	2257
89-31-6	7397
89-37-2	3975
89-39-4	1447
89-54-3	2490
89-55-4	2524
89-56-5	2631
89-57-6	2499
89-58-7	1478
89-59-8	1212
89-60-1	1209
89-61-2	1395
89-62-3	803
89-68-9	9268
89-69-0	2339
89-71-4	2090
89-72-5	9430
89-73-6	613
89-74-7	5809
89-75-8	2893
89-79-2	4544
89-82-7	4608
89-83-8	9418
89-84-9	5803
89-86-1	2569
89-87-2	1479
89-92-9	1062
89-93-0	1815
89-95-2	1897
89-96-3	1161
89-98-5	529
90-00-6	9353
90-01-7	1882
90-02-8	564
90-04-0	766
90-05-1	9379
90-11-9	7739
90-12-0	7915
90-13-1	7780
90-14-2	7902
90-15-3	8017
90-19-7	2815
90-20-0	7874
90-26-6	881
90-27-7	897
90-33-5	2831
90-39-1	7571
90-41-5	3010
90-42-6	2989
90-43-7	3088
90-44-8	433
90-45-9	313
90-47-1	12050
90-59-5	534
90-65-3	6701
90-69-7	5834
90-74-4	6236
90-80-2	6212
90-81-3	1900
90-84-6	10428
90-86-8	1903
90-89-1	9603
90-90-4	7602
90-94-8	7593
90-98-2	7592
90-99-3	1187
91-00-9	1826
91-01-0	1910
91-02-1	7638
91-08-7	1430
91-10-1	9305
91-13-4	983
91-14-5	1403
91-16-7	1436
91-17-8	7789
91-18-9	10747
91-19-0	11333
91-20-3	7732
91-21-4	7388
91-22-5	11214
91-23-6	1935
91-40-7	2715
91-49-6	39
91-53-2	11265
91-56-5	7300
91-57-6	7916
91-58-7	7781
91-59-8	7709
91-60-1	8008
91-62-3	11281
91-63-4	11277
91-64-5	2792
91-65-6	4285
91-66-7	695
91-68-9	9297
91-73-6	1829
91-76-9	11829
91-78-1	11834
91-80-5	5507
91-81-6	5505
91-84-9	5511
91-88-3	5724
91-96-3	3204
92-06-8	11596
92-13-7	6151
92-24-0	7703
92-35-3	11111
92-36-4	782
92-41-1	7869
92-43-3	10844
92-48-8	2835
92-51-3	2988
92-52-4	3008
92-54-6	9620
92-59-1	1811
92-61-5	2829
92-62-6	316
92-66-0	3018
92-67-1	3011
92-69-3	3090
92-70-6	7772
92-71-7	8580
92-82-0	9193
92-83-1	12049
92-84-2	9499
92-85-3	11660
92-86-4	3037
92-87-5	3031
92-91-1	5776
92-92-2	3022
92-93-3	3087
92-94-4	11597
92-99-9	11213
93-01-6	7974
93-02-7	544
93-03-8	1856
93-04-9	7914
93-05-0	1266
93-07-2	2576
93-08-3	5858
93-09-4	7759
93-10-7	11227
93-11-8	7976
93-14-1	10016
93-15-2	1450
93-16-3	1449
93-18-5	7884
93-22-1	7942
93-25-4	921
93-26-5	70
93-28-7	9395
93-35-6	2826
93-37-8	11258
93-39-0	2821
93-42-5	68
93-44-7	8021
93-50-5	667
93-51-6	9385
93-52-7	1299
93-53-8	872
93-54-9	1870
93-55-0	10480
93-56-1	5579
93-58-3	2687
93-59-4	1126
93-60-7	10906
93-61-8	5998
93-71-0	44
93-72-1	11512
93-76-5	256
93-78-7	259
93-79-8	258
93-88-9	1574
93-89-0	2602
93-90-3	5743
93-91-4	3362
93-97-0	2510
93-98-1	626
93-99-2	2716
94-02-0	2141
94-07-5	1887
94-08-6	2692
94-09-7	2497
94-11-1	173
94-12-2	2507
94-13-3	2653
94-14-4	2505
94-20-2	2309
94-25-7	2488
94-26-8	2615
94-30-4	2671
94-33-7	5578
94-36-0	9157
94-41-7	10708
94-44-0	10911
94-46-2	3632
94-49-5	5572
94-53-1	2418
94-58-6	2431
94-59-7	2429
94-62-2	9632
94-65-5	4612
94-67-7	570
94-68-8	746
94-69-9	5996
94-70-2	736
94-71-3	9350
94-74-6	141
94-75-7	169
94-80-4	171
94-81-5	3493
94-82-6	3925
94-96-2	6802
95-00-1	1803
95-01-2	541
95-05-6	11714
95-06-7	4065
95-11-4	2966
95-13-6	7245
95-14-7	2871
95-15-8	2399
95-16-9	2861
95-20-5	7313
95-21-6	2880
95-25-0	2883
95-43-2	3193
95-45-4	3360
95-46-5	1054
95-47-6	1452
95-48-7	9400
95-49-8	1183
95-50-1	1354
95-51-2	658
95-52-3	1726
95-53-4	778
95-54-5	1260
95-55-6	9201
95-56-7	9228
95-57-8	9251
95-59-0	5188
95-63-6	2358
95-64-7	708
95-65-8	9311
95-66-9	1138
95-68-1	705
95-69-2	673
95-70-5	1279
95-71-6	1532
95-72-7	1140
95-73-8	1388
95-74-9	672
95-75-0	1384
95-76-1	689
95-77-2	9286
95-78-3	706
95-79-4	676
95-80-7	1278
95-81-8	669
95-82-9	687
95-83-0	1264
95-84-1	9210
95-85-2	9204
95-86-3	9274
95-87-4	9309
95-92-1	5560
95-93-2	2280
95-94-3	2260
95-95-4	9484
96-00-4	4206
96-01-5	4203
96-02-6	6092
96-04-8	6402
96-05-9	10671
96-09-3	8624
96-11-7	10163
96-12-8	9885
96-14-0	8824
96-18-4	10175
96-19-5	10603
96-21-9	10357
96-22-0	8958
96-23-1	10362
96-24-2	9994
96-26-4	10429
96-27-5	10013
96-29-7	3675
96-31-1	12004
96-32-2	117
96-33-3	10655
96-34-4	138
96-35-5	198
96-37-7	4826
96-40-2	4914
96-41-3	4864
96-43-5	11740
96-45-7	7225
96-47-9	6171
96-48-0	6142
96-49-1	5211
96-50-4	11662
96-54-8	11130
96-76-4	9219
96-77-5	1560
96-80-0	5683
96-82-2	6210
96-88-8	9639
96-91-3	9207
96-97-9	2647
96-99-1	2547
97-00-7	1144
97-02-9	725
97-05-2	2655
97-16-5	9289
97-17-6	9581
97-18-7	9478
97-23-4	9408
97-24-5	9477
97-25-6	6231
97-30-3	6227
97-39-2	6279
97-53-0	9389
97-54-1	9388
97-56-3	798
97-59-6	12009
97-62-1	10298
97-63-2	10658
97-65-4	3328
97-72-3	10291
97-74-5	11715
97-77-8	11721
97-85-8	10304
97-86-9	10666
97-88-1	10650
97-90-5	10656
97-94-9	3118
97-95-0	3620
97-96-1	3183
97-97-2	5472
97-99-4	6124
98-00-0	6118
98-01-1	6041
98-02-2	6117
98-03-3	11733
98-04-4	866
98-05-5	460
98-06-6	1459
98-07-7	2333
98-08-8	2348
98-09-9	2242
98-11-3	2214
98-13-5	11473
98-15-7	1241
98-16-8	856
98-17-9	9491
98-19-1	1460
98-27-1	9333
98-29-3	1516
98-46-4	2061
98-50-0	456
98-51-1	1466
98-54-4	9323
98-56-6	1242
98-58-8	2244
98-59-9	2250
98-60-2	2245
98-61-3	2247
98-62-4	2252
98-66-8	2220
98-67-9	2226
98-71-5	2224
98-72-6	459
98-73-7	2588
98-79-3	9768
98-80-6	3135
98-81-7	1025
98-82-8	1975
98-83-9	1972
98-85-1	1896
98-86-2	5864
98-87-3	1382
98-88-4	2886
98-89-5	4332
98-91-9	1130
98-92-0	10890
98-94-2	4286
98-95-3	2049
98-96-4	10802
98-97-5	10803
98-98-6	10893
99-02-5	5788
99-03-6	5772
99-04-7	2677
99-05-8	2485
99-07-0	9318
99-08-1	2001
99-09-2	823
99-11-6	10900
99-12-7	1477
99-14-9	10165
99-17-2	6223
99-20-7	6220
99-30-9	692
99-31-0	1331
99-32-1	10776
99-33-2	2896

CAS No.	No.
99-34-3	2591
99-35-4	2365
99-36-5	2697
99-45-6	5804
99-48-9	4696
99-50-3	2571
99-51-4	1475
99-52-5	800
99-53-6	9422
99-54-7	1392
99-55-8	801
99-56-9	1281
99-59-2	775
99-61-6	592
99-62-7	993
99-63-8	1319
99-65-0	1492
99-71-8	9431
99-73-0	5777
99-75-2	2698
99-76-3	2637
99-77-4	2709
99-82-1	4438
99-84-3	4675
99-85-4	4275
99-86-5	4273
99-87-6	1990
99-88-7	789
99-89-8	9412
99-90-1	5781
99-91-2	5789
99-92-3	5773
99-93-4	5832
99-94-5	2678
99-96-7	2613
99-97-8	862
99-98-9	1270
99-99-0	2002
100-00-5	1216
100-01-6	824
100-02-7	9437
100-06-1	5841
100-07-2	2901
100-09-4	2666
100-10-7	550
100-12-9	1694
100-13-0	1645
100-14-1	1208
100-15-2	805
100-16-3	7168
100-17-4	1937
100-18-5	994
100-19-6	5861
100-20-9	1320
100-21-0	1329
100-22-1	1287
100-25-4	1493
100-26-5	10929
100-28-7	1790
100-29-8	1664
100-36-7	5502
100-37-8	5706
100-39-0	1053
100-40-3	4658
100-41-4	1668
100-42-5	1624
100-43-6	10954
100-44-7	1182
100-45-8	4629
100-46-9	1795
100-47-0	2740
100-49-2	4419
100-50-5	4632
100-51-6	1836
100-52-7	525
100-53-8	1832
100-54-9	10882
100-55-0	10976
100-60-7	4298
100-61-8	781
100-63-0	7169
100-64-1	4610
100-65-2	757
100-66-3	1915
100-68-5	2041
100-69-6	10952
100-70-9	10881
100-71-0	10956
100-72-1	10784
100-74-3	7694
100-75-4	9688
100-76-5	480
100-79-8	5207
100-80-1	1641
100-81-2	1816
100-83-4	565
100-84-5	1932
100-86-7	1600
100-88-9	11568
100-97-0	11601
101-01-9	6288
101-02-0	9593
101-05-3	360
101-10-0	4141
101-14-4	785
101-17-7	678
101-18-8	9446
101-21-3	3939
101-27-9	510
101-31-5	917
101-38-2	4280
101-39-3	10516
101-40-6	4385
101-41-7	929
101-43-9	10653
101-53-1	9455
101-54-2	1283
101-55-3	1079
101-57-5	2234
101-60-0	9734
101-61-1	786
101-76-8	1963
101-77-9	784
101-80-4	832
101-81-5	1961
101-82-6	11004
101-83-7	4284
101-84-8	2071
101-90-6	8625
101-91-7	3205
101-97-3	899
101-99-5	3965
102-01-2	3207
102-02-3	7234
102-04-5	10433
102-06-7	6282
102-07-8	12010
102-08-9	11805
102-09-0	4001
102-13-6	932
102-14-7	3575
102-20-5	940
102-25-0	2345
102-27-2	747
102-45-4	4810
102-47-6	1357
102-50-1	772
102-51-2	1273
102-52-3	10150
102-53-4	7489
102-54-5	5960
102-56-7	702
102-60-3	10379
102-69-2	9808
102-70-5	10523
102-71-6	5751
102-76-1	10195
102-79-4	5692
102-81-8	5703
102-82-9	3213
102-85-2	9590
102-86-3	6725
102-92-1	10716
102-94-3	10683
102-97-6	1820
103-01-5	6270
103-02-6	10482
103-03-7	7145
103-05-9	2150
103-09-3	187
103-11-7	10636
103-12-8	2209
103-16-2	9453
103-17-3	4075
103-19-5	5246
103-23-1	6781
103-24-2	8132
103-25-3	2139
103-28-6	10307
103-29-7	1587
103-30-0	1619
103-32-2	1827
103-34-4	7688
103-37-7	3585
103-38-8	3560
103-43-5	3298
103-44-6	6405
103-45-7	232
103-48-0	10306
103-49-1	1828
103-50-4	2079
103-55-9	5504
103-63-9	1038
103-65-1	2169
103-67-3	1818
103-69-5	743
103-70-8	5999
103-71-9	1786
103-72-0	1792
103-73-1	1649
103-74-2	10950
103-75-3	10783
103-76-4	9616
103-79-7	10481
103-80-0	971
103-81-1	878
103-82-2	883
103-83-3	1807
103-84-4	86
103-85-5	11808
103-88-8	37
103-89-9	78
103-90-2	67
103-95-7	2117
103-99-1	8232
104-01-8	922
104-03-0	933
104-04-1	83
104-06-3	34
104-07-4	8253
104-09-6	874
104-13-2	655
104-15-4	2229
104-21-2	1892
104-22-3	2213
104-29-0	10000
104-36-9	1314
104-45-0	1945
104-46-1	1942
104-47-2	958
104-50-7	6140
104-51-8	1118
104-52-9	1235
104-53-0	2114
104-57-4	6016
104-61-0	6149
104-63-2	5762
104-65-4	10707
104-66-5	1588
104-67-6	6152
104-72-3	1258
104-75-6	6726
104-76-7	6930
104-78-9	9876
104-80-3	7109
104-81-4	1064
104-82-5	1200
104-83-6	1134
104-84-7	1817
104-85-8	2765
104-87-0	589
104-88-1	531
104-90-5	10966
104-91-6	9438
104-92-7	1052
104-93-8	1933
104-94-9	768
104-96-1	820
105-03-3	8131
105-05-5	1415
105-06-6	1405
105-07-7	2758
105-13-5	1891
105-14-6	8964
105-30-6	8937
105-31-7	7124
105-34-0	154
105-36-2	115
105-37-3	10263
105-38-4	10261
105-39-5	130
105-40-8	3954
105-41-9	6730
105-42-0	6977
105-45-3	3573
105-46-4	214
105-48-6	139
105-50-0	8783
105-53-3	9941
105-54-4	3515
105-55-5	11803
105-56-6	152
105-57-7	5541
105-58-8	3999
105-59-9	5742
105-60-2	493
105-66-8	3588
105-67-9	9308
105-68-0	3639
105-70-4	10190
105-72-6	3512
105-74-8	9155
105-75-9	3774
105-83-9	9875
105-86-2	8338
105-99-7	6784
106-02-5	8562
106-07-0	8259
106-19-4	6787
106-23-0	8490
106-24-1	8333
106-25-2	8334
106-26-3	8318
106-27-4	3544
106-28-5	5333
106-30-9	6462
106-31-0	3467
106-32-1	8428
106-33-2	5312
106-35-4	6511
106-36-5	10326
106-37-6	1290
106-38-7	1056
106-39-8	1006
106-40-1	640
106-41-2	9230
106-42-3	1454
106-43-4	1185
106-44-5	9402
106-45-6	2303
106-46-7	1356
106-47-8	660
106-48-9	9253
106-49-0	780
106-50-3	1262
106-51-4	4257
106-52-5	9717
106-53-6	2292
106-54-7	2295
106-57-0	9612
106-58-1	9608
106-63-8	10673
106-65-0	3318
106-68-3	8472
106-69-4	6868
106-70-7	6898
106-73-0	6466
106-74-1	10634
106-79-6	5046
106-86-5	8555
106-87-6	8557
106-88-7	8615
106-90-1	10680
106-91-2	10668
106-92-3	8626
106-93-4	5518
106-94-5	9836
106-95-6	10531
106-96-7	10728
106-97-8	3229
106-98-9	3704
106-99-0	3146
107-00-6	3897
107-02-8	10509
107-03-9	10156
107-04-0	5456
107-05-1	10536
107-06-2	5529
107-07-3	5693
107-08-4	10054
107-10-8	9803
107-11-9	10522
107-12-0	10090
107-13-1	10583
107-14-2	274
107-15-3	5496
107-16-4	282
107-18-6	10697
107-19-7	10743
107-20-0	12
107-21-1	5569
107-22-2	5491
107-25-5	5908
107-29-9	17
107-30-2	7483
107-31-3	6010
107-34-6	461
107-35-7	5623
107-37-9	11475
107-39-1	9067
107-40-4	9071
107-41-5	8793
107-43-7	7454
107-46-0	5236
107-47-1	10155
107-48-2	8612
107-49-3	5221
107-50-6	4239
107-51-7	11914
107-52-8	6990
107-53-9	11964
107-59-5	128
107-68-6	5624
107-70-0	8978
107-72-2	11472
107-74-4	8392
107-75-5	8347
107-80-2	3266
107-81-3	8711
107-82-4	3239
107-83-5	8823
107-84-6	3252
107-85-7	3220
107-86-8	3700
107-87-9	8957
107-88-0	3345
107-89-1	3184
107-91-5	49
107-92-6	3460
107-93-7	3814
107-94-8	10216
107-95-9	323
107-96-0	10286
107-97-1	6268
107-98-2	10387
108-01-0	5709
108-03-2	10116
108-05-4	179
108-08-7	8762
108-10-1	8981
108-11-2	8942
108-12-3	3693
108-16-7	10376
108-18-9	9821
108-19-0	7232
108-20-3	10120
108-21-4	209
108-22-5	10698
108-23-6	4023
108-24-7	107
108-27-0	11186
108-28-1	6155
108-30-5	6093
108-31-6	6091
108-32-7	5213

CAS	No.	CAS	No.	CAS	No.	CAS	No.	CAS	No.
108-36-1	: 1289	109-90-0	: 5598	111-17-1	: 10329	112-51-6	: 5257	117-37-3	: 7269
108-37-2	: 1005	109-92-2	: 5905	111-18-2	: 6752	112-52-7	: 5295	117-39-5	: 2817
108-38-3	: 1453	109-93-3	: 5912	111-19-3	: 5049	112-53-8	: 5321	117-73-7	: 12064
108-39-4	: 9401	109-94-4	: 6007	111-20-6	: 5042	112-54-9	: 5288	117-78-2	: 372
108-40-7	: 2302	109-95-5	: 8087	111-21-7	: 5714	112-55-0	: 5304	117-79-3	: 381
108-41-8	: 1184	109-96-6	: 11118	111-24-0	: 8741	112-57-2	: 5498	117-80-6	: 7850
108-42-9	: 659	109-97-7	: 11112	111-25-1	: 6732	112-58-3	: 6842	117-81-7	: 1334
108-43-0	: 9252	109-98-8	: 10822	111-26-2	: 6723	112-60-7	: 5756	117-84-0	: 1344
108-44-1	: 779	109-99-9	: 6168	111-27-3	: 6911	112-61-8	: 8266	117-93-1	: 66
108-45-2	: 1261	110-00-9	: 6025	111-28-4	: 6707	112-62-9	: 8307	117-99-7	: 7622
108-46-3	: 1498	110-01-0	: 11783	111-29-5	: 8787	112-63-0	: 8228	118-00-3	: 6289
108-47-4	: 10937	110-02-1	: 11724	111-30-8	: 8734	112-64-1	: 11635	118-08-1	: 7357
108-48-5	: 10939	110-03-2	: 6801	111-31-9	: 6856	112-66-3	: 178	118-10-5	: 4120
108-50-9	: 10807	110-04-3	: 5587	111-34-2	: 3369	112-67-4	: 6665	118-32-1	: 7871
108-52-1	: 11046	110-05-4	: 9153	111-40-0	: 5497	112-70-9	: 11883	118-34-3	: 6225
108-55-4	: 10780	110-06-5	: 5242	111-41-1	: 5676	112-71-0	: 11614	118-41-2	: 2734
108-56-5	: 3340	110-12-3	: 6978	111-42-2	: 5730	112-72-1	: 11630	118-44-5	: 7718
108-57-6	: 1404	110-13-4	: 6805	111-43-3	: 10119	112-73-2	: 3417	118-46-7	: 8020
108-59-8	: 9943	110-14-5	: 3257	111-44-4	: 5611	112-76-5	: 8281	118-55-8	: 2650
108-60-1	: 10123	110-15-6	: 3294	111-45-5	: 5763	112-79-8	: 8297	118-58-1	: 2651
108-64-5	: 3548	110-16-7	: 3765	111-46-6	: 5753	112-80-1	: 8294	118-61-6	: 2619
108-67-8	: 2359	110-17-8	: 3764	111-47-7	: 10151	112-82-3	: 6633	118-69-4	: 1385
108-68-9	: 9312	110-18-9	: 5516	111-48-8	: 5766	112-85-6	: 5276	118-71-8	: 10796
108-69-0	: 709	110-19-0	: 215	111-49-9	: 492	112-86-7	: 5283	118-74-1	: 1749
108-70-3	: 2323	110-22-5	: 9156	111-51-3	: 3263	112-88-9	: 8291	118-75-2	: 4268
108-73-6	: 2368	110-27-0	: 11626	111-55-7	: 5571	112-89-0	: 8235	118-76-3	: 4686
108-75-8	: 11028	110-38-3	: 5063	111-56-8	: 5307	112-91-4	: 8293	118-78-5	: 11077
108-77-0	: 11837	110-39-4	: 3566	111-60-4	: 8260	112-92-5	: 8278	118-79-6	: 9479
108-78-1	: 11835	110-40-7	: 5045	111-61-5	: 8257	112-95-8	: 5350	118-83-2	: 1219
108-82-7	: 6485	110-41-8	: 11936	111-62-6	: 8301	113-00-8	: 6278	118-90-1	: 2676
108-83-8	: 6514	110-42-9	: 5066	111-64-8	: 8481	113-15-5	: 5394	118-91-2	: 2534
108-84-9	: 8947	110-43-0	: 6510	111-65-9	: 8355	113-48-4	: 7660	118-92-3	: 2484
108-85-0	: 4315	110-44-1	: 6698	111-66-0	: 8491	114-07-8	: 5396	118-93-4	: 5830
108-86-1	: 998	110-45-2	: 3636	111-68-2	: 6349	114-25-0	: 3001	118-96-7	: 2048
108-87-2	: 4426	110-46-3	: 8090	111-69-3	: 6777	114-26-1	: 10722	119-13-1	: 2785
108-88-3	: 1947	110-49-6	: 5735	111-70-6	: 6472	114-91-0	: 10979	119-20-0	: 2494
108-89-4	: 10982	110-52-1	: 3267	111-71-7	: 6344	115-07-1	: 10528	119-26-6	: 7156
108-90-7	: 1131	110-53-2	: 8710	111-74-0	: 5501	115-10-6	: 7523	119-27-7	: 1923
108-91-8	: 4283	110-54-3	: 6731	111-75-1	: 5691	115-11-7	: 10578	119-28-8	: 7967
108-93-0	: 4500	110-56-5	: 3279	111-76-2	: 5687	115-17-3	: 20	119-33-5	: 9424
108-94-1	: 4571	110-57-6	: 3750	111-77-3	: 5737	115-18-4	: 3870	119-36-8	: 2634
108-95-2	: 9199	110-58-7	: 8700	111-78-4	: 4733	115-19-5	: 3922	119-41-5	: 225
108-97-4	: 10789	110-59-8	: 8827	111-81-9	: 11978	115-20-8	: 5768	119-44-0	: 10749
108-98-5	: 2289	110-60-1	: 3258	111-82-0	: 5316	115-21-9	: 11463	119-48-2	: 7686
108-99-6	: 10981	110-61-2	: 3251	111-83-1	: 8356	115-24-2	: 9835	119-59-5	: 845
109-01-3	: 9618	110-62-3	: 8690	111-84-2	: 8120	115-25-3	: 4194	119-60-8	: 7606
109-02-4	: 7696	110-63-4	: 3346	111-85-3	: 8359	115-29-7	: 5369	119-61-9	: 7616
109-04-6	: 10875	110-65-6	: 3911	111-86-4	: 8350	115-32-2	: 1851	119-64-2	: 7979
109-06-8	: 10980	110-66-7	: 8844	111-87-5	: 8443	115-37-7	: 7679	119-65-3	: 7373
109-07-9	: 9619	110-67-8	: 10110	111-88-6	: 8418	115-38-8	: 11094	119-67-5	: 2607
109-08-0	: 10810	110-68-9	: 3221	111-90-0	: 5720	115-39-9	: 9366	119-68-6	: 2679
109-09-1	: 10915	110-69-0	: 3188	111-92-2	: 3211	115-40-2	: 9364	119-84-6	: 2798
109-12-6	: 11042	110-70-3	: 5503	111-96-6	: 5613	115-43-5	: 11103	119-90-4	: 3033
109-16-0	: 10657	110-71-4	: 5549	112-04-9	: 11470	115-53-7	: 7676	119-93-7	: 3034
109-19-3	: 3543	110-72-5	: 5509	112-05-0	: 8146	115-56-0	: 11064	120-07-0	: 5760
109-21-7	: 3479	110-73-6	: 5722	112-06-1	: 193	115-61-7	: 11349	120-12-7	: 366
109-43-3	: 5044	110-74-7	: 6018	112-10-7	: 8267	115-63-9	: 9623	120-14-9	: 545
109-49-9	: 7089	110-75-8	: 5887	112-12-9	: 11958	115-69-5	: 9987	120-15-0	: 11298
109-52-4	: 8855	110-76-9	: 5423	112-13-0	: 5078	115-70-8	: 9985	120-18-3	: 7960
109-53-5	: 10035	110-77-0	: 5726	112-14-1	: 222	115-71-9	: 9093	120-20-7	: 1571
109-56-8	: 5741	110-78-1	: 10059	112-15-2	: 5721	115-76-4	: 10004	120-21-8	: 539
109-57-9	: 11809	110-80-5	: 5717	112-16-3	: 5328	115-77-5	: 9989	120-22-9	: 699
109-59-1	: 5740	110-81-6	: 5253	112-17-4	: 157	115-80-0	: 10182	120-23-0	: 217
109-60-4	: 236	110-82-7	: 4305	112-20-9	: 8119	115-84-4	: 9992	120-25-2	: 557
109-61-5	: 4028	110-83-8	: 4626	112-23-2	: 6008	115-86-6	: 9566	120-29-6	: 489
109-63-7	: 3137	110-85-0	: 9602	112-24-3	: 5499	115-90-2	: 5954	120-32-1	: 9270
109-64-8	: 9882	110-86-1	: 10872	112-25-4	: 5728	115-96-8	: 5701	120-33-2	: 2691
109-65-9	: 3231	110-87-2	: 10777	112-26-5	: 5449	115-99-1	: 8337	120-36-5	: 10250
109-66-0	: 8709	110-88-3	: 11905	112-27-6	: 5713	116-06-3	: 334	120-43-4	: 9604
109-67-1	: 9004	110-89-4	: 9629	112-29-8	: 5034	116-09-6	: 10443	120-46-7	: 10027
109-69-3	: 3241	110-91-8	: 7681	112-30-1	: 5070	116-14-3	: 5918	120-47-8	: 2621
109-70-6	: 9839	110-93-0	: 6603	112-31-2	: 5030	116-15-4	: 10572	120-48-9	: 2706
109-73-9	: 3209	110-94-1	: 8770	112-32-3	: 6014	116-16-5	: 10440	120-50-3	: 2701
109-74-0	: 3403	110-95-2	: 9879	112-34-5	: 5688	116-17-6	: 9395	120-51-4	: 2720
109-75-1	: 3803	110-96-3	: 9823	112-36-7	: 5612	116-26-7	: 4250	120-55-8	: 5755
109-76-2	: [illegible]	110-97-4	: 10384	112-37-8	: 11943	116-29-0	: 2325	120-57-0	: 2417
109-77-3	: 9911	110-99-6	: 226	112-38-9	: 11974	116-38-1	: 865	120-61-6	: 1351
109-78-4	: 10106	111-01-3	: 11606	112-39-0	: 6653	116-43-8	: 3578	120-66-1	: 76
109-79-5	: 3437	111-02-4	: 11602	112-40-3	: 5292	116-44-9	: 2202	120-71-8	: 770
109-80-8	: 10031	111-03-5	: 8299	112-41-4	: 5335	116-54-1	: 166	120-72-9	: 7294
109-81-9	: 5512	111-06-8	: 6647	112-42-5	: 11950	116-58-5	: 9757	120-73-0	: 10752
109-82-0	: 285	111-11-5	: 8434	112-43-6	: 11979	116-63-2	: 7966	120-75-2	: 2863
109-83-1	: 5738	111-12-6	: 8542	112-44-7	: 11935	117-08-8	: 7346	120-80-9	: 1497
109-84-2	: 5729	111-13-7	: 8471	112-46-9	: 11981	117-10-2	: 395	120-82-1	: 2322
109-86-4	: 5733	111-14-8	: 6455	112-47-0	: 5047	117-12-4	: 393	120-83-2	: 9283
109-87-5	: 7507	111-15-9	: 5718	112-48-1	: 5527	117-18-0	: 2264	120-85-4	: 9622
109-89-7	: 5427	111-16-0	: 6393	112-49-2	: 11642	117-34-0	: 936	120-87-6	: 8254

CAS NUMBER INDEX OF ORGANIC COMPOUNDS (continued)

CAS No.	No.
120-89-8	7226
120-92-3	4878
120-93-4	7227
120-94-5	11168
120-97-8	1559
121-06-2	9337
121-14-2	1956
121-17-5	1217
121-21-1	4960
121-29-9	4962
121-30-2	1558
121-32-4	555
121-33-5	574
121-34-6	2625
121-43-7	3126
121-44-8	5417
121-45-9	9592
121-46-0	2913
121-47-1	2216
121-57-3	2217
121-59-5	455
121-60-8	2243
121-61-9	31
121-66-4	11666
121-69-7	713
121-71-1	5831
121-72-2	861
121-73-3	1215
121-75-5	3316
121-79-9	2732
121-81-3	614
121-82-4	11833
121-86-8	1210
121-89-1	5860
121-90-4	2905
121-91-5	1328
121-92-6	2704
121-98-2	2675
122-00-9	5853
122-01-0	2892
122-03-2	590
122-04-3	2906
122-07-6	5420
122-09-8	1572
122-11-2	2194
122-14-5	5951
122-20-3	10393
122-25-8	2686
122-28-1	82
122-32-7	8310
122-34-9	11513
122-35-0	10505
122-37-2	9447
122-39-4	836
122-40-7	6347
122-42-9	3966
122-46-3	212
122-48-5	3663
122-51-0	5604
122-52-1	9591
122-57-6	3881
122-59-8	228
122-60-1	8623
122-62-3	5043
122-66-7	7158
122-69-0	10691
122-70-3	10324
122-73-6	1951
122-78-1	868
122-79-2	231
122-80-5	29
122-84-9	10461
122-87-2	6267
122-88-3	146
122-95-2	1408
122-97-4	2148
122-98-5	5759
122-99-6	5758
123-00-2	7701
123-01-3	1566
123-02-4	2342
123-03-5	11030
123-04-6	6368
123-05-7	6719
123-06-8	9914
123-07-9	9355
123-08-0	566
123-11-5	585
123-15-9	8693
123-19-3	6512
123-20-6	3513
123-22-8	9587
123-25-1	3303
123-29-5	8150
123-30-8	9203
123-31-9	1499
123-32-0	10806
123-35-3	8330
123-38-6	9774
123-39-7	5995
123-42-2	8976
123-43-3	238
123-45-5	239
123-51-3	3627
123-54-6	8798
123-56-8	11152
123-62-6	10203
123-63-7	11908
123-66-0	6884
123-72-8	3178
123-73-9	3695
123-75-1	11143
123-76-2	8903
123-77-3	5124
123-80-8	5576
123-82-0	6350
123-83-1	5510
123-84-2	10340
123-86-4	121
123-88-6	11822
123-90-0	11719
123-91-1	5187
123-92-2	3631
123-93-3	240
123-95-5	8251
123-98-8	8138
123-99-9	8130
124-02-7	10526
124-04-9	6779
124-06-1	11624
124-07-2	8422
124-09-4	6749
124-10-7	11625
124-11-8	8182
124-12-9	8411
124-13-0	8346
124-17-4	5689
124-18-5	5033
124-19-6	8117
124-21-0	5425
124-22-1	5291
124-26-5	8231
124-30-1	8233
124-38-9	3992
124-40-3	7448
124-42-5	5671
124-47-0	12021
124-48-1	7493
124-58-3	458
124-63-0	7535
124-68-5	10341
124-70-9	11396
124-73-2	5525
124-80-1	4067
124-83-4	4794
124-87-8	7579
125-15-5	11351
125-20-2	7352
125-24-6	3002
125-28-0	7674
125-29-1	7678
125-33-7	11066
125-55-3	11079
126-00-1	1103
126-07-8	11520
126-11-4	10012
126-14-7	6229
126-17-0	11529
126-18-1	11533
126-19-2	11531
126-29-4	4056
126-30-7	10005
126-33-0	11786
126-54-5	11643
126-68-1	9585
126-71-6	9572
126-72-7	10358
126-73-8	9562
126-81-8	4378
126-84-1	9904
126-98-7	10586
126-99-8	3152
127-00-4	10349
127-06-0	10476
127-07-1	12016
127-08-2	234
127-09-3	237
127-17-3	10314
127-18-4	5917
127-19-5	55
127-25-3	9168
127-27-5	9170
127-29-7	308
127-40-2	4058
127-48-0	8587
127-63-9	2237
127-66-2	1878
127-69-5	2195
127-71-9	603
127-79-7	2199
127-91-3	2925
128-04-1	11514
128-08-5	11153
128-09-6	11154
128-13-2	4098
128-20-1	9745
128-23-4	9743
128-27-8	5381
128-33-6	4104
128-37-0	9224
128-39-2	9220
128-50-7	2969
128-53-0	11122
128-62-1	7356
128-68-7	2979
129-00-0	10853
129-43-1	407
129-44-2	389
129-64-6	7580
129-66-8	2735
129-79-3	5979
130-15-4	7843
130-16-5	11311
130-26-7	11312
130-73-4	9299
130-80-3	9301
130-85-8	7777
130-86-9	3109
130-89-2	4130
130-90-5	4129
130-95-0	4127
131-01-1	12065
131-09-9	388
131-11-3	1343
131-28-2	2586
131-53-3	7619
131-54-4	7595
131-56-6	7609
131-57-7	7620
131-58-8	7631
131-73-7	864
131-89-5	9273
131-91-9	8035
132-57-0	7972
132-60-5	11235
132-64-9	5163
132-65-0	5167
132-66-1	2702
132-75-2	7736
132-86-5	7834
133-06-2	7368
133-07-3	5984
133-08-4	9930
133-11-9	2502
133-13-1	9949
133-14-2	9152
133-26-6	6188
133-32-4	7296
133-37-9	3311
133-53-9	9290
133-58-4	7437
133-59-5	2249
133-67-5	2854
133-90-4	2492
133-91-5	2617
134-04-3	2848
134-11-2	2596
134-20-3	2503
134-31-6	11324
134-32-7	7708
134-49-6	7697
134-55-4	2483
134-58-7	11851
134-62-3	612
134-81-6	5584
134-84-9	7633
134-85-0	7604
134-96-3	569
135-00-2	7641
135-01-3	1413
135-02-4	583
135-09-1	2858
135-19-3	8018
135-20-6	759
135-44-4	8926
135-48-8	8636
135-58-0	11661
135-67-1	9507
135-68-2	3012
135-70-6	11194
135-88-6	7728
135-97-7	486
135-98-8	2035
136-35-6	11826
136-44-7	10188
136-47-0	2531
136-60-7	2532
136-72-1	8678
136-77-6	1519
136-78-7	11358
136-79-8	739
136-80-1	5744
136-95-8	2859
137-00-8	11677
137-04-2	665
137-05-3	10624
137-06-4	2301
137-07-5	2290
137-16-6	6269
137-17-7	858
137-19-9	1510
137-26-8	11722
137-30-4	12068
137-32-6	3626
137-40-6	10327
137-43-9	4777
137-58-6	54
138-41-0	2508
138-52-3	6222
138-53-4	955
138-86-3	4677
138-87-4	4542
138-89-6	720
139-13-9	6256
139-40-2	10508
139-45-7	10198
139-65-1	850
139-66-2	2283
139-85-5	542
139-87-7	5723
139-93-5	9275
140-08-9	5702
140-10-3	10682
140-11-4	233
140-18-1	148
140-22-7	4010
140-25-0	5318
140-28-3	5500
140-29-4	945
140-31-8	9615
140-39-6	213
140-40-9	85
140-55-6	9718
140-57-8	11579
140-65-8	7683
140-67-0	1943
140-76-1	10955
140-77-2	4847
140-80-7	8736
140-82-9	5707
140-87-4	153
140-88-5	10635
140-95-4	11995
141-03-7	3301
141-04-8	6783
141-05-9	3776
141-06-0	8914
141-10-6	11965
141-12-8	8336
141-20-8	5313
141-22-0	8303
141-27-5	8317
141-28-6	6785
141-32-2	10616
141-43-5	5675
141-46-8	15
141-57-1	11476
141-62-8	11646
141-63-9	8997
141-66-2	5172
141-75-3	3681
141-78-6	186
141-79-7	9108
141-82-2	9921
141-83-3	11993
141-84-4	11685
141-86-6	10925
141-90-2	11110
141-91-3	7687
141-93-5	1414
141-97-9	3571
142-04-1	756
142-08-5	11033
142-09-6	10660
142-26-7	65
142-28-9	9893
142-29-0	4903
142-30-3	7119
142-47-2	6248
142-50-7	5332
142-60-9	10313
142-61-0	6986
142-62-1	6870
142-64-3	9607
142-68-7	10800
142-73-4	6260
142-77-8	8298
142-82-5	6355
142-83-6	6675
142-84-7	9826
142-91-6	6654
142-92-7	194
142-96-1	3415
143-07-7	5305
143-08-8	8157
143-10-2	5058
143-13-5	220
143-15-7	5293
143-16-8	6727
143-22-6	5690
143-24-8	8995
143-27-1	6631
143-28-2	8312
143-50-0	7408
143-62-4	4044
143-74-8	9363
144-19-4	8795
144-49-0	189
144-55-8	4004
144-62-7	5553
144-68-3	4057
144-79-6	11380
144-80-9	30
144-83-2	2203
145-13-1	9756
145-73-3	5370
145-94-8	7283
146-06-5	4128
146-48-5	12062
147-71-7	3312
147-73-9	3304
147-82-0	851
147-85-3	9764
147-93-3	2663
148-03-8	2789

148-24-3	: 11308	258-31-1	: 6621	307-34-6	: 8415	353-50-4	: 4008	373-05-7	: 6888
148-25-4	: 7875	260-36-6	: 2473	307-45-9	: 5050	353-54-8	: 7550	373-14-8	: 6829
148-53-8	: 570	260-94-6	: 314	308-24-7	: 4498	353-55-9	: 7494	373-28-4	: 6735
148-75-4	: 7876	261-31-4	: 11811	309-00-2	: 5177	353-58-2	: 7468	373-32-0	: 6936
148-79-8	: 11656	262-30-6	: 11353	311-89-7	: 3228	353-59-3	: 7464	373-44-4	: 8365
148-86-7	: 3091	271-44-3	: 7235	312-40-3	: 11411	353-82-2	: 279	373-49-9	: 6670
149-29-1	: 6183	271-58-9	: 2390	312-84-5	: 11355	353-83-3	: 5667	373-52-4	: 7470
149-30-4	: 2866	271-89-6	: 2438	313-72-4	: 7935	353-85-5	: 289	373-53-5	: 7517
149-32-6	: 3431	271-95-4	: 2389	314-13-6	: 7877	353-89-9	: 11432	373-80-8	: 11584
149-57-5	: 6882	272-16-2	: 2387	314-40-9	: 3139	354-04-1	: 5526	374-07-2	: 5537
149-73-5	: 7561	273-53-0	: 2877	315-30-0	: 10852	354-06-3	: 5457	374-25-4	: 3715
149-74-6	: 11401	274-07-7	: 2412	317-52-2	: 6753	354-11-0	: 5639	374-77-6	: 4343
149-91-7	: 2731	274-09-9	: 2415	319-85-7	: 4406	354-12-1	: 5658	375-01-9	: 3622
150-13-0	: 2486	274-40-8	: 7324	319-86-8	: 4407	354-14-3	: 5638	375-16-6	: 3692
150-19-6	: 9380	275-51-4	: 504	319-88-0	: 2340	354-15-4	: 5660	375-22-4	: 3521
150-25-4	: 6259	278-30-8	: 2991	319-89-1	: 4269	354-21-2	: 5659	375-83-7	: 6432
150-30-1	: 9509	279-23-2	: 2914	320-72-9	: 2566	354-23-4	: 5539	375-96-2	: 8139
150-46-9	: 3125	280-57-9	: 5111	321-28-8	: 1722	354-25-6	: 5485	376-27-2	: 8438
150-50-5	: 7438	280-65-9	: 2990	321-38-0	: 7893	354-26-7	: 5490	376-87-4	: 9027
150-60-7	: 5248	281-23-2	: 11860	321-60-8	: 3074	354-51-8	: 5519	377-41-3	: 4177
150-61-8	: 5508	285-67-6	: 8558	323-09-1	: 7894	354-56-3	: 5618	377-70-8	: 4276
150-68-5	: 11999	286-08-8	: 2915	324-74-3	: 3075	354-58-5	: 5662	377-93-5	: 4199
150-69-6	: 12013	286-18-0	: 478	326-56-7	: 2420	354-92-7	: 10050	378-68-7	: 10251
150-76-5	: 9381	286-20-4	: 8554	326-59-0	: 2416	354-96-1	: 3414	378-75-6	: 10317
150-77-6	: 5515	287-23-0	: 4170	326-62-5	: 952	355-02-2	: 4499	381-61-3	: 3162
150-78-7	: 1438	287-27-4	: 11691	327-54-8	: 2272	355-04-4	: 8854	381-73-7	: 175
150-86-7	: 6672	287-92-3	: 4774	327-57-1	: 8214	355-20-4	: 3287	382-45-6	: 353
151-10-0	: 1437	288-13-1	: 10815	328-39-2	: 7412	355-25-9	: 3255	383-63-1	: 267
151-13-3	: 8304	288-14-2	: 7402	328-42-7	: 3339	355-42-0	: 6843	383-66-4	: 269
151-18-8	: 10092	288-32-4	: 7192	328-50-7	: 8781	355-63-5	: 6560	383-90-4	: 3114
151-19-9	: 8453	288-36-8	: 11846	328-74-5	: 637	355-68-0	: 4381	384-04-3	: 3172
151-56-4	: 498	288-39-1	: 11658	328-96-1	: 723	355-75-9	: 4643	384-22-5	: 2060
151-67-7	: 5458	288-42-6	: 8575	329-71-5	: 9343	355-80-6	: 8953	386-17-4	: 7351
152-16-9	: 5220	288-47-1	: 11667	330-54-1	: 12000	356-16-1	: 5642	388-82-9	: 3042
152-58-9	: 9749	288-92-6	: 11657	330-55-2	: 7415	356-24-1	: 3525	389-08-2	: 8066
152-93-2	: 11107	288-94-8	: 11652	331-39-5	: 10632	356-27-4	: 3524	389-40-2	: 2093
152-95-4	: 2824	288-99-3	: 8563	332-77-4	: 6087	356-28-5	: 3523	392-56-3	: 1757
154-17-6	: 7110	289-80-5	: 10855	333-20-0	: 11711	357-20-0	: 3808	392-69-8	: 1740
154-41-6	: 1838	289-95-2	: 11049	333-41-5	: 9582	357-49-3	: 10721	392-83-6	: 1089
154-42-7	: 10762	290-37-9	: 10801	333-93-7	: 3260	357-57-3	: 11564	392-85-8	: 1737
154-93-8	: 11994	290-67-5	: 5202	334-28-1	: 6463	357-67-5	: 11089	393-52-2	: 2897
156-08-1	: 1575	290-87-9	: 11827	334-42-9	: 6359	357-70-0	: 6200	394-46-7	: 1633
156-10-5	: 829	290-96-0	: 11651	334-43-0	: 6365	358-06-5	: 11410	396-01-0	: 10750
156-38-7	: 906	291-21-4	: 11919	334-48-5	: 5059	358-43-0	: 11484	396-64-5	: 3043
156-43-4	: 737	291-64-5	: 4221	334-56-5	: 5052	358-99-6	: 5883	398-23-2	: 3044
156-59-2	: 5896	292-46-6	: 8998	334-59-8	: 5064	359-01-3	: 3379	399-92-8	: 10328
156-60-5	: 5895	292-64-8	: 4736	334-61-2	: 5035	359-02-4	: 5900	400-52-2	: 266
156-87-6	: 10337	293-55-0	: 4724	334-62-3	: 5037	359-10-4	: 5885	401-56-9	: 133
157-40-4	: 11527	293-96-9	: 4202	334-64-5	: 5074	359-11-5	: 5924	401-78-5	: 1090
180-43-8	: 11537	294-62-2	: 4210	334-88-3	: 7491	359-13-7	: 5708	401-80-9	: 1738
184-97-4	: 5240	295-48-7	: 4759	335-36-4	: 6109	359-25-1	: 116	402-31-3	: 997
189-55-9	: 2850	297-97-2	: 11720	335-44-4	: 3452	359-28-4	: 5661	402-43-7	: 1091
191-07-1	: 4145	298-02-2	: 9513	335-48-8	: 3275	359-29-5	: 5923	402-44-8	: 1739
191-24-2	: 2471	298-04-4	: 5260	335-57-9	: 6418	359-32-0	: 296	402-67-5	: 1734
191-28-6	: 12055	298-12-4	: 223	336-19-6	: 4647	359-35-3	: 5641	403-42-9	: 5815
192-65-4	: 8061	298-45-3	: 2470	336-59-4	: 3522	359-37-5	: 5925	403-43-0	: 2899
192-97-2	: 2436	298-46-4	: 5144	338-65-8	: 5471	359-70-6	: 5436	404-72-8	: 1721
195-19-7	: 2406	298-81-7	: 6187	338-83-0	: 9813	360-68-9	: 4109	404-86-4	: 8181
198-55-0	: 9161	298-96-4	: 11653	339-43-5	: 2192	360-89-4	: 3804	405-50-5	: 901
205-25-4	: 2405	299-84-3	: 9584	343-65-7	: 1102	363-51-9	: 664	405-99-2	: 1635
205-82-3	: 2738	299-86-5	: 4153	344-04-7	: 1078	363-72-4	: 2091	406-74-6	: 190
205-99-2	: 631	300-38-9	: 11927	344-07-0	: 1220	366-18-7	: 3099	407-25-0	: 264
206-44-0	: 5966	300-57-2	: 2165	345-35-7	: 1194	367-11-3	: 1422	407-75-0	: 8879
207-08-9	: 2739	300-62-9	: 1578	346-18-9	: 2857	367-12-4	: 9360	407-97-6	: 8713
207-83-0	: 5148	300-76-5	: 7702	348-51-6	: 1166	367-23-7	: 2346	407-98-7	: 8728
208-96-8	: 3	300-87-8	: 7403	348-52-7	: 1719	367-25-9	: 700	408-16-2	: 6491
212-74-8	: 11644	301-00-8	: 8285	348-54-9	: 752	367-64-6	: 265	408-27-5	: 8460
213-46-7	: 9601	301-12-2	: 8632	348-70-9	: 135	368-34-3	: 132	408-38-8	: 6639
217-59-4	: 11910	301-21-3	: 9399	349-88-2	: 2246	368-43-4	: 2251	420-04-2	: 4155
218-01-9	: 4114	302-17-0	: 5582	350-03-8	: 5868	368-47-8	: 11486	420-05-3	: 4162
222-93-5	: 8996	302-27-2	: 306	350-46-9	: 1735	368-77-4	: 2770	420-12-2	: 11695
224-42-0	: 5143	302-72-7	: 324	350-50-5	: 1725	368-98-9	: 3116	420-24-6	: 463
225-51-4	: 630	302-84-1	: 11356	350-51-6	: 1634	370-14-9	: 9403	420-26-8	: 10047
225-61-6	: 2394	303-04-8	: 3754	350-90-3	: 10637	371-40-4	: 754	420-45-1	: 9906
229-87-8	: 9188	303-38-8	: 2568	352-11-4	: 1195	371-41-5	: 9362	420-46-2	: 5665
230-07-9	: 9191	304-21-2	: 11038	352-32-9	: 1728	371-62-0	: 5727	420-90-6	: 10532
230-27-3	: 2474	304-84-7	: 611	352-33-0	: 1168	372-09-8	: 150	420-97-3	: 9897
230-46-6	: 9190	305-01-1	: 2804	352-34-1	: 1720	372-18-9	: 1423	420-99-5	: 9851
238-84-6	: 2393	305-03-3	: 1101	352-70-5	: 1727	372-19-0	: 753	421-04-5	: 5487
250-84-0	: 11688	305-84-0	: 7133	352-91-0	: 9844	372-20-3	: 9361	421-06-7	: 5466
253-52-1	: 9599	305-85-1	: 9304	352-93-2	: 5644	372-31-6	: 3592	421-83-0	: 7537
253-66-7	: 4136	306-08-1	: 913	352-97-6	: 6253	372-38-3	: 2347	425-23-0	: 3526
253-82-7	: 11195	306-44-5	: 9789	353-25-3	: 8429	372-46-3	: 4403	425-75-2	: 7533
254-04-6	: 2772	306-83-2	: 5540	353-29-7	: 5294	372-54-3	: 6830	428-59-1	: 8629
254-79-5	: 8065	306-94-5	: 7934	353-36-6	: 5593	372-64-5	: 5249	430-40-0	: 445
254-87-5	: 2870	306-98-9	: 4342	353-42-4	: 3138	372-75-8	: 8551	430-51-3	: 10437
255-50-5	: 2434	307-11-9	: 11912	353-49-1	: 4005	372-93-0	: 3621	430-57-9	: 5535

503-66-2 : 10267
503-74-2 : 3539
503-76-4 : 9869
503-80-0 : 452
503-93-5 : 4220
504-02-9 : 4376
504-03-0 : 9650
504-15-4 : 1531
504-20-1 : 6340
504-24-5 : 10859
504-29-0 : 10857
504-31-4 : 10788
504-53-0 : 9000
504-57-4 : 8105
504-61-0 : 3859
504-63-2 : 9983
504-64-3 : 9773
504-78-9 : 11681
505-20-4 : 5261
505-22-6 : 5186
505-23-7 : 5262
505-29-3 : 5263
505-32-8 : 6671
505-34-0 : 10067
505-48-6 : 8386
505-60-2 : 5645
505-84-0 : 10074
505-92-0 : 5338
506-03-6 : 10010
506-12-7 : 6314
506-23-0 : 8284
506-24-1 : 8316
506-30-9 : 5355
506-31-0 : 5364
506-32-1 : 5361
506-33-2 : 5282
506-42-3 : 8311
506-46-7 : 6624
506-48-9 : 8221
506-50-3 : 11820
506-52-5 : 6626
506-68-3 : 4165
506-77-4 : 4166
506-78-5 : 7337
506-80-9 : 4036
506-82-1 : 3929
506-85-4 : 6024
506-89-8 : 12020
506-93-4 : 6286
506-96-7 : 291
507-02-8 : 302
507-09-5 : 5646
507-19-7 : 9846
507-20-0 : 9863
507-25-5 : 7542
507-32-4 : 457
507-36-8 : 3240
507-40-4 : 7183
507-42-6 : 5581
507-45-9 : 3286
507-47-1 : 5680
507-60-8 : 3142
508-02-1 : 8547
508-32-7 : 11867
508-44-1 : 5185
508-54-3 : 7675
508-65-6 : 4068
509-12-6 : 2938
509-14-8 : 7544
509-15-9 : 6203
509-20-6 : 305
509-64-8 : 7670
509-87-5 : 11096
510-15-6 : 4082
510-20-3 : 9939
510-90-7 : 11063
511-07-9 : 5392
511-08-0 : 5391
511-09-1 : 5389
511-10-4 : 5390
511-18-2 : 8210
511-20-6 : 5377
511-21-7 : 5378
511-34-2 : 11535
511-67-1 : 11086
511-70-6 : 3198
511-96-6 : 11530

512-04-9 : 11536
512-12-9 : 8585
512-13-0 : 2941
512-16-3 : 4309
512-48-1 : 9003
512-56-1 : 9564
512-60-7 : 11862
512-69-6 : 6219
512-85-6 : 5184
513-02-0 : 9568
513-08-6 : 9567
513-12-2 : 5725
513-23-5 : 2985
513-31-5 : 10545
513-35-9 : 3801
513-36-0 : 9862
513-37-1 : 10539
513-38-2 : 10057
513-42-8 : 10702
513-44-0 : 10158
513-81-5 : 3159
513-86-0 : 3661
513-88-2 : 10425
513-92-8 : 5919
513-96-2 : 10115
514-10-3 : 9167
514-12-5 : 2953
514-92-1 : 4059
515-12-8 : 4401
515-25-3 : 11178
515-30-0 : 914
515-40-2 : 1143
515-46-8 : 2223
515-49-1 : 2189
515-59-3 : 2200
515-82-2 : 6000
515-83-3 : 5769
515-84-4 : 248
515-88-8 : 2944
515-90-2 : 2964
515-94-6 : 326
515-96-8 : 105
516-05-2 : 9960
516-06-3 : 12040
516-12-1 : 11159
516-78-9 : 5384
516-85-8 : 5380
516-95-0 : 4108
517-04-4 : 5401
517-06-6 : 5407
517-07-7 : 5402
517-09-9 : 5397
517-10-2 : 4111
517-22-6 : 11126
517-23-7 : 6139
517-25-9 : 7562
517-28-2 : 629
517-60-2 : 1748
517-66-8 : 2488
517-82-8 : 7857
517-92-0 : 401
517-97-5 : 5159
518-05-8 : 7805
518-17-2 : 7333
518-28-5 : 6192
518-44-5 : 2572
518-61-6 : 9502
518-69-4 : 5149
518-75-2 : 2774
518-77-4 : 5151
518-82-1 : 419
519-02-8 : 7426
519-05-1 : 2608
519-09-5 : 481
519-34-6 : 7610
519-37-9 : 10758
519-44-8 : 1517
519-72-2 : 1804
519-73-3 : 1982
519-87-9 : 57
520-03-6 : 7366
520-18-3 : 2843
520-26-3 : 2794
520-33-2 : 2800
520-36-5 : 2806
520-45-6 : 10779
520-85-4 : 9752

521-04-0 : 11554
521-11-9 : 350
521-12-0 : 351
521-18-6 : 349
521-24-4 : 7970
521-31-3 : 9600
521-35-7 : 5154
521-74-4 : 11313
521-85-7 : 3108
522-12-3 : 2796
522-27-0 : 5583
522-57-6 : 7611
522-66-7 : 4123
522-75-8 : 2404
523-27-3 : 375
523-31-9 : 1337
523-47-7 : 7898
523-80-8 : 2423
524-15-2 : 6179
524-30-1 : 2823
524-34-5 : 11236
524-36-7 : 10978
524-40-3 : 10883
524-42-5 : 7842
524-63-0 : 4119
524-80-1 : 3982
524-96-9 : 1797
525-06-4 : 5928
525-15-5 : 11041
525-37-1 : 7008
525-64-4 : 5971
525-68-8 : 11253
525-82-6 : 2840
526-08-9 : 2201
526-31-8 : 11923
526-35-2 : 8586
526-55-6 : 7305
526-73-8 : 2357
526-75-0 : 9307
526-84-1 : 3777
526-85-2 : 9493
526-86-3 : 4263
526-95-4 : 6209
526-99-8 : 6193
527-07-1 : 6213
527-17-3 : 4270
527-18-4 : 1552
527-35-5 : 9475
527-52-6 : 7111
527-53-7 : 2279
527-55-9 : 1513
527-60-6 : 9495
527-61-7 : 4264
527-62-8 : 9205
527-69-5 : 6040
527-72-0 : 11737
527-73-1 : 7211
527-84-4 : 1988
527-85-5 : 619
527-89-9 : 1125
527-93-5 : 5153
528-21-2 : 5873
528-23-4 : 4784
528-29-0 : 1491
528-44-9 : 2319
528-45-0 : 2590
528-48-3 : 2813
528-53-0 : 2849
528-71-2 : 9197
528-75-6 : 551
528-76-7 : 2178
528-79-0 : 9420
528-81-4 : 467
529-00-0 : 4602
529-08-8 : 506
529-16-8 : 2968
529-17-9 : 484
529-19-1 : 2763
529-20-4 : 587
529-21-5 : 10964
529-28-2 : 1775
529-33-9 : 8041
529-34-0 : 8046
529-35-1 : 8042
529-36-2 : 8007
529-64-6 : 915
529-65-7 : 62

529-86-2 : 432
529-92-0 : 11216
530-14-3 : 5818
530-22-3 : 2464
530-40-5 : 10892
530-43-8 : 6648
530-44-9 : 7613
530-48-3 : 1636
530-50-7 : 7157
530-55-2 : 4262
530-62-1 : 7193
530-78-9 : 2729
530-85-8 : 10740
530-91-6 : 8043
530-93-8 : 8047
531-02-2 : 3097
531-18-0 : 7584
531-29-3 : 6226
531-37-3 : 9383
531-39-5 : 8886
531-75-9 : 2822
531-81-7 : 2775
531-84-0 : 7774
531-91-9 : 3035
531-95-3 : 2786
532-18-3 : 7727
532-24-1 : 490
532-27-4 : 5790
532-32-1 : 2725
532-48-9 : 5825
532-54-7 : 876
532-82-1 : 1284
532-96-7 : 2717
533-18-6 : 211
533-24-4 : 1544
533-30-2 : 2860
533-32-4 : 3201
533-45-9 : 11670
533-48-2 : 7223
533-50-6 : 3679
533-58-4 : 9371
533-68-6 : 3469
533-70-0 : 701
533-73-3 : 2367
533-75-5 : 4243
533-86-8 : 8300
533-98-2 : 3265
534-00-9 : 3238
534-07-6 : 10426
534-13-4 : 11804
534-15-6 : 5548
534-22-5 : 6130
534-26-9 : 7197
534-52-1 : 9405
534-59-8 : 9029
534-84-9 : 4611
534-85-0 : 1282
535-11-5 : 10207
535-13-7 : 10220
535-15-9 : 164
535-46-6 : 11548
535-75-1 : 9640
535-77-3 : 1989
535-80-8 : 2535
535-83-1 : 11029
535-87-5 : 2554
535-89-7 : 11043
536-08-3 : 2575
536-17-4 : 11684
536-25-4 : 2501
536-33-4 : 10885
536-38-9 : 5778
536-50-5 : 1859
536-57-2 : 2182
536-59-4 : 4664
536-60-7 : 1902
536-66-3 : 2689
536-69-6 : 10898
536-74-3 : 1715
536-75-4 : 10958
536-78-7 : 10957
536-88-9 : 10965
536-89-0 : 7164
536-90-3 : 767
536-95-8 : 2721
537-17-7 : 11830
537-26-8 : 488

537-29-1 : 916
537-45-1 : 4279
537-46-2 : 1573
537-47-3 : 7146
537-49-5 : 11933
537-64-4 : 7432
537-65-5 : 1263
537-91-7 : 5247
537-92-8 : 77
538-08-9 : 4159
538-23-8 : 8440
538-24-9 : 5319
538-32-9 : 12027
538-39-6 : 1590
538-41-0 : 635
538-43-2 : 10026
538-44-3 : 9106
538-49-8 : 11000
538-51-2 : 840
538-56-7 : 10684
538-58-9 : 8683
538-62-5 : 5122
538-64-7 : 3768
538-68-1 : 2107
538-74-9 : 2287
538-75-0 : 4291
538-81-8 : 1095
538-86-3 : 1930
538-93-2 : 2039
539-03-7 : 48
539-08-2 : 9796
539-15-1 : 9320
539-23-1 : 10860
539-30-0 : 1658
539-47-9 : 10638
539-52-6 : 6133
539-74-2 : 10208
539-80-0 : 4242
539-82-2 : 8878
539-86-6 : 10595
539-88-8 : 8907
539-89-9 : 3961
539-90-2 : 3563
539-92-4 : 3997
540-07-8 : 6907
540-08-9 : 6321
540-09-0 : 11857
540-10-3 : 6651
540-18-1 : 3579
540-36-3 : 1424
540-37-4 : 761
540-38-5 : 9373
540-42-1 : 10310
540-47-6 : 4980
540-51-2 : 5885
540-54-5 : 9848
540-61-4 : 272
540-63-6 : 5590
540-67-0 : 5602
540-69-2 : 6003
540-73-8 : 7155
540-80-7 : 8085
540-82-9 : 11576
540-84-1 : 8848
540-88-5 : 177
540-97-6 : 4624
541-01-5 : 6528
541-02-6 : 4901
541-05-9 : 5002
541-16-2 : 9925
541-20-8 : 8738
541-28-6 : 3384
541-31-1 : 3440
541-33-3 : 3276
541-35-5 : 3194
541-41-3 : 4018
541-42-4 : 8092
541-46-8 : 3206
541-47-9 : 3846
541-50-4 : 3568
541-53-7 : 11718
541-58-2 : 11674
541-59-3 : 11121
541-73-1 : 1355
541-88-8 : 124
541-91-3 : 4763
542-05-2 : 8782

CAS No.	No.
542-08-5	3574
542-10-9	5570
542-15-4	821
542-18-7	4339
542-28-9	10798
542-37-0	3546
542-46-1	4218
542-50-7	6303
542-52-9	3998
542-54-1	8831
542-55-2	6013
542-56-3	8094
542-58-5	5695
542-59-6	5577
542-69-8	3381
542-76-7	10097
542-81-4	5481
542-85-8	5600
542-88-1	7525
542-90-5	11703
542-91-6	11409
542-92-7	4764
543-20-4	3366
543-21-5	3902
543-24-8	6251
543-27-1	4024
543-28-2	3959
543-29-3	8078
543-38-4	7135
543-59-9	8719
543-67-9	8098
543-75-9	5203
543-80-6	110
543-83-9	6283
543-86-2	3633
543-87-3	3637
544-00-3	3226
544-01-4	3418
544-02-5	3436
544-10-5	6737
544-13-8	8767
544-16-1	8083
544-25-2	4240
544-35-4	8226
544-40-1	3433
544-62-7	10023
544-63-8	11622
544-65-0	11637
544-66-1	11638
544-72-9	8286
544-73-0	8282
544-76-3	6632
544-77-4	6640
544-85-4	5348
544-97-8	12066
545-06-2	288
545-26-6	4048
545-47-1	7417
545-93-7	11078
546-06-5	4143
546-33-8	5004
546-43-0	8059
546-45-2	5005
546-56-5	4994
546-97-4	5926
547-25-1	6022
547-44-4	2190
547-52-4	2188
547-58-0	2222
547-63-7	10301
547-65-9	6147
547-81-9	5405
547-91-1	11295
548-40-3	8631
548-51-6	2642
548-62-9	7453
548-80-1	7878
548-98-1	4093
550-10-7	5214
550-24-3	4261
550-44-7	7365
550-60-7	8034
550-74-3	10848
550-82-3	9508
550-97-0	2644
551-01-9	9877
551-06-4	7906
551-09-7	5514
551-62-2	2270
551-76-8	9489
551-77-9	9488
551-78-0	9487
551-88-2	8834
551-92-8	7202
552-16-9	2703
552-30-7	7341
552-32-9	81
552-41-0	5823
552-45-4	1198
552-58-9	2812
552-59-0	2828
552-70-5	479
552-72-7	9646
552-79-4	1863
552-82-9	807
552-89-6	591
552-94-3	2616
553-03-7	11327
553-17-3	9384
553-19-5	2397
553-20-8	84
553-26-4	3104
553-39-9	7953
553-60-6	10907
553-69-5	1913
553-79-7	827
553-82-2	1381
553-86-6	2461
553-90-2	5563
553-94-6	1021
553-97-9	4266
554-00-7	686
554-01-8	11106
554-12-1	10296
554-14-3	11769
554-21-2	5212
554-35-8	10104
554-68-7	5418
554-70-1	9521
554-84-7	9436
555-03-3	1936
555-10-2	4673
555-16-8	593
555-21-5	967
555-30-6	11931
555-35-1	6645
555-43-1	8272
555-44-2	6656
555-45-3	11628
555-57-7	1823
555-59-9	3853
555-77-1	5412
555-89-5	1965
556-10-5	12025
556-22-9	6275
556-24-1	3552
556-27-4	333
556-50-3	6265
556-52-5	8618
556-56-9	10573
556-61-6	7521
556-64-9	11705
556-67-2	4993
556-68-3	4741
556-69-4	8482
556-70-7	5079
556-71-8	4726
556-82-1	3869
556-88-7	6287
556-89-8	12024
556-96-7	1019
557-00-6	3562
557-05-1	8277
557-11-9	12028
557-17-5	10069
557-24-4	3817
557-25-5	3505
557-30-2	5492
557-31-3	10566
557-40-4	10592
557-59-5	11607
557-61-9	8222
557-66-4	5433
557-68-6	7471
557-89-1	453
557-91-5	5517
557-93-7	10530
557-98-2	10535
557-99-3	301
558-13-4	7539
558-17-8	10058
558-25-8	7538
558-30-5	8607
558-37-2	3760
558-38-3	10352
558-42-9	10354
558-43-0	10017
559-70-6	8548
560-08-7	3931
560-21-4	8849
560-22-5	9068
560-23-6	9065
560-95-2	7476
561-07-9	311
561-20-6	11352
561-25-1	7664
561-27-3	7667
561-83-1	11090
561-94-4	5393
562-49-2	8763
562-74-3	4699
563-03-1	9387
563-04-2	9570
563-12-2	5931
563-16-6	6775
563-41-7	7143
563-45-1	3800
563-46-2	3799
563-47-3	10540
563-52-0	3723
563-57-5	10553
563-58-6	10547
563-70-2	7473
563-78-0	3759
563-79-1	3761
563-80-4	3668
563-83-7	9798
563-84-8	3606
564-00-1	3007
564-02-3	8847
564-03-4	9069
564-04-5	8966
564-36-3	5387
564-37-4	5388
565-00-4	2921
565-53-7	161
565-59-3	8761
565-60-6	8941
565-63-9	3845
565-64-0	10245
565-67-3	8943
565-69-5	8982
565-75-3	8850
565-76-4	9066
565-77-5	9070
565-80-0	8967
567-18-0	7973
568-02-5	8062
568-21-8	5158
568-93-4	399
568-99-0	383
569-08-4	394
569-31-3	7355
569-34-6	10759
569-41-5	7824
569-51-7	2318
569-57-3	1154
569-64-2	7455
569-65-3	9605
569-77-7	2408
570-08-1	9923
570-22-9	7195
570-24-1	802
571-58-4	7820
571-61-9	7821
572-59-8	4125
572-83-8	386
572-93-0	396
573-17-1	9164
573-20-6	7841
573-26-2	75
573-33-1	7199
573-56-8	9344
573-58-0	7969
573-97-7	8022
573-98-8	7818
574-09-4	5813
574-12-9	2841
574-39-0	3983
574-84-5	2808
575-37-1	7823
575-38-2	7837
575-41-7	7819
575-43-9	7822
575-44-0	7836
575-74-6	607
576-22-7	1020
576-23-8	1017
576-24-9	9282
576-26-1	9310
576-55-6	9468
576-83-0	1094
577-16-2	5851
577-19-5	1075
577-27-5	4989
577-33-3	429
577-37-7	7572
577-55-9	992
577-56-0	2476
577-59-3	5863
577-71-9	9345
577-91-3	2135
578-07-4	312
578-54-1	740
578-57-4	1050
578-58-5	1931
578-66-5	11206
578-67-6	11305
578-68-7	11203
578-76-7	10769
579-04-4	7112
579-07-7	10028
579-10-2	79
579-18-0	2511
579-21-5	5854
579-44-2	5822
579-60-2	9392
579-66-8	694
579-75-9	2664
579-92-0	2656
579-94-2	183
580-02-9	2482
580-13-2	7740
580-15-4	11205
580-16-5	11306
580-20-1	11307
580-22-3	11202
580-48-3	11828
580-51-8	3089
581-08-8	59
581-40-8	7825
581-42-0	7826
581-43-1	7838
581-45-3	11009
581-46-4	3102
581-47-5	3101
581-49-7	3105
581-50-0	3100
581-55-5	7512
581-89-5	7930
582-08-1	5128
582-16-1	7827
582-17-2	7839
582-22-9	1579
582-24-1	5833
582-33-2	2496
582-60-5	2380
582-61-6	2884
582-62-7	3671
582-77-4	622
582-78-5	623
583-03-9	1846
583-04-0	2722
583-05-1	8804
583-33-5	6052
583-39-1	2386
583-48-2	6776
583-50-6	3192
583-53-9	1288
583-55-1	1046
583-58-4	10940
583-61-9	10936
583-68-6	647
583-69-7	1503
583-70-0	1018
583-71-1	1022
583-75-5	650
583-78-8	9284
583-86-8	8276
584-02-1	8917
584-04-3	10371
584-12-3	6033
584-27-0	3331
584-70-3	621
584-79-2	336
584-84-9	1431
584-87-2	2609
584-90-7	5114
584-92-9	8399
584-94-1	6772
584-98-5	3299
584-99-6	3769
585-32-0	1806
585-34-2	9322
585-48-8	10874
585-68-2	6054
585-74-0	5852
585-76-2	2519
585-79-5	1076
585-81-9	2632
586-11-8	9346
586-27-6	4698
586-37-8	5840
586-38-9	2665
586-39-0	1644
586-42-5	2477
586-61-8	1059
586-62-9	4684
586-75-4	2888
586-76-5	2520
586-77-6	645
586-78-7	1077
586-84-5	6129
586-89-0	2478
586-95-8	10977
586-96-9	2059
586-98-1	10975
587-02-0	741
587-03-1	1898
587-04-2	530
587-54-2	2513
587-61-1	10873
587-85-9	7436
587-90-6	11997
588-01-2	11906
588-46-5	87
588-47-6	817
588-52-3	5113
588-63-6	1085
588-67-0	1115
588-72-7	1029
588-73-8	1031
588-96-5	1032
589-08-2	1581
589-09-3	841
589-10-6	1033
589-14-0	6120
589-15-1	1000
589-16-2	742
589-17-3	1008
589-18-4	1899
589-21-9	7141
589-29-7	1435
589-33-3	11113
589-38-8	6960
589-40-2	6012
589-43-5	6773
589-44-6	3465
589-53-7	6424
589-55-9	6475
589-59-3	3555
589-63-9	8473
589-75-3	8427
589-87-7	1048
589-92-4	4598

CAS No.	No.
589-93-5	10938
589-98-0	8445
590-01-2	10214
590-02-3	125
590-11-4	5892
590-12-5	5891
590-13-6	10529
590-16-9	5463
590-18-1	3706
590-19-2	3145
590-26-1	5904
590-35-2	8760
590-36-3	8940
590-42-1	10066
590-50-1	8965
590-60-3	3634
590-66-9	4364
590-67-0	4530
590-73-8	6771
590-86-3	3186
590-90-9	3662
590-92-1	10205
590-93-2	3917
591-07-1	27
591-08-2	33
591-09-3	108
591-11-7	6154
591-12-8	6153
591-17-3	1055
591-18-4	1047
591-19-5	639
591-20-8	9229
591-22-0	10941
591-27-5	9202
591-31-1	584
591-34-4	10309
591-35-5	9287
591-47-9	4672
591-49-1	4670
591-50-4	1766
591-60-6	3569
591-62-8	3936
591-68-4	8862
591-70-4	6834
591-78-6	6959
591-80-0	9073
591-82-2	10065
591-84-4	11707
591-87-7	235
591-93-5	8659
591-95-7	8658
591-96-8	8660
592-01-8	3930
592-13-2	6774
592-20-1	10412
592-22-3	5250
592-27-8	6422
592-34-7	4012
592-35-8	3935
592-41-6	6999
592-42-7	6678
592-44-9	6676
592-45-0	6677
592-46-1	6680
592-49-4	6679
592-50-7	8817
592-51-8	9062
592-55-2	5461
592-57-4	4248
592-65-4	10154
592-76-7	6533
592-80-3	8935
592-82-5	3388
592-84-7	6001
592-88-1	10599
592-90-5	8590
593-03-3	6661
593-08-8	11889
593-12-4	8358
593-14-6	8361
593-45-3	8234
593-49-7	6302
593-50-0	11821
593-51-1	7446
593-52-2	446
593-53-3	7515
593-57-7	442
593-59-9	444
593-60-2	5880
593-61-3	5937
593-63-5	5938
593-66-8	5907
593-70-4	7481
593-71-5	7482
593-74-8	7435
593-75-9	7655
593-77-1	7447
593-79-3	7526
593-81-7	7445
593-88-4	449
593-90-8	3119
593-92-0	5890
593-95-3	4006
593-96-4	5455
593-98-6	7466
594-02-5	5546
594-03-6	5589
594-04-7	7503
594-08-1	4034
594-09-2	9522
594-10-5	11550
594-11-6	4981
594-15-0	7549
594-16-1	9883
594-18-3	7495
594-20-7	9894
594-22-9	7550
594-27-4	11544
594-34-3	9889
594-36-5	3253
594-37-6	9900
594-38-7	3385
594-39-8	3222
594-42-3	7527
594-44-5	5627
594-45-6	5622
594-47-8	94
594-51-4	3274
594-52-5	3236
594-56-9	3812
594-57-0	3247
594-58-1	10228
594-60-5	3618
594-61-6	10278
594-65-0	95
594-70-7	10085
594-71-8	9870
594-72-9	5536
594-73-0	5594
594-81-0	3271
594-82-1	3430
594-83-2	3644
594-89-8	10048
595-30-2	8559
595-37-9	3506
595-38-0	8724
595-39-1	7400
595-40-4	7401
595-41-5	8930
595-44-8	9902
595-49-3	9920
595-89-1	9733
595-90-4	11545
596-27-0	7353
596-29-2	7358
596-43-0	1061
596-51-0	11179
596-75-8	9937
597-04-6	3511
597-05-7	8934
597-08-1	10008
597-12-6	6221
597-32-0	10365
597-33-1	9998
597-35-3	5626
597-49-9	8932
597-50-2	9517
597-52-4	11503
597-55-7	9926
597-67-1	11428
597-71-7	9988
597-72-8	10068
597-76-2	6931
597-77-3	6935
597-90-0	6487
597-93-3	8162
597-96-6	6946
598-01-6	6505
598-02-7	9558
598-03-8	10131
598-04-9	3423
598-05-0	11575
598-09-4	8606
598-16-3	5921
598-17-4	9880
598-19-6	10699
598-21-0	292
598-22-1	10490
598-23-2	3916
598-25-4	3170
598-31-2	10416
598-38-9	5704
598-45-8	10063
598-50-5	12017
598-53-8	10070
598-55-0	3953
598-56-1	5421
598-57-2	7451
598-58-3	8075
598-61-8	4192
598-74-3	3223
598-76-5	9899
598-77-6	10172
598-78-7	10215
598-82-3	10266
598-88-9	5898
598-92-5	5482
598-94-7	12005
598-96-9	9072
598-98-1	10260
598-99-2	253
599-01-9	10333
599-04-2	6143
599-66-6	2240
599-70-2	1704
599-79-1	2654
599-91-7	2233
599-97-3	5657
599-99-5	242
600-00-0	10213
600-05-5	10238
600-06-6	3245
600-07-7	3538
600-11-3	8749
600-13-5	4150
600-14-6	8797
600-15-7	3530
600-17-9	3842
600-18-0	3567
600-22-6	10316
600-25-9	9866
600-30-6	3497
600-36-2	8931
600-40-8	5551
601-75-2	9948
601-76-3	9918
601-77-4	9822
601-84-3	2557
601-88-7	1393
602-03-9	724
602-21-1	1353
602-27-7	1489
602-29-9	2046
602-37-9	7787
602-38-0	7833
602-55-1	428
602-56-2	318
602-60-8	426
602-64-2	415
602-65-3	418
602-82-4	1332
602-92-6	2560
602-96-0	2364
602-99-3	9434
603-26-9	7513
603-32-7	450
603-33-8	3111
603-34-9	728
603-35-0	9523
603-36-1	11551
603-38-3	729
603-39-4	2110
603-45-2	4278
603-50-9	9465
603-52-1	3946
603-54-3	12011
603-69-0	3462
603-71-4	2363
603-76-9	7312
603-81-6	2552
603-83-8	799
603-85-0	9214
604-53-5	3004
604-61-5	2515
604-68-2	6216
604-69-3	6217
604-83-1	9172
604-88-6	1756
605-01-6	2090
605-02-7	7947
605-03-8	11290
605-27-6	411
605-32-3	408
605-37-8	7852
605-39-0	3050
605-42-5	390
605-50-5	1335
605-54-9	1333
605-63-0	7786
605-65-2	7977
605-71-0	7832
605-79-8	7634
605-81-2	6228
605-82-3	421
605-83-4	420
605-85-6	5859
605-99-2	10765
606-17-7	2594
606-18-8	2506
606-20-2	1958
606-21-3	1145
606-22-4	726
606-23-5	7264
606-27-9	2710
606-28-0	2516
606-35-9	1946
606-37-1	7831
606-43-9	11328
606-45-1	2673
606-46-2	696
606-81-5	3063
606-88-2	7173
606-97-3	3093
606-99-5	751
607-00-1	5993
607-31-8	11274
607-32-9	7386
607-34-1	11284
607-42-1	368
607-50-1	7919
607-52-3	7939
607-53-4	8005
607-56-7	12008
607-81-8	9977
607-85-2	2635
607-86-3	2682
607-90-9	2652
607-91-0	2428
607-97-6	3518
607-99-8	2313
608-07-1	7304
608-21-9	2307
608-23-1	1137
608-25-3	1529
608-26-4	9260
608-27-5	685
608-28-6	1772
608-29-7	2349
608-30-0	682
608-31-1	688
608-32-2	2305
608-33-3	9277
608-50-4	2360
608-66-2	6194
608-69-5	3308
608-71-9	9439
608-89-9	3315
608-90-2	2082
608-93-5	2084
608-96-8	2096
608-98-0	1553
609-02-9	9963
609-08-5	9962
609-09-6	9973
609-11-0	3498
609-12-1	3475
609-13-2	3477
609-14-3	3557
609-19-8	9486
609-22-3	9279
609-23-4	9492
609-25-6	2369
609-26-7	8813
609-29-0	8803
609-31-4	3640
609-36-9	9765
609-39-2	6134
609-40-5	11775
609-46-1	2225
609-65-4	2890
609-72-3	860
609-73-4	1783
610-02-6	2730
610-15-1	624
610-16-2	2584
610-17-3	718
610-25-3	2047
610-34-4	2707
610-39-9	1960
610-46-8	377
610-48-0	423
610-50-4	431
610-66-2	965
610-67-3	1662
610-69-5	219
610-71-9	2556
610-72-0	2581
610-89-9	9074
610-91-3	2362
610-97-9	2660
610-99-1	10449
611-00-7	2555
611-01-8	2580
611-03-0	2553
611-06-3	1396
611-07-4	9269
611-10-9	4785
611-12-1	10366
611-13-2	6059
611-14-3	1685
611-15-4	1640
611-19-8	1132
611-20-1	2759
611-21-2	710
611-32-5	11283
611-33-6	11242
611-34-7	11204
611-35-8	11238
611-43-8	3051
611-45-0	7949
611-49-4	7928
611-61-0	3053
611-62-1	3065
611-69-8	1901
611-70-1	10469
611-73-4	934
611-79-0	7589
611-92-7	12006
611-94-9	7625
611-95-0	2512
611-97-2	7590
611-99-4	7597
612-00-0	1677
612-01-1	12018
612-08-8	5109
612-12-4	986
612-13-5	2750
612-15-7	1637
612-16-8	1889
612-17-9	7815
612-19-1	2600
612-22-6	1692
612-23-7	1203
612-25-9	1906
612-28-2	804

CAS NUMBER INDEX OF ORGANIC COMPOUNDS (continued)

612-35-1	: 2718	615-57-6	: 681	618-85-9	: 1957	622-04-8	: 10377	623-96-1	: 4002
612-37-3	: 10766	615-58-7	: 9276	618-87-1	: 727	622-08-2	: 5761	623-97-2	: 5697
612-55-5	: 7903	615-59-8	: 1305	618-89-3	: 2525	622-09-3	: 7693	624-03-3	: 6649
612-57-7	: 11240	615-60-1	: 1141	618-91-7	: 2661	622-15-1	: 7565	624-04-4	: 5311
612-58-8	: 11278	615-62-3	: 9265	618-95-1	: 2711	622-16-2	: 764	624-08-8	: 6319
612-60-2	: 11282	615-65-6	: 668	618-98-4	: 2708	622-24-2	: 1164	624-09-9	: 6464
612-61-3	: 11241	615-67-8	: 1507	619-02-3	: 4601	622-26-4	: 9664	624-13-5	: 8441
612-62-4	: 11237	615-68-9	: 2350	619-12-5	: 9598	622-29-7	: 7452	624-16-8	: 5077
612-64-6	: 750	615-71-4	: 2306	619-15-8	: 1959	622-31-1	: 594	624-17-9	: 8133
612-71-5	: 11598	615-74-7	: 9262	619-20-5	: 2601	622-32-2	: 595	624-20-4	: 6754
612-75-9	: 3057	615-79-2	: 8876	619-23-8	: 1206	622-36-6	: 7166	624-24-8	: 8891
612-76-0	: 3066	615-81-6	: 5556	619-25-0	: 1907	622-37-7	: 976	624-29-3	: 4368
612-78-2	: 3005	615-83-8	: 8860	619-31-8	: 719	622-38-8	: 1705	624-31-7	: 1780
612-81-7	: 3107	615-84-9	: 5494	619-33-0	: 5530	622-39-9	: 11015	624-38-4	: 1429
612-94-2	: 7948	615-85-0	: 5453	619-39-6	: 5068	622-40-2	: 7691	624-41-9	: 3630
612-95-3	: 11291	615-87-2	: 1298	619-41-0	: 5779	622-42-4	: 2052	624-42-0	: 6523
612-96-4	: 11288	615-89-4	: 1514	619-42-1	: 2526	622-45-7	: 155	624-43-1	: 10192
612-97-5	: 11196	615-98-5	: 5565	619-44-3	: 2662	622-47-9	: 926	624-44-2	: 6982
613-03-6	: 2370	615-99-6	: 5564	619-52-3	: 4683	622-57-1	: 748	624-45-3	: 8908
613-08-1	: 369	616-02-4	: 6098	619-58-9	: 2659	622-60-6	: 1661	624-48-6	: 3779
613-12-7	: 424	616-06-8	: 8215	619-60-3	: 9319	622-61-7	: 1158	624-49-7	: 3778
613-13-8	: 365	616-12-6	: 9056	619-73-8	: 1908	622-62-8	: 9352	624-54-4	: 10318
613-21-8	: 11319	616-19-3	: 9901	619-82-9	: 4354	622-63-9	: 1707	624-57-7	: 8617
613-29-6	: 684	616-20-6	: 8721	619-84-1	: 2585	622-78-6	: 1794	624-60-2	: 5435
613-31-0	: 376	616-21-7	: 3277	619-99-8	: 6815	622-79-7	: 980	624-61-3	: 5939
613-33-2	: 3061	616-23-9	: 10359	620-02-0	: 6045	622-80-0	: 843	624-64-6	: 3705
613-37-6	: 3081	616-24-0	: 8702	620-03-1	: 6047	622-82-2	: 7160	624-65-7	: 10729
613-39-8	: 9613	616-25-1	: 9087	620-05-3	: 1777	622-85-5	: 2168	624-66-8	: 10734
613-42-3	: 3094	616-27-3	: 3646	620-11-1	: 8919	622-86-6	: 1159	624-67-9	: 10723
613-43-4	: 5112	616-28-4	: 3912	620-12-2	: 10193	622-93-5	: 10367	624-70-4	: 5477
613-45-6	: 543	616-31-9	: 8846	620-13-3	: 1063	622-96-8	: 1687	624-72-6	: 5545
613-46-7	: 7749	616-38-6	: 4000	620-14-4	: 1686	622-97-9	: 1642	624-73-7	: 5547
613-48-9	: 697	616-39-7	: 5430	620-16-6	: 1162	622-98-0	: 1163	624-74-8	: 5941
613-50-3	: 11285	616-40-0	: 7152	620-17-7	: 9354	623-00-7	: 2746	624-75-9	: 283
613-51-4	: 11286	616-42-2	: 11582	620-18-8	: 9349	623-03-0	: 2749	624-76-0	: 5731
613-59-2	: 7950	616-43-3	: 11132	620-19-9	: 1199	623-05-2	: 1884	624-79-3	: 5599
613-69-4	: 552	616-44-4	: 11770	620-20-2	: 1133	623-08-5	: 712	624-81-7	: 5434
613-70-7	: 9382	616-45-5	: 11180	620-22-4	: 2764	623-09-6	: 1280	624-82-8	: 7566
613-75-2	: 7510	616-47-7	: 7208	620-23-5	: 588	623-10-9	: 758	624-83-9	: 7520
613-77-4	: 11249	616-55-7	: 9223	620-24-6	: 1883	623-12-1	: 1180	624-85-1	: 7184
613-78-5	: 2645	616-91-1	: 5015	620-32-6	: 2241	623-15-4	: 3878	624-89-5	: 5606
613-80-9	: 7940	617-00-5	: 795	620-40-6	: 1798	623-17-6	: 6119	624-90-8	: 7460
613-81-0	: 8006	617-02-7	: 2699	620-42-8	: 9597	623-18-7	: 10640	624-91-9	: 8091
613-84-3	: 577	617-05-0	: 2626	620-43-9	: 3224	623-19-8	: 6123	624-92-0	: 5254
613-88-7	: 956	617-09-4	: 3996	620-47-3	: 2015	623-20-1	: 10639	624-95-3	: 3617
613-89-8	: 5774	617-30-1	: 6797	620-63-3	: 3561	623-21-2	: 3520	624-96-4	: 3284
613-90-1	: 968	617-31-2	: 8883	620-71-3	: 9800	623-22-3	: 10642	624-97-5	: 9100
613-91-2	: 5866	617-33-4	: 160	620-72-4	: 119	623-24-5	: 985	625-01-4	: 4085
613-93-4	: 620	617-35-6	: 10315	620-73-5	: 147	623-25-6	: 988	625-04-7	: 8959
613-94-5	: 2611	617-45-8	: 470	620-80-4	: 3576	623-26-7	: 1316	625-06-9	: 8928
613-97-8	: 749	617-50-5	: 10303	620-81-5	: 5495	623-27-8	: 1323	625-08-1	: 3535
614-00-6	: 806	617-51-6	: 10275	620-82-6	: 5559	623-30-3	: 10512	625-23-0	: 6942
614-16-4	: 2124	617-52-7	: 3330	620-83-7	: 2016	623-34-7	: 3285	625-25-2	: 6495
614-18-6	: 10902	617-53-8	: 3789	620-85-9	: 1696	623-36-9	: 9002	625-27-4	: 9055
614-21-1	: 5862	617-54-9	: 3790	620-86-0	: 2719	623-39-2	: 10014	625-28-5	: 3410
614-27-7	: 2142	617-60-7	: 5309	620-88-2	: 2054	623-40-5	: 8985	625-30-9	: 8701
614-29-9	: 1613	617-65-2	: 6240	620-92-8	: 9407	623-42-7	: 3547	625-31-0	: 9089
614-33-5	: 10196	617-73-2	: 8432	620-93-9	: 797	623-43-8	: 3847	625-36-5	: 10497
614-34-6	: 2700	617-75-4	: 448	620-94-0	: 2286	623-46-1	: 5533	625-37-6	: 3702
614-39-1	: 602	617-78-7	: 8809	620-95-1	: 11005	623-47-2	: 10739	625-38-7	: 3813
614-45-9	: 1128	617-80-1	: 3407	620-99-5	: 5670	623-48-3	: 201	625-44-5	: 10071
614-47-1	: 10709	617-83-4	: 4156	621-08-9	: 2184	623-49-4	: 4030	625-45-6	: 205
614-54-0	: 1881	617-84-5	: 5991	621-14-7	: 3320	623-50-7	: 197	625-46-7	: 10591
614-68-6	: 1787	617-85-6	: 11549	621-15-8	: 6722	623-51-8	: 203	625-47-8	: 5483
614-71-1	: 1659	617-86-7	: 11483	621-23-8	: 2355	623-53-0	: 4003	625-48-9	: 5752
614-72-2	: 1156	617-88-9	: 6075	621-25-0	: 7174	623-56-3	: 6981	625-50-3	: 61
614-75-5	: 904	617-89-0	: 6113	621-26-1	: 7138	623-59-6	: 73	625-52-5	: 12014
614-78-8	: 11806	617-90-3	: 6039	621-27-2	: 9462	623-60-9	: 10351	625-54-7	: 10038
614-96-0	: 7316	617-92-5	: 11124	621-31-8	: 9356	623-65-4	: 6646	625-55-8	: 6011
614-99-3	: 6056	617-94-7	: 1857	621-32-9	: 1660	623-66-5	: 8424	625-56-9	: 7582
615-05-4	: 1274	618-31-5	: 1301	621-34-1	: 9351	623-68-7	: 3818	625-57-0	: 3978
615-09-8	: 6163	618-32-6	: 2885	621-36-3	: 925	623-69-8	: 10372	625-58-1	: 8074
615-10-1	: 6071	618-34-8	: 1148	621-37-4	: 905	623-70-1	: 3839	625-59-2	: 11706
615-11-2	: 6057	618-36-0	: 1814	621-50-1	: 966	623-71-2	: 10221	625-60-5	: 5647
615-13-4	: 7287	618-38-2	: 2910	621-51-2	: 2597	623-72-3	: 10270	625-65-0	: 9035
615-18-9	: 2878	618-40-6	: 7165	621-52-3	: 1663	623-73-4	: 158	625-66-1	: 3282
615-20-3	: 2862	618-41-7	: 2180	621-59-0	: 572	623-76-7	: 12001	625-67-2	: 8750
615-30-5	: 8691	618-42-8	: 9630	621-62-5	: 5469	623-78-9	: 3949	625-68-3	: 3481
615-36-1	: 638	618-45-1	: 9411	621-63-6	: 5705	623-80-3	: 4031	625-69-4	: 8789
615-37-2	: 1778	618-46-2	: 2891	621-64-7	: 9825	623-81-4	: 11581	625-71-8	: 3531
615-38-3	: 4531	618-51-9	: 2658	621-70-5	: 6908	623-84-7	: 10002	625-74-1	: 10084
615-39-4	: 4532	618-62-2	: 1394	621-72-7	: 2385	623-85-8	: 3970	625-76-3	: 7508
615-41-8	: 1171	618-65-5	: 562	621-76-1	: 10078	623-86-9	: 102	625-77-4	: 23
615-42-9	: 1427	618-68-8	: 2144	621-77-2	: 8705	623-87-0	: 10191	625-80-9	: 10152
615-52-1	: 5567	618-71-3	: 2593	621-87-4	: 10479	623-91-6	: 3775	625-82-1	: 11119
615-54-3	: 2308	618-76-8	: 2618	621-95-4	: 731	623-93-8	: 8161	625-84-3	: 11120
615-55-4	: 683	618-80-4	: 9296	622-00-4	: 3306	623-95-0	: 12012	625-86-5	: 6089

CAS NUMBER INDEX OF ORGANIC COMPOUNDS (continued)

625-88-7	: 11753	627-48-5	: 4163	629-39-0	: 8079	634-67-3	: 852	641-36-1	: 5160
625-92-3	: 10926	627-51-0	: 5920	629-40-3	: 8385	634-68-4	: 2273	641-74-7	: 7423
625-95-6	: 1779	627-52-1	: 11572	629-41-4	: 8389	634-89-9	: 2253	641-85-0	: 9740
625-96-7	: 4597	627-53-2	: 5620	629-43-6	: 8088	634-92-4	: 2274	641-91-8	: 8010
625-98-9	: 1167	627-54-3	: 5630	629-45-8	: 5252	634-93-5	: 854	641-96-3	: 11193
626-00-6	: 1428	627-58-7	: 6691	629-46-9	: 8095	634-95-7	: 12002	642-28-4	: 7912
626-01-7	: 760	627-59-8	: 6944	629-49-2	: 5344	635-03-0	: 10139	642-32-0	: 2268
626-02-8	: 9372	627-63-4	: 3795	629-50-5	: 11871	635-10-9	: 2258	642-44-4	: 11060
626-04-0	: 1562	627-67-8	: 3400	629-54-9	: 6629	635-11-0	: 2276	642-84-2	: 608
626-10-8	: 3771	627-70-3	: 10463	629-59-4	: 11613	635-27-8	: 11239	643-28-7	: 788
626-13-1	: 796	627-73-6	: 3323	629-60-7	: 11880	635-46-1	: 11296	643-58-3	: 3082
626-15-3	: 984	627-82-7	: 10025	629-62-9	: 8642	635-65-4	: 3000	643-60-7	: 7393
626-16-4	: 987	627-83-8	: 8256	629-63-0	: 11619	635-67-6	: 1508	643-65-2	: 7632
626-17-5	: 1315	627-88-3	: 8265	629-64-1	: 6430	635-81-4	: 2269	643-75-4	: 10199
626-18-6	: 1434	627-89-4	: 8306	629-65-2	: 6435	635-88-1	: 7918	643-84-5	: 2847
626-19-7	: 1322	627-90-7	: 11946	629-66-3	: 8106	635-90-5	: 11134	643-93-6	: 3083
626-20-0	: 1638	627-91-8	: 6792	629-70-9	: 6662	635-93-8	: 533	644-08-6	: 3084
626-21-1	: 10933	627-93-0	: 6786	629-72-1	: 8643	636-04-4	: 1129	644-15-5	: 7725
626-23-3	: 3227	627-96-3	: 6756	629-73-2	: 6667	636-09-9	: 1350	644-21-3	: 7159
626-26-6	: 3434	627-97-4	: 6548	629-74-3	: 6673	636-10-2	: 7162	644-30-4	: 1468
626-27-7	: 6456	627-98-5	: 6941	629-75-4	: 6674	636-21-5	: 791	644-31-5	: 9151
626-29-9	: 11623	628-02-4	: 6721	629-78-7	: 6306	636-28-2	: 2254	644-32-6	: 5251
626-31-3	: 3297	628-03-5	: 3390	629-79-8	: 6641	636-30-6	: 853	644-35-9	: 9461
626-33-5	: 6524	628-04-6	: 3374	629-80-1	: 6628	636-31-7	: 2275	644-49-5	: 10308
626-35-7	: 218	628-05-7	: 8832	629-82-3	: 8416	636-36-2	: 6891	644-71-3	: 9445
626-36-8	: 3932	628-08-0	: 3862	629-89-0	: 8313	636-41-9	: 11131	644-80-4	: 3770
626-37-9	: 3963	628-09-1	: 10350	629-90-3	: 6304	636-46-4	: 1352	644-90-6	: 8423
626-39-1	: 2309	628-11-5	: 4015	629-92-5	: 8101	636-53-3	: 1346	644-97-3	: 9556
626-41-5	: 9278	628-12-6	: 4019	629-93-6	: 8242	636-60-2	: 3325	644-98-4	: 10984
626-43-7	: 690	628-13-7	: 10969	629-94-7	: 6294	636-72-6	: 11768	645-00-1	: 1784
626-44-8	: 2351	628-16-0	: 6716	629-96-9	: 5358	636-73-7	: 11018	645-09-0	: 625
626-48-2	: 11069	628-17-1	: 8818	629-97-0	: 5268	636-82-8	: 4637	645-12-5	: 6066
626-51-7	: 8779	628-20-6	: 3405	629-98-1	: 5284	636-98-6	: 1785	645-13-6	: 5848
626-53-9	: 6454	628-21-7	: 3288	629-99-2	: 8637	637-03-6	: 447	645-36-3	: 5414
626-55-1	: 10876	628-28-4	: 3391	630-01-3	: 6622	637-07-0	: 10231	645-41-0	: 3225
626-58-4	: 9682	628-29-5	: 3402	630-02-4	: 8220	637-12-7	: 8248	645-45-4	: 2160
626-60-8	: 10916	628-30-8	: 10064	630-03-5	: 8099	637-27-4	: 10323	645-49-8	: 1620
626-61-9	: 10917	628-32-0	: 10037	630-04-6	: 6299	637-47-8	: 7139	645-55-6	: 825
626-62-0	: 4415	628-33-1	: 10733	630-06-8	: 6991	637-50-3	: 2162	645-56-7	: 9463
626-64-2	: 11032	628-34-2	: 5474	630-07-9	: 8999	637-55-8	: 8271	645-59-0	: 2123
626-67-5	: 9679	628-37-5	: 9158	630-08-0	: 4011	637-57-0	: 10679	645-66-9	: 5306
626-70-0	: 6788	628-41-1	: 4249	630-10-4	: 11354	637-59-2	: 1088	645-67-0	: 8912
626-77-7	: 6909	628-44-4	: 8464	630-13-7	: 9908	637-64-9	: 6125	[illegible]	: 3319
626-82-4	: 6880	628-46-6	: [illegible]	630-16-0	: 5631	637-65-0	: 6126	645-96-5	: 2177
626-84-6	: 5736	628-50-2	: 4021	630-17-1	: 9843	637-69-4	: 1639	646-04-8	: 9005
626-85-7	: 11580	628-61-5	: 8360	630-18-2	: 10100	637-78-5	: 10299	646-05-9	: 9111
626-86-8	: 6791	628-62-6	: 6348	630-19-3	: 9783	637-87-6	: 1172	646-06-0	: 5204
626-87-9	: 8740	628-63-7	: 227	630-20-6	: 5634	637-88-7	: 4377	646-07-1	: 8889
626-88-0	: 8716	628-64-8	: 4017	630-25-1	: 5524	637-89-8	: 9378	646-13-9	: 8268
626-89-1	: 8939	628-66-0	: 10003	630-51-3	: 3343	637-92-3	: 10040	646-14-0	: 6841
626-90-4	: 3187	628-67-1	: 3348	630-60-4	: 4040	637-98-9	: 3979	646-19-5	: 6371
626-91-5	: 3395	628-68-2	: 5754	630-72-8	: 290	638-00-6	: 11755	646-20-8	: 6392
626-92-6	: 8746	628-71-7	: 6609	630-76-2	: 1831	638-02-8	: 11756	646-30-0	: 8103
626-94-8	: 7068	628-72-8	: 6329	630-95-5	: 7910	638-04-0	: 4367	646-31-1	: 11603
626-95-9	: 8786	628-73-9	: 6840	631-01-6	: 8546	638-07-3	: 3494	650-69-1	: 435
626-96-0	: 8696	628-76-2	: 8747	631-28-7	: 10164	638-08-4	: 8249	653-37-2	: 596
626-97-1	: 8698	628-77-3	: 8758	631-36-7	: 11450	638-10-8	: 3850	655-48-1	: 5575
626-99-3	: 8676	628-80-8	: 8822	631-40-3	: 9833	638-11-9	: 3549	655-86-7	: 9194
627-00-9	: 3482	628-81-9	: 3372	631-57-2	: 10113	638-17-5	: 5265	659-04-1	: 10678
627-02-1	: 10039	628-83-1	: 11699	631-61-8	: 106	638-20-0	: 40	659-49-4	: 828
627-03-2	: 180	628-89-7	: 5698	631-64-1	: 159	638-21-1	: 9519	659-70-1	: 3551
627-04-3	: 188	628-92-2	: 4244	632-05-3	: 3444	638-23-3	: 5017	659-86-9	: 10735
627-05-4	: 3412	628-96-6	: 5574	632-07-5	: 10233	638-25-5	: 8439	660-68-4	: 5428
627-07-6	: 9824	628-97-7	: 6650	632-14-4	: 12032	638-28-8	: 6738	660-88-8	: 8857
627-08-7	: 10076	629-01-6	: 8349	632-15-5	: 11757	638-29-9	: 8990	661-11-0	: 6417
627-09-8	: 10744	629-03-8	: 6757	632-16-6	: 11754	638-37-9	: 3256	661-19-8	: 5279
627-11-2	: 4013	629-04-9	: 6356	632-21-3	: 10487	638-41-5	: 4025	661-68-7	: 5227
627-12-3	: 3969	629-05-0	: 8529	632-22-4	: 12030	638-45-9	: 6831	661-97-2	: 9898
627-13-4	: 8080	629-06-1	: 6360	632-32-6	: 5385	638-46-0	: 3376	666-99-9	: 8102
627-16-7	: 10730	629-08-3	: 6428	632-46-2	: 2582	638-49-3	: 6015	670-54-2	: 5915
627-18-9	: 10344	629-09-4	: 6769	632-50-8	: 1591	638-51-7	: 8089	670-96-2	: 7213
627-19-0	: 9122	629-11-8	: 6796	632-51-9	: 1623	638-53-9	: 11881	671-56-7	: 3866
627-20-3	: 9006	629-14-1	: 5542	632-56-4	: 5633	638-54-0	: 8364	672-13-9	: 571
627-21-4	: 9123	629-15-2	: 5573	632-77-9	: 412	638-58-4	: 11611	672-15-1	: 7134
627-22-5	: 3151	629-16-3	: 8086	632-80-4	: 7348	638-60-8	: 11640	673-31-4	: 2149
627-26-9	: 3802	629-17-4	: 11702	632-82-6	: 414	638-65-3	: 8243	673-32-5	: 2173
627-27-0	: 3861	629-19-6	: 5259	632-83-7	: 385	638-67-5	: 11855	673-79-0	: 10330
627-30-5	: 10348	629-20-9	: 4742	632-89-3	: 12031	638-68-6	: 11819	674-76-0	: 9058
627-31-6	: 9907	629-23-2	: 11634	632-92-8	: 1490	638-73-3	: 7479	674-82-8	: 8598
627-32-7	: 10385	629-27-6	: 8406	632-99-5	: 634	638-95-9	: 12038	675-09-2	: 10792
627-37-2	: 10524	629-30-1	: 6399	633-03-4	: 5444	639-46-3	: 7668	675-20-7	: 9719
627-39-4	: 9787	629-31-2	: 6346	634-19-5	: 5978	639-58-7	: 11541	677-21-4	: 10608
627-40-7	: 10576	629-32-3	: 6421	634-36-6	: 2354	639-81-6	: 5395	677-68-9	: 10149
627-41-8	: 10736	629-33-4	: 6009	634-41-3	: 7723	640-19-7	: 63	678-26-2	: 8807
627-42-9	: 5480	629-35-6	: 7433	634-43-5	: 7724	640-61-9	: 2211	679-37-8	: 7185
627-44-1	: 7434	629-36-7	: 10122	634-47-9	: 11243	641-06-5	: 50	680-31-9	: 9573
627-45-2	: 5994	629-37-8	: 8412	634-66-2	: 2259	641-16-7	: 9476	681-06-1	: 9550

CAS NUMBER INDEX OF ORGANIC COMPOUNDS (continued)

CAS No.	No.
917-92-0	: 3907
917-93-1	: 9779
918-00-3	: 10488
918-07-0	: 3254
918-79-6	: 8692
918-82-1	: 9128
918-84-3	: 8732
919-12-0	: 9132
919-19-7	: 9530
919-30-2	: 9828
919-86-8	: 5103
919-94-8	: 3375
920-37-6	: 10584
920-46-7	: 10715
921-08-4	: 3487
921-13-1	: 7480
921-47-1	: 6863
922-28-1	: 6389
922-55-4	: 5016
922-59-8	: 9134
922-62-3	: 9057
922-65-6	: 8681
923-06-8	: 3300
923-26-2	: 10662
923-99-9	: 9594
924-16-3	: 3212
924-41-4	: 6706
924-43-6	: 8093
924-50-5	: 3852
924-52-7	: 8077
925-15-5	: 3321
925-78-0	: 8173
926-02-3	: 10036
926-26-1	: 3296
926-55-6	: 9116
926-56-7	: 8670
926-63-6	: 9807
926-64-7	: 281
926-65-8	: 10034
926-82-9	: 6390
927-49-1	: 11962
927-54-8	: 7006
927-62-8	: 3215
927-68-4	: 5686
927-73-1	: 3724
927-74-2	: 3920
927-80-0	: 5942
928-40-5	: 6795
928-45-0	: 8072
928-49-4	: 7116
928-50-7	: 9019
928-51-8	: 3604
928-56-3	: 5886
928-65-4	: 11464
928-68-7	: 6521
928-80-3	: 5076
928-92-7	: 7072
928-94-9	: 7070
928-95-0	: 7069
928-96-1	: 7071
929-55-5	: 8348
929-61-3	: 8402
929-62-4	: 8401
929-77-1	: 5278
930-02-9	: 8240
930-18-7	: 4972
930-21-2	: 496
930-22-3	: 8613
930-27-8	: 6131
930-28-9	: 4789
930-30-3	: 4941
930-36-9	: 10834
930-37-0	: 8620
930-55-2	: 11170
930-66-5	: 4642
930-68-7	: 4703
930-87-0	: 11142
930-89-2	: 4815
930-90-5	: 4816
931-10-2	: 4293
931-20-4	: 9721
931-56-6	: 4425
931-77-1	: 4319
931-87-3	: 4752
931-89-5	: 4751
931-97-5	: 4329
932-01-4	: 4519
932-16-1	: 5849
932-32-1	: 670
932-39-8	: 4797
932-40-1	: 4798
932-43-4	: 4834
932-44-5	: 4835
932-53-6	: 11831
932-56-9	: 4235
932-66-1	: 5793
932-77-4	: 1007
932-96-7	: 675
933-05-1	: 4865
933-21-1	: 4728
933-36-8	: 4616
933-48-2	: 4569
933-67-5	: 7317
933-75-5	: 9483
933-78-8	: 9482
933-88-0	: 2902
933-98-2	: 1670
934-00-9	: 1525
934-34-9	: 2868
934-53-2	: 1189
934-74-7	: 1672
934-80-5	: 1675
935-05-7	: 8097
935-07-9	: 18
935-08-0	: 5134
935-12-6	: 6164
935-13-7	: 6161
935-31-9	: 4208
935-95-5	: 9471
936-48-1	: 10824
936-53-8	: 10823
936-58-3	: 1911
936-72-1	: 2026
937-14-4	: 1127
937-20-2	: 5784
937-32-6	: 2179
937-33-7	: 716
938-16-9	: 10431
938-25-0	: 7795
938-33-0	: 11276
938-39-6	: 11326
938-45-4	: 10446
938-73-8	: 615
939-23-1	: 10999
939-26-4	: 7743
939-27-5	: 7886
939-48-0	: 2693
939-52-6	: 3650
939-55-9	: 5696
939-83-3	: 2766
939-87-7	: 4957
940-41-0	: 11474
940-49-8	: 2033
940-54-5	: 3189
941-69-5	: 11123
941-98-0	: 5857
942-01-8	: 3989
942-06-3	: 7344
942-92-7	: 6983
943-14-6	: 2528
943-15-7	: 1991
943-27-1	: 5808
944-22-9	: 5986
945-51-7	: 2183
947-73-9	: 9162
947-84-2	: 3021
947-91-1	: 877
948-32-3	: 2372
948-65-2	: 7320
948-97-0	: 1966
948-98-1	: 1146
948-99-2	: 1147
949-87-1	: 5139
950-37-8	: 7645
951-86-0	: 1365
952-47-6	: 9426
952-97-6	: 2058
955-83-9	: 6100
957-51-7	: 5219
959-33-1	: 10713
961-11-5	: 6202
961-71-7	: 5506
964-68-1	: 2533
976-71-6	: 9736
982-43-4	: 9657
992-94-9	: 11437
993-00-0	: 11378
993-07-7	: 11489
993-13-5	: 9538
994-05-8	: 3396
994-25-2	: 8726
994-30-9	: 11388
994-49-0	: 5235
996-98-5	: 5562
997-49-9	: 11698
998-29-8	: 11500
998-30-1	: 11477
998-40-3	: 9520
998-41-4	: 11454
998-93-6	: 6358
998-95-8	: 6363
999-21-3	: 3782
999-52-0	: 6362
999-55-3	: 10694
999-78-0	: 9130
999-79-1	: 9127
999-81-5	: 5442
999-97-3	: 11363
1000-49-3	: 11371
1000-50-6	: 11369
1000-86-8	: 8665
1001-53-2	: 28
1001-89-4	: 6361
1002-28-4	: 7126
1002-69-3	: 5036
1002-84-2	: 8645
1003-03-8	: 4771
1003-09-4	: 11726
1003-10-7	: 11794
1003-29-8	: 11114
1003-30-1	: 6107
1003-31-2	: 11732
1003-56-1	: 11036
1003-64-1	: 4391
1003-73-2	: 10988
1003-78-7	: 11784
1003-90-3	: 11141
1004-36-0	: 10793
1004-66-6	: 1919
1004-77-9	: 4599
1005-64-7	: 1109
1005-67-0	: 7684
1006-06-0	: 1874
1006-31-1	: 2203
1007-26-7	: 1484
1007-32-5	: 3677
1008-88-4	: 10998
1008-89-5	: 10997
1009-14-9	: 8986
1009-67-2	: 2137
1009-81-0	: 1754
1009-93-4	: 5001
1010-48-6	: 2130
1012-72-2	: 990
1012-84-6	: 2712
1013-08-7	: 9187
1013-83-8	: 7769
1013-88-3	: 1835
1013-96-3	: 10676
1016-77-9	: 7601
1016-94-0	: 9195
1017-56-7	: 7583
1024-57-3	: 6301
1025-15-6	: 11842
1048-08-4	: 11453
1056-77-5	: 6175
1066-27-9	: 11435
1066-45-1	: 11540
1067-08-9	: 8814
1067-20-5	: 8756
1068-19-5	: 6391
1068-57-1	: 195
1068-87-7	: 8812
1068-90-2	: 9922
1069-53-0	: 6864
1070-14-0	: 6705
1070-32-2	: 6492
1070-34-4	: 3335
1070-71-9	: 10737
1070-78-6	: 10142
1070-83-3	: 3508
1070-87-7	: 8841
1071-26-7	: 6383
1071-73-4	: 8974
1071-81-4	: 6849
1071-83-6	: 6276
1071-94-9	: 6589
1071-98-3	: 3908
1072-05-5	: 6387
1072-16-8	: 8378
1072-21-5	: 6748
1072-36-2	: 8224
1072-39-5	: 5281
1072-43-1	: 11696
1072-44-2	: 500
1072-52-2	: 499
1072-53-3	: 5199
1072-83-9	: 5820
1072-91-9	: 10840
1072-98-6	: 10861
1073-06-9	: 1043
1073-07-0	: 4732
1073-23-0	: 10945
1073-67-2	: 1151
1074-11-9	: 1376
1074-17-5	: 2036
1074-43-7	: 2037
1074-55-1	: 2038
1074-58-4	: 275
1074-92-6	: 1464
1075-22-5	: 7254
1075-38-3	: 1465
1075-74-7	: 2106
1076-26-2	: 8031
1076-55-7	: 9391
1076-61-5	: 7994
1076-66-0	: 1025
1077-16-3	: 1763
1078-04-2	: 7260
1078-19-9	: 8048
1078-21-3	: 2128
1078-30-4	: 11228
1078-71-3	: 1745
1079-71-6	: 427
1080-16-6	: 5130
1080-32-6	: 9545
1081-34-1	: 11600
1081-75-0	: 2121
1081-77-2	: 2064
1083-56-3	: 1099
1086-80-2	: 2472
1088-11-5	: 2411
1093-58-9	: 354
1095-90-5	: 6516
1111-74-6	: 11414
1111-97-3	: 3901
1112-37-4	: 9529
1112-39-6	: 11412
1112-48-7	: 11368
1112-55-6	: 11449
1112-56-7	: 11543
1113-12-8	: 11417
1113-30-0	: 3113
1113-68-4	: 25
1113-74-2	: 8968
1114-09-6	: 3302
1114-51-8	: 9793
1114-71-2	: 3976
1114-76-7	: 3200
1115-11-3	: 3698
1115-12-4	: 9915
1116-00-0	: 5283
1116-24-1	: 58
1116-39-8	: 3123
1116-40-1	: 9820
1116-61-6	: 3121
1117-19-7	: 9945
1117-55-1	: 8431
1117-59-5	: 8882
1118-09-8	: 11491
1118-68-9	: 6261
1118-71-4	: 6404
1118-84-9	: 3577
1119-06-8	: 6465
1119-19-3	: 3895
1119-33-1	: 6245
1119-40-0	: 8775
1119-46-6	: 8866
1119-49-9	: 38
1119-51-3	: 9010
1119-60-4	: 6564
1119-65-9	: 6610
1120-07-6	: 8118
1120-16-7	: 5289
1120-21-4	: 11938
1120-28-1	: 5357
1120-29-2	: 5287
1120-36-1	: 11636
1120-48-5	: 8354
1120-56-5	: 4193
1120-59-8	: 11750
1120-62-3	: 4929
1120-64-5	: 8576
1120-71-4	: 8570
1120-72-5	: 4885
1120-73-6	: 4944
1120-87-2	: 10877
1120-97-4	: 5193
1121-07-9	: 11160
1121-18-2	: 4710
1121-24-0	: 9376
1121-37-5	: 7607
1121-55-7	: 10953
1121-60-4	: 10887
1121-92-2	: 502
1121-95-5	: 4875
1122-20-9	: 4705
1122-25-4	: 4594
1122-39-0	: 11027
1122-42-5	: 1773
1122-54-9	: 5869
1122-60-7	: 4454
1122-62-9	: 5867
1122-69-6	: 10963
1122-81-2	: 11016
1122-82-3	: 4417
1122-83-4	: 844
1122-91-4	: 528
1123-00-8	: 4775
1123-09-7	: 4707
1123-27-9	: 5821
1123-34-8	: 4555
1123-84-8	: 1371
1123-85-9	: 1608
1123-86-0	: 10423
1123-95-1	: 11105
1123-96-2	: 10959
1124-05-6	: 1362
1124-07-8	: 9293
1124-20-5	: 1985
1124-27-2	: 4442
1124-33-0	: 10993
1125-60-6	: 7372
1125-74-2	: 1547
1125-78-6	: 8044
1125-80-0	: 7382
1126-09-6	: 9643
1126-18-7	: 4575
1126-46-1	: 2543
1126-75-6	: 2031
1126-78-9	: 656
1126-79-0	: 1114
1127-76-0	: 7885
1128-08-1	: 4950
1128-54-7	: 10838
1128-76-3	: 2538
1129-47-1	: 10293
1129-50-6	: 3208
1129-89-1	: 4216
1131-16-4	: 10829
1131-18-6	: 10814
1131-60-8	: 9272
1131-62-0	: 5805
1131-63-1	: 7779
1132-21-4	: 2577
1132-39-4	: 2175
1132-66-7	: 1765
1133-80-8	: 5968
1134-23-2	: 4168
1134-62-9	: 7747
1135-12-2	: 839
1140-14-3	: 7614
1140-16-5	: 7628
1142-70-7	: 11080
1142-97-8	: 2236

CAS No.	No.
1143-38-0	430
1144-74-7	7636
1146-98-1	7265
1146-99-2	7266
1148-11-4	11148
1149-24-2	10932
1151-15-1	2667
1154-59-2	610
1165-39-5	6184
1165-48-6	2819
1174-72-7	11509
1184-60-7	10569
1185-33-7	3615
1185-55-3	11487
1186-31-8	6573
1186-53-4	8840
1186-73-8	7553
1187-03-7	12029
1187-13-9	3837
1187-58-2	9799
1188-33-6	7443
1189-71-5	11585
1189-99-7	6448
1190-22-3	3278
1190-34-7	6605
1190-39-2	9938
1190-63-2	8258
1190-83-6	6440
1191-04-4	7049
1191-08-8	3368
1191-41-9	8283
1191-43-1	6812
1191-75-9	6668
1191-95-3	4197
1191-96-4	4974
1191-99-7	6086
1192-18-3	4801
1192-28-5	4892
1192-30-9	6036
1192-37-6	4428
1192-62-7	5817
1192-88-7	4631
1193-02-8	2291
1193-12-0	9651
1193-16-4	4294
1193-17-5	4295
1193-18-6	4711
1193-35-7	6085
1193-46-0	4513
1193-47-1	4580
1193-70-0	4894
1193-72-2	1011
1193-74-4	11044
1193-81-3	4420
1193-82-4	2040
1194-02-1	2756
1194-65-6	2752
1195-31-9	4681
1195-32-0	1986
1195-42-2	4303
1196-31-2	4605
1196-39-0	7383
1196-58-3	1701
1196-67-4	3701
1196-79-8	7299
1197-18-8	4333
1197-37-1	1272
1197-40-6	6132
1197-55-3	886
1198-14-7	11310
1198-37-4	11255
1198-59-0	1398
1198-61-4	1399
1198-62-5	1400
1200-03-9	1003
1200-73-3	5194
1201-26-9	10442
1202-08-0	5850
1202-34-2	10871
1203-39-0	7845
1204-78-0	3015
1205-02-3	327
1205-64-7	808
1205-71-6	679
1205-91-0	1509
1207-69-8	315
1210-35-1	5147
1214-24-0	12053
1241-94-7	9560
1260-17-9	373
1300-94-3	9425
1322-67-4	9178
1330-61-6	10646
1400-62-0	8549
1405-86-3	6230
1405-87-4	508
1407-14-3	6293
1420-07-1	9328
1421-14-3	894
1423-60-5	3923
1435-50-3	1014
1435-53-6	1300
1435-55-8	4124
1436-43-7	11223
1441-87-8	2900
1445-23-4	9324
1445-45-0	5669
1445-73-4	9724
1446-34-0	3006
1447-26-3	6600
1450-14-2	5228
1450-31-3	7547
1450-58-4	6048
1450-72-2	5827
1450-81-3	8871
1453-17-4	4254
1453-24-3	4660
1453-25-4	4245
1453-58-3	10835
1454-80-4	3028
1454-84-8	8104
1454-85-9	6317
1455-20-5	11731
1455-21-6	8145
1457-85-8	224
1458-99-7	9023
1459-00-3	1060
1459-09-2	1751
1459-10-5	2266
1459-14-9	1039
1459-93-4	1347
1460-16-8	4223
1460-88-4	4911
1460-98-6	1948
1462-03-9	4867
1462-12-0	9950
1462-27-7	3652
1462-34-6	5468
1462-84-6	11026
1463-17-8	11259
1467-05-6	1191
1467-79-4	4157
1468-39-9	3542
1470-94-6	7285
1471-03-0	10594
1471-96-1	7858
1472-90-8	8238
1475-13-4	1854
1476-07-9	1666
1476-11-5	3751
1477-55-0	1433
1482-15-1	9144
1482-91-3	8483
1483-60-9	1473
1484-08-8	3984
1484-12-4	3987
1484-19-1	7308
1484-84-0	9663
1484-85-1	2424
1486-75-5	4215
1487-18-9	6102
1489-47-0	8167
1489-57-2	4271
1489-60-7	4201
1490-05-7	7428
1493-01-2	7516
1493-02-3	6019
1493-03-4	7505
1493-27-2	1733
1495-50-7	4167
1498-51-7	9577
1498-65-3	9557
1501-04-8	2100
1501-06-0	8771
1501-27-5	8780
1501-60-6	8668
1502-05-2	4204
1502-06-3	4205
1503-53-3	2481
1504-54-7	3872
1504-58-1	10745
1504-70-7	10614
1504-75-2	10517
1511-62-2	7469
1515-76-0	3176
1515-80-6	6702
1515-95-3	1684
1516-80-9	11431
1517-05-1	5681
1518-75-8	11779
1518-86-1	9215
1519-36-4	402
1519-59-1	11002
1520-21-4	735
1520-45-2	1698
1520-46-3	2021
1521-51-3	4628
1522-22-1	8802
1522-46-9	3461
1526-73-4	7848
1527-89-5	2761
1528-30-9	4827
1529-17-5	11492
1529-41-5	950
1530-03-6	2171
1530-05-8	1746
1530-20-7	1744
1532-97-4	7374
1534-26-5	11890
1538-87-0	5479
1539-42-0	10974
1540-28-9	8856
1540-29-0	6872
1540-34-7	8801
1541-23-7	6328
1547-87-1	99
1550-09-0	6742
1551-06-0	11125
1551-09-3	11139
1551-21-9	10089
1551-27-5	11778
1551-32-2	11765
1551-44-6	3496
1551-88-8	4581
1551-89-9	4582
1552-12-1	4734
1552-67-6	7053
1552-95-0	8680
1555-80-2	887
1556-18-9	4824
1558-17-4	11058
1558-24-3	11459
1558-25-4	11456
1558-31-2	11391
1558-33-4	11390
1559-02-0	4967
1559-81-5	7995
1560-06-1	1111
1560-09-4	1110
1560-11-8	4909
1560-89-0	6310
1561-10-0	6900
1561-20-2	10630
1563-66-2	3991
1563-83-3	24
1563-87-7	26
1565-71-5	1855
1565-86-2	1905
1569-01-3	10401
1569-02-4	10381
1569-43-3	7076
1569-44-4	7084
1569-59-1	9101
1569-60-4	6580
1569-69-3	4480
1570-45-2	10903
1570-64-5	9266
1573-17-7	3913
1573-28-0	6483
1573-58-6	3174
1574-33-0	9117
1574-40-9	9114
1574-41-0	8662
1575-61-7	8992
1576-13-2	9464
1576-95-0	9091
1576-96-1	9090
1577-22-6	7051
1578-21-8	10628
1579-40-4	2078
1582-09-8	11901
1585-06-4	1653
1585-07-5	1037
1587-04-8	2023
1587-20-8	10170
1587-26-4	3753
1587-29-7	3752
1587-41-3	7501
1588-44-9	1762
1589-47-5	10386
1589-60-2	4552
1590-08-5	8049
1591-31-7	3079
1592-23-0	8252
1592-38-7	7909
1595-04-6	1122
1595-05-7	1123
1595-06-8	1994
1595-11-5	1121
1595-16-0	1998
1596-84-5	3334
1599-41-3	10180
1599-67-3	5280
1600-37-9	10593
1601-00-9	5847
1603-40-3	10866
1603-79-8	935
1603-91-4	11665
1604-01-9	11021
1604-02-0	4872
1604-11-1	3327
1606-08-2	4340
1609-47-8	5169
1609-86-5	10062
1609-93-4	10618
1610-24-8	4433
1610-29-3	9094
1611-50-3	3942
1611-52-5	3952
1611-65-0	3386
1613-32-7	11292
1613-34-9	11266
1613-45-2	10044
1613-51-0	11795
1615-14-1	7205
1615-70-9	8673
1615-75-4	5475
1615-80-1	7153
1617-18-1	3841
1617-90-9	5349
1618-26-4	7461
1619-28-9	4947
1619-34-7	485
1619-56-3	3495
1619-57-4	3504
1619-62-1	9942
1620-14-0	10427
1622-10-2	6373
1622-32-8	5629
1623-14-9	9561
1626-09-1	8394
1629-58-9	9104
1630-77-9	5903
1630-94-0	4971
1631-82-9	11381
1631-84-1	11403
1632-16-2	6425
1632-76-4	10856
1632-83-3	2382
1633-90-5	6858
1634-04-4	10072
1634-09-9	7746
1635-02-5	7091
1635-61-6	677
1637-31-6	3173
1638-22-8	9248
1638-26-2	4800
1638-63-7	972
1639-09-4	6436
1640-89-7	4812
1641-40-3	2413
1643-49-8	5970
1646-26-0	5775
1646-88-4	335
1647-08-1	7019
1647-16-1	5024
1647-23-0	3235
1647-26-3	4316
1649-08-7	5531
1653-19-6	3158
1653-31-2	11884
1653-32-3	11632
1653-34-5	8647
1653-40-3	6494
1653-57-2	8084
1654-86-0	5062
1655-70-5	2075
1655-71-6	1929
1656-48-0	10114
1657-52-9	1222
1663-67-8	10029
1666-13-3	5225
1666-96-2	11289
1667-01-2	5876
1667-04-5	1243
1670-46-8	4879
1671-75-6	6526
1672-46-4	4047
1674-10-8	4648
1674-30-2	1852
1674-33-5	8744
1674-37-9	8480
1674-38-0	5327
1674-56-2	10338
1675-69-0	8129
1678-80-4	4492
1678-81-5	4489
1678-82-6	4440
1678-91-7	4390
1678-92-8	4471
1678-93-9	4325
1678-97-3	4488
1678-98-4	4453
1679-06-7	6857
1679-07-8	4852
1679-09-0	3441
1679-36-3	7130
1679-47-6	6145
1680-51-9	7998
1680-53-1	7922
1680-58-6	7927
1681-22-7	7999
1686-41-5	4237
1689-82-3	9448
1689-84-5	2751
1690-75-1	9645
1694-19-5	1939
1695-04-1	1938
1696-20-4	7682
1698-60-8	4080
1702-17-6	4142
1703-46-4	1862
1703-52-2	6106
1704-15-0	10711
1706-11-2	1920
1706-73-6	1546
1708-32-3	11751
1708-39-0	5209
1709-03-1	46
1710-98-1	2895
1711-06-4	2903
1711-07-5	2898
1712-64-7	8076
1717-00-6	5534
1718-50-9	2098
1719-53-5	11393
1719-57-9	11372
1719-58-0	11376
1721-89-7	11254
1721-93-3	7381
1723-94-0	7690
1725-74-2	7035
1726-12-1	2108
1726-14-3	1113
1729-67-5	10241

CAS No.	No.
1731-79-9	5300
1731-84-6	8154
1731-88-0	11882
1731-92-6	6316
1732-08-7	6395
1732-09-8	8388
1732-10-1	8134
1733-12-6	9368
1738-25-6	10101
1738-36-9	284
1739-84-0	7200
1740-97-2	9414
1741-83-9	8826
1745-16-0	2105
1745-81-9	9457
1746-11-8	2443
1746-13-0	2167
1746-77-6	3955
1754-62-7	10686
1755-01-7	7577
1755-05-1	8686
1758-10-7	2027
1758-33-4	8608
1758-85-6	1421
1758-88-9	1674
1759-53-1	4958
1759-58-6	4804
1759-64-4	4652
1759-81-5	4930
1759-88-2	11439
1760-46-9	937
1761-71-3	4299
1768-31-6	10478
1772-10-7	1996
1776 37-0	7241
1777-82-8	1853
1779-81-3	11664
1780-31-0	11052
1781-78-8	4755
1790-22-3	3450
1792-41-2	2379
1792-81-0	4372
1794-84-9	7485
1795-01-3	11760
1795-04-6	11792
1795-05-7	11793
1795-09-1	11787
1795-15-9	4460
1795-16-0	4344
1795-17-1	4382
1795-18-2	4475
1795-20-6	4843
1795-21-7	4790
1795-22-8	4850
1795-26-2	4497
1795-27-3	4496
1795-48-8	10060
1796-05-0	1494
1804-15-5	9784
1805-61-4	9404
1806-29-7	3064
1807-36-9	2837
1809-02-5	3155
1809-05-8	8820
1809-10-5	8712
1809-19-4	9532
1809-26-3	9014
1812-51-7	3072
1814-88-6	10128
1817-47-6	1978
1817-57-8	3924
1817-73-8	646
1818-07-1	2069
1820-09-3	2976
1821-02-9	8902
1821-12-1	1100
1821-36-9	680
1821-39-2	842
1822-74-8	5910
1822-86-2	3149
1823-91-2	959
1824-81-3	10868
1825-30-5	7810
1825-31-6	7809
1825-62-3	11430
1825-82-7	11462
1826-14-8	11676
1830-54-2	8784
1830-95-1	3120
1832-53-7	9543
1833-51-8	11379
1836-75-5	8081
1838-59-1	6017
1838-77-3	7085
1839-88-9	4490
1843-05-6	7621
1845-30-3	2975
1846-70-4	8203
1850-14-2	7543
1855-63-6	4630
1859-76-3	7626
1860-21-5	247
1860-27-1	3399
1861-21-8	11099
1861-32-1	5180
1861-40-1	509
1863-63-4	2509
1864-92-2	738
1866-31-5	10692
1867-37-4	9917
1868-53-7	7498
1871-52-9	4749
1871-57-4	10537
1871-58-5	10178
1871-76-7	975
1871-96-1	5041
1873-25-2	3605
1873-29-6	10061
1873-54-7	11267
1873-92-3	11402
1875-88-3	1597
1877-72-1	2550
1878-68-8	888
1879-06-7	6040
1879-09-0	9325
1883-32-5	1615
1885-14-9	4026
1885-29-6	2741
1885-38-7	10587
1888-57-9	6966
1888-71-7	10571
1888-85-3	3370
1896-62-4	3882
1897-45-6	1317
1899-24-7	6043
1904-98-9	10754
1912-21-6	10319
1912-24-9	476
1917-15-3	6060
1918-00-9	2567
1918-02-1	10896
1918-16-7	9771
1921-70-6	8644
1928-30-9	11856
1928-38-7	172
1929-73-3	170
1929-77-7	12047
1929-82-4	8071
1931-60-8	3367
1932-04-3	9661
1937-54-8	8113
1937-62-8	8308
1938-17-6	1257
1938-22-3	2063
1940-18-7	4522
1942-45-6	8532
1942-46-7	5100
1943-16-4	7487
1943-54-0	9291
1945-53-5	9166
1951-11-7	3684
1967-25-5	11998
1969-43-3	6406
1969-73-9	875
1971-46-6	7322
1973-22-4	1035
1974-04-5	6357
1975-44-6	7767
1975-78-6	5057
1982-47-4	4088
1985-57-5	1455
1985-59-7	7980
1985-97-3	1691
1990-29-0	338
2004-67-3	9102
2004-69-5	9113
2004-70-8	8661
2006-14-6	10675
2008-41-5	11586
2009-74-7	6598
2009-83-8	6916
2016-42-4	11612
2016-57-1	5032
2018-90-8	7907
2021-28-5	2131
2025-40-3	10625
2027-17-0	7921
2027-19-2	7958
2028-63-9	3921
2029-94-9	538
2031-67-6	11479
2032-35-1	5459
2032-65-7	7646
2033-24-1	5190
2035-94-1	1602
2036-41-1	11072
2037-31-2	2291
2038-03-1	7689
2039-82-9	1028
2039-83-0	1368
2039-85-2	1150
2039-86-3	1027
2039-87-4	1149
2039-88-5	1026
2039-89-6	1628
2039-93-2	1968
2040-07-5	5875
2040-14-4	10466
2040-95-1	4778
2040-96-2	4849
2042-37-7	2744
2043-38-1	2856
2043-61-0	4331
2044-08-8	4627
2045-79-6	1577
2047-21-4	5877
2049-66-3	9951
2049-67-4	9041
2049-73-2	1407
2049-74-3	5719
2049-80-1	9980
2049-94-7	1954
2049-95-8	1480
2050-01-3	10302
2050-03-5	1488
2050-04-6	2109
2050-07-9	8983
2050-08-0	2649
2050-14-8	9216
2050-20-6	6394
2050-23-9	8387
2050-24-0	1419
2050-32-0	4256
2050-33-1	4255
2050-43-3	56
2050-46-6	1406
2050-47-7	2072
2050-48-8	2238
2050-60-4	5558
2050-61-5	5557
2050-67-1	8010
2050-68-2	3041
2050-69-3	7807
2050-73-9	7811
2050-74-0	7812
2050-76-2	8026
2050-77-3	5053
2050-78-4	8073
2050-87-5	11917
2050-89-7	3030
2050-92-2	8708
2050-95-5	3635
2051-00-5	5555
2051-04-9	5243
2051-10-7	4116
2051-25-4	5023
2051-30-1	8377
2051-31-2	5072
2051-32-3	8459
2051-33-4	6948
2051-49-2	6874
2051-53-8	793
2051-60-7	3023
2051-61-8	3024
2051-62-9	3025
2051-78-7	3587
2051-84-5	783
2051-90-3	1389
2051-97-0	11136
2051-98-1	4
2052-01-9	10209
2052-06-4	642
2052-07-5	3016
2052-14-4	2614
2052-15-5	8904
2055-40-5	1643
2057-43-4	10996
2057-49-0	11007
2065-23-8	230
2065-70-5	7813
2077-13-6	1190
2078-54-8	9226
2079-95-0	11621
2081-44-9	10787
2082-59-9	8858
2082-76-0	5060
2083-91-2	11361
2084-19-7	8845
2084-45-9	1363
2085-88-3	8622
2086-83-1	2467
2090-14-4	4118
2090-89-3	5748
2091-29-4	6669
2093-20-1	5673
2095-57-0	11065
2100-17-6	9001
2101-86-2	982
2101-88-4	977
2103-64-2	11526
2104-64-5	9551
2106-54-9	6910
2108-92-1	4358
2109-66-2	7692
2110-78-3	10280
2111-75-3	4633
2113-47-5	35
2113-51-1	3077
2113-57-7	3017
2113-58-8	3086
2114-00-3	10419
2114-39-8	1087
2114-42-3	4470
2116-62-3	11003
2116-65-6	11006
2121-66-6	10140
2122-70-5	7735
2123-27-5	1373
2123-28-6	1367
2130-23-6	8504
2131-18-2	2089
2131-38-6	8013
2131-43-3	8002
2131-55-7	1177
2132-80-1	3047
2136-81-4	1238
2136-89-2	1237
2138-05-0	2083
2138-22-9	1505
2138-43-4	1535
2141-62-0	10103
2142-01-0	7367
2142-06-5	11163
2142-63-4	5780
2142-73-6	5810
2146-37-4	4813
2146-38-5	4924
2150-18-7	10962
2150-38-1	2579
2155-30-8	10273
2155-42-2	1370
2157-31-5	8792
2157-50-8	10483
2158-55-6	4228
2159-75-3	5123
2162-92-7	6761
2162-98-3	5040
2162-99-4	8371
2163-00-0	6762
2163-44-2	9934
2163-48-6	9981
2164-08-1	4900
2164-09-2	10520
2164-17-2	12007
2164-19-4	4292
2167-39-7	8594
2168-93-6	3422
2170-03-8	6095
2171-96-2	11436
2173-56-0	8913
2174-64-3	1526
2175-91-9	4768
2176-62-7	10995
2176-98-9	11546
2177-32-4	7093
2177-47-1	7276
2180-43-0	6949
2180-68-9	8674
2180-69-0	8675
2182-66-3	11419
2189-60-8	2067
2196-23-8	6529
2197-37-7	8225
2198-20-1	4207
2198-61-0	6897
2198-72-3	6508
2198-75-6	7808
2199-51-1	11115
2199-52-2	11116
2203-34-1	3607
2203-35-2	3603
2203-80-7	7120
2206-38-4	1253
2206-48-6	1250
2206-89-5	10620
2207-01-4	4365
2207-03-6	4370
2207-04-7	4371
2207-27-4	4770
2207-50-3	8574
2210-28-8	10672
2211-67-8	3281
2211-68-9	3727
2211-69-0	3728
2211 70 3	3722
2212-67-1	7661
2213-08-3	332
2213-23-2	6385
2213-32-3	9030
2213-43-6	9624
2214-14-4	1955
2216-12-8	2053
2216-15-1	698
2216-30-0	6386
2216-32-2	6408
2216-33-3	8409
2216-34-4	8410
2216-36-6	8124
2216-67-3	7722
2216-68-4	7721
2216-69-5	7913
2216-70-8	8328
2216-81-1	10264
2216-87-7	11959
2216-92-4	6271
2216-94-6	10741
2217-06-3	2288
2217-07-4	730
2217-14-3	3310
2217-41-6	7731
2217-45-0	847
2217-60-9	1538
2219-66-1	3904
2219-82-1	9332
2224-37-5	8951
2227-13-6	11648
2230-70-8	4619
2231-57-4	4033
2233-00-3	10604
2233-29-6	11024
2234-13-1	7933
2234-16-4	5799
2234-20-0	1626
2237-30-1	2742
2239-78-3	6657

CAS Number	No.
2243-27-8	8144
2243-30-3	833
2243-32-5	2713
2243-35-8	5771
2243-42-7	2714
2243-43-8	3411
2243-53-0	3855
2243-55-2	7167
2243-61-0	7796
2243-62-1	7797
2243-71-2	400
2243-76-7	2648
2243-80-3	7635
2243-81-4	7757
2243-83-6	7750
2243-89-2	11302
2243-94-9	8016
2243-98-3	11984
2244-07-7	11941
2244-16-8	4715
2254-94-6	2867
2257-35-4	10696
2258-58-4	9086
2259-96-3	2851
2270-20-4	2099
2272-45-9	811
2279-76-7	11542
2288-18-8	1967
2291-80-7	3968
2294-43-1	1083
2294-71-5	900
2294-72-6	11987
2294-94-2	1618
2295-31-0	11683
2295-58-1	10489
2303-16-4	5106
2303-17-5	11823
2305-13-7	2153
2306-88-9	8437
2308-38-5	3509
2310-17-0	9514
2310-98-7	9841
2311-59-3	5067
2312-35-8	10507
2313-65-7	6943
2314-78-5	11157
2314-97-8	7559
2315-68-6	2724
2316-50-9	9234
2316-64-5	1843
2319-29-1	5031
2319-61-1	4392
2320-30-1	4589
2321-07-5	11522
2323-81-1	8868
2328-24-7	924
2338-12-7	2873
2344-70-9	2155
2344-71-0	1106
2344-80-1	11382
2344-83-4	11386
2345-26-8	10295
2345-27-9	11633
2345-28-0	8651
2345-61-1	10617
2346-00-1	11672
2346-81-8	6739
2350-19-8	6367
2351-13-5	5229
2351-97-5	9080
2358-84-1	10669
2362-10-9	5230
2362-12-1	9241
2364-59-2	3581
2365-48-2	204
2366-52-1	3378
2367-82-0	2271
2370-88-9	4999
2371-19-9	6520
2373-51-5	7484
2381-31-9	523
2382-43-6	9831
2384-73-8	6606
2384-75-0	6538
2384-85-2	5098
2384-86-3	5099
2384-90-9	6326
2385-70-8	3468
2385-85-5	7644
2388-14-9	1973
2393-23-9	1813
2396-53-4	2080
2396-60-3	5137
2396-63-6	6343
2396-65-8	8115
2396-78-3	7062
2396-83-0	7054
2396-84-1	6700
2398-09-6	4189
2398-10-9	4190
2398-16-5	4182
2398-37-0	1051
2398-65-4	2068
2398-66-5	2007
2398-68-7	1568
2398-81-4	10909
2399-48-6	10695
2400-01-3	1764
2400-02-4	1695
2400-03-5	2170
2400-04-6	1119
2401-22-1	1173
2401-24-3	666
2401-73-2	5232
2402-06-4	4973
2402-79-1	11019
2404-35-5	4222
2408-20-0	10325
2408-37-9	4617
2409-52-1	3329
2409-55-4	9330
2412-73-9	2551
2415-72-7	4984
2417-04-1	3069
2417-90-5	10095
2418-31-7	3787
2419-73-0	3349
2420-16-8	532
2423-01-0	4906
2423-10-1	8289
2424-92-2	5352
2424-98-8	10297
2425-28-7	1844
2425-54-9	11615
2425-66-3	9867
2425-79-8	8601
2426-02-0	7347
2426-08-6	8602
2426-54-2	10631
2430-00-4	96
2430-22-0	8463
2431-24-5	6540
2431-30-3	7117
2431-50-7	3809
2431-52-9	3420
2431-54-1	3810
2432-12-4	9295
2432-63-5	3784
2436-84-2	6330
2436-85-3	7716
2436-92-2	11260
2436-93-3	11264
2436-96-6	3062
2437-03-8	7719
2437-25-4	5303
2437-56-1	11893
2437-73-2	11300
2437-76-5	1065
2437-88-9	2343
2437-95-8	2972
2437-98-1	88
2438-03-1	2723
2438-04-2	2688
2438-12-2	4668
2438-19-9	10641
2438-72-4	880
2439-01-2	4072
2439-10-3	5347
2439-35-2	10633
2439-99-8	6277
2443-03-0	4837
2443-04-1	4836
2443-91-6	9886
2444-37-3	216
2444-89-5	2073
2445-76-3	10265
2445-82-1	2833
2445-83-2	2836
2447-54-3	2432
2450-71-7	10725
2455-24-5	10674
2459-05-4	3791
2459-09-8	10908
2459-24-7	7766
2459-25-8	7776
2462-85-3	8229
2462-94-4	5363
2463-53-8	8179
2463-63-0	6530
2464-37-1	4079
2467-09-6	3098
2467-10-9	8838
2470-68-0	6352
2473-01-0	8123
2476-37-1	5800
2486-07-9	1495
2489-86-3	7954
2492-22-0	8324
2493-02-9	1049
2495-35-4	10688
2497-21-4	7088
2498-75-1	522
2499-95-8	10643
2505-06-8	4246
2506-41-4	7785
2508-29-4	8920
2512-81-4	9648
2514-91-2	11815
2516-34-9	4169
2516-96-3	2545
2517-04-6	495
2517-43-3	3625
2523-37-7	5974
2523-55-9	4297
2523-56-0	4296
2524-04-1	9576
2524-09-6	9579
2524-37-0	2574
2524-52-9	10901
2524-64-3	9575
2525-02-2	1548
2528-38-3	9565
2528-61-2	6527
2531-80-8	10312
2532-58-3	4803
2536-38-1	8250
2545-59-7	257
2547-61-7	7536
2550-04-1	11482
2550-06-3	11458
2550-21-2	6976
2550-22-3	7097
2550-26-7	3678
2550-28-9	7131
2550-36-9	4318
2553-04-0	7623
2553-19-7	11406
2554-06-5	4996
2555-49-9	229
2562-38-1	4839
2563-97-5	98
2565-39-1	6586
2568-33-4	3354
2568-90-3	3398
2568-91-4	10075
2570-12-9	4747
2570-26-5	8641
2571-39-3	7615
2575-20-4	3658
2579-20-6	4362
2579-22-8	10724
2583-25-7	9979
2586-89-2	6611
2591-86-8	9638
2593-15-9	11599
2594-20-9	3956
2594-21-0	3937
2595-97-3	337
2601-89-0	10099
2605-18-7	1940
2612-33-1	9997
2612-46-6	6993
2613-61-8	6447
2613-65-2	4818
2613-66-3	4817
2613-69-6	4858
2613-72-1	4861
2614-88-2	6711
2620-50-0	2426
2621-46-7	1192
2621-78-5	3967
2622-08-4	9596
2622-21-1	4657
2622-60-8	2383
2622-89-1	3130
2623-23-6	4549
2623-50-9	11263
2623-54-3	2923
2623-82-7	8425
2623-84-9	11945
2623-95-2	5061
2624-01-3	6457
2625-30-1	4455
2625-31-2	4840
2627-95-4	5233
2631-37-0	9769
2639-63-6	3528
2641-89-6	1551
2642-63-9	5801
2642-98-0	4113
2648-61-5	5802
2652-13-3	8178
2655-83-6	1157
2657-65-0	3426
2675-77-6	4084
2676-33-7	10269
2678-54-8	5901
2679-87-0	3373
2681-83-6	6875
2681-92-7	8859
2687-25-4	1275
2688-77-9	7379
2688-84-8	834
2689-59-0	7618
2689-88-5	9946
2690-08-6	8417
2706-56-1	10948
2713-09-9	5943
2717-42-2	8009
2717-44-4	7816
2718-25-4	11342
2719-27-9	4330
2719-52-0	1952
2720-41-4	4481
2721-38-2	4796
2722-36-3	2156
2724-59-6	8263
2725-82-8	1036
2734-70-5	703
2736-37-0	10492
2736-40-5	3691
2738-19-4	7040
2739-12-0	792
2741-16-4	1974
2745-25-7	6031
2745-26-8	6026
2747-53-7	9098
2757-90-6	6247
2758-18-1	4945
2759-28-6	9621
2761-24-2	11480
2761-84-4	5339
2765-04-0	11920
2765-11-9	8640
2765-18-6	7957
2768-41-4	3915
2769-64-4	3387
2777-66-4	11991
2778-68-9	2936
2781-85-3	4990
2782-57-2	11838
2782-91-4	11810
2785-74-2	1511
2785-89-9	9357
2790-09-2	1348
2791-29-9	6636
2798-73-4	3891
2804-50-4	10541
2807-30-9	5764
2807-54-7	3781
2808-71-1	4661
2808-76-6	4649
2808-77-7	4651
2808-79-9	4650
2808-80-2	4434
2809-64-5	7997
2809-67-8	8530
2809-69-0	6717
2810-04-0	11739
2810-69-7	2255
2815-34-1	9610
2816-57-1	4586
2817-71-2	10832
2819-48-9	4314
2819-86-5	9443
2820-37-3	10827
2825-00-5	10797
2825-79-8	8296
2834-05-1	11944
2835-21-4	10574
2835-33-8	11014
2835-81-6	3463
2835-96-3	9213
2835-97-4	9209
2836-00-2	9211
2836-03-5	1267
2836-04-6	1268
2837-89-0	5486
2845-89-8	1179
2847-72-5	5056
2848-25-1	1540
2855-08-5	3246
2856-63-5	949
2858-66-4	10485
2859-67-8	11013
2859-68-9	11012
2859-78-1	1016
2867-05-2	2986
2867-47-2	10654
2867-59-6	3596
2867-63-2	4893
2868-37-3	4963
2869-34-3	11870
2870-04-4	1673
2873-18-9	11728
2873-74-7	8806
2873-89-4	661
2873-97-4	10521
2874-74-0	5314
2876-35-9	7829
2876-44-0	7938
2876-45-1	7896
2876-46-2	7901
2876-51-9	7937
2876-52-0	7895
2876-53-1	7900
2876-62-2	3674
2876-63-3	10473
2882-98-6	4841
2883-02-5	4457
2883-07-0	4341
2883-08-1	4447
2883-12-7	1255
2883-45-6	6337
2885-00-9	8244
2886-52-4	10445
2887-61-8	3665
2890-62-2	5843
2892-18-4	7100
2896-60-8	1518
2899-90-3	8950
2902-69-4	5871
2902-96-7	10394
2905-56-8	9698
2912-62-1	973
2917-26-2	6643
2917-47-7	10410
2918-13-0	6583
2919-05-3	9118
2919-23-5	4196
2921-88-2	4089
2922-51-2	6320
2923-68-4	8442
2933-74-6	5745
2935-44-6	6799

CAS NUMBER INDEX OF ORGANIC COMPOUNDS (continued)

2935-90-2 : 10287
2941-19-7 : 1830
2941-20-0 : 1809
2941-55-1 : 5930
2943-70-6 : 5231
2944-49-2 : 1916
2947-60-6 : 961
2947-61-7 : 962
2955-65-9 : 8395
2955-88-6 : 11165
2959-96-8 : 10781
2961-47-9 : 10880
2969-81-5 : 3470
2971-79-1 : 9644
2973-00-4 : 7497
2973-27-5 : 11224
2977-31-3 : 1617
2978-58-7 : 3896
2979-19-3 : 4587
2980-71-4 : 8185
2983-26-8 : 5523
2983-37-1 : 6885
2984-40-9 : 3806
2987-53-3 : 819
2987-77-1 : 11485
2996-92-1 : 11488
2998-18-7 : 10665
3000-79-1 : 9680
3001-72-7 : 11190
3002-24-2 : 6804
3007-91-8 : 7770
3007-97-4 : 7765
3008-39-7 : 4225
3010-02-4 : 280
3010-04-6 : 273
3012-37-1 : 11710
3016-19-1 : 8487
3017-67-2 : 5086
3017-68-3 : 3714
3017-70-7 : 3718
3017-71-8 : 3713
3017-95-6 : 9840
3017-96-7 : 9838
3018-09-5 : 5881
3018-12-0 : 278
3019-04-3 : 10454
3019-20-3 : 1980
3019-25-8 : 5792
3022-41-1 : 4209
3024-72-4 : 2894
3027-13-2 : 10460
3031-05-8 : 8011
3031-66-1 : 7118
3031-73-0 : 7180
3031-74-1 : 7179
3032-32-4 : 2491
3034-53-5 : 11668
3035-75-4 : 11877
3036-66-6 : 3168
3040-44-6 : 9662
3041-40-5 : 9916
3042-22-6 : 11135
3042-50-0 : 1486
3043-33-2 : 8672
3047-38-9 : 4907
3051-09-0 : 11097
3054-92-0 : 8956
3054-95-3 : 10560
3055-14-9 : 1949
3058-47-7 : 2043
3060-89-7 : 7656
3062-81-5 : 3122
3066-71-5 : 10626
3068-00-6 : 3459
3073-66-3 : 4486
3073-92-5 : 3421
3074-61-1 : 4935
3074-64-4 : 6542
3074-71-3 : 6384
3074-75-7 : 6828
3074-76-8 : 6826
3074-77-9 : 6827
3075-84-1 : 3096
3077-16-5 : 7698
3081-24-1 : 9512
3083-23-6 : 8628
3085-53-8 : 5997
3087-62-5 : 6211
3088-41-3 : 9703
3096-47-7 : 5976
3096-52-4 : 5977
3101-60-8 : 8611
3102-33-8 : 9105
3112-88-7 : 2112
3113-98-2 : 4421
3118-97-6 : 8028
3121-83-3 : 9993
3128-06-1 : 6905
3128-07-2 : 6469
3129-39-3 : 7846
3129-90-6 : 7399
3132-99-8 : 527
3140-92-9 : 11743
3140-93-0 : 11742
3141-12-6 : 440
3141-24-0 : 11790
3141-27-3 : 11744
3142-58-3 : 8799
3142-66-3 : 8972
3142-84-5 : 6608
3144-16-9 : 2929
3144-54-5 : 6962
3147-55-5 : 2558
3152-68-9 : 9110
3153-36-4 : 3485
3153-37-5 : 3489
3156-70-5 : 10589
3156-73-8 : 10396
3156-74-9 : 3641
3163-07-3 : 1541
3163-27-7 : 7742
3167-49-5 : 10897
3172-52-9 : 11748
3173-53-3 : 4416
3173-79-3 : 8597
3174-67-2 : 8785
3174-83-2 : 2157
3175-23-3 : 10174
3176-62-3 : 7240
3178-22-1 : 4369
3178-23-2 : 4429
3178-24-3 : 4466
3178-29-8 : 6434
3179-63-3 : 10375
3180-09-4 : 9246
3187-94-8 : 6073
3188-13-4 : 5478
3189-22-8 : 7311
3194-15-8 : 10438
3195-24-2 : 9944
3196-15-4 : 3473
3199-61-9 : 2440
3202-84-4 : 7695
3208-16-0 : 6105
3208-22-8 : 6173
3209-22-1 : 1391
3211-48-1 : 4183
3212-75-7 : 4753
3213-79-4 : 1269
3217-86-5 : 9802
3218-02-8 : 4418
3221-61-2 : 8408
3228-04-4 : 9416
3228-99-7 : 9895
3229-00-3 : 9884
3230-23-7 : 9670
3232-84-6 : 11850
3234-64-8 : 9121
3238-38-8 : 9474
3238-40-2 : 6081
3238-62-8 : 9677
3238-75-3 : 5712
3240-09-3 : 7099
3243-42-3 : 7952
3244-88-0 : 2218
3252-43-5 : 277
3254-93-1 : 7228
3256-88-0 : 10990
3260-87-5 : 9264
3261-62-9 : 1580
3268-49-3 : 9786
3268-79-9 : 11700
3277-26-7 : 5237
3278-46-4 : 10332
3280-08-8 : 1616
3280-51-1 : 6708
3282-30-2 : 10503
3283-12-3 : 9528
3288-04-8 : 1793
3289-28-9 : 4334
3290-06-0 : 1358
3296-05-7 : 978
3299-38-5 : 3656
3319-15-1 : 1895
3320-83-0 : 1176
3320-86-3 : 1788
3320-87-4 : 1789
3321-50-4 : 4386
3329-48-4 : 7073
3329-56-4 : 3377
3333-23-1 : 7764
3333-44-6 : 6138
3333-52-6 : 3292
3337-71-1 : 474
3338-15-6 : 10444
3344-70-5 : 5296
3347-22-6 : 8056
3350-30-9 : 4725
3350-78-5 : 3888
3351-28-8 : 4117
3351-86-8 : 4055
3352-87-2 : 5290
3354-58-3 : 9427
3355-17-7 : 3900
3358-28-9 : 6174
3368-16-9 : 941
3373-53-3 : 10767
3375-23-3 : 8036
3377-71-7 : 7321
3377-86-4 : 6733
3377-87-5 : 6734
3383-96-8 : 1
3384-04-1 : 1233
3386-33-2 : 8236
3393-64-4 : 3664
3393-78-0 : 2284
3395-79-7 : 1443
3395-91-3 : 10212
3399-21-1 : 4357
3400-45-1 : 4781
3402-74-2 : 2006
3402-76-4 : 7945
3404-57-7 : 6558
3404-58-8 : 7033
3404-61-3 : 7037
3404-63-5 : 9060
3404-67-9 : 6838
3404-71-5 : 9043
3404-72-6 : 9029
3404-73-7 : 9031
3404-77-1 : 7016
3404-78-2 : 7024
3404-80-6 : 6837
3411-95-8 : 9217
3416-24-8 : 6233
3419-71-4 : 4689
3424-93-9 : 618
3426-01-5 : 10843
3430-27-1 : 10870
3433-37-2 : 9674
3433-80-5 : 999
3435-28-7 : 11048
3437-89-6 : 8987
3437-95-4 : 11767
3438-46-8 : 11071
3439-97-2 : 11443
3440-02-6 : 11415
3445-11-2 : 11184
3446-98-8 : 11147
3452-07-1 : 5362
3452-09-3 : 8196
3452-97-9 : 6951
3454-07-7 : 1632
3457-58-7 : 2003
3458-28-4 : 7425
3460-18-2 : 1310
3460-20-6 : 2315
3460-24-0 : 1295
3460-67-1 : 10904
3463-26-1 : 10837
3463-40-9 : 2316
3463-92-1 : 4064
3468-99-3 : 939
3476-89-9 : 11339
3478-88-4 : 7594
3497-00-5 : 9552
3508-00-7 : 6307
3508-78-9 : 8805
3508-94-9 : 930
3510-70-1 : 11496
3512-80-9 : 10864
3521-91-3 : 6566
3522-94-9 : 6861
3524-62-7 : 5839
3524-73-0 : 7039
3524-75-2 : 4848
3528-17-4 : 2869
3535-67-9 : 1925
3536-29-6 : 2158
3538-65-6 : 3529
3540-95-2 : 9658
3544-25-0 : 946
3547-04-4 : 1678
3552-01-0 : 4538
3554-74-3 : 9716
3555-84-8 : 7612
3558-69-8 : 10947
3567-38-2 : 1877
3570-55-6 : 5652
3570-58-9 : 5699
3574-40-1 : 3157
3574-43-4 : 5682
3581-87-1 : 11678
3581-91-7 : 11675
3582-71-6 : 11913
3586-12-7 : 835
3586-14-9 : 2013
3587-57-3 : 9858
3587-60-8 : 1181
3588-30-5 : 3169
3593-00-8 : 6563
3594-37-4 : 3364
3598-14-9 : 287
3600-24-6 : 11916
3607-78-1 : 10052
3609-53-8 : 2479
3610-02-4 : 2403
3613-42-1 : 367
3615-21-2 : 2377
3616-56-6 : 5416
3616-57-7 : 5415
3634-92-2 : 8273
3637-01-2 : 5811
3637-61-4 : 4825
3638-35-5 : 4983
3638-64-0 : 5911
3643-64-9 : 4688
3657-07-6 : 245
3658-79-5 : 8755
3669-80-5 : 11948
3673-79-8 : 10242
3674-09-7 : 10248
3674-10-0 : 10363
3674-13-3 : 10240
3674-75-7 : 9179
3675-68-1 : 3442
3675-69-2 : 3443
3678-62-4 : 10919
3680-02-2 : 5909
3681-78-5 : 5320
3682-35-7 : 11843
3683-19-0 : 7043
3683-22-5 : 7042
3689-24-5 : 11716
3690-04-8 : 9795
3697-27-6 : 4115
3699-01-2 : 2019
3710-30-3 : 8319
3710-43-8 : 6088
3710-84-7 : 5429
3711-34-0 : 7137
3711-37-3 : 7140
3721-95-7 : 4174
3722-73-4 : 2781
3722-74-5 : 2780
3725-05-1 : 6714
3725-07-3 : 8340
3725-30-2 : 4750
3728-43-6 : 11490
3730-60-7 : 6923
3731-51-9 : 10971
3731-52-0 : 10972
3731-53-1 : 10973
3735-90-8 : 3977
3741-00-2 : 4846
3741-38-6 : 5200
3742-38-9 : 4926
3742-42-5 : 4662
3743-22-4 : 9317
3744-02-3 : 9109
3757-32-2 : 8910
3761-94-2 : 4230
3763-39-1 : 4991
3763-55-1 : 4062
3764-01-0 : 11075
3768-43-2 : 9722
3768-60-3 : 11583
3768-63-6 : 11569
3769-23-1 : 7038
3769-41-3 : 9452
3773-93-1 : 5208
3774-03-6 : 5197
3779-29-1 : 4180
3780-58-3 : 6894
3788-32-7 : 4838
3790-71-4 : 5334
3810-26-2 : 4951
3810-74-0 : 11561
3835-64-1 : 1864
3837-38-5 : 404
3839-50-7 : 4765
3840-31-1 : 1914
3848-12-2 : 126
3849-33-0 : 10049
3850-30-4 : 3217
3852-09-3 : 10289
3853-80-3 : 10202
3856-25-5 : 11861
3857-25-8 : 6122
3862-11-1 : 9338
3862-12-2 : 9335
3864-18-4 : 3052
3867-18-3 : 733
3875-51-2 : 4828
3875-52-3 : 4805
3875-68-1 : 12051
3877-15-4 : 10088
3877-19-8 : 7996
3878-55-5 : 3336
3884-88-6 : 8246
3910-35-8 : 7263
3913-71-1 : 5080
3913-85-7 : 5088
3916-64-1 : 6339
3917-15-5 : 10565
3920-53-4 : 6453
3937-45-9 : 4422
3937-46-0 : 4423
3937-56-2 : 8136
3938-95-2 : 10259
3938-96-3 : 206
3943-74-6 : 2627
3943-77-9 : 2578
3944-36-3 : 10390
3955-26-8 : 2324
3958-03-0 : 11781
3958-38-1 : 4727
3958-57-4 : 1067
3959-07-7 : 1799
3962-77-4 : 1926
3963-78-8 : 382
3964-56-5 : 9232
3964-58-7 : 2540
3964-61-2 : 9254
3968-85-2 : 1953
3970-35-2 : 2544
3970-62-5 : 8929
3971-28-6 : 4363
3972-56-3 : 1142
3972-65-4 : 1023
3973-62-4 : 9694
3974-99-0 : 254
3976-34-9 : 3056
3976-36-1 : 3095
3978-81-2 : 10944

CAS NUMBER INDEX OF ORGANIC COMPOUNDS (continued)

CAS No.	No.
6292-82-6	9710
6294-31-1	6855
6294-89-9	7149
6298-37-9	11332
6302-94-9	5264
6304-18-3	10985
6306-30-5	6984
6308-98-1	1585
6309-50-8	5308
6310-09-4	5791
6310-21-0	714
6311-60-0	1309
6314-28-9	2400
6315-03-3	2695
6315-60-2	330
6315-96-4	10474
6319-21-7	5135
6319-40-0	2529
6320-03-2	2293
6320-40-7	2314
6322-01-6	10339
6322-49-2	3648
6326-44-9	9953
6338-45-0	7201
6341-60-2	9970
6341-97-5	9288
6344-28-1	5786
6344-72-5	11338
6346-09-4	3214
6351-10-6	7284
6361-23-5	535
6372-01-6	3831
6374-70-5	5146
6378-11-6	4087
6378-65-0	6890
6380-21-8	9456
6380-24-1	1451
6380-28-5	9419
6380-34-3	464
6382-13-4	8270
6402-36-4	5336
6414-69-3	10283
6423-02-5	6758
6423-29-6	7804
6436-90-4	6272
6443-91-0	5944
6443-92-1	6535
6453-99-2	7642
6482-24-2	5465
6482-34-4	3995
6493-77-2	4640
6493-83-0	10457
6496-48-6	11176
6498-47-1	3868
6513-13-9	12039
6526-72-3	1977
6531-35-7	403
6538-02-9	5379
6553-48-6	4187
6556-12-3	6238
6568-58-7	4731
6573-52-0	4729
6579-55-1	10383
6592-85-4	5217
6597-78-0	7653
6600-40-4	8218
6607-66-5	3456
6609-56-9	2760
6609-57-0	2754
6617-04-5	4311
6622-76-0	3851
6624-76-6	8100
6626-15-9	1502
6626-32-0	9347
6627-55-0	9236
6627-72-1	2942
6627-74-3	4667
6628-00-8	4304
6628-04-2	11210
6628-06-4	9429
6628-18-8	5452
6628-21-3	10246
6628-77-9	10865
6628-79-1	8896
6628-80-4	4505
6628-97-3	7866
6629-96-5	5192
6630-33-7	526
6639-30-1	2336
6639-82-3	11336
6640-27-3	9261
6640-77-3	2685
6641-83-4	8895
6669-13-2	1458
6682-71-9	7253
6688-11-5	4738
6693-29-4	4302
6693-30-7	4301
6693-31-8	4300
6697-12-7	6664
6703-78-2	6298
6703-80-6	6297
6703-82-8	4819
6704-31-0	8596
6705-31-3	10808
6709-22-4	4913
6709-39-3	6331
6728-26-3	6996
6734-98-1	7007
6738-23-4	1917
6739-34-0	4337
6740-88-1	4579
6742-54-7	2374
6748-95-4	436
6753-98-6	5006
6765-39-5	6322
6776-19-8	3840
6780-13-8	5513
6780-49-0	7567
6781-42-6	5865
6781-98-2	1139
6784-18-5	4890
6785-23-5	4863
6789-80-6	6998
6789-88-4	2610
6789-94-2	9626
6790-27-8	5345
6790-47-2	4307
6795-75-1	5935
6802-75-1	9966
6812-38-0	4411
6812-39-1	4822
6812-78-8	8516
6814-58-0	4083
6818-07-1	6901
6829-40-9	9924
6829-41-0	9958
6831-89-6	10831
6832-98-0	6630
6836-38-0	5324
6843-66-9	11413
6846-11-3	7221
6848-84-6	8769
6849-18-9	9082
6850-36-8	4287
6850-57-3	1812
6852-54-6	5437
6861-64-9	7
6871-44-9	4730
6873-15-0	11200
6874-98-2	11346
6876-23-9	4366
6886-16-4	6921
6890-03-5	4520
6891-45-8	9250
6894-69-5	4912
6897-76-3	8014
6897-88-7	8015
6912-05-6	4923
6917-76-6	11377
6921-35-3	8593
6921-40-0	7738
6921-64-8	5826
6923-20-2	10549
6923-22-4	7663
6931-71-1	6800
6933-10-4	651
6938-94-9	6782
6939-75-9	8911
6940-09-6	8275
6940-76-7	9857
6940-78-9	3233
6941-69-1	129
6942-99-0	1313
6943-58-4	10356
6946-29-8	2500
6946-35-6	2668
6946-88-9	3338
6950-43-2	2530
6950-84-1	12022
6951-08-2	2419
6952-59-6	2745
6959-06-4	3624
6963-44-6	8791
6964-04-1	6401
6967-43-7	3623
6969-16-0	11894
6971-51-3	1890
6973-79-1	10206
6974-10-3	7361
6974-77-2	9842
6975-60-6	10439
6975-92-4	7015
6975-98-0	5054
6975-99-1	5346
6976-00-7	11963
6976-72-3	6889
6982-25-8	3347
6983-79-5	5108
6986-48-7	5610
6995-30-8	7905
6995-79-5	4374
6996-92-5	2174
6998-82-9	4352
7005-72-3	1221
7012-16-0	5972
7013-11-8	3744
7021-04-7	889
7035-03-2	957
7036-98-8	8191
7040-43-9	6090
7044-96-4	7654
7045-89-8	11903
7058-01-7	4448
7065-46-5	3690
7069-38-7	10548
7081-78-9	5473
7086-07-9	10147
7087-68-5	9811
7094-26-0	4485
7094-27-1	4487
7094-34-0	7591
7098-07-9	7207
7098-21-7	11918
7101-96-4	11215
7116-86-1	7021
7119-92-8	5431
7119-94-0	9687
7132-64-1	8646
7133-68-8	3358
7137-56-6	1251
7144-05-0	9673
7145-20-2	7022
7146-60-3	8374
7148-74-5	10495
7149-70-4	1066
7149-75-9	674
7152-15-0	8897
7152-88-7	2628
7153-14-2	4576
7154-66-7	2887
7154-73-6	11164
7154-75-8	9135
7154-79-2	8839
7154-80-5	6450
7163-25-9	7773
7164-98-9	7212
7168-55-0	3048
7169-34-8	2462
7170-38-9	10320
7187-01-1	10585
7193-78-4	5844
7200-25-1	437
7205-90-5	1136
7208-05-1	8577
7208-92-6	5554
7209-38-3	9614
7214-50-8	4712
7216-56-0	8488
7220-26-0	6818
7220-81-7	4757
7225-64-1	6313
7225-66-3	11878
7235-40-7	4051
7236-47-7	3703
7239-24-9	9829
7242-17-3	3780
7250-95-5	7712
7251-61-8	11337
7252-83-7	5460
7283-69-4	3767
7283-70-7	3766
7283-96-7	11735
7287-19-6	9770
7287-81-2	1858
7291-33-0	97
7297-25-8	3432
7300-03-0	6555
7307-55-3	11937
7311-34-4	546
7314-06-9	732
7314-37-6	11620
7315-32-4	2425
7318-67-4	6684
7319-00-8	6683
7321-48-4	8500
7321-55-3	9913
7328-33-8	5026
7332-93-6	8695
7335-01-5	9672
7335-02-6	9690
7335-06-0	11166
7335-07-1	11173
7335-25-3	2537
7335-26-4	2669
7335-27-5	2539
7344-34-5	7278
7351-74-8	7803
7357-93-9	3797
7359-72-0	2334
7367-38-6	8184
7369-50-8	1693
7371-67-7	4181
7372-85-2	3055
7372-92-1	7280
7379-12-6	6979
7383-20-2	8208
7383-90-6	3058
7385-78-6	9032
7385-82-2	7044
7391-39-1	6881
7398-82-5	989
7399-49-7	1984
7399-50-0	10967
7400-08-0	10644
7403-66-9	42
7415-31-8	3748
7417-19-8	1599
7417-48-3	8666
7423-69-0	7018
7425-45-8	3483
7425-47-0	3928
7425-48-1	3484
7428-48-0	8261
7433-56-9	5084
7433-78-5	5085
7439-33-0	4185
7443-55-2	4534
7446-43-7	9814
7459-46-3	5654
7462-74-0	9792
7466-38-8	5133
7474-10-4	4720
7474-83-1	7844
7488-99-5	4054
7493-82-5	6470
7495-97-8	9526
7498-57-9	7737
7499-65-2	7744
7521-80-4	11370
7523-44-6	3865
7524-63-2	7988
7525-62-4	1631
7526-26-3	9541
7529-63-7	9697
7539-25-5	364
7554-65-6	10836
7564-63-8	1630
7567-22-8	4743
7568-92-5	1593
7568-93-6	1839
7570-26-5	5552
7570-98-1	7951
7572-29-4	5940
7583-53-1	9678
7611-43-0	472
7611-86-1	3721
7611-87-2	3720
7614-93-9	2022
7615-57-8	11836
7616-22-0	2790
7619-08-1	8227
7623-01-0	4997
7623-10-1	10499
7623-11-2	3683
7623-13-4	10501
7642-04-8	8493
7642-09-3	7003
7642-10-6	6537
7642-15-1	8497
7658-08-4	6235
7661-39-4	11222
7661-55-4	11280
7665-72-7	8610
7667-55-2	4491
7667-58-5	4494
7667-59-6	4493
7667-60-9	4495
7681-84-7	6049
7683-59-2	1522
7688-21-3	7001
7693-52-9	9244
7696-51-7	11129
7697-46-3	7637
7700-17-6	4137
7712-60-9	9026
7712-66-5	11814
7731-28-4	4535
7731-29-5	4536
7735-33-3	8869
7740-33-2	3589
7745-91-7	649
7745-93-9	1072
7756-94-7	10582
7766-23-6	9238
7771-44-0	5360
7773-60-6	9358
7778-87-2	6471
7779-80-8	6467
7781-98-8	2620
7782-21-0	943
7785-26-4	2971
7786-34-7	7659
7786-44-9	8112
7787-72-6	7792
7789-89-1	10171
7789-90-4	9790
7789-92-6	10183
7789-99-3	8945
9005-79-2	6274
10008-73-8	6148
10021-92-8	4353
10022-60-3	6780
10024-74-5	1822
10031-82-0	554
10031-87-5	185
10042-29-2	268
10048-32-5	10791
10061-01-5	10551
10061-02-6	10550
10071-60-0	9020
10075-38-4	3749
10088-95-6	10785
10124-73-9	7005
10130-87-7	2248
10130-91-3	4566
10136-52-4	6324
10138-10-0	3572
10138-89-3	3455
10140-89-3	9781
10143-54-1	10107
10143-60-9	6431
10147-36-1	10133
10151-95-8	80
10152-76-8	10581

60-87-9	: 10731
60-99-3	: 6627
66-08-2	: 873
93-50-7	: 9406
03-08-4	: 537
03-28-8	: 5322
03-30-2	: 5323
03-58-4	: 9969
10-64-7	: 2912
12-74-5	: 3960
20-20-9	: 9659
22-95-4	: 2361
26-29-6	: 6961
29-10-4	: 9142
30-68-9	: 10475
33-13-3	: 5316
36-47-2	: 2795
40-08-1	: 8030
50-47-2	: 8949
57-92-8	: 7318
259-22-0	: 2670
265-92-6	: 7441
273-90-2	: 10991
276-09-2	: 3835
283-15-5	: 9256
289-68-6	: 11885
292-98-5	: 2963
293-06-8	: 2952
297-09-3	: 8541
297-38-8	: 11571
299-30-6	: 6867
302-94-0	: 3772
307-61-6	: 3550
309-50-9	: 2959
311-74-7	: 1622
311-84-9	: 5105
312-83-1	: 16
315-89-6	: 9685
316-00-4	: 7700
317-10-9	: 10353
321-71-8	: 9078
323-40-7	: 3471
324-58-0	: 9692
327-08-9	: 10204
334-26-6	: 2926
339-67-0	: 6602
340-91-7	: 1791
342-83-3	: 10418
343-99-4	: 11246
373-81-6	: 2957
394-35-1	: 2378
403-60-8	: 3451
410-35-2	: 2448
420-90-3	: 6712
420-91-4	: 6715
424-29-0	: 3341
428-64-5	: 3272
435-81-1	: 8419
436-75-6	: 794
439-23-3	: 2186
443-65-9	: 10610
0453-86-8	: 11340
0463-48-6	: 3740
0473-14-0	: 3871
0476-95-6	: 10564
0470-00-1	: 2002
0478-12-3	: 9083
0486-19-8	: 11869
0487-71-5	: 3887
0487-96-4	: 4871
0488-25-2	: 4539
0488-94-5	: 8436
0489-97-1	: 4345
0495-45-1	: 10132
0498-35-8	: 4359
0500-57-9	: 11297
0503-96-5	: 3249
0508-09-5	: 9154
0519-06-9	: 1117
0519-33-2	: 5093
0521-91-2	: 2101
0522-26-6	: 11955
0524-08-0	: 9018
0528-67-3	: 4469
0544-63-5	: 3838
0568-38-4	: 1683
0574-17-1	: 2030
10574-36-4	: 7041
10574-37-5	: 9034
10575-87-8	: 6378
10577-44-3	: 1941
10588-10-0	: 8900
10602-34-3	: 3758
10604-59-8	: 7307
10605-21-7	: 3990
10606-73-2	: 891
11024-24-1	: 6195
11028-42-5	: 4066
12069-69-1	: 511
12122-67-7	: 12067
12321-04-9	: 5961
12427-38-2	: 7420
13012-59-4	: 9864
13014-24-9	: 1401
13019-20-0	: 6522
13019-22-2	: 5092
13023-00-2	: 3500
13024-80-1	: 1750
13027-43-5	: 1650
13030-26-7	: 2778
13038-21-6	: 8190
13045-99-3	: 3798
13047-06-8	: 7600
13054-87-0	: 3595
13057-17-5	: 7472
13057-19-7	: 7519
13063-43-9	: 9912
13065-07-1	: 7989
13071-79-9	: 11592
13073-29-5	: 9423
13073-35-3	: 991
13086-84-5	: 9531
13099-50-8	: 1227
13102-31-3	: 9449
13105-52-7	: 3934
13114-62-0	: 12023
13116-53-5	: 10145
13116-60-4	: 10143
13121-70-5	: 5011
13129-23-2	: 6063
13138-51-7	: 3429
13147-06-3	: 11639
13151-04-7	: 6546
13151-05-8	: 6545
13151-17-2	: 7045
13151-34-3	: 5055
13153-11-2	: 11694
13163-73-0	: 4604
13165-77-0	: 2181
13172-91-3	: 6557
13176-46-0	: 3478
13194-48-4	: 5934
13195-80-7	: 10542
13206-46-7	: 9776
13222-26-9	: 10498
13249-58-6	: 1030
13250-46-9	: 303
13254-34-7	: 6479
13258-63-4	: 10949
13269-52-8	: 7002
13279-87-3	: 3448
13290-74-9	: 1205
13313-91-2	: 261
13325-10-5	: 3597
13327-31-6	: 11273
13329-40-3	: 5837
13347-42-7	: 9255
13351-61-6	: 2151
13351-73-0	: 2872
13356-08-6	: 5950
13360-45-7	: 4076
13360-61-7	: 8653
13360-63-9	: 3218
13360-65-1	: 10809
13361-35-8	: 143
13361-36-9	: 137
13361-63-2	: 9129
13362-52-2	: 11876
13367-29-8	: 2993
13388-93-7	: 5846
13389-42-9	: 8492
13398-94-2	: 1606
13402-32-9	: 1308
13402-35-2	: 4746
13402-36-3	: 4748
13402-37-4	: 4744
13403-37-7	: 8605
13417-43-1	: 3735
13423-15-9	: 6172
13436-48-1	: 7239
13444-24-1	: 9715
13463-39-3	: 8068
13463-40-6	: 7339
13466-40-5	: 8657
13466-41-6	: 11037
13471-42-6	: 6915
13475-76-8	: 5272
13475-77-9	: 5354
13475-78-0	: 6416
13475-79-1	: 8766
13475-81-5	: 6844
13475-82-6	: 6433
13481-70-4	: 9294
13482-62-7	: 9524
13491-79-7	: 4521
13505-34-5	: 6403
13511-38-1	: 10219
13513-82-1	: 1893
13529-27-6	: 6084
13532-18-8	: 10311
13545-04-5	: 3317
13552-31-3	: 9984
13556-55-3	: 7985
13556-58-6	: 7888
13602-13-6	: 3747
13603-04-8	: 11150
13603-21-9	: 9636
13618-93-4	: 7325
13632-93-4	: 1416
13632-94-5	: 1420
13632-95-6	: 1418
13643-05-5	: 8667
13665-04-8	: 5798
13672-18-9	: 7730
13678-54-1	: 11764
13679-46-4	: 6128
13679-70-4	: 11736
13679-74-8	: 5856
13684-56-5	: 5104
13684-63-4	: 9198
13688-75-0	: 11460
13698-16-3	: 3940
13698-87-8	: 1326
13706-86-0	: 6809
13721-54-5	: 7103
13732-80-4	: 1417
13794-14-4	: 3580
13799-90-1	: 1349
13808-64-5	: 10817
13811-01-3	: 4274
13828-34-7	: 4441
13831-30-6	: 101
13835-81-9	: 11467
13855-77-1	: 3964
13862-07-2	: 9665
13865-48-0	: 2018
13877-91-3	: 8485
13898-58-3	: 2514
13898-73-2	: 4678
13905-48-1	: 4320
13909-73-4	: 5874
13937-08-1	: 9957
13939-06-5	: 7662
13956-29-1	: 1537
13963-35-4	: 2017
13979-28-7	: 3836
13984-57-1	: 6906
13993-65-2	: 9500
13997-70-1	: 1232
13997-73-4	: 9271
14007-64-8	: 898
14010-23-2	: 6315
14021-92-2	: 5048
14045-26-2	: 9671
14062-18-1	: 923
14062-78-3	: 627
14064-10-9	: 9931
14064-43-8	: 1223
14064-68-7	: 2768
14072-86-7	: 4656
14073-97-3	: 4606
14077-58-8	: 298
14090-87-0	: 8393
14092-20-7	: 7106
14108-88-4	: 7993
14116-69-9	: 6234
14122-00-0	: 9280
14159-48-9	: 10130
14167-59-0	: 11650
14179-94-3	: 9126
14180-63-3	: 10596
14181-05-6	: 11145
14188-79-5	: 7794
14214-31-4	: 10538
14222-60-7	: 10886
14248-22-7	: 2129
14252-42-7	: 5450
14255-23-3	: 7023
14272-54-9	: 9115
14273-06-4	: 10732
14282-76-9	: 11729
14287-61-7	: 3507
14296-80-1	: 4172
14303-70-9	: 11629
14304-30-4	: 5087
14315-14-1	: 2402
14321-27-8	: 1810
14331-54-5	: 11055
14352-61-5	: 4310
14371-10-9	: 10518
14379-80-7	: 6266
14400-94-3	: 9467
14401-73-1	: 5797
14411-75-7	: 1471
14436-32-9	: 5091
14444-77-0	: 1409
14446-69-6	: 9653
14452-30-3	: 5836
14459-29-1	: 9735
14472-14-1	: 9242
14476-30-3	: 10162
14476-37-0	: 11960
14484-64-1	: 5957
14490-79-0	: 11608
14548-31-3	: 7105
14548-46-0	: 7639
14558-14-6	: 8003
14575-93-0	: 2965
14595-35-8	: 3783
14596-92-0	: 6682
14602-82-1	: 10161
14620-52-1	: 5297
14620-53-2	: 11617
14620-55-4	: 8237
14630-40-1	: 11434
14631-45-9	: 10102
14660-52-7	: 8861
14676-29-0	: 6409
14685-29-1	: 4995
14686-13-6	: 6534
14686-14-7	: 6536
14689-97-5	: 5592
14697-46-2	: 8851
14716-89-3	: 7404
14719-83-6	: 2548
14720-74-2	: 6131
14721-66-5	: 6658
14737-94-1	: 10622
14752-60-4	: 9124
14752-61-5	: 9133
14752-75-1	: 1743
14804-32-1	: 1682
14804-59-2	: 10510
14813-68-4	: 8345
14845-35-3	: 4706
14845-55-7	: 4887
14847-51-9	: 9237
14850-22-7	: 8495
14850-23-8	: 8496
14852-31-4	: 6660
14869-39-7	: 5814
14919-01-8	: 8494
14924-53-9	: 4175
14938-35-3	: 9444
14963-40-7	: 4946
14979-39-6	: 6501
14996-78-2	: 4236
15014-25-2	: 9927
15068-90-3	: 2950
15104-61-7	: 10126
15111-56-5	: 4639
15116-17-3	: 1534
15121-84-3	: 1612
15150-39-7	: 140
15164-44-0	: 582
15192-80-0	: 8489
15220-85-6	: 10580
15232-76-5	: 8531
15232-96-9	: 1247
15240-91-2	: 1800
15250-29-0	: 10670
15268-49-2	: 10086
15299-99-7	: 8067
15301-40-3	: 11293
15320-72-6	: 3833
15329-10-9	: 4708
15333-22-9	: 3829
15339-36-3	: 7421
15356-70-4	: 4548
15358-48-2	: 7574
15361-99-6	: 2277
15409-60-6	: 4591
15411-43-5	: 734
15430-98-5	: 8654
15457-05-3	: 5980
15466-93-0	: 4511
15466-94-1	: 4510
15469-77-9	: 5089
15489-90-4	: 5959
15501-33-4	: 10056
15519-73-0	: 3068
15521-18-3	: 10374
15543-70-1	: 4965
15570-12-4	: 2299
15600-01-8	: 10053
15606-95-8	: 439
15644-80-1	: 11316
15647-08-2	: 9589
15647-11-7	: 4502
15657-96-2	: 10091
15679-02-4	: 4188
15706-73-7	: 3545
15721-05-8	: 4992
15726-15-5	: 6525
15727-65-8	: 7892
15764-04-2	: 8052
15764-16-6	: 547
15797-21-4	: 4795
15802-97-8	: 10841
15802-99-0	: 10839
15825-70-4	: 7424
15826-80-9	: 2981
15840-60-5	: 7047
15843-48-8	: 5568
15854-58-7	: 10647
15869-80-4	: 6407
15869-85-9	: 8143
15869-86-0	: 8404
15869-87-1	: 8373
15869-89-3	: 8376
15869-92-8	: 8380
15869-93-9	: 8381
15869-94-0	: 8382
15869-95-1	: 8383
15869-96-2	: 8384
15870-10-7	: 6543
15874-03-0	: 6637
15877-57-3	: 8694
15890-36-5	: 4541
15890-40-1	: 4859
15901-49-2	: 5003
15909-76-9	: 10254
15918-08-8	: 6426
15922-53-9	: 11849
15932-80-6	: 4607
15950-66-0	: 9481
15953-45-4	: 10821
15953-73-8	: 10819
15971-29-6	: 7755
15972-60-8	: 322
15980-15-1	: 8569
16003-14-8	: 6160
16009-13-5	: 5958
16020-17-0	: 8001
16021-20-8	: 1702

16066-09-4 : 11647
16096-32-5 : 7315
16106-59-5 : 7020
16109-68-5 : 5120
16133-49-6 : 777
16136-84-8 : 10534
16136-85-9 : 10533
16147-07-2 : 1446
16177-46-1 : 4186
16179-36-5 : 8885
16183-46-3 : 3666
16188-57-1 : 9374
16197-90-3 : 3822
16245-79-7 : 831
16246-07-4 : 3947
16292-88-9 : 776
16302-35-5 : 10778
16306-39-1 : 9184
16308-92-2 : 1860
16315-07-4 : 1922
16326-97-9 : 4806
16326-98-0 : 4807
16357-59-8 : 11231
16369-05-4 : 3599
16369-17-8 : 8922
16369-21-4 : 5765
16386-93-9 : 9075
16387-70-5 : 8344
16387-72-7 : 11610
16409-46-4 : 3554
16419-60-6 : 3134
16420-13-6 : 3980
16423-68-0 : 11524
16426-64-5 : 2527
16432-53-4 : 6794
16433-43-5 : 1104
16435-50-0 : 9024
16452-01-0 : 771
16484-17-6 : 7293
16491-15-9 : 4921
16503-53-0 : 2147
16509-46-9 : 8335
16518-62-0 : 644
16520-62-0 : 1124
16532-02-8 : 11365
16538-93-5 : 4737
16545-78-1 : 9061
16560-43-3 : 11272
16580-23-7 : 4436
16580-24-8 : 4437
16582-38-0 : 1207
16587-71-6 : 4592
16605-36-0 : 4435
16619-12-8 : 7291
16629-19-9 : 11780
16631-63-3 : 4833
16650-52-5 : 7762
16661-99-7 : 4349
16672-87-0 : 5929
16674-04-7 : 10230
16694-52-3 : 9868
16696-65-4 : 11940
16714-77-5 : 3654
16726-67-3 : 7761
16736-42-8 : 8325
16737-44-3 : 1983
16744-89-1 : 7083
16745-94-1 : 7017
16746-02-4 : 6836
16746-86-4 : 7013
16746-87-5 : 7014
16747-25-4 : 6859
16747-26-5 : 6860
16747-28-7 : 6862
16747-30-1 : 6865
16747-31-2 : 6866
16747-32-3 : 8810
16747-33-4 : 8811
16747-38-9 : 8842
16747-42-5 : 6848
16747-44-7 : 8836
16747-45-8 : 8837
16747-50-5 : 4814
16749-11-4 : 4361
16751-59-0 : 6351
16752-77-5 : 7650
16754-49-7 : 10080
16789-46-1 : 6825
16789-51-8 : 7034
16790-49-1 : 3197
16805-78-0 : 9578
16813-18-6 : 6318
16837-43-7 : 2118
16841-48-8 : 10597
16874-33-2 : 6072
16883-48-0 : 4860
16889-72-8 : 10294
16898-52-5 : 9701
16932-49-3 : 2753
16939-57-4 : 1097
16957-70-3 : 9076
16980-61-3 : 4512
16995-35-0 : 10486
17005-59-3 : 11383
17015-11-1 : 6914
17021-26-0 : 357
17042-16-9 : 6980
17057-82-8 : 7252
17057-88-4 : 3060
17057-92-0 : 7887
17059-44-8 : 1485
17059-45-9 : 1482
17059-52-8 : 2459
17071-47-5 : 3401
17082-09-6 : 10717
17082-12-1 : 5129
17085-91-5 : 7828
17091-40-6 : 5747
17094-21-2 : 3558
17096-37-6 : 3796
17102-64-6 : 6709
17113-33-6 : 6159
17176-77-1 : 9586
17199-29-0 : 903
17206-52-9 : 4595
17216-08-9 : 8045
17216-62-5 : 9933
17226-28-7 : 3416
17226-34-5 : 7009
17228-64-7 : 10918
17236-58-7 : 11777
17249-75-1 : 11747
17249-76-2 : 11749
17249-77-3 : 11791
17249-79-5 : 11746
17249-82-0 : 11741
17283-45-3 : 2946
17297-04-0 : 1108
17299-41-1 : 4719
17301-94-9 : 8142
17302-01-1 : 6410
17302-02-2 : 6767
17302-04-4 : 6415
17312-63-9 : 8122
17348-59-3 : 10081
17356-19-3 : 4866
17372-87-1 : 11525
17406-45-0 : 6196
17432-37-0 : 598
17447-00-6 : 9085
17447-01-7 : 9084
17478-66-9 : 5138
17484-36-5 : 1934
17496-14-9 : 7289
17534-15-5 : 1561
17556-48-8 : 10409
17577-93-4 : 8523
17587-22-3 : 8398
17609-31-3 : 8484
17618-76-7 : 6554
17627-77-9 : 7395
17639-93-9 : 10222
17640-26-5 : 182
17648-05-4 : 2281
17655-74-2 : 1651
17658-32-1 : 8924
17672-88-7 : 2422
17679-92-4 : 8526
17696-11-6 : 8426
17696-37-6 : 9326
17696-64-9 : 142
17704-30-2 : 165
17715-00-3 : 4473
17742-04-0 : 7528
17745-45-8 : 3136
17773-65-8 : 3737
17788-00-0 : 9292
17789-14-9 : 5205
17804-35-2 : 515
17837-41-1 : 11693
17849-38-6 : 1848
17851-27-3 : 1710
17878-44-3 : 11367
17880-86-3 : 10502
17912-17-3 : 8368
17937-68-7 : 11404
17952-11-3 : 8808
17972-08-6 : 7652
17988-31-7 : 11427
18002-79-4 : 11360
18023-33-1 : 11426
18023-36-4 : 11511
18030-61-0 : 11364
18036-81-2 : 11497
18169-57-8 : 11469
18171-59-0 : 11373
18178-59-1 : 11421
18181-80-1 : 3140
18187-24-1 : 5234
18231-53-3 : 8764
18236-57-2 : 11400
18236-89-0 : 11399
18240-50-1 : 6114
18243-41-9 : 11366
18261-92-2 : 8280
18262-85-6 : 1709
18281-05-5 : 5356
18282-40-1 : 1680
18282-59-2 : 1015
18295-59-5 : 8697
18309-32-5 : 2978
18338-40-4 : 3449
18362-36-2 : 578
18362-97-5 : 8893
18368-63-3 : 10920
18381-45-8 : 3406
18388-45-9 : 11408
18397-07-4 : 4154
18398-36-2 : 7849
18423-69-3 : 6136
18441-43-5 : 2458
18444-79-6 : 11521
18476-57-8 : 8327
18479-51-1 : 8515
18479-57-7 : 8450
18492-37-0 : 2960
18495-23-3 : 8536
18495-26-6 : 6613
18495-30-2 : 10144
18510-29-7 : 11455
18521-07-8 : 8519
18523-48-3 : 109
18530-30-8 : 2917
18530-56-8 : 8212
18589-29-2 : 6420
18598-63-5 : 5019
18649-64-4 : 6108
18668-68-3 : 3147
18685-70-6 : 4073
18699-48-4 : 1338
18719-43-2 : 9932
18720-65-5 : 6502
18729-48-1 : 4870
18747-42-7 : 10470
18748-27-1 : 11578
18752-92-6 : 11445
18758-34-4 : 4998
18773-77-8 : 4160
18786-24-8 : 8572
18787-64-9 : 6663
18791-49-6 : 10282
18794-84-8 : 5331
18812-62-9 : 7079
18829-56-6 : 8180
18839-90-2 : 9359
18840-45-4 : 12042
18854-56-3 : 5586
18858-65-6 : 11441
18869-72-2 : 6311
18904-54-6 : 4146
18905-91-4 : 6973
18913-31-0 : 3175
18917-91-4 : 340
18932-14-4 : 8420
18936-17-9 : 3409
18937-76-3 : 11990
18963-00-3 : 10176
18963-01-4 : 10148
18970-44-0 : 1688
18971-59-0 : 8821
18979-60-7 : 1549
18979-62-9 : 1539
18992-39-7 : 10402
18992-80-8 : 5806
18999-28-5 : 6561
19037-72-0 : 4922
19044-88-3 : 8553
19074-59-0 : 9336
19089-47-5 : 10380
19113-78-1 : 6928
19145-60-9 : 487
19145-96-1 : 4356
19155-24-9 : 7329
19163-05-4 : 3129
19177-04-9 : 1987
19184-10-2 : 10567
19184-67-9 : 8774
19221-28-4 : 3165
19230-27-4 : 1294
19246-38-9 : 4346
19246-39-0 : 3601
19269-28-4 : 6720
19317-11-4 : 5330
19329-89-6 : 10274
19354-27-9 : 6170
19374-46-0 : 4857
19393-45-4 : 3773
19393-92-1 : 1013
19395-58-5 : 11198
19396-83-9 : 2916
19398-47-1 : 3614
19398-53-9 : 8743
19398-61-9 : 1387
19398-75-5 : 4799
19398-77-7 : 6768
19402-87-0 : 3988
19404-07-0 : 5191
19420-29-2 : 7946
19434-65-2 : 9778
19438-10-9 : 2636
19447-29-1 : 8197
19472-74-3 : 948
19481-82-4 : 10094
19482-57-6 : 8535
19489-10-2 : 4397
19493-31-3 : 3026
19493-44-8 : 7376
19523-57-0 : 7753
19549-71-4 : 6480
19549-73-6 : 6481
19549-74-7 : 6482
19549-77-0 : 6484
19549-79-2 : 6486
19549-84-9 : 6515
19549-97-4 : 6617
19550-03-9 : 6922
19550-05-1 : 6924
19550-07-3 : 6927
19550-09-5 : 6929
19550-10-8 : 6964
19550-14-2 : 6967
19550-40-4 : 4655
19550-48-2 : 4920
19564-40-0 : 2115
19594-02-6 : 1652
19613-76-4 : 1024
19617-44-8 : 7231
19653-33-9 : 9704
19655-60-8 : 11268
19666-30-9 : 8565
19686-73-8 : 10345
19689-18-0 : 5083
19694-02-1 : 10854
19700-97-1 : 6693
19718-45-7 : 7956
19735-89-8 : 10849
19752-55-7 : 1012
19780-63-3 : 8933
19780-66-6 : 9046
19780-67-7 : 9049
19780-90-6 : 6798
19780-96-2 : 8164
19781-10-3 : 8451
19781-11-4 : 8458
19781-68-1 : 4976
19781-69-2 : 4977
19781-73-8 : 4404
19781-77-2 : 6572
19812-64-7 : 11618
19814-71-2 : 2077
19819-98-8 : 1609
19821-84-2 : 6612
19832-98-5 : 8050
19860-56-1 : 10256
19866-51-4 : 9077
19879-11-9 : 4508
19879-12-0 : 4509
19889-37-3 : 3516
19890-00-7 : 2956
19891-74-8 : 4060
19902-08-0 : 2924
19915-11-8 : 8689
19942-78-0 : 11708
19947-22-9 : 1700
19961-27-4 : 9810
19967-57-8 : 9775
19987-13-4 : 9865
20009-26-1 : 7955
20024-90-2 : 1703
20027-77-4 : 7991
20030-32-4 : 4690
20063-92-7 : 8183
20089-07-0 : 3439
20098-48-0 : 2338
20177-02-0 : 9028
20184-89-8 : 8198
20184-91-2 : 8199
20185-55-1 : 2694
20193-20-8 : 9809
20213-30-3 : 7917
20231-81-6 : 4045
20232-11-5 : 7279
20237-34-7 : 6681
20244-71-7 : 5888
20247-89-6 : 2008
20266-65-3 : 10022
20278-84-6 : 6446
20278-85-7 : 6443
20278-87-9 : 6449
20278-88-0 : 6451
20278-89-1 : 6452
20279-53-2 : 6062
20281-83-8 : 8938
20281-85-0 : 3616
20281-86-1 : 6912
20281-91-8 : 3619
20291-91-2 : 6816
20291-95-6 : 6439
20327-65-5 : 10568
20333-41-9 : 5245
20354-26-1 : 7643
20357-79-3 : 7247
20359-55-1 : 8979
20395-25-9 : 10173
20442-79-9 : 3078
20474-93-5 : 3856
20485-41-0 : 11669
20589-85-9 : 10598
20599-21-7 : 8534
20599-27-3 : 9011
20634-92-8 : 9812
20637-49-4 : 10185
20668-29-5 : 11261
20699-48-3 : 3261
20731-93-5 : 1875
20734-71-8 : 9386
20739-58-6 : 8543
20748-86-1 : 6400
20754-04-5 : 8478
20769-85-1 : 10491
20809-46-5 : 8479
20820-82-0 : 11022
20830-75-5 : 4041
20850-43-5 : 2421
20893-30-5 : 11725

CAS NUMBER INDEX OF ORGANIC COMPOUNDS (continued)

72-43-4 : 5131
84-33-2 : 11322
20-24-6 : 3899
20-26-8 : 9131
20-27-9 : 9136
40-45-9 : 10706
40-59-5 : 7392
87-29-6 : 1230
87-64-9 : 7658
02-13-6 : 6952
29-27-1 : 4546
79-48-6 : 10239
11-65-4 : 11133
32-53-1 : 361
66-88-6 : 6375
66-90-0 : 6376
93-29-8 : 8679
96-92-4 : 7323
99-26-3 : 4578
26-62-5 : 3611
52-22-7 : 11209
70-71-0 : 8054
91-98-0 : 4635
00-25-9 : 10602
36-44-2 : 2683
72-86-6 : 2335
90-63-1 : 8609
04-43-8 : 10559
531-91-9 : 6793
535-97-7 : 2457
564-91-0 : 7984
564-92-1 : 7986
570-35-4 : 6503
571-91-5 : 8751
617-18-5 : 11250
33-79-4 : 9972
655-48-1 : 9611
693-54-9 : 7992
700-31-2 : 10125
702-79-4 : 2559
725-46-2 : 4161
777-84-4 : 1992
849-97-8 : 7840
856-89-3 : 6970
884-26-4 : 8909
889-88-3 : 6585
890-35-7 : 3148
915-53-7 : 8619
923-23-9 : 4092
948-47-0 : 8372
964-23-8 : 9013
964-44-3 : 8189
968-17-2 : 9827
974-48-1 : 9714
984-93-0 : 6065
987-21-3 : 11886
994-77-4 : 7065
994-78-5 : 7064
2029-76-1 : 3873
2037-28-1 : 6034
2037-73-6 : 3710
2038-57-9 : 3830
2056-82-2 : 14
2077-55-0 : 2176
2104-78-5 : 8510
2118-12-3 : 3682
2128-62-7 : 4014
2128-63-8 : 4016
2139-77-1 : 1545
2144-60-1 : 1912
2160-12-9 : 6893
2160-39-0 : 8887
2160-40-3 : 8888
2160-41-4 : 6895
2202-65-9 : 11989
2224-92-6 : 5947
2227-75-4 : 10556
2232-54-8 : 7194
2253-11-8 : 7259
2287-02-1 : 8176
2319-28-4 : 6592
2331-38-0 : 5351
2349-03-7 : 4456
2364-68-7 : 960
2374-89-6 : 2119
2381-86-8 : 11980
2398-09-0 : 9709
22401-25-8 : 9546
22409-91-2 : 10322
22418-49-1 : 10577
22421-97-2 : 10229
22445-41-6 : 1771
22483-09-6 : 5419
22520-40-7 : 8391
22524-32-9 : 113
22531-06-2 : 8051
22531-20-0 : 7891
22539-65-7 : 1469
22545-12-6 : 1603
22545-13-7 : 1604
22564-83-6 : 7429
22564-99-4 : 8332
22567-21-1 : 4543
22610-63-5 : 8255
22618-23-1 : 9315
22621-74-5 : 2947
22646-79-3 : 4646
22663-64-5 : 8166
22672-74-8 : 5150
22692-70-2 : 11560
22717-57-3 : 2640
22726-00-7 : 605
22734-04-9 : 4905
22767-95-9 : 2140
22767-96-0 : 2145
22768-19-0 : 6550
22768-20-3 : 6549
22781-23-3 : 514
22839-47-0 : 9511
22868-76-4 : 11056
22900-08-9 : 4540
22922-67-4 : 7004
22924-15-8 : 553
22927-13-5 : 558
22936-86-3 : 5013
23010-00-6 : 3733
23010-04-0 : 3283
23012-11-5 : 8578
23048-13-7 : 4943
23063-36-7 : 1390
23074-36-4 : 3709
23079-65-4 : 1116
23007-08-5 : 0018
23103-98-2 : 9728
23132-52-7 : 1467
23135-22-0 : 8568
23144-52-7 : 8448
23184-66-9 : 3144
23204-80-0 : 1229
23287-26-5 : 2638
23308-82-9 : 1894
23378-11-2 : 3732
23399-70-4 : 1174
23418-37-3 : 8469
23418-38-4 : 8170
23422-53-9 : 6001
23431-48-3 : 2028
23438-77-9 : 4722
23444-65-7 : 7854
23470-12-4 : 7127
23505-41-1 : 9729
23511-70-8 : 3582
23564-05-8 : 11723
23578-35-0 : 6751
23580-51-0 : 8544
23580-52-1 : 10960
23617-71-2 : 2589
23619-59-2 : 9459
23676-09-7 : 2599
23708-56-7 : 11954
23733-68-8 : 4717
23769-39-3 : 11921
23843-52-9 : 8230
23947-60-6 : 5932
23950-58-5 : 10746
24127-62-6 : 10615
24152-39-4 : 9637
24247-53-8 : 3696
24264-08-2 : 3877
24280-93-1 : 7052
24306-18-1 : 6323
24307-26-4 : 7430
24331-72-4 : 9794
24332-22-7 : 6584
24342-04-9 : 9139
24346-56-3 : 2132
24356-00-1 : 9017
24393-70-2 : 2943
24395-10-6 : 6571
24398-88-7 : 2522
24410-50-2 : 2449
24410-51-3 : 2446
24410-52-4 : 2452
24425-40-9 : 7244
24425-97-6 : 10051
24438-17-3 : 7328
24446-63-7 : 942
24544-04-5 : 636
24545-81-1 : 2987
24549-06-2 : 745
24557-10-6 : 10627
24579-70-2 : 3941
24580-48-1 : 4550
24580-52-7 : 4551
24602-86-6 : 11898
24612-75-7 : 4477
24618-31-3 : 4317
24622-72-8 : 11169
24634-61-5 : 6704
24644-78-8 : 7290
24677-78-9 : 540
24716-09-4 : 890
24724-07-0 : 11949
24734-68-7 : 2125
24764-97-4 : 3179
24796-87-0 : 4503
24815-24-5 : 12060
24840-05-9 : 10588
24844-28-8 : 1550
24885-45-8 : 9375
24887-75-0 : 345
24910-63-2 : 9036
24932-49-8 : 1297
24948-66-1 : 5095
24964-64-5 : 2757
24965-84-2 : 4596
24973-49-7 : 3020
25006-60-4 : 167
25013-16-5 : 9329
25044-01-3 : 9107
25057-89-0 : 518
25109-28-8 : 1009
25109-57-3 : 3653
25117-74-2 : 2755
25118-28-9 : 6755
25144-04-1 : 4869
25144-05-2 : 4868
25148-87-2 : 3756
25167-70-8 : 9064
25172-42-3 : 4873
25172-43-4 : 4874
25182-84-7 : 10395
25186-47-4 : 1386
25201-40-5 : 6338
25229-42-9 : 3834
25234-83-7 : 8898
25245-33-4 : 1767
25265-71-8 : 10397
25309-64-2 : 1681
25311-71-1 : 7359
25312-34-9 : 3874
25312-65-6 : 4094
25346-31-0 : 8718
25346-32-1 : 8730
25346-33-2 : 8714
25354-97-6 : 5065
25354-98-7 : 8649
25394-75-6 : 10772
25395-22-6 : 103
25095-01-7 : 10189
25414-22-6 : 6127
25415-67-2 : 8892
25415-70-7 : 8894
25419-36-7 : 7990
25419-37-8 : 7987
25542-62-5 : 6879
25611-78-3 : 1584
25641-98-9 : 1340
25659-22-7 : 7087
25695-77-6 : 10689
25790-55-0 : 3150
25871-69-6 : 7587
26073-26-7 : 10500
26106-95-6 : 3602
26116-12-1 : 11167
26118-38-7 : 6963
26120-52-5 : 4253
26153-88-8 : 8703
26171-83-5 : 3344
26186-01-6 : 8107
26186-02-7 : 11897
26190-82-9 : 11256
26198-63-0 : 9892
26225-79-6 : 5933
26259-45-0 : 3143
26260-02-6 : 580
26303-90-2 : 5090
26311-17-1 : 5424
26311-33-1 : 7063
26330-51-8 : 3476
26340-58-9 : 4151
26370-28-5 : 8108
26399-36-0 : 9763
26438-26-6 : 7931
26438-27-7 : 7793
26438-28-8 : 7879
26464-32-4 : 3488
26465-81-6 : 7288
26472-41-3 : 4282
26473-49-4 : 3537
26489-01-0 : 8514
26513-20-2 : 1265
26530-20-1 : 8528
26531-85-1 : 6237
26533-33-5 : 8169
26545-24-0 : 6913
26549-25-7 : 6474
26608-06-0 : 5165
26644-46-2 : 11902
26730-14-3 : 11879
26754-48-3 : 6030
26771-11-9 : 4959
26963-33-7 : 11771
27063-27-0 : 11208
27129-86-8 : 1057
27129-87-9 : 1861
27267-69-2 : 2773
27314-13-2 : 8213
27356-52-1 : 11269
27387-79-7 : 10218
27464-82-0 : 11659
27532-76-9 : 378
27598-85-2 : 9200
27610-88-4 : 8400
27644-49-1 : 8465
27655-41-0 : 7375
27701-01-5 : 3943
27779-01-7 : 4158
27831-13-6 : 1629
27876-94-4 : 5107
27993-42-6 : 8984
28056-54-4 : 4896
28123-70-8 : 8125
28148-04-1 : 9890
28249-77-6 : 11697
28261-97-4 : 5273
28300-74-5 : 11588
28320-88-9 : 8362
28320-89-0 : 6364
28332-44-7 : 7095
28466-26-4 : 10812
28467-92-7 : 3190
28469-92-3 : 1369
28664-03-1 : 4171
28685-60-1 : 3958
28715-26-6 : 2451
28788-62-7 : 2889
28823-41-8 : 6696
28832-55-5 : 3180
28905-12-6 : 6232
28921-35-9 : 6064
28994-41-4 : 9454
28995-89-3 : 8004
29003-73-4 : 163
29022-31-9 : 8201
29030-84-0 : 6689
29040-46-8 : 2445
29072-93-3 : 10087
29091-05-2 : 5181
29091-21-2 : 9761
29106-32-9 : 4936
29110-74-5 : 7236
29113-63-1 : 6055
29136-19-4 : 2062
29172-20-1 : 6111
29204-02-2 : 5365
29232-93-7 : 9730
29276-40-2 : 7707
29289-13-2 : 762
29342-44-7 : 6740
29342-53-8 : 2918
29394-58-9 : 3383
29400-42-8 : 8211
29472-95-5 : 9421
29488-27-5 : 11773
29538-77-0 : 4507
29554-49-2 : 4745
29555-02-0 : 4964
29559-27-1 : 2004
29559-52-2 : 10177
29560-84-7 : 10701
29604-75-9 : 1296
29606-79-9 : 4603
29624-17-7 : 4348
29653-38-1 : 3926
29682-41-5 : 1380
29780-00-5 : 11896
29823-16-3 : 8873
29843-58-1 : 3757
29865-90-5 : 568
29882-57-3 : 8720
29883-15-6 : 954
29897-82-3 : 11172
29916-45-8 : 3269
29921-50-4 : 8063
29943-42-8 : 10799
29949-19-7 : 7511
29972-79-0 : 5341
30090-17-6 : 814
30122-12-4 : 8745
30168-50-4 : 4830
30273-11-1 : 813
30361-33-2 : 5025
30389-18-5 : 4290
30433-57-9 : 10833
30515-28-7 : 6458
30525-89-4 : 8634
30538-80-8 : 9689
30544-34-4 : 6077
30560-19-1 : 10
30600-19-2 : 10619
30614-77-8 : 6082
30673-60-0 : 5069
30685-95-1 : 3886
30689-41-9 : 4916
30752-19-3 : 1045
30926-60-4 : 1188
30964-01-3 : 8207
31053-99-3 : 1092
31059-19-5 : 7944
31116-84-4 : 10459
31161-71-4 : 11627
31202-12-7 : 363
31218-83-4 : 10719
31252-42-3 : 9700
31283-43-9 : 8498
31340-44-0 : 10247
31366-00-4 : 1713
31367-46-1 : 6496
31508-08-4 : 6607
31508-12-0 : 8194
31543-75-6 : 1306
31599-60-7 : 1769
31599-61-8 : 1774
31642-67-8 : 8188
31656-92-5 : 981
31699-35-1 : 8116
31702-33-7 : 10605
31720-69-1 : 4562
31766-07-1 : 2771
31772-05-1 : 11874
31805-48-8 : 11761
31844-95-8 : 9008
31844-96-9 : 3716
31849-78-2 : 3708

CAS NUMBER INDEX OF ORGANIC COMPOUNDS (continued)

Section 4
Properties of the Elements and Inorganic Compounds

THE ELEMENTS

C. R. Hammond

One of the most striking facts about the elements is their unequal distribution and occurrence in nature. Present knowledge of the chemical mposition of the universe, obtained from the study of the spectra of stars and nebulae, indicates that hydrogen is by far the most abundant element d may account for more than 90% of the atoms or about 75% of the mass of the universe. Helium atoms make up most of the remainder. All of the ner elements together contribute only slightly to the total mass.

The chemical composition of the universe is undergoing continuous change. Hydrogen is being converted into helium, and helium is being changed to heavier elements. As time goes on, the ratio of heavier elements increases relative to hydrogen. Presumably, the process is not reversible.

Burbidge, Burbidge, Fowler, and Hoyle, and more recently, Peebles, Penzias, and others have studied the synthesis of elements in stars. To explain of the features of the nuclear abundance curve — obtained by studies of the composition of the earth, meteorites, stars, etc. — it is necessary to stulate that the elements were originally formed by at least eight different processes: (1) hydrogen burning, (2) helium burning, (3) χ process, (4) rocess, (5) *s* process, (6) *r* process, (7) *p* process, and (8) the X process. The X process is thought to account for the existence of light nuclei such D, Li, Be, and B. Common metals such as Fe, Cr, Ni, Cu, Ti, Zn, etc. were likely produced early in the history of our galaxy. It is also probable at most of the heavy elements on earth and elsewhere in the universe were originally formed in supernovae, or in the hot interior of stars.

Studies of the solar spectrum have led to the identification of 67 elements in the sun's atmosphere; however, all elements cannot be identified with e same degree of certainty. Other elements may be present in the sun, although they have not yet been detected spectroscopically. The element helium as discovered on the sun before it was found on earth. Some elements such as scandium are relatively more plentiful in the sun and stars than here earth.

Minerals in lunar rocks brought back from the moon on the Apollo missions consist predominantly of *plagioclase* {$(Ca,Na)(Al,Si)O_4O_8$} and *roxene* {$(Ca,Mg,Fe)_2Si_2O_6$} — two minerals common in terrestrial volcanic rock. No new elements have been found on the moon that cannot be counted for on earth; however, three minerals, *armalcolite* {$(Fe,Mg)Ti_2O_5$}, *pyroxferroite* {$CaFe_6(SiO_3)_7$}, and *tranquillityite* {$Fe_8(Zr,Y)Ti_3Si_3O_2$}, e new. The oldest known terrestrial rocks are about 4 billion years old. One rock, known as the "Genesis Rock," brought back from the Apollo 15 ission, is about 4.15 billion years old. This is only about one-half billion years younger than the supposed age of the moon and solar system. Lunar cks appear to be relatively enriched in refractory elements such as chromium, titanium, zirconium, and the rare earths, and impoverished in volatile ements such as the alkali metals, in chlorine, and in noble metals such as nickel, platinum, and gold.

Even older than the "Genesis Rock" are *carbonaceous chondrites*, a type of meteorite that has fallen to earth and has been studied. These are some the most primitive objects of the solar system yet found. The grains making up these objects probably condensed directly out the gaseous nebula om which the sun and planets were born. Most of the condensation of the grains probably was completed within 50,000 years of the time the disk the nebula was first formed — about 4.6 billion years ago. It is now thought that this type of meteorite may contain a small percentage of presolar ust grains. The relative abundances of the elements of these meteorites are about the same as the abundances found in the solar chromosphere.

The X-ray fluorescent spectrometer sent with the Viking I spacecraft to Mars shows that the Martian soil contains about 12 to 16% iron, 14 to 15% licon, 3 to 8% calcium, 2 to 7% aluminum, and one-half to 2% titanium. The gas chromatograph — mass spectrometer on Viking II found no trace t organic compounds that should be present if life ever existed there.

F. W. Clarke and others have carefully studied the composition of rocks making up the crust of the earth. Oxygen accounts for about 47% of the ust, by weight, while silicon comprises about 28% and aluminum about 8%. These elements, plus iron, calcium, sodium, potassium, and magnesium, ccount for about 99% of the composition of the crust.

Many elements such as tin, copper, zinc, lead, mercury, silver, platinum, antimony, arsenic, and gold, which are so essential to our needs and vilization, are among some of the rarest elements in the earth's crust. These are made available to us only by the processes of concentration in ore odies. Some of the so-called *rare-earth* elements have been found to be much more plentiful than originally thought and are about as abundant as ranium, mercury, lead, or bismuth. The least abundant rare-earth or *lanthanide* element, thulium, is now believed to be more plentiful on earth than lver, cadmium, gold, or iodine, for example. Rubidium, the 16th most abundant element, is more plentiful than chlorine while its compounds are little nown in chemistry and commerce.

It is now thought that at least 24 elements are essential to living matter. The four most abundant in the human body are hydrogen, oxygen, carbon, nd nitrogen. The seven next most common, in order of abundance, are calcium, phosphorus, chlorine, potassium, sulfur, sodium, and magnesium. on, copper, zinc, silicon, iodine, cobalt, manganese, molybdenum, fluorine, tin, chromium, selenium, and vanadium are needed and play a role in ving matter. Boron is also thought essential for some plants, and it is possible that aluminum, nickel, and germanium may turn out to be necessary.

Ninety-one elements occur naturally on earth. Minute traces of plutonium-244 have been discovered in rocks mined in Southern California. This iscovery supports the theory that heavy elements were produced during creation of the solar system. While technetium and promethium have not yet een found naturally on earth, they have been found to be present in stars. Technetium has been identified in the spectra of certain "late" type stars, nd promethium lines have been identified in the spectra of a faintly visible star HR465 in Andromeda. Promethium must have been made very recently ear the star's surface for no known isotope of this element has a half-life longer than 17.7 years.

It has been suggested that californium is present in certain stellar explosions known as supernovae; however, this has not been proved. At present o elements are found elsewhere in the universe that cannot be accounted for here on earth.

All atomic mass numbers from 1 to 238 are found naturally on earth except for masses 5 and 8. About 280 stable and 67 naturally radioactive isotopes ccur on earth totaling 347. In addition, the neutron, technetium, promethium, and the transuranic elements (lying beyond uranium) up to Element 07 have been produced artificially. West German scientists in late 1982 confirmed the existence of Element 107 made by Soviet scientists and nnounced synthesis of Element 109. Laboratory processes have now extended the radioactive element mass numbers beyond 238 to 266. Each element rom atomic number 1 to 107 is known to have at least one radioactive isotope. About 1980 different nuclides (the name given to different kinds of nuclei, whether they are of the same or different elements) are now recognized. Many stable and radioactive isotopes are now produced and distribu by the Oak Ridge National Laboratory, Oak Ridge, Tenn., U.S.A., to customers licensed by the U.S. Department of Energy.

The nucleus of an atom is characterized by the number of protons it contains, denoted by Z, and by the number of neutrons, N. Isotopes of an elem have the same value of Z, but different values of N. The *mass number A*, is the sum of Z and N. For example, Uranium-238 has a mass number of 2 and contains 92 protons and 146 neutrons.

There is evidence that the definition of chemical elements must be broadened to include the electron. Several compounds known as *electrides,* h recently been made of alkaline metal elements and electrons. A relatively stable combination of a positron and electron, known as *positronium,* also been studied.

In addition to the proton, neutron, and electron, there are considerably more than 100 other fundamental particles which have been discoverec hypothesized. The majority of these fall into one of two classes, *leptons* or *hadrons.* The leptons comprise just four known particles, the *electron,* *muon* (μ meson), and two kinds of *neutrinos.* The muon is essentially similar to the electron and has a charge of –1, but it is 200 times heavier. T neutrino is either of two stable particles of small (probably zero) rest mass, carrying no charge. Also there are four *antileptons,* identical to corresponding leptons in some respects, such as mass, but they have properties exactly opposite those of the leptons. The *positron,* for example, is antilepton, with a charge of +1. Leptons cannot be broken into smaller units and are considered to be elementary. On the other hand, hadrons are comp and thought to have internal structure. Protons and neutrons, which make up atomic nuclei, are hadrons.

Elementary particle physics is not yet clearly understood, but groupings and arrangements of these particles have been made resembling the perio table of chemical elements. This has led to the speculation that hadrons are composed of three (or possibly more simpler components called *quar* Quarks are presumed to be elementary particles. There is presently no evidence that quarks exist in isolation. Many physicists now hold that all matter and energy in the universe is controlled by four fundamental natural forces: the electromagnetic force, gravity, a weak nuclear force, and a stro nuclear force. Each of these natural forces is passed back and forth among the basic particles of matter by unique force-carrying particles. T electromagnetic force is carried by the *photon,* the weak nuclear force by the intermediate vector *boson,* and gravity by the *gravitron.* There is n evidence of the existence of a new particle, known as the *gluon,* that binds quarks together by carrying the strong nuclear force.

The available evidence leads to the conclusion that elements 89 (actinium) through 103 (lawrencium) are chemically similar to the rare-earth lanthanide elements (elements 57 to 71, inclusive). These elements therefore have been named *actinides* after the first member of this series. The elements beyond uranium that have been produced artificially have the following names and symbols: neptunium, 93 (Np); plutonium, 94 (P americium, 95 (Am); curium, 96 (Cm); berkelium, 97 (Bk); californium, 98 (Cf); einsteinium, 99 (Es); fermium, 100 (Fm); mendelevium, 101 (M nobelium, 102 (No); and lawrencium, 103 (Lr). It is now claimed that Elements 104, 105, 106, 107, and 109 have been produced and identified. Nam and chemical symbols have been suggested for Elements 104 and 105, but have not been officially adopted. Names for Elements 106, 107, and 1 have not yet been suggested.

Element 104 is expected to have chemical properties similar to those of hafnium and would not be a member of the actinide series. Element 1 probably would have chemical properties similar to those of tantalum, Element 106 similar to tungsten, Element 107 similar to rhenium, and Eleme 109 similar to iridium.

There is still thought to be a possibility of producing elements beyond Element 109 either by bombardment of heavy isotopic targets with hea ions, or by the irradiation of uranium or transuranic elements with the instantaneous high flux of neutrons produced by underground nuclear explosio The limit will be set by the yields of the nuclear reactions and by the half-lives of radioactive decay. It has been suggested that Elements 102 and 1 have abnormally short lives only because they are in a pocket of instability, and that this region of instability might begin to "heal" around Eleme 105. If so, it may be possible to produce heavier isotopes with longer half-lives. It has also been suggested that Element 114, with a mass number 298, and Element 126, with a mass number of 310, may be sufficiently stable to make discovery and identification possible. Calculations indicate th Element 110, a homolog of platinum, may have a half-life of as long as 100 million years. Searches have already been made by workers in a numb of laboratories for Element 110 and its neighboring elements in naturally occurring platinum. Recent studies of the xenon component (^{131}Xe to ^{136}X of certain carbonaceous chronditic meteorites suggest that Elements 113, 114, or 115 may have been its progenitor.

There are many claims in the literature of the existence of various allotropic modifications of the elements, some of which are based on doubt or incomplete evidence. Also, the physical properties of an element may change drastically by the presence of small amounts of impurities. With ne methods of purification, which are now able to produce elements with 99.9999% purity, it has been necessary to restudy the properties of the elemen For example, the melting point of thorium changes by several hundred degrees by the presence of a small percentage of ThO_2 as an impurity. Ordina commercial tungsten is brittle and can be worked only with difficulty. Pure tungsten, however, can be cut with a hacksaw, forged, spun, drawn, extruded. In general, the value of a physical property given here applies to the pure element, when it is known.

Many of the chemical elements and their compounds are toxic and should be handled with due respect and care. In recent years there has been greatly increased knowledge and awareness of the health hazards associated with chemicals, radioactive materials, and other agents. Anyone workin with the elements and certain of their compounds should become thoroughly familiar with the proper safeguards to be taken. Reference should be ma to publications such as the following:

1. *Code of Federal Regulations,* Title 29, Labor, chapter XVII, section 1910.93 of subpart G, redesignated as 1910.1000 at 40 FR (Federal Registe 23072. May 28, 1975; amended at 41 FR 11505, March 19, 1976; 41 FR 35184, August 20, 1976; FR 46784, October 22, 1976; 42 FR 3304, Janua 18, 1977 (corrections and additional amendments and corrections as issued, U.S. Government Printing Office, Supt. of Documents, Washingto D.C.
2. *Code of Federal Regulations,* Title 10, Energy, Chapter 1, Nuclear Regulatory Commission, section 20.103 — 108; 20.201 — 207; 20.301 — 30 20.401 — 409; 20.501 — 502; 20.601; appendices, corrections, and amendments.
3. *Occupational Safety and Health Reporter* (latest edition with amendments and corrections), Bureau of National Affairs, Washington, D.C.
4. *Atomic Energy Law Reporter,* Commerce Clearing House, Chicago, IL.

Nuclear Regulation Reporter, Commerce Clearing House, Chicago, IL.

Maximum Permissible Body Burdens and Maximum Permissible Concentrations of Radionuclides in Air and in Water for Occupational Exposure, with addenda, U.S. Department of Commerce, N.B.S., Handbook No. 69, (NCRP Report No. 22), latest edition, National Council on Radiation Protection and Measurements (NCRP), Bethesda, MD.; also refer to Permissible Quarterly Intakes of Radionuclides. Handbook of Chemistry and Physics, 72nd Edition.

TLVs® Threshold Limit Values for Chemical Substances and Physical Agents in Workroom Environment with Intended Changes, latest edition, American Conference of Governmental Industrial Hygienists, Cincinnati, Ohio.

Actinium — (Gr. *aktis, aktinos*, beam or ray), Ac; at. wt. (227); at. no. 89; m.p. 1051°C, b.p. 3200 ± 300°C (est.); sp. gr. 10.07 (calc.). Discovered y Andre Debierne in 1899 and independently by F. Giesel in 1902. Occurs naturally in association with uranium minerals. Actinium-227, a decay oduct of uranium-235, is a beta emitter with a 21.6-year half-life. Its principal decay products are thorium-227 (18.5-day half-life), radium-223 (11.4-ay half-life), and a number of short-lived products including radon, bismuth, polonium, and lead isotopes. In equilibrium with its decay products, is a powerful source of alpha rays. Actinium metal has been prepared by the reduction of actinium fluoride with lithium vapor at about 1100 to 1300°C. he chemical behavior of actinium is similar to that of the rare earths, particularly lanthanum. Purified actinium comes into equilibrium with its decay oducts at the end of 185 days, and then decays according to its 21.6-year half-life. It is about 150 times as active as radium, making it of value in e production of neutrons.

Aluminum — (L. *alumen, alum*), Al; at. wt. 26.981539; at. no. 13; m.p. 660.32°C; b.p. 2519°C; sp. gr. 2.6989 (20°C); valence 3. The ancient Greeks nd Romans used *alum* in medicine as an astringent, and as a mordant in dyeing. In 1761 de Morveau proposed the name *alumine* for the base in alum, nd Lavoisier, in 1787, thought this to be the oxide of a still undiscovered metal. Wohler is generally credited with having isolated the metal in 1827, though an impure form was prepared by Oersted two years earlier. In 1807, Davy proposed the name *alumium* for the metal, undiscovered at that me, and later agreed to change it to *aluminum*. Shortly thereafter, the name *aluminium* was adopted to conform with the "ium" ending of most elements, nd this spelling is now in use elsewhere in the world. *Aluminium* was also the accepted spelling in the U.S. until 1925, at which time the American hemical Society officially decided to use the name *aluminum* thereafter in their publications. The method of obtaining aluminum metal by the lectrolysis of alumina dissolved in *cryolite* was discovered in 1886 by Hall in the U.S. and at about the same time by Heroult in France. Cryolite, a atural ore found in Greenland, is no longer widely used in commercial production, but has been replaced by an artificial mixture of sodium, aluminum, nd calcium fluorides. *Bauxite*, an impure hydrated oxide ore, is found in large deposits in Jamaica, Australia, Surinam, Guyana, Arkansas, and lsewhere. The Bayer process is most commonly used today to refine bauxite so it can be accommodated in the Hall-Heroult refining process, used) make most aluminum. Aluminum can now be produced from clay, but the process is not economically feasible at present. Aluminum is the most bundant metal to be found in the earth's crust (8.1%), but is never found free in nature. In addition to the minerals mentioned above, it is found in eldspars, granite, and in many other common minerals. Pure aluminum, a silvery-white metal, possesses many desirable characteristics. It is light, ontoxic, has a pleasing appearance, can easily be formed, machined, or cast, has a high thermal conductivity, and has excellent corrosion resistance. is nonmagnetic and nonsparking, stands second among metals in the scale of malleability, and sixth in ductility. It is extensively used for kitchen tensils, outside building decoration, and in thousands of industrial applications where a strong, light, easily constructed material is needed. Although s electrical conductivity is only about 60% that of copper, it is used in electrical transmission lines because of its light weight. Pure aluminum is soft nd lacks strength, but it can be alloyed with small amounts of copper, magnesium, silicon, manganese, and other elements to impart a variety of useful roperties. These alloys are of vital importance in the construction of modern aircraft and rockets. Aluminum, evaporated in a vacuum, forms a highly eflective coating for both visible light and radiant heat. These coatings soon form a thin layer of the protective oxide and do not deteriorate as do silver oatings. They have found application in coatings for telescope mirrors, in making decorative paper, packages, toys, and in many other uses. The ompounds of greatest importance are aluminum oxide, the sulfate, and the soluble sulfate with potassium (alum). The oxide, alumina, occurs naturally s ruby, sapphire, corundum, and emery, and is used in glassmaking and refractories. Synthetic ruby and sapphire have found application in the onstruction of lasers for producing coherent light. In 1852, the price of aluminum was about $545/lb, and just before Hall's discovery in 1886, about 11.00. The price rapidly dropped to 30¢ and has been as low as 15¢/lb.

Americium — (the Americas), Am; at. wt. 243; at. no. 95; m.p. 1176°C; sp. gr. 13.67 (20°C); valence 2, 3, 4, 5, or 6. Americium was the fourth ransuranium element to be discovered; the isotope ^{241}Am was identified by Seaborg, James, Morgan, and Ghiorso late in 1944 at the wartime Metallurgical Laboratory of the University of Chicago as the result of successive neutron capture reactions by plutonium isotopes in a nuclear reactor:

$$^{239}\mathrm{Pu}(n,\gamma)\rightarrow{}^{240}\mathrm{Pu}(n,\gamma)\rightarrow{}^{241}\mathrm{Pu}\xrightarrow{\beta}{}^{241}\mathrm{Am}$$

Since the isotope ^{241}Am can be prepared in relatively pure form by extraction as a decay product over a period of years from strongly neutron-bombarded plutonium, ^{241}Pu, this isotope is used for much of the chemical investigation of this element. Better suited is the isotope ^{243}Am due to its onger half-life (8.8×10^3 years as compared to 470 years for ^{241}Am). A mixture of the isotopes ^{241}Am, ^{242}Am, and ^{243}Am can be prepared by intense neutron irradiation of ^{241}Am according to the reactions ^{241}Am (n, γ) → ^{242}Am (n, γ) → ^{243}Am. Nearly isotopically pure ^{243}Am can be prepared by a equence of neutron bombardments and chemical separations as follows: neutron bombardment of ^{241}Am yields ^{242}Pu by the reactions ^{241}Am (n, γ) → ^{242}Am → ^{242}Pu, after chemical separation the ^{242}Pu can be transformed to ^{243}Am via the reactions ^{242}Pu (n, γ) → ^{243}Pu → ^{243}Am, and the ^{243}Am can be chemically separated. Fairly pure ^{242}Pu can be prepared more simply by very intense neutron irradiation of ^{239}Pu as the result of successive neutron-capture reactions. Americium metal has been prepared by reducing the trifluoride with barium vapor at 1000 to 1200°C or the dioxide by anthanum metal. The luster of freshly prepared americium metal is white and more silvery than plutonium or neptunium prepared in the same manner. It appears to be more malleable than uranium or neptunium and tarnishes slowly in dry air at room temperature. Americium is thought to exist in two forms: an alpha form which has a double hexagonal close-packed structure and a loose-packed cubic beta form. Americium must be handled with great care to avoid personal contamination. As little as 0.03 μCi of ^{241}Am is the maximum permissible total body burden. The alpha activity from ^{241}Am is about three times that of radium. When gram quantities of ^{241}Am are handled, the intense gamma activity makes exposure a serious prob Americium dioxide, AmO_2, is the most important oxide. AmF_3, AmF_4, $AmCl_3$, $AmBr_3$, AmI_3, and other compounds have been prepared. The iso ^{241}Am has been used as a portable source for gamma radiography. It has also been used as a radioactive glass thickness gage for the flat glass indus and as a source of ionization for smoke detectors. Americum-241 is available from the A.E.C. at a cost of $1600/g and Americum-243 at a cost of $ mg.

Antimony — (Gr. *anti* plus *monos* — a metal not found alone), Sb; at. wt. 121.757; at. no. 51; m.p. 630.63°C; b.p. 1587°C; sp. gr. 6.691 (20 valence 0, –3, +3, or +5. Antimony was recognized in compounds by the ancients and was known as a metal at the beginning of the 17th century possibly much earlier. It is not abundant, but is found in over 100 mineral species. It is sometimes found native, but more frequently as the sulf *stibnite* (Sb_2S_3); it is also found as antimonides of the heavy metals, and as oxides. It is extracted from the sulfide by roasting to the oxide, whic reduced by salt and scrap iron; from its oxides it is also prepared by reduction with carbon. Two allotropic forms of antimony exist: the normal sta metallic form, and the amorphous gray form. The so-called explosive antimony is an ill-defined material always containing an appreciable amc of halogen; therefore, it no longer warrants consideration as a separate allotrope. The yellow form, obtained by oxidation of *stibine*, SbH_3, is proba impure, and is not a distinct form. Metallic antimony is an extremely brittle metal of a flaky, crystalline texture. It is bluish white and has a meta luster. It is not acted on by air at room temperature, but burns brilliantly when heated with the formation of white fumes of Sb_2O_3. It is a poor conduc of heat and electricity, and has a hardness of 3 to 3.5. Antimony, available commercially with a purity of 99.999 + %, is finding use in semiconduc technology for making infrared detectors, diodes, and Hall-effect devices. Commercial-grade antimony is widely used in alloys with percenta ranging from 1 to 20. It greatly increases the hardness and mechanical strength of lead. Batteries, antifriction alloys, type metal, small arms and tra bullets, cable sheathing, and minor products use about half the metal produced. Compounds taking up the other half are oxides, sulfides, sodi antimonate, and antimony trichloride. These are used in manufacturing flame-proofing compounds, paints, ceramic enamels, glass, and pottery. Ta emetic (hydrated potassium antimonyl tartate) has been used in medicine. Antimony and many of its compounds are toxic. Exposure to antimony a its compounds should not exceed 0.5 mg/m^3 (8-hour time weight average 40-hour work week).

Argon — (Gr. *argos*, inactive), Ar; at. wt. 39.948; at. no. 18; freezing pt. –189.3°C; b.p. –185.9°C; density 1.7837 g/l. Its presence in air v suspected by Cavendish in 1785, discovered by Lord Rayleigh and Sir William Ramsay in 1894. The gas is prepared by fractionation of liquid a the atmosphere containing 0.94% argon. The atmosphere of Mars contains 1.6% of ^{40}Ar and 5 p.p.m. of ^{36}Ar. Argon is two and one half times as solu in water as nitrogen, having about the same solubility as oxygen. It is recognized by the characteristic lines in the red end of the spectrum. It is us in electric light bulbs and in fluorescent tubes at a pressure of about 400 Pa, and in filling photo tubes, glow tubes, etc. Argon is also used as an in gas shield for arc welding and cutting, as a blanket for the production of titanium and other reactive elements, and as a protective atmosphere for growi silicon and germanium crystals. Argon is colorless and odorless, both as a gas and liquid. It is available in high-purity form. Commercial argor available at a cost of about 3¢ per cubic foot. Argon is considered to be a very inert gas and is not known to form true chemical compounds, as do krypt xenon, and radon. However, it does form a hydrate having a dissociation pressure of 105 atm at 0°C. Ion molecules such as $(ArKr)^+$, $(ArXe)^+$, (NeA have been observed spectroscopically. Argon also forms a clathrate with β-hydroquinone. This clathrate is stable and can be stored for a considera time, but a true chemical bond does not exist. Van der Waals' forces act to hold the argon. Naturally occurring argon is a mixture of three isotop Twelve other radioactive isotopes are now known to exist.

Arsenic — (L. *arsenicum*, Gr. *arsenikon*, yellow orpiment, identified with *arsenikos*, male, from the belief that metals were different sexes; Arab *Az-zernikh*, the orpiment from Persian *zerni-zar*, gold), As; at. wt. 74.92159; at. no. 33; valence –3, 0, +3 or +5. Elemental arsenic occurs in two so modifications: yellow, and gray or metallic, with specific gravities of 1.97, and 5.73, respectively. Gray arsenic, the ordinary stable form, has a m of 817°C (28 atm) and sublimes at 614°C. Several other allotropic forms of arsenic are reported in the literature. It is believed that Albertus Magn obtained the element in 1250 A.D. In 1649 Schroeder published two methods of preparing the element. It is found native, in the sulfides *realgar* a *orpiment*, as arsenides and sulfarsenides of heavy metals, as the oxide, and as arsenates. *Mispickel*, arsenopyrite, (FeSAs) is the most common miner from which on heating the arsenic sublimes leaving ferrous sulfide. The element is a steel gray, very brittle, crystalline, semimetallic solid; it tarnish in air, and when heated is rapidly oxidized to arsenous oxide (As_2O_3) with the odor of garlic. Arsenic and its compounds are poisonous. Exposure arsenic and its compounds should not exceed 0.2 mg/m^3 as elemental As during an 8-hour work day. These values, however, are being studied, a may be lowered. Arsenic is also used in bronzing, pyrotechny, and for hardening and improving the sphericity of shot. The most important compoun are white arsenic (As_2O_3), the sulfide, Paris green $3Cu(AsO_2)_2 \cdot Cu(C_2H_3O_2)_2$, calcium arsenate, and lead arsenate; the last three have been used agricultural insecticides and poisons. Marsh's test makes use of the formation and ready decomposition of arsine (AsH_3). Arsenic is available in hig purity form. It is finding increasing uses as a doping agent in solid-state devices such as transistors. Gallium arsenide is used as a laser material to conv electricity directly into coherent light.

Astatine — (Gr. *astatos*, unstable), At; at. wt. (210); at. no. 85; m.p. 302°C; b.p. 337°C (est.); valence probably 1, 3, 5, or 7. Synthesized in 19 by D. R. Corson, K. R. MacKenzie, and E. Segre at the University of California by bombarding bismuth with alpha particles. The longest-lived isotope ^{210}At, have a half-life of only 8.1 hours. Twenty four isotopes are now known. Minute quantities of ^{215}At, ^{218}At, and ^{219}At exist in equilibrium in natu with naturally occurring uranium and thorium isotopes, and traces of ^{217}At are equilibrium with ^{233}U and ^{239}Np resulting from interaction of thoriu and uranium with naturally produced neutrons. The total amount of astatine present in the earth's crust, however, is less than 1 oz. Astatine can produced by bombarding bismuth with energetic alpha particles to obtain the relatively long-lived $^{209-211}At$, which can be distilled from the target b heating it in air. Only about 0.05 μg of astatine has been prepared to date. The "time of flight" mass spectrometer has been used to confirm that th highly radioactive halogen behaves chemically very much like other halogens, particularly iodine. The interhalogen compounds AtI, AtBr, and AtC are known to form, but it is not yet known if astatine forms diatomic astatine molecules. HAt and CH_3At (methyl astatide) have been detected. Astatir is said to be more metallic that iodine, and, like iodine, it probably accumulates in the thyroid gland. Workers at the Brookhaven National Laborator have recently used reactive scattering in crossed molecular beams to identify and measure elementary reactions involving astatine.

Barium — (Gr. *barys*, heavy), Ba; at. wt. 137.327, at. no. 56; m.p. 727°C; b.p. 1897°C; sp. gr. 3.5 (20°C); valence 2. Baryta was distinguishe from lime by Scheele in 1774; the element was discovered by Sir Humphrey Davy in 1808. It is found only in combination with other elements, chief *arite* or *heavy spar* (sulfate) and *witherite* (carbonate) and is prepared by electrolysis of the chloride. Barium is a metallic element, soft, and when e is silvery white like lead; it belongs to the alkaline earth group, resembling calcium chemically. The metal oxidizes very easily and should be t under petroleum or other suitable oxygen-free liquids to exclude air. It is decomposed by water or alcohol. The metal is used as a "getter" in vacuum es. The most important compounds are the peroxide (BaO_2), chloride, sulfate, carbonate, nitrate, and chlorate. Lithopone, a pigment containing ium sulfate and zinc sulfide, has good covering power, and does not darken in the presence of sulfides. The sulfate, as permanent white or *blanc* , is also used in paint, in X-ray diagnostic work, and in glassmaking. *Barite* is extensively used as a weighing agent in oilwell drilling fluids, and o in making rubber. The carbonate has been used as a rat poison, while the nitrate and chlorate give colors in pyrotechny. The impure sulfide osphoresces after exposure to the light. The compounds and the metal are not expensive. Barium metal (99.5 + % pure) costs about $20.00/lb. All ium compounds that are water or acid soluble are poisonous. Naturally occurring barium is a mixture of seven stable isotopes. Twenty two other ioactive isotopes are known to exist.

Berkelium — (*Berkeley,* home of the University of California), Bk; at. wt. (247); at. no. 97; valence 3 or 4; sp. gr. 14 (est.). Berkelium, the eighth mber of the actinide transition series, was discovered in December 1949 by Thompson, Ghiorso, and Seaborg, and was the fifth transuranium ment synthesized. It was produced by cyclotron bombardment of milligram amounts of ^{241}Am with helium ions at Berkeley, California. The first tope produced had a mass number of 243 and decayed with a half-life of 4.5 hours. Ten isotopes are now known and have been synthesized. The stence of ^{249}Bk, with a half-life of 314 days, makes it feasible to isolate berkelium in weighable amounts so that its properties can be investigated th macroscopic quantities. One of the first visible amounts of a pure berkelium compound, berkelium chloride, was produced in 1962. It weighed illionth of a gram. Berkelium probably has not yet been prepared in elemental form, but it is expected to be a silvery metal, easily soluble in dilute neral acids, and readily oxidized by air or oxygen at elevated temperatures to form the oxide. X-ray diffraction methods have been used to identify following compounds: BkO_2, BkO_3, BkF_3, BkCl, and BkOCl. As with other actinide elements, berkelium tends to accumulate in the skeletal system. e maximum permissible body burden of ^{249}Bk in the human skeleton is about 0.0004 μg. Because of its rarity, berkelium presently has no commercial technological use. Berkelium-249 is available from O.R.N.L. at a cost of $100/mg.

Beryllium — (Gr. *beryllos, beryl;* also called Glucinium or Glucinum, Gr. *glykys,* sweet), Be; at. wt. 9.012182; at no. 4; m.p. 1287°C; b.p. 2471°C; gr. 1.848 (20°C); valence 2. Discovered as the oxide by Vauquelin in beryl and in emeralds in 1798. The metal was isolated in 1828 by Wohler d by Bussy independently by the action of potassium on beryllium chloride. Beryllium is found in some 30 mineral species, the most important of hich are *bertrandite, beryl, chrysoberyl,* and *phenacite. Aquamarine* and *emerald* are precious forms of *beryl. Beryl* ($3BeO \cdot Al_2O_3 \cdot 6SiO_2$) and *rtrandite* ($4BeO \cdot 2SiO_2 \cdot H_2O$) are the most important commercial sources of the element and its compounds. Most of the metal is now prepared reducing beryllium fluoride with magnesium metal. Beryllium metal did not become readily available to industry until 1957. The metal, steel gray color, has many desirable properties. It is one of the lightest of all metals, and has one of the highest melting points of the light metals. Its modulus elasticity is about one third greater than that of steel. It resists attack by concentrated nitric acid, has excellent thermal conductivity, and is onmagnetic. It has a high permeability to X-rays, and when bombarded by alpha particles, as from radium or polonium, neutrons are produced in the tio of about 30 neutrons/million alpha particles. At ordinary temperatures beryllium resists oxidation in air, although its ability to scratch glass is obably due to the formation of a thin layer of the oxide. Beryllium is used as an alloying agent in producing beryllium copper which is extensively ed for springs, electrical contacts, spot-welding electrodes, and nonsparking tools. It has found application as a structural material for high-speed rcraft, missiles, spacecraft, and communication satellites. It is being used in the windshield frame, brake discs, support beams, and other structural omponents of the space shuttle. Because beryllium is relatively transparent to X-rays, ultra-thin Be-foil is finding use in X-ray lithography for production of microminiature integrated circuits.

Beryllium is used in nuclear reactors as a reflector or moderator for it has a low thermal neutron absorption cross section. It is used in gyroscopes, omputer parts, and instruments where lightness, stiffness, and dimensional stability are required. The oxide has a very high melting point and is also sed in nuclear work and ceramic applications. Beryllium and its salts are toxic and should be handled with the greatest of care. Beryllium and its ompounds should not be tasted to verify the sweetish nature of beryllium (as did early experimenters). The metal, its alloys, and its salts can be handled fely if certain work codes are observed, but no attempt should be made to work with beryllium before becoming familiar with proper safeguards. xposure to beryllium dust in air should be limited to 2 μg/m^3 (8-hour time-weighted average — 40-hour week), with a ceiling concentration of 5 μg/ ^{3}. A maximum peak above the acceptable ceiling concentration for an 8-hour shift is 25 μg/m^3 for a maximum duration of 30 min. These values are eing reviewed and studied. Beryllium metal in vacuum cast billet form is priced roughly at $150/lb. Fabricated forms are more expensive.

Bismuth — (Ger. *Weisse Masse,* white mass; later *Wisuth* and *Bisemutum*), Bi; at. wt. 208.98037; at. no. 83; m.p. 271.4°C; b.p. 1564 ± 5°C; sp. r. 9.747 (20°C); valence 3 or 5. In early times bismuth was confused with tin and lead. Claude Geoffroy the Younger showed it to be distinct from ead in 1753. It is a white crystalline, brittle metal with a pinkish tinge. It occurs native. The most important ores are *bismuthinite* or bismuth glance Bi_2S_3) and *bismite* (Bi_2O_3). Peru, Japan, Mexico, Bolivia, and Canada are major bismuth producers. Much of the bismuth produced in the U.S. is btained as a by-product in refining lead, copper, tin, silver, and gold ores. Bismuth is the most diamagnetic of all metals, and the thermal conductivity s lower than any metal, except mercury. It has a high electrical resistance, and has the highest Hall effect of any metal (i.e., greatest increase in electrical esistance when placed in a magnetic field). "Bismanol" is a permanent magnet of high coercive force, made of MnBi, by the U.S. Naval Surface Veapons Center. Bismuth expands 3.32% on solidification. This property makes bismuth alloys particularly suited to the making of sharp castings f objects subject to damage by high temperatures. With other metals such as tin, cadmium, etc., bismuth forms low-melting alloys which are xtensively used for safety devices in fire detection and extinguishing systems. Bismuth is used in producing malleable irons and is finding use as a atalyst for making acrylic fibers. When bismuth is heated in air it burns with a blue flame, forming yellow fumes of the oxide. The metal is also used s a thermocouple material, and has found application as a carrier for U^{235} or U^{233} fuel in atomic reactors. Its soluble salts are characterized by forming nsoluble basic salts on the addition of water, a property sometimes used in detection work. Bismuth oxychloride is used extensively in cosmetics. Bismuth subnitrate and subcarbonate are used in medicine. Bismuth metal costs about $30/lb.

Boron — (Ar. *Buraq,* Pers. *Burah*), B; at. wt. 10.81; at. no. 5; m.p. 2075°C; b.p. 4000°C; sp. gr. of crystals 2.34, of amorphous variety 2.37; valence 3. Boron compounds have been known for thousands of years, but the element was not discovered until 1808 by Sir Humphry Davy and by Gay-Lussac and Thenard. The element is not found free in nature, but occurs as orthoboric acid usually in certain volcanic spring waters and as borates in *bo* and *colemanite. Ulexite,* another boron mineral, is interesting as it is nature's own version of "fiber optics." Important sources of boron are the o *rasorite (kernite)* and *tincal (borax ore).* Both of these ores are found in the Mohave Desert. *Tincal* is the most important source of boron from Mohave. Extensive *borax* deposits are also found in Turkey. Boron exists naturally as 19.78% ^{10}B isotope and 80.22% ^{11}B isotope. High-pu crystalline boron may be prepared by the vapor phase reduction of boron trichloride or tribromide with hydrogen on electrically heated filaments. T impure, or amorphous, boron, a brownish-black powder, can be obtained by heating the trioxide with magnesium powder. Boron of 99.9999% pu has been produced and is available commercially. Elemental boron has an energy band gap of 1.50 to 1.56 eV, which is higher than that of either silic or germanium. It has interesting optical characteristics, transmitting portions of the infrared, and is a poor conductor of electricity at room temperatu but a good conductor at high temperature. Amorphous boron is used in pyrotechnic flares to provide a distinctive green color, and in rockets as an igni By far the most commercially important boron compound in terms of dollar sales is $Na_2B_4O_7 \cdot 5H_2O$. This pentahydrate is used in very large quanti in the manufacture of insulation fiberglass and sodium perborate bleach. Boric acid is also an important boron compound with major markets in tex fiberglass and in cellulose insulation as a flame retardant. Next in order of importance is borax ($Na_2B_4O_7 \cdot 10H_2O$) which is used principally in laun products. Use of borax as a mild antiseptic is minor in terms of dollars and tons. Boron compounds are also extensively used in the manufacture borosilicate glasses. Other boron compounds show promise in treating arthritis. The isotope boron-10 is used as a control for nuclear reactors, a shield for nuclear radiation, and in instruments used for detecting neutrons. Boron nitride has remarkable properties and can be used to make a mater as hard as diamond. The nitride also behaves like an electrical insulator but conducts heat like a metal. It also has lubricating properties similar graphite. The hydrides are easily oxidized with considerable energy liberation, and have been studied for use as rocket fuels. Demand is increasi for boron filaments, a high-strength, lightweight material chiefly employed for advanced aerospace structures. Boron is similar to carbon in that it h a capacity to form stable covalently bonded molecular networks. Carboranes, metalloboranes, phosphacarboranes, and other families compri thousands of compounds. Crystalline boron (99%) costs about $5/g. Amorphous boron costs about $2/g. Elemental boron and the borates are n considered to be toxic, and they do not require special care in handling. However, some of the more exotic boron hydrogen compounds are definite toxic and do require care.

Bromine — (Gr. *bromos,* stench), Br; at. wt. 79.904; at. no. 35; m.p. –7.2°C; b.p. 58.78°C; density of gas 7.59 g/l, liquid 3.12 (20°C); valence 3, 5, or 7. Discovered by Balard in 1826, but not prepared in quantity until 1860. A member of the halogen group of elements, it is obtained from natu brines from wells in Michigan and Arkansas. Little bromine is extracted today from seawater, which contains only about 85 ppm. Bromine is the on liquid nonmetallic element. It is a heavy, mobile, reddish-brown liquid, volatilizing readily at room temperature to a red vapor with a strong disagreeab odor, resembling chlorine, and having a very irritating effect on the eyes and throat; it is readily soluble in water or carbon disulfide, forming a r solution, is less active than chlorine but more so than iodine; it unites readily with many elements and has a bleaching action; when spilled on the sk it produces painful sores. It presents a serious health hazard, and maximum safety precautions should be taken when handling it. Much of the bromi output in the U.S. was used in the production of ethylene dibromide, a lead scavenger used in making gasoline antiknock compounds. Lead in gasolin however, has been drastically reduced, due to environmental considerations. This will greatly affect future production of bromine. Bromine is al used in making fumigants, flameproofing agents, water purification compounds, dyes, medicinals, sanitizers, inorganic bromides for photograph etc. Organic bromides are also important.

Cadmium — (L. *cadmia;* Gr. *kadmeia* — ancient name for calamine, zinc carbonate), Cd; at. wt. 112.411; at. no. 48; m.p. 321.07°C; b.p. 767° sp. gr. 8.65 (20°C); valence 2. Discovered by Stromeyer in 1817 from an impurity in zinc carbonate. Cadmium most often occurs in small quantiti associated with zinc ores, such as *sphalerite* (ZnS). *Greenockite* (CdS) is the only mineral of any consequence bearing cadmium. Almost all cadmiu is obtained as a by-product in the treatment of zinc, copper, and lead ores. It is a soft, bluish-white metal which is easily cut with a knife. It is simil in many respects to zinc. It is a component of some of the lowest melting alloys; it is used in bearing alloys with low coefficients of friction and gre resistance to fatigue; it is used extensively in electroplating, which accounts for about 60% of its use. It is also used in many types of solder, for standa E.M.F. cells, for Ni-Cd batteries, and as a barrier to control atomic fission. Cadmium compounds are used in black and white television phospho and in blue and green phosphors for color TV tubes. It forms a number of salts, of which the sulfate is most common; the sulfide is used as a yello pigment. Cadmium and solutions of its compounds are toxic. Failure to appreciate the toxic properties of cadmium may cause workers to be unwitting exposed to dangerous fumes. Silver solder, for example, which contains cadmium, should be handled with care. Serious toxicity problems have bee found from long-term exposure and work with cadmium plating baths. Exposure to cadmium dust should not exceed 0.01 mg/m^3 (8-hour time-weighte average, 40-hour week). The ceiling concentration (maximum), for a period of 15 min, should not exceed 0.14 mg/m^3. Cadmium oxide fume exposu (8-hour, 40-hour week) should not exceed 0.05 mg/m^3, and the maximum concentration should not exceed 0.05 mg/m^3. These values are presentl being restudied and recommendations have been made to reduce the exposure. In 1927 the International Conference on Weights and Measure redefined the meter in terms of the wavelength of the red cadmium spectral line (i.e. 1 m = 1,553,164.13 wavelengths). This definition has been change (see under Krypton). The current price of cadmium is about $12/lb. It is available in high purity form.

Calcium — (L. *calx,* lime), Ca; at. wt. 40.078; at. no. 20; m.p. 842 ± 2°C; b.p. 1484°C; sp. gr. 1.55 (20°C); valence 2. Though lime was prepare by the Romans in the first century under the name calx, the metal was not discovered until 1808. After learning that Berzelius and Pontin prepare calcium amalgam by electrolyzing lime in mercury, Davy was able to isolate the impure metal. Calcium is a metallic element, fifth in abundance i the earth's crust, of which it forms more than 3%. It is an essential constituent of leaves, bones, teeth, and shells. Never found in nature uncombined it occurs abundantly as *limestone* ($CaCO_3$), *gypsum* ($CaSO_4 \cdot 2H_2O$), and *fluorite* (CaF_2); *apatite* is the fluorophosphate or chlorophosphate of calcium The metal has a silvery color, is rather hard, and is prepared by electrolysis of the fused chloride to which calcium fluoride is added to lower the meltin point. Chemically it is one of the alkaline earth elements; it readily forms a white coating of nitride in air, reacts with water, burns with a yellow-re flame, forming largely the nitride. The metal is used as a reducing agent in preparing other metals such as thorium, uranium, zirconium, etc., and i used as a deoxidizer, desulfurizer, or decarburizer for various ferrous and nonferrous alloys. It is also used as an alloying agent for aluminum, beryllium copper, lead, and magnesium alloys, and serves as a "getter" for residual gases in vacuum tubes, etc. Its natural and prepared compounds are widely used. Quicklime (CaO), made by heating limestone and changed into slaked lime by the careful addition of water, is the great cheap base of chemica ustry with countless uses. Mixed with sand it hardens as mortar and plaster by taking up carbon dioxide from the air. Calcium from limestone is important element in Portland cement. The solubility of the carbonate in water containing carbon dioxide causes the formation of caves with lactites and stalagmites and is responsible for hardness in water. Other important compounds are the carbide (CaC_2), chloride ($CaCl_2$), cyanamide aCN$_2$), hypochlorite ($Ca(OCl)_2$), nitrate ($Ca(NO_3)_2$), and sulfide (CaS).

Californium — (State and University of California), Cf; at. wt. (251); at. no. 98. Californium, the sixth transuranium element to be discovered, s produced by Thompson, Street, Ghioirso, and Seaborg in 1950 by bombarding microgram quantities of ^{242}Cm with 35 MeV helium ions in the rkeley 60-inch cyclotron. Californium (III) is the only ion stable in aqueous solutions, all attempts to reduce or oxidize californium (III) having failed. e isotope ^{249}Cf results from the beta decay of ^{249}Bk while the heavier isotopes are produced by intense neutron irradiation by the reactions:

$$^{249}\text{Bk}(n,\gamma)\rightarrow{}^{250}\text{Bk}\xrightarrow{\beta}{}^{250}\text{Cf and }{}^{249}\text{Cf}(n,\gamma)\rightarrow{}^{250}\text{Cf}$$

llowed by

$$^{250}\text{Cf}(n,\gamma)\rightarrow{}^{251}\text{Cf}(n,\gamma)\rightarrow{}^{252}\text{Cf}$$

ne existence of the isotopes ^{249}Cf, ^{250}Cf, ^{251}Cf, and ^{252}Cf makes it feasible to isolate californium in weighable amounts so that its properties can be vestigated with macroscopic quantities. Californium-252 is a very strong neutron emitter. One microgram releases 170 million neutrons per minute, hich presents biological hazards. Proper safeguards should be used in handling californium. In 1960 a few tenths of a microgram of californium chloride, $CfCl_3$, californium oxychloride, CfOCl, and californium oxide, Cf_2O_3, were first prepared. Reduction of californium to its metallic state s not yet been accomplished. Because californium is a very efficient source of neutrons, many new uses are expected for it. It has already found use neutron moisture gages and in well-logging (the determination of water and oil-bearing layers). It is also being used as a portable neutron source r discovery of metals such as gold or silver by on-the-spot activation analysis. ^{252}Cf is now being offered for sale by the O.R.N.L. at a cost of $10/ g. As of May 1975, more than 63 mg have been produced and sold. It has been suggested that californium may be produced in certain stellar explosions, lled *supernovae*, for the radioactive decay of ^{254}Cf (55-day half-life) agrees with the characteristics of the light curves of such explosions observed rough telescopes. This suggestion, however, is questioned.

Carbon — (L. *carbo*, charcoal), C; at. wt. 12.011; at. no. 6; m.p. ~3550°C, graphite sublimes at 3825°C; triple point; (graphite-liquid-gas), 4492°C a pressure of 10.3 MPa and (graphite-diamond-liquid), 3830—3930° at a pressure of 12—13 GPa; sp. gr. amorphous 1.8 to 2.1, graphite 1.9 to 2.3, iamond 3.15 to 3.53 (depending on variety); gem diamond 3.513 (25°C); valence 2, 3, or 4. Carbon, an element of prehistoric discovery, is very widely istributed in nature. It is found in abundance in the sun, stars, comets, and atmospheres of most plants. Carbon in the form of microscopic diamonds found in some meteorites. Natural diamonds are found in *kimberlite* of ancient volcanic "pipes," such as found in South Africa, Arkansas, and sewhere. Diamonds are now also being recovered from the ocean floor off the Cape of Good Hope. About 30% of all industrial diamonds used in e U.S. are now made synthetically. The energy of the sun and stars can be attributed at least in part to the well-known carbon-nitrogen cycle. Carbon found free in nature in three allotropic forms: amorphous, graphite, and diamond. A fourth form, known as "white" carbon, is now thought to exist. raphite is one of the softest known materials while diamond is one of the hardest. Graphite exists in two forms: alpha and beta. These have identical hysical properties, except for their crystal structure. Naturally occurring graphites are reported to contain as much as 30% of the rhombohedral (beta) orm, whereas synthetic materials contain only the alpha form. The hexagonal alpha type can be converted to the beta by mechanical treatment, and ne beta form reverts to the alpha on heating it above 1000°C. In 1969 a new allotropic form of carbon was produced during the sublimation of pyrolytic raphite at low pressures. Under free-vaporization conditions above ~2550 K, "white" carbon forms as small transparent crystals on the edges of the asal planes of graphite. The interplanar spacings of "white" carbon are identical to those of carbon form noted in the graphitic gneiss from the Ries meteoritic) Crater of Germany. "White" carbon is a transparent birefringent material. Little information is presently available about this allotrope. n combination, carbon is found as carbon dioxide in the atmosphere of the earth and dissolved in all natural waters. It is a component of great rock nasses in the form of carbonates of calcium (limestone), magnesium, and iron. Coal, petroleum, and natural gas are chiefly hydrocarbons. Carbon is nique among the elements in the vast number and variety of compounds it can form. With hydrogen, oxygen, nitrogen, and other elements, it forms very large number of compounds, carbon atom often being linked to carbon atom. There are close to ten million known carbon compounds, many housands of which are vital to organic and life processes. Without carbon, the basis for life would be impossible. While it has been thought that silicon night take the place of carbon in forming a host of similar compounds, it is now not possible to form stable compounds with very long chains of silicon toms. The atmosphere of Mars contains 96.2% CO_2. Some of the most important compounds of carbon are carbon dioxide (CO_2), carbon monoxide CO), carbon disulfide (CS_2), chloroform ($CHCl_3$), carbon tetrachloride (CCl_4), methane (CH_4), ethylene (C_2H_4), acetylene (C_2H_2), benzene (C_6H_6), thyl alcohol (C_2H_5OH), acetic acid (CH_3COOH), and their derivatives. Carbon has seven isotopes. In 1961 the International Union of Pure and Applied Chemistry adopted the isotope carbon-12 as the basis for atomic weights. Carbon-14, an isotope with a half-life of 5715 years, has been widely used o date such materials as wood, archeological specimens, etc. Carbon-13 is now commercially available at a cost of $700/g.

Cerium — (named for the asteroid *Ceres*, which was discovered in 1801 only 2 years before the element), Ce; at. wt. 140.115; at. no. 58; m.p. 799°C; b.p. 3424°C; sp. gr. 6.770 (25°C); valence 3 or 4. Discovered in 1803 by Klaproth and by Berzelius and Hisinger; metal prepared by Hillebrand and Norton in 1875. Cerium is the most abundant of the metals of the so-called rare earths. It is found in a number of minerals including *allanite* (also known as *orthite*), *monazite*, *bastnasite*, *cerite*, and *samarskite*. Monazite and bastnasite are presently the two most important sources of cerium. Large deposits of monazite found on the beaches of Travancore, India, in river sands in Brazil, and deposits of *allanite* in the western United States, and *bastnasite* in Southern California will supply cerium, thorium, and the other rare-earth metals for many years to come. Metallic cerium is prepared by metallothermic reduction techniques, such as by reducing cerous fluoride with calcium, or by electrolysis of molten cerous chloride or other cerous halides. The metallothermic technique is used to produce high-purity cerium. Cerium is especially interesting because of its variable electronic structure. The energy of the inner 4f level is nearly the same as that of the outer or valence electrons, and only small amounts of energy are require to change the relative occupancy of these electronic levels. This gives rise to dual valency states. For example, a volume change of about 10% occu when cerium is subjected to high pressures or low temperatures. It appears that the valence changes from about 3 to 4 when it is cooled or compresse The low temperature behavior of cerium is complex. Four allotropic modifications are thought to exist: cerium at room temperature and at atmospher pressure is known as γ cerium. Upon cooling to –16°C, γ cerium changes to β cerium. The remaining γ cerium starts to change to α cerium when coole to –172°C, and the transformation is complete at –269°C. α Cerium has a density of 8.16; δ cerium exists above 726°C. At atmospheric pressure, liqu cerium is more dense than its solid form at the melting point. Cerium is an iron-gray lustrous metal. It is malleable, and oxidizes very readily at roo temperature, especially in moist air. Except for europium, cerium is the most reactive of the "rare-earth" metals. It slowly decomposes in cold wate and rapidly in hot water. Alkali solutions and dilute and concentrated acids attack the metal rapidly. The pure metal is likely to ignite if scratched wi a knife. Ceric salts are orange red or yellowish; cerous salts are usually white. Cerium is a component of misch metal, which is extensively used the manufacture of pyrophoric alloys for cigarette lighters, etc. While cerium is not radioactive, the impure commercial grade may contain traces thorium, which is radioactive. The oxide is an important constituent of incandescent gas mantles and it is emerging as a hydrocarbon catalyst in "sel cleaning" ovens. In this application it can be incorporated into oven walls to prevent the collection of cooking residues. As ceric sulfate it finds extensiv use as a volumetric oxidizing agent in quantitative analysis. Cerium compounds are used in the manufacture of glass, both as a component and as decolorizer. The oxide is finding increased use as a glass polishing agent instead of rouge, for it is much faster than rouge in polishing glass surface Cerium, with other rare earths, is used in carbon-arc lighting, especially in the motion picture industry. It is also finding use as an important cataly in petroleum refining and in metallurgical and nuclear applications. In small lots, 99.9% cerium costs about $125/kg.

Cesium — (L. *caesius,* sky blue), Cs; at. wt. 132.90543; at. no. 55; m.p. 28.44 ± 0.01°C; b.p. 671°C; sp. gr. 1.873 (20°C); valence 1. Cesium wa discovered spectroscopically by Bunsen and Kirchhoff in 1860 in mineral water from Durkheim. Cesium, an alkali metal, occurs in *lepidolite, pollucit* (a hydrated silicate of aluminum and cesium), and in other sources. One of the world's richest sources of cesium is located at Bernic Lake, Manitob The deposits are estimated to contain 300,000 tons of pollucite, averaging 20% cesium. It can be isolated by electrolysis of the fused cyanide and b a number of other methods. Very pure, gas-free cesium can be prepared by thermal decomposition of cesium azide. The metal is characterized by spectrum containing two bright lines in the blue along with several others in the red, yellow, and green. It is silvery white, soft, and ductile. It is th most electropositive and most alkaline element. Cesium, gallium, and mercury are the only three metals that are liquid at room temperature. Cesiur reacts explosively with cold water, and reacts with ice at temperatures above –116°C. Cesium hydroxide, the strongest base known, attacks glas Because of its great affinity for oxygen the metal is used as a "getter" in electron tubes. It is also used in photoelectric cells, as well as a catalyst i the hydrogenation of certain organic compounds. The metal has recently found application in ion propulsion systems. Cesium is used in atomic clock which are accurate to 5 s in 300 years. Its chief compounds are the chloride and the nitrate. Cesium has 32 isotopes (more than any element) with masse ranging from 114 to 145. The present price of cesium is about $30/g.

Chlorine — (Gr. *chloros,* greenish yellow), Cl; at. wt. 35.4527; at. no. 17; m.p. –101.5°C; b.p. –34.04°C; density 3.214 g/l; sp. gr. 1.56 (–33.6°C valence 1, 3, 5, or 7. Discovered in 1774 by Scheele, who thought it contained oxygen; named in 1810 by Davy, who insisted it was an element. I nature it is found in the combined state only, chiefly with sodium as common salt (NaCl), *carnallite* ($KMgCl_3 \cdot 6H_2O$), and *sylvite* (KCl). It is a membe of the halogen (salt-forming) group of elements and is obtained from chlorides by the action of oxidizing agents and more often by electrolysis; it a greenish-yellow gas, combining directly with nearly all elements. At 10°C one volume of water dissolves 3.10 volumes of chlorine, at 30°C onl 1.77 volumes. Chlorine is widely used in making many everyday products. It is used for producing safe drinking water the world over. Even the smalle water supplies are now usually chlorinated. It is also extensively used in the production of paper products, dyestuffs, textiles, petroleum products medicines, antiseptics, insecticides, foodstuffs, solvents, paints, plastics, and many other consumer products. Most of the chlorine produced is use in the manufacture of chlorinated compounds for sanitation, pulp bleaching, disinfectants, and textile processing. Further use is in the manufactur of chlorates, chloroform, carbon tetrachloride, and in the extraction of bromine. Organic chemistry demands much from chlorine, both as an oxidizin agent and in substitution, since it often brings desired properties in an organic compound when substituted for hydrogen, as in one form of syntheti rubber. Chlorine is a respiratory irritant. The gas irritates the mucous membranes and the liquid burns the skin. As little as 3.5 ppm can be detecte as an odor, and 1000 ppm is likely to be fatal after a few deep breaths. It was used as a war gas in 1915. Exposure to chlorine should not exceed 0. ppm (8-hour time-weighted average — 40 hour week.)

Chromium — (Gr. *chroma,* color), Cr; at. wt. 51.9961; at. no. 24; m.p. 1907°C; b.p. 2671°C; sp. gr. 7.18 to 7.20 (20°C); valence chiefly 2, 3, o 6. Discovered in 1797 by Vauquelin, who prepared the metal the next year, chromium is a steel-gray, lustrous, hard metal that takes a high polish. Th principal ore is *chromite* ($FeCr_2O_4$), which is found in Zimbabwe, U.S.S.R., Transvaal, Turkey, Iran, Albania, Finland, Democratic Republic o Madagascar, and the Philippines. The metal is usually produced by reducing the oxide with aluminum. Chromium is used to harden steel, t manufacture stainless steel, and to form many useful alloys. Much is used in plating to produce a hard, beautiful surface and to prevent corrosion Chromium is used to give glass an emerald green color. It finds wide use as a catalyst. All compounds of chromium are colored; the most importan are the chromates of sodium and potassium (K_2CrO_4) and the dichromates ($K_2Cr_2O_7$) and the potassium and ammonium chrome alums, as $KCr(SO_4)_2 \cdot 12H_2O$. The dichromates are used as oxidizing agents in quantitative analysis, also in tanning leather. Other compounds are of industrial value; lea chromate is chrome yellow, a valued pigment. Chromium compounds are used in the textile industry as mordants, and by the aircraft and other industrie for anodizing aluminum. The refractory industry has found chromite useful for forming bricks and shapes, as it has a high melting point, moderat thermal expansion, and stability of crystalline structure. Chromium compounds are toxic and should be handled with proper safeguards.

Cobalt — (*Kobald,* from the German, goblin or evil spirit, *cobalos,* Greek, mine), Co; at. wt. 58.93320; at. no. 27; m.p. 1495°C; b.p. 2927°C; sp gr. 8.9 (20°C); valence 2 or 3. Discovered by Brandt about 1735. Cobalt occurs in the mineral *cobaltite, smaltite,* and *erythrite,* and is often associated with nickel, silver, lead, copper, and iron ores, from which it is most frequently obtained as a by-product. It is also present in meteorites. Importan ore deposits are found in Zaire, Morocco, and Canada. The U.S. Geological Survey has announced that the bottom of the north central Pacific Ocear may have cobalt-rich deposits at relatively shallow depths in waters close to the Hawaiian Islands and other U.S. Pacific territories. Cobalt is a brittle hard metal, closely resembling iron and nickel in appearance. It has a magnetic permeability of about two thirds that of iron. Cobalt tends to exist as mixture of two allotropes over a wide temperature range; the β-form predominates below 400°C, and the α above that temperature. The transformation sluggish and accounts in part for the wide variation in reported data on physical properties of cobalt. It is alloyed with iron, nickel and other metals make Alnico, an alloy of unusual magnetic strength with many important uses. Stellite alloys, containing cobalt, chromium, and tungsten, are used r high-speed, heavy-duty, high temperature cutting tools, and for dies. Cobalt is also used in other magnet steels and stainless steels, and in alloys ed in jet turbines and gas turbine generators. The metal is used in electroplating because of its appearance, hardness, and resistance to oxidation. ne salts have been used for centuries for the production of brilliant and permanent blue colors in porcelain, glass, pottery, tiles, and enamels. It is e principal ingredient in Sevre's and Thenard's blue. A solution of the chloride ($CoCl_2 \cdot 6H_2O$) is used as sympathetic ink. The cobalt amines are interest; the oxide and the nitrate are important. Cobalt carefully used in the form of the chloride, sulfate, acetate, or nitrate has been found effective correcting a certain mineral deficiency disease in animals. Soils should contain 0.13 to 0.30 ppm of cobalt for proper animal nutrition. Cobalt-60, artificial isotope, is an important gamma ray source, and is extensively used as a tracer and a radiotherapeutic agent. Single compact sources of obalt-60 vary from about $1 to $10/curie, depending on quantity and specific activity. Exposure to cobalt (metal fumes and dust) should be limited 0.05 mg/m^3 (8-hour time-weighted average, 40-hour week).

Columbium — See Niobium.

Copper — (L. *cuprum,* from the island of Cyprus), Cu; at. wt. 63.546; at. no. 29; m.p. 1084.62 ± 0.2°C; b.p. 2562°C; sp. gr. 8.96 (20°C); valence or 2. The discovery of copper dates from prehistoric times. It is said to have been mined for more than 5000 years. It is one of man's most important etals. Copper is reddish colored, takes on a bright metallic luster, and is malleable, ductile, and a good conductor of heat and electricity (second only silver in electrical conductivity). The electrical industry is one of the greatest users of copper. Copper occasionally occurs native, and is found in any minerals such as *cuprite, malachite, azurite, chalcopyrite,* and *bornite*. Large copper ore deposits are found in the U.S., Chile, Zambia, Zaire, eru, and Canada. The most important copper ores are the sulfides, oxides, and carbonates. From these, copper is obtained by smelting, leaching, and y electrolysis. Its alloys, brass and bronze, long used, are still very important; all American coins are now copper alloys; monel and gun metals also ontain copper. The most important compounds are the oxide and the sulfate, blue vitriol; the latter has wide use as an agricultural poison and as an lgicide in water purification. Copper compounds such as Fehling's solution are widely used in analytical chemistry in tests for sugar. High-purity opper (99.999 + %) is available commercially.

Curium — (Pierre and Marie Curie), Cm; at. wt. (247); at. no. 96; m.p. 1345 ± 40°C; sp. gr. 13.51 (calc.); valence 3 and 4. Although curium follows mericium in the periodic system, it was actually known before americium and was the third transuranium element to be discovered. It was identified y Seaborg, James, and Ghiorso in 1944 at the wartime Metallurgical Laboratory in Chicago as a result of helium-ion bombardment of ^{239}Pu in the erkeley, California, 60-inch cyclotron. Visible amounts (30 μg) of ^{242}Cm, in the form of the hydroxide, were first isolated by Werner and Perlman f the University of California in 1947. In 1950, Crane, Wallmann, and Cunningham found that the magnetic susceptibility of microgram samples of CmF$_3$ was of the same magnitude as that of GdF_3. This provided direct experimental evidence for assigning an electronic configuration to Cm^{+3}. In 951, the same workers prepared curium in its elemental form for the first time. Fourteen isotopes of curium are now known. The most stable, ^{247}Cm, with a half-life of 16 million years, is so short compared to the earth's age that any primordial curium must have disappeared long ago from the natural cene. Minute amounts of curium probably exist in natural deposits of uranium, as a result of a sequence of neutron captures and β decays sustained y the very low flux of neutrons naturally present in uranium ores. The presence of natural curium, however, has never been detected. ^{242}Cm and ^{244}Cm re available in multigram quantities. ^{248}Cm has been produced only in milligram amounts. Curium is similar in some regards to gadolinium, its rare-earth homolog, but it has a more complex crystal structure. Curium is silver in color, is chemically reactive, and is more electropositive than aluminum. CmO$_2$, Cm_2O_3, CmF_3, CmF_4, $CmCl_3$, $CmBr_3$, and CmI_3 have been prepared. Most compounds of trivalent curium are faintly yellow in color. ^{242}Cm generates about three watts of thermal energy per gram. This compares to one-half watt per gram of ^{238}Pu. This suggests use for curium as a power ource. ^{244}Cm is now offered for sale at $100/mg. Curium absorbed into the body accumulates in the bones, and is therefore very toxic as its radiation destroys the red-cell forming mechanism. The maximum permissible total body burden of ^{244}Cm (soluble) in a human being is 0.3 μCi (microcurie).

Deuterium, an isotope of hydrogen — see Hydrogen.

Dysprosium — (Gr. *dysprositos,* hard to get at), Dy; at. wt. 162.50; at. no. 66; m.p. 1411°C; b.p. 2561°C; sp. gr. 8.551 (25°C); valence 3. Dysprosium was discovered in 1886 by Lecoq de Boisbaudran, but not isolated. Neither the oxide nor the metal was available in relatively pure form until the development of ion-exchange separation and metallographic reduction techniques by Spedding and associates about 1950. Dysprosium occurs along with other so-called rare-earth or lanthanide elements in a variety of minerals such as *xenotime, fergusonite, gadolinite, euxenite, polycrase,* and *blomstrandine*. The most important sources, however, are from *monazite* and *bastnasite*. Dysprosium can be prepared by reduction of the trifluoride with calcium. The element has a metallic, bright silver luster. It is relatively stable in air at room temperature, and is readily attacked and dissolved, with the evolution of hydrogen, by dilute and concentrated mineral acids. The metal is soft enough to be cut with a knife and can be machined without sparking if overheating is avoided. Small amounts of impurities can greatly affect its physical properties. While dysprosium has not yet found many applications, its thermal neutron absorption cross-section and high melting point suggest metallurgical uses in nuclear control applications and for alloying with special stainless steels. A dysprosium oxide-nickel cermet has found use in cooling nuclear reactor rods. This cermet absorbs neutrons readily without swelling or contracting under prolonged neutron bombardment. In combination with vanadium and other rare earths, dysprosium has been used in making laser materials. Dysprosium-cadmium chalcogenides, as sources of infrared radiation, have been used for studying chemical reactions. The cost of dysprosium metal has dropped in recent years since the development of ion-exchange and solvent extraction techniques, and the discovery of large ore bodies. The metal costs about $300/kg in purities of 99 + %.

Einsteinium — (Albert Einstein), Es; at. wt. (252); at. no. 99. Einsteinium, the seventh transuranic element of the actinide series to be discovered, was identified by Ghiorso and co-workers at Berkeley in December 1952 in debris from the first large thermonuclear explosion, which took place in the Pacific in November 1952. The isotope produced was the 20-day ^{253}Es isotope. In 1961, a sufficient amount of einsteinium was produced to permit separation of a macroscopic amount of ^{253}Es. This sample weighed about 0.01 μg. A special magnetic-type balance was used in making this determination. ^{253}Es so produced was used to produce mendelevium (Element 101). About 3 μg of einsteinium has been produced at Oak Ridge National Laboratories by irradiating for several years kilogram quantities of ^{239}Pu in a reactor to produce ^{242}Pu. This was then fabricated into pellets of plutonium oxide and aluminum powder, and loaded into target rods for an initial 1-year irradiation at the Savannah River Plant, followec irradiation in a HFIR (High Flux Isotopic Reactor). After 4 months in the HFIR the targets were removed for chemical separation of the einstein from californium. Fourteen isotopes of einsteinium are now recognized. ^{254}Es has the longest half-life (275 days). Tracer studies using ^{253}Es show einsteinium has chemical properties typical of a heavy trivalent, actinide element.

Element 104 — In 1964, workers of the Joint Nuclear Research Institute at Dubna (U.S.S.R.) bombarded plutonium with accelerated 113 to MeV neon ions. By measuring fission tracks in a special glass with a microscope, they detected an isotope that decays by spontaneous fission. T suggested that this isotope, which had a half-life of 0.3 ± 0.1 s might be ^{260}104, produced by the following reaction:

$$^{242}_{94}\text{Pu} + ^{22}_{10}\text{Ne} \rightarrow ^{260}104 + 4\text{n}$$

Element 104, the first *transactinide* element, is expected to have chemical properties similar to those of hafnium. It would, for example, form a relativ volatile compound with chlorine (a tetrachloride). The Soviet scientists have performed experiments aimed at chemical identification, and h attempted to show that the 0.3-s activity is more volatile than that of the relatively nonvolatile actinide trichlorides. This experiment does not fu the test of chemically separating the new element from all others, but it provides important evidence for evaluation. New data, reportedly issued Soviet scientists, have reduced the half-life of the isotope they worked with from 0.3 to 0.15 s. The Dubna scientists suggest the name *kurchatovi* and symbol *Ku* for Element 104, in honor of Igor Vasilevich Kurchatov (1903—1960), late Head of Soviet Nuclear Research. In 1969, Ghiorso, Nurm Harris, K. A. Y. Eskola, and P. L. Eskola of the University of California at Berkeley reported they had positively identified two, and possibly thr isotopes of Element 104. The group also indicated that after repeated attempts so far they have been unable to produce isotope ^{260}104 reported by Dubna groups in 1964. The discoveries at Berkeley were made by bombarding a target of ^{249}Cf with ^{12}C nuclei of 71 MeV, and ^{13}C nuclei of 69 Me The combination of ^{12}C with ^{249}Cf followed by instant emission of four neutrons produced Element ^{257}104. This isotope has a half-life of 4 to decaying by emitting an alpha particle into ^{253}No, with a half-life of 105 s. The same reaction, except with the emission of three neutrons, was thou to have produced ^{258}104 with a half-life of about 1/100 s. Element ^{259}104 is formed by the merging of a ^{13}C nuclei with ^{249}Cf, followed by emissi of three neutrons. This isotope has a half-life of 3 to 4 s, and decays by emitting an alpha particle into ^{255}No, which has a half-life of 185 s. Thousar of atoms of ^{257}104 and ^{259}104 have been detected. The Berkeley group believe their identification of ^{258}104 is correct, but they do not attach the sa degree of confidence to this work as to their work on ^{257}104 and ^{259}104. The Berkeley group proposes for the new element the name *rutherfordi* (symbol Rf), in honor of Ernest R. Rutherford, New Zealand physicist. The claims for discovery and the naming of Element 104 are still in questi The International Union of Pure and Applied Physics has proposed using the neutral temporary name, "unnilquadium".

Element 105 — In 1967 G. N. Flerov reported that a Soviet team working at the Joint Institute for Nuclear Research at Dubna may have produc a few atoms of ^{260}105 and ^{261}105 by bombarding ^{243}Am with ^{22}Ne. Their evidence was based on time-coincidence measurements of alpha energi More recently, it was reported that early in 1970 Dubna scientists synthesized Element 105 and that by the end of April 1970 "had investigated all types of decay of the new element and had determined its chemical properties." The Soviet group has not proposed a name for Element 105. In la April 1970, it was announced that Ghiorso, Nurmia, Harris, K. A. Y. Eskola, and P. L. Eskola, working at the University of California at Berkele had positively identified Element 105. The discovery was made by bombarding a target of ^{249}Cf with a beam of 84 MeV nitrogen nuclei in the Hea Ion Linear Accelerator (HILAC). When a ^{15}N nuclear is absorbed by a ^{249}Cf nucleus, four neutrons are emitted and a new atom of ^{260}105 with a ha life of 1.6 s is formed. While the first atoms of Element 105 are said to have been detected conclusively on March 5, 1970, there is evidence that Eleme 105 had been formed in Berkeley experiments a year earlier by the method described. Ghiorso and his associates have attempted to confirm Sovi findings by more sophisticated methods without success. The Berkeley Group proposed the name *hahnium,* after the late German scientist Otto Ha (1879—1968), and *Ha* for the chemical symbol.

In October 1971, it was announced that two new isotopes of Element 105 were synthesized with the heavy ion linear accelerator by A. Ghior and co-workers at Berkeley. Element ^{261}105 was produced both by bombarding ^{250}Cf with ^{15}N and by bombarding ^{249}Bk with ^{16}O. The isotope emi 8.93-MeV α particles and decays to ^{257}Lr with a half-life of about 1.8 s. Element ^{262}105 was produced by bombarding ^{249}Bk with ^{18}O. It emits 8. MeV α particles and decays to ^{258}Lr with a half-life of about 40 s. Seven isotopes of Element 105 (unnilpentium) are now recognized.

Element 106 — In June 1974, members of the Joint Institute for Nuclear Research in Dubna, U.S.S.R., reported their discovery of Element 10 which they claim to have synthesized. In September 1974, workers of the Lawrence Berkeley and Livermore Laboratories also reported creation Element 106 "without any scientific doubt." The LBL and LLL Group used the Super HILAC to accelerate ^{18}O ions onto a ^{249}Cf target. Element 1 was created by the reaction ^{249}Cf(^{18}O, 4n)^{263}X, which decayed by α emission to rutherfordium, and then by α emission to nobelium, which in tu further decayed by α between daughter and granddaughter. The element so identified had α energies of 9.06 and 9.25 MeV with a half-life of 0.9 0.2 s. At Dubna, 280-MeV ions of ^{54}Cr from the 310-cm cyclotron were used to strike targets of ^{206}Pb, ^{207}Pb, and ^{208}Pb, in separate runs. Foils expose to a rotating target disc were used to detect spontaneous fission activities, the foils being etched and examined microscopically to detect the numb of fission tracks and the half-life of the fission activity. Other experiments were made to aid in confirmation of the discovery. Neither the Dubna tea nor the Berkeley-Livermore Group has proposed a name as yet for Element 106 (unnilhexium).

Element 107 — In 1976 Soviet scientists at Dubna announced they had synthesized Element 107 by bombarding ^{204}Bi with heavy nuclei of ^{54}C It is reported that earlier experiments in 1975 had allowed scientists "to glimpse" the new element for 2/1000 s. A rapidly rotating cylinder, coate with a thin layer of the bismuth metal, was used as the target. This was bombarded by a stream of ^{54}Cr ions fired tangentially. The existence of Elemer 107 was confirmed by a team of West German physicists at the Heavy Ion Research Laboratory at Darmstadt, who created and identified six nucle of Element 107.

Element 109 — On August 29, 1982, Element 109 was made and identified by physicists of the Heavy Ion Research Laboratory, Darmstadt, Wes Germany, by bombing a target of Bi-209 with accelerated nuclei of Fe-58. If the combined energy of two nuclei is sufficiently high, the repulsive force between the nuclei can be overcome. In this experiment it took a week of target bombardment to produce a single fused nucleus. The team confirme existence of Element 109 by four independent measurements. The newly formed atom recoiled from the target at a predicted velocity and was rated from smaller, faster nuclei by a newly developed velocity filter. The time of flight to the detector and the striking energy were measured found to match predicted values. The nucleus of ^{266}X started to decay 5 ms after striking the detector. A high-energy α particle was emitted, lucing $^{262}_{107}X$. This in turn emitted an α particle, becoming $^{258}_{105}Ha$, which in turn captured an electron and became $^{258}_{104}Rf$. This in turn decayed other nuclides. This experiment demonstrated the feasibility of using fusion techniques as a method of making new, heavy nuclei.

Element 110 (ununnilium) — In 1987 it was reported that Organessian, et al., in Russia claimed discovery of Element 110 (UUN). Their eriments indicated a spontaneous fissioning nuclide with an atomic number of 110, a mass of 272, and a half-life of 10 ms.

Erbium — (*Ytterby*, a town in Sweden), Er; at. wt. 167.26; at. no. 68; m.p. 1529°C; b.p. 2862°C; sp. gr. 9.066 (25°C); valence 3, Erbium, one of so-called rare-earth elements of the lanthanide series, is found in the minerals mentioned under dysprosium above. In 1842 Mosander separated ria," found in the mineral *gadolinite,* into three fractions which he called *yttria, erbia,* and *terbia.* The names *erbia* and *terbia* became confused is early period. After 1860, Mosander's *terbia* was known as *erbia,* and after 1877, the earlier known *erbia* became *terbia.* The *erbia* of this period later shown to consist of five oxides, now known as *erbia, scandia, holmia, thulia* and *ytterbia.* By 1905 Urbain and James independently succeeded solating fairly pure Er_2O_3. Klemm and Bommer first produced reasonably pure erbium metal in 1934 by reducing the anhydrous chloride with assium vapor. The pure metal is soft and malleable and has a bright, silvery, metallic luster. As with other rare-earth metals, its properties depend certain extent on the impurities present. The metal is fairly stable in air and does not oxidize as rapidly as some of the other rare-earth metals. urally occurring erbium is a mixture of six isotopes, all of which are stable. Nine radioactive isotopes of erbium are also recognized. Recent duction techniques, using ion-exchange reactions, have resulted in much lower prices of the rare-earth metals and their compounds in recent years. e cost of 99+% erbium metal is about $650/kg. Erbium is finding nuclear and metallurgical uses. Added to vanadium, for example, erbium lowers hardness and improves workability. Most of the rare-earth oxides have sharp absorption bands in the visible, ultraviolet, and near infrared. This perty, associated with the electronic structure, gives beautiful pastel colors to many of the rare-earth salts. Erbium oxide gives a pink color and has en used as a colorant in glasses and porcelain enamel glazes.

Europium — (Europe), Eu; at. wt. 151.965; at. no. 63; m.p. 822°C; b.p. 1596°C; sp. gr. 5.244 (25°C); valence 2 or 3. In 1890 Boisbaudran obtained sic fractions from samarium-gadolinium concentrates which had spark spectral lines not accounted for by samarium or gadolinium. These lines bsequently have been shown to belong to europium. The discovery of europium is generally credited to Demarcay, who separated the rare earth in asonably pure form in 1901. The pure metal was not isolated until recent years. Europium is now prepared by mixing Eu_2O_3 with a 10%-excess of thanum metal and heating the mixture in a tantalum crucible under high vacuum. The element is collected as a silvery white metallic deposit on e walls of the crucible. As with other rare-earth metals, except for lanthanum, europium ignites in air at about 150 to 180°C. Europium is about as rd as lead and is quite ductile. It is the most reactive of the rare-earth metals, quickly oxidizing in air. It resembles calcium in its reaction with water. *stnasite* and *monazite* are the principal ores containing europium. Europium has been identified spectroscopically in the sun and certain stars. venteen isotopes are now recognized. Europium isotopes are good neutron absorbers and are being studied for use in nuclear control applications. ropium oxide is now widely used as a phosphor activator and europium-activated yttrium vanadate is in commercial use as the red phosphor in color V tubes. Europium-doped plastic has been used as a laser material. With the development of ion-exchange techniques and special processes, the cost the metal has been greatly reduced in recent years. Europium is one of the rarest and most costly of the rare-earth metals. It is priced at about $7500/ g.

Fermium — (Enrico Fermi), Fm; at. wt. (257); at. no. 100. Fermium, the eighth transuranium element of the actinide series to be discovered, was entified by Ghiorso and co-workers in 1952 in the debris from a thermonuclear explosion in the Pacific in work involving the University of California adiation Laboratory, the Argonne National Laboratory, and the Los Alamos Scientific Laboratory. The isotope produced was the 20-hour ^{255}Fm. uring 1953 and early 1954, while discovery of elements 99 and 100 was withheld from publication for security reasons, a group from the Nobel stitute of Physics in Stockholm bombarded ^{238}U with ^{16}O ions, and isolated a 30-min α-emitter, which they ascribed to $^{250}100$, without claiming scovery of the element. This isotope has since been identified positively, and the 30-min half-life confirmed. The chemical properties of fermium ave been studied solely with tracer amounts, and in normal aqueous media only the (III) oxidation state appears to exist. The isotope ^{254}Fm and heavier otopes can be produced by intense neutron irradiation of lower elements such as plutonium by a process of successive neutron capture interspersed ith beta decays until these mass numbers and atomic numbers are reached. Sixteen isotopes of fermium are known to exist. ^{257}Fm, with a half-life f about 100.5 days, is the longest lived. ^{250}Fm, with a half-life of 30 min, has been shown to be a product of decay of Element $^{254}102$. It was by chemical lentification of ^{250}Fm that production of Element 102 (nobelium) was confirmed.

Fluorine — (L. and F. *fluere,* flow, or flux), F; at. wt. 18.9984032; at. no. 9; m.p. –219.62°C (1 atm); b.p. –188.12°C (1 atm); density 1.696 g/ (0°C, 1 atm); liq. den. at b.p. 1.50 g/cm³; valence 1. In 1529, Georigius Agricola described the use of fluorspar as a flux, and as early as 1670 chwandhard found that glass was etched when exposed to fluorspar treated with acid. Scheele and many later investigators, including Davy, Gay-ussac, Lavoisier, and Thenard, experimented with hydrofluoric acid, some experiments ending in tragedy. The element was finally isolated in 1886 y Moisson after nearly 74 years of continuous effort. Fluorine occurs chiefly in *fluorspar* (CaF_2) and *cryolite* (Na_2AlF_6), but is rather widely distributed n other minerals. It is a member of the halogen family of elements, and is obtained by electrolyzing a solution of potassium hydrogen fluoride in nhydrous hydrogen fluoride in a vessel of metal or transparent fluorspar. Modern commercial production methods are essentially variations on the rocedures first used by Moisson. Fluorine is the most electronegative and reactive of all elements. It is a pale yellow, corrosive gas, which reacts with ractically all organic and inorganic substances. Finely divided metals, glass, ceramics, carbon, and even water burn in fluorine with a bright flame. Jntil World War II, there was no commercial production of elemental fluorine. The atom bomb project and nuclear energy applications, however, made t necessary to produce large quantities. Safe handling techniques have now been developed and it is possible at present to transport liquid fluorine y the ton. Fluorine and its compounds are used in producing uranium (from the hexafluoride) and more than 100 commercial fluorochemicals, ncluding many well-known high-temperature plastics. Hydrofluoric acid is extensively used for etching the glass of light bulbs, etc. Fluorochloro ydrocarbons are extensively used in air conditioning and refrigeration. It has been suggested that fluorine can be substituted for hydrogen wherever t occurs in organic compounds, which could lead to an astronomical number of new fluorine compounds. The presence of fluorine as a soluble fluoride in drinking water to the extent of 2 ppm may cause mottled enamel in teeth, when used by children acquiring permanent teeth; in smaller am however, fluorides are said to be beneficial and used in water supplies to prevent dental cavities. Elemental fluorine has been studied as a propellant as it has an exceptionally high specific impulse value. Compounds of fluorine with rare gases have now been confirmed. Fluorides of x radon, and krypton are among those known. Elemental fluorine and the fluoride ion are highly toxic. The free element has a characteristic pu odor, detectable in concentrations as low as 20 ppb, which is below the safe working level. The recommended maximum allowable concentrati a daily 8-hour time-weighted exposure is 1 ppm.

Francium — (France), Fr; at. no. 87; at. wt. (223); m.p. 27°C; b.p. 677°C; valence 1. Discovered in 1939 by Mlle. Marguerite Perey of the Institute, Paris. Francium, the heaviest known member of the alkali metal series, occurs as a result of an alpha disintegration of actinium. It ca be made artificially by bombarding thorium with protons. While it occurs naturally in uranium minerals, there is probably less than an ounce of fran at any time in the total crust of the earth. It has the highest equivalent weight of any element, and is the most unstable of the first 101 elements periodic system. Thirty-three isotopes of francium are recognized. The longest lived ^{223}Fr(Ac, K), a daughter of ^{227}Ac, has a half-life of 22 min is the only isotope of francium occurring in nature. Because all known isotopes of francium are highly unstable, knowledge of the chemical prop of this element comes from radiochemical techniques. No weighable quantity of the element has been prepared or isolated. The chemical prop of francium most closely resemble cesium.

Gadolinium — (*gadolinite,* a mineral named for Gadolin, a Finnish chemist), Gd; at. wt. 157.25; at. no. 64; m.p. 1314 ± 1°C; b.p. 3264°C gr. 7.901 (25°C); valence 3. Gadolinia, the oxide of gadolinium, was separated by Marignac in 1880 and Lecoq de Boisbaudran independently iso the element from Mosander's "yttria" in 1886. The element was named for the mineral *gadolinite* from which this rare earth was originally obta Gadolinium is found in several other minerals, including *monazite* and *bastnasite,* which are of commercial importance. The element has been iso only in recent years. With the development of ion-exchange and solvent extraction techniques, the availability and price of gadolinium and the rare-earth metals have greatly improved. Seventeen isotopes of gadolinium are now recognized; seven occur naturally. The metal can be prepar the reduction of the anhydrous fluoride with metallic calcium. As with other related rare-earth metals, it is silvery white, has a metallic luster, a malleable and ductile. At room temperature, gadolinium crystallizes in the hexagonal, close-packed α form. Upon heating to 1235°C, α gadoli transforms into the β form, which has a body-centered cubic structure. The metal is relatively stable in dry air, but in moist air it tarnishes wit formation of a loosely adhering oxide film which spalls off and exposes more surface to oxidation. The metal reacts slowly with water and is sol in dilute acid. Gadolinium has the highest thermal neutron capture cross-section of any known element (49,000 barns). Natural gadolinium is a mix of seven isotopes. Two of these, ^{155}Gd and ^{157}Gd, have excellent capture characteristics, but they are present naturally in low concentrations. As a re gadolinium has a very fast burnout rate and has limited use as a nuclear control rod material. It has been used in making gadolinium yttrium gar which have microwave applications. Compounds of gadolinium are used in making phosphors for color TV tubes. The metal has unu superconductive properties. As little as 1% gadolinium has been found to improve the workability and resistance of iron, chromium, and related al to high temperatures and oxidation. Gadolinium ethyl sulfate has extremely low noise characteristics and may find use in duplicating the perform of amplifiers, such as the maser. The metal is ferromagnetic. Gadolinium is unique for its high magnetic moment and for its special Curie tempera (above which ferromagnetism vanishes) lying just at room temperature. This suggests uses as a magnetic component that senses hot and cold. The p of the metal is $485/kg.

Gallium — (L. *Gallia,* France; also from Latin, *gallus,* a translation of Lecoq, a cock), Ga; at. wt. 69.723; at. no. 31; m.p. 29.76°C; b.p. 2204 sp. gr. 5.904 (29.6°C) solid; sp. gr. 6.095 (29.6°C) liquid; valence 2 or 3. Predicted and described by Mendeleev as ekaaluminum, and discove spectroscopically by Lecoq de Boisbaudran in 1875, who in the same year obtained the free metal by electrolysis of a solution of the hydroxide in K Gallium is often found as a trace element in *diaspore, sphalerite, germanite, bauxite,* and *coal.* Some flue dusts from burning coal have been sh to contain as much as 1.5% gallium. It is the only metal, except for mercury, cesium, and rubidium, which can be liquid near room temperatures; makes possible its use in high-temperature thermometers. It has one of the longest liquid ranges of any metal and has a low vapor pressure even at h temperatures. There is a strong tendency for gallium to supercool below its freezing point. Therefore, seeding may be necessary to initiate solidificat Ultra-pure gallium has a beautiful, silvery appearance, and the solid metal exhibits a conchoidal fracture similar to glass. The metal expands 3.1% solidifying; therefore, it should not be stored in glass or metal containers, as they may break as the metal solidifies. Gallium wets glass or porcel and forms a brilliant mirror when it is painted on glass. It is widely used in doping semiconductors and producing solid-state devices such as transist High-purity gallium is attacked only slowly by mineral acids. Magnesium gallate containing divalent impurities such as Mn^{+2} is finding use commercial ultraviolet activated powder phosphors. Gallium arsenide is capable of converting electricity directly into coherent light. Gallium rea alloys with most metals, and has been used as a component in low-melting alloys. Its toxicity appears to be of a low order, but should be handled v care until more data are forthcoming. The metal can be supplied in ultrapure form (99.99999+%). The cost is about $3/g.

Germanium — (L. *Germania,* Germany), Ge; at. wt. 72.61; at. no. 32; m.p. 938.25°C; b.p. 2833°C; sp. gr. 5.323 (25°C); valence 2 and 4. Predic by Mendeleev in 1871 as ekasilicon, and discovered by Winkler in 1886. The metal is found in *argyrodite,* a sulfide of germanium and silver *germanite,* which contains 8% of the element; in zinc ores; in coal; and in other minerals. The element is frequently obtained commercially from f dusts of smelters processing zinc ores, and has been recovered from the by-products of combustion of certain coals. Its presence in coal insures a la reserve of the element in the years to come. Germanium can be separated from other metals by fractional distillation of its volatile tetrachloride. T tetrachloride may then be hydrolyzed to give GeO_2; the dioxide can be reduced with hydrogen to give the metal. Recently developed zone-refini techniques permit the production of germanium of ultra-high purity. The element is a gray-white metalloid, and in its pure state is crystalline and brit retaining its luster in air at room temperature. It is a very important semiconductor material. Zone-refining techniques have led to production crystalline germanium for semiconductor use with an impurity of only one part in 10^{10}. Doped with arsenic, gallium, or other elements, it is used a transistor element in thousands of electronic applications. Its application as a semiconductor element now provides the largest use for germaniu Germanium is also finding many other applications including use as an alloying agent, as a phosphor in fluorescent lamps, and as a catalyst. Germani and germanium oxide are transparent to the infrared and are used in infrared spectroscopes and other optical equipment, including extremely sensiti infrared detectors. Germanium oxide's high index of refraction and dispersion has made it useful as a component of glasses used in wide-angle came ıses and microscope objectives. The field of organogermanium chemistry is becoming increasingly important. Certain germanium compounds have ɔw mammalian toxicity, but a marked activity against certain bacteria, which makes them of interest as chemotherapeutic agents. The cost of rmanium is about \$3/g.

Gold — (Sanskrit *Jval;* Anglo-Saxon *gold*), Au (L. *aurum,* gold); at. wt. 196.96654; at. no. 79; m.p. 1064.18°C; b.p. 2856°C; sp. gr. ~19.3 (20°C); lence 1 or 3. Known and highly valued from earliest times, gold is found in nature as the free metal and in tellurides; it is very widely distributed d is almost always associated with quartz or pyrite. It occurs in veins and alluvial deposits, and is often separated from rocks and other minerals by ıicing and panning operations. About two thirds of the world's gold output comes from South Africa, and about two thirds of the total U.S. production mes from South Dakota and Nevada. The metal is recovered from its ores by cyaniding, amalgamating, and smelting processes. Refining is also quently done by electrolysis. Gold occurs in sea water to the extent of 0.1 to 2 mg/ton, depending on the location where the sample is taken. As yet, method has been found for recovering gold from sea water profitably. It is estimated that all the gold in the world, so far refined, could be placed a single cube 60 ft on a side. Of all the elements, gold in its pure state is undoubtedly the most beautiful. It is metallic, having a yellow color when a mass, but when finely divided it may be black, ruby, or purple. The Purple of Cassius is a delicate test for auric gold. It is the most malleable and ctile metal; 1 oz. of gold can be beaten out to 300 ft^2. It is a soft metal and is usually alloyed to give it more strength. It is a good conductor of heat d electricity, and is unaffected by air and most reagents. It is used in coinage and is a standard for monetary systems in many countries. It is also tensively used for jewelry, decoration, dental work, and for plating. It is used for coating certain space satellites, as it is a good reflector of infrared d is inert. Gold, like other precious metals, is measured in troy weight; when alloyed with other metals, the term *carat* is used to express the amount gold present, 24 carats being pure gold. For many years the value of gold was set by the U.S. at \$20.67/troy ounce; in 1934 this value was fixed law at \$35.00/troy ounce, 9/10th fine. On March 17, 1968, because of a gold crisis, a two-tiered pricing system was established whereby gold was ll used to settle international accounts at the old \$35.00/troy ounce price while the price of gold on the private market would be allowed to fluctuate. nce this time, the price of gold on the free market has fluctuated widely. The price of gold on the free market reached a price of \$620/troy oz. in January 980. The most common gold compounds are auric chloride ($AuCl_3$) and chlorauric acid ($HAuCl_4$), the latter being used in photography for toning e silver image. Gold has 18 isotopes; ^{198}Au, with a half-life of 2.7 days, is used for treating cancer and other diseases. Disodium aurothiomalate is lministered intramuscularly as a treatment for arthritis. A mixture of one part nitric acid with three of hydrochloric acid is called *aqua regia* (because dissolved gold, the King of Metals). Gold is available commercially with a purity of 99.999+%. For many years the temperature assigned to the eezing point of gold has been 1063.0°C; this has served as a calibration point for the International Temperature Scales (ITS-27 and ITS-48) and the ternational Practical Temperature Scale (IPTS-48). In 1968, a new International Practical Temperature Scale (IPTS-68) was adopted, which demands at the freezing point of gold be changed to 1064.43°C. The specific gravity of gold has been found to vary considerably depending on temperature, ɔw the metal is precipitated, and cold-worked. As of January 1990, gold was priced at about \$410/troy oz.

Hafnium — (*Hafinia,* Latin name for Copenhagen), Hf; at. wt. 178.49; at. no. 72; m.p. 2233 ± 20°C; b.p. 4603°C; sp. gr. 13.31 (20°C); valence . Hafnium was thought to be present in various minerals and concentrations many years prior to its discovery, in 1923, credited to D. Coster and G. on Hevesey. On the basis of the Bohr theory, the new element was expected to be associated with zirconium. It was finally identified in *zircon* from orway, by means of X-ray spectroscope analysis. It was named in honor of the city in which the discovery was made. Most zirconium minerals contain to 5% hafnium. It was originally separated from zirconium by repeated recrystallization of the double ammonium or potassium fluorides by von levesey and Jantzen. Metallic hafnium was first prepared by van Arkel and deBoer by passing the vapor of the tetraiodide over a heated tungsten lament. Almost all hafnium metal now produced is made by reducing the tetrachloride with magnesium or with sodium (Kroll Process). Hafnium a ductile metal with a brilliant silver luster. Its properties are considerably influenced by the impurities of zirconium present. Of all the elements, irconium and hafnium are two of the most difficult to separate. Their chemistry is almost identical, however, the density of zirconium is about half ıat of hafnium. Very pure hafnium has been produced, with zirconium being the major impurity. Because hafnium has a good absorption cross section or thermal neutrons (almost 600 times that of zirconium), has excellent mechanical properties, and is extremely corrosion resistant, it is used for reactor ontrol rods. Such rods are used in nuclear submarines. Hafnium has been successfully alloyed with iron, titanium, niobium, tantalum, and other metals. lafnium carbide is the most refractory binary composition known, and the nitride is the most refractory of all known metal nitrides (m.p. 3310°C). lafnium is used in gas-filled and incandescent lamps, and is an efficient "getter" for scavenging oxygen and nitrogen. Finely divided hafnium is yrophoric and can ignite spontaneously in air. Care should be taken when machining the metal or when handling hot sponge hafnium. At 700°C afnium rapidly absorbs hydrogen to form the composition $HfH_{1.86}$. Hafnium is resistant to concentrated alkalis, but at elevated temperatures reacts vith oxygen, nitrogen, carbon, boron, sulfur, and silicon. Halogens react directly to form tetrahalides. Exposure to hafnium should not exceed 0.5 mg/n^3 (8-hour time-weighted average — 40-hour week). The price of the metal is in the broad range of \$100 to \$500/lb, depending on purity and quantity. The yearly demand for hafnium in the U.S. is now in excess of 100,000 lb.

Hahnium — see Element 105 (unnilpentium).

Helium — (Gr. *helios,* the sun), He; at. wt. 4.002602; at. no. 2; m.p. below — 272.2°C (26 atm); b.p. — 268.93°C; density 0.1785 g/l (0°C, 1 atm); iquid density 7.62 lb/ft^3 at. b.p.; valence usually 0. Evidence of the existence of helium was first obtained by Janssen during the solar eclipse of 1868 vhen he detected a new line in the solar spectrum; Lockyer and Frankland suggested the name *helium* for the new element; in 1895, Ramsay discovered ıelium in the uranium mineral *clevite*, and it was independently discovered in cleveite by the Swedish chemists Cleve and Langlet about the same time. Rutherford and Royds in 1907 demonstrated that α particles are helium nuclei. Except for hydrogen, helium is the most abundant element found hroughout the universe. Helium is extracted from natural gas; all natural gas contains at least trace quantities of helium. It has been detected spectroscopically in great abundance, especially in the hotter stars, and it is an important component in both the proton-proton reaction and the carbon cycle, which account for the energy of the sun and stars. The fusion of hydrogen into helium provides the energy of the hydrogen bomb. The helium content of the atmosphere is about 1 part in 200,000. While it is present in various radioactive minerals as a decay product, the bulk of the Free World's supply is obtained from wells in Texas, Oklahoma, and Kansas. The only known helium extraction plants, outside the United States, in 1984 were in Eastern Europe (Poland), the U.S.S.R., and a little in India. The cost of helium fell from \$2500/ft^3 in 1915 to 1.5¢/ft^3 in 1940. The U.S. Bureau of Mines has set the price of Grade A helium at \$37.50/1000 ft^3 in 1986. Helium has the lowest melting point of any element and has found wide use in cryogenic research, as its boiling point is close to absolute zero. Its use in the study of superconductivity is vital. Using liquid helium, Kurti and co-workers, a others, have succeeded in obtaining temperatures of a few microkelvins by the adiabatic demagnetization of copper nuclei, starting from about 0. K. Seven isotopes of helium are known. Liquid helium (He^4) exists in two forms: He^4I and He^4II, with a sharp transition point at 2.174 K (3.83 Hg). He^4I (above this temperature) is a normal liquid, but He^4II (below it) is unlike any other known substance. It expands on cooling; its conductiv for heat is enormous; and neither its heat conduction nor viscosity obeys normal rules. It has other peculiar properties. Helium is the only liquid t cannot be solidified by lowering the temperature. It remains liquid down to absolute zero at ordinary pressures, but it can readily be solidified increasing the pressure. Solid ^3He and ^4He are unusual in that both can readily be changed in volume by more than 30% by application of pressu The specific heat of helium gas is unusually high. The density of helium vapor at the normal boiling point is also very high, with the vapor expandi greatly when heated to room temperature. Containers filled with helium gas at 5 to 10 K should be treated as though they contained liquid helium d to the large increase in pressure resulting from warming the gas to room temperature. While helium normally has a 0 valence, it seems to have a we tendency to combine with certain other elements. Means of preparing helium diflouride have been studied, and species such as HeNe and the molecu ions He^+ and He^{++} have been investigated. Helium is widely used as an inert gas shield for arc welding; as a protective gas in growing silicon a germanium crystals, and in titanium and zirconium production; as a cooling medium for nuclear reactors, and as a gas for supersonic wind tunne A mixture of helium and oxygen is used as an artificial atmosphere for divers and others working under pressure. Different ratios of He/O_2 are us for different depths at which the diver is operating. Helium is extensively used for filling balloons as it is a much safer gas than hydrogen. One of t recent largest uses for helium has been for pressuring liquid fuel rockets. A Saturn booster such as used on the Apollo lunar missions required abc 13 million $ft^3$of helium for a firing, plus more for checkouts. Liquid helium's use in magnetic resonance imaging (MRI) continues to increase as t medical profession accepts and develops new uses for the equipment. This equipment is providing accurate diagnoses of problems where exploratc surgery has previously been required to determine problems. Another medical application that is being developed uses MRI to determine by blo analysis whether a patient has any form of cancer. Lifting gas applications are increasing. Various companies in addition to Goodyear, are now usi "blimps" for advertising. The Navy and the Air Force are investigating the use of airships to provide early warning systems to detect low-flying cru missiles. The Drug Enforcement Agency is using radar-equipped blimps to detect drug smugglers along the southern border of the U.S. In additic NASA is currently using helium-filled balloons to sample the atmosphere in Antarctica to determine what is depleting the ozone layer that protec Earth from harmful U.V. radiation. Research on and development of materials which become superconductive at temperatures well above the boili point of helium could have a major impact on the demand for helium. Less costly refrigerants having boiling points considerably higher could repla the present need to cool such superconductive materials to the boiling point of helium.

Holmium—(L. *Holmia*, for Stockholm), Ho; at. wt. 164.93032; at. no 67; m.p. 1472°C; b.p. 2694°C; sp. gr. 8.795 (25°C); valence +3. The spect absorption bands of holmium were noticed in 1878 by the Swiss chemists Delafontaine and Soret, who announced the existence of an "Element X Cleve, of Sweden, later independently discovered the element while working on erbia earth. The element is named after Cleve's native city. Pu holmia, the yellow oxide, was prepared by Homberg in 1911. Holmium occurs in *gadolinite, monazite,* and in other rare-earth minerals. It commercially obtained from monazite, occurring in that mineral to the extent of about 0.05%. It has been isolated by the reduction of its anhydro chloride or fluoride with calcium metal. Pure holmium has a metallic to bright silver luster. It is relatively soft and malleable, and is stable in dry at room temperature, but rapidly oxidizes in moist air and at elevated temperatures. The metal has unusual magnetic properties. Few uses have yet be found for the element. The element, as with other rare earths, seems to have a low acute toxic rating. The price of 99 + % holmium metal is abo $10/g.

Hydrogen — (Gr. *hydro*, water, and *genes,* forming), H; at. wt. 1.00794; at. no. 1; m.p. –259.34°C; b.p. –252.87°C; density 0.08988 g/l; densi (liquid) 70.8 g/l (–253°C); density (solid) 70.6 g/l (–262°C); valence 1. Hydrogen was prepared many years before it was recognized as a distir substance by Cavendish in 1766. It was named by Lavoisier. Hydrogen is the most abundant of all elements in the universe, and it is thought that t heavier elements were, and still are, being built from hydrogen and helium. It has been estimated that hydrogen makes up more than 90% of all t atoms or three quarters of the mass of the universe. It is found in the sun and most stars, and plays an important part in the proton-proton reaction a carbon-nitrogen cycle, which accounts for the energy of the sun and stars. It is thought that hydrogen is a major component of the planet Jupiter a that at some depth in the planet's interior the pressure is so great that solid molecular hydrogen is converted into solid metallic hydrogen. In 1973, was reported that a group of Russian experimenters may have produced metallic hydrogen at a pressure of 2.8 Mbar. At the transition the densi changed from 1.08 to 1.3 g/cm^3. Earlier, in 1972, a Livermore (California) group also reported on a similar experiment in which they observed pressure-volume point centered at 2 Mbar. It has been predicted that metallic hydrogen may be metastable; others have predicted it would be superconductor at room temperature. On earth, hydrogen occurs chiefly in combination with oxygen in water, but it is also present in organic matt such as living plants, petroleum, coal, etc. It is present as the free element in the atmosphere, but only to the extent of less than 1 ppm by volume. is the lightest of all gases, and combines with other elements, sometimes explosively, to form compounds. Great quantities of hydrogen are require commercially for the fixation of nitrogen from the air in the Haber ammonia process and for the hydrogenation of fats and oils. It is also used in larg quantities in methanol production, in hydrodealkylation, hydrocracking, and hydrodesulfurization. It is also used as a rocket fuel, for welding, f production of hydrochloric acid, for the reduction of metallic ores, and for filling balloons. The lifting power of 1 ft^3 of hydrogen gas is about 0.07 lb at 0°C, 760 mm pressure. Production of hydrogen in the U.S. alone now amounts to about 3 billion cubic feet per year. It is prepared by the actic of steam on heated carbon, by decomposition of certain hydrocarbons with heat, by the electrolysis of water, or by the displacement from acids by certa metals. It is also produced by the action of sodium or potassium hydroxide on aluminum. Liquid hydrogen is important in cryogenics and in the stud of superconductivity, as its melting point is only a 20 degrees above absolute zero. The ordinary isotope of hydrogen, H, is known as *protium*. In 193 Urey announced the preparation of a stable isotope, deuterium (^{2}H or D) with an atomic weight of 2. Two years later an unstable isotope, tritium (H with an atomic weight of 3 was discovered. Tritium has a half-life of about 12.5 years. One atom of deuterium is found in about 6000 ordinary hydroge atoms. Tritium atoms are also present but in much smaller proportion. Tritium is readily produced in nuclear reactors and is used in the productic of the hydrogen bomb. It is also used as a radioactive agent in making luminous paints, and as a tracer. The current price of tritium, to authorize personnel, is about $2/Ci; deuterium gas is readily available, without permit, at about $1/l. Heavy water, deuterium oxide (D_2O), which is used as oderator to slow down neutrons, is available without permit at a cost of 6c to \$1/g, depending on quantity and purity. Quite apart from isotopes, it as been shown that hydrogen gas under ordinary conditions is a mixture of two kinds of molecules, known as *ortho*- and *para*-hydrogen, which differ om one another by the spins of their electrons and nuclei. Normal hydrogen at room temperature contains 25% of the *para* form and 75% of the *ortho* rm. The *ortho* form cannot be prepared in the pure state. Since the two forms differ in energy, the physical properties also differ. The melting and oiling points of *para*hydrogen are about 0.1°C lower than those of normal hydrogen. Consideration is being given to an entire economy based on olar- and nuclear-generated hydrogen. Located in remote regions, power plants would electrolyze sea water; the hydrogen produced would travel to istant cities by pipelines. Pollution-free hydrogen could replace natural gas, gasoline, etc., and could serve as a reducing agent in metallurgy, chemical rocessing, refining, etc. It could also be used to convert trash into methane and ethylene. Public acceptance, high capital investment, and the high resent cost of hydrogen with respect to present fuels are but a few of the problems facing establishment of such an economy.

Indium — (from the brilliant indigo line in its spectrum),In; at. wt. 114.82; at. no. 49; m.p. 156.60°C; b.p. 2072°C; sp. gr. 7.31 (20°C); valence , 2, or 3. Discovered by Reich and Richter, who later isolated the metal. Indium is most frequently associated with zinc materials, and it is from these at most commercial indium is now obtained; however, it is also found in iron, lead, and copper ores. Until 1924, a gram or so constituted the world's upply of this element in isolated form. It is probably about as abundant as silver. About 4 million troy ounces of indium are now produced annually the Free World. Canada is presently producing more than 1,000,000 troy ounces annually. The present cost of indium is about \$1 to \$5/g, depending n quantity and purity. It is available in ultrapure form. Indium is a very soft, silvery-white metal with a brilliant luster. The pure metal gives a high-itched "cry" when bent. It wets glass, as does gallium. It has found application in making low-melting alloys; an alloy of 24% indium-76% gallium liquid at room temperature. It is used in making bearing alloys, germanium transistors, rectifiers, thermistors, and photoconductors. It can be plated nto metal and evaporated onto glass, forming a mirror as good as that made with silver but with more resistance to atmospheric corrosion. There is vidence that indium has a low order of toxicity; however, care should be taken until further information is available.

Iodine — (Gr. *iodes*, violet), I; at. wt. 126.90447; at. no. 53; m.p. 113.7°C; b.p. 184.4°C; density of the gas 11.27 g/l; sp. gr. solid 4.93 (20°C); alence 1, 3, 5, or 7. Discovered by Courtois in 1811. Iodine, a halogen, occurs sparingly in the form of iodides in sea water from which it is assimilated y seaweeds, in Chilean saltpeter and nitrate-bearing earth, known as *caliche* in brines from old sea deposits, and in brackish waters from oil and salt wells. Ultrapure iodine can be obtained from the reaction of potassium iodide with copper sulfate. Several other methods of isolating the element are nown. Iodine is a bluish-black, lustrous solid, volatilizing at ordinary temperatures into a blue-violet gas with an irritating odor; it forms compounds with many elements, but is less active than the other halogens, which displace it from iodides. Iodine exhibits some metallic-like properties. It dissolves eadily in chloroform, carbon tetrachloride, or carbon disulfide to form beautiful purple solutions. It is only slightly soluble in water. Iodine compounds re important in organic chemistry and very useful in medicine. Thirty isotopes are recognized. Only one stable isotope, ^{127}I is found in nature. The rtificial radioisotope ^{131}I, with a half-life of 8 days, has been used in treating the thyroid gland. The most common compounds are the iodides of sodium nd potassium (KI) and the iodates (KIO_3). Lack of iodine is the cause of goiter. Iodides, and thyroxin which contains iodine, are used internally in medicine, and a solution of KI and iodine in alcohol is used for external wounds. Potassium iodide finds use in photography. The deep blue color with starch solution is characteristic of the free element. Care should be taken in handling and using iodine, as contact with the skin can cause lesions; iodine vapor is intensely irritating to the eyes and mucous membranes. The maximum allowable concentration of iodine in air should not exceed 1 mg/m^3 8-hour time-weighted average — 40-hour).

Iridium — (L. *iris*, rainbow), Ir; at. wt. 192.22; at. no. 77; m.p. 2446°C; b.p. 4428°C; sp. gr. 22.42 (17°C); valence 3 or 4. Discovered in 1803 by Tennant in the residue left when crude platinum is dissolved by aqua regia. The name iridium is appropriate, for its salts are highly colored. Iridium, a metal of the platinum family, is white, similar to platinum, but with a slight yellowish cast. It is very hard and brittle, making it very hard to machine, form, or work. It is the most corrosion-resistant metal known, and was used in making the standard meter bar of Paris, which is a 90% platinum-10% iridium alloy. This meter bar was replaced in 1960 as a fundamental unit of length (see under Krypton). Iridium is not attacked by any of the acids nor by aqua regia, but is attacked by molten salts, such as NaCl and NaCN. Iridium occurs uncombined in nature with platinum and other metals of this family in alluvial deposits. It is recovered as a by-product from the nickel mining industry. Iridium has found use in making crucibles and apparatus for use at high temperatures. It is also used for electrical contacts. Its principal use is as a hardening agent for platinum. With osmium, it forms an alloy which is used for tipping pens and compass bearings. The specific gravity of iridium is only very slightly lower than that of osmium, which has been generally credited as being the heaviest known element. Calculations of the densities of iridium and osmium from the space lattices gives values of 22.65 and 22.61 g/cm^3, respectively. These values may be more reliable than actual physical measurements. At present, therefore, we know that either iridium or osmium is the densest known element, but the data do not yet allow selection between the two. Iridium costs about \$500/troy ounce.

Iron — (Anglo-Saxon, *iron*), Fe (L. *ferrum*); at. wt. 55.847 at. no. 26; m.p. 1538°C; b.p. 2861°C; sp. gr. 7.874 (20°C); valence 2, 3, 4, or 6. The use of iron is prehistoric. Genesis mentions that Tubal-Cain, seven generations from Adam, was "an instructor of every artificer in brass and iron." A remarkable iron pillar, dating to about A.D. 400, remains standing today in Delhi, India. This solid shaft of wrought iron is about $7^1/_4$ m high by 40 cm in diameter. Corrosion to the pillar has been minimal although it has been exposed to the weather since its erection. Iron is a relatively abundant element in the universe. It is found in the sun and many types of stars in considerable quantity. Its nuclei are very stable. Iron is found native as a principal component of a class of meteorites known as "siderites, and is a minor constituent of the other two classes. The core of the earth, 2150 miles in radius, is thought to be largely composed of iron with about 10% occluded hydrogen. The metal is the fourth most abundant element, by weight, making up the crust of the earth. The most common ore is *hematite* (Fe_2O_3), which is frequently seen as *black sands* along beaches and banks of streams. *Taconite* is becoming increasingly important as a commercial ore. Common iron is a mixture of four isotopes. Ten other isotopes are known to exist. Iron is a vital constituent of plant and animal life, and appears in hemoglobin. The pure metal is not often encountered in commerce, but is usually alloyed with carbon or other metals. The pure metal is very reactive chemically, and rapidly corrodes, especially in moist air or at elevated temperatures. It has four allotropic forms, or ferrites, known as α, β, γ, and δ, with transition points at 700, 928, and 1530°C. The α form is magnetic, but when transformed into the β form, the magnetism disappears although the lattice remains unchanged. The relations of these forms are peculiar. Pig iron is an alloy containing about 3% carbon with varying amounts of S, Si, Mn, and P. It is hard, brittle, fairly fusible, and is used to produce other alloys, including steel. Wrought iron contains only a few tenths of a percent of carbon, is tough, malleable, less fusible, and has usually a "fibrous" structure.

Carbon steel is an alloy of iron with carbon, with small amounts of Mn, S, P, and Si. Alloy steels are carbon steels with other additives such as ni chromium, vanadium, etc. Iron is the cheapest and most abundant, useful, and important of all metals.

Krypton — (Gr. *kryptos*, hidden), Kr; at. wt. 83.80; at. no. 36; m.p. –157.36°C; b.p. –153.22 ± 0.10°C; density 3.733 g/l (0°C); valence us 0. Discovered in 1898 by Ramsay and Travers in the residue left after liquid air had nearly boiled away. Krypton is present in the air to the exte about 1 ppm. The atmosphere of Mars has been found to contain 0.3 ppm of krypton. It is one of the "noble" gases. It is characterized by its bri green and orange spectral lines. Naturally occurring krypton contains six stable isotopes. Seventeen other unstable isotopes are now recognized. spectral lines of krypton are easily produced and some are very sharp. In 1960 it was internationally agreed that the fundamental unit of length meter, should be defined in terms of the orange-red spectral line of ^{86}Kr. This replaced the standard meter of Paris, which was defined in terms bar made of a platinum-iridium alloy. In October 1983 the meter, which originally was defined as being one ten millionth of a quadrant of the ea polar circumference, was again redefined by the International Bureau of Weights and Measures as being the length of path traveled by light in a vac during a time interval of 1/299,792,458 of a second. Solid krypton is a white crystalline substance with a face-centered cubic structure which is com to all the "rare gases". While krypton is generally thought of as a rare gas that normally does not combine with other elements to form compou it now appears that the existence of some krypton compounds is established. Krypton difluoride has been prepared in gram quantities and can be n by several methods. A higher fluoride of krypton and a salt of an oxyacid of krypton also have been reported. Molecule-ions of $ArKr^+$ and KrH^+ l been identified and investigated, and evidence is provided for the formation of KrXe or $KrXe^+$. Krypton clathrates have been prepared hydroquinone and phenol. ^{85}Kr has found recent application in chemical analysis. By imbedding the isotope in various solids, *kryptonates* are forn The activity of these kryptonates is sensitive to chemical reactions at the surface. Estimates of the concentration of reactants are therefore made possi Krypton is used in certain photographic flash lamps for high-speed photography. Uses thus far have been limited because of its high cost. Kryp gas presently costs about $30/l.

Kurchatovium — see Element 104.

Lanthanum — (Gr. *lanthanein*, to lie hidden), La; at. wt. 138.9055; at. no. 57; m.p. 920°C; b.p. 3455°C; sp. gr. 6.145 (25°C); valence 3. Mosar in 1839 extracted a new earth *lanthana*, from impure cerium nitrate, and recognized the new element. Lanthanum is found in rare-earth minerals s as *cerite, monazite, allanite,* and *bastnasite*. Monazite and bastnasite are principal ores in which lanthanum occurs in percentages up to 25 and 3 respectively. Misch metal, used in making lighter flints, contains about 25% lanthanum. Lanthanum was isolated in relatively pure form in 1923. I exchange and solvent extraction techniques have led to much easier isolation of the so-called "rare-earth" elements. The availability of lanthanum other rare earths has improved greatly in recent years. The metal can be produced by reducing the anhydrous fluoride with calcium. Lanthanum is silv white, malleable, ductile, and soft enough to be cut with a knife. It is one of the most reactive of the rare-earth metals. It oxidizes rapidly when expo to air. Cold water attacks lanthanum slowly, and hot water attacks it much more rapidly. The metal reacts directly with elemental carbon, nitrog boron, selenium, silicon, phosphorus, sulfur, and with halogens. At 310°C, lanthanum changes from a hexagonal to a face-centered cubic struct and at 865°C it again transforms into a body-centered cubic structure. Natural lanthanum is mixture of two stable isotopes, ^{138}La and ^{139}La. Twe three other radioactive isotopes are recognized. Rare-earth compounds containing lanthanum are extensively used in carbon lighting applicatic especially by the motion picture industry for studio lighting and projection. This application consumes about 25% of the rare-earth compou produced. La_2O_3 improves the alkali resistance of glass, and is used in making special optical glasses. Small amounts of lanthanum, as an additi can be used to produce nodular cast iron. There is current interest in hydrogen sponge alloys containing lanthanum. These alloys take up to 400 tin their own volume of hydrogen gas, and the process is reversible. Heat energy is released every time they do so; therefore these alloys have possibilit in energy conservation systems. Lanthanum and its compounds have a low to moderate acute toxicity rating; therefore, care should be taken in handli them. The metal costs about $5/g.

Lawrencium — (Ernest O. Lawrence, inventor of the cyclotron), Lr; at. no. 103; at. mass no. (262); valence + 3(?). This member of the 5f transiti elements (actinide series) was discovered in March 1961 by A. Ghiorso, T. Sikkeland, A. E. Larsh, and R. M. Latimer.A 3-μg californium targ consisting of a mixture of isotopes of mass number 249, 250, 251, and 252, was bombarded with either ^{10}B or ^{11}B. The electrically charged transmutati nuclei recoiled with an atmosphere of helium and were collected on a thin copper conveyor tape which was then moved to place collected atoms front of a series of solid-state detectors. The isotope of element 103 produced in this way decayed by emitting an 8.6-MeV alpha particle with a ha life of 8 s. In 1967, Flerov and associates of the Dubna Laboratory reported their inability to detect an alpha emitter with a half-life of 8 s which w assigned by the Berkeley group to $^{257}103$. This assignment has been changed to ^{258}Lr or ^{259}Lr. In 1965, the Dubna workers found a longer-liv lawrencium isotope, ^{256}Lr, with a half-life of 35 s. In 1968, Ghiorso and associates at Berkeley were able to use a few atoms of this isotope to stu the oxidation behavior of lawrencium. Using solvent extraction techniques and working very rapidly, they extracted lawrencium ions from a buffer aqueous solution into an organic solvent, completing each extraction in about 30 s. It was found that lawrencium behaves differently from dipositi nobelium and more like the tripositive elements earlier in the actinide series.

Lead — (Anglo-Saxon *lead*), Pb (L. *plumbum*); at. wt. 207.2; at. no. 82; m.p. 327.46°C; b.p. 1749°C; sp. gr. 11.35 (20°C); valence 2 or 4. Lo known, mentioned in Exodus. The alchemists believed lead to be the oldest metal and associated it with the planet Saturn. Native lead occurs in natu but it is rare. Lead is obtained chiefly from *galena* (PbS) by a roasting process. *Anglesite* ($PbSO_4$), *cerussite* ($PbCO_3$), and *minim* (Pb_3O_4) are oth common lead minerals. Lead is a bluish-white metal of bright luster, is very soft, highly malleable, ductile, and a poor conductor of electricity. It very resistant to corrosion; lead pipes bearing the insignia of Roman emperors, used as drains from the baths, are still in service. It is used in containe for corrosive liquids (such as sulfuric acid) and may be toughened by the addition of a small percentage of antimony or other metals. Natural lead a mixture of four stable isotopes: ^{204}Pb (1.48%), ^{206}Pb (23.6%), ^{207}Pb (22.6%), and ^{208}Pb (52.3%). Lead isotopes are the end products of each of th three series of naturally occurring radioactive elements: ^{206}Pb for the uranium series, ^{207}Pb for the actinium series, and ^{208}Pb for the thorium serie Twenty seven other isotopes of lead, all of which are radioactive, are recognized. Its alloys include solder, type metal, and various antifriction metal Great quantities of lead, both as the metal and as the dioxide, are used in storage batteries. Much metal also goes into cable covering, plumbin ammunition, and in the manufacture of lead tetraethyl. The metal is very effective as a sound absorber, is used as a radiation shield around X-ra equipment and nuclear reactors, and is used to absorb vibration. White lead, the basic carbonate, sublimed white lead ($PbSO_4$) chrome yellow ($PbCrO_4$

red lead (Pb_3O_4), and other lead compounds are used extensively in paints, although in recent years the use of lead in paints has been drastically curtailed to eliminate or reduce health hazards. Lead oxide is used in producing fine "crystal glass" and "flint glass" of a high index of refraction for achromatic lenses. The nitrate and the acetate are soluble salts. Lead salts such as lead arsenate have been used as insecticides, but their use in recent years has been practically eliminated in favor of less harmful organic compounds. Care must be used in handling lead as it is a cumulative poison. Environmental concern with lead poisoning has resulted in a national program to eliminate the lead in gasoline.

Lithium — (Gr. *lithos*, stone), Li; at. wt. 6.941; at. no. 3; m.p. 180.5°C; b.p. 1342°C; sp. gr. 0.534 (20°C); valence 1. Discovered by Arfvedson in 1817. Lithium is the lightest of all metals, with a density only about half that of water. It does not occur free in nature; combined it is found in small amounts in nearly all igneous rocks and in the waters of many mineral springs. *Lepidolite, spodumene, petalite,* and *amblygonite* are the more important minerals containing it. Lithium is presently being recovered from brines of Searles Lake, in California, and from those in Nevada. Large deposits of spodumene are found in North Carolina. The metal is produced electrolytically from the fused chloride. Lithium is silvery in appearance, much like Na and K, other members of the alkali metal series. It reacts with water, but not as vigorously as sodium. Lithium imparts a beautiful crimson color to a flame, but when the metal burns strongly the flame is a dazzling white. Since World War II, the production of lithium metal and its compounds has increased greatly. Because the metal has the highest specific heat of any solid element, it has found use in heat transfer applications; however, it is corrosive and requires special handling. The metal has been used as an alloying agent, is of interest in synthesis of organic compounds, and has nuclear applications. It ranks as a leading contender as a battery anode material as it has a high electrochemical potential. Lithium is used in special glasses and ceramics. The glass for the 200-inch telescope at Mt. Palomar contains lithium as a minor ingredient. Lithium chloride is one of the most hygroscopic materials known, and it, as well as lithium bromide, is used in air conditioning and industrial drying systems. Lithium stearate is used as an all-purpose and high-temperature lubricant. Other lithium compounds are used in dry cells and storage batteries. The metal is priced at about $100/lb.

Lutetium — (Lutetia, ancient name for Paris, sometimes called *cassiopeium* by the Germans), Lu; at. wt. 174.967; at. no. 71; m.p. 1663°C; b.p. 3393°C; sp. gr. 9.841 (25°C); valence 3. In 1907, Urbain described a process by which Marignac's ytterbium (1879) could be separated into the two elements, ytterbium (neoytterbium)and lutetium. These elements were identical with "aldebaranium" and "cassiopeium," independently discovered by von Welsbach about the same time. Charles James of the University of New Hampshire also independently prepared the very pure oxide, *lutecia*, at this time. The spelling of the element was changed from *lutecium* to *lutetium* in 1949. Lutetium occurs in very small amounts in nearly all minerals containing yttrium, and is present in *monazite* to the extent of about 0.003%, which is a commercial source. The pure metal has been isolated only in recent years and is one of the most difficult to prepare. It can be prepared by the reduction of anhydrous $LuCl_3$ or LuF_3 by an alkali or alkaline earth metal. The metal is silvery white and relatively stable in air. While new techniques, including ion-exchange reactions, have been developed to separate the various rare-earth elements, lutetium is still the most costly of all rare earths. It is priced at about $75/g. ^{176}Lu occurs naturally (2.6%) with ^{175}Lu (97.4%). It is radioactive with a half-life of about 3×10^{10} years. Stable lutetium nuclides, which emit pure beta radiation after thermal neutron activation, can be used as catalysts in cracking, alkylation, hydrogenation, and polymerization. Virtually no other commercial uses have been found yet for lutetium. While lutetium, like other rare-earth metals, is thought to have a low toxicity rating, it should be handled with care until more information is available.

Magnesium — (*Magnesia,* district in Thessaly) Mg; at. wt. 24.3050; at. no. 12; m.p. 650°C; b.p. 1090°C; sp. gr. 1.738 (20°C); valence 2. Compounds of magnesium have long been known. Black recognized magnesium as an element in 1755. It was isolated by Davy in 1808, and prepared in coherent form by Bussy in 1831. Magnesium is the eighth most abundant element in the earth's crust. It does not occur uncombined, but is found in large deposits in the form of *magnesite, dolomite*, and other minerals. The metal is now principally obtained in the U.S. by electrolysis of fused magnesium chloride derived from brines, wells, and sea water. Magnesium is a light, silvery-white, and fairly tough metal. It tarnishes slightly in air, and finely divided magnesium readily ignites upon heating in air and burns with a dazzling white flame. It is used in flashlight photography, flares, and pyrotechnics, including incendiary bombs. It is one third lighter than aluminium, and in alloys is essential for airplane and missile contruction. The metal improves the mechanical, fabrication, and welding characteristics of aluminum when used as an alloying agent. Magnesium is used in producing nodular graphite in cast iron,and is used as an additive to conventional propellants. It is also used as a reducing agent in the production of pure uranium and other metals from their salts. The hydroxide (*milk of magnesia*), chloride, sulfate (*Epsom salts*), and citrate are used in medicine. Dead-burned magnesite is employed for refractory purposes such as brick and liners in furnaces and converters. Organic magnesium compounds (Grignard's reagents) are important. Magnesium is an important element in both plant and animal life. Chlorophylls are magnesium-centered porphyrins. The adult daily requirement of magnesium is about 300 mg/day, but this is affected by various factors. Great care should be taken in handling magnesium metal, especially in the finely divided state, as serious fires can occur. Water should not be used on burning magnesium or on magnesium fires.

Manganese — (L. *magnes*, magnet, from magnetic properties of pyrolusite; It. *manganese*, corrupt form of *magnesia*), Mn; at. wt. 54.93805; at. no. 25; m.p. 1246 ± 3°C; b.p. 2061°C; sp. gr. 7.21 to 7.44, depending on allotropic form; valence 1, 2, 3, 4, 6, or 7. Recognized by Scheele, Bergman, and others as an element and isolated by Gahn in 1774 by reduction of the dioxide with carbon. Manganese minerals are widely distributed; oxides, silicates, and carbonates are the most common. The discovery of large quantities of manganese nodules on the floor of the oceans holds promise as a source of manganese. These nodules contain about 24% manganese together with many other elements in lesser abundance. Most manganese today is obtained from ores found in the U.S.S.R., Brazil, Australia, Republic of So. Africa, Gabon, and India. *Pyrolusite* (MnO_2) and *rhodochrosite* ($MnCO_3$) are among the most common manganese minerals. The metal is obtained by reduction of the oxide with sodium, magnesium, aluminum, or by electrolysis. It is gray-white, resembling iron, but is harder and very brittle. The metal is reactive chemically, and decomposes cold water slowly. Manganese is used to form many important alloys. In steel, manganese improves the rolling and forging qualities, strength, toughness, stiffness, wear resistance, hardness, and hardenability. With aluminum and antimony, especially with small amounts of copper, it forms highly ferromagnetic alloys. Manganese metal is ferromagnetic only after special treatment. The pure metal exists in four allotropic forms. The alpha form is stable at ordinary temperature; gamma manganese, which changes to alpha at ordinary temperatures, is said to be flexible, soft, easily cut, and capable of being bent. The dioxide (pyrolusite) is used as a depolarizer in dry cells, and is used to "decolorize" glass that is colored green by impurities of iron. Manganese by itself colors glass an amethyst color, and is responsible for the color of true amethyst. The dioxide is also used in the preparation of oxygen and chlorine, and in drying black paints. The permanganate is a powerful oxidizing agent and is used in quantitative analysis and in medicine. Manganese is widely distributed throughout the animal kingdom. It is an important trace element and may be essential for utilization of vitamin B_1. Exposure to manganese dusts, fume, and compounds (as Mn) should not exceed the ceiling value of 5 mg/m^3 for even short periods because of the toxicity of the element.

Mendelevium — (Dmitri Mendeleev), Md; at. wt. (258); at. no. 101; valence +2, +3. Mendelevium, the ninth transuranium element of the actinide series to be discovered, was first identified by Ghiorso, Harvey, Choppin, Thompson, and Seaborg early in 1955 as a result of the bombardment of the isotope ^{253}Es with helium ions in the Berkeley 60-inch cyclotron. The isotope produced was ^{256}Md, which has a half-life of 76 min. This first identification was notable in that ^{256}Md was synthesized on a one-atom-at-a-time basis. Fourteen isotopes are now recognized. ^{258}Md has a half-life of 2 months. This isotope has been produced by the bombardment of an isotope of einsteinium with ions of helium. It now appears possible that eventually enough ^{258}Md can be made so that some of its physical properties can be determined. ^{256}Md has been used to elucidate some of the chemical properties of mendelevium in aqueous solution. Experiments seem to show that the element possesses a moderately stable dipositive (II) oxidation state in addition to the tripositive (III) oxidation state,which is characteristic of actinide elements.

Mercury — (Planet *Mercury*), Hg (*hydrargyrum*, liquid silver); at. wt. 200.59; at. no. 80; m.p. –38.83°C; b.p. 356.73°C; sp. gr. 13.546 (20°C); valence 1 or 2. Known to ancient Chinese and Hindus; found in Egyptian tombs of 1500 B.C. Mercury is the only common metal liquid at ordinary temperatures. It only rarely occurs free in nature. The chief ore is *cinnabar* (HgS). Spain and Italy produce about 50% of the world's supply of the metal. The commercial unit for handling mercury is the "flask," which weighs 76 lb. The metal is obtained by heating cinnabar in a current of air and by condensing the vapor. It is a heavy, silvery-white metal; a rather poor conductor of heat, as compared with other metals, and a fair conductor of electricity. It easily forms alloys with many metals, such as gold, silver, and tin, which are called *amalgams*. Its ease in amalgamating with gold is made use of in the recovery of gold from its ores. The metal is widely used in laboratory work for making thermometers, barometers, diffusion pumps, and many other instruments. It is used in making mercury-vapor lamps and advertising signs, etc. and is used in mercury switches and other electrical apparatus. Other uses are in making pesticides, mercury cells for caustic soda and chlorine production, dental preparations, antifouling paint, batteries, and catalysts. The most important salts are mercuric chloride $HgCl_2$ (corrosive sublimate — a violent poison), mercurous chloride Hg_2Cl_2(calomel, occasionally still used in medicine), mercury fulminate ($Hg(ONC)_2$), a detonator widely used in explosives, and mercuric sulfide (HgS, vermillion, a high-grade paint pigment). Organic mercury compounds are important. It has been found that an electrical discharge causes mercury vapor to combine with neon, argon, krypton, and xenon. These products, held together with van der Waals' forces, correspond to HgNe, HgAr, HgKr, and HgXe. Mercury is a virulent poison and is readily absorbed through the respiratory tract, the gastrointestinal tract, or through unbroken skin. It acts as a cumulative poison and dangerous levels are readily attained in air. Air saturated with mercury vapor at 20°C contains a concentration that exceeds the toxic limit many times. The danger increases at higher temperatures. *It is therefore important that mercury be handled with care*. Containers of mercury should be securely covered and spillage should be avoided. If it is necessary to heat mercury or mercury compounds, it should be done in a well-ventilated hood. Methyl mercury is a dangerous pollutant and is now widely found in water and streams. The triple point of mercury, –38.8344°C, is a fixed point on the International Temperature Scale (ITS-90).

Molybdenum — (Gr. *molybdos*, lead), Mo; at. wt. 95.94; at. no. 42; m.p. 2623°C; b.p. 4639°C; sp. gr. 10.22 (20°C); valence 2, 3, 4?, 5?, or 6. Before Scheele recognized molybdenite as a distinct ore of a new element in 1778, it was confused with graphite and lead ore. The metal was prepared in an impure form in 1782 by Hjelm. Molybdenum does not occur native, but is obtained principally from *molybdenite* (MoS_2). *Wulfenite* ($PbMoO_4$) and *Powellite* ($Ca(MoW)O_4$) are also minor commercial ores. Molybdenum is also recovered as a by-product of copper and tungsten mining operations. The metal is prepared from the powder made by the hydrogen reduction of purified molybdic trioxide or ammonium molybdate. The metal is silvery white, very hard, but is softer and more ductile than tungsten. It has a high elastic modulus, and only tungsten and tantalum, of the more readily available metals, have higher melting points. It is a valuable alloying agent, as it contributes to the hardenability and toughness of quenched and tempered steels. It also improves the strength of steel at high temperatures. It is used in certain nickel-based alloys, such as the "Hastelloys®" which are heat-resistant and corrosion-resistant to chemical solutions. Molybdenum oxidizes at elevated temperatures. The metal has found recent application as electrodes for electrically heated glass furnaces and foreheaths. The metal is also used in nuclear energy applications and for missile and aircraft parts. Molybdenum is valuable as a catalyst in the refining of petroleum. It has found application as a filament material in electronic and electrical applications. Molybdenum is an essential trace element in plant nutrition. Some lands are barren for lack of this element in the soil. Molybdenum sulfide is useful as a lubricant, especially at high temperatures where oils would decompose. Almost all ultra-high strength steels with minimum yield points up to 300,000 psi($lb/in.^2$) contain molybdenum in amounts from 0.25 to 8%.

Neodymium — (Gr. *neos*, new, and *didymos*, twin), Nd; at. wt. 144.24; at. no. 60; m.p. 1016°C; b.p. 3066°C; sp. gr. 7.008 (25°C); valence 3. In 1841, Mosander, extracted from *cerite* a new rose-colored oxide, which he believed contained a new element. He named the element *didymium*, as it was *an inseparable twin brother of lanthanum*.. In 1885 von Welsbach separated didymium into two new elemental components, *neodymia* and *praseodymia*, by repeated fractionation of ammonium didymium nitrate. While the free metal is in *misch metal*, long known and used as a pyrophoric alloy for light flints, the element was not isolated in relatively pure form until 1925. Neodymium is present in misch metal to the extent of about 18%. It is present in the minerals *monazite* and *bastnasite*, which are principal sources of rare-earth metals. The element may be obtained by separating neodymium salts from other rare earths by ion-exchange or solvent extraction techniques, and by reducing anhydrous halides such as NdF_3 with calcium metal. Other separation techniques are possible. The metal has a bright silvery metallic luster. Neodymium is one of the more reactive rare-earth metals and quickly tarnishes in air, forming an oxide that spalls off and exposes metal to oxidation. The metal, therefore, should be kept under light mineral oil or sealed in a plastic material. Neodymium exists in two allotropic forms, with a transformation from a double hexagonal to a body-centered cubic structure taking place at 863°C. Natural neodymium is a mixture of seven stable isotopes. Fourteen other radioactive isotopes are recognized. Didymium, of which neodymium is a component, is used for coloring glass to make welder's goggles. By itself, neodymium colors glass delicate shades ranging from pure violet through wine-red and warm gray. Light transmitted through such glass shows unusually sharp absorption bands. The glass has been used in astronomical work to produce sharp bands by which spectral lines may be calibrated. Glass containing neodymium can be used as a laser material to produce coherent light. Neodymium salts are also used as a colorant for enamels. The price of the metal is about \$1/g. Neodymium has a low-to-moderate acute toxic rating. As with other rare earths, neodymium should be handled with care.

Neon — (Gr. *neos*, new), Ne; at. wt. 20.1797; at. no. 10; m.p. –248.59°C; b.p. –246.08°C (1 atm); density of gas 0.89990 g/l (1 atm, 0°C); density of liquid at b.p. 1.207 g/cm^3; valence 0. Discovered by Ramsay and Travers in 1898. Neon is a rare gaseous element present in the atmosphere to the extent of 1 part in 65,000 of air. It is obtained by liquefaction of air and separated from the other gases by fractional distillation. Natural neon is a mixture of three isotopes. Six other unstable isotopes are known. It is very inert element; however, it is said to form a compound with fluorine. It is still questionable if true compounds of neon exist, but evidence is mounting in favor of their existence. The following ions are known from optical and mass spectrometric studies: Ne^+, $(NeAr)^+$, $(NeH)^+$, and $(HeNe^+)$. Neon also forms an unstable hydrate. In a vacuum discharge tube, neon glows reddish orange. Of all the rare gases, the discharge of neon is the most intense at ordinary voltages and currents. Neon is used in making the common neon advertising signs, which accounts for its largest use. It is also used to make high-voltage indicators, lightning arrestors, wave meter tubes, and TV tubes. Neon and helium are used in making gas lasers. Liquid neon is now commercially available and is finding important application as an economical cryogenic refrigerant. It has over 40 times more refrigerating capacity per unit volume than liquid helium and more than three times that of liquid hydrogen. It is compact, inert, and is less expensive than helium when it meets refrigeration requirements. Neon costs about \$2.00/l.

Neptunium — (Planet *Neptune*), Np; at. wt. (237); at. no. 93; m.p. 644°C; b.p. 3902°C (est.); sp. gr. 20.25 (20°C); valence 3, 4, 5, and 6. Neptunium was the first synthetic transuranium element of the actinide series discovered; the isotope ^{239}Np was produced by McMillan and Abelson in 1940 at Berkeley, California, as the result of bombarding uranium with cyclotron-produced neutrons. The isotope ^{237}Np (half-life of 2.14×10^6 years) is currently obtained in gram quantities as a by-product from nuclear reactors in the production of plutonium. Trace quantities of the element are actually found in nature due to transmutation reactions in uranium ores produced by the neutrons which are present. Neptunium is prepared by the reduction of NpF_3 with barium or lithium vapor at about 1200°C. Neptunium metal has a silvery appearance, is chemically reactive, and exists in at least three structural modifications: α-neptunium, orthorhombic, density 20.25 g/cm^3, β-neptunium (above 280°C), tetragonal, density (313°C) 19.36 g/cm^3; γ-neptunium (above 577°C), cubic, density (600°C) 18.0 g/cm^3. Neptunium has four ionic oxidation states in solution: Np^{+3} (pale purple), analogous to the rare earth ion Pm^{+3}, Np^{+4} (yellow green); NpO^+ (green blue); and NpO^{++} (pale pink). These latter oxygenated species are in contrast to the rare earths which exhibit only simple ions of the (II), (III), and (IV) oxidation states in aqueous solution. The element forms tri- and tetrahalides such as NpF_3, NpF_4, $NpCl_4$, $NpBr_3$, NpI_3, and oxides of various compositions such as are found in the uranium-oxygen system, including Np_3O_8 and NpO_2. Fifteen isotopes of neptunium are now recognized. The O.R.N.L. has ^{237}Np available for sale to its licensees and for export. This isotope can be used as a component in neutron detection instruments. It is offered at a price of \$280/g.

Nickel — (Ger. *Nickel*, Satan or Old Nick's and from *kupfernickel*, Old Nick's copper), Ni; at. wt. 58.6934; at. no. 28; m.p. 1455°C; b.p. 2913°C; sp. gr. 8.902 (25°C); valence 0, 1, 2, 3. Discovered by Cronstedt in 1751 in kupfernickel (*niccolite*). Nickel is found as a constituent in most meteorites and often serves as one of the criteria for distinguishing a meteorite from other minerals. Iron meteorites, or *siderites*, may contain iron alloyed with from 5 to nearly 20% nickel. Nickel is obtained commercially from *pentlandite* and *pyrrhotite* of the Sudbury region of Ontario, a district that produces about 30% of the nickel for the Free World. Other deposits are found in New Caledonia, Australia, Cuba, Indonesia, and elsewhere. Nickel is silvery white and takes on a high polish. It is hard, malleable, ductile, somewhat ferromagnetic, and a fair conductor of heat and electricity. It belongs to the iron-cobalt group of metals and is chiefly valuable for the alloys it forms. It is extensively used for making stainless steel and other corrosion-resistant alloys such as Invar®, Monel®, Inconel®, and the Hastelloys®. Tubing made of a copper-nickel alloy is extensively used in making desalination plants for converting sea water into fresh water. Nickel is also now used extensively in coinage and in making nickel steel for armor plate and burglar-proof vaults, and is a component in Nichrome®, Permalloy®, and constantan. Nickel added to glass gives a green color. Nickel plating is often used to provide a protective coating for other metals, and finely divided nickel is a catalyst for hydrogenating vegetable oils. It is also used in ceramics, in the manufacture of Alnico magnets, and in the Edison® storage battery. The sulfate and the oxides are important compounds. Natural nickel is a mixture of five stable isotopes; nine other unstable isotopes are known. Exposure to nickel metal and soluble compounds (as Ni) should not exceed 0.05 mg/m^3 (8-hour time-weighted average — 40-hour week). Nickel sulfide fume and dust is recognized as having carcinogenic potential.

Niobium — (*Niobe*, daughter of Tantalus), Nb; or Columbium (*Columbia*, name for America); at. wt. 92.90638; at. no. 41; m.p. 2477 ± 10°C; b.p. 4744°C, sp. gr. 8.57 (20°C); valence 2, 3, 4?, 5. Discovered in 1801 by Hatchett in an ore sent to England more that a century before by John Winthrop the Younger, first governor of Connecticut. The metal was first prepared in 1864 by Blomstrand, who reduced the chloride by heating it in a hydrogen atmosphere. The name *niobium* was adopted by the International Union of Pure and Applied Chemistry in 1950 after 100 years of controversy. Many leading chemical societies and government organizations refer to it by this name. Most metallurgists, leading metal societies, and all but one of the leading U.S. commercial producers, however, still refer to the metal as "columbium". The element is found in *niobite* (or *columbite*), *niobite-tantalite*, *pyrochlore*, and *euxenite*. Large deposits of niobium have been found associated with *carbonatites* (carbon-silicate rocks), as a constituent of *pyrochlore*. Extensive ore reserves are found in Canada, Brazil, Nigeria, Zaire, and in the U.S.S.R. The metal can be isolated from tantalum, and prepared in several ways. It is a shiny, white, soft, and ductile metal, and takes on a bluish cast when exposed to air at room temperatures for a long time. The metal starts to oxidize in air at 200°C, and when processed at even moderate temperatures must be placed in a protective atmosphere. It is used in arc-welding rods for stabilized grades of stainless steel. Thousands of pounds of niobium have been used in advance air frame systems such as were used in the Gemini space program. The element has superconductive properties; superconductive magnets have been made with Nb-Zr wire, which retains its superconductivity in strong magnetic fields. This type of application offers hope of direct large-scale generation of electric power. Eighteen isotopes of niobium are known. Niobium metal (99.5% pure) is priced at about \$75/lb.

Nitrogen — (L. *nitrum*, Gr. *nitron*, native soda; genes, *forming*, N; at. wt. 14.00674; at. no. 7; m.p. –210.00°C; b.p. –195.8°C; density 1.2506 g/l; sp. gr. liquid 0.808 (–195.8°C), solid 1.026 (–252°C); valence 3 or 5. Discovered by Daniel Rutherford in 1772, but Scheele, Cavendish, Priestley, and others about the same time studied "burnt or dephlogisticated air," as air without oxygen was then called. Nitrogen makes up 78% of the air, by volume. The atmosphere of Mars, by comparison, is 2.6% nitrogen. The estimated amount of this element in our atmosphere is more than 4000 trillion tons. From this inexhaustible source it can be obtained by liquefaction and fractional distillation. Nitrogen molecules give the orange-red, blue-green, blue-violet, and deep violet shades to the aurora. The element is so inert that Lavoisier named it *azote*, meaning without life, yet its compounds are so active as to be most important in foods, poisons, fertilizers, and explosives. Nitrogen can be also easily prepared by heating a water solution of ammonium nitrite. Nitrogen, as a gas, is colorless, odorless, and a generally inert element. As a liquid it is also colorless and odorless, and is similar in appearance to water. Two allotropic forms of solid nitrogen exist, with the transition from the α to the β form taking place at –237°C. When nitrogen is heated, it combines directly with magnesium, lithium, or calcium; when mixed with oxygen and subjected to electric sparks, it forms first nitric oxide (NO) and then the dioxide (NO_2); when heated under pressure with a catalyst with hydrogen, ammonia is formed (Haber process). The ammonia thus formed is of the utmost importance as it is used in fertilizers, and it can be oxidized to nitric acid (Ostwald process). The ammonia industryis the largest consumer of nitrogen. Large amounts of gas are also used by the electronics industry, which uses the gas as a blanketing medium during production of such components as transistors, diodes, etc. Large quantities of nitrogen are used in annealing stainless steel and other steel mill products. The drug industry also uses large quantities. Nitrogen is used as a refrigerant both for the immersion freezing of food products and for transportation of foods. Liquid nitrogen is also used in missile work as a purge for components, insulators for space chambers, etc., and by the oil industry to build up great pressures in wells to force crude oil upward. Sodium and potassium nitrates are formed by the decomposition of organic matter with compounds of the metals present. In certain dry areas of the world these saltpeters are found in quantity. Ammonia, nitric acid, the nitrates, the five oxides (N_2O, NO, N_2O_3, NO_2, and N_2O_5), TNT, the cyanides, etc. are but a few of the important compounds. Nitrogen gas prices vary from 2¢ to \$2.75 per 100 ft^3, depending on purity, etc. Production of elemental nitrogen in the U.S. is more than 9 million short tons per year.

Nobelium — (Alfred Nobel, discoverer of dynamite), No; at. wt. (259); at. no. 102; valence +2, +3. Nobelium was unambiguously discovered and identified in April 1958 at Berkeley by A. Ghiorso, T. Sikkeland, J. R. Walton, and G. T. Seaborg, who used a new double-recoil technique. A heavy-ion linear accelerator (HILAC) was used to bombard a thin target of curium (95% ^{244}Cm and 4.5% ^{246}Cm) with ^{12}C ions to produce 102^{254} according to the ^{246}Cm (^{12}C, 4*n*) reaction. Earlier in 1957 workers of the U.S., Britain, and Sweden announced the discovery of an isotope of Element 102 with a 10-min half-life at 8.5 MeV, as a result of bombarding ^{244}Cm with ^{13}C nuclei. On the basis of this experiment the name *nobelium* was assigned and accepted by the Commission on Atomic Weights of the International Union of Pure and Applied Chemistry. The acceptance of the name was premature, for both Russian and American efforts now completely rule out the possibility of any isotope of Element 102 having a half-life of 10 min in the vicinity of 8.5 MeV. Early work in 1957 on the search for this element, in Russia at the Kurchatov Institute, was marred by the assignment of 8.9 ± 0.4 MeV alpha radiation with a half-life of 2 to 40 sec, which was too indefinite to support claim to discovery. Confirmatory experiments at Berkeley in 1966 have shown the existence of ^{254}102 with a 55-s half-life, ^{252}102 with a 2.3-s half-life, and ^{257}102 with a 23-s half-life. Ten isotopes are now recognized, one of which — ^{255}102 has a half-life of 3 min. In view of the discover's traditional right to name an element, the Berkeley group, in 1967, suggested that the hastily given name *nobelium*, along with the symbol No, be retained.

Osmium — (Gr. *osme*, a smell), Os; at. wt. 190.2; at. no. 76; m.p. 3033 ± 30°C; b.p. 5012 ± 100°C; sp. gr. 22.57; valence 0 to +8, more usually +3, +4, +6, and +8. Discovered in 1803 by Tennant in the residue left when crude platinum is dissolved by *aqua regia*. Osmium occurs in *iridosmine* and in platinum-bearing river sands of the Urals, North America, and South America. It is also found in the nickel-bearing ores of Sudbury, Ontario, region along with other platinum metals. While the quantity of platinum metals in these ores is very small, the large tonnages of nickel ores processed make commercial recovery possible. The metal is lustrous, bluish white, extremely hard, and brittle even at high temperatures. It has the highest melting point and the lowest vapor pressure of the platinum group. The metal is very difficult to fabricate, but the powder can be sintered in a hydrogen atmosphere at a temperature of 2000°C. The solid metal is not affected by air at room temperature, but the powdered or spongy metal slowly gives off osmium tetroxide, which is a powerful oxidizing agent and has a strong smell. The tetroxide is highly toxic, and boils at 130°C (760 mm). Concentrations in air as low as 10^{-7} g/m^3 can cause lung congestion, skin damage, or eye damage. Exposure to osmium tetroxide should not exceed 0.0016 mg/m^3 (8-hour time weighted average — 40-hour work week). The tetroxide has been used to detect fingerprints and to stain fatty tissue for microscope slides. The metal is almost entirely used to produce very hard alloys, with other metals of the platinum group, for fountain pen tips, instrument pivots, phonograph needles, and electrical contacts. The price of 99% pure osmium powder — the form usually supplied commercially — is about \$100/g, depending on quantity and supplier. The measured densities of iridium and osmium seem to indicate that osmium is slightly more dense than iridium, so osmium has generally been credited with being the heaviest known element. Calculations of the density from the space lattice, which may be more reliable for these elements than actual measurements, however, give a density of 22.65 for iridium compared to 226.61 for osmium. At present, therefore, we know either iridium or omium is the heaviest element, but the data do not allow selection between the two.

Oxygen — (Gr. *oxys*, sharp, acid, and *genes*, forming; acid former), O; at. wt. 15.9994; at. no. 8; m.p. –218.79°C; b.p. –182.95°C; valence 2. For many centuries, workers occasionally realized air was composed of more than one component. The behavior of oxygen and nitrogen as components of air led to the advancement of the phlogiston theory of combustion, which captured the minds of chemists for a century. Oxygen was prepared by several workers, including Bayen and Borch, but they did not know how to collect it, did not study its properties, and did not recognize it as an elementary substance. Priestley is generally credited with its discovery, although Scheele also discovered it independently. Oxygen is the third most abundant element found in the sun, and it plays a part in the carbon-nitrogen cycle, one process thought to give the sun and stars their energy. Oxygen under excited conditions is responsible for the bright red and yellow-green colors of the aurora. Oxygen, as a gaseous element, forms 21% of the atmosphere by volume from which it can be obtained by liquefaction and fractional distillation. The atmosphere of Mars contains about 0.15% oxygen. The element and its compounds make up 49.2%, by weight, of the earth's crust. About two thirds of the human body and nine tenths of water is oxygen. In the laboratory it can be prepared by the electrolysis of water or by heating potassium chlorate with manganese dioxide as a catalyst. The gas is colorless, odorless, and tasteless. The liquid and solid forms are a pale blue color and are strongly paramagnetic. Ozone (O_3), a highly active compound, is formed by the action of an electrical discharge or ultraviolet light on oxygen. Ozone's presence in the atmosphere (amounting to the equivalent of a layer 3 mm thick at ordinary pressures and temperatures) is of vital importance in preventing harmful ultraviolet rays of the sun from reaching the earth's surface. There has been recent concern that pollutants in the atmosphere may have a detrimental effect on this ozone layer. Ozone is toxic and exposure should not exceed 0.2 mg/m^3 (8-hour time-weighted average — 40-hour work week). Undiluted ozone has a bluish color. Liquid ozone is bluish black, and solid ozone is violet-black. Oxygen is very reactive and capable of combining with most elements. It is a component of hundreds of thousands of organic compounds. It is essential for respiration of all plants and animals and for practically all combustion. In hospitals it is frequently used to aid respiration of patients. Its atomic weight was used as a standard of comparison for each of the other elements until 1961 when the International

Union of Pure and Applied Chemistry adopted carbon 12 as the new basis. Oxygen has nine isotopes. Natural oxygen is a mixture of three isotopes. Oxygen 18 occurs naturally, is stable, and is available commercially. Water (H_2O with 1.5% ^{18}O) is also available. Commercial oxygen consumption in the U.S. is estimated to be 20 million short tons per year and the demand is expected to increase substantially in the next few years. Oxygen enrichment of steel blast furnaces accounts for the greatest use of the gas. Large quantities are also used in making synthesis gas for ammonia and methanol, ethylene oxide, and for oxy-acetylene welding. Air separation plants produce about 99% of the gas, electrolysis plants about 1%. The gas costs 5¢/ft^3 in small quantities, and about \$15/ton in large quantities.

Palladium — (named after the asteroid *Pallas*, discovered about the same time; Gr. *Pallas*, goddess of wisdom), Pd. at. wt. 106.42 at. no. 46; m.p. 1554.9°C; b.p. 2963°C; sp. gr. 1202 (20°C); valence 2, 3, or 4. Discovered in 1803 by Wollaston. Palladium is found along with platinum and other metals of the platinum group in placer deposits of the U.S.S.R., South and North America, Ethiopia, and Australia. It is also found associated with the nickel-copper deposits of South Africa and Ontario. Its separation from the platinum metals depends upon the type of ore in which it is found. It is a steel-white metal, does not tarnish in air, and is the least dense and lowest melting of the platinum group of metals. When annealed, it is soft and ductile; cold working greatly increases its strength and hardness. Palladium is attacked by nitric and sulfuric acid. At room temperatures the metal has the unusual property of absorbing up to 900 times its own volume of hydrogen, possibly forming Pd_2H. It is not yet clear if this a true compound. Hydrogen readily diffuses through heated palladium and this provides a means of purifying the gas. Finely divided palladium is a good catalyst and is used for hydrogenation and dehydrogenation reactions. It is alloyed and used in jewelry trades. White gold is an alloy of gold decolorized by the addition of palladium. Like gold, palladium can be beaten into leaf as thin as 1/250,000 in. The metal is used in dentistry, watchmaking, and in making surgical instruments and electrical contacts. The metal sells for about \$150/troy oz.

Phosphorus — (Gr. *phosphoros*, light bearing; ancient name for the planet Venus when appearing before sunrise), P; at. wt. 30.973762; at. no. 15; m.p. (white) 44.15°C; sp. gr. (white) 1.82 (red) 2.20, (black) 2.25 to 2.69; valence 3 or 5. Discovered in 1669 by Brand, who prepared it from urine. Phosphorus exists in four or more allotropic forms: white (or yellow), red, and black (or violet). White phosphorus has two modifications: α and β with a transition temperature at –3.8°C. Never found free in nature, it is widely distributed in combination with minerals. *Phosphate* rock, which contains the mineral *apatite*, an impure tri-calcium phosphate, is an important source of the element. Large deposits are found in the U.S.S.R., in Morocco, and in Florida, Tennessee, Utah, Idaho, and elsewhere. Phosphorus in an essential ingredient of all cell protoplasm, nervous tissue, and bones. Ordinary phosphorus is a waxy white solid; when pure it is colorless and transparent. It is insoluble in water, but soluble in carbon disulfide. It takes fire spontaneously in air, burning to the pentoxide. It is very poisonous, 50 mg constituting an approximate fatal dose. Exposure to white phosphorus should not exceed 0.1 mg/m^3 (8-hour time-weighted average — 40-hour work week). White phosphorus should be kept under water, as it is dangerously reactive in air, and it should be handled with forceps, as contact with the skin may cause severe burns. When exposed to sunlight or when heated in its own vapor to 250°C, it is converted to the red variety, which does not phosphoresce in air as does the white variety. This form does not ignite spontaneously and it is not as dangerous as white phosphorus. It should, however, be handled with care as it does convert to the white form at some temperatures and it emits highly toxic fumes of the oxides of phosphorus when heated. The red modification is fairly stable, sublimes with a vapor pressure of 1 atm at 417°C,and is used in the manufacture of safety matches, pyrotechnics, pesticides, incendiary shells, smoke bombs, tracer bullets, etc. White phosphorus may be made by several methods. By one process, tri-calcium phosphate, the essential ingredient of phosphate rock, is heated in the presence of carbon and silica in an electric furnace or fuel-fired furnace. Elementary phosphorus is liberated as vapor and may be collected under water. If desired, the phosphorus vapor and carbon monoxide produced by the reaction can be oxidized at once in the presence of moisture to produce phosphoric acid, an important compound in making super-phosphate fertilizers. In recent years, concentrated phosphoric acids, which may contain as much as 70 to 75% P_2O_5 content, have become of great importance to agriculture and farm production. World-wide demand for fertilizers has caused record phosphate production. Phosphates are used in the production of special glasses, such as those used for sodium lamps. Bone-ash, calcium phosphate, is also used to produce fine chinaware and to produce mono-calcium phosphate used in baking powder. Phosphorus is also important in the production of steels, phosphor bronze, and many other products. Trisodium phosphate is important as a cleaning agent, as a water softener, and for preventing boiler scale and corrosion of pipes and boiler tubes. Organic compounds of phosphorus are important.

Platinum — (Sp. *platina*, silver), Pt; at. wt. 195.08; at. no. 78; m.p. 1768.4°C; b.p. 3825 ± 100°C; sp. gr. 21.45 (20°C); valence 1?, 2, 3, or 4. Discovered in South America by Ulloa in 1735 and by Wood in 1741. The metal was used by pre-Columbian Indians. Platinum occurs native, accompanied by small quantities of iridium, osmium, palladium, ruthenium, and rhodium, all belonging to the same group of metals. These are found in the alluvial deposits of the Ural mountains, of Columbia, and of certain western American states. *Sperrylite* ($PtAs_2$), occurring with the nickel-bearing deposits of Sudbury, Ontario, is the source of a considerable amount of metal. The large production of nickel offsets there being only one part of the platinum metals in two million parts of ore. Platinum is a beautiful silvery-white metal, when pure, and is malleable and ductile. It has a coefficient of expansion almost equal to that of soda-lime-silica glass, and is therefore used to make sealed electrodes in glass systems. The metal does not oxidize in air at any temperature, but is corroded by halogens, cyanides, sulfur, and caustic alkalis. It is insoluble in hydrochloric and nitric acid, but dissolves when they are mixed as *aqua regia*, forming chloroplatinic acid (H_2PtCl_6), an important compound. The metal is extensively used in jewelry, wire, and vessels for laboratory use, and in many valuable instruments including thermocouple elements. It is also used for electrical contacts, corrosion-resistant apparatus, and in dentistry. Platinum-cobalt alloys have magnetic properties. One such alloy made of 76.7% Pt and 23.3% Co, by weight, is an extremely powerful magnet that offers a B-H (max) almost twice that of Alnico V. Platinum resistance wires are used for constructing high-temperature electric furnaces. The metal is used for coating missile nose cones, jet engine fuel nozzles, etc., which must perform reliably for long periods of time at high temperatures. The metal, like palladium, absorbs large volumes, of hydrogen, retaining it at ordinary temperatures but giving it up at red heat. In the finely divided state platinum is an excellent catalyst, having long been used in the contact process for producing sulfuric acid. It is also used as a catalyst in cracking petroleum products. There is also much current interest in the use of platinum as a catalyst in fuel cells and in antipollution devices for automobiles. Platinum anodes are extensively used in cathodic protection systems for large ships and ocean-going vessels, pipelines, steel piers, etc. Fine platinum wire will glow red hot when placed in the vapor of methyl alcohol. It acts here as a catalyst, converting the alcohol to formaldehyde. This phenomenon has been used commercially to produce cigarette lighters and hand warmers. Hydrogen and oxygen explode in the presence of platinum. The price of platinum has varied widely; more than a century ago it was used to adulterate gold. It was nearly eight times as valuable as gold in 1920. The price in January 1990 was about \$500/troy oz.

Plutonium — (Planet *pluto*), Pu; at. wt. (244); at. no. 94; sp. gr. (α modification) 19.84 (25°C); m.p. 640°C; b.p. 3228°C; valence 3, 4, 5, or 6. Plutonium was the second transuranium element of the actinide series to be discovered. The isotope ^{238}Pu was produced in 1940 by Seaborg, McMillan, Kennedy, and Wahl by deuteron bombardment of uranium in the 60-inch cyclotron at Berkeley, California. Plutonium also exists in trace quantities in naturally occurring uranium ores. It is formed in much the same manner as neptunium, by irradiation of natural uranium with the neutrons which are present. By far of greatest importance is the isotope Pu^{239}, with a half-life of 24,100 years, produced in extensive quantities in nuclear reactors from natural uranium:

$$^{238}U(n,\gamma)\rightarrow{}^{239}U\xrightarrow{\beta}{}^{239}Np\xrightarrow{\beta}{}^{239}Pu$$

Fifteen isotopes of plutonium are known. Plutonium has assumed the position of dominant importance among the transuranium elements because of its successful use as an explosive ingredient in nuclear weapons and the place which it holds as a key material in the development of industrial use of nuclear power. One kilogram is equivalent to about 22 million kilowatt hours of heat energy. The complete detonation of a kilogram of plutonium produces an explosion equal to about 20,000 tons of chemical explosive. Its importance depends on the nuclear property of being readily fissionable with neutrons and its availability in quantity. The world's nuclear-power reactors are now producing about 20,000 kg of plutonium/yr. By 1982 it was estimated that about 300,000 kg had accumulated. The various nuclear applications of plutonium are well known. ^{238}Pu has been used in the Apollo lunar missions to power seismic and other equipment on the lunar surface. As with neptunium and uranium, plutonium metal can be prepared by reduction of the trifluoride with alkaline-earth metals. The metal has a silvery appearance and takes on a yellow tarnish when slightly oxidized. It is chemically reactive. A relatively large piece of plutonium is warm to the touch because of the energy given off in alpha decay. Larger pieces will produce enough heat to boil water. The metal readily dissolves in concentrated hydrochloric acid, hydroiodic acid, or perchloric acid with formation of the Pu^{+3} ion. The metal exhibits six allotropic modifications having various crystalline structures. The densities of these vary from 16.00 to 19.86 g/cm^3. Plutonium also exhibits four ionic valence states in aqueous solutions: Pu^{+3}(blue lavender), Pu^{+4} (yellow brown), PuO^{+} (pink?), and PuO^{+2} (pink orange). The ion PuO^{+} is unstable in aqueous solutions, disproportionating into Pu^{+4} and PuO^{+2}. The Pu^{+4} thus formed, however, oxidizes the PuO^{+} into PuO^{+2}, itself being reduced to Pu^{+3}, giving finally Pu^{+3} and PuO^{+2}. Plutonium forms binary compounds with oxygen: PuO, PuO_2, and intermediate oxides of variable composition; with the halides: PuF_3, PuF_4, $PuCl_3$, $PuBr_3$, PuI_3; with carbon, nitrogen, and silicon: PuC, PuN, $PuSi_2$. Oxyhalides are also well known: PuOCl, PuOBr, PuOI. Because of the high rate of emission of alpha particles and the element being specifically absorbed by bone marrow, plutonium, as well as all of the other transuranium elements except neptunium, are radiological poisons and must be handled with very special equipment and precautions. Plutonium is a very dangerous radiological hazard. Precautions must also be taken to prevent the unintentional formation of a critical mass. Plutonium in liquid solution is more likely to become critical than solid plutonium. The shape of the mass must also be considered where criticality is concerned. Plutonium-238 is available from the A.E.C. at a cost of about $700/g (80 to 89% enriched.)

Polonium — (Poland, native country of Mme. Curie), Po; at. wt. (209); at. no. 84; m.p. 254°C; b.p. 962°C; sp. gr. (alpha modification) 9.32; valence −2, 0, +2, +3(?), +4, and +6. Polonium was the first element discovered by Mme. Curie in 1898, while seeking the cause of radioactivity of pitchblende from Joachimsthal, Bohemia. The electroscope showed it separating with bismuth. Polonium is also called Radium F. Polonium is a very rare natural element. Uranium ores contain only about 100 μg of the element per ton. Its abundance is only about 0.2% of that of radium. In 1934, it was found that when natural bismuth (^{209}Bi) was bombarded by neutrons, ^{210}Bi, the parent of polonium, was obtained. Milligram amounts of polonium may now be prepared this way, by using the high neutron fluxes of nuclear reactors. Polonium-210 is a low-melting, fairly volatile metal, 50% of which is vaporized in air in 45 hours at 55°C. It is an alpha emitter with a half-life of 138.39 days. A milligram emits as many alpha particles as 5 g of radium. The energy released by its decay is so large (140 W/g) that a capsule containing about half a gram reaches a temperature above 500°C. The capsule also presents a contact gamma-ray dose rate of 0.012 Gy/h. A few curies (1 curie = 3.7×10^{10} Bq) of polonium exhibit a blue glow, caused by excitation of the surrounding gas. Because almost all alpha radiation is stopped within the solid source and its container, giving up its energy, polonium has attracted attention for uses as a lightweight heat source for thermoelectric power in space satellites. Twenty five isotopes of polonium are known, with atomic masses ranging from 194 to 218. Polonium-210 is the most readily available. Isotopes of mass 209 (half-life 103 years) and mass 208 (half-life 2.9 years) can be prepared by alpha, proton, or deuteron bombardment of lead or bismuth in a cyclotron, but these are expensive to produce. Metallic polonium has been prepared from polonium hydroxide and some other polonium compounds in the presence of concentrated aqueous or anhydrous liquid ammonia. Two allotropic modifications are known to exist. Polonium is readily dissolved in dilute acids, but is only slightly soluble in alkalis. Polonium salts of organic acids char rapidly; halide amines are reduced to the metal. Polonium can be mixed or alloyed with beryllium to provide a source of neutrons. It has been used in devices for eliminating static charges in textile mills, etc.; however, beta sources are more commonly used and are less dangerous. It is also used on brushes for removing dust from photographic films. The polonium for these is carefully sealed and controlled, minimizing hazards to the user. Polonium-210 is very dangerous to handle in even milligram or microgram amounts, and special equipment and strict control is necessary. Damage arises from the complete absorption of the energy of the alpha particle into tissue. The maximum permissible body burden for ingested polonium is only 0.03 μCi, which represents a particle weighing only 6.8×10^{-12} g. Weight for weight it is about 2.5×10^{11} times as toxic as hydrocyanic acid. The maximum allowable concentration for soluble polonium compounds in air is about 2 10^{-11} $\mu Ci/cm^3$. Polonium is available commercially on special order from the Oak Ridge National Laboratory.

Potassium — (English, *potash* — pot ashes; L. *kalium*, Arab. *qali*, alkali), K; at. wt. 39.0983; at. no. 19; m.p. 63.28°C; b.p. 759°C; sp. gr. 0.862 (20°C); valence 1. Discovered in 1807 by Davy, who obtained it from caustic potash (KOH); this was the first metal isolated by electrolysis. The metal is the seventh most abundant and makes up about 2.4% by weight of the earth's crust. Most potassium minerals are insoluble and the metal is obtained from them only with great difficulty. Certain minerals, however, such as *sylvite, carnallite, langbeinite,* and *polyhalite* are found in ancient lake and sea beds and form rather extensive deposits from which potassium and its salts can readily be obtained. Potash is mined in Germany, New Mexico, California, Utah, and elsewhere. Large deposits of potash, found at a depth of some 3000 ft in Saskatchewan, promise to be important in coming years. Potassium is also found in the ocean, but is present only in relatively small amounts, compared to sodium. The greatest demand for potash has been in its use for fertilizers. Potassium is an essential constituent for plant growth and it is found in most soils. Potassium is never found free in nature, but is obtained by electrolysis of the hydroxide, much in the same manner as prepared by Davy. Thermal methods also are commonly used to produce potassium (such as by reduction of potassium compounds with CaC_2, C, Si, or Na). It is one of the most reactive and electropositive of metals. Except for lithium, it is the lightest known metal. It is soft, easily cut with a knife, and is silvery in appearance immediately after a fresh surface is exposed. It rapidly oxidizes in air and must be preserved in a mineral oil such as kerosene. As with other metals of the alkali group, it decomposes in water with the evolution of hydrogen. It catches fire spontaneously on water. Potassium and its salts impart a violet color to flames. Seventeen isotopes of potassium are known. Ordinary potassium is composed of three isotopes, one of which is ^{40}K (0.0118%), a radioactive isotope with a half-life of 1.28×10^9 years. The radioactivity presents no appreciable hazard. An alloy of sodium and potassium (NaK) is used as a heat-transfer medium. Many potassium salts are of utmost importance, including the hydroxide, nitrate, carbonate, chloride, chlorate, bromide, iodide, cyanide, sulfate, chromate, and dichromate. Metallic potassium is available commercially for about $40/lb in small quantities.

Praseodymium — (Gr. *prasios*, green, and *didymos*, twin), Pr; at. wt. 140.90765; at. no. 59; m.p. 931°C; b.p. 3510°C; sp. gr. 6.773; valence 3. In 1841 Mosander extracted the rare earth *didymia* from *lanthana*; in 1879, Lecoq de Boisbaudran isolated a new earth, *samaria*, from didymia obtained from the mineral *samarskite*. Six years later, in 1885, von Welsbach separated didymia into two others, *praseodymia* and *neodymia*, which gave salts of different colors. As with other rare earths, compounds of these elements in solution have distinctive sharp spectral absorption bands or lines, some of which are only a few Angstroms wide. The element occurs along with other rare-earth elements in a variety of minerals. *Monazite* and *bastnasite* are the two principal commercial sources of the rare-earth metals. Ion-exchange and solvent extraction techniques have led to much easier isolation of the rare earths and the cost has dropped greatly in the past few years. Praseodymium can be prepared by several methods, such as by calcium reduction of the anhydrous chloride of fluoride. Misch metal, used in making cigarette lighters, contains about 5% praseodymium metal. Praseodymium is soft, silvery, malleable, and ductile. It was prepared in relatively pure form in 1931. It is somewhat more resistant to corrosion in air than europium, lanthanum, cerium, or neodymium, but it does develop a green oxide coating that spalls off when exposed to air. As with other rare-earth metals it should be kept under a light mineral oil or sealed in plastic. The rare-earth oxides, including Pr_2O_3, are among the most refractory substances known. Along with other rare earths, it is widely used as a core material for carbon arcs used by the motion picture industry for studio lighting and projection. Salts of praseodymium are used to color glasses and enamels; when mixed with certain other materials, praseodymium produces an intense and unusually clean yellow color in glass. Didymium glass, of which praseodymium is a component, is a colorant for welder's goggles. The metal (99 + % pure) is priced at about $70/oz.

Promethium — (*Prometheus*, who, according to mythology, stole fire from heaven), Pm; at. no. 61; at. wt. (145); m.p. 1042°C; b.p. 3000°C (est.), sp. gr. 7.264 (25°C); valence 3. In 1902 Branner predicted the existence of an element between neodymium and samarium, and this was confirmed by Moseley in 1914. In 1941, workers at Ohio State University irradiated neodymium and praseodymium with neutrons, deuterons, and alpha particles, resp., and produced several new radioactivities, which most likely were those of element 61. Wu and Segre, and Bethe, in 1942, confirmed the formation; however, chemical proof of the production of element 61 was lacking because of the difficulty in separating the rare earths from each other at that time. In 1945, Marinsky, Glendenin, and Coryell made the first chemical identification by use of ion-exchange chromatography. Their work was done by fission of uranium and by neutron bombardment of neodymium. Searches for the element on earth have been fruitless, and it now appears that promethium is completely missing from the earth's crust. Promethium, however, has been identified in the spectrum of the star HR[465] in Andromeda. This element is being formed recently near the star's surface, for no known isotope of promethium has a half-life longer than 17.7 years. Seventeen isotopes of promethium, with atomic masses from 134 to 155 are now known. Promethium-147,with a half-life of 2.6 years, is the most generally useful. Promethium-145 is the longest lived, and has a specific activity of 940 Ci/g. It is a soft beta emitter; although no gamma rays are emitted, X-radiation can be generated when beta particles impinge on elements of a high atomic number, and great care must be taken in handling it. Promethium salts luminesce in the dark with a pale blue or greenish glow, due to their high radioactivity. Ion-exchange methods led to the preparation of about 10 g of promethium from atomic reactor fuel processing wastes in early 1963. Little is yet generally known about the properties of metallic promethium. Two allotropic modifications exist. The element has applications as a beta source for thickness gages, and it can be absorbed by a phosphor to produce light. Light produced in this manner can be used for signs or signals that require dependable operation; it can be used as a nuclear-powered battery by capturing light in photocells which convert it into electric current. Such a battery, using ^{147}Pm, would have a useful life of about 5 years. Promethium shows promise as a portable X-ray source, and it may become useful as a heat source to provide auxiliary power for space probes and satellites. More than 30 promethium compounds have been prepared. Most are colored. Promethium-147 is available at a cost of about 50c/Ci.

Protactinium — (Gr. *protos*, first), Pa; at. wt. 231.03588; at. no. 91; m.p. 1572°C; sp. gr. 15.37 (calc.); valence 4 or 5. The first isotope of element 91 to be discovered was ^{234}Pa, also known as UX_2, a short-lived member of the naturally occurring ^{238}U decay series. It was identified by K. Fajans and O. H. Gohring in 1913 and they named the new element *brevium*. When the longer-lived isotope ^{231}Pa was identified by Hahn and Meitner in 1918, the name protoactinium was adopted as being more consistent with the characteristics of the most abundant isotope. Soddy, Cranson, and Fleck were also active in this work. The name *protoactinium* was shortened to *protactinium* in 1949. In 1927, Grosse prepared 2 mg of a white powder, which was shown to be Pa_2O_5. Later, in 1934, from 0.1 g of pure Pa_2O_5 he isolated the element by two methods, one of which was by converting the oxide to an iodide and "cracking" it in a high vacuum by an electrically heated filament by the reaction

$$2PaI_5 \rightarrow 2Pa + 5I_2$$

Protactinium has a bright metallic luster which it retains for some time in air. The element occurs in *pitchblende* to the extent of about 1 part ^{231}Pa to 10 million of ore. Ores from Zaire have about 3 ppm. Protactinium has 20 isotopes, the most common of which is ^{231}Pr with a half-life of 32,700 years. A number of protactinium compounds are known, some of which are colored. The element is superconductive below 1.4 K. The element is a dangerous toxic material and requires precautions similar to those used when handling plutonium. In 1959 and 1961, it was announced that the Great Britain Atomic Energy Authority extracted by a 12-stage process 125 g of 99.9% protactinium, the world's only stock of the metal for many years to come.

The extraction was made from 60 tons of waste material at a cost of about $500,000. Protactinium is one of the rarest and most expensive naturall occurring elements. O.R.N.L. supplies promethium-231 at a cost of about $280/g. The elements is an alpha emitter (5.0 MeV), and is a radiologica hazard similar to polonium.

Radium — (L. *radius*, ray), Ra; at. wt. (226); at. no. 88; m.p. 700°C; b.p. 1140°C; sp. gr. 5; valence 2. Radium was discovered in 1898 by M. an Mme. Curie in the *pitchblende* or *uraninite* of North Bohemia, where it occurs. There is about 1 g of radium in 7 tons of pitchblende. The element wa isolated in 1911 by Mme. Curie and Debierne by the electrolysis of a solution of pure radium chloride, employing a mercury cathode; on distillatio in an atmosphere of hydrogen this amalgam yielded the pure metal. Originally, radium was obtained from the rich pitchblende ore found a Joachimsthal, Bohemia. The *carnotite* sands of Colorado furnish some radium, but richer ores are found in the Republic of Zaire and the Great Bea Lake region of Canada. Radium is present in all uranium minerals, and could be extracted, if desired, from the extensive wastes of uranium processing Large uranium deposits are located in Ontario, New Mexico, Utah, Australia, and elsewhere. Radium is obtained commercially as the bromide o chloride; it is doubtful if any appreciable stock of the isolated element now exists. The pure metal is brilliant white when freshly prepared, but blacken on exposure to air, probably due to formation of the nitride. It exhibits luminescence, as do its salts; it decomposes in water and is somewhat mor volatile than barium. It is a member of the alkaline-earth group of metals. Radium imparts a carmine red color to a flame. Radium emits alpha, beta and gamma rays and when mixed with beryllium produce neutrons. One gram of ^{226}Ra undergoes 3.7×10^{10} disintegrations per s. The *curie (Ci)* is defined as that amount of radioactivity which has the same disintegration rate as 1 g of ^{226}Ra. Twenty five isotopes are now known; radium 226, the common isotope, has a half-life of 1600 years. One gram of radium produces about 0.0001 ml (stp) of emanation, or radon gas, per day. This is pumpec from the radium and sealed in minute tubes, which are used in the treatment of cancer and other diseases. One gram of radium yields about 4186 kJ per year. Radium is used in producing self-luminous paints, neutron sources, and in medicine for the treatment of disease. Some of the more recently discovered radioisotopes, such as ^{60}Co, are now being used in place of radium. Some of these sources are much more powerful, and others are safe to use. Radium loses about 1% of its activity in 25 years, being transformed into elements of lower atomic weight. Lead is a final product o disintegration. Stored radium should be ventilated to prevent build-up of radon. Inhalation, injection, or body exposure to radium can cause cance and other body disorders. The maximum permissible burden in the total body for ^{226}Ra is 7400 becquerel.

Radon — (from *radium*; called *niton* at first, L. *nitens*, shining), Rn; at. wt. (222); at. no. 86; m.p. –71°C; b.p. –61.7°C; density of gas 9.73 g/l; sp. gr. liquid 4.4 at –62°C, solid 4; valence usually 0. The element was discovered in 1900 by Dorn, who called it *radium emanation*. In 1908 Ramsay and Gray, who named it *niton*, isolated the element and determined its density, finding it to be the heaviest known gas. It is essentially inert and occupies the last place in the zero group of gases in the Periodic Table. Since 1923, it has been called radon. Twenty six isotopes are known. Radon-222, coming from radium, has a half-life of 3.823 days and is an alpha emitter; Radon-220, emanating naturally from thorium and called *thoron*, has a half-life of 55.6 s and is also an alpha emitter. Radon-219 emanates from actinium and is called *actinon*. It has a half-life of 3.96 s and is also on alpha emitter. It is estimated that every square mile of soil to a depth of 6 inch contains about 1 g of radium, which releases radon in tiny amounts to the atmosphere. Radon is present in some spring waters, such as those at Hot Springs, Arkansas. On the average, one part of radon is present to 1×10^{21} part of air. At ordinary temperatures radon is a colorless gas; when cooled below the freezing point, radon exhibits a brilliant phosphorescence which becomes yellow as the temperature is lowered and orange-red at the temperature of liquid air. It has been reported that fluorine reacts with radon, forming radon fluoride. Radon clathrates have also been reported. Radon is still produced for therapeutic use by a few hospitals by pumping it from a radium source and sealing it in minute tubes, called seeds or needles, for application to patients. This practice has now been largely discontinued as hospitals can order the seeds directly from suppliers, who make up the seeds with the desired activity for the day of use. Radon is available at a cost of about $4/m Ci. Care must be taken in handling radon, as with other radioactive materials. The main hazard is from inhalation of the element and its solid daughters, which are collected on dust in the air. Good ventilation should be provided where radium, thorium, or actinium is stored to prevent build-up of this element. Radon build-up is a health consideration in uranium mines. Recently radon build-up in homes has been a concern. Many deaths from lung cancer are caused by radon exposure. In the U.S. it is recommended that remedial action be taken if the air in homes exceeds 4 pCi/l.

Rhenium — (L. *Rhenus*, Rhine), Re; at. wt. 186.207; at. no. 75; m.p. 3186°C; b.p. 5596°C; sp. gr. 21.02 (20°C); valence –1, +1, 2, 3, 4, 5, 6, 7. Discovery of rhenium is generally attributed to Noddack, Tacke, and Berg, who announced in 1925 they had detected the element in platinum ores and *columbite*. They also found the element in *gadolinite* and *molybdenite*. By working up 660 kg of molybdenite they were able in 1928 to extract 1 g of rhenium. The price in 1928 was $10,000/g. Rhenium does not occur free in nature or as a compound in a distinct mineral species. It is, however, widely spread throughout the earth's crust to the extent of about 0.001 ppm. Commercial rhenium in the U.S. today is obtained from molybdenite roaster-flue dusts obtained from copper-sulfide ores mined in the vicinity of Miami, Arizona, and elsewhere in Arizona and Utah. Some molybdenites contain from 0.002 to 0.2% rhenium. More than 150,000 troy ounces of rhenium are now being produced yearly in the United States. The total estimated Free World reserve of rhenium metal is 3500 tons. Natural rhenium is a mixture of two stable isotopes. Twenty six other unstable isotopes are recognized. Rhenium metal is prepared by reducing ammonium perrhenate with hydrogen at elevated temperatures. The element is silvery white with a metallic luster; its density is exceeded only by that of platinum, iridium, and osmium, and its melting point is exceeded only by that of tungsten and carbon. It has other useful properties. The usual commercial form of the element is a powder, but it can be consolidated by pressing and resistance-sintering in a vacuum or hydrogen atmosphere. This produces a compact shape in excess of 90% of the density of the metal. Annealed rhenium is very ductile, and can be bent, coiled, or rolled. Rhenium is used as an additive to tungsten and molybdenum-based alloys to impart useful properties. It is widely used for filaments for mass spectrographs and ion gages. Rhenium-molybdenum alloys are superconductive at 10 K. Rhenium is also used as an electrical contact material as it has good wear resistance and withstands arc corrosion. Thermocouples made of Re-W are used for measuring temperatures up to 2200°C, and rhenium wire is used in photoflash lamps for photography. Rhenium catalysts are exceptionally resistant to poisoning from nitrogen, sulfur, and phosphorus, and are used for hydrogenation of fine chemicals, hydrocracking, reforming, and disproportionation of olefins. Rhenium costs about $250/troy oz. Little is known of its toxicity; therefore, it should be handled with care until more data are available.

Rhodium — (Gr. *rhodon*, rose), Rh; at. wt. 102.90550; at. no. 45; m.p. 1964 ± 3°C; b.p. 3695 ± 100°C; sp. gr. 12.41 (20°C); valence 2, 3, 4, 5, and 6. Wollaston discovered rhodium in 1803-4 in crude platinum ore he presumably obtained from South America. Rhodium occurs native with other platinum metals in river sands of the Urals and in North and South America. It is also found with other platinum metals in the copper-nickel sulfide ores of the Sudbury, Ontario region. Although the quantity occurring here is very small, the large tonnages of nickel processed make the recovery commercially feasible. The annual world production of rhodium is only 7 or 8 tons. The metal is silvery white and at red heat slowly changes in air to the sesquioxide. At higher temperatures it converts back to the element. Rhodium has a higher melting point and lower density than platinum. Its major use is as an alloying agent to harden platinum and palladium. Such alloys are used for furnace windings, thermocouple elements, bushings for glass fiber production, electrodes for aircraft spark plugs, and laboratory crucibles. It is useful as an electrical contact material as it has a low electrical resistance, a low and stable contact resistance, and is highly resistant to corrosion. Plated rhodium, produced by electroplating or evaporation, is exceptionally hard and is used for optical instruments. It has a high reflectance and is hard and durable. Rhodium is also used for jewelry, for decoration, and as a catalyst. Exposure to rhodium (metal fume and dust, as Rh) should not exceed 1 mg/m^3 (8-hour time-weighted average, 40-hour wk.). Soluble salts should not exceed 0.01 mg/m^3. Rhodium costs about $1000/troy oz.

Rubidium — (L. *rubidus*, deepest red), Rb; at. wt. 85.4678; at. no. 37; m.p. 39.31°C; b.p. 688°C; sp. gr. (solid) 1.532 (20°C), (liquid) 1.475 (39°C); valence 1, 2, 3, 4. Discovered in 1861 by Bunsen and Kirchoff in the mineral *lepidolite* by use of the spectroscope. The element is much more abundant than was thought several years ago. It is now considered to be the 16th most abundant element in the earth's crust. Rubidium occurs in *pollucite, carnallite, leucite,* and *zinnwaldite*, which contains traces up to 1%, in the form of the oxide. It is found in lepidolite to the extent of about 1.5%, and is recovered commercially from this source. Potassium minerals, such as those found at Searles Lake, California, and potassium chloride recovered from brines in Michigan also contain the element and are commercial sources. It is also found along with cesium in the extensive deposits of *pollucite* at Bernic Lake, Manitoba. Rubidium can be liquid at room temperature. It is a soft, silvery-white metallic element of the alkali group and is the second most electropositive and alkaline element. It ignites spontaneously in air and reacts violently in water, setting fire to the liberated hydrogen. As with other alkali metals, it forms amalgams with mercury and it alloys with gold, cesium, sodium, and potassium. It colors a flame yellowish violet. Rubidium metal can be prepared by reducing rubidium chloride with calcium, and by a number of other methods. It must be kept under a dry mineral oil or in a vacuum or inert atmosphere. Twenty four isotopes of rubidium are known. Naturally occurring rubidium is made of two isotopes, ^{85}Rb and ^{87}Rb. Rubidium-87 is present to the extent of 27.85% in natural rubidium and is a beta emitter with a half-life of 4.9×10^{10} years. Ordinary rubidium is sufficiently radioactive to expose a photographic film in about 30 to 60 days. Rubidium forms four oxides: Rb_2O, Rb_2O_2, Rb_2O_3, Rb_2O_4. Because rubidium can be easily ionized, it has been considered for use in "ion engines" for space vehicles; however, cesium is somewhat more efficient for this purpose. It is also proposed for use as a working fluid for vapor turbines and for use in a thermoelectric generator using the magnetohydrodynamic principle where rubidium ions are formed by heat at high temperature and passed through a magnetic field. These conduct electricity and act like an armature of a generator thereby generating an electric current. Rubidium is used as a getter in vacuum tubes and as a photocell component. It has been used in making special glasses. $RbAg_4I_5$ is important, as it has the highest room conductivity of any known ionic crystal. At 20°C its conductivity is about the same as dilute sulfuric acid. This suggests use in thin film batteries and other applications. The present cost in small quantities in about $25/g.

Ruthenium — (L. *Ruthenia*, Russia), Ru; at. wt. 101.07; at. no. 44, m.p. 2334°C; b.p. 4150°C; sp. gr. 12.41 (20°C); valence 0, 1, 2, 3, 4, 5, 6, 7, 8. Berzelius and Osann in 1827 examined the residues left after dissolving crude platinum from the Ural mountains in *aqua regia*. While Berzelius found no unusual metals, Osann thought he found three new metals, one of which he named ruthenium. In 1844 Klaus, generally recognized as the discoverer, showed that Osann's ruthenium oxide was very impure and that it contained a new metal. Klaus obtained 6 g of ruthenium from the portion of crude platinum that is insoluble in *aqua regia*. A member of the platinum group, ruthenium occurs native with other members of the group in ores found in the Ural mountains and in North and South America. It is also found along with other platinum metals in small but commercial quantities in *pentlandite* of the Sudbury, Ontario, nickel-mining region, and in *pyroxinite* deposits of South Africa. The metal is isolated commercially by a complex chemical process, the final stage of which is the hydrogen reduction of ammonium ruthenium chloride, which yields a powder. The powder is consolidated by powder metallurgy techniques or by argon-arc welding. Ruthenium is a hard, white metal and has four crystal modifications. It does not tarnish at room temperatures, but oxidizes in air at about 800°C. The metal is not attacked by hot or cold acids or *aqua regia*, but when potassium chlorate is added to the solution, it oxidizes explosively. It is attacked by halogens, hydroxides, etc. Ruthenium can be plated by electrodeposition or by thermal decomposition methods. The metal is one of the most effective hardeners for platinum and plalladium, and is alloyed with these metals to make electrical contacts for severe wear resistance. A ruthenium-molybdenum alloy is said to be superconductive at 10.6 K. The corrosion resistance of titanium is improved a hundredfold by addition of 0.1% ruthenium. It is a versatile catalyst. Hydrogen sulfide can be split catalytically by light using an aqueous suspension of CdS particles loaded with ruthenium dioxide. It is thought this may have application to removal of H_2S from oil refining and other industrial processes. Compounds in at least eight oxidation states have been found, but of these, the +2. +3. and +4 states are the most common. Ruthenium tetroxide, like osmium tetroxide, is highly toxic. In addition, it may explode. Ruthenium compounds show a marked resemblance to those of osmium. The metal is priced at about $30/g.

Rutherfordium — See Element 104 (unnilquadium).

Samarium — (*Samarskite* a mineral), Sm; at. wt. 150.36; at. no. 62; m.p. 1072°C; b.p. 1790°C; sp. gr (α) 7.520 (25°C); valence 2 or 3. Discovered spectroscopically by its sharp absorption lines in 1879 by Lecoq de Boisbaudran in the mineral *samarskite*, named in honor of a Russian mine official, Col. Samarski. Samarium is found along with other members of the rare-earth elements in many minerals, including *monazite* and *bastnasite*, which are commercial sources. It occurs in monazite to the extent of 2.8%. While *misch metal* containing about 1% of samarium metal, has long been used, samarium has not been isolated in relatively pure form until recent years. Ion-exchange and solvent extraction techniques have recently simplified separation of the rare earths from one another; more recently, electrochemical deposition, using an electrolytic solution of lithium citrate and a mercury electrode, is said to be a simple, fast, and highly specific way to separate the rare earths. Samarium metal can be produced by reducing the oxide with barium or lanthanum. Samarium has a bright silver luster and is reasonably stable in air. Three crystal modifications of the metal exist, with transformations at 734 and 922°C. The metal ignites in air at about 150°C. Twenty one isotopes of samarium exist. Natural samarium is a mixture of several isotopes, three of which are unstable with long half-lives. Samarium, along with other rare earths, is used for carbon-arc lighting for the motion picture industry. The sulfide has excellent high-temperature stability and good thermoelectric efficiencies up to 1100°C. $SmCo_5$ has been used in making a new permanent magnet material with the highest resistance to demagnetization of any known material. It is said to have an intrinsic coercive force as high as 2200 kA/m. Samarium oxide has been used in optical glass to absorb the infrared. Samarium is used to dope calcium fluoride crysta for use in optical masers or lasers. Compounds of the metal act as sensitizers for phosphors excited in the infrared; the oxide exhibits catalytic properti in the dehydration and dehydrogenation of ethyl alcohol. It is used in infrared absorbing glass and as a neutron absorber in nuclear reactors. The met is priced at about \$5/g. Little is known of the toxicity of samarium; therefore, it should be handled carefully.

Scandium — (L. *Scandia*, Scandinavia), Sc; at. wt. 44.955910; at. no. 21; m.p. 1541°C; b.p. 2830°C; sp. gr. 2.989 (25°C); valence 3. On the bas of the Periodic System, Mendeleev predicted the existence of *ekaboron*, which would have an atomic weight between 40 of calcium and 48 of titaniun The element was discovered by Nilson in 1878 in the minerals *euxenite* and *gadolinite*, which had not yet been found anywhere except in Scandinavi By processing 10 kg of euxenite and other residues of rare-earth minerals, Nilson was able to prepare about 2 g of scandium oxide of high purity. Clev later pointed out that Nilson's scandium was identical with Mendeleev's ekaboron. Scandium is apparently a much more abundant element in the su and certain stars than here on earth. It is about the 23rd most abundant element in the sun, compared to the 50th most abundant on earth. It is widel distributed on earth, occurring in very minute quantities in over 800 mineral species. The blue color of beryl (aquamarine variety) is said to be du to scandium. It occurs as a principal component in the rare mineral *thortveitite*, found in Scandinavia and Malagasy. It is also found in the residue remaining after the extraction of tungsten from Zinnwald *wolframite*, and in *wiikite* and *bazzite*. Most scandium is presently being recovered fro *thortveitite* or is extracted as a by-product from uranium mill tailings. Metallic scandium was first prepared in 1937 by Fischer, Brunger, and Grieneise who electrolyzed a eutectic melt of potassium, lithium, and scandium chlorides at 700 to 800°C. Tungsten wire and a pool of molten zinc served a the electrodes in a graphite crucible. Pure scandium is now produced by reducing scandium fluoride with calcium metal. The production of the fir pound of 99% pure scandium metal was announced in 1960. Scandium is a silver-white metal which develops a slightly yellowish or pinkish cast upo exposure to air. It is relatively soft, and resembles yttrium and the rare-earth metals more than it resembles aluminum or titanium. It is a very light met and has a much higher melting point than aluminum, making it of interest to designers of spacecraft. Scandium is not attacked by a 1:1 mixture of con HNO_3 and 48% HF. Scandium reacts rapidly with many acids. Twelve isotopes of scandium are recognized. The metal is expensive, costing abo \$130/g with a purity of about 99.9%. Scandium oxide costs about \$75/g. About 20 kg of scandium (as Sc_2O_3) are now being used yearly in the U.S to produce high-intensity lights, and the radioactive isotope ^{46}Sc is used as a tracing agent in refinery crackers for crude oil, etc. Scandium iodide adde to mercury vapor lamps produces a highly efficient light source resembling sunlight, which is important for indoor or night-time color TV. Little i yet known about the toxicity of scandium; therefore, it should be handled with care.

Selenium — (Gr. *Selene*, moon), Se; at. wt. 78.96; at. no. 34; m.p. (gray) 221°C; b.p. (gray) 685°C; sp. gr. (gray) 4.79, (vitreous) 4.28; valenc –2, +4, or +6. Discovered by Berzelius in 1817, who found it associated with tellurium, named for the earth. Selenium is found in a few rare mineral such as *crooksite* and *clausthalite*. In years past it has been obtained from flue dusts remaining from processing copper sulfide ores, but the anode mud from electrolytic copper refineries now provide the source of most of the world's selenium. Selenium is recovered by roasting the muds with soda o sulfuric acid, or by smelting them with soda and niter. Selenium exists in several allotropic forms. Three are generally recognized, but as many as si have been claimed. Selenium can be prepared with either an amorphous or crystalline structure. The color of amorphous selenium is either red, i powder form, or black, in vitreous form. Crystalline monoclinic selenium is a deep red; crystalline hexagonal selenium, the most stable variety, is metallic gray. Natural selenium contains six stable isotopes. Fifteen other isotopes have been characterized. The element is a member of the sulfu family and resembles sulfur both in its various forms and in its compounds. Selenium exhibits both photovoltaic action, where light is converted directl into electricity, and photoconductive action, where the electrical resistance decreases with increased illumination. These properties make seleniu useful in the production of photocells and exposure meters for photographic use, as well as solar cells. Selenium is also able to convert a.c. electricit to d.c., and is extensively used in rectifiers. Below its melting point selenium is a p-type semiconductor and is finding many uses in electronic and solid state applications. It is used in Xerography for reproducing and copying documents, letters, etc. It is used by the glass industry to decolorize glass an to make ruby-colored glasses and enamels. It is also used as a photographic toner, and as an additive to stainless steel. Elemental selenium has bee said to be practically nontoxic and is considered to be an essential trace element; however, hydrogen selenide and other selenium compounds ar extremely toxic, and resemble arsenic in their physiological reactions. Hydrogen selenide in a concentration of 1.5 ppm is intolerable to man. Seleniu occurs in some soils in amounts sufficient to produce serious effects on animals feeding on plants, such as locoweed, grown in such soils. Exposur to selenium compounds (as Se) in air should not exceed 0.2 mg/m^3 (8-hour time-weighted average — 40-hour week). Selenium is priced at about \$30 lb. It is also available in high-purity form at a somewhat higher cost.

Silicon — (L. *silex, silicis*, flint), Si; at. wt. 28.0855; at. no. 14; m.p. 1414°C; b.p. 3265°C; sp. gr. 2.33 (25°C); valence 4. Davy in 1800 though silica to be a compound and not an element; later in 1811, Gay Lussac and Thenard probably prepared impure amorphous silicon by heating potassiu with silicon tetrafluoride. Berzelius, generally credited with the discovery, in 1824 succeeded in preparing amorphous silicon by the same genera method as used earlier, but he purified the product by removing the fluosilicates by repeated washings. Deville in 1854 first prepared crystalline silicon the second allotropic form of the element. Silicon is present in the sun and stars and is a principal component of a class of meteorites known a "aerolites". It is also a component of *tektites*, a natural glass of uncertain origin. Silicon makes up 25.7% of the earth's crust, by weight, and is the secon most abundant element, being exceeded only by oxygen. Silicon is not found free in nature, but occurs chiefly as the oxide and as silicates. *Sand, quartz rock crystal, amethyst, agate, flint, jasper,* and *opal* are some of the forms in which the oxide appears. *Granite, hornblende, asbestos, feldspar, cla mica,* etc. are but a few of the numerous silicate minerals. Silicon is prepared commercially by heating silica and carbon in an electric furnace, usin carbon electrodes. Several other methods can be used for preparing the element. Amorphous silicon can be prepared as a brown powder, which ca be easily melted or vaporized. Crystalline silicon has a metallic luster and grayish color. The Czochralski process is commonly used to produce single crystals of silicon used for solid-state or semiconductor devices. Hyperpure silicon can be prepared by the thermal decomposition of ultra-pure trichlorosilane in a hydrogen atmosphere, and by a vacuum float zone process. This product can be doped with boron, gallium, phosphorus, or arsenic to produce silicon for use in transistors, solar cells, rectifiers, and other solid-state devices which are used extensively in the electronics and space-age industries. Hydrogenated amorphous silicon has shown promise in producing economical cells for converting solar energy into electricity. Silicon is a relatively inert element, but it is attacked by halogens and dilute alkali. Most acids except hydrofluoric, do not affect it. Silicones are important products of silicon. They may be prepared by hydrolyzing a silicon organic chloride, such as dimethyl silicon chloride. Hydrolysis and condensation of various substituted chlorosilanes can be used to produce a very great number of polymeric products, or silicones, ranging from liquids to hard, glasslike solids with many useful properties. Elemental silicon transmits more than 95% of all wavelengths of infrared, from 1.3 to 6.7 μm. Silicon is one of man's most useful elements. In the form of sand and clay it is used to make concrete and brick; it is a useful refractory material for high-temperature work, and in the form of silicates it is used in making enamels, pottery, etc. Silica, as sand, is a principal ingredient of glass, one of the most inexpensive of materials with excellent mechanical, optical, thermal, and electrical properties. Glass can be made in a very great variety of shapes, and is used as containers, window glass, insulators, and thousands of other uses. Silicon tetrachloride can be used to iridize glass. Silicon is important in plant and animal life. Diatoms in both fresh and salt water extract silica from the water to build up their cell walls. Silica is present in ashes of plants and in the human skeleton. Silicon is an important ingredient in steel; silicon carbide is one of the most important abrasives and has been used in lasers to produce coherent light of 4560 Å. Regular grade silicon (99%) costs about \$0.50/g. Silicon 99.9% pure costs about \$50/lb; hyperpure silicon may cost as much as \$100/oz. Miners, stonecutters, and other engaged in work where siliceous dust is breathed in large quantities often develop a serious lung disease known as *silicosis*.

Silver — (Anglo-Saxon, *Seolfor siolfur*), Ag (L. argentum), at. wt. 107.8682; at. no. 47; m.p. 961.78°C; b.p. 2162°C; sp. gr. 10.50 (20°C); valence 1, 2. Silver has been known since ancient times. It is mentioned in Genesis. Slag dumps in Asia Minor and on islands in the Aegean Sea indicate that man learned to separate silver from lead as early as 3000 B.C. Silver occurs native and in ores such as *argentite* (Ag_2S) and *horn silver* ($AgCl$); lead, lead-zinc, copper, gold, and copper-nickel ores are principal sources, Mexico, Canada, Peru, and the U.S. are the principal silver producers in the western hemisphere. Silver is also recovered during electrolytic refining of copper. Commercial fine silver contains at least 99.9% silver. Purities of 99.999+% are available commercially. Pure silver has a brilliant white metallic luster. It is a little harder than gold and is very ductile and malleable, being exceeded only by gold and perhaps palladium. Pure silver has the highest electrical and thermal conductivity of all metals, and possesses the lowest contact resistance. It is stable in pure air and water, but tarnishes when exposed to ozone, hydrogen sulfide, or air containing sulfur. The alloys of silver are important. Sterling silver is used for jewelry, silverware, etc. where appearance is paramount. This alloy contains 92.5% silver, the remainder being copper or some other metal. Silver is of utmost importance in photography, about 30% of the U.S. industrial consumption going into this application. It is used for dental alloys. Silver is used in making solder and brazing alloys, electrical contacts, and high capacity silver-zinc and silver-cadmium batteries. Silver paints are used for making printed circuits. It is used in mirror production and may be deposited on glass or metals by chemical deposition, electrodeposition, or by evaporation. When freshly deposited, it is the best reflector of visible light known, but is rapidly tarnishes and loses much of its reflectance. It is a poor reflector of ultraviolet. Silver fulminate ($Ag_2C_2N_2O_2$), a powerful explosive, is sometimes formed during the silvering process. Silver iodide is used in seeding clouds to produce rain. Silver chloride has interesting optical properties as it can be made transparent; it also is a cement for glass. Silver nitrate, or *lunar caustic*, the most important silver compound, is used extensively in photography. While silver itself is not considered to be toxic, most of its salts are poisonous. Exposure to silver (metal and soluble compounds, as Ag) in air should not exceed 0.01 mg/m^3, (8-hour time-weighted average — 40-hour week). Silver compounds can be absorbed in the circulatory system and reduced silver deposited in the various tissues of the body. A condition, known as *argyria*, results, with a greyish pigmentation of the skin and mucous membranes. Silver has germicidal effects and kills many lower organisms effectively without harm to higher animals. Silver for centuries has been used traditionally for coinage by many countries of the world. In recent times, however, consumption of silver has greatly exceeded the output. In 1939, the price of silver was fixed by the U.S. Treasury at 71¢/troy oz., and at 90.5¢/troy oz. in 1946. In November 1961 the U.S. Treasury suspended sales of nonmonetized silver, and the price stabilized for a time at about \$1.29, the melt-down value of silver U.S. coins. The Coinage Act of 1965 authorized a change in the metallic composition of the three U.S. subsidiary denominations to clad or composite type coins. This was the first change in U.S. coinage since the monetary system was established in 1792. Clad dimes and quarters are made of an outer layer of 75% Cu and 25% Ni bonded to a central core of pure Cu. The composition of the one- and five-cent pieces remains unchanged. One-cent coins are 95% Cu and 5% Zn. Five-cent coins are 75% Cu and 25% Ni. Old silver dollars are 90% Ag and 10% Cu. Earlier subsidiary coins of 90% Ag and 10% Cu officially were to circulate alongside the clad coins; however, in practice they have largely disappeared (Gresham's Law), as the value of the silver is now greater than their exchange value. Silver coins of other countries have largely been replaced with coins made of other metals. On June 24, 1968, the U.S. Government ceased to redeem U.S. Silver Certificates with silver. Since that time, the price of silver has fluctuated widely. As of January 1990, the price of silver was about \$5.25/troy oz.; however the price has fluctuated considerably due to market instability.

Sodium — (English, *soda*; Medieval Latin, *sodanum*, headache remedy), Na (L. *natrium*); at. wt. 22.989768; at. no. 11; m.p. 97.72 ± 0.03°C; b.p. 883°C; sp. gr. 0.971 (20°C); valence 1. Long recognized in compounds, sodium was first isolated by Davy in 1807 by electrolysis of caustic soda. Sodium is present in fair abundance in the sun and stars. The D lines of sodium are among the most prominent in the solar spectrum. Sodium is the sixth most abundant element on earth, comprising about 2.6% of the earth's crust; it is the most abundant of the alkali group of metals of which it is a member. The most common compound is sodium chloride, but it occurs in many other minerals, such as *soda niter*, *cryolite*, *amphibole*, *zeolite*, *sodalite*, etc. It is a very reactive element and is never found free in nature. It is now obtained commercially by the electrolysis of absolutely dry fused sodium chloride. This method is much cheaper than that of electrolyzing sodium hydroxide, as was used several years ago. Sodium is a soft, bright, silvery metal which floats on water, decomposing it with the evolution of hydrogen and the formation of the hydroxide. It may or may not ignite spontaneously on water, depending on the amount of oxide and metal exposed to the water. It normally does not ignite in air at temperatures below 115°C. Sodium should be handled with respect, as it can be dangerous when improperly handled. Metallic sodium is vital in the manufacture of sodamide and esters, and in the preparation of organic compounds. The metal may be used to improve the structure of certain alloys, to descale metal, to purify molten metals, and as a heat transfer agent. An alloy of sodium with potassium, NaK, is also an important heat transfer agent. Sodium compounds are important to the paper, glass, soap, textile, petroleum, chemical, and metal industries. Soap is generally a sodium salt of certain fatty acids. The importance of common salt to animal nutrition has been recognized since prehistoric times. Among the many compounds that are of the greatest industrial importance are common salt ($NaCl$), soda ash (Na_2CO_3), baking soda ($NaHCO_3$), caustic soda ($NaOH$), Chile saltpeter ($NaNO_3$), di- and tri-sodium phosphates, sodium thiosulfate (hypo, $Na_2S_2O_3 \cdot 5H_2O$), and borax ($Na_2B_4O_7 \cdot 10H_2O$). Thirteen isotopes of sodium are recognized. Metallic sodium is priced at about 15 to 20¢/lb in quantity. Reagent grade (ACS) sodium in January 1990 cost about \$35/lb. On a volume basis, it is the cheapest of all metals. Sodium metal should be handled with great care. It avoid be maintained in an inert atmosphere and contact with water and other substances with which sodium reacts should be avoided.

Strontium — (*Strontian*, town in Scotland), Sr; at. wt. 87.62; at. no. 38; m.p. 777°C; b.p. 1382°C; sp. gr. 2.54; valence 2. Isolated by Davey electrolysis in 1808; however, Adair Crawford in 1790 recognized a new mineral (strontianite) as differing from other barium minerals (bary Strontium is found chiefly as *celestite* ($SrSO_4$) and *strontianite* ($SrCO_3$). The metal can be prepared by electrolysis of the fused chloride mixed v potassium chloride, or is made by reducing strontium oxide with aluminum in a vacuum at a temperature at which strontium distills off. Three allotr forms of the metal exist, with transition points at 235 and 540°C. Strontium is softer than calcium and decomposes water more vigorously. It does absorb nitrogen below 380°C. It should be kept under kerosene to prevent oxidation. Freshly cut strontium has a silvery appearance, but rapidly tu a yellowish color with the formation of the oxide. The finely divided metal ignites spontaneously in air. Volatile strontium salts impart a beau crimson color to flames, and these salts are used in pyrotechnics and in the production of flares. Natural strontium is a mixture of four stable isoto Sixteen other unstable isotopes are known to exist. Of greatest importance is ^{90}Sr with a half-life of 29 years. It is a product of nuclear fallout and prese a health problem. This isotope is one of the best long-lived high-energy beta emitters known, and is used in SNAP (Systems for Nuclear Auxili Power) devices. These devices hold promise for use in space vehicles, remote weather stations, navigational buoys, etc., where a lightweight, lo lived, nuclear-electric power source is needed. The major use for strontium at present is in producing glass for color television picture tubes. It also found use in producing ferrite magnets and in refining zinc. Strontium titanate is an interesting optical material as it has an extremely high refrac index and an optical dispersion greater than that of diamond. It has been used as a gemstone, but it is very soft. It does not occur naturally. Stronti metal (98% pure) in January 1990 cost about $5/oz.

Sulfur — (Sanskrit, *sulvere*; L. *sulphurium*), S; at. wt. 32.066; at. no. 16; m.p. 115.21°C; b.p. 444.60°C; sp. gr. (rhombic) 2.07, (monoclinic) 1. (20°C); valence 2, 4, or 6. Known to the ancients; referred to in Genesis as *brimstone*. Sulfur is found in meteorites. A dark area near the cra Aristarchus on the moon has been studied by R. W. Wood with ultraviolet light. This study suggests strongly that it is a sulfur deposit. Sulfur occ native in the vicinity of volcanoes and hot springs. It is widely distributed in nature as *iron pyrites, galena, sphalerite, cinnabar, stibnite, gypsum, Eps salts, celestite, barite*,etc. Sulfur is commercially recovered from wells sunk into the salt domes along the Gulf Coast of the U.S. It is obtained fr these wells by the Frasch process, which forces heated water into the wells to melt the sulfur, which is then brought to the surface. Sulfur also occ in natural gas and petroleum crudes and must be removed from these products. Formerly this was done chemically, which wasted the sulfur. N processes now permit recovery, and these sources promise to be very important. Large amounts of sulfur are being recovered from Alberta gas fiel Sulfur is a pale yellow, odorless, brittle solid, which is insoluble in water but soluble in carbon disulfide. In every state, whether gas, liquid or so elemental sulfur occurs in more than one allotropic form or modification; these present a confusing multitude of forms whose relations are not yet fu understood. Amorphous or "plastic" sulfur is obtained by fast cooling of the crystalline form. X-ray studies indicate that amorphous sulfur may ha a helical structure with eight atoms per spiral. Crystalline sulfur seems to be made of rings, each containing eight sulfur atoms, which fit together give a normal X-ray pattern. Eleven isotopes of sulfur exist. Four occur in natural sulfur, none of which is radioactive. A finely divided form of sulf known as *flowers of sulfur*, is obtained by sublimation. Sulfur readily forms sulfides with many elements. Sulfur is a component of black gunpowd and is used in the vulcanization of natural rubber and a fungicide. It is also used extensively is making phosphatic fertilizers. A tremendous tonna is used to produce sulfuric acid, the most important manufactured chemical. It is used in making sulfite paper and other papers, as a fumigant, and the bleaching of dried fruits. The element is a good electrical insulator. Organic compounds containing sulfur are very important. Calcium sulfa ammonium sulfate, carbon disulfide, sulfur dioxide, and hydrogen sulfide are but a few of the many other important compounds of sulfur. Sulfu essential to life. It is a minor constituent of fats, body fluids, and skeletal minerals. Carbon disulfide, hydrogen sulfide, and sulfur dioxide should handled carefully. Hydrogen sulfide in small concentrations can be metabolized, but in higher concentrations it quickly can cause death by respirato paralysis. It is insidious in that it quickly deadens the sense of smell. Sulfur dioxide is a dangerous component in atmospheric air pollution. In 197 University of Pennsylvania scientists reported synthesis of polymeric sulfur nitride, which has the properties of a metal, although it contains no me atoms. The material has unusual optical and electrical properties. High-purity sulfur is commercially available in purities of 99.999+%.

Tantalum — (Gr. *Tantalos*, mythological character, father of *Niobe*), Ta; at. wt. 180.9479; at. no. 73; m.p. 3017°C; b.p. 5458 ± 100°C; sp. 16.654; valence 2?, 3, 4?, or 5. Discovered in 1802 by Ekeberg, but many chemists thought niobium and tantalum were identical elements until Ro in 1844, and Marignac, in1866, showed that niobic and tantalic acids were two different acids. The early investigators only isolated the impure met The first relatively pure ductile tantalum was produced by von Bolton in 1903. Tantalum occurs principally in the mineral *columbite-tantalite* (F Mn)$(Nb, Ta)_2O_6$. Tantalum ores are found in Australia, Brazil, Mozambique, Thailand, Portugal, Nigeria, Zaire, and Canada. Separation of tantalu from niobium requires several complicated steps. Several methods are used to commercially produce the element, including electrolysis of molt potassium fluorotantalate, reduction of potassium fluorotantalate with sodium, or reacting tantalum carbide with tantalum oxide. Twenty five isotop of tantalum are known to exist. Natural tantalum contains two isotopes. Tantalum is a gray, heavy, and very hard metal. When pure, it is ductile ar can be drawn into fine wire, which is used as a filament for evaporating metals such as aluminum. Tantalum is almost completely immune to chemic attack at temperatures below 150°C, and is attacked only by hydrofluoric acid, acidic solutions containing the fluoride ion, and free sulfur trioxid Alkalis attack it only slowly. At high temperatures, tantalum becomes much more reactive. The element has a melting point exceeded only by tungste and rhenium. Tantalum is used to make a variety of alloys with desirable properties such as high melting point, high strength, good ductility, et Scientists at Los Alamos have produced a tantalum carbide graphite composite material, which is said to be one of the hardest materials ever mad The compound has a melting point of 3738°C. Tantalum has good "gettering" ability at high temperatures, and tantalum oxide films are stable ar have good rectifying and dielectric properties. Tantalum is used to make electrolytic capacitors and vacuum furnace parts, which account for abo 60% of its use. The metal is also widely used to fabricate chemical process equipment, nuclear reactors, and aircraft and missile parts. Tantalum completely immune to body liquids and is a nonirritating metal. It has, therefore, found wide use in making surgical appliances. Tantalum oxide used to make special glass with high index of refraction for camera lenses. The metal has many other uses. The price of (99.9%) tantalum in Dec. 198 was about $50/oz.

Technetium — (Gr. *technetos*, artificial), Tc; at. wt. (98); at. no. 43; m.p. 2157°C; b.p. 4265°C; sp. gr. 11.50 (calc.); valence 0, +2, +4, +5, + and +7. Element 43 was predicted on the basis of the periodic table, and was erroneously reported as having been discovered in 1925, at which tim it was named *masurium*. The element was actually discovered by Perrier and Segre in Italy in 1937. It was found in a sample of molybdenum, whic was bombarded by deuterons in the Berkeley cyclotron, and which E. Lawrence sent to these investigators. Technetium was the first element to be produced artificially. Since its discovery, searches for the element in terrestrial materials have been made without success. If it does exist, the concentration must be very small. Technetium has been found in the spectrum of S-, M-, and N-type stars, and its presence in stellar matter is leading to new theories of the production of heavy elements in the stars. Nineteen isotopes of technetium, with atomic masses ranging from 90 to 108, are known. ^{97}Tc has a half-life of 2.6×10^6 years. ^{98}Tc has a half-life of 4.2×10^6 years. The isomeric isotope ^{95m}Tc, with a half-life of 61 days, is useful for tracer work, as it produces energetic gamma rays. Technetium metal has been produced in kilogram quantities. The metal was first prepared by passing hydrogen gas at 1100°C over Tc_2S_7. It is now conveniently prepared by the reduction of ammonium pertechnetate with hydrogen. Technetium is a silvery-gray metal that tarnishes slowly in moist air. Until 1960, technetium was available only in small amounts and the price was as high as $2800/g. It is now commercially available to holders of O.R.N.L. permits at a price of $60/g. The chemistry of technetium is said to be similar to that of rhenium. Technetium dissolves in nitric acid, aqua regia, and conc. sulfuric acid, but is not soluble in hydrochloric acid of any strength. The element is a remarkable corrosion inhibitor for steel. It is reported that mild carbon steels may be effectively protected by as little as 55 ppm of $KTcO_4$ in aerated distilled water at temperatures up to 250°C. This corrosion protection is limited to closed systems, since technetium is radioactive and must be confined. ^{99}Tc has a specific activity of 6.2×10^8 Bq/g. Activity of this level must not be allowed to spread. ^{99}Tc is a contamination hazard and should be handled in a glove box. The metal is an excellent superconductor at 11 K and below.

Tellurium — (L. *tellus*, earth), Te; at. wt. 127.60; at. no. 52; m.p. 449.51 ± 0.3°C; b.p. 988°C; sp. gr. 6.24 (20°C); valence 2, 4, or 6. Discovered by Muller von Reichenstein in 1782; named by Klaproth, who isolated it in 1798. Tellurium is occasionally found native, but is more often found as the telluride of gold (*calaverite*), and combined with other metals. It is recovered commercially from the anode muds produced during the electrolytic refining of blister copper. The U.S., Canada, Peru, and Japan are the largest Free World producers of the element. Crystalline tellurium has a silvery-white appearance, and when pure exhibits a metallic luster. It is brittle and easily pulverized. Amorphous tellurium is formed by precipitating tellurium from a solution of telluric or tellurous acid. Whether this form is truly amorphous, or made of minute crystals, is open to question. Tellurium is a p-type semiconductor, and shows greater conductivity in certain directions, depending on alignment of the atoms. Its conductivity increases slightly with exposure to light. It can be doped with silver, copper, gold, tin, or other elements. In air, tellurium burns with a greenish-blue flame, forming the dioxide. Molten tellurium corrodes iron, copper, and stainless steel. Tellurium and its compounds are probably toxic and should be handled with care. Workmen exposed to as little as 0.01 mg/m^3 of air, or less, develop "tellurium breath," which has a garlic-like odor. Thirty isotopes of tellurium are known, with atomic masses ranging from 108 to 137. Natural tellurium consists of eight isotopes. Tellurium improves the machinability of copper and stainless steel, and its addition to lead decreases the corrosive action of sulfuric acid on lead and improves its strength and hardness. Tellurium is used as a basic ingredient in blasting caps, and is added to cast iron for chill control. Tellurium is used in ceramics. Bismuth telluride has been used in thermoelectric devices. Tellurium costs about $100/lb, with a purity of about 99.5%.

Terbium — (*Ytterby*, village in Sweden), Tb; at. wt. 158.92534; at. no. 65; m.p. 1359°C; b.p. 3221°C; sp. gr. 8.230; valence 3, 4. Discovered by Mosander in 1843. Terbium is a member of the lanthanide or "rare earth" group of elements. It is found in *cerite, gadolinite*, and other minerals along with other rare earths. It is recovered commercially from *monazite* in which it is present to the extent of 0.03%, from *xenotime*, and from *euxenite*, a complex oxide containing 1% of more of terbia. Terbium has been isolated only in recent years with the development of ion-exchange techniques for separating the rare-earth elements. As with other rare earths, it can be produced by reducing the anhydrous chloride or fluoride with calcium metal in a tantalum crucible. Calcium and tantalum impurities can be removed by vacuum remelting. Other methods of isolation are possible. Terbium is reasonably stable in air. It is a silver-gray metal, and is malleable, ductile, and soft enough to be cut with a knife. Two crystal modifications exist, with a transformation temperature of 1289°C. Twenty one isotopes with atomic masses ranging from 145 to 165 are recognized. The oxide is a chocolate or dark maroon color. Sodium terbium borate is used as a laser material and emits coherent light at 0.546 μm. Terbium is used to dope calcium fluoride, calcium tungstate, and strontium molybdate, used in solid-state devices. The oxide has potential application as an activator for green phosphors used in color TV tubes. It can be used with ZrO_2 as a crystal stabilizer of fuel cells which operate at elevated temperature. Few other uses have been found. The element is priced at about $30/g (99.9%). Little is known of the toxicity of terbium. It should be handled with care as with other lanthanide elements.

Thallium — (Gr. *thallos*, a green shoot or twig), Tl; at. wt. 204.3833; at. no. 81; m.p. 304°C; b.p. 1473 ± 10°C; sp. gr. 11.85 (20°C); valence 1, or 3. Thallium was discovered spectroscopically in 1861 by Crookes. The element was named after the beautiful green spectral line, which identified the element. The metal was isolated both by Crookes and Lamy in 1862 about the same time. Thallium occurs in *crooksite, lorandite*, and *hutchinsonite*. It is also present in *pyrites* and is recovered from the roasting of this ore in connection with the production of sulfuric acid. It is also obtained from the smelting of lead and zinc ores. Extraction is somewhat complex and depends on the source of the thallium. Manganese nodules, found on the ocean floor, contain thallium. When freshly exposed to air, thallium exhibits a metallic luster, but soon develops a bluish-gray tinge, resembling lead in appearance. A heavy oxide builds up on thallium if left in air, and in the presence of water the hydroxide is formed. The metal is very soft and malleable. It can be cut with a knife. Twenty five isotopic forms of thallium, with atomic masses ranging from 184 to 210 are recognized. Natural thallium is a mixture of two isotopes. The element and its compounds are toxic and should be handled carefully. Contact of the metal with skin is dangerous, and when melting the metal adequate ventilation should be provided. Exposure to thallium (soluble compounds) — skin, as Tl, should not exceed 0.1 mg/m^3 (8-hour time-weighted average — 40-hour week). Thallium is suspected of carcinogenic potential for man. Thallium sulfate has been widely employed as a rodenticide and ant killer. It is odorless and tasteless, giving no warning of its presence. Its use, however, has been prohibited in the U.S. since 1975 as a household insecticide and rodenticide. The electrical conductivity of thallium sulfide changes with exposure to infrared light, and this compound is used in photocells. Thallium bromide-iodide crystals have been used as infrared optical materials. Thallium has been used, with sulfur or selenium and arsenic, to produce low melting glasses which become fluid between 125 and 150°C. These glasses have properties at room temperatures similar to ordinary glasses and are said to be durable and insoluble in water. Thallium oxide has been used to produce glasses with a high index of refraction. Thallium has been used in treating ringworm and other skin infections; however, its use has been limited because of the narrow margin between toxicity and therapeutic benefits. A mercury-thallium alloy, which forms a eutectic at 8.5% thallium, is reported to freeze at –60°C, some 20° below the freezing point of mercury. Commercial thallium metal (99%) costs about $40/lb.

Thorium — (*Thor*, Scandinavian god of war), Th; at. wt. 232.0381; at. no. 90; m.p. 1750°C; b.p. 4788°C; sp. gr. 11.72; valence +2(?), +3(?), +4.

Discovered by Berzelius in 1828. Thorium occurs in *thorite* ($ThSiO_4$) and in *thorianite* ($ThO_2 + UO_2$). Large deposits of thorium minerals have been reported in New England and elsewhere, but these have not yet been exploited. Thorium is now thought to be about three times as abundant as uranium and about as abundant as lead or molybdenum. The metal is a source of nuclear power. There is probably more energy available for use from thorium in the minerals of the earth's crust than from both uranium and fossil fuels. Any sizable demand for thorium as a nuclear fuel is still several years in the future. Work has been done in developing thorium cycle converter-reactor systems. Several prototypes, including the HTGR (high-temperature gas-cooled reactor) and MSRE (molten salt converter reactor experiment), have operated. While the HTGR reactors are efficient, they are not expected to become important commercially for many years because of certain operating difficulties. Thorium is recovered commercially from the mineral *monazite*, which contains from 3 to 9% ThO_2 along with rare-earth minerals. Much of the internal heat the earth produces has been attributed to thorium and uranium. Several methods are available for producing thorium metal: it can be obtained by reducing thorium oxide with calcium, by electrolysis of anhydrous thorium chloride in a fused mixture of sodium and potassium chlorides, by calcium reduction of thorium tetrachloride mixed with anhydrous zinc chloride, and by reduction of thorium tetrachloride with an alkali metal. Thorium was originally assigned a position in Group IV of the periodic table. Because of its atomic weight, valence, etc., it is now considered to be the second member of the *actinide* series of elements. When pure, thorium is a silvery-white metal which is air-stable and retains its luster for several months. When contaminated with the oxide, thorium slowly tarnishes in air, becoming gray and finally black. The physical properties of thorium are greatly influenced by the degree of contamination with the oxide. The purest specimens often contain several tenths of a percent of the oxide. High-purity thorium has been made. Pure thorium is soft, very ductile, and can be cold-rolled, swaged, and drawn. Thorium is dimorphic, changing at 1400°C from a cubic to a body-centered cubic structure. Thorium oxide has a melting point of 3300°C, which is the highest of all oxides. Only a few elements, such as tungsten, and a few compounds, such as tantalum carbide, have higher melting points. Thorium is slowly attacked by water, but does not dissolve readily in most common acids, except hydrochloric. Powdered thorium metal is often pyrophoric and should be carefully handled. When heated in air, thorium turnings ignite and burn brilliantly with a white light. The principal use of thorium has been in the preparation of the Welsbach mantle, used for portable gas lights. These mantles, consisting of thorium oxide with about 1% cerium oxide and other ingredients, glow with a dazzling light when heated in a gas flame. Thorium is an important alloying element in magnesium, imparting high strength and creep resistance at elevated temperatures. Because thorium has a low work-function and high electron emission, it is sued to coat tungsten wire used in electronic equipment. The oxide is also used to control the grain size of tungsten used for electric lamps; it is also used for high-temperature laboratory crucibles. Glasses containing thorium oxide have a high refractive index and low dispersion. Consequently, they find application in high quality lenses for cameras and scientific instruments. Thorium oxide has also found use as a catalyst in the conversion of ammonia to nitric acid, in petroleum cracking, and in producing sulfuric acid. Twenty five isotopes of thorium are known with atomic masses ranging from 212 to 236. All are unstable. ^{232}Th occurs naturally and has a half-life of 1.4×10^{10} years. It is an alpha emitter. ^{232}Th goes through six alpha and four beta decay steps before becoming the stable isotope ^{208}Pb. ^{232}Th is sufficiently radioactive to expose a photographic plate in a few hours. Thorium disintegrates with the production of "thoron" (^{220}Rn), which is an alpha emitter and presents a radiation hazard. Good ventilation of areas where thorium is stored or handled is therefore essential. Thorium metal (99.9%) costs about $150/oz.

Thulium — (*Thule*, the earliest name for Scandinavia), Tm; at. wt. 168.93421; at. no. 69; m.p. 1545°C; b.p. 1946°C; sp. gr. 9.321 (25°C); valence 3. Discovered in 1879 by Cleve. Thulium occurs in small quantities along with other rare earths in a number of minerals. It is obtained commercially from *monazite*, which contains about 0.007% of the element. Thulium is the least abundant of the rare earth elements, but with new sources recently discovered, it is now considered to be about as rare as silver, gold, or cadmium. Ion-exchange and solvent extraction techniques have recently permitted much easier separation of the rare earths, with much lower costs. Only a few years ago, thulium metal was not obtainable at any cost; in 1985 the oxide sold for $3400/kg. Thulium metal costs $50/g. Thulium can be isolated by reduction of the oxide with lanthanum metal or by calcium reduction of the anhydrous fluoride. The pure metal has a bright, silvery luster. It is reasonably stable in air, but the metal should be protected from moisture in a closed container. The element is silver-gray, soft, malleable, and ductile, and can be cut with a knife. Twenty five isotopes are known, with atomic masses ranging from 152 to 176. Natural thulium, which is 100% ^{169}Tm, is stable. Because of the relatively high price of the metal, thulium has not yet found many practical applications. ^{169}Tm bombarded in a nuclear reactor can be used as a radiation source in portable X-ray equipment. ^{171}Tm is potentially useful as an energy source. Natural thulium also has possible use in *ferrites* (ceramic magnetic materials) used in microwave equipment. As with other lanthanides, thulium has a low-to-moderate acute toxic rating. It should be handled with care.

Tin — (anglo-Saxon, *tin*), Sn (L. *stannum*); at. wt. 118.710; at. no. 50; m.p. 231.9°C; b.p. 2602°C; sp. gr. (gray) 5.75, (white) 7.31; valence 2, 4. Known to the ancients. Tin is found chiefly in *cassiterite* (SnO_2). Most of the world's supply comes from Malaya, Bolivia, Indonesia, Zaire, Thailand, and Nigeria. The U.S. produces almost none, although occurrences have been found in Alaska and California. Tin is obtained by reducing the ore with coal in a reverberatory furnace. Ordinary tin is composed of nine stable isotopes; 18 unstable isotopes are also known. Ordinary tin is a silver-white metal, is malleable, somewhat ductile, and has a highly crystalline structure. Due to the breaking of these crystals, a "tin cry" is heard when a bar is bent. The element has two allotropic forms at normal pressure. On warming, gray, or α tin, with a cubic structure, changes at 13.2°C into white, or β tin, the ordinary form of the metal. White tin has a tetragonal structure. When tin is cooled below 13.2°C, it changes slowly from white to gray. This change is affected by impurities such as aluminum and zinc, and can be prevented by small additions of antimony or bismuth. This change from the α to β form is called the tin pest. There are few if any uses for gray tin. Tin takes a high polish and is used to coat other metals to prevent corrosion or other chemical action. Such tin plate over steel is used in the so-called tin can for preserving food. Alloys of tin are very important. Soft solder, type metal, fusible metal, pewter, bronze, bell metal, Babbitt metal, White metal, die casting alloy, and phosphor bronze are some of the important alloys using tin. Tin resists distilled sea and soft tap water, but is attacked by strong acids, alkalis, and acid salts. Oxygen in solution accelerates the attack. When heated in air, tin forms SnO_2, which is feebly acid, forming stannate salts with basic oxides. The most important salt is the chloride ($SnCl_2 \cdot H_2O$), which is used as a reducing agent and as a mordant in calico printing. Tin salts sprayed onto glass are used to produce electrically conductive coatings. These have been used for panel lighting and for frost-free windshields. Most window glass is now made by floating molten glass on molten tin (float glass) to produce a flat surface (Pilkington process). Of recent interest is a crystalline tin-niobium alloy that is superconductive at very low temperatures. This promises to be important in the construction of superconductive magnets that generate enormous field strengths but use practically no power. Such magnets, made of tin-niobium wire, weigh but a few pounds and produce magnetic fields that, when started with a small battery, are comparable to that of a 100 ton electromagnet operated continuously with a large power supply. The small amount of tin found in canned foods is quite harmless. The agreed limit of tin content in U.S. foods is 300 mg/kg. The trialkyl and triaryl tin compounds are used as biocides and must be handled carefully. Over the past 25 years the price of tin has varied from 50¢/lb to its present price of about $4/lb. as of January 1990.

Titanium — (L. *Titans*, the first sons of the Earth, myth.), Ti; at. wt. 47.88; at. no. 22; m.p. 1668 ± 10°C; b.p. 3287°C; sp. gr. 4.54; valence 2, 3, or 4. Discovered by Gregor in 1791; named by Klaproth in 1795. Impure titanium was prepared by Nilson and Pettersson in 1887; however, the pure metal (99.9%) was not made until 1910 by Hunter by heating $TiCl_4$ with sodium in a steel bomb. Titanium is present in meteorites and in the sun. Rocks obtained during the Apollo 17 lunar mission showed presence of 12.1% TiO_2. Analyses of rocks obtained during earlier Apollo missions show lower percentages. Titanium oxide bands are prominent in the spectra of M-type stars. The element is the ninth most abundant in the crust of the earth. Titanium is almost always present in igneous rocks and in the sediments derived from them. It occurs in the minerals *rutile, ilmenite,* and *sphene*, and is present in titanates and in many iron ores. Titanium is present in the ash of coal, in plants, and in the human body. The metal was a laboratory curiosity until Kroll, in 1946, showed that titanium could be produced commercially by reducing titanium tetrachloride with magnesium. This method is largely used for producing the metal today. The metal can be purified by decomposing the iodide. Titanium, when pure, is a lustrous, white metal. It has a low density, good strength, is easily fabricated, and has excellent corrosion resistance. It is ductile only when it is free of oxygen. The metal burns in air and is the only element that burns in nitrogen. Titanium is resistant to dilute sulfuric and hydrochloric acid, most organic acids, moist chlorine gas, and chloride solutions. Natural titanium consists of five isotopes with atomic masses from 46 to 50. All are stable. Eight other unstable isotopes are known. Natural titanium is reported to become very radioactive after bombardment with deuterons. The emitted radiations are mostly positrons and hard gamma rays. The metal is dimorphic. The hexagonal α form changes to the cubic β form very slowly at about 880°C. The metal combines with oxygen at red heat, and with chlorine at 550°C. Titanium is important as an alloying agent with aluminum, molybdenum, manganese, iron, and other metals. Alloys of titanium are principally used for aircraft and missiles where lightweight strength and ability to withstand extremes of temperature are important. Titanium is as strong as steel, but 45% lighter. It is 60% heavier than aluminum, but twice as strong. Titanium has potential use in desalination plants for converting sea water into fresh water. The metal has excellent resistance to sea water and is used for propeller shafts, rigging, and other parts of ships exposed to salt water. A titanium anode coated with platinum has been used to provide cathodic protection from corrosion by salt water. Titanium metal is considered to be physiologically inert. When pure, titanium dioxide is relatively clear and has an extremely high index of refraction with an optical dispersion higher than diamond. It is produced artificially for use as a gemstone, but it is relatively soft. Star sapphires and rubies exhibit their asterism as a result of the presence of TiO_2. Titanium dioxide is extensively used for both house paint and artist's paint, as it is permanent and has good covering power. Titanium oxide pigment accounts for the largest use of the element. Titanium paint is an excellent reflector of infrared, and is extensively used in solar observatories where heat causes poor seeing conditions. Titanium tetrachloride is used to iridize glass. This compound fumes strongly in air and has been used to produce smoke screens. The price of titanium metal powder (99.95%) is about $100/lb.

Tungsten — (Swedish, *tung sten*, heavy stone); also known as *wolfram* (from *wolframite*, said to be named from *wolf rahm* or *spumi lupi*, because the ore interfered with the smelting of tin and was supposed to devour the tin), W; at. wt. 183.85; at. no. 74; m.p. 3422 ± 20°C; b.p. 5555°C; sp. gr. 19.3 (20°C); valence 2, 3, 4, 5, or 6. In 1779 Peter Woulfe examined the mineral now known as *wolframite* and concluded it must contain a new substance. Scheele, in 1781, found that a new acid could be made from *tung sten* (a name first applied about 1758 to a mineral now known as *scheelite*). Scheele and Berman suggested the possibility of obtaining a new metal by reducing this acid. The de Elhuyar brothers found an acid in *wolframite* in 1783 that was identical to the acid of *tungsten*(tungstic acid) of Scheele, and in that year they succeeded in obtaining the element by reduction of this acid with charcoal. Tungsten occurs in *wolframite*, (Fe, Mn)WO_4; *scheelite*, $CaWO_4$; *huebnerite*, $MnWO_4$; and *ferberite*, $FeWO_4$. Important deposits of tungsten occur in California, Colorado, South Korea, Bolivia, U.S.S.R., and Portugal. China is reported to have about 75% of the world's tungsten resources. Natural tungsten contains five stable isotopes. Twenty one other unstable isotopes are recognized. The metal is obtained commercially by reducing tungsten oxide with hydrogen or carbon. Pure tungsten is a steel-gray to tin-white metal. Very pure tungsten can be cut with a hacksaw, and can be forged, spun, drawn, and extruded. The impure metal is brittle and can be worked only with difficulty. Tungsten has the highest melting point of all metals, and at temperatures over 1650°C has the highest tensile strength. The metal oxidizes in air and must be protected at elevated temperatures. It has excellent corrosion resistance and is attacked only slightly by most mineral acids. The thermal expansion is about the same as borosilicate glass, which makes the metal useful for glass-to-metal seals. Tungsten and its alloys are used extensively for filaments for electric lamps, electron and television tubes, and for metal evaporation work; for electrical contact points for automobile distributors; X-ray targets; windings and heating elements for electrical furnaces; and for numerous spacecraft and high-temperature applications. High-speed tool steels, Hastelloy®, Stellite®, and many other alloys contain tungsten. Tungsten carbide is of great importance to the metal-working, mining, and petroleum industries.Calcium and magnesium tungstates are widely used in fluorescent lighting; other salts of tungsten are used in the chemical and tanning industries. Tungsten disulfide is a dry, high-temperature lubricant, stable to 500°C. Tungsten bronzes and other tungsten compounds are used in paints. Tungsten powder (99.9%) costs about $50/lb.

Uranium — (Planet *Uranus*), U; at. wt. 238.0289; at. no. 92; m.p. 1135°C; b.p. 4131°C; sp. gr. ~18.95; valence 2, 3, 4, 5, or 6. Yellow-colored glass, containing more than 1% uranium oxide and dating back to 79 A.D., has been found near Naples, Italy. Klaproth recognized an unknown element in *pitchblende* and attempted to isolate the metal in 1789. The metal apparently was first isolated in 1841 by Peligot, who reduced the anhydrous chloride with potassium. Uranium is not as rare as it was once thought. It is now considered to be more plentiful than mercury, antimony, silver, or cadmium, and is about as abundant as molybdenum or arsenic. It occurs in numerous minerals such as *pitchblende, uraninite, carnotite, autunite, uranophane, davidite,* and *tobernite*. It is also found in *phosphate rock, lignite, monazite sands*, and can be recovered commercially from these sources. The U.S.D.O.E. purchases uranium in the form of acceptable U_3O_8 concentrates. This incentive program has greatly increased the known uranium reserves. Uranium can be repared by reducing uranium halides with alkali or alkaline earth metals or by reducing uranium oxides by calcium, aluminum, or carbon at high temperatures. The metal can also be produced by electrolysis of KUF_5 or UF_4, dissolved in a molten mixture of $CaCl_2$ and NaCl. High-purity uranium can be prepared by the thermal decomposition of uranium halides on a hot filament. Uranium exhibits three crystallographic modifications as follows:

$$\alpha \xrightarrow{688°C} \beta \xrightarrow{776°C} \gamma$$

Uranium is a heavy, silvery-white metal which is pyrophoric when finely divided. It is a little softer than steel, and is attacked by cold water in a finely divided state. It is malleable, ductile, and slightly paramagnetic. In air, the metal becomes coated with a layer of oxide. Acids dissolve the metal, but it is unaffected by alkalis. Uranium has sixteen isotopes, all of which are radioactive. Naturally occurring uranium nominally contains 99.2830% by weight ^{238}U, 0.7110% ^{235}U, and 0.0054% ^{234}U. Studies show that the percentage weight of ^{235}U in natural uranium varies by as much as 0.1%, depending on the source. The U.S.D.O.E. has adopted the value of 0.711 as being their "official" percentage of ^{235}U in natural uranium. Natural uranium is sufficiently radioactive to expose a photographic plate in an hour or so. Much of the internal heat of the earth is thought to be attributable to the presence of uranium and thorium. ^{238}U with a half-life of 4.51×10^9 years, has been used to estimate the age of igneous rocks. The origin of uranium, the highest member of the naturally occurring elements — except perhaps for traces of neptunium or plutonium — is not clearly understood, although it may be presumed that uranium is a decay product of elements of higher atomic weight, which may have once been present on earth or elsewhere in the universe. These original elements may have been formed as a result of a primordial "creation," known as "the big bang," in a supernova, or in some other stellar processes. Uranium is of great importance as a nuclear fuel. ^{238}U can be converted into fissionable plutonium by the following reactions:

$$^{238}U(n,\gamma) \rightarrow {}^{239}U \xrightarrow{\beta} {}^{239}Np \xrightarrow{\beta} {}^{239}Pu$$

This nuclear conversion can be brought about in "breeder" reactors where it is possible to produce more new fissionable material than the fissionable material used in maintaining the chain reaction. ^{235}U is of even greater importance, for it is the key to the utilization of uranium. ^{235}U, while occurring in natural uranium to the extent of only 0.71%, is so fissionable with slow neutrons that a self-sustaining fission chain reaction can be made to occur in a reactor constructed from natural uranium and a suitable moderator, such as heavy water or graphite, alone. ^{235}U can be concentrated by gaseous diffusion and other physical processes, if desired, and used directly as a nuclear fuel, instead of natural uranium, or used as an explosive. Natural uranium, slightly enriched with ^{235}U by a small percentage, is used to fuel nuclear power reactors for the generation of electricity. Natural thorium can be irradiated with neutrons as follows to produce the important isotope ^{233}U.

$$^{232}Th(n,\gamma) \rightarrow {}^{233}Th \xrightarrow{\beta} {}^{233}Pa \xrightarrow{\beta} {}^{233}U$$

While thorium itself is not fissionable, ^{233}U is, and in this way may be used as a nuclear fuel. One pound of completely fissioned uranium has the fuel value of over 1500 tons of coal. The uses of nuclear fuels to generate electrical power, to make isotopes for peaceful purposes, and to make explosives are well known. The estimated world-wide capacity of the 429 nuclear power reactors in operation in January 1990 amounted to about 311,000 megawatts. Uranium in the U.S.A. is controlled by the U.S. Nuclear Regulatory Commission. New uses are being found for "depleted" uranium, i.e., uranium with the percentage of ^{235}U lowered to about 0.2%. It has found use in inertial guidance devices, gyro compasses, counterweights for aircraft control surfaces, as ballast for missile reentry vehicles, and as a shielding material. Uranium metal is used for X-ray targets for production of high-energy X-rays; the nitrate has been used as photographic toner, and the acetate is used in analytical chemistry. Crystals of uranium nitrate are triboluminescent. Uranium salts have also been used for producing yellow "vaseline" glass and glazes. Uranium and its compounds are highly toxic, both from a chemical and radiological standpoint. Finely divided uranium metal, being pyrophoric, presents a fire hazard. The maximum recommended allowable concentration of soluble uranium compound in air (based on chemical toxicity) is 0.2 mg/m^3 (8-hour time-weighted average — 40-hour week). The maximum permissible total body burden of natural uranium (based on radiotoxicity) is 0.2 μCi for soluble compounds. Recently, the natural presence of uranium in many soils has become of concern to homeowners because of the generation of radon and its daughters (see under Radon).

Unnilquadium, etc. — See under Elements 104 to 110.

Vanadium — (Scandinavian goddess, *Vanadis*), V; at. wt. 50.9415; at. no. 23; m.p. 1910 ± 10°C; b.p. 3407°C; sp. gr. 6.11 (18.7°C); valence 2, 3, 4, or 5. Vanadium was first discovered by del Rio in 1801. Unfortunately, a French chemist incorrectly declared del Rio's new element was only impure chromium; del Rio thought himself to be mistaken and accepted the French chemist's statement. The element was rediscovered in 1830 by Sefstrom, who named the element in honor of the Scandinavian goddess *Vanadis* because of its beautiful multicolored compounds. It was isolated in nearly pure form by Roscoe, in 1867, who reduced the chloride with hydrogen. Vanadium of 99.3 to 99.8% purity was not produced until 1927. Vanadium is found in about 65 different minerals among which are *carnotite, roscoelite, vanadinite,* and *patronite* important sources of the metal. Vanadium is also found in phosphate rock and certain iron ores, and is present in some crude oils in the form of organic complexes. It is also found in small percentages in meteorites. Commercial production from petroleum ash holds promise as an important source of the element. High-purity ductile vanadium can be obtained by reduction of vanadium trichloride with magnesium or with magnesium-sodium mixtures. Much of the vanadium metal being produced is now made by calcium reduction of V_2O_5 in a pressure vessel, an adaption of a process developed by McKechnie and Seybolt. Natural vanadium is a mixture of two isotopes, ^{50}V (0.24%) and ^{51}V (99.76%). ^{50}V is slightly radioactive, having a half-life of $>3.9 \times 10^{17}$ years. Nine other unstable isotopes are recognized. Pure vanadium is a bright white metal, and is soft and ductile. It has good corrosion resistance to alkalis, sulfuric and hydrochloric acid, and salt water, but the metal oxidizes readily above 660°C. The metal has good structural strength and a low fission neutron cross section, making it useful in nuclear applications. Vanadium is used in producing rust resistant, spring, and highspeed tool steels. It is an important carbide stabilizer in making steels. About 80% of the vanadium now produced is used as ferrovanadium or as a steel additive. Vanadium foil is used as a bonding agent in cladding titanium to steel. Vanadium pentoxide is used in ceramics and as a catalyst. It is also used in producing a superconductive magnet with a field of 175,000 gauss. Vanadium and its compounds are toxic and should be handled with care. The maximum allowable concentration of V_2O_5 dust in air is about 0.05 (8-hour time-weighted average — 40-hour week). Ductile vanadium is commercially available. Commercial vanadium metal, of about 95% purity, costs about \$20/lb. Vanadium (99.9%) costs about \$100/oz.

Wolfram — see Tungsten.

Xenon — (Gr. *xenon*, stranger), Xe; at. wt. 131.29; at. no. 54; m.p. –111.75°C; b.p. –108.0°C; density (gas) 5.887 ± 0.009 g/l, sp. gr (liquid) 3.52 (–109°C); valence usually 0. Discovered by Ramsay and Travers in 1898 in the residue left after evaporating liquid air components. Xenon is a member of the so-called noble or "inert" gases. It is present in the atmosphere to the extent of about one part in twenty million. Xenon is present in the Martian atmosphere to the extent of 0.08 ppm. The element is found in the gases evolved from certain mineral springs, and is commercially obtained by extraction from liquid air. Natural xenon is composed of nine stable isotopes. In addition to these, 20 unstable isotopes have been characterized. Before 1962, it had generally been assumed that xenon and other noble gases were unable to form compounds. Evidence has been mounting in the past few years that xenon, as well as other members of the zero valence elements, do form compounds. Among the "compounds" of xenon now reported are xenon hydrate, sodium perxenate, xenon deuterate, difluoride, tetrafluoride, hexafluoride, and $XePtF_6$ and $XeRhF_6$. Xenon trioxide, which is highly explosive, has been prepared. More than 80 xenon compounds have been made with xenon chemically bonded to fluorine and oxygen. Some xenon compounds are colored. Metallic xenon has been produced, using several hundred kilobars of pressure. Xenon in a vacuum tube produces a beautiful blue glow when excited by an electrical discharge. The gas is used in making electron tubes, stroboscopic lamps, bactericidal lamps, and lamps used to excite ruby lasers for generating coherent light. Xenon is used in the atomic energy field in bubble chambers, probes, and other applications where its high molecular weight is of value. The perxenates are used in analytical chemistry as oxidizing agents. ^{133}Xe and ^{135}Xe are produced by neutron irradiation in air cooled nuclear reactors. ^{133}Xe has useful applications as a radioisotope. The element is available in sealed glass containers for about $20/l of gas at standard pressure. Xenon is not toxic, but its compounds are highly toxic because of their strong oxidizing characteristics.

Ytterbium — (Ytterby, village in Sweden), Yb; at. wt. 173.04; at. no. 70; m.p. 824°C; b.p. 1194°C; sp. gr (α) 6.903 (β) 6.966; valence 2, 3. Marignac in 1878 discovered a new component, which he called *ytterbia*, in the earth then known as *erbia*. In 1907, Urbain separated ytterbia into two components, which he called *neoytterbia* and *lutecia*. The elements in these earths are now known as *ytterbium* and *lutetium*, respectively. These elements are identical with *aldebaranium* and *cassiopeium*, discovered independently and at about the same time by von Welsbach. Ytterbium occurs along with other rare earths in a number of rare minerals. It is commercially recovered principally from *monazite sand*, which contains about 0.03%. Ion-exchange and solvent extraction techniques developed in recent years have greatly simplified the separation of the rare earths from one another. The element was first prepared by Klemm and Bonner in 1937 by reducing ytterbium trichloride with potassium. Their metal was mixed, however, with KCl. Daane, Dennison, and Spedding prepared a much purer form in 1953 from which the chemical and physical properties of the element could be determined. Ytterbium has a bright silvery luster, is soft, malleable, and quite ductile. While the element is fairly stable, it should be kept in closed containers to protect it from air and moisture. Ytterbium is readily attacked and dissolved by dilute and concentrated mineral acids and reacts slowly with water. Ytterbium has three allotropic forms with transformation points at –13° and 795°C. The beta form is a room-temperature, face-centered, cubic modification, while the high-temperature gamma form is a body-centered cubic form. Another body-centered cubic phase has recently been found to be stable at high pressures at room temperatures. The beta form ordinarily has metallic-type conductivity, but becomes a semiconductor when the pressure is increased above 16,000 atm. The electrical resistance increases tenfold as the pressure is increased to 39,000 atm and drops to about 80% of its standard temperature-pressure resistivity at a pressure of 40,000 atm. Natural ytterbium is a mixture of seven stable isotopes. Seven other unstable isotopes are known. Ytterbium metal has possible use in improving the grain refinement, strength, and other mechanical properties of stainless steel. One isotope is reported to have been used as a radiation source as a substitute for a portable X ray machine where electricity is unavailable. Few other uses have been found. Ytterbium metal is commercially available with a purity of about 99+% for about $875/kg. Ytterbium has a low acute toxic rating.

Yttrium — (*Ytterby*, village in Sweden near Vauxholm), Y; at. wt. 88.90585; at. no. 39; m.p. 1526°C; b.p. 3336°C; sp. gr. 4.469 (25°C); valence 3. *Yttria*, which is an earth containing yttrium, was discovered by Gadolin in 1794. *Ytterby* is the site of a quarry which yielded many unusually minerals containing rare earths and other elements. This small town, near Stockholm, bears the honor of giving names to *erbium*, *terbium*, and *ytterbium* as well as *ytterium*. In 1843 Mosander showed that yttria could be resolved into the oxides (or earths) of three elements. The name yttria was reserved for the most basic one; the others were named *erbia* and *terbia*. Yttrium occurs in nearly all of the rare-earth minerals. Analysis of lunar rock samples obtained during the Apollo missions show a relatively high yttrium content. It is recovered commercially from *monazite sand*, which contains about 3%, and from *bastnasite*, which contains about 0.2%. Wohler obtained the impure element in 1828 by reduction of the anhydrous chloride with potassium. The metal is now produced commercially by reduction of the fluoride with calcium metal. It can also be prepared by other techniques. Yttrium has a silver-metallic luster and is relatively stable in air. Turnings of the metal, however, ignite in air if their temperature exceeds 400°C, and finely divided yttrium is very unstable in air. Yttrium oxide is one of the most important compounds of yttrium and accounts for the largest use. It is widely used in making YVO_4 europium, and Y_2O_3 europium phosphors to give the red color in color television tubes. Many hundreds of thousands of pounds are now used in this application. Yttrium oxide also is used to produce yttrium-iron-garnets, which are very effective microwave filters. Yttrium iron, aluminum, and gadolinium garnets, with formulas such as $Y_3Fe_5O_{12}$ and $Y_3Al_5O_{12}$, have interesting magnetic properties. Yttrium iron garnet is also exceptionally efficient as both a transmitter and transducer of acoustic energy. Yttrium aluminum garnet, with a hardness of 8.5, is also finding use as a gemstone (simulated diamond). Small amounts of yttrium (0.1 to 0.2%) can be used to reduce the grain size in chromium, molybdenum, zirconium, and titanium, and to increase strength of aluminum and magnesium alloys. Alloys with other useful properties can be obtained by using yttrium as an additive. The metal can be used as a deoxidizer for vanadium and other nonferrous metals. The metal has a low cross section for nuclear capture. ^{90}Y, one of the isotopes of yttrium, exists in equilibrium with its parent ^{90}Sr, a product of atomic explosions. Yttrium has been considered for use as a nodulizer for producing nodular cast iron, in which the graphite forms compact nodules instead of the usual flakes. Such iron has increased ductility. Yttrium is also finding application in laser systems and as a catalyst for ethylene polymerization.It has also potential use in ceramic and glass formulas, as the oxide has a high melting point and imparts shock resistance and low expansion characteristics to glass. Natural yttrium contains but one isotope, ^{89}Y. Nineteen other unstable isotopes have been characterized. Yttrium metal of 99.9% purity is commercially available at a cost of about $75/oz.

Zinc — (Ger. *Zink*, of obscure origin), Zn; at. wt. 65.39; at. no. 30; m.p. 419.53°C; b.p. 907°C; sp. gr. 7.133 (25°C); valence 2. Centuries before zinc was recognized as a distinct element, zinc ores were used for making brass. Tubal-Cain, seven generations from Adam, is mentioned as being an "instructor in every artificer in brass and iron." An alloy containing 87% zinc has been found in prehistoric ruins in Transylvania. Metallic zinc was produced in the 13th century A.D. in India by reducing calamine with organic substances such as wool. The metal was rediscovered in Europe by Marggraf in 1746, who showed that it could be obtained by reducing *calamine* with charcoal. The principal ores of zinc are *sphalerite* or *blende* (sulfide), *smithsonite* (carbonate), *calamine* (silicate), and *franklinite* (zinc, manganese, iron oxide). Zinc can be obtained by roasting its ores to form the oxide and by reduction of the oxide with coal or carbon, with subsequent distillation of the metal. Other methods of extraction are possible. Naturally occurring zinc contains five stable isotopes. Sixteen other unstable isotopes are recognized. Zinc is a bluish-white, lustrous metal. It is brittle at ordinary temperatures but malleable at 100 to 150°C. It is a fair conductor of electricity, and burns in air at high red heat with evolution of white clouds of the oxide. The metal is employed to form numerous alloys with other metals. Brass, nickel silver, typewriter metal, commercial bronze, spring brass, German silver, soft solder, and aluminum solder are some of the more important alloys. Large quantities of zinc are used to produce die castings, used extensively by the automotive, electrical, and hardware industries. An alloy called *Prestal®*, consisting of 78% zinc and 22% aluminum is reported to be almost as strong as steel but as easy to mold as plastic. It is said to be so plastic that it can be molded into form by relatively inexpensive die casts made of ceramics and cement. It exhibits superplasticity. Zinc is also extensively used to galvanize other metals such as iron to prevent corrosion. Neither zinc nor zirconium is ferromagnetic; but $ZrZn_2$ exhibits ferromagnetism at temperatures below 35 K. Zinc oxide is a unique and very useful material to modern civilization. It is widely used in the manufacture of paints, rubber products, cosmetics, pharmaceuticals, floor coverings, plastics, printing inks, soap, storage batteries, textiles, electrical equipment, and other products. It has unusual electrical, thermal, optical, and solid-state properties that have not yet been fully investigated. Lithopone, a mixture of zinc sulfide and barium sulfate, is an important pigment. Zinc sulfide is used in making luminous dials, X-ray and TV screens, and fluorescent lights. The chloride and chromate are also important compounds. Zinc is an essential element in the growth of human beings and animals. Tests show that zinc-deficient animals require 50% more food to gain the same weight as an animal supplied with sufficient zinc. Zinc is not considered to be toxic, but when freshly formed ZnO is inhaled a disorder known as the *oxide shakes* or *zinc chills* sometimes occurs. It is recommended that where zinc oxide is encountered good ventilation be provided to avoid concentrations exceeding 5 mg/m^3, (time-weighted over an 8-hour exposure, 40-hour work week). The price of zinc was roughly $0.70/lb in January 1990.

Zirconium — (Persian *zargun*, gold like), Zr; at. wt. 91.224; at. no. 40; m.p. 1855 ± 2°C; b.p. 4409°C; sp. gr. 6.506 (20°C); valence +2, +3, and +4. The name *zircon* probably originated from the Persian word *zargun*, which describes the color of the gemstone now known as *zircon, jargon, hyacinth, jacinth,* or *ligure*. This mineral, or its variations, is mentioned in biblical writings. The mineral was not known to contain a new element until Klaproth, in 1789, analyzed a jargon from Ceylon and found a new earth, which Werner named zircon (*silex circonius*), and Klaproth called *Zirkonerde (zirconia)*. The impure metal was first isolated by Berzelius in 1824 by heating a mixture of potassium and potassium zirconium fluoride in a small iron tube. Pure zirconium was first prepared in 1914. Very pure zirconium was first produced in 1925 by van Arkel and de Boer by an iodide decomposition process they developed. Zirconium is found in abundance in S-type stars, and has been identified in the sun and meteorites. Analyses of lunar rock samples obtained during the various Apollo missions to the moon show a surprisingly high zirconium oxide content, compared with terrestial rocks. Naturally occurring zirconium contains five isotopes. Fifteen other isotopes are known to exist. *Zircon*, $ZrSiO_4$, the principal ore, is found in deposits in Florida, South Carolina, Australia, and Brazil. *Baddeleyite*, found in Brazil, is an important zirconium mineral. It is principally pure ZrO_2 in crystalline form having a hafnium content of about 1%. Zirconium also occurs in some 30 other recognized mineral species. Zirconium is produced commercially by reduction of the chloride with magnesium (the Kroll Process), and by other methods. It is a grayish-white lustrous metal. When finely divided, the metal may ignite spontaneously in air, especially at elevated temperatures. The solid metal is much more difficult to ignite. The inherent toxicity of zirconium compounds is low. Hafnium is invariably found in zirconium ores, and the separation is difficult. Commercial-grade zirconium contains from 1 to 3% hafnium. Zirconium has a low absorption cross section for neutrons, and is therefore used for nuclear energy applications, such as for cladding fuel elements. Commercial nuclear power generation now takes more than 90% of zirconium metal production. Reactors of the size now being made may use as much as a half-million lineal feet of zirconium alloy tubing. Reactor-grade zirconium is essentially free of hafnium. *Zircaloy®* is an important alloy developed specifically for nuclear applications. Zirconium is exceptionally resistant to corrosion by many common acids and alkalis, by sea water, and by other agents. It is used extensively by the chemical industry where corrosive agents are employed. Zirconium is used as a getter in vacuum tubes, as an alloying agent in steel, in surgical appliances, photoflash bulbs, explosive primers, rayon spinnerets, lamp filaments, etc. It is used in poison ivy lotions in the form of the carbonate as it combines with *urushiol*. With niobium, zirconium is superconductive at low temperatures and is used to make superconductive magnets, which offer hope of direct large-scale generation of electric power. Alloyed with zinc, zirconium becomes magnetic at temperatures below 35 K. Zirconium oxide (zircon) has a high index of refraction and is used as a gem material. The impure oxide, zirconia, is used for laboratory crucibles that will withstand heat shock, for linings of metallurgical furnaces, and by the glass and ceramic industries as a refractory material. Its use as a refractory material accounts for a large share of all zirconium consumed. Zirconium of about 99.6% purity is available at a cost of about $150/kg.

PHYSICAL CONSTANTS OF INORGANIC COMPOUNDS

The compounds in this table were selected on the basis of their laboratory and industrial importance, as well as their value in illustrating trends in the variation of physical properties with position in the periodic table. An effort has been made to include the most frequently encountered inorganic substances; organometallics are not covered, with the exception of metal salts of organic acids. Many, if not most, of the compounds that are solids at ambient temperature can exist in more than one crystalline modification. The information given here applies to the most stable or common crystalline form. In cases where two or more forms are of practical importance, separate entries will be found in the table.

Compounds are arranged primarily in alphabetical order by the most commonly used name. However, adjustments are made in many instances in order to bring closely related compounds together. For example, hydrides of elements such as boron, silicon, and germanium are grouped together immediately following the entry for the parent element, since they would otherwise be scattered throughout the table. Likewise, the oxoacids of an element are given in one group when a strict alphabetical order would separate them (e.g., sulfuric acid and fluorosulfuric acid). Whenever deviations from alphabetical order occur, the names are included in the Synonym Index to provide an easy way to locate these substances.

The data fields in the table are described below:

- **No.:** An identification number used in the indexes.
- **Name:** Common name for the substance. The valence state of a metallic element is indicated by a Roman numeral , e.g., copper in the +2 state is written as copper(II) rather than cupric (but names beginning with cupric, ferrous, etc. are included in the synonym index).
- **Formula:** The simplest descriptive formula is given, but this does not necessarily specify the actual structure of the compound. For example, aluminum chloride is designated as $AlCl_3$, even though a more accurate representation of the structure in the solid phase (and, under some conditions, in the gas phase) is Al_2Cl_6. A few exceptions are made, such as the use of Hg_2^{+2} for the mercury(I) ion.
- **CAS RN:** Chemical Abstracts Service Registry Number. An asterisk * following the CAS RN for a hydrate indicates that the number refers to the anhydrous compound. In most cases the generic CAS RN for the compound is given rather than the number for a specific crystalline form or mineral.
- **Mol. Wt.:** Molecular weight (relative molar mass) as calculated with the 1993 IUPAC Recommended Atomic Weights. The number of decimal places corresponds to the number of places in the atomic weight of the least accurately known element (e.g., one place for lead compounds, two places for compounds of selenium, germanium, etc.); a maximum of three places is given. For compounds of radioactive elements for which IUPAC makes no recommendation, the mass number of the isotope with the longest half-life is used, and the result is rounded to the nearest integer.
- **Physical form:** The crystal system is given ,when available, for compounds that are solid at room temperature, together with color and other descriptive features. Abbreviations are listed below.
- **mp:** Normal melting point in °C. A "tp" following the number indicates a triple point, where solid, liquid, and gas are in equilibrium at a pressure greater than one atmosphere (i.e., the normal melting point does not exist). When available, the triple point pressure is listed.
- **bp:** Normal boiling point in °C (referred to 101.325 kPa or 760 mmHg pressure). An "sp" following the number indicates a sublimation point, where the pressure of the vapor in equilibrium with the solid reaches 101.325 kPa. See Reference 8, p. 23, for further discussion of sublimation points and triple points. Note: The abbreviation "subl" indicates that there is a perceptible sublimation pressure above the solid, but does not imply any quantitative information on the pressure-temperature behavior.
- **den:** Density. Values for solids and liquids are always in grams per cubic centimeter and can be assumed to refer to temperatures near room temperature unless otherwise stated. Values for gases are the calculated ideal gas densities in grams per liter at 25°C and 101.325 kPa; the unit is always specified with a gas value.
- **Solubility:** Qualitative information on the solubility in various solvents is given. The abbreviations are listed below.
- **Other Data:** The codes refer to other tables in this handbook which include an entry for the compound in question. These codes are:

  a. Data on enthalpy of formation, entropy, and heat capacity in the table "Standard Thermodynamic Properties of Chemical Substances" (Section 5).
  b Critical temperature, pressure, and volume in the table "Critical Constants, Boiling Points, and Melting Points of Selected Compounds" (Section 6).
  c. Table of "Enthalpy of Fusion" (Section 6).
  d. Table of "Enthalpy of Vaporization" (Section 6).
  e. Vapor pressure data in one or more of the vapor pressure tables in Section 6.
  f Optical constants and hardness in the table "Physical and Optical Properties of Minerals" (Section 4).

A Synonym Index and a CAS Registry Number Index follow the table.

LIST OF ABBREVIATIONS

Ac	acetyl	gl	glass, glassy	rhom	rhombohedral
ace	acetone	grn	green	s	soluble in
acid	acid solutions	hc	hydrocarbon solvents	silv	silvery
alk	alkaline solutions	hex	hexagonal	sl	slightly soluble in
amorp	amorphous	hp	heptane	soln	solution
anh	anhydrous	hx	hexane	sp	sublimation point
aq	aqueous	hyd	hydrate	stab	stable
blk	black	hyg	hygroscopic	subl	sublimes
brn	brown	i	insoluble in	temp	temperature
bz	benzene	liq	liquid	tetr	tetragonal
chl	chloroform	MeOH	methanol	thf	tetrahydrofuran
col	colorless	mono	monoclinic	tol	toluene
conc	concentrated	octahed	octahedral	tp	triple point
cry	crystals, crystalline	oran	orange	trans	transition, transformation
cub	cubic	orth	orthorhombic	tricl	triclinic
cyhex	cyclohexane	os	organic solvents	trig	trigonal
dec	decomposes	peth	petroleum ether	unstab	unstable
dil	dilute	pow	powder	viol	violet
diox	dioxane	prec	precipitate	visc	viscous
eth	ethyl ether	pur	purple	vs	very soluble in
EtOH	ethanol	py	pyridine	wh	white
exp	explodes, explosive	reac	reacts with	xyl	xylene
flam	flammable	refrac	refractory	yel	yellow

REFERENCES

1. Phillips, S.L., and Perry, D.L., *Handbook of Inorganic Compounds*, CRC Press, Boca Raton, FL, 1995.
2. Trotman-Dickenson, A.F., Executive Editor, *Comprehensive Inorganic Chemistry*, Vol. 1-5, Pergamon Press, Oxford, 1973.
3. Greenwood, N.N., and Earnshaw, A., *Chemistry of the Elements*, Pergamon Press, Oxford, 1984.
4. Budavari, S., Editor, *The Merck Index, Eleventh Edition*, Merck & Co., Rahway, NJ, 1989.
5. *GMELIN Handbook of Inorganic and Organometallic Chemistry*, Springer-Verlag, Heidelberg.
6. Chase, M.W., Davies, C.A., Downey, J.R., Frurip, D.J., McDonald, R.A., and Syverud, A.N.; *JANAF Thermochemical Tables, Third Edition*, *J. Phys. Chem. Ref. Data*, Vol. 14, Suppl. 1, 1985.
7. Donnay, J.D.H., and Ondik, H.M., *Crystal Data Determinative Tables, Third Edition, Volumes 2 and 4, Inorganic Compounds*, Joint Committee on Powder Diffraction Standards, Swarthmore, PA, 1973.
8. Lide, D.R., and Kehiaian, H.V., *CRC Handbook of Thermophysical and Thermochemical Data*, CRC Press, Boca Raton, FL, 1994.
9. *Kirk-Othmer Concise Encyclopedia of Chemical Technology*, Wiley-Interscience, New York, 1985.
10. *Dictionary of Inorganic Compounds*, Chapman & Hall, New York, 1992.
11. Massalski, T.B., Editor, *Binary Alloy Phase Diagrams*, American Society for Metals, Metals Park, Ohio, 1986.
12. *Landolt-Börnstein, Numerical Data and Functional Relationships in Science and Technology, Sixth Edition, II/4, Caloric Quantities of State*, Springer-Verlag, Heidelberg, 1961.
13. Deer, W.A., Howie, R.A., and Zussman, J., *An Introduction to the Rock-Forming Minerals, 2nd Edition*, Longman Scientific & Technical, Harlow, Essex, 1992.
14. Carmichael, R.S., *Practical Handbook of Physical Properties of Rocks and Minerals*, CRC Press, Boca Raton, FL, 1989.
15. Dinsdale, A.T., "SGTE Data for Pure Elements", *CALPHAD*, 15, 317-425, 1991.
16. Madelung, O., *Semiconductors: Group IV Elements and III-IV Compounds*, Springer-Verlag, Heidelberg, 1991.
17. Daubert, T.E., Danner, R.P., Sibul, H.M., and Stebbins, C.C., *Physical and Thermodynamic Properties of Pure Compounds: Data Compilation*, extant 1994 (core with 4 supplements), Taylor & Francis, Bristol, PA.

PHYSICAL CONSTANTS OF INORGANIC COMPOUNDS

No.	Name Formula	CAS RN Mol. Wt.	Physical Form	mp/°C den/g cm^{-3}	bp/°C Other Data	Solubility
1	Actinium Ac	7440-34-8 227	silv metal; cub	1051 10	3198 a	
2	Actinium bromide $AcBr_3$	33689-81-5 467	wh hex cry	 5.85		s H_2O
3	Actinium chloride $AcCl_3$	22986-54-5 333	wh hex cry	 4.81		
4	Actinium fluoride AcF_3	33689-80-4 284	wh hex cry	 7.88		i H_2O
5	Actinium iodide AcI_3	33689-82-6 608	wh cry			s H_2O
6	Actinium oxide Ac_2O_3	12002-61-8 502	wh hex cry	1977 9.19		i H_2O
7	Aluminum Al	7429-90-5 26.982	silv-wh metal; cub cry	660.32 2.70	2519 a,c,d,e	i H_2O; s acid, alk
8	Aluminum ammonium sulfate	7784-25-0 237.148	wh powder			sl H_2O; i EtOH
9	Aluminum ammonium sulfate dodecahydrate $AlNH_4(SO_4)_2 \cdot 12H_2O$	7784-26-1 453.331	col cry or powder	94.5 1.65	dec >280	s H_2O; i EtOH
10	Aluminum antimonide AlSb	25152-52-7 148.742	cub cry	1065 4.26		
11	Aluminum arsenide AlAs	22831-42-1 101.903	oran cub cry; hyg	1740 3.76		
12	Aluminum borate $2Al_2O_3 \cdot B_2O_3$	11121-16-7 273.543	needles	≈1050		i H_2O
13	Aluminum boride AlB_2	12041-50-8 48.604	powder	dec >920 3.19		s dil HCl
14	Aluminum borohydride $Al(BH_4)_3$	16962-07-5 71.510	flam liq	-64.5	44.5 a,d	reac H_2O
15	Aluminum bromate nonahydrate $Al(BrO_3)_3 \cdot 9H_2O$	11126-81-1* 572.826	wh hyg cry	62	dec >100	s H_2O
16	Aluminum bromide $AlBr_3$	7727-15-3 266.694	wh-yel monocl cry; hyg	97.5 3.2	255 a,b,c,d,e	reac H_2O; s bz, tol
17	Aluminum bromide hexahydrate $AlBr_3 \cdot 6H_2O$	7784-11-4 374.785	col-yel hyg cry	93 2.54		s H_2O, EtOH, CS_2
18	Aluminum carbide Al_4C_3	1299-86-1 143.959	yel hex cry	2100 2.36	dec >2200	reac H_2O
19	Aluminum chlorate nonahydrate $Al(ClO_3)_3 \cdot 9H_2O$	15477-33-5 439.472	hyg cry			vs H_2O; s EtOH
20	Aluminum chloride $AlCl_3$	7446-70-0 133.340	wh hex cry or powder; hyg	190 2.48	 a,b,c,e	reac H_2O; s bz, ctc, chl
21	Aluminum chloride hexahydrate $AlCl_3 \cdot 6H_2O$	7784-13-6 241.431	col hyg cry	dec 100 2.398		vs H_2O; s EtOH, eth
22	Aluminum diacetate $Al(OH)(C_2H_3O_2)_2$	142-03-0 162.079	wh amorp powder			i H_2O
23	Aluminum ethoxide $Al(C_2H_5O)_3$	555-75-9 162.165	liq, condenses to wh solid	140		reac H_2O; sl xyl
24	Aluminum fluoride AlF_3	7784-18-1 83.977	wh hex cry	≈2250 tp (220 MPa) 3.10	1276 sp a,c,e	sl H_2O
25	Aluminum fluoride monohydrate $AlF_3 \cdot H_2O$	32287-65-3 101.992	orth cry	 2.17		sl H_2O
26	Aluminum fluoride trihydrate $AlF_3 \cdot 3H_2O$	15098-87-0 138.023	wh hyg cry	 1.914		sl H_2O
27	Aluminum hexafluorosilicate nonahydrate $Al_2(SiF_6)_3 \cdot 9H_2O$	17099-70-6 642.329	hex prisms	dec >500		s H_2O
28	Aluminum hydride AlH_3	7784-21-6 30.006	col hex cry	dec >150	 a	reac H_2O
29	Aluminum hydroxide $Al(OH)_3$	21645-51-2 78.004	wh amorp powder	 2.42	 f	i H_2O; s alk, acid
30	Aluminum hydroxychloride $Al_2(OH)_5Cl \cdot 2H_2O$	1327-41-9 210.483	gl solid			s H_2O
31	Aluminum hypophosphite $Al(H_2PO_2)_3$	7784-22-7 221.948	cry powder	dec 220		i H_2O; s alk, acid
32	Aluminum iodide AlI_3	7784-23-8 407.695	wh leaflets	191 3.98	382 a,b,c,d,e	reac H_2O
33	Aluminum iodide hexahydrate $AlI_3 \cdot 6H_2O$	10090-53-6 515.786	yel hyg cry powder			vs H_2O; s EtOH, eth
34	Aluminum lactate $Al(C_3H_5O_3)_3$	18917-91-4 294.195	powder			vs H_2O
35	Aluminum nitrate nonahydrate $Al(NO_3)_3 \cdot 9H_2O$	7784-27-2 375.134	wh hyg cry; monocl	73 1.72	dec 135	vs H_2O, EtOH; i pyr
36	Aluminum nitride AlN	24304-00-5 40.989	blue-wh hex cry	3000 3.255	 a	reac H_2O
37	Aluminum oleate $Al(C_{18}H_{33}O_2)_3$	688-37-9 871.358	yel solid			i H_2O; s EtOH, bz
38	Aluminum phosphate $AlPO_4$	7784-30-7 121.953	wh rhomb plates	>1460 2.56	 a	i H_2O; sl acid

No.	Name Formula	CAS RN Mol. Wt.	Physical Form	mp/°C den/g cm^{-3}	bp/°C Other Data	Solubility
39	Aluminum metaphosphate $Al(PO_3)_3$	32823-06-6 263.898	col powder; tetr	≈1525 2.78		i H_2O
40	Aluminum oxide Al_2O_3	1344-28-1 101.961	wh powder; hex	2054 3.97	≈3000 a,c,e,f	i H_2O, os; sl alk
41	Aluminum oxyhydroxide $AlO(OH)$	14457-84-2 59.989	ortho cry	 3.44	 f	i H_2O; s acid, alk
42	Aluminum palmitate $Al(C_{15}H_{31}COO)_3$	555-35-1 793.244	wh-yel powder			i H_2O, EtOH; s peth
43	Aluminum perchlorate nonahydrate $Al(ClO_4)_3 \cdot 9H_2O$	14452-39-2 487.470	wh hyg cry	82 dec 2.0		
44	Aluminum phosphide AlP	20859-73-8 57.956	grn or yel cub cry	2550 2.40	 a	reac H_2O
45	Aluminum selenide Al_2Se_3	1302-82-5 290.84	yel-brown powder	960 3.437		reac H_2O
46	Aluminum silicate Al_2SiO_5	12183-80-1 162.046	gray-grn cry	 3.145	 f	
47	Aluminum silicate dihydrate $Al_2O_3 \cdot 2SiO_2 \cdot 2H_2O$	1332-58-7 258.161	wh-yel powder; tricl	 2.59		i H_2O, acid, alk
48	Aluminum stearate $Al(C_{18}H_{35}O_2)_3$	637-12-7 877.406	wh powder	115 1.070		i H_2O, EtOH, eth; s alk
49	Aluminum sulfate $Al_2(SO_4)_3$	10043-01-3 342.154	wh cry	dec 1040		s H_2O; i EtOH
50	Aluminum sulfate octadecahydrate $Al_2(SO_4)_3 \cdot 18H_2O$	7784-31-8 666.429	col monocl cry	dec 86 1.69	 f	s H_2O
51	Aluminum sulfide Al_2S_3	1302-81-4 150.161	yel-gray powder	1100 2.02	 a	
52	Aluminum telluride Al_2Te_3	12043-29-7 436.76	gray-blk hex cry	≈895 4.5		
53	Aluminum thiocyanate $Al(CNS)_3$	538-17-0 201.233	yel powder			s H_2O; i EtOH, eth
54	Americium Am	7440-35-9 243	silv metal; hex or cub	1176 12	2011 a,c	s acid
55	Americium(III) oxide Am_2O_3	12254-64-7 534	tan hex cry	 11.77		s acid
56	Americium(III) bromide $AmBr_3$	14933-38-1 483	wh orth cry	 6.85		s H_2O
57	Americium(III) chloride $AmCl_3$	13464-46-5 349	pink hex cry	500 5.87		
58	Americium(III) fluoride AmF_3	13708-80-0 300	pink hex cry	1393 9.53		
59	Americium(III) iodide AmI_3	13813-47-3 624	yel ortho cry	≈950 6.9		
60	Americium(IV) fluoride AmF_4	15947-41-8 319	tan monocl cry	 7.23		
61	Americium(IV) oxide AmO_2	12005-67-3 275	blk cub cry	dec >1000 11.68		s acid
62	Ammonia NH_3	7664-41-7 17.031	col gas	-77.74 0.747 g/L	-33.33 a,b,c,d,e	vs H_2O; s EtOH, eth
63	Ammonium acetate $NH_4C_2H_3O_2$	631-61-8 77.084	wh hyg cry	114 1.073		vs H_2O
64	Ammonium azide NH_4N_3	12164-94-2 60.059	ortho cry; flam	160 1.346	exp a,e	s H_2O
65	Ammonium benzoate $NH_4C_7H_5O_2$	1863-63-4 139.154	wh cry or powder	198 1.26		s H_2O; sl EtOH
66	Ammonium bimalate $NH_4OOCCH_2CH(OH)COOH$	5972-71-4 151.119	ortho cry	160 1.15		s H_2O; sl EtOH
67	Ammonium borate tetrahydrate $(NH_4)_2B_4O_7 \cdot 4H_2O$	12228-87-4 263.377	tetr cry			s H_2O; i EtOH
68	Ammonium bromide NH_4Br	12124-97-9 97.943	wh hyg tetr cry	542 dec 2.429	396 sp a,e	s H_2O, EtOH, ace; sl eth
69	Ammonium caprylate $NH_4C_8H_{15}O_2$	5972-76-9 161.245	hyg monocl cry	≈75		reac H_2O; s EtOH; i chl, bz
70	Ammonium carbamate NH_2COONH_4	1111-78-0 78.071	cry powder			vs H_2O; s EtOH
71	Ammonium carbonate $(NH_4)_2CO_3$	506-87-6 96.086	col cry powder	dec 58		
72	Ammonium cerium(III) sulfate tetrahydrate $NH_4Ce(SO_4)_2 \cdot 4H_2O$	21995-38-0* 422.342	monocl cry			s H_2O
73	Ammonium cerium(IV) nitrate $(NH_4)_2Ce(NO_3)_6$	16774-21-3 548.222	red-oran cry			vs H_2O
74	Ammonium chlorate NH_4ClO_3	10192-29-7 101.490	wh cry	102 exp 1.80		s H_2O
75	Ammonium chloride NH_4Cl	12125-02-9 53.492	col cub cry	520 tp (dec) 1.519	338 sp a,b,e	s H_2O
76	Ammonium chromate $(NH_4)_2CrO_4$	7788-98-9 152.071	yel cry	dec 185 1.90		s H_2O; sl ace, MeOH; i EtOH

No.	Name Formula	CAS RN Mol. Wt.	Physical Form	mp/°C den/g cm^{-3}	bp/°C Other Data	Solubility
77	Ammonium chromic sulfate dodecahydrate $NH_4Cr(SO_4)_2\cdot 12H_2O$	10022-47-6 478.345	blue-viol cry	94 dec 1.72		s H_2O; sl EtOH
78	Ammonium cobalt(II) phosphate $CoNH_4PO_4$	14590-13-7 171.943	red-viol powder; hyg			i H_2O; s acid
79	Ammonium cobalt(II) sulfate hexahydrate $(NH_4)_2Co(SO_4)_2\cdot 6H_2O$	13586-38-4 395.229	red monocl prisms	 1.90		s H_2O; i EtOH
80	Ammonium copper(II) chloride dihydrate $2NH_4Cl\cdot CuCl_2\cdot 2H_2O$	10060-13-6 277.464	blue-grn tetr cry	≈115 dec 1.99		s H_2O, EtOH
81	Ammonium cyanide NH_4CN	12211-52-8 44.056	col tetr cry	dec 1.10	 a,e	vs H_2O
82	Ammonium dichromate $(NH_4)_2Cr_2O_7$	7789-09-5 252.065	oran-red monocl cry; hyg	dec 180 2.155		vs H_2O
83	Ammonium dihydrogen arsenate $NH_4H_2AsO_4$	13462-93-6 158.975	tetr cry	300 dec 2.311		vs H_2O
84	Ammonium dihydrogen phosphate $NH_4H_2PO_4$	7722-76-1 115.026	wh tetr cry	190 1.80		vs H_2O; sl EtOH; i ace
85	Ammonium dithiocarbamate NH_4NH_2CSS	513-74-6 110.204	yel ortho cry	99 dec 1.45		s H_2O
86	Ammonium ferric chromate $NH_4Fe(CrO_4)_2$	7780-00-4 305.871	red powder			i H_2O
87	Ammonium ferric oxalate trihydrate $(NH_4)_3Fe(C_2O_4)_3\cdot 3H_2O$	13268-42-3 428.065	grn monocl cry; hyg	≈160 dec 1.780		vs H_2O; i EtOH
88	Ammonium ferric sulfate dodecahydrate $NH_4Fe(SO_4)_2\cdot 12H_2O$	10138-04-2 482.194	col to viol cry	≈37 1.71		vs H_2O; i EtOH
89	Ammonium ferricyanide trihydrate $(NH_4)_3Fe(CN)_6\cdot 3H_2O$	14221-48-8* 320.113	red cry			s H_2O; i EtOH
90	Ammonium ferrocyanide trihydrate $(NH_4)_4Fe(CN)_6\cdot 3H_2O$	14481-29-9* 338.151	yel cry	dec		s H_2O; i EtOH
91	Ammonium ferrous sulfate hexahydrate $(NH_4)_2Fe(SO_4)_2\cdot 6H_2O$	10045-89-3 392.141	blue-grn monocl cry	dec ≈100 1.86		s H_2O; i EtOH
92	Ammonium fluoride NH_4F	12125-01-8 37.037	wh hex cry; hyg	dec 1.015	 a	vs H_2O; sl EtOH
93	Ammonium fluoroborate NH_4BF_4	13826-83-0 104.844	wh powder; orth	487 dec 1.871		sl H_2O
94	Ammonium fluorosulfonate NH_4SO_3F	13446-08-7 117.101	col needles	245		s H_2O, EtOH, MeOH
95	Ammonium formate NH_4CHO_2	540-69-2 63.057	hyg cry	116 1.27		vs H_2O; s EtOH
96	Ammonium hexachloroiridate(IV) $(NH_4)_2IrCl_6$	16940-92-4 441.010	blk cry powder	dec 2.856		sl H_2O
97	Ammonium hexachloroosmiate(IV) $(NH_4)_2OsCl_6$	12125-08-5 439.02	red cry or powder	 2.93	subl	s H_2O, EtOH
98	Ammonium hexachloropalladate(IV) $(NH_4)_2PdCl_6$	19168-23-1 355.21	red-brn hyg cry	dec 2.418		
99	Ammonium hexachloroplatinate(IV) $(NH_4)_2PtCl_6$	16919-58-7 443.87	red-oran cub cry	dec 380 3.065		sl H_2O; i EtOH
100	Ammonium hexafluoroaluminate $(NH_4)_3AlF_6$	7784-19-2 195.087	cub cry	 1.78		s H_2O
101	Ammonium hexafluorogallate $(NH_4)_3GaF_6$	14639-94-2 237.828	col cub cry	dec >200 2.10		
102	Ammonium hexafluorogermanate $(NH_4)_2GeF_6$	16962-47-3 222.68	wh cry	380 2.564	subl	s H_2O; i EtOH
103	Ammonium hexafluorophosphate NH_4PF_6	16941-11-0 163.003	wh cub cry	dec 58 2.180		vs H_2O; s ace, EtOH, MeOH
104	Ammonium hexafluorosilicate $(NH_4)_2SiF_6$	16919-19-0 178.153	wh cub or trig cry	dec 2.011	 a	s H_2O; i EtOH, ace
105	Ammonium hydrogen arsenate $(NH_4)_2HAsO_4$	7784-44-3 176.004	wh powder	 1.99		s H_2O
106	Ammonium hydrogen borate trihydrate $NH_4HB_4O_7\cdot 3H_2O$	10135-84-9 228.332	col cry	 ≈2.5		s H_2O
107	Ammonium hydrogen carbonate NH_4HCO_3	1066-33-7 79.056	col or wh prisms	107 dec 1.586	 a	s H_2O; i EtOH, bz
108	Ammonium hydrogen citrate $(NH_4)_2HC_6H_5O_7$	3012-65-5 226.186	col cry	 1.48		vs H_2O; sl EtOH
109	Ammonium hydrogen fluoride NH_4HF_2	1341-49-7 57.044	wh orth cry	125 1.50	240 dec	s H_2O
110	Ammonium hydrogen oxalate monohydrate $NH_4HC_2O_4\cdot H_2O$	5972-72-5* 125.082	col rhomb cry	dec 1.56		sl H_2O, EtOH
111	Ammonium hydrogen phosphate $(NH_4)_2HPO_4$	7783-28-0 132.055	wh cry	155 dec 1.619	 a	s H_2O; i EtOH, ace
112	Ammonium hydrogen sulfate NH_4HSO_4	7803-63-6 115.111	wh hyg cry	147 1.78	 a	vs H_2O; i EtOH, ace, py
113	Ammonium hydrogen sulfide NH_4HS	12124-99-1 51.113	wh tetr or orth cry	dec 1.17	 e	vs H_2O; sl ace; i bz, eth
114	Ammonium hydrogen sulfite NH_4HSO_3	10192-30-0 99.111	col cry	dec 2.03	 a	vs H_2O

PHYSICAL CONSTANTS OF INORGANIC COMPOUNDS (continued)

No.	Name Formula	CAS RN Mol. Wt.	Physical Form	mp/°C den/g cm^{-3}	bp/°C Other Data	Solubility
115	Ammonium hydrogen tartrate $NH_4OOCCH(OH)CH(OH)COOH$	3095-65-6 167.118	wh cry	 1.68		sl H_2O; s alk; i EtOH
116	Ammonium hydroxide NH_4OH	1336-21-6 35.046	exists only in soln		 a	
117	Ammonium hypophosphite $NH_4H_2PO_2$	7803-65-8 83.028	wh hyg cry	dec		vs H_2O; sl EtOH; i ace
118	Ammonium iodate NH_4IO_3	13446-09-8 192.941	wh powder	150 3.3		
119	Ammonium iodide NH_4I	12027-06-4 144.943	wh tetr cry; hyg	551 dec 2.514	405 sp a,c,e	vs H_2O; sl EtOH, MeOH
120	Ammonium lactate $NH_4C_3H_5O_3$	52003-58-4 107.109	col cry	92		s H_2O, EtOH; sl MeOH; i ace, eth
121	Ammonium metatungstate hexahydrate $(NH_4)_6W_7O_{24}$•$6H_2O$	12028-48-7 1887.19	wh cry			s H_2O; i EtOH
122	Ammonium metavanadate NH_4VO_3	7803-55-6 116.979	wh-yel cry	dec 2.326		sl H_2O
123	Ammonium molybdate(IV) tetrahydrate $(NH_4)_6Mo_7O_{24}$•$4H_2O$	12054-85-2 1235.86	col or grn-yel cry	dec 2.498		s H_2O; i EtOH
124	Ammonium nickel chloride hexahydrate NH_4NiCl_3•$6H_2O$	16122-03-5* 291.181	grn hyg cry	 1.65		s H_2O
125	Ammonium nickel sulfate hexahydrate $(NH_4)_2Ni(SO_4)_2$•$6H_2O$	7785-20-8 394.989	blue-grn cry	dec 1.923		sl H_2O; i EtOH
126	Ammonium nitrate NH_4NO_3	6484-52-2 80.043	wh hyg cry; orth	dec 210; exp 1.72	 a,c	vs H_2O; sl MeOH
127	Ammonium nitroferricyanide $(NH_4)_2Fe(CN)_5NO$	14402-70-1 252.017	red-brn cry			s H_2O, EtOH
128	Ammonium oleate $NH_4C_{18}H_{33}O_2$	544-60-5 299.498	yel-brn paste	21		s H_2O; sl ace
129	Ammonium oxalate monohydrate $(NH_4)_2C_2O_4$•H_2O	6009-70-7 142.111	wh orth cry	dec 1.50		sl H_2O, EtOH
130	Ammonium palmitate $NH_4C_{15}H_{31}COO$	593-26-0 273.460	yel-wh powder	22		s H_2O; sl bz, xyl; i ace, EtOH, c
131	Ammonium pentachlorozincate $(NH_4)_3ZnCl_5$	14639-98-6 296.77	hyg orth cry	 1.81		vs H_2O
132	Ammonium perchlorate NH_4ClO_4	7790-98-9 117.490	wh orth cry	dec, exp 1.95	 a	vs H_2O; s MeOH; sl EtOH, ace; eth
133	Ammonium permanganate NH_4MnO_4	13446-10-1 136.975	purp rhomb cry	dec 70 2.22		sl H_2O
134	Ammonium peroxydisulfate $(NH_4)_2S_2O_8$	7727-54-0 228.204	monocl cry or wh powder	dec 1.982		vs H_2O
135	Ammonium perrhenate NH_4ReO_4	13598-65-7 268.244	col powder	 3.97		sl H_2O
136	Ammonium phosphite, dibasic, monohydrate $(NH_4)_2HPO_3$•H_2O	51503-61-8 134.071	hyg cry			s H_2O
137	Ammonium phosphomolybdate monohydrate $(NH_4)_3PO_4$•$12MoO_3$•H_2O	54723-94-3 1894.36	yel cry or powder	dec		sl H_2O
138	Ammonium phosphotungstate dihydrate $(NH_4)_3PO_4$•$12WO_3$•$2H_2O$	1311-90-6 2967.18	cry powder			sl H_2O
139	Ammonium picrate $NH_4C_6H_2N_3O_7$	131-74-8 246.137	yel orth cry	exp 1.72		sl H_2O
140	Ammonium salicylate $NH_4C_7H_5O_3$	528-94-9 155.153	wh cry powder			vs H_2O; s EtOH
141	Ammonium selenate $(NH_4)_2SeO_4$	7783-21-3 179.04	wh monocl cry	dec 2.194		vs H_2O; i EtOH, ace
142	Ammonium selenite $(NH_4)_2SeO_3$	7783-19-9 163.04	wh or red hyg cry	dec		s H_2O
143	Ammonium stearate $NH_4C_{18}H_{35}O_2$	1002-89-7 301.514	yel-wh powder	22 0.89		sl H_2O, bz; s EtOH, MeOH; i ace
144	Ammonium sulfamate $NH_4NH_2SO_3$	7773-06-0 114.125	wh hyg cry	131	dec 160	vs H_2O; sl EtOH
145	Ammonium sulfate $(NH_4)_2SO_4$	7783-20-2 132.141	wh or brn orth cry	dec 280 1.77	 a,f	vs H_2O; i EtOH, ace
146	Ammonium sulfide $(NH_4)_2S$	12135-76-1 68.143	yel-oran cry	dec ≈0		s H_2O, EtOH, alk
147	Ammonium sulfite monohydrate $(NH_4)_2SO_3$•H_2O	7783-11-1 134.156	col cry	dec 1.41		s H_2O; i EtOH, ace
148	Ammonium tartrate $(NH_4)_2C_4H_4O_6$	3164-29-2 184.148	wh cry	dec 1.601		s H_2O
149	Ammonium tellurate $(NH_4)_2TeO_4$	13453-06-0 227.68	wh powder	dec 3.024		
150	Ammonium tetrachloroaluminate NH_4AlCl_4	7784-14-7 186.832	wh hyg solid	304		s H_2O, eth
151	Ammonium tetrachloroplatinate(II) $(NH_4)_2PtCl_4$	13820-41-2 372.97	red cry	dec 2.936		s H_2O; i EtOH
152	Ammonium tetrachlorozincate $(NH_4)_2ZnCl_4$	14639-97-5 243.28	wh orth plates; hyg	150 dec 1.879		vs H_2O

No.	Name Formula	CAS RN Mol. Wt.	Physical Form	mp/°C den/g cm^{-3}	bp/°C Other Data	Solubility
153	Ammonium tetrathiotungstate $(NH_4)_2WS_4$	13862-78-7 348.18	oran cry	dec 2.71		s H_2O
154	Ammonium thiocyanate NH_4SCN	1762-95-4 76.122	col hyg cry	≈149 1.30	dec	vs H_2O, EtOH; s ace; i chl
155	Ammonium thiosulfate $(NH_4)_2S_2O_3$	7783-18-8 148.207	wh cry	dec 150 1.678		vs H_2O; i EtOH, eth
156	Ammonium titanium oxalate monohydrate $(NH_4)_2TiO(C_2O_4)_2 \cdot H_2O$	10580-03-7 293.997	hyg cry			vs H_2O
157	Ammonium uranate(VI) $(NH_4)_2U_2O_7$	7783-22-4 624.131	red-yel amorp powder			i H_2O, akl; s acid
158	Ammonium uranium fluoride $UO_2(NH_4)_3F_5$	18433-40-4 419.135	grn-yel monocl cry			s H_2O; i EtOH
159	Ammonium valerate $NH_4C_4H_9COO$	42739-38-8 119.164	hyg cry	108		vs H_2O, EtOH; s eth
160	Ammonium zirconyl carbonate dihydrate $(NH_4)_3ZrOH(CO_3)_3 \cdot 2H_2O$	12616-24-9* 362.405	prisms; unstable			s H_2O
161	Antimony Sb	7440-36-0 121.760	silv metal; hex	630.63 6.68	1587 a,c,e,f	i dil acid
162	Stibine SbH_3	7803-52-3 124.784	col gas; flam	-88 5.475 g/L	-17 a,d	sl H_2O; s EtOH
163	Antimony arsenide SbAs	12322-34-8 196.682	hex cry	≈680 6.0		
164	Antimony(III) bromide $SbBr_3$	7789-61-9 361.472	yel orth cry; hyg	96.6 4.35	280 a,b,d,e	reac H_2O; s ace, bz, chl
165	Antimony(III) chloride $SbCl_3$	10025-91-9 228.118	col orth cry; hyg	73.4 3.14	220.3 a,b,c,d,e	vs H_2O; s acid, EtOH, bz, ace
166	Antimony(III) fluoride SbF_3	7783-56-4 178.755	wh orth cry; hyg	292 4.38	≈345 a	vs H_2O
167	Antimony(III) iodide SbI_3	7790-44-5 502.473	red rhomb cry	168 4.92	401 a,b,d,e	reac H_2O; s EtOH, ace; i ctc
168	Antimony(III) oxide (senarmontite) Sb_2O_3	1309-64-4 291.518	col cub cry	570 trans 5.58	1425 e,f	sl H_2O; i os
169	Antimony(III) oxide (valentinite) Sb_2O_3	1309-64-4 291.518	wh orth cry	655 5.7	1425 e,f	sl H_2O; i os
170	Antimony(III) oxychloride SbOCl	7791-08-4 173.212	wh momo cry	170 dec		reac H_2O; i EtOH, eth
171	Antimony(III) selenide Sb_2Se_3	1315-05-5 480.40	grn orth cry	611 5.81		sl H_2O
172	Antimony(III) sulfate $Sb_2(SO_4)_3$	7446-32-4 531.711	wh cry powder; hyg	dec 3.62		sl H_2O
173	Antimony(III) sulfide Sb_2S_3	1345-04-6 339.718	gray-blk orth cry	550 4.562	f	i H_2O; s conc HCl
174	Antimony(III) teluride Sb_2Te_3	1327-50-0 626.32	gray cry	620 6.5		
175	Antimony(III,V) oxide Sb_2O_4	1332-81-6 307.518	yel orth cry	6.64	f	
176	Antimony(V) chloride $SbCl_5$	7647-18-9 299.024	col or yel liq	4 2.34	140 dec e	reac H_2O; s chl, ctc
177	Antimony(V) fluoride SbF_5	7783-70-2 216.752	hyg visc liq	8.3 3.10	141	reac H_2O
178	Antimony(V) dichlorotrifluoride $SbCl_2F_3$	7791-16-4 249.660	visc liq			reac H_2O
179	Antimony(V) oxide Sb_2O_5	1314-60-9 323.517	yel powder; cub	dec 3.78	a	sl H_2O
180	Antimony(V) sulfide Sb_2S_5	1315-04-4 403.850	oran-yel powder	75 dec 4.120		i H_2O; s acid, alk
181	Argon Ar	7440-37-1 39.948	col gas	-189.35 1.7[illegible] g/L	-185.85 a,b,c,d,e	sl H_2O
182	Arsenic As	7440-38-2 74.922	gray metal; rhomb	817 tp 5.75	614 sp a,b,c,e,f	i H_2O
183	Arsine AsH_3	7784-42-1 77.946	col gas	-116 3.420 g/L	-62.5 a,b,d,e	sl H_2O
184	Diarsine As_2H_4	15942-63-9 153.875	unstable liq		≈100	
185	Arsenic acid H_3AsO_4	7778-39-4 141.944	exists only in soln		a	
186	Arsenic acid hemihydrate $H_3AsO_4 \cdot 0.5H_2O$	7778-39-4* 150.951	wh hyg cry	35.5 ≈2		vs H_2O, EtOH
187	Arsenious acid H_3AsO_3	13464-58-9 125.944	exists only in soln			
188	Arsenic diiodide As_2I_4	13770-56-4 657.461	red cry	137		reac H_2O; s os
189	Arsenic hemiselenide As_2Se	1303-35-1 228.80	blk cry			i H_2O, os; dec acid, alk
190	Arsenic sulfide As_4S_4	12279-90-2 427.950	red monocl cry	320 3.5	565 f	i H_2O; sl bz; s alk

No.	Name Formula	CAS RN Mol. Wt.	Physical Form	mp/°C den/g cm^{-3}	bp/°C Other Data	Solubility
191	Arsenic(III) bromide $AsBr_3$	7784-33-0 314.634	yel orth cry; hyg	31.1 3.40	221 a,c,d,e	reac H_2O; s hc, ctc; vs eth, bz
192	Arsenic(III) chloride $AsCl_3$	7784-34-1 181.280	col liq	-16 2.150	130 a,b,c,d,e	reac H_2O; vs chl, ctc, eth
193	Arsenic(III) fluoride AsF_3	7784-35-2 131.917	col liq	-5.9 2.7	57.8 a,c,d,e	reac H_2O; s EtOH, eth, bz
194	Arsenic(III) iodide AsI_3	7784-45-4 455.635	red hex cry	140.9 4.73	424 a,d	sl H_2O, EtOH, eth; s bz. tol
195	Arsenic(III) oxide (arsenolite) As_2O_3	1327-53-3 197.841	wh cub cry	274 3.86	460 e,f	sl H_2O
196	Arsenic(III) oxide (claudetite) As_2O_3	1327-53-3 197.841	wh monocl cry	313 3.74	460 e,f	sl H_2O; s dil acid, alk; i EtOH
197	Arsenic(III) selenide As_2Se_3	1303-36-2 386.72	brn-blk solid	260 4.75		i H_2O; s alk
198	Arsenic(III) sulfide As_2S_3	1303-33-9 246.041	yel-oran monocl cry	310 3.46	707 a,f	i H_2O; s alk
199	Arsenic(III) telluride As_2Te_3	12044-54-1 532.64	blk monocl cry	621 6.50		
200	Arsenic(V) chloride $AsCl_5$	22441-45-8 252.186	stable at low temp	≈-50 dec		
201	Arsenic(V) fluoride AsF_5	7784-36-3 169.914	col gas	-79.8 7.456 g/L	-52.8 d,e	reac H_2O; s EtOH, bz, eth
202	Arsenic(V) oxide As_2O_5	1303-28-2 229.840	wh amorp powder	315 4.32	 a	vs H_2O, EtOH
203	Arsenic(V) selenide As_2Se_5	1303-37-3 544.64	blk solid	dec		i H_2O, EtOH, eth; s alk
204	Arsenic(V) sulfide As_2S_5	1303-34-0 310.173	brn-yel amorp solid	dec		i H_2O; s alk
205	Astatine At	7440-68-8 210	cry	302	 a	s HNO_3, os
206	Barium Ba	7440-39-3 137.327	silv-yel metal; cub	727 3.62	1897 a,c,d,e	reac H_2O; sl EtOH
207	Barium acetate $Ba(C_2H_3O_2)_2$	543-80-6 255.417	wh powder	 2.47		vs H_2O
208	Barium acetate monohydrate $Ba(C_2H_3O_2)_2{\cdot}H_2O$	5908-64-5 273.432	wh cry	dec 110 2.19		vs H_2O; sl EtOH
209	Barium aluminate $BaAl_2O_4$	12004-04-5 255.288	hex cry	1827		
210	Barium azide $Ba(N_3)_2$	18810-58-7 221.367	monocl cry; exp	dec ≈120 2.936		s H_2O; sl EtOH; i eth
211	Barium bromate monohydrate $Ba(BrO_3)_2{\cdot}H_2O$	10326-26-8 411.147	wh monocl cry	dec 260 3.99		sl H_2O; i EtOH
212	Barium bromide $BaBr_2$	10553-31-8 297.135	wh orth cry	857 4.781	1835 a,c	vs H_2O
213	Barium bromide dihydrate $BaBr_2{\cdot}2H_2O$	7791-28-8 333.166	wh cry	dec 75 3.7		vs H_2O; s MeOH; i EtOH, ace, diox
214	Barium carbide BaC_2	50813-65-5 161.349	gray tetr cry	dec 3.74		reac H_2O
215	Barium carbonate $BaCO_3$	513-77-9 197.336	wh orth cry	811 4.2865	 a,f	i H_2O; s acid
216	Barium chlorate monohydrate $Ba(ClO_3)_2{\cdot}H_2O$	10294-38-9 322.244	wh monocl cry	414 3.179		s H_2O, acid; sl EtOH, ace
217	Barium chloride $BaCl_2$	10361-37-2 208.232	wh orth cry; hyg	962 3.9	1560 a,c	s H_2O
218	Barium chloride dihydrate $BaCl_2{\cdot}2H_2O$	10326-27-9 244.263	wh monocl cry	dec ≈120 3.097		s H_2O; i EtOH
219	Barium chromate(V) $Ba_3(CrO_4)_2$	12345-14-1 643.968	grn-blk hex cry	 5.25		s H_2O
220	Barium chromate(VI) $BaCrO_4$	10294-40-3 253.321	yel orth cry	dec 4.50		i H_2O; reac acid
221	Barium citrate monohydrate $Ba_3(C_6H_5O_7)_2{\cdot}H_2O$	512-25-4* 808.199	gray-wh cry			s H_2O, acid
222	Barium cyanide $Ba(CN)_2$	542-62-1 189.362	wh cry powder			vs H_2O; s EtOH
223	Barium dichromate dihydrate $BaCr_2O_7{\cdot}2H_2O$	10031-16-0 389.346	brn-red needles	dec		reac H_2O
224	Barium dithionate dihydrate $Ba(SO_3)_2{\cdot}2H_2O$	13845-17-5 333.486	cry	 4.54		s H_2O; sl EtOH
225	Barium ferrocyanide hexahydrate $Ba_2Fe(CN)_6{\cdot}6H_2O$	13821-06-2* 594.696	yel monocl cry	dec 80		i H_2O, EtOH
226	Barium fluoride BaF_2	7787-32-8 175.324	wh cub cry	1368 4.893	2260 a,c	sl H_2O
227	Barium formate $Ba(CHO_2)_2$	541-43-5 227.363	cry	 3.21		s H_2O; i EtOH
228	Barium hexaboride BaB_6	12046-08-1 202.193	blk cub cry	2070 4.36		i H_2O; s acid; i EtOH
229	Barium hexafluorosilicate $BaSiF_6$	17125-80-3 279.403	wh orth needles	dec 300 4.29		i H_2O, EtOH; sl acid

PHYSICAL CONSTANTS OF INORGANIC COMPOUNDS (continued)

No.	Name	Formula	CAS RN	Mol. Wt.	Physical Form	mp/°C	den/g cm^{-3}	bp/°C	Other Data	Solubility
230	Barium hydride	BaH_2	13477-09-3	139.343	gray orth cry	dec 675	4.16		a	reac H_2O
231	Barium hydrogen phosphate	$BaHPO_4$	10048-98-3	233.306	wh cry powder	dec 400	4.16			i H_2O; s dil acid
232	Barium hydrosulfide	$Ba(HS)_2$	25417-81-6	203.475	yel hyg cry					s H_2O
233	Barium hydrosulfide tetrahydrate	$Ba(HS)_2 \cdot 4H_2O$	12230-74-9	275.536	yel rhomb cry	dec 50				s H_2O
234	Barium hydroxide	$Ba(OH)_2$	17194-00-2	171.342	wh powder	408			a,c	sl H_2O
235	Barium hydroxide monohydrate	$Ba(OH)_2 \cdot H_2O$	22326-55-2	189.357	wh powder		3.743			sl H_2O; s acid
236	Barium hydroxide octahydrate	$Ba(OH)_2 \cdot 8H_2O$	12230-71-6	315.464	wh monocl cry	dec 78	2.18			sl H_2O
237	Barium hypophosphite monohydrate	$Ba(H_2PO_2)_2 \cdot H_2O$	14871-79-5*	285.320	monocl plates		2.90			s H_2O; i EtOH
238	Barium iodate	$Ba(IO_3)_2$	10567-69-8	487.132	wh cry powder	dec 476	5.23			sl H_2O
239	Barium iodate monohydrate	$Ba(IO_3)_2 \cdot H_2O$	7787-34-0	505.148	cry	dec 130	5.00			sl H_2O;s acid; i EtOH
240	Barium iodide	BaI_2	13718-50-8	391.136	wh orth cry	711	5.15		a,c	vs H_2O
241	Barium iodide dihydrate	$BaI_2 \cdot 2H_2O$	7787-33-9	427.167	col cry	740 dec	5.0			vs H_2O; s EtOH, ace
242	Barium manganate(VI)	$BaMnO_4$	7787-35-1	256.263	grn-gray hyg cry		4.85			reac H_2O
243	Barium metaborate monohydrate	$Ba(BO_2)_2 \cdot H_2O$	26124-86-7	240.962	wh powder	>900	3.3			sl H_2O
244	Barium molybdate	$BaMoO_4$	7787-37-3	297.27	wh powder	1450	4.975			i H_2O
245	Barium niobate	$Ba(NbO_3)_2$	12009-14-2	419.136	yel orth cry	1455	5.44			i H_2O
246	Barium nitrate	$Ba(NO_3)_2$	10022-31-8	261.336	wh cub cry	590	3.24		a	s H_2O; sl EtOH, ace
247	Barium nitride	Ba_3N_2	12047-79-9	439.994	yel-brn cry		4.78			reac H_2O
248	Barium nitrite	$Ba(NO_2)_2$	13465-94-6	229.338	col hex cry	267	3.234		a	
249	Barium nitrite monohydrate	$Ba(NO_2)_2 \cdot H_2O$	7787-38-4	247.353	yel-wh hex cry	217 dec	3.18			vs H_2O; i EtOH
250	Barium oxalate	BaC_2O_4	516-02-9	225.347	wh powder	400 dec	2.658			i H_2O
251	Barium oxalate monohydrate	$BaC_2O_4 \cdot H_2O$	13463-22-4	243.362	wh cry powder		2.66			i H_2O; s acid
252	Barium oxide	BaO	1304-28-5	153.326	wh-yel powder; cub and hex	2013	5.72(cub)		a,c	sl H_2O; s dil acid, EtOH; i ace
253	Barium perchlorate	$Ba(ClO_4)_2$	13465-95-7	336.227	col hex cry	505	3.20			vs H_2O, EtOH
254	Barium perchlorate trihydrate	$Ba(ClO_4)_2 \cdot 3H_2O$	10294-39-0	390.273	col cry		2.74			s H_2O, MeOH; sl EtOH, ace; i eth
255	Barium permanganate	$Ba(MnO_4)_2$	7787-36-2	375.198	brn-viol cry	dec 200	3.77			sl H_2O; reac EtOH
256	Barium peroxide	BaO_2	1304-29-6	169.326	gray-wh tetr cry	450 dec	4.96			i H_2O; reac dil acid
257	Barium metaphosphate	$Ba(PO_3)_2$	13466-20-1	295.271	wh powder	1560				i H_2O; sl acid
258	Barium potassium chromate	$BaK_2(CrO_4)_2$	27133-66-0	447.511	yel hex cry		3.63			vs H_2O
259	Barium pyrophosphate	$Ba_2P_2O_7$	13466-21-2	448.597	wh powder		3.9			sl H_2O; s acid
260	Barium selenate	$BaSeO_4$	7787-41-9	280.29	wh rhomb cry	dec	4.75			i H_2O
261	Barium selenide	$BaSe$	1304-39-8	216.29	cub cry powder	1780	5.02			reac H_2O
262	Barium selenite	$BaSeO_3$	13718-59-7	264.29	solid					i H_2O
263	Barium disilicate	$BaSi_2O_5$	12650-28-1	273.495	wh orth cry	1420	3.70			
264	Barium metasilicate	$BaSiO_3$	13255-26-0	213.411	col rhomb powder	1605	4.40			i H_2O; s acid
265	Barium silicide	$BaSi_2$	1304-40-1	193.498	gray lumps	1180				reac H_2O
266	Barium sodium niobate	$Ba_2Na(NbO_3)_5$	12323-03-4	1002.167	wh orth cry	1437	5.40			i H_2O
267	Barium stannate	$BaSnO_3$	12009-18-6	304.035	cub cry		7.24			sl H_2O
268	Barium stannate trihydrate	$BaSnO_3 \cdot 3H_2O$	12009-18-6*	358.081	wh cry powder					sl H_2O; s acid

No.	Name Formula	CAS RN Mol. Wt.	Physical Form	mp/°C den/g cm^{-3}	bp/°C Other Data	Solubility
269	Barium stearate $Ba(C_{18}H_{35}O_2)_2$	6865-35-6 704.277	wh powder	160 1.145		i H_2O, alc
270	Barium sulfate $BaSO_4$	7727-43-7 233.391	wh orth cry	1350 4.49	dec 1580 a,c,f	i H_2O, EtOH
271	Barium sulfide BaS	21109-95-5 169.393	col cub cry or gray powder	>2000 4.3	a	sl H_2O
272	Barium sulfite $BaSO_3$	7787-39-5 217.391	wh monocl cry	dec 4.44		i H_2O, EtOH
273	Barium tartrate $BaC_4H_4O_6$	5908-81-6 285.399	wh cry	2.98		s H_2O; i EtOH
274	Barium tetracyanoplatinate(II) tetrahydrate $BaPt(CN)_4•4H_2O$	13755-32-3 508.54	yel powder or cry	2.076		sl H_2O; i EtOH
275	Barium tetraiodomercurate(II) $BaHgI_4$	10048-99-4 845.54	yel-red hyg cry			vs H_2O, EtOH
276	Barium thiocyanate $Ba(SCN)_2$	2092-17-3 253.494	hyg cry			vs H_2O; s ace, MeOH, EtOH
277	Barium thiocyanate dihydrate $Ba(SCN)_2•2H_2O$	2092-17-3* 289.525	hyg wh cry			s H_2O, EtOH
278	Barium thiocyanate trihydrate $Ba(SCN)_2•3H_2O$	68016-36-4 307.540	wh needles; hyg	2.286		s H_2O, EtOH
279	Barium thiosulfate BaS_2O_3	35112-53-9 249.457	wh cry powder	dec 220		sl H_2O; i EtOH
280	Barium thiosulfate monohydrate $BaS_2O_3•H_2O$	7787-40-8 267.473	wh cry powder	dec 3.5		sl H_2O; i EtOH
281	Barium titanate $BaTiO_3$	12047-27-7 233.192	wh tetr cry	1625 6.02		i H_2O
282	Barium tungstate $BaWO_4$	7787-42-0 385.17	wh tetr cry	5.04		i H_2O
283	Barium uranium oxide BaU_2O_7	10380-31-1 725.381	oran-yel powder			i H_2O; s acid
284	Barium orthovanadate $Ba_3(VO_4)_2$	39416-30-3 641.859	hex cry	707 5.14		
285	Barium zirconate $BaZrO_3$	12009-21-1 276.549	gray-wh cub cry	2500 5.52		i H_2O, alk; sl acid
286	Berkelium (α form) Bk	7440-40-6 247	hex	1050 14.78	a	
287	Berkelium (β form) Bk	7440-40-6 247	cub cry	986 13.25	a	
288	Beryllium Be	7440-41-7 9.012	hex	1287 1.85	2471 a,c,e	s acid, alk
289	Beryllium acetate $Be(C_2H_3O_2)_2$	543-81-7 127.102	wh cry	dec 60		i H_2O, EtOH
290	Beryllium acetylacetonate $Be(C_5H_7O_2)_2$	10210-64-7 207.231	monocl cry powder	108 1.168	270	i H_2O, vs EtOH, eth
291	Beryllium aluminate $BeAl_2O_4$	12004-06-7 126.973	orth cry	3.65	f	
292	Beryllium aluminum metasilicate $Be_3Al_2(SiO_3)_6$	1302-52-9 537.502	col or grn-yel cry; hex	2.64	f	
293	Beryllium basic acetate $Be_4O(C_2H_3O_2)_6$	1332-52-1 406.316	wh cry	285 1.25	330	i H_2O; s eth, os
294	Beryllium boride BeB_2	12228-40-9 30.634	refrac solid	>1970		
295	Beryllium borohydride $Be(BH_4)_2$	17440-85-6 36.682	solid	dec 125; flam	subl e	reac H_2O
296	Beryllium bromide $BeBr_2$	7787-46-4 168.820	orth cry; hyg	508 3.465	520 a,c,e	vs H_2O; s EtOH, pyr
297	Beryllium carbide Be_2C	506-66-1 30.035	red cub cry	dec >2100 1.90		reac H_2O
298	Beryllium carbonate tetrahydrate $BeCO_3•4H_2O$	60883-64-9 93.085	unstable solid			
299	Beryllium chloride $BeCl_2$	7787-47-5 79.917	wh-yel orth cry; hyg	399 1.90	482 a,c,d,e	vs H_2O; s EtOH, eth, py; i bz, tol
300	Beryllium fluoride BeF_2	7787-49-7 47.009	tetr cry or gl; hyg	552 2.1	1169 a,c	vs H_2O; sl EtOH
301	Beryllium formate $Be(CHO_2)_2$	1111-71-3 99.048	powder	dec >250		reac H_2O; i os
302	Beryllium hydride BeH_2	7787-52-2 11.028	wh amorp solid	dec 250 0.65		reac H_2O; i eth, tol
303	Beryllium hydroxide $Be(OH)_2$	13327-32-7 43.027	amorp powder or cry	1.92	a	sl H_2O, alk; s acid
304	Beryllium iodide BeI_2	7787-53-3 262.821	hyg needles	480 4.32	487 a,c,d,e	reac H_2O; s EtOH
305	Beryllium nitrate trihydrate $Be(NO_3)_2•3H_2O$	13597-99-4 187.068	yel-wh hyg cry	≈30	dec	vs H_2O; s EtOH
306	Beryllium nitride Be_3N_2	1304-54-7 55.050	gray refrac cry; cub	2200 2.71		reac acid, alk
307	Beryllium oxide BeO	1304-56-9 25.011	wh amorp powder	2507 3.01	a,c	i H_2O; sl acid, alk

No.	Name Formula	CAS RN Mol. Wt.	Physical Form	mp/°C den/g cm^{-3}	bp/°C Other Data	Solubility
308	Beryllium perchlorate tetrahydrate $Be(ClO_4)_2 \cdot 4H_2O$	7787-48-6 279.974	hyg cry			vs H_2O
309	Beryllium selenate tetrahydrate $BeSeO_4 \cdot 4H_2O$	10039-31-3 224.03	orth cry	dec 100 2.03		vs H_2O
310	Beryllium sulfate $BeSO_4$	13510-49-1 105.076	col tetr cry; hyg	 2.5	 a	s H_2O
311	Beryllium sulfate tetrahydrate $BeSO_4 \cdot 4H_2O$	7787-56-6 177.137	col tetr cry	dec ≈100 1.71		vs H_2O; i EtOH
312	Beryllium sulfide BeS	13598-22-6 41.078	col cub cry	dec 2.36	 a	reac hot H_2O
313	Bismuth Bi	7440-69-9 208.980	gray-wh soft metal	271.40 9.79	1564 a,c,d,e,f	s acid
314	Bismuth basic carbonate $(BiO)_2CO_3$	5892-10-4 509.969	wh powder	 6.86		i H_2O; s acid
315	Bismuth bromide $BiBr_3$	7787-58-8 448.692	yel cub cry	218 5.72	453 b,d,e	reac H_2O; s dil acid, ace; i EtOH
316	Bismuth chloride $BiCl_3$	7787-60-2 315.338	yel-wh cub cry; hyg	230 4.75	447 a,b,c,d,e	reac H_2O; s acid, EtOH, ace
317	Bismuth citrate $BiC_6H_5O_7$	813-93-4 398.082	wh powder	 3.458		i H_2O; sl EtOH
318	Bismuth fluoride BiF_3	7787-61-3 265.975	wh-gray cub cry	725 8.3	900	i H_2O
319	Bismuth pentafluoride BiF_5	7787-62-4 303.972	wh tetr needles; hyg	154 5.55	230	reac H_2O
320	Bismuth hydride BiH_3	18288-22-7 212.004	col gas, unstable	-67 9.303 g/L	≈17	
321	Bismuth hydroxide $Bi(OH)_3$	10361-43-0 260.002	wh-yel amorp powder	 4.962	 a	i H_2O; s acid
322	Bismuth iodide BiI_3	7787-64-6 589.693	blk hex cry	408.6 5.778	542 a	i H_2O; s EtOH
323	Bismuth molybdate $Bi_2(MoO_4)_3$	51898-99-8 897.77	monocl cry	 5.95		
324	Bismuth nitrate pentahydrate $Bi(NO_3)_3 \cdot 5H_2O$	10035-06-0 485.071	col tricl cry; hyg	≈75 dec 2.83		reac H_2O; s ace; i EtOH
325	Bismuth oleate $Bi(C_{18}H_{33}O_2)_3$	52951-38-9 1053.356	soft yel-brn solid			i H_2O; s eth; sl bz
326	Bismuth orthovanadate $BiVO_4$	14059-33-7 323.920	oran-yel orth cry	trans 500 6.25		
327	Bismuth oxalate $Bi_2(C_2O_4)_3$	6591-55-5 682.020	wh powder			i H_2O, EtOH; s dil acid
328	Bismuth oxide Bi_2O_3	1304-76-3 465.959	yel monocl cry or powder	817 8.9	1890 a	i H_2O; s acid
329	Bismuth oxybromide BiOBr	7787-57-7 304.883	col tetr cry	 8.08		i H_2O, EtOH; s acid
330	Bismuth oxychloride BiOCl	7787-59-9 260.432	wh tetr cry	 7.72	 a	i H_2O
331	Bismuth oxyiodide BiOI	7787-63-5 351.883	red tetr cry	>300 dec 7.92		i H_2O, EtOH, chl; s HCl
332	Bismuth oxynitrate $BiONO_3$	10361-46-3 286.985	wh powder	260 dec 4.93		i H_2O, EtOH; s acid
333	Bismuth phosphate $BiPO_4$	10049-01-1 303.951	monocl cry	 6.32		sl H_2O, dil acid; i EtOH
334	Bismuth selenide Bi_2Se_3	12068-69-8 654.84	blk hex cry	710 dec 7.5		i H_2O
335	Bismuth stannate pentahydrate $Bi_2(SnO_3)_3 \cdot 5H_2O$	12777-45-6 1008.162	wh cry			i H_2O
336	Bismuth sulfate $Bi_2(SO_4)_3$	7787-68-0 706.152	wh needles or powder	405 dec 5.08	 a	reac H_2O, EtOH
337	Bismuth sulfide Bi_2S_3	1345-07-9 514.159	blk-brn orth cry	850 6.78	 a,f	i H_2O; s acid
338	Bismuth telluride Bi_2Te_3	1304-82-1 800.76	gray hex plates	580 7.74	 f	i H_2O; s EtOH
339	Bismuth tetroxide Bi_2O_4	12048-50-9 481.959	red-oran powder	305 5.6		reac H_2O
340	Bismuth titanate $Bi_4(TiO_4)_3$	12048-51-0 1171.515	wh orth cry	 7.85		
341	Boron B	7440-42-8 10.811	blk rhomb cry	2075 2.34	4000 a,c,d	i H_2O
342	Diborane(6) B_2H_6	19287-45-7 27.670	col gas; flam	-165.5 1.214 g/L	-92.4 a,b,d,e	reac H_2O
343	Tetraborane(10) B_4H_{10}	18283-93-7 53.323	col gas	-121 2.340 g/L	18 a,d,e	reac H_2O
344	Pentaborane(9) B_5H_9	19624-22-7 63.126	flam liq	-46.6 0.60	60 a,e	reac hot H_2O
345	Pentaborane(11) B_5H_{11}	18433-84-6 65.142	col liq; unstable	-123	63 a,d,e	reac H_2O
346	Hexaborane(10) B_6H_{10}	23777-80-2 74.945	col liq	-62.3 0.67	108 dec a	reac hot H_2O

No.	Name Formula	CAS RN Mol. Wt.	Physical Form	mp/°C den/g cm^{-3}	bp/°C Other Data	Solubility
347	Hexaborane(12) B_6H_{12}	12008-19-4 76.961	col liq; unstable	-82	≈80	reac H_2O
348	Nonaborane(15) B_9H_{15}	19465-30-6 112.418	col liq	 2.6		
349	Decaborane(14) $B_{10}H_{14}$	17702-41-9 122.221	wh orth cry	99.6 0.94	≈213 e	sl H_2O; s EtOH, bz, CS_2, ctc
350	Borane carbonyl BH_3CO	13205-44-2 41.845	col gas	-137	-64 e	reac H_2O
351	Borazine $B_3N_3H_6$	6569-51-3 80.501	col liq	-58 0.824	53 a	reac H_2O
352	Boric acid (orthoboric acid) $B(OH)_3$	10043-35-3 61.833	col tricl cry	170.9 1.5	 a	sl H_2O, EtOH
353	Metaboric acid (α form) HBO_2	13460-50-9 43.818	col orth cry; hyg	176 1.784	 a	s H_2O
354	Metaboric acid (β form) HBO_2	13460-50-9 43.818	col monocl cry; hyg	201 2.045	 a	s H_2O
355	Metaboric acid (γ form) HBO_2	13460-50-9 43.818	col cub cry	236 2.487	 a	s H_2O
356	Fluoroboric acid HBF_4	16872-11-0 87.813	col liq	 ≈1.8	130 dec	vs H_2O, EtOH
357	Boron arsenide BAs	12005-69-5 85.733	cub cry	dec 920 5.22		
358	Boron bromide BBr_3	10294-33-4 250.523	col liq; hyg	-45 2.6	91 a,b,d,e	reac H_2O, EtOH
359	Boron carbide B_4C	12069-32-8 55.255	hard blk cry	2350 2.50	>3500	i H_2O, acid
360	Boron chloride BCl_3	10294-34-5 117.169	col liq or gas	-107 5.141 g/L	12.65 a,b,c,d,e	reac H_2O, EtOH
361	Boron chloride B_2Cl_4	13701-67-2 163.433	col liq; flam	-92.6	65 a	reac H_2O
362	Boron fluoride BF_3	7637-07-2 67.806	col gas	-126.8 2.975 g/L	-101 a,b,c,d,e	s H_2O
363	Boron fluoride B_2F_4	13965-73-6 97.616	col gas; flam	-56 4.283 g/L	-34 a,d	reac H_2O
364	Boron iodide BI_3	13517-10-7 391.524	needles	43 3.96	210 a,b,d	
365	Boron nitride BN	10043-11-5 24.818	wh powder; hex or cub cry	≈2975 2.18	 a	i H_2O, acid
366	Boron oxide B_2O_3	1303-86-2 69.620	col gl or hex cry; hyg	450 2.55	 a,c	sl H_2O; s EtOH
367	Boron phosphide BP	20205-91-8 41.785	red cub cry or powder	dec 1125		reac H_2O, acid
368	Boron sulfide B_2S_3	12007-33-9 117.820	yel amorp solid	softens ≈320 ≈1.7	 a	
369	Bromine Br_2	7726-95-6 159.808	red liq	-7.2 3.1028	58.8 a,b,c,d,e	sl H_2O
370	Bromic acid $HBrO_3$	7789-31-3 128.910	stable only in aq soln			s H_2O
371	Bromine oxide Br_2O	21308-80-5 175.807	brn solid	-17.5 dec		
372	Bromine dioxide BrO_2	21255-83-4 111.903	unstable yel cry	dec ≈0	 a	
373	Bromine azide BrN_3	13973-87-0 121.924	red cry; exp	≈45	exp	
374	Bromine chloride BrCl	13863-41-7 115.357	unstable red-brn gas	≈-66 5.062 g/L	≈5 dec a	reac H_2O; s eth, CS_2
375	Bromine fluoride BrF	13863-59-7 98.902	unstable red-brn gas	≈-33 4.340 g/L	≈20 dec a,d	
376	Bromine trifluoride BrF_3	7787-71-5 136.899	col hyg liq	8.77 2.803	125.8 a,d	reac H_2O
377	Bromine pentafluoride BrF_5	7789-30-2 174.896	col liq	-60.5 2.460	40.76 a,c,d,e	reac H_2O (exp)
378	Cadmium Cd	7440-43-9 112.411	silv-wh metal	321.07 8.69	767 a,c,d,e	i H_2O; reac acid
379	Cadmium acetate $Cd(C_2H_3O_2)_2$	543-90-8 230.501	col cry	255 2.34		s H_2O, EtOH
380	Cadmium acetate dihydrate $Cd(C_2H_3O_2)_2 \cdot 2H_2O$	5743-04-4 266.530	wh cry	dec 130 2.01		vs H_2O; s EtOH
381	Cadmium antimonide CdSb	12014-29-8 234.171	orth cry	456 6.92		
382	Cadmium arsenide Cd_3As_2	12006-15-4 487.076	gray tetr cry	721 6.25		
383	Cadmium azide $Cd(N_3)_2$	14215-29-3 196.451	yel-wh orth cry; exp	exp 3.24		
384	Cadmium bromide $CdBr_2$	7789-42-6 272.219	wh hex powder or flakes; hyg	568 5.19	844 a,c,d	s H_2O; sl ace, eth
385	Cadmium bromide tetrahydrate $CdBr_2 \cdot 4H_2O$	13464-92-1 344.281	wh-yel cry			s H_2O, ace, EtOH

No.	Name Formula	CAS RN Mol. Wt.	Physical Form	mp/°C den/g cm^{-3}	bp/°C Other Data	Solubility
386	Cadmium carbonate $CdCO_3$	513-78-0 172.420	wh hex cry	dec 500 4.258	 a	i H_2O; s acid
387	Cadmium chlorate dihydrate $Cd(ClO_3)_2 \cdot 2H_2O$	22750-54-5* 315.343	col hyg cry	80 dec 2.28		vs H_2O
388	Cadmium chloride $CdCl_2$	10108-64-2 183.316	rhomb cry; hyg	564 4.08	960 a,c,d,e	vs H_2O; s ace; sl EtOH; i eth
389	Cadmium chloride hemipentahydrate $CdCl_2 \cdot 2.5H_2O$	7790-78-5 228.354	wh rhomb leaflets	 3.327		vs H_2O; s ace
390	Cadmium chloride monohydrate $CdCl_2 \cdot H_2O$	34330-64-8 201.331	wh cry			vs H_2O
391	Cadmium chromate $CdCrO_4$	14312-00-6 228.405	yel orth cry	 4.5		i H_2O
392	Cadmium cyanide $Cd(CN)_2$	542-83-6 164.446	wh cub cry	 2.23		sl H_2O
393	Cadmium fluoride CdF_2	7790-79-6 150.408	cub cry	1110 6.33	1748 a,c,d,e	sl H_2O; s acid; i EtOH
394	Cadmium hydroxide $Cd(OH)_2$	21041-95-2 146.426	wh trig or hex cry	dec 130 4.79	 a	i H_2O, s dil acid
395	Cadmium iodate $Cd(IO_3)_2$	7790-81-0 462.216	wh powder	 6.48		i H_2O; s HNO_3
396	Cadmium iodide CdI_2	7790-80-9 366.220	hex flakes	387 5.64	742 a,c,d,e	s H_2O, EtOH, eth, ace
397	Cadmium metasilicate $CdSiO_3$	13477-19-5 188.495	grn monocl cry	1252 5.10		
398	Cadmium molybdate $CdMoO_4$	13972-68-4 272.35	col tetr cry	≈900 dec 5.4		i H_2O; s acid
399	Cadmium niobate $Cd_2Nb_2O_7$	12187-14-3 522.631	cub cry	≈1410 6.28		i H_2O
400	Cadmium nitrate $Cd(NO_3)_2$	10325-94-7 236.420	wh cub cry, hyg	350 3.6		vs H_2O; s EtOH
401	Cadmium nitrate tetrahydrate $Cd(NO_3)_2 \cdot 4H_2O$	10022-68-1 308.482	col orth cry, hyg	59.5 2.45		vs H_2O; s EtOH, ace
402	Cadmium oxalate CdC_2O_4	814-88-0 200.431	wh solid	 3.32		i H_2O
403	Cadmium oxalate trihydrate $CdC_2O_4 \cdot 3H_2O$	20712-42-9 254.477	wh amorp powder	340 dec		i H_2O, EtOH; s dil acid
404	Cadmium oxide CdO	1306-19-0 128.410	brn cub cry	 8.15	1559 sp a,e	i H_2O; s dil acid
405	Cadmium perchlorate hexahydrate $Cd(ClO_4)_2 \cdot 6H_2O$	10326-28-0 419.403	wh hex cry	 2.37		vs H_2O
406	Cadmium phosphide Cd_3P_2	12014-28-7 399.181	gr tetr needles	700 5.96		s dil HCl
407	Cadmium selenate dihydrate $CdSeO_4 \cdot 2H_2O$	10060-09-0 291.40	orth cry	dec 100 3.62		vs H_2O
408	Cadmium selenide CdSe	1306-24-7 191.37	wh cub cry	1240 5.81		i H_2O
409	Cadmium sulfate $CdSO_4$	10124-36-4 208.475	col orth cry	1000 4.69	 a	vs H_2O; i EtOH
410	Cadmium sulfate monohydrate $CdSO_4 \cdot H_2O$	7790-84-3 226.490	monocl cry	105 3.79		
411	Cadmium sulfate octahydrate $CdSO_4 \cdot 8H_2O$	15244-35-6 352.597	col monocl cry	dec 40 3.08		vs H_2O
412	Cadmium sulfide CdS	1306-23-6 144.477	yel-oran cub cry	1750 4.83	 a,f	i H_2O; s acid
413	Cadmium telluride CdTe	1306-25-8 240.01	brn-blk cub cry	1042 6.2	 a	i H_2O, dil acid
414	Cadmium titanate $CdTiO_3$	12014-14-1 208.276	orth cry	 6.5		
415	Cadmium tungstate $CdWO_4$	7790-85-4 360.25	wh monocl cry	 8.0		i H_2O, acid; s NH_4OH
416	Calcium Ca	7440-70-2 40.078	silv-wh metal	842 1.54	1484 a,c,e	reac H_2O; i bz
417	Calcium acetate $Ca(C_2H_3O_2)_2$	62-54-4 158.168	wh hyg cry	dec 160 1.50		s H_2O; sl EtOH
418	Calcium acetate monohydrate $Ca(C_2H_3O_2)_2 \cdot H_2O$	5743-26-0 176.183	wh needles or powder	dec ≈150		s H_2O; sl EtOH
419	Calcium aluminate $CaAl_2O_4$	12042-68-1 158.039	wh monocl cry	1605 2.98		reac H_2O
420	Calcium aluminate (β form) $Ca_3Al_2O_6$	12042-78-3 270.193	wh cub cry; refr	1535 3.04		i H_2O
421	Calcium arsenate $Ca_3(AsO_4)_2$	7778-44-1 398.072	wh powder	dec 3.6		sl H_2O; s dil acid
422	Calcium boride CaB_6	12007-99-7 104.944	refrac solid	2235 2.49		
423	Calcium bromide $CaBr_2$	7789-41-5 199.886	rhomb cry; hyg	742 3.38	1815 a,c	vs H_2O; s EtOH, ace
424	Calcium bromide hexahydrate $CaBr_2 \cdot 6H_2O$	13477-28-6 307.977	wh hyg powder	38 dec 2.29		vs H_2O

No.	Name Formula	CAS RN Mol. Wt.	Physical Form	mp/°C den/g cm^{-3}	bp/°C Other Data	Solubility
425	Calcium carbide CaC_2	75-20-7 64.100	gray-blk orth cry	2300 2.22	 a	reac H_2O
426	Calcium carbonate (aragonite) $CaCO_3$	471-34-1 100.087	wh orth cry or powder	825 dec 2.83	 a,c,f	i H_2O; s dil acid
427	Calcium carbonate (calcite) $CaCO_3$	471-34-1 100.087	wh hex cry or powder	1339 2.71	 a,c,f	i H_2O; s dil acid
428	Calcium chlorate $Ca(ClO_3)_2$	10137-74-3 206.979	wh cry	340		vs H_2O
429	Calcium chlorate dihydrate $Ca(ClO_3)_2\cdot 2H_2O$	10035-05-9 243.010	wh monocl cry; hyg	100 dec 2.711		vs H_2O; s EtOH
430	Calcium chloride $CaCl_2$	10043-52-4 110.983	wh cub cry or powder; hyg	772 2.15	1935.5 a,c	vs H_2O, EtOH
431	Calcium chloride dihydrate $CaCl_2\cdot 2H_2O$	10035-04-8 147.014	hyg flakes or powder	175 dec 1.85		vs H_2O, EtOH
432	Calcium chloride hexahydrate $CaCl_2\cdot 6H_2O$	7774-34-7 219.074	wh hex cry; hyg	30 dec 1.71		vs H_2O
433	Calcium chloride monohydrate $CaCl_2\cdot H_2O$	13477-29-7 128.998	wh hyg cry	260 dec 2.24		s H_2O, EtOH
434	Calcium chromate dihydrate $CaCrO_4\cdot 2H_2O$	13765-19-0 192.102	yel orth cry	 2.50		s H_2O
435	Calcium cyanamide $CaCN_2$	156-62-7 80.102	col hex cry	≈1340 2.29	subl	reac H_2O
436	Calcium cyanide $Ca(CN)_2$	592-01-8 92.113	wh rhomb cry; hyg		 a	s H_2O, EtOH
437	Calcium dichromate trihydrate $CaCr_2O_7\cdot 3H_2O$	14307-33-6* 310.112	red-oran cry	dec 100 2.37		vs H_2O; reac EtOH; i eth, ctc
438	Calcium dihydrogen phosphate monohydrate $Ca(H_2PO_4)_2\cdot H_2O$	10031-30-8 252.068	col tricl plates	dec ≈100 2.22		sl H_2O; s dil acid
439	Calcium fluoride CaF_2	7789-75-5 78.075	wh cub cry or powder	1418 3.18	2533.4 a,c,f	i H_2O; sl acid
440	Calcium formate $Ca(CHO_2)_2$	544-17-2 130.114	orth cry	300 dec 2.02		s H_2O; i EtOH
441	Calcium hexafluorosilicate dihydrate $CaSiF_6\cdot 2H_2O$	16925-39-6 218.185	col tetr cry	 2.25		i H_2O, ace; reac hot H_2O
442	Calcium hydride CaH_2	7789-78-8 42.094	gray orth cry or powder	816 1.7	 a	reac H_2O, EtOH
443	Calcium hydrogen phosphate $CaHPO_4$	7757-93-9 136.057	wh tricl cry	dec 2.92	 f	i H_2O, EtOH
444	Calcium hydrogen phosphate dihydrate $CaHPO_4\cdot 2H_2O$	7789-77-7 172.088	monocl cry	dec ≈100 2.31		i H_2O, EtOH; s dil acid
445	Calcium hydroxide $Ca(OH)_2$	1305-62-0 74.093	soft hex cry	 ≈2.2	 a	sl H_2O; s acid
446	Calcium hypophosphite $Ca(H_2PO_2)_2$	7789-79-9 170.055	wh monocl cry	dec 300		s H_2O; i EtOH
447	Calcium iodate $Ca(IO_3)_2$	7789-80-2 389.883	wh monocl cry	 4.52		sl H_2O; s HNO_3; i EtOH
448	Calcium iodide CaI_2	10102-68-8 293.887	hyg hex cry	779 3.96	 a,c	vs H_2O; s MeOH, EtOH, ace; i eth
449	Calcium iodide hexahydrate $CaI_2\cdot 6H_2O$	71626-98-7 401.978	wh hex needles or powder	42 dec 2.55		vs H_2O, EtOH
450	Calcium molybdate $CaMoO_4$	7789-82-4 200.02	wh tetr cry	965 dec 4.35	 f	i H_2O, EtOH; s conc acid
451	Calcium nitrate $Ca(NO_3)_2$	10124-37-5 164.087	wh cub cry; hyg	561 2.5	 a	vs H_2O; s EtOH, MeOH, ace
452	Calcium nitrate tetrahydrate $Ca(NO_3)_2\cdot 4H_2O$	13477-34-4 236.149	wh cry	dec ≈40 1.82		vs H_2O; s EtOH, ace
453	Calcium nitride Ca_3N_2	12013-82-0 148.247	red-brn cub cry	1195 2.67		s H_2O, acid; i EtOH
454	Calcium nitrite $Ca(NO_2)_2$	13780-06-8 132.089	wh-yel hex cry; hyg	 2.23		vs H_2O; sl EtOH
455	Calcium oxalate CaC_2O_4	563-72-4 128.098	wh cry powder	 2.2	 a	i H_2O
456	Calcium oxalate monohydrate $CaC_2O_4\cdot H_2O$	5794-28-5 146.113	cub cry	dec 200 2.2	 f	i H_2O; s dil acid
457	Calcium oxide CaO	1305-78-8 56.077	gray-wh cub cry	2927 3.34	 a,c	reac H_2O; s acid
458	Calcium perchlorate $Ca(ClO_4)_2$	13477-36-6 238.978	wh cry	dec 270 2.65		vs H_2O; s EtOH
459	Calcium permanganate $Ca(MnO_4)_2$	10118-76-0 277.949	purp hyg cry	 2.4		vs H_2O; reac EtOH
460	Calcium peroxide CaO_2	1305-79-9 72.077	wh-yel tetr cry; hyg	dec ≈200 2.9		sl H_2O; s acid
461	Calcium phosphate $Ca_3(PO_4)_2$	7758-87-4 310.177	wh amorp powder	1670 3.14	 a	i H_2O, EtOH; s dil acid
462	Calcium dihydrogen phosphate monohydrate $Ca(H_2PO_4)_2\cdot H_2O$	7758-23-8* 252.068	col hyg cry	dec ≈100 2.22		s H_2O, acid

No.	Name Formula	CAS RN Mol. Wt.	Physical Form	mp/°C den/g cm^{-3}	bp/°C Other Data	Solubility
463	Calcium phosphide Ca_3P_2	1305-99-3 182.182	red-brn hyg cry	≈1600 2.51		reac H_2O; i EtOH, eth
464	Calcium pyrophosphate $Ca_2P_2O_7$	7790-76-3 254.099	wh powder	1353 3.09		i H_2O; s dil acid
465	Calcium selenate dihydrate $CaSeO_4 \cdot 2H_2O$	7790-74-1 219.07	wh monocl cry	 2.75		s H_2O
466	Calcium selenide CaSe	1305-84-6 119.04	wh-brn cub cry	 3.8		reac H_2O
467	Calcium metasilicate $CaSiO_3$	1344-95-2 116.162	wh monocl cry	 2.92	 f	i H_2O
468	Calcium silicide $CaSi_2$	12013-56-8 96.249	gray hex cry	1040 2.50		i cold H_2O; reac hot H_2O; s acid
469	Calcium silicide CaSi	12013-55-7 68.164	orth cry	1324 2.39		
470	Calcium sulfate $CaSO_4$	7778-18-9 136.142	orth cry	1450 2.96	 a,c,f	sl H_2O
471	Calcium sulfate dihydrate $CaSO_4 \cdot 2H_2O$	10101-41-4 172.172	monocl cry or powder	dec 150 2.32	 f	s H_2O; i os
472	Calcium sulfate hemihydrate $CaSO_4 \cdot 0.5H_2O$	10034-76-1 145.149	wh powder			s H_2O
473	Calcium sulfide CaS	20548-54-3 72.144	wh-yel cub cry; hyg	2525 2.59	 a,f	sl H_2O; i EtOH
474	Calcium sulfite dihydrate $CaSO_3 \cdot 2H_2O$	10257-55-3 156.173	wh powder			sl H_2O, EtOH; s acid
475	Calcium telluride CaTe	12013-57-9 167.68	cub cry	 4.87		
476	Calcium tetrahydroaluminate $Ca(AlH_4)_2$	16941-10-9 102.105	gray powder; flam			reac H_2O; s thf; i eth, bz
477	Calcium thiocyanate tetrahydrate $Ca(SCN)_2 \cdot 4H_2O$	2092-16-2 228.307	hygr cry	dec 160		vs H_2O; s EtOH, ace
478	Calcium thiosulfate hexahydrate $CaS_2O_3 \cdot 6H_2O$	10124-41-1 260.300	tricl cry	dec 45 1.87		s H_2O; i EtOH
479	Calcium titanate $CaTiO_3$	12049-50-2 135.943	cub cry	1980 3.98	 f	
480	Calcium tungstate $CaWO_4$	7790-75-2 287.92	wh tetr cry	1620 6.06	 f	i H_2O; s hot acid
481	Californium Cf	7440-71-3 251	hex or cub metal	900 15.1	 a	
482	Carbon (diamond) C	7782-40-3 12.011	col cub cry	4440 (12.4 GPa) 3.513	 e,f	i H_2O
483	Carbon (graphite) C	7440-44-0 12.011	soft blk hex cry	4492 tp (10.3 MPa) 2.2	3825 sp a,c,f	i H_2O
484	Carbon (fullerene-C_{60}) C_{60}	99685-96-8 720.660	yel needles or plates	>280		s os
485	Carbon (fullerene-C_{70}) C_{70}	115383-22-7 840.770	red-brn solid	>280		s bz, tol
486	Fullerene fluoride $C_{60}F_{60}$	134929-59-2 1860.564	col plates	287		vs ace; s thf; i chl
487	Carbon monoxide CO	630-08-0 28.010	col gas	-205 1.229 g/L	-191.5 a,b,c,d,e	sl H_2O; s chl, EtOH
488	Carbon dioxide CO_2	124-38-9 44.010	col gas	-56.57 tp 1.931 g/L	-78.4 sp a,b,c,e	s H_2O
489	Carbon diselenide CSe_2	506-80-9 169.93	yel liq	-45.5 2.6626	125.5 a,e	i H_2O; vs ctc, tol
490	Carbon disulfide CS_2	75-15-0 76.143	col or yel liq	-111.5 1.2555	46 a,b,c,d,e	i H_2O; vs EtOH, bz, os
491	Carbon oxyselenide COSe	1603-84-5 106.97	col gas; unstable	-1[illegible] 4.694 g/L	[illegible]1.3 e	reac H_2O
492	Carbon oxysulfide COS	463-58-1 60.076	col gas	-138.8 2.636 g/L	-50 a,b,e	s H_2O, EtOH
493	Carbon suboxide C_3O_2	504-64-3 68.032	col gas	-111.3 2.985 g/L	6.8 e	reac H_2O
494	Carbon sulfide telluride CSTe	10340-06-4 171.68	red-yel liq; unstable			reac H_2O
495	Carbonyl bromide $COBr_2$	593-95-3 187.818	col liq	 2.5	64.5 a	reac H_2O
496	Carbonyl chloride $COCl_2$	75-44-5 98.915	col gas	-127.9 4.340 g/L	8 a,b,c,e	sl H_2O; s bz, tol
497	Carbonyl fluoride COF_2	353-50-4 66.007	col gas	-111.26 2.896 g/L	-84.57 a	reac H_2O
498	Cyanogen C_2N_2	460-19-5 52.035	col gas	-27.9 2.283 g/L	-21.1 a,b,d,e	sl H_2O, eth; s EtOH
499	Cyanogen bromide BrCN	506-68-3 105.922	wh hyg needles	52 2.005	61.5 a	s H_2O, EtOH, eth
500	Cyanogen chloride ClCN	506-77-4 61.471	col gas	-6.55 2.697 g/L	13 a	s H_2O, EtOH, eth

No.	Name	Formula	CAS RN	Mol. Wt.	Physical Form	mp/°C	den/g cm^{-3}	bp/°C	Other Data	Solubility
501	Cyanogen fluoride	FCN	1495-50-7	45.016	col gas	-82	1.975 g/L	-46	a	
502	Cyanogen iodide	ICN	506-78-5	152.922	col needles	146.7	1.84		a	s H_2O, EtOH, eth
503	Cerium	Ce	7440-45-1	140.115	gray metal; cub or hex	799	8.16	3424	a,c	s dil acid
504	Cerium boride	CeB_6	12008-02-5	204.981	blue refrac solid; hex	2550	4.87			i H_2O, HCl
505	Cerium carbide	CeC_2	12012-32-7	164.137	red hex cry	2250	5.47			reac H_2O
506	Cerium nitride	CeN	25764-08-3	154.122	refrac cub cry	2557	7.89			
507	Cerium silicide	$CeSi_2$	12014-85-6	196.286	tetr cry	1620	5.31			i H_2O
508	Cerium(II) hydride	CeH_2	13569-50-1	142.131	cub cry		5.45			reac H_2O
509	Cerium(II) iodide	CeI_2	19139-47-0	393.924	bronze cry	808				
510	Cerium(II) sulfide	CeS	12014-82-3	172.181	yel cub cry	2445	5.9		a	
511	Cerium(III) bromide	$CeBr_3$	14457-87-5	379.827	wh hex cry; hyg	733		1457		s H_2O
512	Cerium(III) bromide heptahydrate	$CeBr_3 \cdot 7H_2O$	14457-87-5*	505.934	col hyg needles	732				s H_2O, EtOH
513	Cerium(III) carbide	Ce_2C_3	12115-63-8	316.263	yel-brn cub cry	1505	6.9			
514	Cerium(III) carbonate hydrate	$Ce_2(CO_3)_3 \cdot 5H_2O$	72520-94-6	550.334	wh powder					i H_2O; s dil acid
515	Cerium(III) chloride	$CeCl_3$	7790-86-5	246.473	wh hex cry	817	3.97		a	s H_2O, EtOH
516	Cerium(III) chloride heptahydrate	$CeCl_3 \cdot 7H_2O$	18618-55-8	372.580	yel orth cry; hyg	dec 90				vs H_2O, EtOH
517	Cerium(III) fluoride	CeF_3	7758-88-5	197.110	wh hex cry; hyg	1430	6.157			i H_2O
518	Cerium(III) iodide	CeI_3	7790-87-6	520.828	yel orth cry; hyg	766				s H_2O
519	Cerium(III) iodide nonahydrate	$CeI_3 \cdot 9H_2O$	7790-87-6*	682.966	wh-red cry					vs H_2O; s EtOH
520	Cerium(III) oxide	Ce_2O_3	1345-13-7	328.228	yel-grn cub cry	2230	6.2		a	i H_2O; s acid
521	Cerium(III) sulfate octahydrate	$Ce_2(SO_4)_3 \cdot 8H_2O$	13454-94-9	712.543	wh orth cry	dec ≈250	2.87			s H_2O
522	Cerium(III) sulfide	Ce_2S_3	12014-93-6	376.428	red cub cry	2450	5.02			i H_2O
523	Cerium(IV) fluoride	CeF_4	10060-10-3	216.109	wh hyg powder	dec ≈600	4.77			i H_2O
524	Cerium(IV) oxide	CeO_2	1306-38-3	172.114	wh-yel powder; cub	2400	7.65		a	i H_2O, dil acid
525	Cerium(IV) sulfate tetrahydrate	$Ce(SO_4)_2 \cdot 4H_2O$	10294-42-5	404.304	yel-oran orth cry	dec 180	3.91			reac H_2O
526	Cesium	Cs	7440-46-2	132.905	silv-wh metal	28.44	1.93	671	a,c,e	reac H_2O
527	Cesium amide	$CsNH_2$	22205-57-8	148.928	wh tetr cry		3.70		a	
528	Cesium azide	CsN_3	22750-57-8	174.925	hyg tetr cry; exp	326	≈3.5			
529	Cesium bromate	$CsBrO_3$	13454-75-6	260.807	col hex cry		4.11			sl H_2O
530	Cesium bromide	CsBr	7787-69-1	212.809	wh cub cry; hyg	636	4.43	≈1300	a,e	vs H_2O; s EtOH; i ace
531	Cesium carbonate	Cs_2CO_3	534-17-8	325.820	wh monocl cry; hyg	792	4.24		a	s H_2O, EtOH, eth
532	Cesium chlorate	$CsClO_3$	13763-67-2	216.356	col hex cry		3.57			sl H_2O
533	Cesium chloride	CsCl	7647-17-8	168.358	wh cub cry; hyg	645	3.988	1297	a,c,e	vs H_2O; s EtOH
534	Cesium cyanide	CsCN	21159-32-0	158.923	wh cub cry; hyg	350	3.34			vs H_2O
535	Cesium fluoride	CsF	13400-13-0	151.903	wh cub cry; hyg	703	4.64		a,c,e	vs H_2O; s MeOH; i diox, py
536	Cesium hydride	CsH	58724-12-2	133.913	wh cub cry; flam	dec ≈170	3.42		a	reac H_2O
537	Cesium hydrogen fluoride	$CsHF_2$	12280-52-3	171.910	tetr cry	170	3.86		a	
538	Cesium hydroxide	CsOH	21351-79-1	149.912	wh-yel hyg cry	272	3.68		a	vs H_2O; s EtOH
539	Cesium iodide	CsI	7789-17-5	259.809	col cub cry; hyg	621	4.51	≈1280	a,e	vs H_2O; s EtOH, MeOH, ace

No.	Name Formula	CAS RN Mol. Wt.	Physical Form	mp/°C den/g cm^{-3}	bp/°C Other Data	Solubility
540	Cesium metaborate $CsBO_2$	92141-86-1 175.715	cub cry	732 ≈3.7	 a	
541	Cesium nitrate $CsNO_3$	7789-18-6 194.910	wh hex or cub cry	414 3.66	 a	s H_2O, ace; sl EtOH
542	Cesium oxide Cs_2O	20281-00-9 281.810	yel-oran hex cry	490 4.65	 a	vs H_2O
543	Cesium perchlorate $CsClO_4$	13454-84-7 232.356	wh orth cry; hyg	250 3.327	 a	sl H_2O
544	Cesium superoxide CsO_2	12018-61-0 164.904	yel tetr cry	432 3.77	 a	reac H_2O
545	Cesium sulfate Cs_2SO_4	10294-54-9 361.875	wh orth cry or hex prisms; hyg	1005 4.24	 a,c	vs H_2O; i EtOH, ace, py
546	Chlorine Cl_2	7782-50-5 70.905	grn-yel gas	-101.5 3.111 g/L	-34.04 a,b,c,d,e	sl H_2O
547	Hypochlorous acid HOCl	7790-92-3 52.460	grn-yel; stable only in aq soln		 a	s H_2O
548	Perchloric acid $HClO_4$	7601-90-3 100.459	col hyg liq	-112 1.77	dec ≈90 a	s H_2O
549	Chlorine oxide Cl_2O	7791-21-1 86.904	yel-brn gas	-120.6 3.813 g/L	2.2 a,d,e	vs H_2O
550	Chlorine dioxide ClO_2	10049-04-4 67.452	oran-grn gas	-59 2.960 g/L	11 a,d,e	sl H_2O
551	Chlorine trioxide Cl_2O_3	17496-59-2 118.903	dark brn solid	exp <25		
552	Chlorine hexoxide Cl_2O_6	12442-63-6 166.901	red liq	3.5	≈200 e	reac H_2O
553	Chlorine heptoxide Cl_2O_7	10294-48-1 182.901	col oily liq; exp	-91.5 1.9	82 e	reac H_2O
554	Chlorine fluoride ClF	7790-89-8 54.451	col gas	-155.6 2.389 g/L	-101.1 a,d,e	reac H_2O
555	Chlorine trifluoride ClF_3	7790-91-2 92.448	gas	-76.34 4.057 g/L	11.75 a,d,e	reac H_2O
556	Chlorine pentafluoride ClF_5	13637-63-3 130.445	col gas	-103 5.724 g/L	-13.1 b	
557	Chlorine trioxide fluoride (perchloryl fluoride) ClO_3F	7616-94-6 102.449	col gas	-147 4.495 g/L	-46.75 a,b,d	
558	Chlorine perchlorate $ClOClO_3$	27218-16-2 134.903	unstable liq	-117 1.81(0°C)	dec ≈25	
559	Chromium Cr	7440-47-3 51.996	blue-wh metal; cub	1907 7.15	2671 a,c,e	reac dil acid
560	Chromium acetylacetonate $Cr(CH_3COCHCOCH_3)_3$	21679-31-2 349.324	red monocl cry	208 1.34	345	i H_2O; s bz
561	Chromium antimonide CrSb	12053-12-2 173.756	hex cry	1110 7.11		
562	Chromium arsenide Cr_2As	12254-85-2 178.914	tetr cry	 7.04		
563	Chromium boride CrB	12006-79-0 62.807	refrac orth cry	2100 6.1		
564	Chromium boride CrB_2	12007-16-8 73.618	refrac solid; hex	2200 5.22		
565	Chromium boride Cr_5B_3	12007-38-4 292.414	tetr cry	1900 6.10		
566	Chromium carbide Cr_3C_2	12012-35-0 180.010	gray orth cry	1895 6.68		
567	Chromium carbonyl $Cr(CO)_6$	13007-92-6 220.058	col orth cry	dec 130 1.77	subl e	i H_2O, EtOH; s eth, chl
568	Chromium nitride Cr_2N	12053-27-9 117.999	hex cry	1650 6.8		
569	Chromium nitride CrN	24094-93-7 66.003	gray cub cry	dec 1080 5.9		
570	Chromium phosphide CrP	26342-61-0 82.970	orth cry	 5.25		
571	Chromium selenide CrSe	12053-13-3 130.96	hex cry	≈1500 6.1		
572	Chromium silicide Cr_3Si	12018-36-9 184.074	cub cry	1770 6.4		
573	Chromium silicide $CrSi_2$	12018-09-6 108.167	gray hex cry	1490 4.91		
574	Chromium(II) acetate monohydrate $Cr(C_2H_3O_2)_2 \cdot H_2O$	628-52-4* 188.101	red monocl cry	 1.79		sl H_2O
575	Chromium(II) bromide $CrBr_2$	10049-25-9 211.804	wh monocl cry; aq soln blue	842 4.236	 a	s H_2O, EtOH
576	Chromium(II) chloride $CrCl_2$	10049-05-5 122.901	hyg needles; aq soln blue	814 2.88	1300 a,c,d	s H_2O
577	Chromium(II) chloride tetrahydrate $Cr(H_2O)_4Cl_2 \cdot 4H_2O$	13931-94-7 267.023	blue hyg cry	dec 51		s H_2O

No.	Name Formula	CAS RN Mol. Wt.	Physical Form	mp/°C den/g cm^{-3}	bp/°C Other Data	Solubility
578	Chromium(II) fluoride CrF_2	10049-10-2 89.993	blue-grn monocl cry	894 3.79	 a	sl H_2O; i EtOH
579	Chromium(II) iodide CrI_2	13478-28-9 305.805	red-brn cry; hyg	868 5.1	 a	
580	Chromium(II) oxalate monohydrate $CrC_2O_4 \cdot H_2O$	814-90-4* 158.031	yel-grn powder	 2.468		sl H_2O
581	Chromium(II) sulfate pentahydrate $CrSO_4 \cdot 5H_2O$	13825-86-0 238.136	blue cry			s H_2O, dil acid; sl EtOH; i ace
582	Chromium(II,III) oxide Cr_3O_4	12018-34-7 219.986	cub cry	 6.1	 a	
583	Chromium(III) acetate hexahydrate $Cr(C_2H_3O_2)_3 \cdot 6H_2O$	1066-30-4* 285.226	blue needles			s H_2O
584	Chromium(III) bromide $CrBr_3$	10031-25-1 291.708	dark grn hex cry	1130 4.68		s hot H_2O
585	Chromium(III) bromide hexahydrate $Cr(H_2O)_6Br_3$	10031-25-1* 399.799	viol hyg cry			s H_2O; i EtOH, eth
586	Chromium(III) bromide hexahydrate $CrBr_3(H_2O)_4 \cdot 2H_2O$	18721-05-6 399.799	grn hyg cry			s H_2O, EtOH
587	Chromium(III) chloride $CrCl_3$	10025-73-7 158.354	purp hex plates	1152 2.87	dec 1300 a	sl H_2O
588	Chromium(III) chloride hexahydrate $[CrCl_2(H_2O)_4]Cl \cdot 2H_2O$	10060-12-5 266.445	grn monocl cry; hyg			s H_2O, EtOH; sl ace; i eth
589	Chromium(III) fluoride CrF_3	7788-97-8 108.991	grn needles	1400 3.8	 a	i H_2O; EtOH
590	Chromium(III) fluoride trihydrate $CrF_3 \cdot 3H_2O$	16671-27-5 163.037	grn hex cry	 2.2		sl H_2O
591	Chromium(III) hydroxide trihydrate $Cr(OH)_3 \cdot 3H_2O$	1308-14-1 157.063	blue-grn powder			i H_2O; s acid
592	Chromium(III) iodide CrI_3	13569-75-0 432.709	dark grn hex cry	dec 500 5.32	 a	sl H_2O
593	Chromium(III) nitrate $Cr(NO_3)_3$	13548-38-4 238.011	grn hyg powder	dec >60		vs H_2O
594	Chromium(III) nitrate nonahydrate $Cr(NO_3)_3 \cdot 9H_2O$	7789-02-8 400.148	grn-blk monocl cry	66.3 1.80	dec >100	vs H_2O
595	Chromium(III) oxide Cr_2O_3	1308-38-9 151.990	grn hex cry	2330 5.22	≈3000 a,c	i H_2O, EtOH; sl acid, alk
596	Chromium(III) phosphate $CrPO_4$	7789-04-0 146.967	blue orth cry	>1800 4.6		i H_2O, acid, aqua regia
597	Chromium(III) phosphate hemiheptahydrate $CrPO_4 \cdot 3.5H_2O$	84359-31-9 210.021	blue-grn powder	 2.15		i H_2O; s acid
598	Chromium(III) phosphate hexahydrate $CrPO_4 \cdot 6H_2O$	84359-31-9 255.059	viol cry	dec >500 2.121		i H_2O; s acid, alk
599	Chromium(III) potassium sulfate dodecahydrate $CrK(SO_4)_2 \cdot 12H_2O$	7788-99-0 499.405	viol-blk cub cry	89 dec 1.83		s H_2O; i EtOH
600	Chromium(III) sulfate $Cr_2(SO_4)_3$	10101-53-8 392.183	red-brn hex cry	 3.1		i H_2O, acid
601	Chromium(III) sulfide Cr_2S_3	12018-22-3 200.190	brn-blk hex cry	 3.8		
602	Chromium(III) telluride Cr_2Te_3	12053-39-3 486.79	hex cry	≈1300 7.0		
603	Chromium(IV) chloride $CrCl_4$	15597-88-3 193.807	gas, stable at high temp	 8.504 g/L	dec >600	
604	Chromium(IV) fluoride CrF_4	10049-11-3 127.990	grn cry	277		
605	Chromium(IV) oxide CrO_2	12018-01-8 83.995	brn-blk tetr powder	dec ~400 4.89	 a	i H_2O; s acid
606	Chromium(V) fluoride CrF_5	14884-42-5 146.988	red orth cry	34	117	
607	Chromium(VI) fluoride CrF_6	13843-28-2 165.986	yel solid; stable at low temp	dec -100		
608	Chromium(VI) oxide CrO_3	1333-82-0 99.994	red orth cry	197 2.7	dec ≈250	vs H_2O
609	Chromyl chloride CrO_2Cl_2	14977-61-8 154.900	red liq	-96.5 1.91	117 a,d,e	reac H_2O; s ctc, chl, bz
610	Cobalt Co	7440-48-4 58.933	gray metal; hex or cub	1495 8.86	2927 a,c	s dil acid
611	Cobalt antimonide CoSb	12052-42-5 180.69	hex cry	1202 8.8		
612	Cobalt arsenic sulfide CoAsS	12254-82-9 165.92	silv-wh solid	 ≈6.1	 f	
613	Cobalt arsenide CoAs	27016-73-5 133.855	orth cry	1180 8.22		
614	Cobalt arsenide $CoAs_2$	12044-42-7 208.776	monocl cry	 7.2		
615	Cobalt arsenide $CoAs_3$	12256-04-1 283.698	cub cry	942 6.84		

No.	Name Formula	CAS RN Mol. Wt.	Physical Form	mp/°C den/g cm^{-3}	bp/°C Other Data	Solubility
616	Cobalt boride Co_2B	12045-01-1 128.677	refrac solid	1280 8.1		
617	Cobalt boride CoB	12006-77-8 69.744	refrac solid	1460 7.25		reac H_2O, HNO_3
618	Cobalt carbonyl $Co_2(CO)_8$	10210-68-1 341.949	oran cry	51 dec 1.78		i H_2O; s EtOH, eth, CS_2
619	Cobalt phosphide Co_2P	12134-02-0 148.840	gray needles	1386 6.4		i H_2O; s HNO_3
620	Cobalt silicide $CoSi_2$	12017-12-8 115.10	gray cub cry	1326 4.9		s hot HCl
621	Cobalt disulfide CoS_2	12013-10-4 123.07	cub cry	 4.3		
622	Cobalt dodecacarbonyl $Co_4(CO)_{12}$	17786-31-1 571.858	blk cry	dec 60 2.09		
623	Cobalt(II) acetate $Co(C_2H_3O_2)_2$	71-48-7 177.023	pink cry			vs H_2O; s EtOH
624	Cobalt(II) acetate tetrahydrate $Co(C_2H_3O_2)_2 \cdot 4H_2O$	6147-53-1 249.083	red monocl cry	 1.705		s H_2O, EtOH, dil acid
625	Cobalt(II) aluminate $CoAl_2O_4$	13820-62-7 176.894	blue cub cry	 4.37		i H_2O
626	Cobalt(II) arsenate octahydrate $Co_3(AsO_4)_2 \cdot 8H_2O$	24719-19-5 598.760	red monocl needles	dec 1000 3.0		i H_2O; s dil acid
627	Cobalt(II) bromate hexahydrate $Co(BrO_3)_2 \cdot 6H_2O$	13476-01-2 422.829	viol cry	 ≈2.5		vs H_2O
628	Cobalt(II) bromide $CoBr_2$	7789-43-7 218.741	grn hex cry; hyg	678 4.91	 a	vs H_2O; s MeOH, EtOH, ace
629	Cobalt(II) bromide hexahydrate $CoBr_2 \cdot 6H_2O$	13762-12-4 326.832	red hyg cry	47 dec 2.46	dec 100	s H_2O
630	Cobalt(II) carbonate $CoCO_3$	513-79-1 118.942	pink rhomb cry	 4.2		i H_2O, EtOH
631	Cobalt(II) chloride $CoCl_2$	7646-79-9 129.838	blue hyg leaflets	740 3.36	1049 a,c,e	s H_2O, EtOH, eth, ace, py
632	Cobalt(II) chloride dihydrate $CoCl_2 \cdot 2H_2O$	16544-92-6 165.869	viol-blue cry	 2.477		
633	Cobalt(II) chloride hexahydrate $CoCl_2 \cdot 6H_2O$	7791-13-1 237.929	pink red monocl cry	87 dec 1.924		vs H_2O; s EtOH, ace, eth
634	Cobalt(II) chromate $CoCrO_4$	24613-38-5 174.927	yel-brn orth cry	 ≈4.0		i H_2O; s acid
635	Cobalt(II) chromite $CoCr_2O_4$	13455-25-9 226.923	blue-grn cub cry	 5.14		i H_2O, conc acid
636	Cobalt(II) cyanide $Co(CN)_2$	542-84-7 110.968	blue hyg cry	 1.872		i H_2O
637	Cobalt(II) cyanide dihydrate $Co(CN)_2 \cdot 2H_2O$	20427-11-6 146.999	pink-brn needles			i H_2O, acid
638	Cobalt(II) ferricyanide $Co_3[Fe(CN)_6]_2$	14049-81-1 600.703	red needles			i H_2O, HCl; s NH_4OH
639	Cobalt(II) fluoride CoF_2	10026-17-2 96.930	red tetr cry	1127 4.46	≈1400 a,c	sl H_2O; s acid
640	Cobalt(II) fluoride tetrahydrate $CoF_2 \cdot 4H_2O$	13817-37-3 168.992	red orth cry	dec 2.22		s H_2O
641	Cobalt(II) formate dihydrate $Co(CHO_2)_2 \cdot 2H_2O$	6424-20-0 184.999	red cry powder	dec 140 2.13		s H_2O; i EtOH
642	Cobalt(II) hexafluorosilicate hexahydrate $CoSiF_6 \cdot 6H_2O$	12021-68-0 309.100	pale red cry	 2.087		vs H_2O
643	Cobalt(II) hydroxide $Co(OH)_2$	21041-93-0 92.948	blue-grn cry	dec ≈160 3.60	 a	sl H_2O; s acid
644	Cobalt(II) iodate $Co(IO_3)_2$	13455-28-2 408.738	blk-viol needles	dec 200 5.09		sl H_2O
645	Cobalt(II) iodide CoI_2	15238-00-3 312.742	blk hex cry; hyg	520 5.60	 a	vs H_2O
646	Cobalt(II) iodide hexahydrate $CoI_2 \cdot 6H_2O$	15238-00-3* 420.833	red hex prisms	dec 130 2.90		s H_2O, EtOH, eth, ace
647	Cobalt(II) titanate $CoTiO_3$	12017-01-5 154.798	grn rhomb cry	 5.0		
648	Cobalt(II) molybdate $CoMoO_4$	13762-14-6 218.87	blk monocl cry	1040 4.7		
649	Cobalt(II) nitrate $Co(NO_3)_2$	10141-05-6 182.942	pale red powder	dec 100 2.49	 a	vs H_2O
650	Cobalt(II) nitrate hexahydrate $Co(NO_3)_2 \cdot 6H_2O$	10026-22-9 291.034	red monocl cry; hyg	≈55 1.88		vs H_2O; s EtOH
651	Cobalt(II) oxalate CoC_2O_4	814-89-1 146.953	pink powder	250 dec 3.02		i H_2O; s acid, NH_4OH
652	Cobalt(II) oxalate dihydrate $CoC_2O_4 \cdot 2H_2O$	814-89-1* 166.984	pink needles	dec		i H_2O; sl acid; s NH_4OH
653	Cobalt(II) oxide CoO	1307-96-6 74.932	gray cub cry	1830 6.44	 a	i H_2O; s acid
654	Cobalt(II) perchlorate $Co(ClO_4)_2$	13455-31-7 257.833	red needles	 3.33		vs H_2O; i EtOH, ace

No.	Name Formula	CAS RN Mol. Wt.	Physical Form	mp/°C den/g cm^{-3}	bp/°C Other Data	Solubility
655	Cobalt(II) phosphate octahydrate $Co_3(PO_4)_2 \cdot 8H_2O$	10294-50-5 510.865	pink amorp powder	 2.77		i H_2O; s acid
656	Cobalt(II) potassium sulfate hexahydrate $CoK_2(SO_4)_2 \cdot 6H_2O$	10026-20-7 437.349	red monocl cry	dec 75 2.22		vs H_2O
657	Cobalt(II) selenate pentahydrate $CoSeO_4 \cdot 5H_2O$	14590-19-3 291.97	red tricl cry	dec 2.51		vs H_2O
658	Cobalt(II) selenide CoSe	1307-99-9 137.89	yel hex cry	1055 7.65		i H_2O, alk; s aqua regia
659	Cobalt(II) selenite dihydrate $CoSeO_3 \cdot 2H_2O$	19034-13-0 221.92	blue-red powder			i H_2O
660	Cobalt(II) orthosilicate Co_2SiO_4	12017-08-2 209.950	red-viol orth cry	1345 4.63		i H_2O; s dil HCl
661	Cobalt(II) stannate Co_2SnO_4	12139-93-4 300.574	grn-blue cub cry	 6.30		i H_2O; s alk
662	Cobalt(II) sulfate $CoSO_4$	10124-43-3 154.997	red orth cry	>700 3.71	 a	s H_2O
663	Cobalt(II) sulfate heptahydrate $CoSO_4 \cdot 7H_2O$	10026-24-1 281.103	pink monocl cry	41 dec 2.03		vs H_2O; sl EtOH, MeOH
664	Cobalt(II) sulfate monohydrate $CoSO_4 \cdot H_2O$	13455-34-0 173.012	red monocl cry	 3.08		sl H_2O
665	Cobalt(II) sulfide CoS	1317-42-6 91.00	blk amorp powder	1182 5.45	 a	i H_2O; s acid
666	Cobalt(II) telluride CoTe	12017-13-9 186.53	hex cry	 ≈8.8		
667	Cobalt(II) thiocyanate trihydrate $Co(SCN)_2 \cdot 3H_2O$	97126-35-7 229.146	viol rhomb cry			s H_2O, EtOH, eth, ace
668	Cobalt(II) tungstate $CoWO_4$	12640-47-0 306.77	blue monocl cry	 ≈7.8		i H_2O; s hot conc acid
669	Cobalt(II,III) oxide Co_3O_4	1308-06-1 240.798	blk cub cry	dec 900 6.11	 a	i H_2O; s acid, alk
670	Cobalt(III) acetate $Co(C_2H_3O_2)_3$	917-69-1 236.066	grn hyg cry	dec 100		s H_2O, EtOH
671	Cobalt(III) ammonium tetranitrodiammine $NH_4[Co(NH_3)_2(NO_2)_4]$	13600-89-0 295.054	red-brn orth cry	 1.97		s H_2O
672	Cobalt(III) fluoride CoF_3	10026-18-3 115.928	brn hex cry	927 3.88		reac H_2O
673	Cobalt(III) hexammine chloride $Co(NH_3)_6Cl_3$	10534-89-1 267.474	red monocl cry	 1.71		s H_2O; i EtOH
674	Cobalt(III) hydroxide $Co(OH)_3$	1307-86-4 109.955	brn powder	dec ≈4		i H_2O; s acid
675	Cobalt(III) nitrate $Co(NO_3)_3$	15520-84-0 244.948	grn cub cry; hyg	 ≈3.0		s H_2O; reac os
676	Cobalt(III) oxide Co_2O_3	1308-04-9 165.864	gray-blk powder	dec 895 5.18		i H_2O; s conc acid
677	Cobalt(III) oxide monohydrate $Co_2O_3 \cdot H_2O$	12016-80-7 183.880	brn-blk hex cry	dec 150		i H_2O; s acid
678	Cobalt(III) potassium nitrite sesquihydrate $CoK_3(NO_2)_6 \cdot 1.5H_2O$	13782-01-9* 479.284	yel cub cry	 2.6		sl H_2O; reac acid; i EtOH
679	Cobalt(III) sulfide Co_2S_3	1332-71-4 214.064	blk cub cry	 4.8	 a	reac acid
680	Cobalt(III) titanate Co_2TiO_4	12017-38-8 229.731	grn-blk cub cry	 5.1		s conc HCl
681	Copper Cu	7440-50-8 63.546	red metal; cub	1084.62 8.96	2562 a,c,e,f	sl dil acid
682	Copper acetylacetonate $Cu(CH_3COCHCOCH_3)_2$	13395-16-9 261.765	blue powder	284 dec	subl	sl H_2O; s chl
683	Copper nitride Cu_3N	1308-80-1 204.645	cub cry	300 dec 5.84		
684	Copper phosphide CuP_2	12019-11-3 125.494	monocl cry	≈900 4.20		
685	Copper(I) acetate $CuC_2H_3O_2$	598-54-9 122.591	col cry	dec	subl	reac H_2O
686	Copper(I) acetylide Cu_2C_2	1117-94-8 151.114	red amorp powder; exp			
687	Copper(I) azide CuN_3	14336-80-2 105.566	tetr cry; exp			
688	Copper(I) bromide CuBr	7787-70-4 143.450	wh cub cry; hyg	497 4.98	1345 a,e	sl H_2O; i ace
689	Copper(I) chloride CuCl	7758-89-6 98.999	wh cub cry	430 4.14	≈1400 a,c,e,f	sl H_2O; i EtOH, ace
690	Copper(I) cyanide CuCN	544-92-3 89.564	wh powder or grn orth cry	474 2.9	dec a	i H_2O, EtOH; s KCN soln
691	Copper(I) fluoride CuF	13478-41-6 82.544	cub cry	 7.1		
692	Copper(I) hydride CuH	13517-00-5 64.554	red-brn solid	dec 60		

No.	Name Formula	CAS RN Mol. Wt.	Physical Form	mp/°C den/g cm^{-3}	bp/°C Other Data	Solubility
693	Copper(I) iodide CuI	7681-65-4 190.450	wh cub cry	606 5.67	≈1290 a,f	i H_2O, dil acid
694	Copper(I) mercury iodide Cu_2HgI_4	13876-85-2 835.30	red cry powder	trans ≈60 (brn)		i H_2O, EtOH
695	Copper(I) oxide Cu_2O	1317-39-1 143.091	red-brn cub cry	1235 6.0	dec 1800 a,f	i H_2O
696	Copper(I) selenide Cu_2Se	20405-64-5 206.05	blue-blk tetr cry	1113 6.84		i H_2O; s acid
697	Copper(I) sulfide Cu_2S	22205-45-4 159.158	blue-blk orth cry	≈1100 5.6	a,f	i H_2O; sl acid
698	Copper(I) sulfite hemihydrate $Cu_2SO_3{\cdot}0.5H_2O$	13982-53-1* 216.164	wh-yel hex cry			sl H_2O; s acid, alk; i EtOH, eth
699	Copper(I) telluride Cu_2Te	12019-52-2 254.69	blue hex cry	1127 4.6		
700	Copper(I) thiocyanate CuSCN	1111-67-7 121.630	wh-yel amorp powder	1084 2.85		i H_2O, dil acid, EtOH, ace; s eth
701	Copper(I,II) sulfite dihydrate $Cu_2SO_3{\cdot}CuSO_3{\cdot}2H_2O$	13814-81-8 386.797	red prisms or powder			i H_2O, EtOH; s HCl
702	Copper(II) acetate $Cu(C_2H_3O_2)_2$	142-71-2 181.636	blue-grn hyg powder			
703	Copper(II) acetate metaarsenite $Cu(C_2H_3O_2)_2{\cdot}3Cu(AsO_2)_2$	12002-03-8 1013.796	grn cry powder			i H_2O; reac acid
704	Copper(II) acetate monohydrate $Cu(C_2H_3O_2)_2{\cdot}H_2O$	6046-93-1 199.651	grn monocl cry	115 1.88	dec 240	s H_2O, EtOH; sl eth
705	Copper(II) acetylide CuC_2	12540-13-5 87.568	brn-blk solid; exp	exp 100		
706	Copper(II) arsenate $Cu_3(AsO_4)_2$	10103-61-4 468.476	blue-grn cry			i H_2O, EtOH; s dil acid
707	Copper(II) arsenite $CuHAsO_3$	10290-12-7 187.474	yel-grn powder			i H_2O, EtOH; s acid
708	Copper(II) azide $Cu(N_3)_2$	14215-30-6 147.586	brn orth cry; exp	≈2.6		
709	Copper(II) basic acetate $Cu(C_2H_3O_2)_2{\cdot}CuO{\cdot}6H_2O$	52503-64-7 369.272	blue-grn cry or powder			sl H_2O, EtOH; s dil acid, NH_4OH
710	Copper(II) borate $Cu(BO_2)_2$	39290-85-2 149.166	blue-grn powder	3.859		i H_2O; s acid
711	Copper(II) bromide $CuBr_2$	7789-45-9 223.354	blk monocl cry; hyg	498 4.710	900 a	vs H_2O; s EtOH, ace; i bz, eth
712	Copper(II) butanoate monohydrate $Cu(C_4H_7O_2)_2{\cdot}H_2O$	540-16-9 255.758	grn monocl plates			s H_2O, diox, bz; sl EtOH
713	Copper(II) carbonate hydroxide $CuCO_3{\cdot}Cu(OH)_2$	12069-69-1 221.116	grn monocl cry	200 dec 4.0		i H_2O, EtOH; s dil acid
714	Copper(II) chlorate hexahydrate $Cu(ClO_3)_2{\cdot}6H_2O$	14721-21-2 338.539	blue-grn hyg cry	65	dec 100	vs H_2O, EtOH
715	Copper(II) chloride $CuCl_2$	7447-39-4 134.451	yel brn monocl cry; hyg	630 dec 3.4	a,c	s H_2O, EtOH, ace
716	Copper(II) chloride dihydrate $CuCl_2{\cdot}2H_2O$	10125-13-0 170.482	grn-blue orth cry; hyg	100 dec 2.51		vs H_2O, EtOH, MeOH; s ace; i eth
717	Copper(II) chromate $CuCrO_4$	13548-42-0 179.540	red-brn cry			i H_2O; s EtOH
718	Copper(II) chromite $CuCr_2O_4$	12018-10-9 231.536	gray-blk tetr cry	5.4		i H_2O, dil acid
719	Copper(II) citrate hemipentahydrate $Cu_2C_6H_4O_7{\cdot}2.5H_2O$	10402-15-0 360.223	blue-grn cry	dec 100		sl H_2O; s dil acid
720	Copper(II) cyanide $Cu(CN)_2$	14763-77-0 115.581	grn powder			i H_2O; s acid, alk
721	Copper(II) dichromate dihydrate $CuCr_2O_7{\cdot}2H_2O$	13675-47-3 315.565	red-brn tricl cry	2.286		vs H_2O
722	Copper(II) ferrocyanide $Cu_2Fe(CN)_6$	13601-13-3 339.043	red-br cub cry or powder	2.2		i H_2O, acid, os
723	Copper(II) ferrous sulfide $CuFeS_2$	1308-56-1 183.523	yel tetr cry	950 4.2	f	i H_2O, HCl; s HNO_3
724	Copper(II) fluoride CuF_2	7789-19-7 101.543	wh monocl cry	836 4.23	1676 a,c	sl H_2O
725	Copper(II) fluoride dihydrate $CuF_2{\cdot}2H_2O$	13454-88-1 137.574	blue monocl cry	dec 130 2.934		sl H_2O
726	Copper(II) formate $Cu(CHO_2)_2$	544-19-4 153.582	blue cry			s H_2O; i os
727	Copper(II) formate tetrahydrate $Cu(CHO_2)_2{\cdot}4H_2O$	5893-61-8 225.642	blue monocl cry			s H_2O; sl EtOH; i os
728	Copper(II) hexafluorosilicate tetrahydrate $CuSiF_6{\cdot}4H_2O$	12062-24-7 277.684	blue monocl cry	dec 2.56		vs H_2O; sl EtOH
729	Copper(II) hydroxide $Cu(OH)_2$	20427-59-2 97.561	blue-grn powder	3.37	a	i H_2O; s acid, conc alk
730	Copper(II) molybdate $CuMoO_4$	13767-34-5 223.48	grn cry	820 dec 3.4		i H_2O
731	Copper(II) nitrate $Cu(NO_3)_2$	3251-23-8 187.555	blue-grn orth cry; hyg	255	subl a	s H_2O, diox; reac eth

PHYSICAL CONSTANTS OF INORGANIC COMPOUNDS (continued)

No.	Name Formula	CAS RN Mol. Wt.	Physical Form	mp/°C den/g cm^{-3}	bp/°C Other Data	Solubility
732	Copper(II) nitrate hexahydrate $Cu(NO_3)_2 \cdot 6H_2O$	13478-38-1 295.647	blue rhomb cry; hyg	 2.07		vs H_2O; s EtOH
733	Copper(II) nitrate trihydrate $Cu(NO_3)_2 \cdot 3H_2O$	10031-43-3 241.602	blue rhomb cry	114 2.32	170 dec	vs H_2O, EtOH
734	Copper(II) oleate $Cu(C_{18}H_{33}O_2)_2$	1120-44-1 626.464	blue-grn solid			i H_2O; sl EtOH; s eth
735	Copper(II) oxalate CuC_2O_4	814-91-5 151.566	blue-wh powder	dec 310		i H_2O, EtOH, eth; s NH_4OH
736	Copper(II) oxide CuO	1317-38-0 79.545	blk powder or monocl cry	1446 6.31	 a,c	i H_2O, EtOH; s dil acid
737	Copper(II) perchlorate $Cu(ClO_4)_2$	13770-18-8 262.446	grn hyg cry	dec 130		s H_2O, eth, diox; i bz, ctc
738	Copper(II) perchlorate hexahydrate $Cu(ClO_4)_2 \cdot 6H_2O$	10294-46-9 370.538	blue monocl cry; hyg	82 2.22	dec 120	vs H_2O, EtOH, HOAc, ace; sl eth
739	Copper(II) phosphate trihydrate $Cu_3(PO_4)_2 \cdot 3H_2O$	10031-48-8 434.627	blue-grn orth cry			i H_2O; s acid, NH_4OH
740	Copper(II) selenate pentahydrate $CuSeO_4 \cdot 5H_2O$	10031-45-5 296.58	blue tricl cry	dec 80 2.56		s H_2O, acid, NH_4OH; sl ace; i EtOH
741	Copper(II) selenide CuSe	1317-41-5 142.51	blue-blk needles or plates	550 dec 5.99	 a	reac acid
742	Copper(II) selenite dihydrate $CuSeO_3 \cdot 2H_2O$	15168-20-4 226.54	blue orth cry	 3.31		i H_2O; s acid, NH_4OH
743	Copper(II) stearate $Cu(C_{18}H_{35}O_2)_2$	660-60-6 630.496	blue-grn amorp powder	≈250		i H_2O, EtOH, eth; s py
744	Copper(II) sulfate $CuSO_4$	7758-98-7 159.610	wh-grn amorp powder or rhomb cry	560 dec 3.60	 a	s H_2O, i EtOH
745	Copper(II) sulfate pentahydrate $CuSO_4 \cdot 5H_2O$	7758-99-8 249.686	blue tricl cry	dec 110 2.286	 f	vs H_2O; s MeOH; sl EtOH
746	Copper(II) sulfate, basic $Cu_3(OH)_4SO_4$	1332-14-5 200.126	grn rhomb cry	 3.88		i H_2O
747	Copper(II) sulfide CuS	1317-40-4 95.612	blk hex cry	trans 507 4.76	 a,f	i H_2O, EtOH, dil acid, alk
748	Copper(II) tartrate trihydrate $CuC_4H_4O_6 \cdot 3H_2O$	815-82-7 265.664	blue-grn powder			sl H_2O; s acid, alk
749	Copper(II) telluride CuTe	12019-23-7 191.15	yel orth cry	trans ≈400 7.09		
750	Copper(II) tungstate $CuWO_4$	13587-35-4 311.38	yel-brn powder	 7.5	 a	
751	Copper(II) tungstate dihydrate $CuWO_4 \cdot 2H_2O$	13587-35-4* 347.41	grn powder			i H_2O; sl HOAc; reac conc acid
752	Curium Cm	7440-51-9 247	silv metal; hex or cub	1345 13.51	≈3100 a	
753	Dysprosium Dy	7429-91-6 162.50	silv metal; hex	1411 8.55	2561 a,c	s dil acid
754	Dysprosium boride DyB_4	12310-43-9 205.74	tetr cry	2500 6.98		
755	Dysprosium nitride DyN	12019-88-4 176.51	cub cry	 9.93		
756	Dysprosium silicide $DySi_2$	12133-07-2 218.67	orth cry	 5.2		
757	Dysprosium(II) chloride $DyCl_2$	13767-31-2 233.41	blk cry	721 dec		reac H_2O
758	Dysprosium(II) iodide DyI_2	36377-94-3 416.31	purp cry	659		reac H_2O
759	Dysprosium(III) bromide $DyBr_3$	14456-48-5 402.21	wh hyg cry	879		s H_2O
760	Dysprosium(III) chloride $DyCl_3$	10025-74-8 268.86	yel cry	680 3.67	 a	s H_2O
761	Dysprosium(III) fluoride DyF_3	13569-80-7 219.50	grn cry	1154		
762	Dysprosium(III) hydride DyH_3	13537-09-2 165.52	hex cry	 7.1		
763	Dysprosium(III) iodide DyI_3	15474-63-2 543.21	grn cry	978		
764	Dysprosium(III) oxide Dy_2O_3	1308-87-8 373.00	wh cub cry	2408 7.81	 a	s acid
765	Dysprosium(III) sulfide Dy_2S_3	12133-10-7 421.20	red-brn monocl cry	 6.08		
766	Einsteinium Es	7429-92-7 252	metal; cub	860	 a	
767	Erbium Er	7440-52-0 167.26	soft gray metal; hex	1529 9.07	2862 a,c	i H_2O; s acid
768	Erbium boride ErB_4	12310-44-0 210.50	tetr cry	2450 7.0		
769	Erbium bromide $ErBr_3$	13536-73-7 406.97	viol hyg cry	923		s H_2O
770	Erbium chloride $ErCl_3$	10138-41-7 273.62	viol monocl cry; hyg	776 4.1	 a	s H_2O

No.	Name Formula	CAS RN Mol. Wt.	Physical Form	mp/°C den/g cm^{-3}	bp/°C Other Data	Solubility
771	Erbium chloride hexahydrate $ErCl_3 \cdot 6H_2O$	10025-75-9 381.71	pink hyg cry	dec		s H_2O; sl EtOH
772	Erbium fluoride ErF_3	13760-83-3 224.26	pink orth cry	1147 7.8	 a	i H_2O
773	Erbium hydride ErH_3	13550-53-3 170.28	hex cry	 ≈7.6		
774	Erbium iodide ErI_3	13813-42-8 547.97	viol hex cry; hyg	1014 ≈5.5		s H_2O
775	Erbium nitride ErN	12020-21-2 181.27	cub cry	 10.6		
776	Erbium oxide Er_2O_3	12061-16-4 382.52	pink powder	2418 8.64	 a	i H_2O; s acid
777	Erbium silicide $ErSi_2$	12020-28-9 223.43	orth cry	 7.26		
778	Erbium sulfate $Er_2(SO_4)_3$	13478-49-4 622.71	hyg powder	dec 3.68		s H_2O
779	Erbium sulfate octahydrate $Er_2(SO_4)_3 \cdot 8H_2O$	10031-52-4 766.83	pink monocl cry	dec 3.20		s H_2O
780	Erbium sulfide Er_2S_3	12159-66-9 430.72	red-brn monocl cry	1730 6.07		
781	Erbium telluride Er_2Te_3	12020-39-2 717.32	orth cry	1213 7.11		
782	Europium Eu	7440-53-1 151.965	soft silv metal; cub	822 5.24	1596 a,c	reac H_2O
783	Europium boride EuB_6	12008-05-8 216.831	cub cry	≈2600 4.91		
784	Europium nitride EuN	12020-58-5 165.972	cub cry	 8.7		
785	Europium silicide $EuSi_2$	12434-24-1 208.136	tetr cry	1500 5.46		
786	Europium(II) bromide $EuBr_2$	13780-48-8 311.773	wh cry	683		s H_2O
787	Europium(II) chloride $EuCl_2$	13769-20-5 222.870	wh orth cry	731 4.9		s H_2O
788	Europium(II) fluoride EuF_2	14077-39-5 189.962	grn-yel cub cry	≈1380 6.5		
789	Europium(II) iodide EuI_2	22015-35-6 405.774	grn cry	580		s H_2O
790	Europium(II) selenide EuSe	12020-66-5 230.93	brn cub cry	 6.45		
791	Europium(II) sulfide EuS	12020-65-4 184.031	cub cry	 5.7		
792	Europium(II) telluride EuTe	12020-69-8 279.57	blk cub cry	1526 6.48		
793	Europium(III) bromide $EuBr_3$	13759-88-1 391.677	gray cry	dec		s H_2O
794	Europium(III) chloride $EuCl_3$	10025-76-0 258.323	grn-yel needles	623 4.89	 a	
795	Europium(III) chloride hexahydrate $EuCl_3 \cdot 6H_2O$	13759-92-7 366.414	wh-yel hyg cry	850 4.89		s H_2O
796	Europium(III) fluoride EuF_3	13765-25-8 208.960	wh hyg cry	1276		i H_2O
797	Europium(III) nitrate hexahydrate $Eu(NO_3)_3 \cdot 6H_2O$	10031-53-5 446.071	wh-pink hyg cry	dec 85		s H_2O
798	Europium(III) oxide Eu_2O_3	1308-96-9 351.928	pink powder	2350 7.42	 a	i H_2O; s acid
799	Europium(III) sulfate $Eu_2(SO_4)_3$	13537-15-0 592.121	pale pink cry	 4.99		sl H_2O
800	Europium(III) sulfate octahydrate $Eu_2(SO_4)_3 \cdot 8H_2O$	10031-52-4 736.243	pink cry	dec 375		sl H_2O
801	Fermium Fm	7440-72-4 257	metal	1527	 a	
802	Fluorine F_2	7782-41-4 37.997	pale yel gas	-219.66 1.667 g/L	-188.12 a,b,c,d,e	reac H_2O
803	Fluorine monoxide F_2O	7783-41-7 53.996	col gas	-223.8 2.369 g/L	-144.75 a,b,d,e	sl H_2O
804	Fluorine dioxide F_2O_2	7783-44-0 69.996	gas, stable only at low temp	-154 3.071 g/L	-57 a,d	
805	Fluorine nitrate FNO_3	7789-26-6 81.003	col gas	-175 3.554 g/L	-46	reac H_2O, EtOH, eth; s ace
806	Fluorine perchlorate $FOClO_3$	10049-03-3 118.449	col gas; exp	-167.3 5.197 g/L	-16	reac H_2O
807	Francium Fr	7440-73-5 223	short-lived alkali metal	27	 a	
808	Gadolinium Gd	7440-54-2 157.25	col or yel metal; hex	1314 7.90	3264 a,c	s dil acid
809	Gadolinium boride GdB_6	12008-06-9 222.12	blk-brn cub cry	2510 5.31		

No.	Name Formula	CAS RN Mol. Wt.	Physical Form	mp/°C den/g cm^{-3}	bp/°C Other Data	Solubility
810	Gadolinium nitride GdN	25764-15-2 171.26	cub cry	 9.10		
811	Gadolinium silicide $GdSi_2$	12134-75-7 213.42	orth cry	 5.9		
812	Gadolinium(II) iodide GdI_2	13814-72-7 411.06	bronze cry	831		
813	Gadolinium(II) selenide $GdSe$	12024-81-6 236.21	cub cry	2170 8.1		
814	Gadolinium(III) bromide $GdBr_3$	13818-75-2 396.96	wh monocl cry; hyg	770 4.56		
815	Gadolinium(III) chloride $GdCl_3$	10138-52-0 263.61	wh monocl cry; hyg	609 4.52	 a	s H_2O
816	Gadolinium(III) chloride hexahydrate $GdCl_3 \cdot 6H_2O$	19423-81-5 371.70	col hyg cry	 2.424		s H_2O
817	Gadolinium(III) fluoride GdF_3	13765-26-9 214.25	wh cry	1231	 a	
818	Gadolinium(III) iodide GdI_3	13572-98-0 537.96	yel cry	925		
819	Gadolinium(III) nitrate hexahydrate $Gd(NO_3)_3 \cdot 6H_2O$	19598-90-4 451.36	hyg tricl cry	91 dec 2.33		s H_2O, EtOH
820	Gadolinium(III) nitrate pentahydrate $Gd(NO_3)_3 \cdot 5H_2O$	52788-53-1 433.34	wh cry	92 dec 2.41		i H_2O
821	Gadolinium(III) oxide Gd_2O_3	12064-62-9 362.50	wh hyg powder	2420 7.07	 a	i H_2O; s acid
822	Gadolinium(III) sulfate octahydrate $Gd_2(SO_4)_3 \cdot 8H_2O$	13450-87-8 746.81	col monocl cry	dec 400 4.14		s H_2O
823	Gadolinium(III) sulfide Gd_2S_3	12134-77-9 410.70	yel cub cry	 6.1		
824	Gadolinium(III) telluride Gd_2Te_3	12160-99-5 697.30	orth cry	1255 7.7		
825	Gallium Ga	7440-55-3 69.723	silv liq or gray orth cry	29.76 5.91	2204 a,c,d,e	reac alk
826	Gallium antimonide $GaSb$	12064-03-8 191.483	cub cry	712 5.6137	 a,c	
827	Gallium arsenide $GaAs$	1303-00-0 144.645	gray cub cry	1238 5.3176	 a	
828	Gallium nitride GaN	25617-97-4 83.730	gray hex cry	600 6.1	dec >600 a	
829	Gallium phosphide GaP	12063-98-8 100.697	yel cub cry	1457 4.138	 a	
830	Gallium suboxide Ga_2O	12024-20-3 155.445	brn powder	>660 4.77	dec >800 a	
831	Gallium(II) chloride $GaCl_2$	24597-12-4 140.628	wh orth cry	172.4 2.74	535	
832	Gallium(II) selenide $GaSe$	12024-11-2 148.68	hex cry	960 5.03		
833	Gallium(II) sulfide GaS	12024-10-1 101.789	hex cry	965 3.86		
834	Gallium(II) telluride $GaTe$	12024-14-5 197.32	monocl cry	824 5.44		
835	Gallium(III) acetylacetonate $Ga(CH_3COCHCOCH_3)_3$	14405-43-7 367.051	wh powder	193 1.42	subl	
836	Gallium(III) bromide $GaBr_3$	13450-88-9 309.435	wh orth cry	121.5 3.69	279 a,b,c,d	
837	Gallium(III) chloride $GaCl_3$	13450-90-3 176.081	col needles or gl solid	77.9 2.47	201 a,b,c,d,e	
838	Gallium(III) fluoride GaF_3	7783-51-9 126.718	wh powder or col needles	>1000 4.47	 a	i H_2O
839	Gallium(III) fluoride trihydrate $GaF_3 \cdot 3H_2O$	22886-66-4 180.764	wh cry	dec >140		sl H_2O
840	Gallium(III) hydride GaH_3	13572-93-5 72.747	visc liq	-15	dec ≈0	
841	Gallium(III) hydroxide $Ga(OH)_3$	12023-99-3 120.745	unstable prec		 a	
842	Gallium(III) iodide GaI_3	13450-91-4 450.436	monocl cry	210 4.5	340 a,b,c,d	
843	Gallium(III) nitrate $Ga(NO_3)_3$	13494-90-1 255.738	wh cry powder			s H_2O, EtOH, eth
844	Gallium(III) oxide Ga_2O_3	12024-21-4 187.444	wh cry	1725 ≈6.0	 a	s hot acid
845	Gallium(III) oxide hydroxide $GaOOH$	20665-52-5 102.730	orth cry	 5.23		
846	Gallium(III) selenide Ga_2Se_3	12024-24-7 376.33	cub cry	937 4.92		
847	Gallium(III) sulfate $Ga_2(SO_4)_3$	13494-91-2 427.637	hex cry			
848	Gallium(III) sulfate octadecahydrate $Ga_2(SO_4)_3 \cdot 18H_2O$	13780-42-2 751.912	octahed cry			s H_2O, EtOH

No.	Name Formula	CAS RN Mol. Wt.	Physical Form	mp/°C den/g cm^{-3}	bp/°C Other Data	Solubility
849	Gallium(III) sulfide Ga_2S_3	12024-22-5 235.644	monocl cry	1090 3.7		
850	Gallium(III) telluride Ga_2Te_3	12024-27-0 522.25	cub cry	790 5.57		
851	Germanium Ge	7440-56-4 72.61	gray-wh cub cry	938.25 5.3234	2833 a,c,d	i H_2O, dil acid, alk
852	Germanium nitride Ge_3N_4	12065-36-0 273.86	orth cry	dec 900		i H_2O, acid, aqua regia
853	Germane GeH_4	7782-65-2 76.64	col gas; flam	-165 3.363 g/L	-88.1 a,b,d,e	i H_2O
854	Bromogermane GeH_3Br	13569-43-2 155.54	col liq	-32 2.34	52 a	reac H_2O
855	Chlorogermane GeH_3Cl	13637-65-5 111.09	col liq	-52 1.75	28 a	reac H_2O
856	Chlorotrifluorogermane GeF_3Cl	14188-40-0 165.06	gas	-66.2 7.243 g/L	-20.3	
857	Dibromogermane GeH_2Br_2	13769-36-3 234.43	col liq	-15 2.80	89	reac H_2O
858	Dichlorogermane GeH_2Cl_2	15230-48-5 145.53	col liq	-68 1.90	69.5	reac H_2O
859	Dichlorodifluorogermane GeF_2Cl_2	24422-21-7 181.51	col gas	-51.8 7.965 g/L	-2.8	
860	Fluorogermane GeH_3F	13537-30-9 94.63	col gas	 4.152 g/L	 a	reac H_2O
861	Iodogermane GeH_3I	13573-02-9 202.54	liq	-15	≈90 a	reac H_2O
862	Tribromogermane $GeHBr_3$	14779-70-5 313.33	col liq	-25	dec	reac H_2O
863	Trichlorogermane $GeHCl_3$	1184-65-2 179.98	liq	-71 1.93	75.3 e	reac H_2O
864	Trichlorofluorogermane $GeCl_3F$	24422-20-6 197.97	liq	-49.8	37.5	
865	Methylgermane GeH_3CH_3	1449-65-6 90.67	col gas	-158 3.979 g/L	-23	
866	Digermane Ge_2H_6	13818-89-8 151.27	col liq; flam	-109	30.8 a,d,e	
867	Trigermane Ge_3H_8	14691-44-2 225.89	col liq	-105.6 2.20(-105°C)	110.5 a,d,e	i H_2O
868	Germanium(II) bromide $GeBr_2$	24415-00-7 232.42	yel monocl cry	122	dec 150	reac H_2O
869	Germanium(II) chloride $GeCl_2$	10060-11-4 143.51	wh-yel hyg powder	dec		reac H_2O; s eth, bz
870	Germanium(II) fluoride GeF_2	13940-63-1 110.61	wh orth cry; hyg	110 3.64	dec 130	reac H_2O
871	Germanium(II) iodide GeI_2	13573-08-5 326.42	oran-yel hex cry	dec 550 5.4		reac H_2O
872	Germanium(II) oxide GeO	20619-16-3 88.61	blk solid	dec 700	 a	
873	Germanium(II) selenide GeSe	12065-10-0 151.57	gray orth cry or brn powder	667 5.6		
874	Germanium(II) sulfide GeS	12025-32-0 104.68	gray orth cry	615 4.1	 a	
875	Germanium(II) telluride GeTe	12025-39-7 200.21	cub cry	725 6.16		i H_2O; s conc HNO_3
876	Germanium(IV) bromide $GeBr_4$	13450-92-5 392.23	wh cry	26.1 3.132	186.35 a,b,d,e	reac H_2O
877	Germanium(IV) chloride $GeCl_4$	10038-98-9 214.42	col liq	-49.5 1.88	86.55 a,b,d,e	reac H_2O; s bz, eth, EtOH, ctc
878	Germanium(IV) fluoride GeF_4	7783-58-6 148.60	col gas	-15 tp 6.521 g/L	-36.5 sp a	reac H_2O
879	Germanium(IV) iodide GeI_4	13450-95-8 580.23	red-oran cub cry	144 4.4	377 a,b	reac H_2O
880	Germanium(IV) oxide GeO_2	1310-53-8 104.61	wh hex cry	1115 4.25	 a	i H_2O
881	Germanium(IV) selenide $GeSe_2$	12065-11-1 230.53	yel-oran orth ccry	707 dec 4.56		
882	Germanium(IV) sulfide GeS_2	12025-34-2 136.74	blk orth cry	530 3.01		
883	Gold Au	7440-57-5 196.967	soft yel metal	1064.18 19.3	2856 a,c,d,e,f	s aqua regia
884	Bromoauric acid pentahydrate $HAuBr_4 \cdot 5H_2O$	17083-68-0 607.667	red-brn hyg cry	27		s H_2O, EtOH
885	Chloroauric acid tetrahydrate $HAuCl_4 \cdot 4H_2O$	16903-35-8 411.847	yel monocl cry; hyg	 ≈3.9		vs H_2O, EtOH; s eth
886	Gold(I) bromide AuBr	10294-27-6 276.871	yel-gray tetr cry	dec 165 8.20	 a	i H_2O
887	Gold(I) chloride AuCl	10294-29-8 232.420	yel orth cry	dec 289 7.6	 a	i H_2O

No.	Name	Formula	CAS RN	Mol. Wt.	Physical Form	mp/°C	den/g cm^{-3}	bp/°C	Other Data	Solubility
888	Gold(I) cyanide	$AuCN$	506-65-0	222.985	yel hex cry	dec	7.2			i H_2O, EtOH, eth, dil acid
889	Gold(I) iodide	AuI	10294-31-2	323.871	yel-grn powder; tetr	dec 120	8.25		a	i H_2O; s CN soln
890	Gold(I) sulfide	Au_2S	1303-60-2	425.999	brn-blk cub cry; unstable	240 dec	≈11			i H_2O, acid; s aqua regia
891	Gold(III) bromide	$AuBr_3$	10294-28-7	436.679	red-br monocl cry	dec ≈160			a	s H_2O, EtOH
892	Gold(III) chloride	$AuCl_3$	13453-07-1	303.325	red monocl cry	dec >160	4.7		a	
893	Gold(III) cyanide trihydrate	$Au(CN)_3{\cdot}3H_2O$	535-37-5*	329.066	wh hyg cry	dec 50				vs H_2O; sl EtOH
894	Gold(III) fluoride	AuF_3	14720-21-9	253.962	oran-yel hex cry	>300	6.75	subl	a	
895	Gold(III) hydroxide	$Au(OH)_3$	1303-52-2	247.989	brn powder	dec ≈100				i H_2O; s acid
896	Gold(III) iodide	AuI_3	31032-13-0	577.680	unstable grn powder					
897	Gold(III) oxide	Au_2O_3	1303-58-8	441.931	brn powder	dec ≈150				i H_2O; s acid
898	Gold(III) selenate	$Au_2(SeO_4)_3$	10294-32-3	822.81	yel cry					i H_2O; s acid
899	Gold(III) selenide	Au_2Se_3	1303-62-4	630.81	blk amorp solid	dec	4.65			s aqua regia
900	Gold(III) sulfide	Au_2S_3	1303-61-3	490.131	unstable blk powder	dec 200				
901	Hafnium	Hf	7440-58-6	178.49	gray metal; hex	2233	13.3	4603	a,c	s HF
902	Hafnium boride	HfB_2	12007-23-7	200.11	gray hex cry	3100	10.5			
903	Hafnium bromide	$HfBr_4$	13777-22-5	498.11	wh cub cry	424 tp	4.90	323 sp	b	
904	Hafnium carbide	HfC	12069-85-1	190.50	refrac cub cry	≈3000	12.2			
905	Hafnium chloride	$HfCl_4$	13499-05-3	320.30	wh monocl cry	432 tp		317 sp	a,b	reac H_2O
906	Hafnium fluoride	HfF_4	13709-52-9	254.48	wh monocl cry	>970	7.1	970 sp	a	
907	Hafnium hydride	HfH_2	12770-26-2	180.51	refrac tetr cry		11.4			
908	Hafnium iodide	HfI_4	13777-23-6	686.11	yel-oran cub cry	449 tp	5.6	394 sp	b	
909	Hafnium nitride	HfN	25817-87-2	192.50	yel-brn cub cry	3305	13.8			
910	Hafnium oxide	HfO_2	12055-23-1	210.49	wh cub cry	2774	9.68		a	i H_2O
911	Hafnium oxychloride octahydrate	$HfOCl_2{\cdot}8H_2O$	14456-34-9	409.52	wh tetr cry	dec				s H_2O
912	Hafnium phosphide	HfP	12325-59-6	209.46	hex cry		9.78			
913	Hafnium selenide	$HfSe_2$	12162-21-9	336.41	brn hex cry		7.46			
914	Hafnium orthosilicate	$HfSiO_4$	13870-13-8	270.57	tetr cry		7.0			
915	Hafnium silicide	$HfSi_2$	12401-56-8	234.66	gray orth cry	≈1700	7.6			
916	Hafnium sulfate	$Hf(SO_4)_2$	15823-43-5	370.62	wh cry	dec >500				
917	Hafnium sulfide	HfS_2	18855-94-2	242.62	purp-brn hex cry		6.03			
918	Helium	He	7440-59-7	4.003	col gas		0.176 g/L	-268.93	a,b,d,e	sl H_2O; i EtOH
919	Holmium	Ho	7440-60-0	164.930	metal; hex	1472	8.80	2694	a,c	s dil acid
920	Holmium bromide	$HoBr_3$	13825-76-8	404.642	yel hyg cry	919		1470		
921	Holmium chloride	$HoCl_3$	10138-62-2	271.288	yel monocl cry; hyg	718	3.7	1500	a	s H_2O
922	Holmium fluoride	HoF_3	13760-78-6	221.925	pink-yel orth cry; hyg	1143	7.664	>2200	a	s H_2O
923	Holmium iodide	HoI_3	13813-41-7	545.643	yel hex cry	994	5.4			
924	Holmium nitride	HoN	12029-81-1	178.937	cub cry		10.6			
925	Holmium oxide	Ho_2O_3	12055-62-8	377.859	yel cub cry	2415	8.41		a	s acid
926	Holmium silicide	$HoSi_2$	12136-24-2	221.10	hex cry		7.1			

No.	Name	Formula	CAS RN	Mol. Wt.	Physical Form	mp/°C	den/g cm^{-3}	bp/°C	Other Data	Solubility
927	Holmium sulfide	Ho_2S_3	12162-59-3	426.059	yel-oran monocl cry		5.92			
928	Hydrazine	N_2H_4	302-01-2	32.045	col oily liq	1.4	1.0036	113.55	a,b,c,d	vs H_2O, EtOH, MeOH
929	Hydrazine hydrate	$N_2H_4 \cdot H_2O$	7803-57-8	50.060	fuming liq	-51.7	1.030	119		vs H_2O, EtOH; i chl, eth
930	Hydrazine hydrobromide	$N_2H_4 \cdot HBr$	13775-80-9	112.957	wh monocl cry flakes	84	2.3	dec ≈190		s H_2O, EtOH
931	Hydrazine hydrochloride	$N_2H_4 \cdot HCl$	2644-70-4	68.506	wh orth cry	89	1.5	dec 240		s H_2O; i os
932	Hydrazine dihydrochloride	$N_2H_4 \cdot 2HCl$	5341-61-7	104.966	wh orth cry	198 dec	1.42			s H_2O; sl EtOH
933	Hydrazine hydroiodide	$N_2H_4 \cdot HI$	10039-55-1	159.957	hyg cry	125				s H_2O
934	Hydrazine nitrate	$N_2H_4 \cdot HNO_3$	13464-97-6	95.058	monocl cry; exp	70				vs H_2O
935	Hydrazine sulfate	$N_2H_4 \cdot H_2SO_4$	10034-93-2	130.125	col orth cry	254	1.378			sl H_2O; i EtOH
936	Hydrazoic acid	HN_3	7782-79-8	43.028	col liq; exp	-80		35.7	a,d	s H_2O
937	Hydroxylamine	H_2NOH	7803-49-8	33.030	wh orth flakes or needles	33.1	1.21	58	a	vs H_2O, MeOH
938	Hydroxylamine sulfate	$(H_2NOH)_2 \cdot H_2SO_4$	10039-54-0	164.139	cry	170				vs H_2O
939	Hydrogen	H_2	1333-74-0	2.016	col gas; flam	-259.34	0.088 g/L	-252.87	a,b,c,d,e	sl H_2O
940	Hydrogen bromide	HBr	10035-10-6	80.912	col gas	-86.81	3.550 g/L	-66.38	a,b,c,d,e	vs H_2O; s EtOH
941	Hydrogen chloride	HCl	7647-01-0	36.461	col gas	-114.18	1.600 g/L	-85	a,b,c,d,e	vs H_2O
942	Hydrogen cyanide	HCN	74-90-8	27.026	col liq	13.1	0.684	26	a,b,c,e	vs H_2O, EtOH; sl eth
943	Hydrogen fluoride	HF	7664-39-3	20.006	col gas	-83.36	0.878 g/L	20	a,b,c,e	vs H_2O, EtOH; sl eth
944	Hydrogen iodide	HI	10034-85-2	127.912	col or yel gas	-50.77	5.613 g/L	-35.55	a,b,c,d,e	vs H_2O; s os
945	Hydrogen peroxide	H_2O_2	7722-84-1	34.015	col liq	-0.43	1.44	150.2	a,c,d,e	vs H_2O
946	Hydrogen selenide	H_2Se	7783-07-5	80.98	col gas; flam	-65.73	3.553 g/L	-41.25	a,b,d,e	s H_2O
947	Hydrogen sulfide	H_2S	7783-06-4	34.082	col gas; flam	-85.5	1.496 g/L	-59.55	a,b,c,d,e	s H_2O
948	Hydrogen disulfide	H_2S_2	13465-07-1	66.148	col liq		1.334	70.7	a,d,e	
949	Hydrogen telluride	H_2Te	7783-09-7	129.62	col gas	-49	5.687 g/L	-2	a,d,e	s H_2O, EtOH, alk
950	Indium	In	7440-74-6	114.818	soft wh metal	156.60	7.31	2072	a,c	s acid
951	Indium antimonide	$InSb$	1312-41-0	236.578	blk cub cry	525	5.7747		a,c	
952	Indium arsenide	$InAs$	1303-11-3	189.740	gray cub cry	942	5.67		a	i acids
953	Indium nitride	InN	25617-98-5	128.825	hex cry	1100	6.88			
954	Indium phosphide	InP	22398-80-7	145.792	blk cub cry	1062	4.81		a	sl acid
955	Indium(I) bromide	$InBr$	14280-53-6	194.722	oran-red orth cry	290	4.96	656	a,d	reac H_2O
956	Indium(I) chloride	$InCl$	13465-10-6	150.271	yel cub cry	225	4.19	608	a,c	reac H_2O
957	Indium(I) iodide	InI	13966-94-4	241.722	orth cry	351	5.32	712	a,d	
958	Indium(II) bromide	$InBr_2$	21264-43-7	274.626	orth cry					reac H_2O
959	Indium(II) chloride	$InCl_2$	13465-11-7	185.723	col orth cry	235	3.64			reac H_2O
960	Indium(II) sulfide	InS	12030-14-7	146.884	red-brn orth cry	692	5.2		a	
961	Indium(III) bromide	$InBr_3$	13465-09-3	354.530	hyg yel-wh monocl cry	≈436	4.74		a	
962	Indium(III) chloride	$InCl_3$	10025-82-8	221.176	yel monocl cry; hyg	586	4.0		a	vs H_2O; s EtOH
963	Indium(III) fluoride	InF_3	7783-52-0	171.813	wh hex cry; hyg	1170	4.39	>1200		sl H_2O; s dil acid
964	Indium(III) fluoride trihydrate	$InF_3 \cdot 3H_2O$	14166-78-0	225.859	wh cry	dec 100				s H_2O
965	Indium(III) hydroxide	$In(OH)_3$	20661-21-6	165.840	cub cry		4.4			

No.	Name Formula	CAS RN Mol. Wt.	Physical Form	mp/°C den/g cm^{-3}	bp/°C Other Data	Solubility
966	Indium(III) iodide InI_3	13510-35-5 495.531	yel-red monocl cry; hyg	210 4.69	 a	
967	Indium(III) oxide In_2O_3	1312-43-2 277.634	yel cub cry	≈2000 7.18	 a	i H_2O; s hot acid
968	Indium(III) perchlorate octahydrate $In(ClO_4)_3 \cdot 8H_2O$	13465-15-1 557.291	wh cry	≈80	dec 200	
969	Indium(III) phosphate $InPO_4$	14693-82-4 209.789	wh orth cry	 4.9		i H_2O
970	Indium(III) selenide In_2Se_3	1312-42-1 466.52	blk hex cry	660 5.8		
971	Indium(III) sulfate $In_2(SO_4)_3$	13464-82-9 517.827	hyg wh powder	 3.44		s H_2O
972	Indium(III) sulfide In_2S_3	12030-24-9 325.834	oran cub cry	1050 4.45	 a	
973	Indium(III) telluride In_2Te_3	1312-45-4 612.44	blk cub cry	667 5.75		
974	Iodine I_2	7553-56-2 253.809	blue-blk plates	113.7 4.933	184.4 a,b,c,d,e	sl H_2Os; s bz, EtOH, eth, ctc, chl
975	Iodic acid HIO_3	7782-68-5 175.910	col orth cry	110 dec 4.63	 a	vs H_2O; i EtOH, eth
976	Periodic acid dihydrate $HIO_4 \cdot 2H_2O$	10450-60-9 227.940	monocl hyg cry	122 dec		s H_2O, EtOH; sl eth
977	Iodine tetroxide I_2O_4	12399-08-5 317.807	yel cry	dec 85 4.2		sl H_2O
978	Iodine pentoxide I_2O_5	12029-98-0 333.806	hyg wh cry	≈300 dec 4.98		s H_2O; i EtOH, eth, CS_2
979	Iodine nonaoxide I_4O_9	73560-00-6 651.613	hyg yel powder	dec 75		
980	Iodine bromide IBr	7789-33-5 206.808	blk orth cry	40 4.3	116 dec a,b	s H_2O, EtOH, eth
981	Iodine chloride ICl	7790-99-0 162.357	red cry or oily liq	27.38 3.24	100 dec a,c	reac H_2O; s EtOH
982	Iodine trichloride ICl_3	865-44-1 233.262	yel tricl cry; hyg	101 tp (16 atm) 3.2	64 sp (dec)	reac H_2O; s EtOH, bz
983	Iodine fluoride IF	13873-84-2 145.902	disproportionates at room temp		 a	
984	Iodine trifluoride IF_3	22520-96-3 183.899	yel solid, stable at low temp	dec -28		
985	Iodine pentafluoride IF_5	7783-66-6 221.896	yel liq	9.43 3.19	100.5 a,d,e	reac H_2O
986	Iodine heptafluoride IF_7	16921-96-3 259.893	col liq	6.5 tp 2.67	4.8 sp e	s H_2O
987	Iridium Ir	7439-88-5 192.217	silv-wh metal; cub	2446 22.5	4428 a,c	s aqua regia
988	Iridium(III) sulfide Ir_2S_3	12136-42-4 480.632	orth cry	 10.2	 a	
989	Iridium(III) bromide $IrBr_3$	10049-24-8 431.929	red-brn monocl cry	 6.82		i H_2O, acid, alk
990	Iridium(III) bromide tetrahydrate $IrBr_3 \cdot 4H_2O$	10049-24-8* 503.991	grn-brn cry			s H_2O; i EtOH
991	Iridium(III) chloride $IrCl_3$	10025-83-9 298.575	brn monocl cry	dec 763 5.30	 a	i H_2O, acid, alk
992	Iridium(III) fluoride IrF_3	23370-59-4 249.212	blk hex cry	dec 250 ≈8.0		i H_2O, dil acid
993	Iridium(III) iodide IrI_3	7790-41-2 572.930	dark brn monocl cry	 ≈7.4		i H_2O, acid, bz, chl; s alk
994	Iridium(III) oxide Ir_2O_3	1312-46-5 432.432	blue-blk cry	dec 1000		i H_2O; sl hot HCl
995	Iridium(IV) oxide IrO_2	12030-49-8 224.216	brn tetr cry	1100 dec 11.7	 a	
996	Iridium(IV) sulfide IrS_2	12030-51-2 256.349	orth cry	 9.3	 a	
997	Iridium(VI) fluoride IrF_6	7783-75-7 306.207	yel cub cry; hyg	44 4.8	53 a,c,d	reac H_2O
998	Iron Fe	7439-89-6 55.845	silv-wh or gray met	1538 7.87	2861 a,c,e	s dil acid
999	Iron arsenide FeAs	12044-16-5 130.767	gray orth cry	1030 7.85		
1000	Iron boride Fe_2B	12006-86-9 122.501	refr solid; tetr	1389 7.3		
1001	Iron boride FeB	12006-84-7 66.656	refr solid; orth	1650 ≈7		
1002	Iron pentacarbonyl $Fe(CO)_5$	13463-40-6 195.897	yel oily liq; flam	-20 1.490	103 a,e	i H_2O; s eth, bz, ace
1003	Iron nonacarbonyl $Fe_2(CO)_9$	15321-51-4 363.784	oran-yel cry	100 dec 2.85		
1004	Iron dodecacarbonyl $Fe_3(CO)_{12}$	12088-65-2 503.660	blk cry	140 2.00		

No.	Name Formula	CAS RN Mol. Wt.	Physical Form	mp/°C den/g cm^{-3}	bp/°C Other Data	Solubility
1005	Iron hydrocarbonyl $H_2Fe(CO)_4$	17440-90-3 169.903	col liq; unstable	-70	dec	s alk
1006	Iron phosphide Fe_2P	1310-43-6 142.664	gray hex needles	1370 6.8		i H_2O,dil acid, alk
1007	Iron disulfide FeS_2	1317-66-4 119.977	blk cub cry	dec >600 5.02	 a,f	i H_2O
1008	Iron silicide FeSi	12022-95-6 83.931	gray cub cry	1410 6.1		
1009	Iron silicide $FeSi_2$	12022-99-0 112.016	gray tetr cry	1220 4.74		
1010	Iron(II) aluminate $Fe(AlO_2)_2$	12068-49-4 173.806	blk cub cry	 4.3		
1011	Iron(II) bromide $FeBr_2$	7789-46-0 215.653	yel-brn hex cry; hyg	691 4.636	dec a,c	vs H_2O, EtOH
1012	Iron(II) bromide hexahydrate $FeBr_2 \cdot 6H_2O$	13463-12-2 323.744	grn hyg cry	27 dec 4.64		s H_2O, EtOH
1013	Iron(II) carbonate $FeCO_3$	563-71-3 115.854	gray-brn hex cry	 3.9	 f	
1014	Iron(II) chloride $FeCl_2$	7758-94-3 126.750	wh hex cry; hyg	677 3.16	1023 a,c,e	vs H_2O, EtOH, ace; sl bz
1015	Iron(II) chloride dihydrate $FeCl_2 \cdot 2H_2O$	16399-77-2 162.781	wh-grn monocl cry	dec 120 2.39		s H_2O
1016	Iron(II) chloride tetrahydrate $FeCl_2 \cdot 4H_2O$	13478-10-9 198.812	grn monocl cry	dec 105 1.93		s H_2O, EtOH
1017	Iron(II) chromite $FeCr_2O_4$	1308-31-2 223.835	blk cub cry	 5.0	 f	
1018	Iron(II) fluoride FeF_2	7789-28-8 93.842	wh tetr cry	1100 4.09	 a,c	sl H_2O; s dil HF; i EtOH, eth
1019	Iron(II) fluoride tetrahydrate $FeF_2 \cdot 4H_2O$	13940-89-1 165.904	col hex cry	 2.20		
1020	Iron(II) hydroxide $Fe(OH)_2$	18624-44-7 89.860	wh-grn hex cry	 3.4		i H_2O
1021	Iron(II) iodide FeI_2	7783-86-0 309.654	red-viol hex cry; hyg	587 5.3	 a,c	s H_2O, EtOH, eth
1022	Iron(II) iodide tetrahydrate $FeI_2 \cdot 4H_2O$	7783-86-0* 381.716	blk hyg leaflets	dec 90 2.87		s H_2O, EtOH
1023	Iron(II) molybdate $FeMoO_4$	13718-70-2 215.78	brn-yel monocl cry	1115 5.6	 a	i H_2O
1024	Iron(II) nitrate $Fe(NO_3)_2$	14013-86-6 179.854	grn solid			s H_2O
1025	Iron(II) nitrate hexahydrate $Fe(NO_3)_2 \cdot 6H_2O$	13476-08-9 287.946	grn solid	60 dec		vs H_2O
1026	Iron(II) oxalate dihydrate $FeC_2O_4 \cdot 2H_2O$	6047-25-2 179.895	yel cry	dec 150 2.28		sl H_2O; s acid
1027	Iron(II) oxide FeO	1345-25-1 71.844	blk cub cry	1377 6.0	 a,c	i H_2O, alk; s acid
1028	Iron(II) phosphate octahydrate $Fe_3(PO_4)_2 \cdot 8H_2O$	14940-41-1 501.600	gray-blue monocl cry; hyg	 2.58	 f	i H_2O; s acid
1029	Iron(II) selenide FeSe	1310-32-3 134.81	blk hex cry	 6.7		i H_2O
1030	Iron(II) orthosilicate Fe_2SiO_4	10179-73-4 203.774	brn orth cry	 4.30	 a,f	
1031	Iron(II) sulfate $FeSO_4$	7720-78-7 151.909	wh orth cry; hyg	 3.65	 a	s H_2O
1032	Iron(II) sulfate monohydrate $FeSO_4 \cdot H_2O$	17375-41-6 169.924	wh-yel monocl cry	dec 300 3.0		s H_2O
1033	Iron(II) sulfate heptahydrate $FeSO_4 \cdot 7H_2O$	7782-63-0 278.015	blue-grn monocl cry	dec ≈60 1.895	 f	vs H_2O; i EtOH
1034	Iron(II) sulfide FeS	1317-37-9 87.911	col hex or tetr cry; hyg	1188 4.7	dec a,c	i H_2O; reac acid
1035	Iron(II) telluride FeTe	12125-63-2 183.45	tetr cry	914 6.8		
1036	Iron(II) thiocyanate trihydrate $Fe(SCN)_2 \cdot 3H_2O$	6010-09-9 226.058	grn monocl cry			s H_2O, EtOH, eth
1037	Iron(II) titanate $FeTiO_3$	12168-52-4 151.710	blk rhomb cry	≈1470 4.72	 f	
1038	Iron(II) tungstate $FeWO_4$	13870-24-1 303.68	monocl cry	 7.51	 a,f	
1039	Iron(II,III) oxide Fe_3O_4	1317-61-9 231.533	blk cub cry or amorp powder	1597 5.17	 a,c,f	i H_2O; s acid
1040	Iron(III) acetate, basic $FeOH(C_2H_3O_2)_2$	10450-55-2 190.942	brn-red amorp powder			i H_2O; s EtOH, acid
1041	Iron(III) acetylacetonate $Fe(OC(CH_3){=}CHC(O)CH_3)_3$	14024-18-1 353.173	red-oran cry	179 5.24		sl H_2O; s os
1042	Iron(III) arsenate dihydrate $FeAsO_4 \cdot 2H_2O$	10102-49-5 230.795	grn-brn powder	dec 3.18		i H_2O; s dil acid
1043	Iron(III) bromide $FeBr_3$	10031-26-2 295.557	dark red hex cry; hyg	dec 4.5	 a	s H_2O, EtOH, eth

No.	Name Formula	CAS RN Mol. Wt.	Physical Form	mp/°C den/g cm^{-3}	bp/°C Other Data	Solubility
1044	Iron(III) chloride $FeCl_3$	7705-08-0 162.203	grn hex cry; hyg	304 2.90	≈316 a,c,e	s H_2O, EtOH, eth, ace
1045	Iron(III) chloride hexahydrate $FeCl_3 \cdot 6H_2O$	10025-77-1 270.294	yel-oran monocl cry; hyg	dec 37 1.82		s H_2O, EtOH, eth, ace
1046	Iron(III) chromate $Fe_2(CrO_4)_3$	10294-52-7 459.671	yel powder			i H_2O, EtOH; s acid
1047	Iron(III) citrate pentahydrate $FeC_6H_5O_7 \cdot 5H_2O$	3522-50-7 335.023	red-brn cry			s H_2O; i EtOH
1048	Iron(III) dichromate $Fe_2(Cr_2O_7)_3$	10294-53-8 759.654	red-brn solid			s H_2O, acid
1049	Iron(III) ferrocyanide $Fe_4[Fe(CN)_6]_3$	14038-43-8 859.234	dark blue powder	 1.80		i H_2O, dil acid, os
1050	Iron(III) fluoride FeF_3	7783-50-8 112.840	grn hex cry	>1000 3.87		sl H_2O; i EtOH, eth, bz
1051	Iron(III) fluoride trihydrate $FeF_3 \cdot 3H_2O$	15469-38-2 166.886	yel-brn tetr cry	 2.3		
1052	Iron(III) hydroxide $Fe(OH)_3$	1309-33-7 106.867	yel monocl cry	 3.12		
1053	Iron(III) hydroxide oxide $FeO(OH)$	20344-49-4 88.852	red-brn orth cry	 4.26	 f	i H_2O; s acid
1054	Iron(III) nitrate nonahydrate $Fe(NO_3)_3 \cdot 9H_2O$	7782-61-8 403.997	viol-gray hyg cry	47 dec 1.68		vs H_2O, EtOH, ace
1055	Iron(III) oxalate $Fe_2(C_2O_4)_3$	19469-07-9 375.749	yel amorp powder	dec 100		s H_2O, acid; i alk
1056	Iron(III) oxide Fe_2O_3	1309-37-1 159.688	red-brn hex cry	1565 5.25	 a,f	i H_2O; s acid
1057	Iron(III) phosphate dihydrate $FePO_4 \cdot 2H_2O$	10045-86-0 186.847	gray-wh orth cry	 2.87	 f	i H_2O; s HCl
1058	Iron(III) pyrophosphate nonahydrate $Fe_4(P_2O_7)_3 \cdot 9H_2O$	10058-44-3 907.348	yel powder			i H_2O; s acid
1059	Iron(III) hypophosphite $Fe(H_2PO_2)_3$	7783-84-8 250.811	wh-gray powder			i H_2O
1060	Iron(III) sulfate $Fe_2(SO_4)_3$	10028-22-5 399.881	gray-wh rhomb cry; hyg	 3.10		sl H_2O
1061	Iron(III) sulfate nonahydrate $Fe_2(SO_4)_3 \cdot 9H_2O$	13520-56-4 562.018	yel hex cry	dec 400 2.1	 f	vs H_2O
1062	Iron(III) thiocyanate monohydrate $Fe(SCN)_3 \cdot H_2O$	4119-52-2 248.111	red hyg cry	dec		s H_2O, EtOH, ace; i tol, chl
1063	Iron(III) metavanadate $Fe(VO_3)_3$	65842-03-7 352.665	gray-brn powder			i H_2O, EtOH; s acid
1064	Krypton Kr	7439-90-9 83.80	col gas	-157.36 3.677 g/L	-153.22 a,b,c,d,e	sl H_2O
1065	Krypton difluoride KrF_2	13773-81-4 121.80	col tetr cry	dec ≈25 3.24		reac H_2O
1066	Lanthanum La	7439-91-0 138.906	wh metal	920 6.15	3455 a,c	s dil acid
1067	Lanthanum boride LaB_6	12008-21-8 203.772	blk cub cry; refrac	2715 4.76		
1068	Lanthanum bromide $LaBr_3$	13536-79-3 378.618	wh hex cry; hyg	788 5.1		s H_2O
1069	Lanthanum carbide LaC_2	12071-15-7 162.928	tetr cry	2360 5.29		
1070	Lanthanum carbonate octahydrate $La_2(CO_3)_3 \cdot 8H_2O$	6487-39-4 601.961	wh cry powder	 2.6		i H_2O; s dil acid
1071	Lanthanum chloride $LaCl_3$	10099-58-8 245.264	wh hex cry; hyg	859 3.84	 a	s H_2O
1072	Lanthanum chloride heptahydrate $LaCl_3 \cdot 7H_2O$	20211-76-1 371.371	wh tricl cry; hyg	dec 91		s H_2O, EtOH
1073	Lanthanum fluoride LaF_3	13709-38-1 195.901	wh hex cry; hyg	1493 5.9		i H_2O, acid
1074	Lanthanum hydride LaH_3	13864-01-2 141.930	blk cub cry	 5.36		
1075	Lanthanum hydroxide $La(OH)_3$	14507-19-8 189.928	wh amorp solid	dec		i H_2O
1076	Lanthanum iodide LaI_3	13813-22-4 519.619	wh orth cry; hyg	778 5.6		s H_2O
1077	Lanthanum nitrate hexahydrate $La(NO_3)_3 \cdot 6H_2O$	10277-43-7 433.012	wh hyg cry	dec ≈40		vs H_2O, EtOH
1078	Lanthanum nitride LaN	25764-10-7 152.913	cub cry	 6.73		
1079	Lanthanum oxide La_2O_3	1312-81-8 325.809	wh amorp powder	2305 6.51	4200 a	i H_2O; s dil acid
1080	Lanthanum silicide $LaSi_2$	12056-90-5 195.077	gray tetr cry	 5.0		
1081	Lanthanum sulfate nonahydrate $La_2(SO_4)_3 \cdot 9H_2O$	10294-62-9 728.139	hex cry	 2.82		sl H_2O; i EtOH
1082	Lanthanum sulfide La_2S_3	12031-49-1 374.009	red cub cry	2110 4.9		

No.	Name Formula	CAS RN Mol. Wt.	Physical Form	mp/°C den/g cm^{-3}	bp/°C Other Data	Solubility
1083	Lanthanum sulfide LaS	12031-30-0 170.972	yel cub cry	2300 5.61	 a	
1084	Lawrencium Lr	22537-19-5 262	metal	1627	 a	
1085	Lead Pb	7439-92-1 207.2	soft silv-gray metal; cub	327.46 11.3	1749 a,c,d,e	s conc acid
1086	Lead(II) acetate $Pb(C_2H_3O_2)_2$	301-04-2 325.3	wh cry	280 3.25	dec	vs H_2O
1087	Lead(II) acetate trihydrate $Pb(C_2H_3O_2)_2{\cdot}3H_2O$	6080-56-4 427.3	col cry	75 dec 2.55		vs H_2O; sl EtOH
1088	Lead(II) acetate, basic $Pb(C_2H_3O_2)_2{\cdot}2PbO{\cdot}2H_2O$	1335-32-6 807.7	wh powder	dec		
1089	Lead(II) antimonate $Pb_3(SbO_4)_2$	13510-89-9 993.1	oran-yel powder	 6.58		i H_2O, dil acid
1090	Lead(II) arsenate $Pb_3(AsO_4)_2$	3687-31-8 899.4	wh cry	1042 dec 5.8		i H_2O; s HNO_3
1091	Lead(II) arsenite $Pb(AsO_2)_2$	10031-13-7 421.0	wh powder	 5.85		i H_2O; s dil HNO_3
1092	Lead(II) azide $Pb(N_3)_2$	13424-46-9 291.2	col orth needles; exp	exp ≈350 4.7		i H_2O; vs HOAc
1093	Lead(II) borate monohydrate $Pb(BO_2)_2{\cdot}H_2O$	10214-39-8 310.8	wh powder	500 dec 5.6		i H_2O; s dil HNO_3
1094	Lead(II) bromate monohydrate $Pb(BrO_3)_2{\cdot}H_2O$	10031-21-7 481.0	col cry	≈180 dec 5.53		sl H_2O
1095	Lead(II) bromide $PbBr_2$	10031-22-8 367.0	wh orth cry	371 6.69	892 a,c,d,e	sl H_2O; i EtOH
1096	Lead(II) butanoate $Pb(C_4H_7O_2)_2$	819-73-8 381.4	col solid	≈90		i H_2O; s dil HNO_3
1097	Lead(II) carbonate $PbCO_3$	598-63-0 267.2	col orth cry	dec ≈315 6.6	 a,f	i H_2O
1098	Lead(II) carbonate, basic $Pb(OH)_2{\cdot}2PbCO_3$	1319-46-6 775.6	wh hex cry	400 dec ≈6.5		i H_2O, EtOH; s acid
1099	Lead(II) chlorate $Pb(ClO_3)_2$	10294-47-0 374.1	col hyg cry	dec 230 3.9		vs H_2O, EtOH
1100	Lead(II) chloride $PbCl_2$	7758-95-4 278.1	wh orth needles or powder	501 5.98	951 a,c,d,e,f	sl H_2O; s alk
1101	Lead(II) chloride fluoride PbClF	13847-57-9 261.7	tetr cry	 7.05		i H_2O
1102	Lead(II) chromate $PbCrO_4$	7758-97-6 323.2	yel-oran monocl cry	844 6.12	 a,f	i H_2O; s akl, dil acid
1103	Lead(II) chromate(VI) oxide $PbCrO_4{\cdot}PbO$	18454-12-1 546.4	red powder			i H_2O
1104	Lead(II) citrate trihydrate $Pb_3(C_6H_5O_7)_2{\cdot}3H_2O$	512-26-5 1053.8	wh cry powder			s H_2O; sl EtOH
1105	Lead(II) cyanide $Pb(CN)_2$	592-05-2 259.2	wh-yel powder			sl H_2O; reac acid
1106	Lead(II) fluoride PbF_2	7783-46-2 245.2	wh orth cry	830 8.44	1293 a,c,d,e	sl H_2O
1107	Lead(II) fluoroborate $Pb(BF_4)_2$	13814-96-5 380.8	stable only in aq soln			s H_2O
1108	Lead(II) formate $Pb(CHO_2)_2$	811-54-1 297.2	wh prisms or needles	dec 190 4.63		sl H_2O; i EtOH
1109	Lead(II) hydrogen arsenate $PbHAsO_4$	7784-40-9 347.1	wh monocl cry	dec 280 5.943		i H_2O; s HNO_3, alk
1110	Lead(II) hydrogen phosphate $PbHPO_4$	15845-52-0 303.2	wh monocl cry	dec 5.66		
1111	Lead(II) hydroxide $Pb(OH)_2$	19783-14-3 241.2	wh powder	145 dec 7.59		i H_2O; s acid
1112	Lead(II) iodate $Pb(IO_3)_2$	25659-31-8 557.0	wh orth cry	 6.50		i H_2O
1113	Lead(II) iodide PbI_2	10101-63-0 461.0	yel hex cry or powder	410 6.16	872 dec a,c,d,e	sl H_2O; i EtOH
1114	Lead(II) lactate $Pb(C_3H_5O_3)_2$	18917-82-3 385.3	wh cry powder			s H_2O, hot EtOH
1115	Lead(II) molybdate $PbMoO_4$	10190-55-3 367.1	yel tetr cry	≈1060 6.7	 a,f	i H_2O; HNO_3, NaOH
1116	Lead(II) niobate $Pb(NbO_3)_2$	12034-88-7 489.0	rhomb or tetr cry	1343 6.6		i H_2O
1117	Lead(II) nitrate $Pb(NO_3)_2$	10099-74-8 331.2	col cub cry	470 4.53	 a	vs H_2O; sl EtOH
1118	Lead(II) oleate $Pb(C_{18}H_{33}O_2)_2$	1120-46-3 770.1	wax-like solid			i H_2O; s EtOH, bz, eth
1119	Lead(II) oxalate PbC_2O_4	814-93-7 295.2	wh powder	dec 300 5.28	 a	i H_2O; s dil HNO_3
1120	Lead(II) oxide (litharge) PbO	1317-36-8 223.2	red tetr cry	888 9.35	 a,e,f	i H_2O, EtOH; s dil HNO_3
1121	Lead(II) oxide (massicot) PbO	1317-36-8 223.2	yel orth cry	stab >489 9.64	 a,e,f	i H_2O, EtOH; s dil HNO_3

PHYSICAL CONSTANTS OF INORGANIC COMPOUNDS (continued)

No.	Name Formula	CAS RN Mol. Wt.	Physical Form	mp/°C den/g cm^{-3}	bp/°C Other Data	Solubility
1122	Lead(II) oxide hydrate $3PbO \cdot H_2O$	1311-11-1 687.6	wh powder	 7.41		i H_2O; s dil acid
1123	Lead(II) perchlorate $Pb(ClO_4)_2$	13453-62-8 406.1	wh cry			vs H_2O
1124	Lead(II) perchlorate trihydrate $Pb(ClO_4)_2 \cdot 3H_2O$	13637-76-8 460.1	wh cry	100 dec 2.6		vs cold H_2O; s EtOH
1125	Lead(II) phosphate $Pb_3(PO_4)_2$	7446-27-7 811.5	wh hex cry	1014 7.01		i H_2O, EtOH
1126	Lead(II) hypophosphite $Pb(H_2PO_2)_2$	10294-58-3 337.2	hyg cry powder	dec		sl H_2O; i EtOH
1127	Lead(II) metasilicate $PbSiO_3$	10099-76-0 283.3	wh monocl cry powder	764 6.49	 a	i H_2O, os
1128	Lead(II) orthosilicate Pb_2SiO_4	13566-17-1 506.5	monocl cry	743 7.60	 a	
1129	Lead(II) hexafluorosilicate dihydrate $PbSiF_6 \cdot 2H_2O$	1310-03-8 385.3	col cry	dec		vs H_2O
1130	Lead(II) selenate $PbSeO_4$	7446-15-3 350.2	orth cry	 6.37	 a	i H_2O; s conc acid
1131	Lead(II) selenide PbSe	12069-00-0 286.2	gray cub cry	1078 8.1	 a	i H_2O; s HNO_3
1132	Lead(II) selenite $PbSeO_3$	7488-51-9 334.2	wh monocl cry	≈500 7.0		i H_2O
1133	Lead(II) sodium thiosulfate $Na_4Pb(S_2O_3)_3$	10101-94-7 635.6	wh cry			sl H_2O
1134	Lead(II) stearate $Pb(C_{18}H_{35}O_2)_2$	1072-35-1 774.2	wh powder	≈100 1.4		i H_2O, s hot EtOH
1135	Lead(II) sulfate $PbSO_4$	7446-14-2 303.3	orth cry	1087 6.29	 a,f	i H_2O, acid; sl alk
1136	Lead(II) sulfide PbS	1314-87-0 239.3	blk powder or silv cub cry	1118 7.60	 a,c,e,f	i H_2O; s acid
1137	Lead(II) sulfite $PbSO_3$	7446-10-8 287.3	wh powder	dec	 a	i H_2O; s HNO_3
1138	Lead(II) tantalate $PbTa_2O_6$	12065-68-8 665.1	orth cry	 7.9		i H_2O
1139	Lead(II) telluride PbTe	1314-91-6 334.8	gray cub cry	924 8.164	 a,f	i H_2O, acid
1140	Lead(II) thiocyanate $Pb(SCN)_2$	592-87-0 323.4	wh-yel powder	 3.82		sl H_2O
1141	Lead(II) thiosulfate PbS_2O_3	13478-50-7 319.3	wh cry	dec 5.18		i H_2O; s acid
1142	Lead(II) titanate $PbTiO_3$	12060-00-3 303.1	yel tetr cry	 7.9		i H_2O; reac HCl
1143	Lead(II) tungstate (stolzite) $PbWO_4$	7759-01-5 455.0	yel tetr cry	1130 8.24	 f	i H_2O; s alk
1144	Lead(II) tungstate (raspite) $PbWO_4$	7759-01-5 455.0	monocl cry	trans 400 8.46	 f	i H_2O; s alk
1145	Lead(II) metavanadate $Pb(VO_3)_2$	10099-79-3 405.1	yel powder			i H_2O; reac HNO_3
1146	Lead(II) zirconate $PbZrO_3$	12060-01-4 346.4	col orth cry	 ≈8		i H_2O, alk; s acid
1147	Lead(II,IV) oxide Pb_2O_3	1314-27-8 462.4	blk monocl cry or red amorp powder	dec 530 10.05		i H_2O; s alk; reac conc HCl
1148	Lead(II,II,IV) oxide Pb_3O_4	1314-41-6 685.6	red tetr cry	830 8.92	 a,f	i H_2O, EtOH; s hot HCl
1149	Lead(IV) acetate $Pb(C_2H_3O_2)_4$	546-67-8 443.4	col monocl cry	≈175 2.23		reac H_2O, EtOH; s bz, chl
1150	Lead(IV) bromide $PbBr_4$	13701-91-2 526.8	unstable liq			
1151	Lead(IV) chloride $PbCl_4$	13463-30-4 349.0	yel oily liq	-15	dec ≈50 a	
1152	Lead(IV) fluoride PbF_4	7783-59-7 283.2	wh tetr cry; hyg	≈600 6.7		
1153	Lead(IV) oxide PbO_2	1309-60-0 239.2	red tetr cry or brn powder	290 dec 9.64	 a	
1154	Lithium Li	7439-93-2 6.941	soft silv-wh metal	180.5 0.534	1342 a,c,e	reac H_2O
1155	Lithium acetate dihydrate $LiC_2H_3O_2 \cdot 2H_2O$	6108-17-4 102.017	wh rhomb cry	58 dec 1.3		vs H_2O; s EtOH
1156	Lithium aluminum hydride $LiAlH_4$	16853-85-3 37.955	gray-wh monocl cry	dec >125 0.917	 a	reac H_2O, EtOH; s eth, thf
1157	Lithium amide $LiNH_2$	7782-89-0 22.964	tetr cry	380 1.18	 a	reac H_2O
1158	Lithium arsenate Li_3AsO_4	13478-14-3 159.743	col orth cry	 3.07		sl H_2O; s HOAc
1159	Lithium azide LiN_3	19597-69-4 48.961	hyg monocl cry; exp	 1.83		vs H_2O
1160	Lithium borate $LiBO_2$	13453-69-5 49.751	wh monocl cry; hyg	849 2.18	 a	vs H_2O; s EtOH

No.	Name Formula	CAS RN Mol. Wt.	Physical Form	mp/°C den/g cm^{-3}	bp/°C Other Data	Solubility
1161	Lithium borohydride $LiBH_4$	16949-15-8 21.784	wh-gray orth cry or powder	268 0.66	dec 380 a	s alk, eth, thf
1162	Lithium bromide LiBr	7550-35-8 86.845	wh cub cry; hyg	552 3.464	≈1300 a,c,e	s H_2O, EtOH, eth
1163	Lithium carbonate Li_2CO_3	554-13-2 73.891	wh monocl cry	723 2.11	dec 1300 a,c	sl H_2O; s acid; i EtOH
1164	Lithium chloride LiCl	7447-41-8 42.394	wh cub cry or powder; hyg	610 2.07	1383 a,c,e	s H_2O, EtOH, ace, py
1165	Lithium chromate dihydrate $Li_2CrO_4 \cdot 2H_2O$	7789-01-7 165.906	yel orth cry; hyg	dec 75 2.15		vs H_2O; s EtOH
1166	Lithium dichromate dihydrate $Li_2Cr_2O_7 \cdot 2H_2O$	10022-48-7 265.901	yel-red hyg cry	130 dec 2.34		vs H_2O
1167	Lithium fluoride LiF	7789-24-4 25.939	wh cub cry or powder	848.2 2.640	1673 a,c,d,e	sl H_2O; s acid
1168	Lithium formate monohydrate $Li(CHO_2) \cdot H_2O$	6108-23-2 69.974	col-wh cry	 1.46		s H_2O
1169	Lithium hydride LiH	7580-67-8 7.949	gray cub cry or powder; hyg	688.7 0.78	 a,c	reac H_2O, EtOH
1170	Lithium hydroxide LiOH	1310-65-2 23.948	col tetr cry	471.2 1.45	1626 a,c,d	s H_2O; sl EtOH
1171	Lithium hydroxide monohydrate $LiOH \cdot H_2O$	1310-66-3 41.964	wh monocl cry or powder	 1.51		s H_2O; sl EtOH
1172	Lithium iodide LiI	10377-51-2 133.845	wh cub cry; hyg	469 4.06	1171 a,c,e	vs H_2O
1173	Lithium iodide trihydrate $LiI \cdot 3H_2O$	7790-22-9 187.891	wh hyg cry	73 3.48		vs H_2O, EtOH, ace
1174	Lithium niobate $LiNbO_3$	12031-63-9 147.845	wh hex cry	≈1240 4.30		
1175	Lithium nitrate $LiNO_3$	7790-69-4 68.946	col hex cry; hyg	253 2.38	 a,c	vs H_2O; s EtOH
1176	Lithium nitride Li_3N	26134-62-3 34.830	red hex cry	813 1.27		reac H_2O
1177	Lithium phosphate Li_3PO_4	10377-52-3 115.794	wh orth cry	1205 2.46	 a	i H_2O
1178	Lithium oxide Li_2O	12057-24-8 29.881	wh cub cry	1570 2.013	 a	
1179	Lithium perchlorate $LiClO_4$	7791-03-9 106.392	wh orth cry or powder	236 2.428	dec 430 a,c	s H_2O, EtOH, ace, eth
1180	Lithium peroxide Li_2O_2	12031-80-0 45.881	wh hex cry	 2.31	 a	s H_2O; i EtOH
1181	Lithium selenate monohydrate $Li_2SeO_4 \cdot H_2O$	7790-71-8 174.86	monocl cry	 2.56		vs H_2O
1182	Lithium metasilicate Li_2SiO_3	10102-24-6 89.966	wh orth needles	1201 2.52	 a,c	i cold H_2O; reac dil acid
1183	Lithium sulfate Li_2SO_4	10377-48-7 109.946	wh monocl cry; hyg	859 2.21	 a,c	s H_2O
1184	Lithium sulfate monohydrate $Li_2SO_4 \cdot H_2O$	10102-25-7 127.961	col cry	dec 130 2.06		vs H_2O; sl EtOH
1185	Lithium sulfide Li_2S	12136-58-2 45.948	wh cub cry; hyg	1372 1.64	 a	
1186	Lutetium Lu	7439-94-3 174.967	soft silv metal; hex	1663 9.84	3393 a,c	s dil acid
1187	Lutetium boride LuB_4	12688-52-7 218.211	tetr cry	2600 ≈7.0		
1188	Lutetium bromide $LuBr_3$	14456-53-2 414.679	wh hyg cry	1025		vs H_2O
1189	Lutetium chloride $LuCl_3$	10099-66-8 281.325	wh monocl cry; hyg	925 3.98	 a	s H_2O
1190	Lutetium fluoride LuF_3	13760-81-1 231.962	orth cry	1182 8.3	2200	i H_2O
1191	Lutetium iodide LuI_3	13813-45-1 555.680	brn hex cry; hyg	1050 ≈5.6	 a	vs H_2O
1192	Lutetium nitride LuN	12125-25-6 188.974	cub cry	 11.6		
1193	Lutetium oxide Lu_2O_3	12032-20-1 397.932	wh cub cry or powder	2490 9.41	 a	
1194	Lutetium sulfate octahydrate $Lu_2(SO_4)_3 \cdot 8H_2O$	13473-77-3 782.247	wh cry			vs H_2O
1195	Lutetium sulfide Lu_2S_3	12163-20-1 446.132	gray rhomb cry	1750 dec 6.26		
1196	Lutetium telluride Lu_2Te_3	12163-22-3 732.73	orth cry	 7.8		
1197	Magnesium Mg	7439-95-4 24.305	silv-wh metal	650 1.74	1090 a,c,e	s dil acid
1198	Magnesium acetate tetrahydrate $Mg(C_2H_3O_2)_2 \cdot 4H_2O$	16674-78-5 214.455	col monocl cry; hyg	80 dec 1.45		vs H_2O, EtOH
1199	Magnesium amide $Mg(NH_2)_2$	7803-54-5 56.350	wh powder; flam	dec 1.39		reac H_2O

No.	Name Formula	CAS RN Mol. Wt.	Physical Form	mp/°C den/g cm^{-3}	bp/°C Other Data	Solubility
1200	Magnesium antimonide Mg_3Sb_2	12057-75-9 316.435	hex cry	1245 3.99		
1201	Magnesium boride MgB_2	12007-25-9 45.927	hex cry	dec 800 2.57		
1202	Magnesium bromate hexahydrate $Mg(BrO_3)_2 \cdot 6H_2O$	7789-36-8 388.201	col cub cry	dec 200 2.29		vs H_2O
1203	Magnesium bromide $MgBr_2$	7789-48-2 184.113	wh hex cry; hyg	711 3.72	a,c	vs H_2O
1204	Magnesium bromide hexahydrate $MgBr_2 \cdot 6H_2O$	13446-53-2 292.204	col monocl cry	dec 165 2.0		vs H_2O; s EtOH
1205	Magnesium carbonate $MgCO_3$	546-93-0 84.314	wh hex cry	dec 350 3.05	f	i H_2O, EtOH; s acid
1206	Magnesium chlorate hexahydrate $Mg(ClO_3)_2 \cdot 6H_2O$	10326-21-3* 299.298	wh hyg cry	dec $\approx$35 1.80		vs H_2O; sl EtOH
1207	Magnesium chloride $MgCl_2$	7786-30-3 95.210	wh hex leaflets; hyg	714 2.325	1412 a,c,e	vs H_2O
1208	Magnesium chloride hexahydrate $MgCl_2 \cdot 6H_2O$	7791-18-6 203.301	wh hyg cry	dec $\approx$100 1.56		vs H_2O; s EtOH
1209	Magnesium fluoride MgF_2	7783-40-6 62.302	wh tetr cry	1263 3.148	2227 a,c,f	sl H_2O
1210	Magnesium formate dihydrate $Mg(CHO_2)_2 \cdot 2H_2O$	6150-82-9 150.371	wh cry	dec		s H_2O; i EtOH
1211	Magnesium germanide Mg_2Ge	1310-52-7 121.22	cub cry	1117 3.09		
1212	Magnesium hydride MgH_2	60616-74-2 26.321	wh tetr cry	dec 200 1.45		reac H_2O
1213	Magnesium hydrogen phosphate trihydrate $MgHPO_4 \cdot 3H_2O$	7757-86-0 174.331	wh powder	dec 550 2.13	f	sl H_2O; s dil acid
1214	Magnesium hydroxide $Mg(OH)_2$	1309-42-8 58.320	wh hex cry	350 2.37	a,f	i H_2O; s dil acid
1215	Magnesium iodide MgI_2	10377-58-9 278.114	wh hex cry; hyg	634 4.43	a,c	vs H_2O
1216	Magnesium iodide octahydrate $MgI_2 \cdot 8H_2O$	7790-31-0 422.236	wh orth cry; hyg	41 dec 2.10		vs H_2O; s EtOH
1217	Magnesium nitrate $Mg(NO_3)_2$	10377-60-3 148.314	wh cub cry	 $\approx$2.3	a	vs H_2O
1218	Magnesium nitrate dihydrate $Mg(NO_3)_2 \cdot 2H_2O$	15750-45-5 184.345	wh cry	$\approx$100 dec 1.45		s H_2O, EtOH
1219	Magnesium nitrate hexahydrate $Mg(NO_3)_2 \cdot 6H_2O$	13446-18-9 256.406	col monocl cry; hyg	dec $\approx$95 1.46		vs H_2O; s EtOH
1220	Magnesium nitride Mg_3N_2	12057-71-5 100.928	yel cub cry	dec 270 2.71		
1221	Magnesium oxalate dihydrate $MgC_2O_4 \cdot 2H_2O$	6150-88-5 148.355	wh powder			i H_2O, EtOH; s dil acid
1222	Magnesium oxide MgO	1309-48-4 40.304	wh cub cry	2826 3.6	3600 a,c,f	sl H_2O; i EtOH
1223	Magnesium perchlorate $Mg(ClO_4)_2$	10034-81-8 223.205	wh hyg powder	dec 250 2.2		vs H_2O
1224	Magnesium perchlorate hexahydrate $Mg(ClO_4)_2 \cdot 6H_2O$	13446-19-0 331.297	wh hyg cry	dec 190 1.98		vs H_2O; s EtOH
1225	Magnesium permanganate hexahydrate $Mg(MnO_4)_2 \cdot 6H_2O$	10377-62-5 370.268	blue-blk cry	dec 2.18		s H_2O
1226	Magnesium peroxide MgO_2	1335-26-8 56.304	wh cub cry	dec 100 $\approx$3.0		i H_2O; s dil acid
1227	Magnesium phosphate octahydrate $Mg_3(PO_4)_2 \cdot 8H_2O$	13446-23-6 406.980	wh monocl cry	 2.17		i H_2O; s acid
1228	Magnesium pyrophosphate trihydrate $Mg_2P_2O_7 \cdot 3H_2O$	10102-34-8 276.600	wh powder	dec 100 2.56		i H_2O; s acid
1229	Magnesium phosphide Mg_3P_2	12057-74-8 134.863	yel cub cry	 2.06		reac H_2O
1230	Magnesium selenate hexahydrate $MgSeO_4 \cdot 6H_2O$	13446-28-1 275.35	wh monocl cry	 1.928		s H_2O
1231	Magnesium selenide $MgSe$	1313-04-8 103.27	brn cub cry	 4.2		reac H_2O
1232	Magnesium selenite hexahydrate $MgSeO_3 \cdot 6H_2O$	15593-61-0 259.36	col hex cry	 2.09		i H_2O; s dil acid
1233	Magnesium metasilicate $MgSiO_3$	13776-74-4 100.389	wh monocl cry	dec $\approx$1550 3.19	f	i H_2O; sl HF
1234	Magnesium orthosilicate Mg_2SiO_4	26686-77-1 140.694	wh orth cry	1898 3.21	a,c,f	i H_2O
1235	Magnesium hexafluorosilicate hexahydrate $MgSiF_6 \cdot 6H_2O$	60950-56-3 274.472	wh cry	120 dec 1.79		s H_2O; i EtOH
1236	Magnesium silicide Mg_2Si	22831-39-6 76.696	gray cub cry	1102 1.99		reac H_2O
1237	Magnesium stannide Mg_2Sn	1313-08-2 167.320	blue cub cry	771 3.60		s H_2O, dil HCl

No.	Name Formula	CAS RN Mol. Wt.	Physical Form	mp/°C den/g cm^{-3}	bp/°C Other Data	Solubility
1238	Magnesium sulfate $MgSO_4$	7487-88-9 120.369	col orth cry	1127 2.66	a,c	s H_2O
1239	Magnesium sulfate heptahydrate $MgSO_4 \cdot 7H_2O$	10034-99-8 246.475	col orth cry	dec 150 1.67	f	vs H_2O; sl EtOH
1240	Magnesium sulfate monohydrate $MgSO_4 \cdot H_2O$	14168-73-1 138.384	col monocl cry	dec 150 2.57	f	vs H_2O
1241	Magnesium sulfide MgS	12032-36-9 56.371	red-brn cub cry	>2000 dec 2.68	a	reac H_2O
1242	Magnesium sulfite hexahydrate $MgSO_3 \cdot 6H_2O$	13446-29-2 212.461	wh hex cry	dec 200 1.72		sl H_2O; i EtOH
1243	Magnesium sulfite trihydrate $MgSO_3 \cdot 3H_2O$	19086-20-5 158.415	col orth cry	 2.12		sl H_2O
1244	Magnesium thiosulfate hexahydrate $MgS_2O_3 \cdot 6H_2O$	13446-30-5 244.527	col cry	dec 170 1.82		vs H_2O; i EtOH
1245	Magnesium titanate $MgTiO_3$	1312-99-8 120.170	col hex cry	1565 3.85	f	
1246	Magnesium tungstate $MgWO_4$	13573-11-0 272.14	wh monocl cry	 6.89		i H_2O, EtOH
1247	Manganese Mn	7439-96-5 54.938	hard gray metal	1246 7.3	2061 a,c,e	s dil acids
1248	Manganese antimonide $MnSb$	12032-82-5 176.698	hex cry	840 6.9		
1249	Manganese antimonide Mn_2Sb	12032-97-2 231.636	tetr cry	948 7.0		
1250	Manganese boride MnB	12045-15-7 65.749	orth cry	1890 6.45		
1251	Manganese boride MnB_2	12228-50-1 76.560	hex cry	1827 5.3		
1252	Manganese boride Mn_2B	12045-16-8 120.687	red-brn tetr cry	1580 7.20		
1253	Manganese carbide Mn_3C	12266-65-8 176.825	refrac solid	1520 6.89		
1254	Manganese carbonyl $Mn_2(CO)_{10}$	10170-69-1 389.980	yel monocl cry	154 1.75		i H_2O; s os
1255	Manganese phosphide Mn_2P	12333-54-9 140.850	hex cry	1327 6.0		
1256	Manganese phosphide MnP	12032-78-9 85.912	orth cry	1147 5.49		
1257	Manganese(II) acetate tetrahydrate $Mn(C_2H_3O_2)_2 \cdot 4H_2O$	15243-27-3 245.088	red monocl cry	80 1.59		s H_2O, EtOH
1258	Manganese(II) tetraborate octahydrate $MnB_4O_7 \cdot 8H_2O$	12228-91-0 354.300	red solid			i H_2O, EtOH; s dil acid
1259	Manganese(II) bromide $MnBr_2$	13446-03-2 214.746	pink hex cry	698 4.385	a	vs H_2O
1260	Manganese(II) bromide tetrahydrate $MnBr_2 \cdot 4H_2O$	10031-20-6 286.808	red hyg cry	64 dec		vs H_2O
1261	Manganese(II) carbonate $MnCO_3$	598-62-9 114.947	pink hex cry	dec >200 3.70	f	sl H_2O; s dil acid
1262	Manganese(II) chloride $MnCl_2$	7773-01-5 125.843	pink trig cry; hyg	650 2.977	1190 a,c,e	s H_2O, py, EtOH; i eth
1263	Manganese(II) chloride tetrahydrate $MnCl_2 \cdot 4H_2O$	13446-34-9 197.905	red monocl cry; hyg	87.5 1.913		vs H_2O; s EtOH; i eth
1264	Manganese(II) dihydrogen phosphate dihydrate $Mn(H_2PO_4)_2 \cdot 2H_2O$	18718-07-5 284.944	col hyg cry			s H_2O; i EtOH
1265	Manganese(II) fluoride MnF_2	7782-64-1 92.935	red tetr cry	930 3.98		sl H_2O; i EtOH
1266	Manganese(II) hydroxide $Mn(OH)_2$	[illegible] 88.953	pink hex cry	dec 3.26	f	i H_2O
1267	Manganese(II) iodide MnI_2	7790-33-2 308.747	wh hex cry; hyg	638 5.04		s H_2O, EtOH
1268	Manganese(II) iodide tetrahydrate $MnI_2 \cdot 4H_2O$	7790-33-2* 380.809	red cry			vs H_2O; s EtOH
1269	Manganese(II) molybdate $MnMoO_4$	14013-15-1 214.88	yel monocl cry	 4.05		
1270	Manganese(II) nitrate $Mn(NO_3)_2$	10377-93-2 178.948	col orth cry; hyg	 2.2		vs H_2O; s diox, thf
1271	Manganese(II) nitrate hexahydrate $Mn(NO_3)_2 \cdot 6H_2O$	10377-66-9 287.040	rose monocl cry	28 dec 1.8		vs H_2O, EtOH
1272	Manganese(II) nitrate tetrahydrate $Mn(NO_3)_2 \cdot 4H_2O$	20694-39-7 251.010	pink hyg cry	37.1 dec 2.13		vs H_2O; s EtOH
1273	Manganese(II) oxalate dihydrate $MnC_2O_4 \cdot 2H_2O$	6556-16-7 178.988	wh cry powder	dec 150 2.45		sl H_2O; s acid
1274	Manganese(II) oxide MnO	1344-43-0 70.937	gr cub cry or powder	1840 5.37	a,c,f	i H_2O; s acid
1275	Manganese(II) perchlorate hexahydrate $Mn(ClO_4)_2 \cdot 6H_2O$	15364-94-0 361.930	pink hex cry	 2.10		
1276	Manganese(II) pyrophosphate $Mn_2P_2O_7$	53731-35-4 283.819	wh monocl cry	1196 3.71		i H_2O

No.	Name	Formula	CAS RN	Mol. Wt.	Physical Form	mp/°C	den/g cm^{-3}	bp/°C	Other Data	Solubility
1277	Manganese(II) metasilicate	$MnSiO_3$	7759-00-4	131.022	red orth cry	1291	3.48		a,f	i H_2O
1278	Manganese(II) orthosilicate	Mn_2SiO_4	13568-32-6	201.960	orth cry		4.11		a	i H_2O
1279	Manganese(II) selenide	MnSe	1313-22-0	133.90	gray cub cry	1460	5.45		a	i H_2O
1280	Manganese(II) sulfate	$MnSO_4$	7785-87-7	151.002	wh orth cry	700	3.25	dec 850		vs H_2O
1281	Manganese(II) sulfate monohydrate	$MnSO_4 \cdot H_2O$	10034-96-5	169.017	red monocl cry		2.95			vs H_2O; i EtOH
1282	Manganese(II) sulfate tetrahydrate	$MnSO_4 \cdot 4H_2O$	10101-68-5	223.063	red monocl cry	38 dec	2.26			s H_2O; i EtOH
1283	Manganese(II) sulfide (α)	MnS	18820-29-6	87.004	grn cub cry	1610	4.0		a,f	i H_2O; s dil acid
1284	Manganese(II) sulfide (β)	MnS	18820-29-6	87.004	red cub cry		3.3		a,f	i H_2O; s dil acid
1285	Manganese(II) sulfide (γ)	MnS	18820-29-6	87.004	red hex cry		≈3.3		a,f	i H_2O; s dil acid
1286	Manganese(II) telluride	MnTe	12032-88-1	182.54	hex cry	≈1150	6.0			
1287	Manganese(II) titanate	$MnTiO_3$	12032-74-5	150.803	red hex cry	1360	4.55			
1288	Manganese(II) tungstate	$MnWO_4$	13918-22-4	302.78	wh monocl cry		7.2		f	
1289	Manganese(II,III) oxide	Mn_3O_4	1317-35-7	228.812	brn tetr cry	1567	4.84		a,f	i H_2O; s HCl
1290	Manganese(III) fluoride	MnF_3	7783-53-1	111.933	red monocl cry; hyg	dec >600	3.54			reac H_2O
1291	Manganese(III) hydroxide	MnO(OH)	1332-63-4	87.945	blk monocl cry	dec 250	≈4.3		f	i H_2O
1292	Manganese(III) oxide	Mn_2O_3	1317-34-6	157.874	blk cub cry	dec 1080	≈5.0		a	i H_2O
1293	Manganese(IV) oxide	MnO_2	1313-13-9	86.937	blk tetr cry	535 dec	5.08		a,f	i H_2O, HNO_3
1294	Manganese(VII) oxide	Mn_2O_7	12057-92-0	221.872	grn oil; exp	5.9	2.40	exp 95		vs H_2O
1295	Mendelevium	Md	7440-11-1	258	Metal	827			a	
1296	Mercury	Hg	7439-97-6	200.59	heavy silv liq	-38.83	13.5336	356.73	a,b,c,d,e	i H_2O
1297	Mercury(I) acetate	$Hg_2(C_2H_3O_2)_2$	631-60-7	519.27	col scales	dec				sl H_2O; i EtOH, eth
1298	Mercury(I) bromide	Hg_2Br_2	15385-58-7	560.99	wh tetr cry or powder	407	7.307		a	i H_2O, EtOH, eth
1299	Mercury(I) chlorate	$Hg_2(ClO_3)_2$	10294-44-7	568.08	wh cry	dec ≈250	6.41			sl H_2O; s EtOH
1300	Mercury(I) chloride	Hg_2Cl_2	10112-91-1	472.09	wh tetr cry	525 tp	7.16	383 sp	a,f	i H_2O, EtOH, eth
1301	Mercury(I) fluoride	Hg_2F_2	13967-25-4	439.18	yel cub cry	570 dec	8.73	subl		reac H_2O
1302	Mercury(I) iodide	Hg_2I_2	15385-57-6	654.99	yel amorp powder	290	7.70		a,c	i H_2O, EtOH, eth
1303	Mercury(I) nitrate dihydrate	$Hg_2(NO_3)_2 \cdot 2H_2O$	7782-86-7	561.22	col cry	70 dec	4.8			reac H_2O
1304	Mercury(I) oxide	Hg_2O	15829-53-5	417.18	prob mixture of HgO+Hg	100 dec	9.8			i H_2O; s HNO_3
1305	Mercury(I) sulfate	Hg_2SO_4	7783-36-0	497.24	wh-yel cry powder		7.56		a	i H_2O; s dil HNO_3
1306	Mercury(II) acetate	$Hg(C_2H_3O_2)_2$	1600-27-7	318.68	wh-yel cry or powder	179 dec	3.28			s H_2O, EtOH
1307	Mercury(II) amide chloride	$Hg(NH_2)Cl$	10124-48-8	252.07	wh solid		5.38	subl		i H_2O, EtOH; s warm acid
1308	Mercury(II) bromide	$HgBr_2$	7789-47-1	360.40	wh rhomb cry or powder	236	6.05	322	a,b,c,d,e	sl H_2O, chl; s EtOH, MeOH
1309	Mercury(II) chloride	$HgCl_2$	7487-94-7	271.50	wh orth cry	276	5.6	304	a,b,c,d,e	sl H_2O, bz; s EtOH, MeOH, ace, eth
1310	Mercury(II) chromate	$HgCrO_4$	13444-75-2*	316.58	red monocl cry		6.06			sl H_2O
1311	Mercury(II) cyanide	$Hg(CN)_2$	592-04-1	252.63	col tetr cry	dec 320	4.00			s H_2O, EtOH; sl eth
1312	Mercury(II) dichromate	$HgCr_2O_7$	7789-10-8	416.58	red cry powder					i H_2O; s acid
1313	Mercury(II) fluoride	HgF_2	7783-39-3	238.59	wh cub cry; hyg	dec 645	8.95			reac H_2O
1314	Mercury(II) fulminate	$Hg(CNO)_2$	628-86-4	284.62	gray cry	exp	4.42			sl H_2O; s EtOH, NH_4OH
1315	Mercury(II) hydrogen arsenate	$HgHAsO_4$	7784-37-4	340.52	yel powder					i H_2O; s acid

.	Name Formula	CAS RN Mol. Wt.	Physical Form	mp/°C den/g cm^{-3}	bp/°C Other Data	Solubility
16	Mercury(II) iodate $Hg(IO_3)_2$	7783-32-6 550.40	wh powder	dec 175		i H_2O
17	Mercury(II) iodide HgI_2	7774-29-0 454.40	red tetr cry or powder	259 6.28	354 a,b,c,d,e	i H_2O; sl EtOH, ace, eth
18	Mercury(II) nitrate $Hg(NO_3)_2$	10045-94-0 324.60	col hyg cry	79 4.3		s H_2O; i EtOH
19	Mercury(II) nitrate dihydrate $Hg(NO_3)_2 \cdot 2H_2O$	10045-94-0* 360.63	monocl cry	4.78		s H_2O
20	Mercury(II) nitrate monohydrate $Hg(NO_3)_2 \cdot H_2O$	7783-34-8 342.62	wh-yel hyg cry	4.3		s H_2O, dil acid
21	Mercury(II) oxide HgO	21908-53-2 216.59	red or yel orth cry	dec 500 11.14	a,f	i H_2O, EtOH; s dil acid
22	Mercury(II) oxycyanide $Hg(CN)_2 \cdot HgO$	1335-31-5 469.21	wh orth cry	exp 4.44		s H_2O
23	Mercury(II) phosphate $Hg_3(PO_4)_2$	7782-66-3 791.71	wh-yel powder			i H_2O, EtOH; s acid
24	Mercury(II) selenide $HgSe$	20601-83-6 279.55	gray cub cry	8.21	subl	i H_2O
25	Mercury(II) sulfate $HgSO_4$	7783-35-9 296.65	wh monocl cry	6.47	a	reac H_2O
26	Mercury(II) sulfide (black) HgS	1344-48-5 232.66	blk cub cry or powder	583 7.70	a,f	i H_2O; s acid, EtOH
27	Mercury(II) sulfide (red) HgS	1344-48-5 232.66	red hex cry	trans to blk HgO 386 8.17	a,f	i H_2O, acid; s aqua regia
28	Mercury(II) telluride $HgTe$	12068-90-5 328.19	gray cub cry	673 8.63	a	
29	Mercury(II) thiocyanate $Hg(SCN)_2$	592-85-8 316.76	monocl cry	dec ≈165 3.71		i H_2O; s dil HCl
30	Molybdenum Mo	7439-98-7 95.94	gray-blk metal; cub	2623 10.2	4000 a,c,e	i H_2O, dil acid, alk
31	Molybdenum boride Mo_2B	12006-99-4 202.69	refrac tetr cry	2000 9.2		
32	Molybdenum boride Mo_2B_5	12007-97-5 245.94	refrac hex cry	1600 ≈7.2		
33	Molybdenum carbide MoC	12011-97-1 107.95	refrac solid; cub	2577		
34	Molybdenum carbide Mo_2C	12069-89-5 203.89	gray orth cry	2687 9.18		
35	Molybdenum carbonyl $Mo(CO)_6$	13939-06-5 264.00	wh orth cry	dec 150 1.96	subl a,d	i H_2O; s bz; sl eth
36	Molybdenum nitride MoN	12033-19-1 109.95	hex cry	1750 9.20		
37	Molybdenum nitride Mo_2N	12033-31-7 205.89	gray cub cry	790 dec 9.46		
38	Molybdenum phosphide MoP	12163-69-8 126.91	blk hex cry	7.34		
39	Molybdenum silicide $MoSi_2$	12136-78-6 152.11	gray tetr cry	≈1900 6.2		i H_2O; s HF
40	Molybdenum(II) bromide $MoBr_2$	13446-56-5 255.75	yel-red cry	dec 900		
41	Molybdenum(II) chloride $MoCl_2$	13478-17-6 166.85	yel cry	dec 530		
42	Molybdenum(II) iodide MoI_2	14055-74-4 349.75	blk hyg cry	5.278		
43	Molybdenum(III) bromide $MoBr_3$	13446-57-6 335.65	grn hex cry	977 4.89		i H_2O
44	Molybdenum(III) chloride $MoCl_3$	13478-18-7 202.30	dark red monocl cry	1027 3.74		i H_2O
45	Molybdenum(III) fluoride MoF_3	20193-58-2 152.94	brn hex cry	>600 4.64		i H_2O
46	Molybdenum(III) iodide MoI_3	14055-75-5 476.65	blk solid	927		i H_2O
47	Molybdenum(III) oxide Mo_2O_3	1313-29-7 239.88	gray-blk powder			i H_2O; sl acid
48	Molybdenum(IV) bromide $MoBr_4$	13520-59-7 415.56	blk cry	dec		reac H_2O
49	Molybdenum(IV) chloride $MoCl_4$	13320-71-3 237.75	blk cry	dec >170		reac H_2O
350	Molybdenum(IV) fluoride MoF_4	23412-45-5 171.93	grn cry	dec		reac H_2O
351	Molybdenum(IV) oxide MoO_2	18868-43-4 127.94	brn-viol tetr cry	dec ≈1100 6.47	a	sl H_2O
352	Molybdenum(IV) selenide $MoSe_2$	12058-18-3 253.86	gray hex cry	>1200 6.90		
353	Molybdenum(IV) sulfide MoS_2	1317-33-5 160.07	blk powder or hex cry	2375 5.06	a,f	i H_2O; s conc acid

No.	Name Formula	CAS RN Mol. Wt.	Physical Form	mp/°C den/g cm^{-3}	bp/°C Other Data	Solubility
1354	Molybdenum(IV) telluride $MoTe_2$	12058-20-7 351.14	gray hex cry	 7.7		
1355	Molybdenum(V) chloride $MoCl_5$	10241-05-1 273.20	gr-blk monocl cry; hyg	194 2.93	268 b,c,d	s EtOH, eth
1356	Molybdenum(V) fluoride MoF_5	13819-84-6 190.93	yel monocl cry	67 3.5	213	
1357	Molybdenum(V) oxytrichloride $MoOCl_3$	13814-74-9 218.30	blk monocl cry	297	subl	reac H_2O
1358	Molybdenum(VI) acid monohydrate $H_2MoO_4{\cdot}H_2O$	7782-91-4 179.97	wh powder	 3.1		sl H_2O; s alk
1359	Molybdenum(VI) fluoride MoF_6	7783-77-9 209.93	wh cub cry or col liq; hyg	17.5 2.54	34 a,b,c,d,e	reac H_2O
1360	Molybdenum(VI) oxytetrafluoride $MoOF_4$	14459-59-7 187.93	volatile solid	98		
1361	Molybdenum(VI) oxytetrachloride $MoOCl_4$	13814-75-0 253.75	grn hyg powder	101		
1362	Molybdenum(VI) dioxydichloride MoO_2Cl_2	13637-68-8 198.84	yel-oran solid	≈175 3.31		reac H_2O
1363	Molybdenum(VI) oxide MoO_3	1313-27-5 143.94	wh-yel rhomb cry	801 4.70	1155 a,c,d,e	sl H_2O; s conc acid
1364	Molybdenum(VI) metaphosphate $Mo(PO_3)_6$	133863-98-6 569.77	yel powder	 3.28		i H_2O, acid
1365	Neodymium Nd	7440-00-8 144.24	soft silv-wh metal; hex	1016 7.01	3066 a,c	
1366	Neodymium boride NdB_6	12008-23-0 209.11	blk cub cry	2610 4.93		
1367	Neodymium bromide $NdBr_3$	13536-80-6 383.95	viol orth cry; hyg	682 5.3	1540	s H_2O
1368	Neodymium chloride $NdCl_3$	10024-93-8 250.60	viol hex cry	758 4.13	1600 a	vs H_2O, EtOH; i eth, chl
1369	Neodymium chloride hexahydrate $NdCl_3{\cdot}6H_2O$	13477-89-9 358.69	purp cry	dec 124 2.3		vs H_2O; s EtOH
1370	Neodymium fluoride NdF_3	13709-42-7 201.24	viol hex cry; hyg	1377 6.51	2300 a	i H_2O
1371	Neodymium iodide NdI_3	13813-24-6 524.95	grn orth cry; hyg	784 5.85		s H_2O
1372	Neodymium nitrate hexahydrate $Nd(NO_3)_3{\cdot}6H_2O$	14517-29-4 438.35	purp hyg cry			vs H_2O; s EtOH, ace
1373	Neodymium nitride NdN	25764-11-8 158.25	blk cub cry	 7.69		
1374	Neodymium oxide Nd_2O_3	1313-97-9 336.48	blue hex cry; hyg	2320 7.24	 a	i H_2O; s dil acid
1375	Neodymium sulfate $Nd_2(SO_4)_3$	13477-91-3 576.67	pink needles	dec ≈700		s H_2O
1376	Neodymium sulfide Nd_2S_3	12035-32-4 384.68	orth cry	2207 5.46		
1377	Neodymium telluride Nd_2Te_3	12035-35-7 671.28	gray orth cry	1377 7.0		
1378	Neon Ne	7440-01-9 20.180	col gas	-248.59 0.885 g/L	-246.08 a,b,c,d,e	sl H_2O
1379	Neptunium Np	7439-99-8 237	silv metal	644 20.2	 c	s HCl
1380	Neptunium(IV) oxide NpO_2	12035-79-9 269	grn cub cry	2547 11.1		
1381	Nickel Ni	7440-02-0 58.693	wh metal; cub	1455 8.90	2913 a,c,e	i H_2O; sl dil acid
1382	Nickel antimonide NiSb	12035-52-8 180.453	hex cry	1147 8.74	 f	
1383	Nickel arsenide NiAs	27016-75-7 133.615	hex cry	967 7.77	 f	
1384	Nickel boride Ni_3B	12007-02-2 186.891	refrac solid	1156 8.17		
1385	Nickel boride NiB	12007-00-0 69.504	grn refrac solid	1035 7.13		
1386	Nickel boride Ni_2B	12007-01-1 128.198	refrac solid	1125 7.90		
1387	Nickel carbonyl $Ni(CO)_4$	13463-39-3 170.735	col liq	-19.3 1.31	43 (exp ≈60) a,e	i H_2O; s EtOH, bz, ace, ctc
1388	Nickel phosphide Ni_2P	12035-64-2 148.361	hex cry	1100 7.33		
1389	Nickel silicide Ni_2Si	12059-14-2 145.473	orth cry	1255 7.40		
1390	Nickel silicide $NiSi_2$	12201-89-7 114.864	cub cry	993 4.83		
1391	Nickel(II) arsenate octahydrate $Ni_3(AsO_4)_2{\cdot}8H_2O$	7784-48-7 598.040	yel-grn powder	dec 4.98		i H_2O; s acid
1392	Nickel(II) bromide $NiBr_2$	13462-88-9 218.501	yel hex cry; hyg	963 5.10	subl a	vs H_2O

PHYSICAL CONSTANTS OF INORGANIC COMPOUNDS (continued)

No.	Name Formula	CAS RN Mol. Wt.	Physical Form	mp/°C den/g cm^{-3}	bp/°C Other Data	Solubility
	Nickel(II) bromide trihydrate $NiBr_2 \cdot 3H_2O$	13462-88-9* 272.547	yel-grn hyg cry	dec 200		vs H_2O; s EtOH, eth
	Nickel(II) carbonate $NiCO_3$	3333-67-3 118.702	grn rhomb cry	 4.39		i H_2O; s dil acid
5	Nickel(II) chloride $NiCl_2$	7718-54-9 129.598	yel hex cry; hyg	1009 3.51	subl a,c,e	s H_2O, EtOH
6	Nickel(II) chloride hexahydrate $NiCl_2 \cdot 6H_2O$	7791-20-0 237.689	grn monocl cry			vs H_2O; s EtOH
7	Nickel(II) cyanide tetrahydrate $Ni(CN)_2 \cdot 4H_2O$	13477-95-7 182.790	grn plates	dec 200		i H_2O; sl dil acid; s NH_4OH
8	Nickel(II) fluoride NiF_2	10028-18-9 96.690	yel tetr cry	1474 4.7	 a	sl H_2O; i EtOH, eth
9	Nickel(II) hydroxide $Ni(OH)_2$	12054-48-7 92.708	grn hex cry	dec 230 4.1	 a	i H_2O
0	Nickel(II) hydroxide monohydrate $Ni(OH)_2 \cdot H_2O$	36897-37-7 110.723	grn powder			i H_2O; s dil acid
1	Nickel(II) iodide NiI_2	13462-90-3 312.502	blk hex cry; hyg	780 5.22	subl a	vs H_2O
2	Nickel(II) iodide hexahydrate $NiI_2 \cdot 6H_2O$	7790-34-3 420.593	grn monocl cry; hyg			vs H_2O, EtOH
3	Nickel(II) nitrate hexahydrate $Ni(NO_3)_2 \cdot 6H_2O$	13478-00-7 290.794	grn monocl cry; hyg	dec 56 2.05		vs H_2O; s EtOH
4	Nickel(II) oxide NiO	1313-99-1 74.692	grn cub cry	1955 6.72	 f	i H_2O; s acid
5	Nickel(II) selenide $NiSe$	1314-05-2 137.65	yel-grn hex cry	980 7.2		
6	Nickel(II) sulfate $NiSO_4$	7786-81-4 154.757	grn-yel orth cry	dec 840 4.01	 a	vs H_2O
7	Nickel(II) sulfate heptahydrate $NiSO_4 \cdot 7H_2O$	10101-98-1 280.863	grn orth cry	 1.98		s H_2O, EtOH
8	Nickel(II) sulfate hexahydrate $NiSO_4 \cdot 6H_2O$	10101-97-0 262.848	blue-grn tetr cry	dec ≈100 2.07		vs H_2O; sl EtOH
9	Nickel(II) sulfide NiS	16812-54-7 90.759	yel hex cry	976 5.5	 a,c,f	i H_2O
0	Nickel(II) titanate $NiTiO_3$	12035-39-1 154.558	brn hex cry	 5.0		
1	Nickel(II,III) sulfide Ni_3S_4	12137-12-1 304.344	cub cry	995 4.77		
2	Nickel(III) oxide Ni_2O_3	1314-06-3 165.385	gray-blk cub cry	dec ≈600	 a	i H_2O; s hot acid
3	Nickel(III) sulfide Ni_3S_2	12035-72-2 240.212	hex cry	787 5.87		
4	Niobium Nb	7440-03-1 92.906	gray metal; cub	2477 8.57	4744 a,c	i acid
5	Niobium boride NbB	12045-19-1 103.717	gray orth cry	2270 7.5		
6	Niobium boride NbB_2	12007-29-3 114.528	gray hex cry	2900 6.8		
7	Niobium carbide NbC	12069-94-2 104.917	gray cub cry	3608 7.82	4300	i H_2O, acid
8	Niobium carbide Nb_2C	12011-99-3 197.824	refrac hex cry	3080 7.8		i H_2O
9	Niobium nitride NbN	24621-21-4 106.913	gray cry; cub	2300 8.47		i HCl, acid
20	Niobium phosphide NbP	12034-66-1 123.880	tetr cry	 6.5		
21	Niobium silicide $NbSi_2$	12034-80-9 149.077	gray hex cry	1950 5.7		
22	Niobium(II) oxide NbO	12034-57-0 108.905	gray cub cry	1937 7.30	 a,c	
23	Niobium(III) bromide $NbBr_3$	15752-41-7 332.618	dark brn solid		subl	
24	Niobium(III) chloride $NbCl_3$	13569-59-0 199.264	blk solid			
25	Niobium(III) fluoride NbF_3	15195-53-6 149.901	blue cub cry	 4.2		
26	Niobium(IV) chloride $NbCl_4$	13569-70-5 234.717	viol-blk monocl cry	 3.2		
27	Niobium(IV) fluoride NbF_4	13842-88-1 168.900	blk tetr cry; hyg	dec >350 4.01		
28	Niobium(IV) iodide NbI_4	13870-21-8 600.524	gray orth cry	503 5.6		
29	Niobium(IV) oxide NbO_2	12034-59-2 124.905	wh tetr cry or powder	1902 5.9	 a,c	
30	Niobium(IV) selenide $NbSe_2$	12034-77-4 250.83	gray hex cry	>1300 6.3		
31	Niobium(IV) sulfide NbS_2	12136-97-9 157.038	blk rhomb cry	 4.4		

No.	Name Formula	CAS RN Mol. Wt.	Physical Form	mp/°C den/g cm^{-3}	bp/°C Other Data	Solubility
1432	Niobium(IV) telluride $NbTe_2$	12034-83-2 348.11	hex cry	 7.6		
1433	Niobium(V) bromide $NbBr_5$	13478-45-0 492.426	oran orth cry	254 4.36	360	s H_2O, EtOH
1434	Niobium(V) chloride $NbCl_5$	10026-12-7 270.170	yel monocl cry; hyg	204.7 2.78	254.0 a,b,c,d	reac H_2O; s HCl, ctc
1435	Niobium(V) fluoride NbF_5	7783-68-8 187.898	col monocl cry; hyg	80 2.70	229 a,b,c,d,e	reac H_2O; sl CS_2, chl
1436	Niobium(V) iodide NbI_5	13779-92-5 727.428	yel-blk monocl cry	dec ≈200 5.32		
1437	Niobium(V) oxide Nb_2O_5	1313-96-8 265.810	wh orth cry	1512 4.6	 a,c	i H_2O; s HF
1438	Niobium(V) oxybromide $NbOBr_3$	14459-75-7 348.617	yel-brn cry	dec ≈320	subl	
1439	Niobium(V) oxychloride $NbOCl_3$	13597-20-1 215.263	wh tetr cry	 3.72	subl	
1440	Niobium(V) dioxyfluoride NbO_2F	15195-33-2 143.903	wh cub cry	 4.0		
1441	Nitrogen N_2	7727-37-9 28.013	col gas	-210.00 1.229 g/L	-195.79 a,b,c,d,e	sl H_2O; i EtOH
1442	Nitramide NO_2NH_2	7782-94-7 62.028	unstable wh cry	72 dec	 a	s H_2O, EtOH, ace, eth; i chl
1443	Nitric acid HNO_3	7697-37-2 63.013	col liq; hyg	-41.6 1.55	83 a,c,d	vs H_2O
1444	Nitrous acid HNO_2	7782-77-6 47.014	stable only in soln		 a	
1445	Nitrous oxide N_2O	10024-97-2 44.012	col gas	-90.8 1.931 g/L	-88.48 a,b,c,d,e	sl H_2O; s EtOH, eth
1446	Nitric oxide NO	10102-43-9 30.006	col gas	-163.6 1.317 g/L	-151.74 a,b,c,d,e	sl H_2O
1447	Nitrogen dioxide NO_2	10102-44-0 46.006	brn gas; equil with N_2O_4	 2.019 g/L	see N_2O_4 a	reac H_2O
1448	Nitrogen trioxide N_2O_3	10544-73-7 76.011	blue solid or liq (low temp)	-101.1 1.4(2°C)	≈3 dec a	reac H_2O
1449	Nitrogen tetroxide N_2O_4	10544-72-6 92.011	col liq; equil with NO_2	-9.3 1.45(20°C)	21.15 a,b,c,d,e	reac H_2O
1450	Nitrogen pentoxide N_2O_5	10102-03-1 108.010	col hex cry	30 2.0	47 a,e	s chl; sl ctc
1451	Nitrogen tribromide NBr_3	15162-90-0 253.719	unstable solid	exp -100		
1452	Nitrogen trichloride NCl_3	10025-85-1 120.365	yel oily liq; exp	-40 1.653	71 a	i H_2O; s CS_2, bz, ctc
1453	Nitrogen trifluoride NF_3	7783-54-2 71.002	col gas	-206.79 3.116 g/L	-128.75 a,b,d,e	i H_2O
1454	Nitrogen triiodide NI_3	13444-85-4 394.720	unstable blk cry; exp			
1455	Difluoramine NHF_2	10405-27-3 53.012	col gas	 2.326 g/L	-23 a,b	
1456	*cis*-Difluorodiazine N_2F_2	13812-43-6 66.010	col gas	<-195 2.896 g/L	-105.75 a,b	
1457	*trans*-Difluorodiazine N_2F_2	13776-62-0 66.010	col gas	-172 2.896 g/L	-111.45 a,b	
1458	Tetrafluorohydrazine N_2F_4	10036-47-2 104.007	col gas	-164.5 4.564 g/L	-74 a,b,d	
1459	Nitrosyl bromide NOBr	13444-87-6 109.910	red gas	-56 4.823 g/L	≈0 a	reac H_2O
1460	Nitrosyl chloride NOCl	2696-92-6 65.459	yel gas	-59.6 2.872 g/L	-5.5 a,b,d,e	reac H_2O
1461	Nitrosyl fluoride NOF	7789-25-5 49.004	col gas	-132.5 2.150 g/L	-59.9 a,d,e	
1462	Trifluoramine oxide NOF_3	13847-65-9 87.001	col gas	-161 3.818 g/L	-87.5 b	
1463	Nitryl chloride NO_2Cl	13444-90-1 81.459	col gas	-145 3.574 g/L	-15 a,d	
1464	Nitryl fluoride NO_2F	10022-50-1 65.004	col gas	-166 2.852 g/L	-72.4 a,b,d,e	reac H_2O
1465	Nitrogen selenide N_4Se_4	12033-88-4 371.87	red monocl cry; hyg	exp 4.2		i H_2O, eth, EtOH; sl bz, CS_2
1466	Nobelium No	10028-14-5 259	metal	827	 a	
1467	Osmium Os	7440-04-2 190.23	blue-wh metal; hex	3033 22.59	5012 a,c	s aqua regia
1468	Osmium carbonyl $Os_3(CO)_{12}$	15696-40-9 906.82	yel cry	 3.48		
1469	Osmium(III) bromide $OsBr_3$	59201-51-3 429.94	dark gray cry	dec 340		
1470	Osmium(III) chloride $OsCl_3$	13444-93-4 296.59	gray cub cry	dec >450	 a	i H_2O; s HNO_3

	Name Formula	CAS RN Mol. Wt.	Physical Form	mp/°C den/g cm^{-3}	bp/°C Other Data	Solubility
	Osmium(IV) chloride $OsCl_4$	10026-01-4 332.04	red-blk orth cry	 4.38	450 sp	reac H_2O
	Osmium(IV) fluoride OsF_4	54120-05-7 266.22	yel cry	230		
	Osmium(IV) oxide OsO_2	12036-02-1 222.23	yel-brn tetr cry	 11.4		i H_2O, acid
	Osmium(V) fluoride OsF_5	31576-40-6 285.22	blue cry	70		reac H_2O
	Osmium(VI) fluoride OsF_6	13768-38-2 304.22	yel cub cry	33.2 4.1	46 a	reac H_2O
6	Osmium(VIII) oxide OsO_4	20816-12-0 254.23	yel monocl cry	41 5.1	135 a,b,c,e	sl H_2O
7	Oxygen O_2	7782-44-7 31.999	col gas	-218.79 1.404 g/L	-182.95 a,b,c,d,e	sl H_2O, EtOH, os
8	Ozone O_3	10028-15-6 47.998	blue gas	-193 2.106 g/L	-111.35 a,b,e	sl H_2O
9	Palladium Pd	7440-05-3 106.42	silv-wh metal; cub	1554.9 12.0	2963 a,c	s aqua regia
0	Palladium(II) sulfide PdS	12125-22-3 138.49	gray tetr cry	 6.7	 a	
1	Palladium(II) bromide $PdBr_2$	13444-94-5 266.23	red-blk monocl cry; hyg	250 dec ≈5.2		i H_2O
2	Palladium(II) chloride $PdCl_2$	7647-10-1 177.33	red rhomb cry; hyg	679 4.0		s H_2O, EtOH, ace
3	Palladium(II) fluoride PdF_2	13444-96-7 144.42	viol tetr cry; hyg	952 5.76		reac H_2O
4	Palladium(II) iodide PdI_2	7790-38-7 360.23	blk cry	360 dec 6.0		i H_2O, EtOH, eth
5	Palladium(II) nitrate $Pd(NO_3)_2$	10102-05-3 230.43	brn hyg cry	dec		sl H_2O; s dil HNO_3
6	Palladium(II) oxide PdO	1314-08-5 122.42	grn blk tetr cry	750 dec 8.3	 a	i H_2O, acid; sl aqua regia
7	Phosphorus (white) P	7723-14-0 30.974	col waxlike cub cry	44.15 1.823	280.5 a,b,c,d,e	i H_2O; sl bz, EtOH, chl; s CS_2
8	Phosphorus (red) P	7723-14-0 30.974	red-viol amorp powder	597 2.16	280.5 a,b,c,d,e	i H_2O, os
9	Phosphorus (black) P	7723-14-0 30.974	blk orth cry or amorp solid	610 2.69	280.5 a,b,c,d,e	i os
0	Phosphine PH_3	7803-51-2 33.998	col gas; flam	-133 1.492 g/L	-87.75 a,b,d,e	i H_2O; sl EtOH, eth
1	Diphosphine P_2H_4	13445-50-6 65.980	col liq	-99	63.5 dec a,d	reac H_2O
2	Phosphoric acid (orthophosphoric acid) H_3PO_4	7664-38-2 97.995	col visc liq	42.4	407 a,c	s H_2O, EtOH
3	Metaphosphoric acid HPO_3	37267-86-0 79.980	gl solid; hyg			sl H_2O; s EtOH
4	Pyrophosphoric acid $H_4P_2O_7$	2466-09-3 177.975	wh cry	71.5	 a	s H_2O
5	Hypophosphoric acid $H_4P_2O_6$	7803-60-3 161.976	col orth cry	73 dec		vs H_2O
6	Difluorophosphoric acid HPO_2F_2	13779-41-4 101.978	col liq	≈-94 1.583	dec 110	reac H_2O
97	Phosphorous acid H_3PO_3	13598-36-2 81.996	wh hyg cry	74.4 1.65	200 a,c	vs H_2O, EtOH
98	Hypophosphorous acid HPH_2O_2	6303-21-5 65.997	hyg cry or col oily liq	26.5 1.49	130 a,c	vs H_2O, EtOH, eth
99	Phosphonium iodide PH_4I	12125-09-6 161.910	col tetr cry	18.5 2.86	62.5 e	reac H_2O, EtOH
00	Phosphorus heptasulfide P_4S_7	12037-82-0 348.357	pale yel monocl cry	308 2.19	523	sl CS_2
01	Phosphorus nitride P_3N_5	12136-91-3 162.955	wh amorp solid	dec 800		i H_2O; s os
02	Phosphonitrilic chloride trimer $(PNCl_2)_3$	940-71-6 347.657	wh hyg cry	128.8 1.98		reac H_2O
03	Phosphorus(III) bromide PBr_3	7789-60-8 270.686	col liq	-40 2.8	172.95 a,b,d,e	reac H_2O, EtOH; s ace, CS_2
04	Phosphorus(III) chloride PCl_3	7719-12-2 137.332	col liq	-112 1.574	75.95 a,b,c,d,e	reac H_2O, EtOH; s bz, chl, eth
05	Phosphorus(III) chloride difluoride $PClF_2$	14335-40-1 104.424	col gas	 4.582 g/L	-47.25 b,d	
06	Phosphorus(III) dichloride fluoride PCl_2F	15597-63-4 120.877	col gas	 5.304 g/L	14 b,d	
07	Phosphorus(III) fluoride PF_3	7783-55-3 87.969	col gas	-151.5 3.860 g/L	-101.5 a,b,d,e	reac H_2O
08	Phosphorus(III) iodide PI_3	13455-01-1 411.687	red-oran hex cry; hyg	61.5 4.18	227 dec a,d	reac H_2O; s EtOH
09	Phosphorus(III) oxide P_2O_3	1314-24-5 109.946	col monocl cry or liq	23.8 2.13	173 e	reac H_2O

No.	Name	Formula	CAS RN	Mol. Wt.	Physical Form	mp/°C	den/g cm^{-3}	bp/°C	Other Data	Solubility
1510	Phosphorus(III) selenide	P_2Se_3	1314-86-9	298.83	oran-red cry	245	1.31	≈380		reac H_2O; s bz, ctc, CS_2, ace
1511	Phosphorus(III) sulfide	P_2S_3	12165-69-4	158.146	yel solid	290		490		reac H_2O; s EtOH, eth, CS_2
1512	Phosphorus(V) bromide	PBr_5	7789-69-7	430.494	yel orth cry, Donnay	dec >100	3.61		a	reac H_2O, EtOH; s CS_2, ctc
1513	Phosphorus(V) chloride	PCl_5	10026-13-8	208.238	wh-yel tetr cry; hyg	167 tp	2.1	160 sp	a,b,e	reac H_2O; s CS_2, ctc
1514	Phosphorus(V) fluoride	PF_5	7647-19-0	125.966	col gas	-93.8	5.527 g/L	-84.6	a,d	reac H_2O
1515	Phosphorus(V) oxide	P_2O_5	1314-56-3	141.945	wh orth cry; hyg	562	2.30	605	c,e	reac H_2O, EtOH
1516	Phosphorus(V) oxybromide	$POBr_3$	7789-59-5	286.685	faint oran plates	56	2.822	191.7	a,d	reac H_2O; s bz, eth, chl
1517	Phosphorus(V) oxychloride	$POCl_3$	10025-87-3	153.331	col liq	1	1.645	105.5	a,c,d,e	reac H_2O, EtOH
1518	Phosphorus(V) oxyfluoride	POF_3	13478-20-1	103.968	col gas	-39.1 tp	4.562 g/L	-39.7 sp	a	reac H_2O
1519	Phosphorus(V) selenide	P_2Se_5	1314-82-5	456.75	blk-purp amorp solid					reac hot H_2O, ctc; i CS_2
1520	Phosphorus(V) sulfide	P_2S_5	1314-80-3	222.278	grn-yel hyg cry	285	2.03	515		reac H_2O; s CS_2
1521	Phosphorus(V) sulfide trifluoride	PSF_3	2404-52-6	120.035	gas	-148.8	5.267 g/L	-52.25	b,d	
1522	Phosphorus(V) sulfide trichloride	$PSCl_3$	3982-91-0	169.398	fuming liq	-36.2	1.635	125	e	reac H_2O, s bz, ctc, chl, CS_2
1523	Phosphorus(V) sulfide chloride difluoride	$PSClF_2$	2524-02-9	136.490	col gas	-155	5.989 g/L	6.3	b	
1524	Platinum	Pt	7440-06-4	195.08	silv-gray metal; cub	1768.4	21.5	3825	a,c,e,f	i acid; s aqua regia
1525	Hexachloroplatinic acid hexahydrate	$H_2PtCl_6 \cdot 6H_2O$	16941-12-1	517.90	brn-yel hyg cry	60	2.43			vs H_2O, EtOH
1526	Platinum silicide	PtSi	12137-83-6	223.17	orth cry	1229	12.4			
1527	Platinum(II) bromide	$PtBr_2$	13455-12-4	354.89	red-brn powder	dec 250	6.65		a	i H_2O
1528	Platinum(II) chloride	$PtCl_2$	10025-65-7	265.99	grn hex cry	dec 581	6.0		a	i H_2O, EtOH, eth; s HCl
1529	Platinum(II) iodide	PtI_2	7790-39-8	448.89	blk powder	dec 325	6.4			i H_2O
1530	Platinum(II) oxide	PtO	12035-82-4	211.08	blk tetr cry	dec 325	14.1			i H_2O, EtOH; s aqua regia
1531	Platinum(II) sulfide	PtS	12038-20-9	227.15	tetr cry		10.25		a,f	
1532	Platinum(III) bromide	$PtBr_3$	25985-07-3	434.79	grn-blk cry	dec 200			a	
1533	Platinum(III) chloride	$PtCl_3$	25909-39-1	301.44	grn-blk cry	435 dec	5.26		a	
1534	Platinum(IV) bromide	$PtBr_4$	68938-92-1	514.70	brn-blk cry	dec 180			a	i H_2O; sl EtOH, eth
1535	Platinum(IV) chloride	$PtCl_4$	37773-49-2	336.89	red-brn cub cry	327 dec	4.30			sl H_2O
1536	Platinum(IV) chloride pentahydrate	$PtCl_4 \cdot 5H_2O$	13454-96-1	426.97	red cry		2.43			s H_2O, EtOH
1537	Platinum(IV) fluoride	PtF_4	13455-15-7	271.07	red cry	600				
1538	Platinum(IV) iodide	PtI_4	7790-46-7	702.70	brn-blk powder	dec 130			a	s H_2O
1539	Platinum(IV) oxide	PtO_2	1314-15-4	227.08	blk hex cry	450	11.8			i H_2O; s conc acid, dil alk
1540	Platinum(IV) sulfide	PtS_2	12038-21-0	259.21	hex cry		7.85		a	
1541	Platinum(VI) fluoride	PtF_6	13693-05-5	309.07	red cub cry	61.3	≈4.0	69.1	a	
1542	Plutonium	Pu	7440-07-5	244	silv-wh metal; monocl	640	19.7	3228	a,c	
1543	Plutonium nitride	PuN	12033-54-4	258	gray cub cry	2550	14.4			
1544	Plutonium(II) oxide	PuO	12035-83-5	260	cub cry		14.0			
1545	Plutonium(III) bromide	$PuBr_3$	15752-46-2	484	grn orth cry	681	6.75			s H_2O
1546	Plutonium(III) chloride	$PuCl_3$	13569-62-5	350	grn hex cry	760	5.71			s H_2O
1547	Plutonium(III) fluoride	PuF_3	13842-83-6	301	purp hex cry	1425	9.33			i H_2O; sl acid
1548	Plutonium(III) iodide	PuI_3	13813-46-2	625	grn orth cry; hyg	777	6.92			s H_2O

PHYSICAL CONSTANTS OF INORGANIC COMPOUNDS (continued)

No.	Name Formula	CAS RN Mol. Wt.	Physical Form	mp/°C den/g cm^{-3}	bp/°C Other Data	Solubility
1549	Plutonium(III) oxide Pu_2O_3	12036-34-9 536	blk cub cry	 10.5		
1550	Plutonium(IV) fluoride PuF_4	13709-56-3 320	red-brn monocl cry	1027 7.1		
1551	Plutonium(IV) oxide PuO_2	12059-95-9 276	yel-brn cub cry	2400 11.5		
1552	Plutonium(VI) fluoride PuF_6	13693-06-6 358	red-brn orth cry	52 5.08		
1553	Polonium Po	7440-08-6 209	silv metal; cub	254 9.20	962 a	
1554	Polonium(IV) chloride $PoCl_4$	10026-02-5 351	yel hyg cry	≈300	390	s H_2O, EtOH, ace
1555	Polonium(IV) oxide PoO_2	7446-06-2 241	yel cub cry	dec 500 8.9		
1556	Potassium K	7440-09-7 39.098	soft silv-wh metal; cub	63.38 0.89	759 a,c,e	reac H_2O
1557	Potassium acetate $KC_2H_3O_2$	127-08-2 98.143	wh hyg cry	292 1.57	 a	vs H_2O; s EtOH; i eth
1558	Potassium aluminate trihydrate $K_2Al_2O_4 \cdot 3H_2O$	12003-63-3* 250.204	wh orth cry	 2.13		vs H_2O; i EtOH
1559	Potassium aluminum silicate $KAlSi_3O_8$	1327-44-2 278.332	col monocl cry	 2.56		i H_2O
1560	Potassium aluminum sulfate $KAl(SO_4)_2$	10043-67-1 258.207	wh hyg powder			sl H_2O
1561	Potassium aluminum sulfate dodecahydrate $KAl(SO_4)_2 \cdot 12H_2O$	7784-24-9 474.391	col cry	dec ≈100 1.72		sl H_2O
1562	Potassium arsenate K_3AsO_4	7784-41-0 256.215	col cry	 2.8		s H_2O
1563	Potassium azide KN_3	20762-60-1 81.118	tetr cry; exp	 2.04		vs H_2O
1564	Potassium borohydride KBH_4	13762-51-1 53.941	wh cub cry	dec ≈500 1.11	 a	s H_2O
1565	Potassium bromate $KBrO_3$	7758-01-2 167.000	wh hex cry	434 dec 3.27	 a	sl H_2O; i EtOH
1566	Potassium bromide KBr	7758-02-3 119.002	col cub cry; hyg	734 2.74	1435 a,c,e	vs H_2O; sl EtOH
1567	Potassium carbonate K_2CO_3	584-08-7 138.206	wh monocl cry; hyg	898 2.29	dec a,c	s H_2O; i EtOH
1568	Potassium carbonate sesquihydrate $K_2CO_3 \cdot 1.5H_2O$	6381-79-9 165.229	granular cry			vs H_2O
1569	Potassium chlorate $KClO_3$	3811-04-9 122.549	wh monocl cry	368 2.32	dec a	s H_2O
1570	Potassium chloride KCl	7447-40-7 74.551	wh cub cry	771 1.988	 a,c,e,f	s H_2O; i eth, ace
1571	Potassium chromate K_2CrO_4	7789-00-6 194.191	yel orth cry	975 2.73		vs H_2O
1572	Potassium cyanate KCNO	590-28-3 81.115	wh tetr cry	dec ≈700 2.05		s H_2O; sl EtOH
1573	Potassium cyanide KCN	151-50-8 65.116	wh cub cry; hyg	634 1.55	 a	vs H_2O; sl EtOH
1574	Potassium dichromate $K_2Cr_2O_7$	7778-50-9 294.185	oran-red tricl cry	398 2.68	dec ≈500	s H_2O
1575	Potassium dihydrogen arsenate KH_2AsO_4	7784-41-0 180.034	col cry	288 2.87		s H_2O; i EtOH
1576	Potassium dihydrogen phosphate KH_2PO_4	7778-77-0 136.085	wh tetr cry	253 2.34	 a	vs H_2O; sl EtOH
1577	Potassium ferricyanide $K_3Fe(CN)_6$	13746-66-2 329.246	red cry	dec 1.89		vs H_2O
1578	Potassium ferrocyanide trihydrate $K_4Fe(CN)_6 \cdot 3H_2O$	14459-95-1 422.390	yel monocl cry	dec 60 1.85		s H_2O; i EtOH, eth
1579	Potassium fluoride KF	7789-23-3 58.096	wh cub cry	858 2.48	1502 a,c,e	s H_2O
1580	Potassium fluoride dihydrate $KF \cdot 2H_2O$	13455-21-5 94.127	monocl cry	41 dec 2.5		vs H_2O
1581	Potassium fluoroborate KBF_4	14075-53-7 125.903	col orth cry	530 2.505		sl H_2O, EtOH
1582	Potassium formate $KCHO_2$	590-29-4 84.116	col hyg cry	167 1.91	 a	vs H_2O
1583	Potassium hexachloroosmate(IV) K_2OsCl_6	16871-60-6 481.14	red cub cry			vs H_2O; sl EtOH
1584	Potassium hexachloroplatinate K_2PtCl_6	16921-30-5 485.99	yel-oran cub cry	dec 250 3.50		sl H_2O; i EtOH
1585	Potassium hexacyanocobalt $K_3Co(CN)_6$	13963-58-1 332.334	yel monocl cry	dec 1.91		vs H_2O; i EtOH
1586	Potassium hexafluoromanganate(IV) K_2MnF_6	16962-31-5 247.125	yel hex cry			reac H_2O
1587	Potassium hexafluorosilicate K_2SiF_6	16871-90-2 220.273	wh cry	dec 2.27	 a	sl H_2O; i EtOH

PHYSICAL CONSTANTS OF INORGANIC COMPOUNDS (continued)

No.	Name Formula	CAS RN Mol. Wt.	Physical Form	mp/°C den/g cm^{-3}	bp/°C Other Data	Solubility
1588	Potassium hydride KH	7693-26-7 40.106	cub cry	 1.43	 a	reac H_2O
1589	Potassium hydrogen arsenite $KAsO_2 \cdot HAsO_2$	10124-50-2 253.947	wh hyg powder			s H_2O
1590	Potassium hydrogen carbonate $KHCO_3$	298-14-6 100.115	col monocl cry	dec ≈100 2.17	 a	vs H_2O; i EtOH
1591	Potassium hydrogen fluoride KHF_2	7789-29-9 78.103	col tetr cry	238.9 2.37	 a,c	s H_2O; i EtOH
1592	Potassium hydrogen phosphate K_2HPO_4	7758-11-4 174.176	wh hyg cry	dec		vs H_2O; sl EtOH
1593	Potassium hydrogen phosphite K_2HPO_3	13492-26-7 158.177	wh hyg powder	dec		vs H_2O; i EtOH
1594	Potassium hydrogen selenite $KHSeO_3$	7782-70-9 167.06	hyg orth cry	dec >100		s H_2O; sl EtOH
1595	Potassium hydrogen sulfate $KHSO_4$	7646-93-7 136.170	wh monocl cry; hyg	≈200 2.32	 a	vs H_2O
1596	Potassium hydrogen sulfide KHS	1310-61-8 72.172	wh hex cry; hyg	≈450 1.69		s H_2O, EtOH
1597	Potassium hydrogen sulfide hemihydrate $KHS \cdot 0.5H_2O$	1310-61-8* 81.179	wh-yel hyg cry	≈175 1.7		vs H_2O, EtOH
1598	Potassium hydrogen sulfite $KHSO_3$	7773-03-7 120.170	wh cry powder	dec 190		s H_2O; i EtOH
1599	Potassium hydrogen tartrate $KHC_4H_4O_6$	868-14-4 188.178	wh cry	 1.98		sl H_2O; s acid, alk; i EtOH
1600	Potassium hydroxide KOH	1310-58-3 56.105	wh rhomb cry; hyg	406 2.044	1327 a,c,e	vs H_2O; s EtOH
1601	Potassium hypophosphite KH_2PO_2	7782-87-8 104.087	wh hyg cry	dec		vs H_2O; s EtOH
1602	Potassium iodate KIO_3	7758-05-6 214.001	wh monocl cry	560 dec 3.89	 a	sl H_2O
1603	Potassium iodide KI	7681-11-0 166.003	col cub cry	681 3.12	1323 a,c,e	vs H_2O; sl EtOH
1604	Potassium manganate K_2MnO_4	10294-64-1 197.133	grn cry	190 dec		s H_2O; reac HCl
1605	Potassium metabisulfite $K_2S_2O_5$	16731-55-8 222.326	wh powder	dec ≈150 2.3		vs H_2O; reac acid; i EtOH
1606	Potassium metaborate KBO_2	13709-94-9 81.908	wh hex cry	 ≈2.3	 a	
1607	Potassium molybdate K_2MoO_4	13446-49-6 238.14	wh hyg cry	919 2.3		vs H_2O; i EtOH
1608	Potassium niobate $KNbO_3$	12030-85-2 180.002	wh rhomb cry	≈1100 4.64		i H_2O
1609	Potassium nitrate KNO_3	7757-79-1 101.103	col rhomb cry or powder	337 2.11	dec 400 a,c	s H_2O; i EtOH
1610	Potassium nitrite KNO_2	7758-09-0 85.104	wh hyg cry	441 1.915	exp 537 a	vs H_2O; sl EtOH
1611	Potassium oxalate monohydrate $K_2C_2O_4 \cdot H_2O$	6487-48-5 184.232	col cry	dec 160 2.13		vs H_2O
1612	Potassium oxide K_2O	12136-45-7 94.196	gray cub cry	350 dec 2.35	 a	s H_2O, EtOH, eth
1613	Potassium perchlorate $KClO_4$	7778-74-7 138.549	col orth cry; hyg	525 2.52	 a	sl H_2O
1614	Potassium periodate KIO_4	7790-21-8 230.001	col tetr cry	582 3.618	exp a	sl H_2O
1615	Potassium permanganate $KMnO_4$	7722-64-7 158.034	purp orth cry	dec 2.7	 a	s H_2O; reac EtOH
1616	Potassium peroxide K_2O_2	17014-71-0 110.196	yel amorp solid	490	 a	reac H_2O
1617	Potassium persulfate $K_2S_2O_8$	7727-21-1 270.324	col cry	dec ≈100 2.48		sl H_2O
1618	Potassium phosphate K_3PO_4	7778-53-2 212.266	wh orth cry; hyg	1340 2.564	 a	vs H_2O; i EtOH
1619	Potassium pyrophosphate trihydrate $K_4P_2O_7 \cdot 3H_2O$	7320-34-5* 384.383	col hyg cry	1090 2.33		vs H_2O; i EtOH
1620	Potassium pyrosulfate $K_2S_2O_7$	7790-62-7 254.325	col needles	≈325 2.28		s H_2O
1621	Potassium selenate K_2SeO_4	7790-59-2 221.16	wh powder	 3.07		vs H_2O
1622	Potassium selenide K_2Se	1312-74-9 157.16	red cub cry; hyg	800 2.29		s H_2O
1623	Potassium stannate trihydrate $K_2SnO_3 \cdot 3H_2O$	12142-33-5* 298.951	col cry	 3.20		vs H_2O; i EtOH
1624	Potassium sulfate K_2SO_4	7778-80-5 174.261	wh orth cry	1069 2.66	 a,c,f	s H_2O; i EtOH
1625	Potassium sulfide K_2S	1312-73-8 110.263	red-yel cub cry; hyg	948 1.74	 a,c	s H_2O, EtOH; i eth
1626	Potassium sulfide pentahydrate $K_2S \cdot 5H_2O$	1312-73-8* 200.339	col rhomb cry	60		vs H_2O, EtOH; i eth

No.	Name Formula	CAS RN Mol. Wt.	Physical Form	mp/°C den/g cm^{-3}	bp/°C Other Data	Solubility
1627	Potassium sulfite dihydrate $K_2SO_3 \cdot 2H_2O$	7790-56-9 194.292	wh monocl cry	dec		s H_2O; sl EtOH; dec dil acid
1628	Potassium superoxide KO_2	12030-88-5 71.097	yel tetr cry; hyg	380 2.16	 a	reac H_2O
1629	Potassium tellurate(VI) trihydrate $K_2TeO_4 \cdot 3H_2O$	15571-91-2* 323.84	wh cry powder			s H_2O
1630	Potassium tellurite K_2TeO_3	7790-58-1 253.80	wh hyg cry	dec ≈460		vs H_2O
1631	Potassium tetraborate pentahydrate $K_2B_4O_7 \cdot 5H_2O$	1332-77-0 323.513	wh cry powder			s H_2O; sl EtOH
1632	Potassium tetrachloroaurate dihydrate $KAuCl_4 \cdot 2H_2O$	13682-61-6 413.907	yel monocl cry			s H_2O, EtOH, eth
1633	Potassium tetrachloroplatinate K_2PtCl_4	10025-99-7 415.09	pink-red tetr cry	dec 500 3.38		s H_2O; i EtOH
1634	Potassium tetracyanoplatinate(II) trihydrate $K_2Pt(CN)_4 \cdot 3H_2O$	562-76-5* 431.39	col rhomb prisms			s H_2O
1635	Potassium tetraiodomercurate(II) K_2HgI_4	7783-33-7 786.40	yel hyg cry	 4.29		vs H_2O; s EtOH, eth, ace
1636	Potassium thiocyanate KSCN	333-20-0 97.182	col tetr cry; hyg	173 1.88	dec 500 a	vs H_2O; s EtOH
1637	Potassium thiosulfate $K_2S_2O_3$	10294-66-3 190.327	col hyg cry			vs H_2O; i EtOH
1638	Potassium titanate K_2TiO_3	12030-97-6 174.062	wh orth cry	1515 3.1		reac H_2O
1639	Potassium triiodide monohydrate $KI_3 \cdot H_2O$	7790-42-3 437.827	brn monocl cry; hyg	dec 225 3.5		s H_2O; reac EtOH, eth
1640	Potassium thiocarbonate K_2CS_3	26750-66-3 186.406	yel-red hyg cry			vs H_2O
1641	Potassium tungstate K_2WO_4	7790-60-5 326.04	hyg cry	921 3.12		vs H_2O; i EtOH
1642	Potassium uranate $K_2U_2O_7$	7790-63-8 666.251	oran cub cry	 6.12		i H_2O; s acid
1643	Praseodymium Pr	7440-10-0 140.908	yel metal; cub	931 6.77	3510 a,c	
1644	Praseodymium boride PrB_6	12008-27-4 205.774	blk cub cry	2610 4.84		
1645	Praseodymium bromide $PrBr_3$	13536-53-3 380.620	grn hex cry; hyg	693 5.28		s H_2O
1646	Praseodymium chloride $PrCl_3$	10361-79-2 247.266	grn hex needles; hyg	786 4.0	 a	s H_2O, EtOH
1647	Praseodymium chloride heptahydrate $PrCl_3 \cdot 7H_2O$	10025-90-8 373.373	grn cry	dec 110		s H_2O, EtOH
1648	Praseodymium fluoride PrF_3	13709-46-1 197.903	grn hex cry	1395 6.3		
1649	Praseodymium iodide PrI_3	13813-23-5 521.621	orth hyg cry	737 ≈5.8		s H_2O
1650	Praseodymium nitrate hexahydrate $Pr(NO_3)_3 \cdot 6H_2O$	14483-17-1 435.014	ligtt grn cry			vs H_2O
1651	Praseodymium nitride PrN	25764-09-4 154.915	cub cry	 7.46		
1652	Praseodymium oxide Pr_2O_3	12036-32-7 329.813	wh hex cry	2300 6.9	 a	
1653	Praseodymium silicide $PrSi_2$	12066-83-0 197.079	tetr cry	1712 5.46		
1654	Praseodymium sulfide Pr_2S_3	12038-13-0 378.013	cub cry	1765 5.1		
1655	Praseodymium telluride Pr_2Te_3	12038-12-9 664.62	cub cry	1500 ≈7.0		
1656	Promethium Pm	7440-12-2 145	silv metal	1042 7.26	3000 a	
1657	Protactinium Pa	7440-13-3 231.036	shiny metal; tetr or cub	1572 15.4	 a,c	
1658	Protactinium(V) chloride $PaCl_5$	13760-41-3 408.300	yel monocl cry	306 3.74	 a	
1659	Radium Ra	7440-14-4 226	wh metal; cub	700 5	 a	
1660	Radium bromide $RaBr_2$	10031-23-9 386	wh orth cry	728 5.79		s H_2O, EtOH
1661	Radium chloride $RaCl_2$	10025-66-8 297	wh orth cry	1000 4.9		s H_2O, EtOH
1662	Radium fluoride RaF_2	20610-49-5 264	wh cub cry	 6.7		
1663	Radium sulfate $RaSO_4$	7446-16-4 322	wh cry		 a	i H_2O, acid
1664	Radon Rn	10043-92-2 222	col gas	-71 9.741 g/L	-61.7 a,b,e	sl H_2O
1665	Rhenium Re	7440-15-5 186.207	silv-gray metal	3186 20.8	5596 a,c	i HCl

No.	Name Formula	CAS RN Mol. Wt.	Physical Form	mp/°C den/g cm^{-3}	bp/°C Other Data	Solubility
1666	Perrhenic acid $HReO_4$	13768-11-1 251.213	exists only in soln		 a	vs H_2O, os
1667	Rhenium carbonyl $Re_2(CO)_{10}$	14285-68-8 652.518	yel-wh cry	170 dec 2.87		s os
1668	Rhenium(III) bromide $ReBr_3$	13569-49-8 425.919	red-brn monocl cry	 6.10	subl 500 a	s ace, MeOH, EtOH
1669	Rhenium(III) chloride $ReCl_3$	13569-63-6 292.565	red-blk hyg cry	dec 500 4.81	 a	s H_2O
1670	Rhenium(III) iodide ReI_3	15622-42-1 566.920	blk solid	dec		
1671	Rhenium(IV) chloride $ReCl_4$	13569-71-6 328.018	purp-blk cry; hyg	dec 300 4.9		
1672	Rhenium(IV) fluoride ReF_4	15192-42-4 262.201	blue tetr cry	 7.49	subl >300	
1673	Rhenium(IV) oxide ReO_2	12036-09-8 218.206	gray orth cry	dec 900 11.4		
1674	Rhenium(IV) sulfide ReS_2	12038-63-0 250.339	tricl cry	 7.6		
1675	Rhenium(IV) telluride $ReTe_2$	12067-00-4 441.41	orth cry	 8.50		
1676	Rhenium(V) bromide $ReBr_5$	30937-53-2 585.727	brn solid	110 dec		
1677	Rhenium(V) chloride $ReCl_5$	39368-69-9 363.471	brn-blk solid	220 4.9		reac H_2O
1678	Rhenium(V) fluoride ReF_5	30937-52-1 281.199	yel-grn solid	48	≈220	
1679	Rhenium(V) oxide Re_2O_5	12165-05-8 452.411	blue-blk tetr cry	 ≈7		
1680	Rhenium(VI) chloride $ReCl_6$	31234-26-1 398.923	red-grn solid	29		
1681	Rhenium(VI) fluoride ReF_6	10049-17-9 300.197	yel liq or cub cry	18.5 4.06(cry)	33.7	s HNO_3
1682	Rhenium(VI) oxide ReO_3	1314-28-9 234.205	redcub cry	400 dec 6.9		i H_2O, acid, alk
1683	Rhenium(VI) oxytetrachloride $ReOCl_4$	13814-76-1 344.017	brn cry	29.3	223 b	reac H_2O
1684	Rhenium(VI) oxytetrafluoride $ReOF_4$	17026-29-8 278.200	blue solid	108	171	
1685	Rhenium(VII) fluoride ReF_7	17029-21-9 319.196	yel cub cry	48.3 4.32	73.7	
1686	Rhenium(VII) oxide Re_2O_7	1314-68-7 484.410	yel hyg cry	297 6.10	360 a,b,c,e	s H_2O, EtOH, eth, diox, py
1687	Rhenium(VII) trioxychloride ReO_3Cl	7791-09-5 269.658	col liq	4.5 3.87	128	reac H_2O
1688	Rhenium(VII) trioxyfluoride ReO_3F	42246-24-2 253.203	yel solid	147	164	
1689	Rhenium(VII) dioxytrifluoride ReO_2F_3	57246-89-6 275.201	yel solid	90	185	reac H_2O
1690	Rhenium(VII) oxypentafluoride $ReOF_5$	23377-53-9 297.198	cream solid	43.8	73.0	
1691	Rhenium(VII) sulfide Re_2S_7	12038-67-4 596.876	brn-blk tetr cry	 4.87		i H_2O
1692	Rhodium Rh	7440-16-6 102.906	silv-wh metal; cub	1964 12.4	3695 a,c	i acid, sl aqua regia
1693	Rhodium carbonyl chloride $[Rh(CO)_2Cl]_2$	14523-22-9 388.758	red-oran cry	124		s os
1694	Rhodium dodecacarbonyl $Rh_4(CO)_{12}$	19584-30-6 747.747	red hyg cry	 2.52		reac H_2O
1695	Rhodium(III) chloride $RhCl_3$	10049-07-7 209.264	red monocl cry	 5.38	717 a	i H_2O; s alk
1696	Rhodium(III) fluoride RhF_3	60804-25-3 159.901	red hex cry	 5.4		
1697	Rhodium(III) iodide RhI_3	15492-38-3 483.619	blk monocl cry; hyg	 6.4		
1698	Rhodium(III) oxide Rh_2O_3	12036-35-0 253.809	gray hex cry	1100 dec 8.2	 a	
1699	Rhodium(III) sulfate $Rh_2(SO_4)_3$	10489-46-0 494.002	red-yel solid	dec >500		
1700	Rhodium(IV) oxide RhO_2	12137-27-8 134.905	blk tetr cry	 7.2		
1701	Rhodium(VI) fluoride RhF_6	13693-07-7 216.896	blk cub cry	≈70 3.1		
1702	Rubidium Rb	7440-17-7 85.468	soft silv metal; cub	39.31 1.53	688 a,c,e	reac H_2O
1703	Rubidium aluminum sulfate $RbAl(SO_4)_2$	13530-57-9 304.577	hex cry	 ≈3.1		sl H_2O; i EtOH

No.	Name	Formula	CAS RN	Mol. Wt.	Physical Form	mp/°C	den/g cm^{-3}	bp/°C	Other Data	Solubility
1704	Rubidium aluminum sulfate dodecahydrate	$RbAl(SO_4)_2 \cdot 12H_2O$	7784-29-4	520.761	col cub cry	≈100 dec	≈1.9			s H_2O; i EtOH
1705	Rubidium azide	RbN_3	22756-36-1	127.488	tetr cry; exp	317	2.79			
1706	Rubidium bromide	RbBr	7789-39-1	165.372	wh cub cry; hyg	682	3.35	1340	a,c,e	vs H_2O
1707	Rubidium carbonate	Rb_2CO_3	584-09-8	230.945	col monocl cry; hyg	837			a	vs H_2O
1708	Rubidium chloride	RbCl	7791-11-9	120.921	wh cub cry; hyg	715	2.76	1390	a,c,e	s H_2O; sl EtOH
1709	Rubidium cyanide	RbCN	19073-56-4	111.486	wh cub cry		2.3			s H_2O; i EtOH, eth
1710	Rubidium fluoride	RbF	13446-74-7	104.466	wh cub cry; hyg	833	3.2	1410	a,c,e	vs H_2O; i EtOH
1711	Rubidium fluoride	$RbHF_2$	12280-64-7	124.473	tetr cry	188	3.3		a	
1712	Rubidium hydride	RbH	13446-75-8	86.476	wh cub cry; flam	dec ≈170	2.60		a	reac H_2O
1713	Rubidium hydrogen carbonate	$RbHCO_3$	19088-74-5	146.485	wh rhomb cry	dec 175				vs H_2O
1714	Rubidium hydrogen sulfate	$RbHSO_4$	15587-72-1	182.540	col monocl cry	208	2.9		a	s H_2O
1715	Rubidium hydroxide	RbOH	1310-82-3	102.475	gray-wh orth cry; hyg	300	3.2		a	vs H_2O; s EtOH
1716	Rubidium iodide	RbI	7790-29-6	212.372	wh cub cry	642	3.55	1300	a,c,e	s H_2O, EtOH
1717	Rubidium nitrate	$RbNO_3$	13126-12-0	147.473	wh hyg cry	305	3.11		a,c	vs H_2O
1718	Rubidium oxide	Rb_2O	18088-11-4	186.935	yel-brn cub cry; hyg	dec 400	4.0		a	reac H_2O
1719	Rubidium perchlorate	$RbClO_4$	13510-42-4	184.919	wh hyg cry	281	2.8	dec 600	a	sl H_2O
1720	Rubidium peroxide	Rb_2O_2	23611-30-5	202.935	wh orth cry		3.8		a	reac H_2O
1721	Rubidium selenide	Rb_2Se	31052-43-4	249.90	wh cub cry	733	3.22			reac H_2O
1722	Rubidium sulfate	Rb_2SO_4	7488-54-2	267.000	wh orth cry	1050	3.6		a	vs H_2O
1723	Rubidium sulfide	Rb_2S	31083-74-6	203.002	wh cub cry	425	2.91			s H_2O
1724	Rubidium superoxide	RbO_2	12137-25-6	117.467	tetr cry	412	≈3.0		a	
1725	Ruthenium	Ru	7440-18-8	101.07	silv-wh metal; hex	2334	12.1	4150	a,c	i acid, aqua regia
1726	Ruthenium dodecacarbonyl	$Ru_3(CO)_{12}$	15243-33-1	639.34	oran cry	dec 150				
1727	Ruthenium(III) acetylacetonate	$Ru(CH_3COCHCOCH_3)_3$	14284-93-6	398.40	red-brn cry	230				
1728	Ruthenium(III) bromide	$RuBr_3$	14014-88-1	340.78	brn hex cry	dec >400	5.3		a	
1729	Ruthenium(III) chloride	$RuCl_3$	10049-08-8	207.43	brn hex cry	dec >500	3.1		a	i H_2O; s EtOH
1730	Ruthenium(III) fluoride	RuF_3	51621-05-7	158.07	brn rhomb cry	dec >600	5.36			
1731	Ruthenium(III) iodide	RuI_3	13896-65-6	481.78	blk hex cry		6.0		a	
1732	Ruthenium(IV) fluoride	RuF_4	71500-16-8	177.06	yel cry					reac H_2O
1733	Ruthenium(IV) oxide	RuO_2	12036-10-1	133.07	gray-blk tetr cry		7.05		a	i H_2O, acid
1734	Ruthenium(V) fluoride	RuF_5	14521-18-7	196.06	grn monocl cry	86.5	3.90	227		
1735	Ruthenium(VI) fluoride	RuF_6	13693-08-8	215.06	dark brn orth cry	54	3.54			reac H_2O
1736	Ruthenium(VIII) oxide	RuO_4	20427-56-9	165.07	yel monocl prisms	25.4	3.29	40	a	sl H_2O; vs ctc; reac EtOH
1737	Samarium	Sm	7440-19-9	150.36	hard yel metal; rhomb	1072	7.52	1790	a,c	
1738	Samarium boride	SmB_6	12008-29-6	215.23	refrac solid	2580	5.07			
1739	Samarium silicide	$SmSi_2$	12300-22-0	206.53	orth cry		5.14			
1740	Samarium(II) bromide	$SmBr_2$	50801-97-3	310.17	brn cry	669				reac H_2O
1741	Samarium(II) chloride	$SmCl_2$	13874-75-4	221.27	brn cry	855	3.69			reac H_2O
1742	Samarium(II) fluoride	SmF_2	15192-17-3	188.36	purp cry					reac H_2O

No.	Name Formula	CAS RN Mol. Wt.	Physical Form	mp/°C den/g cm^{-3}	bp/°C Other Data	Solubility
1743	Samarium(II) iodide SmI_2	32248-43-4 404.17	grn cry	520		reac H_2O
1744	Samarium(III) bromide $SmBr_3$	13759-87-0 390.07	yel cry	640		reac H_2O
1745	Samarium(III) chloride $SmCl_3$	10361-82-7 256.72	yel cry	682 4.46	 a	reac H_2O
1746	Samarium(III) chloride hexahydrate $SmCl_3 \cdot 6H_2O$	13465-55-9 364.81	yel cry	dec 2.38		reac H_2O
1747	Samarium(III) fluoride SmF_3	13765-24-7 207.36	wh cry	1306	 a	reac H_2O
1748	Samarium(III) iodide SmI_3	13813-25-7 531.07	oran cry	850		reac H_2O
1749	Samarium(III) oxide Sm_2O_3	12060-58-1 348.72	yel-wh cub cry	2335 7.6	 a	
1750	Samarium(III) sulfate octahydrate $Sm_2(SO_4)_3 \cdot 8H_2O$	13465-58-2 733.03	yel cry	 2.93		sl H_2O
1751	Samarium(III) sulfide Sm_2S_3	12067-22-0 396.92	gray-brn cub cry	1720 5.87		
1752	Samarium(III) telluride Sm_2Te_3	12040-00-5 683.52	orth cry	 7.31		
1753	Scandium Sc	7440-20-2 44.956	silv-wh metal; hex	1541 2.99	2830 a,c	
1754	Scandium boride ScB_2	12007-34-0 66.578	refrac solid	2250 3.17		
1755	Scandium bromide $ScBr_3$	13465-59-3 284.668	wh hyg cry	969 9.33	 a	s H_2O
1756	Scandium chloride $ScCl_3$	10361-84-9 151.314	wh hyg cry	967 2.4	 a	s H_2O; i EtOH
1757	Scandium fluoride ScF_3	13709-47-2 101.951	wh powder	1515	 a	sl H_2O
1758	Scandium oxide Sc_2O_3	12060-08-1 137.910	wh cub cry	2485 3.864	 a	s conc acid
1759	Scandium sulfide Sc_2S_3	12166-29-9 186.110	yel orth cry	1775 2.91		
1760	Scandium telluride Sc_2Te_3	12166-44-8 472.71	blk hex cry	 5.29		
1761	Selenium (α form) Se	7782-49-2 78.96	red monocl cry	221 4.39	685 a,b,c,d,e	i H_2O, EtOH; sl eth
1762	Selenium (gray) Se	7782-49-2 78.96	gray metallic cry; hex	221 4.81	685 a,b,c,d,e	i H_2O, CS_2
1763	Selenium (vitreous) Se	7782-49-2 78.96	blk amorp solid	trans to gray Se 180 4.28	685 a,b,c,d,e	i H_2O; sl CS_2
1764	Selenic acid H_2SeO_4	7783-08-6 144.97	wh hyg solid	58 2.95	260 dec a	vs H_2O; reac EtOH
1765	Selenous acid H_2SeO_3	7783-00-8 128.97	wh hyg cry	dec 70 3.0		vs H_2O; s EtOH
1766	Selenium dioxide SeO_2	7446-08-4 110.96	wh tetr needles or powder	340 tp 3.95	315 sp a,e	s H_2O, EtOH, MeOH; sl ace
1767	Selenium trioxide SeO_3	13768-86-0 126.96	wh tetr cry; hyg	118 3.44	subl	s H_2O, os
1768	Selenium bromide Se_2Br_2	7789-52-8 317.73	red liq	 3.60	225 dec	reac H_2O; s CS_2, chl
1769	Selenium chloride Se_2Cl_2	10025-68-0 228.83	yel-brn oily liq	-85 2.774	130 dec	reac H_2O; s CS_2, bz, ctc, chl
1770	Selenium tetrabromide $SeBr_4$	7789-65-3 398.58	oran-red cry	123		reac H_2O; s CS_2, chl
1771	Selenium tetrachloride $SeCl_4$	10026-03-6 220.77	wh-yel cry	305 2.6	subl e	reac H_2O
1772	Selenium tetrafluoride SeF_4	13465-66-2 154.95	col liq	-10 2.75	106 d	reac H_2O, vs EtOH, eth
1773	Selenium hexafluoride SeF_6	7783-79-1 192.95	col gas	-34.6 tp 8.467 g/L	-46.6 sp a,b,e	i H_2O
1774	Selenium oxybromide $SeOBr_2$	7789-51-7 254.77	red-yel solid	41.6 3.38	220 dec	reac H_2O; s CS_2, bz, ctc
1775	Selenium oxychloride $SeOCl_2$	7791-23-3 165.86	col or yel liq	8.5 2.44	177 b,e	reac H_2O; s ctc, chl, bz, tol
1776	Selenium oxyfluoride $SeOF_2$	7783-43-9 132.96	col liq	15 2.8	125	reac H_2O
1777	Selenium dioxydifluoride SeO_2F_2	14984-81-7 148.96	col gas	-99.5 6.536 g/L	-8.4	reac H_2O
1778	Selenium sulfide Se_2S_6	75926-26-0 350.32	oran needles	121.5 2.44		s CS_2; sl bz
1779	Selenium sulfide Se_4S_4	75926-28-2 444.10	red cry	113 dec 3.29		s bz; sl CS_2
1780	Silicon Si	7440-21-3 28.086	gray cry or brn amorp solid	1414 2.3290	3265 a,c,e	i H_2O, acid; s alk
1781	Silane SiH_4	7803-62-5 32.118	col gas; flam	-185 1.409 g/L	-111.9 a,d,e	reac H_2O; i EtOH, bz

PHYSICAL CONSTANTS OF INORGANIC COMPOUNDS (continued)

No.	Name Formula	CAS RN Mol. Wt.	Physical Form	mp/°C den/g cm^{-3}	bp/°C Other Data	Solubility
2	Bromosilane SiH_3Br	13465-73-1 111.014	col gas	-94 4.871 g/L	1.9 a,d,e	
3	Bromotrichlorosilane $SiCl_3Br$	13465-74-2 214.348	col liq	-62 1.826	80.3 a	reac H_2O
4	Chlorosilane SiH_3Cl	13465-78-6 66.563	col gas	-118 2.921 g/L	-30.4 a,d,e	
5	Chlorotrifluorosilane SiF_3Cl	14049-36-6 120.534	col gas	-138 5.289 g/L	-70.0 b,d	reac H_2O
36	Dibromosilane SiH_2Br_2	13768-94-0 189.910	liq	-70.1	66 a,d,e	
37	Dichlorosilane SiH_2Cl_2	4109-96-0 101.007	col gas; flam	-122 4.432 g/L	8.3 a,d	reac H_2O
38	Dichlorodifluorosilane $SiCl_2F_2$	18356-71-3 136.988	col gas	-44 6.011 g/L	-32 b,d,e	reac H_2O
39	Difluorosilane SiH_2F_2	13824-36-7 68.099	col gas	-122 2.988 g/L	-77.8 d,e	
90	Diiodosilane SiH_2I_2	13760-02-6 283.911	col liq	-1	150 e	
91	Fluorosilane SiH_3F	13537-33-2 50.108	col gas	 2.199 g/L	-98.6 a,d,e	
92	Iodosilane SiH_3I	13598-42-0 158.014	col liq	-57	45.6 a,e	
93	Tetrabromosilane $SiBr_4$	7789-66-4 347.702	col fuming liq	5.2 2.8	154 a,b,d	reac H_2O
94	Tetrachlorosilane $SiCl_4$	10026-04-7 169.897	col fuming liq	-68.85 1.5	57.65 a,b,c,d,e	reac H_2O
95	Tetrafluorosilane SiF_4	7783-61-1 104.080	col gas	-90.2 4.567 g/L	-86 a,b,e	reac H_2O
96	Tetraiodosilane SiI_4	13465-84-4 535.704	wh powder	120.5 4.1	287.35 a,b,c,d	
97	Tribromosilane $SiHBr_3$	7789-57-3 268.806	flam liq	70 2.7	109 a,b,d,e	reac H_2O
98	Tribromochlorosilane $SiBr_3Cl$	13465-76-4 303.251	col liq	-20.8 2.497	127 a	reac H_2O
99	Trichlorosilane $SiHCl_3$	10025-78-2 135.452	fuming liq	-128.2 1.331	33 a,b,d,e	reac H_2O
00	Trichlorofluorosilane $SiCl_3F$	14965-52-7 153.442	col gas	 6.733 g/L	12.25 b,e	
01	Trifluorosilane $SiHF_3$	13465-71-9 86.089	col gas	-131 3.778 g/L	-95 a,d,e	
02	Triiodosilane $SiHI_3$	13465-72-0 409.807	liq	8	220 dec	
03	Disilane Si_2H_6	1590-87-0 62.219	col gas; flam	-132.5 2.730 g/L	-14.3 a,d,e	reac H_2O, ctc, chl; s EtOH, bz
304	Trisilane Si_3H_8	7783-26-8 92.321	flam liq	-117.4 0.725(-117°C)	52.9 a,d,e	reac H_2O
305	Tetrasilane Si_4H_{10}	7783-29-1 122.421	col liq	-84.3 0.79(-84°C)	107.4 e	reac H_2O
306	Disiloxane $(SiH_3)_2O$	13597-73-4 78.218	gas	-144 3.432 g/L	-15.2 e	
307	Metasilicic acid H_2SiO_3	7699-41-4 78.100	wh amorp powder		 a,f	i H_2O; s HF
308	Orthosilicic acid H_4SiO_4	10193-36-9 96.116	exists only in soln		 a	
309	Fluorosilicic acid H_2SiF_6	16961-83-4 144.092	stable only in aq soln			s H_2O
310	Silicon carbide SiC	409-21-2 40.097	hard grn-black hex cry	2830 3.16	 a,f	i H_2O, EtOH
811	Silicon nitride Si_3N_4	12033-89-5 140.284	gray refrac solid; hex	1900 3.17	 a	
812	Silicon monoxide SiO	10097-28-6 44.085	blk cub cry, stable >1200	 2.18	 a	
813	Silicon dioxide (α quartz) SiO_2	14808-60-7 60.085	col hex cry	573 trans to quartz 2.648	2950 a,c,f	i H_2O, acid; s HF
814	Silicon dioxide (β quartz) SiO_2	14808-60-7 60.085	col hex cry	867 trans to tridymite 2.533(600°C)	2950 a,c,f	i H_2O, acid; s HF
815	Silicon dioxide (tridymite) SiO_2	15468-32-3 60.085	col hex cry	1470 trans to cristobalite 2.265	2950	i H_2O, acid; s HF
816	Silicon dioxide (cristobalite) SiO_2	14464-46-1 60.085	col hex cry	1713 2.334	2950 e	i H_2O, acid; s HF
817	Silicon dioxide (vitreous) SiO_2	60676-86-0 60.085	col amorp solid	1713 2.196	2950	i H_2O, acid; s HF
818	Silicon monosulfide SiS	50927-81-6 60.152	yel-red hyg powder	≈900 1.85	940	reac H_2O
819	Silicon disulfide SiS_2	13759-10-9 92.218	wh rhomb cry	1090 2.04	subl	reac H_2O, EtOH; i bz

No.	Name Formula	CAS RN Mol. Wt.	Physical Form	mp/°C den/g cm^{-3}	bp/°C Other Data	Solubility
1820	Silver Ag	7440-22-4 107.868	silv metal; cub	961.78 10.5	2162 a,c,e,f	
1821	Silver azide AgN_3	13863-88-2 149.888	orth cry; exp	exp ≈250 4.9		
1822	Silver subfluoride Ag_2F	1302-01-8 234.734	yel hex cry	dec 100 8.6		reac H_2O
1823	Silver(I) acetate $AgC_2H_3O_2$	563-63-3 166.913	wh needles or powder	dec 3.26		s H_2O
1824	Silver(I) acetylide Ag_2C_2	7659-31-6 239.758	wh powder; exp			
1825	Silver(I) acetylide AgC_2H	13092-75-6 132.898	wh powder; exp			
1826	Silver(I) bromate $AgBrO_3$	7783-89-3 235.770	wh tetr cry	dec 5.2	 a	sl H_2O
1827	Silver(I) bromide AgBr	7785-23-1 187.772	yel cub cry	432 6.47	1502 a,c,d,f	i H_2O, acid
1828	Silver(I) carbonate Ag_2CO_3	534-16-7 275.745	yel monocl cry	218 6.077		sl H_2O; s acid
1829	Silver(I) chlorate $AgClO_3$	7783-92-8 191.319	wh tetr cry	230 4.430	dec 270 a	s H_2O; sl EtOH
1830	Silver(I) chloride AgCl	7783-90-6 143.321	wh cub cry	455 5.56	1547 a,c,d,e,f	i H_2O
1831	Silver(I) chromate Ag_2CrO_4	7784-01-2 331.730	brn-red monocl cry	 5.625	 a	i H_2O
1832	Silver(I) citrate $Ag_3C_6H_5O_7$	126-45-4 512.707	wh cry powder			i H_2O; s HNO_3
1833	Silver(I) cyanide AgCN	506-64-9 133.886	wh-gray hex cry	320 dec 3.95	 a	i H_2O, EtOH, dil acid
1834	Silver(I) dichromate $Ag_2Cr_2O_7$	7784-02-3 431.724	red cry	 4.770		sl H_2O
1835	Silver(I) fluoride AgF	7775-41-9 126.866	yel-brn cub cry; hyg	435 5.852	1159 a	s H_2O
1836	Silver(I) hydrogen fluoride $AgHF_2$	12249-52-4 146.873	hyg cry	dec		
1837	Silver(I) iodate $AgIO_3$	7783-97-3 282.770	wh orth cry	>200 5.53	 a	i H_2O
1838	Silver(I) iodide AgI	7783-96-2 234.772	yel powder; hex	558 5.68	1506 a,c,d,e,f	i H_2O, acid
1839	Silver(I) lactate monohydrate $AgC_3H_5O_3 \cdot H_2O$	128-00-7 214.955	gray cry powder			sl H_2O, EtOH
1840	Silver(I) molybdate Ag_2MoO_4	13765-74-7 375.67	yel cub cry	483 6.18		sl H_2O
1841	Silver(I) nitrate $AgNO_3$	7761-88-8 169.873	col rhomb cry	212 4.35	dec 440 a,c	vs H_2O; sl EtOH, ace
1842	Silver(I) nitrite $AgNO_2$	7783-99-5 153.874	yel needles	dec 140 4.453		sl H_2O; i EtOH; reac acid
1843	Silver(I) oxalate $Ag_2C_2O_4$	533-51-7 303.756	wh cry powder	exp 140 5.03		i H_2O
1844	Silver(I) oxide Ag_2O	20667-12-3 231.735	brn-blk cub cry	dec ≈200 7.2	 a	i H_2O, EtOH; s acid, alk
1845	Silver(I) perchlorate $AgClO_4$	7783-93-9 207.319	col cub cry; hyg	dec 486 2.806	 a	vs H_2O; s bz, py, os
1846	Silver(I) perchlorate monohydrate $AgClO_4 \cdot H_2O$	14242-05-8 225.334	hyg wh cry	dec 43		
1847	Silver(I) permanganate $AgMnO_4$	7783-98-4 226.804	viol monocl cry	dec 4.49		sl H_2O; reac EtOH
1848	Silver(I) phosphate Ag_3PO_4	7784-09-0 418.576	yel powder	849 6.37		i H_2O; sl dil acid
1849	Silver(I) picrate monohydrate $AgC_6H_2N_3O_7 \cdot H_2O$	146-84-9 353.981	yel cry			sl H_2O, EtOH; i chl, eth
1850	Silver(I) selenate Ag_2SeO_4	7784-07-8 358.69	orth cry	 5.72		sl H_2O
1851	Silver(I) selenide Ag_2Se	1302-09-6 294.70	gray hex needles	880 8.216		i H_2O
1852	Silver(I) selenite Ag_2SeO_3	7784-05-6 342.69	needles	530 5.930	dec >550	sl H_2O; s acid
1853	Silver(I) sulfate Ag_2SO_4	10294-26-5 311.800	col cry or powder	652 5.45	 a	sl H_2O
1854	Silver(I) sulfide Ag_2S	21548-73-2 247.802	gray-blk orth powder	825 7.23	 a,c,f	i H_2O; s acid
1855	Silver(I) telluride Ag_2Te	12002-99-2 343.34	blk orth cry	955 8.4	 f	
1856	Silver(I) tetraiodomercurate(II) Ag_2HgI_4	7784-03-4 923.94	yel tetr cry	trans to red cub ≈40 6.1		i H_2O, dil acid
1857	Silver(I) thiocyanate AgSCN	14104-20-2 165.952	wh powder	dec		i H_2O

No.	Name Formula	CAS RN Mol. Wt.	Physical Form	mp/°C den/g cm^{-3}	bp/°C Other Data	Solubility
8	Silver(II) oxide AgO	1301-96-8 123.867	gray powder; monocl or cub	dec >100 7.5		i H_2O; s alk; reac acid
9	Silver(II) fluoride AgF_2	7783-95-1 145.865	wh or gray hyg cry	690 4.58	 a	reac H_2O
0	Sodium Na	7440-23-5 22.990	soft silv met; cub	97.72 0.97	883 a,c,e	reac H_2O
1	Sodium acetate $NaC_2H_3O_2$	127-09-3 82.035	col cry	324 1.528	 a	vs H_2O
2	Sodium acetate trihydrate $NaC_2H_3O_2 \cdot 3H_2O$	6131-90-4 136.080	col cry	58 dec 1.45		vs H_2O; sl EtOH
3	Sodium aluminate $NaAlO_2$	1302-42-7 81.971	wh orth cry; hyg	1650 4.63		vs H_2O; i EtOH
4	Sodium aluminum sulfate dodecahydrate $NaAl(SO_4)_2 \cdot 12H_2O$	10102-71-3 458.283	col cry	≈60 1.61		vs H_2O; i EtOH
5	Sodium amide $NaNH_2$	7782-92-5 39.013	wh-grn orth cry	210 1.39	dec 500 a	reac H_2O
6	Sodium ammonium phosphate tetrahydrate $NaNH_4HPO_4 \cdot 4H_2O$	13011-54-6 209.069	monocl cry	≈80 dec 1.54		s H_2O; i EtOH
7	Sodium arsenite $NaAsO_2$	7784-46-5 129.911	wh-gray hyg powder	 1.87		vs H_2O; i EtOH
8	Sodium azide NaN_3	26628-22-8 65.010	col hex cry	dec 300; exp 1.846	 a	vs H_2O; sl EtOH; i eth
9	Sodium borohydride $NaBH_4$	16940-66-2 37.833	wh cub cry; hyg	dec ≈400 1.07	 a	vs H_2O; reac EtOH
0	Sodium bromate $NaBrO_3$	7789-38-0 150.892	col cub cry	381 3.34	 a,c	s H_2O; i EtOH
1	Sodium bromide NaBr	7647-15-6 102.894	wh cub cry	747 3.200	1390 a,c,e	vs H_2O; s EtOH
2	Sodium bromide dihydrate $NaBr \cdot 2H_2O$	13466-08-5 138.925	wh cry	36 dec 2.18		s H_2O; sl EtOH
3	Sodium carbonate Na_2CO_3	497-19-8 105.989	wh hyg powder	858.1 2.54	 a,c	s H_2O; i EtOH
4	Sodium carbonate decahydrate $Na_2CO_3 \cdot 10H_2O$	6132-02-1 286.142	col cry	34 dec 1.46		vs H_2O; i EtOH
5	Sodium carbonate monohydrate $Na_2CO_3 \cdot H_2O$	5968-11-6 124.005	col orth cry	dec 100 2.25	 f	vs H_2O; i EtOH
6	Sodium chlorate $NaClO_3$	7775-09-9 106.441	col cub cry	248 2.5	dec >300 a,c	vs H_2O; sl EtOH
7	Sodium chloride NaCl	7647-14-5 58.443	col cub cry	800.7 2.17	1465 a,c,e,f	s H_2O; sl EtOH
8	Sodium chlorite $NaClO_2$	7758-19-2 90.442	wh hyg cry	dec ≈180	 a	vs H_2O
9	Sodium chromate Na_2CrO_4	7775-11-3 161.974	yel orth cry	792 2.72		vs H_2O; sl EtOH
30	Sodium chromate tetrahydrate $Na_2CrO_4 \cdot 4H_2O$	10034-82-9 234.035	yel hyg cry	dec		vs H_2O; sl EtOH
31	Sodium citrate dihydrate $Na_3C_6H_5O_7 \cdot 2H_2O$	6132-04-3 294.101	wh cry	dec 150		vs H_2O; i EtOH, eth
32	Sodium cyanate NaOCN	917-61-3 65.007	col needles	550 1.89	 a	s H_2O; sl EtOH; i eth
33	Sodium cyanide NaCN	143-33-9 49.008	wh cub cry; hyg	563 1.6	 a,e	vs H_2O; sl EtOH
34	Sodium cyanoborohydride $NaBH_3(CN)$	25895-60-7 62.843	wh hyg powder	240 dec 1.12		vs H_2O; s thf; sl EtOH; i bz, eth
85	Sodium dichromate $Na_2Cr_2O_7$	10588-01-9 261.968	red hyg cry	357	dec 400	vs H_2O
86	Sodium dihydrogen phosphate monohydrate $NaH_2PO_4 \cdot H_2O$	10049-21-5 137.993	wh hyg cry	dec 100		vs H_2O; i EtOH
87	Sodium dihydrogen phosphate dihydrate $NaH_2PO_4 \cdot 2H_2O$	13472-35-0 156.008	col orth cry	60 dec 1.91		vs H_2O; i EtOH
88	Sodium dihydrogen pyrophosphate $Na_2H_2P_2O_7$	7758-16-9 221.939	wh powder	dec 220 ≈1.9		s H_2O
89	Sodium dithionate dihydrate $Na_2S_2O_6 \cdot 2H_2O$	7631-94-9* 242.139	col orth cry	dec 110 2.19		s H_2O; i EtOH
390	Sodium ferricyanide monohydrate $Na_3Fe(CN)_6 \cdot H_2O$	14217-21-1* 298.935	red hyg cry			s H_2O; i EtOH
391	Sodium ferrocyanide decahydrate $Na_4Fe(CN)_6 \cdot 10H_2O$	13601-19-9 484.063	yel monocl cry	dec ≈50 1.46		s H_2O; i os
392	Sodium fluoride NaF	7681-49-4 41.988	col cub or tetr cry	996 2.78	1704 a,c,e,f	sl H_2O; i EtOH
393	Sodium fluoroborate $NaBF_4$	13755-29-8 109.795	wh orth prisms	384 2.47	 a	vs H_2O; sl EtOH
394	Sodium formate $NaCHO_2$	141-53-7 68.008	wh hyg cry	253 1.92	dec a	vs H_2O; sl EtOH
395	Sodium hexafluoroaluminate Na_3AlF_6	13775-53-6 209.941	col monocl cry; trans cub 560	1009 2.97	 f	i H_2O

No.	Name Formula	CAS RN Mol. Wt.	Physical Form	mp/°C den/g cm^{-3}	bp/°C Other Data	Solubility
1896	Sodium hexafluorosilicate Na_2SiF_6	16893-85-9 188.056	wh hex cry	dec 2.7	a	sl H_2O; i EtOH
1897	Sodium hydride NaH	7646-69-7 23.998	silv cub cry; flam	425 dec 1.39	a	reac H_2O, EtOH
1898	Sodium hydrogen arsenate heptahydrate $Na_2HAsO_4{\cdot}7H_2O$	10048-95-0 312.014	wh monocl cry	dec ≈50 1.87		vs H_2O; sl EtOH
1899	Sodium hydrogen carbonate $NaHCO_3$	144-55-8 84.007	wh monocl cry	dec ≈50 2.20	a	s H_2O; i EtOH
1900	Sodium hydrogen fluoride $NaHF_2$	1333-83-1 61.995	wh hex cry	dec >160 2.08	a	s H_2O
1901	Sodium hydrogen phosphate Na_2HPO_4	7558-79-4 141.959	wh hyg powder	 1.7	a	s H_2O
1902	Sodium hydrogen phosphate dodecahydrate $Na_2HPO_4{\cdot}12H_2O$	10039-32-4 358.143	col cry	dec ≈35 ≈1.5		s H_2O; i EtOH
1903	Sodium hydrogen phosphate heptahydrate $Na_2HPO_4{\cdot}7H_2O$	7782-85-6 268.066	col cry	 ≈1.7		s H_2O; i EtOH
1904	Sodium hydrogen sulfate $NaHSO_4$	7681-38-1 120.062	wh hyg cry	≈315 2.43	a	vs H_2O
1905	Sodium hydrogen sulfate monohydrate $NaHSO_4{\cdot}H_2O$	10034-88-5 138.077	wh monocl cry	 2.10		vs H_2O; reac EtOH
1906	Sodium hydrogen sulfide NaHS	16721-80-5 56.064	col rhomb cry	350 1.79		s H_2O, EtOh, eth
1907	Sodium hydrogen sulfide dihydrate $NaHS{\cdot}2H_2O$	16721-80-5 92.095	yel hyg needles	dec 55		vs H_2O, EtOH, eth
1908	Sodium hydrogen sulfite $NaHSO_3$	7631-90-5 104.062	wh cry	 1.48		s H_2O; sl EtOH
1909	Sodium hydroxide NaOH	1310-73-2 39.997	wh orth cry; hyg	323 2.13	1388 a,c,d,e	vs H_2O; s EtOH, MeOH
1910	Sodium hypochlorite NaClO	7681-52-9 74.442	stable in aq soln	anh form exp		vs H_2O
1911	Sodium hypochlorite pentahydrate $NaOCl{\cdot}5H_2O$	7681-52-9* 164.518	pale grn orth cry	18 1.6		s H_2O
1912	Sodium iodate $NaIO_3$	7681-55-2 197.892	wh orth cry	dec 4.28	a	s H_2O; i EtOH
1913	Sodium iodide NaI	7681-82-5 149.894	wh cub cry; hyg	660 3.67	1304 a,c,e	vs H_2O; s EtOH, ace
1914	Sodium bismuthate $NaBiO_3$	12232-99-4 279.968	yel-brn hyg cry			i cold H_2O, reac acid
1915	Sodium metabisulfite $Na_2S_2O_5$	7681-57-4 190.109	wh cry			vs H_2O; sl EtOH
1916	Sodium metaborate $NaBO_2$	7775-19-1 65.800	wh hex cry	966 2.46	1434 a,c	s H_2O
1917	Sodium metasilicate Na_2SiO_3	6834-92-0 122.064	wh amorp solid; hyg	1089 2.61	a,c	s cold H_2O; reac hot H_2O
1918	Sodium molybdate Na_2MoO_4	7631-95-0 205.92	col cub cry	687 ≈3.5	a	vs H_2O
1919	Sodium molybdate dihydrate $Na_2MoO_4{\cdot}2H_2O$	10102-40-6 241.95	cry powder	dec 100 ≈3.5		vs H_2O
1920	Sodium niobate $NaNbO_3$	12034-09-2 163.894	rhomb cry	1422 4.55		i H_2O
1921	Sodium nitrate $NaNO_3$	7631-99-4 84.995	col hex cry; hyg	307 2.26	a,c	vs H_2O; sl EtOH, MeOH
1922	Sodium nitrite $NaNO_2$	7632-00-0 68.996	wh orth cry; hyg	271 2.17	dec >320 a	vs H_2O; sl EtOH; reac acid
1923	Sodium oxalate $Na_2C_2O_4$	62-76-0 134.000	wh powder	dec ≈250 2.34	a	sl H_2O; i EtOH
1924	Sodium oxide Na_2O	1313-59-3 61.979	wh amorp powder	1132 dec 2.27	a,c	reac H_2O
1925	Sodium perchlorate $NaClO_4$	7601-89-0 122.441	wh orth cry; hyg	480 dec 2.52	a	vs H_2O
1926	Sodium perchlorate monohydrate $NaClO_4{\cdot}H_2O$	7791-07-3 140.456	wh hyg cry	dec ≈130 2.02		vs H_2O
1927	Sodium periodate $NaIO_4$	7790-28-5 213.892	wh tetr cry	dec ≈300 3.86	a	s H_2O, acid
1928	Sodium periodate trihydrate $NaIO_4{\cdot}3H_2O$	13472-31-6 267.938	wh hex cry	dec 175 3.22		s H_2O
1929	Sodium permanganate trihydrate $NaMnO_4{\cdot}3H_2O$	10101-50-5* 195.972	red-blk hyg cry	dec 170 2.47		vs H_2O; reac EtOH
1930	Sodium peroxide Na_2O_2	1313-60-6 77.979	yel hyg powder	675 2.805	a	reac H_2O
1931	Sodium persulfate $Na_2S_2O_8$	7775-27-1 238.107	wh hyg cry			vs H_2O; reac EtOH
1932	Sodium phosphate dodecahydrate $Na_3PO_4{\cdot}12H_2O$	7601-54-9 380.124	col hex cry	≈75 1.62		s H_2O; i EtOH
1933	Sodium potassium tartrate tetrahydrate $NaK(C_4H_4O_6){\cdot}4H_2O$	304-59-6 282.221	wh cry	dec ≈70 1.79	anh at 130	vs H_2O; i EtOH

	Name Formula	CAS RN Mol. Wt.	Physical Form	mp/°C den/g cm^{-3}	bp/°C Other Data	Solubility
	Sodium pyrophosphate $Na_4P_2O_7$	7722-88-5 265.902	col cry	988 2.53		sl H_2O
	Sodium selenate decahydrate $Na_2SeO_4 \cdot 10H_2O$	10102-23-5 369.09	wh cry	 1.61		vs H_2O
	Sodium selenide Na_2Se	1313-85-5 124.94	amorp solid	>875 2.62		reac H_2O
	Sodium selenite Na_2SeO_3	10102-18-8 172.94	wh tetr cry			vs H_2O; i EtOH
	Sodium sulfate Na_2SO_4	7757-82-6 142.044	wh orth cry or powder	884 2.7	 a,c,f	s H_2O; i EtOH
	Sodium sulfate decahydrate $Na_2SO_4 \cdot 10H_2O$	7727-73-3 322.197	col monocl cry	32 dec 1.46	 f	s H_2O; i EtOH
	Sodium sulfide Na_2S	1313-82-2 78.046	wh cub cry; hyg	1172 1.856	 a,c	s H_2O; sl EtOH; i eth
	Sodium sulfide nonahydrate $Na_2S \cdot 9H_2O$	1313-84-4 240.184	wh-yel hyg cry	dec ≈50 1.43		vs H_2O; sl EtOH; i eth
	Sodium sulfide pentahydrate $Na_2S \cdot 5H_2O$	1313-83-3 168.122	col orth cry	120 dec 1.58		vs H_2O; s EtOH; i eth
3	Sodium sulfite Na_2SO_3	7757-83-7 126.044	wh hex cry	dec 2.63	 a	s H_2O; i EtOH
4	Sodium sulfite heptahydrate $Na_2SO_3 \cdot 7H_2O$	10102-15-5 252.151	wh monocl cry; unstable	 1.56		vs H_2O; sl EtOH
5	Sodium superoxide NaO_2	12034-12-7 54.989	yel cub cry	552 2.2	 a	reac H_2O
6	Sodium tellurate Na_2TeO_4	10102-20-2 237.58	wh powder			sl H_2O
7	Sodium tetraborate $Na_2B_4O_7$	1330-43-4 201.220	col gl solid; hyg	743 2.4	1575 a	sl H_2O, MeOH
8	Sodium tetraborate decahydrate $Na_2B_4O_7 \cdot 10H_2O$	1303-96-4 381.373	wh monocl cry	75 dec 1.73	 f	sl H_2O; i EtOH
9	Sodium tetraborate pentahydrate $Na_2B_4O_7 \cdot 5H_2O$	12045-88-4 291.296	hex cry	dec 1.88		sl H_2O
0	Sodium tetraborate tetrahydrate $Na_2B_4O_7 \cdot 4H_2O$	12045-87-3 273.281	wh monocl cry	 1.95	 f	sl H_2O
1	Sodium tetrachloroaluminate $NaAlCl_4$	7784-16-9 191.783	orth cry	 2.01		s H_2O
2	Sodium tetrafluoroberyllate Na_2BeF_4	13871-27-7 130.986	orth cry	575 2.47		sl H_2O
3	Sodium thiocyanate NaSCN	540-72-7 81.074	col hyg cry	287		vs H_2O
4	Sodium thiosulfate pentahydrate $Na_2S_2O_3 \cdot 5H_2O$	10102-17-7 248.186	col cry	dec ≈50 1.69		vs H_2O; i EtOH
5	Sodium tungstate dihydrate $Na_2WO_4 \cdot 2H_2O$	10213-10-2 329.85	wh orth cry	dec 100 3.25		vs H_2O; i EtOH
6	Strontium Sr	7440-24-6 87.62	silv-wh metal; cub	777 2.64	1382 a,c,e	reac H_2O; s EtOH
7	Strontium bromate monohydrate $Sr(BrO_3)_2 \cdot H_2O$	14519-18-7 361.44	wh-yel hyg cry	dec 240 3.77		s H_2O
8	Strontium bromide $SrBr_2$	10476-81-0 247.43	wh tetr cry	657 4.216	 a,c	vs H_2O
9	Strontium bromide hexahydrate $SrBr_2 \cdot 6H_2O$	7789-53-9 355.52	col hyg cry	88 dec		vs H_2O; s EtOH; i eth
0	Strontium carbide SrC_2	12071-29-3 111.64	tetr cry	 3.19		
1	Strontium carbonate $SrCO_3$	1633-05-2 147.63	wh orth cry; hyg	dec 1100 3.5	 a,f	i H_2O; s dil acid
2	Strontium chlorate $Sr(ClO_3)_2$	7791-10-8 254.52	col cry	120 dec 3.15		vs H_2O; sl EtOH
63	Strontium chloride $SrCl_2$	10476-85-4 158.53	wh cub cry; hyg	874 3.052	1250 a,c	vs H_2O
64	Strontium chloride hexahydrate $SrCl_2 \cdot 6H_2O$	10025-70-4 266.62	col hyg cry	dec 100 1.96		vs H_2O; s EtOH
65	Strontium chromate $SrCrO_4$	7789-06-2 203.61	yel monocl cry	dec 3.9		sl H_2O; s dil acid
66	Strontium fluoride SrF_2	7783-48-4 125.62	wh cub cry or powder	1477 4.24	2460 a,c	s H_2O, dil acid
67	Strontium hydride SrH_2	13598-33-9 89.64	orth cry	 3.26	 a	reac H_2O
68	Strontium hydroxide $Sr(OH)_2$	18480-07-4 121.64	col orth cry; hyg	510 3.625	dec 710 a,c	s H_2O
69	Strontium iodide SrI_2	10476-86-5 341.43	wh hyg cry	538 4.55	1773 dec a,c	vs H_2O
70	Strontium iodide hexahydrate $SrI_2 \cdot 6H_2O$	73796-25-5 449.52	wh-yel hex cry; hyg	dec 120 4.4		vs H_2O; s EtOH
71	Strontium niobate $SrNb_2O_6$	12034-89-8 369.43	monocl cry	1225 5.11		i H_2O
72	Strontium nitrate $Sr(NO_3)_2$	10042-76-9 211.63	wh cub cry	570 2.99	645 a	vs H_2O; sl EtOH, ace

PHYSICAL CONSTANTS OF INORGANIC COMPOUNDS (continued)

No.	Name Formula	CAS RN Mol. Wt.	Physical Form	mp/°C den/g cm^{-3}	bp/°C Other Data	Solubility
1973	Strontium nitrite $Sr(NO_2)_2$	13470-06-9 179.63	wh-yel hyg needles	dec 240 2.8	 a	vs H_2O
1974	Strontium oxide SrO	1314-11-0 103.62	col cub cry	2665 5.1	 a,c,e	reac H_2O
1975	Strontium peroxide SrO_2	1314-18-7 119.62	wh tetr cry; unstable	215 dec 4.78		reac H_2O
1976	Strontium phosphate $Sr_3(PO_4)_2$	7446-28-8 452.80	wh powder			i H_2O; s acid
1977	Strontium selenate $SrSeO_4$	7446-21-1 230.58	orth cry	 4.25		i H_2O; s hot HCl
1978	Strontium selenide SrSe	1315-07-7 166.58	wh cub cry	1600 4.54	 a	
1979	Strontium orthosilicate Sr_2SiO_4	13597-55-2 267.32	orth cry	 4.5	 a	
1980	Strontium silicide $SrSi_2$	12138-28-2 143.79	silv-gray cub cry	1100 3.35		
1981	Strontium sulfate $SrSO_4$	7759-02-6 183.68	wh orth cry	dec 1580 3.96	 a,f	i H_2O, EtOH; sl acid
1982	Strontium sulfide SrS	1314-96-1 119.69	gray cub cry	2002 3.70	 a	sl H_2O; s acid
1983	Strontium titanate $SrTiO_3$	12060-59-2 183.49	wh cub cry	2080 5.1		i H_2O
1984	Sulfur (α form) S	7704-34-9 32.066	yel orth cry	trans to β form 94.5 2.07	444.60 a,b,c,d,e,f	i H_2O; sl EtOH, bz, eth; s CS_2
1985	Sulfur (β form) S	7704-34-9 32.066	yel monocl needles, stable 94.5-115	115.21 2.07	444.60 a,b,c,d,e,f	i H_2O; sl EtOH, bz, eth; s CS_2
1986	Sulfuric acid H_2SO_4	7664-93-9 98.080	col oily liq	10.31 1.8	337 a,c,e	vs H_2O
1987	Peroxysulfuric acid H_2SO_5	7722-86-3 114.079	wh cry; unstable	dec 45		vs H_2O
1988	Chlorosulfonic acid $SO_2(OH)Cl$	7790-94-5 116.525	col-yel liq	-80 1.75	152	reac H_2O; s py
1989	Fluorosulfuric acid HSO_3F	7789-21-1 100.070	col liq	-89 1.726	163	reac H_2O
1990	Nitrosylsulfuric acid $HNOSO_4$	7782-78-7 127.078	prisms	dec 73		reac H_2O; s H_2SO_4
1991	Sulfurous acid H_2SO_3	7782-99-2 82.080	exists only in soln			soln of SO_2 in H_2O
1992	Sulfur dioxide SO_2	7446-09-5 64.065	col gas	-75.5 2.811 g/L	-10.05 a,b,d,e	s H_2O, EtOh, eth, chl
1993	Sulfur trioxide SO_3	7446-11-9 80.064	col liq	16.8 1.92	45 a,b,d,e	reac H_2O
1994	Sulfur bromide $SSBr_2$	13172-31-1 223.940	red oily liq	-46 2.63	dec >25 a	reac H_2O
1995	Sulfur chloride $SSCl_2$	10025-67-9 135.037	yel-red oily liq	-77 1.69	137 a,e	reac H_2O; s EtOH, bz, eth, ctc
1996	Sulfur fluoride SSF_2	16860-99-4 102.129	col gas	-164.6 4.481 g/L	-10.6	reac H_2O
1997	Sulfur fluoride FSSF	13709-35-8 102.129	col gas	-133 4.481 g/L	15	reac H_2O
1998	Sulfur dichloride SCl_2	10545-99-0 102.971	red visc liq	-122 1.62	59.6 a	reac H_2O
1999	Sulfur tetrafluoride SF_4	7783-60-0 108.060	col gas	-125 4.742 g/L	-40.45 a,b,d	reac H_2O
2000	Sulfur hexafluoride SF_6	2551-62-4 146.056	col gas	-50.7 tp 6.409 g/L	-63.8 sp a,b,c,d,e	sl H_2O; s EtOH
2001	Sulfur bromide pentafluoride SF_5Br	15607-89-3 206.962	col gas	-79 9.081 g/L	3.1	
2002	Sulfur chloride pentafluoride SF_5Cl	13780-57-9 162.511	col gas	-64 7.131 g/L	-19.05 a,b	
2003	Sulfur decafluoride S_2F_{10}	5714-22-7 254.116	liq	-52.7 2.08	30; dec 150	i H_2O
2004	Sulfuryl amide $SO_2(NH_2)_2$	7803-58-9 96.110	orth plates	93	dec 250	s H_2O, hot EtOH, ace
2005	Sulfuryl chloride SO_2Cl_2	7791-25-5 134.970	col liq	-51 1.680	69.4 a,d,e	reac H_2O; s bz, tol, eth
2006	Sulfuryl fluoride SO_2F_2	2699-79-8 102.062	col gas	-135.8 4.478 g/L	-55.4 a	sl H_2O, EtOH; s tol, ctc
2007	Thionyl bromide $SOBr_2$	507-16-4 207.873	yel liq	-50	140 e	reac H_2O
2008	Thionyl chloride $SOCl_2$	7719-09-7 118.970	yel fuming liq	-101 1.631	75.6 a,d,e	reac H_2O; s bz, ctc, chl
2009	Thionyl fluoride SOF_2	7783-42-8 86.062	col gas	-129.5 3.776 g/L	-43.8 a,d	reac H_2O; s bz, eth
2010	Sulfur fluoride hypofluorite F_5SOF	15179-32-5 162.055	col gas	-86 7.111 g/L	-35.1	
2011	Tantalum Ta	7440-25-7 180.948	gray metal; cub	3017 16.4	5458 a,c	reac HF

No.	Name Formula	CAS RN Mol. Wt.	Physical Form	mp/°C den/g cm⁻³	bp/°C Other Data	Solubility
2012	Tantalum aluminide $TaAl_3$	12004-76-1 261.893	gray refrac powder	≈1400 7.02		i H_2O, acid, alk
2013	Tantalum boride TaB	12007-07-7 191.759	refrac orth cry	2040 14.2		
2014	Tantalum boride TaB_2	12007-35-1 202.570	blk hex cry	3140 11.2		i H_2O, acid, alk
2015	Tantalum carbide TaC	12070-06-3 192.959	gold-brown powder; cub	3880 14.3	4780	s HF-HNO_3 mixture
2016	Tantalum carbide Ta_2C	12070-07-4 373.907	refrac hex cry	3327 15.1		
2017	Tantalum nitride TaN	12033-62-4 194.955	blk hex cry	3090 13.7		i H_2O; sl aqua regia; reac alk
2018	Tantalum silicide $TaSi_2$	12039-79-1 237.119	gray powder	2200 9.14		
2019	Tantalum(IV) oxide TaO_2	12036-14-5 212.947	tetr cry	 10.0		
2020	Tantalum(IV) selenide $TaSe_2$	12039-55-3 338.87	hex cry	 6.7		
2021	Tantalum(IV) sulfide TaS_2	12143-72-5 245.080	blk hex cry	>3000 6.86		i H_2O
2022	Tantalum(IV) telluride $TaTe_2$	12067-66-2 436.15	monocl cry	 9.4		
2023	Tantalum(V) bromide $TaBr_5$	13451-11-1 580.468	yel cry powder	265 4.99	349 a,b,c,d	
2024	Tantalum(V) chloride $TaCl_5$	7721-01-9 358.212	yel monocl cry; hyg	216 3.68	239.35 a,b,c,d	reac H_2O; s EtOH
2025	Tantalum(V) fluoride TaF_5	7783-71-3 275.940	wh monocl cry; hyg	95.1 5.0	229.2 a,d,e	s H_2O, eth; sl CS_2, ctc
2026	Tantalum(V) iodide TaI_5	14693-81-3 815.470	blk hex cry; hyg	496 5.80	543	
2027	Tantalum(V) oxide Ta_2O_5	1314-61-0 441.893	wh rhomb cry or powder	1785 8.2	 a,c	i H_2O, EtOH, acid; s HF
2028	Technetium Tc	7440-26-8 98	hex cry	2157 11	4265 a,c	
2029	Technetium(V) fluoride TcF_5	31052-14-9 193	yel solid	50	dec	
2030	Technetium(VI) fluoride TcF_6	13842-93-8 212	yel cub cry	37.4 3.0	55.3	
2031	Tellurium Te	13494-80-9 127.60	gray-wh rhomb cry	449.51 6.24	988 a,c,d,e	i H_2O, bz, CS_2
2032	Telluric(VI) acid H_6TeO_6	7803-68-1 229.64	wh monocl cry	136 3.07		s H_2O
2033	Tellurous acid H_2TeO_3	10049-23-7 177.61	wh cry	dec 40 3.0		sl H_2O; s dil acid, alk
2034	Tellurium dioxide TeO_2	7446-07-3 159.60	wh orth cry	733 5.9	1245 a	i H_2O; s alk, acid
2035	Tellurium trioxide TeO_3	13451-18-8 175.60	yel-oran cry	430 5.07		i H_2O
2036	Tellurium dibromide $TeBr_2$	7789-54-0 287.41	grn-brn hyg cry	210	339	reac H_2O; s eth; sl chl
2037	Tellurium dichloride $TeCl_2$	10025-71-5 198.51	blk amorp solid; hyg	208 6.9	328	reac H_2O; i ctc
2038	Tellurium tetrabromide $TeBr_4$	10031-27-3 447.22	yel-oran monocl cry	388 4.3	≈420 dec a	reac H_2O; s eth
2039	Tellurium tetrachloride $TeCl_4$	10026-07-0 269.41	wh monocl cry; hyg	224 3.0	387 a,b,d,e	reac H_2O; s EtOH, tol
2040	Tellurium tetrafluoride TeF_4	15192-26-4 203.59	col cry	129	dec 195	reac H_2O
2041	Tellurium tetraiodide TeI_4	7790-48-9 635.22	blk orth cry	280 5.05		reac H_2O; sl ace
2042	Tellurium hexafluoride TeF_6	7783-80-4 241.59	col gas	-37.6 tp 10.601 g/L	-38.9 sp a,b,e	reac H_2O
2043	Terbium Tb	7440-27-9 158.925	silv-gray metal	1359 8.23	3221 a,c	
2044	Terbium chloride $TbCl_3$	10042-88-3 265.283	wh orth cry; hyg	588 4.35	 a	s H_2O
2045	Terbium chloride hexahydrate $TbCl_3•6H_2O$	13798-24-8 373.374	hyg cry	 4.35		vs H_2O
2046	Terbium iodide TbI_3	13813-40-6 539.638	hex cry; hyg	957 ≈5.2		s H_2O
2047	Terbium nitride TbN	12033-64-6 172.932	cub cry	 9.55		
2048	Terbium oxide Tb_2O_3	12036-41-8 365.849	wh cub cry	2410 7.91	 a	
2049	Terbium silicide $TbSi_2$	12039-80-4 215.096	orth cry	 6.66		
2050	Terbium sulfide Tb_2S_3	12138-11-3 414.049	cub cry	 6.35		

No.	Name Formula	CAS RN Mol. Wt.	Physical Form	mp/°C den/g cm^{-3}	bp/°C Other Data	Solubility
2051	Thallium Tl	7440-28-0 204.383	soft blue-wh metal	304 11.8	1473 a,c,e	i H_2O; reac acid
2052	Thallium(I) acetate $TlC_2H_3O_2$	563-68-8 263.428	hyg wh cry	131 3.68		s H_2O, EtOH
2053	Thallium(I) bromide $TlBr$	7789-40-4 284.287	yel cub cry	460 7.5	819 a,c,d,e	i H_2O
2054	Thallium(I) carbonate Tl_2CO_3	6533-73-9 468.776	wh monocl cry	272 7.11	a,c	sl H_2O; i EtOH
2055	Thallium(I) chlorate $TlClO_3$	13453-30-0 287.834	col hex cry	5.5		sl H_2O
2056	Thallium(I) chloride $TlCl$	7791-12-0 239.836	wh cub cry	430 7.0	720 a,c,d,e	sl H_2O; i EtOH
2057	Thallium(I) cyanide $TlCN$	13453-34-4 230.401	wh hex plates	6.523		s H_2O, acid, EtOH
2058	Thallium(I) ethoxide TlC_2H_5O	20398-06-5 249.444	cloudy liq	-3 3.49	dec 130	reac H_2O
2059	Thallium(I) fluoride TlF	7789-27-7 223.381	wh orth cry	322 8.36	826 a,c	s H_2O
2060	Thallium(I) formate $TlCHO_2$	992-98-3 249.401	hyg col needles	101 4.97		vs H_2O; s MeOH
2061	Thallium(I) hexafluorophosphate $TlPF_6$	60969-19-9 349.347	wh cub cry	4.6		
2062	Thallium(I) hydroxide $TlOH$	12026-06-1 221.390	yel needles	139 dec 7.44	a	vs H_2O
2063	Thallium(I) iodide TlI	7790-30-9 331.287	yel cry powder	440 7.1	824 a,c,d,e	i H_2O, EtOH
2064	Thallium(I) molybdate Tl_2MoO_4	34128-09-1 568.71	yel-wh cub cry			i H_2O
2065	Thallium(I) nitrate $TlNO_3$	10102-45-1 266.388	wh cry	206 5.55	450 dec a,c	s H_2O; i EtOH
2066	Thallium(I) nitrite $TlNO_2$	13826-63-6 250.389	cub cry	5.7		vs H_2O
2067	Thallium(I) oxalate $Tl_2C_2O_4$	30737-24-7 496.787	wh powder	6.31		
2068	Thallium(I) oxide Tl_2O	1314-12-1 424.766	blk rhomb cry; hyg	596 9.52	≈1080 a	s H_2O, EtOH
2069	Thallium(I) perchlorate $TlClO_4$	13453-40-2 303.834	col orth cry	4.8		s H_2O
2070	Thallium(I) selenate Tl_2SeO_4	7446-22-2 551.73	orth cry	>400 6.875		sl H_2O; i EtOH, eth
2071	Thallium(I) selenide Tl_2Se	15572-25-5 487.73	gray plates	340	a	i H_2O, acid
2072	Thallium(I) sulfate Tl_2SO_4	7446-18-6 504.831	wh rhomb prisms	632 6.77	a,c	sl H_2O
2073	Thallium(I) sulfide Tl_2S	1314-97-2 440.833	blue-blk cry	448 8.39	1367 a,c,d	sl H_2O, alk; s acid
2074	Thallium(III) bromide tetrahydrate $TlBr_3 \cdot 4H_2O$	13701-90-1 516.157	yel orth cry	3.65		s H_2O, EtOH
2075	Thallium(III) chloride $TlCl_3$	13453-32-2 310.741	monocl cry	155 4.7	a	vs H_2O, EtOH, eth
2076	Thallium(III) chloride tetrahydrate $TlCl_3 \cdot 4H_2O$	13453-32-2* 382.803	orth cry	3.00		s H_2O
2077	Thallium(III) fluoride TlF_3	7783-57-5 261.378	wh orth cry; hyg	550 dec 8.65		reac H_2O
2078	Thallium(III) nitrate $Tl(NO_3)_3$	13746-98-0 390.398	col cry			reac H_2O
2079	Thallium(III) oxide Tl_2O_3	1314-32-5 456.765	brn cub cry	717 10.2		i H_2O; reac acid
2080	Thorium Th	7440-29-1 232.038	soft gray-wh metal; cub	1750 11.7	4788 a,c	s acids
2081	Thorium hydride ThH_2	16689-88-6 234.054	tetr cry	9.5	a	
2082	Thorium boride ThB_6	12229-63-9 296.904	refrac solid	2450 6.99		
2083	Thorium bromide $ThBr_4$	13453-49-1 551.654	wh hyg cry	679		reac H_2O
2084	Thorium carbide ThC	12012-16-7 244.049	cub cry	2500 10.6		reac H_2O
2085	Thorium dicarbide ThC_2	12071-31-7 256.060	yel monocl cry	≈2650 9.0		reac H_2O
2086	Thorium chloride $ThCl_4$	10026-08-1 373.849	gray-wh tetr needles; hyg	770 4.59	921 a,c,d	s H_2O, EtOH
2087	Thorium fluoride ThF_4	13709-59-6 308.032	wh monocl cry; hyg	1110 6.1	1680 a,d	
2088	Thorium iodide ThI_4	7790-49-0 739.656	wh-yel monocl cry	570	837	
2089	Thorium nitrate tetrahydrate $Th(NO_3)_4 \cdot 4H_2O$	33088-16-3 552.119	wh hyg cry	dec 500		vs H_2O; s EtOH

No.	Name Formula	CAS RN Mol. Wt.	Physical Form	mp/°C den/g cm^{-3}	bp/°C Other Data	Solubility
2090	Thorium nitride ThN	12033-65-7 246.045	refrac cub cry	2820 11.6		reac H_2O
2091	Thorium oxide ThO_2	1314-20-1 264.037	wh cub cry	3390 10.0	4400 a,f	i H_2O, alk; sl acid
2092	Thorium selenide $ThSe_2$	60763-24-8 389.96	orth cry	 8.5		
2093	Thorium orthosilicate $ThSiO_4$	14553-44-7 324.122	brn tetr cry	 6.7	 f	
2094	Thorium silicide $ThSi_2$	12067-54-8 288.209	tetr cry	1850 7.9		
2095	Thorium sulfate nonahydrate $Th(SO_4)_2{\cdot}9H_2O$	10381-37-0 586.303	wh monocl cry	dec 2.8		s H_2O
2096	Thorium sulfide ThS_2	12138-07-7 296.170	dark brn cry	1905 7.30		i H_2O; s acid
2097	Thulium Tm	7440-30-4 168.934	silv-wh metal	1545 9.32	1946 a,c	s dil acid
2098	Thulium bromide $TmBr_3$	14456-51-0 408.646	wh hyg cry	954		s H_2O
2099	Thulium chloride $TmCl_3$	13537-18-3 275.292	yel hyg cry	824	 a	s H_2O
2100	Thulium chloride heptahydrate $TmCl_3{\cdot}7H_2O$	13778-39-7 401.399	hyg cry			s H_2O, EtOH
2101	Thulium fluoride TmF_3	13760-79-7 225.929	wh cry	1158		s H_2O
2102	Thulium iodide TmI_3	13813-43-9 549.647	yel hyg cry	1021		
2103	Thulium oxide Tm_2O_3	12036-44-1 385.866	grn-wh cub cry	2425 8.6	 a	sl acid
2104	Tin (gray) Sn	7440-31-5 118.710	cub cry	trans to wh Sn 13.2 5.769	2602 a,c,e	
2105	Tin (white) Sn	7440-31-5 118.710	silv tetr cry	231.93 7.265	2602 a,c,e	
2106	Stannane SnH_4	2406-52-2 122.742	unstable col gas	-146 5.386 g/L	-51.8 a,d,e	
2107	Methylstannane SnH_3CH_3	1631-78-3 136.769	col gas	 6.001 g/L	0	reac H_2O
2108	Tin(II) acetate $Sn(C_2H_3O_2)_2$	638-39-1 236.800	wh orth cry	183 2.31	subl	s dil HCl
2109	Tin(II) bromide $SnBr_2$	10031-24-0 278.518	yel powder	215 5.12	639 a,d	s H_2O, EtOH, eth, ace
2110	Tin(II) chloride $SnCl_2$	7772-99-8 189.615	wh orth cry	247 3.90	623 a,c,d,e	s H_2O, EtOH, ace, eth; i xyl
2111	Tin(II) chloride dihydrate $SnCl_2{\cdot}2H_2O$	10025-69-1 225.646	wh monocl cry	37 dec 2.71		s H_2O, EtOH, NaOH; vs HCl
2112	Tin(II) fluoride SnF_2	7783-47-3 156.707	wh monocl cry; hyg	213 4.57	850	s H_2O; i EtOH, eth, chl
2113	Tin(II) hexafluorozirconate $SnZrF_6$	12419-43-1 323.924	cry	 4.21		s H_2O
2114	Tin(II) hydroxide $Sn(OH)_2$	12026-24-3 152.725	wh amorp solid		 a	
2115	Tin(II) iodide SnI_2	10294-70-9 372.519	red-oran powder	320 5.28	714 a,d	sl H_2O; s bz, chl, CS_2
2116	Tin(II) oxalate SnC_2O_4	814-94-8 206.730	wh powder	280 dec 3.56		i H_2O; s dil HCl
2117	Tin(II) oxide SnO	21651-19-4 134.709	blue-blk tetr cry	1080 dec 6.45	 a	i H_2O, EtOH; s acid
2118	Tin(II) pyrophosphate $Sn_2P_2O_7$	15578-26-4 411.363	wh amorp powder	dec 400 4.009		i H_2O; s conc acid
2119	Tin(II) selenide SnSe	1315-06-6 197.67	gray orth cry	861 6.18		i H_2O; s aqua regia
2120	Tin(II) sulfate $SnSO_4$	7488-55-3 214.774	wh orth cry	dec 378 4.15		reac H_2O
2121	Tin(II) sulfide SnS	1314-95-0 150.776	gray orth cry	880 5.08	1210 a	i H_2O; s conc acid
2122	Tin(II) tartrate $SnC_4H_4O_6$	815-85-0 266.782	wh cry powder			s H_2), dil HCl
2123	Tin(II) telluride SnTe	12040-02-7 246.31	gray cub cry	790 6.5		
2124	Tin(IV) bromide $SnBr_4$	7789-67-5 438.326	wh cry	31 3.34	205 a,b,c,d,e	vs H_2O; s EtOH
2125	Tin(IV) chloride $SnCl_4$	7646-78-8 260.521	col fuming liq	-33 2.234	114.15 a,b,c,d,e	reac H_2O; s EtOH, ctc, bz, ace
2126	Tin(IV) chloride pentahydrate $SnCl_4{\cdot}5H_2O$	10026-06-9 350.597	wh-yel cry	dec 56 2.04		vs H_2O; s EtOH
2127	Tin(IV) chromate $Sn(CrO_4)_2$	38455-77-5 350.697	brn-yel cry powder	dec		s H_2O
2128	Tin(IV) fluoride SnF_4	7783-62-2 194.704	wh tetr cry	 4.78	subl 705	reac H_2O

No.	Name Formula	CAS RN Mol. Wt.	Physical Form	mp/°C den/g cm^{-3}	bp/°C Other Data	Solubility
2129	Tin(IV) iodide SnI_4	7790-47-8 626.328	yel-brn cub cry	143 4.46	364.35 a,b,d,e	reac H_2O; s EtOH, bz, chl, eth
2130	Tin(IV) oxide SnO_2	18282-10-5 150.709	gray tetr cry	1630 6.85	 a,f	i H_2O, EtOH; s hot conc alk
2131	Tin(IV) selenide $SnSe_2$	20770-09-6 276.63	red-brn cry	650 ≈5.0		i H_2O; s alk, conc acid
2132	Tin(IV) selenite $Sn(SeO_3)_2$	7446-25-5 372.63	cry powder			i H_2O; s hot HCl
2133	Tin(IV) sulfide SnS_2	1315-01-1 182.842	gold-yel hex cry	dec 600 4.5		i H_2O; s alk, aqua regia
2134	Titanium Ti	7440-32-6 47.867	gray metal; hex	1668 4.506	3287 a,c	
2135	Titanium hydride TiH_2	7704-98-5 49.883	gray-blk powder	dec ≈450 3.75		i H_2O
2136	Titanium boride TiB_2	12045-63-5 69.489	gray refrac solid; hex	3225 4.38		
2137	Titanium carbide TiC	12070-08-5 59.878	cub cry	3067 4.93		i H_2O; s HNO_3
2138	Titanium nitride TiN	25583-20-4 61.874	yel-brn cub cry	3290 5.43		i H_2O; s aqua regia
2139	Titanium phosphide TiP	12037-65-9 78.841	gray hex cry	1990 4.08		
2140	Titanium silicide $TiSi_2$	12039-83-7 104.038	blk orth cry	1500 4.0		i H_2O, acid, alk; s HF
2141	Titanium(II) bromide $TiBr_2$	13783-04-5 207.675	blk powder	 4.0	 a	reac H_2O
2142	Titanium(II) chloride $TiCl_2$	10049-06-6 118.772	blk hex cryc	1035 3.13	1500 a,d	reac H_2O; s EtOH; i chl, eth
2143	Titanium(II) iodide TiI_2	13783-07-8 301.676	blk hex cry	 5.02		reac H_2O
2144	Titanium(II) oxide TiO	12137-20-1 63.866	cub cry	1750 4.95	 a	
2145	Titanium(II) sulfide TiS	12039-07-5 79.933	brn hex cry	1780 3.85		s conc acid
2146	Titanium(III) bromide $TiBr_3$	13135-31-4 287.579	blue-blk hex cry		 a	s H_2O
2147	Titanium(III) chloride $TiCl_3$	7705-07-9 154.225	red-viol hex cry; hyg	425 dec 2.64	 a,d	reac H_2O
2148	Titanium(III) fluoride TiF_3	13470-08-1 104.862	viol hex cry	1200 2.98	1400	i H_2O, dil acid, alk
2149	Titanium(III) oxide Ti_2O_3	1344-54-3 143.732	viol hex cry	1842 4.486	 a	s hot HF
2150	Titanium(III) sulfate $Ti_2(SO_4)_3$	10343-61-0 383.925	grn cry			i H_2O, EtOH; s dil HCl
2151	Titanium(III) sulfide Ti_2S_3	12039-16-6 191.932	blk hex cry	 3.56		
2152	Titanium(III,IV) oxide Ti_3O_5	12065-65-5 223.598	blk monocl cry	1777 4.24	 a	
2153	Titanium(IV) bromide $TiBr_4$	7789-68-6 367.483	yel-oran cub cry; hyg	39 3.37	230 a,b,c,d	reac H_2O
2154	Titanium(IV) chloride $TiCl_4$	7550-45-0 189.678	col or yel liq	-25 1.73	136.45 a,b,c,d,e	reac H_2O; s EtOH
2155	Titanium(IV) fluoride TiF_4	7783-63-3 123.861	wh hyg powder	284 2.798	subl	reac H_2O; s EtOH, py
2156	Titanium(IV) iodide TiI_4	7720-83-4 555.485	red hyg powder	150 4.3	377 a,b,c,d	reac H_2O
2157	Titanium(IV) oxide TiO_2	13463-67-7 79.866	wh tetr cry	1843 4.23	 a,f	i H_2O, dil acid; s conc acid
2158	Titanium(IV) oxysulfate monohydrate $TiOSO_4 \cdot H_2O$	13825-74-6* 177.945	col orth cry	 2.71		reac H_2O
2159	Titanium(IV) sulfide TiS_2	12039-13-3 111.999	yel-brn hex cry; hyg	 3.37		s H_2SO_4
2160	Tungsten W	7440-33-7 183.84	gray-wh metal; cub	3422 19.3	5555 a,c,e	
2161	Tungstic acid H_2WO_4	7783-03-1 249.85	yel amorp powder	dec 100 5.5		i H_2O, acid; s alk
2162	Tungsten boride W_2B	12007-10-2 378.49	refrac blk powder	2670 16.0		i H_2O
2163	Tungsten boride WB	12007-09-9 194.65	blk refrac powder	2665 15.2		i H_2O
2164	Tungsten boride W_2B_5	12007-98-6 421.74	refrac solid	2365 11.0		i H_2O
2165	Tungsten carbide W_2C	12070-13-2 379.69	refrac hex cry	≈2800 14.8		i H_2O
2166	Tungsten carbide WC	12070-12-1 195.85	gray hex cry	2785 15.6		i H_2O; s HNO_3/HF
2167	Tungsten carbonyl $W(CO)_6$	14040-11-0 351.90	wh cry	dec 170 2.65	subl	i H_2O; s os

No.	Name Formula	CAS RN Mol. Wt.	Physical Form	mp/°C den/g cm^{-3}	bp/°C Other Data	Solubility
2168	Tungsten nitride WN_2	60922-26-1 211.85	hex cry	dec 600 7.7		
2169	Tungsten nitride W_2N	12033-72-6 381.69	gray cub cry	dec 17.8		
2170	Tungsten silicide WSi_2	12039-88-2 240.01	blue-gray tetr cry	2160 9.3		i H_2O
2171	Tungsten silicide W_5Si_3	12039-95-1 1003.46	blue-gray refrac solid	2320 14.4		
2172	Tungsten(II) bromide WBr_2	13470-10-5 343.65	yel powder	dec 400		
2173	Tungsten(II) chloride WCl_2	13470-12-7 254.75	yel solid	dec >500		s H_2O
2174	Tungsten(II) iodide WI_2	13470-17-2 437.65	oran cry	 6.79		
2175	Tungsten(III) bromide WBr_3	15163-24-3 423.55	blk hex cry	dec >80		i H_2O
2176	Tungsten(III) chloride WCl_3	20193-56-0 290.20	red solid	550 dec	subl	reac H_2O
2177	Tungsten(IV) bromide WBr_4	14055-81-3 503.46	blk orth cry			reac H_2O
2178	Tungsten(IV) chloride WCl_4	13470-13-8 325.65	blk hyg powder	 4.62		reac H_2O
2179	Tungsten(IV) fluoride WF_4	13766-47-7 259.83	red-brn cry	dec >800		
2180	Tungsten(IV) iodide WI_4	14055-84-6 691.46	blk powder	dec		reac H_2O; s EtOH; i eth chl
2181	Tungsten(IV) oxide WO_2	12036-22-5 215.84	blue monocl cry	dec 1500-1700 10.8	 a	i H_2O, os
2182	Tungsten(IV) selenide WSe_2	12067-46-8 341.76	gray hex cry	 9.2		
2183	Tungsten(IV) sulfide WS_2	12138-09-9 247.97	gray hex cry	1250 dec 7.6		i H_2O, HCl, alk
2184	Tungsten(IV) telluride WTe_2	12067-76-4 439.04	gray orth cry	1020 9.43		
2185	Tungsten(V) bromide WBr_5	13470-11-6 583.36	brn-blk hyg solid	286	333	
2186	Tungsten(V) chloride WCl_5	13470-14-9 361.10	blk hyg cry	242	286	reac H_2O
2187	Tungsten(V) fluoride WF_5	19357-83-6 278.83	yel solid	dec >80		
2188	Tungsten(V) oxytribromide $WOBr_3$	20213-56-3 439.55	dark brn tetr cry	 ≈5.9		
2189	Tungsten(V) oxytrichloride $WOCl_3$	14249-98-0 306.20	grn tetr cry	 ≈4.6		
2190	Tungsten(VI) bromide WBr_6	13701-86-5 663.26	blue-blk cry	309	 a	
2191	Tungsten(VI) chloride WCl_6	13283-01-7 396.56	purp hex cry; hyg	275 3.52	346.75 a,b,c,d	s EtOH, os
2192	Tungsten(VI) dioxydibromide WO_2Br_2	13520-75-7 375.65	red cry			
2193	Tungsten(VI) dioxydichloride WO_2Cl_2	13520-76-8 286.74	yel orth cry	265 4.67		i H_2O
2194	Tungsten(VI) dioxydiiodide WO_2I_2	14447-89-3 469.65	monocl cry	 6.39		
2195	Tungsten(VI) fluoride WF_6	7783-82-6 297.83	col gas	2.3 13.069 g/L	17 a,b,c,d,e	reac H_2O
2196	Tungsten(VI) oxide WO_3	1314-35-8 231.84	yel powder	1472 7.2	 a,b	i H_2O; sl acid; s alk
2197	Tungsten(VI) oxytetrabromide $WOBr_4$	13520-77-9 519.46	red tetr cry	277 ≈5.5	327	reac H_2O
2198	Tungsten(VI) oxytetrachloride $WOCl_4$	13520-78-0 341.65	red hyg cry	211 11.92	227.55 b,c,d	reac H_2O; s bz, CS_2
2199	Tungsten(VI) oxytetrafluoride WOF_4	13520-79-1 275.83	wh monocl cry	106 5.07	186	reac H_2O
2200	Tungsten(VI) sulfide WS_3	12125-19-8 280.04	brn powder			sl H_2O; s alk
2201	Uranium U	7440-61-1 238.029	silv-wh orth cry	1135 19.1	4131 a,c	
2202	Uranium boride UB_2	12007-36-2 259.651	refrac solid	2430 12.7		
2203	Uranium boride UB_4	12007-84-0 281.273	refrac solid	2530 9.32		i H_2O
2204	Uranium carbide UC	12070-09-6 250.040	gray cub cry	2790		
2205	Uranium carbide UC_2	12071-33-9 262.051	gray tetr cry	2350 11.3	4370	reac H_2O; sl EtOH
2206	Uranium carbide U_2C_3	12076-62-9 512.091	gray cub cry	dec ≈1700 12.7		

No.	Name Formula	CAS RN Mol. Wt.	Physical Form	mp/°C den/g cm^{-3}	bp/°C Other Data	Solubility
2207	Uranium nitride UN	25658-43-9 252.036	gray cub cry	2805 14.3		i H_2O
2208	Uranium nitride U_2N_3	12033-83-9 518.078	cub cry	dec 11.3		
2209	Uranium(III) bromide UBr_3	13470-19-4 477.741	red hyg cry	727		s H_2O
2210	Uranium(III) chloride UCl_3	10025-93-1 344.387	grn hyg cry	837 5.51	 a	vs H_2O; i bz, ctc
2211	Uranium(III) fluoride UF_3	13775-06-9 295.024	blk hex cry	dec 8.9	 a	i H_2O; s acid
2212	Uranium(III) hydride UH_3	13598-56-6 241.053	gray-blk cub cry	 11.1	 a	
2213	Uranium(III) iodide UI_3	13775-18-3 618.742	blk hyg cry	766		s H_2O
2214	Uranium(IV) bromide UBr_4	13470-20-7 557.645	brn hyg cry	519		s H_2O, EtOH
2215	Uranium(IV) chloride UCl_4	10026-10-5 379.840	grn octahed cry	590 4.72	791 a,c	reac H_2O; s EtOH
2216	Uranium(IV) fluoride UF_4	10049-14-6 314.023	grn monocl cry	1036 6.7	1417 a	i H_2O; s conc acid, alk
2217	Uranium(IV) iodide UI_4	13470-22-9 745.647	blk hyg cry	506		s H_2O, EtOH
2218	Uranium(IV) oxide UO_2	1344-57-6 270.028	brn cub cry	2827 10.97	 a,f	i H_2O, dil acid; s conc acid
2219	Uranium(IV,V) oxide U_4O_9	12037-15-9 1096.111	cub cry	 11.2	 a	
2220	Uranium(V) bromide UBr_5	13775-16-1 637.549	brn hyg cry			reac H_2O
2221	Uranium(V) chloride UCl_5	13470-21-8 415.293	brn hyg cry	287		reac H_2O
2222	Uranium(V) fluoride UF_5	13775-07-0 333.021	pale blue tetr cry; hyg	348 5.81		s H_2O
2223	Uranium(V,VI) oxide U_3O_8	1344-59-8 842.082	grn-blk orth cry	1300 dec 8.38	 a	
2224	Uranium(VI) chloride UCl_6	13763-23-0 450.745	green hex cry	177 3.6	 a	
2225	Uranium(VI) fluoride UF_6	7783-81-5 352.019	wh monocl solid	64.0 tp 5.09	56.5 sp a,b,c,e	reac H_2O; s ctc, chl
2226	Uranium(VI) oxide UO_3	1344-58-7 286.027	oran-yel cry	 ≈7.3	 a	i H_2O; s acid
2227	Uranium(VI) oxide monohydrate $UO_3 \cdot H_2O$	12326-21-5 304.043	yel orth cry	dec 570 7.05		
2228	Uranyl chloride UO_2Cl_2	7791-26-6 340.933	yel orth cry; hyg	577	 a	vs H_2O; s EtOH, ace; i bz
2229	Uranyl nitrate hexahydrate $UO_2(NO_3)_2 \cdot 6H_2O$	13520-83-7 502.129	yel orth cry; hyg	60 2.81	dec 118	vs H_2O; s EtOH, eth
2230	Uranyl sulfate trihydrate $UO_2SO_4 \cdot 3H_2O$	20910-28-5 420.138	yel cry	 3.28		s H_2O, sl EtOH
2231	Vanadium V	7440-62-2 50.942	gray-wh metal; cub	1910 6.0	3407 a,c	i H_2O; s acid
2232	Vanadium boride VB	12045-27-1 61.753	refrac solid	2250		i H_2O
2233	Vanadium boride VB_2	12007-37-3 72.564	refrac solid	2450		
2234	Vanadium carbide VC	12070-10-9 62.953	refrac blk cry; cub	2810 5.77		i H_2O
2235	Vanadium carbide V_2C	12012-17-8 113.894	hex cry	2167		
2236	Vanadium carbonyl $V(CO)_6$	20644-87-5 219.004	blue-grn cry; flam	dec 60	subl	
2237	Vanadium nitride VN	24646-85-3 64.949	blk powder; cub	2050 6.13		i H_2O; s aqua regia
2238	Vanadium silicide VSi_2	12039-87-1 107.113	metallic prisms	 4.42		s HF
2239	Vanadium silicide V_3Si	12039-76-8 180.911	cub cry	1935 5.70		
2240	Vanadium(II) bromide VBr_2	14890-41-6 210.750	oran-brn hex cry	 4.58	subl 800	reac H_2O
2241	Vanadium(II) chloride VCl_2	10580-52-6 121.847	grn hex plates	 3.23	subl 910	reac H_2O; s EtOH, eth
2242	Vanadium(II) fluoride VF_2	13842-80-3 88.939	blue hyg cry			reac H_2O
2243	Vanadium(II) iodide VI_2	15513-84-5 304.751	red-viol hex cry	 5.44		reac H_2O
2244	Vanadium(II) oxide VO	12035-98-2 66.941	grn cry	1790 5.758	 a,c	s acid
2245	Vanadium(II) sulfate heptahydrate $VSO_4 \cdot 7H_2O$	36907-42-3 273.112	viol cry			

No.	Name Formula	CAS RN Mol. Wt.	Physical Form	mp/°C den/g cm^{-3}	bp/°C Other Data	Solubility
246	Vanadium(III) acetylacetonate $V(CHCOCHCOCH_3)_3$	13476-99-8 348.270	brn cry	≈185 ≈1.0	subl	s MeOH, ace, bz chl
247	Vanadium(III) bromide VBr_3	13470-26-3 290.654	gray-brn hyg cry	 4.00		reac H_2O
248	Vanadium(III) chloride VCl_3	7718-98-1 157.300	red-viol hex cry; hyg	 3.00	 a	reac H_2O; s EtOH, eth
249	Vanadium(III) fluoride VF_3	10049-12-4 107.937	yel-grn hex cry	≈1400 3.363	subl	i H_2O, EtOH
250	Vanadium(III) fluoride trihydrate $VF_3 \cdot 3H_2O$	10049-12-4* 161.983	grn rhomb cry	dec ≈100		sl H_2O
251	Vanadium(III) iodide VI_3	15513-94-7 431.655	brn-blk rhomb cry; hyg	 5.21		reac H_2O
252	Vanadium(III) oxide V_2O_3	1314-34-7 149.881	blk powder	2067 4.87	 a	i H_2O
253	Vanadium(III) sulfate $V_2(SO_4)_3$	13701-70-7 390.074	yel powder	dec ≈400		sl H_2O
254	Vanadium(III) sulfide V_2S_3	1315-03-3 198.081	grn-blk powder	dec 4.7		i H_2O, s hot HCl
255	Vanadium(IV) bromide VBr_4	13595-30-7 370.558	unstable magenta cry	dec -23	 a	
256	Vanadium(IV) chloride VCl_4	7632-51-1 192.753	unstable red liq	-25.7 1.816	148 a,c,d	reac H_2O; s EtOH, eth
257	Vanadium(IV) fluoride VF_4	10049-16-8 126.936	grn hyg powder	dec 325 3.15	subl a	vs H_2O
258	Vanadium(IV) oxide VO_2	12036-21-4 82.941	blue-blk powder	1967 4.339		i H_2O; s acid, alk
259	Vanadium(V) fluoride VF_5	7783-72-4 145.934	col liq	19.5 2.50	48.3 a,c,d	reac H_2O
260	Vanadium(V) oxide V_2O_5	1314-62-1 181.880	yel-brn orth cry	670 3.35	1800 dec a,c	sl H_2O; s conc acid, alk; i EtOH
261	Vanadyl bromide VOBr	13520-88-2 146.845	viol cry	dec 480		
262	Vanadyl chloride VOCl	13520-87-1 102.394	brn orth cry	dec >700 1.72	 a,b	
263	Vanadyl dibromide $VOBr_2$	13520-89-3 226.749	yel-brn cry	dec 180		
264	Vanadyl dichloride $VOCl_2$	10213-09-9 137.846	grn hyg cry	dec 380 2.88		reac H_2O; s EtOH
265	Vanadyl difluoride VOF_2	13814-83-0 104.938	yel cry			
266	Vanadyl selenite hydrate $VOSeO_3 \cdot H_2O$	133578-89-9 211.92	grn tricl plates	 3.506		
267	Vanadyl sulfate dihydrate $VOSO_4 \cdot 2H_2O$	27774-13-6 199.036	blue cry powder			s H_2O
268	Vanadyl tribromide $VOBr_3$	13520-90-6 306.653	deep red liq		dec 180	reac H_2O
269	Vanadyl trichloride $VOCl_3$	7727-18-6 173.299	fuming red liq	-79 1.829	127 a,d,e	reac H_2O; s MeOH, eth, ace
270	Vanadyl trifluoride VOF_3	13709-31-4 123.936	yel hyg powder	300 2.459	480	reac H_2O
271	Water H_2O	7732-18-5 18.015	col liq	0.00 0.9970	100.0 a,b,c,d,e	s EtOH, MeOH, ace
272	Xenon Xe	7440-63-3 131.29	col gas	-111.75 5.761 g/L	-108.04 a,b,c,d,e	sl H_2O
273	Xenon trioxide XeO_3	13776-58-4 179.29	col orth cry	exp ≈25 4.55		s H_2O
274	Xenon tetroxide XeO_4	12340-14-6 195.29	yel solid; exp	-35.9	dec ≈0	
275	Xenon difluoride XeF_2	13709-36-9 169.29	col tetr cry	129.03 tp 4.32	114.35 sp b	sl H_2O
276	Xenon tetrafluoride XeF_4	13709-61-0 207.28	col monocl cry	117.10 tp 4.04	115.75 sp a,b	reac H_2O
277	Xenon hexafluoride XeF_6	13693-09-9 245.28	col monocl cry	49.5 3.56	75.6	reac H_2O
278	Xenon dioxydifluoride XeO_2F_2	13875-06-4 201.29	col orth cry	30.8 exp 4.10		
279	Xenon oxytetrafluoride $XeOF_4$	13774-85-1 223.28	col liq	-46.2 3.17(0°C)		reac H_2O
280	Xenon fluoride hexafluororuthenate $XeFRuF_6$	22527-13-5 365.35	yel-grn monocl cry	110 3.78		
281	Xenon fluoride undecafluoroantimonate $XeFSb_2F_{11}$	15364-10-0 602.79	yel monocl cry	63 3.69		
282	Xenon fluoride hexafluoroarsenate $Xe_2F_3AsF_6$	50432-32-1 508.49	yel-grn monocl cry	99 3.62		reac H_2O
283	Xenon fluoride hexafluoroantimonate XeF_3SbF_6	39797-63-2 424.04	yel-grn monocl cry	≈110 3.92		
284	Xenon trifluoride undecafluoroantimonate $XeF_3Sb_2F_{11}$	35718-37-7 640.79	yel-grn tricl cry	82 3.98		

No.	Name	Formula	CAS RN	Mol. Wt.	Physical Form	mp/°C	den/g cm^{-3}	bp/°C	Other Data	Solubility
2285	Xenon pentafluoride hexafluoroarsenate	XeF_5AsF_6	20328-94-3	415.19	wh monocl cry	130.5	3.51			
2286	Xenon pentafluoride hexafluororuthenate	XeF_5RuF_6	39796-98-0	441.34	grn orth cry	152	3.79			
2287	Ytterbium	Yb	7440-64-4	173.04	silv metal; cub	824	6.90	1194	a,c	s dil acid
2288	Ytterbium silicide	$YbSi_2$	12039-89-3	229.21	hex cry		7.54			
2289	Ytterbium(II) bromide	$YbBr_2$	25502-05-0	332.85	yel cry	673				reac H_2O
2290	Ytterbium(II) chloride	$YbCl_2$	13874-77-6	243.95	grn cry	721	5.27			reac H_2O
2291	Ytterbium(II) iodide	YbI_2	19357-86-9	426.85	blk cry	772				reac H_2O
2292	Ytterbium(III) chloride	$YbCl_3$	10361-91-8	279.40	wh hyg powder	875			a	s H_2O
2293	Ytterbium(III) chloride hexahydrate	$YbCl_3 \cdot 6H_2O$	19423-87-1	387.49	grn hyg cry	dec 150	2.57			vs H_2O
2294	Ytterbium(III) fluoride	YbF_3	13760-80-0	230.04	wh cry	1157	8.2			i H_2O
2295	Ytterbium(III) oxide	Yb_2O_3	1314-37-0	394.08	col cub cry	2435	9.2		a	s dil acid
2296	Ytterbium(III) sulfate octahydrate	$Yb_2(SO_4)_3 \cdot 8H_2O$	10034-98-7	778.39	col cry		3.3			s H_2O
2297	Yttrium	Y	7440-65-5	88.906	silv-wh metal; hex	1526	4.47	3336	a,c	reac H_2O; s dil acid
2298	Yttrium aluminum oxide	$Y_3Al_5O_{12}$	12005-21-9	593.619	grn cub cry		≈4.5			
2299	Yttrium antimonide	YSb	12186-97-9	210.666	cub cry	2310	5.97			
2300	Yttrium arsenide	YAs	12255-48-0	163.828	cub cry		5.59			
2301	Yttrium boride	YB_6	12008-32-1	153.772	refrac solid	2600	3.72			
2302	Yttrium bromide	YBr_3	13469-98-2	328.618	col hyg cry	904				vs H_2O
2303	Yttrium carbide	YC_2	12071-35-1	112.928	refrac solid	≈2400	4.13			
2304	Yttrium carbonate trihydrate	$Y_2(CO_3)_3 \cdot 3H_2O$	5970-44-5	411.886	red-brn powder					i H_2O; s dil acid
2305	Yttrium chloride	YCl_3	10361-92-9	195.264	wh monocl cry; hyg	721	2.61		a	vs H_2O
2306	Yttrium fluoride	YF_3	13709-49-4	145.901	wh hyg powder	≈1150	4.0		a	i H_2O
2307	Yttrium nitrate hexahydrate	$Y(NO_3)_3 \cdot 6H_2O$	13494-98-9	383.012	hyg cry					vs H_2O
2308	Yttrium oxide	Y_2O_3	1314-36-9	225.810	wh cry; cub	2439	5.03		a,c	s dil acid
2309	Yttrium phosphide	YP	12294-01-8	119.880	cub cry		≈4.4			
2310	Yttrium sulfate octahydrate	$Y_2(SO_4)_3 \cdot 8H_2O$	7446-33-5	610.125	red monocl cry		2.6			s H_2O
2311	Yttrium sulfide	Y_2S_3	12039-19-9	274.010	yel cub cry	1925	3.87			
2312	Zinc	Zn	7440-66-6	65.39	blue-wh metal; hex	419.53	7.14	907	a,c,e	s acid, alk
2313	Zinc acetate dihydrate	$Zn(C_2H_3O_2)_2 \cdot 2H_2O$	5970-45-6	219.51	wh powder	237 dec	1.735			vs H_2O; s EtOH
2314	Zinc antimonide	ZnSb	12039-35-9	187.15	silv-wh orth cry	565	6.33			reac H_2O
2315	Zinc arsenate	$Zn_3(AsO_4)_2$	13464-44-3	474.01	wh powder					i H_2O; s acid, alk
2316	Zinc arsenate octahydrate	$Zn_3(AsO_4)_2 \cdot 8H_2O$	13464-45-4	618.13	wh monocl cry		3.33			i H_2O; s acid, alk
2317	Zinc arsenite	$Zn(AsO_2)_2$	10326-24-6	279.23	col powder					i H_2O; s acid
2318	Zinc borate	$3ZnO \cdot 2B_2O_3$	27043-84-1	383.41	wh amorp powder		3.64			sl H_2O; s dil acid
2319	Zinc borate hemiheptahydrate	$2ZnO \cdot 3B_2O_3 \cdot 3.5H_2O$	12513-27-8	434.69	wh cry	980	4.22			i H_2O
2320	Zinc bromate hexahydrate	$Zn(BrO_3)_2 \cdot 6H_2O$	13517-27-6	429.29	wh hyg solid	100	2.57			vs H_2O
2321	Zinc bromide	$ZnBr_2$	7699-45-8	225.20	wh hex cry; hyg	394	4.5	697	a,c,d	vs H_2O, EtOH; s eth
2322	Zinc caprylate	$Zn(C_8H_{15}O_2)_2$	557-09-5	351.80	wh hyg cry	136				sl H_2O
2323	Zinc carbonate	$ZnCO_3$	3486-35-9	125.40	wh rhomb cry	dec	4.4		a,f	i H_2O; s dil acid, alk

No.	Name Formula	CAS RN Mol. Wt.	Physical Form	mp/°C den/g cm^{-3}	bp/°C Other Data	Solubility
	Zinc chlorate $Zn(ClO_3)_2$	10361-95-2 232.29	yel hyg cry	dec 60 2.15		vs H_2O
	Zinc chloride $ZnCl_2$	7646-85-7 136.29	wh hyg cry	290 2.907	732 a,d,e	vs H_2O; s EtOH, ace
	Zinc chromite $ZnCr_2O_4$	12018-19-8 233.38	grn cub cry	 5.29		
	Zinc citrate dihydrate $Zn_3(C_6H_5O_7)_2$•$2H_2O$	546-46-3 610.40	col powder			sl H_2O; s dil acid, alk
	Zinc cyanide $Zn(CN)_2$	557-21-1 117.43	wh powder	 1.852		i H_2O; reac acid
	Zinc fluoride ZnF_2	7783-49-5 103.39	wh tetr needles; hyg	872 4.9	1500 a,d,e	sl H_2O
	Zinc fluoride tetrahydrate ZnF_2•$4H_2O$	13986-18-0 175.45	wh orth cry	 2.30		sl H_2O
	Zinc fluoroborate hexahydrate $Zn(BF_4)_2$•$6H_2O$	27860-83-9 347.09	hex cry	 2.12		vs H_2O; s EtOH
	Zinc formate dihydrate $Zn(CHO_2)_2$•$2H_2O$	5970-62-7 191.46	wh cry	 2.207		s H_2O, i EtOH
3	Zinc hexafluorosilicate hexahydrate $ZnSiF_6$•$6H_2O$	16871-71-9 315.56	wh cry			s H_2O
4	Zinc hydroxide $Zn(OH)_2$	20427-58-1 99.41	col orth cry	125 dec 3.05	 a	sl H_2O
5	Zinc iodate $Zn(IO_3)_2$	7790-37-6 415.20	wh cry powder			sl H_2O
6	Zinc iodide ZnI_2	10139-47-6 319.20	wh hyg cry	446 4.74	625 a	vs H_2O; s EtOH, eth
7	Zinc laurate $Zn(C_{12}H_{23}O_2)_2$	2452-01-9 464.02	wh powder	128		sl H_2O
8	Zinc molybdate $ZnMoO_4$	13767-32-3 225.33	wh tetr cry	>700 4.3		i H_2O
9	Zinc nitrate $Zn(NO_3)_2$	7779-88-6 189.40	wh powder		 a	vs H_2O
0	Zinc nitrate hexahydrate $Zn(NO_3)_2$•$6H_2O$	10196-18-6 297.49	col orth cry	36 dec 2.067		vs H_2O, EtOH
1	Zinc nitride Zn_3N_2	1313-49-1 224.18	blue-gray cub cry			
2	Zinc nitrite $Zn(NO_2)_2$	10102-02-0 157.40	hyg solid			reac H_2O
3	Zinc oleate $Zn(C_{18}H_{33}O_2)_2$	557-07-3 628.31	wh powder	70 dec		i H_2O; s EtOH, eth, bz
4	Zinc oxalate dihydrate ZnC_2O_4•$2H_2O$	547-68-2 189.44	wh powder	100 dec 2.56		i H_2O; s dil acid
45	Zinc oxide ZnO	1314-13-2 81.39	wh powder; hex	1975 5.6	 a,c,f	i H_2O; s dil acid
46	Zinc perchlorate hexahydrate $Zn(ClO_4)_2$•$6H_2O$	10025-64-6 372.38	wh cub cry; hyg	106 dec 2.2		vs H_2O; s EtOH
47	Zinc permanganate hexahydrate $Zn(MnO_4)_2$•$6H_2O$	23414-72-4 411.35	blk orth cry; hyg	 2.45		s H_2O; reac EtOH
48	Zinc peroxide ZnO_2	1314-22-3 97.39	yel-wh powder	dec >150 1.57	exp 212	i H_2O; reac acid, EtOH, ace
49	Zinc phosphate $Zn_3(PO_4)_2$	7779-90-0 386.11	wh monocl cry	900 4.0		i H_2O
50	Zinc phosphate tetrahydrate $Zn_3(PO_4)_2$•$4H_2O$	7543-51-3 458.17	col orth cry	 3.04	 f	i H_2O, EtOH; s dil acid, alk
51	Zinc phosphide Zn_3P_2	1314-84-7 258.12	gray tetr cry	1160 4.55		i H_2O, EtOH; reac acid; s bz
52	Zinc pyrophosphate $Zn_2P_2O_7$	7446-26-6 304.72	wh cry powder	 3.75		i H_2O; s dil acid
53	Zinc selenate pentahydrate $ZnSeO_4$•$5H_2O$	13597-54-1 298.42	tricl cry	dec 50 2.59		s H_2O
354	Zinc selenide $ZnSe$	1315-09-9 144.35	yel-red cub cry	>1100 5.65	subl a	i H_2O; s dil acid
355	Zinc orthosilicate Zn_2SiO_4	13597-65-4 222.86	wh hex cry	1509 4.1	 a,f	i H_2O, dil acid
356	Zinc stearate $Zn(C_{18}H_{35}O_2)_2$	557-05-1 632.34	wh powder	130 1.095		i H_2O, EtOH, eth; s bz
357	Zinc sulfate $ZnSO_4$	7733-02-0 161.45	col orth cry	680 dec 3.8	 a	vs H_2O
358	Zinc sulfate monohydrate $ZnSO_4$•H_2O	7446-19-7 179.47	wh monocl cry	238 dec 3.20		vs H_2O; i EtOH
359	Zinc sulfate heptahydrate $ZnSO_4$•$7H_2O$	7446-20-0 287.56	col orth cry	100 dec 1.97	 f	vs H_2O; i EtOH
360	Zinc sulfide (sphalerite) ZnS	1314-98-3 97.46	gray-wh cub cry	1700 4.04	 a,f	i H_2O, EtOH; s dil acid
361	Zinc sulfide (wurtzite) ZnS	1314-98-3 97.46	wh hex cry	1700 4.09	 a,f	i H_2O; s dil acid
362	Zinc sulfite dihydrate $ZnSO_3$•$2H_2O$	123714-57-8 181.49	wh powder	200 dec		i H_2O, EtOH; reac hot H_2O

No.	Name Formula	CAS RN Mol. Wt.	Physical Form	mp/°C den/g cm^{-3}	bp/°C Other Data	Solubility
2363	Zinc telluride $ZnTe$	1315-11-3 192.99	red cub cry	1239 5.9		i H_2O
2364	Zinc thiocyanate $Zn(SCN)_2$	557-42-6 181.56	wh hyg cry			sl H_2O; s EtOH
2365	Zirconium Zr	7440-67-7 91.224	gray-wh metal; hex	1855 6.52	4409 a,c	s hot conc acid
2366	Zirconium boride ZrB_2	12045-64-6 112.846	gray refrac solid; hex	3245 6.17		
2367	Zirconium bromide $ZrBr_4$	13777-25-8 410.840	wh cub cry	450 tp 3.98	360 sp a,b,e	
2368	Zirconium carbide ZrC	12020-14-3 103.235	gray refrac solid; cub	3532 6.73		s HF
2369	Zirconium chloride $ZrCl_4$	10026-11-6 233.035	wh monocl cry; hyg	437 tp 2.80	331 sp a,b,c,e	reac H_2O; s EtOH, eth
2370	Zirconium fluoride ZrF_4	7783-64-4 167.218	wh monocl cry	932 tp 4.43	912 sp a,c	sl H_2O
2371	Zirconium hydride ZrH_2	7704-99-6 93.240	gray tetr cry	 5.6	 a	i H_2O
2372	Zirconium hydroxide $Zr(OH)_4$	14475-63-9 159.254	wh amorp powder	dec 3.25		i H_2O; s acid
2373	Zirconium iodide ZrI_4	13986-26-0 598.842	oran cub cry; hyg	499 tp 4.85	431 sp a,b,e	vs H_2O
2374	Zirconium nitrate pentahydrate $Zr(NO_3)_4{\cdot}5H_2O$	13746-89-9 429.320	wh hyg cry	dec 100		vs H_2O; s EtOH
2375	Zirconium nitride ZrN	25658-42-8 105.231	yel cub cry	2960 7.09		s conc HF; sl dil acid
2376	Zirconium oxide ZrO_2	1314-23-4 123.223	wh amorp powder	2710 5.68	 a,c,f	i H_2O; sl acid
2377	Zirconium phosphide ZrP_2	12037-80-8 153.172	orth cry	 ≈5.1		
2378	Zirconium orthosilicate $ZrSiO_4$	10101-52-7 183.308	wh tetr cry	dec 1540 4.6	 a,f	i H_2O, acid
2379	Zirconium silicide $ZrSi_2$	12039-90-6 147.395	gray powder	1620 4.88		i H_2O, aqua regia; s HF
2380	Zirconium sulfate tetrahydrate $Zr(SO_4)_2{\cdot}4H_2O$	7446-31-3 355.413	wh tetr cry	dec 100 2.80		vs H_2O
2381	Zirconium sulfide ZrS_2	12039-15-5 155.356	red-brn hex cry	1480 3.82		i H_2O
2382	Zirconyl chloride octahydrate $ZrOCl_2{\cdot}8H_2O$	13520-92-8 322.251	tetr cry	dec 400 1.91		vs H_2O, EtOH

SYNONYM INDEX OF INORGANIC COMPOUNDS

Synonym	No.
Platinum monoxide	1530
Platinum tetrabromide	1534
Platinum tetrachloride	1535
Platinum tetrafluoride	1537
Platinum tetraiodide	1538
Platinum tribromide	1532
Platinum trichloride	1533
Plattnerite	1153
Plessy's green	597
Plutonium dioxide	1551
Plutonium hexafluoride	1552
Plutonium monoxide	1544
Plutonium sesquioxide	1549
Plutonium tetrafluoride	1550
Plutonium tribromide	1545
Plutonium trichloride	1546
Plutonium trifluoride	1547
Plutonium triiodide	1548
Polonium dioxide	1555
Polonium tetrachloride	1554
Polydymite	1411
Portlandite	445
Potassium bicarbonate	1590
Potassium bichromate	1574
Potassium bifluoride	1591
Potassium biselenite	1594
Potassium bisulfate	1595
Potassium bisulfide	1596
Potassium bisulfite	1598
Potassium bitartrate	1599
Potassium chloroplatinate	1633
Potassium cobalticyanide	1585
Potassium dioxide	1628
Potassium hexacyanoferrate(II) trihydrate	1578
Potassium hexacyanoferrate(III)	1577
Potassium hyposulfite	1637
Potassium monoxide	1612
Potassium orthodiuranate	1642
Potassium peroxydisulfate	1617
Potassium phosphate, dibasic	1592
Potassium pyrosulfite	1605
Potassium sodium tartrate	1933
Potassium trithiocarbonate	1640
Potcrate	1569
Powellite	450
Prussic acid	942
Pucherite	326
Pyrite	1007
Pyrochroite	1266
Pyrolusite	1293
Pyrophenite	1287
Pyrophosphoric acid	1494
Quartz	1813-1816
Quartz	1815
Quicklime	457
Quicksilver	1296
[illegible]	1155
Raspite	1144
Realgar	190
Red lead	1148
Red lead chromate	1103
Red phosphorus	1488
Retgersite	1408
Rhenium dioxide	1673
Rhenium disulfide	1674
Rhenium ditelluride	1675
Rhenium heptafluoride	1685
Rhenium heptasulfide	1691
Rhenium heptoxide	1686
Rhenium hexachloride	1680
Rhenium hexafluoride	1681
Rhenium pentabromide	1676
Rhenium pentachloride	1677
Rhenium pentafluoride	1678
Rhenium pentoxide	1679
Rhenium tetrachloride	1671
Rhenium tetrafluoride	1672
Rhenium tribromide	1668
Rhenium trichloride	1669
Rhenium triiodide	1670
Rhenium trioxide	1682
Rhodium dioxide	1700
Rhodium hexafluoride	1701
Rhodium trichloride	1695
Rhodium trifluoride	1696
Rhodium triiodide	1697
Rhodium trioxide	1698
Rhodochrosite	1261
Rhodonite	1277
Rochelle salt	1933
Rock salt	1877
Rohrbach's solution	275
Rubidium alum	1704
Rubidium bicarbonate	1713
Rubidium bifluoride	1711
Rubidium bisulfate	1714
Rubidium dioxide	1724
Ruthenium dioxide	1733
Ruthenium hexafluoride	1735
Ruthenium monoxide	1718
Ruthenium pentafluoride	1734
Ruthenium tetrafluoride	1732
Ruthenium tetroxide	1736
Ruthenium tribromide	1728
Ruthenium trichloride	1729
Ruthenium trifluoride	1730
Ruthenium triiodide	1731
Rutile	2157
Sacchite	1262
Sal ammoniac	75
Sal soda	1874
Salt of tartar	1567
Saltpeter	1609
Samaria	1749
Sanbornite	263
Sassolite	352
Scandia	1758
Scheele's green	707
Scheelite	480
Schultenite	1109
Selenium hexasulfide	1778
Selenium tetrasulfide	1779
Selenium(IV) bromide	1770
Selenium(IV) oxide	1766
Sellaite	1209
Senarmontite	168
Siderite	1013
Silane	1781
Silica	1813
Silicon chloride trifluoride	1785
Silicon decahydride	1805
Silicon octahydride	1804
Silicon tetrabromide	1793
Silicon tetrachloride	1794
Silicon tetrafluoride	1795
Silicon tetraiodide	1796
Silicon(II) oxide	1812
Silver difluoride	1859
Sinjarite	431
Skutterudite	615
Smaltite	614
Smite	1868
Smithsonite	2323
Soda ash	1873
Soda lye	1909
Sodamide	1865
Sodium acid pyrophosphate	1888
Sodium alum	1864
Sodium bicarbonate	1899
Sodium bifluoride	1900
Sodium bisulfate	1904
Sodium bisulfate monohydrate	1905
Sodium bisulfide	1906
Sodium bisulfide dihydrate	1907
Sodium bisulfite	1908
Sodium borate	1947
Sodium borofluoride	1893
Sodium dioxide	1945
Sodium metaarsenite	1867
Sodium metabismuthate	1914
Sodium metaperiodate	1928
Sodium metasilicate	1917
Sodium monobasic phosphate	1886
Sodium monoxide	1924
Sodium peroxydisulfate	1931
Sodium phosphate, dibasic	1901
Sodium prussiate yellow	1891
Sodium pyrosulfite	1915
Sodium tetrafluoroborate	1893
Sphalerite	2360
Spherocobaltite	630
Stannane	2106
Stannic bromide	2124
Stannic chloride	2125
Stannic chloride pentahydrate	2126
Stannic chromate	2127
Stannic fluoride	2128
Stannic iodide	2129
Stannic oxide	2130
Stannic selenide	2131
Stannic selenite	2132
Stannic sulfide	2133
Stannous acetate	2108
Stannous bromide	2109
Stannous chloride	2110
Stannous chloride dihydrate	2111
Stannous fluoride	2112
Stannous hexafluorozirconate	2113
Stannous hydroxide	2114
Stannous iodide	2115
Stannous monoxide	2117
Stannous oxalate	2116
Stannous pyrophosphate	2118
Stannous selenide	2119
Stannous sulfate	2120
Stannous sulfide	2121
Stannous tartrate	2122
Stannous telluride	2123
Stibine	162
Stibium	161
Stibnite	173
Stilleite	2354
Stolzite	1143
Strengite	1057
Strontia	1974
Strontianite	1961
Strontium orthosilicate	1979
Sulfamide	2004
Sulfinyl dibromide	2007
Sulfinyl dichloride	2008
Sulfinyl difluoride	2009
Sulfur chloride	1998
Sulfur decafluoride	2003
Sulfur monobromide	1994
Sulfur monochloride	1995
Sulfur monofluoride	1996, 1997
Sycoporite	665
Sylvite	1570
Szmikite	1281
Szomolnokite	1032
Tantalum boride	2014
Tantalum dioxide	2019
Tantalum diselenide	2020
Tantalum disulfide	2021
Tantalum ditelluride	2022
Tantalum pentabromide	2023

CAS RN	No.
2-54-4	417
2-76-0	1923
1-48-7	623
4-90-8	942
5-15-0	490
5-20-7	425
5-44-5	496
24-38-9	488
26-45-4	1832
27-08-2	1557
27-09-3	1861
28-00-7	1839
31-74-8	139
41-53-7	1894
42-03-0	22
42-71-2	702
43-33-9	1883
44-55-8	1899
46-84-9	1849
51-50-8	1573
56-62-7	435
298-14-6	1590
301-04-2	1086
302-01-2	928
304-59-6	1933
333-20-0	1636
353-50-4	497
409-21-2	1810
460-19-5	498
463-58-1	492
471-34-1	420
471-34-1	427
497-19-8	1873
504-64-3	493
506-64-9	1833
506-65-0	888
506-66-1	297
506-68-3	499
506-77-4	500
506-78-5	502
506-80-9	489
506-87-6	71
507-16-4	2007
512-25-4*	221
512-26-5	1104
513-74-6	85
513-77-9	215
513-78-0	386
513-79-1	630
516-02-9	250
528-94-9	140
533-51-7	1843
534-16-7	1828
534-17-8	531
535-37-5*	893
538-17-0	53
540-16-9	712
540-69-2	95
540-72-7	1953
541-43-5	227
542-62-1	222
542-83-6	392
542-84-7	636
543-80-6	207
543-81-7	289
543-90-8	379
544-17-2	440
544-19-4	726
544-60-5	128
544-92-3	690
546-46-3	2327
546-67-8	1149
546-93-0	1205
547-68-2	2344
554-13-2	1163
555-35-1	42
555-75-9	23
557-05-1	2356
557-07-3	2343
557-09-5	2322
557-21-1	2328
557-42-6	2364
562-76-5*	1634
563-63-3	1823
563-68-8	2052
563-71-3	1013
563-72-4	455
584-08-7	1567
584-09-8	1707
590-28-3	1572
590-29-4	1582
592-01-8	436
592-04-1	1311
592-05-2	1105
592-85-8	1329
592-87-0	1140
593-26-0	130
593-95-3	495
598-54-9	685
598-62-9	1261
598-63-0	1097
628-52-4*	574
628-86-4	1314
630-08-0	487
631-60-7	1297
631-61-8	63
[illegible]	[illegible]
638-39-1	2108
660-60-6	743
688-37-9	37
811-54-1	1108
813-93-4	317
814-88-0	402
814-89-1	651
814-89-1*	652
814-90-4*	580
814-91-5	735
814-93-7	1119
814-94-8	2116
815-82-7	748
815-85-0	2122
819-73-8	1096
865-44-1	982
868-14-4	1599
917-61-3	1882
917-69-1	670
940-71-6	1502
992-98-3	2060
1002-89-7	143
1066-30-4*	583
1066-33-7	107
1072-35-1	1134
1111-67-7	700
1111-71-3	301
1111-78-0	70
1117-94-8	686
1120-44-1	734
1120-46-3	1118
1184-65-2	863
1299-86-1	18
1301-96-8	1858
1302-01-8	1822
1302-09-6	1851
1302-42-7	1863
1302-52-9	292
1302-81-4	51
1302-82-5	45
1303-00-0	827
1303-11-3	952
1303-28-2	202
1303-33-9	198
1303-34-0	204
1303-35-1	189
1303-36-2	197
1303-37-3	203
1303-52-2	895
1303-58-8	897
1303-60-2	890
1303-61-3	900
1303-62-4	899
1303-86-2	366
1303-96-4	1948
1304-28-5	252
1304-29-6	256
1304-39-8	261
1304-40-1	265
1304-54-7	306
1304-56-9	307
1304-76-3	328
1304-82-1	338
1305-62-0	445
1305-78-8	457
1305-79-9	460
1305-84-6	466
1305-99-3	463
1306-19-0	404
1306-23-6	412
1306-24-7	408
1306-25-8	413
1306-38-3	524
1307-86-4	674
1307-96-6	653
1307-99-9	658
1308-04-9	676
1308-06-1	669
1308-14-1	591
1308-31-2	1017
1308-38-9	595
1308-56-1	723
1308-80-1	683
1308-87-8	764
1308-96-9	798
1309-33-7	1052
1309-37-1	1056
1309-42-8	1214
1309-48-4	1222
1309-60-0	1153
1309-64-4	169
1309-64-4	168
1310-03-8	1129
1310-32-3	1029
1310-43-6	1006
1310-52-7	1211
1310-53-8	880
1310-58-3	1600
1310-61-8	1596
1310-61-8*	1597
1310-65-2	1170
1310-66-3	1171
1310-73-2	1909
1310-82-3	1715
1311-11-1	1122
1311-90-6	138
1312-41-0	951
1312-42-1	970
1312-43-2	967
1312-45-4	973
1312-46-5	994
1312-73-8	1625
1312-73-8*	1626
1312-74-9	1622
1312-81-8	1079
1312-99-8	1245
1313-04-8	1231
1313-08-2	1237
1313-13-9	1293
1313-22-0	1279
1313-27-5	1363
1313-29-7	1347
1313-49-1	2341
1313-59-3	1924
1313-60-6	1930
1313-82-2	1940
1313-83-3	1942
1313-84-4	1941
1313-85-5	1936
1313-96-8	1437
1313-97-9	1374
1313-99-1	1404
1314-05-2	1405
1314-06-3	1412
1314-08-5	1486
1314-11-0	1974
1314-12-1	2068
1314-13-2	2345
1314-15-4	1539
1314-18-7	1975
1314-20-1	2091
1314-22-3	2348
1314-23-4	2376
1314-24-5	1509
1314-27-8	1147
1314-28-9	1682
1314-32-5	2079
1314-34-7	2252
1314-35-8	2196
1314-36-9	2308
1314-37-0	2295
1314-41-6	1148
1314-56-3	1515
1314-60-9	179
1314-61-0	2027
1314-62-1	2260
1314-68-7	1686
1314-80-3	1520
1314-82-5	1519
1314-84-7	2351
1314-86-9	1510
1314-87-0	1136
1314-91-6	1139
1314-95-0	2121
1314-96-1	1982
1314-97-2	2073
1314-98-3	2361
1314-98-3	2360
1315-01-1	2133
1315-03-3	2254
1315-04-4	180
1315-05-5	171
1315-06-6	2119
1315-07-7	1978
1315-09-9	2354
1315-11-3	2363
1317-33-5	1353
1317-34-6	1292
1317-35-7	1289
1317-36-8	1120
1317-36-8	1121
1317-37-9	1034
1317-38-0	736
1317-39-1	695
1317-40-4	747
1317-41-5	741
1317-42-6	665
1317-61-9	1039
1317-66-4	1007
1319-46-6	1098
1327-41-9	30
1327-44-2	1559
1327-50-0	174
1327-53-3	196
1327-53-3	195
1330-43-4	1947
1332-14-5	746
1332-52-1	293
1332-58-7	47
1332-63-4	1291
1332-71-4	679
1332-77-0	1631
1332-81-6	175
1333-74-0	939
1333-82-0	608
1333-83-1	1900
1335-26-8	1226
1335-31-5	1322
1335-32-6	1088
1336-21-6	116
1341-49-7	109
1344-28-1	40
1344-43-0	1274
1344-48-5	1327
1344-48-5	1326
1344-54-3	2149
1344-57-6	2218
1344-58-7	2226
1344-59-8	2223
1344-95-2	467
1345-04-6	173
1345-07-9	337
1345-13-7	520
1345-25-1	1027
1449-65-6	865
1495-50-7	501
1590-87-0	1803
1600-27-7	1306
1603-84-5	491
1631-78-3	2107
1633-05-2	1961
1762-95-4	154
1863-63-4	65
2092-16-2	477
2092-17-3	276
2092-17-3*	277
2404-52-6	1521
2406-52-2	2106
2452-01-9	2337
2466-09-3	1494
2524-02-9	1523
2551-62-4	2000
2644-70-4	931
2696-92-6	1460
2699-79-8	2006
3012-65-5	108
3095-65-6	115
3164-29-2	148
3251-23-8	731
3333-67-3	1394
3486-35-9	2323
3522-50-7	1047
3687-31-8	1090
3811-04-9	1569
3982-91-0	1522
4109-96-0	1787
4119-52-2	1062
5341-61-7	932
5714-22-7	2003
5743-04-4	380
5743-26-0	418
5794-28-5	456
5892-10-4	314
5893-61-8	727
5908-64-5	208
5908-81-6	273
5968-11-6	1875
5970-44-5	2304
5970-45-6	2313
5970-62-7	2332
5972-71-4	66
5972-72-5*	110
5972-76-9	69

CAS RN	No.	CAS RN	No.	CAS RN	No.	CAS RN	No.	CAS RN	No.
6009-70-7	129	7440-34-8	1	7580-67-8	1169	7757-86-0	1213	7783-11-1	14
6010-09-9	1036	7440-35-9	54	7601-54-9	1932	7757-93-9	443	7783-18-8	15
6046-93-1	704	7440-36-0	161	7601-89-0	1925	7758-01-2	1565	7783-19-9	142
6047-25-2	1026	7440-37-1	181	7601-90-3	548	7758-02-3	1566	7783-20-2	145
6080-56-4	1087	7440-38-2	182	7616-94-6	557	7758-05-6	1602	7783-21-3	14
6108-17-4	1155	7440-39-3	206	7631-90-5	1908	7758-09-0	1610	7783-22-4	15
6108-23-2	1168	7440-40-6	286	7631-94-9*	1889	7758-11-4	1592	7783-26-8	18
6131-90-4	1862	7440-40-6	287	7631-95-0	1918	7758-16-9	1888	7783-28-0	11
6132-02-1	1874	7440-41-7	288	7631-99-4	1921	7758-19-2	1878	7783-29-1	18
6132-04-3	1881	7440-42-8	341	7632-00-0	1922	7758-23-8*	462	7783-32-6	13
6147-53-1	624	7440-43-9	378	7632-51-1	2256	7758-87-4	461	7783-33-7	16
6150-82-9	1210	7440-44-0	483	7637-07-2	362	7758-88-5	517	7783-34-8	132
6150-88-5	1221	7440-45-1	503	7646-69-7	1897	7758-89-6	689	7783-35-9	132
6303-21-5	1498	7440-46-2	526	7646-78-8	2125	7758-94-3	1014	7783-36-0	13
6381-79-9	1568	7440-47-3	559	7646-79-9	631	7758-95-4	1100	7783-39-3	13
6424-20-0	641	7440-48-4	610	7646-85-7	2325	7758-97-6	1102	7783-40-6	120
6484-52-2	126	7440-50-8	681	7646-93-7	1595	7758-98-7	744	7783-41-7	803
6487-39-4	1070	7440-51-9	752	7647-01-0	941	7758-99-8	745	7783-42-8	200
6487-48-5	1611	7440-52-0	767	7647-10-1	1482	7759-00-4	1277	7783-43-9	177
6533-73-9	2054	7440-53-1	782	7647-14-5	1877	7759-01-5	1143	7783-44-0	804
6556-16-7	1273	7440-54-2	808	7647-15-6	1871	7759-01-5	1144	7783-46-2	110
6569-51-3	351	7440-55-3	825	7647-17-8	533	7759-02-6	1981	7783-47-3	211
6591-55-5	327	7440-56-4	851	7647-18-9	176	7761-88-8	1841	7783-48-4	196
6834-92-0	1917	7440-57-5	883	7647-19-0	1514	7772-99-8	2110	7783-49-5	232
6865-35-6	269	7440-58-6	901	7659-31-6	1824	7773-01-5	1262	7783-50-8	105
7320-34-5*	1619	7440-59-7	918	7664-38-2	1492	7773-03-7	1598	7783-51-9	838
7429-90-5	7	7440-60-0	919	7664-39-3	943	7773-06-0	144	7783-52-0	963
7429-91-6	753	7440-61-1	2201	7664-41-7	62	7774-29-0	1317	7783-53-1	129
7429-92-7	766	7440-62-2	2231	7664-93-9	1986	7774-34-7	432	7783-54-2	145
7439-88-5	987	7440-63-3	2272	7681-11-0	1603	7775-09-9	1876	7783-55-3	150
7439-89-6	998	7440-64-4	2287	7681-38-1	1904	7775-11-3	1879	7783-56-4	166
7439-90-9	1064	7440-65-5	2297	7681-49-4	1892	7775-19-1	1916	7783-57-5	207
7439-91-0	1066	7440-66-6	2312	7681-52-9	1910	7775-27-1	1931	7783-58-6	878
7439-92-1	1085	7440-67-7	2365	7681-52-9*	1911	7775-41-9	1835	7783-59-7	115
7439-93-2	1154	7440-68-8	205	7681-55-2	1912	7778-18-9	470	7783-60-0	199
7439-94-3	1186	7440-69-9	313	7681-57-4	1915	7778-39-4*	186	7783-61-1	179
7439-95-4	1197	7440-70-2	416	7681-65-4	693	7778-39-4	185	7783-62-2	212
7439-96-5	1247	7440-71-3	481	7681-82-5	1913	7778-44-1	421	7783-63-3	215
7439-97-6	1296	7440-72-4	801	7693-26-7	1588	7778-50-9	1574	7783-64-4	237
7439-98-7	1330	7440-73-5	807	7697-37-2	1443	7778-53-2	1618	7783-66-6	985
7439-99-8	1379	7440-74-6	950	7699-41-4	1807	7778-74-7	1613	7783-68-8	143
7440-00-8	1365	7446-06-2	1555	7699-45-8	2321	7778-77-0	1576	7783-70-2	177
7440-01-9	1378	7446-07-3	2034	7704-34-9	1985	7778-80-5	1624	7783-71-3	202
7440-02-0	1381	7446-08-4	1766	7704-34-9	1984	7779-88-6	2339	7783-72-4	225
7440-03-1	1414	7446-09-5	1992	7704-98-5	2135	7779-90-0	2349	7783-75-7	997
7440-04-2	1467	7446-10-8	1137	7704-99-6	2371	7782-40-3	482	7783-77-9	135
7440-05-3	1479	7446-11-9	1993	7705-07-9	2147	7782-41-4	802	7783-79-1	177
7440-06-4	1524	7446-14-2	1135	7705-08-0	1044	7782-44-7	1477	7783-80-4	204
7440-07-5	1542	7446-15-3	1130	7718-54-9	1395	7782-49-2	1761	7783-81-5	222
7440-08-6	1553	7446-16-4	1663	7718-98-1	2248	7782-49-2	1762	7783-82-6	219
7440-09-7	1556	7446-18-6	2072	7719-09-7	2008	7782-49-2	1763	7783-84-8	105
7440-10-0	1643	7446-19-7	2358	7719-12-2	1504	7782-50-5	546	7783-86-0	102
7440-11-1	1295	7446-20-0	2359	7720-78-7	1031	7782-61-8	1054	7783-86-0*	102
7440-12-2	1656	7446-21-1	1977	7720-83-4	2156	7782-63-0	1033	7783-89-3	182
7440-13-3	1657	7446-22-2	2070	7721-01-9	2024	7782-64-1	1265	7783-90-6	183
7440-14-4	1659	7446-25-5	2132	7722-64-7	1615	7782-65-2	853	7783-92-8	182
7440-15-5	1665	7446-26-6	2352	7722-76-1	84	7782-66-3	1323	7783-93-9	1845
7440-16-6	1692	7446-27-7	1125	7722-84-1	945	7782-68-5	975	7783-95-1	1859
7440-17-7	1702	7446-28-8	1976	7722-86-3	1987	7782-70-9	1594	7783-96-2	1838
7440-18-8	1725	7446-31-3	2380	7722-88-5	1934	7782-77-6	1444	7783-97-3	1837
7440-19-9	1737	7446-32-4	172	7723-14-0	1489	7782-78-7	1990	7783-98-4	1847
7440-20-2	1753	7446-33-5	2310	7723-14-0	1487	7782-79-8	936	7783-99-5	1842
7440-21-3	1780	7446-70-0	20	7723-14-0	1488	7782-85-6	1903	7784-01-2	1831
7440-22-4	1820	7447-39-4	715	7726-95-6	369	7782-86-7	1303	7784-02-3	1834
7440-23-5	1860	7447-40-7	1570	7727-15-3	16	7782-87-8	1601	7784-03-4	1856
7440-24-6	1956	7447-41-8	1164	7727-18-6	2269	7782-89-0	1157	7784-05-6	1852
7440-25-7	2011	7487-88-9	1238	7727-21-1	1617	7782-91-4	1358	7784-07-8	1850
7440-26-8	2028	7487-94-7	1309	7727-37-9	1441	7782-92-5	1865	7784-09-0	1848
7440-27-9	2043	7488-51-9	1132	7727-43-7	270	7782-94-7	1442	7784-11-4	17
7440-28-0	2051	7488-54-2	1722	7727-54-0	134	7782-99-2	1991	7784-13-6	21
7440-29-1	2080	7488-55-3	2120	7727-73-3	1939	7783-00-8	1765	7784-14-7	150
7440-30-4	2097	7543-51-3	2350	7732-18-5	2271	7783-03-1	2161	7784-16-9	1951
7440-31-5	2104	7550-35-8	1162	7733-02-0	2357	7783-06-4	947	7784-18-1	24
7440-31-5	2105	7550-45-0	2154	7757-79-1	1609	7783-07-5	946	7784-19-2	100
7440-32-6	2134	7553-56-2	974	7757-82-6	1938	7783-08-6	1764	7784-21-6	28
7440-33-7	2160	7558-79-4	1901	7757-83-7	1943	7783-09-7	949	7784-22-7	31

23-8	32	7789-28-8	1018	7790-92-3	547	10026-24-1	663	10058-44-3	1058
24-9	1561	7789-29-9	1591	7790-94-5	1988	10028-14-5	1466	10060-09-0	407
25-0	8	7789-30-2	377	7790-98-9	132	10028-15-6	1478	10060-10-3	523
26-1	9	7789-31-3	370	7790-99-0	981	10028-18-9	1398	10060-11-4	869
27-2	35	7789-33-5	980	7791-03-9	1179	10028-22-5	1060	10060-12-5	588
29-4	1704	7789-36-8	1202	7791-07-3	1926	10031-13-7	1091	10060-13-6	80
30-7	38	7789-38-0	1870	7791-08-4	170	10031-16-0	223	10090-53-6	33
31-8	50	7789-39-1	1706	7791-09-5	1687	10031-20-6	1260	10097-28-6	1812
33-0	191	7789-40-4	2053	7791-10-8	1962	10031-21-7	1094	10099-58-8	1071
34-1	192	7789-41-5	423	7791-11-9	1708	10031-22-8	1095	10099-66-8	1189
35-2	193	7789-42-6	384	7791-12-0	2056	10031-23-9	1660	10099-74-8	1117
36-3	201	7789-43-7	628	7791-13-1	633	10031-24-0	2109	10099-76-0	1127
37-4	1315	7789-45-9	711	7791-16-4	178	10031-25-1	584	10099-79-3	1145
40-9	1109	7789-46-0	1011	7791-18-6	1208	10031-25-1*	585	10101-41-4	471
41-0	1562	7789-47-1	1308	7791-20-0	1396	10031-26-2	1043	10101-50-5*	1929
41-0	1575	7789-48-2	1203	7791-21-1	549	10031-27-3	2038	10101-52-7	2378
-42-1	183	7789-51-7	1774	7791-23-3	1775	10031-30-8	438	10101-53-8	600
44-3	105	7789-52-8	1768	7791-25-5	2005	10031-43-3	733	10101-63-0	1113
-45-4	104	7789-53-9	1959	7791-26-6	2228	10031-45-5	740	10101-68-5	1282
-46-5	1867	7789-54-0	2036	7791-28-8	213	10031-48-8	739	10101-94-7	1133
-48-7	1391	7789-57-3	1797	7803-49-8	937	10031-52-4	779	10101-97-0	1408
-20-8	125	7789-59-5	1516	7803-51-2	1490	10031-52-4	800	10101-98-1	1407
-23-1	1827	7789-60-8	1503	7803-52-3	162	10031-53-5	797	10102-02-0	2342
-87-7	1280	7789-61-9	164	7803-54-5	1199	10034-76-1	472	10102-03-1	1450
-30-3	1207	7789-65-3	1770	7803-55-6	122	10034-81-8	1223	10102-05-3	1485
-81-4	1406	7789-66-4	1793	7803-57-8	929	10034-82-9	1880	10102-15-5	1944
-32-8	226	7789-67-5	2124	7803-58-9	2004	10034-85-2	944	10102-17-7	1954
-33-9	241	7789-68-6	2153	7803-60-3	1495	10034-88-5	1905	10102-18-8	1937
-34-0	239	7789 69-7	1512	7803-62-5	1781	10034-93-2	935	10102-20-2	1946
-35-1	242	7789-75-5	439	7803-63-6	112	10034-96-5	1281	10102-23-5	1935
-36-2	233	7789-77-7	441	7803-65-8	117	10034-98-7	2296	10102-24-6	1182
-37-3	244	7789-78-8	442	7803-68-1	2032	10034-99-8	1239	10102-25-7	1184
-38-4	249	7789-79-9	446	10022-31-8	246	10035-04-8	431	10102-34-8	1228
-39-5	272	7789-80-2	447	10022-47-6	77	10035-05-9	429	10102-40-6	1919
-40-8	280	7789-82-4	450	10022-48-7	1166	10035-06-0	324	10102-43-9	1446
-41-9	260	7790-21-8	1614	10022-50-1	1464	10035-10-6	940	10102-44-0	1447
-42-0	282	7790-22-9	1173	10022-68-1	401	10036-47-2	1458	10102-45-1	2065
-46-4	296	7790-28-5	1927	10024-93-8	1368	10038-98-9	877	10102-49-5	1042
7-47-5	299	7790-29-6	1716	10024 97-2	1445	10039-31-3	309	10102-68-8	448
7-48-6	308	7790-30-9	2063	10025-64-6	2346	10039-32-4	1902	10102-71-3	1864
7-49-7	300	7790-31-0	1216	10025-65-7	1528	10039-54-0	938	10103-61-4	700
7-52-2	302	7790-33-2	1267	10025-66-8	1661	10039-55-1	933	10108-64-2	388
7-53-3	304	7790-33-2*	1268	10025-67-9	1995	10042-76-9	1972	10112-91-1	1300
7-56-6	311	7790-34-3	1402	10025-68-0	1769	10042-88-3	2044	10118-76-0	459
7-57-7	329	7790-37-6	2335	10025-69-1	2111	10043-01-3	49	10124-36-4	409
7-58-8	315	7790-38-7	1484	10025-70-4	1964	10043-11-5	365	10124-37-5	451
7-59-9	330	7790-39-8	1529	10025-71-5	2037	10043-35-3	352	10124-41-1	478
7-60-2	316	7790-41-2	993	10025-73-7	587	10043-52-4	430	10124-43-3	662
7-61-3	318	7790-42-3	1639	10025-74-8	760	10043-67-1	1560	10124-48-8	1307
7-62-4	319	7790-44-5	167	10025-75-9	771	10043-92-2	1664	10124-50-2	1589
7-63-5	331	7790-46-7	1538	10025-76-0	794	10045-86-0	1057	10125-13-0	716
7-64-6	322	7790-47-8	2129	10025-77-1	1045	10045-89-3	91	10135-84-9	106
7-68-0	336	7790-48-9	2041	10025-78-2	1799	10045-94-0	1318	10137-74-3	428
7-69-1	530	7790-49-0	2088	10025-82-8	962	10045-94-0*	1319	10138-04-2	88
7-70-4	688	7790-56-9	1627	10025-83-9	991	10048-95-0	1898	10138-41-7	770
7-71-5	376	7790-58-1	1630	10025-85-1	1452	10048-98-3	231	10138-52-0	815
8-97-8	589	7790-59-2	1621	10025-87-3	1517	10048-99-4	275	10138-62-2	921
8-98-9	76	7790-60-5	1641	10025-90-8	1647	10049-01-1	333	101[illegible]-47-[illegible]	[illegible]
38-99-0	599	7790-62-7	1620	10025-91-9	165	10049-03-3	806	10141-05-6	649
39-00-6	1571	7790-63-8	1642	10025-93-1	2210	10049-04-4	550	10170-69-1	1254
39-01-7	1165	7790-69-4	1175	10025-99-7	1633	10049-05-5	576	10179-73-4	1030
39-02-8	594	7790-71-8	1181	10026-01-4	1471	10049-06-6	2142	10190-55-3	1115
39-04-0	596	7790-74-1	465	10026-02-5	1554	10049-07-7	1695	10192-29-7	74
39-06-2	1965	7790-75-2	480	10026-03-6	1771	10049-08-8	1729	10192-30-0	114
89-08-4	86	7790-76-3	464	10026-04-7	1794	10049-10-2	578	10193-36-9	1808
39-09-5	82	7790-78-5	389	10026-06-9	2126	10049-11-3	604	10196-18-6	2340
89-10-8	1312	7790-79-6	393	10026-07-0	2039	10049-12-4	2249	10210-64-7	290
39-17-5	539	7790-80-9	396	10026-08-1	2086	10049-12-4*	2250	10210-68-1	618
89-18-6	541	7790-81-0	395	10026-10-5	2215	10049-14-6	2216	10213-09-9	2264
89-19-7	724	7790-84-3	410	10026-11-6	2369	10049-16-8	2257	10213-10-2	1955
39-21-1	1989	7790-85-4	415	10026-12-7	1434	10049-17-9	1681	10214-39-8	1093
89-23-3	1579	7790-86-5	515	10026-13-8	1513	10049-21-5	1886	10241-05-1	1355
89-24-4	1167	7790-87-6	518	10026-17-2	639	10049-23-7	2033	10257-55-3	474
89-25-5	1461	7790-87-6*	519	10026-18-3	672	10049-24-8*	990	10277-43-7	1077
89-26-6	805	7790-89-8	554	10026-20-7	656	10049-24-8	989	10290-12-7	707
89-27-7	2059	7790-91-2	555	10026-22-9	650	10049-25-9	575	10294-26-5	1853

10294-27-6	886	12004-76-1	2012	12019-52-2	699	12035-79-9	1380	12057-71-5	12
10294-28-7	891	12005-21-9	2298	12019-88-4	755	12035-82-4	1530	12057-74-8	12
10294-29-8	887	12005-67-3	61	12020-14-3	2368	12035-83-5	1544	12057-75-9	12
10294-31-2	889	12005-69-5	357	12020-21-2	775	12035-98-2	2244	12057-92-0	12
10294-32-3	898	12006-15-4	382	12020-28-9	777	12036-02-1	1473	12058-18-3	13
10294-33-4	358	12006-77-8	617	12020-39-2	781	12036-09-8	1673	12058-20-7	13
10294-34-5	360	12006-79-0	563	12020-58-5	784	12036-10-1	1733	12059-14-2	13
10294-38-9	216	12006-84-7	1001	12020-65-4	791	12036-14-5	2019	12059-95-9	15
10294-39-0	254	12006-86-9	1000	12020-66-5	790	12036-21-4	2258	12060-00-3	11
10294-40-3	220	12006-99-4	1331	12020-69-8	792	12036-22-5	2181	12060-01-4	11
10294-42-5	525	12007-00-0	1385	12021-68-0	642	12036-32-7	1652	12060-08-1	17
10294-44-7	1299	12007-01-1	1386	12022-95-6	1008	12036-34-9	1549	12060-58-1	17
10294-46-9	738	12007-02-2	1384	12022-99-0	1009	12036-35-0	1698	12060-59-2	19
10294-47-0	1099	12007-07-7	2013	12023-99-3	841	12036-41-8	2048	12061-16-4	77
10294-48-1	553	12007-09-9	2163	12024-10-1	833	12036-44-1	2103	12062-24-7	72
10294-50-5	655	12007-10-2	2162	12024-11-2	832	12037-15-9	2219	12063-98-8	82
10294-52-7	1046	12007-16-8	564	12024-14-5	834	12037-65-9	2139	12064-03-8	82
10294-53-8	1048	12007-23-7	902	12024-20-3	830	12037-80-8	2377	12064-62-9	82
10294-54-9	545	12007-25-9	1201	12024-21-4	844	12037-82-0	1500	12065-10-0	87
10294-58-3	1126	12007-29-3	1416	12024-22-5	849	12038-12-9	1655	12065-11-1	88
10294-62-9	1081	12007-33-9	368	12024-24-7	846	12038-13-0	1654	12065-36-0	852
10294-64-1	1604	12007-34-0	1754	12024-27-0	850	12038-20-9	1531	12065-65-5	21
10294-66-3	1637	12007-35-1	2014	12024-81-6	813	12038-21-0	1540	12065-68-8	11
10294-70-9	2115	12007-36-2	2202	12025-32-0	874	12038-63-0	1674	12066-83-0	16
10325-94-7	400	12007-37-3	2233	12025-34-2	882	12038-67-4	1691	12067-00-4	16
10326-21-3*	1206	12007-38-4	565	12025-39-7	875	12039-07-5	2145	12067-22-0	17
10326-24-6	2317	12007-84-0	2203	12026-06-1	2062	12039-13-3	2159	12067-46-8	21
10326-26-8	211	12007-97-5	1332	12026-24-3	2114	12039-15-5	2381	12067-54-8	209
10326-27-9	218	12007-98-6	2164	12027-06-4	119	12039-16-6	2151	12067-66-2	202
10326-28-0	405	12007-99-7	422	12028-48-7	121	12039-19-9	2311	12067-76-4	21
10340-06-4	494	12008-02-5	504	12029-81-1	924	12039-35-9	2314	12068-49-4	101
10343-61-0	2150	12008-05-8	783	12029-98-0	978	12039-55-3	2020	12068-69-8	334
10361-37-2	217	12008-06-9	809	12030-14-7	960	12039-76-8	2239	12068-90-5	132
10361-43-0	321	12008-19-4	347	12030-24-9	972	12039-79-1	2018	12069-00-0	113
10361-46-3	332	12008-21-8	1067	12030-49-8	995	12039-80-4	2049	12069-32-8	359
10361-79-2	1646	12008-23-0	1366	12030-51-2	996	12039-83-7	2140	12069-69-1	713
10361-82-7	1745	12008-27-4	1644	12030-85-2	1608	12039-87-1	2238	12069-85-1	904
10361-84-9	1756	12008-29-6	1738	12030-88-5	1628	12039-88-2	2170	12069-89-5	133
10361-91-8	2292	12008-32-1	2301	12030-97-6	1638	12039-89-3	2288	12069-94-2	141
10361-92-9	2305	12009-14-2	245	12031-30-0	1083	12039-90-6	2379	12070-06-3	201
10361-95-2	2324	12009-18-6*	268	12031-49-1	1082	12039-95-1	2171	12070-07-4	201
10377-48-7	1183	12009-18-6	267	12031-63-9	1174	12040-00-5	1752	12070-08-5	213
10377-51-2	1172	12009-21-1	285	12031-80-0	1180	12040-02-7	2123	12070-09-6	220
10377-52-3	1177	12011-97-1	1333	12032-20-1	1193	12041-50-8	13	12070-10-9	223
10377-58-9	1215	12011-99-3	1418	12032-36-9	1241	12042-68-1	419	12070-12-1	216
10377-60-3	1217	12012-16-7	2084	12032-74-5	1287	12042-78-3	420	12070-13-2	216
10377-62-5	1225	12012-17-8	2235	12032-78-9	1256	12043-29-7	52	12071-15-7	106
10377-66-9	1271	12012-32-7	505	12032-82-5	1248	12044-16-5	999	12071-29-3	196
10377-93-2	1270	12012-35-0	566	12032-88-1	1286	12044-42-7	614	12071-31-7	208
10380-31-1	283	12013-10-4	621	12032-97-2	1249	12044-54-1	199	12071-33-9	220
10381-37-0	2095	12013-55-7	469	12033-19-1	1336	12045-01-1	616	12071-35-1	230
10402-15-0	719	12013-56-8	468	12033-31-7	1337	12045-15-7	1250	12076-62-9	220
10405-27-3	1455	12013-57-9	475	12033-54-4	1543	12045-16-8	1252	12088-65-2	100
10450-55-2	1040	12013-82-0	453	12033-62-4	2017	12045-19-1	1415	12115-63-8	513
10450-60-9	976	12014-14-1	414	12033-64-6	2047	12045-27-1	2232	12124-97-9	68
10476-81-0	1958	12014-28-7	406	12033-65-7	2090	12045-63-5	2136	12124-99-1	113
10476-85-4	1963	12014-29-8	381	12033-72-6	2169	12045-64-6	2366	12125-01-8	92
10476-86-5	1969	12014-82-3	510	12033-83-9	2208	12045-87-3	1950	12125-02-9	75
10489-46-0	1699	12014-85-6	507	12033-88-4	1465	12045-88-4	1949	12125-08-5	97
10534-89-1	673	12014-93-6	522	12033-89-5	1811	12046-08-1	228	12125-09-6	149
10544-72-6	1449	12016-80-7	677	12034-09-2	1920	12047-27-7	281	12125-19-8	220
10544-73-7	1448	12017-01-5	647	12034-12-7	1945	12047-79-9	247	12125-22-3	148
10545-99-0	1998	12017-08-2	660	12034-57-0	1422	12048-50-9	339	12125-25-6	119
10553-31-8	212	12017-12-8	620	12034-59-2	1429	12048-51-0	340	12125-63-2	103
10567-69-8	238	12017-13-9	666	12034-66-1	1420	12049-50-2	479	12133-07-2	756
10580-03-7	156	12017-38-8	680	12034-77-4	1430	12052-42-5	611	12133-10-7	765
10580-52-6	2241	12018-01-8	605	12034-80-9	1421	12053-12-2	561	12134-02-0	619
10588-01-9	1885	12018-09-6	573	12034-83-2	1432	12053-13-3	571	12134-75-7	811
11121-16-7	12	12018-10-9	718	12034-88-7	1116	12053-27-9	568	12134-77-9	823
11126-81-1*	15	12018-19-8	2326	12034-89-8	1971	12053-39-3	602	12135-76-1	146
12002-03-8	703	12018-22-3	601	12035-32-4	1376	12054-48-7	1399	12136-24-2	926
12002-61-8	6	12018-34-7	582	12035-35-7	1377	12054-85-2	123	12136-42-4	988
12002-99-2	1855	12018-36-9	572	12035-39-1	1410	12055-23-1	910	12136-45-7	1612
12003-63-3*	1558	12018-61-0	544	12035-52-8	1382	12055-62-8	925	12136-58-2	1185
12004-04-5	209	12019-11-3	684	12035-64-2	1388	12056-90-5	1080	12136-78-6	1339
12004-06-7	291	12019-23-7	749	12035-72-2	1413	12057-24-8	1178	12136-91-3	1501

13776-58-4	2273	13940-89-1	1019	14763-77-0	720	16893-85-9	1896	20328-94-3	228
13776-62-0	1457	13963-58-1	1585	14779-70-5	862	16903-35-8	885	20344-49-4	105
13776-74-4	1233	13965-73-6	363	14808-60-7	1813	16919-19-0	104	20398-06-5	205
13777-22-5	903	13966-94-4	957	14808-60-7	1814	16919-58-7	99	20405-64-5	696
13777-23-6	908	13967-25-4	1301	14871-79-5*	237	16921-30-5	1584	20427-11-6	637
13777-25-8	2367	13972-68-4	398	14884-42-5	606	16921-96-3	986	20427-56-9	173
13778-39-7	2100	13973-87-0	373	14890-41-6	2240	16925-39-6	441	20427-58-1	233
13779-41-4	1496	13982-53-1*	698	14933-38-1	56	16940-66-2	1869	20427-59-2	729
13779-92-5	1436	13986-18-0	2330	14940-41-1	1028	16940-92-4	96	20548-54-3	473
13780-06-8	454	13986-26-0	2373	14965-52-7	1800	16941-10-9	476	20601-83-6	132
13780-42-2	848	14013-15-1	1269	14977-61-8	609	16941-11-0	103	20610-49-5	166
13780-48-8	786	14013-86-6	1024	14984-81-7	1777	16941-12-1	1525	20619-16-3	872
13780-57-9	2002	14014-88-1	1728	15098-87-0	26	16949-15-8	1161	20644-87-5	223
13782-01-9*	678	14024-18-1	1041	15162-90-0	1451	16961-83-4	1809	20661-21-6	965
13783-04-5	2141	14038-43-8	1049	15163-24-3	2175	16962-07-5	14	20665-52-5	845
13783-07-8	2143	14040-11-0	2167	15168-20-4	742	16962-31-5	1586	20667-12-3	184
13798-24-8	2045	14049-36-6	1785	15179-32-5	2010	16962-47-3	102	20694-39-7	127
13812-43-6	1456	14049-81-1	638	15192-17-3	1742	17014-71-0	1616	20712-42-9	403
13813-22-4	1076	14055-74-4	1342	15192-26-4	2040	17026-29-8	1684	20762-60-1	156
13813-23-5	1649	14055-75-5	1346	15192-42-4	1672	17029-21-9	1685	20770-09-6	213
13813-24-6	1371	14055-81-3	2177	15195-33-2	1440	17083-68-0	884	20816-12-0	147
13813-25-7	1748	14055-84-6	2180	15195-53-6	1425	17099-70-6	27	20859-73-8	44
13813-40-6	2046	14059-33-7	326	15230-48-5	858	17125-80-3	229	20910-28-5	223
13813-41-7	923	14075-53-7	1581	15238-00-3	645	17194-00-2	234	21041-93-0	643
13813-42-8	774	14077-39-5	788	15238-00-3*	646	17375-41-6	1032	21041-95-2	394
13813-43-9	2102	14104-20-2	1857	15243-27-3	1257	17440-85-6	295	21109-95-5	271
13813-45-1	1191	14166-78-0	964	15243-33-1	1726	17440-90-3	1005	21159-32-0	534
13813-46-2	1548	14168-73-1	1240	15244-35-6	411	17496-59-2	551	21255-83-4	372
13813-47-3	59	14188-40-0	856	15321-51-4	1003	17702-41-9	349	21264-43-7	958
13814-72-7	812	14215-29-3	383	15364-10-0	2281	17786-31-1	622	21308-80-5	371
13814-74-9	1357	14215-30-6	708	15364-94-0	1275	18088-11-4	1718	21351-79-1	538
13814-75-0	1361	14217-21-1*	1890	15385-57-6	1302	18282-10-5	2130	21548-73-2	185
13814-76-1	1683	14221-48-8*	89	15385-58-7	1298	18283-93-7	343	21645-51-2	29
13814-81-8	701	14242-05-8	1846	15468-32-3	1815	18288-22-7	320	21651-19-4	211
13814-83-0	2265	14249-98-0	2189	15469-38-2	1051	18356-71-3	1788	21679-31-2	560
13814-96-5	1107	14280-53-6	955	15474-63-2	763	18433-40-4	158	21908-53-2	132
13817-37-3	640	14284-93-6	1727	15477-33-5	19	18433-84-6	345	21995-38-0*	72
13818-75-2	814	14285-68-8	1667	15492-38-3	1697	18454-12-1	1103	22015-35-6	789
13818-89-8	866	14307-33-6*	437	15513-84-5	2243	18480-07-4	1968	22205-45-4	697
13819-84-6	1356	14312-00-6	391	15513-94-7	2251	18618-55-8	516	22205-57-8	527
13820-41-2	151	14335-40-1	1505	15520-84-0	675	18624-44-7	1020	22326-55-2	235
13820-62-7	625	14336-80-2	687	15571-91-2*	1629	18718-07-5	1264	22398-80-7	954
13821-06-2*	225	14402-70-1	127	15572-25-5	2071	18721-05-6	586	22441-45-8	200
13824-36-7	1789	14405-43-7	835	15578-26-4	2118	18810-58-7	210	22520-96-3	984
13825-74-6*	2158	14447-89-3	2194	15587-72-1	1714	18820-29-6	1285	22527-13-5	228
13825-76-8	920	14452-39-2	43	15593-61-0	1232	18820-29-6	1283	22537-19-5	108
13825-86-0	581	14456-34-9	911	15597-63-4	1506	18820-29-6	1284	22750-54-5*	387
13826-63-6	2066	14456-48-5	759	15597-88-3	603	18855-94-2	917	22750-57-8	528
13826-83-0	93	14456-51-0	2098	15607-89-3	2001	18868-43-4	1351	22756-36-1	170
13842-80-3	2242	14456-53-2	1188	15622-42-1	1670	18917-82-3	1114	22831-39-6	123
13842-83-6	1547	14457-84-2	41	15696-40-9	1468	18917-91-4	34	22831-42-1	11
13842-88-1	1427	14457-87-5	511	15750-45-5	1218	18933-05-6	1266	22886-66-4	839
13842-93-8	2030	14457-87-5*	512	15752-41-7	1423	19034-13-0	659	22986-54-5	3
13843-28-2	607	14459-59-7	1360	15752-46-2	1545	19073-56-4	1709	23370-59-4	992
13845-17-5	224	14459-75-7	1438	15823-43-5	916	19086-20-5	1243	23377-53-9	169
13847-57-9	1101	14459-95-1	1578	15829-53-5	1304	19088-74-5	1713	23412-45-5	135
13847-65-9	1462	14464-46-1	1816	15845-52-0	1110	19139-47-0	509	23414-72-4	2347
13862-78-7	153	14475-63-9	2372	15942-63-9	184	19168-23-1	98	23611-30-5	1720
13863-41-7	374	14481-29-9*	90	15947-41-8	60	19287-45-7	342	23777-80-2	346
13863-59-7	375	14483-17-1	1650	16122-03-5*	124	19357-83-6	2187	24094-93-7	569
13863-88-2	1821	14507-19-8	1075	16399-77-2	1015	19357-86-9	2291	24304-00-5	36
13864-01-2	1074	14517-29-4	1372	16544-92-6	632	19423-81-5	816	24415-00-7	868
13870-13-8	914	14519-18-7	1957	16671-27-5	590	19423-87-1	2293	24422-20-6	864
13870-21-8	1428	14521-18-7	1734	16674-78-5	1198	19465-30-6	348	24422-21-7	859
13870-24-1	1038	14523-22-9	1693	16689-88-6	2081	19469-07-9	1055	24597-12-4	831
13871-27-7	1952	14553-44-7	2093	16721-80-5	1906	19584-30-6	1694	24613-38-5	634
13873-84-2	983	14590-13-7	78	16721-80-5	1907	19597-69-4	1159	24621-21-4	1419
13874-75-4	1741	14590-19-3	657	16731-55-8	1605	19598-90-4	819	24646-85-3	2237
13874-77-6	2290	14639-94-2	101	16774-21-3	73	19624-22-7	344	24719-19-5	626
13875-06-4	2278	14639-97-5	152	16812-54-7	1409	19783-14-3	1111	25152-52-7	10
13876-85-2	694	14639-98-6	131	16853-85-3	1156	20193-56-0	2176	25417-81-6	232
13896-65-6	1731	14691-44-2	867	16860-99-4	1996	20193-58-2	1345	25502-05-0	2289
13918-22-4	1288	14693-81-3	2026	16871-60-6	1583	20205-91-8	367	25583-20-4	2138
13931-94-7	577	14693-82-4	969	16871-71-9	2333	20211-76-1	1072	25617-97-4	828
13939-06-5	1335	14720-21-9	894	16871-90-2	1587	20213-56-3	2188	25617-98-5	953
13940-63-1	870	14721-21-2	714	16872-11-0	356	20281-00-9	542	25658-42-8	2375

-43-9	2207	27218-16-2	558	35112-53-9	279	51621-05-7	1730	65842-03-7	1063
-31-8	1112	27774-13-6	2267	35718-37-7	2284	51898-99-8	323	68016-36-4	278
-08-3	506	27860-83-9	2331	36377-94-3	758	52003-58-4	120	68938-92-1	1534
-09-4	1651	30737-24-7	2067	36897-37-7	1400	52503-64-7	709	71500-16-8	1732
-10-7	1078	30937-52-1	1678	36907-42-3	2245	52788-53-1	820	71626-98-7	449
-11-8	1373	30937-53-2	1676	37267-86-0	1493	52951-38-9	325	72520-94-6	514
-15-2	810	31032-13-0	896	37773-49-2	1535	53731-35-4	1276	73560-00-6	979
-87-2	909	31052-14-9	2029	38455-77-5	2127	54120-05-7	1472	73796-25-5	1970
-60-7	1884	31052-43-4	1721	39290-85-2	710	54723-94-3	137	75926-26-0	1778
-39-1	1533	31083-74-6	1723	39368-69-9	1677	57246-89-6	1689	75926-28-2	1779
-07-3	1532	31234-26-1	1680	39416-30-3	284	58724-12-2	536	84359-31-9	597
-86-7	243	31576-40-6	1474	39796-98-0	2286	59201-51-3	1469	84359-31-9	598
-62-3	1176	32248-43-4	1743	39797-63-2	2283	60616-74-2	1212	92141-86-1	540
-61-0	570	32287-65-3	25	42246-24-2	1688	60676-86-0	1817	97126-35-7	667
-22-8	1868	32823-06-6	39	42739-38-8	159	60763-24-8	2092	99685-96-8	484
-77-1	1234	33088-16-3	2089	50432-32-1	2282	60804-25-3	1696	115383-22-7	485
-66-3	1640	33689-80-4	4	50801-97-3	1740	60883-64-9	298	123714-57-8	2362
-73-5	613	33689-81-5	2	50813-65-5	214	60922-26-1	2168	133578-89-9	2266
-75-7	1383	33689-82-6	5	50927-81-6	1818	60950-56-3	1235	133863-98-6	1364
-84-1	2318	34128-09-1	2064	51503-61-8	136	60969-19-9	2061	134929-59-2	486
-66-0	258	34330-64-8	390						

gistry number refers to the anhydrous compound.

PHYSICAL PROPERTIES OF THE RARE EARTH METALS

K. A. Gschneidner, Jr.

Table 1
DATA FOR THE TRIVALENT IONS OF THE RARE EARTH ELEMENTS

				Electronic configuration for R^{3+}				
Rare earth	**Symbol**	**Atomic no.**	**Atomic wt.**	**No. 4f electrons**	***S***	***L***	***J***	**Spectroscopic ground state**
Scandium	Sc	21	44.955910	0	—	—	—	—
Yttrium	Y	39	88.90585	0	—	—	—	—
Lanthanum	La	57	138.9055	0	—	—	—	—
Cerium	Ce	58	140.115	1	1/2	3	5/2	$^2F_{5/2}$
Praseodymium	Pr	59	140.90765	2	1	5	4	$^3H_4$
Neodymium	Nd	60	144.24	3	3/2	6	9/2	$^4I_{9/2}$
Promethium	Pm	61	(145)	4	2	6	4	$^5I_4$
Samarium	Sm	62	150.36	5	5/2	5	5/2	$^6H_{5/2}$
Europium	Eu	63	151.965	6	3	3	0	$^7F_0$
Gadolinium	Gd	64	157.25	7	7/2	0	7/2	$^8S_{7/2}$
Terbium	Tb	65	158.92534	8	3	3	6	$^7F_6$
Dysprosium	Dy	66	162.50	9	5/2	5	15/2	$^6H_{15/2}$
Holmium	Ho	67	164.93032	10	2	6	8	$^5I_8$
Erbium	Er	68	167.26	11	3/2	6	15/2	$^4I_{15/2}$
Thulium	Tm	69	168.93421	12	1	5	6	$^3H_6$
Ytterbium	Yb	70	173.04	13	1/2	3	7/2	$^2F_{7/2}$
Lutetium	Lu	71	174.967	14	—	—	—	—

Note: For additional information, see Goldschmidt, Z. B., in *Handbook on the Physics and Chemistry of Rare Earths*, Vol. 1, Gschneidner, K. A., Jr. and Eyring, L., Eds., North-Holland Physics, Amsterdam, 1978, 1.

Table 2
CRYSTALLOGRAPHIC DATA FOR THE RARE EARTH METALS AT 24°C (297 K) OR BELOW

Rare earth metal	Crystal structure[a]	Lattice constants(Å) a_0	b_0	c_0	Metallic radius CN = 12 (Å)	Atomic volume $\left(\frac{cm^3}{mol}\right)$	Density (g/cm^3)
αSc	hcp	3.3088	—	5.2680	1.6406	15.039	2.989
αY	hcp	3.6482	—	5.7318	1.8012	19.893	4.469
αLa	dhcp	3.7740	—	12.171	1.8791	22.602	6.146
αCe	fcc	4.58[b]	—	—	1.72	17.2	8.16
βCe	dhcp	3.6810	—	11.857	1.8321	20.947	6.689
γCe	fcc	5.1610	—	—	1.8247	20.696	6.770
αPr	dhcp	3.6721	—	11.8326	1.8279	20.803	6.773
αNd	dhcp	3.6582	—	11.7966	1.8214	20.583	7.008
αPm	dhcp	3.65	—	11.65	1.811	20.24	7.264
αSm	rhomb[c]	3.6290	—	26.207	1.8041	20.000	7.520
Eu	bcc	4.5827	—	—	2.0418	28.979	5.244
αGd	hcp	3.6336	—	5.7810	1.8013	19.903	7.901
α'Tb[d]	ortho	3.605	6.244	5.706	1.858	19.34	8.219
αTb	hcp	3.6055	—	5.6966	1.7833	19.310	8.230
α'Dy[e]	ortho	3.595	6.184	5.678	1.849	19.00	8.551
αDy	hcp	3.5915	—	5.6501	1.7740	19.004	8.551
Ho	hcp	3.5778	—	5.6178	1.7661	18.752	8.795
Er	hcp	3.5592	—	5.5850	1.7566	18.449	9.066
Tm	hcp	3.5375	—	5.5540	1.7462	18.124	9.321
αYb[f]	hcp	3.8799	—	6.3859	1.9451	25.067	6.903
βYb	fcc	5.4848	—	—	1.9392	24.841	6.966
Lu	hcp	3.5052	—	5.5494	1.7349	17.779	9.841

Note: For additional information, see Gschneidner, K. A., Jr. and Calderwood, F. W., in *Handbook on the Physics and Chemistry of Rare Earths,* Vol. 8, Gschneidner, K. A., Jr. and Eyring, L., Eds., North-Holland Physics, Amsterdam, 1986, 1.

[a] dhcp = double-c hexagonal close-packed; fcc = face-centered cubic; bcc = body-centered cubic; hcp = hexagonal close-packed; rhombo = rhombohedral; ortho = orthorhombic.
[b] At 77 K (−196°C).
[c] Rhombohedral is the primitive cell. Lattice parameters given are for the nonprimitive hexagonal cell.
[d] At 220 K (−53°C).
[e] At 86 K (−187°C).
[f] At 23°C.

Table 3
CRYSTALLOGRAPHIC DATA FOR RARE EARTH METALS AT HIGH TEMPERATURE

Rare earth metal	Structure	Lattice parameter (Å)	Temp. (°C)	Metallic radius CN=8 (Å)	CN=12 (Å)	Atomic volume $\left(\frac{cm^3}{mol}\right)$	Density (g/cm^3)
βSc	bcc	3.73 (est.)	1312	1.62	1.66	15.6	2.88
βY	bcc	4.11[a]	24	1.78	1.83	20.9	4.25
		4.40[b]	24	1.75	1.80	19.8	4.48
βLa	fcc	5.303	325	—	1.875	22.45	6.187
γLa	bcc	4.26	887	1.84	1.90	23.3	5.97
δCe	bcc	4.12	757	1.78	1.84	21.1	6.65
βPr	bcc	4.13	821	1.79	1.84	21.2	6.64
βNd	bcc	4.13	883	1.79	1.84	21.2	6.80
βPm	bcc	4.10 (est.)	865	1.78	1.83	20.8	6.99
βSm	hcp	a = 3.6630; c = 5.8448	450[c]	—	1.8176	20.450	7.3527
γSm	bcc	4.10 (est.)	897	1.77	1.82	20.8	7.25
βGd	bcc	4.05[b]	24	1.75	1.80	20.0	7.86
		4.06	1265	1.76	1.81	20.2	7.80
βTb	bcc	4.02[b]	24	1.74	1.79	19.6	8.12
βDy	bcc	3.98[b]	24	1.72	1.77	19.0	8.56
γYb	bcc	4.44	763[d]	1.92	1.98	26.4	6.57

Note: The rare earths Eu, Ho, Er, Tm, and Lu are monomorphic. For additional information, see Gschneidner, K. A., Jr. and Calderwood, F. W., in *Handbook on the Physics and Chemistry of Rare Earths,* Vol. 8, Gschneidner, K. A., Jr. and Eyring, L., Eds., North-Holland Physics, Amsterdam, 1986, 1.

[a] Determined by extrapolation to 0% solute of a vs. composition data for Y–Th alloys.
[b] Determined by extrapolation to 0% solute of a vs. composition data for R–Mg alloys.
[c] The hcp phase was stabilized by impurities and the temperature of measurement was below the equilibrium transition temperature (see Table 4).
[d] The bcc phase was stabilized by impurities and the temperature of measurement was below the equilibrium transition temperature (see Table 4).

Table 4
HIGH TEMPERATURE TRANSITION TEMPERATURES AND MELTING POINT OF RARE EARTH METALS

Rare earth metal	Transition I $(\alpha - \beta)^a$ Temp. (°C)	Transition I Phases	Transition II $(\beta - \gamma)^a$ Temp. (°C)	Transition II Phases	Melting point (°C)
Sc	1337	hcp ⇆ bcc	—	—	1541
Y	1478	hcp ⇆ bcc	—	—	1526
La[b]	310	dhcp → fcc	865	fcc ⇆ bcc	920
Ce[c]	139	dhcp → fcc $(\beta - \gamma)$	726	fcc ⇆ bcc $(\gamma - \delta)$	799
Pr	795	dhcp ⇆ bcc	—	—	931
Nd	863	dhcp ⇆ bcc	—	—	1016
Pm	890	dhcp ⇆ bcc	—	—	1042
Sm[d]	734	rhom → hcp	922	hcp ⇆ bcc	1072
Eu	—	—	—	—	822
Gd	1235	hcp ⇆ bcc	—	—	1314
Tb	1289	hcp ⇆ bcc	—	—	1359
Dy	1381	hcp ⇆ bcc	—	—	1411
Ho	—	—	—	—	1472
Er	—	—	—	—	1545
Tm	—	—	—	—	1545
Yb	795	fcc ⇆ bcc $(\beta - \gamma)$	—	—	824
Lu	—	—	—	—	1663

Note: For additional information, see Gschneidner, K. A., Jr. and Calderwood, F. W., in *Handbook on the Physics and Chemistry of Rare Earths,* Vol. 8, Gschneidner, K. A., Jr. and Eyring, L., Eds., North-Holland Physics, Amsterdam, 1986, 1.

[a] For all the transformations listed, unless otherwise noted.
[b] On cooling, fcc → dhcp $(\beta \rightarrow \alpha)$, 260°C.
[c] On cooling, fcc → dhcp $(\gamma \rightarrow \beta)$, −16°C.
[d] On cooling, hcp → rhomb $(\beta \rightarrow \alpha)$, 727°C.

Table 5
LOW TEMPERATURE TRANSITION TEMPERATURES OF THE RARE EARTH METALS

Cooling

Rare earth metal	Transformation	°C	K
Ce	$\gamma \rightarrow \beta$	−16	257
	$\gamma \rightarrow \alpha$	−172	101
	$\beta \rightarrow \alpha$	−228	45
Tb	$\alpha \rightarrow \alpha'$	−53	220
Dy	$\alpha \rightarrow \alpha'$	−187	86
Yb	$\beta \rightarrow \alpha$	−13	260

Heating

Rare earth metal	Transformation	°C	K
Ce	$\alpha \rightarrow \beta$	−148	125
	$\alpha \rightarrow \beta + \gamma$	−104	169
	$\beta \rightarrow \gamma$	139	412
Yb	$\alpha \rightarrow \beta$	7	280

Note: For additional information, see Beaudry, B. J. and Gschneidner, K. A., Jr., in *Handbook on the Physics and Chemistry of Rare Earths,* Vol. 1, Gschneidner, K. A., Jr. and Eyring, L., Eds., North-Holland Physics, Amsterdam, 1978, 173 and Koskenmaki, D. C. and Gschneidner, K. A., Jr., 1978, in *Handbook on the Physics and Chemistry of Rare Earths,* Vol. 1, Gschneidner, K. A., Jr. and Eyring, L., Eds., North-Holland Physics, Amsterdam, 1978, 337.

Table 6
HEAT CAPACITY, STANDARD ENTROPY, HEATS OF TRANSFORMATION, AND FUSION OF THE RARE EARTH METALS

Rare earth metal	Heat capacity at 298K (J/mol K)	Standard entropy S°_{298} (J/mol K)	Heat of transformation (kJ/mol)				Heat of Fusion (kJ/mol)
			trans.1	ΔH^{1}_{tr}	trans. 2	ΔH^{2}_{tr}	
Sc	25.5	34.6	$\alpha \rightleftarrows \beta$	4.00	—	—	14.1
Y	26.5	44.4	$\alpha \rightleftarrows \beta$	4.99	—	—	11.4
La	27.1	56.9	$\alpha \rightleftarrows \beta$	0.36	$\beta \rightleftarrows \gamma$	3.12	6.20
Ce	26.9	72.0	$\beta \rightleftarrows \gamma$	0.05	$\gamma \rightleftarrows \delta$	2.99	5.46
Pr	27.4	73.9	$\alpha \rightleftarrows \beta$	3.17	—	—	6.89
Nd	27.4	71.1	$\alpha \rightleftarrows \beta$	3.03	—	—	7.14
Pm	27.3[a]	71.6[a]	$\alpha \rightleftarrows \beta$	3.0[a]	—	—	7.7[a]
Sm	29.5	69.5	$\alpha \rightleftarrows \beta$	0.2[a]	$\beta \rightleftarrows \gamma$	3.11	8.62
Eu	27.7	77.8	—	—	—	—	9.21
Gd	37.1	67.9	$\alpha \rightleftarrows \beta$	3.91	—	—	10.0
Tb	28.9	73.3	$\alpha \rightleftarrows \beta$	5.02	—	—	10.79
Dy	27.7	75.6	$\alpha \rightleftarrows \beta$	4.16	—	—	11.06
Ho	27.2	75.0	—	—	—	—	17.0[a]
Er	28.1	73.2	—	—	—	—	19.9
Tm	27.0	74.0	—	—	—	—	16.8
Yb	26.7	59.8	$\beta \rightleftarrows \gamma$	1.75	—	—	7.66
Lu	26.8	51.0	—	—	—	—	22[a]

Note: For additional information, see Hultgren, R., Desai, P. D., Hawkins, D. T., Gleiser, M., Kelley, K. K., and Wagman, D. D., *Selected Values of the Thermodynamic Properties of the Elements,* ASM International, Metals Park, Ohio, 1973 and **Alcock, C. B. and Itkin, V. P.,** *Bull. Alloy Phase Diagrams,* to be published, 1988.

[a] Estimated.

Table 7
VAPOR PRESSURES, BOILING POINTS, AND HEATS OF SUBLIMATION OF RARE EARTH METALS

Rare earth metal	Temperature in °C for a vapor pressure of				Boiling point (°C)	Heat of sublimation at 25°C (kJ/mol)
	10^{-8} atm (0.001 Pa)	10^{-6} atm (0.101 Pa)	10^{-4} atm (10.1 Pa)	10^{-2} atm (1013 Pa)		
Sc	1036	1243	1533	1999	2830	377.8
Y	1220	1458	1809	2356	3336	424.7
La	1301	1566	1938	2506	3455	431.0
Ce	1290	1554	1926	2487	3424	422.6
Pr	1083	1333	1701	2305	3510	355.6
Nd	955	1175	1500	2029	3066	327.6
Pm	—	—	—	—	3000[a]	348[a]
Sm	508	642	835	1150	1790	206.7
Eu	399	515	685	963	1596	175.3
Gd	1167	1408	1760	2306	3264	397.5
Tb	1124	1354	1698	2237	3221	388.7
Dy	804	988	1252	1685	2561	290.4
Ho	845	1036	1313	1771	2694	300.8
Er	908	1113	1405	1896	2862	317.1
Tm	599	748	964	1300	1946	232.2
Yb	301	400	541	776	1194	152.1
Lu	1241	1483	1832	2387	3393	427.6

Note: For additional information, see Hultgren, R., Desai, P. D., Hawkins, D. T., Gleiser, M., Kelley, K. K., and Wagman, D. D., *Selected Values of the Thermodynamic Properties of the Elements,* ASM International, Metals Park, Ohio, 1973 and **Beaudry, B. J. and Gschneidner, K. A., Jr.,** in *Handbook on the Physics and Chemistry of Rare Earths,* Vol. 1, Gschneidner, K. A., Jr. and Eyring, L., Eds., North-Holland Physics, Amsterdam, 1978, 173.

[a] Estimated.

Table 8
MAGNETIC PROPERTIES OF THE RARE EARTH METALS

Rare earth metal	$\chi_A \times 10^6$ at 298 K $\left(\frac{emu}{mol}\right)$	Effective magnetic moment: Paramagnetic at ~298 K, Theory[a]	Paramagnetic at ~298 K, Obs.	Ferromagnetic at ~0 K, Theory[b]	Ferromagnetic at ~0 K, Obs.	Easy axis	Neél Temp. T_N (K): Hex sites	Neél Temp. T_N (K): Cubic sites	Curie temp. T_c (K)	θ_p (K): ‖c	θ_p (K): ⊥c	θ_p (K): Polycryst. or Avg.
αSc	294.6	—	—	—	—	—	—	—	—	—	—	—
αY	187.7	—	—	—	—	—	—	—	—	—	—	—
αLa	95.9	—	—	—	—	—	—	—	—	—	—	—
βLa	105	—	—	—	—	—	—	—	—	—	—	—
γCe	2,270	2.54	2.52	2.14	—	—	—	14.4	—	—	—	−50
βCe	2,500	2.54	2.61	2.14	—	—	13.7	12.5	—	—	—	−41
αPr	5,530	3.58	3.56	3.20	2.7[c]	a	0.03	—	—	—	—	0
αNd	5,930	3.62	3.45	3.27	2.2[c]	b	19.9	7.5	—	0	5	3.3
αPm	—	2.68	—	2.40	—	—	—	—	—	—	—	—
αSm	1,278[d]	0.85	1.74	0.71	0.5[c]	a	109	14.0	—	—	—	—
Eu	30,900	7.94	8.48	7.0	5.9	<110>	—	90.4	—	—	—	100
αGd	185,000[e]	7.94	7.98	7.0	7.63	30° to c	—	—	293.4	317	317	317
αTb	170,000	9.72	9.77	—	—	—	230.0	—	—	195	239	224
α′Tb	—	—	—	9.0	9.34	b	—	—	219.5	—	—	—
αDy	98,000	10.64	10.83	—	—	—	179.0	—	—	121	169	153
α′Dy	—	—	—	10.0	10.33	a	—	—	89.0	—	—	—
Ho	72,900	10.60	11.2	10.0	10.34	b	132	—	20.0	73.0	88.0	83.0
Er	48,000	9.58	9.9	9.0	9.1	30° to c	85	—	20.0	61.7	32.5	42.2
Tm	24,700	7.56	7.61	7.0	7.14	c	58	—	32.0	41.0	−17.0	2.3
βYb	67[d]	—	—	—	—	—	—	—	—	—	—	—
Lu	182.9	—	—	—	—	—	—	—	—	—	—	—

Note: For additional information, see McEwen, K. A., in *Handbook on the Physics and Chemistry of Rare Earths,* Vol. 1, Gschneidner, K. A., Jr. and Eyring, L., Eds., North-Holland Physics, Amsterdam, 1978, 411 and **Legvold, S.,** in *Ferromagnetic Materials,* Vol. 1, Wohlfarth, E. P., Ed., North-Holland Physics, Amsterdam, 1980, 183.

[a] $g[J(J+1)]^{1/2}$.
[b] gJ.
[d] At 38 T and 4.2 K.
[d] At 290 K.
[e] At 350 K.

Table 9
ROOM TEMPERATURE COEFFICIENT OF THERMAL EXPANSION, THERMAL CONDUCTIVITY, ELECTRICAL RESISTANCE, AND HALL COEFFICIENT

Rare earth metal	Expansion ($\alpha_i \times 10^6$) (°C^{-1})			Thermal conductivity $\left(\frac{W}{cm\text{-}K}\right)$	Electrical resistance ($\mu\Omega \cdot cm$)			Hall coefficient ($R_i \times 10^{12}$) ($V \cdot cm/A \cdot Oe$)		
	α_a	α_c	α_{vol}		ρ_a	ρ_c	ρ_{poly}	R_a	R_c	R_{poly}
αSc	7.6	15.3	10.2	0.158	71.0	26.9	56.3[a]	—	—	−0.13
αY	6.0	19.7	10.6	0.172	72.5	35.5	59.6	−0.27	−1.6	—
αLa	4.5	27.2	12.1	0.134	—	—	61.5	—	—	−0.35
βCe	—	—	—	—	—	—	82.8	—	—	—
γCe	6.3	—	6.3	0.113	—	—	74.4	—	—	+1.81
αPr	4.5	11.2	6.7	0.125	—	—	70.0	—	—	+0.709
αNd	7.6	13.5	9.6	0.165	—	—	64.3	—	—	+0.971
αPm	9[b]	16[b]	11[b]	0.15[b]	—	—	75[b]	—	—	—
αSm	9.6	19.0	12.7	0.133	—	—	94.0	—	—	−0.21
Eu	35.0	—	35.0	0.139[b]	—	—	90.0	—	—	+24.4
αGd	9.1[c]	10.0[c]	9.4[c]	0.105	135.1	121.7	131.0	−10	−54	−4.48[d]
αTb	9.3	12.4	10.3	0.111	123.5	101.5	115.0	−1.0	−3.7	—
αDy	7.5	15.6	9.9	0.107	111.0	76.6	92.6	−0.3	−3.7	—
Ho	7.0	19.5	11.2	0.162	101.5	60.5	81.4	+0.2	−3.2	—
Er	7.9	20.9	12.2	0.145	94.5	60.3	86.0	+0.3	−3.6	—
Tm	8.8	22.2	13.3	0.169	88.0	47.2	67.6	—	—	−1.8
βYb	26.3	—	26.3	0.385	—	—	25.0	—	—	+3.77
Lu	4.8	20.0	9.9	0.164	76.6	34.7	58.2	+0.45	−2.6	−0.535

Note: For additional information, see Beaudry, B. J. and Gschneidner, K. A., Jr., in *Handbook on the Physics and Chemistry of Rare Earths,* Vol. 1, Gschneidner, K. A., Jr. and Eyring, L., Eds., North-Holland Physics, Amsterdam, 1978, 173 and **McEwen, K. A.,** in *Handbook on the Physics and Chemistry of Rare Earths,* Vol. 1, Gschneidner, K. A., Jr. and Eyring, L., Eds., North-Holland Physics, Amsterdam, 1978, 411.

[a] Calculated from single crystal values.
[b] Estimated.
[c] At 100°C.
[d] At 77°C.

Table 10
ELECTRONIC SPECIFIC HEAT CONSTANT (γ), ELECTRON-ELECTRON (COULOMB) COUPLING CONSTANT (μ*), ELECTRON-PHONON COUPLING CONSTANT (λ), DEBYE TEMPERATURE AT 0 K(θ_D), AND SUPERCONDUCTING TRANSITION TEMPERATURE

Rare earth metal	γ $\left(\frac{mJ}{mol \cdot K^2}\right)$	μ*	λ	θ_D (K) from		Superconducting transition temperature (K)
				Heat capacity	Elastic constants	
αSc	10.334	0.16	0.30	345.3	—	0.050[a]
αY	7.878	0.15	0.30	244.4	258	1.3[b]
αLa	9.45	0.08	0.76	150	154	5.10
βLa	11.5	—	—	140	—	6.00
αCe	12.8	—	—	179	—	0.022[c]
αPr	20	—	—	—	153	—
αNd	[d]	—	—	[d]	163	—
αPm	—	—	—	—	—	—
αSm	[d]	—	—	[d]	169	—
Eu	[d]	—	—	[d]	118	—
αGd	4.48	—	0.30	169	182	—
α'Tb	3.71	—	—	169.6	177	—
α'Dy	[d]	—	—	[d]	183	—
Ho	6	—	—	—	190	—
Er	10.0	—	—	—	188	—
Tm	[d]	—	—	[d]	200	—
αYb	3.30	—	—	117.6	118	—
βYb	8.36	—	—	109	—	—
Lu	8.194	0.14	0.31	183.2	185	0.022[e]

Note: For additional information, see Sundström, L. J., in *Handbook on the Physics and Chemistry of Rare Earths,* Vol. 1, Gschneidner, K. A., Jr. and Eyring, L., Eds., North-Holland Physics, Amsterdam, 1978, 379, **Scott, T.,** in *Handbook on the Physics and Chemistry of Rare Earths,* Vol. 1, Gschneidner, K. A., Jr. and Eyring, L., Eds., North-Holland Physics, Amsterdam, 1978, 591, **Probst, C. and Wittig, J.,** in *Handbook on the Physics and Chemistry of Rare Earths,* Vol. 1, Gschneidner, K. A., Jr. and Eyring, L., Eds., North-Holland Physics, Amsterdam, 1978, 749, and **Tsang, T.-W. E., Gschneidner, K. A., Jr., Schmidt, F. A., and Thome, D. K.,** *Phys. Rev.,* B, 31, 235, 1985.

[a] At 18.6 GPa.
[b] At 11 GPa.
[c] At 2.2 GPa.
[d] Heat capacity results have been reported, but the resultant γ and θ_D values are unreliable because of the presence of impurities and/or there was no reliable procedure or model to correct for the magnetic contribution to the heat capacity.
[e] At 4.5 GPa.

Table 11
ROOM TEMPERATURE ELASTIC MODULI AND MECHANICAL PROPERTIES

Rare earth metal	Elastic moduli (GPa) Young's (elastic) modulus	Shear modulus	Bulk modulus	Poisson's ratio	Mechanical properties (MPa) Yield strength 0.2% offset	Ultimate tensile strength	Uniform elongation (%)	Reduction in area (%)	Recryst. temp. (°C)
Sc	74.4	29.1	56.6	0.279	173[a]	255[a]	5.0[a]	8.0[a]	550
Y	63.5	25.6	41.2	0.243	42	129	34.0	—	550
αLa	36.6	14.3	27.9	0.280	126[a]	130	7.9[a]	—	300
βCe	—	—	—	—	86	138	—	24.0	—
γCe	33.6	13.5	21.5	0.24	28	117	22.0	30.0	325
αPr	37.3	14.8	28.8	0.281	73	147	15.4	67.0	400
αNd	41.4	16.3	31.8	0.281	71	164	25.0	72.0	400
αPm	46[b]	18[b]	33[b]	0.28[b]	—	—	—	—	400[b]
αSm	49.7	19.5	37.8	0.274	68	156	17.0	29.5	440
Eu	18.2	7.9	8.3	0.152	—	—	—	—	300
αGd	54.8	21.8	37.9	0.259	15	118	37.0	56.0	500
αTb	55.7	22.1	38.7	0.261	—	—	—	—	500
αDy	61.4	24.7	40.5	0.247	43	139	30.0	30.0	550
Ho	64.8	26.3	40.2	0.231	—	—	—	—	520
Er	69.9	28.3	44.4	0.237	60	136	11.5	11.9	520
Tm	74.0	30.5	44.5	0.213	—	—	—	—	600
βYb	23.9	9.9	30.5	0.207	7	58	43.0	92.0	300
Lu	68.6	27.2	47.6	0.261	—	—	—	—	600

Note: For additional information, see Scott, T., in *Handbook on the Physics and Chemistry of Rare Earths,* Vol. 1, Gschneidner, K. A., Jr. and Eyring, L., Eds., North-Holland Physics, Amsterdam, 1978, 591.

[a] Value is questionable.
[b] Estimated.

Table 12
LIQUID METAL PROPERTIES NEAR THE MELTING POINT

Rare earth metal	Density (g/cm³)	Surface tension (N/m)	Viscosity (centipoise)	Heat capacity (J/mol K)	Thermal conductivity (W/cm K)	Magnetic susceptibility $\chi \times 10^4$ $\left(\frac{emu}{mol}\right)$	Electrical resistivity ($\mu\Omega \cdot cm$)	$\Delta V_{L \to S}$[a] (%)	Spectral emittance at λ = 645 nm ϵ (%)	Temp. range (°C)
Sc	2.80	0.954	—	44.2[b]	—	—	—	—	—	—
Y	4.24	0.871	—	43.1	—	—	—	—	37.0	1522—1675
La	5.96	0.718	2.65	34.3	0.238	1.20	133	−0.6	28.2	920—1220
Ce	6.68	0.706	3.20	37.7	0.210	9.37	130	+1.1	30.9	850—1225
Pr	6.59	0.707	2.85	43.0	0.251	17.3	139	−0.02	29.4	931—1225
Nd	6.72	0.687	—	48.8	0.195	18.7	151	−0.9	28.0	1021—1300
Pm	6.9[b]	0.680[b]	—	50[b]	—	—	160[b]	—	—	—
Sm	7.16	0.431	—	50.2[b]	—	18.3	182	−3.6	43.7	1074—1077
Eu	4.87	0.264	—	38.1	—	97	242	−4.8	—	—
Gd	7.4	0.664	—	37.2	0.149	67	195	−2.0	34.2	1313—1600
Tb	7.65	0.669	—	46.5	—	82	193	−3.1	—	—
Dy	8.2	0.648	—	49.9	0.187	95	210	−4.5	29.7	1413—1437
Ho	8.34	0.650	—	43.9	[illegible]	[illegible]	[illegible]	[illegible]	—	—
[illegible]	[illegible]	0.637	—	38.7	—	69	226	−9.0	37.2	1529—1587
Tm	9.0[b]	—	—	41.4	—	41	235[b]	−6.9	—	—
Yb	6.21	0.320	2.67	36.8	—	—	113	−5.1	—	—
Lu	9.3	0.940	—	47.9[b]	—	—	224	−3.6	—	—

Note: For additional information, see Van Zytveld, J., to be published in *Handbook on the Physics and Chemistry of Rare Earths,* Vol. 12, Gschneidner, K. A., Jr. and Eyring, L., Eds., North-Holland Physics, Amsterdam.

[a] Volume change on freezing.
[b] Estimated.

Table 13
IONIZATION POTENTIALS (ELECTRONVOLTS)

Rare earth	I Neutral atom	II Singly ionized	III Doubly ionized	IV Triply ionized	V Quadruply ionized
Sc	6.56	12.80	24.76	73.49	91.65
Y	6.217	12.24	20.52	60.597	77.0
La	5.5770	11.060	19.1773	49.95	61.6
Ce	5.5387	10.85	20.198	36.758	65.55
Pr	5.464	10.55	21.624	38.98	57.53
Nd	5.525	10.73	22.1	40.41	—
Pm	5.554	10.90	22.3	41.1	—
Sm	5.6437	11.07	23.4	41.4	—
Eu	5.6704	11.241	24.92	42.7	—
Gd	6.150	12.09	20.63	44.0	—
Tb	5.8639	11.52	21.91	39.79	—
Dy	5.9389	11.67	22.8	41.47	—
Ho	6.018	11.80	22.84	42.5	—
Er	6.1078	11.93	22.74	42.7	—
Tm	6.1843	12.05	23.68	42.7	—
Yb	6.25416	12.176	25.05	43.56	—
Lu	5.42585	13.9	20.9594	45.25	

Note: For references, see the table "Ionization Potentials of Atoms and Atomic Ions" in Section 10.

Table 14
EFFECTIVE IONIC RADII (Å)[a]

Rare earth ion	R^{2+}		R^{3+}			R^{4+}	
	CN=6	CN=8	CN=6	CN=8	CN=12	CN=6	CN=8
Sc	—	—	0.745	0.87	1.116	—	—
Y	—	—	0.900	1.015	1.220	—	—
La	—	—	1.045	1.18	1.320	—	—
Ce	—	—	1.010	1.14	1.290	0.80	0.97
Pr	—	—	0.997	1.14	1.286	0.78	0.96
Nd	—	—	0.983	1.12	1.276	—	—
Pm	—	—	0.97	1.10	1.267	—	—
Sm	1.19	1.27	0.958	1.09	1.260	—	—
Eu	1.17	1.25	0.947	1.07	1.252	—	—
Gd	—	—	0.938	1.06	1.246	—	—
Tb	—	—	0.923	1.04	1.236	0.76	0.88
Dy	—	—	0.912	1.03	1.228	—	—
Ho	—	—	0.901	1.02	1.221	—	—
Er	—	—	0.890	1.00	1.214	—	—
Tm	—	—	0.880	0.99	1.207	—	—
Yb	1.00	1.07	0.868	0.98	1.199	—	—
Lu	—	—	0.861	0.97	1.194	—	—

Note: For additional information, see Shannon, R. D. and Prewitt, C. T., *Acta Cryst.*, 25, 925, 1969 and **Shannon, R. D. and Prewitt, C. T.,** *Acta Cryst.*, 26, 1046, 1970.

[a] Radius of O^{2-} is 1.40 Å for a coordination number (CN) of 6.

MELTING, BOILING, AND CRITICAL TEMPERATURES OF THE ELEMENTS

This table gives the melting point (T_m), normal boiling point (T_b) at a pressure of 101.325 kPa (1 atmosphere), and critical temperature (T_c) for all the elements for which measurements or reliable estimates are available. Values are given in °C on the ITS-90 scale. A "t" after a value indicates a triple point, and "s" indicates sublimation temperature (i.e., vapor pressure of the solid phase reaches 101.325 kPa, or 1 atmosphere).

Name	T_m/°C	T_b/°C	T_c/°C
Actinium	1051	3198	
Aluminum	660.32	2519	
Americium	1176	2011	
Antimony	630.63	1587	
Argon	-189.35	-185.85	-122.28
Arsenic	817 t	614 s	1400
Astatine	302		
Barium	727	1897	
Berkelium	1050		
Beryllium	1287	2471	
Bismuth	271.40	1564	
Boron	2075	4000	
Bromine	-7.2	58.8	315
Cadmium	321.07	767	
Calcium	842	1484	
Californium	900		
Carbon	4492 t	3642 s	
Cerium	799	3424	
Cesium	28.44	671	
Chlorine	-101.5	-34.04	143.8
Chromium	1907	2671	
Cobalt	1495	2927	
Copper	1084.62	2562	
Curium	1345		
Dysprosium	1411	2561	
Einsteinium	860		
Erbium	1529	2862	
Europium	822	1596	
Fermium	1527		
Fluorine	-219.62	-188.12	-129.02
Francium	27		
Gadolinium	1314	3264	
Gallium	29.76	2204	
Germanium	938.25	2833	
Gold	1064.18	2856	
Hafnium	2233	4603	
Helium		-268.93	-267.96
Holmium	1472	2694	
Hydrogen	-259.34	-252.87	-240.18
Indium	156.60	2072	
Iodine	113.7	184.4	546
Iridium	2446	4428	
Iron	1538	2861	
Krypton	-157.36	-153.22	-63.74
Lanthanum	920	3455	
Lawrencium	1627		
Lead	327.46	1749	
Lithium	180.5	1342	
Lutetium	1663	3393	
Magnesium	650	1090	
Manganese	1246	2061	
Mendelevium	827		
Mercury	-38.83	356.73	1477
Molybdenum	2623	4639	
Neodymium	1016	3066	
Neon	-248.59	-246.08	-228.7
Neptunium	644		
Nickel	1455	2913	
Niobium	2477	4744	
Nitrogen	-210.00	-195.79	-146.94
Nobelium	827		
Osmium	3033	5012	
Oxygen	-218.79	-182.95	-118.56
Palladium	1554.9	2963	
Phosphorus	44.15	280.5	721
Platinum	1768.4	3825	
Plutonium	640	3228	
Polonium	254	962	
Potassium	63.38	759	
Praseodymium	931	3510	
Promethium	1042	3000	
Protactinium	1572		
Radium	700		
Radon	-71	-61.7	104
Rhenium	3186	5596	
Rhodium	1964	3695	
Rubidium	39.31	688	
Ruthenium	2334	4150	
Samarium	1072	1790	
Scandium	1541	2830	
Selenium	221	685	1493
Silicon	1414	3265	
Silver	961.78	2162	
Sodium	97.72	883	
Strontium	777	1382	
Sulfur	115.21	444.60	1041
Tantalum	3017	5458	
Technetium	2157	4265	
Tellurium	449.51	988	
Terbium	1359	3221	
Thallium	304	1473	
Thorium	1750	4788	
Thulium	1545	1946	
Tin	231.93	2602	
Titanium	1668	3287	
Tungsten	3422	5555	
Uranium	1135	4131	
Vanadium	1910	3407	
Xenon	-111.75	-108.04	16.58
Ytterbium	824	1194	
Yttrium	1526	3336	
Zinc	419.53	907	
Zirconium	1855	4409	

HEAT CAPACITY OF THE ELEMENTS AT 25°C

This table gives the specific heat capacity (c_p) in J/g K and the molar heat capacity (C_p) in J/mol K at a temperature of 25°C and a pressure of 100 kPa (1 bar or 0.987 standard atmospheres) for all the elements for which reliable data are available.

Name	c_p J/g K	C_p J/mol K
Actinium	0.120	27.2
Aluminum	0.897	24.200
Antimony	0.207	25.23
Argon	0.520	20.786
Arsenic	0.329	24.64
Barium	0.204	28.07
Beryllium	1.825	16.443
Bismuth	0.122	25.52
Boron	1.026	11.087
Bromine (Br_2)	0.226	36.057
Cadmium	0.232	26.020
Calcium	0.647	25.929
Carbon (graphite)	0.709	8.517
Cerium	0.192	26.94
Cesium	0.242	32.210
Chlorine (Cl_2)	0.479	33.949
Chromium	0.449	23.35
Cobalt	0.421	24.81
Copper	0.385	24.440
Dysprosium	0.173	28.16
Erbium	0.168	28.12
Europium	0.182	27.66
Fluorine (F_2)	0.824	31.304
Gadolinium	0.236	37.03
Gallium	0.371	25.86
Germanium	0.320	23.222
Gold	0.129	25.418
Hafnium	0.144	25.73
Helium	5.193	20.786
Holmium	0.165	27.15
Hydrogen (H_2)	14.304	28.836
Indium	0.233	26.74
Iodine (I_2)	0.145	36.888
Iridium	0.131	25.10
Iron	0.449	25.10
Krypton	0.248	20.786
Lanthanum	0.195	27.11
Lead	0.129	26.650
Lithium	3.582	24.860
Lutetium	0.154	26.86
Magnesium	1.023	24.869
Manganese	0.479	26.32
Mercury	0.140	27.983
Molybdenum	0.251	24.06
Neodymium	0.190	27.45
Neon	1.030	20.786
Nickel	0.444	26.07
Niobium	0.265	24.60
Nitrogen (N_2)	1.040	29.124
Osmium	0.130	24.7
Oxygen (O_2)	0.918	29.378
Palladium	0.244	25.98
Phosphorous (white)	0.769	23.824
Platinum	0.133	25.86
Potassium	0.757	29.600
Praseodymium	0.193	27.20
Radon	0.094	20.786
Rhenium	0.137	25.48
Rhodium	0.243	24.98
Rubidium	0.363	31.060
Ruthenium	0.238	24.06
Samarium	0.197	29.54
Scandium	0.568	25.52
Selenium	0.321	25.363
Silicon	0.705	19.789
Silver	0.235	25.350
Sodium	1.228	28.230
Strontium	0.301	26.4
Sulfur (rhombic)	0.710	22.75
Tantalum	0.140	25.36
Tellurium	0.202	25.73
Terbium	0.182	28.91
Thallium	0.129	26.32
Thorium	0.113	26.230
Thulium	0.160	27.03
Tin (white)	0.228	27.112
Titanium	0.523	25.060
Tungsten	0.132	24.27
Uranium	0.116	27.665
Vanadium	0.489	24.89
Xenon	0.158	20.786
Ytterbium	0.155	26.74
Yttrium	0.298	26.53
Zinc	0.388	25.390
Zirconium	0.278	25.36

VAPOR PRESSURE OF THE METALLIC ELEMENTS

C. B. Alcock

This table gives coefficients in an equation for the vapor pressure of 65 metallic elements in both the solid and liquid state. Vapor pressures in the range 10^{-10} to 10^2 Pa (10^{-15} to 10^{-3} atm) are covered. The equation is:

for p in pascals: $\log(p/\mathrm{Pa}) = 5.006 + A + BT^{-1} + C\log T + DT^{-3}$
for p in atmospheres: $\log(p/\mathrm{atm}) = A + BT^{-1} + C\log T + DT^{-3}$, where T is the temperature in K

This equation reproduces the observed vapor pressures to an accuracy of ±5% or better. Further details are given in:

C. B. Alcock, V. P. Itkin, and M. K. Horrigan, *Canadian Metallurgical Quarterly*, 23, 309, 1984.

Element, state	*A*	*B*	*C*	*D*	Temperature range
Li sol	5.667	−8310			298–m.p.
Li liq	5.055	−8023			m.p.–1000
Na sol	5.298	−5603			298–m.p.
Na liq	4.704	−5377			m.p.–700
K sol	4.961	−4646			298–m.p.
K liq	4.402	−4453			m.p.–600
Rb sol	4.857	−4215			298–m.p.
Rb liq	4.312	−4040			m.p.–550
Cs sol	4.711	−3999			298–m.p.
Cs liq	4.165	−3830			m.p.–550
Be sol	8.042	−17020	−0.4440		298–m.p.
Be liq	5.786	−15731			m.p.–1800
Mg sol	8.489	−7813	−0.8253		298–m.p.
Ca sol	10.127	−9517	−1.4030		298–m.p.
Sr sol	9.226	−8572	−1.1926		298–m.p.
Ba sol	12.405	−9690	−2.2890		298–m.p.
Ba liq	4.007	−8163			m.p.–1200
Al sol	9.459	−17342	−0.7927		298–m.p.
Al liq	5.911	−16211			m.p.–1800
Ga sol	6.657	−14208			298–m.p.
Ga liq	6.754	−13984	−0.3413		m.p.–1600
In sol	5.991	−12548			298–m.p.
In liq	5.374	−12276			m.p.–1500
Tl sol	5.971	−9447			298–m.p.
Tl liq	5.259	−9037			m.p.–1100
Sn sol	6.036	−15710			298–m.p.
Sn liq	5.262	−15332			m.p.–1850
Pb sol	5.643	−10143			298–m.p.
Pb liq	4.911	−9701			m.p.–1200
Sc sol	6.650	−19721	0.2885	−0.3663	298–m.p.
Sc liq	5.795	−17681			m.p.–2000
Y sol	9.735	−22306	−0.8705		298–m.p.
Y liq	5.795	−20341			m.p.–2300
La sol	7.463	−22551	−0.3142		298–m.p.
La liq	5.911	−21855			m.p.–2450
Ti sol	11.925	−24991	−1.3376		298–m.p.
Ti liq	6.358	−22747			m.p.–2400
Zr sol	10.008	−31512	−0.7890		298–m.p.
Zr liq	6.806	−30295			m.p.–2500
Hf sol	9.445	−32482	−0.6735		298–m.p.
V sol	9.744	−27132	−0.5501		298–m.p.
V liq	6.929	−25011			m.p.–2500
Nb sol	8.822	−37818	−0.2575		298–2500
Ta sol	16.807	−41346	−3.2152	0.7437	298–2500
Cr sol	6.800	−20733	0.4391	−0.4094	298–2000
Mo sol	11.529	−34626	−1.1331		298–2500

Element, state	A	B	C	D	Temperature range
W sol	2.945	−44 094	1.3677		298–2350
W sol	−54.527	−57 687	−12.2231		2200–2500
Mn sol	12.805	−15 097	−1.7896		298–m.p.
Re sol	11.543	−40 726	−1.1629		298–2500
Fe sol	7.100	−21 723	0.4536	−0.5846	298–m.p.
Fe liq	6.347	−19 574			m.p.–2100
Ru sol	9.755	−34 154	−0.4723		298–m.p.
Os sol	9.419	−41 198	−0.3896		298–2500
Co sol	10.976	−22 576	−1.0280		298–m.p.
Co liq	6.488	−20 578			m.p.–2150
Rh sol	10.168	−29 010	−0.7068		298–m.p.
Rh liq	6.802	−26 792			m.p.–2500
Ir sol	10.506	−35 099	−0.7500		298–2500
Ni sol	10.557	−22 606	−0.8717		298–m.p.
Ni liq	6.666	−20 765			m.p.–2150
Pd sol	9.502	−19 813	−0.9258		298–m.p.
Pd liq	5.426	−17 899			m.p.–2100
Pt sol	4.882	−29 387	1.1039	−0.4527	298–m.p.
Pt liq	6.386	−26 856			m.p.–2500
Cu sol	9.123	−17 748	−0.7317		298–m.p.
Cu liq	5.849	−16 415			m.p.–1850
Ag sol	9.127	−14 999	−0.7845		298–m.p.
Ag liq	5.752	−13 827			m.p.–1600
Au sol	9.152	−19 343	−0.7479		298–m.p.
Au liq	5.832	−18 024			m.p.–2050
Zn sol	6.102	−6776			298–m.p.
Zn liq	5.378	−6286			m.p.–750
Cd sol	5.939	−5799			298–m.p.
Cd liq	5.242	−5392			m.p.–650
Hg liq	5.116	−3190			298–400
Ce sol	6.139	−21 752			298–m.p.
Ce liq	5.611	−21 200			m.p.–2450
Pr sol	8.859	−18 720	−0.9512		298–m.p.
Pr liq	4.772	−17 315			m.p.–2200
Nd sol	8.996	−17 264	−0.9519		298–m.p.
Nd liq	4.912	−15 824			m.p.–2000
Sm sol	9.988	−11 034	−1.3287		298–m.p.
Eu sol	9.240	−9459	−1.1661		298–m.p.
Gd sol	8.344	−20 861	−0.5775		298–m.p.
Gd liq	5.557	−19 389			m.p.–2250
Tb sol	9.510	−20 457	−0.9247		298–m.p.
Tb liq	5.411	−18 639			m.p.–2200
Dy sol	9.579	−15 336	−1.1114		298–m.p.
Ho sol	9.785	−15 899	−1.1753		298–m.p.
Er sol	9.916	−16 642	−1.2154		298–m.p.
Er liq	4.688	−14 380			m.p.–1900
Tm sol	8.882	−12 270	−0.9564		298–1400
Yb sol	9.111	−8111	−1.0849		298–900
Lu sol	8.793	−22 423	−0.6200		298–m.p.
Lu liq	5.648	−20 302			m.p.–2350
Th sol	8.668	−31 483	−0.5288		298–m.p.
Th liq	−18.453	−24 569	6.6473		m.p.–2500
Pa sol	10.552	−34 869	−1.0075		298–m.p.
Pa liq	6.177	−32 874			m.p.–2500
U sol	0.770	−27 729	2.6982	−1.5471	298–m.p.
U liq	20.735	−28 776	−4.0962		m.p.–2500
Np sol	19.643	−24 886	−3.9991		298–m.p.
Np liq	10.076	−23 378	−1.3250		m.p.–2500
Pu sol	26.160	−19 162	−6.6675		298–600
Pu sol	18.858	−18 460	−4.4720		500–m.p.
Pu liq	3.666	−16 658			m.p.–2450
Am sol	11.311	−15 059	−1.3449		298–m.p.
Cm sol	8.369	−20 364	−0.5770		298–m.p.
Cm liq	5.223	−18 292			m.p.–2200

DENSITY OF LIQUID ELEMENTS

Gernot Lang

Temperatures are in °C and densities in grams per cubic centimeter. t_m is the melting point, and Pt_m is the density at the melting point.

Element	Purity	Pt_m	ρ_{t_1}		Temperature dependence (ρ_t)	Temperature range, equation	Atm.	Method	Ref.
			t_1	ρ_1					
Ag	—	9.32	—	—	$9.32 - 10 \cdot 10^{-4}(t-t_m)$	–1200°C	—	IBF[a]	100
Ag	—	9.33	—	—	$9.33 - 9.8 \cdot 10^{-4}(t-t_m)$	–1300°C	—	DBF[b]	49
Ag	99.9	9.30	—	—	$9.30 - 9.5 \cdot 10^{-4}(t-t_m)$	–1300°C	—	IBF	37
Ag	—	(9.285)	—	—	$9.285 - 9.2 \cdot 10^{-4}(t-t_m)$	–1300°C	—	IBF	36
Ag	—	9.345	1100	9.20	—		H_2	Bubble pressure	66
Ag	—	9.33	—	—	$9.33 - 10.50 \cdot 10^{-4}(t-t_m)$	–1300°C	Ar	Bubble pressure	67
Ag	—	9.346	—	—	$9.346 - 9.12 \cdot 10^{-4}(t-t_m)$	–1300°C	—	DBF	57
Ag	—	9.36	—	—	$9.36 - 11.4 \cdot 10^{-4}(t-t_m)$	–1200°C	—	Bubble pressure	165
Ag	99.95		—	—	$9.318 - 10.8 \cdot 10^{-4}(t-1000)$	—	H_2	Bubble pressure	81
Ag	—	9.33	—	—	$9.33 - 10 \cdot 10^{-4}(t-t_m)$	–1400°C	—	Bubble pressure	69
Ag	99.999	9.31	—	—	$9.31 - 10.51 \cdot 10^{-4}(t-t_m)$	–1150°C	Ar	Sessile drop	122
Ag	99.999	9.320	—	—	$9.320 - 9 \cdot 10^{-4}(t-t_m)$	–1500°C	Ar	DBF	144
Ag	99.999	9.321	—	—	$9.321 - 9.787 \cdot 10^{-4}(T-T_{mp})$	–1400 K	Ar	Pycnometer	188
Ag	99.999	9.37	—	—	$9.37 - 13 \cdot 10^{-4}(t-t_{mp})$	–1210°C	Ar	Bubble pressure	191
Al	99.4	2.41	—	—	—	—	H_2	BF	85
Al	99.4	2.384	—	—	—	—	—	Pycnometer	23
Al	99.996	2.368	—	—	$2.368 - 2.63 \cdot 10^{-4}(t-t_m)$	–900°C	—	BF	34
Al	99.99	2.41	—	—	$2.41 - 2.78 \cdot 10^{-4}(t-t_m)$	—	—	Pycnometer	82
Al	99.997	2.39	—	—	$2.39 - 3.8 \cdot 10^{-4}(t-t_m)$	–912°C	Ar	Bubble pressure	16
Al	—	2.369	—	—	$2.369 - 3.10 \cdot 10^{-4}(t-t_m)$	—	—	Volumetric measurement	132
Al	99.99	2.39	—	—	$2.39 - 3.2 \cdot 10^{-4}(t-t_m)$	—	He	Drop volume	6
Al	99.999	2.365	—	—	$2.365 - 3.2 \cdot 10^{-4}(t-t_m)$	—	He	Sessile drop	116
Al	99.998	2.375	—	—	$2.375 - 5.4 \cdot 10^{-4}(t-t_m)$	—	Ar	Sessile drop	142
Al	99.99	—	1650	2.07	—	—	He	Sessile drop	143
Al	99.999	2.370	—	—	$2.370 - 2.80 \cdot 10^{-4}(T-933)$	–1250K	vac.	Sessile drop	166
Al	99.996	2.369	—	—	$2.369 - 3.11 \cdot 10^{-4}(t-659)$	–1500°C	He	Sessile drop	169
Al	99.99	2.370	—	—	$2.370 - 2.8 \cdot 10^{-4}(T-T_{mp})$	–1473K	vac.	Sessile drop	186
Au	99.96	17.28	—	—	$17.28 - 12.26 \cdot 10^{-4}(t-t_m)$	–1300°C	H_2	IBF	37
Au	—	17.3105	—	—	$17.3105 - 13.43 \cdot 10^{-4}(t-t_m)$		—	DBF	170
Au	—	17.361	—	—	$17.361 - 16.12 \cdot 10^{-4}(t-t_m)$	–1300°C	—	IBF	36
As	—	5.22	—	—	$5.22 - 5.44 \cdot 10^{-4}(t-t_m)$	—	As vapor	Volumetric method	150
B	99.8	2.08 ± 0.03	—	—	—	—	vac.	Weighing and volume determination	111
Ba	—	3.325	—	—	$3.325 - 2.14 \cdot 10^{-4}(t-t_m)$	—	Ar	DBF	1
Ba	—	3.328	—	—	$3.328 - 5.26 \cdot 10^{-4}(t-t_m)$	—	—	Calculated	44
Ba	99.5	3.32	—	—	$3.32 - 2.74 \cdot 10^{-4}(T-997)$	—	—	Bubble pressure	9
Ba	99.9	3.355	—	—	$3.355(\pm 0.006) - 3.24 \cdot 10^{-4}(T-T_{mp})$	–1800K	He	Bubble pressure	189
Be	—	—	1500	1.42 ± 0.04	—	—	—	Calculated	31
Be	—	1.690	—	—	$1.690 - 1.1 \cdot 10^{-4}(t-t_m)$	—	Ar	Bubble pressure	134
Bi	—	10.07	—	—	$10.07 - 12.56 \cdot 10^{-4}(t-t_m)$	–500°C	—	Manometer pressure	46
Bi	—	10.02	—	—	$10.02 - 11.80 \cdot 10^{-4}(t-t_m)$	–800°C	—	Dilatometer	12
Bi	—	10.07	—	—	$10.07 - 12.44 \cdot 10^{-4}(t-t_m)$	–600°C	—	Dilatometer	73
Bi	—	10.05	—	—	$10.05 - 11.44 \cdot 10^{-4}(t-t_m)$	–800°C	—	DBF	50
Bi	—	10.02	—	—	$10.02 - 12.28 \cdot 10^{-4}(t-t_m)$	–700°C	n_2	Bubble pressure	107
Bi	99.94	10.04	—	—	$10.04 - 12.42 \cdot 10^{-4}(t-t_m)$	–700°C	—	DBF	89
Bi	99.90	—	—	—	$9.43 - 14.00 \cdot 10^{-4}(t-800)$	800—1000°C	H_2	Bubble pressure	78
Bi	—	10.02	—	—	$10.02 - 12.66 \cdot 10^{-4}(t-t_m)$	–500°C	—	Dilatometer	45
Bi	99.98	10.06	—	—	$10.06 - 12.27 \cdot 10^{-4}(t-t_m)$	–4000°C	Ar	Bubble pressure	69
Bi	—	10.05	—	—	$10.05 - 12.9 \cdot 10^{-4}(t-t_m)$	–800°C	—	Direct Archimedean	136
Bi	99.9999	10.05	—	—	$10.05 - 14.1 \cdot 10^{-4}(t-t_m)$	—	N_2	Volume determination	51
Bi	99.98	10.04	—	—	$10.04 - 11.1 \cdot 10^{-4}(t-t_m)$	—	N_2	Volume determination	51
Bi	99.999	10.114	—	—	$10.114 - 10.78 \cdot 10^{-4}(t-t_m)$	–414°C	N_2	DBF	123
Bi	—	10.031	—	—	$10.031 - 12.37 \cdot 10^{-4}(t-t_m)$	—	vac.	Pycnometer	18
Bi	99.999	10.049	—	—	$10.049 - 12.4 \cdot 10^{-4}(t-t_m$ ± 0.002	—	—	DBF	121
Bi	99.999	10.049	—	—	$10.049 - 12.22 \cdot 10^{-4}(t-t_m)$	—	—	DBF	147
Bi	—	10.057	700	9.51 ± 0.03	—	—	—	—	19
Bi	99.999	10.040	—	—	$10.040 - 1.29 \cdot 10^{-2}(t-t_{mp})$	–450°C	N_2	Pycnometer	173
Bi	99.99	10.048	—	—	$10.048 - 1.264 \cdot 10^{-3}(t-t_{mp})$	–420°C	—	Pycnometer	174
Bi	99.99	10.0490	—	—	$10.0490 - 12.7 \cdot 10^{-4}(t-t_m)$	–341°C	vac.	Pycnometer	201
Ca	—	1.367	—	—	$1.367 - 2.21 \cdot 10^{-4}(T-1111)$	—	—	Bubble pressure	9
Ca	—	1.406	—	—	$1.406 - 9 \cdot 10^{-4}(t-t_m)$	—	—	Pycnometer	128
Cd	—	8.02	—	—	$8.02 - 11.0 \cdot 10^{-4}(t-t_m)$	—	—	Manometer pressure	46
Cd	99.97	8.03	—	—	$8.03 - 11.78 \cdot 10^{-4}(t-t_m)$	–500°C	N_2	Bubble pressure	41
Cd	—	8.01	—	—	$8.01 - 11.43 \cdot 10^{-4}(t-t_m)$	–560°C	Ar	Bubble pressure	102
Cd	99.999	7.996	—	—	$7.996 - 12.205 \cdot 10^{-4}(t-t_m)$	—	Ar	Pycnometer	126
Cd	—	8.015	—	—	—	—	H_2, Ar	Bubble pressure	155
Cd	99.99	8.0047	—	—	$8.0047 - 124.10^{-4}(t-t_m)$	–364°C	vac.	Pycnometer	201

Element	Purity	Pt_m	ρ_{t_1}: t_1	ρ_{t_1}: ρ_1	Temperature dependence (ρ_t)	Temperature range, equation	Atm.	Method	Ref.
Ce	—	6.68	—	—	$6.68 \quad 2.3 \cdot 10^{-4}(t-t_m)$	—	—	Crit. Rev.	A
Co	—	—	1600	8.08—8.13	—	—	—	Drop volume	80
Co	—	—	1500	8.05	—	—	—	BF	38
Co	—	7.992	—	—	$7.992 - 10.86 \cdot 10^{-4}(t-t_m)$	—	—	Bubble pressure	141
Co	99.99	7.67	—	—	$7.67 - 12.04 \cdot 10^{-4}(t-t_m)$	−1700°C	Ar	Bubble pressure	70
Co	—	—	1500	7.70	—	—	—	—	112
Co	99.53	7.76	—	—	$7.67 - 9.88 \cdot 10^{-4}(T-1768)$	−2200 K	vac.	Sessile drop	99
Co	99.53	7.78	—	—	$7.78 - 10.17 \cdot 10^{-4}(T-1768)$	−2200 K	Ar	Drop volume	154
Co	—	—	1527	7.80	—	—	He	Sessile drop	159
Co	99.67	7.75	—	—	$7.75 - 11.1 \cdot 10^{-4}(T-1768)$	—	Ar	Bubble Pressure	162
Co	99.95	7.75_2	—	—	$7.75 - 7 \cdot 10^{-4}(t-t_m)$	−1750°C	Ar	DBF	145
Co	98.6	7.756	—	—	$7.756 - 1.652 \cdot 10^{-3}(t-t_{mp})$	−1580°C	—	Bubble pressure	171
Co	—	—	1550	7.60	—	—	vac.	Sessile drop	63
Co	—	7.67	—	—	—	—			67
Co	99.99	7.98	—	—	$7.98 - 1.147 \cdot 10^{-3}(t-t_{mp})$	−2200°C	He	Sessile drop	194
Co	—	—	1550	7.73	—		—	Sessile drop	197
Cr	—	—	1950	6.00 ± 0.13	—	—	vac.	Sessile drop	27
Cr	—	6.46	—	—	—	—	—	Calculated	
Cr	99.75	6.3_3	—	—	$6.3_3 - 11._1 \cdot 10^{-4}(t-t_{mp})$	1900—2100°C	Ar	Levitation	195
Cs	—	1.86	—		$\frac{1.84}{1 + 1.1755 \cdot 10^{-4}(t-82.4) + 7.656 \cdot 10^{-8}(t-82.4)^2 (t°F)}$	−704°C	Ar	Pycnometer	120
Cs	—	1.845	—	—	$1.845 - 5.51 \cdot 10^{-4}(t-t_m)$	−376°C	Ar	Dilatometer	160
Cs	—	1.843	—	—	$1.843 - 5.56 \cdot 10^{-4}(t-t_m)$	−510°C	—	Pycnometer	109
Cs	—	1.851	—	—	$1.851 - 5.71 \cdot 10^{-4}(t-t_m)$	−750°C	—	Pycnometer	106
Cu	—	7.99	—	—	$7.99 - 15.36 \cdot 10^{-4}(t-t_m)$	−1300°C	—	IBF	11
Cu	—	7.962	—	—	$7.96 - 15.88 \cdot 10^{-4}(t-t_m)$	−1300°C	—	Dilatometer	119
Cu	—	7.92_4	—	—	$7.92 - 7.46 \cdot 10^{-4}(t-t_m)$	−1500°C	—	DBF	117
Cu	99.99	7.940	—	—	$7.940 - 8.05 \cdot 10^{-4}(t-t_m)$	−1300°C	—	IBF	37
Cu	—	—	1100	8.10	—	—	—	Drop volume	62
Cu	—	7.87_5	—	—	$7.87 - 7.79 \cdot 10^{-4}(t-t_m)$	−1500°C	—	DBF	65
Cu	—	—	1100	7.90	—	—	—	Pycnometer	72
Cu	—	7.99_2	—	—	$7.99 - 8.02 \cdot 10^{-4}(t-t_m)$	−1500°C	Ar	DBF	13
Cu	99.992	8.090	—	—	$8.090 - 9.442 \cdot 10^{-4}(T-1356)$	—	H_2, Ar	Drop volume	130
Cu	—	—	1100	8.07	—	—	—	Bubble pressure	8
Cu	—	8.03	—	—	$8.03 - 7.94 \cdot 10^{-4}(t-t_m)$	−1500°C	Ar	Bubble pressure	69
Cu	pec.p.	7.938	—	—	$7.938 - 7 \cdot 10^{-4}(t-t_m)$	−1600°C	Ar	DBF	144
Cu	—	8.039	—	—	$8.039 - 9.60 \cdot 10^{-4}(t-t_m)$	—	He	Sessile drop	139
Cu	—	7.936	—	—	$7.936 - 7.862 \cdot 10^{-4}(T-1356)$	−1239°C	Ar	Archimedean	180
Cu	99.99	—	1200	7.81	—	—	N_2	Bubble pressure	184
Cu	99.99	7.91	—	—	$7.91 - 7.7 \cdot 10^{-4}(T-T_{mp})$	−1473 K	vac.	Sessile drop	186
Dy	–	8.2	–	–	$8.2 — 14 \cdot 10^{-4}(t-t_m)$	–	–	Crit. Rev.	A
Er	–	8.6	–	–	$8.6 — 16 \cdot 10^{-4}(t-t_m)$	–	–	Crit. Rev.	A
Eu	–	4.87	–	–	$4.87 — 2.4 \cdot 10^{-4}(t-t_m)$	–	–	Crit. Rev.	A
Fe	—	7.13	—	—	—	—	—	DBF	101
Fe	—	7.24	—	—	—	—	vac.	Drop volume	52
Fe	—	—	1550	7.01	—	—	air	Casting	110
Fe	—	7.15	—	—	—	—	—	DBF	60
Fe	—	7.011	—	—	$7.011 - 8.358 \cdot 10^{-4}(T-1809)$	—	Ar	DBF	55
Fe	—	—	1550	7.189	—	—	—	—	39
Fe	—	7.020	—	—	$7.020 - 8.83 \cdot 10^{-4}(T-1809)$	—	Ar	DBF	43
Fe	—	—	1555	6.98	—	—	—	Bubble pressure	8
Fe	—	7.03_5	—	—	$7.03 - 14.19 \cdot 10^{-4}(t-t_m)$	−1700°C	Ar	Bubble pressure	70
Fe	—	—	1550	7.13	—	—	—	—	95
Fe	99.9	7.02	—	—	$7.02 - 8.17 \cdot 10^{-4}(T-1809)$	—	vac.	Sessile drop	99
Fe	—	7.06	—	—	$7.06 - 7.3 \cdot 10^{-4}(t-1550)$	—	H_2, He	Sessile drop	93
Fe	99.9	7.03	—	—	$7.03 - 8.53 \cdot 10^{-4}(T-1809)$	−2150°C	Ar	Drop volume	154
Fe	—	—	1527	7.08	—	—	He	Sessile drop	159
Fe	99.998	7.06	—	—	$7.06 - 7.33 \cdot 10^{-4}(t-t_m)$	−1850°C	He	Sessile drop	157
Fe	99.98	7.05	—	—	$7.05 - 9.58 \cdot 10^{-4}(T-1809)$	—	Ar	Bubble pressure	162
Fe	99.988	—	1650	6.97	—	—	He	Sessile drop	143
Fe	99.96	7.02_2	—	—	$7.02 - 6 \cdot 10^{-4}(t-t_m)$	−1700°C	Ar	DBF	145
Fe	—	—	1550	7.05	—	—	—	Sessile drop	172
Fe	—	6.957			$6.957 - 9.36 \cdot 10^{-4}(t-t_m)$	—	—	Archimedean	177
Fe	—	—	1550	7.01 ± 0.03	—	—	Ar	Sessile drop	63
Fe	—	7.03	—	—	—	—	Ar	Bubble pressure	67
Fe	—	7.05	—	—	$7.05 - 7.3 \cdot 10^{-4}(t-t_m)$	−1850°C	H_2, He	Sessile drop	93
Fe	99.9	7.08	—	—	$7.08 - 6.338 \cdot 10^{-4}(t-t_{mp})$	−2200°C	He	Sessile drop	194
Fe	—	—	1550	7.03	—	—	—	Sessile drop	196
Fr	—	2.29	—	—	—	—	—	Calculated	84
Ga	—	6.20 ± 0.01	—	—	—	—	—	—	10
Ga	—	6.08	—	—	$6.08 - 6 \cdot 10^{-4}(t-t_m)$	303—1773 K	He	Sessile drop	152

DENSITY OF LIQUID ELEMENTS (continued)

Element	Purity	ρ_{t_m}	ρ_{t_1}: t_1	ρ_1	Temperature dependence (ρ_t)	Temperature range, equation	Atm.	Method	Ref.
Ga	—	6.09379	—	—	$6.11564 - 7.37437 \cdot 10^{-4} \cdot t + 1.37767 \cdot 10^{-7} \cdot t^2$	−600°C	—	Pycnometer	61
Ga	99.999	6.12	—	—	$6.12 - 6.2 \cdot 10^{-4}(T\text{-}303)$	−1250 K	vac.	Sessile drop	166
Ga	99.999	6.087	—	—	$6.087 - 5.93 \cdot 10^{-4}(t\text{-}t_m)$	−450°C	vac.	Drop pressure	182
Ga	99.99	6.12	—	—	$6.12 - 6.2 \cdot 10^{-4}(T\text{-}T_{mp})$	−1473 K	vac.	Sessile drop	186
Ga	99.9999	6.05	—	—	$6.05 - 6 \cdot 10^{-4}(t\text{-}t_{mp})$	900—1210°C	Ar	Bubble pressure	191
Ga	99.99	6.0960	—	—	$6.0960 - 7.44 \cdot 10^{-4}(t\text{-}t_m)$	−93°C	vac.	Pycnometer	210
Ge	—	5.52	—	—	$5.52 - 5.10 \cdot 10^{-4}(t\text{-}t_m)$	−1200°C	—	Volumetric measurement	79
Ge	—	5.57_5	—	—	$5.58 - 4.84 \cdot 10^{-4}(t\text{-}t_m)$	−1200°C	N_2	Pycnometer	58
Ge	99.990	5.49	—	—	$5.49 - 4.98 \cdot 10^{-4}(t\text{-}t_m)$	−1600°C	Ar	Bubble pressure	69
Ge	—	5.655	—	—	$5.655 - 6 \cdot 10^{-4}(t\text{-}t_m)$	—	He	Sessile drop	139
Ge	—	5.598	—	—	$5.598 - 6.250 \cdot 10^{-4}(T\text{-}1233)$	−1850 K	vac.	Sessile drop	176
Ge	99.999	5.61	—	—	$5.61 - 7.29 \cdot 10^{-4}(t\text{-}t_m)$	−1600°C	He	Sessile drop	183
Ge	99.999	5.57	—	—	$5.57 - 3.71 \cdot 10^{-4}(T\text{-}T_{mp})$	−1400 K	Ar	Pycnometer	188
Hf	—	12.0	—	—	—	—	—	Calculated	2
Hg	—	13.691	—	—	$13.691 - 24.12 \cdot 10^{-4}(t\text{-}t_m)$	—	—	Absolute displacement	133
Hg	—	13.6873	—	—	$13.6873 - 24.34 \cdot 10^{-4}(t\text{-}t_m)$	−100°C	—	DBF	108
Hg	—	13.53	—	—	$13.53 - 24 \cdot 10^{-4}(t\text{-}t_m)$	—	—	—	192
Hg	—	13.691496	—	—	$13.595080[1 + 1.814401 \cdot 10^{-4} t + 7.016 \cdot 10^{-9} \cdot t^2 + 1.8625 \cdot 10^{-11} \cdot t^3]^{-1}$	—	—	—	202
In	—	7.04	—	—	$7.04 - 7.86 \cdot 10^{-4}(t\text{-}t_m)$	−421°C	Ar	Bubble pressure	102
In	—	7.023	—	—	$7.023 - 6.7 \cdot 10^{-4}(t\text{-}t_m)$	—	—	DBF	151
In	99.999	7.035	—	—	$7.035 - 7.59 \cdot 10^{-4}(t\text{-}t_m)$	−532°C	N_2	DBF	123
In	99.999	7.016	—	—	$7.016 - 8.362 \cdot 10^{-4}(t\text{-}t_m)$	—	Ar	Pycnometer	126
In	9.99	7.035	—	—	$7.035 - 4.3 \cdot 10^{-4}(T\text{-}T_{mp})$	−1473 K	vac.	Sessile drop	186
In	99.9995	6.95	—	—	$6.95 - 5 \cdot 10^{-4}(t\text{-}t_{mp})$	595—890°C	Ar	Bubble pressure	191
Ir	—	20.0	—	—	—	—	—	Calculated	2
Ir	—	19.39	—	—	—	—	—	—	181
Ir	—	—	2450	19.23	—	—	He	Sessile drop	203
K	—	0.826	—	—	$0.826 - 2.22 \cdot 10^{-4}(t\text{-}t_m)$	—	—	—	96
K	—	0.819	—	—	$0.819 - 2.38 \cdot 10^{-4}(t\text{-}t_m)$	−1400°C	—	Pycnometer	40
K	—	0.828	—	—	$0.828 - 2.32 \cdot 10^{-4}(t\text{-}t_m)$	−510°C	—	Pycnometer	109
K	—	0.8288	—	—	$0.8288 - 2.42 \cdot 10^{-4}(t\text{-}t_m)$	—	—	Pycnometer	158
La	—	5.955	—	—	$5.955 - 2.42 \cdot 10^{-4}(t\text{-}t_m)$	—	—	Pycnometer	163
Li	—	0.510	—	—	$0.510 - 0.75 \cdot 10^{-4}(t\text{-}t_m)$	−800°C	—	—	7
Li	—	0.512	—	—	$0.512 - 0.52 \cdot 10^{-4}(t\text{-}t_m)$	−285°C	—	—	4
Li	—	0.515	—	—	$0.515 - 1.01 \cdot 10^{-4}(t\text{-}t_m)$	−1600°C	—	Pycnometer	40
Li	—	—	—	—	$0.5368 - 1.021 \cdot 10^{-4} \cdot t$	400—1125°C	—	Pycnometer	106
Mg	—	1.585	—	—	$1.585 - 2.0 \cdot 10^{-4}(t\text{-}t_m)$	−800°C	—	DBF	90
Mg	99.5	1.584	—	—	$1.584 - 2.34 \cdot 10^{-4}(t\text{-}t_m)$	−900°C	—	IBF	35
Mg	—	1.588	—	—	$1.588 - 2.67 \cdot 10^{-4}(t\text{-}t_m)$	—	Ar	DBF	74
Mg	99.5	—	—	—	$2.03 - 5.31 \cdot 10^{-4} \cdot t$	959—1053°C	—	Bubble pressure	9
Mg	—	1.587	—	—	—	—	Ar	Direct Archimedean	167
Mg	99.9	—	800	1.50	—	—	vac.	Sessile drop	187
Mn	—	—	1440	5.84 ± 2%	—	—	He	Volumetric method	115
Mn	—	6.43	—	—	—	—	—	—	95
Mn	99.999	5.3_8	—	—	$5.3_8 - 9.0_2(t\text{-}t_m)$	1700—1900°C	Ar	Levitation	195
Mo	99.7	9.33	—	—	—	—	vac.	Drop volume	87
Mo	—	9.35	—	—	—	—	—	Calculated	2
Mo	—	9.1	—	—	—	—	vac.	Pendant drop + drop weight	129
Na	—	0.938_5	—	—	$0.938_5 - 2.60 \cdot 10^{-4}(t\text{-}t_m)$	—	—	—	96
Na	—	0.945	—	—	$0.945 - 3.25 \cdot 10^{-4}(t\text{-}t_m)$	−450°C	—	Bubble pressure	113
Na	—	0.927	—	—	$0.927 - 2.24 \cdot 10^{-4}(t\text{-}t_m)$	—	—	DBF	135
Na	—	0.927	—	—	$0.927 - 2.38 \cdot 10^{-4}(t\text{-}t_m)$	—	—	Pycnometer	40
Na	—	0.929	—	—	$0.929 - 2.44 \cdot 10^{-4}(t\text{-}t_m)$	−810°C	—	Pycnometer	109
Nb	—	7.83	—	—	—	—	—	Calculated	2
Nb	—	7.6	—	—	—	—	vac.	Pendant drop + drop weight	129
Nd	—	6.688	—	—	$6.688 - 5.27 \cdot 10^{-4}(t\text{-}t_m)$	—	vac.	Pycnometer	153
Ni	—	—	1500	8.04	—	—	—	BF	38
Ni	—	7.906	—	—	$7.906 - 11.598 \cdot 10^{-4}(T\text{-}1726)$	—	Ar	DBF	43
Ni	99.85	7.77	—	—	$7.77 - 11.77 \cdot 10^{-4}(T\text{-}1726)$	−1700°C	Ar	Bubble pressure	70
Ni	99.99	7.78	—	—	$7.78 - 60 \cdot 10^{-4}(t\text{-}t_m)$	—	vac.	Sessile drop	30
Ni	—	—	1500	7.78	—	—	—	—	112
Ni	99.95	7.78	—	—	$7.78 - 10 \cdot 10^{-4}(t\text{-}t_m)$	—	vac.	Sessile drop	99
Ni	—	7.81	—	—	$7.81 - 8.7 \cdot 10^{-4}(t\text{-}t_m)$	—	H_2, He	Sessile drop	99
Ni	—	7.65	—	—	—	—	vac.	Pendant drop + drop weight	129
Ni	99.95	7.95	—	—	$7.95 - 10.8 \cdot 10^{-4}(T\text{-}1726)$	−2150°C	Ar	Drop volume	154
Ni	99.997	7.81	—	—	$7.81 - 8.33 \cdot 10^{-4}(t\text{-}t_m)$	−1850°C	He	Sessile drop	157
Ni	99.85	7.78_5	—	—	$7.79 - 6 \cdot 10^{-4}(t\text{-}t_m)$	−1700°C	Ar	DBF	145
Ni	—	—	1550	7.64	—	—	Ar	Sessile drop	63

Element	Purity	Pt_m	ρ_{t_1}: t_1	ρ_{t_1}: ρ_1	Temperature dependence (ρ_t)	Temperature range, equation	Atm.	Method	Ref.
Ni	—	7.75	—	—	—	—	Ar	Bubble pressure	67
Ni	—	7.81	—	—	$7.81 - 8.7\cdot10^{-4}(t-t_m)$	−1850°C	H_2, He	Sessile drop	93
Ni	99.99	7.82	—	—	$7.82 - 6.376\cdot10^{-4}(t-t_m)$	−2200°C	He	Sessile drop	194
Ni	—	—	1550	7.78	—		—	Sessile drop	197
Ni	99.99	—	1600	7.74 ± 0.06	—	—	Ar, H_2	Sessile drop	199
Os	—	20.1	—	—	—	—	—	Calculated	2
Pb	—	10.71	—	—	$10.71 - 13.9\cdot10^{-4}(t-t_m)$	—	—	Manometer pressure	46
Pb	99.98	10.59	—	—	$10.59 - 13.95\cdot10^{-4}(t-t_m)$	−440°C	N_2	Bubble pressure	41
Pb	—	10.66	—	—	$10.66 - 11.77\cdot10^{-4}(t-t_m)$	−720°C	—	DBF	64
Pb	—	10.65	—	—	$10.65 - 13.0\cdot10^{-4}(t-t_m)$	−700°C	N_2	Bubble pressure	107
Pb	—	10.65	—	—	$10.65 - 13\cdot10^{-4}(t-t_m)$	—	H_2	Bubble pressure	155
Pb	—	10.66	—	—	$10.66 - 13\cdot10^{-4}(t-t_m)$	—	Ar	Bubble pressure	155
Pb	—	10.785 ± 0.0173	—	—	—	—	—	Pycnometer	59
Pb	—	10.67	—	—	$10.67 - 13.34\cdot10^{-4}(t-t_m)$	−705°C	Ar	Bubble pressure	102
Pb	99.9923	10.678	—	—	$10.678 - 13.174\cdot10^{-4}(t-t_m)$	—	Ar	DBF	56
Pb	99.997	10.660	—	—	$10.660 - 12.220\cdot10^{-4}(t-t_m)$	—	—	Pycnometer	161
Pb	—	10.66	—	—	$10.66 - 12\cdot10^{-4}(t-t_m)$	−740°C	—	Direct Archimedean	136
Pb	99.999	—	—	—	$10.650 - 9.8\cdot10^{-4}\cdot t$	1000—1600°C	He	Sessile drop	116
Pb	99.99	10.687	—	—	$10.687 - 12.2\cdot10^{-4}(t-t_m)$	−700°C	—	Electronic densimeter	148
Pb	99.999	10.662	—	—	$10.662 - 13.1\cdot10^{-4}(t-t_m)$	−440°C	—	Pycnometer	174
Pb	—	10.683	—	—	$10.683 - 12.53\cdot10^{-4}(t-t_m)$	—	—	—	175
Pb	99.999	10.665	—	—	$10.665 - 12.64\cdot10^{-4}(t-t_m)$ $10.08\cdot10^{-8}(t-t_m)^2$	−1550°C	Ar	Direct Archimedean	144
Pb	99.9	10.760	—	—	$10.760 - 16.03\cdot10^{-4}(t-t_m)$	−900°C	—	IBF	35
Pb	99.999	10.62	—	—	$10.62 - 13\cdot10^{-4}(t-t_m)$	−450°C	vac.	Drop pressure	193
Pb	99.99	10.6604	—	—	$10.6604 - 13.9\cdot10^{-4}(t-t_m)$	−381°C	vac.	Pycnometer	101
Pd	—	10.7	—	—	—	—	—	Calculated	29
Pd	—	10.7	—	—	—	—	—	Calculated	2
Pd	99.95	10.49	—	—	$10.49 - 12.26\cdot10^{-4}(t-t_m)$	−1800°C	Ar	Bubble pressure	70
Pd	99.998	10.379	—	—	$10.379 - 11.69\cdot10^{-4}(t-t_m)$	−1700°C	He	Sessile drop	116
Pd	—	10.52	—	—	—	—	—	—	181
Pr	—	6.61	—	—	$6.61 - 2.5\cdot10^{-4}(t-t_m)$	—	—	Pycnometer	163
Pt	99.84	19.7 ± 0.25	—	—	—	—	vac.	Drop volume	28
Pt	99.999	—	1800	18.82 ± 0.02	—	—	Ar	Sessile drop	63
Pt	—	18.91	—	—	$18.91 - 28.82\cdot10^{-4}(t-t_m)$	−1875°C	Ar	Bubble pressure	68, 70
Pt	—	18.81	—	—	—	—	—	—	181
Pt	99.99	19.77	—	—	$19.77 - 24\cdot10^{-4}(t-t_m)$	−2200°C	vac.	Sessile drop	185
Pu	—	16.64	—	—	$16.64 - 14.5\cdot10^{-4}(t-t_m)$	−960°C	—	—	118
Pu	—	16.66	—	—	$16.66 - 17.32\cdot10^{-4}(t-t_m)$	—	—	Pycnometer	156
Pu	99.95	16.628	—	—	$16.628 - 14.19\cdot10^{-4}(t-t_m)$	−950°C	—	Pycnometer	48
Pu	99.97	—	700	15.90	—	—	vac.	Pycnometer	164
Rb	99.4	1.385			$1.55643 - 2.6511\cdot10^{-4}\cdot t\frac{6.26779}{t}$ (t°F)	−1076°C	Ar	Dilatometer	160
Rb	—	1.484			$\frac{1.472}{1 + 1.3309\cdot10^{-4}(t-102) + 5.2106\cdot10^{-8}(t-102)^2}$ (t°F)	−730°C	Ar	Pycnometer	120
Rb	—	1.463	—	—	$1.463 - 4.51\cdot10^{-4}(t-t_m)$	−800°C	—	Pycnometer	106
Rb	—	1.457	—	—	$1.457 - 9.47\cdot10^{-4}(t-t_m)$	−50°C	vac.	Pycnometer	47
Re	99.4	18.9	—	—	—	—	vac.	Drop volume	87
Re	—	18.7	—	—	—	—	—	Calculated	2
Rh	—	10.65	—	—	—	—	—	Calculated	29
Rh	—	11.1	—	—	—	—	—	Calculated	2
Rh	99.99	12.20	—	—	$12.20 - 50\cdot10^{-4}(t-t_m)$	−2200°C	vac.	Sessile drop	185
Rh	99.99	10.7	—	—	$10.7 - 8.955\cdot10^{-4}(t-t_{mp})$	−2200°C	He	Sessile drop	194
Ru	—	10.9	—	—	—	—	—	Calculated	2
S	—	1.819	—	—	$1.819 - 8.00\cdot10^{-4}(t-t_m)$	−160°C	N_2	Bubble pressure	83
Sb	—	6.49	—	—	$6.49 - 5.97\cdot10^{-4}(t-t_m)$	−1000°C	—	IBF	12
Sb	99.52	6.55	—	—	$6.55 - 7.27\cdot10^{-4}(t-t_m)$	−800°C	N_2	Bubble pressure	41
Sb	—	6.43	—	—	$6.43 - 3.0\cdot10^{-4}(t-t_m)$	−900°C	H_2	Bubble pressure	107
Sb	—	6.53	—	—	$6.53 - 4.0\cdot10^{-4}(t-t_m)$	−700°C	H_2	Bubble pressure	32
Sb	—	6.50	—	—	$6.50 - 6.1\cdot10^{-4}(t-t_m)$	—	Ar	Bubble pressure	102, 155
Sb	—	6.50	—	—	$6.50 - 6.34\cdot10^{-4}(t-t_m)$	−917°C	Ar	Bubble pressure	102
Sb	99.9	6.50	—	—	$6.50 - 7.35\cdot10^{-4}(t-t_m)$	−700°C	—	DBF	89
Sb	99.992	6.483	—	—	$6.596 + 2.022\cdot10^{-4}\cdot T - 3.629\cdot10^{-7}\cdot T^2$	—	—	Bubble pressure, DBF	54
Sb	99.6	6.46_5	—	—	$6.46_5 - 5.90\cdot10^{-4}(t-t_m)$	−1200°C	Ar	Bubble pressure	69
Sb	99.999	6.452	—	—	$6.452 - 5.8\cdot10^{-4}(t-t_m)$	−1600°C	He	Sessile drop	116

Element	Purity	Pt_m	ρ_{t_1}		Temperature dependence (ρ_t)	Temperature range, equation	Atm.	Method	Ref.
			t_1	ρ_1					
SB	99.999	6.535	—	—	$6.535 - 6.73 \cdot 10^{-4}(t-t_m)$	–745°C	N_2	DBF	123
Sb	—	6.493	—	—	$6.493 - 6.486 \cdot 10^{-4}(t-t_m)$	–746°C	Ar	Pycnometer	127
Sb	99.9999	6.48	—	—	$6.48 - 6 \cdot 10^{-4}(t-t_{mp})$	670—1010°C	Ar	Bubble pressure	191
Se	—	3.987	—	—	$3.987 - 16 \cdot 10^{-4}(T-490)$	—	—	—	21
Se	—	3.985	—	—	$3.985 - 15.5 \cdot 10^{-4}(T-490)$	—	—	Pycnometer	15
Se	—	4.06	—	—	$4.06 - 5 \cdot 10^{-4}(T-490)$	—	—	—	5
Se	—	3.984	—	—	—	—	—	—	104
Se	—	4.01	—	—	$4.01 - 12.02 \cdot 10^{-4}(t-t_m)$	—	—	DBF	71
Se	—	4.011	—	—	$4.011 - 10 \cdot 10^{-4}(t-t_m)$	–400°C	Ar	DBF	144
Se	99.995	4.00	—	—	$3.75 - 7.5 \cdot 10^{-4}(t-t_m)$ (nonlinear from t_{mp} to 350°C)	350—600°C	Ar	Bubble pressure	191
Si	—	—	1550	2.54	—	—	—	Sessile drop	22
Si	99.9	2.52_5	—	—	$2.52_5 - 3.51 \cdot 10^{-4}(t-t_m)$	–1650°C	Ar	Bubble pressure	69
Si	99.9999	—	1500	2.46	—	—	Ar	Sessile drop	24
Si	99.999	2.57	—	—	$2.57 - 9.36 \cdot 10^{-4}(t-t_m)$	–1500°C	He	Bubble pressure	131
Sn	—	6.988	—	—	$6.988 - 6.11 \cdot 10^{-4}(t-t_m)$	–800°C	—	DBF	20
Sn	—	6.98	—	—	—	—	—	DBF	86
Sn	—	7.01	—	—	$7.01 - 7.4 \cdot 10^{-4}(t-t_m)$	—	—	Manometer pressure	46
Sn	—	6.97	—	—	$6.97 - 7.03 \cdot 10^{-4}(t-t_m)$	–800°C	—	IBF	100
Sn	—	6.93	—	—	$6.93 - 5.69 \cdot 10^{-4}(t-t_m)$	–600°C	—	Dilatometer	73
Sn	—	6.96	—	—	$6.96 - 6.36 \cdot 10^{-4}(t-t_m)$	–1200°C	—	DBF	117
Sn	—	6.983	—	—	$6.98 - 7.17 \cdot 10^{-4}(t-t_m)$	–500°C	—	Pycnometer	98
Sn	—	7.00	—	—	$7.00 - 7.14 \cdot 10^{-4}(t-t_m)$	–600°C	N_2	Bubble pressure	88
Sn	—	—	625	6.67	—	—	N_2	Bubble pressure	107
Sn	—	6.99	—	—	$6.99 - 7.13 \cdot 10^{-4}(t-t_m)$	–400°C	H_2	Bubble pressure	32
Sn	—	6.993	—	—	$6.99 - 7.4 \cdot 10^{-4}(t-t_m)$	—	—	Manometer pressure	92
Sn	—	6.972	—	—	$6.972 - 6.60 \cdot 10^{-4}(t-t_m)$	–500°C	—	Dilatometer	45
Sn	—	7.000	—	—	$7.000 - 6.13 \cdot 10^{-4}(t-t_m)$	2480°C	—	DBF	53
Sn	—	6.986			$6.986 - 8.02 \cdot 10^{-4}(t-t_m)$	–500°C	Ar	Bubble pressure	114
Sn	—	6.978	—	—	$6.978 - 7.13 \cdot 10^{-4}(t-t_m)$	–1600°C	Ar	Bubble pressure	69
Sn	99.999	6.973	—	—	$6.973 - 7.125 \cdot 10^{-4}(t-t_m)$	—	—	Pycnometer	161
Sn	99.999	6.974	—	—	$6.974 - 7.14 \cdot 10^{-4}(t-t_m)$	–400°C	vac.	Pycnometer	17
Sn	99.999	7.01	—	—	$7.01 - 6.3 \cdot 10^{-4}(t-t_m)$	—	—	Bubble pressure	103
Sn	99.999	6.981	—	—	$6.981 - 6.63 \cdot 10^{-4}(t-t_m)$	–438°C	N_2	DBF	123
Sn	99.999	6.986	—	—	$6.986 - 6 \cdot 10^{-4}(t-t_m)$	–1500°C	Ar	DBF	18
Sn	99.999	6.95	—	—	$6.95 - 1.94 \cdot 10^{-4}(t-T_M)$	–950 K	vac.	Sessile drop	166
Sn	—	6.83	—	—	6.83	(350—700°C)	Ar	Bubble pressure	168
Sn	99.999		—	—	$7.390 + 1.035 \cdot 10^{-7}T^2 - 8.604 \cdot 10^{-4}T$	(K)	vac.	Pycnometer	170
Sn	99.999	6.983	—	—	$6.983 - 6.88 \cdot 10^{-4}(t-t_m)$ (non linear from 800 to 1500°C)	232—800°C	Ar	Direct Archimedean	144
Sn	99.99	6.964	—	—	$6.964 - 6.74 \cdot 10^{-4}(t-t_m)$ (non linear from 800 to 1300°C)	232—800°C	—	IBF	179
Sn	99.99	6.95	—	—	$6.95 - 1.94 \cdot 10^{-4}(T-T_m)$	–1473 K	vac.	Sessile drop	186
Sn	99.999	—	800	6.56	—	—	vac.	Sessile drop	187
Sn	99.99	6.954	—	—	$6.954 - 6.389 \cdot 10^{-4}(t-t_m)$	–500°C	Ar	Pycnometer	190
Sn	99.99	6.9801	—	—	$6.9801 - 7.5 \cdot 10^{-4}(t-t_m)$	–341°C	vac.	Pycnometer	201
Sr	99.5	2.375	—	—	$2.375 - 2.62 \cdot 10^{-4}(T-1041)$	—	—	Bubble pressure	9
Ta	—	15.0	—	—	—	—	—	Calculated	2
Te	—	5.75	—	—	$5.75 - 3.45 \cdot 10^{-4}(t-t_m)$	–600°C	—	Pycnometer	58
Te	—	5.58	—	—	$5.58 - 3.57 \cdot 10^{-4}(t-t_m)$	–600°C	N_2	Bubble pressure	105
Te	99.9999	5.797	—	—	$5.797 - 5.35 \cdot 10^{-4}(t-t_m)$	–700°C	Ar	Bubble pressure	69, 71
Te	—	5.71	—	—	$5.71 - 3.6 \cdot 10^{-4}(t-t_m)$		—	Bubble pressure	165
Te	99.7	5.86	—	—	$5.86 - 7.3 \cdot 10^{-4}(t-t_m)$	—	N_2	Volume determination	51
Te	99.999	5.66	—	—	$5.66 - 7.5 \cdot 10^{-4}(t-t_{mp})$	470—900°C	Ar	Bubble pressure	191
Te	99.9999	5.68	—	—	$5.68 - 7 \cdot 10^{-4}(t-t_{mp})$	470—900°C	Ar	Bubble pressure	191
Ti	98.7	4.11 ± 0.08	—	—	—	—	—	Capillary method	25
Ti	—	4.15	—	—	—	—	—	Calculated	2
Ti	—	4.11	—	—	$4.11 - 6.99 \cdot 10^{-4}(t-t_m)$	—	—	3 different methods	149
Ti	—	4.10	—	—	—	—	vac.	Pendant drop + drop weight	129
Tl	—	11.29	—	—	$11.29 - 14.85 \cdot 10^{-4}(t-t_m)$	–651°C	Ar	Bubble pressure	102
Tl	—	11.255	—	—	$11.255 - 13.88 \cdot 10^{-4}(t-t_m)$	—	—	DBF	138
Tl	99.999	11.222	—	—	$11.222 - 14.39 \cdot 10^{-4}(t-t_m)$	—	Ar	Pycnometer	125
Tl	99.999	11.350	—	—	$11.350 - 13 \cdot 10^{-4}(t-t_m)$	–905°C	—	Electronic densimeter	148
Tl	99.999	11.31	—	—	$11.31 - 17 \cdot 10^{-4}(t-t_m)$	500—900°C	Ar	Bubble pressure	191
Tl	99.99	11.2166	—	—	$11.2166 - 14.8 \cdot 10^{-4}(t-t_m)$	–346°C	vac.	Pycnometer	201
U	—	17.905	—	—	$17.905 - 10.328 \cdot 10^{-4}(T-1405)$	—	Ar	DBF	42
U	—	17.27	—	—	$17.27 - 16.01 \cdot 10^{-4}(T-1405)$	—	vac.	Pycnometer	97
U	—	16.95	—	—	$16.95 - 12.9 \cdot 10^{-4}(T-1407)$	(K) –1707 K	vac.	Gamma densitometry	200

Element	Purity	Pt_m	ρ_{t_1} t_1	ρ_{t_1} ρ_1	Temperature dependence (ρ_t)	Temperature range, equation	Atm.	Method	Ref.
V	—	5.55	—	—	—	—	—	Calculated	2
V	—	—	1935	5.734	—	—	—	Bubble pressure	149
V	—	5.3	—	—	—	—	vac.	Pendant drop + drop weight	129
W	—	17.6	—	—	—	—	—	Estimated	14
W	99.8	17.7	—	—	—	—	vac.	Drop volume	87
W	—	17.5	—	—	—	—	—	Calculated	2
Zn	—	6.59	—	—	$6.59 - 9.7{\cdot}10^{-4}(t\text{-}t_m)$	—	—	Manometer pressure	46
Zn	—	—	600	6.35	—	—	—	IBF	12
Zn	—	6.55	—	—	$6.55 - 9.68{\cdot}10^{-4}(t\text{-}t_m)$	–700°C	—	Dilatometer	73
Zn	—	6.55	—	—	$6.55 - 9.25{\cdot}10^{-4}(t\text{-}t_m)$	–700°C	—	Pycnometer	98
Zn	—	6.64_5	—	—	$6.64 - 10.38{\cdot}10^{-4}(t\text{-}t_m)$	–800°C	—	DBF	90
Zn	—	6.66	—	—	$6.66 - 11.27{\cdot}10^{-4}(t\text{-}t_m)$	–800°C	—	DBF	91
Zn	—	6.64	—	—	$6.64 - 7.41{\cdot}10^{-4}(t\text{-}t_m)$	–700°C	N_2	Bubble pressure	88
Zn	—	—	500	6.55	—	—	H_2	Bubble pressure	107
Zn	—	6.64_5	—	—	—	—	—	Pycnometer	59
Zn	99.995	6.562	—	—	$6.562 - 10.76{\cdot}10^{-4}(t\text{-}t_m)$	–700°C	—	IBF	34
Zn	—	6.57_5	—	—	$6.57_5 - 11.30{\cdot}10^{-4}(t\text{-}t_m)$	–700°C	Ar	Bubble pressure	114
Zn	99.999	6.57_7 ± 00.012	—	—	$6.57_7 - 11.11{\cdot}10^{-4}(t\text{-}t_m)$	–700°C	Ar	Bubble pressure	69
Zn	—	6.576	—	—	$6.576 - 9.80{\cdot}10^{-4}(T\text{-}692)$	—	—	DBF	121, 147
Zr	—	5.80	—	—	—	—	—	Calculated	2
Zr	—	—	1835	6.06	—	—	—	3 different methods	149
Zr	—	5.60	—	—	—	—	vac.	Pendant drop + drop weight	129

[a] Indirect buoyancy force.
[b] Direct buoyancy force.

REFERENCES

1. **Addison and Pulham,** *J. Chem. Soc.*, 3873, 1962.
2. **Allen,** *Trans. AIME*, 227, 1175, 1963.
3. **Allen,** *Trans. AIME*, 230, 1537, 1964.
4. **Andrade and Dobbs,** *Proc. Roy. Soc.*, 211, 12, 1952.
5. **Astakhov, Penin, and Dobkina,** *Zh. Fiz. Khim.*, 20, 403, 1946.
6. **Ayushina, Levin, and Geld,** *Zh. Fiz. Khim.*, 42, 2799, 1968.
7. **Been, Edwards, Teeter, and Chalkins,** NEPA-1585, US-AEC., 1950.
8. **Beer,** in *Lucas Mém. Sci. Rev. Mét.*, 61, 1, 97, 1964.
9. **Bohdansky and Schins,** *J. Inorg. Nucl. Chem.*, 30, 2331, 1968.
10. **Bosio,** *C. R., Paris*, 259, 4545, 1964.
11. **Bornemann and Sauerwald,** *Z. Metallkunde*, 14, 145, 1922.
12. **Bornemann and Sauerwald,** *Z. Metallkunde*, 14, 254, 1922.
13. **Cahill and Kirshenbaum,** *J. Phys. Chem.*, 66, 1080, 1962.
14. **Calverley,** *Proc. Phys. Soc.*, 70, 1040, 1957.
15. **Campbell and Epstein,** *J. Am. Chem. Soc.*, 64, 2679, 1942.
16. **Coy and Mateer,** *Trans. ASM*, 58, 99, 1965.
17. **Crawley,** *Trans. AIME*, 245, 1655, 1969.
18. **Crawley and Kiff,** *Metall. Trans.*, 2, 609, 1971.
19. **Cubicciotti,** *J. Phys. Chem.*, 68, 537, 1964.
20. **Day, Sosman, and Hostetter,** *Am. J. Sci.*, 187, 1, 1914.
21. **Dobinsky and Veselovsky,** *Bull. Int. Acad. Pol.*, A, 446, 1936.
22. **Dshemilev, Popel, and Zarevski,** *Fiz. Met. i Met.*, 18, 83, 1964.
23. **Edwards and Moorman,** *Chem. Met. Engin.*, 24, 61, 1921.
24. **Eljutin, Kostikov, and Levin,** *Izv. Vys. Uch. Sav. Tsvetn. Met.*, 2, 131, 1970.
25. **Eljutin and Maurakh,** *Izv. A.N., OTN*, 4, 129, 1956.
26. **Eremenko,** *Ukr. Khim. Zh.*, 28, 427, 1962.
27. **Eremenko and Naidich,** *Izv. A.N., OTN*, 2, 111, 1959.
28. **Eremenko and Naidich,** *Izv. A.N., OTN*, 6, 129, 1959.
29. **Eremenko and Naidich,** *Izv. A.N., OTN*, 6, 100, 1961.
30. **Eremenko and Nishenko,** *Ukr. Khim. Zh.*, 30, 125, 1964.
31. **Eremenko, Nishenko, and Taj-Shou-Wej,** *Izv. A.N., OTN*, 3, 116, 1960.
32. **Fisher and Phillips,** *J. Metals*, 6, 1060, 1954.
34. **Gebhardt, Becker, and Dorner,** *Aluminium*, 31, 315, 1955.
35. **Gebhardt, Becker, and Trägner,** *Z. Metallkunde*, 46, 90, 1955.
36. **Gebhardt, and Dorner,** *Z. Metallkunde*, 42, 353, 1951.
37. **Gebhardt and Wörwag,** *Z. Metallkunde*, 42, 358, 1951.
38. **Geld and Vertman,** *Fiz. Met. i Met.*, 10, 793, 1960.
39. **Gogiberidse and Kekelidse,** *Izv. A.N.*, 3, 125, 1963.
40. **Golchova,** *Teplofiz. Vysok. Temp.*, 4, 360, 1966.

41. **Greenaway,** *J. Inst. Met.*, 74, 133, 1947.
42. **Grosse, Cahill, and Kirshenbaum,** *J. Am. Chem. Soc.*, 83, 4665, 1961.
43. **Grosse and Kirshenbaum,** *J. Inorg. Nucl. Chem.*, 25, 331, 1963.
44. **Grosse and McGonigal,** *J. Phys. Chem.*, 68, 414, 1964.
45. **Herczynska,** *Naturwiss.*, 47, 200, 1960.
46. **Hogness,** *J. Am. Chem. Soc.*, 43, 1621, 1921.
47. **Jakimovich and Saars,** *Teplofiz. Vysok. Temp.*, 5, 532, 1967.
48. **Jones, Ofte, Rohr, and Wittenberg,** *Trans. ASM*, 55, 819, 1962.
49. **Jouniaux,** *Bull. Soc. Chim. France*, 47, 524, 1930.
50. **Jouniaux,** *Bull. Soc. Chim. France*, 51, 677, 1932.
51. **Keskar and Hruska,** *Metall. Trans.*, 1, 2357, 1970.
52. **Kingery and Humenik,** *J. Phys. Chem.*, 57, 359, 1953.
53. **Kirshenbaum and Cahill,** *Trans. ASM*, 55, 844, 1962.
54. **Kirshenbaum and Cahill,** *Trans. ASM*, 55, 849, 1962.
55. **Kirshenbaum and Cahill,** *Trans. AIME*, 224, 816, 1962.
56. **Kirshenbaum, Cahill, and Grosse,** *J. Inorg. Nucl. Chem.*, 22, 33, 1961.
57. **Kirshenbaum, Cahill, and Grosse,** *J. Inorg. Nucl. Chem.*, 24, 333, 1962.
58. **Klemm et al.,** *Monatsh. Chem.*, 83, 629, 1952.
59. **Knappwost and Restle,** *Z. Elektrochem.*, 58, 112, 1954.
60. **Königer and Nagel,** *Gießerei TWB*, 13, 57, 1960.
61. **Koster, Hensel, and Franck,** *Ber. Bunsenges.*, 74, 43, 1970.
62. **Kozakevitch, Châtel, Urbain, and Sage,** *Rev. Mét.*, 52, 139, 1955.
63. **Kozakevitch and Urbain,** *C. R., Paris*, 253, 2229, 1961.
64. **Kubaschewski and Hörnle,** *Z. Metallkunde*, 42, 129, 1951.
65. **Leng,** WADC-Techn. Rep. 57-488, 1957.
66. **Lauermann and Metzger,** *Z. Phys. Chem.*, 216, 37, 1961.
67. **Lucas,** *C. R., Paris*, 250, 1850, 1960.
68. **Lucas,** *C. R., Paris*, 253, 2526, 1961.
69. **Lucas,** *Mém. Sci. Rev. Mét.*, 61, 1, 1964.
70. **Lucas,** *Mém. Sci. Rev. Mét.*, 61, 97, 1964.
71. **Lucas and Urbain,** *C. R., Paris*, 258, 6403, 1964.
72. **Malmberg,** *J. Inst. Metals* 89, 137, 1960.
73. **Matuyama,** *Sci. Rep. RITU*, 18, 19, 1929.
74. **McGonigal, Kirshenbaum and Grosse,** *J. Phys. Chem.*, 66, 737, 1962.
75. **Melik-Gajkazan, Voronchikhina, and Sakharova,** *Elektrokhim.*, 4, 1420, 1968.
76. *Metals Handbook: Properties and Selection of Metals*, 8th ed., Vol. 1, 1961.
77. *Metals References Book*, 4th ed., Vol. 1, Smithells, C. J., Ed., 1967, 688.
78. **Metzger,** *Z. Phys. Chem.*, 211, 1, 1959.
79. **Mokrovski and Regel,** *J. Phys. Tech.*, 22, 1281, 1952.
80. **Monma and Suto,** *J. Jpn. Inst. Met.*, 1, 69, 1960.
81. **Nagamori,** *Trans. AIME*, 245, 1897, 1969.
82. **Naidich and Eremenko,** *Fiz. Met. i Met.*, 6, 62, 1961.
83. **Ono and Matsushima,** *Sci. Rep. RITU*, 9, 309, 1957.
84. **Osminin,** *Zh. Fiz. Khim.*, 43, 2610, 1969.
85. **Pascal and Jouniaux,** *C. R., Paris*, 158, 414, 1914.
86. **Pascal and Jouniaux,** *Z. Elektrochem.*, 22, 72, 1916.
87. **Pekarev,** *Izv. Vys. Uch. Sav., Tsvetn. Met.*, 6, 111, 1963.
88. **Pelzel,** *Berg-u. Hütt. Mon. Hefte, Leoben*, 93, 248, 1948.
89. **Pelzel,** *Z. Metallkunde*, 50, 392, 1959.
90. **Pelzel and Sauerwald,** *Z. Metallkunde*, 33, 229, 1941.
91. **Pelzel and Schneider,** *Z. Metallkunde*, 35, 121, 1943.
92. **Pokrovski and Saidov,** *Zh. Fiz. Khim.*, 29, 1601, 1955.
93. **Popel, Shergin, and Zarevski,** *Zh. Fiz. Khim.*, 43, 2365, 1969.
94. **Popel, Smirnov, Zarevski, Dshemilev, and Pastuknov,** *Izv.A.N.*, 1, 62, 1965.
95. **Popel, Zarevski, and Dshemilev,** *Fiz. Met. i Met.*, 18, 468, 1964.
96. **Rink,** *C. R., Paris*, 189, 39, 1929.
97. **Rohr and Wittenberg,** *J. Phys. Chem.*, 74, 1151, 1970.
98. **Saeger and Ash,** *J. Res. Natl. Bur. Stand.*, 8, 37, 1932.
99. **Saito and Sakuma,** *J. Jpn. Inst. Met.*, 31, 1140, 1967.
100. **Sauerwald,** *Z. Metallkunde*, 14, 457, 1922.
101. **Sauerwald and Widawski,** *Z. Anorg. Allg. Chem.*, 155, 1, 1926.
102. **Schneider and Heymer,** *Z. Anorg. Allg. Chem.*, 286, 118, 1956.
103. **Schwaneke and Falke,** US-Bur. Min., Invest. Rep. No. 7372, 1970.
104. **Shirai, Hamada, and Kobayashi,** *J. Chem. Soc. Jpn.*, 84, 968, 1963.
105. **Smith and Spitzer,** *J. Phys. Chem.*, 66, 946, 1962.
106. **Spilrajn and Jakimovich,** *Teplofiz. Vysok. Temp.*, 5, 239, 1967.
107. **Stauffer,** Thesis Göttingen, 1953.
108. *Stoffhütte: Taschenbuch der Werkstoffkunde*, 4th ed., 1967, 1059.
109. **Stone, Ewing, Spann, Steinkuller, Williams, and Miller,** *J. Chem. Eng. Data*, 11, 320, 1966.
110. **Stott and Rendall,** *J. Iron Steel Inst.*, 175, 374, 1953.
111. **Tavadse, Bairamashvili, Khantadse, and Zagareishvili,** *Doklady A.N.*, 150, 544, 1963.
112. **Tavadse, Bairamashvili, and Khantadse,** *Doklady A.N.*, 162, 67, 1965.
113. **Taylor,** *J. Inst. Met.*, 83, 143, 1954.
114. **Übelacker and Lucas,** *C. R., Paris*, 254, 1622, 1962.
115. **Vatolin and Exin,** *Fiz. Met. I Met.*, 16, 936, 1963.

116. **Vatolin, Esin, Ukhov, and Dubinin,** *Trudy Inst. Met. Sverdlovsk,* 18, 73, 1969.
117. **Widawski and Sauerwald,** *Z. Anorg. Allg. Chem.,* 192, 145, 1930.
118. **Wilkinson,** *Extractive and Physical Metallurgy of Plutonium and its Alloys,* 1960.
119. **Zimmermann and Esser,** *Arch. f. Eisenh.,* 2, 867, 1929.
120. **Achener,** HTLMHTTM., Vol. I, Oak Ridge, Nov. 1964, 5.
121. **Bedon and Desré,** *C. R., Paris,* 274, 40, 1972.
122. **Bernard and Lupis,** *Metall. Trans.,* 2, 555, 1971.
123. **Berthou and Tougas,** *Metall. Trans.,* 1, 2978, 1970.
124. **Cohen,** *Nucl. Sci. Eng.,* 2, 530, 1957.
125. **Crawley,** *Trans. AIME,* 242, 2309, 1968.
126. **Crawley,** *Trans. AIME,* 242, 2237, 1968.
127. **Crawley and Kiff,** *Metall. Trans.,* 3, 158, 1972.
128. **Culpin,** *Proc. Phys. Soc.,* 70, 1669, 1957.
129. **Eljutin, Kostikov and Penkov,** *Poroshk. Met.,* 9, 46, 1970.
130. **El-Mehairy and Ward,** *Trans. Met. Soc. AIME,* 227, 1226, 1963.
131. **Freeman,** Ann. Rep. INCRA, Proj. No. 175A, 1972.
132. **Golchova,** *Teplofis. Vysok. Temp.,* 3, 483, 1965.
133. **Grosse,** US-AEC, Contr. AT 30-1-2082, 1965.
134. **Grosse and Cahill,** A.S.M. Trans., 57, 739, 1964.
135. **Grusdev,** *Shidkie Met., Sb. Statej,* 256, 1963.
136. **Hesson, Shimotake, and Tralmer,** *J. Met.,* 6, 1968.
137. **Kanda,** US-AEC, REp. TID 15836, 1961.
138. **Kanda,** US-AEC, Rep. TID 20849, 1964.
139. **Khilya, Ivashchenko, and Eremenko,** *A.N. Ukr. SSR, Fis. Khim. Pov. Yavl. v Raspl.,* 149, 1971.
141. **Kirshenbaum and Cahill,** *A.S.M. Trans.,* 56, 281, 1963.
142. **Körber and Löhberg,** *Giessereiforschung,* 23, 173, 1971.
143. **Levin and Ayushina,** *Izv. Vyss. Uch. Sav., Chern. Met.,* 2, 15, 1972.
144. **Lucas,** Mém. Sci. Rev. Mét., 69, 395, 1972.
145. **Lucas** Mém. Sci. Rev. Mét., 69, 479, 1972.
146. **Lucas and Urbain,** *C. R., Paris,* 255, 2414, 340, 1962.
147. **Martin-Garin, Bedon, and Desré,** *J. Chim. Phys.,* 70, 112, 1973.
148. **Martinez and Walls,** *Met. Trans.,* 4, 1419, 1973.
149. **Maurakh,** *Trans. Indian Inst. Met.,* 14, 209, 1964.
150. **McGonigal and Grosse,** *J. Phys. Chem.,* 67, 924, 1963.
151. **McGonigal, Cahill, and Kirshenbaum,** *J. Inorg. Nucl. Chem.,* 24, 1012, 1962.
152. **Nishenko, Skljarenko, and Eremenko,** *Ukr. Khim. Sh.,* 31, 559, 1965.
153. **Rohr,** *J. Less-Common Metals,* 10, 389, 1966.
154. **Saito and Sakuma,** *Sci. Rep. RITU.,* 22, 57, 1970.
155. **Schneider, Stauffer, and Heymer,** *Naturwiss.,* 14, 326, 1954.
156. **Serpan and Wittenberg,** *Trans. Met. Soc. AIME,* 221, 1017, 1961.
157. **Shergin, Popel, and Zarevski,** *A.N. Ukr. SSR, Fis. Khim. Pov. Yavl. v Raspl.,* 161, 1971.
158. **Stokes,** *J. Phys. Chem. Solids,* 27, 51, 1966.
159. **Tavadse, Khantadse, and Tsertsvadse,** *A.N. Ukr. SSR, Fis. Khim. Pov. Yavl. v Raspl.,* 169, 1971.
160. **Tepper, Murchison, Zelenak, and Roehlich,** HTLMHTTM, Vol. I, Oak Ridge, Nov. 1964, 26.
161. **Thresh, Crawley, and White,** *Trans. AIME,* 242, 819, 1968.
162. **Watanabe,** *Trans. Jpn. Inst. Met.,* 12, 17, 1971.
163. **Wittenberg, Ofte, and Rohr,** Proc. 3rd Rare Earth Conf. 1963, Vol. II, 1964.
164. **Wittenberg, Ofte, Rohr, and Rigney,** *Met. Trans.,* 2, 287, 1971.
165. **Wobst and Rentzsch,** *Z. Phys. Chem.,* 240, 36, 1969.
166. **Yatsenko, Kononenko, and Shukman,** *Teplofis. Vysok. Temp.,* 10, 66, 1972.
167. **Kanda and Keller,** Syracuse University, Contr. No. AT(30-1)-2731, 1963.
168. **Abdel-Aziz, Kirshah, and Aref,** *Z. Metallkunde,* 66, 183, 1975.
169. **Lewin, Ajuschina, and Gold,** *Teplofis. Vysok. Temp.,* 6, 432, 1968.
170. **Bedon,** Private communication, 1974.
171. **Frohberg and Weber,** Arch. Eisenh., 35, 877, 1964.
172. **Kawai, Moni, Kishimoto, Ishikura, and Shimada,** *Tetsu-to-Hagane,* 60, 29, 1974.
173. **Kenesha, Jr. and Cubicciotti,** *J. Phys. Chem.,* 62, 843, 1958.
174. **Nücker,** *Z. Angew. Physik,* 27, 33, 1969.
175. **Strauss, Richards, and Brown,** *Nucl. Sci. Eng.,* 7, 422, 1960.
176. **Tavadse, Khantadse, and Tsertsvadse,** *Vopr. Metalloved. i Korros. Met., Isdat. Metsniereba,* Tiflis, 1968.
177. **Adachi, Morita, Kitaura, and Demukai,** Techn. Rep. Osaka Univ., 20, 67, 1970.
179. **Gebhardt, Becker, and Sebastian,** Z. Metallkunde, 46, 669, 1955.
180. **Gomex, Martin-Garin, Ebert, Bedon, and Desré,** *Z. Metallkunde,* 67, 131, 1976.
181. **Matsenyuk and Iwaschtschenko,** *Ukr. Chim. Sh.,* 40, 431, 1974.
182. **Naleyev and Ibragimov,** *Zh. Fiz. Khim.,* 48, 1289, 1974.
183. **Lewin and Geld,** *Sbornik Trud., Klyuch. Zav. Ferrosplav.,* 4, 140, 1969.
184. **Sikora, Zielinski, and Orecki,** *Rudy Met. Niezel.,* 17, 20, 1972.
184. **Dubinin, Vlasov, Timofejev, Safonov, and Chegodajev,** *Izv. Vyss. Uchebn. Saved., Tsvetn, Met.,* 4, 160, 1975.
186. **Bykova and Shevchenko,** *Akad. Nauk SSSR, Uralskij Nauchnyj Zentr., Fis.-Chim. Issled. Shidk. Met. i Splav.,* 42, 1974.
187. **Eremenko, Ivashchenko, and Khilya,** *Izv. A.N., Met.,* 3, 38, 1977.
188. **Martin-Garin, Gomez, Bedon, and Desré,** *J. Less-Common Met.,* 41, 65, 1975.
189. **Spilrajn, Fomin, Kagan, Sokol, Kachalov, and Ulyanov,** *High Temp.-High Pressures,* 9, 49, 1977.
190. **Kucharski,** *Arch. Hutn.,* 22, 181, 1977.
191. **Wobst,** *Wiss. Z. Techn. Hochsch. Karl-Marx-Stadt,* 12, 393, 1970.
192. **Vukalovich and Ivanov,** *Teplofisicheskie svojstva rtyti met., Isdat. Stand.,* Moskva, 1971.

193. **Ibragimov and Savvin,** *Izv. Vyss. Uchebn. Saved., Tsvetn. Met.*, 4, 148, 1976.
194. **Mitko, Dubinin, Timofejev, and Chegodajev,** *Izv. Vyss. Uchebn. Saved., Tsvetn. Met.*, 3, 84, 1978.
195. **Park and Kim,** *J. Korean Inst. Met.*, 16, 463, 1978.
196. **Nishenko and Floka,** *V. Kn. poverchnostnye javlenija v raspl. k "Naukova Dumka"*, 130, 1968.
197. **Naidich, Perevertailo, and Nevodnuk,** *Izv. A.N. SSSR, Met.*, 3, 240, 1972.
198. **Lukin, Shukhov, Vatolin, and Koslov,** *J. Less-Common Met.*, 67, 407, 1979.
199. **Ogino and Taimatsu,** *J. Jpn. Inst. Het.*, 9, 871, 1979.
200. **Drotnig,** *High Temp.-High Pressures,* 14, 253, 1982.
201. **Mathiak, Nistler, Waschkowski, and Koester,** *Z. Metallkunde,* 74, 793, 1983.
202. **Cook,** *Phil. Trans. Roy. Soc. London,* A254, 125, 1961.
203. **Apollova, Dubinin, Mitko, Chegodajev, and Besukladnikova,** *Izv. A.N., SSSR, Met.*, 6, 55, 1982.

PHYSICAL AND OPTICAL PROPERTIES OF MINERALS

The chemical formula, crystal system, density, hardness, and index of refraction of some common minerals are given in this table. Entries are arranged alphabetically by mineral name. The columns are:

- **Formula:** Chemical formula for a typical sample of the mineral. Composition often varies considerably with the origin of the sample.
- **Crystal system:** tricl = triclinic; monocl = monoclinic; orth = orthorhombic; tetr = tetragonal; hex = hexagonal; rhomb = rhombohedral; cub = cubic.
- **Density:** Typical density in g/cm^3. Individual samples may vary by a few percent.
- **Hardness:** On the Mohs' scale (range of 1 to 10, with talc = 1 and diamond = 10).
- **Index of refraction:** Values are given for the three coordinate axes in the order of least, intermediate, and greatest index. For cubic crystals there is only a single value. See Reference 1 for details on the axis systems. Variations of several percent, depending on the origin and exact composition of the sample, are common.

REFERENCES

1. Deer, W.A., Howie, R.A., and Zussman, J., *An Introduction to the Rock-Forming Minerals*, 2nd Edition, Longman Scientific & Technical, Harlow, Essex, 1992.
2. Carmichael, R.S., *Practical Handbook of Physical Properties of Rocks and Minerals*, CRC Press, Boca Raton, FL, 1989.
3. Donnay, J.D.H., and Ondik, H.M., *Crystal Data Determinative Tables, Third Edition, Volume 2, Inorganic Compounds*, Joint Committee on Powder Diffraction Standards, Swarthmore, PA, 1973.

Name	Formula	Crystal system	Density g/cm^3	Hard-ness	Index of refraction n_α	n_β	n_γ
Acanthite	Ag_2S	orth	7.2	2.3			
Actinolite	$Ca_2(Mg,Fe)_5Si_8O_{22}(OH,F)_2$	monocl	3.23	5.5	1.624	1.655	1.664
Aegirine	$NaFe(SiO_3)_2$	monocl	3.58	6	1.763	1.800	1.815
Akermanite	$Ca_2MgSi_2O_7$	tetr	2.94	5.5	1.632	1.640	
Alabandite	MnS	cub	4.0	3.8			
Albite	$NaAlSi_3O_8$	tricl	2.63	6.3	1.527	1.531	1.538
Allanite	$(Ca,Mn,Ce,La,Y,Th)_2(Fe,Ti)(Al,Fe)O{\cdot}OH(Si_2O_7)(SiO_4)$	monocl	3.8	5.8	1.75	1.78	1.80
Allemontite	SbAs	hex	6.0	3.5			
Almandine	$Fe_3Al_2Si_3O_{12}$	cub	4.32	6.8	1.830		
Altaite	PbTe	cub	8.16	3			
Aluminite	$Al_2(SO_4)(OH)_4{\cdot}7H_2O$	monocl	1.74	1.5	1.459	1.464	1.470
Alunite	$(K,Na)Al_3(SO_4)_2(OH)_6$	rhomb	2.8	3.8	1.572	1.592	
Alunogen	$Al_2(SO_4)_3{\cdot}18H_2O$	monocl	1.69	1.8	1.467	1.47	1.478
Amblygonite	$(Li,Na)Al(PO_4)(F,OH)$	tricl	3.1	5.8	1.591	1.604	1.613
Analcite	$NaAlSi_2O{\cdot}6H20$	cub	2.27	5.5	1.486		
Anatase	TiO_2	tetr	4.23	5.8	2.488	2.561	
Andalusite	Al_2OSiO_4	orth	3.15	7.5	1.635	1.639	1.644
Andesine	$([NaSi]_{0.7\text{-}0.5}[CaAl]_{0.3\text{-}0.5}AlSi_2O_8$	tricl	2.67	6.3	1.550	1.553	1.557
Andorite	$PbAgSb_3S_6$	rhomb	5.35	3.3			
Andradite	$Ca_3(Fe,Ti)_2Si_3O_{12}$	cub	3.86	6.8	1.887		
Anglesite	$PbSO_4$	orth	6.29	2.8	1.877	1.883	1.894
Anhydrite	$CaSO_4$	orth	2.96	3.5	1.570	1.575	1.614
Ankerite	$Ca(Fe,Mg,Mn)(CO_3)_2$	rhomb	3.0	3.8	1.529	1.720	
Anorthite	$CaAl_2Si_2O_8$	tricl	2.76	6.3	1.577	1.585	1.590
Anorthoclase	$(Na,K)AlSi_3O_8$	tricl	2.58	6	1.523	1.528	1.529
Anthophyllite	$(Mg,Fe)_7Si_8O_{22}(OH,F)_2$	rhomb	3.21	5.8	1.645	1.658	1.668
Apatite	$Ca_5(PO_4)_3(OH,F,Cl)$	hex	3.2	5	1.645	1.648	
Apophyllite	$KFCa_4Si_8O_{20}{\cdot}8H_2O$	tetr	2.35	4.8	1.535	1.536	
Aragonite	$CaCO_3$	orth	2.83	3.5	1.531	1.680	1.686
Arcanite	K_2SO_4	orth	2.66		1.494	1.494	1.497
Argentite	Ag_2S	orth	7.2	2.3			
Arsenolite	As_2O_3	cub	3.86	1.5	1.755		
Arsenopyrite	FeAsS	monocl	6.1	5.8			
Atacamite	$Cu_2(OH)_3Cl$	rhomb	3.76	3.3	1.831	1.861	1.880
Augelite	$Al_2(PO_4)(OH)_3$	monocl	2.70	4.8	1.574	1.576	1.588
Augite	$(Ca,Mg,Fe,Ti,Al)_2(Si,Al)_2O_6$	monocl	3.38	6	1.703	1.707	1.738
Autunite	$Ca(UO_{22})(PO_4)_2{\cdot}10H20$	tetr	3.2	2.3	1.553	1.577	
Axinite	$(Ca,Mn,Fe)_3Al_2BO_3Si_4O_{12}(OH)$	tricl	3.31	6.8	1.684	1.691	1.694

PHYSICAL AND OPTICAL PROPERTIES OF MINERALS (continued)

Name	Formula	Crystal system	Density g/cm³	Hard-ness	Index of refraction n_α	n_β	n_γ
Azurite	$Cu_3(OH)_2(CO_3)_2$	monocl	3.77	3.8	1.730	1.758	1.838
Baddeleyite	ZrO_2	monocl	5.7	6.5	2.13	2.19	2.20
Barite	$BaSO_4$	orth	4.49	3.3	1.636	1.637	1.648
Benitoite	$BaTi(SiO_3)_3$	rhomb	3.65	6.3	1.757	1.804	
Bertrandite	$Be_4Si_2O_7(OH)_2$	rhomb	2.6	6	1.589	1.602	1.613
Beryl	$Be_3Al_2(SiO_3)_6$	hex	2.64	7.8	1.582	1.589	
Beryllonite	$NaBe(PO)_4$	monocl	2.81	5.8	1.552	1.558	1.561
Biotite	$K(Mg,Fe)_3AlSi_3O_{10}(OH,F)_2$	monocl	3.0	2.8	1.595	1.651	1.651
Bismuthinite	Bi_2S_3	orth	6.78	2			
Bixbyite	$(Mn,Fe)_2O_3$	cub	4.95	6.3			
Bloedite	$Na_2Mg(SO_4)_2 \cdot 4H_2O$	monocl	2.25	2.8	1.483	1.486	1.487
Boehmite	$AlO(OH)$	orth	3.44	3.8	1.64	1.65	1.66
Boracite	$Mg_3B_7O_{13}Cl$	rhomb	2.94	7.3	1.66	1.66	1.67
Borax	$Na_2B_4O_7 \cdot 10H_2O$	monocl	1.73	2.3	1.447	1.469	1.472
Bornite	Cu_5FeS_4	cub	5.07	3			
Boulangerite	$Pb_5Sb_4S_{11}$	monocl	6.1	2.8			
Bournonite	$PbCuSbS_3$	rhomb	5.83	2.8			
Braggite	PtS	tetr	10.2				
Braunite	$(Mn,Si)_2O_3$	tetr	4.78	6.3			
Bravoite	$(Ni,Fe)S_2$	cub	4.62	5.8			
Breithauptite	$NiSb$	hex	≈8.7	5.5			
Brochantite	$Cu_4(SO_4)(OH)_6$	monocl	3.79	3.8	1.728	1.771	1.800
Bromyrite	$AgBr$	cub	6.47	2.5	2.253		
Brookite	TiO_2	tetr	4.23	5.8	2.583	2.584	2.700
Brucite	$Mg(OH)_2$	hex	2.37	2.5	1.575	1.59	
Bunsenite	NiO	cub	6.72	5.5			
Cacoxenite	$Fe_4(PO_4)_3(OH)_3 \cdot 12H_20$	hex	2.3	3.5	1.580	1.646	
Calcite	$CaCO_3$	hex	2.71	3	1.486	1.658	
Caledonite	$Cu_2Pb_5(SO_4)_3(CO_3)(OH)_6$	rhomb	5.76	2.8	1.818	1.866	1.909
Calomel	Hg_2Cl_2	tetr	7.16	1.5	1.973	2.656	
Cancrinite	$(Na,Ca,K)_7[Al_6Si_6O_{24}]$ $(CO_3,SO_4,Cl,OH)_2 \cdot H_20$	hex	2.42	5.5	1.495	1.509	
Carnalite	$KMgCl_3 \cdot 6H_20$	rhomb	1.60	2.5	1.466	1.475	1.494
Carnotite	$K_2(UO_2)_2(VO_4)_2 \cdot 3H_2O$	rhomb		1.5	1.75	1.92	1.95
Cassiterite	SnO_2	tetr	6.85	6.5	2.006	2.097	
Celestite	$SrSO_4$	orth	3.96	3.3	1.622	1.624	1.631
Celsian	$BaAl_2Si_2O_8$	monocl	3.25	6.3	1.583	1.588	1.594
Cerargyrite	$AgCl$	cub	5.56	2.5	2.071		
Cerussite	$PbCO_3$	orth	6.6	3.3	1.804	2.076	2.079
Cervantite	Sb_2O_4	orth	6.64	4.5			
Chabazite	$Ca[Al_2Si_4O_{12}] \cdot 6H_2O$	trig	2.08	4.5	1.482		
Chalcanthite	$CuSO_4 \cdot 5H_2O$	tricl	2.29	2.5	1.514	1.537	1.543
Chalcocite	Cu_2S	orth	5.6	2.8			
Chalcopyrite	$CuFeS_2$	tetr	4.2	3.8			
Chiolite	$Na_5Al_3F_{14}$	tetr	3.00	3.8	1.342	1.349	
Chlorite	$(Mg,Al,Fe)_{12}(Si,Al)_8O_{20}(OH)_{16}$	monocl	3.0	2.5	1.61	1.62	1.62
Chloritoid	$(Fe,Mg,Mn)_2(Al,Fe)Al_3O_2(OH)_4$ $(SiO_4)_2$	monocl	3.66	6.5	1.717	1.721	1.726
Chondrodite	$Mg(OH,F)_2 \cdot 2Mg_2SiO_4$	monocl	3.21	6.5	1.604	1.615	1.634
Chromite	$FeCr_2O_4$	cub	5.0	5.5	2.16		
Chrysoberyl	$BeAl_2O_4$	orth	3.65	8.5	1.746	1.748	1.756
Chrysocolla	$CuSiO_3 \cdot 2H_2O$	rhomb	2.4	2	1.575	1.597	1.598
Cinnabar	HgS	hex	8.17	2.3	2.814	3.143	
Claudetite	As_2O_3	monocl	3.74	2.5	1.87	1.92	2.01
Clinohumite	$Mg(OH,F)_2 \cdot 4Mg_2SiO_4$	monocl	3.21	6	1.633	1.647	1.668
Clinozoisite	$Ca_2Al_3Si_3O_{12}(OH)$	monocl	3.30	6.5	1.693	1.700	1.712
Cobaltite	$CoAsS$	cub	≈6.1	5.5			
Colemanite	$Ca_2B_6O_{11} \cdot 5H_2O$	monocl	2.42	4.5	1.586	1.592	1.614
Columbite	$(Fe,Mn)(Nb,Ta)_2O_6$	rhomb	5.20	6			

PHYSICAL AND OPTICAL PROPERTIES OF MINERALS (continued)

Name	Formula	Crystal system	Density g/cm³	Hard-ness	Index of refraction n_α	n_β	n_γ
Connellite	$Cu_{19}(SO_4)Cl_4(OH)_{32}\cdot 3H_2O$	hex	3.36	3	1.731	1.752	
Copiapite	$(Fe,Mg)Fe_4(SO_4)_6(OH)_2\cdot 20H_2O$	tricl	2.13	2.8	1.52	1.54	1.59
Coquimbite	$Fe_2(SO_4)_3\cdot 9H_2O$	hex	2.1	2.5	1.54	1.56	
Cordierite	$Al_3(Mg,Fe)_2Si_5AlO_{18}$	rhomb	2.66	7	1.540	1.549	1.553
Corundum	Al_2O_3	hex	3.97	9	1.761	1.769	
Cotunnite	$PbCl_2$	orth	5.98	2.5	2.199	2.217	2.260
Covellite	CuS	hex	4.8	1.8			
Cristobalite	SiO_2	hex	2.33	6.5	1.484	1.487	
Crocoite	$PbCrO_4$	monocl	6.12	2.8	2.29	2.36	2.66
Cryolite	Na_3AlF_6	monocl	2.97	2.5	1.338	1.338	1.339
Cryolithionite	$Na_3Li_3Al_2F_{12}$	cub	2.77	2.8	1.340		
Cubanite	$CuFe_2S_3$	rhomb	4.11	3.5			
Cummingtonite	$(Mg,Fe)_7Si_8O_{22}(OH)_2$	monocl	3.4	5.5	1.650	1.660	1.676
Cuprite	Cu_2O	cub	6.0	3.8			
Danburite	$CaSi_2B_2O_8$	rhomb	3.0	7	1.63	1.63	1.63
Datolite	$CaBSiO_4(OH)$	monocl	2.98	5.3	1.624	1.652	1.668
Daubreelite	Cr_2FeS_4	cub	3.81				
Derbylite	$Fe_6Ti_6Sb_2O_{23}$	rhomb	4.53	5	2.45	2.45	2.51
Diamond	C	cub	3.51	10	2.418		
Diaspore	$AlO(OH)$	orth	3.4	6.8	1.694	1.715	1.741
Digenite	$Cu_{2-x}S$	cub	5.55	2.8			
Diopside	$CaMgSi_2O_6$	monocl	3.30	6	1.680	1.687	1.708
Dioptase	$Cu_6Si_6O_{18}\cdot 6H_2O$	rhomb	3.5	5	1.65	1.70	
Dolomite	$CaMg(CO_3)_2$	rhomb	2.86	3.5	1.500	1.679	
Douglasite	$K_2FeCl_4\cdot 2H_2O$	orth	2.16		1.488	1.500	
Dyscrasite	Ag_3Sb	rhomb	9.74	3.8			
Eddingtonite	$BaAl_2Si_3O_{10}\cdot 4H_2O$	rhomb	2.8		1.541	1.553	1.557
Eglestonite	Hg_4OCl_2	cub	8.4	2.5	2.49		
Emplectite	$CuBiS_2$	rhomb	6.38	2			
Enargite	Cu_3AsS_4	rhomb	4.5	3			
Enstatite	$MgSiO_3$	monocl	3.19	5.5	1.656	1.662	1.669
Epidote	$Ca_2Al_2O(Al,Fe)OH(Si_2O_7)(SiO_4)$	monocl	3.44	6	1.733	1.755	1.765
Epsomite	$MgSO_4\cdot 7H_2O$	orth	1.67	2.3	1.433	1.455	1.461
Erythrite	$(Co,Ni)_3(AsO_4)_2\cdot 8H_2O$	monocl	3.06	2	1.626	1.661	1.699
Eucairite	$CuAgSe$	orth	7.7	2.5			
Euclasite	$BeAlSiO_4(OH)$	monocl	3.1	7.5	1.651	1.655	1.671
Eudialite	$(Na,Ca,Ce)_5(Fe,Mn)(Zr,Ti)(Si_3O_9)_2(OH,Cl)$	hex	3.0	5.5	1.623	1.600	1.615
Eulytite	$Bi_4Si_3O_{12}$	cub	6.6	4.5	2.05		
Euxenite	$(Y,Ca,Ce,U,Th)(Nb,Ta,Ti)_2O_6$	rhomb	5.5	6	2.2		
Fayalite	Fe_2SiO_4	orth	4.30	6.5	1.827	1.869	1.879
Ferberite	$FeWO_4$	monocl	7.51	4.3			
Fergussonite	$(Y,Er,Ce,Fe)(Nb,Ta,Ti)O_4$	tetr	5.7	6	2.1		
Fluorite	CaF_2	cub	3.18	4	1.434		
Forsterite	Mg_2SiO_4	orth	3.21	7	1.635	1.651	1.670
Franklinite	$ZnFe_2O_4$	cub	5.21	6	2.36		
Gahnite	$ZnAl_2O_4$	cub	4.62	7.8	1.805		
Galaxite	$MnAl_2O_4$	cub	4.04	7.8	1.92		
Galena	PbS	cub	7.60	2.5	3.91		
Galenabismuthite	$PbBi_2S_4$	rhomb	7.04	3			
Ganomalite	$(Ca,Pb)_{10}(OH,Cl)_2(Si_2O_7)_3$	hex	5.6	3.5	1.910	1.945	
Gaylussite	$Na_2Ca(CO_3)_2\cdot 5H_2O$	monocl	1.99	2.8	1.444	1.516	1.523
Gehlenite	$Ca_2Al_2SiO_7$	tetr	3.04	5.5	1.658	1.669	
Geikielite	$MgTiO_3$	hex	3.85	5.5	1.95	2.31	
Gibbsite	$Al(OH)_3$	monocl	2.42	3	1.57	1.57	1.59
Glauberite	$Na_2Ca(SO_4)_2$	monocl	2.80	2.8	1.515	1.535	1.536
Glauconite	$(K,Na,Ca)_{1.6}(Fe,Al,Mg)_{4.0}Si_{7.3}Al_{0.7}O_{20}(OH)_4$	monocl	2.7	2	1.60	1.63	1.63
Glaucophane	$Na_2Mg_3Al_2Si_8O_{22}(OH)_2$	monocl	3.19	6	1.634	1.645	1.648

Name	Formula	Crystal system	Density g/cm³	Hard-ness	Index of refraction n_α	n_β	n_γ
Gmelinite	$(Ca,Na_2)[Al_2Si_4O_{12}]\cdot 6H_2O$	hex	2.10	4.5	1.477	1.485	
Goethite	$FeO(OH)$	orth	4.3	5.3	2.268	2.401	2.457
Goslarite	$ZnSO_4\cdot 7H_2O$	orth	1.97	2.3	1.457	1.480	1.484
Greenockite	CdS	cub	4.8	3.3	2.506	2.529	
Grossularite	$Ca_3Al_2Si_3O_{12}$	cub	3.59	6.8	1.734		
Gummite	$UO_3\cdot H_2O$	orth	7.05	3.8			
Gypsum	$CaSO_4\cdot 2H_2O$	monocl	2.32	2	1.520	1.525	1.530
Halite	$NaCl$	cub	2.17	2	1.544		
Hambergite	$Be_2(OH)(BO_3)$	rhomb	2.36	7.5	1.56	1.59	1.63
Hanksite	$Na_{22}K(SO_4)_9(CO_3)_2Cl$	hex	2.56	3.3	1.461	1.481	
Harmotome	$Ba[Al_2Si_6O_{16}]\cdot 6H_2O$	monocl	2.44	4.5	1.506	1.507	1.511
Hausmannite	Mn_3O_4	tetr	4.84	5.5	2.15	2.46	
Haüyne	$(Na,Ca)_{4\text{-}8}Al_6Si_6O_{24}(SO_4,S)_{1\text{-}2}$	cub	2.47	5.8	1.502		
Hedenbergite	$CaFeSi_2O_6$	monocl	3.53	6	1.721	1.727	1.746
Helvite	$Mn_4Be_3Si_3O_{12}S$	cub	3.32	6	1.739		
Hematite	Fe_2O_3	hex	5.25	6	2.91	3.19	
Hemimorphite	$Zn_4Si_2O_7(OH)_2\cdot H_2O$	rhomb	3.45	5	1.614	1.617	1.636
Hercynite	$Fe(AlO_2)_2$	cub	4.3	7.8	1.835		
Herderite	$CaBe(PO_4)(Fe,OH)$	monocl	2.98	5.3	1.592	1.612	1.621
Hessite	Ag_2Te	orth	8.4	2.5			
Heulandite	$(Ca,Na_2,K_2)[Al_2Si_7O_{18}]\cdot 6H_2O$	monocl	2.2	3.8	1.498	1.498	1.506
Hopeite	$Zn_3(PO_4)_2\cdot 4H_2O$	orth	3.0	3.2	1.58	1.59	1.59
Hornblende	$Ca_2(Mg,Fe)_4Al(Si_7AlO_{22})(OH)_2$	monocl	3.24	5.5	1.67	1.67	1.69
Huebnerite	$MnWO_4$	monocl	7.2	4.3	2.17	2.22	2.32
Humite	$Mg(OH,F)_2\cdot 3Mg_2SiO_4$	orth	3.3	6	1.625	1.636	1.657
Huntite	$Mg_3Ca(CO_3)_4$	trig	2.70				
Hydrogrossularite	$Ca_3Al_2Si_2O_8(SiO_4)_{1\text{-}m}(OH)_{4m}$	cub	3.4	6.8	1.70		
Hydromagnesite	$3MgCO_3\cdot Mg(OH)_2\cdot 3H_2O$	monocl	2.24	3.5	1.523	1.527	1.545
Illite	$KAl_4[Si_7AlO_{20}](OH)_4$	monocl	2.8	1.5	1.56	1.59	1.59
Ilmenite	$FeTiO_3$	rhomb	4.72	5.5			
Iodyrite	AgI	hex	5.68	1.5	2.21	2.22	
Jacobsite	$MnFe_2O_4$	cub	4.87	7.8	2.3		
Jadeite	$NaAlSi_2O_6$	monocl	3.34	6	1.649	1.654	1.663
Jamesonite	$Pb_4FeSb_6S_{14}$	monocl	5.63	2.5			
Jarosite	$KFe_3(SO_4)_2(OH)_6$	rhomb	3.09	3	1.715	1.820	
Kainite	$KMg(SO_4)Cl\cdot 3H_2O$	monocl	2.15	2.8	1.494	1.505	1.516
Kaliophyllite	$KAlSiO_4$	hex	2.61	6	1.532	1.537	
Kaolinite	$Al_4Si_4O_{10}(OH)_8$	tricl	2.65	2.3	1.549	1.564	1.565
Kernite	$Na_2B_4O_7\cdot 4H_2O$	monocl	1.95	2.5	1.454	1.472	1.488
Kieserite	$MgSO_4\cdot H_2O$	monocl	2.57	3.5	1.520	1.533	1.584
Kyanite	Al_2OSiO_4	tricl	3.59	6.3	1.715	1.722	1.731
Lanarkite	$Pb_2(SO_4)O$	monocl	6.92	2.3	1.928	2.007	2.036
Lanthanite	$(La,Ce)_2(CO_3)_3\cdot 8H_2O$	rhomb	2.72	2.8	1.52	1.587	1.613
Laumontite	$Ca_4[Al_8Si_{16}O_{48}]\cdot 16H_2O$	monocl	2.3	3.3	1.508	1.517	1.519
Laurionite	$Pb(OH)Cl$	rhomb	6.24	3.3	2.08	2.12	2.16
Lawsonite	$CaAl_2(OH)_2Si_2O_7\cdot H_2O$	rhomb	3.08	6	1.655	1.675	1.685
Lazulite	$(Mg,Fe)Al_2(PO_4)_2(OH)_2$	monocl	3.23	5.8	1.615	1.64	1.650
Lazurite	$Na_4SSi_3Al_3O_{12}$	cub	2.42	5.3	1.500		
Leadhillite	$Pb_4(SO_4)(CO_3)_2(OH)_2$	monocl	6.55	2.8	1.87	2.00	2.01
Lepidocrocite	$FeO(OH)$	orth	4.26	5	1.94	2.20	2.51
Lepidolite	$K_2(Li,Al)_{5\text{-}6}[Si_{6\text{-}7}Al_{2\text{-}1}O_{20}](OH,F)_4$	monocl	2.85	3.3	1.536	1.565	1.566
Leucite	$KAlSi_2O_6$	tetr	2.49	5.8	1.510		
Levyne	$(Ca,Na_2)Al_2Si_4O_{12}\cdot 6H_2O$	rhomb	2.10	4.5	1.496	1.501	
Litharge	PbO	tetr	9.35	2	2.535	2.665	
Loellingite	$FeAs_2$	rhomb	7.40	5.3			
Maghemite	Fe_2O_3	cub	4.88	7.8	2.63		
Magnesite	$MgCO_3$	hex	3.05	4	1.536	1.741	
Magnetite	Fe_3O_4	cub	5.17	6	2.42		
Malachite	$Cu_2(OH)_2(CO_3)$	monocl	4.05	3.8	1.655	1.875	1.909

Name	Formula	Crystal system	Density g/cm³	Hard-ness	Index of refraction		
					n_α	n_β	n_γ
Manganite	$MnO(OH)$	monocl	≈4.3	4	2.25	2.25	2.5[illegible]
Manganosite	MnO	cub	5.37	5.5			
Marcasite	FeS_2	cub	5.02	6.3			
Marialite	$Na_4Al_3Si_9O_{24}Cl$	tetr	2.56	5.5	1.541	1.548	
Marshite	CuI	cub	5.67	2.5	2.346		
Mascagnite	$(NH_4)_2SO_4$	orth	1.77	2.3	1.520	1.523	1.53[illegible]
Matlockite	$PbClF$	tetr	7.05	2.8	2.006	2.145	
Meionite	$Ca_4Al_6Si_6O_{24}CO_3$	tetr	2.78	5.5	1.559	1.595	
Melanterite	$FeSO_4{\cdot}7H_2O$	monocl	1.89	2	1.47	1.48	1.4[illegible]
Melilite	$(Ca,Na)_2(Mg,Fe,Al,Si)_3O_7$	tetr	3.00	5.5	1.639	1.645	
Mellite	$Al_2C_{12}O_{12}{\cdot}18H_2O$	tetr	1.64	2.3	1.511	1.539	
Mendipite	$Pb_3O_2Cl_2$	rhomb	7.24	2.5	2.24	2.27	2.3[illegible]
Mesolite	$Na_2Ca_2(Al_2Si_3O_{10})_3{\cdot}8H_2O$	orth	2.26	5	1.506		
Metacinnabar	HgS	cub	7.70	3			
Microcline	$KAlSi_3O_8$	monocl	2.56	6.3	1.522	1.526	1.53[illegible]
Miersite	AgI	hex	5.68	2.5	2.20		
Millerite	NiS	hex	5.5	3.3			
Mimetite	$Pb_5(AsO_4,PO_4)_3Cl$	hex	7.24	3.8	2.128	2.147	
Minium	Pb_3O_4	tetr	8.9	2.5			
Mirabilite	$Na_2SO_4{\cdot}10H_2O$	monocl	1.46	1.8	1.394	1.396	1.39[illegible]
Moissanite	SiC	hex	3.16	9.5	2.648	2.691	
Molybdenite	MoS_2	hex	5.06	1.3			
Monazite	$(Ce,La,Th)PO_4$	monocl	5.2	5	1.787	1.789	1.84[illegible]
Monetite	$CaHPO_4$	tricl	2.92	3.5	1.587	1.61	1.64[illegible]
Monticellite	$Ca(Mg,Fe)SiO_4$	orth	3.18	5.5	1.647	1.655	1.66[illegible]
Montmorillonite	$(0.5Ca,Na)_{0.7}(Al,Mg,Fe)_4$ $[(Si,Al)_8O_{20}](OH)_4{\cdot}nH_2O$	monocl	2.5	1.5	1.55	1.57	1.5[illegible]
Montroydite	HgO	orth	11.14	2.5	2.37	2.50	2.6[illegible]
Mordenite	$(Na,K,Ca)[Al_2Si_{10}O_{24}]{\cdot}7H_2O$	orth	2.13	3.5	1.478	1.480	1.48[illegible]
Muscovite	$KAl_2Si_3AlO_{10}(OH,F)_2$	monocl	2.83	2.8	1.563	1.596	1.60[illegible]
Nantokite	$CuCl$	cub	4.14	2.5	1.930		
Natrolite	$Na_2Al_2Si_3O_{10}{\cdot}2H_2O$	orth	2.23	5	1.478	1.481	1.49[illegible]
Nepheline	$Na_3KAl_4Si_4O_{16}$	hex	2.61	5.8	1.534	1.538	
Newberyite	$MgHPO_4{\cdot}3H_2O$	orth	2.13	3.3	1.514	1.517	1.53[illegible]
Niccolite	$NiAs$	hex	7.77	5.3			
Norbergite	$Mg(OH,F)_2{\cdot}Mg_2SiO_4$	orth	3.21	6.5	1.565	1.573	1.59[illegible]
Nosean	$Na_8Al_6Si_6O_{24}SO_4$	cub	2.35	5.5	1.495		
Oldhamite	CaS	cub	2.59	4	2.137		
Oligoclase	$([NaSi]_{0.9\text{-}0.7}[CaAl]_{0.1\text{-}0.3})AlSi_2O_8$	tricl	2.64	6.3	1.539	1.543	1.54[illegible]
Olivenite	$Cu_2(AsO_4)(OH)$	rhomb	4.2	3	1.77	1.80	1.8[illegible]
Olivine	$(Mg,Fe)SiO_4$	rhomb	3.81	6.8	1.73	1.76	1.7[illegible]
Opal	$SiO_2{\cdot}nH_2O$	amorp	1.9	5	1.44		
Orpiment	As_2S_3	monocl	3.46	1.8	2.40	2.81	3.0[illegible]
Orthoclase	$KAlSi_3O_8$	monocl	2.56	6	1.523	1.527	1.53[illegible]
Orthopyroxene	$(Mg,Fe)SiO_3$	rhomb	3.6	5.5	1.709	1.712	1.72[illegible]
Paragonite	$NaAl_2Si_3AlO_{10}(OH)_2$	monocl	2.85	2.5	1.572	1.602	1.60[illegible]
Parisite	$(Ce,La,Na)FCO_3{\cdot}CaCO_3$	hex	4.42	4.5	1.672	1.771	
Pectolite	$Ca_2NaH(SiO_3)_3$	tricl	2.88	4.8	1.603	1.610	1.63[illegible]
Penfieldite	$Pb_4Cl_6(OH)_2$	hex	6.6		2.13	2.21	
Pentlandite	$(Fe,Ni)_9S_8$	cub	4.8	3.8			
Percylite	$PbCuCl_2(OH)_2$	cub		2.5	2.05		
Periclase	MgO	cub	3.6	5.5	1.735		
Perovskite	$CaTiO_3$	cub	3.98	5.5	2.34		
Petalite	$LiAlSi_4O_{10}$	monocl	2.42	6.5	1.506	1.511	1.51[illegible]
Pharmacosiderite	$Fe_3(AsO_4)_2(OH)_3{\cdot}5H_2O$	cub	2.80	2.5	1.690		
Phenakite	Be_2SiO_4	rhomb	2.98	7.5	1.654	1.670	
Phillipsite	$K(Ca_{0.5},Na)_2[Al_3Si_5O_{16}]{\cdot}6H_2O$	monocl	2.2	4.3	1.494	1.497	1.50[illegible]
Phlogopite	$KMg_3AlSi_3O_{10}(OH,F)_2$	monocl	2.83	2.3	1.560	1.597	1.59[illegible]
Phosgenite	$Pb_2(CO_3)Cl_2$	tetr	6.13	2.5	2.118	2.145	

Name	Formula	Crystal system	Density g/cm³	Hard-ness	Index of refraction n_α	n_β	n_γ
Piemontite	$Ca_2(Mn,Fe,Al)_3O(Si_2O_7)(SiO_4)(OH)$	monocl	3.49	6	1.762	1.773	1.796
Pigeonite	$(Mg,Fe,Ca)(Mg,Fe)Si_2O_6$	monocl	3.38	6	1.702	1.703	1.728
Pollucite	$CsAlSi_2O_6$	tetr	2.9	6.5	1.517		
Polybasite	$(Ag,Cu)_{16}Sb_2S_{11}$	monocl	6.1	2.5			
Powellite	$Ca(Mo,W)O_4$	tetr	4.35	3.8	1.971	1.980	
Prehnite	$Ca_2Al_2Si_3O_{10}(OH)_2$	rhomb	2.93	6.3	1.622	1.628	1.648
Proustite	Ag_3AsS_3	rhomb	5.57	2.3	2.792	3.088	
Pseudobrookite	Fe_2TiO_5	rhomb	4.36	6	2.38	2.39	2.42
Psilomelane	$BaMn_9O_{16}(OH)_4$	rhomb	4.71	5.5			
Pumpellyite	$Ca_2Al_2(Al,Fe,Mg)[Si_2(O,OH)_7](SiO_4)(OH,O)_3$	monocl	3.21	5.5	1.688	1.695	1.705
Pyrargyrite	Ag_3SbS_3	rhomb	5.85	2.5	2.88	3.08	
Pyrite	FeS_2	cub	5.02	6.3			
Pyrochlore	$NaCaNb_2O_6F$	cub	5.3	5.3			
Pyrochroite	$Mn(OH)_2$	hex	3.26	2.5	1.68	1.72	
Pyrolusite	MnO_2	tetr	5.08	6.3			
Pyromorphite	$Pb_5(PO_4,AsO_4)_3Cl$	hex	7.04	3.8	2.048	2.058	
Pyrope	$Mg_3Al_2Si_3O_{12}$	cub	3.58	6.8	1.714		
Pyrophyllite	$Al_2Si_4O_{10}(OH)_2$	monocl	2.78	1.5	1.545	1.579	1.599
Pyrrhotite	Fe_7S_8	hex	4.62	4			
Quartz	SiO_2	hex	2.65	7	1.544	1.553	
Rammelsbergite	$NiAs_2$	orth	7.1	5.8			
Raspite	$PbWO_4$	monocl	8.46	2.8	1.27	1.27	1.30
Realgar	As_4S_4	monocl	3.5	1.8	2.538	2.684	2.704
Rhodochrosite	$MnCO_3$	hex	3.70	3.8	1.597	1.816	
Rhodonite	$(Mn,Fe,Ca)SiO_3$	orth	3.48	6	1.725	1.729	1.737
Riebeckite	$Na_2Fe_5(Si_8O_{22})(OH)_2$	monocl	3.3	5	1.675	1.683	1.694
Rutile	TiO_2	tetr	4.23	6.2	2.609	2.900	
Safflorite	$(Co,Fe)As_2$	rhomb	7.3	4.8			
Samarskite	$(Y,Er,Ce,U,Ca,Fe,Pb,Th)(Nb,Ta,Ti,Sn)_2O_6$	rhomb	5.69	5.5	2.200		
Sapphirine	$(Mg,Fe)_2Al_4O_6SiO_4$	monocl	3.49	7.5	1.709	1.712	1.715
Scapolite	$(Na,Ca)_4Al_3(Al,Si)_3Si_6O_{24}(Cl,F,OH,CO_3,SO_4)$	tetr	2.64	5.5	1.551	1.573	
Scheelite	$CaWO_4$	tetr	6.06	4.8	1.920	1.936	
Scolecite	$CaAl_2Si_3O_{10}\cdot 3H_2O$	monocl	2.27	5	1.510	1.518	1.519
Scorodite	$Fe(AsO_4)\cdot 2H_2O$	rhomb	3.28	3.8	1.784	1.795	1.814
Sellaite	MgF_2	tetr	3.15	5	1.378	1.390	
Senarmontite	Sb_2O_3	cub	5.58	2.3	2.087		
Serpentine	$Mg_3Si_2O_5(OH)_4$	monocl	2.55	3	1.55	1.56	1.56
Siderite	$FeCO_3$	hex	3.9	4.3	1.635	1.875	
Sillimanite	Al_2OSiO_4	rhomb	3.25	7	1.658	1.660	1.660
Skutterudite	$(Co,Ni)As_3$	cub	6.8	5.8			
Smithsonite	$ZnCO_3$	rhomb	4.4	4.3	1.621	1.848	
Sodalite	$Na_8Al_6Si_6O_{24}Cl_2$	cub	2.30	5.8	1.485		
Sperrylite	$PtAs_2$	cub	10.58	6.5			
Spessartite	$Mn_3Al_2Si_3O_{12}$	cub	4.19	6.8	1.800		
Sphalerite	ZnS	cub	4.0	3.8	2.369		
Sphene	$CaTiSiO_4(O,OH,F)$	monocl	3.50	5	1.90	1.95	2.03
Spinel	$MgAl_2O_4$	cub	3.55	7.8	1.719		
Spodumene	$LiAlSi_2O_6$	monocl	3.13	6.8	1.656	1.662	1.671
Stannite	Cu_2FeSn_4	tetr	4.4	4			
Staurolite	$(Fe,Mg,Zn)_2(Al,Fe,Ti)_9O_6[(Si,Al)O_4]_4(O,OH)_2$	monocl	3.79	7.5	1.743	1.747	1.755
Stercorite	$Na(NH_4)H(PO_4)\cdot 4H_2O$	tricl	1.62	2	1.439	1.442	1.469
Stibiotantalite	$Sb(Ta,Nb)O_4$	rhomb	6.6	5.5	2.38	2.41	2.46
Stibnite	Sb_2S_3	orth	4.56	2			
Stilbite	$NaCa_2[Al_5Si_{13}O_{36}]\cdot 14H_2O$	monocl	2.2	3.8	1.492	1.499	1.503

Name	Formula	Crystal system	Density g/cm^3	Hardness	Index of refraction		
					n_α	n_β	n_γ
Stilpnomelane	$(K,Na,Ca)_{0.6}(Fe,Mg)_6Si_8Al(O,OH)_{27}\cdot 2H_2O$	monocl	2.8	3.5	1.585	1.665	1.665
Stolzite	$PbWO_4$	tetr	8.2	2.8	2.19	2.27	
Strengite	$FePO_4\cdot 2H_2O$	orth	2.87	4	1.707	1.719	1.741
Strontianite	$SrCO_3$	orth	3.5	3.5	1.518	1.666	1.668
Struvite	$Mg(NH_4)(PO_4)\cdot 6H_2O$	rhomb	1.71	2	1.495	1.496	1.504
Sulfur	S	orth	2.07	2	1.958	2.038	2.245
Sylvanite	$(Ag,Au)Te_2$	monocl	8.16	1.8			
Sylvite	KCl	cub	1.99	2	1.490		
Talc	$Mg_3Si_4O_{10}(OH)_2$	monocl	2.71	1	1.545	1.592	1.595
Tantalite	$(Fe,Mn)(Ta,Nb)_2O_6$	rhomb	7.95	6.5	2.26	2.32	2.43
Tapiolite	$FeTa_2O_6$	tetr	7.9	6.3	2.27	2.42	
Tellurobismuthite	Bi_2Te_3	hex	7.74	1.8			
Terlinguaite	Hg_2OCl	monocl	8.73	2.5	2.35	2.64	2.66
Tetrahedrite	$(Cu,Fe)_{12}Sb_4S_{13}$	cub	4.9	3.8			
Thenardite	Na_2SO_4	orth	2.7	2.8	1.468	1.475	1.483
Thermonatrite	$Na_2CO_3\cdot H_2O$	orth	2.25	1.3	1.420	1.506	1.524
Thomsenolite	$NaCaAlF_6\cdot H_2O$	monocl	2.98	2	1.407	1.414	1.415
Thorianite	ThO_2	cub	10.0	6.5	2.200		
Thorite	$ThSiO_4$	tetr	6.7	4.8	1.8		
Topaz	$Al_2SiO_3(OH,F)_2$	rhomb	3.53	8	1.618	1.620	1.627
Torbernite	$Cu(UO_2)_2(PO_4)_2\cdot 8H_2O$	tetr	3.22	2.3	1.582	1.592	
Tourmaline	$Na(Mg,Fe,Mn,Li,Al)_3Al_6Si_6O_{18}(BO_3)_3$	rhomb	3.14	7	1.62	1.65	
Tremolite	$Ca_2Mg_5Si_8O_{22}(OH,F)_2$	monocl	3.0	5.5	1.599	1.612	1.622
Trevorite	$NiFe_2O_4$	cub	5.33	7.8	2.3		
Tridymite	SiO_2	hex	2.27	7	1.475	1.476	1.479
Triphyllite-Lithiophyllite	$Li(Fe,Mn)PO_4$	rhomb	3.46	4.5	1.68	1.68	1.69
Troegerite	$(UO_2)_3(AsO_4)_2\cdot 12H_2O$	tetr		2.5	1.59	1.630	
Troilite	FeS	hex	4.7	4			
Trona	$Na_3H(CO_3)_2\cdot 2H_2O$	monocl	2.14	2.8	1.412	1.492	1.540
Turquois	$Cu(Al,Fe)_6(PO_4)_4(OH)_8\cdot 4H_2O$	tricl	2.9	5.3	1.70	1.73	1.75
Ullmannite	$NiSbS$	cub	6.65	5.3			
Uraninite	UO_2	cub	11.0	5.5			
Uvarovite	$Ca_3Cr_2Si_3O_{12}$	cub	3.83	6.8	1.865		
Valentinite	Sb_2O_3	orth	5.7	2.8	2.18	2.35	2.35
Vanadinite	$Pb_5(VO_4)_3Cl$	hex	6.8	2.9	2.350	2.416	
Variseite-Strengite	$(Al,Fe)(PO_4)\cdot 2H_2O$	rhomb	2.72	4	1.635	1.654	1.668
Vaterite	$CaCO_3$	hex	2.71		1.550	1.645	
Vermiculite	$(Mg,Ca)_{0.7}(Mg,Fe,Al)_6[(Al,Si)_8O_{20}](OH)_4\cdot 8H_2O$	monocl	2.3	1.5	1.542	1.556	1.556
Vesuvianite	$Ca_{10}(Mg,Fe)_2Al_4(Si_2O_7)_2(SiO_4)_5(OH,F)_4$	tetr	3.33	6.5	1.72	1.73	
Villiaumite	NaF	cub	2.78	2.3	1.327		
Vivianite	$Fe_3(PO_4)_2\cdot 8H_2O$	monocl	2.58	1.8	1.598	1.629	1.652
Wagnerite	$Mg_2(PO_4)F$	monocl	3.15	5.3	1.568	1.572	1.582
Wavellite	$Al_3(OH)_3(PO_4)_2\cdot 5H_2O$	rhomb	2.36	3.6	1.527	1.535	1.553
Whewellite	$CaC_2O_4\cdot H_2O$	cub	2.2	2.8	1.491	1.554	1.650
Willemite	Zn_2SiO_4	hex	4.1	5.5	1.691	1.719	
Witherite	$BaCO_3$	orth	4.29	3.5	1.529	1.676	1.677
Wolframite	$(Fe,Mn)WO_4$	monocl	7.3	4.3	2.26	2.32	2.42
Wollastonite	$CaSiO_3$	monocl	2.92	4.8	1.628	1.639	1.642
Wulfenite	$PbMoO_4$	tetr	6.7	2.9	2.283	2.403	
Wurtzite	ZnS	hex	4.09	3.8	2.356	2.378	
Xenotime	YPO_4	tetr	4.8	4.5	1.721	1.816	
Zeunerite	$Cu(UO_2)_2(AsO_4)_2\cdot 10H_2O$	tetr			1.606		
Zincite	ZnO	hex	5.6	4	2.013	2.029	
Zircon	$ZrSiO_4$	tetr	4.6	7.5	1.94	1.99	
Zoisite	$Ca_2Al_3Si_3O_{12}(OH)$	rhomb	3.26	6	1.695	1.699	1.711

X-RAY CRYSTALLOGRAPHIC DATA ON INORGANIC SUBSTANCES AND MINERALS

From U.S. Geological Survey Bulletin 1248 by
Richard A. Robie, Philip M. Bethke and Keith M. Beardsley

An extensive list of references and the bases for the calculations and the selection of data are given in the above referenced Bulletin. Bulletin 1248 may be obtained from the Superintendent of Documents, U.S. Government Printing Office, Washington, D.C., 20402.

Z; The number of gram formula weights per unit cell.

r; Indicates the data were obtained at an unspecified room temperature and may be taken as 25°+5°C.

*; Indicates the measurements were made on a natural specimen which may have deviated slightly from the listed formula. Densities for these minerals were calculated using the formula weight for the stoichiometric phase.

hex-R; Rhombohedral symmetry. To distinguish from true hexagonal symmetry.

X-Ray Crystallographic Data of Minerals

	Name and formula	Crystal system	Space group	Structure type	Z	a_o	b_o	c_o	α_o	β_o	γ_o	Cell volume 10^{-24} cm^3	Molar volume cm^3	Molar volume cal bar^{-1}	X-Ray density grams cm^{-3}	Temp. °C	
	Elements																
1	Silver Ag	cubic	Fm3m(225)	face-centered cubic	4	4.0862 ±.0002						68.227 ± .010	10.272 ± .002	.24556 ±.00008	10.501 ±.002	25	1
2	Arsenic As	hex-R	R$\bar{3}$m(166)	arsenic	6	3.760 ±.001		10.555 ±.003				129.23 ± .08	12.972 ± .002	.31007 ±.00023	5.776 ±.004	26	2
3	Gold Au	cubic	Fm3m(225)	face-centered cubic	4	4.0786 ±.0002						67.847 ± .010	10.215 ± .002	.24420 ±.00008	19.282 ±.003	25	3
4	Bismuth Bi	hex-R	R$\bar{3}$m(166)	arsenic	6	4.5459 ±.0010		11.8622 ±.0030				212.29 ± .11	21.309 ± .011	.50934 ±.00030	9.8071 ±.0050	26	4
5	Diamond C*	cubic	Fd3m(227)	diamond	8	3.5670 ±.0001						45.385 ± .004	3.4166 ± .0003	.08170 ±.00005	3.5155 ±.0003	25	5
6	Graphite C*	hex.	C6/mmc(194)	graphite	4	2.4612 ±.0001		6.7079 ±.0010				35.189 ± .006	5.2982 ± .0009	.12668 ±.00007	2.2670 ±.0004	15	6
7	Copper Cu	cubic	Fm3m(225)	face-centered cubic	4	3.6150 ±.0005						47.242 ± .020	7.1128 ± .0030	.17005 ±.00012	8.9331 ±.0037	25	7
8	α-Iron Fe	cubic	Im3m(229)	body-centered cubic	2	2.8664 .0005						23.551 = .012	7.0918 ± .0037	.16954 ±.00013	7.8748 ±.0041	25	8
9	Nickel Ni	cubic	Fm3m(225)	face-centered cubic	4	3.5238 ±.0005						43.756 = .019	6.5880 ± .0028	.15750 ±.00011	8.9117 ±.0038	25	9
10	Lead Pb	cubic	Fm3m(225)	face-centered cubic	4	4.9505 ±.0005						121.32 = .04	18.267 ± .006	.43663 ±.00018	11.342 ±.003	25	10
11	Platinum Pt	cubic	Fm3m(225)	face-centered cubic	4	3.9231 ±.0005						60.379 = .023	9.0909 ± .0035	.21732 ±.00013	21.460 ±.008	25	11
12	orthorhombic Sulfur S	orth.	Fddd(70)	S_8 ring molecules	128	10.4646 ±.0020	12.8660 ±.0020	24.4860 ±.0040				3296.73 = .97	15.511 ± .005	.37078 ±.00015	2.0671 ±.0006	25	12
13	monoclinic Sulfur S	mon.	P2$_1$/c(14)	S_8 ring molecules	48	11.04 ±.03	10.98 ±.03	10.92 ±.03		96.73 ±.50		1314.6 ± 6.4	16.49 ± .08	.3943 ±.0020	1.944 ±.009	103	13
14	rhombohedral Sulfur S	hex-R	R$\bar{3}$(148)	S_6 ring molecules	18	10.818 ±.002		4.280 ±.001				433.78 ± .19	14.514 ± .006	.34693 ±.00020	2.2092 ±.0010	r	14
15	Antimony Sb	hex-R	R$\bar{3}$m(166)	arsenic	6	4.310 ±.001		11.279 ±.003				181.45 ± .09	18.213 ± .010	.43535 ±.00028	6.685 ±.004	26	15
16	Selenium Se	hex.	P3$_1$21(152) P3$_2$21(154)		3	4.3642 ±.0008		4.9588 ±.0008				81.793 ± .033	16.420 ± .007	.39249 ±.00020	4.8088 ±.0019	26	16
17	Silicon Si	cubic	Fd3m(227)	diamond	8	5.4305 ±.0003						160.15 ± .03	12.056 ± .002	.28819 ±.00009	2.3296 ±.0004	25	17
18	β-Tin (white) Sn	tet.	I4$_1$/amd(141)		4	5.8315 ±.0008		3.1813 ±.0006				108.18 ± .04	16.289 ± .005	.38935 ±.00017	7.2867 ±.0024	26	18
19	Tellurium Te	hex.	P3$_1$21(152) P3$_2$21(154)		3	4.4570 ±.0008		5.9290 ±.0010				102.00 ± .04	20.476 ± .008	.48944 ±.00024	6.2316 ±.0025	25	19

X-Ray Crystallographic Data of Minerals

	Name and formula	Crystal system	Space group	Structure type	Z	a_o	b_o	c_o	α_o	β_o	γ_o	Cell volume 10^{-24} cm^3	Molar volume cm^3	Molar volume cal bar^{-1}	X-Ray density grams cm^{-3}	Temp. °C	
20	Zinc Zn	hex.	$P6_3/mmc$(194)	hexagonal close packed	2	2.665 ±.001		4.947 ±.001				30.428 ± .024	9.162 ± .007	.2190 ±.0002	7.134 ±.006	25	20
	Sulfides, arsenides, tellurides, selenides, and sulfosalts																
21	Shandite β-$Ni_3Pb_2S_2$*	hex-R	$R\bar{3}m$(166)		3	5.576 ±.010		13.658 ±.010				367.76 ± 1.35	73.83 ± .27	1.765 ±.007	8.867 ±.033	r	21
22	High-Argentite Ag_2S I	cubic			4	6.269 ±.020						246.4 ± 2.4	37.09 ± .36	.8866 ±.0085	6.680 ±.064	600	22
23	Argentite Ag_2S II	cubic			2	4.870 ±.008						115.5 ± .6	34.78 ± .17	.8313 ±.0041	7.125 ±.035	189	23
24	Acanthite Ag_2S III	mon.	$P2_1/c$(14)		4	4.228 ±.002	6.928 ±.005	7.862 ±.003		99.58 ±.30		227.08 ± .29	34.19 ± .04	.8172 ±.0011	7.248 ±.009	25	24
25	High-Naumanite Ag_2Se	cubic			2	4.993 ±.016						124.48 ± 1.20	37.48 ± .36	.8959 ±.0087	7.862 ±.076	170	25
26	Ag_2Te I	cubic			2	5.29 ±.01						148.0 ± .8	44.58 ± .26	1.065 ±.006	7.702 ±.044	825	26
27	Ag_2Te II	cubic			4	6.585 ±.010						285.54 ± 1.30	42.99 ± .20	1.028 ±.005	7.986 ±.036	250	27
28	Hessite Ag_2 Te III	mon.	$P2_1/c$(14)		4	8.09 ±.02	4.48 ±.01	8.96 ±.02		123.33 ±.30		271.33 ± 1.43	40.85 ± .22	.9764 ±.0052	8.405 ±.044	r	28
29	$Ag_{1.55}Cu_{.45}S$ I	cubic			4	6.110 ±.010						228.10 ± 1.12	34.34 ± .17	.8209 ±.0041	6.635 ±.033	300	29
30	$Ag_{1.55}Cu_{.45}S$ II	cubic			2	4.825 ±.005						112.33 ± .35	33.83 ± .11	.8085 ±.0026	6.736 ±.021	116	30
31	Jalpaite $Ag_{1.55}Cu_{.45}S$ III	tet.			16	8.673 ±.004		11.756 ±.006				884.30 ± .93	33.286 ± .035	.79559 ±.00088	6.8455 ±.0072	r	31
32	$Ag_{.93}Cu_{1.07}S$ I	cubic			4	5.961 ±.009						211.82 ± .96	31.89 ± .14	.7623 ±.0035	6.283 ±.029	196	32
33	$Ag_{.93}Cu_{1.07}S$ II	hex.			2	4.138 ±.004		7.105 ±.007				105.36 ± .23	31.73 ± .07	.7583 ±.0017	6.316 ±.014	100	33
34	Stromeyerite $Ag_{.93}Cu_{1.07}S$ III	orth.	Cmcm(63)		4	4.066 ±.002	6.628 ±.003	7.972 ±.004				214.84 ± .18	32.35 ± .03	.7732 ±.0007	6.194 ±.005	r	34
35	Eucairite AgCuSe	orth.	pseudo P4/nmm(129)		10	4.105 ±.010	20.35 ±.02	6.31 ±.01				527.12 ± 1.62	31.75 ± .10	.7588 ±.0024	7.887 ±.024	r	35
36	Petzite Ag_3AuTe_2*	cubic	$I4_132$(214)		8	10.38 ±.02						1118.4 ± 6.5	84.19 ± .49	2.012 ±.012	9.214 ±.053	r	36
37	Maldonite Au_2Bi	cubic	Fd3m(227)	Cu_2Mg	8	7.958 ±.002						503.98 ± .38	37.94 ± .03	.9068 ±.0007	15.891 ±.012	r	37
38	High-Digenite Cu_2S I	cubic			4	5.725 ±.010						187.64 ± .98	28.25 ± .15	.6753 ±.0036	5.633 ±.030	465	38
39	High-Chalcocite Cu_2S II	hex.			2	3.961 ±.004		6.722 ±.007				91.34 ± .21	27.50 ± .06	.6574 ±.0015	5.786 ±.013	152	39
40	Chalcocite Cu_2S III	orth.	Ab2m(39)		96	11.881 ±.004	27.323 ±.010	13.491 ±.004				4379.5 ± 2.5	27.475 ± .016	.65671 ±.00043	5.7924 ±.0034	r	40
41	Digenite $Cu_{1.79}S$ (Cu rich side)	cubic		deformed fluorite	4	5.5695 ±.0010						172.76 ± .09	26.012 ± .014	.6217 ±.0004	5.605 ±.003	25	41
42	Digenite $Cu_{1.77}S$ (S rich side)	cubic		deformed fluorite	4	5.5542 ±.0010						171.34 ± .09	25.798 ± .014	.6166 ±.0004	5.602 ±.003	25	42
43	Berzelianite Cu_2Se	cubic			4	5.85 ±.01						200.2 ± 1.0	30.14 ± .15	.7205 ±.0037	6.835 ±.035	170	43
44	High-Bornite Cu_5FeS_4*	cubic			1	5.50 ±.01						166.4 ± .9	100.2 ± .5	2.395 ±.013	5.008 ±.027	240	44

X-Ray Crystallographic Data of Minerals

	Name and formula	Crystal system	Space group	Structure type	Z	a_o	b_o	c_o	α_o	β_o	γ_o	Cell volume 10^{-24} cm³	Molar volume cm³	Molar volume cal bar⁻¹	X-Ray density grams cm⁻³	Temp. °C	
45	Metastable Bornite Cu_5FeS_4	cubic			8	10.94 ± .02						1309.34 ± 7.18	98.57 ± .54	2.356 ± .013	5.091 ± .028	r	45
46	Low-Bornite Cu_5FeS_4*	tet.	P$\bar{4}$2₁c(144)		16	10.94 ± .02		21.88 ± .04				2618.7 ±10.7	98.57 ± .40	2.356 ± .010	5.091 ± .021	r	46
47	Umangite Cu_3Se_2	tet.	P4/mmm(123)		2	6.402 ± .010		4.276 ± .010				175.25 ± .68	52.77 ± .21	1.261 ± .005	6.604 ± .026	r	47
48	Heazelwoodite Ni_3S_2	hex-R	R32(155)		3	5.746 ± .001		7.134 ± .002				203.98 ± .09	40.95 ± .02	.9788 ± .0005	5.867 ± .003	r	48
49	Maucherite $Ni_{11}As_8$	tet.	P4₁2₁2(92)		4	6.870 ± .001		21.81 ± .01				1029.36 ± .56	154.98 ± .08	3.7043 ± .0021	8.0343 ± .0044	r	49
50	Pentlandite $Fe_{5.25}Ni_{3.75}S_8$	cubic	Fm3m(225)		4	10.196 ± .010						1059.96 ± 3.12	159.59 ± .47	3.8144 ± .0113	4.823 ± .014	r	50
51	Pentlandite $Fe_{4.75}Ni_{5.25}S_8$	cubic	Fm3m(225)		4	10.095 ± .010						1028.77 ± 3.06	154.89 ± .46	3.702 ± .011	4.998 ± .015	r	51
52	Sternbergite $AgFe_2S_3$*	orth.	Ccmm(63)		8	11.60 ± .02	12.675 ± .020	6.63 ± .01				974.81 ± 2.71	73.39 ± .20	1.754 ± .005	4.303 ± .012	r	52
53	Argentopyrite $AgFe_2S_3$*	orth.	Pmmm(47)		4	6.64 ± .01	11.47 ± .02	6.45 ± .02				491.2 ± 1.9	73.96 ± .29	1.768 ± .007	4.269 ± .017	r	53
54	Realgar AsS*	mon.	P2₁/m(11)		16	9.29 ± .05	13.53 ± .05	6.57 ± .03		106.55 ± .30		791.6 ± 6.4	29.80 ± .24	.7122 ± .0058	3.591 ± .029	r	54
55	Oldhamite CaS	cubic	Fm3m(225)	rock salt	4	5.689 ± .006						184.12 ± .58	27.722 ± .088	.6626 ± .0021	2.602 ± .008	r	55
56	Greenockite CdS	hex.	P6₃mc(186)	zincite	2	4.1354 ± .0010		6.7120 ± .0010				99.407 ± .050	29.934 ± .015	.71549 ± .00041	4.8261 ± .0024	r	56
57	Hawleyite CdS	cubic	F$\bar{4}$3m(216)	sphalerite	4	5.833 ± .002						198.46 ± .20	29.88 ± .03	.7142 ± .0008	4.835 ± .005	r	57
58	(hypothetical) CdS	cubic	Fm3m(225)	rock salt	4	5.516 ± .002						167.83 ± .18	25.27 ± .03	.6040 ± .0007	5.717 ± .006	r	58
59	Cadmoselite CdSe	hex.	P6₃mc(186)	zincite	2	4.2977 ± .0010		7.0021 ± .0010				112.00 = .05	33.727 ± .016	.80614 ± .00044	5.6738 ± .0028	r	59
60	CdTe	cubic	F$\bar{4}$3m(216)	sphalerite	4	6.4805 ± .0006						272.16 = .08	40.977 ± .012	.97943 ± .00032	5.8569 ± .0016	25	60
61	(hypothetical) CoS	cubic	F$\bar{4}$3m(216)	sphalerite	4	5.339 ± .001						152.19 = .09	22.91 ± .02	.5477 ± .0004	3.971 ± .002	r	61
62	Chalcopyrite ($CuFeS_2$) $CuFeS_{1.90}$	tet.	I42d(122)		4	5.2988 ± .0010		10.434 ± .005				292.96 ± .18	44.109 ± .027	1.0543 ± .0007	4.0878 ± .0025	r	62
63	Cubanite $CuFe_2S_3$*	orth.	Pcmn(62)		4	6.46 ± .01	11.12 ± .01	6.23 ± .01				447.53 ± 1.08	67.38 ± .16	1.611 ± .004	4.026 ± .010	r	63
64	Covellite CuS	hex.	P6₃/mmc(194)		6	3.792 ± .001		16.34 ± .01				203.48 ± .16	20.42 ± .02	.4882 ± .0005	4.682 ± .001	r	64
65	Klockmannite CuSe	hex.		deformed covellite	78	14.206 ± .010		17.25 ± .05				3014.8 ± 9.7	23.28 ± .08	.5564 ± .0018	6.122 ± .020	r	65
66	Troilite FeS	hex.	P6₃/mmc(194)	niccolite	2	3.445 ± .003		5.877 ± .001				60.439 ± .106	18.20 ± .03	.4350 ± .0008	4.830 ± .009	28	66
67	Pyrrhotite $Fe_{.980}S$	hex.	P6₃/mmc(194)	defect niccolite	2	3.445 ± .001		5.848 ± .002				60.14 ± .04	18.11 ± .01	.4329 ± .0003	4.793 ± .003	28	67
68	Pyrrhotite $Fe_{.885}S$	hex.	P6₃/mmc(194)	defect niccolite	2	3.440 ± .001		5.709 ± .003				58.507 ± .046	17.62 ± .02	.4211 ± .0004	4.625 ± .004	28	68
69	(hypothetical) FeS	cubic	F$\bar{4}$3m(216)	sphalerite	4	5.455 ± .001						162.32 ± .09	24.44 ± .01	.5842 ± .0004	3.597 ± .002	r	69
70	(hypothetical) FeS	hex.	P6₃mc(186)	zincite	2	3.872 ± .001		6.345 ± .002				82.38 ± .05	24.81 ± .02	.5930 ± .0004	3.544 ± .002	r	70

X-Ray Crystallographic Data of Minerals

	Name and formula	Crystal system	Space group	Structure type	Z	a_o	b_o	c_o	α_o	β_o	γ_o	Cell volume 10^{-24} cm³	Molar volume cm³	Molar volume cal bar⁻¹	X-Ray density grams cm⁻³	Temp. °C	
71	Cinnabar HgS	hex.	$P3_121(152)$ $P3_21(154)$	cinnabar	3	4.149 ±.001		9.495 ±.002				141.55 ± .07	28.416 ± .015	.6792 ±.0004	8.187 ±.004	r	71
72	Metacinnabar HgS	cubic	$F\bar{4}3m(216)$	sphalerite	4	5.8517 ±.0010						200.38 ± .10	30.169 ± .016	.7211 ±.0004	7.712 ±.004	r	72
73	Tiemannite $HgSe$	cubic	$F\bar{4}3m(216)$	sphalerite	4	6.0853 ±.0050						225.34 ± .56	33.928 ± .084	.8110 ±.0020	8.239 ±.020	r	73
74	Coloradoite $HgTe$	cubic	$F\bar{4}3m(216)$	sphalerite	4	6.4600 ±.0006						269.59 ± .08	40.590 ± .011	.97016 ±.00032	8.0855 ±.0023	r	74
75	Alabandite MnS	cubic	Fm3m(225)	rock salt	4	5.2234 ±.0005						142.51 ± .04	21.457 ± .006	.51289 ±.00019	4.0546 ±.0012	r	75
76	(hypothetical) MnS	cubic	$F\bar{4}3m(216)$	sphalerite	4	5.611 ±.002						176.65 ± .19	26.60 ± .03	.6357 ±.0007	3.271 ±.004	r	76
77	(hypothetical) MnS	hex.	$P6_3mc(186)$	zincite	2	3.986 ±.001		6.465 ±.002				88.96 ± .05	26.79 ± .02	.6403 ±.0004	3.248 ±.002	r	77
78	Niccolite $NiAs$	hex.	$P6_3/mmc(194)$	niccolite	2	3.618 ±.001		5.034 ±.001				57.07 ± .03	17.18 ± .01	.4108 ±.0003	7.776 ±.005	r	78
79	Millerite NiS	hex-R	R3m(160)		9	9.616 ±.001		3.152 ±.001				252.41 ± .10	16.891 ± .006	.40374 ±.00020	5.3743 ±.0020	r	79
80	Breithauptite $NiSb$	hex.	$P6_3/mmc(194)$	niccolite	2	3.942 ±.001		5.155 ±.001				69.37 ± .04	20.89 ± .01	.4994 ±.0004	8.639 ±.005	r	80
81	Galena PbS	cubic	Fm3m(225)	rock salt	4	5.9360 ±.0005						209.16 ± .05	31.492 ± .008	.75272 ±.00024	7.5973 ±.0019	26	81
82	Clausthalite $PbSe$	cubic	Fm3m(225)	rock salt	4	6.1255 ±.0005						229.84 ± .06	34.605 ± .009	.82713 ±.00025	8.2690 ±.0020	r	82
83	Teallite $PbSnS_2$	orth.	Pbnm(62)	GeS	2	4.266 ±.003	11.419 ±.007	4.090 ±.002				199.24 ± .21	59.996 ± .063	1.4340 ±.0016	6.501 ±.007	r	83
84	Altaite $PbTe$	cubic	Fm3m(225)	rock salt	4	6.4606 ±.0005						269.66 ± .06	40.601 ± .009	.97043 ±.00027	8.2459 ±.0019	r	84
85	Cooperite PtS	tet.	$P4_2/mmc(131)$		2	3 4699 ±.0006		6.1098 ±.0010				73.563 ± .028	22.152 ± .008	.5295 ±.0003	10.254 ±.004	r	85
86	Herzenbergite SnS	orth.	Pbnm(62)	GeS	4	4.328 ±.002	11.190 ±.004	3.978 ±.001				192.66 ± .12	29.01 ± .02	.6933 ±.0005	5.197 ±.003	r	86
87	Sphalerite ZnS	cubic	$F\bar{4}3m(216)$	sphalerite	4	5.4093 ±.0005						158.28 ± .04	23.831 ± .007	.56962 ±.00020	4.0885 ±.0011	r	87
88	Wurtzite ZnS	hex.	$P6_3mc(186)$	zincite	2	3.8230 ±.0010		6.2565 ±.0010				79.190 ± .043	23.846 ± .013	.56998 ±.00036	4.0859 ±.0022	r	88
89	Stilleite $ZnSe$	cubic	$F\bar{4}3m(216)$	sphalerite	4	5.6685 ±.0005						182.14 ± .05	27.424 ± .007	.65548 ±.00022	5.2630 ±.0014	r	89
90	$ZnTe$	cubic	$F\bar{4}3m(216)$	sphalerite	4	6.1020 ±.0006						227.20 ± .07	34.209 ± .010	.81765 ±.00029	5.6410 ±.0017	r	90
91	Orpiment As_2S_3*	mon.	$P2_1/n(14)$		4	11.49 ±.02	9.59 ±.02	4.25 ±.01		90.45 ±.30		468.3 ± 1.7	70.51 ± .25	1.685 ±.006	3.490 ±.013	r	91
92	Bismuthinite Bi_2S_3	orth.	Pbnm(62)	stibnite	4	11.150 ±.004	11.300 ±.004	3.981 ±.001				501.59 ± .28	75.520 ± .043	1.8050 ±.0011	6.8081 ±.0038	26	92
93	Tellurobismuthite Bi_2Te_3	hex-R	$R\bar{3}m(166)$	Bi_2Te_2S	3	4.3835 ±.0020		30.487 ±.003				507.33 ± .47	101.85 ± .09	2.4342 ±.0023	7.862 ±.007	25	93
94	Stibnite Sb_2S_3	orth.	Pbnm(62)	stibnite	4	11.229 ±.004	11.310 ±.004	3.8389 ±.0010				487.54 ± .28	73.406 ± .042	1.7545 ±.0010	4.6276 ±.0026	25	94
95	Linnaeite Co_3S_4	cubic	Fd3m(227)	spinel	8	9.401 ±.001						830.85 ± .27	62.548 ± .020	1.4950 ±.0005	4.8772 ±.0016	r	95
96	Greigite Fe_3S_4	cubic	Fd3m(227)	spinel	8	9.876 ±.002						963.26 ± .59	72.52 ± .04	1.733 ±.001	4.079 ±.003	r	96

	Name and formula	Crystal system	Space group	Structure type	Z	a_o	b_o	c_o	α_o	β_o	γ_o	Cell volume 10^{-24} cm³	Molar volume cm³	Molar volume cal bar⁻¹	X-Ray density grams cm⁻³	Temp. °C	
97	Daubreeite $FeCr_2S_4$	cubic	Fd3m(227)	spinel	8	9.966 ±.005						989.83 ± 1.49	74.52 ± .11	1.781 ±.003	3.866 ±.006	r	97
98	Violarite $FeNi_2S_4$	cubic	Fd3m(227)	spinel	8	9.464 ±.005						847.66 ± 1.34	63.81 ± .10	1.525 ±.002	4.725 ±.008	r	98
99	Polymidite Ni_3S_4	cubic	Fd3m(227)	spinel	8	9.480 ±.001						851.97 ± .27	64.138 ± .020	1.5330 ±.0005	4.7458 ±.0015	r	99
100	Co-Safflorite $CoAs_2$	mon.		deformed marcasite	2	5.049 ±.002	5.872 ±.002	3.127 ±.001		90.45 ±.20		92.706 ± .057	27.92 ± .02	.6672 ±.0005	7.479 ±.005	26	100
101	Safflorite $(CO_{.5}Fe_{.5})As_2$	orth.	Pnnm(58)	marcasite	2	5.231 ±.002	5.953 ±.002	2.962 ±.002				92.237 ± .078	27.775 ± .024	.6639 ±.0006	7.461 ±.006	26	101
102	Cobaltite CoAsS*	cubic	P2₁3(198)	NiSbS	4	5.60 ±.05						175.62 ± 4.70	26.44 ± .71	.6320 ±.0170	6.275 ±.168	r	102
103	Glaucodot (Co,Fe)AsS*	orth.	Cmmm(65)		24	6.64 ±.05	28.39 ±.10	5.64 ±.05				1063.2 ±12.9	26.68 ± .32	.6377 ±.0078	6.161 ±.075	r	103
104	Cattierite CoS_2	cubic	Pa3(205)	pyrite	4	5.5345 ±.0005						169.53 ± .05	25.524 ± .007	.61009 ±.00021	4.8213 ±.0013	r	104
105	Trogtalite $CoSe_2$	cubic	Pa3(205)	pyrite	4	5.8588 ±.0010						201.11 ± .10	30.279 ± .016	.72374 ±.00042	7.1618 ±.0037	r	105
106	Loellingite $FeAs_2$	orth.	Pnnm(58)	marcasite	2	5.300 ±.002	5.981 ±.002	2.882 ±.001				91.357 ± .056	27.51 ± .02	.6576 ±.0005	7.477 ±.005	26	106
107	Arsenopyrite FeAsS*	tri.	P$\bar{1}$(2)		4	5.760 ±.010	5.690 ±.005	5.785 ±.005	90.00 ±.20	112.23 ±.20	90.00 ±.20	175.51 ± .44	26.42 ± .07	.6316 ±.0016	6.162 ±.015	r	107
108	Gudmundite FeSbS*	mon.	B2₁/d(14)		8	10.09 ±.05	5.93 ±.03	6.73 ±.03		90.00 ± 50		399.09 ± 3.35	30.04 ± .25	.7181 ±.0061	6.978 ±.059	r	108
109	Pyrite FeS_2	cubic	Pa3(205)	pyrite	4	5.4175 ±.0005						159.00 ± .04	23.940 ± .007	.57221 ±.00020	5.0116 ±.0014	r	109
110	Marcasite FeS_2*	orth.	Pnnm(58)	marcasite	2	4.443 ±.002	5.423 ±.002	3.3876 ±.0015				81.622 ± .060	24.579 ± .018	.58749 ±.00047	4.8813 ±.0036	25	110
111	Ferroselite $FeSe_2$	orth.	Pnnm(58)	marcasite	2	4.801 ±.005	5.778 ±.005	3.587 ±.004				99.50 ± .17	29.96 ± .05	.7162 ±.0013	7.134 ±.013	r	111
112	Frohbergite $FeTe_2$	orth.	Pnnm(58)	marcasite	2	5.265 ±.005	6.265 ±.005	3.869 ±.002				127.62 ± .17	38.43 ± .05	.9185 ±.0013	8.094 ±.011	r	112
113	Hauerite MnS_2	cubic	Pa3(205)	pyrite	4	6.1014 ±.0006						227.14 ± .07	34.198 ± .010	.81741 ±.00029	3.4816 ±.0010	28	113
114	Molybdenite MoS_2	hex.	P6₃/mmc(194)	molybdenite	2	3.1604 ±.0010		12.295 ±.002				106.35 ± .07	32.025 ± .021	.76547 ±.00055	4.9982 ±.0033	26	114
115	Rammelsbergite $NiAs_2$	orth.	Pnnm(58)	marcasite	2	4.757 ±.002	5.797 ±.004	3.542 ±.002				97.645 ± .096	29.41 ± .03	.7030 ±.0007	7.091 ±.007	26	115
116	Pararammelsbergite $NiAs_2$	orth.	Pbca(61)		8	5.75 ±.01	5.82 ±.01	11.428 ±.02				382.42 ± 1.15	28.79 ± .09	.6882 ±.0021	7.244 ±.022	r	116
117	Gersdorfite NiAsS	cubic	P2₁3(198)		4	5.693 ±.001						184.51 ± .10	27.78 ± .01	.6640 ±.0004	5.964 ±.003	26	117
118	Vaesite NiS_2	cubic	Pa3(205)	pyrite	4	5.6873 ±.0005						183.96 ± .05	27.697 ± .007	.66203 ±.00022	4.4350 ±.0012	r	118
119	$NiSe_2$	cubic	Pa3(205)	pyrite	4	5.9604 ±.0010						211.75 ± .11	31.882 ± .016	.76204 ±.00043	6.7948 ±.0034	20	119
120	Melonite $NiTe_2$	hex.	P$\bar{3}$ml(164)	cadmium iodide	1	3.869 ±.010		5.308 ±.010				68.81 ± .38	41.44 ± .23	.9905 ±.0055	7.575 ±.042	84	120
121	Sperrylite $PtAs_2$	cubic	Pa3(205)	pyrite	4	5.968 ±.005						212.56 ± .53	32.00 ± .08	.7650 ±.0020	10.778 ±.027	r	121
122	Laurite RuS_2	cubic	Pa3(205)	pyrite	4	5.60 ±.02						175.6 ± 1.9	26.44 ± .28	.6320 ±.0068	6.248 ±.067	r	122

X-Ray Crystallographic Data of Minerals

No.	Name and formula	Crystal system	Space group	Structure type	Z	a_o	b_o	c_o	α_o	β_o	γ_o	Cell volume 10^{-24} cm^3	Molar volume cm^3	Molar volume cal bar^{-1}	X-Ray density grams cm^{-3}	Temp. °C	No.
123	Tungstenite WS_2	hex.	$P6_3/mmc$(194)	molybdenite	2	3.154 ±.001		12.362 ±.004				106.50 ± .08	32.069 ± .023	.76652 ±.00059	7.7325 ±.0055	26	123
124	Co-Skutterudite $CoAs_{3-x}$ $CoAs_{2.95}$	cubic	Im3(204)		8	8.2060 ±.0010						552.58 ± .20	41.599 ± .015	.99428 ±.00041	6.7298 ±.0025	r	124
125	Fe-Skutterudite $FeAs_{3-x}$ $FeAs_{2.95}$	cubic	Im3(204)		8	8.1814 ±.0010						547.62 ± .20	41.226 ± .015	.98537 ±.00041	6.7158 ±.0025	r	125
126	Ni-Skutterudite $NiAs_{3-x}$ $NiAs_{2.95}$	cubic	Im3(204)		8	8.3300 ±.0010						578.01 ± .21	43.513 ± .016	1.0400 ±.0004	6.4286 ±.0023	r	126
127	Tennantite $Cu_{12}As_4S_{13}$	cubic	I$\bar{4}$3m(217)	tetrahedrite	2 2	10.190 ±.004						1058.09 ± 1.25	318.62 ± .38	7.1652 ±.0090	4.642 ±.006	r	127
128	Tetrahedrite $Cu_{12}Sb_4S_{13}$	cubic	I$\bar{4}$3m(217)	tetrahedrite	2 2	10.327 ±.004						1101.3 ± 1.3	331.64 ± .39	7.9266 ±.0094	5.024 ±.006	r	128
129	Enargite Cu_3AsS_4	orth.	Pnn2(34)		2	6.426 ±.005	7.422 ±.005	6.144 ±.005				293.03 ± .38	88.24 ± .12	2.109 ±.003	4.463 ±.006	26	129
130	Luzonite Cu_3AsS_4*	tet.	I$\bar{4}$2m(121)		2	5.289 ±.005		10.440 ±.008				292.04 ± .60	87.94 ± .18	2.1019 ±.0043	4.478 ±.009	26	130
131	Famatimite Cu_3SbS_4*	tet.	I$\bar{4}$m(121)		2	5.384 ±.005		10.770 ±.008				312.19 ± .62	94.01 ± .19	2.2469 ±.0045	4.687 ±.009	26	131
132	Proustite Ag_3AsS_3	hex-R	R3c(161)		6	10.816 ±.001		8.6948 ±.0013				880.89 ± .21	88.420 ± .021	2.1133 ±.0006	5.595 ±.001	26	132
133	Pyrargyrite Ag_3SbS_3	hex-R	R3c(161)		6	11.052 ±.002		8.7177 ±.0020				922.18 ± .40	92.564 ± .040	2.2124 ±.0010	5.8506 ±.0025	26	133
134	Miargyrite $AgSbS_2$*	mon.	Cc(9)		8	12.862 ±.013	4.111 ±.004	13.220 ±.010		98.63 ±.15		691.10 ± 1.14	52.027 ± .086	1.244 ±.002	5.646 ±.009	r	134
	Oxides and hydroxides																
135	Corundum Al_2O_3	hex-R	R$\bar{3}$c(167)	corundum	6	4.7591 ±.0004		12.9894 ±.0030				254.78 ± .07	25.575 ± .007	.61128 ±.00022	3.9869 ±.0011	25	135
136	Boehmite AlO(OH)*	orth.	Cmcm(63)	lepidocrocite	4	2.868 ±.003	12.227 ±.003	3.700 ±.003				129.75 ± .17	19.535 ± .026	.46695 ±.00067	3.071 ±.004	26	136
137	Diaspore AlO(OH)*	orth.	Pbnm(62)		4	4.401 ±.005	9.421 ±.005	2.845 ±.002				117.96 ± .17	17.760 ± .026	.4245 ±.0007	3.378 ±.005	r	137
138	Gibbsite $Al(OH)_3$	mon.	$P2_1/n$(14)		8	9.719 ±.002	5.0705 ±.0010	8.6412 ±.0010		94.57 ±.25		424.49 ± .20	31.956 ± .015	.7638 ±.0004	2.441 ±.001	r	138
139	Arsenolite As_2O_3	cubic	Fd3m(227)	diamond	16	11.074 ±.005						1358.0 ± 1.8	51.118 ± .069	1.2218 ±.0017	3.870 ±.005	25	139
140	Claudetite As_2O_3	mon.	$P2_1/n$(14)		4	5.339 ±.002	12.984 ±.005	4.5405 ±.0010		94.27 ±.10		313.88 ± .19	47.259 ± .028	1.1296 ±.0007	4.1863 ±.0025	25	140
141	Bromellite BeO	hex.	$P6_3mc$(186)	zincite	2	2.6979 ±.0005		4.3772 ±.0005				27.592 ± .011	8.3086 ± .0032	.19862 ±.00012	3.0104 ±.0012	26	141
142	Bismite α-Bi_2O_3	mon.	$P2_1/c$(14)	pseudo orthorhombic	8	8.166 ±.005	13.827 ±.010	5.850 ±.004		90.00 ±.20		660.53 ± .77	49.73 ± .06	1.1885 ±.0014	9.371 ±.011	25	142
143	Lime CaO	cubic	Fm3m(225)	rock salt	4	4.8108 ±.0005						111.34 ± .03	16.764 ± .005	.40071 ±.00017	3.3453 ±.0010	26	143
144	Portlandite $Ca(OH)_2$	hex.	P$\bar{3}$ml(164)	CdI_2	1	3.5933 ±.0005		4.9086 ±.0020				54.888 ± .027	33.056 ± .016	.79011 ±.00043	2.2415 ±.0011	26	144
145	Monteponite CdO	cubic	Fm3m(225)	rock salt	4	4.6953 ±.0010						103.51 ± .07	15.585 ± .010	.37254 ±.00028	8.2386 ±.0053	27	145
146	Cerianite CeO_2	cubic	Fm3m(225)	fluorite	4	5.4110 ±.0020						158.43 ± .18	23.853 ± .026	.57016 ±.00068	7.216 ±.008	26	146
147	CoO	cubic	Fm3m(225)	rock salt	4	4.260 ±.002						77.31 ± .11	11.64 ± .02	.2782 ±.0004	6.438 ±.009	26	147

X-Ray Crystallographic Data of Minerals

	Name and formula	Crystal system	Space group	Structure type	Z	a_o	b_o	c_o	α_o	β_o	γ_o	Cell volume 10^{-24} cm³	Molar volume cm³	Molar volume cal bar⁻¹	X-Ray density grams cm⁻³	Temp. °C	
148	Eskolaite Cr_2O_3	hex-R	R$\bar{3}$c(167)	corundum	6	4.9607 ±.0020		13.599 ±.010				289.82 ± .32	29.090 ± .032	.6953 ±.0008	5.225 ±.006	r	148
149	Tenorite CuO	mon.	C2/c(15)		4	4.684 ±.005	3.425 ±.005	5.129 ±.005		99.47 ±.17		81.16 ± .17	12.22 ± .03	.2921 ±.0007	6.509 ±.014	26	149
150	Cuprite Cu_2O	cubic	Pn3m(224)		2	4.2696 ±.0010						77.833 ± .055	23.437 ± .016	.56021 ±.00044	6.1047 ±.0043	26	150
151	Wustite $Fe_{.953}O$	cubic	Fm3m(225)	defect rock salt	4	4.3088 ±.0003						79.996 ± .017	12.044 ± .003	.28791 ±.00011	5.7471 ±.0012	17	151
152	Hematite Fe_2O_3	hex-R	R$\bar{3}$c(167)	corundum	6	5.0329 ±.0010				13.7492 ±.0010		301.61 ± .12	30.274 ± .012	.72361 ±.00034	5.2749 ±.0021	25	152
153	Magnetite Fe_3O_4	cubic	Fd3m(227)	spinel	8	8.3940 ±.0005						591.43 ± .11	44.524 ± .008	1.0642 ±.0002	5.2003 ±.0009	22	153
154	Goethite α-FeO(OH)*	orth.	Pbnm(62)		4	4.596 ±.005	9.957 ±.010	3.021 ±.003				138.2 ± .2	20.82 ± .04	.4975 ±.0009	4.269 ±.008	r	154
155	Lepidocrocite γ-FeO(OH)*	orth.	Amam(63)		4	3.868 ±.010	12.525 ±.010	3.066 ±.003				148.54 ± .43	22.364 ± .064	.5346 ±.0016	3.973 ±.011	r	155
156	α-Ga_2O_3	hex-R	R$\bar{3}$c(167)	corundum	6	4.9793 ±.0010		13.429 ±.003				288.34 ± .13	28.943 ± .013	.69179 ±.00036	6.4762 ±.0030	24	156
157	Low-germania GeO_2	tet.	P4/mnm(136)	rutile	2	4.3963 ±.0010		2.8626 ±.0010				55.327 ± .032	16.660 ± .010	.39824 ±.00027	6.2777 ±.0036	25	157
158	High-germania GeO_2	hex.	$P3_121(152)$ $P3_221(154)$	α-quartz	3	4.987 ±.002		5.652 ±.002				121.73 ± .11	24.438 ± .021	.58413 ±.00056	4.2797 ±.0038	26	158
159	Ice H_2O	hex.	$P6_3/mmc(194)$		4	4.5212 ±.0010		7.3666 ±.0010				130.41 ± .06	19.635 ± .009	.46932 ±.00026	.9175 ±.0004	0	159
160	Hafnia HfO_2	mon.	$P2_1/c(14)$	baddeleyite	4	5.1156 ±.0010	5.1722 ±.0010	5.2948 ±.0010		99.18 ±.08		138.30 ± .06	20.823 ± .008	.49772 ±.00025	10.108 ±.004	r	160
161	Montroydite HgO	orth.	Pnma(62)		4	6.608 ±.003	5.518 ±.003	3.519 ±.003				128.3 ± .1	19.32 ± .02	.4618 ±.0006	11.21 ±.01	25	161
162	Periclase MgO	cubic	Fm3m(225)	rock salt	4	4.2117 ±.0005						74.709 ± .027	11.248 ± .004	.26889 ±.00014	3.5837 ±.0013	25	162
163	Brucite $Mg(OH)_2$	hex.	P$\bar{3}$ml(164)	CdI_2	1	3.147 ±.004		4.769 ±.004				40.90 ± .11	24.63 ± .07	.5888 ±.0016	2.368 ±.006	26	163
164	Manganosite MnO	cubic	Fm3m(225)	rock salt	4	4.4448 ±.0005						87.813 ± .030	13.221 ± .004	.31604 ±.00015	5.3653 ±.0018	26	164
165	Pyrolusite MnO_2	tet.	P4/mnm(136)	rutile	2	4.388 ±.003		2.865 ±.002				55.16 ± .08	16.61 ± .02	.3971 ±.0007	5.234 ±.008	r	165
166	Bixbyite Mn_2O_3	cubic	Ia3(206)	Tl_2O_3	16	9.411 ±.005						833.5 ± 1.3	31.37 ± .05	.7499 ±.0012	5.032 ±.008	25	166
167	Hausmanite Mn_3O_4	tet.	$I4_1/amd(141)$		8	8.136 ±.005		9.422 ±.005				623.68 ± .84	46.95 ± .06	1.1222 ±.0016	4.873 ±.007	20	167
168	Molybdite MoO_3	orth.	Pbnm(62)		4	3.962 ±.002	13.858 ±.005	3.697 ±.004				202.98 ± .25	30.56 ± .04	.7305 ±.0010	4.710 ±.006	26	168
169	Bunsenite NiO	cubic	Fm3m(225)	rock salt	4	4.177 ±.002						72.88 ± .10	10.97 ± .02	.2623 ±.0004	6.809 ±.010	26	169
170	Litharge PbO red	tet.	P4/nmm(129)		2	3.9759 ±.0040		5.023 ±.004				79.40 ± .17	23.91 ± .05	.5715 ±.0013	9.334 ±.020	27	170
171	Massicot PbO yellow	orth.	Pb2a(32)		4	5.489 ±.003	4.755 ±.004	5.891 ±.004				153.8 ± .2	23.15 ± .03	.5533 ±.0007	9.641 ±.012	27	171
172	Minium Pb_3O_4	tet.	$P4_2/mbc(135)$		4	8.815 ±.005		6.565 ±.003				510.13 ± .62	76.81 ± .09	1.836 ±.002	8.926 ±.009	25	172
173	Senarmontite Sb_2O_3	cubic	Fm3m(225)	arsenic trioxide	16	11.152 ±.003						1336.9 ± 1.1	52.206 ± .042	1.2478 ±.0011	5.5837 ±.0045	26	173

X-Ray Crystallographic Data of Minerals

	Name and formula	Crystal system	Space group	Structure type	Z	a_o	b_o	c_o	α_o	β_o	γ_o	Cell volume 10^{-24} cm³	Molar volume cm³	Molar volume cal bar⁻¹	X-Ray density grams cm⁻³	Temp. °C	
174	Valentinite Sb_2O_3	orth.	Pccn(56)	antimony trioxide	4	4.914 ±.002	12.468 ±.005	5.421 ±.004				332.13 ± .31	50.007 ± .047	1.1952 ±.0012	5.8292 ±.0054	25	174
175	Cervantite Sb_2O_4	cubic	Fd3m(227)		8	10.305 ±.005						1094.3 ± 1.6	82.38 ± .12	1.9690 ±.0029	3.733 ±.005	26	175
176	Selenolite SeO_2	tet.	$P4_2/mbc$(135) $P4_2bc$(106)		8	8.35 ±.01		5.08 ±.01				354.2 ± 1.1	26.66 ± .08	.6373 ±.0020	4.161 ±.013	26	176
177	α-Quartz SiO_2*	hex.	$P3_121$(152) $P3_221$(154)		3	4.9136 ±.0001		5.4051 ±.0001				113.01 ± .01	22.688 ± .001	.54229 ±.00007	2.6483 ±.0001	25	177
178	β-Quartz SiO_2*	hex.	$P6_422$(181) $P6_222$(180)		3	4.999 ±.001		5.4592 ±.0020				118.15 ± .06	23.718 ± .013	.5669 ±.0004	2.533 ±.002	575	178
179	α-Cristobalite SiO_2	tet.	$P4_12_12$(92) $P4_32_12$(96)		4	4.971 ±.003		6.918 ±.003				170.95 ± .22	25.739 ± .033	.61521 ±.00083	2.3344 ±.0030	25	179
180	β-Cristobalite SiO_2	cubic	Fd3m(227)		8	7.1382 ±.0010						363.72 ± .15	27.381 ± .012	.65447 ±.00032	2.1944 ±.0009	405	180
181	Keatite SiO_2	tet.	$P4_12_12$(92) $P4_32_12$(96)		12	7.456 ±.003		8.604 ±.005				478.3 ± .5	24.01 ± .02	.5738 ±.0006	2.503 ±.003	r	181
182	β-Tridymite SiO_2	hex.	$P\bar{6}2c$(172) $P6_3/mmc$(194)		4	5.0463 ±.0020		8.2563 ±.0030				182.08 ± .16	27.414 ± .024	.65527 ±.00062	2.1917 ±.0019	405	182
183	Coesite SiO_2*	mon.	B2/b(15)		16	7.152 ±.001	12.379 ±.002	7.152 ±.001		120.00 ±.17		548.37 ± .95	20.641 ± .036	.49338 ±.00090	2.9110 ±.0050	25	183
184	Stishovite SiO_2*	tet.	P4/mnm(136)	rutile	2	4.1790 ±.0010		2.6649 ±.0010				46.540 ± .028	14.014 ± .009	.33500 ±.00025	4.2874 ±.0026	r	184
185	Melanophlogite SiO_2*	cubic	Pm3n(223)	clathrate type	46	13.402 ±.004						2407.2 ± 2.2	31.516 ± .028	.75325 ±.00072	1.9065 ±.0017	r	185
186	Cassiterite SnO_2	tet.	P4/mnm(136)	rutile	2	4.738 ±.003		3.188 ±.003				71.57 ± .11	21.55 ± .03	.5151 ±.0009	6.992 ±.011	26	186
187	Tellurite TeO_2*	orth.	Pbca(61)	tellurite	8	5.607 ±.003	12.034 ±.005	5.463 ±.003				368.61 ± .32	27.750 ± .024	.66328 ±.00062	5.7514 ±.0050	25	187
188	Paratellurite TeO_2	tet.	$P4_12_12$(92) $P4_32_12$(96)		4	4.810 ±.002		7.613 ±.002				176.14 ± .15	26.52 ± .02	.6339 ±.0006	6.018 ±.005	25	188
189	Thorianite ThO_2	cubic	Fm3m(225)	fluorite	4	5.5952 ±.0005						175.16 ± .05	26.373 ± .007	.63038 ±.00021	10.012 ±.003	25	189
190	Rutile TiO_2	tet.	P4/mnm(136)		2	4.5937 ±.0005		2.9618 ±.0010				62.500 ± .025	18.820 ± .008	.44986 ±.00023	4.2453 ±.0017	25	190
191	Anatase TiO_2	tet.	$I4_1/amd$(141)		4	3.785 ±.002		9.514 ±.006				136.30 ± .17	20.522 ± .025	.4905 ±.0007	3.893 ±.005	r	191
192	Brookite TiO_2*	orth.	Pcab(61)		8	5.456 ±.002	9.182 ±.005	5.143 ±.003				257.6 ± .2	19.40 ± .02	.4636 ±.0005	4.119 ±.004	r	192
193	Titanium sesquioxide Ti_2O_3	hex-R	$R\bar{3}c$(167)	corundum	6	5.149 ±.002		13.642 ±.010				313.2 ± .3	31.44 ± .03	.7515 ±.0009	4.574 ±.005	r	193
194	Uraninite UO_2	cubic	Fm3m(225)	fluorite	4	5.4682 ±.0010						163.51 ± .09	24.618 ± .014	.58843 ±.00037	10.969 ±.006	26	194
195	Karelianite V_2O_3	hex-R	$R\bar{3}c$(167)	corundum	6	4.952 ±.002		14.002 ±.010				297.36 ± .32	29.848 ± .032	.71342 ±.00081	5.0216 ±.0054	r	195
196	Zincite ZnO	hex.	$P6_3mc$(186)	zincite	2	3.2495 ±.0005		5.2069 ±.0005				47.615 ± .015	14.338 ± .005	.34273 ±.00016	5.6750 ±.0018	25	196
197	Baddeleyite ZrO_2	mon.	$P2_1/c$(14)	baddeleyite	4	5.1454 ±.0010	5.2075 ±.0010	5.3107 ±.0010		99.23 ±.08		140.46 ± .06	21.148 ± .009	.50548 ±.00025	5.8267 ±.0023	r	197
	Multiple oxides																
198	Spinel $MgAl_2O_4$	cubic	Fd3m(227)	spinel	8	8.080 ±.002						527.5 ± .4	39.71 ± .03	.9492 ±.0008	3.583 ±.003	26	198

X-Ray Crystallographic Data of Minerals

	Name and formula	Crystal system	Space group	Structure type	Z	a_o	b_o	c_o	α_o	β_o	γ_o	Cell volume 10^{-24} cm³	Molar volume cm³	Molar volume cal bar⁻¹	X-Ray density grams cm⁻³	Temp. °C	
199	Hercynite $FeAl_2O_4$	cubic	Fd3m(227)	spinel	8	8.150 ±.004						541.3 ± .6	40.75 ± .05	.9740 ±.0011	4.265 ±.005	25	199
200	Galaxite $MnAl_2O_4$	cubic	Fd3m(227)	spinel	8	8.258 ±.002						563.2 ± .4	42.39 ± .03	1.013 ±.001	4.078 ±.003	25	200
201	Gahnite $ZnAl_2O_4$	cubic	Fd3m(227)	spinel	8	8.0848 ±.0020						528.45 ± .39	39.783 ± .030	.95088 ±.00075	4.6083 ±.0034	26	201
202	Magnetite $FeFe_2O_4$	cubic	Fd3m(227)	spinel	8	8.3940 ±.0005						591.43 ± .11	44.524 ± .008	1.0642 ±.0002	5.2003 ±.0009	22	202
203	Jacobsite $MnFe_2O_4$	cubic	Fd3m(227)	spinel	8	8.499 ±.002						613.9 ± .4	46.22 ± .03	1.105 ±.001	4.990 ±.004	25	203
204	Trevorite $NiFe_2O_4$	cubic	Fd3m(227)	spinel	8	8.339 ±.003						579.9 ± .6	43.65 ± .05	1.043 ±.001	5.370 ±.006	25	204
205	Picrochromite $MgCr_2O_4$	cubic	Fd3m(227)	spinel	8	8.333 ±.003						578.6 ± .6	43.56 ± .05	1.041 ±.001	4.415 ±.005	26	205
206	Ilmenite $FeTiO_3$	hex-R	R$\bar{3}$(148)	ilmenite	6	5.093 ±.005		14.055 ±.020				315.73 ± .75	31.69 ± .08	.7574 ±.0019	4.788 ±.012	r	206
207	Geikielite $MgTiO_3$	hex-R	R$\bar{3}$(148)	ilmenite	6	5.054 ±.005		13.898 ±.010				307.44 ± .65	30.86 ± .07	.7376 ±.0016	3.896 ±.008	26	207
208	Pyrophanite $MnTiO_3$	hex-R	R$\bar{3}$(148)	ilmenite	6	5.155 ±.005		14.18 ±.01				326.3 ± .7	32.76 ± .07	.7829 ±.0017	4.605 ±.010	r	208
209	Cobalt Titanate $CoTiO_3$	hex-R	R$\bar{3}$(148)	ilmenite	6	5.066 ±.001		13.918 ±.005				309.34 ± .17	31.05 ± .02	.7422 ±.0004	4.986 ±.003	r	209
210	Perovskite $CaTiO_3$	orth.	Pcmn(62)	perovskite	4	5.3670 ±.0010	7.6438 ±.0010	5.443 ±.001				223.33 ± .07	33.626 ± .010	.80371 ±.00028	4.0439 ±.0012	r	210
211	Chrysoberyl $BeAl_2O_4$	orth.	Pmnb(62)	olivine	4	5.4756 ±.0020	9.4041 ±.0030	4.426 ±.002				227.94 ± .15	34.320 ± .023	.82031 ±.00059	3.6997 ±.0025	25	211
	Halides																
212	Halite $NaCl$	cubic	Fm3m(225)	rock salt	4	5.6402 ±.0002						179.43 ± .02	27.015 ± .003	.64571 ±.00011	2.1634 ±.0002	26	212
213	Sylvite KCl	cubic	Fm3m(225)	rock salt	4	6.2931 ±.0002						249.23 ± .02	37.524 ± .004	.89690 ±.00013	1.9868 ±.0002	25	213
214	Villiaumite NaF	cubic	Fm3m(225)	rock salt	4	4.6342 ±.0005						99.523 ± .032	14.984 ± .005	.35818 ±.00016	2.8021 ±.0009	25	214
215	Chlorargyrite $AgCl$	cubic	Fm3m(225)	rock salt	4	5.5491 ±.0005						170.87 ± .05	25.727 ± .007	.61493 ±.00021	5.5710 ±.0015	26	215
216	Bromargyrite $AgBr$	cubic	Fm3m(225)	rock salt	4	5.7745 ±.0005						192.55 ± .05	28.991 ± .008	.69294 ±.00022	6.4772 ±.0017	26	216
217	Nantockite $CuCl$	cubic	F$\bar{4}$3m(216)	sphalerite	4	5.416 ±.003						158.87 ± .26	23.92 ± .04	.5717 ±.0010	4.139 ±.007	25	217
218	Marshite CuI	cubic	F$\bar{4}$3m(216)	sphalerite	4	6.0507 ±.0010						221.52 ± .11	33.353 ± .017	.7972 ±.0004	5.710 ±.003	26	218
219	Miersite AgI	cubic	F$\bar{4}$3m(216)	sphalerite	4	6.4963 ±.0010						274.16 ± .10	41.278 ± .020	.9866 ±.0004	5.688 ±.003	r	219
220	Iodargyrite AgI	hex.	P6_3mc(186)	zincite	2	4.5955 ±.0010		7.5005 ±.0033				137.18 ± .10	41.308 ± .030	.9873 ±.0009	5.683 ±.004	25	220
221	Calomel $HgCl$	tet.	14/mm(139)		4	4.478 ±.005		10.910 ±.005				218.77 ± .50	32.939 ± .075	.7873 ±.0018	7.166 ±.016	26	221
222	Fluorite CaF_2	cubic	Fm3m(225)	fluorite	4	5.4638 ±.0004						163.11 ± .04	24.558 ± .005	.58701 ±.00017	3.1792 ±.0007	25	222
223	Sellaite MgF_2	tet.	P4_2/mnm(136)	rutile	2	4.621 ±.001		3.050 ±.001				65.13 ± .04	19.61 ± .01	.4688 ±.0003	3.177 ±.002	18	223

X-Ray Crystallographic Data of Minerals

	Name and formula	Crystal system	Space group	Structure type	Z	a_o	b_o	c_o	α_o	β_o	γ_o	Cell volume 10^{-24} cm^3	Molar volume cm^3	Molar volume cal bar^{-1}	X-Ray density grams cm^{-3}	Temp. °C	
224	Chloromagnesite $MgCl_2$	hex-R	R3m(166)		3	3.632 ± .004		17.795 ± .016				203.29 ± .48	40.81 ± .10	.9754 ± .0024	2.333 ± .006	r	224
225	Lawrencite $FeCl_2$	hex-R	R$\bar{3}$m(166)		3	3.593 ± .003		17.58 ± .09				196.55 ± 1.06	39.46 ± .21	.9431 ± .0051	3.212 ± .017	r	225
226	Scacchite $MnCl_2$	hex-R	R$\bar{3}$m(166)		3	3.711 ± .002		17.59 ± .07				209.79 ± .86	42.11 ± .17	1.007 ± .004	2.988 ± .012	r	226
227	Cotunnite $PbCl_2$	orth.	Pnmb(62)		4	4.535 ± .005	7.62 ± .01	9.05 ± .01				312.74 ± .64	47.09 ± .10	1.1254 ± .0023	5.906 ± .012	26	227
228	Matlockite PbFCl	tet.	P4/nmm(129)		2	4.106 ± .005		7.23 ± .01				121.89 ± .34	36.70 ± .10	.8773 ± .0025	9.853 ± .028	26	228
229	Cryolite Na_3AlF_6*	mon.	P2$_1$/n(14)		2	5.40 ± .01	5.60 ± .01	7.776 ± .010		90.18 ± .25		235.1 ± .7	70.81 ± .20	1.692 ± .005	2.965 ± .009	r	229
230	Neighborite $NaMgF_3$	orth.	Pcmn(62)	perovskite	4	5.363 ± .001	7.676 ± .001	5.503 ± .001				226.54 ± .07	34.11 ± .01	.8152 ± .0003	3.058 ± .001	18	230
	Carbonates and nitrates																
231	Calcite $CaCO_3$	hex-R	R$\bar{3}$c(167)	calcite	6	4.9899 ± .0010		17.064 ± .002				367.96 ± .15	36.934 ± .015	.88278 ± .00041	2.7100 ± .0011	26	231
232	Otavite $CdCO_3$	hex-R	R$\bar{3}$c(167)	calcite	6	4.9204 ± .0010		16.298 ± .003				341.72 ± .15	34.300 ± .015	.81983 ± .00041	5.0265 ± .0022	26	232
233	Cobalticalcite $CoCO_3$	hex-R	R$\bar{3}$c(167)	calcite	6	4.6581 ± .0010		14.958 ± .003				281.07 ± .13	28.213 ± .013	.67435 ± .00036	4.2159 ± .0020	26	233
234	Siderite $FeCO_3$	hex-R	R$\bar{3}$c(167)	calcite	6	4.6887 ± .0010		15.373 ± .003				292.68 ± .14	29.378 ± .014	.70219 ± .00037	3.9436 ± .0018	26	234
235	Magnesite $MgCO_3$	hex-R	R$\bar{3}$c(167)	calcite	6	4.6330 ± .0010		15.016 ± .003				279.13 ± .13	28.018 ± .013	.66969 ± .00036	3.0095 ± .0014	26	235
236	Rhodochrosite $MnCO_3$	hex-R	R$\bar{3}$c(167)	calcite	6	4.7771 ± .0010		15.664 ± .003				309.57 ± .14	31.073 ± .014	.74272 ± .00039	3.6992 ± .0017	26	236
237	Nickelous Carbonate $NiCO_3$	hex-R	R$\bar{3}$c(167)	calcite	6	4.5975 ± .0010		14.723 ± .002				269.51 ± .12	27.052 ± .012	.64660 ± .00034	4.3886 ± .0020	26	237
238	Smithsonite $ZnCO_3$	hex-R	R$\bar{3}$(167)	calcite	6	4.6528 ± .0010		15.025 ± .003				281.69 ± .13	28.275 ± .013	.67583 ± .00037	4.4343 ± .0021	26	238
239	Dolomite $CaMg(CO_3)_2$*	hex-R	R$\bar{3}$(148)	calcite	3	4.8079 ± .0010		16.010 ± .003				320.50 ± .15	64.341 ± .029	1.5378 ± .0008	2.8661 ± .0013	26	239
240	Huntite $Mg_3Ca(CO_3)_4$*	hex-R	R32(155)	calcite	3	9.498 ± .003		7.816 ± .004				610.63 ± .50	122.58 ± .10	2.9299 ± .0024	2.880 ± .002	26	240
241	Norsethite $BaMg(CO_3)_2$*	hex-R	R32(155)	calcite	3	5.020 ± .005		16.75 ± .02				365.6 ± .8	73.39 ± .17	1.754 ± .004	3.838 ± .009	r	241
242	Vaterite $CaCO_3$	hex.			6	7.135 ± .005		8.524 ± .007				375.80 ± .61	37.72 ± .06	.9016 ± .0015	2.653 ± .004	r	242
243	Witherite $BaCO_3$	orth.	Pnam(62)	aragonite	4	6.430 ± .005	8.904 ± .005	5.314 ± .005				304.24 ± .41	45.81 ± .06	1.095 ± .002	4.308 ± .006	26	243
244	Aragonite $CaCO_3$	orth.	Pnam(62)	aragonite	4	5.741 ± .005	7.968 ± .005	4.959 ± .005				226.85 ± .33	34.15 ± .05	.8164 ± .0012	2.930 ± .004	26	244
245	Cerussite $PbCO_3$	orth.	Pnam(62)	aragonite	4	6.152 ± .005	8.436 ± .005	5.195 ± .005				269.61 ± .38	40.59 ± .06	.9702 ± .0014	6.582 ± .009	26	245
246	Strontianite $SrCO_3$	orth.	Pnam(62)	aragonite	4	6.029 ± .005	8.414 ± .005	5.107 ± .005				259.07 ± .37	39.01 ± .06	.9323 ± .0014	3.785 ± .005	26	246
247	Shortite $Na_2Ca_2(CO_3)_3$	orth.	Amm2(38)		2	4.961 ± .005	11.03 ± .02	7.12 ± .01				389.6 ± 1.0	117.3 ± .3	2.804 ± .007	2.610 ± .007	r	247
248	Malachite $Cu_2(OH)_2CO_3$	mon.	P2$_1$/a(14)		4	9.502 ± .007	11.974 ± .007	3.240 ± .003		98.75 ± .25		364.35 ± .54	54.86 ± .08	1.311 ± .002	4.030 ± .006	25	248

X-Ray Crystallographic Data of Minerals

	Name and formula	Crystal system	Space group	Structure type	Z	a_o	b_o	c_o	α_o	β_o	γ_o	Cell volume 10^{-24} cm³	Molar volume cm³	Molar volume cal bar⁻¹	X-Ray density grams cm⁻³	Temp. °C	
249	Azurite $Cu_3(OH)_2(CO_3)_2$	mon	$P2_1/a$(14)		2	5.008 ±.005	5.844 ±.005	10.336 ±.005		92.45 ±.25		302.22 ± .43	91.01 ± .13	2.1752 ±.0031	3.787 ±.005	25	249
250	Niter KNO_3	orth.	Pnam(62)	aragonite	4	5.431 ±.005	9.164 ±.005	5.414 ±.005				319.07 ± .42	48.04 ± .06	1.148 ±.002	2.105 ±.003	26	250
251	Soda Niter $NaNO_3$	hex-R	$R\bar{3}c$(167)	calcite	6	5.0696 ±.0010		16.829 ±.005				374.57 ± .19	37.598 ± .019	.89866 ±.00049	2.2606 ±.0011	25	251
252	Gerhardite $Cu_2(NO_3)(OH)_3$	orth.	$P2_12_12_1$(19)		4	6.075 ±.004	13.812 ±.008	5.592 ±.004				469.21 ± .53	70.65 ± .08	1.689 ±.002	3.399 ±.004	r	252
	Sulfates and borates																
253	Barite $BaSO_4$	orth.	Pnma(62)	barite	4	8.878 ±.005	5.450 ±.005	7.152 ±.003				346.05 ± .40	52.10 ± .06	1.245 ±.002	4.480 ±.005	26	253
254	Anhydrite $CaSO_4$	orth.	Amma(63) Ccmm(63)	anhydrite	4	6.991 ±.005	6.996 ±.005	6.238 ±.005				305.09 ± .39	45.94 ± .06	1.098 ±.002	2.964 ±.004	26	254
255	Anglesite $PbSO_4$	orth.	Pnma(62)	barite	4	8.480 ±.005	5.398 ±.005	6.958 ±.003				318.50 ± .38	47.95 ± .06	1.146 ±.002	6.324 ±.008	25	255
256	Celestite $SrSO_4$	orth.	Pnma(62)	barite	4	8.359 ±.005	5.352 ±.005	6.866 ±.005				307.17 ± .41	46.25 ± .06	1.105 ±.002	3.972 ±.005	26	256
257	Zinkosite $ZnSO_4$	orth.	Pnma(62)	barite	4	8.588 ±.008	6.740 ±.006	4.770 ±.005				276.10 ± .46	41.57 ± .07	.9936 ±.0017	3.883 ±.006	25	257
258	Arcanite KS_2SO_4	orth.	Pnma(62)	arcanite	4	5.772 ±.005	10.072 ±.005	7.483 ±.004				435.03 ± .49	65.50 ± .07	1.566 ±.002	2.661 ±.003	25	258
259	Mascagnite $(NH_4)_2SO_4$	orth.	Pnma(62)	arcanite	4	7.782 ±.005	5.993 ±.005	10.636 ±.005				496.04 ± .57	74.68 ± .09	1.7851 ±.0021	1.7693 ±.0020	25	259
260	Thenardite Na_2SO_4	orth.	Fddd(70)	thenardite	8	5.863 ±.005	12.304 ±.005	9.821 ±.005				708.47 ± .76	53.33 ± .06	1.275 ±.002	2.663 ±.003	25	260
261	Gypsum $CaSO_4.2H_2O$*	mon.	C2/c(15)		4	5.68 ±.01	15.18 ±.01	6.29 ±.01		113.83 ±.22		496.1 ± 1.5	74.69 ± .22	1.785 ±.005	2.305 ±.007	r	261
262	Epsomite $MgSO_4.7H_2O$	orth.	$P2_12_12_1$(19)		4	11.86 ±.01	11.99 ±.01	6.858 ±.007				975.22 ± 1.53	146.83 ± .23	3.5094 ±.0055	1.679 ±.003	25	262
263	Goslarite $ZnSO_4.7H_2O$	orth.	$P2_12_12_1$(19)	epsomite	4	11.779 ±.005	12.050 ±.005	6.822 ±.003				968.29 ± .72	145.79 ± .11	3.4845 ±.0026	1.9723 ±.0015	25	263
264	Mirabilite $Na_2SO_4.10H_2O$	mon.	$P2_1/c$(14)		4	11.51 ±.01	10.38 ±.01	12.83 ±.01		107.75 ±.17		1459.9 ± 2.6	219.8 ± .4	5.253 ±.009	1.466 ±.003	24	264
265	Chalcanthite $CuSO_4.5H_2O$	tri.	$P\bar{1}$(2)		2	6.1045 ±.0050	10.72 ±.01	5.949 ±.007	97.57 ±.17	107.28 ±.17	77.43 ±.17	361.88 ± .72	108.97 ± .22	2.6045 ±.0052	2.2912 ±.0046	r	265
266	Brochantite $Cu_4SO_4(OH)_6$*	mon.	$P2_1/c$(14)		4	13.066 ±.010	9.85 ±.01	6.022 ±.010		103.27 ±.25		754.3 ± 1.8	113.6 ± .2	2.715 ±.006	3.982 ±.009	r	266
267	Syngenite $K_2Ca(SO_4)_2.H_2O$	mon.	$P2_1/m$(11)		2	9.775 ±.005	7.156 ±.005	6.251 ±.005		104.00 ±.25		424.27 ± .68	127.76 ± .20	3.0535 ±.0049	2.5707 ±.0041	r	267
268	Alunite $KAl_3(SO_4)_2(OH)_6$	hex-R	R3m(160)		3	6.982 ±.005		17.32 ±.01				731.2 ± 1.1	146.8 ± .2	3.508 ±.005	2.822 ±.004	r	268
269	Natroalunite $NaAl_3(SO_4)_2(OH)_6$	hex-R	R3m(160)		3	6.974 ±.005		16.69 ±.01				702.99 ± 1.09	141.1 ± .2	3.373 ±.005	2.821 ±.004	r	269
270	Hexahydrite $MgSO_4.6H_2O$	mon.	C2/c(15)		8	10.110 ±.005	7.212 ±.004	24.41 ±.01		98.30 ±.10		1731.2 ± 1.6	132.58 ± .12	3.1689 ±.0029	1.7232 ±.0015	r	270
271	Leonhardtite $MgSO_4.4H_2O$	mon.	$P2_1/n$(14)		4	5.922 ±.006	13.604 ±.004	7.905 ±.005		90.85 ±.20		636.78 ± .78	95.88 ± .12	2.2915 ±.0029	2.0071 ±.0025	r	271
272	Melanterite $FeSO_4.7H_2O$	mon.	$P2_1/c$(14)		4	14.072 ±.010	6.503 ±.007	11.041 ±.010		105.57 ±.15		973.29 ± 1.69	146.54 ± .25	3.5025 ±.0061	1.8972 ±.0033	r	272
273	Vanthoffite $MgSO_4.3Na_2SO_4$	mon.	$P2_1/c$(14)		2	9.797 ±.003	9.217 ±.003	8.199 ±.003		113.50 ±.10		678.96 ± .65	204.45 ± .20	4.8866 ±.0047	2.6730 ±.0025	r	273

X-Ray Crystallographic Data of Minerals

No.	Name and formula	Crystal system	Space group	Structure type	Z	a_o	b_o	c_o	α_o	β_o	γ_o	Cell volume 10^{-24} cm^3	Molar volume cm^3	Molar volume cal bar^{-1}	X-Ray density grams cm^{-3}	Temp. °C	No.
274	Dolerophanite $Cu_2O(SO_4)$	mon.	C2/m(15)		4	9.355 ±.010	6.312 ±.005	7.628 ±.005		122.29 ±.10		380.77 ± .70	57.33 ± .11	1.3703 ±.0026	4.171 ±.008	r	274
275	Retgersite $NiSO_4.4H_2O$	tet.	$P4_12_12$(92) $P4_32_1$(96)		4	6.782 ±.004		18.28 ±.01				840.80 ± 1.09	126.59 ± .16	3.0257 ±.0040	2.076 ±.003	25	275
276	Colemanite $CaB_3O_4(OH)_3.H_2O$*	mon.	$P2_1$/a(14)		4	8.743 ±.004	11.264 ±.002	6.102 ±.003		110.12 ±.08		564.26 ± .49	84.957 ± .073	2.0306 ±.0018	2.4194 ±.0021	r	276
277	Borax $Na_2B_4O_7.10H_2O$	mon.	C2/c(15)		4	11.858 ±.005	10.674 ±.005	12.197 ±.005		106.68 ±.03		1478.8 ± 1.1	222.66 ± .17	5.3217 ±.0041	1.7128 ±.0013	r	277
278	Kernite $Na_2B_4O_7.4H_2O$	mon.	$P2_1$/c(14)		4	7.022 ±.003	9.151 ±.004	15.676 ±.008		108.83 ±.25		953.40 ± 1.61	143.55 ± .24	3.4309 ±.0058	1.9038 ±.0032	r	278
279	Hambergite $Be_2BO_3.(OH,F)$*	orth.	Pbca(61)		8	9.755 ±.001	12.201 ±.001	4.426 ±.001				526.79 ± .14	39.658 ± .011	.9479 ±.0003	2.3663 ±.0006	r	279
	Phosphates, molybdates, and tungstates																
280	Berlinite $AlPO_4$	hex.	$P3_121$(152) $P3_221$(154)	α-quartz	3	4.942 ±.005		10.97 ±.007				232.03 ± .50	46.58 ± .10	1.113 ±.002	2.618 ±.006	25	280
281	Xenotime YPO_4	tet.	$I4_1$/amd(141)	zircon	4	6.885 ±.005		5.982 ±.005				283.57 ± .48	42.69 ± .07	1.020 ±.002	4.307 ±.008	26	281
282	Hydroxylapatite $Ca_5(PO_4)_3OH$	hex.	$P6_3$/m(176)	apatite	2	9.418 ±.003		6.883 ±.003				528.7 ± .5	159.2 ± .2	3.805 ±.004	3.155 ±.004	r	282
283	Fluorapatite $Ca_5(PO_4)_3F$	hex.	$P6_3$/m(176)	apatite	2	9.3684 ±.0030		6.8841 ±.0030				523.25 ± .41	157.56 ± .12	3.7659 ±.0030	3.2007 ±.0025	25	283
284	Chlorapatite $Ca_5(PO_4)_3Cl$	hex.	$P6/_3m$(176)	apatite	2	9.629 ±.005		6.777 ±.003				544.16 ± .61	163.86 ± .19	3.916 ±.004	3.178 ±.004	r	284
285	Carbonate-apatite $Ca_{10}(PO_4)_6CO_3H_2O$	hex.	$P6_3$/m(176)	apatite	1	9.436 ±.010		6.883 ±.010				530.74 ± 1.36	319.6 ± .8	7.640 ±.020	3.281 ±.008	r	285
286	Turquois $CuAl_6(PO_4)_4(OH)_8.4H_2O$*	tri.	$P\bar{1}$(2)		1	7.424 ±.008	7.629 ±.008	9.910 ±.010	68.61 ±.20	69.71 ±.20	65.08 ±.20	461.40 ± 1.12	277.9 ± .7	6.6416 ±.0162	2.927 ±.007	r	286
287	Powellite $CaMoO_4$	tet.	$I4_1$/a(100)	scheelite	4	5.226 ±.005		11.43 ±.007				312.17 ± .63	47.00 ± .09	1.1234 ±.0023	4.256 ±.009	25	287
288	Wulfenite $PbMoO_4$	tet.	$I4_1$/a(100)	scheelite	4	5.435 ±.005		12.110 ±.007				357.72 ± .69	53.859 ± .104	1.2873 ±.0025	6.816 ±.013	25	288
289	Scheelite $CaWO_4$	tet.	$I4_1$/a(100)	scheelite	4	5.242 ±.005		11.372 ±.005				312.49 ± .61	47.049 ± .092	1.1245 ±.0023	6.120 ±.012	25	289
290	Stolzite $PbWO_4$	tet.	$I4_1$/a(100)	scheelite	4	5.4616 ±.0030		12.046 ±.005				359.32 ± .42	54.100 ± .064	1.2931 ±.0016	8.4110 ±.0099	25	290
291	Ferberite $FeWO_4$	mon.	P2/c(13)	wolframite	2	4.732 ±.004	5.708 ±.003	4.965 ±.004		90.00 ±.05		134.11 ± .17	40.38 ± .05	.9652 ±.0013	7.520 ±.010	r	291
292	Huebnerite $MnWO_4$	mon.	P2/c(13)	wolframite	2	4.834 ±.004	5.758 ±.005	4.999 ±.004		91.18 ±.10		139.11 ± .20	41.89 ± .06	1.001 ±.002	7.228 ±.010	r	292
293	Wolframite $Fe_{.5}Mn_{.5}WO_4$	mon.	P2/c(13)	wolframite	2	4.782 ±.004	5.731 ±.004	4.982 ±.004		90.57 ±.10		136.53 ± .18	41.11 ± .06	.9826 ±.0014	7.376 ±.010	r	293
294	Sanmartinite $ZnWO_4$	mon.	P2/c(13)	wolframite	2	4.691 ±.003	5.720 ±.003	4 925 ±.003		89.36 ±.20		132.14 ± .14	39.79 ± .04	.9511 ±.0010	7.872 ±.008	25	294
	Ortho and ring structure silicates																
295	Forsterite Mg_2SiO_4	orth.	Pbnm(62)	olivine	4	4.758 ±.002	10.214 ±.003	5.984 ±.002				290.81 ± .18	43.786 ± .027	1.0465 ±.0007	3.2136 ±.0020	25	295
296	Fayalite Fe_2SiO_4	orth.	Pbnm(62)	olivine	4	4.817 ±.005	10.477 ±.005	6.105 ±.010				308.11 ± .62	46.389 ± .093	1.1088 ±.0023	4.3928 ±.0088	r	296
297	Tephroite Mn_2SiO_4*	orth.	Pbnm(62)	olivine	4	4.871 ±.005	10.636 ±.005	6.232 ±.005				322.87 ± .45	48.612 ± .067	1.1619 ±.0017	4.1545 ±.0058	r	297

	Name and formula	Crystal system	Space group	Structure type	Z	a_o	b_o	c_o	α_o	β_c	γ_o	Cell volume 10^{-24} cm³	Molar volume cm³	Molar volume cal bar⁻¹	X-Ray density grams cm⁻³	Temp. °C	
298	Lime Olivine γCa_2SiO_4	orth.	Pbnm(62)	olivine	4	5.091 ±.010	11.371 ±.020	6.782 ±.010				392.61 ± 1.19	59.11 ± .18	1.4129 ±.0043	2.914 ±.009	r	298
299	Nickel Olivine Ni_2SiO_4	orth.	Pbnm(62)	olivine	4	4.727 ±.002	10.121 ±.005	5.915 ±.002				282.98 ± .21	42.61 ± .03	1.0184 ±.0008	4.917 ±.004	r	299
300	Cobalt Olivine Co_2SiO_4	orth.	Pbnm(62)	olivine	4	4.782 ±.002	10.301 ±.005	6.003 ±.002				295.70 ± .21	44.52 ± .03	1.0642 ±.0008	4.716 ±.003	r	300
301	Monticellite $CaMgSiO_4$	orth.	Pbnm(62)	olivine	4	4.827 ±.005	11.084 ±.005	6.376 ±.005				341.13 ± .47	51.362 ± .071	1.2276 ±.0017	3.046 ±.004	r	301
302	Kerschsteinite $CaFeSiO_4$	orth.	Pbnm(62)	olivine	4	4.886 ±.005	11.146 ±.005	6.434 ±.010				350.39 ± .67	52.756 ± .101	1.2609 ±.0025	3.564 ±.007	r	302
303	Knebelite $MnFeSiO_4$*	orth.	Pbnm(62)	olivine	4	4.854 ±.010	10.602 ±.010	6.162 ±.010				317.11 ± .88	47.74 ± .13	1.1412 ±.0032	4.249 ±.012	r	303
304	Glauchroite $CaMnSiO_4$	orth.	Pbnm(62)	olivine	4	4.944 ±.004	11.19 ±.01	6.529 ±.005				361.2 ± .9	54.38 ± .14	1.2997 ±.0032	3.441 ±.009	r	304
305	Fluor-Norbergite $Mg_2SiO_4.MgF_2$	orth.	Pnmb(62)		4	8.727 ±.005	10.271 ±.010	4.709 ±.002				422.09 ± .51	63.551 ± .077	1.5190 ±.0019	3.194 ±.004	25	305
306	Chondrodite $2Mg_2SiO_4.MgF_2$*	mon.	$P2_1/c(14)$		2	7.89 ±.03	4.743 ±.020	10.29 ±.03		109.03 ±.30		364.0 ± 2.4	109.6 ± .7	2.620 ±.017	3.136 ±.021	r	306
307	Fluor-Humite $3Mg_2SiO_4.MgF_2$	orth.	Pnma(62)		4	10.243 ±.005	20.72 ±.02	4.735 ±.002				1004.9 ± 1.2	151.31 ± .18	3.6163 ±.0042	3.2017 ±.0037	25	307
308	Clinohumite $4Mg_2SiO_4.MgF_2$*	mon.	$P2_1/c(14)$		2	13.68 ±.04	4.75 ±.02	10.27 ±.02		100.83 ±.50		655.5 ± 3.8	197.4 ± 1.1	4.717 ±.027	3.167 ±.018	r	308
309	Grossularite $Ca_3Al_2Si_3O_{12}$	cubic	Ia3d(230)	garnet	8	11.851 ±.001						1664.43 ± .42	125.30 ± .03	2.9948 ±.0008	3.595 ±.001	25	309
310	Uvarovite $Ca_3Cr_2Si_3O_{12}$	cubic	Ia3d(230)	garnet	8	11.999 ±.002						1727.57 ± .86	130.05 ± .07	3.1084 ±.0016	3.848 ±.002	26	310
311	Andradite $Ca_3Fe_2Si_3O_{12}$	cubic	Ia3d(230)	garnet	8	12.048 ±.001						1748.82 ± .44	131.65 ± .03	3.1466 ±.0008	3.860 ±.001	25	311
312	Goldmanite $Ca_3V_2Si_3O_{12}$	cubic	Ia3d(230)	garnet	8	12.070 ±.005						1758.42 2.19	132.38 ± .16	3.1639 ±.0040	3.765 ±.005	r	312
313	Almandite $Fe_3Al_2Si_3O_{12}$	cubic	Ia3d(230)	garnet	8	11.526 ±.001						1531.21 ± .40	115.27 ± .04	2.7551 ±.0008	4.318 ±.001	25	313
314	Pyrope $Mg_3Al_2Si_3O_{12}$	cubic	Ia3d(230)	garnet	8	11.459 ±.001						1504.67 ± .39	113.27 ± .03	2.7074 ±.0008	3.559 ±.001	25	314
315	Spessartite $Mn_3Al_2Si_3O_{12}$	cubic	Ia3d(230)	garnet	8	11.621 ±.001						1569.39 ± .41	118.15 ± .03	2.8238 ±.0008	4.190 ±.001	25	315
316	Zircon $ZrSiO_4$*	tet.	I4/amd(141)	zircon	4	6.604 ±.005		5.979 ±.005				260.76 ± .45	39.261 ± .068	.9384 ±.0017	4.669 ±.008	25	316
317	Thorite $ThSiO_4$	tet.	I4/amd(141)	zircon	4	7.143 ±.004		6.327 ±.003				322.82 ± .39	48.60 ± .06	1.1617 ±.0015	6.668 ±.008	r	317
318	Coffinite $USiO_4$	tet.	I4/amd(141)	zircon	4	6.995 ±.004		6.263 ±.005				306.45 ± .43	46.140 ± .064	1.103 ±.002	7.155 ±.010	r	318
319	Kyanite Al_2SiO_5*	tri.	$P\bar{1}(2)$		4	7.123 ±.001	7.848 ±.002	5.564 ±.008	89.92 ±.15	101.25 ±.08	105.97 ±.08	292.83 ± .45	44.09 ± .07	1.054 ±.002	3.675 ±.006	25	319
320	Andalusite Al_2SiO_5*	orth.	Pnnm(58)		4	7.7959 ±.0050	7.8983 ±.0020	5.5583 ±.0020				342.25 ± .27	51.530 ± .040	1.2316 ±.0010	3.145 ±.002	25	320
321	Sillimanite Al_2SiO_5*	orth.	Pbnm(62) Pnma(62)		4	7.4843 ±.0030	7.6730 ±.0030	5.7711 ±.0040				331.42 ± .30	49.899 ± .044	1.1927 ±.0011	3.248 ±.003	25	321
322	3.2 Mullite $3AL_2O_3.2SiO_2$	orth.			3/4	7.557 ±.002	7.6876 ±.0020	2.8842 ±.0010				167.56 ± .09	134.55 ± .07	3.2159 ±.0016	3.166 ±.002	r	322
323	2.1 Mullite $2Al_2O_3.SiO_2$	orth.	Pbam(55)		6/5	7.5788 ±.0020	7.6909 ±.0020	2.8883 ±.0010				168.35 ± .09	84.492 ± .043	2.0195 ±.0011	3.125 ±.002	r	323

X-Ray Crystallographic Data of Minerals

	Name and formula	Crystal system	Space group	Structure type	Z	a_o	b_o	c_o	α_o	β_o	γ_o	Cell volume 10^{-24} cm³	Molar volume cm³	Molar volume cal bar⁻¹	X-Ray density grams cm⁻³	Temp. °C	
324	Staurolite $Fe_2Al_9Si_4O_{22}(OH)_2$*	mon.	C2/m(15)		2	7.90 ±.10	16.65 ±.15	5.63 ±.10		90.00 ±.25		740.5 ±17.5	223.0 ± 5.3	5.330 ±.126	3.825 ±.090	r	324
325	Topaz $Al_2(SiO_4)(OH)$*	orth.	Pmnb(62)		4	8.394 ±.005	8.792 ±.007	4.649 ±.003				343.10 ± .41	51.66 ± .06	1.2347 ±.0015	3.563 ±.005	26	325
326	Phenacite Be_2SiO_4*	hex-R	R$\bar{3}$(148)	phenacite	18	12.472 ±.005		8.252 ±.005				1111.6 ± 1.1	37.194 ± .037	.8890 ±.0009	2.960 ±.003	25	326
327	Willemite Zn_2SiO_4	hex-R	R$\bar{3}$(148)	phenacite	18	13.94 ±.01		9.309 ±.003				1566.6 ± 2.3	52.42 ± .08	1.253 ±.002	4.251 ±.006	25	327
328	Dioptase CuH_2SiO_4*	hex-R	R$\bar{3}$(148)	phenacite	18	14.61 ±.02		7.80 ±.01				1441.9 ± 4.4	48.24 ± .15	1.153 ±.004	3.247 ±.010	r	328
329	Larnite β-Ca_2SiO_4*	mon.	$P2_1/n$(14)		4	5.48 ±.02	6.76 ±.02	9.28 ±.02		94.55 ±.33		342.7 ± 1.8	51.60 ± .27	1.233 ±.006	3.338 ±.017	r	329
330	Akermanite $Ca_2MgSi_2O_7$	tet.	P$\bar{4}2_1$m(113)	melilite	2	7.8435 ±.0030		5.010 ±.003				308.22 ± .30	92.812 ± .090	2.2183 ±.0022	2.9375 ±.0029	r	330
331	Gehlenite $Ca_2Al_2SiO_7$	tet.	P$\bar{4}2_1$m(113)	melilite	2	7.690 ±.003		5.0675 ±.0030				299.67 ± .29	90.239 ± .088	2.1568 ±.0022	3.0387 ±.0030	r	331
332	Fe-Gehlenite $Ca_2Fe_2SiO_7$	tet.	P$\bar{4}2_1$m(113)	melilite	2	7.54 ±.01		4.855 ±.005				276.01 ± .79	83.12 ± .24	1.9865 ±.0057	3.994 ±.011	r	332
333	Hardystonite $Ca_2ZnSi_2O_7$*	tet.	P$\bar{4}2_1$m(113)	melilite	2	7.87 ±.03		5.01 ±.02				310.3 ± 2.7	93.44 ± .80	2.233 ±.019	3.357 ±.029	r	333
334	Sodium Melilite $NaCaAlSi_2O_7$	tet.	P$\bar{4}2_1$m(113)	melilite	2	8.511 ±.005		4.809 ±.003				348.35 ± .46	104.90 ± .14	2.507 ±.003	2.462 ±.003	r	334
335	Beryl $Be_3Al_2(Si_6O_{18})$*	hex.	P6/mmc(192)	beryl	2	9.215 ±.005		9.192 ±.005				675.98 ± .82	203.55 ± .25	4.8651 ±.0060	2.641 ±.003	25	335
336	Indialite high Cordierite $Mg_2Al_3(AlSi_5O_{18})$	hex.	P6/mmc(192)	beryl	2	9.7698 ±.0030		9.3517 ±.0030				773.02 ± .54	232.78 ± .16	5.5636 ±.0039	2.513 ±.002	25	336
337	Low Cordierite $Mg_2Al_3(AlSi_5O_{18})$	orth.	Cccm(66)		4	9.721 ±.003	17.062 ±.006	9.339 ±.003				1548.96 ± .88	233.22 ± .13	5.5741 ±.0032	2.508 ±.001	25	337
338	Fe-Indialite $Fe_2Al_3(AlSi_5O_{18})$	hex.	P6/mmc(192)	beryl	2	9.860 ±.010		9.285 ±.010				781.75 ± 1.80	235.40 ± .54	5.6264 ±.0130	2.753 ±.006	r	338
339	Fe-Cordierite $Fe_2Al_3(AlSi_5O_{18})$	orth.	Cccm(66)	cordierite	4	9.726 ±.010	17.065 ±.010	9.287 ±.010				1541.40 ± 2.47	232.08 ± .37	5.5468 ±.0089	2.792 ±.005	r	339
340	Mn-Indialite $Mn_2Al_3(AlSi_5O_{18})$	hex.	P6/mmc(192)	beryl	2	9.925 ±.010		9.297 ±.010				793.11 ± .81	238.8 ± .5	5.708 ±.013	2.706 ±.006	r	340
341	Sapphirine $Mg_2Al_4O_6SiO_4$*	mon.	$P2_1/c$(14)		8	11.26 ±.03	14.46 ±.03	9.95 ±.02		125.33 ±.50		1321.7 ± 9.7	99.50 ± .73	2.378 ±.017	3.464 ±.025	r	341
342	Elbaite $NaLiAl_{7.67}B_3Si_6O_{27}(OH)_4$*	hex-R	R3m(160)	tourmaline	3	15.842 ±.010		7.009 ±.010				1523.4 ± 2.9	305.82 ± .58	7.3093 ±.0140	3.271 ±.006	r	342
343	Schorl $NaFe_3Al_6B_3Si_6O_{27}(OH)_4$*	hex-R	R3m(160)	tourmaline	3	16.032 ±.010		7.149 ±.010				1591.3 ± 3.0	319.45 ± .60	7.635 ±.014	3.297 ±.006	r	343
344	Dravite $NaMg_3Al_6B_3Si_6O_{27}(OH)_4$	hex-R	R3m(160)	tourmaline	3	15.942 ±.010		7.224 ±.010				1589.99 ± 2.97	319.19 ± .60	7.629 ±.014	3.004 ±.006	r	344
345	Uvite $CaMg_4Al_5B_3Si_6O_{27}(OH)_4$	hex-R	R3m(160)	tourmaline	3	15.86 ±.01		7.19 ±.01				1566.3 ± 2.9	314.4 ± .6	7.515 ±.014	3.095 ±.006	r	345
346	Sphene $CaTiSiO_5$*	mon.	A2/a(15)		4	7.07 ±.01	8.72 ±.01	6.56 ±.01		113.95 ±.25		369.61 ± 1.13	55.65 ± .17	1.330 ±.004	3.523 ±.011	r	346
347	Datolite $CaBSiO_4(OH)$*	mon.	$P2_1/c$(14)		4	9.62 ±.03	7.60 ±.03	4.84 ±.02		90.15 ±.25		353.9 ± 2.3	53.28 ± .35	1.273 ±.008	3.003 ±.020	r	347
348	Euclase $AlBeSiO_4(OH)$*	mon.	$P2_1/a$(14)		4	4.763 ±.005	14.29 ±.02	4.618 ±.005		100.25 ±.10		309.30 ± .64	46.57 ± .10	1.113 ±.002	3.116 ±.007	r	348
349	Chloritoid $H_2FeAl_2SiO_7$*	mon.	C2/c(15)		8	9.48 ±.01	5.48 ±.01	18.18 ±.01		101.77 ±.25		924.6 ± 2.2	69.61 ± .16	1.664 ±.004	3.619 ±.008	r	349

X-Ray Crystallographic Data of Minerals

	Name and formula	Crystal system	Space group	Structure type	Z	a_o	b_o	c_o	α_o	β_o	γ_o	Cell volume 10^{-24} cm³	Molar volume cm³	Molar volume cal bar⁻¹	X-Ray density grams cm⁻³	Temp. °C	
350	Hemimorphite $Zn_4(OH)_2Si_2O_7.H_2O$*	orth.	Imm2(35)		2	8.370 ±.005	10.719 ±.005	5.120 ±.005				459.36 ± .57	138.32 ± .17	3.306 ±.004	3.482 ±.004	25	350
351	Zoisite $Ca_2Al_3(SiO_4)_3OH$	orth.	Pnma(62)		4	16.15 ±.01	5.581 ±.005	10.06 ±.01				906.74 ± 1.34	136.52 ± .20	3.263 ±.005	3.328 ±.005	r	351
352	Clinozoisite $Ca_2Al_3(SiO_4)_3OH$	mon.	$P2_1/m(11)$		2	8.387 ±.007	5.581 ±.005	10.14 ±.01		115.93 ±.33		452.30 ± 1.45	136.20 ± .44	3.255 ±.010	3.336 ±.011	r	352
353	Epidote $Ca_2Al_{1.5}Fe_{1.5}(SiO_4)_3OH$*	mon.	$P2_1/m(11)$		2	8.89 ±.02	5.63 ±.01	10.19 ±.02		115.40 ±.30		460.72 ± 1.97	138.7 ± .6	3.316 ±.014	3.587 ±.015	r	353
354	Piemontite $Ca_2Al_{1.5}Mn_{1.5}(SiO_4)_3OH$*	mon.	$P2_1/m(11)$		2	8.95 ±.02	5.70 ±.01	9.41 ±.02		115.70 ±.50		432.56 ± 2.38	130.3 ± .7	3.113 ±.017	3.810 ±.021	r	354
355	Lawsonite $CaAl_2Si_2O_7(OH)_2.H_2O$	orth.	Cccm(63)		4	8.787 ±.005	5.836 ±.005	13.123 ±.008				672.96 ± .80	101.32 ± .12	2.4217 ±.0029	3.101 ±.004	r	355
	Chain and band structure silicates																
356	Enstatite $MgSiO_3$*	orth.	Pcab(61)		16	8.829 ±.010	18.22 ±.01	5.192 ±.005				835.21 ± 1.32	31.44 ± .05	.7514 ±.0012	3.194 ±.005	r	356
357	Clinoenstatite $MgSiO_3$	mon.	$P2_1/c(15)$		8	9.620 ±.005	8.825 ±.005	5.188 ±.005		108.33 ±.17		418.10 ± .66	31.47 ± .05	.7523 ±.0012	3.190 ±.005	r	357
358	Protoenstatite $MgSiO_3$	orth.	Pbcn(60)		8	9.25 ±.01	8.74 ±.01	5.32 ±.01				430.10 ± 1.05	32.38 ± .08	.7739 ±.0019	3.101 ±.008	r	358
359	High Clinoenstatite $MgSiO_3$	tri.			8	10.000 ±.005	8.934 ±.004	5.170 ±.003	88.27 ±.05	70.03 ±.04	91.01 ±.04	433.72 ± .40	32.65 ± .03	.7804 ±.0008	3.075 ±.003		359
360	Clinoferrosilite $FeSiO_3$	mon.	$P2_1/c(14)$		8	9.7085 ±.0010	9.0872 ±.0011	5.2284 ±.004		108.43 ±.05		437.60 ± .15	32.943 ± .011	.7874 ±.0003	4.005 ±.002	r	360
361	Orthoferrosilite $FeSiO_3$	orth.	Pcab(61)	enstatite	16	9.080 ±.002	18.431 ±.004	5.238 ±.001				876.6 ± .54	33.00 ± .02	.7887 ±.0008	3.998 ±.004	r	361
362	Diopside $CaMg(SiO_3)_2$	mon.	C2/c(15)	diopside	4	9.743 ±.005	8.923 ±.005	5.251 ±.003		105.93 ±.25		438.97 ± .69	66.09 ± .10	1.580 ±.003	3.277 ±.005	r	362
363	Hedenbergite $CaFe(SiO_3)_2$*	mon.	C2/c(15)	diopside	4	9.854 ±.010	9.024 ±.010	5.263 ±.010		104.23 ±.33		453.64 ± 1.28	68.30 ± .19	1.632 ±.005	3.632 ±.010	r	363
364	Johannsenite $CaMn(SiO_3)_2$*	mon.	C2/c(15)	diopside	4	9.83 ±.03	9.04 ±.03	5.27 ±.02		105.00 ±.50		452.35 ± 2.87	68.11 ± .43	1.628 ±.010	3.629 ±.023	r	364
365	Ureyite $NaCr(SiO_3)_2$	mon.	C2/c(15)	diopside	4	9.550 ±.016	8.712 ±.007	5.273 ±.008		107.44 ±.16		418.6 ± 1.1	63.02 ± .16	1.506 ±.004	3.605 ±.009	r	365
366	Jadeite $NaAl(SiO_3)_2$*	mon.	C2/c(15)	diopside	4	9.409 ±.005	8.564 ±.005	5.220 ±.005		107.50 ±.20		401.15 ± .67	60.40 ± .10	1.444 ±.002	3.347 ±.006	r	366
367	Acmite (Aegirine) $NaFe(SiO_3)_2$	mon.	C2/c(15)	diopside	4	9.658 ±.005	8.795 ±.005	5.294 ±.005		107.42 ±.20		429.06 ± .70	64.60 ± .11	1.544 ±.003	4.411 ±.007	r	367
368	Ca Tschermak Molecule $CaAl_2SiO_6$	mon.	C2/c(15)	diopside	4	9.615 ±.005	8.661 ±.005	5.272 ±.003		106.12 ±.20		421.77 ± .59	63.50 ± .09	1.518 ±.002	3.435 ±.005	r	368
369	Spodumene $LiAl(SiO_3)_2$	mon.	C2/c(15)	diopside	4	9.451 ±.002	8.387 ±.002	5.208 ±.001		110.07 ±.03		387.7 ± .1	58.37 ± .02	1.395 ±.001	3.188 ±.001	r	369
370	β-Spodumene $LiAl(SiO_3)_2$	tet.	$P4_32_12(96)$ $P4_12_12(92)$		4	7.5332 ±.0008		9.1540 ±.0008				519.48 ± .12	78.215 ± .018	1.8694 ±.0005	2.379 ±.001	r	370
371	Pectolite $Ca_2NaH(SiO_3)_3$*	tri.	$P\bar{1}(2)$		2	7.99 ±.01	7.04 ±.01	7.02 ±.01	90.05 ±.25	95.27 ±.25	102.47 ±.25	383.84 ± .99	115.58 ± .30	2.763 ±.007	2.876 ±.007	r	371
372	Wollastonite $CaSiO_3$*	tri.	$P\bar{1}(2)$		6	7.94 ±.01	7.32 ±.01	7.07 ±.01	90.03 ±.25	95.37 ±.25	103.43 ±.25	397.82 ± 1.03	39.93 ± .10	.9544 ±.0025	2.909 ±.008	r	372
373	Parawollastonite $CaSiO_3$*	mon.	$P2_1(4)$		12	15.417 ±.004	7.321 ±.002	7.066 ±.003		95.40 ±.10		793.98 ± .47	39.85 ± .02	.9524 ±.0006	2.915 ±.002	r	373
374	Pseudowollastonite $CaSiO_3$*	tri.			24	6.90 ±.02	11.78 ±.02	19.65 ±.02	90.00 ±.30	90.80 ±.30	90.00 ±.30	1597.0 ± 5.6	40.08 ± .14	.9579 ±.0034	2.899 ±.010	r	374

X-Ray Crystallographic Data of Minerals

	Name and formula	Crystal system	Space group	Structure type	Z	a_o	b_o	c_o	α_o	β_o	γ_o	Cell volume 10^{-24} cm^3	Molar volume cm^3	Molar volume cal bar^{-1}	X-Ray density grams cm^{-3}	Temp. °C	
375	Rhodonite $MnSiO_3$*	tri.	P$\bar{1}$(2)		10	7.682 ±.002	11.818 ±.003	6.707 ±.002	92.36 ±.05	93.95 ±.05	105.66 ±.05	583.77 ± .31	35.158 ± .019	.8403 ±.0005	3.727 ±.002	r	375
376	Bustamite $CaMn(SiO_3)_2$*	tri.	A$\bar{1}$(2)		6	7.736 ±.003	7.157 ±.003	13.824 ±.010	90.52 ±.25	94.58 ±.25	103.87 ±.25	740.38 ± 1.08	74.32 ± .11	1.776 ±.003	3.326 ±.005	r	376
377	Pyroxmangite $MnFe(SiO_3)_2$*	tri.	P$\bar{1}$(2)		7	7.56 ±.02	17.45 ±.05	6.67 ±.02	84.00 ±.30	94.30 ±.30	113.70 ±.30	800.77 ± 4.29	68.90 ± .36	1.647 ±.009	3.817 ±.020	r	377
378	Tremolite $Ca_2Mg_5[Si_8O_{22}](OH)_2$*	mon.	C2/m(12)	tremolite	2	9.840 ±.010	18.052 ±.020	5.275 ±.010		104.70 ±.25		906.34 ± 2.43	272.92 ± .73	6.523 ±.018	2.977 ±.008	r	378
379	Fluor-tremolite $Ca_2Mg_5[Si_8O_{22}]F_2$	mon.	C2/m(12)	tremolite	2	9.781 ±.007	18.01 ±.01	5.267 ±.005		104.52 ±.25		898.18 ± 1.56	270.46 ± .47	6.464 ±.011	3.018 ±.005	20	379
380	Ferrotremolite $Ca_2Fe_5[Si_8O_{22}](OH)_2$	mon.	C2/m(12)	tremolite	2	9.97 ±.01	18.34 ±.02	5.30 ±.01		104.50 ±.10		938.24 ± 2.92	282.53 ± .69	6.753 ±.017	3.434 ±.008	r	380
381	Grunerite $Fe_7[Si_8O_{22}](OH)_2$	mon.	C2/m(12)	tremolite	2	9.572 ±.005	18.44 ±.01	5.342 ±.007		101.77 ±.25		923.08 ± 1.63	277.96 ± .49	6.644 ±.012	3.603 ±.006	r	381
382	Cummingtonite (hypo.) $Mg_7[Si_8O_{22}](OH)_2$	mon.	C2/m(12)	tremolite	2	9.476 ±.010	17.935 ±.010	5.292 ±.005		102.23 ±.25		878.97 ± 1.58	264.68 ± .47	6.326 ±.011	2.950 ±.005	r	382
383	Riebeckite $Na_2Fe_3Fe_2[Si_8O_{22}](OH)_2$	mon.	C2/m(12)	tremolite	2	9.729 ±.020	18.065 ±.020	5.334 ±.010		103.31 ±.25		912.29 ± 2.89	274.71 ± .87	6.566 ±.021	3.407 ±.011	r	383
384	Magnesioriebeckite $Na_2Mg_3Fe_2[Si_8O_{22}](OH)_2$	mon.	C2/m(12)	tremolite	2	9.733 ±.010	17.946 ±.020	5.299 ±.010		103.30 ±.25		900.74 ± 2.37	271.24 ± .71	6.483 ±.017	3.102 ±.008	r	384
385	Gaucophane I $Na_2Mg_3Al_2[Si_8O_{22}](OH)_2$	mon.	C2/m(12)	tremolite	2	9.748 ±.010	17.915 ±.020	5.273 ±.010		102.78 ±.25		898.04 ± 2.35	270.42 ± .71	6.463 ±.017	2.898 ±.008	r	385
386	Glaucophane II $Na_2Mg_3Al_2[Si_8O_{22}](OH)_2$	mon.	C2/m(12)	tremolite	2	9.663 ±.010	17.696 ±.020	5.277 ±.010		103.67 ±.10		876.79 ± 2.17	264.02 ± .65	6.310 ±.016	2.968 ±.007	r	386
387	Fluor-edenite $NaCa_2Mg_5[AlSi_7O_{22}]F_2$	mon.	C2/m(12)	tremolite	2	9.847 ±.005	18.00 ±.01	5.282 ±.005		104.83 ±.25		905.03 ± 1.51	272.53 ± .46	6.514 ±.011	3.076 ±.005	r	387
388	Fluor-richterite $Na_2CaMg_5[Si_8O_{22}]F_2$	mon.	C2/m(12)	tremolite	2	9.823 ±.005	17.96 ±.01	5.268 ±.005		104.33 ±.25		900.47 ± 1.48	271.15 ± .45	6.481 ±.011	3.033 ±.005	r	388
389	Anthophyllite $Mg_7[Si_8O_{22}](OH)_2$	orth.	Pnma(62)		4	18.61 ±.02	18.01 ±.06	5.24 ±.01				1756.3 ± 7.0	264.4 ± 1.1	6.320 ±.025	2.953 ±.012	r	389
	Framework structure silicates																
390	Microcline $KAlSi_3O_8$	tri.	C$\bar{1}$(2)		4	8.582 ±.002	12.964 ±.005	7.222 ±.002	90.62 ±.10	115.92 ±.10	87.68 ±.10	722.06 ± .67	108.72 ± .10	2.5984 ±.0025	2.560 ±.002	r	390
391	High Sanidine $KAlSi_3O_8$	mon.	C2/m(12)		4	8.615 ±.002	13.031 ±.003	7.177 ±.002		115.98 ±.10		724.28 ± .69	109.05 ± .10	2.6064 ±.0025	2.552 ±.002	r	391
392	Orthoclase $KAlSi_3O_8$*	mon.	C2/m(12)		4	8.562 ±.003	12.996 ±.004	7.193 ±.003		116.02 ±.15		719.25 ± 1.02	108.29 ± .15	2.5883 ±.0037	2.570 ±.004	r	392
393	Fe-Sanidine $KFeSi_3O_8$	mon.	C2/m(12)		4	8.689 ±.008	13.12 ±.01	7.319 ±.007		116.10 ±.30		749.28 ± 2.24	112.81 ± .34	2.6964 ±.0081	2.723 ±.008	r	393
394	Fe-Microcline $KFeSi_3O_8$	tri.	C$\bar{1}$(2)		4	8.68 ±.01	13.10 ±.01	7.340 ±.007	90.75 ±.25	116.05 ±.25	86.23 ±.25	748.09 ± 1.92	112.63 ± .29	2.692 ±.007	2.727 ±.007	r	394
395	Low Albite $NaAlSi_3O_8$	tri.	C$\bar{1}$(2)		4	8.139 ±.002	12.788 ±.003	7.160 ±.002	94.27 ±.10	116.57 ±.10	87.68 ±.10	664.65 ± .60	100.07 ± .09	2.3918 ±.0022	2.620 ±.002	26	395
396	High Albite (Analbite) $NaAlSi_3O_8$	tri.	C$\bar{1}$(2)		4	8.160 ±.002	12.870 ±.003	7.106 ±.002	93.54 ±.10	116.36 ±.10	90.19 ±.10	667.00 ± .60	100.43 ± .09	2.4003 ±.0022	2.611 ±.002	r	396
397	Anorthite $CaAl_2Si_2O_8$	tri.	P$\bar{1}$(2)	primitive cell	8	8.177 ±.002	12.877 ±.003	14.169 ±.003	93.17 ±.02	115.85 ±.02	91.22 ±.02	1338.9 ± .6	100.79 ± .04	2.4090 ±.0011	2.760 ±.001	r	397
398	Synthetic $CaAl_2Si_2O_8$	hex.	P6_3/mcm(193)		2	5.10 ±.02		14.72 ±.02				331.57 ± 2.64	99.85 ± .79	2.386 ±.019	2.786 ±.022	r	398
399	Synthetic $CaAl_2Si_2O_8$	orth.	P2_12_12(18)		2	8.22 ±.02	8.60 ±.02	4.83 ±.01				341.44 ± 1.35	102.82 ± .41	2.457 ±.010	2.706 ±.011	r	399

X-Ray Crystallographic Data of Minerals

	Name and formula	Crystal system	Space group	Structure type	Z	a_0	b_0	c_0	α_0	β_0	γ_0	Cell volume 10^{-24} cm³	Molar volume cm³	Molar volume cal bar⁻¹	X-Ray density grams cm⁻³	Temp. °C	
400	Celsian $BaAl_2Si_2O_8$*	mon.	$I2_1/c(15)$		8	8.627 ±.010	13.045 ±.010	14.408 ±.020		115.20 ±.25		1467.1 ± 4.2	110.45 ± .31	2.640 ±.008	3.400 ±.010	r	400
401	Paracelsian $BaAl_2Si_2O_8$*	mon.	$P2_1/a(14)$		4	8.58 ±.02	9.583 ±.020	9.08 ±.02		90.00 ±.50		746.6 ± 2.9	112.4 ± .4	2.687 ±.010	3.340 ±.013	r	401
402	Banalsite $BaNa_2Al_4Si_4O_{16}$*	orth.			4	8.50 ±.02	9.97 ±.02	16.72 ±.03				1416.9 ± 5.1	213.3 ± .8	5.099 ±.018	3.092 ±.011	r	402
403	Danburite $CaB_2Si_2O_8$*	orth.	Pnam(62)		4	8.04 ±.02	8.77 ±.02	7.74 ±.02				545.8 ± 2.3	82.17 ± .35	1.964 ±.008	2.992 ±.013	r	403
404	Low Nepheline $NaAlSiO_4$	hex.	$C6_3(178)$		8	9.986 ±.005		8.330 ±.004				719.38 ± .80	54.16 ± .06	1.294 ±.002	2.623 ±.003	r	404
405	High Carnegeite $NaAlSiO_4$	cubic			4	7.325 ±.004						393.03 ± .64	59.18 ± .10	1.414 ±.002	2.401 ±.004	750	405
406	Kaliophilite natural $KAlSiO_4$*	hex.	$P6_322(182)$		54	26.930 ±.010		8.522 ±.004				5352.4 ± 4.7	59.69 ± .05	1.427 ±.001	2.650 ±.002	r	406
407	Kaliophilite synthetic $KAlSiO_4$	hex.	$P6_3(173)$ $P6_322(182)$		2	5.180 ±.002		8.559 ±.004				198.89 ± .18	59.89 ± .05	1.431 ±.001	2.641 ±.002	r	407
408	Kalsilite $KAlSiO_4$	hex.	$P6_3(173)$		2	5.1597 ±.0020		8.7032 ±.0030				200.66 ± .17	60.424 ± .051	1.4442 ±.0031	2.618 ±.002	r	408
409	Leucite $KAlSi_2O_6$	tet.	$I4_1/a(100)$		16	13.074 ±.003		13.738 ±.003				2348.23 ± 1.19	88.389 ± .045	2.1126 ±.0011	2.469 ±.001	25	409
410	High Leucite $KAlSi_2O_6$*	cubic	Ia3d(230)		16	13.43 ±.05						2422.3 ±27.1	91.18 ± 1.02	2.179 ±.024	2.394 ±.027	625	410
411	Fe-Leucite $KFeSi_2O_6$	tet.	$I4_1/a(100)$		16	13.205 ±.002		13.970 ±.003				2435.98 ± .91	91.692 ± .034	2.1915 ±.0009	2.695 ±.001	25	411
412	Petalite $LiAlSi_4O_{10}$*	mon.	$P2_1/n(14)$		2	11.32 ±.03	5.14 ±.01	7.62 ±.01		105.90 ±.20		426.41 ± 1.57	128.4 ± .5	3.069 ±.011	2.385 ±.009	r	412
413	Marialite $Na_4Al_3Si_9O_{24}Cl$	tet.	I4/m(87) P4/m(83)		2	12.064 ±.008		7.514 ±.004				1093.6 ± 1.6	329.3 ± .5	7.871 ±.011	2.566 ±.004	r	413
414	Meionite $Ca_4Al_6Si_6O_{24}CO_3$	tet.	I4/m(87) P4/m(83)		2	12.174 ±.008		7.652 ±.015				1134.07 ± 2.68	341.5 ± .8	8.162 ±.019	2.737 ±.007	r	414
	Sheet structure silicates																
415	Muscovite $KAl_2[AlSi_3O_{10}](OH)_2$*	mon.	C2/c(15)	$2M_2$ mica	4	5.203 ±.005	8.995 ±.005	20.030 ±.010		94.47 ±.33		934.57 ± 1.21	140.71 ± .18	3.363 ±.004	2.831 ±.004	r	415
416	Paragonite $NaAl_2[AlSi_3O_{10}](OH)_2$*	mon.	C2/c(15)	$2M_1$ mica	4	5.13 ±.03	8.89 ±.05	19.32 ±.10		95.17 ±.50		877.52 ± 8.47	132.1 ± 1.3	3.158 ±.031	2.893 ±.028	r	416
417	Lepidolite $K_2Al_3Li_2[AlSi_7O_{20}](OH)_4$*	mon.	C2/c(15)	$2M_2$ mica	2	9.2 ±.1	5.3 ±.1	20.0 ±.2		98.00 ±.50		965.7 ±23.2	290.8 ± 7.0	6.950 ±.167	2.698 ±.065	r	417
418	Phlogopite $KMg_3[AlSi_3O_{10}](OH)_2$	mon.	Cm(8)	1M mica	2	5.326 ±.010	9.210 ±.010	10.311 ±.010		100.17 ±.10		497.83 ± 1.19	149.91 ± .36	3.5830 ±.0086	2.784 ±.007	r	418
419	Fluor-phlogopite $KMg_3[AlSi_3O_{10}]F_2$	mon.	Cm(8)	1M mica	2	5.299 ±.005	9.188 ±.005	10.135 ±.005		99.92 ±.10		486.07 ± .60	146.37 ± .18	3.498 ±.004	2.878 ±.004	r	419
420	Annite $KFe_3[AlSi_3O_{10}](OH)_2$	mon.	Cm(8)	1M mica	2	5.391 ±.010	9.350 ±.005	10.313 ±.020		99.70 ±.25		512.40 ± 1.45	154.30 ± .44	3.688 ±.010	3.318 ±.009	r	420
421	Ferriannite $KFe_3[FeSi_3O_{10}](OH)_2$	mon.	C2/m(12)		2	5.430 ±.002	9.404 ±.003	10.341 ±.006		100.07 ±.20		519.92 ± .51	156.56 ± .15	3.7419 ±.0037	3.454 ±.003	r	421
422	Margarite $CaAl_2[Al_2Si_2O_{10}](OH)_2$*	mon.	C2/c(15)	2M mica	4	5.13 ±.02	8.92 ±.03	19.50 ±.05		95.00 ±.50		888.9 ± 5.2	133.8 ± .8	3.199 ±.019	2.975 ±.017	r	422
423	Talc $Mg_3Si_4O_{10}(OH)_2$*	mon.	C2/c(15)	$2M_1$	4	5.287 ±.007	9.158 ±.010	18.95 ±.01		99.50 ±.20		904.94 ± 1.71	136.25 ± .26	3.2565 ±.0062	2.784 ±.005	r	423
424	Pyrophyllite $Al_2Si_4O_{10}(OH)_2$*	mon.	C2/c(15)	$2M_1$	4	5.14 ±.02	8.90 ±.02	18.55 ±.03		99.92 ±.20		835.9 ± 4.0	125.9 ± .6	3.008 ±.015	2.863 ±.014	r	424

X-Ray Crystallographic Data of Minerals

	Name and formula	Crystal system	Space group	Structure type	Z	a_o	b_o	c_o	α_o	β_o	γ_o	Cell volume 10^{-24} cm^3	Molar volume cm^3	Molar volume cal bar^{-1}	X-Ray density grams cm^{-3}	Temp. °C	
425	Minnesotaite $Fe_3Si_4O_{10}(OH)_2$*	mon.	C2/c(15)		4	5.4 ±.1	9.42 ±.04	19.4 ±.1		100.00 ±.50		971.8 ±19.2	146.3 ± 2.9	3.497 ±.069	3.239 ±.064	r	425
426	Dickite $Al_2Si_2O_5(OH)_4$*	mon.	Cc(9)		4	5.150 ±.002	8.940 ±.003	14.736 ±.005		103.58 ±.10		659.49 ± .49	99.30 ± .07	2.3733 ±.0018	2.600 ±.002	r	426
427	Kaolinite $Al_2Si_2O_5(OH)_4$*	tri.	P1(1)		2	5.155 ±.007	8.959 ±.010	7.407 ±.008	91.68 ±.35	104.87 ±.35	89.93 ±.35	330.48 ± .86	99.52 ± .26	2.3785 ±.0062	2.594 ±.007	r	427
428	Nacrite $Al_2Si_2O_5(OH)_4$*	mon.	Cc(9)		4	8.909 ±.010	5.146 ±.010	15.697 ±.020		113.70 ±.25		658.9 ± 2.1	99.21 ± .32	2.3713 ±.0076	2.602 ±.008	r	428
	Zeolites																
429	Analcite $NaAlSi_2O_6.H_2O$	cubic	Ia3d(230)		16	13.733 ±.005						2589.98 ± 2.83	97.49 ± .11	2.3301 ±.0026	2.258 ±.003	r	429
430	Natrolite $Na_2Al_2Si_3O_{10}.2H_2O$*	orth.	Fdd2(43)		8	18.30 ±.02	18.63 ±.02	6.60 ±.01				2250.1 ± 4.8	169.39 ± .37	4.049 ±.009	2.245 ±.005	r	430

Section 5
Thermochemistry, Electrochemistry, and Kinetics

CODATA KEY VALUES FOR THERMODYNAMICS

The Committee on Data for Science and Technology (CODATA) has conducted a project to establish internationally agreed values for the ermodynamic properties of key chemical substances. This table presents the final results of the project. Use of these recommended, internally nsistent values is encouraged in the analysis of thermodynamic measurements, data reduction, and preparation of other thermodynamic tables.

The table includes the standard enthalpy of formation at 298.15 K, the entropy at 298.15 K, and the quantity $H°$ (298.15 K)–$H°$ (0). A value of n the $\Delta_f H°$ column for an element indicates the reference state for that element. The standard state pressure is 100000 Pa (1 bar). See the reference information on the dependence of gas-phase entropy on the choice of standard state pressure.

Substances are listed in alphabetical order of their chemical formulas when written in the most common form.

The table is reprinted with permission of CODATA.

REFERENCE

x, J. D., Wagman, D. D., and Medvedev, V. A., *CODATA Key Values for Thermodynamics,* Hemisphere Publishing Corp., New York, 1989.

Substance	State	$\Delta_f H°$ (298.15 K) kJ·mol^{-1}	$S°$ (298.15 K) J·K^{-1}·mol^{-1}	$H°$ (298.15 K)–$H°$ (0) kJ·mol^{-1}
Ag	cr	0	42.55 ± 0.20	5.745 ± 0.020
Ag	g	284.9 ± 0.8	172.997 ± 0.004	6.197 ± 0.001
Ag^+	aq	105.79 ± 0.08	73.45 ± 0.40	
AgCl	cr	-127.01 ± 0.05	96.25 ± 0.20	12.033 ± 0.020
Al	cr	0	28.30 ± 0.10	4.540 ± 0.020
Al	g	330.0 ± 4.0	164.554 ± 0.004	6.919 ± 0.001
Al^{+3}	aq	-538.4 ± 1.5	-325 ± 10	
AlF_3	cr	-1510.4 ± 1.3	66.5 ± 0.5	11.62 ± 0.04
Al_2O_3	cr, corundum	-1675.7 ± 1.3	50.92 ± 0.10	10.016 ± 0.020
Ar	g	0	154.846 ± 0.003	6.197 ± 0.001
B	cr, rhombic	0	5.90 ± 0.08	1.222 ± 0.008
B	g	565 ± 5	153.436 ± 0.015	6.316 ± 0.002
BF_3	g	-1136.0 ± 0.8	254.42 ± 0.20	11.650 ± 0.020
B_2O_3	cr	-1273.5 ± 1.4	53.97 ± 0.30	9.301 ± 0.040
Be	cr	0	9.50 ± 0.08	1.950 ± 0.020
Be	g	324 ± 5	136.275 ± 0.003	6.197 ± 0.001
BeO	cr	-609.4 ± 2.5	13.77 ± 0.04	2.837 ± 0.008
Br	g	111.87 ± 0.12	175.018 ± 0.004	6.197 ± 0.001
Br^-	aq	121.41 ± 0.15	82.55 ± 0.20	
Br_2	l	0	152.21 ± 0.30	24.52 ± 0.01
Br_2	g	30.91 ± 0.11	245.468 ± 0.005	9.725 ± 0.001
C	cr, graphite	0	5.74 ± 0.10	1.050 ± 0.020
C	g	716.68 ± 0.45	158.100 ± 0.003	6.536 ± 0.001
CO	g	-110.53 ± 0.17	197.660 ± 0.004	8.671 ± 0.001
CO_2	g	-393.51 ± 0.13	213.785 ± 0.010	9.365 ± 0.003
CO_2	aq, undissoc.	-413.26 ± 0.20	119.36 ± 0.60	
CO_3^{-2}	aq	-675.23 ± 0.25	-50.0 ± 1.0	
Ca	cr	0	41.59 ± 0.40	5.736 ± 0.040
Ca	g	177.8 ± 0.8	154.887 ± 0.004	6.197 ± 0.001
Ca^{+2}	aq	-543.0 ± 1.0	-56.2 ± 1.0	
CaO	cr	-634.92 ± 0.90	38.1 ± 0.4	6.75 ± 0.06
Cd	cr	0	51.80 ± 0.15	6.247 ± 0.015
Cd	g	111.80 ± 0.20	167.749 ± 0.004	6.197 ± 0.001
Cd^{+2}	aq	-75.92 ± 0.60	-72.8 ± 1.5	
CdO	cr	-258.35 ± 0.40	54.8 ± 1.5	8.41 ± 0.08
$CdSO_4·8/3H_2O$	cr	-1729.30 ± 0.80	229.65 ± 0.40	35.56 ± 0.04
Cl	g	121.301 ± 0.008	165.190 ± 0.004	6.272 ± 0.001
Cl^-	aq	-167.080 ± 0.10	56.60 ± 0.20	
ClO_4^-	aq	-128.10 ± 0.40	184.0 ± 1.5	
Cl_2	g	0	223.081 ± 0.010	9.181 ± 0.001
Cs	cr	0	85.23 ± 0.40	7.711 ± 0.020
Cs	g	76.5 ± 1.0	175.601 ± 0.003	6.197 ± 0.001
Cs^+	aq	-258.00 ± 0.50	132.1 ± 0.5	

Substance	State	$\Delta_f H°$ (298.15 K) kJ·mol^{-1}	$S°$ (298.15 K) J·K^{-1}·mol^{-1}	$H°$ (298.15 K)–$H°$ (0) kJ·mol^{-1}
Cu	cr	0	33.15 ± 0.08	5.004 ± 0.008
Cu	g	337.4 ± 1.2	166.398 ± 0.004	6.197 ± 0.001
Cu^{+2}	aq	64.9 ± 1.0	-98 ± 4	
$CuSO_4$	cr	-771.4 ± 1.2	109.2 ± 0.4	16.86 ± 0.08
F	g	79.38 ± 0.30	158.751 ± 0.004	6.518 ± 0.001
F^-	aq	-335.35 ± 0.65	-13.8 ± 0.8	
F_2	g	0	202.791 ± 0.005	8.825 ± 0.001
Ge	cr	0	31.09 ± 0.15	4.636 ± 0.020
Ge	g	372 ± 3	167.904 ± 0.005	7.398 ± 0.001
GeF_4	g	-1190.20 ± 0.50	301.9 ± 1.0	17.29 ± 0.10
GeO_2	cr, tetragonal	-580.0 ± 1.0	39.71 ± 0.15	7.230 ± 0.020
H	g	217.998 ± 0.006	114.717 ± 0.002	6.197 ± 0.001
H^+	aq	0	0	
HBr	g	-36.29 ± 0.16	198.700 ± 0.004	8.648 ± 0.001
HCO_3^-	aq	-689.93 ± 0.20	98.4 ± 0.5	
HCl	g	-92.31 ± 0.10	186.902 ± 0.005	8.640 ± 0.001
HF	g	-273.30 ± 0.70	173.779 ± 0.003	8.599 ± 0.001
HI	g	26.50 ± 0.10	206.590 ± 0.004	8.657 ± 0.001
HPO_4^{-2}	aq	-1299.0 ± 1.5	-33.5 ± 1.5	
HS^-	aq	-16.3 ± 1.5	67 ± 5	
HSO_4^-	aq	-886.9 ± 1.0	131.7 ± 3.0	
H_2	g	0	130.680 ± 0.003	8.468 ± 0.001
H_2O	l	-285.830 ± 0.040	69.95 ± 0.03	13.273 ± 0.020
H_2O	g	-241.826 ± 0.040	188.835 ± 0.010	9.905 ± 0.005
$H_2PO_4^-$	aq	-1302.6 ± 1.5	92.5 ± 1.5	
H_2S	g	-20.6 ± 0.5	205.81 ± 0.05	9.957 ± 0.010
H_2S	aq, undissoc.	-38.6 ± 1.5	126 ± 5	
H_3BO_3	cr	-1094.8 ± 0.8	89.95 ± 0.60	13.52 ± 0.04
H_3BO_3	aq, undissoc.	-1072.8 ± 0.8	162.4 ± 0.6	
He	g	0	126.153 ± 0.002	6.197 ± 0.001
Hg	l	0	75.90 ± 0.12	9.342 ± 0.008
Hg	g	61.38 ± 0.04	174.971 ± 0.005	6.197 ± 0.001
Hg^{+2}	aq	170.21 ± 0.20	-36.19 ± 0.80	
HgO	cr, red	-90.79 ± 0.12	70.25 ± 0.30	9.117 ± 0.025
Hg_2^{+2}	aq	166.87 ± 0.50	65.74 ± 0.80	
Hg_2Cl_2	cr	-265.37 ± 0.40	191.6 ± 0.8	23.35 ± 0.20
Hg_2SO_4	cr	-743.09 ± 0.40	200.70 ± 0.20	26.070 ± 0.030
I	g	106.76 ± 0.04	180.787 ± 0.004	6.197 ± 0.001
I^-	aq	-56.78 ± 0.05	106.45 ± 0.30	
I_2	cr	0	116.14 ± 0.30	13.196 ± 0.040
I_2	g	62.42 ± 0.08	260.687 ± 0.005	10.116 ± 0.001
K	cr	0	64.68 ± 0.20	7.088 ± 0.020
K	g	89.0 ± 0.8	160.341 ± 0.003	6.197 ± 0.001
K^+	aq	-252.14 ± 0.08	101.20 ± 0.20	
Kr	g	0	164.085 ± 0.003	6.197 ± 0.001
Li	cr	0	29.12 ± 0.20	4.632 ± 0.040
Li	g	159.3 ± 1.0	138.782 ± 0.010	6.197 ± 0.001
Li^+	aq	-278.47 ± 0.08	12.24 ± 0.15	
Mg	cr	0	32.67 ± 0.10	4.998 ± 0.030
Mg	g	147.1 ± 0.8	148.648 ± 0.003	6.197 ± 0.001
Mg^{+2}	aq	-467.0 ± 0.6	-137 ± 4	
MgF_2	cr	-1124.2 ± 1.2	57.2 ± 0.5	9.91 ± 0.06
MgO	cr	-601.60 ± 0.30	26.95 ± 0.15	5.160 ± 0.020
N	g	472.68 ± 0.40	153.301 ± 0.003	6.197 ± 0.001
NH_3	g	-45.94 ± 0.35	192.77 ± 0.05	10.043 ± 0.010
NH_4^+	aq	-133.26 ± 0.25	111.17 ± 0.40	
NO_3^-	aq	-206.85 ± 0.40	146.70 ± 0.40	

CODATA KEY VALUES FOR THERMODYNAMICS (continued)

Substance	State	$\Delta_f H°$ (298.15 K) kJ·mol⁻¹	$S°$ (298.15 K) J·K⁻¹·mol⁻¹	$H°$ (298.15 K)–$H°$ (0) kJ·mol⁻¹
N_2	g	0	191.609 ± 0.004	8.670 ± 0.001
Na	cr	0	51.30 ± 0.20	6.460 ± 0.020
Na	g	107.5 ± 0.7	153.718 ± 0.003	6.197 ± 0.001
Na^+	aq	-240.34 ± 0.06	58.45 ± 0.15	
Ne	g	0	146.328 ± 0.003	6.197 ± 0.001
O	g	249.18 ± 0.10	161.059 ± 0.003	6.725 ± 0.001
OH^-	aq	-230.015 ± 0.040	-10.90 ± 0.20	
O_2	g	0	205.152 ± 0.005	8.680 ± 0.002
P	cr, white	0	41.09 ± 0.25	5.360 ± 0.015
P	g	316.5 ± 1.0	163.199 ± 0.003	6.197 ± 0.001
P_2	g	144.0 ± 2.0	218.123 ± 0.004	8.904 ± 0.001
P_4	g	58.9 ± 0.3	280.01 ± 0.50	14.10 ± 0.20
Pb	cr	0	64.80 ± 0.30	6.870 ± 0.030
Pb	g	195.2 ± 0.8	175.375 ± 0.005	6.197 ± 0.001
Pb^{+2}	aq	0.92 ± 0.25	18.5 ± 1.0	
$PbSO_4$	cr	-919.97 ± 0.40	148.50 ± 0.60	20.050 ± 0.040
Rb	cr	0	76.78 ± 0.30	7.489 ± 0.020
Rb	g	80.9 ± 0.8	170.094 ± 0.003	6.197 ± 0.001
Rb^+	aq	-251.12 ± 0.10	121.75 ± 0.25	
S	cr, rhombic	0	32.054 ± 0.050	4.412 ± 0.006
S	g	277.17 ± 0.15	167.829 ± 0.006	6.657 ± 0.001
SO_2	g	-296.81 ± 0.20	248.223 ± 0.050	10.549 ± 0.010
SO_4^{-2}	aq	-909.34 ± 0.40	18.50 ± 0.40	
S_2	g	128.60 ± 0.30	228.167 ± 0.010	9.132 ± 0.002
Si	cr	0	18.81 ± 0.08	3.217 ± 0.008
Si	g	450 ± 8	167.981 ± 0.004	7.550 ± 0.001
SiF_4	g	-1615.0 ± 0.8	282.76 ± 0.50	15.36 ± 0.05
SiO_2	cr, alpha quartz	-910.7 ± 1.0	41.46 ± 0.20	6.916 ± 0.020
Sn	cr, white	0	51.18 ± 0.08	6.323 ± 0.008
Sn	g	301.2 ± 1.5	168.492 ± 0.004	6.215 ± 0.001
Sn^{+2}	aq	-8.9 ± 1.0	-16.7 ± 4.0	
SnO	cr, tetragonal	-280.71 ± 0.20	57.17 ± 0.30	8.736 ± 0.020
SnO_2	cr, tetragonal	-577.63 ± 0.20	49.04 ± 0.10	8.384 ± 0.020
Th	cr	0	51.8 ± 0.5	6.35 ± 0.05
Th	g	602 ± 6	190.17 ± 0.05	6.197 ± 0.003
ThO_2	cr	-1226.4 ± 3.5	65.23 ± 0.20	10.560 ± 0.020
Ti	cr	0	30.72 ± 0.10	4.824 ± 0.015
Ti	g	473 ± 3	180.298 ± 0.010	7.539 ± 0.002
$TiCl_4$	g	-763.2 ± 3.0	353.2 ± 4.0	21.5 ± 0.5
TiO_2	cr, rutile	-944.0 ± 0.8	50.62 ± 0.30	8.68 ± 0.05
U	cr	0	50.20 ± 0.20	6.364 ± 0.020
U	g	533 ± 8	199.79 ± 0.10	6.499 ± 0.020
UO_2	cr	-1085.0 ± 1.0	77.03 ± 0.20	11.280 ± 0.020
UO_2^{+2}	aq	-1019.0 ± 1.5	-98.2 ± 3.0	
UO_3	cr, gamma	-1223.8 ± 1.2	96.11 ± 0.40	14.585 ± 0.050
U_3O_8	cr	-3574.8 ± 2.5	282.55 ± 0.50	42.74 ± 0.10
Xe	g	0	169.685 ± 0.003	6.197 ± 0.001
Zn	cr	0	41.63 ± 0.15	5.657 ± 0.020
Zn	g	130.40 ± 0.40	160.990 ± 0.004	6.197 ± 0.001
Zn^{+2}	aq	-153.39 ± 0.20	-109.8 ± 0.5	
ZnO	cr	-350.46 ± 0.27	43.65 ± 0.40	6.933 ± 0.040

STANDARD THERMODYNAMIC PROPERTIES OF CHEMICAL SUBSTANCES

This table gives the standard state chemical thermodynamic properties of 1750 individual substances in various states of aggregation, representing a total of about 2500 entries. Substances are listed by molecular formula in a modified Hill order; all compounds not containing carbon appear first, followed by those that contain carbon. See Appendix B for an index to molecular formula by compound name.

The properties listed are:

$\Delta_f H°$ Standard enthalpy of formation at 298.15 K in J/mol
$\Delta_f G°$ Standard Gibbs energy of formation at 298.15 K in J/mol
$S°$ Standard entropy at 298.15 K in J/mol K
C_p Heat capacity at constant pressure at 298.15 K in J/mol K

The standard state pressure is 100 kPa (1 bar). An entry of 0.0 for $\Delta_f H°$ for an element indicates the reference state of that element. See References 1 and 2 for further information on reference states.

REFERENCES

1. Cox, J. D., Wagman, D. D., and Medvedev, V. A., *CODATA Key Values for Thermodynamics,* Hemisphere Publishing, New York, 1989.
2. Wagman, D. D. et al., *The NBS Tables of Chemical Thermodynamic Properties, J. Phys. Chem. Ref. Data*, 11, Suppl. 2, 1982.
3. Chase, M. W. et al., *JANAF Thermochemical Tables, Third Edition, J. Phys. Chem. Ref. Data*, 14, Suppl. 1, 1985.
4. Daubert, T. E., Danner, R. P., Sibul, H. M., and Stebbins, C. C., *Physical and Thermodynamic Properties of Pure Compounds: Data Compilation*, extant 1994 (core with 4 supplements), Taylor & Francis, Bristol, PA (also available as database).
5. Pedley, J. B., Naylor, R. D., and Kirby, S. P., *Thermochemical Data of Organic Compounds, Second Edition*, Chapman & Hall, London, 1986.
6. Domalski, E. S., Evans, W. H., and Hearing, E. D., *Heat Capacities and Entropies of Organic Compounds in the Condensed Phase, J. Phys. Chem. Ref. Data*, 13, Suppl. 1, 1984.

Molecular formula	Name	State	$\Delta_f H°$ kJ/mol	$\Delta_f G°$ kJ/mol	$S°$ J/mol K	C_p J/mol K
Compounds not containing carbon:						
Ac	Actinium	cry	0.0		56.5	27.2
		gas	406.0	366.0	188.1	20.8
Ag	Silver	cry	0.0		42.6	25.4
		gas	284.9	246.0	173.0	20.8
AgBr	Silver bromide	cry	–100.4	–96.9	107.1	52.4
$AgBrO_3$	Silver bromate	cry	–10.5	71.3	151.9	
AgCl	Silver chloride	cry	–127.0	–109.8	96.3	50.8
$AgClO_3$	Silver chlorate	cry	–30.3	64.5	142.0	
$AgClO_4$	Silver perchlorate	cry	–31.1			
AgF	Silver fluoride	cry	–204.6			
AgF_2	Silver fluoride (AgF_2)	cry	–360.0			
AgI	Silver iodide	cry	–61.8	–66.2	115.5	56.8
$AgIO_3$	Silver iodate	cry	–171.1	–93.7	149.4	102.9
$AgNO_3$	Silver nitrate	cry	–124.4	–33.4	140.9	93.1
Ag_2	Silver (Ag_2)	gas	410.0	358.8	257.1	37.0
Ag_2CrO_4	Silver chromate	cry	–731.7	–641.8	217.6	142.3
Ag_2O	Silver oxide (Ag_2O)	cry	–31.1	–11.2	121.3	65.9
Ag_2O_2	Silver oxide (Ag_2O_2)	cry	–24.3	27.6	117.0	88.0
Ag_2O_3	Silver oxide (Ag_2O_3)	cry	33.9	121.4	100.0	
Ag_2O_4S	Silver sulfate	cry	–715.9	–618.4	200.4	131.4
Ag_2S	Silver sulfide (argentite)	cry	–32.6	–40.7	144.0	76.5
Al	Aluminum	cry	0.0		28.3	24.4
		gas	330.0	289.4	164.6	21.4
AlB_3H_{12}	Aluminum borohydride	liq	–16.3	145.0	289.1	194.6
		gas	13.0	147.0	379.2	
AlBr	Aluminum bromide (AlBr)	gas	–4.0	–42.0	239.5	35.6
$AlBr_3$	Aluminum tribromide	cry	–527.2			101.7
		gas	–425.1			
AlCl	Aluminum chloride (AlCl)	gas	–47.7	–74.1	228.1	35.0
$AlCl_2$	Aluminum chloride ($AlCl_2$)	gas	–331.0			

STANDARD THERMODYNAMIC PROPERTIES OF CHEMICAL SUBSTANCES (continued)

Molecular formula	Name	State	$\Delta_f H^\circ$ kJ/mol	$\Delta_f G^\circ$ kJ/mol	S° J/mol K	C_p J/mol K
$AlCl_3$	Aluminum trichloride	cry	–704.2	–628.8	110.7	91.8
		gas	–583.2			
AlF	Aluminum fluoride (AlF)	gas	–258.2	–283.7	215.0	31.9
AlF_3	Aluminum trifluoride	cry	–1510.4	–1431.1	66.5	75.1
		gas	–1204.6	–1188.2	277.1	62.6
AlF_4Na	Sodium tetrafluoroaluminate	gas	–1869.0	–1827.5	345.7	105.9
AlH	Aluminum hydride (AlH)	gas	259.2	231.2	187.9	29.4
AlH_3	Aluminum hydride (AlH_3)	cry	–46.0			
AlH_4K	Potassium tetrahydroaluminate	cry	–183.7			
AlH_4Li	Lithium tetrahydroaluminate	cry	–116.3	–44.7	78.7	83.2
AlI	Aluminum iodide (AlI)	gas	65.5			36.0
AlI_3	Aluminum triiodide	cry	–313.8	–300.8	159.0	98.7
		gas	–207.5			
AlN	Aluminum nitride (AlN)	cry	–318.0	–287.0	20.2	30.1
AlO	Aluminum oxide (AlO)	gas	91.2	65.3	218.4	30.9
AlO_4P	Aluminum phosphate ($AlPO_4$)	cry	–1733.8	–1617.9	90.8	93.2
AlP	Aluminum phosphide (AlP)	cry	–166.5			
AlS	Aluminum sulfide (AlS)	gas	200.9	150.1	230.6	33.4
Al_2	Aluminum (Al_2)	gas	485.9	433.3	233.2	36.4
Al_2Br_6	Aluminum hexabromide	gas	–970.7			
Al_2Cl_6	Aluminum hexachloride	gas	–1290.8	1220.4	490.0	
Al_2F_6	Aluminum hexafluoride	gas	–2628.0			
Al_2I_6	Aluminum hexaiodide	gas	–516.7			
Al_2O	Aluminum oxide (Al_2O)	gas	–130.0	–159.0	259.4	45.7
Al_2O_3	Aluminum oxide (Al_2O_3)	cry	–1675.7	–1582.3	50.9	79.0
Al_2S_3	Aluminum sulfide (Al_2S_3)	cry	–724.0			
Am	Americium	cry	0.0			
Ar	Argon	gas	0.0		154.8	20.8
As	Arsenic (gray)	cry	0.0		35.1	24.6
	Arsenic (yellow)	cry	14.6			
		gas	302.5	261.0	174.2	20.8
$AsBr_3$	Arsenic tribromide	cry	–197.5			
		gas	–130.0	–159.0	363.9	79.2
$AsCl_3$	Arsenic trichloride	liq	–305.0	–259.4	216.3	
		gas	–261.5	–248.9	327.2	75.7
AsF_3	Arsenic trifluoride	liq	–821.3	–774.2	181.2	126.6
		gas	–785.8	–770.8	289.1	65.6
AsGa	Gallium arsenide (GaAs)	cry	–71.0	–67.8	64.2	46.2
AsH_3	Arsine	gas	66.4	68.9	222.8	38.1
AsH_3O_4	Arsenic acid (H_3AsO_4)	cry	–906.3			
AsI_3	Arsenic triiodide	cry	–58.2	–59.4	213.1	105.8
		gas			388.3	80.6
AsIn	Indium arsenide (InAs)	cry	–58.6	–53.6	75.7	47.8
AsO	Arsenic oxide (AsO)	gas	70.0			
As_2	Arsenic (As_2)	gas	222.2	171.9	239.4	35.0
As_2O_5	Arsenic pentoxide (As_2O_5)	cry	–924.9	–782.3	105.4	116.5
As_2S_3	Arsenic trisulfide (As_2S_3)	cry	–169.0	–168.6	163.6	116.3
At	Astatine	cry	0.0			
Au	Gold	cry	0.0		47.4	25.4
		gas	366.1	326.3	180.5	20.8
AuBr	Gold bromide (AuBr)	cry	–14.0			
$AuBr_3$	Gold bromide ($AuBr_3$)	cry	–53.3			
AuCl	Gold chloride (AuCl)	cry	–34.7			
$AuCl_3$	Gold chloride ($AuCl_3$)	cry	–117.6			
AuF_3	Gold fluoride (AuF_3)	cry	–363.6			
AuH	Gold hydride (AuH)	gas	295.0	265.7	211.2	29.2
AuI	Gold iodide (AuI)	cry	0.0			

Molecular formula	Name	State	$\Delta_f H°$ kJ/mol	$\Delta_f G°$ kJ/mol	$S°$ J/mol K	C_p J/mol K
Au_2	Gold (Au_2)	gas	515.1			36.9
B	Boron (rhombic)	cry	0.0		5.9	11.1
		gas	565.0	521.0	153.4	20.8
BBr	Bromoborane (BBr)	gas	238.1	195.4	225.0	32.9
BBr_3	Boron tribromide	liq	–239.7	–238.5	229.7	
		gas	–205.6	–232.5	324.2	67.8
BCl	Chloroborane (BCl)	gas	149.5	120.9	213.2	31.7
BClO	Chloroxyborane (ClBO)	gas	–314.0			
BCl_3	Boron trichloride	liq	–427.2	–387.4	206.3	106.7
		gas	–403.8	–388.7	290.1	62.7
$BCsO_2$	Cesium metaborate	cry	–972.0	–915.0	104.4	80.6
BF	Fluoroborane (BF)	gas	–122.2	–149.8	200.5	29.6
BFO	Fluorooxyborane (FBO)	gas	–607.0			
BF_3	Boron trifluoride	gas	–1136.0	–1119.4	254.4	
BF_3H_3N	Aminetrifluoroboron	cry	–1353.9			
BF_3H_3P	Trihydro(phosprorous trifluoride) boron	gas	–854.0			
BF_4Na	Sodium tetrafluoroborate	cry	–1844.7	–1750.1	145.3	120.3
BH	Borane (BH)	gas	449.6	419.6	171.9	29.2
BHO_2	Metaboric acid (monoclinic)	cry	–794.3	–723.4	38.0	
		gas	–561.9	–551.0	240.1	42.2
BH_3	Borane (BH_3)	gas	100.0			
BH_3O_3	Boric acid (H_3BO_3)	cry	–1094.3	–968.9	88.8	81.4
		gas	–994.1			
BH_4K	Potassium borohydride	cry	–227.4	–160.3	106.3	96.1
BH_4Li	Lithium borohydride	cry	–190.8	–125.0	75.9	82.6
BH_4Na	Sodium borohydride	cry	–188.6	–123.9	101.3	86.8
BI_3	Boron triiodide	gas	71.1	20.7	349.2	70.8
BKO_2	Potassium metaborate	cry	–981.6	–923.4	80.0	66.7
$BLiO_2$	Lithium metaborate	cry	–1032.2	–976.1	51.5	59.8
BN	Boron nitride (BN)	cry	–254.4	–228.4	14.8	19.7
		gas	647.5	614.5	212.3	29.5
$BNaO_2$	Sodium metaborate	cry	–977.0	–920.7	73.5	65.9
BO	Boron oxide (BO)	gas	25.0	–4.0	203.5	29.2
BO_2	Boron oxide (BO_2)	gas	–300.4	–305.9	229.6	43.0
BO_2Rb	Rubidium metaborate	cry	–971.0	–913.0	94.3	74.1
BS	Boron sulfide (BS)	gas	342.0	288.8	216.2	30.0
B_2	Boron (B_2)	gas	830.5	774.0	201.9	30.5
B_2Cl_4	Tetrachlorodiborane	liq	–523.0	–464.8	262.3	137.7
		gas	–490.4	–460.6	357.4	95.4
B_2F_4	Tetrafluorodiborane	gas	–1440.1	–1410.4	317.3	79.1
B_2H_6	Diborane	gas	35.6	86.7	232.1	56.9
B_2O_3	Boron oxide (B_2O_3)	cry	–1273.5	–1194.3	54.0	
		gas	–843.8	–832.0	279.8	66.9
B_2S_3	Boron sulfide (B_2S_3)	cry	–240.6			
		gas	67.0			
$B_3H_6N_3$	s–Triazaborane	liq	–541.0	–392.7	199.6	
B_4H_{10}	Tetraborane	gas	66.1			
$B_4Na_2O_7$	Sodium tetraborate	cry	–3291.1	–3096.0	189.5	186.8
B_5H_9	Pentaborane	liq	42.7	171.8	184.2	151.1
		gas	73.2	175.0	275.9	96.8
B_5H_{11}	Pentaborane	liq	73.2			
		gas	103.3			
B_6H_{10}	Hexaborane	liq	56.3			
		gas	94.6			
Ba	Barium	cry	0.0		62.8	28.1
		gas	180.0	146.0	170.2	20.8

Molecular formula	Name	State	$\Delta_f H°$ kJ/mol	$\Delta_f G°$ kJ/mol	$S°$ J/mol K	C_p J/mol K
$BaBr_2$	Barium bromide	cry	–757.3	–736.8	146.0	
$BaCl_2$	Barium chloride	cry	–858.6	–810.4	123.7	75.1
BaF_2	Barium fluoride	cry	–1207.1	–1156.8	96.4	71.2
BaH_2	Barium hydride (BaH_2)	cry	–178.7			
BaH_2O_2	Barium hydroxide	cry	–944.7			
BaI_2	Barium iodide	cry	–602.1			
BaN_2O_4	Barium nitrite	cry	–768.2			
BaN_2O_6	Barium nitrate	cry	–992.1	–796.6	213.8	151.4
BaO	Barium oxide	cry	–553.5	–525.1	70.4	47.8
BaO_4S	Barium sulfate	cry	–1473.2	–1362.2	132.2	101.8
BaS	Barium sulfide	cry	–460.0	–456.0	78.2	49.4
Be	Beryllium	cry	0.0		9.5	16.4
		gas	324.0	286.6	136.3	20.8
$BeBr_2$	Beryllium bromide	cry	–353.5			
$BeCl_2$	Beryllium chloride	cry	–490.4	–445.6	82.7	64.8
BeF_2	Beryllium fluoride	cry	–1026.8	–979.4	53.4	51.8
BeH_2O_2	Beryllium hydroxide	cry	–902.5	–815.0	51.9	
BeI_2	Beryllium iodide	cry	–192.5			
BeO	Beryllium oxide	cry	–609.4	–580.1	13.8	
BeO_4S	Beryllium sulfate	cry	–1205.2	–1093.8	77.9	85.7
BeS	Beryllium sulfide	cry	–234.3			
Bi	Bismuth	cry	0.0		56.7	25.5
		gas	207.1	168.2	187.0	20.8
BiClO	Bismuth oxychloride	cry	–366.9	–322.1	120.5	
$BiCl_3$	Bismuth trichloride	cry	–379.1	–315.0	177.0	105.0
		gas	–265.7	–256.0	358.9	79.7
BiH_3O_3	Bismuth hydroxide	cry	–711.3			
BiI_3	Bismuth triodide	cry		–175.3		
Bi_2	Bismuth (Bi_2)	gas	219.7			36.9
$Bi_2O_{12}S_3$	Bismuth sulfate	cry	–2544.3			
Bi_2O_3	Bismuth oxide (Bi_2O_3)	cry	–573.9	–493.7	151.5	113.5
Bi_2S_3	Bismuth sulfide (Bi_2S_3)	cry	–143.1	–140.6	200.4	122.2
Bk	Berkelium	cry	0.0			
Br	Bromine	gas	111.9	82.4	175.0	20.8
BrCl	Bromine chloride	gas	14.6	–1.0	240.1	35.0
$BrCl_3Si$	Bromotrichlorosilane	gas			350.1	90.9
BrCs	Cesium bromide	cry	–405.8	–391.4	113.1	52.9
BrCu	Copper bromide (CuBr)	cry	–104.6	–100.8	96.1	54.7
BrF	Bromine fluoride	gas	–93.8	–109.2	229.0	33.0
BrF_3	Bromine trifluoride	liq	–300.8	–240.5	178.2	124.6
		gas	–255.6	–229.4	292.5	66.6
BrF_5	Bromine pentafluoride	liq	–458.6	–351.8	225.1	
		gas	–428.9	–350.6	320.2	99.6
BrGe	Germanium bromide (GeBr)	gas	235.6			37.1
$BrGeH_3$	Bromogermane	gas			274.8	56.4
BrH	Hydrogen bromide	gas	–36.3	–53.4	198.7	29.1
BrHSi	Bromosilylene (HSiBr)	cry	–464.4			
BrH_3Si	Bromosilane	gas			262.4	52.8
BrH_4N	Ammonium bromide	cry	–270.8	–175.2	113.0	96.0
BrI	Iodine bromide	gas	40.8	3.7	258.8	36.4
BrIn	Indium bromide (InBr)	cry	–175.3	–169.0	113.0	
		gas	–56.9	–94.3	259.5	36.7
BrK	Potassium bromide	cry	–393.8	–380.7	95.9	52.3
$BrKO_3$	Potassium bromate	cry	–360.2	–271.2	149.2	105.2
$BrKO_4$	Potassium perbromate	cry	–287.9	–174.4	170.1	120.2
BrLi	Lithium bromide	cry	–351.2	–342.0	74.3	
BrNO	Nitrosyl bromide	gas	82.2	82.4	273.7	45.5

Molecular formula	Name	State	$\Delta_f H°$ kJ/mol	$\Delta_f G°$ kJ/mol	$S°$ J/mol K	C_p J/mol K
BrNa	Sodium bromide	cry	–361.1	–349.0	86.8	51.4
		gas	–143.1	–177.1	241.2	36.3
$BrNaO_3$	Sodium bromate	cry	–334.1	–242.6	128.9	
BrO	Bromine oxide (BrO)	gas	125.8	108.2	237.6	32.1
BrO_2	Bromine superoxide (BrOO)	cry	48.5			
BrRb	Rubidium bromide	cry	–394.6	–381.8	110.0	52.8
BrSi	Bromosilyldyne	gas	209.0			38.6
BrTl	Thallium bromide (TlBr)	cry	–173.2	–167.4	120.5	
		gas	–37.7			
Br_2	Bromine (Br_2)	liq	0.0		152.2	75.7
		gas	30.9	3.1	245.5	36.0
Br_2Ca	Calcium bromide	cry	–682.8	–663.6	130.0	
Br_2Cd	Cadmium bromide	cry	–316.2	–296.3	137.2	76.7
Br_2Co	Cobalt bromide ($CoBr_2$)	cry	–220.9			79.5
Br_2Cr	Chromium bromide ($CrBr_2$)	cry	–302.1			
Br_2Cu	Copper bromide ($CuBr_2$)	cry	–141.8			
Br_2Fe	Iron bromide ($FeBr_2$)	cry	–249.8	–238.1	140.6	
Br_2H_2Si	Dibromosilane	gas			309.7	65.5
Br_2Hg	Mercury bromide ($HgBr_2$)	cry	–170.7	–153.1	172.0	
Br_2Hg_2	Mercury bromide (Hg_2Br_2)	cry	–206.9	–181.1	218.0	
Br_2Mg	Magnesium bromide	cry	–524.3	–503.8	117.2	
Br_2Mn	Manganese bromide ($MnBr_2$)	cry	–384.9			
Br_2Ni	Nickel bromide ($NiBr_2$)	cry	–212.1			
Br_2Pb	Lead bromide ($PbBr_2$)	cry	–278.7	–261.9	161.5	80.1
Br_2Pt	Platinum bromide ($PtBr_2$)	cry	–82.0			
Br_2S_2	Sulfur bromide (S_2Br_2)	liq	–13.0			
Br_2Se	Selenium bromide ($SeBr_2$)	gas	–21.0			
Br_2Sn	Tin bromide ($SnBr_2$)	cry	–243.5			
Br_2Sr	Strontium bromide	cry	–717.6	–697.1	135.1	75.3
Br_2Ti	Titanium bromide ($TiBr_2$)	cry	–402.0			
Br_2Zn	Zinc bromide	cry	–328.7	–312.1	138.5	
Br_3ClSi	Tribromochlorosilane	gas			377.1	95.3
Br_3Fe	Iron bromide ($FeBr_3$)	cry	–268.2			
Br_3Ga	Gallium bromide ($GaBr_3$)	cry	–386.6	–359.8	180.0	
Br_3HSi	Tribromosilane	liq	–355.6	–336.4	248.1	
		gas	–317.6	–328.5	348.6	80.8
Br_3In	Indium bromide ($InBr_3$)	cry	–428.9			
		gas	–282.0			
Br_3OP	Phosphoryl bromide	cry	–458.6			
		gas			359.8	89.9
Br_3P	Phosphorus tribromide	liq	–184.5	–175.7	240.2	
		gas	–139.3	–162.8	348.1	76.0
Br_3Pt	Platinum bromide ($PtBr_3$)	cry	–120.9			
Br_3Re	Rhenium bromide ($ReBr_3$)	cry	–167.0			
Br_3Ru	Ruthenium bromide ($RuBr_3$)	cry	–138.0			
Br_3Sb	Antimony tribromide	cry	–259.4	–239.3	207.1	
		gas	–194.6	–223.9	372.9	80.2
Br_3Sc	Scandium bromide ($ScBr_3$)	cry	–743.1			
Br_3Ti	Titanium bromide ($TiBr_3$)	cry	–548.5	–523.8	176.6	101.7
Br_4Ge	Germanium tetrabromide	liq	–347.7	–331.4	280.7	
		gas	–300.0	–318.0	396.2	101.8
Br_4Pa	Protactinium bromide ($PaBr_4$)	cry	–824.0	–787.8	234.0	
Br_4Pt	Platinum bromide ($PtBr_4$)	cry	–156.5			
Br_4Si	Silicon tetrabromide	liq	–457.3	–443.9	277.8	
		gas	–415.5	–431.8	377.9	97.1
Br_4Sn	Tin bromide ($SnBr_4$)	cry	–377.4	–350.2	264.4	
Br_4Te	Tellurium bromide ($TeBr_4$)	cry	–190.4			

STANDARD THERMODYNAMIC PROPERTIES OF CHEMICAL SUBSTANCES (continued)

Molecular formula	Name	State	$\Delta_f H°$ kJ/mol	$\Delta_f G°$ kJ/mol	$S°$ J/mol K	C_p J/mol K
Br_4Ti	Titanium bromide ($TiBr_4$)	cry	–616.7	–589.5	243.5	131.5
		gas	–549.4	–568.2	398.4	100.8
Br_4V	Vanadium bromide (VBr_4)	gas	–336.8			
Br_4Zr	Zirconium bromide ($ZrBr_4$)	cry	–760.7			
Br_5P	Phosphorus pentabromide	cry	–269.9			
Br_5Ta	Tantalum bromide ($TaBr_5$)	cry	–598.3			
Br_6W	Tungsten bromide (WBr_6)	cry	–348.5			
Ca	Calcium	cry	0.0		41.6	25.9
		gas	177.8	144.0	154.9	20.8
$CaCl_2$	Calcium chloride	cry	–795.4	–748.8	108.4	72.9
CaF_2	Calcium fluoride	cry	–1228.0	–1175.6	68.5	67.0
CaH_2	Calcium hydride (CaH_2)	cry	–181.5	–142.5	41.4	41.0
CaH_2O_2	Calcium hydroxide	cry	–985.2	–897.5	83.4	87.5
CaI_2	Calcium iodide	cry	–533.5	–528.9	142.0	
CaN_2O_6	Calcium nitrate	cry	–938.2	–742.8	193.2	149.4
CaO	Calcium oxide	cry	–634.9	–603.3	38.1	42.0
CaO_4S	Calcium sulfate	cry	–1434.5	–1322.0	106.5	99.7
CaS	Calcium sulfide	cry	–482.4	–477.4	56.5	47.4
$Ca_3O_8P_2$	Calcium phosphate	cry	–4120.8	–3884.7	236.0	227.8
Cd	Cadmium	cry	0.0		51.8	26.0
		gas	111.8		167.7	20.8
$CdCl_2$	Cadmium chloride	cry	–391.5	–343.9	115.3	74.7
CdF_2	Cadmium fluoride	cry	–700.4	–647.7	77.4	
CdH_2O_2	Cadmium hydroxide	cry	–560.7	–473.6	96.0	
CdI_2	Cadmium iodide	cry	–203.3	–201.4	161.1	80.0
CdO	Cadmium oxide	cry	–258.4	–228.7	54.8	43.4
CdO_4S	Cadmium sulfate	cry	–933.3	–822.7	123.0	99.6
CdS	Cadmium sulfide	cry	–161.9	–156.5	64.9	
CdTe	Cadmium telluride	cry	–92.5	–92.0	100.0	
Ce	Cerium	cry	0.0		72.0	26.9
		gas	423.0	385.0	191.8	23.1
$CeCl_3$	Cerium chloride ($CeCl_3$)	cry	–1053.5	–977.8	151.0	87.4
CeO_2	Cerium oxide (CeO_2)	cry	–1088.7	–1024.6	62.3	61.6
CeS	Cerium sulfide (CeS)	cry	–459.4	–451.5	78.2	50.0
Ce_2O_3	Cerium oxide (Ce_2O_3)	cry	–1796.2	–1706.2	150.6	114.6
Cf	Californium	cry	0.0			
Cl	Chlorine	gas	121.3	105.3	165.2	21.8
ClCs	Cesium chloride	cry	–443.0	–414.5	101.2	52.5
$ClCsO_4$	Cesium perchlorate	cry	–443.1	–314.3	175.1	108.3
ClCu	Copper chloride (CuCl)	cry	–137.2	–119.9	86.2	48.5
ClF	Chlorine fluoride	gas	–50.3	–51.8	217.9	32.1
$ClFO_3$	Perchloryl fluoride	gas	–23.8	48.2	279.0	64.9
ClGe	Germanium chloride (GeCl)	gas	155.2	124.2	247.0	36.9
ClF_3	Chlorine trifluoride	liq	–189.5			
		gas	–163.2	–123.0	281.6	63.9
ClF_5S	Sulfur chloride pentafluoride	liq	–1065.7			
$ClGeH_3$	Chlorogermane	gas			263.7	54.7
ClH	Hydrogen chloride	gas	–92.3	–95.3	186.9	29.1
ClHO	Hypochlorous acid (HOCl)	gas	–78.7	–66.1	236.7	37.2
$ClHO_4$	Perchloric acid	liq	–40.6			
ClH_3Si	Chlorosilane	gas			250.7	51.0
ClH_4N	Ammonium chloride	cry	–314.4	–202.9	94.6	84.1
ClH_4NO_4	Ammonium perchlorate	cry	–295.3	–88.8	186.2	
ClH_4P	Phosphonium chloride	cry	–145.2			
ClI	Iodine chloride	liq	–23.9	–13.6	135.1	
		gas	17.8	–5.5	247.6	35.6
ClIn	Indium chloride (InCl)	cry	–186.2			

Molecular formula	Name	State	$\Delta_f H°$ kJ/mol	$\Delta_f G°$ kJ/mol	$S°$ J/mol K	C_p J/mol K
		gas	–75.0			
ClK	Potassium chloride	cry	–436.5	–408.5	82.6	51.3
$ClKO_3$	Potassium chlorate	cry	–397.7	–296.3	143.1	100.3
$ClKO_4$	Potassium perchlorate	cry	–432.8	–303.1	151.0	112.4
ClLi	Lithium chloride	cry	–408.6	–384.4	59.3	48.0
$ClLiO_4$	Lithium perchlorate	cry	–381.0			
ClNO	Nitrosyl chloride	gas	51.7	66.1	261.7	44.7
$ClNO_2$	Nitryl chloride	gas	12.6	54.4	272.2	53.2
ClNa	Sodium chloride	cry	–411.2	–384.1	72.1	50.5
$ClNaO_2$	Sodium chlorite	cry	–307.0			
$ClNaO_3$	Sodium chlorate	cry	–365.8	–262.3	123.4	
$ClNaO_4$	Sodium perchlorate	cry	–383.3	–254.9	142.3	
ClO	Chlorine oxide (ClO)	gas	101.8	98.1	226.6	31.5
ClOV	Vanadium oxychloride	cry	–607.0	–556.0	75.0	
ClO_2	Chlorine dioxide (ClO_2)	gas	102.5	120.5	256.8	42.0
ClO_2	Chlorine superoxide (ClOO)	gas	89.1	105.0	263.7	46.0
ClO_4Rb	Rubidium perchlorate	cry	–437.2	–306.9	161.1	
ClRb	Rubidium chloride	cry	–435.4	–407.8	95.9	52.4
ClSi	Chlorosilylidyne	gas	189.9			36.9
ClTl	Thallium chloride (TlCl)	cry	–204.1	–184.9	111.3	50.9
		gas	–67.8			
Cl_2	Chlorine (Cl_2)	gas	0.0		223.1	33.9
Cl_2Co	Cobalt chloride ($CoCl_2$)	cry	–312.5	–269.8	109.2	78.5
Cl_2Cr	Chromium chloride ($CrCl_2$)	cry	–395.4	–356.0	115.3	71.2
Cl_2CrO_2	Chromyl chloride (CrO_2Cl_2)	liq	–579.5	–510.8	221.8	
		gas	–538.1	–501.6	329.8	84.5
Cl_2Cu	Copper chloride ($CuCl_2$)	cry	–220.1	–175.7	108.1	71.9
Cl_2Fe	Iron chloride ($FeCl_2$)	cry	–341.8	–302.3	118.0	76.7
Cl_2H_2Si	Dichlorosilane	gas			285.7	60.5
Cl_2Hg	Mercury chloride ($HgCl_2$)	cry	–224.3	–178.6	146.0	
Cl_2Hg_2	Mercury chloride (Hg_2Cl_2)	cry	–265.4	–210.7	191.6	
Cl_2Mg	Magnesium chloride	cry	–641.3	–591.8	89.6	71.4
Cl_2Mn	Manganese chloride ($MnCl_2$)	cry	–481.3	–440.5	118.2	72.9
Cl_2Ni	Nickel chloride ($NiCl_2$)	cry	–305.3	–259.0	97.7	71.7
Cl_2O	Oxygen dichloride	gas	80.3	97.9	266.2	45.4
Cl_2OS	Thionyl chloride	liq	–245.6			121.0
		gas	–212.5	–198.3	309.8	66.5
Cl_2O_2S	Sulfuryl chloride	liq	–394.1			134.0
		gas	–364.0	–320.0	311.9	77.0
Cl_2O_2U	Dichlorodioxouranium	cry	–1243.9	–1146.4	150.5	107.9
Cl_2Pb	Lead chloride ($PbCl_2$)	cry	–359.4	–314.1	136.0	
Cl_2Pt	Platinum chloride ($PtCl_2$)	cry	–123.4			
Cl_2S	Sulfur chloride (SCl_2)	liq	–50.0			
Cl_2S_2	Sulfur chloride (S_2Cl_2)	liq	–59.4			
Cl_2Sn	Tin chloride ($SnCl_2$)	cry	–325.1			
Cl_2Sr	Strontium chloride	cry	–828.9	–781.1	114.9	75.6
Cl_2Ti	Titanium chloride ($TiCl_2$)	cry	–513.8	–464.4	87.4	69.8
Cl_2Zn	Zinc chloride	cry	–415.1	–369.4	111.5	71.3
		gas	–266.1			
Cl_2Zr	Zirconium chloride ($ZrCl_2$)	cry	–502.0			
Cl_3Cr	Chromium chloride ($CrCl_3$)	cry	–556.5	–486.1	123.0	91.8
Cl_3Dy	Dysprosium chloride ($DyCl_3$)	cry	–1000.0			
Cl_3Er	Erbium chloride ($ErCl_3$)	cry	–998.7			100.0
Cl_3Eu	Europium chloride ($EuCl_3$)	cry	–936.0			
Cl_3Fe	Iron chloride ($FeCl_3$)	cry	–399.5	–334.0	142.3	96.7
Cl_3Ga	Gallium chloride ($GaCl_3$)	cry	–524.7	–454.8	142.0	
Cl_3Gd	Gadolinium chloride ($GdCl_3$)	cry	–1008.0			88.0

Molecular formula	Name	State	$\Delta_f H°$ kJ/mol	$\Delta_f G°$ kJ/mol	$S°$ J/mol K	C_p J/mol K
Cl_3HSi	Trichlorosilane	liq	–539.3	–482.5	227.6	
		gas	–513.0	–482.0	313.9	75.8
Cl_3Ho	Holmium chloride ($HoCl_3$)	cry	–1005.4			88.0
Cl_3In	Indium chloride ($InCl_3$)	cry	–537.2			
		gas	–374.0			
Cl_3La	Lanthanum chloride ($LaCl_3$)	cry	–1071.1			108.8
Cl_3Lu	Lutetium chloride ($LuCl_3$)	cry	–945.6			
		gas	–649.0			
Cl_3N	Nitrogen trichloride	liq	230.0			
Cl_3Nd	Neodymium chloride ($NdCl_3$)	cry	–1041.0			113.0
Cl_3OP	Phosphoryl chloride	liq	–597.1	–520.8	222.5	138.8
		gas	–558.5	–512.9	325.5	84.9
Cl_3OV	Vanadium oxytrichloride	liq	–734.7	–668.5	244.3	
		gas	–695.6	–659.3	344.3	89.9
Cl_3Os	Osmium chloride ($OsCl_3$)	cry	–190.4			
Cl_3P	Phosphorus trichloride	liq	–319.7	–272.3	217.1	
		gas	–287.0	–267.8	311.8	71.8
Cl_3Pr	Praseodymium chloride ($PrCl_3$)	cry	–1056.9			100.0
Cl_3Pt	Platinum chloride ($PtCl_3$)	cry	–182.0			
Cl_3Re	Rhenium chloride ($ReCl_3$)	cry	–264.0	–188.0	123.8	92.4
Cl_3Rh	Rhodium chloride ($RhCl_3$)	cry	–299.2			
Cl_3Ru	Ruthenium chloride ($RuCl_3$)	cry	–205.0			
Cl_3Sb	Antimony trichloride	cry	–382.2	–323.7	184.1	107.9
Cl_3Sc	Scandium chloride ($ScCl_3$)	cry	–925.1			
Cl_3Sm	Samarium chloride ($SmCl_3$)	cry	–1025.9			
Cl_3Tb	Terbium chloride ($TbCl_3$)	cry	–997.0			
Cl_3Ti	Titanium chloride ($TiCl_3$)	cry	–720.9	–653.5	139.7	97.2
Cl_3Tl	Thallium chloride ($TlCl_3$)	cry	–315.1			
Cl_3Tm	Thullium chloride ($TmCl_3$)	cry	–986.6			
Cl_3U	Uranium chloride (UCl_3)	cry	–866.5	–799.1	159.0	102.5
Cl_3V	Vanadium chloride (VCl_3)	cry	–580.7	511.2	131.0	93.2
Cl_3Y	Yttrium chloride (YCl_3)	cry	–1000.0			
		gas	–750.2			75.0
Cl_3Yb	Ytterbium chloride ($YbCl_3$)	cry	–959.8			
Cl_4Ge	Germanium tetrachloride	liq	–531.8	–462.7	245.6	
		gas	–495.8	–457.3	347.7	96.1
Cl_4Hf	Hafnium chloride ($HfCl_4$)	cry	–990.4	–901.3	190.8	120.5
		gas	–884.5			
Cl_4Pa	Protactinium chloride ($PaCl_4$)	cry	–1043.0	–953.0	192.0	
Cl_4Pb	Lead chloride ($PbCl_4$)	liq	–329.3			
Cl_4Pt	Platinum chloride ($PtCl_4$)	cry	–231.8			
Cl_4Si	Silicon tetrachloride	liq	–687.0	–619.8	239.7	145.3
		gas	–657.0	–617.0	330.7	90.3
Cl_4Sn	Tin chloride ($SnCl_4$)	liq	–511.3	–440.1	258.6	165.3
		gas	–471.5	–432.2	365.8	98.3
Cl_4Te	Tellurium chloride ($TeCl_4$)	cry	–326.4			138.5
Cl_4Th	Thorium chloride ($ThCl_4$)	cry	–1186.6	–1094.5	190.4	
Cl_4Ti	Titanium chloride ($TiCl_4$)	liq	–804.2	–737.2	252.3	145.2
		gas	–763.2	–726.3	353.2	95.4
Cl_4U	Uranium chloride (UCl_4)	cry	–1019.2	–930.0	197.1	122.0
		gas	–809.6	–786.6	419.0	
Cl_4V	Vanadium chloride (VCl_4)	liq	–569.4	–503.7	255.0	
		gas	–525.5	–492.0	362.4	96.2
Cl_4Zr	Zirconium chloride ($ZrCl_4$)	cry	–980.5	–889.9	181.6	119.8
Cl_5Nb	Niobium chloride ($NbCl_5$)	cry	–797.5	–683.2	210.5	148.1
		gas	–703.7	–646.0	400.6	120.8
Cl_5P	Phosphorus pentachloride	cry	–443.5			

Molecular formula	Name	State	$\Delta_f H°$ kJ/mol	$\Delta_f G°$ kJ/mol	$S°$ J/mol K	C_p J/mol K
		gas	–374.9	–305.0	364.6	112.8
Cl_5Pa	Protactinium chloride ($PaCl_5$)	cry	–1145.0	–1034.0	238.0	
Cl_5Ta	Tantalum chloride ($TaCl_5$)	cry	–859.0			
Cl_6U	Uranium chloride (UCl_6)	cry	–1092.0	–962.0	285.8	175.7
		gas	–1013.0	–928.0	431.0	
Cl_6W	Tungsten chloride (WCl_6)	cry	–602.5			
		gas	–513.8			
Cm	Curium	cry	0.0			
Co	Cobalt	cry	0.0		30.0	24.8
		gas	424.7	380.3	179.5	23.0
CoF_2	Cobalt fluoride (CoF_2)	cry	–692.0	–647.2	82.0	68.8
CoH_2O_2	Cobalt hydroxide ($Co(OH)_2$)	cry	–539.7	–454.3	79.0	
CoI_2	Cobalt iodide (CoI_2)	cry	–88.7			
CoN_2O_6	Cobalt nitrate ($Co(NO_3)_2$)	cry	–420.5			
CoO	Cobalt oxide (CoO)	cry	–237.9	–214.2	53.0	55.2
CoO_4S	Cobalt sulfate ($CoSO_4$)	cry	–888.3	–782.3	118.0	
CoS	Cobalt sulfide (CoS)	cry	–82.8			
Co_2S_3	Cobalt sulfide (Co_2S_3)	cry	–147.3			
Co_3O_4	Cobalt oxide (Co_3O_4)	cry	–891.0	–774.0	102.5	123.4
Cr	Chromium	cry	0.0		23.8	23.4
		gas	396.6	351.8	174.5	20.8
CrF_2	Chromium fluoride (CrF_2)	cry	–778.0			
CrF_3	Chromium fluoride (CrF_3)	cry	–1159.0	–1088.0	93.9	78.7
CrI_2	Chromium iodide (CrI_2)	cry	–156.9			
CrI_3	Chromium iodide (CrI_3)	cry	–205.0			
CrO_2	Chromium oxide (CrO_2)	cry	–598.0			
CrO_4Pb	Lead chromate ($PbCrO_4$)	cry	–930.9			
Cr_2FeO_4	Chromium iron oxide ($FeCr_2O_4$)	cry	–1444.7	–1343.8	146.0	133.6
Cr_2O_3	Chromium oxide (Cr_2O_3)	cry	–1139.7	–1058.1	81.2	118.7
Cr_3O_4	Chromium oxide (Cr_3O_4)	cry	–1531.0			
Cs	Cesium	cry	0.0		85.2	32.2
		gas	76.5	49.6	175.6	20.8
CsF	Cesium fluoride	cry	–553.5	–525.5	92.8	51.1
CsF_2H	Cesium hydrogen fluoride ($CsHF_2$)	cry	–923.8	–858.9	135.2	87.3
CsH	Cesium hydride	cry	–54.2			
CsHO	Cesium hydroxide	cry	–417.2			
$CsHO_4S$	Cesium hydrogen sulfate	cry	–1158.1			
CsH_2N	Cesium amide	cry	–118.4			
CsI	Cesium iodide	cry	–346.6	–340.6	123.1	52.8
$CsNO_3$	Cesium nitrate	cry	–506.0	–406.5	155.2	
CsO_2	Cesium superoxide (CsO_2)	cry	–286.2			
Cs_2O	Cesium oxide (Cs_2O)	cry	–345.8	–308.1	146.9	76.0
Cs_2O_3S	Cesium sulfite	cry	–1134.7			
Cs_2O_4S	Cesium sulfate	cry	–1443.0	–1323.6	211.9	134.9
Cs_2S	Cesium sulfide (Cs_2S)	cry	–359.8			
Cu	Copper	cry	0.0		33.2	24.4
		gas	337.4	297.7	166.4	20.8
CuF_2	Copper fluoride (CuF_2)	cry	–542.7			
CuH_2O_2	Copper hydroxide ($Cu(OH)_2$)	cry	–449.8			
CuI	Copper iodide (CuI)	cry	–67.8	–69.5	96.7	54.1
CuN_2O_6	Copper nitrate ($Cu(NO_3)_2$)	cry	–302.9			
CuO	Copper oxide (CuO)	cry	–157.3	–129.7	42.6	42.3
CuO_4S	Copper sulfate ($CuSO_4$)	cry	–771.4	–662.2	109.2	
CuO_4W	Copper tungstate ($CuWO_4$)	cry	–1105.0			
CuS	Copper sulfide (CuS)	cry	–53.1	–53.6	66.5	47.8
CuSe	Copper selenide (CuSe)	cry	–39.5			
Cu_2	Copper (Cu_2)	gas	484.2	431.9	241.6	36.6

Molecular formula	Name	State	$\Delta_f H°$ kJ/mol	$\Delta_f G°$ kJ/mol	$S°$ J/mol K	C_p J/mol K
Cu_2O	Copper oxide (Cu_2O)	cry	–168.6	–146.0	93.1	63.6
Cu_2S	Copper sulfide (Cu_2S)	cry	–79.5	–86.2	120.9	76.3
Dy	Dysprosium	cry	0.0		74.8	28.2
		gas	290.4	254.4	196.6	20.8
Dy_2O_3	Dysprosium oxide (Dy_2O_3)	cry	–1863.1	–1771.5	149.8	116.3
Er	Erbium	cry	0.0		73.2	28.1
		gas	317.1	280.7	195.6	20.8
ErF_3	Erbium fluoride (ErF_3)	cry	–1711.0			
Er_2O_3	Erbium oxide (Er_2O_3)	cry	–1897.9	–1808.7	155.6	108.5
Eu	Europium	cry	0.0		77.8	27.7
		gas	175.3	142.2	188.8	20.8
Eu_2O_3	Europium oxide (Eu_2O_3)	cry	–1651.4	–1556.8	146.0	122.2
Eu_3O_4	Europium oxide (Eu_3O_4)	cry	–2272.0	–2142.0	205.0	
F	Fluorine	gas	79.4	62.3	158.8	22.7
FGa	Gallium fluoride (GaF)	gas	–251.9			33.3
FGe	Germanium fluoride (GeF)	gas	–33.4			34.7
$FGeH_3$	Fluorogermane	gas			252.8	51.6
FH	Hydrogen fluoride	liq	–299.8			
		gas	–273.3	–275.4	173.8	
FH_3Si	Fluorosilane	gas			238.4	47.4
FH_4N	Ammonium fluoride	cry	–464.0	–348.7	72.0	65.3
FI	Iodine fluoride	gas	–95.7	–118.5	236.2	33.4
FIn	Indium fluoride (InF)	gas	–203.4			
FK	Potassium fluoride	cry	–567.3	–537.8	66.6	49.0
FLi	Lithium fluoride	cry	–616.0	–587.7	35.7	41.6
FNO	Nitrosyl fluoride	gas	–66.5	–51.0	248.1	41.3
FNO_2	Nitryl fluoride	gas			260.4	49.8
FNS	Thionitrosyl fluoride (NSF)	gas			259.8	44.1
FNa	Sodium fluoride	cry	–576.6	–546.3	51.1	46.9
FO	Fluorine oxide (FO)	gas	109.0	105.0	216.8	30.5
FRb	Rubidium fluoride	cry	–557.7			
FSi	Fluorosilylidyne	gas	7.1	–24.3	225.8	32.6
FTl	Thallium fluoride (TlF)	cry	–324.7			
		gas	–182.4			
F_2	Fluorine (F_2)	gas	0.0		202.8	31.3
F_2Fe	Iron fluoride (FeF_2)	cry	–711.3	–668.6	87.0	68.1
F_2HK	Potassium hydrogen fluoride (KHF_2)	cry	–927.7	–859.7	104.3	76.9
F_2HN	Difluoramine	gas			252.8	43.4
F_2HNa	Sodium hydrogen fluoride ($NaHF_2$)	cry	–920.3	–852.2	90.9	75.0
F_2HRb	Rubidium fluoride ($RbHF_2$)	cry	–922.6	–855.6	120.1	79.4
F_2Mg	Magnesium fluoride	cry	–1124.2	–1071.1	57.2	61.6
F_2N	Difluoroamidogen (NF_2)	gas	43.1	57.8	249.9	41.0
F_2N_2	*cis*–Difluorodiazine	gas	69.5			
F_2N_2	*trans*–Difluorodiazine	gas	82.0			
F_2Ni	Nickel fluoride (NiF_2)	cry	–651.4	–604.1	73.6	64.1
F_2O	Oxygen difluoride	gas	24.7	41.9	247.4	43.3
F_2OS	Thionyl fluoride	gas			278.7	56.8
F_2O_2	Dioxygen difluoride (FOOF)	gas	18.0			
F_2O_2S	Sulfuryl fluoride	gas			284.0	66.0
F_2O_2U	Uranyl fluoride	cry	–1648.1	–1551.8	135.6	103.2
F_2Pb	Lead fluoride (PbF_2)	cry	–664.0	–617.1	110.5	
F_2Si	Difluorosilylene (SiF_2)	gas	–619.0	–628.0	252.7	43.9
F_2Sr	Strontium fluoride	cry	–1216.3	–1164.8	82.1	70.0
F_2Zn	Zinc fluoride	cry	–764.4	–713.3	73.7	65.7
F_3Ga	Gallium fluoride (GaF_3)	cry	–1163.0	–1085.3	84.0	
F_3Gd	Gadolinium fluoride (GdF_3)	gas	–1297.0			
F_3HSi	Trifluorosilane	gas			271.9	60.5

Molecular formula	Name	State	$\Delta_f H°$ kJ/mol	$\Delta_f G°$ kJ/mol	$S°$ J/mol K	C_p J/mol K
F_3Ho	Holmium fluoride (HoF_3)	cry	–1707.0			
F_3N	Nitrogen trifluoride	gas	–132.1	–90.6	260.8	53.4
F_3Nd	Neodymium fluoride (NdF_3)	cry	–1657.0			
F_3OP	Phosphoryl fluoride	gas	–1254.3	–1205.8	285.4	68.8
F_3P	Phosphorus trifluoride	gas	–958.4	–936.9	273.1	58.7
F_3Sb	Antimony trifluoride	cry	–915.5			
F_3Sc	Scandium fluoride (ScF_3)	cry	–1629.2	–1555.6	92.0	
		gas	–1247.0	–1234.0	300.5	67.8
F_3Sm	Samarium fluoride (SmF_3)	cry	–1778.0			
F_3Th	Thorium fluoride (ThF_3)	gas	–1182.0	–1176.5	339.4	73.2
F_3U	Uranium fluoride (UF_3)	cry	–1502.1	–1433.4	123.4	95.1
F_3Y	Yttrium fluoride (YF_3)	cry	–1718.8	–1644.7	100.0	
		gas	–1288.7	–1277.8	311.8	70.3
F_4Ge	Germanium tetrafluoride	gas	–1190.2	–1150.0	301.9	
F_4Hf	Hafnium fluoride (HfF_4)	cry	–1930.5	–1830.4	113.0	
		gas	–1669.8			
F_4N_2	Tetrafluorohydrazine	gas	–8.4	79.9	301.2	79.2
F_4Pb	Lead fluoride (PbF_4)	cry	–941.8			
F_4S	Sulfur fluoride (SF_4)	gas	–763.2	–722.0	299.6	77.6
F_4Si	Silicon tetrafluoride	gas	–1615.0	–1572.8	282.8	73.6
F_4Th	Thorium fluoride (ThF_4)	cry	–2091.6	–1997.0	142.1	110.5
F_4U	Uranium fluoride (UF_4)	cry	–1914.2	–1823.3	151.7	116.0
		gas	–1598.7	–1572.7	368.0	91.2
F_4V	Vanadium fluoride (VF_4)	cry	–1403.3			
F_4Xe	Xenon tetrafluoride	cry	–261.5			
F_4Zr	Zirconium fluoride (ZrF_4)	cry	–1911.3	–1809.9	104.6	103.7
F_5I	Iodine pentafluoride	liq	–864.8			
		gas	–822.5	–751.7	327.7	99.2
F_5Nb	Niobium fluoride (NbF_5)	cry	–1813.8	–1699.0	160.2	134.7
		gas	–1739.7	–1673.6	321.9	97.1
F_5P	Phosphorus pentafluoride	gas	–1594.4	–1520.7	300.8	84.8
F_5Ta	Tantalum fluoride (TaF_5)	cry	–1903.6			
F_5V	Vanadium fluoride (VF_5)	liq	–1480.3	–1373.1	175.7	
		gas	–1433.9	–1369.8	320.9	98.6
$F_6H_8N_2Si$	Ammonium hexafluorosilicate	cry	–2681.7	–2365.3	280.2	228.1
F_6Ir	Iridium fluoride (IrF_6)	cry	–579.7	–461.6	247.7	
		gas	–544.0	–460.0	357.8	121.1
F_6K_2Si	Potassium hexafluorosilicate	cry	–2956.0	–2798.6	226.0	
F_6Mo	Molybdenum fluoride (MoF_6)	liq	–1585.5	–1473.0	259.7	169.8
		gas	–1557.7	–1472.2	350.5	120.6
F_6Na_2Si	Sodium hexafluorosilicate	cry	–2909.6	–2754.2	207.1	187.1
F_6Os	Osmium fluoride (OsF_6)	cry			246.0	
		gas			358.1	120.8
F_6Pt	Platinum fluoride (PtF_6)	cry			235.6	
		gas			348.3	122.8
F_6S	Sulfur fluoride (SF_6)	gas	–1220.5	–1116.5	291.5	97.0
F_6Se	Selenium fluoride (SeF_6)	gas	–1117.0	–1017.0	313.9	110.5
F_6Te	Tellurium fluoride (TeF_6)	gas	–1318.0			
F_6U	Uranium fluoride (UF_6)	cry	–2197.0	–2068.5	227.6	166.8
		gas	–2147.4	–2063.7	377.9	129.6
F_6W	Tungsten fluoride (WF_6)	liq	–1747.7	–1631.4	251.5	
		gas	–1721.7	–1632.1	341.1	119.0
Fe	Iron	cry	0.0		27.3	25.1
		gas	416.3	370.7	180.5	25.7
FeI_2	Iron iodide (FeI_2)	cry	–113.0			
FeI_3	Iron iodide (FeI_3)	gas	71.0			
$FeMoO_4$	Iron molybdate ($FeMoO_4$)	cry	–1075.0	–975.0	129.3	118.5

Molecular formula	Name	State	$\Delta_f H°$ kJ/mol	$\Delta_f G°$ kJ/mol	$S°$ J/mol K	C_p J/mol K
FeO	Iron oxide (FeO)	cry	–272.0			
FeO_4S	Iron sulfate ($FeSO_4$)	cry	–928.4	–820.8	107.5	100.6
FeO_4W	Iron tungstate ($FeWO_4$)	cry	–1155.0	–1054.0	131.8	114.6
FeS	Iron sulfide (FeS)	cry	–100.0	–100.4	60.3	50.5
FeS_2	Iron sulfide (FeS_2)	cry	–178.2	–166.9	52.9	62.2
Fe_2O_3	Iron oxide (Fe_2O_3)	cry	–824.2	–742.2	87.4	103.9
Fe_2O_4Si	Iron silicate (Fe_2SiO_4)	cry	–1479.9	–1379.0	145.2	132.9
Fe_3O_4	Iron oxide (Fe_3O_4)	cry	–1118.4	–1015.4	146.4	143.4
Fm	Fermium	cry	0.0			
Fr	Francium	cry	0.0		95.4	
Ga	Gallium	cry	0.0		40.9	25.9
		liq	5.6			
		gas	277.0	238.9	169.1	25.4
GaH_3O_3	Gallium hydroxide ($Ga(OH)_3$)	cry	–964.4	–831.3	100.0	
GaN	Gallium nitride (GaN)	cry	–110.5			
GaO	Gallium oxide (GaO)	gas	279.5	253.5	231.1	32.1
GaP	Gallium phosphide (GaP)	cry	–88.0			
GaSb	Gallium antimonide (GaSb)	cry	–41.8	–38.9	76.1	48.5
Ga_2	Gallium (Ga_2)	gas	438.5			
Ga_2O	Gallium oxide (Ga_2O)	cry	–356.0			
Ga_2O_3	Gallium oxide (Ga_2O_3)	cry	–1089.1	–998.3	85.0	92.1
Gd	Gadolinium	cry	0.0		68.1	37.0
		gas	397.5	359.8	194.3	27.5
Gd_2O_3	Gadolinium oxide (Gd_2O_3)	cry	–1819.6			106.7
Ge	Germanium	cry	0.0		31.1	23.3
		gas	372.0	331.2	167.9	30.7
GeH_3I	Iodogermane	gas			283.2	57.5
GeH_4	Germane	gas	90.8	113.4	217.1	45.0
GeI_4	Germanium tetraiodide	cry	–141.8	–144.3	271.1	
		gas	–56.9	–106.3	428.9	104.1
GeO	Germanium oxide (GeO) (brown)	cry	–261.9	–237.2	50.0	
		gas	–46.2	–73.2	224.3	30.9
GeO_2	Germanium dioxide (tetragonal)	cry	–580.0	–521.4	39.7	52.1
GeP	Germanium phosphide (GeP)	cry	–21.0	–17.0	63.0	
GeS	Germanium sulfide (GeS)	cry	–69.0	–71.5	71.0	
		gas	92.0	42.0	234.0	33.7
Ge_2	Germanium (Ge_2)	gas	473.1	416.3	252.8	35.6
Ge_2H_6	Digermane	liq	137.3			
		gas	162.3			
Ge_3H_8	Trigermane	liq	193.7			
		gas	226.8			
H	Hydrogen	gas	218.0	203.3	114.7	20.8
HI	Hydrogen iodide	gas	26.5	1.7	206.6	29.2
HIO_3	Iodic acid	cry	–230.1			
HK	Potassium hydride	cry	–57.7			
HKO	Potassium hydroxide	cry	–424.8	–379.1	78.9	64.9
HKO_4S	Potassium hydrogen sulfate	cry	–1160.6	–1031.3	138.1	
HLi	Lithium hydride	cry	–90.5	–68.3	20.0	27.9
HLiO	Lithium hydroxide	cry	–484.9	–439.0	42.8	49.7
HN	Imidogen (NH)	gas	351.5	345.6	181.2	29.2
HNO_2	Nitrous acid (HONO)	gas	–79.5	–46.0	254.1	45.6
HNO_3	Nitric acid	liq	–174.1	–80.7	155.6	109.9
		gas	–135.1	–74.7	266.4	53.4
HN_3	Hydrazoic acid	liq	264.0	327.3	140.6	
		gas	294.1	328.1	239.0	43.7
HNa	Sodium hydride	cry	–56.3	–33.5	40.0	36.4
HNaO	Sodium hydroxide	cry	–425.6	–379.5	64.5	59.5

STANDARD THERMODYNAMIC PROPERTIES OF CHEMICAL SUBSTANCES (continued)

Molecular formula	Name	State	$\Delta_f H°$ kJ/mol	$\Delta_f G°$ kJ/mol	$S°$ J/mol K	C_p J/mol K
$HNaO_4S$	Sodium hydrogen sulfate	cry	–1125.5	–992.8	113.0	
HNa_2O_4P	Disodium hydrogen phosphate	cry	–1748.1	–1608.2	150.5	135.3
HO	Hydroxyl (OH)	gas	39.0	34.2	183.7	29.9
HORb	Rubidium hydroxide	cry	–418.2			
HOTl	Thallium hydroxide (TlOH)	cry	–238.9	–195.8	88.0	
HO_2	Hydroperoxy (HOO)	gas	10.5	22.6	229.0	34.9
HO_3P	Metaphosphoric acid	cry	–948.5			
HO_4RbS	Rubidium hydrogen sulfate	cry	–1159.0			
HO_4Re	Perrhenic acid	cry	–762.3	–656.4	158.2	
HRb	Rubidium hydride	cry	–52.3			
HS	Mercapto (SH)	gas	142.7	113.3	195.7	32.3
HSi	Silylidyne (SiH)	gas	361.0			
HTa_2	Tantalum hydride (Ta_2H)	cry	–32.6	–69.0	79.1	90.8
H_2	Hydrogen (H_2)	gas	0.0		130.7	28.8
H_2KN	Potassium amide	cry	–128.9			
H_2KO_4P	Potassium dihydrogen phosphate	cry	–1568.3	–1415.9	134.9	116.6
H_2LiN	Lithium amide	cry	–179.5			
H_2Mg	Magnesium hydride	cry	–75.3	–35.9	31.1	35.4
H_2MgO_2	Magnesium hydroxide	cry	–924.5	–833.5	63.2	77.0
H_2N	Amidogen (NH_2)	gas	184.9	194.6	195.0	33.9
H_2NNa	Sodium amide	cry	–123.8	–64.0	76.9	66.2
H_2NRb	Rubidium amide	cry	–113.0			
$H_2N_2O_2$	Nitramide	cry	–89.5			
H_2NiO_2	Nickel hydroxide ($Ni(OH)_2$)	cry	–529.7	–447.2	88.0	
H_2O	Water	liq	–285.8	–237.1	70.0	75.3
		gas	–241.8	–228.6	188.8	33.6
H_2O_2	Hydrogen peroxide	liq	–187.8	–120.4	109.6	89.1
		gas	–136.3	–105.6	232.7	43.1
H_2O_2Sn	Tin hydroxide ($Sn(OH)_2$)	cry	–561.1	–491.6	155.0	
H_2O_2Sr	Strontium hydroxide	cry	–959.0			
H_2O_2Zn	Zinc hydroxide	cry	–641.9	–553.5	81.2	
H_2O_3Si	Metasilicic acid (H_2SiO_3)	cry	–1188.7	–1092.4	134.0	
H_2O_4S	Sulfuric acid	liq	–814.0	–690.0	156.9	138.9
H_2O_4Se	Selenic acid	cry	–530.1			
H_2S	Hydrogen sulfide	gas	–20.6	–33.4	205.8	34.2
H_2S_2	Hydrogen sulfide (H_2S_2)	liq	–18.1			84.1
		gas	15.5			51.5
H_2Se	Hydrogen selenide	gas	29.7	15.9	219.0	34.7
H_2Sr	Strontium hydride	cry	–180.3			
H_2Te	Hydrogen telluride	gas	99.6			
H_2Th	Thorium hydride (ThH_2)	cry	–139.7	–100.0	50.7	36.7
H_2Zr	Zirconium hydride (ZrH_2)	cry	–169.0	–128.8	35.0	31.0
H_3ISi	Iodosilane	gas			270.9	54.4
H_3N	Ammonia	gas	–45.9	–16.4	192.8	35.1
H_3NO	Hydroxylamine (NH_2OH)	cry	–114.2			
H_3O_2P	Hypophosphorous acid (H_3PO_2)	cry	–604.6			
		liq	–595.4			
H_3O_3P	Phosphorous acid	cry	–964.4			
H_3O_4P	Phosphoric acid	cry	–1284.4	–1124.3	110.5	106.1
		liq	–1271.7	–1123.6	150.8	145.0
H_3P	Phosphine	gas	5.4	13.4	210.2	37.1
H_3Sb	Stibine	gas	145.1	147.8	232.8	41.1
H_3U	Uranium hydride (UH_3)	cry	–127.2	–72.8	63.7	49.3
H_4IN	Ammonium iodide	cry	–201.4	–112.5	117.0	
H_4N_2	Hydrazine	liq	50.6	149.3	121.2	98.9
		gas	95.4	159.4	238.5	49.6
$H_4N_2O_2$	Ammonium nitrite	cry	–256.5			

Molecular formula	Name	State	$\Delta_f H°$ kJ/mol	$\Delta_f G°$ kJ/mol	$S°$ J/mol K	C_p J/mol K
$H_4N_2O_3$	Ammonium nitrate	cry	–365.6	–183.9	151.1	139.3
H_4N_4	Ammonium azide	cry	115.5	274.2	112.5	
H_4O_4Si	Silicic acid (H_4SiO_4)	cry	–1481.1	–1332.9	192.0	
$H_4O_7P_2$	Diphosphoric acid ($H_4P_2O_7$)	cry	–2241.0			
		liq	–2231.7			
H_4P_2	Diphosphine (P_2H_4)	liq	–5.0			
		gas	20.9			
H_4Si	Silane	gas	34.3	56.9	204.6	42.8
H_4Sn	Stannane	gas	162.8	188.3	227.7	49.0
H_5NO	Ammonium hydroxide	liq	–361.2	–254.0	165.6	154.9
H_5NO_3S	Ammonium hydrogen sulfite	cry	–768.6			
H_5NO_4S	Ammonium hydrogen sulfate	cry	–1027.0			
H_6Si_2	Disilane	gas	80.3	127.3	272.7	80.8
$H_8N_2O_4S$	Ammonium sulfate	cry	–1180.9	–901.7	220.1	187.5
H_8Si_3	Trisilane	liq	92.5			
		gas	120.9			
$H_9N_2O_4P$	Diammonium hydrogen phosphate	cry	–1566.9			188.0
$H_{12}N_3O_4P$	Ammonium phosphate	cry	1671.9			
He	Helium	gas	0.0		126.2	20.8
Hf	Hafnium	cry	0.0		43.6	25.7
		gas	619.2	576.5	186.9	20.8
HfO_2	Hafnium oxide (HfO_2)	cry	–1144.7	–1088.2	59.3	60.3
Hg	Mercury	liq	0.0		75.9	28.0
		gas	61.4	31.8	175.0	20.8
HgI_2	Mercury iodide (HgI_2) (red)	cry	–105.4	–101.7	180.0	
HgO	Mercury oxide (HgO) (red)	cry	–90.8	–58.5	70.3	44.1
HgO_4S	Mercury sulfate ($HgSO_4$)	cry	–707.5			
HgS	Mercury sulfide (HgS)	cry	–58.2	–50.6	82.4	48.4
Hg_2	Mercury (Hg_2)	gas	108.8	68.2	288.1	37.4
Hg_2I_2	Mercury iodide (Hg_2I_2)	cry	–121.3	–111.0	233.5	
Hg_2O_4S	Mercury sulfate (Hg_2SO_4)	cry	–743.1	–625.8	200.7	132.0
Ho	Holmium	cry	0.0		75.3	27.2
		gas	300.8	264.8	195.6	20.8
Ho_2O_3	Holmium oxide (Ho_2O_3)	cry	–1880.7	–1791.1	158.2	115.0
I	Iodine	gas	106.8	70.2	180.8	20.8
IIn	Indium iodide (InI)	cry	–116.3	–120.5	130.0	
		gas	7.5	–37.7	267.3	36.8
IK	Potassium iodide	cry	–327.9	–324.9	106.3	52.9
IKO_3	Potassium iodate	cry	–501.4	–418.4	151.5	106.5
IKO_4	Potassium periodate	cry	–467.2	–361.4	175.7	
ILi	Lithium iodide	cry	–270.4	–270.3	86.8	51.0
INa	Sodium iodide	cry	–287.8	–286.1	98.5	52.1
$INaO_3$	Sodium iodate	cry	–481.8			92.0
$INaO_4$	Sodium periodate	cry	–429.3	–323.0	163.0	
IO	Iodine oxide (IO)	gas	175.1	149.8	245.5	32.9
IRb	Rubidium iodide	cry	–333.8	–328.9	118.4	53.2
ITl	Thallium iodide (TlI)	cry	–123.8	–125.4	127.6	
		gas	7.1			
I_2	Iodine (I_2) (rhombic)	cry	0.0		116.1	54.4
		gas	62.4	19.3	260.7	36.9
I_2Mg	Magnesium iodide	cry	–364.0	–358.2	129.7	
I_2Ni	Nickel iodide (NiI_2)	cry	–78.2			
I_2Pb	Lead iodide (PbI_2)	cry	–175.5	–173.6	174.9	77.4
I_2Sn	Tin iodide (SnI_2)	cry	–143.5			
I_2Sr	Strontium iodide	cry	–558.1			81.6
I_2Zn	Zinc iodide	cry	–208.0	–209.0	161.1	

STANDARD THERMODYNAMIC PROPERTIES OF CHEMICAL SUBSTANCES (continued)

Molecular formula	Name	State	$\Delta_f H°$ kJ/mol	$\Delta_f G°$ kJ/mol	$S°$ J/mol K	C_p J/mol K
I_3In	Indium iodide (InI_3)	cry	–238.0			
		gas	–120.5			
I_3Lu	Lutecium iodide (LuI_3)	cry	–548.0			
I_3P	Phosphorus triiodide	cry	–45.6			
		gas			374.4	78.4
I_3Ru	Ruthenium iodide (RuI_3)	cry	–65.7			
I_3Sb	Antimony triiodide	cry	–100.4			
I_4Pt	Platinum iodide (PtI_4)	cry	–72.8			
I_4Si	Silicon tetraiodide	cry	–189.5			
I_4Sn	Tin iodide (SnI_4)	cry				84.9
		gas			446.1	105.4
I_4Ti	Titanium iodide (TiI_4)	cry	–375.7	–371.5	249.4	125.7
		gas	–277.8			
I_4V	Vanadium iodide (VI_4)	gas	–122.6			
I_4Zr	Zirconium iodide (ZrI_4)	cry	–481.6			
In	Indium	cry	0.0		57.8	26.7
		gas	243.3	208.7	173.8	20.8
InO	Indium oxide (InO)	gas	387.0	364.4	236.5	32.6
InP	Indium phosphide (InP)	cry	–88.7	–77.0	59.8	45.4
InS	Indium sulfide (InS)	cry	–138.1	–131.8	67.0	
		gas	238.0			
InSb	Indium antimonide (InSb)	cry	–30.5	–25.5	86.2	49.5
		gas	344.3			
In_2	Indium (In_2)	gas	380.9			
In_2O_3	Indium oxide (In_2O_3)	cry	–925.8	–830.7	104.2	92.0
In_2S_3	Indium sulfide (In_2S_3)	cry	–427.0	–412.5	163.6	118.0
Ir	Iridium	cry	0.0		35.5	25.1
		gas	665.3	617.9	193.6	20.8
IrO_2	Iridium oxide (IrO_2)	cry	–274.1			57.3
IrS_2	Iridium sulfide (IrS_2)	cry	–138.0			
Ir_2S_3	Iridium sulfide (Ir_2S_3)	cry	–234.0			
K	Potassium	cry	0.0		64.7	29.6
		gas	89.0	60.5	160.3	20.8
$KMnO_4$	Potassium permanganate	cry	–837.2	–737.6	171.7	117.6
KNO_2	Potassium nitrite	cry	–369.8	–306.6	152.1	107.4
KNO_3	Potassium nitrate	cry	–494.6	–394.9	133.1	96.4
KNa	Potassium sodium (KNa)	liq	6.3			
KO_2	Potassium superoxide (KO_2)	cry	–284.9	–239.4	116.7	77.5
K_2	Potassium (K_2)	gas	123.7	87.5	249.7	37.9
K_2O	Potassium oxide (K_2O)	cry	–361.5			
K_2O_2	Potassium peroxide (K_2O_2)	cry	–494.1	–425.1	102.1	
K_2O_4S	Potassium sulfate	cry	–1437.8	–1321.4	175.6	131.5
K_2S	Potassium sulfide (K_2S)	cry	–380.7	–364.0	105.0	
K_3O_4P	Potassium phosphate	cry	–1950.2			
Kr	Krypton	gas	0.0		164.1	20.8
La	Lanthanum	cry	0.0		56.9	27.1
		gas	431.0	393.6	182.4	22.8
LaS	Lanthanum sulfide (LaS)	cry	–456.0	–451.5	73.2	59.0
La_2O_3	Lanthanum oxide (La_2O_3)	cry	–1793.7	–1705.8	127.3	108.8
Li	Lithium	cry	0.0		29.1	24.8
		gas	159.3	126.6	138.8	20.8
$LiNO_2$	Lithium nitrite	cry	–372.4	–302.0	96.0	
$LiNO_3$	Lithium nitrate	cry	–483.1	–381.1	90.0	
Li_2	Lithium (Li_2)	gas	215.9	174.4	197.0	36.1
Li_2O	Lithium oxide (Li_2O)	cry	–597.9	–561.2	37.6	54.1
Li_2O_2	Lithium peroxide (Li_2O_2)	cry	–634.3			
Li_2O_3Si	Lithium metasilicate	cry	–1648.1	–1557.2	79.8	99.1

Molecular formula	Name	State	$\Delta_f H°$ kJ/mol	$\Delta_f G°$ kJ/mol	$S°$ J/mol K	C_p J/mol K
Li_2O_4S	Lithium sulfate	cry	–1436.5	–1321.7	115.1	117.6
Li_2S	Lithium sulfide (Li_2S)	cry	–441.4			
Li_3O_4P	Lithium phosphate	cry	–2095.8			
Lr	Lawrencium	cry	0.0			
Lu	Lutetium	cry	0.0		51.0	26.9
		gas	427.6	387.8	184.8	20.9
Lu_2O_3	Lutetium oxide (Lu_2O_3)	cry	–1878.2	–1789.0	110.0	101.8
Md	Mendelevium	cry	0.0			
Mg	Magnesium	cry	0.0		32.7	24.9
		gas	147.1	112.5	148.6	20.8
MgN_2O_6	Magnesium nitrate	cry	–790.7	–589.4	164.0	141.9
MgO	Magnesium oxide	cry	–601.6	–569.3	27.0	37.2
MgO_4S	Magnesium sulfate	cry	–1284.9	–1170.6	91.6	96.5
MgO_4Se	Magnesium selenate	cry	–968.5			
MgS	Magnesium sulfide	cry	–346.0	–341.8	50.3	45.6
Mg_2	Magnesium (Mg_2)	gas	287.7			
Mg_2O_4Si	Magnesium silicate	cry	–2174.0	–2055.1	95.1	118.5
Mn	Manganese	cry	0.0		32.0	26.3
		gas	280.7	238.5	173.7	20.8
MnN_2O_6	Manganese nitrate ($Mn(NO_3)_2$)	cry	–576.3			
$MnNa_2O_4$	Sodium permanganate	cry	–1156.0			
MnO	Manganese oxide (MnO)	cry	–385.2	–362.9	59.7	45.4
MnO_2	Manganese oxide (MnO_2)	cry	–520.0	–465.1	53.1	54.1
MnO_3Si	Manganese metasilicate ($MnSiO_3$)	cry	–1320.9	–1240.5	89.1	86.4
MnS	Manganese sulfide (MnS)	cry	–214.2	–218.4	78.2	50.0
MnSe	Manganese selenide (MnSe)	cry	–106.7	–111.7	90.8	51.0
Mn_2O_3	Manganese oxide (Mn_2O_3)	cry	–959.0	–881.1	110.5	107.7
Mn_2O_4Si	Manganese silicate (Mn_2SiO_4)	cry	–1730.5	–1632.1	163.2	129.9
Mn_3O_4	Manganese oxide (Mn_3O_4)	cry	1387.8	–1283.2	155.6	139.7
Mo	Molybdenum	cry	0.0		28.7	24.1
		gas	658.1	612.5	182.0	20.8
$MoNa_2O_4$	Sodium molybdate	cry	–1468.1	–1354.3	159.7	141.7
MoO_2	Molybdenum oxide (MoO_2)	cry	–588.9	533.0	46.3	56.0
MoO_3	Molybdenum oxide (MoO_3)	cry	–745.1	–668.0	77.7	75.0
MoO_4Pb	Lead molybdate ($PbMO_4$)	cry	–1051.9	–951.4	166.1	119.7
MoS_2	Molybdenum sulfide (MoS_2)	cry	–235.1	–225.9	62.6	63.6
N	Nitrogen	gas	472.7	455.5	153.3	20.8
$NNaO_2$	Sodium nitrite	cry	–358.7	–284.6	103.8	
$NNaO_3$	Sodium nitrate	cry	–467.9	–367.0	116.5	92.9
NO_2	Nitrogen dioxide	gas	33.2	51.3	240.1	37.2
NO_2Rb	Rubidium nitrite	cry	–367.4	–306.2	172.0	
NO_3Rb	Rubidium nitrate	cry	–495.1	–395.8	147.3	102.1
NO_3Tl	Thallium nitrate	cry	–243.9	–152.4	160.7	99.5
NP	Phosphorus nitride (PN)	cry	–63.0			
		gas	109.9	87.7	211.2	29.7
N_2	Nitrogen (N_2)	gas	0.0		191.6	29.1
N_2O	Nitrous oxide	gas	82.1	104.2	219.9	38.5
N_2O_3	Nitrogen trioxide	liq	50.3			
		gas	83.7	139.5	312.3	65.6
N_2O_4	Nitrogen tetroxide	liq	–19.5	97.5	209.2	142.7
		gas	9.2	97.9	304.3	77.3
N_2O_4Sr	Strontium nitrite	cry	–762.3			
N_2O_5	Nitrogen pentoxide	cry	–43.1	113.9	178.2	143.1
		gas	11.3	115.1	355.7	84.5
N_2O_6Pb	Lead nitrate ($Pb(NO_3)_2$)	cry	–451.9			
N_2O_6Ra	Radium nitrate	cry	–992.0	–796.1	222.0	
N_2O_6Sr	Strontium nitrate	cry	–978.2	–780.0	194.6	149.9

Molecular formula	Name	State	$\Delta_f H°$ kJ/mol	$\Delta_f G°$ kJ/mol	$S°$ J/mol K	C_p J/mol K
N_2O_6Zn	Zinc nitrate	cry	–483.7			
N_3Na	Sodium azide	cry	21.7	93.8	96.9	76.6
N_4Si_3	Silicon nitride (Si_3N_4)	cry	–743.5	–642.6	101.3	
Na	Sodium	cry	0.0		51.3	28.2
		gas	107.5	77.0	153.7	20.8
NaO_2	Sodium superoxide (NaO_2)	cry	–260.2	–218.4	115.9	72.1
Na_2	Sodium (Na_2)	gas	142.1	103.9	230.2	37.6
Na_2O	Sodium oxide (Na_2O)	cry	–414.2	–375.5	75.1	69.1
Na_2O_2	Sodium peroxide (Na_2O_2)	cry	–510.9	–447.7	95.0	89.2
Na_2O_3S	Sodium sulfite	cry	–1100.8	–1012.5	145.9	120.3
Na_2O_3Si	Sodium metasilicate	cry	–1554.9	–1462.8	113.9	
Na_2O_4S	Sodium sulfate	cry	–1387.1	–1270.2	149.6	128.2
Na_2S	Sodium sulfide (Na_2S)	cry	–364.8	–349.8	83.7	
Nb	Niobium	cry	0.0		36.4	24.6
		gas	725.9	681.1	186.3	30.2
NbO	Niobium oxide (NbO)	cry	–405.8	–378.6	48.1	41.3
NbO_2	Niobium oxide (NbO_2)	cry	–796.2	–740.5	54.5	57.5
Nb_2O_5	Niobium oxide (Nb_2O_5)	cry	–1899.5	–1766.0	137.2	132.1
Nd	Neodymium	cry	0.0		71.5	27.5
		gas	327.6	292.4	189.4	22.1
Nd_2O_3	Neodymium oxide	cry	–1807.9	–1720.8	158.6	111.3
Ne	Neon	gas	0.0		146.3	20.8
Ni	Nickel	cry	0.0		29.9	26.1
		gas	429.7	384.5	182.2	23.4
NiO_4S	Nickel sulfate ($NiSO_4$)	cry	–872.9	–759.7	92.0	138.0
NiS	Nickel sulfide (NiS)	cry	–82.0	–79.5	53.0	47.1
Ni_2O_3	Nickel oxide (Ni_2O_3)	cry	–489.5			
No	Nobelium	cry	0.0			
O	Oxygen	gas	249.2	231.7	161.1	21.9
OP	Phosphorus oxide (PO)	gas	–28.5	–51.9	222.8	31.8
OPb	Lead oxide (PbO) (yellow)	cry	–217.3	–187.9	68.7	45.8
	Lead oxide (PbO) (red)	cry	–219.0	–188.9	66.5	45.8
OPd	Palladium oxide (PdO)	cry	–85.4			31.4
		gas	348.9	325.9	218.0	
ORa	Radium oxide	cry	–523.0			
ORb_2	Rubidium oxide (Rb_2O)	cry	–339.0			
ORh	Rhodium oxide (RhO)	gas	385.0			
OS	Sulfur oxide (SO)	gas	6.3	–19.9	222.0	30.2
OSe	Selenium oxide (SeO)	gas	53.4	26.8	234.0	31.3
OSi	Silicon oxide (SiO)	gas	–99.6	–126.4	211.6	29.9
OSn	Tin oxide (SnO) (tetragonal)	cry	–280.7	–251.9	57.2	44.3
		gas	15.1	–8.4	232.1	31.6
OSr	Strontium oxide	cry	–592.0	–561.9	54.4	45.0
OTi	Titanium oxide (TiO)	cry	–519.7	–495.0	50.0	40.0
OTl_2	Thallium oxide (Tl_2O)	cry	–178.7	–147.3	126.0	
OU	Uranium oxide (UO)	gas	21.0			
OV	Vanadium oxide (VO)	cry	–431.8	–404.2	38.9	45.4
OZn	Zinc oxide	cry	–350.5	–320.5	43.7	40.3
O_2	Oxygen (O_2)	gas	0.0		205.2	29.4
O_2P	Phosphorous oxide (PO_2)	gas	–279.9	–281.6	252.1	39.5
O_2Pb	Lead oxide (PbO_2)	cry	–277.4	–217.3	68.6	64.6
O_2Rb	Rubidium superoxide (RbO_2)	cry	–278.7			
O_2Rb_2	Rubidium peroxide (Rb_2O_2)	cry	–472.0			
O_2Ru	Ruthenium oxide (RuO_2)	cry	–305.0			
O_2S	Sulfur dioxide	liq	–320.5			
		gas	–296.8	–300.1	248.2	39.9
O_2Se	Selenium dioxide	cry	–225.4			

Molecular formula	Name	State	$\Delta_f H°$ kJ/mol	$\Delta_f G°$ kJ/mol	$S°$ J/mol K	C_p J/mol K
O_2Si	Silicon dioxide (α–quartz)	cry	–910.7	–856.3	41.5	44.4
		gas	–322.0			
O_2Sn	Tin oxide (SnO_2) (tetragonal)	cry	–577.6	–515.8	49.0	52.6
O_2Te	Tellurium dioxide	cry	–322.6	–270.3	79.5	
O_2Th	Thorium oxide (ThO_2)	cry	–1226.4	–1169.2	65.2	61.8
O_2Ti	Titanium oxide (TiO_2) (rutile)	cry	–944.0	–888.8	50.6	55.0
O_2U	Uranium oxide (UO_2)	cry	–1085.0	–1031.8	77.0	63.6
		gas	–465.7	–471.5	274.6	51.4
O_2W	Tungsten oxide (WO_2)	cry	–589.7	–533.9	50.5	56.1
O_2Zr	Zirconium oxide (ZrO_2)	cry	–1100.6	–1042.8	50.4	56.2
O_3	Ozone	gas	142.7	163.2	238.9	39.2
O_3PbS	Lead sulfite ($PbSO_3$)	cry	–669.9			
O_3PbSi	Lead metasilicate ($PbSiO_3$)	cry	–1145.7	–1062.1	109.6	90.0
O_3Pr_2	Praseodymium oxide (Pr_2O_3)	cry	–1809.6			117.4
O_3Rh_2	Rhodium oxide (Rh_2O_3)	cry	–343.0			103.8
O_3S	Sulfur trioxide	cry	–454.5	–374.2	70.7	
		liq	–441.0	–373.8	113.8	
		gas	–395.7	–371.1	256.8	50.7
O_3Sc_2	Scandium oxide (Sc_2O_3)	cry	–1908.8	–1819.4	77.0	94.2
O_3SiSr	Strontium metasilicate	cry	–1633.9	–1549.7	96.7	88.5
O_3Sm_2	Samarium oxide (Sm_2O_3)	cry	–1823.0	–1734.6	151.0	114.5
O_3Tb_2	Terbium oxide (Tb_2O_3)	cry	–1865.2			115.9
O_3Ti_2	Titanium oxide (Ti_2O_3)	cry	–1520.9	–1434.2	78.8	97.4
O_3Tm_2	Thullium oxide (Tm_2O_3)	cry	–1888.7	–1794.5	139.7	116.7
O_3U	Uranium oxide (UO_3)	cry	–1223.8	–1145.7	96.1	81.7
O_3V_2	Vanadium oxide (V_2O_3)	cry	–1218.8	–1139.3	98.3	103.2
O_3W	Tungsten oxide (WO_3)	cry	–842.9	–764.0	75.9	73.8
O_3Y_2	Yttrium oxide (Y_2O_3)	cry	–1905.3	–1816.6	99.1	102.5
O_3Yb_2	Ytterbium oxide (Yb_2O_3)	cry	–1814.6	1726.7	133.1	115.4
O_4Os	Osmium oxide (OsO_4)	cry	–394.1	–304.9	143.9	
		gas	–337.2	–292.8	293.8	74.1
O_4PbS	Lead sulfate ($PbSO_4$)	cry	–920.0	–813.0	148.5	103.2
O_4PbSe	Lead selenate ($PbSeO_4$)	cry	–609.2	–504.9	167.8	
O_4Pb_2Si	Lead silicate (Pb_2SiO_4)	cry	–1363.1	–1252.6	186.6	137.2
O_4Pb_3	Lead oxide (Pb_3O_4)	cry	–718.4	–601.2	211.3	146.9
O_4RaS	Radium sulfate	cry	–1471.1	–1365.6	138.0	
O_4Rb_2S	Rubidium sulfate	cry	–1435.6	–1316.9	197.4	134.1
O_4Ru	Ruthenium oxide (RuO_4)	cry	–239.3	–152.2	146.4	
O_4SSr	Strontium sulfate	cry	–1453.1	–1340.9	117.0	
O_4STl_2	Thallium sulfate (Tl_2SO_4)	cry	–931.8	–830.4	230.5	
O_4SZn	Zinc sulfate	cry	–982.8	–871.5	110.5	99.2
O_4SiSr_2	Strontium silicate	cry	–2304.5	–2191.1	153.1	134.3
O_4SiZn_2	Zinc silicate	cry	–1636.7	–1523.2	131.4	123.3
O_4SiZr	Zirconium silicate ($ZrSiO_4$)	cry	–2033.4	–1919.1	84.1	98.7
O_5Ta_2	Tantalum oxide (Ta_2O_5)	cry	–2046.0	–1911.2	143.1	135.1
O_5Ti_3	Titanium oxide (Ti_3O_5)	cry	–2459.4	–2317.4	129.3	154.8
O_5V_2	Vanadium oxide (V_2O_5)	cry	–1550.6	–1419.5	131.0	127.7
O_5V_3	Vanadium oxide (V_3O_5)	cry	–1933.0	–1803.0	163.0	
O_7Re_2	Rhenium oxide (Re_2O_7)	cry	–1240.1	–1066.0	207.1	166.1
		gas	–1100.0	–994.0	452.0	
O_7U_3	Uranium oxide (U_3O_7)	cry	–3427.1	–3242.9	250.5	215.5
O_8S_2Zr	Zirconium sulfate	cry	–2217.1			172.0
O_8U_3	Uranium oxide (U_3O_8)	cry	–3574.8	–3369.5	282.6	238.4
O_9U_4	Uranium oxide (U_4O_9)	cry	–4510.4	–4275.1	334.1	293.3
Os	Osmium	cry	0.0		32.6	24.7
		gas	791.0	745.0	192.6	20.8
P	Phosphorus (white)	cry	0.0		41.1	23.8

STANDARD THERMODYNAMIC PROPERTIES OF CHEMICAL SUBSTANCES (continued)

Molecular formula	Name	State	$\Delta_f H°$ kJ/mol	$\Delta_f G°$ kJ/mol	$S°$ J/mol K	C_p J/mol K
	Phosphorus (red)	cry	–17.6		22.8	21.2
	Phosphorus (black)	cry	–39.3			
		gas	316.5	280.1	163.2	20.8
P_2	Phosphorus (P_2)	gas	144.0	103.5	218.1	32.1
P_4	Phosphorus (P_4)	gas	58.9	24.4	280.0	67.2
Pa	Protactinium	cry	0.0		51.9	
		gas	607.0	563.0	198.1	22.9
Pb	Lead	cry	0.0		64.8	26.4
		gas	195.2	162.2	175.4	20.8
PbS	Lead sulfide (PbS)	cry	–100.4	–98.7	91.2	49.5
PbSe	Lead selenide (PbSe)	cry	–102.9	–101.7	102.5	50.2
PbTe	Lead telluride (PbTe)	cry	–70.7	–69.5	110.0	50.5
Pd	Palladium	cry	0.0		37.6	26.0
		gas	378.2	339.7	167.1	20.8
PdS	Palladium sulfide (PdS)	cry	–75.0	–67.0	46.0	
Pm	Promethium	cry	0.0			
		gas			187.1	24.3
Po	Polonium	cry	0.0			
Pr	Praseodymium	cry	0.0		73.2	27.2
		gas	355.6	320.9	189.8	21.4
Pt	Platinum	cry	0.0		41.6	25.9
		gas	565.3	520.5	192.4	25.5
PtS	Platinum sulfide (PtS)	cry	–81.6	–76.1	55.1	43.4
PtS_2	Platinum sulfide (PtS_2)	cry	–108.8	–99.6	74.7	65.9
Pu	Plutonium	cry	0.0			
Ra	Radium	cry	0.0		71.0	
		gas	159.0	130.0	176.5	20.8
Rb	Rubidium	cry	0.0		76.8	31.1
		gas	80.9	53.1	170.1	20.8
Re	Rhenium	cry	0.0		36.9	25.5
		gas	769.9	724.6	188.9	20.8
Rh	Rhodium	cry	0.0		31.5	25.0
		gas	556.9	510.8	185.8	21.0
Rn	Radon	gas	0.0		176.2	20.8
Ru	Ruthenium	cry	0.0		28.5	24.1
		gas	642.7	595.8	186.5	21.5
S	Sulfur (rhombic)	cry	0.0		32.1	22.6
	Sulfur (monoclinic)	cry	0.3			
		gas	277.2	236.7	167.8	23.7
SSi	Silicon sulfide (SiS)	gas	112.5	60.9	223.7	32.3
SSn	Tin sulfide (SnS)	cry	–100.0	–98.3	77.0	49.3
SSr	Strontium sulfide	cry	–472.4	–467.8	68.2	48.7
STl_2	Thallium sulfide (Tl_2S)	cry	–97.1	–93.7	151.0	
SZn	Zinc sulfide (wurtzite)	cry	–192.6			
	Zinc sulfide (sphalerite)	cry	–206.0	–201.3	57.7	46.0
S_2	Sulfur (S_2)	gas	128.6	79.7	228.2	32.5
Sb	Antimony	cry	0.0		45.7	25.2
		gas	262.3	222.1	180.3	20.8
Sb_2	Antimony (Sb_2)	gas	235.6	187.0	254.9	36.4
Sc	Scandium	cry	0.0		34.6	25.5
		gas	377.8	336.0	174.8	22.1
Se	Selenium	cry	0.0		42.4	25.4
		gas	227.1	187.0	176.7	20.8
SeSr	Strontium selenide	cry	–385.8			
$SeTl_2$	Thallium selenide (Tl_2Se)	cry	–59.0	–59.0	172.0	
SeZn	Zinc selenide	cry	–163.0	–163.0	84.0	
Se_2	Selenium (Se_2)	gas	146.0	96.2	252.0	35.4

STANDARD THERMODYNAMIC PROPERTIES OF CHEMICAL SUBSTANCES (continued)

Molecular formula	Name	State	$\Delta_f H°$ kJ/mol	$\Delta_f G°$ kJ/mol	$S°$ J/mol K	C_p J/mol K
Si	Silicon	cry	0.0		18.8	20.0
		gas	450.0	405.5	168.0	22.3
Si_2	Silicon (Si_2)	gas	594.0	536.0	229.9	34.4
Sm	Samarium	cry	0.0		69.6	29.5
		gas	206.7	172.8	183.0	30.4
Sn	Tin (white)	cry	0.0		51.2	27.0
	Tin (gray)	cry	–2.1	0.1	44.1	25.8
		gas	301.2	266.2	168.5	21.3
Sr	Strontium	cry	0.0		52.3	26.4
		gas	164.4	130.9	164.6	20.8
Ta	Tantalum	cry	0.0		41.5	25.4
		gas	782.0	739.3	185.2	20.9
Tb	Terbium	cry	0.0		73.2	28.9
		gas	388.7	349.7	203.6	24.6
Tc	Technetium	cry	0.0			
		gas	678.0		181.1	20.8
Te	Tellurium	cry	0.0		49.7	25.7
		gas	196.7	157.1	182.7	20.8
Te_2	Tellurium (Te_2)	gas	168.2	118.0	268.1	36.7
Th	Thorium	cry	0.0		51.8	27.3
		gas	602.0	560.7	190.2	20.8
Ti	Titanium	cry	0.0		30.7	25.0
		gas	473.0	428.4	180.3	24.4
Tl	Thallium	cry	0.0		64.2	26.3
		gas	182.2	147.4	181.0	20.8
Tm	Thulium	cry	0.0		74.0	27.0
		gas	232.2	197.5	190.1	20.8
U	Uranium	cry	0.0		50.2	27.7
		gas	533.0	488.4	199.8	23.7
V	Vanadium	cry	0.0		28.9	24.9
		gas	514.2	754.4	182.3	26.0
W	Tungsten	cry	0.0		32.6	24.3
		gas	849.4	807.1	174.0	21.3
Xe	Xenon	gas	0.0		169.7	20.8
Y	Yttrium	cry	0.0		44.4	26.5
		gas	421.3	381.1	179.5	25.9
Yb	Ytterbium	cry	0.0		59.9	26.7
		gas	152.3	118.4	173.1	20.8
Zn	Zinc	cry	0.0		41.6	25.4
		gas	130.4	94.8	161.0	20.8
Zr	Zirconium	cry	0.0		39.0	25.4
		gas	608.8	566.5	181.4	26.7
Compounds containing carbon:						
C	Carbon (graphite)	cry	0.0		5.7	8.5
	Carbon (diamond)	cry	1.9	2.9	2.4	6.1
		gas	716.7	671.3	158.1	20.8
CAgN	Silver cyanide	cry	146.0	156.9	107.2	66.7
CAg_2O_3	Silver carbonate	cry	–505.8	–436.8	167.4	112.3
$CBaO_3$	Barium carbonate	cry	–1216.3	–1137.6	112.1	85.3
$CBeO_3$	Beryllium carbonate	cry	–1025.0			
$CBrClF_2$	Bromochlorodifluoromethane	gas			318.5	74.6
$CBrCl_2F$	Bromodichlorofluoromethane	gas			330.6	80.0
$CBrCl_3$	Bromotrichloromethane	gas	–41.8			85.3
$CBrF_3$	Bromotrifluoromethane	gas	–648.3			69.3
CBrN	Cyanogen bromide	cry	140.5			

Molecular formula	Name	State	$\Delta_f H°$ kJ/mol	$\Delta_f G°$ kJ/mol	$S°$ J/mol K	C_p J/mol K
		gas	186.2	165.3	248.3	46.9
CBr_2ClF	Dibromochlorofluoromethane	gas			342.8	82.4
CBr_2Cl_2	Dibromodichloromethane	gas			347.8	87.1
CBr_2F_2	Dibromodifluoromethane	gas			325.3	77.0
CBr_2O	Carbonyl bromide	liq	–127.2			
		gas	–96.2	–110.9	309.1	61.8
CBr_3Cl	Tribromochloromethane	gas			357.8	89.4
CBr_3F	Tribromofluoromethane	gas			345.9	84.4
CBr_4	Tetrabromomethane	cry	18.8	47.7	212.5	144.3
		gas	79.0	67.0	358.1	91.2
$CCaO_3$	Calcium carbonate (calcite)	cry	–1207.6	–1129.1	91.7	83.5
	Calcium carbonate (aragonite)	cry	–1207.8	–1128.2	88.0	82.3
$CCdO_3$	Cadmium carbonate	cry	–750.6	–669.4	92.5	
$CClFO$	Carbonyl chloride fluoride	gas			276.7	52.4
$CClF_3$	Chlorotrifluoromethane	gas	–706.3			66.9
$CClN$	Cyanogen chloride	liq	112.1			
		gas	138.0	131.0	236.2	45.0
CCl_2F_2	Dichlorodifluoromethane	gas	–477.4	–439.4	300.8	72.3
CCl_2O	Carbonyl chloride	gas	–219.1	–204.9	283.5	57.7
CCl_3	Trichloromethyl	gas	59.0			
CCl_3F	Trichlorofluoromethane	liq	–301.3	–236.8	225.4	121.6
		gas	–268.3			78.1
CCl_4	Tetrachloromethane	liq	–128.2			130.7
		gas	–95.8			83.3
$CCoO_3$	Cobalt carbonate ($CoCO_3$)	cry	–713.0			
$CCsHO_3$	Cesium hydrogen carbonate	cry	–966.1			
CCs_2O_3	Cesium carbonate	cry	–1139.7	–1054.3	204.5	123.9
$CCuN$	Copper cyanide (CuCN)	cry	96.2	111.3	84.5	
CFN	Cyanogen fluoride	gas			224.7	41.8
CF_2O	Carbonyl fluoride	gas	–639.8			46.8
CF_3	Trifluoromethyl	gas	–477.0	–464.0	264.5	49.6
CF_3I	Trifluoroiodomethane	gas	–587.8		307.4	70.9
CF_4	Tetrafluoromethane	gas	–933.6		261.6	61.1
CFe_3	Iron carbide (Fe_3C)	cry	25.1	20.1	104.6	105.9
$CFeO_3$	Iron carbonate ($FeCO_3$)	cry	–740.6	–666.7	92.9	82.1
CHg_2O_3	Mercury carbonate (Hg_2CO_3)	cry	–553.5	–468.1	180.0	
CIN	Cyanogen iodide	cry	166.2	185.0	96.2	
		gas	225.5	196.6	256.8	48.3
CI_4	Tetraiodomethane	gas			391.9	95.9
CKN	Potassium cyanide	cry	–113.0	–101.9	128.5	66.3
$CKNS$	Potassium thiocyanate	cry	–200.2	–178.3	124.3	88.5
CK_2O_3	Potassium carbonate	cry	–1151.0	–1063.5	155.5	114.4
CLi_2O_3	Lithium carbonate	cry	–1215.9	–1132.1	90.4	99.1
$CMgO_3$	Magnesium carbonate	cry	–1095.8	–1012.1	65.7	75.5
$CMnO_3$	Manganese carbonate ($MnCO_3$)	cry	–894.1	–816.7	85.8	81.5
CN	Cyanide (CN)	gas	437.6	407.5	202.6	29.2
$CNNa$	Sodium cyanide	cry	–87.5	–76.4	115.6	70.4
$CNNaO$	Sodium cyanate	cry	–405.4	–358.1	96.7	86.6
CN_4O_8	Tetranitromethane	liq	38.4			
CNa_2O_3	Sodium carbonate	cry	–1130.7	–1044.4	135.0	112.3
CO	Carbon monoxide	gas	–110.5	–137.2	197.7	29.1
COS	Carbon oxysulfide	gas	–142.0	–169.2	231.6	41.5
CO_2	Carbon dioxide	gas	–393.5	–394.4	213.8	37.1
CO_3Pb	Lead carbonate ($PbCO_3$)	cry	–699.1	–625.5	131.0	87.4
CO_3Rb_2	Rubidium carbonate	cry	–1136.0	–1051.0	181.3	117.6
CO_3Sr	Strontium carbonate	cry	–1220.1	–1140.1	97.1	81.4
CO_3Tl_2	Thallium carbonate (Tl_2CO_3)	cry	–700.0	–614.6	155.2	

STANDARD THERMODYNAMIC PROPERTIES OF CHEMICAL SUBSTANCES (continued)

Molecular formula	Name	State	$\Delta_f H°$ kJ/mol	$\Delta_f G°$ kJ/mol	$S°$ J/mol K	C_p J/mol K
CO_3Zn	Zinc carbonate	cry	–812.8	–731.5	82.4	79.7
CS	Carbon sulfide	gas	234.0	184.0	210.6	29.8
CS_2	Carbon disulfide	liq	89.0	64.6	151.3	76.4
		gas	116.6	67.1	237.8	45.4
CSe_2	Carbon diselenide	liq	164.8			
CSi	Silicon carbide (cubic)	cry	–65.3	–62.8	16.6	26.9
CSi	Silicon carbide (hexagonal)	cry	–62.8	–60.2	16.5	26.7
CH	Methylidyne	gas	595.8			
CHBrClF	Bromochlorofluoromethane	gas			304.3	63.2
$CHBrCl_2$	Bromodichloromethane	gas			316.4	67.4
$CHBrF_2$	Bromodifluoromethane	gas			295.1	58.7
$CHBr_2Cl$	Chlorodibromomethane	gas			327.7	69.2
$CHBr_2F$	Dibromofluoromethane	gas			316.8	65.1
$CHBr_3$	Tribromomethane	liq	–28.5	–5.0	220.9	130.7
		gas	17.0	8.0	330.9	71.2
$CHClF_2$	Chlorodifluoromethane	gas	–482.6		280.9	55.9
$CHCl_2F$	Dichlorofluoromethane	gas			293.1	60.9
$CHCl_3$	Trichloromethane	liq	–134.5	–73.7	201.7	114.2
		gas	–103.1	6.0	295.7	65.7
CHFO	Formyl fluoride	gas			246.6	39.9
CHF_3	Trifluoromethane	gas	–695.4		259.7	51.0
CHI_3	Triiodomethane	cry	141.0			
		gas			356.2	75.0
$CHKO_2$	Potassium formate	cry	–679.7			
$CHKO_3$	Potassium hydrogen carbonate	cry	–963.2	–863.5	115.5	
CHN	Hydrogen cyanide	liq	108.9	125.0	112.8	70.6
		gas	135.1	124.7	201.8	35.9
CHNO	Isocyanic acid (HNCO)	gas			238.0	44.9
CHNS	Isothiocyanic acid	gas	127.6	113.0	247.8	46.9
$CHNaO_2$	Sodium formate	cry	–666.5	–599.9	103.8	82.7
$CHNaO_3$	Sodium hydrogen carbonate	cry	–950.8	–851.0	101.7	87.6
CHO	Oxomethyl (HCO)	gas	43.1	28.0	224.7	34.6
CH_2	Methylene	gas	390.4	372.9	194.9	33.8
CH_2BrCl	Bromochloromethane	gas			287.6	52.7
CH_2BrF	Bromofluoromethane	gas			276.3	49.2
CH_2Br_2	Dibromomethane	gas			293.2	54.7
CH_2ClF	Chlorofluoromethane	gas			264.4	47.0
CH_2Cl_2	Dichloromethane	liq	–124.1		177.8	101.2
		gas	–95.6		270.2	51.0
CH_2F_2	Difluoromethane	gas	–452.2		246.7	42.9
CH_2I_2	Diiodomethane	liq	66.9	90.4	174.1	134.0
		gas	113.0	95.8	309.7	57.7
CH_2N_2	Cyanamide	cry	58.8			
CH_2N_2	Diazomethane	gas			242.9	52.5
CH_2O	Formaldehyde	gas	–108.6	–102.5	218.8	35.4
CH_2O_2	Formic acid	liq	–424.7	–361.4	129.0	99.0
		gas	–378.6			
CH_3	Methyl	gas	145.7	147.9	194.2	38.7
CH_3BO	Carbonyltrihydroboron (BH_3CO)	gas	–111.2	–92.9	249.4	59.5
CH_3Br	Bromomethane	liq	–59.4			
		gas	–35.5	–26.3	246.4	42.4
CH_3Cl	Chloromethane	gas	–81.9		234.6	40.8
CH_3F	Fluoromethane	gas			222.9	37.5
CH_3I	Iodomethane	liq	–12.3		163.2	126.0
		gas	14.7		254.1	44.1
CH_3NO	Formamide	liq	–254.0			
CH_3NO_2	Nitromethane	liq	–113.1	–14.4	171.8	106.6

Molecular formula	Name	State	$\Delta_f H°$ kJ/mol	$\Delta_f G°$ kJ/mol	$S°$ J/mol K	C_p J/mol K
		gas	–74.7	–6.8	275.0	57.3
CH_3NO_3	Methyl nitrate	liq	–159.0	–43.4	217.1	157.3
CH_4	Methane	gas	–74.4	–50.3	186.3	35.3
CH_4N_2	Ammonium cyanide	cry	0.4			134.0
CH_4N_2O	Urea	cry	–333.6			
CH_4O	Methanol	liq	–239.1	–166.6	126.8	81.1
		gas	–201.5	–162.6	239.8	43.9
CH_4S	Methanethiol	liq	–46.4	–7.7	169.2	90.5
		gas	–22.3	–9.3	255.2	50.3
CH_5N	Methylamine	liq	–47.3	35.7	150.2	102.1
		gas	–22.5	32.7	242.9	50.1
CH_5NO_3	Ammonium hydrogen carbonate	cry	–849.4	–665.9	120.9	
CH_6N_2	Methylhydrazine	liq	54.0	180.0	165.9	134.9
		gas	94.3	187.0	278.8	71.1
CH_6Si	Methylsilane	gas			256.5	65.9
C_2	Carbon (C_2)	gas	831.9	775.9	199.4	43.2
$C_2Br_2ClF_3$	1,2–Dibromo–1–chloro–1,2,2-trifluoroethane	gas	–656.6			
$C_2Br_2F_4$	1,2–Dibromotetrafluoroethane	gas	–789.1			
C_2Br_4	Tetrabromoethylene	gas			387.1	102.7
C_2Br_6	Hexabromoethane	gas			441.9	139.3
C_2Ca	Calcium carbide	cry	–59.8	–64.9	70.0	62.7
C_2CaN_2	Calcium cyanide	cry	–184.5			
C_2CaO_4	Calcium oxalate	cry	–1360.6			
C_2ClF_3	Chlorotrifluoroethylene	gas	–555.2	–523.8	322.1	83.9
$C_2Cl_2F_4$	1,2–Dichlorotetrafluoroethane	liq	–939.7			111.7
		gas	–916.3			
$C_2Cl_3F_3$	1,1,2–Trichlorotrifluoroethane	liq	–805.8			170.1
		gas	–777.3			
C_2Cl_3N	Trichloroacetonitrile	gas			336.6	96.1
C_2Cl_4	Tetrachloroethylene	liq	–50.6	3.0	266.9	143.4
$C_2Cl_4F_2$	1,1,1,2–Tetrachloro–2,2–difluoroethane	gas	–489.9	–407.0	382.9	123.4
C_2Cl_4O	Trichloroacetyl chloride	liq	–280.8			
C_2Cl_6	Hexachloroethane	cry	–202.8		237.3	198.2
C_2F_3N	Trifluoroacetonitrile	gas	–497.9		298.1	77.9
C_2F_4	Tetrafluoroethylene	cry	–820.5			
		gas	–658.9		300.1	80.5
C_2F_6	Hexafluoroethane	gas	–1344.2		332.3	106.7
C_2HgO_4	Mercury oxalate (HgC_2O_4)	cry	–678.2			
C_2I_2	Diiodoacetylene	gas			313.1	70.3
C_2I_4	Tetraiodoethylene	cry	305.0			
$C_2K_2O_4$	Potassium oxalate	gas	–1346.0			
C_2MgO_4	Magnesium oxalate	cry	–1269.0			
C_2N_2	Cyanogen	gas	306.7		241.9	56.8
		liq	285.9			
$C_2Na_2O_4$	Sodium oxalate	gas	–1318.0			
C_2O_4Pb	Lead oxalate (PbC_2O_4)	cry	–851.4	–750.1	146.0	105.4
C_2HBr	Bromoacetylene	gas			253.7	55.7
C_2HCl	Chloroacetylene	gas			242.0	54.3
C_2HClF_2	1–Chloro–2,2–difluoroethylene	gas	–315.5	–289.1	303.0	72.1
C_2HCl_2F	1,1–Dichloro–2–fluoroethylene	gas			313.9	76.5
$C_2HCl_2F_3$	2,2–Dichloro–1,1,1–trifluoroethane	gas			352.8	102.5
C_2HCl_3	Trichloroethylene	liq	–43.6		228.4	124.4
		gas	–8.1		324.8	80.3
C_2HCl_3O	Trichloroacetaldehyde	liq	–236.2			151.0
		gas	–196.6			

Molecular formula	Name	State	$\Delta_f H°$ kJ/mol	$\Delta_f G°$ kJ/mol	$S°$ J/mol K	C_p J/mol K
C_2HCl_3O	Dichloroacetyl chloride	liq	–280.4			
$C_2HCl_3O_2$	Trichloroacetic acid	cry	–503.3			
C_2HCl_5	Pentachloroethane	liq	–187.6			173.8
C_2HF	Fluoroacetylene	gas			231.7	52.4
C_2HF_3	Trifluoroethylene	gas	–490.4			
$C_2HF_3O_2$	Trifluoroacetic acid	liq	–1069.9			
		gas	–1031.4			
C_2H_2	Acetylene	gas	228.2	210.7	200.9	43.9
$C_2H_2Br_2$	*cis*–1,2–Dibromoethylene	gas			311.3	68.8
$C_2H_2Br_2$	*trans*–1,2–Dibromoethylene	gas			313.5	70.3
$C_2H_2ClF_3$	2–Chloro–1,1,1–trifluoroethane	gas			326.5	89.1
$C_2H_2Cl_2$	1,1–Dichloroethylene	liq	–23.9	24.1	201.5	111.3
		gas	2.6	25.4	289.0	67.1
$C_2H_2Cl_2$	*cis*–1,2–Dichloroethylene	liq	–26.4		198.4	116.4
		gas	4.6		289.6	65.1
$C_2H_2Cl_2$	*trans*–1,2–Dichloroethylene	liq	–23.1	27.3	195.9	116.8
		gas	6.2	28.6	290.0	66.7
$C_2H_2Cl_2O$	Chloroacetyl chloride	liq	–283.7			
$C_2H_2Cl_4$	1,1,1,2–Tetrachloroethane	gas			356.0	102.7
$C_2H_2Cl_4$	1,1,2,2–Tetrachloroethane	liq	–195.0		246.9	162.3
		gas	–149.2		362.8	100.8
$C_2H_2F_2$	1,1–Difluoroethylene	gas	–335.0		266.2	60.1
$C_2H_2F_2$	*cis*–1,2–Difluoroethylene	gas			268.3	58.2
C_2H_2O	Ketene	liq	–67.9			
		gas	–47.5	–48.3	247.6	51.8
$C_2H_2O_2$	Glyoxal	gas	–212.0			
$C_2H_2O_4$	Oxalic acid	cry	–821.7		109.8	91.0
		gas	–723.7			
$C_2H_2O_4Sr$	Strontium formate	cry	–1393.3			
C_2H_3Br	Bromoethylene	gas	79.2	81.8	275.8	55.5
C_2H_3BrO	Acetyl bromide	liq	–223.4			
C_2H_3Cl	Chloroethylene	cry	–94.1			59.4
		liq	14.6			
		gas	37.3	53.6	264.0	53.7
$C_2H_3ClF_2$	1–Chloro–1,1–difluoroethane	gas			307.2	82.5
C_2H_3ClO	Acetyl chloride	liq	–273.8	–208.0	200.8	117.0
		gas	–243.5	–205.8	295.1	67.8
$C_2H_3ClO_2$	Chloroacetic acid	cry	–510.5			
$C_2H_3Cl_2F$	1,1–Dichloro–1–fluoroethane	gas			320.2	88.7
$C_2H_3Cl_3$	1,1,1–Trichloroethane	liq	–177.4		227.4	144.3
		gas	–144.6		323.1	93.3
$C_2H_3Cl_3$	1,1,2–Trichloroethane	liq	–191.5		232.6	150.9
		gas	–151.2		337.2	89.0
C_2H_3F	Fluoroethylene	gas	–138.8			
C_2H_3FO	Acetyl fluoride	liq	–467.2			
		gas	–442.1			
$C_2H_3F_3$	1,1,1–Trifluoroethane	gas	–744.6		279.9	78.2
$C_2H_3F_3$	1,1,2–Trifluoroethane	gas	–730.9			
C_2H_3I	Iodoethylene	gas			285.0	57.9
C_2H_3IO	Acetyl iodide	liq	–162.5			
$C_2H_3KO_2$	Potassium acetate	cry	–723.0			
C_2H_3N	Acetonitrile	liq	31.4	77.2	149.6	91.4
		gas	64.3	81.7	245.1	52.2
C_2H_3N	Isocyanomethane	liq	117.2	159.5	159.0	
		gas	149.0	165.7	246.9	52.9
C_2H_3NO	Methylisocyanate	liq	–92.0			
$C_2H_3NaO_2$	Sodium acetate	cry	–708.8	–607.2	123.0	79.9

Molecular formula	Name	State	$\Delta_f H°$ kJ/mol	$\Delta_f G°$ kJ/mol	$S°$ J/mol K	C_p J/mol K
C_2H_4	Ethylene	gas	52.5	68.4	219.6	43.6
$C_2H_4Br_2$	Dibromoethane	gas			327.7	80.8
$C_2H_4Br_2$	1,2–Dibromoethane	liq	–79.2		223.3	136.0
C_2H_4ClF	1–Chloro–1–fluoroethane	gas	–313.4			
$C_2H_4Cl_2$	1,1–Dichloroethane	liq	–158.4	–73.8	211.8	126.3
		gas	–127.7	–70.8	305.1	76.2
$C_2H_4Cl_2$	1,2–Dichloroethane	liq	–167.4			128.4
		gas	–126.9		308.4	78.7
C_2H_4O	Acetaldehyde	liq	–191.8	–127.6	160.2	89.0
		gas	–166.2	–132.8	263.7	55.3
C_2H_4O	Ethylene oxide	liq	–77.8	–11.8	153.9	88.0
		gas	–52.6	–13.0	242.5	47.9
$C_2H_4O_2$	Acetic acid	liq	–484.5	–389.9	159.8	123.3
		gas	–432.8	–374.5	282.5	66.5
$C_2H_4O_2$	Methyl formate	liq	–386.1			119.1
		gas	–355.5		285.3	64.4
C_2H_4Si	Ethynylsilane	gas			269.4	72.6
C_2H_5Br	Bromoethane	liq	–90.1	–25.8	198.7	100.8
		gas	–61.9	–23.9	286.7	64.5
C_2H_5Cl	Chloroethane	liq	–136.5	–59.3	190.8	104.3
		gas	–112.2	–60.4	276.0	62.8
C_2H_5F	Fluoroethane	gas			264.5	58.6
C_2H_5I	Iodoethane	liq	–40.2	14.7	211.7	115.1
		gas	–7.7	19.2	306.0	66.9
C_2H_5N	Ethyleneimine	liq	91.9			
		gas	126.5			
C_2H_5NO	Acetamide	cry	–317.0		115.0	91.3
$C_2H_5NO_2$	Nitroethane	liq	–143.9			134.4
$C_2H_5NO_2$	Glycine	cry	–528.5			
		gas	–392.1			
C_2H_6	Ethane	gas	–83.8	–31.9	229.6	52.6
C_2H_6Cd	Dimethyl cadmium	liq	63.6	139.0	201.9	132.0
		gas	101.6	146.9	303.0	
C_2H_6Hg	Dimethyl mercury	liq	59.8	140.3	209.0	
		gas	94.4	146.1	306.0	83.3
C_2H_6O	Dimethyl ether	gas	–184.1	–112.6	266.4	64.4
C_2H_6O	Ethanol	liq	–277.7	–174.8	160.7	112.3
		gas	–235.1	–168.5	282.7	65.4
C_2H_6OS	Dimethyl sulfoxide	liq	–204.2	–99.9	188.3	153.0
$C_2H_6O_2$	Ethylene glycol	liq	–455.3		163.2	148.6
		gas	–387.5		303.8	82.7
$C_2H_6O_2S$	Dimethyl sulfone	cry	–451.0	–302.4	142.0	
		gas	–371.1	–272.7	310.6	100.0
C_2H_6S	Dimethyl sulfide	liq	–65.4		196.4	118.1
		gas	–37.5		286.0	74.1
C_2H_6S	Ethanethiol	liq	–73.6	–5.5	207.0	117.9
		gas	–45.3	–4.8	296.2	72.7
$C_2H_6S_2$	Dimethyl disulfide	liq	–62.6		235.4	146.1
C_2H_6Zn	Dimethyl zinc	liq	23.4		201.6	129.2
		gas	53.0			
C_2H_7N	Dimethylamine	liq	–43.9	70.0	182.3	137.7
		gas	–18.5	68.5	273.1	70.7
C_2H_7N	Ethylamine	liq	–74.1			130.0
		gas	–47.5	36.3	283.8	71.5
$C_2H_8N_2$	1,1–Dimethylhydrazine	liq	48.9	206.4	198.0	164.1
		gas	83.9			
$C_2H_8N_2$	1,2–Ethanediamine	liq	–63.0			172.6

Molecular formula	Name	State	$\Delta_f H°$ kJ/mol	$\Delta_f G°$ kJ/mol	$S°$ J/mol K	C_p J/mol K
$C_2H_8N_2O_4$	Ammonium oxalate	cry	–1123.0			226.0
C_3F_8	Perfluoropropane	gas	–1783.2			
$C_3H_3F_3$	3,3,3–Trifluoropropene	gas	–614.2			
C_3H_3N	Propenenitrile	liq	147.1			
		gas	180.6			
C_3H_3NO	Oxazole	liq	–48.0			
		gas	–15.5			
C_3H_3NO	Isoxazole	liq	42.1			
		gas	78.6			
C_3H_4	Propyne	gas	184.9			
C_3H_4	Allene	gas	190.5			
C_3H_4	Cyclopropene	gas	277.1			
$C_3H_4Cl_2$	2,3–Dichloropropene	liq	–73.3			
$C_3H_4N_2$	Imidazole	cry	58.5			
$C_3H_4O_2$	Propenoic acid	liq	–383.8			145.7
$C_3H_4O_2$	2–Oxetanone	liq	–329.9		175.3	122.1
C_3H_5Br	3–Bromopropene	liq	12.2			
		gas	45.2			
C_3H_5Cl	2–Chloropropene	gas	–21.0			
C_3H_5ClO	Epichlorohydrin	liq	–148.4			131.6
		gas	–107.8			
$C_3H_5Cl_3$	1,2,3–Trichloropropane	liq	–230.6			183.6
C_3H_5N	Propanenitrile	liq	15.5			119.3
		gas	51.5			
$C_3H_5N_3O_9$	Trinitroglycerol	liq	–370.9			
		gas	–270.9			
C_3H_6	Propene	liq	1.7			
		gas	20.0			
C_3H_6	Cyclopropane	gas	53.3			
$C_3H_6Br_2$	1,2–Dibromopropane	gas	–71.5			
$C_3H_6Cl_2$	1,2–Dichloropropane	liq	–198.8			149.1
		gas	–162.8			
$C_3H_6Cl_2$	1,3–Dichloropropane	liq	–200.0			
C_3H_6O	Acetone	liq	–248.1		199.8	126.3
		gas	–217.3		297.6	75.0
C_3H_6O	Allyl alcohol	liq	–171.8			138.9
		gas	–124.5			
C_3H_6O	Propanal	liq	–215.3			
		gas	–185.6		304.5	80.7
C_3H_6O	Methyloxirane	liq	–122.6		196.5	120.4
		gas	–94.7		286.9	72.6
$C_3H_6O_2$	Methyl acetate	liq	–445.8			141.9
		gas	–411.9		324.4	86.0
$C_3H_6O_2$	Propanoic acid	liq	–510.7		191.0	152.8
$C_3H_6O_2$	1,3–Dioxolane	liq	–333.5			118.0
		gas	–298.0			
$C_3H_6O_3$	Trioxane	cry	–522.5		133.0	111.4
C_3H_6S	Thiacyclobutane	liq	24.7		184.9	
		gas	60.6			
C_3H_7Br	1–Bromopropane	liq	–121.8			
		gas	–87.0			
C_3H_7Br	2–Bromopropane	liq	–130.5			
		gas	–99.4			
C_3H_7Cl	1–Chloropropane	liq	–160.6			
		gas	–131.9			
C_3H_7Cl	2–Chloropropane	liq	–172.1			
		gas	–144.9			

STANDARD THERMODYNAMIC PROPERTIES OF CHEMICAL SUBSTANCES (continued)

Molecular formula	Name	State	$\Delta_f H°$ kJ/mol	$\Delta_f G°$ kJ/mol	$S°$ J/mol K	C_p J/mol K
C_3H_7F	1–Fluoropropane	gas	–285.9			
C_3H_7F	2–Fluoropropane	gas	–293.5			
C_3H_7I	1–Iodopropane	liq	–66.0			
C_3H_7I	2–Iodopropane	liq	–74.8			
		gas	–40.3			
C_3H_7N	Cyclopropylamine	liq	45.8		187.7	147.1
		gas	77.0			
C_3H_7NO	*N,N*–Dimethylformamide	liq	–239.3			150.6
		gas	–191.7			
$C_3H_7NO_2$	1–Nitropropane	liq	–167.2			
$C_3H_7NO_2$	2–Nitropropane	liq	–180.3			170.3
$C_3H_7NO_2$	l–Alanine	cry	–604.0			
		gas	–465.9			
$C_3H_7NO_2$	d–Alanine	cry	–561.2			
$C_3H_7NO_2S$	l–Cysteine	cry	–515.5			
$C_3H_7NO_3$	l–Serine	cry	–732.7			
C_3H_8	Propane	gas	–104.7			
C_3H_8O	1–Propanol	liq	–302.6		193.6	143.9
		gas	–255.1		322.6	85.6
C_3H_8O	2–Propanol	liq	–318.1		181.1	156.5
		gas	–272.8		309.2	89.3
C_3H_8O	Ethyl methyl ether	gas	–216.4		309.2	93.3
$C_3H_8O_2$	1,2–Propylene glycol	liq	–485.7			190.8
$C_3H_8O_2$	1,3–Propylene glycol	liq	–464.9			
$C_3H_8O_2$	Dimethoxymethane	liq	–377.7		244.0	162.0
		gas	–348.4			
$C_3H_8O_3$	Glycerol	liq	–668.5		206.3	218.9
		gas	–582.7			
C_3H_8S	Ethyl methyl sulfide	liq	–91.6		239.1	144.6
		gas	–59.6			
C_3H_8S	1–Propanethiol	liq	–99.9		242.5	144.6
		gas	–67.9			
C_3H_8S	2–Propanethiol	liq	–105.9		233.5	145.3
		gas	–76.2			
C_3H_9Al	Trimethyl aluminum	liq	–136.4	–9.9	209.4	155.6
C_3H_9B	Trimethylborane	liq	–143.1	–32.1	238.9	
		gas	–124.3	–35.9	314.7	88.5
C_3H_9ClSi	Trimethylchlorosilane	liq	–382.8	–246.4	278.2	
		gas	–352.8	–243.5	369.1	
C_3H_9N	Propylamine	liq	–101.5			164.1
		gas	–70.1	39.9	325.4	91.2
C_3H_9N	Isopropylamine	liq	–112.3		218.3	163.8
		gas	–83.7	32.2	312.2	97.5
C_3H_9N	Trimethylamine	liq	–45.7		208.5	137.9
		gas	–23.7		287.1	91.8
$C_3H_{10}Si$	Trimethylsilane	gas			331.0	117.9
$C_3H_{12}BN$	Trimethylamine borane	cry	–142.5	70.7	187.0	
$C_3H_{12}BN$	Aminetrimethylboron	cry	–284.1	–79.3	218.0	
C_4F_8	Perfluorocyclobutane	gas	–1542.6			
C_4NiO_4	Nickel carbonyl	liq	–633.0	–588.2	313.4	204.6
		gas	–602.9	–587.2	410.6	145.2
$C_4H_2O_3$	Maleic anhydride	cry	–469.8			
$C_4H_4N_2$	Pyridazine	liq	224.8			
		gas	278.3			
$C_4H_4N_2$	Pyrimidine	liq	145.9			
$C_4H_4N_2$	Succinonitrile	liq	139.7		191.6	145.6
		gas	209.7			

Molecular formula	Name	State	$\Delta_f H°$ kJ/mol	$\Delta_f G°$ kJ/mol	$S°$ J/mol K	C_p J/mol K
$C_4H_4N_2O_2$	Uracil	cry	–429.4			120.5
		gas	–302.9			
C_4H_4O	Furan	liq	–62.3		177.0	115.3
		gas	–34.9		267.2	65.4
$C_4H_4O_2$	Diketene	liq	–233.1			
$C_4H_4O_3$	Succinic anhydride	cry	–607.8			
$C_4H_4O_4$	Fumaric acid	cry	–811.7		168.0	142.0
		gas	–675.8			
$C_4H_4O_4$	Maleic acid	cry	–789.4		160.8	137.0
		gas	–679.4			
C_4H_4S	Thiophene	liq	80.2		181.2	123.8
		gas	114.9			
C_4H_5N	Cyclopropanecarbonitrile	liq	140.8			
		gas	181.8			
C_4H_5N	Pyrrole	liq	63.1		156.4	127.7
C_4H_5NS	4–Methylthiazole	liq	67.9			
$C_4H_5N_3O$	Cytosine	cry	–221.3			132.6
C_4H_6	1,2–Butadiene	liq	139.0			
		gas	162.3			
C_4H_6	1,3–Butadiene	liq	87.9		199.0	123.6
		gas	110.0			
C_4H_6	1–Butyne	liq	141.9			
		gas	165.2			
C_4H_6	2–Butyne	liq	119.1			
		gas	145.7			
C_4H_6	Cyclobutene	gas	156.7			
C_4H_6O	*trans*–2–Butenal	liq	–138.7			
		gas	–100.6			
C_4H_6O	Divinyl ether	liq	–39.8			
		gas	–13.6			
$C_4H_6O_2$	Methyl acrylate	liq	–362.2		239.5	158.8
		gas	–333.0			
$C_4H_6O_2$	Vinyl acetate	liq	–280.1			
		gas	–314.9			
$C_4H_6O_3$	Acetic anhydride	liq	–624.4			
		gas	–572.5			
$C_4H_6O_4$	Succinic acid	cry	–940.5		167.3	153.1
		gas	–823.0			
$C_4H_6O_4$	Dimethyl oxalate	cry	–756.3			
		gas	–708.9			
C_4H_6S	2,3–Dihydrothiophene	liq	52.9			
C_4H_6S	2,5–Dihydrothiophene	liq	47.0			
C_4H_7N	Butanenitrile	liq	–5.8			
C_4H_7N	2–Methylpropanenitrile	liq	–13.8			
C_4H_7NO	2–Pyrrolidone	liq	–286.2			
$C_4H_7NO_4$	l–Aspartic acid	cry	–973.3			
C_4H_8	1–Butene	liq	–20.5		227.0	118.0
		gas	0.1			
C_4H_8	*cis*–2–Butene	liq	–29.7		219.9	127.0
		gas	–7.1			
C_4H_8	*trans*–2–Butene	liq	–33.0			
		gas	–11.4			
C_4H_8	Isobutene	liq	–37.5			
		gas	–16.9			
C_4H_8	Cyclobutane	liq	3.7			
		gas	28.4			
C_4H_8	Methylcyclopropane	liq	1.7			

Molecular formula	Name	State	$\Delta_f H°$ kJ/mol	$\Delta_f G°$ kJ/mol	$S°$ J/mol K	C_p J/mol K
$C_4H_8Br_2$	1,4–Dibromobutane	liq	–140.1			
		gas	–87.0			
$C_4H_8N_2O_3$	l–Asparagine	cry	–789.4			
C_4H_8O	1,2–Epoxybutane	liq	–168.9		230.9	147.0
C_4H_8O	Butanal	liq	–239.2		246.6	163.7
		gas	–204.8		343.7	103.4
C_4H_8O	Isobutanal	liq	–247.4			
		gas	–215.8			
C_4H_8O	2–Butanone	liq	–273.3		239.1	158.7
		gas	–238.7		339.9	101.7
C_4H_8O	Tetrahydrofuran	liq	–216.2		204.3	124.0
		gas	–184.2		302.4	76.3
C_4H_8O	Ethyl vinyl ether	liq	–167.4			
		gas	–140.8			
$C_4H_8O_2$	1,3–Dioxane	liq	–379.7			143.9
$C_4H_8O_2$	1,4–Dioxane	liq	–353.9		270.2	152.1
		gas	–315.8			
$C_4H_8O_2$	Ethyl acetate	liq	–479.3		257.7	170.7
		gas	–444.1			
$C_4H_8O_2$	Propyl formate	liq	–500.3			
$C_4H_8O_2$	Butanoic acid	liq	–533.8		222.2	178.6
C_4H_8S	Tetrahydrothiophene	liq	–72.9			
C_4H_9Br	1–Bromobutane	liq	–143.8			
C_4H_9Br	2–Bromobutane	liq	–154.8			
		gas	–120.3			
C_4H_9Br	2–Bromo–2–methylpropane	liq	–163.8			
		gas	–132.4			
C_4H_9Cl	1–Chlorobutane	liq	–188.1			
		gas	–154.6			
C_4H_9Cl	1–Chloro–2–methylpropane	liq	–191.1			
		gas	–159.4			
C_4H_9Cl	2–Chloro–2–methylpropane	liq	–211.2			
		gas	–182.2			
C_4H_9I	2–Iodo–2–methylpropane	liq	–107.4			
C_4H_9N	Pyrrolidine	liq	–41.0		204.1	156.6
		gas	–3.4			
C_4H_9NO	*N,N*–Dimethylacetamide	liq	–278.3			175.6
$C_4H_9NO_3$	l–Threonine	cry	–807.2			
C_4H_{10}	Butane	liq	–146.6			140.9
		gas	–125.6			
C_4H_{10}	Isobutane	liq	–153.5			
		gas	–134.2			
$C_4H_{10}Hg$	Diethyl mercury	liq	30.1			182.8
		gas	75.3			
$C_4H_{10}O$	1–Butanol	liq	–327.3		225.8	177.2
$C_4H_{10}O$	2–Butanol	liq	–342.6		214.9	196.9
		gas	–292.9		359.5	112.7
$C_4H_{10}O$	2–Methyl–2–propanol	liq	–359.2		193.3	218.6
		gas	–312.5		326.7	113.6
$C_4H_{10}O$	2–Methyl–1–propanol	liq	–334.7		214.7	181.5
$C_4H_{10}O$	Diethyl ether	liq	–279.3		172.4	175.6
		gas	–252.1		342.7	119.5
$C_4H_{10}O$	Methyl propyl ether	liq	–266.0		262.9	165.4
		gas	–238.2			
$C_4H_{10}O$	Isopropyl methyl ether	liq	–278.7		253.8	161.9
		gas	–252.0			
$C_4H_{10}O_2$	1,3–Butanediol	liq	–501.0			

Molecular formula	Name	State	$\Delta_f H°$ kJ/mol	$\Delta_f G°$ kJ/mol	$S°$ J/mol K	C_p J/mol K
		gas	–433.2			
$C_4H_{10}O_2$	1,4–Butanediol	liq	–503.3		223.4	200.1
		gas	–426.7			
$C_4H_{10}O_3$	Diethylene glycol	liq	–628.5			244.8
		gas	–571.2			
$C_4H_{10}S$	1–Butanethiol	liq	–124.7			171.2
		gas	–88.1			
$C_4H_{10}S$	2–Butanethiol	liq	–131.0			
		gas	–96.9			
$C_4H_{10}S$	2–Methyl–1–propanethiol	liq	–132.0			
$C_4H_{10}S$	2–Methyl–2–propanethiol	liq	–140.5			
		gas	–109.6			
$C_4H_{10}S$	Diethyl sulfide	liq	–119.4		269.3	171.4
		gas	–83.6		368.1	117.0
$C_4H_{10}S$	Methyl propyl sulfide	liq	–118.5		272.5	171.6
		gas	–82.3			
$C_4H_{10}S$	Isopropyl methyl sulfide	liq	–124.7		263.1	172.4
		gas	–90.5			
$C_4H_{10}S_2$	Diethyl disulfide	liq	–120.1		269.3	171.4
$C_4H_{11}N$	Butylamine	liq	–127.7			179.2
		gas	–92.0			
$C_4H_{11}N$	Isobutylamine	liq	–132.6			183.2
		gas	98.7			
$C_4H_{11}N$	*sec*–Butylamine	liq	–137.5			
		gas	–104.9			
$C_4H_{11}N$	*tert*–Butylamine	liq	–150.6			192.1
		gas	–120.9			
$C_4H_{11}N$	Diethylamine	liq	–103.7			169.2
		gas	–72.5			
$C_4H_{12}Pb$	Tetramethyl lead	liq	97.9			
		gas	135.9			
$C_4H_{12}Si$	Tetramethylsilane	liq	–264.0	–100.0	277.3	204.1
		gas	–239.1	–99.9	359.0	143.9
$C_4H_{12}Sn$	Tetramethylstannane	liq	–52.3			
		gas	–18.8			
C_5FeO_5	Iron pentacarbonyl	liq	–774.0	–705.3	338.1	240.6
$C_5H_4N_4O$	Hypoxanthine	cry	–110.8		145.6	134.5
$C_5H_4N_4O_2$	Xanthine	cry	–379.6		161.1	151.3
$C_5H_4N_4O_3$	Uric acid	cry	–618.8		173.2	166.1
$C_5H_4O_2$	Furfural	liq	–201.6			163.2
$C_5H_5N_5$	Adenine	cry	96.0			147.0
		gas	204.8			
C_5H_5N	Pyridine	liq	100.2			132.7
		gas	140.4			
$C_5H_5N_5O$	Guanine	cry	–183.9			
C_5H_6	*cis*–3–Penten–1–yne	gas	81.4			
C_5H_6	*trans*–3–Penten–1–yne	liq	228.2			
C_5H_6	1,3–Cyclopentadiene	liq	105.9			
		gas	134.3			
$C_5H_6N_2O_2$	Thymine	cry	–462.8			150.8
		gas	–328.7			
$C_5H_6O_2$	Furfuryl alcohol	liq	–276.2			204.0
C_5H_6S	2–Methylthiophene	liq	44.6		218.5	149.8
C_5H_6S	3–Methylthiophene	liq	43.1			
C_5H_7N	Cyclobutanecarbonitrile	liq	103.0			
C_5H_8	2–Methyl–1,3–butadiene	liq	48.2		229.3	152.6
		gas	75.5			

Molecular formula	Name	State	$\Delta_f H°$ kJ/mol	$\Delta_f G°$ kJ/mol	$S°$ J/mol K	C_p J/mol K
C_5H_8	*cis*–1,3–Pentadiene	gas	81.4			
C_5H_8	*trans*–1,3–Pentadiene	gas	76.1			
C_5H_8	1,4–Pentadiene	gas	105.6			
C_5H_8	Cyclopentene	liq	4.4		201.2	122.4
		gas	33.9			
C_5H_8	Spiropentane	liq	157.7		193.7	134.5
		gas	185.2			
C_5H_8O	Cyclopentanone	liq	–235.7			
$C_5H_8O_2$	2,4–Pentanedione	liq	–423.8			
$C_5H_8O_4$	Glutaric acid	cry	–960.0			
C_5H_9N	Pentanenitrile	liq	–33.1			
C_5H_9N	2,2–Dimethylpropanenitrile	liq	–39.8		232.0	179.4
C_5H_9NO	*N*–Methyl–2–pyrrolidone	liq	–262.2			307.8
$C_5H_9NO_2$	l–Proline	cry	–512.2			
		gas	–366.2			
$C_5H_9NO_4$	l–Glutamic acid	cry	–1009.7			
C_5H_{10}	1–Pentene	liq	–46.9		262.6	154.0
		gas	–21.3			
C_5H_{10}	*cis*–2–Pentene	liq	–53.7		258.6	151.7
		gas	–27.6			
C_5H_{10}	*trans*–2–Pentene	liq	–58.2		256.5	157.0
		gas	–31.9			
C_5H_{10}	2–Methyl–1–butene	liq	–61.0		254.0	157.2
		gas	–35.3			
C_5H_{10}	2–Methyl–2–butene	liq	–68.6		251.0	152.8
		gas	–41.8			
C_5H_{10}	3–Methyl–1–butene	liq	–51.5		253.3	156.1
		gas	–27.6			
C_5H_{10}	Cyclopentane	liq	–105.1		204.5	128.8
		gas	–76.4			
$C_5H_{10}N_2O_3$	l–Glutamine	cry	–826.4			
$C_5H_{10}O$	Cyclopentanol	liq	–300.1		206.3	184.1
$C_5H_{10}O$	2–Pentanone	liq	–297.3			184.1
$C_5H_{10}O$	3–Pentanone	liq	–296.5		266.0	190.9
$C_5H_{10}O$	3–Methyl–2–butanone	liq	–299.4		268.5	179.9
		gas	–262.5			
$C_5H_{10}O$	3,3–Dimethyloxetane	liq	–182.2			
		gas	–148.2			
$C_5H_{10}O$	Tetrahydropyran	liq	–258.3			
		gas	–223.4			
$C_5H_{10}O$	Pentanal	liq	–267.3			
		gas	–288.5			
$C_5H_{10}O_2$	Isopropyl acetate	liq	–518.9			199.4
		gas	–481.7			
$C_5H_{10}O_2$	Ethyl propanoate	liq	–502.7			
		gas	–463.6			
$C_5H_{10}O_2$	Pentanoic acid	liq	–559.4		259.8	210.3
$C_5H_{10}O_2$	3–Methylbutanoic acid	liq	–561.6			
$C_5H_{10}O_2$	2–Methylbutanoic acid	liq	–554.5			
$C_5H_{10}O_2$	Tetrahydrofurfuryl alcohol	liq	–435.7			
		gas	–369.2			
$C_5H_{10}O_3$	Diethyl carbonate	liq	–681.5			
		gas	–637.9			
$C_5H_{10}S$	Cyclopentanethiol	liq	–89.5		256.9	165.2
$C_5H_{10}S$	Thiacyclohexane	liq	–106.3		218.2	163.3
$C_5H_{11}Br$	1–Bromopentane	liq	–170.2			
$C_5H_{11}Cl$	1–Chloropentane	liq	–213.2			

Molecular formula	Name	State	$\Delta_f H°$ kJ/mol	$\Delta_f G°$ kJ/mol	$S°$ J/mol K	C_p J/mol K
$C_5H_{11}Cl$	1–Chloro–3–methylbutane	liq	–216.0			
		gas	–179.2			
$C_5H_{11}N$	Piperidine	liq	–86.4		210.0	179.9
$C_5H_{11}N$	Cyclopentylamine	liq	–95.1		241.0	181.2
		gas	–54.9			
$C_5H_{11}NO_2$	l–Valine	cry	–617.9			
		gas	–455.1			
C_5H_{12}	Pentane	liq	–173.5			167.2
		gas	–146.9			
C_5H_{12}	Isopentane	liq	–178.5		260.4	164.8
		gas	–153.7			
C_5H_{12}	Neopentane	liq	–190.2			
		gas	–168.1			
$C_5H_{12}O$	Butyl methyl ether	liq	–290.6		295.3	192.7
		gas	–258.1			
$C_5H_{12}O$	*tert*–Butyl methyl ether	liq	–313.6		265.3	187.5
		gas	–283.5			
$C_5H_{12}O$	Ethyl propyl ether	liq	–303.6		295.0	197.2
		gas	–272.2			
$C_5H_{12}O$	1–Pentanol	liq	–351.6			208.1
$C_5H_{12}O$	2–Pentanol	liq	–365.2			
$C_5H_{12}O$	3–Pentanol	liq	–368.9			239.7
$C_5H_{12}O$	2–Methyl–1–butanol	liq	–356.6			
		gas	–302.0			
$C_5H_{12}O$	2–Methyl–2–butanol	liq	–379.5			247.1
$C_5H_{12}O$	3–Methyl–1–butanol	liq	–356.4			
$C_5H_{12}O$	3–Methyl–2–butanol	liq	–366.6			
		gas	–315.2			
$C_5H_{12}O$	2,2–Dimethyl–1–propanol	liq	–399.4			
$C_5H_{12}O_2$	1,5–Pentanediol	liq	–531.5			
		gas	–449.1			
$C_5H_{12}O_2$	Diethoxymethane	liq	–450.4			
		gas	–414.8			
$C_5H_{12}O_4$	Pentaerythritol	cry	–920.6			
		gas	–776.7			
$C_5H_{12}S$	Butyl methyl sulfide	liq	–142.9		307.5	200.9
		gas	–121.3			
$C_5H_{12}S$	Ethyl propyl sulfide	liq	–144.8		309.5	198.4
$C_5H_{12}S$	Ethyl isopropyl sulfide	liq	–156.1			
$C_5H_{12}S$	1–Pentanethiol	liq	–151.3			
$C_5H_{12}S$	2–Methyl–1–butanethiol	liq	–154.4			
$C_5H_{12}S$	3–Methyl–1–butanethiol	liq	–154.3			
$C_5H_{12}S$	2–Methyl–2–butanethiol	liq	–162.8		290.1	198.1
		gas	–127.1			
C_6ClF_5	Chloropentafluorobenzene	cry	–858.7			
C_6Cl_6	Hexachlorobenzene	cry	–127.6		260.2	201.2
		gas	–35.5			
C_6F_6	Hexafluorobenzene	liq	–991.3		280.8	221.6
		gas	–955.4			
C_6F_{10}	Perfluorocyclohexene	liq	–1963.5			
		gas	–1932.7			
C_6F_{12}	Perfluorocyclohexane	liq	–2406.3			
		gas	–2370.4			
C_6MoO_6	Molybdenum hexacarbonyl	cry	–982.8	–877.7	325.9	242.3
		gas	–912.1	–856.0	490.0	205.0
C_6HF_5	Pentafluorobenzene	cry	–852.7			
		liq	–841.8			

Molecular formula	Name	State	$\Delta_f H°$ kJ/mol	$\Delta_f G°$ kJ/mol	$S°$ J/mol K	C_p J/mol K
		gas	–806.5			
C_6HF_5O	Pentafluorophenol	cry	–1024.1			
		liq	–1007.7			
$C_6H_2F_4$	1,2,4,5–Tetrafluorobenzene	liq	–683.7			
$C_6H_4Cl_2$	*o*–Dichlorobenzene	liq	–17.5			162.4
		gas	30.2			
$C_6H_4Cl_2$	*m*–Dichlorobenzene	liq	–20.7			
		gas	25.7			
$C_6H_4Cl_2$	*p*–Dichlorobenzene	cry	–42.3		175.4	147.8
		gas	22.5			
$C_6H_4F_2$	*o*–Difluorobenzene	liq	–330.0		222.6	159.0
		gas	–293.8			
$C_6H_4F_2$	*m*–Difluorobenzene	liq	–343.9		223.8	159.1
		gas	–309.2			
$C_6H_4F_2$	*p*–Difluorobenzene	liq	–342.3			157.5
		gas	–306.7			
$C_6H_4O_2$	*p*–Benzoquinone	cry	–185.7			129.0
		gas	–122.9			
C_6H_5Br	Bromobenzene	liq	60.9		219.2	154.3
C_6H_5Cl	Chlorobenzene	liq	11.0			150.1
C_6H_5ClO	*m*–Chlorophenol	cry	–206.4			
		liq	–189.3			
C_6H_5ClO	*p*–Chlorophenol	cry	–197.7			
		liq	–181.3			
C_6H_5F	Fluorobenzene	liq	–150.6		205.9	146.4
C_6H_5I	Iodobenzene	liq	117.2		205.4	158.7
$C_6H_5NO_2$	Nitrobenzene	liq	12.5			185.8
		gas	67.5			
C_6H_6	Benzene	liq	49.0			136.3
		gas	82.6			
$C_6H_6N_2O_2$	*o*–Nitroaniline	cry	–26.1			166.0
		liq	–9.4			
		gas	63.8			
$C_6H_6N_2O_2$	*m*–Nitroaniline	cry	–38.3			158.8
		liq	–14.4			
		gas	58.4			
$C_6H_6N_2O_2$	*p*–Nitroaniline	cry	–42.0			167.0
		liq	–20.7			
		gas	58.8			
C_6H_6O	Phenol	cry	–165.1		144.0	127.4
		gas	–96.4			
$C_6H_6O_2$	*p*–Hydroquinone	cry	–364.5			136.0
		gas	–265.3			
C_6H_6S	Benzenethiol	liq	63.7		222.8	173.2
C_6H_7N	Aniline	liq	31.3			191.9
		gas	87.5	–7.0	317.9	107.9
C_6H_7N	2–Methylpyridine	liq	56.7			158.6
C_6H_7N	3–Methylpyridine	liq	61.9		216.3	158.7
C_6H_7N	4–Methylpyridine	liq	59.2		209.1	159.0
C_6H_7N	1–Cyclopentenecarbonitrile	liq	111.5			
		gas	156.5			
$C_6H_8N_2$	Adiponitrile	liq	85.1			128.7
$C_6H_8N_2$	Phenylhydrazine	liq	141.0			217.0
$C_6H_8N_2$	*o*–Phenylenediamine	cry	–0.3			
$C_6H_8N_2$	*m*–Phenylenediamine	cry	–7.8		154.5	159.6
$C_6H_8N_2$	*p*–Phenylenediamine	cry	3.1			
C_6H_9N	Cyclopentanecarbonitrile	liq	0.7			

Molecular formula	Name	State	$\Delta_f H°$ kJ/mol	$\Delta_f G°$ kJ/mol	$S°$ J/mol K	C_p J/mol K
		gas	43.0			
$C_6H_9NO_3$	Triacetamide	liq	–610.5			
		gas	–550.1			
$C_6H_9N_3O_2$	l–Histidine	cry	–466.7			
C_6H_{10}	Cyclohexene	liq	–38.5		214.6	148.3
		gas	–5.0			
C_6H_{10}	1,5–Hexadiene	liq	54.1			
		gas	84.1			
C_6H_{10}	3,3–Dimethyl–1–butyne	liq	78.4			
$C_6H_{10}O$	Cyclohexanone	liq	–271.2			182.2
$C_6H_{10}O_2$	Methyl cyclobutanecarboxylate	liq	–395.0			
$C_6H_{10}O_3$	Propanoic anhydride	liq	–679.1			
		gas	–626.5			
$C_6H_{10}O_4$	Adipic acid	cry	–994.3			
$C_6H_{10}O_4$	Diethyl oxalate	liq	–805.5			
		gas	–742.0			
$C_6H_{11}Cl$	Chlorocyclohexane	liq	–207.2			
		gas	–163.7			
$C_6H_{11}NO$	Caprolactam	cry	–329.4			156.8
		gas	–246.2			
C_6H_{12}	Cyclohexane	liq	–156.4			154.9
		gas	–123.4			
C_6H_{12}	Methylcyclopentane	liq	–137.9			
		gas	–106.2			
C_6H_{12}	Ethylcyclobutane	liq	–59.0			
		gas	–26.3			
C_6H_{12}	1–Hexene	liq	–74.2		295.2	183.3
		gas	–43.5			
C_6H_{12}	*cis*–2–Hexene	liq	–83.9			
		gas	–52.3			
C_6H_{12}	*trans*–2–Hexene	liq	–85.5			
		gas	–53.9			
C_6H_{12}	2–Methyl–1–pentene	liq	–90.0			
		gas	–59.4			
C_6H_{12}	2–Methyl–2–pentene	liq	–98.5			
		gas	–66.9			
C_6H_{12}	4–Methyl–1–pentene	liq	–80.0			
		gas	–51.3			
C_6H_{12}	4–Methyl–*cis*–2–pentene	liq	–87.0			
		gas	–57.5			
C_6H_{12}	4–Methyl–*trans*–2–pentene	liq	–91.5			
		gas	–61.5			
C_6H_{12}	2,3–Dimethyl–1–butene	liq	–93.3			
		gas	62.6			
C_6H_{12}	2,3–Dimethyl–2–butene	liq	–101.5		270.2	174.7
		gas	–68.2			
C_6H_{12}	2–Ethyl–1–butene	liq	–87.1			
		gas	–56.0			
$C_6H_{12}O$	Cyclohexanol	liq	–348.2			208.2
$C_6H_{12}O$	2–Hexanone	liq	–322.0			213.3
$C_6H_{12}O$	3–Hexanone	liq	–320.2		305.3	216.9
$C_6H_{12}O$	2–Methyl–3–pentanone	liq	–325.9			
$C_6H_{12}O$	3,3–Dimethyl–2–butanone	liq	–328.6			
$C_6H_{12}O_2$	Butyl acetate	liq	–529.2			227.8
$C_6H_{12}O_2$	Methyl pentanoate	liq	–514.2			229.3
$C_6H_{12}O_2$	Methyl 2,2–dimethylpropanoate	liq	–530.0			257.9
$C_6H_{12}O_2$	Hexanoic acid	liq	–583.8			

Molecular formula	Name	State	$\Delta_f H°$ kJ/mol	$\Delta_f G°$ kJ/mol	$S°$ J/mol K	C_p J/mol K
$C_6H_{12}S$	Cyclohexanethiol	liq	–140.7		255.6	192.6
$C_6H_{13}Br$	1–Bromohexane	liq	–194.2		453.0	203.5
		gas	–148.1			
$C_6H_{13}N$	Cyclohexylamine	liq	–147.7			
$C_6H_{13}NO_2$	l–Leucine	cry	–637.4			200.1
		gas	–486.8			
$C_6H_{13}NO_2$	l–Isoleucine	cry	–637.9			
C_6H_{14}	Hexane	liq	–198.7			195.6
		gas	–167.1			
C_6H_{14}	2–Methylpentane	liq	–204.6		290.6	193.7
		gas	–174.8			
C_6H_{14}	3–Methylpentane	liq	–202.4		292.5	190.7
		gas	–172.1			
C_6H_{14}	2,2–Dimethylbutane	liq	–213.8		272.5	191.9
		gas	–186.1			
C_6H_{14}	2,3–Dimethylbutane	liq	–207.4		287.8	189.7
		gas	–178.3			
$C_6H_{14}N_2$	Azopropane	liq	11.5			
		gas	51.5			
$C_6H_{14}N_2O_2$	Lysine	cry	–678.7			
$C_6H_{14}N_4O_2$	d–Arginine	cry	–623.5		250.6	232.0
$C_6H_{14}O$	Dipropyl ether	liq	–328.8		323.9	221.6
		gas	–292.9			
$C_6H_{14}O$	Diisopropyl ether	liq	–351.5			216.8
		gas	–319.2			
$C_6H_{14}O$	1–Hexanol	liq	–377.5		287.4	240.4
$C_6H_{14}O$	2–Hexanol	liq	–392.0			
$C_6H_{14}O$	4–Methyl–2–pentanol	liq	–394.7			273.0
$C_6H_{14}O_2$	1,1–Diethoxyethane	liq	–491.4			
$C_6H_{14}O_2$	1,2–Diethoxyethane	liq	–451.4			259.4
$C_6H_{14}O_2$	1,6–Hexanediol	cry	–569.9			
		liq	–544.4			
		gas	–461.2			
$C_6H_{14}O_3$	Trimethylolpropane	cry	–750.9			
$C_6H_{14}O_4$	Triethylene glycol	liq	–804.2			
		gas	–725.0			
$C_6H_{14}S$	Methyl pentyl sulfide	liq	–167.1			
$C_6H_{14}S$	Butyl ethyl sulfide	liq	–172.3			
$C_6H_{14}S$	Diisopropyl sulfide	liq	–181.6		313.0	232.0
$C_6H_{15}B$	Triethylborane	liq	–194.6	9.4	336.7	241.2
		gas	–157.7	16.1	437.8	
$C_6H_{15}N$	Triethylamine	liq	–127.7			219.9
		gas	–92.8			
$C_6H_{15}N$	Dipropylamine	liq	–156.1			
$C_6H_{15}N$	Diisopropylamine	liq	–178.5			
		gas	–144.0			
$C_6H_{18}OSi_2$	Hexamethyldisiloxane	liq	–815.0	–541.5	433.8	311.4
		gas	–777.7	–534.5	535.0	238.5
C_7F_8	Perfluorotoluene	liq	–1311.1		355.5	262.3
C_7F_{14}	Perfluoromethylcyclohexane	liq	–2931.1			353.1
		gas	–2897.2			
C_7F_{16}	Perfluoroheptane	liq	–3420.0		561.8	419.0
		gas	–3383.6			
$C_7H_3F_5$	2,3,4,5,6–Pentafluorotoluene	liq	–883.8		306.4	225.8
		gas	–842.9			
$C_7H_4Cl_2O_2$	*m*–Chlorobenzoyl chloride	liq	–189.7			
C_7H_5ClO	Benzoyl chloride	liq	–158.0			

Molecular formula	Name	State	$\Delta_f H°$ kJ/mol	$\Delta_f G°$ kJ/mol	$S°$ J/mol K	C_p J/mol K
		gas	–103.2			
$C_7H_5ClO_2$	*o*–Chlorobenzoic acid	cry	–428.9			163.2
		gas	–341.0			
C_7H_5N	Benzonitrile	liq	163.2		209.1	165.2
		gas	215.7			
C_7H_6O	Benzaldehyde	liq	–87.0		221.2	172.0
		gas	–36.7			
$C_7H_6O_2$	Benzoic acid	cry	–385.2		167.6	146.8
		gas	–294.1			
$C_7H_6O_3$	Salicylic acid	cry	–589.9			
		gas	–494.8			
C_7H_7Cl	(Chloromethyl)benzene	liq	–32.6			
		gas	18.9			
C_7H_7NO	Benzamide	cry	–202.6			
$C_7H_7NO_2$	*o*–Nitrotoluene	liq	–9.7			
$C_7H_7NO_2$	*m*–Nitrotoluene	liq	–31.5			
$C_7H_7NO_2$	*p*–Nitrotoluene	cry	–48.1			172.3
		gas	31.0			
C_7H_8	Toluene	liq	12.4			157.3
		gas	50.4			
C_7H_8O	*o*–Cresol	cry	–204.6		165.4	154.6
		gas	–128.6			
C_7H_8O	*m*–Cresol	liq	–194.0		212.6	224.9
		gas	–132.3			
C_7H_8O	*p*–Cresol	cry	–199.3		167.3	150.2
		gas	–125.4			
C_7H_8O	Benzyl alcohol	liq	–160.7		216.7	217.9
		gas	–100.4			
C_7H_8O	Anisole	liq	–114.8			
		gas	–67.9			
C_7H_9N	*o*–Methyl aniline	gas	56.4	167.6	351.0	130.2
		liq	–6.3			
C_7H_9N	*m*–Methyl aniline	gas	54.6	165.4	352.5	125.5
		liq	–8.1			
C_7H_9N	*p*–Methyl aniline	cry	–23.5			
		gas	55.3	167.7	347.0	126.2
C_7H_9N	1–Cyclohexenecarbonitrile	liq	48.1			
		gas	101.6			
C_7H_9N	2,3–Dimethylpyridine	liq	19.4		243.7	189.5
		gas	68.3			
C_7H_9N	2,4–Dimethylpyridine	liq	16.2		248.5	184.8
		gas	63.9			
C_7H_9N	2,5–Dimethylpyridine	liq	18.7		248.8	184.7
		gas	66.5			
C_7H_9N	2,6–Dimethylpyridine	liq	12.7		244.2	185.2
		gas	58.7			
C_7H_9N	3,4–Dimethylpyridine	liq	18.3		240.7	191.8
		gas	70.1			
C_7H_9N	3,5–Dimethylpyridine	liq	22.5		241.7	184.5
		gas	72.8			
$C_7H_{11}N$	Cyclohexanecarbonitrile	liq	–47.2			
		gas	4.8			
C_7H_{12}	1–Methylbicyclo(3,1,0)hexane	liq	–33.2			
		gas	1.5			
C_7H_{14}	Cycloheptane	liq	–156.6			
		gas	–118.1			
C_7H_{14}	Methylcyclohexane	liq	–190.1			184.8

Molecular formula	Name	State	$\Delta_f H°$ kJ/mol	$\Delta_f G°$ kJ/mol	$S°$ J/mol K	C_p J/mol K
		gas	–154.7			
C_7H_{14}	Ethylcyclopentane	liq	–163.4		279.9	
		gas	–126.9			
C_7H_{14}	*cis*–1,3–Dimethylcyclopentane	liq	–170.1			
		gas	–135.9			
C_7H_{14}	*cis*–1,2–Dimethylcyclopentane	liq	–165.3		269.2	
		gas	–129.5			
C_7H_{14}	*trans*–1,2–Dimethylcyclopentane	liq	–171.2			
		gas	–136.6			
C_7H_{14}	1–Heptene	liq	–97.9		327.6	211.8
		gas	–62.3			
C_7H_{14}	*cis*–2–Heptene	liq	–105.1			
C_7H_{14}	*cis*–3–Heptene	liq	–104.3			
$C_7H_{14}O$	2,4–Dimethyl–3–pentanone	liq	–352.9		318.0	233.7
		gas	–311.3			
$C_7H_{14}O$	2,2–Dimethyl–3–pentanone	liq	–356.1			
		gas	–313.7			
$C_7H_{14}O$	1–Heptanal	liq	–311.5		335.4	230.1
		gas	–263.8			
$C_7H_{14}O$	*cis*–2–Methylcyclohexanol	liq	–390.2			
		gas	–327.0			
$C_7H_{14}O$	*trans*–2–Methylcyclohexanol	liq	–415.7			
		gas	–352.5			
$C_7H_{14}O$	*cis*–3–Methylcyclohexanol	liq	–416.1			
		gas	–350.9			
$C_7H_{14}O$	*trans*–3–Methylcyclohexanol	liq	–394.4			
		gas	–329.1			
$C_7H_{14}O$	*cis*–4–Methylcyclohexanol	liq	–413.2			
		gas	–347.5			
$C_7H_{14}O$	*trans*–4–Methylcyclohexanol	liq	–433.3			
		gas	–367.2			
$C_7H_{14}O_2$	Ethyl 2,2–dimethylpropanoate	liq	–577.2			
		gas	–536.0			
$C_7H_{14}O_2$	Methyl hexanoate	liq	–540.2			
		gas	–492.6			
		gas	–506.9			
$C_7H_{14}O_2$	Ethyl 3–methylbutanoate	liq	–570.9			
		gas	–527.0			
$C_7H_{14}O_2$	Heptanoic acid	liq	–610.2			265.4
		gas	–536.2			
$C_7H_{15}Br$	1–Bromoheptane	liq	–218.4			
		gas	–167.8			
C_7H_{16}	Heptane	liq	–224.2			224.7
		gas	–187.7			
C_7H_{16}	2–Methylhexane	liq	–229.5		323.3	222.9
		gas	–194.6			
C_7H_{16}	3–Methylhexane	liq	–226.4			
		gas	–191.3			
C_7H_{16}	2,2–Dimethylpentane	liq	–238.3		300.3	221.1
		gas	–205.9			
C_7H_{16}	2,3–Dimethylpentane	liq	–233.1			
		gas	–198.9			
C_7H_{16}	2,4–Dimethylpentane	liq	–234.6		303.2	224.2
		gas	–201.7			
C_7H_{16}	3,3–Dimethylpentane	liq	–234.2			
		gas	–201.2			
C_7H_{16}	3–Ethylpentane	liq	–224.8		314.5	219.6

Molecular formula	Name	State	$\Delta_f H°$ kJ/mol	$\Delta_f G°$ kJ/mol	$S°$ J/mol K	C_p J/mol K
		gas	–189.6			
C_7H_{16}	2,2,3–Trimethylbutane	liq	–236.5		292.2	213.5
		gas	–204.5			
$C_7H_{16}O$	1–Heptanol	liq	–403.3			272.1
		gas	–336.4			
$C_8H_4O_3$	Phthalic anhydride	cry	–460.1		180.0	160.0
		gas	–371.4			
$C_8H_6O_4$	Phthalic acid	cry	–782.0		207.9	188.1
$C_8H_6O_4$	Isophthalic acid	cry	–803.0			
		gas	–696.3			
$C_8H_6O_4$	Terephthalic acid	cry	–816.1			
		gas	–717.9			
C_8H_8	Styrene	liq	103.8			182.0
		gas	147.9			
C_8H_8O	Acetophenone	liq	–142.5			
		gas	–86.7			
$C_8H_8O_2$	*o*–Toluic acid	cry	–416.5			174.9
$C_8H_8O_2$	*m*–Toluic acid	cry	–426.1			163.6
$C_8H_8O_2$	*p*–Toluic acid	cry	–429.2			169.0
$C_8H_8O_2$	Methyl benzoate	liq	–343.5			221.3
		gas	–287.9			
C_8H_{10}	Ethylbenzene	liq	–12.3			183.2
		gas	29.9			
C_8H_{10}	*o*–Xylene	liq	–24.4			186.1
		gas	19.1			
C_8H_{10}	*m*–Xylene	liq	–25.4			183.0
		gas	17.3			
C_8H_{10}	*p*–Xylene	liq	–24.4			181.5
		gas	18.0			
$C_8H_{10}O$	*o*–Ethylphenol	liq	–208.8			
		gas	–145.2			
$C_8H_{10}O$	*m*–Ethylphenol	liq	–214.3			
		gas	–146.1			
$C_8H_{10}O$	*p*–Ethylphenol	cry	–224.4			206.9
		gas	–144.1			
$C_8H_{10}O$	2,3–Xylenol	cry	–241.1			
		gas	–157.2			
$C_8H_{10}O$	2,4–Xylenol	liq	–228.7			
		gas	–162.9			
$C_8H_{10}O$	2,5–Xylenol	cry	–246.6			
		gas	–161.6			
$C_8H_{10}O$	2,6–Xylenol	cry	–237.4			
		gas	–161.8			
$C_8H_{10}O$	3,4–Xylenol	cry	–242.3			
		gas	–156.6			
$C_8H_{10}O$	3,5–Xylenol	cry	–244.4			
		gas	–161.5			
$C_8H_{10}O$	Phenetole	liq	–152.6			228.5
		gas	–101.6			
$C_8H_{11}N$	*N,N*–Dimethylaniline	liq	47.7			
		gas	100.5			
$C_8H_{11}N$	*N*–Ethylaniline	liq	4.0			
		gas	56.3			
$C_8H_{15}N$	Octanenitrile	liq	–107.3			
		gas	–50.5			
C_8H_{16}	Cyclooctane	liq	–167.7			
		gas	–124.4			

Molecular formula	Name	State	$\Delta_f H°$ kJ/mol	$\Delta_f G°$ kJ/mol	$S°$ J/mol K	C_p J/mol K
C_8H_{16}	Propylcyclopentane	liq	–188.8		310.8	216.3
		gas	–147.7			
C_8H_{16}	1–Ethyl–1–methylcyclopentane	liq	–193.8			
C_8H_{16}	Ethylcyclohexane	liq	–211.9		280.9	211.8
		gas	–171.7			
C_8H_{16}	1,1–Dimethylcyclohexane	liq	–218.7		267.2	209.2
		gas	–180.9			
C_8H_{16}	*cis*–1,2–Dimethylcyclohexane	liq	–211.8		274.1	210.2
		gas	–172.1			
C_8H_{16}	*trans*–1,2–Dimethylcyclohexane	liq	–218.2		273.2	209.4
		gas	–179.9			
C_8H_{16}	*cis*–1,3–Dimethylcyclohexane	liq	–222.9		272.6	209.4
		gas	–184.6			
C_8H_{16}	*trans*–1,3–Dimethylcyclohexane	liq	–215.7		276.3	212.8
		gas	–176.5			
C_8H_{16}	*cis*–1,4–Dimethylcyclohexane	liq	–215.6		271.1	212.1
		gas	–176.6			
C_8H_{16}	*trans*–1,4–Dimethylcyclohexane	liq	–222.4		268.0	210.2
		gas	–184.5			
$C_8H_{16}O$	2,2,4–Trimethyl–3–pentanone	liq	–381.6			
		gas	–338.3			
$C_8H_{16}O$	2–Ethylhexanal	liq	–348.5			
		gas	–299.6			
$C_8H_{16}O_2$	Octanoic acid	liq	–636.0			297.9
		gas	–554.3			
$C_8H_{16}O_2$	2–Ethylhexanoic acid	liq	–635.1			
		gas	–559.5			
$C_8H_{16}O_2$	Methyl heptanoate	liq	–567.1			285.1
		gas	–515.9			
$C_8H_{17}Br$	1–Bromooctane	liq	–245.1			
		gas	–189.7			
$C_8H_{17}Cl$	1–Chlorooctane	liq	–291.3			
		gas	–238.9			
C_8H_{18}	Octane	liq	–250.1			254.6
		gas	–208.6			
C_8H_{18}	2–Methylheptane	liq	–255.0		356.4	252.0
		gas	–215.4			
C_8H_{18}	3–Methylheptane	liq	–252.3		362.6	250.2
		gas	–212.5			
C_8H_{18}	4–Methylheptane	liq	–251.6			251.1
		gas	–212.0			
C_8H_{18}	2,2–Dimethylhexane	liq	–261.9			
		gas	–224.6			
C_8H_{18}	2,3–Dimethylhexane	liq	–252.6			
		gas	–213.8			
C_8H_{18}	2,4–Dimethylhexane	liq	–257.0			
		gas	–219.2			
C_8H_{18}	2,5–Dimethylhexane	liq	–260.4			249.2
		gas	–222.5			
C_8H_{18}	3,3–Dimethylhexane	liq	–257.5			246.6
		gas	–220.0			
C_8H_{18}	3,4–Dimethylhexane	liq	–251.8			
		gas	–212.8			
C_8H_{18}	3–Ethylhexane	liq	–250.4			
		gas	–210.7			
C_8H_{18}	3–Ethyl–2–methylpentane	liq	–249.6			
		gas	–211.0			

Molecular formula	Name	State	$\Delta_f H°$ kJ/mol	$\Delta_f G°$ kJ/mol	$S°$ J/mol K	C_p J/mol K
C_8H_{18}	3–Ethyl–3–methylpentane	liq	–252.8			
		gas	–214.8			
C_8H_{18}	2,2,3–Trimethylpentane	liq	–256.9			
		gas	–220.0			
C_8H_{18}	2,2,4–Trimethylpentane	liq	–259.2			239.1
		gas	–224.0			
C_8H_{18}	2,3,3–Trimethylpentane	liq	–253.5			245.6
		gas	–216.3			
C_8H_{18}	2,3,4–Trimethylpentane	liq	–255.0		329.3	247.3
		gas	–217.3			
C_8H_{18}	2,2,3,3–Tetramethylbutane	cry	–268.9		273.7	239.2
		gas	–225.6			
$C_8H_{18}N_2$	Azobutane	liq	–40.1			
		gas	9.2			
$C_8H_{18}O$	1–Octanol	liq	–426.5			305.2
		gas	–355.5			
$C_8H_{18}O$	2–Ethyl–1–hexanol	liq	–432.8		347.0	317.5
		gas	–365.3			
$C_8H_{18}O$	Dibutyl ether	liq	–377.9			278.2
$C_8H_{18}O$	Di–*sec*–butyl ether	liq	–401.5			
		gas	–360.9			
$C_8H_{18}O$	Di–*tert*–butyl ether	liq	–399.6			276.1
		gas	–362.0			
$C_8H_{18}O_5$	Tetraethylene glycol	liq	–981.7			428.8
		gas	–883.0			
$C_8H_{18}S$	Dibutyl sulfide	liq	–220.7		405.1	284.3
		gas	–167.4			
$C_8H_{18}S$	Diisobutyl sulfide	liq	–229.2			
		gas	–179.5			
$C_8H_{18}S$	Di–*tert*–butyl sulfide	liq	–232.6			
		gas	–188.9			
$C_8H_{19}N$	Dibutylamine	liq	–206.0			292.9
		gas	–156.6			
$C_8H_{19}N$	Diisobutylamine	liq	–218.5			
		gas	–179.2			
$C_8H_{20}Pb$	Tetraethyl lead	liq	52.7		464.6	307.4
		gas	109.6			
C_9H_7N	Isoquinoline	liq	144.5		216.0	196.2
C_9H_8	Indene	liq	110.6		215.3	186.9
		gas	163.4			
C_9H_{10}	Cyclopropylbenzene	liq	100.3			
		gas	150.5			
C_9H_{10}	Indan	liq	11.5		56.0	190.2
		gas	60.7			
$C_9H_{11}NO_2$	l–Phenylalanine	cry	–466.9		213.6	203.0
		gas	–312.9			
$C_9H_{11}NO_3$	l–Tyrosine	cry	–685.1		214.0	216.4
C_9H_{12}	Cumene	liq	–41.1			210.7
		gas	4.0			
C_9H_{12}	*o*–Ethyltoluene	liq	–46.4			
		gas	1.3			
C_9H_{12}	*m*–Ethyltoluene	liq	–48.7			
		gas	–1.8			
C_9H_{12}	*p*–Ethyltoluene	liq	–49.8			
		gas	–3.2			
C_9H_{12}	Propylbenzene	liq	–38.3		287.8	214.7
		gas	7.9			

Molecular formula	Name	State	$\Delta_f H°$ kJ/mol	$\Delta_f G°$ kJ/mol	$S°$ J/mol K	C_p J/mol K
C_9H_{12}	1,2,3–Trimethylbenzene	liq	–58.5		267.9	216.4
		gas	–9.5			
C_9H_{12}	1,2,4–Trimethylbenzene	liq	–61.8			215.0
		gas	–13.8			
C_9H_{12}	Mesitylene	liq	–63.4			209.3
		gas	–15.9			
$C_9H_{14}O_6$	Triacetin	liq	–1330.8		458.3	384.7
		gas	–1248.8			
C_9H_{18}	Propylcyclohexane	liq	–237.4		311.9	242.0
		gas	–192.5			
$C_9H_{18}O$	2–Nonanone	liq	–397.2			
		gas	–340.7			
$C_9H_{18}O$	5–Nonanone	liq	–398.2		401.4	303.6
		gas	–344.9			
$C_9H_{18}O$	2,6–Dimethyl–4–heptanone	liq	–408.5			297.3
		gas	–357.6			
$C_9H_{18}O_2$	Nonanoic acid	liq	–659.7			362.4
		gas	–577.3			
$C_9H_{18}O_2$	Methyl octanoate	liq	–590.3			
		gas	–533.8			
C_9H_{20}	Nonane	liq	–274.7			284.4
		gas	–228.2			
C_9H_{20}	2,2–Dimethylheptane	liq	–288.2			
C_9H_{20}	2,2,5–Trimethylhexane	liq	–293.3			
C_9H_{20}	2,3,5–Trimethylhexane	liq	–284.0			
C_9H_{20}	2,2,3,3–Tetramethylpentane	liq	–278.3			271.5
		gas	–237.1			
C_9H_{20}	2,2,3,4–Tetramethylpentane	liq	–277.7			
		gas	–236.9			
C_9H_{20}	3,3–Diethylpentane	liq	–275.4			278.2
		gas	–232.3			
C_9H_{20}	2,2,4,4–Tetramethylpentane	liq	–280.0			266.3
		gas	–241.6			
C_9H_{20}	2,3,3,4–Tetramethylpentane	liq	–277.9			
		gas	–236.1			
$C_9H_{20}O$	1–Nonanol	liq	–456.5			
$C_{10}H_7Cl$	1–Chloronaphthalene	liq	54.6			212.6
		gas	119.8			
$C_{10}H_8$	Azulene	cry	212.3			
		gas	289.1			
$C_{10}H_8$	Naphthalene	cry	77.9		167.4	165.7
		gas	150.3			
$C_{10}H_8O$	1–Naphthol	cry	–121.0			166.9
		gas	–29.9			
$C_{10}H_8O$	2–Naphthol	liq	–124.2			
		gas	–30.0			
$C_{10}H_{10}O_4$	Dimethyl isophthalate	cry	–730.9			
$C_{10}H_{10}O_4$	Dimethyl terephthalate	cry	–732.6			261.1
$C_{10}H_{12}$	1,2,3,4–Tetrahydronaphthalene	liq	–29.2			217.5
		gas	26.0			
$C_{10}H_{14}$	Butylbenzene	liq	–63.2		321.2	243.4
		gas	–13.1			
$C_{10}H_{14}$	Isobutylbenzene	liq	–69.8			
		gas	–21.5			
$C_{10}H_{14}$	*sec*–Butylbenzene	liq	–66.4			
		gas	–17.4			
$C_{10}H_{14}$	*tert*–Butylbenzene	liq	–70.7			

Molecular formula	Name	State	$\Delta_f H°$ kJ/mol	$\Delta_f G°$ kJ/mol	$S°$ J/mol K	C_p J/mol K
		gas	–22.6			
$C_{10}H_{14}$	*o*–Cymene	liq	–73.3			
$C_{10}H_{14}$	*m*–Cymene	liq	–78.6			
$C_{10}H_{14}$	*p*–Cymene	liq	–78.0			236.4
$C_{10}H_{14}$	*o*–Diethylbenzene	liq	–68.5			
$C_{10}H_{14}$	*m*–Diethylbenzene	liq	–73.5			
$C_{10}H_{14}$	*p*–Diethylbenzene	liq	–72.8			
$C_{10}H_{14}$	1,2,4,5–Tetramethylbenzene	cry	–119.9		245.6	215.1
$C_{10}H_{14}$	3–Ethyl–*o*–xylene	liq	–80.5			
$C_{10}H_{14}O$	Thymol	cry	–309.7			
		gas	–218.5			
$C_{10}H_{16}$	α–Pinene	liq	–16.4			
		gas	28.3			
$C_{10}H_{16}$	beta–Pinene	liq	–7.7			
		gas	38.7			
$C_{10}H_{16}O$	Camphor	cry	–319.4			271.2
		gas	–267.5			
$C_{10}H_{18}$	*cis*–Decahydronaphthalene	liq	–219.4		265.0	232.0
		gas	–169.2			
$C_{10}H_{18}$	*trans*–Decahydronaphthalene	liq	–230.6		264.9	228.5
		gas	–182.1			
$C_{10}H_{18}O_4$	Sebacic acid	cry	–1082.6			
		gas	–921.9			
$C_{10}H_{19}N$	Decanenitrile	liq	–158.4			
		gas	–91.5			
$C_{10}H_{20}$	Butylcyclohexane	liq	–263.1		345.0	271.0
		gas	–213.3			
$C_{10}H_{20}$	1–Decene	liq	–173.8		425.0	300.8
		gas	–123.4			
$C_{10}H_{20}O_2$	Decanoic acid	cry	–713.7			
		liq	–684.3			
		gas	–594.9			
$C_{10}H_{22}$	Decane	liq	–300.9			314.4
		gas	–249.5			
$C_{10}H_{22}$	2–Methylnonane	liq	–309.8		420.1	313.3
		gas	–259.9			
$C_{10}H_{22}$	5–Methylnonane	liq	–307.9		423.8	314.4
		gas	–258.6			
$C_{10}H_{22}O$	1–Decanol	liq	–478.1			370.6
		gas	–396.4			
$C_{10}H_{22}S$	1–Decanethiol	cry	–309.9			
		liq	–276.5		476.1	350.4
		gas	–211.5			
$C_{10}H_{22}S$	Diisopentylsulfide	liq	–281.8			
		gas	–221.5			
$C_{11}H_{10}$	1–Methylnaphthalene	liq	56.3		254.8	224.4
$C_{11}H_{10}$	2–Methylnaphthalene	cry	44.9		220.0	196.0
		gas	106.7			
$C_{11}H_{12}N_2O_2$	l–Tryptophan	cry	–415.3		251.0	238.1
$C_{11}H_{24}$	Undecane	liq	–327.2			344.9
		gas	–270.9			
$C_{12}H_8$	Acenaphthylene	cry	186.7			166.4
		gas	259.7			
$C_{12}H_9N$	Carbazole	cry	125.1			
		gas	209.6			
$C_{12}H_{10}$	Acenaphthene	cry	70.3		188.9	190.4

Molecular formula	Name	State	$\Delta_f H°$ kJ/mol	$\Delta_f G°$ kJ/mol	$S°$ J/mol K	C_p J/mol K
$C_{12}H_{10}$	Biphenyl	cry	99.4		209.4	198.4
		gas	181.4			
$C_{12}H_{10}O$	Diphenyl ether	cry	–32.1		233.9	216.6
		gas	52.0			
$C_{12}H_{11}N$	Diphenylamine	cry	130.2			
		gas	219.3			
$C_{12}H_{12}N_2$	*p*–Benzidine	cry	70.7			
$C_{12}H_{14}O_4$	Diethyl phthalate	liq	–776.6		425.1	366.1
		gas	–688.4			
$C_{12}H_{16}$	Cyclohexylbenzene	liq	–76.6			
		gas	–16.7			
$C_{12}H_{18}$	5,7–Dodecadiyne	liq	181.5			
$C_{12}H_{18}$	3,9–Dodecadiyne	liq	197.8			
$C_{12}H_{18}$	Hexamethylbenzene	cry	–161.5		306.3	245.6
		gas	–86.8			
$C_{12}H_{22}$	Cyclohexylcyclohexane	liq	–273.7			
		gas	–215.7			
$C_{12}H_{24}$	1–Dodecene	liq	–226.2		484.8	360.7
		gas	–165.4			
$C_{12}H_{24}O_2$	Dodecanoic acid	cry	–774.6			404.3
		liq	–737.9			
		gas	–642.0			
$C_{12}H_{26}$	Dodecane	liq	–350.9			375.8
		gas	–289.7			
$C_{12}H_{26}O$	1–Dodecanol	liq	–528.5			438.1
		gas	–436.6			
$C_{12}H_{27}N$	Tributylamine	liq	–281.6			
$C_{13}H_{10}O$	Benzophenone	cry	–34.5			224.8
		gas	54.9			
$C_{13}H_{12}$	Diphenylmethane	cry	71.5		239.3	
		liq	89.7			
		gas	139.0			
$C_{13}H_{26}O_2$	Methyl dodecanoate	liq	–693.0			
		gas	–614.8			
$C_{13}H_{28}O$	1–Tridecanol	cry	–599.4			
$C_{14}H_{10}$	Anthracene	cry	129.2		207.5	210.5
		gas	230.9			
$C_{14}H_{10}$	Phenanthrene	cry	116.2		215.1	220.6
		gas	207.5			
$C_{14}H_{10}$	Diphenylacetylene	cry	312.4			225.9
$C_{14}H_{12}$	*cis*–Stilbene	liq	183.3			
		gas	252.3			
$C_{14}H_{12}$	*trans*–Stilbene	cry	136.9			
		gas	236.1			
$C_{14}H_{14}$	1,1–Diphenylethane	liq	48.7			
$C_{14}H_{14}$	1,2–Diphenylethane	cry	51.5			
		gas	142.9			
$C_{14}H_{27}N$	Tetradecanenitrile	liq	–260.2			
		gas	–174.9			
$C_{14}H_{28}O_2$	Tetradecanoic acid	cry	–833.5			432.0
		liq	–788.8			
		gas	–693.7			
$C_{14}H_{30}O$	1–Tetradecanol	cry	–629.6			388.0
		liq	–580.6			
$C_{15}H_{30}O_2$	Pentadecylic acid	cry	–861.7			443.3
		liq	–811.7			
		gas	–699.0			

Molecular formula	Name	State	$\Delta_f H°$ kJ/mol	$\Delta_f G°$ kJ/mol	$S°$ J/mol K	C_p J/mol K
$C_{16}H_{10}$	Fluoranthene	cry	189.9		230.6	230.2
		gas	289.0			
$C_{16}H_{10}$	Pyrene	cry	125.5		224.9	229.7
		gas	225.7			
$C_{16}H_{22}O_4$	Dibutyl phthalate	liq	–842.6			
		gas	–750.9			
$C_{16}H_{26}$	Decylbenzene	liq	–218.3			
		gas	–138.6			
$C_{16}H_{32}$	1–Hexadecene	liq	–328.7		587.9	488.9
		gas	–248.5			
$C_{16}H_{32}O_2$	Hexadecanoic acid	cry	–891.5		452.4	460.7
		liq	–838.1			
		gas	–737.1			
$C_{16}H_{34}$	Hexadecane	liq	–456.1			501.6
		gas	–374.8			
$C_{16}H_{34}O$	1–Hexadecanol	cry	–686.5			422.0
		gas	–517.0			
$C_{17}H_{34}O_2$	Margaric acid	cry	–924.4			475.7
		liq	–865.6			
$C_{18}H_{12}$	Chrysene	cry	145.3			
		gas	269.8			
$C_{18}H_{36}O_2$	Stearic acid	cry	–947.7			501.5
		liq	–884.7			
		gas	–781.2			
$C_{18}H_{38}$	Octadecane	cry	–567.4		480.2	485.6
		gas	–414.6			
$C_{19}H_{36}O_2$	Methyl oleate	liq	–734.5			
		gas	–649.9			
$C_{20}H_{12}$	Perylene	cry	182.8		264.6	274.9
$C_{20}H_{40}O_2$	Arachidic acid	cry	–1011.9			545.1
		liq	–940.0			
		gas	–812.4			
$C_{22}H_{42}O_2$	Brassidic acid	cry	–960.7			

THERMODYNAMIC PROPERTIES AS A FUNCTION OF TEMPERATURE

L. V. Gurvich, V. S. Iorish, V. S. Yungman, and O. V. Dorofeeva

The thermodynamic properties $C_p°(T)$, $S°(T)$, $H°(T)-H°(T_r)$, $-[G°(T)-H°(T_r)]/T$ and formation properties $\Delta_f H°(T)$, $\Delta_f G°(T)$, $\log K_f°(T)$ are tabulated as functions of temperature in the range 298.15 to 1500 K for 80 substances in the standard state. The reference temperature, T_r, is equal to 298.15 K. The standard state pressure is taken as 1 bar (100,000 Pa). The tables are presented in the JANAF Thermochemical Tables format (Reference 2). The numerical data are extracted from IVTANTHERMO databases except for C_2H_4O, C_3H_6O, C_6H_6, C_6H_6O, $C_{10}H_8$, and CH_5N, which are based upon TRC Tables. See the references for information on standard states and other details.

REFERENCES

1. Gurvich, L. V., Veyts, I. V., and Alcock, C. B., Eds., *Thermodynamic Properties of Individual Substances, 4th ed.*, Hemisphere Publishing Corp., New York, 1989.
2. Chase, M. W., et al., *JANAF Thermochemical Tables, 3rd ed., J. Phys. Chem. Ref. Data*, 14, Suppl. 1, 1985.

Order of Listing of Tables

No.	Formula	Name	State	No.	Formula	Name	State
1	Ar	Argon	g	41	$CuCl_2$	Copper dichloride	cr, l
2	Br	Bromine	g	42	$CuCl_2$	Copper dichloride	g
3	Br_2	Dibromine	g	43	F	Fluorine	g
4	BrH	Hydrogen bromide	g	44	F_2	Difluorine	g
5	C	Carbon (graphite)	cr	45	FH	Hydrogen fluoride	g
6	C	Carbon (diamond)	cr	46	Ge	Germanium	cr, l
7	C_2	Dicarbon	g	47	Ge	Germanium	g
8	C_3	Tricarbon	g	48	GeO_2	Germanium dioxide	cr, l
9	CO	Carbon oxide	g	49	$GeCl_4$	Germanium tetrachloride	g
10	CO_2	Carbon dioxide	g	50	H	Hydrogen	g
11	CH_4	Methane	g	51	H_2	Dihydrogen	g
12	C_2H_2	Acetylene	g	52	HO	Hydroxyl	g
13	C_2H_4	Ethylene	g	53	H_2O	Water	l
14	C_2H_6	Ethane	g	54	H_2O	Water	g
15	C_3H_6	Cyclopropane	g	55	I	Iodine	g
16	C_3H_8	Propane	g	56	I_2	Diiodine	cr, l
17	C_6H_6	Benzene	l	57	I_2	Diiodine	g
18	C_6H_6	Benzene	g	58	IH	Hydrogen iodide	g
19	$C_{10}H_8$	Naphthalene	cr, l	59	K	Potassium	cr, l
20	$C_{10}H_8$	Naphthalene	g	60	K	Potassium	g
21	CH_2O	Formaldehyde	g	61	K_2O	Dipotassium oxide	cr, l
22	CH_4O	Methanol	g	62	KOH	Potassium hydroxide	cr, l
23	C_2H_4O	Acetaldehyde	g	63	KOH	Potassium hydroxide	g
24	C_2H_6O	Ethanol	g	64	KCl	Potassium chloride	cr, l
25	$C_2H_4O_2$	Acetic acid	g	65	KCl	Potassium chloride	g
26	C_3H_6O	Acetone	g	66	N_2	Dinitrogen	g
27	C_6H_6O	Phenol	g	67	NO	Nitric oxide	g
28	CF_4	Carbon tetrafluoride	g	68	NO_2	Nitrogen dioxide	g
29	CHF_3	Trifluoromethane	g	69	NH_3	Ammonia	g
30	$CClF_3$	Chlorotrifluoromethane	g	70	O	Oxygen	g
31	CCl_2F_2	Dichlorodifluoromethane	g	71	O_2	Dioxygen	g
32	$CHClF_2$	Chlorodifluoromethane	g	72	S	Sulfur	cr, l
33	CH_5N	Methylamine	g	73	S	Sulfur	g
34	Cl	Chlorine	g	74	S_2	Disulfur	g
35	Cl_2	Dichlorine	g	75	S_8	Octasulfur	g
36	ClH	Hydrogen chloride	g	76	SO_2	Sulfur dioxide	g
37	Cu	Copper	cr, l	77	Si	Silicon	cr
38	Cu	Copper	g	78	Si	Silicon	g
39	CuO	Copper oxide	cr	79	SiO_2	Silicon dioxide	cr
40	Cu_2O	Dicopper oxide	cr	80	$SiCl_4$	Silicon tetrachloride	g

	J/K·mol			kJ/mol			
T/K	C_p°	S°	$-(G^\circ\text{-}H^\circ(T_r))/T$	$H^\circ\text{-}H^\circ(T_r)$	$\Delta_f H^\circ$	$\Delta_f G^\circ$	Log K_f
1. ARGON Ar (g)							
298.15	20.786	154.845	154.845	0.000	0.000	0.000	0.000
300	20.786	154.973	154.845	0.038	0.000	0.000	0.000
400	20.786	160.953	155.660	2.117	0.000	0.000	0.000
500	20.786	165.591	157.200	4.196	0.000	0.000	0.000
600	20.786	169.381	158.924	6.274	0.000	0.000	0.000
700	20.786	172.585	160.653	8.353	0.000	0.000	0.000
800	20.786	175.361	162.322	10.431	0.000	0.000	0.000
900	20.786	177.809	163.909	12.510	0.000	0.000	0.000
1000	20.786	179.999	165.410	14.589	0.000	0.000	0.000
1100	20.786	181.980	166.828	16.667	0.000	0.000	0.000
1200	20.786	183.789	168.167	18.746	0.000	0.000	0.000
1300	20.786	185.453	169.434	20.824	0.000	0.000	0.000
1400	20.786	186.993	170.634	22.903	0.000	0.000	0.000
1500	20.786	188.427	171.773	24.982	0.000	0.000	0.000
2. BROMINE Br (g)							
298.15	20.786	175.017	175.017	0.000	111.870	82.379	–14.432
300	20.786	175.146	175.018	0.038	111.838	82.196	–14.311
400	20.787	181.126	175.833	2.117	96.677	75.460	–9.854
500	20.798	185.765	177.373	4.196	96.910	70.129	–7.326
600	20.833	189.559	179.097	6.277	97.131	64.752	–5.637
700	20.908	192.776	180.827	8.364	97.348	59.338	–4.428
800	21.027	195.575	182.499	10.461	97.568	53.893	–3.519
900	21.184	198.061	184.093	12.571	97.796	48.420	–2.810
1000	21.365	200.302	185.604	14.698	98.036	42.921	–2.242
1100	21.559	202.347	187.034	16.844	98.291	37.397	–1.776
1200	21.752	204.231	188.390	19.010	98.560	31.850	–1.386
1300	21.937	205.980	189.676	21.195	98.844	26.279	–1.056
1400	22.107	207.612	190.900	23.397	99.141	20.686	–0.772
1500	22.258	209.142	192.065	25.615	99.449	15.072	–0.525
3. DIBROMINE Br_2 (g)							
298.15	36.057	245.467	245.467	0.000	30.910	3.105	–0.544
300	36.074	245.690	245.468	0.067	30.836	2.933	–0.511
332.25	36.340	249.387	245.671	1.235		pressure = 1 bar	
400	36.729	256.169	246.892	3.711	0.000	0.000	0.000
500	37.082	264.406	249.600	7.403	0.000	0.000	0.000
600	37.305	271.188	252.650	11.123	0.000	0.000	0.000
700	37.464	276.951	255.720	14.862	0.000	0.000	0.000
800	37.590	281.962	258.694	18.615	0.000	0.000	0.000
900	37.697	286.396	261.530	22.379	0.000	0.000	0.000
1000	37.793	290.373	264.219	26.154	0.000	0.000	0.000
1100	37.883	293.979	266.763	29.938	0.000	0.000	0.000
1200	37.970	297.279	269.170	33.730	0.000	0.000	0.000
1300	38.060	300.322	271.451	37.532	0.000	0.000	0.000
1400	38.158	303.146	273.615	41.343	0.000	0.000	0.000
1500	38.264	305.782	275.673	45.164	0.000	0.000	0.000
4. HYDROGEN BROMIDE HBr (g)							
298.15	29.141	198.697	198.697	0.000	–36.290	–53.360	9.348
300	29.141	198.878	198.698	0.054	–36.333	–53.466	9.309
400	29.220	207.269	199.842	2.971	–52.109	–55.940	7.305
500	29.454	213.811	202.005	5.903	–52.484	–56.854	5.939
600	29.872	219.216	204.436	8.868	–52.844	–57.694	5.023
700	30.431	223.861	206.886	11.882	–53.168	–58.476	4.363
800	31.063	227.965	209.269	14.957	–53.446	–59.214	3.866

	J/K·mol			kJ/mol			
T/K	C_p°	S°	$-(G^\circ-H^\circ(T_r))/T$	$H^\circ-H^\circ(T_r)$	Δ_fH°	Δ_fG°	Log K_f
4. HYDROGEN BROMIDE HBr (g) (continued)							
900	31.709	231.661	211.555	18.095	–53.677	–59.921	3.478
1000	32.335	235.035	213.737	21.298	–53.864	–60.604	3.166
1100	32.919	238.145	215.816	24.561	–54.012	–61.271	2.909
1200	33.454	241.032	217.799	27.880	–54.129	–61.925	2.696
1300	33.938	243.729	219.691	31.250	–54.220	–62.571	2.514
1400	34.374	246.261	221.499	34.666	–54.291	–63.211	2.358
1500	34.766	248.646	223.230	38.123	–54.348	–63.846	2.223
5. CARBON (GRAPHITE) C (cr; graphite)							
298.15	8.536	5.740	5.740	0.000	0.000	0.000	0.000
300	8.610	5.793	5.740	0.016	0.000	0.000	0.000
400	11.974	8.757	6.122	1.054	0.000	0.000	0.000
500	14.537	11.715	6.946	2.385	0.000	0.000	0.000
600	16.607	14.555	7.979	3.945	0.000	0.000	0.000
700	18.306	17.247	9.113	5.694	0.000	0.000	0.000
800	19.699	19.785	10.290	7.596	0.000	0.000	0.000
900	20.832	22.173	11.479	9.625	0.000	0.000	0.000
1000	21.739	24.417	12.662	11.755	0.000	0.000	0.000
1100	22.452	26.524	13.827	13.966	0.000	0.000	0.000
1200	23.000	28.502	14.968	16.240	0.000	0.000	0.000
1300	23.409	30.360	16.082	18.562	0.000	0.000	0.000
1400	23.707	32.106	17.164	20.918	0.000	0.000	0.000
1500	23.919	33.749	18.216	23.300	0.000	0.000	0.000
6. CARBON (DIAMOND) C (cr; diamond)							
298.15	6.109	2.362	2.362	0.000	1.850	2.857	–0.501
300	6.201	2.400	2.362	0.011	1.846	2.863	–0.499
400	10.321	4.783	2.659	0.850	1.645	3.235	–0.422
500	13.404	7.431	3.347	2.042	1.507	3.649	–0.381
600	15.885	10.102	4.251	3.511	1.415	4.087	–0.356
700	17.930	12.709	5.274	5.205	1.361	4.537	–0.339
800	19.619	15.217	6.361	7.085	1.338	4.993	–0.326
900	21.006	17.611	7.479	9.118	1.343	5.450	–0.316
1000	22.129	19.884	8.607	11.277	1.372	5.905	–0.308
1100	23.020	22.037	9.731	13.536	1.420	6.356	–0.302
1200	23.709	24.071	10.842	15.874	1.484	6.802	–0.296
1300	24.222	25.990	11.934	18.272	1.561	7.242	–0.291
1400	24.585	27.799	13.003	20.714	1.646	7.675	–0.286
1500	24.824	29.504	14.047	23.185	1.735	8.103	–0.282
7. DICARBON C_2 (g)							
298.15	43.548	197.095	197.095	0.000	830.457	775.116	–135.795
300	43.575	197.365	197.096	0.081	830.506	774.772	–134.898
400	42.169	209.809	198.802	4.403	832.751	755.833	–98.700
500	39.529	218.924	201.959	8.483	834.170	736.423	–76.933
600	37.837	225.966	205.395	12.342	834.909	716.795	–62.402
700	36.984	231.726	208.758	16.078	835.148	697.085	–52.016
800	36.621	236.637	211.943	19.755	835.020	677.366	–44.227
900	36.524	240.943	214.931	23.411	834.618	657.681	–38.170
1000	36.569	244.793	217.728	27.065	834.012	638.052	–33.328
1100	36.696	248.284	220.349	30.728	833.252	618.492	–29.369
1200	36.874	251.484	222.812	34.406	832.383	599.006	–26.074
1300	37.089	254.444	225.133	38.104	831.437	579.596	–23.288
1400	37.329	257.201	227.326	41.824	830.445	560.261	–20.903
1500	37.589	259.785	229.405	45.570	829.427	540.997	–18.839

	J/K·mol			kJ/mol			
T/K	C_p°	S°	$-(G^\circ-H^\circ(T_r))/T$	$H^\circ-H^\circ(T_r)$	$\Delta_f H^\circ$	$\Delta_f G^\circ$	Log K_f
8. TRICARBON C_3 (g)							
298.15	42.202	237.611	237.611	0.000	839.958	774.249	–135.643
300	42.218	237.872	237.611	0.078	839.989	773.841	–134.736
400	43.383	250.164	239.280	4.354	841.149	751.592	–98.147
500	44.883	260.003	242.471	8.766	841.570	729.141	–76.172
600	46.406	268.322	246.104	13.331	841.453	706.659	–61.519
700	47.796	275.582	249.807	18.042	840.919	684.230	–51.057
800	48.997	282.045	253.440	22.884	840.053	661.901	–43.217
900	50.006	287.876	256.948	27.835	838.919	639.698	–37.127
1000	50.844	293.189	260.310	32.879	837.572	617.633	–32.261
1100	51.535	298.069	263.524	37.999	836.059	595.711	–28.288
1200	52.106	302.578	266.593	43.182	834.420	573.933	–24.982
1300	52.579	306.768	269.524	48.417	832.690	552.295	–22.191
1400	52.974	310.679	272.326	53.695	830.899	530.793	–19.804
1500	53.307	314.346	275.006	59.010	829.068	509.421	–17.739
9. CARBON OXIDE CO (g)							
298.15	29.141	197.658	197.658	0.000	–110.530	–137.168	24.031
300	29.142	197.838	197.659	0.054	–110.519	–137.333	23.912
400	29.340	206.243	198.803	2.976	–110.121	–146.341	19.110
500	29.792	212.834	200.973	5.930	–110.027	–155.412	16.236
600	30.440	218.321	203.419	8.941	–110.157	–164.480	14.319
700	31.170	223.067	205.895	12.021	–110.453	–173.513	12.948
800	31.898	227.277	208.309	15.175	–110.870	–182.494	11.915
900	32.573	231.074	210.631	18.399	–111.378	–191.417	11.109
1000	33.178	234.538	212.851	21.687	–111.952	–200.281	10.461
1100	33.709	237.726	214.969	25.032	–112.573	–209.084	9.928
1200	34.169	240.679	216.990	28.426	–113.228	–217.829	9.482
1300	34.568	243.430	218.920	31.864	–113.904	–226.518	9.101
1400	34.914	246.005	220.763	35.338	–114.594	–235.155	8.774
1500	35.213	248.424	222.527	38.845	–115.291	–243.742	8.488
10. CARBON DIOXIDE CO_2 (g)							
298.15	37.135	213.783	213.783	0.000	–393.510	–394.373	69.092
300	37.220	214.013	213.784	0.069	–393.511	–394.379	68.667
400	41.328	225.305	215.296	4.004	–393.586	–394.656	51.536
500	44.627	234.895	218.280	8.307	–393.672	–394.914	41.256
600	47.327	243.278	221.762	12.909	–393.791	–395.152	34.401
700	49.569	250.747	225.379	17.758	–393.946	–395.367	29.502
800	51.442	257.492	228.978	22.811	–394.133	–395.558	25.827
900	53.008	263.644	232.493	28.036	–394.343	–395.724	22.967
1000	54.320	269.299	235.895	33.404	–394.568	–395.865	20.678
1100	55.423	274.529	239.172	38.893	–394.801	–395.984	18.803
1200	56.354	279.393	242.324	44.483	–395.035	–396.081	17.241
1300	57.144	283.936	245.352	50.159	–395.265	–396.159	15.918
1400	57.818	288.196	248.261	55.908	–395.488	–396.219	14.783
1500	58.397	292.205	251.059	61.719	–395.702	–396.264	13.799
11. METHANE CH_4 (g)							
298.15	35.695	186.369	186.369	0.000	–74.600	–50.530	8.853
300	35.765	186.590	186.370	0.066	–74.656	–50.381	8.772
400	40.631	197.501	187.825	3.871	–77.703	–41.827	5.462
500	46.627	207.202	190.744	8.229	–80.520	–32.525	3.398
600	52.742	216.246	194.248	13.199	–82.969	–22.690	1.975
700	58.603	224.821	198.008	18.769	–85.023	–12.476	0.931
800	64.084	233.008	201.875	24.907	–86.693	–1.993	0.130

T/K	C_p° (J/K·mol)	S° (J/K·mol)	$-(G^\circ-H^\circ(T_r))/T$ (J/K·mol)	$H^\circ-H^\circ(T_r)$ (kJ/mol)	Δ_fH° (kJ/mol)	Δ_fG° (kJ/mol)	Log K_f
11. METHANE CH_4 (g) (continued)							
900	69.137	240.852	205.773	31.571	–88.006	8.677	–0.504
1000	73.746	248.379	209.660	38.719	–88.996	19.475	–1.017
1100	77.919	255.607	213.511	46.306	–89.698	30.358	–1.442
1200	81.682	262.551	217.310	54.289	–90.145	41.294	–1.797
1300	85.067	269.225	221.048	62.630	–90.367	52.258	–2.100
1400	88.112	275.643	224.720	71.291	–90.390	63.231	–2.359
1500	90.856	281.817	228.322	80.242	–90.237	74.200	–2.584
12. ACETYLENE C_2H_2 (g)							
298.15	44.036	200.927	200.927	0.000	227.400	209.879	–36.769
300	44.174	201.199	200.927	0.082	227.397	209.770	–36.524
400	50.388	214.814	202.741	4.829	227.161	203.928	–26.630
500	54.751	226.552	206.357	10.097	226.846	198.154	–20.701
600	58.121	236.842	210.598	15.747	226.445	192.452	–16.754
700	60.970	246.021	215.014	21.704	225.968	186.823	–13.941
800	63.511	254.331	219.418	27.931	225.436	181.267	–11.835
900	65.831	261.947	223.726	34.399	224.873	175.779	–10.202
1000	67.960	268.995	227.905	41.090	224.300	170.355	–8.898
1100	69.909	275.565	231.942	47.985	223.734	164.988	–7.835
1200	71.686	281.725	235.837	55.067	223.189	159.672	–6.950
1300	73.299	287.528	239.592	62.317	222.676	154.400	–6.204
1400	74.758	293.014	243.214	69.721	222.203	149.166	–5.565
1500	76.077	298.218	246.709	77.264	221.774	143.964	–5.013
13. ETHYLENE C_2H_4 (g)							
298.15	42.883	219.316	219.316	0.000	52.400	68.358	–11.976
300	43.059	219.582	219.317	0.079	52.341	68.457	–11.919
400	53.045	233.327	221.124	4.881	49.254	74.302	–9.703
500	62.479	246.198	224.864	10.667	46.533	80.887	–8.450
600	70.673	258.332	229.441	17.335	44.221	87.982	–7.659
700	77.733	269.770	234.393	24.764	42.278	95.434	–7.121
800	83.868	280.559	239.496	32.851	40.655	103.142	–6.734
900	89.234	290.754	244.630	41.512	39.310	111.036	–6.444
1000	93.939	300.405	249.730	50.675	38.205	119.067	–6.219
1100	98.061	309.556	254.756	60.280	37.310	127.198	–6.040
1200	101.670	318.247	259.688	70.271	36.596	135.402	–5.894
1300	104.829	326.512	264.513	80.599	36.041	143.660	–5.772
1400	107.594	334.384	269.225	91.223	35.623	151.955	–5.669
1500	110.018	341.892	273.821	102.107	35.327	160.275	–5.581
14. ETHANE C_2H_6 (g)							
298.15	52.487	229.161	229.161	0.000	–84.000	–32.015	5.609
300	52.711	229.487	229.162	0.097	–84.094	–31.692	5.518
400	65.459	246.378	231.379	5.999	–88.988	–13.473	1.759
500	77.941	262.344	235.989	13.177	–93.238	5.912	–0.618
600	89.188	277.568	241.660	21.545	–96.779	26.086	–2.271
700	99.136	292.080	247.835	30.972	–99.663	46.800	–3.492
800	107.936	305.904	254.236	41.334	–101.963	67.887	–4.433
900	115.709	319.075	260.715	52.525	–103.754	89.231	–5.179
1000	122.552	331.628	267.183	64.445	–105.105	110.750	–5.785
1100	128.553	343.597	273.590	77.007	–106.082	132.385	–6.286
1200	133.804	355.012	279.904	90.131	–106.741	154.096	–6.708
1300	138.391	365.908	286.103	103.746	–107.131	175.850	–7.066
1400	142.399	376.314	292.178	117.790	–107.292	197.625	–7.373
1500	145.905	386.260	298.121	132.209	–107.260	219.404	–7.640

	J/K·mol			kJ/mol			
T/K	C_p°	S°	$-(G^\circ-H^\circ(T_r))/T$	$H^\circ-H^\circ(T_r)$	$\Delta_f H^\circ$	$\Delta_f G^\circ$	Log K_f
15. CYCLOPROPANE C_3H_6 (g)							
298.15	55.571	237.488	237.488	0.000	53.300	104.514	–18.310
300	55.941	237.832	237.489	0.103	53.195	104.832	–18.253
400	76.052	256.695	239.924	6.708	47.967	122.857	–16.043
500	93.859	275.637	245.177	15.230	43.730	142.091	–14.844
600	108.542	294.092	251.801	25.374	40.405	162.089	–14.111
700	120.682	311.763	259.115	36.854	37.825	182.583	–13.624
800	130.910	328.564	266.755	49.447	35.854	203.404	–13.281
900	139.658	344.501	274.516	62.987	34.384	224.441	–13.026
1000	147.207	359.616	282.277	77.339	33.334	245.618	–12.830
1100	153.749	373.961	289.965	92.395	32.640	266.883	–12.673
1200	159.432	387.588	297.538	108.060	32.249	288.197	–12.545
1300	164.378	400.549	304.967	124.257	32.119	309.533	–12.437
1400	168.689	412.892	312.239	140.915	32.215	330.870	–12.345
1500	172.453	424.662	319.344	157.976	32.507	352.193	–12.264
16. PROPANE C_3H_8 (g)							
298.15	73.597	270.313	270.313	0.000	–103.847	–23.458	4.110
300	73.931	270.769	270.314	0.136	–103.972	–22.959	3.997
400	94.014	294.739	273.447	8.517	–110.33	15.029	–0.657
500	112.591	317.768	280.025	18.872	–115.658	34.507	–3.605
600	128.700	339.753	288.162	30.955	–119.973	64.961	–5.655
700	142.674	360.668	297.039	44.540	–123.384	96.065	–7.168
800	154.766	380.528	306.245	59.427	–126.016	127.603	–8.331
900	165.352	399.381	315.555	75.444	–127.982	159.430	–9.253
1000	174.598	417.293	324.841	92.452	–129.380	191.444	–10.000
1100	182.673	434.321	334.026	110.325	–130.296	223.574	–10.617
1200	189.745	450.526	343.064	128.954	–130.802	255.770	–11.133
1300	195.853	465.961	351.929	148.241	–130.961	287.993	–11.572
1400	201.209	480.675	360.604	168.100	–130.829	320.217	–11.947
1500	205.895	494.721	369.080	188.460	–130.445	352.422	–12.272
17. BENZENE C_6H_6 (l)							
298.15	135.950	173.450	173.450	0.000	49.080	124.521	–21.815
300	136.312	174.292	173.453	.252	49.077	124.989	–21.762
400	161.793	216.837	179.082	15.102	48.978	150.320	–19.630
500	207.599	257.048	190.639	33.204	50.330	175.559	–18.340
18. BENZENE C_6H_6 (g)							
298.15	82.430	269.190	269.190	0.000	82.880	129.750	–22.731
300	83.020	269.700	269.190	0.153	82.780	130.040	–22.641
400	113.510	297.840	272.823	10.007	77.780	146.570	–19.140
500	139.340	326.050	280.658	22.696	73.740	164.260	–17.160
600	160.090	353.360	290.517	37.706	70.490	182.680	–15.903
700	176.790	379.330	301.360	54.579	67.910	201.590	–15.042
800	190.460	403.860	312.658	72.962	65.910	220.820	–14.418
900	201.840	426.970	324.084	92.597	64.410	240.280	–13.945
1000	211.430	448.740	335.473	113.267	63.340	259.890	–13.575
1100	219.580	469.280	346.710	134.827	62.620	277.640	–13.184
1200	226.540	488.690	357.743	157.137	62.200	299.320	–13.029
1300	232.520	507.070	368.534	180.097	62.000	319.090	–12.821
1400	237.680	524.490	379.056	203.607	61.990	338.870	–12.643
1500	242.140	541.040	389.302	227.607	62.110	358.640	–12.489

	J/K·mol			kJ/mol			
T/K	C_p°	S°	$-(G^\circ-H^\circ(T_r))/T$	$H^\circ-H^\circ(T_r)$	$\Delta_f H^\circ$	$\Delta_f G^\circ$	Log K_f

19. NAPHTHALENE $C_{10}H_8$ (cr, l)

T/K	C_p°	S°	$-(G^\circ-H^\circ(T_r))/T$	$H^\circ-H^\circ(T_r)$	$\Delta_f H^\circ$	$\Delta_f G^\circ$	Log K_f
298.15	165.720	167.390	167.390	0.000	78.530	201.585	–35.316
300	167.001	168.419	167.393	0.308	78.466	202.349	–35.232
353.43	208.722	198.948	169.833	10.290	96.099	224.543	–33.186
PHASE TRANSITION: $\Delta_{trs}H$ = 18.980 kJ/mol, $\Delta_{trs}S$ = 53.702 J/K·mol, cr–l							
353.43	217.200	252.650	169.833	29.270	96.099	224.543	–33.186
400	241.577	280.916	181.124	39.917	96.067	241.475	–31.533
470	276.409	322.712	199.114	58.091	97.012	266.859	–29.658

20. NAPHTHALENE $C_{10}H_8$ (g)

T/K	C_p°	S°	$-(G^\circ-H^\circ(T_r))/T$	$H^\circ-H^\circ(T_r)$	$\Delta_f H^\circ$	$\Delta_f G^\circ$	Log K_f
298.15	131.920	333.150	333.150	0.000	150.580	224.100	–39.260
300	132.840	333.970	333.157	0.244	150.450	224.560	–39.098
400	180.070	378.800	338.950	15.940	144.190	250.270	–32.681
500	219.740	423.400	351.400	36.000	139.220	277.340	–28.973
600	251.530	466.380	367.007	59.624	135.350	305.330	–26.581
700	277.010	507.140	384.146	86.096	132.330	333.950	–24.919
800	297.730	545.520	401.935	114.868	130.050	362.920	–23.696
900	314.850	581.610	419.918	145.523	128.430	392.150	–22.759
1000	329.170	615.550	437.806	177.744	127.510	421.700	–22.027
1100	341.240	647.500	455.426	211.281	127.100	450.630	–21.398
1200	351.500	677.650	472.707	245.932	126.960	480.450	–20.913
1300	360.260	706.130	489.568	281.531	127.060	509.770	–20.482
1400	367.780	733.110	506.009	317.941	127.390	539.740	–20.137
1500	374.270	758.720	522.019	355.051	127.920	568.940	–19.812

21. FORMALDEHYDE H_2CO (g)

T/K	C_p°	S°	$-(G^\circ-H^\circ(T_r))/T$	$H^\circ-H^\circ(T_r)$	$\Delta_f H^\circ$	$\Delta_f G^\circ$	Log K_f
298.15	35.387	218.760	218.760	0.000	–108.700	–102.667	17.987
300	35.443	218.979	218.761	0.066	–108.731	–102.630	17.869
400	39.240	229.665	220.192	3.789	–110.438	–100.340	13.103
500	43.736	238.900	223.028	7.936	–112.073	–97.623	10.198
600	48.181	247.270	226.381	12.534	–113.545	–94.592	8.235
700	52.280	255.011	229.924	17.560	–114.833	–91.328	6.815
800	55.941	262.236	233.517	22.975	–115.942	–87.893	5.739
900	59.156	269.014	237.088	28.734	–116.889	–84.328	4.894
1000	61.951	275.395	240.603	34.792	–117.696	–80.666	4.213
1100	64.368	281.416	244.042	41.111	–118.382	–76.929	3.653
1200	66.453	287.108	247.396	47.655	–118.966	–73.134	3.183
1300	68.251	292.500	250.660	54.392	–119.463	–69.294	2.784
1400	69.803	297.616	253.833	61.297	–119.887	–65.418	2.441
1500	71.146	302.479	256.915	68.346	–120.249	–61.514	2.142

22. METHANOL CH_3OH (g)

T/K	C_p°	S°	$-(G^\circ-H^\circ(T_r))/T$	$H^\circ-H^\circ(T_r)$	$\Delta_f H^\circ$	$\Delta_f G^\circ$	Log K_f
298.15	44.101	239.865	239.865	0.000	–201.000	–162.298	28.434
300	44.219	240.139	239.866	0.082	–201.068	–162.057	28.216
400	51.713	253.845	241.685	4.864	–204.622	–148.509	19.393
500	59.800	266.257	245.374	10.442	–207.750	–134.109	14.010
600	67.294	277.835	249.830	16.803	–210.387	–119.125	10.371
700	73.958	288.719	254.616	23.873	–212.570	–103.737	7.741
800	79.838	298.987	259.526	31.569	–214.350	–88.063	5.750
900	85.025	308.696	264.455	39.817	–215.782	–72.188	4.190
1000	89.597	317.896	269.343	48.553	–216.916	–56.170	2.934
1100	93.624	326.629	274.158	57.718	–217.794	–40.050	1.902
1200	97.165	334.930	278.879	67.262	–218.457	–23.861	1.039
1300	100.277	342.833	283.497	77.137	–218.936	–7.624	0.306
1400	103.014	350.367	288.007	87.304	–219.261	8.644	–0.322
1500	105.422	357.558	292.405	97.729	–219.456	24.930	–0.868

	J/K·mol			kJ/mol			
T/K	C_p°	S°	$-(G^\circ - H^\circ(T_r))/T$	$H^\circ - H^\circ(T_r)$	$\Delta_f H^\circ$	$\Delta_f G^\circ$	Log K_f
23. ACETALDEHYDE C_2H_4O (g)							
298.15	55.318	263.840	263.840	0.000	–166.190	–133.010	23.302
300	55.510	264.180	263.837	0.103	–166.250	–132.800	23.122
400	66.282	281.620	266.147	6.189	–169.530	–121.130	15.818
500	76.675	297.540	270.850	13.345	–172.420	–108.700	11.356
600	85.942	312.360	276.550	21.486	–174.870	–95.720	8.334
700	94.035	326.230	282.667	30.494	–176.910	–82.350	6.145
800	101.070	339.260	288.938	40.258	–178.570	–68.730	4.487
900	107.190	351.520	295.189	50.698	–179.880	–54.920	3.187
1000	112.490	363.100	301.431	61.669	–180.850	–40.930	2.138
1100	117.080	374.040	307.537	73.153	–181.560	–27.010	1.283
1200	121.060	384.400	313.512	85.065	–182.070	–12.860	0.560
1300	124.500	394.230	319.350	97.344	–182.420	1.240	0.050
1400	127.490	403.570	325.031	109.954	–182.640	15.470	–0.577
1500	130.090	412.460	330.571	122.834	–182.750	29.580	–1.030
24. ETHANOL C_2H_5OH (g)							
298.15	65.652	281.622	281.622	0.000	–234.800	–167.874	29.410
300	65.926	282.029	281.623	0.122	–234.897	–167.458	29.157
400	81.169	303.076	284.390	7.474	–239.826	–144.216	18.832
500	95.400	322.750	290.115	16.318	–243.940	–119.820	12.517
600	107.656	341.257	297.112	26.487	–247.260	–94.672	8.242
700	118.129	358.659	304.674	37.790	–249.895	–69.023	5.151
800	127.171	375.038	312.456	50.065	–251.951	–43.038	2.810
900	135.049	390.482	320.276	63.185	–253.515	–16.825	0.976
1000	141.934	405.075	328.033	77.042	–254.662	9.539	–0.498
1100	147.958	418.892	335.670	91.543	–255.454	36.000	–1.709
1200	153.232	431.997	343.156	106.609	–255.947	62.520	–2.721
1300	157.849	444.448	350.473	122.168	–256.184	89.070	–3.579
1400	161.896	456.298	357.612	138.160	–256.206	115.630	–4.314
1500	165.447	467.591	364.571	154.531	–256.044	142.185	–4.951
25. ACETIC ACID $C_2H_4O_2$ (g)							
298.15	63.438	283.470	283.470	0.000	–432.249	–374.254	65.567
300	63.739	283.863	283.471	0.118	–432.324	–373.893	65.100
400	79.665	304.404	286.164	7.296	–436.006	–353.840	46.206
500	93.926	323.751	291.765	15.993	–438.875	–332.950	34.783
600	106.181	341.988	298.631	26.014	–440.993	–311.554	27.123
700	116.627	359.162	306.064	37.169	–442.466	–289.856	21.629
800	125.501	375.331	313.722	49.287	–443.395	–267.985	17.497
900	132.989	390.558	321.422	62.223	–443.873	–246.026	14.279
1000	139.257	404.904	329.060	75.844	–443.982	–224.034	11.702
1100	144.462	418.429	336.576	90.039	–443.798	–202.046	9.594
1200	148.760	431.189	343.933	104.707	–443.385	–180.086	7.839
1300	152.302	443.240	351.113	119.765	–442.795	–158.167	6.355
1400	155.220	454.637	358.105	135.146	–442.071	–136.299	5.085
1500	157.631	465.432	364.903	150.793	–441.247	–114.486	3.987
26. ACETONE C_3H_6O (g)							
298.15	74.517	295.349	295.349	0.000	–217.150	–152.716	26.757
300	74.810	295.809	295.349	0.138	–217.233	–152.339	26.521
400	91.755	319.658	298.498	8.464	–222.212	–129.913	16.962
500	107.864	341.916	304.988	18.464	–226.522	–106.315	11.107
600	122.047	362.836	312.873	29.978	–230.120	–81.923	7.133
700	134.306	382.627	321.470	42.810	–233.049	–56.986	4.252
800	144.934	401.246	330.265	56.785	–235.350	–31.673	2.068

	J/K·mol			kJ/mol			
T/K	C_p°	S°	$-(G^\circ-H^\circ(T_r))/T$	$H^\circ-H^\circ(T_r)$	Δ_fH°	Δ_fG°	Log K_f
26. ACETONE C_3H_6O (g) (continued)							
900	154.097	418.860	339.141	71.747	–237.149	–6.109	0.353
1000	162.046	435.513	347.950	87.563	–238.404	19.707	–1.030
1100	168.908	451.286	356.617	104.136	–239.283	45.396	–2.157
1200	174.891	466.265	365.155	121.332	–239.827	71.463	–3.110
1300	180.079	480.491	373.513	139.072	–240.120	97.362	–3.912
1400	184.556	493.963	381.596	157.314	–240.203	123.470	–4.607
1500	188.447	506.850	389.533	175.975	–240.120	149.369	–5.202
27. PHENOL C_6H_6O (g)							
298.15	103.220	314.810	314.810	0.000	–96.400	–32.630	5.720
300	103.860	315.450	314.810	0.192	–96.490	–32.230	5.610
400	135.790	349.820	319.278	12.217	–100.870	–10.180	1.330
500	161.910	383.040	328.736	27.152	–104.240	12.970	–1.360
600	182.480	414.450	340.430	44.412	–106.810	36.650	–3.190
700	198.840	443.860	353.134	63.508	–108.800	60.750	–4.530
800	212.140	471.310	366.211	84.079	–110.300	85.020	–5.550
900	223.190	496.950	379.327	105.861	–111.370	109.590	–6.360
1000	232.490	520.960	392.302	128.658	–111.990	134.280	–7.010
1100	240.410	543.500	405.033	152.314	–112.280	158.620	–7.530
1200	247.200	564.720	417.468	176.703	–112.390	183.350	–7.980
1300	253.060	584.740	429.568	201.723	–112.330	208.070	–8.360
1400	258.120	603.680	441.331	227.288	–112.120	233.050	–8.700
1500	262.520	621.650	452.767	253.325	–111.780	257.540	–8.970
28. CARBON TETRAFLUORIDE CF_4 (g)							
298.15	61.050	261.455	261.455	0.000	–933.200	–888.518	155.663
300	61.284	261.833	261.456	0.113	–933.219	–888.240	154.654
400	72.399	281.057	264.001	6.822	–933.986	–873.120	114.016
500	80.713	298.153	269.155	14.499	–934.372	–857.852	89.618
600	86.783	313.434	275.284	22.890	–934.490	–842.533	73.348
700	91.212	327.162	281.732	31.801	–934.431	–827.210	61.726
800	94.479	339.566	288.199	41.094	–934.261	–811.903	53.011
900	96.929	350.842	294.542	50.670	–934.024	–796.622	46.234
1000	98.798	361.156	300.695	60.460	–933.745	–781.369	40.814
1100	100.250	370.643	306.629	70.416	–933.442	–766.146	36.381
1200	101.396	379.417	312.334	80.500	–933.125	–750.952	32.688
1300	102.314	387.571	317.811	90.687	–932.800	–735.784	29.564
1400	103.059	395.181	323.069	100.957	–932.470	–720.641	26.887
1500	103.671	402.313	328.116	111.295	–932.137	–705.522	24.568
29. TRIFLUOROMETHANE CHF_3 (g)							
298.15	51.069	259.675	259.675	0.000	–696.700	–662.237	116.020
300	51.258	259.991	259.676	0.095	–696.735	–662.023	115.267
400	61.148	276.113	261.807	5.722	–698.427	–650.186	84.905
500	69.631	290.700	266.149	12.275	–699.715	–637.969	66.647
600	76.453	304.022	271.368	19.593	–700.634	–625.528	54.456
700	81.868	316.230	276.917	27.519	–701.253	–612.957	45.739
800	86.201	327.455	282.542	35.930	–701.636	–600.315	39.196
900	89.719	337.818	288.116	44.732	–701.832	–587.636	34.105
1000	92.617	347.426	293.572	53.854	–701.879	–574.944	30.032
1100	95.038	356.370	298.879	63.240	–701.805	–562.253	26.699
1200	97.084	364.730	304.022	72.849	–701.629	–549.574	23.922
1300	98.833	372.571	308.997	82.647	–701.368	–536.913	21.573
1400	100.344	379.952	313.804	92.607	–701.033	–524.274	19.561
1500	101.660	386.921	318.449	102.709	–700.635	–511.662	17.817

	J/K·mol			kJ/mol			
T/K	C_p°	S°	$-(G^\circ-H^\circ(T_r))/T$	$H^\circ-H^\circ(T_r)$	$\Delta_f H^\circ$	$\Delta_f G^\circ$	Log K_f
30. CHLOROTRIFLUOROMETHANE $CClF_3$ (g)							
298.15	66.886	285.419	285.419	0.000	–707.800	–667.238	116.896
300	67.111	285.834	285.421	0.124	–707.810	–666.986	116.131
400	77.528	306.646	288.187	7.383	–708.153	–653.316	85.313
500	85.013	324.797	293.734	15.532	–708.170	–639.599	66.818
600	90.329	340.794	300.271	24.314	–707.975	–625.901	54.489
700	94.132	355.020	307.096	33.547	–707.654	–612.246	45.686
800	96.899	367.780	313.897	43.106	–707.264	–598.642	39.087
900	98.951	379.317	320.536	52.903	–706.837	–585.090	33.957
1000	100.507	389.827	326.947	62.880	–706.396	–571.586	29.856
1100	101.708	399.465	333.108	72.993	–705.950	–558.126	26.503
1200	102.651	408.357	339.013	83.213	–705.505	–544.707	23.710
1300	103.404	416.604	344.668	93.517	–705.064	–531.326	21.349
1400	104.012	424.290	350.084	103.889	–704.628	–517.977	19.326
1500	104.512	431.484	355.273	114.316	–704.196	–504.660	17.574
31. DICHLORODIFLUOROMETHANE CCl_2F_2 (g)							
298.15	72.476	300.903	300.903	0.000	–486.000	–447.030	78.317
300	72.691	301.352	300.905	0.134	–486.002	–446.788	77.792
400	82.408	323.682	303.883	7.919	–485.945	–433.716	56.637
500	89.063	342.833	309.804	16.514	–485.618	–420.692	43.949
600	93.635	359.500	316.729	25.663	–485.136	–407.751	35.497
700	96.832	374.189	323.909	35.196	–484.576	–394.897	29.467
800	99.121	387.276	331.027	44.999	–483.984	–382.126	24.950
900	100.801	399.053	337.942	55.000	–483.388	–369.429	21.441
1000	102.062	409.742	344.596	65.146	–482.800	–356.799	18.637
1100	103.030	419.517	350.969	75.402	–482.226	–344.227	16.346
1200	103.786	428.515	357.061	85.745	–481.667	–331.706	14.439
1300	104.388	436.847	362.882	96.154	–481.121	–319.232	12.827
1400	104.874	444.602	368.445	106.618	–480.588	–306.799	11.447
1500	105.270	451.851	373.767	117.126	–480.065	–294.404	10.252
32. CHLORODIFLUOROMETHANE $CHClF_2$ (g)							
298.15	55.853	280.915	280.915	0.000	–475.000	–443.845	77.759
300	56.039	281.261	280.916	0.104	–475.028	–443.652	77.246
400	65.395	298.701	283.231	6.188	–476.390	–432.978	56.540
500	73.008	314.145	287.898	13.123	–477.398	–422.001	44.086
600	78.940	328.003	293.448	20.733	–478.103	–410.851	35.767
700	83.551	340.533	299.294	28.867	–478.574	–399.603	29.818
800	87.185	351.936	305.172	37.411	–478.870	–388.299	25.353
900	90.100	362.379	310.956	46.280	–479.031	–376.967	21.878
1000	92.475	371.999	316.586	55.413	–479.090	–365.622	19.098
1100	94.433	380.908	322.033	64.761	–479.068	–354.276	16.823
1200	96.066	389.196	327.289	74.289	–478.982	–342.935	14.927
1300	97.438	396.941	332.352	83.966	–478.843	–331.603	13.324
1400	98.601	404.206	337.228	93.769	–478.661	–320.283	11.950
1500	99.593	411.044	341.923	103.681	–478.443	–308.978	10.759
33. METHYLAMINE CH_5N (g)							
298.15	50.053	242.881	242.881	0.000	–22.529	32.734	–5.735
300	50.227	243.196	242.893	0.091	–22.614	33.077	–5.759
400	60.171	258.986	244.975	5.604	–26.846	52.294	–6.829
500	70.057	273.486	249.244	12.121	–30.431	72.510	–7.575
600	78.929	287.063	254.431	19.579	–33.364	93.382	–8.129
700	86.711	299.826	260.008	27.873	–35.712	114.702	–8.559
800	93.545	311.865	265.749	36.893	–37.548	136.316	–8.900

T/K	C_p° (J/K·mol)	S° (J/K·mol)	$-(G^\circ-H^\circ(T_r))/T$ (J/K·mol)	$H^\circ-H^\circ(T_r)$ (kJ/mol)	$\Delta_f H^\circ$ (kJ/mol)	$\Delta_f G^\circ$ (kJ/mol)	Log K_f
33. METHYLAMINE CH_5N (g) (continued)							
900	99.573	323.239	271.511	46.555	–38.949	158.138	–9.178
1000	104.886	334.006	277.220	56.786	–39.967	180.098	–9.407
1100	109.576	344.233	282.861	67.509	–40.681	201.822	–9.584
1200	113.708	353.944	288.374	78.685	–41.136	224.240	–9.761
1300	117.341	363.190	293.775	90.239	–41.376	246.364	–9.899
1400	120.542	372.012	299.061	102.131	–41.451	268.504	–10.018
1500	123.353	380.426	304.209	114.326	–41.381	290.639	–10.121
34. CHLORINE Cl (g)							
298.15	21.838	165.190	165.190	0.000	121.302	105.306	–18.449
300	21.852	165.325	165.190	0.040	121.311	105.207	–18.318
400	22.467	171.703	166.055	2.259	121.795	99.766	–13.028
500	22.744	176.752	167.708	4.522	122.272	94.203	–9.841
600	22.781	180.905	169.571	6.800	122.734	88.546	–7.709
700	22.692	184.411	171.448	9.074	123.172	82.813	–6.179
800	22.549	187.432	173.261	11.337	123.585	77.019	–5.029
900	22.389	190.079	174.986	13.584	123.971	71.175	–4.131
1000	22.233	192.430	176.615	15.815	124.334	65.289	–3.410
1100	22.089	194.542	178.150	18.031	124.675	59.368	–2.819
1200	21.959	196.458	179.597	20.233	124.996	53.416	–2.325
1300	21.843	198.211	180.963	22.423	125.299	47.439	–1.906
1400	21.742	199.826	182.253	24.602	125.587	41.439	–1.546
1500	21.652	201.323	183.475	26.772	125.861	35.418	–1.233
35. DICHLORINE Cl_2 (g)							
298.15	33.949	223.079	223.079	0.000	0.000	0.000	0.000
300	33.981	223.290	223.080	0.063	0.000	0.000	0.000
400	35.296	233.263	224.431	3.533	0.000	0.000	0.000
500	36.064	241.229	227.021	7.104	0.000	0.000	0.000
600	36.547	247.850	229.956	10.736	0.000	0.000	0.000
700	36.874	253.510	232.926	14.408	0.000	0.000	0.000
800	37.111	258.450	235.815	18.108	0.000	0.000	0.000
900	37.294	262.832	238.578	21.829	0.000	0.000	0.000
1000	37.442	266.769	241.203	25.566	0.000	0.000	0.000
1100	37.567	270.343	243.692	29.316	0.000	0.000	0.000
1200	37.678	273.617	246.052	33.079	0.000	0.000	0.000
1300	37.778	276.637	248.290	36.851	0.000	0.000	0.000
1400	37.872	279.440	250.416	40.634	0.000	0.000	0.000
1500	37.961	282.056	252.439	44.426	0.000	0.000	0.000
36. HYDROGEN CHLORIDE HCl (g)							
298.15	29.136	186.902	186.902	0.000	–92.310	–95.298	16.696
300	29.137	187.082	186.902	0.054	–92.314	–95.317	16.596
400	29.175	195.468	188.045	2.969	–92.587	–96.278	12.573
500	29.304	201.990	190.206	5.892	–92.911	–97.164	10.151
600	29.576	207.354	192.630	8.835	–93.249	–97.983	8.530
700	29.988	211.943	195.069	11.812	–93.577	–98.746	7.368
800	30.500	215.980	197.435	14.836	–93.879	–99.464	6.494
900	31.063	219.604	199.700	17.913	–94.149	–100.145	5.812
1000	31.639	222.907	201.858	21.049	–94.384	–100.798	5.265
1100	32.201	225.949	203.912	24.241	–94.587	–101.430	4.816
1200	32.734	228.774	205.867	27.488	–94.760	–102.044	4.442
1300	33.229	231.414	207.732	30.786	–94.908	–102.645	4.124
1400	33.684	233.893	209.513	34.132	–95.035	–103.235	3.852
1500	34.100	236.232	211.217	37.522	–95.146	–103.817	3.615

T/K	C_p° (J/K·mol)	S° (J/K·mol)	$-(G^\circ-H^\circ(T_r))/T$ (J/K·mol)	$H^\circ-H^\circ(T_r)$ (kJ/mol)	$\Delta_f H^\circ$ (kJ/mol)	$\Delta_f G^\circ$ (kJ/mol)	Log K_f
37. COPPER Cu (cr, l)							
298.15	24.440	33.150	33.150	0.000	0.000	0.000	0.000
300	24.460	33.301	33.150	0.045	0.000	0.000	0.000
400	25.339	40.467	34.122	2.538	0.000	0.000	0.000
500	25.966	46.192	35.982	5.105	0.000	0.000	0.000
600	26.479	50.973	38.093	7.728	0.000	0.000	0.000
700	26.953	55.090	40.234	10.399	0.000	0.000	0.000
800	27.448	58.721	42.322	13.119	0.000	0.000	0.000
900	28.014	61.986	44.328	15.891	0.000	0.000	0.000
1000	28.700	64.971	46.245	18.726	0.000	0.000	0.000
1100	29.553	67.745	48.075	21.637	0.000	0.000	0.000
1200	30.617	70.361	49.824	24.644	0.000	0.000	0.000
1300	31.940	72.862	51.501	27.769	0.000	0.000	0.000
1358	32.844	74.275	52.443	29.647	0.000	0.000	0.000
PHASE TRANSITION: $\Delta_{trs} H$ = 13.141 kJ/mol, $\Delta_{trs} S$ = 9.676 J/K·mol, cr–l							
1358	32.800	83.951	52.443	42.788	0.000	0.000	0.000
1400	32.800	84.950	53.403	44.166	0.000	0.000	0.000
1500	32.800	87.213	55.583	47.446	0.000	0.000	0.000
38. COPPER Cu (g)							
298.15	20.786	166.397	166.397	0.000	337.600	297.873	–52.185
300	20.786	166.525	166.397	0.038	337.594	297.626	–51.821
400	20.786	172.505	167.213	2.117	337.179	284.364	–37.134
500	20.786	177.143	168.752	4.196	336.691	271.215	–28.333
600	20.786	180.933	170.476	6.274	336.147	258.170	–22.475
700	20.786	184.137	172.205	8.353	335.554	245.221	–18.298
800	20.786	186.913	173.874	10.431	334.913	232.359	–15.171
900	20.786	189.361	175.461	12.510	334.219	219.581	–12.744
1000	20.786	191.551	176.963	14.589	333.463	206.883	–10.806
1100	20.788	193.532	178.380	16.667	332.631	194.265	–9.225
1200	20.793	195.341	179.719	18.746	331.703	181.726	–7.910
1300	20.803	197.006	180.986	20.826	330.657	169.270	–6.801
1400	20.823	198.548	182.186	22.907	316.342	157.305	–5.869
1500	20.856	199.986	183.325	24.991	315.146	145.987	–5.084
39. COPPER OXIDE CuO (cr)							
298.15	42.300	42.740	42.740	0.000	–162.000	–134.277	23.524
300	42.417	43.002	42.741	0.078	–161.994	–134.105	23.349
400	46.783	55.878	44.467	4.564	–161.487	–124.876	16.307
500	49.190	66.596	47.852	9.372	–160.775	–115.803	12.098
600	50.827	75.717	51.755	14.377	–159.973	–106.883	9.305
700	52.099	83.651	55.757	19.526	–159.124	–98.102	7.320
800	53.178	90.680	59.691	24.791	–158.247	–89.444	5.840
900	54.144	97.000	63.491	30.158	–157.356	–80.897	4.695
1000	55.040	102.751	67.134	35.617	–156.462	–72.450	3.784
1100	55.890	108.037	70.615	41.164	–155.582	–64.091	3.043
1200	56.709	112.936	73.941	46.794	–154.733	–55.812	2.429
1300	57.507	117.507	77.118	52.505	–153.940	–47.601	1.913
1400	58.288	121.797	80.158	58.295	–166.354	–39.043	1.457
1500	59.057	125.845	83.070	64.163	–165.589	–29.975	1.044
40. DICOPPER OXIDE Cu_2O (cr)							
298.15	62.600	92.550	92.550	0.000	–173.100	–150.344	26.339
300	62.721	92.938	92.551	0.116	–173.102	–150.203	26.152
400	67.587	111.712	95.078	6.654	–173.036	–142.572	18.618
500	70.784	127.155	99.995	13.580	–172.772	–134.984	14.101

T/K	J/K·mol C_p°	J/K·mol S°	J/K·mol $-(G^\circ-H^\circ(T_r))/T$	kJ/mol $H^\circ-H^\circ(T_r)$	kJ/mol $\Delta_f H^\circ$	kJ/mol $\Delta_f G^\circ$	Log K_f
40. DICOPPER OXIDE Cu_2O (cr) (continued)							
600	73.323	140.291	105.643	20.789	–172.389	–127.460	11.096
700	75.552	151.764	111.429	28.235	–171.914	–120.009	8.955
800	77.616	161.989	117.121	35.894	–171.363	–112.631	7.354
900	79.584	171.245	122.629	43.755	–170.750	–105.325	6.113
1000	81.492	179.729	127.920	51.809	–170.097	–98.091	5.124
1100	83.360	187.584	132.992	60.052	–169.431	–90.922	4.317
1200	85.202	194.917	137.850	68.480	–168.791	–83.814	3.648
1300	87.026	201.808	142.507	77.092	–168.223	–76.756	3.084
1400	88.836	208.324	146.978	85.885	–194.030	–68.926	2.572
1500	90.636	214.515	151.276	94.858	–193.438	–60.010	2.090
41. COPPER DICHLORIDE $CuCl_2$ (cr, l)							
298.15	71.880	108.070	108.070	0.000	–218.000	–173.826	30.453
300	71.998	108.515	108.071	0.133	–217.975	–173.552	30.218
400	76.338	129.899	110.957	7.577	–216.494	–158.962	20.758
500	78.654	147.204	116.532	15.336	–214.873	–144.765	15.123
600	80.175	161.687	122.884	23.282	–213.182	–130.901	11.396
675	81.056	171.183	127.732	29.329	–211.185	–120.693	9.340
PHASE TRANSITION: $\Delta_{trs}H$ = 0.700 kJ/mol, $\Delta_{trs}S$ = 1.037 J/K·mol, crII–crI							
675	82.400	172.220	127.732	30.029	–211.185	–120.693	9.340
700	82.400	175.216	129.375	32.089	–210.719	–117.350	8.757
800	82.400	186.219	135.808	40.329	–208.898	–104.137	6.799
871	82.400	193.226	140.207	46.179	–192.649	–94.893	5.691
PHASE TRANSITION: $\Delta_{trs}H$ = 15.001 kJ/mol, $\Delta_{trs}S$ = 17.221 J/K·mol, crI–l							
871	100.000	210.447	140.207	61.180	–192.649	–94.893	5.691
900	100.000	213.723	142.523	64.080	–191.640	–91.655	5.319
1000	100.000	224.259	150.179	74.080	–188.212	–80.730	4.217
1100	100.000	233.790	157.353	84.080	–184.873	–70.144	3.331
1130.75	100.000	236.547	159.470	87.155	–183.867	–66.951	3.093
42. COPPER DICHLORIDE $CuCl_2$ (g)							
298.15	56.814	278.418	278.418	0.000	–43.268	–49.883	8.739
300	56.869	278.769	278.419	0.105	–43.271	–49.924	8.692
400	58.992	295.456	280.679	5.911	–43.428	–52.119	6.806
500	60.111	308.752	285.010	11.871	–43.606	–54.271	5.670
600	60.761	319.774	289.911	17.918	–43.814	–56.385	4.909
700	61.168	329.173	294.865	24.015	–44.060	–58.462	4.362
800	61.439	337.360	299.677	30.147	–44.349	–60.500	3.950
900	61.630	344.608	304.274	36.301	–44.688	–62.499	3.627
1000	61.776	351.109	308.638	42.471	–45.088	–64.457	3.367
1100	61.900	357.003	312.771	48.655	–45.566	–66.372	3.152
1200	62.022	362.394	316.685	54.851	–46.139	–68.239	2.970
1300	62.159	367.364	320.395	61.060	–46.829	–70.053	2.815
1400	62.325	371.976	323.916	67.284	–60.784	–71.404	2.664
1500	62.531	376.283	327.265	73.526	–61.613	–72.133	2.512
43. FLUORINE F (g)							
298.15	22.746	158.750	158.750	0.000	79.380	62.280	–10.911
300	22.742	158.891	158.750	0.042	79.393	62.173	–10.825
400	22.432	165.394	159.639	2.302	80.043	56.332	–7.356
500	22.100	170.363	161.307	4.528	80.587	50.340	–5.259
600	21.832	174.368	163.161	6.724	81.046	44.246	–3.852
700	21.629	177.717	165.008	8.897	81.442	38.081	–2.842
800	21.475	180.595	166.780	11.052	81.792	31.862	–2.080
900	21.357	183.117	168.458	13.193	82.106	25.601	–1.486

43. FLUORINE F (g) (continued)

T/K	C_p° (J/K·mol)	S° (J/K·mol)	$-(G^\circ-H^\circ(T_r))/T$ (J/K·mol)	$H^\circ-H^\circ(T_r)$ (kJ/mol)	$\Delta_f H^\circ$ (kJ/mol)	$\Delta_f G^\circ$ (kJ/mol)	Log K_f
1000	21.266	185.362	170.039	15.324	82.391	19.308	–1.009
1100	21.194	187.386	171.525	17.447	82.654	12.986	–0.617
1200	21.137	189.227	172.925	19.563	82.897	6.642	–0.289
1300	21.091	190.917	174.245	21.675	83.123	0.278	–0.011
1400	21.054	192.479	175.492	23.782	83.335	–6.103	0.228
1500	21.022	193.930	176.673	25.886	83.533	–12.498	0.435

44. DIFLUORINE F_2 (g)

T/K	C_p°	S°	$-(G^\circ-H^\circ(T_r))/T$	$H^\circ-H^\circ(T_r)$	$\Delta_f H^\circ$	$\Delta_f G^\circ$	Log K_f
298.15	31.304	202.790	202.790	0.000	0.000	0.000	0.000
300	31.337	202.984	202.790	0.058	0.000	0.000	0.000
400	32.995	212.233	204.040	3.277	0.000	0.000	0.000
500	34.258	219.739	206.453	6.643	0.000	0.000	0.000
600	35.171	226.070	209.208	10.117	0.000	0.000	0.000
700	35.839	231.545	212.017	13.669	0.000	0.000	0.000
800	36.343	236.365	214.765	17.279	0.000	0.000	0.000
900	36.740	240.669	217.409	20.934	0.000	0.000	0.000
1000	37.065	244.557	219.932	24.625	0.000	0.000	0.000
1100	37.342	248.103	222.334	28.346	0.000	0.000	0.000
1200	37.588	251.363	224.619	32.093	0.000	0.000	0.000
1300	37.811	254.381	226.794	35.863	0.000	0.000	0.000
1400	38.019	257.191	228.866	39.654	0.000	0.000	0.000
1500	38.214	239.820	230.843	43.466	0.000	0.000	0.000

45. HYDROGEN FLUORIDE HF (g)

T/K	C_p°	S°	$-(G^\circ-H^\circ(T_r))/T$	$H^\circ-H^\circ(T_r)$	$\Delta_f H^\circ$	$\Delta_f G^\circ$	Log K_f
298.15	29.137	173.776	173.776	0.000	–273.300	–275.399	48.248
300	29.137	173.956	173.776	0.054	–273.302	–275.412	47.953
400	29.149	182.340	174.919	2.968	–273.450	–276.096	36.054
500	29.172	188.846	177.078	5.884	–273.679	–276.733	28.910
600	29.230	194.169	179.496	8.804	–273.961	–277.318	24.142
700	29.350	198.683	181.923	11.732	–274.277	–277.852	20.733
800	29.549	202.614	184.269	14.676	–274.614	–278.340	18.174
900	29.827	206.110	186.505	17.645	–274.961	–278.785	16.180
1000	30.169	209.270	188.626	20.644	–275.309	–279.191	14.583
1100	30.558	212.163	190.636	23.680	–275.652	–279.563	13.275
1200	30.974	214.840	192.543	26.756	–275.988	–279.904	12.184
1300	31.403	217.336	194.355	29.875	–276.315	–280.217	11.259
1400	31.831	219.679	196.081	33.037	–276.631	–280.505	10.466
1500	32.250	221.889	197.729	36.241	–276.937	–280.771	9.777

46. GERMANIUM Ge (cr, l)

T/K	C_p°	S°	$-(G^\circ-H^\circ(T_r))/T$	$H^\circ-H^\circ(T_r)$	$\Delta_f H^\circ$	$\Delta_f G^\circ$	Log K_f
298.15	23.222	31.090	31.090	0.000	0.000	0.000	0.000
300	23.249	31.234	31.090	0.043	0.000	0.000	0.000
400	24.310	38.083	32.017	2.426	0.000	0.000	0.000
500	24.962	43.582	33.798	4.892	0.000	0.000	0.000
600	25.452	48.178	35.822	7.414	0.000	0.000	0.000
700	25.867	52.133	37.876	9.980	0.000	0.000	0.000
800	26.240	55.612	39.880	12.586	0.000	0.000	0.000
900	26.591	58.723	41.804	15.227	0.000	0.000	0.000
1000	26.926	61.542	43.639	17.903	0.000	0.000	0.000
1100	27.252	64.124	45.386	20.612	0.000	0.000	0.000
1200	27.571	66.509	47.048	23.353	0.000	0.000	0.000
1211.4	27.608	66.770	47.232	23.668	0.000	0.000	0.000
PHASE TRANSITION: $\Delta_{trs}H$ = 37.030 kJ/mol, $\Delta_{trs}S$ = 30.568 J/K·mol, cr–l							
1211.4	27.600	97.338	47.232	60.698	0.000	0.000	0.000
1300	27.600	99.286	50.714	63.143	0.000	0.000	0.000

	J/K·mol			kJ/mol			
T/K	C_p°	S°	-(G°-H° (T_r))/T	H°-H° (T_r)	$\Delta_f H^\circ$	$\Delta_f G^\circ$	Log K_f
46. GERMANIUM Ge (cr, l) (continued)							
1400	27.600	101.331	54.258	65.903	0.000	0.000	0.000
1500	27.600	103.236	57.460	68.663	0.000	0.000	0.000
47. GERMANIUM Ge (g)							
298.15	30.733	167.903	167.903	0.000	367.800	327.009	–57.290
300	30.757	168.094	167.904	0.057	367.814	326.756	–56.893
400	31.071	177.025	169.119	3.162	368.536	312.959	–40.868
500	30.360	183.893	171.415	6.239	369.147	298.991	–31.235
600	29.265	189.334	173.965	9.222	369.608	284.914	–24.804
700	28.102	193.758	176.487	12.090	369.910	270.773	–20.205
800	27.029	197.439	178.882	14.845	370.060	256.598	–16.754
900	26.108	200.567	181.122	17.501	370.073	242.414	–14.069
1000	25.349	203.277	183.205	20.072	369.969	228.234	–11.922
1100	24.741	205.664	185.141	22.575	369.763	214.069	–10.165
1200	24.264	207.795	186.941	25.025	369.471	199.928	–8.703
1300	23.898	209.722	188.621	27.432	332.088	188.521	–7.575
1400	23.624	211.483	190.192	29.807	331.704	177.492	–6.622
1500	23.426	213.105	191.666	32.159	331.296	166.491	–5.798
48. GERMANIUM DIOXIDE GeO_2 (cr, l)							
298.15	50.166	39.710	39.710	0.000	–580.200	–521.605	91.382
300	50.475	40.021	39.711	0.093	–580.204	–521.242	90.755
400	61.281	56.248	41.850	5.759	–579.893	–501.610	65.503
500	66.273	70.519	46.191	12.164	–579.013	–482.134	50.368
600	69.089	82.872	51.299	18.943	–577.915	–462.859	40.295
700	70.974	93.671	56.597	25.952	–576.729	–443.776	33.115
800	72.449	103.247	61.841	33.125	–575.498	–424.866	27.741
900	73.764	111.857	66.928	40.436	–574.235	–406.113	23.570
1000	75.049	119.696	71.819	47.877	–572.934	–387.502	20.241
1100	76.378	126.910	76.504	55.447	–571.582	–369.024	17.523
1200	77.796	133.616	80.987	63.155	–570.166	–350.671	15.264
1300	79.332	139.903	85.279	71.010	–605.685	–329.732	13.249
1308	79.460	140.390	85.615	71.646	–584.059	–328.034	13.100
PHASE TRANSITION: $\Delta_{trs} H$ = 21.500 kJ/mol, $\Delta_{trs} S$ = 16.437 J/K·mol, crII–crI							
1308	80.075	156.827	85.615	93.146	–584.059	–328.034	13.100
1388	81.297	161.617	89.858	99.601	–565.504	–312.415	11.757
PHASE TRANSITION: $\Delta_{trs} H$ = 17.200 kJ/mol, $\Delta_{trs} S$ = 12.392 J/K·mol, crI–l							
1388	78.500	174.009	89.858	116.801	–565.504	–312.415	11.757
1400	78.500	174.685	90.582	117.743	–565.328	–310.228	11.575
1500	78.500	180.100	96.372	125.593	–563.882	–292.057	10.170
49. GERMANIUM TETRACHLORIDE $GeCl_4$ (g)							
298.15	95.918	348.393	348.393	0.000	–500.000	–461.582	80.866
300	96.041	348.987	348.395	0.178	–499.991	–461.343	80.326
400	100.750	377.342	352.229	10.045	–499.447	–448.540	58.573
500	103.206	400.114	359.604	20.255	–498.845	–435.882	45.536
600	104.624	419.067	367.980	30.652	–498.234	–423.347	36.855
700	105.509	435.266	376.463	41.162	–497.634	–410.914	30.662
800	106.096	449.396	384.715	51.744	–497.057	–398.565	26.023
900	106.504	461.917	392.611	62.375	–496.509	–386.287	22.419
1000	106.799	473.155	400.113	73.041	–495.993	–374.068	19.539
1100	107.020	483.344	407.224	83.733	–495.512	–361.899	17.185
1200	107.189	492.664	413.961	94.444	–495.067	–349.772	15.225
1300	107.320	501.249	420.349	105.169	–531.677	–334.973	13.459
1400	107.425	509.206	426.416	115.907	–531.265	–319.857	11.934
1500	107.509	516.621	432.185	126.654	–530.861	–304.771	10.613

T/K	C_p°	S°	$-(G^\circ-H^\circ(T_r))/T$	$H^\circ-H^\circ(T_r)$	$\Delta_f H^\circ$	$\Delta_f G^\circ$	Log K_f
	J/K·mol			kJ/mol			
50. HYDROGEN H (g)							
298.15	20.786	114.716	114.716	0.000	217.998	203.276	–35.613
300	20.786	114.845	114.716	0.038	218.010	203.185	–35.377
400	20.786	120.824	115.532	2.117	218.635	198.149	–25.875
500	20.786	125.463	117.071	4.196	219.253	192.956	–20.158
600	20.786	129.252	118.795	6.274	219.867	187.639	–16.335
700	20.786	132.457	120.524	8.353	220.476	182.219	–13.597
800	20.786	135.232	122.193	10.431	221.079	176.712	–11.538
900	20.786	137.680	123.780	12.510	221.670	171.131	–9.932
1000	20.786	139.870	125.282	14.589	222.247	165.485	–8.644
1100	20.786	141.852	126.700	16.667	222.806	159.781	–7.587
1200	20.786	143.660	128.039	18.746	223.345	154.028	–6.705
1300	20.786	145.324	129.305	20.824	223.864	148.230	–5.956
1400	20.786	146.864	130.505	22.903	224.360	142.393	–5.313
1500	20.786	148.298	131.644	24.982	224.835	136.522	–4.754
51. DIHYDROGEN H_2 (g)							
298.15	28.836	130.680	130.680	0.000	0.000	0.000	0.000
300	28.849	130.858	130.680	0.053	0.000	0.000	0.000
400	29.181	139.217	131.818	2.960	0.000	0.000	0.000
500	29.260	145.738	133.974	5.882	0.000	0.000	0.000
600	29.327	151.078	136.393	8.811	0.000	0.000	0.000
700	29.440	155.607	138.822	11.749	0.000	0.000	0.000
800	29.623	159.549	141.172	14.702	0.000	0.000	0.000
900	29.880	163.052	143.412	17.676	0.000	0.000	0.000
1000	30.204	166.217	145.537	20.680	0.000	0.000	0.000
1100	30.580	169.113	147.550	23.719	0.000	0.000	0.000
1200	30.991	171.791	149.460	26.797	0.000	0.000	0.000
1300	31.422	174.288	151.275	29.918	0.000	0.000	0.000
1400	31.860	176.633	153.003	33.082	0.000	0.000	0.000
1500	32.296	178.846	154.653	36.290	0.000	0.000	0.000
52. HYDROXYL OH (g)							
298.15	29.886	183.737	183.737	0.000	39.349	34.631	–6.067
300	29.879	183.922	183.738	0.055	39.350	34.602	–6.025
400	29.604	192.476	184.906	3.028	39.384	33.012	–4.311
500	29.495	199.067	187.104	5.982	39.347	31.422	–3.283
600	29.513	204.445	189.560	8.931	39.252	29.845	–2.598
700	29.655	209.003	192.020	11.888	39.113	28.287	–2.111
800	29.914	212.979	194.396	14.866	38.945	26.752	–1.747
900	30.265	216.522	196.661	17.874	38.763	25.239	–1.465
1000	30.682	219.731	198.810	20.921	38.577	23.746	–1.240
1100	31.135	222.677	200.848	24.012	38.393	22.272	–1.058
1200	31.603	225.406	202.782	27.149	38.215	20.814	–0.906
1300	32.069	227.954	204.621	30.332	38.046	19.371	–0.778
1400	32.522	230.347	206.374	33.562	37.886	17.941	–0.669
1500	32.956	232.606	208.048	36.836	37.735	16.521	–0.575
53. WATER H_2O (l)							
298.15	75.300	69.950	69.950	0.000	–285.830	–237.141	41.546
300	75.281	70.416	69.951	0.139	–285.771	–236.839	41.237
373.21	76.079	86.896	71.715	5.666	–283.454	–225.160	31.513
54. WATER H_2O (g)							
298.15	33.598	188.832	188.832	0.000	–241.826	–228.582	40.046

	J/K·mol			kJ/mol			
T/K	C_p°	S°	$-(G^\circ-H^\circ(T_r))/T$	$H^\circ-H^\circ(T_r)$	$\Delta_f H^\circ$	$\Delta_f G^\circ$	Log K_f

54. WATER H_2O (g) (continued)

T/K	C_p°	S°	$-(G^\circ-H^\circ(T_r))/T$	$H^\circ-H^\circ(T_r)$	$\Delta_f H^\circ$	$\Delta_f G^\circ$	Log K_f
300	33.606	189.040	188.833	0.062	–241.844	–228.500	39.785
400	34.283	198.791	190.158	3.453	–242.845	–223.900	29.238
500	35.259	206.542	192.685	6.929	–243.822	–219.050	22.884
600	36.371	213.067	195.552	10.509	–244.751	–214.008	18.631
700	37.557	218.762	198.469	14.205	–245.620	–208.814	15.582
800	38.800	223.858	201.329	18.023	–246.424	–203.501	13.287
900	40.084	228.501	204.094	21.966	–247.158	–198.091	11.497
1000	41.385	232.792	206.752	26.040	–247.820	–192.603	10.060
1100	42.675	236.797	209.303	30.243	–248.410	–187.052	8.882
1200	43.932	240.565	211.753	34.574	–248.933	–181.450	7.898
1300	45.138	244.129	214.108	39.028	–249.392	–175.807	7.064
1400	46.281	247.516	216.374	43.599	–249.792	–170.132	6.348
1500	47.356	250.746	218.559	48.282	–250.139	–164.429	5.726

55. IODINE I (g)

T/K	C_p°	S°	$-(G^\circ-H^\circ(T_r))/T$	$H^\circ-H^\circ(T_r)$	$\Delta_f H^\circ$	$\Delta_f G^\circ$	Log K_f
298.15	20.786	180.787	180.787	0.000	106.760	70.172	–12.294
300	20.786	180.915	180.787	0.038	106.748	69.945	–12.178
400	20.786	186.895	181.602	2.117	97.974	58.060	–7.582
500	20.786	191.533	183.142	4.196	75.988	50.202	–5.244
600	20.786	195.323	184.866	6.274	76.190	45.025	–3.920
700	20.786	198.527	186.594	8.353	76.385	39.816	–2.971
800	20.787	201.303	188.263	10.432	76.574	34.579	–2.258
900	20.789	203.751	189.851	12.510	76.757	29.319	–1.702
1000	20.795	205.942	191.352	14.589	76.936	24.038	–1.256
1100	20.806	207.924	192.770	16.669	77.109	18.740	–0.890
1200	20.824	209.735	194.110	18.751	77.277	13.426	–0.584
1300	20.851	211.403	195.377	20.835	77.440	8.098	–0.325
1400	20.889	212.950	196.577	22.921	77.596	2.758	–0.103
1500	20.936	214.392	197.717	25.013	77.745	–2.592	0.090

56. DIIODINE I_2 (cr, l)

T/K	C_p°	S°	$-(G^\circ-H^\circ(T_r))/T$	$H^\circ-H^\circ(T_r)$	$\Delta_f H^\circ$	$\Delta_f G^\circ$	Log K_f
298.15	54.440	116.139	116.139	0.000	0.000	0.000	0.000
300	54.518	116.476	116.140	0.101	0.000	0.000	0.000
386.75	61.531	131.039	117.884	5.088	0.000	0.000	0.000
		PHASE TRANSITION: $\Delta_{trs}H$ = 15.665 kJ/mol, $\Delta_{trs}S$ = 40.504 J/K·mol, cr–l					
386.75	79.555	171.543	117.884	20.753	0.000	0.000	0.000
400	79.555	174.223	119.706	21.807	0.000	0.000	0.000
457.67	79.555	184.938	127.266	26.395	0.000	0.000	0.000

57. DIIODINE I_2 (g)

T/K	C_p°	S°	$-(G^\circ-H^\circ(T_r))/T$	$H^\circ-H^\circ(T_r)$	$\Delta_f H^\circ$	$\Delta_f G^\circ$	Log K_f
298.15	36.887	260.685	260.685	0.000	62.420	19.324	–3.385
300	36.897	260.913	260.685	0.068	62.387	19.056	–3.318
400	37.256	271.584	262.138	3.778	44.391	5.447	–0.711
457.67	37.385	276.610	263.652	5.931		pressure = 1 bar	
500	37.464	279.921	264.891	7.515	0.000	0.000	0.000
600	37.613	286.765	267.983	11.269	0.000	0.000	0.000
700	37.735	292.573	271.092	15.037	0.000	0.000	0.000
800	37.847	297.619	274.099	18.816	0.000	0.000	0.000
900	37.956	302.083	276.965	22.606	0.000	0.000	0.000
1000	38.070	306.088	279.681	26.407	0.000	0.000	0.000
1100	38.196	309.722	282.249	30.220	0.000	0.000	0.000
1200	38.341	313.052	284.679	34.047	0.000	0.000	0.000
1300	38.514	316.127	286.981	37.890	0.000	0.000	0.000
1400	38.719	318.989	289.166	41.751	0.000	0.000	0.000
1500	38.959	321.668	291.245	45.635	0.000	0.000	0.000

	J/K·mol			kJ/mol			
T/K	C_p°	S°	$-(G^\circ - H^\circ(T_r))/T$	$H^\circ - H^\circ(T_r)$	$\Delta_f H^\circ$	$\Delta_f G^\circ$	Log K_f

58. HYDROGEN IODIDE HI (g)

T/K	C_p°	S°	$-(G^\circ - H^\circ(T_r))/T$	$H^\circ - H^\circ(T_r)$	$\Delta_f H^\circ$	$\Delta_f G^\circ$	Log K_f
298.15	29.157	206.589	206.589	0.000	26.500	1.700	–0.298
300	29.158	206.769	206.589	0.054	26.477	1.546	–0.269
400	29.329	215.176	207.734	2.977	17.093	–6.289	0.821
500	29.738	221.760	209.904	5.928	–5.481	–9.946	1.039
600	30.351	227.233	212.348	8.931	–5.819	–10.806	0.941
700	31.070	231.965	214.820	12.002	–6.101	–11.614	0.867
800	31.807	236.162	217.230	15.145	–6.323	–12.386	0.809
900	32.511	239.950	219.548	18.362	–6.489	–13.133	0.762
1000	33.156	243.409	221.763	21.646	–6.608	–13.865	0.724
1100	33.735	246.597	223.878	24.991	–6.689	–14.586	0.693
1200	34.249	249.555	225.896	28.391	–6.741	–15.302	0.666
1300	34.703	252.314	227.823	31.839	–6.775	–16.014	0.643
1400	35.106	254.901	229.666	35.330	–6.797	–16.723	0.624
1500	35.463	257.336	231.430	38.858	–6.814	–17.432	0.607

59. POTASSIUM K (cr, l)

T/K	C_p°	S°	$-(G^\circ - H^\circ(T_r))/T$	$H^\circ - H^\circ(T_r)$	$\Delta_f H^\circ$	$\Delta_f G^\circ$	Log K_f
298.15	29.600	64.680	64.680	0.000	0.000	0.000	0.000
300	29.671	64.863	64.681	0.055	0.000	0.000	0.000
336.86	32.130	68.422	64.896	1.188	0.000	0.000	0.000
PHASE TRANSITION: $\Delta_{trs} H$ = 2.321 kJ/mol, $\Delta_{trs} S$ = 6.891 J/K·mol, cr–l							
336.86	32.129	75.313	64.896	3.509	0.000	0.000	0.000
400	31.552	80.784	66.986	5.519	0.000	0.000	0.000
500	30.741	87.734	70.469	8.632	0.000	0.000	0.000
600	30.158	93.283	73.824	11.675	0.000	0.000	0.000
700	29.851	97.905	76.943	14.673	0.000	0.000	0.000
800	29.838	101.887	79.818	17.655	0.000	0.000	0.000
900	30.130	105.415	82.470	20.651	0.000	0.000	0.000
1000	30.730	108.618	84.927	23.691	0.000	0.000	0.000
1039.4	31.053	109.812	85.847	24.908	0.000	0.000	0.000

60. POTASSIUM K (g)

T/K	C_p°	S°	$-(G^\circ - H^\circ(T_r))/T$	$H^\circ - H^\circ(T_r)$	$\Delta_f H^\circ$	$\Delta_f G^\circ$	Log K_f
298.15	20.786	160.340	160.340	0.000	89.000	60.479	–10.596
300	20.786	160.468	160.340	0.038	88.984	60.302	–10.499
400	20.786	166.448	161.155	2.117	85.598	51.332	–6.703
500	20.786	171.086	162.695	4.196	84.563	42.887	–4.480
600	20.786	174.876	164.419	6.274	83.599	34.643	–3.016
700	20.786	178.080	166.148	8.353	82.680	26.557	–1.982
800	20.786	180.856	167.817	10.431	81.776	18.601	–1.215
900	20.786	183.304	169.404	12.510	80.859	10.759	–0.624
1000	20.786	185.494	170.905	14.589	79.897	3.021	–0.158
1039.4	20.786	186.297	171.474	15.408		pressure = 1 bar	
1100	20.786	187.175	172.323	16.667	0.000	0.000	0.000
1200	20.786	189.284	173.662	18.746	0.000	0.000	0.000
1300	20.789	190.948	174.929	20.825	0.000	0.000	0.000
1400	20.793	192.489	176.129	22.904	0.000	0.000	0.000
1500	20.801	193.923	177.268	24.983	0.000	0.000	0.000

61. DIPOTASSIUM OXIDE K_2O (cr, l)

T/K	C_p°	S°	$-(G^\circ - H^\circ(T_r))/T$	$H^\circ - H^\circ(T_r)$	$\Delta_f H^\circ$	$\Delta_f G^\circ$	Log K_f
298.15	72.000	96.000	96.000	0.000	–361.700	–321.171	56.267
300	72.130	96.446	96.001	0.133	–361.704	–320.920	55.876
400	79.154	118.158	98.914	7.698	–366.554	–306.416	40.013
500	86.178	136.575	104.647	15.964	–366.043	–291.423	30.444
590	92.500	151.348	110.662	24.005	–364.204	–278.079	24.619
PHASE TRANSITION: $\Delta_{trs} H$ = 0.700 kJ/mol, $\Delta_{trs} S$ = 1.186 J/K·mol, crIII–crII							

	J/K·mol			kJ/mol			
T/K	C_p°	S°	$-(G^\circ - H^\circ(T_r))/T$	$H^\circ - H^\circ(T_r)$	$\Delta_f H^\circ$	$\Delta_f G^\circ$	Log K_f
61. DIPOTASSIUM OXIDE K_2O (cr, l) (continued)							
590	100.000	152.534	110.662	24.705	–364.204	–278.079	24.619
600	100.000	154.215	111.374	25.705	–363.968	–276.621	24.082
645	100.000	161.447	114.618	30.205	–358.901	–270.109	21.874
PHASE TRANSITION: $\Delta_{trs}H$ = 4.000 kJ/mol, $\Delta_{trs}S$ = 6.202 J/K·mol, crII–crI							
645	100.000	167.649	114.618	34.205	–358.901	–270.109	21.874
700	100.000	175.832	119.111	39.705	–357.592	–262.592	19.595
800	100.000	189.185	127.054	49.705	–355.224	–249.183	16.270
900	100.000	200.963	134.625	59.705	–352.919	–236.067	13.701
1000	100.000	211.499	141.794	69.705	–350.732	–223.202	11.659
1013	100.000	212.791	142.697	71.005	–323.459	–221.546	11.424
PHASE TRANSITION: $\Delta_{trs}H$ = 27.000 kJ/mol, $\Delta_{trs}S$ =26.654 J/K·mol, crI–l							
1013	100.000	239.444	142.697	98.005	–323.459	–221.546	11.424
1100	100.000	247.684	150.679	106.705	–479.439	–203.633	9.670
1200	100.000	256.385	159.131	116.705	–475.371	–178.740	7.780
1300	100.000	264.389	166.924	126.705	–471.321	–154.185	6.195
1400	100.000	271.800	174.154	136.705	–467.287	–129.941	4.848
1500	100.000	278.699	180.896	146.705	–463.268	–105.986	3.691
62. POTASSIUM HYDROXIDE KOH (cr, l)							
298.15	64.900	78.870	78.870	0.000	–424.580	–378.747	66.354
300	65.038	79.272	78.871	0.120	–424.569	–378.463	65.895
400	72.519	99.007	81.512	6.998	–426.094	–362.765	47.372
500	80.000	115.993	86.745	14.624	–424.572	–347.093	36.260
520	81.496	119.159	87.931	16.239	–417.725	–344.002	34.555
PHASE TRANSITION: $\Delta_{trs}H$ = 6.450 kJ/mol, $\Delta_{trs}S$ = 12.404 J/K·mol, crII–crI							
520	79.000	131.563	87.931	22.689	–417.725	–344.002	34.555
600	79.000	142.868	94.520	29.009	–416.274	–332.766	28.969
678	79.000	152.523	100.649	35.171	–405.464	–321.998	24.807
PHASE TRANSITION: $\Delta_{trs}H$ = 9.400 kJ/mol, $\Delta_{trs}S$ = 13.865 J/K·mol, crI–l							
678	83.000	166.388	100.649	44.571	–405.464	–321.998	24.807
700	83.000	169.038	102.757	46.397	–404.981	–319.297	23.826
800	83.000	180.121	111.750	54.697	–402.808	–307.206	20.058
900	83.000	189.897	119.901	62.997	–400.694	–295.383	17.143
1000	83.000	198.642	127.345	71.297	–398.668	–283.791	14.824
1100	83.000	206.553	134.192	79.597	–475.618	–267.780	12.716
1200	83.000	213.775	140.527	87.897	–472.711	–249.014	10.839
1300	83.000	220.418	146.421	96.197	–469.843	–230.490	9.261
1400	83.000	226.569	151.929	104.497	–467.011	–212.184	7.917
1500	83.000	232.296	157.098	112.797	–464.217	–194.080	6.758
63. POTASSIUM HYDROXIDE KOH (g)							
298.15	49.184	238.283	238.283	0.000	–227.989	–229.685	40.239
300	49.236	238.588	238.284	0.091	–228.007	–229.696	39.993
400	51.178	253.053	240.243	5.124	–231.377	–229.667	29.991
500	52.178	264.591	243.998	10.296	–232.309	–229.129	23.937
600	52.804	274.163	248.251	15.547	–233.145	–228.413	19.885
700	53.296	282.340	252.551	20.853	–233.934	–227.562	16.981
800	53.758	289.487	256.730	26.206	–234.708	–226.599	14.795
900	54.229	295.846	260.730	31.605	–235.495	–225.538	13.090
1000	54.713	301.585	264.533	37.052	–236.322	–224.388	11.721
1100	55.203	306.823	268.143	42.548	–316.077	–218.535	10.377
1200	55.686	311.647	271.570	48.092	–315.925	–209.674	9.127
1300	56.153	316.122	274.827	53.684	–315.764	–200.826	8.069
1400	56.598	320.300	277.927	59.322	–315.595	–191.991	7.163
1500	57.016	324.220	280.884	65.003	–315.420	–183.169	6.378

64. POTASSIUM CHLORIDE KCl (cr, l)

	J/K·mol			kJ/mol			
T/K	C_p°	S°	$-(G^\circ-H^\circ(T_r))/T$	$H^\circ-H^\circ(T_r)$	$\Delta_f H^\circ$	$\Delta_f G^\circ$	Log K_f
298.15	51.300	82.570	82.570	0.000	–436.490	–408.568	71.579
300	51.333	82.887	82.571	0.095	–436.481	–408.395	71.107
400	52.977	97.886	84.605	5.312	–438.463	–398.651	52.058
500	54.448	109.867	88.498	10.685	–437.990	–388.749	40.612
600	55.885	119.921	92.919	16.201	–437.332	–378.960	32.991
700	57.425	128.649	97.413	21.865	–436.502	–369.295	27.557
800	59.205	136.430	101.812	27.694	–435.505	–359.760	23.490
900	61.361	143.523	106.058	33.719	–434.337	–350.360	20.334
1000	64.032	150.121	110.138	39.983	–432.981	–341.100	17.817
1044	65.405	152.908	111.882	42.830	–485.450	–336.720	16.847
	PHASE TRANSITION: $\Delta_{trs}H$ = 26.320 kJ/mol, $\Delta_{trs}S$ = 25.210 J/K·mol, cr–l						
1044	72.000	178.118	111.882	69.150	–485.450	–336.720	16.847
1100	72.000	181.880	115.351	73.182	–483.633	–328.790	15.613
1200	72.000	188.145	121.160	80.382	–480.393	–314.856	13.705
1300	72.000	193.908	126.537	87.582	–477.158	–301.192	12.102
1400	72.000	199.244	131.542	94.782	–473.928	–287.778	10.737
1500	72.000	204.211	136.223	101.982	–470.704	–274.594	9.562

65. POTASSIUM CHLORIDE KCl (g)

T/K	C_p°	S°	$-(G^\circ-H^\circ(T_r))/T$	$H^\circ-H^\circ(T_r)$	$\Delta_f H^\circ$	$\Delta_f G^\circ$	Log K_f
298.15	36.505	239.091	239.091	0.000	–214.575	–233.320	40.876
300	36.518	239.317	239.092	0.068	–214.594	–233.436	40.644
400	37.066	249.904	240.532	3.749	–218.112	–239.107	31.224
500	37.384	258.212	243.267	7.473	–219.287	–244.219	25.513
600	37.597	265.048	246.344	11.222	–220.396	–249.100	21.686
700	37.769	270.857	249.441	14.991	–221.461	–253.799	18.938
800	37.907	275.910	252.441	18.775	–222.509	–258.347	16.868
900	38.041	280.382	255.302	22.572	–223.568	–262.764	15.250
1000	38.162	284.397	258.014	26.383	–224.667	–267.061	13.950
1100	38.279	288.039	260.581	30.205	–304.696	–266.627	12.661
1200	38.401	291.375	263.010	34.039	–304.821	–263.161	11.455
1300	38.518	294.454	265.312	37.885	–304.941	–259.684	10.434
1400	38.639	297.313	267.496	41.743	–305.053	–256.199	9.559
1500	38.761	299.983	269.574	45.613	–305.159	–252.706	8.800

66. DINITROGEN N_2 (g)

T/K	C_p°	S°	$-(G^\circ-H^\circ(T_r))/T$	$H^\circ-H^\circ(T_r)$	$\Delta_f H^\circ$	$\Delta_f G^\circ$	Log K_f
298.15	29.124	191.608	191.608	0.000	0.000	0.000	0.000
300	29.125	191.788	191.608	0.054	0.000	0.000	0.000
400	29.249	200.180	192.752	2.971	0.000	0.000	0.000
500	29.580	206.738	194.916	5.911	0.000	0.000	0.000
600	30.109	212.175	197.352	8.894	0.000	0.000	0.000
700	30.754	216.864	199.812	11.936	0.000	0.000	0.000
800	31.433	221.015	202.208	15.046	0.000	0.000	0.000
900	32.090	224.756	204.509	18.222	0.000	0.000	0.000
1000	32.696	228.169	206.706	21.462	0.000	0.000	0.000
1100	33.241	231.311	208.802	24.759	0.000	0.000	0.000
1200	33.723	234.224	210.801	28.108	0.000	0.000	0.000
1300	34.147	236.941	212.708	31.502	0.000	0.000	0.000
1400	34.517	239.485	214.531	34.936	0.000	0.000	0.000
1500	34.842	241.878	216.275	38.404	0.000	0.000	0.000

67. NITRIC OXIDE NO (g)

T/K	C_p°	S°	$-(G^\circ-H^\circ(T_r))/T$	$H^\circ-H^\circ(T_r)$	$\Delta_f H^\circ$	$\Delta_f G^\circ$	Log K_f
298.15	29.862	210.745	210.745	0.000	91.277	87.590	–15.345
300	29.858	210.930	210.746	0.055	91.278	87.567	–15.247
400	29.954	219.519	211.916	3.041	91.320	86.323	–11.272
500	30.493	226.255	214.133	6.061	91.340	85.071	–8.887

T/K	C_p° (J/K·mol)	S° (J/K·mol)	$-(G^\circ - H^\circ(T_r))/T$ (J/K·mol)	$H^\circ - H^\circ(T_r)$ (kJ/mol)	$\Delta_f H^\circ$ (kJ/mol)	$\Delta_f G^\circ$ (kJ/mol)	Log K_f
67. NITRIC OXIDE NO (g) (continued)							
600	31.243	231.879	216.635	9.147	91.354	83.816	–7.297
700	32.031	236.754	219.168	12.310	91.369	82.558	–6.160
800	32.770	241.081	221.642	15.551	91.386	81.298	–5.308
900	33.425	244.979	224.022	18.862	91.405	80.036	–4.645
1000	33.990	248.531	226.298	22.233	91.426	78.772	–4.115
1100	34.473	251.794	228.469	25.657	91.445	77.505	–3.680
1200	34.883	254.811	230.540	29.125	91.464	76.237	–3.318
1300	35.234	257.618	232.516	32.632	91.481	74.967	–3.012
1400	35.533	260.240	234.404	36.170	91.495	73.697	–2.750
1500	35.792	262.700	236.209	39.737	91.506	72.425	–2.522
68. NITROGEN DIOXIDE NO_2 (g)							
298.15	37.178	240.166	240.166	0.000	34.193	52.316	–9.165
300	37.236	240.397	240.167	0.069	34.181	52.429	–9.129
400	40.513	251.554	241.666	3.955	33.637	58.600	–7.652
500	43.664	260.939	244.605	8.167	33.319	64.882	–6.778
600	46.383	269.147	248.026	12.673	33.174	71.211	–6.199
700	48.612	276.471	251.575	17.427	33.151	77.553	–5.787
800	50.405	283.083	255.107	22.381	33.213	83.893	–5.478
900	51.844	289.106	258.555	27.496	33.334	90.221	–5.236
1000	53.007	294.631	261.891	32.741	33.495	96.534	–5.042
1100	53.956	299.729	265.102	38.090	33.686	102.828	–4.883
1200	54.741	304.459	268.187	43.526	33.898	109.105	–4.749
1300	55.399	308.867	271.148	49.034	34.124	115.363	–4.635
1400	55.960	312.994	273.992	54.603	34.360	121.603	–4.537
1500	56.446	316.871	276.722	60.224	34.604	127.827	–4.451
69. AMMONIA NH_3 (g)							
298.15	35.630	192.768	192.768	0.000	–45.940	–16.407	2.874
300	35.678	192.989	192.769	0.066	–45.981	–16.223	2.825
400	38.674	203.647	194.202	3.778	–48.087	–5.980	0.781
500	41.994	212.633	197.011	7.811	–49.908	4.764	–0.498
600	45.229	220.578	200.289	12.174	–51.430	15.846	–1.379
700	48.269	227.781	203.709	16.850	–52.682	27.161	–2.027
800	51.112	234.414	207.138	21.821	–53.695	38.639	–2.523
900	53.769	240.589	210.516	27.066	–54.499	50.231	–2.915
1000	56.244	246.384	213.816	32.569	–55.122	61.903	–3.233
1100	58.535	251.854	217.027	38.309	–55.589	73.629	–3.496
1200	60.644	257.039	220.147	44.270	–55.920	85.392	–3.717
1300	62.576	261.970	223.176	50.432	–56.136	97.177	–3.905
1400	64.339	266.673	226.117	56.779	–56.251	108.975	–4.066
1500	65.945	271.168	228.971	63.295	–56.282	120.779	–4.206
70. OXYGEN O (g)							
298.15	21.911	161.058	161.058	0.000	249.180	231.743	–40.600
300	21.901	161.194	161.059	0.041	249.193	231.635	–40.331
400	21.482	167.430	161.912	2.207	249.874	225.677	–29.470
500	21.257	172.197	163.511	4.343	250.481	219.556	–22.937
600	21.124	176.060	165.290	6.462	251.019	213.319	–18.571
700	21.040	179.310	167.067	8.570	251.500	206.997	–15.446
800	20.984	182.115	168.777	10.671	251.932	200.610	–13.098
900	20.944	184.584	170.399	12.767	252.325	194.171	–11.269
1000	20.915	186.789	171.930	14.860	252.686	187.689	–9.804
1100	20.893	188.782	173.372	16.950	253.022	181.173	–8.603
1200	20.877	190.599	174.733	19.039	253.335	174.628	–7.601
1300	20.864	192.270	176.019	21.126	253.630	168.057	–6.753

	J/K·mol			kJ/mol			
T/K	C_p°	S°	$-(G^\circ-H^\circ(T_r))/T$	$H^\circ-H^\circ(T_r)$	$\Delta_f H^\circ$	$\Delta_f G^\circ$	Log K_f
70. OXYGEN O (g) (continued)							
1400	20.853	193.815	177.236	23.212	253.908	161.463	–6.024
1500	20.845	195.254	178.389	25.296	254.171	154.851	–5.392
71. DIOXYGEN O_2 (g)							
298.15	29.378	205.148	205.148	0.000	0.000	0.000	0.000
300	29.387	205.330	205.148	0.054	0.000	0.000	0.000
400	30.109	213.873	206.308	3.026	0.000	0.000	0.000
500	31.094	220.695	208.525	6.085	0.000	0.000	0.000
600	32.095	226.454	211.045	9.245	0.000	0.000	0.000
700	32.987	231.470	213.612	12.500	0.000	0.000	0.000
800	33.741	235.925	216.128	15.838	0.000	0.000	0.000
900	34.365	239.937	218.554	19.244	0.000	0.000	0.000
1000	34.881	243.585	220.878	22.707	0.000	0.000	0.000
1100	35.314	246.930	223.096	26.217	0.000	0.000	0.000
1200	35.683	250.019	225.213	29.768	0.000	0.000	0.000
1300	36.006	252.888	227.233	33.352	0.000	0.000	0.000
1400	36.297	255.568	229.162	36.968	0.000	0.000	0.000
1500	36.567	258.081	231.007	40.611	0.000	0.000	0.000
72. SULFUR S (cr, l)							
298.15	22.690	32.070	32.070	0.000	0.000	0.000	0.000
300	22.737	32.210	32.070	0.042	0.000	0.000	0.000
368.3	24.237	37.030	32.554	1.649	0.000	0.000	0.000
PHASE TRANSITION: $\Delta_{trs}H$ = 0.401 kJ/mol, $\Delta_{trs}S$ = 1.089 J/K·mol, crII–crI							
368.3	24.773	38.119	32.553	2.050	0.000	0.000	0.000
388.36	25.180	39.444	32.875	2.551	0.000	0.000	0.000
PHASE TRANSITION: $\Delta_{trs}H$ = 1.722 kJ/mol, $\Delta_{trs}S$ = 4.431 J/K·mol, crI–l							
388.36	31.710	43.875	32.872	4.273	0.000	0.000	0.000
400	32.369	44.824	33.206	4.647	0.000	0.000	0.000
500	38.026	53.578	36.411	8.584	0.000	0.000	0.000
600	34.371	60.116	39.842	12.164	0.000	0.000	0.000
700	32.451	65.278	43.120	15.511	0.000	0.000	0.000
800	32.000	69.557	46.163	18.715	0.000	0.000	0.000
882.38	32.000	72.693	48.496	21.351	0.000	0.000	0.000
73. SULFUR S (g)							
298.15	23.673	167.828	167.828	0.000	277.180	236.704	–41.469
300	23.669	167.974	167.828	0.044	277.182	236.453	–41.170
400	23.233	174.730	168.752	2.391	274.924	222.962	–29.115
500	22.741	179.860	170.482	4.689	273.286	210.145	–21.953
600	22.338	183.969	172.398	6.942	271.958	197.646	–17.206
700	22.031	187.388	174.302	9.160	270.829	185.352	–13.831
800	21.800	190.314	176.125	11.351	269.816	173.210	–11.309
900	21.624	192.871	177.847	13.522	215.723	162.258	–9.417
1000	21.489	195.142	179.465	15.677	216.018	156.301	–8.164
1100	21.386	197.185	180.985	17.821	216.284	150.317	–7.138
1200	21.307	199.043	182.413	19.955	216.525	144.309	–6.282
1300	21.249	200.746	183.759	22.083	216.743	138.282	–5.556
1400	21.209	202.319	185.029	24.206	216.940	132.239	–4.934
1500	21.186	203.781	186.231	26.325	217.119	126.182	–4.394
74. DISULFUR S_2 (g)							
298.15	32.505	228.165	228.165	0.000	128.600	79.696	–13.962
300	32.540	228.366	228.165	0.060	128.576	79.393	–13.823
400	34.108	237.956	229.462	3.398	122.703	63.380	–8.276
500	35.133	245.686	231.959	6.863	118.296	49.031	–5.122

	J/K·mol			kJ/mol			
T/K	C_p°	S°	$-(G^\circ - H^\circ(T_r))/T$	$H^\circ - H^\circ(T_r)$	$\Delta_f H^\circ$	$\Delta_f G^\circ$	Log K_f
74. DISULFUR S_2 (g) (continued)							
600	35.815	252.156	234.800	10.413	114.685	35.530	–3.093
700	36.305	257.715	237.686	14.020	111.599	22.588	–1.685
800	36.697	262.589	240.501	17.671	108.841	10.060	–0.657
882.38	36.985	266.200	242.734	20.706		pressure = 1 bar	
900	37.045	266.932	243.201	21.358	0.000	0.000	0.000
1000	37.377	270.852	245.773	25.079	0.000	0.000	0.000
1100	37.704	274.430	248.218	28.833	0.000	0.000	0.000
1200	38.030	277.725	250.541	32.620	0.000	0.000	0.000
1300	38.353	280.781	252.751	36.439	0.000	0.000	0.000
1400	38.669	283.635	254.856	40.290	0.000	0.000	0.000
1500	38.976	286.314	256.865	44.173	0.000	0.000	0.000
75. OCTASULFUR S_8 (g)							
298.15	156.500	432.536	432.536	0.000	101.277	48.810	–8.551
300	156.768	433.505	432.539	0.290	101.231	48.484	–8.442
400	167.125	480.190	438.834	16.542	80.642	32.003	–4.179
500	173.181	518.176	451.022	33.577	66.185	21.409	–2.237
600	177.936	550.180	464.951	51.137	55.101	13.549	–1.180
700	182.441	577.948	479.152	69.157	46.349	7.343	–0.548
800	186.764	602.596	493.071	87.620	39.177	2.263	–0.148
900	190.595	624.821	506.495	106.494	–392.062	6.554	–0.380
1000	193.618	645.067	519.355	125.712	–387.728	50.614	–2.644
1100	195.684	663.625	531.639	145.185	–383.272	94.233	–4.475
1200	196.825	680.707	543.359	164.817	–378.786	137.444	–5.983
1300	197.195	696.480	554.539	184.524	–374.356	180.283	–7.244
1400	196.988	711.089	565.206	204.237	–370.048	222.785	–8.312
1500	196.396	724.662	575.389	223.909	–365.905	264.984	–9.227
76. SULFUR DIOXIDE SO_2 (g)							
298.15	39.842	248.219	248.219	0.000	–296.810	–300.090	52.574
300	39.909	248.466	248.220	0.074	–296.833	–300.110	52.253
400	43.427	260.435	249.828	4.243	–300.240	–300.935	39.298
500	46.490	270.465	252.978	8.744	–302.735	–300.831	31.427
600	48.938	279.167	256.634	13.520	–304.699	–300.258	26.139
700	50.829	286.859	260.413	18.513	–306.308	–299.386	22.340
800	52.282	293.746	264.157	23.671	–307.691	–298.302	19.477
900	53.407	299.971	267.796	28.958	–362.075	–295.987	17.178
1000	54.290	305.646	271.301	34.345	–362.012	–288.647	15.077
1100	54.993	310.855	274.664	39.810	–361.934	–281.314	13.358
1200	55.564	315.665	277.882	45.339	–361.849	–273.989	11.926
1300	56.033	320.131	280.963	50.920	–361.763	–266.671	10.715
1400	56.426	324.299	283.911	56.543	–361.680	–259.359	9.677
1500	56.759	328.203	286.735	62.203	–361.605	–252.053	8.777
77. SILICON Si (cr)							
298.15	19.789	18.810	18.810	0.000	0.000	0.000	0.000
300	19.855	18.933	18.810	0.037	0.000	0.000	0.000
400	22.301	25.023	19.624	2.160	0.000	0.000	0.000
500	23.610	30.152	21.231	4.461	0.000	0.000	0.000
600	24.472	34.537	23.092	6.867	0.000	0.000	0.000
700	25.124	38.361	25.006	9.348	0.000	0.000	0.000
800	25.662	41.752	26.891	11.888	0.000	0.000	0.000
900	26.135	44.802	28.715	14.478	0.000	0.000	0.000
1000	26.568	47.578	30.464	17.114	0.000	0.000	0.000
1100	26.974	50.130	32.138	19.791	0.000	0.000	0.000
1200	27.362	52.493	33.737	22.508	0.000	0.000	0.000

T/K	J/K·mol			kJ/mol			
	C_p°	S°	$-(G^\circ-H^\circ(T_r))/T$	$H^\circ-H^\circ(T_r)$	$\Delta_f H^\circ$	$\Delta_f G^\circ$	Log K_f
77. SILICON Si (cr) (continued)							
1300	27.737	54.698	35.265	25.263	0.000	0.000	0.000
1400	28.103	56.767	36.728	28.055	0.000	0.000	0.000
1500	28.462	58.719	38.130	30.883	0.000	0.000	0.000
78. SILICON Si (g)							
298.15	22.251	167.980	167.980	0.000	450.000	405.525	–71.045
300	22.234	168.117	167.980	0.041	450.004	405.249	–70.559
400	21.613	174.416	168.843	2.229	450.070	390.312	–50.969
500	21.316	179.204	170.456	4.374	449.913	375.388	–39.216
600	21.153	183.074	172.246	6.497	449.630	360.508	–31.385
700	21.057	186.327	174.032	8.607	449.259	345.682	–25.795
800	21.000	189.135	175.748	10.709	448.821	330.915	–21.606
900	20.971	191.606	177.375	12.808	448.329	316.205	–18.352
1000	20.968	193.815	178.911	14.904	447.791	301.553	–15.751
1100	20.989	195.815	180.358	17.002	447.211	286.957	–13.626
1200	21.033	197.643	181.723	19.103	446.595	272.416	–11.858
1300	21.099	199.329	183.014	21.209	445.946	257.927	–10.364
1400	21.183	200.895	184.236	23.323	445.268	243.489	–9.085
1500	21.282	202.360	185.396	25.446	444.563	229.101	–7.978
79. SILICON DIOXIDE SiO_2 (cr)							
298.15	44.602	41.460	41.460	0.000	–910.700	–856.288	150.016
300	44.712	41.736	41.461	0.083	–910.708	–855.951	149.032
400	53.477	55.744	43.311	4.973	–910.912	–837.651	109.385
500	60.533	68.505	47.094	10.705	–910.540	–819.369	85.598
600	64.452	79.919	51.633	16.971	–909.841	–801.197	69.749
700	68.234	90.114	56.414	23.590	–908.958	–783.157	58.439
800	76.224	99.674	61.226	30.758	–907.668	–765.265	49.966
848	82.967	104.298	63.533	34.569	–906.310	–756.747	46.613
PHASE TRANSITION: $\Delta_{trs}H$ = 0.411 kJ/mol, $\Delta_{trs}S$ = 0.484 J/K·mol, crII–crII′							
848	67.446	104.782	63.532	34.980	–906.310	–756.747	46.613
900	67.953	108.811	66.033	38.500	–905.922	–747.587	43.388
1000	68.941	116.021	70.676	45.345	–905.176	–730.034	38.133
1100	69.940	122.639	75.104	52.289	–904.420	–712.557	33.836
1200	70.947	128.768	79.323	59.333	–901.382	–695.148	30.259
PHASE TRANSITION: $\Delta_{trs}H$ = 2.261 kJ/mol, $\Delta_{trs}S$ = 1.883 J/K·mol, crII′–crI							
1200	71.199	130.651	79.323	61.594	–901.382	–695.148	30.259
1300	71.743	136.372	83.494	68.742	–900.574	–677.994	27.242
1400	72.249	141.707	87.463	75.941	–899.782	–660.903	24.658
1500	72.739	146.709	91.248	83.191	–899.004	–643.867	22.421
80. SILICON TETRACHLORIDE $SiCl_4$ (g)							
298.15	90.404	331.446	331.446	0.000	–662.200	–622.390	109.039
300	90.562	332.006	331.448	0.167	–662.195	–622.143	108.323
400	96.893	359.019	335.088	9.572	–661.853	–608.841	79.505
500	100.449	381.058	342.147	19.456	–661.413	–595.637	62.225
600	102.587	399.576	350.216	29.616	–660.924	–582.527	50.713
700	103.954	415.500	358.432	39.948	–660.417	–569.501	42.496
800	104.875	429.445	366.455	50.392	–659.912	–556.548	36.338
900	105.523	441.837	374.155	60.914	–659.422	–543.657	31.553
1000	105.995	452.981	381.490	71.491	–658.954	–530.819	27.727
1100	106.349	463.101	388.456	82.109	–658.515	–518.027	24.599
1200	106.620	472.366	395.068	92.758	–658.107	–505.274	21.994
1300	106.834	480.909	401.347	103.431	–657.735	–492.553	19.791
1400	107.003	488.833	407.316	114.123	–657.400	–479.860	17.904
1500	107.141	496.220	413.000	124.830	–657.104	–467.189	16.269

GIBBS ENERGY OF FORMATION OF METAL OXIDES

C. B. Alcock

This table gives standard Gibbs energies of formation as a function of temperature for metallic oxides in units of joules per mole (J/mol). The data are presented in two forms:

1. The coefficients of a two-term equation
$$\Delta G^\circ = A + BT$$
2. The coefficients of a three-term equation
$$\Delta G^\circ = A + BT + CT \log T$$

The first equation is able to fit the original data to ±500 J/mol over a restricted temperature range, i.e., about 300 K. The second equation is considerably more accurate, and usually fits the original data to ±500 J/mol over a temperature range of 1500 K.

The original data to which these equations have been fitted are mainly drawn from the JANAF and the IVTAN tabulations cited below.

The following notation is used in the reactions:

[ ] indicates solid
{ } indicates liquid
( ) indicates gas

For each reaction, the first line gives the coefficients *A* and *B* of Equation 1, while the second line gives *A*, *B*, and *C* of Equation 2. The unit of *T* is K, and the unit of ΔG° is J/mol for the reaction as written.

REFERENCES

1. M. W. Chase, et al., JANAF Thermochemical Tables, Third Edition, *J. Phys. Chem. Ref. Data*, 14, Supplement No. 1, 1985.
2. L. V. Gurvich, I. V. Veyts, and C. B. Alcock, *Thermodynamic Properties of Individual Substances*, Fourth Edition, Vol. 1, Hemisphere Publishing Corp., New York, 1989.

Reaction	Temp. range (K)	*A*	*B*	*C*
2 [Li] + $^1/_2O_2 \rightarrow$ [Li_2O]	300—450	–598424	124.50	—
		—	—	—
2{Li} + $^1/_2O_2 \rightarrow$ [Li_2O]	500—1550	–603513	137.41	—
		–611755	206.72	–8.77
2[Na] + $^1/_2O_2 \rightarrow$ [Na_2O]	300—370	–414650	130.34	—
2{Na} + $^1/_2O_2 \rightarrow$ [Na_2O]	400—1200	–418282	141.97	—
		–426588	228.19	–11.26
2{K} + $^1/_2O_2 \rightarrow$ [K_2O]	400—1000	–360427	138.40	—
		–382654	392.41	–33.74
2{Cs} + $^1/_2O_2 \rightarrow$ [Cs_2O]	350—750	–349702	139.05	—
		–353361	189.63	–6.93
2[Cu] + $^1/_2O_2 \rightarrow$ [Cu_2O]	300—1250	–169881	74.43	—
		–172866	109.28	–4.62
2{Cu} + $^1/_2O_2 \rightarrow$ [Cu_2O]	1375—1500	–190241	88.50	—
[Cu_2O] + $^1/_2O_2 \rightarrow$ 2CuO	300—1100	–136825	100.17	—
		–144564	192.82	–12.33
[Be] + $^1/_2O_2 \rightarrow$ [BeO]	300—1500	–608357	96.88	—
		–610659	120.04	–2.99
{Be} + $^1/_2O_2 \rightarrow$ [BeO]	1500—2000	–618864	103.47	—
		–606471	43.13	7.13
[Mg] + $^1/_2O_2 \rightarrow$ [MgO]	300—900	–600751	107.11	—
		–602369	128.73	–2.93
{Mg} + $^1/_2O_2 \rightarrow$ [MgO]	900—1380	–608234	115.09	—
		–604628	89.33	3.21

GIBBS ENERGY OF FORMATION OF METAL OXIDES (continued)

Reaction	Temp. range (K)	*A*	*B*	*C*
[Ca] + $^1/_2O_2 \rightarrow$ [CaO]	300—1100	–634096	103.38	—
		–635000	114.21	–1.44
{Ca} + $^1/_2O_2 \rightarrow$ [CaO]	1125—1750	–633646	102.96	—
		–635996	116.74	–1.67
[Sr] + $^1/_2O_2 \rightarrow$ [SrO]	300—1000	–590509	98.76	—
		–593069	131.066	–4.34
{Sr} + $^1/_2O_2 \rightarrow$ [SrO]	1050—1650	–596363	104.28	—
		–601767	137.62	–4.07
[Ba] + $^1/_2O_2 \rightarrow$ [BaO]	300—1100	–547818	93.81	—
		–546873	81.88	1.60
[Zn] + $^1/_2O_2 \rightarrow$ [ZnO]	300—675	–349962	99.15	—
		–351584	124.16	–3.49
{Zn} + $^1/_2O_2 \rightarrow$ [ZnO]	700—1150	–356005	107.81	—
		–358467	129.00	–2.71
[Cd] + $^1/_2O_2 \rightarrow$ [CdO]	300—600	–257481	97.54	—
		–259474	130.12	–4.59
{Cd} + $^1/_2O_2 \rightarrow$ [CdO]	595—1040	–261633	104.60	—
		–266457	151.04	–6.03
2[B] + $^3/_2O_2 \rightarrow B_2O_3$	300—700	–1273294	265.0	—
2[B] + $^3/_2O_2 \rightarrow$ {B_2O_3}	700—2000	–1255420	243.74	—
		–1288814	462.28	26.72
2[Al] + $^3/_2O_2 \rightarrow$ [Al_2O_3]	300—900	–1675432	313.20	—
		–1678712	357.00	–5.95
2{Al} + $^3/_2O_2 \rightarrow$ [Al_2O_3]	1000—1800	–1689572	328.66	—
		–1706832	432.76	–12.64
2[Y] + $^3/_2O_2 \rightarrow$ [Y_2O_3]	300—1800	–1897878	281.88	—
		–1908170	375.02	–11.81
2[Sc] + $^3/_2O_2 \rightarrow$ [Sc_2O_3]	300—1600	–190362	289.32	—
		–1912414	374.94	–10.98
2[La] + $^3/_2O_2 \rightarrow$ [La_2O_3]	300—1150	–178949	279.86	—
		–1798877	407.47	–16.91
2[Ce] + $^3/_2O_2 \rightarrow$ [Ce_2O_3]	300—975	–1789061	285.11	—
		–1804375	482.26	–26.57
[Ce_2O_3] + $^1/_2O_2 \rightarrow$ 2[CeO_2]	300—975	–383476	132.05	—
		–378627	69.56	8.42
2{Ga} + $^3/_2O_2 \rightarrow$ [Ga_2O_3]	300—1000	–1097904	336.87	—
		–1105260	429.69	–12.47
2{In} + $^3/_2O_2 \rightarrow$ [In_2O_3]	430—1400	–923702	318.56	—
		–941468	478.91	–20.54
2[Tl] + $^1/_2O_2 \rightarrow$ [Tl_2O]	300—550	–167711	81.96	—
2{Tl} + $^1/_2O_2 \rightarrow$ [Tl_2O]	500—850	–173558	92.18	—
[Tl_2O] + $O_2 \rightarrow$ [Tl_2O_3]	300—850	–223779	186.84	—
3[Ti_2O_3] + $^1/_2O_2 \rightarrow$ 2[Ti_3O_5]	300—1200	–360548	82.79	—
		–339218	–160.28	32.08
[Zr] + $O_2 \rightarrow$ [ZrO_2]	300—1800	–1105994	185.57	—
		–1100032	245.21	–7.56
[Hf] + $O_2 \rightarrow$ [HfO_2]	300—1800	–1110438	174.89	—
		–1122308	282.32	–13.62

Reaction	Temp. range (K)	*A*	*B*	*C*
$[Th] + O_2 \rightarrow [ThO_2]$	300—1800	–1220280	178.99	—
		–1229326	260.87	–10.38
$[U] + O_2 \rightarrow [UO_2]$	300—1400	–1082037	170.66	—
		–1083049	181.33	–1.39
$\{Sn\} + O_2 \rightarrow [SnO_2]$	550—1400	–581581	208.12	—
		–579665	345.82	–17.52
$\{Pb\} + {}^1/_2O_2 \rightarrow [PbO]$	625—1125	–215343	94.10	—
		–222991	163.81	–8.98
$\{Pb\} + {}^1/_2O_2 \rightarrow \{PbO\}$	1200—1600	–209494	88.96	—
		–225003	180.82	–11.15
$3[PbO] + {}^1/_2O_2 \rightarrow [Pb_3O_4]$	650—1150	–52719	74.23	—
		–68566	214.96	–18.06
$3\{PbO\} + {}^1/_2O_2 \rightarrow [Pb_3O_4]$	1200—1600	–41503	64.58	—
		–83535	313.72	–30.23
$[Nb] + {}^1/_2O_2 \rightarrow [NbO]$	300—1800	–403087	85.28	—
		–408074	130.43	–5.72
$[NbO] + {}^1/_2O_2 \rightarrow [NbO_2]$	300—1800	–383422	83.87	—
		–398354	219.04	–17.13
$2[NbO_2] + {}^1/_2O_2 \rightarrow [Nb_2O_5]$	300—1700	–311239	79.62	—
		–297089	–52.93	16.92
$[Ta] + 2.5O_2 \rightarrow [Ta_2O_5]$	300—1600	–1036040	511.36	—
		–1015664	312.96	25.36
$2[Bi] + {}^3/_2O_2 \rightarrow [Bi_2O_3]$	300—525	–571731	263.41	—
		–580157	410.25	–20.94
$2\{Bi\} + {}^3/_2O_2 \rightarrow [Bi_2O_3]$	600—1125	–585420	289.9	—
		–605540	478.3	–24.36
$2[Cr] + {}^3/_2O_2 \rightarrow [Cr_2O_3]$	300—1600	–1131960	253.90	—
		–1140545	331.58	–9.85
$[Mo] + O_2 \rightarrow [MoO_2]$	300—1800	–581245	172.81	—
		–596245	308.58	–17.21
$[W] + O_2 \rightarrow [WO_2]$	300—1800	–582425	173.05	—
		–595492	291.31	–14.99
$[Mn] + {}^1/_2O_2 \rightarrow [MnO]$	300—1500	–384852	–73.32	—
		–383209	56.68	2.15
$\{Mn\} + {}^1/_2O_2 \rightarrow [MnO]$	1500—2000	–402913	85.62	—
		–389544	20.52	7.79
$3[MnO] + {}^1/_2O_2 \rightarrow [Mn_3O_4]$	300—1500	–230432	123.42	—
		–237011	190.03	–8.60
$[Mn_3O_4] + O_2 \rightarrow 3[MnO_2]$	300—800	–171164	198.83	—
		–175346	258.27	–8.16
$[Re] + O_2 \rightarrow [ReO_2]$	300—1200	–445594	186.35	—
		–453691	278.62	–12.18
$[ReO_2] + {}^1/_2O_2 \rightarrow [ReO_3]$	300—1200	–136914	73.62	—
		–143984	154.17	–10.63
$[Fe] + {}^1/_2O_2 \rightarrow [FeO]$	300—1800	–269493	62.70	—
		–270590	72.64	–1.26
$\{Fe\} + {}^1/_2O_2 \rightarrow [FeO]$	1800—2000	–279330	68.00	—
		–259034	–23.42	10.69

GIBBS ENERGY OF FORMATION OF METAL OXIDES (continued)

Reaction	Temp. range (K)	*A*	*B*	*C*
$3[FeO] + {}^1/_2O_2 \rightarrow [Fe_3O_4]$	300—1800	–290306	116.27	—
		–309467	289.69	–21.98
$2[Fe_3O_4] + {}^1/_2O_2 \rightarrow 3[Fe_2O_3]$	300—1500	–240360	141.07	—
		–233957	76.24	8.37
$[Co] + {}^1/_2O_2 \rightarrow [CoO]$	300—1800	–234566	71.03	—
		–236285	86.59	–1.97
$\{Co\} + {}^1/_2O_2 \rightarrow [CoO]$	1850—2000	–248302	78.68	—
		–261213	136.14	–6.71
$3[CoO] + {}^1/_2O_2 \rightarrow [Co_3O_4]$	300—1100	–212553	172.91	—
		–205988	94.33	10.46
$[Ni] + {}^1/_2O_2 \rightarrow [NiO]$	300—1700	–235969	86.08	—
		–240826	131.63	–5.81
$[Ni] + {}^1/_2O_2 \rightarrow [NiO]$	300—600	–238410	90.70	—
		–243380	168.75	–10.92
	650—1400	–235020	85.53	—
		–239864	124.56	–4.93
$\{Ni\} + {}^1/_2O_2 \rightarrow [NiO]$	1750—2000	–248161	92.64	—
		–259364	143.71	–5.98
$[Ru] + O_2 \rightarrow [RuO_2]$	300—1800	–307942	261.08	–28.87
$[Rh] + O_2 \rightarrow (RhO_2)$	1473—1773	–188866	20.67	
$[Pd] + {}^1/_2O_2 \rightarrow [PdO]$	300—1100	–93722	154.3	–24.06
$[Os] + O_2 \rightarrow [OsO_2]$	300—1600	–284748	172.2	—
		–300392	293.38	–15.54
$[Ir] + O_2 \rightarrow [IrO_2]$	300—1300	–218405	323.02	–52.93
$[Pt] + O_2 \rightarrow (PtO_2)$	1373—1823	–164306	3.89	—

ENTHALPY OF COMBUSTION OF SELECTED ORGANIC COMPOUNDS

This table gives the standard molar enthalpy (heat) of combustion at 298.15 K (25°C) of selected compounds containing carbon, hydrogen, nitrogen, and oxygen. The products of combustion are taken to be CO_2 (gas), N_2 (gas), and H_2O (liquid) in their standard states. The molar mass is included in the table to facilitate conversion from a molar to mass basis.

Note that the quantity tabulated here is the *negative* of the enthalpy of combustion; thus the tabulated number gives the heat released upon combustion of one mole of the compound.

Substances are arranged by molecular formula in the Hill order. See Appendix B for an index to molecular formula by compound name. For references see the table "Standard Thermodynamic Properties of Chemical Substances" in this Section.

Mol. form.	Name	M g/mol	$-\Delta_c H°_{298.15}$/(kJ/mol) Crystal	Liquid	Gas
C	Carbon	12.011	393.5		1110.2
CN_4O_8	Tetranitromethane	196.033		431.9	475.5
CO	Carbon monoxide	28.010			283.0
CHN	Hydrogen cyanide	27.026		645.3	671.5
CH_2N_2	Cyanamide	42.040	738.1		
CH_2O	Formaldehyde	30.026			570.7
CH_2O_2	Formic acid	46.026		254.6	300.7
CH_3NO	Formamide	45.041		568.3	
CH_3NO_2	Nitromethane	61.040		709.2	747.6
CH_3NO_3	Methyl nitrate	77.040		663.3	697.6
CH_4	Methane	16.043			890.8
CH_4N_2	Ammonium cyanide	44.056	965.6		
CH_4N_2O	Urea	60.056	631.6		719.4
CH_4O	Methanol	32.042		726.1	763.7
CH_5N	Methylamine	31.057		1060.8	1085.6
CH_5NO_3	Ammonium hydrogen carbonate	79.056	258.7		
CH_6N_2	Methylhydrazine	46.072		1305.0	1345.3
C_2N_2	Cyanogen	52.036		1072.9	1093.7
C_2H_2	Acetylene	26.038			1301.1
C_2H_2O	Ketene	42.037		1005.0	1025.4
$C_2H_2O_2$	Glyoxal	58.037			860.8
$C_2H_2O_4$	Oxalic acid	90.036	251.1		349.1
C_2H_3N	Acetonitrile	41.053		1247.2	1280.1
C_2H_3N	Isocyanomethane	41.053		1333.0	1364.8
C_2H_3NO	Methylisocyanate	57.052		1123.8	
C_2H_4	Ethylene	28.054			1411.2
C_2H_4O	Acetaldehyde	44.053		1166.9	1192.5
C_2H_4O	Ethylene oxide	44.053		1280.9	1306.1
$C_2H_4O_2$	Acetic acid	60.053		874.2	925.9
$C_2H_4O_2$	Methyl formate	60.053		972.6	1003.2
C_2H_5N	Ethyleneimine	43.068		1593.5	1628.1
C_2H_5NO	Acetamide	59.068	1184.6		1263.3
$C_2H_5NO_2$	Nitroethane	75.067		1357.7	1399.3
$C_2H_5NO_2$	Glycine	75.067	973.1		1109.5
C_2H_6	Ethane	30.070			1560.7
C_2H_6O	Dimethyl ether	46.069			1460.4
C_2H_6O	Ethanol	46.069		1366.8	1409.4
$C_2H_6O_2$	Ethylene glycol	62.068		1189.2	1257.0
C_2H_7N	Dimethylamine	45.084		1743.5	1768.9
C_2H_7N	Ethylamine	45.084		1713.3	1739.9
$C_2H_8N_2$	1,1-Dimethylhydrazine	60.099		1979.2	2014.2
$C_2H_8N_2$	1,2-Ethanediamine	60.099		1867.3	1912.7
$C_2H_8N_2O_4$	Ammonium oxalate	124.097	807.3		
C_3H_3N	Propenenitrile	53.064		1756.4	1789.9
C_3H_3NO	Oxazole	69.063		1561.3	1593.8
C_3H_3NO	Isoxazole	69.063		1651.4	1687.9
C_3H_4	Propyne	40.065			1937.1
C_3H_4	Allene	40.065			1942.7
C_3H_4	Cyclopropene	40.065			2029.3

Mol. form.	Name	M g/mol	$-\Delta_c H°_{298.15}$/(kJ/mol) Crystal	Liquid	Gas
$C_3H_4N_2$	Imidazole	68.078	1810.7		
$C_3H_4O_2$	Propenoic acid	72.064		1368.4	
$C_3H_4O_2$	2-Oxetanone	72.064		1422.3	1469.3
C_3H_5N	Propanenitrile	55.079		1910.6	1946.6
$C_3H_5N_3O_9$	Trinitroglycerol	227.088		1524.2	1624.2
C_3H_6	Propene	42.081		2039.7	2058.0
C_3H_6	Cyclopropane	42.081			2091.3
C_3H_6O	Acetone	58.080		1789.9	1820.7
C_3H_6O	Allyl alcohol	58.080		1866.2	1913.5
C_3H_6O	Propanal	58.080		1822.7	1852.4
C_3H_6O	Methyloxirane	58.080		1915.4	1943.3
$C_3H_6O_2$	Methyl acetate	74.079		1592.2	1626.1
$C_3H_6O_2$	Propanoic acid	74.079		1527.3	1584.5
$C_3H_6O_2$	1,3-Dioxolane	74.079		1704.5	1740.0
$C_3H_6O_3$	Trioxane	90.079	1515.5		1572.1
C_3H_7N	Cyclopropylamine	57.095		2226.7	2257.9
C_3H_7NO	*N,N*-Dimethylformamide	73.095		1941.6	1989.2
$C_3H_7NO_2$	1-Nitropropane	89.094		2013.7	2057.1
$C_3H_7NO_2$	2-Nitropropane	89.094		2000.6	2041.9
$C_3H_7NO_2$	l-Alanine	89.094	1576.9		1715.0
$C_3H_7NO_2$	d-Alanine	89.094	1619.7		
$C_3H_7NO_3$	l-Serine	105.094	1448.2		
C_3H_8	Propane	44.097			2219.2
C_3H_8O	1-Propanol	60.096		2021.3	2068.8
C_3H_8O	2-Propanol	60.096		2005.8	2051.1
C_3H_8O	Ethyl methyl ether	60.096			2107.5
$C_3H_8O_2$	1,2-Propylene glycol	76.095		1838.2	1902.6
$C_3H_8O_2$	1,3-Propylene glycol	76.095		1859.0	1931.8
$C_3H_8O_2$	Dimethoxymethane	76.095		1946.2	1975.5
$C_3H_8O_3$	Glycerol	92.095		1655.4	1741.2
C_3H_9N	Propylamine	59.111		2365.3	2396.7
C_3H_9N	Isopropylamine	59.111		2354.5	2383.1
C_3H_9N	Trimethylamine	59.111		2421.1	2443.1
$C_4H_2O_3$	Maleic anhydride	98.058	1390.1		1461.6
$C_4H_4N_2$	Pyridazine	80.089		2370.5	2424.0
$C_4H_4N_2$	Pyrimidine	80.089		2291.6	2341.6
$C_4H_4N_2$	Succinonitrile	80.089		2285.4	2355.4
$C_4H_4N_2O_2$	Uracil	112.088	1716.3		1842.8
C_4H_4O	Furan	68.075		2083.4	2110.8
$C_4H_4O_2$	Diketene	84.075		1912.6	1955.5
$C_4H_4O_3$	Succinic anhydride	100.074	1537.9		
$C_4H_4O_4$	Fumaric acid	116.073	1334.0		1469.9
$C_4H_4O_4$	Maleic acid	116.073	1356.3		1466.3
C_4H_5N	Cyclopropanecarbonitrile	67.090		2429.4	2470.4
C_4H_5N	Pyrrole	67.090		2351.7	2396.9
$C_4H_5N_3O$	Cytosine	111.103	2067.3		
C_4H_6	1,2-Butadiene	54.092		2570.5	2593.8
C_4H_6	1,3-Butadiene	54.092		2519.4	2541.5
C_4H_6	1-Butyne	54.092		2573.4	2596.7
C_4H_6	2-Butyne	54.092		2550.6	2577.2
C_4H_6	Cyclobutene	54.092			2588.2
C_4H_6O	*trans*-2-Butenal	70.091		2292.8	2330.9
C_4H_6O	Divinyl ether	70.091		2391.7	2417.9
$C_4H_6O_2$	Methyl acrylate	86.090		2069.3	2098.5
$C_4H_6O_2$	Vinyl acetate	86.090		2151.4	2116.6
$C_4H_6O_3$	Acetic anhydride	102.090		1807.1	1859.0
$C_4H_6O_4$	Succinic acid	118.089	1491.0		1608.5

Mol. form.	Name	M g/mol	$-\Delta_c H°_{298.15}$/(kJ/mol) Crystal	Liquid	Gas
$C_4H_6O_4$	Dimethyl oxalate	118.089	1675.2		1722.6
C_4H_7N	Butanenitrile	69.106		2568.6	2608.0
C_4H_7N	2-Methylpropanenitrile	69.106		2560.6	2597.7
C_4H_7NO	2-Pyrrolidone	85.106		2288.2	
$C_4H_7NO_4$	l-Aspartic acid	133.104	1601.1		
C_4H_8	1-Butene	56.107		2696.9	2717.5
C_4H_8	*cis*-2-Butene	56.107		2687.7	2710.3
C_4H_8	*trans*-2-Butene	56.107		2684.4	2706.0
C_4H_8	Isobutene	56.107		2679.9	2700.5
C_4H_8	Cyclobutane	56.107		2721.1	2745.8
C_4H_8	Methylcyclopropane	56.107		2719.1	
$C_4H_8N_2O_3$	l-Asparagine	132.119	1928.0		
C_4H_8O	1,2-Epoxybutane	72.107		2548.5	
C_4H_8O	Butanal	72.107		2478.2	2512.6
C_4H_8O	Isobutanal	72.107		2470.0	2501.6
C_4H_8O	2-Butanone	72.107		2444.1	2478.7
C_4H_8O	Tetrahydrofuran	72.107		2501.2	2533.2
C_4H_8O	Ethyl vinyl ether	72.107		2550.0	2576.6
$C_4H_8O_2$	1,3-Dioxane	88.106		2337.7	2375.1
$C_4H_8O_2$	1,4-Dioxane	88.106		2363.5	2401.6
$C_4H_8O_2$	Ethyl acetate	88.106		2238.1	2273.3
$C_4H_8O_2$	Propyl formate	88.106		2217.1	
$C_4H_8O_2$	Butanoic acid	88.106		2183.6	2241.6
C_4H_9N	Pyrrolidine	71.122		2819.3	2856.9
C_4H_9NO	*N,N*-Dimethylacetamide	87.122		2582.0	
$C_4H_9NO_3$	l-Threonine	119.120	2053.1		
C_4H_{10}	Butane	58.123		2856.6	2877.6
C_4H_{10}	Isobutane	58.123		2849.7	2869.0
$C_4H_{10}O$	1-Butanol	74.123		2675.9	2728.2
$C_4H_{10}O$	2-Butanol	74.123		2660.6	2710.3
$C_4H_{10}O$	2-Methyl-2-propanol	74.123		2644.0	2690.7
$C_4H_{10}O$	2-Methyl-1-propanol	74.123		2668.5	2719.3
$C_4H_{10}O$	Diethyl ether	74.123		2723.9	2751.1
$C_4H_{10}O$	Methyl propyl ether	74.123		2737.2	2765.0
$C_4H_{10}O$	Isopropyl methyl ether	74.123		2724.5	2751.2
$C_4H_{10}O_2$	1,3-Butanediol	90.122		2502.2	2570.0
$C_4H_{10}O_2$	1,4-Butanediol	90.122		2499.9	2576.5
$C_4H_{10}O_3$	Diethylene glycol	106.122		2374.7	2432.0
$C_4H_{11}N$	Butylamine	73.138		3018.4	3054.1
$C_4H_{11}N$	Isobutylamine	73.138		3013.5	3047.4
$C_4H_{11}N$	*sec*-Butylamine	73.138		3008.6	3041.2
$C_4H_{11}N$	*tert*-Butylamine	73.138		2995.5	3025.2
$C_4H_{11}N$	Diethylamine	73.138		3042.4	3073.6
$C_5H_4N_4O$	Hypoxanthine	136.113	2428.4		
$C_5H_4N_4O_2$	Xanthine	152.113	2159.6		
$C_5H_4N_4O_3$	Uric acid	168.112	1920.4		
$C_5H_4O_2$	Furfural	96.086		2337.6	2388.2
$C_5H_5N_5$	Adenine	135.128	2778.1		2886.9
C_5H_5N	Pyridine	79.101		2782.3	2822.5
$C_5H_5N_5O$	Guanine	151.128	2498.2		
C_5H_6	*cis*-3-Penten-1-yne	66.103			2906.4
C_5H_6	*trans*-3-Penten-1-yne	66.103		3053.2	
C_5H_6	1,3-Cyclopentadiene	66.103		2930.9	2959.3
$C_5H_6N_2O_2$	Thymine	126.115	2362.2		2496.3
$C_5H_6O_2$	Furfuryl alcohol	98.101		2548.8	2613.2
C_5H_7N	Cyclobutanecarbonitrile	81.117		3071.0	3111.1
C_5H_8	2-Methyl-1,3-butadiene	68.119		3159.1	3186.4

Mol. form.	Name	*M* g/mol	$-\Delta_c H°_{298.15}$/(kJ/mol) Crystal	Liquid	Gas
C_5H_8	*cis*-1,3-Pentadiene	68.119			3192.3
C_5H_8	*trans*-1,3-Pentadiene	68.119			3187.0
C_5H_8	1,4-Pentadiene	68.119			3216.5
C_5H_8	Cyclopentene	68.119		3115.3	3144.8
C_5H_8	Spiropentane	68.119		3268.6	3296.1
C_5H_8O	Cyclopentanone	84.118		2875.2	2918.8
$C_5H_8O_2$	2,4-Pentanedione	100.117		2687.1	2730.3
$C_5H_8O_4$	Glutaric acid	132.116	2150.9		
C_5H_9N	Pentanenitrile	83.133		3220.7	3264.3
C_5H_9N	2,2-Dimethylpropanenitrile	83.133		3214.0	3251.3
C_5H_9NO	*N*-Methyl-2-pyrrolidone	99.133		2991.6	
$C_5H_9NO_2$	l-Proline	115.132	2741.6		2887.6
$C_5H_9NO_4$	l-Glutamic acid	147.131	2244.1		
C_5H_{10}	1-Pentene	70.134		3349.8	3375.4
C_5H_{10}	*cis*-2-Pentene	70.134		3343.0	3369.1
C_5H_{10}	*trans*-2-Pentene	70.134		3338.5	3364.8
C_5H_{10}	2-Methyl-1-butene	70.134		3335.7	3361.4
C_5H_{10}	2-Methyl-2-butene	70.134		3328.1	3354.9
C_5H_{10}	3-Methyl-1-butene	70.134		3345.2	3369.1
C_5H_{10}	Cyclopentane	70.134		3291.6	3320.3
$C_5H_{10}N_2O_3$	l-Glutamine	146.146	2570.3		
$C_5H_{10}O$	Cyclopentanol	86.134		3096.6	3154.1
$C_5H_{10}O$	2-Pentanone	86.134		3099.4	3137.7
$C_5H_{10}O$	3-Pentanone	86.134		3100.2	3138.8
$C_5H_{10}O$	3-Methyl-2-butanone	86.134		3097.3	3134.2
$C_5H_{10}O$	3,3-Dimethyloxetane	86.134		3214.5	3248.5
$C_5H_{10}O$	Tetrahydropyran	86.134		3138.4	3173.3
$C_5H_{10}O$	Pentanal	86.134		3129.4	3108.2
$C_5H_{10}O_2$	Isopropyl acetate	102.133		2877.8	2915.0
$C_5H_{10}O_2$	Ethyl propanoate	102.133		2894.0	2933.1
$C_5H_{10}O_2$	Pentanoic acid	102.133		2837.3	2904.8
$C_5H_{10}O_2$	3-Methylbutanoic acid	102.133		2835.1	2886.7
$C_5H_{10}O_2$	2-Methylbutanoic acid	102.133		2842.2	
$C_5H_{10}O_2$	Tetrahydrofurfuryl alcohol	102.133		2961.0	3027.5
$C_5H_{10}O_3$	Diethyl carbonate	118.133		2715.2	2758.8
$C_5H_{11}N$	Piperidine	85.149		3453.2	3492.4
$C_5H_{11}N$	Cyclopentylamine	85.149		3444.5	3484.7
$C_5H_{11}NO_2$	l-Valine	117.148	2921.7		3084.5
C_5H_{12}	Pentane	72.150		3509.0	3535.6
C_5H_{12}	Isopentane	72.150		3504.0	3528.8
C_5H_{12}	Neopentane	72.150		3492.3	3514.4
$C_5H_{12}O$	Butyl methyl ether	88.150		3391.9	3424.4
$C_5H_{12}O$	*tert*-Butyl methyl ether	88.150		3368.9	3399.0
$C_5H_{12}O$	Ethyl propyl ether	88.150		3378.9	3410.3
$C_5H_{12}O$	1-Pentanol	88.150		3330.9	3387.8
$C_5H_{12}O$	2-Pentanol	88.150		3317.3	3369.8
$C_5H_{12}O$	3-Pentanol	88.150		3313.6	3365.3
$C_5H_{12}O$	2-Methyl-1-butanol	88.150		3325.9	3380.5
$C_5H_{12}O$	2-Methyl-2-butanol	88.150		3303.0	3351.7
$C_5H_{12}O$	3-Methyl-1-butanol	88.150		3326.1	3381.2
$C_5H_{12}O$	3-Methyl-2-butanol	88.150		3315.9	3367.3
$C_5H_{12}O$	2,2-Dimethyl-1-propanol	88.150		3283.1	
$C_5H_{12}O_2$	1,5-Pentanediol	104.149		3151.0	3233.4
$C_5H_{12}O_2$	Diethoxymethane	104.149		3232.1	3267.7
$C_5H_{12}O_4$	Pentaerythritol	136.148	2761.9		2905.8
$C_6H_4O_2$	*p*-Benzoquinone	108.097	2747.0		2809.8
$C_6H_5NO_2$	Nitrobenzene	123.111		3088.1	3143.1

Mol. form.	Name	*M* g/mol	$-\Delta_c H°_{298.15}$/(kJ/mol) Crystal	Liquid	Gas
C_6H_6	Benzene	78.114		3267.6	3301.2
$C_6H_6N_2O_2$	*o*-Nitroaniline	138.126	3192.5	3209.2	3282.4
$C_6H_6N_2O_2$	*m*-Nitroaniline	138.126	3180.3	3204.2	3277.0
$C_6H_6N_2O_2$	*p*-Nitroaniline	138.126	3176.6	3197.9	3277.4
C_6H_6O	Phenol	94.113	3053.5		3122.2
$C_6H_6O_2$	*p*-Hydroquinone	110.112	2854.1		2953.3
C_6H_7N	Aniline	93.128		3392.8	3449.0
C_6H_7N	2-Methylpyridine	93.128		3418.2	3460.7
C_6H_7N	3-Methylpyridine	93.128		3423.4	3467.9
C_6H_7N	4-Methylpyridine	93.128		3420.7	3465.6
C_6H_7N	1-Cyclopentenecarbonitrile	93.128		3473.0	3518.0
$C_6H_8N_2$	Adiponitrile	108.143		3589.5	3653.9
$C_6H_8N_2$	Phenylhydrazine	108.143		3645.4	3707.3
$C_6H_8N_2$	*o*-Phenylenediamine	108.143	3504.1		
$C_6H_8N_2$	*m*-Phenylenediamine	108.143	3496.6		
$C_6H_8N_2$	*p*-Phenylenediamine	108.143	3507.5		
C_6H_9N	Cyclopentanecarbonitrile	95.144		3648.0	3690.3
$C_6H_9NO_3$	Triacetamide	143.142		3036.8	3097.2
$C_6H_9N_3O_2$	l-Histidine	155.157	3180.6		
C_6H_{10}	Cyclohexene	82.145		3751.7	3785.2
C_6H_{10}	1,5-Hexadiene	82.145		3844.3	3874.3
C_6H_{10}	3,3-Dimethyl-1-butyne	82.145		3868.6	
$C_6H_{10}O$	Cyclohexanone	98.145		3519.0	3564.1
$C_6H_{10}O_2$	Methyl cyclobutanecarboxylate	114.144		3395.2	3434.9
$C_6H_{10}O_3$	Propanoic anhydride	130.144		3111.1	3163.7
$C_6H_{10}O_4$	Adipic acid	146.143	2795.9		2925.2
$C_6H_{10}O_4$	Diethyl oxalate	146.143		2984.7	3048.2
$C_6H_{11}NO$	Caprolactam	113.160	3603.7		3686.9
C_6H_{12}	Cyclohexane	84.161		3919.6	3952.6
C_6H_{12}	Methylcyclopentane	84.161		3938.1	3969.8
C_6H_{12}	Ethylcyclobutane	84.161		4017.0	4049.7
C_6H_{12}	1-Hexene	84.161		4001.8	4032.5
C_6H_{12}	*cis*-2-Hexene	84.161		3992.1	4023.7
C_6H_{12}	*trans*-2-Hexene	84.161		3990.5	4022.1
C_6H_{12}	2-Methyl-1-pentene	84.161		3986.0	4016.6
C_6H_{12}	2-Methyl-2-pentene	84.161		3977.5	4009.1
C_6H_{12}	4-Methyl-1-pentene	84.161		3996.0	4024.7
C_6H_{12}	4-Methyl-*cis*-2-pentene	84.161		3989.0	4018.5
C_6H_{12}	4-Methyl-*trans*-2-pentene	84.161		3984.5	4014.5
C_6H_{12}	2,3-Dimethyl-1-butene	84.161		3982.7	4013.4
C_6H_{12}	2,3-Dimethyl-2-butene	84.161		3974.5	4007.8
C_6H_{12}	2-Ethyl-1-butene	84.161		3988.9	4020.0
$C_6H_{12}O$	Cyclohexanol	100.161		3727.8	3789.8
$C_6H_{12}O$	2-Hexanone	100.161		3754.0	3796.2
$C_6H_{12}O$	3-Hexanone	100.161		3755.8	3797.7
$C_6H_{12}O$	2-Methyl-3-pentanone	100.161		3750.1	3789.9
$C_6H_{12}O$	3,3-Dimethyl-2-butanone	100.161		3747.4	3785.3
$C_6H_{12}O_2$	Butyl acetate	116.160		3546.8	3590.4
$C_6H_{12}O_2$	Methyl pentanoate	116.160		3561.8	3604.8
$C_6H_{12}O_2$	Methyl 2,2-dimethylpropanoate	116.160		3546.0	3581.7
$C_6H_{12}O_2$	Hexanoic acid	116.160		3492.2	3564.1
$C_6H_{13}N$	Cyclohexylamine	99.176		4071.3	4114.1
$C_6H_{13}NO_2$	l-Leucine	131.175	3581.6		3732.2
$C_6H_{13}NO_2$	d-Leucine	131.175	3581.7		
$C_6H_{13}NO_2$	l-Isoleucine	131.175	3581.1		
C_6H_{14}	Hexane	86.177		4163.2	4194.8
C_6H_{14}	2-Methylpentane	86.177		4157.3	4187.1

ENTHALPY OF COMBUSTION OF SELECTED ORGANIC COMPOUNDS (continued)

Mol. form.	Name	M g/mol	$-\Delta_c H°_{298.15}$/(kJ/mol) Crystal	Liquid	Gas
C_6H_{14}	3-Methylpentane	86.177		4159.5	4189.8
C_6H_{14}	2,2-Dimethylbutane	86.177		4148.1	4175.8
C_6H_{14}	2,3-Dimethylbutane	86.177		4154.5	4183.6
$C_6H_{14}N_2$	Azopropane	114.191		4373.4	4413.4
$C_6H_{14}N_2O_2$	Lysine	146.189	3683.2		
$C_6H_{14}N_4O_2$	d-Arginine	174.203	3738.4		
$C_6H_{14}O$	Dipropyl ether	102.177		4033.1	4069.0
$C_6H_{14}O$	Diisopropyl ether	102.177		4010.4	4042.7
$C_6H_{14}O$	1-Hexanol	102.177		3984.4	4046.1
$C_6H_{14}O$	2-Hexanol	102.177		3969.9	
$C_6H_{14}O$	4-Methyl-2-pentanol	102.177		3967.2	
$C_6H_{14}O_2$	1,1-Diethoxyethane	118.176		3870.5	3908.4
$C_6H_{14}O_2$	1,2 Diethoxyethane	118.176		3910.5	3953.7
$C_6H_{14}O_2$	1,6-Hexanediol	118.176	3792.0	3817.5	3900.7
$C_6H_{14}O_3$	Trimethylolpropane	134.175	3611.0		
$C_6H_{14}O_4$	Triethylene glycol	150.175		3557.7	3636.9
$C_6H_{15}N$	Triethylamine	101.192		4377.1	4412.0
$C_6H_{15}N$	Dipropylamine	101.192		4348.7	4388.7
$C_6H_{15}N$	Diisopropylamine	101.192		4326.3	4360.8
C_7H_5N	Benzonitrile	103.123		3632.3	3684.8
C_7H_6O	Benzaldehyde	106.124		3525.1	3575.4
$C_7H_6O_2$	Benzoic acid	122.123	3226.9		3318.0
$C_7H_6O_3$	Salicylic acid	138.123	3022.2		3117.3
C_7H_7NO	Benzamide	121.139	3552.4		
$C_7H_7NO_2$	*o*-Nitrotoluene	137.138		3745.3	
$C_7H_7NO_2$	*m*-Nitrotoluene	137.138		3723.5	
$C_7H_7NO_2$	*p*-Nitrotoluene	137.138	3706.9		3786.0
C_7H_8	Toluene	92.141		3910.3	3948.3
C_7H_8O	*o*-Cresol	108.140	3693.3		3769.3
C_7H_8O	*m*-Cresol	108.140		3703.9	3765.6
C_7H_8O	*p*-Cresol	108.140	3698.6		3772.5
C_7H_8O	Benzyl alcohol	108.140		3737.2	3797.5
C_7H_8O	Anisole	108.140		3783.1	3830.0
C_7H_9N	*o*-Methyl aniline	107.155		4034.5	4097.2
C_7H_9N	*m*-Methyl aniline	107.155		4032.7	4095.4
C_7H_9N	*p*-Methyl aniline	107.155	4017.3		4096.1
C_7H_9N	1-Cyclohexenecarbonitrile	107.155		4088.9	4142.4
C_7H_9N	2,3-Dimethylpyridine	107.155		4060.2	4109.1
C_7H_9N	2,4-Dimethylpyridine	107.155		4057.0	4104.7
C_7H_9N	2,5-Dimethylpyridine	107.155		4059.5	4107.3
C_7H_9N	2,6-Dimethylpyridine	107.155		4053.5	4099.5
C_7H_9N	3,4-Dimethylpyridine	107.155		4059.1	4110.9
C_7H_9N	3,5-Dimethylpyridine	107.155		4063.3	4113.6
$C_7H_{11}N$	Cyclohexanecarbonitrile	109.171		4279.4	4331.4
C_7H_{12}	1-Methylbicyclo(3,1,0)hexane	96.172		4436.4	4471.1
C_7H_{14}	Cycloheptane	98.188		4598.8	4637.3
C_7H_{14}	Methylcyclohexane	98.188		4565.3	4600.7
C_7H_{14}	Ethylcyclopentane	98.188		4592.0	4628.5
C_7H_{14}	*cis*-1,3-Dimethylcyclopentane	98.188		4585.3	4619.5
C_7H_{14}	*cis*-1,2-Dimethylcyclopentane	98.188		4590.1	4625.9
C_7H_{14}	*trans*-1,2-Dimethylcyclopentane	98.188		4584.2	4618.8
C_7H_{14}	1-Heptene	98.188		4657.5	4693.1
C_7H_{14}	*cis*-2-Heptene	98.188		4650.3	
C_7H_{14}	*cis*-3-Heptene	98.188		4651.1	
$C_7H_{14}O$	2,4-Dimethyl-3-pentanone	114.188		4402.5	4444.1
$C_7H_{14}O$	2,2-Dimethyl-3-pentanone	114.188		4399.3	4441.7
$C_7H_{14}O$	1-Heptanal	114.188		4443.9	4491.6

Mol. form.	Name	M g/mol	$-\Delta_c H°_{298.15}$/(kJ/mol) Crystal	Liquid	Gas
$C_7H_{14}O$	*cis*-2-Methylcyclohexanol	114.188		4365.2	4428.4
$C_7H_{14}O$	*trans*-2-Methylcyclohexanol	114.188		4339.7	4402.9
$C_7H_{14}O$	*cis*-3-Methylcyclohexanol	114.188		4339.3	4404.5
$C_7H_{14}O$	*trans*-3-Methylcyclohexanol	114.188		4361.0	4426.3
$C_7H_{14}O$	*cis*-4-Methylcyclohexanol	114.188		4342.2	4407.9
$C_7H_{14}O$	*trans*-4-Methylcyclohexanol	114.188		4322.1	4388.2
$C_7H_{14}O_2$	Ethyl 2,2-dimethylpropanoate	130.187		4178.2	4219.4
$C_7H_{14}O_2$	Methyl hexanoate	130.187		4215.2	4262.8
$C_7H_{14}O_2$	Ethyl pentanoate	130.187		4202.4	4248.5
$C_7H_{14}O_2$	Ethyl 3-methylbutanoate	130.187		4184.5	4228.4
$C_7H_{14}O_2$	Heptanoic acid	130.187		4145.2	4219.2
C_7H_{16}	Heptane	100.204		4817.0	4853.5
C_7H_{16}	2-Methylhexane	100.204		4811.7	4846.6
C_7H_{16}	3-Methylhexane	100.204		4814.8	4849.9
C_7H_{16}	2,2-Dimethylpentane	100.204		4802.9	4835.3
C_7H_{16}	2,3-Dimethylpentane	100.204		4808.1	4842.3
C_7H_{16}	2,4-Dimethylpentane	100.204		4806.6	4839.5
C_7H_{16}	3,3-Dimethylpentane	100.204		4807.0	4840.0
C_7H_{16}	3-Ethylpentane	100.204		4816.4	4851.6
C_7H_{16}	2,2,3-Trimethylbutane	100.204		4804.7	4836.7
$C_7H_{16}O$	1-Heptanol	116.203		4637.9	4704.8
$C_8H_4O_3$	Phthalic anhydride	148.118	3259.6		3348.3
$C_8H_6O_4$	Phthalic acid	166.133	3223.6		
$C_8H_6O_4$	Isophthalic acid	166.133	3202.6		3309.3
$C_8H_6O_4$	Terephthalic acid	166.133	3189.5		3287.7
C_8H_8	Styrene	104.152		4395.2	4439.3
C_8H_8O	Acetophenone	120.151		4148.9	4204.7
$C_8H_8O_2$	*o*-Toluic acid	136.150	3874.9		
$C_8H_8O_2$	*m*-Toluic acid	136.150	3865.3		
$C_8H_8O_2$	*p*-Toluic acid	136.150	3862.2		
$C_8H_8O_2$	Methyl benzoate	136.150		3947.9	4003.5
C_8H_{10}	Ethylbenzene	106.167		4564.9	4607.1
C_8H_{10}	*o*-Xylene	106.167		4552.8	4596.3
C_8H_{10}	*m*-Xylene	106.167		4551.8	4594.5
C_8H_{10}	*p*-Xylene	106.167		4552.8	4595.2
$C_8H_{10}O$	*o*-Ethylphenol	122.167		4368.4	4432.0
$C_8H_{10}O$	*m*-Ethylphenol	122.167		4362.9	4431.1
$C_8H_{10}O$	*p*-Ethylphenol	122.167	4352.8		4433.1
$C_8H_{10}O$	2,3-Xylenol	122.167	4336.1		4420.0
$C_8H_{10}O$	2,4-Xylenol	122.167		4348.5	4414.3
$C_8H_{10}O$	2,5-Xylenol	122.167	4330.6		4415.6
$C_8H_{10}O$	2,6-Xylenol	122.167	4339.8		4415.4
$C_8H_{10}O$	3,4-Xylenol	122.167	4334.9		4420.6
$C_8H_{10}O$	3,5-Xylenol	122.167	4332.8		4415.7
$C_8H_{10}O$	Phenetole	122.167		4424.6	4475.6
$C_8H_{11}N$	*N,N*-Dimethylaniline	121.182		4767.8	4820.6
$C_8H_{11}N$	*N*-Ethylaniline	121.182		4724.1	4776.4
$C_8H_{15}N$	Octanenitrile	125.214		5184.5	5241.3
C_8H_{16}	Cyclooctane	112.215		5267.0	5310.3
C_8H_{16}	Propylcyclopentane	112.215		5245.9	5287.0
C_8H_{16}	1-Ethyl-1-methylcyclopentane	112.215		5240.9	
C_8H_{16}	Ethylcyclohexane	112.215		5222.8	5263.0
C_8H_{16}	1,1-Dimethylcyclohexane	112.215		5216.0	5253.8
C_8H_{16}	*cis*-1,2-Dimethylcyclohexane	112.215		5222.9	5262.6
C_8H_{16}	*trans*-1,2-Dimethylcyclohexane	112.215		5216.5	5254.8
C_8H_{16}	*cis*-1,3-Dimethylcyclohexane	112.215		5211.8	5250.1
C_8H_{16}	*trans*-1,3-Dimethylcyclohexane	112.215		5219.0	5258.2

Mol. form.	Name	*M* g/mol	$-\Delta_c H°_{298.15}$/(kJ/mol) Crystal	Liquid	Gas
C_8H_{16}	*cis*-1,4-Dimethylcyclohexane	112.215		5219.1	5258.1
C_8H_{16}	*trans*-1,4-Dimethylcyclohexane	112.215		5212.3	5250.2
$C_8H_{16}O$	2,2,4-Trimethyl-3-pentanone	128.214		5053.1	5096.4
$C_8H_{16}O$	2-Ethylhexanal	128.214		5086.2	5135.1
$C_8H_{16}O_2$	Octanoic acid	144.214		4798.7	4880.4
$C_8H_{16}O_2$	2-Ethylhexanoic acid	144.214		4799.6	4875.2
$C_8H_{16}O_2$	Methyl heptanoate	144.214		4867.6	4918.8
C_8H_{18}	Octane	114.231		5470.5	5512.0
C_8H_{18}	2-Methylheptane	114.231		5465.6	5505.2
C_8H_{18}	3-Methylheptane	114.231		5468.3	5508.1
C_8H_{18}	4-Methylheptane	114.231		5469.0	5508.6
C_8H_{18}	2,2-Dimethylhexane	114.231		5458.7	5496.0
C_8H_{18}	2,3-Dimethylhexane	114.231		5468.0	5506.8
C_8H_{18}	2,4-Dimethylhexane	114.231		5463.6	5501.4
C_8H_{18}	2,5-Dimethylhexane	114.231		5460.2	5498.1
C_8H_{18}	3,3-Dimethylhexane	114.231		5463.1	5500.6
C_8H_{18}	3,4-Dimethylhexane	114.231		5468.8	5507.8
C_8H_{18}	3-Ethylhexane	114.231		5470.2	5509.9
C_8H_{18}	3-Ethyl-2-methylpentane	114.231		5471.0	5509.6
C_8H_{18}	3-Ethyl-3-methylpentane	114.231		5467.8	5505.8
C_8H_{18}	2,2,3-Trimethylpentane	114.231		5463.7	5500.6
C_8H_{18}	2,2,4-Trimethylpentane	114.231		5461.4	5496.6
C_8H_{18}	2,3,3-Trimethylpentane	114.231		5467.1	5504.3
C_8H_{18}	2,3,4-Trimethylpentane	114.231		5465.6	5503.3
C_8H_{18}	2,2,3,3-Tetramethylbutane	114.231	5451.7		5495.0
$C_8H_{18}N_2$	Azobutane	142.244		5680.5	5729.8
$C_8H_{18}O$	1-Octanol	130.230		5294.1	5365.1
$C_8H_{18}O$	2-Ethyl-1-hexanol	130.230		5287.8	5355.3
$C_8H_{18}O$	Dibutyl ether	130.230		5342.7	5387.2
$C_8H_{18}O$	Di-*sec*-butyl ether	130.230		5319.1	5359.7
$C_8H_{18}O$	Di-*tert*-butyl ether	130.230		5321.0	5358.6
$C_8H_{18}O_5$	Tetraethylene glycol	194.228		4738.9	4837.6
$C_8H_{19}N$	Dibutylamine	129.246		5657.5	5706.9
$C_8H_{19}N$	Diisobutylamine	129.246		5645.0	5684.3
C_9H_7N	Isoquinoline	129.161		4686.5	
C_9H_8	Indene	116.163		4795.5	4848.3
C_9H_{10}	Cyclopropylbenzene	118.178		5071.0	5121.2
C_9H_{10}	Indan	118.178		4982.2	5031.4
$C_9H_{11}NO_2$	l-Phenylalanine	165.192	4646.8		4800.8
$C_9H_{11}NO_3$	l-Tyrosine	181.191	4428.6		
C_9H_{12}	Cumene	120.194		5215.5	5260.6
C_9H_{12}	*o*-Ethyltoluene	120.194		5210.2	5257.9
C_9H_{12}	*m*-Ethyltoluene	120.194		5207.9	5254.8
C_9H_{12}	*p*-Ethyltoluene	120.194		5206.8	5253.4
C_9H_{12}	Propylbenzene	120.194		5218.3	5264.5
C_9H_{12}	1,2,3-Trimethylbenzene	120.194		5198.1	5247.1
C_9H_{12}	1,2,4-Trimethylbenzene	120.194		5194.8	5242.8
C_9H_{12}	Mesitylene	120.194		5193.2	5240.7
$C_9H_{14}O_6$	Triacetin	218.207		4211.6	4293.6
C_9H_{18}	Propylcyclohexane	126.242		5876.7	5921.6
$C_9H_{18}O$	2-Nonanone	142.241		5716.9	5773.4
$C_9H_{18}O$	5-Nonanone	142.241		5715.9	5769.2
$C_9H_{18}O$	2,6-Dimethyl-4-heptanone	142.241		5705.6	5756.5
$C_9H_{18}O_2$	Nonanoic acid	158.241		5454.4	5536.8
$C_9H_{18}O_2$	Methyl octanoate	158.241		5523.8	5580.3
C_9H_{20}	Nonane	128.258		6125.2	6171.7
C_9H_{20}	2,2-Dimethylheptane	128.258		6111.7	

Mol. form.	Name	M g/mol	$-\Delta_c H°_{298.15}$/(kJ/mol) Crystal	Liquid	Gas
C_9H_{20}	2,2,5-Trimethylhexane	128.258		6106.6	
C_9H_{20}	2,3,5-Trimethylhexane	128.258		6115.9	
C_9H_{20}	2,2,3,3-Tetramethylpentane	128.258		6121.6	6162.8
C_9H_{20}	2,2,3,4-Tetramethylpentane	128.258		6122.2	6163.0
C_9H_{20}	3,3-Diethylpentane	128.258		6124.5	6167.6
C_9H_{20}	2,2,4,4-Tetramethylpentane	128.258		6119.9	6158.3
C_9H_{20}	2,3,3,4-Tetramethylpentane	128.258		6122.0	6163.8
$C_9H_{20}O$	1-Nonanol	144.257		5943.4	
$C_{10}H_8$	Azulene	128.174	5290.7		5367.5
$C_{10}H_8$	Naphthalene	128.174	5156.3		5228.7
$C_{10}H_8O$	1-Naphthol	144.173	4957.4		5048.5
$C_{10}H_8O$	2-Naphthol	144.173		4954.2	5048.4
$C_{10}H_{10}O_4$	Dimethyl isophthalate	194.187	4633.4		
$C_{10}H_{10}O_4$	Dimethyl terephthalate	194.187	4631.7		
$C_{10}H_{12}$	1,2,3,4-Tetrahydronaphthalene	132.205		5620.9	5676.1
$C_{10}H_{14}$	Butylbenzene	134.221		5872.7	5922.8
$C_{10}H_{14}$	Isobutylbenzene	134.221		5866.1	5914.4
$C_{10}H_{14}$	*sec*-Butylbenzene	134.221		5869.5	5918.5
$C_{10}H_{14}$	*tert*-Butylbenzene	134.221		5865.2	5913.3
$C_{10}H_{14}$	*o*-Cymene	134.221		5862.6	
$C_{10}H_{14}$	*m*-Cymene	134.221		5857.3	
$C_{10}H_{14}$	*p*-Cymene	134.221		5857.9	
$C_{10}H_{14}$	*o*-Diethylbenzene	134.221		5867.4	
$C_{10}H_{14}$	*m*-Diethylbenzene	134.221		5862.4	
$C_{10}H_{14}$	*p*-Diethylbenzene	134.221		5863.1	
$C_{10}H_{14}$	1,2,4,5-Tetramethylbenzene	134.221	5816.0		
$C_{10}H_{14}$	3-Ethyl-*o*-xylene	134.221		5855.4	
$C_{10}H_{14}O$	Thymol	150.221	5626.2		5717.4
$C_{10}H_{16}$	α-Pinene	136.237		6205.3	6250.0
$C_{10}H_{16}$	β-Pinene	136.237		6214.0	6260.4
$C_{10}H_{16}O$	Camphor	152.236	5902.3		5954.2
$C_{10}H_{18}$	*cis*-Decahydronaphthalene	138.253		6288.2	6338.4
$C_{10}H_{18}$	*trans*-Decahydronaphthalene	138.253		6277.0	6325.5
$C_{10}H_{18}O_4$	Sebacic acid	202.251	5425.0		5585.7
$C_{10}H_{19}N$	Decanenitrile	153.268		6492.1	6559.0
$C_{10}H_{20}$	Butylcyclohexane	140.269		6530.3	6580.1
$C_{10}H_{20}$	1-Decene	140.269		6619.6	6670.0
$C_{10}H_{20}O_2$	Decanoic acid	172.268	6079.7	6109.1	6198.5
$C_{10}H_{22}$	Decane	142.285		6778.3	6829.7
$C_{10}H_{22}$	2-Methylnonane	142.285		6769.4	6819.3
$C_{10}H_{22}$	5-Methylnonane	142.285		6771.3	6820.6
$C_{10}H_{22}O$	1-Decanol	158.284		6601.1	6682.8
$C_{11}H_{10}$	1-Methylnaphthalene	142.200		5814.1	
$C_{11}H_{10}$	2-Methylnaphthalene	142.200	5802.7		5864.5
$C_{11}H_{12}N_2O_2$	l-Tryptophan	204.229	5628.3		
$C_{11}H_{24}$	Undecane	156.312		7431.4	7487.7
$C_{12}H_8$	Acenaphthylene	152.196	6052.1		6125.1
$C_{12}H_9N$	Carbazole	167.210	6133.5		6218.0
$C_{12}H_{10}$	Acenaphthene	154.211	6221.6		6307.3
$C_{12}H_{10}$	Biphenyl	154.211	6250.7		6332.7
$C_{12}H_{10}O$	Diphenyl ether	170.211	6119.2		6203.3
$C_{12}H_{11}N$	Diphenylamine	169.226	6424.4		6513.5
$C_{12}H_{12}N_2$	*p*-Benzidine	184.241	6507.8		
$C_{12}H_{14}O_4$	Diethyl phthalate	222.241		5946.3	6034.5
$C_{12}H_{16}$	Cyclohexylbenzene	160.259		6932.2	6992.1
$C_{12}H_{18}$	5,7-Dodecadiyne	162.275		7476.1	
$C_{12}H_{18}$	3,9-Dodecadiyne	162.275		7492.4	

ENTHALPY OF COMBUSTION OF SELECTED ORGANIC COMPOUNDS (continued)

Mol. form.	Name	*M* g/mol	$-\Delta_c H°_{298.15}$/(kJ/mol) Crystal	Liquid	Gas
$C_{12}H_{18}$	Hexamethylbenzene	162.275	7133.1		7207.8
$C_{12}H_{22}$	Cyclohexylcyclohexane	166.307		7592.6	7650.6
$C_{12}H_{24}$	1-Dodecene	168.323		7925.9	7986.7
$C_{12}H_{24}O_2$	Dodecanoic acid	200.321	7377.5	7414.2	7510.1
$C_{12}H_{26}$	Dodecane	170.338		8087.0	8148.2
$C_{12}H_{26}O$	1-Dodecanol	186.338		7909.4	8001.3
$C_{12}H_{27}N$	Tributylamine	185.353		8299.2	
$C_{13}H_{10}O$	Benzophenone	182.222	6510.3		6599.7
$C_{13}H_{12}$	Diphenylmethane	168.238	6902.1	6920.3	6969.6
$C_{13}H_{26}O_2$	Methyl dodecanoate	214.348		8138.4	8216.6
$C_{13}H_{28}O$	1-Tridecanol	200.365	8517.9		
$C_{14}H_{10}$	Anthracene	178.233	7067.5		7169.2
$C_{14}H_{10}$	Phenanthrene	178.233	7054.5		7145.8
$C_{14}H_{10}$	Diphenylacetylene	178.233	7250.7		
$C_{14}H_{10}O_2$	Benzil	210.232	6784.4		6882.8
$C_{14}H_{12}$	*cis*-Stilbene	180.249		7407.4	7476.4
$C_{14}H_{12}$	*trans*-Stilbene	180.249	7361.0		7460.2
$C_{14}H_{14}$	1,1-Diphenylethane	182.265		7558.7	
$C_{14}H_{14}$	1,2-Diphenylethane	182.265	7561.5		7652.9
$C_{14}H_{27}N$	Tetradecanenitrile	209.375		9107.6	9192.9
$C_{14}H_{28}O_2$	Tetradecanoic acid	228.375	8677.3	8722.0	8817.1
$C_{14}H_{30}O$	1-Tetradecanol	214.392	9167.0	9216.0	
$C_{15}H_{30}O_2$	Pentadecylic acid	242.402	9328.4	9378.4	9491.1
$C_{16}H_{10}$	Fluoranthene	202.255	7915.2		8014.3
$C_{16}H_{10}$	Pyrene	202.255	7850.8		7951.0
$C_{16}H_{22}O_4$	Dibutyl phthalate	278.348		8597.7	8689.4
$C_{16}H_{26}$	Decylbenzene	218.382		9793.7	9873.4
$C_{16}H_{32}$	1-Hexadecene	224.430		10540.7	10620.9
$C_{16}H_{32}O_2$	Hexadecanoic acid	256.429	9977.9	10031.3	10132.3
$C_{16}H_{34}$	Hexadecane	226.446		10699.2	10780.5
$C_{16}H_{34}O$	1-Hexadecanol	242.445	10468.8		10638.3
$C_{17}H_{34}O_2$	Margaric acid	270.456	10624.4	10683.2	
$C_{18}H_{12}$	Chrysene	228.293	8943.5		9068.0
$C_{18}H_{36}O_2$	Stearic acid	284.483	11280.4	11343.4	11446.9
$C_{18}H_{38}$	Octadecane	254.500	11946.6		12099.4
$C_{19}H_{36}O_2$	Methyl oleate	296.494		11887.1	11971.7
$C_{20}H_{12}$	Perylene	252.315	9768.0		
$C_{20}H_{40}O_2$	Arachidic acid	312.536	12574.9	12646.8	12774.4
$C_{22}H_{42}O_2$	Brassidic acid	338.574	13699.0		

STANDARD SOLUTIONS FOR CALIBRATING CONDUCTIVITY CELLS

Grams of KCl	κ/(S/m)			
per kg of H_2O	0°C	18°C	25°C	25°C (Ref. 2)
76.5829	6.514	9.781	11.131	
7.47458	0.7134	1.1163	1.2852	1.2854
0.745819	0.07733	0.12201	0.14083	0.14086
7.45510 (0.1 m)				1.28217
0.745510 (0.01 m)				0.14079

Data are from Reference 1 except last column.

REFERENCES

1. K. N. Marsh, Ed., *Recommended Reference Materials for the Realization of Physicochemical Properties*, Blackwell, Oxford, 1987.
2. Y. C. Wu, K. W. Pratt, and W. F. Koch, *J. Solution Chem.*, 18, 515,1989.

EQUIVALENT CONDUCTIVITY OF AQUEOUS SOLUTIONS OF HYDROHALOGEN ACIDS

From NSRDS-NBS 33
Walter J. Hamer and Harold J. DeWane

One may wish to refer to the above document which defines terms relating to conductance of electrolytic solutions and also discusses some general considerations of the migration of ions under applied potential gradients. In NSRDS-NBS 33 conductance equations are given and some treatment of the Debye–Hückel–Onsager–Fuoss theories is presented.

Equivalent conductances (Ω^{-1} cm^2 equiv^{-1}) of aqueous solutions of HF at 0, 16, 18, 20, and 25°C

c	0°C	16°C	18°C	20°C	25°C
mol l^{-1}					
0.004	106.7	—	—	—	140.5
0.005	97.7	—	—	—	128.1
0.006	90.9	112.9	114.6	116.3	118.8
0.007	85.5	105.7	107.3	108.9	111.4
0.008	81.1	99.9	101.3	102.9	105.4
0.009	77.3	95.0	96.4	97.8	100.4
0.01	74.2	90.8	92.1	93.5	96.1
0.02	56.2	67.7	68.8	69.8	72.2
0.05	39.3	46.8	47.6	48.3	50.1
0.07	34.7	41.3	41.9	42.6	44.3
0.10	30.7	36.4	37.0	37.7	39.1
0.20	24.8	29.6	30.1	30.7	31.7
0.50	20.4	—	—	—	26.3
0.70	19.5	—	—	—	25.1
1.0	18.8	—	—	—	24.3

Equivalent conductances (Ω^{-1} cm^2 equiv^{-1}) of aqueous solutions of HCl from −20 to 65°C

c	−20°C	−10°C	0°C	5°C	10°C	15°C	20°C	25°C	30°C	35°C	40°C	45°C	50°C	55°C	65°C
mol l^{-1}															
0.0001	—	—	—	296.4	—	360.8	—	424.5	—	487.0	—	547.9	577.7	606.6	662.9
0.0005	—	—	—	295.2	—	359.2	—	422.6	—	484.7	—	545.2	575.1	603.5	660.0
0.001	—	—	—	294.3	—	358.0	—	421.2	—	483.1	—	543.2	573.1	601.3	657.8
0.005	—	—	—	291.0	—	353.5	—	415.7	—	476.7	—	535.5	564.4	592.6	647.3
0.01	—	—	—	288.6	—	350.3	—	411.9	—	472.2	—	530.3	558.7	586.5	641.2
0.05	—	—	—	280.3	—	339.9	—	398.9	—	456.7	—	512.4	—	565.6	616.9
0.1	—	—	—	275.0	—	333.3	—	391.1	—	446.8	—	501.1	—	552.8	602.8
0.5	—	—	228.7	254.8	283.0	308.1	336.4	360.7	386.8	411.9	436.9	461.1	482.4	508.0	552.3
1.0	—	—	211.7	235.2	261.6	283.9	312.2	332.2	359.0	379.4	402.9	424.8	445.3	468.1	509.3
1.5	—	—	196.2	216.9	241.5	261.5	287.5	305.8	331.1	349.4	371.6	391.5	410.8	431.7	469.9
2.0	—	—	182.0	199.9	222.7	240.7	262.9	281.4	303.3	321.6	342.4	360.6	378.2	398.0	433.6
2.5	—	131.7	168.5	184.3	205.1	221.4	239.8	258.9	277.0	295.8	315.2	332.0	347.6	366.7	399.9
3.0	—	120.8	154.6	169.5	188.5	203.4	219.3	237.6	253.3	271.5	289.3	304.8	319.0	336.9	368.0
3.5	85.5	111.3	139.6	155.6	172.2	186.5	201.6	218.3	232.9	248.6	263.9	279.4	292.1	308.6	337.2
4.0	79.3	102.7	129.2	143.4	158.1	171.5	185.6	200.0	214.2	228.4	242.2	256.6	268.2	283.6	310.1
4.5	73.7	94.9	119.5	132.0	145.4	157.4	170.6	183.1	196.6	209.5	222.5	235.2	246.7	260.2	284.7
5.0	68.5	87.8	110.3	121.3	133.5	144.4	156.6	167.4	180.2	191.9	204.1	215.4	226.5	238.4	261.0
5.5	63.6	81.1	101.7	111.4	122.5	132.3	143.6	152.9	165.0	175.6	187.1	197.1	207.7	218.3	239.1
6.0	58.9	74.9	93.7	102.2	112.3	121.2	131.5	139.7	151.0	160.6	171.3	180.2	190.3	199.7	218.9
6.5	54.4	69.1	86.2	93.8	103.0	111.0	120.4	127.7	138.2	146.8	156.9	164.8	174.3	182.7	200.4
7.0	50.2	63.7	79.3	86.0	94.4	101.7	110.2	116.9	126.4	134.3	143.3	150.8	159.7	167.2	183.5
7.5	46.3	58.6	73.0	78.9	86.5	93.3	100.9	107.0	115.7	122.9	131.6	138.1	146.2	153.1	168.1
8.0	42.7	54.0	67.1	72.4	79.4	85.6	92.4	98.2	106.1	112.6	120.6	126.7	134.0	140.4	154.1
8.5	39.4	49.8	61.7	66.5	72.9	78.7	84.7	90.3	97.3	103.2	110.7	116.4	123.0	128.8	141.5
9.0	36.4	45.9	56.8	61.2	67.1	72.5	77.8	83.1	89.4	94.8	101.7	107.1	112.9	118.3	130.1
9.5	33.6	42.3	52.3	56.4	61.8	66.8	71.5	76.6	82.3	87.3	93.6	98.7	103.9	108.8	119.7
10.0	31.2	39.1	48.2	52.0	57.0	61.6	65.8	70.7	75.9	80.5	86.3	91.1	95.7	—	—
10.5	28.9	36.1	44.5	48.0	52.7	56.9	60.7	65.3	70.1	74.3	79.6	84.1	88.4	—	—
11.0	26.8	33.4	41.1	44.4	48.8	52.6	56.1	60.2	64.9	68.7	73.6	77.7	81.7	—	—
11.5	24.9	31.0	38.0	41.1	45.3	48.5	51.9	55.3	60.1	63.6	68.0	71.7	75.6	—	—
[illegible]	[illegible]	[illegible]	[illegible]	[illegible]	[illegible]	—	[illegible]		[illegible]		[illegible]		[illegible]		
12.5	21.4	26.7	32.7	—	39.0	—	44.4	—	51.4	—	57.9	—	64.8	—	—
13.0	20.0	24.9	—	—	—	—	—	—	—	—	—	—	—	—	—

Equivalent conductances (Ω^{-1} cm^2 equiv^{-1}) of aqueous solutions of HBr at 25°C

c	Λ	*c*	Λ	*c*	Λ	*c*	Λ	*c*	Λ	*c*	Λ
mol l^{-1}		mol l^{-1}		mol l^{-1}		mol l^{-1}		mol l^{-1}		mol l^{-1}	
0.0002	425.5	0.003	419.7	0.04	402.7	0.30	374.0	0.80	345.3	4.0	199.4
0.0003	425.0	0.004	418.5	0.05	400.4	0.35	370.8	0.85	342.6	4.5	182.4
0.0004	424.6	0.005	417.6	0.06	398.4	0.40	367.7	0.90	339.9	5.0	166.5
0.0005	424.3	0.006	416.7	0.07	396.5	0.45	364.7	0.95	337.2	5.5	151.8
0.0006	424.0	0.007	415.8	0.08	394.9	0.50	361.9	1.0	334.5	6.0	138.2
0.0007	423.7	0.008	415.1	0.09	393.4	0.55	359.0	1.5	307.6	6.5	125.7
0.0008	423.4	0.009	414.4	0.10	391.9	0.60	356.2	2.0	281.7	7.0	114.2
0.0009	423.2	0.01	413.7	0.15	386.0	0.65	353.5	2.5	257.8	7.5	103.8
0.001	422.9	0.02	408.9	0.20	381.4	0.70	350.8	3.0	236.8	8.0	94.4
0.002	421.1	0.03	405.4	0.25	377.5	0.75	348.0	3.5	217.5	8.5	85.8

EQUIVALENT CONDUCTIVITY OF AQUEOUS SOLUTIONS OF HYDROHALOGEN ACIDS (continued)

Equivalent conductances (Ω^{-1} cm^2 $equiv^{-1}$) of aqueous solutions of HBr from −20 to 50°C

c	−20°C	−10°C	0°C	10°C	20°C	30°C	40°C	50°C
mol l^{-1}								
0.50	—	—	240.9	295.9	347.0	398.9	453.6	496.8
0.75	—	—	234.7	284.9	339.0	387.2	433.8	480.6
1.00	—	—	229.6	276.0	329.0	380.4	418.6	465.2
1.25	—	—	221.7	265.8	314.9	362.8	401.8	442.9
1.50	—	—	209.5	254.9	298.9	340.6	381.8	421.4
1.75	—	—	198.3	243.1	284.6	327.3	366.2	404.8
2.00	—	150.8	188.6	231.3	271.8	314.1	350.5	387.4
2.25	—	143.4	180.1	219.6	258.3	296.9	332.3	367.0
2.50	—	136.8	171.7	208.3	244.8	281.7	316.0	349.1
2.75	—	131.1	164.1	198.3	232.9	267.6	301.2	333.2
3.00	—	125.7	157.2	189.5	222.2	255.0	287.8	318.6
3.25	—	120.6	150.8	181.7	212.4	244.3	275.4	304.9
3.50	—	116.1	144.1	174.6	203.2	234.4	263.7	291.9
3.75	87.1	112.0	137.6	167.4	194.7	224.2	252.2	279.4
4.00	84.0	107.5	132.3	160.2	186.8	214.2	239.7	266.9
4.25	80.9	103.1	127.7	153.2	179.0	204.5	228.8	254.6
4.50	78.0	99.0	123.0	146.4	171.2	195.1	218.8	242.6
4.75	75.1	95.1	117.7	139.9	163.1	186.1	209.3	231.3
5.00	72.3	91.4	112.6	134.0	155.7	178.2	199.6	221.3
5.25	69.6	87.8	107.8	128.3	148.7	170.5	190.0	211.4
5.50	67.0	84.2	103.1	122.7	142.1	162.8	181.4	201.8
5.75	64.4	80.6	98.6	117.3	135.7	155.3	173.2	192.4
6.00	61.8	77.2	94.3	112.0	129.6	148.0	165.4	183.4
6.25	59.3	73.8	90.1	106.9	123.7	140.9	157.8	174.7
6.50	56.8	70.7	86.0	102.0	118.0	134.1	150.5	166.3
6.75	54.4	67.7	82.1	97.2	112.5	127.6	143.3	158.4
7.00	51.9	64.6	78.4	92.6	107.1	121.4	136.3	150.8

Equivalent conductances (Ω^{-1} cm^2 $equiv^{-1}$) of aqueous solutions of HI at 25°C

c	Λ	c	Λ	c	Λ	c	Λ	c	Λ
mol l^{-1}		mol l^{-1}		mol l^{-1}		mol l^{-1}		mol l^{-1}	
0.00045	423.2								
0.00050	423.0	0.0035	417.9	0.025	406.6	0.10	394.0	0.90	349.2
0.00055	422.9	0.0040	417.3	0.030	405.2	0.15	389.5	0.95	346.6
0.00060	422.7	0.0045	416.8	0.035	403.9	0.20	385.9	1.0	343.9
0.00065	422.6	0.0050	416.4	0.040	402.8	0.25	382.9	1.5	316.4
0.00070	422.4	0.0055	415.9	0.045	401.7	0.30	380.1	2.0	288.9
0.00075	422.3	0.0060	415.5	0.050	400.8	0.40	374.9	2.5	262.5
0.00080	422.2	0.0065	415.1	0.055	399.9	0.45	372.3	3.0	237.9
0.00085	422.0	0.0070	414.8	0.060	399.1	0.50	369.8	3.5	215.4
0.00090	421.9	0.0075	414.4	0.065	398.3	0.55	367.3	4.0	195.1
0.00095	421.8	0.0080	414.1	0.070	397.6	0.60	364.8	4.5	176.8
0.0010	421.7	0.0090	413.4	0.075	396.9	0.65	362.2	5.0	160.4
0.0015	420.7	0.0095	413.1	0.080	396.3	0.70	359.7	5.5	145.5
0.0020	419.8	0.010	412.8	0.085	395.6	0.75	357.1	6.0	131.7
0.0025	419.1	0.015	410.3	0.090	395.1	0.80	354.5	6.5	118.6
0.0030	418.5	0.020	408.3	0.095	394.5	0.85	351.9	7.0	105.7

Equivalent conductances (Ω^{-1} cm^2 $equiv^{-1}$) of aqueous solutions of HI from −20to 50°C

c	−20°C	−10°C	0°C	10°C	20°C	30°C	40°C	50°C
mol l^{-1}								
0.4	—	—	253.9	300.7	354.5	411.9	453.6	501.8
0.6	—	—	249.1	293.6	344.6	401.1	441.5	488.4
0.8	—	—	242.4	285.5	333.9	389.1	429.3	474.8
1.0	—	—	234.8	276.9	322.7	376.3	416.5	460.8
1.2	—	—	226.6	267.9	311.4	363.1	403.4	446.3
1.4	—	—	218.2	258.8	300.2	349.8	390.0	431.4
1.6	—	—	209.9	249.7	289.1	336.6	376.5	416.3
1.8	—	—	201.8	240.7	278.4	323.7	363.0	401.2
2.0	—	—	194.0	231.9	268.1	311.1	349.6	386.1
2.2	—	147.7	186.6	223.4	258.2	299.0	336.3	371.2
2.4	—	143.5	179.5	215.1	248.8	287.3	323.3	356.5
2.6	—	138.8	172.8	207.2	239.7	276.1	310.5	342.2
2.8	—	133.9	166.5	199.5	231.0	265.4	298.1	328.2
3.0	99.9	129.0	160.5	192.0	222.7	255.1	286.1	314.7
3.2	97.4	124.1	154.7	184.8	214.6	245.2	274.5	301.7
3.4	94.2	119.5	149.1	177.8	206.7	235.7	263.4	289.2
3.6	90.6	115.0	143.7	171.0	198.9	226.4	252.6	277.3
3.8	86.9	110.9	138.3	164.3	191.2	217.4	242.3	266.0
4.0	83.6	106.9	132.8	157.6	183.5	208.4	232.5	255.3
4.2	80.9	103.0	127.3	151.0	175.6	199.6	223.1	245.3
4.4	79.3	99.2	121.6	144.3	167.5	190.7	214.1	235.9

EQUIVALENT CONDUCTIVITY OF ELECTROLYTES IN AQUEOUS SOLUTION

Petr Vanýsek

This table gives the equivalent (molar) conductivity Λ at 25°C for some common electrolytes in aqueous solution at concentrations up to 0.01 mol/L. The units of Λ are 10^{-4} m^2 S mol^{-1}.

For very dilute solutions, the equivalent conductivity for any electrolyte of concentration c can be approximately calculated using the Debye-Hückel-Onsager equation, which can be written for a symmetrical (equal charge on cation and anion) electrolyte as

$$\Lambda = \Lambda° - (A + B\Lambda°)c^{1/2}$$

For a solution at 25°C and both cation and anion with charge $|1|$, the constants are $A = 60.20$ and $B = 0.229$. $\Lambda°$ can be found from the next table, "Ionic Conductivity and Diffusion at Infinite Dilution". The equation is reliable for $c < 0.001$ mol/L; with higher concentration the error increases.

		Concentration (mol/L)						
Compound	Infinite dilution	0.0005	0.001	0.005	0.01	0.02	0.05	0.01
	$\Lambda°$				Λ			
$AgNO_3$	133.29	131.29	130.45	127.14	124.70	121.35	115.18	109.09
$1/2BaCl_2$	139.91	135.89	134.27	127.96	123.88	119.03	111.42	105.14
$1/2CaCl_2$	135.77	131.86	130.30	124.19	120.30	115.59	108.42	102.41
$1/2Ca(OH)_2$	258	—	—	233	226	214	—	—
$1/2CuSO_4$	133.6	121.6	115.20	94.02	83.08	72.16	59.02	50.55
HCl	425.95	422.53	421.15	415.59	411.80	407.04	398.89	391.13
KBr	151.9	149.8	148.9	146.02	143.36	140.41	135.61	131.32
KCl	149.79	147.74	146.88	143.48	141.20	138.27	133.30	128.90
$KClO_4$	139.97	138.69	137.80	134.09	131.39	127.86	121.56	115.14
$1/3K_3Fe(CN)_6$	174.5	166.4	163.1	150.7	—	—	—	—
$1/4K_4Fe(CN)_6$	184	—	167.16	146.02	134.76	122.76	107.65	97.82
$KHCO_3$	117.94	116.04	115.28	112.18	110.03	107.17	—	—
KI	150.31	148.2	143.32	144.30	142.11	139.38	134.90	131.05
KIO_4	127.86	125.74	124.88	121.18	118.45	114.08	106.67	98.2
KNO_3	144.89	142.70	141.77	138.41	132.75	132.34	126.25	120.34
$KMnO_4$	134.8	132.7	131.9	—	126.5	—	—	113
KOH	271.5	—	234	230	228	—	219	213
$KReO_4$	128.20	126.03	125.12	121.31	118.49	114.49	106.40	97.40
$1/3LaCl_3$	145.9	139.6	137.0	127.5	121.8	115.3	106.2	99.1
LiCl	114.97	113.09	112.34	109.35	107.27	104.60	100.06	95.81
$LiClO_4$	105.93	104.13	103.39	100.52	98.56	96.13	92.15	88.52
$1/2MgCl_2$	129.34	125.55	124.15	118.25	114.49	109.99	103.03	97.05
NH_4Cl	149.6	147.5	146.7	134.4	141.21	138.25	133.22	128.69
NaCl	126.39	124.44	123.68	120.59	118.45	115.70	111.01	106.69
$NaClO_4$	117.42	115.58	114.82	111.70	109.54	106.91	102.35	98.38
NaI	126.88	125.30	124.19	121.19	119.18	116.64	112.73	108.73
$NaOOCCH_3$	91.0	89.2	88.5	85.68	83.72	81.20	76.88	72.76
NaOH	247.7	245.5	244.6	240.7	237.9	—	—	—
Na picrate	80.45	78.7	78.6	75.7	73.7	—	66.3	61.8
$1/2Na_2SO_4$	129.8	125.68	124.09	117.09	112.38	106.73	97.70	89.94
$1/2SrCl_2$	135.73	131.84	130.27	124.18	120.23	115.48	108.20	102.14
$1/2ZnSO_4$	132.7	121.3	114.47	95.44	84.87	74.20	61.17	52.61

IONIC CONDUCTIVITY AND DIFFUSION AT INFINITE DILUTION

Petr Vanýsek

This table gives the molar (equivalent) conductivity λ for common ions at infinite dilution. All values refer to aqueous solutions at 25°C. It also lists the diffusion coefficient D of the ion in dilute aqueous solution, which is related to λ through the equation

$$D = \left(RT / F^2\right)\left(\lambda / |z|\right)$$

where R is the molar gas constant, T the temperature, F the Faraday constant, and z the charge on the ion. The variation with temperature is fairly sharp; for typical ions, λ and D increase by 2 to 3% per degree as the temperature increases from 25°C.

The diffusion coefficient for a salt, $D_{salt,}$ may be calculated from the D_+ and D_- values of the constituent ions by the relation

$$D_{salt} = \frac{\left(z_+ + |z_-|\right) D_+ D_-}{z_+ D_+ + |z_-| D_-}$$

For solutions of simple, pure electrolytes (one positive and one negative ionic species), such as NaCl or $CaCl_2$, equivalent ionic conductivity $\Lambda°$, which is the conductivity per unit concentration of charge, is defined as

$$\Lambda° = \lambda_+ + \lambda_-$$

where λ_+ and λ_- are equivalent ionic conductivities of the cation and anion.

The ions are arranged in four groups, Inorganic Cations, Inorganic Anions, Organic Cations, and Organic Anions. Within each group, entries are listed in alphabetical order either by chemical formula (for inorganic ions) or name (for organic ions).

REFERENCES

1. Gray, D. E., Ed., *American Institute of Physics Handbook,* McGraw-Hill, New York, 1972, 2—226.
2. Robinson, R. A., and Stokes, R. H., *Electrolyte Solutions*, Butterworths, London, 1959.
3. Lobo, V. M. M., and Quaresma, J. L., *Handbook of Electrolyte Solutions*, Physical Science Data Series 41, Elsevier, Amsterdam, 1989.

Ion	λ 10^{-4} m² S mol⁻¹	D 10^{-5} cm² s⁻¹
Inorganic Cations		
Ag^+	61.9	1.648
$1/3Al^{3+}$	61	0.541
$1/2Ba^{2+}$	63.6	0.847
$1/2Be^{2+}$	45	0.599
$1/2Ca^{2+}$	59.47	0.792
$1/2Cd^{2+}$	54	0.719
$1/3Ce^{3+}$	69.8	0.620
$1/2Co^{2+}$	55	0.732
$1/3[Co(NH_3)_6]^{3+}$	101.9	0.904
$1/3[Co(en)_3]^{3+}$	74.7	0.663
$1/6[Co_2(trien)_3]^{6+}$	69	0.306
$1/3Cr^{3+}$	67	0.595
Cs^+	77.2	2.056
$1/2Cu^{2+}$	53.6	0.714
D^+	249.9	6.655
$1/3Dy^{3+}$	65.6	0.582
$1/3Er^{3+}$	65.9	0.585
$1/3Eu^{3+}$	67.8	0.602
$1/2Fe^{2+}$	54	0.719
$1/3Fe^{3+}$	68	0.604
$1/3Gd^{3+}$	67.3	0.597
H^+	349.65	9.311
$1/2Hg^{2+}$	68.6	0.913
$1/2Hg^{2+}$	63.6	0.847
$1/3Ho^{3+}$	66.3	0.589
K^+	73.48	1.957
$1/3La^{3+}$	69.7	0.619
Li^+	38.66	1.029
$1/2Mg^{2+}$	53.0	0.706
$1/2Mn^{2+}$	53.5	0.712
NH_4^+	73.5	1.957
$N_2H_5^+$	59	1.571
Na^+	50.08	1.334
$1/3Nd^{3+}$	69.4	0.616
$1/2Ni^{2+}$	49.6	0.661
$1/4[Ni_2(trien)_3]^{4+}$	52	0.346
$1/2Pb^{2+}$	71	0.945
$1/3Pr^{3+}$	69.5	0.617
$1/2Ra^{2+}$	66.8	0.889
Rb^+	77.8	2.072
$1/3Sc^{3+}$	64.7	0.574
$1/3Sm^{3+}$	68.5	0.608
$1/2Sr^{2+}$	59.4	0.791
Tl^+	74.7	1.989
$1/3Tm^{3+}$	65.4	0.581
$1/2UO_2^{2+}$	32	0.426
$1/3Y^{3+}$	62	0.550
$1/3Yb^{3+}$	65.6	0.582
$1/2Zn^{2+}$	52.8	0.703
Inorganic Anions		
$Au(CN)_2^-$	50	1.331
$Au(CN)_4^-$	36	0.959

IONIC CONDUCTIVITY AND DIFFUSION AT INFINITE DILUTION (continued)

Ion	λ 10^{-4} m^2 S mol^{-1}	*D* 10^{-5} cm^2 s^{-1}
$B(C_6H_5)_4^-$	21	0.559
Br^-	78.1	2.080
Br_3^-	43	1.145
BrO_3^-	55.7	1.483
CN^-	78	2.077
CNO^-	64.6	1.720
$1/2CO_3^{2-}$	69.3	0.923
Cl^-	76.31	2.032
ClO_2^-	52	1.385
ClO_3^-	64.6	1.720
ClO_4^-	67.3	1.792
$1/3[Co(CN)_6]^{3-}$	98.9	0.878
$1/2CrO_4^{2-}$	85	1.132
F^-	55.4	1.475
$1/4[Fe(CN)_6]^{4-}$	110.4	0.735
$1/3[Fe(CN)_6]^{3-}$	100.9	0.896
$H_2AsO_4^-$	34	0.905
HCO_3^-	44.5	1.185
HF_2^-	75	1.997
$1/2HPO_4^{2-}$	33	0.439
$H_2PO_4^-$	33	0.879
$H_2PO_3^-$	46	1.225
HS^-	65	1.731
HSO_3^-	50	1.331
HSO_4^-	50	1.331
$H_2SbO_4^-$	31	0.825
I^-	76.8	2.045
IO_3^-	40.5	1.078
IO_4^-	54.5	1.451
MnO_4^-	61.3	1.632
MoO_4^-	74.5	1.984
$N(CN)_2^-$	54.5	1.451
NO_2^-	71.8	1.912
NO_3^-	71.42	1.902
$NH_2SO_3^-$	48.3	1.286
N_3^-	69	1.837
OCN^-	64.6	1.720
OH^-	198	5.273
PF_6^-	56.9	1.515
$1/2PO_3F^{2-}$	63.3	0.843
$1/3PO_4^{3-}$	69.0	0.612
$1/4P_2O_7^{4-}$	96	0.639
$1/3P_3O_9^{3-}$	83.6	0.742
$1/5P_3O_{10}^{5-}$	109	0.581
ReO_4^-	54.9	1.462
SCN^-	66	1.758
$1/2SO_3^{2-}$	79.9	1.064
$1/2SO_4^{2-}$	80.0	1.065
$1/2S_2O_3^{2-}$	85.0	1.132
$1/2S_4O_4^{2-}$	66.5	0.885
$1/2S_2O_6^{2-}$	93	1.238
$1/2S_2O_8^{2-}$	86	1.145
$Sb(OH)_6^-$	31.9	0.849
$SeCN^-$	64.7	1.723
$1/2SeO_4^{2-}$	75.7	1.008
$1/2WO_4^{2-}$	69	0.919

Ion	λ 10^{-4} m^2 S mol^{-1}	*D* 10^{-5} cm^2 s^{-1}
Organic Cations		
Benzyltrimethylammonium$^+$	34.6	0.921
Isobutylammonium$^+$	38	1.012
Butyltrimethylammonium$^+$	33.6	0.895
Decylpyridinium$^+$	29.5	0.786
Decyltrimethylammonium$^+$	24.4	0.650
Diethylammonium$^+$	42.0	1.118
Dimethylammonium$^+$	51.8	1.379
Dipropylammonium$^+$	30.1	0.802
Dodecylammonium$^+$	23.8	0.634
Dodecyltrimethylammonium$^+$	22.6	0.602
Ethanolammonium$^+$	42.2	1.124
Ethylammonium$^+$	47.2	1.257
Ethyltrimethylammonium$^+$	40.5	1.078
Hexadecyltrimethylammonium$^+$	20.9	0.557
Hexyltrimethylammonium$^+$	29.6	0.788
Histidyl$^+$	23.0	0.612
Hydroxyethyltrimethylarsonium$^+$	39.4	1.049
Methylammonium$^+$	58.7	1.563
Octadecylpyridinium$^+$	20	0.533
Octadecyltributylammonium$^+$	16.6	0.442
Octadecyltriethylammonium$^+$	17.9	0.477
Octadecyltrimethylammonium$^+$	19.9	0.530
Octadecyltripropylammonium$^+$	17.2	0.458
Octyltrimethylammonium$^+$	26.5	0.706
Pentylammonium$^+$	37	0.985
Piperidinium$^+$	37.2	0.991
Propylammonium$^+$	40.8	1.086
Pyrilammonium$^+$	24.3	0.647
Tetrabutylammonium$^+$	19.5	0.519
Tetradecyltrimethylammonium$^+$	21.5	0.573
Tetraethylammonium$^+$	32.6	0.868
Tetramethylammonium$^+$	44.9	1.196
Tetraisopentylammonium$^+$	17.9	0.477
Tetrapentylammmonium$^+$	17.5	0.466
Tetrapropylammonium$^+$	23.4	0.623
Triethylammonium$^+$	34.3	0.913
Triethylsulfonium$^+$	36.1	0.961
Trimethylammonium$^+$	47.23	1.258
Trimethylhexylammonium$^+$	34.6	0.921
Trimethylsulfonium$^+$	51.4	1.369
Tripropylammonium$^+$	26.1	0.695
Organic Anions		
Acetate$^-$	40.9	1.089
p-Anisate$^-$	29.0	0.772
1/2Azelate^{2-}	40.6	0.541
Benzoate$^-$	32.4	0.863
Bromoacetate$^-$	39.2	1.044
Bromobenzoate$^-$	30	0.799
Butyrate$^-$	32.6	0.868
Chloroacetate$^-$	42.2	1.124
m-Chlorobenzoate$^-$	31	0.825
o-Chlorobenzoate$^-$	30.2	0.804

Ion	λ 10^{-4} m^2 S mol^{-1}	*D* 10^{-5} cm^2 s^{-1}
1/3Citrate^{3-}	70.2	0.623
Crotonate$^-$	33.2	0.884
Cyanoacetate$^-$	43.4	1.156
Cyclohexane carboxylate$^-$	28.7	0.764
1/2 1,1-Cyclopropanedicarboxylate^{2-}	53.4	0.711
Decylsulfate$^-$	26	0.692
Dichloroacetate$^-$	38.3	1.020
1/2Diethylbarbiturate^{2-}	26.3	0.350
Dihydrogencitrate$^-$	30	0.799
1/2Dimethylmalonate^{2-}	49.4	0.658
3,5-Dinitrobenzoate$^-$	28.3	0.754
Dodecylsulfate$^-$	24	0.639
Ethylmalonate$^-$	49.3	1.313
Ethylsulfate$^-$	39.6	1.055
Fluoroacetate$^-$	44.4	1.182
Fluorobenzoate$^-$	33	0.879
Formate$^-$	54.6	1.454
1/2Fumarate^{2-}	61.8	0.823
1/2Glutarate^{2-}	52.6	0.700
Hydrogenoxalate$^-$	40.2	1.070
Isovalerate$^-$	32.7	0.871
Iodoacetate$^-$	40.6	1.081
Lactate$^-$	38.8	1.033
1/2Malate^{2-}	58.8	0.783
1/2Maleate^{2-}	61.9	0.824
1/2Malonate^{2-}	63.5	0.845
Methylsulfate$^-$	48.8	1.299
Naphthylacetate$^-$	28.4	0.756
1/2Oxalate^{2-}	74.11	0.987
Octylsulfate$^-$	29	0.772
Phenylacetate$^-$	30.6	0.815
1/2*o*-Phthalate^{2-}	52.3	0.696
1/2*m*-Phthalate^{2-}	54.7	0.728
Picrate$^-$	30.37	0.809
Pivalate$^-$	31.9	0.849
Propionate$^-$	35.8	0.953
Propylsulfate$^-$	37.1	0.988
Salicylate$^-$	36	0.959
1/2Suberate^{2-}	36	0.479
1/2Succinate^{2-}	58.8	0.783
1/2Tartarate^{2-}	59.6	0.794
Trichloroacetate$^-$	36.6	0.975

ACTIVITY COEFFICIENTS OF ACIDS, BASES, AND SALTS

Petr Vanýsek

This table gives mean activity coefficients at 25°C for molalities in the range 0.1 to 1.0. See the following table for definitions, references, and data over a wider concentration range.

	0.1	0.2	0.3	0.4	0.5	0.6	0.7	0.8	0.9	1.0
$AgNO_3$	0.734	0.657	0.606	0.567	0.536	0.509	0.485	0.464	0.446	0.429
$AlCl_3$	0.337	0.305	0.302	0.313	0.331	0.356	0.388	0.429	0.479	0.539
$Al_2(SO_4)_3$	0.035	0.0225	0.0176	0.0153	0.0143	0.014	0.0142	0.0149	0.0159	0.0175
$BaCl_2$	0.500	0.444	0.419	0.405	0.397	0.391	0.391	0.391	0.392	0.395
$BeSO_4$	0.150	0.109	0.0885	0.0769	0.0692	0.0639	0.0600	0.0570	0.0546	0.0530
$CaCl_2$	0.518	0.472	0.455	0.448	0.448	0.453	0.460	0.470	0.484	0.500
$CdCl_2$	0.2280	0.1638	0.1329	0.1139	0.1006	0.0905	0.0827	0.0765	0.0713	0.0669
$Cd(NO_3)_2$	0.513	0.464	0.442	0.430	0.425	0.423	0.423	0.425	0.428	0.433
$CdSO_4$	0.150	0.103	0.0822	0.0699	0.0615	0.0553	0.0505	0.0468	0.0438	0.0415
$CoCl_2$	0.522	0.479	0.463	0.459	0.462	0.470	0.479	0.492	0.511	0.531
$CrCl_3$	0.331	0.298	0.294	0.300	0.314	0.335	0.362	0.397	0.436	0.481
$Cr(NO_3)_3$	0.319	0.285	0.279	0.281	0.291	0.304	0.322	0.344	0.371	0.401
$Cr_2(SO_4)_3$	0.0458	0.0300	0.0238	0.0207	0.0190	0.0182	0.0181	0.0185	0.0194	0.0208
CsBr	0.754	0.694	0.654	0.626	0.603	0.586	0.571	0.558	0.547	0.538
CsCl	0.756	0.694	0.656	0.628	0.606	0.589	0.575	0.563	0.553	0.544
CsI	0.754	0.692	0.651	0.621	0.599	0.581	0.567	0.554	0.543	0.533
$CsNO_3$	0.733	0.655	0.602	0.561	0.528	0.501	0.478	0.458	0.439	0.422
CsOH	0.795	0.761	0.744	0.739	0.739	0.742	0.748	0.754	0.762	0.771
CsOAc	0.799	0.771	0.761	0.759	0.762	0.768	0.776	0.783	0.792	0.802
Cs_2SO_4	0.456	0.382	0.338	0.311	0.291	0.274	0.262	0.251	0.242	0.235
$CuCl_2$	0.508	0.455	0.429	0.417	0.411	0.409	0.409	0.410	0.413	0.417
$Cu(NO_3)_2$	0.511	0.460	0.439	0.429	0.426	0.427	0.431	0.437	0.445	0.455
$CuSO_4$	0.150	0.104	0.0829	0.0704	0.0620	0.0559	0.0512	0.0475	0.0446	0.0423
$FeCl_2$	0.5185	0.473	0.454	0.448	0.450	0.454	0.463	0.473	0.488	0.506
HBr	0.805	0.782	0.777	0.781	0.789	0.801	0.815	0.832	0.850	0.871
HCl	0.796	0.767	0.756	0.755	0.757	0.763	0.772	0.783	0.795	0.809
$HClO_4$	0.803	0.778	0.768	0.766	0.769	0.776	0.785	0.795	0.808	0.823
HI	0.818	0.807	0.811	0.823	0.839	0.860	0.883	0.908	0.935	0.963

ACTIVITY COEFFICIENTS OF ACIDS, BASES, AND SALTS (continued)

	0.1	0.2	0.3	0.4	0.5	0.6	0.7	0.8	0.9	1.0
HNO_3	0.791	0.754	0.735	0.725	0.720	0.717	0.717	0.718	0.721	0.724
H_2SO_4	0.2655	0.2090	0.1826	—	0.1557	—	0.1417	—	—	0.1316
KBr	0.772	0.722	0.693	0.673	0.657	0.646	0.636	0.629	0.622	0.617
KCl	0.770	0.718	0.688	0.666	0.649	0.637	0.626	0.618	0.610	0.604
$KClO_3$	0.749	0.681	0.635	0.599	0.568	0.541	0.518	—	—	—
K_2CrO_4	0.456	0.382	0.340	0.313	0.292	0.276	0.263	0.253	0.243	0.235
KF	0.775	0.727	0.700	0.682	0.670	0.661	0.654	0.650	0.646	0.645
$K_3Fe(CN)_6$	0.268	0.212	0.184	0.167	0.155	0.146	0.140	0.135	0.131	0.128
$K_4Fe(CN)_6$	0.139	0.0993	0.0808	0.0693	0.0614	0.0556	0.0512	0.0479	0.0454	—
KH_2PO_4	0.731	0.653	0.602	0.561	0.529	0.501	0.477	0.456	0.438	0.421
KI	0.778	0.733	0.707	0.689	0.676	0.667	0.660	0.654	0.649	0.645
KNO_3	0.739	0.663	0.614	0.576	0.545	0.519	0.496	0.476	0.459	0.443
KOAc	0.796	0.766	0.754	0.750	0.751	0.754	0.759	0.766	0.774	0.783
KOH	0.798	0.760	0.742	0.734	0.732	0.733	0.736	0.742	0.749	0.756
KSCN	0.769	0.716	0.685	0.663	0.646	0.633	0.623	0.614	0.606	0.599
K_2SO_4	0.441	0.360	0.316	0.286	0.264	0.246	0.232	—	—	—
LiBr	0.796	0.766	0.756	0.752	0.753	0.758	0.767	0.777	0.789	0.803
LiCl	0.790	0.757	0.744	0.740	0.739	0.743	0.748	0.755	0.764	0.774
$LiClO_4$	0.812	0.794	0.792	0.798	0.808	0.820	0.834	0.852	0.869	0.887
LiI	0.815	0.802	0.804	0.813	0.824	0.838	0.852	0.870	0.888	0.910
$LiNO_3$	0.788	0.752	0.736	0.728	0.726	0.727	0.729	0.733	0.737	0.743
LiOH	0.760	0.702	0.665	0.638	0.617	0.599	0.585	0.573	0.563	0.554
LiOAc	0.784	0.742	0.721	0.709	0.700	0.691	0.689	0.688	0.688	0.689
Li_2SO_4	0.468	0.398	0.361	0.337	0.319	0.307	0.297	0.289	0.282	0.277
$MgCl_2$	0.529	0.489	0.477	0.475	0.481	0.491	0.506	0.522	0.544	0.570
$MgSO_4$	0.150	0.107	0.0874	0.0756	0.0675	0.0616	0.0571	0.0536	0.0508	0.0485
$MnCl_2$	0.516	0.469	0.450	0.442	0.440	0.443	0.448	0.455	0.466	0.479
$MnSO_4$	0.150	0.105	0.0848	0.0725	0.0640	0.0578	0.0530	0.0493	0.0463	0.0439
NH_4Cl	0.770	0.718	0.687	0.665	0.649	0.636	0.625	0.617	0.609	0.603
NH_4NO_3	0.740	0.677	0.636	0.606	0.582	0.562	0.545	0.530	0.516	0.504
$(NH_4)_2SO_4$	0.439	0.356	0.311	0.280	0.257	0.240	0.226	0.214	0.205	0.196
NaBr	0.782	0.741	0.719	0.704	0.697	0.692	0.689	0.687	0.687	0.687
NaCl	0.778	0.735	0.710	0.693	0.681	0.673	0.667	0.662	0.659	0.657
$NaClO_3$	0.772	0.720	0.688	0.664	0.645	0.630	0.617	0.606	0.597	0.589
$NaClO_4$	0.775	0.729	0.701	0.683	0.668	0.656	0.648	0.641	0.635	0.629
Na_2CrO_4	0.464	0.394	0.353	0.327	0.307	0.292	0.280	0.269	0.261	0.253
NaF	0.765	0.710	0.676	0.651	0.632	0.616	0.603	0.592	0.582	0.573
NaH_2PO_4	0.744	0.675	0.629	0.593	0.563	0.539	0.517	0.499	0.483	0.468
NaI	0.787	0.751	0.735	0.727	0.723	0.723	0.724	0.727	0.731	0.736
$NaNO_3$	0.762	0.703	0.666	0.638	0.617	0.599	0.583	0.570	0.558	0.548
NaOAc	0.791	0.757	0.744	0.737	0.735	0.736	0.740	0.745	0.752	0.757
NaOH	0.766	0.727	0.708	0.697	0.690	0.685	0.681	0.679	0.678	0.678
NaSCN	0.787	0.750	—	0.720	0.715	0.712	0.710	0.710	0.711	0.712
Na_2SO_4	0.445	0.365	0.320	0.289	0.266	0.248	0.233	0.221	0.210	0.201
$NiCl_2$	0.522	0.479	0.463	0.460	0.464	0.471	0.482	0.496	0.515	0.563
$NiSO_4$	0.150	0.105	0.0841	0.0713	0.0627	0.0562	0.0515	0.0478	0.0448	0.0425
$Pb(NO_3)_2$	0.395	0.308	0.260	0.228	0.205	0.187	0.172	0.160	0.150	0.141
RbBr	0.763	0.706	0.673	0.650	0.632	0.617	0.605	0.595	0.586	0.578
RbCl	0.764	0.709	0.675	0.652	0.634	0.620	0.608	0.599	0.590	0.583
RbI	0.762	0.705	0.671	0.647	0.629	0.614	0.602	0.591	0.583	0.575
$RbNO_3$	0.734	0.658	0.606	0.565	0.534	0.508	0.485	0.465	0.446	0.430
RbOAc	0.796	0.767	0.756	0.753	0.755	0.759	0.766	0.773	0.782	0.792
Rb_2SO_4	0.451	0.374	0.331	0.301	0.279	0.263	0.249	0.238	0.228	0.219
$SrCl_2$	0.511	0.462	0.442	0.433	0.430	0.431	0.434	0.441	0.449	0.461
$TlClO_4$	0.730	0.652	0.599	0.559	0.527	—	—	—	—	—
$TlNO_3$	0.702	0.606	0.545	0.500	—	—	—	—	—	—
UO_2Cl_2	0.544	0.510	0.520	0.505	0.517	0.532	0.549	0.571	0.595	0.620
UO_2SO_4	0.150	0.102	0.0807	0.0689	0.0611	0.0566	0.0515	0.0483	0.0458	0.0439
$ZnCl_2$	0.515	0.462	0.432	0.411	0.394	0.380	0.369	0.357	0.348	0.339
$Zn(NO_3)_2$	0.531	0.489	0.474	0.469	0.473	0.480	0.489	0.501	0.518	0.535
$ZnSO_4$	0.150	0.140	0.0835	0.0714	0.0630	0.0569	0.0523	0.0487	0.0458	0.0435

MEAN ACTIVITY COEFFICIENTS OF ELECTROLYTES AS A FUNCTION OF CONCENTRATION

The mean activity coefficient γ of an electrolyte X_aY_b is defined as

$$\gamma = \left(\gamma_+{}^a + \gamma_-{}^b\right)^{1/(a+b)}$$

where γ_+ and γ_- are activity coefficients of the individual ions (which cannot be directly measured). This table gives the mean activity coefficients of about 100 electrolytes in aqueous solution as a function of concentration, expressed in molality terms. All values refer to a temperature of 25°C. Substances are arranged in alphabetical order by formula.

REFERENCES

1. Hamer, W. J., and Wu, Y. C., *J. Phys. Chem. Ref. Data*, 1, 1047, 1972.
2. Staples, B. R., *J. Phys. Chem. Ref. Data*, 6, 385, 1977; 10, 767, 1981; 10, 779, 1981.
3. Goldberg, R. N. et al., *J. Phys. Chem. Ref. Data*, 7, 263, 1978; 8, 923, 1979; 8, 1005, 1979; 10, 1, 1981; 10, 671, 1981.

Mean Activity Coefficient at 25°C

m/mol kg^{-1}	$AgNO_3$	$BaBr_2$	$BaCl_2$	BaI_2	$CaBr_2$	$CaCl_2$	CaI_2
0.001	0.964	0.881	0.887	0.890	0.890	0.888	0.890
0.002	0.950	0.850	0.849	0.853	0.853	0.851	0.853
0.005	0.924	0.785	0.782	0.792	0.791	0.787	0.791
0.010	0.896	0.727	0.721	0.737	0.735	0.727	0.736
0.020	0.859	0.661	0.653	0.678	0.674	0.664	0.677
0.050	0.794	0.573	0.559	0.600	0.594	0.577	0.600
0.100	0.732	0.517	0.492	0.551	0.540	0.517	0.552
0.200	0.656	0.463	0.436	0.520	0.502	0.469	0.524
0.500	0.536	0.435	0.391	0.536	0.500	0.444	0.554
1.000	0.430	0.470	0.393	0.664	0.604	0.495	0.729
2.000	0.316	0.654		1.242	1.125	0.784	
5.000	0.181				18.7	5.907	
10.000	0.108					43.1	
15.000	0.085						

m/mol kg^{-1}	$Cd(NO_2)_2$	$Cd(NO_3)_2$	$CoBr_2$	$CoCl_2$	CoI_2	$Co(NO_3)_2$	$CsBr$
0.001	0.881	0.888	0.890	0.889	0.887	0.888	0.965
0.002	0.837	0.851	0.854	0.852	0.849	0.850	0.951
0.005	0.759	0.787	0.794	0.789	0.783	0.786	0.925
0.010	0.681	0.728	0.740	0.732	0.724	0.728	0.898
0.020	0.589	0.664	0.681	0.670	0.661	0.663	0.864
0.050	0.451	0.576	0.605	0.586	0.582	0.576	0.806
0.100	0.344	0.515	0.556	0.528	0.540	0.516	0.752
0.200	0.247	0.465	0.523	0.483	0.527	0.469	0.691
0.500	0.148	0.428	0.538	0.465	0.596	0.446	0.605
1.000	0.098	0.437	0.685	0.532	0.845	0.492	0.540
2.000	0.069	0.517	1.421	0.864	2.287	0.722	0.485
5.000	0.054		13.9		55.3	3.338	0.454
10.000					196		

m/mol kg^{-1}	$CsCl$	CsF	CsI	$CsNO_3$	$CsOH$	Cs_2SO_4	$CuBr_2$
0.001	0.965	0.965	0.965	0.964	0.966	0.885	0.889
0.002	0.951	0.952	0.951	0.951	0.953	0.845	0.853
0.005	0.925	0.929	0.925	0.924	0.930	0.775	0.791
0.010	0.898	0.905	0.898	0.897	0.906	0.709	0.735
0.020	0.864	0.876	0.863	0.860	0.878	0.634	0.674
0.050	0.805	0.830	0.804	0.796	0.836	0.526	0.594
0.100	0.751	0.792	0.749	0.733	0.802	0.444	0.541

MEAN ACTIVITY COEFFICIENTS OF ELECTROLYTES AS A FUNCTION OF CONCENTRATION (continued)

m/mol kg^{-1}	CsCl	CsF	CsI	$CsNO_3$	CsOH	Cs_2SO_4	$CuBr_2$
0.200	0.691	0.755	0.688	0.655	0.772	0.369	0.504
0.500	0.607	0.721	0.601	0.529	0.755	0.285	0.503
1.000	0.546	0.726	0.534	0.421	0.782	0.233	0.591
2.000	0.496	0.803	0.470				0.859
5.000	0.474						
10.000	0.508						

m/mol kg^{-1}	$CuCl_2$	$Cu(ClO_4)_2$	$Cu(NO_3)_2$	$FeCl_2$	HBr	HCl	$HClO_4$
0.001	0.887	0.890	0.888	0.888	0.966	0.965	0.966
0.002	0.849	0.854	0.851	0.850	0.953	0.952	0.953
0.005	0.783	0.795	0.787	0.785	0.930	0.929	0.929
0.010	0.722	0.741	0.729	0.725	0.907	0.905	0.906
0.020	0.654	0.685	0.664	0.659	0.879	0.876	0.878
0.050	0.561	0.613	0.577	0.570	0.837	0.832	0.836
0.100	0.495	0.572	0.516	0.509	0.806	0.797	0.803
0.200	0.441	0.553	0.466	0.462	0.783	0.768	0.776
0.500	0.401	0.617	0.431	0.443	0.790	0.759	0.769
1.000	0.405	0.892	0.456	0.500	0.872	0.811	0.826
2.000	0.453	2.445	0.615	0.782	1.167	1.009	1.055
5.000	0.601		2.083		3.800	2.380	3.100
10.000					33.4	10.4	30.8
15.000							323

m/mol kg^{-1}	HF	HI	HNO_3	H_2SO_4	KBr	KCNS	KCl
0.001	0.551	0.966	0.965	0.804	0.965	0.965	0.965
0.002	0.429	0.953	0.952	0.740	0.952	0.951	0.951
0.005	0.302	0.931	0.929	0.634	0.927	0.927	0.927
0.010	0.225	0.909	0.905	0.542	0.902	0.901	0.901
0.020	0.163	0.884	0.875	0.445	0.870	0.869	0.869
0.050	0.106	0.847	0.829	0.325	0.817	0.815	0.816
0.100	0.0766	0.823	0.792	0.251	0.771	0.768	0.768
0.200	0.0550	0.811	0.756	0.195	0.772	0.716	0.717
0.500	0.0352	0.845	0.725	0.146	0.658	0.647	0.649
1.000	0.0249	0.969	0.730	0.125	0.617	0.598	0.604
2.000	0.0175	1.363	0.788	0.119	0.593	0.556	0.573
5.000	0.0110	4.760	1.063	0.197	0.626	0.525	0.593
10.000	0.0085	49.100	1.644	0.527			
15.000	0.0077		2.212	1.077			
20.000	0.0075		2.607	1.701			

m/mol kg^{-1}	$KClO_3$	K_2CrO_4	KF	KH_2PO_4*	K_2HPO_4**	KI	KNO_3
0.001	0.965	0.886	0.965	0.964	0.886	0.965	0.964
0.002	0.951	0.847	0.952	0.950	0.847	0.952	0.950
0.005	0.926	0.779	0.927	0.924	0.779	0.927	0.924
0.010	0.899	0.715	0.902	0.896	0.715	0.902	0.896
0.020	0.865	0.643	0.870	0.859	0.643	0.871	0.860
0.050	0.805	0.539	0.818	0.793	0.538	0.820	0.797
0.100	0.749	0.460	0.773	0.730	0.457	0.776	0.735
0.200	0.681	0.385	0.726	0.652	0.379	0.731	0.662
0.500	0.569	0.296	0.670	0.529	0.283	0.676	0.546
1.000		0.239	0.645	0.422		0.646	0.444

MEAN ACTIVITY COEFFICIENTS OF ELECTROLYTES AS A FUNCTION OF CONCENTRATION (continued)

m/mol kg^{-1}	$KClO_3$	K_2CrO_4	KF	KH_2PO_4*	K_2HPO_4**	KI	KNO_3
2.000		0.199	0.658			0.638	0.332
5.000			0.871				
10.000			1.715				
15.000			3.120				

m/mol kg^{-1}	KOH	K_2SO_4	LiBr	LiCl	$LiClO_4$	LiI	$LiNO_3$
0.001	0.965	0.885	0.965	0.965	0.966	0.966	0.965
0.002	0.952	0.844	0.952	0.952	0.953	0.953	0.952
0.005	0.927	0.772	0.929	0.928	0.931	0.930	0.928
0.010	0.902	0.704	0.905	0.904	0.908	0.908	0.904
0.020	0.871	0.625	0.877	0.874	0.882	0.882	0.874
0.050	0.821	0.511	0.832	0.827	0.843	0.843	0.827
0.100	0.779	0.424	0.797	0.789	0.815	0.817	0.788
0.200	0.740	0.343	0.767	0.756	0.795	0.802	0.753
0.500	0.710	0.251	0.754	0.739	0.806	0.824	0.726
1.000	0.733		0.803	0.775	0.887	0.912	0.743
2.000	0.860		1.012	0.924	1.161	1.197	0.837
5.000	1.697		2.696	2.000			1.298
10.000	6.110		20.0	9.600			2.500
15.000	19.9		147	30.9			3.960
20.000	46.4		486				4.970

m/mol kg^{-1}	LiOH	Li_2SO_4	$MgBr_2$	$MgCl_2$	MgI_2	$MnBr_2$	$MnCl_2$
0.001	0.964	0.887	0.889	0.889	0.889	0.889	0.888
0.002	0.950	0.847	0.852	0.852	0.853	0.853	0.850
0.005	0.923	0.780	0.790	0.790	0.791	0.791	0.786
0.010	0.895	0.716	0.733	0.734	0.736	0.735	0.727
0.020	0.858	0.645	0.672	0.672	0.677	0.674	0.662
0.050	0.794	0.544	0.593	0.590	0.602	0.595	0.574
0.100	0.735	0.469	0.543	0.535	0.556	0.543	0.513
0.200	0.668	0.400	0.512	0.493	0.535	0.508	0.464
0.500	0.579	0.325	0.540	0.485	0.594	0.519	0.437
1.000	0.522	0.284	0.715	0.577	0.858	0.650	0.477
2.000	0.484	0.270	1.590	1.065	2.326	1.224	0.661
5.000	0.493		36.1	14.40	109.8	6.697	1.539

m/mol kg^{-1}	$Mn(ClO_4)_2$	NH_4Cl	NH_4ClO_4	$(NH_4)_2HPO_4$**	NH_4NO_3	NaBr	$NaBrO_3$
0.001	0.892	0.965	0.964	0.882	0.964	0.965	0.965
0.002	0.858	0.952	0.950	0.839	0.951	0.952	0.951
0.005	0.801	0.927	0.924	0.763	0.925	0.928	0.926
0.010	0.752	0.901	0.895	0.688	0.897	0.903	0.900
0.020	0.700	0.869	0.859	0.600	0.862	0.873	0.867
0.050	0.637	0.816	0.794	0.469	0.801	0.824	0.811
0.100	0.604	0.769	0.734	0.367	0.744	0.783	0.759
0.200	0.596	0.718	0.663	0.273	0.678	0.742	0.698
0.500	0.686	0.649	0.560	0.171	0.582	0.697	0.605
1.000	1.030	0.603	0.479	0.114	0.502	0.687	0.528
2.000	3.072	0.569	0.399	0.074	0.419	0.730	0.449
5.000		0.563			0.303	1.083	
10.000					0.220		
15.000					0.179		
20.000					0.154		

MEAN ACTIVITY COEFFICIENTS OF ELECTROLYTES AS A FUNCTION OF CONCENTRATION (continued)

m/mol kg^{-1}	Na_2CO_3	NaCl	$NaClO_3$	$NaClO_4$	Na_2CrO_4	NaF	Na_2HPO_4*
0.001	0.887	0.965	0.965	0.965	0.887	0.965	0.887
0.002	0.847	0.952	0.952	0.952	0.849	0.951	0.848
0.005	0.780	0.928	0.927	0.928	0.783	0.926	0.780
0.010	0.716	0.903	0.902	0.903	0.722	0.901	0.717
0.020	0.644	0.872	0.870	0.872	0.653	0.868	0.644
0.050	0.541	0.822	0.818	0.821	0.554	0.813	0.539
0.100	0.462	0.779	0.771	0.777	0.479	0.764	0.456
0.200	0.385	0.734	0.719	0.729	0.406	0.710	0.373
0.500	0.292	0.681	0.646	0.668	0.318	0.633	0.266
1.000	0.229	0.657	0.590	0.630	0.261	0.573	0.191
2.000	0.182	0.668	0.537	0.608	0.231		0.133
5.000		0.874		0.648			

m/mol kg^{-1}	NaI	$NaNO_3$	NaOH	Na_2SO_3	Na_2SO_4	Na_2WO_4	$NiBr_2$
0.001	0.965	0.965	0.965	0.887	0.886	0.886	0.889
0.002	0.952	0.951	0.952	0.847	0.846	0.846	0.853
0.005	0.928	0.926	0.927	0.779	0.777	0.777	0.791
0.010	0.904	0.900	0.902	0.716	0.712	0.712	0.735
0.020	0.874	0.866	0.870	0.644	0.637	0.638	0.675
0.050	0.827	0.810	0.819	0.540	0.529	0.534	0.596
0.100	0.789	0.759	0.775	0.462	0.446	0.457	0.546
0.200	0.753	0.701	0.731	0.386	0.366	0.388	0.514
0.500	0.722	0.617	0.685	0.296	0.268	0.320	0.535
1.000	0.734	0.550	0.674	0.237	0.204	0.291	0.692
2.000	0.823	0.480	0.714	0.196	0.155	0.291	1.476
5.000	1.402	0.388	1.076				
10.000	4.011	0.329	3.258				
15.000			9.796				
20.000			19.410				

m/mol kg^{-1}	$NiCl_2$	$Ni(ClO_4)_2$	$Ni(NO_3)_2$	$Pb(ClO_4)_2$	$Pb(NO_3)_2$	RbBr	RbCl
0.001	0.889	0.891	0.889	0.889	0.882	0.965	0.965
0.002	0.852	0.855	0.851	0.851	0.840	0.951	0.951
0.005	0.789	0.797	0.787	0.787	0.764	0.926	0.926
0.010	0.732	0.745	0.730	0.729	0.690	0.900	0.900
0.020	0.669	0.690	0.666	0.666	0.604	0.866	0.867
0.050	0.584	0.621	0.581	0.580	0.476	0.811	0.811
0.100	0.527	0.582	0.524	0.522	0.379	0.760	0.761
0.200	0.482	0.567	0.481	0.476	0.291	0.705	0.707
0.500	0.465	0.639	0.467	0.438	0.195	0.630	0.633
1.000	0.538	0.946	0.528	0.516	0.136	0.578	0.583
2.000	0.915	2.812	0.797	0.799		0.535	0.546
5.000	4.785			4.043		0.514	0.544
10.000				33.8			

m/mol kg^{-1}	RbF	RbI	$RbNO_3$	Rb_2SO_4	$SrBr_2$	$SrCl_2$	SrI_2
0.001	0.965	0.965	0.964	0.886	0.889	0.888	0.890
0.002	0.952	0.951	0.950	0.845	0.852	0.850	0.854
0.005	0.927	0.926	0.924	0.776	0.790	0.785	0.793
0.010	0.902	0.900	0.896	0.710	0.734	0.725	0.740
0.020	0.871	0.866	0.859	0.635	0.673	0.659	0.681

MEAN ACTIVITY COEFFICIENTS OF ELECTROLYTES AS A FUNCTION OF CONCENTRATION (continued)

m/mol kg^{-1}	RbF	RbI	$RbNO_3$	Rb_2SO_4	$SrBr_2$	$SrCl_2$	SrI_2
0.050	0.821	0.810	0.795	0.526	0.591	0.569	0.606
0.100	0.780	0.759	0.733	0.443	0.535	0.506	0.557
0.200	0.739	0.703	0.657	0.365	0.492	0.455	0.526
0.500	0.701	0.627	0.536	0.274	0.476	0.421	0.542
1.000	0.697	0.574	0.430	0.217	0.545	0.451	0.686
2.000	0.724	0.532	0.320		0.921	0.650	
5.000		0.517					

m/mol kg^{-1}	UO_2Cl_2	$UO_2(NO_3)_2$	$ZnBr_2$	$ZnCl_2$	ZnI_2
0.001	0.888	0.888	0.890	0.887	0.893
0.002	0.851	0.849	0.854	0.847	0.859
0.005	0.787	0.784	0.794	0.781	0.804
0.010	0.729	0.726	0.741	0.719	0.757
0.020	0.666	0.663	0.683	0.652	0.708
0.050	0.583	0.583	0.606	0.561	0.644
0.100	0.529	0.535	0.553	0.499	0.601
0.200	0.493	0.509	0.515	0.447	0.574
0.500	0.501	0.532	0.516	0.384	0.635
1.000	0.601	0.673	0.558	0.330	0.836
2.000	0.948	1.223	0.578	0.283	1.062
5.000		3.020	0.788	0.342	1.546
10.000			2.317	0.876	4.698
15.000			5.381	1.914	
20.000			7.965	2.968	

* The anion is $H_2PO_4^-$.

** The anion is HPO_4^{-2}.

ENTHALPY OF DILUTION OF ACIDS

The quantity given in this table is $-\Delta_{dil}H$, the negative of the enthalpy (heat) of dilution to infinite dilution for aqueous solutions of several common acids; i.e., the negative of the enthalpy change when a solution of molality m at a temperature of 25°C is diluted with an infinite amount of water. The tabulated numbers thus represent the heat produced (or, if the value is negative, the heat absorbed) when the acid is diluted. The initial molality m is given in the first column. The second column gives the dilution ratio, which is the number of moles of water that must be added to one mole of the acid to produce a solution of the molality in the first column.

REFERENCE

Parker, V. B., *Thermal Properties of Aqueous Uni-Univalent Electrolytes*, Natl. Stand. Ref. Data Ser. - Natl. Bur. Stand. (U.S.) 2, U.S. Government Printing Office, 1965.

$-\Delta_{dil}H$ in kJ/mol at 25°C

m	Dil. ratio	HF	HCl	$HClO_4$	HBr	HI	HNO_3	CH_2O_2	$C_2H_4O_2$
55.506	1.0		45.61		48.83		19.73	0.046	2.167
20	2.775	14.88	19.87	13.81	19.92	21.71	9.498	0.038	2.075
15	3.700	14.34	15.40	7.920	14.29	14.02	6.883	0.109	1.962
10	5.551	13.87	10.24	2.013	8.694	7.615	3.933	0.205	1.824
9	6.167	13.81	9.213	1.280	7.719	6.569	3.368	0.230	1.782
8	6.938	13.77	8.201	0.611	6.786	5.607	2.791	0.255	1.724
7	7.929	13.73	7.217	0.046	5.925	4.728	2.251	0.272	1.648
6	9.251	13.69	6.268	-0.351	5.004	3.975	1.749	0.280	1.540
5.5506	10	13.66	5.841	-0.490	4.590	3.577	1.540	0.285	1.477
5	11.10	13.62	5.318	0.620	4.113	3.197	1.310	0.289	1.393
4.5	12.33	13.58	4.899	-0.732	3.711	2.828	1.109	0.289	1.310
4	13.88	13.53	4.402	-0.787	3.330	2.460	0.958	0.289	1.218
3.5	15.86	13.47	3.958	-0.820	2.966	2.105	0.791	0.289	1.121
3	18.50	13.45	3.506	-0.782	2.611	1.787	0.665	0.289	1.025
2.5	22.20	13.43	3.063	-0.724	2.301	1.527	0.582	0.285	0.912
2	27.75	13.40	2.623	-0.623	1.996	1.318	0.527	0.276	0.803
1.5	37.00	13.36	2.167	-0.431	1.665	1.125	0.506	0.259	0.678
1	55.51	13.30	1.695	-0.201	1.314	0.933	0.506	0.226	0.544
0.5551	100	13.22	1.234	0.050	0.983	0.736	0.502	0.184	0.423
0.5	111.0	13.20	1.172	0.075	0.941	0.711	0.498	0.176	0.406
0.2	277.5	13.09	0.761	0.247	0.649	0.536	0.439	0.146	0.331
0.1	555.1	12.80	0.556	0.272	0.498	0.439	0.372	0.134	0.289
0.0925	600	12.79	0.540	0.272	0.481	0.427	0.368	0.134	0.285
0.0793	700	12.70	0.502	0.272	0.452	0.402	0.351	0.134	0.285
0.0694	800	12.61	0.473	0.268	0.427	0.385	0.339	0.130	0.280
0.0617	900	12.50	0.448	0.264	0.406	0.368	0.326	0.126	0.276
0.05551	1000	12.42	0.427	0.259	0.385	0.351	0.318	0.121	0.272
0.05	1110	12.24	0.406	0.259	0.372	0.339	0.305	0.121	0.272
0.02775	2000	11.29	0.310	0.226	0.285	0.264	0.247	0.117	0.264
0.01850	3000	10.66	0.251	0.197	0.234	0.218	0.213	0.117	0.259
0.01388	4000	10.25	0.226	0.180	0.205	0.192	0.192	0.113	0.259
0.01110	5000	9.874	0.197	0.167	0.184	0.172	0.176	0.109	0.255
0.00555	10000	8.912	0.142	0.126	0.130	0.121	0.130	0.105	0.243
0.00278	20000	7.531	0.105	0.092	0.092	0.084	0.096	0.096	0.230
0.00111	50000	5.439	0.067	0.059	0.054	0.050	0.063	0.084	0.222
0.000555	100000	3.766	0.042	0.042	0.038	0.038	0.046	0.054	0.209
0.000111	500000	1.255	0.021	0.021	0.021	0.021	0.021	0.038	0.167
0	∞	0	0	0	0	0	0	0	0

ENTHALPY OF SOLUTION OF ELECTROLYTES

This table gives the molar enthalpy (heat) of solution at infinite dilution for some common uni-univalent electrolytes. This is the enthalpy change when 1 mol of solute in its standard state is dissolved in an infinite amount of water. Values are given in kilojoules per mole at 25°C.

REFERENCE

Parker, V. B., *Thermal Properties of Uni-Univalent Electrolytes*, Natl. Stand. Ref. Data Series — Natl. Bur. Stand.(U.S.), No.2, 1965.

Solute	State	$\Delta_{sol}H°$ kJ/mol
HF	g	–61.50
HCl	g	–74.84
$HClO_4$	l	–88.76
$HClO_4 \cdot H_20$	c	–32.95
HBr	g	–85.14
HI	g	–81.67
HIO_3	c	8.79
HNO_3	l	–33.28
HCOOH	l	–0.86
CH_3COOH	l	–1.51
NH_3	g	–30.50
NH_4Cl	c	14.78
NH_4ClO_4	c	33.47
NH_4Br	c	16.78
NH_4I	c	13.72
NH_4IO_3	c	31.80
NH_4NO_2	c	19.25
NH_4NO_3	c	25.69
$NH_4C_2H_3O_2$	c	–2.38
NH_4CN	c	17.57
NH_4CNS	c	22.59
CH_3NH_3Cl	c	5.77
$(CH_3)_3NHCl$	c	1.46
$N(CH_3)_4Cl$	c	4.08
$N(CH_3)_4Br$	c	24.27
$N(CH_3)_4I$	c	42.07
$AgClO_4$	c	7.36
$AgNO_2$	c	36.94
$AgNO_3$	c	22.59
LiOH	c	–23.56
$LiOH \cdot H_2O$	c	–6.69
LiF	c	4.73
LiCl	c	–37.03
$LiCl \cdot H_2O$	c	–19.08
$LiClO_4$	c	–26.55
$LiClO_4 \cdot 3H_2O$	c	32.61
LiBr	c	–48.83
$LiBr \cdot H_2O$	c	–23.26
$LiBr \cdot 2H_2O$	c	–9.41
$LiBrO_3$	c	1.42
LiI	c	–63.30
$LiI \cdot H_2O$	c	–29.66
$LiI \cdot 2H_2O$	c	–14.77
$LiI \cdot 3H_2O$	c	0.59
$LiNO_2$	c	–11.00
$LiNO_2 \cdot H_2O$	c	7.03
$LiNO_3$	c	–2.51
NaOH	c	–44.51
$NaOH \cdot H_2O$	c	–21.41
NaF	c	0.91
NaCl	c	3.88
$NaClO_2$	c	0.33
$NaClO_2 \cdot 3H_2O$	c	28.58
$NaClO_3$	c	21.72
$NaClO_4$	c	13.88
$NaClO_4 \cdot H_2O$	c	22.51
NaBr	c	–0.60
$NaBr \cdot 2H_2O$	c	18.64
$NaBrO_3$	c	26.90
NaI	c	–7.53
$NaI \cdot 2H_2O$	c	16.13
$NaIO_3$	c	20.29
$NaNO_2$	c	13.89
$NaNO_3$	c	20.50
$NaC_2H_3O_2$	c	–17.32
$NaC_2H_3O_2 \cdot 3H_2O$	c	19.66
NaCN	c	1.21
$NaCN \cdot 0.5H_2O$	c	3.31
$NaCN \cdot 2H_2O$	c	18.58
NaCNO	c	19.20
NaCNS	c	6.83
KOH	c	–57.61
$KOH \cdot H_2O$	c	–14.64
$KOH \cdot 1.5H_2O$	c	–10.46
KF	c	–17.73
$KF \cdot 2H_2O$	c	6.97
KCl	c	17.22
$KClO_3$	c	41.38
$KClO_4$	c	51.04
KBr	c	19.87
$KBrO_3$	c	41.13
KI	c	20.33
KIO_3	c	27.74
KNO_2	c	13.35
KNO_3	c	34.89
$KC_2H_3O_2$	c	–15.33
KCN	c	11.72
KCNO	c	20.25
KCNS	c	24.23
$KMnO_4$	c	43.56
RbOH	c	–62.34
$RbOH \cdot H_2O$	c	–17.99
$RbOH \cdot 2H_2O$	c	0.88
RbF	c	–26.11
$RbF \cdot H_2O$	c	–0.42
$RbF \cdot 1.5H_2O$	c	1.34
RbCl	c	17.28
$RbClO_3$	c	47.74
$RbClO_4$	c	56.74
RbBr	c	21.88
$RbBrO_3$	c	48.95
RbI	c	25.10
$RbNO_3$	c	36.48
CsOH	c	–71.55
$CsOH \cdot H_2O$	c	–20.50
CsF	c	–36.86
$CsF \cdot H_2O$	c	–10.46
$CsF \cdot 1.5H_2O$	c	–5.44
CsCl	c	17.78
$CsClO_4$	c	55.44
CsBr	c	25.98
$CsBrO_3$	c	50.46
CsI	c	33.35
$CsNO_3$	c	40.00

LOWERING OF VAPOR PRESSURE BY SALTS IN AQUEOUS SOLUTIONS

The table gives the reduction of the vapor pressure in mmHg due to the presence of the number of moles of salt per liter of water given at the head of the columns, at the temperature 100° C, at which temperature the vapor pressure of pure water is 760 mmHg.

(From Smithsonian Tables.)

Substance	0.5	1.0	2.0	3.0	4.0	5.0	6.0	8.0	10.0
$Al_2(SO_4)_3$	12.8	36.5							
$AlCl_3$	22.5	61.0	179.0	318.0					
BaS_2O_6	6.6	15.4	34.4						
$Ba(OH)_2$	12.3	22.5	39.0						
$Ba(NO_3)_2$	13.5	27.0							
$Ba(ClO_3)_2$	15.8	33.3	70.5	108.2					
$BaCl_2$	16.4	36.7	77.6						
$BaBr_2$	16.8	38.8	91.4	150.0	204.7				
CaS_2O_3	9.9	23.0	56.0	106.0					
$Ca(NO_3)_2$	16.4	34.8	74.6	139.3	161.7	205.4			
$CaCl_2$	17.0	39.8	95.3	166.6	241.5	319.5			
$CaBr_2$	17.7	44.2	105.8	191.0	283.3	368.5			
$CdSO_4$	4.1	8.9	18.1						
CdI_2	7.6	14.8	33.5	52.7					
$CdBr_2$	8.6	17.8	36.7	55.7	80.0				
$CdCl_2$	9.6	18.8	36.7	57.0	77.3	99.0			
$Cd(NO_3)_2$	15.9	30.1	78.0	122.2					
$Cd(ClO_3)_2$	17.5								
$CoSO_4$	5.5	10.7	22.9	45.5					
$CoCl_2$	15.0	34.8	83.0	136.0	186.4				
$Co(NO_3)_2$	17.3	39.2	89.0	152.0	218.7	282.0	332.0		
$FeSO_4$	5.8	10.7	24.0	42.4					
H_3BO_3	6.0	12.3	25.1	38.0	51.0				
H_3PO_4	6.6	14.0	28.6	45.2	62.0	81.5	103.0	146.9	189.5
H_3AsO_4	7.3	15.0	30.2	46.4	64.9				
H_2SO_4	12.9	26.5	62.8	104.0	148.0	198.4	247.0	343.2	
KH_2PO_4	10.2	19.5	33.3	47.8	60.5	73.1	85.2		
KNO_3	10.3	21.1	40.1	57.6	74.5	88.2	102.1	126.3	148.0
$KClO_3$	10.6	21.6	42.8	62.1	80.0				
$KBrO_3$	10.9	22.4	45.0						
$KHSO_4$	10.9	21.9	43.3	65.3	85.5	107.8	129.9	170.0	
KNO_2	11.1	22.8	44.8	67.0	90.0	110.5	130.7	167.0	198.8
$KClO_4$	11.5	22.3							
KCl	12.2	24.4	48.8	74.1	100.9	128.5	152.2		
$KHCO_3$	11.6	23.6	59.0	77.6	104.2	132.0	160.0	210.0	255.0
KI	12.5	25.3	52.2	82.6	112.2	141.5	171.8	225.5	278.5
$K_2C_2O_4$	13.9	28.3	59.8	94.2	131.0				
K_2WO_4	13.9	33.0	75.0	123.8	175.4	226.4			
K_2CO_3	14.4	31.0	68.3	105.5	152.0	209.0	258.5	350.0	
KOH	15.0	29.5	64.0	99.2	140.0	181.8	223.0	309.5	387.8
K_2CrO_4	16.2	29.5	60.0						
$LiNO_3$	12.2	25.9	55.7	88.9	122.2	155.1	188.0	253.4	309.2
$LiCl$	12.1	25.5	57.1	95.0	132.5	175.5	219.5	311.5	393.5
$LiBr$	12.2	26.2	60.0	97.0	140.0	186.3	241.5	341.5	438.0
Li_2SO_4	13.3	28.1	56.8	89.0					
$LiHSO_4$	12.8	27.0	57.0	93.0	130.0	168.0			
LiI	13.6	28.6	64.7	105.2	154.5	206.0	264.0	357.0	445.0
Li_2SiFl_6	15.4	34.0	70.0	106.0					

Substance	0.5	1.0	2.0	3.0	4.0	5.0	6.0	8.0	10.0
$LiOH$	15.9	37.4	78.1						
Li_2CrO_4	16.4	32.6	74.0	120.0	171.0				
$MgSO_4$	6.5	12.0	24.5	47.5					
$MgCl_2$	16.8	39.0	100.5	183.3	277.0	377.0			
$Mg(NO_3)_2$	17.6	42.0	101.0	174.8					
$MgBr_2$	17.9	44.0	115.8	205.3	298.5				
$MgH_2(SO_4)_2$	18.3	46.0	116.0						
$MnSO_4$	6.0	10.5	21.0						
$MnCl_2$	15.0	34.0	76.0	122.3	107.0	209.0			
NaH_2PO_4	10.5	20.0	36.5	51.7	66.8	82.0	96.5	126.7	157.1
$NaHSO_4$	10.9	22.1	47.3	75.0	100.2	126.1	148.5	189.7	231.4
$NaNO_3$	10.6	22.5	46.2	68.1	90.3	111.5	131.7	167.8	198.8
$NaClO_3$	10.5	23.0	48.4	73.5	98.5	123.3	147.5	196.5	223.5
$(NaPO_3)_6$	11.6								
$NaOH$	11.8	22.8	48.2	77.3	107.5	139.1	172.5	243.3	314.0
$NaNO_2$	11.6	24.4	50.0	75.0	98.2	122.5	146.5	189.0	226.2
Na_2HPO_4	12.1	23.5	43.0	60.0	78.7	99.8	122.1		
$NaHCO_3$	12.9	24.1	48.2	77.6	102.2	127.8	152.0	198.0	239.4
Na_2SO_4	12.6	25.0	48.9	74.2					
$NaCl$	12.3	25.2	52.1	80.0	111.0	143.0	176.5		
$NaBrO_3$	12.1	25.0	54.1	81.3	108.8	136.0			
$NaBr$	12.6	25.9	57.0	89.2	124.2	159.5	197.5	268.0	
NaI	12.1	25.6	60.2	99.5	136.7	177.5	221.0	301.5	370.0
$Na_4P_2O_7$	13.2	22.0							
Na_2CO_3	14.3	27.3	53.5	80.2	111.0				
$Na_2C_2O_4$	14.5	30.0	65.8	105.8	146.0				
Na_2WO_4	14.8	33.6	71.6	115.7	162.6				
Na_3PO_4	16.5	30.0	52.5						
$(NaPO_3)_3$	17.1	36.5							
NH_4NO_3	12.8	22.0	42.1	62.7	82.9	103.8	121.0	152.2	180.0
$(NH_4)_2SiFl_6$	11.5	25.0	44.5						
NH_4Cl	12.0	23.7	45.1	69.3	94.2	118.5	138.2	179.0	213.8
NH_4HSO_4	11.5	22.0	46.8	71.0	94.5	118.	139.0	181.2	218.0
$(NH_4)_2SO_4$	11.0	24.0	46.5	69.5	93.0	117.0	141.8		
NH_4Br	11.9	23.9	48.8	74.1	99.4	121.5	145.5	190.2	228.5
NH_4I	12.9	25.1	49.8	78.5	104.5	132.3	156.0	200.0	243.5
$NiSO_4$	5.0	10.2	21.5						
$NiCl_2$	16.1	37.0	86.7	147.0	212.8				
$Ni(NO_3)_2$	16.1	37.3	91.3	156.2	235.0				
$Pb(NO_3)_2$	12.3	23.5	45.0	63.0					
$Sr(SO_3)_2$	7.2	20.3	47.0						
$Sr(NO_3)_2$	15.8	31.0	64.0	97.4	131.4				
$SrCl_2$	16.8	38.8	91.4	156.8	223.3	281.5			
$SrBr_2$	17.8	42.0	101.1	179.0	267.0				
$ZnSO_4$	4.9	10.4	21.5	42.1	66.2				
$ZnCl_2$	9.2	18.7	46.2	75.0	107.0	153.0	195.0		
$Zn(NO_3)_2$	16.6	39.0	93.5	157.5	223.8				

KINETIC AND PHOTOCHEMICAL DATA FOR ATMOSPHERIC CHEMISTRY

R. Atkinson, D. L. Baulch, R. A. Cox, R. F. Hampson, Jr., J. A. Kerr, and J. Troe
IUPAC Subcommittee on Gas Kinetic Data Evaluation for Atmospheric Chemistry

The International Union of Pure and Applied Chemistry (IUPAC) maintains a Subcommittee on Gas Kinetic Data evaluation for Atmospheric Chemistry which issues periodic recommendations on rate constants and related data for chemical reactions that are important in the chemistry of the atmosphere. These recommended values are based on the latest (1992) evaluation of all relevant data reported in the literature. The table below provides a summary of the recommended values and their assigned uncertainties. Full details of the data and evaluation process may be found in the references.

This table is reprinted with permission of IUPAC and of the American Institute of Physics and American Chemical Society, who are the publishers of the *Journal of Physical and Chemical Reference Data*.

REFERENCE

1. Atkinson, R., et al., *Journal of Physical and Chemical Reference Data*, 21, 1125-1568, 1992; 18, 881-1097, 1989; 13, 1259-1380, 1984; 11, 327-496, 1982; 9, 295-471, 1980.

Reaction	k_{298} cm³ molecule⁻¹ s⁻¹	$\Delta\log k_{298}$	Temp. dependence of k/cm³ molecule⁻¹ s⁻¹	Temp. range/K	$\Delta(E/R)$/K
O_x Reactions					
$O + O_2 + M \rightarrow O_3 + M$	6.0×10^{-34} [O_2] (k_0)	±0.05	$6.0 \times 10^{-34}(T/300)^{-2.8}[O_2]$	100–300	Δn = ±0.5
	5.6×10^{-34} [N_2] (k_0)	±0.05	$5.6 \times 10^{-34}(T/300)^{-2.8}[N_2]$	100–300	Δn = ±0.5
$O + O_2 \rightarrow O_3^*$	See previous evaluation				
$O_3^* + M \rightarrow O_3 + M$	See previous evaluation				
$O + O_3 \rightarrow 2\,O_2$	8.0×10^{-15}	±0.08	$8.0 \times 10^{-12}\exp(-2060/T)$	200–400	±200
$O + O_3^* \rightarrow$ products	See previous evaluation				
$O(^1D) + O_2 \rightarrow O(^3P) + O_2$	4.0×10^{-11}	±0.05	$3.2 \times 10^{-11}\exp(67/T)$	200–350	±100
$O(^1D) + O_3 \rightarrow O_2 + 2\,O(^3P)$	1.2×10^{-10}	±0.1	2.4×10^{-10}	100–400	$\Delta\log k$ = ±0.05
$\rightarrow 2\,O_2(^3\Sigma_g^-)$	1.2×10^{-10}	±0.1			
$O_2^* + O_3 \rightarrow O + 2\,O_2$	See data sheet				
$O_2(^1\Delta_g) + M \rightarrow O_2(^3\Sigma_g^-) + M$	1.6×10^{-18} (M = O_2)	±0.2	$3.0 \times 10^{-18}\exp(-200/T)$	100–450	±200
	$\leqslant 1.4 \times 10^{-19}$ (M = N_2)				
	5×10^{-18} (M = H_2O)	±0.3			
	$\leqslant 2 \times 10^{-20}$ (M = CO_2)				
$O_2(^1\Sigma_g^+) + M \rightarrow O_2(^3\Sigma_g^-) + M$	See previous evaluation				
$O_2(^1\Sigma_g^+) + O_3 \rightarrow$ products	See previous evaluation				
$O_2(^1\Sigma_g^+)^* + O_2 \rightarrow O_2(^1\Sigma_g^+) + O_2$	See previous evaluation				
$O_2 + h\nu \rightarrow$ products	See data sheets				
$O_3 + h\nu \rightarrow$ products	See data sheets				
HO_x Reactions					
$H + HO_2 \rightarrow H_2 + O_2$	5.6×10^{-12}	±0.5	5.6×10^{-12}	245–300	$\Delta\log k$ = ±0.5
$\rightarrow 2\,HO$	7.2×10^{-11}	±0.1	7.2×10^{-11}	245–300	$\Delta\log k$ = ±0.1
$\rightarrow H_2O + O$	2.4×10^{-12}	±0.5	2.4×10^{-12}	245–300	$\Delta\log k$ = ±0.5
$H + O_2 + M \rightarrow HO_2 + M$	$6.2 \times 10^{-32}[N_2]$ (k_0)	±0.05	$6.2 \times 10^{-32}(T/300)^{-1.6}[N_2]$	200–600	Δn = ±0.6
	7.5×10^{-11} (k_∞)	±0.3	7.5×10^{-11}	200–300	Δn = ±0.6
	$F_c = 0.55$	ΔF_c = ±0.15	$F_c = \exp(-T/498)$	200–300	
$H + O_3 \rightarrow HO + O_2$	See previous evaluation				
$H + O_3 \rightarrow HO^* + O_2$	See previous evaluation				
$O + H_2 \rightarrow HO + H$	See previous evaluation				
$O + HO \rightarrow O_2 + H$	3.3×10^{-11}	±0.1	$2.3 \times 10^{-11}\exp(110/T)$	220–500	±100
$O + HO_2 \rightarrow HO + O_2$	5.8×10^{-11}	±0.08	$2.7 \times 10^{-11}\exp(224/T)$	200–400	±100
$O + H_2O_2 \rightarrow HO + HO_2$	1.7×10^{-15}	±0.3	$1.4 \times 10^{-12}\exp(-2000/T)$	250–390	±1000
$O(^1D) + H_2 \rightarrow HO + H$	1.1×10^{-10}	±0.1	1.1×10^{-10}	200–350	±100
$O(^1D) + H_2O \rightarrow 2\,HO$	2.2×10^{-10}	±0.1	2.2×10^{-10}	200–350	±100
$HO + H_2 \rightarrow H_2O + H$	6.7×10^{-15}	±0.1	$7.7 \times 10^{-12}\exp(-2100/T)$	200–450	±200
$HO + H_2(v=1) \rightarrow H_2O + H$	See previous evaluation				
$HO + HO \rightarrow H_2O + O$	1.9×10^{-12}	±0.15	$4.2 \times 10^{-12}\exp(-240/T)$	250–500	±240
$HO + HO + M \rightarrow H_2O_2 + M$	$8 \times 10^{-31}[N_2]$ (k_0)	±0.3	$8 \times 10^{-31}(T/300)^{-0.8}[N_2]$	200–300	Δn = ±0.5
	3×10^{-11} (k_∞)	±0.3	3×10^{-11}	200–300	Δn = ±0.5
	$F_c = 0.5$		$F_c = 0.5$	200–300	
$HO + HO_2 \rightarrow H_2O + O_2$	1.1×10^{-10}	±0.1	$4.8 \times 10^{-11}\exp(250/T)$	250–400	±200
$HO + H_2O_2 \rightarrow H_2O + HO_2$	1.7×10^{-12}	±0.1	$2.9 \times 10^{-12}\exp(-160/T)$	240–460	±100
$HO + O_3 \rightarrow HO_2 + O_2$	6.7×10^{-14}	±0.15	$1.9 \times 10^{-12}\exp(-1000/T)$	220–450	±300
$HO^* + M \rightarrow HO + M$	See previous evaluation				
$HO^* + O_3 \rightarrow$ products	See previous evaluation				
$HO_2 + HO_2 \rightarrow H_2O_2 + O_2$	1.6×10^{-12}	±0.15	$2.2 \times 10^{-13}\exp(600/T)$	230–420	±200
$HO_2 + HO_2 + M \rightarrow H_2O_2 + O_2 + M$	5.2×10^{-32} [N_2]	±0.15	$1.9 \times 10^{-33}\exp(980/T)[N_2]$	230–420	±300
	4.5×10^{-32} [O_2]	±0.15			
	See data sheets for effect of H_2O				
$HO_2 + O_3 \rightarrow HO + 2\,O_2$	2.0×10^{-15}	±0.2	$1.4 \times 10^{-14}\exp(-600/T)$	250–350	+500 −100
$H_2O + h\nu \rightarrow HO + H$	See data sheets				
$H_2O_2 + h\nu \rightarrow 2\,HO$	See data sheets				

Reaction	k_{298} cm³ molecule⁻¹ s⁻¹		$\Delta\log k_{298}$	Temp. dependence of k/cm³ molecule⁻¹ s⁻¹	Temp. range/K	$\Delta(E/R)$/K
NO_x Reactions						
$O + NO + M \rightarrow NO_2 + M$	1.0×10^{-31} $[N_2]$	(k_0)	± 0.1	$1.0 \times 10^{-31}(T/300)^{-1.6}[N_2]$	200–300	$\Delta n = \pm 0.3$
	3.0×10^{-11}	(k_∞)	± 0.3	$3.0 \times 10^{-11}(T/300)^{0.3}$	200–1500	$\Delta n = \pm 0.3$
	$F_c = 0.85$		$\Delta F_c = \pm 0.1$	$F_c = \exp(-T/1850)$	200–300	
$O + NO_2 \rightarrow O_2 + NO$	9.7×10^{-12}		± 0.06	$6.5 \times 10^{-12}\exp(120/T)$	230–350	± 120
$O + NO_2 + M \rightarrow NO_3 + M$	9.0×10^{-32} $[N_2]$	(k_0)	± 0.1	$9.0 \times 10^{-32}(T/300)^{-2.0}[N_2]$	200–400	$\Delta n = \pm 1$
	2.2×10^{-11}	(k_∞)	± 0.2	2.2×10^{-11}	200–400	$\Delta n = \pm 0.5$
	$F_c = 0.8$		$\Delta F_c = \pm 0.1$	$F_c = \exp(-T/1300)$	200–400	
$O + NO_3 \rightarrow O_2 + NO_2$	1.7×10^{-11}		± 0.3			
$O + N_2O_5 \rightarrow$ products	See previous evaluation					
$O(^1D) + N_2 \rightarrow O(^3P) + N_2$	2.6×10^{-11}		± 0.1	$1.8 \times 10^{-11}\exp(107/T)$	200–350	± 100
$O(^1D) + N_2O \rightarrow N_2 + O_2$	4.4×10^{-11}		± 0.15	4.4×10^{-11}	200–350	± 100
$\rightarrow 2\ NO$	7.2×10^{-11}		± 0.15	7.2×10^{-11}	200–350	± 100
$N + HO \rightarrow NO + H$	See previous evaluation					
$N + O_2 \rightarrow NO + O$	See previous evaluation					
$N + O_2(^1\Delta_g) \rightarrow NO + O$	See previous evaluation					
$N + O_3 \rightarrow NO + O_2$	See previous evaluation					
$N + NO \rightarrow N_2 + O$	See previous evaluation					
$N + NO_2 \rightarrow N_2O + O$	See previous evaluation					
$HO + NH_3 \rightarrow H_2O + NH_2$	1.6×10^{-13}		± 0.1	$3.5 \times 10^{-12}\exp(-925/T)$	230–450	± 200
$HO + HONO \rightarrow H_2O + NO_2$	4.9×10^{-12}		± 0.3	$1.8 \times 10^{-11}\exp(-390/T)$	280–340	± 400
$HO + HONO_2 \rightarrow H_2O + NO_3$	1.5×10^{-13}	(1 bar)	± 0.1	See data sheets		
$HO + HO_2NO_2 \rightarrow$ products	5.0×10^{-12}		± 0.2	$1.5 \times 10^{-12}\exp(360/T)$	240–340	$^{+300}_{-600}$
$HO + NO + M \rightarrow HONO + M$	7.4×10^{-31} $[N_2]$	(k_0)	± 0.1	$7.4 \times 10^{-31}(T/300)^{-2.4}[N_2]$	200–300	$\Delta n = \pm 0.5$
	3.2×10^{-11}	(k_∞)	± 0.3	3.2×10^{-11}	200 400	$\Delta\log k = \pm 0.3$
	$F_c = 0.8$					
$HO + NO_2 + M \rightarrow HONO_2 + M$	2.6×10^{-30} $[N_2]$	(k_0)	± 0.1	$2.6 \times 10^{-30}(T/300)^{-2.9}[N_2]$	200–300	$\Delta n = \pm 0.5$
	6.0×10^{-11}	(k_∞)	± 0.1	6.0×10^{-11}	200–300	$\Delta n = \pm 0.5$
	$F_c = 0.43$					
$HO + NO_3 \rightarrow HO_2 + NO_2$	2.3×10^{-11}		± 0.2			
$HO_2 + NO \rightarrow HO + NO_2$	8.3×10^{-12}		± 0.1	$3.7 \times 10^{-12}\exp(240/T)$	230–500	± 100
$HO_2 + NO_2 + M \rightarrow HO_2NO_2 + M$	1.8×10^{-31} $[N_2]$	(k_0)	± 0.1	$1.8 \times 10^{-31}(T/300)^{[illegible]}[N_2]$	200–300	$\Delta n = \pm 1$
	4.7×10^{-12}	(k_∞)	± 0.2	4.7×10^{-12}	200–300	$\Delta n = \pm 1$
	$F_c = 0.6$			$F_c = 0.6$		
$HO_2NO_2 + M \rightarrow HO_2 + NO_2 + M$	1.3×10^{-20} $[N_2]$	(k_0/s^{-1})	± 0.3	$5 \times 10^{-6}\exp(-10000/T)[N_2]$	260–300	± 500
	0.34	(k_∞/s^{-1})	± 0.5	$2.6 \times 10^{15}\exp(-10900/T)$	260–300	± 500
	$F_c = 0.6$					
$HO_2 + NO_3 \rightarrow O_2 + HONO_2$ } $\rightarrow HO + NO_2 + O_2$ }	4.3×10^{-12}		± 0.2			
$NH_2 + HO \rightarrow$ products	See previous evaluation					
$NH_2 + HO_2 \rightarrow$ products	See previous evaluation					
$NH_2 + O_2 \rightarrow$ products	$<3 \times 10^{-18}$					
$NH_2 + O_3 \rightarrow$ products	1.7×10^{-13}		± 0.5	$4.9 \times 10^{-12}\exp(-1000/T)$	250–380	± 500
$NH_2 + NO \rightarrow$ products	1.6×10^{-11}		± 0.2	$1.6 \times 10^{-11}(T/298)^{-1.5}$	210–500	$\Delta n = \pm 0.5$
$NH_2 + NO_2 \rightarrow$ products	2.0×10^{-11}		± 0.2	$2.0 \times 10^{-11}(T/298)^{-2.0}$	250–500	$\Delta n = \pm 0.7$
$2\ NO + O_2 \rightarrow 2\ NO_2$	2.0×10^{-38} (cm⁶ molecule⁻² s⁻¹)		± 0.1	$3.3 \times 10^{-39}\exp(530/T)$	273 600	± 400
$NO + O_3 \rightarrow NO_2 + O_2$	1.8×10^{-14}		± 0.08	$1.8 \times 10^{-12}\exp(-1370/T)$	195–304	± 200
$NO + NO_3 \rightarrow 2\ NO_2$	2.6×10^{-11}		± 0.1	$1.8 \times 10^{-11}\exp(110/T)$	220–400	± 100
$NO_2 + O_3 \rightarrow NO_3 + O_2$	3.2×10^{-17}		± 0.06	$1.2 \times 10^{-13}\exp(-2450/T)$	230–360	± 150
$NO_2 + NO_3 + M \rightarrow N_2O_5 + M$	$2.7 \times 10^{-30}[N_2]$	(k_0)	± 0.1	$2.7 \times 10^{-30}(T/300)^{-3.4}[N_2]$	200–400	$\Delta n = \pm 0.5$
	2.0×10^{-12}	(k_∞)	± 0.2	$2.0 \times 10^{-12}(T/300)^{0.2}$	200–500	$\Delta n = \pm 0.6$
	$F_c = 0.33$			$F_c = [\exp(-T/250) + \exp(-1050/T)]$	200–500	
$N_2O_5 + M \rightarrow NO_2 + NO_3 + M$	$1.6 \times 10^{-19}[N_2]$	(k_0/s^{-1})	± 0.2	$2.2 \times 10^{-3}(T/300)^{-4.4}\exp(-11080/T)[N_2]$	220–300	$\Delta n = \pm 0.5$
	6.9×10^{-2}	(k_∞/s^{-1})	± 0.3	$9.7 \times 10^{14}(T/300)^{0.1}\exp(-11080/T)$	200–300	$\Delta n = \pm 0.2$
	$F_c = 0.33$			$F_c = [\exp(-T/250) + \exp(-1050/T)]$	200–300	
$N_2O_5 + H_2O \rightarrow 2\ HONO_2$	$<2 \times 10^{-21}$					
$HONO + h\nu \rightarrow$ products	See data sheets					
$HONO_2 + h\nu \rightarrow$ products	See data sheets					
$HO_2NO_2 + h\nu \rightarrow$ products	See data sheets					
$NO + h\nu \rightarrow$ products	See previous evaluation					
$NO_2 + h\nu \rightarrow$ products	See data sheets					
$NO_3 + h\nu \rightarrow$ products	See data sheets					
$N_2O + h\nu \rightarrow$ products	See data sheets					
$N_2O_5 + h\nu \rightarrow$ products	See data sheets					
Organic Reactions						
$O + CH_3 \rightarrow HCHO + H$	1.4×10^{-10}		± 0.1	1.4×10^{-10}	200–900	± 100
$O + CN \rightarrow CO + N(^2D)$	See previous evaluation					
$\rightarrow CO + N(^4S)$	See previous evaluation					
$O(^1D) + CH_4 \rightarrow HO + CH_3$	1.35×10^{-10}		± 0.1	1.35×10^{-10}	200–300	± 100
$\rightarrow HCHO + H_2$	1.5×10^{-11}		± 0.1	1.5×10^{-11}	200–300	± 100
$HO + CH_4 \rightarrow H_2O + CH_3$	7.0×10^{-15}		± 0.10	$3.9 \times 10^{-12}\exp(-1885/T)$	240–300	± 100
$HO + C_2H_2 + M \rightarrow C_2H_2OH + M$	$5 \times 10^{-30}[N_2]$	(k_0)	± 0.1	$5 \times 10^{-30}(T/300)^{-1.5}[N_2]$	200–300	$\Delta n = \pm 1.5$
	9.0×10^{-13}	(k_∞)	± 0.1	$9.0 \times 10^{-13}(T/300)^2$	200–300	$\Delta n = \pm 1$
	$F_c = 0.62$			$F_c = \exp(-T/623)$	200–300	

Reaction	k_{298} cm^3 molecule^{-1} s^{-1}		Δlog k_{298}	Temp. dependence of k/cm^3 molecule^{-1} s^{-1}	Temp. range/K	$\Delta(E/R)$/ K
$HO + C_2H_4 + M \rightarrow C_2H_4OH + M$	$7 \times 10^{-29}[N_2]$	(k_0)	±0.3	$7 \times 10^{-29}(T/300)^{-3.1}[N_2]$	200–300	$\Delta n = \pm 2$
	9×10^{-12}	(k_∞)	±0.3	9×10^{-12}	200–300	$\Delta n = \pm 0.5$
	$F_c = 0.7$					
$HO + C_2H_6 \rightarrow H_2O + C_2H_5$	2.5×10^{-13}		±0.10	$7.8 \times 10^{-12}\exp(-1020/T)$	240–300	±100
$HO + C_3H_6 + M \rightarrow C_3H_6OH + M$	$8 \times 10^{-27}[N_2]$	(k_0)	±1	$8 \times 10^{-27}(T/300)^{-3.5}[N_2]$	200–300	$\Delta n = \pm 1$
	3.0×10^{-11}	(k_∞)	±0.1	3.0×10^{-11}	200–300	$\Delta n = \pm 1$
	$F_c = 0.5$					
$HO + C_3H_8 \rightarrow H_2O + C_3H_7$	1.14×10^{-12}		±0.10	$9.8 \times 10^{-12}\exp(-640/T)$	~300	±150
$HO + CO \rightarrow H + CO_2$	1.5×10^{-13} (1 + 0.6 P/bar)		±0.1	1.5×10^{-13}(1 + 0.6 P/bar)	200–300	±300
$HO + HCHO \rightarrow H_2O + HCO$	9.6×10^{-12}		±0.10	$8.8 \times 10^{-12}\exp(25/T)$	240–300	±150
$HO + CH_3CHO \rightarrow H_2O + CH_3CO$	1.6×10^{-11}		±0.10	$5.6 \times 10^{-12}\exp(310/T)$	240–530	±200
$HO + C_2H_5CHO \rightarrow$ products	2.0×10^{-11}		±0.15			
$HO + (CHO)_2 \rightarrow$ products	1.1×10^{-11}		±0.3			
$HO + HOCH_2CHO \rightarrow H_2O + HOCH_2CO$	8.0×10^{-12}		±0.3			
$\rightarrow H_2O + HOCHCHO$	2.0×10^{-12}		±0.3			
$HO + CH_3COCHO \rightarrow H_2O + CH_3COCO$	1.7×10^{-11}		±0.3			
$HO + CH_3COCH_3 \rightarrow H_2O + CH_2COCH_3$	2.3×10^{-13}		±0.2	$1.7 \times 10^{-12}\exp(-600/T)$	240–440	±300
$HO + CH_3OH \rightarrow H_2O + CH_2OH$	7.8×10^{-13}		±0.15 }	$3.3 \times 10^{-12}\exp(-380/T)$	240–300	±200
$\rightarrow H_2O + CH_3O$	1.4×10^{-13}		±0.15 }			
$HO + C_2H_5OH \rightarrow H_2O + CH_2CH_2OH$	1.6×10^{-13}		±0.15 }			
$\rightarrow H_2O + CH_3CHOH$	2.9×10^{-12}		±0.15 }	$4.1 \times 10^{-12}\exp(-70/T)$	270–340	±200
$\rightarrow H_2O + CH_3CH_2O$	1.6×10^{-13}		±0.15 }			
$HO + n\text{-}C_3H_7OH \rightarrow$ products	5.5×10^{-12}		±0.2			
$HO + i\text{-}C_3H_7OH \rightarrow$ products	5.7×10^{-12}		±0.2	5.7×10^{-12}	240–440	±200
$HO + CH_3COCH_2OH \rightarrow$ products	3.0×10^{-12}		±0.3			
$HO + CH_3OOH \rightarrow H_2O + CH_2OOH$	1.9×10^{-12}		±0.2	$1.0 \times 10^{-12}\exp(190/T)$	220–430	±150
$\rightarrow H_2O + CH_3OO$	3.6×10^{-12}		±0.2	$1.9 \times 10^{-12}\exp(190/T)$	220–430	±150
$HO + HCOOH \rightarrow$ products	4.5×10^{-13}		±0.15	4.5×10^{-13}	290–450	±250
$HO + CH_3COOH \rightarrow$ products	8×10^{-13}		±0.3			
$HO + CH_3ONO_2 \rightarrow$ products	3.5×10^{-13}	(1 bar)	±0.10	$1.0 \times 10^{-14}\exp(1060/T)$ (1 bar)	290–400	±500
$HO + C_2H_5ONO_2 \rightarrow$ products	4.9×10^{-13}	(1 bar)	±0.15	$4.4 \times 10^{-14}\exp(720/T)$ (1 bar)	290–380	±500
$HO + n\text{-}C_3H_7ONO_2 \rightarrow$ products	7.3×10^{-13}	(1 bar)	±0.15	7.3×10^{-13} (1 bar)	290–370	±500
$HO + i\text{-}C_3H_7ONO_2 \rightarrow$ products	4.9×10^{-13}	(1 bar)	±0.25			
$HO + CH_3CO_3NO_2 \rightarrow$ products	1.1×10^{-13}		±0.2	$9.5 \times 10^{-13}\exp(-650/T)$	270–300	±400
$HO + HCN \rightarrow$ products	3×10^{-14}	(1 bar)	±0.5	$1.2 \times 10^{-13}(-400/T)$ (1 bar)	290–440	±300
$HO + CH_3CN \rightarrow$ products	2.2×10^{-14}	(1 bar)	±0.15	$8.1 \times 10^{-13}\exp(-1080/T)$ (1 bar)	250–390	±200
$HO_2 + CH_3O_2 \rightarrow O_2 + CH_3O_2H$	5.2×10^{-12}		±0.3	$3.8 \times 10^{-13}\exp(780/T)$	225–580	±500
$HO_2 + HOCH_2O_2 \rightarrow O_2 + HOCH_2O_2H$ $\rightarrow O_2 + HCO_2H + H_2O$ }	1.2×10^{-11}		±0.3	$5.6 \times 10^{-15}\exp(2300/T)$	275–335	±1500
$HO_2 + C_2H_5O_2 \rightarrow O_2 + C_2H_5O_2H$	5.8×10^{-12}		±0.2	$6.5 \times 10^{-13}\exp(650/T)$	240–380	±200
$HO_2 + CH_3CO_3 \rightarrow O_2 + CH_3CO_3H$	4.2×10^{-12}		±0.3	$1.3 \times 10^{-13}\exp(1040/T)$	250–370	±500
$\rightarrow O_3 + CH_3CO_2H$	1.0×10^{-11}		±0.3	$3.0 \times 10^{-13}\exp(1040/T)$	250–370	±500
$HO_2 + HOCH_2CH_2O_2 \rightarrow$ products	1.0×10^{-11}		±0.3			
$HO_2 + HCHO \rightarrow HOCH_2OO$	7.9×10^{-14}		±0.3	$9.7 \times 10^{-15}\exp(625/T)$	275–333	±600
$HOCH_2OO \rightarrow HO_2 + HCHO$	1.5×10^{2}	(k/s^{-1})	±0.3	$2.4 \times 10^{12}\exp(-7000/T)$	275–333	±2000
$NO_3 + C_2H_2 \rightarrow$ products	$<1 \times 10^{-16}$					
$NO_3 + C_2H_4 \rightarrow$ products	2.1×10^{-16}		±0.2	$3.3 \times 10^{-12}\exp(-2880/T)$	270–330	±500
$NO_3 + C_3H_6 \rightarrow$ products	9.4×10^{-15}		±0.2			
$NO_3 + HCHO \rightarrow HNO_3 + HCO$	5.8×10^{-16}		±0.3			
$NO_3 + CH_3CHO \rightarrow HNO_3 + CH_3CO$	2.7×10^{-15}		±0.2	$1.4 \times 10^{-12}\exp(-1860/T)$	260–370	±500
$NO_3 + CH_3OH \rightarrow$ products	2.4×10^{-16}		±0.5	$1.3 \times 10^{-12}\exp(-2560/T)$	290–480	±700
$NO_3 + C_2H_5OH \rightarrow$ products	$<2 \times 10^{-15}$					
$NO_3 + i\text{-}C_3H_7OH \rightarrow$ products	$<5 \times 10^{-15}$					
$CH_3 + O_2 + M \rightarrow CH_3O_2 + M$	$1.0 \times 10^{-30}[N_2]$	(k_0)	±0.2	$1.0 \times 10^{-30}(T/300)^{-3.3}[N_2]$	200–300	$\Delta n = \pm 1$
	2.2×10^{-12}	(k_∞)	±0.3	$2.2 \times 10^{-12}(T/300)^{1.0}$	200–300	$\Delta n = \pm 1$
	$F_c = 0.27$					
$C_2H_5 + O_2 \rightarrow C_2H_4 + HO_2$	3.8×10^{-15}	(1 bar air)	±0.5			
	1.9×10^{-14}	(0.133 bar air)	±0.5			
$C_2H_5 + O_2 + M \rightarrow C_2H_5O_2 + M$	$5.9 \times 10^{-29}[N_2]$	(k_0)	±0.3	$5.9 \times 10^{-29}(T/300)^{-3.8}[N_2]$	200–300	$\Delta n = \pm 1$
	7.8×10^{-12}	(k_∞)	±0.2	7.8×10^{-12}	200–300	Δlog k = ±0.2
	$F_c = 0.54$			$F_c = \{0.58\exp(-T/1250) + 0.42\exp(-T/183)\}$	200–300	
$n\text{-}C_3H_7 + O_2 + M \rightarrow n\text{-}C_3H_7O_2 + M$	8×10^{-12}	(k_∞)	±0.2	8×10^{-12}	200–300	Δlog k = ±0.2
$i\text{-}C_3H_7 + O_2 + M \rightarrow i\text{-}C_3H_7O_2 + M$	1.1×10^{-11}	(k_∞)	±0.3	1.1×10^{-11}	200–300	Δlog k = ±0.3
$CH_3COCH_2 + O_2 + M \rightarrow CH_3COCH_2O_2 + M$	1.5×10^{-12}	(k_∞)	±0.5			
$HCO + O_2 \rightarrow CO + HO_2$	5.5×10^{-12}		±0.15	5.5×10^{-12}	200–400	±150
$CH_3CO + O_2 + M \rightarrow CH_3CO_3 + M$	5.0×10^{-12}	(k_∞)	±0.5	5.0×10^{-12}	200–300	Δlog k = ±0.5
$CH_2OH + O_2 \rightarrow HCHO + HO_2$	9.4×10^{-12}		±0.12			
$CH_3CHOH + O_2 \rightarrow CH_3CHO + HO_2$	1.9×10^{-11}		±0.3			
$CH_2CH_2OH + O_2 \rightarrow$ products	3.0×10^{-12}		±0.3			
$CH_3O + O_2 \rightarrow HCHO + HO_2$	1.9×10^{-15}		±0.2	$7.2 \times 10^{-14}\exp(-1080/T)$	298–610	±300
$C_2H_5O + O_2 \rightarrow CH_3CHO + HO_2$	9.5×10^{-15}		±0.2	$6.0 \times 10^{-14}\exp(-550/T)$	295–425	±300
$n\text{-}C_3H_7O + O_2 \rightarrow C_2H_5CHO + HO_2$	8×10^{-15}		±0.5			
$i\text{-}C_3H_7O + O_2 \rightarrow CH_3COCH_3 + HO_2$	8×10^{-15}		±0.3	$1.5 \times 10^{-14}\exp(-200/T)$	290–390	±200
$CH_3 + O_3 \rightarrow$ products	2.5×10^{-12}		±0.3	$5.1 \times 10^{-12}\exp(-210/T)$	240–400	±200
$CH_3O + NO + M \rightarrow CH_3ONO + M$	$1.6 \times 10^{-29}[N_2]$	(k_0)	±0.1	$1.6 \times 10^{-29}(T/300)^{-3.5}[N_2]$	200–400	$\Delta n = \pm 0.5$
	3.6×10^{-11}	(k_∞)	±0.5	$3.6 \times 10^{-11}(T/300)^{-0.6}$	200–400	$\Delta n = \pm 0.5$
	$F_c = 0.6$					
$CH_3O + NO \rightarrow HCHO + HNO$	4×10^{-12}			$4 \times 10^{-12}(T/300)^{-0.7}$	200–400	$\Delta n = \pm 0.5$
$C_2H_5O + NO + M \rightarrow C_2H_5ONO + M$	4.4×10^{-11}	(k_∞)	±0.3	4.4×10^{-11}	200–300	$\Delta n = \pm 0.5$
$C_2H_5O + NO \rightarrow CH_3CHO + HNO$	1.3×10^{-11}					
$i\text{-}C_3H_7O + NO + M \rightarrow i\text{-}C_3H_7ONO + M$	3.4×10^{-11}	(k_∞)	±0.3	3.4×10^{-11}	200–300	$\Delta n = \pm 0.5$

Reaction	k_{298} cm³ molecule⁻¹ s⁻¹		Δlog k_{298}	Temp. dependence of k/cm³ molecule⁻¹ s⁻¹	Temp. range/K	Δ(E/R)/ K
$i\text{-}C_3H_7O + NO \rightarrow CH_3COCH_3 + HNO$	6.5×10^{-12}		±0.5			
$CH_3O + NO_2 + M \rightarrow CH_3ONO_2 + M$	$2.8 \times 10^{-29}[N_2]$	(k_0)	±0.3	$2.8 \times 10^{-29}(T/300)^{-4.5}[N_2]$	200–400	$\Delta n = \pm 1$
	2×10^{-11}	(k_∞)	±0.3	2×10^{-11}	200–400	$\Delta n = \pm 0.5$
	$F_c = 0.44$					
$CH_3O + NO_2 \rightarrow HCHO + HONO$	See data sheets					
$C_2H_5O + NO_2 + M \rightarrow C_2H_5ONO_2 + M$	2.8×10^{-11}	(k_∞)	±0.3	2.8×10^{-11}	200–300	$\Delta n = \pm 0.5$
$C_2H_5O + NO_2 \rightarrow CH_3CHO + HONO$	See data sheets					
$i\text{-}C_3H_7O + NO_2 + M \rightarrow i\text{-}C_3H_7ONO_2 + M$	3.5×10^{-11}	(k_∞)	±0.3	3.5×10^{-11}	200–300	$\Delta n = \pm 0.5$
$i\text{-}C_3H_7O + NO_2 \rightarrow CH_3COCH_3 + HONO$	See data sheets					
$CH_3O_2 + NO \rightarrow CH_3O + NO_2$	7.6×10^{-12}		±0.1	$4.2 \times 10^{-12}\exp(180/T)$	240–360	±180
$C_2H_5O_2 + NO \rightarrow C_2H_5O + NO_2$	8.9×10^{-12}		±0.3			
$C_2H_5O_2 + NO\ (+M) \rightarrow C_2H_5ONO_2\ (+M)$	$\leq 1.3 \times 10^{-13}$	(1 bar)				
$n\text{-}C_3H_7O_2 + NO \rightarrow n\text{-}C_3H_7O + NO_2$	8.7×10^{-12}		±0.3			
$n\text{-}C_3H_7O_2 + NO\ (+M) \rightarrow n\text{-}C_3H_7ONO_2$ $(+M)$	1.8×10^{-13}	(1 bar)	±0.5			
$i\text{-}C_3H_7O_2 + NO \rightarrow i\text{-}C_3H_7O + NO_2$	8.5×10^{-12}		±0.5			
$i\text{-}C_3H_7O_2 + NO\ (+M) \rightarrow i\text{-}C_3H_7ONO_2$ $(+M)$	3.7×10^{-13}	(1 bar)	±0.5			
$CH_3CO_3 + NO \rightarrow CH_3 + CO_2 + NO_2$	2.0×10^{-11}		±0.2	2.0×10^{-11}	280–325	±600
$CH_3O_2 + NO_2 + M \rightarrow CH_3O_2NO_2 + M$	$2.5 \times 10^{-30}[N_2]$	(k_0)	±0.3	$2.5 \times 10^{-30}(T/300)^{-5.5}[N_2]$	250–350	$\Delta n = \pm 1$
	7.5×10^{-12}	(k_∞)	±0.3	7.5×10^{-12}	250–350	$\Delta n = \pm 0.5$
	$F_c = 0.4$					
$CH_3O_2NO_2 + M \rightarrow CH_3O_2 + NO_2 + M$	$6.8 \times 10^{-19}[N_2]$	(k_0/s⁻¹)	±0.3	$9 \times 10^{-5}\exp(-9690/T)[N_2]$	250–300	±500
	4.5	(k_∞/s⁻¹)	±0.3	$1.1 \times 10^{16}\exp(-10560/T)$	250–300	±500
	$F_c = 0.4$					
$C_2H_5O_2 + NO_2 + M \rightarrow C_2H_5O_2NO_2 + M$	$1.3 \times 10^{-29}[N_2]$	(k_0)	±0.3	$1.3 \times 10^{-29}(T/300)^{-6.2}[N_2]$	200–300	$\Delta n = \pm 1$
	8.8×10^{-12}	(k_∞)	±0.3	8.8×10^{-12}	200–300	$\Delta \log k = \pm 0.3$
	$F_c = 0.31$			$F_c = 0.31$	250–300	
$C_2H_5O_2NO_2 + M \rightarrow C_2H_5O_2 + NO_2 + M$	$1.4 \times 10^{-17}[N_2]$	(k_0/s⁻¹)	±0.5	$4.8 \times 10^{-4}\exp(-9285/T)[N_2]$	250–300	±1000
	5.4	(k_∞/s⁻¹)	±0.5	$8.8 \times 10^{15}\exp(-10440/T)$	250–300	±1000
	$F_c = 0.31$			$F_c = 0.31$	250–300	
$CH_3CO_3 + NO_2 + M \rightarrow CH_3CO_3NO_2 + M$	$2.7 \times 10^{-28}[N_2]$	(k_0)	±0.4	$2.7 \times 10^{-28}(T/300)^{-7.1}[N_2]$	250–300	$\Delta n = \pm 2$
	1.2×10^{-11}	(k_∞)	±0.2	$1.2 \times 10^{-11}(T/300)^{-0.9}$	250–300	$\Delta n = \pm 1$
	$F_c = 0.3$					
$CH_3CO_3NO_2 + M \rightarrow CH_3CO_3 + NO_2 + M$	$1.1 \times 10^{-20}[N_2]$	(k_0/s⁻¹)	±0.4	$4.9 \times 10^{-3}\exp(-12100/T)[N_2]$	300–330	±1000
	6.1×10^{-4}	(k_∞/s⁻¹)	±0.2	$4.0 \times 10^{16}\exp(-13600/T)$	280–330	±200
	$F_c = 0.3$					
$CH_3O_2 + NO_3 \rightarrow$ products	No recommendation (see data sheets)					
$CH_3O_2 + CH_3O_2 \rightarrow CH_3OH + HCHO + O_2$ $\rightarrow 2\ CH_3O + O_2$ $\rightarrow CH_3OOCH_3 + O_2$	3.7×10^{-13}		±0.12	$1.1 \times 10^{-13}\exp(365/T)$	200–400	±200
$CH_3O_2 + CH_3CO_3 \rightarrow CH_3O + CH_3CO_2 + O_2$	5.5×10^{-12}		±0.5			
$\rightarrow CH_3CO_2H + HCHO + O_2$	5.5×10^{-12}		±0.5			
$HOCH_2O_2 + HOCH_2O_2 \rightarrow HCOOH + CH_2(OH)_2 + O_2$	7.0×10^{-13}		±0.3	$5.7 \times 10^{-14}\exp(750/T)$	275–325	±750
$\rightarrow 2\ HOCH_2O + O_2$	5.5×10^{-12}		±0.3			
$C_2H_5O_2 + C_2H_5O_2 \rightarrow C_2H_5OH + CH_3CHO + O_2$ $\rightarrow 2\ C_2H_5O + O_2$ $\rightarrow C_2H_5OOC_2H_5 + O_2$	6.8×10^{-14}		±0.12	$9.8 \times 10^{-14}\exp(-110/T)$	250–450	$^{+300}_{-100}$
$CH_3CO_3 + CH_3CO_3 \rightarrow 2\ CH_3CO_2 + O_2$	1.6×10^{-11}		±0.5	$2.8 \times 10^{-12}\exp(530/T)$	250–370	±500
$HOCH_2CH_2O_2 + HOCH_2CH_2O_2$						
$\rightarrow HOCH_2CH_2OH + HOCH_2CHO + O_2$	1.5×10^{-12}		±0.3			
$\rightarrow 2\ HOCH_2CH_2O + O_2$	8.3×10^{-13}		±0.3			
$n\text{-}C_3H_7O_2 + n\text{-}C_3H_7O_2 \rightarrow n\text{-}C_3H_7OH + C_2H_5CHO + O_2$ $\rightarrow 2\ n\text{-}C_3H_7O + O_2$	3×10^{-13}		±0.5			
$i\text{-}C_3H_7O_2 + i\text{-}C_3H_7O_2 \rightarrow i\text{-}C_3H_7OH + CH_3COCH_3 + O_2$	4.4×10^{-16}		±0.3	$1.6 \times 10^{-12}\exp(-2200/T)$	300–400	±300
$\rightarrow 2\ i\text{-}C_3H_7O + O_2$	5.6×10^{-16}		±0.3			
$CH_3COCH_2O_2 + CH_3COCH_2O_2$ $\rightarrow CH_3COCH_2OH + CH_3COCHO + O_2$ $\rightarrow 2\ CH_3COCH_2O + O_2$	$\leq 1 \times 10^{-11}$					
$RCHOO + H_2O \rightarrow RCOOH + H_2O$ $RCHOO + NO_2 \rightarrow RCHO + NO_3$ $RCHOO + SO_2 \rightarrow$ products $RCHOO + HCHO \rightarrow$ products	No recommendations (see data sheets)					
$CN + O_2 \rightarrow$ products	See previous evaluation					
$O_3 + C_2H_2 \rightarrow$ products	1×10^{-20}		±1.0			
$O_3 + C_2H_4 \rightarrow$ products	1.7×10^{-18}		±0.10	$1.2 \times 10^{-14}\exp(-2630/T)$	180–360	±100
$O_3 + C_3H_6 \rightarrow$ products	1.2×10^{-17}		±0.15	$6.5 \times 10^{-15}\exp(-1880/T)$	230–370	±400
$HCHO + h\nu \rightarrow$ products	See data sheets					
$CH_3CHO + h\nu \rightarrow$ products	See data sheets					
$C_2H_5CHO + h\nu \rightarrow$ products	See data sheets					
$(CHO)_2 + h\nu \rightarrow$ products	See data sheets					
$CH_3COCHO + h\nu \rightarrow$ products	See data sheets					
$CH_3COCH_3 + h\nu \rightarrow$ products	See data sheets					
$CH_3OOH + h\nu \rightarrow$ products	See data sheets					
$CH_3ONO_2 + h\nu \rightarrow$ products	See data sheets					
$C_2H_5ONO_2 + h\nu \rightarrow$ products	See data sheets					
$n\text{-}C_3H_7ONO_2 + h\nu \rightarrow$ products	See data sheets					
$i\text{-}C_3H_7ONO_2 + h\nu \rightarrow$ products	See data sheets					
$CH_3O_2NO_2 + h\nu \rightarrow$ products	See data sheets					
$CH_3CO_3NO_2 + h\nu \rightarrow$ products	See data sheets					

Reaction	k_{298} cm^3 molecule^{-1} s^{-1}		$\Delta\log k_{298}$	Temp. dependence of k/cm^3 molecule^{-1} s^{-1}	Temp. range/K	$\Delta(E/R)$/ K
SO_x Reactions						
$O + H_2S \rightarrow HO + HS$	See previous evaluation					
$O + CS \rightarrow CO + S$	2.1×10^{-11}		±0.1	$2.7 \times 10^{-10}\exp(-760/T)$	150–300	±250
$O + CH_3SCH_3 \rightarrow CH_3SO + CH_3$	5.0×10^{-11}		±0.1	$1.3 \times 10^{-11}\exp(409/T)$	270–560	±100
$O + CS_2 \rightarrow SO + CS$	3.6×10^{-12}		±0.2	$3.2 \times 10^{-11}\exp(-650/T)$	200–500	±100
$O + CH_3SSCH_3 \rightarrow CH_3SO + CH_3S$	1.3×10^{-10}		±0.3	$5.5 \times 10^{-11}\exp(250/T)$	290–570	±100
$O + OCS \rightarrow SO + CO$	1.2×10^{-14}		±0.2	$1.6 \times 10^{-11}\exp(-2150/T)$	220–500	±150
$O + SO_2 + M \rightarrow SO_3 + M$	$1.4 \times 10^{-33}[N_2]$	(k_0)	±0.3	$4.0 \times 10^{-32}\exp(-1000/T)[N_2]$	200–400	±200
$S + O_2 \rightarrow SO + O$	2.1×10^{-12}		±0.2	2.1×10^{-12}	230–400	±200
$S + O_3 \rightarrow SO + O_2$	1.2×10^{-11}		±0.3			
$Cl + H_2S \rightarrow HCl + HS$	5.7×10^{-11}		±0.3	5.7×10^{-11}	210–350	±100
$HO + H_2S \rightarrow H_2O + HS$	4.8×10^{-12}		±0.08	$6.3 \times 10^{-12}\exp(-80/T)$	200–300	±80
$HO + SO_2 + M \rightarrow HOSO_2 + M$	$4.0 \times 10^{-31}[N_2]$	(k_0)	±0.3	$4.0 \times 10^{-31}(T/300)^{-3.3}[N_2]$	300–400	$\Delta n = \pm1$
	$F_c = 0.8$					
	2×10^{-12}	(k_∞)	±0.3	2×10^{-12}	200–300	$\Delta\log k = \pm0.3$
	$F_c = 0.45$					
$HOSO_2 + O_2 \rightarrow HO_2 + SO_3$	4.0×10^{-13}		±0.1	$1.3 \times 10^{-12}\exp(-330/T)$	290–420	±200
$HO + OCS \rightarrow$ products	2.0×10^{-15}		±0.3	$1.1 \times 10^{-13}\exp(-1200/T)$	250–500	±500
$HO + CS_2 + M \rightarrow HOCS_2 + M$	$8 \times 10^{-31}[N_2]$	(k_0)	±0.5	$8 \times 10^{-31}[N_2]$	270–300	$\Delta\log k = \pm0.5$
	8×10^{-12}	(k_∞)	±0.5	8×10^{-12}	250–300	$\Delta\log k = \pm0.5$
	$F_c = 0.8$					
$HO + CS_2 \rightarrow HS + OCS$	$<2 \times 10^{-15}$					
$HOCS_2 + M \rightarrow HO + CS_2 + M$	$4.8 \times 10^{-14}[N_2]$	(k_0/s^{-1})	±0.5	$1.6 \times 10^{-6}\exp(-5160/T)[N_2]$	250–300	±500
	4.8×10^{5}	(k_∞/s^{-1})	±0.5	$1.6 \times 10^{13}\exp(-5160/T)$	250–300	±500
	$F_c = 0.8$					
$HOCS_2 + O_2 \rightarrow$ products	2.8×10^{-14}		±0.3	2.8×10^{-14}	240–300	$\Delta\log k = \pm0.3$
$HO + CH_3SH \rightarrow$ products	3.3×10^{-11}		±0.10	$9.9 \times 10^{-12}\exp(356/T)$	240–430	±100
$HO + CH_3SCH_3 \rightarrow H_2O + CH_2SCH_3$	4.4×10^{-12}		±0.10	$9.6 \times 10^{-12}\exp(-234/T)$	250–400	±300
$\rightarrow CH_3S(OH)CH_3$	See data sheets					
$HO + CH_3SSCH_3 \rightarrow$ products	2.0×10^{-10}		±0.10	$6.0 \times 10^{-11}\exp(380/T)$	250–370	±300
$HO_2 + SO_2 \rightarrow$ products	$\leq 1 \times 10^{-18}$					
$NO_3 + H_2S \rightarrow$ products	$<1 \times 10^{-15}$					
$NO_3 + CS_2 \rightarrow$ products	$<1 \times 10^{-15}$					
$NO_3 + OCS \rightarrow$ products	$<1 \times 10^{-16}$					
$NO_3 + SO_2 \rightarrow$ products	$<1 \times 10^{-19}$					
$NO_3 + CH_3SH \rightarrow$ products	9.2×10^{-13}		±0.15	9.2×10^{-13}	250–370	±400
$NO_3 + CH_3SCH_3 \rightarrow$ products	1.1×10^{-12}		±0.15	$1.9 \times 10^{-13}\exp(520/T)$	250–380	±200
$NO_3 + CH_3SSCH_3 \rightarrow$ products	7×10^{-13}		±0.3	7×10^{-13}	300–380	±500
$CH_3O_2 + SO_2 \rightarrow CH_3O + SO_3$	See previous evaluation					
$\rightarrow CH_3O_2SO_2$	See previous evaluation					
$HS + O_2 \rightarrow$ products	$\leq 4 \times 10^{-19}$					
$HS + O_3 \rightarrow HSO + O_2$	3.7×10^{-12}		±0.2	$9.5 \times 10^{-12}\exp(-280/T)$	290–450	±250
$HS + NO + M \rightarrow HSNO + M$	$2.4 \times 10^{-31}[N_2]$	(k_0)	±0.3	$2.4 \times 10^{-31}(T/300)^{-2.5}[N_2]$	200–300	$\Delta n = \pm1$
	2.7×10^{-11}	(k_∞)	±0.5	2.7×10^{-11}	200–300	$\Delta\log k = \pm0.5$
	$F_c = 0.6$					
$HS + NO_2 \rightarrow HSO + NO$	5.8×10^{-11}		±0.3	$2.6 \times 10^{-11}\exp(240/T)$	220–450	±200
$HSO + O_2 \rightarrow$ products	$\leq 2.0 \times 10^{-17}$					
$HSO + O_3 \rightarrow$ products	1.1×10^{-13}		±0.3			
$HSO + NO \rightarrow$ products	$\leq 1.0 \times 10^{-15}$					
$HSO + NO_2 \rightarrow$ products	9.6×10^{-12}		±0.3			
$HSO_2 + O_2 \rightarrow$ products	3.0×10^{-13}		±0.8			
$SO + O_2 \rightarrow SO_2 + O$	6.7×10^{-17}		±0.15	$1.4 \times 10^{-13}\exp(-2280/T)$	230–420	±500
$SO + O_3 \rightarrow SO_2 + O_2$	8.9×10^{-14}		±0.10	$4.5 \times 10^{-12}\exp(-1170/T)$	230–420	±150
$SO + NO_2 \rightarrow SO_2 + NO$	1.4×10^{-11}		±0.1	1.4×10^{-11}	210–360	±100
$SO_3 + H_2O \rightarrow$ products	$<6 \times 10^{-15}$					
$CS + O_2 \rightarrow$ products	2.9×10^{-19}		±0.6			
$CS + O_3 \rightarrow OCS + O_2$	3.0×10^{-16}		±0.5			
$CS + NO_2 \rightarrow OCS + NO$	7.6×10^{-17}		±0.5			
$CH_3S + O_2 \rightarrow$ products	$<2.5 \times 10^{-18}$					
$CH_3S + O_3 \rightarrow$ products	4.1×10^{-12}		±0.5			
$CH_3S + NO + M \rightarrow CH_3SNO + M$	$3.2 \times 10^{-29}[N_2]$	(k_0)	±0.3	$3.2 \times 10^{-29}(T/298)^{-4}[N_2]$	250–450	$\Delta n = \pm2$
	4×10^{-11}	(k_∞)	±0.5	4×10^{-11}	250–450	$\Delta\log k = \pm0.5$
	$F_c = 0.60$			$F_c = \exp(-T/580)$		
$CH_3S + NO_2 \rightarrow CH_3SO + NO$	5.6×10^{-11}		±0.2			
$CH_3SO + O_3 \rightarrow$ products	1×10^{-12}		±0.7			
$CH_3SO + NO_2 \rightarrow$ products	1.2×10^{-11}		±0.5			
$O_3 + CH_3SCH_3 \rightarrow$ products	$<1 \times 10^{-18}$					
$OCS + h\nu \rightarrow$ products	See data sheets					
$CS_2 + h\nu \rightarrow$ products	See data sheets					
$CH_3SSCH_3 + h\nu \rightarrow$ products	See data sheets					
$CH_3SNO + h\nu \rightarrow$ products	See data sheets					
FO_x Reactions						
$O + FO \rightarrow O_2 + F$	5×10^{-11}		±0.5			
$O + FO_2 \rightarrow O_2 + FO$	5×10^{-11}		±0.7			
$O(^1D) + HF \rightarrow HO + F$; $\rightarrow O(^3P) + HF$	1×10^{-10}		±0.5			
$O(^1D) + COF_2 \rightarrow CO_2 + F_2$	2.2×10^{-11}		±0.2			
$\rightarrow O(^3P) + COF_2$	5.2×10^{-11}		±0.2			
$O(^1D) + CH_3F \rightarrow$ products	1.4×10^{-10}		±0.5			
$O(^1D) + CH_2F_2 \rightarrow$ products	9×10^{-11}		±0.5			
$O(^1D) + CHF_3 \rightarrow$ products	8.4×10^{-12}		±0.5			
$O(^1D) + CH_3CHF_2 \rightarrow$ products	1×10^{-10}		±0.7			

Reaction	k_{298} cm^3 molecule^{-1} s^{-1}	$\Delta\log k_{298}$	Temp. dependence of k/cm^3 molecule^{-1} s^{-1}	Temp. range/K	$\Delta(E/R)$/K
$O(^1D) + CH_3CF_3 \rightarrow$ products	1×10^{-10}	±0.5			
$O(^1D) + CH_2FCF_3 \rightarrow$ products	1×10^{-10}	±0.7			
$O(^1D) + CHF_2CF_3 \rightarrow$ products	5×10^{-11}	±0.5			
$F + H_2 \rightarrow HF + H$	2.6×10^{-11}	±0.1	$1.4 \times 10^{-10}\exp(-500/T)$	200–375	±200
$F + H_2O \rightarrow HF + HO$	1.4×10^{-11}	±0.1	1.4×10^{-11}	240–370	±200
$F + O_2 + M \rightarrow FO_2 + M$	$3.7 \times 10^{-33}[N_2]$ (k_0)	±0.3	$3.7 \times 10^{-33}(T/300)^{-1}[N_2]$	300–400	$\Delta n = \pm0.5$
$FO_2 + M \rightarrow F + O_2 + M$	No recommendation (see data sheets)				
$F + O_3 \rightarrow FO + O_2$	1.3×10^{-11}	±0.3	$2.8 \times 10^{-11}\exp(-230/T)$	250–365	±200
$F + HONO_2 \rightarrow HF + NO_3$	2.3×10^{-11}	±0.1	$6.0 \times 10^{-12}\exp(400/T)$	260–320	±200
$F + NO_2 + M \rightarrow FONO + M$	See previous evaluation				
$F + CH_4 \rightarrow HF + CH_3$	8×10^{-11}	±0.2	$3.0 \times 10^{-10}\exp(-400/T)$	250–450	±200
$HO + CH_3F \rightarrow H_2O + CH_2F$	1.7×10^{-14}	±0.10	$3.7 \times 10^{-12}\exp(-1600/T)$	270–340	±300
$HO + CH_2F_2 \rightarrow H_2O + CHF_2$	1.1×10^{-14}	±0.10	$2.0 \times 10^{-12}\exp(-1545/T)$	240–300	±200
$HO + CHF_3 \rightarrow H_2O + CF_3$	2.4×10^{-16}	±0.5	$1.0 \times 10^{-12}\exp(-2490/T)$	270–340	±500
$HO + CH_3CH_2F \rightarrow$ products	2.3×10^{-13}	±0.3			
$HO + CH_3CHF_2 \rightarrow$ products	3.6×10^{-14}	±0.10	$1.0 \times 10^{-12}\exp(-990/T)$	240–300	±200
$HO + CH_3CF_3 \rightarrow H_2O + CH_2CF_3$	1.3×10^{-15}	±0.15	$1.05 \times 10^{-12}\exp(-1990/T)$	240–300	±300
$HO + CH_2FCHF_2 \rightarrow$ products	1.8×10^{-14}	±0.3			
$HO + CH_2FCF_3 \rightarrow H_2O + CHFCF_3$	4.9×10^{-15}	±0.2	$8.4 \times 10^{-13}\exp(-1535/T)$	240–300	±300
$HO + CHF_2CHF_2 \rightarrow H_2O + CF_2CHF_2$	5.7×10^{-15}	±0.3			
$HO + CHF_2CF_3 \rightarrow H_2O + CF_2CF_3$	1.9×10^{-15}	±0.2	$4.9 \times 10^{-13}\exp(-1655/T)$	240–300	±300
$HO + CF_3CHO \rightarrow H_2O + CF_3CO$	1.1×10^{-12}	±0.3			
$FO + O_3 \rightarrow$ products	No recommendation (see data sheets)				
$FO + NO \rightarrow F + NO_2$	2.6×10^{-11}	±0.3			
$FO + NO_2 + M \rightarrow FONO_2 + M$	See previous evaluation				
$FO + FO \rightarrow$ products	1.5×10^{-11}	±0.3			
$COF_2 + h\nu \rightarrow$ products	See data sheets				
$HCOF + h\nu \rightarrow$ products	See data sheets				
$CF_3COF + h\nu \rightarrow$ products	See data sheets				
ClO_x Reactions					
$O + HCl \rightarrow HO + Cl$	See previous evaluation				
$O + HOCl \rightarrow HO + ClO$	No recommendation (see data sheets)				
$O + ClO \rightarrow Cl + O_2$	3.8×10^{-11}	±0.1	3.8×10^{-11}	200–300	±250
$O + OClO \rightarrow O_2 + ClO$	1.0×10^{-13}	±0.3	$2.5 \times 10^{-12}\exp(-950/T)$	240–400	±300
$O + OClO + M \rightarrow ClO_3 + M$	$1.8 \times 10^{-31}[N_2]$ (k_0)	±0.3	$1.8 \times 10^{-31}(T/298)^{-1}[N_2]$	250–300	$\Delta n = \pm0.5$
	3.1×10^{-11} (k_∞)	±0.3	$3.1 \times 10^{-11}(T/298)^{1}$	250–300	$\Delta n = \pm1$
	$F_c = 0.48$				
$O + Cl_2O \rightarrow ClO + ClO$	3.5×10^{-12}	±0.15	$2.9 \times 10^{-11}\exp(-630/T)$	235–300	±200
$O + ClONO_2 \rightarrow$ products	2.0×10^{-13}	±0.1	$3.0 \times 10^{-12}\exp(-800/T)$	213–295	±200
$O(^1D) + CHF_2Cl \rightarrow$ products	9.5×10^{-11}	±0.2	9.5×10^{-11}	175–340	$\Delta\log k = \pm0.2$
$O(^1D) + CHFCl_2 \rightarrow$ products	1.9×10^{-10}	±0.2	1.9×10^{-10}	175–340	$\Delta\log k = \pm0.2$
$O(^1D) + CH_3CF_2Cl \rightarrow$ products	1.4×10^{-10}	±0.3			
$O(^1D) + CH_3CFCl_2 \rightarrow$ products	1.5×10^{-10}	±0.5			
$O(^1D) + CH_2ClCF_3 \rightarrow$ products	1.5×10^{-10}	±0.3			
$O(^1D) + CH_2ClCF_2Cl \rightarrow$ products	1.6×10^{-10}	±0.3			
$O(^1D) + CHFClCF_3 \rightarrow$ products	1.0×10^{-10}	±0.5			
$O(^1D) + CHCl_2CF_3 \rightarrow$ products	2.2×10^{-10}	±0.3			
$O(^1D) + CF_2Cl_2 \rightarrow$ products	1.4×10^{-10}	±0.1			
$O(^1D) + CFCl_3 \rightarrow$ products	2.3×10^{-10}	±0.1			
$O(^1D) + CCl_4 \rightarrow$ products	3.3×10^{-10}	±0.1			
$O(^1D) + COFCl \rightarrow$ products	1.9×10^{-10}	±0.3			
$O(^1D) + COCl_2 \rightarrow$ products	3.6×10^{-10}	±0.3			
$Cl + H_2 \rightarrow HCl + H$	1.6×10^{-14}	±0.1	$3.7 \times 10^{-11}\exp(-2300/T)$	200–300	±200
$Cl + HO_2 \rightarrow HCl + O_2$	3.2×10^{-11}	±0.2	$1.8 \times 10^{-11}\exp(170/T)$	250–420	±250
$\quad\rightarrow ClO + HO$	9.1×10^{-12}	±0.3	$4.1 \times 10^{-11}\exp(-450/T)$	250–420	±250
$Cl + H_2O_2 \rightarrow HCl + HO_2$	4.1×10^{-13}	±0.2	$1.1 \times 10^{-11}\exp(-980/T)$	265–424	±500
$Cl + O_2 + M \rightarrow ClOO + M$	$1.4 \times 10^{-33}[N_2]$ (k_0)	±0.2	$1.4 \times 10^{-33}(T/300)^{-3.9}[N_2]$	160–300	$\Delta n = \pm1$
	$1.6 \times 10^{-33}[O_2]$ (k_0)	±0.2	$1.6 \times 10^{-33}(T/300)^{-2.9}[O_2]$	160–300	$\Delta n = \pm1$
$ClOO + M \rightarrow Cl + O_2 + M$	$6.2 \times 10^{-13}[N_2]$ (k_0/s^{-1})	±0.3	$2.8 \times 10^{-10}\exp(-1820/T)[N_2]$	200–300	±200
$Cl + O_3 \rightarrow ClO + O_2$	1.2×10^{-11}	±0.06	$2.9 \times 10^{-11}\exp(-260/T)$	205–298	±100
$Cl + HONO_2 \rightarrow HCl + NO_3$	$<2.0 \times 10^{-16}$				
$Cl + NO_3 \rightarrow ClO + NO_2$	2.6×10^{-11}	±0.3	2.6×10^{-11}	200–300	±400
$Cl + OClO \rightarrow ClO + ClO$	5.8×10^{-11}	±0.1	$3.4 \times 10^{-11}\exp(160/T)$	298–450	±200
$Cl + Cl_2O \rightarrow Cl_2 + ClO$	9.8×10^{-11}	±0.1	9.8×10^{-11}	200–300	±250
$Cl + Cl_2O_2 \rightarrow Cl_2 + ClOO$	1.0×10^{-10}	±0.3	1.0×10^{-10}	230–298	±300
$Cl + ClONO_2 \rightarrow Cl_2 + NO_3$	1.2×10^{-11}	±0.12	$6.8 \times 10^{-12}\exp(160/T)$	219–298	±200
$Cl + CH_4 \rightarrow HCl + CH_3$	1.0×10^{-13}	±0.08	$9.6 \times 10^{-12}\exp(-1350/T)$	200–300	±250
$Cl + C_2H_2 + M \rightarrow C_2H_2Cl + M$	$6 \times 10^{-30}[N_2]$ (k_0)	±0.3	$6 \times 10^{-30}(T/300)^{-3.5}[N_2]$	200–300	$\Delta n = \pm1$
	2.3×10^{-10} (k_∞)	±0.5	2.3×10^{-10}	200–300	$\Delta n = \pm1$
	$F_c = 0.6$				
$Cl + C_2H_4 + M \rightarrow C_2H_4Cl + M$	$1.6 \times 10^{-29}[air]$ (k_0)	±0.5	$1.6 \times 10^{-29}(T/298)^{-3.5}[air]$	250–300	$\Delta n = \pm1$
	3×10^{-10} (k_∞)	±0.3	3×10^{-10}	250–300	$\Delta n = \pm1$
	$F_c = 0.6$				
$Cl + C_2H_6 \rightarrow HCl + C_2H_5$	5.9×10^{-11}	±0.06	$8.2 \times 10^{-11}\exp(-100/T)$	220–600	±100
$Cl + C_3H_8 \rightarrow HCl + C_3H_7$	1.4×10^{-10}	±0.12	$1.2 \times 10^{-10}\exp(40/T)$	220–600	±200
$Cl + HCHO \rightarrow HCl + HCO$	7.3×10^{-11}	±0.06	$8.2 \times 10^{-11}\exp(-34/T)$	200–500	±100
$Cl + CH_3CHO \rightarrow HCl + CH_3CO$	7.2×10^{-11}	±0.15	7.2×10^{-11}	210–340	±300
$Cl + C_2H_5CHO \rightarrow$ products	1.2×10^{-10}	±0.3			
$Cl + CH_3COCH_3 \rightarrow HCl + CH_3COCH_2$	3.5×10^{-12}	±0.3			
$Cl + CH_3OH \rightarrow HCl + CH_2OH$	5.3×10^{-11}	±0.15	5.3×10^{-11}	200–500	±200
$Cl + C_2H_5OH \rightarrow$ products	9.4×10^{-11}	±0.2			
$Cl + n\text{-}C_3H_7OH \rightarrow$ products	1.5×10^{-10}	±0.2			
$Cl + i\text{-}C_3H_7OH \rightarrow$ products	8.4×10^{-11}	±0.3			
$Cl + CH_3OOH \rightarrow$ products	5.9×10^{-11}	±0.5			

Reaction	k_{298} cm^3 molecule^{-1} s^{-1}	$\Delta \log k_{298}$	Temp. dependence of k/cm^3 molecule^{-1} s^{-1}	Temp. range/K	$\Delta(E/R)$/K
$Cl + HCOOH \rightarrow$ products	2.0×10^{-13}	±0.2			
$Cl + CH_3COOH \rightarrow$ products	2.8×10^{-14}	±0.3			
$Cl + CH_3ONO_2 \rightarrow$ products	2.4×10^{-13}	±0.3			
$Cl + C_2H_5ONO_2 \rightarrow$ products	4.7×10^{-12}	±0.2			
$Cl + n\text{-}C_3H_7ONO_2 \rightarrow$ products	2.7×10^{-11}	±0.2			
$Cl + i\text{-}C_3H_7ONO_2 \rightarrow$ products	5.8×10^{-12}	±0.3			
$Cl + CH_3CO_3NO_2 \rightarrow$ products	$<2 \times 10^{-14}$				
$Cl + CH_3CN \rightarrow$ products	$\leq 2 \times 10^{-15}$				
$Cl + HCOCl \rightarrow HCl + ClCO$	7.8×10^{-13}	±0.15	$1.2 \times 10^{-11}\exp(-815/T)$	265–325	±300
$Cl + CH_3Cl \rightarrow HCl + CH_2Cl$	4.9×10^{-13}	±0.15	$3.3 \times 10^{-11}\exp(-1250/T)$	233–322	±300
$Cl + CH_2Cl_2 \rightarrow HCl + CHCl_2$	4.1×10^{-13}	±0.25	$8.7 \times 10^{-12}\exp(-910/T)$	270–330	±400
$Cl + CHCl_3 \rightarrow HCl + CCl_3$	7.6×10^{-14}	±0.3	$4.9 \times 10^{-12}\exp(-1240/T)$	240–330	±400
$Cl + CH_3CCl_3 \rightarrow HCl + CH_2CCl_3$	$<4 \times 10^{-14}$				
$HO + HCl \rightarrow H_2O + Cl$	8.1×10^{-13}	±0.1	$2.4 \times 10^{-12}\exp(-330/T)$	200–300	±150
$HO + HOCl \rightarrow ClO + H_2O$	5.0×10^{-13}	±0.5	$3.0 \times 10^{-12}\exp(-500/T)$	200–300	±500
$HO + ClO \rightarrow HO_2 + Cl$ $\rightarrow HCl + O_2$	1.7×10^{-11}	±0.2	$1.1 \times 10^{-11}\exp(120/T)$	200–373	±150
$HO + OClO \rightarrow HOCl + O_2$	7.0×10^{-12}	±0.3	$4.5 \times 10^{-13}\exp(800/T)$	290–480	±200
$HO + ClNO_2 \rightarrow HOCl + NO_2$	3.5×10^{-14}	±0.3			
$HO + ClONO_2 \rightarrow$ products	3.9×10^{-13}	±0.2	$1.2 \times 10^{-12}\exp(-330/T)$	246–387	±200
$HO + CH_3Cl \rightarrow H_2O + CH_2Cl$	4.3×10^{-14}	±0.10	$1.8 \times 10^{-12}\exp(-1115/T)$	240–300	±200
$HO + CH_2FCl \rightarrow H_2O + CHFCl$	4.4×10^{-14}	±0.10	$2.0 \times 10^{-12}\exp(-1135/T)$	240–300	±200
$HO + CHF_2Cl \rightarrow H_2O + CF_2Cl$	4.6×10^{-15}	±0.10	$7.8 \times 10^{-13}\exp(-1530/T)$	240–300	±200
$HO + CHFCl_2 \rightarrow H_2O + CFCl_2$	3.0×10^{-14}	±0.10	$8.8 \times 10^{-13}\exp(-1010/T)$	240–300	±200
$HO + CH_2Cl_2 \rightarrow H_2O + CHCl_2$	1.4×10^{-13}	±0.10	$4.4 \times 10^{-12}\exp(-1030/T)$	240–300	±250
$HO + CHCl_3 \rightarrow H_2O + CCl_3$	1.0×10^{-13}	±0.10	$3.3 \times 10^{-12}\exp(-1030/T)$	240–300	±100
$HO + CFCl_3 \rightarrow HOCl + CFCl_2$	$<5 \times 10^{-18}$		$<1 \times 10^{-12}\exp(-3650/T)$	250–480	
$HO + CF_2Cl_2 \rightarrow HOCl + CF_2Cl$	$<7 \times 10^{-18}$		$<1 \times 10^{-12}\exp(-3540/T)$	250–478	
$HO + CCl_4 \rightarrow HOCl + CCl_3$	$<5 \times 10^{-16}$		$<1 \times 10^{-12}\exp(-2260/T)$	250–300	
$HO + C_2HCl_3 \rightarrow$ products	2.2×10^{-12}	±0.10	$5.0 \times 10^{-13}\exp(445/T)$	230–420	±200
$HO + C_2Cl_4 \rightarrow$ products	1.7×10^{-13}	±0.10	$9.4 \times 10^{-12}\exp(-1200/T)$	300–420	±200
$HO + CH_3CF_2Cl \rightarrow H_2O + CH_2CF_2Cl$	3.0×10^{-15}	±0.10	$9.2 \times 10^{-13}\exp(-1705/T)$	240–300	±200
$HO + CH_3CFCl_2 \rightarrow H_2O + CH_2CFCl_2$	5.9×10^{-15}	±0.2	$7.0 \times 10^{-13}\exp(-1425/T)$	240–300	±300
$HO + CH_3CCl_3 \rightarrow H_2O + CH_2CCl_3$	9.5×10^{-15}	±0.10	$1.2 \times 10^{-12}\exp(-1440/T)$	240–300	±200
$HO + CH_2ClCF_3 \rightarrow H_2O + CHClCF_3$	1.3×10^{-14}	±0.2	$5.2 \times 10^{-13}\exp(-1100/T)$	260–380	±250
$HO + CH_2ClCF_2Cl \rightarrow H_2O + CHClCF_2Cl$	1.6×10^{-14}	±0.3	$3.2 \times 10^{-12}\exp(-1580/T)$	250–350	±500
$HO + CHFClCF_3 \rightarrow H_2O + CFClCF_3$	9.5×10^{-15}	±0.10	$5.4 \times 10^{-13}\exp(-1205/T)$	240–300	±200
$HO + CHCl_2CF_3 \rightarrow H_2O + CCl_2CF_3$	3.6×10^{-14}	±0.15	$5.5 \times 10^{-13}\exp(-815/T)$	240–300	±200
$HO + CHCl_2CF_2CF_3 \rightarrow H_2O + CCl_2CF_2CF_3$	2.5×10^{-14}	±0.15	$1.1 \times 10^{-12}\exp(-1130/T)$	270–400	±300
$HO + CHFClCF_2CF_2Cl \rightarrow H_2O + CFClCF_2CF_2Cl$	8.9×10^{-15}	±0.10	$5.5 \times 10^{-13}\exp(-1230/T)$	290–400	±300
$HO + CH_3CF_2CFCl_2 \rightarrow H_2O + CH_2CF_2CFCl_2$	2.4×10^{-15}	±0.3	$7.0 \times 10^{-13}\exp(-1690/T)$	290–370	±300
$HO + HCOCl \rightarrow H_2O + ClCO$	$<5 \times 10^{-13}$				
$HO + COCl_2 \rightarrow$ products	$<5 \times 10^{-15}$				
$HO + CH_2ClCHO \rightarrow$ products	3.0×10^{-12}	±0.3			
$HO + CHCl_2CHO \rightarrow$ products	2.4×10^{-12}	±0.3			
$HO + CCl_3CHO \rightarrow H_2O + CCl_3CO$	1.4×10^{-12}	±0.3			
$HO + CH_3COCl \rightarrow H_2O + CH_2COCl$	9×10^{-15}	±1.0			
$HO + CHF_2OCHClCF_3 \rightarrow$ products	2.1×10^{-14}	±0.5			
$HO + CHF_2OCF_2CHFCl \rightarrow$ products	1.6×10^{-14}	±0.5	$6.1 \times 10^{-13}\exp(-1080/T)$	300–430	±500
$NO_3 + C_2HCl_3 \rightarrow$ products	2.9×10^{-16}	±0.3			
$NO_3 + C_2Cl_4 \rightarrow$ products	$<1 \times 10^{-16}$				
$ClO + HO_2 \rightarrow HOCl + O_2$ $\rightarrow HCl + O_3$	5.0×10^{-12}	±0.15	$4.6 \times 10^{-13}\exp(710/T)$	200–300	±300
$ClO + O_2(^1\Delta_g) \rightarrow sym\text{-}ClO_3$	See previous evaluation				
$ClO + O_3 \rightarrow ClOO + O_2$	$<1.5 \times 10^{-17}$				
$\rightarrow OClO + O_2$	$<1 \times 10^{-18}$				
$ClO + NO \rightarrow Cl + NO_2$	1.7×10^{-11}	±0.1	$6.2 \times 10^{-12}\exp(294/T)$	202–415	±100
$ClO + NO_2 + M \rightarrow ClONO_2 + M$	$1.6 \times 10^{-31}[N_2]$ (k_0)	±0.1	$1.6 \times 10^{-31}(T/300)^{-3.4}[N_2]$	200–300	$\Delta n = \pm 1$
	2×10^{-11} (k_∞)	±0.3	2×10^{-11}	200–300	$\Delta \log k = \pm 0.3$
	$F_c = 0.5$		$F_c = \exp(-T/430)$	200–300	
$ClO + NO_3 \rightarrow ClOO + NO_2$ $\rightarrow OClO + NO_2$	4.0×10^{-13}	±0.3			
$ClO + HCHO \rightarrow$ products	See previous evaluation				
$ClO + ClO \rightarrow Cl + ClOO$	3.4×10^{-15}	±0.3			
$\rightarrow Cl + OClO$	1.7×10^{-15}	±0.3			
$\rightarrow Cl_2 + O_2$	4.9×10^{-15}	±0.3			
$ClO + ClO + M \rightarrow Cl_2O_2 + M$	$1.7 \times 10^{-32}[N_2]$ (k_0)	±0.1	$1.7 \times 10^{-32}(T/300)^{-4}[N_2]$	200–260	$\Delta n = \pm 1.5$
	5.4×10^{-12} (k_∞)	±0.3	5.4×10^{-12}	200–300	$\Delta \log k = \pm 0.3$
	$F_c = 0.6$				
$Cl_2O_2 + M \rightarrow ClO + ClO + M$	$2.7 \times 10^{-18}[N_2]$ (k_0/s^{-1})	±0.3	$1.35 \times 10^{-5}(T/300)^{-5}$ $\exp(-8720/T)[N_2]$	200–300	±900
	8.7×10^{2} (k_∞/s^{-1})	±0.3	$1.8 \times 10^{15}\exp(-8450/T)$	200–300	±900
	$F_c = 0.6$				
$ClO + OClO + M \rightarrow Cl_2O_3 + M$	$2.8 \times 10^{-31}[N_2]$ (k_0; 226 K)	±0.5 (226 K)			
$Cl_2O_3 + M \rightarrow ClO + OClO + M$	$2.8 \times 10^{-18}[N_2]$ (k_0/s^{-1}; 226 K)	±0.5 (226 K)			
$ClO + CH_3O_2 \rightarrow ClOO + CH_3O$	$<4 \times 10^{-12}$ (200 K)				
$\rightarrow OClO + CH_3O$	$<1 \times 10^{-15}$ (200 K)				
$OClO + O_3 \rightarrow ClO_3 + O_2$	3.0×10^{-19}	±0.4	$2.1 \times 10^{-12}\exp(-4700/T)$	262–298	±1000
$OClO + NO \rightarrow NO_2 + ClO$	3.4×10^{-13}	±0.3			
$Cl_2O_2 + O_3 \rightarrow ClO + ClOO + O_2$	$<1 \times 10^{-19}$ (200 K)				

Reaction	k_{298} cm^3 molecule^{-1} s^{-1}		$\Delta\log k_{298}$	Temp. dependence of k/cm^3 molecule^{-1} s^{-1}	Temp. range/K	$\Delta(E/R)$/K
$CF_3 + O_2 + M \rightarrow CF_3O_2 + M$	$1.9 \times 10^{-29}[N_2]$	(k_0)	±0.2	$1.9 \times 10^{-29}(T/300)^{-4.7}[N_2]$	200–300	Δn = ±1
	1.0×10^{-11}	(k_∞)	±0.3	1.0×10^{-11}	200–400	$\Delta\log k$ = ±0.3
	$F_c = 0.6$					
$CF_2Cl + O_2 + M \rightarrow CF_2ClO_2 + M$	$1.4 \times 10^{-29}[N_2]$	(k_0)	±0.5	$1.4 \times 10^{-29}(T/300)^{-5}[N_2]$	200–300	Δn = ±2
	9×10^{-12}	(k_∞)	±0.5	9×10^{-12}	200–300	$\Delta\log k$ = ±0.5
	$F_c = 0.6$					
$CFCl_2 + O_2 + M \rightarrow CFCl_2O_2 + M$	$5.5 \times 10^{-30}[N_2]$	(k_0)	±0.3	$5.5 \times 10^{-30}(T/300)^{-6}[N_2]$	200–300	Δn = ±2
	9×10^{-12}	(k_∞)	±0.5	9×10^{-12}	200–300	Δn = ±1
	$F_c = 0.6$					
$CCl_3 + O_2 + M \rightarrow CCl_3O_2 + M$	$1.6 \times 10^{-30}[N_2]$	(k_0)	±0.3	$1.6 \times 10^{-30}(T/300)^{-6}[N_2]$	200–300	Δn = ±2
	3.6×10^{-12}	(k_∞)	±0.5	3.6×10^{-12}	200–300	$\Delta\log k$ = ±0.5
	$F_c = 0.6$					
$CF_3O \rightarrow COF_2 + F$	$<10^{-5}$	(k_∞/s^{-1})				
$CF_2ClO \rightarrow COF_2 + Cl$	7×10^5	(k/s^{-1})	±1.0	$3 \times 10^{13}\exp(-5250/T)$	220–300	±1000
$CFCl_2O \rightarrow COFCl + Cl$	7×10^5	(k/s^{-1})	±1.0	$3 \times 10^{13}\exp(-5250/T)$	220–300	±1000
$CCl_3O \rightarrow COCl_2 + Cl$	8×10^6	(k/s^{-1})	±1.0	$4 \times 10^{13}\exp(-4600/T)$	220–300	±1000
$CF_3O_2 + NO \rightarrow CF_3O + NO_2$	1.6×10^{-11}		±0.2	$1.6 \times 10^{-11}(T/300)^{-1.2}$	230–430	$\Delta\log k$ = ±0.2
$CF_2ClO_2 + NO \rightarrow CF_2ClO + NO_2$	1.6×10^{-11}		±0.3	$1.6 \times 10^{-11}(T/300)^{-1.5}$	230–430	$\Delta\log k$ = ±0.3
$CFCl_2O_2 + NO \rightarrow CFCl_2O + NO_2$	1.5×10^{-11}		±0.2	$1.5 \times 10^{-11}(T/300)^{-1.3}$	230–430	$\Delta\log k$ = ±0.2
$CCl_3O_2 + NO \rightarrow CCl_3O + NO_2$	1.8×10^{-11}		±0.2	$1.8 \times 10^{-11}(T/300)^{-1.0}$	230–430	$\Delta\log k$ = ±0.2
$CF_3O_2 + NO_2 + M \rightarrow CF_3O_2NO_2 + M$	$4.5 \times 10^{-29}[N_2]$	(k_0)	±0.3	$4.5 \times 10^{-29}(T/300)^{-6.4}[N_2]$	220–300	Δn = ±1
	7.5×10^{-12}	(k_∞)	±0.5	7.5×10^{-12}	200–300	$\Delta\log k$ = ±0.5
	$F_c = 0.28$			$F_c = 0.28$	220–300	
$CF_3O_2NO_2 + M \rightarrow CF_3O_2 + NO_2 + M$	$3.6 \times 10^{-19}[N_2]$	(k_0/s^{-1})	±0.4	$5 \times 10^{-1}(T/300)^{-6}$ $\exp(-12460/T)[N_2]$	233–373	±500
	5.6×10^{-2}	(k_∞/s^{-1})	±0.5	$1.2 \times 10^{17}\exp(-12580/T)$	233–373	±500
	$F_c = 0.28$			$F_c = 0.28$	220–300	
$CF_2ClO_2 + NO_2 + M \rightarrow CF_2ClO_2NO_2 + M$	$1.4 \times 10^{-28}[N_2]$	(k_0)	±0.5	$1.4 \times 10^{-28}(T/300)^{-6.4}[N_2]$	200–300	Δn = ±2
	7.5×10^{-12}	(k_∞)	±0.3	7.5×10^{-12}	200–300	$\Delta\log k$ = ±0.3
	$F_c = 0.26$			$F_c = 0.26$	220–300	
$CF_2ClO_2NO_2 + M \rightarrow CF_2ClO_2 + NO_2 + M$	$9.0 \times 10^{-19}[N_2]$	(k_0/s^{-1})	±0.3	$1.8 \times 10^{-3}\exp(-10500/T)[N_2]$	260–300	±500
	5.4×10^{-2}	(k_∞/s^{-1})	±0.3	$1.6 \times 10^{16}\exp(-11990/T)$	260–300	±500
	$F_c = 0.26$			$F_c = 0.26$	250–300	
$CFCl_2O_2 + NO_2 + M \rightarrow CFCl_2O_2NO_2 + M$	$1.7 \times 10^{-28}[N_2]$	(k_0)	±0.3	$1.7 \times 10^{-28}(T/300)^{-6.7}[N_2]$	230–300	Δn = ±2
	7.5×10^{-12}	(k_∞)	±0.3	7.5×10^{-12}	250–300	$\Delta\log k$ = ±0.3
	$F_c = 0.23$			$F_c = 0.23$	230–300	
$CFCl_2O_2NO_2 + M \rightarrow CFCl_2O_2 + NO_2 + M$	$1.5 \times 10^{-18}[N_2]$	(k_0/s^{-1})	±0.3	$1.0 \times 10^{-2}\exp(-10860/T)$	250–300	±500
	9.6×10^{-2}	(k_∞/s^{-1})	±0.3	$6.6 \times 10^{16}\exp(-12240/T)$	250–300	±500
	$F_c = 0.23$			$F_c = 0.23$	250–300	
$CCl_3O_2 + NO_2 + M \rightarrow CCl_3O_2NO_2 + M$	$3.2 \times 10^{-28}[N_2]$	(k_0)	±0.5	$3.2 \times 10^{-28}(T/300)^{-7.7}[N_2]$	230–300	Δn = ±3
	7.5×10^{-12}	(k_∞)	±0.3	7.5×10^{-12}	250–300	$\Delta\log k$ = ±0.3
	$F_c = 0.21$			$F_c = 0.21$	250–300	
$CCl_3O_2NO_2 + M \rightarrow CCl_3O_2 + NO_2 + M$	$7.6 \times 10^{-18}[N_2]$	(k_0/s^{-1})	±0.3	$6.3 \times 10^{-3}\exp(-10235/T)[N_2]$	250–300	±500
	0.29	(k_∞/s^{-1})	±0.3	$4.8 \times 10^{16}\exp(-11820/T)$	250–300	±500
	$F_c = 0.20$			$F_c = 0.20$	250–300	
$O_3 + C_2HCl_3 \rightarrow$ products	$<5 \times 10^{-20}$					
$O_3 + C_2Cl_4 \rightarrow$ products	$<10^{-21}$					
$HCl + h\nu \rightarrow$ products	See data sheets					
$HOCl + h\nu \rightarrow$ products	See data sheets					
$OClO + h\nu \rightarrow$ products	See data sheets					
$Cl_2O + h\nu \rightarrow$ products	See data sheets					
$Cl_2O_2 + h\nu \rightarrow$ products	See data sheets					
$Cl_2O_3 + h\nu \rightarrow$ products	See data sheets					
$ClNO + h\nu \rightarrow$ products	See data sheets					
$ClONO + h\nu \rightarrow$ products	See data sheets					
$ClNO_2 + h\nu \rightarrow$ products	See data sheets					
$ClONO_2 + h\nu \rightarrow$ products	See data sheets					
$Cl_2 + h\nu \rightarrow$ products	See data sheets					
$CH_3Cl + h\nu \rightarrow$ products	See data sheets					
$CHF_2Cl + h\nu \rightarrow$ products	See data sheets					
$CF_2Cl_2 + h\nu \rightarrow$ products	See data sheets					
$CFCl_3 + h\nu \rightarrow$ products	See data sheets					
$CCl_4 + h\nu \rightarrow$ products	See data sheets					
$CH_3CF_2Cl + h\nu \rightarrow$ products	See data sheets					
$CH_3CFCl_2 + h\nu \rightarrow$ products	See data sheets					
$CH_3CCl_3 + h\nu \rightarrow$ products	See data sheets					
$CF_3CHFCl + h\nu \rightarrow$ products	See data sheets					
$CF_3CHCl_2 + h\nu \rightarrow$ products	See data sheets					
$CF_2ClCFCl_2 + h\nu \rightarrow$ products	See data sheets					
$CF_2ClCF_2Cl + h\nu \rightarrow$ products	See data sheets					
$CF_3CF_2Cl + h\nu \rightarrow$ products	See data sheets					
$CF_3CF_2CHCl_2 + h\nu \rightarrow$ products	See data sheets					
$CF_2ClCF_2CHFCl + h\nu \rightarrow$ products	See data sheets					
$HCOCl + h\nu \rightarrow$ products	See data sheets					
$COFCl + h\nu \rightarrow$ products	See data sheets					
$COCl_2 + h\nu \rightarrow$ products	See data sheets					
$CCl_3CHO + h\nu \rightarrow$ products	See data sheets					
$CF_3COCl + h\nu \rightarrow$ products	See data sheets					

BrO$_x$ Reactions

Reaction	k_{298} cm^3 molecule^{-1} s^{-1}		$\Delta\log k_{298}$	Temp. dependence of k/cm^3 molecule^{-1} s^{-1}	Temp. range/K	$\Delta(E/R)$/K
$O + HBr \rightarrow HO + Br$	See previous evaluation					
$O + Br_2 \rightarrow BrO + Br$	See previous evaluation					
$O + BrO \rightarrow O_2 + Br$	3×10^{-11}		±0.5			

Reaction	k_{298} cm^3 molecule^{-1} s^{-1}	$\Delta\log k_{298}$	Temp. dependence of k/cm^3 molecule^{-1} s^{-1}	Temp. range/K	$\Delta(E/R)$/K
$Br + HO_2 \rightarrow HBr + O_2$	2.0×10^{-12}	±0.3	$1.4 \times 10^{-11}\exp(-590/T)$	260–390	±200
$Br + H_2O_2 \rightarrow HBr + HO_2$ $\rightarrow HOBr + HO$	$<5 \times 10^{-16}$				
$Br + O_3 \rightarrow BrO + O_2$	1.2×10^{-12}	±0.08	$1.7 \times 10^{-11}\exp(-800/T)$	195–392	±200
$Br + NO_2 + M \rightarrow BrNO_2 + M$	4.2×10^{-31} [N_2] (k_0)	±0.3	$4.2 \times 10^{-31}(T/300)^{-2.4}$[$N_2$]	200–300	$\Delta n = \pm 1$
	2.7×10^{-11} (k_∞)	±0.4	2.7×10^{-11}	200–300	$\Delta\log k = \pm 0.4$
	$F_c = 0.55$				
$Br + OClO \rightarrow BrO + ClO$	3.4×10^{-13}	±0.3	$2.6 \times 10^{-11}\exp(-1300/T)$	200–450	±300
$Br + Cl_2O \rightarrow BrCl + ClO$	3.8×10^{-12}	±0.3	$2.1 \times 10^{-11}\exp(-520/T)$	220–298	±300
$Br + Cl_2O_2 \rightarrow BrCl + ClOO$	3.0×10^{-12}	±0.3			
$Br + HCHO \rightarrow HBr + HCO$	1.0×10^{-12}	±0.15	$1.7 \times 10^{-11}\exp(-800/T)$	223–480	±250
$Br + CH_3CHO \rightarrow HBr + CH_3CO$	3.9×10^{-12}	±0.2	$1.3 \times 10^{-11}\exp(-360/T)$	250–400	±200
$HO + HBr \rightarrow H_2O + Br$	1.1×10^{-11}	±0.1	1.1×10^{-11}	249–416	±250
$HO + Br_2 \rightarrow HOBr + Br$	4.5×10^{-11}	±0.15	$1.2 \times 10^{-11}\exp(400/T)$	260–360	±400
$HO + CH_3Br \rightarrow H_2O + CH_2Br$	3.0×10^{-14}	±0.10	$1.9 \times 10^{-12}\exp(-1240/T)$	240–300	±200
$HO + CHF_2Br \rightarrow H_2O + CF_2Br$	9.5×10^{-15}	±0.2	$7.7 \times 10^{-13}\exp(-1310/T)$	240–300	±200
$HO + CF_3Br \rightarrow$ products	$<1 \times 10^{-16}$				
$HO + CF_2ClBr \rightarrow$ products	$<1 \times 10^{-16}$				
$HO + CF_2Br_2 \rightarrow$ products	$<5 \times 10^{-16}$				
$HO + CF_3CHFBr \rightarrow H_2O + CF_3CFBr$	1.7×10^{-14}	±0.3	$1.1 \times 10^{-12}\exp(-1250/T)$	270–430	±500
$HO + CF_3CHClBr \rightarrow H_2O + CF_3CClBr$	5.8×10^{-14}	±0.3			
$HO + CF_2BrCF_2Br \rightarrow$ products	$<1.3 \times 10^{-16}$				
$BrO + HO_2 \rightarrow HOBr + O_2$ $\rightarrow HBr + O_3$	3.3×10^{-11}	±0.5	$6.2 \times 10^{-12}\exp(500/T)$	200–300	±500
$BrO + O_3 \rightarrow Br + 2O_2$	$<5 \times 10^{-15}$				
$BrO + NO \rightarrow Br + NO_2$	2.1×10^{-11}	±0.1	$8.7 \times 10^{-12}\exp(260/T)$	224–425	±100
$BrO + NO_2 + M \rightarrow BrONO_2 + M$	4.7×10^{-31} [N_2] (k_0)	±0.1	$4.7 \times 10^{-31}(T/300)^{-3.1}$[$N_2$]	200–300	$\Delta n = \pm 1$
	1.7×10^{-11} (k_∞)	±0.1	$1.7 \times 10^{-11}(T/298)^{-0.6}$	200–300	$\Delta n = \pm 1$
	$F_c = 0.40$		$F_c = \exp(-T/327)$	200–300	
$BrO + ClO \rightarrow Br + OClO$	6.8×10^{-12}	±0.1	$1.6 \times 10^{-12}\exp(430/T)$	220–400	±200
$\rightarrow Br + ClOO$	6.1×10^{-12}	±0.1	$2.9 \times 10^{-12}\exp(220/T)$	220–400	±200
$\rightarrow BrCl + O_2$	1.0×10^{-12}	±0.1	$5.8 \times 10^{-13}\exp(170/T)$	220–400	±200
$BrO + BrO \rightarrow 2Br + O_2$	2.1×10^{-12}	±0.1	$1.1 \times 10^{-12}\exp(250/T)$	223–400	±200
$\rightarrow Br_2 + O_2$	4.1×10^{-13}	±0.2			
$HOBr + h\nu \rightarrow$ products	See data sheets				
$BrO + h\nu \rightarrow$ products	See data sheets				
$BrONO_2 + h\nu \rightarrow$ products	See data sheets				
$CH_3Br + h\nu \rightarrow$ products	See data sheets				
$CF_3Br + h\nu \rightarrow$ products	See data sheets				
$CF_2ClBr + h\nu \rightarrow$ products	See data sheets				
$CF_2Br_2 + h\nu \rightarrow$ products	See data sheets				
$CHBr_3 + h\nu \rightarrow$ products	See data sheets				
$CF_2BrCF_2Br + h\nu \rightarrow$ products	See data sheets				
IO_x Reactions					
$O + I_2 \rightarrow IO + I$	1.4×10^{-10}	±0.3	1.4×10^{-10}	200–400	±250
$O + IO \rightarrow O_2 + I$	3×10^{-11}	±0.5			
$I + HO_2 \rightarrow HI + O_2$	3.8×10^{-13}	±0.3	$1.5 \times 10^{-11}\exp(-1090/T)$	250–350	±500
$I + O_3 \rightarrow IO + O_2$	1.0×10^{-12}	±0.2	$2.0 \times 10^{-11}\exp(-890/T)$	200–350	±300
$I + NO + M \rightarrow INO + M$	1.8×10^{-32} [N_2] (k_0)	±0.1	$1.8 \times 10^{-32}(T/300)^{-1.0}$[$N_2$]	200–400	$\Delta n = \pm 0.5$
	1.7×10^{-11} (k_∞)	±0.3	1.7×10^{-11}	200–400	$\Delta n = \pm 0.5$
	$F_c = 0.75$		$F_c = [\exp(-T/1040) + \exp(-4160/T)]$	200–400	
$I + NO_2 + M \rightarrow INO_2 + M$	3.0×10^{-31} [N_2] (k_0)	±0.2	$3.0 \times 10^{-31}(T/300)^{-1}$[$N_2$]	200–400	$\Delta n = \pm 1$
	6.6×10^{-11} (k_∞)	±0.3	6.6×10^{-11}	200–400	$\Delta\log k = \pm 0.3$
	$F_c = 0.63$		$F_c = [\exp(-T/650) + \exp(-2600/T)]$	200–400	
$HO + HI \rightarrow H_2O + I$	3.0×10^{-11}	±0.3			
$HO + I_2 \rightarrow HOI + I$	1.8×10^{-10}	±0.3			
$HO + CH_3I \rightarrow H_2O + CH_2I$	7.2×10^{-14}	±0.5	$3.1 \times 10^{-12}\exp(-1120/T)$	270–430	±500
$NO_3 + HI \rightarrow HNO_3 + I$	No recommendation (see data sheets)				
$IO + HO_2 \rightarrow HOI + O_2$	6.4×10^{-11}	±0.3			
$IO + IO \rightarrow$ products	5.2×10^{-11}	±0.3	$1.7 \times 10^{-12}\exp(1020/T)$	250–373	±500
$IO + NO \rightarrow I + NO_2$	2.2×10^{-11}	±0.3	$7.3 \times 10^{-12}\exp(330/T)$	200–400	±150
$IO + NO_2 + M \rightarrow IONO_2 + M$	7.7×10^{-31} [N_2] (k_0)	±0.3	$7.7 \times 10^{-31}(T/300)^{-5}$[$N_2$]	250–350	$\Delta n = \pm 2$
	1.6×10^{-11} (k_∞)	±0.3	1.6×10^{-11}	250–350	$\Delta\log k = \pm 0.3$
	$F_c = 0.4$				
$IO + CH_3SCH_3 \rightarrow$ products	1.2×10^{-14}	±0.3			
$INO + INO \rightarrow I_2 + 2\,NO$	1.3×10^{-14}	±0.4	$8.4 \times 10^{-11}\exp(-2620/T)$	300–450	±600
$INO_2 + INO_2 \rightarrow I_2 + 2\,NO_2$	4.7×10^{-15}	±0.5	$2.9 \times 10^{-11}\exp(-2600/T)$	298–400	±1000
$HOI + h\nu \rightarrow$ products	See data sheets				
$IO + h\nu \rightarrow$ products	See data sheets				
$INO + h\nu \rightarrow$ products	See data sheets				
$INO_2 + h\nu \rightarrow$ products	See data sheets				
$IONO_2 + h\nu \rightarrow$ products	See data sheets				

*No data sheet or recommendation presented in this article. See our earlier evaluation; J. Phys. Chem. Ref. Data **18**, 881 (1989) for our most recent recommendation.

KINETIC DATA FOR COMBUSTION MODELLING

D. L. Baulch, C. J. Cobos, R. A. Cox, C. Esser, P. Frank, Th. Just, J. A. Kerr, M. J. Pilling, J. Troe, R. W. Walker, and J. Warnatz

The following tables present evaluated rate constants and other chemical kinetic data required for modelling the combustion of hydrocarbons. The compilation was prepared as part of the project "Kinetics and Mechanisms of Chemical Processes in Combustion", which is one of the projects in the third European Community Energy Research and Development Program. The tables are reprinted from the *Journal of Physical and Chemical Reference Data* by permission of the authors and the American Institute of Physics.

Table 1 lists all the reactions studied and gives the recommended rate constant *k* for every bimolecular reaction, as well as the applicable temperature range and the associated error limits. Where more than one set of products is possible, rate constants or branching ratios are given for all channels considered feasible. The data for decomposition reactions and combination reactions are given in Tables 2 and 3, respectively. The reference includes a detailed data sheet for each reaction listed here, covering the thermodynamic data, kinetic measurements, and reliability assessments.

REFERENCE

Baulch, D. L., et al., *J. Phys. Chem. Ref. Data*, 21, 411-734, 1992.

Table 1
BIMOLECULAR REACTIONS

Reaction	k/cm^3 molecule^{-1}s^{-1}	Temp/K	Error limits ($\Delta \log k$)
O Atom Reactions			
$O + H_2 \rightarrow OH + H$	$8.5 \times 10^{-20} T^{2.67} \exp(-3160/T)$	300–2500	± 0.5 at 300 K falling to ± 0.2 for $T > 500$ K
$O + OH \rightarrow O_2 + H$	$2.0 \times 10^{-11} \exp(112/T)$	220–500	± 0.2
	$2.4 \times 10^{-11} \exp(-353/T)$	1000–2000	± 0.1
$O + HO_2 \rightarrow OH + O_2$	5.3×10^{-11}	300–1000	± 0.3 at 300 K rising to ± 0.5 at 1000 K.
$O + H_2O_2 \rightarrow OH + HO_2$	$1.1 \times 10^{-12} \exp(-2000/T)$	300–500	± 0.3
$O + NH_3 \rightarrow OH + NH_2$	$1.6 \times 10^{-11} \exp(-3670/T)$	500–2500	± 0.5
$O + CH \rightarrow CO + H$	6.6×10^{-11}	300–2000	± 0.5
$\rightarrow CHO^+ + e$	$4.2 \times 10^{-13} \exp(-850/T)$	300–2500	± 0.5
$O + {}^3CH_2 \rightarrow CO + 2H$	2×10^{-10}	300–2500	± 0.2 at 300 K rising to ± 0.7 at 2500 K.
$\rightarrow CO + H_2$	$k_1/k = 0.6 \pm 0.3$ over whole range		
$O + CH_3 \rightarrow HCHO + H$	1.4×10^{-10}	300–2500	± 0.2
$O + CH_4 \rightarrow OH + CH_3$	$1.5 \times 10^{-15} T^{1.56} \exp(-4270/T)$	300–2500	± 0.3 at 300 K falling to ± 0.15 at 2500 K.
$O + CHO \rightarrow OH + CO$	5.0×10^{-11}	300–2500	± 0.3
$\rightarrow CO_2 + H$	5.0×10^{-11}	300–2500	± 0.3
$O + HCHO \rightarrow OH + CHO$	$6.9 \times 10^{-13} T^{0.57} \exp(-1390/T)$	250–2200	± 0.1 at 250 K rising to ± 0.3 at 2200 K.
$O + CH_3O \rightarrow O_2 + CH_3$	2.5×10^{-11}	300–1000	± 0.3 at 300 K rising to ± 0.7 at 1000 K.
$\rightarrow OH + HCHO$	$k_2/k = (0.12 \pm 0.1)$ at 300 K		
$O + CN \rightarrow CO + N(^4S)$	1.7×10^{-11}	300–5000	± 0.2 at 300 K rising to ± 0.6 × 5000 K.
$\rightarrow CO + N(^2D)$			
$O + NCO \rightarrow NO + CO$	7.0×10^{-11}	1450–2600	± 0.8
$\rightarrow O_2 + CN$			

Table 1
BIMOLECULAR REACTIONS (continued)

Reaction	k/cm^3 molecule^{-1}s^{-1}	Temp/K	Error limits (Δ log k)
$O + HCN \rightarrow NCO + H$ $\rightarrow CO + NH$ $\rightarrow OH + CN$	$2.3 \times 10^{-18}\ T^{2.1} \exp(-3075/T)$	450–2500	± 0.2 at 450 K rising to ± 0.3 at 2500 K.
$O + CH_3OOH \rightarrow OH + CH_2COOH$ $\rightarrow OH + CH_3O_2$	$6.9 \times 10^{-13}\ T^{0.57} \exp(-1390/T)$ [estimate]	250–2200	± 0.1 at 250 K rising to ± 0.3 at 2200 K.
$O + C_2H \rightarrow CO + CH$	1.7×10^{-11}	300–2500	± 1.0
$O + C_2H_2 \rightarrow CO + {}^3CH_2$ $\rightarrow CHCO + H$	$3.6 \times 10^{-20}\ T^{2.8} \exp(-250/T)$ $k_1/k = 0.5 \pm 0.3$ over whole range.	300–2500	± 0.2
$O + C_2H_3 \rightarrow OH + C_2H_2$ $\rightarrow CO + CH_3$ $\rightarrow HCO + CH_2$	5×10^{-11}	300–2000	± 0.5
$O + C_2H_4 \rightarrow CH_2CHO + H$ $\rightarrow HCO + CH_3$ $\rightarrow HCHO + CH_2$ $\rightarrow CH_2CO + H_2$	$5.75 \times 10^{-18}\ T^{2.08}$ $k_1/k = 0.35 \pm 0.05$ at p> 3 Torr $k_2/k = 0.6 \pm 0.10$	300–2000 over whole temperature range	± 0.1 for $T < 1000$ K rising to ± 0.3 at 2000 K.
$O + C_2H_5 \rightarrow CH_3CHO + H$ $\rightarrow HCHO + CH_3$	1.1×10^{-10} $k_2/k = 0.17 \pm 0.2$ at 300 K	300–2500	± 0.3 from 300 to 1000 K ± 0.5 from 1000 to 2500 K
$O + C_2H_6 \rightarrow OH + C_2H_5$	$1.66 \times 10^{-15}\ T^{1.5} \exp(-2920/T)$	300–1200	± 0.3 at 300 K falling to ± 0.15 at 1200 K.
$O + CHCO \rightarrow 2CO + H$	1.6×10^{-10}	300–2500	± 0.3
$O + CH_2CO \rightarrow CH_2O + CO$ $\rightarrow HCO + H + CO$ $\rightarrow HCO + HCO$	$3.8 \times 10^{-12} \exp(-680/T)$	230–500	± 0.3
$O + CH_3CHO \rightarrow OH + CH_3CO$ $\rightarrow OH + CH_2CHO$	$9.7 \times 10^{-12} \exp(-910/T)$	300–1500	± 0.05 at 300 K rising to ± 0.5 at 1500 K.
$O + C_2H_5OOH \rightarrow OH + C_2H_4OOH$ $\rightarrow OH + C_2H_5OO$	$6.9 \times 10^{-13}\ T^{0.57} \exp(-1390/T)$ [estimate]	250–2200	± 0.1 at 150 K rising to ± 0.3 at 2200 K.
$O + C_6H_6 \rightarrow OH + C_6H_5$ $\rightarrow C_6H_5OH$	$1.2 \times 10^{-22}\ T^{3.7} \exp(-570/T)$	300–1000	± 0.5
$O + C_6H_5CH_2 \rightarrow HCO + C_6H_6$ $\rightarrow C_6H_5CH + H$ $\rightarrow CH_2O + C_6H_5$	5.5×10^{-10} No recommendation	300	±0.3
$O + C_6H_5CH_3 \rightarrow$ products	$5.3 \times 10^{-15}\ T^{1.21} \exp(-1260/T)$	300–2800	± 0.1 at 300 K rising to ± 0.4 at 2800 K
$O + p\text{-}C_6H_4(CH_3)_2 \rightarrow$ products	$2.6 \times 10^{-11} \exp(-1409/T)$	300–600	± 0.3
$O + C_6H_5C_2H_5 \rightarrow$ products	1.0×10^{-13}	298	± 0.3
O_2 Reactions			
$O_2 + CH_4 \rightarrow HO_2 + CH_3$	$6.6 \times 10^{-11} \exp(-28630/T)$	500–2000	± 0.5 at 500 K rising to ± 1.0 at 2000 K.

Table 1

BIMOLECULAR REACTIONS (continued)

Reaction	k/cm³ molecule⁻¹s⁻¹	Temp/K	Error limits ($\Delta \log k$)
$O_2 + C_2H_6 \rightarrow HO_2 + C_2H_5$	$1.0 \times 10^{-10} \exp(-26100/T)$	500–2000	± 0.5 at 500 K rising to ± 1.0 at 2000 K
$O_2 + HCHO \rightarrow HO_2 + HCO$	$1.0 \times 10^{-10} \exp(-20460/T)$	700–1000	± 0.5
$O_2 + CH_3CHO \rightarrow HO_2 + CH_3CO$	$5.0 \times 10^{-11} \exp(-19700/T)$	600–1100	± 0.5 at 600 K rising to ± 1.0 at 1100 K.
H Atom Reactions			
$H + O_2 \rightarrow OH + O$	$3.3 \times 10^{-10} \exp(-8460/T)$	300–2500	± 0.1 at 300 K rising to ± 0.2 at 2500 K.
$H + O_2 + Ar \rightarrow HO_2 + Ar$	See Table 3		
$H + O_2 + H_2 \rightarrow HO_2 + H_2$	See Table 3		
$H + O_2 + N_2 \rightarrow HO_2 + N_2$	See Table 3		
$H + O_2 + H_2O \rightarrow HO_2 + H_2O$	See Table 3		
$H + H + Ar \rightarrow H_2 + Ar$	See Table 3		
$H + H + H_2 \rightarrow H_2 + H_2$	See Table 3		
$H + OH + H_2O \rightarrow H_2O + H_2O$	See Table 3		
$H + OH + Ar \rightarrow H_2O + Ar$	See Table 3		
$H + HO + N_2 \rightarrow H_2O + N_2$	See Table 3		
$H + HO_2 \rightarrow H_2 + O_2$ $\rightarrow 2\,OH$ $\rightarrow H_2O + O$	$7.1 \times 10^{-11} \exp(-710/T)$ $2.8 \times 10^{-10} \exp(-440/T)$ $5.0 \times 10^{-11} \exp(-866/T)$	300–1000 300–1000 300–1000	± 0.3 ± 0.3 ± 0.3
$H + H_2O \rightarrow OH + H_2$	$7.5 \times 10^{-16}\, T^{1.6} \exp(-9270/T)$	300–2500	±0.2
$H + H_2O_2 \rightarrow H_2 + HO_2$ $\rightarrow OH + H_2O$	$2.8 \times 10^{-12} \exp(-1890/T)$ $1.7 \times 10^{-11} \exp(-1800/T)$	300–1000 300–1000	± 0.3 ± 0.3
$H + NH \rightarrow H_2 + N$	1.7×10^{-11}	1500–2500	± 1.0
$H + NH_2 \rightarrow H_2 + NH$	1.0×10^{-11}	2000–3000	± 1.0
$H + {}^3CH_2 \rightarrow H_2 + CH$	$1.0 \times 10^{-11} \exp(900/T)$	300–3000	± 0.7
$H + CH_3 \rightarrow H_2 + {}^1CH_2$ $\rightarrow CH_4$	$1.0 \times 10^{-10} \exp(-7600/T)$ See Table 3	300–2500	± 1.0
$H + CH_4 \rightarrow H_2 + CH_3$	$2.2 \times 10^{-20}\, T^{3.0} \exp(-4045/T)$	300–2500	± 0.2
$H + CHO \rightarrow H_2 + CO$	1.5×10^{-10}	300–2500	± 0.3
$H + HCHO \rightarrow H_2 + HCO$	$3.8 \times 10^{-14}\, T^{1.05} \exp(-1650/T)$	300–2200	± 0.1 at 300 K rising to ± 0.5 at 2200 K
$H + CH_3O \rightarrow H_2 + HCHO$	3.0×10^{-11}	300–1000	± 0.5
$H + HNCO \rightarrow NH_2 + CO$ $\rightarrow H_2 + NCO$	No recommendation $3.4 \times 10^{-10}\, T^{-0.27} \exp(-10190/T)$	 500–1000	 ± 1.0
$H + NCO \rightarrow NH + CO$ $\rightarrow HCN + O$]	8.7×10^{-11}	1400–1500	± 0.5

Table 1
BIMOLECULAR REACTIONS (continued)

Reaction	k/cm^3 molecule^{-1}s^{-1}	Temp/K	Error limits (Δ log k)
$H + C_2H_2 \rightarrow H_2 + C_2H$	$1.0 \times 10^{-10} \exp(-14000/T)$	1000–3000	± 1.0
$\rightarrow C_2H_3$	See Table 3		
$H + C_2H_3 \rightarrow H_2 + C_2H_2$	2.0×10^{-11}	300–2500	± 0.5
$\rightarrow C_2H_4$	See Table 3		
$H + C_2H_4 \rightarrow C_2H_3 + H_2$	$9.0 \times 10^{-10} \exp(-7500/T)$	700–2000	± 0.5
$\rightarrow C_2H_5$	See Table 3		
$H + C_2H_5 \rightarrow 2CH_3$	6.0×10^{-11}	300–2000	± 0.3
$\rightarrow C_2H_6$	See Table 3		
$H + C_2H_6 \rightarrow H_2 + C_2H_5$	$2.4 \times 10^{-15} T^{1.5} \exp(-3730/T)$	300–2000	± 0.15 at 300 K rising to ± 0.3 at 2000 K
$H + CHCO \rightarrow CH_2 + CO$]			
$\rightarrow H_2 + C_2O$]	2.5×10^{-10}	300–2500	± 0.4
$\rightarrow HCCOH$]			
$H + CH_2CO \rightarrow CH_3 + CO$	$3.0 \times 10^{-11} \exp(-1700/T)$	200–2000	± 0.5 at 200 K rising to ± 1.0 at 2000 K.
$\rightarrow CH_2CHO$	k_2/k very small		
$H + CH_3CHO \rightarrow H_2 + CH_3CO$]	$6.8 \times 10^{-15} T^{1.16} \exp(-1210/T)$	300–2000	± 0.1 at 300 rising to ± 0.4 at 2000 K.
$\rightarrow H_2 + CH_2CHO$]			
$H + C_6H_5 + M \rightarrow C_6H_6 + M$	See Table 3		
$H + C_6H_6 \rightarrow H_2 + C_6H_5$	No recommendation		
$\rightarrow C_6H_7$	See Table 3		
$H + C_6H_5O + M \rightarrow C_6H_5OH + M$	See Table 3		
$H + C_6H_5OH \rightarrow C_6H_5O + H_2$	$1.9 \times 10^{-10} \exp(-6240/T)$	1000–1150	± 0.3
$\rightarrow C_6H_6 + OH$	$3.7 \times 10^{-11} \exp(-3990/T)$	1000–1150	± 0.3
$H + C_6H_5CH_2 + M \rightarrow C_6H_5CH_3 + M$	See Table 3		
$H + C_6H_5CH_3 \rightarrow H_2 + C_6H_5CH_2$	$6.6 \times 10^{-22} T^{3.44} \exp(-1570/T)$	600–2800	± 0.3 at 600 K rising to ± 0.5 at 2800 K.
$\rightarrow H_2 + C_6H_4CH_3$	No recommendation		
$\rightarrow C_6H_6 + CH_3$	No recommendation		
$\rightarrow C_6H_6CH_3$	See Table 3		
$H + p\text{-}C_6H_4(CH_3)_2 \rightarrow$ products	5.8×10^{-13}	298	± 0.1
$H + C_6H_5C_2H_5 \rightarrow H_2 + C_6H_5C_2H_4$	2.4×10^{-12}	773	± 0.1
$\rightarrow C_6H_6C_2H_5$	See Table 3		
H_2 Reactions			
$H_2 + Ar \rightarrow 2H + Ar$	See Table 2		
$H_2 + H_2 \rightarrow 2H + H_2$	See Table 2		
OH Radical Reactions			
$OH + H_2 \rightarrow H_2O + H$	$1.7 \times 10^{-16} T^{1.6} \exp(-1660/T)$	300–2500	± 0.1 at 300 K rising to ± 0.3 at 2500 K
$OH + OH \rightarrow H_2O + O$	$2.5 \times 10^{-15} T^{1.14} \exp(-50/T)$	250–2500	± 0.2
$OH + OH + M \rightarrow H_2O_2 + M$	See Table 3		

Table 1
BIMOLECULAR REACTIONS (continued)

Reaction	k/cm^3 molecule^{-1}s^{-1}	Temp/K	Error limits (Δ log k)
$OH + HO_2 \rightarrow H_2O + O_2$	$4.8 \times 10^{-11} \exp(250/T)$	300–2000	± 0.2 at 300 K rising to ± 0.5 at 2000 K.
$OH + H_2O_2 \rightarrow H_2O + HO_2$	$1.3 \times 10^{-11} \exp(-670/T)$	300–1000	± 0.2
$OH + NH \rightarrow NO + H_2$] $\rightarrow H_2O + N$]	8.0×10^{-11}	300–1000	± 0.5
$OH + NH_2 \rightarrow O + NH_3$	$3.3 \times 10^{-14} T^{0.405} \exp(-250/T)$	500–2500	± 0.5
$\rightarrow H_2O + NH$	No recommendation		
$OH + CO \rightarrow H + CO_2$	$1.05 \times 10^{-17} T^{1.5} \exp(250/T)$	300–2000	± 0.2 at 300 K rising to ± 0.5 at 2000 K.
$OH + CH_3 \rightarrow H + CH_2OH$] $\rightarrow H + CH_3O$] $\rightarrow H_2O + {}^1CH_2$]	6.0×10^{-11}	300–2000	± 0.7
$\rightarrow CH_3OH$	See Table 3		
$OH + CH_4 \rightarrow H_2O + CH_3$	$2.6 \times 10^{-17} T^{1.83} \exp(-1400/T)$	250–2500	± 0.07 at 250 K rising to ± 0.15 at 1200 K.
$OH + CHO \rightarrow H_2O + CO$	1.7×10^{-10}	300–2500	± 0.3
$OH + HCHO \rightarrow H_2O + CHO$	$5.7 \times 10^{-15} T^{1.18} \exp(225/T)$	300–3000	± 0.1 at 300 K rising to ± 0.7 at 3000 K.
$OH + CN \rightarrow O + HCN$] $\rightarrow NCO + H$]	1.0×10^{-10}	1500–3000	± 0.5
$OH + HCN \rightarrow H_2O + CN$	$1.5 \times 10^{-11} \exp(-5400/T)$	1500–2500	± 0.5
$\rightarrow HOCN + H$] $\rightarrow HNCO + H$]	No recommendation		
$OH + CH_3OOH \rightarrow H_2O + CH_3OO$	$1.2 \times 10^{-12} \exp(130/T)$	300–1000	± 0.2 at 300 K rising to ± 0.4 at 1000 K
$\rightarrow H_2O + CH_2OOH$	$1.8 \times 10^{-12} \exp(220/T)$	300–1000	± 0.1 at 300 K rising to ±0.3 at 1000 K.
$OH + C_2H_2 \rightarrow H_2O + C_2H$] $\rightarrow H + CH_2CO$]	$1.0 \times 10^{-10} \exp(-6500/T)$	1000–2000	± 1.0
$\rightarrow C_2H_2OH$	See Table 3		
$OH + C_2H_4 \rightarrow H_2O + C_2H_3$	$3.4 \times 10^{-11} \exp(-2990/T)$	650–1500	± 0.5
$OH + C_2H_6 \rightarrow H_2O + C_2H_5$	$1.2 \times 10^{-17} T^{2.0} \exp(-435/T)$	250–2000	± 0.07 at 250 K rising to ± 0.15 at 2000 K.
$OH + CH_2CO \rightarrow CH_2OH + CO$] $\rightarrow H_2CO + HCO$]	1.7×10^{-11}	300–2000	± 1.0
$OH + CH_3CHO \rightarrow H_2O + CH_3CO$] $\rightarrow H_2O + CH_2CHO$]	$3.9 \times 10^{-14} T^{0.73} \exp(560/T)$	250–1200	± 0.1 at 250 K rising to ± 0.3 at 1200 K.
$OH + C_2H_5OOH \rightarrow H_2O + C_2H_5OO$] $\rightarrow H_2O + C_2H_4OOH$]	$3.0 \times 10^{-12} \exp(190/T)$ [estimate]	250–1000	± 0.3 at 250 K rising to ± 0.7 at 1000 K

KINETIC DATA FOR COMBUSTION MODELLING (continued)
Table 1
BIMOLECULAR REACTIONS (continued)

Reaction	k/cm^3 molecule^{-1}s^{-1}	Temp/K	Error limits (Δ log k)
$OH + C_6H_6 \rightarrow H_2O + C_6H_5$	$2.7 \times 10^{-16}\ T^{1.42} \exp(-730/T)$	400–1500	± 0.3
$\rightarrow H + C_6H_5OH$	$2.2 \times 10^{-11} \exp(-5330/T)$	1000–1150	± 0.3
$\rightarrow C_6H_6OH$	See Table 3		
$OH + C_6H_5OH \rightarrow C_6H_5(OH)_2$	See Table 3		
$\rightarrow H_2O + C_6H_5O$]			
$\rightarrow H_2O + C_6H_4OH$]	1.0×10^{-11}	1000–1150	± 0.5
$OH + C_6H_5CH_3 \rightarrow H_2O + C_6H_5CH_2$	$8.6 \cdot 10^{-15}\ T \exp(-1440/T)$ See Table 3	400–1200	± 0.5 at 400 K reducing to ± 0.3 at 1200 K.
$OH + p\text{-}C_6H_4(CH_3)_2 \rightarrow C_6H_4CH_2CH_3 + H_2O$	$6.4 \times 10^{-11} \exp(-1440/T)$	500–960	± 0.1
$\rightarrow p\text{-}C_6H_4(CH_3)_2OH$	See Table 3		
$OH + C_6H_5C_2H_5 \rightarrow HOC_6H_5C_2H_5$	See Table 3		
$\rightarrow H_2O + C_6H_5C_2H_4$]	8.7×10^{-12}	773	± 0.1
$\rightarrow H_2O + C_6H_4C_2H_5$]			
H_2O Reactions			
$H_2O + M \rightarrow H + OH + M$	See Table 2		
HO_2 Radical Reactions			
$HO_2 + HO_2 \rightarrow H_2O_2 + O_2$	$3.1 \times 10^{-12} \exp(-775/T)$	550–1250	± 0.15 at 550 K rising to ± 0.3 at 1250 K.
$HO_2 + NH_2 \rightarrow NH_3 + O_2$]			
$\rightarrow HNO + H_2O$]	2.6×10^{-11}	300–400	± 0.4
$HO_2 + CH_3 \rightarrow OH + CH_3O$	3×10^{-11}	300–2500	± 0.7
$\rightarrow O_2 + CH_4$	No recommendation		
$HO_2 + CH_4 \rightarrow H_2O_2 + CH_3$	$1.5 \times 10^{-11} \exp(-12400/T)$	600–1000	± 0.2 at 600 K rising to ± 0.3 at 1000 K.
$HO_2 + HCHO \rightarrow H_2O_2 + CHO$	$5.0 \times 10^{-12} \exp(-6580/T)$	600–1000	± 0.5
$HO_2 + C_2H_4 \rightarrow OH + C_2H_4O$	$3.7 \times 10^{-12} \exp(-8650/T)$	600–900	± 0.15 at 600 K rising to ± 0.25 at 900 K.
$HO_2 + C_2H_6 \rightarrow H_2O_2 + C_2H_5$	$2.2 \times 10^{-11} \exp(-10300/T)$	500–1000	± 0.2 at 500 K rising to ± 0.3 at 1000 K.
$HO_2 + CH_3CHO \rightarrow H_2O_2 + CH_3CO$	$5.0 \times 10^{-12} \exp(-6000/T)$	900–1200	± 0.7
H_2O_2 Reactions			
$H_2O_2 + M \rightarrow 2OH + M$	See Table 2		
N Atom Reactions			
$N + CN \rightarrow N_2 + C$	3×10^{-10}	300–2500	± 1.0
$N + NCO \rightarrow NO + CN$	No recommendation		
$\rightarrow N_2 + CO$	3.3×10^{-11}	1700	± 0.5

Table 1
BIMOLECULAR REACTIONS (continued)

Reaction	k/cm^3 molecule^{-1}s^{-1}	Temp/K	Error limits ($\Delta \log k$)
NH Radical Reactions			
$NH + O_2 \rightarrow NO + OH$ $\rightarrow NO_2 + H$ $\rightarrow HNO + O$	$1.26 \times 10^{-13} \exp(-770/T)$	270–550	$\pm$ 0.2 at 270 K rising to $\pm$ 0.5 at 550 K.
$NH + NO \rightarrow N_2O + H$ $\rightarrow HN_2 + O$ $\rightarrow N_2 + OH$	5.0×10^{-11}	270–380	$\pm$ 0.2
NH_2 Radical Reactions			
$NH_2 + O_2 \rightarrow$ products	$<3 \times 10^{-18}$	298	
$NH_2 + NO \rightarrow N_2 + H_2O$ $\rightarrow N_2 + H + OH$ $\rightarrow N_2H + OH$ $\rightarrow N_2O + H_2$	$1.8 \times 10^{-12} \exp(650/T)$ $(k_2 + k_3)/k \approx 0.12$ at 298 K.	220–2000	$\pm$ 0.5
$^1C_2$ and $^3C_2$ Radical Reactions	See data sheets.		
CH Radical Reactions			
$CH + O_2 \rightarrow CHO + O$ $\rightarrow CO + OH$	5.5×10^{-11}	300–2000	$\pm$ 0.3 at 300 K rising to $\pm$ 0.5 at 2000 K.
$CH + H_2 \rightarrow CH_2 + H$ $\rightarrow CH_3$	$2.4 \times 10^{-10} \exp(-1760/T)$	300–1000	$\pm$ 0.3
$CH + H_2O \rightarrow$ products	$9.5 \times 10^{-12} \exp(380/T)$	300–1000	$\pm$ 1.0
$CH + CO \rightarrow$ products	$4.6 \times 10^{-13} \exp(860/T)$	300–1000	$\pm$ 1.0
$CH + CO_2 \rightarrow$ products	$5.7 \times 10^{-12} \exp(-345/T)$	300–1000	$\pm$ 1.0
$CH + CH_4 \rightarrow$ products	$5.0 \times 10^{-11} \exp(200/T)$	200–700	$\pm$ 1.0
$CH + C_2H_2 \rightarrow$ products	$3.5 \times 10^{-10} \exp(61/T)$	200–700	$\pm$ 1.0
$CH + C_2H_4 \rightarrow$ products	$2.2 \times 10^{-10} \exp(173/T)$	200–700	$\pm$ 1.0
$CH + C_2H_6 \rightarrow$ products	$1.8 \times 10^{-10} \exp(132/T)$	200–700	$\pm$ 1.0
$CH + C_3H_8 \rightarrow$ products	$1.9 \times 10^{-10} \exp(240/T)$	300–700	$\pm$ 1.0
$CH + n\text{-}C_4H_{10} \rightarrow$ products	$4.4 \times 10^{-10} \exp(28/T)$	250–700	$\pm$ 1.0
$CH + i\text{-}C_4H_{10} \rightarrow$ products	$2.0 \times 10^{-10} \exp(240/T)$	300–700	$\pm$ 1.0
$CH + \text{neo-}C_5H_{12} \rightarrow$ products	$1.6 \times 10^{-10} \exp(340/T)$	300–700	$\pm$ 1.0
$CH + CH_3C_2H \rightarrow$ products	No recommendation		
$CH + CH_2O \rightarrow$ products	$1.6 \times 10^{-10} \exp(260/T)$	300–700	$\pm$ 1.0

Table 1

BIMOLECULAR REACTIONS (continued)

Reaction	k/cm³ molecule⁻¹s⁻¹	Temp/K	Error limits (Δ log k)
$^3CH_2$ Radical Reactions			
$^3CH_2 + O_2 \rightarrow CO + H + OH$ $\rightarrow CO_2 + H + H$ $\rightarrow CO + H_2O$ $\rightarrow CO_2 + H_2$ $\rightarrow HCHO + O$	$4.1 \times 10^{-11} \exp(-750/T)$	300–1000	± 0.3 at 300 K rising to ± 0.5 at 1000 K.
$^3CH_2 + {}^3CH_2 \rightarrow C_2H_2 + H_2$ $\rightarrow C_2H_2 + 2H$	$2.0 \times 10^{-10} \exp(-400/T)$ $k_2/k = 0.9 \pm 0.1$ over range 300–3000 K.	300–3000	± 0.5
$^3CH_2 + CH_3 \rightarrow C_2H_4 + H$	7.0×10^{-11}	300–3000	± 0.3 at 300 K rising to ± 0.5 at 3000 K.
$^3CH_2 + C_2H_2 \rightarrow C_3H_4$	See Table 3		
$^3CH_2 + C_2H_4 \rightarrow C_3H_6$ $\rightarrow c\text{-}C_3H_6$ $\rightarrow CH_2CHCH_2 + H$	See Table 3		
$^1CH_2$ Radical Reactions			
$^1CH_2 + Ar \rightarrow {}^3CH_2 + Ar$	6.0×10^{-12}	300–2000	± 0.3
$^1CH_2 + N_2 \rightarrow {}^3CH_2 + N_2$	1.0×10^{-11}	300–2000	± 0.3
$^1CH_2 + CH_4 \rightarrow {}^3CH_2 + CH_4$	1.2×10^{-11}	300–2000	± 0.4
$^1CH_2 + C_2H_2 \rightarrow {}^3CH_2 + C_2H_2$	8.0×10^{-11}	300–2000	± 0.4
$^1CH_2 + C_2H_4 \rightarrow {}^3CH_2 + C_2H_4$	2.3×10^{-11}	300–2000	± 0.4
$^1CH_2 + C_2H_6 \rightarrow {}^3CH_2 + C_2H_6$	3.6×10^{-11}	300–2000	± 0.4
$^1CH_2 + O_2 \rightarrow CO + H + OH$ $\rightarrow CO_2 + H_2$ $\rightarrow CO + H_2O$ $\rightarrow {}^3CH_2 + O_2$	5.2×10^{-11}	300–1000	± 0.3 at 300 K rising to ± 0.5 at 1000 K
$^1CH_2 + H_2 \rightarrow CH_3 + H$	1.2×10^{-10}	300–1000	± 0.1 at 300 K rising to ± 0.3 at 1000 K
$^1CH_2 + C_2H_2 \rightarrow CH_2CCH_2$ $\rightarrow CH_3CCH$ $\rightarrow CH_2CCH + H$ $\rightarrow {}^3CH_2 + C_2H_2$	See Table 3 See earlier entry		
$^1CH_2 + C_2H_4 \rightarrow C_3H_6$ $\rightarrow {}^3CH_2 + C_2H_4$	See Table 3 See earlier entry		
CH_3 Radical Reactions			
$CH_3 + M \rightarrow CH_2 + H + M$	See Table 2		
$CH_3 + O_2 \rightarrow CH_3O + O$ $\rightarrow HCHO + OH$ $\rightarrow CH_3O_2$	$2.2 \times 10^{-10} \exp(-15800/T)$ $5.5 \times 10^{-13} \exp(-4500/T)$ See Table 3	300–2500 1000–2500	± 0.5 ± 0.5

Table 1
BIMOLECULAR REACTIONS (continued)

Reaction	$k/cm^3\ molecule^{-1}s^{-1}$	Temp/K	Error limits ($\Delta \log k$)
$CH_3 + H_2 \rightarrow CH_4 + H$	$1.14 \times 10^{-20}\ T^{2.74} \exp(-4740/T)$	300–2500	± 0.15 in the range 300–700 K. ± 0.3 in the range 700–2500 K.
$CH_3 + CH_3 \rightarrow C_2H_5 + H$	$5 \times 10^{-11} \exp(-6800/T)$	1300–2500	± 0.6
$\rightarrow C_2H_4 + H_2$	No recommendation (see data sheets)		
$\rightarrow C_2H_6$	See Table 3		
$CH_3 + HCHO \rightarrow CH_4 + HCO$	$6.8 \times 10^{-12} \exp(-4450/T)$	300–1000	± 0.3
$CH_3 + C_2H_2 + M \rightarrow C_3H_5 + M$	See Table 3		
$\rightarrow CH_4 + C_2H$	No recommendation		
$CH_3 + C_2H_4 \rightarrow CH_4 + C_2H_3$	$6.9 \times 10^{-12} \exp(-5600/T)$	400–3000	± 0.5
$\rightarrow n\text{-}C_3H_7$	See Table 3		
$CH_3 + C_2H_5 \rightarrow CH_4 + C_2H_4$	1.9×10^{-12}	300–800	± 0.4
$\rightarrow C_3H_8$	See Table 3		
$CH_3 + C_2H_6 \rightarrow CH_4 + C_2H_5$	$2.5 \times 10^{-31}\ T^{6.0} \exp(-3043/T)$	300–1500	± 0.1 at 300 K rising to ± 0.2 at 1500 K.
$CH_3 + CH_3CHO \rightarrow CH_4 + CH_3CO$	$3.3 \times 10^{-30}\ T^{5.64} \exp(-1240/T)$	300–1250	± 0.3
$\rightarrow CH_4 + CH_2CHO$	No recommendation (see data sheets)		
CH_4 Reactions			
$CH_4 + M \rightarrow CH_3 + H + M$	See Table 2		
CHO Radical Reactions			
$CHO + O_2 \rightarrow CO + HO_2$; $\rightarrow OH + CO_2$; $\rightarrow HCO_3$	5.0×10^{-12}	300–2500	± 0.3
$CHO + CHO \rightarrow HCHO + CO$	5.0×10^{-11}	300	± 0.3
HCHO Reactions			
$HCHO + M \rightarrow H + CHO + M$; $\rightarrow H_2 + CO + M$	See Table 2		
CH_2OH Reactions			
$CH_2OH + O_2 \rightarrow CH_2O + HO_2$	$2.6 \times 10^{-9}\ T^{-1.0} + 1.2 \times 10^{-10} \exp(-1800/T)$	300–1200	± 0.1 at 300 K rising to ± 0.3 at 1200 K.
CH_3O Radical Reactions			
$CH_3O + M \rightarrow HCHO + H + M$	See Table 2		
$CH_3O + O_2 \rightarrow HCHO + HO_2$	$6.7 \times 10^{-14} \exp(-1070/T)$	300–1000	± 0.2 at 500 K rising to ± 0.3 at 300 K and 1000 K.

KINETIC DATA FOR COMBUSTION MODELLING (continued)

Table 1
BIMOLECULAR REACTIONS (continued)

Reaction	k/cm³ molecule⁻¹s⁻¹	Temp/K	Error limits (Δ log k)
CH_3OOH Reactions			
$CH_3OOH + M \rightarrow CH_3O + OH + M$	See Table 2		
CN Radical Reactions			
$CN + O_2 \rightarrow NCO + O$	$1.1 \times 10^{-11} \exp(205/T)$	300–2500	± 0.25 at 300 K rising to ± 0.5 at 2500 K.
$CN + H_2O \rightarrow HCN + OH$ $\rightarrow HOCN + H$	$1.3 \times 10^{-11} \exp(-3750/T)$	500–3000	± 0.3 at 500 K rising to ± 0.5 at 3000 K.
$CN + CH_4 \rightarrow HCN + CH_3$	$1.5 \times 10^{-11} \exp(-940/T)$	260–400	± 0.3
NCO Radical Reactions			
$NCO + M \rightarrow N + CO + M$	See Table 2		
$NCO + NO \rightarrow N_2O + CO$ $\rightarrow N_2 + CO_2$ $\rightarrow N_2 + CO + O$	$1.7 \times 10^{-11} \exp(200/T)$	300–600	± 0.5
C_2H Radical Reactions			
$C_2H + O_2 \rightarrow CO_2 + CH$ $\rightarrow 2CO + H$ $\rightarrow C_2HO + O$ $\rightarrow CO + HCO$	3.0×10^{-11}	300	± 0.5
$C_2H + H_2 \rightarrow C_2H_2 + H$	$2.5 \times 10^{-11} \exp(-1560/T)$	300–2500	± 0.3 at 300 K rising to ± 0.7 at 2500 K
$C_2H + C_2H_2 \rightarrow C_4H_2 + H$	5.0×10^{-11}	300–2700	± 0.3
$C_2H + CH_4 \rightarrow$ products	2.0×10^{-12}	298	± 1
$C_2H + C_2H_6 \rightarrow$ products	No recommendation		
C_2H_3 Radical Reactions			
$C_2H_3 + M \rightarrow C_2H_2 + H + M$	See Table 2		
$C_2H_3 + O_2 \rightarrow HCHO + CHO$	9.0×10^{-12}	300–2000	± 0.3 at 300 K rising to ± 0.5 at 2000 K
C_2H_5 Radical Reactions			
$C_2H_5 + O_2 \rightarrow C_2H_4 + HO_2$	$1.7 \times 10^{-14} \exp(1100/T)$	600–1200	± 0.3
$C_2H_5 + C_2H_5 \rightarrow C_2H_6 + C_2H_4$ $\rightarrow n\text{-}C_4H_{10}$	2.4×10^{-12} See Table 3	300–1200	± 0.4
C_2H_6 Reactions			
$C_2H_6 + M \rightarrow CH_3 + CH_3 + M$	See Table 2		

KINETIC DATA FOR COMBUSTION MODELLING (continued)
Table 1
BIMOLECULAR REACTIONS (continued)

Reaction	k/cm^3 molecule^{-1}s^{-1}	Temp/K	Error limits ($\Delta \log k$)
CHCO Reactions			
$CHCO + O_2 \rightarrow CO_2 + HCO$ $\rightarrow 2CO + OH$ $\rightarrow C_2O + HO_2$ $\rightarrow CHO_2CO$	$2.7 \times 10^{-12} \exp(430/T)$ M = He, 2 Torr	300–550	± 0.7
CH_2CHO Radical Reactions			
$CH_2CHO + O_2 \rightarrow HO_2 + CH_2CHO$ $\rightarrow HCHO + CO + OH$ $\rightarrow O_2CH_2CHO$	$k_\infty = 2.6 \times 10^{-13}$ $k_2 = 3.0 \times 10^{-14}$	250–1000 300	± 0.2 ± 0.3
CH_3CO Radical Reactions			
$CH_3CO + O_2 + M \rightarrow CH_3CO_3 + M$	See Table 3		
CH_3CHO Reactions			
$CH_3CHO + M \rightarrow CH_3 + HCO + M$	See Table 2		
C_2H_5O Reactions			
$C_2H_5O + M \rightarrow HCHO + CH_3 + M$ $\rightarrow CH_3CHO + H + M$	See Table 2		
$C_2H_5O + O_2 \rightarrow CH_3CHO + HO_2$	$1.0 \times 10^{-13} \exp(-830/T)$	300–1000	± 0.3 at 300 K rising to ± 0.5 at 1000 K
C_2H_5OOH Reactions			
$C_2H_5OOH + M \rightarrow C_2H_5O + OH + M$	See Table 2		
C_6H_5 Radical Reactions			
$C_6H_5 + M \rightarrow C_2H_2 + C_4H_3 + M$ $\rightarrow C_2H_3 + C_4H_2 + M$ $\rightarrow$ linear-$C_6H_5 + M$	See Table 2		
C_6H_6 Reactions			
$C_6H_6 + M \rightarrow C_6H_5 + H + M$ $\rightarrow C_4H_4 + C_2H_2 + M$	See Table 2		
C_6H_5O Radical Reactions			
$C_6H_5O + M \rightarrow C_5H_5 + CO + M$	See Table 2		
$C_6H_5CH_2$ Radical Reactions			
$C_6H_5CH_2 + M \rightarrow C_3H_3 + 2C_2H_2 + M$ $\rightarrow C_4H_4 + C_3H_3 + M$ $\rightarrow C_5H_5 + C_2H_2 + M$ $\rightarrow C_7H_7(BCH) + M$	See Table 2		

Table 1
BIMOLECULAR REACTIONS (continued)

Reaction	k/cm^3 molecule^{-1}s^{-1}	Temp/K	Error limits (Δ log k)
$C_6H_5CH_3$ Reactions			
$C_6H_5CH_3$ + M → $C_6H_5CH_2$ + H + M → C_6H_5 + CH_3 + M	See Table 2		
p-$C_6H_4(CH_3)_2$ Reactions			
p-$C_6H_4(CH_3)_2$ + M → $C_6H_4CH_2CH_3$ + H + M	See Table 2		
$C_6H_5C_2H_5$ Reactions			
$C_6H_5C_2H_5$ + M → $C_6H_5CH_2$ + CH + M → C_6H_6 + C_2H_4 + M → $C_6H_5CHCH_2$ + H_2 + M → C_6H_5 + C_2H_5 + M → $C_6H_5CHCH_3$ + H + M	See Table 2		

Table 2
DECOMPOSITION REACTIONS

Reaction	k_∞/s^{-1} k_0/cm^3 molecule^{-1} s^{-1} F_c $k/\text{s}^{-1} = \dfrac{k_0 k_\infty [\text{M}]}{k_0[\text{M}] + k_\infty} F$	Temp/K	Error limits (Δ log k)
H_2 + Ar → 2H + Ar	$k_0 = 3.7 \times 10^{-10} \exp(-48350/T)$	2500–8000	± 0.3
H_2 + H_2 → 2H + H_2	$k_0 = 1.5 \times 10^{-9} \exp(-48350/T)$	2500–8000	± 0.5
H_2O + N_2 → H + OH + N_2	$k_0 = 5.8 \times 10^{-9} \exp(-52920/T)$	2000–6000	± 0.5
H_2O_2 + M → 2 OH + M	$k_0(\text{Ar}) = 3 \times 10^{-8} \exp(-21600/T)$ $k_0(\text{N}_2) = 2 \times 10^{-7} \exp(-22900/T)$ $k_\infty = 3 \times 10^{14} \exp(-24400/T)$ $F_c(\text{Ar}) = 0.5$	1000–1500 700–1500 1000–1500 700–1500	± 0.2 ± 0.2 ± 0.5 ΔF_c = ± 0.1
CH_3 + M → CH_2 + H + M	$k_0 = 1.7 \times 10^{-8} \exp(-45600/T)$	1500–3000	± 0.5
CH_4 + M → CH_3 + H + M	$k_0(\text{Ar}) = 1.2 \times 10^{-6} \exp(-47000/T)$ $k_0(\text{CH}_4) = 1.4 \times 10^{-5} \exp(-48100/T)$ $k_\infty = 2.4 \times 10^{16} \exp(-52800/T)$ $F_c\ (\text{Ar}) = \exp(-0.45 - T/3231)$ $F_c\ (\text{CH}_4) = \exp(-0.37 - T/2210)$	1000–3000 1000–2000 1000–3000 1000–3000 1000–2000	± 0.3 ± 0.3 ± 0.5 ΔF_c = ± 0.1 ΔF_c = ± 0.1
HCHO + M → H + CHO + M → H_2 + CO + M	$k_0(1) = 2.1 \times 10^{-8} \exp(-39200/T)$ $k_1/k_2 = 0.5$ at 2200 K	1500–2500	± 0.3
CH_3O + M → HCHO + H + M	$k_0 = 3.16 \times 10^2\, T^{-2.7} \exp(-15400/T)$ [estimate]	300–1000	± 1.0
CH_3OOH + M → CH_3O + OH + M	$k_\infty = 4 \times 10^{15} \exp(-21600/T)$	400–1000	± 0.5 at 600 K rising to ± 1.0 at 400 and 1000 K
NCO + Ar → N + CO + Ar	$k_0 = 1.7 \times 10^{-9} \exp(-23500/T)$	1450–2600	± 0.4

Table 2
DECOMPOSITION REACTIONS (continued)

Reaction	k_∞/s^{-1} $k_0/cm^3\ molecule^{-1}\ s^{-1}$ F_c $k/s^{-1} = \frac{k_0 k_\infty [M]}{k_0[M] + k_\infty} F$	Temp/K	Error limits ($\Delta \log k$)
$C_2H_3 + M \rightarrow C_2H_2 + H + M$	$k_0 = 6.9 \times 10^{17}\ T^{-7.5} \exp(-22900/T)$ $k_\infty = 2 \times 10^{14} \exp(-20000/T)$ $F_c = 0.35$	500–2500 500–2500 500–2500	± 0.5 ± 0.5 $\Delta F_c = \pm 0.1$
$C_2H_6 + M \rightarrow 2CH_3 + M$	$k_0(Ar) = 1.1 \times 10^{25}\ T^{-8.24} \exp(-47090/T)$ $k_0(C_2H_6) = 4.5 \times 10^{-2} \exp(-41930/T)$ $k_\infty = 1.8 \times 10^{21}\ T^{-1.24} \exp(-45700/T)$ $F_c(Ar) = 0.38 \exp(-T/73) + 0.62 \exp(-T/1180)$ $F_c(C_2H_6) = 0.54 \exp(-T/1250)$	300–2000 800–1000 300–2000 300–2000 800–1000	± 0.5 ± 0.5 ± 0.3 $\Delta F_c = \pm 0.1$ $\Delta F_c = \pm 0.1$
$CH_3CHO + M \rightarrow CH_3 + CHO + M$	$k(1\ atm.) = 7 \times 10^{15} \exp(-41100/T)$ (pressure dependent region)	750–1200	± 0.4
$C_2H_5O + M \rightarrow HCHO + CH_3 + M$	$k_\infty = 8 \times 10^{13} \exp(-10830/T)$ [estimate]	300–600	± 1.0
$C_2H_5OOH + M \rightarrow C_2H_5O + OH + M$	$k_\infty\ 4 \times 10^{15} \exp(-21600/T)$	400–1000	± 1.0
$C_6H_5 + M \rightarrow C_2H_2 + C_4H_3 + M$ $\rightarrow C_2H_3 + C_4H_2 + M$	No recommendation		
$\rightarrow$ linear-CH_5 + M	$4.0 \times 10^{13} \exp(-36700/T)$	1450–1900	± 0.4
$C_6H_6 + M \rightarrow C_6H_5 + H + M$ $\rightarrow C_4H_4 + H_2 + M$	$9.0 \times 10^{15} \exp(-54060/T)$	1200–2500	± 0.4 at 1200 K reducing to ± 0.3 at 2500 K
$C_6H_5O + M \rightarrow C_5H_5 + CO + M$	$2.5 \times 10^{11} \exp(-22100/T)$	1000–1580	± 0.2
$C_6H_5CH_2 + M \rightarrow C_3H_3 + 2C_2H_2 + M$ $\rightarrow C_4H_4 + C_3H_3 + M$ $\rightarrow C_5H_5 + C_2H_2 + M$ $\rightarrow C_7H_7$ (BCH) + M	$5.1 \times 10^{13} \exp(-36370/T)$	1350–1900	± 0.3 at 1350 K rising to ± 0.5 1900 K
$C_6H_5CH_3 + M \rightarrow C_6H_5CH_2 + H + M$ $\rightarrow C_6H_5 + CH_3 + M$	$3.1 \times 10^{15} \exp(-44890/T)$ No recommendation	920–2200	± 0.3 at 900 K rising to ± 0.5 at 2200 K
p-$C_6H_4(CH_3)_2 + M \rightarrow p$-$C_6H_4CH_2CH_3$ + H + M	$4.0 \times 10^{15} \exp(-42600/T)$	1400–1800	± 0.5
$C_6H_5C_2H_5 + M \rightarrow C_6H_5CH_2 + CH + M$	$6.1 \times 10^{15} \exp(-37800/T)$	770–1800	± 0.1 at 770 K rising to ± 0.4 at
$\rightarrow C_6H_6 + C_2H_4 + M$ $\rightarrow C_6H_5CHCH_2 + H_2 + M$ $\rightarrow C_6H_5 + C_2H_5 + M$ $\rightarrow C_6H_5CHCH_3 + H + M$	No recommendations		

Table 3
COMBINATION REACTIONS

Reaction	$k_\infty/cm^3\ molecule^{-1}s^{-1}$ $k_0/cm^6\ molecule^{-2}\ s^{-1}$ F_c $k/cm^3\ molecule^{-1}s^{-1} = \frac{k_0 k_\infty [M]}{k_0[M] + k_\infty} F$	Temp/K	Error limits ($\Delta \log k$)
$H + O_2 + Ar \rightarrow HO_2 + Ar$	$k_0 = 1.7 \times 10^{-30}\ T^{-0.8}$	300–2000	± 0.5

Table 3
COMBINATION REACTIONS (continued)

Reaction	k_∞/cm³ molecule⁻¹s⁻¹ k_0/cm⁶ molecule⁻² s⁻¹ F_c k/cm³ molecule⁻¹s⁻¹ $= \frac{k_0 k_\infty [M]}{k_0[M] + k_\infty} F$	Temp/K	Error limits ($\Delta \log k$)
$H + O_2 + H_2 \rightarrow HO_2 + H_2$	$k_0 = 5.8 \times 10^{-30} T^{-0.8}$	300–2000	± 0.5
$H + O_2 + N_2 \rightarrow HO_2 + N_2$	$k_0 = 3.9 \times 10^{-30} T^{-0.8}$	300–2000	± 0.5
$H + O_2 + H_2O \rightarrow HO_2 + H_2O$	$k_0 = 4.3 \times 10^{-30} T^{-0.8}$	300–2000	± 0.5
$H + H + Ar \rightarrow H_2 + Ar$	$k_0 = 1.8 \times 10^{-30} T^{-1.0}$	300–2500	± 0.5
$H + H + H_2 \rightarrow H_2 + H_2$	$k_0 = 2.7 \times 10^{-31} T^{-0.6}$	100–5000	± 0.5
$H + OH + H_2O \rightarrow H_2O + H_2O$	$k_0 = 3.9 \times 10^{-25} T^{-2.0}$	300–3000	± 0.3
$H + OH + Ar \rightarrow H_2O + Ar$	$k_0 = 2.3 \times 10^{-26} T^{-2.0}$	300–3000	± 0.3
$H + OH + N_2 \rightarrow H_2O + N_2$	$k_0 = 6.1 \times 10^{-26} T^{-2.0}$	300–3000	± 0.3
$H + CH_3 + M \rightarrow CH_4 + M$	$k_0(He) = 6.2 \times 10^{-29} (T/3000)^{-1.8}$ $k_0(Ar) = 6 \times 10^{-29} (T/300)^{-1.8}$ $k_0(C_2H_6) = 3 \times 10^{-28} (T/300)^{-1.8}$ $k_\infty = 3.5 \times 10^{-10}$ $F_c(He,Ar) = \exp(-0.45 - T/3231)$ $F_c(C_2H_6) = \exp(-0.34 - T/3053)$	300–1000 300–1000 300–1000 300–1000 300–1000 300–1000	± 0.3 ± 0.5 ± 0.5 ± 0.3 ΔF_c = ± 0.1 ΔF_c = ± 0.1
$H + C_2H_2 + He \rightarrow C_2H_3 + He$	$k_\infty = 1.4 \times 10^{-11} \exp(-1300/T)$ $k_0 = 3.3 \times 10^{-30} \exp(-740/T)$ $F_c = 0.44$	200–400 200–400 200–400	± 0.3 ± 0.5 ΔF_c = ± 0.1
$H + C_2H_3 + M \rightarrow C_2H_4 + M$	No recommendation		
$H + C_2H_4 + M \rightarrow C_2H_5 + M$	No recommendation		
$H + C_2H_5 + M \rightarrow C_2H_6 + M$	No recommendation		
$H + C_6H_5 + M \rightarrow C_6H_6 + M$	$k_\infty = 1.3 \times 10^{-10}$	1400–1700	± 0.5
$H + C_6H_6 + M \rightarrow C_6H_7 + M$	$k_\infty = 6.7 \times 10^{-11} \exp(-2170/T)$	300–1000	± 0.2
$H + C_6H_5O + M \rightarrow C_6H_5OH + M$	$k_\infty = 4.2 \times 10^{-10}$	1000	± 0.3
$H + C_6H_5CH_2 + M \rightarrow C_6H_5CH_3 + M$	$k_\infty = 5.5 \times 10^{-10}$	300–2000	± 0.2 at 300 K rising to ± 0.7 at 2000 K.
$H + C_6H_5CH_3 + M \rightarrow C_6H_6CH_3 + M$	$k_\infty = 1.2 \times 10^{-13}$	298	± 0.2
$H + C_6H_5C_2H_5 + M \rightarrow C_6H_6C_2H_5 + M$	$k_\infty = 3.3 \times 10^{-13}$	298	± 0.1
$OH + OH + M \rightarrow H_2O_2 + M$	$k_0(N_2) = 8 \times 10^{-31} (T/300)^{-0.76}$ $k_0(H_2O) = 4 \times 10^{-30}$ $k_\infty = 1.5 \times 10^{-11} (T/300)^{-0.37}$ $F_c(N_2) = 0.5$	250–1400 300–400 200–1500 200–1500	± 0.4 ± 0.5 ΔF_c = ± 0.2
$OH + CH_3 + M \rightarrow CH_3OH + M$	No data available for this channel (See Table 1)		

KINETIC DATA FOR COMBUSTION MODELLING (continued)
Table 3
COMBINATION REACTIONS (continued)

Reaction	k_∞/cm³ molecule⁻¹s⁻¹ k_0/cm⁶ molecule⁻² s⁻¹ F_c k/cm³ molecule⁻¹s⁻¹ = $\frac{k_0 k_\infty [M]}{k_0[M] + k_\infty} F$	Temp/K	Error limits (Δ log k)
$OH + C_2H_2 + M \rightarrow C_2H_2OH + M$	See data sheet		
$OH + C_6H_6 + M \rightarrow C_6H_6OH + M$	$k_\infty = 3.8 \times 10^{-12} \exp(-340/T)$	240–340	± 0.2
$OH + C_6H_5OH + M \rightarrow C_6H_5(OH)_2 + M$	$k_\infty = 2.8 \times 10^{-11}$	298	± 0.1
$OH + C_6H_5CH_3 + M \rightarrow HOC_6H_5CH_3 + M$	$k_\infty = 3.8 \times 10^{-12} \exp(180/T)$	200–300	± 0.4
$OH + C_6H_4(CH_3)_2 + M \rightarrow C_6H_4(CH_3)_2OH + M$	$k_\infty = 1.4 \times 10^{-11}$	300–320	± 0.1
$OH + C_6H_5C_2H_5 + M \rightarrow HOC_6H_5C_2H_5 + M$	7.5×10^{-12} at $p \leqslant 1$ atm.	298	± 0.1
$^3CH_2 + C_2H_2 + M \rightarrow C_3H_4 + M$	$2.0 \times 10^{-11} \exp(-3330/T)$ at $p = \leqslant 10$ Torr.	300–1000	± 0.3
$^3CH_2 + C_2H_4 + M \rightarrow C_3H_6 + M$ $\rightarrow c\text{-}C_3H_6 + M$ $\rightarrow C_3H_5 + H + M$	$5.3 \times 10^{-12} \exp(-2660/T)$	300–1000	± 0.2 at 300 K rising to ± 0.3 at 1000 K
$^1CH_2 + C_2H_2 + M \rightarrow CH_2CCH_2 + M$ $\rightarrow CH_3CCH + M$ $\rightarrow CH_2CCH + H + M$	3.7×10^{-10} independent of p	300–1000	± 0.3 at 300 K rising to ± 0.7 at 1000 K
$^1CH_2 + C_2H_4 + M \rightarrow C_3H_6$	1.1×10^{-10} independent of p	300–1000	± 0.2 at 300 K rising to ± 0.5 at 1000 K.
$CH_3 + O_2 + M \rightarrow CH_3O_2 + M$	$k_0(Ar) = 1.5 \times 10^{-22} T^{-3.3}$ $k_0(N_2) = 1.6 \times 10^{-22} T^{-3.3}$ $k_\infty = 1.3 \times 10^{-15} T^{-1.2}$ $F_c = 0.466 - 1.30 \times 10^{-4} T$	300–800 300–800 300–800 300–800	± 0.3 ± 0.3 ± 0.3
$CH_3 + CH_3 + Ar \rightarrow C_2H_6 + Ar$	$k_\infty = 6 \times 10^{-11}$	300–2000	± 0.05 at 300 K rising to ± 0.3 at 2000 K
	$k_0 = 3.5 \times 10^{-7} T^{-7.0} \exp(-1390/T)$	300–2000	± 0.3
	$F_c = 0.38 \exp(-T/73)$ $+ 0.62\exp(-T/1180)$	300–2000	$\Delta F_c = \pm 0.1$
$CH_3 + C_2H_2 + M \rightarrow C_3H_5 + M$	$k_\infty = 1 \times 10^{-12} \exp(-3900/T)$	300–600	± 0.5
$CH_3 + C_2H_4 + M \rightarrow n\text{-}C_3H_7 + M$	$3.5 \times 10^{-13} \exp(-3700/T)$	300–600	± 0.3
$CH_3 + C_2H_5 + M \rightarrow C_3H_8 + M$	$k_\infty = 4.7 \times 10^{-11}$	300–800	± 0.3
$C_2H_5 + C_2H_5 + M \rightarrow n\text{-}C_4H_{10} + M$	$k_\infty = 1.9 \times 10^{-11}$	300–1200	± 0.3
$CH_3CO + O_2 + M \rightarrow CH_3CO_3 + M$	2×10^{-12} for $p = 1\text{–}4$ Torr.	300	± 0.3

Section 6
Fluid Properties

Section 6
Fluid Properties

THERMODYNAMIC PROPERTIES OF AIR

P(boil) is the pressure at which boiling begins.
P(con) is the pressure at which condensation begins.

References:

A. A. Vasserman and V. A. Rabinovich, *Thermophysical Properties of Liquid Air and its Components*, Izdatel'stvo Komiteta, Standartov, Moscow, 1968
A. A. Vasserman et al., *Thermophysical Properties of Air and Air Components*, Izdatel'stvo Nauka, Moscow, 1966

Properties in the saturation state:

T K	*P*(boil) bar	*P*(con) bar	ρ (liq) g/cm³	ρ (gas) g/L
65	0.1468	0.0861	0.939	0.464
70	0.3234	0.2052	0.917	1.033
75	0.6366	0.4321	0.894	2.048
80	1.146	0.8245	0.871	3.709
85	1.921	1.453	0.845	6.258
90	3.036	2.397	0.819	9.980
95	4.574	3.748	0.792	15.21
100	6.621	5.599	0.763	22.39
110	12.59	11.22	0.699	45.15
120	21.61	20.14	0.622	87.34
130	34.16	33.32	0.487	184.33
132.55	37.69	37.69	0.313	312.89

Properties of liquid air:

P bar	*T* K	ρ g/cm³	*H* J/g	*S* J/g K	C_p J/g K
1	75	0.8935	–131.7	2.918	1.843
5	75	0.8942	–131.4	2.916	1.840
5	80	0.8718	–122.3	3.031	1.868
5	85	0.8482	–112.9	3.143	1.901
5	90	0.8230	–103.3	3.250	1.941
5	95	0.7962	–93.5	3.356	1.991
10	75	0.8952	–131.1	2.913	1.836
10	80	0.8729	–122.0	3.028	1.863
10	90	0.8245	–103.1	3.246	1.932
10	100	0.7695	–83.2	3.452	2.041
50	75	0.9025	–128.2	2.892	1.806
50	100	0.7859	–81.8	3.415	1.939
50	125	0.6222	–28.3	3.889	2.614
50	150	0.1879	91.9	4.764	2.721
100	75	0.9111	–124.5	2.867	1.774
100	100	0.8033	–79.4	3.376	1.852
100	125	0.6746	–31.4	3.805	2.062
100	150	0.4871	32.8	4.271	2.832

Properties of air in the gaseous state:

P bar	*T* K	ρ g/L	*H* J/g	*S* J/g K	C_p J/g K
1	100	3.556	98.3	5.759	1.032
1	200	1.746	199.7	6.463	1.007
1	300	1.161	300.3	6.871	1.007

THERMODYNAMIC PROPERTIES OF AIR (continued)

P bar	*T* K	ρ g/L	*H* J/g	*S* J/g K	C_p J/g K
1	500	0.696	503.4	7.389	1.030
1	1000	0.348	1046.6	8.138	1.141
10	200	17.835	195.2	5.766	1.049
10	300	11.643	298.3	6.204	1.021
10	500	6.944	502.9	6.727	1.034
10	1000	3.471	1047.2	7.477	1.142
100	200	213.950	148.8	4.949	1.650
100	300	116.945	279.9	5.486	1.158
100	500	66.934	499.0	6.048	1.073
100	1000	33.613	1052.4	6.812	1.151

SOLUBILITY OF SELECTED GASES IN WATER

L. H. Gevantman

The values in this table are taken almost exclusively from the International Union of Pure and Applied Chemistry "Solubility Data Series". Unless noted, they comprise evaluated data fitted to a smoothing equation. The data at each temperature are then derived from the smoothing equation which expresses the mole fraction solubility X_1 of the gas in solution as:

$$\ln X_1 = A + B/T^* + C \ln T^*$$

where

$$T^* = T/100 \text{ K}$$

All values refer to a partial pressure of the gas of 101.325 kPa (one atmosphere).

The equation constants, the standard deviation for $\ln X_1$ (except where noted), and the temperature range over which the equation applies are given in the column headed Equation constants. There are two exceptions. The equation for methane has an added term, DT^*. The equation for H_2Se and H_2S takes the form,

$$\ln X_1 = A + B/T + C \ln T + DT$$

where T is the temperature in kelvin.

Solubilities given for those gases which react with water, namely ozone, nitrogen oxides, chlorine and its oxides, carbon dioxide, hydrogen sulfide, hydrogen selenide and sulfur dioxide, are recorded as bulk solubilities; i.e., all chemical species of the gas and its reaction products with water are included.

Gas	T/K	Solubility (X_1)	Equation constants	Ref.
Hydrogen (H_2)	288.15	1.510×10^{-5}	$A = -48.1611$	1
$M_r = 2.01588$	293.15	1.455×10^{-5}	$B = 55.2845$	
	298.15	1.411×10^{-5}	$C = 16.8893$	
	303.15	1.377×10^{-5}	Std. dev. = ± 0.54%	
	308.15	1.350×10^{-5}	Temp.range = 273.15—353.15	
Deuterium (D_2)	283.15	1.675×10^{-5} ± 0.57%	Averaged experimental	1
$M_r = 4.0282$	288.15	1.595×10^{-5} ± 0.57%	values	
	293.15	1.512×10^{-5} ± 0.78%	Temp. range = 278.15—303.15	
	298.15	1.460×10^{-5} ± 0.52%		
	303.15	1.395×10^{-5} ± 0.37%		
Helium (He)	288.15	7.123×10^{-6}	$A = -41.4611$	2
$A_r = 4.0026$	293.15	7.044×10^{-6}	$B = 42.5962$	
	298.15	6.997×10^{-6}	$C = 14.0094$	
	303.15	6.978×10^{-6}	Std. dev. = ±0.54%	
	308.15	6.987×10^{-6}	Temp.range = 273.15—348.15	
Neon (Ne)	288.15	8.702×10^{-6}	$A = -52.8573$	2
$A_r = 20.1797$	293.15	8.395×10^{-6}	$B = 61.0494$	
	298.15	8.152×10^{-6}	$C = 18.9157$	
	303.15	7.966×10^{-6}	Std. dev. = ±0.47%	
	308.15	7.829×10^{-6}	Temp.range = 273.15—348.15	
Argon (Ar)	288.15	3.025×10^{-5}	$A = -57.6661$	3
$A_r = 39.948$	293.15	2.748×10^{-5}	$B = 74.7627$	
	298.15	2.519×10^{-5}	$C = 20.1398$	
	303.15	2.328×10^{-5}	Std. dev. = ±0.26%	
	308.15	2.169×10^{-5}	Temp.range = 273.15—348.15	
Krypton (Kr)	288.15	5.696×10^{-5}	$A = -66.9928$	4
$A_r = 83.80$	293.15	5.041×10^{-5}	$B = 91.0166$	
	298.15	4.512×10^{-5}	$C = 24.2207$	

Gas	*T*/K	Solubility (X_1)	Equation constants	Ref.
	303.15	4.079×10^{-5}	Std. dev. = ±0.32%	
	308.15	3.725×10^{-5}	Temp.range = 273.15—353.15	
Xenon (Xe)	288.15	10.519×10^{-5}	$A = -74.7398$	4
$A_r = 131.29$	293.15	9.051×10^{-5}	$B = 105.210$	
	298.15	7.890×10^{-5}	$C = 27.4664$	
	303.15	6.961×10^{-5}	Std. dev. = ±0.35%	
	308.15	6.212×10^{-5}	Temp.range = 273.15—348.15	
Radon-222(^{222}Rn)	288.15	2.299×10^{-4}	$A = -90.5481$	
$A_r = 222$	293.15	1.945×10^{-4}	$B = 130.026$	
	298.15	1.671×10^{-4}	$C = 35.0047$	
	303.15	1.457×10^{-4}	Std. dev. = ±1.02%	
	308.15	1.288×10^{-4}	Temp.range = 273.15—373.15	
Oxygen (O_2)	288.15	2.756×10^{-5}	$A = -66.7354$	5
$M_r = 31.9988$	293.15	2.501×10^{-5}	$B = 87.4755$	
	298.15	2.293×10^{-5}	$C = 24.4526$	
	303.15	2.122×10^{-5}	Std. dev. = ±0.36%	
	308.15	1.982×10^{-5}	Temp.range = 273.15—348.15	
Ozone (O_3)	293.15	$1.885 \times 10^{-6} \pm 10\%$	Experimental value derived	5
$M_r = 47.9982$		pH = 7.0	from Henry's Law Constant	
Nitrogen (N_2)	288.15	1.386×10^{-5}	$A = -67.3877$	6
$M_r = 28.0134$	293.15	1.274×10^{-5}	$B = 86.3213$	
	298.15	1.183×10^{-5}	$C = 24.7981$	
	303.15	1.108×10^{-5}	Std. dev. = ±0.72%	
	308.15	1.047×10^{-5}	Temp.range = 273.15—348.15	
Nitrous oxide (N_2O)	288.15	5.948×10^{-4}	$A = -60.7467$	7
$M_r = 44.0129$	293.15	5.068×10^{-4}	$B = 88.8280$	
	298.15	4.367×10^{-4}	$C = 21.2531$	
	303.15	3.805×10^{-4}	Std. dev. = ±1.2%	
	308.15	3.348×10^{-4}	Temp.range = 273.15—313.15	
Nitric oxide (NO)	288.15	4.163×10^{-5}	$A = -62.8086$	7
$M_r = 30.0061$	293.15	3.786×10^{-5}	$B = 82.3420$	
	298.15	3.477×10^{-5}	$C = 22.8155$	
	303.15	3.222×10^{-5}	Std. dev. = ±0.76%	
	308.15	3.012×10^{-5}	Temp.range = 273.15—358.15	
Carbon monoxide (CO)	288.15	2.095×10^{-5}	Derived from Henry's	8
$M_r = 28.0104$	293.15	1.918×10^{-5}	Law Constant Equation	
	298.15	1.774×10^{-5}	Std. dev. = ±0.043%	
	303.15	1.657×10^{-5}	Temp.range = 273.15—328.15	
	308.15	1.562×10^{-5}		
Carbon dioxide (CO_2)	288.15	8.21×10^{-4}	Derived from Henry's	9
$M_r = 44.0098$	293.15	7.07×10^{-4}	Law Constant Equation	
	298.15	6.15×10^{-4}	Std. dev. = ±1.1%	
	303.15	5.41×10^{-4}	Temp.range = 273.15—353.15	
	308.15	4.80×10^{-4}		
Hydrogen selenide (H_2Se)	288.15	1.80×10^{-3}	$A = 9.15$	10
$M_r = 80.976$	298.15	1.49×10^{-3}	$B = 974$	
	308.15	1.24×10^{-3}	$C = -3.542$	
			$D = 0.0042$	

Gas	T/K	Solubility (X_1)	Equation constants	Ref.
			Std. dev. = ±2.3 × 10^{-5}	
			Temp. range = 288.15—343.15	
Hydrogen sulfide (H_2S)	288.15	2.335 × 10^{-3}	A = –24.912	10
M_r = 34.082	293.15	2.075 × 10^{-3}	B = 3477	
	298.15	1.85 × 10^{-3}	C = 0.3993	
	303.15	1.66 × 10^{-3}	D = 0.0157	
	308.15	1.51 × 10^{-3}	Std. dev. = ±6.5 × 10^{-5}	
			Temp. range = 283.15—603.15	
Sulfur dioxide (SO_2)	288.15	3.45 × 10^{-2}	A = –25.2629	11
M_r = 64.0648	293.15	2.90 × 10^{-2}	B = 45.7552	
	298.15	2.46 × 10^{-2}	C = 5.6855	
	303.15	2.10 × 10^{-2}	Std. dev. = ±1.8%	
	308.15	1.80 × 10^{-2}	Temp.range = 278.15—328.15	
Chlorine (Cl_2)	283.15	2.48 × 10^{-3} ± 2%	Experimental data	11
M_r = 70.9054	293.15	1.88 × 10^{-3} ± 2%	Temp.range = 283.15—333.15	
	303.15	1.50 × 10^{-3} ± 2%		
	313.15	1.23 × 10^{-3} ± 2%		
Chlorine monoxide (Cl_2O)	273.15	5.25 × 10^{-1} ± 1%	Experimental data	11
M_r = 86.9048	276.61	4.54 × 10^{-1} ± 1%	Temp. range = 273.15—293.15	
	283.15	4.273 × 10^{-1} ± 1%		
	293.15	3.353 × 10^{-1} ± 1%		
Chlorine dioxide (ClO_2)	288.15	2.67 × 10^{-2}	A = 7.9163	11
M_r = 67.4515	293.15	2.20 × 10^{-2}	B = 0.4791	
	298.15	1.823 × 10^{-2}	C = 11.0593	
	303.15	1.513 × 10^{-2}	Std. dev. = ±4.6%	
	308.15	1.259 × 10^{-2}	Temp.range = 283.15—333.15	
Methane (CH_4)	288.15	3.122 × 10^{-5}	A = –115.6477	12
M_r = 16.0428	293.15	2.806 × 10^{-5}	B = 155.5756	
	298.15	2.552 × 10^{-5}	C = 65.2553	
	303.15	2.346 × 10^{-5}	D = –6.6170	
	308.15	2.180 × 10^{-5}	Std. dev. = ±0.056%	
			Temp.range = 273.15—328.15	
Ethane (C_2H_6)	288.15	4.556 × 10^{-5}	A = –90.8225	13
M_r = 30.0696	293.15	3.907 × 10^{-5}	B = 126.9559	
	298.15	3.401 × 10^{-5}	C = 34.7413	
	303.15	3.002 × 10^{-5}	Std. dev. = ±0.13%	
	308.15	2.686 × 10^{-5}	Temp.range = 273.15—323.15	
Propane (C_3H_8)	288.15	3.813 × 10^{-5}	A = –102.044	14
M_r = 44.097	293.15	3.200 × 10^{-5}	B = 144.345	
	298.15	2.732 × 10^{-5}	C = 39.4740	
	303.15	2.370 × 10^{-5}	Std. dev. = ±0.012%	
	308.15	2.088 × 10^{-5}	Temp.range = 273.15—347.15	
Butane (C_4H_{10})	288.15	3.274 × 10^{-5}	A = –102.029	14
M_r = 58.123	293.15	2.687 × 10^{-5}	B = 146.040	
	298.15	2.244 × 10^{-5}	C = 38.7599	
	303.15	1.906 × 10^{-5}	Std. dev. = ±0.026%	
	308.15	1.645 × 10^{-5}	Temp.range = 273.15—349.15	
2-Methyl propane (Isobutane)	288.15	2.333 × 10^{-5}	A = –129.714	14

SOLUBILITY OF SELECTED GASES IN WATER (continued)

Gas	T/K	Solubility (X_1)	Equation constants	Ref.
(C_4H_{10})	293.15	1.947×10^{-5}	B = 183.044	
M_r = 58.123	298.15	1.659×10^{-5}	C = 53.4651	
	303.15	1.443×10^{-5}	Std. dev. = ±0.034%	
	308.15	1.278×10^{-5}	Temp.range = 278.15—318.15	

REFERENCES

1. C. L. Young, Ed., *IUPAC Solubility Data Series*, Vol. 5/6, Hydrogen and Deuterium, Pergamon Press, Oxford, England, 1981.
2. H. L. Clever, Ed., *IUPAC Solubility Data Series*, Vol. 1, Helium and Neon, Pergamon Press, Oxford, England, 1979.
3. H. L. Clever, Ed., *IUPAC Solubility Data Series*, Vol. 4, Argon, Pergamon Press, Oxford, England, 1980.
4. H. L. Clever, Ed., *IUPAC Solubility Data Series*, Vol. 2, Krypton, Xenon and Radon, Pergamon Press, Oxford, England, 1979.
5. R. Battino, Ed., *IUPAC Solubility Data Series*, Vol. 7, Oxygen and Ozone, Pergamon Press, Oxford, England, 1981.
6. R. Battino, Ed., *IUPAC Solubility Data Series*, Vol. 10, Nitrogen and Air, Pergamon Press, Oxford, England, 1982.
7. C. L. Young, Ed., *IUPAC Solubility Data Series*, Vol. 8, Oxides of Nitrogen, Pergamon Press, Oxford, England, 1981.
8. R. W. Cargill, Ed., *IUPAC Solubility Data Series*, Vol. 43, Carbon Monoxide, Pergamon Press, Oxford, England, 1990.
9. R. Crovetto, Evaluation of Solubility Data for the System CO_2-H_2O, *J. Phys. Chem. Ref. Data*, 20, 575, 1991.
10. P. G. T. Fogg and C. L. Young, Eds., *IUPAC Solubility Data Series*, Vol. 32, Hydrogen Sulfide, Deuterium Sulfide, and Hydrogen Selenide, Pergamon Press, Oxford, England, 1988.
11. C. L. Young, Ed., *IUPAC Solubility Data Series*, Vol. 12, Sulfur Dioxide, Chlorine, Fluorine and Chlorine Oxides, Pergamon Press, Oxford, England, 1983.
12. H. L. Clever and C. L. Young, Eds., *IUPAC Solubility Data Series*, Vol. 27/28, Methane, Pergamon Press, Oxford, England, 1987.
13. W. Hayduk, Ed., *IUPAC Solubility Data Series*, Vol. 9, Ethane, Pergamon Press, Oxford, England, 1982.
14. W. Hayduk, Ed., *IUPAC Solubility Data Series*, Vol. 24, Propane, Butane and 2-Methylpropane, Pergamon Press, Oxford, England, 1986.

SOLUBILITY OF CARBON DIOXIDE IN WATER AT VARIOUS TEMPERATURES AND PRESSURES

The solubility of CO_2 in water, expressed as mole fraction of CO_2 in the liquid phase, is given for pressures up to atmospheric and temperatures of 0 to 100°C. Note that 1 standard atmosphere equals 101.325 kPa. The references give data over a wider range of temperature and pressure. The estimated accuracy is about 2%.

REFERENCES

1. Carroll, J. J., Slupsky, J. D., and Mather, A. E., *J. Phys. Chem. Ref. Data*, 20, 1201, 1991.
2. Fernandez-Prini, R. and Crovetto, R., *J. Phys. Chem. Ref. Data*, 18, 1231, 1989.
3. Crovetto, R., *J. Phys. Chem. Ref. Data*, 20, 575,1991

1000 × mole fraction of CO_2 in liquid phase

	Partial pressure of CO_2 in kPa						
t/°C	5	10	20	30	40	50	100
0	0.067	0.135	0.269	0.404	0.538	0.671	1.337
5	0.056	0.113	0.226	0.338	0.451	0.564	1.123
10	0.048	0.096	0.191	0.287	0.382	0.477	0.950
15	0.041	0.082	0.164	0.245	0.327	0.409	0.814
20	0.035	0.071	0.141	0.212	0.283	0.353	0.704
25	0.031	0.062	0.123	0.185	0.247	0.308	0.614
30	0.027	0.054	0.109	0.163	0.218	0.271	0.541
35	0.024	0.048	0.097	0.145	0.193	0.242	0.481
40	0.022	0.043	0.087	0.130	0.173	0.216	0.431
45	0.020	0.039	0.078	0.117	0.156	0.196	0.389
50	0.018	0.036	0.071	0.107	0.142	0.178	0.354
55	0.016	0.033	0.065	0.098	0.131	0.163	0.325
60	0.015	0.030	0.060	0.090	0.121	0.150	0.300
65	0.014	0.028	0.056	0.084	0.112	0.140	0.279
70	0.013	0.026	0.052	0.079	0.105	0.131	0.261
75	0.012	0.025	0.049	0.074	0.099	0.123	0.245
80	0.012	0.023	0.047	0.070	0.093	0.116	0.232
85	0.011	0.022	0.044	0.067	0.089	0.111	0.221
90	0.011	0.021	0.042	0.064	0.085	0.106	0.211
95	0.010	0.020	0.041	0.061	0.082	0.102	0.203
100	0.010	0.020	0.039	0.059	0.079	0.098	0.196

OXYGEN SOLUBILITY IN AQUEOUS ELECTROLYTE SOLUTIONS

The Bunsen coefficients, α, and the standard deviation among a number of tests performed under identical conditions are presented. The Bunsen coefficient is defined as the gas volume at STP (0.101 325 MPa and 273.15 K) absorbed per unit volume of pure liquid at the temperature of the measurement. All data refer to a temperature of 310.2 K and a pressure of 0.101 325 MPa.

These data are from *Industrial & Engineering Chemistry Fundamentals*, 25, 778—779, 1986, Werner Lang and Rolf Zander. They are reproduced with permission of the copyright owner, the American Chemical Society.

Electrolyte	Concn of solution (mol/dm³)	$10^4\alpha$
HCl	2.005	211 ± 4
	3.050	198 ± 7
	3.910	191 ± 4
	4.000	185 ± 2
$AlCl_3$	0.501	173 ± 1
	0.745	144 ± 3
	1.006	119 ± 2
	1.081	113 ± 3
	2.009	63 ± 3
	2.503	41 ± 2
$BaCl_2$	0.507	178 ± 3
	0.997	131 ± 2
	1.509	96 ± 2
$CaCl_2$	0.494	184 ± 2
	0.747	162 ± 2
	0.987	144 ± 1
	1.021	141 ± 1
	1.421	110 ± 2
	1.490	107 ± 1
	2.985	51 ± 2
	3.557	37 ± 1
	3.894	34 ± 1
	4.477	23 ± 1
$CdCl_2$	0.260	213 ± 4
	0.479	196 ± 1
	0.505	194 ± 3
	0.523	193 ± 3
	0.747	179 ± 3
	0.760	178 ± 2
	0.966	164 ± 1
	0.997	163 ± 2
	1.025	161 ± 3
	1.046	164 ± 3
	1.264	149 ± 3
	1.457	141 ± 3
	1.491	141 ± 3
	1.503	138 ± 3
	1.922	120 ± 1
	1.996	116 ± 2
	2.041	116 ± 2
	2.029	115 ± 1
$CeCl_3$	0.498	159 ± 7
	0.979	111 ± 2
	1.974	51 ± 1
	2.462	33 ± 4
$CoCl_2$	0.501	184 ± 2
	0.749	162 ± 2
	0.993	144 ± 3
	1.494	114 ± 2
CsCl	0.515	212 ± 5
	1.001	192 ± 2
	1.028	187 ± 4
	1.505	170 ± 3
	1.993	152 ± 3
	2.203	138 ± 2
	2.517	132 ± 3
	3.002	119 ± 2
	3.382	105 ± 1
	4.003	97 ± 1
	4.028	98 ± 1
	5.003	78 ± 2
	6.003	64 ± 2
	6.122	60 ± 2
$CuCl_2$	0.497	192 ± 2
	0.519	189 ± 1
	0.763	169 ± 1
	0.994	155 ± 1
	1.002	156 ± 1
	1.499	122 ± 1
	1.517	124 ± 1
$FeCl_3$	0.500	186 ± 3
	1.016	148 ± 2
	2.064	84 ± 7
KCl	0.503	208 ± 1
	1.002	178 ± 2
	1.502	158 ± 5
	1.992	134 ± 3
	2.976	105 ± 1
$LaCl_3$	0.514	162 ± 2
	0.992	112 ± 1
	1.993	50 ± 1
	2.478	37 ± 2
LiCl	0.985	194 ± 4
	1.069	189 ± 8
	1.482	174 ± 6
	1.993	159 ± 7
	2.336	150 ± 3
	3.978	109 ± 3
$MgCl_2$	0.503	190 ± 3
	0.523	187 ± 4
	0.982	153 ± 1
	0.998	149 ± 3
	1.463	120 ± 2
	1.592	113 ± 2
	1.745	106 ± 3
	2.126	91 ± 1
	2.879	64 ± 2
$MnCl_2$	0.840	159 ± 2
	1.230	134 ± 3
	1.756	106 ± 2
	2.127	88 ± 1
NaCl	1.001	177 ± 4
	1.265	164 ± 8
	1.503	152 ± 2
	2.016	130 ± 3
	2.989	100 ± 2
	3.030	98 ± 3
	4.017	71 ± 2
NH_4Cl	1.002	200 ± 3
	2.008	172 ± 1
	2.215	160 ± 1
	3.001	144 ± 2
$NiCl_2$	0.744	167 ± 2
	0.991	145 ± 1
	1.490	117 ± 2
RbCl	0.990	187 ± 3
	1.014	184 ± 7
	1.359	166 ± 2
	1.984	143 ± 1
	2.957	112 ± 2
$ZnCl_2$	0.250	213 ± 7
	0.493	190 ± 3
	0.762	176 ± 3
	0.816	174 ± 2
	1.017	164 ± 4
	1.069	158 ± 3
	1.250	154 ± 1
	1.460	149 ± 1
	1.939	132 ± 3
	2.005	132 ± 2
	2.959	108 ± 2
HNO_3	1.000	236 ± 4
	2.000	230 ± 2
	4.000	221 ± 4
$Al(NO_3)_3$	0.308	202 ± 5
	0.602	171 ± 3
	1.095	127 ± 7
$Ba(NO_3)_2$	0.158	222 ± 3
	0.298	207 ± 4
$Ca(NO_3)_2$	0.497	195 ± 4
	0.990	155 ± 6
	1.961	101 ± 3
$Cd(NO_3)_2$	0.551	194 ± 5
	0.740	181 ± 3
	1.548	133 ± 2
	2.046	110 ± 3
$Ce(NO_3)_3$	0.997	137 ± 3
	1.862	85 ± 2
	2.952	50 ± 3
$Co(NO_3)_2$	0.502	194 ± 4
	0.753	177 ± 4
	1.002	157 ± 2
	1.487	126 ± 2
$CsNO_3$	0.314	226 ± 2
	0.325	228 ± 2
	0.615	215 ± 2
	0.624	209 ± 1
	0.930	198 ± 2
$Cu(NO_3)_2$	0.298	213 ± 5
	0.497	198 ± 1
	0.613	188 ± 3
	0.987	160 ± 1
	1.267	144 ± 1
$Fe(NO_3)_3$	0.498	189 ± 5
	0.739	169 ± 1
	0.999	147 ± 3
	1.037	147 ± 1
KNO_3	0.746	201 ± 1
	1.013	191 ± 3
	1.510	169 ± 2
	1.852	161 ± 6
$La(NO_3)_3$	0.493	184 ± 2
	0.966	141 ± 2
	2.227	69 ± 2
$LiNO_3$	0.955	201 ± 4
	1.916	169 ± 4
	2.070	164 ± 3
	2.869	140 ± 3
$Mg(NO_3)_2$	0.861	175 ± 6
	0.992	162 ± 2
	1.663	124 ± 2
	2.238	99 ± 3
$Mn(NO_3)_2$	0.506	198 ± 2
	1.011	161 ± 3
	1.482	132 ± 5
	1.936	112 ± 1

Electrolyte	Concn of solution (mol/dm³)	$10^4\alpha$
$NaNO_3$	0.762	200 ± 3
	1.017	189 ± 4
	1.533	166 ± 3
	2.078	151 ± 4
NH_4NO_3	0.998	210 ± 4
	1.991	185 ± 4
	2.050	181 ± 4
	3.120	158 ± 2
$Ni(NO_3)_2$	0.752	173 ± 1
	0.852	167 ± 2
	1.385	135 ± 1
	2.062	99 ± 1
$RbNO_3$	0.487	219 ± 3
	0.522	216 ± 5
	0.958	193 ± 3
	1.006	192 ± 3
	1.058	192 ± 5
	1.360	177 ± 4
	1.989	158 ± 1
	2.017	155 ± 7
$Th(NO_3)_4$	0.488	174 ± 1
	0.741	145 ± 2
	0.974	121 ± 3
	1.385	92 ± 1
$Zn(NO_3)_2$	0.499	196 ± 3
	0.756	178 ± 3
	0.995	159 ± 5
	1.239	146 ± 3
	1.506	128 ± 4
	1.651	119 ± 1
H_2SO_4	1.000	209 ± 7
	1.500	189 ± 5
	1.600	190 ± 1
	2.000	177 ± 5
	2.500	163 ± 4
$Al_2(SO_4)_3$	0.197	175 ± 2
	0.298	151 ± 3
	0.396	129 ± 2
$CdSO_4$	1.012	132 ± 2
	1.583	96 ± 3
	1.989	70 ± 2
	2.510	50 ± 2
	2.735	45 ± 1
	3.026	36 ± 1
$CoSO_4$	0.497	174 ± 2
	0.767	150 ± 1
	1.001	128 ± 2
	1.494	96 ± 1
Cs_2SO_4	0.500	171 ± 1
	1.015	120 ± 1
	1.525	86 ± 1
	1.909	66 ± 1
$CuSO_4$	0.496	177 ± 1
	0.591	168 ± 4
	0.741	155 ± 3
	0.959	139 ± 1
	1.000	133 ± 1
$Fe_2(SO_4)_3$	0.487	148 ± 2
	0.723	115 ± 2
	0.940	94 ± 2
K_2SO_4	0.247	200 ± 3
	0.297	190 ± 2
	0.397	177 ± 2
	0.405	174 ± 3
	0.500	166 ± 3
	0.582	157 ± 2
Li_2SO_4	0.510	177 ± 2
	1.007	127 ± 2
	1.903	71 ± 1
$MgSO_4$	0.497	174 ± 3
	0.750	148 ± 4
	0.996	128 ± 1
	1.196	112 ± 2
	1.202	113 ± 2
	1.500	89 ± 2
	2.003	66 ± 3
	2.013	66 ± 2
$MnSO_4$	0.503	179 ± 2
	0.691	161 ± 4
	0.998	132 ± 3
	1.386	108 ± 2
	1.407	108 ± 1
	1.740	88 ± 1
Na_2SO_4	0.498	165 ± 2
	0.748	136 ± 2
	0.883	120 ± 2
	0.993	113 ± 2
	1.325	87 ± 1
	1.487	76 ± 2
	1.762	63 ± 1
$(NH_4)_2SO_4$	1.016	140 ± 2
	1.047	137 ± 2
	1.988	86 ± 2
	2.007	82 ± 2
	2.498	66 ± 2
	2.984	51 ± 1
	3.499	44 ± 1
$NiSO_4$	0.499	176 ± 2
	0.749	150 ± 1
	1.181	116 ± 3
	1.489	95 ± 3
Rb_2SO_4	0.401	181 ± 1
	0.803	138 ± 2
	1.199	104 ± 1
$ZnSO_4$	0.501	170 ± 2
	0.719	148 ± 5
	1.006	122 ± 4
	1.481	98 ± 2
	1.499	95 ± 3
	1.783	82 ± 1
	2.151	66 ± 1
$KHSO_4$	0.565	192 ± 2
	0.771	176 ± 1
	1.011	162 ± 3
	1.484	141 ± 2
	1.522	137 ± 3
	1.638	133 ± 2
	1.798	128 ± 3
	1.805	127 ± 2
	2.017	115 ± 2
$NaHSO_4$	0.489	194 ± 2
	0.519	193 ± 1
	0.776	175 ± 3
	0.995	159 ± 2
	1.018	159 ± 2
	1.503	135 ± 1
	1.574	132 ± 2
	1.939	118 ± 3
	2.628	92 ± 2
NH_4HSO_4	1.031	176 ± 3
	1.503	153 ± 2
	1.982	135 ± 2
	2.643	116 ± 1
	3.010	108 ± 2
$CsOH$	0.890	174 ± 1
	2.142	108 ± 3
	2.325	100 ± 1
KOH	0.330	169 ± 1
	1.030	164 ± 2
	1.844	122 ± 1
	2.135	108 ± 2
	2.311	102 ± 1
	2.329	100 ± 1
	3.077	79 ± 1
	3.502	67 ± 1
	4.848	42 ± 1
	4.871	37 ± 1
$LiOH$	1.015	175 ± 1
	1.856	135 ± 1
	3.075	95 ± 2
	4.059	65 ± 4
$NaOH$	1.000	164 ± 1
	1.139	155 ± 2
	2.000	114 ± 1
	2.105	104 ± 1
	2.122	106 ± 1
	3.035	74 ± 1
	4.071	49 ± 1
$RbOH$	1.112	157 ± 2
	2.070	109 ± 3
	3.187	76 ± 3

PROPERTIES OF WATER IN THE RANGE 0 — 100 °C

This table summarizes the best available values of the density, specific heat capacity at constant pressure (C_p), vapor pressure, viscosity, thermal conductivity, dielectric constant, and surface tension for liquid water in the range 0 — 100 °C. All values (except vapor pressure) refer to a pressure of 100 kPa (1 bar). The temperature scale is IPTS-68.

t °C	Density g/cm³	C_p J/g K	Vap. pres. kPa	Visc. μPa s	Ther. cond. mW/K m	Diel. const.	Surf. ten. mN/m
0	0.99984	4.2176	0.6113	1793	561.0	87.90	75.64
10	0.99970	4.1921	1.2281	1307	580.0	83.96	74.23
20	0.99821	4.1818	2.3388	1002	598.4	80.20	72.75
30	0.99565	4.1784	4.2455	797.7	615.4	76.60	71.20
40	0.99222	4.1785	7.3814	653.2	630.5	73.17	69.60
50	0.98803	4.1806	12.344	547.0	643.5	69.88	67.94
60	0.98320	4.1843	19.932	466.5	654.3	66.73	66.24
70	0.97778	4.1895	31.176	404.0	663.1	63.73	64.47
80	0.97182	4.1963	47.373	354.4	670.0	60.86	62.67
90	0.96535	4.2050	70.117	314.5	675.3	58.12	60.82
100	0.95840	4.2159	101.325	281.8	679.1	55.51	58.91
Ref.	1—3	2	1, 3	3	3	4	5

REFERENCES

1. L. Harr, J. S. Gallagher, and G. S. Kell, *NBS/NRC Steam Tables*, Hemisphere Publishing Corp., 1984.
2. K. N. Marsh, Ed., *Recommended Reference Materials for the Realization of Physicochemical Properties*, Blackwell Scientific Publications, Oxford, 1987.
3. J. V. Sengers and J. T. R. Watson, Improved international formulations for the viscosity and thermal conductivity of water substance, *J. Phys. Chem. Ref. Data*, 15, 1291, 1986.
4. D. G. Archer and P. Wang, The dielectric constant of water and debye-huckel limiting law slopes, *J. Phys. Chem. Ref. Data*, 19, 371, 1990.
5. N. B. Vargaftik, et al., International tables of the surface tension of water, *J. Phys. Chem. Ref. Data*, 12, 817, 1983.

ENTHALPY OF VAPORIZATION OF WATER

The enthalpy (heat) of vaporization of water is tabulated as a function of temperature on the IPTS-68 scale.

REFERENCE

Marsh, K. N., Ed., *Recommended Reference Materials for the Realization of Physicochemical Properties,* Blackwell, Oxford, 1987.

t °C	$\Delta_{vap}H$ kJ/mol	*t* °C	$\Delta_{vap}H$ kJ/mol
0	45.054	200	34.962
25	43.990	220	33.468
40	43.350	240	31.809
60	42.482	260	29.930
80	41.585	280	27.795
100	40.657	300	25.300
120	39.684	320	22.297
140	38.643	340	18.502
160	37.518	360	12.966
180	36.304	374	2.066

FIXED POINT PROPERTIES OF H_2O AND D_2O

	Unit	H_2O	D_2O
Molar mass	g/mol	18.01528	20.02748
Melting point (101.325 kPa)	°C	0.00	3.82
Boiling point (101.325 kPa)	°C	100.00	101.42
Triple point temperature	°C	0.01	3.82
Triple point pressure	Pa	611.73	661
Triple point density (l)	g/cm^3	0.99978	1.1055
Triple point density (g)	mg/L	4.885	5.75
Critical temperature	°C	373.99	370.74
Critical pressure	MPa	22.064	21.671
Critical density	g/cm^3	0.322	0.356
Critical specific volume	cm^3/g	3.11	2.81
Maximum density (saturated liquid)	g/cm^3	0.99995	1.1053
Temperature of maximum density	°C	4.0	11.2

REFERENCES

L. Haar, J. S. Gallagher, and G. S. Kell, *NBS/NRC Steam Tables*, Hemisphere Publishing Corp., 1984.

J. M. H. Levelt Sengers, J. Straub, K. Watanabe, and P. G. Hill, Assessment of critical parameter values for H_2O and D_2O, *J. Phys. Chem. Ref. Data*, 14, 193, 1985.

J. Kestin, et. al., Thermophysical properties of fluid D_2O, *J. Phys. Chem. Ref. Data*, 13, 601, 1984.

J. Kestin, et. al., Thermophysical properties of fluid H_2O, *J. Phys. Chem. Ref. Data*, 13, 175, 1984.

P. G. Hill, R. D. C. MacMillan, and V. Lee, A fundamental equation of state for heavy water, *J. Phys. Chem. Ref. Data*, 11, 1, 1982.

THERMAL CONDUCTIVITY OF SATURATED H_2O AND D_2O

This table gives the thermal conductivity λ for water (H_2O or D_2O) in equilibrium with its vapor. Values for the liquid (λ_l) and vapor (λ_v) are listed, as well as the vapor pressure.

REFERENCES

1. J. V. Sengers and J. T. R. Watson, Improved international formulations for the viscosity and thermal conductivity of water substance, *J. Phys. Chem. Ref. Data*, 15, 1291, 1986.
2. N. Matsunaga and A. Nagashima, Transport properties of liquid and gaseous D_2O over a wide range of temperature and pressure, *J. Phys. Chem. Ref. Data*, 12, 933, 1983.

	H_2O			D_2O		
t/°C	*P*/kPa	λ_l/(mW/K m)	λ_v/(mW/K m)	*P*/kPa	λ_l/(mW/K m)	λ_v/(mW/K m)
0	0.6	561.0	16.49			
10	1.2	580.0	17.21	1.0	575	17.0
20	2.3	598.4	17.95	2.0	589	17.8
30	4.2	615.4	18.70	3.7	600	18.5
40	7.4	630.5	19.48	6.5	610	19.3
50	12.3	643.5	20.28	11.1	618	20.2
60	19.9	654.3	21.10	18.2	625	21.0
70	31.2	663.1	21.96	28.8	629	21.9
80	47.4	670.0	22.86	44.2	633	22.8
90	70.1	675.3	23.80	66.1	635	23.8
100	101.3	679.1	24.79	96.2	636	24.8
150	476	682.1	30.77	465	625	30.8
200	1555	663.4	39.10	1546	592	39.0
250	3978	621.4	51.18	3995	541	52.0
300	8593	547.7	71.78	8688	473	75.2
350	16530	447.6	134.59	16820	391	143.0

STANDARD DENSITY OF WATER

This table gives the density of standard mean ocean water (SMOW), free from dissolved gases, at a pressure of 101325 Pa. SMOW is a standard water sample of high purity and known isotopic composition. Methods of correcting for different isotopic compositions are discussed in the reference. The table below is reprinted with the permission of IUPAC. Note that the temperature scale is IPTS-68.

REFERENCE

Marsh, K. N., Ed., *Recommended Reference Materials for the Realization of Physicochemical Properties,* Blackwell Scientific Publications, Oxford, 1987.

	ρ/kg m^{-3}									
t_{68}/°C	0.0	0.1	0.2	0.3	0.4	0.5	0.6	0.7	0.8	0.9
0	999.8426	8493	8558	8622	8683	8743	8801	8857	8912	8964
1	999.9015	9065	9112	9158	9202	9244	9284	9323	9360	9395
2	999.9429	9461	9491	9519	9546	9571	9595	9616	9636	9655
3	999.9672	9687	9700	9712	9722	9731	9738	9743	9747	9749
4	999.9750	9748	9746	9742	9736	9728	9719	9709	9696	9683
5	999.9668	9651	9632	9612	9591	9568	9544	9518	9490	9461
6	999.9430	9398	9365	9330	9293	9255	9216	9175	9132	9088
7	999.9043	8996	8948	8898	8847	8794	8740	8684	8627	8569
8	999.8509	8448	8385	8321	8256	8189	8121	8051	7980	7908
9	999.7834	7759	7682	7604	7525	7444	7362	7279	7194	7108
10	999.7021	6932	6842	6751	6658	6564	6468	6372	6274	6174
11	999.6074	5972	5869	5764	5658	5551	5443	5333	5222	5110
12	999.4996	4882	4766	4648	4530	4410	4289	4167	4043	3918
13	999.3792	3665	3536	3407	3276	3143	3010	2875	2740	2602
14	999.2464	2325	2184	2042	1899	1755	1609	1463	1315	1166
15	999.1016	0864	0712	0558	0403	0247	0090	9932*	9772*	9612*
16	998.9450	9287	9123	8957	8791	8623	8455	8285	8114	7942
17	998.7769	7595	7419	7243	7065	6886	6706	6525	6343	6160
18	998.5976	5790	5604	5416	5228	5038	4847	4655	4462	4268
19	998.4073	3877	3680	3481	3282	3081	2880	2677	2474	2269
20	998.2063	1856	1649	1440	1230	1019	0807	0594	0380	0164
21	997.9948	9731	9513	9294	9073	8852	8630	8406	8182	7957
22	997.7730	7503	7275	7045	6815	6584	6351	6118	5883	5648
23	997.5412	5174	4936	4697	4456	4215	3973	3730	3485	3240
24	997.2994	2747	2499	2250	2000	1749	1497	1244	0990	0735
25	997.0480	0223	9965*	9707*	9447*	9186*	8925*	8663*	8399*	8135*
26	996.7870	7604	7337	7069	6800	6530	6259	5987	5714	5441
27	996.5166	4891	4615	4337	4059	3780	3500	3219	2938	2655
28	996.2371	2087	1801	1515	1228	0940	0651	0361	0070	9778*
29	995.9486	9192	8898	8603	8306	8009	7712	7413	7113	6813
30	995.6511	6209	5906	5602	5297	4991	4685	4377	4069	3760
31	995.3450	3139	2827	2514	2201	1887	1572	1255	0939	0621
32	995.0302	9983*	9663*	9342*	9020*	8697*	8373*	8049*	7724*	7397*
33	994.7071	6743	6414	6085	5755	5423	5092	4759	4425	4091
34	994.3756	3420	3083	2745	2407	2068	1728	1387	1045	0703
35	994.0359	0015	9671*	9325*	8978*	8631*	8283*	7934*	7585*	7234*
36	993.6883	6531	6178	5825	5470	5115	4759	4403	4045	3687
37	993.3328	2968	2607	2246	1884	1521	1157	0793	0428	0062
38	992.9695	9328	8960	8591	8221	7850	7479	7107	6735	6361
39	992.5987	5612	5236	4860	4483	4105	3726	3347	2966	2586
40	992.2204									

* The leading figure decreases by 1.

VOLUMETRIC PROPERTIES OF AQUEOUS SODIUM CHLORIDE SOLUTIONS

This table gives the following properties of aqueous solutions of NaCl as a function of temperature and concentration:

Specific volume v (reciprocal of density) in cm^3/g
Isothermal compressibility $\kappa_T = -(1/v)(\partial v/\partial P)_T$ in GPa^{-1}
Cubic expansion coefficient $\alpha_v = (1/v)(\partial v/\partial T)_P$ in kK^{-1}

All data refer to a pressure of 100 kPa (1 bar). The reference gives properties over a wider range of temperature and pressure.

REFERENCE

Rogers, P. S. Z., and Pitzer, K. S., *J. Phys. Chem. Ref. Data*, 11, 15, 1982.

	Molality in mol/kg								
T/°C	0.100	0.250	0.500	0.750	1.000	2.000	3.000	4.000	5.000
					Specific volume v in cm^3/g				
0	0.995732	0.989259	0.978889	0.968991	0.959525	0.925426	0.896292	0.870996	0.848646
10	0.995998	0.989781	0.979804	0.970256	0.961101	0.927905	0.899262	0.874201	0.851958
20	0.997620	0.991564	0.981833	0.972505	0.963544	0.930909	0.902565	0.877643	0.855469
25	0.998834	0.992832	0.983185	0.973932	0.965038	0.932590	0.904339	0.879457	0.857301
30	1.000279	0.994319	0.984735	0.975539	0.966694	0.934382	0.906194	0.881334	0.859185
40	1.003796	0.997883	0.988374	0.979243	0.970455	0.938287	0.910145	0.885276	0.863108
50	1.008064	1.002161	0.992668	0.983551	0.974772	0.942603	0.914411	0.889473	0.867241
60	1.0130	1.0071	0.9976	0.9885	0.9797	0.9474	0.9191	0.8940	0.8716
70	1.0186	1.0127	1.0031	0.9939	0.9851	0.9526	0.9240	0.8987	0.8762
80	1.0249	1.0188	1.0092	0.9999	0.9909	0.9581	0.9293	0.9037	0.8809
90	1.0317	1.0256	1.0157	1.0063	0.9972	0.9640	0.9348	0.9089	0.8858
100	1.0391	1.0329	1.0228	1.0133	1.0040	0.9703	0.9406	0.9144	0.8910
					Compressibility κ_T in GPa^{-1}				
0	0.503	0.492	0.475	0.459	0.443	0.389	0.346	0.315	0.294
10	0.472	0.463	0.449	0.436	0.423	0.377	0.341	0.313	0.294
20	0.453	0.446	0.433	0.422	0.411	0.371	0.338	0.313	0.294
25	0.447	0.440	0.428	0.417	0.407	0.369	0.337	0.313	0.294
30	0.443	0.436	0.425	0.414	0.404	0.367	0.337	0.313	0.294
40	0.438	0.432	0.421	0.411	0.401	0.367	0.338	0.315	0.296
50	0.438	0.431	0.421	0.411	0.402	0.369	0.340	0.317	0.299
60	0.44	0.44	0.43	0.42	0.41	0.38	0.35	0.32	0.30
70	0.45	0.44	0.43	0.42	0.42	0.38	0.36	0.33	0.31
80	0.46	0.45	0.44	0.43	0.43	0.39	0.37	0.34	0.32
90	0.47	0.47	0.46	0.45	0.44	0.41	0.38	0.35	0.33
100	0.49	0.48	0.47	0.46	0.45	0.42	0.39	0.37	0.34
					Cubic expansion coefficient α_V in kK^{-1}				
0	-0.058	-0.026	0.024	0.069	0.110	0.237	0.313	0.355	
10	0.102	0.123	0.156	0.186	0.213	0.297	0.349	0.380	
20	0.218	0.232	0.254	0.274	0.292	0.349	0.384	0.406	
25	0.267	0.278	0.296	0.312	0.327	0.373	0.401	0.420	
30	0.311	0.320	0.334	0.347	0.359	0.395	0.418	0.433	
40	0.389	0.394	0.402	0.410	0.417	0.438	0.451	0.460	
50	0.458	0.460	0.464	0.467	0.470	0.479	0.484	0.486	
60	0.52	0.52	0.52	0.52	0.52	0.52	0.52	0.52	
70	0.58	0.58	0.58	0.57	0.57	0.56	0.55	0.54	
80	0.64	0.63	0.63	0.62	0.61	0.60	0.58	0.56	
90	0.69	0.68	0.67	0.67	0.66	0.63	0.61	0.59	
100	0.74	0.73	0.72	0.71	0.70	0.66	0.64	0.61	

DENSITY OF D_2O

Density of liquid D_2O in g/cm^3 at a pressure of 100 kPa (1 bar).

t/°C	3.8	5	10	15	20	25	30
Density	1.1053	1.1055	1.1057	1.1056	1.105	1.1044	1.1034
t/°C	35	40	45	50	55	60	65
Density	1.1019	1.1001	1.0979	1.0957	1.0931	1.0905	1.0875
t/°C	70	75	80	85	90	95	100
Density	1.0847	1.0815	1.0783	1.0748	1.0712	1.0673	1.0635

REFERENCE

V. A. Kirillin, Ed., *Heavy Water: Thermophysical Properties*, Gosudarstvennoe Energeticheskoe Izdatel'stvo, Moscow, 1963.

VAPOR PRESSURE OF ICE

T/°C	*P*/Pa	*T*/°C	*P*/Pa	*T*/°C	*P*/Pa
0	611	−15	165	−30	38
−1	563	−16	150	−31	34
−2	518	−17	137	−32	31
−3	476	−18	125	−33	28
−4	437	−19	113	−34	25
−5	402	−20	103	−35	22
−6	369	−21	94	−36	20
−7	338	−22	85	−37	18
−8	310	−23	77	−38	16
−9	284	−24	70	−39	14
−10	260	−25	63	−40	13
−11	237	−26	57	−50	3.9
−12	217	−27	52	−60	1.1
−13	198	−28	47	−70	0.3
−14	181	−29	42	−80	0.1

REFERENCE

G. Fischer, Ed., *Landolt-Börnstein, Numerical Data and Functional Relationships in Science and Technology*, Volume 4, Subvolume b, Physical and Chemical Properties of Air, Springer-Verlag, Heidelberg, 1988.

VAPOR PRESSURE OF WATER FROM 0 TO 370° C

This table gives the vapor pressure of water at intervals of 1° C from the melting point to the critical point.

T/°C	*P*/kPa	*T*/°C	*P*/kPa	*T*/°C	*P*/kPa	*T*/°C	*P*/kPa
0	0.61129	55	15.752	110	143.24	165	700.29
1	0.65716	56	16.522	111	148.12	166	717.83
2	0.70605	57	17.324	112	153.13	167	735.70
3	0.75813	58	18.159	113	158.29	168	753.94
4	0.81359	59	19.028	114	163.58	169	772.52
5	0.87260	60	19.932	115	169.02	170	791.47
6	0.93537	61	20.873	116	174.61	171	810.78
7	1.0021	62	21.851	117	180.34	172	830.47
8	1.0730	63	22.868	118	186.23	173	850.53
9	1.1482	64	23.925	119	192.28	174	870.98
10	1.2281	65	25.022	120	198.48	175	891.80
11	1.3129	66	26.163	121	204.85	176	913.03
12	1.4027	67	27.347	122	211.38	177	934.64
13	1.4979	68	28.576	123	218.09	178	956.66
14	1.5988	69	29.852	124	224.96	179	979.09
15	1.7056	70	31.176	125	232.01	180	1001.9
16	1.8185	71	32.549	126	239.24	181	1025.2
17	1.9380	72	33.972	127	246.66	182	1048.9
18	2.0644	73	35.448	128	254.25	183	1073.0
19	2.1978	74	36.978	129	262.04	184	1097.5
20	2.3388	75	38.563	130	270.02	185	1122.5
21	2.4877	76	40.205	131	278.20	186	1147.9
22	2.6447	77	41.905	132	286.57	187	1173.8
23	2.8104	78	43.665	133	295.15	188	1200.1
24	2.9850	79	45.487	134	303.93	189	1226.9
25	3.1690	80	47.373	135	312.93	190	1254.2
26	3.3629	81	49.324	136	322.14	191	1281.9
27	3.5670	82	51.342	137	331.57	192	1310.1
28	3.7818	83	53.428	138	341.22	193	1338.8
29	4.0078	84	55.585	139	351.09	194	1368.0
30	4.2455	85	57.815	140	361.19	195	1397.6
31	4.4953	86	60.119	141	371.53	196	1427.8
32	4.7578	87	62.499	142	382.11	197	1458.5
33	5.0335	88	64.958	143	392.92	198	1489.7
34	5.3229	89	67.496	144	403.98	199	1521.4
35	5.6267	90	70.117	145	415.29	200	1553.6
36	5.9453	91	72.823	146	426.85	201	1586.4
37	6.2795	92	75.614	147	438.67	202	1619.7
38	6.6298	93	78.494	148	450.75	203	1653.6
39	6.9969	94	81.465	149	463.10	204	1688.0
40	7.3814	95	84.529	150	475.72	205	1722.9
41	7.7840	96	87.688	151	488.61	206	1758.4
42	8.2054	97	90.945	152	501.78	207	1794.5
43	8.6463	98	94.301	153	515.23	208	1831.1
44	9.1075	99	97.759	154	528.96	209	1868.4
45	9.5898	100	101.32	155	542.99	210	1906.2
46	10.094	101	104.99	156	557.32	211	1944.6
47	10.620	102	108.77	157	571.94	212	1983.6
48	11.171	103	112.66	158	586.87	213	2023.2
49	11.745	104	116.67	159	602.11	214	2063.4
50	12.344	105	120.79	160	617.66	215	2104.2
51	12.970	106	125.03	161	633.53	216	2145.7
52	13.623	107	129.39	162	649.73	217	2187.8
53	14.303	108	133.88	163	666.25	218	2230.5
54	15.012	109	138.50	164	683.10	219	2273.8

VAPOR PRESSURE OF WATER FROM 0 TO 370° C (continued)

T/°C	P/kPa	T/°C	P/kPa	T/°C	P/kPa	T/°C	P/kPa
220	2317.8	259	4613.7	298	8344.5	336	13876
221	2362.5	260	4689.4	299	8463.5	337	14053
222	2407.8	261	4766.1	300	8583.8	338	14232
223	2453.8	262	4843.7	301	8705.4	339	14412
224	2500.5	263	4922.3	302	8828.3	340	14594
225	2547.9	264	5001.8	303	8952.6	341	14778
226	2595.9	265	5082.3	304	9078.2	342	14964
227	2644.6	266	5163.8	305	9205.1	343	15152
228	2694.1	267	5246.3	306	9333.4	344	15342
229	2744.2	268	5329.8	307	9463.1	345	15533
230	2795.1	269	5414.3	308	9594.2	346	15727
231	2846.7	270	5499.9	309	9726.7	347	15922
232	2899.0	271	5586.4	310	9860.5	348	16120
233	2952.1	272	5674.0	311	9995.8	349	16320
234	3005.9	273	5762.7	312	10133	350	16521
235	3060.4	274	5852.4	313	10271	351	16725
236	3115.7	275	5943.1	314	10410	352	16931
237	3171.8	276	6035.0	315	10551	353	17138
238	3228.6	277	6127.9	316	10694	354	17348
239	3286.3	278	6221.9	317	10838	355	17561
240	3344.7	279	6317.0	318	10984	356	17775
241	3403.9	280	6413.2	319	11131	357	17992
242	3463.9	281	6510.5	320	11279	358	18211
243	3524.7	282	6608.9	321	11429	359	18432
244	3586.3	283	6708.5	322	11581	360	18655
245	3648.8	284	6809.2	323	11734	361	18881
246	3712.1	285	6911.1	324	11889	362	19110
247	3776.2	286	7014.1	325	12046	363	19340
248	3841.2	287	7118.3	326	12204	364	19574
249	3907.0	288	7223.7	327	12364	365	19809
250	3973.6	289	7330.2	328	12525	366	20048
251	4041.2	290	7438.0	329	12688	367	20289
252	4109.6	291	7547.0	330	12852	368	20533
253	4178.9	292	7657.2	331	13019	369	20780
254	4249.1	293	7768.6	332	13187	370	21030
255	4320.2	294	7881.3	333	13357	371	21286
256	4392.2	295	7995.2	334	13528	372	21539
257	4465.1	296	8110.3	335	13701	373	21803
258	4539.0	297	8226.8				

REFERENCE

L. Haar, J. S. Gallagher, and G. S. Kell, *NBS/NRC Steam Tables*, Hemisphere Publishing Corp., New York, 1984.

BOILING POINT OF WATER AT VARIOUS PRESSURES

Data based on the equation of state recommended by the International Association for the Properties of Steam in 1984, as presented in Haar, Gallagher, and Kell, "NBS-NRC Steam Tables" (Hemisphere Publishing Corp., New York, 1984). The temperature scale is IPTS-68.
Note that: 1 mbar = 100 Pa = 0.000986923 atmos = 0.750062 mmHg.

P/mbar	*T*/°C	*P*/mbar	*T*/°C	*P*/mbar	*T*/°C	*P*/mbar	*T*/°C
50	32.88	915	97.17	1013.25	100.00	1200	104.81
100	45.82	920	97.32	1015	100.05	1250	105.99
150	53.98	925	97.47	1020	100.19	1300	107.14
200	60.07	930	97.62	1025	100.32	1350	108.25
250	64.98	935	97.76	1030	100.46	1400	109.32
300	69.11	940	97.91	1035	100.60	1450	110.36
350	72.70	945	98.06	1040	100.73	1500	111.38
400	75.88	950	98.21	1045	100.87	1550	112.37
450	78.74	955	98.35	1050	101.00	1600	113.33
500	81.34	960	98.50	1055	101.14	1650	114.26
550	83.73	965	98.64	1060	101.27	1700	115.18
600	85.95	970	98.78	1065	101.40	1750	116.07
650	88.02	975	98.93	1070	101.54	1800	116.94
700	89.96	980	99.07	1075	101.67	1850	117.79
750	91.78	985	99.21	1080	101.80	1900	118.63
800	93.51	990	99.35	1085	101.93	1950	119.44
850	95.15	995	99.49	1090	102.06	2000	120.24
900	96.71	1000	99.63	1095	102.19	2050	121.02
905	96.87	1005	99.77	1100	102.32	2100	121.79
910	97.02	1010	99.91	1150	103.59	2150	122.54

STEAM TABLES

This table gives properties of compressed water and superheated steam at selected pressures and temperatures. It was generated from the formulation approved by the International Association for the Properties of Steam in 1984. This formulation, as well as a much more extensive set of tables, is given in:

L. Haar, J. S. Gallagher, and G. S. Kell, *NBS/NRC Steam Tables*, Hemisphere Press, New York, 1984

The reference state for these tables is the liquid at the triple point, at which the internal energy and entropy are taken as zero. A duplicate entry in the temperature column indicates a phase transition (liquid-vapor) at that temperature; property values are then given for both phases.

Pressure MPa	Temp. K	Density kg/m³	Enthalpy J/g	Entropy J/g K	C_p J/g K
0.1	273.15	999.83	0.06	–0.00015	4.2282
0.1	300	996.57	112.58	0.3928	4.1831
0.1	372.78	958.66	417.51	1.3027	4.2166
0.1	372.78	0.59021	2675.1	7.3589	2.0427
0.1	373.15	0.58958	2675.9	7.3609	2.0418
0.1	400	0.54765	2730.0	7.5010	1.9973
0.1	500	0.43517	2927.9	7.9427	1.9816
0.1	600	0.36186	3128.2	8.3076	2.0272
0.1	700	0.30988	3333.8	8.6245	2.0869
0.1	800	0.27102	3545.8	8.9074	2.1526
0.1	900	0.24085	3764.4	9.1649	2.2216
0.1	1000	0.21673	3990.1	9.4026	2.2921
0.1	1100	0.19701	4222.8	9.6244	2.3621
0.1	1200	0.18058	4462.5	9.8329	2.4302
0.1	1300	0.16668	4708.8	10.030	2.4954
0.1	1400	0.15477	4961.4	10.217	2.5568
0.1	1500	0.14445	5220.0	10.396	2.6143
0.1	1600	0.13542	5484.2	10.566	2.6676
0.1	1700	0.12745	5753.4	10.729	2.7166
0.1	1800	0.12037	6027.3	10.886	2.7617
0.1	1900	0.11403	6305.6	11.036	2.8031
0.1	2000	0.10833	6587.9	11.181	2.8412
1	273.15	1000.30	0.98	–0.0001	4.2233
1	300	996.97	113.41	0.3926	4.1806
1	373.15	958.81	419.74	1.3062	4.2150
1	400	937.92	533.47	1.6005	4.2593
1	453.07	887.15	762.88	2.1388	4.4030
1	453.07	5.1445	2777.7	6.5859	2.5569
1	500	4.5348	2890.0	6.8220	2.2760
1	600	3.6877	3108.0	7.2199	2.1332
1	700	3.1307	3321.1	7.5483	2.1386
1	800	2.7266	3537.0	7.8365	2.1821
1	900	2.4175	3758.0	8.0968	2.2402
1	1000	2.1724	3985.2	8.3361	2.3046
1	1100	1.9730	4219.0	8.5588	2.3710
1	1200	1.8074	4459.4	8.7680	2.4368
1	1300	1.6677	4706.3	8.9656	2.5004
1	1400	1.5481	4959.3	9.1531	2.5608
1	1500	1.4445	5218.3	9.3317	2.6175
1	1600	1.3540	5482.7	9.5024	2.6702
1	1700	1.2742	5752.2	9.6657	2.7188
1	1800	1.2034	6026.3	9.8224	2.7636
1	1900	1.1399	6304.8	9.9729	2.8047
1	2000	1.0829	6587.2	10.1180	2.8426

Pressure MPa	Temp. K	Density kg/m³	Enthalpy J/g	Entropy J/g K	C_p J/g K
10	273.15	1004.80	10.10	0.0004	4.1765
10	300	1001.00	121.66	0.3900	4.1564
10	373.15	962.98	426.52	1.2992	4.1948
10	400	942.47	539.67	1.5921	4.2360
10	500	838.17	977.04	2.5666	4.5903
10	584.18	688.63	1407.3	3.3591	6.1244
10	584.18	55.477	2724.5	5.6139	6.8973
10	600	49.830	2817.8	5.7716	5.1447
10	700	35.392	3176.1	6.3279	2.8769
10	800	29.128	3442.6	6.6844	2.5324
10	900	25.132	3690.6	6.9764	2.4463
10	1000	22.243	3934.5	7.2334	2.4395
10	1100	20.015	4179.5	7.4670	2.4654
10	1200	18.227	4428.0	7.6832	2.5059
10	1300	16.751	4680.9	7.8856	2.5529
10	1400	15.508	4938.6	8.0766	2.6018
10	1500	14.443	5201.3	8.2577	2.6503
10	1600	13.520	5468.6	8.4303	2.6969
10	1700	12.710	5740.6	8.5951	2.7410
10	1800	11.994	6016.8	8.7530	2.7823
10	1900	11.356	6296.9	8.9044	2.8206
10	2000	10.784	6580.8	9.0500	2.8563
100	273.15	1045.30	95.40	–0.0085	3.9092
100	300	1037.20	201.35	0.3614	3.9846
100	373.15	999.70	495.00	1.2371	4.0392
100	400	981.74	603.77	1.5186	4.0633
100	500	899.23	1015.3	2.4365	4.1854
100	600	791.46	1447.6	3.2237	4.5025
100	700	651.37	1925.2	3.9588	5.0954
100	800	482.17	2466.3	4.6806	5.6041
100	900	343.48	2997.0	5.3065	4.8280
100	1000	265.84	3431.3	5.7649	3.9368
100	1100	220.96	3798.6	6.1154	3.4632
100	1200	191.59	4131.4	6.4051	3.2174
100	1300	170.56	4445.8	6.6568	3.0848
100	1400	154.56	4750.3	6.8825	3.0129
100	1500	141.85	5049.5	7.0890	2.9759
100	1600	131.43	5346.2	7.2804	2.9600
100	1700	122.68	5641.9	7.4597	2.9570
100	1800	115.18	5937.8	7.6289	2.9620
100	1900	108.67	6234.5	7.7893	2.9720
100	2000	102.94	6532.4	7.9420	2.9852

THERMOPHYSICAL PROPERTIES OF FLUIDS

These tables were generated from equations of state discussed in the references below. Please consult these references for information on the uncertainties and reference states for *E*, *H*, and *S*.

B. A. Younglove, Thermophysical properties of fluids: Part I, *J. Phys. Chem. Ref. Data*, 11, Suppl. 1, 1982; erratum, 14, 619, 1985; B. A. Younglove and J. F. Ely, Part II, 16, 577, 1987.

R. D. McCarty, Thermodynamic properties of helium-4, *J. Phys. Chem. Ref. Data*, 2, 923, 1973.

Nitrogen (N_2)

T K	ρ mol/L	*E* J/mol	*H* J/mol	*S* J/mol K	C_v J/mol K	C_p J/mol K	η μPa s	λ mW/m K	*D*
P = 0.1 MPa (1 bar)									
70	30.017	–3828	–3824	73.8	28.5	57.2	203.9	143.5	1.45269
77.25	28.881	–3411	–3407	79.5	27.8	57.8	152.2	133.8	1.43386
77.25	0.163	1546	2161	151.6	21.6	31.4	5.3	7.6	1.00215
100	0.123	2041	2856	159.5	21.1	30.0	6.8	9.6	1.00162
200	0.060	4140	5800	179.9	20.8	29.2	12.9	18.4	1.00079
300	0.040	6223	8717	191.8	20.8	29.2	18.0	25.8	1.00053
400	0.030	8308	11635	200.2	20.9	29.2	22.2	32.3	1.00040
500	0.024	10414	14573	206.7	21.2	29.6	26.1	38.5	1.00032
600	0.020	12563	17554	212.2	21.8	30.1	29.5	44.5	1.00026
700	0.017	14770	20593	216.8	22.4	30.7	32.8	50.5	1.00023
800	0.015	17044	23698	221.0	23.1	31.4	35.8	56.3	1.00020
900	0.013	19383	26869	224.7	23.7	32.0	38.7	62.0	1.00017
1000	0.012	21786	30103	228.1	24.3	32.6	41.5	67.7	1.00016
1500	0.008	34530	47004	241.8	26.4	34.7	54.0	93.3	1.00010
P = 1 MPa									
70	30.070	–3838	–3805	73.6	28.9	56.9	205.9	144.1	1.45355
80	28.504	–3267	–3232	81.3	27.8	57.7	139.5	130.7	1.42760
90	26.721	–2685	–2648	88.2	26.7	59.4	100.1	115.3	1.39824
100	24.634	–2073	–2032	94.6	26.2	64.4	73.1	98.5	1.36417
103.75	23.727	–1828	–1786	97.1	26.2	67.8	64.8	91.8	1.34947
103.75	1.472	1788	2467	138.1	24.1	45.0	7.6	12.5	1.01954
200	0.614	4048	5675	160.3	21.0	30.4	13.2	19.3	1.00812
300	0.402	6171	8661	172.5	20.9	29.6	18.1	26.3	1.00529
400	0.300	8273	11609	180.9	20.9	29.5	22.4	32.7	1.00395
500	0.240	10389	14563	187.5	21.3	29.7	26.1	38.8	1.00315
600	0.200	12544	17554	193.0	21.8	30.2	29.6	44.8	1.00262
700	0.171	14756	20600	197.7	22.4	30.8	32.8	50.7	1.00224
800	0.150	17032	23709	201.8	23.1	31.4	35.9	56.5	1.00196
900	0.133	19374	26884	205.6	23.7	32.1	38.8	62.2	1.00174
1000	0.120	21778	30121	209.0	24.3	32.7	41.5	67.8	1.00157
1500	0.080	34527	47029	222.7	26.4	34.8	54.0	93.4	1.00104
P = 10 MPa									
65.32	31.120	–4176	–3855	68.6	31.8	53.8	275.7	153.8	1.47067
100	26.201	–2328	–1946	92.0	27.4	56.3	90.2	112.3	1.38942
200	7.117	3037	4442	136.4	22.7	45.5	17.6	30.4	1.09698
300	3.989	5667	8174	151.7	21.4	33.4	20.1	31.9	1.05347
400	2.898	7941	11392	161.0	21.3	31.3	23.7	36.7	1.03860
500	2.302	10148	14492	167.9	21.5	30.8	27.1	42.0	1.03055
600	1.918	12361	17575	173.5	21.9	30.9	30.4	47.4	1.02538
700	1.647	14613	20683	178.3	22.5	31.3	33.5	53.0	1.02175
800	1.445	16919	23837	182.5	23.2	31.8	36.4	58.6	1.01904
900	1.288	19283	27046	186.3	23.8	32.4	39.3	64.1	1.01694
1000	1.162	21705	30308	189.8	24.4	32.9	42.0	69.6	1.01526
1500	0.783	34504	47283	203.5	26.5	34.8	54.3	94.7	1.01020

Oxygen (O_2)

T K	ρ mol/L	*E* J/mol	*H* J/mol	*S* J/mol K	C_v J/mol K	C_p J/mol K	η μPa s	λ mW/m K	*D*
P = 0.1 MPa (1 bar)									
60	40.049	–5883	–5880	72.4	34.9	53.4	425.2	188.2	1.55619
80	37.204	–4814	–4812	87.7	31.0	53.6	251.7	166.1	1.51114
100	0.123	2029	2840	172.9	21.4	30.5	7.5	9.3	1.00146
120	0.102	2458	3442	178.4	21.0	29.8	9.0	11.2	1.00121
140	0.087	2881	4035	182.9	20.9	29.5	10.5	13.1	1.00103
160	0.076	3301	4624	186.9	20.9	29.4	11.9	15.0	1.00090
180	0.067	3720	5210	190.3	20.8	29.3	13.3	16.7	1.00080
200	0.060	4138	5796	193.4	20.8	29.3	14.6	18.4	1.00072
220	0.055	4556	6381	196.2	20.8	29.3	15.9	20.1	1.00065
240	0.050	4974	6966	198.8	20.9	29.3	17.2	21.7	1.00060
260	0.046	5393	7552	201.1	20.9	29.3	18.4	23.2	1.00055
280	0.043	5812	8138	203.3	21.0	29.4	19.5	24.8	1.00051
300	0.040	6234	8726	205.3	21.1	29.4	20.6	26.3	1.00048
320	0.038	6657	9316	207.2	21.2	29.5	21.7	27.8	1.00045
340	0.035	7082	9908	209.0	21.3	29.7	22.8	29.3	1.00042
360	0.033	7510	10503	210.7	21.5	29.8	23.8	30.8	1.00040
380	0.032	7941	11100	212.3	21.6	30.0	24.8	32.2	1.00038
P = 1 MPa									
60	40.084	–5887	–5863	72.3	34.9	53.3	428.5	188.4	1.55671
80	37.254	–4822	–4795	87.6	31.0	53.5	253.8	166.4	1.51192
100	34.153	–3741	–3712	99.7	28.5	55.2	155.6	137.9	1.46381
120	1.198	2163	2997	156.7	24.0	40.6	9.4	13.9	1.01429
140	0.950	2683	3735	162.4	22.2	34.4	10.8	14.9	1.01133
160	0.802	3151	4398	166.8	21.5	32.2	12.2	16.3	1.00955
180	0.698	3598	5030	170.5	21.2	31.2	13.5	17.7	1.00831
200	0.620	4035	5647	173.8	21.1	30.6	14.8	19.3	1.00738
220	0.559	4466	6255	176.7	21.0	30.3	16.1	20.8	1.00665
240	0.509	4894	6858	179.3	21.0	30.1	17.3	22.3	1.00606
260	0.468	5321	7458	181.7	21.0	29.9	18.5	23.8	1.00556
280	0.433	5748	8056	183.9	21.1	29.9	19.6	25.2	1.00515
300	0.403	6174	8654	186.0	21.1	29.9	20.7	26.7	1.00479
320	0.377	6602	9252	187.9	21.2	29.9	21.8	28.2	1.00448
340	0.355	7032	9851	189.7	21.4	30.0	22.8	29.6	1.00421
360	0.335	7463	10452	191.4	21.5	30.1	23.9	31.1	1.00397
380	0.317	7898	11056	193.1	21.7	30.2	24.9	32.6	1.00376
P = 10 MPa									
60	40.419	–5931	–5684	71.5	35.1	53.0	461.8	189.9	1.56210
80	37.727	–4893	–4628	86.7	31.6	52.7	274.4	168.6	1.51936
100	34.881	–3856	–3570	98.5	29.1	53.4	171.0	141.2	1.47500
120	31.721	–2796	–2481	108.4	27.3	55.9	113.0	115.1	1.42677
140	27.890	–1662	–1304	117.5	26.2	62.9	76.3	91.8	1.36972
160	22.379	-322	125	127.0	26.1	84.8	48.6	71.2	1.29037
180	13.232	1489	2245	139.5	26.6	105.9	26.2	46.8	1.16560
200	8.666	2681	3835	147.9	24.0	60.6	21.2	34.0	1.10650
220	6.868	3424	4880	152.9	22.6	46.4	20.5	30.8	1.08380
240	5.836	4029	5742	156.6	22.0	40.6	20.8	30.1	1.07090
260	5.134	4573	6521	159.7	21.8	37.6	21.4	30.2	1.06219
280	4.613	5086	7254	162.5	21.6	35.8	22.1	30.8	1.05575
300	4.205	5581	7959	164.9	21.6	34.7	22.9	31.6	1.05073
320	3.874	6063	8645	167.1	21.7	33.9	23.7	32.6	1.04667
340	3.598	6538	9318	169.1	21.8	33.4	24.6	33.7	1.04329
360	3.363	7009	9982	171.0	21.9	33.0	25.4	34.9	1.04043
380	3.161	7477	10641	172.8	22.0	32.8	26.3	36.1	1.03796

Hydrogen (H_2)

T K	ρ mol/L	*E* J/mol	*H* J/mol	*S* J/mol K	C_v J/mol K	C_p J/mol K	v_s m/s	*D*
P = 0.1 MPa (1 bar)								
15	37.738	–605	–603	11.2	9.7	14.4	1319	1.24827
20	35.278	–524	–521	15.8	11.3	19.1	1111	1.23093
40	0.305	491	818	75.6	12.5	21.3	521	1.00186
60	0.201	748	1244	84.3	13.1	21.6	636	1.00122
80	0.151	1030	1694	90.7	15.3	23.7	714	1.00091
100	0.120	1370	2202	96.4	18.7	27.1	773	1.00073
120	0.100	1777	2776	101.6	21.8	30.2	827	1.00061
140	0.086	2237	3401	106.4	23.8	32.2	883	1.00052
160	0.075	2723	4054	110.8	24.6	33.0	940	1.00046
180	0.067	3216	4714	114.7	24.6	32.9	998	1.00041
200	0.060	3703	5367	118.1	24.1	32.4	1054	1.00037
220	0.055	4179	6009	121.2	23.4	31.8	1110	1.00033
240	0.050	4641	6638	123.9	22.8	31.2	1163	1.00030
260	0.046	5093	7256	126.4	22.3	30.6	1214	1.00028
280	0.043	5535	7865	128.6	21.9	30.2	1263	1.00026
300	0.040	5970	8466	130.7	21.6	29.9	1310	1.00024
400	0.030	8093	11421	139.2	21.0	29.3	1518	1.00018
P = 1 MPa								
15	38.109	–609	–583	10.9	10.1	14.1	1315	1.25089
20	35.852	–532	–504	15.5	11.4	18.4	1155	1.23496
40	3.608	399	676	54.1	12.9	28.4	498	1.02209
60	2.098	697	1173	64.3	13.2	23.5	635	1.01280
80	1.523	994	1651	71.1	15.4	24.7	719	1.00928
100	1.204	1343	2174	77.0	18.8	27.7	779	1.00733
120	0.999	1756	2758	82.3	21.9	30.6	835	1.00608
140	0.854	2219	3390	87.1	23.9	32.5	891	1.00520
160	0.747	2709	4048	91.5	24.7	33.2	949	1.00454
180	0.663	3204	4712	95.4	24.6	33.1	1006	1.00404
200	0.597	3693	5368	98.9	24.1	32.5	1063	1.00363
220	0.543	4170	6012	102.0	23.5	31.9	1118	1.00330
240	0.498	4634	6643	104.7	22.9	31.2	1171	1.00303
260	0.460	5087	7263	107.2	22.3	30.7	1222	1.00279
280	0.427	5530	7873	109.5	21.9	30.3	1271	1.00259
300	0.399	5966	8475	111.5	21.6	30.0	1317	1.00242
400	0.299	8091	11433	120.1	21.0	29.4	1525	1.00182
P = 10 MPa								
20	39.669	–568	–316	13.0	10.9	15.0	1458	1.26198
40	31.344	–209	110	27.3	13.2	27.0	1171	1.20354
60	21.273	255	725	39.7	13.8	32.5	931	1.13527
80	14.830	686	1360	48.8	15.9	31.1	886	1.09303
100	11.417	1110	1986	55.8	19.3	31.9	904	1.07109
120	9.357	1571	2640	61.8	22.4	33.5	941	1.05801
140	7.969	2068	3323	67.0	24.3	34.6	989	1.04925
160	6.963	2583	4020	71.7	25.0	34.9	1042	1.04294
180	6.195	3099	4713	75.7	24.9	34.4	1096	1.03814
200	5.588	3604	5393	79.3	24.4	33.6	1150	1.03436
220	5.094	4094	6057	82.5	23.7	32.8	1203	1.03129
240	4.683	4569	6704	85.3	23.1	32.0	1254	1.02874
260	4.336	5030	7336	87.8	22.6	31.3	1302	1.02659
280	4.038	5481	7958	90.1	22.1	30.8	1349	1.02475
300	3.780	5924	8570	92.3	21.8	30.4	1394	1.02315
400	2.869	8073	11559	100.9	21.2	29.6	1592	1.01753

Helium (He-4)

T K	ρ mol/L	*E* J/mol	*H* J/mol	*S* J/mol K	C_v J/mol K	C_p J/mol K	v_s m/s	η μPa s	*D*
P = 0.1 MPa (1 bar)									
3	35.794	−39	−36	9.8	7.6	9.4	222	3.85	1.05646
4	32.477	−27	−24	13.3	9.1	16.3	185	3.33	1.05114
5	2.935	52	86	39.1	12.7	27.1	120	1.39	1.00456
10	1.238	120	201	55.2	12.5	21.7	185	2.26	1.00192
20	0.602	247	413	69.9	12.5	21.0	264	3.58	1.00093
50	0.240	623	1039	89.0	12.5	20.8	417	6.36	1.00037
100	0.120	1247	2079	103.4	12.5	20.8	589	9.78	1.00019
200	0.060	2494	4158	117.8	12.5	20.8	833	15.14	1.00009
300	0.040	3741	6237	126.3	12.5	20.8	1020	19.93	1.00006
400	0.030	4988	8315	132.3	12.5	20.8	1177	24.29	1.00005
500	0.024	6236	10394	136.9	12.5	20.8	1316	28.36	1.00004
600	0.020	7483	12472	140.7	12.5	20.8	1441	32.22	1.00003
700	0.017	8730	14551	143.9	12.5	20.8	1557	35.89	1.00003
800	0.015	9977	16630	146.7	12.5	20.8	1664	39.43	1.00002
900	0.013	11224	18708	149.1	12.5	20.8	1765	42.85	1.00002
1000	0.012	12471	20787	151.3	12.5	20.8	1861	46.16	1.00002
1500	0.008	18707	31179	159.7	12.5	20.8	2279	61.55	1.00001
P = 1 MPa									
3	39.703	−42	−16	8.6	7.1	7.8	300	5.63	1.06274
4	38.210	−34	−7	11.2	8.3	10.9	290	5.01	1.06034
5	35.818	−22	6	14.0	9.7	15.1	269	4.38	1.05650
10	15.378	78	143	32.2	12.3	30.5	198	3.07	1.02402
20	6.067	228	393	49.8	12.6	22.9	274	3.94	1.00943
50	2.353	617	1042	69.8	12.5	21.1	428	6.53	1.00365
100	1.186	1245	2089	84.3	12.5	20.9	597	9.89	1.00184
200	0.597	2495	4170	98.7	12.5	20.8	838	15.21	1.00093
300	0.399	3742	6249	107.1	12.5	20.8	1024	19.96	1.00062
400	0.300	4990	8327	113.1	12.5	20.8	1180	24.32	1.00046
500	0.240	6237	10406	117.8	12.5	20.8	1319	28.38	1.00037
600	0.200	7485	12484	121.5	12.5	20.8	1444	32.23	1.00031
700	0.172	8732	14562	124.7	12.5	20.8	1559	35.91	1.00027
800	0.150	9979	16641	127.5	12.5	20.8	1666	39.44	1.00023
900	0.133	11227	18719	130.0	12.5	20.8	1767	42.86	1.00021
1000	0.120	12474	20798	132.2	12.5	20.8	1862	46.17	1.00019
1500	0.080	18710	31190	140.6	12.5	20.8	2280	61.55	1.00012
P = 10 MPa									
4	51.978	−24	169	6.7	6.0	7.3	586	24.27	1.08262
5	51.118	−18	177	8.5	7.9	9.3	576	18.16	1.08122
10	46.872	23	236	16.6	11.0	14.5	546	9.31	1.07432
20	37.092	154	423	29.5	12.6	20.7	498	6.99	1.05854
50	19.192	572	1093	49.9	12.9	22.4	541	8.07	1.03003
100	10.525	1231	2181	65.0	12.8	21.3	674	10.93	1.01640
200	5.605	2500	4284	79.6	12.6	20.9	889	15.82	1.00871
300	3.829	3755	6367	88.0	12.6	20.8	1063	20.25	1.00595
400	2.908	5006	8445	94.0	12.6	20.8	1212	24.54	1.00452
500	2.344	6256	10522	98.6	12.5	20.8	1346	28.56	1.00364
600	1.963	7505	12599	102.4	12.5	20.8	1467	32.38	1.00305
700	1.689	8754	14676	105.6	12.5	20.8	1580	36.04	1.00262
800	1.481	10003	16753	108.4	12.5	20.8	1685	39.56	1.00230
900	1.320	11252	18830	110.9	12.5	20.8	1784	42.96	1.00205
1000	1.189	12500	20907	113.0	12.5	20.8	1877	46.26	1.00185
1500	0.797	18742	31294	121.5	12.5	20.8	2289	61.62	1.00124

Argon (Ar)

T K	ρ mol/L	*E* J/mol	*H* J/mol	*S* J/mol K	C_v J/mol K	C_p J/mol K	v_s m/s	η μPa s	λ mW/m K
P = 0.1 MPa (1 bar)									
85	35.243	–4811	–4808	53.6	23.1	44.7	820	278.8	132.4
90	0.138	1077	1802	129.4	13.1	22.5	174	7.5	6.0
100	0.123	1211	2024	131.8	12.9	21.9	184	8.2	6.6
120	0.102	1471	2456	135.7	12.6	21.4	203	9.8	7.8
140	0.087	1727	2881	139.0	12.6	21.1	220	11.4	9.0
160	0.076	1980	3302	141.8	12.5	21.0	235	13.0	10.2
180	0.067	2232	3722	144.3	12.5	21.0	250	14.5	11.4
200	0.060	2483	4141	146.5	12.5	20.9	263	16.0	12.5
220	0.055	2734	4559	148.5	12.5	20.9	276	17.5	13.7
240	0.050	2984	4976	150.3	12.5	20.9	289	18.9	14.8
260	0.046	3234	5394	152.0	12.5	20.9	300	20.3	15.8
280	0.043	3484	5811	153.5	12.5	20.8	312	21.6	16.9
300	0.040	3734	6227	155.0	12.5	20.8	323	22.9	17.9
320	0.038	3984	6644	156.3	12.5	20.8	333	24.2	18.9
340	0.035	4234	7060	157.6	12.5	20.8	344	25.4	19.9
360	0.033	4484	7477	158.7	12.5	20.8	354	26.6	20.8
380	0.032	4734	7893	159.9	12.5	20.8	363	27.8	21.7
P = 1 MPa									
85	35.307	–4820	–4792	53.5	23.1	44.6	823	281.3	133.0
90	34.542	–4598	–4569	56.1	21.6	44.7	808	242.7	124.2
100	32.909	–4145	–4115	60.9	19.9	46.2	753	185.0	109.2
120	1.181	1210	2057	114.3	14.7	30.1	189	10.3	9.3
140	0.945	1544	2603	118.5	13.5	25.4	212	11.8	10.1
160	0.799	1838	3089	121.8	13.0	23.6	231	13.3	11.1
180	0.697	2116	3551	124.5	12.8	22.7	247	14.8	12.1
200	0.619	2384	3999	126.9	12.7	22.2	262	16.3	13.2
220	0.559	2648	4438	128.9	12.6	21.8	275	17.7	14.2
240	0.509	2908	4873	130.8	12.6	21.6	288	19.1	15.3
260	0.468	3167	5304	132.6	12.6	21.5	301	20.4	16.3
280	0.433	3423	5732	134.2	12.6	21.4	312	21.8	17.3
300	0.403	3679	6159	135.6	12.5	21.3	324	23.1	18.3
320	0.377	3934	6583	137.0	12.5	21.2	334	24.3	19.2
340	0.355	4188	7007	138.3	12.5	21.2	345	25.5	20.2
360	0.335	4441	7429	139.5	12.5	21.1	355	26.7	21.1
380	0.317	4694	7851	140.6	12.5	21.1	365	27.9	22.0
P = 10 MPa									
90	35.208	–4694	–4410	55.0	21.9	43.2	846	265.2	129.5
100	33.744	–4271	–3974	59.6	20.4	44.0	800	205.0	115.1
120	30.525	–3396	–3069	67.8	18.8	46.9	672	131.2	92.1
140	26.609	–2447	–2072	75.5	17.6	54.1	526	85.9	71.7
160	20.816	–1279	–799	83.9	17.4	78.6	357	51.3	52.8
180	12.296	228	1042	94.8	17.3	83.6	257	27.8	32.0
200	8.442	1118	2302	101.4	15.3	48.6	268	23.3	23.6
220	6.776	1661	3137	105.4	14.2	36.8	284	22.8	21.6
240	5.787	2087	3815	108.4	13.7	31.6	300	23.2	21.3
260	5.105	2458	4416	110.8	13.4	28.8	314	23.9	21.4
280	4.596	2798	4974	112.9	13.2	27.1	327	24.8	21.8
300	4.195	3119	5503	114.7	13.1	25.9	339	25.7	22.3
320	3.869	3427	6012	116.3	13.0	25.0	350	26.7	22.9
340	3.596	3726	6506	117.8	13.0	24.4	361	27.7	23.5
360	3.364	4017	6989	119.2	12.9	23.9	372	28.7	24.2
380	3.164	4303	7464	120.5	12.9	23.5	381	29.7	24.9

Methane (CH_4)

T K	ρ mol/L	*E* J/mol	*H* J/mol	*S* J/mol K	C_v J/mol K	C_p J/mol K	η μPa s	λ mW/m K	*D*
P = 0.1 MPa (1 bar)									
100	27.370	–5258	–5254	73.0	33.4	54.1	156.3	208.1	1.65504
125	0.099	3026	4039	156.5	25.4	34.6	5.0	13.4	1.00193
150	0.081	3667	4896	162.7	25.2	34.0	5.9	16.2	1.00159
175	0.069	4301	5743	168.0	25.2	33.8	6.9	19.1	1.00136
200	0.061	4935	6587	172.5	25.3	33.8	7.8	21.9	1.00119
225	0.054	5571	7434	176.5	25.5	34.0	8.7	24.8	1.00105
250	0.048	6216	8288	180.1	26.0	34.4	9.6	27.8	1.00095
275	0.044	6875	9156	183.4	26.6	35.0	10.4	30.9	1.00086
300	0.040	7552	10042	186.4	27.5	35.9	11.2	34.1	1.00079
325	0.037	8252	10951	189.4	28.5	36.9	12.0	37.6	1.00073
350	0.034	8979	11887	192.1	29.7	38.0	12.8	41.2	1.00068
375	0.032	9737	12853	194.8	30.9	39.3	13.5	45.1	1.00063
400	0.030	10528	13852	197.4	32.3	40.7	14.3	49.1	1.00059
425	0.028	11354	14886	199.9	33.7	42.1	15.0	53.3	1.00056
450	0.027	12215	15956	202.3	35.2	43.5	15.7	57.6	1.00053
500	0.024	14047	18204	207.1	38.0	46.4	17.0	66.5	1.00047
600	0.020	18111	23101	216.0	42.9	51.3	19.4	84.1	1.00039
P = 1 MPa									
100	27.413	–5268	–5231	72.9	33.4	54.0	158.1	208.9	1.65617
125	25.137	–3882	–3842	85.3	32.4	57.4	89.2	168.2	1.59261
150	0.969	3282	4315	140.9	27.9	45.2	6.2	18.4	1.01911
175	0.765	4041	5348	147.3	26.4	38.9	7.1	20.6	1.01507
200	0.644	4736	6289	152.3	25.9	36.8	8.0	23.1	1.01268
225	0.560	5410	7197	156.6	25.9	36.0	8.9	25.8	1.01102
250	0.497	6081	8093	160.4	26.2	35.8	9.7	28.7	1.00979
275	0.448	6758	8991	163.8	26.8	36.1	10.6	31.7	1.00882
300	0.408	7449	9901	167.0	27.6	36.7	11.4	34.9	1.00803
325	0.375	8160	10829	169.9	28.6	37.6	12.1	38.3	1.00738
350	0.347	8897	11781	172.8	29.7	38.6	12.9	41.9	1.00683
375	0.323	9662	12760	175.5	31.0	39.8	13.6	45.7	1.00636
400	0.302	10460	13770	178.1	32.4	41.1	14.4	49.6	1.00595
425	0.284	11291	14814	180.6	33.8	42.4	15.1	53.8	1.00559
450	0.268	12157	15892	183.1	35.2	43.8	15.7	58.1	1.00527
500	0.241	13997	18153	187.8	38.1	46.6	17.0	66.9	1.00474
600	0.200	18073	23070	196.8	43.0	51.4	19.5	84.5	1.00394
P = 10 MPa									
100	27.815	–5362	–5003	72.0	33.8	53.2	175.4	217	1.66668
125	25.754	–4036	–3648	84.1	32.7	55.3	100.4	178.8	1.60895
150	23.441	–2655	–2229	94.4	31.4	50.6	65.7	144.6	1.54553
175	20.613	–1175	–689	103.9	30.3	65.5	44.9	113.4	1.47021
200	16.602	542	1144	113.6	30.1	84.7	29.4	85.8	1.36789
225	10.547	2680	3628	125.3	30.8	102.2	17.6	61.0	1.22352
250	7.013	4289	5714	134.1	29.3	67.4	14.3	47.6	1.14481
275	5.530	5387	7195	139.8	28.7	53.4	13.8	44.1	1.11297
300	4.685	6320	8454	144.2	28.9	48.0	13.9	44.6	1.09513
325	4.115	7192	9622	147.9	29.6	45.8	14.3	46.6	1.08322
350	3.695	8047	10753	151.3	30.5	44.9	14.7	49.2	1.07450
375	3.366	8903	11874	154.4	31.7	44.8	15.2	52.3	1.06773
400	3.101	9774	12999	157.3	32.9	45.2	15.8	55.7	1.06227
425	2.880	10666	14138	160.0	34.3	46.0	16.3	59.4	1.05775
450	2.692	11584	15298	162.7	35.7	46.9	16.9	63.3	1.05392
500	2.389	13507	17692	167.7	38.5	48.9	18.0	71.6	1.04775
600	1.963	17700	22795	177.0	43.3	52.9	20.2	88.3	1.03911

Ethane (C_2H_6)

T K	ρ mol/L	E J/mol	H J/mol	S J/mol K	C_v J/mol K	C_p J/mol K	v_s m/s	D
P = 0.1 MPa (1 bar)								
95	21.50	–14555	–14550	80.2	47.2	68.7	1970	1.93480
100	21.32	–14210	–14205	83.8	47.1	69.3	1943	1.92500
125	20.41	–12468	–12463	99.3	45.0	69.8	1775	1.87634
150	19.47	–10717	–10712	112.1	43.4	70.4	1587	1.82726
175	18.49	–8938	-8933	123.1	42.7	72.1	1396	1.77671
200	0.062	5503	7123	210.1	34.5	43.8	258	1.00208
225	0.054	6401	8238	215.4	36.5	45.5	273	1.00183
250	0.049	7349	9401	220.3	38.9	47.7	287	1.00164
275	0.044	8360	10624	224.9	41.6	50.2	300	1.00148
300	0.040	9439	11914	229.4	44.5	53.1	312	1.00136
325	0.037	10592	13278	233.8	47.6	56.1	324	1.00125
350	0.035	11823	14719	238.1	50.7	59.2	335	1.00116
375	0.032	13133	16240	242.3	54.0	62.4	345	1.00108
400	0.030	14525	17841	246.4	57.2	65.6	355	1.00101
450	0.027	17548	21282	254.5	63.6	72.0	375	1.00090
500	0.024	20883	25035	262.4	69.7	78.1	393	1.00081
600	0.020	28429	33415	277.6	80.9	89.3	428	1.00067
P = 1 MPa								
95	21.514	–14562	–14515	80.2	47.3	68.7	1972	1.93537
100	21.334	–14217	–14170	83.7	47.2	69.3	1946	1.92560
125	20.427	–12478	–12429	99.2	45.0	69.8	1778	1.87709
150	19.494	–10731	–10679	112.0	43.4	70.3	1592	1.82823
175	18.515	–8957	–8903	123.0	42.7	72.0	1402	1.77800
200	17.464	–7127	–7070	132.7	42.9	74.9	1209	1.72513
225	16.288	–5199	–5137	141.8	43.8	80.2	1008	1.66733
250	0.564	6762	8534	198.7	41.6	57.5	260	1.01909
275	0.489	7902	9949	204.1	43.2	56.2	280	1.01650
300	0.435	9063	11363	209.0	45.5	57.2	297	1.01467
325	0.393	10273	12815	213.7	48.3	59.1	311	1.01327
350	0.360	11546	14321	218.1	51.3	61.5	325	1.01214
375	0.333	12889	15893	222.5	54.4	64.2	337	1.01121
400	0.310	14306	17534	226.7	57.5	67.1	349	1.01043
450	0.272	17367	21038	234.9	63.8	73.0	370	1.00917
500	0.244	20730	24836	242.9	69.9	78.9	390	1.00819
600	0.201	28313	33278	258.3	81.0	89.8	427	1.00677
P = 10 MPa								
95	21.624	–14626	–14163	79.5	47.4	68.5	2000	1.94104
100	21.448	–14286	–13819	83.0	47.4	69.1	1974	1.93146
125	20.570	–12572	–12086	98.5	45.5	69.3	1814	1.88436
150	19.678	–10858	–10350	111.1	43.9	69.6	1637	1.83753
175	18.758	–9130	–8596	121.9	43.3	70.8	1459	1.79010
200	17.793	–7363	–6801	131.5	43.5	73.0	1284	1.74134
225	16.760	–5535	–4938	140.3	44.3	76.4	1110	1.69017
250	15.620	–3609	–2969	148.6	45.8	81.5	935	1.63488
275	14.301	–1539	–839	156.7	47.9	89.4	758	1.57249
300	12.666	757	1547	165.0	50.8	102.7	577	1.49740
325	10.398	3443	4404	174.1	54.7	129.1	399	1.39745
350	7.292	6643	8015	184.8	58.8	150.1	290	1.26832
375	5.182	9419	11349	194.1	60.0	115.7	289	1.18570
400	4.182	11577	13968	200.8	61.4	96.9	310	1.14797
450	3.204	15379	18500	211.5	65.8	87.5	347	1.11193
500	2.677	19135	22870	220.7	71.2	88.0	378	1.09288
600	2.076	27160	31978	237.3	81.8	94.7	427	1.07142

Propane (C_3H_8)

T K	ρ mol/L	*E* J/mol	*H* J/mol	*S* J/mol K	C_v J/mol K	C_p J/mol K	v_s m/s	*D*
P = 0.1 MPa (1 bar)								
90	16.526	–21486	–21426	87.3	59.2	84.5	2126	2.07988
100	16.295	–20639	–20577	96.2	59.6	85.2	2041	2.05806
125	15.726	–18495	–18432	115.4	59.2	86.5	1856	2.00674
150	15.156	–16319	–16253	131.3	58.9	88.0	1685	1.95796
175	14.577	–14096	–14028	145.0	59.5	90.3	1521	1.91036
200	13.982	–11806	–11735	157.3	61.0	93.5	1359	1.86300
225	13.339	–9395	–9387	168.5	63.4	97.9	1197	1.81487
250	0.050	9194	11213	257.6	57.2	66.8	228	1.00238
275	0.045	10691	12930	264.1	61.6	70.7	239	1.00215
300	0.041	12297	14752	270.5	66.2	75.1	249	1.00195
325	0.037	14019	16689	276.7	71.1	79.8	259	1.00179
350	0.035	15862	18744	282.8	76.0	84.6	269	1.00166
375	0.032	17827	20921	288.8	80.9	89.5	278	1.00154
400	0.030	19912	23217	294.7	85.7	94.3	286	1.00144
450	0.027	24441	28166	306.4	95.2	103.6	303	1.00128
500	0.024	29428	33573	317.7	104.1	112.6	318	1.00115
600	0.020	40677	45658	339.7	120.4	128.8	347	1.00095
P = 1 MPa								
90	16.526	–21486	–21426	87.2	59.3	84.5	2128	2.08034
100	16.295	–20639	–20577	96.2	59.7	85.2	2043	2.05856
125	15.726	–18495	–18432	115.3	59.2	86.4	1859	2.00736
150	15.156	–16319	–16253	131.2	59.0	88.0	1690	1.95873
175	14.577	–14096	–14028	144.9	59.6	90.2	1526	1.91132
200	13.982	–11806	–11735	157.2	61.1	93.4	1365	1.86421
225	13.361	–9424	–9349	168.4	63.4	97.7	1205	1.81642
250	12.696	–6919	–6840	179.0	66.4	103.3	1045	1.76672
275	11.962	–4252	–4169	189.1	70.0	110.8	881	1.71316
300	11.102	–1360	–1270	199.2	74.1	121.9	708	1.65216
325	0.428	13278	15614	255.2	74.1	89.6	233	1.02067
350	0.383	15259	17869	261.9	78.0	91.2	248	1.01846
375	0.349	17318	20183	268.3	82.2	94.2	261	1.01678
400	0.322	19472	22582	274.4	86.7	97.8	272	1.01544
450	0.279	24092	27672	286.4	95.7	105.9	293	1.01337
500	0.248	29137	33172	298.0	104.4	114.1	312	1.01184
600	0.203	40455	45374	320.2	120.5	129.7	344	1.00968
P = 10 MPa								
90	16.590	–21553	–20951	86.5	59.9	84.4	2146	2.08489
100	16.364	–20714	–20103	95.4	60.1	85.1	2068	2.06350
125	15.810	18595	17962	114.5	59.6	86.1	1895	2.01342
150	15.259	–16448	–15793	130.3	59.3	87.5	1733	1.96617
175	14.705	–14261	–13581	144.0	59.9	89.5	1577	1.92048
200	14.141	–12016	–11309	156.1	61.4	92.4	1425	1.87557
225	13.562	–9692	–8955	167.2	63.7	96.1	1277	1.83076
250	12.960	–7268	–6496	177.5	66.7	100.7	1133	1.78529
275	12.322	–4721	–3909	187.4	70.2	106.4	991	1.73826
300	11.631	–2027	–1167	196.9	74.1	113.2	851	1.68849
325	10.860	843	1764	206.3	78.4	121.5	715	1.63437
350	9.973	3924	4927	215.7	82.9	132.0	582	1.57361
375	8.905	7270	8393	225.2	87.7	146.1	455	1.50271
400	7.561	10957	12279	235.3	93.0	165.7	339	1.41671
450	4.614	18845	21013	255.8	101.8	167.8	249	1.24060
500	3.241	25567	28652	272.0	107.8	142.7	276	1.16439
600	2.242	38131	42591	297.4	121.7	140.5	332	1.11122

VIRIAL COEFFICIENTS OF SELECTED GASES

Henry V. Kehiaian

This table gives second virial coefficients of about 110 inorganic and organic gases as a function of temperature. Selected data from the literature have been fitted by least squares to the equation

$$B/\text{cm}^3\,\text{mol}^{-1} = \sum_{i=1}^{n} a(i)\left[(T_o/T) - 1\right]^{i-1}$$

where T_o = 298.15 K. The table gives the coefficients $a(i)$ and values of B at fixed temperature increments, as calculated from this smoothing equation.

The equation may be used with the tabulated coefficients for interpolation within the indicated temperature range. It should not be used for extrapolation beyond this range.

Compounds are listed in the modified Hill order (see Introduction), with carbon-containing compounds following those compounds not containing carbon.

A useful compilation of virial coefficient data from the literature may be found in:

J. H. Dymond and E. B. Smith, *The Virial Coefficients of Pure Gases and Mixtures, A Critical Compilation*, Oxford University Press, Oxford, 1980.

Compounds Not Containing Carbon

Mol. form.	Name	T/K	B/cm³ mol⁻¹
Ar	Argon	100	–184
		120	–131
		140	–98
	a(1) = –16	160	–76
	a(2) = –60	80	–60
	a(3) = –10	200	–48
		300	–16
		400	–1
		500	7
		600	12
		700	15
		800	18
		900	20
		1000	22
BF_3	Boron trifluoride	200	–338
		240	–202
		280	–129
	a(1) = –106	320	–85
	a(2) = –330	360	–56
	a(3) = –251	400	–37
	a(4) = –80	440	–23
ClH	Hydrogen chloride	190	–451
		230	–269
		270	–181
	a(1) = –144	310	–132
	a(2) = –325	350	–102
	a(3) = –277	390	–81
	a(4) = –170	430	–66
		470	–54
Cl_2	Chlorine	210	–508
		220	–483

Mol. form.	Name	*T*/K	*B*/cm³ mol⁻¹
		230	–457
	a(1) = –303	240	–432
	a(2) = –555	250	–407
	a(3) = 9	260	–383
	a(4) = 329	270	–360
	a(5) = 68	280	–339
		290	–318
		300	–299
		350	–221
		400	–166
		450	–126
		500	–97
		600	–59
		700	–36
		800	–22
		900	–12
F_2	Fluorine	80	–386
		110	–171
		140	–113
	a(1) = –25	170	–73
	a(2) = 21	200	–47
	a(3) = –185	230	–32
	a(4) = 113	260	25
F_4Si	Silicon tetrafluoride	210	–268
		240	–213
		270	–170
	a(1) = –138	300	–136
	a(2) = –312	330	–108
		360	–84
		390	–64
		420	–47
		450	–32
F_5I	Iodine pentafluoride	320	–2540
		330	–2344
		340	–2172
	a(1) = –3077	350	–2021
	a(2) = –8474	360	–1890
	a(3) = –9116	370	–1775
		380	–1674
		390	–1587
		400	–1510
		410	–1443
F_5P	Phosphorus pentafluoride	320	–162
		340	–143
		360	–127
	a(1) = –186	380	–112
	a(2) = –345	400	–98
		420	–86
		440	–75
		460	–64
F_6Mo	Molybdenum hexafluoride	300	–896
		310	–810
		320	–737

Mol. form.	Name	*T*/K	*B*/cm³ mol⁻¹
	a(1) = –914	330	–677
	a(2) = –2922	340	–627
	a(3) = –4778	350	–586
		360	–553
		370	–527
		380	–506
		390	–491
F_6S	Sulfur hexafluoride	200	–685
		250	–416
		300	–275
	a(1) = –279	350	–190
	a(2) = –647	400	–135
	a(3) = –335	450	–96
	a(4) = –72	500	–68
F_6U	Uranium hexafluoride	320	–1030
		340	–905
		360	–805
	a(1) = –1204	380	–724
	a(2) = –2690	400	–658
	a(3) = –2144	420	–604
		440	–560
F_6W	Tungsten hexafluoride	320	–641
		340	–578
		360	–523
	a(1) = –719	380	–473
	a(2) = –1143	400	–428
		420	–387
		440	–350
		460	–317
H_2	Hydrogen	15	–230
		20	–151
		25	–108
	a(1) = 15.4	30	–82
	a(2) = –9.0	35	–64
	a(3) = –0.2	40	–52
		45	–42
		50	–35
		60	–24
		70	–16
		80	–11
		90	–7
		100	–3
		200	11
		300	15
		400	18
H_2O	Water	300	–1126
		320	–850
		340	–660
	a(1) = –1158	360	–526
	a(2) = –5157	380	–428
	a(3) = –10301	400	–356
	a(4) =–10597	420	–301
	a(5) = –4415	440	–258

Mol. form.	Name	*T*/K	*B*/cm³ mol⁻¹
		460	–224
		480	–197
		500	–175
		600	–104
		700	–67
		800	–44
		900	–30
		1000	–20
		1100	–14
		1200	–11
H_3N	Ammonia	290	–302
		300	–265
		310	–236
	a(1) = –271	320	–213
	a(2) = –1022	330	–194
	a(3) = –2715	340	–179
	a(4) = –4189	350	–166
		360	–154
		370	–144
		380	–135
		400	–118
		420	–101
H_3P	Phosphine	190	–457
		200	–404
		210	–364
	a(1) = –146	220	–332
	a(2) = –733	230	–305
	a(3) = 1022	240	–281
	a(4) = –1220	250	–258
		260	–235
		270	–213
		280	–190
		290	–166
He	Helium	2	–172
		6	–48
		10	–24
	a(1) = 12	14	–13
	a(2) = –1	18	–7
		22	–3
		26	–1
		30	1
		50	6
		70	8
		90	10
		110	10
		150	11
		250	12
		650	13
		700	13
Kr	Krypton	110	–363
		120	–307
		130	–263
	a(1) = –51	140	–229
	a(2) = –118	150	–201

VIRIAL COEFFICIENTS OF SELECTED GASES (continued)

Mol. form.	Name	T/K	B/cm^3 mol^{-1}
	a(3) = –29	160	–178
	a(4) = –5	170	–159
		180	–143
		190	–129
		200	–117
		250	–75
		300	–51
		400	–23
		500	–8
		600	2
		700	8
NO	Nitric oxide	120	–232
		130	–176
		140	–138
	a(1) = –12	150	–113
	a(2) = –119	160	–96
	a(3) = 89	170	–83
	a(4) = –73	180	–73
		190	–65
		200	–58
		210	–52
		230	–42
		250	–32
		270	–24
N_2	Nitrogen	75	–274
		100	–161
		125	–104
	a(1) = –4	150	–71
	a(2) = –56	175	–49
	a(3) = –12	200	–34
		225	–24
		250	–15
		300	–4
		400	9
		500	16
		600	21
		700	24
N_2O	Nitrous oxide	240	–219
		260	–181
		280	–151
	a(1) = –130	300	–128
	a(2) = –307	320	–110
	a(3) = –248	340	–96
		360	–85
		380	–76
		400	–68
Ne	Neon	60	–25
		80	–13
		100	–6
	a(1) = 10.8	120	–1
	a(2) = –7.5	140	2
	a(3) = 0.4	160	4
		180	6
		200	7

Mol. form.	Name	T/K	B/cm³ mol⁻¹
		300	11
		400	13
		500	14
		600	15
O_2	Oxygen	90	–241
		110	–161
		130	–117
	a(1) = –16	150	–88
	a(2) = –62	170	–69
	a(3) = –8	190	–55
	a(4) = –3	210	–44
		230	–36
		250	–29
		270	–23
		290	–18
		310	–14
		330	–10
		350	–7
		400	–1
O_2S	Sulfur dioxide	290	–465
		320	–354
		350	–276
	a(1) = –430	380	–221
	a(2) = –1193	410	–181
	a(3) = 1029	440	–153
		470	–132
Xe	Xenon	160	–421
		170	–377
		180	–340
	a(1) = –130	190	–307
	a(2) = –262	200	–280
	a(3) = –87	210	–255
		220	–234
		230	–215
		240	–199
		250	–184
		300	–129
		350	–93
		400	–69
		500	–39
		600	–21
		650	–14

Compounds Containing Carbon

Mol. form.	Name	T/K	B/cm³ mol⁻¹
$CClF_3$	Chlorotrifluoromethane	240	–369
		290	–237
		340	–165
	a(1) = –223	390	–119
	a(2) = –504	440	–86
	a(3) = –340	490	–60
	a(4) = –291	540	–39

VIRIAL COEFFICIENTS OF SELECTED GASES (continued)

Mol. form.	Name	T/K	B/cm³ mol⁻¹
CCl_2F_2	Dichlorodifluoromethane	250	–769
		280	–570
		310	–441
	a(1) = –486	340	–353
	a(2) = –1217	370	–289
	a(3) = –1188	400	–241
	a(4) = –698	430	–204
		460	–174
CCl_3F	Trichlorofluoromethane	240	–1140
		280	–879
		320	–689
	a(1) = –786	360	–545
	a(2) = –1428	400	–431
	a(3) = –142	440	–340
		480	–265
CCl_4	Tetrachloromethane	320	–1345
		340	–1171
		360	–1040
	a(1) = –1600	380	–942
	a(2) = –4059	400	–868
	a(3) = –4653	420	–814
CF_4	Tetrafluoromethane	250	–137
		300	–87
		350	–55
	a(1) = –88	400	–32
	a(2) = –238	450	–16
	a(3) = –70	500	–4
		600	14
		700	25
		800	33
$CHClF_2$	Chlorodifluoromethane	300	–343
		325	–298
		350	–257
	a(1) = –347	375	–221
	a(2) = –575	400	–188
	a(3) = 187	425	–158
$CHCl_2F$	Dichlorofluoromethane	250	–728
		275	–634
		300	–557
	a(1) = –562	325	–491
	a(2) = –862	350	–434
		375	–385
		400	–343
		425	–305
		450	–271
$CHCl_3$	Trichloromethane	320	–1001
		330	–926
		340	–858
	a(1) = –1193	350	–797
	a(2) = –2936	360	–740
	a(3) = –1751	370	–689
		380	–642

Mol. form.	Name	T/K	B/cm³ mol⁻¹
		390	–599
		400	–559
CHF_3	Trifluoromethane	200	–433
		220	–350
		240	–288
	a(1) = –177	260	–241
	a(2) = –399	280	–204
	a(3) = –250	300	–174
		320	–151
		340	–132
		360	–116
		380	–103
		400	–91
CH_2Cl_2	Dichloromethane	320	–706
		330	–634
		340	–574
	a(1) = –913	350	–524
	a(2) = –3371	360	–482
	a(3) = –5013	370	–447
		380	–420
		400	–380
		420	357
CH_2F_2	Difluoromethane	280	–375
		290	–343
		300	–316
	a(1) = –321	310	–294
	a(2) = –754	320	–275
	a(3) = –1300	330	–260
		340	–248
		350	–238
CH_3Br	Bromomethane	280	–645
		290	–596
		300	–551
	a(1) = –559	310	–509
	a(2) = –1324	320	–469
		340	–396
		360	–332
		380	–274
CH_3Cl	Chloromethane	280	–466
		300	–402
		320	–348
	a(1) = –407	340	–304
	a(2) = –887	360	–266
	a(3) = –385	380	–234
		400	–206
		420	–182
		440	–161
		460	–142
		480	–126
		500	–112
		600	–58
CH_3F	Fluoromethane	280	–244

Mol. form.	Name	*T*/K	*B*/cm³ mol⁻¹
		300	–205
		320	–174
	a(1) = –209	340	–150
	a(2) = –525	360	–129
	a(3) = –365	380	–112
		400	–99
		420	–87
CH_3I	Iodomethane	310	–725
		320	–646
		330	–582
	a(1) = –844	340	–531
	a(2) = –3353	350	–492
	a(3) = –6590	360	–462
		370	–441
		380	–427
CH_4	Methane	110	–328
		120	–276
		130	–237
	a(1) = –43	140	–206
	a(2) = –114	150	–181
	a(3) = –19	160	–160
	a(4) = –7	170	–143
		180	–128
		190	–116
		200	–105
		250	–66
		300	–43
		350	–27
		400	–16
		500	0
		600	10
CH_4O	Methanol	320	–1431
		330	–1299
		340	–1174
	a(1) = –1752	350	–1056
	a(2) = –4694	360	–945
		370	–840
		380	–741
		390	–646
		400	–557
CH_5N	Methylamine	300	–451
		325	–367
		350	–304
	a(1) = –459	375	–257
	a(2) = –1191	400	–220
	a(3) = –995	425	–192
		450	–170
		500	–140
		550	–122
CO	Carbon monoxide	210	–36
		240	–24
		270	–15
	a(1) = –9	300	–8

Mol. form.	Name	*T*/K	*B*/cm³ mol⁻¹
	a(2) = –58	330	–3
	a(3) = –18	360	1
		420	7
		480	11
CO_2	Carbon dioxide	220	–244
		240	–204
		260	–172
	a(1) = –127	280	–146
	a(2) = –288	300	–126
	a(3) = –118	320	–108
		340	–94
		360	–81
		380	71
		400	–62
		500	–30
		600	–13
		700	–1
		800	7
		900	12
		1000	16
		1100	19
CS_2	Carbon disulfide	280	–932
		310	–740
		340	–603
	a(1) = –807	370	–504
	a(2) = –1829	400	–431
	a(3) = –1371	430	–375
$C_2Cl_2F_4$	1,2-Dichloro-1,1,2,2-tetrafluoroethane	300	–801
		320	–695
		340	–608
	a(1) = –812	360	–536
	a(2) = –1773	380	–475
	a(3) = –963	400	–423
		420	–379
		440	–341
		460	–307
		480	–279
		500	–253
$C_2Cl_3F_3$	1,1,2-Trichloro-1,2,2-trifluoroethane	290	–1041
		310	–943
		330	–830
	a(1) = –999	350	–780
	a(2) = –1479	370	–712
		390	–651
		410	–596
		430	–546
		450	–500
C_2H_2	Ethyne	200	–573
		210	–500
		220	–440
	a(1) = –216	230	–390
	a(2) = –375	240	–349
	a(3) = –716	250	–315

Mol. form.	Name	*T*/K	*B*/cm³ mol⁻¹
		260	–287
		270	–263
C_2H_3N	Ethanenitrile	330	–3468
		340	–2971
		350	–2563
	a(1) = –5840	360	–2233
	a(2) = –29175	370	–1970
	a(3) = –47611	380	–1765
		390	–1610
		400	–1499
		410	–1425
C_2H_4	Ethene	240	–218
		270	–172
		300	–139
	a(1) = –140	330	–113
	a(2) = –296	360	–92
	a(3) = –101	390	–76
		420	–63
		450	–52
$C_2H_4Cl_2$	1,2-Dichloroethane	370	–812
		390	–716
		410	–635
	a(1) = –1362	430	–566
	a(2) = –3240	450	–508
	a(3) = –2100	470	–458
		490	–416
		510	–379
		530	–347
		550	–319
		570	–295
C_2H_4O	Ethanal	290	–1352
		320	–927
		350	–654
	a(1) = –1217	380	–482
	a(2) = –4647	410	–375
	a(3) = –5725	440	–314
		470	–283
$C_2H_4O_2$	Methyl methanoate	320	–821
		330	–744
		340	–677
	a(1) = –1035	350	–620
	a(2) = –3425	360	–571
	a(3) = –4203	370	–528
		380	–492
		390	–461
		400	–435
C_2H_5Cl	Chloroethane	320	–634
		360	–450
		400	–330
	a(1) = –777	440	–249
	a(2) = –2205	480	–195
	a(3) = –1764	520	–157

VIRIAL COEFFICIENTS OF SELECTED GASES (continued)

Mol. form.	Name	*T*/K	*B*/cm³ mol⁻¹
		560	–131
		600	–114
C_2H_6	Ethane	200	–409
		220	–337
		240	–284
	a(1) = –184	260	–242
	a(2) = –376	280	–209
	a(3) = –143	300	–181
	a(4) = –54	320	–159
		340	–140
		360	–123
		380	–109
		400	–96
		500	–52
		600	–24
C_2H_6O	Ethanol	320	–2710
		330	–2135
		340	–1676
	a(1) = –4475	350	–1317
	a(2) = –29719	360	–1043
	a(3) = –56716	370	–843
		380	–705
		390	–622
C_2H_6O	Dimethyl ether	275	–536
		280	–517
		285	–499
	a(1) = –455	290	–482
	a(2) = –965	295	–465
		300	–449
		305	–433
		310	–418
C_2H_7N	Dimethylamine	310	–606
		320	–563
		330	–523
	a(1) = –662	340	–487
	a(2) = –1504	350	–454
	a(3) = –667	360	–423
		370	–395
		380	–369
		390	–345
		400	322
C_2H_7N	Ethylamine	300	–773
		310	–710
		320	–654
	a(1) = –785	330	–604
	a(2) = –2012	340	–558
	a(3) = –1397	350	–517
		360	–480
		370	–447
		380	–416
		390	–389
		400	–363
C_3H_6	Cyclopropane	300	–383

Mol. form.	Name	*T*/K	*B*/cm³ mol⁻¹
		310	–356
		320	–332
	a(1) = –388	330	–310
	a(2) = –861	340	–290
	a(3) = –538	350	–272
		360	–256
		370	–241
		380	–227
		390	–215
		400	–204
C_3H_6	Propene	280	–395
		300	–342
		320	–299
	a(1) = –347	340	–262
	a(2) = –727	360	–232
	a(3) = –325	380	–205
		400	–183
		420	–163
		440	–146
		460	–131
		480	–118
		500	–106
C_3H_6O	2-Propanone	300	–1996
		320	–1522
		340	–1198
	a(1) = –2051	360	–971
	a(2) = –8903	380	–806
	a(3) = –18056	400	–683
	a(4) = –16448	420	–586
		440	–506
		460	–437
		480	–375
C_3H_6O	Ethyl methanoate	330	–1003
		340	–916
		350	–839
	a(1) = –1371	360	–771
	a(2) = –4231	370	–712
	a(3) = –4312	380	–660
		390	–614
C_3H_6O	Methyl ethanoate	320	–1320
		330	–1186
		340	–1074
	a(1) = –1709	350	–980
	a(2) = –6348	360	–903
	a(3) = –9650	370	–840
		380	–789
		390	–749
C_3H_7Cl	1-Chloropropane	310	–1001
		340	–772
		370	–614
	a(1) = –1121	400	–501
	a(2) = –3271	430	–417
	a(3) = –3786	460	–352
	a(4) = –1974	490	–302

Mol. form.	Name	T/K	B/cm³ mol⁻¹
		520	–261
		550	–227
		580	–198
C_3H_8	Propane	240	–641
		260	–527
		280	–444
	a(1) = –386	300	–381
	a(2) = –844	320	–331
	a(3) = –720	340	–292
	a(4) = –574	360	–259
		380	–232
		400	–208
		440	–169
		480	–138
		520	–112
		560	–90
C_3H_8O	1-Propanol	380	–873
		385	–826
		390	–783
	a(1) = –2690	395	–744
	a(2) = –12040	400	–709
	a(3) = –16738	405	–679
		410	–651
		415	–627
		420	–606
C_3H_8O	2-Propanol	380	–821
		385	–766
		390	–717
	a(1) = –3165	395	–674
	a(2) = –16092	400	–636
	a(3) = –24197	405	–604
		410	–576
		415	–552
		420	–533
C_3H_9N	Trimethylamine	310	–675
		320	–628
		330	–585
	a(1) = –737	340	–547
	a(2) = –1669	350	–512
	a(3) = –986	360	–480
		370	–450
C_4H_8	1-Butene	300	–624
		320	–539
		340	–470
	a(1) = –633	360	–413
	a(2) = –1442	380	–366
	a(3) = –932	400	–327
		420	–294
C_4H_8O	2-Butanone	310	–2056
		320	–1878
		330	–1712
	a(1) = –2282	340	–1555

Mol. form.	Name	T/K	B/cm³ mol⁻¹
	a(2) = –5907	350	–1407
		360	–1267
		370	–1135
$C_4H_8O_2$	Propyl methanoate	330	–1496
		340	–1354
		350	–1231
	a(1) = –2118	360	–1126
	a(2) = –7299	370	–1035
	a(3) = –8851	380	–957
		390	–890
		400	–834
$C_4H_8O_2$	Ethyl ethanoate	330	–1543
		340	–1385
		350	–1254
	a(1) = –2272	360	–1144
	a(2) = –8818	370	–1055
	a(3) = –13130	380	–982
		390	–923
		400	–878
$C_4H_8O_2$	Methyl propanoate	330	–1588
		340	–1444
		350	–1319
	a(1) = –2216	360	–1211
	a(2) = –7339	370	–1117
	a(3) = –8658	380	–1037
		390	–968
		400	–908
C_4H_9Cl	1-Chlorobutane	330	–1224
		370	–898
		410	–691
	a(1) = –1643	450	–551
	a(2) = –4897	490	–449
	a(3) = –6178	530	–371
	a(4) = –3718	570	–309
C_4H_{10}	Butane	250	–1170
		280	–863
		310	–668
	a(1) = –735	340	–536
	a(2) = –1835	370	–442
	a(3) = –1922	400	–371
	a(4) = –1330	430	–315
		460	–270
		490	–232
		520	–199
		550	–171
C_4H_{10}	2-Methylpropane	270	–900
		300	–697
		330	–553
	a(1) = –707	360	–450
	a(2) = –1719	390	–374
	a(3) = –1282	420	–317
		450	–273

Mol. form.	Name	T/K	B/cm³ mol⁻¹
		480	–240
		510	–215
$C_4H_{10}O$	1-Butanol	350	–1693
		360	–1544
		370	–1402
	a(1) = –2629	380	–1268
	a(2) = –6315	390	–1141
		400	–1021
		420	–796
		440	–593
$C_4H_{10}O$	2-Methyl-1-propanol	390	–1076
		400	–979
		410	–887
	a(1) = –2269	420	–800
	a(2) = –5065	430	–716
		440	–636
$C_4H_{10}O$	2-Butanol	380	–1110
		390	–1005
		400	–906
	a(1) = –2232	410	–811
	a(2) = –5209	420	–721
$C_4H_{10}O$	2-Methyl-2-propanol	380	–924
		390	–827
		400	–736
	a(1) = –1952	410	–649
	a(2) = –4775	420	–567
$C_4H_{10}O$	Diethyl ether	280	–1550
		300	–1199
		320	–954
	a(1) = –1226	340	–776
	a(2) = –4458	360	–638
	a(3) = –7746	380	–525
	a(4) = –10005	400	–428
		420	–340
$C_4H_{11}N$	Diethylamine	320	–1228
		330	–1134
		340	–1056
	a(1) = –1522	350	–988
	a(2) = –5204	360	–926
	a(3) = –15047	370	–868
	a(4) = –28835	380	–812
		390	–755
		400	–697
C_5H_5N	Pyridine	350	–1257
		360	–1176
		370	–1099
	a(1) = –1765	380	–1026
	a(2) = –3431	390	–957
		400	–892
		420	–770
		440	–659

Mol. form.	Name	*T*/K	*B*/cm³ mol⁻¹
C_5H_{10}	Cyclopentane	300	–1049
		305	–1015
		310	–981
	a(1) = –1062	315	–949
	a(2) = –2116	320	–918
C_5H_{10}	1-Pentene	310	–966
		320	–898
		330	–836
	a(1) = –1055	340	–780
	a(2) = –2377	350	–729
	a(3) = –1189	360	–681
		370	–638
		380	–598
		390	–561
		400	–527
		410	–495
$C_5H_{10}O$	2-Pentanone	330	–2850
		340	–2420
		350	–2076
	a(1) = –4962	360	–1804
	a(2) = –26372	370	–1595
	a(3) = –46537	380	–1440
		390	–1332
C_5H_{12}	Pentane	300	–1234
		310	–1130
		320	–1038
	a(1) = –1254	330	–957
	a(2) = –3345	340	–884
	a(3) = –2726	350	–818
		400	–579
		450	–436
		500	–348
		550	–294
C_5H_{12}	2-Methylbutane	280	–1263
		290	–1166
		300	–1079
	a(1) = –1095	310	–1001
	a(2) = –2503	320	–931
	a(3) = –1534	330	–867
		340	–810
		350	–757
		400	–557
		450	–424
C_5H_{12}	2,2-Dimethylpropane	300	–916
		310	–843
		320	–780
	a(1) = –931	330	–724
	a(2) = –2387	340	–674
	a(3) = –2641	350	–629
	a(4) = –1810	360	–590
		370	–554
		380	–521
		390	–492

Mol. form.	Name	*T*/K	*B*/cm³ mol⁻¹
		400	–464
		450	–357
		500	–279
		550	–218
C_6H_6	Benzene	290	–1588
		300	–1454
		310	–1335
	a(1) = –1477	320	–1231
	a(2) = –3851	330	–1139
	a(3) = –3683	340	–1056
	a(4) = –1423	350	–983
		400	–712
		450	–542
		500	–429
		550	–349
		600	–291
C_6H_7N	2-Methylpyridine	360	–1656
		370	–1523
		380	–1404
	a(1) = –2940	390	–1297
	a(2) = –8813	400	–1202
	a(3) = –7809	410	–1117
		420	–1040
		430	–972
C_6H_7N	3-Methylpyridine	380	–1819
		390	–1612
		400	–1448
	a(1) = –6304	410	–1322
	a(2) = –30415	420	–1230
	a(3) = –44549	430	–1166
C_6H_7N	4-Methylpyridine	380	–1787
		390	–1578
		400	–1417
	a(1) = –6553	410	–1297
	a(2) = –32873	420	–1214
	a(3) = –49874	430	–1163
C_6H_{12}	Cyclohexane	300	–1698
		320	–1391
		340	–1170
	a(1) = –1733	360	1007
	a(2) = –5618	380	–883
	a(3) = –9486	400	–786
	a(4) = –7936	420	–707
		440	–641
		460	–584
		480	–534
		500	–488
		520	–446
		540	–406
		560	–368
C_6H_{12}	Methylcyclopentane	305	–1447
		315	–1357

VIRIAL COEFFICIENTS OF SELECTED GASES (continued)

Mol. form.	Name	*T*/K	*B*/cm³ mol⁻¹
		325	–1272
	a(1) = –1512	335	–1192
	a(2) = –2910	345	–1117
C_6H_{14}	Hexane	300	–1920
		310	–1724
		320	–1561
	a(1) = –1961	330	–1424
	a(2) = –6691	340	–1309
	a(3) = –13167	350	–1209
	a(4) = –15273	360	–1123
		370	–1046
		380	–978
		390	–916
		400	–859
		410	–806
		430	–707
		450	–616
$C_6H_{15}N$	Triethylamine	330	–1562
		340	–1444
		350	–1340
	a(1) = –2061	360	–1249
	a(2) = –5735	370	–1169
	a(3) = –5899	380	–1099
		390	–1037
		400	–983
C_7H_8	Toluene	350	–1641
		360	–1511
		370	–1394
	a(1) = –2620	380	–1289
	a(2) = –7548	390	–1195
	a(3) = –6349	400	–1110
		410	–1034
		420	–965
		430	–903
C_7H_{14}	1-Heptene	340	–1781
		350	–1651
		360	–1532
	a(1) = –2491	370	–1424
	a(2) = –6230	380	–1324
	a(3) = –3780	390	–1233
		400	–1150
		410	–1073
C_7H_{16}	Heptane	300	–2782
		320	–2297
		340	–1928
	a(1) = –2834	360	–1641
	a(2) = –8523	380	–1415
	a(3) = –10068	400	–1233
	a(4) = –5051	420	–1085
		440	–963
		460	–862
		480	–775
		500	–702

VIRIAL COEFFICIENTS OF SELECTED GASES (continued)

Mol. form.	Name	T/K	B/cm³ mol⁻¹
		540	–583
		580	–490
		620	–416
		660	–355
		700	–304
C_8H_{10}	1,2-Dimethylbenzene	380	–2046
		390	–1848
		400	–1681
	a(1) = –5632	410	–1543
	a(2) = –22873	420	–1428
	a(3) = –28900	430	–1335
		440	–1261
C_8H_{10}	1,3-Dimethylbenzene	380	–2082
		390	–1865
		400	–1679
	a(1) = –5808	410	–1521
	a(2) = –23244	420	–1388
	a(3) = –27607	430	–1276
		440	–1184
C_8H_{10}	1,4-Dimethylbenzene	380	–2043
		390	–1851
		400	–1680
	a(1) = –4921	410	–1529
	a(2) = –16843	420	–1395
	a(3) = –16159	430	–1276
		440	–1171
C_8H_{16}	1-Octene	360	–2147
		370	–2000
		380	–1861
	a(1) = –3273	390	–1729
	a(2) = –6557	400	–1604
		410	–1485
C_8H_{18}	Octane	300	–4042
		350	–2511
		400	–1704
	a(1) = –4123	450	–1234
	a(2) = –13120	500	–936
	a(3) = –16408	550	–732
	a(4) = –8580	600	–583
		650	–468
		700	–375

VAN DER WAALS' CONSTANTS FOR GASES

The van der Waals' equation of state for a real gas is

$$(P + n^2a/V^2)(V - nb) = nRT$$

where P is the pressure, V the volume, T the temperature, n the amount of substance (in moles), and R the gas constant. The van der Waals constants a and b are characteristic of the substance and are independent of temperature. They are related to the critical temperature and pressure, T_c and P_c, by

$$a = 27R^2T_c^2/64P_c \qquad b = RT_c/8P_c$$

This table gives values of a and b for over 200 common gases. Most of the values have been calculated from the critical constants.

To convert the van der Waals constants to SI units, note that

$$1 \text{ bar L}^2/\text{mol}^2 = 0.1 \text{ Pa m}^6/\text{mol}^2$$
$$1 \text{ L/mol} = 0.001 \text{ m}^3/\text{mol}$$

Substances are listed by molecular formula in the modified Hill order, with substances not containing carbon preceding those that do contain carbon.

REFERENCE

Reid, R.C, Prausnitz, J. M., and Poling,B. E., *The Properties of Gases and Liquids*, Fourth Edition, McGraw-Hill, New York, 1987.

Molecular formula	Name	a bar L²/mol	b L/mol
$AlCl_3$	Aluminum trichloride	42.63	0.2450
Ar	Argon	1.355	0.03201
BCl_3	Boron trichloride	15.60	0.1222
BF_3	Boron trifluoride	3.98	0.05443
B_2H_6	Diborane	6.048	0.07437
BrH	Hydrogen bromide	4.500	0.04415
Br_2	Bromine	9.75	0.0591
$ClFO_3$	Perchloryl fluoride	7.371	0.07130
ClF_5	Chlorine pentafluoride	9.58	0.08204
ClH	Hydrogen chloride	3.700	0.04061
ClH_4N	Ammonium chloride	2.380	0.00734
ClH_4P	Phosphonium chloride	4.111	0.04545
Cl_2	Chlorine	6.343	0.05422
Cl_3FSi	Trichlorofluorosilane	15.67	0.1273
Cl_4Ge	Germanium tetrachloride	23.12	0.1489
Cl_4Si	Silicon tetrachloride	20.96	0.1470
Cl_4Sn	Tin chloride ($SnCl_4$)	27.25	0.1641
Cl_4Ti	Titanium chloride ($TiCl_4$)	25.47	0.1423
FH	Hydrogen fluoride	9.565	0.0739
F_2	Fluorine	1.171	0.02896
F_2Xe	Xenon difluoride	12.46	0.07037
F_3N	Nitrogen trifluoride	3.58	0.05453
F_3P	Phosphorus trifluoride	4.954	0.06510
F_4N_2	Tetrafluorohydrazine	7.426	0.08564
F_4Si	Silicon tetrafluoride	5.259	0.07236
F_4Xe	Xenon tetrafluoride	15.52	0.09035
F_6S	Sulfur fluoride (SF_6)	7.857	0.08786
F_6U	Uranium fluoride (UF_6)	16.01	0.1128
F_6W	Tungsten fluoride (WF_6)	13.25	0.1063
GeH_4	Germane	5.743	0.06555
HI	Hydrogen iodide	6.309	0.05303
H_2	Hydrogen	0.2453	0.02651
H_2O	Water	5.537	0.03049
H_2S	Hydrogen sulfide	4.544	0.04339
H_2Se	Hydrogen selenide	5.523	0.0479

Molecular formula	Name	*a* bar L²/mol	*b* L/mol
H_3N	Ammonia	4.225	0.03713
H_3P	Phosphine	4.696	0.05157
H_4N_2	Hydrazine	8.46	0.0462
H_4Si	Silane	4.38	0.0579
He	Helium	0.0346	0.0238
Hg	Mercury	5.193	0.01057
Kr	Krypton	2.325	0.0396
NO	Nitric oxide	1.46	0.0289
NO_2	Nitrogen dioxide	5.36	0.0443
N_2	Nitrogen	1.370	0.0387
N_2O	Nitrous oxide	3.852	0.04435
Ne	Neon	0.208	0.01672
O_2	Oxygen	1.382	0.03186
O_2S	Sulfur dioxide	6.865	0.05679
O_3	Ozone	3.570	0.0487
P	Phosphorus	53.6	0.157
Rn	Radon	6.601	0.06239
S	Sulfur	24.3	0.0660
Se	Selenium	33.4	0.0675
Xe	Xenon	4.192	0.05156
$CClF_3$	Chlorotrifluoromethane	6.873	0.08110
CCl_3F	Trichlorofluoromethane	14.68	0.1111
CCl_4	Tetrachloromethane	20.01	0.1281
CF_4	Tetrafluoromethane	4.040	0.06325
CO	Carbon monoxide	1.472	0.03948
COS	Carbon oxysulfide	6.975	0.06628
CO_2	Carbon dioxide	3.658	0.04286
CS_2	Carbon disulfide	11.25	0.07262
$CHCl_3$	Trichloromethane	15.34	0.1019
CHF_3	Trifluoromethane	5.378	0.06403
CHN	Hydrogen cyanide	11.29	0.08806
CH_2Cl_2	Dichloromethane	12.44	0.08689
CH_2F_2	Difluoromethane	6.184	0.06268
CH_3Cl	Chloromethane	7.566	0.06477
CH_3F	Fluoromethane	5.009	0.05617
CH_3NO_2	Nitromethane	17.18	0.1041
CH_4	Methane	2.300	0.04301
CH_4O	Methanol	9.472	0.06584
CH_4S	Methanethiol	8.911	0.06756
CH_5N	Methylamine	7.106	0.05879
$C_2Cl_3F_3$	1,1,2-Trichlorotrifluoroethane	20.25	0.1481
C_2F_4	Tetrafluoroethylene	6.954	0.08085
C_2N_2	Cyanogen	7.803	0.06952
C_2H_2	Acetylene	4.516	0.05220
$C_2H_2F_2$	1,1-Difluoroethylene	6.000	0.07058
$C_2H_3Cl_3$	1,1,1-Trichloroethane	20.15	0.1317
C_2H_3F	Fluoroethylene	5.984	0.06504
$C_2H_3F_3$	1,1,1-Trifluoroethane	9.302	0.09572
C_2H_3N	Acetonitrile	17.89	0.1169
C_2H_4	Ethylene	4.612	0.05821
$C_2H_4Cl_2$	1,1-Dichloroethane	15.73	0.1072
$C_2H_4Cl_2$	1,2-Dichloroethane	17.0	0.108
C_2H_4O	Ethylene oxide	8.922	0.06779
$C_2H_4O_2$	Acetic acid	17.71	0.1065
$C_2H_4O_2$	Methyl formate	11.54	0.08442
C_2H_5Br	Bromoethane	11.89	0.08406
C_2H_5Cl	Chloroethane	11.7	0.090

VAN DER WAALS' CONSTANTS FOR GASES (continued)

Molecular formula	Name	*a* bar L²/mol	*b* L/mol
C_2H_5F	Fluoroethane	8.170	0.07758
C_2H_6	Ethane	5.570	0.06499
C_2H_6O	Dimethyl ether	8.690	0.07742
C_2H_6O	Ethanol	12.56	0.08710
C_2H_6S	Dimethyl sulfide	13.34	0.09453
C_2H_6S	Ethanethiol	13.23	0.09447
C_2H_7N	Dimethylamine	10.44	0.08510
C_2H_7N	Ethylamine	10.79	0.08433
C_3F_8	Perfluoropropane	12.96	0.1338
C_3H_5N	Propanenitrile	21.57	0.1369
C_3H_6	Propene	8.438	0.08242
C_3H_6	Cyclopropane	8.293	0.07420
C_3H_6O	Acetone	16.02	0.1124
C_3H_6O	Propanal	14.08	0.09947
$C_3H_6O_2$	Ethyl formate	15.91	0.1115
$C_3H_6O_2$	Methyl acetate	15.75	0.1108
$C_3H_6O_2$	Propanoic acid	23.49	0.1386
C_3H_7Cl	1-Chloropropane	16.11	0.1141
C_3H_8	Propane	9.385	0.09044
C_3H_8O	1-Propanol	16.26	0.1080
C_3H_8O	2-Propanol	15.82	0.1109
C_3H_8O	Ethyl methyl ether	12.70	0.1034
C_3H_8S	Ethyl methyl sulfide	19.45	0.1300
C_3H_9N	Propylamine	15.26	0.1094
C_3H_9N	Trimethylamine	13.37	0.1101
C_4H_4O	Furan	12.74	0.0926
C_4H_4S	Thiophene	17.21	0.1058
C_4H_5N	Pyrrole	18.82	0.1049
C_4H_6	1,3-Butadiene	12.17	0.1020
$C_4H_6O_3$	Acetic anhydride	26.8	0.157
C_4H_7N	Butanenitrile	25.76	0.1568
C_4H_8	1-Butene	12.76	0.1084
C_4H_8	Cyclobutane	12.39	0.0960
C_4H_8O	2-Butanone	19.97	0.1326
C_4H_8O	Tetrahydrofuran	16.39	0.1082
$C_4H_8O_2$	1,4-Dioxane	19.29	0.1171
$C_4H_8O_2$	Ethyl acetate	20.57	0.1401
$C_4H_8O_2$	Methyl propanoate	20.51	0.1377
$C_4H_8O_2$	Propyl formate	20.79	0.1377
$C_4H_8O_2$	Butanoic acid	28.18	0.1609
C_4H_9N	Pyrrolidine	16.84	0.1056
C_4H_{10}	Butane	13.93	0.1168
C_4H_{10}	Isobutane	13.36	0.1168
$C_4H_{10}O$	1-Butanol	20.90	0.1323
$C_4H_{10}O$	2-Methyl-2-propanol	18.81	0.1324
$C_4H_{10}O$	2-Methyl-1-propanol	20.35	0.1324
$C_4H_{10}O$	Diethyl ether	17.46	0.1333
$C_4H_{10}S$	Diethyl sulfide	22.85	0.1462
$C_4H_{11}N$	Butylamine	19.41	0.1301
$C_4H_{11}N$	Diethylamine	19.40	0.1383
$C_4H_{12}Si$	Tetramethylsilane	20.81	0.1653
$C_5H_4O_2$	Furfural	22.23	0.1182
C_5H_5N	Pyridine	19.77	0.1137
C_5H_8	Cyclopentene	15.61	0.1097
C_5H_{10}	1-Pentene	17.86	0.1370
C_5H_{10}	2-Methyl-1-butene	16.9	0.129
C_5H_{10}	2-Methyl-2-butene	17.26	0.1279

VAN DER WAALS' CONSTANTS FOR GASES (continued)

Molecular formula	Name	*a* bar L²/mol	*b* L/mol
C_5H_{10}	Cyclopentane	16.94	0.1180
$C_5H_{10}O$	Tetrahydropyran	20.02	0.1247
$C_5H_{10}O_2$	Isobutyl formate	22.82	0.1476
$C_5H_{10}O_2$	Propyl acetate	26.23	0.1700
$C_5H_{10}O_2$	Ethyl propanoate	25.86	0.1688
$C_5H_{10}O_2$	Methyl butanoate	25.83	0.1661
$C_5H_{10}O_2$	Methyl isobutanoate	24.87	0.1639
$C_5H_{11}N$	Piperidine	20.84	0.1250
C_5H_{12}	Pentane	19.13	0.1451
C_5H_{12}	Isopentane	18.29	0.1415
C_5H_{12}	Neopentane	17.17	0.1410
$C_5H_{12}O$	1-Pentanol	25.81	0.1564
C_6H_5Br	Bromobenzene	28.96	0.1541
C_6H_5Cl	Chlorobenzene	25.80	0.1454
C_6H_5F	Fluorobenzene	20.10	0.1279
C_6H_5I	Iodobenzene	33.54	0.1658
C_6H_6	Benzene	18.82	0.1193
C_6H_6O	Phenol	22.93	0.1177
C_6H_7N	Aniline	29.14	0.1486
$C_6H_{10}O$	Cyclohexanone	31.1	0.170
$C_6H_{11}N$	Hexanenitrile	35.50	0.1996
C_6H_{12}	Cyclohexane	21.95	0.1413
$C_6H_{12}O$	Cyclohexanol	28.93	0.1586
$C_6H_{12}O_2$	Pentyl formate	27.97	0.1730
$C_6H_{12}O_2$	Isobutyl acetate	29.05	0.1845
$C_6H_{12}O_2$	Ethyl butanoate	30.53	0.1922
$C_6H_{12}O_2$	Ethyl 2-methylpropanoate	29.05	0.1872
$C_6H_{12}O_2$	Methyl pentanoate	29.39	0.1847
C_6H_{14}	Hexane	24.97	0.1753
C_6H_{14}	2,3-Dimethylbutane	23.29	0.1660
$C_6H_{14}O$	1-Hexanol	31.35	0.1829
$C_6H_{15}N$	Triethylamine	27.59	0.1836
$C_6H_{15}N$	Dipropylamine	24.82	0.1591
C_7H_5N	Benzonitrile	33.89	0.1727
C_7H_6O	Benzaldehyde	30.30	0.1553
C_7H_8	Toluene	24.89	0.1499
C_7H_8O	*o*-Cresol	28.33	0.1447
C_7H_8O	*m*-Cresol	31.86	0.1609
C_7H_8O	*p*-Cresol	28.11	0.1422
C_7H_8O	Benzyl alcohol	34.7	0.173
C_7H_8O	Anisole	28.60	0.1579
C_7H_{16}	Heptane	30.89	0.2038
$C_7H_{16}O$	1-Heptanol	37.22	0.2097
C_8H_{10}	Ethylbenzene	30.86	0.1782
C_8H_{10}	*o*-Xylene	31.06	0.1756
C_8H_{10}	*m*-Xylene	31.41	0.1814
C_8H_{10}	*p*-Xylene	31.54	0.1824
$C_8H_{10}O$	Phenetole	35.70	0.1966
$C_8H_{11}N$	*N,N*-Dimethylaniline	37.92	0.1967
C_8H_{18}	Octane	37.86	0.2372
C_8H_{18}	2,5-Dimethylhexane	35.49	0.2299
$C_8H_{18}O$	1-Octanol	43.42	0.2371
C_9H_7N	Quinoline	36.70	0.1672
C_9H_{12}	Cumene	36.20	0.2044
C_9H_{12}	Propylbenzene	37.14	0.2073
C_9H_{12}	1,2,4-Trimethylbenzene	38.03	0.2088
C_9H_{12}	Mesitylene	37.87	0.2118

Molecular formula	Name	*a* bar L²/mol	*b* L/mol
C_9H_{20}	Nonane	45.11	0.2702
$C_9H_{20}O$	1-Nonanol	50.00	0.2654
$C_{10}H_8$	Naphthalene	40.32	0.1920
$C_{10}H_{14}$	Butylbenzene	44.07	0.2378
$C_{10}H_{14}$	Isobutylbenzene	40.40	0.2215
$C_{10}H_{14}$	*o*-Cymene	42.7	0.234
$C_{10}H_{14}$	*p*-Cymene	45.27	0.2478
$C_{10}H_{14}$	*p*-Diethylbenzene	45.03	0.2439
$C_{10}H_{14}$	1,2,4,5-Tetramethylbenzene	45.8	0.24
$C_{10}H_{22}$	Decane	52.88	0.3051
$C_{10}H_{22}O$	1-Decanol	57.45	0.2971
$C_{11}H_{24}$	Undecane	60.88	0.3396
$C_{12}H_{10}$	Biphenyl	47.16	0.2130
$C_{12}H_{26}$	Dodecane	69.14	0.3741
$C_{12}H_{26}O$	1-Dodecanol	72.69	0.3598
$C_{13}H_{12}$	Diphenylmethane	60.46	0.2798
$C_{13}H_{28}$	Tridecane	77.94	0.4109
$C_{13}H_{28}O$	1-Tridecanol	81.20	0.3942
$C_{14}H_{30}O$	1-Tetradecanol	89.91	0.4289
$C_{15}H_{32}$	Pentadecane	96.50	0.4857

INFLUENCE OF PRESSURE ON FREEZING POINTS

This table illustrates the variation of the freezing point of representative types of liquids with pressure. Substances are listed in alphabetical order. Note that 1 MPa = 0.01 kbar = 9.87 atm.

REFERENCES

1. Isaacs, N.S., *Liquid Phase High Pressure Chemistry*, John Wiley, New York, 1981.
2. Merrill, L., *J. Phys. Chem. Ref. Data*, 6, 1205, 1977; 11, 1005, 1982.

Substance	Molecular formula	Freezing point in °C at: 0.1 MPa	100 MPa	1000 MPa
Acetic acid	$C_2H_4O_2$	16.6	37	
Acetophenone	C_8H_8O	20.0	41.2	
Aniline	C_6H_7N	–6.0	13.5	140
Benzene	C_6H_6	5.5	33.4	
Benzonitrile	C_7H_5N	–12.8	7.6	
Benzyl alcohol	C_7H_8O	–15.2	0.2	
Bromobenzene	C_6H_5Br	–30.6	–12	108
Bromoethane	C_2H_5Br	–118.6	–108	
1-Bromonaphthalene	$C_{10}H_7Br$	–1.8	6.1	
1-Bromopropane	C_3H_7Br	–110	–98	
p-Bromotoluene	C_7H_7Br	28.0	56.7	
Butanoic acid	$C_4H_8O_2$	–5.7	13.8	
1-Butanol	$C_4H_{10}O$	–89.8	–77.2	
Carbon disulfide	CS_2	–111.5	–98	
Chlorobenzene	C_6H_5Cl	–45.2	–28	84
p-Chlorotoluene	C_7H_7Cl	6.9	33.1	
o-Cresol	C_7H_8O	29.8	47.7	
m-Cresol	C_7H_8O	11.8	25.6	
p-Cresol	C_7H_8O	35.8	56.2	
Cyclohexane	C_6H_{12}	6.6	32.5	
Cyclohexanol	$C_6H_{12}O$	25.5	62.3	
1,2-Dibromoethane	$C_2H_4Br_2$	9.9	34.0	
p-Dichlorobenzene	$C_6H_4Cl_2$	52.7	79.1	
Dichloromethane	CH_2Cl_2	–95.1	–83	
N,N-Dimethylaniline	$C_8H_{11}N$	2.5	26.3	
1,4-Dioxane	$C_4H_8O_2$	11	23	
Ethanol	C_2H_6O	–114.1	–108	
Formamide	CH_3NO	–15.5	10.8	
Formic acid	CH_2O_2	8.3	20.6	
Furan	C_4H_4O	–85.6	–73	
Hexamethyldisiloxane	$C_6H_{18}OSi_2$	–66	–37	
Menthol	$C_{10}H_{20}O$	42	60	
Methyl benzoate	$C_8H_8O_2$	–15	31.8	
2-Methyl-2-butanol	$C_5H_{12}O$	–8.8	13.4	
2-Methyl-2-propanol	$C_4H_{10}O$	25.4	58.1	
Naphthalene	$C_{10}H_8$	78.2	115.7	
Nitrobenzene	$C_6H_5NO_2$	5.7	13.5	
m-Nitrotoluene	$C_7H_7NO_2$	15.5	40.6	
Pentachloroethane	C_2HCl_5	–29.0	–6.3	
Potassium	K	63.7	78	170
Potassium chloride	ClK	771		945
Propanoic acid	$C_3H_6O_2$	–20.7	–1.2	
Silver chloride	AgCl	455		545
Sodium	Na	97.8	106	167
Sodium chloride	ClNa	800.7		997
Sodium fluoride	FNa	996		1115
Tetrachloromethane	CCl_4	–23.0	14.2	
Tribromomethane	$CHBr_3$	8.1	31.5	
Trichloromethane	$CHCl_3$	–63.6	–45.2	
Water	H_2O	0.0	–9.0	
o-Xylene	C_8H_{10}	–25.2	–3.5	
m-Xylene	C_8H_{10}	–47.8	–25.2	
p-Xylene	C_8H_{10}	13.2	46.0	

CRITICAL CONSTANTS, BOILING POINTS, AND MELTING POINTS OF SELECTED COMPOUNDS

Douglas Ambrose

This table gives the critical constants, boiling points, and melting points of about 600 inorganic and organic substances. The properties tabulated are:

T_m: Melting point in K; a "t" following the value indicates a triple point
T_b: Normal boiling point in K at a pressure of 101.325 kPa (1 atmosphere); an "s" following the value indicates a sublimation point (temperature at which the solid is in equilibrium with the gas at 101.325 kPa)
T_c: Critical temperature in K
P_c: Critical pressure in MPa
V_c: Critical molar volume in cm^3/mol

The number of digits given for T_m, T_b, T_c, and P_c indicates the estimated accuracy of these quantities: however, values of T_c greater than 750 K may be in error by 10 K or more. Although V_c values are given to three figures, they cannot be assumed accurate to better than a few percent.

Compounds are listed by molecular formula in the modified Hill order, with compounds not containing carbon preceeding those that do contain carbon.

REFERENCES

1. Stephenson, R. M., and Malanowski, S., *Handbook of the Thermodynamics of Organic Compounds*, Elsevier, New York, 1987.
2. Reid, R. C., Prausnitz, J. M., and Poling, B. E., *The Properties of Gases and Liquids*, 4th ed., McGraw-Hill, New York, 1987.
3. Daubert, T. E., Danner, R. P., Sibul, H. M., and Stebbins, C. C., *Physical and Thermodynamic Properties of Pure Compounds: Data Compilation*, extant 1994 (core with 4 supplements), Taylor & Francis, Bristol, PA (also available as database).

Molecular formula	Name	T_m/K	T_b/K	T_c/K	P_c/MPa	V_c/ cm^3/mol
$AlBr_3$	Aluminum tribromide	370.6	528	763	2.89	310
$AlCl_3$	Aluminum trichloride	463		620	2.63	257
AlI_3	Aluminum triiodide	464	655	983		408
Ar	Argon	83.80	87.30	150.87	4.898	75
As	Arsenic	1090 t	887 s	1673		35
$AsCl_3$	Arsenic trichloride	257	403	654		252
AsH_3	Arsine	157	210.6	373.1		
BBr_3	Boron tribromide	228	364	581		272
BCl_3	Boron trichloride	166	285.80	455	3.87	239
BF_3	Boron trifluoride	146.3	172	260.8	4.98	115
BI_3	Boron triiodide	316	483	773		356
B_2H_6	Diborane	107.6	180.7	289.8	4.05	
$BiBr_3$	Bismuth tribromide	491	726	1220		301
$BiCl_3$	Bismuth trichloride	503	720	1179	12.0	261
BrH	Hydrogen bromide	186.34	206.77	363.2	8.55	
BrI	Iodine bromide	313	389	719		139
Br_2	Bromine (Br_2)	265.9	331.9	588	10.34	127
Br_2Hg	Mercury bromide ($HgBr_2$)	509	595	1012		
Br_3Ga	Gallium bromide ($GaBr_3$)	394.6	552	806.7		303
Br_3HSi	Tribromosilane	200	382	610.0		305
Br_3P	Phosphorus tribromide	233	446.10	711		300
Br_3Sb	Antimony tribromide	369.7	553	904		300
Br_4Ge	Germanium tetrabromide	299.2	459.50	718		392
Br_4Hf	Hafnium bromide ($HfBr_4$)	697 t	596 s	746		415
Br_4Si	Silicon tetrabromide	278.3	427	663		382
Br_4Sn	Tin bromide ($SnBr_4$)	304	478	744		417
Br_4Ti	Titanium bromide ($TiBr_4$)	312	503	795.7		391
Br_4Zr	Zirconium bromide ($ZrBr_4$)	723	633 s	805		424
Br_5Ta	Tantalum bromide ($TaBr_5$)	538	622	974		461
$ClFO_3$	Perchloryl fluoride	126	226.40	368.4	5.37	161
ClF_2N	Nitrogen chloride difluoride		207	337.5	5.15	
ClF_2P	Phosphorus chloride difluoride		225.90	362.4	4.52	

CRITICAL CONSTANTS, BOILING POINTS, AND MELTING POINTS OF SELECTED COMPOUNDS (continued)

Molecular formula	Name	T_m/K	T_b/K	T_c/K	P_c/MPa	V_c/ cm^3/mol
ClF_2PS	Thiophosphoryl chloride difluoride	118	279.4	439.2	4.14	
ClF_3Si	Chlorotrifluorosilane	135	203.10	307.7	3.46	
ClF_5	Chlorine pentafluoride	170	260.1	416	5.27	233
ClF_5S	Sulfur chloride pentafluoride	209	254.10	390.9		
ClH	Hydrogen chloride	158.97	188	324.7	8.31	81
ClH_4N	Ammonium chloride	793	611 s	1155	163.5	
ClH_4P	Phosphonium chloride			322.3	7.37	
$ClNO$	Nitrosyl chloride	213.6	267.60	440		
$ClOV$	Vanadium oxychloride		400	636		171
Cl_2	Chlorine (Cl_2)	171.6	239.11	416.9	7.991	123
Cl_2FP	Phosphorus dichloride fluoride		287	463.0	4.96	
Cl_2F_2Si	Dichlorodifluorosilane	229	241	369.0	3.5	
Cl_2Hg	Mercury chloride ($HgCl_2$)	549	577	973		174
Cl_2OSe	Selenium oxychloride	281.6	450	730	7.09	235
Cl_3FSi	Trichlorofluorosilane		285.40	438.6	3.58	
Cl_3Ga	Gallium chloride ($GaCl_3$)	351.0	474	694		263
Cl_3HSi	Trichlorosilane	144.9	306	479		268
Cl_3P	Phosphorus trichloride	161	349.10	563		264
Cl_3Sb	Antimony trichloride	346.5	493.4	794		272
Cl_4Ge	Germanium tetrachloride	223.6	359.70	553.2	3.861	330
Cl_4Hf	Hafnium chloride ($HfCl_4$)	705 t	590 s	725.7	5.42	314
Cl_4ORe	Rhenium oxide tetrachloride	302.4	496	781		362
Cl_4OW	Tungsten oxide tetrachloride	484	500.70	782		338
Cl_4Si	Silicon tetrachloride	204.30	330.80	508.1	3.593	326
Cl_4Sn	Tin chloride ($SnCl_4$)	240	387.30	591.9	3.75	351
Cl_4Te	Tellurium chloride ($TeCl_4$)	497	660	1002	8.56	310
Cl_4Ti	Titanium chloride ($TiCl_4$)	248	409.60	638	4.66	339
Cl_4Zr	Zirconium chloride ($ZrCl_4$)	710	604 s	778	5.77	319
Cl_5Mo	Molybdenum pentachloride	467	541	850		369
Cl_5Nb	Niobium chloride ($NbCl_5$)	477.8	527.20	803.5	4.88	397
Cl_5P	Phosphorus pentachloride	440 t	433	646		
Cl_5Ta	Tantalum chloride ($TaCl_5$)	489	512.50	767		402
Cl_6W	Tungsten chloride (WCl_6)	548	619.90	923		422
FH	Hydrogen fluoride	189.79	293	461	6.48	69
FNO_2	Nitryl fluoride	107	200.7	349.5		
F_2	Fluorine (F_2)	53.49	85.03	144.13	5.172	66
F_2HN	Difluoramine		250	403		
F_2N_2	*cis*-Difluorodiazine		167.40	272	7.09	
F_2N_2	*trans*-Difluorodiazine	101	161.70	260	5.57	
F_2O	Oxygen difluoride	49.3	128.40	215		
F_2Xe	Xenon difluoride	402.18 t	387.50	631	9.32	148
F_3N	Nitrogen trifluoride	66.36	144.40	234.0	4.46	126
F_3NO	Trifluoramine oxide	112	185.60	303	6.43	147
F_3P	Phosphorus trifluoride	121.6	171.6	271.2	4.33	
F_3PS	Thiophosphoryl trifluoride	124.3	220.90	346.0	3.82	
F_4N_2	Tetrafluorohydrazine	108.6	199	309	3.75	
F_4S	Sulfur fluoride (SF_4)	148	232.70	364		
F_4Si	Silicon tetrafluoride	182.9	187	259.0	3.72	
F_4Xe	Xenon tetrafluoride	390.25 t	388.90	612	7.04	188
F_5Nb	Niobium fluoride (NbF_5)	353	502	737	6.28	155
F_6Mo	Molybdenum hexafluoride	290.6	307	473	4.75	226
F_6S	Sulfur fluoride (SF_6)	222.4		318.69	3.77	199
F_6Se	Selenium fluoride (SeF_6)	238.5 t	226.5 t	345.5		
F_6Te	Tellurium fluoride (TeF_6)	235.5 t	234	356		
F_6U	Uranium fluoride (UF_6)	337.20	329.6	505.8	4.66	250
F_6W	Tungsten fluoride (WF_6)	275.4	290	444	4.34	233

CRITICAL CONSTANTS, BOILING POINTS, AND MELTING POINTS OF SELECTED COMPOUNDS (continued)

Molecular formula	Name	T_m/K	T_b/K	T_c/K	P_c/MPa	V_c/ cm³/mol
GaI_3	Gallium iodide (GaI_3)	483	613	951		395
GeH_4	Germane	108	185.0	312.2	4.95	147
GeI_4	Germanium tetraiodide	417	650	973		500
HI	Hydrogen iodide	222.38	237.60	424.0	8.31	
H_2	Hydrogen (H_2)	13.81	20.28	32.97	1.293	65
H_2O	Water	273.15	373.15	647.14	22.06	56
H_2S	Hydrogen sulfide	187.6	213.60	373.2	8.94	99
H_2Se	Hydrogen selenide	207.42	231.90	411	8.92	
H_3N	Ammonia	195.41	239.82	405.5	11.35	72
H_3P	Phosphine	140	185.40	324.5	6.54	
H_4N_2	Hydrazine	274.5	386.70	653	14.7	
He	Helium		4.22	5.19	0.227	57
HfI_4	Hafnium iodide (HfI_4)	722	667 s	916		528
Hg	Mercury	234.32	629.88	1750	172.00	43
HgI_2	Mercury iodide (HgI_2)	532	627	1072		
I_2	Iodine (I_2)	386.8	457.5	819		155
I_3Sb	Antimony triiodide	441	674	1102		
I_4Si	Silicon tetraiodide	393.6	560.50	944		558
I_4Sn	Tin iodide (SnI_4)	416	637.50	968		531
I_4Ti	Titanium iodide (TiI_4)	423	650	1040		505
I_4Zr	Zirconium iodide (ZrI_4)	772	704	960		530
Kr	Krypton	115.79	119.93	209.41	5.50	91
NO	Nitric oxide	109.5	121.41	180	6.48	58
N_2	Nitrogen (N_2)	63.15	77.36	126.21	3.39	90
N_2O	Nitrous oxide	182.3	184.67	309.57	7.255	97
N_2O_4	Nitrogen tetroxide	263.8	294.30	431	10.1	167
Ne	Neon	24.56	27.07	44.4	2.76	42
O_2	Oxygen (O_2)	54.36	90.20	154.59	5.043	73
O_2S	Sulfur dioxide	197.6	263.10	430.8	7.884	122
O_3	Ozone	80	161.80	261.1	5.57	89
O_3S	Sulfur trioxide	289.9	318	491.0	8.2	127
O_4Os	Osmium oxide (OsO_4)	314	408	678		
O_7Re_2	Rhenium oxide (Re_2O_7)	570	633	942		334
P	Phosphorus	317.30	553.6	994		
Rn	Radon	202	211.4	377	6.28	
S	Sulfur	388.36	717.75	1314	20.7	
Se	Selenium	494	958	1766	27.2	
Xe	Xenon	161.40	165.11	289.73	5.84	118
$CBrClF_2$	Bromochlorodifluoromethane	113.6	269.43	426.88	4.254	246
$CBrF_3$	Bromotrifluoromethane	101	215.26	340.2	3.97	196
CBr_2F_2	Dibromodifluoromethane	163.0	298	471.3		
$CClF_3$	Chlorotrifluoromethane	92	191.7	302	3.870	180
CCl_2F_2	Dichlorodifluoromethane	115	243.3	384.95	4.136	217
CCl_2O	Carbonyl chloride	145.2	281	455	5.67	190
CCl_3F	Trichlorofluoromethane	162.04	296.90	471.2	4.41	248
CCl_4	Tetrachloromethane	250	349.9	556.6	4.516	276
CF_4	Tetrafluoromethane	89.56	145.13	227.6	3.74	140
CO	Carbon monoxide	68	81.6	132.91	3.499	93
COS	Carbon oxysulfide	134.3	223	375	5.88	137
CO_2	Carbon dioxide	216.58		304.14	7.375	94
CS_2	Carbon disulfide	161.6	319	552	7.90	173
$CHClF_2$	Chlorodifluoromethane	115.73	232.40	369.3	4.99	169
$CHCl_2F$	Dichlorofluoromethane	138	282.10	451.58	5.18	196
$CHCl_3$	Trichloromethane	209.5	334.32	536.4	5.47	239
CHF_3	Trifluoromethane	117.97	191.0	299.3	4.858	133
CHN	Hydrogen cyanide	259.7	299	456.7	5.39	139

CRITICAL CONSTANTS, BOILING POINTS, AND MELTING POINTS OF SELECTED COMPOUNDS (continued)

Molecular formula	Name	T_m/K	T_b/K	T_c/K	P_c/MPa	V_c/ cm^3/mol
CH_2Cl_2	Dichloromethane	178.01	313	510	6.10	
CH_2F_2	Difluoromethane	137	221.46	351.6	5.830	121
CH_2O_2	Formic acid	281.4	374	588		
CH_3Cl	Chloromethane	175.4	249.06	416.25	6.679	139
CH_3Cl_3Si	Methyltrichlorosilane	183	338.80	517	3.28	348
CH_3F	Fluoromethane	131.3	194.74	317.8	5.88	113
CH_3I	Iodomethane	206.70	315.70	528		
CH_3NO_2	Nitromethane	244.60	374.34	588	5.87	173
CH_4	Methane	90.69	111.67	190.53	4.604	99
CH_4O	Methanol	175.47	337.7	512.64	8.092	118
CH_4S	Methanethiol	150	279.11	470.0	7.23	145
CH_5ClSi	Chloromethylsilane		280	517.8		
CH_5N	Methylamine	179.71	266.83	430.7	7.614	
CH_6N_2	Methylhydrazine	220.7	360.6	567	8.24	271
CH_6Si	Methylsilane	116.6	215.60	352.5		
$C_2Br_2ClF_3$	1,2-Dibromo-1-chloro-1,2,2-trifluoroethane		366	560.7	3.61	368
$C_2Br_2F_4$	1,2-Dibromotetrafluoroethane	162.7	320.50	487.8	3.393	341
C_2ClF_3	Chlorotrifluoroethylene	115	245.30	379	4.05	212
C_2ClF_5	Chloropentafluoroethane	173.71	235.20	353.2	3.229	252
$C_2Cl_2F_4$	1,1-Dichlorotetrafluoroethane	216.5	277	418.6	3.30	294
$C_2Cl_2F_4$	1,2-Dichlorotetrafluoroethane	179	276.9	418.78	3.252	297
$C_2Cl_3F_3$	1,1,2-Trichlorotrifluoroethane	238	320.8	487.3	3.42	325
C_2Cl_4	Tetrachloroethylene	250.80	394.4	620.2		
$C_2Cl_4F_2$	Tetrachloro-1,2-difluoroethane	299	366	551		
C_2F_3N	Trifluoroacetonitrile		204.33	311.11	3.618	202
C_2F_4	Tetrafluoroethylene	130.6	197.20	306.5	3.94	172
C_2F_6	Hexafluoroethane	172.4	195.0	293		222
C_2N_2	Cyanogen	245.2	252.0	400	5.98	
C_2HClF_2	1-Chloro-2,2-difluoroethylene	134.6	254.60	400.6	4.46	197
C_2HClF_4	1-Chloro-1,1,2,2-tetrafluoroethane	156	263	399.9	3.72	244
C_2HCl_3	Trichloroethylene	188.40	360.36	544.2	5.02	
$C_2HF_3O_2$	Trifluoroacetic acid	257.90	346	491.3	3.258	204
C_2H_2	Acetylene	192.40	188.43	308.33	6.139	113
$C_2H_2Cl_2$	*cis*-1,2-Dichloroethylene	193	333.34	544.2		
$C_2H_2Cl_2$	*trans*-1,2-Dichloroethylene	223.3	321.88	516.5	5.51	
$C_2H_2Cl_4$	1,1,2,2-Tetrachloroethane	229.3	419.6	661.15		
$C_2H_2F_2$	1,1-Difluoroethylene	129	187.4	302.9	4.46	154
$C_2H_2F_4$	1,1,1,2-Tetrafluoroethane	172	246.6	374.3	4.065	198
$C_2H_3ClF_2$	1-Chloro-1,1-difluoroethane	142.3	263.40	410.29	4.041	225
$C_2H_3Cl_3$	1,1,1-Trichloroethane	242.7	347.24	545	4.30	
C_2H_3F	Fluoroethylene	112.6	201	327.9	5.24	144
$C_2H_3F_3$	1,1,1-Trifluoroethane	161.8	225.60	346.3	3.76	194
C_2H_3N	Acetonitrile	229.32	354.80	545.5	4.85	173
C_2H_4	Ethylene	104	169.38	282.34	5.041	131
$C_2H_4Br_2$	1,2-Dibromoethane	283.08	404.7	583.0	7.2	
$C_2H_4Cl_2$	1,1-Dichloroethane	176.19	330.5	523	5.07	236
$C_2H_4Cl_2$	1,2-Dichloroethane	237.6	356.6	561	5.4	225
$C_2H_4F_2$	1,1-Difluoroethane	156	248.20	386.7	4.50	181
C_2H_4O	Acetaldehyde	150	293.2	466		154
C_2H_4O	Ethylene oxide	161.4	283.7	469	7.19	140
$C_2H_4O_2$	Acetic acid	289.7	391.0	592.71	5.786	171
$C_2H_4O_2$	Methyl formate	174	304.8	487.2	5.998	172
C_2H_5Br	Bromoethane	154.5	311.6	503.9	6.23	215
C_2H_5Cl	Chloroethane	134.4	286.2	460.4	5.3	
C_2H_5F	Fluoroethane	129.9	235.50	375.31	5.028	

CRITICAL CONSTANTS, BOILING POINTS, AND MELTING POINTS OF SELECTED COMPOUNDS (continued)

Molecular formula	Name	T_m/K	T_b/K	T_c/K	P_c/MPa	V_c/ cm³/mol
C_2H_6	Ethane	90.3	184.5	305.4	4.884	148
$C_2H_6Cl_2Si$	Dichlorodimethylsilane	257	343.4	520.4	3.49	350
C_2H_6O	Dimethyl ether	131.6	248.3	400.0	5.37	190
C_2H_6O	Ethanol	159.0	351.44	513.92	6.132	167
$C_2H_6O_2$	Ethylene glycol	260	470.49	718		
C_2H_6S	Dimethyl sulfide	174.8	310.48	503.0	5.53	201
C_2H_6S	Ethanethiol	125.26	308.2	499	5.49	207
C_2H_7N	Dimethylamine	180.9	280.03	437.22	5.340	
C_2H_7N	Ethylamine	192.62	289.70	456	5.62	182
C_3ClF_5O	Chloropentafluoroacetone		281	410.6	2.878	
C_3F_6O	Perfluoroacetone	148	245.70	357.14	2.84	329
C_3F_8	Perfluoropropane	125.46	236.50	345.1	2.680	299
$C_3H_3F_3$	3,3,3-Trifluoropropene	256	244	376.2	3.80	211
$C_3H_3F_5$	1,1,1,2,2-Pentafluoropropane		255.71	380.11	3.137	273
C_3H_3NO	Isoxazole		368	552.0		
C_3H_4	Propyne	170.4	249.9	402.38	5.628	164
C_3H_4	Allene	136.87	238.70	393		
C_3H_5Cl	3-Chloropropene	138.6	318.30	514		
C_3H_5N	Propanenitrile	180.26	370.29	561.3	4.26	229
C_3H_6	Propene	87.90	225.46	364.85	4.601	181
C_3H_6	Cyclopropane	145.7	240.34	398.3	5.579	162
C_3H_6O	Acetone	178.3	329.20	508.1	4.700	209
C_3H_6O	Propanal	193	321	504.4	5.27	204
C_3H_6O	Methyloxirane	161.22	308	482.2	4.92	186
$C_3H_6O_2$	Ethyl formate	193.5	327.5	508.5	4.74	229
$C_3H_6O_2$	Methyl acetate	175	330.02	506.55	4.75	228
$C_3H_6O_2$	Propanoic acid	252.4	414.30	604	4.53	222
C_3H_7Cl	1-Chloropropane	150.3	319.6	503	4.58	
C_3H_7NO	*N,N*-Dimethylformamide	212.72	426	649.6		262
C_3H_8	Propane	85.46	231.0	369.82	4.250	203
C_3H_8O	1-Propanol	147.0	370.3	536.78	5.168	219
C_3H_8O	2-Propanol	183.6	355.4	508.3	4.762	220
C_3H_8O	Ethyl methyl ether	160	280.60	437.8	4.40	221
C_3H_8S	Ethyl methyl sulfide	167.2	339.8	533	4.26	
C_3H_8S	1-Propanethiol	159.8	340.9	536.6		286
$C_3H_9BO_3$	Trimethylborate	243.8	340.6	501.7	3.59	
C_3H_9ClSi	Trimethylchlorosilane	233	333	497.8	3.20	366
C_3H_9N	Propylamine	190	320.37	497.0	4.72	
C_3H_9N	Isopropylamine	178.01	304.91	471.8	4.54	221
C_3H_9N	Trimethylamine	156.0	276.02	432.79	4.087	254
$C_4Br_2F_8$	1,4-Dibromooctafluorobutane		370	532.5	2.39	
C_4F_8	Perfluorocyclobutane	232.96	267.16	388.46	2.784	324
C_4F_{10}	Perfluorobutane	144.9	271.20	386.4	2.323	378
C_4F_{10}	Perfluoroisobutane		273	395.4		
C_4H_4O	Furan	187.5	304.6	490.2	5.50	218
C_4H_4S	Thiophene	233.7	357.1	579.4	5.69	219
C_4H_5N	Pyrrole	249.73	402.94	639.7	6.34	200
C_4H_6	1,3-Butadiene	164.2	268.74	425	4.33	221
C_4H_6	1-Butyne	147.43	281.23	463.7		
C_4H_6	2-Butyne	240.79	300.12	488.7		
$C_4H_6O_3$	Acetic anhydride	200	412.70	606	4.0	
C_4H_7N	Butanenitrile	161.2	390.7	585.4	3.88	
C_4H_8	1-Butene	87.80	266.89	419.57	4.023	240
C_4H_8	*cis*-2-Butene	134.2	276.86	435.58	4.197	234
C_4H_8	*trans*-2-Butene	167.62	274.03	428.63	3.985	238
C_4H_8	Isobutene	132.7	266.20	417.9	4.000	239

CRITICAL CONSTANTS, BOILING POINTS, AND MELTING POINTS OF SELECTED COMPOUNDS (continued)

Molecular formula	Name	T_m/K	T_b/K	T_c/K	P_c/MPa	V_c/ cm³/mol
C_4H_8	Cyclobutane	182.48	285.7	460.0	4.98	210
C_4H_8O	Butanal	174	347.9	537.2	4.32	258
C_4H_8O	2-Butanone	186.48	352.74	536.78	4.207	267
C_4H_8O	Tetrahydrofuran	164.76	338	540.1	5.19	224
C_4H_8O	Ethyl vinyl ether	157.3	308.70	475	4.07	
$C_4H_8O_2$	1,4-Dioxane	284.9	374.6	587	5.21	238
$C_4H_8O_2$	Ethyl acetate	189.5	350.26	523.3	3.882	286
$C_4H_8O_2$	Methyl propanoate	185.6	352.9	530.6	4.004	282
$C_4H_8O_2$	Propyl formate	180.2	354.0	538.0	4.06	285
$C_4H_8O_2$	Butanoic acid	267.4	436.90	624	4.03	290
$C_4H_8O_2$	2-Methylpropanoic acid	227	427.60	605	3.7	292
C_4H_8S	Tetrahydrothiophene	176.99	394.1	632.0		
C_4H_9N	Pyrrolidine	215.31	359.71	568.2	5.59	238
C_4H_{10}	Butane	134.86	272.6	425.14	3.784	255
C_4H_{10}	Isobutane	134.8	261.42	407.85	3.630	257
$C_4H_{10}O$	1-Butanol	183.3	390.88	563.05	4.423	275
$C_4H_{10}O$	2-Butanol	158.4	372.66	536.05	4.179	269
$C_4H_{10}O$	2-Methyl-2-propanol	298.5	355.5	506.21	3.973	275
$C_4H_{10}O$	2-Methyl-1-propanol	165	381.04	547.78	4.300	273
$C_4H_{10}O$	Diethyl ether	156.8	307.6	466.74	3.638	280
$C_4H_{10}O$	Methyl propyl ether		312.2	476.25	3.801	
$C_4H_{10}O$	Isopropyl methyl ether		303.92	464.48	3.762	
$C_4H_{10}O_2$	1,2-Dimethoxyethane	215	358	536	3.87	271
$C_4H_{10}S$	1-Butanethiol	157.4	371.6	570.1		324
$C_4H_{10}S$	Diethyl sulfide	169.20	365.2	557	3.96	318
$C_4H_{11}N$	Butylamine	224.05	350.15	531.9	4.25	277
$C_4H_{11}N$	Isobutylamine	186.4	340.90	519	4.07	278
$C_4H_{11}N$	*sec*-Butylamine	168.6	335.88	514.3	4.20	278
$C_4H_{11}N$	*tert*-Butylamine	206.20	317.19	483.9	3.84	292
$C_4H_{11}N$	Diethylamine	223.3	328.6	499.99	3.758	
$C_4H_{12}Si$	Tetramethylsilane	174.11	299.80	448.64	2.821	362
$C_4H_{12}Sn$	Tetramethylstannane	218.3	351	521.8	2.981	
C_5F_{12}	Perfluoropentane	263	302.40	420.59	2.045	473
$C_5H_2F_6O_2$	Hexafluoroacetylacetone		327.30	485.1	2.767	
$C_5H_4O_2$	Furfural	236.6	434.90	670	5.89	
C_5H_5N	Pyridine	231.49	388.38	620.0	5.67	243
$C_5H_6N_2$	2-Methylpyrazine	244	410	634.3	5.01	283
C_5H_6O	2-Methylfuran		338	527	4.72	247
C_5H_8	Cyclopentene	138.0	317.39	507.0	4.802	245
C_5H_8	1-Pentyne	183	313.33	493.5		
C_5H_8O	Cyclopentanone	221.8	403.72	624.5	4.60	
C_5H_8O	3,4-Dihydro-2H-pyran		359	561.7	4.56	268
C_5H_9N	Pentanenitrile	176.9	414.4	610.3	3.58	
C_5H_9NO	*N*-Methyl-2-pyrrolidone	249	475	721.8		311
C_5H_{10}	1-Pentene	107.9	303.11	464.78	3.527	293
C_5H_{10}	*cis*-2-Pentene	121.7	310.08	475	3.69	
C_5H_{10}	*trans*-2-Pentene	132.9	309.49	471	3.52	
C_5H_{10}	2-Methyl-1-butene	135.58	304.3	470	3.8	
C_5H_{10}	2-Methyl-2-butene	139.39	311.71	481	3.91	
C_5H_{10}	Cyclopentane	179.3	322.4	511.7	4.508	260
$C_5H_{10}O$	Cyclopentanol	254	413.57	619.5	4.90	
$C_5H_{10}O$	2-Pentanone	196.2	375.41	561.08	3.694	301
$C_5H_{10}O$	3-Pentanone	234	375.11	561.46	3.729	336
$C_5H_{10}O$	3-Methyl-2-butanone	181	367.48	553.4	3.85	310
$C_5H_{10}O$	Tetrahydropyran	228	361	572.2	4.77	263
$C_5H_{10}O$	2-Methyltetrahydrofuran		351	537	3.76	267

CRITICAL CONSTANTS, BOILING POINTS, AND MELTING POINTS OF SELECTED COMPOUNDS (continued)

Molecular formula	Name	T_m/K	T_b/K	T_c/K	P_c/MPa	V_c/ cm^3/mol
$C_5H_{10}O$	Pentanal	181.6	376	566.1	3.97	313
$C_5H_{10}O$	Allyl ethyl ether		340.80	518		
$C_5H_{10}O_2$	Isobutyl formate	177.3	371.40	551	3.88	352
$C_5H_{10}O_2$	Propyl acetate	180	374.69	549.7	3.36	345
$C_5H_{10}O_2$	Isopropyl acetate	199.7	361.7	531		
$C_5H_{10}O_2$	Ethyl propanoate	199.2	372.2	546.0	3.362	345
$C_5H_{10}O_2$	Methyl butanoate	187.3	375.9	554.4	3.47	340
$C_5H_{10}O_2$	Methyl isobutanoate	188.4	365.6	540.8	3.43	339
$C_5H_{10}O_2$	Pentanoic acid	239	459.30	643	3.58	340
$C_5H_{10}O_2$	3-Methylbutanoic acid	243.8	449.70	629	3.40	
$C_5H_{11}N$	Piperidine	262.12	379.37	594.1	4.94	288
C_5H_{12}	Pentane	143.4	309.21	469.69	3.364	311
C_5H_{12}	Isopentane	113.2	301.03	460.43	3.381	306
C_5H_{12}	Neopentane	256.5	282.63	433.8	3.197	307
$C_5H_{12}O$	Butyl methyl ether	157.6	343.31	512.78	3.371	329
$C_5H_{12}O$	*tert*-Butyl methyl ether	164.5	328.3	497.1	3.430	
$C_5H_{12}O$	Ethyl propyl ether	145.6	336.36	500.23	3.370	339
$C_5H_{12}O$	1-Pentanol	194.2	411.13	588.15	3.909	326
$C_5H_{12}O$	2-Pentanol	200	392.4	560.4		
$C_5H_{12}O$	3-Pentanol	204	389.40	559.6		
$C_5H_{12}O$	2-Methyl-2-butanol	264.3	375.5	545		
$C_5H_{12}O$	3-Methyl-1-butanol	155.9	404.2	579.4		
$C_5H_{12}O_2$	2-Propoxyethanol		422.9	614.6		364
$C_5H_{12}S$	3-Methyl-1-butanethiol		393	604		
C_6BrF_5	Bromopentafluorobenzene	242	410	601	3.0	
C_6ClF_5	Chloropentafluorobenzene		391.11	570.81	3.238	376
$C_6Cl_2F_4$	1,2-Dichlorotetrafluorobenzene			626	5.32	
$C_6Cl_3F_3$	1,3,5-Trichlorofluorobenzene		471.52	684.8	3.27	448
C_6F_6	Hexafluorobenzene	278.50	353.41	516.73	3.273	335
C_6F_{10}	Perfluorocyclohexene		325.20	461.8		
C_6F_{12}	Perfluorocyclohexane	321.6	323.76	457.2	2.43	
C_6F_{12}	Perfluoro-1-hexene		330.20	454.4		
C_6F_{14}	Perfluorohexane	186.0	329.80	448.77	1.868	606
C_6F_{14}	Perfluoro-2-methylpentane		330.81	455.3	1.923	532
C_6F_{14}	Perfluoro-3-methylpentane		331.52	450	1.69	
C_6F_{14}	Perfluoro-2,3-dimethylbutane		332.93	463	1.87	525
C_6HF_5	Pentafluorobenzene	225.8	358.89	530.97	3.531	324
C_6HF_5O	Pentafluorophenol	305.9	418.79	609	4.0	348
C_6HF_{11}	Undecafluorocyclohexane		335.20	477.7		
$C_6H_2F_4$	1,2,3,4-Tetrafluorobenzene		367.51	550.83	3.791	313
$C_6H_2F_4$	1,2,3,5-Tetrafluorobenzene	225	357.61	535.25	3.747	
$C_6H_2F_4$	1,2,4,5-Tetrafluorobenzene	277.6	363.41	543.35	3.801	
$C_6H_4F_2$	*p*-Difluorobenzene	260	362	556	4.40	
C_6H_5Br	Bromobenzene	242.5	429.21	670	4.52	324
C_6H_5Cl	Chlorobenzene	227.9	404.87	632.4	4.52	308
C_6H_5F	Fluorobenzene	230.94	357.88	560.09	4.551	269
C_6H_5I	Iodobenzene	241.8	461.60	721	4.52	351
C_6H_6	Benzene	278.68	353.24	562.16	4.898	259
C_6H_6O	Phenol	314.0	455.02	694.2	6.13	
C_6H_7N	Aniline	267.13	457.32	699	4.89	287
C_6H_7N	2-Methylpyridine	206.47	402.53	621.0	4.60	292
C_6H_7N	3-Methylpyridine	255.01	417.29	645.0	4.48	288
C_6H_7N	4-Methylpyridine	276.81	418.51	645.7	4.70	292
C_6H_{10}	Cyclohexene	169.6	356.13	560.48		
C_6H_{10}	1,5-Hexadiene	132.4	332.60	507		
$C_6H_{10}O$	Cyclohexanone	242	428.58	653.0	4.0	

CRITICAL CONSTANTS, BOILING POINTS, AND MELTING POINTS OF SELECTED COMPOUNDS (continued)

Molecular formula	Name	T_m/K	T_b/K	T_c/K	P_c/MPa	V_c/ cm³/mol
$C_6H_{10}O_2$	Ethyl *trans*-2-butenoate		411	599		
$C_6H_{10}S$	Diallylsulfide	188	411.7	653		
$C_6H_{11}N$	Hexanenitrile	192.8	436.80	633.8	3.30	
C_6H_{12}	Cyclohexane	279.7	353.88	553.5	4.07	308
C_6H_{12}	Methylcyclopentane	130.6	344.9	532.73	3.784	319
C_6H_{12}	1-Hexene	133.39	336.63	504.1	3.206	348
$C_6H_{12}O$	Cyclohexanol	298.61	433.99	650.0	4.26	
$C_6H_{12}O$	2-Hexanone	217.6	400.7	587.0	3.32	
$C_6H_{12}O$	3-Hexanone	217.6	396.6	582.82	3.320	
$C_6H_{12}O$	4-Methyl-2-pentanone	189	389.6	571	3.27	
$C_6H_{12}O$	Hexanal	217	404	591	3.46	
$C_6H_{12}O_2$	Pentyl formate	199.6	403.60	576	3.46	
$C_6H_{12}O_2$	Isopentyl formate	179.6	396.70	578		
$C_6H_{12}O_2$	Butyl acetate	195	399.2	579		
$C_6H_{12}O_2$	Isobutyl acetate	174.30	389.70	561	3.16	
$C_6H_{12}O_2$	Propyl propanoate	197.2	395.6	578		
$C_6H_{12}O_2$	Ethyl butanoate	175	394.6	566	3.06	421
$C_6H_{12}O_2$	Ethyl 2-methylpropanoate	184.9	383.2	553	3.07	421
$C_6H_{12}O_2$	Methyl pentanoate		400.5	567	3.19	
$C_6H_{12}O_2$	Hexanoic acid	270	478.40	662	3.20	
$C_6H_{12}O_3$	Paraldehyde	285.7	397.50	563		
$C_6H_{12}O_3$	2-Ethoxyethyl acetate	211.4	429.5	607.3	3.166	443
C_6H_{14}	Hexane	177.8	341.88	507.7	3.010	370
C_6H_{14}	2-Methylpentane	119.4	333.41	497.7	3.031	367
C_6H_{14}	3-Methylpentane	110.2	336.42	504.5	3.126	367
C_6H_{14}	2,2-Dimethylbutane	174	322.88	488.8	3.090	359
C_6H_{14}	2,3-Dimethylbutane	144.3	331.13	500.0	3.131	358
$C_6H_{14}O$	Methyl pentyl ether		372	546.53	3.042	391
$C_6H_{14}O$	Dipropyl ether	147.0	363.23	530.6	3.028	
$C_6H_{14}O$	Diisopropyl ether	186.3	341.66	500.32	2.832	386
$C_6H_{14}O$	1-Hexanol	228.5	430.7	610.7	3.47	381
$C_6H_{14}O$	2-Hexanol		411	586.2		
$C_6H_{14}O$	4-Methyl-1-pentanol		425.0	603.5		
$C_6H_{14}O$	2-Methyl-2-pentanol	170	394.2	559.5		
$C_6H_{14}O$	4-Methyl-2-pentanol	183	404.80	574.4		
$C_6H_{14}O_2$	1,1-Diethoxyethane	173	375.40	527		
$C_6H_{14}O_2$	2-Butoxyethanol	198.3	441.5	633.9		424
$C_6H_{15}N$	Triethylamine	158.4	362	535.6	3.032	389
$C_6H_{15}N$	Dipropylamine	210	382.4	555.8	3.63	
$C_6H_{15}N$	Diisopropylamine	212	357.0	523.1	3.02	
C_7F_8	Perfluorotoluene	207.5	377.73	534.47	2.705	428
C_7F_{14}	Perfluoro-1-heptene		354.20	478.2		
C_7F_{14}	Perfluoromethylcyclohexane	220.4	349.50	485.91	2.019	570
C_7F_{16}	Perfluoroheptane	195	355.66	474.8	1.62	664
C_7HF_{15}	1H-Pentadecafluoroheptane		369.20	495.8		
$C_7H_3F_5$	2,3,4,5,6-Pentafluorotoluene	243.3	390.6	566.52	3.126	384
C_7H_5N	Benzonitrile	260.40	464.30	699.4	4.21	
C_7H_6O	Benzaldehyde	247	452.20	695	4.65	
C_7H_8	Toluene	178.16	383.78	591.79	4.104	316
C_7H_8O	*o*-Cresol	302.9	464.19	697.6	5.01	
C_7H_8O	*m*-Cresol	284.9	475.42	705.8	4.56	309
C_7H_8O	*p*-Cresol	308.9	475.13	704.6	5.15	
C_7H_8O	Benzyl alcohol	257.9	478.46	715	4.3	
C_7H_8O	Anisole	235.6	426.8	645.6	4.25	
C_7H_9N	*N*-Methylaniline	216	469.40	701	5.20	
C_7H_9N	*o*-Methylaniline	256.80	473.49	707	4.37	
C_7H_9N	*m*-Methylaniline	241.90	476.52	707	4.28	

CRITICAL CONSTANTS, BOILING POINTS, AND MELTING POINTS OF SELECTED COMPOUNDS (continued)

Molecular formula	Name	T_m/K	T_b/K	T_c/K	P_c/MPa	V_c/ cm^3/mol
C_7H_9N	*p*-Methylaniline	316.90	473.57	706	4.58	
C_7H_9N	2,3-Dimethylpyridine		434.41	655.4		
C_7H_9N	2,4-Dimethylpyridine		431.6	647		
C_7H_9N	2,5-Dimethylpyridine	257	430.16	644.2		
C_7H_9N	2,6-Dimethylpyridine	267.0	417.2	623.8		
C_7H_9N	3,4-Dimethylpyridine		452.28	683.8		
C_7H_9N	3,5-Dimethylpyridine	266.5	445	667.2		
C_7H_{14}	Cycloheptane	265.12	391.63	604.2	3.81	353
C_7H_{14}	Methylcyclohexane	146.5	374.08	572.2	3.471	368
C_7H_{14}	Ethylcyclopentane	134.71	376.6	569.5	3.397	375
C_7H_{14}	1-Heptene	153.4	366.79	537.3	2.921	402
$C_7H_{14}O$	2-Heptanone	238	424.20	611.5	3.436	
$C_7H_{14}O_2$	Isopentyl acetate	194.6	415.70	599		
$C_7H_{14}O_2$	Ethyl pentanoate	181.9	419.2	570		
$C_7H_{14}O_2$	Propyl butanoate	177.9	416.20	600		
$C_7H_{14}O_2$	Propyl 2-methylpropanoate		408.60	589		
$C_7H_{14}O_2$	Isobutyl propanoate	201.7	410	592		
$C_7H_{14}O_2$	Ethyl 3-methylbutanoate	173.8	408.20	588		
$C_7H_{14}O_2$	Heptanoic acid	265.6	495.40	679	2.90	
C_7H_{16}	Heptane	182.5	371.6	540.3	2.756	428
C_7H_{16}	2-Methylhexane	154.9	363.19	530.4	2.734	421
C_7H_{16}	3-Methylhexane	153.7	364.99	535.3	2.814	404
C_7H_{16}	2,2-Dimethylpentane	149.3	352.3	520.5	2.773	416
C_7H_{16}	2,3-Dimethylpentane		362.93	537.4	2.908	393
C_7H_{16}	2,4-Dimethylpentane	153.2	353.64	519.8	2.737	418
C_7H_{16}	3,3-Dimethylpentane	138.2	359.21	536.4	2.946	414
C_7H_{16}	3-Ethylpentane	154.5	366.6	540.7	2.891	416
C_7H_{16}	2,2,3-Trimethylbutane	248	354.01	531.2	2.954	398
$C_7H_{16}O$	1-Heptanol	239	449.60	632.5	3.135	435
$C_7H_{16}O$	2-Heptanol	243	432	611.4		
$C_7H_{16}O$	3-Heptanol	203	430	605.4		
$C_7H_{16}O$	4-Heptanol	231.9	429	602.6		
$C_8F_{16}O$	Perfluoro-2-butyltetrahydrofuran		375.80	500.2	1.607	588
C_8F_{18}	Perfluorooctane		379.0	502	1.66	
C_8H_7N	*p*-Tolunitrile	302.6	490.20	723		
C_8H_8O	Acetophenone	293	475	709.5		386
$C_8H_8O_3$	Methyl salicylate	265	496.10	709		
C_8H_{10}	Ethylbenzene	178.20	409.34	617.2	3.600	374
C_8H_{10}	*o*-Xylene	247.9	417.6	630.3	3.730	369
C_8H_{10}	*m*-Xylene	225.3	412.27	617.05	3.535	376
C_8H_{10}	*p*-Xylene	286.3	411.52	616.2	3.511	379
$C_8H_{10}O$	*o*-Ethylphenol	291	477.67	703.0		
$C_8H_{10}O$	*m*-Ethylphenol	269	491.57	718.8		
$C_8H_{10}O$	*p*-Ethylphenol	318.23	491.13	716.4		
$C_8H_{10}O$	2,3-Xylenol	345.9	490.07	722.8		
$C_8H_{10}O$	2,4-Xylenol	297.68	484.13	707.6		
$C_8H_{10}O$	2,5-Xylenol	347.9	484.33	706.9		
$C_8H_{10}O$	2,6-Xylenol	318.8	474.22	701.0		
$C_8H_{10}O$	3,4-Xylenol	333.9	500	729.8		
$C_8H_{10}O$	3,5-Xylenol	336.7	494.89	715.6		
$C_8H_{10}O$	Phenetole	243.63	442.96	647	3.42	
$C_8H_{11}N$	*N,N*-Dimethylaniline	275.60	467.30	687	3.63	
$C_8H_{11}N$	*N*-Ethylaniline	209.6	476.20	698		
$C_8H_{14}O_4$	Ethyl succinate	252	490.90	663		
$C_8H_{15}N$	Octanenitrile	227.5	478.40	674.4	2.85	
C_8H_{16}	Cyclooctane	287.98	422	647.2	3.56	410
C_8H_{16}	*trans*-1,4-Dimethylcyclohexane	236.2	392.5	587.7		

CRITICAL CONSTANTS, BOILING POINTS, AND MELTING POINTS OF SELECTED COMPOUNDS (continued)

Molecular formula	Name	T_m/K	T_b/K	T_c/K	P_c/MPa	V_c/ cm³/mol
C_8H_{16}	1-Octene	171.4	394.44	566.7	2.675	464
$C_8H_{16}O_2$	Octanoic acid	289.4	512	695	2.64	
$C_8H_{16}O_2$	2-Ethylhexanoic acid			674.6	2.778	528
$C_8H_{16}O_2$	Propyl 3-methylbutanoate		429.10	609		
$C_8H_{16}O_2$	Isobutyl butanoate		430.10	611		
$C_8H_{16}O_2$	Isobutyl 2-methylpropanoate	192.4	421.80	602		
$C_8H_{16}O_2$	Isopentyl propanoate		433.40	611		
C_8H_{18}	Octane	216.3	398.82	568.9	2.493	492
C_8H_{18}	2-Methylheptane	164.16	390.81	559.7	2.484	488
C_8H_{18}	3-Methylheptane	152.6	392.09	563.7	2.546	464
C_8H_{18}	4-Methylheptane	152	390.87	561.8	2.542	476
C_8H_{18}	2,2-Dimethylhexane	151.97	380.01	549.9	2.529	478
C_8H_{18}	2,3-Dimethylhexane		388.77	563.5	2.628	468
C_8H_{18}	2,4-Dimethylhexane		382.6	553.6	2.556	472
C_8H_{18}	2,5-Dimethylhexane	182	382.27	550.1	2.487	482
C_8H_{18}	3,3-Dimethylhexane	147.0	385.12	562.1	2.654	443
C_8H_{18}	3,4-Dimethylhexane		390.88	568.9	2.692	466
C_8H_{18}	3-Ethylhexane		391.7	565.5	2.608	455
C_8H_{18}	3-Ethyl-2-methylpentane	158.20	388.81	567.1	2.700	443
C_8H_{18}	3-Ethyl-3-methylpentane	182.2	391.42	576.6	2.808	455
C_8H_{18}	2,2,3-Trimethylpentane	160.89	383	563.5	2.730	436
C_8H_{18}	2,2,4-Trimethylpentane	165.8	372.37	544.0	2.568	468
C_8H_{18}	2,3,3-Trimethylpentane	172.22	387.9	573.6	2.820	455
C_8H_{18}	2,3,4-Trimethylpentane	163.9	386.6	566.5	2.730	461
C_8H_{18}	2,2,3,3-Tetramethylbutane	373.8	379.60	567.8	2.87	461
$C_8H_{18}O$	1-Octanol	257.6	468.31	652.5	2.86	490
$C_8H_{18}O$	2-Octanol	241	453.03	638	2.9	
$C_8H_{18}O$	4-Methyl-3-heptanol	150	443	623.5		
$C_8H_{18}O$	5-Methyl-3-heptanol	181.9	445	621.2		
$C_8H_{18}O$	2-Ethyl-1-hexanol	203	457.77	641	2.8	
$C_8H_{18}O$	Dibutyl ether	177.9	413.43	584.1	3.01	
$C_8H_{18}O$	Di-*tert*-butyl ether		380.38	550		
$C_8H_{19}N$	Dibutylamine	211	432.7	607.5	3.11	
$C_8H_{19}N$	Diisobutylamine	199.6	412.84	584.4	3.20	
$C_8H_{20}Si$	Tetraethylsilane		427.90	603.7	2.602	
C_9F_{20}	Perfluorononane		398.50	524	1.56	
C_9H_7N	Quinoline	258.37	510.31	782	4.86	437
C_9H_7N	Isoquinoline	299.62	516.37	803	5.10	374
C_9H_{10}	Indan	221.7	451.12	684.9	3.95	
C_9H_{12}	Cumene	177.14	425.56	631.1	3.209	
C_9H_{12}	Propylbenzene	173.59	432.39	638.32	3.200	440
C_9H_{12}	1,2,3-Trimethylbenzene	247.7	449.27	664.47	3.454	
C_9H_{12}	1,2,4-Trimethylbenzene	229.3	442.53	649.17	3.232	
C_9H_{12}	Mesitylene	228.4	437.89	637.25	3.127	
$C_9H_{13}N$	*N,N*-Dimethyl-*o*-toluidine	213	467.30	668	3.12	
C_9H_{18}	*trans*-1,3,5-Trimethylcyclohexane	165.7	413.70	602.2		
$C_9H_{18}O$	5-Nonanone	267.2	461.60	640		
$C_9H_{18}O_2$	Nonanoic acid	285.50	527.70	711	2.40	
$C_9H_{18}O_2$	Isopentyl butanoate		452	619		
$C_9H_{18}O_2$	Isobutyl 3-methylbutanoate		441.70	621		
C_9H_{20}	Nonane	219.6	423.97	594.9	2.288	
C_9H_{20}	2-Methyloctane	192.78	416.43	587.0	2.310	
C_9H_{20}	2,2-Dimethylheptane	160	405.8	576.8	2.350	
C_9H_{20}	2,2,5-Trimethylhexane	167.39	397.24	568		
C_9H_{20}	2,2,3,3-Tetramethylpentane	263.3	413.44	607.7	2.741	
C_9H_{20}	2,2,3,4-Tetramethylpentane	152.06	406.18	592.7	2.602	
C_9H_{20}	2,2,4,4-Tetramethylpentane	206.61	395.44	574.7	2.485	

CRITICAL CONSTANTS, BOILING POINTS, AND MELTING POINTS OF SELECTED COMPOUNDS (continued)

Molecular formula	Name	T_m/K	T_b/K	T_c/K	P_c/MPa	V_c/ cm³/mol
C_9H_{20}	2,3,3,4-Tetramethylpentane	171.0	414.72	607.7	2.716	
$C_9H_{20}O$	1-Nonanol	268	486.52	671.5	2.63	
$C_{10}F_8$	Perfluoronaphthalene	360.6	482	673.1		
$C_{10}F_{18}$	Perfluorodecalin	263	415	566	1.52	
$C_{10}F_{22}$	Perfluorodecane		417.40	542	1.45	
$C_{10}H_8$	Naphthalene	353.3	491.14	748.4	4.051	413
$C_{10}H_{12}$	1,2,3,4-Tetrahydronaphthalene	237.40	480.77	719.9		408
$C_{10}H_{14}$	Butylbenzene	185.2	456.46	660.5	2.887	497
$C_{10}H_{14}$	Isobutylbenzene	221.70	445.94	650	3.05	
$C_{10}H_{14}$	*p*-Cymene	204.2	450.28	651	2.73	
$C_{10}H_{14}$	*p*-Diethylbenzene	230.32	456.94	657.88	2.803	
$C_{10}H_{14}$	1,2,4,5-Tetramethylbenzene	352.4	469.99	675	2.9	
$C_{10}H_{14}O$	Thymol	324.6	505.70	698		
$C_{10}H_{18}$	*cis*-Decahydronaphthalene	230.20	468.96	702.3	3.20	
$C_{10}H_{18}$	*trans*-Decahydronaphthalene	242.79	460.46	687.1		
$C_{10}H_{20}$	1-Decene		443.6	616.4	2.218	584
$C_{10}H_{20}O$	Menthol	315	489.50	694		
$C_{10}H_{20}O$	Decanal	268	481.6	674.2		
$C_{10}H_{20}O_2$	Decanoic acid	305.14	541.90	726	2.23	
$C_{10}H_{20}O_2$	Ethyl octanoate	230.0	481.70	659		
$C_{10}H_{22}$	Decane	243.4	447.30	617.65	2.104	
$C_{10}H_{22}$	3,3,5-Trimethylheptane		428.8	609.7	2.317	
$C_{10}H_{22}$	2,2,3,3-Tetramethylhexane	219	433.48	623.2	2.510	
$C_{10}H_{22}$	2,2,5,5-Tetramethylhexane	260.5	410.63	581.6	2.186	
$C_{10}H_{22}O$	1-Decanol	280.0	504.2	689	2.41	
$C_{10}H_{22}S$	Diisopentylsulfide		484	664		
$C_{11}H_{10}$	1-Methylnaphthalene	242.67	517.89	772		
$C_{11}H_{10}$	2-Methylnaphthalene	307.5	514.26	761		
$C_{11}H_{22}O_2$	Ethyl nonanoate	236.4	500.20	674		
$C_{11}H_{24}$	Undecane	247.5	469.08	638.85	1.955	
$C_{12}H_9N$	Carbazole	519.3	627.85	901.8	3.93	502
$C_{12}H_{10}$	Biphenyl	342	529.2	789	3.85	502
$C_{12}H_{10}O$	Diphenyl ether	300.02	531.20	766.8		
$C_{12}H_{18}$	Hexamethylbenzene	439.6	536.60	758		
$C_{12}H_{24}$	1-Dodecene	237.9	486.9	657.6	1.930	
$C_{12}H_{26}$	Dodecane	263.5	489.47	658.65	1.830	
$C_{12}H_{26}O$	1-Dodecanol	297	532	720	2.08	
$C_{13}H_{12}$	Diphenylmethane	298.39	538.20	770	2.86	
$C_{13}H_{26}O_2$	Methyl dodecanoate	278.3	540	712		
$C_{13}H_{28}$	Tridecane	267.76	508.62	676	1.71	
$C_{13}H_{28}O$	1-Tridecanol	305.6		734	1.935	
$C_{14}H_{10}$	Phenanthrene	372.39	613	869		554
$C_{14}H_{30}$	Tetradecane	279.01	526.73	693	1.61	
$C_{14}H_{30}O$	1-Tetradecanol	312.6	562	747	1.81	
$C_{15}H_{32}$	Pentadecane	283.0	543.7	708	1.515	
$C_{16}H_{34}$	Hexadecane	291.34	560.01	722	1.435	
$C_{16}H_{34}O$	1-Hexadecanol	322.4	607	770	1.61	
$C_{17}H_{36}$	Heptadecane	295	575.1	735	1.37	
$C_{17}H_{36}O$	1-Heptadecanol	326.9	606	780	1.50	
$C_{18}H_{14}$	*o*-Terphenyl	329.3	605	891.0	3.90	753
$C_{18}H_{14}$	*m*-Terphenyl	360	636	924.9	3.51	768
$C_{18}H_{14}$	*p*-Terphenyl	483.2	649	926.0	3.32	763
$C_{18}H_{38}$	Octadecane	301.3	589.50	746	1.30	
$C_{18}H_{38}O$	1-Octadecanol	332.6		790	1.44	
$C_{19}H_{40}$	Nonadecane	305.2	603.0	758	1.23	
$C_{20}H_{42}$	Eicosane	309.9	616	769	1.16	
$C_{20}H_{42}O$	1-Eicosanol	339.2		809	1.30	

SUBLIMATION PRESSURE OF SOLIDS

This table gives the sublimation pressure of some representative solids as a function of temperature. The last entry for each substance is the solid-liquid-gas triple point. Substances are listed by molecular formula in the Hill order. Note that the pressure is given in kilopascals (kPa), except for water and naphthalene, for which the pressure unit is pascals. For conversion, 1 kPa = 7.506 Torr = 0.0098692 atm.

Substance									
Ar	*T*/K	55	60	65	70	75	80	83.81	
Argon	*P*/kPa	0.2	0.8	2.8	7.7	18.7	40.7	68.8	
BrH	*T*/K	135	140	150	160	170	180	185.1	
Hydrogen bromide	*P*/kPa	0.1	0.3	1.1	3.3	8.7	20.1	27.4	
ClH	*T*/K	120	130	140	150	155	159.0		
Hydrogen chloride	*P*/kPa	0.1	0.5	1.9	5.8	9.5	13.5		
F_4Si	*T*/K	130	140	150	160	170	175	180	186.3
Silicon tetrafluoride	*P*/kPa	0.2	0.9	3.9	14.0	43.8	74.2	122.4	220.8
F_6S	*T*/K	150	165	180	190	200	210	220	223.1
Sulfur hexafluoride	*P*/kPa	0.4	2.6	11.3	25.9	54.5	106.1	195.1	232.7
HI	*T*/K	160	170	180	190	200	210	220	222.4
Hydrogen iodide	*P*/kPa	0.2	0.8	2.2	5.3	11.7	23.6	44.1	49.3
H_2O	*T*/K	190	210	225	240	250	260	270	273.16
Water	*P*/Pa	0.032	0.702	4.942	27.28	76.04	195.8	470.1	611.66
H_2S	*T*/K	140	150	160	165	170	175	180	187.6
Hydrogen sulfide	*P*/kPa	0.2	0.6	1.9	3.2	5.2	8.3	12.7	22.7
H_3N	*T*/K	160	170	180	190	195	195.4		
Ammonia	*P*/kPa	0.1	0.4	1.2	3.5	5.8	6.12		
Kr	*T*/K	80	90	95	100	105	110	115.8	
Krypton	*P*/kPa	0.4	2.7	6.0	12.1	22.8	40.4	73.1	
NO	*T*/K	85	90	95	100	105	109.5		
Nitric oxide	*P*/kPa	0.1	0.4	1.3	3.8	10.0	21.9		
Xe	*T*/K	110	120	130	140	150	155	160	161.4
Xenon	*P*/kPa	0.3	1.5	4.9	14.0	34.2	51.1	74.2	81.7
CO	*T*/K	50	55	60	65	68.13			
Carbon monoxide	*P*/kPa	0.1	0.6	2.6	8.2	15.4			
CO_2	*T*/K	130	140	155	170	185	194.7	205	216.58
Carbon dioxide	*P*/kPa	0.032	0.187	1.674	9.987	44.02	101.3	227.1	518.0
CHN	*T*/K	200	210	220	230	240	250	255	259.83
Hydrogen cyanide	*P*/kPa	0.2	0.4	1.0	2.2	4.8	9.7	13.6	18.62
CH_4	*T*/K	65	70	75	80	85	90.69		
Methane	*P*/kPa	0.1	0.3	0.8	2.1	4.9	11.70		
C_2H_2	*T*/K	130	140	150	160	170	180	190	192.4
Acetylene	*P*/kPa	0.2	0.7	2.6	7.8	20.6	49.0	106.3	126.0
$C_2H_4O_2$	*T*/K	250	260	270	280	289.7			
Acetic acid	*P*/kPa	0.092	0.199	0.406	0.79	1.29			
C_5H_{12}	*T*/K	200	210	220	230	240	250	255	256.58
Neopentane	*P*/kPa	0.7	1.6	3.6	7.3	13.9	24.8	32.4	35.8
$C_{10}H_8$	*T*/K	250	270	280	290	300	310	330	353.43
Naphthalene	*P*/Pa	0.036	0.514	1.662	4.918	13.43	34.15	182.9	999.6

VAPOR PRESSURE OF FLUIDS AT TEMPERATURES BELOW 300 K

This table gives vapor pressures of 67 important fluids in the temperature range 2 to 300 K. Helium (^4He), hydrogen (H_2), and neon (Ne) are covered on this page. The remaining fluids are listed on subsequent pages by molecular formula in the Hill order (see Introduction). The data have been taken from evaluated sources; references are listed at the end of the table.

Pressures are given in kilopascals (kPa). Note that:

1 kPa = 7.50062 Torr
100 kPa = 1 bar
101.325 kPa = 1 atmos

s following an entry indicates that the compound is solid at that temperature.

Helium		Hydrogen		Neon	
T/K	*P*/kPa	*T*/K	*P*/kPa	*T*/K	*P*/kPa
2.2	5.3	14.0	7.90	25.0	51.3
2.3	6.7	14.5	10.38	26.0	71.8
2.4	8.3	15.0	13.43	27.0	98.5
2.5	10.2	15.5	17.12	28.0	132.1
2.6	12.4	16.0	21.53	29.0	173.5
2.7	14.8	16.5	26.74	30.0	223.8
2.8	17.5	17.0	32.84	31.0	284.0
2.9	20.6	17.5	39.92	32.0	355.2
3.0	24.0	18.0	48.08	33.0	438.6
3.1	27.8	18.5	57.39	34.0	535.2
3.2	32.0	19.0	67.96	35.0	646.2
3.3	36.5	19.5	79.89	36.0	772.8
3.4	41.5	20.0	93.26	37.0	916.4
3.5	47.0	20.5	108.2	38.0	1078
3.6	52.9	21.0	124.7	39.0	1260
3.7	59.3	21.5	143.1	40.0	1462
3.8	66.1	22.0	163.2	41.0	1688
3.9	73.5	22.5	185.3	42.0	1939
4.0	81.5	23.0	209.4	43.0	2216
4.1	90.0	23.5	235.7	44.0	2522
4.2	99.0	24.0	264.2		
4.3	108.7	24.5	295.1		
4.4	119.0	25.0	328.5		
4.5	129.9	25.5	364.3		
4.6	141.6	26.0	402.9		
4.7	153.9	26.5	444.3		
4.8	167.0	27.0	488.5		
4.9	180.8	27.5	535.7		
5.0	195.4	28.0	586.1		
5.1	210.9	28.5	639.7		
		29.0	696.7		
		29.5	757.3		
		30.0	821.4		
		30.5	889.5		
		31.0	961.5		
		31.5	1038.0		
		32.0	1119.0		
		32.5	1204.0		
Ref.	17,18		1		13

VAPOR PRESSURE OF FLUIDS AT TEMPERATURES BELOW 300 K (continued)

T/K	Ar Argon	BCl_3 Boron trichloride	BF_3 Boron trifluoride	BrH Hydrogen bromide	Br_2 Bromine	ClF Chlorine fluoride	ClH Hydrogen chloride
50	0.1 s						
55	0.2 s						
60	0.8 s						
65	2.8 s						
70	7.7 s						
75	18.7 s						
80	40.7 s						
85	79.0						
90	134						
95	213						
100	324						
105	473						
110	666						
115	910					0.1	
120	1214					0.3	0.1 s
125	1584					0.6	0.3 s
130	2027					1.2	0.5 s
135	2553			0.1 s		2.1	1.0 s
140	3170			0.3 s		3.6	1.9 s
145	3892		7.7	0.6 s		6.0	3.4 s
150	4736		13.4	1.1 s		9.5	5.8 s
155			22.3	1.9 s		14.6	9.5 s
160			35.2	3.3 s		21.8	14.7
165			53.7	5.4 s		31.7	22.0
170			79.1	8.7 s		44.8	31.9
175			113	13.4 s		62.0	45.1
180		0.1	157	20.1 s		84.2	62.5
185		0.2	214	29.5 s		112	84.7
190		0.3	285	37.9		147	113
195		0.5	372	51.8		190	148
200		0.8	479	69.5		242	190
205		1.2	608	91.8		304	242
210		1.8	762	119		378	304
215		2.6	944	153		464	377
220		3.8	1160	194	0.1 s	564	463
225		5.2	1413	242	0.2 s	680	563
230		7.2	1709	299	0.3 s	812	678
235		9.7	2056	366	0.4 s	961	811
240		12.9	2460	443	0.7 s	1130	961
245		17.0	2913	532	1.1 s	1319	1132
250		22.0	3481	633	1.7 s	1529	1325
255		28.1	4123	748	2.6 s	1762	1542
260		35.0	4874	878	3.8 s	2019	1784
265		44.5		1023	5.5 s	2301	2054
270		55.1		1185	7.3	2608	2354
275		67.6		1364	9.5	2941	2686
280		82.2		1562	12.3	3303	3053
285		99.1		1780	15.6	3693	3457
290		119		2018	19.7	4111	3901
295		141		2278	24.6	4560	4388
300		166		2561	30.5	5039	4921
Ref.	8,15	12	12	12	12	12	12

VAPOR PRESSURE OF FLUIDS AT TEMPERATURES BELOW 300 K (continued)

T/K	ClO_2 Chlorine dioxide	Cl_2 Chlorine	Cl_4Si Silicon tetrachloride	FH Hydrogen fluoride	F_2 Fluorine	F_2O Difluorine oxide	F_3N Nitrogen trifluoride
50							
55					0.4		
60					1.5		
65					4.8		
70					12.3		
75					27.6	0.1	
80					55.3	0.2	
85					101	0.5	0.1
90					172	1.2	0.2
95					276	2.6	0.4
100					420	5.3	0.9
105					615	10.1	2.0
110					870	18.0	4.0
115					1196	30.5	7.3
120					1605	49.3	12.8
125					2108	76.7	21.1
130					2721	115	33.5
135					3458	168	51.1
140					4339	237	75.4
145						328	108
150						444	150
155						588	205
160						766	273
165						981	357
170						1238	459
175		1.8				1541	581
180		2.8				1895	726
185		4.2				2303	896
190		6.1		0.3		2771	1092
195	0.1	8.7		0.5		3302	1319
200	0.3	12.3		0.8		3899	1578
205	0.5	16.9		1.2		4567	1871
210	0.9	22.9	0.1	1.7		5308	2203
215	1.4	30.5	0.2	2.3			2577
220	2.3	40.1	0.3	3.2			2995
225	3.5	51.9	0.5	4.4			3464
230	5.3	66.4	0.7	5.9			3991
235	7.6	84.0	1.0	7.9			
240	10.8	105	1.5	10.3			
245	14.9	130	2.0	13.4			
250	20.1	160	2.8	17.2			
255	26.6	194	3.8	21.8			
260	34.6	234	5.0	27.4			
265	44.4	280	6.6	34.2			
270	56.1	332	8.6	42.2			
275	69.9	392	11.1	51.8			
280	86.2	459	14.2	63.1			
285	105	535	17.9	76.3			
290	127	619	22.3	91.7			
295	151	714	27.7	110			
300	179	818	34.0	130			
Ref.	12	5	12	12	12	12	1

VAPOR PRESSURE OF FLUIDS AT TEMPERATURES BELOW 300 K (continued)

T/K	F_3P Phosphorous trifluoride	F_4Si Silicon tetrafluoride	F_6S Sulfur hexafluoride	HI Hydrogen iodide	H_2S Hydrogen sulfide	H_3N Ammonia	H_3P Phosphine
50							
55							
60							
65							
70							
75							
80							
85							
90							
95							
100							
105	0.1						
110	0.2						0.1
115	0.5						0.2
120	1.0						0.4
125	1.9	0.1 s					0.7
130	3.5	0.2 s					1.3
135	5.9	0.4 s			0.1 s		2.3
140	9.5	0.9 s	0.1 s		0.2 s		3.9
145	14.9	1.9 s	0.2 s		0.3 s		6.2
150	22.5	3.8 s	0.4 s		0.6 s		9.6
155	33.1	7.5 s	0.8 s	0.1 s	1.1 s		14.5
160	47.3	14.0 s	1.5 s	0.2 s	1.9 s	0.1 s	21.1
165	66.0	25.2 s	2.6 s	0.4 s	3.2 s	0.2 s	30.0
170	90.1	43.8 s	4.4 s	0.8 s	5.2 s	0.3 s	41.6
175	121	74.2 s	7.1 s	1.3 s	8.3 s	0.6 s	56.6
180	159	122 s	11.3 s	2.2 s	12.7 s	1.2 s	75.6
185	206	197 s	17.3 s	3.4 s	18.9 s	2.1 s	99.2
190	262	280	25.9 s	5.3 s	26.6	3.5 s	128
195	330	376	38.0 s	8.0 s	36.7	5.8 s	163
200	410	488	54.4 s	11.7 s	49.8	8.7	205
205	503	618	76.6 s	16.8 s	66.4	12.6	254
210	611	766	106 s	23.6 s	87.1	17.9	312
215	736	932	145 s	32.5 s	113	24.9	379
220	877	1117	195 s	44.0 s	144	34.1	456
225	1037	1324	249	56.2	182	45.9	544
230	1217	1555	305	71.4	227	60.8	644
235	1418	1816	371	89.7	281	79.6	756
240	1640	2111	448	112	344	103	881
245	1885	2449	536	137	416	131	1019
250	2154	2841	636	168	500	165	1172
255	2448	3301	750	203	597	207	1341
260	2767		870	244	706	256	1525
265	3112		1021	290	830	313	1725
270			1181	343	969	381	1942
275			1358	404	1124	460	2176
280			1554	472	1297	552	2428
285			1768	548	1488	655	2699
290			2003	633	1698	774	2987
295			2258	727	1929	909	3295
300			2534	831	2181	1062	3621
Ref.	12	12	12,15	12	12,15	11	12

T/K	H_4Si Silane	Kr Krypton	NO Nitric oxide	N_2 Nitrogen	N_2O Nitrous oxide	O_2 Oxygen	O_2S Sulfur dioxide
50				0.4 s			
55				1.8 s		0.2	
60				6.3 s		0.7	
65				17.4		2.3	
70				38.6		6.3	
75		0.1 s		76.1		14.5	
80		0.4 s		137		30.1	
85		1.1 s	0.1 s	229		56.8	
90		2.7 s	0.4 s	361		99.3	
95	0.1	6.0 s	1.3 s	541		163	
100	0.2	12.1 s	3.8 s	779		254	
105	0.4	22.8 s	10.0 s	1084		379	
110	1.0	40.4 s	23.5	1467		543	
115	1.9	68.0 s	46.8	1939	0.1	756	
120	3.5	103	86.5	2513	0.1	1022	
125	6.1	150	151	3209	0.3	1351	
130	10.0	211	248		0.7	1749	
135	15.8	290	391		1.3	2225	
140	24.1	388	592		2.5	2788	
145	35.3	509	867		4.3	3448	
150	50.3	655	1231		7.1	4219	
155	69.8	830	1703		11.4		
160	94.6	1037	2302		17.6		
165	126	1278	3050		26.4		
170	164	1557	3971		38.5		0.1
175	210	1877	5089		54.7		0.2
180	265	2241	6433		75.9		0.3
185	331	2655			103		0.5
190	408	3120			138		0.8
195	498	3641			181		1.3
200	602	4223			234		2.0
205	722	4870			298		3.0
210	859				374		4.4
215	1017				465		6.3
220	1196				571		9.0
225	1398				694		12.6
230	1628				835		17.3
235	1888				996		23.3
240	2180				1179		31.1
245	2509				1385		40.9
250	2880				1615		53.2
255	3296				1870		68.3
260	3763				2152		86.7
265	4288				2462		109
270					2802		136
275					3172		168
280					3573		205
285					4006		249
290					4473		300
295					4973		359
300					5508		426
Ref.	12	13, 15	12, 15	1	12	3	12

VAPOR PRESSURE OF FLUIDS AT TEMPERATURES BELOW 300 K (continued)

T/K	O_3 Ozone	Ra Radon	Xe Xenon	$CBrF_3$ Bromotri-fluoromethane	$CClF_3$ Chlorotri-fluoromethane	CCl_2F_2 Dichlorodi-fluoromethane	CCl_3F Trichloro-fluoromethane
50							
55							
60							
65							
70							
75							
80							
85							
90							
95							
100	0.1		0.1 s				
105	0.2		0.1 s				
110	0.4		0.3 s				
115	1.0		0.7 s		0.1		
120	2.0		1.5 s		0.2		
125	3.8		2.7 s		0.3		
130	6.8	0.1	4.9 s		0.6		
135	11.5	0.3	8.5 s	0.1	1.1		
140	18.7	0.5	14.0 s	0.3	2.0		
145	29.1	0.9	22.2 s	0.5	3.3		
150	43.7	1.5	34.2 s	0.9	5.3		
155	63.6	2.4	51.1 s	1.5	8.3	0.1	
160	89.9	3.8	74.2 s	2.5	12.6	0.3	
165	124	5.8	101	3.9	18.6	0.5	
170	168	8.6	134	5.9	26.8	0.8	
175	222	12.5	173	8.8	37.6	1.3	
180	289	17.7	222	12.8	51.7	2.1	
185	367	24.5	280	18.1	69.7	3.2	
190	468	33.2	348	25.1	92.3	4.8	0.2
195	584	44.4	428	34.1	120	6.9	0.3
200	721	58.2	521	45.6	155	9.9	0.4
205	881	75.3	628	60.0	196	13.7	0.6
210	1068	96	750	77.8	246	18.8	1.0
215	1285	121	889	99.5	304	25.2	1.4
220	1536	151	1045	126	372	33.3	2.0
225	1824	185	1220	157	451	43.3	2.9
230	2155		1416	194	542	55.5	4.1
235	2534		1633	237	646	70.4	5.6
240	2968		1872	287	763	88.1	7.6
245	3464		2136	344	896	109	10.1
250	4031		2425	410	1044	134	13.3
255	4678		2742	485	1210	163	17.2
260	5417		3087	570	1394	196	22.1
265			3462	665	1598	234	28.0
270			3869	771	1823	278	35.1
275			4310	889	2071	327	43.7
280			4786	1021	2343	383	53.8
285			5299	1166	2641	445	65.7
290				1325	2968	515	79.6
295				1501	3325	593	95.6
300				1692	3716	679	114.1
Ref.	12	15	12,13	12	12	12	12

VAPOR PRESSURE OF FLUIDS AT TEMPERATURES BELOW 300 K (continued)

T/K	CCl_4 Tetrachloro-methane	CF_4 Tetrafluoro-methane	CO Carbon monoxide	COS Carbon oxysulfide	CO_2 Carbon dioxide	$CHClF_2$ Chlorodifluo-methane	$CHCl_3$ Trichloro-methane
50			0.1 s				
55			0.6 s				
60			2.6 s				
65			8.2 s				
70			21.0				
75			44.4				
80			83.7				
85			147				
90		0.1	239				
95		0.3	371				
100		0.8	545				
105		1.7	771				
110		3.4	1067				
115		6.5	1428				
120		11.5	1877				
125		19.3	2400				
130		30.8	3064				
135		47.4			0.1		
140		70.2		0.1	0.2		
145		101		0.2	0.4		
150		141		0.4	0.8	0.1	
155		191		0.8	1.7	0.3	
160		254		1.3	3.1	0.5	
165		332		2.2	5.7	0.8	
170		425		3.4	9.9	1.4	
175		537		5.2	16.8	2.3	
180		669		7.8	27.6	3.6	
185		824		11.3	44.0	5.5	
190		1005		15.9	68.4	8.1	
195		1216		22.1	104	11.8	
200		1460		30.0	155	16.7	
205		1743		40.1	227	23.1	
210		2073		52.7	327	31.5	
215		2457		68.2	465	42.1	0.1
220		2907		87.2	600	55.3	0.2
225		3438		110	735	71.7	0.3
230				137	894	91.6	0.4
235				169	1075	116	0.7
240				207	1283	144	1.0
245				250	1519	178	1.4
250				301	1786	218	2.0
255	1.5			358	2085	264	2.7
260	2.1			423	2419	317	3.7
265	2.8			497	2790	377	5.0
270	3.7			580	3203	446	6.6
275	4.9			673	3658	525	8.7
280	6.4			777	4160	613	11.3
285	8.2			892	4712	711	14.4
290	10.5			1019	5315	821	18.3
295	13.2			1159	5984	944	22.9
300	16.5			1313	6710	1080	28.5
Ref.	12	12	9	12	6	12	12

T/K	CHF_3 Trifluoro-methane	CHN Hydrogen cyanide	CH_2Cl_2 Dichloro-methane	CH_2F_2 Difluoro methane	CH_2O Formaldehyde	CH_3Cl Chloromethane	CH_3F Fluoromethane
50							
55							
60							
65							
70							
75							
80							
85							
90							
95							
100							
105							
110							
115							
120	0.1						
125	0.2						
130	0.4						
135	0.7						0.6
140	1.4			0.1			1.2
145	2.5			0.2			2.1
150	4.3			0.3			3.6
155	7.1			0.6			5.9
160	11.1			1.0			9.3
165	17.0			1.7			14.1
170	25.3			2.8			20.9
175	36.5			4.4			29.9
180	51.4			6.8			42.0
185	70.9			10.2	1.3	2.1	57.6
190	95.8			14.8	2.0	3.1	77.4
195	127			21.2	3.0	4.6	102
200	166	0.1 s	0.1	29.5	4.4	6.7	133
205	214	0.2 s	0.2	40.5	6.4	9.5	171
210	271	0.4 s	0.3	54.5	9.1	13.1	216
215	340	0.6 s	0.4	72.1	12.7	17.9	270
220	421	1 s	0.6	94.1	17.4	24.0	333
225	516	1.5 s	0.9	121	23.4	31.8	408
230	626	2.2 s	1.4	154	31.0	41.4	495
235	754	3.3 s	2.0	193	40.6	53.3	595
240	900	4.7 s	2.8	240	52.5	67.7	711
245	1067	6.8 s	3.8	295	67.0	85.1	843
250	1257	9.7 s	5.3	360	84.6	106	993
255	1472	13.6 s	7.1	434	106	131	1163
260	1713	18.8	9.5	521	131	159	1355
265	1984	24.1	12.4	620	161	193	1571
270	2287	30.5	16.1	732	196	232	1813
275	2624	38.3	20.7	860	236	277	2084
280	3000	47.7	26.3	1004	283	327	2387
285	3418	58.8	33.0	1165	337	385	2724
290	3881	72.1	41.1	1346	399	450	3099
295	4393	87.6	50.8	1547	470	524	3516
300		105.9	62.1	1770	549	606	3978
Ref.	12	12,16	12	12	12	12	12

VAPOR PRESSURE OF FLUIDS AT TEMPERATURES BELOW 300 K (continued)

T/K	CH_4 Methane	CH_4O Methanol	C_2H_2 Acetylene	C_2H_4 Ethylene	C_2H_6 Ethane	C_2H_6O Dimethyl ether	C_3H_4 Propadiene
50							
55							
60							
65	0.1						
70	0.3						
75	0.8						
80	2.1						
85	4.9						
90	10.6						
95	20.0						
100	34.5						
105	57.0						
110	88.4			0.3			
115	133			0.8	0.1		
120	192			1.4	0.4		
125	269			2.7	0.7		
130	368		0.1 s	4.5	1.3		
135	491		0.3 s	7.7	2.2		
140	642		0.7 s	11.9	3.8		
145	824		1.3 s	18.3	6.0		
150	1041		2.6 s	27.5	9.7		0.1
155	1297		4.6 s	39.9	15.0	0.1	0.2
160	1594		7.8 s	56.4	21.5	0.2	0.3
165	1937		12.8 s	77.9	31.0	0.3	0.6
170	2331		20.6 s	105	42.9	0.5	1.0
175	2779		32.2 s	140	59.0	0.9	1.7
180	3288		49.0 s	182	78.7	1.4	2.7
185	3865		72.9 s	234	104	2.1	4.1
190	4520		106 s	296	135	3.2	6.1
195			146	369	172	4.7	8.9
200			190	456	217	6.8	12.5
205			244	557	271	9.6	17.4
210			309	673	334	13.3	23.7
215			385	806	407	18.1	31.6
220			475	958	492	24.3	41.4
225			579	1128	590	32.1	53.5
230		0.1	699	1321	700	41.9	68.2
235		0.2	837	1535	826	53.9	85.8
240		0.4	993	1774	967	68.6	107
245		0.5	1170	2039	1125	86.3	131
250		0.8	1370	2331	1301	108	160
255		1.2	1593	2652	1496	133	193
260		1.7	1843	3005	1712	162	230
265		2.4	2121	3391	1949	197	273
270		3.3	2429	3813	2210	237	322
275		4.5	2771	4275	2495	283	376
280		6.2	3150		2806	335	438
285		8.3	3567		3146	395	506
290		11	4028		3515	463	582
295		14.4	4535		3917	538	666
300		18.7	5093		4355	623	759
Ref.	2,16	12	12,16	4	2	12	12

VAPOR PRESSURE OF FLUIDS AT TEMPERATURES BELOW 300 K (continued)

T/K	C_3H_6 Propylene	C_3H_8 Propane	C_4H_6 Buta-1,3-diene	C_4H_{10} *n*-Butane	C_4H_{10} Isobutane	C_5H_{12} *n*-Pentane	C_5H_{12} Neopentane
50							
55							
60							
65							
70							
75							
80							
85							
90							
95							
100							
105							
110							
115							
120							
125							
130							
135							
140	0.1						
145	0.2						
150	0.4						
155	0.7						
160	1.2	0.8			0.1		
165	2.0	1.4			0.1		
170	3.1	2.2	0.1	0.1	0.3		
175	4.7	3.3	0.2	0.2	0.4		
180	7.0	5.0	0.4	0.3	0.7		
185	10.1	7.3	0.6	0.5	1.1		0.1 s
190	14.2	10.5	1.0	0.8	1.7		0.2 s
195	19.7	15.0	1.5	1.3	2.5		0.4 s
200	26.9	20.1	2.3	1.9	3.7		0.7 s
205	35.9	27.0	3.4	2.8	5.3		1.1 s
210	47.3	36.0	4.8	4.0	7.4		1.6 s
215	61.3	47.0	6.7	5.7	10.2		2.4 s
220	78.5	60.0	9.2	7.8	13.8	1.0	3.6 s
225	99.2	77.0	12.5	10.6	18.3	1.5	5.2 s
230	124	97.0	16.7	14.1	24.0	2.1	7.3 s
235	153	120	21.9	18.5	31.1	3.0	10.2 s
240	188	148	28.4	24.1	39.8	4.2	13.9 s
245	228	180	36.3	30.9	50.3	5.7	18.7 s
250	274	218	46.0	39.1	62.9	7.6	24.8 s
255	327	261	57.6	49.1	77.8	10.0	32.4 s
260	387	311	71.1	61.0	93.4	13.0	41.6
265	456	367	87.6	75.0	116	16.6	51.4
270	533	431	107	91.5	140	21.1	63.0
275	619	502	129	111	167	26.6	76.6
280	715	582	154	133	198	33.1	92.3
285	822	671	184	159	234	40.8	111
290	940	769	217	188	274	50.0	131
295	1069	878	255	221	319	60.7	155
300	1212	998	297	258	370	73.2	182
Ref.	7	2	12	2	2	14	12,16

REFERENCES

1. B. A. Younglove, Thermophysical properties of fluids. I. Ethylene, parahydrogen, nitrogen trifluoride, and oxygen, *J. Phys. Chem. Ref. Data*, 11, Supp. 1, 1982.
2. B. A. Younglove and J. F. Ely, Thermophysical properties of fluids. II. Methane, ethane, propane, isobutane, and normal butane, *J. Phys. Chem. Ref. Data*, 16, 577, 1987.
3. W. Wagner, et al., *International Tables for the Fluid State: Oxygen*, Blackwell Scientific Publications, Oxford, 1987.
4. R. T. Jacobsen, et al., *International Tables for the Fluid State: Ethylene*, Blackwell Scientific Publications, Oxford, 1988.
5. S. Angus, et al., *International Tables for the Fluid State: Chlorine*, Pergamon Press, Oxford, 1985.
6. S. Angus, et al., *International Tables for the Fluid State: Carbon Dioxide*, Pergamon Press, Oxford, 1976.
7. S. Angus, et al., *International Tables for the Fluid State: Propylene*, Pergamon Press, Oxford, 1980.
8. R. B. Stewart and R. T. Jacobsen, Thermophysical properties of argon, *J. Phys. Chem. Ref. Data*, 18, 639, 1989.
9. R. D. Goodwin, Carbon monoxide thermophysical properties, *J. Phys. Chem. Ref. Data,* 14, 849, 1985.
10. R. D. Goodwin, Methanol thermophysical properties, *J. Phys. Chem. Ref. Data*, 16, 799, 1987.
11. L. Haar, Thermodynamic properties of ammonia, *J. Phys. Chem. Ref. Data*, 7, 635, 1978.
12. DIPPR Data Compilation of Pure Compound Properties, Design Institute for Physical Properties Data, American Institute of Chemical Engineers, 1987.
13. V. A. Rabinovich, et al., *Thermophysical Properties of Neon, Argon, Krypton, and Xenon*, Hemisphere Publishing Corp., New York, 1987.
14. K. N. Marsh, *Recommended Reference Methods for the Realization of Physicochemical Properties*, Blackwell Scientific Publications, Oxford, 1987.
15. TRC Thermodynamic Tables: Non-Hydrocarbons, Thermodynamic Research Center, Texas A & M University, College Station, Texas, 1985.
16. R. M. Stevenson and S. Malanowski, *Handbook of the Thermodynamics of Organic Compounds*, Elsevier, New York, 1987.
17. S. Angus and K. M. de Reuck, *International Tables of the Fluid State: Helium-4*, Pergamon Press, Oxford, 1977.
18. R. D. McCarty, *J. Phys. Chem. Ref. Data*, 2, 923, 1973.

VAPOR PRESSURE IN THE TEMPERATURE RANGE −25°C TO 150°C

This table gives the vapor pressure of 1277 substances near ambient temperature, specifically, in the range from −25°C to 150°C. More volatile substances are covered in the preceding table, *Vapor Pressure at Temperatures Below 300 K*. Substances in the present table that are also included in the preceding table are indicated by an asterisk (*).

All vapor pressures are given in kilopascal (kPa) units. The correspondence between kPa and other common pressure units is:

kPa	mmHg	atm
0.1	0.750	0.987×10^{-3}
1	7.501	0.987×10^{-2}
10	75.01	0.0987
100	750.1	0.987
1000	7501	9.87

If a value is given at a temperature below the normal melting point of a compound, it refers to the sublimation pressure of the solid.

REFERENCES

1. Lide, D. R., and Kehiaian, H. V., *Handbook of Thermophysical and Thermochemical Data*, CRC Press, Boca Raton, FL, 1994.
2. Riddick, J. A., Bunger, W. B., and Sakano, T. K., *Organic Solvents, Fourth Edition*, John Wiley & Sons, New York, 1986.
3. Daubert, T. E., Danner, R. P., Sibul, H. M., and Stebbins, C. C., *Physical and Thermodynamic Properties of Pure Compounds: Data Compilation*, extant 1994 (core with 4 supplements), Taylor & Francis, Bristol, PA (also available as database).
4. *TRC Thermodynamic Tables*, Thermodynamic Research Center, Texas A & M University, College Station, TX.
5. Stevenson, R. M., and Malanowski, S., *Handbook of the Thermodynamics of Organic Compounds*, Elsevier, New York, 1987.
6. Ohe, S., *Computer Aided Data Book of Vapor Pressure*, Data Book Publishing Co., Tokyo, 1976.
7. Boublik, T., Fried, V., and Hala, E., *The Vapor Pressure of Pure Substances, Second Edition*, Elsevier, Amsterdam, 1984.

Molecular formula	Name	Vapor pressure in kPa							
		−25°C	0°C	25°C	50°C	75°C	100°C	125°C	150°C
$AsCl_3$	Arsenic trichloride			1.29	5.38	16.7	41.6	88.8	
BBr_3	Boron tribromide		2.46	8.94	25.7	62.2			
$BiCl_3$	Bismuth trichloride								0.016
Br_2	Bromine*	1.46	8.80	28.7	75.3	168			
Br_3Sb	Antimony tribromide								1.81
Cl_2	Chlorine*	148	370	780			3927		
Cl_4Si	Silicon tetrachloride*			31.3	78.8				
FH	Hydrogen fluoride*	15.5	48.3	123					
H_2O	Water		0.611	3.17	12.3	38.6	101	232	476
H_2O_2	Hydrogen peroxide				1.32	5.14	16.2	43.1	101
H_3N	Ammonia*	152	429	1003	2033	3709	6253	9963	
H_4N_2	Hydrazine			1.91	7.59				
Hg	Mercury				0.002	0.009	0.037	0.129	0.383
O_2S	Sulfur dioxide*	49.5	155						
$CBrClF_2$	Bromochlorodifluoromethane	41.7	118						
$CBrCl_3$	Bromotrichloromethane		1.47	5.35	15.8	39.7	87.8	175	
CBrN	Cyanogen bromide		2.90	15.9	66.6				

VAPOR PRESSURE IN THE TEMPERATURE RANGE –25°C TO 150°C (continued)

Molecular formula	Name	Vapor pressure in kPa							
		–25°C	0°C	25°C	50°C	75°C	100°C	125°C	150°C
CBr_2F_2	Dibromodifluoromethane	13.0	41.9	110					
CBr_4	Tetrabromomethane			0.096					
CClN	Cyanogen chloride	13.2							
CCl_2F_2	Dichlorodifluoromethane	123	308	651	1216	2076	3332		
CCl_2O	Carbonyl chloride	23.8	75.1						
CCl_3F	Trichlorofluoromethane*	12.1	40.3	106	237		824		
CCl_3NO_2	Trichloronitromethane		0.765	3.18					
CCl_4	Tetrachloromethane	0.959	4.49	15.2	41.6	96.3	197	364	
$CHBrF_2$	Bromodifluoromethane	67.1	190						
$CHBr_3$	Tribromomethane			0.726	2.83	8.77	22.7	51.2	103
$CHClF_2$	Chlorodifluoromethane*	201	498	1044	1944	3317			
$CHCl_2F$	Dichlorofluoromethane	13.7	67.2	187					
$CHCl_3$	Trichloromethane*	1.75	8.02	26.2	69.3	156	308		
CHI_3	Triiodomethane				0.001	0.006	0.030		
CHN	Hydrogen cyanide*	8.53	35.3	98.8	233	492	944	1677	2791
CHNO	Cyanic acid (HOCN)	9.37							
CH_2BrCl	Bromochloromethane		5.26	19.5	54.6				
CH_2Br_2	Dibromomethane		1.53	6.12	19.1	49.6	111		
CH_2ClF	Chlorofluoromethane	51.6	140	326	647	1158			
CH_2Cl_2	Dichloromethane*	4.77	19.2	58.2	145				
CH_2I_2	Diiodomethane			0.172	0.783	2.75	7.93	19.6	42.8
CH_2O	Formaldehyde*	77.6	220						
CH_2O_2	Formic acid			5.75	17.4	44.4	99.6		
CH_3AsF_2	Methyldifluoroarsine		2.80	11.6	37.0	96.4			
CH_3Br	Bromomethane	28.8	88.0	217					
CH_3Cl	Chloromethane*	97.8	259	574	1115	1965	3220		
CH_3Cl_3Si	Methyltrichlorosilane			22.5	59.1	132			
CH_3I	Iodomethane	4.97	18.6	53.9	130				
CH_3NO	Formamide							4.52	12.1
CH_3NO_2	Nitromethane			4.79		42.1	97.6	201	
CH_3NO_3	Methyl nitrate		5.84	21.2					
CH_4Cl_2Si	Dichloromethylsilane	5.59	19.7	20.1	54.3				
CH_4O	Methanol*		4.03	16.9	55.5	151	353	735	1391
CH_4S	Methanethiol	25.4	79.8	202					
CH_5N	Methylamine	40.4		353					
CH_6N_2	Methylhydrazine			6.61					
CIN	Cyanogen iodide								90.2
CO_2	Carbon dioxide*	1684	3483						
CS_2	Carbon disulfide	4.68	16.9	48.2	114				
CSe_2	Carbon diselenide			2.38	7.74				
$C_2Br_2F_4$	1,2-Dibromotetrafluoroethane	3.79	14.3	43.4					
C_2Br_4	Tetrabromoethylene	0.181	1.29	6.58					

VAPOR PRESSURE IN THE TEMPERATURE RANGE –25°C TO 150°C (continued)

Molecular formula	Name	Vapor pressure in kPa –25°C	0°C	25°C	50°C	75°C	100°C	125°C	150°C
C_2ClF_3	Chlorotrifluoroethylene	116	689	1293	2240	3634			
$C_2Cl_2F_4$	1,1-Dichlorotetrafluoroethane	30.2	90.2	218	452	842	1443	2322	
$C_2Cl_2F_4$	1,2-Dichlorotetrafluoroethane	29.4	88.3	215	447	830	1413	2262	
$C_2Cl_3F_3$	1,1,1-Trichlorotrifluoroethane	2.74	14.9	48.5	115				
$C_2Cl_3F_3$	1,1,2-Trichlorotrifluoroethane	3.88	15.1	44.8	110	232	438	757	1225
C_2Cl_3N	Trichloroacetonitrile			9.89	29.2	72.3			
C_2Cl_4	Tetrachloroethylene			2.42	8.27	22.9	54.2	113	
$C_2Cl_4F_2$	1,1,1,2-Tetrachloro-2,2-difluoro-ethane			7.36	23.6	60.2	129		
$C_2Cl_4F_2$	Tetrachloro-1,2-difluoroethane		1.73	7.51	23.5	58.7	124		
C_2Cl_4O	Trichloroacetyl chloride			2.77	9.29	25.4	59.2		
C_2Cl_6	Hexachloroethane			0.047	0.339	1.47	4.89	13.8	34.2
$C_2F_4N_2O_4$	1,1,2,2-Tetrafluoro-1,2-dinitro-ethane		7.04	24.9	70.5				
$C_2HBrClF_3$	1,1,1-Trifluoro-2-bromo-2-chloro-ethane	3.13	12.8	39.9	101				
C_2HBr_3O	Tribromoacetaldehyde			0.196	0.866	3.04	8.90	22.5	50.8
C_2HClF_4	1-Chloro-1,1,2,2-tetrafluoro-ethane	56.6	161	377	770	1410	2337	3632	
C_2HCl_3	Trichloroethylene		2.73	9.91	28.6	69.3			
C_2HCl_3O	Trichloroacetaldehyde	0.362	1.80	6.66	19.7	49.4	108		
$C_2HCl_3O_2$	Trichloroacetic acid				0.101	0.588	2.47	8.08	22.0
C_2HCl_5	Pentachloroethane			0.478	1.99	6.39	16.9	38.7	78.6
$C_2HF_3O_2$	Trifluoroacetic acid		4.00	15.1	45.1	113			
$C_2H_2Br_2$	*cis*-1,2-Dibromoethylene			2.53	9.05	26.0			
$C_2H_2Br_2$	*trans*-1,2-Dibromoethylene		1.26	4.43	13.9	35.5			
$C_2H_2Br_2Cl_2$	1,2-Dibromo-1,1-dichloro-ethane					2.97	8.68	21.6	47.3
$C_2H_2Br_2Cl_2$	1,2-Dibromo-1,2-dichloro-ethane				0.491	1.67	4.87		
$C_2H_2Br_4$	1,1,2,2-Tetrabromoethane			0.003					6.02
$C_2H_2Cl_2$	1,1-Dichloroethylene	8.22	28.9	80.0					
$C_2H_2Cl_2$	*cis*-1,2-Dichloroethylene			26.8	70.6	159	320		
$C_2H_2Cl_2$	*trans*-1,2-Dichloroethylene	3.73	14.7	44.2	110	234			
$C_2H_2Cl_2F_2$	1,2-Dichloro-1,1-difluoroethane		14.3	44.8	114	247	477	842	1383
$C_2H_2Cl_2O$	Chloroacetyl chloride	0.088	0.676	3.33	12.0	34.6	83.8		
$C_2H_2Cl_4$	1,1,1,2-Tetrachloroethane			1.60	5.83	16.9	41.2	87.9	169
$C_2H_2Cl_4$	1,1,2,2-Tetrachloroethane			0.622	2.78	9.25	24.9	57.0	
C_2H_3Br	Bromoethylene	16.9	54.5	141	307				
C_2H_3BrO	Acetyl bromide			16.2	39.7				
$C_2H_3Br_3$	1,1,2-Tribromoethane						4.90	13.6	32.5
C_2H_3Cl	Chloroethylene	62.6	170	355					

VAPOR PRESSURE IN THE TEMPERATURE RANGE –25°C TO 150°C (continued)

Molecular formula	Name	Vapor pressure in kPa							
		–25°C	0°C	25°C	50°C	75°C	100°C	125°C	150°C
$C_2H_3ClF_2$	1-Chloro-1,1-difluoroethane	52.9	147	351	711	1298	2192	3512	
C_2H_3ClO	Acetyl chloride		12.1	38.4	98.8				
$C_2H_3ClO_2$	Chloroacetic acid						3.31	11.0	29.4
$C_2H_3Cl_2F$	1,1-Dichloro-1-fluoroethane	8.80	30.1	80.3	179				
$C_2H_3Cl_2F$	1,2-Dichloro-1-fluoroethane			16.8	46.1	105			
$C_2H_3Cl_3$	1,1,1-Trichloroethane		4.80	16.5	45.1	104	211		
$C_2H_3Cl_3$	1,1,2-Trichloroethane			3.10	10.1	28.2	66.6		
C_2H_3FO	Acetyl fluoride	36.6	69.1	117					
$C_2H_3F_3$	1,1,1-Trifluoroethane	264	627	1267					
$C_2H_3F_3O$	2,2,2-Trifluoroethanol			9.87	37.0				
C_2H_3I	Iodoethylene				83.7				
C_2H_3IO	Acetyl iodide		1.04	4.32					
C_2H_3N	Acetonitrile			11.9	33.9	82.1			
C_2H_3NO	Methylisocyanate	3.98	17.7	57.7	151				
C_2H_3NS	Methylthiocyanate		0.371	1.64	5.59	15.7	38.0	81.2	
C_2H_4BrCl	1-Bromo-2-chloroethane	0.177	1.03	4.24	13.6	36.2	83.0		
$C_2H_4Br_2$	1,1-Dibromoethane		0.738	3.40	11.8	33.1	78.9	166	
$C_2H_4Br_2$	1,2-Dibromoethane			1.55	5.79	16.9	40.7	85.5	
C_2H_4ClF	1-Chloro-1-fluoroethane	18.1	55.4	137					
$C_2H_4Cl_2$	1,1-Dichloroethane	2.25	9.55	30.5	79.2				
$C_2H_4Cl_2$	1,2-Dichloroethane	0.541	2.84	10.6	31.4	77.2	162		
$C_2H_4N_2O_6$	Ethylene glycol dinitrate			0.009	0.092				
C_2H_4O	Ethylene oxide	19.8	65.8						
C_2H_4O	Acetaldehyde	12.8	44.3	120					
$C_2H_4O_2$	Acetic acid			2.07	7.62	22.7	57.0	125	
$C_2H_4O_2$	Methyl formate	6.63	26.0	78.1	193	410	779	1351	
$C_2H_4O_3$	Peroxyacetic acid		0.376	1.93	7.67	25.0	69.6		
$C_2H_4O_3$	Glycolic acid					30.3	101		
$C_2H_5AsF_2$	Ethyldifluoroarsine	0.256	1.47	6.08	19.8	53.4	125		
C_2H_5Br	Bromoethane	5.86	21.7	62.5	149	308	571	970	
C_2H_5Cl	Chloroethane	19.5	62.3	160					
C_2H_5ClO	2-Chloroethanol				4.45	14.2	38.6	92.7	
C_2H_5ClO	Chloromethyl methyl ether			24.9	67.7				
$C_2H_5Cl_3OSi$	Trichloroethoxysilane	0.242	1.33	5.25	16.3	42.0	94.1		
$C_2H_5Cl_3Si$	Trichloroethylsilane			6.29	19.0	47.7	104		
C_2H_5F	Fluoroethane	172	425	890					
C_2H_5FO	2-Fluoroethanol		0.563	2.87	11.4				
C_2H_5I	Iodoethane	1.22	5.46	18.2	48.5	110			
C_2H_5N	Ethyleneimine		7.90	28.9					
C_2H_5NO	*N*-Methylformamide						3.04	8.94	22.6
$C_2H_5NO_2$	Nitroethane			2.79	9.95	28.1	66.6		
$C_2H_5NO_3$	Ethyl nitrate		2.17	8.56	26.3				

VAPOR PRESSURE IN THE TEMPERATURE RANGE –25°C TO 150°C (continued)

Molecular formula	Name	Vapor pressure in kPa							
		–25°C	0°C	25°C	50°C	75°C	100°C	125°C	150°C
$C_2H_6Cl_2Si$	Dichlorodimethylsilane		5.73	18.9	50.7	116			
C_2H_6Hg	Dimethyl mercury		2.27	8.30	24.1	58.9			
$C_2H_6N_2O$	*N*-Nitrosodimethylamine				2.65	8.01	20.9	48.1	101
C_2H_6O	Dimethyl ether*	101	273						
C_2H_6O	Ethanol		1.50	7.87	29.5	88.8	224	495	976
C_2H_6OS	Dimethyl sulfoxide			0.084	0.431	1.67	5.27	14.1	33.0
$C_2H_6O_2$	Ethylene glycol			0.010	0.092	0.512	2.14	7.18	20.2
$C_2H_6O_2$	Ethyl hydroperoxide	0.096	0.634	3.06	11.6	36.3			
$C_2H_6O_2S$	Dimethyl sulfone							2.04	5.51
C_2H_6S	Dimethyl sulfide		22.3	64.4					
C_2H_6S	Ethanethiol		24.5	70.3	167				
$C_2H_6S_2$	Dimethyl disulfide		0.893	3.82	12.5	33.2	76.0		
$C_2H_7BO_2$	Dimethoxyborane	10.2	36.4						
C_2H_7N	Dimethylamine	21.1	75.0	203					
C_2H_7NO	Ethanolamine			0.050		1.72	6.56	20.0	51.2
$C_2H_8N_2$	1,1-Dimethylhydrazine	1.05	5.53						
$C_2H_8N_2$	1,2-Ethanediamine			1.62	6.77	21.8	57.4		
C_2N_2	Cyanogen	84.6							
C_3Cl_6	1,1,2,3,3,3-Hexachloro-1-propene						2.73	7.51	17.9
C_3F_6	Perfluoropropene	128	342	768					
C_3F_6O	Perfluoroacetone	113							
C_3HN	Cyanoacetylene	2.51	16.4	54.3	131				
$C_3H_2F_6O$	1,1,1,3,3,3-Hexafluoro-2-propanol			21.2	74.1				
C_3H_3N	Propenenitrile		3.97	14.1	39.7	93.9	194		
C_3H_3NS	Thiazole					23.6	57.5		
C_3H_4	Propyne	92.7	255	581	153	2059	3385	5209	
C_3H_4	Allene*	145							
$C_3H_4ClF_3$	3-Chloro-1,1,1-trifluoropropane			46.4	117				
$C_3H_4Cl_2O$	1,1-Dichloroacetone			3.59	10.9	27.7	60.9	120	
$C_3H_4Cl_2O_2$	Methyl dichloroacetate				2.63	8.86	24.6	58.8	124
$C_3H_4Cl_4$	1,1,1,2-Tetrachloropropane				2.68	8.45	22.1	50.1	101
$C_3H_4F_4O$	2,2,3,3-Tetrafluoro-1-propanol			1.78	8.14	27.9	77.0		
C_3H_4O	2-Propenal	2.86	11.7	36.2	91.3	198			
$C_3H_4O_2$	2-Propenoic acid				2.45	8.75	25.2	61.3	131
$C_3H_4O_2$	Vinyl formate	2.11	11.0	40.6	116	276			
$C_3H_4O_2$	2-Oxetanone				1.28	4.41	12.9	33.0	75.4
$C_3H_4O_3$	1,3-Dioxolan-2-one			0.003					
C_3H_5Br	*cis*-1-Bromopropene	2.24	9.49	30.2	78.0	172	337		

VAPOR PRESSURE IN THE TEMPERATURE RANGE –25°C TO 150°C (continued)

Molecular formula	Name	Vapor pressure in kPa							
		–25°C	0°C	25°C	50°C	75°C	100°C	125°C	150°C
C_3H_5Br	2-Bromopropene			43.0	107				
C_3H_5Br	3-Bromopropene			18.6	50.8				
C_3H_5Cl	*cis*-1-Chloropropene	7.62	27.3	76.5	178				
C_3H_5Cl	*trans*-1-Chloropropene	6.11	22.6	64.7	154	317			
C_3H_5Cl	2-Chloropropene	12.3	41.4	110	247				
C_3H_5Cl	3-Chloropropene	4.24	16.4	48.9	120				
C_3H_5ClO	Epichlorohydrin			2.20	8.41	24.5	60.8		
$C_3H_5ClO_2$	Methyl chloroacetate				4.36	14.4	38.5	88.3	
$C_3H_5Cl_3$	1,1,3-Trichloropropane				3.20	9.87	25.4	56.9	114
$C_3H_5Cl_3$	1,2,3-Trichloropropane			0.492		6.64	17.8	41.1	84.6
$C_3H_5Cl_3Si$	Trichloro-2-propenylsilane				8.71	25.2	60.6	127	
C_3H_5I	Allyl iodide						95.8		
C_3H_5N	Propanenitrile	0.276	1.66	6.14					
C_3H_5NO	Hydracrylonitrile			0.010	0.083	0.345	1.21	3.66	9.87
C_3H_5NO	Propenamide					0.014	0.060	0.211	
C_3H_5NS	Ethylthiocyanate					8.34	23.6	56.9	121
C_3H_5NS	Ethylisothiocyanate			1.52					
$C_3H_5N_3O_9$	Trinitroglycerol				0.001	0.010	0.043		
C_3H_6BrCl	1-Bromo-3-chloropropane				3.36	10.4	26.8	60.2	121
$C_3H_6Br_2$	1,2-Dibromopropane			1.07	2.96	9.86	27.1	64.1	135
$C_3H_6Br_2$	1,3-Dibromopropane					3.47	10.6	27.2	61.5
$C_3H_6Cl_2$	1,1-Dichloropropane		2.40	9.09	27.0	66.7	143		
$C_3H_6Cl_2$	1,2-Dichloropropane	0.299	1.68	6.62	20.3	51.4	113		
$C_3H_6Cl_2$	2,2-Dichloropropane		5.74	19.4	52.6	121	245		
$C_3H_6Cl_2O$	1,3-Dichloro-2-propanol			0.125	0.602	2.30	7.26	19.8	47.5
$C_3H_6N_2O_4$	1,1-Dinitropropane				0.231	1.11	4.10		
C_3H_6O	Acetone	2.19	9.35	30.8	82.0	186	372	676	1136
C_3H_6O	Allyl alcohol			3.14	13.3	42.8	113	254	
C_3H_6O	Propanal		13.5	42.2					
C_3H_6O	Methyloxirane	6.34	24.4	71.7					
C_3H_6O	Methyl vinyl ether	35.4	86.5	180					
$C_3H_6O_2$	Ethyl formate	2.06	9.58	32.3	87.0	197	393	708	1177
$C_3H_6O_2$	Methyl acetate		8.37	28.8	79.2	184	371		
$C_3H_6O_2$	Propanoic acid			0.553	2.21	7.98	23.7	59.8	132
$C_3H_6O_2$	1,3-Dioxolane		3.51	14.6	41.6	99.2			
$C_3H_6O_3$	1,3,5-Trioxane					25.5	64.0		
C_3H_6S	Thiacyclobutane		1.80	7.01	21.3	53.7	117		
C_3H_7Br	1-Bromopropane		5.56	18.6	50.1				
C_3H_7Br	2-Bromopropane		9.23	28.9	74.1				
C_3H_7Cl	1-Chloropropane		15.1	45.8					
C_3H_7Cl	2-Chloropropane		25.1	68.9					
C_3H_7ClO	2-Chloro-1-propanol				5.04	16.5	43.7	98.1	

VAPOR PRESSURE IN THE TEMPERATURE RANGE –25°C TO 150°C (continued)

Molecular formula	Name	Vapor pressure in kPa –25°C	0°C	25°C	50°C	75°C	100°C	125°C	150°C
C_3H_7F	1-Fluoropropane	37.9	112						
C_3H_7I	1-Iodopropane		1.48	5.75	17.4	43.6	94.5	183	
C_3H_7I	2-Iodopropane		2.65	9.36	26.7	64.4	137		
C_3H_7N	Allylamine		9.77	33.1					
C_3H_7NO	*N,N*-Dimethylformamide			0.439	1.99	6.82			
$C_3H_7NO_2$	1-Nitropropane			1.36		15.3	38.4	84.4	
$C_3H_7NO_2$	2-Nitropropane			2.30	8.20	23.2	55.4	116	
$C_3H_7NO_3$	Propyl nitrate		0.660	3.12	10.9				
C_3H_8	Propane*	203	472	939	1665	2702			
C_3H_8O	1-Propanol		0.445	2.76	12.2	40.9	113	265	545
C_3H_8O	2-Propanol		1.11	6.02	23.9	75.3	198	452	
$C_3H_8O_2$	1,2-Propanediol			0.020	0.175	0.829	3.15	10.00	27.5
$C_3H_8O_2$	1,3-Propanediol							3.35	10.0
$C_3H_8O_2$	Dimethoxymethane		16.7	53.1	133				
C_3H_8S	Ethyl methyl sulfide			21.3	57.5	131	265		
C_3H_8S	1-Propanethiol			20.6	55.5	127	256		
C_3H_8S	2-Propanethiol		11.9	36.9	93.1	201	386		
$C_3H_8S_2$	1,3-Propanedithiol						11.3	26.3	55.2
C_3H_9As	Trimethylarsine	3.45	13.0						
$C_3H_9BO_3$	Trimethylborate			17.2	53.6	124			
C_3H_9BS	Dimethyl(methylthio) borane	1.41	5.72	18.4					
C_3H_9N	Propylamine		12.4	42.1	112	251			
C_3H_9N	Isopropylamine			78.0	192				
C_3H_9N	Trimethylamine	30.5	90.7	215					
C_3H_9NO	1-Amino-2-propanol			0.165	0.829	3.29	10.8	30.6	76.6
$C_3H_9O_3P$	Trimethyl phosphite								295
$C_3H_9O_4P$	Trimethyl phosphate			0.110	0.505	1.82	5.41	13.8	31.3
C_3H_9P	Trimethylphosphine	5.82	21.0						
C_3H_9Sb	Trimethylstibine	0.925	3.95						
$C_3H_{10}N_2$	1,2-Propanediamine	0.030	0.266						
C_3N_2O	Carbonyl dicyanide	0.787	4.16	16.6					
C_4Cl_6	Hexachloro-1,3-butadiene					0.516	1.99	6.18	16.3
$C_4F_6O_3$	Trifluoroacetic acid anhydride		15.2	54.8					
C_4F_8	Perfluorocyclobutane	42.8	129						
C_4F_{10}	Perfluorobutane	36.8	111						
$C_4H_2Cl_2O_2$	*trans*-2-Butenedioyl dichloride			0.303	1.26	4.32	12.6	32.2	74.1
$C_4H_2Cl_2S$	2,5-Dichlorothiophene				3.29	8.11	17.7		
$C_4H_2O_3$	Maleic anhydride					1.07	3.32	8.99	21.7
C_4H_3ClS	2-Chlorothiophene				9.33	23.4	52.0		
C_4H_3IS	2-Iodothiophene				2.67	5.81	11.4		

VAPOR PRESSURE IN THE TEMPERATURE RANGE –25°C TO 150°C (continued)

Molecular formula	Name	Vapor pressure in kPa –25°C	0°C	25°C	50°C	75°C	100°C	125°C	150°C
C_4H_4	1-Buten-3-yne	25.8	81.9						
C_4H_4O	Furan	7.37	27.7	80.0	191	394	730		
$C_4H_4O_2$	Diketene			1.41	5.38	16.5	42.8		
$C_4H_4O_3$	Succinic anhydride							1.19	3.30
$C_4H_4O_4$	Fumaric acid							0.001	
C_4H_4S	Thiophene		2.85	10.6	31.1	75.8	161	307	
C_4H_5Cl	2-Chloro-1,3-butadiene		9.88	29.5	74.2				
C_4H_5ClO	2-Methyl-2-propenoyl chloride				18.2	47.3	106		
$C_4H_5Cl_3O_2$	Ethyl trichloroacetate			0.196	0.901	3.28	9.95	26.1	60.5
C_4H_5N	1*H*-Pyrrole			1.10		14.4	38.2	87.7	179
C_4H_5N	Methylacrylonitrile		2.13	8.26	24.9	62.2	135		
C_4H_5N	Allyl cyanide		0.552	2.47	8.48	23.8	56.9	121	233
$C_4H_5NO_2$	Methyl cyanoacetate			0.019			2.20	6.94	18.3
C_4H_5NS	4-Methylthiazole					0.001	0.005	0.025	
C_4H_5NS	Allyl isothiocyanate			0.677					
C_4H_6	1,2-Butadiene	21.2	66.3	167					
C_4H_6	1,3-Butadiene*	42.2	120						
C_4H_6	1-Butyne	22.6	73.1						
C_4H_6	2-Butyne	9.39	33.7	94.3					
C_4H_6O	*trans*-2-Butenal			4.92	15.8	41.3	93.2		
C_4H_6O	Divinyl ether	8.45	31.2	89.3	211				
C_4H_6O	Cyclobutanone		1.31	5.50					
C_4H_6O	3-Buten-2-one			12.0	34.3	82.6			
$C_4H_6O_2$	γ-Butyrolactone			0.430				8.30	19.7
$C_4H_6O_2$	*cis*-Crotonic acid				0.435	2.04	7.36	21.6	54.3
$C_4H_6O_2$	*trans*-Crotonic acid					1.04	3.88	12.0	31.6
$C_4H_6O_2$	Methacrylic acid				0.703	2.98	10.1	28.3	69.1
$C_4H_6O_2$	Methyl acrylate			11.0	34.2	85.1			
$C_4H_6O_2$	Vinyl acetate			15.4	45.0				
$C_4H_6O_2$	2,3-Butanedione		1.80	7.45	24.8	69.4			
$C_4H_6O_2$	3-Butenoic acid					2.25	7.80	22.4	55.2
$C_4H_6O_3$	Acetic anhydride			0.680		9.98	27.4	64.4	
$C_4H_6O_3$	4-Methyl-1,3-dioxolan-2-one				0.131	0.324	0.710		
$C_4H_6O_4$	Dimethyl oxalate			0.213	0.974	3.55	10.8	28.5	66.8
C_4H_7Br	*trans*-1-Bromo-1-butene	0.406	1.98	7.27	21.5	53.8	118		
C_4H_7Br	2-Bromo-1-butene		3.40	12.3	35.0	83.9	176	331	
C_4H_7Br	*cis*-2-Bromo-2-butene	0.417	2.06	7.57	22.3	55.5	121		
C_4H_7Br	*trans*-2-Bromo-2-butene	0.647	3.05	10.7	30.3	72.7			
$C_4H_7Br_3$	1,2,3-Tribromobutane							4.80	12.6
$C_4H_7Br_3$	1,2,4-Tribromobutane							5.69	14.7

VAPOR PRESSURE IN THE TEMPERATURE RANGE –25°C TO 150°C (continued)

Molecular formula	Name	Vapor pressure in kPa –25°C	0°C	25°C	50°C	75°C	100°C	125°C	150°C
C_4H_7Cl	3-Chloro-1-butene				64.6				
C_4H_7Cl	*cis*-2-Chloro-2-butene				59.0	129			
C_4H_7Cl	*trans*-2-Chloro-2-butene				67.6				
C_4H_7Cl	3-Chloro-2-methylpropene			16.9	47.4	112			
$C_4H_7ClO_2$	Ethyl chloroacetate		0.121	0.640	2.57	8.37	23.1	55.6	120
C_4H_7N	Butanenitrile			2.55		25.1	59.5	125	
C_4H_7N	2-Methylpropanenitrile						90.9		
C_4H_8	1-Butene	46.4	128	296	593	1074	1796	2816	
C_4H_8	*cis*-2-Butene	29.2	87.7	214	447	835	1431	2292	3485
C_4H_8	*trans*-2-Butene	33.6	98.1	234	483	896	1528	2444	3719
C_4H_8	Isobutene	47.6	131	300	602	1091	1829	2877	
C_4H_8	Cyclobutane	20.2	62.6	157	336	642	1123	1833	2833
C_4H_8	Methylcyclopropane	28.3	85.0						
$C_4H_8Br_2$	1,2-Dibromobutane		0.097	0.456	1.68	5.11	13.4	31.0	65.0
$C_4H_8Br_2$	1,4-Dibromobutane						3.66	10.4	25.5
$C_4H_8Cl_2$	1,1-Dichlorobutane			3.04	10.3	28.4	66.7	138	
$C_4H_8Cl_2$	1,2-Dichlorobutane	0.130	0.710	2.82	8.80	23.0	52.0	105	
$C_4H_8Cl_2$	1,4-Dichlorobutane					7.38	19.6	44.8	91.5
$C_4H_8Cl_2$	2,2-Dichlorobutane			5.45	16.9	43.0	94.2		
$C_4H_8Cl_2O$	Bis(2-chloroethyl) ether			0.143	0.681	2.51	7.58	19.6	44.8
C_4H_8O	Butanal			15.7	43.2	102			
C_4H_8O	Isobutanal		6.66	23.0	62.5				
C_4H_8O	2-Butanone		3.51	12.6	35.6	87.4	184		
C_4H_8O	Tetrahydrofuran			21.6	58.6	135	272		
C_4H_8O	Ethyl vinyl ether	6.23	23.1	68.8					
C_4H_8O	1,2-Epoxybutane	4.22	12.5	31.7	70.5	142			
$C_4H_8O_2$	1,3-Dioxane		1.16						
$C_4H_8O_2$	1,4-Dioxane			4.95	15.8	42.2	97.2	200	
$C_4H_8O_2$	Butanoic acid			0.221	0.895	2.81	9.39	26.8	65.5
$C_4H_8O_2$	2-Methylpropanoic acid					4.04	16.5	44.0	90.3
$C_4H_8O_2$	Propyl formate		2.76	10.9	33.4	83.8	180		
$C_4H_8O_2$	Isopropyl formate			18.2	53.1	127			
$C_4H_8O_2$	Ethyl acetate		3.25	12.6	37.9	94.5	204		
$C_4H_8O_2$	Methyl propanoate		2.86	11.5	35.1	87.8	188		
$C_4H_8O_2S$	Sulfolane								1.83
C_4H_8S	Tetrahydrothiophene			2.45	8.32	23.1	54.5	113	214
C_4H_9Br	1-Bromobutane	0.232	1.41	5.26					
C_4H_9Br	1-Bromo-2-methylpropane		2.58	9.20	26.0	61.7	128	239	
C_4H_9Br	2-Bromobutane		2.61	9.32	26.3	62.3	129	241	
C_4H_9Br	2-Bromo-2-methylpropane		5.58	17.7	46.8				
C_4H_9Cl	1-Chlorobutane		3.82	13.7	38.6	91.1	188		
C_4H_9Cl	1-Chloro-2-methylpropane	1.30	5.90	19.9	53.5	122			

Molecular formula	Name	Vapor pressure in kPa –25°C	0°C	25°C	50°C	75°C	100°C	125°C	150°C
C_4H_9Cl	2-Chlorobutane		6.51	21.2	55.6				
C_4H_9Cl	2-Chloro-2-methylpropane			42.7	99.2				
$C_4H_9Cl_3Si$	Butyltrichlorosilane					9.17	23.0	51.4	105
C_4H_9F	1-Fluorobutane	7.54	27.3	77.1	181				
C_4H_9F	2-Fluorobutane	11.0	37.6	101					
C_4H_9I	1-Iodobutane			1.85	6.37	17.7	42.0	87.4	165
C_4H_9I	1-Iodo-2-methylpropane	0.075	0.468	2.10	7.36	21.3	52.8	116	
C_4H_9I	2-Iodobutane						57.8	115	
C_4H_9I	2-Iodo-2-methylpropane	0.287							
C_4H_9N	Pyrrolidine			8.40	26.7	69.0	153		
C_4H_9NO	*N,N*-Dimethylacetamide			0.075	0.639	3.16	10.9	29.1	65.0
C_4H_9NO	Morpholine			1.34		15.7	40.4	90.3	
C_4H_9NO	Methylpropylamine				0.271	0.708			
C_4H_9NO	2-Butanone oxime			0.396	1.94	7.23	21.7	55.5	
$C_4H_9NO_3$	Isobutyl nitrate		0.381	1.78	6.53				
C_4H_{10}	Butane	36.9	103	242					
C_4H_{10}	Isobutane	58.7	156	348	677	1190	1928	2929	
$C_4H_{10}O$	1-Butanol		0.103	0.860	4.52	17.2	51.9	130	283
$C_4H_{10}O$	2-Butanol			2.32	11.1	37.2	103	240	495
$C_4H_{10}O$	2-Methyl-2-propanol			5.52	23.3	75.2	191		
$C_4H_{10}O$	2-Methyl-1-propanol			1.39	7.10	26.1	75.4	184	
$C_4H_{10}O$	Diethyl ether	6.61	24.9	71.7	171				
$C_4H_{10}O$	Methyl propyl ether		19.4	60.9	142	274			
$C_4H_{10}O$	Isopropyl methyl ether		29.2	82.1	192				
$C_4H_{10}O_2$	1,3-Butanediol			0.008			1.37	4.59	13.4
$C_4H_{10}O_2$	1,4-Butanediol							1.61	5.30
$C_4H_{10}O_2$	2,3-Butanediol						3.59	11.9	33.6
$C_4H_{10}O_2$	1,2-Dimethoxyethane		2.62	9.93	29.4	72.3	155		
$C_4H_{10}O_2$	2-Ethoxyethanol			0.710		10.7	29.8	72.1	
$C_4H_{10}O_2$	Dimethylacetal		6.62	22.9	61.9				
$C_4H_{10}O_2$	Diethylperoxide	2.33	8.46	24.7	61.2				
$C_4H_{10}O_2S$	Bis(2-hydroxyethyl) sulfide						0.726	1.26	2.04
$C_4H_{10}O_3$	Diethylene glycol			0.001			0.273	0.935	3.72
$C_4H_{10}O_4S$	Diethyl sulfate								17.8
$C_4H_{10}S$	1-Butanethiol		1.51	6.07	18.8	48.1	106		
$C_4H_{10}S$	2-Butanethiol		2.97	10.8	30.9	74.3	156	294	
$C_4H_{10}S$	2-Methyl-1-propanethiol		2.49	9.30	27.2	66.5	141	269	
$C_4H_{10}S$	2-Methyl-2-propanethiol		7.45	24.2	63.2	141	276		
$C_4H_{10}S$	Diethyl sulfide			7.78	23.5	58.9	128		
$C_4H_{10}S$	Methyl propyl sulfide		1.71	6.78	20.8	52.7	116		
$C_4H_{10}S$	Isopropyl methyl sulfide		2.94	10.8	31.0	74.7	157		
$C_4H_{10}S_2$	Diethyl disulfide		0.097	0.560	2.30	7.36	19.5	44.7	91.3

VAPOR PRESSURE IN THE TEMPERATURE RANGE –25°C TO 150°C (continued)

Molecular formula	Name	Vapor pressure in kPa							
		–25°C	0°C	25°C	50°C	75°C	100°C	125°C	150°C
$C_4H_{11}N$	Butylamine		3.17	12.2	36.8	92.1	200		
$C_4H_{11}N$	Isobutylamine	0.961	5.14	19.0	54.2	128			
$C_4H_{11}N$	*sec*-Butylamine		6.62	23.0	63.9	150	308		
$C_4H_{11}N$	*tert*-Butylamine			48.4	123				
$C_4H_{11}N$	Diethylamine			30.1	83.6				
$C_4H_{11}NO$	2-(Dimethylamino)ethanol					12.0	32.7		
$C_4H_{12}BN$	(Dimethylamino)dimethylborane	1.60	6.98	23.0	61.5				
$C_4H_{12}Cl_2OSi_2$	1,3-Dichloro-1,1,3,3-tetramethyl-disiloxane			1.07	4.18	12.8	32.6	72.1	
$C_4H_{12}O_4Si$	Tetramethyl silicate		0.408	1.84	6.55	19.5	50.0		
$C_4H_{12}Si$	Tetramethylsilane			94.2	215	429	769	1271	1964
$C_4H_{12}Sn$	Tetramethylstannane	1.04	4.43	14.6	39.5	92.1			
$C_4H_{13}N_3$	Bis(2-amimoethyl)amine			0.030			2.74	8.29	
C_5F_{12}	Perfluoropentane	7.87	29.8	87.0					
C_5H_4ClN	2-Chloropyridine			0.311	1.24	4.03	11.1	26.7	57.7
$C_5H_4O_2$	Furfural			0.290		4.57	13.7	34.1	73.4
C_5H_5N	Pyridine			2.76	9.57	26.8	63.7	133	252
C_5H_6	1,3-Cyclopentadiene			58.5	140				
$C_5H_6N_2$	Pentanedinitrile			0.001					
C_5H_6O	2-Methylfuran		6.74	23.0	61.8	138	282		
$C_5H_6O_2$	Furfuryl alcohol			0.097	0.484	1.95	6.60	19.4	50.4
C_5H_6S	2-Methylthiophene			3.45	11.0	29.4	69.0		
C_5H_6S	3-Methylthiophene			2.98	9.78	26.7	63.1		
$C_5H_7NO_2$	Ethyl cyanoacetate					0.177	0.885	3.51	11.6
C_5H_8	Spiropentane		21.1	61.0	146	303			
C_5H_8	Cyclopentene	4.59	17.4	50.7	122				
C_5H_8	1,2-Pentadiene	2.79	11.4	35.4	89.4				
C_5H_8	1,4-Pentadiene	10.7	36.5	97.9					
C_5H_8	2,3-Pentadiene	3.41	13.9	42.7	108				
C_5H_8	3-Methyl-1,2-butadiene	5.09	19.4	56.8	138				
C_5H_8	2-Methyl-1,3-butadiene		26.4	73.4					
C_5H_8	1-Pentyne			58.1	140				
C_5H_8	2-Pentyne				82.7				
C_5H_8	3-Methyl-1-butyne			87.7					
C_5H_8O	Cyclopentanone			1.55	5.82	17.0	41.4	87.5	
C_5H_8O	Cyclopropyl methyl ketone						68.9		
C_5H_8O	3-Methyl-3-buten-2-one				18.6	48.5	109		
C_5H_8O	3,4-Dihydro-2H-pyran		3.48	11.4	31.1	73.6			
$C_5H_8O_2$	2,4-Pentanedione			1.02	4.35	13.6	33.9	72.1	
$C_5H_8O_2$	Vinyl propanoate		2.14						
$C_5H_8O_2$	Ethyl acrylate	0.193	1.20	5.14	16.8	44.8	103		
$C_5H_8O_2$	Methyl methacrylate			5.10	16.0	43.4	100		

VAPOR PRESSURE IN THE TEMPERATURE RANGE –25°C TO 150°C (continued)

Molecular formula	Name	Vapor pressure in kPa –25°C	0°C	25°C	50°C	75°C	100°C	125°C	150°C
$C_5H_8O_2$	4-Pentenoic acid						3.57	11.4	30.1
$C_5H_8O_2$	Tetrahydro-2H-pyran-2-one				0.260	1.03	3.30	8.87	20.8
$C_5H_8O_3$	Methyl acetoacetate			0.241	0.995	3.35	9.57	24.0	53.9
$C_5H_8O_4$	Glutaric acid								0.081
$C_5H_8O_4$	Dimethyl malonate			0.069	0.376	1.56	5.28	15.2	38.0
$C_5H_9ClO_2$	Isopropyl chloroacetate		0.118	0.569	2.19	7.02	19.5	47.7	106
$C_5H_9ClO_2$	Ethyl 2-chloropropanoate		0.090	0.504	2.13	7.20	20.5	50.9	113
C_5H_9N	Pentanenitrile			0.943		11.2	28.7	63.8	
C_5H_9N	3-Methylbutanenitrile					9.93	22.7	46.8	88.4
C_5H_9N	2,2-Dimethylpropanenitrile				14.8	38.7	86.9		
C_5H_9NO	*N*-Methyl-2-pyrrolidone			0.040			3.28	8.88	21.4
C_5H_{10}	1-Pentene		31.4	85.0	193	385	695	1162	1829
C_5H_{10}	*cis*-2-Pentene		23.2	66.0	156	318	586	994	1580
C_5H_{10}	*trans*-2-Pentene		23.9	67.4	159	324	597	1012	1607
C_5H_{10}	2-Methyl-1-butene	8.41	29.6	81.4	187	375	679	1134	1779
C_5H_{10}	2-Methyl-2-butene	6.01	21.9	62.1	147	305	568	973	1566
C_5H_{10}	3-Methyl-1-butene	14.7	47.0	120	262				
C_5H_{10}	*cis*-1,2-Dimethylcyclopropane			67.5					
C_5H_{10}	*trans*-1,2-Dimethylcyclopropane			90.9					
C_5H_{10}	Ethylcyclopropane			69.5					
C_5H_{10}	Cyclopentane	3.69	14.2	42.3	104	220	418	725	1175
$C_5H_{10}Br_2$	1,5-Dibromopentane							4.45	11.7
$C_5H_{10}Cl_2$	1,2-Dichloropentane				2.91	9.05	23.5	52.8	106
$C_5H_{10}Cl_2$	1,5-Dichloropentane					2.93	8.54	21.2	46.1
$C_5H_{10}N_2$	3-(Dimethylamino)propanenitrile				0.944	3.22	9.31	23.6	53.5
$C_5H_{10}O$	Cyclopentanol		0.035	0.294	1.65				
$C_5H_{10}O$	2-Pentanone		1.25	4.97	15.7	41.4	94.6	193	360
$C_5H_{10}O$	3-Pentanone		1.18	4.72	15.6	41.7	95.5	195	363
$C_5H_{10}O$	3-Methyl-2-butanone		1.62	6.99	22.0	55.5	119		
$C_5H_{10}O$	Tetrahydropyran		2.61	9.54	27.5				
$C_5H_{10}O$	2-Methyltetrahydrofuran		3.53	12.6	36.0	86.3			
$C_5H_{10}O$	Pentanal			4.58	15.1	40.4	92.5	188	
$C_5H_{10}O$	Allyl ethyl ether	1.30	6.00	20.4	55.4	128	259	474	
$C_5H_{10}O_2$	Butyl formate			3.53	15.0	41.1	86.8		
$C_5H_{10}O_2$	Isobutyl formate			5.34	17.7	47.3	107		
$C_5H_{10}O_2$	Propyl acetate			4.49	15.2	41.6	96.6	198	
$C_5H_{10}O_2$	Ethyl propanoate			4.97	16.6	45.1	104		
$C_5H_{10}O_2$	Methyl butanoate	0.135	0.939	4.30	14.6	40.1	93.2	191	354
$C_5H_{10}O_2$	Methyl isobutanoate						127	250	453
$C_5H_{10}O_2$	Pentanoic acid		0.002	0.024	0.172	0.924	3.86	12.3	32.5
$C_5H_{10}O_2$	3-Methylbutanoic acid		0.006	0.067	0.414	1.77	6.23	18.0	44.4
$C_5H_{10}O_2$	Tetrahydrofurfuryl alcohol			0.100				20.1	45.8

VAPOR PRESSURE IN THE TEMPERATURE RANGE –25°C TO 150°C (continued)

Molecular formula	Name	Vapor pressure in kPa –25°C	0°C	25°C	50°C	75°C	100°C	125°C	150°C
$C_5H_{10}O_3$	Diethyl carbonate			1.63	5.91	17.5	44.0	97.3	
$C_5H_{10}O_3$	2-Methoxyethyl acetate			0.670		8.25	23.0	55.3	119
$C_5H_{10}S$	Cyclopentanethiol					15.9	38.9	83.2	160
$C_5H_{10}S$	Thiacyclohexane				3.94	11.7	29.3	63.9	125
$C_5H_{11}Br$	1-Bromopentane			1.68	6.01	17.3	41.9	89.3	171
$C_5H_{11}Br$	2-Bromopentane			3.10	10.1	26.9	61.4	124	228
$C_5H_{11}Br$	3-Bromopentane			2.94	9.65	25.9	59.3	120	222
$C_5H_{11}Br$	1-Bromo-3-methylbutane			2.72	9.01	24.3	56.2	115	212
$C_5H_{11}Cl$	1-Chloropentane			4.36	13.8	35.7	79.6	158	
$C_5H_{11}Cl$	2-Chloropentane			6.79	20.5	51.5	112		
$C_5H_{11}Cl$	3-Chloropentane			6.23	19.2	49.1	108	212	
$C_5H_{11}Cl$	2-Chloro-2-methylbutane		3.41	11.5	31.5	73.5	151		
$C_5H_{11}Cl$	1-Chloro-2,2-dimethylpropane		2.92	10.8	31.2	75.6	160	303	
$C_5H_{11}F$	1-Fluoropentane	1.67	7.42	24.6	65.5	148			
$C_5H_{11}I$	1-Iodopentane				2.32	7.24	18.8	42.2	84.8
$C_5H_{11}I$	1-Iodo-3-methylbutane		0.162	0.805	3.02	9.13	23.4	52.5	106
$C_5H_{11}N$	*N*-Methylpyrrolidine		3.89	13.5	38.2				
$C_5H_{11}N$	Piperidine			4.28	13.5	36.6	84.1	171	
$C_5H_{11}N$	Cyclopentylamine				12.1	33.4			
$C_5H_{11}NO_3$	Isopentyl nitrate			0.530	2.24	7.49	20.9	50.6	
C_5H_{12}	Pentane		24.5	68.3	159				
C_5H_{12}	Isopentane		34.6	91.7			723		
C_5H_{12}	Neopentane		71.0	171					
$C_5H_{12}O$	Butyl methyl ether			18.5	50.9				
$C_5H_{12}O$	*tert*-Butyl methyl ether			33.6	85.3	186			
$C_5H_{12}O$	Ethyl propyl ether			24.2	64.7				
$C_5H_{12}O$	1-Pentanol			0.259	1.66	7.25	24.0	64.6	149
$C_5H_{12}O$	2-Pentanol			0.804	4.35	16.7	49.9	124	
$C_5H_{12}O$	3-Pentanol			1.10	5.47	19.9	57.7	141	
$C_5H_{12}O$	2-Methyl-1-butanol			0.416	2.59	10.9	34.5	89.4	198
$C_5H_{12}O$	2-Methyl-2-butanol			2.19	10.1	34.3	94.0		
$C_5H_{12}O$	3-Methyl-1-butanol			0.315	2.17	9.65	31.7	83.7	188
$C_5H_{12}O$	3-Methyl-2-butanol			1.20	6.32	23.3	66.8	159	
$C_5H_{12}O$	2,2-Dimethyl-1-propanol				5.77	20.6	59.8	148	
$C_5H_{12}O_2$	1,5-Pentanediol								3.36
$C_5H_{12}O_2$	Diethoxymethane		2.36	8.97	26.7	66.5			
$C_5H_{12}O_2$	2-Propoxyethanol					6.22	18.2	45.9	102
$C_5H_{12}O_3$	2-(2-Methoxyethoxy)ethanol			0.024			3.65	10.4	26.3
$C_5H_{12}S$	Butyl methyl sulfide					20.9	50.2	106	201
$C_5H_{12}S$	Ethyl propyl sulfide			2.58	8.88	24.7	58.4	121	
$C_5H_{12}S$	Ethyl isopropyl sulfide			4.26	13.7	36.1	81.6	164	
$C_5H_{12}S$	1-Pentanethiol						45.6	96.8	185

VAPOR PRESSURE IN THE TEMPERATURE RANGE –25°C TO 150°C (continued)

Molecular formula	Name	Vapor pressure in kPa							
		–25°C	0°C	25°C	50°C	75°C	100°C	125°C	150°C
$C_5H_{12}S$	2-Pentanethiol			3.55	11.6	30.9	70.5	143	
$C_5H_{12}S$	3-Pentanethiol			1.83	6.57	18.8	45.6	96.8	185
$C_5H_{12}S$	3-Methyl-1-butanethiol			2.71	9.19	25.2	58.9	122	227
$C_5H_{12}S$	2-Methyl-2-butanethiol		1.65	6.39	19.2	48.0	104	201	355
$C_5H_{13}N$	Pentylamine			4.00	13.6	37.5	88.3	184	
C_6BrF_5	Bromopentafluorobenzene								145
C_6ClF_5	Chloropentafluorobenzene			2.37	8.44	24.1	58.3	124	
$C_6Cl_3F_3$	1,3,5-Trichloro-2,4,6-trifluoro-benzene						4.21	11.3	26.4
C_6F_6	Hexafluorobenzene			11.3	34.1	85.3	184		
C_6F_{12}	Perfluorocyclohexane					222	425	747	1225
C_6F_{14}	Perfluorohexane		8.57	29.5	79.1				
C_6F_{14}	Perfluoro-2-methylpentane	1.81	8.57	29.1	78.1				
C_6HF_5	Pentafluorobenzene				28.0	71.2	156		
C_6HF_5O	Pentafluorophenol						21.5	53.3	115
$C_6H_2F_4$	1,2,3,4-Tetrafluorobenzene			6.50	20.6	53.7	120		
$C_6H_2F_4$	1,2,3,5-Tetrafluorobenzene			9.80	29.7	74.5	162		
$C_6H_2F_4$	1,2,4,5-Tetrafluorobenzene			7.56	23.7	61.3	136		
$C_6H_3Cl_3O$	2,4,6-Trichlorophenol						0.495	1.67	4.79
$C_6H_3F_3$	1,3,5-Trifluorobenzene			13.8	39.5				
$C_6H_4Br_2$	*m*-Dibromobenzene								15.6
$C_6H_4ClNO_2$	*p*-Chloronitrobenzene			0.003				3.27	8.16
$C_6H_4Cl_2$	*o*-Dichlorobenzene			0.180			8.39	20.7	45.0
$C_6H_4Cl_2$	*m*-Dichlorobenzene			0.252			10.9	26.1	55.3
$C_6H_4Cl_2$	*p*-Dichlorobenzene			0.235			10.4	25.1	53.5
$C_6H_4O_2$	1,4-Benzoquinone			0.020	0.169	1.05	5.13	16.7	
$C_6H_5AsCl_2$	Dichlorophenylarsine		0.001	0.005	0.028				
C_6H_5Br	Bromobenzene			0.556	2.27	7.20	18.9	43.0	87.0
C_6H_5Cl	Chlorobenzene			1.60		16.3	39.5	84.3	
C_6H_5ClO	*o*-Chlorophenol			0.308	1.24	4.00	10.8	25.4	53.5
C_6H_5ClO	*m*-Chlorophenol				0.192	0.770	2.51	6.92	16.7
C_6H_5ClO	*p*-Chlorophenol				0.136	0.561	1.88	5.33	13.2
$C_6H_5Cl_3Si$	Trichlorophenylsilane				0.317	1.28	4.10	10.9	25.1
C_6H_5F	Fluorobenzene		2.84	10.4	30.3	74.2	156		
C_6H_5I	Iodobenzene			0.133	0.632	2.28	6.67	16.7	36.7
$C_6H_5NO_2$	Nitrobenzene			0.030					18.8
$C_6H_5NO_3$	*p*-Nitrophenol					0.001			
C_6H_6	Benzene	0.485	3.29	12.7	36.2	86.4	180	338	583
C_6H_6	1,5-Hexadien-3-yne	0.519	2.81	10.6	31.3	76.3			
C_6H_6ClN	*o*-Chloroaniline			0.034				7.82	19.3
C_6H_6ClN	*m*-Chloroaniline					0.203	1.40		
$C_6H_6N_2O_2$	*p*-Nitroaniline						0.003		

VAPOR PRESSURE IN THE TEMPERATURE RANGE –25°C TO 150°C (continued)

Molecular formula	Name	Vapor pressure in kPa –25°C	0°C	25°C	50°C	75°C	100°C	125°C	150°C
C_6H_6O	Phenol			0.055		1.53	5.47	15.8	38.8
$C_6H_6O_3$	1,2,3-Benzenetriol								0.576
C_6H_7N	Aniline			0.090			6.10	16.4	38.2
C_6H_7N	2-Methylpyridine			1.50	5.58	16.5	41.1	89.4	
C_6H_7N	3-Methylpyridine			0.795	3.18	9.96	26.0	58.7	
C_6H_7N	4-Methylpyridine			0.759	3.04	9.56	25.0	56.7	
C_6H_8	*cis*-1,3,5-Hexatriene			12.0	34.0				
C_6H_8	1,3-Cyclohexadiene			13.0	36.5	85.9	177		
C_6H_8	1,4-Cyclohexadiene			9.01	26.2				
$C_6H_8N_2$	Adiponitrile							0.351	1.06
$C_6H_8N_2$	Phenylhydrazine								5.97
$C_6H_8N_2$	*m*-Phenylenediamine						0.137	0.496	1.52
$C_6H_8O_4$	Dimethyl maleate							9.14	23.1
C_6H_8S	2,5-Dimethylthiophene				4.67	13.5	33.9		
C_6H_{10}	Cyclohexene			11.8	33.4	79.2	165		
C_6H_{10}	*trans*-1,3-Hexadiene			17.3	47.8				
C_6H_{10}	*trans*-1,4-Hexadiene			23.3	61.1				
C_6H_{10}	1,5-Hexadiene		9.57	29.7	74.4				
C_6H_{10}	*cis,cis*-2,4-Hexadiene					87.3			
C_6H_{10}	*trans,trans*-2,4-Hexadiene					87.3			
C_6H_{10}	*trans,cis*-2,4-Hexadiene					87.3			
C_6H_{10}	*trans*-2-Methyl-1,3-pentadiene					98.4			
C_6H_{10}	2,3-Dimethyl-1,3-butadiene		6.13	20.2	54.0	124			
C_6H_{10}	1-Hexyne				48.8	114			
C_6H_{10}	2-Hexyne		2.89	10.7	31.1	75.2	158		
C_6H_{10}	4-Methyl-1-pentyne		8.10	26.5	69.5				
C_6H_{10}	4-Methyl-2-pentyne		4.89	16.9	46.4	107			
$C_6H_{10}Cl_2$	1,1-Dichlorocyclohexane					4.73	12.7	29.3	59.8
$C_6H_{10}Cl_2$	*cis*-1,2-Dichlorocyclohexane						3.86	9.76	22.0
$C_6H_{10}O$	Cyclohexanone			0.530			18.6	42.7	87.4
$C_6H_{10}O$	4-Methyl-4-penten-2-one			1.97	7.26	21.2	52.2	112	
$C_6H_{10}O$	Mesityl oxide			1.47	5.40	16.2	40.5	88.5	
$C_6H_{10}O_2$	Ethyl methacrylate		0.640	2.62	8.68	24.2	58.8	128	
$C_6H_{10}O_2$	Vinyl butanoate						62.3		
$C_6H_{10}O_2$	Allyl glycidyl ether				1.77	6.24	17.8	43.2	92.5
$C_6H_{10}O_3$	Ethyl acetoacetate			0.095	0.494	1.95	6.26	17.0	40.5
$C_6H_{10}O_3$	Propanoic anhydride				2.44	8.87	25.1	59.1	121
$C_6H_{10}O_4$	Diethyl oxalate			0.030	0.146	0.769	3.12	10.3	29.0
$C_6H_{10}O_4$	Ethylene glycol diacetate			0.030	0.275	1.19	4.07	11.7	29.2
$C_6H_{10}S$	Diallylsulfide		0.251	1.21	4.39	12.8	31.9	69.4	
$C_6H_{11}Br$	Bromocyclohexane				17.7	45.2	101		
$C_6H_{11}Cl$	Chlorocyclohexane					11.5	28.7	62.5	122

VAPOR PRESSURE IN THE TEMPERATURE RANGE –25°C TO 150°C (continued)

Molecular formula	Name	Vapor pressure in kPa –25°C	0°C	25°C	50°C	75°C	100°C	125°C	150°C
$C_6H_{11}N$	Hexanenitrile			0.355			13.8	33.2	70.5
$C_6H_{11}N$	Isopentyl cyanide				3.82	9.93	22.7	46.8	88.4
C_6H_{12}	Cyclohexane			13.0	36.3	85.0	175	324	553
C_6H_{12}	Methylcyclopentane		5.50	18.3	49.1				
C_6H_{12}	Ethylcyclobutane					115			
C_6H_{12}	1,1,2-Trimethylcyclopropane				93.9				
C_6H_{12}	1-Ethyl-1-methylcyclopropane				81.7				
C_6H_{12}	Isopropylcyclopropane				77.7				
C_6H_{12}	1-Hexene		7.68	24.8	64.7		284		
C_6H_{12}	*cis*-2-Hexene	1.37	6.04	20.0	53.6	122			
C_6H_{12}	*trans*-2-Hexene	1.37	6.20	20.7	55.5	126			
C_6H_{12}	*cis*-3-Hexene	1.50	6.66	22.0	58.4	132			
C_6H_{12}	*trans*-3-Hexene	1.41	6.37	21.2	56.9	129			
C_6H_{12}	2-Methyl-1-pentene	1.88	8.10	26.0	67.7	150			
C_6H_{12}	3-Methyl-1-pentene	3.03	12.0	36.0	88.5	188			
C_6H_{12}	4-Methyl-1-pentene	3.03	12.0	36.1	89.3	191			
C_6H_{12}	2-Methyl-2-pentene	1.40	6.30	21.0	56.5	129			
C_6H_{12}	*cis*-3-Methyl-2-pentene	1.42	6.34	21.0	56.0	127			
C_6H_{12}	*trans*-3-Methyl-2-pentene		5.47	18.6	50.7	117			
C_6H_{12}	4-Methyl-*cis*-2-pentene	2.55	10.5	32.5	82.1	178			
C_6H_{12}	4-Methyl-*trans*-2-pentene	2.26	9.41	29.7	76.1	167			
C_6H_{12}	2,3-Dimethyl-1-butene	2.72	11.0	33.6	84.3	182			
C_6H_{12}	3,3-Dimethyl-1-butene	5.59	20.4	57.4	134				
C_6H_{12}	2-Ethyl-1-butene	1.65	7.16	23.4	61.8	139			
C_6H_{12}	2,3-Dimethyl-2-butene			16.7	46.0				
$C_6H_{12}Cl_2$	1,2-Dichlorohexane					3.78	10.8	26.2	56.1
$C_6H_{12}Cl_2O$	2,2'-Dichlorodiisopropyl ether			0.085	0.458	1.85	6.00	16.3	38.7
$C_6H_{12}O$	Cyclohexanol			0.100			10.4	30.6	72.7
$C_6H_{12}O$	2-Hexanone			1.54	5.80	17.3	43.2	94.2	184
$C_6H_{12}O$	3-Hexanone						49.1	106	
$C_6H_{12}O$	2-Methyl-3-pentanone					27.4	66.2		
$C_6H_{12}O$	4-Methyl-2-pentanone		0.539	2.64	9.37	26.3	62.0	128	
$C_6H_{12}O$	3,3-Dimethyl-2-butanone			4.27	13.9	37.1	84.5	171	
$C_6H_{12}O$	Butyl vinyl ether		1.76	6.65	20.5	53.8	124		
$C_6H_{12}O$	Isobutyl vinyl ether		2.34	9.30	29.9	81.2			
$C_6H_{12}O$	Hexanal			1.48	5.61	16.8	42.2	92.2	181
$C_6H_{12}O_2$	Isopentyl formate			1.79	6.59	19.4	47.9	104	
$C_6H_{12}O_2$	Butyl acetate			1.66		19.3	46.4	98.4	
$C_6H_{12}O_2$	Isobutyl acetate			2.39	8.56	24.7	60.3	129	
$C_6H_{12}O_2$	Propyl propanoate			1.88	6.84	20.1	49.8	109	
$C_6H_{12}O_2$	Ethyl butanoate			2.01	7.84	22.9	54.4	111	202
$C_6H_{12}O_2$	Ethyl 2-methylpropanoate	0.096	0.693	3.25	11.3	31.5	74.2	154	

VAPOR PRESSURE IN THE TEMPERATURE RANGE –25°C TO 150°C (continued)

Molecular formula	Name	Vapor pressure in kPa –25°C	0°C	25°C	50°C	75°C	100°C	125°C	150°C
$C_6H_{12}O_2$	Methyl pentanoate			1.41	5.32	16.1	41.4	93.1	
$C_6H_{12}O_2$	Methyl isopentanoate	0.089	0.560	2.50	8.63	24.5	59.8	129	
$C_6H_{12}O_2$	Hexanoic acid			0.005				5.28	15.6
$C_6H_{12}O_2$	4-Methylpentanoic acid					0.224	1.63	6.72	19.4
$C_6H_{12}O_2$	Diethylacetic acid						2.41	7.81	22.1
$C_6H_{12}O_2$	Diacetone alcohol			0.224	0.989	3.53	10.6		
$C_6H_{12}O_3$	Paraldehyde				5.70	17.3	45.2	105	
$C_6H_{12}O_3$	2-Ethoxyethyl acetate			0.240	1.40	5.50	16.5	40.5	85.8
$C_6H_{12}S$	Cyclohexanethiol					6.72	17.6	39.9	80.8
$C_6H_{12}S$	*cis*-Tetrahydro-2,5-dimethyl-thiophene-				3.77	11.3	28.5	62.4	123
$C_6H_{12}S$	Tetrahydro-3-methyl-2H-thiopyran				2.22	6.93	18.0	40.8	82.4
$C_6H_{13}Br$	1-Bromohexane				2.15	6.96	18.6	42.9	88.1
$C_6H_{13}Cl$	1-Chlorohexane			1.25	4.71	14.1	35.1	76.5	149
$C_6H_{13}Cl$	3-Chloro-2,2-dimethylbutane			25.9	84.9	228	523	1064	1964
$C_6H_{13}F$	1-Fluorohexane		1.96	7.71	23.6	59.6	131		
$C_6H_{13}I$	1-Iodohexane				0.860	3.01	8.58	20.9	44.8
$C_6H_{13}N$	Cyclohexylamine			1.20		14.3	36.1	78.9	
C_6H_{14}	Hexane	1.34	6.05	20.2	54.1	123	246	446	749
C_6H_{14}	2-Methylpentane	2.16	8.99	28.2	72.2	158	308	548	906
C_6H_{14}	3-Methylpentane	1.86	7.93	25.3	65.3	144	282	503	834
C_6H_{14}	2,2-Dimethylbutane	3.88	14.6	42.5	102	213	397	684	1101
C_6H_{14}	2,3-Dimethylbutane	2.56	10.3	31.3	78.2	168	323	567	929
$C_6H_{14}O$	1-Hexanol			0.110	0.611	3.12	11.4	32.9	79.9
$C_6H_{14}O$	2-Hexanol			0.346	1.76	7.26	23.2	61.1	140
$C_6H_{14}O$	3-Hexanol			0.684	2.75	9.82	28.7	71.8	161
$C_6H_{14}O$	2-Methyl-1-pentanol			0.236	1.29	5.24	16.9	45.8	108
$C_6H_{14}O$	4-Methyl-1-pentanol			0.111	0.833	4.03	14.3	40.3	95.8
$C_6H_{14}O$	2-Methyl-2-pentanol			0.845	4.83	17.9	49.8	113	
$C_6H_{14}O$	3-Methyl-2-pentanol			0.431	2.40	9.47	29.2	74.5	165
$C_6H_{14}O$	4-Methyl-2-pentanol			0.698	3.26	11.5	33.2	81.5	
$C_6H_{14}O$	2-Methyl-3-pentanol			0.709	3.64	13.5	39.5	96.7	
$C_6H_{14}O$	3-Methyl-3-pentanol			1.16	5.60	19.0	50.5	112	
$C_6H_{14}O$	2-Ethyl-1-butanol			0.206	1.35	5.86	19.1	50.2	113
$C_6H_{14}O$	3,3-Dimethyl-1-butanol				1.65	6.38	20.1	54.1	128
$C_6H_{14}O$	2,3-Dimethyl-2-butanol			1.18	5.52	19.1	52.9	124	
$C_6H_{14}O$	Dipropyl ether			8.35	25.1	62.7	135		
$C_6H_{14}O$	Diisopropyl ether			19.9	54.0				
$C_6H_{14}O$	Butyl ethyl ether			7.46	23.0	58.3	127		
$C_6H_{14}O$	Ethyl *tert*-butyl ether	0.975	4.69	16.5	46.0	108			
$C_6H_{14}O_2$	2-Butoxyethanol			0.150			8.77	23.3	54.4
$C_6H_{14}O_2$	Hexylene glycol						2.06	6.68	18.9

VAPOR PRESSURE IN THE TEMPERATURE RANGE –25°C TO 150°C (continued)

Molecular formula	Name	Vapor pressure in kPa –25°C	0°C	25°C	50°C	75°C	100°C	125°C	150°C
$C_6H_{14}O_2$	1,1-Diethoxyethane		0.777	3.68	13.2	38.1	93.9		
$C_6H_{14}O_2$	1,2-Diethoxyethane	0.226	1.18	4.33	12.5	29.8	62.0	116	
$C_6H_{14}O_3$	Trimethylolpropane								0.383
$C_6H_{14}O_3$	2-(2-Ethoxyethoxy)ethanol			0.017	0.189	0.761	2.57	7.48	19.3
$C_6H_{14}O_3$	1,2,6-Hexanetriol							0.023	0.128
$C_6H_{14}O_3$	Dipropylene glycol								6.04
$C_6H_{14}O_3$	Diethylene glycol dimethyl ether		0.054	0.315	1.37	4.73	13.6	34.1	76.0
$C_6H_{14}O_4$	Triethylene glycol							0.247	0.885
$C_6H_{14}S$	Butyl ethyl sulfide				3.23	10.1	26.2	58.8	118
$C_6H_{14}S$	Dipropyl sulfide				3.42	10.6	27.4	61.2	122
$C_6H_{14}S$	Isopropyl propyl sulfide			1.47	5.39	15.8	38.7	83.3	161
$C_6H_{14}S$	Diisopropyl sulfide			2.57	8.74	24.0	56.2	116	
$C_6H_{14}S$	2-Hexanethiol				4.04	12.4	31.3	68.8	135
$C_6H_{14}S$	1-Hexanethiol				2.34	7.55	20.1	46.2	94.5
$C_6H_{15}N$	Hexylamine			1.17	4.54	13.9	35.8	80.1	160
$C_6H_{15}N$	Butylethylamine			3.19					
$C_6H_{15}N$	Triethylamine			7.70	25.8	65.8			
$C_6H_{15}N$	Dipropylamine			3.21	11.2	31.7	76.0	160	306
$C_6H_{15}N$	Diisopropylamine		3.00	10.7	31.0	75.3	160	306	
$C_6H_{15}NO$	2-Diethylaminoethanol							30.0	71.3
$C_6H_{16}N_2$	Hexamethylenediamine					0.953	3.12	8.82	22.0
$C_6H_{16}O_2Si$	Diethoxydimethylsilane	0.071	0.537	2.64	9.53	27.5	66.6		
$C_6H_{18}Cl_2O_2Si_3$	1,5-Dichloro-1,1,3,3,5,5-hexamethyltrisiloxane			0.121	0.585	2.17	6.59	17.1	38.9
$C_6H_{18}OSi_2$	Hexamethyldisiloxane				17.6	45.2	99.8	196	
C_7F_{14}	Perfluoromethylcyclohexane			14.1	40.6	97.3	202		
C_7F_{16}	Perfluoroheptane		2.52	10.2	31.3	79.2	172		
C_7HF_{15}	1H-Pentadecafluoroheptane			5.90	18.8				
$C_7H_3ClF_3NO_2$	1-Chloro-2-nitro-4-(trifluoromethyl)benzene					0.370	1.44	4.52	11.9
$C_7H_3F_5$	2,3,4,5,6-Pentafluorotoluene				8.50	24.4	59.1	126	241
$C_7H_4ClF_3$	1-Chloro-2-(trifluoromethyl)benzene				2.33	7.54	20.1	46.4	95.3
$C_7H_4ClF_3$	1-Chloro-3-(trifluoromethyl)benzene			1.05	4.06	12.4	31.8	70.6	
$C_7H_4ClF_3$	1-Chloro-4-(trifluoromethyl)benzene			1.05	3.99	12.1	30.8	68.7	
$C_7H_4Cl_2O_2$	*o*-Chlorobenzoyl chloride						1.37	4.04	
$C_7H_4Cl_2O_2$	*m*-Chlorobenzoyl chloride						1.71	4.64	
$C_7H_4F_3NO_2$	1-Nitro-3-(trifluoromethyl)benzene					0.938	3.23	9.15	22.3
$C_7H_4F_4$	1-Fluoro-4-(trifluoromethyl)benzene			5.33	16.2	41.6	93.3		
C_7H_5BrO	Benzoyl bromide				0.158	0.640	2.10	5.87	14.4
C_7H_5ClO	Benzoyl chloride			0.084	0.402	1.49	4.53	11.8	27.3
$C_7H_5Cl_3$	(Trichloromethyl)benzene				0.180	0.721	2.36	6.59	16.1
$C_7H_5F_3$	(Trifluoromethyl)benzene			5.14	16.2	42.4	95.4	190	

VAPOR PRESSURE IN THE TEMPERATURE RANGE –25°C TO 150°C (continued)

Molecular formula	Name	Vapor pressure in kPa –25°C	0°C	25°C	50°C	75°C	100°C	125°C	150°C
C_7H_5N	Benzonitrile			0.110	0.496	1.80	5.41	14.0	32.2
C_7H_5NS	Phenylisothiocyanate			0.196	0.330	0.801			
$C_7H_6Cl_2$	2,4-Dichlorotoluene			0.055		1.43	4.61	12.2	27.7
$C_7H_6Cl_2$	3,4-Dichlorotoluene						3.11	8.53	20.2
C_7H_6O	Benzaldehyde			0.169	0.776	2.83	8.37	21.0	46.3
$C_7H_6O_2$	Salicylaldehyde						4.42	11.7	27.3
C_7H_7Br	*o*-Bromotoluene					2.88	8.32	20.4	44.1
C_7H_7Br	*m*-Bromotoluene			0.145	0.649	2.30	6.79	17.3	39.2
C_7H_7Br	*p*-Bromotoluene					2.52	7.41	18.5	40.6
C_7H_7Br	Benzyl bromide			0.098	0.423	1.48	4.37	11.2	25.8
C_7H_7Cl	*o*-Chlorotoluene			0.482		6.29	16.8	38.8	79.8
C_7H_7Cl	*m*-Chlorotoluene						15.1	35.3	73.3
C_7H_7Cl	*p*-Chlorotoluene					5.64	15.3	35.7	74.0
C_7H_7Cl	Benzyl chloride			0.164	0.747	2.66	7.83	19.8	44.4
C_7H_7ClO	1-Chloro-2-methoxybenzene							9.93	23.4
C_7H_7F	*o*-Fluorotoluene			3.13	10.5	28.4	66.1		
C_7H_7F	*m*-Fluorotoluene			2.83	9.59	26.4	61.8	128	
C_7H_7F	*p*-Fluorotoluene			3.00		26.3	61.7	128	240
$C_7H_7NO_2$	*o*-Nitrotoluene								13.5
$C_7H_7NO_2$	*m*-Nitrotoluene					0.493	1.58	4.31	10.5
$C_7H_7NO_3$	2-Nitroanisole								2.42
C_7H_8	Toluene			3.79			74.6	149	274
C_7H_8	Bicyclo[2.2.1]hepta-2,5-diene			8.97	25.7	61.9			
$C_7H_8Cl_2Si$	Dichloromethylphenylsilane			0.060	0.304	1.17	3.65	9.65	22.4
C_7H_8O	*o*-Cresol			0.041			4.23		30.3
C_7H_8O	*m*-Cresol			0.019			2.50		20.6
C_7H_8O	*p*-Cresol			0.017			2.45		20.7
C_7H_8O	Benzyl alcohol			0.015			2.27		18.1
C_7H_8O	Anisole			0.472			18.5	43.9	91.9
C_7H_8S	3-Methylbenzenethiol					1.58	4.95	13.0	29.7
C_7H_9N	*N*-Methylaniline				0.306	1.28	4.23	11.6	27.4
C_7H_9N	*o*-Methylaniline			0.043				9.81	23.9
C_7H_9N	*m*-Methylaniline			0.036				8.65	21.5
C_7H_9N	*p*-Methylaniline	0.191	0.638	1.74	4.07	8.41	15.8	27.4	44.4
C_7H_9N	Benzylamine			0.096	0.489	1.90	5.97	15.9	37.3
C_7H_9N	2-Ethylpyridine				2.59	8.36	22.1		
C_7H_9N	3-Ethylpyridine					4.78	13.2		
C_7H_9N	4-Ethylpyridine				1.29	4.42	12.4		
C_7H_9N	2,4-Dimethylpyridine								136
C_7H_9N	2,5-Dimethylpyridine						17.1	40.2	83.9
C_7H_9N	2,6-Dimethylpyridine			0.746	3.05	9.71	25.6	58.5	
$C_7H_{10}N_2$	Toluene-2,4-diamine						0.098	0.378	1.24

VAPOR PRESSURE IN THE TEMPERATURE RANGE –25°C TO 150°C (continued)

Molecular formula	Name	Vapor pressure in kPa –25°C	0°C	25°C	50°C	75°C	100°C	125°C	150°C
C_7H_{12}	Cycloheptene	0.147	0.811	3.36					
C_7H_{12}	1-Methylcyclohexene				12.6	33.2	75.1		
C_7H_{12}	4-Methylcyclohexene			5.43	16.7	42.6	93.6	184	329
C_7H_{12}	1,2-Dimethylcyclopentene			4.89	15.2	38.9	85.8	169	303
C_7H_{12}	1,5-Dimethylcyclopentene		1.42	5.61	17.2	43.6	95.7	187	335
C_7H_{12}	1-Ethylcyclopentene			4.78	14.9	38.2	84.6	167	300
C_7H_{12}	Bicyclo[4.1.0]heptane				10.0	27.0	62.1		
C_7H_{12}	Methylenecyclohexane				16.5	41.9	91.9		
C_7H_{12}	1-Heptyne						101		
C_7H_{12}	3-Heptyne						82.7	166	
C_7H_{12}	5-Methyl-1-hexyne			7.85	23.7	59.3	129		
C_7H_{12}	5-Methyl-2-hexyne			6.96	21.2	53.5	117		
C_7H_{12}	2-Methyl-3-hexyne			6.96	21.2	53.5	117		
C_7H_{12}	4,4-Dimethyl-1-pentyne		4.39	15.3	42.2	98.0	200		
C_7H_{12}	4,4-Dimethyl-2-pentyne		3.29	11.8	33.4	79.2	164		
$C_7H_{12}O$	Cycloheptanone				0.815	2.93	8.55	21.2	46.3
$C_7H_{12}O_2$	Butyl acrylate		0.140	0.731	2.84	8.85	23.2	53.0	109
$C_7H_{12}O_2$	Propyl methacrylate			0.941	3.45	10.5	27.5	64.0	
$C_7H_{12}O_3$	Ethyl levulinate				0.139	0.658	2.41	7.23	18.6
$C_7H_{12}O_4$	Diethyl malonate				0.237	0.932	3.07	8.73	22.0
$C_7H_{13}ClO$	Heptanoyl chloride			0.069	0.503	2.75	11.8	42.2	129
C_7H_{14}	Cycloheptane			2.90	9.55	25.6	58.8		
C_7H_{14}	Methylcyclohexane			6.18	18.4	45.8	98.7		
C_7H_{14}	Ethylcyclopentane		1.34	5.32	16.4	41.7	91.7		
C_7H_{14}	1,1-Dimethylcyclopentane		2.80	10.1	28.7	68.5	143		
C_7H_{14}	*cis*-1,2-Dimethylcyclopentane		1.63	6.30	19.0	47.4	103		
C_7H_{14}	*trans*-1,2-Dimethylcyclopentane		2.31	8.54	24.8	60.3	128		
C_7H_{14}	*cis*-1,3-Dimethylcyclopentane		2.44	8.96	25.9	62.5	132		
C_7H_{14}	*trans*-1,3-Dimethylcyclopentane		2.33	8.61	25.0	60.6	128		
C_7H_{14}	1-Heptene	0.355	1.94	7.52	22.6	56.4	122	236	418
C_7H_{14}	*cis*-2-Heptene		1.66	6.45	19.5	48.9	106		
C_7H_{14}	*trans*-2-Heptene		1.69	6.56	19.8	49.6	108		
C_7H_{14}	*cis*-3-Heptene				21.3	52.5	115		
C_7H_{14}	*trans*-3-Heptene				21.4	52.6	115		
C_7H_{14}	2-Methyl-1-hexene				24.2	59.6	128	245	
C_7H_{14}	4-Methyl-1-hexene				28.8	70.2	149		
C_7H_{14}	2-Methyl-2-hexene				22.4	55.4	119	229	
C_7H_{14}	*cis*-3-Methyl-2-hexene				20.3	50.6	110	212	
C_7H_{14}	*trans*-4-Methyl-2-hexene				28.0	68.4	145		
C_7H_{14}	*trans*-5-Methyl-2-hexene				27.5	67.2	143		
C_7H_{14}	*trans*-2-Methyl-3-hexene				22.6	56.4	122		
C_7H_{14}	2,3-Dimethyl-1-pentene				31.3	75.7	160		

VAPOR PRESSURE IN THE TEMPERATURE RANGE –25°C TO 150°C (continued)

Molecular formula	Name	Vapor pressure in kPa							
		–25°C	0°C	25°C	50°C	75°C	100°C	125°C	150°C
C_7H_{14}	2,4-Dimethyl-1-pentene		3.58	12.6	35.2	82.8	170		
C_7H_{14}	3,3-Dimethyl-1-pentene			13.9	39.4	93.7	195		
C_7H_{14}	4,4-Dimethyl-1-pentene	1.26	5.60	18.4	48.6	109			
C_7H_{14}	3-Ethyl-1-pentene				31.5	76.1	160		
C_7H_{14}	2,3-Dimethyl-2-pentene				20.2	50.4	109	211	
C_7H_{14}	2,4-Dimethyl-2-pentene		3.03	11.2	32.4	78.1	164		
C_7H_{14}	*cis*-3,4-Dimethyl-2-pentene				26.5	64.9	138		
C_7H_{14}	*trans*-3,4-Dimethyl-2-pentene				24.6	60.5	129	248	
C_7H_{14}	*cis*-4,4-Dimethyl-2-pentene		3.89	13.4	37.0	85.9	175		
C_7H_{14}	*trans*-4,4-Dimethyl-2-pentene		4.23	14.8	41.1	96.0	197		
C_7H_{14}	2,3,3-Trimethyl-1-butene		4.38	14.9	40.4	92.8	188		
$C_7H_{14}O$	2-Heptanone			0.490	2.17	7.37	20.2	47.4	98.5
$C_7H_{14}O$	3-Heptanone					6.98	19.8	49.1	110
$C_7H_{14}O$	4-Heptanone			0.164	0.988	4.46	16.1	48.6	127
$C_7H_{14}O$	5-Methyl-2-hexanone			0.691	2.89	9.33	24.8	57.0	116
$C_7H_{14}O$	2,4-Dimethyl-3-pentanone				6.87	19.6	47.4	101	
$C_7H_{14}O$	Heptanal				2.04	6.90	19.0	44.9	93.9
$C_7H_{14}O_2$	Pentyl acetate			0.600	2.33	7.40	20.1	48.3	105
$C_7H_{14}O_2$	Isopentyl acetate			0.728	2.94	9.49	25.7	60.6	128
$C_7H_{14}O_2$	Isobutyl propanoate				3.27	11.5			
$C_7H_{14}O_2$	Propyl butanoate			0.618	2.95	10.1	27.0	60.8	120
$C_7H_{14}O_2$	Propyl 2-methylpropanoate		0.164	1.03	4.35	13.8	35.7	78.9	
$C_7H_{14}O_2$	Isopropyl isobutanoate	0.063	0.448	2.13	7.62	21.9	53.6	115	
$C_7H_{14}O_2$	Ethyl 3-methylbutanoate			1.07	4.09	12.6	33.0	75.5	155
$C_7H_{14}O_2$	Methyl hexanoate				2.09	6.97	19.5		
$C_7H_{14}O_2$	Heptanoic acid					0.123	0.655	2.60	8.25
$C_7H_{14}O_2$	4-Methoxy-4-methyl-2-pentanone					5.29	14.9	35.9	76.0
$C_7H_{15}Br$	1-Bromoheptane				0.784	2.86	8.41	21.1	46.3
$C_7H_{15}Cl$	1-Chloroheptane				1.67	5.65	15.6	36.8	76.8
$C_7H_{15}F$	1-Fluoroheptane			2.45	8.62	24.4	58.6	124	236
$C_7H_{15}I$	1-Iodoheptane					1.26	3.96	10.4	23.9
C_7H_{16}	Heptane		1.52	6.09	18.9	48.2	106		
C_7H_{16}	2-Methylhexane		2.34	8.78	25.8	63.3	135		
C_7H_{16}	3-Methylhexane		2.17	8.21	24.3	59.9	128		
C_7H_{16}	2,2-Dimethylpentane		4.08	14.0	38.5	89.2	182		
C_7H_{16}	2,3-Dimethylpentane		2.48	9.18	26.6	64.3	135		
C_7H_{16}	2,4-Dimethylpentane		3.71	13.1	36.6	85.7	175		
C_7H_{16}	3,3-Dimethylpentane		3.13	11.0	30.8	72.5	149		
C_7H_{16}	3-Ethylpentane		2.03	7.74	23.0	57.0	122		
C_7H_{16}	2,2,3-Trimethylbutane		9.13	28.5	72.2	157			
$C_7H_{16}O$	1-Heptanol					1.37	5.53	17.2	44.2
$C_7H_{16}O$	2-Heptanol			0.100	0.700	3.54	12.5	34.1	77.7

VAPOR PRESSURE IN THE TEMPERATURE RANGE –25°C TO 150°C (continued)

Molecular formula	Name	Vapor pressure in kPa –25°C	0°C	25°C	50°C	75°C	100°C	125°C	150°C
$C_7H_{16}O$	3-Heptanol				0.721	3.73	13.3	36.4	83.1
$C_7H_{16}O$	4-Heptanol				0.939	4.29	14.4	38.5	87.4
$C_7H_{16}O$	2,2-Dimethyl-3-pentanol				2.81	11.0	31.6	73.6	
$C_7H_{16}S$	1-Heptanethiol					3.05	8.99	22.4	49.0
$C_7H_{17}N$	Heptylamine				1.70	5.82	16.3	39.3	83.6
$C_7H_{18}N_2$	*N,N*-Diethyl-1,3-propanediamine				0.996	3.44	10.1	25.7	58.8
C_8F_{18}	Perfluorooctane				12.6	35.6	84.2		
$C_8H_4O_3$	Phthalic anhydride								2.53
C_8H_6O	Benzofuran				1.12	3.86	10.9	26.6	
C_8H_7Cl	*o*-Chlorostyrene						6.63	16.5	36.4
C_8H_7N	*o*-Tolunitrile				0.303	1.15	3.57	9.45	22.0
C_8H_7N	*p*-Tolunitrile				0.195	0.829	2.70	7.23	16.6
C_8H_7N	Phenylacetonitrile					0.318	1.13	3.38	8.77
$C_8H_7NO_4$	Methyl 2-nitrobenzoate								1.48
C_8H_8	Styrene			0.810	3.18	9.85	25.5	57.3	
C_8H_8	1,3,5,7-Cyclooctatetraene		0.205	1.04	3.92	11.8			
C_8H_8O	Acetophenone			0.049			0.172	0.640	2.03
$C_8H_8O_2$	Methyl benzoate			0.052				11.5	26.3
$C_8H_8O_2$	4-Methoxybenzaldehyde					0.161	0.602	1.89	5.17
$C_8H_8O_2$	Phenyl acetate				0.292	1.16	3.76	10.4	25.5
$C_8H_8O_3$	Methyl salicylate			0.015			1.74	5.24	13.4
C_8H_9Cl	1-Chloro-2-ethylbenzene					2.94	8.65	21.6	47.2
C_8H_9Cl	1-Chloro-4-ethylbenzene					2.36	7.10	18.0	40.0
$C_8H_9NO_2$	1-Ethyl-4-nitrobenzene						0.565	1.86	5.21
C_8H_{10}	Ethylbenzene			1.28		13.8	34.2	74.3	
C_8H_{10}	*o*-Xylene			0.880		10.4	26.5	58.9	
C_8H_{10}	*m*-Xylene			1.13		12.4	31.1	68.3	
C_8H_{10}	*p*-Xylene			1.19		12.8	32.1	69.9	
$C_8H_{10}O$	*o*-Ethylphenol						2.41	7.01	17.7
$C_8H_{10}O$	*m*-Ethylphenol						1.35	4.43	12.2
$C_8H_{10}O$	*p*-Ethylphenol						1.12	3.72	10.5
$C_8H_{10}O$	2,3-Xylenol						1.60	5.11	13.6
$C_8H_{10}O$	2,4-Xylenol			0.022			1.93	6.12	16.2
$C_8H_{10}O$	2,5-Xylenol			0.022			1.96	6.17	16.2
$C_8H_{10}O$	2,6-Xylenol			0.019			3.41	9.69	23.6
$C_8H_{10}O$	2-Phenylethanol							4.76	12.7
$C_8H_{10}O$	Phenetole			0.204				26.6	58.6
$C_8H_{10}O_2$	1,3-Dimethoxybenzene					0.467	2.12	6.67	16.5
$C_8H_{10}O_2$	2-Phenoxyethanol					0.094	0.439	1.60	4.84
$C_8H_{11}N$	*N,N*-Dimethylaniline					1.59	4.90	12.9	29.7
$C_8H_{11}N$	*N*-Ethylaniline				0.222	0.931	3.11	8.68	21.1
$C_8H_{11}N$	*p*-Ethylaniline								14.6

VAPOR PRESSURE IN THE TEMPERATURE RANGE –25°C TO 150°C (continued)

Molecular formula	Name	Vapor pressure in kPa							
		–25°C	0°C	25°C	50°C	75°C	100°C	125°C	150°C
$C_8H_{11}N$	2,4-Xylidine						1.82	5.66	15.0
$C_8H_{11}N$	2,6-Xylidine						2.38	6.36	15.1
$C_8H_{11}N$	5-Ethyl-2-picoline		0.022						
$C_8H_{11}NO$	*o*-Phenetidine						0.898	2.86	7.96
C_8H_{12}	4-Vinylcyclohexene			1.87	6.49	18.2			
C_8H_{12}	1,5-Cyclooctadiene					8.13	21.0		
$C_8H_{12}O_4$	Diethyl maleate				0.087	0.383	1.36	4.07	10.6
C_8H_{14}	1-Octyne						45.3	97.8	
C_8H_{14}	2-Octyne						31.0	69.3	
C_8H_{14}	3-Octyne						36.0	79.9	
C_8H_{14}	4-Octyne						37.6	83.3	
C_8H_{14}	2,5-Dimethyl-1,5-hexadiene				9.61	28.4	65.7		
C_8H_{14}	1-Ethylcyclohexene					13.4	33.4	72.6	
$C_8H_{14}O_2$	Butyl methacrylate					4.41	13.2		
$C_8H_{14}O_2$	Cyclohexyl acetate						8.84	22.7	51.6
$C_8H_{14}O_3$	Butanoic anhydride					1.25	3.88	10.5	25.2
$C_8H_{14}O_4$	Ethyl succinate				0.093	0.425	1.55	4.76	12.6
$C_8H_{14}O_4$	Dipropyl oxalate				0.101	0.476	1.77	5.42	14.3
$C_8H_{15}Br$	(2-Bromoethyl)cyclohexane				0.239	0.984	3.17	8.47	19.5
$C_8H_{15}ClO$	Octanoyl chloride					1.02	5.74		
$C_8H_{15}N$	Octanenitrile						3.28	9.05	21.7
C_8H_{16}	Cyclooctane						22.4	49.5	98.4
C_8H_{16}	Propylcyclopentane			1.64	5.86	16.8	40.6	86.1	
C_8H_{16}	Isopropylcyclopentane			1.66	5.88	16.8	40.4	85.5	
C_8H_{16}	1-Ethyl-1-methylcyclopentane			2.64	8.73	23.6	54.4	111	
C_8H_{16}	Ethylcyclohexane			1.71	5.96	16.8	40.1	84.3	
C_8H_{16}	1,1-Dimethylcyclohexane			3.02	9.74	25.7	58.2	117	
C_8H_{16}	*cis*-1,2-Dimethylcyclohexane			1.93	6.59	18.2	43.0	89.3	
C_8H_{16}	*trans*-1,2-Dimethylcyclohexane			2.58	8.46	22.7	52.0	106	
C_8H_{16}	*cis*-1,3-Dimethylcyclohexane			2.87	9.35	24.9	57.0	115	
C_8H_{16}	*trans*-1,3-Dimethylcyclohexane			2.34	7.88	21.5	50.0	103	
C_8H_{16}	*cis*-1,4-Dimethylcyclohexane			2.39	7.98	21.7	50.3	103	
C_8H_{16}	*trans*-1,4-Dimethylcyclohexane			3.02	9.75	25.8	58.4	118	
C_8H_{16}	1,1,2-Trimethylcyclopentane			3.71	11.8	30.6	68.5	137	
C_8H_{16}	1,1,3-Trimethylcyclopentane		1.35	5.30	16.1	40.5	88.2		
C_8H_{16}	1α,2α,4β-1,2,4-Trimethyl-cyclopentane			3.25	10.5	27.6	62.7	126	
C_8H_{16}	1α,2β,4α-1,2,4-Trimethyl-cyclopentane			4.36	13.6	35.0	77.7	154	
C_8H_{16}	*cis*-1-Ethyl-2-methylcyclopentane			1.95	6.74	18.8	44.6	93.3	
C_8H_{16}	1-Octene			2.30	8.07	22.6	53.7		
C_8H_{16}	*cis*-2-Octene					19.3	46.9	99.5	

VAPOR PRESSURE IN THE TEMPERATURE RANGE –25°C TO 150°C (continued)

Molecular formula	Name	Vapor pressure in kPa –25°C	0°C	25°C	50°C	75°C	100°C	125°C	150°C
C_8H_{16}	*trans*-2-Octene					19.6	47.7	101	
C_8H_{16}	*cis*-3-Octene			2.35	8.00	22.1	51.8	107	
C_8H_{16}	*trans*-3-Octene					20.9	50.4	106	
C_8H_{16}	*cis*-4-Octene					21.8	51.8	108	
C_8H_{16}	*trans*-4-Octene					21.7	51.9	109	
C_8H_{16}	2-Methyl-1-heptene			2.69	9.06	24.8	57.6	119	
C_8H_{16}	2,3-Dimethyl-2-hexene			2.44	8.30	22.8	53.5	111	
C_8H_{16}	*cis*-2,2-Dimethyl-3-hexene			4.90	15.3	39.3	86.7	170	
C_8H_{16}	2,3,3-Trimethyl-1-pentene			4.03	13.1	34.9	79.3	160	
C_8H_{16}	2,4,4-Trimethyl-1-pentene		1.54	5.96	17.9	44.9	97.2	188	
C_8H_{16}	2,3,4-Trimethyl-2-pentene			3.00	10.0	27.2	62.8	129	
C_8H_{16}	2,4,4-Trimethyl-2-pentene			4.80	15.2	39.4	87.8	174	
$C_8H_{16}O$	2-Octanone				0.528	2.20	7.34	20.5	50.0
$C_8H_{16}O$	3-Octanone			0.286	1.13	3.79			
$C_8H_{16}O$	2,2,4-Trimethyl-3-pentanone			0.298	1.67	7.07	24.1	69.3	
$C_8H_{16}O$	Octanal			0.321	1.25	3.98	10.9	26.2	57.0
$C_8H_{16}O$	1-Propylcyclopentanol					1.53	6.63	20.5	50.2
$C_8H_{16}O_2$	Octanoic acid							1.44	5.14
$C_8H_{16}O_2$	Methyl heptanoate				0.788	2.97	9.07	23.5	
$C_8H_{16}O_2$	Ethyl hexanoate			0.158	0.798	3.08	9.69	25.9	60.9
$C_8H_{16}O_2$	Propyl 3-methylbutanoate		0.088	0.453	1.80	5.83	16.1	39.0	85.0
$C_8H_{16}O_2$	Isobutyl butanoate		0.096	0.500	1.95	6.20	16.7	39.5	83.7
$C_8H_{16}O_2$	Isobutyl 2-methylpropanoate		0.097	0.552	2.31	7.67	21.2	50.9	109
$C_8H_{16}O_2$	Hexyl acetate			0.185	0.854	3.16	9.81		
$C_8H_{16}O_4$	Diethylene glycol monoethyl-ether acetate			0.029	0.145	0.575	1.90	5.41	13.6
$C_8H_{17}Br$	1-Bromooctane					1.18	3.85	10.5	24.6
$C_8H_{17}Cl$	1-Chlorooctane					2.30	7.05	18.1	40.7
$C_8H_{17}Cl$	3-(Chloromethyl)heptane						9.89	24.2	53.1
$C_8H_{17}F$	1-Fluorooctane				3.20	10.2	26.8	61.3	125
$C_8H_{17}I$	1-Iodooctane						1.83	5.24	12.9
C_8H_{18}	Octane	0.058	0.386	1.86	6.71	19.3	46.8	99.5	190
C_8H_{18}	2-Methylheptane				9.32	25.7	60.1		
C_8H_{18}	3-Methylheptane				8.91	24.6	57.8		
C_8H_{18}	4-Methylheptane			2.74	9.31	25.6	60.0	124	231
C_8H_{18}	2,2-Dimethylhexane		1.10	4.54	14.4	37.3	83.1	165	298
C_8H_{18}	2,3-Dimethylhexane			3.13	10.4	28.0	64.3	131	241
C_8H_{18}	2,5-Dimethylhexane			4.06	13.1	34.4	77.7	156	284
C_8H_{18}	3,3-Dimethylhexane			3.82	12.2	31.9	71.8	144	261
C_8H_{18}	3,4-Dimethylhexane			2.89	9.63	26.1	60.3	123	229
C_8H_{18}	3-Ethylhexane			2.68	9.11	25.1	58.6	121	226
C_8H_{18}	3-Ethyl-2-methylpentane			3.19	10.5	28.1	64.3	130	240

VAPOR PRESSURE IN THE TEMPERATURE RANGE –25°C TO 150°C (continued)

Molecular formula	Name	Vapor pressure in kPa –25°C	0°C	25°C	50°C	75°C	100°C	125°C	150°C
C_8H_{18}	3-Ethyl-3-methylpentane			3.07	9.93	26.3	60.0	121	223
C_8H_{18}	2,2,3-Trimethylpentane			4.28	13.4	34.5	76.5	151	273
C_8H_{18}	2,2,4-Trimethylpentane			6.50		48.8	103	198	351
C_8H_{18}	2,3,3-Trimethylpentane			3.60	11.4	29.7	66.6	133	241
C_8H_{18}	2,3,4-Trimethylpentane				11.6	30.4	68.7		
C_8H_{18}	2,2,3,3-Tetramethylbutane		0.555	2.76	10.6	32.6	84.5	165	
$C_8H_{18}O$	1-Octanol			0.010	0.079	0.554	2.55	8.72	24.0
$C_8H_{18}O$	2-Octanol				0.232	1.39	5.59	16.9	41.8
$C_8H_{18}O$	3-Methyl-3-heptanol					3.43	11.8	32.2	73.9
$C_8H_{18}O$	4-Methyl-3-heptanol				1.83	5.90	16.3	39.5	86.4
$C_8H_{18}O$	5-Methyl-3-heptanol				1.83	6.50	18.4	44.0	92.1
$C_8H_{18}O$	4-Methyl-4-heptanol					3.34	11.3	31.0	72.3
$C_8H_{18}O$	2-Ethyl-1-hexanol			0.019			4.19	13.4	34.7
$C_8H_{18}O$	2,4,4-Trimethyl-2-pentanol				2.02	8.00	23.3	54.9	111
$C_8H_{18}O$	2,2,4-Trimethyl-3-pentanol				1.31	6.76	21.5	50.9	99.0
$C_8H_{18}O$	Dibutyl ether			0.898			29.1	65.5	
$C_8H_{18}O$	Di-*tert*-butyl ether			4.34	13.6	35.1	78.2		
$C_8H_{18}O_2$	Di-*tert*-butyl peroxide		0.751	3.43	11.6	31.7	73.2		
$C_8H_{18}O_2$	Ethylene glycol monohexyl ether						1.99	6.00	15.9
$C_8H_{18}O_3$	Diethylene glycol diethyl ether				0.453	1.65	5.05	13.4	31.8
$C_8H_{18}O_3$	Diethylene glycol monobutyl ether				0.029	0.173	0.768	2.70	7.91
$C_8H_{18}O_5$	Tetraethylene glycol								0.093
$C_8H_{18}S$	Dibutyl sulfide							15.3	34.8
$C_8H_{18}S$	1-Octanethiol						4.10	11.1	26.1
$C_8H_{19}N$	Dibutylamine			0.340		5.01	14.5	35.6	77.3
$C_8H_{19}N$	Diisobutylamine		0.194	0.972	3.68	11.2	29.1	66.2	
$C_8H_{20}O_4Si$	Ethyl silicate		0.335	1.17	3.37	8.34	18.3	36.3	66.4
$C_8H_{20}Si$	Tetraethylsilane		0.157	0.739	2.69	8.00	20.4	45.9	93.3
$C_9H_6N_2O_2$	Toluene-2,4-diisocyanate						0.499	1.67	4.79
C_9H_8	Indene			0.220	0.860	2.79	7.74	19.0	42.0
C_9H_{10}	Indan						9.16	22.4	48.3
C_9H_{10}	Isopropenylbenzene					4.88	13.5	32.1	67.5
C_9H_{10}	*cis*-1-Propenylbenzene					4.11	11.6	28.0	59.6
C_9H_{10}	*trans*-1-Propenylbenzene			0.217	0.918	3.09	8.70	21.3	46.3
$C_9H_{10}O$	2,4-Dimethylbenzaldehyde					0.370	1.41	4.47	12.3
$C_9H_{10}O_2$	Benzyl acetate				0.124	0.537	1.89	5.65	14.8
$C_9H_{10}O_2$	Ethyl benzoate					0.774	2.51	6.91	16.7
$C_9H_{11}Br$	1-Bromo-4-isopropylbenzene							6.17	15.1
$C_9H_{11}Cl$	1-Chloro-2-isopropylbenzene						5.61	14.6	33.1
$C_9H_{11}Cl$	1-Chloro-4-isopropylbenzene						4.43	11.8	27.2
C_9H_{12}	Cumene			0.610	2.49	7.86	20.7	46.9	95.1
C_9H_{12}	*o*-Ethyltoluene					4.90	13.6	32.2	67.7

VAPOR PRESSURE IN THE TEMPERATURE RANGE –25°C TO 150°C (continued)

Molecular formula	Name	Vapor pressure in kPa –25°C	0°C	25°C	50°C	75°C	100°C	125°C	150°C
C_9H_{12}	*m*-Ethyltoluene					5.60	15.3	35.9	75.0
C_9H_{12}	*p*-Ethyltoluene					5.55	15.1	35.4	73.8
C_9H_{12}	Propylbenzene						16.6	38.5	79.5
C_9H_{12}	1,2,3-Trimethylbenzene						9.44	23.2	50.3
C_9H_{12}	1,2,4-Trimethylbenzene			0.300			11.7	28.3	60.3
C_9H_{12}	Mesitylene			0.330			13.5	32.2	68.2
$C_9H_{12}O$	Benzyl ethyl ether			0.136	0.608	2.16	6.42	16.5	37.5
$C_9H_{12}O$	Propoxybenzene						5.86	14.9	33.5
$C_9H_{12}O$	Isopropoxybenzene					2.76	8.60	22.1	48.6
$C_9H_{13}N$	*N,N*-Dimethyl-*o*-toluidine				0.545	2.11	6.51	16.9	38.4
$C_9H_{13}N$	2,4,6-Trimethylbenzenamine					0.184	0.804	2.81	8.24
$C_9H_{13}N$	Amphetamine					1.30			
$C_9H_{14}O$	Isophorone			0.060	0.274	0.995	2.99	7.73	17.7
$C_9H_{14}O_6$	Triacetin				0.003				
C_9H_{16}	1-Nonyne			1.73	6.33	18.5	45.5	97.9	190
$C_9H_{16}O_4$	Diethyl glutarate					0.236	0.856	2.63	7.05
$C_9H_{17}N$	Nonanenitrile					0.435	1.57	4.63	11.8
C_9H_{18}	1,1,2-Trimethylcyclohexane								114
C_9H_{18}	1,1,3-Trimethylcyclohexane					14.7	35.2	74.2	
C_9H_{18}	1α,2β,4β-1,2,4-Trimethylcyclohexane							67.3	125
C_9H_{18}	*trans*-1,3,5-Trimethylcyclohexane							67.3	125
C_9H_{18}	*trans*-1-Ethyl-4-methylcyclohexane								106
C_9H_{18}	Propylcyclohexane				2.23	7.05	18.5	42.1	85.2
C_9H_{18}	Isopropylcyclohexane				2.49	7.75	20.0	44.9	90.1
C_9H_{18}	1,1,3,3-Tetramethylcyclopentane						62.2	121	
C_9H_{18}	*cis*-1-Methyl-2-propylcyclopentane								95.2
C_9H_{18}	*trans*-1-Methyl-2-propylcyclopentane								111
C_9H_{18}	Isobutylcyclopentane								104
C_9H_{18}	1-Nonene			0.714	2.90	9.17	24.0	54.5	110
C_9H_{18}	2-Methyl-1-octene							59.8	116
$C_9H_{18}O$	2-Nonanone						4.33	12.2	29.2
$C_9H_{18}O$	5-Nonanone						0.049	0.302	1.32
$C_9H_{18}O$	2,6-Dimethyl-4-heptanone			0.230		4.00	11.6	28.6	61.8
$C_9H_{18}O$	Nonanal			0.085	0.421	1.64	5.25	14.4	34.8
$C_9H_{18}O_2$	Heptyl acetate					1.30	4.36	12.2	29.4
$C_9H_{18}O_2$	Isopentyl butanoate			0.160	0.751	2.71	8.01	20.3	45.3
$C_9H_{18}O_2$	Isobutyl 3-methylbutanoate			0.253	1.10	3.78	10.8	26.9	59.5
$C_9H_{18}O_2$	Propyl hexanoate				0.420	1.69	5.60	15.9	
$C_9H_{18}O_2$	Methyl octanoate						31.9	89.8	215
$C_9H_{18}O_2$	Nonanoic acid						0.127	0.637	2.41

VAPOR PRESSURE IN THE TEMPERATURE RANGE –25°C TO 150°C (continued)

Molecular formula	Name	Vapor pressure in kPa –25°C	0°C	25°C	50°C	75°C	100°C	125°C	150°C
$C_9H_{19}Cl$	1-Chlorononane					0.940	3.21	8.99	21.6
C_9H_{20}	Nonane			0.570	2.41	7.85	21.0	48.5	99.2
C_9H_{20}	2-Methyloctane				84.9	198	400	729	1220
C_9H_{20}	4-Methyloctane				3.56	11.0	28.0	62.3	124
C_9H_{20}	2,2-Dimethylheptane				5.30	15.5	38.1	81.9	158
C_9H_{20}	2,3-Dimethylheptane				3.99	12.0	30.2	66.0	130
C_9H_{20}	2,6-Dimethylheptane				4.66	14.0	35.0	76.2	149
C_9H_{20}	2,2,4-Trimethylhexane			2.00	6.97	19.6	46.5	97.2	183
C_9H_{20}	2,2,5-Trimethylhexane			2.21	7.62	21.2	50.0	104	
C_9H_{20}	2,3,3-Trimethyl-hexane			1.31	4.75	13.8	33.6	71.8	138
C_9H_{20}	2,3,5-Trimethylhexane			1.57	5.68	16.4	39.9	85.1	163
C_9H_{20}	2,4,4-Trimethylhexane			1.78	6.21	17.4	41.5	87.0	165
C_9H_{20}	3,3,4-Trimethylhexane				4.32	12.6	31.0	66.6	129
C_9H_{20}	2,2,3,3-Tetramethylpentane					13.0	31.5	67.3	
C_9H_{20}	2,2,3,4-Tetramethylpentane					16.4	39.0	81.7	
C_9H_{20}	2,2,4,4-Tetramethylpentane				8.75	23.4	53.7		
C_9H_{20}	2,3,3,4-Tetramethylpentane					12.4	30.2	64.8	
C_9H_{20}	2,3-Diethylpentane				4.05	12.1	30.3	66.2	130
C_9H_{20}	3,3-Diethylpentane					10.5	26.2	57.1	
C_9H_{20}	3-Ethyl-2,4-dimethylpentane			1.33	4.85	14.1	34.4	73.7	142
$C_9H_{20}S$	1-Nonanethiol						1.88	5.57	14.0
$C_9H_{21}BO_3$	Triisopropyl borate					10.8	28.2	65.3	
$C_9H_{21}N$	Nonylamine					0.984	3.33	9.37	22.7
$C_9H_{21}N$	Tripropylamine					5.63	16.1	39.2	84.4
$C_{10}F_8$	Perfluoronaphthalene			0.010	0.119				
$C_{10}H_7Br$	1-Bromonaphthalene			0.001		0.074	0.289	0.930	2.57
$C_{10}H_7Cl$	1-Chloronaphthalene					0.132	0.534	1.73	4.71
$C_{10}H_8$	Azulene			0.001	0.015	0.134	0.859	2.51	6.47
$C_{10}H_8$	Naphthalene			0.011		0.768	2.50	6.84	16.2
$C_{10}H_8O$	1-Naphthol								1.75
$C_{10}H_8O$	2-Naphthol								1.51
$C_{10}H_{10}$	*m*-Divinylbenzene			0.087	0.397	1.43	4.29	11.1	25.6
$C_{10}H_{10}O_4$	Dimethyl phthalate						0.151	0.548	1.71
$C_{10}H_{10}O_4$	Dimethyl isophthalate							0.813	2.39
$C_{10}H_{10}O_4$	Dimethyl terephthalate						0.061	0.413	1.69
$C_{10}H_{12}$	1,2,3,4-Tetrahydronaphthalene			0.050	0.289	1.11	3.48	9.21	21.4
$C_{10}H_{12}$	2-Ethylstyrene						6.36	16.2	
$C_{10}H_{12}$	3-Ethylstyrene					1.79	5.60	14.7	33.6
$C_{10}H_{12}$	4-Ethylstyrene					1.87	5.68	14.5	32.5
$C_{10}H_{12}O$	Estragole				0.111	0.501	1.80	5.40	14.0
$C_{10}H_{12}O$	Cuminaldehyde				0.077	0.341	1.21	3.59	9.27
$C_{10}H_{12}O_2$	Propyl benzoate				0.099	0.420	1.43	4.11	10.3

VAPOR PRESSURE IN THE TEMPERATURE RANGE –25°C TO 150°C (continued)

Molecular formula	Name	Vapor pressure in kPa –25°C	0°C	25°C	50°C	75°C	100°C	125°C	150°C
$C_{10}H_{12}O_2$	2-Phenylethyl acetate								9.23
$C_{10}H_{12}O_2$	Ethyl phenylacetate							3.80	9.94
$C_{10}H_{12}O_2$	4-Allyl-2-methoxyphenol							1.50	4.20
$C_{10}H_{12}O_2$	Isoeugenol							1.01	2.83
$C_{10}H_{14}$	Butylbenzene			0.150			7.41	18.7	41.3
$C_{10}H_{14}$	Isobutylbenzene			0.257		3.87	10.9	26.2	55.6
$C_{10}H_{14}$	*sec*-Butylbenzene			0.230		3.78	10.7	25.7	54.8
$C_{10}H_{14}$	*tert*-Butylbenzene			0.280	1.28	4.36	12.2	29.0	61.2
$C_{10}H_{14}$	*o*-Cymene					3.17	8.90	21.7	47.1
$C_{10}H_{14}$	*m*-Cymene					3.53	10.0	24.4	52.1
$C_{10}H_{14}$	*p*-Cymene			0.190			9.20	22.6	49.1
$C_{10}H_{14}$	*o*-Diethylbenzene					2.45	7.33	18.5	41.1
$C_{10}H_{14}$	*m*-Diethylbenzene					2.63	7.84	19.8	43.6
$C_{10}H_{14}$	*p*-Diethylbenzene					2.45	7.30	18.4	40.8
$C_{10}H_{14}$	3-Ethyl-*o*-xylene					1.62	5.07	13.3	30.5
$C_{10}H_{14}$	4-Ethyl-1,2-dimethylbenzene					1.90	5.85	15.1	34.3
$C_{10}H_{14}$	2-Ethyl-1,3-dimethylbenzene					1.89	5.81	15.0	34.0
$C_{10}H_{14}$	2-Ethyl-1,4-dimethylbenzene					2.20	6.60	16.8	37.4
$C_{10}H_{14}$	1-Ethyl-2,4-dimethylbenzene					2.03	6.19	15.9	35.7
$C_{10}H_{14}$	1-Ethyl-3,5-dimethylbenzene					2.41	7.22	18.3	40.6
$C_{10}H_{14}$	1-Methyl-2-propylbenzene					2.36	7.05	17.9	39.6
$C_{10}H_{14}$	1-Methyl-3-propylbenzene					2.61	7.75	19.5	42.9
$C_{10}H_{14}$	1-Methyl-4-propylbenzene					2.54	7.51	18.8	41.5
$C_{10}H_{14}$	1,2,3,4-Tetramethylbenzene						3.46	9.43	22.2
$C_{10}H_{14}$	1,2,4,5-Tetramethylbenzene					1.41	4.48	11.9	27.6
$C_{10}H_{14}$	Isodurene					1.37	4.36	11.6	27.0
$C_{10}H_{14}O$	Butoxybenzene							7.95	19.0
$C_{10}H_{14}O$	2-Butylphenol							3.07	8.27
$C_{10}H_{15}N$	*N*-Butylaniline								6.73
$C_{10}H_{15}N$	*N*,*N*-Diethylaniline				0.145	0.621	2.12	6.05	15.0
$C_{10}H_{15}N$	2-Methyl-5-isopropylaniline						0.638		
$C_{10}H_{16}$	(±)-1-Methyl-4-isopropenyl-cyclohexene			0.259	1.07	3.56	9.91	24.0	52.0
$C_{10}H_{16}$	Terpinolene			0.090	0.438	1.68	5.31	14.5	34.9
$C_{10}H_{16}$	β-Phellandrene			0.189	0.843	2.95			
$C_{10}H_{16}$	(+)-α-Pinene			0.582	2.35	7.37	19.2	43.2	86.6
$C_{10}H_{16}$	β-Myrcene			0.280	1.16	3.83	10.7	25.8	56.0
$C_{10}H_{16}O$	Pulegone				0.011	0.242	1.69	6.48	17.4
$C_{10}H_{18}$	*cis*-Decahydronaphthalene			0.100		1.87	5.56	14.0	30.8
$C_{10}H_{18}$	*trans*-Decahydronaphthalene			0.164		2.62	7.46	18.1	38.9
$C_{10}H_{18}$	1-Decyne					3.42	9.87	24.3	52.8
$C_{10}H_{18}O$	Eucalyptol			0.260	1.08	3.59	9.90	23.7	50.8

VAPOR PRESSURE IN THE TEMPERATURE RANGE –25°C TO 150°C (continued)

Molecular formula	Name	Vapor pressure in kPa –25°C	0°C	25°C	50°C	75°C	100°C	125°C	150°C
$C_{10}H_{18}O$	*trans*-Geraniol					0.205	0.806	2.64	7.46
$C_{10}H_{18}O_4$	Sebacic acid							0.001	
$C_{10}H_{18}O_4$	Dipropyl succinate					0.119	0.471	1.55	4.39
$C_{10}H_{18}O_4$	Diethyl adipate					0.118	0.446	1.47	4.33
$C_{10}H_{19}N$	Decanenitrile							2.47	6.80
$C_{10}H_{20}$	Butylcyclohexane					2.87	8.34	20.6	44.7
$C_{10}H_{20}$	Isobutylcyclohexane					4.22	11.7	27.7	58.2
$C_{10}H_{20}$	*tert*-Butylcyclohexane					4.37	11.9	27.9	58.1
$C_{10}H_{20}$	1-Decene			0.210			10.9	26.8	58.1
$C_{10}H_{20}O$	Decanal				0.120	0.544	1.97	6.01	15.9
$C_{10}H_{20}O_2$	Decanoic acid							0.319	1.33
$C_{10}H_{20}O_2$	Ethyl octanoate				0.173	0.719	2.46	7.22	18.7
$C_{10}H_{20}O_2$	Isopentyl isopentanoate					1.80	5.25	13.3	30.1
$C_{10}H_{20}O_2$	2-Ethylhexyl acetate				0.619	2.58	7.68	18.2	36.4
$C_{10}H_{20}O_2$	Octyl acetate					1.53	4.63	11.9	27.1
$C_{10}H_{20}O_4$	Diethylene glycol monobutyl - ether acetate							1.75	4.90
$C_{10}H_{21}Br$	1-Bromodecane							2.66	7.19
$C_{10}H_{21}Cl$	1-Chlorodecane						1.48	4.52	11.7
$C_{10}H_{21}F$	1-Fluorodecane					1.82	5.81	15.5	36.1
$C_{10}H_{22}$	Decane			0.170			9.56	23.9	52.5
$C_{10}H_{22}$	2-Methylnonane					4.28	12.3	29.9	64.1
$C_{10}H_{22}$	3-Methylnonane					4.23	12.1	29.4	62.8
$C_{10}H_{22}$	4-Methylnonane				1.33	4.62	13.1	31.4	66.7
$C_{10}H_{22}$	5-Methylnonane				1.36	4.73	13.3	32.0	67.6
$C_{10}H_{22}$	2,4-Dimethyloctane				1.98	6.58	17.9	41.9	86.6
$C_{10}H_{22}$	2,7-Dimethyloctane		0.093	0.501	1.99	5.67	15.7	37.2	77.9
$C_{10}H_{22}$	2,2,6-Trimethylheptane				2.72	8.65	22.7	51.6	104
$C_{10}H_{22}$	3,3,5-Trimethylheptane				2.30	7.29	19.1	43.3	87.6
$C_{10}H_{22}$	2,2,3,3-Tetramethylhexane				2.08	6.54	17.1	38.6	78.0
$C_{10}H_{22}$	2,2,5,5-Tetramethylhexane				4.41	13.2	32.9	71.7	140
$C_{10}H_{22}$	2,2,3,3,4-Pentamethylpentane				1.83	5.72	14.9	33.6	67.8
$C_{10}H_{22}$	2,2,3,4,4-Pentamethylpentane				2.26	6.86	17.4	38.5	76.1
$C_{10}H_{22}$	2,4-Dimethyl-3-isopropylpentane				2.29	7.19			
$C_{10}H_{22}O$	1-Decanol			0.009			0.584	2.37	7.44
$C_{10}H_{22}O$	4-Decanol						1.50		
$C_{10}H_{22}O$	Dipentyl ether						6.35	16.2	36.6
$C_{10}H_{22}O$	Diisopentyl ether			0.210	0.922	3.21	9.32	23.4	52.3
$C_{10}H_{22}O_5$	Tetraethylene glycol dimethyl ether								1.62
$C_{10}H_{22}S$	1-Decanethiol							2.81	7.63
$C_{10}H_{22}S$	Diisopentylsulfide		0.088	0.150	0.296	0.743			
$C_{10}H_{23}N$	Dipentylamine							9.02	22.1

VAPOR PRESSURE IN THE TEMPERATURE RANGE –25°C TO 150°C (continued)

Molecular formula	Name	Vapor pressure in kPa –25°C	0°C	25°C	50°C	75°C	100°C	125°C	150°C
$C_{10}H_{30}O_3Si_4$	Decamethyltetrasiloxane					1.49	4.60	12.6	29.7
$C_{10}H_{30}O_5Si_5$	Decamethylcyclopentasiloxane						2.50	7.38	18.3
$C_{11}H_{10}$	1-Methylnaphthalene			0.009					7.33
$C_{11}H_{10}$	2-Methylnaphthalene								8.15
$C_{11}H_{12}O_3$	Myristicin						0.183	0.631	1.87
$C_{11}H_{14}$	4-Isopropylstyrene				0.338	1.26	3.87	10.2	23.6
$C_{11}H_{14}$	1,2,3,4-Tetrahydro-5-methyl-naphthalene								8.95
$C_{11}H_{14}$	1,2,3,4-Tetrahydro-6-methyl-naphthalene								10.9
$C_{11}H_{14}O_2$	Butyl benzoate					0.153	0.600	1.98	
$C_{11}H_{16}$	*p-tert*-Butyltoluene			0.090		1.75	5.41	14.1	32.3
$C_{11}H_{16}$	1,3-Diethyl-5-methylbenzene			0.074	0.352	1.31	4.01	10.5	24.5
$C_{11}H_{16}$	Pentylbenzene						3.47	9.43	22.2
$C_{11}H_{16}$	2-Ethyl-1,3,5-trimethylbenzene				0.241	0.965	3.11	8.47	20.1
$C_{11}H_{16}$	1-Ethyl-2,4,5-trimethylbenzene				0.188	0.797	2.70	7.63	18.8
$C_{11}H_{20}$	2-Undecyne						3.39	9.17	21.6
$C_{11}H_{20}$	1-Undecyne					1.54	4.81	12.7	29.2
$C_{11}H_{20}O_4$	Ethyl diethylmalonate							3.17	10.3
$C_{11}H_{21}N$	Undecanenitrile						0.355	1.25	3.66
$C_{11}H_{22}$	Pentylcyclohexane						3.77	10.2	23.7
$C_{11}H_{22}$	Hexylcyclopentane						3.68	10.0	23.6
$C_{11}H_{22}$	1-Undecene					1.55	5.00	13.4	31.1
$C_{11}H_{22}$	*cis*-2-Undecene				0.300	1.29	4.30	11.8	
$C_{11}H_{22}$	*trans*-2-Undecene				0.317	1.36	4.50	12.2	
$C_{11}H_{22}$	*cis*-4-Undecene				0.381	1.56	5.02	13.4	
$C_{11}H_{22}$	*trans*-4-Undecene				0.365	1.52	4.94	13.3	
$C_{11}H_{22}$	*cis*-5-Undecene				0.390	1.59	5.10	13.6	
$C_{11}H_{22}$	*trans*-5-Undecene				0.371	1.53	4.95	13.3	
$C_{11}H_{22}O_2$	Heptyl butanoate							2.87	8.19
$C_{11}H_{22}O_2$	Propyl octanoate					0.354	1.36	4.35	12.0
$C_{11}H_{22}O_2$	Methyl decanoate							3.14	8.62
$C_{11}H_{24}$	Undecane						4.38	12.0	28.1
$C_{11}H_{24}$	2-Methyldecane					1.80	5.69	15.0	34.4
$C_{11}H_{24}$	3-Methyldecane					1.87	5.49	14.0	32.0
$C_{11}H_{24}S$	Undecyl mercaptan							1.43	4.18
$C_{12}F_{27}N$	Trinonafluorobutylamine			0.073	0.441	1.92	6.51	18.3	44.3
$C_{12}H_8$	Acenaphthylene			0.001	0.010				
$C_{12}H_{10}$	Acenaphthene						0.281	0.949	
$C_{12}H_{10}$	Biphenyl					0.149	0.586	1.87	5.05
$C_{12}H_{10}S$	Phenyl sulfide						0.180	0.586	1.65
$C_{12}H_{12}$	1,2-Dimethylnaphthalene							1.10	3.26

VAPOR PRESSURE IN THE TEMPERATURE RANGE –25°C TO 150°C (continued)

Molecular formula	Name	Vapor pressure in kPa							
		–25°C	0°C	25°C	50°C	75°C	100°C	125°C	150°C
$C_{12}H_{12}$	1-Ethylnaphthalene							1.67	4.62
$C_{12}H_{12}$	2-Ethylnaphthalene							1.73	4.74
$C_{12}H_{14}O_4$	Diethyl phthalate								0.980
$C_{12}H_{16}$	Cyclohexylbenzene								8.50
$C_{12}H_{16}$	*p*-Isopropenylisopropylbenzene							5.24	13.0
$C_{12}H_{16}O_2$	Isoamyl benzoate					0.158	0.521	1.51	3.90
$C_{12}H_{18}$	1,2-Diisopropylbenzene							9.67	22.9
$C_{12}H_{18}$	1,3-Diisopropylbenzene							9.82	23.3
$C_{12}H_{18}$	1,4-Diisopropylbenzene							4.35	10.4
$C_{12}H_{18}$	Hexylbenzene						1.62	4.80	12.1
$C_{12}H_{18}$	1,5,9-Cyclododecatriene								11.8
$C_{12}H_{20}O_2$	Geranyl acetate					0.154	0.591	1.91	5.34
$C_{12}H_{20}O_4$	Dibutyl maleate			0.002	0.010	0.038	0.134	0.431	1.28
$C_{12}H_{22}$	1-Dodecyne						2.37	6.69	16.3
$C_{12}H_{22}O_2$	Methyl 10-undecenoate							1.50	4.31
$C_{12}H_{24}$	Hexylcyclohexane						1.74	5.09	12.7
$C_{12}H_{24}$	Heptylcyclopentane						1.66	4.95	12.6
$C_{12}H_{24}$	1-Dodecene			0.019	0.134	0.637	2.30	6.73	16.8
$C_{12}H_{24}O$	Dodecanal					0.131	0.475	1.47	3.98
$C_{12}H_{24}O_2$	Dodecanoic acid								0.406
$C_{12}H_{24}O_2$	Decyl acetate						0.460	1.61	4.85
$C_{12}H_{24}O_2$	Ethyl decanoate						0.556	1.86	5.38
$C_{12}H_{25}Br$	1-Bromododecane							0.701	2.17
$C_{12}H_{25}Cl$	1-Chlorododecane							1.15	3.45
$C_{12}H_{26}$	Dodecane			0.016			2.01		15.2
$C_{12}H_{26}O_3$	Diethylene glycol dibutyl ether			0.005	0.029	0.132	0.488	1.53	4.21
$C_{12}H_{27}N$	Tributylamine			0.010		0.881	2.69	7.12	16.8
$C_{12}H_{27}N$	Triisobutylamine			0.075	0.413	1.73	5.81	16.5	40.8
$C_{12}H_{36}O_6Si_6$	Dodecamethylcyclohexasiloxane							2.28	6.41
$C_{13}H_{12}$	Diphenylmethane								3.69
$C_{13}H_{13}N$	Methyldiphenylamine							0.422	1.36
$C_{13}H_{14}$	1-Isopropylnaphthalene							1.09	3.24
$C_{13}H_{20}$	Heptylbenzene								6.50
$C_{13}H_{24}O_2$	Ethyl 10-undecenoate							0.991	3.16
$C_{13}H_{26}$	Heptylcyclohexane							2.52	6.79
$C_{13}H_{26}$	Octylcyclopentane							2.47	6.75
$C_{13}H_{28}$	Tridecane			0.005					8.31
$C_{13}H_{28}O$	1-Tridecanol								1.69
$C_{14}H_{12}$	*cis*-Stilbene						0.199	0.765	2.51
$C_{14}H_{12}$	*trans*-Stilbene								0.784
$C_{14}H_{14}$	1,1-Diphenylethane					0.066	0.278	0.984	
$C_{14}H_{16}$	1-Butylnaphthalene								2.27

VAPOR PRESSURE IN THE TEMPERATURE RANGE –25°C TO 150°C (continued)

Molecular formula	Name	Vapor pressure in kPa –25°C	0°C	25°C	50°C	75°C	100°C	125°C	150°C
$C_{14}H_{18}$	2-Butylnaphthalene								1.66
$C_{14}H_{22}$	Octylbenzene					0.075	0.342	1.22	3.55
$C_{14}H_{27}N$	Tetradecanenitrile							0.187	0.664
$C_{14}H_{28}$	Octylcyclohexane						0.357	1.27	
$C_{14}H_{28}$	Nonylcyclopentane								3.67
$C_{14}H_{28}$	1-Tetradecene							1.74	5.01
$C_{14}H_{28}O_2$	Tetradecanoic acid								0.121
$C_{14}H_{30}O$	1-Tetradecanol							0.118	0.785
$C_{14}H_{42}O_5Si_6$	Tetradecamethylhexasiloxane							1.43	3.92
$C_{15}H_{18}$	1-Pentylnaphthalene								1.46
$C_{15}H_{24}$	Nonylbenzene								1.92
$C_{15}H_{30}$	Nonylcyclohexane								2.00
$C_{15}H_{30}$	Decylcyclopentane								2.02
$C_{16}H_{32}$	1-Hexadecene								1.52
$C_{16}H_{34}O$	1-Hexadecanol					0.001	0.011		
$C_{18}H_{38}$	Octadecane								0.466
$C_{18}H_{38}O$	1-Octadecanol							0.014	0.084
$C_{19}H_{36}O_2$	Methyl oleate								0.102
$C_{20}H_{60}O_8Si_9$	Eicosamethylnonasiloxane								0.172
$C_{21}H_{21}O_4P$	Tris(4-methylphenyl) phosphate								0.002
$C_{22}H_{42}O_2$	Butyl *cis*-9-octadecenoate						0.001	0.011	
$C_{22}H_{44}O_2$	Behenic acid								0.001
$C_{22}H_{44}O_2$	Butyl stearate						0.001	0.007	
$C_{24}H_{38}O_4$	Dioctyl phthalate								0.004
$C_{24}H_{38}O_4$	Bis(2-ethylhexyl) phthalate							0.001	0.008
$C_{30}H_{62}$	Squalane						0.056	0.566	3.94

VAPOR PRESSURE AT ELEVATED TEMPERATURES

This table provides vapor pressure information for about 200 substances of low volatility. The data are presented in the form of the temperature at which the vapor pressure reaches the following fixed pressures in kilopascals (kPa): 0.13 (1 mmHg), 0.67 (5 mmHg), 2.66 (20 mmHg), 8.0 (60 mmHg), 26.7 (200 mmHg), and 53.3 (400 mmHg). An "s" following an entry indicates the substance is a solid at that temperature.

Substances are listed in a modified Hill order, with substances not containing carbon preceding those that contain carbon.

REFERENCES

1. Lide, D.R., and Kehiaian, H.V., *CRC Handbook of Thermophysical and Thermochemical Data*, CRC Press, Boca Raton, FL, 1994.
2. Stull, Daniel R., *Ind. Eng. Chem.*, 39, 517-550, 1947
3. Trotman-Dickenson, A.F., Executive Editor, *Comprehensive Inorganic Chemistry*, Vol. 1-5, Pergamon Press, Oxford, 1973.
4. Marsh, K.N., Wilhoit, R.C., and Yin, D., *TRC Databases in Chemistry and Engineering — Vapor Pressures, Ver 2.2*, Thermodynamics Research Center, College Station, TX, 1994.

Molecular formula	Name	Temperature in °C for the indicated pressure 0.13 kPa 1 mmHg	0.67 kPa 5 mmHg	2.66 kPa 20 mmHg	8.0 kPa 60 mmHg	26.7 kPa 200 mmHg	53.3 kPa 400 mmHg
	SUBSTANCES NOT CONTAINING CARBON						
Ag	Silver	1357	1500	1658	1795	1971	2090
AgCl	Silver(I) chloride	912	1019	1134	1242	1379	1467
AgI	Silver(I) iodide	820	927	1045	1152	1297	1400
Al	Aluminum	1561	1726	1894	2050	2250	2382
$AlCl_3$	Aluminum chloride	100.0 s	116.4 s	131.8 s	145.4 s	161.8 s	171.6 s
AlF_3	Aluminum fluoride	1238	1298	1350	1398	1457	1496
AlI_3	Aluminum iodide	178.0 s	207.7	244.2	277.8	322.0	354.0
Al_2O_3	Aluminum oxide	2148	2306	2465	2599	2766	2874
As	Arsenic	354 s	402 s	450 s	493 s	546 s	580 s
As_2O_3	Arsenic trioxide	211 s	234 s	277 s	302	373	423
Au	Gold	1869	2059	2256	2431	2657	2807
Ba	Barium	912	1046	1188			
Be	Beryllium	1547	1707	1871	2021	2212	2339
$BeBr_2$	Beryllium bromide	289 s	325 s	361 s	390 s	427 s	451 s
$BeCl_2$	Beryllium chloride	291 s	328 s	365 s	395 s	435	461
BeI_2	Beryllium iodide	283 s	322 s	361 s	394 s	435 s	461 s
Bi	Bismuth	1021	1099	1177	1240	1319	1370
$BiBr_3$	Bismuth bromide	s	261	305	340	392	425
$BiCl_3$	Bismuth chloride	s	242	287	324	372	405
BrCs	Cesium bromide	748	838	938	1026	1140	1221
BrCu	Copper(I) bromide	572	666	777	887	1052	1189
BrH_4N	Ammonium bromide	198.3 s	234.5 s	270.6 s	303.8 s	345.3 s	370.9 s
BrK	Potassium bromide	795	940	994	1087	1212	1297
BrLi	Lithium bromide	748	840	939	1028	1147	1226
BrNa	Sodium bromide	806	903	1005	1099	1220	1304
BrRb	Rubidium bromide	781	876	975	1066	1186	1267
BrTl	Thallium(I) bromide	s	490	559	621	703	759
Br_2Hg	Mercury(II) bromide	136.5 s	165.3 s	191.3 s	221.0 s	262.7	290.0
Br_2Pb	Lead(II) bromide	513	578	646	711	796	856
Br_4Ge	Germanium(IV) bromide	s	43.3	71.8	98.8	135.4	161.6
Br_4Sn	Tin(IV) bromide	s	58.3	88.1	116.2	152.8	177.7
Br_4Zr	Zirconium bromide	207 s	237 s	266 s	289 s	318 s	337 s
Ca	Calcium	s	926	1046	1152	1288	1388
Cd	Cadmium	394	455	516	578	658	711
$CdCl_2$	Cadmium chloride	s	618	695	762	847	908
CdF_2	Cadmium fluoride	1112	1231	1344	1436	1561	1651
CdI_2	Cadmium iodide	416	481	546	608	688	742
CdO	Cadmium oxide	1000 s	1100 s	1200 s	1295 s	1409 s	1484 s
ClCs	Cesium chloride	744	837	934	1023	1139	1217
ClCu	Copper(I) chloride	546	645	766	886	1077	1249
ClH_4N	Ammonium chloride	160.4 s	193.8 s	226.1 s	256.2 s	293.2 s	316.5 s
ClK	Potassium chloride	821	919	1020	1115	1239	1322
ClLi	Lithium chloride	783	880	987	1081	1203	1290

Molecular formula	Name	Temperature in °C for the indicated pressure					
		0.13 kPa 1 mmHg	0.67 kPa 5 mmHg	2.66 kPa 20 mmHg	8.0 kPa 60 mmHg	26.7 kPa 200 mmHg	53.3 kPa 400 mmHg
ClNa	Sodium chloride	865	967	1072	1169	1296	1379
ClRb	Rubidium chloride	792	887	990	1084	1207	1294
ClTl	Thallium(I) chloride	s	487	550	612	694	748
Cl_2Co	Cobalt(II) chloride	s	s	s	801	904	974
Cl_2Fe	Iron(II) chloride			737	805	897	961
Cl_2Hg	Mercury(II) chloride	136.2 s	166.0 s	195.8 s	222.2 s	256.5 s	275.5 s
Cl_2Mg	Magnesium chloride	778	877	988	1088	1223	1316
Cl_2Mn	Manganese(II) chloride	s	736	825	913	1028	1108
Cl_2Ni	Nickel(II) chloride	671 s	731 s	789 s	840 s	904 s	945 s
Cl_2Pb	Lead(II) chloride	547	615	684	750	833	893
Cl_2Sn	Tin(II) chloride	316	366	420	467	533	577
Cl_2Zn	Zinc chloride	428	481	536	584	648	689
Cl_3Fe	Iron(III) chloride	194.0 s	221.8 s	246.0 s	263.7 s	285.0 s	298.0 s
Cl_4Te	Tellurium tetrachloride	s	s	253	287	330	360
Cl_4Zr	Zirconium chloride	190 s	217 s	243 s	268 s	295 s	312 s
Cr	Chromium	1616	1768	1928	2067	2243	2361
Cs	Cesium	279	341	409	474	561	624
CsF	Cesium fluoride	712	798	893	980	1092	1170
CsI	Cesium iodide	738	828	923	1009	1124	1200
Cu	Copper	1612	1801	1950	2107	2306	2434
CuI	Copper(I) iodide	s	610	716	836	1018	1158
FK	Potassium fluoride	885	988	1096	1193	1323	1411
FLi	Lithium fluoride	1047	1156	1270	1372	1503	1591
FNa	Sodium fluoride	1077	1186	1300	1403	1531	1617
FRb	Rubidium fluoride	921	982	1052	1123	1239	1322
F_2Pb	Lead(II) fluoride	s	861	950	1036	1144	1219
F_2Zn	Zinc fluoride	1243	1328	1402	1480	1602	1690
F_5Nb	Niobium(V) fluoride	s	s	103.0	133.2	172.2	198.0
F_5Ta	Tantalum(V) fluoride	s	s	s	110.3	159.9	194.0
Fe	Iron	1851	2031	2213	2379	2588	2724
Ga	Gallium	1349	1478	1608	1725	1874	1974
HKO	Potassium hydroxide	719	814	918	1013	1142	1233
HNaO	Sodium hydroxide	739	843	953	1057	1192	1286
H_2O_4S	Sulfuric acid	145.8	178.0	211.5	241.5	279.8	305.0
H_4IN	Ammonium iodide	210.9 s	247.0 s	282.8 s	316.0 s	355.8 s	381.0 s
Hg	Mercury	126.2	164.8	204.6	242.0	290.7	323.0
HgI_2	Mercury(II) iodide	157.5 s	189.2 s	220.0 s	249.0 s	291.0	324.2
IK	Potassium iodide	745	840	938	1030	1152	1238
ILi	Lithium iodide	723	802	883	955	1049	1110
INa	Sodium iodide	767	857	952	1039	1150	1225
IRb	Rubidium iodide	748	839	935	1026	1141	1223
ITl	Thallium(I) iodide	440	502	567	631	712	763
I_2Pb	Lead(II) iodide	479	540	605	668	750	807
I_3Sb	Antimony(III) iodide	163.6 s	203.8	244.8	282.5	333.8	368.5
I_4Sn	Tin(IV) iodide	s	156.0	196.2	234.2	283.5	315.5
I_4Zr	Zirconium iodide	264 s	297 s	329 s	355 s	389 s	409 s
K	Potassium	341	408	483	550	643	708
Li	Lithium	723	828	940	1042	1178	1273
Mg	Magnesium	595 s	678	763	843	947	1017
Mn	Manganese	1292	1434	1583	1720	1900	2029
Mo	Molybdenum	3102	3393	3690	3964	4322	4553
MoO_3	Molybdenum(VI) oxide	734 s	785 s	851	917	1014	1082
Na	Sodium	439	511	589	662	758	823
Ni	Nickel	1810	1979	2143	2289	2473	2603
OPb	Lead(II) oxide	943	1039	1134	1222	1330	1402
OSr	Strontium oxide	2068 s	2198 s	2333 s			
O_2Se	Selenium dioxide	157.0 s	187.7 s	217.5 s	244.6 s	277.0 s	297.7 s
O_2Si	Silicon dioxide	s	s	1798	1911	2053	2141

VAPOR PRESSURE AT ELEVATED TEMPERATURES (continued)

Molecular formula	Name	Temperature in °C for the indicated pressure 0.13 kPa 1 mmHg	0.67 kPa 5 mmHg	2.66 kPa 20 mmHg	8.0 kPa 60 mmHg	26.7 kPa 200 mmHg	53.3 kPa 400 mmHg
O_3Sb_2	Antimony trioxide	574 s	626 s	729	873	1085	1242
O_5P_2	Phosphorus pentoxide	384 s	424 s	462 s	493 s	532 s	556 s
O_7Re_2	Rhenium(VII) oxide	212.5 s	237.5 s	261.0 s	280.0 s	307.0	336.0
P	Phosphorus (white)	76.6	111.2	146.2	179.8	222.7	251.0
P	Phosphorus (red)	237 s	271 s	306 s	334 s	370 s	391 s
P	Phosphorus (black)	290 s	323 s	354 s	381 s	413 s	432 s
Pb	Lead	973	1099	1234	1358	1519	1630
PbS	Lead(II) sulfide	852 s	928 s	1005 s	1074 s	1160	1221
Pt	Platinum	2730	3007	3302	3574	3923	4169
Rb	Rubidium	297	358	422	482	563	620
S	Sulfur	183.8	223.0	264.7	305.5	359.7	399.6
Sb	Antimony	732	858	996	1131	1315	1445
Se	Selenium	356	413	473	527	594	637
Si	Silicon	1724	1835	1942	2036	2151	2220
Sn	Tin	1492	1634	1777	1903	2063	2169
Sr	Strontium	s	847	953	1057	1192	1285
Te	Tellurium	520	605	697	789	910	997
Tl	Thallium	825	931	1040	1143	1274	1364
W	Tungsten	3990	4337	4690	5007	5403	5666
Zn	Zinc	487	558	632	700	788	844
	SUBSTANCES CONTAINING CARBON						
C	Carbon (graphite)	2802 s	2971 s	3132 s	3271 s	3438 s	3541 s
CBr_4	Tetrabromomethane	s	s	s	106.3	139.7	163.5
CNNa	Sodium cyanide	817	928	1046	1156	1302	1401
C_2H_5NO	Acetamide	65.0 s	92.0	120.0	145.8	178.3	200.0
$C_3H_8O_3$	Glycerol	125.5	153.8	182.2	208.0	240.0	263.0
$C_4H_8O_2$	*cis*-2-Butene-1,4-diol	81.5	109.6	137.6	162.9	194.5	214.8
$C_6H_{10}O_4$	Dimethyl succinate	34.4	53.0	91.1	117.4	151.1	173.2
$C_6H_{15}O_4P$	Triethyl phosphate	39.6	67.8	97.8	126.3	163.7	187.0
$C_7H_6Cl_2$	Benzal chloride	35.4	64.0	94.3	123.4	160.7	187.0
$C_7H_{12}O_4$	Dimethyl glutarate		79.8	108.2	134.1	166.8	188.2
$C_8H_{10}O$	3,4-Xylenol	66.2	93.8	122.0	148.0	181.5	203.6
$C_8H_{10}O$	3,5-Xylenol	62.0 s	89.2	117.0	143.5	176.2	197.8
C_9H_7N	Quinoline	59.7	89.6	119.8	148.1	186.2	212.3
C_9H_7N	Isoquinoline	63.5	92.7	123.7	152.0	190.0	214.5
$C_{10}H_{10}N_2$	Azobenzene	103.5	135.7	168.3	199.8	240.0	266.1
$C_{10}H_{10}O_4$	Dimethyl phthalate	100.3	131.8	164.0	194.0	232.7	257.8
$C_{10}H_{16}$	Camphene	s	s	60.4	85.0	117.5	138.7
$C_{10}H_{16}$	*d*-Limonene	14.0	40.4	68.2	94.6	128.5	151.4
$C_{10}H_{16}$	α-Pinene	-1.0	24.6	51.4	76.8	110.2	132.3
$C_{10}H_{16}$	β-Pinene	4.2	30.0	58.1	81.2	114.1	136.1
$C_{10}H_{16}O$	*d*-Camphor	41.5 s	68.6 s	97.5 s	124.0 s	157.9 s	182.0
$C_{10}H_{18}O$	α-Terpineol	52.8	80.4	109.8	136.3	171.2	194.3
$C_{10}H_{18}O_4$	Sebacic acid	183.0	215.7	250.0	279.8	313.2	332.8
$C_{10}H_{20}O_2$	Decanoic acid	125.0	142.0	165.0	189.8	217.1	240.3
$C_{10}H_{22}O_2$	Ethylene glycol dibutyl ether	48	71	96.8	121.4	154.9	178.3
$C_{11}H_8O_2$	1-Naphthoic acid	156.0	184.0	211.2	234.5	263.5	281.4
$C_{11}H_8O_2$	2-Naphthoic acid	160.8	189.7	216.9	241.3	270.3	289.5
$C_{11}H_{12}O_2$	Ethyl *trans*-cinnamate	87.6	108.5	150.3	181.2	219.3	245.0
$C_{11}H_{20}O_2$	10-Undecenoic acid	114.0	142.8	172.0	199.5	232.8	254.0
$C_{11}H_{22}O_2$	Undecanoic acid	101.4	133.1	166.0	197.2	237.8	262.8
$C_{12}H_9N$	Carbazole	s	s	s	248.2	292.5	323.0
$C_{12}H_{10}$	Acenaphthene	s	114.8	148.7	181.2	222.1	250.0
$C_{12}H_{10}O$	1-Acetonaphthone	115.6	146.3	178.4	208.6	246.7	270.5
$C_{12}H_{10}O$	2-Acetonaphthone	120.2	152.3	185.7	214.7	251.6	275.8
$C_{12}H_{10}S_2$	Diphenyl disulfide	131.6	164.0	197.0	226.2	262.6	285.8

Molecular formula	Name	0.13 kPa 1 mmHg	0.67 kPa 5 mmHg	2.66 kPa 20 mmHg	8.0 kPa 60 mmHg	26.7 kPa 200 mmHg	53.3 kPa 400 mmHg
		Temperature in °C for the indicated pressure					
$C_{12}H_{11}N$	Diphenylamine	108.3	141.7	175.2	206.9	247.5	274.1
$C_{12}H_{12}$	1-Ethylnaphthalene	70.0	101.4	133.8	164.1	204.6	230.8
$C_{12}H_{14}O_4$	Diethyl phthalate	108.8	140.7	173.6	204.1	243.0	267.5
$C_{12}H_{22}O4$	Dimethyl sebacate	104.0	139.8	175.8	208.0	245.0	269.6
$C_{12}H_{24}O_2$	Lauric acid	121.0	150.6	183.6	212.7	249.8	273.8
$C_{12}H_{27}O_4P$	Tributyl phosphate			167	198	237.2	262.8
$C_{13}H_{10}$	Fluorene	s	129.3	164.2	197.8	240.3	268.6
$C_{13}H_{10}O_2$	Phenyl benzoate	106.8	141.5	177.0	210.8	254.0	283.5
$C_{13}H_{10}O_3$	Phenyl salicylate	117.8	150.7	185.6	217.2	257.8	284.8
$C_{13}H_{26}O_2$	Tridecanoic acid	137.8	166.3	195.8	222.0	255.2	276.5
$C_{14}H_{10}$	Anthracene	145.0 s	173.5 s	201.9 s	231.8	279.0	310.2
$C_{14}H_{10}$	Phenanthrene	118.2	154.3	193.7	229.9	277.1	308.0
$C_{14}H_{10}O_2$	Benzil	128.4	165.2	202.8	238.2	283.5	314.3
$C_{14}H_{12}O_2$	Benzoin	135.6	170.2	207.0	241.7	284.4	313.5
$C_{14}H_{15}N$	Dibenzylamine	118.3	149.8	182.2	212.2	249.8	274.3
$C_{14}H_{26}O_4$	Diethyl sebacate	125.3	156.2	189.8	218.4	255.8	280.3
$C_{14}H_{28}O_2$	Tetradecanoic acid	142.0	174.1	207.6	237.2	272.3	294.6
$C_{14}H_{31}N$	Tetradecylamine	102.6	135.8	170.0	200.2	239.8	264.6
$C_{15}H_{30}O_3$	Methyl tetradecanoate	115.0	145.7	177.8	207.5	245.3	269.8
$C_{16}H_{22}O_4$	Dibutyl phthalate	148.2	182.1	216.2	247.8	287.0	313.5
$C_{16}H_{32}$	1-Hexadecene	101.6	131.7	162.0	190.8	226.8	250.0
$C_{16}H_{32}O_2$	Hexadecanoic acid	153.6	188.1	223.8	256.0	298.7	326.0
$C_{16}H_{34}$	Hexadecane	105.3	135.2	164.7	193.2	231.7	258.3
$C_{16}H_{34}O$	1-Hexadecanol	122.7	158.3	197.8	234.3	280.2	312.7
$C_{16}H_{36}N$	Hexadecylamine	123.6	157.8	195.7	228.8	272.2	300.4
$C_{17}H_{10}O$	Benzanthrone	225.0	274.5	322.5	368.8	426.5	
$C_{17}H_{34}O_2$	Methyl hexadecanoate	134.3	166.8	202.0			
$C_{17}H_{36}$	Heptadecane	115.0	145.2	177.7	207.3	247.8	274.5
$C_{18}H_9N$	Acridine	129.4	165.8	203.5	238.7	284.0	314.3
$C_{18}H_{30}$	Hexaethylbenzene	s	134.3	168.0	199.7	241.7	268.5
$C_{18}H_{34}O_2$	Oleic acid	176.5	208.5	240.0	269.8	309.8	334.7
$C_{18}H_{34}O_2$	Elaidic acid	171.3	206.7	242.3	273.0	312.4	337.0
$C_{18}H_{36}O$	Stearaldehyde	140.4	174.6	210.6	244.2	285.0	313.8
$C_{18}H_{36}O_2$	Stearic acid	173.7	209.0	243.4	275.5	316.5	343.0
$C_{18}H_{38}$	Octadecane	119.6	152.1	187.5	219.7	260.6	288.0
$C_{18}H_{38}O$	1-Octadecanol	150.3	185.6	220.0	252.7	293.5	320.3
$C_{19}H_{16}$	Triphenylmethane	169.7	188.4	206.8	221.2	239.7	249.8
$C_{19}H_{40}$	Nonadecane	133.2	166.3	200.8	232.8	271.8	299.8
$C_{21}H_{21}O_4P$	Tricresyl phosphate	154.6	184.2	213.2	239.8	271.8	292.7
$C_{21}H_{44}$	Heneicosane	152.6	188.0	223.2	255.3	296.5	323.8
$C_{22}H_{42}O_2$	Erucic acid	206.7	239.7	270.6	300.2	336.5	358.8
$C_{22}H_{42}O_2$	Brassidic acid	209.6	241.7	272.9	301.5	336.8	359.6
$C_{22}H_{46}$	Docosane	157.8	195.4	233.5	268.3	314.2	343.5
$C_{22}H_{48}$	Tricosane	170.0	206.3	242.0	273.8	313.5	339.8
$C_{24}H_{50}$	Tetracosane	183.8	219.6	255.3	288.4	330.3	358.0
$C_{25}H_{52}$	Pentacosane	194.2	230.0	266.1	298.4	339.0	365.4
$C_{26}H_{54}$	Hexacosane	204.0	240.0	275.8	307.8	348.4	374.6
$C_{27}H_{56}$	Heptacosane	211.7	248.6	284.6	318.3	359.4	385.0
$C_{28}H_{58}$	Octacosane	226.5	260.3	295.4	326.8	364.8	388.9
$C_{29}H_{60}$	Nonacosane	234.2	269.8	303.6	334.8	373.2	397.2

IUPAC RECOMMENDED DATA FOR VAPOR PRESSURE CALIBRATION

Vapor pressures are given in kPa
1 kPa = 0.0098692 atmos = 7.5006 Torr

Source: K. N. Marsh, Ed., *Recommended Reference Materials for the Realization of Physicochemical Properties*, Blackwell Scientific Publications,Oxford (1987).

T/K	CO_2(s)	H_2O(s)	$C_{10}H_8$(s)	n-C_5H_{12}	C_6H_6	C_6F_6	H_2O	Hg
180	27.62							
190	68.44							
200	155.11	0.0002						
210	327.17	0.0007						
220		0.0026						
230		0.0089						
240		0.0273						
250		0.0760		7.60				
260		0.1958	0.0001	12.98				
270		0.4701	0.0005	21.15			0.485	
280			0.0017	33.11	5.148	4.322	0.991	
290			0.0049	50.01	8.606	7.463	1.919	
300			0.0134	73.17	13.816	12.328	3.535	
310			0.0341	104.07	21.389	19.576	6.228	
320			0.0814	144.3	32.054	30.009	10.540	
330			0.1829	195.7	46.656	44.578	17.202	
340			0.3899	260.1	66.152	64.380	27.167	
350			0.7920	339.4	91.609	90.664	41.647	
360				435.9	124.192	124.816	62.139	
370				551.5	165.2	168.4	90.453	
380				688.8	215.9	223.0	128.74	
390				850.2	277.7	290.4	179.48	
400				1038	353.2	372.6	245.54	0.138
410				1256	441.0	471.5	330.15	0.215
420				1507	545.5	589.3	436.89	0.329
430				1793	667.6	728.3	569.73	0.493
440				2120	808.8	890.9	732.99	0.724
450				2490	971.1	1080	931.36	1.045
460				2910	1156	1297	1169.9	1.485
470					1366	1547	1453.9	2.078
480					1602	1833	1789.0	2.866
490					1868	2159	2181.4	3.899
500					2164	2530	2637.3	5.239
510					2494	2954	3163.3	6.955
520					2861		3766.4	9.131
530					3267		4453.9	11.861
540					3717		5233.5	15.256
550					4216		6113.4	19.438
560					4770		7102.0	25.547
570							8208.6	30.74
580							9443.0	38.19
590							10816	47.09
600							12339	57.64

ENTHALPY OF VAPORIZATION

The molar enthalpy (heat) of vaporization, $\Delta_{vap}H$, of about 700 inorganic and organic substances is tabulated here. Values are given, when available, both at the normal boiling point T_b, referred to a pressure of 101.325 kPa (760 mmHg), and at 25°C. Substances are listed by molecular formula in the modified Hill order (see Preface).

See Reference 4 for a discussion of the accuracy of the data and methods of estimating enthalpy of vaporization at other temperatures.

REFERENCES

1. Chase, M.W., et al., *JANAF Thermochemical Tables, Third Edition, J. Phys. Chem. Ref. Data*, 14, Suppl. 1, 1985.
2. Daubert, T. E., Danner, R. P., Sibul, H. M., and Stebbins, C. C., *Physical and Thermodynamic Properties of Pure Compounds: Data Compilation*, extant 1994 (core with 4 supplements), Taylor & Francis, Bristol, PA (also available as database).
3. Bartels, J., et al., *Landolt-Bornstein Numerical Values and Functions of Physics, Chemistry, Astronomy, Geophysics, and Technology*, Vol. 2, Part 4: Thermal Properties, Springer, Heidelberg, 1961.
4. Majer, V., and Svoboda, V., *Enthalpies of Vaporization of Organic Compounds*, Blackwell Scientific Publications, Oxford, 1985.

Molecular formula	Name	T_b °C	$\Delta_{vap}H$ (T_b) kJ/mol	$\Delta_{vap}H$ (25°C) kJ/mol
Compounds not containing carbon:				
AgBr	Silver bromide	1502	198	
AgCl	Silver chloride	1547	199	
AgI	Silver iodide	1506	143.9	
Al	Aluminum	2519	294	
AlB_3H_{12}	Aluminum borohydride	44.5	30	
$AlBr_3$	Aluminum tribromide	255	23.5	
AlI_3	Aluminum triiodide	382	32.2	
Ar	Argon	-185.85	6.43	
$AsBr_3$	Arsenic tribromide	221	41.8	
$AsCl_3$	Arsenic trichloride	130	35.01	
AsF_3	Arsenic trifluoride	57.8	29.7	
AsF_5	Arsenic pentafluoride	-52.8	20.8	
AsH_3	Arsine	-62.5	16.69	
AsI_3	Arsenic triiodide	424	59.3	
Au	Gold	2856	324	
B	Boron	4000	480	
BBr_3	Boron tribromide	91	30.5	
BCl_3	Boron trichloride	12.65	23.77	23.1
BF_3	Boron trifluoride	-101	19.33	
BI_3	Boron triiodide	210	40.5	
B_2F_4	Tetrafluorodiborane(4)	-34	28	
B_2H_6	Diborane	-92.4	14.28	
B_4H_{10}	Tetraborane	18	27.1	
B_5H_{11}	Pentaborane(11)	63	31.8	
Ba	Barium	1897	140	
$BeCl_2$	Beryllium chloride	482	105	
BeI_2	Beryllium iodide	487	70.5	
Bi	Bismuth	1564	151	
$BiBr_3$	Bismuth tribromide	453	75.4	
$BiCl_3$	Bismuth trichloride	447	72.61	
BrF	Bromine fluoride	20	25.1	
BrF_3	Bromine trifluoride	125.8	47.57	
BrF_5	Bromine pentafluoride	40.76	30.6	
BrH	Hydrogen bromide	-66.38		12.69
BrH_3Si	Bromosilane	1.9	24.4	
BrIn	Indium bromide (InBr)	656	92	
BrTl	Thallium bromide (TlBr)	819	99.56	
Br_2	Bromine (Br_2)	58.8	29.96	30.91
Br_2Cd	Cadmium bromide	844	115	
Br_2H_2Si	Dibromosilane	66	31	

ENTHALPY OF VAPORIZATION (continued)

Molecular formula	Name	T_b °C	$\Delta_{vap}H$ (T_b) kJ/mol	$\Delta_{vap}H$ (25°C) kJ/mol
Br_2Hg	Mercury bromide ($HgBr_2$)	322	58.89	
Br_2Pb	Lead bromide ($PbBr_2$)	892	133	
Br_2Sn	Tin bromide ($SnBr_2$)	639	102	
Br_2Zn	Zinc bromide	697	118	
Br_3Ga	Gallium bromide ($GaBr_3$)	279	38.9	
Br_3HSi	Tribromosilane	109	34.8	
Br_3OP	Phosphoryl bromide	191.7	38	
Br_3P	Phosphorus tribromide	172.95	38.8	
Br_3Sb	Antimony tribromide	280	59	
Br_4Ge	Germanium tetrabromide	186.35	41.4	
Br_4Si	Silicon tetrabromide	154	37.9	
Br_4Sn	Tin bromide ($SnBr_4$)	205	43.5	
Br_4Ti	Titanium bromide ($TiBr_4$)	230	44.37	
Br_5Ta	Tantalum bromide ($TaBr_5$)	349	62.3	
Cd	Cadmium	767	99.87	
$CdCl_2$	Cadmium chloride	960	124.3	
CdF_2	Cadmium fluoride	1748	214	
CdI_2	Cadmium iodide	742	115	
ClF	Chlorine fluoride	-101.1	24	
$ClFO_3$	Perchloryl fluoride	-46.75	19.33	
ClF_2P	Phosphorus chloride difluoride	-47.25	17.6	
ClF_3	Chlorine trifluoride	11.75	27.53	
ClF_3Si	Chlorotrifluorosilane	-70.0	18.7	
ClH	Hydrogen chloride	-85	16.15	9.08
ClH_3Si	Chlorosilane	-30.4	21	
ClNO	Nitrosyl chloride	-5.5	25.78	
$ClNO_2$	Nitryl chloride	-15	25.7	
ClO_2	Chlorine dioxide (ClO_2)	11	30	
ClTl	Thallium chloride (TlCl)	720	102.2	
Cl_2	Chlorine (Cl_2)	-34.04	20.41	17.65
Cl_2Cr	Chromium chloride ($CrCl_2$)	1300	197	
Cl_2CrO_2	Chromyl chloride (CrO_2Cl_2)	117	35.1	
Cl_2FP	Phosphorus dichloride fluoride	14	24.9	
Cl_2F_2Si	Dichlorodifluorosilane	-32	21.2	
Cl_2H_2Si	Dichlorosilane	8.3	25	24.2
Cl_2Hg	Mercury chloride ($HgCl_2$)	304	58.9	
Cl_2O	Oxygen dichloride	2.2	25.9	
Cl_2OS	Thionyl chloride	75.6	31.7	31
Cl_2O_2S	Sulfuryl chloride	69.4	31.4	30.1
Cl_2Pb	Lead chloride ($PbCl_2$)	951	127	
Cl_2Sn	Tin chloride ($SnCl_2$)	623	86.8	
Cl_2Ti	Titanium chloride ($TiCl_2$)	1500	232	
Cl_2Zn	Zinc chloride	732	126	
Cl_3Ga	Gallium chloride ($GaCl_3$)	201	23.9	
Cl_3HSi	Trichlorosilane	33		25.7
Cl_3OP	Phosphoryl chloride	105.5	34.35	38.6
Cl_3OV	Vanadium oxytrichloride	127	36.78	
Cl_3P	Phosphorus trichloride	75.95	30.5	32.1
Cl_3Sb	Antimony trichloride	220.3	45.19	
Cl_3Ti	Titanium chloride ($TiCl_3$)	960	124	
Cl_4Ge	Germanium tetrachloride	86.55	27.9	
Cl_4OW	Tungsten oxide tetrachloride	227.55	67.8	
Cl_4Si	Silicon tetrachloride	57.65	28.7	29.7
Cl_4Sn	Tin chloride ($SnCl_4$)	114.15	34.9	
Cl_4Te	Tellurium chloride ($TeCl_4$)	387	77	
Cl_4Th	Thorium chloride ($ThCl_4$)	921	146.4	
Cl_4Ti	Titanium chloride ($TiCl_4$)	136.45	36.2	

Molecular formula	Name	T_b °C	$\Delta_{vap}H$ (T_b) kJ/mol	$\Delta_{vap}H$ (25°C) kJ/mol
Cl_4V	Vanadium chloride (VCl_4)	148	41.4	42.5
Cl_5Mo	Molybdenum pentachloride	268	62.8	
Cl_5Nb	Niobium chloride ($NbCl_5$)	254.0	52.7	
Cl_5Ta	Tantalum chloride ($TaCl_5$)	239.35	54.8	
Cl_6W	Tungsten chloride (WCl_6)	346.75	52.7	
FH_3Si	Fluorosilane	-98.6	18.8	
FLi	Lithium fluoride	1673	147	
FNO	Nitrosyl fluoride	-59.9	19.28	
FNO_2	Nitryl fluoride	-72.4	18.05	
FNS	Thionitrosyl fluoride (NSF)	4.8	22.2	
F_2	Fluorine (F_2)	-188.12	6.62	
F_2H_2Si	Difluorosilane	-77.8	16.3	
F_2O	Oxygen difluoride	-144.75	11.09	
F_2OS	Thionyl fluoride	-43.8	21.8	
F_2O_2	Dioxygen difluoride (FOOF)	-57	19.1	
F_2Pb	Lead fluoride (PbF_2)	1293	160.4	
F_2Zn	Zinc fluoride	1500	190.1	
F_3HSi	Trifluorosilane	-95	16.2	
F_3N	Nitrogen trifluoride	-128.75	11.56	
F_3P	Phosphorus trifluoride	-101.5	16.5	
F_3PS	Thiophosphoryl trifluoride	-52.25	19.6	
F_4N_2	Tetrafluorohydrazine	-74	13.27	
F_4S	Sulfur fluoride (SF_4)	-40.45	26.44	
F_4Se	Selenium fluoride (SeF_4)	106	47.2	
F_4Th	Thorium fluoride (ThF_4)	1680	258	
F_5I	Iodine pentafluoride	100.5	41.3	
F_5Nb	Niobium fluoride (NbF_5)	229	52.3	
F_5P	Phosphorus pentafluoride	-84.6	17.2	
F_5Ta	Tantalum fluoride (TaF_5)	229.2	56.9	
F_5V	Vanadium fluoride (VF_5)	48.3	44.52	
F_6Ir	Iridium fluoride (IrF_6)	53	36	
F_6Mo	Molybdenum hexafluoride	34	27.2	28
F_6S	Sulfur fluoride (SF_6)			8.99
F_6W	Tungsten fluoride (WF_6)	17	27	26.6
Ga	Gallium	2204	254	
GaI_3	Gallium iodide (GaI_3)	340	56.5	
Ge	Germanium	2833	334	
GeH_4	Germane	-88.1	14.06	
Ge_2H_6	Digermane	30.8	25.1	
Ge_3H_8	Trigermane	110.5	32.2	
HI	Hydrogen iodide	-35.55	19.76	17.36
HLiO	Lithium hydroxide	1626	188	
HNO_3	Nitric acid	83		39.1
HN_3	Hydrazoic acid	35.7	30.5	
HNaO	Sodium hydroxide	1388	175	
H_2	Hydrogen (H_2)	-252.87	0.898	
H_2O	Water	100.0	40.65	43.98
H_2O_2	Hydrogen peroxide	150.2		51.6
H_2S	Hydrogen sulfide	-59.55	18.67	14.08
H_2S_2	Hydrogen sulfide (H_2S_2)	70.7		33.78
H_2Se	Hydrogen selenide	-41.25	19.7	
H_2Te	Hydrogen telluride	-2	19.2	
H_3N	Ammonia	-33.33	23.33	19.86
H_3P	Phosphine	-87.75	14.6	
H_3Sb	Stibine	-17	21.3	
H_4N_2	Hydrazine	113.55	41.8	44.7
H_4P_2	Diphosphine (P_2H_4)	63.5	28.8	

Molecular formula	Name	T_b °C	$\Delta_{vap}H$ (T_b) kJ/mol	$\Delta_{vap}H$ (25°C) kJ/mol
H_4Si	Silane	-111.9	12.1	
H_4Sn	Stannane	-51.8	19.05	
H_6Si_2	Disilane	-14.3	21.2	
H_8Si_3	Trisilane	52.9	28.5	
He	Helium	-268.93	0.0829	
Hg	Mercury	356.73	59.11	
HgI_2	Mercury iodide (HgI_2)	354	59.2	
IIn	Indium iodide (InI)	712	90.8	
ITl	Thallium iodide (TlI)	824	104.7	
I_2	Iodine (I_2)	184.4	41.57	
I_2Pb	Lead iodide (PbI_2)	872	104	
I_2Sn	Tin iodide (SnI_2)	714	105	
I_3P	Phosphorus triiodide	227	43.9	
I_3Sb	Antimony triiodide	401	68.6	
I_4Si	Silicon tetraiodide	287.35	50.2	
I_4Sn	Tin iodide (SnI_4)	364.35	56.9	
I_4Ti	Titanium iodide (TiI_4)	377	58.4	
Kr	Krypton	-153.22	9.08	
MoO_3	Molybdenum oxide (MoO_3)	1155	138	
NO	Nitric oxide	-151.74	13.83	
N_2	Nitrogen (N_2)	-195.79	5.57	
N_2O	Nitrous oxide	-88.48	16.53	
N_2O_4	Nitrogen tetroxide	21.15	38.12	
Ne	Neon	-246.08	1.71	
O_2	Oxygen (O_2)	-182.95	6.82	
O_2S	Sulfur dioxide	-10.05	24.94	22.92
O_3S	Sulfur trioxide	45	40.69	43.14
P	Phosphorus	280.5	12.4	14.2
Pb	Lead	1749	179.5	
S	Sulfur	444.60	45	
STl_2	Thallium sulfide (Tl_2S)	1367	154	
Se	Selenium	685	95.48	
Te	Tellurium	988	114.1	
Xe	Xenon	-108.04	12.62	

Compounds containing carbon:

Molecular formula	Name	T_b °C	$\Delta_{vap}H$ (T_b) kJ/mol	$\Delta_{vap}H$ (25°C) kJ/mol
CCl_4	Tetrachloromethane	76.8	29.82	32.43
CN_4O_8	Tetranitromethane	126.1	40.74	49.93
CO	Carbon monoxide	-191.5	6.04	
CS_2	Carbon disulfide	46	26.74	27.51
$CHBr_3$	Tribromomethane	149.1	39.66	46.05
$CHCl_3$	Trichloromethane	61.17	29.24	31.28
CH_2Br_2	Dibromomethane	97	32.92	36.97
CH_2Cl_2	Dichloromethane	40	28.06	28.82
CH_2O_2	Formic acid	101	22.69	20.10
CH_3Br	Bromomethane	3.5	23.91	22.81
CH_3Cl	Chloromethane	-24.09	21.40	18.92
CH_3I	Iodomethane	42.55	27.34	27.97
CH_3NO	Formamide	220		60.15
CH_3NO_2	Nitromethane	101.19	33.99	38.27
CH_4	Methane	-161.48	8.19	
CH_4O	Methanol	64.6	35.21	37.43
CH_5N	Methylamine	-6.32	25.60	23.37
CH_6N_2	Methylhydrazine	87.5	36.12	40.37
$C_2Br_2ClF_3$	1,2-Dibromo-1-chloro-1,2,2-trifluoroethane	93	31.17	35.04
$C_2Br_2F_4$	1,2-Dibromotetrafluoroethane	47.35	27.03	28.39

Molecular formula	Name	T_b °C	$\Delta_{vap}H$ (T_b) kJ/mol	$\Delta_{vap}H$ (25°C) kJ/mol
C_2ClF_5	Chloropentafluoroethane	-37.95	19.41	
$C_2Cl_3F_3$	1,1,1-Trichlorotrifluoroethane	46.1	26.85	28.08
$C_2Cl_3F_3$	1,1,2-Trichlorotrifluoroethane	47.7	27.04	28.40
C_2Cl_4	Tetrachloroethylene	121.3	34.68	39.68
C_2F_6	Hexafluoroethane	-78.1	16.15	
C_2N_2	Cyanogen	-21.1	23.33	19.75
$C_2HBrClF_3$	1,1,1-Trifluoro-2-bromo-2-chloroethane	50.2	28.08	29.61
C_2HCl_3	Trichloroethylene	87.21	31.40	34.54
$C_2H_2Cl_2$	1,1-Dichloroethylene	31.6	26.14	26.48
$C_2H_2Cl_4$	1,1,2,2-Tetrachloroethane	146.5	37.64	45.71
$C_2H_3Cl_3$	1,1,1-Trichloroethane	74.09	29.86	32.50
$C_2H_3Cl_3$	1,1,2-Trichloroethane	113.8	34.82	40.24
C_2H_3N	Acetonitrile	81.65	29.75	32.94
C_2H_4	Ethylene	-103.77	13.53	
$C_2H_4Br_2$	1,2-Dibromoethane	131.6	34.77	41.73
$C_2H_4Cl_2$	1,1-Dichloroethane	57.4	28.85	30.62
$C_2H_4Cl_2$	1,2-Dichloroethane	83.5	31.98	35.16
$C_2H_4F_2$	1,1-Difluoroethane	-24.95	21.56	19.08
C_2H_4O	Acetaldehyde	20.1	25.76	25.47
C_2H_4O	Ethylene oxide	10.6	25.54	24.75
$C_2H_4O_2$	Acetic acid	117.9	23.70	23.36
$C_2H_4O_2$	Methyl formate	31.7	27.92	28.35
C_2H_5Br	Bromoethane	38.5	27.04	28.03
C_2H_5Cl	Chloroethane	12.3	24.65	
C_2H_5I	Iodoethane	72.5	29.44	31.93
C_2H_5NO	*N*-Methylformamide	199.51		56.19
C_2H_6	Ethane	-88.6	14.69	5.16
C_2H_6O	Dimethyl ether	-24.8	21.51	18.51
C_2H_6O	Ethanol	78.29	38.56	42.32
C_2H_6S	Dimethyl sulfide	37.33	27.0	27.65
C_2H_6S	Ethanethiol	35.1	26.79	27.30
$C_2H_6S_2$	Dimethyl disulfide	109.8	33.78	37.86
$C_2H_6S_2$	1,2-Ethanedithiol	146.1	37.93	44.68
C_2H_7N	Dimethylamine	6.88	26.40	25.05
C_2H_7NO	Ethanolamine	171	49.83	
$C_2H_8N_2$	1,1-Dimethylhydrazine	63.9	32.55	35.0
$C_2H_8N_2$	1,2-Ethanediamine	117	37.98	44.98
$C_3Cl_2F_6$	1,2-Dichlorohexafluoropropane	34.1	26.28	26.93
$C_3H_3Cl_3O_2$	Methyl trichloroacetate	153.8		48.33
$C_3H_4Cl_2O_2$	Methyl dichloroacetate	142.9	39.28	47.72
$C_3H_4O_2$	2-Oxetanone	162		47.03
C_3H_5Br	3-Bromopropene	70.1	30.24	32.73
$C_3H_5ClO_2$	Methyl chloroacetate	129.5	39.23	46.73
C_3H_5N	Propanenitrile	97.14	31.81	36.03
C_3H_6	Propene	-47.69	18.42	14.24
C_3H_6	Cyclopropane	-32.81	20.05	16.93
$C_3H_6Br_2$	1,2-Dibromopropane	141.9	35.61	41.67
$C_3H_6Br_2$	1,3-Dibromopropane	167.3		47.45
$C_3H_6Cl_2$	1,3-Dichloropropane	120.9	35.18	40.75
C_3H_6O	Acetone	56.05	29.10	30.99
C_3H_6O	Propanal	48	28.31	29.62
C_3H_6O	Methyloxirane	35	27.35	27.89
C_3H_6O	Oxetane	47.6	28.67	29.85
$C_3H_6O_2$	Ethyl formate	54.4	29.91	31.96
$C_3H_6O_2$	Methyl acetate	56.87	30.32	32.29
$C_3H_6O_2$	Propanoic acid	141.15		32.14
C_3H_6S	Thiacyclobutane	95	32.32	35.97

Molecular formula	Name	T_b °C	$\Delta_{vap}H$ (T_b) kJ/mol	$\Delta_{vap}H$ (25°C) kJ/mol
C_3H_7Br	1-Bromopropane	71.1	29.84	32.01
C_3H_7Br	2-Bromopropane	59.5	28.33	30.17
C_3H_7Cl	1-Chloropropane	46.5	27.18	28.35
C_3H_7Cl	2-Chloropropane	35.7	26.30	26.90
C_3H_7I	1-Iodopropane	102.6	32.08	36.25
C_3H_7I	2-Iodopropane	89.5	30.68	34.06
C_3H_7NO	*N*-Ethylformamide	198		58.44
C_3H_7NO	*N,N*-Dimethylformamide	153		46.89
C_3H_8	Propane	-42.1	19.04	14.79
C_3H_8O	1-Propanol	97.2	41.44	47.45
C_3H_8O	2-Propanol	82.3	39.85	45.39
$C_3H_8O_2$	2-Methoxyethanol	124.1	37.54	45.17
C_3H_8S	Ethyl methyl sulfide	66.7	29.53	31.85
C_3H_8S	1-Propanethiol	67.8	29.54	31.89
C_3H_8S	2-Propanethiol	52.6	27.91	29.45
$C_3H_8S_2$	1,3-Propanedithiol	172.9		49.66
C_3H_9N	Propylamine	47.22	29.55	31.27
C_3H_9N	Isopropylamine	31.76	27.83	28.36
C_3H_9N	Trimethylamine	2.87	22.94	21.66
$C_3H_{10}N_2$	1,3-Propanediamine	139.8	40.85	50.16
$C_4H_4N_2$	Pyridazine	208		53.47
$C_4H_4N_2$	Pyrimidine	123.8	43.09	49.79
C_4H_4O	Furan	31.5	27.10	27.45
$C_4H_4O_2$	Diketene	126.1	36.80	42.89
C_4H_4S	Thiophene	84.0	31.48	34.70
$C_4H_5Cl_3O_2$	Ethyl trichloroacetate	167.5		50.97
C_4H_5N	Cyclopropanecarbonitrile	135.1	35.55	41.94
C_4H_5N	Pyrrole	129.79	38.75	45.09
C_4H_5NS	4-Methylthiazole	133.3	37.58	43.85
C_4H_6	1,2-Butadiene	10.9	24.02	23.21
C_4H_6	1,3-Butadiene	-4.41	22.47	20.86
C_4H_6	1-Butyne	8.08	24.52	23.35
$C_4H_6Cl_2O_2$	Ethyl dichloroacetate			50.60
C_4H_6S	2,3-Dihydrothiophene	112.1	33.24	37.74
C_4H_6S	2,5-Dihydrothiophene	122.4	34.83	39.95
$C_4H_7ClO_2$	Ethyl chloroacetate	144.3	40.43	49.47
C_4H_7N	Butanenitrile	117.6	33.68	39.33
C_4H_7N	2-Methylpropanenitrile	103.9	32.39	37.13
C_4H_8	1-Butene	-6.26	22.07	20.22
C_4H_8	*cis*-2-Butene	3.71	23.34	22.16
C_4H_8	*trans*-2-Butene	0.88	22.72	21.40
C_4H_8	Cyclobutane	12.6	24.19	23.51
$C_4H_8Br_2$	1,4-Dibromobutane	197		53.09
$C_4H_8Cl_2$	1,2-Dichlorobutane	124.1	33.90	39.58
$C_4H_8Cl_2$	1,4-Dichlorobutane	161		46.36
C_4H_8O	2-Butanone	79.59	31.30	34.79
C_4H_8O	Tetrahydrofuran	65	29.81	31.99
$C_4H_8O_2$	1,3-Dioxane	106.1	34.37	39.09
$C_4H_8O_2$	1,4-Dioxane	101.5	34.16	38.60
$C_4H_8O_2$	Ethyl acetate	77.11	31.94	35.60
$C_4H_8O_2$	Methyl propanoate	79.8	32.24	35.85
$C_4H_8O_2$	Propyl formate	80.9	33.61	37.53
$C_4H_8O_2$	Butanoic acid	163.75		40.45
$C_4H_8O_2$	2-Methylpropanoic acid	154.45		35.30
C_4H_8S	Tetrahydrothiophene	121.0	34.66	39.43
C_4H_9Br	1-Bromobutane	101.6	32.51	36.64
C_4H_9Br	1-Bromo-2-methylpropane	91.1	31.33	34.82

Molecular formula	Name	T_b °C	$\Delta_{vap}H$ (T_b) kJ/mol	$\Delta_{vap}H$ (25°C) kJ/mol
C_4H_9Br	2-Bromobutane	91.4	30.77	34.41
C_4H_9Br	2-Bromo-2-methylpropane	73.3	29.23	31.81
C_4H_9Cl	1-Chlorobutane	78.6	30.39	33.51
C_4H_9Cl	1-Chloro-2-methylpropane	68.5	29.22	31.67
C_4H_9Cl	2-Chlorobutane	68.3	29.17	31.53
C_4H_9Cl	2-Chloro-2-methylpropane	50.9	27.55	28.98
C_4H_9I	1-Iodobutane	130.6	34.66	40.63
C_4H_9I	1-Iodo-2-methylpropane	121.1	33.54	38.83
C_4H_9I	2-Iodobutane	120.1	33.27	38.46
C_4H_9I	2-Iodo-2-methylpropane	100.1	31.43	35.41
C_4H_9N	Pyrrolidine	86.56	33.01	37.52
C_4H_9NO	*N*-Ethylacetamide	205		64.89
C_4H_9NO	*N,N*-Dimethylacetamide	165		50.24
C_4H_{10}	Butane	-0.5	22.44	21.02
C_4H_{10}	Isobutane	-11.73	21.30	19.23
$C_4H_{10}O$	1-Butanol	117.73	43.29	52.35
$C_4H_{10}O$	2-Butanol	99.51	40.75	49.72
$C_4H_{10}O$	2-Methyl-2-propanol	82.4	39.07	46.69
$C_4H_{10}O$	2-Methyl-1-propanol	107.89	41.82	50.82
$C_4H_{10}O$	Diethyl ether	34.5	26.52	27.10
$C_4H_{10}O$	Methyl propyl ether	39.1	26.75	27.60
$C_4H_{10}O$	Isopropyl methyl ether	30.77	26.05	26.41
$C_4H_{10}O_2$	2-Ethoxyethanol	135	39.22	48.21
$C_4H_{10}O_2$	1,2-Dimethoxyethane	85	32.42	36.39
$C_4H_{10}S$	1-Butanethiol	98.5	32.23	36.63
$C_4H_{10}S$	2-Butanethiol	85	30.59	33.99
$C_4H_{10}S$	2-Methyl-1-propanethiol	88.5	31.01	34.63
$C_4H_{10}S$	2-Methyl-2-propanethiol	64.3	28.45	30.78
$C_4H_{10}S$	Diethyl sulfide	92.1	31.77	35.80
$C_4H_{10}S$	Methyl propyl sulfide	95.6	32.08	36.24
$C_4H_{10}S$	Isopropyl methyl sulfide	84.8	30.71	34.15
$C_4H_{10}S_2$	1,4-Butanedithiol			55.10
$C_4H_{10}S_2$	Diethyl disulfide	154.1	37.58	45.18
$C_4H_{11}N$	Butylamine	77.00	31.81	35.72
$C_4H_{11}N$	Isobutylamine	67.75	30.61	33.85
$C_4H_{11}N$	*sec*-Butylamine	62.73	29.92	32.85
$C_4H_{11}N$	*tert*-Butylamine	44.04	28.27	29.64
$C_4H_{11}N$	Diethylamine	55.5	29.06	31.31
$C_4H_{11}N$	Isopropylmethylamine	50.4	28.71	30.69
$C_5H_2F_6O_2$	Hexafluoroacetylacetone	54.15	27.05	30.58
C_5H_5N	Pyridine	115.23	35.09	40.21
C_5H_6S	2-Methylthiophene	112.6	33.90	38.87
C_5H_6S	3-Methylthiophene	115.5	34.24	39.43
C_5H_7N	*trans*-3-Pentenenitrile	142.6	37.09	44.77
C_5H_7N	Cyclobutanecarbonitrile	149.6	36.88	44.34
C_5H_8	Spiropentane	39	26.76	27.49
C_5H_8O	Cyclopentanone	130.57	36.35	42.72
C_5H_8O	Cyclopropyl methyl ketone	111.3	34.07	39.41
$C_5H_8O_2$	2,4-Pentanedione	138	34.30	41.77
$C_5H_8O_2$	Methyl cyclopropanecarboxylate	114.9	35.25	41.27
C_5H_9N	Pentanenitrile	141.3	36.09	43.60
C_5H_9N	3-Methylbutanenitrile	127.5	35.10	41.64
C_5H_9N	2,2-Dimethylpropanenitrile	106.1	32.40	37.35
C_5H_{10}	1-Pentene	29.96	25.20	25.47
C_5H_{10}	*cis*-2-Pentene	36.93		26.86
C_5H_{10}	*trans*-2-Pentene	36.34		26.76
C_5H_{10}	2-Methyl-1-butene	31.2	25.50	25.86

Molecular formula	Name	T_b °C	$\Delta_{vap}H$ (T_b) kJ/mol	$\Delta_{vap}H$ (25°C) kJ/mol
C_5H_{10}	2-Methyl-2-butene	38.56	26.31	27.06
C_5H_{10}	3-Methyl-1-butene	20.1		23.80
C_5H_{10}	Cyclopentane	49.3	27.30	28.52
$C_5H_{10}Cl_2$	1,2-Dichloropentane	148.3	36.45	43.89
$C_5H_{10}Cl_2$	1,5-Dichloropentane	179		50.71
$C_5H_{10}O$	Cyclopentanol	140.42		57.60
$C_5H_{10}O$	2-Pentanone	102.26	33.44	38.40
$C_5H_{10}O$	3-Pentanone	101.96	33.45	38.52
$C_5H_{10}O$	3-Methyl-2-butanone	94.33	32.35	36.78
$C_5H_{10}O$	3,3-Dimethyloxetane	80.6	30.85	33.94
$C_5H_{10}O$	Tetrahydropyran	88	31.17	34.58
$C_5H_{10}O_2$	Butyl formate	106.1	36.58	41.11
$C_5H_{10}O_2$	Propyl acetate	101.54	33.92	39.72
$C_5H_{10}O_2$	Isopropyl acetate	88.6	32.93	37.20
$C_5H_{10}O_2$	Ethyl propanoate	99.1	33.88	39.21
$C_5H_{10}O_2$	Methyl butanoate	102.8	33.79	39.28
$C_5H_{10}O_2$	Methyl isobutanoate	92.5	32.61	37.32
$C_5H_{10}O_2$	2-Methylbutanoic acid			46.91
$C_5H_{10}O_3$	Diethyl carbonate			43.60
$C_5H_{10}S$	Cyclopentanethiol	132.1	35.32	41.42
$C_5H_{10}S$	Thiacyclohexane	141.8	35.96	42.58
$C_5H_{11}Br$	1-Bromopentane	129.8	35.01	41.28
$C_5H_{11}Cl$	1-Chloropentane	107.8	33.15	38.24
$C_5H_{11}Cl$	1-Chloro-3-methylbutane	98.9	32.02	36.24
$C_5H_{11}Cl$	2-Chloropentane	97.1	31.79	36.03
$C_5H_{11}I$	1-Iodopentane	155		45.27
$C_5H_{11}N$	Piperidine	106.22		39.29
C_5H_{12}	Pentane	36.06	25.79	26.43
C_5H_{12}	Isopentane	27.88	24.69	24.85
C_5H_{12}	Neopentane	9.48	22.74	21.84
$C_5H_{12}O$	Butyl methyl ether	70.16	29.55	32.37
$C_5H_{12}O$	Isobutyl methyl ether	58.6	28.02	30.13
$C_5H_{12}O$	*sec*-Butyl methyl ether	59.1	28.09	30.23
$C_5H_{12}O$	*tert*-Butyl methyl ether	55.2	27.94	29.82
$C_5H_{12}O$	Ethyl propyl ether	63.21	28.94	31.43
$C_5H_{12}O$	Ethyl isopropyl ether	54.1	28.21	30.08
$C_5H_{12}O$	1-Pentanol	137.98	44.36	57.02
$C_5H_{12}O$	2-Pentanol	119.3	41.40	54.21
$C_5H_{12}O$	3-Pentanol	116.25		54.0
$C_5H_{12}O$	2-Methyl-1-butanol	128		55.16
$C_5H_{12}O$	2-Methyl-2-butanol	102.4	39.04	50.10
$C_5H_{12}O$	3-Methyl-1-butanol	131.1	44.07	55.61
$C_5H_{12}O$	3-Methyl-2-butanol			53.0
$C_5H_{12}O_2$	1-Ethoxy-2-methoxyethane	102.1	34.33	39.83
$C_5H_{12}O_2$	Diethoxymethane	88	31.33	35.65
$C_5H_{12}O_2$	2-Propoxyethanol	149.8	41.40	52.12
$C_5H_{12}S$	Butyl methyl sulfide	123.5	34.47	40.46
$C_5H_{12}S$	*tert*-Butyl methyl sulfide	99	31.47	35.84
$C_5H_{12}S$	Ethyl propyl sulfide	118.6	34.24	39.97
$C_5H_{12}S$	Ethyl isopropyl sulfide	107.5	32.74	37.78
$C_5H_{12}S$	1-Pentanethiol	126.6	34.88	41.24
$C_5H_{12}S$	2-Methyl-1-butanethiol	119.1	33.79	39.45
$C_5H_{12}S$	2-Methyl-2-butanethiol	99.1	31.37	35.67
$C_5H_{13}N$	Ethylisopropylamine	69.6	29.94	33.13
$C_5H_{13}N$	Pentylamine	104.3	34.01	40.08
C_6ClF_5	Chloropentafluorobenzene	117.96	34.76	41.07
C_6F_6	Hexafluorobenzene	80.26	31.66	35.71

Molecular formula	Name	T_b °C	$\Delta_{vap}H$ (T_b) kJ/mol	$\Delta_{vap}H$ (25°C) kJ/mol
C_6MoO_6	Molybdenum hexacarbonyl	701.2	72.51	
C_6HF_5	Pentafluorobenzene	85.74	32.15	36.27
$C_6H_4Cl_2$	*o*-Dichlorobenzene	180	39.66	50.21
$C_6H_4Cl_2$	*m*-Dichlorobenzene	173	38.62	48.58
$C_6H_4Cl_2$	*p*-Dichlorobenzene	174	38.79	49.0
$C_6H_4F_2$	*o*-Difluorobenzene	94	32.21	36.18
$C_6H_4F_2$	*m*-Difluorobenzene	82.6	31.10	34.59
$C_6H_4F_2$	*p*-Difluorobenzene	89	31.77	35.54
C_6H_5Br	Bromobenzene	156.06		44.54
C_6H_5Cl	Chlorobenzene	131.72	35.19	40.97
C_6H_5F	Fluorobenzene	84.73	31.19	34.58
$C_6H_5NO_2$	Nitrobenzene	210.8		55.01
C_6H_6	Benzene	80.09	30.72	33.83
C_6H_6O	Phenol	181.87	45.69	57.82
C_6H_6S	Benzenethiol	169.1	39.93	47.56
C_6H_7N	Aniline	184.17	42.44	55.83
C_6H_7N	2-Methylpyridine	129.38	36.17	42.48
C_6H_7N	3-Methylpyridine	144.14	37.35	44.44
C_6H_7N	4-Methylpyridine	145.36	37.51	44.56
C_6H_7N	1-Cyclopentenecarbonitrile			44.98
C_6H_9N	Cyclopentanecarbonitrile			43.43
$C_6H_9NO_3$	Triacetamide			60.41
C_6H_{10}	Cyclohexene	82.98	30.46	33.47
$C_6H_{10}O$	Cyclohexanone	155.43		45.06
$C_6H_{10}O_2$	Methyl cyclobutanecarboxylate	135.5	37.13	44.72
$C_6H_{10}O_4$	Ethylene glycol diacetate	190		61.44
$C_6H_{11}N$	Hexanenitrile	163.65		47.91
C_6H_{12}	Cyclohexane	80.73	29.97	33.01
C_6H_{12}	Methylcyclopentane	71.8	29.08	31.64
C_6H_{12}	Ethylcyclobutane	70.8	28.67	31.24
C_6H_{12}	1-Hexene	63.48		30.61
C_6H_{12}	*cis*-2-Hexene	68.8		32.19
C_6H_{12}	*trans*-2-Hexene	67.9		31.60
C_6H_{12}	2-Methyl-1-pentene	62.1		30.48
C_6H_{12}	2-Methyl-2-pentene	67.3		31.60
C_6H_{12}	4-Methyl-1-pentene	53.9		28.71
C_6H_{12}	4-Methyl-*cis*-2-pentene	56.3		29.48
C_6H_{12}	4-Methyl-*trans*-2-pentene	58.6		29.97
C_6H_{12}	2,3-Dimethyl-1-butene	55.6		29.18
C_6H_{12}	2,3-Dimethyl-2-butene	73.3	29.64	32.53
C_6H_{12}	2-Ethyl-1-butene	64.7		31.13
$C_6H_{12}Cl_2$	1,2-Dichlorohexane	173		48.16
$C_6H_{12}O$	Cyclohexanol	160.84		62.01
$C_6H_{12}O$	2-Hexanone	127.6	36.35	43.14
$C_6H_{12}O$	3-Hexanone	123.5	35.36	42.47
$C_6H_{12}O$	2-Methyl-3-pentanone	113.5	33.84	39.79
$C_6H_{12}O$	3-Methyl-2-pentanone	117.5	34.16	40.53
$C_6H_{12}O$	4-Methyl-2-pentanone	116.5	34.49	40.61
$C_6H_{12}O$	3,3-Dimethyl-2-butanone	106.1	33.39	37.91
$C_6H_{12}O$	Butyl vinyl ether	94	31.58	36.17
$C_6H_{12}O_2$	Butyl acetate	126.1	36.28	43.86
$C_6H_{12}O_2$	*tert*-Butyl acetate	95.1	33.07	38.03
$C_6H_{12}O_2$	Propyl propanoate	122.5	35.54	43.45
$C_6H_{12}O_2$	Ethyl butanoate	121.5	35.47	42.68
$C_6H_{12}O_2$	Ethyl 2-methylpropanoate	110.1	33.67	39.83
$C_6H_{12}O_2$	Methyl pentanoate	127.4	35.36	43.10
$C_6H_{12}O_2$	Methyl 2,2-dimethylpropanoate	101.1	33.42	38.76

Molecular formula	Name	T_b °C	$\Delta_{vap}H$ (T_b) kJ/mol	$\Delta_{vap}H$ (25°C) kJ/mol
$C_6H_{12}S$	Cyclohexanethiol	158.9	37.06	44.57
$C_6H_{13}Br$	1-Bromohexane	155.3		45.89
$C_6H_{13}Cl$	1-Chlorohexane	135	35.67	42.83
$C_6H_{13}I$	1-Iodohexane	181		49.75
$C_6H_{13}N$	Cyclohexylamine	134	36.14	43.67
C_6H_{14}	Hexane	68.73	28.85	31.56
C_6H_{14}	2-Methylpentane	60.26	27.79	29.89
C_6H_{14}	3-Methylpentane	63.27	28.06	30.28
C_6H_{14}	2,2-Dimethylbutane	49.73	26.31	27.68
C_6H_{14}	2,3-Dimethylbutane	57.98	27.38	29.12
$C_6H_{14}N_2$	Azopropane			39.88
$C_6H_{14}O$	Methyl pentyl ether	99	32.02	36.85
$C_6H_{14}O$	Butyl ethyl ether	92.3	31.63	36.32
$C_6H_{14}O$	Dipropyl ether	90.08	31.31	35.69
$C_6H_{14}O$	Diisopropyl ether	68.51	29.10	32.12
$C_6H_{14}O$	1-Hexanol	157.6	44.50	61.61
$C_6H_{14}O$	2-Hexanol	138	41.01	58.46
$C_6H_{14}O$	4-Methyl-1-pentanol	151.9	44.46	60.47
$C_6H_{14}O$	2-Methyl-2-pentanol	121.1	39.59	54.77
$C_6H_{14}O_2$	1,1-Diethoxyethane	102.25	36.28	43.20
$C_6H_{14}O_2$	1,2-Diethoxyethane	119.4	36.28	43.20
$C_6H_{14}O_2$	2-Butoxyethanol	168.4		56.59
$C_6H_{14}O_3$	Bis(ethoxymethyl)ether	140.6	36.17	44.69
$C_6H_{14}O_3$	Diethylene glycol dimethyl ether	162	36.17	44.69
$C_6H_{14}S$	Methyl pentyl sulfide	145.1	37.41	45.24
$C_6H_{14}S$	Butyl ethyl sulfide	144.3	37.01	44.51
$C_6H_{14}S$	Dipropyl sulfide	142.9	36.60	44.21
$C_6H_{14}S$	Isopropyl propyl sulfide	132.1	35.11	41.78
$C_6H_{14}S$	Diisopropyl sulfide	120.1	33.80	39.60
$C_6H_{15}N$	Hexylamine	132.8	36.54	45.10
$C_6H_{15}N$	Butylethylamine	107.5	33.97	40.15
$C_6H_{15}N$	Isopropylpropylamine	96.9	32.14	37.23
$C_6H_{15}N$	Triethylamine	89	31.01	34.84
$C_6H_{15}N$	Dipropylamine	109.3	33.47	40.04
$C_6H_{15}N$	Diisopropylamine	83.9	30.40	34.61
$C_7H_3F_5$	2,3,4,5,6-Pentafluorotoluene	117.5	34.75	41.12
$C_7H_5F_3$	(Trifluoromethyl)benzene	102.1	32.63	37.60
C_7H_7F	*p*-Fluorotoluene	116.6	34.08	39.42
C_7H_8	Toluene	110.63	33.18	38.01
C_7H_8O	*o*-Cresol	191.04	45.19	
C_7H_8O	*m*-Cresol	202.27	47.40	61.71
C_7H_8O	*p*-Cresol	201.98	47.45	
C_7H_8O	Benzyl alcohol	205.31	50.48	
C_7H_8O	Anisole	153.7	38.97	46.90
C_7H_9N	1-Cyclohexenecarbonitrile			53.55
C_7H_9N	2,3-Dimethylpyridine	161.26	39.08	47.73
C_7H_9N	2,4-Dimethylpyridine	158.5	38.53	47.48
C_7H_9N	2,6-Dimethylpyridine	144.1	37.46	45.36
C_7H_9N	3,4-Dimethylpyridine	179.13	39.99	50.53
C_7H_9N	3,5-Dimethylpyridine	172	39.46	49.48
C_7H_9N	Benzylamine	185		60.16
$C_7H_{10}O$	Dicyclopropyl ketone	161		53.70
$C_7H_{11}N$	Cyclohexanecarbonitrile			51.92
C_7H_{12}	1-Methylbicyclo(3,1,0)hexane	93.1	31.07	34.77
C_7H_{14}	Methylcyclohexane	100.93	31.27	35.36
C_7H_{14}	Ethylcyclopentane	103.5	31.96	36.40
C_7H_{14}	*cis*-1,3-Dimethylcyclopentane	90.8	30.40	34.20

Molecular formula	Name	T_b °C	$\Delta_{vap}H$ (T_b) kJ/mol	$\Delta_{vap}H$ (25°C) kJ/mol
C_7H_{14}	1-Heptene	93.64		35.49
$C_7H_{14}O$	2-Heptanone	151.05		47.24
$C_7H_{14}O$	2,4-Dimethyl-3-pentanone	125.4	34.64	41.51
$C_7H_{14}O$	2,2-Dimethyl-3-pentanone	125.6	36.09	42.34
$C_7H_{14}O_2$	Ethyl 2,2-dimethylpropanoate	118.4	34.51	41.25
$C_7H_{14}O_2$	Methyl hexanoate	149.5	38.55	48.04
$C_7H_{14}O_2$	Ethyl pentanoate	146.1	36.96	47.01
$C_7H_{15}Br$	1-Bromoheptane	179		50.60
$C_7H_{15}Cl$	1-Chloroheptane	159		47.66
C_7H_{16}	Heptane	98.5	31.77	36.57
C_7H_{16}	2-Methylhexane	90.04	30.62	34.87
C_7H_{16}	3-Methylhexane	91.84	30.89	35.06
C_7H_{16}	2,2-Dimethylpentane	79.2	29.23	32.42
C_7H_{16}	2,3-Dimethylpentane	89.78	30.46	34.26
C_7H_{16}	2,4-Dimethylpentane	80.49	29.55	32.88
C_7H_{16}	3,3-Dimethylpentane	86.06	29.62	33.03
C_7H_{16}	3-Ethylpentane	93.5	31.12	35.22
C_7H_{16}	2,2,3-Trimethylbutane	80.86	28.90	32.05
$C_7H_{16}O$	Hexyl methyl ether	126.1	34.93	42.07
$C_7H_{16}O$	Ethyl pentyl ether	117.6	34.41	41.01
$C_7H_{16}O$	Butyl propyl ether	118.1	33.72	40.22
$C_7H_{16}O$	1-Heptanol	176.45		66.81
$C_7H_{17}N$	Heptylamine	156		49.96
C_8F_{18}	Perfluorooctane	105.9	33.38	41.13
$C_8H_8O_2$	Methyl benzoate	199		55.57
C_8H_{10}	Ethylbenzene	136.19	35.57	42.24
C_8H_{10}	*o*-Xylene	144.5	36.24	43.43
C_8H_{10}	*m*-Xylene	139.12	35.66	42.65
C_8H_{10}	*p*-Xylene	138.37	35.67	42.40
$C_8H_{10}O$	2,4-Xylenol	210.98		64.96
$C_8H_{10}O$	2,6-Xylenol	201.07		75.31
$C_8H_{10}O$	3,4-Xylenol	227		85.03
$C_8H_{10}O$	3,5-Xylenol	221.74		82.01
$C_8H_{10}O$	Phenetole	169.81		51.04
$C_8H_{11}N$	*N,N*-Dimethylaniline	194.15		52.83
$C_8H_{11}N$	2,3,6-Trimethylpyridine	171.6	39.95	50.61
$C_8H_{11}N$	2,4,6-Trimethylpyridine	170.6	39.87	50.33
C_8H_{14}	1-Octyne	126.3	35.83	42.30
C_8H_{14}	2-Octyne	137.6	37.26	44.49
C_8H_{14}	3-Octyne	133.1	36.94	43.92
C_8H_{14}	4-Octyne	131.6	36.0	42.73
$C_8H_{15}N$	Octanenitrile	205.25		56.80
C_8H_{16}	Propylcyclopentane	131	34.70	41.08
C_8H_{16}	Isopropylcyclopentane	126.5	33.56	39.44
C_8H_{16}	1-Ethyl-1-methylcyclopentane	121.6	33.20	38.85
C_8H_{16}	Ethylcyclohexane	131.9	34.04	40.56
C_8H_{16}	1,1-Dimethylcyclohexane	119.6	32.51	37.92
C_8H_{16}	*cis*-1,2-Dimethylcyclohexane	129.8	33.47	39.70
C_8H_{16}	*trans*-1,2-Dimethylcyclohexane	123.5	32.96	38.36
C_8H_{16}	*cis*-1,3-Dimethylcyclohexane	120.1	32.91	38.26
C_8H_{16}	*trans*-1,3-Dimethylcyclohexane	124.5	33.39	39.16
C_8H_{16}	*cis*-1,4-Dimethylcyclohexane	124.4	33.28	39.02
C_8H_{16}	*trans*-1,4-Dimethylcyclohexane	119.4	32.56	37.90
C_8H_{16}	1-Octene	121.29	34.07	40.39
$C_8H_{16}O$	2,2,4-Trimethyl-3-pentanone	135.1	35.64	43.30
$C_8H_{16}O_2$	2-Ethylhexanoic acid			75.60
$C_8H_{16}O_2$	Methyl heptanoate	172		51.62

Molecular formula	Name	T_b °C	$\Delta_{vap}H$ (T_b) kJ/mol	$\Delta_{vap}H$ (25°C) kJ/mol
$C_8H_{16}O_2$	Ethyl hexanoate	167		51.72
$C_8H_{17}Br$	1-Bromooctane	200		55.77
$C_8H_{17}Cl$	1-Chlorooctane	181.5		52.42
$C_8H_{17}F$	1-Fluorooctane	142.4	40.43	49.65
C_8H_{18}	Octane	125.67	34.41	41.49
C_8H_{18}	2-Methylheptane	117.66	33.26	39.67
C_8H_{18}	3-Methylheptane	118.94	33.66	39.83
C_8H_{18}	4-Methylheptane	117.72	33.35	39.69
C_8H_{18}	2,2-Dimethylhexane	106.86	32.07	37.28
C_8H_{18}	2,3-Dimethylhexane	115.62	33.17	38.78
C_8H_{18}	2,4-Dimethylhexane	109.5	32.51	37.76
C_8H_{18}	2,5-Dimethylhexane	109.12	32.54	37.85
C_8H_{18}	3,3-Dimethylhexane	111.97	32.31	37.53
C_8H_{18}	3,4-Dimethylhexane	117.73	33.24	38.97
C_8H_{18}	3-Ethylhexane	118.6	33.59	39.64
C_8H_{18}	3-Ethyl-2-methylpentane	115.66	32.93	38.52
C_8H_{18}	3-Ethyl-3-methylpentane	118.27	32.78	37.99
C_8H_{18}	2,2,3-Trimethylpentane	110	31.94	36.91
C_8H_{18}	2,2,4-Trimethylpentane	99.22	30.79	35.14
C_8H_{18}	2,3,3-Trimethylpentane	114.8	32.12	37.27
C_8H_{18}	2,3,4-Trimethylpentane	113.5	32.36	37.75
C_8H_{18}	2,2,3,3-Tetramethylbutane	106.45		42.90
$C_8H_{18}N_2$	Azobutane			49.31
$C_8H_{18}O$	1-Octanol	195.16		70.98
$C_8H_{18}O$	Dibutyl ether	140.28	36.49	44.97
$C_8H_{18}O$	Di-*sec*-butyl ether	121.1	34.06	40.84
$C_8H_{18}O$	Di-*tert*-butyl ether	107.23	32.15	37.61
$C_8H_{18}O_2$	1,2-Dipropoxyethane			50.62
$C_8H_{18}O_3$	Diethylene glycol diethyl ether	188		58.40
$C_8H_{18}S$	Dibutyl sulfide	185		52.96
$C_8H_{18}S$	Diisobutyl sulfide			48.71
$C_8H_{18}S$	Di-*tert*-butyl sulfide	149.1	33.26	43.76
$C_8H_{19}N$	Dibutylamine	159.6	38.44	49.45
C_9H_7N	Quinoline	237.16		53.90
C_9H_7N	Isoquinoline	243.22		60.26
C_9H_{10}	Cyclopropylbenzene			50.22
C_9H_{10}	Indan	177.97	39.63	48.79
C_9H_{12}	Cumene	152.41		45.13
C_9H_{12}	Propylbenzene	159.24		46.22
C_9H_{12}	1,2,3-Trimethylbenzene	176.12		49.05
C_9H_{12}	1,2,4-Trimethylbenzene	169.38		47.93
C_9H_{12}	Mesitylene	164.74		47.50
$C_9H_{14}O_6$	Triacetin	259		85.74
C_9H_{18}	Butylcyclopentane	156.6	36.16	45.89
C_9H_{18}	Propylcyclohexane	156.7		45.08
C_9H_{18}	Isopropylcyclohexane	154.8		44.02
$C_9H_{18}O$	2-Nonanone	195.3		56.44
$C_9H_{18}O$	5-Nonanone	188.45		53.30
$C_9H_{18}O$	2,6-Dimethyl-4-heptanone	169.4		50.92
$C_9H_{18}O_2$	Methyl octanoate	192.9		56.41
C_9H_{20}	Nonane	150.82	36.91	46.41
C_9H_{20}	2,2,5-Trimethylhexane	124.09	33.65	40.16
C_9H_{20}	2,3,5-Trimethylhexane	131.4	34.43	41.41
C_9H_{20}	3,3-Diethylpentane	146.3	34.61	42.0
C_9H_{20}	2,2,4,4-Tetramethylpentane	122.29	32.51	38.49
$C_9H_{20}O$	1-Nonanol	213.37		76.86
$C_{10}H_{14}$	Butylbenzene	183.31	38.87	51.36

Molecular formula	Name	T_b °C	$\Delta_{vap}H$ (T_b) kJ/mol	$\Delta_{vap}H$ (25°C) kJ/mol
$C_{10}H_{14}$	Isobutylbenzene	172.79		47.86
$C_{10}H_{14}$	*sec*-Butylbenzene			47.98
$C_{10}H_{14}$	*tert*-Butylbenzene	169.1		47.71
$C_{10}H_{19}N$	Decanenitrile			66.84
$C_{10}H_{20}$	Butylcyclohexane	180.9		49.36
$C_{10}H_{20}$	1-Decene	170.5		50.43
$C_{10}H_{22}$	Decane	174.15	38.75	51.38
$C_{10}H_{22}$	2-Methylnonane	167.1	38.23	49.63
$C_{10}H_{22}$	3-Methylnonane	167.9	38.26	49.71
$C_{10}H_{22}$	5-Methylnonane	165.1	38.14	49.34
$C_{10}H_{22}$	2,4-Dimethyloctane	156	36.47	47.13
$C_{10}H_{22}O$	1-Decanol	231.1		81.50
$C_{10}H_{22}S$	1-Decanethiol	240.6		65.48
$C_{11}H_{21}N$	Undecanenitrile	253		71.14
$C_{11}H_{22}$	Pentylcyclohexane	203.7		53.88
$C_{11}H_{24}$	Undecane	195.93		56.43
$C_{11}H_{24}$	2-Methyldecane	189.3	40.25	54.28
$C_{11}H_{24}$	4-Methyldecane	187	40.70	53.76
$C_{11}H_{24}$	2,4,7-Trimethyloctane	168.1	38.22	49.91
$C_{12}H_{16}$	Cyclohexylbenzene	240.1		59.94
$C_{12}H_{22}$	Cyclohexylcyclohexane			57.98
$C_{12}H_{23}N$	Dodecanenitrile	277		76.12
$C_{12}H_{24}$	1-Dodecene	213.8		60.78
$C_{12}H_{26}$	Dodecane	216.32		61.51
$C_{12}H_{26}$	2,2,4,6,6-Pentamethylheptane	177.8		48.97
$C_{12}H_{26}O$	1-Dodecanol	259		91.96
$C_{13}H_{26}O_2$	Methyl dodecanoate	267		77.17
$C_{13}H_{28}$	Tridecane	235.47		66.43
$C_{14}H_{10}$	Phenanthrene	340		75.50
$C_{14}H_{27}N$	Tetradecanenitrile			85.29
$C_{14}H_{30}$	Tetradecane	253.58		71.30
$C_{14}H_{30}O$	1-Tetradecanol	289		102.20
$C_{15}H_{32}$	Pentadecane	270.6		76.11
$C_{16}H_{32}$	1-Hexadecene	284.9		80.25
$C_{16}H_{34}$	Hexadecane	286.86		81.38
$C_{17}H_{36}$	Heptadecane	302.0		86.02

ENTHALPY OF FUSION

This table lists the molar enthalpy (heat) of fusion of about 550 inorganic and organic compounds. Values of $\Delta_{fus}H$ are given in kJ/mol at the normal melting point T_m, except for substances whose triple-point pressures are greater than normal atmospheric pressure (i.e., the sublimation pressure of the solid exceeds 1 atm at temperatures for which the liquid phase does not exist). In such cases the entries are designated by "t" and refer to the triple-point temperature. Substances are listed by molecular formula in the Hill order, with compounds not containing carbon preceding those that do contain carbon. Data for the organic compounds were compiled by William E. Acree, Jr.

REFERENCES

1. Chase, M. W., et al., *JANAF Thermochemical Tables, Third Edition, J. Phys. Chem. Ref. Data*, 14, Suppl. 1, 1985.
2. Daubert, T. E., Danner, R. P., Sibul, H. M., and Stebbins, C. C., *Physical and Thermodynamic Properties of Pure Compounds: Data Compilation*, extant 1994 (core with 4 supplements), Taylor & Francis, Bristol, PA (also available as database).
3. Janz, G. J., et al., *Physical Properties Data Compilations Relevant to Energy Storage. II. Molten Salts*, Natl. Stand. Ref. Data Sys.- Natl. Bur. Standards (U.S.), No. 61, Part 2, 1979.
4. Dinsdale, A. T., *CALPHAD*, 15, 317, 1991.
5. *Landolt-Bornstein, Numerical Values and Functions for Physics, Chemistry, Astronomy, Geophysics, and Technology, Sixth Edition*, Vol. 2, Part 4, Springer-Verlag, Heidelberg, 1961.

Molecular formula	Name	T_m °C	$\Delta_{fus}H$ kJ/mol
Ag	Silver	961.78	11.30
AgBr	Silver bromide	432	9.12
AgCl	Silver chloride	455	13.20
AgI	Silver iodide	558	9.41
$AgNO_3$	Silver nitrate	212	11.50
Ag_2S	Silver sulfide	825	14.10
Al	Aluminum	660.32	10.71
$AlBr_3$	Aluminum tribromide	97.5	11.25
$AlCl_3$	Aluminum trichloride	190	35.40
AlF_3	Aluminum trifluoride	2250 t	98.00
AlI_3	Aluminum triiodide	191	15.90
Al_2O_3	Aluminum oxide (Al_2O_3)	2054	111.10
Am	Americium	1176	14.39
Ar	Argon	-189.35	1.12
As	Arsenic	817 t	24.44
$AsBr_3$	Arsenic tribromide	31.1	11.70
$AsCl_3$	Arsenic trichloride	-16	10.10
AsF_3	Arsenic trifluoride	-5.9	10.40
Au	Gold	1064.18	12.55
B	Boron	2075	50.20
BCl_3	Boron trichloride	-107	2.10
BF_3	Boron trifluoride	-126.8	4.20
$BNaO_2$	Sodium metaborate	966	36.20
Ba	Barium	727	7.12
$BaBr_2$	Barium bromide	857	31.96
$BaCl_2$	Barium chloride	962	16.00
BaF_2	Barium fluoride	1368	23.36
BaH_2O_2	Barium hydroxide	408	16.70
BaI_2	Barium iodide	711	26.53
BaO	Barium oxide	2013	59.00
BaO_4S	Barium sulfate	1350	40.60
Be	Beryllium	1287	7.90
$BeBr_2$	Beryllium bromide	508	9.80
$BeCl_2$	Beryllium chloride	399	8.66
BeF_2	Beryllium fluoride	552	4.76
BeI_2	Beryllium iodide	480	21.00
BeO	Beryllium oxide	2507	85.00
Bi	Bismuth	271.40	11.30
$BiCl_3$	Bismuth trichloride	230	10.90

Molecular formula	Name	T_m °C	$\Delta_{fus}H$ kJ/mol
BrF_5	Bromine pentafluoride	-60.5	5.67
BrH	Hydrogen bromide	-86.81	2.41
BrK	Potassium bromide	734	25.50
BrLi	Lithium bromide	552	17.60
BrNa	Sodium bromide	747	26.11
$BrNaO_3$	Sodium bromate	381	28.11
BrRb	Rubidium bromide	682	15.50
BrTl	Thallium bromide (TlBr)	460	25.10
Br_2	Bromine (Br_2)	-7.2	10.57
Br_2Ca	Calcium bromide	742	29.08
Br_2Cd	Cadmium bromide	568	20.90
Br_2Fe	Iron bromide ($FeBr_2$)	691	50.20
Br_2Hg	Mercury bromide ($HgBr_2$)	236	17.90
Br_2Mg	Magnesium bromide	711	39.30
Br_2Pb	Lead bromide ($PbBr_2$)	371	16.44
Br_2Sr	Strontium bromide	657	10.12
Br_2Zn	Zinc bromide	394	16.70
Br_3Ga	Gallium bromide ($GaBr_3$)	121.5	11.70
Br_4Sn	Tin bromide ($SnBr_4$)	31	12.00
Br_4Ti	Titanium bromide ($TiBr_4$)	39	12.90
Br_5Ta	Tantalum bromide ($TaBr_5$)	265	45.60
Ca	Calcium	842	8.54
$CaCl_2$	Calcium chloride	772	28.54
CaF_2	Calcium fluoride	1418	29.71
CaI_2	Calcium iodide	779	41.80
CaO	Calcium oxide	2927	59.00
CaO_4S	Calcium sulfate	1450	28.03
Cd	Cadmium	321.07	6.19
$CdCl_2$	Cadmium chloride	564	48.58
CdF_2	Cadmium fluoride	1110	22.60
CdI_2	Cadmium iodide	387	15.30
Ce	Cerium	799	5.46
ClCs	Cesium chloride	645	15.90
ClCu	Copper chloride (CuCl)	430	10.20
ClH	Hydrogen chloride	-114.18	2.00
ClI	Iodine chloride	27.38	11.60
ClIn	Indium chloride (InCl)	225	17.20
ClK	Potassium chloride	771	26.53
ClLi	Lithium chloride	610	19.90
$ClLiO_4$	Lithium perchlorate	236	29.00
ClNa	Sodium chloride	800.7	28.16
$ClNaO_3$	Sodium chlorate	248	22.10
ClRb	Rubidium chloride	715	18.40
ClTl	Thallium chloride (TlCl)	430	17.80
Cl_2	Chlorine (Cl_2)	-101.5	6.40
Cl_2Co	Cobalt chloride ($CoCl_2$)	740	45.00
Cl_2Cr	Chromium chloride ($CrCl_2$)	814	32.20
Cl_2Cu	Copper chloride ($CuCl_2$)	630	20.40
Cl_2Fe	Iron chloride ($FeCl_2$)	677	43.01
Cl_2Hg	Mercury chloride ($HgCl_2$)	276	19.41
Cl_2Mg	Magnesium chloride	714	43.10
Cl_2Mn	Manganese chloride ($MnCl_2$)	650	30.70
Cl_2Ni	Nickel chloride ($NiCl_2$)	1009	71.20
Cl_2Pb	Lead chloride ($PbCl_2$)	501	21.90
Cl_2Sn	Tin chloride ($SnCl_2$)	247	12.80
Cl_2Sr	Strontium chloride	874	16.20
Cl_3Fe	Iron chloride ($FeCl_3$)	304	43.10
Cl_3Ga	Gallium chloride ($GaCl_3$)	77.9	10.90

ENTHALPY OF FUSION (continued)

Molecular formula	Name	T_m °C	$\Delta_{fus}H$ kJ/mol
Cl_3OP	Phosphoryl chloride	1	13.10
Cl_3P	Phosphorus trichloride	-112	7.10
Cl_3Sb	Antimony trichloride	73.4	12.70
Cl_4OW	Tungsten oxide tetrachloride	211	45.00
Cl_4Si	Silicon tetrachloride	-68.85	7.60
Cl_4Sn	Tin chloride ($SnCl_4$)	-33	9.20
Cl_4Th	Thorium chloride ($ThCl_4$)	770	40.20
Cl_4Ti	Titanium chloride ($TiCl_4$)	-25	9.97
Cl_4U	Uranium chloride (UCl_4)	590	45.00
Cl_4V	Vanadium chloride (VCl_4)	-25.7	2.30
Cl_4Zr	Zirconium chloride ($ZrCl_4$)	437	50.00
Cl_5Mo	Molybdenum pentachloride	194	19.00
Cl_5Nb	Niobium chloride ($NbCl_5$)	204.7	33.90
Cl_5Ta	Tantalum chloride ($TaCl_5$)	216	35.10
Cl_6W	Tungsten chloride (WCl_6)	275	6.60
Co	Cobalt	1495	16.20
CoF_2	Cobalt fluoride (CoF_2)	1127	59.00
Cr	Chromium	1907	21.00
Cr_2O_3	Chromium oxide (Cr_2O_3)	2330	130.00
Cs	Cesium	28.44	2.10
CsF	Cesium fluoride	703	21.70
Cs_2O_4S	Cesium sulfate	1005	35.70
Cu	Copper	1084.62	13.26
CuF_2	Copper fluoride (CuF_2)	836	55.00
CuO	Copper oxide (CuO)	1446	11.80
Dy	Dysprosium	1411	10.78
Er	Erbium	1529	19.90
Eu	Europium	822	9.21
FH	Hydrogen fluoride	-83.36	4.58
FK	Potassium fluoride	858	27.20
FLi	Lithium fluoride	848.2	27.09
FNa	Sodium fluoride	996	33.35
FRb	Rubidium fluoride	833	17.30
FTl	Thallium fluoride (TlF)	322	14.00
F_2	Fluorine (F_2)	-219.66	0.51
F_2Fe	Iron fluoride (FeF_2)	1100	52.00
F_2HK	Potassium hydrogen fluoride (KHF_2)	238.9	6.62
F_2Mg	Magnesium fluoride	1263	58.70
F_2Pb	Lead fluoride (PbF_2)	830	14.70
F_2Sr	Strontium fluoride	1477	29.70
F_4Zr	Zirconium fluoride (ZrF_4)	932	64.20
F_5Nb	Niobium fluoride (NbF_5)	80	36.00
F_5V	Vanadium fluoride (VF_5)	19.5	49.96
F_6Ir	Iridium fluoride (IrF_6)	44	8.40
F_6Mo	Molybdenum hexafluoride	17.5	4.33
F_6S	Sulfur fluoride (SF_6)	-50.7	5.02
F_6U	Uranium fluoride (UF_6)	64.0 t	19.19
F_6W	Tungsten fluoride (WF_6)	2.3	4.10
Fe	Iron	1538	13.81
FeI_2	Iron iodide (FeI_2)	587	45.00
FeO	Iron oxide (FeO)	1377	24.00
FeS	Iron sulfide (FeS)	1188	31.50
Fe_3O_4	Iron oxide (Fe_3O_4)	1597	138.00
Ga	Gallium	29.76	5.59
GaI_3	Gallium iodide (GaI_3)	210	16.30
GaSb	Gallium antimonide (GaSb)	712	25.10
Gd	Gadolinium	1314	9.81
Ge	Germanium	938.25	36.94

ENTHALPY OF FUSION (continued)

Molecular formula	Name	T_m °C	$\Delta_{fus}H$ kJ/mol
HI	Hydrogen iodide	-50.77	2.87
HKO	Potassium hydroxide	406	8.60
HLi	Lithium hydride	688.7	22.59
HLiO	Lithium hydroxide	471.2	20.88
HNO_3	Nitric acid	-41.6	10.50
HNaO	Sodium hydroxide	323	6.60
H_2	Hydrogen (H_2)	-259.34	0.12
H_2O	Water	0.00	6.01
H_2O_2	Hydrogen peroxide	-0.43	12.50
H_2O_2Sr	Strontium hydroxide	510	21.00
H_2O_4S	Sulfuric acid	10.31	10.71
H_2S	Hydrogen sulfide	-85.5	23.80
H_3N	Ammonia	-77.74	5.66
H_3O_2P	Hypophosphorous acid (H_3PO_2)	26.5	9.70
H_3O_3P	Phosphorous acid	74.4	12.80
H_3O_4P	Phosphoric acid	42.4	13.40
H_4IN	Ammonium iodide	551	21.00
H_4N_2	Hydrazine	1.4	12.60
$H_4N_2O_3$	Ammonium nitrate	210	6.40
Hf	Hafnium	2233	27.20
Hg	Mercury	-38.83	2.29
HgI_2	Mercury iodide (HgI_2)	259	18.90
Hg_2I_2	Mercury iodide (Hg_2I_2)	290	27.00
Ho	Holmium	1472	11.76
IK	Potassium iodide	681	24.00
ILi	Lithium iodide	469	14.60
INa	Sodium iodide	660	23.60
IRb	Rubidium iodide	642	12.50
ITl	Thallium iodide (TlI)	440	13.10
I_2	Iodine (I_2)	113.7	15.52
I_2Mg	Magnesium iodide	634	29.00
I_2Pb	Lead iodide (PbI_2)	410	23.40
I_2Sr	Strontium iodide	538	19.70
I_4Si	Silicon tetraiodide	120.5	19.70
I_4Ti	Titanium iodide (TiI_4)	150	19.80
In	Indium	156.60	3.28
InSb	Indium antimonide (InSb)	525	25.50
Ir	Iridium	2446	41.12
K	Potassium	63.38	2.32
KNO_3	Potassium nitrate	337	10.10
K_2O_4S	Potassium sulfate	1069	36.40
K_2S	Potassium sulfide (K_2S)	948	16.15
Kr	Krypton	-157.36	1.37
La	Lanthanum	920	6.20
Li	Lithium	180.5	3.00
$LiNO_3$	Lithium nitrate	253	24.90
Li_2O_3Si	Lithium metasilicate	1201	28.00
Li_2O_4S	Lithium sulfate	859	7.50
Lu	Lutetium	1663	18.65
Mg	Magnesium	650	8.48
MgO	Magnesium oxide	2826	78.00
MgO_4S	Magnesium sulfate	1127	14.60
Mg_2O_4Si	Magnesium silicate	1898	71.00
Mn	Manganese	1246	12.91
MnO	Manganese oxide (MnO)	1840	54.40
Mo	Molybdenum	2623	37.48
MoO_3	Molybdenum oxide (MoO_3)	801	48.00
$NNaO_3$	Sodium nitrate	307	15.00

Molecular formula	Name	T_m °C	$\Delta_{fus}H$ kJ/mol
NO	Nitric oxide	-163.6	2.30
NO_3Rb	Rubidium nitrate	305	5.60
NO_3Tl	Thallium nitrate ($TlNO_3$)	206	9.60
N_2	Nitrogen (N_2)	-210.00	0.71
N_2O	Nitrous oxide	-90.8	6.54
N_2O_4	Nitrogen tetroxide	-9.3	14.65
Na	Sodium	97.72	2.60
Na_2O	Sodium oxide (Na_2O)	1132	48.00
Na_2O_3Si	Sodium metasilicate	1089	52.00
Na_2O_4S	Sodium sulfate	884	23.60
Na_2S	Sodium sulfide (Na_2S)	1172	19.00
Nb	Niobium	2477	30.00
NbO	Niobium oxide (NbO)	1937	85.00
NbO_2	Niobium oxide (NbO_2)	1902	92.00
Nb_2O_5	Niobium oxide (Nb_2O_5)	1512	104.30
Nd	Neodymium	1016	7.14
Ne	Neon	-248.59	0.34
Ni	Nickel	1455	17.48
NiS	Nickel sulfide (NiS)	976	30.10
Np	Neptunium	644	3.20
OSr	Strontium oxide	2665	75.00
OV	Vanadium oxide (VO)	1790	63.00
OZn	Zinc oxide	1975	52.30
O_2	Oxygen (O_2)	-218.79	0.44
O_2Si	Silicon dioxide	1713	8.51
O_2Zr	Zirconium oxide (ZrO_2)	2710	87.00
O_3S	Sulfur trioxide	16.8	8.60
O_3W	Tungsten oxide (WO_3)	1472	73.00
O_3Y_2	Yttrium oxide (Y_2O_3)	2439	105.00
O_4Os	Osmium oxide (OsO_4)	41	9.80
O_4STl_2	Thallium sulfate (Tl_2SO_4)	632	23.00
O_5P_2	Phosphorus pentoxide	562	27.20
O_5Ta_2	Tantalum oxide (Ta_2O_5)	1785	120.00
O_5V_2	Vanadium oxide (V_2O_5)	670	64.50
O_7Re_2	Rhenium oxide (Re_2O_7)	297	64.20
Os	Osmium	3033	57.85
P	Phosphorus (white)	44.15	0.66
Pa	Protactinium	1572	12.34
Pb	Lead	327.46	4.77
PbS	Lead sulfide (PbS)	1118	19.00
Pd	Palladium	1554.9	16.74
Pr	Praseodymium	931	6.89
Pt	Platinum	1768.4	22.17
Pu	Plutonium	640	2.82
Rb	Rubidium	39.31	2.19
Re	Rhenium	3186	60.43
Rh	Rhodium	1964	26.59
Ru	Ruthenium	2334	38.59
S	Sulfur	115.21	1.72
STl_2	Thallium sulfide (Tl_2S)	448	12.00
Sb	Antimony	630.63	19.87
Sc	Scandium	1541	14.10
Se	Selenium	221	6.69
Si	Silicon	1414	50.21
Sm	Samarium	1072	8.62
Sn	Tin	231.93	7.03
Sr	Strontium	777	7.43
Ta	Tantalum	3017	36.57

ENTHALPY OF FUSION (continued)

Molecular formula	Name	T_m °C	$\Delta_{fus}H$ kJ/mol
Tb	Terbium	1359	10.15
Tc	Technetium	2157	33.29
Te	Tellurium	449.51	17.49
Th	Thorium	1750	13.81
Ti	Titanium	1668	14.15
Tl	Thallium	304	4.14
Tm	Thulium	1545	16.84
U	Uranium	1135	9.14
V	Vanadium	1910	21.50
W	Tungsten	3422	52.31
Xe	Xenon	-111.75	1.81
Y	Yttrium	1526	11.40
Yb	Ytterbium	824	7.66
Zn	Zinc	419.53	7.32
Zr	Zirconium	1855	21.00
C	Carbon	4492 t	117
$CBrCl_3$	Bromotrichloromethane	-5.7	2.54
$CCaO_3$	Calcium carbonate	825	53.10
CCl_2O	Carbonyl chloride	-127.9	5.74
CCl_4	Tetrachloromethane	-23	3.28
CK_2O_3	Potassium carbonate	898	27.60
CLi_2O_3	Lithium carbonate	723	41.00
CNa_2O_3	Sodium carbonate	858.1	29.70
CO	Carbon monoxide	-205	0.83
CO_2	Carbon dioxide	-56.57	9.02
CO_3Tl_2	Thallium carbonate (Tl_2CO_3)	272	18.40
CS_2	Carbon disulfide	-111.5	4.40
$CHCl_3$	Trichloromethane	-63.6	8.80
CHN	Hydrogen cyanide	-13.4	8.41
CH_2Cl_2	Dichloromethane	-95.14	6.00
CH_2N_2	Cyanamide	44	8.76
CH_2O_2	Formic acid	8.3	12.72
CH_3Br	Bromomethane	-93.7	5.98
CH_4	Methane	-182.46	0.94
CH_4O	Methanol	-97.68	3.18
CH_4S	Methanethiol	-123	5.91
CH_5N	Methylamine	-93.44	6.13
$C_2Br_2F_4$	1,2-Dibromotetrafluoroethane	-110.4	7.04
$C_2Cl_2F_4$	1,2-Dichlorotetrafluoroethane	-94	1.51
$C_2Cl_3F_3$	1,1,2-Trichlorotrifluoroethane	-35	2.47
$C_2Cl_4F_2$	Tetrachloro-1,2-difluoroethane	26	3.70
$C_2HCl_3O_2$	Trichloroacetic acid	57.5	5.88
$C_2H_3ClO_2$	Chloroacetic acid	63	12.28
$C_2H_3Cl_3$	1,1,1-Trichloroethane	-30.4	2.73
$C_2H_3Cl_3$	1,1,2-Trichloroethane	-36.6	11.54
$C_2H_3F_3$	1,1,1-Trifluoroethane	-111.3	6.19
$C_2H_4Br_2$	1,2-Dibromoethane	9.93	10.84
$C_2H_4Cl_2$	1,2-Dichloroethane	-35.5	8.83
$C_2H_4O_2$	Acetic acid	16.6	11.54
C_2H_5Cl	Chloroethane	-138.7	4.45
C_2H_6	Ethane	-182.8	2.86
C_2H_6O	Dimethyl ether	-141.5	4.94
C_2H_6O	Ethanol	-114.1	5.02
$C_2H_6O_2$	Ethylene glycol	-13	11.23
C_2H_6S	Dimethyl sulfide	-98.3	7.99
C_2H_6S	Ethanethiol	-147.89	4.98
$C_2H_6S_2$	Dimethyl disulfide	-85	9.19
C_2H_7N	Dimethylamine	-92.2	5.94

Molecular formula	Name	T_m °C	$\Delta_{fus}H$ kJ/mol
$C_2H_8N_2$	1,2-Ethanediamine	11.1	22.58
C_3H_3N	Propenenitrile	-83.5	6.23
$C_3H_4O_2$	Propenoic acid	12.3	11.16
$C_3H_5N_3O_9$	Trinitroglycerol	13.5	21.87
C_3H_6	Propene	-185.25	3.00
C_3H_6	Cyclopropane	-127.4	5.44
$C_3H_6Br_2$	1,3-Dibromopropane	-34.2	13.60
$C_3H_6Cl_2$	1,2-Dichloropropane	-100.5	6.40
C_3H_6O	Acetone	-94.8	5.69
$C_3H_6O_3$	1,3,5-Trioxane	60.2	15.11
C_3H_7Cl	2-Chloropropane	-117.2	7.39
C_3H_7N	Cyclopropylamine	-35.4	13.18
C_3H_8	Propane	-187.6	3.53
C_3H_8O	1-Propanol	-126.1	5.20
C_3H_8O	2-Propanol	-89.5	5.37
$C_3H_8O_3$	Glycerol	18.2	8.48
C_3H_9N	Propylamine	-83	10.97
C_3H_9N	Isopropylamine	-95.14	7.33
C_3H_9N	Trimethylamine	-117.1	6.55
$C_4H_4N_2$	Succinonitrile	54.5	3.92
$C_4H_4O_3$	Succinic anhydride	119	20.41
C_4H_4S	Thiophene	-39.4	4.97
C_4H_5N	Pyrrole	23.42	7.91
C_4H_6	1,3-Butadiene	-108.9	7.98
C_4H_6	2-Butyne	-32.36	9.23
$C_4H_6O_2$	γ-Butyrolactone	-43.37	9.57
$C_4H_6O_2$	*cis*-Crotonic acid	15	12.57
$C_4H_6O_2$	*trans*-Crotonic acid	72	9.12
$C_4H_6O_4$	Succinic acid	188	32.95
$C_4H_6O_4$	Dimethyl oxalate	54.35	21.07
C_4H_8	*cis*-2-Butene	-138.9	7.58
C_4H_8	Isobutene	-140.4	5.93
C_4H_8O	2-Butanone	-86.67	8.44
C_4H_8O	Tetrahydrofuran	-108.3	8.54
$C_4H_8O_2$	1,4-Dioxane	11.8	12.85
$C_4H_8O_2$	Ethyl acetate	-83.6	10.48
$C_4H_8O_2$	Butanoic acid	-5.7	11.08
C_4H_9Br	2-Bromobutane	-112.7	6.89
C_4H_9N	Pyrrolidine	-57.84	8.58
C_4H_{10}	Isobutane	-138.3	4.66
$C_4H_{10}O$	1-Butanol	-89.8	9.28
$C_4H_{10}O$	2-Methyl-2-propanol	25.4	6.79
$C_4H_{10}O$	Diethyl ether	-116.3	7.27
$C_4H_{11}N$	*tert*-Butylamine	-66.95	0.88
$C_4H_{12}Si$	Tetramethylsilane	-99.04	6.88
C_5H_5N	Pyridine	-41.66	8.28
C_5H_8	2-Methyl-1,3-butadiene	-145.9	4.79
C_5H_8	1,4-Pentadiene	-148.8	6.14
C_5H_8	Cyclopentene	-135.1	3.36
C_5H_8	Spiropentane	-134.6	5.76
$C_5H_8O_3$	Levulinic acid	33	9.22
$C_5H_8O_4$	Glutaric acid	97.8	20.90
C_5H_{10}	1-Pentene	-165.2	5.81
C_5H_{10}	*cis*-2-Pentene	-151.4	7.12
C_5H_{10}	*trans*-2-Pentene	-140.2	8.36
C_5H_{10}	Cyclopentane	-93.8	0.61
$C_5H_{10}O$	2-Pentanone	-76.9	10.63
$C_5H_{10}O$	3-Pentanone	-39	11.59

Molecular formula	Name	T_m °C	$\Delta_{fus}H$ kJ/mol
$C_5H_{11}N$	Piperidine	-11.03	14.85
$C_5H_{11}N$	Cyclopentylamine	-82.7	8.31
C_5H_{12}	Pentane	-129.7	8.42
C_5H_{12}	Isopentane	-159.9	5.15
C_5H_{12}	Neopentane	-16.6	3.10
$C_5H_{12}O$	1-Pentanol	-78.9	9.83
C_6Cl_6	Hexachlorobenzene	231.8	23.85
C_6F_6	Hexafluorobenzene	5.35	11.59
C_6HF_5	Pentafluorobenzene	-47.3	10.85
C_6HF_5O	Pentafluorophenol	32.8	12.85
$C_6H_3Cl_3$	1,3,5-Trichlorobenzene	63.5	18.20
$C_6H_4ClNO_2$	*m*-Chloronitrobenzene	44.4	19.37
$C_6H_4ClNO_2$	*p*-Chloronitrobenzene	83.5	20.77
$C_6H_4Cl_2$	*o*-Dichlorobenzene	-16.7	12.93
$C_6H_4Cl_2$	*m*-Dichlorobenzene	-24.8	12.64
$C_6H_4Cl_2$	*p*-Dichlorobenzene	52.7	17.15
$C_6H_4F_2$	*m*-Difluorobenzene	-69.09	8.58
$C_6H_4O_2$	*p*-Benzoquinone	115.7	18.53
C_6H_5Br	Bromobenzene	-30.6	10.62
C_6H_5Cl	Chlorobenzene	-45.2	9.61
C_6H_5ClO	*o*-Chlorophenol	9.8	12.52
C_6H_5ClO	*m*-Chlorophenol	32.6	14.91
C_6H_5ClO	*p*-Chlorophenol	42.7	14.07
C_6H_5F	Fluorobenzene	-42.21	11.31
C_6H_5I	Iodobenzene	-31.3	9.76
$C_6H_5NO_2$	Nitrobenzene	5.7	11.59
$C_6H_5NO_3$	*o*-Nitrophenol	44.8	17.44
$C_6H_5NO_3$	*m*-Nitrophenol	96.8	19.20
$C_6H_5NO_3$	*p*-Nitrophenol	113.8	18.25
C_6H_6	Benzene	5.53	9.95
$C_6H_6N_2O_2$	*o*-Nitroaniline	71.2	16.11
$C_6H_6N_2O_2$	*m*-Nitroaniline	114	23.68
$C_6H_6N_2O_2$	*p*-Nitroaniline	147	21.10
C_6H_6O	Phenol	40.9	11.29
$C_6H_6O_2$	*p*-Hydroquinone	172.3	27.11
C_6H_6S	Benzenethiol	-14.9	11.48
C_6H_7N	Aniline	-6.02	10.56
C_6H_7N	2-Methylpyridine	-66.68	9.72
C_6H_7N	3-Methylpyridine	-18.14	14.18
$C_6H_8N_2$	Phenylhydrazine	19.6	16.43
C_6H_{10}	Cyclohexene	-103.5	3.29
$C_6H_{10}O_4$	Adipic acid	153.2	34.85
C_6H_{12}	Cyclohexane	6.6	2.63
C_6H_{12}	Methylcyclopentane	-142.5	6.93
C_6H_{12}	2,3-Dimethyl-2-butene	-74.6	5.46
$C_6H_{12}O$	Cyclohexanol	25.46	1.76
$C_6H_{12}O$	2-Hexanone	-55.5	14.90
$C_6H_{12}O$	3-Hexanone	-55.5	13.49
C_6H_{14}	Hexane	-95.3	13.08
C_6H_{14}	2-Methylpentane	-153.7	6.27
C_6H_{14}	2,2-Dimethylbutane	-99	0.58
C_6H_{14}	2,3-Dimethylbutane	-128.8	0.80
$C_6H_{14}O$	Dipropyl ether	-126.1	8.83
$C_6H_{14}O$	Diisopropyl ether	-86.8	11.03
C_7F_8	Perfluorotoluene	-65.6	11.58
$C_7H_3F_5$	2,3,4,5,6-Pentafluorotoluene	-29.8	12.99
$C_7H_5ClO_2$	*o*-Chlorobenzoic acid	140.2	25.73
$C_7H_6O_2$	Benzoic acid	122.4	18.06

ENTHALPY OF FUSION (continued)

Molecular formula	Name	T_m °C	$\Delta_{fus}H$ kJ/mol
C_7H_7NO	Benzamide	129.1	18.49
$C_7H_7NO_2$	*p*-Nitrotoluene	51.6	16.81
C_7H_8	Toluene	-94.99	6.85
C_7H_8O	*o*-Cresol	29.8	13.94
C_7H_8O	*m*-Cresol	11.8	9.41
C_7H_8O	*p*-Cresol	35.5	11.89
C_7H_8O	Benzyl alcohol	-15.2	8.97
C_7H_9N	*p*-Methylaniline	43.75	18.22
C_7H_{14}	Cycloheptane	-8.03	1.88
C_7H_{14}	Methylcyclohexane	-126.6	6.75
C_7H_{14}	1-Heptene	-119.7	12.66
C_7H_{16}	Heptane	-90.6	14.16
C_7H_{16}	2-Methylhexane	-118.2	8.87
C_7H_{16}	2,2-Dimethylpentane	-123.8	5.86
C_7H_{16}	2,4-Dimethylpentane	-119.9	6.69
C_7H_{16}	3,3-Dimethylpentane	-134.9	7.07
C_7H_{16}	3-Ethylpentane	-118.6	9.55
C_7H_{16}	2,2,3-Trimethylbutane	-25	2.20
$C_8H_8O_2$	*o*-Toluic acid	103.7	20.17
$C_8H_8O_2$	*m*-Toluic acid	108.75	15.72
$C_8H_8O_2$	*p*-Toluic acid	179.6	22.73
$C_8H_8O_2$	Phenylacetic acid	76.7	14.49
C_8H_{10}	*o*-Xylene	-25.2	13.61
C_8H_{10}	*m*-Xylene	-47.8	11.55
C_8H_{10}	*p*-Xylene	13.2	16.81
$C_8H_{10}O$	2,3-Xylenol	72.8	21.02
$C_8H_{10}O$	2,5-Xylenol	74.8	23.38
$C_8H_{10}O$	2,6-Xylenol	45.7	18.90
$C_8H_{10}O$	3,4-Xylenol	60.8	18.13
$C_8H_{10}O$	3,5-Xylenol	63.6	18.00
C_8H_{16}	Cyclooctane	14.83	2.41
C_8H_{16}	Ethylcyclohexane	-111.3	8.33
C_8H_{16}	1,1-Dimethylcyclohexane	-33.3	2.06
C_8H_{16}	*cis*-1,2-Dimethylcyclohexane	-49.9	1.64
C_8H_{16}	*trans*-1,2-Dimethylcyclohexane	-90	10.49
C_8H_{16}	*cis*-1,3-Dimethylcyclohexane	-75.6	10.82
C_8H_{16}	*trans*-1,3-Dimethylcyclohexane	-90.1	9.86
C_8H_{16}	*cis*-1,4-Dimethylcyclohexane	-87.4	9.31
C_8H_{16}	*trans*-1,4-Dimethylcyclohexane	-36.9	12.33
$C_8H_{16}O_2$	Octanoic acid	16.3	21.36
C_8H_{18}	Octane	-56.8	20.65
C_8H_{18}	3-Methylheptane	-120.5	11.38
C_8H_{18}	4-Methylheptane	-121	10.84
C_8H_{18}	2,2,4-Trimethylpentane	-107.3	9.04
C_9H_7N	Quinoline	-14.78	10.66
C_9H_7N	Isoquinoline	26.47	13.54
C_9H_{12}	Propylbenzene	-99.56	9.27
C_9H_{12}	1,2,3-Trimethylbenzene	-25.4	8.37
C_9H_{12}	1,2,4-Trimethylbenzene	-43.8	3.76
C_9H_{12}	Mesitylene	-44.7	9.51
C_9H_{18}	Propylcyclohexane	-94.9	10.37
$C_9H_{18}O$	5-Nonanone	-5.9	24.93
$C_9H_{18}O_2$	Nonanoic acid	12.35	20.28
C_9H_{20}	Nonane	-53.5	15.47
C_9H_{20}	2,2,3,3-Tetramethylpentane	-9.8	2.33
C_9H_{20}	3,3-Diethylpentane	-33.1	10.09
C_9H_{20}	2,2,4,4-Tetramethylpentane	-66.54	9.75
$C_{10}H_7Br$	1-Bromonaphthalene	-0.9	15.16

Molecular formula	Name	T_m °C	$\Delta_{fus}H$ kJ/mol
$C_{10}H_7Cl$	1-Chloronaphthalene	-2.5	12.90
$C_{10}H_8$	Naphthalene	80.2	19.06
$C_{10}H_8O$	1-Naphthol	95	23.33
$C_{10}H_8O$	2-Naphthol	123	17.51
$C_{10}H_{14}$	Butylbenzene	-87.9	11.22
$C_{10}H_{14}$	*p*-Cymene	-68.9	9.60
$C_{10}H_{14}$	1,2,4,5-Tetramethylbenzene	79.3	21.00
$C_{10}H_{14}O$	Thymol	51.5	17.27
$C_{10}H_{18}O_4$	Sebacic acid	130.8	40.80
$C_{10}H_{20}$	Butylcyclohexane	-74.73	14.16
$C_{10}H_{20}O_2$	Decanoic acid	31.99	28.02
$C_{10}H_{22}$	Decane	-29.7	28.78
$C_{11}H_{10}$	2-Methylnaphthalene	34.4	11.97
$C_{11}H_{24}$	Undecane	-25.6	22.32
$C_{12}H_9N$	Carbazole	246.2	26.90
$C_{12}H_{10}$	Acenaphthene	93.4	21.54
$C_{12}H_{10}$	Biphenyl	69	18.60
$C_{12}H_{10}N_2$	Azobenzene	67.1	22.04
$C_{12}H_{10}N_2O$	Azoxybenzene	36	17.93
$C_{12}H_{10}O$	Diphenyl ether	26.87	17.22
$C_{12}H_{11}N$	Diphenylamine	52.98	17.86
$C_{12}H_{16}$	Cyclohexylbenzene	7.3	15.30
$C_{12}H_{24}O_2$	Dodecanoic acid	43.22	36.64
$C_{12}H_{26}$	Dodecane	-9.6	36.58
$C_{13}H_{10}$	Fluorene	114.8	19.58
$C_{13}H_{10}O$	Benzophenone	47.88	18.19
$C_{14}H_{10}$	Anthracene	215.0	28.83
$C_{14}H_{10}$	Phenanthrene	99.24	16.46
$C_{14}H_{10}O_2$	Benzil	94.86	23.54
$C_{14}H_{12}$	*trans*-Stilbene	123	27.40
$C_{14}H_{12}O_2$	Diphenylacetic acid	147	31.27
$C_{14}H_{28}O_2$	Tetradecanoic acid	53.96	45.38
$C_{16}H_{10}$	Fluoranthene	107.8	18.87
$C_{16}H_{10}$	Pyrene	151.2	17.11
$C_{16}H_{32}O_2$	Hexadecanoic acid	61.82	42.04
$C_{16}H_{34}O$	1-Hexadecanol	49.3	34.29
$C_{18}H_{12}$	Chrysene	258.2	26.15
$C_{18}H_{14}$	*p*-Terphenyl	210.1	35.50
$C_{18}H_{36}O_2$	Stearic acid	68.82	56.59
$C_{18}H_{38}$	Octadecane	28.2	61.39
$C_{19}H_{40}$	Nonadecane	32.1	45.82
$C_{20}H_{12}$	Perylene	274	31.75
$C_{20}H_{42}$	Eicosane	36.8	69.88
$C_{24}H_{12}$	Coronene	437.3	19.20

ISOTHERMAL COMPRESSIBILITY OF LIQUIDS

J. C. McGowan

The isothermal compressibility $\kappa_T = -(1/V)(\partial V/\partial P)_T$ is given in this table at pressures of 0.101325 MPa (1 atm) and 101.325 MPa (1000 atm) as a function of temperature.

	T/°C	κ_T/Pa		Ref.
		1 atm	1000 atm	
Acetic acid	15	8.75	–	1
	20	9.08	–	1
	30	9.72	–	1
	40	10.37	–	1
	50	11.11	–	1
	60	11.91	–	1
	70	12.77	–	1
	80	13.68	–	1
Acetic acid, ethyl ester	0	9.78	–	1
	10	10.36	–	1
	20	11.32	–	1
	30	12.37	–	1
	40	13.52	–	1
	50	14.78	–	1
	60	16.21	–	1
	70	17.90	–	1
Acetone	20	12.75	–	3
	20	12.29	–	10
	25	12.39	6.02	4
	30	13.34	–	10
	40	15.61		3
	40	14.64	–	10
	50	16.03	–	10
Aniline	0	4.08	–	1
	10	4.30	–	1
	20	4.53	–	1
	25	4.67	3.23	5
	40	5.04	–	1
	45	5.22	3.48	5
	50	5.33	–	1
	60	5.64	–	1
	65	5.84	3.76	5
	70	5.97	–	1
	80	6.32	–	1
	85	6.56	4.04	5
	90	6.70	–	1
Anisole	21	6.67	–	2
	30	7.04	–	2
	45	7.72	–	2
	60	8.50	–	2
	81	9.79	–	2
	100	11.25	–	2
	120	13.07	–	2
	140	15.45	–	2
Benzene	0	8.09	–	1
	10	8.73	–	1
	10	8.64	–	3
	20	9.44	–	1
	20	9.37	–	3
	20	9.54	–	10
	25	[illegible]	[illegible]	6
	25	9.7	–	15
	30	10.27	–	10
	30	10.18	–	1
	30	10.12	–	3
	35	10.43	5.28	6
	39.5	10.91	–	15
	40	11.05	–	10
	40	11.00	–	1
	40	10.96	–	3
	45	11.32	5.50	6
	50	11.90	–	10
	50	11.89	–	1
	50	11.83	–	3
	50.1	11.91	–	15
	55	12.29	5.73	6
	60	12.78	–	10
	60	12.83	–	1
	60	12.96	–	15
	65	13.39	5.98	6
	70	13.72	–	10
	70	14.13	–	1
	75.9	14.95	–	15
	80	15.44	–	1
Benzene, bromo-	25	6.68	4.09	5
	45	7.52	4.39	5
	65	8.50	4.72	5
	85	9.65	5.06	5
Benezene, chloro-	0	6.61	–	1
	10	7.02	–	1
	20	7.45	–	1
	20	7.38	–	3
	25	7.51	4.39	5
	30	7.89	–	1
	30	7.84	–	3
	40	8.39	–	1
	40	8.32	–	3
	45	8.55	4.73	5
	50	8.92	–	1
	50	8.83	–	3
	60	9.50	–	1
	65	9.76	5.10	5
	70	10.13	–	1
	80	10.79	–	1
	85	11.23	5.49	5
Benezene, nitro-	0	4.41	–	1
	10	4.67	–	1
	20	4.93	–	1
	25	5.03	3.39	5
	30	5.23	–	1
	40	5.49	–	1
	45	5.59	3.64	5
	65	6.24	3.91	5
	85	6.99	4.20	5
n-Butyl alcohol	0	8.10	–	11
Carbon disulphide	0	8.04	–	1
	0	7.95	–	3
	10	8.64	–	1
	10	8.54	–	3
	20	9.13	–	10
	20	9.26	–	1
	20	9.19	–	3
	30	9.72	–	10
	30	9.92	–	1
	30	9.96	–	3
	40	10.65	–	1
	40	10.57	–	10
	40	10.89	–	3
	50	11.48	–	1
	50	11.95	–	3
Carbon tetrachloride	[illegible]	7.01		12
	−13.1	8.31	–	12
	−3.1	8.87	–	12
	0	8.98	–	1
	0	8.85	–	3
	6.9	9.59	–	12
	10	9.70	–	1
	10	9.57	–	3
	10	9.45	–	16
	16.9	10.35	–	12
	20	10.46	–	1
	20	10.34	–	3
	20	10.40	–	10
	25	10.67	5.30	7
	25	10.77	–	15
	25	10.58	–	16
	26.9	11.15	–	12
	30	11.28	–	10
	30	11.29	–	1
	30	11.18	–	3

ISOTHERMAL COMPRESSIBILITY OF LIQUIDS (continued)

	T/°C	κ_T/Pa 1 atm	1000 atm	Ref.
	35	11.95	5.52	7
	37.5	11.99	–	15
	40	12.20	–	10
	40	12.23	–	1
	40	12.16	–	3
	40	11.96	–	16
	45	12.54	5.75	7
	50	13.20	–	10
	50	13.32	–	1
	50	13.26	–	3
	50.3	13.28	–	15
	55	13.63	5.97	7
	55	13.51	–	15
	60	14.26	–	10
	60	14.52	–	1
	62.6	14.84	–	15
	65	14.87	6.22	7
	70	15.43	–	10
	70	15.77	–	1
	75	16.70	–	15
Chloroform	−33.1	7.13	–	12
	−23.1	7.61	–	12
	−13.1	8.12	–	12
	−3.1	8.66	–	12
	0	8.48	–	1
	0	8.55	–	3
	6.9	9.30	–	12
	10	9.17	–	1
	10	9.19	–	3
	16.9	10.35	–	12
	20	9.98	–	1
	20	9.94	–	3
	20	10.15	–	10
	25	9.74	5.34	4
	26.9	11.15	–	12
	30	10.86	–	1
	30	10.81	–	3
	30	10.99	–	10
	40	11.84	–	1
	40	11.79	–	3
	40	11.94	–	10
	50	12.90	–	1
	50	12.99	–	10
	60	14.06	–	1
Cyclohexane	25	11.20	–	16
	25	11.40	–	15
	35	12.19	–	16
	37.6	12.67	–	15
	40	12.56	–	16
	45	13.31	–	16
	50.1	14.15	–	15
	55	14.35	–	16
	60	15.20	–	16
	62.4	15.76	–	15
	75	17.84	–	15
n-Decane	25	10.95	–	16
	35	11.76	–	16
	45	12.65	–	16
	60	14.11	–	16
Dodecane	25	9.87	–	16
	35	10.52	–	16
	37.8	9.9	–	16
	45	11.27	–	16
	60	12.51	–	16
	79.4	12.8	–	16
	98.9	14.4	–	16
	115.0	16.1	–	16
	135.0	18.3	–	16
Ethane 1,1,2,2-tetrachloro-	25	6.17	3.88	4
Ethyl alcohol	0	9.87	–	1
	0	9.87	–	11
	0	9.63	–	3
	10	10.49	–	1
	10	10.30	–	3
	20	11.19	–	1
	20	10.98	–	3
	30	11.91	–	1
	30	11.80	–	3
	40	12.74	–	1
	40	12.61	–	3
	50	13.70	–	1
	50	13.60	–	3
	60	14.74	–	1
	70	15.93	–	1
	75	16.67	–	1
Ethyl bromide	0	10.76	–	1
	10	11.78	–	1
	20	12.94	–	1
	30	14.23	–	1
	40	15.52	–	1
Ethylene, 1,2-dichloro-(trans)	25	11.19	5.62	4
Ethylene, tetrachloro-	25	7.56	4.45	4
Ethylene, trichloro-	25	8.57	4.99	4
Ethylene, chloride	0	6.91	–	1
	10	7.42	–	1
	20	7.97	–	1
	20	7.82	–	10
	25	7.78	4.54	4
	30	8.41	–	10
	30	8.58	–	1
	40	9.09	–	10
	40	9.25	–	1
	50	9.86	–	10
	50	9.99	–	1
	60	10.66	–	10
	60	10.83	–	1
	70	11.54	–	10
	70	11.76	–	1
	80	12.79	–	1
Ethyl ether	0	15.10	–	1
	0	15.07	–	3
	10	16.81	–	1
	10	16.52	–	3
	20	18.65	–	1
	20	18.44	–	3
	30	20.90	–	1
	30	20.80	–	3
	35	24.15	–	1
Ethyl iodide	0	8.45	–	1
	10	9.12	–	1
	20	9.82	–	1
	30	10.59	–	1
	40	11.44	–	1
	50	12.38	–	1
	60	13.40	–	1
	70	14.49	–	1
Glycol	25	3.72	2.73	7
	45	4.00	2.89	7
	65	4.32	3.05	7
	85	4.70	3.24	7
	105	5.14	3.44	7
n-Hendecane	25	10.81	–	17
	35	11.50	–	17
	50	12.47	–	17
n-Heptane	25	14.40	6.18	8
	35	15.69	–	17
	45	17.12	–	8
	60	19.62	–	8
1-Heptanol	0	7.05	–	11
n-Hexadecane	25	8.67	–	17
	35	9.43	–	17
	50	10.29	–	17
n-Hexane	25	16.72	6.51	8
	35	18.44	–	8
	45	20.33	–	8
	60	23.84	–	8
1-Hexanol	0	7.47	–	11
Mercury	0	0.40	–	14
	20	0.40	0.39	14
	40	0.41	–	14
	80	0.42	–	14
	120	0.44	–	14

	T/°C	κ_T/Pa 1 atm	1000 atm	Ref.
	160	0.46	–	14
Methanol	0	10.62	–	1
	0	10.68	–	11
	0	10.78	–	3
	10	11.34	–	1
	10	11.45	–	3
	20	12.11	–	1
	20	12.18	–	3
	30	12.93	–	1
	30	12.98	–	3
	40	13.85	–	1
	40	13.82	–	3
	50	14.76	–	3
Methylene bromide	-33.1	4.95	–	12
	-23.1	5.11	–	12
	-13.1	5.50	–	12
	-3.1	5.81	–	12
	6.9	6.13	–	12
	16.9	6.47	–	12
	26.9	6.85	–	12
Methylene chloride	25	9.74	5.31	4
Methyl iodide	-33.1	6.86	–	12
	-23.1	7.41	–	12
	-13.1	7.97	–	12
	-3.1	8.55	–	12
	6.9	9.13	–	12
	16.9	9.71	–	12
	26.9	10.33	–	12
n-Nonane	25	11.77	–	8
	35	12.68	–	8
	45	13.66	–	8
	60	15.33	–	8
nOctadecane	60	9.4	5.1	13
	79.4	10.4	5.5	13
	98.9	11.6	5.8	13
	115.0	12.8	6.1	13
	135	14.4	6.4	13
n-Octane	25	12.80	–	8
	35	13.86	–	8
	45	15.04	–	8
	60	15.33	–	8
1-Octanol	0	6.82	–	11
n-Pentadecane	37.8	9.1	–	13
	60	10.2	5.2	13
	79.4	11.7	5.5	13
	98.9	13.2	5.8	13
	115	14.7	6.1	13
	135	16.8	6.4	13
Pentanol	0	7.71	–	11
Phenol	46	5.61	–	2
	60	6.05	–	2
	80	6.78	–	2
	110	8.12	–	2
	125	8.88	–	2
	150	10.30	–	2
	175	12.35	–	2
n-Propyl alcohol	0	8.43	–	11
Toluene	-59.3	5.27	–	9
	-41.1	5.93	–	9
	-19.2	6.87	–	9
	0	7.83	–	1
	0	7.97	–	3
	0	7.84	–	9
	10	8.38	–	1
	10	8.44	–	3
	20	8.96	–	1
	20	8.94	–	3
	30	9.60	–	1
	30	9.49	–	3
	40	10.33	–	1
	40	10.14	–	3
	50	11.13	–	1
	50	10.90	–	3
	60	11.99	–	1
	70	12.95	–	1
Water	0	5.01	–	1
	0	5.04	–	9
	10	4.78	–	1
	20	4.58	–	1
	25	4.57	3.48	7
	30	4.46	–	1
	34.8	4.44	–	9
	35	4.48	3.42	7
	40	4.41	–	1
	45	4.41	3.40	7
	50	4.40	–	1
	55	4.44	3.40	7
	60	4.43	–	1
	65	4.48	3.42	7
	70	4.49	–	1
	75	4.55	3.47	7
	80	4.57	–	1
	85	4.65	3.53	7
	90	4.68	–	1
	100	4.80	–	1
m-Xylene	0	7.44	–	1
	10	7.94	–	1
	20	8.46	–	1
	30	9.03	–	1
	40	9.63	–	1
	50	10.25	–	1
	60	11.01	–	1
	70	11.77	–	1
	80	12.56	–	1

REFERENCES

1. Tyrer, D., *J. Chem. Soc.*, 105, 2534, 1914.
2. Lutskii, A. E. and Solonko, V. N., *Russ. J. Phys. Chem.*, 38, 602, 1964.
3. Fryer, E. B., Hubbard, J. C., and Andrews, D. H., *J. Am. Chem. Soc.*, 51, 759, 1929.
4. Newitt, D. M. and Weale, K. E., *J. Chem. Soc.*, p. 3092, 1951.
5. Gibson, R. E. and Loeffler, O. H., *J. Am. Chem. Soc.*, 61, 2515, 1939.
6. Gibson, R. E. and Kincaid, J. F., *J. Am. Chem. Soc.*, 60, 511, 1938.
7. Gibson, R. E. and Loeffler, O. H., *J. Am. Chem. Soc.*, 63, 898, 1941.
8. Aicart, E., Tardajos, G., and Diaz Pena, M., *J. Chem. Eng. Data*, 26, 22, 1981.
9. Marshall, J. G., Staveley, L. A. K., and Hart, K. R., *Trans. Faraday Soc.*, 52, 23, 1956.
10. Staveley, L. A. K., Tupman, W. I., and Hart, K. R., *Trans. Faraday Soc.*, 51, 323, 1955.
11. McKinney, W. P., Skinner, G. F., and Staveley, L. A. K., *J. Chem. Soc.*, p. 2415, 1959.
12. Harrison, D. and Moelwyn-Hughes, E. A., *Proc. R. Soc. London Ser. A*, 239, 230, 1957.
13. Cutler, W. G., McMickle, R. H., Webb, W., and Schiessler, R. W., *J. Chem. Phys.*, 29, 727, 1958.
14. Moelwyn-Hughes, E. A., *J. Phys. Colloid Chem.*, 55, 1246, 1951.
15. Holder, G. A. and Walley, E., *Trans. Faraday Soc.*, 58, 2095, 1962.
16. Aicart, E., Tardajos, G., and Diaz Pena, M., *J. Chem. Eng. Data*, 25, 140, 1980.
17. Blinowska, A. and Brostow, W., *J. Chem. Thermodyn.*, 7, 787, 1975.

PROPERTIES OF CRYOGENIC FLUIDS

This table gives physical and thermodynamic properties of eight cryogenic fluids. The properties are:

M	Molar mass in grams per mole
T_t	Triple point temperature in kelvins
P_t	Triple point pressure in kilopascals
ρ_t (l)	Liquid density at the triple point in grams per milliliter
$\Delta_{fus}H$ @ T_t	Enthalpy of fusion at the triple point in joules per gram
T_b	Normal boiling point in kelvins at a pressure of 101325 pascals (760 mmHg)
$\Delta_{vap}H$ @ T_b	Enthalpy of vaporization at the normal boiling point in joules per gram
ρ (l) @ T_b	Liquid density at the normal boiling point in grams per milliliter
ρ (g) @ T_b	Vapor density at the normal boiling point in grams per liter
C_p (l) @ T_b	Liquid heat capacity at constant pressure at the normal boiling point in joules per gram kelvin
C_p (g) @ T_b	Vapor heat capacity at constant pressure at the normal boiling point in joules per gram kelvin
T_c	Critical temperature in kelvins
P_c	Critical pressure in megapascals
ρ_c	Critical density in grams per milliliter

In the case of air, the value given for the triple point temperature is the incipient solidification temperature, and the normal boiling point value is the incipient boiling (bubble) point. See Reference 3 for more details.

REFERENCES

1. Younglove, B. A., *J. Phys. Chem. Ref. Data*, 11, Suppl. 1, 1982.
2. Daubert, T. E., Danner, R. P., Sibul, H. M., and Stebbins, C. C., *Physical and Thermodynamic Properties of Pure Compounds: Data Compilation*, extant 1994 (core with 4 supplements), Taylor & Francis, Bristol, PA (also available as database).
3. Sytchev, V. V., et al., *Thermodynamic Properties of Air*, Hemisphere Publishing, New York, 1987.
4. Jacobsen, R. T., Stewart, R. B., and Jahangiri, M., *J. Phys. Chem. Ref. Data*, 15, 735, 1986. [Nitrogen]
5. Stewart, R. B., Jacobsen, R. T., and Wagner, W., *J. Phys. Chem. Ref. Data*, 20, 917, 1991. [Oxygen]
6. McCarty, R. D., *J. Phys. Chem. Ref. Data*, 2, 923, 1973. [Helium] Also, Donnelly, R. J., private communication.
7. Stewart, R. B. and Jacobsen, R. T., *J. Phys. Chem. Ref. Data*, 18, 639, 1989. [Argon]
8. Setzmann, U. and Wagner, W., *J. Phys. Chem. Ref. Data*, 20, 1061, 1991. [Methane]
9. Vargaftik, N. B., *Thermophysical Properties of Liquids and Gases*, 2nd ed., John Wiley, New York, 1975.

Property	Units	Air	N_2	O_2	H_2	He	Ne	Ar	Kr	Xe	CH_4
M	g/mol	28.96	28.014	31.999	2.0159	4.0026	20.180	39.948	83.800	131.290	16.043
T_t	K	59.75	63.15	54.3584	13.8		24.5561	83.8058	115.8	161.4	90.694
P_t	kPa		12.463	0.14633	7.042		50	68.95	72.92	81.59	11.696
ρ_t (l)	g/mL	0.959	0.870	1.306	0.0770		1.251	1.417	2.449	2.978	0.4515
$\Delta_{fus}H$ @ T_t	J/g		25.3	13.7	59.5		16.8	28.0	16.3	13.8	58.41
T_b	K	78.67	77.35	90.188	20.28	4.2221	27.07	87.293	119.92	165.10	111.668
$\Delta_{vap}H$ @ T_b	J/g	198.7	198.8	213.1	445	20.7	84.8	161.0	108.4	96.1	510.83
ρ (l) @ T_b	g/mL	0.8754	0.807	1.141	0.0708	0.124901	1.204	1.396	2.418	2.953	0.4224
ρ (g) @ T_b	g/L	3.199	4.622	4.467	1.3390	16.89	9.51	5.79	8.94		1.816
C_p (l) @ T_b	J/g K	1.865	2.042	1.699	9.668	4.545	1.877	1.078	0.533	0.340	3.481
C_p (g) @ T_b	J/g K		1.341	0.980	12.24	9.78		0.570	0.248	0.158	2.218
T_c	K	132.5	126.20	154.581	32.98	5.1953	44.40	150.663	209.40	289.73	190.56
P_c	MPa	3.766	3.390	5.043	1.293	0.227460	2.760	4.860	5.500	5.840	4.592
ρ_c	g/mL	0.316	0.313	0.436	0.031	0.06964	0.484	0.531	0.919	1.110	0.1627

HALOCARBON REFRIGERANTS

Halogen derivatives of the lower aliphatic hydrocarbons, especially fluorocarbons and chlorocarbons, are widely used as refrigerants, solvents, and cleaning agents. This table lists the most important halocarbons in current use or being considered as possible substitutes for those compounds that are potential environmental hazards. The code number for each compound appears in the first column; this number is frequently used in expressions like R 11 or CFC 11 or HFC 134.

In addition to name, molecular formula, CAS Registry Number, and molecular weight, the table lists the following properties:

T_m: normal melting point
T_b: normal boiling point (at 101.324 kPa)
T_c: critical temperature
TLV: Threshold Limit Value, which is the maximum safe concentration in air in the workplace, expressed as the time-weighted average (TWA) in parts per million by volume over an 8-hr workday and 40-hr workweek (see Reference 3). An * following the TLV indicates that the substance in a confirmed or suspected human carcinogen.

REFERENCES

1. *1989 ASHRAE Handbook: Fundamentals*, American Society of Heating, Refrigerating, and Air Conditioning Engineers, Atlanta, GA, 1989.
2. Platzer, B., Polt, A., and Mauer,G.,*Thermophysical Properties of Refrigerants*, Springer, Berlin, 1990.
3. 1991-1992 Threshold Limit Values and Biological Exposure Indexes, American Conference of Government Industrial Hygienists, Cincinnati, OH, 1991.

Code no.	Name	Molecular formula	CAS reg. no.	Molecular weight	T_m/°C	T_b/°C	T_c/°C	TLV ppm
10	Tetrachloromethane	CCl_4	56-23-5	153.822	-23	76.8	283.5	5*
11	Trichlorofluoromethane	CCl_3F	75-69-4	137.368	-111.11	23.75	198.1	1000
12	Dichlorodifluoromethane	CCl_2F_2	75-71-8	120.913	-158	-29.8	111.80	1000
12B1	Bromochlorodifluoromethane	$CBrClF_2$	353-59-3	165.365	-159.5	-3.72	153.73	
13	Chlorotrifluoromethane	$CClF_3$	75-72-9	104.459	-181	-81.4	29	
13B1	Bromotrifluoromethane	$CBrF_3$	75-63-8	148.910	-172	-57.89	67.1	
14	Tetrafluoromethane	CF_4	75-73-0	88.005	-183.59	-128.02	-45.5	
20	Trichloromethane	$CHCl_3$	67-66-3	119.377	-63.6	61.17	263.3	10*
21	Dichlorofluoromethane	$CHCl_2F$	75-43-4	102.923	-135	8.95	178.43	10
22	Chlorodifluoromethane	$CHClF_2$	75-45-6	86.468	-157.42	-40.75	96.2	1000
23	Trifluoromethane	CHF_3	75-46-7	70.014	-155.18	-82.1	26.2	
30	Dichloromethane	CH_2Cl_2	75-09-2	84.932	-95.14	40	237	50*
31	Chlorofluoromethane	CH_2ClF	593-70-4	68.478	-133	-9.1		
32	Difluoromethane	CH_2F_2	75-10-5	52.024	-136	-51.69	78.5	
40	Chloromethane	CH_3Cl	74-87-3	50.488	-97.7	-24.09	143.10	50*
41	Fluoromethane	CH_3F	593-53-3	34.033	-141.8	-78.41	44.7	
110	Hexachloroethane	C_2Cl_6	67-72-1	236.738	187	187		1*
111	Pentachlorofluoroethane	C_2Cl_5F	354-56-3	220.284				
112	Tetrachloro-1,2-difluoroethane	$C_2Cl_4F_2$	76-12-0	203.830	26	93	278	500
112a	1,1,1,2-Tetrachloro-2,2-difluoroethane	$C_2Cl_4F_2$	76-11-9	203.830	40.6	91.5		
113	1,1,2-Trichlorotrifluoroethane	$C_2Cl_3F_3$	76-13-1	187.375	-35	47.7	214.2	1000
113a	1,1,1-Trichlorotrifluoroethane	$C_2Cl_3F_3$	354-58-5	187.375	14.2	46.1		
114	1,2-Dichlorotetrafluoroethane	$C_2Cl_2F_4$	76-14-2	170.921	-94	3.8	145.63	1000
114a	1,1-Dichlorotetrafluoroethane	$C_2Cl_2F_4$	374-07-2	170.921	-56.6	4	145.5	

HALOCARBON REFRIGERANTS (continued)

Code no.	Name	Molecular formula	CAS reg. no.	Molecular weight	T_m/°C	T_b/°C	T_c/°C	TLV ppm
114B2	1,2-Dibromotetrafluoroethane	$C_2Br_2F_4$	124-73-2	259.824	-110.4	47.35	214.7	
115	Chloropentafluoroethane	C_2ClF_5	76-15-3	154.467	-99.44	-37.95	80.1	1000
116	Hexafluoroethane	C_2F_6	76-16-4	138.012	-100.7	-78.1	20	
120	Pentachloroethane	C_2HCl_5	76-01-7	202.293	-29	159.88		
121	1,1,2,2-Tetrachloro-1-fluoroethane	C_2HCl_4F	354-14-3	185.839	-82.6	116.6		
121a	1,1,1,2-Tetrachloro-2-fluoroethane	C_2HCl4_F	354-11-0	185.840	-95.35	116.5		
122	1,2,2-Trichloro-1,1-difluoroethane	$C_2HCl_3F_2$	354-21-2	169.385	-140	71.9		
122a	1,1,2-Trichloro-1,2-difluoroethane	$C_2HCl_3F_2$	354-15-4	169.385		72.5		
122b	1,1,1-Trichloro-2,2-difluoroethane	$C_2HCl_3F_2$	354-12-1	169.385		73		
123	2,2-Dichloro-1,1,1-trifluoroethane	$C_2HCl_2F_3$	306-83-2	152.931	-107	27.1		
123a	1,2-Dichloro-1,2,2-trifluoroethane	$C_2HCl_2F_3$	354-23-4	152.931		28.2		
124	2-Chloro-1,1,1,2-tetrafluoroethane	C_2HClF_4	2837-89-0	136.476		-12		
124a	1-Chloro-1,1,2,2-tetrafluoroethane	C_2HClF_4	354-25-6	136.476	-117	-10.2	126.8	
125	Pentafluoroethane	C_2HF_5	354-33-6	120.022	-103	-48.5		
130	1,1,2,2-Tetrachloroethane	$C_2H_2Cl_4$	79-34-5	167.849	-43.8	146.5	388.00	1*
131	1,1,2-Trichloro-2-fluoroethane	$C_2H_2Cl_3F$	359-28-4	151.394		102.5		
132	1,2-Dichloro-1,2-difluoroethane	$C_2H_2Cl_2F_2$	431-06-1	134.940	-155	59		
132b	1,2-Dichloro-1,1-difluoroethane	$C_2H_2Cl_2F_2$	1649-08-7	134.940				
133	1-Chloro-1,2,2-trifluoroethane	$C_2H_2ClF_3$	431-07-2	118.486		17		
133a	2-Chloro-1,1,1-trifluoroethane	$C_2H_2ClF_3$	75-88-7	118.486	-105.5	6.1		
133b	1-Chloro-1,1,2-trifluoroethane	$C_2H_2ClF_3$	421-04-5	118.486		12		
134	1,1,2,2-Tetrafluoroethane	$C_2H_2F_4$	359-35-3	102.032	-89	-19.9	119	
134a	1,1,1,2-Tetrafluoroethane	$C_2H_2F_4$	811-97-2	102.032	-101	-26.5	101.2	
140	1,1,2-Trichloroethane	$C_2H_3Cl_3$	79-00-5	133.404	-36.6	113.8		10
140a	1,1,1-Trichloroethane	$C_2H_3Cl_3$	71-55-6	133.404	-30.4	74.09	272	350
141	1,2-Dichloro-1-fluoroethane	$C_2H_3Cl_2F$	430-57-9	116.950		75.7		
141b	1,1-Dichloro-1-fluoroethane	$C_2H_3Cl_2F$	1717-00-6	116.950	-103.5	32.11		
142	2-Chloro-1,1-difluoroethane	$C_2H_3ClF_2$	338-65-8	100.495		35.1		
142b	1-Chloro-1,1-difluoroethane	$C_2H_3ClF_2$	75-68-3	100.495	-130.8	-9.75	137.14	
143	1,1,2-Trifluoroethane	$C_2H_3F_3$	430-66-0	84.041	-84	5		
143a	1,1,1-Trifluoroethane	$C_2H_3F_3$	420-46-2	84.041	-111.3	-47.55	73.2	
150	1,2-Dichloroethane	$C_2H_4Cl_2$	107-06-2	98.959	-35.5	83.5	288	10*
150a	1,1-Dichloroethane	$C_2H_4Cl_2$	75-34-3	98.959	-96.96	57.4	250	100
151	1-Chloro-2-fluoroethane	C_2H_4ClF	762-50-5	82.505		53.2		
151a	1-Chloro-1-fluoroethane	C_2H_4ClF	1615-75-4	82.505		16.1		
152	1,2-Difluoroethane	$C_2H_4F_2$	624-72-6	66.051		30.7		
152a	1,1-Difluoroethane	$C_2H_4F_2$	75-37-6	66.051	-117	-24.95	113.6	
160	Chloroethane	C_2H_5Cl	75-00-3	64.514	-138.7	13.1	187.3	1000
161	Fluoroethane	C_2H_5F	353-36-6	48.060	-143.2	-37.65	102.16	
218	Perfluoropropane	C_3F_8	76-19-7	188.020	-147.69	-36.65	72.0	
1112a	1,1-Dichloro-2,2-difluoroethylene	$C_2Cl_2F_2$	79-35-6	132.924				
1113	Chlorotrifluoroethylene	C_2ClF_3	79-38-9	116.470	-158	-27.85	106	

HALOCARBON REFRIGERANTS (continued)

Code no.	Name	Molecular formula	CAS reg. no.	Molecular weight	T_m/°C	T_b/°C	T_c/°C	TLV ppm
1114	Tetrafluoroethylene	C_2F_4	116-14-3	100.016	-142.5	-75.95	33.4	
1120	Trichloroethylene	C_2HCl_3	79-01-6	131.388	-84.75	87.21	271.1	50*
1130	*cis*-1,2-Dichloroethylene	$C_2H_2Cl_2$	156-59-2	96.943	-80	60.19	271.1	200
1130	*trans*-1,2-Dichloroethylene	$C_2H_2Cl_2$	156-60-5	96.943	-49.8	48.73	243.4	200
1132a	1,1-Difluoroethylene	$C_2H_2F_2$	75-38-7	64.035	-144	-85.7	29.8	
1140	Chloroethylene	C_2H_3Cl	75-01-4	62.499	-153.79	-13.37		5*
1141	Fluoroethylene	C_2H_3F	75-02-5	46.044	-160.5	-72	54.8	
C316	1,2-Dichlorohexafluorocyclobutane	$C_4Cl_2F_6$	356-18-3	232.940				
C317	Chloroheptafluorocyclobutane	C_4ClF_7	377-41-3	216.486				
C318	Perfluorocyclobutane	C_4F_8	115-25-3	200.031	-40.19	-5.99	115.31	

* Confirmed or suspected human carcinogen.

HEAT CAPACITY OF LIQUIDS AND GASES AT 25°C

The molar heat capacities of liquids at 25°C and ideal gases at 25°C and 100 kPa (0.9869 atm) pressure are given in this table. The data on liquids are given at a nominal pressure of 1 atmosphere except for a few substances whose normal boiling points are somewhat below 25°C. In such cases the applicable pressure is the vapor pressure of the liquid at 25°C. Substances are arranged by molecular formula in the modified Hill order.

The units are joules per mole kelvin. Molar masses are listed in order to facilitate conversion to specific heat capacity in J/g K units.

References may be found in the table in Section 5, "Standard State Thermodynamic Properties of Chemical Substances".

Molecular formula	Name	Molar mass g/mol	C_p (liq) J/mol K	C_p (ideal gas) J/mol K
AlB_3H_{12}	Aluminum borohydride	71.510	194.6	
Ar	Argon	39.948		20.8
$AsCl_3$	Arsenic trichloride	181.280		75.7
AsF_3	Arsenic trifluoride	131.917	126.6	65.6
AsH_3	Arsine	77.945		38.1
BBr_3	Boron tribromide	250.523		67.8
BCl_3	Boron trichloride	117.169	106.7	62.7
B_2Cl_4	Tetrachlorodiborane	163.433	137.7	95.4
B_2F_4	Tetrafluorodiborane	97.616		79.1
B_2H_6	Diborane	27.670		56.9
B_5H_9	Pentaborane	63.127	151.1	96.8
BrCl	Bromine chloride	115.357		35.0
$BrCl_3Si$	Bromotrichlorosilane	214.348		90.9
BrF	Bromine fluoride	98.902		33.0
BrF_3	Bromine trifluoride	136.899	124.6	66.6
BrF_5	Bromine pentafluoride	174.896		99.6
$BrGeH_3$	Bromogermane	155.538		56.4
BrH	Hydrogen bromide	80.912		29.1
BrH_3Si	Bromosilane	111.013		52.8
BrI	Iodine bromide	206.809		36.4
BrNO	Nitrosyl bromide	109.910		45.5
Br_2	Bromine (Br_2)	159.808	75.7	36.0
Br_2H_2Si	Dibromosilane	189.909		65.5
Br_3ClSi	Tribromochlorosilane	303.250		95.3
Br_3HSi	Tribromosilane	268.805		80.8
Br_3OP	Phosphoryl bromide	286.685		89.9
Br_3P	Phosphorous tribromide	270.686		76.0
Br_4Ge	Germanium tetrabromide	392.226		101.8
Br_4Si	Silicon tetrabromide	347.702		97.1
ClF	Chlorine fluoride	54.451		32.1
$ClFO_3$	Perchloryl fluoride	102.449		64.9
ClF_3	Chlorine trifluoride	92.448		63.9
$ClGeH_3$	Chlorogermane	111.087		54.7
ClH	Hydrogen chloride	36.461		29.1
ClH_3Si	Chlorosilane	66.562		51.0
ClI	Iodine chloride	162.357		35.6
ClNO	Nitrosyl chloride	65.459		44.7
$ClNO_2$	Nitryl chloride	81.458		53.2
ClO_2	Chlorine dioxide (ClO_2)	67.452		42.0
Cl_2	Chlorine (Cl_2)	70.905		33.9
Cl_2CrO_2	Chromyl chloride (CrO_2Cl_2)	154.900		84.5
Cl_2H_2Si	Dichlorosilane	101.007		60.5
Cl_2O	Oxygen dichloride	86.905		45.4
Cl_2OS	Thionyl chloride	118.971	121.0	66.5
Cl_2O_2S	Sulfuryl chloride	134.970	134.0	77.0
Cl_3HSi	Trichlorosilane	135.452		75.8
Cl_3OP	Phosphoryl chloride	153.331	138.8	84.9
Cl_3P	Phosphorous trichloride	137.332		71.8
Cl_4Ge	Germanium tetrachloride	214.421		96.1
Cl_4Si	Silicon tetrachloride	169.896	145.3	90.3
Cl_4Sn	Tin chloride ($SnCl_4$)	260.521	165.3	98.3
Cl_4Ti	Titanium chloride ($TiCl_4$)	189.691	145.2	95.4

Molecular formula	Name	Molar mass g/mol	C_p (liq) J/mol K	C_p (ideal gas) J/mol K
FI	Iodine fluoride	145.903		33.4
FNO	Nitrosyl fluoride	49.005		41.3
FNO_2	Nitryl fluoride	65.004		49.8
F_2	Fluorine (F_2)	37.997		31.3
F_2HN	Difluoramine	53.012		43.4
F_2O	Oxygen difluoride	53.996		43.3
F_2OS	Thionyl fluoride	86.062		56.8
F_2O_2S	Sufuryl fluoride	102.062		66.0
F_3HSi	Trifluorosilane	86.089		60.5
F_3N	Nitrogen trifluoride	71.002		53.4
F_3OP	Phosphoryl fluoride	103.968		68.8
F_3P	Phosphorous trifluoride	87.969		58.7
F_4N_2	Tetrafluorohydrazine	104.007		79.2
F_4S	Sulfur fluoride (SF_4)	108.060		77.6
F_4Si	Silicon tetrafluoride	104.079		73.6
F_6Mo	Molybdenum fluoride (MoF_6)	209.930	169.8	120.6
F_6S	Sulfur fluoride (SF_6)	146.056		97.0
F_6Se	Selenium fluoride (SeF_6)	192.950		110.5
F_6W	Tungsten fluoride (WF_6)	297.840		119.0
GeH_4	Germane	76.642		45.0
HI	Hydrogen iodide	127.912		29.2
HNO_2	Nitrous acid (HONO)	47.014		45.6
HNO_3	Nitric acid	63.013	109.9	53.4
HN_3	Hydrazoic acid	43.028		43.7
H_2	Hydrogen (H_2)	2.016		28.8
H_2O	Water	18.015	75.3	33.6
H_2O_2	Hydrogen peroxide	34.015	89.1	43.1
H_2O_4S	Sulfuric acid	98.080	138.9	
H_2S	Hydrogen sulfide	34.082		34.2
H_2S_2	Hydrogen sulfide (H_2S_2)	66.148	84.1	51.5
H_2Se	Hydrogen selenide	80.976		34.7
H_3ISi	Iodosilane	158.014		54.4
H_3N	Ammonia	17.031		35.1
H_3O_4P	Phosphoric acid	97.995	145.0	
H_3P	Phosphine	33.998		37.1
H_3Sb	Stibine	124.781		41.1
H_4N_2	Hydrazine	32.045	98.9	49.6
H_4Si	Silane	32.117		42.8
H_4Sn	Stannane	122.742		49.0
H_5NO	Ammonium hydroxide	35.046	154.9	
H_6Si_2	Disilane	62.219		80.8
He	Helium	4.003		20.8
Hg	Mercury	200.590	28.0	20.8
I_2	Iodine (I_2)	253.809		36.9
Kr	Krypton	83.800		20.8
NO	Nitric oxide	30.006		29.8
NO_2	Nitrogen dioxide	46.006		37.2
N_2	Nitrogen (N_2)	28.014		29.1
N_2O	Nitrous oxide	44.013		38.5
N_2O_3	Nitrogen trioxide	76.012		65.6
N_2O_4	Nitrogen tetroxide	92.011	142.7	77.3
N_2O_5	Nitrogen pentoxide	108.011		84.5
O_2	Oxygen (O_2)	31.999		29.4
O_2S	Sulfur dioxide	64.065		39.9
O_3	Ozone	47.998		39.2
Rn	Radon			20.8
$CBrClF_2$	Bromochlorodifluoromethane	165.365		74.6
$CBrCl_2F$	Bromodichlorofluoromethane	181.819		80.0
$CBrCl_3$	Bromotrichloromethane	198.273		85.3

Molecular formula	Name	Molar mass g/mol	C_p (liq) J/mol K	C_p (ideal gas) J/mol K
$CBrF_3$	Bromotrifluoromethane	148.910		69.3
CBrN	Cyanogen bromide	105.922		46.9
CBr_2ClF	Dibromochlorofluoromethane	226.270		82.4
CBr_2Cl_2	Dibromodichloromethane	242.724		87.1
CBr_2F_2	Dibromodifluoromethane	209.816		77.0
CBr_2O	Carbonyl bromide	187.818		61.8
CBr_3Cl	Tribromochloromethane	287.176		89.4
CBr_3F	Tribromofluoromethane	270.721		84.4
CClFO	Carbonyl chloride fluoride	82.462		52.4
$CClF_3$	Chlorotrifluoromethane	104.459		66.9
CClN	Cyanogen chloride	61.470		45.0
CCl_2F_2	Dichlorodifluoromethane	120.913		72.3
CCl_2O	Carbonyl chloride	98.916		57.7
CCl_3F	Trichlorofluoromethane	137.368	121.6	78.1
CCl_4	Tetrachloromethane	153.822	130.7	83.3
CFN	Cyanogen fluoride	45.016		41.8
CF_2O	Carbonyl fluoride	66.007		46.8
CF_3I	Trifluoroiodomethane	195.911		70.9
CF_4	Tetrafluoromethane	88.005		61.1
CIN	Cyanogen iodide	152.922		48.3
CI_4	Tetraiodomethane	519.629		95.9
CO	Carbon monoxide	28.010		29.1
COS	Carbon oxysulfide	60.076		41.5
CO_2	Carbon dioxide	44.010		37.1
CS_2	Carbon disulfide	76.143	76.4	45.4
CHBrClF	Bromochlorofluoromethane	147.374		63.2
$CHBrCl_2$	Bromodichloromethane	163.828		67.4
$CHBrF_2$	Bromodifluoromethane	130.920		58.7
$CHBr_2Cl$	Chlorodibromomethane	208.280		69.2
$CHBr_2F$	Dibromofluoromethane	191.825		65.1
$CHBr_3$	Tribromomethane	252.731	130.7	71.2
$CHClF_2$	Chlorodifluoromethane	86.468		55.9
$CHCl_2F$	Dichlorofluoromethane	102.923		60.9
$CHCl_3$	Trichloromethane	119.377	114.2	65.7
CHFO	Formyl fluoride	48.017		39.9
CHF_3	Trifluoromethane	70.014		51.0
CHN	Hydrogen cyanide	27.026	70.6	35.9
CHNS	Isothiocyanic acid	59.092		46.9
CH_2BrCl	Bromochloromethane	129.384		52.7
CH_2BrF	Bromofluoromethane	112.929		49.2
CH_2Br_2	Dibromomethane	173.835		54.7
CH_2ClF	Chlorofluoromethane	68.478		47.0
CH_2Cl_2	Dichloromethane	84.932	101.2	51.0
CH_2F_2	Difluoromethane	52.024		42.9
CH_2I_2	Diiodomethane	267.836	134.0	57.7
CH_2N_2	Diazomethane	42.040		52.5
CH_2O	Formaldehyde	30.026		35.4
CH_2O_2	Formic acid	46.026	99.0	
CH_3Br	Bromomethane	94.939		42.4
CH_3Cl	Chloromethane	50.488		40.8
CH_3F	Fluoromethane	34.033		37.5
CH_3I	Iodomethane	141.939	126.0	44.1
CH_3NO_2	Nitromethane	61.040	106.6	57.3
CH_3NO_3	Methyl nitrate	77.040	157.3	
CH_4	Methane	16.043		35.3
CH_4O	Methanol	32.042	81.1	43.9
CH_5N	Methylamine	31.057	102.1	50.1
CH_6N_2	Methylhydrazine	46.072	134.9	71.1
CH_6Si	Methylsilane	46.144		65.9

HEAT CAPACITY OF LIQUIDS AND GASES AT 25°C (continued)

Molecular formula	Name	Molar mass g/mol	C_p (liq) J/mol K	C_p (ideal gas) J/mol K
C_2Br_4	Tetrabromoethylene	343.638		102.7
C_2Br_6	Hexabromoethane	503.446		139.3
C_2ClF_3	Chlorotrifluoroethylene	116.470		83.9
$C_2Cl_2F_4$	1,2-Dichlorotetrafluoroethane	170.921	111.7	
$C_2Cl_3F_3$	1,1,2-Trichlorotrifluoroethane	187.375	170.1	
C_2Cl_3N	Trichloroacetonitrile	144.387		96.1
C_2Cl_4	Tetrachloroethylene	165.833	143.4	
$C_2Cl_4F_2$	1,1,1,2-Tetrachloro-2,2-difluoroethane	203.830		123.4
$C_2Cl_4F_2$	Tetrachloro-1,2-difluoroethane	203.830	173.6	132.3
C_2F_3N	Trifluoroacetonitrile	95.024		77.9
C_2F_4	Tetrafluoroethylene	100.016		80.5
C_2F_6	Hexafluoroethane	138.012		106.7
C_2N_2	Cyanogen	52.036		56.8
C_2HBr	Bromoacetylene	104.934		55.7
C_2HCl	Chloroacetylene	60.483		54.3
C_2HClF_2	1-Chloro-2,2-difluoroethylene	98.479		72.1
C_2HCl_2F	1,1-Dichloro-2-fluoroethylene	114.934		76.5
$C_2HCl_2F_3$	2,2-Dichloro-1,1,1-trifluoroethane	152.931		102.5
C_2HCl_3	Trichloroethylene	131.388	124.4	80.3
C_2HCl_3O	Trichloroacetaldehyde	147.387	151.0	
C_2HCl_5	Pentachloroethane	202.293	173.8	
C_2HF	Fluoroacetylene	44.028		52.1
C_2H_2	Acetylene	26.038		43.9
$C_2H_2Br_2$	*cis*-1,2-Dibromoethylene	185.846		68.8
$C_2H_2Br_2$	*trans*-1,2-Dibromoethylene	185.846		70.3
$C_2H_2ClF_3$	2-Chloro-1,1,1-trifluoroethane	118.486		89.1
$C_2H_2Cl_2$	1,1-Dichloroethylene	96.943	111.3	67.1
$C_2H_2Cl_2$	*cis*-1,2-Dichloroethylene	96.943	116.4	65.1
$C_2H_2Cl_2$	*trans*-1,2-Dichloroethylene	96.943	116.8	66.7
$C_2H_2Cl_4$	1,1,1,2-Tetrachloroethane	167.849	153.8	102.7
$C_2H_2Cl_4$	1,1,2,2-Tetrachloroethane	167.849	162.3	100.8
$C_2H_2F_2$	1,1-Difluoroethylene	64.035		60.1
$C_2H_2F_2$	*cis*-1,2-Difluoroethylene	64.035		58.2
C_2H_2O	Ketene	42.037		51.8
C_2H_3Br	Bromoethylene	106.950		55.5
C_2H_3Cl	Chloroethylene	62.499		53.7
$C_2H_3ClF_2$	1-Chloro-1,1-difluoroethane	100.495	130.1	82.5
C_2H_3ClO	Acetyl chloride	78.498	117.0	67.8
$C_2H_3Cl_2F$	1,1-Dichloro-1-fluoroethane	116.950		88.7
$C_2H_3Cl_3$	1,1,1-Trichloroethane	133.404	144.3	93.3
$C_2H_3Cl_3$	1,1,2-Trichloroethane	133.404	150.9	89.0
$C_2H_3F_3$	1,1,1-Trifluoroethane	84.041		78.2
C_2H_3I	Iodoethylene	153.950		57.9
C_2H_3N	Acetonitrile	41.053	91.4	52.2
C_2H_3N	Isocyanomethane	41.053		52.9
C_2H_4	Ethylene	28.054		43.6
$C_2H_4Br_2$	Dibromoethane	187.862		80.8
$C_2H_4Br_2$	1,2-Dibromoethane	187.862	136.0	
$C_2H_4Cl_2$	1,1-Dichloroethane	98.959	126.3	76.2
$C_2H_4Cl_2$	1,2-Dichloroethane	98.959	128.4	78.7
$C_2H_4F_2$	1,1-Difluoroethane	66.051		67.8
C_2H_4O	Acetaldehyde	44.053	89.0	55.3
C_2H_4O	Ethylene oxide	44.053	88.0	47.9
$C_2H_4O_2$	Acetic acid	60.053	123.3	66.5
$C_2H_4O_2$	Methyl formate	60.053	119.1	64.4
C_2H_5Br	Bromoethane	108.966	100.8	64.5
C_2H_5Cl	Chloroethane	64.514	104.3	62.8
C_2H_5F	Fluoroethane	48.060		58.6
C_2H_5I	Iodoethane	155.966	115.1	66.9

Molecular formula	Name	Molar mass g/mol	C_p (liq) J/mol K	C_p (ideal gas) J/mol K
$C_2H_5NO_2$	Nitroethane	75.067	134.4	
C_2H_6	Ethane	30.070		52.6
C_2H_6Cd	Dimethyl cadmium	142.481	132.0	
C_2H_6Hg	Dimethyl mercury	230.660		83.3
C_2H_6O	Dimethyl ether	46.069		64.4
C_2H_6O	Ethanol	46.069	112.3	65.4
C_2H_6OS	Dimethyl sulfoxide	78.135	153.0	
$C_2H_6O_2$	Ethylene glycol	62.068	148.6	82.7
C_2H_6S	Dimethyl sulfide	62.136	118.1	74.1
C_2H_6S	Ethanethiol	62.136	117.9	72.7
$C_2H_6S_2$	Dimethyl disulfide	94.202	146.1	
C_2H_6Zn	Dimethyl zinc	95.460	129.2	
C_2H_7N	Dimethylamine	45.084	137.7	70.7
C_2H_7N	Ethylamine	45.084	130.0	71.5
C_2H_7NO	Ethanolamine	61.084	195.5	
$C_2H_8N_2$	1,1-Dimethylhydrazine	60.099	164.1	
$C_2H_8N_2$	1,2-Ethanediamine	60.099	172.6	
$C_3H_4O_2$	Propenoic acid	72.064	145.7	
$C_3H_4O_2$	2-Oxetanone	72.064	122.1	
C_3H_5ClO	Epichlorohydrin	92.525	131.6	
$C_3H_5Cl_3$	1,2,3-Trichloropropane	147.431	183.6	
C_3H_5N	Propanenitrile	55.079	119.3	
$C_3H_6Cl_2$	1,2-Dichloropropane	112.986	149.1	
C_3H_6O	Acetone	58.080	126.3	75.0
C_3H_6O	Allyl alcohol	58.080	138.9	
C_3H_6O	Propanal	58.080		80.7
C_3H_6O	Methyloxirane	58.080	120.4	72.6
$C_3H_6O_2$	Ethyl formate	74.079	149.3	
$C_3H_6O_2$	Methyl acetate	74.079	141.9	86.0
$C_3H_6O_2$	Propanoic acid	74.079	152.8	
$C_3H_6O_2$	1,3-Dioxolane	74.079	118.0	
C_3H_7N	Cyclopropylamine	57.095	147.1	
C_3H_7NO	*N,N*-Dimethylformamide	73.095	150.6	
$C_3H_7NO_2$	2-Nitropropane	89.094	170.3	
C_3H_8O	1-Propanol	60.096	143.9	85.6
C_3H_8O	2-Propanol	60.096	156.5	89.3
C_3H_8O	Ethyl methyl ether	60.096		93.3
$C_3H_8O_2$	1,2-Propylene glycol	76.095	190.8	
$C_3H_8O_2$	2-Methoxyethanol	76.095	171.1	
$C_3H_8O_2$	Dimethoxymethane	76.095	162.0	
$C_3H_8O_3$	Glycerol	92.095	218.9	
C_3H_8S	Ethyl methyl sulfide	76.163	144.6	
C_3H_8S	1-Propanethiol	76.163	144.6	
C_3H_8S	2-Propanethiol	76.163	145.3	
C_3H_9Al	Trimethyl aluminum	72.086	155.6	
C_3H_9B	Trimethylborane	55.916		88.5
C_3H_9N	Propylamine	59.111	164.1	91.2
C_3H_9N	Isopropylamine	59.111	163.8	97.5
C_3H_9N	Trimethylamine	59.111	137.9	91.8
$C_3H_{10}Si$	Trimethylsilane	74.198		117.9
C_4NiO_4	Nickel carbonyl	170.735	204.6	145.2
$C_4H_4N_2$	Succinonitrile	80.089	145.6	
C_4H_4O	Furan	68.075	115.3	65.4
C_4H_4S	Thiophene	84.142	123.8	
C_4H_5N	Pyrrole	67.090	127.7	
C_4H_6	1,3-Butadiene	54.092	123.6	
$C_4H_6O_2$	Methyl acrylate	86.090	158.8	
C_4H_8	1-Butene	56.107	118.0	
C_4H_8	*cis*-2-Butene	56.107	127.0	

HEAT CAPACITY OF LIQUIDS AND GASES AT 25°C (continued)

Molecular formula	Name	Molar mass g/mol	C_p (liq) J/mol K	C_p (ideal gas) J/mol K
C_4H_8O	1,2-Epoxybutane	72.107	147.0	
C_4H_8O	Butanal	72.107	163.7	103.4
C_4H_8O	2-Butanone	72.107	158.7	101.7
C_4H_8O	Tetrahydrofuran	72.107	124.0	76.3
$C_4H_8O_2$	1,3-Dioxane	88.106	143.9	
$C_4H_8O_2$	1,4-Dioxane	88.106	152.1	
$C_4H_8O_2$	Ethyl acetate	88.106	170.7	
$C_4H_8O_2$	Butanoic acid	88.106	178.6	
C_4H_9N	Pyrrolidine	71.122	156.6	
C_4H_9NO	*N,N*-Dimethylacetamide	87.122	175.6	
C_4H_{10}	Butane	58.123	140.9	
$C_4H_{10}Hg$	Diethyl mercury	258.713	182.8	
$C_4H_{10}O$	1 Butanol	74.123	177.2	
$C_4H_{10}O$	2-Butanol	74.123	196.9	112.7
$C_4H_{10}O$	2-Methyl-2-propanol	74.123	218.6	113.6
$C_4H_{10}O$	2-Methyl-1-propanol	74.123	181.5	111.3
$C_4H_{10}O$	Diethyl ether	74.123	175.6	119.5
$C_4H_{10}O$	Methyl propyl ether	74.123	165.4	
$C_4H_{10}O$	Isopropyl methyl ether	74.123	161.9	
$C_4H_{10}O_2$	1,4-Butanediol	90.122	200.1	
$C_4H_{10}O_2$	2-Ethoxyethanol	90.122	210.8	
$C_4H_{10}O_3$	Diethylene glycol	106.122	244.8	
$C_4H_{10}S$	1-Butanethiol	90.189	171.2	
$C_4H_{10}S$	Diethyl sulfide	90.189	171.4	117.0
$C_4H_{10}S$	Methyl propyl sulfide	90.189	171.6	
$C_4H_{10}S$	Isopropyl methyl sulfide	90.189	172.4	
$C_4H_{10}S_2$	Diethyl disulfide	122.255	171.4	
$C_4H_{11}N$	Butylamine	73.138	179.2	
$C_4H_{11}N$	Isobutylamine	73.138	183.2	
$C_4H_{11}N$	*tert*-Butylamine	73.138	192.1	
$C_4H_{11}N$	Diethylamine	73.138	169.2	
$C_4H_{12}Si$	Tetramethylsilane	88.225	204.1	143.9
C_5FeO_5	Iron pentacarbonyl	195.899	240.6	
$C_5H_4O_2$	Furfural	96.086	163.2	
C_5H_5N	Pyridine	79.101	132.7	
$C_5H_6O_2$	Furfuryl alcohol	98.101	204.0	
C_5H_6S	2-Methylthiophene	98.169	149.8	
C_5H_8	2-Methyl-1,3-butadiene	68.119	152.6	
C_5H_8	Cyclopentene	68.119	122.4	
C_5H_8	Spiropentane	68.119	134.5	
C_5H_9N	2,2-Dimethylpropanenitrile	83.133	179.4	
C_5H_9NO	*N*-Methyl-2-pyrrolidone	99.133	307.8	
C_5H_{10}	1-Pentene	70.134	154.0	
C_5H_{10}	*cis*-2-Pentene	70.134	151.7	
C_5H_{10}	*trans*-2-Pentene	70.134	157.0	
C_5H_{10}	2-Methyl-1-butene	70.134	157.2	
C_5H_{10}	2-Methyl-2-butene	70.134	152.8	
C_5H_{10}	3-Methyl-1-butene	70.134	156.1	
C_5H_{10}	Cyclopentane	70.134	128.8	
$C_5H_{10}O$	Cyclopentanol	86.134	184.1	
$C_5H_{10}O$	Mesityl oxide	86.134	212.5	
$C_5H_{10}O$	2-Pentanone	86.134	184.1	
$C_5H_{10}O$	3-Pentanone	86.134	190.9	
$C_5H_{10}O$	3-Methyl-2-butanone	86.134	179.9	
$C_5H_{10}O_2$	Propyl acetate	102.133	196.2	
$C_5H_{10}O_2$	Isopropyl acetate	102.133	199.4	
$C_5H_{10}O_2$	Pentanoic acid	102.133	210.3	
$C_5H_{10}O_3$	2-Methoxyethyl acetate	118.133	310.0	
$C_5H_{10}S$	Cyclopentanethiol	102.200	165.2	

HEAT CAPACITY OF LIQUIDS AND GASES AT 25°C (continued)

Molecular formula	Name	Molar mass g/mol	C_p (liq) J/mol K	C_p (ideal gas) J/mol K
$C_5H_{10}S$	Thiacyclohexane	102.200	163.3	
$C_5H_{11}N$	Piperidine	85.149	179.9	
$C_5H_{11}N$	Cyclopentylamine	85.149	181.2	
C_5H_{12}	Pentane	72.150	167.2	
C_5H_{12}	Isopentane	72.150	164.8	
$C_5H_{12}O$	Butyl methyl ether	88.150	192.7	
$C_5H_{12}O$	*tert*-Butyl methyl ether	88.150	187.5	
$C_5H_{12}O$	Ethyl propyl ether	88.150	197.2	
$C_5H_{12}O$	1-Pentanol	88.150	208.1	
$C_5H_{12}O$	3-Pentanol	88.150	239.7	
$C_5H_{12}O$	2-Methyl-2-butanol	88.150	247.1	
$C_5H_{12}S$	Butyl methyl sulfide	104.216	200.9	
$C_5H_{12}S$	*tert*-Butyl methyl sulfide	104.216	199.9	
$C_5H_{12}S$	Ethyl propyl sulfide	104.216	198.4	
$C_5H_{12}S$	2-Methyl-2-butanethiol	104.216	198.1	
C_6F_6	Hexafluorobenzene	186.056	221.6	
C_6MoO_6	Molybdenum hexacarbonyl	264.002	205.0	
$C_6H_4Cl_2$	*o*-Dichlorobenzene	147.003	162.4	
$C_6H_4F_2$	*o*-Difluorobenzene	114.095	159.0	
$C_6H_4F_2$	*m*-Difluorobenzene	114.095	159.1	
$C_6H_4F_2$	*p*-Difluorobenzene	114.095	157.5	
C_6H_5Br	Bromobenzene	157.010	154.3	
C_6H_5Cl	Chlorobenzene	112.558	150.1	
C_6H_5F	Fluorobenzene	96.104	146.4	
C_6H_5I	Iodobenzene	204.010	158.7	
$C_6H_5NO_2$	Nitrobenzene	123.111	185.8	
C_6H_6	Benzene	78.114	136.3	
C_6H_6S	Benzenethiol	110.180	173.2	
C_6H_7N	Aniline	93.128	191.9	107.9
C_6H_7N	2-Methylpyridine	93.128	158.6	
C_6H_7N	3-Methylpyridine	93.128	158.7	
C_6H_7N	4-Methylpyridine	93.128	159.0	
$C_6H_8N_2$	Adiponitrile	108.143	128.7	
$C_6H_8N_2$	Phenylhydrazine	108.143	217.0	
C_6H_{10}	Cyclohexene	82.145	148.3	
$C_6H_{10}O$	Cyclohexanone	98.145	182.2	
C_6H_{12}	Cyclohexane	84.161	154.9	
C_6H_{12}	1-Hexene	84.161	183.3	
C_6H_{12}	2,3-Dimethyl-2-butene	84.161	174.7	
$C_6H_{12}O$	Cyclohexanol	100.161	208.2	
$C_6H_{12}O$	2-Hexanone	100.161	213.3	
$C_6H_{12}O$	3-Hexanone	100.161	216.9	
$C_6H_{12}O$	4-Methyl-2-pentanone	100.161	213.3	
$C_6H_{12}O_2$	Butyl acetate	116.160	227.8	
$C_6H_{12}O_2$	Isobutyl acetate	116.160	233.8	
$C_6H_{12}O_2$	Methyl pentanoate	116.160	229.3	
$C_6H_{12}O_2$	Methyl 2,2-dimethylpropanoate	116.160	257.9	
$C_6H_{12}O_2$	Diacetone alcohol	116.160	221.3	
$C_6H_{12}O_3$	2-Ethoxyethyl acetate	132.160	376.0	
$C_6H_{12}S$	Cyclohexanethiol	116.227	192.6	
$C_6H_{13}Br$	1-Bromohexane	165.073	203.5	
C_6H_{14}	Hexane	86.177	195.6	
C_6H_{14}	2-Methylpentane	86.177	193.7	
C_6H_{14}	3-Methylpentane	86.177	190.7	
C_6H_{14}	2,2-Dimethylbutane	86.177	191.9	
C_6H_{14}	2,3-Dimethylbutane	86.177	189.7	
$C_6H_{14}O$	Dipropyl ether	102.177	221.6	
$C_6H_{14}O$	Diisopropyl ether	102.177	216.8	
$C_6H_{14}O$	1-Hexanol	102.177	240.4	

HEAT CAPACITY OF LIQUIDS AND GASES AT 25°C (continued)

Molecular formula	Name	Molar mass g/mol	C_p (liq) J/mol K	C_p (ideal gas) J/mol K
$C_6H_{14}O$	4-Methyl-2-pentanol	102.177	273.0	
$C_6H_{14}O_2$	1,2-Diethoxyethane	118.176	259.4	
$C_6H_{14}O_2$	2-Butoxyethanol	118.176	281.0	
$C_6H_{14}O_2$	Hexylene glycol	118.176	336.0	
$C_6H_{14}S$	Diisopropyl sulfide	118.243	232.0	
$C_6H_{15}B$	Triethylborane	97.996	241.2	
$C_6H_{15}N$	Triethylamine	101.192	219.9	
$C_6H_{15}NO_3$	Triethanolamine	149.190	389.0	
$C_6H_{18}OSi_2$	Hexamethyldisiloxane	162.379	311.4	238.5
C_7F_8	Perfluorotoluene	236.064	262.3	
C_7F_{14}	Perfluoromethylcyclohexane	350.055	353.1	
C_7F_{16}	Perfluoroheptane	388.052	419.0	
$C_7H_3F_5$	2,3,4,5,6 Pentafluorotoluene	182.093	225.8	
C_7H_5N	Benzonitrile	103.123	165.2	
C_7H_6O	Benzaldehyde	106.124	172.0	
C_7H_8	Toluene	92.141	157.3	
C_7H_8O	*m*-Cresol	108.140	224.9	
C_7H_8O	Benzyl alcohol	108.140	217.9	
C_7H_9N	*o*-Methyl aniline	107.155	130.2	
C_7H_9N	*m*-Methyl aniline	107.155	125.5	
C_7H_9N	*p* Methyl aniline	107.155	126.2	
C_7H_9N	2,3-Dimethylpyridine	107.155	180.5	
C_7H_9N	2,4-Dimethylpyridine	107.155	184.8	
C_7H_9N	2,5-Dimethylpyridine	107.155	184.7	
C_7H_9N	2,6-Dimethylpyridine	107.155	185.2	
C_7H_9N	3,4-Dimethylpyridine	107.155	191.8	
C_7H_9N	3,5-Dimethylpyridine	107.155	184.5	
C_7H_{14}	1-Heptene	98.188	211.8	
$C_7H_{14}O$	2-Heptanone	114.188	232.6	
$C_7H_{14}O$	2,4-Dimethyl-3-pentanone	114.188	233.7	
$C_7H_{14}O$	1-Heptanal	114.188	230.1	
$C_7H_{14}O_2$	Pentyl acetate	130.187	261.0	
$C_7H_{14}O_2$	Isopentyl acetate	130.187	248.5	
$C_7H_{14}O_2$	Heptanoic acid	130.187	265.4	
C_7H_{16}	Heptane	100.204	224.7	
C_7H_{16}	2-Methylhexane	100.204	222.9	
C_7H_{16}	2,2-Dimethylpentane	100.204	221.1	
C_7H_{16}	2,4-Dimethylpentane	100.204	224.2	
C_7H_{16}	3-Ethylpentane	100.204	219.6	
C_7H_{16}	2,2,3-Trimethylbutane	100.204	213.5	
$C_7H_{16}O$	1-Heptanol	116.203	272.1	
C_8H_8	Styrene	104.152	182.0	
$C_8H_8O_2$	Methyl benzoate	136.150	221.3	
C_8H_{10}	Ethylbenzene	106.167	183.2	
C_8H_{10}	*o*-Xylene	106.167	186.1	
C_8H_{10}	*m*-Xylene	106.167	183.0	
C_8H_{10}	*p*-Xylene	106.167	181.5	
$C_8H_{10}O$	Phenetole	122.167	228.5	
C_8H_{16}	Propylcyclopentane	112.215	216.3	
C_8H_{16}	Ethylcyclohexane	112.215	211.8	
C_8H_{16}	1,1-Dimethylcyclohexane	112.215	209.2	
C_8H_{16}	*cis*-1,2-Dimethylcyclohexane	112.215	210.2	
C_8H_{16}	*trans*-1,2-Dimethylcyclohexane	112.215	209.4	
C_8H_{16}	*cis*-1,3-Dimethylcyclohexane	112.215	209.4	
C_8H_{16}	*trans*-1,3-Dimethylcyclohexane	112.215	212.8	
C_8H_{16}	*cis*-1,4-Dimethylcyclohexane	112.215	212.1	
C_8H_{16}	trans-1,4-Dimethylcyclohexane	112.215	210.2	
$C_8H_{16}O_2$	Octanoic acid	144.214	297.9	
$C_8H_{16}O_2$	Methyl heptanoate	144.214	285.1	

Molecular formula	Name	Molar mass g/mol	C_p (liq) J/mol K	C_p (ideal gas) J/mol K
C_8H_{18}	Octane	114.231	254.6	
C_8H_{18}	2-Methylheptane	114.231	252.0	
C_8H_{18}	3-Methylheptane	114.231	250.2	
C_8H_{18}	4-Methylheptane	114.231	251.1	
C_8H_{18}	2,5-Dimethylhexane	114.231	249.2	
C_8H_{18}	3,3-Dimethylhexane	114.231	246.6	
C_8H_{18}	2,2,4-Trimethylpentane	114.231	239.1	
C_8H_{18}	2,3,3-Trimethylpentane	114.231	245.6	
C_8H_{18}	2,3,4-Trimethylpentane	114.231	247.3	
$C_8H_{18}O$	1-Octanol	130.230	305.2	
$C_8H_{18}O$	2-Ethyl-1-hexanol	130.230	317.5	
$C_8H_{18}O$	Dibutyl ether	130.230	278.2	
$C_8H_{18}O$	Di-*tert*-butyl ether	130.230	276.1	
$C_8H_{18}O_5$	Tetraethylene glycol	194.228	428.8	
$C_8H_{18}S$	Dibutyl sulfide	146.297	284.3	
$C_8H_{19}N$	Dibutylamine	129.246	292.9	
$C_8H_{20}Pb$	Tetraethyl lead	323.447	307.4	
C_9H_7N	Isoquinoline	129.161	196.2	
C_9H_8	Indene	116.163	186.9	
C_9H_{10}	Indane	118.178	190.2	
C_9H_{12}	Cumene	120.194	210.7	
C_9H_{12}	Propylbenzene	120.194	214.7	
C_9H_{12}	1,2,3-Trimethylbenzene	120.194	216.4	
C_9H_{12}	1,2,4-Trimethylbenzene	120.194	215.0	
C_9H_{12}	Mesitylene	120.194	209.3	
$C_9H_{14}O$	Isophorone	138.210	253.5	
$C_9H_{14}O_6$	Triacetin	218.207	384.7	
C_9H_{18}	Propylcyclohexane	126.242	242.0	
$C_9H_{18}O$	5-Nonanone	142.241	303.6	
$C_9H_{18}O$	2,6-Dimethyl-4-heptanone	142.241	297.3	
$C_9H_{18}O_2$	Nonanoic acid	158.241	362.4	
C_9H_{20}	Nonane	128.258	284.4	
C_9H_{20}	2,2,3,3-Tetramethylpentane	128.258	271.5	
C_9H_{20}	3,3-Diethylpentane	128.258	278.2	
C_9H_{20}	2,2,4,4-Tetramethylpentane	128.258	266.3	
$C_{10}H_7Cl$	1-Chloronaphthalene	162.618	212.6	
$C_{10}H_{12}$	1,2,3,4-Tetrahydronaphthalene	132.205	217.5	
$C_{10}H_{14}$	Butylbenzene	134.221	243.4	
$C_{10}H_{14}$	*p*-Cymene	134.221	236.4	
$C_{10}H_{18}$	*cis*-Decahydronaphthalene	138.253	232.0	
$C_{10}H_{18}$	*trans*-Decahydronaphthalene	138.253	228.5	
$C_{10}H_{20}$	Butylcyclohexane	140.269	271.0	
$C_{10}H_{20}$	1-Decene	140.269	300.8	
$C_{10}H_{22}$	Decane	142.285	314.4	
$C_{10}H_{22}$	2-Methylnonane	142.285	313.3	
$C_{10}H_{22}$	5-Methylnonane	142.285	314.4	
$C_{10}H_{22}O$	1-Decanol	158.284	370.6	
$C_{10}H_{22}S$	1-Decanethiol	174.351	350.4	
$C_{11}H_{10}$	1-Methylnaphthalene	142.200	224.4	
$C_{11}H_{24}$	Undecane	156.312	344.9	
$C_{12}H_{14}O_4$	Diethyl phthalate	222.241	366.1	
$C_{12}H_{24}$	1-Dodecene	168.323	360.7	
$C_{12}H_{26}$	Dodecane	170.338	375.8	
$C_{12}H_{26}O$	1-Dodecanol	186.338	438.1	
$C_{13}H_{28}$	Tridecane	184.365	406.7	
$C_{14}H_{30}$	Tetradecane	198.392	438.3	
$C_{15}H_{32}$	Pentadecane	212.419	469.9	
$C_{16}H_{32}$	1-Hexadecene	224.430	488.9	
$C_{16}H_{34}$	Hexadecane	226.446	501.6	

DENSITY AND SPECIFIC VOLUME OF MERCURY

The data in this table have been adjusted to the ITS-90 temperature scale. The uncertainty in density values is 0.0003 g/mL between –20 and –10°C; 0.0001 or less between –10 and 200°C; and 0.0002 between 200 and 300°C.

REFERENCE

Ambrose, D., *Metrologia*, 27, 245, 1990.

t/°C	ρ/(g/mL)	*v*/(mL/kg)
–20	13.64461	73.2890
–19	13.64212	73.3024
–18	13.63964	73.3157
–17	13.63716	73.3291
–16	13.63468	73.3424
–15	13.63220	73.3558
–14	13.62972	73.3691
–13	13.62724	73.3824
–12	13.62476	73.3958
–11	13.62228	73.4091
–10	13.61981	73.4225
–9	13.61733	73.4358
–8	13.61485	73.4492
–7	13.61238	73.4625
–6	13.60991	73.4759
–5	13.60743	73.4892
–4	13.60496	73.5026
–3	13.60249	73.5160
–2	13.60002	73.5293
–1	13.59755	73.5427
0	13.59508	73.5560
1	13.59261	73.5694
2	13.59014	73.5827
3	13.58768	73.5961
4	13.58521	73.6095
5	13.58275	73.6228
6	13.58028	73.6362
7	13.57782	73.6495
8	13.57535	73.6629
9	13.57289	73.6763
10	13.57043	73.6896
11	13.56797	73.7030
12	13.56551	73.7164
13	13.56305	73.7297
14	13.56059	73.7431
15	13.55813	73.7565
16	13.55567	73.7698
17	13.55322	73.7832
18	13.55076	73.7966
19	13.54831	73.8100
20	13.54585	73.8233
21	13.54340	73.8367
22	13.54094	73.8501
23	13.53849	73.8635
24	13.53604	73.8769
25	13.53359	73.8902
26	13.53114	73.9036
27	13.52869	73.9170
28	13.52624	73.9304
29	13.52379	73.9438
30	13.52134	73.9572
31	13.51889	73.9705
32	13.51645	73.9839
33	13.51400	73.9973
34	13.51156	74.0107
35	13.50911	74.0241
36	13.50667	74.0375
37	13.50422	74.0509
38	13.50178	74.0643
39	13.49934	74.0777
40	13.49690	74.0911
41	13.49446	74.1045
42	13.49202	74.1179
43	13.48958	74.1313
44	13.48714	74.1447
45	13.48470	74.1581
46	13.48226	74.1715
47	13.47982	74.1850
48	13.47739	74.1984
49	13.47495	74.2118
50	13.47251	74.2252
51	13.47008	74.2386
52	13.46765	74.2520
53	13.46521	74.2655
54	13.46278	74.2789
55	13.46035	74.2923
56	13.45791	74.3057
57	13.45548	74.3192
58	13.45305	74.3326
59	13.45062	74.3460
60	13.44819	74.3594
61	13.44576	74.3729
62	13.44333	74.3863
63	13.44090	74.3998
64	13.43848	74.4132
65	13.43605	74.4266
66	13.43362	74.4401
67	13.43120	74.4535
68	13.42877	74.4670
69	13.42635	74.4804
70	13.42392	74.4939
71	13.42150	74.5073
72	13.41908	74.5208
73	13.41665	74.5342
74	13.41423	74.5477
75	13.41181	74.5612
76	13.40939	74.5746
77	13.40697	74.5881
78	13.40455	74.6016
79	13.40213	74.6150
80	13.39971	74.6285
81	13.39729	74.6420
82	13.39487	74.6554
83	13.39245	74.6689
84	13.39003	74.6824
85	13.38762	74.6959
86	13.38520	74.7094
87	13.38278	74.7229
88	13.38037	74.7364
89	13.37795	74.7498
90	13.37554	74.7633
91	13.37313	74.7768
92	13.37071	74.7903
93	13.36830	74.8038
94	13.36589	74.8173
95	13.36347	74.8308
96	13.36106	74.8443
97	13.35865	74.8579
98	13.35624	74.8714
99	13.35383	74.8849
100	13.35142	74.8984
110	13.3273	75.0337
120	13.3033	75.1693
130	13.2793	75.3052
140	13.2553	75.4413
150	13.2314	75.5778
160	13.2075	75.7147
170	13.1836	75.8519
180	13.1597	75.9895
190	13.1359	76.1274
200	13.1120	76.2659
210	13.0882	76.4047
220	13.0644	76.5440
230	13.0406	76.6838
240	13.0167	76.8241
250	12.9929	76.9650
260	12.9691	77.1064
270	12.9453	77.2484
280	12.9214	77.3909
290	12.8975	77.5341
300	12.8736	77.6779

THERMAL PROPERTIES OF MERCURY

Lev R. Fokin

The first of these tables gives the molar heat capacity at constant pressure of liquid and gaseous mercury as a function of temperature. To convert to specific heat in units of J/g K, divide these values by 200.59, the atomic weight of mercury.

REFERENCE

Douglas, T. B., Ball, A. T., and Ginnings, D. C., *J. Res. Natl. Bur. Stands.*, 46, 334, 1951.

T/°C	C_p/(J/mol K) Liquid	C_p/(J/mol K) Gas	T/°C	C_p/(J/mol K) Liquid	C_p/(J/mol K) Gas	T/°C	C_p/(J/mol K) Liquid	C_p/(J/mol K) Gas
–38.84	28.2746	20.786	140	27.3675	20.786	340	27.1500	20.836
–20	28.1466	20.786	160	27.3090	20.786	356.73	27.1677	20.849
0	28.0190	20.786	180	27.2588	20.790	360	27.1709	20.853
20	27.9002	20.786	200	27.2169	20.790	380	27.1981	20.870
25	27.8717	20.786	220	27.1834	20.794	400	27.2324	20.891
40	27.7897	20.786	240	27.1583	20.794	420	27.2738	20.916
60	27.6880	20.786	260	27.1412	20.799	440	27.3207	20.941
80	27.5952	20.786	280	27.1320	20.807	460	27.3742	20.974
100	27.5106	20.786	300	27.1303	20.815	480	27.4332	21.008
120	27.4349	20.786	320	27.1366	20.824	500	27.4985	21.046

The second table gives the molar heat capacity of solid mercury in its rhombohedral (α-mercury) form.

REFERENCES

1. Busey and Giaque, *J. Am. Chem. Soc.*, 75, 806, 1953.
2. Amitin, Lebedeva, and Paukov, *Rus. J. Phys. Chem.*, 2666, 1979.

T/°C	C_p/J mol^{-1}	T/°C	C_p/J mol^{-1}	T/°C	C_p/J mol^{-1}	T/°C	C_p/J mol^{-1}
–268.99	0.99*	–248.15	12.74	–193.15	23.16	–113.15	26.15
–268.99	0.97**	–243.15	14.78	–183.15	23.76	–93.15	26.69
–268.15	1.6	–233.15	17.90	–173.15	24.24	–73.15	27.28
–263.15	4.6	–223.15	19.94	–153.15	25.00	–53.15	27.96
–258.15	7.6	–213.15	21.40	–133.15	25.61	–38.87	28.5
–253.15	10.33	–203.15	22.42				

* Superconducting state
** Normal state

The final table gives the cubic thermal expansion coefficient α, the isothermal compressibility coefficient κ_T, and the speed of sound U for liquid mercury as a function of temperature. These properties are defined as follows:

$$\alpha = \frac{1}{v}\left(\frac{\partial v}{\partial T}\right)_p \qquad \kappa_T = -\frac{1}{v}\left(\frac{\partial v}{\partial P}\right)_T \qquad U^2 = \left(\frac{\partial P}{\partial \rho}\right)_S \qquad \rho = v^{-1}$$

where v is the specific volume (given in the table on the preceding page).

REFERENCE

Vukalovich, M. P., et al., *Thermophysical Properties of Mercury,* Moscow Standard Press, 1971.

T/°C	$\alpha \times 10^6$/K^{-1}	$\kappa_T \times 10^6$/bar^{-1} At 1 bar	$\kappa_T \times 10^6$/bar^{-1} At 1000 bar	U/m s^{-1}	T/°C	$\alpha \times 10^6$/K^{-1}	$\kappa_T \times 10^6$/bar^{-1} At 1 bar	$\kappa_T \times 10^6$/bar^{-1} At 1000 bar	U/m s^{-1}
–20	1.818	3.83		1470	120	1.8058	4.513	4.33	1404.7
0	1.8144	3.918	3.78	1460.8	140	1.8074	4.622		1395.4
20	1.8110	4.013	3.87	1451.4	160	1.8100	4.731	4.53	1386.1
40	1.8083	4.109	3.96	1442.0	180	1.8136	4.844		1376.7
60	1.8064	4.207		1432.7	200	1.818	4.96		1367
80	1.8053	4.308	4.14	1423.4	250	1.834	5.26		1344
100	1.8051	4.410		1414.1	300	1.856	5.59		1321

SURFACE TENSION OF COMMON LIQUIDS

The surface tension γ of about 200 liquids is tabulated here as a function of temperature. Values of γ are given in units of millinewtons per meter (mN/m), which is equivalent to dyn/cm in cgs units. The values refer to a nominal pressure of one atmosphere (about 100 kPa) except in cases where the indicated temperature is above the normal boiling point of the substance; in those cases, the applicable pressure is the saturation vapor pressure at the temperature in question.

The uncertainty of the values is 0.1 to 0.2 mN/m or less in most cases. Values at temperatures between the points tabulated can be obtained by linear interpolation to a good approximation.

Substances are listed by molecular formula in the modified Hill order, with substances not containing carbon appearing before those that do contain carbon. A more extensive compilation of surface tension may be found in the Reference.

REFERENCE

Jasper, J. J., *J. Phys. Chem. Ref. Data*, 1, 841, 1972.

Mol. form.	**Name**	**γ in mN/m**				
		10°C	**25°C**	**50°C**	**75°C**	**100°C**
Br_2	Bromine	43.68	40.95	36.40		
Cl_2O_2S	Sulfuryl chloride		28.78			
Cl_3OP	Phosphoryl chloride		32.03	28.85	25.66	
Cl_3P	Phosphorus trichloride		27.98	24.81		
Cl_4Si	Silicon tetrachloride	19.78	18.29	15.80		
H_2O	Water	74.23	71.99	67.94	63.57	58.91
H_4N_2	Hydrazine		66.39			
Hg	Mercury	488.55	485.48	480.36	475.23	470.11
CCl_4	Tetrachloromethane		26.43	23.37	20.31	17.25
CS_2	Carbon disulfide	33.81	31.58	27.87		
$CHBr_3$	Tribromomethane		44.87	41.60	38.33	
$CHCl_3$	Trichloromethane		26.67	23.44	20.20	
CH_2Br_2	Dibromomethane		39.05	35.33	31.61	
CH_2Cl_2	Dichloromethane		27.20			
CH_2O_2	Formic acid		37.13	34.38	31.64	
CH_3I	Iodomethane	32.19	30.34			
CH_3NO	Formamide		57.03	54.92	52.82	50.71
CH_3NO_2	Nitromethane	39.04	36.53	32.33		
CH_4O	Methanol	23.23	22.07	20.14		
CH_5N	Methylamine		19.15			
C_2HCl_5	Pentachloroethane		34.15	31.20	28.26	
$C_2HF_3O_2$	Trifluoroacetic acid		13.53	11.42		
$C_2H_2Cl_4$	1,1,2,2-Tetrachloroethane		35.58	32.41	29.24	26.07
$C_2H_3Cl_3$	1,1,1-Trichloroethane		25.18	22.07		
$C_2H_3Cl_3$	1,1,2-Trichloroethane		34.02	30.65	27.27	23.89
C_2H_3N	Acetonitrile		28.66	25.51		
$C_2H_4Br_2$	1,2-Dibromoethane		39.55	36.25	32.95	
$C_2H_4Cl_2$	1,1-Dichloroethane		24.07			
$C_2H_4Cl_2$	1,2-Dichloroethane		31.86	28.29	24.72	
C_2H_4O	Acetaldehyde	22.54	20.50	17.10		
$C_2H_4O_2$	Acetic acid		27.10	24.61	22.13	
$C_2H_4O_2$	Methyl formate	26.72	24.36	20.43	16.50	12.57
C_2H_5Br	Bromoethane	25.36	23.62			
C_2H_5I	Iodoethane	30.38	28.46	25.24		
$C_2H_5NO_2$	Nitroethane	34.02	32.13	29.00		
C_2H_6O	Ethanol	23.22	21.97	19.89		
C_2H_6OS	Dimethyl sulfoxide		42.92	40.06		
$C_2H_6O_2$	Ethylene glycol		47.99	45.76	43.54	41.31
C_2H_6S	Dimethyl sulfide	25.27	24.06			
C_2H_6S	Ethanethiol		23.08			
$C_2H_6S_2$	Dimethyl disulfide		33.39	30.04		
C_2H_7N	Dimethylamine		26.34			
C_2H_7N	Ethylamine		19.20			

SURFACE TENSION OF COMMON LIQUIDS (continued)

Mol. form.	Name	γ in mN/m 10°C	25°C	50°C	75°C	100°C
C_2H_7NO	Ethanolamine		48.32	45.53	42.73	
C_3H_5Br	3-Bromopropene		26.31	23.17		
C_3H_5Cl	3-Chloropropene		23.14			
C_3H_5ClO	Epichlorohydrin	38.40	36.36	32.96	29.56	26.16
C_3H_5N	Propanenitrile		26.75	23.87		
$C_3H_6Cl_2$	1,2-Dichloropropane		28.32	25.22	22.12	
C_3H_6O	Acetone		23.46	20.66		
C_3H_6O	Allyl alcohol	26.63	25.28	23.02	20.77	
$C_3H_6O_2$	Ethyl formate	25.16	23.18			
$C_3H_6O_2$	Methyl acetate	26.66	24.73	21.51		
$C_3H_6O_2$	Propanoic acid		26.20	23.72	21.23	
C_3H_7Br	1-Bromopropane	27.08	25.26	22.21		
C_3H_7Br	2-Bromopropane	25.03	23.25	20.30		
C_3H_7Cl	1-Chloropropane	23.16	21.30			
C_3H_7Cl	2-Chloropropane	20.49	19.16			
$C_3H_7NO_2$	2-Nitropropane	31.02	29.29	26.39		
C_3H_8O	1-Propanol	24.48	23.32	21.38	19.43	
C_3H_8O	2-Propanol	22.11	20.93	18.96	16.98	
$C_3H_8O_2$	2-Methoxyethanol	32.32	30.84	28.38	25.92	23.46
C_3H_8S	1-Propanethiol		24.20	21.02		
C_3H_8S	2-Propanethiol		21.33	18.39		
C_3H_9N	Propylamine		21.75			
C_3H_9N	Trimethylamine		13.41			
$C_4H_4N_2$	Pyridazine	49.51	47.96	45.37	42.78	40.19
$C_4H_4N_2$	Pyrimidine		30.33	27.80	25.28	22.75
C_4H_4S	Thiophene		30.68	27.36		
C_4H_5N	Pyrrole	38.71	37.06	34.31		
$C_4H_6O_3$	Acetic anhydride	34.08	31.93	28.34	24.75	21.16
C_4H_7N	Butanenitrile		26.92	24.33	21.73	
C_4H_8O	2-Butanone		23.97	21.16		
$C_4H_8O_2$	1,4-Dioxane		32.75	29.28	25.80	22.32
$C_4H_8O_2$	Ethyl acetate	25.13	23.39	20.49	17.58	14.68
$C_4H_8O_2$	Methyl propanoate	26.32	24.44	21.29		
$C_4H_8O_2$	Butanoic acid		26.05	23.75	21.45	
C_4H_9Br	1-Bromobutane	27.58	25.90	23.08	20.27	17.45
C_4H_9Cl	1-Chlorobutane	24.85	23.18	20.39		
C_4H_9I	1-Iodobutane	29.79	28.24	25.67	23.09	20.51
C_4H_9N	Pyrrolidine	30.58	29.23	26.98		
$C_4H_{10}O$	1-Butanol	26.28	24.93	22.69	20.44	18.20
$C_4H_{10}O$	2-Butanol	23.74	22.54	20.56	18.57	16.58
$C_4H_{10}O$	2-Methyl-2-propanol		19.96	17.71		
$C_4H_{10}O$	Diethyl ether		16.65			
$C_4H_{10}O_2$	2-Ethoxyethanol		28.35	26.11	23.86	21.62
$C_4H_{10}O_3$	Diethylene glycol		44.77	42.57	40.37	38.17
$C_4H_{10}S$	Diethyl sulfide	26.22	24.57	21.80		
$C_4H_{11}N$	Butylamine		23.44	20.63		
$C_4H_{11}N$	Isobutylamine		21.75	19.02		
$C_4H_{11}N$	*tert*-Butylamine		16.87			
$C_4H_{11}N$	Diethylamine		19.85			
$C_5H_4O_2$	Furfural	45.08	43.09	39.78	36.46	33.14
C_5H_5N	Pyridine		36.56	33.29	30.03	
C_5H_8	Cyclopentene	24.45	22.20			
C_5H_8O	Cyclopentanone	34.45	32.80	30.05	27.30	24.55
C_5H_{10}	1-Pentene	17.10	15.45			
C_5H_{10}	2-Methyl-2-butene	18.61	17.15			
C_5H_{10}	Cyclopentane	24.07	21.88	18.22		
$C_5H_{10}O$	2-Pentanone		23.25	21.62		

SURFACE TENSION OF COMMON LIQUIDS (continued)

Mol. form.	Name	γ in mN/m				
		10°C	25°C	50°C	75°C	100°C
$C_5H_{10}O$	3-Pentanone		24.74	22.13		
$C_5H_{10}O$	Pentanal	26.95	25.44	22.91		
$C_5H_{10}O_2$	Butyl formate	26.05	24.52	21.95	19.39	16.82
$C_5H_{10}O_2$	Propyl acetate	25.48	23.80	21.00	18.20	15.40
$C_5H_{10}O_2$	Isopropyl acetate	23.37	21.76	19.08	16.40	
$C_5H_{10}O_2$	Ethyl propanoate	25.55	23.80	20.88	17.96	
$C_5H_{10}O_2$	Methyl butanoate	26.34	24.62	21.76	18.89	16.03
$C_5H_{11}Cl$	1-Chloropentane	26.01	24.40	21.71	19.02	16.33
$C_5H_{11}N$	Piperidine	30.64	28.91	26.03	23.14	20.26
C_5H_{12}	Pentane	17.15	15.49			
$C_5H_{12}O$	1-Pentanol	26.67	25.36	23.17	20.99	18.80
$C_5H_{12}O$	2-Pentanol	24.96	23.45	20.94	18.43	15.92
$C_5H_{12}O$	3 Methyl-1-butanol	24.94	23.71	21.66	19.61	17.56
$C_5H_{13}N$	Pentylamine		24.69	22.14	19.58	
$C_6H_4Cl_2$	*m*-Dichlorobenzene	37.15	35.43	32.57	29.70	26.83
C_6H_5Br	Bromobenzene	36.98	35.24	32.34	29.44	26.54
C_6H_5Cl	Chlorobenzene	34.78	32.99	30.02	27.04	24.06
C_6H_5ClO	*o*-Chlorophenol		39.70	36.89	34.09	31.28
C_6H_5ClO	*m*-Chlorophenol		41.18	38.66	36.13	33.61
C_6H_5F	Fluorobenzene	28.47	26.66	23.65	20.64	
C_6H_5I	Iodobenzene	40.40	38.71	35.91	33.10	30.29
$C_6H_5NO_2$	Nitrobenzene			40.56	37.66	34.77
C_6H_6	Benzene		28.22	25.00	21.77	
C_6H_6O	Phenol			38.20	35.53	32.86
C_6H_7N	Aniline		42.12	39.41	36.69	
C_6H_7N	2-Methylpyridine		33.00	29.90	26.79	
$C_6H_8N_2$	Adiponitrile		45.45	43.02	40.58	
C_6H_{10}	Cyclohexene	28.01	26.17	23.12		
$C_6H_{10}O$	Cyclohexanone	36.43	34.57	31.46	28.36	25.25
$C_6H_{11}N$	Hexanenitrile		27.37	25.11	22.84	
C_6H_{12}	Cyclohexane	26.43	24.65	21.68		
C_6H_{12}	Methylcyclopentane	23.47	21.72	18.82		
C_6H_{12}	1-Hexene	19.44	17.90	15.33		
$C_6H_{12}O$	Cyclohexanol		32.92	30.50	28.09	25.67
$C_6H_{12}O$	2-Hexanone		25.45	22.72		
$C_6H_{12}O_2$	Butyl acetate	26.48	24.88	22.21	19.54	16.87
$C_6H_{12}O_2$	Isobutyl acetate	24.58	23.06	20.53	17.99	15.46
$C_6H_{12}O_2$	Ethyl butanoate	25.51	23.94	21.33	18.71	16.10
$C_6H_{12}O_3$	Paraldehyde	27.22	25.63	22.97	20.32	17.66
$C_6H_{13}Cl$	1-Chlorohexane	27.28	25.73	23.13	20.54	17.94
$C_6H_{13}N$	Cyclohexylamine		31.22	28.25	25.28	
C_6H_{14}	Hexane	19.42	17.89	15.33		
C_6H_{14}	2-Methylpentane	18.37	16.88	14.39		
C_6H_{14}	3-Methylpentane	19.20	17.61	14.96		
$C_6H_{14}O$	Diisopropyl ether		17.27	14.65		
$C_6H_{14}O$	1-Hexanol		25.81	23.81	21.80	19.80
$C_6H_{14}O_2$	1,1-Diethoxyethane		20.89	18.31	15.74	
$C_6H_{14}O_2$	2-Butoxyethanol	27.36	26.14	24.10	22.06	20.02
$C_6H_{15}N$	Triethylamine		20.22	17.74		
$C_6H_{15}N$	Dipropylamine		22.31	19.75	17.20	
$C_6H_{15}N$	Diisopropylamine		19.14	16.45		
C_7H_5N	Benzonitrile		38.79	35.90	33.00	
C_7H_6O	Benzaldehyde	39.63	38.00	35.27	32.55	29.82
C_7H_8	Toluene	29.71	27.93	24.96	21.98	19.01
C_7H_8O	*o*-Cresol		36.90	34.38	31.85	29.32
C_7H_8O	*m*-Cresol		35.69	33.38	31.07	28.76
C_7H_8O	Benzyl alcohol				27.89	24.44

Mol. form.	Name	γ in mN/m 10°C	25°C	50°C	75°C	100°C
C_7H_8O	Anisole		35.10	32.09	29.08	
C_7H_9N	*N*-Methylaniline		36.90	34.47	32.05	
C_7H_9N	2,3-Dimethylpyridine		32.71	30.04	27.36	
C_7H_9N	Benzylamine		39.30	36.27	33.23	
C_7H_{14}	Methylcyclohexane	24.98	23.29	20.46		
C_7H_{14}	1-Heptene	21.29	19.80	17.33	14.85	
$C_7H_{14}O$	2-Heptanone		26.12	23.48		
$C_7H_{14}O_2$	Pentyl acetate	26.67	25.17	22.69	20.20	17.72
$C_7H_{14}O_2$	Heptanoic acid		27.76	25.64		
C_7H_{16}	Heptane	21.12	19.65	17.20	14.75	
C_7H_{16}	3-Methylhexane	20.76	19.31	16.88	14.46	
C_8H_8O	Acetophenone		39.04	36.15	33.27	
$C_8H_8O_2$	Methyl benzoate		37.17	34.25	31.32	
$C_8H_8O_3$	Methyl salicylate	40.98	39.22	36.28	33.35	30.41
C_8H_{10}	Ethylbenzene	30.39	28.75	26.01	23.28	20.54
C_8H_{10}	*o*-Xylene	31.41	29.76	27.01	24.25	21.50
C_8H_{10}	*m*-Xylene	30.13	28.47	25.71	22.95	20.19
C_8H_{10}	*p*-Xylene		28.01	25.32	22.64	19.95
$C_8H_{10}O$	Phenetole		32.41	29.65	26.89	
$C_8H_{11}N$	*N,N*-Dimethylaniline		35.52	32.90	30.27	
$C_8H_{11}N$	*N*-Ethylaniline		36.33	33.65	30.98	
C_8H_{16}	Ethylcyclohexane	26.73	25.15	22.51		
C_8H_{18}	Octane	22.57	21.14	18.77	16.39	14.01
C_8H_{18}	2,5-Dimethylhexane	20.77	19.40	17.12	14.84	12.56
$C_8H_{18}O$	1-Octanol	28.30	27.10	25.12		
$C_8H_{19}N$	Dibutylamine		24.12	21.74	19.36	
$C_8H_{19}N$	Diisobutylamine		21.72	19.44	17.16	
C_9H_7N	Quinoline	44.19	42.59	39.94	37.28	34.62
C_9H_{12}	Cumene	29.27	27.69	25.05	22.42	19.78
C_9H_{12}	1,2,4-Trimethylbenzene	30.74	29.20	26.64	24.07	21.51
C_9H_{12}	Mesitylene	28.89	27.55	25.31	23.07	20.82
$C_9H_{18}O$	5-Nonanone		26.28	23.85		
C_9H_{20}	Nonane	23.79	22.38	20.05	17.71	15.37
$C_9H_{20}O$	1-Nonanol	29.03	27.89	26.00	24.10	22.20
$C_{10}H_{12}$	1,2,3,4-Tetrahydronaphthalene		33.17	30.78	28.40	
$C_{10}H_{22}$	Decane	24.75	23.37	21.07	18.77	16.47
$C_{10}H_{22}O$	1-Decanol	29.61	28.51	26.68	24.85	23.02
$C_{11}H_{24}$	Undecane	25.56	24.21	21.96	19.70	17.45
$C_{12}H_{10}O$	Diphenyl ether		26.75	24.80		
$C_{12}H_{27}N$	Tributylamine		24.39	22.32	20.24	
$C_{13}H_{28}$	Tridecane	26.86	25.55	23.37	21.19	19.01
$C_{14}H_{12}O_2$	Benzyl benzoate	44.47	42.82	40.06	37.31	34.55
$C_{14}H_{30}$	Tetradecane	27.43	26.13	23.96	21.78	19.61
$C_{16}H_{34}$	Hexadecane		27.05	24.91	22.78	20.64
$C_{18}H_{38}$	Octadecane		27.87	25.77	23.66	21.55

PERMITTIVITY (DIELECTRIC CONSTANT) OF LIQUIDS

Christian Wohlfarth

The permittivity of a substance (often called the dielectric constant) is the ratio of the electric displacement **D** to the electric field strength **E** when an external field is applied to the substance. The quantity tabulated here is the relative permittivity, which is the ratio of the actual permittivity to the permittivity of a vacuum; it is a dimensionless number. The table gives the static permittivity ε, measured in static fields or at relatively low frequencies where no relaxation effects occur.

This table gives the static permittivity at a specified temperature, usually at 293.15 or 298.15 K. Otherwise, the temperature closest to 293.15 K was chosen or (as is the case for about half the substances included here) a measurement was available at only one temperature, and that value is given in the table. The temperature dependence of the permittivity of many of these substances is given in the table that follows this one.

The static permittivities refer to nominal atmospheric pressure as long as the corresponding temperature is below the normal boiling point. Otherwise, at temperatures above the normal boiling point, the pressure is the saturated vapor pressure of the substance considered.

Substances are listed by molecular formula in the modified Hill order, with substances not containing carbon preceding those that do contain carbon. An asterisk following the substance name indicates that the isomer was not specified in the original reference.

REFERENCES

1. Wohlfarth, Ch., Static Dielectric Constants of Pure Liquids and Binary Liquid Mixtures, *Landolt-Börstein, Numerical Data and Functional Relationships in Science and Technology*, New Series, Editor-in-Chief O. Madelung, Group IV, Macroscopic and Technical Properties of Matter, Volume 6, Springer Verlag, Berlin, Heidelberg, New York, 1991.
2. Marsh, K. N., Ed., *Recommended Reference Materials for the Realization of Physicochemical Properties*, Blackwell Scientific Publications, Oxford, 1987.

Molecular formula	Name	T/K	ε
Ar	Argon	140.00	1.3247
$AlBr_3$	Aluminum tribromide	373.2	3.38
AsH_3	Arsine	200.9	2.40
BBr_3	Boron tribromide	273.2	2.58
B_2H_6	Diborane	180.66	1.8725
B_5H_9	Pentaborane (9)	298.2	21.1
BrF_3	Bromine trifluoride	298.2	106.8
BrF_5	Bromine pentafluoride	297.7	7.91
BrH	Hydrogen bromide	186.8	8.23
BrNO	Nitrosyl bromide	288.4	13.4
Br_2	Bromine	297.9	3.1484
Br_2OS	Thionyl bromide	293.2	9.06
Br_3OV	Vanadium oxytribromide	298.2	3.6
Br_4Ge	Germanium tetrabromide	299.9	2.955
Br_4Sn	Tin tetrabromide	303.45	3.169
$ClFO_3$	Perchloryl fluoride	150.2	2.194
ClF_3	Chlorine trifluoride	293.2	4.394
ClF_5	Chlorine pentafluoride	193.2	4.28
ClH	Hydrogen chloride	158.9	14.3
ClNO	Nitrosyl chloride	285.2	18.2
Cl_2	Chlorine	208.0	2.147
Cl_2F_3P	Phosphorus dichloride trifluoride	228.63	2.8129
Cl_2OS	Thionyl chloride	298.2	8.675
Cl_2OSe	Selenium oxychloride	293.2	46.2
Cl_2O_2S	Sulfuryl chloride	293.2	9.1
Cl_2S	Sulfur chloride (SCl_2)	298.2	2.915
Cl_2S_2	Sulfur chloride (S_2Cl_2)	288.2	4.79
Cl_3F_2P	Phosphorus trichloride difluoride	268.0	2.3752
Cl_3OP	Phosphoryl chloride	293.2	14.1
Cl_3OV	Vanadium oxytrichloride	298.2	3.4
Cl_3P	Phosphorus trichloride	290.2	3.498
Cl_3PS	Phosphorus thiochloride	298.2	4.94
Cl_4FP	Phosphorus tetrachloride fluoride	272.64	2.6499
Cl_4Ge	Germanium tetrachloride	273.2	2.463
Cl_4Si	Silicon tetrachloride	273.2	2.248

Molecular formula	Name	T/K	ε
Cl_4Sn	Tin tetrachloride	273.2	3.014
Cl_4Ti	Titanium tetrachloride	257.4	2.843
Cl_4V	Vanadium tetrachloride	298.2	3.05
Cl_5P	Phosphorus pentachloride	433.2	2.85
Cl_4Pb	Lead tetrachloride	293.2	2.78
Cl_5Sb	Antimony pentachloride	293.0	3.222
DH	Hydrogen-d_1	16.783	1.2690
D_2	Hydrogen-d_2	17.759	1.2896
D_2O	Water-d_2	293.2	79.754
FH	Hydrogen fluoride	273.2	83.6
F_2	Fluorine	53.48	1.4913
F_5I	Iodine pentafluoride	293.2	37.13
F_5IO	Iodine oxypentafluoride	296.2	1.97
F_6S	Sulfur hexafluoride	223.2	1.81
F_6Xe	Xenon hexafluoride	398.2	4.10
F_7I	Iodine heptafluoride	298.2	1.75
$F_{10}S_2$	Disulfur decafluoride	293.2	2.0202
HI	Hydrogen iodide	220.2	3.87
H_2	Hydrogen	13.52	1.2792
H_2O	Water	293.2	80.100
H_2O_2	Hydrogen peroxide	290.2	74.6
H_2S	Hydrogen sulfide	283.2	5.93
H_3N	Ammonia	293.2	16.61
H_4N_2	Hydrazine	298.2	51.7
He	Helium	2.055	1.0555
I_2	Iodine	391.25	11.08
Kr	Krypton	119.80	1.664
Mn_2O_7	Manganese oxide (Mn_2O_7)	293.2	3.28
NO	Nitric oxide		1.997
N_2	Nitrogen	63.15	1.4680
N_2O_3	Nitrogen trioxide	203.2	31.13
N_2O_4	Nitrogen tetroxide	293.2	2.44
Ne	Neon	26.11	1.1907
O_2	Oxygen	54.478	1.5684
O_2S	Sulfur dioxide	298.2	16.3
O_3	Ozone	90.2	4.75
O_3S	Sulfur trioxide	291.2	3.11
P	Phosphorus	307.2	4.096
S	Sulfur	407.2	3.4991
Se	Selenium	510.65	5.44
Xe	Xenon	161.35	1.880
$CBrClF_2$	Bromochlorodifluoromethane	123.2	3.920
$CBrCl_3$	Bromotrichloromethane	293.2	2.405
$CBrF_3$	Bromotrifluoromethane	123.2	3.730
CBr_2Cl_2	Dibromodichloromethane	298.2	2.542
CBr_2F_2	Dibromodifluoromethane	273.2	2.939
CBr_3Cl	Tribromochloromethane	333.2	2.601
CBr_3F	Tribromofluoromethane	293.2	3.00
CBr_3NO_2	Tribromonitromethane	298.2	9.034
$CClF_3$	Chlorotrifluoromethane	123.2	3.010
CCl_2F_2	Dichlorodifluoromethane	123.2	3.500
CCl_2NOP	Phosphorous dichloride isocyanate	293.2	4.16
CCl_2O	Carbonyl chloride	295.2	4.30
CCl_3D	Trichloromethane-d_1	298.2	4.67
CCl_3F	Trichlorofluoromethane	293.2	3.00
CCl_3NO_2	Trichloronitromethane	293.2	7.319
CCl_4	Tetrachloromethane	293.2	2.2379
CF_4	Tetrafluoromethane	126.3	1.685

PERMITTIVITY (DIELECTRIC CONSTANT) OF LIQUIDS (continued)

Molecular formula	Name	T/K	ε
CN_4O_8	Tetranitromethane	293.2	2.317
COS	Carbon oxysulfide	185.0	4.47
COSe	Carbon oxyselenide	283.2	3.47
CO_2	Carbon dioxide	295.0	1.4492
CS_2	Carbon disulfide	293.2	2.6320
$CHBr_3$	Tribromomethane	283.2	4.404
$CHCl_3$	Trichloromethane	293.2	4.8069
CHF_3	Trifluoromethane	294.0	5.2
CHN	Hydrogen cyanide	293.2	114.9
CH_2Br_2	Dibromomethane	283.2	7.77
CH_2Cl_2	Dichloromethane	298.0	8.93
CH_2F_2	Difluoromethane	152.2	53.74
CH_2I_2	Diiodomethane	298.2	5.32
CH_2O_2	Formic acid	298.2	51.1
CH_3Br	Bromomethane	275.7	9.71
CH_3Cl	Chloromethane	295.2	10.0
CH_3ClO_2S	Methanesulfonyl chloride	293.2	34.0
$CH_3Cl_2NO_2S_2$	(Methylsulfonyl)imidosulfurous dichloride	313.2	31.6
CH_3DO	Methan-d_1-ol	297.5	31.68
CH_3F	Fluoromethane	131.0	51.0
CH_3I	Iodomethane	293.2	6.97
CH_3NO	Formamide	293.2	111.0
CH_3NO_2	Methyl nitrite	200.0	20.77
CH_3NO_2	Nitromethane	293.2	37.27
CH_3NO_3	Methyl nitrate	293.2	23.9
CH_4	Methane	91.0	1.6761
CH_4O	Methanol	293.2	33.0
CH_5N	Methylamine	215.2	16.7
$C_2Br_2F_4$	1,2-Dibromotetrafluoroethane	298.2	2.34
$C_2Cl_2F_4$	1,2-Dichlorotetrafluoroethane	273.2	2.4842
$C_2Cl_2O_2$	Oxalyl chloride	294.35	3.470
C_2Cl_3N	Trichloroacetonitrile	292.2	7.85
C_2Cl_4	Tetrachloroethylene	303.2	2.268
$C_2Cl_4F_2$	Tetrachloro-1,2-difluoroethane	308.2	2.52
C_2HBr_3O	Tribromoacetaldehyde	293.2	7.6
C_2HCl_2NO	Dichloromethyl isocyanate	288.2	7.36
C_2HCl_3	Trichloroethylene	301.5	3.390
$C_2HCl_3F_2$	1,2,2-Trichloro-1,1-difluoroethane	303.2	4.01
C_2HCl_3O	Trichloroacetaldehyde	298.2	6.8
$C_2HCl_3O_2$	Trichloroacetic acid	333.2	4.34
C_2HCl_5	Pentachloroethane	298.2	3.716
$C_2HF_3O_2$	Trifluoroacetic acid	293.2	8.42
C_2H_2	Acetylene	195.0	2.4841
C_2H_2BrCl	*cis*-1-Bromo-2-chloroethene	290.2	7.31
C_2H_2BrCl	*trans*-1-Bromo-2-chloroethene	290.2	2.50
$C_2H_2Br_2$	*cis*-1,2-Dibromoethylene	298.2	7.08
$C_2H_2Br_2$	*trans*-1,2-Dibromoethylene	298.2	2.88
$C_2H_2Br_4$	1,1,2,2-Tetrabromoethane	303.2	6.72
$C_2H_2Cl_2$	1,1-Dichloroethylene	293.2	4.60
$C_2H_2Cl_2$	*cis*-1,2-Dichloroethylene	298.2	9.20
$C_2H_2Cl_2$	*trans*-1,2-Dichloroethylene	293.2	2.14
$C_2H_2Cl_2O_2$	Dichloroacetic acid	293.2	8.33
$C_2H_2Cl_4$	1,1,1,2-Tetrachloroethane	207.2	9.22
$C_2H_2Cl_4$	1,1,2,2-Tetrachloroethane	293.2	8.50
$C_2H_2I_2$	*cis*-1,2-Diiodoethene	345.65	4.46
$C_2H_2I_2$	*trans*-1,2-Diiodoethene	350.0	3.19
C_2H_3ClO	Acetyl chloride	295.2	15.8
$C_2H_3ClO_2$	Chloroacetic acid	338.2	12.35

Molecular formula	Name	*T*/K	ε
$C_2H_3Cl_2NO_2$	1,1-Dichloro-1-nitroethane	303.2	16.3
$C_2H_3Cl_3$	1,1,1-Trichloroethane	293.2	7.243
$C_2H_3Cl_3$	1,1,2-Trichloroethane	298.2	7.1937
$C_2H_3F_3O$	2,2,2-Trifluoroethanol	293.2	27.68
$C_2H_3F_3O_2S$	Methyl trifluoromethyl sulfone	293.2	32.0
C_2H_3N	Acetonitrile	293.2	36.64
C_2H_3NO	Methylisocyanate	288.7	21.75
C_2H_4	Ethylene	270.0	1.4833
C_2H_4BrCl	1-Bromo-2-chloroethane	283.2	7.41
$C_2H_4Br_2$	1,2-Dibromoethane	293.2	4.9612
$C_2H_4Cl_2$	1,1-Dichloroethane	298.2	10.10
$C_2H_4Cl_2$	1,2-Dichloroethane	293.2	10.42
$C_2H_4Cl_2O$	Bis(chloromethyl) ether	293.2	3.51
$C_2H_4N_2O_6$	1,2-Ethanediol dinitrate	293.2	28.26
C_2H_4O	Acetaldehyde	291.2	21.0
C_2H_4O	Ethylene oxide	293.2	12.42
C_2H_4OS	Thioacetic acid	298.2	14.30
$C_2H_4O_2$	Methyl formate	288.2	9.20
$C_2H_4O_2$	Acetic acid	293.2	6.20
$C_2H_4O_3S$	Ethylene sulfite	298.2	39.6
C_2H_5Br	Bromoethane	298.2	9.01
C_2H_5Cl	Chloroethane	293.2	9.45
C_2H_5ClO	2-Chloroethanol	293.2	25.80
C_2H_5I	Iodoethane	293.2	7.82
C_2H_5N	Ethyleneimine	298.2	18.3
C_2H_5NO	Acetaldehyde oxime	298.2	4.70
C_2H_5NO	Acetamide	363.7	67.6
C_2H_5NO	*N*-Methylformamide	293.2	189.0
$C_2H_5NO_2$	Methyl carbamate	328.2	18.48
$C_2H_5NO_2$	Nitroethane	288.2	29.11
$C_2H_5NO_3$	Ethyl nitrate	293.2	19.7
C_2H_6	Ethane	95.0	1.9356
C_2H_6O	Dimethyl ether	258.0	6.18
C_2H_6O	Ethanol	293.2	25.3
C_2H_6OS	Dimethyl sulfoxide	293.2	47.24
$C_2H_6O_2$	Ethylene glycol	293.2	41.4
$C_2H_6O_2S$	Dimethyl sulfone	383.2	47.39
$C_2H_6O_4S$	Dimethyl sulfate	298.2	55.0
C_2H_6S	Dimethyl sulfide	294.2	6.70
C_2H_6S	Ethanethiol	298.2	6.667
$C_2H_6S_2$	1,2-Ethanedithiol	293.2	7.26
$C_2H_6S_2$	Dimethyl disulfide	298.2	9.6
C_2H_7N	Ethylamine	273.2	8.7
C_2H_7NO	Ethanolamine	293.2	31.94
$C_2H_7NO_2S$	*N*-Methyl methanesulfonamide	298.2	104.4
$C_2H_8N_2$	1,2-Ethanediamine	293.2	13.82
C_3Cl_6O	Hexachloroacetone	291.9	3.925
C_3F_6O	Hexafluoroacetone	202.2	2.104
$C_3N_3O_3P$	Phosphorus triisocyanate	293.2	5.30
$C_3N_3O_3PS$	Phosphorus triisocyanate sulfide	293.2	6.00
$C_3N_3O_4P$	Phosphorus triisocyanate oxide	298.2	4.40
C_3HN	Cyanoacetylene	291.9	72.3
$C_3H_2F_6O$	1,1,1,3,3,3-Hexafluoro-2-propanol	293.2	16.70
$C_3H_2Cl_2O_3$	4,5-Dichloro-1,3-dioxolan-2-one	313.2	31.8
$C_3H_3ClO_3$	4-Chloro-1,3-dioxolan-2-one	313.2	62.0
C_3H_3NO	Vinyl isocyanate	298.2	10.62
C_3H_3N	Propenenitrile	293.2	33.0
$C_3H_3NO_2$	Cyanoacetic acid	277.2	33.4

Molecular formula	Name	*T*/K	ε
C_3H_4	Allene	269.0	2.025
C_3H_4	Propyne	246.0	3.218
$C_3H_4ClF_3$	3-Chloro-1,1,1-trifluoropropane	295.2	7.32
C_3H_4ClNO	1-Chloro-2-isocyanatoethane	288.2	29.1
$C_3H_4Cl_2O$	1,1-Dichloroacetone	293.2	14.6
$C_3H_4F_4O$	2,2,3,3-Tetrafluoro-1-propanol	298.2	21.03
$C_3H_4N_2O_2$	3-Methylsydnone	313.2	144.0
C_3H_4O	Propargyl alcohol	293.2	20.8
$C_3H_4O_3$	1,3-Dioxolan-2-one	313.2	89.78
C_3H_5Br	3-Bromopropene	293.2	7.0
$C_3H_5BrO_2$	2-Bromopropanoic acid	294.2	11.0
$C_3H_5Br_3$	1,2,3-Tribromopropane	303.2	6.00
C_3H_5Cl	2-Chloropropene	299.25	8.92
C_3H_5Cl	3-Chloropropene	293.2	8.2
$C_3H_5ClN_2O_6$	3-Chloro-1,2-propanediol dinitrate	293.2	17.50
C_3H_5ClO	Epichlorohydrin	293.2	22.6
$C_3H_5ClO_2$	Ethyl chlorocarbonate	308.7	9.736
$C_3H_5ClO_2$	Methyl chloroacetate	293.2	12.0
$C_3H_5Cl_2NO_2S$	Ethoxycarbonylimidosulfurous dichloride	294.2	19.7
$C_3H_5Cl_2NO_3$	1,3-Dichloroisopropyl nitrate	293.2	13.28
$C_3H_5Cl_3$	1,2,3-Trichloropropane	293.2	7.5
C_3H_5I	3-Iodopropene	292.2	6.1
C_3H_5N	Propanenitrile	293.2	29.7
C_3H_5NO	Ethyl isocyanate	293.2	19.7
$C_3H_5NO_2$	Allyl nitrite	298.2	9.12
C_3H_5NS	Ethyl rhodanide	293.2	29.7
C_3H_5NS	Ethyl isothiocyanate	293.2	19.6
$C_3H_5N_3O_9$	Trinitroglycerol	293.2	19.25
C_3H_6	Propene	220.0	2.1365
$C_3H_6Br_2$	1,2-Dibromopropane	283.2	4.60
$C_3H_6Br_2$	1,3-Dibromopropane	293.2	9.482
$C_3H_6ClNO_2$	2-Chloro-2-nitropropane	250.4	31.90
C_3H_6ClNO	*N*-Methyl-2-chloroacetamide	323.2	92.3
$C_3H_6Cl_2$	1,2-Dichloropropane	293.2	8.37
$C_3H_6Cl_2$	1,3-Dichloropropane	303.2	10.27
$C_3H_6Cl_2$	2,2-Dichloropropane	293.2	11.37
$C_3H_6N_2O_4$	2,2-Dinitropropane	325.1	42.4
$C_3H_6N_2O_6$	1,2-Propanediol dinitrate	293.2	26.80
$C_3H_6N_2O_6N$	1,3-Propanediol dinitrate	293.2	18.97
C_3H_6O	Acetone	293.2	21.01
C_3H_6O	Allyl alcohol	293.2	19.7
C_3H_6O	Propanal	290.2	18.5
$C_3H_6O_2$	Ethyl formate	288.2	8.57
$C_3H_6O_2$	Methyl acetate	288.2	7.07
$C_3H_6O_2$	Propanoic acid	298.2	3.44
$C_3H_6O_3$	Dimethyl carbonate	298.2	3.087
$C_3H_6O_3$	3-Hydroxypropanoic acid	296.2	30.0
$C_3H_6O_3$	1,3,5-Trioxane	338.2	15.55
$C_3H_6O_3S$	1,2-Oxathiolane-2,2-dioxide	308.2	86.3
C_3H_7Br	1-Bromopropane	293.2	8.09
C_3H_7Br	2-Bromopropane	293.2	9.46
C_3H_7Cl	1-Chloropropane	293.2	8.588
C_3H_7ClO	3-Chloro-1-propanol	215.2	36.0
C_3H_7ClO	1-Chloro-2-propanol	153.2	59.0
$C_3H_7ClO_2$	3-Chloro-1,2-propanediol	293.2	31.0
C_3H_7I	1-Iodopropane	293.2	7.07
C_3H_7I	2-Iodopropane	298.2	8.19
C_3H_7NO	*N*,*N*-Dimethylformamide	293.2	38.25

Molecular formula	Name	*T*/K	ε
C_3H_7NO	*N*-Ethylformamide	298.2	102.7
C_3H_7NO	*N*-Methylacetamide	303.2	179.0
$C_3H_7NO_2$	Ethyl carbamate	328.2	14.14
$C_3H_7NO_2$	1-Nitropropane	288.2	24.70
$C_3H_7NO_2$	2-Nitropropane	288.2	26.74
$C_3H_7NO_2$	Propyl nitrite	250.0	12.35
$C_3H_7NO_2$	Isoropyl nitrite	260.0	13.92
C_3H_7NS	*N,N*-Dimethylthioformamide	298.2	47.5
C_3H_8	Propane	293.19	1.6678
C_3H_8O	1-Propanol	293.2	20.8
C_3H_8O	2-Propanol	293.2	20.18
$C_3H_8O_2$	Dimethoxymethane	293.2	2.644
$C_3H_8O_2$	2-Methoxyethanol	298.2	17.2
$C_3H_8O_2$	1,2-Propanediol	303.2	27.5
$C_3H_8O_2$	1,3-Propanediol	293.2	35.1
$C_3H_8O_3$	Glycerol	293.2	46.53
C_3H_8S	1-Propanethiol	288.2	5.937
C_3H_8S	2-Propanethiol	298.2	5.952
$C_3H_8S_2$	1,2-Propanedithiol	293.2	7.24
$C_3H_8S_2$	1,3-Propanedithiol	303.2	8.11
$C_3H_9BO_3$	Trimethylborate	293.2	2.2762
C_3H_9ClSi	Trimethylchlorosilane	273.2	10.21
C_3H_9N	Propylamine	296.2	5.08
C_3H_9N	Isopropylamine	293.2	5.6268
C_3H_9N	Trimethylamine	298.2	2.440
$C_3H_9NO_2S$	*N,N*-Dimethyl methanesulfonamide	323.2	80.4
$C_3H_9O_3P$	Dimethyl methanephosphate	293.2	22.3
$C_3H_9O_4P$	Trimethyl phosphate	293.2	20.6
C_3H_9PS	Trimethylphosphine sulfide	293.2	71.6
C_4Cl_6	Hexachloro-1,3-butadiene	293.2	2.55
$C_4Cl_6O_3$	Trichloroacetic anhydride	298.2	5.0
$C_4F_6Cl_4$	1,1,3,4-Tetrachlorohexafluorobutane	293.2	2.86
$C_4F_6O_3$	Trifluoroacetic anhydride	298.2	2.7
$C_4H_2Cl_4O_3$	Dichloroacetic anhydride	298.2	15.8
$C_4H_2O_3$	Maleic anhydride	326.2	52.75
$C_4H_3F_7O$	2,2,3,3,4,4,4-Heptafluoro-1-butanol	298.2	14.4
$C_4H_4N_2$	Succinonitrile	298.2	62.6
$C_4H_4N_2$	Pyrazine	323.2	2.80
C_4H_4O	Furan	277.1	2.88
C_4H_4S	Thiophene	293.2	2.739
C_4H_5Cl	2-Chloro-1,3-butadiene	293.2	4.914
$C_4H_5ClO_3$	4-Chloromethyl-1,3-dioxolan-2-one	313.2	97.5
$C_4H_5Cl_3O$	4,4,4-Trichlorobutanal	291.2	10.0
$C_4H_5Cl_3O_2$	Ethyl trichloroacetate	293.2	8.428
C_4H_5N	1*H*-Pyrrole	293.0	8.00
C_4H_5NO	Allyl isocynate	288.2	15.15
$C_4H_5N_2O_3P$	Diisocyanato ethyl phosphite	298.2	6.93
$C_4H_5N_2O_3PS$	Diisocyanato *O*-ethyl thionphosphate	293.2	13.4
$C_4H_5N_2O_4P$	Diisocyanato ethyl phosphate	293.2	14.1
C_4H_6	1,3-Butadiene	265.0	2.050
$C_4H_6Cl_2N_2S$	(1-Cyano-1-isopropyl)imido sulfurous dichloride	293.2	21.8
$C_4H_6N_2OS$	2-Methyl-2-(sulfinylamino)-propanenitrile	293.2	16.14
C_4H_6O	Cyclobutanone	298.2	14.27
C_4H_6O	Divinyl ether	288.2	3.94
C_4H_6O	Ethoxyacetylene	298.2	8.05
$C_4H_6O_2$	Methyl acrylate	303.2	7.03
$C_4H_6O_2$	2,3-Butanedione	298.2	4.04
$C_4H_6O_2$	γ-Butyrolactone	293.2	39.0

Molecular formula	Name	*T*/K	ε
$C_4H_6O_3$	Acetic anhydride	293.2	22.45
$C_4H_6O_3$	4-Methyl-1,3-dioxolan-2-one	293.0	66.14
C_4H_7Br	*trans*-2-Bromo-2-butene	293.2	6.76
C_4H_7Br	*cis*-2-Bromo-2-butene	293.2	5.38
$C_4H_7BrO_2$	Ethyl 2-bromoacetate	303.2	9.75
$C_4H_7BrO_2$	Methyl 2-bromopropanoate	303.2	10.35
$C_4H_7BrO_2$	Methyl 3-bromopropanoate	303.2	5.81
$C_4H_7BrO_2$	2-Bromobutanoic acid	293.2	7.2
$C_4H_7ClO_2$	Propyl chlorocarbonate	293.2	11.2
$C_4H_7ClO_2$	Methyl 2-chloropropanoate	303.2	11.45
C_4H_7N	Butanenitrile	293.2	24.83
C_4H_7N	2-Methylpropanenitrile	293.2	24.42
C_4H_7NO	Pyrrolidine-2-one	298.2	28.18
$C_4H_7NO_2$	3-Methyl oxazolidine-2-one	298.2	77.5
$C_4H_7NO_2$	5-Methyl oxazolidine-2-one	298.2	84.0
C_4H_8	1-Butene	220.0	2.2195
C_4H_8	*cis*-2-Butene	296.0	1.960
C_4H_8	Isobutene	288.7	2.1225
C_4H_8BrCl	3-Bromo-1-chloro-2-methylpropane	303.2	8.90
$C_4H_8Br_2$	1,2-Dibromobutane	293.2	4.74
$C_4H_8Br_2$	1,3-Dibromobutane	293.2	9.14
$C_4H_8Br_2$	1,4-Dibromobutane	303.2	8.68
$C_4H_8Br_2$	2,3-Dibromobutane	293.2	6.36
$C_4H_8Br_2$	*meso*-2,3-Dibromobutane	298.2	6.245
$C_4H_8Br_2$	(±)-2,3-Dibromobutane	298.2	5.758
$C_4H_8Br_2$	1,2-Dibromo-2-methylpropane	293.2	4.1
C_4H_8ClNO	2-Chloro-*N*,*N*-dimethylacetamide	298.2	49.4
$C_4H_8Cl_2$	1,2-Dichlorobutane	293.2	7.74
$C_4H_8Cl_2$	1,4-Dichlorobutane	308.2	9.30
$C_4H_8Cl_2$	1,2-Dichloro-2-methylpropane	296.0	7.15
$C_4H_8Cl_2O$	Bis(2-chloroethyl) ether	293.2	21.20
$C_4H_8N_2O_6$	1,3-Butanediol dinitrate	293.2	18.85
$C_4H_8N_2O_6$	2,3-Butanediol dinitrate	293.2	28.85
C_4H_8O	Butanal	298.2	13.45
C_4H_8O	2-Butanone	293.2	18.56
C_4H_8O	Tetrahydrofuran	295.2	7.52
C_4H_8OS	Tetrahydrothiophene-*S*-oxide	298.2	42.96
$C_4H_8O_2$	Propyl formate	303.2	6.92
$C_4H_8O_2$	Butanoic acid	287.2	2.98
$C_4H_8O_2$	1,4-Dioxane	293.2	2.2189
$C_4H_8O_2$	Ethyl acetate	293.2	6.0814
$C_4H_8O_2$	2-Methylpropanoic acid	293.2	2.58
$C_4H_8O_2$	Methyl propanoate	293.2	6.200
$C_4H_8O_2S$	Tetrahydrothiophene-*S*,*S*-dioxide	303.2	43.36
$C_4H_8O_3$	2-Hydroxybutanoic acid	296.2	37.7
$C_4H_8O_3$	3-Hydroxybutanoic acid	296.2	31.5
$C_4H_8O_3$	1,2-Ethanediol monoacetate	303.2	12.95
$C_4H_8O_3$	Ethyl methyl carbonate	293.2	2.985
C_4H_9Br	1-Bromobutane	283.2	7.315
C_4H_9Br	1-Bromo-2-methylpropane	273.2	7.70
C_4H_9Br	2-Bromobutane	298.2	8.64
C_4H_9Br	2-Bromo-2-methylpropane	293.0	10.98
C_4H_9Cl	1-Chlorobutane	293.2	7.276
C_4H_9Cl	1-Chloro-2-methylpropane	293.2	7.027
C_4H_9Cl	2-Chlorobutane	293.2	8.564
C_4H_9Cl	2-Chloro-2-methylpropane	293.2	9.663
$C_4H_9Cl_2NS$	(2-Methyl-2-propyl)imido-sulfurous dichloride	298.2	17.0
C_4H_9I	1-Iodobutane	293.2	6.27

Molecular formula	Name	*T*/K	ε
C_4H_9I	2-Iodobutane	293.2	7.873
C_4H_9I	2-Iodo-2-methylpropane	283.2	6.65
C_4H_9N	Pyrrolidine	293.0	8.30
C_4H_9NO	*N,N*-Dimethylacetamide	294.2	38.85
C_4H_9NO	*N*-Ethylacetamide	293.2	135.0
C_4H_9NO	*N*-Methylpropanamide	293.2	170.0
C_4H_9NO	*N*-Isopropylformamide	298.2	65.7
C_4H_9NO	2-Butanone oxime	293.2	3.4
C_4H_9NO	Morpholine	298.2	7.42
C_4H_9NOS	2-Methyl-2-sulfinylaminopropane	298.2	7.22
$C_4H_9NO_2$	*tert*-Butyl nitrite	298.2	11.47
$C_4H_9NO_2$	Ethyl-*N*-methyl carbamate	298.2	21.10
$C_4H_9NO_2$	*N*-(2-Hydroxyethyl)acetamide	298.2	96.6
$C_4H_9NO_2$	Propyl carbamate	338.2	12.06
$C_4H_9NO_3$	Butyl nitrate	293.2	13.10
C_4H_{10}	Butane	295.0	1.7697
C_4H_{10}	Isobutane	295.0	1.7518
$C_4H_{10}Cl_2NO_3P$	Diethyl *N,N*-dichloroamidophosphate	295.2	12.85
$C_4H_{10}Cl_2NO_3PS$	(Diethoxyphosphinyl)imido-sulfurous dichloride	293.2	19.15
$C_4H_{10}O$	1-Butanol	293.2	17.84
$C_4H_{10}O$	2-Butanol	293.2	17.26
$C_4H_{10}O$	2-Methyl-1-propanol	293.2	17.93
$C_4H_{10}O$	2-Methyl-2-propanol	298.2	12.47
$C_4H_{10}O$	Diethyl ether	293.2	4.2666
$C_4H_{10}O_2$	1,2-Butanediol	298.2	22.4
$C_4H_{10}O_2$	1,3-Butanediol	298.2	28.8
$C_4H_{10}O_2$	1,4-Butanediol	298.2	31.9
$C_4H_{10}O_2$	1,2-Dimethoxyethane	296.7	7.30
$C_4H_{10}O_2$	2-Ethoxyethanol	298.2	13.38
$C_4H_{10}O_2S$	Bis(2-hydroxyethyl) sulfide	293.2	28.61
$C_4H_{10}O_3$	Diethylene glycol	293.2	31.82
$C_4H_{10}O_3S$	Diethyl sulfite	293.2	15.6
$C_4H_{10}O_4$	1,2,3,4-Butanetetrol	393.2	28.2
$C_4H_{10}O_4S$	Diethyl sulfate	293.2	29.2
$C_4H_{10}S$	1-Butanethiol	288.2	5.204
$C_4H_{10}S$	2-Butanethiol	288.2	5.645
$C_4H_{10}S$	2-Methyl-1-propanethiol	298.2	4.961
$C_4H_{10}S$	2-Methyl-2-propanethiol	293.2	5.475
$C_4H_{10}S$	Diethyl sulfide	298.2	5.723
$C_4H_{10}Zn$	Diethyl zinc	293.2	2.55
$C_4H_{11}N$	Butylamine	293.2	4.71
$C_4H_{11}N$	Diethylamine	293.2	3.680
$C_4H_{11}NO_2$	Diethanolamine	293.2	25.75
$C_4H_{11}OP$	Ethyldimethylphosphine oxide	363.2	34.4
$C_4H_{11}O_3P$	Ethanephosphonic acid dimethyl ester	313.2	15.89
$C_4H_{12}Ge$	Tetramethyl germanium	297.2	1.817
$C_4H_{12}NO_3P$	Phosphoramidic acid diethyl ester	313.2	25.3
$C_4H_{12}N_2O$	2-(2-Aminoethylamino)ethanol	293.2	21.81
$C_4H_{12}OSi$	Methoxytrimethylsilane	298.2	3.248
$C_4H_{12}O_2Si$	Dimethoxydimethylsilane	298.2	3.663
$C_4H_{12}O_3Si$	Trimethoxymethylsilane	298.2	4.9
$C_4H_{12}O_4Si$	Tetramethyl silicate	293.2	6.0
$C_4H_{12}Si$	Butylsilane	293.2	2.537
$C_4H_{12}Si$	Isobutylsilane	293.2	2.497
$C_4H_{12}Si$	Diethylsilane	293.2	2.544
$C_4H_{12}Si$	Tetramethylsilane	293.2	1.921
$C_4H_{13}N_3$	Bis(2-amimoethyl)amine	293.2	12.62
C_5FeO_5	Iron pentacarbonyl	293.2	2.602

Molecular formula	Name	*T*/K	ε
$C_5H_4F_8O$	2,2,3,3,4,4,5,5-Octafluoro-1-pentanol	298.2	15.30
C_5H_4ClN	2-Chloropyridine	298.2	27.32
C_5H_4BrN	2-Bromopyridine	298.2	23.18
$C_5H_4O_2$	Furfural	293.2	42.1
C_5H_5N	Pyridine	293.2	13.260
C_5H_5NO	Pyridine-1-oxide	343.0	35.94
C_5H_6O	2-Methylfuran	293.2	2.76
$C_5H_6O_2$	Furfuryl alcohol	298.2	16.85
$C_5H_7Cl_3O_2$	Propyl trichloroacetate	298.2	8.32
$C_5H_7NO_2$	Ethyl 2-cyanoacetate	263.2	31.62
$C_5H_7N_2O_3P$	Diisocyanato isopropyl phosphite	293.0	7.25
$C_5H_7N_2O_3PS$	Diisocyanato *O*-isopropyl thionphosphate	288.0	12.8
$C_5H_7N_2O_4P$	Diisocyanato isopropyl phosphate	298.0	13.2
C_5H_8	Cyclopentene	295.0	2.083
C_5H_8	2-Methyl-1,3-butadiene	293.2	2.098
C_5H_8	1,3-Pentadiene*	298.2	2.319
C_5H_8	1,4-Pentadiene	294.0	2.054
$C_5H_8N_2OS$	2-Methyl-2-(sulfinylamino)butanenitrile	293.2	14.70
$C_5H_8N_2O_2$	3-Isopropylsydnone	333.2	66.0
C_5H_8O	Cyclopentanone	298.2	13.58
C_5H_8O	2,3-Dihydropyran	308.2	5.1364
C_5H_8O	2-Methyl-1-butene-3-one	303.2	10.30
$C_5H_8O_2$	2,4-Pentanedione	303.2	26.524
$C_5H_8O_2$	Ethyl acrylate	303.2	6.05
$C_5H_8O_2$	Methyl crotonate	293.2	6.6645
$C_5H_8O_2$	Methyl methacrylate	303.2	6.32
$C_5H_8O_4$	Dimethyl malonate	293.2	9.82
$C_5H_9BrO_2$	2-Bromo-3-methyl butanoic acid	293.2	6.5
$C_5H_9BrO_2$	Ethyl 2-bromopropanoate	293.2	9.4
$C_5H_9ClO_2$	Isobutyl chlorocarbonate	293.2	9.1
$C_5H_9ClO_2$	Methyl 4-chlorobutanoate	303.2	9.51
$C_5H_9ClO_2$	Ethyl 2-chloropropanoate	303.2	11.95
$C_5H_9ClO_2$	Ethyl 3-chloropropanoate	303.2	10.19
$C_5H_9IO_2$	Ethyl 2-iodopropanoate	293.2	8.6
C_5H_9N	2,2-Dimethylpropane nitrile	293.2	21.1
C_5H_9N	Pentanenitrile	293.2	20.04
C_5H_9NO	Butyl isocyanate	293.2	12.290
C_5H_9NO	Isobutyl isocyanate	293.2	11.638
C_5H_9NO	1-Methyl-pyrrolidine-2-one	293.2	32.55
$C_5H_9NO_2$	4,4-Dimethyl-oxazolidine-2-one	333.2	39.2
$C_5H_9NO_2$	4-Ethyl-oxazolidine-2-one	298.2	42.6
$C_5H_9NO_2$	3-Ethyl-oxazolidine-2-one	298.2	66.8
C_5H_{10}	Ethylcyclopropane	293.2	1.933
C_5H_{10}	Cyclopentane	293.2	1.9687
C_5H_{10}	2-Methyl-1-butene	293.2	2.180
C_5H_{10}	2-Methyl-2-butene	296.0	1.979
C_5H_{10}	1-Pentene	293.2	2.011
$C_5H_{10}Br_2$	1,2-Dibromopentane	298.2	4.39
$C_5H_{10}Br_2$	(±)-*erythro*-2,3-Dibromopentane	298.2	5.430
$C_5H_{10}Br_2$	(±)-*threo*-2,3-Bibromopentane	298.2	6.507
$C_5H_{10}Br_2$	1,4-Dibromopentane	293.2	9.05
$C_5H_{10}Br_2$	1,5-Dibromopentane	303.2	9.14
$C_5H_{10}Cl_2$	1,2-Dichloropentane	293.2	6.89
$C_5H_{10}Cl_2$	1,5-Dichloropentane	298.2	9.92
$C_5H_{10}N_2O$	1,3-Dimethyl-imidazolidin-2-one	298.2	37.60
$C_5H_{10}O$	Cyclopentanol	288.2	18.5
$C_5H_{10}O$	3-Pentanone	293.2	17.00
$C_5H_{10}O$	2-Pentanone	293.2	15.45

Molecular formula	Name	*T*/K	ε
$C_5H_{10}O$	3-Methyl-2-butanone	293.2	10.37
$C_5H_{10}O$	2-Methyltetrahydrofuran	298.2	6.97
$C_5H_{10}O$	Tetrahydropyran	293.2	5.66
$C_5H_{10}O$	2,2-Dimethylpropanal	293.2	9.051
$C_5H_{10}O$	Pentanal	293.2	10.00
$C_5H_{10}O_2$	Butyl formate	303.2	6.10
$C_5H_{10}O_2$	Isobutyl formate	293.2	6.41
$C_5H_{10}O_2$	Methyl butanoate	301.2	5.48
$C_5H_{10}O_2$	Propyl acetate	293.2	5.62
$C_5H_{10}O_2$	Ethyl propanoate	293.2	5.76
$C_5H_{10}O_2$	Tetrahydrofurfuryl alcohol	303.2	13.48
$C_5H_{10}O_2$	Pentanoic acid	294.4	2.661
$C_5H_{10}O_2S$	3-Methyl sulfolane	298.2	29.4
$C_5H_{10}O_2S$	Methyl 3-(methylthio)propanoate	303.2	8.66
$C_5H_{10}O_3$	Diethyl carbonate	297.2	2.820
$C_5H_{10}O_3$	Ethyl lactate	303.2	15.4
$C_5H_{10}O_4$	1,2,3-Propanetriol 1-acetate	242.2	38.57
$C_5H_{11}Br$	2-Bromo-2-methylbutane	298.2	9.21
$C_5H_{11}Br$	1-Bromopentane	299.2	6.31
$C_5H_{11}Br$	1-Bromo-3-methylbutane	291.5	6.33
$C_5H_{11}Br$	3-Bromopentane	298.2	8.37
$C_5H_{11}Cl$	1-Chloropentane	293.2	6.654
$C_5H_{11}Cl$	1-Chloro-3-methylbutane	292.0	6.10
$C_5H_{11}Cl$	2-Chloro-2-methylbutane	222.75	12.31
$C_5H_{11}F$	1-Fluoropentane	293.2	3.931
$C_5H_{11}F$	2-Fluoro-2-methylbutane	293.2	5.89
$C_5H_{11}I$	1-Iodopentane	293.2	5.78
$C_5H_{11}I$	2-Iodo-2-methylbutane	293.2	8.192
$C_5H_{11}I$	3-Iodopentane	293.2	7.432
$C_5H_{11}I$	1-Iodo-3-methylbutane	292.2	5.6
$C_5H_{11}N$	Piperidine	293.0	4.33
$C_5H_{11}N$	*N*-Methylpyrrolidine	298.2	32.2
$C_5H_{11}NO$	*N,N*-Diethylformamide	293.2	29.6
$C_5H_{11}NO$	*N,N*-Dimethylpropanamide	293.2	34.6
$C_5H_{11}NO$	2,2-Dimethylpropionamide	298.2	20.13
$C_5H_{11}NO$	*N*-Ethylpropanamide	298.2	126.8
$C_5H_{11}NO$	*N*-Propylacetamide	298.2	117.8
$C_5H_{11}NO$	2-Pentanone oxime	293.2	3.3
$C_5H_{11}NO_2$	*N*-(2-Methoxyethyl)acetamide	298.2	80.7
$C_5H_{11}NO_2$	Pentyl nitrite	298.2	7.21
$C_5H_{11}NO_2$	*tert*-Pentyl nitrite	298.2	10.88
$C_5H_{11}NO_3$	Pentyl nitrate	291.2	9.0
C_5H_{12}	Isopentane	293.2	1.845
C_5H_{12}	Neopentane	296.0	1.769
C_5H_{12}	Pentane	293.2	1.8371
$C_5H_{12}N_2O$	Tetramethyl urea	293.2	23.10
$C_5H_{12}O$	2,2-Dimethyl-1-propanol	333.2	8.35
$C_5H_{12}O$	2-Methyl-1-butanol	298.2	15.63
$C_5H_{12}O$	2-Methyl-2-butanol	298.2	5.78
$C_5H_{12}O$	3-Methyl-1-butanol	293.2	15.63
$C_5H_{12}O$	3-Methyl-2-butanol	298.2	12.1
$C_5H_{12}O$	1-Pentanol	298.2	15.13
$C_5H_{12}O$	2-Pentanol	298.2	13.71
$C_5H_{12}O$	3-Pentanol	298.2	13.35
$C_5H_{12}O_2$	1,2-Pentanediol	296.8	17.31
$C_5H_{12}O_2$	1,4-Pentanediol	295.7	26.74
$C_5H_{12}O_2$	1,5-Pentanediol	293.2	26.2
$C_5H_{12}O_2$	2,3-Pentanediol	296.9	17.37

Molecular formula	Name	*T*/K	ε
$C_5H_{12}O_2$	2,4-Pentanediol	294.2	24.69
$C_5H_{12}O_2$	Diethoxymethane	293.2	2.527
$C_5H_{12}O_4$	Tetramethoxymethane	293.2	2.40
$C_5H_{12}O_5$	Xylitol	293.2	40.0
$C_5H_{12}S$	2-Methyl-2-butanethiol	293.2	5.087
$C_5H_{12}S$	1-Pentanethiol	293.2	4.847
$C_5H_{12}S_4$	Tetrakis(methylthio)methane	343.2	2.818
$C_5H_{13}ClNO_3P$	Diethyl *N*-chloro-*N*-methylamidophosphate	293.2	8.47
$C_5H_{13}N$	Pentylamine	293.2	4.27
$C_5H_{13}N_3$	1,1,3,3-Tetramethylguanidine	298.2	11.5
$C_5H_{13}O_3P$	Diethyl methanephosphate	313.2	13.405
$C_5H_{14}OSi$	Ethoxytrimethylsilane	298.2	3.013
$C_5H_{15}NSi$	Ethyltrimethylsilazane	303.2	2.2751
$C_5H_{18}B_{10}$	Isopropyl carborane	293.2	45.0
$C_5H_{20}O_5Si_5$	Pentamethyl cyclopentasiloxane	293.2	2.740
C_6F_6	Hexafluorobenzene	298.2	2.029
$C_6F_9Cl_5$	Pentachloro-1,1,2,3,3,4,5,5,6-nonafluorohexane	293.2	2.750
C_6F_{14}	Perfluorohexane	298.2	1.76
$C_6F_{15}OP$	Tris(perfluoroethyl)phosphine oxide	313.2	2.34
$C_6H_3Cl_2NO_2$	2,4-Dichloro-1-nitrobenzene	301.2	13.06
$C_6H_3N_3O_7$	2,4,6-Trinitrophenol	294.2	4.0
$C_6H_4BrClO_2S$	4-Bromobenzene sulfonylchloride	353.2	10.90
$C_6H_4BrCl_2NO_2S_2$	[(4-Bromophenyl)sulfonyl]-imidosulfurous dichloride	351.7	11.08
C_6H_4BrF	1-Bromo-2-fluorobenzene	298.2	4.72
C_6H_4BrF	1-Bromo-3-fluorobenzene	298.2	4.85
C_6H_4BrF	1-Bromo-4-fluorobenzene	298.2	2.60
C_6H_4BrNOS	4-Bromo-*N*-sulfinylaniline	354.2	3.46
$C_6H_4BrNO_2$	3-Bromonitrobenzene	328.2	20.2
$C_6H_4Br_2$	*o*-Dibromobenzene	293.2	7.86
$C_6H_4Br_2$	*m*-Dibromobenzene	293.2	4.81
$C_6H_4Br_2$	*p*-Dibromobenzene	368.2	2.57
C_6H_4ClF	2-Chlorofluorobenzene	298.2	6.10
C_6H_4ClF	3-Chlorofluorobenzene	298.2	4.96
C_6H_4ClF	4-Chlorofluorobenzene	298.2	3.34
$C_6H_4ClFO_2S$	4-Fluorobenzene sulfonylchloride	313.2	12.65
C_6H_4ClNOS	3-Chloro-*N*-sulfinylaniline	293.2	7.00
C_6H_4ClNOS	4-Chloro-*N*-sulfinylaniline	333.2	4.17
$C_6H_4ClNO_2$	*o*-Chloronitrobenzene	323.2	37.7
$C_6H_4ClNO_2$	*m*-Chloronitrobenzene	323.2	20.9
$C_6H_4ClNO_2$	*p*-Chloronitrobenzene	393.2	8.09
$C_6H_4ClNO_4S$	4-Nitrobenzene sulfonylchloride	355.0	3.74
$C_6H_4Cl_2$	*o*-Dichlorobenzene	293.2	10.12
$C_6H_4Cl_2$	*m*-Dichlorobenzene	293.2	5.02
$C_6H_4Cl_2$	*p*-Dichlorobenzene	328.2	2.3943
$C_6H_4Cl_2N_2O_2S$	(3-Nitrophenyl)imidosulfurous dichloride	294.2	21.2
$C_6H_4Cl_2N_2O_4S_2$	[(3-Nitrophenyl)sulfonyl]-imidosulfurous dichloride	348.2	25.55
$C_6H_4Cl_2N_2O_4S_2$	[(4-Nitrophenyl)sulfonyl]-imodosulfurous dichloride	356.7	6.85
$C_6H_4Cl_2O_2S$	4-Chlorobenzene sulfonylchloride	333.2	11.80
$C_6H_4Cl_3NO_2S_2$	[(4-Chlorophenyl)sulfonyl]-imidosulfurous dichloride	341.2	12.50
C_6H_4FI	2-Fluoroiodobenzene	298.2	8.22
C_6H_4FI	*m*-Fluoroiodobenzene	298.2	4.62
C_6H_4FI	4-Fluoroiodobenzene	298.2	3.12
$C_6H_4F_2$	*o*-Difluorobenzene	301.2	13.38
$C_6H_4F_2$	*m*-Difluorobenzene	301.2	5.01
$C_6H_4I_2$	1,2-Diiodobenzene	323.2	5.41
$C_6H_4I_2$	1,3-Diiodobenzene	323.2	4.11
$C_6H_4I_2$	1,4-Diiodobenzene	393.2	2.88
$C_6H_4N_2$	2-Cyanopyridine	303.2	93.77

Molecular formula	Name	*T*/K	ε
$C_6H_4N_2$	3-Cyanopyridine	323.2	20.54
$C_6H_4N_2$	4-Cyanopyridine	353.2	5.23
$C_6H_4N_2O_2S_2$	*N,N′*-Disulfinyl-1,2-benzene diamine	339.7	6.99
$C_6H_4N_2O_2S_2$	*N,N′*-Disulfinyl-1,4-benzene diamine	298.2	7.8
$C_6H_4N_2O_3S$	3-Nitro-*N*-sulfinylaniline	343.0	19.49
$C_6H_4N_2O_3S$	4-Nitro-*N*-sulfinylaniline	348.2	14.35
$C_6H_4N_2O_4$	*m*-Dinitrobenzene	365.2	22.9
C_6H_5Br	Bromobenzene	293.2	5.45
C_6H_5Cl	Chlorobenzene	293.2	5.6895
C_6H_5ClO	*o*-Chlorophenol	296.2	7.40
C_6H_5ClO	*m*-Chlorophenol	293.2	6.255
C_6H_5ClO	*p*-Chlorophenol	314.2	11.18
$C_6H_5ClO_2S$	Benzenesulfonyl chloride	323.2	28.90
C_6H_5ClS	4-Chlorothiophenol	338.2	3.59
$C_6H_5Cl_2OP$	Benzenephosphonic acid dichloride	293.2	26.0
$C_6H_5Cl_2NO_2S_2$	Phenylsulfonylimidosulfurous dichloride	298.2	29.7
C_6H_5F	Fluorobenzene	293.2	5.465
$C_6H_5F_2OP$	Benzenephosphonic acid difluoride	293.2	27.9
C_6H_5I	Iodobenzene	293.2	4.59
C_6H_5NOS	*N*-Sulfinylaniline	298.2	6.97
$C_6H_5NO_2$	Nitrobenzene	293.0	35.6
$C_6H_5NO_3$	*o*-Nitrophenol	323.2	16.50
$C_6H_5NO_3$	*m*-Nitrophenol	373.2	35.45
$C_6H_5NO_3$	*p*-Nitrophenol	393.2	42.20
C_6H_6	Benzene	293.2	2.2825
C_6H_6BrN	3-Bromoaniline	293.2	13.0
C_6H_6ClN	*o*-Chloroaniline	293.2	13.40
C_6H_6ClN	*m*-Chloroaniline	293.2	13.3
$C_6H_6Cl_6$	α-Hexachlorocyclohexane*	429.2	4.77
$C_6H_6N_2O_2$	*o*-Nitroaniline	353.0	47.3
$C_6H_6N_2O_2$	*m*-Nitroaniline	398.0	35.6
$C_6H_6N_2O_2$	*p*-Nitroaniline	428.0	78.5
C_6H_6O	Phenol	303.2	12.40
$C_6H_6O_2$	1,2-Benzenediol	388.2	17.57
$C_6H_6O_2$	1,3-Benzenediol	393.2	13.55
$C_6H_6O_2S$	Methyl thiophene-2-carboxylate	293.2	8.81
$C_6H_6O_3$	Methyl furan-2-carboxylate	293.2	11.01
C_6H_6S	Benzenethiol	303.2	4.26
C_6H_7N	Aniline	293.2	7.06
C_6H_7N	2-Methylpyridine	293.2	10.18
C_6H_7N	3-Methylpyridine	303.0	11.10
C_6H_7N	4-Methylpyridine	293.0	12.2
C_6H_7NO	2-Methylpyridine 1-oxide	323.2	36.4
C_6H_7NO	3-Methylpyridine 1-oxide	318.2	28.26
C_6H_8	1,3-Cyclohexadiene	184.2	2.68
C_6H_8	1,4-Cyclohexadiene	296.0	2.211
C_6H_8	1,3,5-Hexatriene(*cis*/*trans* = 0.58)	297.0	2.276
$C_6H_8N_2$	Phenylhydrazine	293.2	7.15
$C_6H_8N_2$	2,5-Dimethylpyrazine	293.2	2.436
$C_6H_8N_2$	2,6-Dimethylpyrazine	308.2	2.653
$C_6H_8O_2$	1,4-Cyclohexanedione	351.2	4.40
$C_6H_9ClO_2$	*cis*-Ethyl 3-chlorocrotonate	349.2	7.67
$C_6H_9ClO_2$	*trans*-Ethyl 3-chlorocrotonate	327.2	4.70
$C_6H_9Cl_3O_2$	Butyl trichloroacetate	293.2	7.480
$C_6H_9Cl_3O_2$	Isobutyl trichloroacetate	293.2	7.667
C_6H_9N	Cyclopentanecarbonitrile	293.2	22.68
C_6H_{10}	Cyclohexene	293.2	2.2176
C_6H_{10}	2,3-Dimethyl-1,3-butadiene	293.2	2.102

Molecular formula	Name	*T*/K	ε
C_6H_{10}	2-Methyl-1,3-pentadiene	298.2	2.422
C_6H_{10}	3-Methyl-1,3-pentadiene	298.2	2.426
C_6H_{10}	4-Methyl-1,3-pentadiene	293.2	2.599
C_6H_{10}	1,5-Hexadiene	294.0	2.125
C_6H_{10}	2,4-Hexadiene	298.2	2.207
C_6H_{10}	*cis*, *cis*-2,4-Hexadiene	297.0	2.163
C_6H_{10}	*trans*, *trans*-2,4-Hexadiene	297.0	2.123
C_6H_{10}	1-Hexyne	296.0	2.621
$C_6H_{10}N_2O_2$	3-Butylsydnone	298.2	52.8
$C_6H_{10}N_2O_2$	3-Propyl-4-methylsydnone	298.2	66.4
$C_6H_{10}O$	Cyclohexanone	293.0	16.1
$C_6H_{10}O$	Mesityl oxide	273.2	15.6
$C_6H_{10}O$	Butoxyacetylene	298.2	6.62
$C_6H_{10}O_2$	Ethyl methacrylate	303.2	5.68
$C_6H_{10}O_2$	Ethyl crotonate	293.2	5.4
$C_6H_{10}O_2S$	Methyl tetrahydrothiophene-2-carboxylate	293.2	7.30
$C_6H_{10}O_3$	Ethyl acetoacetate	293.2	14.0
$C_6H_{10}O_3$	Propanoic anhydride	293.2	18.30
$C_6H_{10}O_4$	Ethylene glycol diacetate	290.2	7.7
$C_6H_{10}O_4$	Diethyl oxalate	293.2	8.266
$C_6H_{10}O_4$	Monomethyl glutarate	293.2	8.37
$C_6H_{10}O_4$	Dimethyl succinate	293.2	7.19
$C_6H_{11}Br$	Bromocyclohexane	303.2	8.0026
$C_6H_{11}BrO_2$	Ethyl 2-bromobutanoate	303.2	8.57
$C_6H_{11}BrO_2$	Ethyl 2-bromo-2-methylpropanoate	303.2	8.55
$C_6H_{11}BrO_2$	(±)-*erythro*-2-Acetoxy-3-bromobutane	298.2	7.268
$C_6H_{11}BrO_2$	(±)-*threo*-2-Acetoxy-3-bromobutane	298.2	7.414
$C_6H_{11}Cl$	Chlorocyclohexane	303.2	7.9505
$C_6H_{11}ClO_2$	3-Methylbutyl chlorocarbonate	293.2	7.8
$C_6H_{11}N$	Hexanenitrile	298.2	17.26
$C_6H_{11}N$	4-Methylpentanenitrile	295.2	17.5
$C_6H_{11}NO$	Cyclohexanone oxime	362.2	3.04
$C_6H_{11}NO_2$	Cyclohexyl nitrite	298.2	9.33
$C_6H_{11}NO_2$	3-Isopropyloxazolidin-2-one	298.2	51.0
$C_6H_{14}O_2$	Hexylene glycol	303.2	23.4
C_6H_{12}	Cyclohexane	293.2	2.0243
C_6H_{12}	Methylcyclopentane	293.2	1.9853
C_6H_{12}	Ethylcyclobutane	293.2	1.965
C_6H_{12}	1-Hexene	294.0	2.077
C_6H_{12}	*trans*-2-Hexene	295.0	1.978
C_6H_{12}	*cis*-3-Hexene	296.0	2.069
C_6H_{12}	*trans*-3-Hexene	293.2	1.954
$C_6H_{12}Br_2$	*meso*-3,4-Dibromohexane	298.2	4.67
$C_6H_{12}Br_2$	(±)-3,4-Dibromohexane	298.2	6.732
$C_6H_{12}Br_2$	1,6-Dibromohexane	298.2	8.52
$C_6H_{12}ClNO$	2-Chloro-*N*,*N*-diethylacetamide	298.2	39.2
$C_6H_{12}Cl_2$	1,6-Dichlorohexane	308.2	8.60
$C_6H_{12}Cl_2O$	Bis(3-chloropropyl) ether	293.2	10.10
$C_6H_{12}N_2O$	1,3-Dimethyl-2-oxohexahydropyrimidine	298.2	36.12
$C_6H_{12}N_3OP$	Tri(1-aziridinyl)phosphine oxide	298.2	14.62
$C_6H_{12}O$	Isobutyl vinyl ether	293.2	3.34
$C_6H_{12}O$	Cyclohexanol	293.2	16.40
$C_6H_{12}O$	*cis*-2,5-Dimethyltetrahydrofuran	296.2	5.03
$C_6H_{12}O$	2-Hexanone	293.2	14.56
$C_6H_{12}O$	4-Methyl-2-pentanone	293.2	13.11
$C_6H_{12}O$	3,3-Dimethyl-2-butanone	293.2	12.73
$C_6H_{12}O$	1-Methylcyclopentanol	310.1	7.11
$C_6H_{12}O_2$	Diacetone alcohol	298.2	18.2

Molecular formula	Name	T/K	ε
$C_6H_{12}O_2$	Pentyl formate	292.2	5.7
$C_6H_{12}O_2$	Isopentyl formate	288.2	5.44
$C_6H_{12}O_2$	Ethyl butanoate	301.2	5.18
$C_6H_{12}O_2$	Hexanoic acid	298.2	2.600
$C_6H_{12}O_2$	*tert*-Butylacetic acid	296.2	2.85
$C_6H_{12}O_2$	Butyl acetate	293.2	5.07
$C_6H_{12}O_2$	Isobutyl acetate	293.2	5.068
$C_6H_{12}O_2$	*sec*-Butyl acetate	293.2	5.135
$C_6H_{12}O_2$	*tert*-Butyl acetate	293.2	5.672
$C_6H_{12}O_2$	2-Ethylbutanoic acid	296.2	2.72
$C_6H_{12}O_2$	Propyl propanoate	293.2	5.249
$C_6H_{12}O_2$	Methyl pentanoate	293.2	4.992
$C_6H_{12}O_2S$	2,4-Dimethyltetrahydrothiophene-*S*,*S*-dioxide	293.2	29.69
$C_6H_{12}O_3$	2-Ethoxyethyl acetate	303.2	7.567
$C_6H_{12}O_3$	2,4,6-Trimethyl-1,3,5-trioxan	292.2	14.70
$C_6H_{12}S$	Cyclohexanethiol	298.2	5.420
$C_6H_{13}Br$	1-Bromohexane	298.2	5.82
$C_6H_{13}Cl$	1-Chlorohexane	293.2	6.104
$C_6H_{13}ClO$	6-Chloro-1-hexanol	242.2	21.6
$C_6H_{13}I$	1-Iodohexane	293.3	5.35
$C_6H_{13}N$	Cyclohexylamine	293.2	4.547
$C_6H_{13}NO$	*N*-Butylacetamide	293.2	104.0
$C_6H_{13}NO$	*N*-Isobutylacetamide	293.2	111.0
$C_6H_{13}NO$	*N*-*sec*-Butylacetamide	293.2	100.0
$C_6H_{13}NO$	*N*,*N*-Diethylacetamide	293.2	32.1
$C_6H_{13}NO$	*N*,*N*-Dimethylbutanamide	293.2	29.7
$C_6H_{13}NO$	*N*-Ethylbutanamide	298.2	107.0
$C_6H_{13}NO$	*N*-Methylpentanamide	286.0	131.0
$C_6H_{13}NO$	*N*-Methyl-3-methylbutanamide	299.0	114.0
$C_6H_{13}NO$	*N*-Methyl-2-methylbutanamide	307.0	123.0
$C_6H_{13}NO$	*N*-Propylpropanamide	298.2	118.1
C_6H_{14}	2,2-Dimethylbutane	293.2	1.869
C_6H_{14}	2,3-Dimethylbutane	293.2	1.889
C_6H_{14}	2-Methylpentane	293.2	1.886
C_6H_{14}	3-Methylpentane	293.2	1.886
C_6H_{14}	Hexane	293.2	1.8865
$C_6H_{14}Cl_2NO_3PS$	Dipropoxyphosphinylimidosulfurous dichloride	293.2	17.65
$C_6H_{14}N_3OP$	*P*,*P*-Bis(1-aziridinyl)-*N*,*N*-dimethylphosphinic amide	298.2	8.58
$C_6H_{14}O$	Dipropyl ether	297.0	3.38
$C_6H_{14}O$	Diisopropyl ether	303.2	3.805
$C_6H_{14}O$	Ethyl *tert*-butyl ether	298.2	7.07
$C_6H_{14}O$	2,2-Dimethyl-1-butanol	293.2	10.5
$C_6H_{14}O$	2-Ethyl-1-butanol	362.2	6.19
$C_6H_{14}O$	1-Hexanol	293.2	13.03
$C_6H_{14}O$	(±)-2-Hexanol	298.2	11.06
$C_6H_{14}O$	3-Hexanol	298.2	9.66
$C_6H_{14}O$	3-Methyl-1-pentanol	298.2	15.2
$C_6H_{14}O$	3-Methyl-3-pentanol	293.2	4.322
$C_6H_{14}OS$	Dipropyl sulfoxide	303.2	30.37
$C_6H_{14}O_2$	2-Butoxyethanol	298.2	9.43
$C_6H_{14}O_2$	1,2-Diethoxyethane	293.2	3.90
$C_6H_{14}O_2$	Hexylene glycol	293.2	25.86
$C_6H_{14}O_2S$	Dipropyl sulfone	303.2	32.62
$C_6H_{14}O_3$	Bis(2-methoyxethyl) ether	298.2	7.23
$C_6H_{14}O_3$	Bis(2-hydroxypropyl) ether	293.2	20.38
$C_6H_{14}O_3$	1,2,6-Hexanetriol	285.3	31.5
$C_6H_{14}O_4$	Triethylene glycol	293.2	23.69
$C_6H_{14}O_6$	Sorbitol	353.2	35.5

Molecular formula	Name	*T*/K	ε
$C_6H_{14}O_6$	D-Mannitol	443.2	24.6
$C_6H_{14}S$	1-Hexanethiol	293.2	4.436
$C_6H_{15}Al$	Triethyl aluminium	293.2	2.9
$C_6H_{15}B$	Triethylborane	293.2	1.974
$C_{10}H_{22}O_2$	Dipropylamine	293.2	2.923
$C_6H_{15}N$	Hexylamine	293.2	4.08
$C_6H_{15}N$	Triethylamine	293.2	2.418
$C_6H_{15}NO$	2-Dimethylamino-2-methyl-1-propanol	298.2	12.36
$C_6H_{15}OP$	Triethylphosphine oxide	323.2	35.5
$C_6H_{15}O_3P$	Diethyl ethanephosphonate	305.2	10.653
$C_6H_{15}O_3PS$	Triethyl thionphosphate	298.2	10.4
$C_6H_{15}O_4P$	Triethyl phosphate	298.2	13.20
$C_6H_{15}O_4V$	Triethyl orthovanadate	298.2	3.333
$C_6H_{15}PS$	Triethylphosphine sulfide	371.2	39.0
$C_6H_{16}NO_3P$	*N*,*N*-Dimethylphosphoramidic acid diethyl ester	293.2	11.75
$C_6H_{16}N_3OP$	*P*-1-Aziridinyl-*N*,*N*,*N*′,*N*′-tetramethylphosphonic diamide	298.2	14.69
$C_6H_{16}OSi$	Trimethylpropoxysilane	298.2	2.850
$C_6H_{16}OSi$	Trimethylisopropoxysilane	298.2	2.851
$C_6H_{16}O_2Si$	Diethoxydimethylsilane	298.2	3.216
$C_6H_{16}Si$	Triethylsilane	293.2	2.323
$C_6H_{17}N_2O_2P$	*N*,*N*,*N*′,*N*′-Tetramethylphosphordiamic acid ethyl ester	298.2	14.73
$C_6H_{18}N_3OP$	Tris(dimethylamino)phosphine oxide	293.2	31.3
$C_6H_{18}N_3P$	Tris(dimethylamino)phosphine	293.2	3.65
$C_6H_{18}N_3PS$	Tris(dimethylamino)phosphine sulfide	303.2	39.5
$C_6H_{18}N_4$	Triethylenetetramine	293.2	10.76
$C_6H_{18}OSi_2$	Hexamethyldisiloxane	293.2	2.179
$C_6H_{18}O_3Si_3$	Hexamethylcyclotrisiloxane	343.2	2.139
$C_6H_{19}NSi_2$	Hexamethyldisilazane	294.2	2.273
C_7F_{14}	Perfluoromethylcyclohexane	298.2	1.82
C_7F_{16}	Perfluoroheptane	289.2	1.847
$C_7H_3ClF_3NO_2$	2-Chloro-1-trifluoromethyl-5-nitrobenzene	303.2	9.8
$C_7H_3ClF_3NO_2$	4-Chloro-1-trifluoromethyl-3-nitrobenzene	303.2	12.8
$C_7H_3Cl_5$	2,3,4,5,6-Pentachlorotoluene	293.2	4.8
$C_7H_4BrCl_2NOS$	(4-Bromobenzoyl)imidosulfurous dichloride	353.2	7.80
$C_7H_4ClF_3O_2S$	4-(Trifluoromethylsulfonyl)-1-chlorobenzene	333.2	10.3
$C_7H_4ClF_3O_4S_2$	4-(Trifluoromethylsulfonyl)-benzenesulfonyl chloride	298.2	4.72
C_7H_4ClNO	4-Chlorophenyl isocyanate	288.2	3.177
$C_7H_4Cl_2N_2O_3S$	(4-Nitrobenzoyl)imidosulfurous dichloride	366.2	6.02
$C_7H_4Cl_3NOS$	(4-Chlorobenzoyl)imidosulfurous dichloride	298.2	8.25
$C_7H_4F_3NO_2$	1-Trifluoromethyl-3-nitrobenzene	303.2	17.0
$C_7H_4F_3NO_3S_2$	4-(Trifluoromethylsulfonyl)-*N*-sulfinyl aniline	349.2	15.05
$C_7H_4F_3NO_4S$	4-(Trifluoromethylsulfonyl)-1-nitrobenzene	373.2	3.62
$C_7H_4F_4O_2S$	4-(Trifluoromethylsulfonyl)-1-fluorobenzene	293.2	12.2
C_7H_5BrO	Benzoyl bromide	293.2	21.33
C_7H_5ClO	Benzoyl chloride	293.2	23.0
$C_7H_5Cl_2NOS$	Benzoylimidosulfurous dichloride	313.2	31.6
$C_7H_5Cl_3$	2,4-Dichloro-1-chloromethylbenzene	298.2	6.290
C_7H_5FO	Benzoyl fluoride	293.2	22.7
$C_7H_5F_3$	(Trifluoromethyl)benzene	298.2	9.22
$C_7H_5F_3O_2S$	Trifluoromethylphenyl sulfone	293.2	31.0
$C_7H_5F_3O_2S_2$	4-(Trifluoromethylsulfonyl)-benzenethiol	298.2	28.5
$C_7H_5F_3O_3S$	4-(Trifluoromethylsulfonyl)-phenol	298.2	42.2
C_7H_5N	Benzonitrile	293.2	25.9
C_7H_5NO	Phenyl isocyanate	293.2	8.940
$C_7H_6ClNO_2$	4-Chloro-3-nitrotoluene	301.2	28.07
$C_7H_6Cl_2$	2,6-Dichlorotoluene	301.2	3.36
$C_7H_6Cl_2$	2,4-Dichlorotoluene	301.2	5.68
$C_7H_6Cl_2$	3,4-Dichlorotoluene	301.2	9.39

Molecular formula	Name	T/K	ε
$C_7H_6Cl_2$	(Dichloromethyl)benzene	293.2	6.9
$C_7H_6F_3NO_2S$	4-(Trifluoromethylsulfonyl)aniline	298.2	77.0
C_7H_6O	Benzaldehyde	293.2	17.85
$C_7H_6O_2$	Salicylaldehyde	293.2	18.35
C_7H_7Br	(Bromomethyl)benzene	293.2	6.658
C_7H_7Br	2-Bromotoluene	293.2	4.641
C_7H_7Br	3-Bromotoluene	293.2	5.566
C_7H_7Br	4-Bromotoluene	293.2	5.503
C_7H_7BrO	2-Bromoanisole	303.2	8.96
C_7H_7BrO	4-Bromoanisole	303.2	7.40
C_7H_7BrS	4-Bromo-3-methylthiophenol	303.2	3.70
C_7H_7Cl	Benzyl chloride	293.2	6.854
C_7H_7Cl	*o*-Chlorotoluene	293.2	4.721
C_7H_7Cl	*m*-Chlorotoluene	293.2	5.763
C_7H_7Cl	*p*-Chlorotoluene	293.2	6.25
$C_7H_7Cl_2NO_2S_2$	[(4-Methylphenyl)sulfonyl] imidosulfurous dichloride	327.2	11.08
C_7H_7ClO	4-Chloroanisole	293.2	7.84
$C_7H_7ClO_2S$	*p*-Toluenesulfonyl chloride	343.2	22.6
$C_7H_7ClO_3S$	4-Methoxybenzenesulfonyl chloride	314.2	27.2
C_7H_7ClS	4-Chlorothioanisole	298.2	5.95
C_7H_7ClS	4-Chloro-2-methylthiophenol	303.2	5.50
C_7H_7F	*o*-Fluorotoluene	298.2	4.23
C_7H_7F	*m*-Fluorotoluene	298.2	5.41
C_7H_7F	*p*-Fluorotoluene	298.2	5.88
C_7H_7I	4-Iodotoluene	308.2	4.4
C_7H_7N	2-Vinylpyridine	293.2	9.126
C_7H_7N	4-Vinylpyridine	293.2	10.50
$C_7H_7NO_2$	Benzyl nitrite	298.2	7.78
$C_7H_7NO_2$	*o*-Nitrotoluene	293.0	26.26
$C_7H_7NO_2$	*m*-Nitrotoluene	303.2	24.95
$C_7H_7NO_2$	*p*-Nitrotoluene	331.2	22.2
$C_7H_7NO_2$	Methyl pyridine-4-carboxylate	293.2	9.51
$C_7H_7NO_2S$	4-Nitrothioanisole	346.0	21.7
$C_7H_7NO_2S$	4-Methyl-*N*-sulfinyl aniline	293.2	8.75
$C_7H_7NO_3$	2-Nitroanisole	293.2	45.75
$C_7H_7NO_3$	3-Nitroanisole	318.2	25.7
$C_7H_7NO_3$	4-Nitroanisole	338.2	26.95
$C_7H_7NO_3S$	4-Methoxy-*N*-sulfinylaniline	293.2	11.85
C_7H_8	Toluene	296.35	2.379
$C_7H_8Cl_2N_2S$	Cyanocyclohexenylimidosulfurous dichloride	298.2	23.2
C_7H_8O	Anisole	294.2	4.30
C_7H_8O	Benzyl alcohol	303.2	11.916
C_7H_8O	*o*-Cresol	298.2	6.76
C_7H_8O	*m*-Cresol	298.2	12.44
C_7H_8O	*p*-Cresol	298.2	13.05
$C_7H_8O_2$	2-Methoxyphenol	298.2	11.95
$C_7H_8O_2$	3-Methoxyphenol	298.2	11.59
$C_7H_8O_2$	4-Methoxyphenol	333.7	11.05
$C_7H_8O_2S$	Methyl phenyl sulfone	373.2	37.9
$C_7H_8O_2S$	Ethyl thiophene-2-carboxylate	293.2	7.50
$C_7H_8O_3$	2-Furfuryl acetate	293.2	5.85
$C_7H_8O_3$	Ethyl furan-2-carboxylate	293.2	9.02
C_7H_8S	Benzylthiol	298.2	4.705
C_7H_8S	4-Methylthiophenol	323.2	4.74
C_7H_8S	Thioanisole	303.2	4.88
C_7H_9N	Benzylamine	293.2	5.18
C_7H_9N	2,4-Dimethylpyridine	293.2	9.60
C_7H_9N	2,6-Dimethylpyridine	293.2	7.33

Molecular formula	Name	*T*/K	ε
C_7H_9N	2-Ethylpyridine	293.2	8.33
C_7H_9N	4-Ethylpyridine	293.2	10.98
C_7H_9N	*N*-Methylaniline	293.2	5.96
C_7H_9N	2-Methylaniline	298.2	6.138
C_7H_9N	3-Methylaniline	298.2	5.816
C_7H_9N	4-Methylaniline	333.2	5.058
C_7H_9NO	2-Methoxyaniline	303.2	5.230
C_7H_9NO	3-Methoxyaniline	298.2	8.76
C_7H_9NO	4-Methoxyaniline	333.2	7.85
C_7H_9NO	2,6-Dimethylpyridine-1-oxide	298.2	46.11
$C_7H_9NO_2S$	*N*-Methylbenzenesulfonamide	303.2	67.1
$C_7H_{10}N_2$	1-Methyl-1-phenylhydrazine	292.2	7.3
$C_7H_{11}F_3$	Trifluoromethylcyclohexane	188.2	11.9
$C_7H_{11}Cl_3O_2$	3-Methyl-1-butyl trichloroacetate	293.2	7.287
C_7H_{12}	Cycloheptene	295.0	2.265
C_7H_{12}	1,6-Heptadiene	293.0	2.161
$C_7H_{12}O$	2-Methylcyclohexanone	293.2	14.0
$C_7H_{12}O$	3-Methylcyclohexanone	293.2	12.4
$C_7H_{12}O$	4-Methylcyclohexanone	293.2	12.35
$C_7H_{12}O$	Cycloheptanone	298.2	13.16
$C_7H_{12}O_2$	Butyl acrylate	301.2	5.25
$C_7H_{12}O_2$	Cyclohexyl formate	293.2	6.47
$C_7H_{12}O_2$	Cyclohexanecarboxylic acid	304.2	2.67
$C_7H_{12}O_2S$	Ethyl tetrahydrothiophene-2-carboxylate	293.2	6.18
$C_7H_{12}O_3$	2-Tetrahydrofurfuryl acetate	293.2	9.65
$C_7H_{12}O_4$	Monomethyl adipate	293.2	6.69
$C_7H_{12}O_4$	Dimethyl glutarate	293.2	7.87
$C_7H_{12}O_4$	Diethyl malonate	304.2	7.550
$C_7H_{12}O_5$	1,2,3-Propanetriol 1,2-diacetate	244.7	18.16
$C_7H_{12}O_5$	1,2,3-Propanetriol 1,3-diacetate	288.2	9.80
$C_7H_{13}ClO_2$	3-Methylbutyl chloroacetate	293.2	7.8
$C_7H_{13}NO$	*N*-Methyl 6-aminohexanoic lactam	293.2	31.0
$C_7H_{13}NO$	3-Pentanoneoxime O-vinyl ether	293.2	2.15
$C_7H_{13}NO_2$	2-Butoxyethyl isocyanate	293.2	9.355
$C_7H_{13}NO_2$	3-Butyloxazolidine-2-one	298.2	52.4
$C_7H_{13}NO_2$	3-*tert*-Butyloxazolidine-2-one	298.2	57.6
$C_7H_{13}NO_2$	Methyl piperidine-4-carboxylate	293.2	8.24
C_7H_{14}	Cycloheptane	293.2	2.0784
C_7H_{14}	3-Ethyl-2-pentene	293.2	2.051
C_7H_{14}	1-Heptene	293.2	2.092
C_7H_{14}	Methylcyclohexane	293.2	2.024
C_7H_{14}	2-Methyl-2-hexene	293.2	2.962
$C_7H_{14}Br_2$	1,2-Dibromoheptane	298.2	3.77
$C_7H_{14}Br_2$	2,3-Dibromoheptane	298.2	5.08
$C_7H_{14}Br_2$	3,4-Dibromoheptane	298.2	4.70
$C_7H_{14}Cl_2$	1,7-Dichloroheptane	298.2	8.34
$C_7H_{14}O$	2-Heptanone	293.2	11.95
$C_7H_{14}O$	3-Heptanone	293.2	12.7
$C_7H_{14}O$	4-Heptanone	293.2	12.60
$C_7H_{14}O$	5-Methyl-2-hexanone	293.2	13.53
$C_7H_{14}O$	Heptanal	295.2	9.07
$C_7H_{14}O$	Cyclohexyl methanol	333.2	9.70
$C_7H_{14}O$	2-Methylcyclohexanol*	293.2	9.375
$C_7H_{14}O$	3-Methylcyclohexanol*	293.2	13.79
$C_7H_{14}O$	4-Methylcyclohexanol*	293.2	13.45
$C_7H_{14}O_2$	Methyl hexanoate	293.2	4.615
$C_7H_{14}O_2$	Ethyl pentanoate	291.2	4.71
$C_7H_{14}O_2$	2,2-Dimethyl pentanoic acid	296.2	2.58

Molecular formula	Name	*T*/K	ε
$C_7H_{14}O_2$	Propyl butanoate	293.2	4.3
$C_7H_{14}O_2$	Pentyl acetate	293.2	4.79
$C_7H_{14}O_2$	Isopentyl acetate	293.2	4.72
$C_7H_{14}O_2$	Ethyl 3-methylbutanoate	293.2	4.71
$C_7H_{14}O_2$	Heptanoic acid	288.2	3.04
$C_7H_{14}O_2$	Butyl propanoate	293.2	4.838
$C_7H_{15}Br$	1-Bromoheptane	303.2	5.255
$C_7H_{15}Br$	2-Bromoheptane	295.2	6.46
$C_7H_{15}Br$	3-Bromoheptane	295.2	6.93
$C_7H_{15}Br$	4-Bromoheptane	295.2	6.81
$C_7H_{15}BrO$	1-Bromo-2-ethoxypentane	298.2	6.45
$C_7H_{15}BrO$	2-Bromo-3-ethoxypentane	298.2	6.40
$C_7H_{15}BrO$	3-Bromo-2-ethoxypentane	298.2	8.24
$C_7H_{15}Cl$	1-Chloroheptane	293.2	5.521
$C_7H_{15}Cl$	2-Chloroheptane	295.2	6.52
$C_7H_{15}Cl$	3-Chloroheptane	295.2	6.70
$C_7H_{15}Cl$	4-Chloroheptane	295.2	6.54
$C_7H_{15}I$	1-Iodoheptane	298.2	4.92
$C_7H_{15}I$	3-Iodoheptane	295.2	6.39
$C_7H_{15}NO$	*N*-Butylpropamide	298.2	100.6
$C_7H_{15}NO$	*N,N*-Dimethylpentanamide	293.2	26.4
$C_7H_{15}NO$	*N,N*-Dipropylformamide	293.2	23.5
$C_7H_{15}NO$	*N,N*-Diisopropylformamide	298.2	24.2
$C_7H_{15}NO$	*N*-Pentylacetamide	298.2	89.9
$C_7H_{15}NO$	*N*-Propylbutanamide	313.2	90.8
C_7H_{16}	2,2-Dimethylpentane	293.2	1.915
C_7H_{16}	2,3-Dimethylpentane	293.2	1.929
C_7H_{16}	2,4-Dimethylpentane	293.2	1.902
C_7H_{16}	3,3-Dimethylpentane	291.3	1.9419
C_7H_{16}	3-Ethylpentane	293.2	1.942
C_7H_{16}	Heptane	293.2	1.9209
C_7H_{16}	2-Methylhexane	293.2	1.9221
C_7H_{16}	3-Methylhexane	293.2	1.920
C_7H_{16}	2,2,3-Trimethylbutane	293.2	1.930
$C_7H_{16}O$	Ethyl pentyl ether	296.2	3.6
$C_7H_{16}O$	Ethyl 3-methylbutyl ether	293.2	3.955
$C_7H_{16}O$	2,2-Dimethyl-1-pentanol	293.2	6.020
$C_7H_{16}O$	3-Ethyl-3-pentanol	293.2	3.158
$C_7H_{16}O$	1-Heptanol	293.2	11.75
$C_7H_{16}O$	(±)-2-Heptanol	293.7	9.72
$C_7H_{16}O$	*D*-(+)-2-Heptanol	293.4	9.67
$C_7H_{16}O$	*L*-(-)-2-Heptanol	293.3	9.65
$C_7H_{16}O$	3-Heptanol	296.1	7.07
$C_7H_{16}O$	4-Heptanol	296.2	6.18
$C_7H_{16}O$	2-Methyl-2-hexanol	297.0	3.257
$C_7H_{16}O$	3-Methyl-2-hexanol	297.2	4.990
$C_7H_{16}O$	3-Methyl-3-hexanol	298.2	3.248
$C_7H_{16}N_2O$	*N,N*-Diethyl-*N′,N′*-dimethyl urea	298.2	17.89
$C_7H_{16}O_3$	Triethoxymethane	293.2	4.779
$C_7H_{16}O_7$	Glucoheptitol	293.2	27.4
$C_7H_{16}S$	1-Heptanethiol	293.2	4.194
$C_7H_{17}N$	Heptylamine	293.2	3.81
$C_7H_{17}O_3P$	Diisopropyl methanephosphonate	298.2	7.72
$C_7H_{17}O_3P$	Diethyl propanephosphonate	303.2	9.542
$C_7H_{17}O_3P$	Diethyl 1-methylethylphosphonate	303.2	8.481
$C_7H_{18}OSi$	Butoxytrimethylsilane	298.2	2.758
$C_7H_{18}OSi$	2-Butoxytrimethylsilane	298.2	2.722
$C_7H_{18}O_3Si$	Triethoxymethylsilane	298.2	3.845

Molecular formula	Name	T/K	ε
$C_7H_{19}NSi$	Isobutyltrimethyl silazane	303.2	2.2523
$C_8Cl_6F_{12}$	Hexachloro-1,1,2,3,3,4,5,5,6,7,7,8-dodecafluorooctane	293.2	2.860
C_8F_{16}	Perflouro-1,3-dimethylcyclohexane	298.2	1.91
$C_8H_2Cl_2F_6$	1,2-Dichloro-3,5-bis(trifluoromethyl)benzene	303.2	3.12
$C_8H_3ClF_6$	2-Chloro-1,3-bis(trifluoromethyl)benzene	303.2	3.20
$C_8H_3ClF_6$	4-Chloro-1,3-bis(trifluoromethyl)benzene	303.2	5.44
$C_8H_4F_3NO$	3-Trifluoromethylphenyl isocyanate	293.2	11.665
$C_8H_4F_6$	1,3-Bis(trifluoromethyl)benzene	303.2	5.98
$C_8H_4F_6O_2S_2$	1-Trifluoromethylsulfonyl-4-(trifluoromethylthio)-benzene	298.2	8.34
$C_8H_4F_6O_4S_2$	1,4-Bis(trifluoromethyl sulfonyl)benzene	298.2	4.11
$C_8H_5N_2O_3P$	Phosphorodiisocyanatidous acid phenyl ester	293.2	5.45
$C_8H_5N_2O_3PS$	Phosphorodiisocyanatidothioic acid O-phenyl ester	293.2	9.92
$C_8H_5N_2O_4P$	Phosphorodiisocyanatidic acid phenyl ester	293.2	8.82
C_8H_6	Phenylacetylene	298.2	2.98
$C_8H_6Cl_2$	2,5-Dichlorostyrene	298.2	2.58
$C_8H_6Cl_4$	3,4,5,6-Tetrachloro-1,2-dimethylbenzene	293.2	8.0
$C_8H_6Cl_4$	2,4,5,6-Tetrachloro-1,3-dimethylbenzene	293.2	5.4
C_8H_6O	Phenoxyacetylene	298.2	4.76
$C_8H_7Cl_2NO_2S$	4-Methoxybenzoylimidosulfurous dichloride	298.2	17.7
$C_8H_7Cl_3$	2,5-Dichloro-2-chloroethylbenzene	297.2	5.20
$C_8H_7F_3O_2S$	4-(Trifluoromethylsulfonyl)toluene	313.2	23.4
$C_8H_7F_3O_3S$	4-(Trifluoromethylsulfonyl)anisole	298.2	29.1
C_8H_7N	Phenylacetonitrile	299.2	17.87
$C_8H_7NO_2$	4-Methoxyphenyl isocyanate	333.2	10.26
$C_8H_7NO_4$	Methyl 2-nitrobenzoate	300.1	27.76
C_8H_8	Styrene	293.2	2.4737
C_8H_8O	Acetophenone	298.2	17.44
$C_8H_8O_2$	Benzyl formate	303.2	6.34
$C_8H_8O_2$	Methyl benzoate	302.7	6.642
$C_8H_8O_2$	Phenyl acetate	298.2	5.403
$C_8H_8O_2$	4-Methoxybenzaldehyde	303.2	22.0
$C_8H_8O_2$	2-Hydroxyacetophenone	298.2	21.33
$C_8H_8O_2$	Phenylacetic acid	353.2	3.47
$C_8H_8O_3$	Methyl salicylate	314.4	8.80
C_8H_9Br	1-Bromo-2-ethylbenzene	298.2	5.55
C_8H_9Br	1-Bromo-3-ethylbenzene	298.2	5.56
C_8H_9Br	1-Bromo-4-ethylbenzene	298.2	5.42
C_8H_9BrO	2-Bromophenetole	313.2	7.04
C_8H_9Cl	2-Chloro-ethylbenzene	298.2	4.36
C_8H_9Cl	3-Chloro-ethylbenzene	298.2	5.18
C_8H_9Cl	4-Chloro-ethylbenzene	298.2	5.16
$C_8H_9NO_2$	1-Phenylethyl nitrite	298.2	7.72
$C_8H_9NO_2$	Ethyl 4-pyridinecarboxylate	293.2	8.95
$C_8H_9NO_2$	2-Nitro-ethylbenzene	273.4	21.9
$C_8H_9NO_2$	Methyl 2-aminobenzoate	298.2	21.9
C_8H_{10}	Ethylbenzene	293.2	2.4463
C_8H_{10}	*o*-Xylene	293.2	2.562
C_8H_{10}	*m*-Xylene	293.2	2.359
C_8H_{10}	*p*-Xylene	293.2	2.2735
$C_8H_{10}N_2OS$	*N,N*-Dimethyl-*N′*-sulfinyl-1,4-benzenediamine	298.2	34.2
$C_8H_{10}O$	2-Methylanisole	293.2	3.502
$C_8H_{10}O$	3-Methylanisole	293.2	3.967
$C_8H_{10}O$	4-Methylanisole	293.2	3.914
$C_8H_{10}O$	Phenetole	293.2	4.216
$C_8H_{10}O$	1-Phenylethanol	293.2	8.77
$C_8H_{10}O$	2-Phenylethanol	293.2	12.31
$C_8H_{10}O$	2,3-Xylenol	343.2	4.81
$C_8H_{10}O$	2,4-Xylenol	303.2	5.060

Molecular formula	Name	*T*/K	ε
$C_8H_{10}O$	2,5-Xylenol	338.2	5.36
$C_8H_{10}O$	2,6-Xylenol	313.2	4.90
$C_8H_{10}O$	3,4-Xylenol	333.2	9.02
$C_8H_{10}O$	3,5-Xylenol	323.2	9.06
$C_8H_{10}OS$	4-Methoxythioanisole	298.2	6.9
$C_8H_{10}O_2$	1,2-Dimethoxybenzene	293.2	4.45
$C_8H_{10}O_2$	1,3-Dimethoxybenzene	298.2	5.363
$C_8H_{10}O_2$	1,4-Dimethoxybenzene	333.7	5.60
$C_8H_{10}O_2S$	Ethyl phenyl sulfone	348.2	39.0
$C_8H_{10}O_2S$	Propyl 2-thiophenecarboxylate	293.2	7.03
$C_8H_{10}O_3$	Propyl 2-furancarboxylate	293.2	8.37
$C_8H_{10}O_3$	2-Furfuryl propanoate	293.2	5.45
$C_8H_{10}S$	4-Methylthioanisole	298.2	4.74
$C_8H_{10}S$	Thiophenetole	298.2	4.95
$C_8H_{11}N$	*N,N*-Dimethylaniline	298.2	4.90
$C_8H_{11}N$	*N*-Ethylaniline	293.2	5.87
$C_8H_{11}N$	4-Ethylaniline	298.2	4.84
$C_8H_{11}N$	*N*-Benzylmethylamine	292.2	4.4
$C_8H_{11}N$	2,4,6-Trimethylpyridine	298.2	7.807
$C_8H_{11}NO$	4-Ethoxyaniline	298.2	7.43
$C_8H_{11}NO_2S$	*N,N*-Dimethylbenzenesulfonamide	323.2	48.6
$C_8H_{11}NO_2S_2$	*S,S*-Dimethyl-*N*-(phenylsulfonyl)sulfilimine	381.2	66.0
$C_8H_{12}N_2O_2$	1,6-Diisocyanatohexane	288.2	14.41
$C_8H_{12}O_4$	Diethyl fumarate	296.2	6.56
$C_8H_{12}O_4$	Diethyl maleate	298.2	7.560
C_8H_{14}	*cis*-Cyclooctene	296.0	2.306
C_8H_{14}	1,2-Dimethylcyclohexene	296.0	2.144
C_8H_{14}	1,3-Dimethylcyclohexene	296.0	2.182
C_8H_{14}	1,7-Octadiene	293.0	2.186
$C_8H_{14}O_2$	Methyl cyclohexanecarboxylate	293.2	4.87
$C_8H_{14}O_2$	Cyclohexyl acetate	293.2	5.08
$C_8H_{14}O_2S$	Propyl tetrahydrothiophene-2-carboxylate	293.2	5.60
$C_8H_{14}O_3$	Butanoic anhydride	293.2	12.8
$C_8H_{14}O_3$	2-Methylpropanoic anhydride	292.2	13.6
$C_8H_{14}O_3$	Tetrahydrofurfuryl propanoate	293.2	8.70
$C_8H_{14}O_4$	Dimethyl adipate	293.2	6.84
$C_8H_{14}O_4$	Diisopropyl oxalate	293.2	6.403
$C_8H_{14}O_4$	Diethyl succinate	293.2	6.098
$C_8H_{14}O_4$	*meso*-2,3-Diacetoxybutane	298.2	6.644
$C_8H_{14}O_4$	(±)-2,3-Diacetoxybutane	298.2	5.10
$C_8H_{15}NO_2$	*N,N*-Diethylacetoacetamide	298.2	40.8
$C_8H_{15}NO_2$	Ethyl piperidine-4-carboxylate	293.2	7.60
$C_8H_{15}N$	Octanenitrile	293.2	13.90
C_8H_{16}	Cyclooctane	295.0	2.116
C_8H_{16}	1-Octene	293.2	2.113
C_8H_{16}	*cis*-3-Octene	298.2	2.062
C_8H_{16}	*trans*-3-Octene	298.2	2.002
C_8H_{16}	*cis*-4-Octene	298.2	2.053
C_8H_{16}	*trans*-4-Octene	298.2	2.004
C_8H_{16}	3-Methyl-2-heptene	293.2	2.436
C_8H_{16}	2,5-Dimethyl-2-hexene	293.2	2.431
C_8H_{16}	2,2,4-Trimethyl-4-pentene	298.2	2.0908
$C_8H_{16}Cl_2$	1,8-Dichlorooctane	298.2	7.64
$C_8H_{16}Br_2$	1,8-Dibromooctane	298.2	7.43
$C_8H_{16}O$	2-Octanone	293.2	9.51
$C_8H_{16}O$	3-Octanone	303.2	10.50
$C_8H_{16}O_2$	Propyl pentanoate	292.2	4.0
$C_8H_{16}O_2$	Butyl butanoate	298.2	4.39

Molecular formula	Name	*T*/K	ε
$C_8H_{16}O_2$	Ethyl hexanoate	293.2	4.45
$C_8H_{16}O_2$	Octanoic acid	288.2	2.85
$C_8H_{16}O_2$	Hexyl acetate	293.2	4.42
$C_8H_{16}O_2$	2-Ethylhexanoic acid	296.2	2.64
$C_8H_{16}O_2$	Methyl heptanoate	293.2	4.355
$C_8H_{16}O_2$	Pentyl propanoate	293.2	4.552
$C_8H_{16}O_3$	3-Methyl-1-butyl lactate	273.2	11.2
$C_8H_{16}O_2$	Isopentyl propanoate	273.2	5.21
$C_8H_{17}Br$	3-Bromomethylheptane	298.2	6.00
$C_8H_{17}Br$	1-Bromooctane	293.2	5.0957
$C_8H_{17}Br$	2-Bromooctane	293.2	5.44
$C_8H_{17}Br$	3-Bromooctane	293.2	6.08
$C_8H_{17}Br$	4-Bromooctane	293.2	5.95
$C_8H_{17}Cl$	1-Chlorooctane	298.2	5.05
$C_8H_{17}Cl$	2-Chlorooctane	293.2	5.42
$C_8H_{17}Cl$	3-Chlorooctane	293.2	6.17
$C_8H_{17}Cl$	4-Chlorooctane	293.2	5.83
$C_8H_{17}F$	1-Fluorooctane	293.2	3.89
$C_8H_{17}F$	2-Fluorooctane	293.2	4.17
$C_8H_{17}F$	3-Fluorooctane	293.2	4.28
$C_8H_{17}F$	4-Fluorooctane	293.2	4.16
$C_8H_{17}I$	1-Iodooctane	293.2	4.67
$C_8H_{17}I$	2-Iodooctane	293.2	5.48
$C_8H_{17}I$	3-Iodooctane	293.2	5.23
$C_8H_{17}I$	4-Iodooctane	293.2	5.42
$C_8H_{17}NO$	*N,N*-Dimethylhexanamide	293.2	22.7
$C_8H_{17}NO$	*N,N*-Dipropylacetamide	293.2	24.5
$C_8H_{17}NO$	*N,N*-Diisopropylacetamide	298.2	23.2
$C_8H_{17}NO_2$	1-Nitrooctane	293.2	11.46
$C_8H_{17}NO_2$	2-Nitrooctane	293.2	14.0
$C_8H_{17}NO_2$	3-Nitrooctane	293.2	15.85
$C_8H_{17}NO_2$	4-Nitrooctane	293.2	15.7
C_8H_{18}	2,2-Dimethylhexane	293.2	1.9498
C_8H_{18}	2,5-Dimethylhexane	293.95	1.9619
C_8H_{18}	3,3-Dimethylhexane	293.2	1.9645
C_8H_{18}	3,4-Dimethylhexane	292.1	1.9814
C_8H_{18}	3-Ethylhexane	293.2	1.9617
C_8H_{18}	3-Ethyl-3-methylpentane	291.49	1.9869
C_8H_{18}	2-Methylheptane	293.2	1.9519
C_8H_{18}	Octane	293.2	1.948
C_8H_{18}	2,2,4-Trimethylpentane	293.2	1.943
C_8H_{18}	2,2,3-Trimethylpentane	293.2	1.960
C_8H_{18}	2,3,4-Trimethylpentane	293.2	1.9738
C_8H_{18}	2,3,3-Trimethylpentane	293.2	1.9780
$C_8H_{18}Cl_2O_3NPS$	Dibutoxyphosphinylimidosulfurous dichloride	293.2	12.21
$C_8H_{18}N_2S$	Thiodi-*tert*-butylimine	298.2	3.41
$C_8H_{18}O$	Dibutyl ether	293.2	3.0830
$C_8H_{18}O$	2,2-Dimethyl-1-hexanol	293.2	4.50
$C_8H_{18}O$	2-Ethyl-1-hexanol	298.2	7.58
$C_8H_{18}O$	2-Isopropyl-1-pentanol	298.2	8.50
$C_8H_{18}O$	2-Methyl-1-heptanol	293.1	5.16
$C_8H_{18}O$	3-Methyl-1-heptanol	290.3	2.884
$C_8H_{18}O$	4-Methyl-1-heptanol	290.6	4.63
$C_8H_{18}O$	5-Methyl-1-heptanol	290.4	7.68
$C_8H_{18}O$	6-Methyl-1-heptanol	290.3	10.54
$C_8H_{18}O$	2-Methyl-2-heptanol	292.2	3.43
$C_8H_{18}O$	3-Methyl-2-heptanol	289.6	7.47
$C_8H_{18}O$	4-Methyl-2-heptanol	290.0	3.59

Molecular formula	Name	*T*/K	ε
$C_8H_{18}O$	5-Methyl-2-heptanol	278.5	7.5
$C_8H_{18}O$	6-Methyl-2-heptanol	290.1	6.41
$C_8H_{18}O$	2-Methyl-3-heptanol	293.2	3.260
$C_8H_{18}O$	3-Methyl-3-heptanol	293.2	3.013
$C_8H_{18}O$	4-Methyl-3-heptanol	293.2	3.312
$C_8H_{18}O$	5-Methyl-3-heptanol	293.2	3.832
$C_8H_{18}O$	6-Methyl-3-heptanol	293.2	4.992
$C_8H_{18}O$	2-Methyl-4-heptanol	296.3	3.338
$C_8H_{18}O$	3-Methyl-4-heptanol	290.0	7.46
$C_8H_{18}O$	4-Methyl-4-heptanol	296.2	2.902
$C_8H_{18}O$	3-Methyl-2-isopropyl-1-butanol	298.2	9.83
$C_8H_{18}O$	1-Octanol	293.2	10.30
$C_8H_{18}O$	2-Octanol	293.2	8.13
$C_8H_{18}O$	3-Octanol	293.2	5.55
$C_8H_{18}O$	4-Octanol	293.2	4.48
$C_8H_{18}OS$	Dibutyl sulfoxide	313.2	24.73
$C_8H_{18}O_2$	2-Ethyl-1,3-hexanediol	293.2	18.73
$C_8H_{18}O_2S$	Dibutyl sulfone	323.2	25.72
$C_8H_{18}O_4$	Triethylene glycol dimethyl ether	298.2	7.62
$C_8H_{18}O_5$	Tetraethylene glycol	293.2	20.44
$C_8H_{18}S$	Dibutyl sulfide	298.2	4.29
$C_8H_{18}S$	1-Octanethiol	293.2	3.949
$C_8H_{19}N$	Dibutylamine	293.2	2.765
$C_8H_{19}N$	Octylamine	293.2	3.58
$C_8H_{19}OP$	Dimethylhexylphosphine oxide	323.2	25.8
$C_8H_{20}Ge$	Tetraethylgermane	274.2	1.971
$C_8H_{20}O_2Si$	Dimethyldipropoxysilane	298.2	2.992
$C_8H_{20}O_2Si$	Dimethylbis(1-methylethoxy)silane	298.2	3.050
$C_8H_{20}O_4Si$	2-Ethylhexyl silicate	293.2	4.1
$C_8H_{20}O_4Si$	Tetraethoxysilane	293.2	2.50
$C_8H_{20}Si$	Dimethyldipropylsilane	293.2	2.054
$C_8H_{20}Si$	Tetraethylsilane	293.2	2.090
$C_8H_{20}Sn$	Tetraethyltin	293.2	2.241
$C_8H_{23}NOSi_2$	*N*-Trimethylsilyl-2-aminoethyl trimethylsilyl ether	293.2	3.045
$C_8H_{23}N_5$	Tetraethylenepentamine	293.2	9.40
$C_8H_{24}O_4Si_4$	Octamethylcyclotetrasiloxane	296.2	2.390
$C_9H_6F_3ClO_2$	Trifluoroethyl 4-chlorobenzoate	298.2	3.96
$C_9H_6N_2O_2$	Toluene-2,4-diisocyanate	293.2	8.433
$C_9H_6N_2O_2$	Toluene-2,6-diisocyanate	293.2	5.147
$C_9H_6O_2$	Coumarin	343.2	34.04
$C_9H_7F_3O_2$	2,2,2-Trifluoroethyl benzoate	298.2	6.98
C_9H_7N	Quinoline	293.2	9.16
C_9H_7N	Isoquinoline	298.2	11.0
C_9H_8O	Cinnamaldehyde	305.8	17.72
$C_9H_8O_4$	Acetylsalicylic acid	333.2	6.55
$C_9H_9Cl_3$	1,2,3-Trichloro-4,5,6-trimethylbenzene	293.2	8.6
$C_9H_9Cl_3$	1,2,5-Trichloro-3,4,6-trimethylbenzene	293.2	6.4
C_9H_9N	2,6-Dimethylbenzonitrile	373.2	16.8
C_9H_{10}	1-Phenylpropene	293.2	2.73
C_9H_{10}	2-Phenylpropene	293.2	2.28
C_9H_{10}	3-Phenylpropene	293.2	2.63
$C_9H_{10}F_3NO_2S$	4-(Trifluoromethylsulfonyl)-*N*,*N*-dimethylaniline	432.2	30.1
$C_9H_{10}O_2$	4-Acetylanisole	313.2	17.3
$C_9H_{10}O_2$	Ethyl benzoate	293.2	6.20
$C_9H_{10}O_2$	Benzyl acetate	303.2	5.34
$C_9H_{10}O_2$	Methyl 2-phenylacetate	297.2	5.30
$C_9H_{10}O_2$	Phenyl propanoate	293.2	4.77
$C_9H_{10}O_2$	Methyl 4-methylbenzoate	306.2	4.3

Molecular formula	Name	T/K	ε
$C_9H_{10}O_2S$	Methyl 4-(methylthio)benzoate	357.2	6.39
$C_9H_{10}O_3$	Ethyl salicylate	308.2	8.48
$C_9H_{10}O_3$	Methyl 2-methoxybenzoate	294.2	7.7
$C_9H_{10}OS$	4-Acetylthioanisole	355.2	11.34
$C_9H_{11}Br$	1-Bromo-3-phenylpropane	302.2	5.41
$C_9H_{11}NO$	*N,N*-Dimethylbenzamide	318.2	20.77
$C_9H_{11}NO$	*N*-Ethylbenzamide	352.7	42.6
$C_9H_{11}NO_2$	Propyl nicotinate	293.2	9.7
$C_9H_{11}NO_2$	Propyl pyridine-4-carboxylate	293.2	8.24
$C_9H_{11}NO_2$	Ethyl 2-aminobenzoate	298.2	4.14
C_9H_{12}	Propylbenzene	293.2	2.370
C_9H_{12}	Isopropylbenzene	293.2	2.381
C_9H_{12}	1,2,3-Trimethylbenzene	293.2	2.656
C_9H_{12}	1,2,4-Trimethylbenzene	293.2	2.377
C_9H_{12}	Mesitylene	293.2	2.279
C_9H_{12}	2-Ethyl-1-methylbenzene	293.2	2.595
C_9H_{12}	3-Ethyl-1-methylbenzene	293.2	2.365
C_9H_{12}	*p*-Ethyltoluene	293.2	2.265
$C_9H_{12}O$	2,6-Dimethylanisole	293.2	3.780
$C_9H_{12}O$	3,5-Dimethylanisole	293.2	3.711
$C_9H_{12}O$	Benzyl ethyl ether	298.2	3.90
$C_9H_{12}O$	1-Phenyl-1-propanol	293.2	6.68
$C_9H_{12}O$	1-Phenyl-2-propanol	293.2	9.35
$C_9H_{12}O$	2-Phenyl-2-propanol	303.2	5.61
$C_9H_{12}O$	3-Phenyl-1-propanol	293.2	11.97
$C_9H_{12}O_2S$	1-Butyl thiophene-2-carboxylate	293.2	6.40
$C_9H_{12}O_3$	2-Furfuryl butanoate	293.2	4.93
$C_9H_{12}O_3$	Butyl furan-2-carboxylate	293.2	7.62
$C_9H_{12}S$	3-Phenyl-1-propanethiol	303.2	4.36
$C_9H_{13}N$	3-Phenylpropylamine	303.2	4.83
$C_9H_{13}N$	*N*-Propylaniline	293.2	5.48
$C_9H_{13}N$	*N*-Ethylbenzylamine	293.2	4.3
$C_9H_{13}N$	*N,N*-Dimethyl-*o*-toluidine	293.2	3.4
$C_9H_{13}N$	*N,N*-Dimethyl-*p*-toluidine	293.2	3.9
$C_9H_{14}OSi$	Trimethylphenoxysilane	298.2	3.3953
$C_9H_{14}O_6$	Triacetin	293.6	7.11
$C_9H_{14}Si$	Trimethylphenylsilane	298.2	2.3533
$C_9H_{15}BO_3$	Triallyl borate	293.2	2.38
$C_9H_{16}O_2$	Ethyl cyclohexanecarboxylate	293.2	4.64
$C_9H_{16}O_2$	2-Nonenoic acid	296.2	2.5
$C_9H_{16}O_2$	Cyclohexyl propanoate	293.2	4.82
$C_9H_{16}O_2S$	Butyl tetrahydrothiophene-2-carboxylate	293.2	5.40
$C_9H_{16}O_3$	2-Tetrahydrofurfuryl butanoate	293.2	8.13
$C_9H_{16}O_4$	Diethyl glutarate	303.2	6.659
$C_9H_{16}O_4$	(±)-*erythro*-2,3-Diacetoxypentane	298.2	6.734
$C_9H_{16}O_4$	(±)-*threo*-2,3-Diacetoxypentane	298.2	5.228
$C_9H_{17}NO_2$	Propyl piperidine-4-carboxylate	293.2	6.90
$C_9H_{17}N$	1-Cyanooctane	293.2	12.08
$C_9H_{17}N$	2-Cyanooctane	293.2	13.76
$C_9H_{17}N$	3-Cyanooctane	293.2	15.10
$C_9H_{17}N$	4-Cyanooctane	293.2	14.83
C_9H_{18}	1-Nonene	293.2	2.180
C_9H_{18}	4-Ethyl-3-heptene	293.2	2.475
C_9H_{18}	2,6-Dimethyl-2-heptene	293.2	2.606
C_9H_{18}	3,6-Dimethyl-3-heptene	293.2	2.343
$C_9H_{18}Br_2$	1,9-Dibromononane	293.2	7.153
$C_9H_{18}O$	2-Nonanone	295.2	9.14
$C_9H_{18}O$	5-Nonanone	293.2	10.6

Molecular formula	Name	T/K	ε
$C_9H_{18}O$	2,6-Dimethyl-4-heptanone	293.2	9.91
$C_9H_{18}O$	2,2,4,4-Tetramethyl-3-pentanone	287.65	10.0
$C_9H_{18}O_2$	Pentyl butanoate	301.2	4.08
$C_9H_{18}O_2$	2-Methylbutyl butanoate	292.2	4.1
$C_9H_{18}O_2$	3-Methylbutyl butanoate	293.2	4.0
$C_9H_{18}O_2$	2-Methyl-1-propyl pentanoate	292.2	3.8
$C_9H_{18}O_2$	Methyl octanoate	293.2	4.101
$C_9H_{18}O_2$	Heptyl acetate	293.2	4.2
$C_9H_{18}O_2$	2-Ethyl-heptanoic acid	293.2	1.98
$C_9H_{18}O_2$	2-Methyl-octanoic acid	293.2	2.39
$C_9H_{18}O_2$	2-Ethyl-2-methyl hexanoic acid	296.2	2.7
$C_9H_{18}O_2$	Nonanoic acid	294.9	2.475
$C_9H_{18}O_2$	3-Propylhexanoic acid	293.2	2.38
$C_9H_{18}O_2$	2,3,4-Trimethylhexanoic acid	296.2	3.40
$C_9H_{18}O_2$	2,3,5-Trimethylhexanoic acid	296.2	2.60
$C_9H_{19}Br$	1-Bromononane	298.2	4.74
$C_9H_{19}Cl$	1-Chlorononane	293.2	4.803
$C_9H_{19}NO$	*N,N*-Dibutylformamide	293.2	18.4
$C_9H_{19}NO$	*N,N*-Diethyl-3-aminopropyl vinyl ether	293.2	2.12
$C_9H_{19}NO$	*N,N*-Dimethylenanthamide	293.2	20.0
C_9H_{20}	Nonane	293.2	1.9722
C_9H_{20}	4-Methyloctane	293.2	1.967
C_9H_{20}	2-Methyloctane	293.2	1.967
C_9H_{20}	2,6-Dimethylheptane	293.2	1.987
C_9H_{20}	2,4-Dimethylheptane	293.2	1.89
C_9H_{20}	2,5-Dimethylheptane	293.2	1.89
$C_9H_{20}O$	1-Nonanol	293.2	8.83
$C_9H_{20}O$	2-Nonanol	298.2	6.66
$C_9H_{20}O$	3-Nonanol	298.2	4.49
$C_9H_{20}O$	4-Nonanol	298.2	3.69
$C_9H_{20}O$	5-Nonanol	298.2	3.54
$C_9H_{20}N_2O$	Tetraethyl urea	296.8	14.29
$C_9H_{20}S$	2-Methyl-2-octanethiol	293.2	4.069
$C_9H_{21}B$	Tripropylborane	293.2	2.026
$C_9H_{21}N$	Nonylamine	293.2	3.42
$C_9H_{21}N$	Tripropylamine	293.2	2.380
$C_9H_{21}O_4P$	Tripropyl phosphate	293.2	10.93
$C_9H_{21}O_4V$	Tripropyl orthovanadate	298.2	2.961
$C_9H_{21}O_4V$	Triisopropyl orthovanadate	298.2	3.299
$C_9H_{21}PS$	Tripropyl phosphine sulfide	293.2	32.8
$C_9H_{22}OSi$	Trimethylhexoxy silane	298.2	2.690
$C_{10}H_6ClF_5O_2$	Pentafluoropropyl 4-chlorobenzoate	298.2	3.80
$C_{10}H_7Br$	1-Bromonaphthalene	298.2	4.768
$C_{10}H_7Cl$	1-Chloronaphthalene	298.2	5.04
$C_{10}H_7ClF_4O_2$	Tetrafluoropropyl 4-chlorobenzoate	298.2	5.21
$C_{10}H_7F_5O_2$	2,2,3,3,3-Pentafluoropropyl benzoate	298.2	6.86
$C_{10}H_7NO_2$	1-Nitronaphthalene	333.2	19.68
$C_{10}H_8$	Naphthalene	363.2	2.54
$C_{10}H_8F_4O_2$	2,2,3,3-Tetrafluoropropyl benzoate	298.2	9.78
$C_{10}H_8N_2O_2$	1,3-Xylylene diisocyanate	293.2	13.787
$C_{10}H_8N_2O_2$	1,4-Xylylene diisocyanate	323.2	15.438
$C_{10}H_8O$	1-Naphthol	373.0	5.03
$C_{10}H_8O$	2-Naphthol	413.0	4.95
$C_{10}H_9N$	2-Methylquinoline	293.2	7.24
$C_{10}H_9N$	4-Methylquinoline	293.2	9.31
$C_{10}H_9N$	6-Methylquinoline	293.2	8.48
$C_{10}H_9N$	8-Methylquinoline	293.2	6.58
$C_{10}H_9N$	1-Naphthylamine	333.2	5.20

Molecular formula	Name	T/K	ε
$C_{10}H_9N$	2-Naphthylamine	393.0	5.26
$C_{10}H_{10}O_4$	Methyl *o*-acetylsalicylate	328.9	5.31
$C_{10}H_{10}O_4$	Dimethyl phthalate	293.2	8.66
$C_{10}H_{12}$	Dicyclopentadiene	313.2	2.43
$C_{10}H_{12}$	4-Ethylstyrene	298.2	3.350
$C_{10}H_{12}$	1,2,3,4-Tetrahydronaphthalene	298.2	2.771
$C_{10}H_{12}O$	Tetrahydro-2-naphthol	293.2	11.70
$C_{10}H_{12}O$	4-Isopropylbenzaldehyde	288.2	10.68
$C_{10}H_{12}O_2$	Propyl benzoate	303.2	5.78
$C_{10}H_{12}O_2$	Phenyl butanoate	293.2	4.48
$C_{10}H_{12}O_2$	2-Phenylethyl acetate	297.2	4.93
$C_{10}H_{12}O_2$	4-Allyl-2-methoxyphenol	293.2	9.55
$C_{10}H_{12}O_2$	Ethyl 2-phenylacetate	293.2	5.320
$C_{10}H_{12}O_2$	Benzyl propanoate	303.0	5.11
$C_{10}H_{13}NO_2$	Butyl pyridine-4-carboxylate	293.2	7.80
$C_{10}H_{13}N$	*N*-Vinyl-*N*-ethylaniline	293.2	7.8
$C_{10}H_{14}$	Butylbenzene	293.2	2.359
$C_{10}H_{14}$	Isobutylbenzene	293.2	2.318
$C_{10}H_{14}$	*sec*-Butylbenzene	293.2	2.357
$C_{10}H_{14}$	*tert*-Butylbenzene	293.2	2.359
$C_{10}H_{14}$	4-Isopropyltoluene	298.2	2.2322
$C_{10}H_{14}$	*o*-Diethylbenzene	293.2	2.594
$C_{10}H_{14}$	*m*-Diethylbenzene	293.2	2.369
$C_{10}H_{14}$	*p*-Diethylbenzene	293.2	2.259
$C_{10}H_{14}$	1-Ethyl-3,5-dimethylbenzene	293.2	2.275
$C_{10}H_{14}$	1,2,3,4-Tetramethylbenzene	296.0	2.538
$C_{10}H_{14}$	1,2,4,5-Tetramethylbenzene	356.0	2.223
$C_{10}H_{14}N_2$	Nicotine	293.2	8.937
$C_{10}H_{14}O$	Butyl phenyl ether	293.2	3.734
$C_{10}H_{14}O$	2-Hydroxy-4-isopropyltoluene	293.2	8.10
$C_{10}H_{14}O$	2-Methyl-1-phenyl-2-propanol	298.2	5.71
$C_{10}H_{14}O$	Thymol	333.2	4.259
$C_{10}H_{14}O_2$	(±)-Camphandione	476.2	16.3
$C_{10}H_{14}O_2S$	Pentyl thiophene-2-carboxylate	293.2	5.85
$C_{10}H_{14}O_3$	Pentyl furan-2-carboxylate	293.2	6.97
$C_{10}H_{14}O_3$	2-Furfuryl pentanoate	293.2	4.73
$C_{10}H_{15}N$	*N,N*-Diethylaniline	303.2	5.15
$C_{10}H_{16}$	2,6-Dimethyl-2,4,6-octatriene, a-form	298.2	2.5574
$C_{10}H_{16}$	Δ-3-Carene	298.2	2.1988
$C_{10}H_{16}$	(R)-(+)-Limonene	298.2	2.3746
$C_{10}H_{16}$	(±)-Limonene	298.2	2.381
$C_{10}H_{16}$	(S)-(-)-Limonene	298.2	2.3738
$C_{10}H_{16}$	Myrcene	298.2	2.3
$C_{10}H_{16}$	α-Pinene	298.2	2.1787
$C_{10}H_{16}$	(±)-α-Pinene	298.2	2.3695
$C_{10}H_{16}$	(1R)-(+)-α-Pinene	298.2	2.3684
$C_{10}H_{16}$	(1S)-(-)-α-Pinene	298.2	2.3701
$C_{10}H_{16}$	β-Pinene	298.2	2.4970
$C_{10}H_{16}$	α-Terpinene	298.2	2.4526
$C_{10}H_{16}$	γ-Terpinene	298.2	2.2738
$C_{10}H_{16}$	Terpinolene	298.2	2.2918
$C_{10}H_{16}O$	Carvenone	293.2	18.8
$C_{10}H_{16}O$	(±)-Fenchone	294.2	12.8
$C_{10}H_{16}O$	Thujone	273.2	10.8
$C_{10}H_{17}Cl$	(±)-2-Chlorobornane	368.2	5.21
$C_{10}H_{18}$	Decahydronaphthalene	293.2	2.1662
$C_{10}H_{18}$	*cis*-Decahydronaphthalene	293.2	2.219
$C_{10}H_{18}$	*trans*-Decahydronaphthalene	293.2	2.184

Molecular formula	Name	T/K	ε
$C_{10}H_{18}$	Pinane	298.2	2.1456
$C_{10}H_{18}O$	1,8-Cineole	298.2	4.57
$C_{10}H_{18}O_2$	Cyclohexyl butanoate	293.2	4.58
$C_{10}H_{18}O_2$	Propyl cyclohexanecarboxylate	293.2	4.44
$C_{10}H_{18}O_2$	6,7-Epoxy-3,7-dimethyl-1-octen-3-ol	297.85	5.78
$C_{10}H_{18}O_2S$	Pentyl tetrahydrothiophene-2-carboxylate	293.2	4.93
$C_{10}H_{18}O_3$	Tetrahydro-2-furfuryl pentanoate	293.2	7.49
$C_{10}H_{18}O_4$	Diethyl adipate	293.2	6.109
$C_{10}H_{19}NO_2$	Butyl piperidine-4-carboxylate	293.2	6.33
$C_{10}H_{20}$	1-Decene	293.2	2.136
$C_{10}H_{20}$	*cis*-5-Decene	298.2	2.071
$C_{10}H_{20}$	*trans*-5-Decene	298.2	2.030
$C_{10}H_{20}$	5-Methyl-4-nonene	293.2	2.175
$C_{10}H_{20}$	2,4,6-Trimethyl-3-heptene	293.2	2.293
$C_{10}H_{20}Br_2$	1,10-Dibromodecane	303.2	6.56
$C_{10}H_{20}Cl_2$	1,10-Dichlorodecane	308.2	6.68
$C_{10}H_{20}O$	2-Decanone	287.2	8.3
$C_{10}H_{20}O$	Menthol	309.3	3.90
$C_{10}H_{20}O_2$	2,2-Dimethyloctanoic acid	296.2	2.8
$C_{10}H_{20}O_2$	Octyl acetate	288.2	4.18
$C_{10}H_{20}O_2$	2-Methylheptyl acetate	288.2	4.27
$C_{10}H_{20}O_2$	1-Pentyl pentanoate	305.6	4.076
$C_{10}H_{20}O_2$	3-Methylbutyl pentanoate	292.2	3.6
$C_{10}H_{20}O_2$	3-Methylbutyl 3-methylbutanoate	288.2	4.39
$C_{10}H_{20}O_2$	Methyl nonanoate	293.2	3.943
$C_{10}H_{20}O_3$	2-Methyl-3-hydroxynonanonic acid	296.2	7.45
$C_{10}H_{21}Br$	1-Bromodecane	298.2	4.44
$C_{10}H_{21}Cl$	1-Chlorodecane	293.2	4.581
$C_{10}H_{21}NO$	*N,N*-Dibutylacetamide	293.2	19.1
$C_{10}H_{21}NO$	*N,N*-Dimethyloctanamide	293.2	17.4
$C_{10}H_{22}$	Decane	293.2	1.9853
$C_{10}H_{22}$	2,7-Dimethyloctane	293.2	1.98
$C_{10}H_{22}$	4-Propylheptane	293.2	1.9955
$C_{10}H_{22}Cl_2O_3NPS$	Di(pentoxy)phosphinylimidosulfurous dichloride	298.2	11.03
$C_{10}H_{22}O$	Dipentyl ether	298.2	2.798
$C_{10}H_{22}O$	Bis(3-methylbutyl) ether	293.2	2.817
$C_{10}H_{22}O$	1-Decanol	293.2	7.93
$C_{10}H_{22}O$	2-Decanol	298.2	5.82
$C_{10}H_{22}O$	(±)-3-Decanol	298.2	4.05
$C_{10}H_{22}O$	4-Decanol	298.2	3.42
$C_{10}H_{22}O$	5-Decanol	298.2	3.24
$C_{10}H_{22}O$	2,2-Dimethyl-1-octanol	293.2	7.86
$C_{10}H_{22}OS$	Dipentyl sulfoxide	348.2	18.8
$C_{10}H_{22}O_2S$	Dipentyl sulfone	348.2	20.0
$C_{10}H_{22}O_5$	Tetraethylene glycol dimethyl ether	298.2	7.68
$C_{10}H_{22}O_6$	Pentaethylene glycol	293.2	18.16
$C_{10}H_{22}S$	Dipentyl sulfide	298.2	3.826
$C_{10}H_{23}N$	Decylamine	293.2	3.31
$C_{10}H_{23}N$	Bis(3-methylbutyl)amine	291.2	2.5
$C_{10}H_{23}O_3P$	Dimethyl octanephosphonate	293.2	7.45
$C_{10}H_{24}O_2Si$	Dibutoxydimethylsilane	298.2	2.836
$C_{10}H_{24}O_2Si$	Bis(1-methylpropoxy)dimethylsilane	298.2	2.966
$C_{10}H_{24}O_3Si$	Tripropoxymethylsilane	298.2	3.339
$C_{10}H_{24}O_3Si$	Tris(1-methylethoxy)methylsilane	298.2	3.255
$C_{10}H_{27}NO_2Si_2$	Bis(trimethylsilaneoxyethyl)amine	313.2	3.572
$C_{10}H_{30}O_3Si_4$	Decamethyltetrasiloxane	293.2	2.370
$C_{10}H_{30}O_5Si_5$	Decamethylcyclopentasiloxane	293.2	2.50
$C_{11}F_{20}$	Perfluoro-1-methyldecalin	298.2	2.0

PERMITTIVITY (DIELECTRIC CONSTANT) OF LIQUIDS (continued)

Molecular formula	Name	*T*/K	ε
$C_{11}H_6ClF_7O_2$	2,2,3,3,4,4,4-Heptafluorobutyl 4-chlorobenzoate	298.2	3.56
$C_{11}H_7F_7O_2$	2,2,3,3,4,4,4-Heptafluorobutyl benzoate	298.00	6.08
$C_{11}H_{10}$	1-Methylnaphthalene	293.2	2.915
$C_{11}H_{10}$	2-Methylnaphthalene	313.2	2.747
$C_{11}H_{10}O$	1-Methoxynaphthalene	293.2	4.020
$C_{11}H_{10}O$	2-Methoxynaphthalene	353.2	3.563
$C_{11}H_{12}O_2$	Ethyl *trans*-cinnamate	293.2	5.63
$C_{11}H_{12}O_3$	Ethyl benzoylacetate	303.2	13.50
$C_{11}H_{14}O_2$	Butyl benzoate	303.2	5.52
$C_{11}H_{14}O_2$	(2-Methylpropyl) benzoate	291.2	5.39
$C_{11}H_{14}O_2$	Phenyl pentanoate	293.2	4.30
$C_{11}H_{14}O_2$	Benzyl butanoate	301.2	4.55
$C_{11}H_{15}Cl$	Chloropentamethylbenzene	293.2	5.8
$C_{11}H_{15}NO_2$	Pentyl pyridine-4-carboxylate	293.2	7.00
$C_{11}H_{16}$	3-*tert*-Butyl-1-methylbenzene	293.2	2.330
$C_{11}H_{16}$	4-*tert*-Butyl-1-methylbenzene	293.2	2.250
$C_{11}H_{16}$	3,5-Diethyltoluene	293.2	2.264
$C_{11}H_{16}$	Pentamethylbenzene	334.0	2.358
$C_{11}H_{16}O_2S$	Hexyl thiophene-2-carboxylate	293.2	5.57
$C_{11}H_{16}O_3$	2-Furfuryl hexanoate	293.2	4.54
$C_{11}H_{16}O_3$	Hexyl furan-2-carboxylate	293.2	6.4
$C_{11}H_{17}N$	*N*-Ethyl-*N*-propylaniline	293.2	4.90
$C_{11}H_{17}O_3P$	Diethyl 4-methylbenzene phosphonate	303.2	11.18
$C_{11}H_{18}N_2O_2$	2,2,4-Trimethylhexamethylene diisocyanate	293.2	10.731
$C_{11}H_{20}O_2$	Butyl cyclohexanecarboxylate	293.2	4.26
$C_{11}H_{20}O_2$	Cyclohexyl pentanoate	293.2	4.32
$C_{11}H_{20}O_2S$	Hexyl tetrahydrothiophene-2-carboxylate	293.2	4.65
$C_{11}H_{20}O_3$	Tetrahydro-2-furfuryl hexanoate	293.2	6.91
$C_{11}H_{20}O_4$	(±)-erythro-3,4-Diacetoxyheptane	298.2	6.684
$C_{11}H_{20}O_4$	(±)-threo-3,4-Diacetoxyheptane	298.2	5.029
$C_{11}H_{21}NO_2$	Pentyl piperidine-4-carboxylate	293.2	5.95
$C_{11}H_{22}$	1-Undecene	293.2	2.137
$C_{11}H_{22}O$	2-Undecanone	285.3	8.3
$C_7H_{14}O_2$	Methyl hexanoate	293.2	3.785
$C_{11}H_{22}O_2$	Pentyl hexanoate	288.2	4.22
$C_{11}H_{22}O_2$	Nonyl acetate	293.2	3.87
$C_{11}H_{22}N_2O_2$	*N,N,N′,N′*-Tetramethylpimelamide	298.2	37.6
$C_{11}H_{22}O_3$	3-Hydroxy-2-methyl decanoic acid	296.2	5.5
$C_{11}H_{23}Br$	1-Bromoundecane	272.6	4.61
$C_{11}H_{23}NO$	*N,N*-Dimethylpelargonamide	293.2	15.8
$C_{11}H_{24}$	Undecane	293.2	1.9972
$C_{11}H_{24}O$	1-Undecanol	313.2	5.98
$C_{11}H_{25}N$	Undecylamine	293.2	3.25
$C_{12}F_{27}OP$	Tris(perfluorobutyl)phosphine oxide	293.2	2.21
$C_{12}F_{27}N$	Tris(perfluorobutyl)amine	293.2	2.15
$C_{12}H_4Cl_6$	Hexachlorodiphenyl*	298.6	4.22
$C_{12}H_5Cl_5$	Pentachlorodiphenyl*	282.2	5.26
$C_{12}H_8F_8O_2$	1*H*,1*H*,5*H*-Octafluoropentyl benzoate	293.2	8.19
$C_{12}H_8Br_2N_2S$	Thiobis(*p*-bromophenyl)imine	298.2	7.7
$C_{12}H_8O$	Dibenzofuran	373.2	3.00
$C_{12}H_{10}$	Biphenyl	348.2	2.53
$C_{12}H_{10}O$	2-Acetonaphthone	333.2	13.03
$C_{12}H_{10}O$	Diphenyl ether	283.2	3.726
$C_{12}H_{10}N_2O$	Azoxybenzene	311.2	5.2
$C_{12}H_{10}OS$	Diphenyl sulfoxide	344.7	16.6
$C_{12}H_{10}O_2S$	Diphenyl sulfone	406.2	21.1
$C_{12}H_{10}N_2S$	Thiobis(*N*-phenyl)imine	333.2	3.93
$C_{12}H_{10}S$	Diphenyl sulfide	298.2	5.43

Molecular formula	Name	T/K	ε
$C_{12}H_{11}NO$	*N*-1-naphthylenylacetamide	433.2	24.3
$C_{12}H_{11}N$	Diphenylamine	323.2	3.73
$C_{12}H_{12}$	1,6-Dimethylnaphthalene	293.2	2.7250
$C_{12}H_{12}O$	1-Ethoxynaphthalene	292.2	3.3
$C_{12}H_{14}F_8O_4$	(1*H*,1*H*,3*H*-Tetrafluoropropyl) 3-methylglutarate	293.2	8.85
$C_{12}H_{14}O_2$	Propyl cinnamate	293.2	5.45
$C_{12}H_{14}O_4$	Diethyl phthalate	293.2	7.86
$C_{12}H_{16}O$	2-Cyclohexylphenol	328.2	3.97
$C_{12}H_{16}O$	4-Cyclohexylphenol	404.2	4.42
$C_{12}H_{16}O_2$	Pentyl benzoate	293.2	5.07
$C_{12}H_{16}O_2$	Phenyl hexanoate	293.2	4.10
$C_{12}H_{16}O_3$	Pentyl salicylate	301.2	6.25
$C_{12}H_{16}O_3$	3-Methylbutyl salicylate	293.12	7.26
$C_{12}H_{17}NO$	*N*-Butylacetanilide	298.2	11.66
$C_{12}H_{17}NO_2$	Hexyl pyridine-4-carboxylate	293.2	6.75
$C_{12}H_{18}$	Hexamethylbenzene	449.0	2.172
$C_{12}H_{18}$	Hexylbenzene	293.2	2.3
$C_{12}H_{18}$	1,3,5-Triethylbenzene	293.2	2.256
$C_{12}H_{18}N_2O$	*N*,*N*-Dipropyl nicotinamide	298.2	2.2770
$C_{12}H_{18}O_2S$	Heptyl thiophene-2-carboxylate	293.2	5.25
$C_{12}H_{18}O_3$	Heptyl furan-2-carboxylate	293.2	5.8
$C_{12}H_{18}O_3$	2-Furfuryl heptanoate	293.2	4.28
$C_{12}H_{20}O_2$	*l*-Bornyl acetate	303.2	4.46
$C_{12}H_{20}O_6$	1,2,3-Propanetriol tripropanoate	212.7	9.34
$C_{12}H_{22}$	6-Dodecyne	298.2	2.171
$C_{12}H_{22}O$	Cyclododecanone	303.2	11.4
$C_{12}H_{22}O$	Dicyclohexyl ether	293.2	3.45
$C_{12}H_{22}O_2$	Cyclohexyl hexanoate	293.2	4.15
$C_{12}H_{22}O_2$	Pentyl cyclohexanecarboxylate	293.2	4.10
$C_{12}H_{22}O_2S$	Heptyl tetrahydrothiophene-2-carboxylate	293.2	4.53
$C_{12}H_{22}O_2Si_2$	1,2-Bis-(trimethylsiloxy)benzene	298.2	2.9767
$C_{12}H_{22}O_2Si_2$	1,3-Bis-(trimethylsiloxy)benzene	298.2	3.7847
$C_{12}H_{22}O_3$	Tetrahydrofurfuryl heptanoate	293.2	6.33
$C_{12}H_{22}O_6$	Dibutyl tartrate	314.2	9.4
$C_{12}H_{24}$	1-Dodecene	293.2	2.152
$C_{12}H_{24}O_2$	Ethyl decanoate	293.2	3.75
$C_{12}H_{24}O_2$	5,9-Dimethyldecanoic acid	296.2	3.0
$C_{12}H_{24}O_2$	Decyl acetate	293.2	3.75
$C_{12}H_{24}O_2$	Methyl undecanoate	293.2	3.671
$C_{12}H_{25}Br$	1-Bromododecane	298.2	4.07
$C_{12}H_{25}Cl$	1-Chlorododecane	298.2	4.17
$C_{12}H_{25}I$	1-Iodododecane	298.2	3.91
$C_{12}H_{25}NO$	*N*,*N*-Dimethyldecanamide	293.2	13.8
$C_{12}H_{25}NO$	*N*,*N*-Dipentylacetamide	293.2	15.8
$C_{12}H_{26}$	Dodecane	293.2	2.0120
$C_{12}H_{26}O$	2-Butyl-1-octanol	363.2	3.28
$C_{12}H_{26}O$	2,2-Dimethyl-1-decanol	293.2	3.27
$C_{12}H_{26}O$	1-Dodecanol	303.2	5.82
$C_{12}H_{26}O_7$	Hexaethylene glycol	293.2	16.00
$C_{12}H_{27}BO_3$	Tributyl borate	293.2	2.23
$C_{12}H_{27}N$	Dodecylamine	303.2	3.07
$C_{12}H_{27}N$	Tributylamine	293.2	2.340
$C_{12}H_{27}OP$	Tributyl phosphine oxide	323.2	26.4
$C_{12}H_{27}O_3P$	Diethyl octanephosphonate	305.2	6.291
$C_{12}H_{27}O_3PS$	Tributyl thiophosphate	298.2	6.92
$C_{12}H_{27}O_4P$	Tributyl phosphate	293.2	8.34
$C_{12}H_{27}O_4V$	Tributyl orthovanadate	298.2	2.780
$C_{12}H_{27}O_4V$	Trisisobutyl orthovandate	298.2	2.761

Molecular formula	Name	*T*/K	ε
$C_{12}H_{27}O_4V$	Tri-*sec*-butyl orthovanadate	298.2	2.699
$C_{12}H_{27}PS$	Tributyl phosphine sulfide	293.2	30.8
$C_{12}H_{28}Ge$	Tetrapropyl germanium	297.2	1.921
$C_{12}H_{28}O_4Si$	Tetrapropoxysilane	298.2	3.21
$C_{12}H_{28}Sn$	Tetrapropyltin	293.2	2.267
$C_{12}H_{30}OSi_2$	Hexaethyl disiloxane	298.2	2.259
$C_{12}H_{30}O_7Si_2$	Hexaethoxy disiloxane	298.2	5.18
$C_{12}H_{36}O_4Si_5$	Dodecamethyl pentasiloxane	293.2	2.476
$C_{12}H_{36}O_6Si_6$	Dodecamethyl cyclohexasiloxane	293.2	2.59
$C_{13}H_9ClN_2OS$	*N*-Benzoyl-*N′*-(*p*-chlorophenyl) thiodiimine	366.2	7.92
$C_{13}H_9FN_2OS$	*N*-Benzoyl-*N′*-(*p*-fluorophenyl) thiodiimine	340.2	7.25
$C_{13}H_{10}O$	Benzophenone	300.2	12.62
$C_{13}H_{10}NO_2P$	Diphenylphosphinic isocyanate	298.2	22.9
$C_{13}H_{10}O_3$	Phenyl salicylate	290.2	6.92
$C_{13}H_{12}$	Diphenylmethane	303.2	2.540
$C_{13}H_{12}O$	Benzyl phenyl ether	313.2	3.748
$C_{13}H_{16}O_2$	Butyl cinnamate	293.2	5.13
$C_{13}H_{18}O_2$	Hexyl benzoate	293.2	4.80
$C_{13}H_{18}O_2$	Phenyl heptanoate	293.2	3.90
$C_{13}H_{18}O_2$	Pentyl 2-phenylacetate	303.2	4.43
$C_{13}H_{19}NO_2$	Heptyl pyridine-4-carboxylate	293.2	6.07
$C_{13}H_{20}$	Heptylbenzene	293.2	2.26
$C_{13}H_{20}O$	α-Ionone	292.4	10.78
$C_{13}H_{20}O$	β-Ionone	297.65	11.66
$C_{13}H_{20}O_2S$	Octyl thiophene-2-carboxylate	293.2	4.93
$C_{13}H_{20}O_3$	2-Furfuryl octanoate	293.2	4.02
$C_{13}H_{20}O_3$	Octyl furan-2-carboxylate	293.2	5.67
$C_{13}H_{24}O$	Cyclotridecanone	303.2	9.34
$C_{13}H_{24}O_2$	Hexyl cyclohexanecarboxylate	293.2	3.94
$C_{13}H_{24}O_2$	Cyclohexyl heptanoate	293.2	3.95
$C_{13}H_{24}O_2S$	Octyl tetrahydrothiophene-2-carboxylate	293.2	4.20
$C_{13}H_{24}O_3$	Tetrahydro-2-furfuryl octanoate	293.2	5.89
$C_{13}H_{24}O_4$	Diethyl azelate	303.2	5.133
$C_{13}H_{25}NO_2$	Heptyl piperidine-4-carboxylate	293.2	5.20
$C_{13}H_{26}$	1-Tridecene	293.2	2.139
$C_{13}H_{26}O$	7-Tridecanone	303.2	7.6
$C_{13}H_{26}O_2$	2,2-Dimethylundecanoic acid	296.2	2.76
$C_{13}H_{26}O_2$	Methyl dodecanoate	293.2	3.539
$C_{13}H_{26}O_2$	Ethyl undecanoate	293.2	3.55
$C_{13}H_{26}O_2$	2,5,9-Trimethyldecanoic acid	296.2	3.04
$C_{13}H_{26}O_3$	3,5,9-Trimethyl-3-hydroxydecanoic acid	296.2	8.75
$C_{13}H_{27}Br$	1-Bromotridecane	281.15	4.19
$C_{13}H_{28}$	5-Butylnonane	293.2	2.0319
$C_{13}H_{28}$	Tridecane	293.2	2.0213
$C_{13}H_{28}O$	1-Tridecanol	333.2	4.02
$C_{13}H_{28}O_4Si_2$	Bis(trimethylsilylmethyl) glutarate	293.2	4.17
$C_{13}H_{30}O_3Si$	Tris(butoxymethyl)silane	298.2	3.138
$C_{13}H_{30}O_3Si$	Tris(*sec*-butoxymethyl)silane	298.2	2.95
$C_{14}H_9Cl_5$	1,1,1-Trichloro-2,2-bis(4-chlorophenyl)ethane	377.2	2.900
$C_{14}H_9F_3O_3N_2S_2$	*N*-Benzoyl-*N′*-(*p*-trifluoro-methylsulfonyl)phenylthiodiimine	343.2	15.9
$C_{14}H_{10}$	Anthracene	502.0	2.649
$C_{14}H_{10}$	Phenanthrene	383.2	2.72
$C_{14}H_{10}ClNO$	1-Chloro-1,1-diphenylmethyl isocyanate	288.2	20.3
$C_{14}H_{10}O_2$	Benzil	368.2	13.04
$C_{14}H_{12}N_2OS$	*N*-Benzoyl-*N′*-(*p*-methylphenyl) thiodiimine	341.2	13.1
$C_{14}H_{12}O_2$	Benzyl benzoate	303.2	5.26
$C_{14}H_{12}O_3$	Benzyl salicylate	301.2	4.12
$C_{14}H_{14}$	1,2-Diphenylethane	331.2	2.47

Molecular formula	Name	T/K	ε
$C_{14}H_{14}$	3,3′-Dimethylbiphenyl	298.15	2.519
$C_{14}H_{14}O$	Dibenzyl ether	293.2	3.821
$C_{14}H_{14}O_3N_2$	4-Azoxyanisole	408.6	5.702
$C_{14}H_{15}N$	Dibenzylamine	293.2	3.446
$C_{14}H_{16}O_2Si$	Dimethyldiphenoxysilane	298.2	3.500
$C_{14}H_{18}O$	Pentyl-3-phenyl-2-propenal*	293.2	11.00
$C_{14}H_{18}O_2$	Pentyl cinnamate	293.2	4.89
$C_{14}H_{20}O_2$	Heptyl benzoate	293.2	4.50
$C_{14}H_{20}O_2$	Phenyl octanoate	293.2	3.72
$C_{14}H_{21}NO_2$	Octyl pyridine-4-carboxylate	293.2	5.80
$C_{14}H_{22}$	Octylbenzene	293.2	2.26
$C_{14}H_{22}O_2S$	Nonyl thiophene-2-carboxylate	293.2	4.77
$C_{14}H_{22}O_3$	Nonyl furan-2-carboxylate	293.2	5.1
$C_{14}H_{26}O_2$	Cyclohexyl octanoate	293.2	3.70
$C_{14}H_{26}O_2$	Heptyl cyclohexanecarboxylate	293.2	3.82
$C_{14}H_{26}O_2S$	Nonyl tetrahydrothiophene-2-carboxylate	293.2	4.02
$C_{14}H_{26}O_3$	Tetrahydro-2-furfuryl pelargonate	293.2	5.67
$C_{14}H_{26}O_4$	Diisobutyl adipate	293.2	5.19
$C_{14}H_{26}O_4$	Diethyl sebacate	303.2	4.995
$C_{14}H_{27}NO_2$	Octyl piperidine-4-carboxylate	293.2	5.05
$C_{14}H_{28}O_2$	Dodecyl acetate	293.2	3.6
$C_{14}H_{28}O_2$	Ethyl laurate	273.2	3.94
$C_{14}H_{28}O_2$	Methyl tridecanoate	293.2	3.442
$C_{14}H_{29}Br$	1-Bromotetradecane	293.2	3.84
$C_{14}H_{30}$	Tetradecane	293.2	2.0343
$C_{14}H_{30}O$	1-Tetradecanol	318.2	4.42
$C_{14}H_{30}OS$	Diheptyl sulfoxide	347.2	13.0
$C_{14}H_{30}O_5Si_2$	Bis(isobutoxymethyl) tetramethyldisiloxane	293.2	4.25
$C_{14}H_{30}O_8$	Heptaethylene glycol	293.2	14.85
$C_{14}H_{31}O_3P$	Diethyl decanephosphonate	293.2	6.06
$C_{14}H_{31}N$	Tetradecylamine	312.55	2.90
$C_{14}H_{42}O_5Si_6$	Tetradecamethyl hexasiloxane	293.2	2.50
$C_{14}H_{42}O_7Si_7$	Tetradecamethyl cycloheptasiloxane	293.2	2.68
$C_{15}F_{33}OP$	Tris(perfluoroisopentyl)phosphine oxide	293.2	2.12
$C_{15}H_9F_3ClNO$	Chlorophenyl(*p*-trifluoromethylphenyl)methyl isocyanate	303.2	11.40
$C_{15}H_{10}N_2O_2$	1,1′-Diphenylmethane diisocyanate	313.2	7.300
$C_{15}H_{12}F_{16}O_4$	Bis(1*H*,1*H*,5*H*-octafluoropentyl) glutarate	293.2	7.48
$C_{15}H_{12}ClNO$	Chlorophenyl(*m*-methylphenyl)methyl isocyanate	303.2	18.59
$C_{15}H_{12}ClNO$	Chlorophenyl(*p*-methylphenyl)methyl isocyanate	288.2	19.4
$C_{15}H_{12}ClNO_2$	Chlorophenyl(*p*-methoxyphenyl)methyl isocyanate	288.2	24.45
$C_{15}H_{12}O_4$	Phenyl acetylsalicylate	384.2	4.33
$C_{15}H_{13}NO$	*N*-Vinylbenzanilide	293.2	3.6
$C_{15}H_{14}F_{12}O_6$	Tris(1*H*,1*H*,3*H*-tetrafluoropropyl) carballate	293.2	10.8
$C_{15}H_{14}O_2$	Benzyl 2-phenylacetate	303.2	4.54
$C_{15}H_{16}O_2S$	Tolyl xylyl sulfone*	297.2	23.5
$C_{15}H_{18}O_2$	Hexyl cinnamate	293.2	4.73
$C_{15}H_{22}O_2$	Octyl benzoate	293.2	4.34
$C_{15}H_{23}NO_2$	Nonyl pyridine-4-carboxylate	293.2	5.55
$C_{15}H_{24}O_2S$	Nonyl phenyl sulfone	323.2	17.6
$C_{15}H_{24}O_2S$	Decyl thiophene-2-carboxylate	293.2	4.57
$C_{15}H_{24}O_3$	Decyl furan-2-carboxylate	293.2	4.93
$C_{15}H_{26}O_6$	Tributyrin	282.8	5.72
$C_{15}H_{28}O_2$	Octyl cyclohexanecarboxylate	293.2	3.70
$C_{15}H_{28}O_3$	Tetrahydro-2-furfuryl decanoate	293.2	5.35
$C_{15}H_{30}O_2$	Methyl myristate	293.2	3.352
$C_{15}H_{31}Br$	1-Bromopentadecane	293.35	3.88
$C_{15}H_{32}$	Pentadecane	293.2	2.0391
$C_{15}H_{32}O$	1-Pentadecanol	333.2	3.70

Molecular formula	Name	T/K	ε
$C_{15}H_{33}O_4V$	Tris(1,1-dimethylpropyl) orthovanandate	298.2	2.764
$C_{15}H_{33}N$	Pentadecylamine	313.25	2.85
$C_{15}H_{33}N$	Tris(3-methyl-1-butyl)amine	294.2	2.29
$C_{15}H_{33}PS$	Tripentylphosphine sulfide	293.2	23.3
$C_{15}H_{39}O_3NSi_3$	Tris-(trimethylsilanoxyethyl)amine	293.2	3.333
$C_{16}H_{14}F_{16}O_4$	Bis(1*H*,1*H*,5*H*-octafluoropentyl) 3-methylglutarate	293.2	7.44
$C_{16}H_{18}N_2O_3$	4-Azoxyphenetole	416.2	5.02
$C_{16}H_{22}O_2$	Heptyl cinnamate	293.2	4.58
$C_{16}H_{22}O_4$	Dibutyl phthalate	293.2	6.58
$C_{16}H_{24}O_2$	Nonyl benzoate	293.2	4.20
$C_{16}H_{24}O_2$	Phenyl decanoate	293.2	3.52
$C_{16}H_{25}NO_2$	Decyl pyridine-4-carboxylate	293.2	5.33
$C_{16}H_{26}O_2S$	Undecyl thiophene-2-carboxylate	293.2	4.40
$C_{16}H_{26}O_3$	Undecyl furan-2-carboxylate	303.2	4.73
$C_{16}H_{30}O_2$	Nonyl cyclohexanecarboxylate	293.2	3.60
$C_{16}H_{30}O_2$	Cyclohexyl decanoate	293.2	3.70
$C_{16}H_{30}O_3$	Tetrahydro-2-furfuryl undecanoate	293.2	5.15
$C_{16}H_{32}O_2$	Tetradecyl acetate	293.2	3.41
$C_{16}H_{32}O_2$	Ethyl myristate	293.2	3.50
$C_{16}H_{32}O_2$	Hexadecanoic acid	338.2	2.417
$C_{16}H_{32}O_2$	Methyl pentadecanoate	293.2	3.296
$C_{16}H_{33}Br$	1-Bromohexadecane	298.2	3.68
$C_{16}H_{33}I$	1-Iodohexadecane	293.2	3.57
$C_{16}H_{34}$	Hexadecane	293.2	2.0460
$C_{16}H_{34}$	6-Pentylundecane	293.2	2.0540
$C_{16}H_{34}O$	1-Hexadecanol	333.2	3.69
$C_{16}H_{34}OS$	Dioctyl sulfoxide	360.2	11.8
$C_{16}H_{34}O_2S$	Dioctyl sulfone	348.2	14.4
$C_{16}H_{34}O_2Si$	Trimethylsilylmethyl laurate	293.2	2.92
$C_{16}H_{34}O_5$	Tetraethylene glycol dibutyl ether	298.2	5.15
$C_{16}H_{35}N$	Hexadecylamine	328.35	2.71
$C_{16}H_{35}O_3P$	Diethyl dodecanephosphonate	305.2	5.162
$C_{16}H_{36}Ge$	Tetrabutyl germanium	297.2	2.334
$C_{16}H_{36}O_4Si$	Tetra(1-methylpropoxy)silane	298.2	2.59
$C_{16}H_{36}O_4Si$	Tetra(2-methylpropoxy)silane	298.2	2.81
$C_{16}H_{36}Sn$	Tetrabutyl tin	293.2	9.74
$C_{16}H_{40}O_{10}Si_3$	Octaethoxytrisiloxane	298.2	4.76
$C_{16}H_{48}O_6Si_7$	Hexadecamethyl heptasiloxane	293.2	2.589
$C_{16}H_{48}O_8Si_8$	Hexadecamethyl cycloheptasiloxane	293.2	2.74
$C_{17}H_{12}O_3$	2-Naphthyl salicylate	293.0	6.30
$C_{17}H_{18}F_{16}O_2$	1*H*,1*H*,9*H*-Hexadecafluorononyl 2-ethylhexanoate	293.2	4.66
$C_{17}H_{24}O_2$	Octyl cinnamate	293.2	4.38
$C_{17}H_{26}O_2$	Decyl benzoate	293.2	4.05
$C_{17}H_{27}NO_2$	Undecyl pyridine-4-carboxylate	295.2	5.15
$C_{17}H_{28}O_2S$	Dodecyl thiophene-2-carboxylate	293.2	4.25
$C_{17}H_{32}O_2$	Decyl cyclohexanecarboxylate	293.2	3.52
$C_{17}H_{32}O_3$	Tetrahydro-2-furfuryl laurate	293.2	4.88
$C_{17}H_{34}O$	9-Heptadecanone	328.2	5.43
$C_{17}H_{34}O_2$	Methyl palmitate	313.2	3.124
$C_{17}H_{36}$	Heptadecane	293.2	2.0578
$C_{17}H_{36}N_2O$	Tetrabutyl urea	298.2	9.41
$C_{17}H_{36}O$	1-Heptadecanol	333.2	3.41
$C_{17}H_{36}O_4Si_2$	Bis(trimethylsilylmethyl) azelate	293.2	4.32
$C_{18}H_{10}F_{16}O_4$	Bis(1*H*,1*H*,5*H*-octafluoropentyl) isophthalate	293.2	7.87
$C_{18}H_{10}F_{16}O_4$	Bis(1*H*,1*H*,5*H*-octafluoropentyl) phthalate	293.2	8.67
$C_{18}H_{12}F_{22}O_4$	Bis(1*H*,1*H*-undecafluorohexyl) 3-methylglutarate	293.2	4.2
$C_{18}H_{14}O_2$	1,3-Diphenoxybenzene	343.2	3.82
$C_{18}H_{15}NO_4S_2$	Phenylsulfonylimidosulfurous acid diphenyl ester	339.2	23.0

Molecular formula	Name	T/K	ε
$C_{18}H_{26}O_4$	Dipentyl phthalate	293.2	6.00
$C_{18}H_{28}O_2$	Phenyl laurate	293.2	3.28
$C_{18}H_{29}NO_2$	Dodecyl pyridine-4-carboxylate	308.2	4.33
$C_{18}H_{30}O$	1-Phenyl-1-dodecanol	333.2	3.30
$C_{18}H_{30}O_2$	Dodecyl benzoate	293.2	3.90
$C_{18}H_{30}O_2$	Linolenic acid	293.2	2.825
$C_{18}H_{30}O_4$	Dicyclohexyl adipate	308.2	4.84
$C_{18}H_{32}O_2$	Linoleic acid	293.2	2.754
$C_{18}H_{32}O_6$	Tributyl tricarballate	293.2	5.7
$C_{18}H_{34}O_2$	Undecyl cyclohexanecarboxylate	293.2	3.45
$C_{18}H_{34}O_2$	Cyclohexyl laurate	293.2	3.50
$C_{18}H_{34}O_2$	Oleic acid	293.2	2.336
$C_{18}H_{34}O_3$	*cis*-Ricinoleic acid	373.2	3.547
$C_{18}H_{34}O_3$	*trans*-Ricinoleic acid	373.2	3.520
$C_{18}H_{34}O_4$	Dibutyl sebacate	293.2	4.54
$C_{18}H_{36}O$	9-Octadecene-1-ol	295.2	3.85
$C_{18}H_{36}O_2$	Cetyl acetate	308.2	3.19
$C_{18}H_{36}O_2$	Methyl heptadecanoate	313.2	3.07
$C_{18}H_{36}O_2$	Ethyl palmitate	303.2	3.07
$C_{18}H_{36}O_2$	Stearic acid	293.2	2.314
$C_{18}H_{37}Br$	1-Bromooctadecane	303.35	3.53
$C_{18}H_{37}NO$	*N,N*-Dioctylacetamide	293.2	11.5
$C_{18}H_{38}O$	1-Octadecanol	333.2	3.38
$C_{18}H_{38}OS$	Dinonyl sulfoxide	359.2	10.0
$C_{18}H_{38}O_2S$	Dinonyl sulfone	358.2	12.1
$C_{18}H_{39}BO_3$	Trihexyl borate	293.2	2.22
$C_{18}H_{39}N$	1-Octadecylamine	326.35	2.67
$C_{18}H_{39}O_3P$	Diethyl tetradecanephosphonate	305.2	4.631
$C_{19}H_{14}F_6O_4$	2,2,3,3,4,4-Hexafluoro-1,5-pentanediol dibenzoate	293.2	6.73
$C_{19}H_{16}$	Triphenylmethane	367.2	2.46
$C_{19}H_{18}O_3Si$	Methyltriphenoxysilane	298.2	3.628
$C_{19}H_{28}O_2$	Decyl cinnamate	293.2	4.05
$C_{19}H_{32}O$	1-Phenyl-1-tridecanol	333.2	3.20
$C_{19}H_{32}O_2$	Methyl linolenate	293.2	3.466
$C_{19}H_{34}O_2$	Methyl linoleate	293.2	3.355
$C_{19}H_{36}O_2$	Dodecyl cyclohexanecarboxylate	293.2	3.37
$C_{19}H_{36}O_2$	Methyl oleate	293.2	3.211
$C_{19}H_{37}NO$	Octadecyl isocyanate	303.2	5.137
$C_{19}H_{38}O$	10-Nonadecanone	353.2	5.37
$C_{19}H_{38}O_2$	Methyl stearate	313.2	3.021
$C_{19}H_{38}O_4$	1,2,3-Propanetriol monopalmitate	340.3	5.34
$C_{19}H_{40}$	7-Hexyltridecane	293.2	2.0715
$C_{19}H_{40}$	Nonadecane	293.2	2.0706
$C_{19}H_{42}O_3Si$	Trihexoxymethylsilane	298.2	2.853
$C_{20}H_{12}F_{12}O_3$	1*H*,1*H*,7*H*-Dodecafluoroheptyl 2-phenoxybenzoate	293.2	7.60
$C_{20}H_{14}F_8O_4$	Bis(1*H*,1*H*,3*H*-tetrafluoropropyl) 2,2′-biphenyldicarboxylate	293.2	9.7
$C_{20}H_{14}F_{24}O_4$	Bis(1*H*,1*H*,7*H*-dodecafluoroheptyl) 3-methylglutarate	293.2	5.94
$C_{20}H_{18}F_{24}O_2$	Bis(1*H*,1*H*,7*H*-dodecafluoroheptoxy)hexane	293.2	9.4
$C_{20}H_{26}O_5Si_2$	Bis(benzoyloxymethyl)tetramethyl disiloxane	293.2	4.88
$C_{20}H_{30}O_4$	Dihexyl phthalate	293.2	5.62
$C_{20}H_{34}O$	1-Phenyl-1-tetradecanol	333.2	3.10
$C_{20}H_{38}O_2$	Ethyl oleate	301.2	3.17
$C_{20}H_{38}O_3$	*cis*-Ethyl ricinoleate	373.2	3.475
$C_{20}H_{38}O_3$	*trans*-Ethyl ricinoleate	373.2	3.775
$C_{20}H_{40}O_2$	Octadecyl acetate	308.2	3.07
$C_{20}H_{40}O_2$	Methyl nonadecanoate	313.2	2.982
$C_{20}H_{40}O_2$	Ethyl stearate	313.2	2.958
$C_{20}H_{42}O$	Didecyl ether	293.2	2.644

Molecular formula	Name	*T*/K	ε
$C_{20}H_{42}O$	1-Eicosanol	338.2	3.13
$C_{20}H_{42}O_2$	Octadecoxy ethanol	328.2	3.56
$C_{20}H_{43}O_3P$	Dihexyl octanephosphonate	298.2	5.15
$C_{20}H_{43}O_3P$	Diethyl hexadecanephosphonate	305.2	4.28
$C_{20}H_{44}O_4Si$	Tetrapentoxysilane	298.2	2.82
$C_{20}H_{44}Ge$	Tetrapentylgermanium	297.2	2.299
$C_{20}H_{60}O_8Si_9$	Eicosamethyl nonasiloxane	293.2	2.645
$C_{21}H_{14}F_{24}O_6$	Tris(1*H*,1*H*,5*H*-octafluoropentyl) tricarballate	293.2	7.81
$C_{21}H_{21}O_4P$	Tricresyl phosphate*	298.2	6.7
$C_{21}H_{32}O_2$	Dodecyl cinnamate	293.2	3.82
$C_{21}H_{36}O$	1-Phenyl-1-pentadecanol	333.2	3.06
$C_{21}H_{38}O_6$	1,2,3-Propanetriyl hexanoate	293.2	4.476
$C_{21}H_{38}O_6$	Tris(3 methyl 1 butyl)tricarballate	293.2	5.1
$C_{21}H_{42}O_3$	2-Methoxyethyl stearate	323.2	3.387
$C_{21}H_{42}O_4$	1,2,3-Propanetriol monostearate	353.2	4.84
$C_{21}H_{44}O_2$	Octadecoxypropanol	328.2	3.86
$C_{21}H_{45}BO_3$	Triheptyl borate	293.2	2.20
$C_{21}H_{45}OP$	Triheptylphosphine oxide	323.2	30.0
$C_{21}H_{45}PS$	Triheptylphosphine sulfide	293.2	20.4
$C_{22}H_{12}F_{30}O_4$	Bis(1*H*,1*H*-pentadecafluorooctyl) 3-methylglutarate	293.2	3.81
$C_{22}H_{30}O_4Si_2$	Bis(trimethylsilylmethyl) 2,2′-biphenyldicarboxylate	293.2	4.50
$C_{22}H_{30}O_5Si_2$	Bis(2-phenylacetoxymethyl)tetramethyldisiloxane	293.2	4.10
$C_{22}H_{34}O_4$	Diheptyl phthalate	293.2	5.22
$C_{22}H_{38}O$	1-Phenyl-1-hexadecanol	333.2	3.00
$C_{22}H_{42}O_2$	*cis*-13-Docosene carboxylic acid	307.8	2.7028
$C_{22}H_{42}O_2$	Butyl oleate	298.2	4.00
$C_{22}H_{42}O_3$	Isobutyl 12-hydroxy-9-octadecenoate	294.2	4.7
$C_{22}H_{44}O_2$	Butyl stearate	298.2	3.120
$C_{22}H_{44}O_2$	2,9,13,17-Tetramethyloctadecanoic acid	296.2	2.72
$C_{22}H_{45}Br$	1-Bromodocosane	315.85	3.20
$C_{22}H_{46}$	Docosane	293.2	2.0840
$C_{22}H_{46}$	8-Heptylpentadecane	293.2	2.1017
$C_{22}H_{46}O$	1-Docosanol	348.2	2.94
$C_{22}H_{46}O_2$	Eicosanoxyethanol	338.2	3.38
$C_{22}H_{47}O_3P$	Dihexyl decanephosphonate	293.2	4.74
$C_{22}H_{47}O_3P$	Diethyl octadecanephosphonate	305.2	4.052
$C_{23}H_{40}O$	1-Phenyl-1-heptadecanol	333.2	2.98
$C_{23}H_{44}O_2$	Methyl *cis*-13-docosenecarboxylate	293.2	3.043
$C_{23}H_{44}O_4$	Dinonyl glutarate	293.2	4.0
$C_{23}H_{46}O_2$	2,5,9,13,17-Pentamethyloctadecanoic acid	296.2	2.76
$C_{23}H_{48}O_2$	Eicosanoxypropanol	333.2	3.63
$C_{23}H_{48}O_4$	Dimethoxydidecaoxy methane	293.2	2.39
$C_{24}H_{12}F_{24}O_6$	Tris(1*H*,1*H*,5*H*-octafluoropentyl) hemimellate	293.2	8.98
$C_{24}H_{14}F_{16}O_4$	Bis(1*H*,1*H*,5*H*-octafluoropentyl) 2,2′-biphenyldicarboxylate	293.2	8.0
$C_{24}H_{15}F_{24}O_2$	Bis(1*H*,1*H*,7*H*-dodecafluoroheptyl) 2-phenylsuccinate	293.2	5.60
$C_{24}H_{20}O_4Si$	Tetraphenoxysilane	333.2	3.4915
$C_{24}H_{30}O_4$	Dibenzyl sebacate	298.2	4.61
$C_{24}H_{38}O_4$	Bis(2-ethylhexyl) phthalate	293.2	4.96
$C_{24}H_{38}O_4$	Dioctyl phthalate	293.2	5.22
$C_{24}H_{42}O$	1-Phenyl 1-octadecanol	333.2	2.93
$C_{24}H_{50}O_2$	Docosanoxyethanol	338.2	3.27
$C_{24}H_{51}BO_3$	Trioctyl borate	293.2	2.16
$C_{24}H_{51}O_3P$	Diheptyl decanephosphonate	293.2	4.77
$C_{24}H_{51}PS$	Trioctylphosphine sulfide	293.2	17.2
$C_{24}H_{72}O_{10}Si_{11}$	Tetracosamethyl undecasiloxane	293.2	2.669
$C_{25}H_{16}F_{24}O_4$	Bis(1*H*,1*H*,7*H*-dodecafluoroheptyl) 2-phenylglutarate	293.2	5.34
$C_{25}H_{44}O$	1-Phenyl-1-nonadecanol	333.2	2.88
$C_{25}H_{52}$	9-Octylheptadecane	293.2	2.1061

Molecular formula	Name	T/K	ε
$C_{25}H_{52}O_2$	Docosanoxypropanol	338.2	3.47
$C_{26}H_{42}O_4$	Bis(3,3,5-trimethylhexyl) phthalate	293.2	4.72
$C_{26}H_{50}O_4$	Bis(2-ethylhexyl) sebacate	298.2	4.03
$C_{26}H_{50}O_4$	Dioctyl sebacate	299.2	4.01
$C_{27}H_{50}O_6$	1,2,3-Propanetriyl octanoate	293.2	3.931
$C_{27}H_{57}BO_3$	Trinonyl borate	293.2	2.20
$C_{27}H_{57}OP$	Trinonylphosphine oxide	323.2	15.4
$C_{28}H_{18}F_{32}O_8$	Tetra(1*H*,1*H*,5*H*-octofluoropentyl) 1,2,3,4-cyclobutanetetracarboxylate	350.2	5.24
$C_{28}H_{56}O_2$	Decyl stearate	313.2	2.81
$C_{28}H_{58}$	10-Nonylnonadecane	293.2	2.102
$C_{29}H_{60}O_4Si_2$	Bis(methyldipentylsilylmethyl)glutarate	293.2	3.19
$C_{30}H_{58}O_4$	Ethylene dimyristate	343.2	2.98
$C_{30}H_{60}O_2$	Tetradecyl palmitate	323.2	2.66
$C_{30}H_{62}$	Triacontane	373.2	1.9112
$C_{30}H_{62}$	Squalane	373.2	1.9106
$C_{30}H_{63}BO_3$	Tridecyl borate	293.2	2.21
$C_{30}H_{63}PS$	Tridecylphosphine sulfide	343.2	11.0
$C_{31}H_{64}$	11-Decylheneicosane	293.2	2.1255
$C_{32}H_{64}O_2$	Tetradecyl stearate	323.2	2.67
$C_{33}H_{62}O_6$	1,2,3-Propanetriol tridecanoate	313.2	3.480
$C_{34}H_{66}$	Tetratriacontadiene*	298.2	2.82
$C_{34}H_{66}O_4$	Ethylene dipalmitate	348.2	2.89
$C_{34}H_{68}O_2$	Cetyl stearate	333.2	2.61
$C_{34}H_{70}$	12-Undecyltricosane	293.2	2.111
$C_{35}H_{68}O_5$	1,2,3-Propanetriol-1,3-dipalmitate	345.2	3.52
$C_{37}H_{76}$	13-Dodecylpentacosane	293.2	2.1203
$C_{38}H_{74}O_4$	Ethylene distearate	353.2	2.79
$C_{39}H_{74}O_6$	Trilaurin	313.2	3.287
$C_{39}H_{76}O_5$	1,2,3-Propanetriol 1,3-distearate	351.2	3.32
$C_{41}H_{84}O_4$	Tetradecoxymethane	293.2	2.44
$C_{48}H_{99}PS$	Trihexadecylphosphine sulfide	378.2	7.7
$C_{51}H_{98}O_6$	Tripalmitin	328.2	2.901
$C_{57}H_{98}O_6$	1,2,3-Propanetriol trilinoleate	293.2	3.470
$C_{57}H_{104}O_6$	*trans, trans, trans*-1,2,3-Propanetriyl 9-octadecenoate	313.2	2.980
$C_{57}H_{104}O_6$	Triolein	293.2	3.109
$C_{57}H_{110}O_6$	Tristearin	353.2	2.740

* Unspecified isomer.

TEMPERATURE DEPENDENCE OF THE PERMITTIVITY (DIELECTRIC CONSTANT) OF LIQUIDS

Christian Wohlfarth

This table describes the temperature dependence of the permittivity (dielectric constant) of about 900 pure liquids (see the preceding table for permittivity values at fixed temperature). The table gives the coefficients of a simple polynomial fitting of permittivity to temperature with an equation of the form

$$\varepsilon(T) = a + bT + cT^2 + dT^3$$

where T is the absolute temperature in K. The temperature range of the fit and the sum of the squares of the deviations divided by the number of points, ESQS/N, are also given. The parameter d was used in only a few cases where the quadratic fit was not satisfactory. When ESQS/N is blank, the number of available data points was equal to the number of fitting parameters.

The coefficients of the fitting equation can be used to calculate dielectric constants within the fitted temperature range. They must not be used for extrapolation without care.

The reader who needs dielectric constant data with more accuracy than can be provided by the fit is referred to Reference 1, which gives the original data together with their literature source.

The static permittivities are given here at nominal atmospheric pressure as long as the corresponding temperature is below the normal boiling point. Otherwise, at temperatures above the normal boiling point, the pressure is the saturated vapor pressure of the substance considered.

Substances are listed by molecular formula in the modified Hill order, with substances not containing carbon preceding those that do contain carbon.

REFERENCES

1. Wohlfarth, Ch., Static Dielectric Constants of Pure Liquids and Binary Liquid Mixtures, *Landolt-Börstein, Numerical Data and Functional Relationships in Science and Technology*, New Series, Editor-in-Chief O. Madelung, Group IV, Macroscopic and Technical Properties of Matter, Volume 6, Springer Verlag Berlin, Heidelberg, New York, 1991.
2. Marsh, K. N., Ed., *Recommended Reference Materials for the Realization of Physicochemical Properties*, Blackwell Scientific Publications, Oxford, 1987.

Molecular formula	Name	*a*	*b*	*c*	*d*	ESQS/N	Range *T*/K
Ar	Argon	0.12408E+01	0.68755E-02	-0.45344E-04		0.5213E-04	87.0-149.1
AsH_3	Arsine	0.37674E+01	-0.97454E-02	0.14537E-04		0.6566E-04	157.1-200.9
B_2H_6	Diborane	0.23848E+01	-0.29501E-02	0.64189E-06		0.1731E-06	108.3-180.7
B_5H_9	Pentaborane(9)	0.40952E+03	-0.24414E+01	0.38225E-02		0.4185E-01	226.2-298.2
BrF_5	Bromine pentafluoride	0.11428E+02	-0.11822E-01			0.7246E-04	261.5-297.7
Br_2	Bromine	0.32701E+01	-0.12535E-03			0.8507E-01	273.2-327.0
Br_3OV	Vanadium oxytribromide	0.61112E+01	-0.84211E-02				203.2-298.2
Br_4Ge	Germanium tetrabromide	0.34450E+01	-0.16083E-02			0.3991E-04	299.9-316.0
Br_4Sn	Tin tetrabromide	0.50001E+01	-0.60383E-02			0.1724E-05	303.5-316.0
$ClFO_3$	Perchloryl fluoride	0.23808E+01	-0.38629E-03	-0.57143E-05		0.2286E-06	125.2-150.2
ClF_3	Chlorine trifluoride	0.96716E+01	-0.18000E-01				273.2-313.2
ClF_5	Chlorine pentafluoride	0.78192E+01	-0.20860E-01	0.13132E-04		0.1498E-03	193.2-256.2
ClH	Hydrogen chloride	0.47316E+02	-0.28455E+00	0.48650E-03		0.3369E-01	158.9-258.2
Cl_2	Chlorine	0.29440E+01	-0.44649E-02	0.30388E-05		0.3874E-05	208.0-240.0
Cl_2F_3P	Phosphorus dichloride trifluoride	0.46501E+01	-0.80358E-02				171.7-228.6
Cl_3F_2P	Phosphorus trichloride difluoride	0.28905E+01	-0.19228E-02				214.9-268.0
Cl_3P	Phosphorus trichloride	0.59098E+01	-0.83322E-02			0.3091E-04	290.2-333.0
Cl_4FP	Phosphorus tetrachloride fluoride	0.33503E+01	-0.29651E-02				244.1-272.6
Cl_4Ge	Germanium tetrachloride	-0.55078E+01	0.64881E-01	-0.13091E-03		0.1299E-03	245.7-273.2
Cl_4Si	Silicon tetrachloride	0.58041E+01	-0.27129E-01	0.51678E-04		0.3275E-04	206.7-273.2
Cl_4Sn	Tin tetrachloride	0.43951E+01	-0.48805E-02			0.1190E-02	233.7-273.2
Cl_4Ti	Titanium tetrachloride	0.33668E+01	-0.19675E-02			0.1033E-03	236.7-257.4
Cl_5Sb	Antimony pentachloride	0.45413E+01	-0.45078E-02			0.3465E-04	275.8-320.4
DH	Hydrogen-d_1	0.12837E+01	-0.10700E-02			0.6236E-04	16.8- 22.2
D_2	Hydrogen-d_2	0.13063E+01	-0.15143E-02			0.3609E-04	17.8- 22.7
D_2O	Water-d_2	0.25232E+03	-0.81067E+00	0.75784E-03		0.7063E-03	277.2-373.2
FH	Hydrogen fluoride	0.50352E+03	-0.19297E+01	0.14372E-02			200.2-273.2
F_2	Fluorine	0.14144E+01	0.26387E-02	-0.28356E-04		0.2438E-03	53.5-144.3
F_5I	Iodine pentafluoride	0.95184E+02	-0.19800E+00				273.2-313.2
$F_{10}S_2$	Sulfur hexafluoride	0.26509E+01	-0.21500E-02			0.9000E-07	263.2-293.2
HI	Hydrogen iodide	0.51557E+03	-0.44552E+01	0.96795E-02			220.2-236.2
H_2	Hydrogen	0.13327E+01	-0.51946E-02			0.3687E-04	13.5- 19.4
H_2O	Water	0.24921E+03	-0.79069E+00	0.72997E-03		0.1184E-02	273.3-372.2

Molecular formula	Name	*a*	*b*	*c*	*d*	ESQS/N	Range *T*/K
H_2O_2	Hydrogen peroxide	0.48511E+03	-0.23145E+01	0.31020E-02		0.3479E-01	233.2-303.2
H_2S	Hydrogen sulfide	0.14736E+02	-0.33675E-01	0.96740E-05		0.1459E-02	212.0-363.2
H_3N	Ammonia	0.66756E+02	-0.24696E+00	0.25913E-03		0.9088E-04	238.2-323.2
H_4N_2	Hydrazine	0.22061E+03	-0.89633E+00	0.11066E-02		0.2980E-03	278.2-323.2
He	Helium	0.10640E+01	-0.35584E-02			0.4670E-06	2.1- 4.2
I_2	Iodine	0.64730E+02	-0.29266E+00	0.39759E-03		0.3503E-03	391.3-440.9
Mn_2O_7	Manganese oxide (Mn_2O_7)	0.37655E+01	-0.16463E-02			0.6028E-05	283.2-312.2
N_2	Nitrogen	0.12550E+01	0.67949E-02	-0.56704E-04		0.2390E-03	63.2-126.2
N_2O_3	Nitrogen trioxide	0.92287E+02	-0.43306E+00	0.65000E-03		0.8000E-05	203.2-243.2
N_2O_4	Nitrogen tetroxide	0.28212E+01	-0.13000E-02			0.6000E-05	253.2-293.2
Ne	Neon	0.12667E+01	-0.29064E-02			0.4057E-08	26.1- 29.0
O_2	Oxygen	0.15434E+01	0.14615E-02	-0.21964E-04		0.1085E-03	54.5-154.0
O_2S	Sulfur dioxide	0.52045E+02	-0.16125E+00	0.11042E-03		0.1637E+00	213.2-449.2
O_3	Ozone	0.86344E+01	-0.54807E-01	0.12596E-03		0.9018E-03	90.2-185.2
P	Phosphorus	0.79018E+00	0.23911E-01	-0.42826E-04		0.2475E-05	307.2-358.2
S	Sulfur	0.51651E+01	-0.77381E-02	0.89120E-05		0.5433E-04	407.2-479.2
Se	Selenium	0.67569E+01	-0.25829E-02			0.4163E-04	510.7-574.6
$CBrClF_2$	Bromochlorodifluoromethane	0.52442E+01	-0.11000E-01			0.1922E-02	123.2-223.2
$CBrCl_3$	Bromotrichloromethane	0.29249E+01	-0.17650E-02			0.3208E-04	273.2-333.2
$CBrF_3$	Bromotrifluoromethane	0.54154E+01	-0.13680E-01				123.2-173.2
CBr_2Cl_2	Dibromodichloromethane	0.32330E+01	-0.23162E-02			0.1126E-06	298.2-333.2
CBr_2F_2	Dibromodifluoromethane	0.67296E+01	-0.22133E-01	0.30213E-04		0.3198E-04	138.5-273.2
CBr_3F	Tribromofluoromethane	0.53203E+01	-0.11061E-01	0.10688E-04		0.8687E-05	205.7-323.2
CBr_3NO_2	Tribromonitromethane	0.16079E+02	-0.23630E-01			0.2011E-03	298.2-328.2
$CClF_3$	Chlorotrifluoromethane	0.43677E+01	-0.11020E-01				123.2-173.2
CCl_2F_2	Dichlorodifluoromethane	0.46984E+01	-0.97600E-02			0.3200E-04	123.2-223.2
CCl_2NOP	Phosphorous dichloride isocyanate	0.69421E+01	-0.95000E-02			0.2222E-04	293.2-333.2
CCl_3F	Trichlorofluoromethane	0.53203E+01	-0.11061E-01	0.10688E-04		0.8687E-05	205.7-323.2
CCl_3NO_2	Trichloronitromethane	0.14403E+02	-0.24178E-01			0.9025E-03	276.2-333.2
CCl_4	Tetrachloromethane	0.28280E+01	-0.20339E-02	0.71795E-07		0.1033E-08	283.2-333.2
CF_4	Tetrafluoromethane	0.20350E+01	-0.27616E-02			0.1176E-05	126.3-141.5
COS	Carbon oxysulfide	0.84702E+01	-0.21488E-01			0.1371E-02	143.0-185.0
COSe	Carbon oxyselenide	0.48740E+01	-0.49425E-02			0.1515E-04	219.2-283.2
CO_2	Carbon dioxide	0.79062E+00	0.10639E-01	-0.28510E-04		0.4024E-04	220.0-300.0
CS_2	Carbon disulfide	0.45024E+01	-0.12054E-01	0.19147E-04		0.7347E-03	154.2-319.4
$CHBr_3$	Tribromomethane	0.71707E+01	-0.98000E-02			0.1502E-03	283.2-343.2
$CHCl_3$	Trichloromethane	0.15115E+02	-0.51830E-01	0.56803E-04		0.5163E-04	218.2-323.2
CHF_3	Trifluoromethane	0.11442E+03	-0.75600E+00	0.13562E-02		0.2121E+00	130.0-263.0
CHN	Hydrogen cyanide	0.37331E+04	-0.23180E+02	0.36963E-01		0.6864E+00	258.2-298.9
CH_2Br_2	Dibromomethane	0.18060E+02	-0.36333E-01				283.2-313.2
CH_2Cl_2	Dichloromethane	0.40452E+02	-0.17748E+00	0.23942E-03		0.4335E-02	184.1-306.0
CH_2F_2	Difluoromethane	0.19428E+03	-0.12939E+01	0.24280E-02		0.1725E-01	152.2-224.2
CH_2O_2	Formic acid	0.14040E+03	-0.24673E+00	-0.17151E-03		0.8184E-01	286.7-358.2
CH_3Br	Bromomethane	0.40580E+02	-0.18418E+00	0.26219E-03		0.1898E-03	194.6-275.7
CH_3Cl	Chloromethane	0.42775E+02	-0.16175E+00	0.17108E-03		0.3439E-01	190.2-392.2
CH_3ClO_2S	Methanesulfonyl chloride	0.10384E+03	-0.33838E+00	0.34156E-03		0.5282E-03	293.2-373.2
CH_3DO	Methan-d_1-ol	0.20839E+03	-0.10318E+01	0.14740E-02		0.5753E-01	176.3-297.5
CH_3F	Fluoromethane	0.11338E+03	-0.63979E+00	0.96983E-03		0.2596E+00	150.0-299.0
CH_3I	Iodomethane	0.24264E+02	-0.93914E-01	0.11926E-03		0.1538E-02	223.2-303.2
CH_3NO	Formamide	0.26076E+03	-0.61145E+00	0.34296E-03		0.2016E-02	278.2-333.2
CH_3NO_2	Methyl nitrite	0.11071E+03	-0.73428E+00	0.14054E-02		0.6757E+00	110.0-260.0
CH_3NO_2	Nitromethane	0.11227E+03	-0.35591E+00	0.34206E-03		0.2039E-01	288.2-343.2
CH_4	Methane	0.15996E+01	0.27434E-02	-0.22086E-04		0.6061E-04	91.0-184.0
CH_4O	Methanol	0.19341E+03	-0.92211E+00	0.12839E-02		0.6618E-01	177.2-293.2
CH_5N	Methylamine	0.34398E+02	-0.73630E-01	-0.41279E-04		0.3172E-01	198.2-258.2
$C_2Cl_2F_4$	1,2-Dichlorotetrafluoroethane	0.36663E+01	-0.42271E-02	-0.36255E-06		0.2399E-07	193.2-273.2
C_2HCl_2NO	Dichloromethyl isocyanate	0.14757E+02	-0.25667E-01				288.2-348.2
C_2HCl_3	Trichloroethylene	0.58319E+01	-0.80828E-02			0.1025E-04	301.5-337.6
$C_2HCl_3F_2$	1,2,2-Trichloro-1,1-difluoroethane	0.75423E+01	-0.11667E-01			0.5000E-04	303.2-333.2
$C_2HCl_3O_2$	Trichloroacetic acid	0.13412E+01	0.90000E-02	-0.24130E-14			333.2-393.2
C_2HCl_5	Pentachloroethane	0.65972E+01	-0.96800E-02			0.5392E-04	298.2-338.2
$C_2HF_3O_2$	Trifluoroacetic acid	0.21652E+02	-0.68146E-01	0.78571E-04		0.2041E-04	263.2-323.2
$C_2H_2Br_4$	1,1,2,2-Tetrabromoethane	0.16246E+02	-0.31500E-01			0.5375E-03	303.2-333.2
$C_2H_2Cl_2O_2$	Dichloroacetic acid	0.11014E+02	-0.10859E-01	0.49242E-05		0.8608E-05	284.0-363.2

TEMPERATURE DEPENDENCE OF THE PERMITTIVITY (DIELECTRIC CONSTANT) OF LIQUIDS (continued)

Molecular formula	Name	*a*	*b*	*c*	*d*	ESQS/N	Range *T*/K
$C_2H_2Cl_4$	1,1,1,2-Tetrachloroethane	0.19606E+02	-0.49847E-01			0.3148E-02	207.2-233.2
$C_2H_3ClO_2$	Chloroacetic acid	0.17310E+02	-0.14674E-01			0.7689E-05	338.2-393.2
$C_2H_3Cl_2NO_2$	1,1-Dichloro-1-nitroethane	0.37576E+02	-0.70400E-01			0.4305E-02	303.2-333.2
$C_2H_3Cl_3$	1,1,1-Trichloroethane	0.27705E+02	-0.10621E+00	0.12424E-03		0.1899E-05	258.2-318.2
$C_2H_3Cl_3$	1,1,2-Trichloroethane	0.17147E+02	-0.33371E-01			0.9082E-05	288.2-318.2
$C_2H_3F_3O$	2,2,2-Trifluoroethanol	0.90593E+02	-0.21421E+00			0.6122E-01	293.2-318.2
C_2H_3N	Acetonitrile	0.29724E+03	-0.15508E+01	0.22591E-02		0.7082E-01	288.2-333.2
C_2H_4	Ethylene	0.13546E+01	0.62614E-02	-0.21374E-04		0.6934E-05	200.0-270.0
C_2H_4BrCl	1-Bromo-2-chloroethane	0.19493E+02	-0.59054E-01	0.58036E-04		0.1224E-03	263.2-363.2
$C_2H_4Br_2$	1,2-Dibromoethane	0.67142E+01	-0.59800E-02			0.2689E-06	293.2-313.2
$C_2H_4Cl_2$	1,1-Dichloroethane	0.24429E+02	-0.48000E-01			0.2250E-03	288.2-318.2
$C_2H_4Cl_2$	1,2-Dichloroethane	0.24404E+02	-0.47892E-01			0.2261E-02	293.2-343.2
C_2H_4O	Ethylene oxide	0.52661E+02	-0.21337E+00	0.25947E-03		0.2614E-03	293.2-243.2
$C_2H_4O_2$	Methyl formate	0.19699E+02	-0.36429E-01				288.2-302.2
$C_2H_4O_2$	Acetic acid	-0.15731E+02	0.12662E+00	-0.17738E-03		0.2367E-02	293.2-363.2
$C_2H_4O_3S$	Ethylene sulfite	0.85483E+02	-0.15400E+00			0.3000E-02	298.2-328.2
C_2H_5Br	Bromoethane	0.28473E+02	-0.85495E-01	0.67971E-04		0.9485E-04	243.2-308.2
C_2H_5Cl	Chloroethane	0.60693E+02	-0.31290E+00	0.47154E-03		0.7954E-03	237.2-293.2
C_2H_5ClO	2-Chloroethanol	0.11155E+03	0.30149E+00			0.2877E-01	140.2-175.2
C_2H_5I	Iodoethane	0.25598E+02	-0.94367E-01	0.11424E-03		0.3382E-02	183.2-343.2
C_2H_5N	Ethyleneimine	0.61405E+02	-0.14474E+00			0.4298E-02	273.2-298.2
C_2H_5NO	Acetamide	-0.20055E+03	0.15515E+01	-0.22392E-02		0.1046E+00	363.7-448.2
C_2H_5NO	*N*-Methylformamide	0.10383E+04	-0.43165E+01	0.48398E-02		0.5075E+00	276.2-353.2
$C_2H_5NO_2$	Methyl carbamate	0.36773E+02	-0.55700E-01			0.1998E-02	328.2-368.2
$C_2H_5NO_2$	Nitroethane	0.57406E+02	-0.97657E-01			0.1890E-01	276.2-333.2
C_2H_6	Ethane	0.20185E+01	-0.51493E-03	-0.48148E-05		0.5398E-04	95.0-295.0
C_2H_6O	Dimethyl ether	0.22389E+02	-0.86524E-01	0.91291E-04		0.1023E-01	155.0-258.0
C_2H_6O	Ethanol	0.15145E+03	-0.87020E+00	0.19570E-02	-0.15512E-05	0.1369E+00	163.2-523.2
C_2H_6OS	Dimethyl sulfoxide	0.38478E+02	0.16939E+00	-0.47423E-03		0.3291E-01	288.2-343.2
$C_2H_6O_2$	Ethylene glycol	0.14355E+03	-0.48573E+00	0.46703E-03		0.5617E-02	293.2-423.2
$C_2H_6O_2S$	Dimethyl sulfone	0.10830E+03	-0.15900E+00			0.1875E-03	383.2-398.2
$C_2H_6S_2$	1,2-Ethanedithiol	0.11228E+02	-0.13500E-01			0.2000E-03	293.2-333.2
$C_2H_6S_2$	Dimethyl disulfide	0.19109E+02	-0.32000E-01			0.8889E-03	298.2-323.2
C_2H_7N	Ethylamine	0.30163E+02	-0.79000E-01			0.1420E-01	233.2-273.2
C_2H_7NO	Ethanolamine	0.14890E+03	-0.62491E+00	0.77143E-03		0.5449E-02	253.2-293.2
$C_2H_7NO_2S$	*N*-Methyl methanesulfonamide	0.54340E+03	-0.22958E+01	0.27619E-02		0.4422E-02	298.2-328.2
$C_2H_8N_2$	1,2-Ethanediamine	0.48922E+02	-0.17021E+00	0.17262E-03		0.1150E-02	273.2-333.2
C_3Cl_6O	Hexachloroacetone	0.76423E+01	-0.15838E-01	0.10618E-04		0.1350E-05	268.7-303.3
C_3F_6O	Hexafluoroacetone	0.34809E+01	-0.92883E-02	0.12282E-04		0.1605E-04	150.6-238.1
C_3HN	Cyanoacetylene	0.91803E+03	-0.49149E+01	0.69104E-02		0.1980E-01	280.8-313.6
$C_3H_2F_6O$	1,1,1,3,3,3-Hexafluoro-2-propanol	0.51961E+03	0.31421E+01	0.48667E-02			283.2-313.2
C_3H_3N	Propenenitrile	0.11109E+03	-0.36806E+00	0.34879E-03		0.3168E-01	233.2-413.2
C_3H_4	Allene	0.26049E+01	-0.44147E-03	-0.63420E-05		0.1019E-04	156.0-269.0
C_3H_4	Propyne	0.60871E+01	-0.11730E-01			0.4998E-03	185.0-246.0
$C_3H_4ClF_3$	3-Chloro-1,1,1-trifluoropropane	0.22361E+02	-0.68840E-01	0.60594E-04		0.2374E-03	275.2-313.2
C_3H_4ClNO	1-Chloro-2-isocyanatoethane	0.64311E+02	-0.12217E+00				288.2-403.2
$C_3H_4N_2O_2$	3-Methylsydnone	0.75278E+03	-0.30204E+01	0.34389E-02		0.5346E-01	313.2-398.2
C_3H_4O	Propargyl alcohol	0.99895E+02	-0.38911E+00	0.40776E-03		0.5853E-02	213.2-293.2
$C_3H_4O_3$	1,3-Dioxolan-2-one	0.20746E+03	-0.37610E+00			0.1402E-01	313.2-343.2
$C_3H_5Br_3$	1,2,3-Tribromopropane	0.11024E+02	-0.16596E-01			0.3222E-03	303.2-358.2
$C_3H_5ClO_2$	Ethyl chlorocarbonate	0.15356E+02	-0.18250E-01			0.1105E-03	308.7-349.2
C_3H_5N	Propanenitrile	0.82222E+02	-0.22937E+00	0.17424E-03		0.4097E-01	213.2-473.2
C_3H_6	Propene	0.29623E+01	-0.37564E-02			0.2492E-06	220.0-250.0
$C_3H_6Br_2$	1,2-Dibromopropane	0.54973E+01	-0.31695E-02			0.1144E-03	283.2-333.2
$C_3H_6Br_2$	1,3-Dibromopropane	0.29193E+02	-0.94450E-01	0.92800E-04		0.7220E-04	293.2-368.2
C_3H_6ClNO	*N*-Methyl-2-chloroacetamide	0.30695E+03	-0.66475E+00			0.5042E-01	323.2-348.2
$C_3H_6Cl_2$	1,2-Dichloropropane	0.18915E+02	-0.35907E-01			0.1981E-02	281.2-323.2
$C_3H_6Cl_2$	1,3-Dichloropropane	0.21609E+02	-0.37333E-01			0.8000E-03	303.2-333.2
$C_3H_6Cl_2$	2,2-Dichloropropane	0.32421E+02	-0.72188E-01			0.8805E-02	244.7-293.2
C_3H_6O	Acetone	0.88157E+02	-0.34300E+00	0.38925E-03		0.9183E-03	273.2-323.2
C_3H_6O	Allyl alcohol	0.62714E+02	-0.14771E+00	0.37879E-05		0.7439E-03	213.2-303.2
$C_3H_6O_2$	Ethyl formate	0.15884E+02	-0.25333E-01			0.3556E-03	288.2-318.2
$C_3H_6O_2$	Methyl acetate	0.13190E+02	-0.21226E-01			0.2439E-03	276.2-318.2
$C_3H_6O_2$	Propanoic acid	0.18793E+01	0.46841E-02	0.19983E-05		0.4416E-03	289.2-408.2

Molecular formula	Name	*a*	*b*	*c*	*d*	ESQS/N	Range *T*/K
$C_3H_6O_3S$	1,2-Oxathiolane-2,2-dioxide	0.24989E+03	-0.72897E+00	0.64286E-03		0.2286E-03	308.2-348.2
C_3H_7Br	1-Bromopropane	0.17769E+02	-0.32599E-01			0.5212E-02	274.2-328.2
C_3H_7Br	2-Bromopropane	0.26195E+02	-0.72995E-01	0.55454E-04		0.9752E-02	186.3-328.2
C_3H_7Cl	1-Chloropropane	0.21214E+02	-0.43130E-01			0.3194E-03	273.2-313.2
C_3H_7ClO	3-Chloro-1-propanol	0.12436E+03	-0.60841E+00	0.92060E-03		0.1844E+00	145.2-215.2
C_3H_7ClO	1-Chloro-2-propanol	-0.19169E+02	0.13605E+01	-0.55567E-02		0.1472E+00	153.2-177.2
C_3H_7I	1-Iodopropane	0.13744E+02	-0.22745E-01			0.4062E-04	293.2-323.3
C_3H_7NO	*N,N*-Dimethylformamide	0.15364E+03	-0.60367E+00	0.71505E-03		0.3503E-01	213.2-353.2
C_3H_7NO	*N*-Ethylformamide	0.64764E+03	-0.28499E+01	0.34286E-02		0.3657E-02	298.2-338.2
C_3H_7NO	*N*-Methylacetamide	0.15975E+04	-0.90451E+01	0.18345E-01	-0.12998E-04	0.3900E-01	303.2-473.2
$C_3H_7NO_2$	Ethyl carbamate	0.32431E+02	-0.65097E-01	0.28571E-04		0.5531E-03	328.2-368.2
$C_3H_7NO_2$	1-Nitropropane	0.94999E+02	-0.38358E+00	0.48480E-03		0.2608E-02	276.2-333.2
$C_3H_7NO_2$	2-Nitropropane	0.60138E+02	-0.11566E+00			0.2164E-02	276.2-303.2
$C_3H_7NO_2$	Propyl nitrite	0.70552E+02	-0.40362E+00	0.66687E-03		0.4240E+00	110.0-310.0
$C_3H_7NO_2$	Isoropyl nitrite	0.74578E+02	-0.38283E+00	0.57071E-03		0.5473E-01	150.0-300.0
C_3H_8	Propane	0.22883E+01	-0.23276E-02	0.84710E-06		0.2970E-04	90.0-300.0
C_3H_8O	1-Propanol	0.98045E+02	-0.36860E+00	0.36422E-03		0.2544E+00	193.2-493.2
C_3H_8O	2-Propanol	0.10416E+03	-0.41011E+00	0.42049E-03		0.1587E+00	193.2-493.2
$C_3H_8O_2$	Dimethoxymethane	0.25877E+01	-0.93019E-03	0.38472E-05		0.6436E-05	170.7-293.2
$C_3H_8O_2$	2-Methoxyethanol	0.11803E+03	-0.58000E+00	0.81001E-03		0.2102E-01	253.6-318.4
$C_3H_8O_2$	1,2-Propanediol	0.24546E+03	-0.15738E+01	0.38068E-02	-0.32544E-05	0.2557E-01	193.2-403.2
$C_3H_8O_2$	1,3-Propanediol	0.11365E+03	-0.36680E+00	0.33766E-03		0.1060E-01	288.2-328.2
$C_3H_8O_3$	Glycerol	0.77503E+02	-0.37984E-01	-0.23107E-03		0.7294E-01	288.2-343.2
C_3H_8S	1-Propanethiol	0.11602E+02	-0.19580E-01			0.2421E-03	273.2-318.2
$C_3H_8S_2$	1,2-Propanedithiol	0.14667E+02	-0.32660E-01	0.25000E-04			293.2-333.2
$C_3H_8S_2$	1,3-Propanedithiol	0.66607E+01	0.31310E-01	-0.87500E-04			303.2-343.2
C_3H_9ClSi	Trimethylchlorosilane	-0.19492E+02	0.29806E+00	-0.69284E-03		0.2640E-02	223.2-273.2
C_3H_9N	Propylamine	0.17719E+02	-0.59022E-01	0.54780E-04		0.3082E-02	204.2-296.2
C_3H_9N	Isopropylamine	0.40429E+02	-0.21441E+00	0.32634E-03		0.7288E-03	213.2-298.2
C_3H_9N	Trimethylamine	0.39745E+01	-0.51331E-02			0.1833E-04	273.2-298.2
$C_3H_9O_3P$	Dimethyl methanephosphate	0.74253E+02	-0.24790E+00	0.24107E-03		0.3143E-03	293.2-373.2
$C_4F_6Cl_4$	1,1,3,4-Tetrachlorohexafluorobutane	0.52319E+01	-0.11882E-01	0.12891E-04		0.4905E-04	203.7-313.2
$C_4H_4N_2$	Succinonitrile	0.17724E+03	-0.54654E+00	0.54046E-03		0.1137E+00	235.5-351.4
C_4H_4O	Furan	0.13636E+01	0.12864E-01	-0.22701E-04		0.9970E-02	187.5-277.1
C_4H_4S	Thiophene	0.32941E+01	-0.19019E-02			0.1433E-04	252.7-293.2
C_4H_5N	1H-Pyrrole	0.12672E+02	-0.14075E-01	-0.62671E-05		0.7743E-04	293.0-356.7
C_4H_5NO	Allyl isocynate	0.34299E+02	-0.66444E-01				288.2-333.2
$C_4H_5N_2O_3P$	Diisocyanato ethyl phosphite	0.11820E+02	-0.16400E-01				298.2-323.2
$C_4H_5N_2O_3PS$	Diisocyanato *O*-ethyl thionphosphate	0.39572E+02	-0.12459E+00	0.12054E-03		0.6786E-03	293.2-393.2
$C_4H_5N_2O_4P$	Diisocyanato ethyl phosphate	0.24312E+02	-0.35000E-01			0.2500E-02	293.2-353.2
C_4H_6	1,3-Butadiene	0.27674E+01	-0.26738E-02			0.5654E-04	185.0-265.0
$C_4H_6Cl_2N_2S$	(1-Cyano-1-isopropyl)imido sulfurous dichloride	0.38834E+02	-O.58095E-01				293.2-398.2
$C_4H_6N_2OS$	2-Methyl-2-(sulfinylamino)-propanenitrile	0.35288E+02	-0.86192E-01	0.71875E-04		0.6122E-02	293.2-392.2
C_4H_6O	Cyclobutanone	0.43974E+02	-0.15712E+00	0.19264E-03		0.2857E-02	220.2-317.2
$C_4H_6O_2$	Methyl acrylate	0.11968E+02	-0.16500E-01			0.3588E-02	303.2-333.2
$C_4H_6O_2$	2,3-Butanedione	0.46907E+01	-0.22302E-02			0.8205E-04	278.2-348.2
$C_4H_6O_3$	4-Methyl-1,3-dioxolan-2-one	0.15940E+03	-0.39530E+00	0.26284E-03		0.1181E-02	273.0-333.0
$C_4H_7BrO_2$	Ethyl 2-bromoacetate	0.15627E+02	-0.19600E-01			0.3780E-02	303.2-333.2
$C_4H_7BrO_2$	Methyl 2-bromopropanoate	0.19850E+02	-0.31250E-01			0.1250E-02	303.2-343.2
$C_4H_7BrO_2$	Methyl 3-bromopropanoate	0.36001E+01	0.72500E-02			0.2722E-03	303.2-343.2
$C_4H_7ClO_2$	Methyl 2-chloropropanoate	0.22449E+02	-0.36250E-01			0.1389E-03	303.2-343.2
C_4H_7N	Butanenitrile	0.53884E+02	-0.99257E-01			0.1554E-02	293.2-333.2
C_4H_7N	2-Methylpropanenitrile	0.52554E+02	-0.96000E-01			0.3556E-03	293.2-313.2
C_4H_7NO	Pyrrolidine-2-one	0.11054E+03	-0.47945E+00	0.68182E-03		0.8182E-03	298.2-338.2
C_4H_8	1-Butene	0.29354E+01	-0.32580E-02			0.7206E-06	220.0-250.0
C_4H_8	*cis*-2-Butene	0.28802E+01	-0.31064E-02			0.5256E-04	197.0-296.0
C_4H_8	Isobutene	0.33701E+01	-0.43295E-02			0.8003E-05	220.0-288.7
C_4H_8BrCl	3-Bromo-1-chloro-2-methylpropane	0.32838E+02	-0.10590E+00	0.88889E-04			303.2-333.2
$C_4H_8Br_2$	1,2-Dibromobutane	0.11199E+03	-0.63334E+00	0.91250E-03			293.2-333.2
$C_4H_8Br_2$	1,3-Dibromobutane	0.34031E+02	-0.13254E+00	0.16250E-03			293.2-333.2
$C_4H_8Br_2$	1,4-Dibromobutane	0.20944E+02	-0.55620E-01	0.50000E-04		0.4500E-04	303.2-333.2
$C_4H_8Br_2$	2,3-Dibromobutane	0.23849E+02	-0.96300E-01	0.12500E-03			293.2-333.2

Molecular formula	Name	*a*	*b*	*c*	*d*	ESQS/N	Range *T*/K
$C_4H_8Cl_2$	1,2-Dichlorobutane	0.31925E+02	-0.13232E+00	0.17007E-03		0.6993E-03	293.2-356.2
$C_4H_8Cl_2$	1,4-Dichlorobutane	0.59766E+01	0.49300E-01	-0.12500E-03		0.3125E-04	308.2-338.2
$C_4H_8Cl_2$	1,2-Dichloro-2-methylpropane	0.39429E+02	-0.20028E+00	0.30917E-03		0.1837E-01	165.2-296.0
C_4H_8O	2-Butanone	0.15457E+02	0.90152E-01	-0.27100E-03		0.1405E-02	293.2-333.2
C_4H_8O	Tetrahydrofuran	0.30739E+02	-0.12946E+00	0.17195E-03		0.2233E-03	224.2-295.2
C_4H_8OS	Tetrahydrothiophene-*S*-oxide	0.11842E+03	-0.34315E+00	0.30071E-03		0.5846E-02	298.2-398.2
$C_4H_8O_2$	Butanoic acid	0.15010E+01	0.50046E-02			0.9836E-03	287.2-403.2
$C_4H_8O_2$	1,4-Dioxane	0.27299E+01	-0.17440E-02			0.6960E-07	293.2-313.2
$C_4H_8O_2$	Ethyl acetate	0.15646E+02	-0.44066E-01	0.39137E-04		0.1251E-03	293.2-433.2
$C_4H_8O_2$	Methyl propanoate	0.12798E+02	-0.22540E-01			0.1420E-02	293.2-333.2
$C_4H_8O_2S$	Tetrahydrothiophene-*S,S*-dioxide	0.10981E+03	-0.30973E+00	0.29809E-03		0.3962E-02	303.2-353.2
C_4H_9Br	1-Bromobutane	0.22542E+02	-0.79306E-01	0.89867E-04		0.1719E-02	183.2-363.2
C_4H_9Br	1-Bromo-2-methylpropane	0.37558E+02	-0.20571E+00	0.35496E-03		0.2884E-01	111.9-273.2
C_4H_9Br	2-Bromobutane	0.18461E+02	-0.32933E-01			0.4880E-05	274.2-328.2
C_4H_9Br	2-Bromo-2-methylpropane	0.35085E+02	-0.14075E+00	0.19960E-03		0.4872E-03	258.0-293.0
C_4H_9Cl	1-Chlorobutane	0.13565E+02	-0.10161E-01	-0.38750E-04		0.4762E-03	273.2-323.2
C_4H_9Cl	1-Chloro-2-methylpropane	0.14945E+02	-0.33747E-01	0.23036E-04		0.4888E-04	273.2-323.2
C_4H_9Cl	2-Chlorobutane	0.30376E+02	-0.11377E+00	0.13429E-03		0.1653E-03	273.2-323.2
C_4H_9Cl	2-Chloro-2-methylpropane	0.35077E+02	-0.12867E+00	0.14304E-03		0.4425E-03	273.2-323.2
C_4H_9I	1-Iodobutane	0.16493E+02	-0.50262E-01	0.52485E-04		0.3476E-05	293.2-323.2
C_4H_9I	2-Iodobutane	0.10883E+02	-0.14680E-02	-0.30000E-04		0.8000E-06	293.2-323.2
C_4H_9I	2-Iodo-2-methylpropane	0.76780E+01	0.69900E-02	-0.37500E-04			283.2-323.2
C_4H_9N	Pyrrolidine	0.38191E+02	-0.15462E+00	0.17941E-03		0.6758E-04	274.0-333.0
C_4H_9NO	*N,N*-Dimethylacetamide	0.15420E+03	-0.57506E+00	0.61911E-03		0.3225E-01	294.2-433.2
$C_4H_9NO_2$	Ethyl-*N*-methyl carbamate	0.11477E+03	-0.47568E+00	0.54127E-03		0.1989E-02	298.2-373.2
$C_4H_9NO_2$	*N*-(2-Hydroxyethyl)acetamide	0.37016E+03	-0.13113E+01	0.13214E-02		0.1190E-02	298.2-348.2
$C_4H_9NO_2$	Propyl carbamate	0.24356E+02	-0.36400E-01			0.9440E-03	338.2-378.2
C_4H_{10}	Butane	0.22379E+01	-0.13884E-02	-0.66711E-06		0.3929E-06	135.0-303.2
C_4H_{10}	Isobutane	0.23295E+01	-0.19953E-02	0.14197E-06		0.2146E-05	115.0-303.2
$C_4H_{10}O$	1-Butanol	0.10578E+03	-0.50587E+00	0.84733E-03	-0.48841E-06	0.1053E-01	193.2-553.2
$C_4H_{10}O$	2-Butanol	0.13850E+03	-0.75146E+00	0.14086E-02	-0.89512E-06	0.4298E-01	172.2-533.2
$C_4H_{10}O$	2-Methyl-1-propanol	0.10762E+03	-0.51398E+00	0.83702E-03	-0.45299E-06	0.8274E-01	173.2-533.2
$C_4H_{10}O$	2-Methyl-2-propanol	0.22541E+03	-0.14990E+01	0.34050E-02	-0.25968E-05	0.2272E-01	298.2-503.2
$C_4H_{10}O$	Diethyl ether	0.79725E+01	-0.12519E-01			0.2822E-02	283.2-301.2
$C_4H_{10}O_2$	1,2-Butanediol	0.63702E+02	-0.13807E+00			0.1069E-01	278.2-323.2
$C_4H_{10}O_2$	1,3-Butanediol	0.72883E+02	-0.14770E+00			0.8115E-02	278.2-323.2
$C_4H_{10}O_2$	1,4-Butanediol	0.13079E+03	-0.46985E+00	0.46320E-03		0.1417E-02	288.2-328.2
$C_4H_{10}O_2$	1,2-Dimethoxyethane	0.48832E+02	-0.24218E+00	0.34413E-03		0.3811E-02	255.6-318.4
$C_4H_{10}O_2S$	Bis(2-hydroxyethyl) sulfide	0.13128E+03	-0.52719E+00	0.60465E-03		0.2327E-02	253.2-333.2
$C_4H_{10}O_3$	Diethylene glycol	0.13973E+03	-0.54725E+00	0.61149E-03		0.1135E-02	288.2-343.2
$C_4H_{10}S$	1-Butanethiol	0.11201E+02	-0.20767E-01			0.8438E-04	273.2-318.2
$C_4H_{10}S$	2-Butanethiol	0.10866E+02	-0.17993E-01			0.6267E-03	273.2-318.2
$C_4H_{10}S$	2-Methyl-2-propanethiol	0.10597E+02	-0.17500E-01			0.1935E-03	283.2-313.2
$C_4H_{11}N$	Butylamine	0.13322E+02	-0.44176E-01	0.50250E-04		0.3162E-03	223.2-333.2
$C_4H_{11}N$	Diethylamine	0.26462E+02	-0.13750E+00	0.20373E-03		0.8237E-03	243.2-323.2
$C_4H_{11}NO_2$	Diethanolamine	0.73435E+02	-0.21377E+00	0.17500E-03		0.7870E-02	273.2-323.2
$C_4H_{11}OP$	Ethyldimethylphosphine oxide	0.76085E+02	-0.11500E+00			0.1389E-01	363.2-383.2
$C_4H_{12}NO_3P$	Phosphoramidic acid diethyl ester	-0.24973E+02	0.40569E+00	-0.78311E-03		0.6669E-02	313.2-383.2
$C_4H_{12}N_2O$	2-(2-Aminoethylamino)ethanol	0.10819E+03	-0.48865E+00	0.65979E-03		0.6209E-01	223.2-323.2
$C_4H_{13}N_3$	Bis(2-amimoethyl)amine	0.57840E+02	-0.23873E+00	0.28841E-03		0.1541E-02	213.2-333.2
C_5H_4ClN	2-Chloropyridine	0.98702E+02	-0.34237E+00	0.34502E-03		0.1033E-02	298.2-398.2
C_5H_4BrN	2-Bromopyridine	0.73391E+02	-0.23678E+00	0.22930E-03		0.2277E-03	298.2-398.2
C_5H_5N	Pyridine	0.43991E+02	-0.15150E+00	0.15925E-03		0.7021E-05	293.2-323.2
C_5H_5NO	Pyridine-1-oxide	0.20878E+02	0.16450E+00	-0.35269E-03		0.9213E-01	343.0-398.0
$C_5H_7N_2O_4P$	Diisocyanato isopropyl phosphate	0.13872E+02	-0.30859E-01	0.28201E-04		0.9850E-05	293.0-398.0
$C_5H_7N_2O_3PS$	Diisocyanato *O*-isopropyl-thionphosphate	0.41564E+02	-0.14562E+00	0.15873E-03		0.2286E-03	288.0-348.0
$C_5H_7N_2O_4P$	Diisocyanato isopropyl phosphate	0.40521E+02	-0.13936E+00	0.16000E-03			298.0-373.0
C_5H_8	Cyclopentene	0.28177E+01	-0.27597E-02	0.89346E-06		0.2447E-05	171.0-319.0
C_5H_8	2-Methyl-1,3-butadiene	0.28170E+01	-0.23147E-02	-0.43975E-06		0.9675E-05	198.2-293.2
C_5H_8	1,4-Pentadiene	0.29994E+01	-0.34578E-02	0.85300E-06		0.2110E-04	178.0-294.0
$C_5H_8N_2OS$	2-Methyl-2-(sulfinylamino)-butanenitrile	0.37682E+02	-0.10926E+00	0.10536E-03		0.2011E-03	293.2-373.2
$C_5H_8N_2O_2$	3-Isopropylsydnone	0.20651E+03	-0.58628E+00	0.49403E-03		0.4210E-03	333.2-398.2

TEMPERATURE DEPENDENCE OF THE PERMITTIVITY (DIELECTRIC CONSTANT) OF LIQUIDS (continued)

Molecular formula	Name	*a*	*b*	*c*	*d*	ESQS/N	Range *T*/K
C_5H_8O	Cyclopentanone	0.24083E+02	-0.30286E-01	-0.16802E-04		0.2072E-02	219.2-298.2
$C_5H_8O_2$	Ethyl acrylate	0.47827E+02	-0.24394E+00	0.35000E-03		0.2560E-03	303.2-343.2
$C_5H_8O_2$	Methyl crotonate	0.16248E+02	-0.45167E-01	0.42571E-04		0.2684E-05	293.2-343.2
$C_5H_8O_2$	Methyl methacrylate	0.32098E+02	-0.14568E+00	0.20000E-03		0.1120E-03	303.2-343.2
$C_5H_8O_4$	Dimethyl malonate	0.26470E+02	-0.76656E-01	0.67888E-04		0.1435E-01	293.2-433.2
$C_5H_9ClO_2$	Methyl 4-chlorobutanoate	0.17127E+02	-0.25000E-01			0.2689E-02	303.2-343.2
$C_5H_9ClO_2$	Ethyl 2-chloropropanoate	0.25965E+02	-0.46250E-01			0.1389E-03	303.2-343.2
$C_5H_9ClO_2$	Ethyl 3-chloropropanoate	0.21951E+02	-0.38750E-01			0.2722E-03	303.2-343.2
C_5H_9N	2,2-Dimethylpropane nitrile	0.58418E+02	-0.16884E+00	0.14131E-03		0.5891E-03	293.2-453.2
C_5H_9N	Pentanenitrile	0.55793E+02	-0.15750E+00	0.12432E-03		0.5378E-01	182.7-333.2
C_5H_9NO	Butyl isocyanate	0.50564E+02	-0.19487E+00	0.21940E-03		0.4483E-05	293.2-353.2
C_5H_9NO	Isobutyl isocyanate	0.38026E+02	-0.12714E+00	0.12679E-03		0.1083E-02	293.2-353.2
C_4H_9NO	*N*-Ethylacetamide	0.74494E+03	-0.31400E+01	0.36131E-02		0.6774E-01	213.2-353.2
C_5H_{10}	Cyclopentane	0.24287E+01	-0.15304E-02	-0.13095E-06		0.6057E-08	278.2-313.2
C_5H_{10}	2-Methyl-2-butene	0.26064E+01	-0.19578E-02	-0.53908E-06		0.1541E-05	225.0-296.0
C_5H_{10}	1-Pentene	-0.11438E+01	0.25420E-01	-0.50000E-04			273.2-293.2
$C_5H_{10}Br_2$	1,4-Dibromopentane	0.26443E+02	-0.88640E-01	0.10000E-03			293.2-333.2
$C_5H_{10}Br_2$	1,5-Dibromopentane	0.38192E+02	-0.15648E+00	0.20000E-03		0.2450E-03	303.2-333.2
$C_5H_{10}Cl_2$	1,2-Dichloropentane	0.19016E+02	-0.57954E-01	0.56801E-04		0.5883E-03	293.2-356.2
$C_5H_{10}N_2O$	1,3-Dimethyl-imidazolidin-2-one	0.12262E+03	-0.40480E+00	0.40121E-03		0.3515E-03	298.2-373.2
$C_5H_{10}O$	Cyclopentanol	0.10565E+03	-0.44244E+00	0.48657E-03		0.3078E-02	258.2-323.2
$C_5H_{10}O$	3-Pentanone	0.12690E+02	0.95177E-01	-0.27321E-03		0.1063E+00	233.2-353.2
$C_5H_{10}O$	2-Pentanone	0.40893E+02	-0.10423E+00	0.60557E-04		0.1952E-01	204.2-353.2
$C_5H_{10}O$	3-Methyl-2-butanone	0.30695E+02	-0.10962E+00	0.13810E-03		0.2095E-02	293.2-328.2
$C_5H_{10}O$	Tetrahydropyran	0.19793E+02	-0.76071E-01	0.94852E-04		0.1003E-02	234.0-333.2
$C_5H_{10}O$	2,2-Dimethylpropanal	0.18645E+02	-0.32395E-01	-0.16157E-05		0.8880E-03	280.2-333.2
$C_5H_{10}O_2$	Butyl formate	0.21532E+02	-0.84106E-01	0.10952E-03			288.2-323.2
$C_5H_{10}O_2$	Methyl butanoate	0.38604E+02	-0.19171E+00	0.27128E-03		0.2243E-03	301.2-343.2
$C_5H_{10}O_2$	Propyl acetate	0.17677E+02	-0.61404E-01	0.69196E-04		0.1821E-04	253.2-353.2
$C_5H_{10}O_2$	Pentanoic acid	0.33491E+01	-0.75156E-02	0.17820E-04		0.2526E-03	250.4-343.5
$C_5H_{10}O_2S$	3-Methyl sulfolane	0.53158E+02	-0.93730E-01	0.47275E-04		0.1758E-03	298.2-398.2
$C_5H_{10}O_2S$	Methyl 3-(methylthio)propanoate	0.18059E+02	-0.31000E-01	-0.23141E-13			303.2-343.2
$C_5H_{10}O_3$	Ethyl lactate	0.31225E+02	-0.43531E-01	-0.28571E-04			273.2-373.2
$C_5H_{10}O_4$	1,2,3-Propanetriol 1-acetate	0.10653E+03	-0.26439E+00	-0.62371E-04		0.4024E-01	215.1-242.2
$C_5H_{11}Br$	1-Bromopentane	0.20954E+02	-0.78743E-01	0.98908E-04		0.2956E-02	182.9-328.2
$C_5H_{11}Br$	1-Bromo-3-methylbutane	0.27743E+02	-0.13927E+00	0.22627E-03		0.7417E-02	122.5-291.5
$C_5H_{11}Cl$	1-Chloropentane	0.18626E+02	-0.54719E-01	0.47143E-04		0.1366E-03	273.2-323.2
$C_5H_{11}Cl$	1-Chloro-3-methylbutane	0.22228E+02	-0.93189E-01	0.12991E-03		0.7145E-03	170.7-296.8
$C_5H_{11}Cl$	2-Chloro-2-methylbutane	0.55104E+02	-0.29866E+00	0.47840E-03		0.1025E-03	201.3-222.8
$C_5H_{11}I$	1-Iodopentane	0.15753E+02	-0.50543E-01	0.56401E-04		0.5103E-05	293.2-323.2
$C_5H_{11}N$	Piperidine	0.82317E+01	-0.11229E-01	-0.71429E-05		0.4571E-05	293.0-333.0
$C_5H_{11}NO$	2,2-Dimethylpropionamide	0.10400E+03	-0.46017E+00	0.60000E-03			298.2-328.2
$C_5H_{11}NO$	*N*-Ethylpropanamide	0.10118E+04	-0.48000E+01	0.61429E-02		0.2714E-01	298.2-328.2
$C_5H_{11}NO$	*N*-Propylacetamide	0.66775E+03	-0.27964E+01	0.31905E-02		0.4728E-01	298.2-328.2
$C_5H_{11}NO_2$	*N*-(2-Methoxyethyl)acetamide	0.40999E+03	-0.16529E+01	0.18393E-02		0.3726E-02	298.2-348.2
C_5H_{12}	Isopentane	0.22384E+01	-0.12985E-02	-0.16182E-06		0.4882E-05	142.7-293.2
C_5H_{12}	Neopentane	0.10949E+02	-0.63057E-01	0.10835E-03		0.1509E-03	251.0-296.0
$C_5H_{12}O$	2,2-Dimethyl-1-propanol	0.92350E+02	-0.41870E+00	0.50000E-03			333.2-373.2
$C_5H_{12}O$	2-Methyl-1-butanol	0.14020E+02	0.13948E+00	-0.45000E-03		0.6050E-03	288.2-318.2
$C_5H_{12}O$	2-Methyl-2-butanol	0.11662E+03	-0.69756E+00	0.10920E-02		0.5226E-02	268.2-318.2
$C_5H_{12}O$	3-Methyl-1-butanol	0.79733E+02	-0.31272E+00	0.32014E-03		0.5792E-01	173.2-513.2
$C_5H_{12}O$	1-Pentanol	0.73397E+02	-0.28165E+00	0.28427E-03		0.4917E-01	213.2-513.2
$C_5H_{12}O$	2-Pentanol	0.16437E+03	-0.86506E+00	0.11955E-02		0.7775E-02	273.2-323.2
$C_5H_{12}O$	3-Pentanol	0.12838E+03	-0.60980E+00	0.75000E-03		0.8000E-02	288.2-318.2
$C_5H_{12}O_2$	1,2-Pentanediol	0.18436E+03	-0.10682E+01	0.17037E-02		0.5330E-01	197.2-296.8
$C_5H_{12}O_2$	1,4-Pentanediol	0.13568E+03	-0.59198E+00	0.75398E-03		0.3712E-01	193.2-317.8
$C_5H_{12}O_2$	1,5-Pentanediol	0.11858E+03	-0.45920E+00	0.49341E-03		0.1002E-01	243.2-343.2
$C_5H_{12}O_2$	2,3-Pentanediol	0.95876E+02	-0.46463E+00	0.67434E-03			237.8-296.9
$C_5H_{12}O_2$	2,4-Pentanediol	0.11914E+03	-0.52569E+00	0.69607E-03		0.1773E-01	223.7-294.2
$C_5H_{12}O_2$	Diethoxymethane	0.25294E+01	0.73988E-04	-0.28331E-06		0.1583E-05	227.2-293.2
$C_5H_{12}S$	2-Methyl-2-butanethiol	0.15116E+02	-0.50700E-01	0.56250E-04		0.2000E-06	273.2-333.2
$C_5H_{12}S$	1-Pentanethiol	0.71131E+01	-0.30228E-02	-0.16414E-04		0.5249E-03	273.2-333.2
$C_5H_{13}ClNO_3P$	Diethyl *N*-chloro-*N*-methyl-amidophosphate	0.18399E+02	-0.48520E-01	0.50000E-04		0.5000E-05	293.2-353.2

Molecular formula	Name	*a*	*b*	*c*	*d*	ESQS/N	Range *T*/K
$C_5H_{13}N$	Pentylamine	0.11274E+02	-0.34965E-01	0.37706E-04		0.1098E-03	223.2-353.2
$C_5H_{18}B_{10}$	Isopropyl carborane	0.77985E+02	-0.11250E+00				293.2-373.2
C_6F_6	Hexafluorobenzene	0.24041E+01	-0.83086E-03	-0.14286E-05		0.1829E-06	298.2-338.2
$C_6F_9Cl_5$	Pentachloro-1,1,2,3,3,4,5,5,6-nonafluorohexane	0.53413E+01	-0.13644E-01	0.16309E-04		0.5296E-04	203.2-323.2
$C_6F_{15}OP$	Tris(perfluoroethyl)phosphine oxide	0.34362E+01	-0.35000E-02				313.2-353.2
$C_6H_4Br_2$	*o*-Dibromobenzene	-0.81849E-02	0.62671E-01	-0.12222E-03			293.2-353.2
$C_6H_4Br_2$	*m*-Dibromobenzene	0.93214E+01	-0.20273E-01	0.16667E-04			293.2-353.2
C_6H_4ClNOS	3-Chloro-*N*-sulfinylaniline	0.18958E+02	-0.58646E-01	0.60714E-04		0.3181E-03	293.2-393.2
$C_6H_4ClNO_2$	*o*-Chloronitrobenzene	0.16800E+03	-0.59708E+00	0.59957E-03		0.4105E-02	323.2-436.2
$C_6H_4ClNO_2$	*m*-Chloronitrobenzene	0.77193E+02	-0.25118E+00	0.23798E-03		0.2184E-02	323.2-433.2
$C_6H_4Cl_2$	*o*-Dichlorobenzene	0.13629E+02	0.10622E-02	-0.44444E-04			293.2-353.2
$C_6H_4Cl_2$	*m*-Dichlorobenzene	0.77565E+01	-0.93333E-02	-0.26880E-14			293.2-353.2
$C_6H_4Cl_2$	*p*-Dichlorobenzene	0.26999E+01	-0.35325E-03	-0.17619E-05		0.3961E-07	328.2-363.2
$C_6H_4F_2$	*o*-Difluorobenzene	0.59107E+02	-0.23611E+00	0.27987E-03			273.2-323.2
$C_6H_4F_2$	*m*-Difluorobenzene	0.14448E+02	-0.46982E-01	0.51948E-04			273.2-323.2
$C_6H_4I_2$	1,2-Diiodobenzene	0.31150E+02	-0.14428E+00	0.20000E-03			323.2-353.2
$C_6H_4N_2$	2-Cyanopyridine	0.45596E+03	-0.17746E+01	0.19105E-02		0.3398E-01	303.2-398.2
$C_6H_4N_2$	3-Cyanopyridine	0.60484E+02	-0.17280E+00	0.15218E-03		0.7473E-04	323.2-398.2
$C_6H_4N_2$	4-Cyanopyridine	0.12533E+02	-0.30115E-01	0.26674E-04		0.6924E-04	353.2-398.2
$C_6H_4N_2O_3S$	3-Nitro-*N*-sulfinylaniline	0.65583E+02	-0.19656E+00	0.18125E-03		0.1012E-03	343.0-403.0
$C_6H_4N_2O_4$	*m*-Dinitrobenzene	0.10406E+03	-0.34133E+00	0.32609E-03			365.2-413.2
C_6H_5Br	Bromobenzene	0.94100E+01	-0.12537E-01	-0.31127E-05		0.1294E-03	234.2-333.2
$C_6H_4BrNO_2$	3-Bromonitrobenzene	0.81413E+02	-0.27645E+00	0.27367E-03		0.2068E-02	328.2-413.2
C_6H_5Cl	Chlorobenzene	0.19471E+02	-0.70786E-01	0.82466E-04		0.1658E-01	293.2-430.2
C_6H_5ClO	*o*-Chlorophenol	0.29755E+02	-0.11256E+00	0.12390E-03		0.7405E-02	296.2-448.2
C_6H_5ClO	*p*-Chlorophenol	0.31997E+02	-0.94241E-01	0.88392E-04		0.2430E-02	314.2-453.2
$C_6H_5ClO_2S$	Benzenesulfonyl chloride	0.83886E+02	-0.23405E+00	0.19713E-03		0.2732E-02	323.2-473.2
$C_6H_5Cl_2NO_2S_2$	Phenylsulfonylimidosulfurous dichloride	0.57134E+02	-0.92000E-01				298.2-373.2
C_6H_5I	Iodobenzene	0.89442E+01	-0.20008E-01	0.17641E-04		0.6898E-05	243.2-323.2
$C_6H_5NO_2$	Nitrobenzene	0.11212E+03	-0.35211E+00	0.31128E-03		0.1831E+00	278.7-533.0
$C_6H_5NO_3$	*o*-Nitrophenol	0.33827E+02	-0.62123E-01	0.26774E-04		0.2244E-02	323.2-453.2
$C_6H_5NO_3$	*m*-Nitrophenol	0.18967E+03	-0.66144E+00	0.66532E-03		0.2395E-01	373.2-458.2
$C_6H_5NO_3$	*p*-Nitrophenol	0.22901E+03	-0.74264E+00	0.68006E-03		0.2344E-02	393.2-463.2
C_6H_6	Benzene	0.26706E+01	-0.91648E-03	-0.14257E-05		0.2276E-04	293.2-513.2
$C_6H_6N_2O_2$	*o*-Nitroaniline	0.18900E+03	-0.56977E+00	0.47484E-03		0.3427E-01	353.0-468.0
$C_6H_6N_2O_2$	*m*-Nitroaniline	0.20352E+03	-0.66582E+00	0.61310E-03		0.1090E-01	398.0-468.0
$C_6H_6N_2O_2$	*p*-Nitroaniline	0.48673E+03	-0.15040E+01	0.12857E-02		0.3314E-02	428.0-468.0
C_6H_6O	Phenol	0.63391E+02	-0.24988E+00	0.26930E-03		0.1433E-02	303.2-433.2
$C_6H_6O_2$	1,2-Benzenediol	0.74930E+02	-0.22142E+00	0.18919E-03		0.5003E-02	388.2-463.2
$C_6H_6O_2$	1,3-Benzenediol	0.30252E+02	-0.56443E-01	0.35578E-04		0.2084E-03	393.2-463.2
C_6H_6S	Benzenethiol	0.57155E+01	-0.70336E-02	0.73617E-05		0.9383E-06	303.2-358.2
C_6H_7N	Aniline	0.89534E+01	0.38990E-02	-0.36310E-04		0.1847E-01	293.2-413.2
C_6H_7N	2-Methylpyridine	0.34560E+02	-0.11980E+00	0.12500E-03			293.2-333.2
C_6H_7N	3-Methylpyridine	0.19643E+03	-0.11167E+01	0.16667E-02			303.0-333.0
C_6H_7N	4-Methylpyridine	0.33765E+02	-0.10113E+00	0.93860E-04		0.3153E-04	274.0-333.0
C_6H_7NO	2-Methylpyridine 1-oxide	0.11705E+03	-0.35301E+00	0.32000E-03		0.1805E-02	323.2-398.2
C_6H_7NO	3-Methylpyridine 1-oxide	0.59851E+02	-0.12682E+00	0.86622E-04		0.4261E-04	318.2-398.2
C_6H_8	1,4-Cyclohexadiene	0.27459E+01	-0.16975E-02	-0.36461E-06		0.3818E-06	232.0-356.0
C_6H_8	1,3,5-Hexatriene(*cis*/*trans*=0.58)	0.40452E+01	-0.95442E-02	0.12064E-04		0.1330E-04	275.0-373.0
C_6H_9N	Cyclopentanecarbonitrile	0.69830E+02	-0.25303E+00	0.31491E-03		0.3209E-02	200.7-293.2
C_6H_{10}	Cyclohexene	0.30598E+01	-0.39841E-02	0.37554E-05		0.5680E-05	141.2-313.2
C_6H_{10}	2,3-Dimethyl-1,3-butadiene	0.26258E+01	-0.17990E-02	0.12035E-06		0.1485E-04	223.2-323.2
C_6H_{10}	4-Methyl-1,3-pentadiene	0.51328E+01	-0.12774E-01	0.14215E-04		0.4665E-04	198.2-323.2
C_6H_{10}	1,5-Hexadiene	0.30014E+01	-0.28668E-02	-0.31026E-06		0.1227E-03	151.0-294.0
C_6H_{10}	*cis*, *cis*-2,4-Hexadiene	0.27284E+01	-0.17178E-02	-0.62926E-06		0.3438E-07	234.0-351.0
C_6H_{10}	*trans*, *trans*-2,4-Hexadiene	0.26774E+01	-0.16977E-02	-0.55637E-06		0.1285E-05	232.0-353.0
C_6H_{10}	1-Hexyne	0.58591E+01	-0.17099E-01	0.20856E-04		0.3800E-04	184.0-296.0
$C_6H_{10}O$	Cyclohexanone	0.41577E+02	-0.11463E+00	0.92454E-04		0.8572E-02	253.0-423.0
$C_6H_{10}O_2$	Ethyl methacrylate	0.40962E+02	-0.20520E+00	0.29286E-03		0.1997E-02	303.2-343.2
$C_6H_{10}O_4$	Ethylene glycol diacetate	0.25093E+02	-0.95171E-01	0.12224E-03		0.4611E-02	223.2-290.2
$C_6H_{10}O_4$	Diethyl oxalate	0.21938E+02	-0.66226E-01	0.66800E-04		0.2761E-04	293.2-368.2
$C_6H_{10}O_4$	Monomethyl glutarate	0.16779E+02	-0.39839E-01	0.38095E-04		0.4940E-04	293.2-363.2

TEMPERATURE DEPENDENCE OF THE PERMITTIVITY (DIELECTRIC CONSTANT) OF LIQUIDS (continued)

Molecular formula	Name	*a*	*b*	*c*	*d*	ESQS/N	Range *T*/K
$C_6H_{10}O_4$	Dimethyl succinate	0.13551E+02	-0.23109E-01	0.55440E-05		0.4924E-02	293.2-433.2
$C_6H_{11}BrO_2$	Ethyl 2-bromobutanoate	0.49005E+02	-0.23193E+00	0.32500E-03		0.3613E-03	303.2-333.2
$C_6H_{11}BrO_2$	Ethyl 2-bromo-2-methylpropanoate	0.77044E+02	-0.40784E+00	0.60000E-03		0.1250E-03	303.2-333.2
C_6H_{12}	Cyclohexane	0.24293E+01	-0.12095E-02	-0.58741E-06		0.5248E-08	283.2-333.2
C_6H_{12}	Methylcyclopentane	0.21587E+01	-0.22450E-03	-0.12500E-05		0.2813E-07	293.2-323.2
C_6H_{12}	1-Hexene	0.31476E+01	-0.50003E-02	0.46673E-05		0.1654E-04	149.0-294.0
C_6H_{12}	*trans*-2-Hexene	0.24338E+01	-0.11323E-02	-0.13720E-05		0.5306E-05	157.0-295.0
C_6H_{12}	*cis*-3-Hexene	0.30691E+01	-0.45458E-02	0.39898E-05		0.1842E-04	155.0-296.0
$C_6H_{12}Br_2$	1,6-Dibromohexane	-0.55185E+01	0.11746E+00	-0.23658E-03		0.1744E-02	274.2-328.2
$C_6H_{12}Cl_2$	1,6-Dichlorohexane	0.11277E+02	0.67200E-02	-0.50000E-04		0.5000E-05	308.2-338.2
$C_6H_{12}N_2O$	1,3-Dimethyl-2-oxohexa-hydropyrimidine	0.12648E+03	-0.43593E+00	0.44574E-03		0.1850E-03	298.2-373.2
$C_6H_{12}O$	Isobutyl vinyl ether	0.48060E+01	-0.50000E-02	-0.41495E-14			293.2-323.2
$C_6H_{12}O$	Cyclohexanol	0.10173E+03	-0.43072E+00	0.47926E-03		0.1373E-01	293.2-423.2
$C_6H_{12}O$	*cis*-2,5-Dimethyltetrahydrofuran	0.23827E+02	-0.10811E+00	0.15086E-03		0.6833E-03	207.7-302.2
$C_6H_{12}O$	2-Hexanone	0.70378E+02	-0.29385E+00	0.35289E-03		0.5283E-04	243.2-293.2
$C_6H_{12}O$	4-Methyl-2-pentanone	0.36341E+02	-0.97119E-01	0.61896E-04		0.3696E-01	204.2-373.2
$C_6H_{12}O$	3,3-Dimethyl-2-butanone	0.66857E+02	-0.28552E+00	0.34422E-03		0.2113E-03	243.2-293.2
$C_6H_{12}O$	1-Methylcyclopentanol	0.75444E+02	-0.36617E+00	0.47021E-03			310.1-332.8
$C_6H_{12}O_2$	Isopentyl formate	0.29257E+02	-0.14028E+00	0.20000E-03			288.2-323.2
$C_6H_{12}O_2$	Ethyl butanoate	0.48698E+02	-0.25660E+00	0.37237E-03		0.7392E-03	301.2-343.2
$C_6H_{12}O_2$	Hexanoic acid	0.21730E+01	0.14840E-02	-0.16526E-06		0.1204E-04	298.2-433.2
$C_6H_{12}O_2$	Butyl acetate	0.13825E+02	-0.43994E-01	0.48214E-04		0.8571E-05	253.2-353.2
$C_6H_{12}O_2$	Isobutyl acetate	0.14323E+02	-0.46048E-01	0.49286E-04		0.4791E-04	273.2-323.2
$C_6H_{12}O_2$	*sec*-Butyl acetate	0.12427E+02	-0.32035E-01	0.24286E-04		0.5109E-04	273.2-323.2
$C_6H_{12}O_2$	*tert*-Butyl acetate	0.55435E+02	-0.30494E+00	0.46107E-03		0.6024E-02	273.2-323.2
$C_6H_{12}O_2S$	2,4-Dimethyltetrahydrothio-phene-S,S-dioxide	0.47328E+02	-0.54486E-01	-0.17599E-04		0.7803E-02	293.2-398.2
$C_6H_{12}O_3$	2-Ethoxyethyl acetate	0.23290E+02	-0.71566E-01	0.65000E-04			303.2-323.2
$C_6H_{12}O_3$	2,4,6-Trimethyl-1,3,5-trioxan	0.86106E+02	-0.36553E+00	0.41440E-03		0.8110E-03	292.2-393.2
$C_6H_{13}Br$	1-Bromohexane	0.15233E+02	-0.44385E-01	0.43039E-04		0.1695E-04	274.2-328.2
$C_6H_{13}Cl$	1-Chlorohexane	0.15994E+02	-0.43647E-01	0.33393E-04		0.5041E-03	273.2-323.2
$C_6H_{13}ClO$	6-Chloro-1-hexanol	-0.73364E+01	0.46377E+00	-0.14202E-02		0.2098E-01	195.2-242.2
$C_6H_{13}I$	1-Iodohexane	0.16685E+02	-0.61309E-01	0.77262E-04		0.2052E-05	293.3-323.2
$C_6H_{13}NO$	*N*-Butylacetamide	0.70739E+03	-0.37369E+01	0.71585E-02	-0.48716E-05	0.9343E-01	253.2-493.2
$C_6H_{13}NO$	*N*-Isobutylacetamide	0.84645E+03	-0.46747E+01	0.92801E-02	-0.64581E-05	0.5325E-01	253.2-493.2
$C_6H_{13}NO$	*N*-*sec*-Butylacetamide	0.40410E+03	-0.14289E+01	0.13341E-02		0.4917E+00	273.2-493.2
$C_6H_{13}NO$	*N*-Ethylbutanamide	0.67036E+03	-0.29402E+01	0.35238E-02		0.2014E-01	298.2-328.2
$C_6H_{13}NO$	*N*-Methylpentanamide	0.81195E+03	-0.35899E+01	0.42135E-02		0.6648E+00	271.0-306.0
$C_6H_{13}NO$	*N*-Methyl-3-methylbutanamide	0.13713E+04	-0.69930E+01	0.93240E-02			286.0-308.0
$C_6H_{13}NO$	*N*-Methyl-2-methylbutanamide	0.11578E+04	-0.54895E+01	0.68981E-02		0.1917E+01	307.0-341.0
$C_6H_{13}NO$	*N*-Propylpropanamide	0.58846E+03	-0.22012E+01	0.20870E-02		0.1640E+01	298.2-328.2
C_6H_{14}	2,2-Dimethylbutane	0.22740E+01	-0.96229E-03	-0.14286E-05		0.5029E-06	273.2-313.2
C_6H_{14}	2,3-Dimethylbutane	0.24305E+01	-0.20081E-02	0.53571E-06		0.6107E-06	273.2-323.2
C_6H_{14}	2-Methylpentane	0.20745E+01	0.50871E-03	-0.39286E-05		0.5857E-06	273.2-323.2
C_6H_{14}	3-Methylpentane	0.24739E+01	-0.23190E-02	0.10714E-05		0.3952E-06	273.2-323.2
C_6H_{14}	Hexane	0.19768E+01	0.70933E-03	-0.34470E-05		0.8445E-05	293.2-473.2
$C_6H_{14}O$	Dipropyl ether	0.14600E+02	-0.72670E-01	0.11742E-03		0.8795E-03	161.0-297.0
$C_6H_{14}O$	2,2-Dimethyl-1-butanol	0.14054E+03	-0.72925E+00	0.97821E-03		0.1192E+00	243.2-393.2
$C_6H_{14}O$	1-Hexanol	0.62744E+02	-0.24214E+00	0.24704E-03		0.4950E-01	233.3-513.2
$C_6H_{14}OS$	Dipropyl sulfoxide	0.84868E+02	-0.23486E+00	0.18198E-03		0.2756E-03	303.2-373.2
$C_6H_{14}O_2$	1,2-Diethoxyethane	0.99099E+01	-0.33403E-01	0.44048E-04		0.4794E-04	223.2-303.2
$C_6H_{14}O_2$	Hexylene glycol	0.14531E+03	-0.65285E+00	0.83503E-03		0.8870E-01	203.2-333.2
$C_6H_{14}O_2S$	Dipropyl sulfone	0.70195E+02	-0.15008E+00	0.86506E-04		0.3052E-03	303.2-398.2
$C_6H_{14}O_3$	Bis(2-methoyxethyl) ether	0.28291E+02	-0.11236E+00	0.14000E-03		0.6228E-04	298.2-333.2
$C_6H_{14}O_3$	Bis(2-hydroxypropyl) ether	0.97446E+02	-0.38029E+00	0.40285E-03		0.4342E-01	213.2-423.2
$C_6H_{14}O_3$	1,2,6-Hexanetriol	0.26127E+03	-0.14552E+01	0.22765E-02		0.1011E+00	261.4-285.3
$C_6H_{14}O_4$	Triethylene glycol	0.91845E+02	-0.33827E+00	0.36062E-03		0.3385E-02	253.2-333.2
$C_6H_{14}S$	1-Hexanethiol	0.11774E+02	-0.37298E-01	0.41875E-04		0.3613E-05	273.2-333.2
$C_{10}H_{22}O_2$	Dipropylamine	0.11376E+02	-0.49796E-01	0.71792E-04		0.2072E-03	243.2-323.2
$C_6H_{15}N$	Hexylamine	0.80244E+01	-0.16627E-01	0.10874E-04		0.7781E-04	253.2-373.2
$C_6H_{15}N$	Triethylamine	0.29205E+01	-0.14007E-02	-0.13469E-05		0.1704E-03	233.2-323.2
$C_6H_{15}O_4P$	Triethyl phosphate	0.61230E+02	-0.26047E+00	0.33333E-03			298.2-333.2
$C_6H_{16}NO_3P$	*N*,*N*-Dimethylphosphoramidic acid diethyl ester	0.22012E+02	-0.35000E-01				293.2-393.2

TEMPERATURE DEPENDENCE OF THE PERMITTIVITY (DIELECTRIC CONSTANT) OF LIQUIDS (continued)

Molecular formula	Name	*a*	*b*	*c*	*d*	ESQS/N	Range *T*/K
$C_6H_{18}N_3OP$	Tris(dimethylamino)-phosphine oxide	0.95666E+02	-0.29769E+00	0.26407E-03		0.1112E-01	283.2-363.2
$C_6H_{18}N_4$	Triethylenetetramine	0.50699E+02	-0.21730E+00	0.27582E-03		0.2501E-02	213.2-333.2
$C_6H_{18}OSi_2$	Hexamethyldisiloxane	0.34537E+01	-0.61530E-02	0.61544E-05		0.1541E-06	213.2-313.2
$C_6H_{19}NSi_2$	Hexamethyldisilazane	0.23358E+01	0.16127E-02	-0.62078E-05			294.2-333.2
C_7H_4ClNO	4-Chlorophenyl isocyanate	0.40896E+01	-0.31667E-02				288.2-348.2
C_7H_5BrO	Benzoyl bromide	0.84231E+02	-0.31089E+00	0.32857E-03		0.1429E-04	283.2-313.2
$C_7H_5F_3O_2S$	Trifluoromethylphenyl sulfone	0.96137E+02	-0.30867E+00	0.29464E-03		0.1190E-02	293.2-393.2
$C_7H_5F_3O_3S$	4-(Trifluoromethylsulfonyl)-phenol	0.88287E+02	-0.18630E+00	0.10650E-03		0.5804E-01	298.2-423.2
C_7H_5N	Benzonitrile	0.57605E+02	-0.13354E+00	0.87767E-04		0.6307E-02	273.2-453.2
C_7H_5NO	Phenyl isocyanate	0.17541E+02	-0.29790E-01	0.15476E-05		0.3340E-05	293.2-353.2
C_7H_6O	Benzaldehyde	0.35046E+02	-0.61271E-01	0.16222E-04		0.1256E-02	300.7-345.7
$C_7H_6O_2$	Salicylaldehyde	0.51315E+02	0.15379E+00	0.14111E-03		0.3279E-02	289.2-453.2
C_7H_7Br	(Bromomethyl)benzene	0.18482E+02	-0.57207E-01	0.57321E-04		0.2730E-03	273.2-323.2
C_7H_7Br	2-Bromotoluene	0.10229E+02	-0.25050E-01	0.20357E-04		0.2511E-04	273.2-323.2
C_7H_7Br	3-Bromotoluene	0.11522E+02	-0.24946E-01	0.15714E-04		0.3025E-04	273.2-323.2
C_7H_7Br	4-Bromotoluene	0.10014E+02	-0.13918E-01	-0.50000E-05			273.2-293.2
C_7H_7BrO	2-Bromoanisole	0.12023E+02	-0.59116E-02	-0.13787E-04		0.4660E-04	303.2-358.2
C_7H_7BrO	4-Bromoanisole	0.74367E+01	0.12648E-01	-0.42128E-04		0.1043E-04	303.2-358.2
C_7H_7BrS	4-Bromo-3-methylthiophenol	0.62893E+01	-0.12330E-01	0.12500E-04			303.2-343.2
C_7H_7Cl	Benzyl chloride	0.17108E+02	-0.45285E-01	0.35000E-04		0.1033E-03	273.2-323.2
C_7H_7Cl	*o*-Chlorotoluene	0.11507E+02	-0.31148E-01	0.27143E-04		0.6825E-04	273.2-323.2
C_7H_7Cl	*m*-Chlorotoluene	0.13921E+02	-0.37186E-01	0.31786E-04		0.5979E-04	273.2-323.2
C_7H_7Cl	*p*-Chlorotoluene	0.20265E+01	0.40060E-01	-0.87500E-04			293.2-333.2
C_7H_7ClO	4-Chloroanisole	0.64019E+01	0.30560E-01	-0.87500E-04			293.2-333.2
C_7H_7ClS	4-Chloro-2-methylthiophenol	0.48451E+01	0.17320E-01	-0.50000E-04			303.2-343.2
$C_7H_7NO_2$	*o*-Nitrotoluene	0.10420E+03	-0.41726E+00	0.51607E-03		0.6147E-02	273.0-323.0
$C_7H_7NO_2$	*m*-Nitrotoluene	0.62492E+02	-0.16235E+00	0.12844E-03		0.3424E-01	303.2-403.2
$C_7H_7NO_3$	2-Nitroanisole	0.16684E+03	-0.58196E+00	0.57382E-03		0.3134E-01	293.2-423.2
$C_7H_7NO_3$	3-Nitroanisole	0.65402E+02	-0.16460E+00	0.12560E-03		0.1628E-02	318.2-443.2
$C_7H_7NO_3$	4-Nitroanisole	0.59811E+02	-0.10955E+00	0.36042E-04		0.2498E-01	338.2-443.2
$C_7H_7NO_3S$	4-Methoxy-*N*-sulfinylaniline	0.33176E+02	-0.10572E+00	0.11250E-03			293.2-333.2
C_7H_8	Toluene	0.32584E+01	-0.34410E-02	0.15937E-05		0.1176E-05	206.7-316.0
C_7H_8O	Anisole	0.10887E+02	-0.32372E-01	0.33629E-04		0.3854E-03	294.2-413.2
C_7H_8O	Benzyl alcohol	0.13661E+03	-0.72127E+00	0.10225E-02		0.5281E-04	303.2-333.2
C_7H_8O	*o*-Cresol	0.21633E+02	-0.71069E-01	0.70590E-04		0.8393E-03	298.2-453.2
C_7H_8O	*m*-Cresol	0.81716E+02	-0.35039E+00	0.39878E-03		0.8561E-01	273.7-463.2
C_7H_8O	*p*-Cresol	0.70253E+02	-0.28870E+00	0.31979E-03		0.4488E-01	298.2-453.2
$C_7H_8O_2$	2-Methoxyphenol	0.31751E+02	-0.88173E-01	0.72953E-04		0.1702E-02	291.2-448.2
$C_7H_8O_2$	3-Methoxyphenol	0.37279E+02	-0.12113E+00	0.11698E-03		0.4976E-03	298.2-432.7
$C_7H_8O_2$	4-Methoxyphenol	0.39483E+02	-0.12142E+00	0.10841E-03		0.1556E-02	333.7-453.2
C_7H_8S	Benzylthiol	0.16628E+02	-0.68276E-01	0.94636E-04		0.2513E-03	298.2-358.2
C_7H_8S	4-Methylthiophenol	0.87052E+01	-0.15347E-01	0.95238E-05			323.2-358.2
C_7H_8S	Thioanisole	0.21841E+02	-0.97630E-01	0.13750E-03			303.2-343.2
C_7H_9N	2,4-Dimethylpyridine	0.25895E+02	-0.73900E-01	0.62500E-04			293.2-333.2
C_7H_9N	2,6-Dimethylpyridine	0.17714E+02	-0.39080E-01	0.12500E-04			293.2-333.2
C_7H_9N	2-Ethylpyridine	0.36397E+02	-0.15070E+00	0.18750E-03			293.2-333.2
C_7H_9N	4-Ethylpyridine	-0.73831E+01	0.14326E+00	-0.27500E-03			293.2-333.2
C_7H_9N	2-Methylaniline	0.10988E+02	-0.18976E-01	0.91958E-05		0.1131E-03	298.2-398.2
C_7H_9N	3-Methylaniline	0.13477E+02	-0.35551E-01	0.33135E-04		0.3507E-04	298.2-398.2
C_7H_9N	4-Methylaniline	0.78897E+01	-0.10196E-01	0.51190E-05		0.1207E-04	333.2-403.2
C_7H_9NO	2-Methoxyaniline	0.79911E+01	-0.92183E-02	0.37879E-06		0.4470E-05	303.2-393.2
C_7H_9NO	3-Methoxyaniline	0.28179E+02	-0.97840E-01	0.11027E-03		0.5895E-03	289.2-393.2
C_7H_9NO	4-Methoxyaniline	0.30149E+02	-0.10523E+00	0.11467E-03		0.3966E-03	333.2-453.2
C_7H_9NO	2,6-Dimethylpyridine-1-oxide	0.22765E+03	-0.90760E+00	0.10011E-02		0.1289E-01	298.2-398.2
$C_7H_9NO_2S$	*N*-Methylbenzenesulfonamide	0.46369E+02	0.45809E+00	-0.12857E-02		0.6524E-02	303.2-328.2
C_7H_{12}	Cycloheptene	0.32309E+01	-0.42373E-02	0.32572E-05		0.4158E-05	227.0-363.0
C_7H_{12}	1,6-Heptadiene	0.30815E+01	-0.36095E-02	0.16354E-05		0.1583E-04	184.0-293.0
$C_7H_{12}O$	Cycloheptanone	0.17511E+03	-0.11221E+01	0.19417E-02			258.2-298.2
$C_7H_{12}O_2$	Butyl acrylate	0.38296E+02	-0.19109E+00	0.27006E-03		0.3702E-03	301.2-343.2
$C_7H_{12}O_4$	Monomethyl adipate	0.11962E+02	-0.23973E-01	0.20608E-04		0.1881E-03	293.2-433.2
$C_7H_{12}O_4$	Dimethyl glutarate	0.20697E+02	-0.57794E-01	0.48405E-04		0.8804E-03	293.2-433.2
$C_7H_{12}O_4$	Diethyl malonate	0.14809E+02	-0.31207E-01	0.24066E-04		0.6969E-04	304.2-393.2

Molecular formula	Name	*a*	*b*	*c*	*d*	ESQS/N	Range *T*/K
$C_7H_{12}O_5$	1,2,3-Propanetriol 1,2-diacetate	0.43342E+02	-0.83764E-01	-0.78234E-04		0.8062E-03	214.9-244.7
$C_7H_{12}O_5$	1,2,3-Propanetriol 1,3-diacetate	0.28321E+02	-0.89073E-01	0.86891E-04		0.7246E-03	257.7-373.7
$C_7H_{13}NO$	*N*-Methyl 6-aminohexanoic lactam	0.11976E+03	-0.44832E+00	0.49745E-03		0.3082E-02	253.2-393.2
$C_7H_{13}NO_2$	2-Butoxyethyl isocyanate	0.18674E+02	-0.31786E-01	0.21944E-14		0.3306E-05	293.2-353.2
C_7H_{14}	Cycloheptane	0.25136E+01	-0.15089E-02	0.84915E-07		0.1369E-08	278.2-333.2
C_7H_{14}	1-Heptene	0.21755E+01	0.13896E-02	-0.57049E-05		0.5053E-05	273.2-323.2
$C_7H_{14}O$	2-Heptanone	0.38348E+02	-0.12531E+00	0.12005E-03		0.1079E-02	253.2-413.2
$C_7H_{14}O$	4-Heptanone	0.41520E+02	-0.13839E+00	0.13497E-03		0.8429E-03	253.2-393.2
$C_7H_{14}O$	5-Methyl-2-hexanone	0.52353E+02	-0.17695E+00	0.15195E-03		0.3638E-03	293.2-333.2
$C_7H_{14}O$	Cyclohexyl methanol	0.10164E+03	-0.45839E+00	0.54762E-03			333.2-368.2
$C_7H_{14}O$	2-Methylcyclohexanol*	0.17315E+03	-0.98794E+00	0.14634E-02		0.2420E-02	273.2-323.2
$C_7H_{14}O$	3-Methylcyclohexanol*	0.65896E+02	-0.21954E+00	0.14107E-03		0.8784E-02	273.2-323.2
$C_7H_{14}O$	4-Methylcyclohexanol*	0.65021E+02	-0.22896E+00	0.17946E-03		0.5192E-02	273.2-323.2
$C_7H_{14}O_2$	Pentyl acetate	0.12091E+02	-0.36536E-01	0.39732E-04		0.1345E-04	253.2-353.2
$C_7H_{14}O_2$	Heptanoic acid	0.36423E+01	-0.31996E-02	0.39362E-05		0.5705E-04	288.2-423.2
$C_7H_{15}Br$	1-Bromoheptane	0.15289E+02	-0.50621E-01	0.57753E-04		0.4005E-04	203.2-343.2
$C_7H_{15}Cl$	1-Chloroheptane	0.14279E+02	-0.39431E-01	0.32321E-04		0.2586E-03	273.2-323.2
$C_7H_{15}I$	1-Iodoheptane	0.11856E+02	-0.33493E-01	0.34368E-04		0.9935E-05	293.5-323.2
$C_7H_{15}NO$	*N*-Butylpropamide	0.72659E+03	-0.33492E+01	0.41905E-02		0.9320E-02	298.2-328.2
$C_7H_{15}NO$	*N*-Pentylacetamide	0.39500E+03	-0.14209E+01	0.13333E-02		0.6259E-02	298.2-328.2
$C_7H_{15}NO$	*N*-Propylbutanamide	-0.78077E+01	0.12542E+01	-0.30000E-02		0.2813E-01	313.2-328.2
C_7H_{16}	2,2-Dimethylpentane	0.23414E+01	-0.14362E-02	-0.51322E-07		0.1471E-06	153.2-353.2
C_7H_{16}	2,3-Dimethylpentane	0.25637E+01	-0.26328E-02	0.16071E-05		0.5631E-06	273.2-323.2
C_7H_{16}	2,4-Dimethylpentane	0.23979E+01	-0.17436E-02	0.17857E-06		0.6786E-07	273.2-323.2
C_7H_{16}	3,3-Dimethylpentane	0.24007E+01	-0.16802E-02	0.36069E-06		0.4382E-09	291.3-322.2
C_7H_{16}	3-Ethylpentane	0.23771E+01	-0.15140E-02	0.10093E-06		0.3033E-06	163.2-363.2
C_7H_{16}	Heptane	0.24740E+01	-0.22577E-02	0.12428E-05		0.1552E-04	273.2-373.2
C_7H_{16}	2-Methylhexane	0.24759E+01	-0.22535E-02	0.12500E-05		0.1225E-05	293.2-323.2
C_7H_{16}	3-Methylhexane	0.27089E+01	-0.37908E-02	0.37500E-05		0.5155E-06	273.2-323.2
$C_7H_{16}O$	Ethyl 3-methylbutyl ether	0.66541E+01	-0.55450E-02	-0.12500E-04		0.7813E-05	293.2-323.2
$C_7H_{16}O$	2,2-Dimethyl-1-pentanol	0.37318E+02	-0.17095E+00	0.22022E-03		0.3098E-01	283.2-393.2
$C_7H_{16}O$	1-Heptanol	0.60662E+02	-0.24049E+00	0.25155E-03		0.8574E-01	239.2-513.2
$C_7H_{16}O$	(±)-2-Heptanol	0.10050E+03	-0.49793E+00	0.64504E-03		0.5121E-01	207.4-365.1
$C_7H_{16}O$	D-(+)-2-Heptanol	0.10076E+03	-0.50063E+00	0.65038E-03		0.3983E-01	208.0-368.5
$C_7H_{16}O$	L-(-)-2-Heptanol	0.99744E+02	-0.49045E+00	0.62725E-03		0.5421E-01	207.0-352.0
$C_7H_{16}O$	3-Heptanol	0.19586E+03	-0.11465E+01	0.17175E-02		0.1375E+00	247.7-348.8
$C_7H_{16}O$	4-Heptanol	0.28995E+03	-0.18499E+01	0.30109E-02		0.1681E-02	270.0-301.1
$C_7H_{16}O$	3-Methyl-2-hexanol	0.59724E+02	-0.32417E+00	0.47058E-03		0.1191E+00	244.0-372.2
$C_7H_{16}S$	1-Heptanethiol	0.71333E+01	-0.97320E-02	-0.12500E-05		0.2592E-03	273.2-333.2
$C_7H_{17}N$	Heptylamine	0.87794E+01	-0.24363E-01	0.25325E-04		0.4459E-04	253.2-373.2
$C_7H_{17}O_3P$	Diisopropyl methanephosphonate	0.11157E+02	-0.72657E-02	-0.14286E-04			298.2-333.2
$C_8Cl_6F_{12}$	Hexachloro-1,1,2,3,3,4,5,5,6,7,7,8-dodecafluorooctane	0.52291E+01	-0.11867E-01	0.12876E-04		0.4572E-04	203.7-313.2
$C_8H_4F_3NO$	3-Trifluoromethylphenyl isocyanate	0.31492E+02	-0.95578E-01	0.95357E-04		0.9204E-05	293.2-353.2
$C_8H_5N_2O_3PS$	Phosphorodiisocyanatidothioic acid O-phenyl ester	0.25982E+02	-0.74947E-01	0.68750E-04		0.3395E-03	293.2-393.2
$C_8H_5N_2O_4P$	Phosphorodiisocyanatidic acid phenyl ester	0.78505E+01	0.16110E-01	-0.43656E-04		0.1177E-03	283.2-393.2
C_8H_7N	Phenylacetonitrile	0.82175E+02	-0.37416E+00	0.53220E-03			299.2-343.2
$C_8H_7NO_2$	4-Methoxyphenyl isocyanate	0.20780E+02	-0.31571E-01				333.2-403.2
C_8H_8	Styrene	0.44473E+01	-0.11422E-01	0.16000E-04			293.2-313.2
C_8H_8O	Acetophenone	0.26099E+02	0.64048E-02	-0.11905E-03		0.1338E-02	298.2-333.2
$C_8H_8O_2$	Benzyl formate	0.26162E+02	-0.11026E+00	0.14787E-03		0.1278E-02	303.2-358.2
$C_8H_8O_2$	Methyl benzoate	0.17486E+02	-0.51027E-01	0.50222E-04		0.4216E-04	302.7-393.2
$C_8H_8O_2$	Phenyl acetate	0.11327E+02	-0.26707E-01	0.22938E-04		0.4466E-06	298.2-404.2
$C_8H_8O_2$	2-Hydroxyacetophenone	0.42286E+02	-0.69215E-01	-0.35714E-05		0.1131E-04	298.2-368.2
$C_8H_8O_2$	Phenylacetic acid	0.24104E+01	0.30000E-02				353.2-393.2
$C_8H_8O_3$	Methyl salicylate	0.20501E+02	-0.39045E-01	0.68298E-05		0.6832E-02	223.2-398.2
C_8H_9BrO	2-Bromophenetole	0.23146E+02	-0.75753E-01	0.77778E-04		0.4512E-03	313.2-358.2
C_8H_{10}	Ethylbenzene	0.35969E+01	-0.53169E-02	0.47500E-05		0.6125E-08	293.2-323.2
C_8H_{10}	*o*-Xylene	0.36163E+01	-0.40177E-02	0.14286E-05		0.1086E-05	273.2-323.2
C_8H_{10}	*m*-Xylene	0.28421E+01	-0.10191E-02	-0.21429E-05		0.2952E-06	273.2-323.2
C_8H_{10}	*p*-Xylene	0.23140E+01	0.97221E-03	-0.37500E-05		0.4622E-05	293.2-363.2
$C_8H_{10}O$	2-Methylanisole	0.50825E+01	-0.62297E-02	0.28571E-05		0.8114E-06	293.2-333.2

Molecular formula	Name	*a*	*b*	*c*	*d*	ESQS/N	Range *T*/K
$C_8H_{10}O$	3-Methylanisole	0.12830E+02	-0.49701E-01	0.66429E-04		0.4023E-05	293.2-333.2
$C_8H_{10}O$	4-Methylanisole	0.86608E+01	-0.23510E-01	0.25000E-04		0.1376E-04	293.2-333.2
$C_8H_{10}O$	Phenetole	-0.15043E+02	0.13752E+00	-0.24500E-03			293.2-313.2
$C_8H_{10}O$	1-Phenylethanol	0.32971E+02	-0.12042E+00	0.12809E-03		0.2179E-02	293.2-423.2
$C_8H_{10}O$	2-Phenylethanol	0.12170E+03	-0.63124E+00	0.87776E-03		0.2728E-01	278.2-333.2
$C_8H_{10}O$	2,3-Xylenol	0.14399E+02	-0.41438E-01	0.39244E-04		0.3926E-04	343.2-433.2
$C_8H_{10}O$	2,4-Xylenol	0.22125E+02	-0.85543E-01	0.96548E-04		0.2163E-04	303.2-363.2
$C_8H_{10}O$	2,5-Xylenol	0.18049E+02	-0.54991E-01	0.51656E-04		0.2805E-05	338.2-455.2
$C_8H_{10}O$	2,6-Xylenol	0.12284E+02	-0.32996E-01	0.29867E-04		0.2147E-03	313.2-453.2
$C_8H_{10}O$	3,4-Xylenol	0.54423E+02	-0.21153E+00	0.22508E-03		0.4946E-02	333.2-453.2
$C_8H_{10}O$	3,5-Xylenol	0.54251E+02	-0.21647E+00	0.23542E-03		0.8998E-02	323.2-453.2
$C_8H_{10}O_2$	1,2-Dimethoxybenzene	0.74604E+01	-0.13445E-01	0.10737E-04		0.1615E-03	293.2-443.2
$C_8H_{10}O_2$	1,3-Dimethoxybenzene	0.11911E+02	-0.30804E-01	0.29643E-04		0.2280E-04	298.2-358.2
$C_8H_{10}O_2$	1,4-Dimethoxybenzene	0.11289E+02	-0.20765E-01	0.11987E-04		0.6627E-02	333.7-463.2
$C_8H_{11}N$	*N,N*-Dimethylaniline	0.84052E+01	-0.13549E-01	0.62835E-05		0.6669E-03	289.2-453.2
$C_8H_{11}N$	2,4,6-Trimethylpyridine	0.20990E+02	-0.57419E-01	0.44286E-04		0.2100E-04	298.2-358.2
$C_8H_{12}N_2O_2$	1,6-Diisocyanatohexane	0.26715E+02	-0.42696E-01				288.2-403.2
$C_8H_{12}O_4$	Diethyl maleate	0.13953E+02	-0.21969E-01	0.17817E-05		0.9522E-05	298.2-343.2
C_8H_{14}	*cis*-Cyclooctene	0.31115E+01	-0.32058E-02	0.16713E-05		0.2864E-05	269.0-406.0
C_8H_{14}	1,2-Dimethylcyclohexene	0.26443E+01	-0.17973E-02	0.35815E-06		0.9402E-06	211.0-374.0
C_8H_{14}	1,3-Dimethylcyclohexene	0.29951E+01	-0.34615E-02	0.24026E-05		0.1855E-05	213.0-373.0
C_8H_{14}	1,7-Octadiene	0.28376E+01	-0.17442E-02	-0.16141E-05		0.1505E-04	214.0-293.0
$C_8H_{14}O_4$	Dimethyl adipate	0.11739E+02	-0.17281E-01	0.11447E-05		0.3617E-02	293.2-433.2
$C_8H_{14}O_4$	Diisopropyl oxalate	0.10709E+02	-0.16328E-01	0.56000E-05		0.8000E-06	293.2-368.2
$C_8H_{14}O_4$	Diethyl succinate	0.80213E+01	0.11810E-02	-0.26400E-04			293.2-343.2
C_8H_{16}	Cyclooctane	0.25036E+01	-0.12460E-02	-0.23175E-06		0.2104E-06	295.0-411.0
C_8H_{16}	1-Octene	0.24348E+01	0.34200E-03	-0.50000E-05		0.2520E-04	273.2-323.2
$C_8H_{16}Br_2$	1,8-Dibromooctane	0.94117E+00	0.61520E-01	-0.13333E-03			298.2-328.2
$C_8H_{16}O$	2-Octanone	-0.16219E+02	0.18799E+00	-0.34156E-03		0.1305E-02	293.2-333.2
$C_8H_{16}O_2$	Butyl butanoate	0.79684E+01	-0.12000E-01	0.15266E-13			298.2-318.2
$C_8H_{16}O_2$	Ethyl hexanoate	0.11007E+02	-0.32800E-01	0.35714E-04		0.9524E-04	253.2-353.2
$C_8H_{16}O_2$	Octanoic acid	0.29391E+01	-0.38721E-03			0.2468E-03	288.2-423.2
$C_8H_{16}O_3$	3-Methyl-1-butyl lactate	0.48649E+02	-0.21253E+00	0.27619E-03			273.2-373.2
$C_8H_{16}O_2$	Isopentyl propanoate	0.17665E+02	-0.71718E-01	0.95635E-04			273.2-373.2
$C_8H_{17}Br$	1-Bromooctane	0.12404E+02	-0.35050E-01	0.34542E-04		0.6355E-06	283.2-353.2
$C_8H_{17}Cl$	1-Chlorooctane	0.11346E+02	-0.25120E-01	0.13450E-04		0.1655E-05	274.2-328.2
$C_8H_{17}I$	1-Iodooctane	0.12452E+02	-0.41229E-01	0.50108E-04		0.2450E-04	233.2-313.2
C_8H_{18}	2,5-Dimethylhexane	0.25821E+01	-0.26804E-02	0.19404E-05		0.3626E-09	294.0-323.7
C_8H_{18}	3,4-Dimethylhexane	0.26849E+01	-0.33712E-02	0.32949E-05		0.2463E-07	292.1-323.7
C_8H_{18}	3-Ethyl-3-methylpentane	0.25983E+01	-0.28027E-02	0.24195E-05		0.1888E-07	291.5-324.1
C_8H_{18}	Octane	0.22590E+01	-0.84212E-03	-0.75758E-06		0.5993E-05	233.2-393.2
C_8H_{18}	2,2,4-Trimethylpentane	0.23677E+01	-0.14768E-02	0.94261E-07		0.1993E-06	173.2-373.2
$C_8H_{18}O$	Dibutyl ether	0.65383E+01	-0.16172E-01	0.14969E-04		0.1305E-06	293.2-313.8
$C_8H_{18}O$	2,2-Dimethyl-1-hexanol	0.91244E+01	-0.21785E-01	0.21018E-04		0.1124E-02	283.2-393.2
$C_8H_{18}O$	2-Ethyl-1-hexanol	0.86074E+02	-0.42636E+00	0.55078E-03		0.8583E-01	208.2-318.2
$C_8H_{18}O$	2-Isopropyl-1-pentanol	0.46429E+02	-0.16764E+00	0.13590E-03		0.2950E-02	208.2-318.2
$C_8H_{18}O$	2-Methyl-1-heptanol	0.61698E+02	-0.33647E+00	0.49066E-03		0.1467E-01	235.6-328.1
$C_8H_{18}O$	3-Methyl-1-heptanol	0.84087E+01	-0.33712E-01	0.49799E-04		0.1699E-01	211.2-315.6
$C_8H_{18}O$	4-Methyl-1-heptanol	0.48612E+02	-0.26773E+00	0.39972E-03		0.1079E-01	236.9-332.2
$C_8H_{18}O$	5-Methyl-1-heptanol	0.54581E+02	-0.24772E+00	0.29734E-03		0.4591E-02	235.1-328.2
$C_8H_{18}O$	6-Methyl-1-heptanol	0.57997E+02	-0.23517E+00	0.24663E-03		0.9944E-03	265.1-328.2
$C_8H_{18}O$	3-Methyl-2-heptanol	0.39178E+02	-0.17976E+00	0.24218E-03		0.1914E-02	228.8-328.8
$C_8H_{18}O$	4-Methyl-2-heptanol	0.39715E+02	-0.23115E+00	0.36771E-03		0.1400E-01	239.8-333.2
$C_8H_{18}O$	5-Methyl-2-heptanol	0.68568E+02	-0.40706E+00	0.67433E-03		0.7625E-04	230.2-278.5
$C_8H_{18}O$	6-Methyl-2-heptanol	0.77520E+02	-0.41724E+00	0.59448E-03		0.5279E-01	239.0-328.7
$C_8H_{18}O$	2-Methyl-3-heptanol	-0.59739E+01	0.56700E-01	-0.83125E-04		0.2761E-04	343.2-403.2
$C_8H_{18}O$	3-Methyl-3-heptanol	-0.38440E+01	0.42327E-01	-0.61250E-04		0.7200E-05	343.2-403.2
$C_8H_{18}O$	4-Methyl-3-heptanol	-0.48003E+01	0.50740E-01	-0.75000E-04		0.1445E-04	343.2-403.2
$C_8H_{18}O$	5-Methyl-3-heptanol	0.61967E+01	-0.63750E-02			0.1561E-03	343.2-383.2
$C_8H_{18}O$	6-Methyl-3-heptanol	0.23037E+02	-0.98029E-01	0.12479E-03		0.2760E-02	283.2-383.2
$C_8H_{18}O$	2-Methyl-4-heptanol	0.42102E+00	0.10427E-01	-0.20438E-05		0.4771E-03	229.5-332.7
$C_8H_{18}O$	3-Methyl-4-heptanol	0.33354E+02	-0.14077E+00	0.17750E-03		0.3057E-03	230.1-329.7
$C_8H_{18}O$	3-Methyl-2-isopropyl-1-butanol	0.66911E+02	-0.29898E+00	0.36326E-03		0.3318E-01	208.2-318.2
$C_8H_{18}O$	1-Octanol	0.51647E+02	-0.20371E+00	0.21320E-03		0.6005E-01	258.2-513.2

Molecular formula	Name	*a*	*b*	*c*	*d*	ESQS/N	Range *T*/K
$C_8H_{18}O$	2-Octanol	0.63760E+02	-0.27643E+00	0.31075E-03		0.6774E+00	213.2-513.2
$C_8H_{18}O$	3-Octanol	0.12505E+03	-0.70646E+00	0.10245E-02		0.4787E+00	223.2-383.2
$C_8H_{18}O$	4-Octanol	0.51049E+02	-0.26664E+00	0.37280E-03		0.4124E+00	243.2-403.2
$C_8H_{18}OS$	Dibutyl sulfoxide	0.67156E+02	-0.16448E+00	0.92275E-04		0.3399E-02	313.2-393.2
$C_8H_{18}O_2$	2-Ethyl-1,3-hexanediol	0.57919E+02	-0.17128E+00	0.12949E-03		0.2224E-02	233.2-333.2
$C_8H_{18}O_2S$	Dibutyl sulfone	0.66248E+02	-0.16417E+00	0.12001E-03		0.1001E-03	323.2-398.2
$C_8H_{18}O_5$	Tetraethylene glycol	0.83547E+02	-0.31691E+00	0.34689E-03		0.2482E-02	253.2-333.2
$C_8H_{18}S$	1-Octanethiol	0.63667E+01	-0.87920E-02	0.18750E-05		0.6125E-06	273.2-333.2
$C_8H_{19}N$	Dibutylamine	0.52504E+01	-0.10538E-01	0.71485E-05		0.1637E-03	243.2-323.2
$C_8H_{19}N$	1-Octylamine	0.77931E+01	-0.20015E-01	0.19347E-04		0.3729E-04	273.2-373.2
$C_8H_{19}OP$	Dimethylhexylphosphine oxide	0.77256E+02	-0.21595E+00	0.17489E-03		0.1768E-02	323.2-473.2
$C_8H_{23}NOSi_2$	*N*-Trimethylsilyl-2-amino-ethyl trimethylsilyl ether	0.69934E+01	-0.20076E-01	0.22500E-04		0.1496E-04	293.2-353.2
$C_8H_{23}N_5$	Tetraethylenepentamine	0.40553E+02	-0.16681E+00	0.20659E-03		0.3780E-03	213.2-333.2
$C_8H_{24}O_4Si_4$	Octamethylcyclotetrasiloxane	0.36286E+01	-0.56885E-02	0.50874E-05			296.2-333.2
$C_9H_6N_2O_2$	Toluene-2,4-diisocyanate	0.22174E+02	-0.66982E-01	0.68571E-04		0.5714E-05	293.2-353.2
$C_9H_6N_2O_2$	2,6-Toluene diisocyanate	0.10834E+02	-0.27987E-01	0.29286E-04		0.4184E-05	293.2-353.2
$C_9H_6O_2$	Coumarin	0.11311E+03	-0.33804E+00	0.31324E-03		0.6791E-02	343.2-423.2
C_9H_7N	Quinoline	0.33432E+02	-0.13497E+00	0.17788E-03		0.3077E-04	258.2-323.2
C_9H_7N	Isoquinoline	0.14412E+03	-0.79935E+00	0.11839E-02		0.1413E-02	298.2-323.2
C_9H_8O	Cinnamaldehyde	0.41837E+02	-0.11060E+00	0.10401E-03		0.2188E-02	305.8-353.5
$C_9H_8O_4$	Acetylsalicylic acid	0.69994E+01	-0.14553E-02			0.3666E-03	333.2-416.2
C_9H_9N	2,6-Dimethylbenzonitrile	0.70509E+02	-0.21399E+00	0.18750E-03		0.1571E-02	373.2-473.2
$C_9H_{10}O_2$	Ethyl benzoate	0.18216E+02	-0.62361E-01	0.72884E-04		0.2257E-04	288.2-343.2
$C_9H_{10}O_2$	Benzyl acetate	0.11727E+02	-0.30869E-01	0.32340E-04		0.1915E-05	303.2-358.2
$C_9H_{10}O_2$	Methyl 2-phenylacetate	0.17413E+02	-0.64222E-01	0.78865E-04		0.5303E-04	297.2-343.2
$C_9H_{10}O_3$	Ethyl salicylate	0.18910E+02	-0.35623E-01	0.46529E-05		0.3547E-02	225.2-321.2
$C_9H_{11}Br$	1-Bromo-3-phenylpropane	0.11360E+02	-0.27471E-01	0.25775E-04		0.2860E-04	302.2-358.2
$C_9H_{11}NO$	*N,N*-Dimethylbenzamide	0.76725E+02	-0.26908E+00	0.29409E-03		0.2506E-01	318.2-443.2
$C_9H_{11}NO$	*N*-Ethylbenzamide	-0.20109E+03	0.17866E+01	-0.31065E-02		0.1512E-04	352.7-388.7
C_9H_{12}	Propylbenzene	0.26933E+01	0.21679E-03	-0.44643E-05		0.1304E-04	273.2-323.2
C_9H_{12}	Isopropylbenzene	0.31149E+01	-0.30801E-02	0.19643E-05		0.2107E-06	273.2-323.2
C_9H_{12}	1,2,3-Trimethylbenzene	0.76006E+01	-0.29118E-01	0.41786E-04		0.8694E-04	273.2-323.2
C_9H_{12}	1,2,4-Trimethylbenzene	0.31517E+01	-0.30634E-02	0.14286E-05		0.1390E-05	273.2-323.2
C_9H_{12}	Mesitylene	0.38998E+01	-0.88072E-02	0.11149E-04		0.3592E-04	288.2-358.2
$C_9H_{12}O$	2,6-Dimethylanisole	0.76700E+01	-0.18298E-01	0.17143E-04		0.5851E-05	293.2-333.2
$C_9H_{12}O$	3,5-Dimethylanisole	0.54981E+01	-0.56651E-02	-0.14286E-05		0.2946E-04	293.2-333.2
$C_9H_{12}O$	1-Phenyl-1-propanol	0.44520E+02	-0.21505E+00	0.29443E-03		0.6928E-01	233.2-373.2
$C_9H_{12}O$	1-Phenyl-2-propanol	0.10762E+03	-0.56026E+00	0.76915E-03		0.4330E-01	233.2-373.2
$C_9H_{12}O$	2-Phenyl-2-propanol	0.57072E+01	0.86568E-02	-0.29580E-04		0.9596E-05	303.2-373.2
$C_9H_{12}O$	3-Phenyl-1-propanol	0.94482E+02	-0.45540E+00	0.59307E-03		0.3936E-03	213.2-303.2
$C_9H_{12}S$	3-Phenyl-1-propanethiol	0.82411E+01	-0.15034E-01	0.73617E-05		0.9383E-06	303.2-358.2
$C_9H_{13}N$	3-Phenylpropylamine	0.86941E+01	-0.12750E-01			0.5556E-05	303.2-343.2
$C_9H_{14}O_6$	Triacetin	0.17819E+02	-0.53656E-01	0.57759E-04		0.9260E-03	218.8-303.7
$C_9H_{14}Si$	Trimethylphenylsilane	0.21463E+01	0.32711E-02	-0.86264E-05		0.8667E-06	288.2-323.2
C_9H_{18}	1-Nonene	0.22710E+01	0.15797E-02	-0.64286E-05		0.2152E-05	273.2-323.2
$C_9H_{18}Br_2$	1,9-Dibromononane	0.18931E+02	-0.57764E-01	0.60000E-04			293.2-343.2
$C_9H_{18}O$	2,6-Dimethyl-4-heptanone	0.33178E+02	-0.11290E+00	0.11454E-03		0.1258E-03	273.2-393.2
$C_9H_{18}O_2$	Pentyl butanoate	0.59029E+01	-0.49905E-02	-0.34292E-05		0.3163E-03	301.2-343.2
$C_9H_{18}O_2$	Nonanoic acid	0.25039E+01	0.67274E-03	-0.24180E-05		0.6371E-04	294.9-364.7
$C_9H_{19}Br$	1-Bromononane	0.79870E+01	-0.10488E-01	-0.13450E-05		0.1655E-07	274.2-328.2
$C_9H_{19}Cl$	1-Chlorononane	0.95528E+01	-0.16200E-01	-0.16365E-13			293.2-323.2
C_9H_{20}	Nonane	0.23894E+01	-0.14830E-02	0.14881E-06		0.9003E-05	253.0-393.0
$C_9H_{20}O$	1-Nonanol	0.97467E+02	-0.51103E+00	0.71429E-03		0.4418E-01	288.2-343.2
$C_9H_{20}O$	2-Nonanol	0.10136E+03	-0.55612E+00	0.80000E-03			288.2-308.2
$C_9H_{20}O$	3-Nonanol	0.55214E+02	-0.31920E+00	0.50000E-03			288.2-308.2
$C_9H_{20}O$	4-Nonanol	0.27954E+01	0.30000E-02	-0.52375E-13			288.2-308.2
$C_9H_{20}O$	5-Nonanol	-0.25463E+01	0.35320E-01	-0.50000E-04			288.2-308.2
$C_9H_{20}N_2O$	Tetraethyl urea	0.52820E+02	-0.18790E+00	0.19580E-03		0.1444E-01	205.0-411.4
$C_9H_{20}S$	2-Methyl-2-octanethiol	0.61521E+01	-0.50260E-02	-0.68750E-05		0.1891E-03	273.2-333.2
$C_9H_{21}N$	Nonylamine	0.53575E+01	-0.71982E-02	0.19481E-05		0.1426E-04	293.2-373.2
$C_9H_{21}N$	Tripropylamine	0.33380E+01	-0.86332E-02	0.18322E-04		0.1626E-04	243.2-293.2
$C_9H_{21}O_4P$	Tripropyl phosphate	0.33166E+02	-0.10514E+00	0.10000E-03		0.6400E-04	293.2-373.2
$C_9H_{21}PS$	Tripropyl phosphine sulfide	0.80045E+02	-0.22665E+00	0.22378E-03		0.5040E-02	273.2-373.2

TEMPERATURE DEPENDENCE OF THE PERMITTIVITY (DIELECTRIC CONSTANT) OF LIQUIDS (continued)

Molecular formula	Name	*a*	*b*	*c*	*d*	ESQS/N	Range *T*/K
$C_{10}H_7Br$	1-Bromonaphthalene	0.10561E+02	-0.27671E-01	0.27655E-04		0.2793E-07	293.2-323.2
$C_{10}H_7Cl$	1-Chloronaphthalene	0.84861E+01	-0.12357E-01	0.26899E-05		0.6622E-07	274.2-328.2
$C_{10}H_7NO_2$	1-Nitronaphthalene	0.36267E+02	-0.41283E-01	-0.25595E-04		0.3519E-03	333.2-403.2
$C_{10}H_8N_2O_2$	1,3-Xylylene diisocyanate	0.19264E+02	-0.18354E-01	-0.11310E-05		0.6459E-05	293.2-363.2
$C_{10}H_8N_2O_2$	1,4-Xylylene diisocyanate	0.48117E+02	-0.84299E-01	-0.52143E-04		0.6084E-03	323.2-363.2
$C_{10}H_8O$	1-Naphthol	0.16489E+02	-0.46700E-01	0.42857E-04		0.9143E-05	373.0-453.0
$C_{10}H_8O$	2-Naphthol	0.92865E+01	-0.10500E-01	0.42501E-15			413.0-453.0
$C_{10}H_9N$	2-Methylquinoline	0.11688E+02	-0.78400E-02	-0.25000E-04			293.2-333.2
$C_{10}H_9N$	4-Methylquinoline	0.17788E+02	-0.32580E-01	0.12500E-04			293.2-333.2
$C_{10}H_9N$	6-Methylquinoline	0.21696E+02	-0.63400E-01	0.62500E-04			293.2-333.2
$C_{10}H_9N$	8-Methylquinoline	0.19356E+02	-0.61900E-01	0.62500E-04			293.2-333.2
$C_{10}H_9N$	1-Naphthylamine	0.10577E+02	-0.22114E-01	0.17857E-04		0.4898E-04	333.2-453.2
$C_{10}H_9N$	2-Naphthylamine	0.19722E+02	-0.60679E-01	0.60714E-04		0.1006E-03	393.0-473.0
$C_{10}H_{10}O_4$	Methyl o-acetylsalicylate	0.19579E+02	0.69970E-01	0.80889E-04		0.4132E-04	328.9-370.7
$C_{10}H_{12}$	Dicyclopentadiene	0.30564E+01	-0.20000E-02	0.82443E-15			313.2-373.2
$C_{10}H_{12}$	1,2,3,4-Tetrahydronaphthalene	0.29172E+01	0.12832E-02	-0.59453E-05		0.1324E-05	298.2-343.2
$C_{10}H_{12}O$	Tetrahydro-2-naphthol	0.98978E+02	-0.48267E+00	0.63008E-03		0.3708E-02	293.2-363.2
$C_{10}H_{12}O_2$	Propyl benzoate	0.10927E+02	-0.20535E-01	0.11745E-04		0.4902E-05	303.2-358.2
$C_{10}H_{12}O_2$	4-Allyl-2-methoxyphenol	0.52377E+02	-0.24380E+00	0.33333E-03			273.2-323.2
$C_{10}H_{12}O_2$	Ethyl 2-phenylacetate	0.13334E+02	-0.39485E-01	0.41440E-04		0.2553E-05	293.2-353.2
$C_{10}H_{12}O_2$	Benzyl propanoate	0.42301E+01	0.13962E-01	-0.36426E-04		0.2788E-03	303.0-358.0
$C_{10}H_{14}$	Isobutylbenzene	0.28055E+01	-0.92614E-03	-0.25000E-05		0.4143E-06	273.2-323.2
$C_{10}H_{14}$	*sec*-Butylbenzene	0.28348E+01	-0.68586E-03	-0.32143E-05		0.2714E-06	273.2-323.2
$C_{10}H_{14}$	*tert*-Butylbenzene	0.27924E+01	-0.38350E-03	-0.37500E-05		0.5917E-06	273.2-323.2
$C_{10}H_{14}$	4-Isopropyltoluene	0.25266E+01	-0.25121E-03	-0.24867E-05		0.2735E-05	277.2-333.2
$C_{10}H_{14}$	1,2,3,4-Tetramethylbenzene	0.33822E+01	-0.33630E-02	0.17475E-05		0.2476E-05	273.0-412.0
$C_{10}H_{14}$	1,2,4,5-Tetramethylbenzene	0.26834E+01	-0.10327E-02	-0.73533E-06		0.3294E-06	356.0-430.0
$C_{10}H_{14}N_2$	Nicotine	0.21347E+02	-0.57177E-01	0.50655E-04		0.1625E-04	293.2-363.2
$C_{10}H_{14}O$	2-Methyl-1-phenyl-2-propanol	0.21922E+02	-0.84231E-01	0.99475E-04		0.2998E-02	298.2-423.2
$C_{10}H_{15}N$	*N,N*-Diethylaniline	0.50773E+01	0.15399E-01	-0.50000E-04		0.3536E-04	303.2-328.2
$C_{10}H_{18}$	Decahydronaphthalene	0.25521E+01	-0.13651E-02	0.16667E-06		0.1190E-08	283.2-353.2
$C_{10}H_{18}$	*cis*-Decahydronaphthalene	0.25410E+01	-0.11420E-02	0.15092E-06		0.9482E-07	293.2-373.2
$C_{10}H_{18}$	*trans*-Decahydronaphthalene	0.26615E+01	-0.21241E-02	0.16864E-05		0.9216E-06	293.2-373.2
$C_{10}H_{18}O_4$	Diethyl adipate	0.14824E+02	-0.40749E-01	0.37600E-04			293.2-343.2
$C_{10}H_{20}$	1-Decene	0.19091E+01	0.33442E-02	-0.87500E-05		0.1515E-05	273.2-323.2
$C_{10}H_{20}Br_2$	1,10-Dibromodecane	0.17350E+02	-0.50328E-01	0.48633E-04		0.1554E-05	303.2-368.2
$C_{10}H_{20}Cl_2$	1,10-Dichlorodecane	-0.57423E+01	0.94220E-01	-0.17500E-03		0.2113E-03	308.2-338.2
$C_{10}H_{20}O$	Menthol	0.68202E+01	-0.15894E-01	0.20837E-04		0.4054E-05	309.3-357.6
$C_{10}H_{20}O_2$	Octyl acetate	-0.34691E+01	0.58106E-01	-0.10952E-03			288.2-323.2
$C_{10}H_{20}O_2$	2-Methylheptyl acetate	0.23285E+02	-0.11538E+00	0.17143E-03			288.2-323.2
$C_{10}H_{20}O_2$	1-Pentyl pentanoate	0.77641E+01	-0.14335E-01	0.73740E-05		0.8203E-04	305.6-393.2
$C_{10}H_{20}O_2$	3-Methylbutyl 3-methylbutanoate	0.14698E+02	-0.57726E-01	0.76190E-04			288.2-323.2
$C_{10}H_{21}Br$	1-Bromodecane	0.11202E+02	-0.33491E-01	0.36314E-04		0.1207E-04	274.2-328.2
$C_{10}H_{21}Cl$	1-Chlorodecane	0.68741E+01	-0.12210E-02	-0.22500E-04		0.3613E-05	293.2-323.2
$C_{10}H_{22}$	Decane	0.24054E+01	-0.15445E-02	0.44643E-06		0.1019E-04	253.2-393.2
$C_{10}H_{22}O$	Bis(3-methylbutyl) ether	0.44690E+01	-0.63710E-02	0.25000E-05		0.5513E-05	293.2-323.2
$C_{10}H_{22}O$	1-Decanol	0.47195E+02	-0.20740E+00	0.24942E-03		0.1102E-01	293.2-343.2
$C_{10}H_{22}O$	2-Decanol	0.13621E+03	0.01000E+00	0.12500E-02			288.2-308.2
$C_{10}H_{22}O$	(±)-3-Decanol	0.52090E+02	-0.31020E+00	0.50000E-03			288.2-308.2
$C_{10}H_{22}O$	4-Decanol	-0.11260E+02	0.93960E-01	-0.15000E-03			288.2-308.2
$C_{10}H_{22}O$	5-Decanol	-0.25832E+01	0.31456E-01	-0.40000E-04			288.2-308.2
$C_{10}H_{22}O$	2,2-Dimethyl-1-octanol	0.69536E+02	-0.34596E+00	0.46250E-03			293.2-333.2
$C_{10}H_{22}O_6$	Pentaethylene glycol	0.66659E+02	-0.23729E+00	0.24524E-03		0.5164E-03	253.2-333.2
$C_{10}H_{23}N$	Decylamine	0.61497E+01	-0.12801E-01	0.10606E-04		0.1286E-04	293.2-373.2
$C_{11}H_{10}$	1-Methylnaphthalene	0.45126E+01	-0.76480E-02	0.75000E-05			293.2-333.2
$C_{11}H_{10}O$	1-Methoxynaphthalene	0.71885E+01	-0.14838E-01	0.13750E-04			293.2-333.2
$C_{11}H_{10}O$	2-Methoxynaphthalene	0.56702E+01	-0.69754E-02	0.28571E-05		0.4571E-07	353.2-373.2
$C_{11}H_{12}O_3$	Ethyl benzoylacetate	0.93644E+01	0.74280E-01	-0.20000E-03			303.2-323.2
$C_{11}H_{14}O_2$	Butyl benzoate	0.77854E+01	-0.34972E-02	-0.13149E-04		0.9117E-04	303.2-358.2
$C_{11}H_{16}$	Pentamethylbenzene	0.30196E+01	-0.22619E-02	0.83831E-06		0.3007E-06	334.0-413.0
$C_{11}H_{18}N_2O_2$	2,2,4-Trimethylhexamethylene diisocyanate	0.32920E+02	-0.10646E+00	0.10500E-03		0.1224E-06	293.2-353.2
$C_{11}H_{22}$	1-Undecene	0.22132E+01	0.13121E-02	-0.53571E-05		0.1462E-05	273.2-323.2
$C_{11}H_{22}O_2$	Pentyl hexanoate	0.83503E+01	-0.18449E-01	0.14286E-04			288.2-323.2

TEMPERATURE DEPENDENCE OF THE PERMITTIVITY (DIELECTRIC CONSTANT) OF LIQUIDS (continued)

Molecular formula	Name	*a*	*b*	*c*	*d*	ESQS/N	Range *T*/K
$C_{11}H_{24}$	Undecane	0.23637E+01	-0.12500E-02	-0.85869E-16			283.2-363.2
$C_{11}H_{25}N$	Undecylamine	0.54945E+01	-0.96161E-02	0.66017E-05		0.3774E-04	293.2-373.2
$C_{12}H_4Cl_6$	Hexachlorodiphenyl*	0.53843E+01	0.85521E-03	-0.15822E-04		0.1952E-04	291.2-303.2
$C_{12}H_5Cl_5$	Pentachlorodiphenyl*	0.92450E+02	-0.60858E+00	0.10617E-02		0.2755E-07	273.2-282.2
$C_{12}H_{10}$	Biphenyl	0.26869E+01	0.63072E-03	-0.30995E-05		0.6116E-05	348.2-428.2
$C_{12}H_{10}O$	2-Acetonaphthone	0.14538E+03	-0.73040E+00	0.10000E-02			333.2-363.2
$C_{12}H_{10}N_2S$	Thio-bis(*N*-phenyl)imine	0.80384E+01	-0.20660E-01	0.25000E-04			333.2-373.2
$C_{12}H_{11}NO$	*N*-1-naphthylenylacetamide	0.84739E+02	-0.12391E+00	-0.35714E-04		0.7238E-02	433.2-533.2
$C_{12}H_{16}O_3$	3-Methylbutyl salicylate	0.13129E+02	-0.19190E-01	-0.36060E-05		0.6386E-03	225.2-397.2
$C_{12}H_{18}$	Hexamethylbenzene	0.35710E+01	-0.46912E-02	0.35088E-05			449.0-489.0
$C_{12}H_{20}O_2$	*l*-Bornyl acetate	0.60791E+01	0.98200E-02	-0.50000E-04			303.2-323.2
$C_{12}H_{22}O$	Cyclododecanone	0.39327E+02	-0.13248E+00	0.13298E-03		0.5512E-03	303.2-423.2
$C_{12}H_{22}O$	Dicyclohexyl ether	0.95324E+01	-0.31740E-01	0.37500E-04			293.2-333.2
$C_{12}H_{24}$	1-Dodecene	0.22581E+01	0.11106E-02	-0.50000E-05		0.1429E-05	273.2-323.2
$C_{12}H_{24}O_2$	Ethyl decanoate	0.70969E+01	-0.15080E-01	0.12500E-04			293.2-353.2
$C_{12}H_{25}Br$	1-Bromododecane	0.86103E+01	-0.20891E-01	0.18994E-04		0.9535E-07	274.2-328.2
$C_{12}H_{25}Cl$	1-Chlorododecane	0.10002E+02	-0.27798E-01	0.27559E-04		0.3442E-04	274.2-328.2
$C_{12}H_{25}I$	1-Iodododecane	0.34641E+01	0.97404E-02	-0.27602E-04		0.5381E-05	293.1-323.0
$C_{12}H_{26}$	Dodecane	0.23697E+01	-0.12200E-02	-0.36375E-16			283.2-363.2
$C_{12}H_{26}O$	2,2-Dimethyl-1-decanol	0.16410E+02	-0.92839E-01	0.16429E-03		0.2333E-02	203.2-293.2
$C_{12}H_{26}O$	1-Dodecanol	0.18518E+02	-0.44859E-01	0.99900E-05		0.9640E-04	303.2-358.2
$C_{12}H_{26}O_7$	Hexaethylene glycol	0.59773E+02	-0.22186E+00	0.24713E-03		0.1306E-02	253.2-333.2
$C_{12}H_{27}N$	Dodecylamine	0.27999E+01	0.44810E-02	-0.11905E-04		0.1756E-04	303.2-373.2
$C_{12}H_{27}N$	Tributylamine	0.19846E+01	0.28108E-02	-0.54545E-05		0.1633E-05	233.2-293.2
$C_{12}H_{27}OP$	Tributyl phosphine oxide	0.94866E+02	-0.29542E+00	0.25788E-03		0.1161E-01	323.2-473.2
$C_{12}H_{27}O_3PS$	Tributyl thiophosphate	0.87018E+01	0.17968E-02	-0.25857E-04		0.7545E-03	298.2-363.2
$C_{12}H_{27}O_4P$	Tributyl phosphate	0.26304E+02	-0.88480E-01	0.92857E-04		0.3314E-04	293.2-373.2
$C_{12}H_{27}PS$	Tributyl phosphine sulfide	0.92150E+02	-0.30005E+00	0.30653E-03		0.5216E-01	273.2-373.2
$C_{12}H_{30}OSi_2$	Hexaethyl disiloxane	0.36559E+01	-0.72406E-02	0.85714E-05			298.2-333.2
$C_{12}H_{36}O_4Si_5$	Dodecamethyl pentasiloxane	0.47894E+01	-0.12286E-01	0.15000E-04		0.1800E-05	293.2-323.2
$C_{13}H_{10}O$	Benzophenone	0.34130E+02	-0.10249E+00	0.10268E-03		0.7143E-04	300.2-420.2
$C_{13}H_{10}O_3$	Phenyl salicylate	0.26545E+02	-0.11180E+00	0.15220E-03		0.5236E-03	290.2-358.2
$C_{13}H_{12}$	Diphenylmethane	0.30638E+01	-0.17286E-02			0.9524E-07	303.2-333.2
$C_{13}H_{18}O_2$	Pentyl 2-phenylacetate	0.16491E+02	-0.66310E-01	0.87500E-04			303.2-343.2
$C_{13}H_{24}O$	Cyclotridecanone	0.10446E+03	-0.58666E+00	0.90000E-03		0.1901E-02	303.2-318.2
$C_{13}H_{26}$	1-Tridecene	0.14154E+01	0.66514E-02	-0.14286E-04		0.2152E-05	273.2-323.2
$C_{13}H_{28}$	Tridecane	0.23731E+01	-0.12000E-02	-0.21841E-15			283.2-363.2
$C_{13}H_{28}O_4Si_2$	Bis(trimethylsilylmethyl) glutarate	0.66133E+01	-0.83333E-02				293.2-413.2
$C_{14}H_9Cl_5$	1,1,1-Trichloro-2,2-bis (4-chlorophenyl)ethane	0.76748E+01	-0.12659E-01				377.2-418.2
$C_{14}H_{10}$	Anthracene	0.20571E+02	-0.69169E-01	0.66667E-04		0.8491E-07	502.0-516.0
$C_{14}H_{10}O_2$	Benzil	-0.23599E+02	0.22715E+00	-0.34667E-03			368.2-393.2
$C_{14}H_{12}O_2$	Benzyl benzoate	0.76856E+01	-0.80000E-02	-0.80361E-15			303.2-358.2
$C_{14}H_{14}$	1,2-Diphenylethane	0.31178E+01	-0.21572E-02	0.59800E-06		0.3427E-05	331.2-451.2
$C_{14}H_{14}O$	Dibenzyl ether	0.80154E+01	-0.20536E-01	0.21250E-04			293.2-333.2
$C_{14}H_{16}O_2Si$	Dimethyldiphenoxysilane	0.51669E+01	-0.77001E-02	0.70156E-05		0.7283E-04	283.2-353.2
$C_{14}H_{18}O$	Pentyl-3-phenyl-2-propenal*	-0.31153E+00	0.11822E+00	-0.27143E-03		0.7273E-02	293.2-333.2
$C_{14}H_{26}O_4$	Diethyl sebacate	0.39143E+02	-0.20965E+00	0.32000E-03			303.2-313.2
$C_{14}H_{29}Br$	1-Bromotetradecane	0.10058E+02	-0.33905E-01	0.43528E-04		0.2320E-03	274.2-328.2
$C_{14}H_{30}$	Tetradecane	0.23832E+01	-0.11900E-02	-0.51229E-16			283.2-363.2
$C_{14}H_{30}O$	1-Tetradecanol	0.12272E+02	-0.24667E-01	-0.13168E-13		0.1481E-04	318.2-358.2
$C_{14}H_{30}O_5Si_2$	Bis(isobutoxymethyl) tetramethyldisiloxane	0.65712E+01	-0.79167E-02				293.2-413.2
$C_{14}H_{30}O_8$	Heptaethylene glycol	0.52101E+02	-0.18670E+00	0.20345E-03		0.1891E-02	253.2-333.2
$C_{14}H_{31}O_3P$	Diethyl decanephosphonate	0.14446E+02	-0.39313E-01	0.36607E-04		0.7286E-04	293.2-393.2
$C_{15}H_9F_3ClNO$	Chlorophenyl(p-trifluoromethylphenyl)methyl isocyanate	0.42140E+02	-0.15199E+00	0.16698E-03		0.1781E-02	288.2-403.2
$C_{15}H_{10}N_2O_2$	1,1′-Diphenylmethane diisocyanate	0.20279E+02	-0.60902E-01	0.62143E-04		0.9611E-05	313.2-353.2
$C_{15}H_{12}ClNO$	Chlorophenyl(m-methylphenyl)methyl isocyanate	0.19764E+02	0.36339E-01	-0.13228E-03		0.1330E-02	288.2-378.2
$C_{15}H_{14}O_2$	Benzyl 2-phenylacetate	0.11247E+02	-0.33490E-01	0.37500E-04			303.2-343.2
$C_{15}H_{16}O_2S$	Tolyl xylyl sulfone*	0.37584E+02	-0.25882E-01	-0.72420E-04		0.1386E-03	273.2-297.2
$C_{15}H_{26}O_6$	Tributyrin	0.13152E+02	-0.36684E-01	0.36795E-04		0.1298E-06	199.4-282.8

TEMPERATURE DEPENDENCE OF THE PERMITTIVITY (DIELECTRIC CONSTANT) OF LIQUIDS (continued)

Molecular formula	Name	*a*	*b*	*c*	*d*	ESQS/N	Range *T*/K
$C_{15}H_{32}$	Pentadecane	0.23792E+01	-0.11600E-02	-0.71069E-16			283.2-363.2
$C_{15}H_{33}PS$	Tripentylphosphine sulfide	0.71435E+02	-0.23105E+00	0.22960E-03		0.8459E-02	273.2-373.2
$C_{15}H_{39}O_3NSi_3$	Tris-(trimethylsilanoxyethyl)amine	0.60295E+01	-0.12415E-01	0.10952E-04		0.1244E-04	293.2-353.2
$C_{16}H_{22}O_4$	Dibutyl phthalate	0.12444E+02	-0.20000E-01				293.2-333.2
$C_{16}H_{32}O_2$	Ethyl myristate	0.52642E+01	-0.60000E-02	-0.47358E-15		0.1250E-03	293.2-353.2
$C_{16}H_{33}Br$	1-Bromohexadecane	0.58668E+01	-0.73333E-02	-0.52666E-14			298.2-328.2
$C_{16}H_{33}I$	1-Iodohexadecane	0.79531E+01	-0.22859E-01	0.26955E-04		0.3825E-05	293.2-323.2
$C_{16}H_{34}$	Hexadecane	0.23861E+01	-0.11600E-02	0.25555E-15			293.2-363.2
$C_{16}H_{34}O$	1-Hexadecanol	0.85935E+01	-0.14714E-01	-0.45533E-13		0.6122E-05	333.2-363.2
$C_{16}H_{34}O_2Si$	Trimethylsilylmethyl laurate	0.38240E+01	-0.30833E-02				293.2-413.2
$C_{16}H_{36}Sn$	Tetrabutyl tin	0.56115E+02	-0.24812E+00	0.30682E-03			293.2-313.2
$C_{16}H_{48}O_6Si_7$	Hexadecamethyl heptasiloxane	-0.21789E+01	0.36057E-01	-0.67500E-04		0.7813E-05	293.2-323.2
$C_{17}H_{12}O_3$	2-Naphthyl salicylate	0.11229E+02	-0.18857E-01	0.70332E-05		0.3590E-03	293.0-353.0
$C_{17}H_{34}O$	9-Heptadecanone	0.44176E+02	-0.21183E+00	0.28571E-03			328.2-363.2
$C_{17}H_{36}$	Heptadecane	0.23627E+01	-0.10400E-02	-0.10397E-12			293.2-308.2
$C_{17}H_{36}O_4Si_2$	Bis(trimethylsilylmethyl) azelate	0.70077E+01	-0.91667E-02				293.2-413.2
$C_{18}H_{30}O_2$	Linolenic acid	0.33867E+01	-0.19181E-02			0.1227E-05	274.2-368.2
$C_{18}H_{32}O_2$	Linoleic acid	0.32073E+01	-0.15477E-02			0.5542E-06	275.2-368.2
$C_{18}H_{34}O_2$	Oleic acid	0.25385E+01	-0.69448E-03			0.1339E-05	275.2-368.2
$C_{18}H_{36}O$	9-Octadecene-1-ol	0.48099E+01	-0.22728E-02	-0.33404E-05		0.6876E-05	295.2-333.2
$C_{18}H_{36}O_2$	Cetyl acetate	0.47310E+01	-0.50000E-02	0.41338E-14			308.2-348.2
$C_{18}H_{36}O_2$	Ethyl palmitate	0.57938E+01	-0.12294E-01	0.10919E-04			303.2-455.2
$C_{18}H_{36}O_2$	Stearic acid	0.27159E+01	-0.13300E-02			0.6291E-04	293.2-373.2
$C_{18}H_{37}Br$	1-Bromooctadecane	0.46790E+01	-0.30355E-02	-0.24798E-05			303.4-331.6
$C_{18}H_{38}O$	1-Octadecanol	0.73784E+01	-0.12000E-01	-0.22871E-13			333.2-363.2
$C_{19}H_{16}$	Triphenylmethane	0.40201E+01	-0.66507E-02	0.65329E-05		0.1018E-04	367.2-448.2
$C_{19}H_{37}NO$	Octadecyl isocyanate	0.11826E+02	-0.30441E-01	0.27619E-04		0.2361E-05	303.2-363.2
$C_{20}H_{26}O_5Si_2$	Bis(benzoyloxymethyl)tetra-methyl disiloxane	0.77631E+01	-0.98333E-02				293.2-413.2
$C_{20}H_{38}O_2$	Ethyl oleate	0.57033E+01	-0.11223E-01	0.93447E-05		0.1951E-04	301.2-423.2
$C_{20}H_{40}O_2$	Octadecyl acetate	0.44569E+01	-0.45000E-02	0.33923E-14			308.2-348.2
$C_{20}H_{40}O_2$	Ethyl stearate	0.70930E+01	-0.19081E-01	0.19555E-04		0.2137E 04	331.2-440.2
$C_{20}H_{42}O$	Didecyl ether	0.41465E+01	-0.62240E-02	0.37500E-05			293.2-333.2
$C_{20}H_{42}O$	1-Eicosanol	0.21700E+01	0.12497E 01	-0.28571E-04		0.3810E-05	338.2-363.2
$C_{20}H_{42}O_2$	Octadecoxy ethanol	0.67914E+01	-0.12966E-01	0.95238E-05		0.5442E-05	328.2-358.2
$C_{20}H_{60}O_8Si_9$	Eicosamethyl nonasiloxane	0.57840E+01	-0.16568E-01	0.20000E-04		0.1800E-05	293.2-323.2
$C_{21}H_{44}O_2$	Octadecoxypropanol	0.85781E-01	0.30251E-01	-0.57143E-04		0.2286E-05	328.2-348.2
$C_{21}H_{45}OP$	Triheptylphosphine oxide	0.68784E+02	-0.12000E+00				323.2-343.2
$C_{21}H_{45}PS$	Triheptylphosphine sulfide	0.11983E+03	-0.49569E+00	0.53571E-03		0.8363E-02	273.2-343.2
$C_{22}H_{42}O_2$	*cis*-13-Docosene carboxylic acid	0.32678E+01	-0.18506E-02			0.9529E-05	307.8-343.9
$C_{22}H_{44}O_2$	Butyl stearate	0.73894E+02	-0.46261E+00	0.75500E-03		0.4466E-01	298.2-343.2
$C_{22}H_{45}Br$	1-Bromodocosane	0.19444E+02	-0.94747E-01	0.13714E-03			315.9-333.4
$C_{22}H_{46}O$	1-Docosanol	0.82062E+01	-0.25069E-01	0.28571E-04		0.3810E-05	348.2-373.2
$C_{22}H_{46}O_2$	Eicosanoxyethanol	0.95252E+01	-0.30250E-01	0.35714E-04		0.3929E-05	338.2-363.2
$C_{22}H_{47}O_3P$	Dihexyl decanephosphonate	0.81703E+01	-0.14300E-01	0.89286E-05		0.5029E-04	293.2-373.2
$C_{23}H_{48}O_2$	Eicosanoxypropanol	0.85266E+01	-0.24211E-01	0.28571E-04		0.3810E-05	333.2-358.2
$C_{24}H_{50}O_2$	Docosanoxyethanol	0.77097E+01	-0.21176E-01	0.23810E-04		0.3401E-05	338.2-368.2
$C_{24}H_{51}O_3P$	Diheptyl decanephosphonate	0.98028E+01	-0.23707E-01	0.22321E-04		0.1095E-04	293.2-393.2
$C_{24}H_{51}PS$	Trioctylphosphine sulfide	0.65165E+02	-0.24748E+00	0.28680E-03		0.8228E-02	273.2-353.2
$C_{25}H_{52}O_2$	Docosanoxypropanol	0.54992E+01	-0.60000E-02	-0.63054E-13			338.2-363.2
$C_{27}H_{57}OP$	Trinonylphosphine oxide	0.76825E+02	-0.27965E+00	0.27749E-03		0.1002E-01	323.2-473.2
$C_{28}H_{56}O_2$	Decyl stearate	0.40628E+01	-0.40000E-02	0.11614E-14			313.2-353.2
$C_{29}H_{60}O_4Si_2$	Bis(methyldipentylsilylmethyl)-glutarate	0.43872E+01	-0.40833E-02				293.2-413.2
$C_{30}H_{63}PS$	Tridecylphosphine sulfide	-0.11749E+02	0.15210E+00	-0.25000E-03		0.1250E-03	343.2-373.2
$C_{51}H_{98}O_6$	Tripalmitin	-0.29131E+01	0.32206E-01	-0.44154E-04			328.2-393.2

* Unspecified isomer.

PERMITTIVITY (DIELECTRIC CONSTANT) OF GASES

This table gives the permittivity ε (often called the dielectric constant) of some common gases at a temperature of 20°C and pressure of one atmosphere (101.325 kPa). Values of the permanent dipole moment μ in Debye Units (1 D = 3.33564×10^{-30} C m) are also included.

The density dependence of the permittivity is given by the equation

$$\frac{\varepsilon - 1}{\varepsilon + 2} = \rho_m \left(\frac{4\pi N \alpha}{3} + \frac{4\pi N \mu^2}{9kT} \right)$$

where ρ_m is the molar density, N is Avogadro's number, k is the Boltzmann constant, T is the temperature, and α is the molecular polarizability. Therefore, in regions where the gas can be considered ideal, ε – 1 is approximately proportional to the pressure at constant temperature. For nonpolar gases (μ = 0), ε –1 is inversely proportional to temperature at constant pressure.

The number of significant figures indicates the accuracy of the values given. The values of ε for air, Ar, H_2, He, N_2, O_2, and CO_2 are recommended as reference values; these are accurate to 1 ppm or better.

The second part of the table gives the permittivity of water vapor in equilibrium with liquid water as a function of temperature (derived from Reference 4).

REFERENCE

1. A. A. Maryott and F. Buckley, *Table of Dielectric Constants and Electric Dipole Moments of Substances in the Gaseous State*, National Bureau of Standards Circular 537, 1953.
2. B. A. Younglove, *J. Phys. Chem. Ref. Data*, 11, Suppl. 1, 1982; 16, 577, 1987 (for data on N_2, H_2, O_2, and hydrocarbons over a range of pressure and temperature).
3. Landolt-Börnstein, *Numerical Data and Functional Relationships in Science and Technology*, New Series, Group IV, Vol. 4, Springer-Verlag, Heidelberg, 1980 (for data at high pressures).
4. G. Birnbaum and S. K. Chatterjee, *J. Appl. Phys.*, 23, 220, 1952 (for data on water vapor).

Mol. form.	Name	ε	μ/D
	Compounds not containing carbon		
	Air (dry,CO_2 free)	1.0005364	
Ar	Argon	1.0005172	0
BF_3	Boron trifluoride	1.0011	0
BrH	Hydrogen bromide	1.00279	0.827
ClH	Hydrogen chloride	1.00390	1.109
F_3N	Nitrogen trifluoride	1.0013	0.235
F_6S	Sulfur hexafluoride	1.00200	0
HI	Hydrogen iodide	1.00214	0.448
H_2	Hydrogen	1.0002538	0
H_2S	Hydrogen sulfide	1.00344	0.97
H_3N	Ammonia	1.00622	1.471
He	Helium	1.0000650	0
Kr	Krypton	1.00078	0
NO	Nitric oxide	1.00060	0.159
N_2	Nitrogen	1.0005480	0
N_2O	Nitrous oxide	1.00104	0.161
Ne	Neon	1.00013	0
O_2	Oxygen	1.0004947	0
O_2S	Sulfur dioxide	1.00825	1.633
O_3	Ozone	1.0017	0.534
Xe	Xenon	1.00126	0
	Compounds containing carbon		
CF_4	Tetrafluoromethane	1.00121	0
CO	Carbon monoxide	1.00262	0.110
CO_2	Carbon dioxide	1.000922	0

PERMITTIVITY (DIELECTRIC CONSTANT) OF GASES (continued)

Mol. form.	Name	ε	μ/D
CH_3Br	Bromomethane	1.01028	1.822
CH_3Cl	Chloromethane	1.01080	1.892
CH_3F	Fluoromethane	1.00973	1.858
CH_3I	Iodomethane	1.00914	1.62
CH_4	Methane	1.00081	0
C_2H_2	Acetylene	1.00124	0
C_2H_3Cl	Chloroethylene	1.0075	1.45
C_2H_4	Ethylene	1.00134	0
C_2H_5Cl	Chloroethane	1.01325	2.05
C_2H_6	Ethane	1.00140	0
C_2H_6O	Dimethyl ether	1.0062	1.30
C_3H_6	Propene	1.00228	0.366
C_3H_6	Cyclopropane	1.00178	0
C_3H_8	Propane	1.00200	0.084
C_4H_{10}	Butane	1.00258	0
C_4H_{10}	Isobutane	1.00260	0.132

PERMITTIVITY OF SATURATED WATER VAPOR

T/°C	ε	*T*/°C	ε
0	1.00007	60	1.00144
10	1.00012	70	1.00213
20	1.00022	80	1.00305
30	1.00037	90	1.00428
40	1.00060	100	1.00587
50	1.00095		

AZEOTROPES

Zdzislaw M. Kurtyka

GENERAL CLASSIFICATION OF AZEOTROPES AND THEIR SYMBOLISM

Although the first reported observations of the appearance of minimum vapor pressure in binary mixtures were in 1802 by Dalton,[1] the term "azeotrope" was not introduced until 1911 by Wade and Merriman.[2]

Azeotrope from the Greek "not to boil with change" or "to boil unchanged" means literally the same as the English "constant-boiling", i.e., the vapor boiling from a liquid has the same composition as the liquid.

A liquid mixture of two or more components which can be separated by distillation was termed a "zeotrope" by Swietoslawski.[3] Such a mixture is also called a "nonazeotrope" or a "non-azeotropic" system in many countries. However, some objection may be placed against the use of the terms "nonazeotrope" and "non-azeotropic" system due to the use of two negatives in the same word. In Greek "a" means "non". Therefore, accuracy indicates that the word nonazeotrope should be replaced by non-non-azeotrope. It is obvious the latter is awkward and that it is reasonable to replace it with Swietoslawski's suggestion, zeotrope.

Azeotropic systems may be classified broadly in relation to the character of the extremum (maximum or minimum), the number of components in the system and whether they form one or more liquid phases.

Positive and Negative Azeotropes

In 1926 Lecat[4] proposed to divide azeotropes into positive and negative. Positive azeotropes are characterized by a minimum boiling temperature at constant preassure, i.e. a maximum in the vapor pressure at constant temperature. Negative azeotropes, on the other hand, have a maximum boiling temperature and a mininum vapor pressure.

In Anglo-Saxon literature a positive azeotrope is equivalent to a pressure-maximum azeotrope, while a negative one — to a pressure-minimum azeotrope. This description of azeotropes cannot be extended to systems exhibiting neither a minimum nor a maximum in either boiling temperature or vapor pressure (saddle or positive-negative azeotropes).

One may say at this point that neither azeotrope nor zeotrope, thus described, gives any indication whether the liquid phase consists of one, two or more liquid phases. To make this perfectly clear, the terms homo- and heteroazeotrope, and homo- and heterozeotrope were introduced. For practical purposes negative azeotropes are divided into three groups.[5]

In view of the fact that the number of different types of azeotropes continue to increase, symbols are given below for positive and negative azeotropes and zeotropes.

Type of homoazeotrope	Type of zeotrope
1. (A,B) binary positive	$(A,B)_z$ binary positive
2. A,B,C) ternary positive	$(A,B,C)_z$ ternary positive
3. (A,B,C,D)quaternary positive	$(A,B,C,D)_z$quaternary positive
4. [(−)A,B]binary negative	$[(-)A,B]_z$ binary negative

Symbols are also used for heteroazeotropes. For example, there is a combination of letters, dots, and dashes. There is one letter for each of the components in the system, one dash for each of the phases in the azeotrope and one less dot than there are dashes. The system may be illustrated by the following two examples. The system benzene-ethanol-water forms a two-phase heteroazeotrope. Symbolically this is written as (B,E,W, -·-). The system nitromethane-water-n-paraffin forms three liquid phases at the boiling temperature. This applies for n-paraffins from heptane to tridecane. If the symbol H_i is assigned to the n-parafffin, the azeotrope will be designated by the symbols (N,W,H_i,-·-·-).

Saddle or Positive-Negative Azeotropes

In spite of the fact that a ternary saddle azeotrope was predicted by Ostwald at the end of the last century, the first azeotrope of this type was found in 1945 by Ewell and Welch[7] in the system acetone-chloroform-methanol.

Saddle azeotropic systems, also called positive-negative systems, exhibit a hyperbolic point which is neither a minimum nor a maximum in either boiling temperature or vapor pressure, and are characterized by the presence of a "top-ridge" line. They also exhibit some peculiar properties called distillation anomalies.[7-10]

In general, ternary saddle azeotropes are classified according to the number of binary negative systems forming such an azeotrope. From this point of view, ternary saddle azeotropes may be divided into bipositive-negative and binegative-positive azeotropes. All possible types of ternary bipositive-negative azeotropes were found.[11-14] The bipositive-negative azeotropes may be designated by the symbols [(−)A,B(±)H], where, A, B, and H are the components forming these azeotropes. For the ternary binegative-positive azeotropes with B, E, and C as the components, the symbols are [(+)B,E(−)C] (two binary negative azeotropes [(−)C,B] and [(−)C,E] occur in this system). This type of a ternary saddle azeotrope is rather a rare phenomenon.[15,16]

Although the terms "saddle azeotrope" and "positive-negative" azeotrope are equivalent in the context of ternary systems, the latter is more preferable when discussing multicomponent systems.

THE COMMONNESS OF THE PHENOMENON OF AZEOTROPY

Azeotropes occur in organic and inorganic systems, although most of the known azeotropes are formed by organic compounds. This is because the very large number of organic compounds boil without decomposition in the easily accessible ranges of temperature and pressure.

In the last 2 decades of the 19th century the appearance of an extremum vapor pressure and boiling temperature was believed to be a rare phenomenon. For this reason Ostwald[17] used the term "ausgezeichnete Lösungen" to emphasize the phenomenon was not often encountered. This view survived in certain circles for many years despite convincing evidence the phenomenon of azeotropy is definitely a common one.

The first investigator who provided evidence of the commonness of azeotropy was Ryland.[18] His word dealt with 80 systems, 45 of which were azeotropic; 80 further new azeotropes were described in Lecats doctoral dissertation, published in 1908 to 1909. Lecat was able to list 1000 azeotropes in his monograph 10 years later.[19] Among them were numerous binary and ternary heteroazeotropes. This monograph showed that the appearance of maximum or miminum vapor pressures of binary and ternary mixtures should not be regarded as a rare phenomenon. From then on many physical chemists and technologists began to take an interest in the theoretical and practical application of azeotropy.

In 1949 Lecat[20] published the azeotropic data for 13,290 binary systems. The number of azeotropes reached 6287 or 47% of the systems examined.

In 1973 Horsley[21] published his Azeotropic Data-III. In this volume 15,823 binary, 725 ternary, 21 quaternary, and 2 quinary systems are reported. The number of the azeotropic systems is as follows: binary 7945 (52%), ternary 371 (51%), quaternary 9 (43%), and quinary 1.

It is interesting to see that 119 binary azeotropes (47% of the systems examined), including 32 negative, occur in the systems composed either of two inorganic compounds (elements) or an inorganic-organic compound system; 768 binary systems contain water as one component (among them 665 [86%] are azeotropic). The ternary positive-negative (saddle) azeotropes occur in 40 systems; 267 (72%) ternary azeotropic systems contain water as one component. There are also 4 ternary negative azeotropes. As far as the quaternary systems are concerned, 8 systems form positive azeotropes and one, a positive-negative.

In the last decade, studies of vapor-liquid equilibria were developed rapidly, mostly at a constant temperature, with the aim of correlating and predicting the equilibrium parameters. Unfortunately, a large number of investigators did not report whether an azeotrope or a zeotrope is formed in a particular system. In view of the fact that the number of the systems examined in that period of time is small, compared to that of the systems examined previously, it becomes clear that the current situation regarding the commonness of the phenomenon of azeotropy remains essentially unchanged.

From the existing experimental work one might conclude that the frequency of occurence of azeotropes diminishes as the number of components in the system increases, and that multicomponent azeotropes should be expected to be very rare. Such a point of view, however, would be wrong. Although the conditions of the formation of multicomponent azeotropes are complex, they are not difficult to fulfil in practice.[22-24]

THE AZEOTROPIC RANGE

The idea of relating a certain "azeotropic ability" to the chemical character of a substance is already apparent in the first works of Lecat,[19] where he arranged the experimental data on azeotropes according to the chemical character of the components.

The term "azeotropic range" is due to Swietoslawski.[25] There is also another closely related term known, namely, the "relative azeotropic effect",[26] but its use is very limited. Malesinski[27] developed the concept of Swietoslawski's azeotropic range on a general assumption that the components of the system form a regular solution.

The symmetrical azeotropic range, Z, is given by the formula

$$(\delta_1^{1/2} + \delta_2^{1/2}) = Z_{12} = 0.5Z \qquad (1)$$

where z_{12} is the half-value of the symmetrical azeotropic range and the quantities δ_1 and δ_2, known otherwise as azeotropic deviations, for positive azeotropes are defined by $\delta_1 = T_1 - T^{Az}$ and $\delta_2 = T_2 - T^{Az}$; T_1, T_2 are the boiling temperatures of the pure components and T^{Az} is the boiling temperature of the azeotrope. For negative azeotropes we have $\delta_1 = T^{Az} - T_1$ and $\delta_2 = T^{Az} - T_2$. The symmetrical azeotropic range means that both parts of the range (the upper and the lower part) are equal.

Equation 1 which is also valid at high pressures[28] enables us to compute z_{12} of any binary azeotropic system, provided that the boiing temperature of the pure components and of the azeotrope are known. It should be remembered that by convention the z_{12} values are positive for positive azeotropic systems, and negative for negative ones.

The knowledge of the azeotropic ranges usually leads to a better understanding of the distillation course of complex azeotropic mixtures, e.g. high- and low-temperature coal tars, petroleum and synthetic gasoline. In addition, the azeotropic ranges appear in the equations for calculating the boiling temperatures and compositions of ternary homoazeotropes. These equations were found useful in predicting the existence of a large number of ternary azeotropes.[24]

Empirical Correlations

Lecat[19] first observed that the composition of a binary azeotrope is related to the difference between the boiling temperatures of the components. He used a power series to relate the above quantities for the systems formed by a common substance with members of a homologous series.

His relation may be written in the form

$$x_1 = A_0 + A_1\Delta + A_2\Delta^2 + A_3\Delta^3 + \cdots \tag{2}$$

where x_1 is the weight fraction of component 1 in the azeotrope and Δ is the absolute difference between the boiling temperatures of the components.

Lecat[29] proposed also a relation between Δ and the azeotropic deviation, δ, i.e. the absolute difference between the boiling temperatures of the azeotrope and that of the more volatile component, for positive azeotropes, and the less volatile component, for negative azeotropes. The Δ—δ relation, otherwise known as Lecat's rule is

$$\delta = C_0 + C_1\Delta + C_2\Delta^2 + C_3\Delta^3 + \cdots \tag{3}$$

For most cases the terms with Δ higher than Δ^2 may be neglected to give

$$\delta = C_0 + C_1\Delta + C_2\Delta^2 \tag{4}$$

The existence of an approximate relation of this kind was expected for a series of binary regular solutions.[30] For instance, the constants of Equation 4 for the azeotropes of ethanol with aliphatic halogen derivatives are $C_o = 12$, $C_1 = -0.5$ and $C_2 = 0.00526$. The general agreement with Equation 4 for that series is satisfactory, despite that these alcohol solutions show large deviations from regularity.

In the 1940s several graphical correlations of empirical nature for the composition of azeotropes within a series of organic compounds were reported, notably by Mair et al.,[31] Horsley,[32] Meissner and Greenfield,[33] Skolnik,[26] and by Seymour.[34] A decade later Johnson and Madonis[35] described another correlation that included a number of other series of compounds.

The empirical treatment suffered mainly due to many variations in the form of the equations and the number of constants required for their evaluation.

A further attempt in improving the situation in this field was made by Seymour et al.[36] They proposed a correlation between x and Δt in the form of a master equation to correlate, among other things, the data from other series already reported. This equation is

$$\log(10 \frac{x_1}{x_2} = mf(\tau)\,\Delta t + b \tag{5}$$

where x_1 and x_2 are the mole fractions of component 1 and 2 in the azeotrope, respectively, Δt is the difference between the boiling temperatures of the two components, m and b are constants and $f(\tau) = \tau_1/\tau_2$. The quantity τ is defined as the ratio of the boiling temperatures (°K) of a compound and a hypothetical n-paraffin of the same molecular weight. The differences between the calculated and observed azeotropic compositions are reported for 15 series of organic compounds.

An approximate linear relation between the logarithm of the mole fraction, $\log x_i$, of the main and the secondary azeotropic agent, and the average condensation temperature, and was found for several ternary homo- and heteropolyazeotropic mixtures.[37] Empirical correlations for the azeotropic composition apply to both homo- and heteroazeotropes.

Composition of a Binary Homoazeotrope

The theory of regular solutions offers methods for calculating the composition of a binary (positive or negative) azeotrope from certain properties of the pure components and of the azeotrope.

Generally speaking there are three known methods serving this purpose. Two of them involve the boiling temperatures of the pure components, T_1, T_2, and that of the azeotrope, T^{Az}, and were developed by Prigogine[30] and Malesinski.[23] The third method is based on the activity coefficients of the components at the azeotropic point, and is due to Kireev.[38]

The Prigogine equation for equal molar vaporization entropies of the components, $\Delta S_1^o = \Delta S_2^o$, may be written in the form

$$x_2 = \alpha(1 + \alpha)^{-1} \tag{6}$$

were α is the square root of the ratio of azeotropic deviations, i.e., $(\delta_1/\delta_2)^{1/2}$, and x_2 is themole fraction of components 2 in the azeotrope.

When $\Delta S_1^o \neq \Delta S_2^o$, Equation 6 takes the form

$$x_2 = \alpha'(1 + \alpha')^{-1} \tag{7}$$

In this case $\alpha' = c\alpha$ and $c = (\Delta S_1^o/\Delta S_2^o)^{1/2}$.

The Malesinski equation for the components having equal molar vaporization entropies is given by

$$x_2 = 0.5 + \frac{T_1 - T_2}{2z_{12}} \qquad (8)$$

where z_{12} is the half-value of the symmetrical azeotropic range. For systems with unequal molar vaporization entropies, Malesinski's equation requires evaluation of z_u (the upper part of the azeotropic range), the quantity which is not easily available.

Recently, Equations 6, 7, and 8 were evaluated with regard to their usefulness for the calculation of the composition of binary azeotropes.[39]

The simplest and most suitable form of Kireev's equation for the case in which $\Delta S_1^\circ = \Delta S_2^\circ$, is the expression

$$x_2 = (1 + b)^{-1} \qquad (9)$$

where b is the square root of the ratio of the logarithms of the activity coefficients of the components γ_2 and γ_1, i.e. $(\ln \gamma_2/\ln\gamma_2)^{1/2}$. At the azeotropic point the composition of the liquid and the vapor are equal, $x_2 = y_2$, and in the case of an ideal vapor phase the expressions for γ_1 and γ_2 are given by

$$\gamma_1 = P/p_1^0 \text{ and } \alpha_2 = P/p_2^0$$

In these relations p_1° and p_2° are the vapor pressures of the pure components at the boiling temperature of the aceotrope, and P is the total pressure of the mixture at equilibrium.

Equation 9 for $\Delta S_1^\circ \neq \Delta S_2^\circ$ becomes

$$x_2 = (1 + b')^{-1} \qquad (10)$$

where $b' = cb$ and $c = (\Delta S_2^\circ/\Delta S_1^\circ)^{1/2}$.

The possible error in the composition of the binary azeotrope under isobaric conditions is due not only to the deviations of the system from regularity but also to the change of the regular solutions constant, A_{12}, with temperature, and to the differences in the vaporization entropies of the pure components. One of the causes of the deviations of the system from regularity is the nonideality of the vapor phase.

The effect of the nonideality of the vapor phase and the differences in the vaporization entropies of the components on the azeotropic composition were studied by Kurtyka and Kurtyka[40] on the systems of acetic acid with n-paraffins. Acetic acid is associated (dimerized) in the vapor phase and its vaporization entropy is 14.85 cal/g-mole, while those of the hydrocarbons are ~20 cal/g-mole.

To get a general idea about those effects on the azeotropic composition, the results obtained for some systems of acetic acid with n-paraffins are reproduced in part from this paper. The system, $\Delta x_2 = x_{2(calcd.)} - x_{2(obsd.)}$, with x_2 computed by Equation 9; Δx_2, with x_2 computed with the corrections for the dimerization of the acid in the vapor phase, and Δx_2, with x_2 computed by Equation 10 are: acetic acid — n-heptane, 12.3, 1.8, 9.3; acetic acid — n-octane, 12.8, 3.8, 9.2, and acetic acid — n-nonane, 9.8 mole %, 4.3 mole %, and 7.0 mole %, respectively.

These results show that the Δx_2 values are reduced to a reasonable magnitude when the corrections for the dimerization of the acid are taken into account. This simply means that the dimerization of acetic acid in the vapor phase is the dominant factor contributing to the large differences in Δx_2. The effect of the differences in the vaporization entropies of the components on the azeotropic composition as exemplified by the systems containing acetic acid and n-paraffins is small compared to the deviations from regularity caused, among other things, by the dimerization of acetic acid in the vapor phase.

Equation 9 has a built-in unfavorable factor because it involves the vapor pressure, p_i°, of each pure component at the boiling temperature of the azeotrope. p_i° values are almost exclusively computed by means of the Antoine equation. Therefore, the accuracy of the computed p_i° is related to the constants of that equation. However, the effect of the differences in p_i° values on the azeotropic composition is usually small, except in cases where the boiling temperature of the azeotrope is close to the boiling temperature of one component of the system.

The Azeotropic Data of a Binary Heteroazeotrope

Many systems of two or more components exhibit a limited solubility and most of them form positive heteroazeotropes. In binary systems it is easy to predict from the data on critical solution temperatures (CST) whether an azeotrope that occurs at a certain temperature is a homoazeotrope or heteroazeotrope.[41]

To describe the vapor-liquid equilibrium in a heterogeneous system, it is necessary to introduce certain simplifying assumptions.

The simplest case is obtained upon the assumption that liquids are completely immiscible in each other and that the vapor phase is ideal.

Then the composition of a binary heteroazeotrope, y_1, is given by the formula

$$x_1 = y_1 = \frac{p_1^0 (T^{Az})}{p_1^0 (T^{Az}) + p_2^0 (T^{Az})} \qquad (11)$$

where p_1° and p_2° are the vapor pressures of pure components 1 and 2, respectively, expressed as functions of the boiling temperature of the azeotrope, T^{Az}, or simply, at the boiling temperature of the azeotrope.

And for the boiling temperature of the azeotrope, T^{Az} we have the relation

$$P = p_1^0 (T^{Az}) + p_2^0 (T^{Az}) \quad (12)$$

where P is the total pressure at equilibrium.

In the cases in which the condition of complete immiscibility is not satisfied, we introduce the correction factors in terms of the activities of the components, α_1 and α_2 in the liquid phase.

Accordingly, the expression for the composition of an azeotrope takes the form

$$y_1 = \frac{p_1^0 (T^{Az}) \alpha_1}{p_1^0 (T^{Az}) \alpha_1 + p_2^0 (T^{Az}) \alpha_2} \quad (13)$$

And the boiling temperature of the azeotrope may be calculated from

$$P = p_1^0 (T^{Az}) \alpha_1 + p_2^0 (T^{Az}) \alpha_2 \quad (14)$$

There is also another possibility, namely to assume that the activities of both components in the liquid phase are equal, i.e., $\alpha_1 = \alpha_2 = \alpha$. The systems behavior of which is well described by a common activity, are those of nitromethane with n-paraffins.[42] The computation of the activities, α_1 and α_2 of the components is a straightforward procedure involving the solution of Equations 13 and 14.

Equations 11 and 12 were found to be satisfied for the systems of aromatic and n-paraffin hydrocarbons with water. Examples of the systems in which considerable miscibility of the components occurs are: n-butanol-water, aniline-water, and acetonitrile-n-paraffins. For instance, the solubility of aniline in water at 90°C is 6.4 g/100 mℓ.

For calculating p_1^o and p_2^o at the boiling temperatures of the respective azeotropes, the Antoine equation is recommended.[43] The constants of the Antoine equation, viz. $\log p^o = A - B/(C + t)$, where p^o is expressed in millimeters Hg and t is in this case the boiling temperature of the azeotrope (°C), are compiled for quite a large number of organic compounds.[44,45]

For example, the activities of the components, α_1 and α_2 in the systems aniline-water[2] and n-butanol-water were found to be: 0.315, 1.00, and 0.649 and 0.9824, respectively.

The relations, thus described, can be easily extended to ternary and multicomponent heteroazeotropes.

THE PREDICTION OF AZEOTROPIC DATA IN TERNARY SYSTEMS

To separate nonideal liquid mixtures by fractional distillation, it is important to establish, among other things, whether the mixtures to be separated form azeotropes. In view that the experimental methods for determining the azeotropic composition are rather difficult and time-consuming operations, expecially in the case of azeotropes containing three and more components, several computational methods were developed to predict the composition of ternary and multicomponent azeotropes.

In general, two approaches in this area may be distinguished. The first one, originated by Haase[46,47] and developed by Malesinski,[48] is based on the theory of regular solutions and is restricted to homoazeotropes. The Malesinski method makes it possible to predict the appearance of ternary azeotropes of various types and to calculate their composition and boiling temperatures. The second approach, which is not limited only to the azeotropic points, is based on the use of various equations of empirical and semiempirical nautre, that relate the liquid-phase activity coefficients to the composition of the liquid phase.

The Azeotropic Data of a Ternary Homoazeotrope from the Binary Azeotropic Data

The composition of a ternary homoazeotrope (1,2,3) under isobaric conditions, when the vaporization entropies of the components are equal, is related to the two pairs of binary azeotropes by the equations.

Pairs (1,2) and (2,3):

$$x_1 = \frac{x_1^{(1,2)} + a x_3^{(2,3)}}{1 - ab} \quad (15)$$

$$x_3 = \frac{x_3^{(2,3)} + b x_1^{(1,2)}}{1 - ab} \quad (16)$$

where

$$a = \frac{z_{13} - z_{23} - z_{12}}{2z_{12}} \quad \text{and} \quad b = \frac{z_{13} - z_{23} - z_{12}}{2z_{23}}$$

$x_1^{(1,2)}$ and $x_3^{(2,3)}$ are the mole fractions of components 1 and 3 in the binary azeotropes (1,2) and (2,3), respectively; x_1 and x_3 are the mole fractions of the above components in the ternary azeotrope and z_{12}, z_{13}, and z_{23} are the half-values of the symmetrical azeotropic range.

When the vaporization entropies of the components are not equal, the regular solution constants, A_{12}, A_{13}, and A_{23} take the place of z_{12}, $z_{13}z$, and z_{23}, respectively.

Pairs (1,2), and (2,3):

$$x_1 = \frac{x_1^{(1,3)} + cx_2^{(2,3)}}{1 - cd} \qquad (17)$$

$$x_2 = \frac{x_2^{(2,3)} + dx_1^{(1,3)}}{1 - cd} \qquad (18)$$

where

$$c = \frac{z_{12} - z_{23} - z_{13}}{2z_{13}} \quad \text{and} \quad d = \frac{z_{12} - z_{23} - z_{13}}{2z_{23}}$$

Pairs (1,2) and (1,3):

$$x_2 = \frac{x_2^{(1,2)} + ex_3^{(1,3)}}{1 - ef} \qquad (19)$$

$$x_3 = \frac{x_3^{(1,3)} + fx_2^{(1,2)}}{1 - ef} \qquad (20)$$

where

$$e = \frac{z_{23} - z_{13} - z_{12}}{2z_{12}} \quad \text{and} \quad f = \frac{z_{23} - z_{13} - z_{12}}{2z_{13}}$$

The values of z_{12}, z_{13}, and z_{23} should be computed from Equation 1.

The ternary system is zeotropic if, for example, the composition of one component of the system is zero or takes a negative value.

In the case when one binary system constituting the ternary system is zeotropic, the respective z_{ij} value may be estimated from that of any close member of a homologous series, its isomers or closely related substances.

The sources of error in the calculated composition of the ternary azeotrope are similar to those of the binary azeotrope.

The boiling temperature of a ternary homoazeotrope, T, with component 2 as the reference component, can be computed from the equation

$$T = T_2 - \frac{\delta_2^{(1,2)} + \left(\frac{\delta_2^{(1,2)}}{z_{12}} \cdot \frac{\delta_2^{(2,3)}}{z_{23}}\right)^{1/2} (z_{13} - z_{23} - z_{12}) + \delta_2^{(2,3)}}{1 - \frac{(z_{13} - z_{23} - z_{12})^2}{4z_{12}z_{23}}} \qquad (21)$$

where $\delta_2^{(1,2)}$ and $\delta_2^{(2,3)}$ are the azeotropic depressions or elevations in the azeotropes (1,2) and (2,3) in relation to component 2, e.g. $\delta_2^{(1,2)} = T_2 - T^{(1,2)}$: $T^{(1,2)}$ is the boiling temperature of the azeotrope (1,2).

By convention, the $\delta_i^{(i,j)}$ and z_{ij} are positive for a positive azeotrope and negative for a negative one.

By interchanging the components, the boiling temperature of the ternary homoazeotrope may be computed from the two remaining sets of the pairs of the components.

For mixtures which exactly fulfil the requirements for regular solutions, the result is independent of the choice of the reference component. If there are deviations from regularity, then the values of z_{ij} or $\delta_i^{(i,j)}$ depend on which experimentally determined quantity was used in the calculations.

For positive azeotropes Equations 15 to 21 usually give good results for the calculated azeotropic data.[23] In the case of positive-negative (saddle) azeotropes the agreement between the calculated and observed azeotropic data is less satisfactory. But this is understandable in view that these systems are complex mixtures, which often contain polar and associated components, e.g. alcohols and low-molecular fatty acids.

For the series of ternary saddle systems acetic acid-pyridine(2-picoline)-n-paraffins, the results obtained for the calculated azeotropic compositions improved considerably, when the corrections for the association (dimerization) of acetic acid in the vapor phase were taken into account.[49]

Equations 15 to 21 made it possible to predict the existence of a large number of ternary saddle (positive-negative) azeotropes that may appear in the course of fractional distillation of certain fractions of coal tar.[24]

It has been found, for instance, that ternary saddle azeotropes are formed in the series of ternary systems aniline-phenol-n-paraffins (ranging from nonane to tetradecane). Only one saddle azeotrope of this type was examined in the system aniline(1)-phenol(2)-tridecane(3).[50] For this system the differences between the calculated and observed azeotropic compositions, Δx_1, Δx_2, and Δx_3, and that of the boiling temperatures of the azeotrope, ΔT, are: 1.0, −2.1, and 1.0 mole %, and −0.11°C, respectively.

In general, good results for ternary homoazeotropes are obtained for systems which show moderate deviations from regularity, any of the binary azeotropes is close to what is called a tangent azeotrope,[23] and the differences between the boiling temperature of the ternary azeotrope and those of the components are not large.[24]

It should be also mentioned that two empirical correlations for the azeotropic composition in the series of ternary saddle azeotropes were proposed by Zeiborak.[51]

Composition of a Ternary Homoazeotrope from Vapor-Liquid Equilibrium Data of the Binary Systems

A general method for predicting the vapor-liquid equilibrium data in ternary and multicomponent systems can be restricted to the azeotropic point by the use of the relative volatility, α_{ij}, defined as

$$\alpha_{ij} = \frac{y_i x_j}{x_i y_j} = \frac{p_i^0 \gamma_i}{p_j^0 \gamma_j} = 1 \qquad (22)$$

In relation (Equation 22) p_i^0 and p_j^0 are the vapor pressures of components i and j at the boiling temperature of the azeotrope, and γ_i and γ_j are the liquid-phase activity coefficients of these components.

The method involves minimization of the function, f, which for a ternary azeotrope becomes

$$f = |\alpha_{13} - 1| + |\alpha_{23} - 1| \qquad (23)$$

α_{13} and α_{23} are the relative volatilities of the pairs of components (1,3) and (2,3), respectively.

The value of the function, f, sufficiently close to zero, corresponds to the azeotropic composition.

This method was used by Aristovicz and Stepanova[52] for the calculation of the azeotropic composition in 19 ternary systems and 1 quaternary system. The results obtained were good. In this procedure the liquid-phase activity coefficients were correlated by the Wilson equation.[53,54]

For ternary and multicomponent heteroazeotropes the above procedure remains essentially unchanged but requires the use of an equation that is applicable to partially miscible systems, e.g. the NRTL (Non-Random, Two-Liquid) equation.[55]

The Azeotropic Data of a Ternary Heteroazeotrope

On the basis of the arguments similar to those described previously, which are applicable to heteroazeotropic systems of any number of components, we can obtain the expressions for the composition and the boiling temperature of a ternary heteroazeotrope.

For the case in which the components are immiscible in each other, the expressions for y_1, y_2, and T^{Az} of a ternary heteroazeotrope are

$$y_1 = \frac{p_1^0}{p_1^0 + p_2^0 + p_3^0} \qquad (24)$$

$$y_2 = \frac{p_2^0}{p_1^0 + p_2^0 + p_3^0} \qquad (25)$$

and

$$P = p_1^0 + p_2^0 + p_3^0 \qquad (26)$$

where p_1^0, p_2^0 and p_3^0 are the vapor pressures of the pure components at the boiling temperature of the azeotrope, P is the total pressure of the mixture at equilibrium, and y_1 and y_2 are the mole fractions of component 1 and 2 in the vapor of the azeotrope, respectively.

Other cases, in which the condition of complete immiscibility of the components is not fulfilled, may be described by introducing the activities of the components, α_1, α_2, and α_3.

The procedure, which is straightforward, and involves the multiplication of each p^0_i by α_i in expressions (24-26), will not be reproduced here.

The activities, α_1, α_2, and α_3 are related to P, p_1^0, p_2^0, and p_3^0 by

$$\alpha_1 = \frac{P - p_2^0\,\alpha_2 - p_3^0\,\alpha_3}{p_1^0} \qquad (27)$$

$$\alpha_2 = \frac{Py_2}{p_2^0} \qquad (28)$$

and

$$\alpha_3 = \frac{Py_3}{p_3^0} \qquad (29)$$

Only one series of ternary heteroazeotropes with three liquid phases has been investigated to date.[6] The heteroazeotropes occur in the systems nitromethane-water-n-paraffins. For this series satisfactory agreement was obtained between the calculated (using Equations 24 and 25) and observed azeotropic compositions, but the calculated boiling temperatures of the azeotropes by Equation 26 were found to be much lower than those observed. Low calculated boiling temperatures are due to the relatively high miscibility of nitromethane in water at the respective boiling temperatures of the azeotropes. The difference is lower for the systems with higher-boiling hydrocarbons and tends to the difference between the calculated and observed boiling temperatures for the nitromethane-water system.

It is interesting to note that in some cases the behavior of a series of ternary heteroazeotropes, in which two components remain unchanged for the series, may be described by the activities α_1, α_2, and α_3, which are common for each component within the whole series. This case is exemplified by the series of ternary heteroazeotropes water-pyridine-n-paraffins.[56]

Studies of ternary heteroazeotropes based on the theory of regular solutions were made by Malesinska and Malesinski[57] and by Stecki.[58]

All the examined heteroazeotropes were found to be positive. The existence of negative heteroazeotropes is rather doubtful.

REFERENCES

1. **Dalton, J.,** *Mem. Manchester Phil. Soc.,* 5, 585, 1802; *Ann. Phil.,* 9, 186, 1817.
2. **Wade, J. and Merriman, R. W.,** *J. Chem. Soc. Trans.,* 99, 997, 1911.
3. **Swictoslawski, W.,** *Ebulliometric Measurements,* Reinhold, New York, 1945.
4. **Lecat, M.,** *Compt. Rend.,* 183, 880, 1926.
5. **Swietoslawski, W.,** *Rocz. Chem.* 26, 632, 1952; *Bull. Acad. Polon. Sci. Cl. III,* 1, 63, 1953.
6. **Malesinska, B. and Malesinski, W.,** *Bull. Acad. Polon. Sci. Ser. Sci. Chim.,* 11, 475, 1963.
7. **Ewell, R. L. and Welch, L. M.,** *Ind. Eng. Chem.,* 37, 1244, 1945.
8. **Lang, H.,** *Z. Physik. Chem.,* 196, 278, 1950.
9. **Swietoslawski, W. and Trabczynski, W.,** *Bull. Acad. Polon. Sci. Cl. III,* 3, 333, 1955.
10. **Galska-Krajewska, A.,** *Bull. Acad. Polon. Sci. Ser. Sci. Chim.,* 10, 45, 51, 1962.
11. **Zieborak, K.,** *Bull. Acad. Polon. Sci. Cl. III,* 3, 53, 1955.
12. **Kurtyka, Z. M.,** *J. Chem. Eng. Data,* 16, 310, 1971.
13. **Kurtyka, Z.,** *Bull. Acad. Polon. Sci. Ser. Sci. Chim.,* 9, 741, 1961.
14. **Zieborak, K. and Wyrzykowska-Stankiewicz, D.,** *Bull. Acad. Polon. Sci. Ser. Sci. Chim.,* 8, 137, 1960.
15. **Orszagh, A., Lelakowska, J., and Beldowicz, M.,** *Bull. Acad. Polon. Sci. Ser. Sci. Chim. Geol. et Geogr.,* 6, 419, 1958.
16. **Orszagh, A., Lelakowska, J., and Radecki, A.,** *Bull. Acad. Polon. Sci. Ser. Sci. Chim. Geol. et Geogr.,* 6, 605, 1958.
17. **Ostwald, W.,** *Lehrbuch der Allgemeinen Chimie,* Vol. 2, Engelman, Leipzig, 1899.
18. **Ryland, G.,** *Am. Chem. J.,* 22, 384, 1899; *Chem. News,* 81, 15, 42, 50, 1900.
19. **Lecat, M.,** *L'Azeotropisme,* Lamartin, Bruxelles, 1918.
20. **Lecat, M.,** *Tables Azeotropiques,* Vol. 1, L'Auteur, Bruxelles, 1949.
21. **Horsley, L. H.,** *Azeotropic Data-III,* American Chemical Society, Washington, D.C., 1973.
22. **Swietoslawski, W.,** *Azeotropy and Polyazeotrophy,* Pergamon Press, Oxford, London, 1963.
23. **Malesinski, W.,** *Azeotropy and Other Theoretical Problems of Vapour-Liquid Equilibrium,* Wiley-Interscience, New York, 1965.
24. **Kurtyka, Z. M.,** *Azeotropy and Its Applications,* to be published.
25. **Swietoslawski, W.,** *Bull. Acad. Sci. Polon. Ser. A,* 19, 29, 1950, *Przem. Chem.,* 7, 363, 1951.
26. **Skolnik, H.,** *Ind. Eng. Chem.,* 40, 442, 1948.
27. **Malesinski, W.,** *Bull. Acad. Polon. Sci. Cl. III,* 3, 601, 1955; 4, 295, 1956.
28. **Zawisza, A.,** *Bull. Acad. Polon. Sci. Ser. Sci. Chim.,* 9, 141, 1961.
29. **Lecat, M.,** *Azeotropisme et Distillation, Traite de Chimie Organique, Vol. I,* V. Grignard, Ed., Mason et Cie., Paris, 1935.
30. **Prigogine, I., and Defay, R.,** *Chemical Thermodynamics,* translated by D. H. Everett, Longmans, Green, London, 1954.
31. **Mair, B. J., Glasgow, A. R., and Rossini, F. D.,** *J. Res. Bur. Std.,* 27, 39, 1941.
32. **Horsley, L. H.,** *Anal. Chem.,* 19, 508, 1947.
33. **Meissner, H. P. and Greenfield, S. H.,** *Ind. Eng. Chem.,* 40, 438, 1948.
34. **Seymour, K. M.,** Abstracts 50-I, 110th National Meeting of the American Chemical Sciety, Chicago, September 1946.
35. **Johnson, A. I. and Madonis, J. A.,** *Can. J. Chem. Eng.,* 37, 71, 1959.
36. **Seymour, K. M., Carmichael, R. H., Carter, J., Ely, J., Isaacs, E., King, J., Taylor, R., and Northern, T.,** *Ind. Eng. Chem. Fundam.,* 16, 200, 1977.
37. **Orszagh, A.,** *Rocz. Chem.,* 29, 623, 636, 1955.
38. **Kireev, V. A.** *Acta Physicochim. URSS,* 14, 371, 1941.
39. **Kurtyka, Z. M. and Kurtyka, A.,** *Ind. Eng. Chem. Fundam.,* 19, 225, 1980.
40. **Kurtyka, Z. M. and Kurtyka, Z.,** *Ind. Eng. Chem. Fundam.,* in press.
41. **Francis, A. W.,** *Critical Solution Temperatures,* American Chemical Society, Washington, D.C. 1961.

42. **Malesinska, B . and Malesinski, W.,** *Bull. Acad. Polon. Sci. Ser. Sci. Chim.*, 11, 469, 1963.
43. **Antoine, C.,** *Compt. Rend.*, 107, 681, 836, 1143, 1888.
44. **Hala, E., Wichterle, I., Polak J., and Boublik, T.,** *Vapour-Liquid Equilibrium Data at Normal Pressures,* Pergamon Press, Oxford, London, 1968.
45. **Hirata, M., Ohe, S., and Nagahama, K.,** *Computer-Aided Data Book of Vapour-Liquid Equilibria,* Kodansha-Elsevier, Tokyo, Amsterdam, 1975.
46. **Haase, R.,** *Z. Physik. Chem.*, 195, 362, 1950.
47. **Haase, R.,** *Termodynamik des Mischphasen,* Springer-Verlag, Berlin, 1956.
48. **Malesinski, W.,** *Bull. Acad. Polon. Sci. Cl. III,* 4, 701, 709, 1956; 5, 177, 183, 1957.
49. **Zeiborak, K. and Wyrzykowska-Stankiewicz, D.,** *Bull. Acad. Polon. Sci. Ser. Sci. Chim. Geol. et Geogr.*, 6, 755, 1958.
50. **Stadnicki, J. S.,** *Bull. Acad. Polon. Sci. Ser. Sci. Chim.*, 10, 357, 1962.
51. **Zieborak, K.,** *Bull. Acad. Polon. Sci. Cl. III,* 3, 531, 1955.
52. **Aristovicz, V. Y. and Stepanova, E. I.,** *Zhur. Prikl. Khim.*, 43, 2192, 1970.
53. **Wilson, G. M.,** *J. Am. Chem. Soc.*, 86, 127, 1964.
54. **Prausnitz, J. M.,** *Molecular Thermodynamics of Fluid-Phase Equilibria,* Prentice Hall, Englewood Cliffs, New Jersey, 1969.
55. **Renon, H. and Prausnitz, J. M.,** *A. I. Ch. E. Journal,* 14, 135, 1968.
56. **Trabczynski, W.,** *Bull. Acad. Polon. Sci. Ser. Sci. Chim. Geol. et Geogr.*, 6, 269, 1958.
57. **Malesinska, B. and Malesinski, W.,** *Bull. Acad. Polon. Sci. Ser. Sci. Chim.*, 12, 861, 867, 1964.
58. **Stecki, J.,** *Bull. Acad. Polon. Sci. Cl. III,* 5, 421, 1957; *Ser. Sci. Chim. Geol. et Geogr.*, 6, 47, 1958.

TABLES OF AZEOTROPES AND ZEOTROPES

In Tables 1, 2, and 3 the different types of azeotropes are identified as:

1. Homoazeotrope, positive; no marking
2. Homoazeotrope, negative; N
3. Homoazeotrope, saddle or positive-negative; S
4. Heteroazeotrope, positive, with two phases; H
5. Heteroazeotrope, positive, with three phases; H-3

Throughout Tables 1,2, and 3 the azeotropic composition is expressed as weight percent (wt %). In Table 1 only the weight percent of component 2, x_2, the variable component, is listed. However, in Tables 2, and 3 the weight percents of all of the components are listed.

Compounds are listed in the following tables according to the empirical formula convention employed by Chemical Abstracts. As further assistance in locating particular systems the following index has been arranged in alphabetical order with respect to component X_1 of each system. The entry number for the particular system is in the second column of the list.

A much more extensive compilation has been done by Horsley (Reference 21).

Table 1
BINARY SYSTEMS

Component X_1	Entry No.	Component X_1	Entry No.
Acetal	1364	Benzyl phenyl ether	1743
Acetaldehyde	377	Borneol	1717
Acetic acid	383	Boron fluoride	2
Acetone	544	Boron hydride	10
Acetonitrile	369	Bromoacetic acid	349
Acetophenone	1529	Bromodichloromethane	217
Acrylonitrile	525	Bromoform	224
Allyl alcohol	554	Bromomethane	271
Aluminum chloride	1	1-Butanethiol	1032
n-Amyl alcohol	1143	1-Butanol	908
p-tert-Amyl alcohol	1729	2-Butanol	939
Aniline	1255	2-Butanone	770
o-Anisidine	1511	2-Butoxyethanol	1368
Benzaldehyde	1408	Butyl acetate	1335
Benzene	1180	Butyl alcohol	908
Benzoic acid	1415	Butyl formate	1100
Benzonitrile	1400	Butyl nitrite	893
Benzyl alcohol	1448	Butyraldehyde	782

Table 1 (Continued)
BINARY SYSTEMS

Component X_1	Entry No.
Butyronotile	764
Camphene	1715
Capric acid	1722
Caproic acid	1338
Capronitrile	1325
Caprylic acid	1622
Carbon disulfide	204
Carbon tetrachloride	185
Carvacol	1696
Carvone	1703
Chloroacetic acid	354
Chloroform	232
m-Cresol	1473
o-Cresol	1462
p-Cresol	1485
Cumene	1655
Cyclohexanol	1327
Cyclohexanone	1297
Diethylene glycol	1008
Doxane	811
Dipropylene glycol	1386
Enanthic acid	1514
Ethanol	460
2-Ethoxyethanol	994
Ethyl	
acetate	821
acetoacetate	1301
alcohol	460
aniline	1592
benzene	1551
bromoacetate	750
carbamate	631
chloroacetate	756
formate	563
fumarate	1613
lactate	1136
maleate	1616
methyl sulfide	712
nitrate	455
oxalate	1319
phenol, para	1554
pyruvate	1071
salicylate	1651
succinate	1619
Ethylene glycol	176
Ethylene sulfide	411
Ethylenediamine	522
Ethylidine diacetate	1313
Formic acid	242
2-Furaldehyde	1044
Glycol diacetate	1324
Glycol monoacetate	865
Glycerol	687
Guaiacol	1497
1-Heptanol	1521
n-Heptyl alcohol	1521
Hexachloroethane	337
1-Hexanol	1357
n-Hexyl alcohol	1357
Hydrogen chloride	14
Hydrogen cyanide	11
Hydrogen fluoride	15
Indole	1527
Iodobenzene	1168
Iodoethane	420
Iodomethane	272
Isoamyl alcohol	1148
Isoamyl benzoate	1735
Isoamyl formate	1345
Isoamyl oxalate	1739
Isobutyl nitrate	902
Isobutyronitrile	767
Isobutyric acid	829
Isopropyl alcohol	663
Isopropyl lactate	1354
Isopropyl methyl sulfide	1038
Isovaleric acid	1105
Levulinic acid	1074
2,4-Lutidine	1501
Menthol	301
Mesitol	1662
Mesitylene	1656
Methanol	301
Methyl	
acetate	567
acetoacetate	1085
acetophenone, para	1643
alcohol	301
aniline	1503
anisole, para	1562
butyrate	1120
chloroacetate	534
disulfide	520
formate	414
fumarate	1284
lactate	876
maleate	1290
1-Methylnaphthalene	1724
2-Methylnaphthalene	1728
2-Methyl-2-propanol	949
2-Methylthiophene	1063
3-Methylthiophene	1067
2-Octanol	1630
sec-Octyl alcohol	1630
Naphthalene	1679
Nitrobenzene	1159
Nitroethane	453
Nitromethane	276
1-Nitropropane	654
o-Nitrotoluene	1426
p-Nitrotoluene	1437
Pelargonic acid	1669
2-Pentanone	1096
Perfluorobutyric acid	724
Phenethyl alcohol	1568
o-Phenetidine	1600
p-Phenetidine	1607
Phenol	1190

Table 1 (Continued)
BINARY SYSTEMS

Component X_1	Entry No.	Component X_1	Entry No.
Phenyl acetate	1538	Pyridine	1056
Phenyl benzoate	1741	Pyrogallol	1252
Phenyl ether	1732	Pyrrol	734
o-Phenylenediamine	1278	Pyruvic acid	527
3-Phenylpropanol	1664	Qunialdine	1683
Phosphorus oxychloride	16	Quinoline	1635
Phosphorus trichloride	17	Resorcinol	1238
2-Picoline	1273	Silicon tetrachloride	20
3-Picoline	1276	Tetrachlorothylene	327
Pinacol	1377	Tetrahydrothiophene	888
1-Propanethiol	716	Thioacetic acid	380
1-Propanol	673	Thiophene	727
Propioamide	593	Thymol	1707
Propionic acid	574	Tin chloride	26
Propionitrile	541	Toluene	1442
Propiophenone	1648	o-Toluidine	1505
Propyl		Trichloroacetic acid	346
acetate	1123	Trichloroethylene	342
alcohol	673	Trichlorofluoromethane	173
benzene	1659	Trinitromethane	176
benzoate	1690	Triethylene glycol	1391
formate	858	Valeric acid	1128
isovalerate	1626	Water	28
lactate	1350	2,4-Xylenol	1576
nitrite	661	3,4-Xylenol	1582
propionate	1348	2,4-Xylidine	1595
Pseudocumene	1660		

Table 2
TERNARY SYSTEMS

Component X_1	Entry No.	Component X_1	Entry No.
Acetic acid	108	Hydrogen chloride	5
Acetone	141	Hydrogen cyanide	3
Acetonitrile	106	Hydrogen fluoride	7
Aniline	167	Isobutyl alcohol	161
Argon	1	Isobutyl lactate	176
1-Butanol	158	Isopropyl alcohol	151
2-Butanone	154	Methanol	101
Butyric acid	156	Methyl formate	
Carbon tetrachloride	92	Nitromethane	129
Chlorine trifluoride	4	Phenol	163
Chloroform	93	1-Propanol	152
m-Cresol	175	1-Propanol, 2-methyl	161
m,p-Cresol (mixture)	172	2-Propanol	151
p-Dioxane	157	Propionic acid	145
Ethanol	130	Propyl lactate	171
Ethyl benzene	166, 177	Pyridine	162
Ethylene glycol	135	Silicon tetrafluoride	8
Hydrogen bromide	2	Trichloroethylene	107

Table 3
QUATERNARY AND QUINARY SYSTEMS

Component X_1	Entry No.	Component X_1	Entry No.
Acetic acid	18	Hydrogen cyanide	1
Acetone	20	Isopropyl alcohol	21
Chloroform	16	Water	3

BINARY SYSTEMS

No.	System	B.P. (°C)	Azeotropic data: Compn. X_2 (wt. %)	Azeotropic data: B.P. (°C)
	Aluminum chloride	183		
1	-tantalum chloride (2)	242	90.4	235
	Boron fluoride	−100		
2	-boron hydride	−92	22.8	−106
3	-acetonitrile, N	81.6	38	101
4	-methyl formate, N	31.9	47	91
5	-methyl ether, N	−21	40	127
6	-ethyl formate, N	54.1	52	102
7	-methyl acetate, N	57.1	52	110
8	-ethyl ether, N	34.5	52	125
9	-ethyl propionate, N	99.15	60	116
	Boron hydride	−92.5		
10	-hydrogen chloride	−85	36	−94
	Hydrogen cyanide	26		
11	-methyl alcohol	64.7	Zeotropic	
12	-methyl formate	31.7	48	24
13	-ethyl nitrite	17.4	85	16.5
	Hydrogen chloride	−85		
14	-methyl ether, N	−22	62	−2
	Hydrogen fluoride	19.4		
15	-ethyl ether, N	34.5	60	74
	Phosphorus oxychloride	107.2		
16	-titanium tetrachloride, N	136.5	53.4	143.2
	Phosphorus trichloride	76		
17	-cyclohexane	80.75	Zeotropic	
18	-2,3-dimethylpentane	80.5	27	74.2
19	-2,2,3-trimethylbutane	80.9	23	74.5
	Silicon tetrachloride	56.9		
20	-carbon tetrachloride	76.75	Zeotropic	
21	-chloroform	61.0	30	55.6
22	-nitromethane	101	6	53.0
23	-acetonitrile	82	9.4	49.0
24	-acrylonitrile	79	11	51.2
25	-propionitrile	97	8	55.6
	Tin chloride	113.85		
26	-toluene	110.7	48	109.15
27	-2,5-dimethylhexane	109.4	60	107.5
	Water	100		
28	-hydrogen chloride, N	−85	20.22	108.58
29	-hydrogen bromide, N	−73	47.5	126
30	-hydrogen iodide, N	−34	57	127
31	-hydrogen fluoride, N	19.4	35.6	111.35
32	-nitric acid, N	86	67.4	120.7
33	-hydrogen peroxide	152.1	Zeotropic	
34	-hydrazine, N	113.8	67.7	120
35	-carbon tetrachloride, H	76.75	95.9	66
36	-carbon disulfide, H	46.25	97.2	42.6
37	-chloroform, H	61	97.2	56.1
38	-formic acid, N	100.75	77.4	107.2
39	-nitromethane, H	101.2	76.4	83.59
40	-tetrachloroethylene, H	121	82.8	88.5
41	-trichloroethylene	86.2	83	73.4
42	-acetonitrile, H	80.1	83.7	76.5
43	-acetic acid	118.1	Zeotropic	
44	-acetamide, H	221.2	Zeotropic	
45	-nitroethane, H	114.07	71.5	87.22
46	-ethyl nitrate	87.68	78	74.35
47	-ethyl alcohol	78.32	96	78.17
48	-methyl sulfate	189.1	27	98.6
49	-acrylonitrile, H	77.2	85.7	70.6
50	-acrolein, H	52.8	97.4	52.4
51	-acetone	56.1	Zeotropic	
52	-allyl alcohol	96.9	72.3	88.9
53	-propionaldehyde, H	47.9	98	47.5
54	-ethyl formate	54.2	95	52.6
55	-propionic acid	141.1	17.7	99.9
56	-trioxane	114.5	70	91.4
57	-l-chloropropane	46.6	97.8	44
58	-isopropyl alcohol	82.3	87.4	80.3
59	-propyl alcohol, 740 mm	97.3	71.7	87
60	-perfluorobutyric acid	122.0	29	97
61	-crotonic acid	189	2.2	99.9
62	-methyl acrylate	80	92.8	71
63	-ethyl chloroacetate	143.5	54.9	95.2
64	-butyronitrile, H	117.6	67.5	88.7
65	-isobutyronitrile, H	103	77	82.5
66	-ethyl vinyl ether, H	35.5	98.5	34.6
67	-butyric acid	163.5	3	99.4
68	-ethyl acetate	77.15	91.53	70.38
69	-isopropyl formate	68.8	97	65.0
70	-propyl formate	80.9	97.7	71.6
71	-methyl lactate	144.8	20	99
72	-butyl alcohol, H	117.4	57.5	92.7
73	-*sec*-butyl alcohol	99.5	73.2	87.0
74	-pyridine	115.5	58.7	93.6
75	-furfuryl alcohol	169.35	20	98.5
76	-furfurylamine	144	26	99
77	-isoprene	34.1	99.86	32.4
78	-cyclopentanone	130.8	57.6	94.6
79	-allyl acetate	104.1	85.3	83
80	-cyclopentanol	140.85	42	96.25
81	-valeraldehyde, H	103.3	81	83
82	-butyl formate	106.6	85.5	83.8
83	-isopropyl acetate	88.6	89.4	76.6
84	-isovaleric acid	176.5	18.4	99.5
85	-methyl butyrate	102.65	88.5	82.7
86	-methyl isobutyrate	92.3	93.2	77.7
87	-valeric acid	188.5	11	99.8
88	-piperidine	105.8	65	92.8
89	-n-pentane, H	36.1	98.6	34.6
90	-n-amyl alcohol, H	137.8	45.6	95.8
91	-*tert*-amyl alcohol	102.25	72.5	87.35
92	-2-pentanol	119.3	63.5	91.7
93	-N-methylbutylamine	91.1	85	82.7
94	-chlorobenzene	131.8	71.6	90.2
95	-nitrobenzene, H	210.85	—	98.6
96	-benzene, H	80.1	91.17	69.25
97	-phenol	182	9.2	99.52
98	-aniline, H, 742 mm	184.3	19.2	98.6
99	-2-picoline	129.5	52	93.5
100	-3-picoline	144.1	40	97
101	-4-picoline	144.3	37.2	97.35
102	-2,5-dimethylfuran	93.3	88.3	77.0
103	-cyclohexene, H	82.75	91.07	70.8
104	-ethyl crotonate	137.8	62	93.5
105	-ethylene glycol diacetate	190.8	15.4	99.7
106	-butyl chloroacetate	181.9	24.5	98.12
107	-cyclohexane, H	80.8	91.6	69.5
108	-amyl formate	132	71.6	91.6
109	-butyl acetate	126.2	71.3	90.2
110	-ethyl butyrate	120.1	78.5	87.9
111	-isoamyl formate	124.2	79	90.2
112	-isobutyl acetate	117.2	83.5	87.4
113	-isopropyl propionate	110.3	80.1	85.2
114	-propyl propionate	122.1	77	88.9
115	-paraldehyde, H	124	71.5	90
116	-n-hexane, H	68.7	94.4	61.6
117	-butyl ethyl ether, H	92.2	88.1	76.6
118	-n-hexyl alcohol, H	157.1	32.8	97.8
119	-acetal	103.6	85.5	82.6
120	-pinacol	174.35	Zeotropic	
121	-toluene, H	110.7	86.5	84.1
122	-anisole	153.85	59.5	95.5
123	-benzyl alcohol, H	205.2	9	99.9
124	-guaiacol	205.0	12.5	99.5
125	-2,6-lutidine, H	144.0	48.2	96.02
126	-o-toluidine	199.7	84.6	—
127	-p-toluidine, H	200.4	86.2	—
128	-2-heptanone	149	52	95
129	-3-heptanone	147.6	57.8	94.6
130	-4-heptanone	143.7	59.5	94.3
131	-ethyl valerate	145.45	60	91.5
132	-isoamyl acetate	142	63.7	93.8
133	-isobutyl propionate	136.85	47.8	92.75
134	-n-heptane, H	98.4	87.1	79.2
135	-benzyl formate	202.3	20	99.2
136	-methyl benzoate	199.45	20.8	99.08
137	-phenyl acetate	195.7	24.9	98.9
138	-ethylbenzene, H	136.2	67.0	92.0
139	-m-xylene, H	139	64.2	92
140	-N-ethylaniline	204.8	16.1	99.2
141	-1-octene, H	121.28	71.3	88.0
142	-hexyl acetate	171.0	39	97.4
143	-isoamyl propionate	160.3	51.5	96.55
144	-isobutyl butyrate	156.8	54	96.3
145	-isobutyl isobutyrate	147.3	60.6	95.5
146	-propyl isovalerate	155.8	54.8	96.2
147	-n-octane, H	125.7	75.5	89.6
148	-isooctane, H	99.3	88.9	78.8
149	-butyl ether, H	142.6	67	92.9
150	-n-octyl alcohol, H	195.15	10	99.4
151	-dibutylamine, H	159.6	49.5	97

No.	System	B.P. (°C)	Azeotropic data Compn. X_2 (wt. %)	B.P. (°C)
152	-quinoline, H	237.3	3.4	—
153	-ethyl benzoate	212.4	16.0	99.4
154	-cumene, H	152.4	56.2	95
155	-mesitylene, H	164.6	—	96.5
156	-triallylamine, H	151.1	62	95
157	-isoamyl butyrate	178.5	36.5	98.05
158	-isobutyl carbonate	190.3	26	98.6
159	-n-nonane, H	150.8	18	94.8
160	-naphthalene, H	218.0	16	98.8
161	-methyl phthalate, H	283.2	2.5	99.95
162	-nicotine, H	—	2.5	99.85
163	-camphene, H	159.6	—	96.0
164	-n-decane, H	173.3	—	97.2
165	-n-undecane, H	194.5	4.0	98.85
166	-o-phenyl phenol	—	1.25	99.95
167	-phenyl ether, H	259.3	3.25	99.33
168	-ethyl phthalate, H	298.5	2.0	99.98
169	-isoamyl benzoate, H	262.3	4.4	99.9
170	-n-dodecane, H	214.5	2	99.45
171	-dihexylamine, H	239.8	7.2	99.8
172	-tributylamine, H	213.9	20.3	99.65
	Trichlorofluoromethane	24.9		
173	-acetaldehyde	20.2	45	15.6
174	-methyl formate	32	18	20.0
175	-2-methylbutane	27	8	23.16
	Trichloronitromethane	111.9		
176	-acetic acid	118.1	19.5	107.65
177	-ethyl alcohol	78.3	66	77.5
178	-isopropyl alcohol	82.4	65	81.95
179	-propyl alcohol	97.2	41.5	94.05
180	-isoamyl alcohol	131.9	7	111.15
181	-n-pentanol	119.8	17	108.0
182	-toluene	110.75	Zeotropic	
183	-methylcyclohexane	101.15	73	100.8
184	-n-heptane	98.4	93	98.32
	Carbon tetrachloride	76.75		
185	-carbon disulfide	46.25	Zeotropic	
186	-chloroform	62.1	Zeotropic	
187	-formic acid	100.7	18.5	66.65
188	-nitromethane	101.2	17	71.3
189	-methyl alcohol	64.7	20.56	55.7
190	-acetonitrile	81.6	17	65.1
191	-acetic acid	118.1	1.54	76
192	-ethyl alcohol	78.3	15.8	65.04
193	-acrylonitrile	77.3	21	66.2
194	-acetone	56.15	88.5	56.08
195	-propyl alcohol	97.25	7.9	73.4
196	-thiophene	84	Zeotropic	
197	-butyl nitrite	78.2	30	75.3
198	-butyl alcohol	117.75	2.4	76.55
199	-ethyl ether	34.6	Zeotropic	
200	-pyridine	115.5	Zeotropic	
201	-benzene	80.1	Zeotropic	
202	-n-heptane	98.4	Zeotropic	
203	-o-xylene	143.6	Zeotropic	
	Carbon disulfide	46.25		
204	-chloroform	61.2	Zeotropic	
205	-formic acid	100.75	17	42.55
206	-nitromethane	101.2	81.4	41.2
207	-methyl alcohol	64.7	29	39.8
208	-acetic acid	118.1	Zeotropic	
209	-propyl nitrite	47.75	38	40.15
210	-ethyl alcohol	78.3	9	42.6
211	-acetone	56.15	33	39.25
212	-propyl alcohol	97.1	5.5	45.65
213	-ethyl acetate	76.7	3	46.1
214	-n-pentane	36.15	89	35.7
215	-n-hexane	68.95	Zeotropic	
216	-toluene	110.7	Zeotropic	
	Bromodichloromethane	90.2		
217	-nitromethane	101.2	25	87.3
218	-methyl alchol	64.7	40	63.8
219	-ethyl alcohol	78.3	28	75.5
220	-ethyl acetate	77.1	12	90.55
221	-benzene	80.2	Zeotropic	
222	-cyclohexane	80.75	Zeotropic	
223	-n-hexane	68.8	Zeotropic	
	Bromoform	149.5		
224	-formic acid	100.75	48	97.4
225	-acetamide	221.15	Zeotropic	
226	-butyric acid	162.45	6.8	146.8
227	-phenol	182.2	Zeotropic	
228	-aniline	184.35	Zeotropic	
229	-toluene	110.65	Zeotropic	
230	-o-cresol	191.1	Zeotropic	
231	-α-pinene	155.8	25	146.5
	Chloroform	61.2		
232	-formic acid	100.75	15	59.15
233	-methyl alcohol	64.7	12.6	53.43
234	-ethyl alcohol	78.3	7	59.35
235	-acetone, N	56.5	21.9	64.4
236	-propyl alcohol	97.2	Zeotropic	
	Chloroform	61.2		
237	-p-dioxane	101	Zeotropic	
238	-cyclohexane	80.75	Zeotropic	
239	-methylcyclopentane	72.0	20	60.5
240	-n-hexane	68.7	16.5	60.4
241	-toluene	110.65	Zeotropic	
	Formic acid	100.75		
242	-nitromethane	101.22	54.5	97.07
243	-trichloroethylene	86.95	75	74.1
244	-tetrachloroethylene	121.1	50	88.15
245	-acetic acid	118.1	Zeotropic	
246	-nitroethane	114.2	Zeotropic	
247	-ethyl ether	34.6	Zeotropic	
248	-ethyl sulfide	92.2	65	82.2
249	-pyridine, N	115.5	38.6	127.43
250	-2-methylbutane	27.95	96	27.2
251	-n-pentane	36.15	80	34.2
252	-bromobenzene	156.1	32	98.1
253	-chlorobenzene	131.75	41	93.7
254	-fluorobenzene	84.9	73	73.0
255	-benzene	80.2	69	71.05
256	-aniline	184.35	Zeotropic	
257	-2-picoline, N	129	75	158.0
258	-cyclohexane	80,75	30	70.7
259	-methylcyclopentane	72.0	71	63.3
260	-n-hexane	68.95	72	60.6
261	-propyl sulfide	141.5	17	98.0
262	-isopropyl sulfide	120.5	38	93.5
263	-toluene	110.7	50	85.8
264	-o-chlorotoluene	159.3	17	100.2
265	-methylcyclohexane	101.1	53.5	80.2
266	-n-heptane	98.45	43.5	78.2
267	-styrene	145.8	27	97.75
268	-o-xylene	143.6	26	95.5
269	-m-xylene	139.0	28.2	92.8
270	-n-octane	125.8	37	90.5
	Bromomethane	3.65		
271	-methyl alcohol	64.7	0.45	3.55
	Iodomethane	42.5		
272	-methyl alcohol	64.7	4.5	37.8
273	-ethyl alcohol	78.3	3.2	41.2
274	-acetone	56.15	5	42.4
275	-n-hexane	68.85	Zeotropic	
	Nitromethane	101.2		
276	-methyl alcohol	64.7	90.9	64.4
277	-acetic acid	118.1	4	101.2
278	-ethyl alcohol	78.3	71	76.05
279	-propyl alcohol	97.15	51.6	89.09
280	-p-dioxane	101.35	43.5	100.55
281	-n-butyl alcohol	117.73	28.6	98.0
282	-n-pentane, H	36.07	99	35
283	-cyclohexane	80.75	73.5	69.5
284	-methylcyclopentane	72.0	77	64.2
285	-n-hexane, H	68.74	81.5	61.7
286	-toluene	110.75	45	96.5
287	-n-heptane, 748 mm Hg, H	98.4	64.4	79.7
288	-styrene	145.8	Zeotropic	
289	-o-xylene	144.3	Zeotropic	
290	-n-octane, 748 mm; H	125.75	44.8	90.23
291	-cumene	152.8	Zeotropic	
292	-mesitylene	164.6	Zeotropic	
293	-n-nonane, 748 mm; H	150.85	28.4	96.14
294	-n-decane, 748 mm; H	174.12	16.1	98.81
295	-n-undecane, 748 mm; H	194.5	9.3	100.01
296	-n-dodecane, 748 mm; H	216.0	4.2	100.60
	Methyl nitrate	64.8		
297	-methyl alcohol	64.65	27	52.5
298	-cyclohexane	80.75	23	61.0
299	-n-hexane	68.8	44	56.0
300	-n-heptane	98.4	Zeotropic	

No.	System	B.P. (°C)	Azeotropic data: Compn. X_2 (wt. %)	Azeotropic data: B.P. (°C)
	Methyl alcohol	64.7		
301	-trichloroethylene	87	62	59.3
302	-bromoethane	38	94.7	34.9
303	-acetic acid	118.1	Zeotropic	
304	-acetone	56.15	88	55.5
305	-methyl acetate	57.1	81	53.5
306	-thiophene	84	83.6	59.71
307	-methyl acrylate	80	46	62.5
308	-p-dioxane	101.05	Zeotropic	
309	-ethyl sulfide	92.2	38	61.2
310	-pyridine	115.4	Zeotropic	
311	-cyclopentane	49.4	86	38.8
312	-isobutyl formate	97.9	5	64.6
313	-piperidine	106.4	Zeotropic	
314	-n-pentane	36.15	93	30.85
315	-chlorobenzene	132.0	Zeotropic	
316	-fluorobenzene	85.15	68	59.7
317	-benzene	80.1	60.9	57.5
318	-cyclohexane	80.7	63.6	53.9
319	-toluene	110.6	27.5	63.5
320	-methylcyclohexane	100.8	46	59.2
321	-n-heptane, H	98.45	48.5	59.1
322	-o-xylene	143.6	Zeotropic	
323	-n-octane	125.75	32.5	62.75
324	-n-nonane	150.7	16.6	64.1
325	-n-decane	173.8	Zeotropic	
326	-methyl *tert*-butyl ether	55.06	85.7	51.27
	Tetrachloroethylene	121.1		
327	-acetic acid	118.1	38.5	107.35
328	-acetamide	221.2	2.6	120.45
329	-ethylene glycol	197.4	6	119.1
330	-acetone	56.1	Zeotropic	
331	-propionic acid	140.9	8.5	119.1
332	-propyl alcohol	97.25	48	94.05
333	-n-butyl alcohol	117.7	32	110.0
334	-pyridine	115.4	48.5	112.85
335	-n-amyl alcohol	138.2	15	117.0
336	-toluene	110.75	Zeotropic	
	Hexachloroethane	185		
337	-trichloroacetic acid	196	15	181
338	-phenol	182.2	30	173.7
339	aniline	184.35	34	176.75
340	-benzyl alcohol	205.15	12	182.0
341	-p-cresol	201.7	10	183.0
	Trichloroethylene	86.9		
342	-acetic acid	118.1	3.8	86.5
343	-benzene	80.2	Zeotropic	
344	-cyclohexane	80.7	83.4	80.5
345	-n-heptane	98.45	Zeotropic	
	Trichloroacetic acid	197.55		
346	-pentachloroethane	161.95	96.5	161.8
347	-naphthalene	218.05	Zeotropic	
348	-butylbenzene	183.1	80	181.3
	Bromoacetic acid	205.1		
349	-o-dichlorobenzene	179.5	84	177.0
350	-o-bromotoluene	181.5	82	179.0
351	-acetophenone	202.0	30	206.5
352	-butylbenzene	183.1	75	179.5
353	-cymene	176.7	85	174.7
	Chloroacetic acid	189.35		
354	-bromobenzene	156.1	89	154.3
355	-phenol	181.5	Zeotropic	
356	-m-bromotoluene	183.8	70	174
357	-p-bromotoluene	185.0	66	174.1
358	-styrene	145.8	86	144.8
359	-o-xylene	144.3	88	143.5
360	-m-xylene	139.2	93	139.05
361	-n-octane	125.75	Zeotropic	
362	-cumene	152.8	79	150.8
363	-mesitylene	164.6	83	162.0
364	-pseudocumene	168.2	66	162.8
365	-naphthalene	218.05	22	187.1
366	-cymene	176.7	58	169.0
367	-n-decane	173.3	58	165.2
368	-1,3,5-triethylbenzene	215.5	25	185.5
	Acetonitrile	81.6		
369	-acetic acid	118.1	Zeotropic	
370	-ethyl alcohol	78.3	56	72.5
371	-pyridine	115.5	Zeotropic	
372	-isoprene	34.1	97.6	33.7
373	-isopropyl acetate	89.5	40.0	79.5
374	-toluene	110.7	20	81.4
375	-ethylbenzene	136.2	Zeotropic	
376	-n-undecane	195.4	Zeotropic	
	Acetaldehyde	20.4		
377	-acetone	56.15	Zeotropic	
378	-ethyl ether	34.5	23.5	18.9
379	-benzene	80.1	Zeotropic	
	Thioacetic acid	89.5		
380	-benzene	80.15	Zeotropic	
381	-cyclohexane	80.75	Zeotropic	
382	-methylcyclopentane	72.0	Zeotropic	
	Acetic acid	118.1		
383	-nitroethane	114.2	70	112.4
384	-dioxane	101.35	23	119.5
385	-acetone	56.1	Zeotropic	
386	-pyridine, N	115.5	48.9	138.1
387	-2-picoline, N	129.3	59.6	144.12
388	-3-picoline, N	144	69.6	152.5
389	4 picoline, N	144.3	69.7	154.3
390	-benzene	80.2	98.0	80.05
391	-cyclohexane	80.75	90.4	78.8
392	-n-hexane	68.6	94.0	68.25
393	-isopropyl sulfide	120	52	111.5
394	-toluene	110.7	71.9	100.6
395	-triethylamine, N	89	33	163
396	-2,6-lutidine, N	144.0	77.1	148.1
397	-methylcyclohexane	101.1	69	96.3
398	-n-heptane	98.25	67	91.72
399	-styrene	145.2	14.3	116.8
400	-ethylbenzene	136.15	34	114.65
401	-o-xylene	143.6	22	116.6
402	-m-xylene	139.0	27.5	115.35
403	-p-xylene	138.4	28	115.25
404	-ethylcyclohexane	131.8	—	107.9
405	-n-octane	125.75	46.3	105.7
406	-cumene	152.8	16	116.0
407	-mesitylene	164.6	Zeotropic	
408	-n-nonane	150.8	31	112.9
409	-n-decane	173.3	20.5	116.75
410	-n-undecane	194.5	5	117.87
	Ethylene sulfide	55.7		
411	-acetone	56.15	43	51.5
412	-n-hexane	68.8	Zeotropic	
413	-2,3-dimethylbutane	58.0	35	54.0
	Methyl formate	31.7		
414	-ethyl ether	34.6	45	28.4
415	-isoprene	34.1	50	22.5
416	-2-methylbutane	27.95	53	17.05
417	-n-pentane	36.15	47	21.8
418	-n-hexane	69.0	Zeotropic	
419	-2,3-dimethylbutane	58.0	15	30.5
	Iodoethane	72.3		
420	-ethyl alcohol	78.3	14	63
421	-propyl alcohol	97.2	7	70
422	-n-hexane	68.85	24	68.0
	Acetamide	221.2		
423	-benzaldehyde	179.2	93.5	178.6
424	-methylaniline	196.25	86	193.8
425	-m-cresol	202.1	Zeotropic	
426	-styrene	145.8	88	144
427	-o-xylene	144.3	89	142.6
428	-m-xylene	139.0	90	138.4
429	-p-xylene	138.2	92	137.75
430	-2,4-xylenol	210.5	Zeotropic	
431	-3,4-xylenol	226.8	4	221.1
432	-ethylaniline	205.5	82	199.0
433	-quinoline	237.3	Zeotropic	
434	-indene	183.0	82.5	177.2
435	-naphthalene	218.05	73	199.55
436	-safrol	235.9	68	208.8
437	-eugenol	255.0	12	220.8
438	-p-cymene	176.7	81	170.5
439	-diethylaniline	217.05	76	198.05
440	-camphene	159.6	88	155.5
441	-dipentene	177.7	82	169.15
442	-camphor	209.1	77	199.8
443	-isoamyl valerate	192.7	84	184.85
444	-isoamyl sulfide	214.8	83	199.5
445	-1-methylnaphthalene	245.1	56.2	209.8
446	-2-methylnaphthalene	241.15	60	208.25
447	-acenaphthene	277.9	35.8	217.1
448	-biphenyl	255.9	49.5	212.95
449	-phenyl ether	259.3	48	214.55

No.	System	B.P. (°C)	Azeotropic data Compn. X_2 (wt. %)	B.P. (°C)
450	-diphenylmethane	265.6	43.5	215.15
451	-1,2-diphenylethane	284	32	218.2
452	-benzyl ether	297	Zeotropic	
	Nitroethane	114.2		
453	-n-hexane, H	68.74	89.4	59.4
454	-toluene	110.75	75	106.2
	Ethyl nitrate	87.68		
455	-thiophene	84.7	Zeotropic	
456	-benzene	80.15	88	80.03
457	-cyclohexane	80.75	64	74.5
458	-n-hexane	68.8	76	66.25
459	-n-heptane	98.4	37	82.6
	Ethyl alcohol	78.3		
460	-acrylonitrile	77.3	59	70.8
461	-acetone	56.1	Zeotropic	
462	-ethyl sulfide	92.2	44	72.6
463	-pyridine	115.4	Zeotropic	
464	-cyclopentane	49.4	92.5	44.7
465	-n-pentane	36.15	95	34.3
466	-fluorobenzene	85.15	25	70.0
467	-benzene	80.1	68.3	67.9
468	-cyclohexane	80.8	70.8	64.8
469	-n-hexane	68.95	79	58.68
470	-propyl ether	90.4	56	74.4
471	-toluene	110.7	32	76.7
472	-ethylbenzene	136.15	Zeotropic	
473	-p-xylene	138.3	Zeotropic	
474	-n-octane	125.6	22	77.0
	Ethylene glycol	197.4		
475	-pyridine	115.5	Zeotropic	
476	-benzene	80.2	Zeotropic	
477	-phenol	182.2	Zeotropic	
478	-aniline	184.35	76	180.55
479	-o-bromotoluene	181.75	75	166.8
480	-o-nitrotoluene	221.75	51.5	188.55
481	-toluene	110.6	97.7	110.1
482	-m-toluidine	200.3	58	188.55
483	-o-cresol	191.1	73	189.6
484	-m-cresol	202.1	40	195.2
485	-2,6-lutidine	144.0	Zeotropic	
486	-n-heptane	98.45	97	97.9
487	-styrene	145.8	83.5	139.5
488	-m-xylene	139.1	93.45	135.1
489	-p-xylene	138.4	93.6	134.5
490	-3,4-xylenol	226.8	11	197.2
491	-2,4,6-collidine	171.3	90.3	170.5
492	-2,4-xylidine	214.0	53	188.6
493	-butyl ether	142.1	93.6	139.5
494	-quinoline	237.3	20.5	196.35
495	-indene	183.0	74	168.4
496	-cumene	152.8	82	147.0
497	-mesitylene	164.6	87	156
498	-propylbenzene	158.8	81	152
499	-cymene	176.7	74.5	163.2
500	-camphene	159.5	80	152.5
501	-camphor	209.1	60	186.15
502	-menthol	216.3	48.5	188.55
503	-n-decane	173.3	77	161.0
504	-naphthalene	218.05	49	183.9
505	-1-methylnaphthalene	245.1	40.0	190.25
506	-2-methylnaphthalene	241.15	42.8	189.1
507	-acenaphthene	277.9	25.8	194.65
508	-biphenyl	256.1	33.5	192.25
509	-fluorene	296.4	18	196.0
510	-diphenylmethane	265.6	31.5	193.3
511	-benzyl phenyl ether	286.5	13	195.5
512	-n-tridecane	234.0	45	188.0
513	-anthracene	340	1.7	197
514	-stilbene	306.4	13	196.8
	Methyl sulfide	37.3		
515	-acetone	56.15	Zeotropic	
516	-isoprene	34.3	65	32.5
517	-cyclopentane	49.35	12.5	37.1
518	-n-pentane	36.15	53.4	31.8
519	-2,2-dimethylbutane	49.7	20.2	36.5
	Methyl disulfide	109.44		
520	-n-heptane	98.4	73.7	96.44
521	-2,3-dimethylhexane	109.15	51.8	102.84
	Ethylenediamine	116.5		
522	-n-butyl alcohol	117.7	64.3	124.7
523	-benzene	80.1	Zeotropic	
524	-toluene	110.7	69.2	104
	Acrylonitrile	77.3		
525	-isopropyl alcohol	82.55	44	71.7
526	-benzene	80.2	53	73.3
	Pyruvic acid	166.8		
527	-propionic acid	141.3	Zeotropic	
528	-benzene	80.15	Zeotropic	
529	-toluene	110.75	92.5	110.05
530	-o-xylene	144.3	72	137.0
531	-ethylbenzene	136.15	78	130.5
532	-mesitylene	164.6	60	151.2
533	-propylbenzene	159.3	63	147.6
	Methyl chloroacetate	129.95		
534	-isobutyl alcohol	107.85	88	107.55
535	-cyclopentanol	140.85	23	127.5
536	-amyl alcohol	138.2	30	126.8
537	-isoamyl alcohol	131.3	39.5	124.9
538	-ethylbenzene	136.15	37.5	127.2
539	-m-xylene	139.2	10	128.25
540	-p-xylene	138.45	15	128.3
	Propionitrile	97.2		
541	-propyl alcohol	97.2	50	90.5
542	-n-hexane	68.8	91	63.5
543	-ethylbenzene	136.15	Zeotropic	
	Acetone	56.15		
544	-methyl acetate	57	51.7	55.8
545	-diethylamine	55.5	61.8	51.4
546	-pyridine	115.4	Zeotropic	
547	-cyclopentane	49.3	64	41.0
548	-n-pentane	36.15	80	32.5
549	-benzene	80.1	Zeotropic	
550	-cyclohexane	80.75	32.5	53.0
551	-n-hexane	68.95	41	49.8
552	-isopropyl ether	69.0	39	54.2
553	-n-heptane	98.4	10.5	55.85
	Allyl alcohol	96.95		
554	-ethyl sulfide	92.1	55	85.1
555	-pyridine	115.4	Zeotropic	
556	-benzene	80.2	82.64	76.75
557	-cyclohexane	80.8	42	74.0
558	-n-hexane	68.95	95.5	65.5
559	-methylcyclohexane	101.1	58	85.0
560	-m-xylene	139.0	Zeotropic	
561	-2,5-dimethylhexane	109.4	50	89.3
562	-n-octane	125.75	32	93.4
	Ethyl formate	54.1		
563	-n-pentane	36.2	70	32.5
564	-benzene	80.2	Zeotropic	
565	-methylcyclopentane	72.0	25	51.2
566	-n-hexane	68.95	33	49.0
	Methyl acetate	56.95		
567	-cyclopentane	49.3	62.1	43.2
568	-benzene	80.2	0.3	56.7
569	-cyclohexane	80.7	22.0	55.5
570	-n-hexane	68.95	39.3	51.75
571	-n-heptane	98.45	3.55	56.65
572	-2-methylhexane	90.0	11.4	56.0
573	-2,2,3-trimethylbutane	80.9	25.8	55.1
	Propionic acid	141.0		
574	-pyridine, N	115.5	32.8	148.6
575	-2-picoline, N	129.3	45.0	154.5
576	-chlorobenzene	132.0	82	128.9
577	-benzene	80.15	Zeotropic	
578	-o-xylene	143.6	57	135.4
579	-p-xylene	138.2	66	132.5
580	-n-hexane	68.85	Zeotropic	
581	-n-heptane	98.15	98	97.82
582	-n-octane	125.12	78.5	120.89
583	-n-nonane	150.67	46.0	134.27
584	-n-decane	174.06	19.5	139.76
585	-n-undecane	193.85	Zeotropic	
586	-propyl sulfide	141.5	55	136.5
587	-quinoline	237.5	Zeotropic	
588	-cumene	152.8	35	139.0
589	-mesitylene	164.0	23	139.3
590	-propylbenzene	158.0	25	139.5
591	-camphene	159.6	35	138.0
592	-α-pinene	155.8	41.5	136.4
	Propionamide	222.2		
593	-p-bromochlorobenzene	196.4	84	189.5
594	-p-dibromobenzene	220.25	78	204.9

No.	System	B.P. (°C)	Azeotropic data Compn. X_2 (wt. %)	Azeotropic data B.P. (°C)
595	-iodobenzene	188.45	90	183.5
596	-nitrobenzene	210.75	76	205.4
597	-o-nitrophenol	217.25	75.2	211.15
598	-phenol	182.2	Zeotropic	
599	-p-bromotoluene	185.0	90	181.0
600	-m-nitrotoluene	230.8	56	214.5
601	-toluene	110.75	Zeotropic	
602	-o-cresol	191.1	Zeotropic	
603	-m-cresol	202.2	Zeotropic	
604	-o-toluidine	200.35	97.5	200.25
605	-m-toluidine	203.1	Zeotropic	
606	-acetophenone	202	85	200.35
607	-methyl salicylate	222.35	66	210.55
608	-o-xylene	144.3	98	144.0
609	-dimethylaniline	194.15	84.5	190.5
610	-3,4-xylidine	225.5	72	217.2
611	-quinoline	237.3	Zeotropic	
612	-indene	182.6	88	179.5
613	-ethyl benzoate	212.6	75	205.0
614	-cumene	152.8	96	151.8
615	-mesitylene	164.6	90	162.3
616	-naphthalene	218.05	68.5	204.65
617	-cymene	176.7	85	172.8
618	-carvone	231.0	52	214.5
619	-camphene	159.6	87	156.5
620	-camphor	209.1	83	203.5
621	-borneol	213.4	78	209.2
622	-n-decane	173.3	88.2	168
623	-1-methylnaphthalene	245.1	48	213.8
624	-2-methylnaphthalene	241.15	50	213.0
625	-n-undecane	194.5	79	183
626	-acenaphthene	277.9	25	220.8
627	-biphenyl	256.1	45	216.0
628	-n-dodecane	216.0	68.4	193.0
629	-fluorene	295	10	221.5
630	-diphenylmethane	265.6	40	218.2
	Ethyl carbamate	185.25		
631	-bromobenzene	156.1	90.2	153.95
632	-iodobenzene	188.45	67	174.5
633	-nitrobenzene	210.75	12	184.95
634	-phenol	182.2	46.5	190.75
635	-benzonitrile	191.1	43	182.1
636	-anisole	153.85	95	153.5
637	-2,4-xylenol	210.5	Zeotropic	
638	-n-octyl alcohol	195.2	27.5	183.5
639	-isobutyl sulfide	172.0	77	166.5
640	-indene	182.6	65	172.65
641	-cumene	152.8	94	151.5
642	-mesitylene	164.6	78	159.0
643	-propylbenzene	159.3	85	157.0
644	-pseudocumene	168.2	75	161.4
645	-naphthalene	218.0	23	184.05
646	-butylbenzene	183.1	63	172.0
645	-camphene	159.6	85	157.0
648	-limonene	177.6	68	168.07
649	-camphor	209.1	16	184.85
650	-2-methylnaphthalene	241.15	Zeotropic	
651	-amyl ether	187.4	63	171.0
652	-isoamyl ether	173.35	73	163.15
653	-methyl pelargonate	213.8	15	184.3
	1-Nitropropane	131		
654	-propyl alcohol	97.15	91.2	96.95
655	-n-butyl alcohol	117.73	67.8	115.3
656	-isobutyl alcohol	107.89	84.8	105.28
657	-n-heptane	98.43	86.5	96.6
658	-ethylbenzene	136.19	44.0	129.0
659	-n-octane	125.66	65.8	115.8
660	-n-nonane	150.8	38.4	126.6
	Propyl nitrite	47.75		
661	-n-pentane	36.15	91	35.8
662	-cyclopentane	49.3	46	45.5
	Isopropyl alcohol	82.45		
663	-butylamine	77.8	40	74.7
664	-n-pentane	36.15	94	35.5
665	-fluorobenzene	85.15	70	74.5
666	-benzene	80.2	66.3	71.74
667	-cyclohexane	80.7	68	69.4
668	-n-hexane	68.85	77	62.7
669	-toluene	110.6	31	80.6
670	-n-heptane	98.45	49.5	76.4
671	-o-xylene	144.3	Zeotropic	
672	-n-octane	124.75	16	81.6
	Propyl alcohol	97.2		
673	-dioxane	101.35	45	95.3
674	-butyl formate	106.8	36	95.5
675	-chlorobenzene	132	20	96.5
676	-fluorobenzene	85.15	82	80.2
677	-benzene	80.2	83.1	77.12
678	-cyclohexane	80.75	81.5	74.69
679	-toluene	110.6	48.8	92.5
680	-methylcyclohexane	100.8	65.2	87.0
681	-n-heptane	98.4	65.3	84.6
682	-styrene	145.8	92	97.0
683	-o-xylene	143.6	Zeotropic	
684	-m-xylene	139.2	6	97.08
685	-p-xylene	138.4	7.8	96.88
686	-n-octane	125.6	30	93.9
	Glycerol	290.5		
687	-p-chloronitrobenzene	239.1	87	235.6
688	-triethylene glycol	288.7	63	285.1
689	-m-nitrotoluene	230.8	87	228.8
690	-p-cresol	201.7	Zeotropic	
691	-methyl salicylate	222.35	92.5	221.4
269	-3,4-xylenol	226.8	Zeotropic	
693	-o-xylene	143.6	Zeotropic	
694	-quinoline	237.3	Zeotropic	
695	-ethyl salicylate	233.7	89.7	230.5
696	-naphthalene	218.05	90	215.2
697	-safrol	235.9	85.5	231.3
698	-methyl phthalate	283.2	69	271.5
699	-estragol	215.6	92.5	213.5
700	-eugenol	254.5	96	251.3
701	-propyl benzoate	230.85	92	228.8
702	-carvone	231.0	97	230.85
703	-2-methylnaphthalene	241.15	83.5	233.7
704	-acenaphthene	277.9	71	259.1
705	-biphenyl	254.9	75	246.1
706	-phenyl ether	259.3	78	247.9
707	-1,3,5-triethylbenzene	215.5	92	212.9
708	-bornyl acetate	227.7	90	226.0
709	-diphenylmethane	265.6	73	250.8
710	-benzyl phenyl ether	286.5	70	264.5
711	-benzyl ether	297.0	64	269.5
	Ethyl methyl sulfide	66.61		
712	-cyclohexane	80,75	Zeotropic	
713	-methylcyclopentane	71.85	35.9	65.6
714	-n-hexane	68.75	43.4	63.94
715	-2,2-dimethylpentane	79.2	11.8	66.37
	1-Propanethiol	67.3		
716	-thiophene	84.7	Zeotropic	
717	-cyclohexane	80.75	2.4	67.77
718	-2,3-dimethylbutane	58.0	83.7	57.54
719	-n-hexane	68.75	47.4	64.35
720	-2-methylpentane	60.27	76.1	59.2
721	-isopropyl ether	68.3	35	66.0
722	-2,2-dimethylpentane	79.2	18.7	67.2
723	-2,2,3-trimethylbutane	80.97	12.6	67.57
	Perfluorobutyric acid	122.0		
724	-ethyl-benzene	136.15	20	115.4
725	-m-xylene	139.0	17	117.5
726	-p-xylene	138.4	18	117.6
	Thiophene	84.7		
727	-benzene	80.15	Zeotropic	
728	-cyclohexane	80.8	38.8	77.9
729	-methylcyclopentane	71.85	86	71.47
730	-n-hexane	68.75	88.8	68.46
731	-2,3-dimethylpentane	89.9	36	80.9
732	-2,4-dimethylpentane	80.55	57.3	76.58
733	-n-heptane	98.4	16.8	83.09
	Pyrrol	129.2		
734	-chlorobenzene	131.75	57	124.5
735	-isopropyl sulfide	120.5	80	117.5
736	-propyl sulfide	140.8	35	127.5
737	-toluene	110.75	Zeotropic	
	Methyl pyruvate	137.5		
738	-isoamyl acetate	142.1	35	135.0
739	-m-xylene	139.2	50	130.0
	Methyl oxalate	163.3		
740	-p-dichlorobenzene	174.35	35	162.05
741	-pinacol	174.35	19	163.15
742	-o-bromotoluene	181.5	2	164.1
743	-butyl butyrate	166.4	42	160.5
744	-ethyl caproate	167.7	40	161.0
745	-indene	182.6	17	163.6

No.	System	B.P. (°C)	Azeotropic data Compn. X_2 (wt. %)	B.P. (°C)
746	-mesitylene	164.0	50.2	154.8
747	-naphthalene	218.0	Zeotropic	
748	-2,7-dimethyloctane	160.6	55	147.0
749	-1,3,5-triethylbenzene	215.5	Zeotropic	
	Ethyl bromoacetate	158.8		
750	-butyric acid	164.0	16	157.4
751	-isobutyric acid	154.6	60	153.0
752	-bromobenzene	156.1	72	155.3
753	-cyclohexanol	160.8	35	155.5
754	-o-chlorotoluene	159.3	48	156.2
755	-propylbenzene	159.3	50	155.8
	Ethyl chloroacetate	143.55		
756	-isoamyl acetate	142.1	60	141.7
757	-isoamyl alcohol	131.3	77	131.0
758	-allyl sulfide	139.35	78	138.5
759	-propyl butyrate	142.8	53	141.7
760	-ethylbenzene	136.15	82	135.3
761	-o-xylene	144.3	42	140.2
762	-m-xylene	139.0	68	137.45
763	-butyl ether	142.4	55	139.8
	Butyronitrile	117.9		
764	-n-butyl alcohol	117.8	50	113.0
765	-toluene	110.75	73	107.0
766	-methylcyclohexane	101.15	80	90.5
	Isobutyronitrile	103.85		
767	-benzene	80.15	Zeotropic	
768	-methylcyclohexane	101.15	60	85.5
769	-n-heptane	98.4	62	80.5
	2-Butanone	79.6		
770	-methyl propionate	79.85	40	79.0
771	-ethyl acetate	77.1	88.2	77.05
772	-1-chlorobutane	78.5	62	77.0
773	-butyl nitrite	78.2	70	76.7
774	-*tert*-butyl alcohol	82.45	31	78.7
775	-butylamine	77.8	65	74.0
776	-fluorobenzene	84.9	25	79.3
777	-benzene	80.1	56	78.33
778	-cyclohexane	80.75	60	71.8
779	-n-hexane	68.8	71.4	64.2
780	-n-heptane	98.5	30	77.0
781	-2,5-dimethylhexane	109.4	5	79.0
	Butyraldehyde	74.8		
782	-benzene	80.1	Zeotropic	
783	-n-hexane	68.7	74	60.0
	Butyric acid	164.0		
784	-iodobutane	130.4	97.5	129.8
785	-2-furaldehyde	161.45	57.5	159.4
786	-pyridine	115.5	8.0	163.2
787	-propyl chloroacetate	162.5	60	160.5
788	-isoamyl nitrate	149.75	88	147.85
789	-p-dichlorobenzene	174.4	43	162.0
790	-chlorobenzene	132.0	97.2	131.75
791	-o-bromotoluene	181.5	28	163.0
792	-m-bromotoluene	184.3	20.5	163.62
793	-p-bromotoluene	185.0	25	161.5
794	-anisole	153.85	88	152.85
795	-n-heptane	98.4	Zeotropic	
796	-styrene	145.8	85	143.5
797	-ethylbenzene	136.15	96	135.8
798	-o-xylene	144.3	90	143.0
799	-m-xylene	139.0	94	138.5
800	-p-xylene	138.45	94.5	137.8
801	-indene	182.6	16	163.65
802	-cumene	152.8	80	149.5
803	-mesitylene	164.8	62	158.0
804	-propylbenzene	158.9	72	154.5
805	-pseudocumene	169	55	159.5
806	-naphthalene	218.1	Zeotropic	
807	-butylbenzene	183.1	25	162.5
808	-cymene	176.7	40	161.0
809	-camphene	159.6	97.2	152.3
810	-n-undecane	194.5	15.5	162.4
	Dioxane	101.35		
811	-ethyl acetate	77.1	Zeotropic	
812	-1-bromobutane	101.5	53	98.0
813	-pyridine	115.5	Zeotropic	
814	-piperidine	106.4	Zeotropic	
815	-*tert*-amyl alcohol	102.35	20	100.65
816	-benzene	80.15	Zeotropic	
817	-cyclohexane	80.75	75.4	79.5
818	-ethyl borate	118.6	8	100.7
819	-toluene	110.75	Zeotropic	
820	-n-heptane	98.4	56	91.85
	Ethyl acetate	77.1		
821	-butyl nitrite	78.2	29	76.3
822	-isobutyl nitrite	67.1	Zeotropic	
823	-*tert*-butyl alcohol	82.45	27	76.0
824	-benzene	80.15	Zeotropic	
825	-cyclohexane	80.75	44	71.6
826	-methylcyclopentane	72.0	62	67.2
827	-n-hexane	68.7	60.1	65.15
828	-methylcyclohexane	101.1	Zeotropic	
	Isobutyric acid	154.6		
829	-iodobutane	130.4	93	128.8
830	-ethyl pyruvate	155.5	40	153.0
831	-bromobenzene	156.15	65	148.6
832	-chlorobenzene	132.0	92	131.2
833	-phenol	182.2	Zeotropic	
834	-1-bromohexane	156.5	65	148.0
835	-o-bromotoluene	181.5	15	153.9
836	-toluene	110.75	Zeotropic	
837	-anisole	153.85	58	149.0
838	-styrene	145.8	73	142.0
839	-o-xylene	144.3	78	141.0
840	-m-xylene	139.0	85	136.9
841	-p-xylene	138.4	87	136.4
842	-cumene	152.8	65	146.8
843	-propylbenzene	158.9	51	149.3
844	-pseudocumene	168.2	37	152.3
845	-cymene	176.7	20	153.4
846	-camphene	159.6	55	148.1
847	-d-limonene	177.8	22	152.5
848	-2,7-dimethyloctane	160.2	52	148.55
849	-isoamyl ether	173.2	7	154.2
	Methyl propionate	79.85		
850	-1-chlorobutane	78.05	62	76.8
851	-butyl nitrite	78.2	88	77.7
852	-n-butyl alcohol	117.8	Zeotropic	
853	-benzene	80.2	48	79.45
854	-cyclohexane	80.75	48	75.0
855	-methylcyclopentane	72.0	72	69.5
856	-propyl ether	90.5	Zeotropic	
857	-methylcyclohexane	101.1	11.5	79.3
	Propyl formate	80.85		
858	-1-chlorobutane	78.5	62	76.1
859	-butyl nitrite	78.2	65	76.8
860	-*tert*-butyl alcohol	82.6	60	78.0
861	-benzene	80.2	53	78.5
862	-cyclohexane	80.75	52	75.0
863	-n-hexane	68.95	70.5	63.6
864	-n-heptane	98.5	29	78.2
	Glycol monoacetate	190.9		
865	-phenol, N	182.2	35	197.5
866	-m-bromotoluene	184.3	68	182.0
867	-o-cresol, N	191.1	49	199.45
868	-m-cresol, N	202.2	69	206.5
869	-p-cresol, N	201.7	67	206.0
870	-n-octyl alcohol	195.2	29	189.5
871	-indene	182.6	80	180.0
872	-naphthalene	218.0	Zeotropic	
873	-amyl ether	187.5	58	180.8
874	-isoamyl ether	173.2	72	170.2
875	-1,3,5-triethylbenzene	215.5	Zeotropic	
	Methyl lactate	143.8		
876	-phenol	182.2	Zeotropic	
877	-anisole	153.85	18	142.8
878	-4-heptanone	143.55	53	142.7
879	-ethyl valerate	145.45	42	140.0
880	-methyl caproate	149.8	30	141.7
881	-m-xylene	139.0	57.5	131.2
882	-p-xylene	138.2	60	130.8
883	-n-octane	125.8	70	120.3
884	-butyl ether	142.8	58	137.0
885	-cumene	152.8	38	137.8
886	-camphene	159.6	15	140.0
887	-2,7-dimethyloctane	160.1	32	137.8
	Tetrahydrothiophene	118.8		
888	-pyridine	115.4	55	113.5
889	-1-methylpyrrol	112.8	82	111.5
890	-ethylcyclohexane	131.85	19.3	117.46
891	-2-methylheptane	117.70	61.8	113.96
892	-n-octane	125.7	39.7	117.79
	Butyl nitrite	78.2		
893	-benzene	80.15	25	77.95
894	-cyclohexane	80.75	37	76.5

No.	System	B.P. (°C)	Azeotropic data: Compn. X_2 (wt. %)	Azeotropic data: B.P. (°C)
895	-n-hexane	68.8	82	68.5
896	-methylcyclohexane	101.15	Zeotropic	
897	-n-heptane	98.4	Zeotropic	
	Isobutyl nitrite	67.1		
898	-benzene	80.15	Zeotropic	
899	-cyclohexane	80.75	Zeotropic	
900	-methylcyclopentane	72.0	32	65.9
901	-n-hexane	68.8	46	65.0
	Isobutyl nitrate	123.5		
902	-n-butyl alcohol	117.8	55	112.8
903	-isobutyl alcohol	107.85	64	105.6
904	-chlorobenzene	131.75	Zeotropic	
905	-propyl sulfide	141.5	Zeotropic	
906	-toluene	110.75	Zeotropic	
907	-ethylbenzene	136.15	Zeotropic	
	n-Butyl alcohol	117.75		
908	-pyridine	115.5	31	118.6
909	-butyl formate	106.6	76.4	105.8
910	-ethyl carbonate	125.9	37	116.5
911	-chlorobenzene	132.0	44	115.3
912	-fluorobenzene	84.9	Zeotropic	
913	-benzene	80.1	Zeotropic	
914	-2-picoline	129.4	Zeotropic	
915	-cyclohexene	82.7	95	82.0
916	-cyclohexane	80.75	90.5	79.8
917	-hexaldehyde	128.3	22.9	116.8
918	-ethyl isobutyrate	110.1	83	109.2
919	-isoamyl formate	123.8	31	115.9
920	-isobutyl acetate	117.2	50	114.5
921	-methyl isoyalerate	116.3	60	113.5
922	-paraldehyde	123.9	48	115.75
923	-n-hexane	68.95	96.8	68.2
924	-acetal	103.55	87	101.0
925	-isopropyl sulfide	120.5	55	112.0
926	-ethyl borate	118.6	48	113.0
927	-toluene	110.7	72.2	105.5
928	-methylcyclohexane	100.8	80	95.3
929	-n-heptane	98.4	82	93.85
930	-ethylbenzene	136.15	34.9	115.85
931	-o-xylene	143.6	25	116.8
932	-m-xylene	139.0	28.5	116.5
933	-p-xylene	138.3	32	115.7
934	-n-octane	125.75	54.8	108.45
935	-butyl ether	142.1	17.5	117.65
936	-isobutyl ether	122.3	52	113.5
937	-n-nonane	150.7	28.5	115.9
938	-2,7-dimethyloctane	160.2	Zeotropic	
	***sec*-Butyl alcohol**	99.5		
939	-butyl formate	106.8	32	98.0
940	-ethyl propionate	99.15	53	95.7
941	-benzene	80.15	84.6	78.5
942	-cyclohexane	80.75	82	76.0
943	-methylcyclopentane	72.0	88.5	69.7
944	-propyl ether	90.4	78	87.0
945	-toluene	110.7	45	95.3
946	-methylcyclohexane	101.5	61.8	89.7
947	-n-heptane	98.4	63.3	88.1
948	-isooctane	99.3	66.2	88.0
	***tert*-Butyl alcohol**	82.9		
949	-ethyl sulfide	92.1	30	79.8
950	-flurobenzene	85.15	69	76.0
951	-benzene	80.2	63.4	73.95
952	-cyclohexane	80.7	65.8	71.2
953	-methylcyclopentane	72.0	74	66.6
954	-n-hexane	68.85	78	63.7
955	-isopropyl ether	68.3	92.1	67.3
956	-propyl ether	90.4	48	79.0
957	-toluene	110.7	Zeotropic	
958	-methylcyclohexane	100.8	34	78.8
959	-n-heptane	98.45	38	78.0
960	-p-xylene	138.45	Zeotropic	
961	-2,5-dimethylhexane	109.2	23	81.5
	Ethyl ether	34.6		
962	-isoprene	34.3	52	33.2
963	-2-methyl-2-butene	37.1	15	34.2
964	-n-pentane	36.16	44	33.7
965	-benzene	80.2	Zeotropic	
966	-n-hexane	68.85	Zeotropic	
	Isobutyl alcohol	108.0		
967	-2-pentanone	102.35	81	101.8
968	-3-pentanone	102.05	80	101.7
969	-butyl formate	106.8	60	103.0
970	-methyl butyrate	102.65	75	101.3
971	-propyl acetate	101.6	83	101.0
972	-n-pentane	36.15	Zeotropic	
973	-chlorobenzene	132.0	37	107.1
974	-fluorobenzene	84.9	91	84.0
975	-benzene	80.1	92.6	79.3
976	-cyclohexene	82.7	85.8	80.5
977	-cyclohexane	80.75	86	78.3
978	-methylcyclopentane	72.0	95	71.0
979	-isobutyl vinyl ether	83.0	93.8	82.7
980	-ethyl isobutyrate	110.1	48	105.5
981	-n-hexane	68.9	97.5	68.3
982	-propyl ether	90.55	90	89.5
983	-acetal	103.55	80	98.2
984	-isopropyl sulfide	100.5	27	105.8
985	-toluene	110.7	55	101.2
986	-methylcyclohexane	100.8	68	92.6
987	-n-heptane	98.45	73	90.8
988	-ethylbenzene	136.15	20	107.2
989	-p-xylene	138.4	11.4	107.1
990	-1,3-dimethylcyclohexane	120.7	44	102.2
991	-2,5-dimethylhexane	109.2	58	98.7
992	-2,2,4-trimethylpentane	99.3	73	92.0
993	-butyl ether	142.4	Zeotropic	
	2-Ethoxyethanol	135.3		
994	-toluene	110.75	89.2	110.15
995	-methylcyclohexane	101.15	85	98.6
996	-propyl butyrate	143.7	28	133.5
997	-n-heptane	98.4	86	96.5
998	-styrene	145.8	45	130.0
999	-ethylbenzene	136.15	52	127.8
1000	-p-xylene	138.45	50	128.6
1001	-n-octane	125.75	62	116.0
1002	-cumene	152.8	33	133.2
1003	-propylbenzene	159.3	20	134.6
1004	-camphene	159.6	35	131.0
	Methyl propyl ether	38.95		
1005	-2-methyl-2-butene	37.15	75	36.3
1006	-n-pentane	36.2	78	35.6
1007	-isoprene	34.3	Zeotropic	
	Diethylene glycol	245.5		
1008	-p-dibromobenzene	220.25	87	212.85
1009	-nitrobenzene	210.75	90	210.0
1010	-o-nitrophenol	217.2	89.5	216.0
1011	-pyrocatechol, N	245.9	54	259.5
1012	-m-nitrotoluene	230.8	75	224.2
1013	-methyl salicylate	222.95	85	220.55
1014	-p-cresol	202.0	Zeotropic	
1015	-ethyl fumarate	217.85	90	217.1
1016	-quinoline	237.3	71	233.6
1017	-benzyl acetate	215.0	93	214.85
1018	-naphthalane	218.0	78	212.6
1019	-isosafrol	252.0	54	233.5
1020	-safrol	235.9	67	225.5
1021	-methyl phthalate	283.7	3.7	245.4
1022	-thymol	232.9	87	232.25
1023	-1-methylnaphthalene	244.6	55	277.0
1024	-2-methylnaphthalene	241.15	61	225.45
1025	-biphenyl	256.1	52	232.65
1026	-acenaphthene	277.9	38	239.6
1027	-1,3,5-triethylbenzene	215.5	78	210.0
1028	-bornyl acetate	227.6	[illegible]	[illegible]
1029	-fluorene	295.0	20	243.0
1030	-diphenylmethane	265.4	48	236.0
1031	-benzyl phenyl ether	286.5	20	241.5
	1-Butanethiol	97.8		
1032	-benzene	80.15	Zeotropic	
1033	-pyridine	115.4	Zeotropic	
1034	-n-heptane	98.42	50.6	95.45
1035	-2-methylhexane	90.05	84.6	89.74
1036	-3-methylhexane	91.95	77.2	91.2
1037	-2,5-dimethylhexane	109.1	12	98.22
	Isopropyl methyl sulfide	84.76		
1038	-cyclohexane	80.85	70	79.76
1039	-3-methylhexane	91.6	17.6	84.38
1040	-2,4-dimethylpentane	80.55	70.3	79.39
	Methyl propyl sulfide	95.47		
1041	-ethylcyclopentane	103.45	9.3	95.41
1042	-methylcyclohexane	101.05	22.0	95.06
1043	-3-methylhexane	91.6	67.05	90.53
	2-Furaldehyde	161.45		
1044	-n-heptane	98.4	94.7	98.3

No.	System	B.P. (°C)	Azeotropic data: Compn. X_2 (wt. %)	Azeotropic data: B.P. (°C)
1045	-ethylbenzene	136.15	Zeotropic	
1046	-o-xylene	143.6	87	140.5
1047	-m-xylene	139.0	88	138.4
1048	-p-xylene	138.4	80	138.0
1049	-cumene	152.8	73	148.5
1050	-mesitylene	164.6	40	155.2
1051	-pseudocumene	168.2	33	157.0
1052	-propylbenzene	159.2	58	151.4
1053	-cymene	176.7	32	157.8
1054	-camphene	159.5	60	146.75
1055	-cineol	176.35	41	157.25
	Pyridine	115.4		
1056	-piperidine	105.8	92	106.1
1057	-phenol, N	181.4	86.9	183.1
1058	-toluene	110.75	77.8	110.1
1059	-n-heptane	98.4	74.7	95.6
1060	-n-octane	125.75	43.9	109.5
1061	-n-nonane	150.7	10.1	115.1
1062	-n-decane	173.3	Zeotropic	
	2-Methylthiophene	111.92		
1063	-n-heptane	98.4	97.8	97.77
1064	-2-methylheptane	117.7	32.2	109.77
1065	-2,2-dimethylhexane	106.85	66.8	104.62
1066	-2,5-dimethylhexane	109.15	60.4	106.12
	3-Methylthiophene	114.96		
1067	-ethylcyclopentane	103.45	96.1	102.82
1068	-n-octane	125.75	18.0	114.15
1069	-2-methylheptane	117.7	41.2	111.86
1070	-2,5-dimethylhexane	109.15	68.3	107.12
	Ethyl pyruvate	155.1		
1071	-bromobenzene	156.1	52	149.5
1072	-m-xylene	139.2	70	137.2
1073	-cumene	152.8	55	146.2
	Levulinic acid	252.0		
1074	-m-nitrotoluene	230.8	85	229.5
1075	-p-nitrotoluene	238.9	78	236.4
1076	-methyl salicylate	222.95	94	222.75
1077	-3,4-xylenol	226.8	Zeotropic	
1078	-ethyl salicylate	233.8	82	230.5
1079	-naphthalene	218.0	89	216.7
1080	-safrol	235.9	83	232.5
1081	-1-methylnaphthalene	244.6	64	237.0
1082	-2-methylnaphthalene	241.15	71	234.55
1083	-isobutyl benzoate	241.9	75	238.6
1084	-1,3,5-triethylbenzene	215.5	89	214.0
	Methyl acetoacetate	169.5		
1085	-siobutyl sulfide	172.0	42	166.0
1086	-mesitylene	164.6	57	159.5
1086	-cymene	176.7	44	165.0
1088	-camphene	159.6	60	152.8
1089	-isoamyl ether	173.2	40	160.5
	Methyl malonate	181.4		
1090	-acetophenone	202.0	61	201.0
1091	-naphthalene	218.0	Zeotropic	
1092	-butylbenzene	183.2	48	173.0
1093	-cymene	176.7	60	169.0
1094	-camphene	159.6	74	154.6
1095	-d-limonene	177.8	52	167.3
	2-Pentanone	102.25		
1096	-methyl butyrate	102.65	50	101.9
1097	-toluene	110.7	Zeotropic	
1098	-methylcyclohexane	101.15	60	95.2
1099	-n-heptane	98.4	66	93.2
	Butyl formate	106.8		
1100	-*tert*-amyl alcohol	102.35	65	101.0
1101	-benzene	80.15	Zeotropic	
1102	-pinacolone	106.2	62	106.0
1103	-methylcyclohexane	101.15	65	96.0
1104	-n-heptane	98.45	60	90.7
	Isovaleric acid	176.5		
1105	-ethyl acetoacetate	180.4	23	176.1
1106	-ethyl oxalate	185.65	16	176.3
1107	-o-xylene	144.3	95	143.8
1108	-butyl sulfide	185.0	27	175.0
1109	-indene	183.0	40	173.0
1110	-cumene	152.8	88	152.0
1111	-mesitylene	164.6	81	162.5
1112	-pseudocumene	168.2	77	165.7
1113	-naphthalene	218.05	Zeotropic	
1114	-butylbenzene	183.1	50	173.0
1115	-cymene	175.3	62	170.8
1116	-camphene	159.6	83	156.5

No.	System	B.P. (°C)	Azeotropic data: Compn. X_2 (wt. %)	Azeotropic data: B.P. (°C)
1117	-cineol	176.3	57.5	175.0
1118	-n-decane	173.3	67	167.0
1119	-n-tridecane	234.0	Zeotropic	
	Methyl butyrate	102.65		
1120	-methylcyclohexane	101.1	55	97.0
1121	-n-heptane	98.45	65	95.1
1122	-n-octane	125.8	Zeotropic	
	Propyl acetate	101.6		
1123	-*tert*-amyl alcohol	102.0	42	99.5
1124	-benzene	80.2	Zeotropic	
1125	-cyclohexane	80.75	Zeotropic	
1126	-n-hexane	69.0	Zeotropic	
1127	-acetal	103.55	32	101.25
	Valeric acid	186.35		
1128	-phenol	182.2	Zeotropic	
1129	-indene	182.6	70	178.5
1130	-mesitylene	164.6	90	164.0
1131	-naphthalene	218.0	4	186.0
1132	-cymene	176.7	78	176.5
1133	-camphene	159.6	92	158.5
1134	-amyl ether	187.5	55	181.5
1135	-isoamyl ether	173.2	87.5	171.8
	Ethyl lactate	154.1		
1136	-toluene	110.75	Zeotropic	
1137	-o-xylene	144.3	70	140.2
1138	-p-xylene	138.45	83	136.6
1139	-cumene	152.8	52	143.5
1140	-mesitylene	164.9	27	150.05
1141	-pseudocumene	168.2	27	152.4
1142	-camphene	159.5	45	144.95
	n-Amyl alcohol	138.2		
1143	-benzene	80.2	Zeotropic	
1144	-phenol	182.2	Zeotropic	
1145	-amyl formate	132.0	57	131.4
1146	-ethylbenzene	136.15	60	129.8
1147	-p-xylene	138.45	58.1	130.9
	Isoamyl alcohol	131.9		
1148	-bromobenzene	156.15	15	131.65
1149	-butyl acetate	126.0	83.5	125.85
1150	-paraldehyde	124.0	78.0	123.5
1151	-o-fluorotoluene	114.0	86.0	112.1
1152	-toluene	110.7	90	109.7
1153	-n-heptane	98.45	93	97.7
1154	-ethylbenzene	136.15	51	125.7
1155	-n-octane	125.8	70	117.0
1156	-butyl ether	142.1	35	129.8
1157	-cumene	152.8	6	131.6
1158	-camphene	159.6	76	130.9
	Nitrobenzene	210.75		
1159	-aniline	184.35	Zeotropic	
1160	-methyl maleate	204.05	93	203.9
1161	-benzyl alcohol	205.25	62	204.2
1162	-3,4-xylenol	226.8	Zeotropic	
1163	-ethyl benzoate	212.5	19	210.6
1164	-camphor	208.9	65	208.4
1165	-borneol	215.0	42	207.8
1166	-1,3,5-triethylbenzene	215.5	Zeotropic	
1167	-ethyl bornyl ether	204.9	70	203.0
	Iodobenzene	188.55		
1168	-nitrobenzene	210.75	Zeotropic	
1169	-phenol	181.5	47	177.7
1170	-ethyl oxalate	185.65	52	181.0
1171	-caproic acid	205.15	12	186.8
1172	-isocaproic acid	199.5	15	185.5
1173	-benzyl alcohol	205.2	12	187.75
1174	-p-cresol	201.7	10	188.1
1175	-o-toluidine	200.35	Zeotropic	
1176	-isobutyl lactate	182.15	70	180.5
1177	-indene	182.6	Zeotropic	
1178	-isoamyl butyrate	178.5	Zeotropic	
1179	-butylbenzene	183.1	Zeotropic	
	Benzene	80.15		
1180	-aniline	184.35	Zeotropic	
1181	-cyclohexene	82.1	35.3	78.9
1182	-cyclohexane	80.75	48.1	77.56
1183	-methylcyclopentane	71.85	84	71.7
1184	-n-hexane	69.0	95.3	68.5
1185	-2,2-dimethylpentane	79.1	53.7	75.85
1186	-2,3-dimethylpentane	89.79	21.2	79.4
1187	-2,4-dimethylpentane	80.8	51.7	75.2
1188	-n-heptane	98.4	0.7	80.1
1189	-2,2,4-trimethylpentane	99.2	2.3	80.1

BINARY SYSTEMS (Continued)

No.	System	B.P. (°C)	Azeotropic data: Compn. X_2 (wt. %)	Azeotropic data: B.P. (°C)
	Phenol	182.2		
1190	-aniline, N	183.91	58.1	185.84
1191	-2-picoline, N	129.2	24.6	185.5
1192	-3-picoline, N	143.5	29.8	188.93
1194	-4-picoline, N	144.8	32.5	190.0
1195	-ethylene diacetate, N	189.86	60.8	195.53
1196	-benzaldehyde	179.2	49.0	175.6
1197	-o-cresol	191.1	Zeotropic	
1198	-2,4-lutidine, N	159.0	43.0	193.4
1199	-2,6-lutidine, N	144.0	27.5	185.5
1200	-o-toluidine	200.35	Zeotorpic	
1201	-2,4,6-collidine, N	171.0	47.7	195.23
1202	-n-octyl alcohol	195.15	87	195.4
1203	-*sec*-octyl alcohol	179.0	50	184.5
1204	-indene	182.2	53	177.8
1205	-mesitylene	164.5	79	163.5
1206	-pseudocumene	168.2	75	166.0
1207	-naphthalene	218.1	Zeotropic	
1208	-butylbenzene	183.1	54	175.0
1209	-camphene	159.6	78	156.1
1210	-n-decane	173.3	65	168.0
1211	-2,7-dimethyloctane	160.25	94	159.5
1212	-amyl ether	187.5	22	180.2
1213	-isoamyl ether	173.2	85	172.2
1214	-isoamyl sulfide	214.8	Zeotropic	
1215	-1,3,5-triethylbenzene	215.5	Zeotropic	
1216	-n-tridecane	235.42	16.9	180.56
	Pyrocatechol	245.9		
1217	-indole	253.5	85.0	255.0
1218	-o-phenetidine, N	232.5	8	246.0
1219	-p-phenetidine, N	249.9	66	253.8
1220	-quinoline, N	237.4	39	257.9
1221	-naphthalene	218.05	88.5	217.45
1222	-quinaldine, N	246.5	52.0	252.5
1223	-safrole	235.9	77.0	233.55
1224	-isosafrole	252.0	30	243.0
1225	-eugenol	254.8	1.5	245.85
1226	-carvone	231.0	29	248.3
1227	-thymol	232.9	83	232.2
1228	-1-methylnapthalene	244.9	60	235.1
1229	-2-methylnaphthalene	241.15	63	233.25
1230	-acenaphthene	277.9	16	245.25
1231	-biphenyl	255.9	43.5	239.85
1232	-phenyl ether	259.3	40.7	242.0
1233	-1,3,5-triethylbenzene	215.5	91.1	214.7
1234	-fluorene	295.0	Zeotropic	
1235	-diphenyl methane	265.6	35.0	243.05
1236	-n-tridecane	234.0	70.0	229.7
1237	-1,2-diphenylethane	284.9	Zeotropic	
	Resorcinol	281.4		
1238	-naphthalene	218.05	Zeotropic	
1239	-1-naphthol	288.0	30	280.2
1240	-2-naphthol	295.0	15	280.8
1241	-methyl phthalate	283.7	62	287.5
1242	-1-methylnaphthalene	244.6	85.5	243.1
1243	-2-methylnaphthalene	241.15	89.5	240.05
	Resorcinol	281.4		
1244	-p-*tert*-amylphenol	266.5	85	265.8
1245	-acenaphthene	277.9	59	266.2
1246	-biphenyl	255.9	79	252.15
1247	-phenyl ether	259.3	77	255.65
1248	-fluorene	295.0	52	274.0
1249	-n-tridecane	234.0	88	233.25
1250	-stilbene	306.5	44	277.5
1251	-1,2-diphenylethane	284.9	53	269.7
	Pyrogallol	309.0		
1252	-2-naphthol	295.0	22	293.5
1253	-acenaphthene	277.9	80	272.8
1254	-biphenyl	256.1	90	253.5
	Aniline	184.35		
1255	-o-cresol, N	191.1	92	191.25
1256	-n-octyl alcohol	195.2	17	183.95
1257	-o-xylene	144.3	Zeotropic	
1258	-indene	182.6	58.5	179.75
1259	-mesitylene	164.7	88.0	164.35
1260	-pseudocumene	169.35	86.5	168.64
1261	-naphthalene	218.0	Zeotropic	
1262	-butylbenzene	183.1	54	177.8
1263	-n-nonane	150.7	86.5	149.2
1264	-n-decane	174.6	64	167.28
1265	-amyl ether	187.5	45	177.5
1266	-isoamyl ether	173.2	72	169.35
1267	-2-methylnaphthalene	241.15	Zeotropic	
1268	-n-undecane	194.5	42.5	175.31
1269	-1,3,5-triethylbenzene	215.5	Zeotropic	
1270	-n-dodecane	216.5	28.5	180.37
1271	-n-tridecane	235.4	13.8	182.94
1272	-n-tetradecane	252.5	4.8	183.90
	2-Picoline	129.3		
1273	-n-octane	125.75	58.0	121.12
1274	-n-nonane	150.7	15.9	129.2
1275	-n-decane	174.6	Zeotropic	
	3-Picoline	144.0		
1276	-allyl sulfide	139.35	70	135.5
1277	-2,6-lutidine	144.06	27.3	143.5
	o-Phenylenediamine	258.6		
1278	-isosafrole	252.0	70	249.2
1279	-isafrole	235.9	Zeotropic	
1280	-biphenyl	256.1	63	249.7
1281	-phenyl ether	259.0	54	251.2
1282	-diphenylmethane	265.4	30	254.0
1283	-1,2-diphenylethane	284.5	Zeotropic	
	Methyl fumarate	193.25		
1284	-m-bromotoluene	184.3	84	183.65
1285	-o-cresol, N	191.1	40	197.8
1286	-m-cresol, N	202.2	28	204.3
1287	-benzyl ethyl ether	185.0	68	183.5
1288	-naphthalene	218.0	Zeotropic	
1289	-dipentene	177.7	30	172.5
	Methyl maleate	204.05		
1290	-caproic acid	205.15	37	201.5
1291	-o-cresol, N	191.1	22	204.65
1292	-m-cresol, N	202.2	15	208.75
1293	-p-cresol, N	201.7	44	208.6
1294	-naphthalene	218.0	13	203.7
1295	-borneol	215.0	22	202.95
1296	-isoamyl sulfide	214.8	18	203.0
	Cyclohexanone	155.7		
1297	-n-hexyl alcohol	157.85	6	155.65
1298	-cumene	152.8	35	152.0
1299	-camphene	159.6	42.5	150.55
1300	-2,7-dimethyloctane	160.1	45	151.5
	Ethyl acetoacetate	180.4		
1301	-phenetole	170.45	76	169.8
1302	-isobutyl sulfide	172.0	90	171.0
1303	-indene	182.6	32	177.15
1304	-propylbenzene	159.3	76	158.3
1305	-pseudocumene	168.2	63	165.2
1306	-butylbenzene	183.1	48	172.0
1307	-cymene	176.7	59	170.5
1308	-camphene	159.6	70	156.15
1309	-dipentene	177.7	57	169.05
1310	-d-limonene	177.8	57	169.05
1311	-2,7-dimethyloctane	160.1	76	156.0
1312	-amyl ether	187.5	30	174.5
	Ethylidene diacetate	168.5		
1313	-phenetole	170.45	44	164.5
1314	-butyl butyrate	166.4	63	163.5
1315	-ethyl caproate	167.7	55	164.0
1316	-*sec*-octyl alcohol	180.4	6.5	168.3
1317	-cineole	176.35	34	164.95
1318	-isoamyl ether	173.2	43	161.5
	Ethyl oxalate	185.65		
1319	-o-cresol, N	191.1	64	194.1
1320	-camphene	159.6	84	158.5
1321	-2,7-dimethyloctane	160.1	78	159.5
1322	-amyl ether	187.5	46	177.7
1323	-isoamyl ether	173.2	71	170.15
	Glycol diacetate	186.3		
1324	-o-cresol, N	191.1	65	194.5
	Capronitrile	163.9		
1325	-cumene	152.8	82	150.8
1326	-camphene	159.6	65	143.0
	Cyclohexanol	160.8		
1327	-o-xylene	143.6	86	143.0
1328	-m-xylene	139.0	95	138.9
1329	-indene	181.7	25	160.0
1330	-propylbenzene	158.8	60	153.8
1331	-naphthalene	218.05	Zeotropic	
1332	-cymene	176.7	28	159.5
1333	-camphene	159.5	59	151.9
1334	-cineole	176.35	8	160.55
	Butyl acetate	126.0		
1335	-paraldehyde	124.35	91	124.25

No.	System	B.P. (°C)	Azeotropic data	
			Compn. X_2 (wt. %)	B.P. (°C)
1336	-n-octane	125.8	48	119.0
1337	-butyl ether	142.1	5	125.9
	Caproic acid	205.3		
1338	-m-cresol	202.2	87	201.9
1339	-guaiacol	205.05	58	200.8
1340	-acetophenone	202.0	68	200.5
1341	-naphthalene	218.05	39	203.75
1342	-1-methylnaphthalene	244.6	Zeotropic	
1343	-2-methylnaphthalene	241.15	Zeotropic	
1344	-1,3,5-triethylbenzene	215.5	37	202.0
	Isoamyl formate	123.8		
1345	-paraldehyde	124.1	44	123.0
1346	-ethylbenzene	136.15	Zeotropic	
1347	-isobutyl ether	122.3	35	121.5
	Propyl propionate	122.5		
1348	-toluene	110.75	Zeotropic	
1349	-n-octane	125.8	40	118.2
	Propyl lactate	171.7		
1350	-o-cresol	191.1	Zeotropic	
1351	-isobutyl sulfide	172.0	52	169.0
1352	-mesitylene	164.6	72	160.5
1353	-isoamyl ether	173.2	47	167.5
	Isopropyl lactate	166.8		
1354	-o-cresol	191.1	Zeotropic	
1355	-mesitylene	164.6	40	159.5
1356	-camphene	159.6	70	154.2
	n-Hexyl alcohol	157.8		
1357	-o-cresol	191.1	zeotropic	
1358	-anisole	153.85	63.5	151.0
1359	-m-xylene	139.0	85	138.3
1360	-cumene	152.8	65	149.5
1361	-mesitylene	164.6	45	153.5
1362	-pesudocumene	168.2	32	156.3
1363	-propylbenzene	158.8	45	152.5
	Acetal	103.55		
1364	-methylcyclohexane	101.15	60	99.65
1365	-n-heptane	98.45	72	97.75
1366	-2,5-dimethylhexane	109.3	25	103.0
1367	-n-octane	125.75	Zeotropic	
	2-Butoxyethanol	171.15		
1368	-benzaldehyde	179.2	9	170.95
1369	-o-cresol, N	191.1	85	191.55
1370	-phenetole	170.45	48	167.1
1371	-isobutyl sulfide	172.0	58	163.8
1372	-mesitylene	164.6	68	162.0
1373	-butylbenzene	183.4	26.6	169.6
1374	-camphene	159.6	70	154.5
1375	-dipentene	177.7	47	164.0
1376	-cineole	176.35	41.5	168.9
	Pinacol	174.35		
1377	-o-cresol, N	191.1	92	191.5
1378	-p-cresol	201.7	Zeotropic	
1379	-n-octane	125.75	Zeotropic	
1380	-pseudocumene	168.2	62	162.9
1381	-propylbenzene	159.3	72	156.3
1382	-naphthalene	218.05	Zeotropic	
1383	-p-cymene	176.7	50	167.7
1384	-cineole	176.35	55	168.5
1385	-isoamyl ether	173.4	60	167.2
	Dipropylene glycol	229.2		
1386	-p-cresol	201.7	Zeotropic	
1387	-methyl salicylate	222.95	65	213.0
1388	-isosafrole	252.0	40	225.5
1389	-safrole	235.9	50	222.0
1390	-2-methylnaphthalene	241.1	Zeotropic	
	Triethylene glycol	288.7		
1391	-methyl phthalate	283.2	67	277.0
1392	-1-methylnaphthalene	244.6	Zeotropic	
1393	-acenaphthene	277.9	65	271.5
1394	-biphenyl	256.1	90	255.3
1395	-fluorene	294.0	Zeotropic	
1396	-phenyl benzoate	315.0	20	286.0
1397	-diphenylmethane	265.4	80	263.0
1398	-stilbene	306.5	40	284.5
1399	-1,2-diphenylethane	284.5	58	275.5
	Benzonitrile	191.1		
1400	-o-cresol, N	191.1	51	195.95
1401	-m-cresol, N	202.2	89	202.5
1402	-p-cresol, N	201.7	86	202.1
1403	-o-toluidine	200.35	Zeotropic	
1404	-isoamyl butyrate	181.05	92	180.85
1405	-cineole	176.35	86	175.6
1406	-amyl ether	187.5	58	180.5
1407	-isoamyl ether	173.2	84	171.4
	Benzaldehyde	179.2		
1408	-o-cresol, N	191.1	77	192.0
1409	-p-cresol	2017	Zeotropic	
1410	-naphthalene	218.0	Zeotropic	
1411	-p-cymene	175.3	72	171.0
1412	-d-limonene	177.8	57	171.2
1413	-camphene	159.6	84.5	158.45
1414	-isoamyl ether	173.2	62.5	168.6
	Benzoic acid	250.8		
1415	-p-nitrotoluene	238.9	89	237.4
1416	-3,4-xylenol	226.8	Zeotropic	
1417	-propyl succinate	250.5	57	248.0
1418	-naphthalene	218.05	95	217.7
1419	-1-methylnaphthalene	244.6	73	239.6
1420	-2-methylnaphthalene	241.15	75	237.25
1421	-biphenyl	277.9	49.5	246.05
1422	-phenyl ether	259.3	41	247.3
1423	-fluorene	295.0	Zeotropic	
1424	-diphenylmethane	265.6	18	248.95
1425	-1,2-diphenylethane	284.0	Zeotropic	
	o-Nitrotoluene	221.75		
1426	-benzyl alcohol	205.2	91	204.75
1427	-methyl salicylate	222.95	14	221.65
1428	-3,4-xylenol	226.8	Zeotropic	
1429	-2,4-xylidine	214.0	Zeotropic	
1430	-naphthalene	218.0	Zeotropic	
1431	-diethylaniline	217.05	88	216.85
1432	-geraniol	229.6	19	220.7
1433	-menthol	216.3	66	214.65
1434	-n-decyl alcohol	232.8	15	221.0
1435	-2-methylnaphthalene	241.15	Zeotropic	
1436	-bornyl acetate	227.6	27	221.15
	p-Nitrotoluene	238.9		
1437	-quinoline	237.3	92	237.2
1438	-safrole	235.9	82	234.5
1439	-geraniol	229.6	75	228.8
1440	-n-decyl alcohol	232.8	67	231.5
1441	-bornyl acetate	227.6	90	227.45
	Toluene	110.7		
1442	-2,6-lutidine	144.0	Zeotropic	
1443	-ethylcyclopentane	103.5	93	103.0
1444	-n-heptane	98.4	Zeotropic	
1445	-2,5-dimethylhexane	109.4	65	107.0
1446	-2-methylheptane	117.6	18	110.3
1447	-2,3,4-trimethylpentane	113.5	40	109.5
	Benzyl alcohol	205.2		
1448	-o-cresol	191.1	Zeotropic	
1449	-m-cresol, N	202.2	39	207.1
1450	-p-cresol, N	201.7	38	206.8
1451	-methylaniline	196.25	70	195.8
1452	-o-toluidine	200.35	Zeotropic	
1453	-3,4-xylenol	226.8	Zeotropic	
1454	-dimethylaniline	194.05	93.5	193.9
1455	-ethylaniline	205.5	50.0	202.8
1456	-2,4-xylidine	214.0	Zeotropic	
1457	-naphthalene	218.05	40	204.1
1458	-diethylaniline	217.05	28	20.42
1459	-d-limonene	177.8	89	176.4
1460	-borneol	215.0	14.2	205.07
1461	-1,3,5-triethylbenzene	215.5	43	203.2
	o-Cresol	191.1		
1462	-benzylamine, N	185.0	33	201.45
1463	-phenyl acetate, N	195.7	64	198.5
1464	-2,4,6-collidine, N	171.3	37	197.2
1465	-n-octyl alcohol	195.15	62	196.9
1466	-butyl sulfide	185.0	75	183.8
1467	-indene	183.0	91	182.9
1468	-naphthalene	218.05	Zeotropic	
1469	-terpinene	181.5	72	177.8
1470	-terpinolene	184.6	66	179.5
1471	-thymene	179.7	27	176.6
1472	-camphor, N	209.1	85	209.85
	m-Cresol	202.2		
1473	-o-toluidine, N	200.35	38.5	203.65
1474	-m-toluidine, N	203.1	47	205.5
1475	-p-toluidine, N	200.55	38	204.3
1476	-phenyl acetate, N	195.7	30	204.4
1477	-2,4,6-collidine, N	171.3	27	206.2
1478	-isoamyl lactate, N	202.4	50	207.6
1479	-n-octyl alcohol, N	195.15	38	203.3

BINARY SYSTEMS (Continued)

No.	System	B.P. (°C)	Azeotropic data: Compn. X_2 (wt. %)	Azeotropic data: B.P. (°C)
1480	-propiophenone, N	217.7	83	218.6
1481	-phorone, N	197.8	45	206.5
1482	-naphthalene	218.05	Zeotropic	
1483	-camphor, N	209.1	63.5	213.35
1484	-1,3,5-triethylbenzene	215.5	Zeotropic	
	p-Cresol	201.6		
1485	-o-toluidine, N	200.35	43	203.5
1486	-m-toluidine, N	203.1	53	204.9
1487	-p-toluidine, N	200.55	43	204.05
1488	-o-anisidine	219.0	Zeotropic	
1489	-acetophenone, N	202.0	53.5	208.4
1490	-benzyl formate, N	202.4	58.0	207.0
1491	-methyl benzoate, N	199.4	60.0	204.35
1492	-phenyl acetate, N	195.7	32	204.3
1493	-isoamyl lactate, N	202.4	52	207.25
1494	-n-octyl alcohol, N	195.2	30	202.25
1495	-camphor, N	209.1	69.5	213.5
1496	-ethyl caprylate, N	208.35	75	209.5
	Guaiacol	205.05		
1497	-acetophenone	202.0	32.5	205.25
1498	-m-toluidine	203.1	Zeotropic	
1499	-ethylaniline	205.5	45	204.4
1500	-ethyl caprylate, N	208.35	85	208.9
	2,4-Lutidine	159.0		
1501	-n-nonane	150.7	67.75	148.3
1502	-n-undecane	195.4	Zeotropic	
	Methylaniline	196.25		
1503	-n-octyl alcohol	195.2	55	193.0
1504	-d-limonene	177.8	97	174.5
	o-Toluidine	200.35		
1505	-acetophenone, N	202.0	68	203.65
1506	-n-octyl alcohol	195.2	77	194.7
1507	-n-decane	174.6	87	173.76
1508	-n-undecane	195.5	60.3	188.25
1509	-n-dodecane	216.5	37	195.75
1510	-n-tridecane	234.6	14.5	199.45
	o-Anisidine	219.0		
1511	-naphthalene	218.0	50	217.0
1512	-2-methylnaphthalene	241.15	Zeotropic	
1513	-1,3,5-triethylbenzene	215.5	65	214.5
	Enanthic acid	222.0		
1514	-ethyl fumarate	217.85	78	216.4
1515	-ethyl maleate	223.3	50	220.0
1516	-ethyl succinate	217.85	80	216.0
1517	-propiophenone	217.7	80	216.5
1518	-naphthalene	218.0	70	214.2
1519	-biphenyl	256.1	Zeotropic	
1520	1,3,5-triethylbenzene	215.5	73	211.0
	n-Heptyl alcohol	176.15		
1521	-benzyl methyl ether	167.8	80	167.0
1522	-p-methylanisole	177.05	48	173.3
1523	-phenetole	170.45	72	169.0
1524	-p-cymene	176.0	52	173.0
1526	-isoamyl ether	173.35	63	170.35
	Indole	253.5		
1527	-carvacrol	237.85	12	254.5
1528	-p-*tert*-amylphenol	266.5	88	268.0
	Acetophenone	202.		
1529	-p-ethylphenol N	218.8	85	219.5
1530	-2,4-xylenol, N	210.5	70	213.0
1531	-3,4-xylenol	226.8	Zeotropic	
1532	-dimethylaniline	194.15	Zeotropic	
1533	-ethylaniline	205.5	Zeotropic	
1534	-2,4-xylidine	214.0	Zeotropic	
1535	-n-octyl alcohol	195.2	87.5	194.95
1536	-naphthalene	218.0	Zeotropic	
1537	-1,3,5-triethylbenzene	215.5	Zeotropic	
	Phenyl acetate	195.7		
1538	-2,4-xylenol	210.5	Zeotropic	
1539	-n-octyl alcohol	195.15	47	192.4
1540	-indene	182.6	Zeotropic	
1541	-naphthalene	218.05	Zeotropic	
1542	-thymene	179.7	82	179.3
1543	-linalool	198.6	39	193.5
	Methyl salicylate	222.95		
1544	-phenethyl alcohol	219.4	57	218.0
1545	-3,4-xylenol	226.8	Zeotropic	
1546	-ethyl maleate	223.3	40	221.95
1547	-quinoline	237.3	Zeotropic	
1548	-geraniol	229.7	3	222.2
1549	-menthol	216.4	85	216.25
1550	-n-tridecane	234.0	Zeotropic	

No.	System	B.P. (°C)	Azeotropic data: Compn. X_2 (wt. %)	Azeotropic data: B.P. (°C)
	Ethylbenzene	136.15		
1551	-ethylcyclohexane	131.8	85	131.2
1552	-n-octane	125.75	Zeotropic	
1553	-n-nonane	150.7	Zeotropic	
	p-Ethylphenol	218.8		
1554	-ethyl fumarate, N	217.85	52	223.0
1555	-ethyl maleate, N	223.3	62	226.3
1556	-p-methylacetophenone, N	226.35	70	229.5
1557	-benzyl acetate, N	215.0	40	221.0
1558	-ethyl benzoate, N	212.5	20	219.8
1559	-naphthalene	218.0	55	215.0
1560	-diethylaniline	217.05	40	214.0
1561	-1,3,5-triethylbenzene	215.5	60	212.0
	p-Methylanisole	177.05		
1562	-*sec*-octyl alcohol	180.4	21	176.3
1563	-pseudocumene	169.0	Zeotropic	
1564	-butyl isovalerate	177.6	42	176.4
1565	-butylbenzene	183.2	Zeotropic	
1566	-cineole	176.35	65	175.35
1567	-isoamyl ether	173.2	70.5	172.5
	Phenethyl alcohol	219.4		
1568	-3,4-xylenol	226.8	Zeotropic	
1569	-2,4-xylidine	214.0	Zeotropic	
1570	-naphthalene	218.05	56	214.2
1571	-diethylaniline	217.05	60	213.95
1572	-borneol	213.4	80	213.0
1573	-menthol	216.3	70	215.05
1574	-1-methylnaphthalene	244.9	Zeotropic	
1575	-biphenyl	256.1	Zeotropic	
	2,4-Xylenol	210.5		
1576	-ethyl fumarate, N	217.85	68	219.65
1577	-quinoline, N	237.3	92	239.0
1578	-p-methylacetophenone, N	226.35	15	227.0
1579	-propiophenone, N	217.7	35	221.0
1580	-benzyl acetate, N	215.0	64	216.8
1581	-camphor, N	209.1	50	217.0
	3,4-Xylenol	226.8		
1582	-o-phenetidine, N	232.5	92	232.65
1583	-ethyl fumarate, N	217.85	35	228.2
1584	-ethyl maleate, N	223.3	45	230.0
1585	-quinoline, N	237.3	65	241.95
1586	-p-methylacetophenone, N	226.35	49	231.35
1587	-propiophenone, N	217.7	33	228.5
1588	-naphthalene	218.0	84	217.6
1589	-diethylaniline	217.05	92	217.0
1590	-camphor, N	209.1	27	227.55
1591	-n-tridecane	234.0	42	223.5
	Ethylaniline	205.5		
1592	-n-octyl alcohol	195.2	85	194.9
1593	-naphthalene	218.0	Zeotropic	
1594	-camphor	209.1	Zeotropic	
	2,4-Xylidine	217.4		
1595	-menthol	216.3	30	213.5
1596	-n-undecane	195.5	88	194.98
1597	-n-dodecane	216.5	63	209.8
1598	-n-tridecane	234.6	29	215.28
1599	-n-tetradecane	252.0	2.5	217.38
	o-Phenetidine	232.5		
1600	-ethyl salicylate	233.8	18	232.2
1601	-naphthalene	218.0	Zeotropic	
1602	-safrole	235.9	14	232.38
1603	-anethole	235.7	25	232.25
1604	-carvacrol	237.85	87	238.0
1605	-thymol, N	232.9	54.9	234.3
1606	-2-methylnaphthalene	241.15	Zeotropic	
	p-Phenetidine	249.9		
1607	-safrole	235.9	Zeotropic	
1608	-isosafrole	252.0	36	248.8
1609	-1-methylnaphthalene	244.6	73	243.95
1610	-2-methylnaphthalene	241.15	85	240.85
1611	-biphenyl	256.1	10	249.5
1612	-phenyl ether	259.0	15	249.75
	Ethyl fumarate	217.85		
1613	-naphthalene	218.0	42	216.7
1614	-thymol, N	232.9	87.5	233.35
1615	-menthol	216.3	70	216.0
	Ethyl maleate	223.3		
1616	-p-methylacetophenone	226.35	12	223.15
1617	-naphthalene	218.0	77	217.65
1618	-thymol, N	232.9	73	234.9
	Ethyl succinate	217.25		

No.	System	B.P. (°C)	Azeotropic data: Compn. X_2 (wt. %)	Azeotropic data: B.P. (°C)
1619	-propiophenone	217.7	33	216.7
1620	-naphthalene	218.05	38.5	216.3
1621	-2-methylnaphthalene	241.15	Zeotropic	
	Caprylic acid	238.5		
1622	-naphthalene	218.05	94	216.2
1623	-carvacrol	237.85	75	237.6
1624	-1-methylnaphthalene	244.6	48	233.5
1625	-2-methylnaphthalene	241.15	52	235.0
	Propyl isovalerate	155.7		
1626	-cumene	152.8	Zeotropic	
1627	-propylbenzene	158.9	Zeotropic	
1628	-camphene	159.6	35	145.0
1629	-nopinene	163.8	25	155.0
	***sec*-Octyl alcohol**	179.0		
1630	butylbenzene	183.1	50	178.2
1631	-p-cymene	176.7	56	174.0
1632	-thymene	179.7	48	176.0
1633	-cineole	176.35	73.5	175.85
1634	-amyl ether	187.5	14	178.8
	Quinoline	237.3		
1635	-mesitol, N	220.5	15	240.4
1636	-safrole	235.9	73	235.15
1637	-carvacrol, N	237.85	52	244.3
1638	-thymol, N	232.9	45	243.1
1639	-1-methylnaphthalene	244.6	Zeotropic	
1640	2-methylnaphththalene	241.15	7	237.25
1641	-p-*tert*-amylphenol, N	266.5	94	267.5
1642	-biphenyl	256.1	Zeotropic	
	p-Methylacetophenone	226.35		
1643	-thymol, N	232.9	68	234.9
1644	-geraniol	229.6	5	226.25
1645	-citronellol	224.4	68	223.7
1646	-2-methylnaphthalene	241.15	Zeotropic	
1647	-bornyl acetate	227.6	40	225.8
	Propiophenone	217.7		
1648	-benzyl acetate	215.0	Zeotropic	
1649	-borneol	215.0	Zeotropic	
1650	-1,3,5-triethylbenzene	215.5	75	215.4
	Ethylsalicylate	233.8		
1651	-safrole	235.9	12	233.65
1652	-geraniol	229.7	60	228.5
1653	-n-decyl alcohol	232.9	52	230.5
1654	-2-methylnaphthalene	241.15	Zeotropic	
	Cumene	152.8		
1655	-n-nonane	150.75	77	148.0
	Mesitylene	164.6		
1656	-propylbenzene	159.3	Zeotropic	
1657	-camphene	159.6		
1658	2,7-dimethyloctane	160.1	72	158.6
	Propylbenzene	159.3		
1659	-camphene	159.6	53	158.0
	Pseudocumene	168.2		
1660	-p-cymene	176.7	Zeotropic	
1661	-n-decane	173.3	25	166.5
	Mesitol	230.5		
1662	-naphthalene	218.0	63	215.5
1663	-1,3,5-triethylbenzene	215.5	70	213.0
	3-Phenylpropanol	235.6		
1664	-naphthalene	218.05	~80	217.8
1665	-safrole	235.9	53	233.8
1666	-anethole	235.7	52	234.0
1667	-thymol, N	232.9	38	237.5
1668	-biphenyl	254.9	—	235.4
	Pelargonic acid	254.0		
1669	-naphthalene	218.0	Zeotropic	
1670	-isosafrole	252.0	65	249.5
1671	-eugenol	254.8	48	250.5
1672	-thymol	232.9	Zeotropic	
1673	-1-methylnaphthalene	244.6	82	243.0
1674	-2-methylnaphthalene	241.15	90	240.2
1675	-biphenyl	256.1	55	250.0
1676	-phenyl ether	259.0	45	250.5
1677	-1,3,5-triethylbenzene	215.5	Zeotropic	
1678	-diphenylmethane	265.4	25	252.7
	Naphthalene	218.0		
1679	-borneol	213.4	65	213.0
1680	-citronellol	224.5	30	217.8
1681	-menthol	216.4	74.5	215.5
1682	-n-tridecane	234.0	Zeotropic	
	Quinaldine	246.5		
1683	-safrole	235.9	Zeotropic	
1684	-carvacrol, N	237.85	33	250.8
1685	-thymol, N	232.9	20	250.0
	Methyl phthalate	283.2		
1686	-acenaphthene	277.9	66.5	276.35
1687	-biphenyl	255.9	Zeotropic	
1688	-diphenylmethane	265.6	Zeotropic	
1689	-1,2-diphenylethane	284.0	47	280.5
	Propyl benzoate	230.85		
1690	-carvacrol, N	237.85	82	238.85
1691	-thymol, N	232.8	55	235.5
1692	-2-methylnaphthalene	241.15	Zeotropic	
	p-Cymene	176.7		
1693	-dipentene	177.7	40	175.8
1694	-d-limonene	177.8	25	174.5
1695	-cineoloe	176.35	55	176.2
	Carvacrol	237.85		
1696	-carvenone, N	234.5	45	243.0
1697	-menthenone, N	222.5	25	239.5
1698	-propyl succinate, N	250.5	75	251.5
1699	-n-decyl alcohol	232.8	Zeotropic	
1700	-isobutyl benzoate, N	241.9	67	243.85
1701	-biphenyl	256.1	Zeotropic	
1702	-bornyl acetate, N	227.6	25	238.2
	Carvone	230.95		
1703	-thymol, N	232.9	52	238.65
1704	-geraniol	229.6	60	229.2
1705	-n-decyl alcohol	232.8	19	230.85
1706	-2-methylnaphthalene	241.15	Zeotropic	
	Thymol	232.9		
1707	-carvenone, N	234.5	50	241.0
1708	-pulegone, N	223.8	35	235.3
1709	-geraniol	229.6	42.5	225.6
1710	-menthone, N	209.5	8	233.2
1711	-2-methylnaphthalene	241.15	Zeotropic	
1712	-isobutyl benzoate, N	242.15	80	243.2
1713	-1,3,5-triethylbenzene	215.5	Zeotropic	
1714	-bornyl acetate, N	227.7	40	235.6
	Camphene	159.6		
1715	-dipentene	177.7	Zeotropic	
1716	-2,7-dimethyloctane	160.25	38	158.0
	Borneol	211.8		
1717	-methol	216.4	Zeotropic	
1718	-1,3,5-triethylbenzene	215.5	38	212.2
	Menthol	216.3		
1719	-2-methylnaphthalene	241.15	Zeotropic	
1720	-terpineol methyl ether	216.2	50	215.3
1721	-1,3,5-triethylbenzene	215.5	~45	214.0
	Capric acid	268.8		
1722	-1-methylnaphthalene	244.6	Zeotropic	
1723	-diphenylmethane	265.4	72	262.5
	1-Methylnaphthalene	244.6		
1724	-2-methylnaphthalene	241.15	Zeotropic	
1725	-biphenyl	256.1	Zeotropic	
1726	-phenyl ether	259.0	Zeotropic	
1727	-diphenylmethane	265.4	Zeotropic	
	2-Methylnaphthalene	241.15		
1728	-isobutyl benzoate	241.9	40	240.8
	p-*tert*-Amylphenol	266.5		
1729	-acenaphthene	277.9	Zeotropic	
1730	-fluorene	295.0	Zeotropic	
1731	-diphenylmethane	265.4	60	263.0
	Phenyl ether	259.0		
1732	-isoamyl benzoate	262.05	10	258.9
1733	-isoamyl oxalate	268.0	Zeotropic	
1734	-diphenylmethane	265.6	Zeotropic	
	Isoamyl benzoate	262.0		
1735	-isoamyl oxalate	268.0	Zeotropic	
1736	-diphenylmethane	265.6		
	1,3,5-Triethylbenzene	215.5		
1737	-bornyl acetate	227.2	Zeotropic	
1738	-bornyl ethyl ether	204.9	Zeotropic	
	Isoamyl oxalate	268.0		
1739	-diphenylmethane	265.4	86	265.25
1740	-1,2-diphenylethane	284.5	Zeotropic	
	Phenyl benzoate	315.0		
1741	-stilbene	306.5	Zeotropic	
1742	-benzyl ether	297.0	Zeotropic	
	Benzyl phenyl ether	286.5		
1743	-1,2-diphenylethane	284.5	Zeotropic	

TERNARY SYSTEMS

			Azeotropic data			
No.	System	B.P. (°C)	Compn. (wt. %)	B.P. (°C)	Type of azeotrope	Ref.
1	Argon	−186			—	1
	Nitrogen	−195	Zeotropic			
	Oxygen	−183	90—120°K			
2	Hydrogen bromide	−67	10.4		H	2
	Water	100	11.0	105		
	Chlorobenzene	131.8	78.6			
3	Hydrogen cyanide	26			—	3
	Acetonitrile	81.6	Zeotropic			
	Acrolein	52.45				
4	Chlorine trifluoride	—			—	4
	Hydrogen fluoride	19.4	Zeotropic			
	Uranium hexafluoride	56				
5	Hydrogen chloride	−80	15.8		H	5
	Water	100	64.8	107.33		
	Phenol	182	19.4			
6	Hydrogen chloride	−80	5.3		H	5
	Water	100	20.2	96.9		
	Chlorobenzene	131.8	74.5			
7	Hydrogen fluoride	19.4	10		N	6
	Fluosilicic acid	—	36	116.1		
	Water	100	54			
8	Silicon tetrafluoride	—	24.6		—	7
	Hexafluoroethane	−78	32.7	−104		
	Ethane	−88	42.7			
9	Water	100	4.5		H	8
	Carbon tetrachloride	76.75	85.5	62		
	Ethyl alcohol	78.3	10			
10	Water	100	5		H	8
	Carbon tetrachloride	76.75	84	65.15		
	Allyl alcohol	96.95	11			
11	Water	100	4.05		H	9
	Carbon tetrachloride	76.7	91.0	65		
	sec-Butyl alcohol	99	4.95			
12	Water	100	1.6		H	10
	Carbon disulfide	46.25	93.4	41.3		
	Ethyl alcohol	78.3	5.0			
13	Water	100			—	11
	Carbon disulfide	46.25	Zeotropic			
	Dioxane	101.4				
14	Water	100			—	12
	Chloroform	61	Zeotropic			
	Formic acid	100.75				
15	Water	100	1.3		—	13
	Chloroform	61	90.5	52.3		
	Methyl alcohol	64.7	8.2			
16	Water	100	2.3		—	14
	Chloroform	61	94.2	55.3		
	Ethyl alcohol	78.3	3.5			
17	Water	100	18.6		S	15
	Formic acid	100.8	71.9	107.2		
	Propionic acid	140.7	9.5			
18	Water	100	19.5		S	16
	Formic acid	100.8	75.9	107.62		
	Butyric acid	162.4	4.6			
19	Water	100	15.5		S	16
	Formic acid	100.8	66.8	107.02		
	Isobutyric acid	154	17.7			
20	Water	100	21.3		S	16
	Formic acid	100.8	76.3	107.64		
	Isovaleric acid	176.5	2.4			
21	Water	100			—	16
	Formic acid	100.8	Zeotropic			
	Valeric acid	186				
22	Water	100	2.1		H-3	17
	Nitromethane	101.2	6.5	33.1		
	n-Pentane	36.07	91.4			
23	Water	100	7.88		H-3; 748 mm Hg	18
	Nitromethane	101.2	29.73	71.43		
	n-Heptane	98.43	62.39			
24	Water	100	12.4		H-3; 748 mm Hg	18
	Nitromethane	101.2	44.25	77.35		
	n-Octane	125.7	43.35			
25	Water	100	17.4		H-3; 748 mm Hg	18
	Nitromethane	101.2	58.3	80.72		
	n-Nonane	150.8	24.3			
26	Water	100	19.1		H-3; 748 mm Hg	18
	Nitromethane	101.2	68.1	82.35		
	n-Decane	174.12	12.8			

No.	System	B.P. (°C)	Azeotropic data: Compn. (wt. %)	Azeotropic data: B.P. (°C)	Type of azeotrope	Ref.
27	Water	100	20.6		H-3; 748 mm Hg	18
	Nitromethane	101.2	73.3	82.82		
	n-Undecane	194.5	6.1			
28	Water	100	21.5		H-3; 748 mm Hg	18
	Nitromethane	101.2	75.3	83.13		
	n-Dodecane	214.5	3.2			
29	Water	100	22.8		H-3; 748 mm Hg	18
	Nitromethane	101.2	75.4	83.21		
	n-Tridecane	234	1.8			
30	Water	100			—	19
	Methyl alcohol	64.7	Zeotropic			
	Ethyl alcohol	78.3				
31	Water	100	5.26		—	20
	Methyl alcohol	64.7	81.20	67.85		
	Methyl chloroacetate	131.4	13.54			
32	Water	100			—	21
	Methyl alcohol	64.7	Zeotropic			
	Methyl acetate	57.1				
33	Water	100	0.6		—	22
	Methyl alcohol	64.7	5.4	30.2		
	Isoprene	34.0	94.0			
34	Water	100	12.45		—	23
	Tetrachloroethylene	120.8	66.75	81.18		
	n-Propyl alcohol	97.2	20.8			
35	Water	100	6.4		—	24
	Trichloroethylene	86.95	73.1	67		
	Acetonitrile	81.6	20.5			
36	Water	100	5.5		—	25
	Trichloroethylene	86.95	78.4	67.0		
	Ethyl alcohol	78.3	16.1			
37	Water	100	7		—	26
	Trichloroethylene	86.95	73	69.4		
	Isopropyl alcohol	82.45	20			
38	Water	100			—	24
	Acetonitrile	81.6	Zeotropic			
	Acetone	56.4				
39	Water	100	1		—	26
	Acetonitrile	81.6	44	72.9		
	Ethyl alcohol	78.3	55			
40	Water	100			—	27
	Acetonitrile	81.6	Zeotropic			
	Diethylamine	55.5				
41	Water	100	8.2		—	24
	Acetonitrile	81.6	23.3	66		
	Benzene	80.2	68.5			
42	Water	100	3.5		—	13
	Acetonitrile	81.6	9.6	68.6		
	Triethylamine	89.7	86.9			
43	Water	100			—	8
	Acetic acid	118.1	Zeotropic			
	Toluene	110.7				
44	Water	100	8.4		H	17
	Nitroethane	114.07	9.3	59.5		
	n-Hexane	68.74				
45	Water	100	11.5		H	17
	Nitroethane	114.07	24.5	75.1		
	n-Heptane	98.43	64.0			
46	Water	100	8.7		—	26
	Ethyl alcohol	78.3	20.3	69.5		
	Acrylonitrile	77.2	71.0			
47	Water	100	4.8		—	26
	Ethyl alcohol	78.3	87.9	78.0		
	Crotonaldehyde	102.4	7.3			
48	Water	100	9.0		—	28
	Ethyl alcohol	78.3	8.4	70.23		
	Ethyl acetate	77.05	82.6			
49	Water	100	7.5		—	26
	Ethyl alcohol	78.3	42.5	81.8		
	Butylamine	77.8	50.0			
50	Water	100	6.3		—	26
	Ethyl alcohol	78.3	8.6	62		
	Butyl methyl ether	70.3	85.1			
51	Water	100	7.4		H	29
	Ethyl alcohol	78.3	18.5	64.86		
	Benzene	80.2	74.1			
52	Water	100	4.8		H	30
	Ethyl alcohol	78.3	19.7	62.60		
	Cyclohexane	80.75	75.5			

No.	System	B.P. (°C)	Azeotropic data: Compn. (wt. %)	Azeotropic data: B.P. (°C)	Type of azeotrope	Ref.
53	Water	100	9		—	31
	Ethyl alcohol	78.3	13	74.7		
	Triethylamine	89.4	78			
54	Water	100	12		H	26
	Ethyl alcohol	78.3	37	74.4		
	Toluene	110.6	51			
55	Water	100	3		H	26
	Ethyl alcohol	78.3	12	56.0		
	n-Hexane	68.7	85			
56	Water	100	6.1		H	26
	Ethyl alcohol	78.3	33.0	68.8		
	n-Heptane	98.45	60.9			
57	Water	100	0.4		—	32
	Acetone	56.7	7.6	32.5		
	Isoprene	34.7	92.0			
58	Water	100	8.5		H	33
	Allyl alcohol	96.95	5.1	59.7		
	n-Hexane	68.95	86.4			
59	Water	100	12.5		H	26
	Isopropyl alcohol	82.3	40.5	83		
	Butylamine	77.8	47.0			
60	Water	100	7.5		H	34
	Isopropyl alcohol	82.45	19.0	66.3		
	Benzene	80.2				
61	Water	100	7.5		H	8
	Isopropyl alcohol	82.45	18.5	64.3		
	Cyclohexane	80.75	74.0			
62	Water	100	13.1		H	26
	Isopropyl alcohol	82.3	38.2	76.3		
	Toluene	110.6	48.7			
63	Water	100	17.0		—	35
	Propyl alcohol	97.2	10.0	82.15		
	Propyl acetate	101.6	73.0			
64	Water	100	7.6		740mm Hg; H	26
	Propyl alcohol	97.2	10.1	67		
	Benzene	80.1	82.3			
65	Water	100	8.5		H	8
	Propyl alcohol	97.2	10.0	66.55		
	Cyclohexane	80.75	81.5			
66	Water	100	19.2		H	36
	n-Butyl alcohol	117.75	2.9	61.5		
	n-Hexane	68.95	77.9			
67	Water	100	41.4		H	36
	n-Butyl alcohol	117.75	7.6	78.1		
	n-Heptane	98.4	51.0			
68	Water	100	60.0		H	36
	n-Butyl alcohol	117.75	14.6	86.1		
	n-Octane	125.75	25.4			
69	Water	100	69.9		H	36
	n-Butyl alcohol	117.75	18.3	90.0		
	n-Nonane	150.7	11.8			
70	Water	100	29.9		H	26
	n-Butyl alcohol	117.75	34.6	90.6		
	Butyl ether	142.1	35.5			
71	Water	100	5		H	37
	2-Butanone	79.6	35	63.6		
	Cyclohexane	80.7	60			
72	Water	100	4		H	37
	Butyraldehyde	74.8	21	55.0		
	n-Hexane	68.7	75			
73	Water	100	8.9		H	14
	sec-Butyl alcohol	99.6	10.8	69.7		
	Cyclohexane	80.75	80.3			
74	Water	100	9		H	13
	sec-Butyl alcohol	99.4	19	76.3		
	Isooctane	99.0	72			
75	Water	100	8.1		H	38
	tert-Butyl alcohol	82.55	21.4	67.3		
	Benzene	80.2	70.5			
76	Water	100			—	22
	tert-Butyl alcohol	82.55	Zeotropic			
	Isoprene	34.0				
77	Water	100	8		H	8
	tert-Butyl alcohol	82.55	21	65.0		
	Cyclohexane	80.75	71			
78	Water	100			H	38
	Isobutyl alcohol	108	Zeotropic			
	Benzene	80.2				
79	Water	100	17.9		H	39
	Isobutyl alcohol	108	16.4	81.3		
	Toluene	110.7	65.7			

TERNARY SYSTEMS (Continued)

No.	System	B.P. (°C)	Azeotropic data Compn. (wt. %)	B.P. (°C)	Type of azeotrope	Ref.
80	Water	100			—	37
	Ethyl acrylate	99.3	Zeotropic			
	Isopropyl ether	68.3				
81	Water	100			—	40
	Toluene	110.7	Zeotropic			
	Benzyl alcohol	204.7				
82	Water	100			—	2
	Pyridine	115.5	Zeotropic			
	Benzene	80.1				
83	Water	100	14.0		H	41
	Pyridine	115.5	15.5	78.6		
	n-Heptane	98.4	70.5			
84	Water	100	22.5		H	41
	Pyridine	115.5	25.5	86.7		
	n-Octane	125.75	52.0			
85	Water	100	30.5		H	41
	Pyridine	115.5	37.0	90.5		
	n-Nonane	150.7	32.5			
86	Water	100	35.5		H	41
	Pyridine	115.5	45.5	92.3		
	N- Decane	173.3	19.0			
87	Water	100	38.5		H	41
	Pyridine	115.5	51.0	93.1		
	n-Undecane	194.5	10.5			
88	Water	100	40.5		H	41
	Pyridine	115.5	54.5	93.5		
	n-Dodecane	216.0	5.0			
89	Water	100	32.4		H	42
	Isoamyl alcohol	131.5	19.6	89.8		
	Isoamyl formate	124.2	48.0			
90	Water	100	44.8		—	43
	Isoamyl alcohol	131.5	31.2	93.6		
	Isoamyl acetate	142.0	24.0			
91	Water	100			—	37
	2-Picoline	129.2	Zeotropic			
	Paraldehyde	124.5				
92	Carbon tetrachloride	76.8			—	44
	Methyl alcohol	64.7	Zeotropic			
	Benzene	80.1				
93	Chloroform	61			—	45
	Formic Acid	100.75	Zeotropic			
	Acetic acid	118.1				
94	Chloroform	61	47		S	46
	Methyl alcohol	64.7	23	57.5		
	Acetone	56.1	30			
95	Chloroform	61	65.3		S	47
	Ethyl alcohol	78.3	10.4	63.2		
	Acetone	56.1	24.3			
96	Chloroform	61.0	56.1		—	48
	Ethyl alcohol	78.3	9.5	57.3		
	n-Hexane	68.7	34.4			
97	Chloroform	61.2	68.8		—	48
	Acetone	56.5	3.6	60.79		
	n-Hexane	68.7	27.6			
98	Chloroform	61.0			—	49
	Acetone	56.4	Zeotropic			
	Toluene	110.7				
99	Chloroform	61.2	79.70		S	50
	Ethyl formate	54.1	5.3	61.97		
	Isopropyl bromide	59.4	15.7			
100	Chloroform	61.2			—	51
	2-Bromopropane	59.4	Zeotropic			
	Isopropyl formate	68.8				
101	Methyl alcohol	64.7	17.4		—	26
	Acetone	56.1	5.8	53.7		
	Methyl acetate	56.3	76.8			
102	Methyl alcohol	64.7	14.6		—	52
	Acetone	56.25	30.8	47		
	n-Hexane	68.95	59.6			
103	Methyl alcohol	64.7	17.8		—	53
	Methyl acetate	57	48.6	50.8		
	Cyclohexane	80.75	33.6			
104	Methyl alcohol	64.7	14.6		—	14
	Methyl acetate	56.3	36.8	47.4		
	n-Hexane	68.7	48.6			
105	Methyl alcohol	64.7			—	8
	Benzene	80.1	Zeotropic			
	Cyclohexane	80.75				
106	Acetonitrile	81.6	34		—	26
	Ethyl alcohol	78.3	8	70.1		
	Triethylamine	89.7	58			

TERNARY SYSTEMS (Continued)

No.	System	B.P. (°C)	Azeotropic data Compn. (wt. %)	B.P. (°C)	Type of azeotrope	Ref.
107	Trichloroethylene	87.2			—	54
	Benzene	80.1	Zeotropic			
	Cyclohexane	80.7				
108	Acetic acid	118.1	23		S	55
	Acetic anhydride	139.6	55	134.4		
	Pyridine	115.5	22			
109	Acetic acid	118.1	3.4		S	56
	Pyridine	115.5	10.6	98.5		
	n-Heptane	98.4	86.0			
110	Acetic acid	118.1	10.4		S	56
	Pyridine	115.5	20.1	115.7		
	n-Octane	125.75	69.5			
111	Acetic acid	118.1	20.7		S	57
	Pyridine	115.5	29.4	128.0		
	n-Nonane	150.7	49.9			
112	Acetic acid	118.1	31.4		S	56
	Pyridine	115.5	38.2	134.1		
	n-Decane	173.3	30.4			
113	Acetic acid	118.1	37.5		S	58
	Pyridine	115.5	43.5	137.1		
	n-Undecane	194.5	19.0			
114	Acetic acid	118.1	13.5		S	59
	Pyridine	115.5	25.2	129.08		
	Ethylbenzene	136.5	61.3			
115	Acetic acid	118.1	17.7		S	57
	Pyridine	115.5	30.5	132.2		
	o-Xylene	143.6	51.8			
116	Acetic acid	118.1	10.2		S	60
	Pyridine	115.5	22.5	129.22		
	p-Xylene	138.4	67.3			
117	Acetic acid	118.1	15		—	61
	Isoamyl alcohol	132	54	132		
	Isoamyl acetate	142	31			
118	Acetic acid	118.1	7.6		—	62
	Benzene	80.1	34.4	77.2		
	Cyclohexane	80.75	58.0			
119	Acetic acid	118.1	3.6		S	63
	2-Picoline	129.45	24.8	121.3		
	n-Octane	125.75	71.6			
120	Acetic acid	118.1	12.8		S	63
	2-Picoline	129.45	38.4	135.0		
	n-Nonane	150.7	48.8			
121	Acetic acid	118.1	19.9		—	63
	2-Picoline	129.45	46.8	141.3		
	n-Decane	173.3	33.3			
122	Acetic acid	118.1	30.5		—	63
	2-Picoline	129.45	55.2	143.4		
	n-Undecane	194.5	14.3			
123	Acetic acid	118.1			—	64
	2,6-Lutidine	144.0	Zeotropic			
	n-Octane	125.75				
124	Acetic acid	118.1	12.6		S	64
	2,6-Lutidine	144.0	74.3	147.0		
	n-Decane	173.3	13.1			
125	Acetic acid	118.1	75.0		S	58
	2,6-Lutidine	144.0	13.8	163.0		
	n-Undecane	194.5	11.2			
126	Acetic acid	118.1			—	65
	Ethylbenzene	136.15	Zeotropic			
	n-Nonane	150.7				
127	Acetic acid	118.1			—	66
	Acetic anhydride	139.6	Zeotropic			
	Methylene diacetate	164.0				
128	Methyl formate	31.9			—	14
	Ethyl ether	34.6	Zeotropic			
	n-Pentane	36.15				
129	Nitroethane	114.2	31.7		S	67
	p-Dioxane	101.3	17.7	102.87		
	Isobutyl alcohol	108	50.6			
130	Ethyl alcohol	78.3	29.6		—	68
	Benzene	80.1	12.8	64.7		
	Cyclohexane	80.75	57.6			
131	Ethyl alcohol	78.3			—	69
	Benzene	80.1	Zeotropic			
	n-Hexane	68.7				
132	Ethyl alcohol	78.3			—	70
	Aniline	184.35	Zeotropic			
	Toluene	110.7				
133	Ethyl alcohol	78.3			—	70
	Aniline	184.35	Zeotropic			
	n-Heptane	98.4				

No.	System	B.P. (°C)	Azeotropic data: Compn. (wt. %)	Azeotropic data: B.P. (°C)	Type of azeotrope	Ref.
134	Ethyl alcohol	78.3			—	70
	Toluene	110.7	Zeotropic			
	n-Heptane	98.4				
135	Ethylene glycol	197.4			—	71
	Pyridine	115.5	Zeotropic			
	Phenol	181.4				
136	Ethylene glycol	197.4	5.9		S	71
	Phenol	181.4	79.1	185.01		
	2-Picoline	128.8	15.0			
137	Ethylene glycol	197.4	15.9		S	71
	Phenol	181.4	67.7	186.41		
	3-Picoline	143.5	16.4			
138	Ethylene glycol	197.4	8.7		S	71
	Phenol	181.4	74.6	185.04		
	2,6-Lutidine	144.0	16.7			
139	Ethylene glycol	197.4	29.5		S	71
	Phenol	181.4	54.8	188.55		
	2,4,6-Collidine	171.0	15.7			
140	Ethylene glycol	197.45	33.6		S	72
	o-Cresol	191.0	62.4	189.65		
	2,4,6-Collidine	171.3	4.0			
141	Acetone	56.1	51.1		—	14
	Methyl acetate	56.3	5.6	49.7		
	n-Hexane	68.7	43.3			
142	Acetone	56.1			—	73
	2-Butanone	79.6	Zeotropic			
	Ethyl acetate	77.0				
143	Acetone	56.4			—	74
	Benzene	80.1	Zeotropic			
	Cyclohexane	80.75				
144	Acetone	56.1			—	75
	Benzene	80.1	Zeotropic			
	Toluene	110.7				
145	Propionic acid	140.7	55.5		S	58
	Pyridine	115.5	26.4	147.1		
	n-Undecane	194.5	18.1			
146	Propionic acid	141.05	4.5		S	76
	2-Picoline	129.3	10.5	123.7		
	n-Octane	125.4	85.0			
147	Propionic acid	141.05	16.5		S	76
	2-Picoline	129.3	21.5	140.1		
	n-Nonane	150.6				
148	Propionic acid	141.05	29.5		S	76
	2-Picoline	129.3	32.0	149.33		
	n-Decane	174.0	38.5			
149	Propionic acid	141.05	43.0		S	76
	2-Picoline	129.3	40.0	153.4		
	n-Undecane	194.8	17.0			
151	Propionic acid	141.05			S	76
	2-Picoline	129.3	Zeotropic			
	n-Dodecane	216.1				
151	Isopropyl alcohol	82.3	31.1		—	76
	Benzene	80.1	15.0	69.1		
	Cyclohexane	80.75	53.9			
152	Propyl alcohol	97.2	15.5		—	77
	Benzene	80.1	30.4	73.81		
	Cyclohexane	80.75	54.2			
153	Propyl alcohol	97.2			—	78
	Benzene	80.1	Zeotropic			
	n-Heptane	98.4				
154	2-Butanone	79.6			—	79
	Ethyl acetate	77.1	Zeotropic			
	n-Hexane	68.7				
155	2-Butanone	79.6			—	80
	Benzene	80.1	Zeotropic			
	Cyclohexane	80.75				
156	Butyric acid	162.45			—	58
	Pyridine	115.5	Zeotropic			
	n-Undecane	194.5				
157	p-Dioxane	101.1	44.3		S	81
	Isobutyl alcohol	107.0	26.7	101.8		
	Toluene	110.7				
158	n-Butyl alcohol	117.75	11.9		—	82
	Pyridine	115.5	20.7	108.7		
	Toluene	110.7	67.4			
159	n-Butyl alcohol	117.75	4		—	83
	Benzene	80.1	48	77.42		
	Cyclohexane	80.75	48			
160	n-Butyl alcohol	117.7			—	84
	Benzene	80.1	Zeotropic			
	n-Heptane	98.4				

TERNARY SYSTEMS (Continued)

No.	System	B.P. (°C)	Azeotropic data Compn. (wt. %)	Azeotropic data B.P. (°C)	Type of azeotrope	Ref.
161	Isobutyl alcohol	107.0	43.2		—	85
	Benzene	80.1	47.0	77.2		
	Cyclohexane	80.75				
162	Pyridine	115.5	8.6		S	86
	Isoamyl alcohol	131.0	4.1	110.79		
	Toluene	110.7				
163	Phenol	181.4	33.5		—	87
	Aniline	183.95	48.5	184.45		
	n-Tridecane	234.0	18.0			
164	Phenol	182.0	26.4		S	88
	Ethylene diacetate	186.0	34.4	194.45		
	Phenyl acetate	195.7	39.2			
165	Phenol	181.4	19.88		S	89
	2,4-Lutidine	159.0	21.52	181.78		
	n-Undecane	194.5	58.60			
166	Ethylbenzene	136.15			—	59
	Pyridine	115.5	Zeotropic			
	n-Nonane	150.7				
167	Aniline	184.35			—	70
	Toluene	110.7	Zeotropic			
	n-Heptane	98.4				
168	Aniline	184.35			—	8
	Benzyl alcohol	205.5	Zeotropic			
	d-Limonene	177.8				
169	Aniline	184.35			—	8
	sec-Octyl alcohol	178.7	Zeotropic			
	d-Limonene	177.8				
170	Aniline	184.35			—	8
	o-Bromotoluene	181.75	Zeotropic			
	sec-Octyl alcohol	178.7				
171	Propyl lactate	171.7	31		—	8
	Phenetole	171.5	33	163.0		
	Menthene	170.8	36			
172	m-,p-Cresol (mixt.)	202	81		S	90
	Pyridine bases (mixt.)	143	9	202.81		
	Naphthalene	218.1	10			
173	m-, p-Cresol (mixt.)	202	65.5		S	90
	Pyridine bases (mixt.)	157	16.5	202.03		
	Naphthalene	218.1				
174	m-, p-Cresol (mixt.)	202	62		S	90
	Pyridine bases (mixt.)	163	17	202.39		
	Naphthalene	218.1	21			
175	m-Cresol	202.8	61.5		S	91
	2,4,6-Collidine	171.3	20.8	205.82		
	Naphthalene	217.9	17.7			
176	Isobutyl lactate	182.15			—	8
	sec-Octyl alcohol	178.7	Zeotropic			
	Terpinene	180.5				
177	Ethylbenzene	136.1			—	92
	Isopropylbenzene	152.8	Zeotropic			
	Butylbenzene	183.1				

REFERENCES

1. **Narinskii,** *Tr. Vses. Nauch. Issled. Inst. Kisloror. Mashinostr.,* 11, 3, 1967.
2. Dow Chemical Co., unpublished data.
3. **Sokolov, Sevryogova, Zhavoronskov,** *Rev. Chim. (Bucharest),* 20, 169, 1969.
4. **Ellis, Johnson,** *J. Inorg. Nucl. Chem.,* 6, 194, 199, 1958.
5. **Prahl, Mathes,** *Angew. Chem.,* 47, 11, 1934.
6. **Munter, Aepli, Kossatz,** *Ind. Eng. Chem.,* 39, 427, 1947.
7. **Calfee, Fukuhara, Bigelow,** *J. Amer. Chem. Soc.,* 61, 3552, 1939.
8. **Lecat,** *L'Azeotropisme,* Lamartin, Bruxelles, 1918.
9. **Marinichev, Susarev,** *Zhur. Fiz. Khim.,* 43, 1132, 1969.
10. **Ghysels,** *Bull. Soc. Chim. Belges,* 33, 57, 1924.
11. **De Mol,** *Ingr. Chim.,* 22, 262, 1938.
12. **Conti, Othmer, Gilmont,** *J. Chem. Eng. Data,* 5, 301, 1960.
13. **Kudryavtseva, Susarev, Eisen,** *Zhur. Fiz. Khim.,* 43, 437, 1969.
14. **Kudryavtseva, Eisen, Susarev,** *Zhur. Fiz. Khim.* 40, 1285, 1652, 1966.
15. **Kushner, Tatsievskaya, Serafimov,** *Zhur. Fiz. Khim.,* 41, 237, 1967.
16. **Kushner, Tatsievskaya, Serafimov,** *Zhur. Fiz. Khim.,* 42, 2248, 1968.
17. **Riddick,** Commercial Solvents Corp., unpublished data.
18. **Malesinska, Malesinski,** *Bull. Acad. Polon. Sci. Ser. Sci. Chim.,* 11, 475, 1963.

19. Delzenne, *J. Chem. Eng. Data*, 3, 224, 1958.
20. Calices, Hannotte, *Ingr. Chim.*, 20, 1, 1936.
21. Balashov, Serafimov, Bessonova, *Zhur. Fiz. Khim.*, 40, 2294, 1966.
22. Lesteva, Kachalova, Morozova, Ogorodnikov, Trenke, *Zhur. Prikl. Khim.*, 40, 1808, 1967.
23. Malesinska, *Bull. Acad. Polon. Sci. Ser. Sci. Chim.*, 12, 853, 1964.
24. Pratt, Preprint, Trans. Inst. Chem. Engrs. (London), March 1947.
25. Licht, Denzler, *Chem. Eng. Progr.*, 44, 627, 1948.
26. Union Carbide Chemicals, *Alcohols*, 1961.
27. Union Carbide Chemicals, unpublished data.
28. Merriman, *J. Chem. Soc. Trans.*, 103, 1790, 1801, 1913.
29. Young, Fortey, *J. Chem. Soc. Trans.*, 81, 717, 1902.
30. Zieborak, Galska, *Bull. Acad. Polon. Sci. Cl. III*, 3, 383, 1955.
31. Tyerman, Br. Pat. 590.713, 1947.
32. Patterson, U.S. Pat. 2,407.997, 1946.
33. Kogan, Tolstova, *Zhur. Fiz. Khim.*, 33, 276, 1959.
34. Yorizane, Yoshimura, *Hiroshima Daigaku Kogakuba Kenkya Hokoku*, 13, 41, 1965.
35. Smirnova, Moraczevskii, Storonkin, *Vest. Leningrad Univ.*, 14, 70, 1959.
36. Kogan, Fridman, Deizenrot, *Zhur. Prikl. Khim.* 30, 1339, 1957.
37. Union Carbide Chemicals Co., unpublished data.
38. Young, Fortey, *J. Chem. Soc. Trans.*, 81, 739, 1902.
39. Frolov, Loginova, Nazarova, *Zhur. Fiz. Khim.* 43, 2632, 1969.
40. Susarev, Gorbunov, *Zhur. Prikl. Khim.*, 36, 459, 1963.
41. Trabczynski, *Bull. Acad. Polon. Sci. Ser. Sci. Chim. Geol. Geogr.*, 6, 269, 1958.
42. Hannotte, *Bull. Soc. Chim. Belges*, 35, 85, 1926.
43. Hyatt, U.S. Patent 2,176.500, 1939.
44. Hirata, Hirose, *Mem. Fac. Technol. Tokyo Metrop. Univ.*, No. 11, 876, 1961.
45. Conti, Othmer, Gilmont, *J. Chem. Eng. Data*, 5, 301, 1960.
46. Ewell, Welch, *Ind. Eng. Chem.*, 37, 1224, 1945.
47. Morachevskii, Leontev, *Zhur. Fiz. Khim.*, 34, 2347, 1960.
48. Kidryavtseva, Susarev, *Zhur. Prikl. Khim.*, 36, 1231, 1471, 1710, 2025, 1963.
49. Satapathy et al., *J. Appl. Chem. (London)*, 6, 261, 1956.
50. Orszagh, Lelakowska, Beldowicz, *Bull. Acad. Polon. Sci. Ser. Sci. Chim. Geol. Geogr.*, 6, 419, 1958.
51. Lelakowska, *Bull. Acad. Polon. Sci. Ser. Sci. Chim. Geol. Geogr.*, 6, 645, 1958.
52. Forman, U.S. Patent 2,581.789, 1952.
53. Fisher, U.S. Patent 2,341.433, 1944.
54. Rao, Dakshinamurty, Rao, *J. Sci. Ind. Res. (India)*, 20B, 218, 1961.
55. Jones, *J. Chem. Eng. Data*, 7, 13, 1962.
56. Zieborak, *Bull. Acad. Polon. Sci. Cl. III*, 3, 531, 1955.
57. Zieborak, Wyrzykowska-Stankiewicz, *Bull. Acad. Polon. Sci. Ser. Sci. Chim. Geol. Geogr.*, 7, 247, 1959.
58. Zieborak, Wyrzykowska-Stankiewicz, *Bull. Acad. Polon. Sci. Ser. Sci. Chim. Geol. Geogr.*, 6, 517, 1958.
59. Galska-Krajewska, Zieborak, *Rocz. Chem.*, 36, 119, 1962.
60. Galska-Krajewska, *Bull. Acad. Polon. Sci. Ser. Sci. Chim.*, 9, 455, 1961.
61. Krokhin, *Zhur. Fiz. Khim.*, 43, 442, 1969.
62. Baradarajan, Satyanarayana, *J. Chem. Eng. Data*, 13, 148, 1968.
63. Zieborak, Wyrzykowska-Stankiewicz, *Bull. Acad. Polon. Sci. Ser. Sci. Chim. Geol. Geogr.*, 6, 377, 1958.
64. Zieborak, Kaczorowana-Badyoczek, Maczynska, *Rocz. Chem.*, 29, 783, 1955.
65. Zieborak, Galska-Krajewska, *Bull. Acad. Polon. Sci. Ser. Sci. Chim. Geol. Geogr.*, 7, 253, 1959.
66. Tatscheff et al., *Z. Phys. Chem. (Leipzig)*, 237, 52, 1968.
67. Malesinska, Malinsinski, *Bull. Acad. Polon. Sci. Ser. Sci. Chim.*, 8, 191, 1960.
68. Morachevskii, Zharov, *Zhur. Prikl. Khim.*, 36, 2771, 1963.
69. Yuan, Ho, Keshpande, Lu, *J. Chem. Eng. Data*, 8, 549, 1963.
70. Hollo, Ember, Lengyel, Weig, *Acta Chim. Acad. Sci. Hung.*, 13, 307, 1957.
71. Razniewska, *Rocz. Cham.*, 38, 851, 1964.
72. Kurtyka, *Bull. Acad. Polon. Sci. Cl. III*, 4, 49, 1956.
73. Babicz, Ivanchikova, Serafimov, *Zhur. Prikl. Khim.*, 42, 1354, 1969.
74. Kurmanadharao, Krishnamurty, Rao, *Rec. Trav. Chim.*, 76, 769, 1957.
75. Vitman, Zharov, *Zhur. Prikl. Khim.*, 42, 2858, 1969.
76. Trabczynski, *Bull. Acad. Polon. Sci. Ser. Sci. Chim.*, 12, 335, 1965.
76. Nagata, *Can. J. Chem. Eng.*, 42, 82, 1964.
77. Moraczevskii, Cheng, *Zhur. Fiz. Khim.*, 35, 2535, 1961.
78. Fu, Lu, *J. Chem. Eng. Data*, 13, 6, 1968.
79. Gorbunova, Lutigina, Malenko, *Zhur. Prikl. Khim.*, 38, 374, 622, 1965.
80. Donald, Ridgway, *J. Appl. Chem. (London)*, 8, 403, 408, 1958.

81. **Wyrzykowska-Stankiewicz, Zieborak,** *Bull. Acad. Polon. Sci. Ser. Sci. Chim.*, 8, 655, 1960.
82. **Hollo, Lengyel,** *Ind. Eng. Chem.*, 51, 957, 1959.
83. **Zieborak, Galska-Krajewska,** *Bull. Acad. Polon. Sci. Ser. Sci. Chim. Geol. Geogr.*, 6, 763, 1958.
84. **Vijayaraghavan, Deshpande, Kuloor,** *J. Chem. Eng. Data*, 12, 13, 1967.
85. **Nataraj, Rao,** *Trans. Indian Inst. Chem. Eng.*, 95, 1968.
86. **Zieborak, Wyrzykowska-Stankiewicz,** *Bull. Acad. Polon. Sci. Ser. Sci. Chim.*, 8, 137, 1960.
87. **Stadnicki,** *Bull. Acad. Polon. Sci. Ser. Sci. Chim.*, 10, 357, 1962.
88. **Orszagh, Lelakowska, Radecki,** *Bull. Acad. Polon. Sci. Ser. Sci. Chim. Geol. Geogr.*, 6, 605, 1958.
89. **Fahmy, Assal,** *Bull. Acad. Polon. Sci. Ser. Sci. Chim.*, 14, 773, 1966.
90. **Zieborak, Markowska-Majewska,** *Bull. Acad. Polon. Sci. Cl. III*, 2, 341, 1954.
91. **Kurtyka,** *Bull. Acad. Polon. Sci. Ser. Sci. Chim.*, 9, 741, 1961.
92. **Linek, Fried, Pick,** *Coll. Czech. Chem. Commun.*, 30, 1358, 1965.

QUATERNARY AND QUINARY SYSTEMS

No.	System	B.P. (°C)	Azeotropic data: Compn. (wt. %)	Azeotropic data: B.P. (°C)	Type	Ref.
1	Hydrocyanic acid	26			—	1
	Water	100.0	Zeotropic			
	Acrylonitrile	77.3				
	Acrolein	52.4				
2	Hydrocyanic acid	26			—	1
	Acetonitrile	81.6	Zeotropic			
	Acrylonitrile	77.3				
	Acrolein	52.4				
3	Water	100.0			—	2
	Formic acid	100.8	Zeotropic			
	Acetic acid	118.1				
	Butyric acid	162.4				
4	Water	100.0	7.38		H	3
	Nitromethane	101.2	20.65	76.88		
	Tetrachloroethylene	120.8	59.45			
	n-Propyl alcohol	97.2	12.52			
5	Water	100.0	9.86		H	3
	Nitromethane	101.2	34.40	77.06		
	Tetrachloroethylene	120.8	32.60			
	n-Octane	125.75	23.14			
6	Water	100.0	—		H	3
	Tetrachloroethylene	120.8	—	80.98		
	n-Propyl alcohol	97.2	—			
	n-Octane	125.75	—			
7	Water	100.0	9.98		H	3
	Nitromethane	101.2	41.00	76.34		
	n-Propyl alcohol	97.2	12.42			
	n-Octane	125.75	36.60			
8	Water	100.0			—	4
	Acetonitrile	81.6	Zeotropic			
	Ethyl alcohol	78.3				
	Triethylamine	89.7				
9	Water	100.0	8.7		—	5
	Ethyl alcohol	78.3	11.1	70		
	Crotonaldehyde	102.2	0.1			
	Ethyl acetate	77.1	80.1			
10	Water	100.0	6.1		H	6
	Ethyl alcohol	78.3	19.2	62.14		
	Benzene	80.1	20.4			
	Cyclohexane	80.75	54.3			
11	Water	100.0			—	7
	Ethyl alcohol	78.3	Zeotropic			
	Benzene	80.1				
	n-Hexane	68.95				
12	Water	100.0			—	7
	Ethyl alcohol	78.3	Zeotropic			
	Benzene	80.1				
	Methylcyclohexane	100.88				
13	Water	100.0	6.8		H	8
	Ethyl alcohol	78.3	18.7	64.97		
	Benzene	80.1	62.4			
	n-Heptane	98.4	12.1			
14	Water	100.0	6.7		H	8
	Ethyl alcohol	78.3	17.7	64.69		
	Benzene	80.1	61.4			
	Isooctane	99.3	14.1			
15	Water	100.0			—	9
	1-Chlorobutane	78.44	Zeotropic			
	n-Butyl alcohol	117.73				
	Butyl ether	—				

No.	System	B.P. (°C)	Azeotropic data Compn. (wt. %)	B.P. (°C)	Type	Ref.
16	Chloroform	61.2			—	10
	Methyl alcohol	64.6	Zeotropic			
	Methyl acetate	56.9				
	Benzene	80.1				
17	Acetic acid	118.1	17		S	11
	Pyridine	115.4	27	127.9		
	Ethylbenzene	136.4	18			
	n-Nonane	150.8	38			
18	Acetic acid	118.1			—	12
	Pyridine	115.4	Zeotropic			
	p-Xylene	138.4				
	n-Nonane	150.8				
19	Acetone	56.1			—	13
	Isopropyl alcohol	82.3	Zeotropic			
	Benzene	80.1				
	Toluene	110.7				
20	Acetone	56.1			—	14
	Benzene	80.1	Zeotropic			
	Cyclohexane	80.9				
	Toluene	110.7				
21	Isopropyl alcohol	82.3			—	15
	2-Butanone	79.6	Zeotropic			
	Benzene	80.1				
	Cyclohexane	80.9				
22	Water	100.0	9.45		H	3
	Nitromethane	101.2	37.30			
	Tetrachloroethylene	120.8	21.15	76.5		
	n-Propyl alcohol	97.2	10.58			
	n-Octane	125.75	21.52			
23	Chloroform	61.2			—	10
	Methyl alcohol	64.6				
	Acetone	56.15	Zeotropic			
	Methyl acetate	56.9				
	Benzene	80.1				

REFERENCES

1. **Sokolov, Sevryoguva, Zhavoronkov,** *Teor. Osn. Khim. Tekhnol.*, 3, 288, 1969.
2. **Kushner, Lebedeva, Tatsievskaya, Serafimov,** *Shur. Prikl. Khim.*, 42, 1104, 1968.
3. **Malesinska,** *Bull. Acad. Polon. Sci. Ser. Sci. Chim.*, 12, 853, 1964.
4. Union Carbide Chemicals, unpublished data.
5. Eastman Chemical Products, unpublished data.
6. **Zieborak, Galska,** *Bull. Acad. Polon. Sci. Cl. III*, 3, 383, 1955.
7. **Swietoslawski, Zieborak, Galska-Krajewska,** *Bull. Acad. Polon. Sci. Ser. Sci. Chim. Geol. Geogr.*, 7, 43, 1959.
8. **Swietoslawski, Zieborak,** *Bull. Acad. Polon. Sci. Ser. A*, 9, 13, 1950.
9. **Riddick,** Commercial Solvents, unpublished data.
10. **Hudson, van Winkle,** *J. Chem. Eng. Data*, 14, 310, 1969.
11. **Zieborak, Galska-Krajewska,** *Bull. Acad. Polon. Sci. Ser. Sci. Chim. Geol. Geogr.*, 7, 253, 1959.
12. **Galska-Krajewska,** *Bull. Acad. Polon. Sci. Ser. Sci. Chim.*, 9, 455, 1961.
13. **Vitman, Zharov,** *Zhur. Prikl. Khim.*, 42, 2858, 1969.
14. **Vitman, Markova,** *Zhur. Prikl. Khim.*, 42, 2360, 1969.
15. **Lutugina, Kolbina,** *Zhur. Prikl. Khim.*, 41, 2766, 1968.

VISCOSITY OF GASES

The following table gives the viscosity of some common gases as a function of temperature. Unless otherwise noted, the viscosity values refer to a pressure of 100 kPa (1 bar). The notation P=0 indicates the low pressure limiting value is given. The difference between the viscosity at 100 kPa and the limiting value is generally less than 1%. Viscosity is given in units of μPa s; note that 1 μPa s = 10^{-5} poise. Substances are listed in the modified Hill order (see Introduction).

		Viscosity in micropascal seconds (μPa s)						
		100 K	**200 K**	**300 K**	**400 K**	**500 K**	**600 K**	**Ref.**
	Air	7.1	13.3	18.6	23.1	27.1	30.8	1
Ar	Argon	8.0	15.9	22.9	28.8	34.2	39.0	2,8
BF_3	Boron trifluoride		12.3	17.1	21.7	26.1	30.2	13
ClH	Hydrogen chloride			14.6	19.7	24.3		13
F_6S	Sulfur hexafluoride (P=0)			15.3	19.8	23.9	27.7	10
H_2	Hydrogen (P=0)	4.2	6.8	9.0	10.9	12.7	14.4	4
D_2	Deuterium (P=0)	5.9	9.6	12.6	15.4	17.9	20.3	11
H_2O	Water			10.0	13.3	17.3	21.4	6
D_2O	Deuterium oxide			11.1	13.7	17.7	22.0	7
He	Helium (P=0)	9.7	15.3	20.0	24.4	28.4	32.3	8
Kr	Krypton (P=0)	8.8	17.1	25.6	33.1	39.8	45.9	8
NO	Nitric oxide		13.8	19.2	23.8	28.0	31.9	13
N_2	Nitrogen (P=0)		12.9	17.9	22.2	26.1	29.6	12
N_2O	Nitrous oxide		10.0	15.0	19.4	23.6	27.4	13
Ne	Neon (P=0)	14.4	24.3	32.1	38.9	45.0	50.8	8
O_2	Oxygen (P=0)	7.5	14.6	20.8	26.1	30.8	35.1	12
O_2S	Sulfur dioxide		8.6	12.9	17.5	21.7		13
Xe	Xenon (P=0)	8.3	15.4	23.2	30.7	37.6	44.0	8
CO	Carbon monoxide	6.7	12.9	17.8	22.1	25.8	29.1	13
CO_2	Carbon dioxide		10.0	15.0	19.7	24.0	28.0	9,10
$CHCl_3$	Chloroform			10.2	13.7	16.9	20.1	13
CH_4	Methane		7.7	11.2	14.3	17.0	19.4	10
CH_4O	Methanol				13.2	16.5	19.6	13
C_2H_2	Acetylene			10.4	13.5	16.5		13
C_2H_4	Ethylene		7.0	10.4	13.6	16.5	19.1	3
C_2H_6	Ethane		6.4	9.5	12.3	14.9	17.3	5
C_2H_6O	Ethanol				11.6	14.5	17.0	13
C_3H_8	Propane			8.3	10.9	13.4	15.8	5
C_4H_{10}	*n*-Butane			7.5	10.0	12.3	14.6	5
C_4H_{10}	Isobutane			7.6	10.0	12.3	14.6	5
$C_4H_{10}O$	Diethyl ether			7.6	10.1	12.4		13
C_5H_{12}	*n*-Pentane			6.7	9.2	11.4	13.4	13
C_6H_{14}	*n*-Hexane				8.6	10.8	12.8	13

REFERENCES

1. K. Kadoya, N. Matsunaga, and A. Nagashima, Viscosity and thermal conductivity of dry air in the gaseous phase, *J. Phys. Chem. Ref. Data*, 14, 947, 1985.
2. B. A. Younglove and H. J. M. Hanley, The viscosity and thermal conductivity coefficients of gaseous and liquid argon, *J. Phys. Chem. Ref. Data*, 15, 1323, 1986.
3. P. M. Holland, B. E. Eaton, and H. J. M. Hanley, A Correlation of the viscosity and thermal conductivity data of gaseous and liquid ethylene, *J. Phys. Chem. Ref. Data*, 12, 917, 1983.
4. M. J. Assael, S. Mixafendi, and W. A. Wakeham, The viscosity and thermal conductivity of normal hydrogen in the limit zero density, *J. Phys. Chem. Ref. Data*, 15, 1315, 1986.
5. B. A. Younglove and J. F. Ely, Thermophysical properties of fluids. II. Methane, ethane, propane, isobutane, and normal butane, *J. Phys. Chem. Ref. Data*, 16, 577, 1987.

6. J. V. Sengers and J. T. R. Watson, Improved international formulations for the viscosity and thermal conductivity of water substance, *J. Phys. Chem. Ref. Data*, 15, 1291, 1986.
7. N. Matsunaga and A. Nagashima, Transport properties of liquid and gaseous D_2O over a wide range of temperature and pressure, *J. Phys. Chem. Ref. Data*, 12, 933, 1983.
8. J. Kestin, et al., Equilibrium and transport properties of the noble gases and their mixtures at low density, *J. Phys. Chem. Ref. Data*, 13, 299, 1984.
9. V. Vescovic, et al., The transport properties of carbon dioxide, *J. Phys. Chem. Ref. Data*, 19, 1990.
10. R. D. Trengove and W. A. Wakeham, The viscosity of carbon dioxide, methane, and sulfur hexafluoride in the limit of zero density, *J. Phys. Chem. Ref. Data*, 16, 175, 1987.
11. M. J. Assael, S. Mixafendi, and W. A. Wakeham, The viscosity of normal deuterium in the limit of zero density, *J. Phys. Chem. Ref. Data*, 16, 189, 1987.
12. W. A. Cole and W. A. Wakeham, The viscosity of nitrogen, oxygen, and their binary mixtures in the limit of zero density, *J. Phys. Chem. Ref. Data*, 14, 209, 1985.
13. C. Y. Ho, Ed., *Properties of Inorganic and Organic Fluids, CINDAS Data Series on Materials Properties*, Vol. V-1, Hemisphere Publishing Corp., New York, 1988.

MEAN FREE PATH IN GASES

$t = 20°C$ for data at pressures below 760 mmHg
$t = 0°C$ for data at 760 mmHg

Gas	Pressure				
	1 mmHg	**0.1 mmHg**	**0.01 mmHg**	**0.001 mmHg**	**760 mmHg**
Argon	4.73×10^{-5} m	4.73×10^{-4} m	4.73×10^{-3} m	4.73×10^{-2} m	6.3×10^{-8} m
Helium	13.32	13.32	13.32	13.32	17.4
Hydrogen	8.81	8.81	8.81	8.81	11.1
Krypton	3.63	3.63	3.63	3.63	4.8
Neon	9.4	9.4	9.4	9.4	12.4
Nitrogen	4.5	4.5	4.5	4.5	5.9
Oxygen	4.82	4.82	4.82	4.82	6.3
Xenon	2.62	2.62	2.62	2.62	3.5

Gas	Molecular diameter, cm		
	From viscosity	**From van der Waal's equation**	**From heat conductivity**
Ammonia	2.97×10^{-8}	3.08×10^{-8}	—
Argon	2.88	2.94	2.86×10^{-8}
Carbon monoxide	3.19	3.12	—
Carbon dioxide	3.34	3.23	3.40
Helium	1.90	2.65	2.30
Hydrogen	2.40	2.34	2.32
Krypton	—	(3.69)	3.14
Mercury	—	3.01	—
Nitrogen	3.15	3.15	3.53
Oxygen	2.98	2.92	—
Xenon	—	4.02	3.42

VISCOSITY OF LIQUIDS

The absolute viscosity of some common liquids at temperatures between –25 and 100°C is given in this table. Values were derived by fitting experimental data to suitable expressions for the temperature dependence. The substances are arranged by molecular formula in the modified Hill order (see Preface). All values are given in units of millipascal seconds (mPa s); this unit is identical to centipoise (cp).

Viscosity values correspond to a nominal pressure of 1 atmosphere. If a value is given at a temperature above the normal boiling point, the applicable pressure is understood to be the vapor pressure of the liquid at that temperature. A few values are given at a temperature slightly below the normal freezing point; these refer to the supercooled liquid.

The accuracy ranges from 1% in the best cases to 5 to 10% in the worst cases. Additional significant figures are included in the table to facilitate interpolation.

REFERENCES

1. Viswanath, D. S. and Natarajan, G., *Data Book on the Viscosity of Liquids*, Hemisphere Publishing Corp., New York, 1989.
2. Daubert, T. E., Danner, R. P., Sibul, H. M., and Stebbins, C. C., *Physical and Thermodynamic Properties of Pure Compounds: Data Compilation*, extant 1994 (core with 4 supplements), Taylor & Francis, Bristol, PA (also available as database).
3. Ho, C. Y., Ed., *CINDAS Data Series on Material Properties*, Vol. V-1, *Properties of Inorganic and Organic Fluids*, Hemisphere Publishing Corp., New York, 1988.
4. Stephan, K. and Lucas, K., *Viscosity of Dense Fluids*, Plenum Press, New York, 1979.
5. Vargaftik, N. B., *Tables of Thermophysical Properties of Liquids and Gases*, 2nd ed., John Wiley, New York, 1975.

Molecular formula	Name	Viscosity in mPa s					
		–25°C	0°C	25°C	50°C	75°C	100°C
	Compounds not containing carbon						
Br_2	Bromine		1.252	0.944	0.746		
Cl_3HSi	Trichlorosilane		0.415	0.326			
Cl_3P	Phosphorous trichloride	0.870	0.662	0.529	0.439		
Cl_4Si	Tetrachlorosilane			99.4	96.2		
H_2O	Water		1.793	0.890	0.547	0.378	0.282
H_4N_2	Hydrazine			0.876	0.628	0.480	0.384
Hg	Mercury			1.526	1.402	1.312	1.245
NO_2	Nitrogen dioxide		0.532	0.402			
	Compounds containing carbon						
CCl_3F	Trichlorofluoromethane	0.740	0.539	0.421			
CCl_4	Tetrachloromethane		1.321	0.908	0.656	0.494	
CS_2	Carbon disulfide		0.429	0.352			
$CHBr_3$	Tribromomethane			1.857	1.367	1.029	
$CHCl_3$	Trichloromethane	0.988	0.706	0.537	0.427		
CHN	Hydrogen cyanide		0.235	0.183			
CH_2Br_2	Dibromomethane	1.948	1.320	0.980	0.779	0.652	
CH_2Cl_2	Dichloromethane	0.727	0.533	0.413			
CH_2O_2	Formic acid			1.607	1.030	0.724	0.545
CH_3I	Iodomethane		0.594	0.469			
CH_3NO	Formamide		7.114	3.343	1.833		
CH_3NO_2	Nitromethane	1.311	0.875	0.630	0.481	0.383	0.317
CH_4O	Methanol	1.258	0.793	0.544			
CH_5N	Methylamine	0.319	0.231				
$C_2Cl_3F_3$	1,1,2-Trichlorotrifluoro-ethane	1.465	0.945	0.656	0.481		
C_2Cl_4	Tetrachloroethylene		1.114	0.844	0.663	0.535	0.442
C_2HCl_3	Trichloroethylene		0.703	0.545	0.444	0.376	
C_2HCl_5	Pentachloroethane		3.761	2.254	1.491	1.061	
$C_2HF_3O_2$	Trifluoroacetic acid			0.808	0.571		
$C_2H_2Cl_2$	*cis*-1,2-Dichloroethylene	0.786	0.575	0.445			
$C_2H_2Cl_2$	*trans*-1,2-Dichloroethylene	0.522	0.398	0.317	0.261		
$C_2H_2Cl_4$	1,1,1,2-Tetrachloroethane	3.660	2.200	1.437	1.006	0.741	0.570
$C_2H_3ClF_2$	1-Chloro-1,1-difluoro-ethane	0.477	0.376				

Molecular formula	Name	Viscosity in mPa s					
		–25°C	0°C	25°C	50°C	75°C	100°C
C_2H_3ClO	Acetyl chloride			0.368	0.294		
$C_2H_3Cl_3$	1,1,1-Trichloroethane	1.847	1.161	0.793	0.578	0.428	
C_2H_3N	Acetonitrile		0.400	0.369	0.284	0.234	
$C_2H_4Br_2$	1,2-Dibromoethane			1.595	1.116	0.837	0.661
$C_2H_4Cl_2$	1,1-Dichloroethane			0.464	0.362		
$C_2H_4Cl_2$	1,2-Dichloroethane		1.125	0.779	0.576	0.447	
$C_2H_4O_2$	Acetic acid			1.056	0.786	0.599	0.464
$C_2H_4O_2$	Methyl formate		0.424	0.325			
C_2H_5Br	Bromoethane	0.635	0.477	0.374			
C_2H_5Cl	Chloroethane	0.416	0.319				
C_2H_5I	Iodoethane		0.723	0.556	0.444	0.365	
C_2H_5NO	*N*-Methylformamide		2.549	1.678	1.155	0.824	0.606
$C_2H_5NO_2$	Nitroethane	1.354	0.940	0.688	0.526	0.415	0.337
C_2H_6O	Ethanol	3.262	1.786	1.074	0.694	0.476	
C_2H_6OS	Dimethyl sulfoxide			1.987	1.290		
$C_2H_6O_2$	Ethylene glycol			16.1	6.554	3.340	1.975
C_2H_6S	Dimethyl sulfide		0.356	0.284			
C_2H_6S	Ethanethiol		0.364	0.287			
C_2H_7N	Dimethylamine	0.300	0.232				
C_2H_7NO	Ethanolamine			21.1	8.560	3.935	1.998
C_3H_5Br	3-Bromopropene		0.620	0.471	0.373		
C_3H_5Cl	3-Chloropropene		0.408	0.314			
C_3H_5ClO	Epichlorohydrin	2.492	1.570	1.073	0.781	0.597	0.474
C_3H_5N	Propanenitrile			0.294	0.240	0.202	
C_3H_6O	Acetone	0.540	0.395	0.306	0.247		
C_3H_6O	Allyl alcohol			1.218	0.759	0.505	
C_3H_6O	Propanal			0.321	0.249		
$C_3H_6O_2$	Ethyl formate		0.506	0.380	0.300		
$C_3H_6O_2$	Methyl acetate		0.477	0.364	0.284		
$C_3H_6O_2$	Propanoic acid		1.499	1.030	0.749	0.569	0.449
C_3H_7Br	1-Bromopropane		0.645	0.489	0.387		
C_3H_7Br	2-Bromopropane		0.612	0.458	0.359		
C_3H_7Cl	1-Chloropropane		0.436	0.334			
C_3H_7Cl	2-Chloropropane		0.401	0.303			
C_3H_7I	1-Iodopropane		0.970	0.703	0.541	0.436	0.363
C_3H_7I	2-Iodopropane		0.883	0.653	0.506	0.407	
C_3H_7NO	*N,N*-Dimethylformamide		1.176	0.794	0.624		
$C_3H_7NO_2$	1-Nitropropane	1.851	1.160	0.798	0.589	0.460	0.374
C_3H_8O	1-Propanol	8.645	3.815	1.945	1.107	0.685	
C_3H_8O	2-Propanol		4.619	2.038	1.028	0.576	
$C_3H_8O_2$	1,2-Propylene glycol		248	40.4	11.3	4.770	2.750
$C_3H_8O_3$	Glycerol			934	152	39.8	14.8
C_3H_8S	1-Propanethiol		0.503	0.385			
C_3H_8S	2-Propanethiol		0.477	0.357	0.280		
C_3H_9N	Propylamine			0.376			
C_3H_9N	Isopropylamine		0.454	0.325			
C_4H_4O	Furan	0.661	0.475	0.361			
C_4H_5N	Pyrrole		2.085	1.225	0.828	0.612	
$C_4H_6O_3$	Acetic anhydride		1.241	0.843	0.614	0.472	0.377
C_4H_7N	Butanenitrile			0.553	0.418	0.330	0.268
C_4H_8O	2-Butanone	0.720	0.533	0.405	0.315	0.249	
C_4H_8O	Tetrahydrofuran	0.849	0.605	0.456	0.359		
$C_4H_8O_2$	1,4-Dioxane			1.177	0.787	0.569	
$C_4H_8O_2$	Ethyl acetate		0.578	0.423	0.325	0.259	
$C_4H_8O_2$	Methyl propionate		0.581	0.431	0.333	0.266	
$C_4H_8O_2$	Propyl formate		0.669	0.485	0.370	0.293	
$C_4H_8O_2$	Butanoic acid		2.215	1.426	0.982	0.714	0.542

Molecular formula	Name	Viscosity in mPa s					
		–25°C	0°C	25°C	50°C	75°C	100°C
$C_4H_8O_2$	2-Methylpropanoic acid		1.857	1.226	0.863	0.639	0.492
$C_4H_8O_2S$	Sulfolane				6.280	3.818	2.559
C_4H_8S	Tetrahydrothiophene			0.973	0.912		
C_4H_9Br	1-Bromobutane		0.815	0.606	0.471	0.379	
C_4H_9Cl	1-Chlorobutane		0.556	0.422	0.329	0.261	
C_4H_9N	Pyrrolidine	1.914	1.071	0.704	0.512		
C_4H_9NO	*N,N*-Dimethylacetamide			1.956	1.279	0.896	0.661
C_4H_9NO	Morpholine			2.021	1.247	0.850	0.627
$C_4H_{10}O$	1-Butanol	12.19	5.185	2.544	1.394	0.833	0.533
$C_4H_{10}O$	2-Butanol			3.096	1.332	0.698	0.419
$C_4H_{10}O$	2-Methyl-2-propanol			4.312	1.421	0.678	
$C_4H_{10}O$	Diethyl ether		0.283	0.224			
$C_4H_{10}O_3$	Diethylene glycol			30.200	11.130	4.917	2.505
$C_4H_{10}S$	Diethyl sulfide		0.558	0.422	0.331	0.267	
$C_4H_{11}N$	Butylamine		0.830	0.574	0.409	0.298	
$C_4H_{11}N$	Isobutylamine		0.770	0.571	0.367		
$C_4H_{11}N$	Diethylamine			0.319	0.239		
$C_4H_{11}NO_2$	Diethanolamine				109.5	28.7	9.100
$C_5H_4O_2$	Furfural		2.501	1.587	1.143	0.906	0.772
C_5H_5N	Pyridine		1.361	0.879	0.637	0.497	0.409
C_5H_{10}	1-Pentene	0.313	0.241	0.195			
C_5H_{10}	2-Methyl-2-butene		0.255	0.203			
C_5H_{10}	Cyclopentane		0.555	0.413	0.321		
$C_5H_{10}O$	Mesityl oxide	1.291	0.838	0.602	0.465	0.381	0.326
$C_5H_{10}O$	2-Pentanone		0.641	0.470	0.362	0.289	0.238
$C_5H_{10}O$	3-Pentanone		0.592	0.444	0.345	0.276	0.227
$C_5H_{10}O_2$	Butyl formate		0.937	0.644	0.472	0.362	0.289
$C_5H_{10}O_2$	Propyl acetate		0.768	0.544	0.406	0.316	0.255
$C_5H_{10}O_2$	Ethyl propanoate		0.691	0.501	0.380	0.299	0.242
$C_5H_{10}O_2$	Methyl butanoate		0.759	0.541	0.406	0.318	0.257
$C_5H_{10}O_2$	Methyl isobutanoate		0.672	0.488	0.373	0.296	
$C_5H_{11}N$	Piperidine			1.573	0.958	0.649	0.474
C_5H_{12}	Pentane	0.351	0.274	0.224			
C_5H_{12}	Isopentane	0.376	0.277	0.214			
$C_5H_{12}O$	1-Pentanol	25.4	8.512	3.619	1.820	1.035	0.646
$C_5H_{12}O$	2-Pentanol			3.470	1.447	0.761	0.465
$C_5H_{12}O$	3-Pentanol			4.149	1.473	0.727	0.436
$C_5H_{12}O$	2-Methyl-1-butanol			4.453	1.963	1.031	0.612
$C_5H_{12}O$	3-Methyl-1-butanol		8.627	3.692	1.842	1.031	0.631
$C_5H_{13}N$	Pentylamine		1.030	0.702	0.493	0.356	
C_6F_6	Hexafluorobenzene			2.789	1.730	1.151	
$C_6H_4Cl_2$	*o*-Dichlorobenzene		1.958	1.324	0.962	0.739	0.593
$C_6H_4Cl_2$	*m*-Dichlorobenzene		1.492	1.044	0.787	0.628	0.525
C_6H_5Br	Bromobenzene		1.560	1.074	0.798	0.627	0.512
C_6H_5Cl	Chlorobenzene	1.703	1.058	0.753	0.575	0.456	0.369
C_6H_5ClO	*o*-Chlorophenol			3.589	1.835	1.131	0.786
C_6H_5ClO	*m*-Chlorophenol				4.041		
C_6H_5F	Fluorobenzene		0.749	0.550	0.423	0.338	
C_6H_5I	Iodobenzene		2.354	1.554	1.117	0.854	0.683
$C_6H_5NO_2$	Nitrobenzene		3.036	1.863	1.262	0.918	0.704
C_6H_6	Benzene			0.604	0.436	0.335	
C_6H_6ClN	*o*-Chloroaniline			3.316	1.913	1.248	0.887
C_6H_6O	Phenol				3.437	1.784	1.099
C_6H_7N	Aniline			3.847	2.029	1.247	0.850
$C_6H_8N_2$	Phenylhydrazine			13.0	4.553	1.850	0.848
C_6H_{10}	Cyclohexene		0.882	0.625	0.467	0.364	
$C_6H_{10}O$	Cyclohexanone			2.017	1.321	0.919	0.671

VISCOSITY OF LIQUIDS (continued)

Molecular formula	Name	Viscosity in mPa s –25°C	0°C	25°C	50°C	75°C	100°C
$C_6H_{11}N$	Hexanenitrile			0.912	0.650	0.488	0.382
C_6H_{12}	Cyclohexane			0.894	0.615	0.447	
C_6H_{12}	Methylcyclopentane	0.927	0.653	0.479	0.364		
C_6H_{12}	1-Hexene	0.441	0.326	0.252	0.202		
$C_6H_{12}O$	Cyclohexanol			57.5	12.3	4.274	1.982
$C_6H_{12}O$	2-Hexanone	1.300	0.840	0.583	0.429	0.329	0.262
$C_6H_{12}O$	4-Methyl-2-pentanone			0.545	0.406		
$C_6H_{12}O_2$	Butyl acetate		1.002	0.685	0.500	0.383	0.305
$C_6H_{12}O_2$	Isobutyl acetate			0.676	0.493	0.370	0.286
$C_6H_{12}O_2$	Ethyl butanoate			0.639	0.453		
$C_6H_{12}O_2$	Diacetone alcohol	28.7	6.621	2.798	1.829	1.648	
$C_6H_{12}O_3$	Paraldehyde			1.079	0.692	0.485	0.362
$C_6H_{13}N$	Cyclohexylamine			1.944	1.169	0.782	0.565
C_6H_{14}	Hexane		0.405	0.300	0.240		
C_6H_{14}	2-Methylpentane		0.372	0.286	0.226		
C_6H_{14}	3-Methylpentane		0.395	0.306			
$C_6H_{14}O$	Dipropyl ether		0.542	0.396	0.304	0.242	
$C_6H_{14}O$	1-Hexanol			4.578	2.271	1.270	0.781
$C_6H_{15}N$	Triethylamine		0.455	0.347	0.273	0.221	
$C_6H_{15}N$	Dipropylamine		0.751	0.517	0.377	0.288	0.228
$C_6H_{15}N$	Diisopropylamine			0.393	0.300	0.237	
$C_6H_{15}NO_3$	Triethanolamine			609	114	31.5	11.7
C_7H_5N	Benzonitrile			1.267	0.883	0.662	0.524
C_7H_7Cl	*o*-Chlorotoluene		1.390	0.964	0.710	0.547	0.437
C_7H_7Cl	*m*-Chlorotoluene		1.165	0.823	0.616	0.482	0.391
C_7H_7Cl	*p*-Chlorotoluene			0.837	0.621	0.483	0.390
C_7H_8	Toluene	1.165	0.778	0.560	0.424	0.333	0.270
C_7H_8O	*o*-Cresol				3.035	1.562	0.961
C_7H_8O	*m*-Cresol			12.9	4.417	2.093	1.207
C_7H_8O	Benzyl alcohol			5.474	2.760	1.618	1.055
C_7H_8O	Anisole			1.056	0.747	0.554	0.427
C_7H_9N	*N*-Methylaniline		4.120	2.042	1.222	0.825	0.606
C_7H_9N	*o*-Methyl aniline		10.3	3.823	1.936	1.198	0.839
C_7H_9N	*m*-Methyl aniline		8.180	3.306	1.679	1.014	0.699
C_7H_9N	Benzylamine			1.624	1.080	0.769	0.577
C_7H_{14}	Methylcyclohexane		0.991	0.679	0.501	0.390	0.316
C_7H_{14}	1-Heptene		0.441	0.340	0.273	0.226	
$C_7H_{14}O$	2-Heptanone			0.714	0.407	0.297	
$C_7H_{14}O_2$	Heptanoic acid			3.840	2.282	1.488	1.041
C_7H_{16}	Heptane	0.757	0.523	0.387	0.301	0.243	
C_7H_{16}	3-Methylhexane			0.350			
$C_7H_{16}O$	1-Heptanol			5.810	2.603	1.389	0.849
$C_7H_{16}O$	2-Heptanol			3.955	1.799	0.987	0.615
$C_7H_{16}O$	3-Heptanol				1.957	0.976	0.584
$C_7H_{16}O$	4-Heptanol			4.207	1.695	0.882	0.539
$C_7H_{17}N$	Heptylamine			1.314	0.865	0.600	0.434
C_8H_8	Styrene		1.050	0.695	0.507	0.390	0.310
C_8H_8O	Acetophenone			1.681			0.634
$C_8H_8O_2$	Methyl benzoate			1.857			
$C_8H_8O_3$	Methyl salicylate					1.102	0.815
C_8H_{10}	Ethylbenzene		0.872	0.631	0.482	0.380	0.304
C_8H_{10}	*o*-Xylene		1.084	0.760	0.561	0.432	0.345
C_8H_{10}	*m*-Xylene		0.795	0.581	0.445	0.353	0.289
C_8H_{10}	*p*-Xylene			0.603	0.457	0.359	0.290
$C_8H_{10}O$	Phenetole			1.197	0.817	0.594	0.453
$C_8H_{11}N$	*N,N*-Dimethylaniline		1.996	1.300	0.911	0.675	0.523
$C_8H_{11}N$	*N*-Ethylaniline		3.981	2.047	1.231	0.825	0.596

Molecular formula	Name	Viscosity in mPa s					
		–25°C	0°C	25°C	50°C	75°C	100°C
C_8H_{16}	Ethylcyclohexane		1.139	0.784	0.579		
$C_8H_{16}O_2$	Octanoic acid			5.020	2.656	1.654	1.147
C_8H_{18}	Octane		0.700	0.508	0.385	0.302	0.243
$C_8H_{18}O$	1-Octanol			7.288	3.232	1.681	0.991
$C_8H_{18}O$	4-Methyl-3-heptanol		1.904	1.085	0.702	0.497	0.375
$C_8H_{18}O$	5-Methyl-3-heptanol		2.052	1.178	0.762	0.536	0.401
$C_8H_{18}O$	2-Ethyl-1-hexanol		20.7	6.271	2.631	1.360	0.810
$C_8H_{18}O$	Dibutyl ether	1.417	0.918	0.637	0.466	0.356	0.281
$C_8H_{19}N$	Dibutylamine		1.509	0.918	0.619	0.449	0.345
$C_8H_{19}N$	Diisobutylamine		1.115	0.723	0.511	0.384	0.303
C_9H_7N	Quinoline			3.337	1.892	1.201	0.833
C_9H_{10}	Indane		2.230	1.357	0.931	0.692	0.545
C_9H_{12}	Cumene		1.075	0.737	0.547		
$C_9H_{14}O$	Isophorone		4.201	2.329	1.415	0.923	0.638
$C_9H_{18}O$	5-Nonanone			1.199	0.834	0.619	0.484
$C_9H_{18}O_2$	Nonanoic acid			7.011	3.712	2.234	1.475
C_9H_{20}	Nonane		0.964	0.665	0.488	0.375	0.300
$C_9H_{20}O$	1-Nonanol			9.123	4.032		
$C_{10}H_{10}O_4$	Dimethyl phthalate		63.2	14.4	5.309	2.824	1.980
$C_{10}H_{14}$	Butylbenzene			0.950	0.683	0.515	
$C_{10}H_{18}$	*cis*-Decahydronaphthalene	12.8	5.645	3.042	1.875	1.271	0.924
$C_{10}H_{18}$	*trans*-Decahydronaphthalene	6.192	3.243	1.940	1.289	0.917	0.689
$C_{10}H_{20}O_2$	Decanoic acid				4.327	2.651	
$C_{10}H_{22}$	Decane	2.188	1.277	0.838	0.598	0.453	0.359
$C_{10}H_{22}O$	1-Decanol			10.9	4.590		
$C_{11}H_{24}$	Undecane		1.707	1.098	0.763	0.562	0.433
$C_{12}H_{10}O$	Diphenyl ether				2.130	1.407	1.023
$C_{12}H_{26}$	Dodecane		2.277	1.383	0.930	0.673	0.514
$C_{13}H_{12}$	Diphenylmethane					1.265	0.929
$C_{13}H_{28}$	Tridecane		2.909	1.724	1.129	0.796	0.594
$C_{14}H_{30}$	Tetradecane			2.128	1.376	0.953	0.697
$C_{16}H_{22}O_4$	Dibutyl phthalate	483	66.4	16.6	6.470	3.495	2.425
$C_{16}H_{34}$	Hexadecane			3.032	1.879	1.260	0.899
$C_{18}H_{38}$	Octadecane				2.487	1.609	1.132

VISCOSITY OF AQUEOUS SOLUTIONS

This table gives the absolute viscosity of aqueous solutions of several common compounds as a function of concentration expresed in mass percent. Viscosity values are in units of millipascal seconds (mPa s), which is equivalent to centipoises (cp).

REFERENCE

Borchers, H., Ed., *Landoldt-Börnstein Numerical Data and Functional Relationships in Physics, Chemistry, Astronomy, Geophysics, and Technology*, 6th ed., Vol.4, Part 1, *Material Values and Mechanical Behavior of Non-Metals*, 1955.

		Concentration in mass percent (%)							
Solute	***t*/°C**	**0**	**5**	**10**	**20**	**40**	**60**	**80**	**100**
Ethanol	25	0.89		1.32	1.81	2.35	2.24	1.75	1.07
Glycerol	25	0.89	1.01	1.15	1.54	3.18	8.82	45.9	934
Sucrose	20	1.00	1.13	1.31	1.92	5.98	58.5		
Pyridine	25	0.89		1.14	1.35	1.83	2.19	1.91	0.88
Sodium chloride	20	1.00	1.08	1.19	1.53				
Nitric acid	20	1.00		1.04	1.14	1.55	2.02	1.84	0.88
Sulfuric acid	25	0.89	1.01	1.12	1.40	2.51	5.37	17.4	24.2

THERMAL CONDUCTIVITY OF GASES

This table gives the thermal conductivity of several gases as a function of temperature. Unless otherwise noted, the values refer to a pressure of 100 kPa (1 bar) or to the saturation vapor pressure if that is less than 100 kPa. The notation $P = 0$ indicates the low pressure limiting value is given. In general, the $P = 0$ and $P = 100$ kPa values differ by less than 1%. Units are milliwatts per meter kelvin. Substances are listed in the modified Hill order.

MF	Name	Thermal conductivity in mW/m K						Ref.
		100 K	200 K	300 K	400 K	500 K	600 K	
	Air	9.4	18.4	26.2	33.3	39.7	45.7	1
Ar	Argon	6.2	12.4	17.9	22.6	26.8	30.6	2,8
BF_3	Boron trifluoride			19.0	24.6			11
ClH	Hydrogen chloride		9.2	14.5	19.5	24.0	28.1	11
F_6S	Sulfur hexafluoride ($P = 0$)			13.0	20.6	27.5	33.8	16
H_2	Hydrogen ($P = 0$)	68.6	131.7	186.9	230.4			4
H_2O	Water			18.7	27.1	35.7	47.1	6
	Deuterium oxide				27.0	36.5	47.6	7
H_2S	Hydrogen sulfide			14.6	20.5	26.4	32.4	11
H_3N	Ammonia			24.4	37.4	51.6	66.8	11
He	Helium ($P = 0$)	75.5	119.3	156.7	190.6	222.3	252.4	8
Kr	Krypton ($P = 0$)	3.3	6.4	9.5	12.3	14.8	17.1	8
NO	Nitric oxide		17.8	25.9	33.1	39.6	46.2	11
N_2	Nitrogen	9.8	18.7	26.0	32.3	38.3	44.0	12
N_2O	Nitrous oxide		9.8	17.4	26.0	34.1	41.8	11
Ne	Neon ($P = 0$)	22.3	37.6	49.8	60.3	69.9	78.7	8
O_2	Oxygen	9.3	18.1	26.3	33.7	41.0	48.1	10
O_2S	Sulfur dioxide			9.6	14.3	20.0	25.6	11
Xe	Xenon ($P = 0$)	2.0	3.6	5.5	7.3	8.9	10.4	8
CCl_2F_2	Dichlorodifluoromethane			9.9	15.0	20.1	25.2	13
CF_4	Tetrafluoromethane ($P = 0$)			16.0	24.1	32.2	39.9	16
CO	Carbon monoxide ($P = 0$)			25.0	32.3	39.2	45.7	14
CO_2	Carbon dioxide		9.6	16.8	25.1	33.5	41.6	9
$CHCl_3$	Trichloromethane			7.5	11.1	15.1		11
CH_4	Methane		22.5	34.1	49.1	66.5	84.1	5,15
CH_4O	Methanol				26.2	38.6	53.0	11
$C_2Cl_2F_4$	1,2-Dichlorotetrafluoro-ethane			10.25	15.7	21.1		13
$C_2Cl_3F_3$	1,1,2-Trichlorotrifluoro-ethane			9.0	13.6	18.3		13
C_2H_2	Acetylene			21.4	33.3	45.4	56.8	11
C_2H_4	Ethylene		11.1	20.5	34.6	49.9	68.6	3
C_2H_6	Ethane		11.0	21.3	35.4	52.2	70.5	5
C_2H_6O	Ethanol			14.4	25.8	38.4	53.2	11
C_3H_6O	Acetone			11.5	20.2	30.6	42.7	11
C_3H_8	Propane			18.0	30.6	45.5	61.9	5
C_4F_8	Perfluorocyclobutane			12.5	19.5			13
C_4H_{10}	Butane			16.4	28.4	43.0	59.1	5
C_4H_{10}	Isobutane			16.1	27.9	42.1	57.6	5
$C_4H_{10}O$	Diethyl ether			15.1	25.0	37.1		11
C_5H_{12}	Pentane			14.4	24.9	37.8	52.7	11
C_6H_{14}	Hexane				23.4	35.4	48.7	11

REFERENCES

1. Kadoya, K. Matsunaga, N., and Nagashima, A., Viscosity and thermal conductivity of dry air in the gaseous phase, *J. Phys. Chem. Ref. Data*, 14, 947, 1985.
2. Younglove, B. A. and Hanley, H. J. M., The viscosity and thermal conductivity coefficients of gaseous and liquid argon, *J. Phys. Chem. Ref. Data*, 15, 1323, 1986.
3. Holland, P. M., Eaton, B. E., and Hanley, H. J. M., A correlation of the viscosity and thermal conductivity data of gaseous and liquid ethylene,

J. Phys. Chem. Ref. Data, 12, 917, 1983.

4. Assael, M. J., Mixafendi, S., and Wakeham, W. A., The viscosity and thermal conductivity of normal hydrogen in the limit of zero density, *J. Phys. Chem. Ref. Data*, 15, 1315, 1986.
5. Younglove, B. A. and Ely, J. F., Thermophysical properties of fluids. II. Methane, ethane, propane, isobutane, and normal butane, *J. Phys. Chem. Ref. Data*, 16, 577, 1987.
6. Sengers, J. V. and Watson, J. T. R., Improved international formulations for the viscosity and thermal conductivity of water substance, *J. Phys. Chem. Ref. Data*, 15, 1291, 1986.
7. Matsunaga, N. and Nagashima, A., Transport properties of liquid and gaseous D_2O over a wide range of temperature and pressure, *J. Phys. Chem. Ref. Data*, 12, 933, 1983.
8. Kestin, J. et al., Equilibrium and transport properties of the noble gases and their mixtures at low density, *J. Phys. Chem. Ref. Data*, 13, 229, 1984.
9. Vescovic, V. et al., The transport properties of carbon dioxide, *J. Phys. Chem. Ref. Data*, 19, 1990.
10. Younglove, B. A., Thermophysical properties of fluids. I. Argon, ethylene, parahydrogen, nitrogen, nitrogen trifluoride, and oxygen, *J. Phys. Chem. Ref. Data*, 11, Suppl. 1, 1982.
11. Ho, C. Y., Ed., *Properties of Inorganic and Organic Fluids, CINDAS Data Series on Materials Properties*, Volume V-1, Hemisphere Publishing Corp., New York, 1988.
12. Stephen, K., Krauss, R., and Laesecke, A., Viscosity and thermal conductivity of nitrogen for a wide range of fluid states, *J. Phys. Chem. Ref. Data*, 16, 993, 1987.
13. Krauss, R. and Stephan, K., Thermal conductivity of refrigerants in a wide range of temperature and pressure, *J. Phys. Chem. Ref. Data*, 18, 43, 1989.
14. Millat, J. and Wakeham, W. A., The thermal conductivity of nitrogen and carbon monoxide in the limit of zero density, *J. Phys. Chem. Ref. Data*, 18, 565, 1989.
15. Friend, D. G., Ely, J. F., and Ingham, H., Thermophysical properties of methane, *J. Phys. Chem. Ref. Data*, 18, 583, 1989.
16. Uribe, F. J., Mason, E. A., and Kestin, J., Thermal conductivity of nine polyatomic gases at low density, *J. Phys. Chem. Ref. Data*, 19, 1123, 1990.

THERMAL CONDUCTIVITY OF LIQUIDS

This table gives the thermal conductivity of some common liquids at temperatures between -25 and 100°C. All values are given in units of watts per meter kelvin (W/m K). Values refer to nominal atmospheric pressure (about 100 kPa); when an entry is given at a temperature above the normal boiling point of the substance, the pressure is understood to be the saturation vapor pressure at that temperature.

Substances are arranged by molecular formula in the modified Hill order, with compounds not containing carbon preceding those that do contain carbon.

The values for water, benzene, toluene, heptane, and dimethyl phthalate are particularly well determined and can be used for calibration purposes.

REFERENCES

1. Daubert, T. E., Danner, R. P., Sibul, H. M., and Stebbins, C. C., *Physical and Thermodynamic Properties of Pure Compounds: Data Compilation*, extant 1994 (core with 4 supplements), Taylor & Francis, Bristol, PA (also available as database).
2. Marsh, K. N., Ed., *Recommended Reference Materials for the Realization of Physicochemical Properties*, Blackwell Scientific Publications, Oxford, 1987.

Molecular formula	Name	Thermal conductivity in W/m K					
		−25°C	0°C	25°C	50°C	75°C	100°C
Cl_4Si	Silicon tetrachloride			0.099	0.096		
H_2O	Water		0.5610	0.6071	0.6435	0.6668	0.6791
Hg	Mercury	7.25	7.77	8.25	8.68	9.07	9.43
CCl_4	Tetrachloromethane		0.104	0.099	0.093	0.088	
CS_2	Carbon disulfide		0.154	0.149			
$CHCl_3$	Trichloromethane	0.127	0.122	0.117	0.112	0.107	0.102
CH_2Br_2	Dibromomethane	0.120	0.114	0.108	0.103	0.097	
CH_4O	Methanol	0.214	0.207	0.200	0.193		
C_2Cl_4	Tetrachloroethylene		0.117	0.110	0.104	0.097	0.091
C_2HCl_3	Trichloroethylene	0.133	0.124	0.116	0.108	0.100	
$C_2H_3Cl_3$	1,1,1-Trichloroethane		0.106	0.101	0.096		
C_2H_3N	Acetonitrile	0.208	0.198	0.188	0.178	0.168	
$C_2H_4O_2$	Acetic acid			0.158	0.153	0.149	0.144
C_2H_5Cl	Chloroethane	0.145	0.132	0.119	0.106	0.093	
C_2H_5NO	*N*-Methylformamide			0.203	0.201	0.199	0.196
C_2H_6O	Ethanol		0.176	0.169	0.162		
$C_2H_6O_2$	Ethylene glycol		0.256	0.256	0.256	0.256	0.256
C_2H_7NO	Ethanolamine			0.299	0.286	0.274	0.261
C_3H_5ClO	Epichlorohydrin	0.142	0.137	0.131	0.125	0.119	0.114
C_3H_6O	Acetone		0.169	0.161			
$C_3H_6O_2$	Methyl acetate	0.174	0.164	0.153	0.143	0.133	0.122
C_3H_7NO	*N,N*-Dimethylformamide			0.184	0.178	0.171	0.165
C_3H_8O	1-Propanol	0.162	0.158	0.154	0.149	0.145	0.141
C_3H_8O	2-Propanol	0.146	0.141	0.135	0.129	0.124	0.118
$C_3H_8O_2$	1,2-Propanediol		0.202	0.200	0.199	0.198	0.197
$C_3H_8O_3$	Glycerol			0.292	0.295	0.297	0.300
C_3H_9N	Trimethylamine	0.143	0.133				
C_4H_4O	Furan	0.142	0.134	0.126			
C_4H_4S	Thiophene			0.199	0.195	0.191	0.186
C_4H_6	2-Butyne	0.137	0.129	0.121			
C_4H_8O	2-Butanone	0.158	0.151	0.145	0.139	0.133	
C_4H_8O	Tetrahydrofuran	0.132	0.126	0.120	0.114		
$C_4H_8O_2$	1,4-Dioxane			0.159	0.147	0.135	0.123
$C_4H_8O_2$	Ethyl acetate	0.162	0.153	0.144	0.135	0.126	
$C_4H_{10}O$	1-Butanol		0.158	0.154	0.149		
$C_4H_{10}O$	Diethyl ether	0.150	0.140	0.130	0.120	0.110	0.100
C_5H_5N	Pyridine		0.169	0.165	0.161	0.158	
C_5H_8	Cyclopentene	0.143	0.136	0.129			
C_5H_{10}	1-Pentene	0.131	0.124	0.116			
C_5H_{10}	Cyclopentane	0.140	0.133	0.126			
C_5H_{12}	Pentane	0.132	0.122	0.113	0.103	0.095	0.087
$C_5H_{12}O$	1-Pentanol		0.157	0.153	0.149	0.145	
C_6H_5Cl	Chlorobenzene	0.136	0.131	0.127	0.122	0.117	0.112

Molecular formula	Name	Thermal conductivity in W/m K					
		–25°C	0°C	25°C	50°C	75°C	100°C
C_6H_6	Benzene			0.1411	0.1329	0.1247	
C_6H_6O	Phenol				0.156	0.153	0.151
C_6H_{10}	Cyclohexene	0.142	0.136	0.130	0.124	0.118	
$C_6H_{10}O$	Mesityl oxide	0.170	0.163	0.156	0.149	0.142	0.134
C_6H_{12}	Cyclohexane			0.123	0.117	0.111	
C_6H_{12}	1-Hexene	0.137	0.129	0.121	0.113		
$C_6H_{12}O$	Cyclohexanol			0.134	0.131		
$C_6H_{12}O$	2-Hexanone	0.151	0.145	0.139	0.133	0.127	0.121
C_6H_{14}	Hexane	0.137	0.128	0.120	0.111	0.102	0.093
$C_6H_{14}O$	1-Hexanol	0.159	0.154	0.150	0.145	0.141	0.137
C_7H_6O	Benzaldehyde			0.151	0.141	0.131	0.121
C_7H_8	Toluene	0.1461	0.1386	0.1311	0.1236	0.1161	
C_7H_8O	Anisole	0.170	0.163	0.156	0.150	0.143	0.136
C_7H_{16}	Heptane	0.1378	0.1303	0.1228	0.1152	0.1077	
$C_7H_{16}O$	1-Heptanol		0.166	0.159	0.153	0.147	0.141
C_8H_8	Styrene	0.148	0.142	0.137	0.131	0.126	0.120
C_8H_{10}	Ethylbenzene			0.130	0.124	0.118	0.112
C_8H_{10}	*o*-Xylene			0.131	0.126	0.120	0.114
C_8H_{10}	*m*-Xylene			0.130	0.124	0.118	0.113
C_8H_{10}	*p*-Xylene			0.130	0.124	0.118	0.112
C_8H_{18}	Octane	0.143	0.135	0.128	0.120	0.113	0.106
$C_8H_{18}O$	1-Octanol		0.168	0.161	0.154	0.147	0.141
C_9H_{12}	Cumene			0.128	0.120	0.112	0.107
C_9H_{12}	Mesitylene	0.147	0.141	0.136	0.130	0.124	0.118
C_9H_{20}	Nonane	0.144	0.138	0.131	0.124	0.118	0.111
$C_9H_{20}O$	1-Nonanol		0.166	0.161	0.155	0.149	0.143
$C_{10}H_{10}O_4$	Dimethyl phthalate		0.1501	0.1473	0.1443	0.1409	0.1373
$C_{10}H_{14}$	*p*-Cymene	0.132	0.127	0.122	0.117	0.112	0.107
$C_{10}H_{22}$	Decane	0.144	0.138	0.132	0.126	0.119	0.113
$C_{10}H_{22}O$	1-Decanol			0.162	0.156	0.150	0.145
$C_{11}H_{24}$	Undecane			0.140	0.135	0.129	0.123
$C_{12}H_{10}O$	Diphenyl ether				0.139	0.135	0.131
$C_{12}H_{26}$	Dodecane		0.157	0.152	0.146	0.140	0.135
$C_{12}H_{26}O$	1-Dodecanol			0.146	0.142	0.139	0.135
$C_{13}H_{28}$	Tridecane			0.137	0.132	0.127	0.122
$C_{14}H_{30}$	Tetradecane			0.136	0.131	0.126	0.121
$C_{14}H_{30}O$	1-Tetradecanol				0.167	0.162	0.157
$C_{16}H_{22}O_4$	Dibutyl phthalate	0.144	0.140	0.136	0.133	0.129	0.125
$C_{16}H_{34}$	Hexadecane			0.140	0.135	0.130	0.125
$C_{18}H_{38}$	Octadecane				0.146	0.142	0.137

DIFFUSION IN GASES

This table gives binary diffusion coefficients D_{12} for a number of common gases as a function of temperature. Values refer to atmospheric pressure. The diffusion coefficient is inversely proportional to pressure as long as the gas is in a regime where binary collisions dominate. See Reference 1 for a discussion of the dependence of D_{12} on temperature and composition.

The first part of the table gives data for several gases in the presence of a large excess of air. The remainder applies to equimolar mixtures of gases. Each gas pair is ordered alphabetically according to the most common way of writing the formula. The listing of pairs then follows alphabetical order by the first constituent.

REFERENCES

1. Marrero, T. R., and Mason, E. A., *J. Phys. Chem. Ref. Data*, 1, 1, 1972.
2. Kestin, J., et al., *J. Phys. Chem. Ref. Data*, 13, 229, 1984.

$D_{12}/cm^2\ s^{-1}$ for p = 101.325 kPa and the Specified T/K

System	200	273.15	293.15	373.15	473.15	573.15	673.15
Large Excess of Air							
Ar-air		0.167	0.148	0.289	0.437	0.612	0.810
CH_4-air			0.106	0.321	0.485	0.678	0.899
CO-air			0.208	0.315	0.475	0.662	0.875
CO_2-air			0.160	0.252	0.390	0.549	0.728
H_2-air		0.668	0.627	1.153	1.747	2.444	3.238
H_2O-air			0.242	0.399	0.638	0.873	1.135
He-air		0.617	0.580	1.057	1.594	2.221	2.933
SF_6-air				0.150	0.233	0.329	0.438
Equimolar Mixture							
Ar-CH_4				0.306	0.467	0.657	0.876
Ar-CO		0.168	0.187	0.290	0.439	0.615	0.815
Ar-CO_2		0.129	0.078	0.235	0.365	0.517	0.689
Ar-H_2		0.698	0.794	1.228	1.876	2.634	3.496
Ar-He	0.381	0.645	0.726	1.088	1.617	2.226	2.911
Ar-Kr	0.064	0.117	0.134	0.210	0.323	0.456	0.605
Ar-N_2		0.168	0.190	0.290	0.439	0.615	0.815
Ar-Ne	0.160	0.277	0.313	0.475	0.710	0.979	1.283
Ar-O_2		0.166	0.189	0.285	0.430	0.600	0.793
Ar-SF_6				0.128	0.202	0.290	0.389
Ar-Xe	0.052	0.095	0.108	0.171	0.264	0.374	0.498
CH_4-H_2			0.782	1.084	1.648	2.311	3.070
CH_4-He			0.723	0.992	1.502	2.101	2.784
CH_4-N_2			0.220	0.317	0.480	0.671	0.890
CH_4-O_2			0.210	0.341	0.523	0.736	0.978
CH_4-SF_6				0.167	0.257	0.363	0.482
CO-CO_2			0.162	0.250	0.384		
CO-H_2	0.408	0.686	0.772	1.162	1.743	2.423	3.196
CO-He	0.365	0.619	0.698	1.052	1.577	2.188	2.882
CO-Kr		0.131	0.581	0.227	0.346	0.485	0.645
CO-N_2	0.133	0.208	0.231	0.336	0.491	0.673	0.878
CO-O_2			0.202	0.307	0.462	0.643	0.849
CO-SF_6				0.144	0.226	0.323	0.432
CO_2-C_3H_8			0.084	0.133	0.209		
CO_2-H_2	0.315	0.552	0.412	0.964	1.470	2.066	2.745
CO_2-H_2O			0.162	0.292	0.496	0.741	1.021
CO_2-He	0.300	0.513	0.400	0.878	1.321		
CO_2-N_2			0.160	0.253	0.392	0.553	0.733
CO_2-N_2O	0.055	0.099	0.113	0.177	0.276		
CO_2-Ne	0.131	0.227	0.199	0.395	0.603	0.847	

System	200	273.15	293.15	373.15	473.15	573.15	673.15
CO_2-O_2			0.159	0.248	0.380	0.535	0.710
CO_2-SF_6				0.099	0.155		
D_2-H_2	0.631	1.079	1.219	1.846	2.778	3.866	5.103
H_2-He	0.775	1.320	1.490	2.255	3.394	4.726	6.242
H_2-Kr	0.340	0.601	0.682	1.053	1.607	2.258	2.999
H_2-N_2	0.408	0.686	0.772	1.162	1.743	2.423	3.196
H_2-Ne	0.572	0.982	0.317	1.684	2.541	3.541	4.677
H_2-O_2		0.692	0.756	1.188	1.792	2.497	3.299
H_2-SF_6			0.208	0.649	0.998	1.400	1.851
H_2-Xe		0.513	0.122	0.890	1.349	1.885	2.493
H_2O-N_2			0.242	0.399			
H_2O-O_2			0.244	0.403	0.645	0.882	1.147
He-Kr	0.330	0.559	0.629	0.942	1.404	1.942	2.550
He-N_2	0.365	0.619	0.698	1.052	1.577	2.188	2.882
He-Ne	0.563	0.948	1.066	1.592	2.362	3.254	4.262
He-O_2		0.641	0.697	1.092	1.640	2.276	2.996
He-SF_6			1.109	0.592	0.871	1.190	1.545
He-Xe	0.282	0.478	0.538	0.807	1.201	1.655	2.168
Kr-N_2		0.131	0.149	0.227	0.346	0.485	0.645
Kr-Ne	0.131	0.228	0.258	0.392	0.587	0.812	1.063
Kr-Xe	0.035	0.064	0.073	0.116	0.181	0.257	0.344
N_2-Ne			0.258	0.483	0.731	1.021	1.351
N_2-O_2			0.202	0.307	0.462	0.643	0.849
N_2-SF_6				0.148	0.231	0.328	0.436
N_2-Xe		0.107	0.708	0.188	0.287	0.404	0.539
Ne-Xe	0.111	0.193	0.219	0.332	0.498	0.688	0.901
O_2-SF_6			0.097	0.154	0.238	0.334	0.441

DIFFUSION COEFFICIENTS IN LIQUIDS AT INFINITE DILUTION

This table lists diffusion coefficients D_{AB} at infinite dilution for some binary liquid mixtures. Although values are given to two decimal places, measurements in the literature are often in poor agreement. Therefore most values in the table cannot be relied upon to better than 10%.

Solvents are listed in alphabetical order, as are the solutes within each solvent group.

REFERENCE

Landolt-Börnstein, *Numerical Data and Functional Relationships in Science and Technology*, Sixth Edition, Vol. II/5a, 1969.

Solute	Solvent	T/°C	D_{AB} 10^{-5} cm^2 s^{-1}
Acetic acid	Acetone	25	3.31
Benzoic acid	Acetone	25	2.62
Formic acid	Acetone	25	3.77
Nitrobenzene	Acetone	20	2.94
Tetrachloromethane	Acetone	25	3.29
Trichloromethane	Acetone	25	3.64
Water	Acetone	25	4.56
Acetic acid	Benzene	25	2.09
Aniline	Benzene	25	1.96
Benzoic acid	Benzene	25	1.38
Bromobenzene	Benzene	8	1.45
2-Butanone	Benzene	30	2.09
Chloroethylene	Benzene	8	1.77
Cyclohexane	Benzene	25	2.25
Ethanol	Benzene	25	3.02
Formic acid	Benzene	25	2.28
Heptane	Benzene	25	1.78
Methanol	Benzene	25	3.80
Toluene	Benzene	25	1.85
1,2,4-Trichlorobenzene	Benzene	8	1.34
Trichloromethane	Benzene	25	2.26
Adipic acid	1-Butanol	30	0.40
Benzene	1-Butanol	25	1.00
Biphenyl	1-Butanol	25	0.63
Butyric acid	1-Butanol	30	0.51
p-Dichlorobenzene	1-Butanol	25	0.82
Methanol	1-Butanol	30	0.59
Oleic acid	1 Butanol	30	0.25
Propane	1-Butanol	25	1.57
Water	1-Butanol	25	0.56
Benzene	Cyclohexane	25	1.41
Tetrachloromethane	Cyclohexane	25	1.49
Toluene	Cyclohexane	25	1.57
Allyl alcohol	Ethanol	20	0.98
Benzene	Ethanol	25	1.81
Iodine	Ethanol	25	1.32
Iodobenzene	Ethanol	20	1.00
3-Methyl-1-butanol	Ethanol	20	0.81
Pyridine	Ethanol	20	1.10
Tetrachloromethane	Ethanol	25	1.50
Water	Ethanol	25	1.24
Acetic acid	Ethyl acetate	20	2.18
Acetone	Ethyl acetate	20	3.18
2-Butanone	Ethyl acetate	30	2.93
Ethyl benzoate	Ethyl acetate	20	1.85
Nitrobenzene	Ethyl acetate	20	2.25
Water	Ethyl acetate	25	3.20
Benzene	Heptane	25	3.91
Toluene	Heptane	25	3.72
Bromobenzene	Hexane	8	2.60
2-Butanone	Hexane	30	3.74
Dodecane	Hexane	25	2.73
Iodine	Hexane	25	4.45
Methane	Hexane	25	0.09
Propane	Hexane	25	4.87
Tetrachloromethane	Hexane	25	3.70
Toluene	Hexane	25	4.21
Acetone	Tetrachloromethane	25	1.75
Benzene	Tetrachloromethane	25	1.42
Cyclohexane	Tetrachloromethane	25	1.30
Ethanol	Tetrachloromethane	25	1.90
Iodine	Tetrachloromethane	30	1.63
Trichloromethane	Tetrachloromethane	25	1.66
Acetic acid	Toluene	25	2.26
Benzene	Toluene	25	2.54
Benzoic acid	Toluene	25	1.49
Cyclohexane	Toluene	25	2.42
Formic acid	Toluene	25	2.65
Water	Toluene	25	6.19
Acetone	Trichloromethane	25	2.55
Benzene	Trichloromethane	25	2.89
2-Butanone	Trichloromethane	25	2.13
Butyl acetate	Trichloromethane	25	1.71
Diethyl ether	Trichloromethane	25	2.15
Ethanol	Trichloromethane	15	2.20
Ethyl acetate	Trichloromethane	25	2.02
Acetic acid	Water	25	1.29
Acetone	Water	25	1.28
Acetonitrile	Water	15	1.26
Alanine	Water	25	0.91
Allyl alcohol	Water	15	0.90
Aniline	Water	20	0.92
Arabinose	Water	20	0.69
Benzene	Water	20	1.02
1-Butanol	Water	25	0.56
Caprolactam	Water	25	0.87
Chloroethylene	Water	25	1.34
Cyclohexane	Water	20	0.84
Diethylamine	Water	20	0.97
Ethanol	Water	25	1.24
Ethanolamine	Water	25	1.08
Ethyl acetate	Water	20	1.00
Ethylbenzene	Water	20	0.81
Ethylene glycol	Water	25	1.16
Glucose	Water	25	0.67
Glycerol	Water	25	1.06
Glycine	Water	25	1.05
Lactose	Water	15	0.38
Maltose	Water	15	0.38
Mannitol	Water	15	0.50
Methane	Water	25	1.49
Methanol	Water	15	1.28
3-Methyl-1-butanol	Water	10	0.69
Methylcyclopentane	Water	20	0.85
Phenol	Water	20	0.89
1-Propanol	Water	15	0.87
Propene	Water	25	1.44
Pyridine	Water	25	0.58
Raffinose	Water	15	0.33
Sucrose	Water	25	0.52
Toluene	Water	20	0.85
Urea	Water	25	1.38
Urethane	Water	15	0.80

DIFFUSIVITIES OF METALLIC SOLUTES IN MOLTEN METALS

A. K. Roy and R. P. Chhabra

Solute	Solvent	Temperature (K)	Diffusivity × 10^5 cm^2/s
Au	Ag	1253	2.46
		1300	2.80
		1350	3.19
		1400	3.59
		1450	4.00
		1500	4.44
		1533	4.73
In		1253	3.81
		1300	4.20
		1350	4.63
		1400	5.06
		1450	5.50
		1500	5.95
		1533	6.25
Sb		1253	4.11
		1300	4.47
		1350	4.85
		1400	5.23
		1450	5.62
		1500	6.00
		1533	6.25
Sn		1248	3.87
		1298	4.25
		1348	4.63
		1398	5.02
		1448	5.41
		1498	5.80
		1548	6.20
		1623	6.78
Ag	Al	1183	11.00
		1183	20.00
Co		980	2.82
		1080	3.85
		1180	4.98
		1280	6.20
		1320	6.70
Cu		973	7.20
		1000	6.00
		1050	6.87
		1100	7.78
		1150	8.71
		1200	9.67
		1250	10.63
		1273	15.00
		1300	11.61
Fe		973	1.40
		1273	20.00
Ga		1000	7.70
		1050	8.67
		1100	9.64
		1150	10.63
		1200	11.62
		1250	12.62
		1300	13.61
Mg		973	2.70
		973	7.54
		1073	6.40
Ni		980	3.86
		1080	5.22
		1180	6.70
		1280	8.27
		1320	8.92
Te		980	2.94
		1030	4.00
		1080	5.30
		1130	6.83
		1180	8.62
		1230	10.68
		1280	13.01
		1320	15.07
Zn		970	6.20
		1073	14.00
Ag	Bi	573	2.24
		673	5.17
		773	9.61
		873	16.00
		973	22.63
Au		773	5.22
Sn		773	4.90
		773	6.50
		873	5.50
		873	8.20
Ag	Cu	1373	2.33
		1473	2.82
		1573	3.32
Au		1373	2.42
		1473	2.79
		1573	3.17
Co		1373	3.61
		1423	4.18
		1473	4.80
		1523	5.45
		1573	6.14
Ga		1373	4.63
		1423	5.13
		1473	5.65
		1523	6.18
		1573	6.72
In		1373	2.56
		1473	3.28
		1573	4.08
Ni		1373	3.62
		1423	4.15
		1473	4.70
		1523	5.29
		1573	5.90
Sb		1373	2.25
		1473	3.07
		1573	4.03
In	Ga	473	3.67
		523	4.33
		573	4.97
		623	5.58
		673	6.16
		723	6.70
Ga	In	473	3.08
		523	3.75
		573	4.41
		623	5.05
		673	5.67
Ag	Pb	623	1.74
		673	1.96
		723	2.16
		773	2.36

DIFFUSIVITIES OF METALLIC SOLUTES IN MOLTEN METALS (continued)

Solute	Solvent	Temperature (K)	Diffusivity × 10^5 cm^2/s
Au		773	3.70
Bi		773	4.93
		773	5.60
		823	6.23
		873	8.30
Cd		723	3.90
		773	5.00
		823	6.00
		873	6.80
Cu		723	3.43
		773	3.92
		823	4.41
		873	4.89
		923	5.37
		973	5.83
		1040	6.44
Na		625	2.75
		673	1.21
		773	1.42
		853	5.25
		873	1.79
		903	5.91
Sb		723	3.10
		773	4.10
		823	5.50
		873	6.40
Sn		723	2.60
		773	3.52
		783	1.75
		823	4.30
		873	5.50
Ag	Sn	560	3.02
		600	3.71
		650	4.57
		700	5.43
		750	6.29
		773	4.80
		943	10.20
		1108	11.80
Al		560	1.76
		600	2.41
		650	3.37
		700	4.50
		750	5.77
Au		600	2.98
		700	4.75
		773	5.37
		800	6.52
		900	8.29
		1000	10.06
		1100	11.83
		1300	15.37
		1500	18.91
		1683	22.10
		1853	26.10
Bi		723	3.60
		773	4.60
		823	5.80
		873	6.60
Co		623	2.70
Cu		560	4.13
		600	5.31
		650	6.97
		700	8.79
		750	10.75
Na		550	1.24
		600	1.67
		625	1.90
		700	2.66
		853	6.39
		903	6.98
Ni		560	3.92
		600	5.15
		650	6.89
		700	8.86
		750	11.00
Pb		773	3.68
		823	3.68
Sb		600	3.15
		700	4.14
		800	5.36
		900	6.42
		1000	7.48
		1100	8.54
		1300	10.66
		1500	12.78
Zn		560	7.72
		600	10.34
		650	14.20
		700	18.56
		750	23.46
Mn	Zn	973	3.28
		1023	3.75
		1073	4.22

VISCOSITY OF LIQUID METALS*

R. P. Chhabra

Temp. (°C)	Viscosity (mPa s)
Aluminium[1]	
662	1.379
669	1.364
769	1.339
685	1.324
689	1.317
700	1.286
718	1.250
768	1.175
806	1.102
833	1.058
Calcium[2]	
812	1.222
816	1.209
833	1.153
851	1.101
867	1.057
883	1.012
Cobalt[3]	
1450	4.46
1495	4.18
1500	4.14
1550	3.85
1600	3.61
1650	3.40
1700	3.20
1750	3.03
Gallium[4]	
52.9	1.894
97.7	1.612
102	1.604
149	1.406
200	1.266
203	1.243
301	1.03
402	0.878
402	0.886
500	0.811
500	0.814
600	0.769
600	0.770
604	0.749
806	0.652
1010	0.592
1100	0.578
Gold[3]	
1100	5.13
1200	4.64
1300	4.24
1364	3.58
Indium[5]	
250	1.35
300	1.22
350	1.12
400	1.05
450	0.98
Lithium[6]	
187	0.589
207	0.558
227	0.5306
247	0.506
267	0.484
287	0.4637
307	0.445
327	0.429
347	0.413
367	0.399
387	0.386
407	0.374
427	0.362
447	0.352
467	0.342
487	0.333
507	0.324
527	0.316
547	0.308
567	0.301
587	0.294
607	0.288
627	0.281
647	0.276
667	0.270
687	0.265
707	0.260
727	0.255
747	0.250
767	0.246
787	0.241
807	0.237
827	0.2334
847	0.230
867	0.226
887	0.223
907	0.219
957	0.211
1007	0.204
1057	0.198
1107	0.191
1157	0.186
1207	0.180
1257	0.176
1307	0.171
1357	0.167
1407	0.163
1457	0.159
1507	0.155
1557	0.152
1607	0.149
1657	0.146
1707	0.143
1807	0.137
1907	0.132
2007	0.128
2107	0.124
2207	0.120
2307	0.116
2407	0.113
2507	0.110
2607	0.107
2707	0.105
2807	0.102
2907	0.0997
3007	0.0975
3107	0.0954
3127	0.0950
Magnesium[3]	
700	1.10
800	0.84
900	0.67
Nickel[3]	
1454	4.61
1500	4.35
1550	4.09
1600	3.66
1700	3.49
1750	3.32
Potassium[6]	
67	0.522
87	0.470
107	0.428
127	0.393
147	0.363
167	0.338
187	0.316
207	0.297
227	0.280
247	0.265
267	0.252
287	0.241
307	0.230
327	0.220
347	0.212
367	0.204
387	0.197
407	0.190
427	0.184
447	0.178
467	0.173
487	0.168
507	0.163
527	0.159
547	0.155
567	0.151
587	0.148
607	0.145
627	0.141
647	0.138
667	0.136
687	0.133
707	0.130
727	0.128
747	0.126
767	0.124
787	0.122
807	0.119
827	0.118
847	0.116
867	0.114
887	0.112
907	0.111
957	0.107
1007	0.104
1057	0.100
1107	0.0976
1157	0.0949
1207	0.0925
1257	0.0902
1307	0.0881
1357	0.0861
1407	0.0842
1457	0.0825
1507	0.0809
1557	0.0793
1607	0.0778
1657	0.0765
1707	0.0751
1727	0.0746
Rubidium[6]	
47	0.550
67	0.500
87	0.458
107	0.424
127	0.394
147	0.369
167	0.347
187	0.327

* Note that 1 millipascal second (mPa s) = 1 centipoise (cP).

Temp. (°C)	Viscosity (mPa s)	Temp. (°C)	Viscosity (mPa s)	Temp. (°C)	Viscosity (mPa s)	Temp. (°C)	Viscosity (mPa s)
207	0.310	857	0.132	127	0.600	727	0.181
227	0.295	907	0.128	147	0.550	747	0.177
247	0.282	957	0.124	167	0.508	767	0.174
267	0.270	1007	0.120	187	0.473	787	0.171
287	0.259	1057	0.117	207	0.442	807	0.168
307	0.249	1107	0.113	227	0.415	857	0.162
327	0.239	1157	0.110	247	0.392	907	0.156
347	0.231	1207	0.108	267	0.371	957	0.150
367	0.223	1257	0.105	287	0.352	1007	0.145
387	0.216	1307	0.103	307	0.336	1057	0.141
407	0.210	1357	0.100	327	0.321	1107	0.137
427	0.204	1407	0.0981	347	0.307	1157	0.133
447	0.198	1457	0.0961	367	0.295	1207	0.129
467	0.193	1507	0.0942	387	0.284	1257	0.126
487	0.188	1557	0.0924	407	0.274	1307	0.123
507	0.183	1607	0.0907	427	0.264	1357	0.120
527	0.179	1627	0.0900	447	0.256	1407	0.118
547	0.175			467	0.248	1457	0.115
567	0.171	**Silver[3]**		487	0.240	1507	0.113
587	0.167			507	0.233	1557	0.111
607	0.164	1020	3.69	527	0.227	1607	0.109
627	0.161	1100	3.34	547	0.221	1657	0.107
647	0.158	1190	2.89	567	0.215	1707	0.105
667	0.155	1255	2.52	587	0.210	1757	0.103
687	0.152	1320	2.41	607	0.205	1807	0.101
707	0.149	1420	2.18	627	0.201	1857	0.0998
727	0.147			647	0.196	1907	0.0983
747	0.144	**Sodium[6]**		667	0.192	1957	0.0968
767	0.142			687	0.188	2007	0.0955
787	0.140	107	0.658	707	0.184	2027	0.0949
807	0.137						

REFERENCES

1. E. Rothwell, *J. Inst. Metals,* 90, 389, 1961—1962.
2. M. F. Culpin, *Proc. Phys. Soc.,* 70, 1079, 1957.
3. *Landolt–Börnstein Tables*, 5 Teil, Band a, Springer Verlag, 1969, p. 123.
4. K. E. Spells, *Proc. Phys. Soc.,* 48, 299, 1936.
5. H. Walsdorfer, I. Arpshofen, B. Predel, *Z. Met.,* 79, 503, 1988.
6. E. E. Shipilrain, K. A. Yakimovich, V. A. Fomin, S. N. Skovoredjko, and M. S. Mozgovoi, in *Handbook of Thermodynamic and Transport Properties of Alkali Metals,* R. W. Ohse, Ed., Blackwell Scientific Publishers, Oxford, 1985, p. 753.

THE LIMITS OF SUPERHEAT OF PURE LIQUIDS

C. T. Avedisian

The limit of superheat of a liquid, also known as the homogenous nucleation limit, represents the deepest possible penetration of a liquid into the domain of metastable states. At constant pressure (P) it is the highest temperature (T) below the critical point that a liquid can sustain without undergoing a phase transition; at constant temperature, it is the lowest pressure. The importance of this limit resides in the consequences of the phase transition that eventually occurs when the limit is reached. The phase transition may be explosive or nonexplosive as it relates to such processes as contact vapor explosions in postulated nuclear reactor accidents, spills of liquid natural gas during transport, and burning of fuel droplets.

The perspective upon which the concept of a superheat limit emerges is based on random molecular density fluctuations within the bulk of a liquid producing hole-like regions of molecular dimensions that act as bubbles. A phase transition occurs when a bubble, formed by these molecular processes, grows to a size such that it is in unstable equilibrium (i.e., a "critical size nucleus") with the surrounding liquid. The "nucleation rate", J, refers to the mean rate of forming such vapor nuclei in units of nuclei/(volume•time).

According to classical homogeneous nucleation theory, the nucleation rate depends exponentially on the energy (Φ) of forming a critical size nucleus as follows:

$$J = \Gamma k_n N_o \exp[-\Phi/(kT)] \qquad (1)$$

where

$$\Phi = 16\pi\sigma^3/[3(P_b - P)^2] \qquad (2)$$

and k is the Boltzmann constant, k_n is the molecular evaporation rate for a nucleus containing n molecules, N_o is the number density of liquid molecules, P is the ambient pressure, P_b is the gas pressure within the critical size nucleus, σ is surface tension, and Γ is a kinetic pre-factor which accounts for the possibility that nuclei larger than the critical size may decay; if every critical size nucleus continues to grow, then $\Gamma = 1$. From Equations (1) and (2), it is evident that 1) a unique superheat limit does not exist because to each nucleation rate corresponds its own superheat limit at a given ambient pressure, 2) if J increases so should T at constant pressure, and 3) J depends very strongly on T because of the variation of physical properties (surface tension, vapor pressure, and density) with temperature. This strong dependence implies the existence of a threshold temperature above which J will be large and below which it will be negligibly small. The limit of superheat is defined as the mean temperature in this range of large change of J.

The limits of superheat given below are organized as follows. All substances are listed in alphabetical order by chemical formula. Inorganic substances are listed first, followed by organic compounds in the order of increasing carbon number. The second and third columns list the nucleation pressures and temperatures at the corresponding nucleation rate. The nucleation rates listed are characteristic of the various experimental methods used to perform the measurements.

This table is reprinted from C.T. Avedisian, *J. Phys. Chem. Ref. Data,* 14, 695-729, 1985, by permission of the copyright holders, American Chemical Society and American Institute of Physics, and the author.

Substance	*P* [MPa]	*T* [K]	*J* [nuclei/(cm³·s)]
Ar	–1.220	85.0	10^2
Argon	0.101	130.8	10^2
	0.190	131.2	10^2
	0.260	131.5	10^5
	0.360	131.8	10
	0.410	131.9	10^5
	0.600	132.8	10^5
	0.810	133.5	10^3
	1.100	134.3	10
	1.150	135.1	10^5
	1.400	135.3	10
	1.420	136.0	10^5
	1.720	137.1	10^5
	2.140	138.6	10^5
	2.450	139.5	10^5
	2.710	141.3	10^5
H_2	0.076	27.8	10^{-2}
Hydrogen	0.149	27.9	10^{-2}
	0.381	29.4	10^{-2}
	0.751	30.6	10^{-2}
	0.834	30.8	10^{-2}

THE LIMITS OF SUPERHEAT OF PURE LIQUIDS (CONTINUED)

P Substance	*T* [MPa]	*J* [K]	[nuclei/(cm^3·s)]
H_2O	–27.700	283.2	10^3
Water	0.101	553.0	10^6
	0.101	575.2	10^{15}
	1.293	580.4	10^{21}
	2.519	584.9	10^{21}
	2.710	588.3	10^{21}
	5.000	593.6	10^{21}
	6.808	600.4	10^{21}
	8.500	606.5	10^{21}
	9.731	607.2	10^{21}
	10.746	610.3	10^{21}
	11.978	615.6	10^{21}
	12.873	616.7	10^{21}
	13.731	620.2	10^{21}
	15.789	627.0	10^{21}
	17.556	632.3	10^{21}
	20.113	642.2	10^{21}
He I	0.012	4.05	10^7
Helium I	0.017	4.12	10^7
	0.037	4.22	10^7
	0.054	4.31	10^7
	0.066	4.37	10^7
	0.081	4.45	10^7
	0.100	4.55	10^7
	0.112	4.62	10^7
	0.129	4.70	10^7
	0.143	4.76	10^7
He II	–0.06	2.09	10^5
Helium II			
Kr	0.400	182.5	10^5
Krypton	0.820	184.3	10^5
	1.200	187.0	10^5
	1.410	187.6	10^5
	1.630	189.1	10^5
	1.900	189.9	10^5
	2.200	192.1	10^5
	2.430	192.9	10^5
	2.800	194.8	10^5
	3.140	196.6	10^5
	3.460	198.0	10^5
	3.800	199.4	10^5
N_2	–1.010	75.0	10^2
Nitrogen	0.101	110.0	1.0
	0.410	111.4	10^5
	0.520	112.0	10^5
	0.610	112.1	10^5
	0.700	112.7	10^5
	0.820	113.2	10^5
	0.940	113.8	10^5
	1.060	114.2	10^5
	1.210	114.8	10^5
	1.240	115.2	10^5
	1.330	115.5	10^5

THE LIMITS OF SUPERHEAT OF PURE LIQUIDS (CONTINUED)

P Substance	*T* [MPa]	*J* [K]	[nuclei/(cm^3·s)]
	1.360	115.6	10^5
	1.460	116.2	10^5
	1.590	116.8	10^5
	1.620	117.0	10^5
	1.730	117.6	10^5
	1.770	117.7	10^5
	1.870	118.3	10^5
	1.920	118.4	10^5
	2.070	119.1	10^5
N_2O_4	0.154	395.6	10^2
Nitrogen tetroxide	0.554	396.2	10^2
	0.980	398.2	10
	2.000	401.5	10
	3.040	405.2	10
	3.920	408.1	10
	4.500	410.2	10
	5.000	412.5	10
	5.500	414.5	10
	6.000	416.4	10
O_2	–1.520	75.0	10^2
Oxygen	0.101	134.1	1.0
	0.400	135.4	10^5
	0.500	136.2	10^5
	0.680	136.5	10^5
	0.920	137.4	10^5
	1.060	137.5	10^5
	1.180	138.3	10^5
	1.350	138.9	10^5
	1.480	139.3	10^5
	1.740	140.7	10^5
	2.030	141.9	10^5
	2.260	142.8	10^5
	2.500	143.6	10^5
	2.700	144.5	10^5
	2.970	145.9	10^5
SO_2 Sulfur dioxide	0.101	323.2	10^2
Xe	0.500	254.1	10^5
Xenon	0.830	256.3	10^5
	1.070	257.2	10^5
	1.260	258.2	10^5
	1.470	259.6	10^5
	1.550	260.3	10^5
	1.680	261.0	10^5
	1.750	261.6	10^5
	1.860	261.9	10^5
	1.970	262.8	10^5
	2.070	263.4	10^5
	2.170	263.8	10^5
	2.370	265.2	10^5
	2.480	266.1	10^5
	2.630	266.9	10^5
	2.750	267.5	10^5

THE LIMITS OF SUPERHEAT OF PURE LIQUIDS (CONTINUED)

P Substance	*T* [MPa]	*J* [K]	[nuclei/(cm^3·s)]
	2.850	267.8	10^5
	2.970	269.1	10^5
	3.050	269.7	10^5
	3.130	270.0	10^5
	3.450	272.0	10^5
	3.630	273.0	10^5
CCl_2F_2	0.221	342.5	10^6
Dichlorodifluoromethane	0.427	344.3	10^6
	0.462	344.7	10^6
	0.655	346.6	10^6
	0.896	348.8	10^6
	0.931	349.0	10^6
	1.227	351.7	10^6
	1.489	354.4	10^6
	1.917	358.8	10^6
	2.399	363.7	10^6
	2.910	369.0	10^6
	3.289	373.0	10^6
	3.323	373.4	10^6
	3.585	376.2	10^6
	3.634	376.9	10^6
CCl_4	–27.600	268.2	10^3
Carbon tetrachloride			
$CHClF_2$	0.101	327.8	10^4
Chlorodifluoromethane	0.236	328.2	10^4
	0.280	329.4	10^4
	0.510	330.8	10^4
	0.560	331.5	10^4
	0.710	332.4	10^4
	0.810	332.9	10^4
	0.910	334.2	10^4
$CHCl_3$	–31.700	258.2	10^3
Trichloromethane	0.101	466.2	10^2
CH_2Cl_2	0.101	394.8	10
Dichloromethane			
CH_3Cl	0.101	366.2	10^5
Chloromethane			
CH_4	0.400	167.6	10^5
Methane	0.620	168.3	10^5
	0.820	169.3	10^5
	1.030	170.5	10^5
	1.230	171.4	10^5
	1.430	172.1	10^5
	1.630	173.1	10^5
	1.830	174.0	10^5
	2.030	175.2	10^5
	2.220	176.4	10^5
	2.430	177.6	10^5
	2.630	178.6	10^5
	2.820	180.0	10^5

P Substance	*T* [MPa]	*J* [K]	[nuclei/(cm³·s)]
CH_4O	0.101	458.4	10
Methanol	0.101	461.2	10^5
	0.101	466.2	10^{18}
	0.600	469.2	10^{19}
	1.050	471.2	10^{20}
	2.030	476.7	10^{16}
	2.030	478.2	10^{20}
	3.000	482.2	10^{21}
	4.000	488.7	10^{22}
	4.980	494.7	10^{22}
	5.970	501.2	10^{23}
	6.960	507.7	10^{23}
C_2H_3Cl	0.101	374.1	10^5
Chloroethane			
C_2H_3F	0.101	290.1	10^5
Fluoroethene			
C_2H_3N	0.101	497.0	10^6
Acetonitrile			
$C_2H_4F_2$	0.101	343.6	10^5
1,1–Difluoroethane			
$C_2H_4O_2$	–28.800	292.7	10^3
Acetic acid	0.101	526.2	10^6
$C_2H_4O_2$	0.101	423.2	10
Methyl formate			
C_2H_5Br	0.101	422.2	10
Ethyl bromide			
C_2H_5Cl	0.101	399.2	10
Ethyl chloride			
C_2H_6	0.101	269.2	10^5
Ethane			
C_2H_6O	0.101	464.1	10
Ethanol	0.101	466.0	10^4
	0.101	471.5	10^{17}
	0.580	474.2	10^{19}
	0.980	471.0	10^2
	1.070	477.2	10^{20}
	1.540	481.7	10^{20}
	2.030	484.2	10^{21}
	2.520	486.7	10^{21}
	3.010	490.2	10^{21}
	3.500	494.2	10^{22}
C_3H_3N	0.101	489.0	10^5
Acrylonitrile			
C_3H_4	0.101	346.2	10^5
Propadiene			

THE LIMITS OF SUPERHEAT OF PURE LIQUIDS (CONTINUED)

P Substance	*T* [MPa]	*J* [K]	[nuclei/(cm³·s)]
C_3H_4 Propyne	0.101	356.8	10^5
C_3H_6 Cyclopropane	0.101	350.7	10^5
C_3H_6 Propene	0.101	325.6	10^5
C_3H_6O Acetone	0.101	454.5	10
	0.101	456.4	10^2
	0.101	458.7	10^{13}
	0.101	462.7	10^{18}
	0.980	462.6	10
$C_3H_6O_2$ Ethyl formate	0.101	428.5	10
$C_3H_6O_2$ Methyl acetate	0.101	416.6	10
C_3H_8 Propane	0.101	326.4	10^6
	0.302	332.8	10^4
	0.491	336.8	10^4
	0.715	339.1	10^4
	0.907	343.2	10^4
C_3H_8O 1-Propanol	0.101	487.4	10^4
	0.101	493.0	10^{15}
	0.101	495.7	10^{18}
C_3H_8O Isopropanol	0.101	473.0	10^6
C_4H_6 1,3-Butadiene	0.101	377.3	10^5
C_4H_8 1-Butene	0.101	371.0	10^5
C_4H_8 *cis*-2-Butene	0.101	385.4	10^5
C_4H_8 *trans*-2-Butene	0.101	379.7	10^5
C_4H_8 2-Methylpropene	0.101	369.6	10^5
C_4H_{10} Butane	0.101	377.6	10^5
C_4H_{10} 2-Methylpropane	0.101	361.0	10^5
$C_4H_{10}O$ 1-Butanol	0.101	509.6	10^2
	0.101	511.9	10^4

THE LIMITS OF SUPERHEAT OF PURE LIQUIDS (CONTINUED)

P Substance	*T* [MPa]	*J* [K]	[nuclei/(cm^3·s)]
	0.101	513.2	10^{13}
	0.101	516.2	10^{16}
	0.101	518.2	10^{18}
	0.908	519.4	10^{2}
$C_4H_{10}O$	–1.75	293.	10^{2}
Diethyl ether	–1.520	402.7	10^{4}
	–1.220	407.6	10^{4}
	–1.120	409.2	10^{4}
	–1.000	410.2	10^{4}
	–0.740	413.4	10^{4}
	0.101	417.5	10^{2}
	0.101	425.7	10^{19}
	0.211	419.4	10
	0.415	420.3	10
	0.480	427.7	10^{18}
	0.500	421.1	10^{2}
	0.641	424.3	10^{2}
	0.777	426.3	10
	0.880	432.7	10^{18}
	1.000	428.4	10
	1.280	436.7	10
	1.366	433.6	10
	1.442	435.1	10^{2}
	1.575	437.2	10
	1.660	440.7	10^{19}
	1.865	441.2	10
	2.089	443.3	10
	2.450	450.7	10^{21}
	2.850	455.7	10^{20}
$C_4H_{10}O$	0.101	437.2	10
Isobutanol			
$C_4H_{11}N$	0.101	408.5	10
Diethylamine			
C_5F_{12}	0.101	381.5	10^{6}
Perfluoropentane	0.300	385.4	10^{6}
	0.500	388.7	10^{6}
	0.700	392.2	10^{6}
	0.890	396.4	10^{6}
	1.090	399.0	10^{6}
	1.280	403.1	10^{6}
	1.480	407.4	10^{6}
C_5H_8	0.101	451.4	10^{6}
Cyclopentene			
C_5H_{10}	0.101	455.1	10^{6}
Cyclopentane			
C_5H_{10}	0.101	417.2	10^{6}
1-Pentene			
C_5H_{12}	0.101	386.1	10^{6}
2,2-Dimethylpropane			

THE LIMITS OF SUPERHEAT OF PURE LIQUIDS (CONTINUED)

P Substance	*T* [MPa]	*J* [K]	[nuclei/(cm³·s)]
C_5H_{12}	0.101	409.2	10
Isopentane	0.101	411.7	10^{7}
C_5H_{12}	0.101	418.8	10^{4}
Pentane	0.101	426.2	10^{18}
	0.490	423.7	10^{2}
	0.880	429.1	10^{2}
	1.280	435.3	10^{2}
	2.600	451.2	10^{6}
$C_5H_{12}O$	0.101	551.7	10^{4}
1-Pentanol			
C_6F_6	0.101	464.8	10
Hexafluorobenzene	0.101	467.9	10^{6}
	0.500	469.9	10^{2}
	0.570	474.1	10^{6}
	1.000	477.1	10^{2}
	1.050	480.4	10^{6}
	1.540	486.0	10^{6}
	2.030	494.2	10^{6}
C_6F_{14}	0.101	409.8	10^{6}
Perfluorohexane	0.300	414.4	10^{6}
	0.500	418.5	10^{6}
	0.700	422.3	10^{6}
	0.880	425.6	10^{6}
	1.050	430.3	10^{6}
	1.240	434.6	10^{6}
C_6H_5Br	0.101	534.2	10^{2}
Bromobenzene			
C_6H_5Cl	0.101	523.2	10^{2}
Chlorobenzene			
C_6H_6	–15.000	291.2	10^{3}
Benzene	0.101	498.9	10^{2}
	0.101	510.2	10^{18}
	0.490	502.2	10^{2}
	0.580	514.2	10^{19}
	0.980	509.2	10^{2}
	1.070	516.7	10^{18}
	1.470	513.8	10^{2}
	1.540	520.7	10^{19}
	2.030	525.7	10^{19}
	2.520	532.2	10^{20}
	3.010	537.7	10^{17}
	3.500	544.7	10^{18}
C_6H_7N	–30.000	272.2	10^{3}
Aniline	0.101	535.2	10^{2}
C_6H_{10}	0.101	465.2	10^{6}
1–Hexyne			

THE LIMITS OF SUPERHEAT OF PURE LIQUIDS (CONTINUED)

P Substance	*T* [MPa]	*J* [K]	[nuclei/(cm^3·s)]
C_6H_{12}	0.101	490.8	10^6
Cyclohexane	0.300	493.1	10^2
	0.420	495.2	10^6
	0.720	499.7	10^6
	0.950	501.7	10^6
	0.980	502.1	10^2
	1.110	504.2	10^6
	1.350	506.2	10^6
	1.700	512.2	10^6
	2.160	518.2	10^6
	2.370	519.2	10^6
	2.550	523.2	10^6
C_6H_{12}	0.101	476.1	10^6
Methylcyclopentane			
C_6H_{14}	0.101	446.4	10^6
2,3-Dimethylbutane			
C_6H_{14}	0.101	453.5	10^2
Hexane	0.101	454.9	10^5
	0.101	459.2	10^{13}
	0.101	463.7	10^{20}
	0.290	465.2	10^{15}
	0.420	461.7	10^6
	0.490	459.3	10^2
	0.490	468.2	10^{22}
	0.760	466.7	10^6
	0.980	467.0	10^2
	0.980	475.2	10^{23}
	1.080	471.7	10^6
	1.120	478.2	10^{15}
	1.280	474.7	10^6
	1.420	475.7	10^6
	1.590	479.7	10^6
	1.600	486.2	10^{16}
	1.720	481.7	10^6
	1.960	487.7	10^{17}
	2.060	493.2	10^{16}
	2.390	496.7	10^6
	2.570	501.2	10^{16}
$C_6H_{14}O$	0.101	551.7	10^4
1-Hexanol			
C_7F_8	0.101	485.3	10
Octafluorotoluene	0.490	489.7	10
	0.980	499.8	10
C_7F_{16}	0.101	434.8	10^6
Perfluoroheptane	0.230	436.9	10^6
	0.400	440.5	10^6
	0.570	444.4	10^6
	0.770	448.3	10^6
	0.920	452.7	10^6

P Substance	*T* [MPa]	*J* [K]	[nuclei/(cm³·s)]
	1.070	456.1	10^6
	1.150	459.0	10^6
	1.280	461.3	10^6
C_7H_8 Toluene	0.101	526.7	10^2
C_7H_{14} Methylcyclohexane	0.101	510.4	10^6
C_7H_{16}	0.101	486.9	10^6
Heptane	0.101	493.7	10^{18}
	0.294	489.2	10^6
	0.392	490.7	10^6
	0.490	493.7	10^6
	0.589	494.2	10^6
	0.736	498.7	10^6
	0.952	500.7	10^6
	1.275	505.2	10^6
	1.373	509.7	10^6
	1.570	512.7	10^6
	1.736	515.2	10^6
	1.805	516.7	10^6
	2.001	519.7	10^6
$C_7H_{16}O$ 1-Heptanol	0.101	566.3	10^4
C_8F_{18}	0.101	457.0	10^6
Perfluorooctane	0.300	461.1	10^6
	0.500	467.1	10^6
	0.700	471.2	10^6
	0.890	476.9	10^6
	1.090	482.8	10^6
	1.190	484.1	10^6
C_8H_{10} Cyclooctane	0.101	560.7	10^6
C_8H_{10} 2,3-dimethylbenzene	0.101	508.2	10^2
C_8H_{16} 1-Octene	0.101	510.3	10^6
C_8H_{18}	0.101	513.8	10^6
Octane	0.377	519.3	10^4
	0.653	525.2	10^4
	0.929	528.6	10^4
	1.204	532.4	10^4
C_8H_{18} 2,2,4-Trimethylpentane	0.101	488.5	10^6
$C_8H_{18}O$ 1-Octanol	0.101	586.0	10^4

THE LIMITS OF SUPERHEAT OF PURE LIQUIDS (CONTINUED)

P	*T*	*J*	
Substance	**[MPa]**	**[K]**	**[nuclei/($cm^3 \cdot s$)]**
C_9F_{20}	0.101	478.5	10^6
Perfluorononane	0.300	484.4	10^6
	0.500	489.3	10^6
	0.700	493.3	10^6
	0.890	499.7	10^6
	1.090	505.7	10^6
C_9H_{20}	0.101	538.5	10^6
Nonane			
$C_{10}F_{22}$	0.101	497.1	10^6
Perfluorodecane	0.300	503.2	10^6
	0.500	508.6	10^6
	0.700	515.6	10^6
	0.890	521.2	10^6
	1.090	527.7	10^6
$C_{10}H_{22}$	0.101	558.3	10^6
Decane			
$C_{12}H_{10}O$	0.101	703.2	10^{17}
Diphenylether	0.101	708.7	10^{19}

Section 7
Biochemistry

PROPERTIES OF COMMON AMINO ACIDS

This table gives selected properties of 20 α-amino acids commonly found in proteins. The structures of these amino acids are given in a separate table. The compounds are listed in alphabetical order by the three-letter symbols. Dissociation constants refer to aqueous solutions at 25° C.

M_r — Molecular weight
T_m — Melting point
pK_a — Negative of the logarithm of the dissociation constant for the α-COOH group
pK_b — Negative of the logarithm of the dissociation constant for the α-NH_3^+ group
pK_x — Negative of the logarithm of the dissociation constant for any other group present in the molecule
pI — pH at the isoelectronic point
S — Solubility in water at 25° C in units of grams per kilogram of water

Symbol	Name	Mol. form.	M_r	T_m/°C	pK_a	pK_b	pK_x	pI	S
Ala	Alanine	$C_3H_7NO_2$	89.09	297	2.34	9.69		6.00	167
Arg	Arginine	$C_6H_{14}N_4O_2$	174.20	238	2.17	9.04	12.48	10.76	181
Asn	Asparagine	$C_4H_8N_2O_3$	132.12	236	2.02	8.80		5.41	25
Asp	Aspartic acid	$C_4H_7NO_4$	133.10	270	1.88	9.60	3.65	2.77	5
Cys	Cysteine	$C_3H_7NO_2S$	121.16	178	1.96	10.28	8.18	5.07	
Gln	Glutamine	$C_5H_{10}N_2O_3$	146.15	185	2.17	9.13		5.65	42
Glu	Glutamic acid	$C_5H_9NO_4$	147.13	249	2.19	9.67	4.25	3.22	
Gly	Glycine	$C_2H_5NO_2$	75.07	290	2.34	9.60		5.97	251
His	Histidine	$C_6H_9N_3O_2$	155.16	277	1.82	9.17	6.00	7.59	43
Ile	Isoleucine	$C_6H_{13}NO_2$	131.17	284	2.36	9.60		6.02	34
Leu	Leucine	$C_6H_{13}NO_2$	131.17	337	2.36	9.60		5.98	23
Lys	Lysine	$C_6H_{14}N_2O_2$	146.19	224—225	2.18	8.95	10.53	9.74	6
Met	Methionine	$C_5H_{11}NO_2S$	149.21	283	2.28	9.21		5.74	56
Phe	Phenylalanine	$C_9H_{11}NO_2$	165.19	284	1.83	9.13		5.48	29
Pro	Proline	$C_5H_9NO_2$	115.13	222	1.99	10.60		6.30	1622
Ser	Serine	$C_3H_7NO_3$	105.09	228	2.21	9.15		5.68	422
Thr	Threonine	$C_4H_9NO_3$	119.12	253	2.09	9.10		5.60	97
Trp	Tryptophan	$C_{11}H_{12}N_2O_2$	204.23	282	2.83	9.39		5.89	12
Tyr	Tyrosine	$C_9H_{11}NO_3$	181.19	344	2.20	9.11	10.07	5.66	0.5
Val	Valine	$C_5H_{11}NO_2$	117.15	292-295	2.32	9.62		5.96	58

REFERENCES

G. D. Fasman, Ed., *Practical Handbook of Biochemistry and Molecular Biology*, CRC Press, Boca Raton, FL, 1989.
E. L. Smith, et al., *Principles of Biochemistry*, 7th ed., McGraw Hill, New York, 1983.
H. J. Hinz, Ed., *Thermodynamic Data for Biochemistry and Biotechnology*, Springer-Verlag, Heidelberg, 1986.

STRUCTURES OF COMMON AMINO ACIDS

Alanine (Ala, A)	$H_3C-CH(NH_2)-COOH$
Arginine (Arg, R)	$H_2N-C(=NH)-NH-CH_2-CH_2-CH_2-CH(NH_2)-COOH$
Asparagine (Asn, N)	$H_2N-C(=O)-CH_2-CH(NH_2)-COOH$
Aspartic acid (Asp, D)	$HOOC-CH_2-CH(NH_2)-COOH$
Cysteine (Cys, C)	$HS-CH_2-CH(NH_2)-COOH$
Glutamine (Gln, Q)	$H_2N-C(=O)-CH_2-CH_2-CH(NH_2)-COOH$
Glutamic acid (Glu, E)	$HOOC-CH_2-CH_2-CH(NH_2)-COOH$
Glycine (Gly, G)	$H-CH(NH_2)-COOH$
Histidine (His, H)	[imidazolyl]$-CH_2-CH(NH_2)-COOH$
Isoleucine (Ile, I)	$CH_3-CH_2-CH(CH_3)-CH(NH_2)-COOH$
Leucine (Leu, L)	$(CH_3)_2CH-CH_2-CH(NH_2)-COOH$
Lysine (Lys, K)	$H_2N-CH_2-CH_2-CH_2-CH_2-CH(NH_2)-COOH$
Methionine (Met, M)	$CH_3-S-CH_2-CH_2-CH(NH_2)-COOH$
Phenylalanine (Phe, F)	[phenyl]$-CH_2-CH(NH_2)-COOH$
Proline (Pro, P)	[ring: $CH_2-CH_2-CH(COOH)-NH-CH_2$]
Serine (Ser, S)	$HO-CH_2-CH(NH_2)-COOH$
Threonine (Thr, T)	$CH_3-CH(OH)-CH(NH_2)-COOH$
Tryptophan (Trp, W)	[indolyl]$-CH_2-CH(NH_2)-COOH$
Tyrosine (Tyr, Y)	$HO-$[phenylene]$-CH_2-CH(NH_2)-COOH$
Valine (Val, V)	$(CH_3)_2CH-CH(NH_2)-COOH$

PROPERTIES OF PURINE AND PYRIMIDINE BASES

This table lists some of the important purine and pyrimidine bases that occur in nucleic acids. The last column gives the aqueous solubility at the indicated temperature in units of grams per 100 milliliters of water.

The numbering system in the rings is:

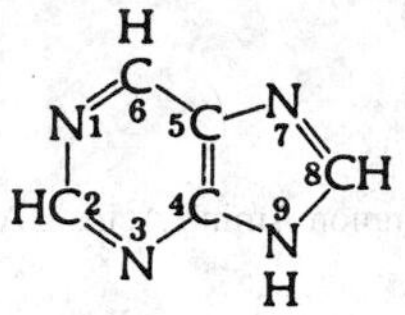

Purine

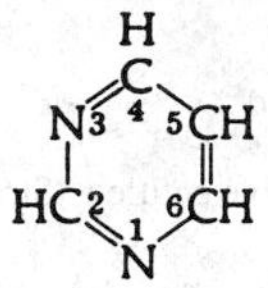

Pyrmidine

REFERENCES

1. R. M. C. Dawson, et al., *Data for Biochemical Research*, 3rd Ed., Clarendon Press, Oxford, 1986.
2. S. Budavari, Ed., *The Merck Index*, 11th Ed., Merk and Co., Rahway, NJ., 1989.

Common name	Systematic name	Mol form.	Mol. wt.	pK_a values			Sol. (temp.)
		Pyrimidines					
Cytosine	4-Amino-2-hydroxypyrimidine	$C_4H_5N_3O$	111.10	4.5	12.2		0.77 (25°C)
5-Methylcytosine	4-Amino-2-hydroxy-5-methylpyrimidine	$C_5H_7N_3O$	125.13	4.6	12.4		0.45 (25°C)
5-Hydroxymethyl-cytosine	4-Amino-2-hydroxy-5-hydroxy-methylpyrimidine	$C_5H_7N_3O_2$	141.13	4.3	13		
Uracil	2,4-Dihydroxypyrimidine	$C_4H_4N_2O_2$	112.09	0.5	9.5	>13	0.36 (25°C)
Thymine	5-Methyluracil	$C_5H_6N_2O_2$	126.11	9.9	>13		0.4 (25°C)
Orotic acid	Uracil-6-carboxylic acid	$C_5H_4N_2O_4$	156.10	2.4	9.5	>13	0.18 (18°C)
		Purines					
Adenine	6-Aminopurine	$C_5H_5N_5$	135.14	<1	4.1	9.8	0.09 (25°C)
Guanine	2-Amino-6-hydroxypurine	$C_5H_5N_5O$	151.13	3.3	9.2	12.3	0.004 (40°C)
7-Methylguanine	7-Methyl-2-amino-6-hydroxypurine	$C_6H_7N_5O$	165.16	3.5	9.9		
Isoguanine	6-Amino-2-hydroxypurine	$C_5H_5N_5O$	151.13	4.5	9.0		0.006 (25°C)
Xanthine	2,6-Dioxopurine	$C_5H_4N_4O_2$	152.11	0.8	7.4	11.1	0.05 (20°C)
Hypoxanthine	6-Hydroxypurine	$C_5H_4N_4O$	136.11	2.0	8.9	12.1	0.07 (19°C)
Uric acid	2,6,8-Trihydroxypurine	$C_5H_4N_4O_3$	168.11	5.4	11.3		0.002 (20°C)

THE GENETIC CODE

This table gives the correspondence between a messenger RNA codon and the amino acid which it specifies. The symbols for bases in the codon are:

U: uracil
C: Cytosine
A: adenine
G: guanine

The amino acid symbols are given in the table entitled "Structures of Common Amino Acids". A chain-initiating codon is indicated by **init** and a chain-terminating codon by **term**.

Example: UCA codes for **Ser**, UAC codes for **Tyr**, etc.

	Second position				
First position	**U**	**C**	**A**	**G**	**Third position**
U	Phe	Ser	Tyr	Cys	U
	Phe	Ser	Tyr	Cys	C
	Leu	Ser	**term**	**term**	A
	Leu	Ser	**term**	Trp	G
C	Leu	Pro	His	Arg	U
	Leu	Pro	His	Arg	C
	Leu	Pro	Gln	Arg	A
	Leu	Pro	Gln	Arg	G
A	Ile	Thr	Asn	Ser	U
	Ile	Thr	Asn	Ser	C
	Ile	Thr	Lys	Arg	A
	Met (**init**)	Thr	Lys	Arg	G
G	Val	Ala	Asp	Gly	U
	Val	Ala	Asp	Gly	C
	Val	Ala	Glu	Gly	A
	Val (**init**)	Ala	Glu	Gly	G

STEROID HORMONES AND OTHER STEROIDAL SYNTHETICS

Erwin DiCyan

The field of steroids has expanded considerably and rapidly in degree and in kind, because synthetic steroids have been synthesized which though resembling the hormones in the body have no natural counterpart, but exert an effect comparable to those of the natural hormones.

In fact, the term *steroid hormone* thus becomes a misnomer when applied to the newer synthetically prepared steroids which do not have a counterpart in the body of man or other animals—as prednisone. (A hormone, by definition, is a material with certain functions and characteristics, *secreted by the ductless glands*. That part of the definition cannot be met by prednisone or by similar steroids as these are not secreted by the ductless, or endocrine glands.)

All the hormones as well as the synthetic analogues have in common the cyclopentanophenanthrene nucleus. Although chemically very similar, a comparatively slight structural change is in many instances productive of substances which have physiologically dissimilar effects, often acting upon different physiologic systems. But in many cases a small change in structure will result merely in an accentuation of certain effects.

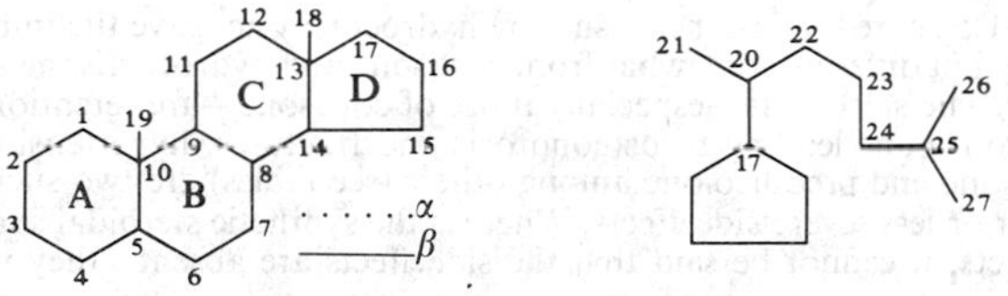

The Cyclopentanophenanthrene Nucleus

Classification. Classification becomes a bizarre problem by reason of the (a) overlapping uses to which these substances are put, and (b) the multiple purposes for which the hormones or synthetic substances are used. Indeed, the steroids may be classified by structure; that however would be uninformative to the student as to their use. Classification by origin, as adrenal, would also be unsuitable because, for example, a number of the adrenal corticosteroids are not found in the adrenal cortex at all, but merely resemble the natural hormones found in the adrenal cortex.

For these reasons the hormonal or hormonelike entries in the tables are classified by-and-large, by their predominant pharmacologic effects. Even that classification has its disparities as for example, the use of male sex hormones, i.e. the androgens, is neither limited to men, nor to uses which entail their effect upon male sex characteristics.

Uses. Originally, the use of steroid hormones was largely based upon one or more of the following predicates:

(a) To supplement the progressively declining secretion of a specific hormone due to natural biologic aging of the organism; in the menopause as an example of such declining secretion, a female sex hormone is used for such supplementation;

(b) To make available to the body a specific hormone, the natural secretion of which is inhibited because of a congenital or developmental anomaly; the underdevelopment of male secondary sex characteristics is an example of such an inhibited secretion, in which a male sex hormone is used—and correspondingly, female sex hormones in underdevelopment in females;

(c) To cause a reversal of hormonal balance in the treatment of diseases peculiar to a sex; for example, in the case of cancer of the female breast, a male sex hormone is administered, and in cancer of the prostate, a female sex hormone is used;

(d) To mimic a natural function, as menstruation, by the administration of estrogens—on withdrawal of which bleeding occurs; or by the alternate use of estrogenic and progestational—both female sex hormones.

(e) To delay a function, as ovulation, as in oral contraceptives, or *birth control pills*.

Since the finding that cortisone ameliorates the symptoms of rheumatoid arthritis (1949) the adrenal corticosteroid hormones and especially the synthetically prepared steroid analogues which have no natural counterpart in the body, have been successfully employed in the treatment of diseases not related to sex or sex function.

Androgens and Anabolic Agents. The agents listed in the tables under this classification have the effect of male sex hormones (androgens) i.e., to stimulate sexual maturation, in the "male climacteric," etc. But all androgens have in greater or lesser degree the ability to stimulate muscle development, i.e., an anabolic effect. Among the synthetically prepared agents which have no counterpart in the body (Methandrostenolone or Oxymetholone) are those which have a lessened androgenic, but a heightened anabolic effect. These qualities are determined by biological tests on animals but principally confirmed by clinical use in man. The anabolic effect includes remineralization of bone, which may be partially demineralized (osteoporosis) by age, or by certain drugs, as the adrenal corticosteroids (q.v.).

Anabolic agents are used for muscle and bone nutrition in men as well as women. The reason for the high interest in synthetic steroidal substances for anabolic use, is based on the need for materials, which within a given effective dose have a greater anabolic-to-androgenic ratio than such androgens as methyl testosterone. Otherwise, the administration of androgens to women produces manifestations of virilism, such as growth of hair on the face, a deepening of the voice, etc. Androgens are also used in the female in the suppression of excessive bleeding and in the treatment of cancer of the breast and cervix. (For other androgen-like agents, see also Progestogens and Progestins.)

Estrogens. Estrogenic agents hasten sexual maturation in the female. Therefore, they are used in underdevelopment in the female. The widest use of estrogens is in the treatment of the menopause, in which they supplement from without, the secretion of natural estrogens by the ovary, which begins to decline at about the 40th year. The menopause is usually a slow process, and the declining secretion gives rise to various symptoms during the time that the secretion declines, until adjustment to the new status takes place. The menopause, a period of physical and psychological stress, is made less precipitous by estrogens.

Frequently, a menopause must be quickly induced, as in cancer of the ovary or in uterine hemorrhage. This is done by radiation or by the removal of the uterus. Severe vasomotor symptoms occur when the menopause is thus suddenly induced. Estrogens —among other drugs—are used in the amelioration of these symptoms.

Estrogens (especially diethylstilbestrol which though not a hormone has an estrogenic effect) are also used in the control of cancer of the prostate in the male. Note the inverse correspondence to the use of male sex hormones in cancer of the breast in the female.

Progestogens and Progestins (Including 19-Norsteroid Compounds). The agents under that listing include progesterone, a female sex hormone, as well as progestins, i.e., synthetic progesterone-like compounds which have no natural counterpart in the body. Their use includes a variety of conditions: functional uterine bleeding, absence of menstruation (amenorrhea) used at times with estrogens, painful menstruation (dysmenorrhea), infertility, habitual abortion in order to maintain pregnancy, and in fact, to suppress ovulation hence their use as antifertility drugs. Certain progestins—as norethindrone combined with an estrogen, are the principal components of birth control pills—suppressing ovulation, there is no egg to fertilize, hence conception does not take place.

Adrenal Corticosteroids, Including Antiinflammatory, Antiallergic and Antirheumatic Agents. The adrenal cortex secretes a large number of hormones. They usually differ from each other in the accentuation of some phases of their properties. Virtually all of the cortical hormones are catabolic, thus having an effect in this respect, diametrically opposed to the androgens which are anabolic. Nearly all the cortical hormones—differing in degree from each other—cause retention of sodium and water by the body and hasten the excretion of potassium. These effects are utilized in the treatment of adrenal insufficiency or Addison's disease, in which conversely, there is an undue excretion of sodium and a strong retention of potassium. Desoxycorticosterone is used in Addison's disease because it has a particularly strong sodium retaining and potassium excreting effect.

Since the finding in 1949 of the usefulness of cortisone in profoundly reducing the symptoms of rheumatoid arthritis, the adrenal corticosteroids, including hydrocortisone, a natural hormone secreted by the adrenal cortex, and particularly the synthetic analogues not found in the body, as prednisone, have been used in the treatment of a wide variety of inflammatory diseases—especially diseases of collagen tissue. The same antiinflammatory effect is also brought into use in the reduction of inflammations associated with diseases of the skin, allergy, asthma, and in such systematic diseases as disseminated lupus erythematosus, also a collagen disease.

The drawbacks of cortisone, also shared in lesser measure by hydrocortisone, gave the impetus to the synthesis of steroidal substances not native to the body but differing somewhat from cortisone and hydrocortisone, in order to reduce the drawbacks attendant to the use of the latter. The sideeffects—especially those of cortisone—are retention of water and sodium, excretion of potassium, loss of mineral from bone leading to osteoporosis and fractures, hypertension, at times diabetes, personality changes or gastric ulcer. Prednisone and prednisolone among others (see tables) are two such steroidal synthetics which have the effects of cortisone, but fewer or less severe sideeffects. Whereas the synthetic steroidal substances are superior to cortisone with respect to lessened sideeffects, it cannot be said that the sideeffects are absent—they vary in degree from substance to substance.

Diuretic, Antidiuretic and Local Anesthetic Agents. Aldosterone, a natural hormone of the adrenal cortex promotes retention in the body of sodium and water, and facilitates excretion of potassium. Hence its effect is almost diametrically opposed to diuretics—especially the thiazide diuretics. Aldosterone is much more active in this respect than desoxycorticosterone, and is used in the treatment of Addison's disease, a hypofunction of the adrenal glands.

Spironolactone is an antagonist to aldosterone—the latter when elaborated in the body in excessive amounts gives rise to a syndrome called aldosteronism. Spironolactone, a synthetically produced steroid does not have a natural counterpart in the body, is diuretic when mercurial or thiazide diuretics are ineffective; it prevents sodium retention and potassium excretion—effects opposite to aldosterone. Hence spironolactone is used in aldosteronism, against edema, in the treatment of congestive heart failure and in other conditions in which an accumulation of water, and water-retaining salt, is to be corrected.

Doses. The amount of substance which comprises a dose of steroid hormones, or of the steroidal synthetics varies from substance to substance—from 0.1 mg for an estradiol ester, to 50 mg for a 19-norsteroid compound. The dose is conditioned upon the order of activity of the substance, the purpose for which it is administered, as well as the patient's response. However, as additional steroids for hormonal use are synthesized—especially those with adrenocortical activity, their average dose is usually smaller than the previously available steroid. The smaller effective dose of the more recent steroid is cited as an advantage over the previously available steroid.

However, a smaller dose cannot be claimed as an inherent advantage of a new steroid in comparison with an existing one, unless the lower dosage exhibits either greater or more prolonged activity or lesser sideeffects. One cannot meaningfully compare a dose, milligram for milligram, without taking into consideration if a heightened effect of the smaller dose produces fewer sideeffects. For example, it does not make any difference if a given effect and the same accompanying sideeffects are produced by a 50 mg or a 5 mg dose.

ADRENAL CORTICOSTEROIDS, INCLUDING ANTIINFLAMMATORY, ANTIALLERGIC AND ANTIRHEUMATIC AGENTS

Names & synonyms:	BETAMETHASONE; 9α-fluoro-16β-methylprednisolone; 16β-methyl-11β,17α,21-trihydroxy-9α-fluoro-1,4-pregnadiene-3,20-dione.	BETAMETHASONE ACETATE; 9α-fluoro-16β-methylprednisolone-21-acetate.	BETAMETHASONE DISODIUM PHOSPHATE; 9α-fluoro-16β-methylprednisolone-21-disodium phosphate.
Formulae:	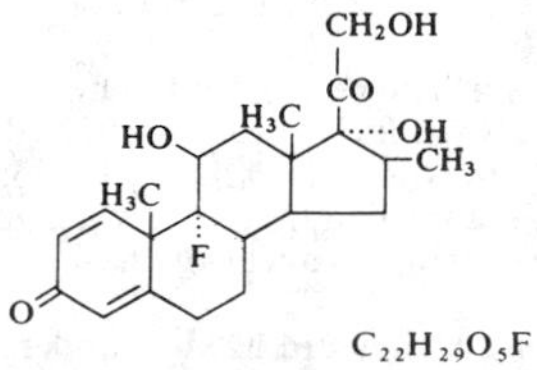 $C_{22}H_{29}O_5F$	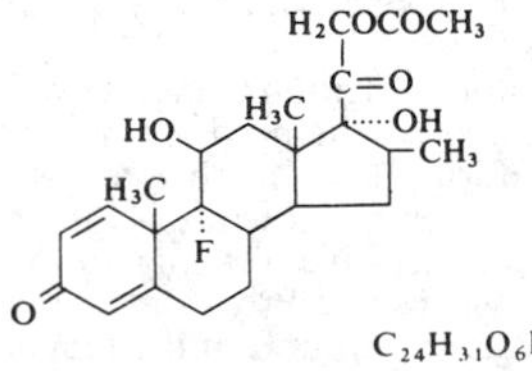 $C_{24}H_{31}O_6F$	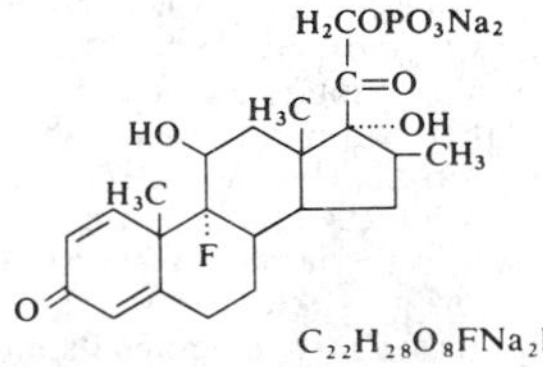 $C_{22}H_{28}O_8FNa_2P$
Molecular weight	392.5	434.5	516.4
Melting point (°C)	240 (dec.)	200 to 220 (dec.)	decomposes
Specific rotation	$(\alpha)^{25}_{D}$ +112 to +120 (100 mg. in 10 ml. dioxane)	$(\alpha)^{25}_{D}$ +120 to +128 (100 mg. in 10 ml. dioxane)	$(\alpha)^{25}_{D}$ +99 to +105 (100 mg. in 10 ml. water)
Absorption max.	239 mμ, E(1 %, 1 cm) 390, methanol	239 mμ, methanol	241 mμ, water

Names & synonyms:	CHLOROPREDNISONE ACETATE; 6α-chloroprednisone acetate; 6α-chloro-$\Delta^{1,4}$-pregnadien-17β,21-diol-3,11,20-trione 21-acetate.	CORTICOSTERONE; 11,21-dihydroxyprogesterone; Δ^4-pregnene-11β,21-diol-3,20-dione; 11β,21-dihydroxy-4-pregnene-3,20-dione; Kendall compound B; Reichstein substance H.	CORTISONE; 17-hydroxy-11-dehydrocorticosterone; 17α,21-dihydroxy-4-pregnene-3,11,20-trione; Δ^4-pregnene-17α,21-diol-3,11,20-trione; Kendall compound E; Wintersteiner compound F.
Formulae:	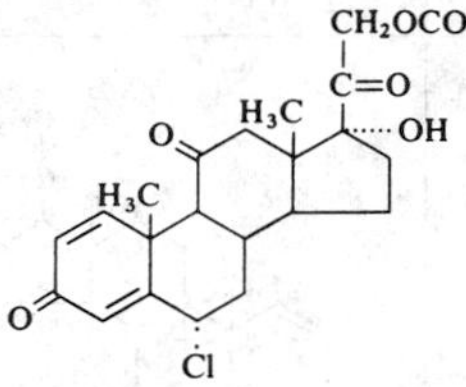 CH_2OCOCH_3, C=O, ····OH, O, H_3C, H_3C, O, Cl	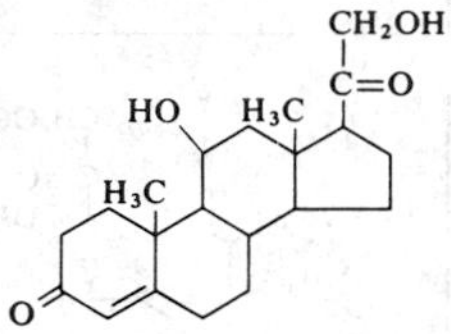 CH_2OH, C=O, HO, H_3C, H_3C, O; $C_{21}H_{30}O_4$	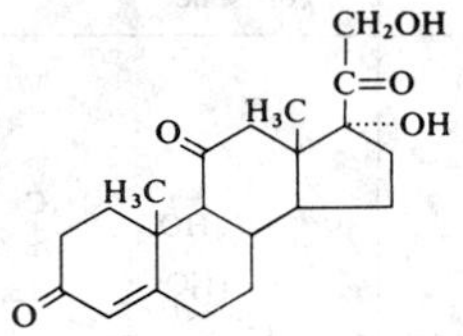 CH_2OH, C=O, ····OH, O, H_3C, H_3C, O; $C_{21}H_{28}O_5$
Molecular weight	436.6	346.40	360.4
Melting point (°C)	207–213	180–182	220–224
Specific rotation	$(\alpha)\frac{25}{D}$ +137 to +142 (100 mg. in 10 ml. chloroform)	$(\alpha)\frac{15}{D}$ +222 (110 mg. in 10 ml. alcohol)	$(\alpha)\frac{25}{D}$ +209 (120 mg. in 10 ml. alcohol)
Absorption max.		240 mμ	237 mμ

Names & synonyms:	DESOXYCORTICOSTERONE; deoxycorticosterone; 11-desoxycorticosterone; 21-hydroxyprogesterone; 4-pregnen-21-ol-3,20-dione; Kendall desoxy compound B; Reichstein substance Q.	DESOXYCORTICOSTERONE ACETATE; DCA; 11-desoxycorticosterone acetate.	DESOXYCORTICOSTERONE PIVALATE; desoxycorticosterone trimethylacetate; 21-hydroxy-4-pregnene-3,20-dione pivalate.
Formulae:	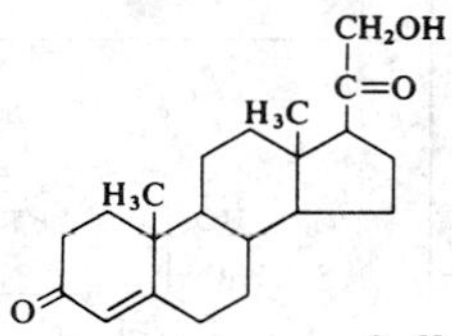 CH_2OH, C=O, H_3C, H_3C, O; $C_{21}H_{30}O_3$	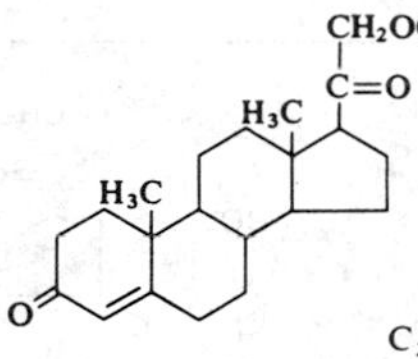 CH_2OOCCH_3, C=O, H_3C, H_3C, O; $C_{23}H_{32}O_4$	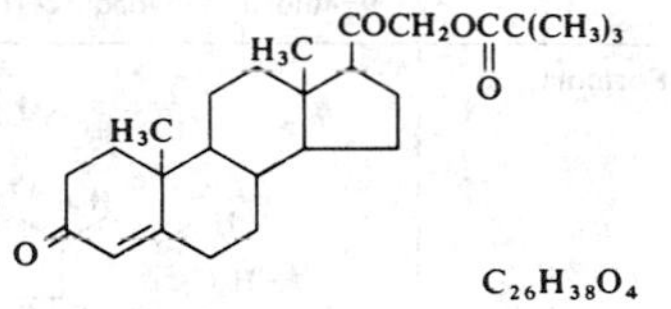 $COCH_2OCC(CH_3)_3$, O, H_3C, H_3C, O; $C_{26}H_{38}O_4$
Molecular weight	330.2	372.4	414.6
Melting point (°C)	140–142	154–160	198–204
Specific rotation	$(\alpha)\frac{22}{D}$ +176 – +178 (100 mg. in 10 ml. alcohol)	$(\alpha)\frac{20}{D}$ +168 – +178 (100 mg. in 10 ml. dioxane)	$(\alpha)\frac{25}{D}$ +157 ±4 (1 % in dioxane)
Absorption max.	240 mμ		240 mμ (in ethanol)

STEROID HORMONES AND OTHER STEROIDAL SYNTHETICS (Continued)

Names & synonyms:	DEXAMETHASONE; hexadecadrol; 9α-fluoro-16α-methyl prednisolone; 9α-fluoro-11β,17α-21-trihydroxy-16α-methyl-1,4-pregnadiene-3,20-dione; 16α-methyl-9α-fluoro-1,4-pregnadiene-11β,17α-21-triol-3,20-dione; 16α-methyl-9α-fluoro-Δ^1-hydrocortisone; 1-dehydro-16α-methyl-9α-fluorohydrocortisone.	DICHLORISONE ACETATE; 9α-11β-dichloro-1,4-pregnadiene-17α,21-diol-3,20-dione-21-acetate	FLUOCINOLONE ACETONIDE; 6α,9α-difluoro-16α hydroxyprednisolone-16,17-acetonide; 6α,9α-difluoro-16α,17α-isopropylidenediosy-1,4-pregnadiene-3,20-dione.
Formulae:	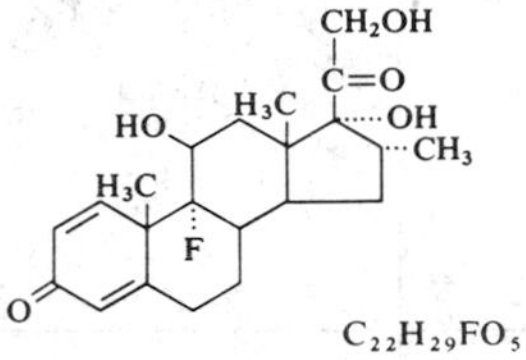 $C_{22}H_{29}FO_5$	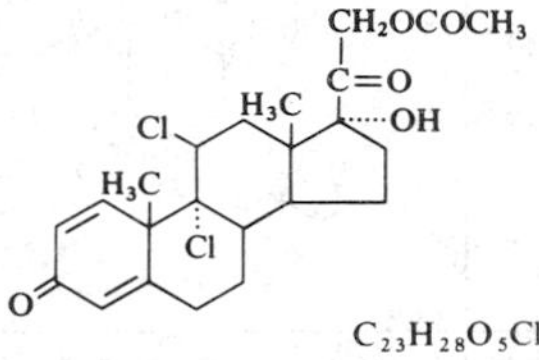 $C_{23}H_{28}O_5Cl$	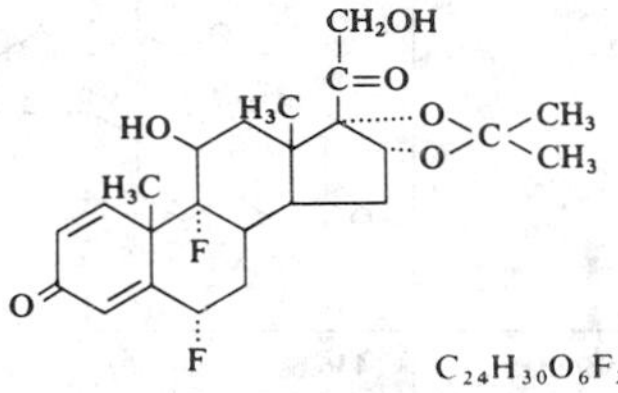 $C_{24}H_{30}O_6F_2$
Molecular weight	392.4	455.3	452.50
Melting point (°C)	262–264	235 (dec.)	255–266
Specific rotation	$(\alpha)\frac{25}{D}$ +78 (100 mg. in 10 ml. dioxane)	$(\alpha)\frac{25}{D}$ +160 – 168 (100 mg. in 10 ml. dioxane)	not less than +95° and not more than +105°C at 25°C.
Absorption max.		237 mμ – 316 – 337 (ε_1^1)	237 mμ ± 1 mμ

Names & synonyms:	FLUOROHYDROCORTISONE; fludrocortisone; 9α-fluorohydrocortisone; 9α-fluorocortisol; fluohydrisone; 9α-fluoro-11β,17α,21-trihydroxy-4-pregnene-3,20-dione; 9α-fluoro-17-hydroxycorticosterone.	FLUOROMETHOLONE; 9α-fluoro-11β,17α-dihydroxy-6α-methyl-1,4-pregnadiene-3,20-dione; 21-desoxy-9α-fluoro-6α-methyl-prednisolone.	FLUPREDNISOLONE; 6α-fluoroprednisolone; 6α-fluoro-1-dehydrohydrocortisone; 6α-fluoro-11β,17α,21-trihydroxy-1,4-pregnadiene-3,20-dione.
Formulae:	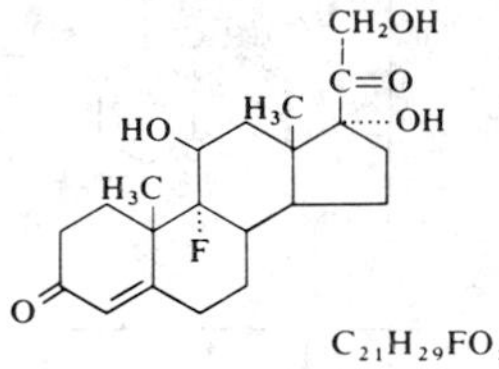 $C_{21}H_{29}FO_5$	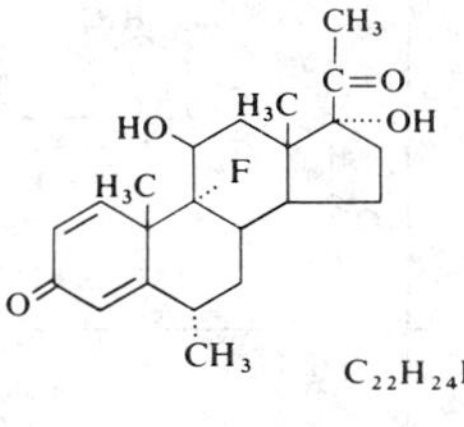 $C_{22}H_{24}FO_4$	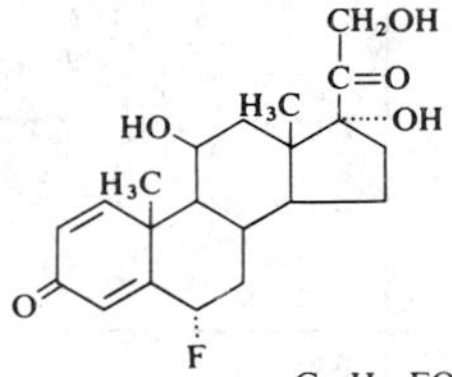 $C_{21}H_{27}FO_5$
Molecular weight	380.4	376.4	378.4
Melting point (°C)	260–262 (dec.)	290 (dec.)	205–210
Specific rotation	$(\alpha)\frac{23}{D}$ +139 (55 mg. in 10 ml. alcohol)	$(\alpha)\frac{25}{D}$ +56 (pyridine)	$(\alpha)_D$ +88 (dioxane)
Absorption max.		239 mu (a_M = 15,050) methanol	λ_{max} 241.5 mu (ε 16,000)

STEROID HORMONES AND OTHER STEROIDAL SYNTHETICS (Continued)

Names & synonyms:	FLURANDRENOLONE; 6-fluoro-16α-hydroxyhydrocortisone-16,17-acetonide; 6α-fluoro-11β,21-dehydroxy-16α,17α-isopropylidenedioxy-pregna-4-ene-3,20-dione.	HYDROCORTISONE; cortisol; 17-hydroxycorticosterone; hydrocortisone free alcohol; 11β,17α,21-trihydroxy-4-pregnene-3,20-dione; 4-pregnene-11β,17α,21-triol-3,20-dione; Kendall compound F; Reichstein substance M.	HYDROCORTISONE ACETATE; cortisol acetate; hydrocortisone-21-acetate; 17-hydroxycorticosterone-21-acetate.
Formulae:	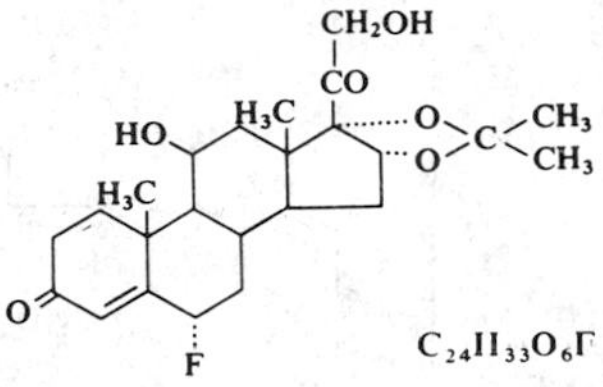 $C_{24}H_{33}O_6F$	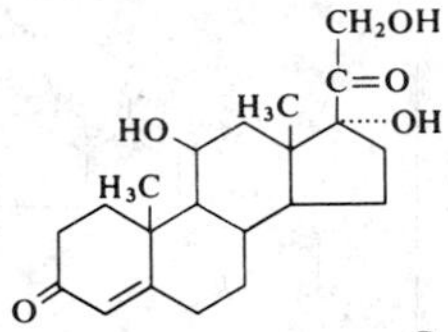 $C_{21}H_{30}O_5$	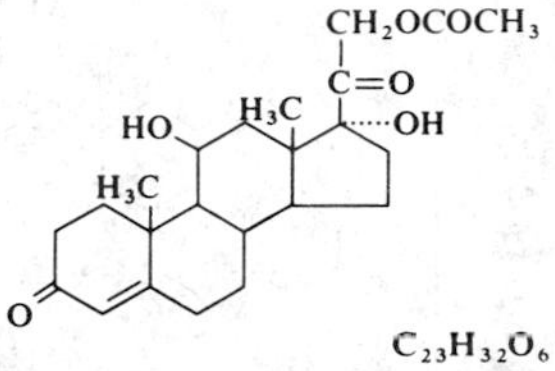 $C_{23}H_{32}O_6$
Molecular weight	436.5	362.5	404.5
Melting point (°C)	240–250	215–220 (dec.)	223 (dec.)
Specific rotation	$(\alpha)\frac{25}{D} = +145$ (1 % in $CHCl_3$)	$(\alpha)\frac{25}{D} +150 - +156$ (100 mg. in 10 ml.) dioxane)	$(\alpha)\frac{25}{D} +158 - +165$ (100 mg. in 10 ml. dioxane)
Absorption max.	236 mμ (methanol)	242 mμ	242 mμ (methanol)

Names & synonyms:	HYDROCORTISONE SODIUM SUCCINATE; 11β, 17α, 21-trihydroxy-4-pregnene-3,20-dione, 21 hydrogen succinate, sodium salt; hydrocortisone, 21 hydrogen succinate, sodium salt.	METHYLPREDNISOLONE; 6α-methylprednisolone; Δ^1-6α-methylhydrocortisone; 1-dehydro-6α-methylhydrocortisone; 11β, 17α,21-trihydroxy-6α-methyl-1,4-pregnadiene-3,20-dione.	METHYLPREDNISOLONE SODIUM SUCCINATE; 1-dehydro-6α-methylhydrocortisone, 21-hydrogen succinate, sodium salt; 6α-methylprednisolone 21-hydrogen succinate, sodium salt; 11β, 17α, 21-trihydroxy-6α-methyl-1,4-pregnadiene-3, 20-dione, 21-hydrogen succinate, sodium salt.
Formulae:	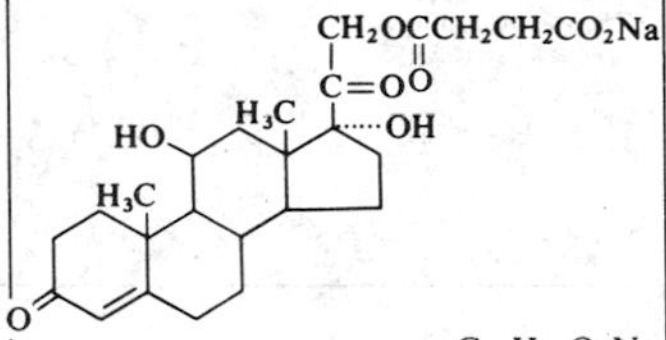 $C_{25}H_{33}O_8Na$	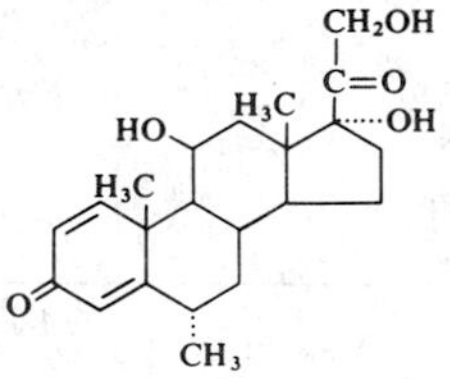 $C_{22}H_{30}O_5$	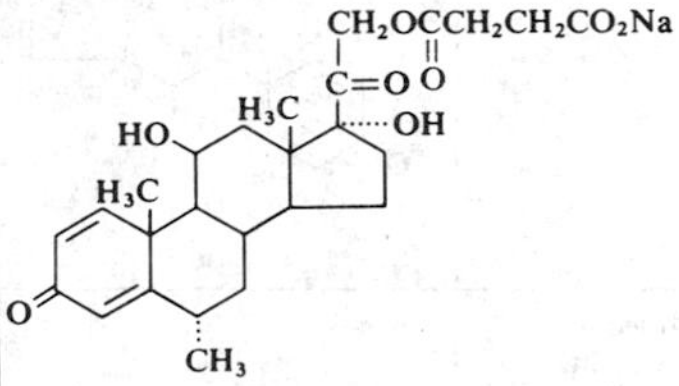 $C_{26}H_{33}O_8Na$
Molecular weight	484.5	374.5	496.5
Melting point (°C)	decomposes	230–240 (dec.)	decomposes
Specific rotation	$(\alpha)_D +140 \pm 5$ (alcohol)	$(\alpha)\frac{25}{D} +85$ (dioxane)	$(\alpha)_D +100 \pm 4$ (alcohol)
Absorption max.	λ 242 mμ (ε 15,700)	243 mμ	λ_{max} 242 mμ (ε 14,500)

	PARAMETHASONE	PARAMETHASONE ACETATE	PREDNISOLONE
Names & synonyms:	PARAMETHASONE; 6α-fluoro-16α-methylprednisolone; 6α-fluoro-11β-17α,21-trihydroxy-16α-methyl-1,4-pregnadiene-3,20-dione.	PARAMETHASONE ACETATE; 6α-fluoro-16α-methylprednisolone-21-acetate; 6α-fluoro-16α-methylpregna-1,4-diene-11β,21-diol-3,20-dione-21-acetate; 6α-fluoro-17β,17α,21-trihydroxy-16α-methyl-1,4-pregnadiene-3,20-dione-21-acetate.	PREDNISOLONE; metacortandralone; Δ[1]-dehydrocortisol; delta F; Δ[1]-hydrocortisone; Δ[1]-dehydrohydrocortisone; 1,4-pregnadiene-3,20-dione-11β,17α,21-triol; 11β,17α,21-trihydroxy-1,4-pregnadiene-3,20-dione.
Formulae:	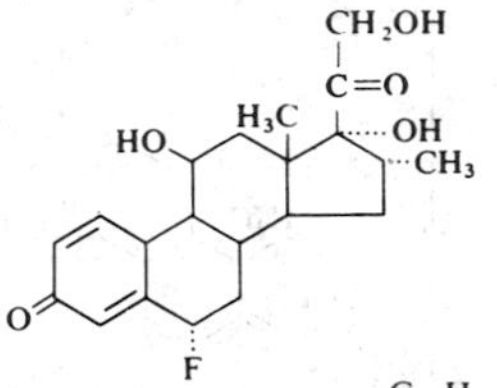 $C_{22}H_{30}O_5$	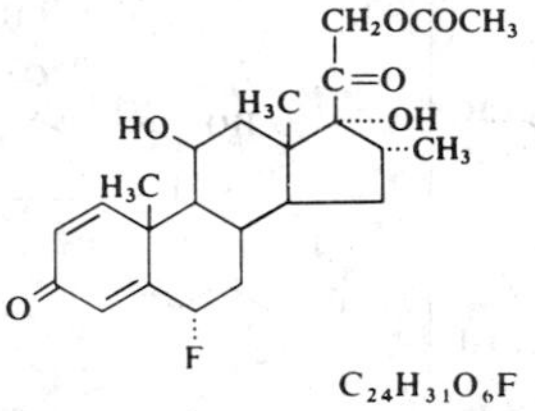 $C_{24}H_{31}O_6F$	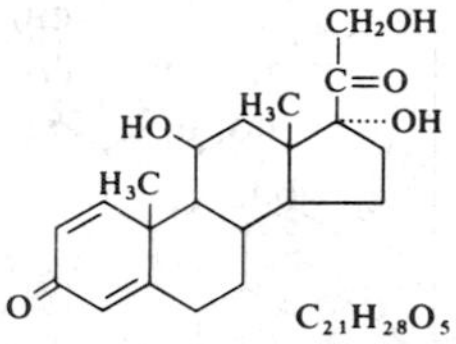 $C_{21}H_{28}O_5$
Molecular weight	392.45	434.5	360.4
Melting point (°C)	228–241	233–246	240 (dec.)
Specific rotation	+59 to +69 at 25°C	$(\alpha)\frac{25}{D}$ +72 (1 % in $CHCl_3$)	$(\alpha)\frac{25}{D}$ +97 – +103 (100 mg. in 10 ml. dioxane)
Absorption max.	242 mμ	242 mμ (methanol)	242 mμ (ε = 15,000) methanol

	PREDNISOLONE PHOSPHATE SODIUM	PREDNISOLONE PIVALATE
Names & synonyms:	PREDNISOLONE PHOSPHATE SODIUM; disodium prednisolone 21-phosphate.	PREDNISOLONE PIVALATE; prednisolone trimethylacetate; 11β,17α,21-trihydroxy-1,4-pregnadiene-3,20-dione 21-pivalate.
Formulae:	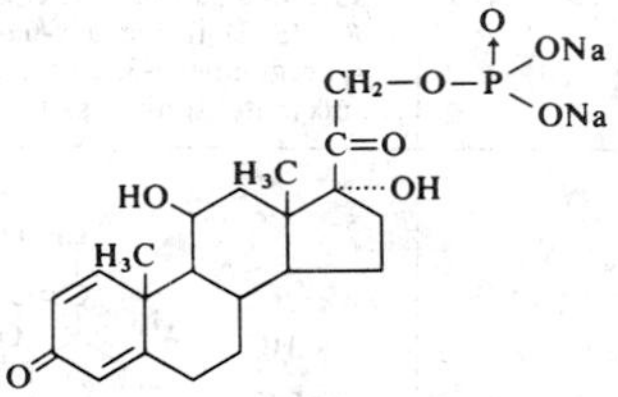 $C_{21}H_{27}Na_2O_8P$	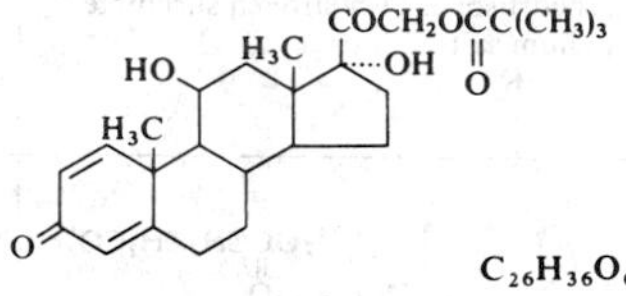 $C_{26}H_{36}O_6$
Molecular weight	484.4	444.6
Melting point (°C)		229
Specific rotation	$(\alpha)\frac{25}{D}$ +102.5 (100 mg. in 10 ml. H_2O)	+108 ±4 (1 % in dioxane)
Absorption max.	243 mμ	240 and 263 mμ (in absolute ethanol)

Names & synonyms:	PREDNISONE; metacortandricin; Δ^1-dehydrocortisone; delta E; Δ^1-cortisone; 1,4-pregnadiene-17α,21-diol-3,11,20-trione; 17α,21-dihydroxy-1,4-pregnadiene-3,11,20-trione.	TRIAMCINOLONE; 9α-fluoro-16α-hydroxyprednisolone; 9α-fluoro-11β,16α,17α,21-tetrahydroxy-1,4-pregnadiene-3,20-dione.
Formulae:	$C_{21}H_{26}O_5$	$C_{21}H_{27}FO_6$
Molecular weight	358.4	394.4
Melting point (°C)	225 (dec.)	260–262.5 (dec.)
Specific rotation	$(\alpha)\frac{25}{D}$ +167 – +175 (100 mg. in 10 ml. dioxane)	$(\alpha)\frac{25}{D}$ +75 (200 mg. in 100 ml. acetone)
Absorption max.	239 mμ (ε = 15,500) methanol	238 mμ (ε = 15,800)

Names & synonyms:	TRIAMCINOLONE ACETONIDE; 9α-fluoro-11β,21-dihydroxy-16α,17α-isopropylidene-dioxy-1,4-pregnadiene-3,20-dione; 9α-fluoro-16α-hydroxyprednisolone 16,17-acetonide.	TRIAMCINOLONE DIACETATE; 16α,21-diacetoxy-9α-fluoro-11β,17α-dihydroxy-1,4-pregnadiene-3,20-dione; 9α-fluoro-16α-hydroxyprednisolone 16,21-diacetate.
Formulae:	$C_{24}H_{31}FO_6$	$C_{25}H_{31}FO_8$
Molecular weight	434.4	478.49
Melting point (°C)	274–278 (dec.); 292–294	variable: 158–235
Specific rotation	$(\alpha)\frac{25}{D}$ +109 – +112 (53.7 mg. in 10 ml. chloroform)	$(\alpha)\frac{25}{D}$ +22 (78.8 mg. in 10 ml. chloroform)
Absorption max.	238–239 mμ (ε = 14,600)	239 mμ (ε = 15,200)

ANDROGENS AND ANABOLIC AGENTS

Names & synonyms:	ANDROSTERONE; cis-androsterone; 3α-hydroxy-17-androstanone; androstane-3α-ol-17-one.	FLUOXYMESTERONE; 9α-fluoro-11β-hydroxy-17α-methyltestosterone 9α-fluoro-11β,17β-dihydroxy-17α-methyl-4-androsten-3-one.	ALDOSTERONE; electrocortin; 18-oxocorticosterone; 18-formyl-11β,21-dihydroxy-4-pregnene-3,20-dione.
Formulae:	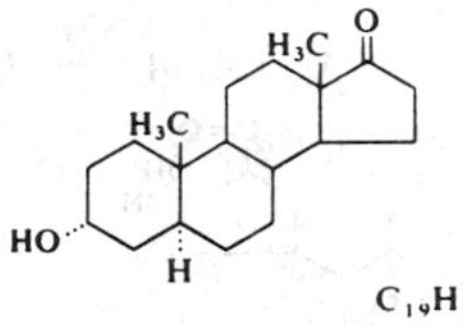 $C_{19}H_{30}O_2$	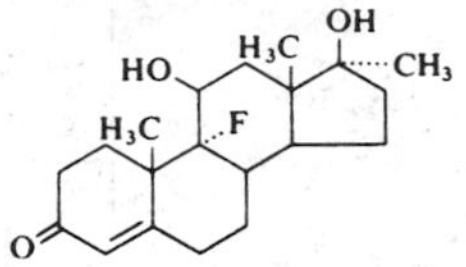 $C_{20}H_{29}FO_3$	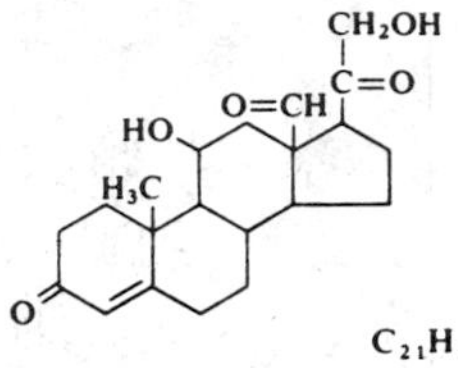 $C_{21}H_{28}O_5$
Molecular weight	290.4	336.4	360.4
Melting point (°C)	185–185.5	270 (dec.)	108–112 (hydrate); 164 (anhydrous)
Specific rotation	$(\alpha)^{15}_{D}$ +85 – +90 (150 mg. in 10 ml. dioxane)	$(\alpha)^{25}_{D}$ +107 – +109 (alcohol)	$(\alpha)^{25}_{D}$ +161 (10 mg. in 10 ml. chloroform)
Absorption max.		240 mμ (ε = 16,700) alcohol	240 mμ (log ε = 4.20 monohydr.; ε mol. 15,000 anhydr.)

Names & synonyms:	HYDROXYDIONE SODIUM; 21-hydroxypregnane-3,20-dione-21-sodium hemisuccinate.	SPIRONOLACTONE; 3-(3-oxo-7α-acetylthio-17β-hydroxy-4-androsten-17α-yl)-propionic acid γ lactone.
Formulae:	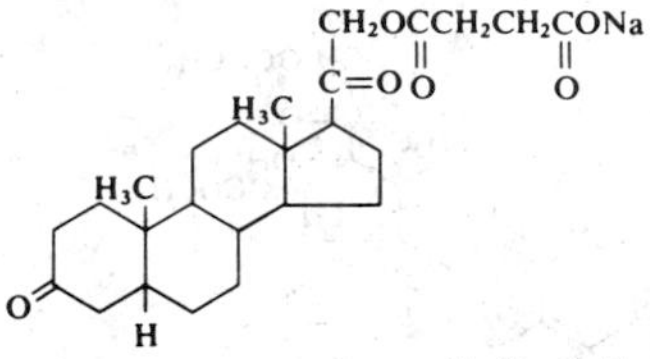 $C_{25}H_{35}O_6Na$	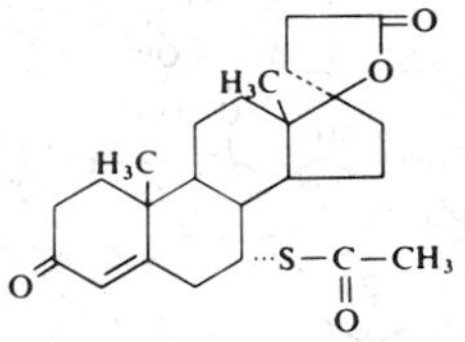 $C_{24}H_{32}O_4S$
Molecular weight	454.5	416.5
Melting point (°C)	193–203 (dec.)	135 (preliminary)–202 (dec.)
Specific rotation	$(\alpha)^{25}_{D}$ +95 (chloroform) for free acid.	$(\alpha)^{25}_{D}$ – 34 (chloroform)
Absorption max.	280 mμ (ε = 93.2)	ε^{238} = 20,200

Names & synonyms:	METHANDROSTENOLONE; 17α-methyl-17β-hydroxy-1,4-androstadien-3-one.	METHYLANDROSTENEDIOL; MAD; methandriol; 17α-methyl-5-androsten-3β,17β-diol.	METHYL TESTOSTERONE; 17-methyl testosterone; 17α-methyl-Δ^4-androsten-17-β-ol-3-one; 17(β)-hydroxy-17(α-methyl-4-androsten-3-one.
Formulae:	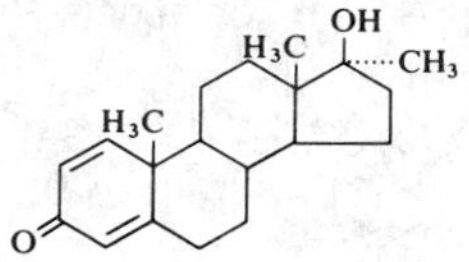 $C_{20}H_{28}O_2$	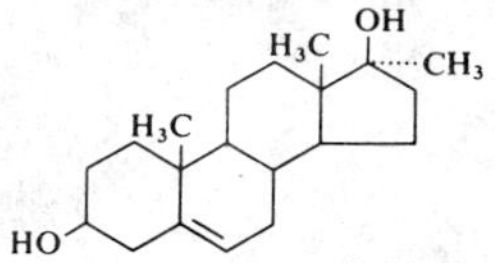 $C_{20}H_{32}O_2$	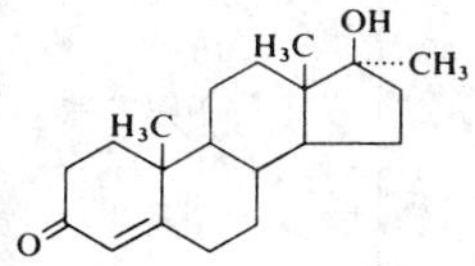 $C_{20}H_{30}O_2$
Molecular weight	300.4	304.4	302.4
Melting point (°C)	166–167	205–207	161–166
Specific rotation	$(\alpha)\frac{20}{D}$ +9 – +17 (100 mg. in 10 ml. alcohol)	$(\alpha)\frac{20}{D}$ −73 (100 mg. in 10 ml. alcohol)	$(\alpha)\frac{25}{D}$ +69 – +75 (100 mg. in 10 ml. dioxane)
Absorption max.			

Names & synonyms:	NORETHANDROLONE; 17α-ethyl-19-nortestosterone; 17α-ethyl-17-hydroxy-4-norandrosten-3-one; 17α-ethyl-17-hydroxy-19-norandrost-4-en-3-one.	OXANDROLONE; 17β-hydroxy-17α-methyl-2-oxa-5α-androstane-3-one
Formulae:	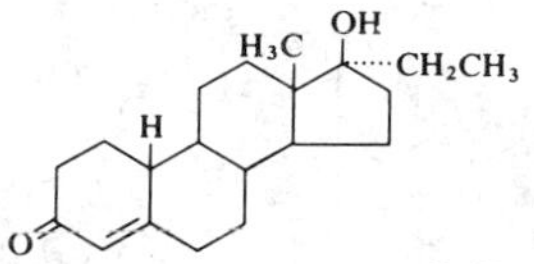 $C_{20}H_{30}O_2$	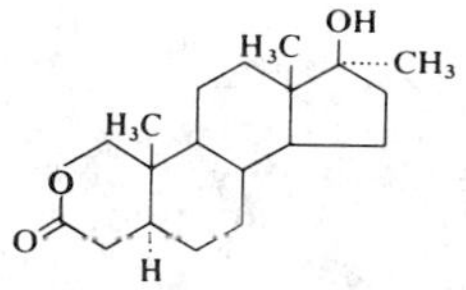 $C_{19}H_{30}O_3$
Molecular weight	302.4	306.4
Melting point (°C)	130–136	230–233
Specific rotation	$(\alpha)\frac{25}{D}$ +21 (dioxane)	$(\alpha)\frac{25}{D}$ −21 (1 % in chloroform)
Absorption max.	240 mμ (ε = 16,500)	None

Names & synonyms:	OXYMETHOLONE; 17β-hydroxy-2-hydroxymethylene-17α-methyl-3-androstanone; 2-hydroxymethylene-17-α-methyl dihydrotestosterone.	PROMETHOLONE: 2α-methyl-dihydro-testosterone propionate; 2α-methyl-5α-androstane-17β-ol-3-one-propionate.
Formulae:	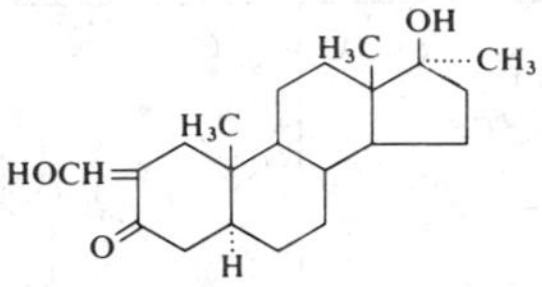 $C_{21}H_{32}O_3$	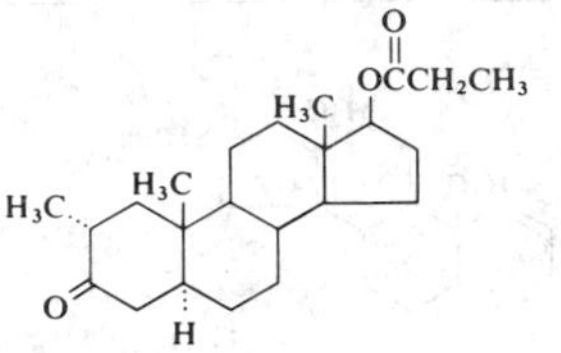
Molecular weight	332.4	360.5
Melting point (°C)	182	124–130
Specific rotation	$(\alpha)\frac{25}{D} = +36$ (200 mg. in 10 ml. dioxane)	$(\alpha)\frac{25}{D} +22 - +29$ (200 mg. in 10 ml. chloroform)
Absorption max.	$E_1^1 = 547$ at 315 mμ (in alkaline methanol made 0.01 N with NaOH)	without significant absorption from 220–300 mμ (methanol)

Names & synonyms:	TESTOSTERONE; trans-testosterone; Δ^4-androsten-17-β-ol-3-one; 17β-hydroxy-4-androsten-3-one.	TESTOSTERONE CYPIONATE; testosterone cyclopentylpropionate; 17β-hydroxy-4-androsten-3-one, cyclopentanepropionate.
Formulae:	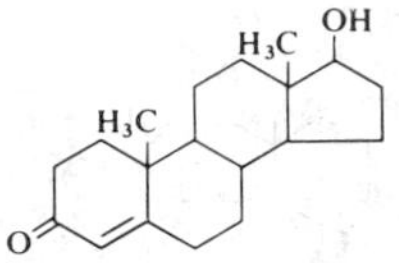 $C_{19}H_{28}O_2$	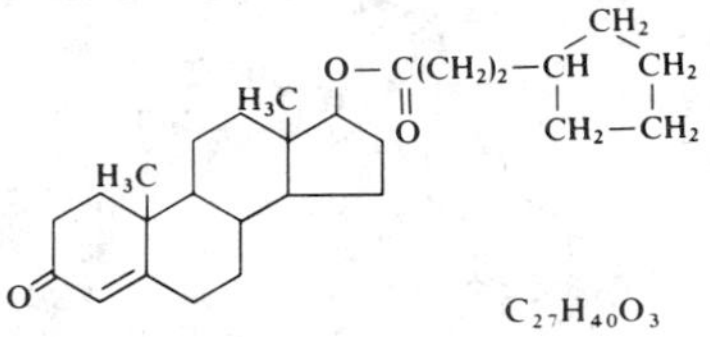 $C_{27}H_{40}O_3$
Molecular weight	288.4	412.6
Melting point (°C)	151–156	100–102
Specific rotation	$(\alpha)\frac{24}{D} +109$ (400 mg. in 10 ml. alcohol)	$(\alpha)_D +88.5 \pm 3.5$ ($CHCl_3$)
Absorption max.	238 mμ	λ_{max} 241 mμ (ε 16,125)

Names & synonyms:	TESTOSTERONE ENANTHATE; testosterone heptanoate; 17β-hydroxyandrost-4-en-3-one-17-enanthate.	TESTOSTERONE PHENYLACETATE; 17β-hydroxy-4-androsten-3-one phenylacetate; testosterone α-toluate.	TESTOSTERONE PROPIONATE: Δ^4-androstene-17-β-propionate-3-one.
Formulae:	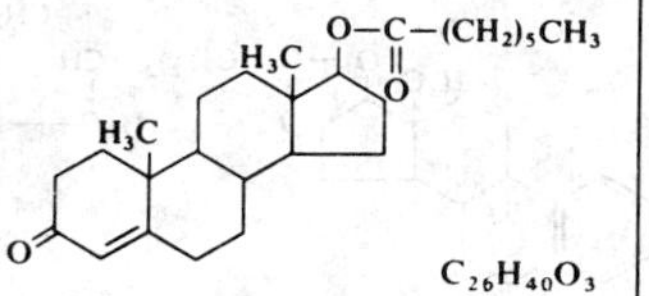 $C_{26}H_{40}O_3$	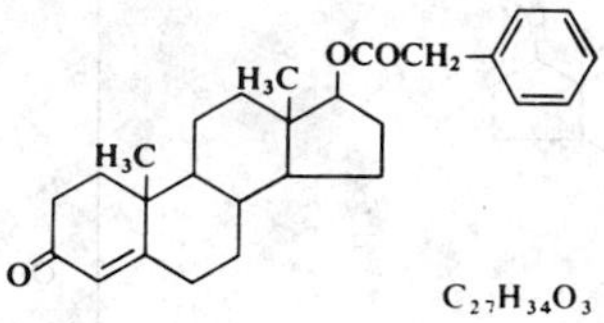 $C_{27}H_{34}O_3$	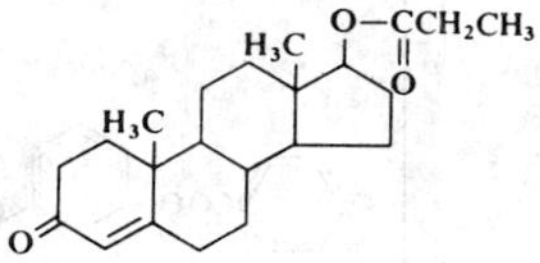 $C_{22}H_{32}O_3$
Molecular weight	400.6	406.5	344.4
Melting point (°C)	34–39	129–131	118–122
Specific rotation	$(\alpha)\frac{25}{D}$ +77 – +82 (2 % in dioxane)	$(\alpha)\frac{25}{D}$ +101 ±3 (1 % in chloroform)	$(\alpha)\frac{25}{D}$ +83 – +90 (100 mg. in 10 ml. dioxane)
Absorption max.	241 mμ (in ethanol)	241 mμ (in ethanol)	

ESTROGENS

Names & synonyms:	EQUILENIN; 3-hydroxy-17-keto-$\Delta^{1,3,5-10,6,8}$ estrapentaene; 1,3,5–10,6,8-estrapentaen-3-ol-17-one.	EQUILIN; 3-hydroxy-17-keto-$\Delta^{1,3,5-10,7}$ estratetraene; 1,3,5,7-estratetraen-3-ol-17-one.	ESTRADIOL (formerly called α-estradiol); β-estradiol; dihydrofolliculin; dihydroxyestrin; 1,3,5-estratriene-3,17β-diol; 3,17-dihydroxy-$\Delta^{1,3,5-10}$-estratriene; 3,17-epidihydroxyestratriene.
Formulae:	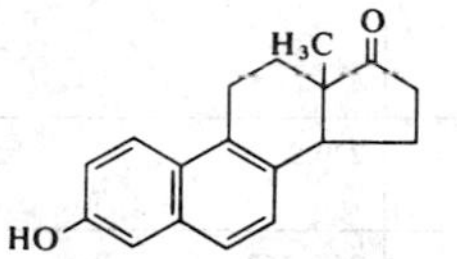 $C_{18}H_{18}O_2$	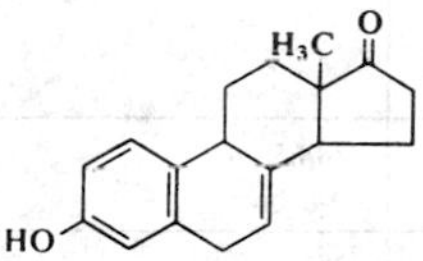 $C_{18}H_{20}O_2$	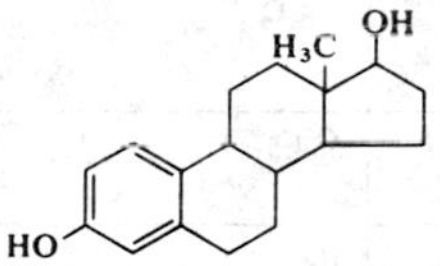 $C_{18}H_{24}O_2$
Molecular weight	266.3	268.3	272.3
Melting point (°C)	258–259	236–240	173–179
Specific rotation	$(\alpha)\frac{25}{D}$ +89 (dioxane)	$(\alpha)\frac{25}{D}$ +308 (200 mg. in 10 ml. dioxane); +325 (200 mg. in 10 ml. alcohol).	$(\alpha)\frac{25}{D}$ +76 – +83 (100 mg. in 10 ml. dioxane)
Absorption max.	231, 270, 282, 292, 325, 340 mμ	283–285 mμ	225, 280 mμ

Names & synonyms:	ESTRADIOL BENZOATE; β-estradiol-3-benzoate; estradiol monobenzoate.	ESTRADIOL CYPIONATE; estradiol cyclopentylpropionate; β-estradiol 17-cyclopentanepropionate; 1,3,5(10)-estratriene-3,17β-diol,17-cyclopentanepropionate.
Formulae:	$C_{25}H_{28}O_3$	$C_{26}H_{36}O_3$
Molecular weight	376.4	396.6
Melting point (°C)	191–196	151–154
Specific rotation	$(\alpha)\frac{25}{D}$ +58 – +63 (200 mg. in 10 ml. dioxane)	$(\alpha)_D$ +41.5 ±3.5 (dioxane)
Absorption max.		223 mμ

Names & synonyms:	ESTRADIOL DIPROPIONATE; α-estradiol dipropionate; 17β-estradiol dipropionate.	ESTRIOL; trihydroxyestrin; $\Delta^{1,3,5-10}$-estratriene-3-16-cis-17-trans-diol; 1,3,5-estratriene-3,16α,17β-triol.	ESTRONE; folliculin; ketohydroxyestrin; 1,3,5-estratrien-3-ol-17-one.
Formulae:	$C_{24}H_{32}O_4$	$C_{18}H_{24}O_3$	$C_{18}H_{22}O_2$
Molecular weight	384.5	288.3	270.3
Melting point (°C)	104–109	282	258–262
Specific rotation	$(\alpha)\frac{25}{D}$ +39 ±2 (1 % in dioxane)	$(\alpha)\frac{25}{D}$ +53 – +63 (40 mg. in 1 ml. dioxane)	$(\alpha)\frac{25}{D}$ +158 – +168 (100 mg. in 10 ml. dioxane)
Absorption max.	268 mμ	280 mμ	283–285 mμ

STEROID HORMONES AND OTHER STEROIDAL SYNTHETICS (Continued)

Names & synonyms:	ESTRONE BENZOATE	ETHYNYL ESTRADIOL; 17-ethinyl estradiol; 17α-ethynyl-1,3,5-estratriene-3,17β-diol.	MESTRANOL; ethynylestradiol 3-methyl ether; 3-methoxy-17α-ethynyl-1,3,5(10)-estratriene-17β-ol; 17α-ethynyl-estradiol-3-methyl ether; 3-methoxy-19-nor-17α-pregna-1,3,5,trien-20-yn-17-ol.
Formulae:	$C_{25}H_{26}O_3$	$C_{20}H_{24}O_2$	$C_{21}H_{26}O_2$
Molecular weight	374.4	296.4	310.4
Melting point (°C)	220	141–146	148–154
Specific rotation	$(\alpha)\frac{25}{D}$ + 120 (dioxane)	$(\alpha)\frac{25}{D}$ + 1 – + 10 (100 mg. in 10 ml. dioxane)	$(\alpha)\frac{25}{D}$ + 2 to + 8 (200 mg. in 10 ml. dioxane)
Absorption max.		248 mμ	278 to 287 mμ (methanol)

PROGESTOGENS AND PROGESTINS (INCLUDING 19-NORSTEROID COMPOUNDS)

Names & synonyms:	ACETOXYPREGNENOLONE; 21-acetoxypregnenolone; prebediolone acetate; Δ^5-pregnene-3β,21-diol-20-one-21-monoacetate; 21-acetoxy-5-pregnene-3-ol-20-one; 3-hydroxy-21-acetoxy-5-pregnen-20-one.	ANAGESTONE ACETATE; 6α-methyl-4-pregnen-17α-ol-20-one acetate; 17α-acetoxy-6α-methylpregn-4-en-20-one; 17α-acetoxy-6α-methyl-4-pregnen-20-one.	CHLORMADINONE ACETATE; 6-chloro-Δ^6-dehydro-17α-acetoxyprogesterone; 6-chloro-$\Delta^{4,6}$-pregnadiene-17α-ol-3,20-dioneacetate.
Formulae:	$C_{23}H_{34}O_4$	$C_{24}H_{36}O_3$	$C_{23}H_{29}ClO_4$
Molecular weight	374.5	372.6	404.9
Melting point (°C)	184–185	172–178	204–212
Specific rotation	$(\alpha)\frac{20}{D}$ + 37 – + 43 (dioxane)	$(\alpha)\frac{25}{D}$ + 40 to + 45 (10 mg. in 10 ml. chloroform)	$(\alpha)\frac{25}{D}$ 0 to – 6 (200 mg. in 10 ml. chloroform)
Absorption max.			284 mμ (methanol) Log ε = 4.34 ± 0.02

Names & synonyms:	DIMETHISTERONE; 6α,21-dimethylethisterone; 6α,21-dimethyl-17β-hydroxy-17α-pregn-4-en-20-yn-3-one; 6α-methyl-17α-propynylandrost-4-en-17β-ol-3-one; 17β-hydroxy-6α-methyl-17α-(prop-1-ynyl)-androst-4-ene-3-one.	ETHISTERONE; anhydrohydroxyprogesterone; ethinyl testosterone; pregneninolone; 17α ethynyl testosterone; 17α-ethynyl-17β-hydroxy-4-androsten-3-one.	ETHYNODIOL DIACETATE; 17α-ethynyl-4-estrene-3β,17β-diol-17-diacetate; 19-nor-17α-pregn-4-en-20-yne-3β,17-diol diacetate.
Formulae:	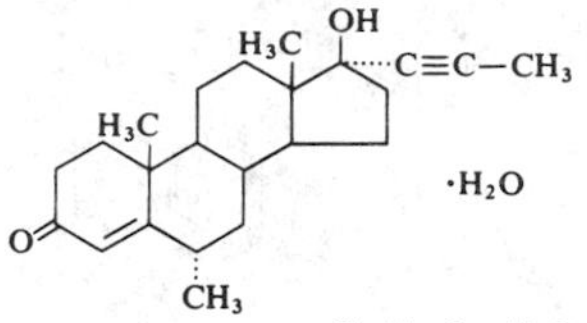 $\cdot H_2O$ $C_{23}H_{32}O_2.H_2O$	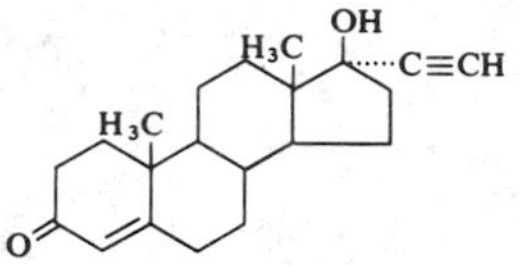 $C_{21}H_{28}O_2$	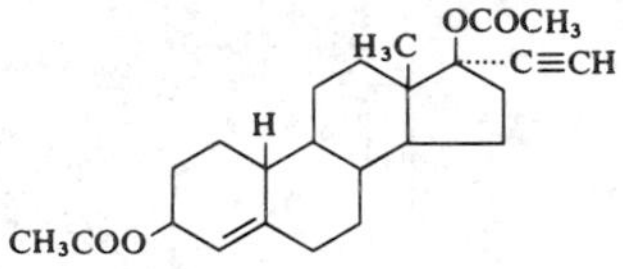 $C_{24}H_{32}O_4$
Molecular weight	358.5	312.4	384.5
Melting point (°C)	App. 100 (dec.)	266–273	126–132
Specific rotation	(α) $\frac{20}{D}$ +16.5 to +18.5 (2% solution in chloroform) (calculated to the anhydrous basis)	(α) $\frac{25}{D}$ −32° (100 mg. in 10 ml. pyridine)	(α) $\frac{25}{D}$ −74 (1% in chloroform)
Absorption max.	App. 240 mμ (anhydrous ethanol) $E_{1}^{1\%}$ cm = 443	241 mμ (methanol)	None

Names & synonyms:	FLUROGESTONE ACETATE; 17α-acetoxy-9α-fluoro-11β-hydroxy-4-pregnene-3,20-dione.	HYDROXYMETHYLPROGESTERONE; medroxyprogesterone; 17α-hydroxy-6α-methylprogesterone; 17α-hydroxy-6α-methyl-4-pregnene-3,20-dione.	HYDROXYMETHYLPROGESTERONE ACETATE; medroxyprogesterone acetate; 17α-hydroxy-6α-methylprogesterone acetate; 17α-hydroxy-6α-methyl-4-pregnene-3,20-dione acetate.
Formulae:	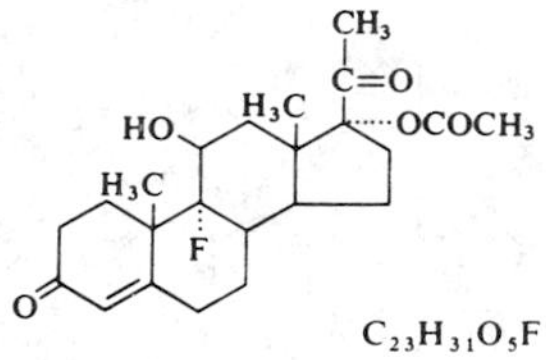 $C_{23}H_{31}O_5F$	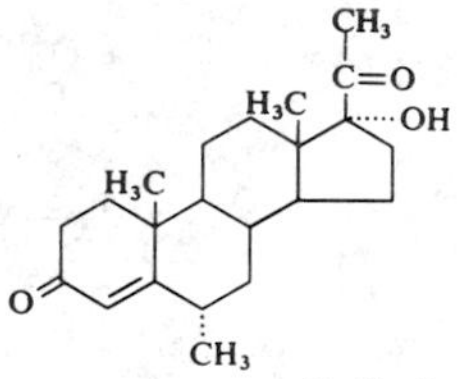 $C_{22}H_{32}O_3$	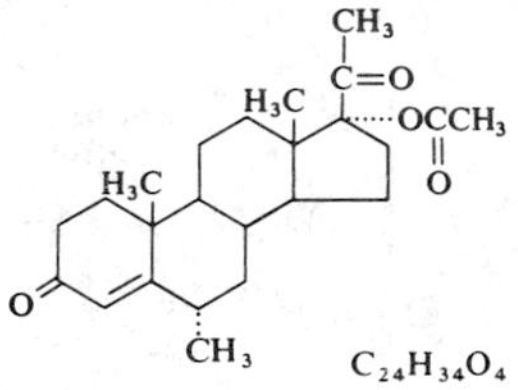 $C_{24}H_{34}O_4$
Molecular weight	406.5	344.5	386.5
Melting point (°C)	250–251	220–223.5	202–207
Specific rotation	(α) $\frac{25}{D}$ +78	(α) $\frac{25}{D}$ +75	(α) $\frac{25}{D}$ +51 (dioxane)
Absorption max.	238 mμ (ε = 17,100)	241 mμ (ε = 16,150)	241 mμ ($α_M$ = 16,500) ethanol

Names & synonyms:	HYDROXYPROGESTERONE; 17α-hydroxyprogesterone; 17α-hydroxy-4-pregene-3,20 dione; 4-pregnen-17α-ol-3,20-dione.	HYDROXYPROGESTERONE ACETATE; 17α-acetoxyprogesterone; 17α-hydroxyprogesterone acetate; 17α-hydroxy-4-pregnene-3,20 dione acetate.
Formulae:	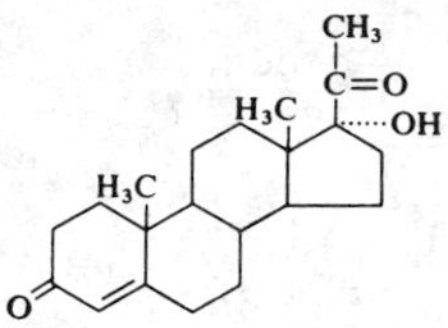 $C_{21}H_{30}O_3$	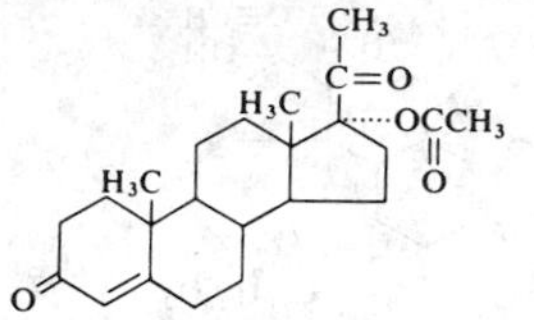 $C_{23}H_{32}O_4$
Molecular weight	330.4	372.5
Melting point (°C)	276	249–250
Specific rotation	$(\alpha)\frac{17}{D}$ +105 (104 mg. in 10 ml. chloroform)	$(\alpha)\frac{25}{D}$ +72 (chloroform)
Absorption max.		240 mμ (a_M = 16,875) ethanol

Names & synonyms:	HYDROXYPROGESTERONE CAPROATE; 17α-hydroxyprogesterone caproate; 17α-hydroxy-4-pregnene-3,20-dione caproate.	MELENGESTROL ACETATE: MGA; 17α-hydroxy-6-methyl-16-methylene-4,6-pregnadiene-3, 20-dione acetate; 6-dehydro-17-hydroxy-6-methyl-16-methylene-progesterone acetate.
Formulae:	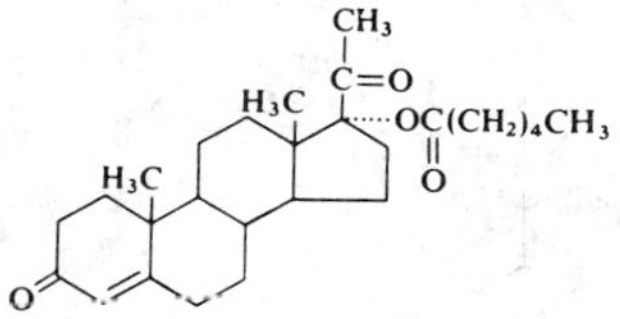 $C_{27}H_{40}O_4$	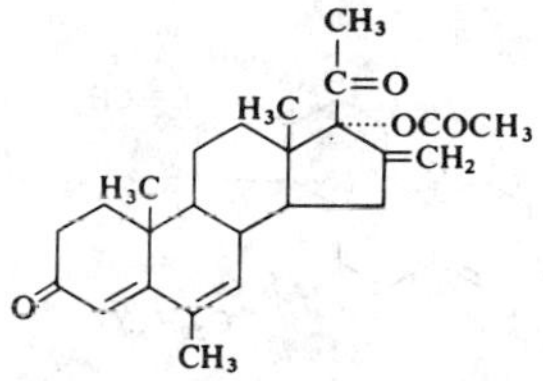 $C_{25}H_{32}O_4$
Molecular weight	428.6	396.51
Melting point (°C)	121–123	215–227
Specific rotation	$(\alpha)\frac{25}{D}$ +57 (chloroform)	$(\alpha)_D$ −127 to −135 (in $CHCl_3$)
Absorption max.		288 mμ (ε_1^1 = 24,000) (ethanol)

Names & synonyms:	NORETHINDRONE; Norethisterone; 17α-ethynyl-19-nortestosterone; 17α-ethynyl-17-hydroxy-19-nor-17α-4-en-20-yn-3-one.	NORETHINDRONE ACETATE; 17α-ethinyl-19-nortestosterone acetate.
Formulae:	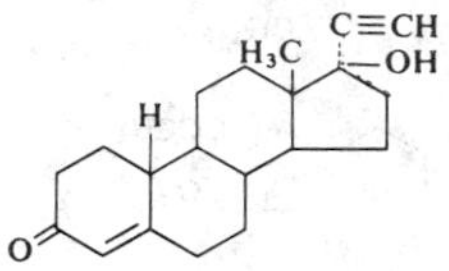 $C_{20}H_{26}O_2$	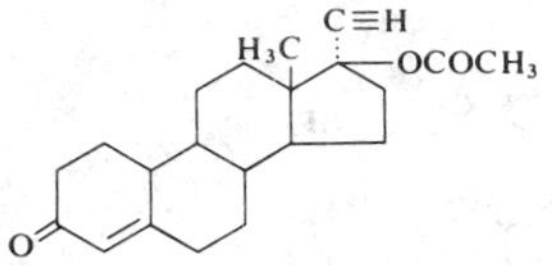 $C_{22}H_{28}O_3$
Molecular weight	298.4	340.4
Melting point (°C)	202 and 208	157–163
Specific rotation	$(\alpha)\frac{25}{D}$ −30 − −35 (200 mg. in 10 ml. dioxane)	$(\alpha)\frac{25}{D}$ −32 − −35 (200 mg. in 10 ml. dioxane)
Absorption max.	α (1 %, 1 cm) λ 240 = 535 ± 15	α (1 %, 1 cm.) λ 240 = 490 to 520 (505 ± 15) (ethanol)

Names & synonyms:	NORETHISTERONE; norethindrone; 19-norethisterone; 17α-ethynyl-19-nor-Δ⁴-androstan-17β-ol-3-one; 17α-ethynyl-19-nor-testosterone; 17-hydroxy-3-oxo-19-nor-17α-pregn-4-ene-20-yne; 17-hydroxy-19-nor-17α-pregn-4-en-20-yn-3-one.	NORETHYNODREL; 17α-ethynyl-17β-hydroxy-5(10)-estren-3-one.
Formulae:	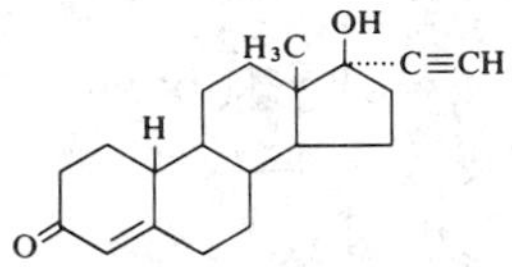 $C_{20}H_{26}O_2$	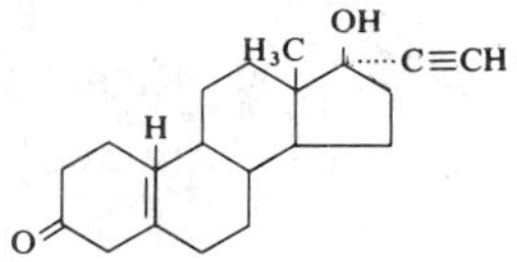 $C_{20}H_{26}O_2$
Molecular weight	298.4	298.4
Melting point (°C)	200–207	174–184
Specific rotation	$(\alpha)\frac{25}{D}$ −30 − −38 (200 mg. in 10 ml. dioxane)	$(\alpha)\frac{25}{D}$ +125 (dioxane)
Absorption max.	240 mμ (ε^1_1 = 576)	

STEROID HORMONES AND OTHER STEROIDAL SYNTHETICS (Continued)

Names & synonyms:	NORMETHISTERONE; 19-normethisterone; normethandrolone; metalutin; normetandrone; 17α-methyl-19-nor-Δ⁴-androsten-17β-ol-3-one; 17α-methyl-19-nor-testosterone; 17β-hydroxy-3-oxo-17α-methyl-estra-4-ene; 17β-hydroxy-17-methyl-estr-4-en-3-one.	PREGNENOLONE; Δ⁵-pregnenolone; Δ⁵-pregnen-3β-ol-20-one; 17β(1-ketoethyl)-Δ⁵-androstene-3β-ol.	PROGESTERONE; progestin; progestone; pregnendione; Δ⁴-pregnene-3,20-dione.
Formulae:	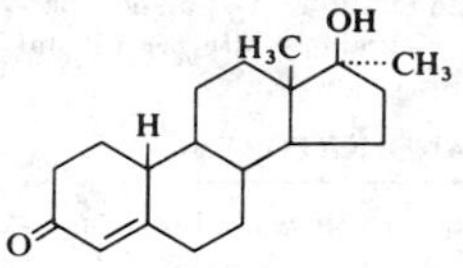 $C_{19}H_{28}O_2$	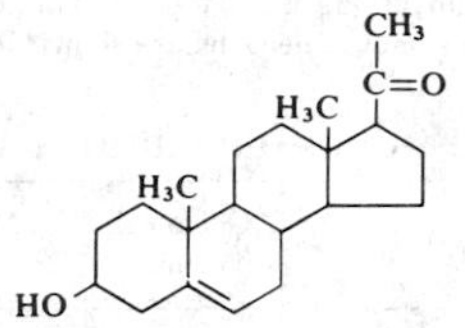 $C_{21}H_{32}O_2$	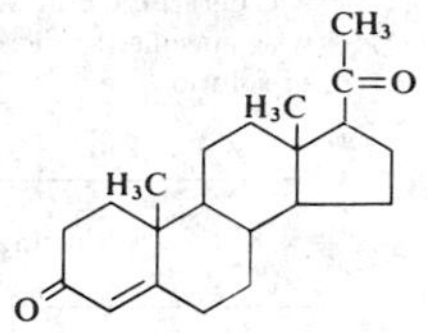 $C_{21}H_{30}O_2$
Molecular weight	288.4	308.4	314.4
Melting point (°C)	153–158	193	(β) isomer 121; (α) isomer 127–131
Specific rotation	$(\alpha)\frac{25}{D}$ +25 to +29 (200 mg. in 10 ml. chloroform)	$(\alpha)\frac{20}{D}$ +28 – +30 (alcohol)	$(\alpha)\frac{20}{D}$ +172 – +182 (200 mg. in 10 ml. dioxane)
Absorption max.	241 mμ – 565 ± 15		240 mμ

DIURETIC, ANTIDIURETIC AND LOCAL ANESTHETIC AGENTS

Names & synonyms:	ALDOSTERONE; electrocortin; 18-oxocorticosterone; 18-formyl-11β,21-dihydroxy-4-pregnene-3,20-dione.	HYDROXYDIONE SODIUM; 21-hydroxypregnane-3,20-dione-21-sodium hemisuccinate.	SPIRONOLACTONE; 3-(3-oxo-7α-acetylthio-17β-hydroxy-4-androsten-17α-yl)-propionic acid γ lactone.
Formulae:	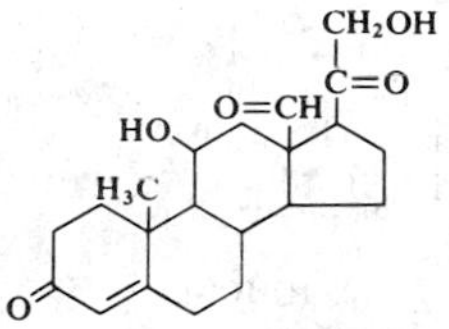 $C_{21}H_{28}O_5$	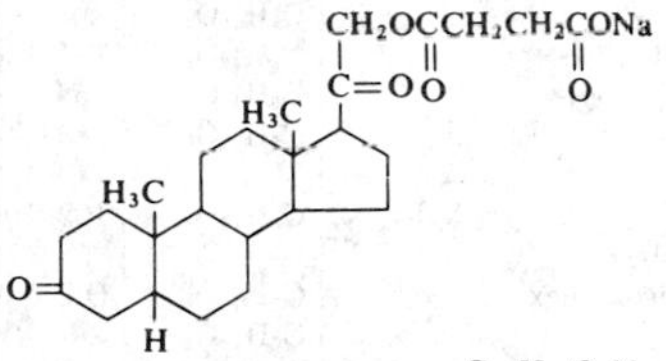 $C_{25}H_{35}O_6Na$	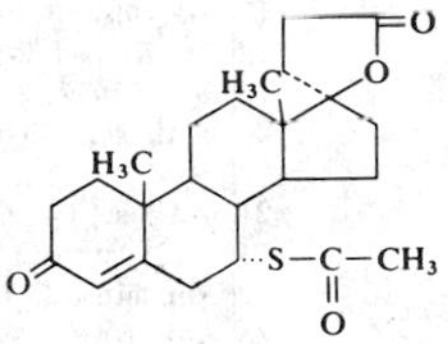 $C_{24}H_{32}O_4S$
Molecular weight	360.4	454.5	416.5
Melting point (°C)	108–112 (hydrate); 164 (anhydrous)	193–203 (dec.)	135; 202 (dec.)
Specific rotation	$(\alpha)\frac{25}{D}$ +161 (10 mg. in 10 ml. chloroform)	$(\alpha)\frac{25}{D}$ +95 (chloroform) for free acid.	$(\alpha)\frac{25}{D}$ – 34 (chloroform)
Absorption max.	240 mμ (log ε = 4.20 monohydr.; ε mol. 15,000 anhydr.)	280 mμ (ε = 93.2)	ε^{238} = 20,200

PROPERTIES OF CARBOHYDRATES

These data for carbohydrates were compiled originally for the Biology Data Book by M. L. Wolfram, G. G. Maher and R. G. Pagnucco (1964). Data are reproduced here by permission of the copyright owners of the above publication, the Federation of American Societies for Experimental Biology, Washington, D.C. pp. 351–359.

All data are for crystalline substances, unless otherwise specified. Selection of substances was restricted to natural carbohydrates found free (or in chemical combination and released on hydrolysis) and to biological oxidation products of the natural carbohydrates. The nomenclature conforms with that of the British-American report as published in the *Journal of Organic Chemistry,* 28:281 (1963). Substances have been arranged alphabetically under the name of the parent sugar within groups formulated according to increasing carbon content (excluding carbon in substituents), with synonymous common names in parentheses. **Melting Point:** b.p. = boiling point; d. = decomposes; s. = sinters. **Specific Rotation** was determined in water at concentrations of 1–5 g per 100 ml. of solution and at 20°–25°C, unless otherwise specified; other temperatures or wavelengths are shown in brackets; *c* = grams solute per 100 ml of solution.

Part I. NATURAL MONOSACCHARIDES: ALDOSES AND KETOSES

	Substance (Synonym)	Chemical Formula	Melting Point °C	Specific Rotation $[\alpha]_D$
	(A)	(B)	(C)	(D)
	Aldoses			
1	D-Glyceraldehyde	$C_3H_6O_3$	....	+13.5 ± 0.5 (syrup)
2	D-Glyceraldehyde, 3-deoxy-3,3-*C*-bis-(hydroxymethyl)- (Cordycepose)	$C_5H_{10}O_4$	....	−26 (*c* 0.6, C_2H_5OH)
3	D-Glyceraldehyde, 3,3-bis(*C*-hydroxy-methyl)- (Apiose)	$C_5H_{10}O_5$	....	+5.6 (*c* 10) [15°] syrup
4	β-D-Arabinose	$C_5H_{10}O_5$	155	−175 → −103
5	D-Arabinose, 2-*O*-methyl-	$C_6H_{12}O_5$	Syrup	−102
6	α-L-Arabinose	$C_5H_{10}O_5$	158 amorphous	+55.4 → +105
7	β-L-Arabinose	$C_5H_{10}O_5$	160	+190.6 → +104.5
8	DL-Arabinose	$C_5H_{10}O_5$	163.5–164.5	None
9	α-L-Lyxose	$C_5H_{10}O_5$	105	+5.8 → +13.5
10	L-Lyxose, 5-deoxy-3-*C*-formyl- (Streptose)	$C_6H_{10}O_5$	....	....
11	L-Lyxose, 3-*C*-formyl- (Hydroxy-streptose)	$C_6H_{10}O_6$	....	....
12	Pentose, 4,5-anhydro-5-deoxy-D-*erythro*-	$C_5H_8O_3$	....	....
13	Pentose, 2-deoxy-D-*erythro*-	$C_5H_{10}O_4$	96–98	−91 → −58
14	D-Ribose	$C_5H_{10}O_5$	87	−23.1 → −23.7
15	D-Ribose, 2-*C*-hydroxymethyl- (Hamamelose)	$C_6H_{12}O_6$	....	−7.1 [λ578]
16	α-D-Xylose	$C_5H_{10}O_5$	145	+93.6 → +18.8
17	D-Xylose, 5-deoxy-	$C_5H_{10}O_4$	....	+16
18	β-D-Xylose, 2-*O*-methyl-	$C_6H_{12}O_5$	137–138	−21 → +34
19	α-D-Xylose, 3-*O*-methyl-	$C_6H_{12}O_5$	95	+45 → +19
20	D-Allose, 6-deoxy-	$C_6H_{12}O_5$	140–143 146–148	+1.6 [18°] (*c* 0.6) −4.7 → 0
21	D-Allose, 6-deoxy-2,3-di-*O*-methyl- (Mycinose)	$C_8H_{16}O_5$	102–106	−46 → −29
22	Amicetose (a trideoxy hexose)	$C_6H_{12}O_3$	Oil, b.p. 65–70	+28.6 ($CHCl_3$)
23	Antiarose	$C_6H_{12}O_5$	....	Levo
24	α-D-Galactose	$C_6H_{12}O_6$	167	+150.7 → +80.2
25	β-D-Galactose	$C_6H_{12}O_6$	143–145	+52.8 → +80.2
26	D-Galactose, 3,6-anhydro-	$C_6H_{10}O_5$	....	+21.3 [10°]
27	α-D-Galactose, 6-deoxy- (D-Fucose; Rhodeose)	$C_6H_{12}O_5$	140–145	+127 → +76.3 (*c* 10)
28	D-Galactose, 6-deoxy-3-*O*-methyl- (Digitalose)	$C_7H_{14}O_5$	106[1], 119[2]	+106
29	D-Galactose, 6-deoxy-4-*O*-methyl-	$C_7H_{14}O_5$	131–132	+82
30	D-Galactose, 6-deoxy-2,3-di-*O*-methyl-	$C_8H_{16}O_5$	....	+73
31	α-D-Galactose, 3-*O*-methyl-	$C_7H_{14}O_6$	144–147	+150.6 → +108.6
32	α-D-Galactose, 6-*O*-methyl-	$C_7H_{14}O_6$	122–123	+117 → +77.3
33	L-Galactose	$C_6H_{12}O_6$	....	*See* D-Galactose
34	α-L-Galactose, 3,6-anhydro-	$C_6H_{10}O_5$	....	−39.4 → −25.2
35	α-L-Galactose, 6-deoxy- (L-Fucose)	$C_6H_{12}O_5$	145	−124.1 → −76.4
36	L-Galactose, 6-deoxy-2-*O*-methyl-	$C_7H_{14}O_5$	149–150	−75 ± 4 (*c* 0.5)
37	L-Galactose, 6-sulfate	$C_6H_{12}O_9S$	....	−47 (*c* 0.2) (Na salt)
38	DL-Galactose	$C_6H_{12}O_6$	143–144, 163	None (racemic)
39	α-D-Glucose	$C_6H_{12}O_6$	146, 83 (H_2O)	+112 → +52.7
40	β-D-Glucose	$C_6H_{12}O_6$	148–150	+18.7 → +52.7
41	D-Glucose, 6-acetate	$C_7H_{14}O_7$	135	+48
42	D-Glucose, 2,3-di-*O*-methyl-	$C_8H_{16}O_6$	85–86, 121	+50
43	D-Glucose, 6-*O*-benzoyl- (Vaccinin)	$C_{13}H_{16}O_7$	Amorphous	+48 (C_2H_5OH)
44	α-D-Glucose, 6-deoxy- (Chinovose; Epirhamnose; Glucomethylose; Isorhamnose; Isorhodeose; Quinovose)	$C_6H_{12}O_5$	139–140	+73.3 → +29.7 (*c* 8)
45	α-D-Glucose, 6-deoxy-3-*O*-methyl- (D-Thevetose)	$C_7H_{14}O_5$	116	+84 → +33

PROPERTIES OF CARBOHYDRATES (continued)

Part I. NATURAL MONOSACCHARIDES: ALDOSES AND KETOSES (Continued)

	Substance (Synonym)	Chemical Formula	Melting Point °C	Specific Rotation $[\alpha]_D$
	(A)	(B)	(C)	(D)
	Aldoses (Con't)			
46	D-Glucose, 6-sulfonic acid, 6-deoxy- (6-Sulfoquinovose)	$C_6H_{12}O_8S$	173–174	+87[3]
47	D-Glucose, 3-*O*-methyl-	$C_7H_{14}O_6$	162–167	+98 → +59.5
48	α-L-Glucose	$C_6H_{12}O_6$	141–143	−95.5 → −51.4
49	L-Glucose, 6-deoxy-3-*O*-methyl- (L-Thevetose)	$C_7H_{14}O_5$	126–129	−36.9 ± 2
50	D-Gulose, 6-deoxy-	$C_6H_{12}O_5$	....	....
51	Hexose, 2-deoxy-D-*arabino*-[4]	$C_6H_{12}O_5$	148	+46.6 [18°]
52	Hexose, 2,6-dideoxy-3-*O*-methyl-D-*arabino*- (D-Oleandrose)	$C_7H_{14}O_4$	....	−11
53	Hexose, 3,6-dideoxy-D-*arabino*- (Tyvelose)	$C_6H_{12}O_4$	....	+24 ± 2
54	Hexose, 2,6-dideoxy-3-*O*-methyl-L-*arabino*- (L-Oleandrose)	$C_7H_{14}O_4$	62–63	+11.9 ± 2.5
55	Hexose, 3,6 dideoxy-L-*arabino*- (Ascarylose)	$C_6H_{12}O_4$	....	−24 ± 2
56	Hexose, 2,6-dideoxy-3-*O*-methyl-D-*lyxo*- (Diginose)	$C_7H_{14}O_4$	90–92	+56 ± 4
57	Hexose, 2,6-dideoxy-L-*lyxo*- (L-Fucose, 2-deoxy-)	$C_6H_{12}O_4$	103–106	−61.6
58	Hexose, 2,6-dideoxy-3-*O*-methyl-L-*lyxo*-	$C_7H_{14}O_4$	78–85	−65
59	Hexose, 2,6-dideoxy-D-*ribo*- (Digitoxose; D-Altrose, 2,6-dideoxy-)	$C_6H_{12}O_4$	110	+46.4
60	Hexose, 2,6-dideoxy-3-*O*-methyl-D-*ribo*- (Cymarose)	$C_7H_{14}O_4$	93	+52
61	Hexose, 3,6-dideoxy-D-*ribo*- (Paratose)	$C_6H_{12}O_4$	....	+10 ± 2 (*c* 0.9)
62	Hexose, 4,6-dideoxy-3-*O*-methyl-D-*ribo*- (D-Gulose, 4,6-dideoxy-3-*O*-methyl-; Chalcose)	$C_7H_{14}O_4$	96–99	+120 → +76
63	Hexose, 2,6-dideoxy-D-*xylo*- (Boivinose)	$C_6H_{12}O_4$	96–98	−3.9 → +3.9
64	Hexose, 2,6-dideoxy-3-*O*-methyl-D-*xylo*- (Sarmentose)	$C_7H_{14}O_4$	78–79	+12 → +15.8
65	Hexose, 3,6-dideoxy-D-*xylo*- (Abequose)	$C_6H_{12}O_4$	....	−3.2 ± 0.6
66	Hexose, 2,6-dideoxy-3-*C*-methyl-L-*xylo*- (Mycarose)	$C_7H_{14}O_4$	129–129	−31.1
67	Hexose, 2,6-dideoxy-3-*C*-methyl-3-*O*-methyl-L-*xylo*-(Cladinose)	$C_8H_{16}O_4$	oil, b.p. 120–132 (0.25 mm)	−23.1
68	Hexose, 3,6-dideoxy-L-*xylo*- (Colitose)	$C_6H_{12}O_4$	....	+4 (H_2O); −51 ± 2 (CH_3OH)
69	D-Idose[5]	$C_6H_{12}O_6$	....	....
70	L-Idose, 1,6-anhydro-	$C_6H_{10}O_5$	....	....
71	α-D-Mannose	$C_6H_{12}O_6$	133	+29.3 → +14.5
72	β-D-Mannose	$C_6H_{12}O_6$	132	−16.3 → +14.5
73	D-Mannose, 6-deoxy- (D-Rhamnose)	$C_6H_{12}O_5$	86–90	−7.0
74	α-L-Mannose, 6-deoxy-monohydrate (L-Rhamnose)	$C_6H_{14}O_6$	93–94	−8.6 → +8.2
75	β-L-Mannose, 6-deoxy-	$C_6H_{12}O_5$	123–125	+38.4 → +8.9
76	L-Mannose, 6-deoxy-2-*O*-methyl-	$C_7H_{14}O_5$	....	....
77	L-Mannose, 6-deoxy-3-*O*-methyl- (L-Acofriose)	$C_7H_{14}O_5$	114–115	+30 [18°]
78	L-Mannose, 6-deoxy-2,4-di-*O*-methyl-	$C_8H_{16}O_5$	82	−19 [16°]
79	L-Mannose, 6-deoxy-5-*C*-methyl-4-*O*-methyl-(Noviose)	$C_8H_{16}O_5$	128–130	+19.9 (50% C_2H_5OH)
80	Rhodinose (a 2,3,6-trideoxyhexose)	$C_6H_{12}O_3$	....	−11 ± 1.6
81	D-Talose	$C_6H_{12}O_6$	128–132	+16.9
82	D-Talose, 6-deoxy- (D-Talomethylose)	$C_6H_{12}O_5$	129–131	+20.6
83	L-Talose, 6-deoxy- (L-Talomethylose)	$C_6H_{12}O_5$	116–118	−19.5 ± 2 [18°]
84	L-Talose, 6-deoxy-2-*O*-methyl- (L-Acovenose)	$C_7H_{14}O_5$	....	−19.4
85	Heptose, D-*glycero*-D-*galacto*-	$C_7H_{14}O_7$	139–140	+47 → +64 (*c* 0.5)
86	Heptose, D-*glycero*-D-*manno*-	$C_7H_{14}O_7$	....	....
87	Heptose, D-*glycero*-L-*manno*-	$C_7H_{14}O_7$	....	....
	Ketoses			
88	Dihydroxyacetone	$C_3H_6O_3$	80 (dimer)	None
89	Tetrulose, L-*glycero*-[8] (L-Erythrulose; Ketoerythritol; L-Threulose)	$C_4H_8O_4$	Syrup	+12
90	Pentulose, D-*erythro*- (Adonose; D-Ribulose)	$C_5H_{10}O_5$	Syrup	+16.6 [27°]
91	Pentulose, L-*erythro*- (L-Ribulose)	$C_5H_{10}O_5$	....	−16.6

PROPERTIES OF CARBOHYDRATES (continued)

Part I. NATURAL MONOSACCHARIDES: ALDOSES AND KETOSES (Continued)

	Substance (Synonym)	Chemical Formula	Melting Point °C	Specific Rotation $[\alpha]_D$
	(A)	(B)	(C)	(D)
	Ketoses (Con't)			
92	Pentulose, D-*threo*- (D-Xylulose)	$C_5H_{10}O_5$	....	−33
93	Pentulose, 5-deoxy-D-*threo*-	$C_5H_{10}O_4$	....	−5 ± 1 (CH_3OH)
94	Pentulose, L-*threo*- (L-Xylulose; L-Lyxulose; Xyloketose)	$C_5H_{10}O_5$	Syrup	+33.1
95	Hexulose, β-D-*arabino*-(β-D-Fructose; Levulose)	$C_6H_{12}O_6$	102–104[7]	−133.5 → −92
96	Hexulose, 6-deoxy-D-*arabino*- (D-Rhamnulose)	$C_6H_{12}O_5$	....	−13 ± 2
97	Hexulose, D-*lyxo*- (D-Tagatose)	$C_6H_{12}O_6$	131–132	+2.7 → −4, −5
98	5-Hexulose, D-*lyxo*	$C_6H_{12}O_6$	158	−86.6
99	Hexulose, 6-deoxy-L-*lyxo*- (L-Fuculose)	$C_6H_{12}O_5$	....	....
100	Hexulose, D-*ribo*- (D-Psicose)	$C_6H_{12}O_6$	Amorphous	+4.7
101	Hexulose, L-*xylo*- (L-Sorbose)	$C_6H_{12}O_6$	159–161	−43.1
102	Hexulose, 6-deoxy-L-*xylo*-	$C_6H_{12}O_5$	88	−25 ± 2 (*c* 0.7)
103	Heptulose, D-*altro*- (Sedoheptulose; Sedoheptose)	$C_7H_{14}O_7$	Amorphous	+2.5 (*c* 10)
104	Heptulose·hemihydrate, L-*galacto*- (Perseulose)	$C_7H_{14}O_7\cdot\frac{1}{2}H_2O$	110–115	−90 → −80
105	Heptulose, L-*gulo*-	$C_7H_{14}O_7$	....	−28
106	Heptulose, D-*ido*-	$C_7H_{14}O_7$	172	−34 ± 8 (*c* 0.3)
107	Heptulose, D-*manno*- (Mannoketo-heptose; D-Mannotagatoheptose)	$C_7H_{14}O_7$	152	+29.4
108	Heptulose, D-*talo*-	$C_7H_{14}O_7$	....	....
109	Octulose, D-*glycero*-L-*galacto*-	$C_8H_{16}O_8$	....	−57, −43.4 → −13.4
110	Octulose, D-*glycero*-D-*manno*-	$C_8H_{16}O_8$	....	+20 (CH_3OH)

[1] Original melting point. [2] Melting point after four-months' storage. [3] As a methyl glycoside cyclohexylamine salt. [4] Included because of speculations concerning it in biological processes. [5] Either D-idose or L-altrose is in the polysaccharide varianose. [6] Early literature refers to this as D-erythrose. [7] The $\cdot\frac{1}{2}H_2O$ and $\cdot 2H_2O$ forms also exist.

Part II. NATURAL MONOSACCHARIDES: AMINO SUGARS

	Substance (Synonym)	Chemical Formula	Melting Point °C	Specific Rotation $[\alpha]_D$
	(A)	(B)	(C)	(D)
	Aldosamines			
1	D-Ribose, 3-amino-3-deoxy-	$C_5H_{11}NO_4$	158–158.5 d.	−24.6 (hydrochloride)
2	D-Galactose, 2-amino-2-deoxy- (Galactosamine; Chondrosamine)	$C_6H_{13}NO_5$	185	+121 → +80 (hydrochloride)
3	α-L-Galactose, 2-amino-2,6-dideoxy- (L-Fucosamine)	$C_6H_{13}NO_4$	192–193 d.	−119 → −92 [27°] (hydrochloride)
4	α-D-Glucose, 2-amino-2-deoxy- (Glucosamine; Chitosamine)	$C_6H_{13}NO_5$	88	+100 → +47.5
5	β-D-Glucose, 2-amino-2-deoxy-	$C_6H_{13}NO_5$	110–111	+28 → +47.5
6	D-Glucose, 3-amino-3-deoxy- (Kanosamine)	$C_6H_{13}NO_5$	128 d.	+19 [14°]
7	D-Glucose, 6-amino-6-deoxy-	$C_6H_{13}NO_5$	161–162 d.	+23 → +50.1 (hydrochloride)
8	D-Glucose, 2,6-diamino-2,6-dideoxy- (Neosamine C)	$C_6H_{14}N_2O_4$	>230	+61.5 (dihydrochloride)
9	D-Glucose, 3,6-dideoxy-3-dimethyl-amino- (Mycaminose)	$C_8H_{17}NO_4$	115–116	+31 (hydrochloride)
10	D-Glucose, 4,6-dideoxy-4-dimethyl-amino-	$C_8H_{17}NO_4$	192–193	+45.5 (hydrochloride)
11	L-Glucose, 2-deoxy-2-methylamino-	$C_7H_{15}NO_5$	130–132	−64
12	D-Gulose, 2-amino-1,6-anhydro-2-deoxy-	$C_6H_{11}NO_4$	250–260 d.	+41 ± 2 (hydrochloride)
13	D-Gulose, 2-amino-2-deoxy-	$C_6H_{13}NO_5$	152–162 d.	+5.6 → −18.7 (hydrochloride)
14	Hexose, 3,4,6-trideoxy-3-dimethyl-amino-D-*xylo*- (Desosamine; Picrocine)	$C_8H_{17}NO_3$	189–191 d.	+49.5 (*c* 10) (hydrochloride)
15	Hexose, a 4-acetamido-2-amino-2,4,6-trideoxy-	$C_8H_{16}N_2O_4$	216–219	+115 → +94 [26°] (*c* 0.05)
16	Hexose, an amino-deoxy-3-*O*-carboxy-ethyl-	$C_9H_{17}NO_7$	....	....
17	Hexose, a 2,6-diamino-2,6-dideoxy- (Neosamine B; Paramose)	$C_6H_{14}N_2O_4$	135–150 d.	+17.5 (*c* 0.9 (hydrochloride)

Part II. NATURAL MONOSACCHARIDES: AMINO SUGARS (Continued)

	Substance (Synonym)	Chemical Formula	Melting Point °C	Specific Rotation $[\alpha]_D$
	(A)	(B)	(C)	(D)
	Aldosamines (Con't)			
18	Hexose, a 3-dimethylamino-2,3,6-trideoxy- (Rhodosamine)	$C_8H_{17}NO_3$	....	....
19	D-Mannose, 2-amino-2-deoxy- (Mannosamine)	$C_6H_{13}NO_5$	142 d.	−4.3 (*c* 9) (hydrochloride)
20	D-Mannose, 3-amino-3,6-dideoxy- (Mycosamine)	$C_6H_{13}NO_4$	162	−11.5 (hydrochloride)
21	D-Talose, 2-amino-2-deoxy- (Talosamine)	$C_6H_{13}NO_5$	151–153	+3.4 → −5.7 (*c* 0.9) (hydrochloride)
22	L-Talose, 2-amino-2,6-dideoxy- (Pneumosamine)	$C_6H_{13}NO_4$	162–163	+6.9 → +10.4 (hydrochloride)
	Ketosamines			
23	Pentulose, 1-(*o*-carboxyanilino)-1-deoxy-D-*erythro*-	$C_{12}H_{14}NO_6$	....	....
24	Hexulose, 1-(*o*-carboxyanilino)-1-deoxy-D-*arabino*-	$C_{13}H_{16}NO_7$	....	....
25	Hexulose, 5-amino-5-deoxy-L-*xylo*-	$C_6H_{13}NO_5$	174–176	−62
26	Hexulose, 6-deoxy-6-(*N*-methylacetamido)-L-*xylo*-	$C_9H_{17}NO_6$	....	....

Part III. NATURAL ALDITOLS AND INOSITOLS (with Inososes and Inosamines)

	Substance (Synonym)	Chemical Formula	Melting Point °C	Specific Rotation $[\alpha]_D$
	(A)	(B)	(C)	(D)
	Alditols			
1	Glycerol	$C_3H_8O_3$	20	None
2	Glycerol, 1-deoxy- (1,2-Propane-diol)[1]	$C_3H_8O_2$	Oil, b.p. 188–189	None (racemic)
3	Erythritol	$C_4H_{10}O_4$	118–120	None (meso)
4	Erythritol, 1,4-dideoxy- (2,3-Butylene-glycol)	$C_4H_{10}O_2$	25, 34	None (meso)
5	D-Threitol, 1,4-dideoxy-	$C_4H_{10}O_2$	19	−13.0
6	L-Threitol, 1,4-dideoxy-	$C_4H_{10}O_2$	....	+10.2
7	DL-Threitol, 1,4-dideoxy-	$C_4H_{10}O_2$	7.6	None (racemic)
8	D-Arabinitol	$C_5H_{12}O_5$	103	+7.82 (*c* 8, borax solution)
9	L-Arabinitol	$C_5H_{12}O_5$	101–102	−32 (*c* 0.4, 5% molybdate)
10	Ribitol (Adomitol)	$C_5H_{12}O_5$	102	None (meso)
11	Galactitol (Dulcitol)	$C_6H_{14}O_6$	186–188	None (meso)
12	D-Glucitol (Sorbitol)	$C_6H_{14}O_6$	112	−1.8 [15°]
13	D-Glucitol, 1,5-anhydro- (Polygalitol)	$C_6H_{12}O_5$	140–141	+42.4
14	L-Iditol	$C_6H_{14}O_6$	73.5	−3.5 (*c* 10)
15	D-Mannitol	$C_6H_{14}O_6$	166	−0.21
16	D-Mannitol, 1,5-anhydro- (Styracitol)	$C_6H_{12}O_5$	157	−49.9
17	Heptitol, D-*glycero*-D-*galacto*- (Heptitol, L-*glycero*-D-*manno*-; Perseitol)	$C_7H_{16}O_7$	183–185, 188	−1.1
18	Heptitol, D-*glycero*-D-*gluco*- (Heptitol, L-*glycero*-D-*talo*-; β-Sedoheptitol)	$C_7H_{16}O_7$	131–132	+46 (5% NH_4 molybdate)
19	Heptitol, D-*glycero*-D-*manno*- (Heptitol, D-*glycero*-D-*talo*-; Volemitol)	$C_7H_{16}O_7$	153	+2.65
20	Octitol, D-*erythro*-D-*galacto*-	$C_8H_{18}O_8 \cdot H_2O$	169–170	−11 (5% NH_4 molybdate)
	Inositols			
21	Betitol (a dideoxy inositol)	$C_6H_{12}O_4$	224	....
22	Bioinosose (*scyllo*-Inosose; *myo*-Inosose-2; a deoxy keto inositol)	$C_6H_{10}O_6$	198–200	None (meso)
23	*h*-Bornesitol (a *myo*-inositol monomethyl ether)	$C_7H_{14}O_6$	200	+31.6
24	*l*-Bornesitol (a *myo*-inositol monomethyl ether)	$C_7H_{14}O_6$	205–206	−32.1
25	Conduritol (a 2,3-dehydro-2,3-dideoxyinositol)	$C_6H_{10}O_4$	142–143	None (meso)
26	Cordycepic acid (a tetrahydroxycyclohexanecarboxylic acid)[2]	$C_7H_{12}O_6$	....	....
27	Dambonitol (a *myo*-inositol dimethyl ether)	$C_8H_{16}O_6$	206	None (meso)
28	DL-Inositol	$C_6H_{12}O_6$	253	None (racemic)

PROPERTIES OF CARBOHYDRATES (continued)

Part III. NATURAL ALDITOLS AND INOSITOLS (with Inososes and Inosamines) (Continued)

	Substance (Synonym)	Chemical Formula	Melting Point °C	Specific Rotation $[\alpha]_D$
	(A)	(B)	(C)	(D)
	Inositols (Con't)			
29	*d*-Inositol	$C_6H_{12}O_6$	....	+60
30	*l*-Inositol	$C_6H_{12}O_6$	240	−65
31	Laminitol (a *C*-methyl *myo*-inositol)	$C_7H_{14}O_6$	266–269	−3
32	Liriodendritol (a *myo*-inositol dimethyl ether)	$C_8H_{16}O_6$	224	−25
33	*muco*-Inositol monomethyl ether	$C_7H_{14}O_6$	322–325	....
34	*myo*-Inositol (*meso*-Inositol)	$C_6H_{12}O_6$	217–218	None (meso)
35	*d*-*myo*-Inosose-1 (a deoxy keto inositol)	$C_6H_{10}O_6$	138–139	+19.6
36	Mytilitol (a *C*-methyl *scyllo*-inositol)	$C_7H_{14}O_6$	259	None (meso)
37	*neo*-Inosamine-2 (a deoxy amino inositol)	$C_6H_{13}O_5N$	239–241 d.	None (meso)
38	*d*-Ononitol (a *myo*-inositol monomethyl ether)	$C_7H_{14}O_6$	172	+6.6
39	*h*-Pinitol (a *dextro*-inositol monomethyl ether)	$C_7H_{14}O_6$	186	+65.5
40	*l*-Pinitol (a *levo*-inositol monomethyl ether)	$C_7H_{14}O_6$	186	−65
41	*l*-Quebrachitol (a *levo*-inositol monomethyl ether)	$C_7H_{14}O_6$	190–191	−80.2 [28°]
42	*d*-Quercitol (a deoxy *dextro*-inositol)	$C_6H_{12}O_5$	235	+24.2
43	*d*-Quinic acid (a trideoxy carboxy *dextro*-inositol)	$C_7H_{12}O_6$	164	+44 (*c* 10)
44	*l*-Quinic acid (a trideoxy carboxy *levo*-inositol)	$C_7H_{12}O_6$	162	−42.1
45	Quinic acid, 5-dehydro-	$C_7H_{10}O_6$	140–142 (138 s.)	−82.4 [28°]
46	Scyllitol (*scyllo*-Inositol; Cocositol)	$C_6H_{12}O_6$	352–353	None (meso)
47	Sequoyitol (a *myo*-inositol monomethyl ether)	$C_7H_{14}O_6$	234–235	None (meso)
48	Shikimic acid (a 3,4-anhydro-quinic acid)	$C_7H_{10}O_5$	183–184	−200 [16°]
49	Shikimic acid, 5-dehydro-	$C_7H_8O_5$	150–152	−57.5 [28°] (EtOH)
50	Streptamine (2,4-diaminodideoxy-scyllitol)	$C_6H_{14}O_4N_2$	88, 210–250 d.	None (meso)
51	Streptamine, 2-deoxy-	$C_6H_{14}O_3N_2$	....	None (meso)
52	Streptadine (1,3-Dideoxy-1,3-diguanidino-scyllitol)	$C_8H_{18}N_6O_4$	....	None (meso)
53	Viburnitol (a deoxy *levo*-inositol)[3]	$C_6H_{12}O_5$	174	−73.9

[1] The 1-phosphate ester of this diol is said to occur in brain tissue and sea-urchin eggs. [2] Strong evidence that cordycepic acid is really D-mannitol. [3] Not an enantiomorph of *d*-quercitol; other isomeric relationship is involved.

Part IV. NATURAL ALDONIC, URONIC, AND ALDARIC ACIDS

	Substance (Synonym)	Chemical Formula	Melting Point °C	Specific Rotation $[\alpha]_D$
	(A)	(B)	(C)	(D)
	Aldonic Acids			
1	D-Glyceric acid	$C_3H_6O_4$	Gum	Dextro
2	L-Glyceric acid	$C_3H_6O_4$	Gum	Levo
3	D-Arabinonic acid	$C_5H_{10}O_6$	114–116	+10.5 (*c* 6)
4	L-Arabinonic acid	$C_5H_{10}O_6$	118–119	−9.6 → −41.7[1]
5	L-Arabinonic-1,4-lactone	$C_5H_8O_5$	97–99	−72
6	D-Ribonic acid	$C_5H_{10}O_6$	112–113	−17.0
7	D-Xylonic acid	$C_5H_{10}O_6$	....	−2.9 → +20.1[1]
8	L-Xylonic acid	$C_5H_{10}O_6$	....	−91.8[1]
9	D-Altronic acid	$C_6H_{12}O_7$	....	+11.5 → +24.8[1] (Ca salt, *N* HCl)
10	D-Galactonic acid	$C_6H_{12}O_7$	122	−11.2 → +57.6[1]
11	D-Gluconic acid	$C_6H_{12}O_7$	130–132 (110–112 s.)	−6.7 → +11.9[1]
12	L-Gulonic acid	$C_6H_{12}O_7$	Exists only in soln.	[ca. 0°]
13	Hexsonic acid, 2-deoxy-D-*arabino*-	$C_6H_{12}O_6$	93–95	+68 (lactone)
14	2-Hexulosonic acid, D-*arabino*-	$C_6H_{10}O_7$	....	−81.7 (Na salt)
15	2-Hexulosonic acid, 3-deoxy-D-*erythro*-	$C_6H_{10}O_6$	....	−29.2 (*c* 6, Ca salt)
16	2-Hexulosonic acid, D-*lyxo*-	$C_6H_{10}O_7$	169	−5
17	5-Hexulosonic acid, D-*arabino*-	$C_6H_{10}O_7$	108–109	....
18	5-Hexulosonic acid, D-*xylo*-	$C_6H_{10}O_7$	....	−14.5
19	D-Mannonic acid	$C_6H_{12}O_7$	....	−15.6
20	D-Gluconic acid, *O*-β-D-galactopyranosyl- (1 → 4)- (Lactobionic acid)	$C_{12}H_{22}O_{12}$	....	+25.1 (Ca salt)

Part IV. NATURAL ALDONIC, URONIC, AND ALDARIC ACIDS (Continued)

	Substance (Synonym)	Chemical Formula	Melting Point °C	Specific Rotation $[\alpha]_D$
	(A)	(B)	(C)	(D)
	Uronic Acids			
21	L-Lyxuronic acid	$C_5H_8O_6$	....	....
22	β-D-Galacturonic acid	$C_6H_{10}O_7$	160	+27 → +55.6
23	α-D-Galacturonic acid·monohydrate	$C_6H_{12}O_8$	159–160 (110–115 s.)	+97.9 → +50.9
24	D-Galacturonic acid, 2-amino-2-deoxy-	$C_6H_{11}O_6N$	160 d.	+84.5 (pH 2 HCl)
25	β-D-Glucuronic acid	$C_6H_{10}O_7$	156	+11.7 → +36.3
26	D-Glucuronic acid, 2-amino-2-deoxy-	$C_6H_{11}O_6N$	120–172 d.	+55
27	D-Glucuronic acid, 3-*O*-methyl-	$C_7H_{12}O_7$	Syrup	+6
28	L-Guluronic acid	$C_6H_{10}O_7$	....	....
29	L-Iduronic acid	$C_6H_{10}O_7$	....	+30
30	β-D-Mannuronic acid	$C_6H_{10}O_7$	165–167	−47.9 → −23.9
31	α-D-Mannuronic acid·monohydrate	$C_6H_{12}O_8$	110 s., 120–130 d.	+16 → −6.1 (*c* 6.8)
	Aldaric Acids			
32	D-Tartaric acid	$C_4H_6O_6$	170	−15
33	L-Tartaric acid	$C_4H_6O_6$	170	+15 [15°]
34	L-Malic acid	$C_4H_6O_5$	100	−2.3 (*c* 8.4)

[1] Equilibrates with the lactone.

PROPERTIES OF SELECTED FATTY ACIDS

This table gives the systematic names and selected properties of the more important fatty acids of five or more carbon atoms. Compounds are listed first by degree of saturation and, secondly, by number of carbon atoms.

M_r: Molecular weight
T_m: Melting point in °C
S: Aqueous solubility at 20°C in units of grams of solute per 100 grams of water

REFERENCES

1. R. C. M. Dawson, et al., *Data for Biochemical Research*, 3rd ed., Clarendon Press, Oxford, 1986.
2. G. D. Fasman, Ed., *Practical Handbook of Biochemistry and Molecular Biology*, CRC Press, Boca Raton, FL, 1989.

Common name	Systematic name	Mol. form.	M_r	T_m/°C	S
		Saturated			
Isovaleric acid	3-Methylbutanoic acid	$C_5H_{10}O_2$	102.13	–29	4.0
Valeric acid	Pentanoic acid	$C_5H_{10}O_2$	102.13	–34	3.2
Caproic acid	Hexanoic acid	$C_6H_{12}O_2$	116.16	–3	0.97
Enanthic acid	Heptanoic acid	$C_7H_{14}O_2$	130.19	–8	0.24
Caprylic acid	Octanoic acid	$C_8H_{16}O_2$	144.21	16	0.068
Pelargonic acid	Nonanoic acid	$C_9H_{18}O_2$	158.24	12	0.026
Capric acid	Decanoic acid	$C_{10}H_{20}O_2$	172.27	32	0.015
Lauric acid	Dodecanoic acid	$C_{12}H_{24}O_2$	200.32	43	0.0055
Tridecylic acid	Tridecanoic acid	$C_{13}H_{26}O_2$	214.35	42	0.0033
Myristic acid	Tetradecanoic acid	$C_{14}H_{28}O_2$	228.38	54	0.0020
Pentadecylic acid	Pentadecanoic acid	$C_{15}H_{30}O_2$	242.40	52	0.0012
Palmitic acid	Hexadecanoic acid	$C_{16}H_{32}O_2$	256.43	62	0.00072
Margaric acid	Heptadecanoic acid	$C_{17}H_{34}O_2$	270.46	61	0.00042
Stearic acid	Octadecanoic acid	$C_{18}H_{36}O_2$	284.48	69	0.00029
Arachidic acid	Eicosanoic acid	$C_{20}H_{40}O_2$	312.54	75	
Phytanic acid	3,7,11,15-Tetramethyl-hexadecanoic acid	$C_{20}H_{40}O_2$	312.54	–65	
Behenic acid	Docosanoic acid	$C_{22}H_{40}O_2$	340.59	81	
Lignoceric acid	Tetracosanoic acid	$C_{24}H_{48}O_2$	368.64	84	
Cerotic acid	Hexacosanoic acid	$C_{26}H_{52}O_2$	396.70	89	
Montanic acid	Octacosanoic acid	$C_{28}H_{56}O_2$	424.75	91	
		Monounsaturated			
Caproleic acid	9-Decenoic acid	$C_{10}H_{18}O_2$	170.25		
Palmitoleic acid	*cis*-9-Hexadecenoic acid	$C_{16}H_{30}O_2$	254.41	0	
Oleic acid	*cis*-9-Octadecanoic acid	$C_{18}H_{34}O_2$	282.47	13	
Vaccenic acid	*trans*-11-Octadecenoic acid	$C_{18}H_{34}O_2$	282.47	7	
Elaidic acid	*trans*-9-Octadecanoic acid	$C_{18}H_{34}O_2$	282.47	44	
Brassidic acid	*trans*-13-Docosenoic acid	$C_{22}H_{42}O_2$	338.57	60	
Erucic acid	*cis*-13-Docosenoic acid	$C_{22}H_{42}O_2$	338.57	34	
Nervonic acid	*cis*-15-Tetracosenoic acid	$C_{24}H_{46}O_2$	366.63	39	
		Diunsaturated			
Linoleic acid	*cis,cis*-9,12-Octadeca-dienoic acid	$C_{18}H_{32}O_2$	280.45	–9	
		Triunsaturated			
Eleosteric acid	*cis,trans,trans*-9,11,13-Octa-decatrienoic acid	$C_{18}H_{30}O_2$	278.44	49	
Linolenic acid	*cis,cis,cis*-9,12,15-Octa-decatrienoic acid	$C_{18}H_{30}O_2$	278.44	–17	
		Tetraunsaturated			
Arachidonic acid	5,8,11,14-Eicosatetraenoic acid	$C_{20}H_{32}O_2$	304.47	–50	

FATS AND OILS

These data for fats and oils were compiled originally for the *Biology Data Book* by H. J. Harwood, and R. P. Geyer, 1964. Data are reproduced here by permission of the copyright owners of the above publication, the Federation of American Societies for Experimental Biology, Washington, D.C. pp. 380—82.

Values are typical rather than average, and frequently were derived from specific analyses for particular samples (especially the constituent fatty acids). Extreme variations may occur, depending on a number of variables such as source, treatment, and age of a fat or oil. **Specific Gravity** (column D) was calculated at the specified temperature (degrees centigrade) and referred to water at the same temperature, unless otherwise specified. **Density,** shown in parentheses (column D), was measured at the specified temperature (degrees centigrade). **Refractive Index** (column E) was measured at 50°C, unless otherwise specified.

		Constants					Constituent fatty acids (g/100 g total fatty acids)										
							Saturated						Unsaturated				
Fat or oil	Source	Melting (or solidification) Point (°C)	Specific gravity (or density)	Refractive index $\left(n_D^{50°C}\right)$	Iodine value	Saponification value	Lauric	Myristic	Palmitic	Stearic	Arachidic	Other	Palmitoleic	Oleic	Linoleic	Linolenic	Other
(A)	(B)	(C)	(D)	(E)	(F)	(G)	(H)	(I)	(J)	(K)	(L)	(M)	(N)	(O)	(P)	(Q)	(R)
Land animals																	
Butterfat	*Bos taurus*	32.2	$0.911^{40°/15°}$	1.4548	36.1	227	2.5	11.1	29.0	9.2	2.4	2.0;[a] 0.5;[b] 2.3[c]	4.6	26.7	3.6	—	3.6;[d] 0.1;[e] 0.1;[f] 0.9;[g] 1.4;[h] 1.0;[i] 1.0;[j] 0.4[k]
Depot fat	*Homo sapiens*	(15)	$0.918^{15°}$	1.4602	67.6	196.2	—	2.7	24.0	8.4	—	—	5	46.9	10.2	—	2.5[h]
Lard oil	*Sus scrofa*	(30.5)	$0.919^{15°}$	1.4615	58.6	194.6	—	1.3	28.3	11.9	—	—	2.7	47.5	6	—	0.2;[g] 2.1[h]
Neat's-foot oil	*B. taurus*	—	$0.910^{25°}$	$1.464^{25°}$	69—76	190—199	—	—	17—18	2—3	—	—	—	74—76	—	—	—
Tallow, beef	*B. taurus*	—	—	—	49.5	197	—	6.3	27.4	14.1	—	—	—	49.6	2.5	—	—
Tallow, mutton	*Ovis aries*	(42.0)	$0.945^{15°}$	1.4565	40	194	—	4.6	24.6	30.5	—	—	—	36.0	4.3	—	—
Marine animals																	
Cod-liver oil	*Gadus morhua*	—	0.925^{25}	$1.481^{25°}$	165	186	—	5.8	8.4	0.6	—	—	20.0	← 29.1 →		—	25.4;[l] 9.6[m]
Herring oil	*Clupea harengus*	—	$0.900^{60°}$	$1.4610^{60°}$	140	192	—	7.3	13.0	Trace	—	—	4.9	—	—	20.7	30.1;[l] 23.2[m]
Menhaden oil	*Brevoortia tyrannus*	—	$0.903^{60°}$	$1.4645^{60°}$	170	191	—	5.9	16.3	0.6	0.6	—	15.5	—	—	29.6	19.0;[l] 11.7[m] 0.8[n]
Sardine oil	*Sardinops caerulea*	—	$0.905^{60°}$	$1.4660^{60°}$	185	191	—	5.1	14.6	3.2	—	—	11.8	← 17.8 →		—	18.1;[l] 11.0;[m] trace;[g] 15.4[o]
Sperm oil(body)	*Physeter macrocephalus*	—	—	—	76—88	122—130	1	5	6.5	—	—	—	26.5	37	19	—	1;[m] 4;[g] 19[p]
Sperm oil(head)	*P. macrocephalus*	—	—	—	70	140—144	16	14	8	2	—	3.5[c]	15	17	6.5	—	4;[f] 14;[g] 6.5[p]
Whale oil	*Balaena mysticetus*	—	0.892^{60}	$1.460^{60°}$	120	195	0.2	9.3	15.6	2.8	—	—	14.4	35.2	—	—	13.6;[l] 5.9;[m] 2.5;[g] 0.2[q]
Plants																	
Babassu oil	*Attalea funifera*	22—26	$(0.893^{60°})$	$1.443^{60°}$	15.5	247	44.1	15.4	8.5	2.7	0.2	0.2;[a] 4.8;[b] 6.6[c]	—	16.1	1.4	—	—
Castor oil	*Ricinus communis*	(−18.0)	$0.961^{15°}$	1.4770	85.5	180.3	←		2.4	→	—	—	—	7.4	3.1	—	87[r]
Cocoa butter	*Theobroma cacao*	34.1	$0.964^{15°}$	1.4568	36.5	193.8	—	—	24.4	35.4	—	—	—	38.1	2.1	—	—
Coconut oil	*Cocos nucifera*	25.1	$0.924^{15°}$	1.4493	10.4	268	45.4	18.0	10.5	2.3	0.4[19]	0.8;[a] 5.4;[b] 8.4[c]	0.4	7.5	Trace	—	—
Corn oil	*Zea mays*	(−20.0)	$0.922^{15°}$	1.4734	122.6	192.0	—	1.4	10.2	3.0	—	—	1.5	49.6	34.3	—	—
Cotton seed oil	*Gossypium hirsutum*	(−1.0)	$0.917^{25°}$	1.4735	105.7	194.3	—	1.4	23.4	1.1	1.3	—	2.0	22.9	47.8	—	—
Linseed oil	*Linum usitatissimum*	(−24.0)	$0.938^{15°}$	$1.4782^{25°}$	178.7	190.3	—	—	6.3	2.5	0.5	—	—	19.0	24.1	47.4	0.2[n]
Mustard oil	*Brassica hirta*	—	$0.9145^{15°}$	1.475	102	174	—	1.3[20]	—	—	—	—	—	27.2[t]	16.6[t]	401.8[t]	1.1;[n] 1.0;[u] 51.0[v]
Neem oil	*Melia azadirachta*	−3	$0.017^{15°}$	1.4615	71	194.5	—	2.6[20]	14.1[20]	24.0[20]	0.8[20]	—	—	58.5[t]	—	—	—
Niger-seed oil	*Guizotia abyssinica*	—	$0.925^{15°}$	1.471	128.5	190	—	3.3[20]	8.2[20]	4.8[20]	0.5[20]	—	—	30.3[t]	57.3[t]	—	—
Oiticica oil	*Licania rigida*	—	$0.974^{25°}$	—	140—180	—	←		11.3[23]		→	—	—	6.2	—	—	82.5[x]
Olive oil	*Olea europaea sativa*	(−6.0)	$0.918^{15°}$	1.4679	81.1	189.7	—	Trace	6.9	2.3	0.1	—	—	84.4	4.6	—	—
Palm oil	*Elaeis guineensis*	35.0	$0.915^{15°}$	1.4578	54.2	199.1	—	1.4	40.1	5.5	—	—	—	42.7	10.3	—	—
Palm-kernel oil	*E. guineensis*	24.1	$0.923^{15°}$	1.4569	37.0	219.9	46.9	14.1	8.8	1.3	—	2.7;[b] 7.0[c]	—	18.5	0.7	—	—
Peanut oil	*Arachis hypogaea*	(3.0)	$0.914^{15°}$	1.4691	93.4	192.1	—	—	8.3	3.1	2.4	—	—	56.0	26.0	—	3.1;[n] 1.1[u]
Perilla oil	*Perilla frutescens*	—	$(0.935^{15°})$	$1.481^{25°}$	195	192	←		9.6[23]		→	—	—	17.8	—	17.5	—
Poppy-seed oil	*Papaver somniferum*	(−15)	$0.925^{15°}$	1.4685	135	194	—	—	4.8[20]	2.9[20]	—	—	—	30.1[t]	62.2[t]	—	—
Rapeseed oil	*Brassica campestris*	(−10)	$0.915^{15°}$	1.4706	98.6	174.7	—	—	1	—	—	—	—	32	15	1	50[v]
Safflower oil	*Carthamus tinctorius*	—	$(0.900^{60°})$	$1.462^{60°}$	145	192	←		6.8[23]		→	—	—	18.6	70.1	3.4	—
Sesame oil	*Sesamum indicum*	(−6.0)	$0.919^{25°}$	1.4646	106.6	187.9	—	—	9.1	4.3	0.8	—	—	45.4	40.4	—	—
Soybean oil	*Glycine soja*	(−16.0)	$0.927^{15°}$	1.4729	130.0	190.6	0.2	0.1	9.8	2.4	0.9	—	0.4	28.9	50.7	6.5	0.1[g]
Sunflower-seed oil	*Helianthus annuus*	(−17.0)	$0.923^{15°}$	1.4694	125.5	188.7	—	—	5.6	2.2	0.9	—	—	25.1	66.2	—	—
Tung oil	*Aleurites fordi*	(−2.5)	$0.934^{15°}$	$1.5174^{25°}$	168.2	193.1	←		4.6[23]		→	—	—	4.1	0.6	—	90.7[y]
Wheat-germ oil	*Triticum aestivum*	—	—	—	125	—	←		16.0[23]			—	—	28.1	52.3	3.6	—

[a] Caproic
[b] Capryli
[c] Capric
[d] Butyric
[e] Decenoic
[f] C_{12} monoethenoic
[g] C_{14} monoethenoic
[h] Gadoleic plus erucic
[i] C_{12} n-pentadecanoic.
[j] C_{17} margaric
[k] 12-Methyl tetradecanoic
[l] C_{20} polyethenoic
[m] C_{22} polyethenoic
[n] Behenic
[o] C_{14} polyethenoic
[p] Gadoleic
[q] C_{24} polyethenoic
[r] Ricinoleic
[s] Includes behenic and lignoceric
[t] Percent by weight
[u] Lignoceric
[v] Erucic
[w] Includes behenic
[x] Licanic.
[y] Eleostearic

BIOLOGICAL BUFFERS

This table of frequently used buffers gives the pK_a value at 25°C and the useful pH range of each buffer. The buffers are listed in order of increasing pH.

The table is reprinted with permission of Sigma Chemical Company, St. Louis, Mo.

Acronym	Name	Mol. wt.	pK_a	Useful pH range
MES	2-(*N*-Morpholino)ethanesulfonic acid	195.2	6.1	5.5—6.7
BIS TRIS	*Bis*(2-hydroxyethyl)iminotris(hydroxymethyl)methane	209.2	6.5	5.8—7.2
ADA	*N*-(2-Acetamido)-2-iminodiacetic acid	190.2	6.6	6.0—7.2
ACES	2-[(2-Amino-2-oxoethyl)amino]ethanesulfonic acid	182.2	6.8	6.1—7.5
PIPES	Piperazine-*N,N′-bis*(2-ethanesulfonic acid)	302.4	6.8	6.1—7.5
MOPSO	3-(*N*-Morpholino)-2-hydroxypropanesulfonic acid	225.3	6.9	6.2—7.6
BIS TRIS PROPANE	1,3-*Bis*[*tris*(hydroxymethyl)methylamino]propane	282.3	6.8[a]	6.3—9.5
BES	*N,N-Bis*(2-hydroxyethyl)-2-aminoethanesulfonic acid	213.2	7.1	6.4—7.8
MOPS	3-(*N*-Morpholino)propanesulfonic acid	209.3	7.2	6.5—7.9
HEPES	*N*-(2-Hydroxyethyl)piperazine-*N′*-(2-ethanesulfonic acid)	238.3	7.5	6.8—8.2
TES	*N-Tris*(hydroxymethyl)methyl-2-aminoethanesulfonic acid	229.2	7.5	6.8—8.2
DIPSO	3-[*N,N-Bis*(2-hydroxyethyl)amino]-2-hydroxypropanesulfonic acid	243.3	7.6	7.0—8.2
TAPSO	3-[*N-Tris*(hydroxymethyl)methylamino)-2-hydroxypropanesulfonic acid	259.3	7.6	7.0—8.2
TRIZMA	Tris(hydroxymethyl)aminomethane	121.1	8.1	7.0—9.1
HEPPSO	*N*-(2-hydroxyethyl)piperazine-*N′*-(2-hydroxypropanesulfonic acid)	268.3	7.8	7.1—8.5
POPSO	Piperazine-*N,N′-bis*(2-hydroxypropanesulfonic acid)	362.4	7.8	7.2—8.5
EPPS	*N*-(2-Hydroxyethyl)piperazine-*N′*-(3-propanesulfonic acid)	252.3	8.0	7.3—8.7
TEA	Triethanolamine	149.2	7.8	7.3—8.3
TRICINE	*N-Tris*(hydroxymethyl)methylglycine	179.2	8.1	7.4—8.8
BICINE	*N,N-Bis*(2-hydroxyethyl)glycine	163.2	8.3	7.6—9.0
TAPS	*N-Tris*(hydroxymethyl)methyl-3-aminopropanesulfonic acid	243.3	8.4	7.7—9.1
AMPSO	3-[(1,1-Dimethyl-2-hydroxyethyl)amino]-2-hydroxypropanesulfonic acid	227.3	9.0	8.3—9.7
CHES	2-(*N*-Cyclohexylamino)ethanesulfonic acid	207.3	9.3	8.6—10.0
CAPSO	3-(Cyclohexylamino)-2-hydroxy-1-propanesulfonic acid	237.3	9.6	8.9—10.3
AMP	2-Amino-2-methyl-1-propanol	89.1	9.7	9.0—10.5
CAPS	3-(Cyclohexylamino)-1-propanesulfonic acid	221.3	10.4	9.7—11.1

[a] pK_a = 9.0 for the second dissociation stage.

APPROXIMATE pH VALUES OF BIOLOGICAL MATERIALS AND FOODS

The following table give approximate pH values for a number of substances of biological importance. All values are rounded off to the nearest tenth and are based on measurements made at 25° C.

Biological Materials

Material	pH
Blood, plasma, human	7.3–7.5
Spinal fluid, human	7.3–7.5
Blood, whole, dog	6.9–7.2
Saliva, human	6.5–7.5
Gastric contents, human	1.0–3.0
Duodenal contents, human	4.8–8.2
Feces, human	4.6–8.4
Urine, human	4.8–8.4
Milk, human	6.6–7.6
Bile, human	6.8–7.0

Foods

Food	pH
Apples	2.9–3.3
Apricots	3.6–4.0
Asparagus	5.4–5.8
Bananas	4.5–4.7
Beans	5.0–6.0
Beers	4.0–5.0
Beets	4.9–5.5
Blackberries	3.2–3.6
Bread, white	5.0–6.0
Butter	6.1–6.4
Cabbage	5.2–5.4
Carrots	4.9–5.3
Cheese	4.8–6.4
Cherries	3.2–4.0
Cider	2.9–3.3
Corn	6.0–6.5
Crackers	6.5–8.5
Dates	6.2–6.4
Eggs, fresh white	7.6–8.0
Flour, wheat	5.5–6.5
Gooseberries	2.8–3.0
Grapefruit	3.0–3.3
Grapes	3.5–4.5
Hominy (lye)	6.8–8.0
Jams, fruit	3.5–4.0
Jellies, fruit	2.8–3.4
Lemons	2.2–2.4
Limes	1.8–2.0
Maple syrup	6.5–7.0
Milk, cows	6.3–6.6
Olives	3.6–3.8
Oranges	3.0–4.0
Oysters	6.1–6.6
Peaches	3.4–3.6
Pears	3.6–4.0
Peas	5.8–6.4
Pickles, dill	3.2–3.6
Pickles, sour	3.0–3.4
Pimento	4.6–5.2
Plums	2.8–3.0
Potatoes	5.6–6.0
Pumpkin	4.8–5.2
Raspberries	3.2–3.6
Rhubarb	3.1–3.2
Salmon	6.1–6.3
Sauerkraut	3.4–3.6
Shrimp	6.8–7.0
Soft drinks	2.0–4.0
Spinach	5.1–5.7
Squash	5.0–5.4
Strawberries	3.0–3.5
Sweet potatoes	5.3–5.6
Tomatoes	4.0–4.4
Tuna	5.9–6.1
Turnips	5.2–5.6
Vinegar	2.4–3.4
Water, drinking	6.5–8.0
Wines	2.8–3.8

RECOMMENDED DAILY DIETARY ALLOWANCES

Designed for the Maintenance of Good Nutrition of Practically all Healthy People in the United States[a]

1989 Recommendations of the Food and Nutrition Board, National Academy of Sciences

Category	Age (years) or Condition	Weight[b] (kg)	Weight[b] (lb)	Height[b] (cm)	Height[b] (in)	Protein (g)	Fat-Soluble Vitamins: Vitamin A (μg RE)[c]	Vitamin D (μg)[d]	Vitamin E (mg α-TE)[e]	Vitamin K (μg)	Water-Soluble Vitamins: Vitamin C (mg)	Thiamin (mg)	Riboflavin (mg)	Niacin (mg NE)[f]	Vitamin B6 (mg)	Folate (μg)	Vitamin B12 (μg)	Minerals: Calcium (mg)	Phosphorus (mg)	Magnesium (mg)	Iron (mg)	Zinc (mg)	Iodine (μg)	Selenium (μg)
Infants	0.0–0.5	6	13	60	24	13	375	7.5	3	5	30	0.3	0.4	5	0.3	25	0.3	400	300	40	6	5	40	10
	0.5–1.0	9	20	71	28	14	375	10	4	10	35	0.4	0.5	6	0.6	35	0.5	600	500	60	10	5	50	15
Children	1–3	13	29	90	35	16	400	10	6	15	40	0.7	0.8	9	1.0	50	0.7	800	800	80	10	10	70	20
	4–6	20	44	112	44	24	500	10	7	20	45	0.9	1.1	12	1.1	75	1.0	800	800	120	10	10	90	20
	7–10	28	62	132	52	28	700	10	7	30	45	1.0	1.2	13	1.4	100	1.4	800	800	170	10	10	120	30
Males	11–14	45	99	157	62	45	1,000	10	10	45	50	1.3	1.5	17	1.7	150	2.0	1,200	1,200	270	12	15	150	40
	15–18	66	145	176	69	59	1,000	10	10	65	60	1.5	1.8	20	2.0	200	2.0	1,200	1,200	400	12	15	150	50
	19–24	72	160	177	70	58	1,000	10	10	70	60	1.5	1.7	19	2.0	200	2.0	1,200	1,200	350	10	15	150	70
	25–50	79	174	176	70	63	1,000	5	10	80	60	1.5	1.7	19	2.0	200	2.0	800	800	350	10	15	150	70
	51+	77	170	173	68	63	1,000	5	10	80	60	1.2	1.4	15	2.0	200	2.0	800	800	350	10	15	150	70
Females	11–14	46	101	157	62	46	800	10	8	45	50	1.1	1.3	15	1.4	150	2.0	1,200	1,200	280	15	12	150	45
	15–18	55	120	163	64	44	800	10	8	55	60	1.1	1.3	15	1.5	180	2.0	1,200	1,200	300	15	12	150	50
	19–24	58	128	164	65	46	800	10	8	60	60	1.1	1.3	15	1.6	180	2.0	1,200	1,200	280	15	12	150	55
	25–50	63	138	163	64	50	800	5	8	65	60	1.1	1.3	15	1.6	180	2.0	800	800	280	15	12	150	55
	51+	65	143	160	63	50	800	5	8	65	60	1.0	1.2	13	1.6	180	2.0	800	800	280	10	12	150	55
Pregnant						60	800	10	10	65	70	1.5	1.6	17	2.2	400	2.2	1,200	1,200	320	30	15	175	65
Lactating	1st 6 months					65	1,300	10	12	65	95	1.6	1.8	20	2.1	280	2.6	1,200	1,200	355	15	19	200	75
	2nd 6 months					62	1,200	10	11	65	90	1.6	1.7	20	2.1	260	2.6	1,200	1,200	340	15	16	200	75

[a] The allowances, expressed as average daily intakes over time, are intended to provide for individual variations among most normal persons as they live in the United States under usual environmental stresses. Diets should be based on a variety of common foods in order to provide other nutrients for which human requirements have been less well defined.

[b] Weights and heights of Reference Adults are actual medians for the U.S. population of the designated age. The use of these figures does not imply that the height-to-weight ratios are ideal.

[c] Retinol equivalents. 1 retinol equivalent = 1μg retinol or 6 μg β-carotene.

[d] As cholecalciferol. 10 μg cholecalciferol = 400 IU of vitamin D.

[e] α-Tocopherol equivalents. 1 mg d-α tocopherol = 1 α-TE.

[f] 1 NE (niacin equivalent) is equal to 1 mg of niacin or 60 mg of dietary tryptophan.

RECOMMENDED DAILY DIETARY ALLOWANCES (continued)

Estimated Safe and Adequate Daily Dietary Intakes of Selected Vitamins and Minerals[a]

		Vitamins	
Category	Age (years)	Biotin (μg)	Pantothenic Acid (mg)
Infants	0–0.5	10	2
	0.5–1	15	3
Children and adolescents	1–3	20	3
	4–6	25	3–4
	7–10	30	4–5
	11+	30–100	4–7
Adults		30–100	4–7

		Trace Elements[b]				
Category	Age (years)	Copper (mg)	Manganese (mg)	Fluoride (mg)	Chromium (μg)	Molybdenum (μg)
Infants	0–0.5	0.4–0.6	0.3–0.6	0.1–0.5	10–40	15–30
	0.5–1	0.6–0.7	0.6–1.0	0.2–1.0	20–60	20–40
Children and adolescents	1–3	0.7–1.0	1.0–1.5	0.5–1.5	20–80	25–50
	4–6	1.0–1.5	1.5–2.0	1.0–2.5	30–120	30–75
	7–10	1.0–2.0	2.0–3.0	1.5–2.5	50–200	50–150
	11+	1.5–2.5	2.0–5.0	1.5–2.5	50–200	75–250
Adults		1.5–3.0	2.0–5.0	1.5–4.0	50–200	75–250

[a] Because there is less information on which to base allowances, these figures are not given in the main table of RDA and are provided here in the form of ranges of recommended intakes.

[b] Since the toxic levels for many trace elements may be only several times usual intakes, the upper levels for the trace elements given in this table should not be habitually exceeded.

Section 8
Analytical Chemistry

PREPARATION OF REAGENTS

The following pages present directions for the preparation of various reagents. The collection has been prepared with the active collaboration of W. D. Bonner, R. K. Carleton, L. L. Carrick, Giles B. Cooke, E. J. Cragoe, Thos. De Vries, James L. Kassner, Thos. W. Mason, F. C. Mathers, M. G. Mellon, W. C. Pierce, J. H. Reedy, Arthur A. Vernon and S. R. Wood. Many others have contributed valuable suggestions.

Volumes have been stated in milliliters (ml) and liters (l). One milliliter is equivalent to one cubic centimeter (cm^3 or cc.). Masses are indicated in grams (g).

The relation to molar solution *(M)* or normal solution *(N)* is indicated in many cases.

Distilled water should be used.

LABORATORY REAGENTS FOR GENERAL USE

DILUTE ACIDS, 3 molar. Use the amount of concentrated acid indicated and dilute to one liter.

Acetic acid, 3 *N*. Use 172 ml of 17.4 *M* acid (99-100%).

Hydrochloric acid, 3 *N*. Use 258 ml of 11.6 *M* acid (36% HCl).

Nitric acid, 3 *N*. Use 195 ml of 15.4 *M* acid (69% HNO_3).

Phosphoric acid, 9 *N*. Use 205 ml of 14.6 *M* acid (85% H_3PO_4).

Sulfuric acid, 6 *N*. Use 168 ml of 17.8 *M* acid (95% H_2SO_4).

DILUTE BASES.

Ammonium hydroxide, 3 *M*, 3 *N*. Dilute 200 ml of concentrated solution (14.8 *M*, 28% NH_3) to 1 liter.

Barium hydroxide, 0.2 *M*, 0.4 *N*. Saturated solution, 63 g per liter of $Ba(OH)_2 \cdot 8H_2O$. Use some excess, filter off $BaCO_3$ and protect from CO_2 of the air with soda lime or ascarite in a guard tube.

Calcium hydroxide, 0.02 *M*, 0.04 *N*. Saturated solution, 1.5 g per liter of $Ca(OH)_2$. Use some excess, filter off $CaCO_3$ and protect from CO_2 of the air.

Potassium hydroxide, 3 *M*, 3 *N*. Dissolve 176 g of the sticks (95%) in water and dilute to 1 liter.

Sodium hydroxide, 3 *M*, 3 *N*. Dissolve 126 g of the sticks (95%) in water and dilute to 1 liter.

GENERAL REAGENTS (See also Decinormal Solutions of Salts and Other Reagents.)

Aluminum chloride, 0.167 *M*, 0.5 *N*. Dissolve 22 g of $AlCl_3$ in 1 liter of water.

Aluminum nitrate, 0.167 *M*, 0.5 *N*. Dissolve 58 g of $Al(NO_3)_3 \cdot 7.5H_2O$ in 1 liter of water.

Aluminum sulfate, 0.083 *M*, 0.5 *N*. Dissolve 56 g of $Al_2(SO_4)_3 \cdot 18H_2O$ in 1 liter of water.

Ammonium acetate, 3 *M*, 3 *N*. Dissolve 230 g of $NH_4C_2H_3O_2$ in water and dilute to 1 liter.

Ammonium carbonate, 1.5 *M* Dissolve 144 g of the commercial salt (mixture of $(NH_4)_2CO_3 \cdot H_2O$ and $NH_4CO_2NH_2$) in 500 ml of 3 *N* NH_4OH and dilute to 1 liter.

Ammonium chloride, 3 *M*, 3 *N*. Dissolve 160 g of NH_4Cl in water. Dilute to 1 liter.

Ammonium molybdate.

1. 0.5 *M*, 1 *N*. Mix well 72 g of pure MoO_3 (or 81 g of H_2MoO_4) with 200 ml of water, and add 60 ml of conc. ammonium hydroxide. When solution is complete, filter and pour filtrate, very slowly **and with rapid stirring**, into a mixture of 270 ml of conc. HNO_3 and 400 ml of water. Allow to stand over night, filter and dilute to 1 liter.

2. The reagent is prepared as two solutions which are mixed as needed, thus always providing fresh reagent of proper strength and composition. Since ammonium molybdate is an expensive reagent, and since an acid solution of this reagent as usually prepared keeps for only a few days, the method proposed will avoid loss of reagent and provide more certain results for quantitative work.

Solution 1. Dissolve 100 g of ammonium molybdate (C.P. grade) in 400 ml of water and 80 ml of 15 *M* NH_4OH. Filter if necessary, though this seldom has to be done.

Solution 2. Mix 400 ml of 16 *M* nitric acid with 600 ml of water.

For use, mix the calculated amount of solution 1 with twice its volume of solution 2, adding solution 1 to solution 2 slowly with vigorous stirring. Thus, for amounts of phosphorus up to 20 mg, 10 ml of solution 1 to 20 ml of solution 2 is adequate. Increase amount as needed.

Ammonium nitrate, 1 *M*, 1 *N*. Dissolve 80 g of NH_4NO_3 in 1 liter of water.

Ammonium oxalate, 0.25 *M*, 0.5 *N*. Dissolve 35.5 g of $(NH_4)_2C_2O_4 \cdot H_2O$ in water. Dilute to 1 liter.

Ammonium sulfate, 0.25 *M*, 0.5 *N*, Dissolve 33 g of $(NH_4)_2SO_4$ in 1 liter of water.

Ammonium sulfide, colorless.

1. 3 *M*. Treat 200 ml of conc. NH_4OH with H_2S until saturated, keeping the solution cold. Add 200 ml of conc. NH_4OH and dilute of 1 liter.

2. 6 *N*. Saturate 6 *N* ammonium hydroxide (40 ml conc. ammonia solution + 60 ml H_2O) with washed H_2S gas. The ammonium hydroxide bottle must be completely full and must be kept surrounded by ice while being saturated (about 48 hours for two liters). The reagent is best preserved in brown, completely filled, glass-stoppered bottles.

Ammonium sulfide, yellow, Treat 150 ml of conc. NH_4OH with H_2S until saturated, keeping the solution cool. Add 250 ml of conc. NH_4OH and 10 g of powdered sulfur. Shake the mixture until the sulfur is dissolved and dilute to 1 liter with water. In the solution the concentration of $(NH_4)_2S_2$, $(NH_4)_2S$ and NH_4OH are 0.625, 0.4 and 1.5 normal respectively. On standing, the concentration of $(NH_4)_2S_2$ increases and that of $(NH_4)_2S$ and NH_4OH decreases.

Antimony pentachloride, 0.1 *M*, 0.5 *N*. Dissolve 30 g of $SbCl_5$ in 1 liter of water.

Antimony trichloride, 0.167 *M*, 0.5 *N*. Dissolve 38 g of $SbCl_3$ in 1 liter of water.

Aqua regia. Mix 1 part concentrated HNO_3 with 3 parts of concentrated HCl. This formula should include one volume of water if the aqua regia is to be stored for any length of time. Without water, objectionable quantities of chlorine and other gases are evolved.

Barium chloride, 0.25 *M*, 0.5 *N*. Dissolve 61 g of $BaCl_2 \cdot 2H_2O$ in water. Dilute to 1 liter.

Barium hydroxide, 0.1 *M*, about 0.2 *N*. Dissolve 32 g of $Ba(OH)_2 8H_2O$ in 1 liter of water.

Barium nitrate, 0.25 *M*, 0.5 *N*. Dissolve 65 g of $Ba(NO_3)_2$ in 1 liter of water.

Bismuth cloride, 0.167 *M*, 0.5 *N*. Dissolve 53 g of $BiCl_3$ in 1 liter of dilute HCl. Use 1 part HCl to 5 parts water.

Bismuth nitrate, 0.083 *M*, 0.25 *N*. Dissolve 40 g of $Bi(NO_3)_3 \cdot 5H_2O$ in 1 liter of dilute HNO_3. U se 1 part of HNO_3 to 5 parts of water.

Cadmium chloride, 0.25 *M*, 0.5 *N*. Dissolve 46 g of $CdCl_2$ in 1 liter of water.

Cadmium nitrate, 0.25 *M*, - 0.5 *N*. Dissolve 77 g of $Cd(NO_3)_2 \cdot 4H_2O$ in 1 liter of water.

Cadmium sulfate, 0.25 *M*, 0.5 *N*. Dissolve 70 g of $CdSO_4 \cdot 4H_2O$ in 1 liter of water.

Calcium chloride, 0.25 *M*, 0.5 *N*. Dissolve 55 g of $CaCl_2 \cdot 6H_2O$ in water. Dilute to 1 liter.

Calcium nitrate, 0.25 *M*, 0.5 *N*. Dissolve 41 g of $Ca(NO_3)_2$ in 1 liter of water.

Chloroplatinic acid.

1. 0.0512 *M*, 0.102 *N*. Dissolve 26.53 g of $H_2PtCl_6 \cdot 6H_2O$ in water. Dilute to 100 ml. Contains 0.100 g Pt per ml.

2. Make a 10% solution by dissolving 1 g of $H_2PtCl_6 \cdot 6H_2O$ in 9 ml of water. Shake thoroughly to insure complete mixing. Keep in a dropping bottle.

Chromic chloride, 0.167 *M*, 0.5 *N*. Dissolve 26 g of $CrCl_3$ in 1 liter of water.

Chromic nitrate, 0.167 *M*, 0.5 *N*. Dissolve 40 g of $Cr(NO_3)_3$ in 1 liter of water.

Chromic sulfate, 0.083 *M*, -.5 *N*. Dissolve 60 g of $Cr_2(SO_4)_3 \cdot 18H_2O$ in 1 liter of water.

Cobaltous nitrate, 0.25 *M*, 0.5 *N*. Dissolve 73 g of $Co(NO_3)_2 \cdot 6H_2O$ in 1 liter of water.

Cobaltous sulfate, 0.25 *M*, 0.5 *N*. Dissolve 70 g of $CoSO_4 \cdot 7H_2O$ in 1 liter of water.

Cupric chloride, 0.25 *M*, 0.5 *N*. Dissolve 43 g of $CuCl_2 \cdot 2H_2O$ in 1 liter of water.

Cupric nitrate, 0.25 *M*, 0.5 *N*. Dissolve 74 g of $Cu(NO_3)_2 \cdot 6H_2O$ in 1 liter of water.

Cupric sulfate, 0.5 *M*, 1 *N*. Dissolve 124.8 g of $CuSO_4 \cdot 5H_2O$ in water to which 5 ml of H_2SO_4 has been added. Dilute to 1 liter.

Ferric chloride, 0.5 *M*, 1.5 *N*. Dissolve 135.2 g of $FeCl_3 \cdot 6H_2O$ in water containing 20 ml of conc. HCl. Dilute to 1 liter.

Ferric nitrate, 0.167 *M*, 0.5 *N*. Dissolve 67 g of $Fe(NO_3)_3 \cdot 9H_2O$ in 1 liter of water.

Ferric sulfate, 0.25 *M*, 0.5 *N*. Dissolve 140.5 g of $Fe_2(SO_4)_3 \cdot 9H_2O$ in water containing 100 ml of conc. H_2SO_4. Dilute to 1 liter.

Ferrous ammonium sulfate, 0.5 *M*, 1 *N*. Dissolve 196 g of $Fe(NH_4SO_4)_2 \cdot 6H_2O$ in water containing 10 ml of conc. H_2SO_4. Dilute to 1 liter. Prepare fresh solutions for best results.

Ferrous sulfate, 0.5 *M*, 1 *N*. Dissolve 139 g of $FeSO_4 \cdot 7H_2O$ in water containing 10 ml of conc. H_2SO_4. Dilute to 1 liter. Solution does not keep well.

Lead acetate, 0.5 *M*, 1 *N*. Dissolve 190 g of $Pb(C_2H_3O_2)_2 \cdot 3H_2O$ in water. Dilute to 1 liter.

Lead nitrate, 0.25 *M*, 0.5 *N*. Dissolve 83 g of $Pb(NO_3)_2$ in water. Dilute to one liter.

Lime water, See Calcium hydroxide.

Magnesium chloride, 0.25 *M*, 0.5 *N*. Dissolve 51 g of $MgCl_2 \cdot 6H_2O$ in 1 liter of water.

Magnesium chloride reagent. Dissolve 50 g of $MgCl_2 \cdot 6H_2O$ and 100 g of NH_4Cl in 500 ml of water. Add 10 ml of conc. NH_4OH, allow to stand over night and filter if a precipitate has formed. Make acid to methyl red with dilute HCl. Dilute to 1 liter. Solution contains 0.25 *M* $MgCl_2$ and 2 *M* NH_4Cl. Solution may also be diluted with 133 ml of conc. NH_4OH and water to make 1 liter. Such a solution will contain 2 *M* NH_4OH.

Magnesium nitrate, 0.25 *M*, 0.5 *N*. Dissolve 64 g of $Mg(NO_3)_2 \cdot 6H_2O$ in 1 liter of water.

Magnesium sulfate, 0.25 *M*, 0.5 *N*. Dissolve 62 g of $MgSO_4 \cdot 7H_2O$ in 1 liter of water.

Manganous chloride, 0.25 *M*, 0.05 *N*. Dissolve 50 g of $MnCl_2 \cdot 4H_2O$ in 1 liter of water.

Manganous nitrate, 0.25 *M*, 0.5 *N*. Dissolve 72 g of $Mn(NO_3)_2 \cdot 6H_2O$ in 1 liter of water.

Manganous sulfate, 0.25 *M*, 0.5 *N*. Dissolve 69 g of $MnSO_4 \cdot 7H_2O$ in 1 liter of water.

Mercuric chloride, 0.25 *M*, 0.5 *N*. Dissolve 68 g of $HgCl_2$ in water. Dilute to 1 liter.

Mercuric nitrate, 0.25 *M*, 0.5 *N*. Dissolve 81 g of $Hg(NO_3)_2$ in 1 liter of water.

Mercuric sulfate, 0.25 *M*, 0.5 *N*. Dissolve 74 g of $HgSO_4$ in 1 liter of water.

Mercurous nitrate. Use 1 part $HgNO_3$, 20 parts water and 1 part HNO_3.

Nickel chloride, 0.25 *M*, 0.5 *N*. Dissolve 59 g of $NiCl_2 \cdot 6H_2O$ in 1 liter of water.

Nickel nitrate, 0.25 *M*, 0.5 *N*. Dissolve 73 g of $Ni(NO_3)_2 \cdot 6H_2O$ in 1 liter of water.

Nickel sulfate, 0.25 *M*, 0.5 *N*. Dissolve 66 g of $NiSO_4 \cdot 6H_2O$ in 1 liter of water.

Potassium bromide, 0.5 *M*, 0.5 *N*. Dissolve 60 g of KBr in 1 liter of water.

Potassium carbonate, 1.5 *M*, 3 *N*. Dissolve 207 g of K_2CO_3 in 1 liter of water.

Potassium chloride, 0.5 *M*, 0.5 *N*. Dissolve 37 g of KCl in 1 liter of water.

Potassium chromate, 0.25 *M*, 0.5 *N*. Dissolve 49 g of K_2CrO_4 in 1 liter of water.

Potassium cyanide, 0.5 *M*, 0.5 *N*. Dissolve 33 g of KCN in 1 liter of water.

Potassium dichromate, 0.125 *M*. Dissolve 37 g of $K_2Cr_2O_7$ in 1 liter of water.

Potassium ferricyanide, 0.167 *M*, 0.5 *N*. Dissolve 55 g of $K_3Fe(CN)_6$ in 1 liter of water.

Potassium ferrocyanide, 0.5 *M*, 2 *N*. Dissolve 211 g of $K_4Fe(CN)_6 \cdot 3H_2O$ in water. Dilute to 1 liter.

Potassium iodide, 0.5 *M*, 0.5 *N*. Dissolve 83 g of KI in 1 liter of water.

Potassium nitrate, 0.5 *M*, 0.5 *N*. Dissolve 51 g of KNO_3 in 1 liter of water.

Potassium sulfate, 0.25 *M*, 0.5 *N*. Dissolve 44 g of K_2SO_4 in 1 liter of water.

Silver nitrate, 0.5 *M*, 0.5 *N*. Dissolve 85 g of $AgNO_3$ in water. Dilute to 1 liter.

Sodium acetate, 3 *M*, 3 *N*. Dissolve 408 g of $NaC_2H_3O_2 \cdot 3H_2O$ in water. Dilute to 1 liter.

Sodium carbonate, 1.5 *M*, 3 *N*. Dissolve 159 g of Na_2CO_3, or 430 g of $Na_2CO_3 \cdot 10H_2O$ in water. Dilute to 1 liter.

Sodium chloride, 0.5 *M*, 0.5 *N*. Dissolve 29 g of NaCl in 1 liter of water.

Sodium cobaltinitrite, 0.08 *M* (reagent for potassium). Dissolve 25 g of $NaNO_2$ in 75 ml of water, add 2 ml of glacial acetic acid and then 2.5 g of $Co(NO_3)_2 \cdot 6H_2O$. Allow to stand for several days, filter and dilute to 100 ml. Reagent is somewhat unstable.

Sodium hydrogen phosphate, 0.167 *M*, 0.5 *N*. Dissolve 60 g of $Na_2HPO_4 \cdot 12H_2O$ in 1 liter of water.

Sodium nitrate, 0.5 *M*, 0.5 *N*. Dissolve 43 g of $NaNO_5$ in 1 liter of water.

Sodium sulfate, 0.25 *M*, 0.5 *N*. Dissolve 36 g of Na_2SO_4 in 1 liter of water.

Sodium sulfide, 0.5 *M*, 1 *N*. Dissolve 120 g of $Na_2S \cdot 9H_2O$ in water and dilute to 1 liter. Or, saturate 500 ml of 1 *M* NaOH (21 g of 95% NaOH sticks) with H_2S, keeping the solution cool, and dilute with 500 ml of 1 *M* NaOH.

Stannic chloride, 0.125 *M*, 0.5 *N*. Dissolve 33 g of $SnCl_4$ in 1 liter of water.

Stannous chloride, 0.5 *M*, 1 *N*. Dissolve 113 g of $SnCl_2 \cdot 2H_2O$ in 170 ml of conc. HCl, using heat if necessary. Dilute with water to 1 liter. Add a few pieces of tin foil. Prepare solution fresh at frequent intervals.

Stannous chloride (for Bettendorf test). Dissolve 113 g of $SnCl_2 \cdot 2H_2O$ in 75 ml of conc. HCl. Add a few pieces of tin foil.

Strontium chloride, 0.25 *M*, 0.5 *N*. Dissolve 67 g of $SrCl_2 \cdot 6H_2O$ in 1 liter of water.

Zinc nitrate, 0.25 *M*, 0.5 *N*. Dissolve 74 g of $Zn(NO_3)_2 \cdot 6H_2O$ in 1 liter of water.

Zinc sulfate, 0.25 *M*, 0.5 *N*. Dissolve 72 g of $ZnSO_4 \cdot 7H_2O$ in 1 liter of water.

SPECIAL SOLUTIONS AND REAGENTS

Aluminon (qualitative test for aluminum). Aluminon is a trade name for the ammonium salt of aurin tricarboxylic acid. Dissolve 1 g of the salt in 1 liter of distilled water. Shake the solution well to insure thorough mixing.

Bang's reagent (for glucose estimation). Dissolve 100 g of K_2CO_3, 66 g of KCl and 160 g of $KHCO_3$ in the order given in about 700 ml of water at 30° C. Add 4.4 g of $CuSO_4$ and dilute to 1 liter after the CO_2 is evolved. This solution should be shaken only in such a manner as not to allow entry of air. After 24 hours 300 ml are diluted to 1 liter with saturated KCl solution, shaken gently and used after 24 hours; 50 ml equivalent to 10 mg glucose.

Barfoed's reagent (test for glucose). See Cupric acetate.

Baudisch's reagent. See Cupferron.

Benedict's solution (qualitative reagent for glucose). With the aid of heat, dissolve 173 g of sodium citrate and 100 g of Na_2CO_3 in 800 ml of water. Filter, if necessary, and dilute to 850 ml. Dissolve 17.3 g of $CuSO_4 \cdot 5H_2O$ in 100 ml of water. Pour the latter solution, with constant stirring, into the arbonate-citrate solution, and make up to 1 liter.

Benzidine hydrochloride solution (for sulfite determination). Make a paste of 8 g of benzidine hydrochloride ($C_{12}H_8(NH_2)_2 \cdot 2HCl$) and 20 ml of water, add 20 ml of HCl (sp. gr. 1.12) and dilute to 1 liter with water. Each ml of this solution is equivalent to 0.00357 g of H_2SO_4.

Bertrand's reagent (glucose estimation). Consists of the following solutions:

(a) Dissolve 200 g of Rochelle salts and 150 g of NaOH in sufficient water to make 1 liter of solution.
(b) Dissolve 40 g of $CuSO_4$ in enough water to make 1 liter of solution.
(c) Dissolve 50 g of $Fe_2(SO_4)_3$ and 200 g of H_2SO_4 (sp. gr. 1.84) in sufficient water to make 1 liter of solution.
(d) Dissolve 5 g of $KMnO_4$ in sufficient water to make 1 liter of solution.

Bial's reagent (for pentose). Dissolve 1 g of orcinol ($CH_3 \cdot C_6H_3(OH)_2$) in 500 ml of 30% HCl to which 30 drops of a 10% solution of $FeCl_3$ has been added.

Boutron-Boudet soap solution.
(a) Dissolve 100 g of pure castile soap in about 2500 ml of 56% ethyl alcohol.
(b) Dissolve 0.59 g of $Ba(NO_3)_2$ in 1 liter of water.
Adjust the castile soap solution so that 2.4 ml of it will give a permanent lather with 40 ml of solution (b). When adjusted, 2.4 ml of soap solution is equivalent to 220 parts per million of hardness (as $CaCO_3$) for a 40 ml sample.
See also Soap solution.

Brucke's reagent (protein precipitation). See Potassium iodide-mercuric iodide.

Clarke's soap solution (or A.P.H.A., standard method). Estimation of hardness in water.
(a) Dissolve 100 g of pure powdered castile soap in 1 liter of 80% ethyl alcohol and allow to stand over night.
(b) Prepare a standard solution of $CaCl_2$ by dissolving 0.5 g of $CaCO_3$ in HCl (sp. gr. 1.19), neutralize with NH_4OH and make slightly alkaline to litmus, and dilute to 500 ml. One ml is equivalent to 1 mg of $CaCO_3$.
Titrate (a) against (b) and dilute (a) with 80% ethyl alcohol until 1 ml of the resulting solution is equivalent to 1 ml of (b) after making allowance for the lather factor (the amount of standard soap solution required to produce a permanent lather in 50 ml of distilled water). One ml of the adjusted solution after subtracting the lather factor is equivalent to 1 mg of $CaCO_3$.
See also Soap solution.

Cobalticyanide paper (Rinnmann's test for Zn). Dissolve 4 g of $K_3Co(CN)_6$ and 1 g of $KClO_3$ in 100 ml of water. Soak filter paper in solution and dry at 100°C. Apply drop of zinc solution and burn in an evaporating dish. A green disk is obtained if zinc is present.

Cochineal, Extract 1 g of cochineal for four days with 20 ml of alcohol and 60 ml of distilled water. Filter.

Congo red. Dissolve 0.5 g of congo red in 90 ml of distilled water and 10 ml of alcohol.

Cupferron (Baudisch's reagent for iron analysis). Dissolve 6 g of the ammonium salt of nitroso-phenyl-hydroxyl-amine (cupferron) in 100 ml of H_2O. Reagent good for one week only and must be kept in the dark.

Cupric acetate (Barfoed's reagent for reducing monosaccharides). Dissolve 66 g of cupric acetate and 10 ml of glacial acetic acid in water and dilute to 1 liter.

Cupric oxide, ammoniacal; Schweitzer's reagent (dissolves cotton, linen and silk, but not wool).
1. Dissolve 5 g of cupric sulfate in 100 ml of boiling water, and add sodium hydroxide until precipitation is complete. Wash the precipitate well, and dissolve it in a minimum quantity of ammonium hydroxide.
2. Bubble a slow stream of air through 300 ml of strong ammonium hydroxide containing 50 g of fine copper turnings. Continue for one hour.

Cupric sulfate in glycerin-potassium hydroxide (reagent for silk). Dissolve 10 g of cupric sulfate, $CuSO_4 \cdot 5H_2O$, in 100 ml of water and add 5 g of glucerin. Add KOH solution slowly until a deep blue solution is obtained.

Cupron (benzoin oxime). Dissolve 5 g in 100 ml of 95% alcohol.

Cuprous chloride, acidic (reagent for CO in gas analysis).
1. Cover the bottom of a two-liter flask with a layer of cupric oxide about one-half inch deep, suspend a bunch of copper wire so as to reach from the bottom to the top of the solution, and fill the flask with hydrochloric acid (sp. gr. 1.10). Shake occasionally. When the solution becomes nearly colorless, transfer to reagent bottles, which should also contain copper wire. The stock bottle may be refilled with dilute hydrochloric acid until either the cupric oxide or the copper wire is used up.
Copper sulfate may be substituted for copper oxide in the above procedure.
2. Dissolve 340 g of $CuCl_2 \cdot 2H_2O$ in 600 ml of conc. HCl and reduce the cupric chloride by adding 190 ml ofa saturated solution of stannous chloride or until the solution is colorless. The stannous chloride is prepared by treating 300 g of metallic tin in a 500 ml flask with conc. HCl until no more tin goes into solution.
3. (Winkler method). Add a mixture of 86 g of CuO and 17 g of finely divided metallic Cu, made by the reduction of CuO with hydrogen, to a solution of HCl, made by diluting 650 ml of conc. HCl with 325 ml of water. After the mixture has been added slowly and with frequent stirring, a spiral of copper wire is suspended in the bottle, reaching all the way to the bottom. Shake occasionally, and when the solution becomes colorless, it is ready for use.

Cuprous chloride, ammoniacal (reagent for CO in gas analysis).
1. The acid solution of cuprous chloride as prepared above is neutralized with ammonium hydroxide until an ammonia odor persists. An excess of metallic copper must be kept in the solution.
2. Pour 800 ml of acidic cuprous chloride, prepared by the Winkler method, into about 4 liters of water. Transfer the precipitate to a 250 ml graduate. After several hours, siphon off the liquid above the 50 ml mark and refill with 7.5% NH_4OH solution which may be prepared by diluting 50 ml of conc. NH_4OH with 150 ml of water. The solution is well shaken and allowed to stand for several hours. It should have a faint odor of ammonia.

Dichlorofluorescein indicator. Dissolve 1 g in 1 liter of 70% alcohol or 1 g of the sodium salt in 1 liter of water.

Dimethylglyoxime (diacetyl dioxime), 0.01 *N*. Dissolve 0.6 g of dimethylglyoxime, $(CH_3CNOH)_2$, in 500 ml of 95% ethyl alcohol. This is an especially sensitive test for nickel, a very definite crimson color being produced.

Diphenylamine (reagent for rayon). Dissolve 0.2 g in 100 ml of concentrated sulfuric acid.

Diphenylamine sulfonate (for titration of iron with $K_2Cr_2O_7$). Dissolve 0.32 g of the barium salt of diphenylamine sulfonic acid in 100 ml of water, add 0.5 g of sodium sulfate and filter off the precipitate of $BaSO_4$

Diphenylcarbazide. Dissolve 0.2 g of diphenylcarbazide in 10 ml of glacial acetic acid and dilute to 100 ml with 95% ethyl alcohol.

Esbach's reagent (estimation of protein). To a water solution of 10 g of picric acid and 20 g of citric acid, add sufficient water to make one liter of solution.

Eschka's compound. Two parts of calcined ("light") magnesia are thoroughly mixed with one part of anhydrous sodium carbonate.

Fehling's solution (reagent for reducing sugars.)
(a) Copper sulfate solution. Dissolve 34.66 g of $CuSO_4 \cdot 5H_2O$ in water and dilute to 500 ml.
(b) Alkaline tartrate solution. Dissolve 173 g of potassium sodium tartrate (Rochelle salts, $KNaC_4H_4O_6 \cdot 4H_2O$) and 50 g of NaOH in water and dilute when cold to 500 ml.
For use, mix equal volumes of the two solutions at the time of using.

Ferric-alum indicator. Dissolve 140 g of ferric-ammonium sulfate cyrstals in 400 ml of hot water. When cool, filter, and make up to a volume of 500 ml with dilute (6 *N*) nitric acid.

Folin's mixture (for uric acid). To 650 ml of water add 500 g of $(NH_4)_2SO_4$, 5 g of uranium acetate and 6 g of glacial acetic acid. Dilute to 1 liter.

Formaldehyde-sulfuric acid (Marquis' reagent for alkaloids). Add 10 ml of formaldehyde solution to 50 ml of sulfuric acid.

Froehde's reagent. See Sulfomolybdic acid.

Fuchsin (reagent for linen). Dissolve 1 g of fuchsin in 100 ml of alcohol.

Fuchsin-sulfurous acid (Schiff's reagent for aldehydes). Dissolve 0.5 g of fuchsin and 9 g of sodium bisulfite in 500 ml of water, and add 10 ml of HCl. Keep in well-stoppered bottles and protect from light.

Gunzberg's reagent (detection of HCl in gastric juice). Prepare as needed a solution containing 4 g of phloroglucinol and 2 g of vanillin in 100 ml of absolute ethyl alcohol.

Hager's reagent. See Picric acid.

Hanus solution (for iodine number). Dissolve 13.2 g of resublimed iodine in one liter of glacial acetic acid which will pass the dichromate test for reducible matter. Add sufficient bromine to double the halogen content, determined by titration (3 ml is about the proper amount). The iodine may be dissolved by the aid of heat, but the solution should be cold when the bromine is added.

Iodine, tincture of. To 50 ml of water add 70 g of I_2 and 50 g of KI. Dilute to 1 liter with alcohol.

Iodo-potassium iodide (Wagner's reagent for alkaloids). Dissolve 2 g of iodine and 6 g of KI in 100 ml of water.

Litmus (indicator). Extract litmus powder three times with boiling alcohol, each treatment consuming an hour. Reject the alcoholic extract. Treat residue with an equal weight of cold water and filter; then exhaust with five times its weight of boiling water, cool and filter. Combine the aqueous extracts.

Magnesia mixture (reagent for phosphates and arsenates). Dissolve 55 g of magnesium chloride and 105 g of ammonium chloride in water, barely acidify with hydrochloric acid, and dilute to 1 liter. The ammonium hydroxide may be omitted until just previous to use. The reagent, if completely mixed and stored for any period of time, becomes turbid.

Magnesium reagent. See S and O reagent.

Magnesium uranyl acetate. Dissolve 100 g of $UO_2(C_2H_3O_2)_2 \cdot 2H_2O$ in 60 ml of glacial acetic acid and dilute to 500 ml. Dissolve 330 g of $Mg(C_2H_3O_2)_2 \cdot 4H_2O$ in 60 ml of glacial acetic acid and dilute to 200 ml. Heat solutions to te boiling point until clear, pour the magnesium solution into the uranyl solution, cool and dilute to 1 liter. Let stand over night and filter if necessary.

Marme's reagent. See Potassium-cadmium iodide.

Marquis' reagent. See Formaldehyde-sulfuric acid.

Mayer's reagent (white precipitate with most alkaloids in slightly acid solutions). Dissolve 1.358 g of $HgCl_2$ in 60 ml of water and pour into a solution of 5 g of KI in 10 ml of H_2O. Add sufficient water to make 100 ml.

Methyl orange indicator. Dissolve 1 g of methyl orange in 1 liter of water. Filter, if necessary.

Methyl orange, modified. Dissolve 2 g of methyl orange and 2.8 g of xylene cyanole FF in 1 liter of 50% alcohol.

Methyl red indicator. Dissolve 1 g of methyl red in 600 ml of alcohol and dilute with 400 ml of water.

Methyl red, modified. Dissolve 0.50 g of methyl red and 1.25 g of xylene cyanole FF in 1 liter of 90% alcohol. Or, dissolve 1.25 g of methyl red and 0.825 g of methylene blue in 1 liter of 90% alcohol.

Millon's reagent (for albumins and phenols). Dissolve 1 part of mercury in 1 part of cold fuming nitric acid. Dilute with twice the volume of water and decant the clear solution after several hours.

Mixed indicator. Prepared by adding about 1.4 g of xylene cyanole FF to 1 g of methyl orange. The dye is seldom pure enough for these proportions to be satisfactory. Each new lot of dye should be tsted by adding additional amounts of the dye until a test portion gives the proper color change. The acid color of this indicator is like that of permanganate; the neutral color is gray; and the alkaline color is green. Described by Hickman and Linstead, J. Chem. Soc. (Lon.), 121, 2502 (1922).

Molisch's reagent. See α-Naphthol.

α-Naphthol (Molisch's reagent for wool). Dissolve 15 g of α-naphthol in 100 ml of alcohol or chloroform.

Nessler's reagent (for ammonia). Dissolve 50 g of KI the smallest possible quantity of cold water (50 ml). Add a saturated solution of mercuric chloride (about 22 g in 350 ml of water will be needed) until an excess is indicated by the formation of a precipitate. Then add 200 ml of 5 *N* NaOH and dilute to 1 liter. Let settle, and draw off the clear liquid.

Nickel oxide, ammoniacal (reagent for silk). Dissolve 5 g of nickel sulfate in 100 ml of water, and add sodium hydroxide solution until nickel hydroxide is completely precipitated. Wash the precipitate well and dissolve in 25 ml of concentrated ammonium hydroxide and 25 ml of water.

ρ-Nitrobenzene-azo-resorcinol (reagent for magnesium). Dissolve 1 g of the dye in 10 ml of *N*NaOH and dilute to 1 liter.

Nitron (detection of nitrate radical). Dissolve 10 g of nitron ($C_{20}H_{16}N_4$, 4, 5-dihydro-1, 4-diphenyl-3, 5-phenylimino-1, 2, 4-triazole) in 5 ml of glacial acetic acid and 95 ml of water. The solution may be filtered with slight suction through an alumdum crucible and kept in a dark bottle.

α-Nitroso-β-naphthol. Make a saturated solution in 50% acetic acid (1 part of glacial acetic acid with 1 part of water). Does not keep well.

Nylander's solution (carbohydrates). Dissolve 20 g of bismuth subnitrate and 40 g of Rochelle salts in 1 liter of 8% NaOH solution. Cool and filter.

Obermayer's reagent (for indoxyl in urine). Dissolve 4 g of $FeCl_3$ in one liter of HCl (sp. gr. 1.19).

Oxine. Dissolve 14 g of HC_9H_6ON in 30 ml of glacial acetic acid. Warm slightly, if necessary. Dilute to 1 liter.

Oxygen absorbent. Dissolve 300 g of ammonium chloride in one liter of water and add one liter of concentrated ammonium hydroxide solution. Shake the solution thorughly. For use as n oxygen absorbent, a bottle half full of copper turnings is filled nearly full with the NH_4Cl-NH_4OH solution and the gas passed through.

Pasteur's salt solution. To one liter of distilled water add 2.5 g of potassium phosphate, 0.25 g of calcium phosphate, 0.25 g of magnesium sulfate and 12.00 g of ammonium tartrate.

Pavy's solution (glucose reagent). To 120 ml of Fehling's solution, add 300 ml of NH_4OH (sp. gr. 0.88) and dilute to 1 liter with water.

Phenathroline ferrous ion indicator. Dissolve 1.485 g of phenanthroline monohydrate in 100 ml of 0.025 *M* ferrous sulfate solution.

Phenolphthalein. Dissolve 1 g of phenolphthalein in 50 ml of alcohol and add 50 ml of water.

Phenolsulfonic acid (determination of nitrogen as nitrate). Dissolve 25 g of phenol in 150 ml of conc. H_2SO_4, add 75 ml of fuming H_2SO_4 (15% SO_3), stir well and heat for two hours at 100° C.

Phloroglucinol solution (pentosans). Make a 3% phloroglucinol solution in alcohol. Keep in a dark bottle.

Phosphomolybdic acid (Sonnenschein's reagent for alkaloids).

1. Prepare ammonium phosphomolybdate and after washing with water, boil with nitric acid and expel NH_3; evaporate to dryness and dissolve in 2 *N* nitric acid.

2. Dissolve ammonium molybdate in HNO_3 and treat with phosphoric acid. Filter, wash the precipitate, and boil with aqua regia until the ammonium salt is decomposed. Evaporate to dryness. The residue dissolved in 10% HNO_3 constitutes Sonnenschein's reagent.

Phosphoric acid—sulfuric acid mixture. Dilute 150 ml of conc. H_2SO_4 and 100 ml of conc. H_3PO_4 (85%) with water to a volume of 1 liter.

Phosphotungstic acid (Scheibler's reagent for alkaloids).

1. Dissolve 20 g of sodium tungstate and 15 g of sodium phosphate in 100 ml of water containing a little nitric acid.

2. The reagent is a 10% solution of phosphotungstic acid in water. The phosphotungstic acid is prepared by evaporating a mixture of 10 g of sodium tungstate dissolved in 5 g of phosphoric acid (sp. gr. 1.13) and enough boiling water to effect solution. Crystals of phosphotungstic acid separate.

Picric acid (Hager's reagent for alkaloids, wool and silk). Dissolve 1 g of picric acid in 100 ml of water.

Potassium antimonate (reagent for sodium). Boil 22 g of potassium antimonate with 1 liter of water until nearly all of the salt has dissolved, cool quickly, and add 35 ml of 10% potassium hydroxide. Filter after standing over night.

Potassium-cadmium iodide (Marme's reagent for alkaloids). Add 2 g of CdI_2 to a boiling solution of 4 g of KI in 12 ml of water, and then mix with 12 ml of saturated KI solution.

Potassium hydroxide (for CO_2 absorption). Dissolve 360 gof KOH in water and dilute to 1 liter.

Potassium iodide-mercric iodide (Brucke's reagent for proteins). Dissolve 50 g of KI in 500 ml of water, and saturate with mercuric iodide (about 120 g). Dilute to 1 liter.

Potassium pyrogallate (for oxygen absorption). For mixtures of gases containing less than 28% oxygen, add 100 ml of KOH solution (50 g of KOH to 100 ml of water) to 5 g of pyrogallol. For mixtures containing more than 28% orygen the KOH solution should contain 120 g of KOH to 100 ml of water.

Pyrogallol, alkaline.

(a) Dissolve 75 g of pyrogallic acid in 75 ml of water.

(b) Dissolve 500 g of KOH in 250 ml of water. When cool, adjust until sp. gr. is 1.55.

For use, add 270 ml of solution (b) to 30 ml of solution (a).

Rosolic acid (indicator). Dissolve 1 g of rosolic acid in 10 ml of alcohol and add 100 ml of water.

S and O reagent (Suitsu and Okuma's test for Mg). Dissolve 0.5 g of the dye (*o-p*-dihydroxy-monazo-*p*-nitrobenzene) in 100 ml of 0.25 *N* NaOH.

Scheibler's reagent. See Phosphotungstic acid.

Schiff's reagent. See Fuchsin-sulfurous acid.

Schweitzer's reagent. See Cupric oxide, ammoniacal.

Soap solution (reagent for hardness in water). D issolve 100 g of dry castile soap in 1 liter of 80% alcohol (5 parts alcohol to 1 part water). Allow to stand several days and dilute with 70% to 80% alcohol until 6.4 ml produces a permanent lather with 20 ml of standard calcium solution. The latter solution is made by dissolving 0.2 g of $CaCO_3$ in a small amount of dilute HCl, evaporating to dryness and making up to 1 liter.

Sodium bismuthate (oxidation of manganese). Heat 20 parts of NaOH nearly to redness in an iron or nickel crucible and add slowly 10 parts of basic bismuth nitrate which has been previously dried. Add two parts of sodium peroxide, and pour the brownish-yellow fused mass on an iron plate to cool. When cold, break up in a mortar, extract with water, and collect on an asbestos filter.

Sodium hydroxide (for CO_2 absorption). Dissolve 330 g of NaOH in water and dilute to 1 liter.

Sodium nitroprusside (reagent for hydrogen sulfide and wool). Use a freshly prepared solution of 1 g of sodium nitroprusside in 10 ml of water.

PREPARATION OF REAGENTS (continued)

Sodium oxalate, according to Sorensen (primary standard). Dissolve 30 g of the commercial salt in 1 liter of water, make slightly alkaline with sodium hydroxide, and let stand until perfectly clear. Filter and evaporate the filtrate to 100 ml. Cool and filter. Pulverize the residue and wash it several times with small volumes of water. The procedure is repeated until the mother liquor is free from sulfate and is neutral to phenolphthalein.

Sodium plumbite (reagent for wool). Dissolve 5 g of sodium hydroxide in 100 ml of water. Add 5 g of litharge and boil until dissolved.

Sodium polysulfide. Dissolve 480 g of $Na_2S \cdot 9H_2O$ in 50G ml of water, add 40 g of NaOH and 18 g of sulfur. Stir thoroughly and dilute to 1 liter with water.

Sonnenschein's reagent. See Phosphomolybdic acid.

Starch solution.

1. Make a paste with 2 g of soluble starch and 0.01 g of HgI_2 with a small amount of water. Add the mixture slowly to 1 liter of boiling water and boil for a few minutes. Keep in a glass stoppered bottle. If other than soluble starch is used, the solution will not clear on boiling; it should be allowed to stand and the clear liquid decanted.

2. A solution of starch which keeps indefinitely is made as follows: Mix 500 ml of saturated NaCl solution (filtered), 80 ml of glacial acetic acid, 20 ml of water and 3 g of starch. Bring slowly to a boil and boil for two minutes.

3. Make a paste with 1 g of soluble starch and 5 mg of HgI_2, using as little cold water as possible. Then pour about 200 ml of boiling water on the paste and stir immediately. This will give a clear solution if the paste is prepared correctly and the water actually boiling. Cool and add 4 g of KI. Starch solution decomposes on standing due to bacterial action, but this solution will keep a long time if stored under a layer of toluene.

Stoke's reagent. Dissolve 30 g of $FeSO_4$ and 20 g of tartaric acid in water and dilute to 1 liter. Just before using, add concentrated NH_4OH until the precipitate first formed is redissolved.

Sulfanilic acid (reagent for nitrites). Dissolve 0.5 g of sulfanilic acid in a mixture of 15 ml of glacial acetic acid and 135 ml of recently boiled water.

Sulfomolybdic acid (Froehde's reagent for alkaloids and glucosides). Dissolve 10 g of molybdic acid or sodium molybdate in 100 ml of conc. H_2SO_4.

Tannic acid (reagent for albumen, alkaloids and gelatin). Dissolve 10 g of tannic acid in 10 ml of alcohol and dilute with water to 100 ml.

Titration mixture. (residual chlorine in water analysis). Prepare 1 liter of dilute HCl (100 ml of HCl (sp. gr. 1.19) in sufficient water to make 1 liter). Dissolve 1 g of o-tolidine in 100 ml of the dilute HCl and dilute to 1 liter with dilute HCl solution.

Trinitrophenol solution. See Picric acid.

Turmeric paper. Impregnant white, unsized paper with the tincture, and dry.

Turmeric tincture (reagent for borates). Digest ground turmeric root with several quantities of water which are discarded. Dry the residue and digest it several days with six times its weight of alcohol. Filter.

Uffelmann's reagent (turns yellow in presence of a lactic acid). To a 2% solution of pure phenol in water, add a water solution of $FeCl_3$ until the phenol solution becomes a violet in color.

Wagner's reagent. See Iodo-potassium iodide.

Wagner's solution (used in phosphate rock analysis to prevent precipitation of iron and aluminum). Dissolve 25 g of citric acid and 1 g of salicylic acid in water and dilute to 1 liter. Use 50 ml of the reagent.

Wij's iodine monochloride solution (for iodine number). Dissolve 13 g of resublimed iodine in 1 liter of glacial acetic acid which will pass the dichromate test for reducible matter. Set aside 25 ml of this solution. Pass into the remainder of the solution dry chlorine gas (dried and washed by passing through H_2SO_4 (sp. gr. 1.84)) until the characteristic color of free iodine has been discharged. Now add the iodine solution which was reserved, until all free chlorine has been destroyed. A slight excess of iodine does little or no harm, but an excess of chlorine must be avoided. Preserve in well stoppered, amber colored bottles. Avoid use of solutions which have been prepared for more than 30 days.

Wij's special solution (for iodine number—Analyst 58, 523-7, 1933). To 200 ml of glacial acetic acid that will pass the dichromate test for reducible matter, add 12 g of dichloroamine T (paratoluene-sulfonedichloroamide), and 16.6 g of dry KI (in small quantities with continual shaking until all the KI has dissolved). Make up to 1 liter with the same quality of acetic acid used above and preserve in a dark colored bottle.

Zimmermann-Reinhardt reagent (determination of iron). Dissolve 70 g of $MnSO_4 \cdot 4H_2O$ in 500 ml of water, add 125 ml of conc. H_2SO_4 and 125 ml of 85% H_3PO_4, and dilute to 1 liter.

Zinc chloride solution, basic (reagent for silk). Dissolve 1000 g of zinc chloride in 850 ml of water, and add 40 g of zinc oxide. Heat until solution is complete.

Zinc uranyl acetate (reagent for sodium). Dissolve 10 g of $UO_2(C_2H_3O_2)_2 \cdot 2H_2O$ in 6 g of 30% acetic acid with heat, if necessary, and dilute to 50 ml. Dissolve 30 g of $Zn(C_2H_3O_2)_2 \cdot 2H_2O$ in 3 g of 30% acetic acid and dilute to 50 ml. Mix the two solutions, add 50 mg of NcCl, allow to stand over night and filter.

STANDARD SOLUTIONS OF ACIDS, BASES, AND SALTS

For each compound listed, the last column of this table gives the mass in grams which is contained in 1 liter of a solution whose amount-of-substance concentration divided by the equivalence factor of the compound equals 0.1 mol/L. In the older literature such a solution is often referred to as a "decinormal solution" (0.1 N).

REFERENCE

Compendium of Analytical Nomenclature (IUPAC), Pergamon Press, Oxford, 1978.

Name	Formula	Atomic or molecular weight	Equivalence factor	Mass in grams
Acetic acid	$HC_2H_3O_2$	60.0530	1	6.0053
Ammonia	NH_3	17.0306	1	1.7031
Ammonium ion	NH_4^+	18.0386	1	1.8039
Ammonium chloride	NH_4Cl	53.4916	1	5.3492
Ammonium sulfate	$(NH_4)_2SO_4$	132.1388	1/2	6.6069
Ammonium thiocyanate	NH_4CNS	76.1204	1	7.6120
Barium	Ba	137.34	1/2	6.867
Barium carbonate	$BaCO_3$	197.3494	1/2	9.8675
Barium chloride hydrate	$BaCl_2 \cdot 2H_2O$	244.2767	1/2	12.2138
Barium hydroxide	$Ba(OH)_2$	171.3547	1/2	8.5677
Barium oxide	BaO	153.3394	1/2	7.6670
Bromine	Be	79.909	1	7.9909
Calcium	Ca	40.08	1/2	2.004
Calcium carbonate	$CaCO_3$	100.0894	1/2	5.0045
Calcium chloride	$CaCl_2$	110.9860	1/2	5.5493
Calcium chloride hydrate	$CaCl_2 \cdot 6H_2O$	219.0150	1/2	10.9508
Calcium hydroxide	$Ca(OH)_2$	74.0947	1/2	3.7047
Calcium oxide	CaO	56.0794	1/2	2.8040
Chlorine	Cl	35.453	1	3.5453
Citric acid	$C_6H_8O_7 \cdot H_2O$	210.1418	1/3	7.0047
Cobalt	Co	58.9332	1/2	2.9466
Copper	Cu	63.54	1/2	3.177
Copper oxide (cupric)	CuO	79.5394	1/2	3.9770
Copper sulfate hydrate	$CuSO_4 \cdot 5H_2O$	249.6783	1/2	12.4839
Hydrochloric acid	HCl	36.4610	1	3.6461
Hydrocyanic acid	HCN	27.0258	1	2.7026
Iodine	I	126.9044	1	12.6904
Lactic acid	$C_3H_6O_3$	90.0795	1	9.0080
Malic acid	$C_4H_6O_5$	134.0894	1/2	6.7045
Magnesium	Mg	24.312	1/2	1.2156
Magnesium carbonate	$MgCO_3$	84.3214	1/2	4.2161
Magnesium chloride	$MgCl_2$	95.2180	1/2	4.7609
Magnesium chloride hydrate	$MgCl_2 \cdot 6H_2O$	203.2370	1/2	10.1623
Magnesium oxide	MgO	40.3114	1/2	2.0156
Manganese	Mn	54.938	1/2	2.7469
Manganese sulfate	$MnSO_4$	150.9996	1/2	7.5500
Mercuric chloride	$HgCl_2$	271.4960	1/2	13.5748
Nickel	Ni	58.71	1/2	2.9356
Nitric acid	HNO_3	63.0129	1	6.3013
Oxalic acid	$H_2C_2O_4$	90.0358	1/2	4.5018
Oxalic acid hydrate	$H_2C_2O_4 \cdot 2H_2O$	126.0665	1/2	6.3033
Oxalic acid anhydride	C_2O_3	72.0205	1/2	3.6010
Phosphoric acid	H_3PO_4	97.9953	1/3	3.2665
Potassium	K	39.102	1	3.9102
Potassium bicarbonate	$KHCO_3$	100.1193	1	10.0119
Potassium carbonate	K_2CO_3	138.2134	1/2	6.9106
Potassium chloride	KCl	74.5550	1	7.4555

STANDARD SOLUTIONS OF ACIDS, BASES, AND SALTS (continued)

Name	Formula	Atomic or molecular weight	Equivalence factor	Mass in grams
Potassium cyanide	KCN	65.1199	1	6.5120
Potassium hydroxide	KOH	56.1094	1	5.6109
Potassium oxide	K_2O	94.2034	1/2	4.7102
Potassium tartrate	$K_2H_4C_4O_6$	226.2769	1/2	11.3139
Silver	Ag	107.87	1	10.787
Silver nitrate	$AgNO_3$	169.8749	1	16.9875
Sodium	Na	22.9898	1	2.2990
Sodium bicarbonate	$NaHCO_3$	84.0071	1	8.4007
Sodium carbonate	Na_2CO_3	105.9890	1/2	5.2995
Sodium chloride	$NaCl$	58.4428	1	5.8443
Sodium hydroxide	$NaOH$	39.9972	1	3.9997
Sodium oxide	Na_2O	61.9790	1/2	3.0990
Sodium sulfide	Na_2S	78.0436	1/2	3.9022
Succinic acid	$H_2C_4H_4O_4$	118.0900	1/2	5.9045
Sulfuric acid	H_2SO_4	98.0775	1/2	4.9039
Tartaric acid	$C_4H_6O_6$	150.0888	1/2	7.5044
Zinc	Zn	65.37	1/2	3.269
Zinc sulfate hydrate	$ZnSO_4 \cdot 7H_2O$	287.5390	1/2	14.3769

STANDARD SOLUTIONS OF OXIDATION AND REDUCTION REAGENTS

For each reagent listed, the last column of this table gives the mass in grams which is contained in a solution whose amount-of-substance concentration divided by the equivalence factor of the compound equals 0.1 mol/L. The equivalence factor given refers to the most common reactions of the reagent. In the older literature such a solution is often called a "decinormal solution" (0.1 N).

REFERENCE

Compendium of Analytical Nomenclature (IUPAC), Pergamon Press, Oxford, 1978.

Name	Formula	Atomic or molecular weight	Equivalence factor	Mass in grams
Antimony	Sb	121.75	1/2	6.0875
Arsenic	As	74.9216	1/2	3.7461
Arsenic trisulfide	As_2S_3	246.0352	1/4	6.1509
Arsenous oxide	As_2O_3	197.8414	1/4	4.9460
Barium peroxide	BaO_2	169.3388	1/2	8.4669
Barium peroxide hydrate	$BaO_2 \cdot 8H_2O$	313.4615	1/2	15.6730
Calcium	Ca	40.08	1/2	2.004
Calcium carbonate	$CaCO_3$	100.0894	1/2	5.0045
Calcium hypochlorite	$Ca(OCl)_2$	142.9848	1/4	3.5746
Calcium oxide	CaO	56.0794	1/2	2.8040
Chlorine	Cl	35.453	1	3.5453
Chromium trioxide	CrO_3	99.9942	1/3	3.3331
Ferrous ammonium sulfate	$FeSO_4(NH_4)SO_4 \cdot 6H_2O$	392.0764	1	39.2076
Hydroferrocyanic acid	$H_4Fe(CN)_6$	215.9860	1	21.5986
Hydrogen peroxide	H_2O_2	34.0147	1/2	1.7007
Hydrogen sulfide	H_2S	34.0799	1/2	1.7040
Iodine	I	126.9044	1	12.6904
Iron	Fe	55.847	1	5.5847
Iron oxide (ferrous)	FeO	71.8464	1	7.1846
Iron oxide (ferric)	Fe_2O_3	159.6922	1/2	7.9846
Lead peroxide	PbO_2	239.1888	1/2	11.9594
Manganese dioxide	MnO_2	86.9368	1/2	4.3468
Nitric acid	HNO_3	63.0129	1/3	2.1004
Nitrogen trioxide	N_2O_3	76.0116	1/4	1.9002
Nitrogen pentoxide	N_2O_5	108.0104	1/6	1.8001
Oxalic acid	$C_2H_2O_4$	90.0358	1/2	4.5018
Oxalic acid hydrate	$C_2H_2O_4 \cdot 2H_2O$	126.0665	1/2	6.3033
Oxygen	O	15.9994	1/2	0.8000
Potassium dichromate	$K_2Cr_2O_7$	294.1918	1/6	4.9032
Potassium chlorate	$KClO_3$	122.5532	1/6	2.0425
Potassium chromate	K_2CrO_4	194.1076	1/3	6.4733
Potassium ferrocyanide	$K_4Fe(CN)_6$	368.3621	1	36.8362
Potassium ferrocyanide hydrate	$K_4Fe(CN)_6 \cdot 3H_2O$	422.4081	1	42.2408
Potassium iodide	KI	166.0064	1	16.6006
Potassium nitrate	KNO_3	101.1069	1/3	3.3702
Potassium perchlorate	$KClO_4$	138.5526	1/8	1.7319
Potassium permanganate	$KMnO_4$	158.0376	1/5	3.1608
Sodium chlorate	$NaClO_3$	106.4410	1/6	1.7740
Sodium nitrate	$NaNO_3$	84.9947	1/3	2.8332
Sodium thiosulfate hydrate	$Na_2S_2O_3 \cdot 5H_2O$	248.1825	1	24.8183
Stannous chloride	$SnCl_2$	189.5960	1/2	9.4798
Stannous oxide	SnO	134.6894	1/2	6.7345
Sulfur dioxide	SO_2	64.0628	1/2	3.2031
Tin	Sn	118.69	1/2	5.935

ORGANIC ANALYTICAL REAGENTS FOR THE DETERMINATION OF INORGANIC SUBSTANCES

G. Ackermann, L. Sommer, and D. Thorburn Burns

Determination	Reagents	Ref.
Aluminium	Alizarin Red S	Onishi, Part II a, p 28. (5), Snell, *Metals I*, p 587. (7)
	Aluminon	Fries/Getrost, p 16. (2), Onishi, IIa, p 21. (5), Snell, *Metals I*, p 590. (7)
	Aluminon + Cetyltrimethylammonium bromide	Huaxue Shiji, *8*, 85, (1986)
	Chrome Azurol S	Onishi, Part IIa, p 26. (5), Snell, *Metals I*, p 605. (7)
	Chrome Azurol S + Cetyltrimethylammonium bromide	Marczenko, p 133 (3), Snell, *Metals I*, p 606. (7)
	Chromazol KS + Cetylpyridinium bromide	*Analyst*, *107*, 428, (1982).
	Eriochrome Cyanine R	Fries/Getrost, p 19 (2), Onishi, Part IIa p 25. (5), Snell, *Metals I*, p 611. (7)
	Eriochrome Cyanine R + Cetyltrimethylammonium bromide	Snell, *Metals I*, p 613. (7), *Analyst, 107*, 1431, (1982).
	8-Hydroxyquinoline	Fries/Getrost, p 22 (2), Marczenko, p 131 (3), Onishi, Part IIa, p 31. (5), Snell, *Metals I , p* 622 (7)
Ammonia	Phenol + Sodium hypochlorite	Boltz, p 210 (1), Marczenko, p 413 (3), Snell, *Nonmetals*, p 604 (9)
Antimony	Brilliant Green	Onishi, Part IIa, p 102. (5), Snell, *Metals I*, p 304. (7)
	Bromopyrogallol Red	*Talanta*, *13*, 507, (1966).
	Rhodamine B	Fries/Getrost, p 32, (2), Marczenko, p 141. (3), Onishi, Part IIa, p 93. (5), Snell, *Metals I*, p 404. (7)
	Silver diethyldithiocarbamate	Fries/Getrost, p 36. (2)
Arsenic	Silver diethyldithiocarbamate	Fries/Getrost, p 41. (2), Marczenko, p 153. (3), Onishi, Part IIa, p 153. (5), Snell, *Metals I*, p 370. (7)
Barium	Sulfonazo III	Fries/Getrost, p 46. (2), Snell, *Metals II*, p 1782. (8), Onishi, Part IIa, p 202. (5)
Beryllium	Beryllon II	Snell, *Metals I*, p 667. (7)
	Chrome Azurol S	Marczenko, p 163. (3), Snell, *Metals I*, p 672. (7)
	Chrome Azurol S + Cetyltrimethylammonium bromide	Marczenko, p 164. (3), Snell, *Metals I*, p 673. (7)
	Eriochrome Cyanine R	Snell, *Metals I*, p 675. (7), *Talanta*, *31*, 249, (1984).
	Eriochrome Cyanine R + Cetyltrimethylammonium bromide	Zh, *Anal. Khim.*, *33*, 1298, (1978).
Bismuth	Dithizone	Onishi, Part IIa, p 262. (5), Snell, *Metals I*, p 303. (7)
	Pyrocatechol Violet	Fres. Z. *Anal. Chem.*, *186*, 418, (1962).
	Pyrocatechol Violet + Cetyltrimethylammonium bromide	Zh. *Anal. Khim.*, *38*, 216, (1983).
	Thiourea	Onishi, Part IIa, p 260. (5), Snell, *Metals I*, p 317. (7)
	Xylenol Orange	Friez/Getrost, p 57. (2), Marczenko, p 172. (3), Snell, *Metals I*, p 320. (7)
Boron	Azomethine H	Snell, Nonmetals, p 165. (9)
	Carminic acid	Boltz, p 14. (1), Fries/Getrost, p 65. (2), Snell, Nonmetals, p 170. (9), Williams, p 35. (11)
	Curcumin	Boltz, p 8. (1), Fries/Getrost, p 68. (2), Marczenko, p 180. (3), Snell, Nonmetals, p 180. (9), Fres. Z. *Anal. Chem.*, *323*, 266, (1986).

Determination	Reagents	Ref.
	Methylene Blue	Boltz, p 21. (1), Marczenko, p 183. (3), Snell, Nonmetals, p 205. (9), T*alanta, 31*, 547, (1984).
Bromide	Fluorescein	Boltz, p 48. (1), Snell, Nonmetals, p 276., Fres. Z. *Anal. Chem.*, *301*, 28 (1980).
	Phenol Red	Boltz, p 44. (1), Marczenko, p 190. (3), Snell, Nonmetals, p 28. (9)
Cadmium	2-(5-Bromo-2-pyridylazo)-5-diethylaminophenol	Marczenko, p 197. (3)
	Cadion	Onishi, Part IIa, p 323. (5)
	Dithizone	Fries/Getrost, p 78. (2), Onishi, Part IIa, p 315. (5), Snell, *Metals I*, p 279. (7), West, p 25. (10).
	4–(2-Pyridylazo)resorcinol	Fres. Z. *Anal. Chem.*, *310*, 51, (1982).
Calcium	Chlorophosphonazo III	Marczenko, p 207. (3), Snell, *Metals II*, p 1744. (8)
	Glyoxal-bis(2-hydroxyanil)	Fries/Getrost, p 86. (2), Onishi, Part IIa, p 352. (5), Snell, *Metals I*, p 1762. (8)
	Murexide	Onishi, Part IIa, p 357. (5), Snell, *Metals II*, p 1769. (8)
	Phthalein Purple	*Anal. Chim. Acta*, *34*, 71 (1966).
Cerium	*N*-benzoyl-*N*-phenylhydroxylamine	*Anal. Chim. Acta*, *48*, 155, (1969).
	8-Hydroxyquinoline	Fries/Getrost, p 93. (2), Marczenko, p 220. (3), Onishi, Part IIa, p 383. (7)
Chlorine	*N,N*-Diethyl-1,4-phenylenediamine	Boltz, p 92. (1), Fries/Getrost, p 101. (2), Snell, Nonmetals, p 225. (9), *Analyst*, *90*, 187, (1965).
Chromium	1,5-Diphenylcarbazide	Fries/Getrost, p 105. (2), Onishi, Part IIa, p 412. (5), Snell, *Metals I*, p 714. (7), West, p 12. (10)
	4-(2-Pyridylazo)resorcinol	Snell, *Metals I*, p 736. (7), West, p 17. (10)
	4-(2-Pyridylazo)resorcinol + Tetradecyldimethyl-benzylammonium chloride	West, p 17. (10), *Anal. Chim. Acta*, *67*, 297, (1973).
	4-(2-Pyridylazo)resorcinol + Hydrogen peroxide	Fres. Z. *Anal. Chem.*, *304*, 382, (1980).
Cobalt	Nitroso-R salt	Fries/Getrost, p 118. (2), Onishi, Part IIa, p 454. (5), Snell, *Metals I*, p 953. (7)
	1-Nitroso-2-naphthol	Fries/Getrost, p 111. (2), Marczenko, p 246. (3), Snell, *Metals I*, p 947. (5)
	2-Nitroso-1-naphthol	Fries/Getrost, p 113. (2), Onishi, Part IIa, p 459. (5), Snell, *Metals I*, p 949. (7), West, p 45. (10)
	4-(2-Pyridylazo)resorcinol	Snell, *Metals I*, p 969. (7), West, p 44. (10)
	4-(2-Pyridylazo)resorcinol + Diphenylguanidine	Zh. *Anal. Khim.*, *35*, 1306, (1980).
Copper	Bathocuproine	Fries/Getrost, p 135. (2), Snell, *Metals I*, p 148. (7)
	Bathocuproine disulfonic acid	Fries/Getrost, p 137. (2), West, p 52. (10)
	Dithizone	Marczenko, p 258. (3), Onishi, Part IIa, p 529. (5), Snell, *Metals I*, p 199. (7)
	Neocuproine	Snell, *Metals I*, p 217. (5), West, p 51. (10)
	Cuprizone	Onishi, Part IIa, p 534. (5), Snell, *Metals I*, p 157. (7), West, p 53. (10)
	4-(2-pyridylazo)resorcinol + Tetradecyldimethyl-benzylammonium chloride	*Anal. Chim.* Acta, *138*, 321, (1982).
Cyanide	Barbituric Acid + Pyridine	Fries/Getrost, p 153. (2), Snell, Nonmetals, p 653. (9)
	Barbituric Acid + Pyridine-4-carboxylic acid	*Anal. Chim. Acta*, *99*, 197, (1978).

ORGANIC ANALYTICAL REAGENTS FOR THE DETERMINATION OF INORGANIC SUBSTANCES (continued)

Determination	Reagents	Ref.
Fluoride	Alizarin Fluorine blue + Lanthanum(III) ion	Boltz, p 129. (1), Fries/Getrost, p 158. (2), Snell, *Nonmetals*, p 333. (9), Williams, p 354. (11)
	Eriochrome Cyanine R + Zirconium(IV) ion	Boltz, p 119. (1), Snell, *Nonmetals*, p 359. (2), Williams, p 357. (10)
Gallium	Pyrocatechol violet + Diphenylguanidine	Snell, *Metals I*, p 500. (7)
	8-Hydroxyquinoline	Onishi Pt IIa, p 582. (5), Snell, *Metals I*, p 505. (7)
	1-(2-Pyridylazo)-2-naphthol	Snell, *Metals I*, p 512. (7)
	4-(2-Pyridylazo)resorcinol	Snell, *Metals I*, p 513. (7)
	Rhodamine B	Marczenko, p 284. (3), Onishi, Part IIa, p 578. (5), Snell, *Metals I*, p 515. (7)
	Xylenol Orange	Fries/Getrost, p 166. (2), Snell, *Metals I*, p 523. (7)
	Xylenol Orange + 8-Hydroxyquinoline	*Zh. Anal. Khim.*, *26*, 75, (1971).
Germanium	Brilliant Green + Molybdate	Snell, *Metals I*, p 562. (7)
	Phenylfluorone	Fries/Getrost, p 168. (2), Marczenko, p 292. (3), Onishi, Part IIa, p 607. (5), Snell, *Metals I*, p 570. (7)
Gold	5-(4-Diethylaminobenzylidene) rhodanine	Fries/Getrost, p 173. (2), Onishi, Part IIa, p 631. (5), Snell, *Metals II*, p 1516. (8)
	Rhodamine B	Fries/Getrost, p 175. (2), Marczenko, p 301. (3), Onishi, Part IIa, p 637. (5), Snell, *Metals II*, p 513. (8)
Hafnium	Arsenazo III	Snell, *Metals II*, p 1184. (8), *Talanta*, *19*, 807, (1972).
Indium	Bromopyrogallol Red	Snell, *Metals I*, p 469. (7)
	Chrome Azurol S	Snell, *Metals I*, p 474. (7)
	Chrome Azurol S + Cetyltrimethylammonium bromide	*Anal. Chim. Acta*, *67*, 107, (1973).
	Dithizone	Fries/Getrost, p 179. (2), Onishi, Part IIa, p 672. (5), Snell, *Metals I*, p 474. (7)
	8-Hydroxyquinoline	Onishi, Part IIa, p 670. (5), Snell, *Metals I*, p 475. (7)
	1-(2-Pyridylazo)-2-naphthol	Snell, *Metals I*, p 480. (7)
	4-(2-Pyridylazo)resorcinol	Marczenko, p 309. (3), Snell, Metals I, p 480. (7)
Iodide	Neocuproine + Copper(II)	*Anal. Chim. Acta*, *69*, 321, (1974).
Iodine	Starch	Boltz, p 162. (1), Marczenko, p 316. (3), Snell, *Nonmetals*, p 307. (9)
Iridium	Rhodamine 6G + Tin(II)	Marczenko, p 323. (3)
	N,N-Dimethyl-4-nitrosoaniline	*Anal. Chem.*, *27*, 1776, (1955).
Iron	Bathophenanthroline	Fries/Getrost, p 189. (2), Onishi, Part IIa, p 729. (5), Snell, *Metals I*, p 763. (7)
	Bathophenanthroline disulfonic acid	Fries/Getrost, p 191. (2), Snell, *Metals I*, p 772. (7)
	2,2′-Bipyridyl	Snell, *Metals I*, p 750. (7)
	Chrome Azurol S + Cetyltrimethylammonium bromide	Snell, *Metals I*, p 757. (7), *Coll. Czech. Chem. Comm.*, *45*, 2656, (1980).
	1,10-Phenanthroline	Fries/Getrost, p 199. (2), Marczenko, p 331. (3), Onishi, Part IIa, p 725. (5), Snell, *Metals I*, p 795. (7)
	1,10-Phenanthroline + Bromothymol Blue	*Zh. Anal. Khim.*, *25*, 1348, (1970).
	Ferrozine	Onishi, Part IIa, p 730. (5), Snell, *Metals I*, p 783. (7)
Lanthanum	Arsenazo III	Marczenko, p 468. (3), Snell, *Metals II*, p 1910. (8)

Determination	Reagents	Ref.
Lead	Dithizone	Fries/Getrost, p 207. (2), Onishi, Part IIa, p 824. (5), Snell, *Metals I,* p 2. (7), West, p 34. (10)
	Sodium diethyldithiocarbamate	Fries/Getrost, p 214. (2), Snell, *Metals I,* p 27. (7)
	4-(2-Pyridylazo)resorcinol	Fries/Getrost, p 220. (2), Marczenko, p 347. (3), Snell, *Metals I,* p 34. (7)
Lithium	Thoron	Onishi, Part IIa, p 863. (5), Snell, *Metals II,* p 1726. (8), *Talanta, 30*, 587, (1983).
Magnesium	Eriochrome Black T	Fries/Getrost, p 226. (2), Marczenko, p 355. (3), Onishi, Part IIb, p 13. (6), Snell, *Metals II,* p 1932. (8)
	8-Hydroxyquinoline	Onishi, Part IIb, p 11. (6), Snell, *Metals II,* p 1938. (8)
	8-Hydroxyquinoline + Butylamine	Fries/Getrost, p 228. (2), Snell, *Metals II,* p 1938. (8)
	Titan Yellow	Fries/Getrost, p 234. (2), Marczenko, p 352. (3), Snell, *Metals II,* p 1945. (8)
	Xylidyl Blue	Fries/Getrost, p 231. (2), Onishi, Part IIb, p 14. (6), Snell, *Metals II,* p 1950. (8)
Manganese	Formaldoxime	Fries/Getrost, p 236. (2), Marczenko, p 364. (3), Onishi Part IIb, p 38. (6), Snell, *Metals II,* 1010. (8)
Mercury	Dithizone	Fries/Getrost, p 243. (2), Marczenko, p 373. (3), Onishi, Part IIb, p 66. (6), Snell, *Metals I,* p 107. (7), West, p 29. (10)
	Michler's thioketone	Marczenko, p 375. (3), Snell, *Metals I,* p 126. (7)
	Xylenol Orange	*Talanta, 16*, 1023, (1969)
Molybdenum	Bromopyrogallol Red + Cetylpyridium chloride	West, p 58. (10)
	Phenylfluorone	Snell, *Metals II*, p 1311., *Microchem. J.*, *31*, 56, (1985).
	Toluene-3,4-dithiol	Fries/Getrost, p 251. (2), Marczenko, p 384. (3), Onishi, Part IIb, p 96. (6), Snell, *Metals II,* p 1301. (8)
Nickel	2-(5-Bromo-2-pyridylazo)-5-diethylaminophenol	Marczenko, p 397. (3), *Talanta 28*, 189, (1981).
	Dimethylglyoxime	Fries/Getrost, p 263. (2), Marczenko, p 393. (3), Onishi, Part IIb, p 125. (6), Snell, *Metals I,* p 887. (7)
	Dimethylglyoxime + Oxidant	Fries/Getrost, p 263. (2), Onishi, Part IIb, p 125. (6), Snell, *Metals I,* p 887. (7)
	2,2′-Furildioxime	Marczenko, p 396. (3), Snell, *Metals I,* p 904. (7)
	2-(2-Pyridylazo)-2-naphthol	Snell, *Metals I,* p 910. (7)
	4-(2-Pyridylazo)resorcinol	Snell, *Metals I,* p 911. (7), West, p 39. (10), *Anal. Chim. Acta, 82*, 431, (1976).
Niobium	*N*-Benzoyl-*N*-phenylhydroxylamine	Snell, *Metals II,* p 1425. (8)
	Pyrocatechol + EDTA or 2,2′Bipyridyl or 1-(2-thenoyl)-3,3,3,-trifluoroacetone	Snell, *Metals II,* p 1427. (8)
	Bromopyrogallol red	Marczenko, p 407. (3), Snell, *Metals II,* p 1426. (8)
	Bromopyrogallol red + Cetylpyridinium chloride	*Talanta*, *32*, 189, (1985).
	4-(2-Pyridylazo)resorcinol	Fries/Getrost, p 274. (2), Marczenko, p 406. (3), Onishi, Part IIb, p 160. (7), Snell, *Metals II,* p 1447. (8)

ORGANIC ANALYTICAL REAGENTS FOR THE DETERMINATION OF INORGANIC SUBSTANCES (continued)

Determination	Reagents	Ref.
	Sulfochlorophenol S	Onishi, Part IIb, p 161. (7), Snell, *Metals II,* p 1430. (8)
	Xylenol Orange	Onishi, Part IIb, p 164. (7)
Nitrate	Brucine	Boltz, p 227. (1), Fries/Getrost, p 280. (2), Snell, *Nonmetals,* p 546. (9)
	Chromotropic acid	Boltz, p 229. (1), Fries/Getrost, p 281. (2), Snell, *Nonmetals,* p 548. (9), Williams, p 132. (11), *Fres. Z. Anal. Chem.*, *320*, 490, (1985).
	Sulfanilamide + *N*-(1-Naphthyl)ethylenediamine dihydrochloride	Fries/Getrost, p 279. (2), Snell, *Nonmetals,* p 559. (9)
Nitrite	Sulfanilamide + *N*-(1-Naphthyl)ethylenediamine dihydrochlorine	Boltz, p 241. (1), Snell, *Nonmetals, p* 585. (8), *Analyst, 109,* 1281, (1984).
	Sulfanilic acid + 1-Naphthylamine	Boltz, p 237. (1), Fries/Getrost, p 285. (2), Marczenko, p 419. (3), Snell, *Nonmetals,* p 586. (9)
Osmium	1,5-Diphenylcarbazide	Marczenko, p 428. (3)
Palladium	2-(5 Bromo-2-pyridylazo)-5-diethylaminophenol	*Talanta*, *33*, 939, (1986).
	Dithizone	Marczenko, p 440. (3), Onishi, Part IIb, p 227. (6), Snell, *Metals II,* p 1577. (8)
	2-Nitroso-1-naphthol	Fries/Getrost, p 294. (2), Onishi, Part IIb, p 226. (6), Snell, *Metals II,* p 1581. (8)
	4-(2-Pyridylazo)resorcinol	Snell, *Metals II,* p 1583. (8) *Analyst, 107,* 708, (1982).
Phosphate	Rhodamine B + Molybdate	Snell, *Nonmetals*, p 103. (9)
	Malachite Green + Molybdate	Snell, *Nonmetals*, p 12. (9), Analyst, *108,* 361, (1983).
Platinum	Sulfochlorophenolazorhodamine	Onishi, Part IIb, p 253. (6), *Talanta*, *34*, 87, (1987).
	Dithizone	Fries/Getrost, p 300. (2), Onishi, Part IIb, p 253. (6), Snell, *Metals II,* p 1534. (8)
	2-Mercaptobenzothiazole	Fries/Getrost, p 302. (2), *Zh. Anal. Khim.*, *24*, 1172, (1969).
Rare Earths	Arsenazo I	Marczenko, p 470. (3), Onishi, Part IIa, p 785. (5), Snell, *Metals II,* p 1857. (8)
	Arsenazo III	Fries/Getrost, p 309. (2), Marczenko, p 468. (3), Onishi, Part IIa, p 786. (5), Snell, *Metals II,* p 1862. (8)
	Xylenol Orange	Onishi, Part IIa, p 787. (5), Snell, *Metals II,* p 1874. (8)
Rhenium	2,2′-Furildioxime	Fries/Getrost, p 310. (2), Marczenko, p 181. (3), Onishi, Part IIb, p 288. (6), Snell, *Metals II,* p 1659. (8)
Rhodium	1-(2-Pyridylazo)-2-naphthol	Fries/Getrost, p 311. (2), Snell, *Metals II,* p 1553. (8)
Ruthenium	1,10-Phenanthroline	Onishi, Part IIb, p 331. (6), Snell, *Metals II,* p 1623. (8)
	Thiourea	Fries/Getrost, p 318. (2), Onishi, Part IIb, 329. (6), Snell, *Metals II,* p 1626. (8)
	1,4-Diphenylthiosemicarbazide	Marczenko, p 493. (3), Onishi, Part IIb, p 330. (8)
Scandium	Alizarin red S	Fries/Getrost, p 319. (2), Onishi, Part IIb, p 360. (6), Snell, *Metals I,* p 536. (7)
	Arsenazo III	Onishi, Part IIb, p 359. (6), Snell, *Metals I,* p 539. (7)

Determination	Reagents	Ref.
	Chrome Azurol S	Snell, Metals I, p 551. (7), *Anal. Chim. Acta, 159*, 309, (1984).
	Xylenol Orange	Marczenko, p 501. (3), Onishi, Part IIb, p 357. (6), Snell, *Metals I,* p 547. (7)
Selenium	3,3′-Diaminobenzidine	Boltz, p 391. (1), Fries/Getrost, p 323. (2), Marczenko, p 508. (3), Snell, *Nonmetals,* p 490. (9), West, p 4. (10).
	2,3-Diaminonaphthaline	Snell, *Nonmetals,* p 501. (9)
Silver	Dithizone	Fries/Getrost, p 328. (2), Marczenko, p 524. (3), Onishi, Part IIb, p 379. (6), Snell, *Metals I,* p 82. (7)
	Eosin + 1,10-Phenanthroline	Snell, *Metals I,* p 93. (7)
Sulfate	Methylthymol blue + Barium (II)	Snell, *Nonmetals,* p 457. (9)
Sulfide	*N,N,*-Dimethyl-1,4-phenylenediamine	Boltz, p 483. (1), Fries/Getrost, p 344. (2), Snell, *Nonmetals,* p 400. (9), Williams, p 578. (11)
Sulfite	Pararosaniline + Formaldehyde	Boltz, p 478. (1), Marczenko, p 540. (3), Snell, *Nonmetals,* p 430. (9), Williams, p 591. (11)
Tantalum	Methyl Violet	Marczenko, p 551. (3), Snell, *Metals II,* p 1485. (8)
	4-(2-Pyridylazo)resorcinol	Snell, *Metals II,* p 1488. (8)
	Phenylfluorone	Onishi, Part IIb, p 166. (6), Snell, *Metals II,* p 1486. (8)
Tellurium	Diethyldithiocarbamate	Boltz, p 402. (1), Fries/Getrost, p 348. (2), Snell, *Nonmetals,* p 533. (9), Williams, p 220. (10)
	Bismuthiol II	Boltz, p 401. (1), Marczenko, p 557. (3), Snell, *Nonmetals,* p 524. (9)
Thallium	Brilliant green	Fries/Getrost, p 352. (2), Marczenko, p 567. (3), Onishi, Part IIb, p 426. (6), Snell, *Metals I,* p 45. (7)
	Dithizone	Fries/Getrost, p 355. (2), Onishi, Part IIb, p 426. (6), Snell, *Metals I,* p 54. (7)
	Rhodamine B	Fries/Getrost, p 354. (2), Marczenko, p 566. (3), Onishi, Part IIb, p 424. (6), Snell, *Metals I,* p 63. (7)
Thorium	Arsenazo III	Fries/Getrost, p 360. (2), Marczenko, p 575. (3), Onishi, Part IIb, p 460. (6), Snell, *Metals II,* p 1820. (8)
	Thoron	Marczenko, p 574. (3), Onishi, Part IIb, p 463. (6), Snell, *Metals I,* p 1835. (7)
	Xylenol Orange	Snell, *Metals I,* p 1852. (7)
	Xylenol Orange + Cetyltrimethylammonium bromide	*Talanta, 26*, 499, (1979).
Tin	Pyrocatechol violet (and + Cetyltrimethylammonium bromide)	Marczenko, p 585. (3), Onishi, Part IIb, p 501. (6), Snell, *Metals I, p* 422. (7)
	Gallein	Onishi, Part IIb, p 507, 510. (6), Snell, *Metals I,* p 432. (7)
	Phenylfluorone	Fries/Getrost, p 368. (2), Marczenko, p 582. (3), Onishi, Part IIb, p 497. (6), Snell, *Metals I,* p 444. (7)
	Toluene-3,4-dithiol + Dispersant	Fries/Getrost, p 366. (2), Onishi, Part IIb, p 502. (6), Snell, *Metals I,* p 427. (7)
Titanium	Chromotropic acid	Marczenko, p 593. (3), Onishi, Part IIb, p 551. (6), Snell, *Metals II,* p 1080. (8)

ORGANIC ANALYTICAL REAGENTS FOR THE DETERMINATION OF INORGANIC SUBSTANCES (continued)

Determination	Reagents	Ref.
	Diantipyrinylmethane	Onishi, Part IIb, p 545. (6), Snell, *Metals II,* 1085. (8)
	Tiron	Fries/Getrost, p 376. (2), Onishi, Part IIb, p 549. (6), Snell, *Metals II,* p 1114. (8)
Tungsten	Pyrocatechol Violet	Snell, Metals II, p 1265. (8)
	Tetraphenylarsonium chloride + Thiocyanate	Onishi, Part IIb, p 596. (6), Snell, *Metals II,* p 1278. (8)
	Toluene-3,5-dithiol	Marczenko, p 605. (3), Onishi, Part IIb, p 590. (6), Snell, *Metals II,* p 1267. (8)
Uranium	Arsenazo III	Marczenko, p 611. (3), Onishi, Part IIb, p 627. (6), Snell, *Metals II,* p 1356. (8)
	2-(5-Bromo-2-pyridylazo)diethylaminophenol	Fries/Getrost, p 388. (2), Onishi, Part IIb, p 625. (6)
	Chlorophosphonazo III	Snell, Metals II, p 1367. (8), *Fres. Z. Anal. Chem., 306*, 110, (1981).
	1-(2-Pyridylazo)-2-naphthol	Fries/Getrost, p 386. (2), Onishi, Part IIb, p 625. (6), Snell, *Metals II,* p 1387. (8)
Vanadium	*N*-Benzoyl-*N*-phenylhydroxylamine	Fries/Getrost, p 395. (2), Marczenko, p 625. (3), Snell, *Metals II,* p 1196. (8)
	8-Hydroxyquinoline	Marczenko, p 623. (3), Snell, *Metals II,* p 1209. (8)
	4-(2-pyridylazo)resorcinol	Fries/Getrost, p 404. (23), Marczenko, p 628. (3), Onishi, Part IIb, p 625. (6), Snell, *Metals II,* p 1226. (8)
Yttrium	Alizarin Red S	Fries/Getrost, p 406. (2), Onishi, Part IIa, p 784. (5), Snell, *Metals II,* p 1919. (8)
	Arsenazo III	Marczenko, p 468. (3), Onishi, Part IIa, p 786. (5), Snell, *Metals II,* p 1921. (8)
	Xylenol Orange	Fries/Getrost, p 406. (2), Onishi, Part IIa, p 787. (5), Snell, *Metals II,* p 1923. (8)
Zinc	Dithizone	Fries/Getrost, p 408. (2), Marczenko, p 637. (3), Onishi, Part IIb, p 708. (6), Snell, *Metals II,* p 1042. (8)
	1-(2-Pyridylazo)-2-naphthol	Marczenko, p 639. (3), Onishi, Part IIb, p 719. (6), Snell, *Metals II,* p 1056. (8)
	Xylenol Orange	Fries/Getrost, p 417. (2), Snell, *Metals II,* p 1062. (8), Talanta, *26*, 693, (1979).
	Zircon	Fries/Getrost, p 412. (2), Onishi, Part IIb, p 719. (6), Snell, *Metals II,* p 1063. (8), West, p 23. (10)
Zirconium	Alizarin Red S	Fries/Getrost, p 421. (2), Marczenko, p 647. (3), Onishi, Part IIb, p 763. (6), Snell, *Metals II,* p 1136. (8)
	Arsenazo III	Fries/Getrost, p 421. (2), Onishi, Part IIb, p 770. (6), Snell, *Metals II,* p 1143. (8)
	Pyrocatechol Violet	Onishi, Part IIb, p 771. (6), Snell, *Metals II,* p 1149. (8)
	Morin	Fries/Getrost, p 424. (2), Onishi, Part IIb, p 765. (6), Snell, *Metals II,* p 1158. (8)
	Xylenol Orange	Fries/Getrost, p 419. (2), Marczenko, p 648. (3), Onishi, Part IIb, p 767. (6), Snell, *Metals II,* p 1167. (8)

REVIEWS

Sommer, L, Ackermann, G., Thorburn Burns, D., and Savvin, S. B., *Pure and Applied Chem.*, 62, 2147, 1990.

Sommer, L., Ackermann, G., and Thorburn Burns, D., *Pure and Applied Chem.*, 62, 2323, 1990)

Sommer, L., Komarek, J., and Thorburn Burns, D., *Pure and Applied Chem., 64,* 213, 1992.

Savvin, S. B., *Crit. Rev. Anal. Chem.,* 8, 55, 1979.

MONOGRAPHS

1. Boltz, D. F., and Howell, J. A., *Colorimetric Determination of Nonmetals,* 2nd ed, Wiley, New York, 1978.
2. Fries, J. and Getrost, H., *Organic Reagents for Trace Analysis,* E Merck, Darmstadt, 1977.
3. Marczenko, Z., *Separation and Spectrophotometric Determination of Elements,* Ellis Horwood, Chichester, 1986.
4. Sandell, E. B. and Onishi, H., *Photometric Determination of Traces of Metals. General Aspects, Part I,* 4th ed, J. Wiley, New York, 1978.
5. Onishi, H., *Photometric Determination of Traces of Metals. Part IIa: Individual Metals, Aluminium to Lithium,* 4th ed, J. Wiley, New York, 1986.
6. Onishi, H., *Photometric Determination of Traces of Metals. Part IIb: Individual Metals, Magnesium to Zinc,* 4th ed, J. Wiley, New York, 1989.
7. Snell, F. D., *Photometric and Fluorimetric Methods of Analysis, Metals Part 1,* J. Wiley, New York, 1978.
8. Snell, F. D., *Photometric and Fluorimetric Methods of Analysis, Metals Part 2,* J. Wiley, New York, 1978.
9. Snell, F. D., *Photometric and Fluorimetric Methods of Analysis, Nonmetals,* J. Wiley, New York, 1981.
10. West, T. S. and Nürnberg, H. W., Eds., *The Determination of Trace Metals in Natural Waters,* Blackwell, Oxford, 1988.
11. Williams, W. J.,, *Handbook of Anion Determination,* Butterworth, London, 1979.
12. Townshend, A., Burns, D. T., Guilbault, G. G., Lobinski, R., Mavenzenko, Z., Newman, E., and Onishi, H., *Dictionary of Analytical Reagents,* Chapman & Hall, London, 1993.

ACID-BASE INDICATORS

A. K. Covington

Indicator	p*K* (20° C)	Approx. pH range	Color change	Preparation
Methyl Violet	—	0.0—1.6	Y-B	0.01—0.05% in water
Crystal Violet	—	0.0—1.8	Y-B	0.02% in water
Ethyl Violet	—	0.0—2.4	Y-B	0.1 g in 50 ml 50% v/v methanol-water
Malachite Green	1.3	0.2—1.8	Y-B/G	water
Methyl Green	—	0.2—1.8	Y-B	0.1% in water
2-(*p*-dimethylaminophenylazo)pyridine	—	0.2—1.8	Y-B	0.1% in ethanol
	—	4.4—5.6	R-Y	
o-cresolsulfonephthalein (Cresol Red)	—	0.4—1.8	Y-R	0.1 g in 26.2 ml 0.01 M NaOH + 222.8 ml water
	8.46	7.0—8.8	Y-R	
Quinaldine Red	2.63	1.4—3.2	C-R	1% in ethanol
p-(*p*-dimethylaminophenylazo)benzoic acid, Na salt (Paramethyl Red)	—	1.0—3.0	R-Y	ethanol
4′-aniline azobenzene *m*-sulfonic acid, Na salt (Metanil Yellow)	—	1.2—2.4	R-Y	0.01% in water
4-phenylazodiphenylamine	—	1.2—2.6	R-Y	0.01 g in 1 ml 1 M HCl + 50 ml ethanol + 49 ml water
thymolsulfonephthalein (Thymol Blue)	1.65	1.2—2.8	R-Y	0.1 g in 21.5 ml 0.01 M NaOH + 229.5 ml water
	—	8.0—9.6	R-B	
m-cresolsulfonephthalein (Metacresol Purple)	1.51	1.2—2.8	R-Y	0.1 g in 26.2 ml 0.01 M NaOH + 223.8 ml water
	—	7.4—9.0	Y-P	
p-(*p*-anilinophenylazo)benzenesulfonic acid, Na salt (Orange IV)	—	1.4—2.8	R-Y	0.01% in water
4-*o*-tolylazo-*o*-toluidine	—	1.4—2.8	O-Y	water
erythrosin, disodium salt	—	2.2—3.6	O-R	0.1% in water
Benzopurpurine 48	—	2.2—4.2	V-R	0.1 % in water
N,N-dimethyl-*p*-(*m*-tolylazo)aniline	—	2.6—4.8	R-Y	0.1% in water
4,4′-bis(2-amino-1-naphthylazo) 2,2′-stilbenedisulfonic acid	—	3.0—4.0	P-R	0.1 g in 5.9 ml 0.05 M NaOH + 94.1 ml water
tetrabromophenolphthaleinethyl ester, K salt	—	3.0—4.2	Y-B	0.1% in ethanol
3′,3″,5′,5″ tetrabromophenolsulfone-phthalein (Bromophenol Blue)	4.10	3.0—4.6	Y-B	0.1 g in 14.9 ml 0.01 M NaOH + 235.1 ml water
2,4-dinitrophenol	4.10	2.0—4.7	C-Y	sat. solution in water
p-dimethylaminoazobenzene	—	2.8—4.4	R-Y	0.1 g in 100 ml 90% v/v ethanol-water
diphenyl bis(1-amino-2-naphthyl-azo-4-sulfonic acid) (Congo Red)	—	3.0—5.0	B-R	0.1% in water
4′-dimethylaminoazobenzene-4-sulfonic acid, Na salt (Methyl Orange)	3.46	3.2—4.4	R-Y	0.01% in water
Ethyl Orange	4.34	3.4—4.8	R-Y	0.05—0.2% in water or aqueous ethanol
4-(4-dimethylamino-1-naphylazo)-3-methoxybenzenesulfonic acid	—	3.5—4.8	V-Y	0.1% in 60% ethanol-water
3′,3″,5′,5″-tetrabromo-*m*-cresol-sulfonephthalein (Bromocresol Green)	4.90	3.8—5.4	Y-B	0.1 g in 14.3 ml 0.01 M NaOH + 235.7 ml water
Resazurin	—	3.8—6.4	O-V	water
4-phenylazo-1-naphthylamine	—	4.0—5.6	R-Y	0.1% in ethanol
Ethyl Red	5.42	4.0—5.8	C-R	0.1 g in 100 ml 50% v/v methanol-water
2-(*p*-dimethylaminophenylazo)pyridine	—	0.2—1.8	Y-R	0.1% in ethanol
	—	4.4—5.6	R-Y	
4-(*p*-ethoxyphenylazo)-*m*-phenylene-diamine monohydrochloride	—	4.4—5.8	O-Y	0.1% in water

ACID-BASE INDICATORS (continued)

Indicator	pK (20° C)	Approx. pH range	Color change	Preparation
Resorcin Blue	—	4.4—6.2	R-B	0.2% in ethanol
1,2-dihydroxyanthroquinone-3-sulfonic acid (Alizarin Red S)	—	4.6—6.0	Y-R	water
4-dimethylaminoazobenzene-2-carboxylic acid (Methyl Red)	5.00	4.8—6.0	R-Y	0.02 g in 100 ml 60% v/v ethanol-water
Propyl Red	5.48	4.8—6.6	R-Y	ethanol
5′,5″-dibromo-*o*-cresolsulfonephthalein (Bromocresol Purple)	6.40	5.2—6.8	Y-P	0.1 g in 18.5 ml 0.01 M NaOH + 231.5 ml water
3′,3″-dichlorophenolsulfonephthalein (Chlorophenol Red)	6.25	5.2—6.8	Y-R	0.1 g in 23.6 ml 0.01 M NaOH + 226.4 ml water
p-nitropheno	17.15	5.4—6.6	C-Y	0.1% in water
alizarin	—	5.6—7.2	Y-R	0.1% in methanol
	—	11.0—12.4	R-P	
2-(2,4-dinitrophenylazo)-1-naphthol-3,6-disulfonic acid, disodium salt	—	6.0—7.0	Y-B	0.1% in water
3′,3″-dibromothymolsulfonephthalein (Bromothymol Blue)	7.30	6.0—7.6	Y-B	0.1 g in 16 ml 0.01 M NaOH + 234 ml water
6,8-dinitro-2,4-(1H)quinazolinedione	—	6.4—8.0	C-Y	25 g in 115 ml 1 M NaOH + 50 ml water at 100° C
Brilliant Yellow	—	6.6—7.8	Y-R	1% in water
phenolsulfonephthalein (Phenol Red)	8.00	6.6—8.0	Y-R	0.1 g in 28.2 ml 0.01 M NaOH + 221.8 ml water
3-amino-6-dimethylamino-2-methyl-phenazinium chloride (Neutral Red)	—	6.8—8.0	R-A	0.01 g in 100 ml 50%v/v ethanol-water
m-nitrophenol	8.35	6.8—8.6	C-Y	0.3% in water
o-cresolsulfonephthalein (Cresol Red)	—	0.0—1.0	R-Y	0.1 g in 26.2 ml 0.01 M NaOH + 223.8 ml water
	8.46	7.0—8.8	Y-R	
Turmaric (Curcumin)	—	7.4—8.6	Y-R	ethanol
m-cresolsulfonephthalein (Metacresol Purple)	1.5	1.2—2.8	R-Y	0.1 g in 26.2 ml 0.01 M NaOH + 223.8 ml water
	8.3	7.4—9.0	Y-P	
4,4′-bis(4-amino-1-naphthylazo)-2,2′-stilbenedisulfonic acid	—	8.0—9.0	B-R	0.1 g in 5.9 ml 0.05 M NaOH + 94.1 ml water
thymolsulfonephthalein (Thymol Blue)	1.65	1.2—2.8	R-Y	0.1 g in 21.5 ml 0.01 M NaOH + 228.5 ml water
	9.20	8.0—9.6		
o-cresolphthalein	—	8.2—9.8	C-R	0.04% in ethanol
p-naphtholbenzene	—	8.2—10.0	O-B	1% in dil. alkali
di(*p*-dioxydiphenyl)phthalide (Phenolphthalein)	9.5	8.2—10.0	C-Pk	0.05 g in 100 ml 50% v/v ethanol-water
ethyl-bis(2,4—dimethylphenyl)ethanoate	—	8.4—9.6	C-B	sat. solution in 50% acetone-alcohol
thymolphthalein	—	9.4—10.6	C-B	0.04 g in 100 ml in 50% v/v ethanol-water
5-(*p*-nitrophenylazo)salicylic acid, Na salt (Alizarin Yellow R)	—	10.1—12.0	Y-R	0.01% in water
p-(2,4-dihydroxyphenylazo)benzene-sulfonic acid, Na salt	—	11.4—12.6	Y-O	0.1% in water
5,5′-indigodisulfonic acid, disodium salt	—	11.4—13.0	B-Y	water
2,4,6-trinitrotoluene	—	11.5—13.0	C-O	0.1—0.5% in ethanol
1,3,5-trinitrobenzene	—	12.0—14.0	C-O	0.1—0.5% in ethanol
Clayton Yellow	—	12.2—13.2	Y-A	0.1% in water

Note: C = colorless; Pk = pink; R = red; O = orange; A = amber; Y = yellow; B = blue; B/G = blue/ green; V = violet; P = purple.

Useful reference: E. Bishop, Ed., *Indicators*, Pergamon, Oxford, 1972.

FLUORESCENT INDICATORS

Jack DeMent

Fluorescent indicators are substances which show definite changes in fluorescence with change in pH. Some fluorescent materials are not suitable for indicators since their change in fluorescence is too gradual. Fluorescent indicators find greatest utility in the titration of opaque, highly turbid or deeply colored solutions. A long wavelength ultraviolet ("black light") lamp in a dimly lighted room provides the best environment for titrations involving fluorescent indicators, although bright daylight is sometimes sufficient to evoke a response in the bright green, yellow and orange fluorescent indicators. Titrations are carried out in non-fluorescent glassware. One should check the glassware prior to use to make certain that it does not fluoresce due to the wavelengths of light involved in the titration. The meniscus of the liquid in the burette can be followed when a few particles of an insoluble fluorescent solid are dropped onto its surface.

In this table the indicators are arranged by approximate pH range covered. In the case of some of the dyestuffs the end point may vary slightly with the source or manufacturer.

pH 0 to 2

Indicator	C.I.	From pH	To pH
Benzoflavine	—	0.3, yellow fl.	1.7, green fl.
3,6-Dioxyphthalimide	—	0, blue fl.	2.4, green fl.
Eosine YS	768	0, yellow colored	3.0, yellow fl.
Erythrosine	772	0, yellow colored	3.6, yellow fl.
Esculin	—	1.5, colorless	2, blue fl.
4-Ethoxyacridone	—	1.2, green fl.	3.2, blue fl.
3,6-Tetramethyldiaminooxanthone	—	1.2, green fl.	3.4, blue fl.

pH 2 to 4

Indicator	C.I.	From pH	To pH
Chromotropic acid	—	3.5, colorless	4.5, blue fl.
Fluorescein	766	4, colorless	4.5, green fl.
Magdala Red	—	3.0, purple colored	4.0, fl.
α-Naphthylamine	—	3.4, colorless	4.8, blue fl.
β-Naphthylamine	—	2.8, colorless	4.4, violet fl.
Phloxine	774	3.4, colorless	5.0, bright yellow fl.
Salicylic acid	—	2.5, colorless	3.5, blue fl.

pH 4 to 6

Indicator	C.I.	From pH	To pH
Acridine	788	4.9, green fl.	5.1, violet colored
Dichlorofluorescein	—	4.0, colorless	5.0, green fl.
3,6-Dioxyxanthone	—	5.4, colorless	7.6, blue-violet fl.
Erythrosine	772	4.0, colorless	4.5, yellow-green fl.
β-Methylesculetin	—	4.0, colorless	6.2, blue fl.
Neville-Winther acid	—	6.0, colorless	6.5, blue fl.
Resorufin	—	4.4, yellow fl.	6.4, weak orange fl.
Quininic acid	—	4.0, yellow colored	5.0, blue fl.
Quinine [first end point]	—	5.0, blue fl.	6.1, violet fl.

pH 6 to 8

Indicator	C.I.	From pH	To pH
Acid R Phosphine	—	(claimed for range pH 6.0–7.0)	
Brilliant Diazol Yellow	—	6.5, colorless	7.5, violet fl.
Cleves acid	—	6.5, colorless	7.5, green fl.
Coumaric acid	—	7.2, colorless	9.0, green fl.
3,6-Dioxyphthalic dinitrile	—	5.8, blue fl.	8.2, green fl.
Magnesium 8-hydroxyquinolinate	—	6.5, colorless	7.5, golden fl.
β-Methylumbelliferone	—	7.0, colorless	7.5, blue fl.
1-Naphthol-4-sulfonic acid	—	6.0, colorless	6.5, blue fl.
Orcinaurine	—	6.5, colorless	8.0, green fl.
Patent Phosphine	789	(for the range pH 6.0–7.0, green-yellow fl.)	
Thioflavine	816	(for the region pH 6.5–7.0, yellow fl.)	
Umbelliferone	—	6.5, colorless	7.6, blue fl.

FLUORESCENT INDICATORS (Continued)

pH 8 to 10

Indicator	C.I.	From pH	To pH
Acridine Orange	788	8.4, orange colored	10.4, green fl.
Ethoxyphenylnaphthostilbazonium chloride	—	9, green fl.	11, non-fl.
G Salt	—	9.0, dull blue fl.	9.5, bright blue fl.
Naphthazol derivatives	—	8.2, colorless	10.0, yellow or green fl.
α-Naphthionic acid	—	9, blue fl.	11, green fl.
2-Naphthol-3,6-disulfonic acid	—	9.5, dark blue fl.	Light blue fl. at higher pH
β-Naphthol	—	8.6, colorless	Blue fl. at higher pH
α-Naphtholsulfonic acid	—	8.0, dark blue fl.	9.0, bright violet fl.
1,4-Naphtholsulfonic acid	—	8.2, dark blue fl.	Light blue fl. at higher pH
Orcinsulfonphthalein	—	8.6, yellow colored	10.0 fl.
Quinine [second end point]	—	9.5, violet fl.	10.0, colorless
R-Salt	—	9.0, dull blue fl.	9.5, bright blue fl.
Sodium 1-naphthol-2-sulfonate	—	9.0, dark blue fl.	10.0, bright violet fl.

pH 10 to 12

Indicator	C.I.	From pH	To pH
Coumarin	—	9.8, deep green fl.	12, light green fl.
Eosine BN	771	10.5, colorless	14.0, yellow fl.
Papaverine (permanganate oxidized)	—	9.5, yellow fl.	11.0, blue fl.
Schaffers Salt	—	5.0, violet fl.	11.0, green-blue fl.
SS-Acid (sodium salt)	—	10.0, violet fl.	12.0, yellow colored

pH 12 to 14

Indicator	C.I.	From pH	To pH
Cotarnine	—	12.0, yellow fl.	13.0, white fl.
α-Naphthionic acid	—	12, blue fl.	13, green fl.
β-Naphthionic acid	—	12, blue fl.	13, violet fl.

CONVERSION FORMULAS FOR CONCENTRATION OF SOLUTIONS

A = Weight per cent of solute
B = Molecular weight of solvent
E = Molecular weight of solute
F = Grams of solute per liter of solution
G = Molality
M = Molarity
N = Mole fraction
R = Density of solution in grams per milliliter

Concentration of solute—SOUGHT	Concentration of solute—GIVEN: A	N	G	M	F
A	—	$\frac{100N \times E}{N \times E + (1 - N)B}$	$\frac{100G \times E}{1000 + G \times E}$	$\frac{M \times E}{10R}$	$\frac{F}{10R}$
N	$\frac{\frac{A}{E}}{\frac{A}{E} + \frac{100 - A}{B}}$	—	$\frac{B \times G}{B \times G + 1000}$	$\frac{B \times M}{M(B - E) + 1000R}$	$\frac{B \times F}{F(B - E) + 1000R \times E}$
G	$\frac{1000A}{E(100 - A)}$	$\frac{1000N}{B - N \times B}$	—	$\frac{1000M}{1000R - (M \times E)}$	$\frac{1000F}{E(1000R - F)}$
M	$\frac{10R \times A}{E}$	$\frac{1000R \times N}{N \times E + (1 - N)B}$	$\frac{1000R \times G}{1000 + E \times G}$	—	$\frac{F}{E}$
F	$10AR$	$\frac{1000R \times N \times E}{N \times E + (1 - N)B}$	$\frac{1000R \times G \times E}{1000 + G \times E}$	$M \times E$	—

ELECTROCHEMICAL SERIES

Petr Vanýsek

There are three tables for this electrochemical series. Each table lists standard reduction potentials, $E°$ values, at 298.15 K (25°C), and at a pressure of 101.325 kPa (1 atm). Table 1 is an alphabetical listing of the elements, according to the symbol of the elements. Thus, data for silver (Ag) precedes those for aluminum (Al). Table 2 lists only those reduction reactions which have $E°$ values positive in respect to the standard hydrogen electrode. In Table 2, the reactions are listed in the order of increasing positive potential, and they range form 0.0000 V to + 3.4 V. Table 3 lists only those reduction potentials which have $E°$ negative with respect to the standard hydrogen electrode. In Table 3, the reactions are listed in the order of decreasing potential and range from 0.0000 V to –4.10 V. The reliability of the potentials is not the same for all the data. Typically, the values with fewer significant figures have lower reliability. The values of reduction potentials, in particular those of less common reactions, are not definite; they are subject to occasional revisions.

Abbreviations: bipy = 2,2′-dipyridine, or bipyridine; en = ethylenediamine; phen = 1,10-phenanthroline.

REFERENCES

1. G. Milazzo, S. Caroli, and V. K. Sharma, *Tables of Standard Electrode Potentials*, Wiley, Chichester, 1978.
2. A. J. Bard, R. Parsons, and J. Jordan, *Standard Potentials in Aqueous Solutions*, Marcel Dekker, New York, 1985.
3. S. G. Bratsch, *J. Phys. Chem. Ref. Data*, 18, 1—21, 1989.

TABLE 1
Alphabetical Listing

Reaction	$E°$/V
$Ac^{3+} + 3e \rightleftharpoons Ac$	–2.20
$Ag^{+} + e \rightleftharpoons Ag$	0.7996
$Ag^{2+} + e \rightleftharpoons Ag^{+}$	1.980
$Ag(ac) + e \rightleftharpoons Ag + (ac)^{-}$	0.643
$AgBr + e \rightleftharpoons Ag + Br^{-}$	0.07133
$AgBrO_3 + e \rightleftharpoons Ag + BrO_3^{-}$	0.546
$Ag_2C_2O_4 + 2e \rightleftharpoons 2Ag + C_2O_4^{2-}$	0.4647
$AgCl + e \rightleftharpoons Ag + Cl^{-}$	0.22233
$AgCN + e \rightleftharpoons Ag + CN^{-}$	–0.017
$Ag_2CO_3 + 2e \rightleftharpoons 2Ag + CO_3^{2-}$	0.47
$Ag_2CrO_4 + 2e \rightleftharpoons 2Ag + CrO_4^{2-}$	0.4470
$AgF + e \rightleftharpoons Ag + F^{-}$	0.779
$Ag_4[Fe(CN)_6] + 4e \rightleftharpoons 4Ag + [Fe(CN)_6]^{4-}$	0.1478
$AgI + e \rightleftharpoons Ag + I^{-}$	–0.15224
$AgIO_3 + e \rightleftharpoons Ag + IO_3^{-}$	0.354
$Ag_2MoO_4 + 2e \rightleftharpoons 2Ag + MoO_4^{2-}$	0.4573
$AgNO_2 + e \rightleftharpoons Ag + 2NO_2^{-}$	0.564
$Ag_2O + H_2O + 2e \rightleftharpoons 2Ag + 2OH^{-}$	0.342
$Ag_2O_3 + H_2O + 2e \rightleftharpoons 2AgO + 2OH^{-}$	0.739
$Ag^{3+} + 2e \rightleftharpoons Ag^{+}$	1.9
$Ag^{3+} + e \rightleftharpoons Ag^{2+}$	1.8
$Ag_2O_2 + 4H^{+} + e \rightleftharpoons 2Ag + 2H_2O$	1.802
$2AgO + H_2O + 2e \rightleftharpoons Ag_2O + 2OH^{-}$	0.607
$AgOCN + e \rightleftharpoons Ag + OCN^{-}$	0.41
$Ag_2S + 2e \rightleftharpoons 2Ag + S^{2+}$	–0.691
$Ag_2S + 2H^{+} + 2e \rightleftharpoons 2Ag + H_2S$	–0.0366
$AgSCN + e \rightleftharpoons Ag + SCN^{-}$	0.08951
$Ag_2SeO_3 + 2e \rightleftharpoons 2Ag + SeO_4^{2-}$	0.3629
$Ag_2SO_4 + 2e \rightleftharpoons 2Ag + SO_4^{2-}$	0.654
$Ag_2WO_4 + 2e \rightleftharpoons 2Ag + WO_4^{2-}$	0.4660
$Al^{3+} + 3e \rightleftharpoons Al$	–1.662
$Al(OH)_3 + 3e \rightleftharpoons Al + 3OH^{-}$	–2.31
$Al(OH)_4 + 3e \rightleftharpoons Al + 4OH^{-}$	–2.328
$H_2AlO_3^{-} + H_2O + 3e \rightleftharpoons Al + 4OH^{-}$	–2.33
$AlF_6^{3-} + 3e \rightleftharpoons Al + 6F^{-}$	–2.069
$Am^{4+} + e \rightleftharpoons Am^{3+}$	2.60
$Am^{2+} + 2e \rightleftharpoons Am$	–1.9
$Am^{3+} + 3e \rightleftharpoons Am$	–2.048
$Am^{3+} + e \rightleftharpoons Am^{2+}$	–2.3
$As + 3H^{+} + 3e \rightleftharpoons AsH_3$	–0.608
$As_2O_3 + 6H^{+} + 6e \rightleftharpoons 2As + 3H_2O$	0.234
$HAsO_2 + 3H^{+} + 3e \rightleftharpoons As + 2H_2O$	0.248
$AsO_2^{-} + 2H_2O + 3e \rightleftharpoons As + 4OH^{-}$	–0.68
$H_3AsO_4 + 2H^{+} + 2e^{-} \rightleftharpoons HAsO_2 + 2H_2O$	0.560
$AsO_4^{3-} + 2H_2O + 2e \rightleftharpoons AsO_2^{-} + 4OH^{-}$	–0.71
$At_2 + 2e \rightleftharpoons 2At^{-}$	0.3
$Au^{+} + e \rightleftharpoons Au$	1.692
$Au^{3+} + 2e \rightleftharpoons Au^{+}$	1.401
$Au^{3+} + 3e \rightleftharpoons Au$	1.498
$Au^{2+} + e^{-} \rightleftharpoons Au^{+}$	1.8
$AuOH^{2+} + H^{+} + 2e \rightleftharpoons Au^{+} + H_2O$	1.32
$AuBr_2^{-} + e \rightleftharpoons Au + 2Br^{-}$	0.959
$AuBr_4^{-} + 3e \rightleftharpoons Au + 4Br^{-}$	0.854
$AuCl_4^{-} + 3e \rightleftharpoons Au + 4Cl^{-}$	1.002
$Au(OH)_3 + 3H^{+} + 3e \rightleftharpoons Au + 3H_2O$	1.45
$H_2BO_3^{-} + 5H_2O + 8e \rightleftharpoons BH_4^{-} + 8OH^{-}$	–1.24
$H_2BO_3^{-} + H_2O + 3e \rightleftharpoons B + 4OH^{-}$	–1.79
$H_3BO_3 + 3H^{+} + 3e \rightleftharpoons B + 3H_2O$	–0.8698
$B(OH)_3 + 7H^{+} + 8e \rightleftharpoons BH_4^{-} + 3H_2O$	–0.481
$Ba^{2+} + 2e \rightleftharpoons Ba$	–2.912
$Ba^{2+} + 2e \rightleftharpoons Ba(Hg)$	–1.570
$Ba(OH)_2 + 2e \rightleftharpoons Ba + 2OH^{-}$	–2.99
$Be^{2+} + 2e \rightleftharpoons Be$	–1.847
$Be_2O_3^{2-} + 3H_2O + 4e \rightleftharpoons 2Be + 6OH^{-}$	–2.63

TABLE 1
Alphabetical Listing (continued)

Reaction	E°/V
p–benzoquinone $+ 2H^+ + 2e \rightleftharpoons$ hydroquinone	0.6992
$Bi^+ + e \rightleftharpoons Bi$	0.5
$Bi^{3+} + 3e \rightleftharpoons Bi$	0.308
$Bi^{3+} + 2e \rightleftharpoons Bi^+$	0.2
$Bi + 3H^+ + 3e \rightleftharpoons BiH_3$	–0.8
$BiCl_4^- + 3e \rightleftharpoons Bi + 4Cl^-$	0.16
$Bi_2O_3 + 3H_2O + 6e \rightleftharpoons 2Bi + 6OH^-$	–0.46
$Bi_2O_4 + 4H^+ + 2e \rightleftharpoons 2BiO^+ + 2H_2O$	1.593
$BiO^+ + 2H^+ + 3e \rightleftharpoons Bi + H_2O$	0.320
$BiOCl + 2H^+ + 3e \rightleftharpoons Bi + Cl^- + H_2O$	0.1583
$Bk^{4+} + e \rightleftharpoons Bk^{3+}$	1.67
$Bk^{2+} + 2e \rightleftharpoons Bk$	–1.6
$Bk^{3+} + e \rightleftharpoons Bk^{2+}$	–2.8
$Br_2(aq) + 2e \rightleftharpoons 2Br^-$	1.0873
$Br_2(\ell) + 2e \rightleftharpoons 2Br^-$	1.066
$HBrO + H^+ + 2e \rightleftharpoons Br^- + H_2O$	1.331
$HBrO + H^+ + e \rightleftharpoons 1/2\,Br_2(aq) + H_2O$	1.574
$HBrO + H^+ + e \rightleftharpoons 1/2\,Br_2(\ell) + H_2O$	1.596
$BrO^- + H_2O + 2e \rightleftharpoons Br^- + 2OH^-$	0.761
$BrO_3^- + 6H^+ + 5e \rightleftharpoons 1/2\,Br_2 + 3H_2O$	1.482
$BrO_3^- + 6H^+ + 6e \rightleftharpoons Br^- + 3H_2O$	1.423
$BrO_3^- + 3H_2O + 6e \rightleftharpoons Br^- + 6OH^-$	0.61
$(CN)_2 + 2H^+ + 2e \rightleftharpoons 2HCN$	0.373
$2HCNO + 2H^+ + 2e \rightleftharpoons (CN)_2 + 2H_2O$	0.330
$(CNS)_2 + 2e \rightleftharpoons 2CNS^-$	0.77
$CO_2 + 2H^+ + 2e \rightleftharpoons HCOOH$	–0.199
$Ca^+ + e \rightleftharpoons Ca$	–3.80
$Ca^{2+} + 2e \rightleftharpoons Ca$	–2.868
$Ca(OH)_2 + 2e \rightleftharpoons Ca + 2OH^-$	–3.02
Calomel electrode, 1 molal KCl	0.2800
Calomel electrode, 1 molar KCl (NCE)	0.2801
Calomel electrode, 0.1 molar KCl	0.3337
Calomel electrode, saturated KCl (SCE)	0.2412
Calomel electrode, saturated NaCl (SSCE)	0.2360
$Cd^{2+} + 2e \rightleftharpoons Cd$	–0.4030
$Cd^{2+} + 2e \rightleftharpoons Cd(Hg)$	–0.3521
$Cd(OH)_2 + 2e \rightleftharpoons Cd(Hg) + 2OH^-$	–0.809
$CdSO_4 + 2e \rightleftharpoons Cd + SO_4^{2-}$	–0.246
$Cd(OH)_4^{2-} + 2e \rightleftharpoons Cd + 4OH^-$	–0.658
$CdO + H_2O + 2e \rightleftharpoons Cd + 2OH^-$	–0.783
$Ce^{3+} + 3e \rightleftharpoons Ce$	–2.336
$Ce^{3+} + 3e \rightleftharpoons Ce(Hg)$	–1.4373
$Ce^{4+} + e \rightleftharpoons Ce^{3+}$	1.72
$CeOH^{3+} + H^+ + e \rightleftharpoons Ce^{3+} + H_2O$	1.715
$Cf^{4+} + e \rightleftharpoons Cf^{3+}$	3.3
$Cf^{3+} + e \rightleftharpoons Cf^{2+}$	–1.6
$Cf^{3+} + 3e \rightleftharpoons Cf$	–1.94
$Cf^{2+} + 2e \rightleftharpoons Cf$	–2.12
$Cl_2(g) + 2e \rightleftharpoons 2Cl^-$	1.35827
$HClO + H^+ + e \rightleftharpoons 1/2\,Cl_2 + H_2O$	1.611
$HClO + H^+ + 2e \rightleftharpoons Cl^- + H_2O$	1.482
$ClO^- + H_2O + 2e \rightleftharpoons Cl^- + 2OH^-$	0.81
$ClO_2 + H^+ + e \rightleftharpoons HClO_2$	1.277
$HClO_2 + 2H^+ + 2e \rightleftharpoons HClO + H_2O$	1.645
$HClO_2 + 3H^+ + 3e \rightleftharpoons 1/2\,Cl_2 + 2H_2O$	1.628

Reaction	E°/V
$HClO_2 + 3H^+ + 4e \rightleftharpoons Cl^- + 2H_2O$	1.570
$ClO_2^- + H_2O + 2e \rightleftharpoons ClO^- + 2OH^-$	0.66
$ClO_2^- + 2H_2O + 4e \rightleftharpoons Cl^- + 4OH^-$	0.76
$ClO_2(aq) + e \rightleftharpoons ClO_2^-$	0.954
$ClO_3^- + 2H^+ + e \rightleftharpoons ClO_2 + H_2O$	1.152
$ClO_3^- + 3H^+ + 2e \rightleftharpoons HClO_2 + H_2O$	1.214
$ClO_3^- + 6H^+ + 5e \rightleftharpoons 1/2\,Cl_2 + 3H_2O$	1.47
$ClO_3^- + 6H^+ + 6e \rightleftharpoons Cl^- + 3H_2O$	1.451
$ClO_3^- + H_2O + 2e \rightleftharpoons ClO_2^- + 2OH^-$	0.33
$ClO_3^- + 3H_2O + 6e \rightleftharpoons Cl^- + 6OH^-$	0.62
$ClO_4^- + 2H^+ + 2e \rightleftharpoons ClO_3^- H_2O$	1.189
$ClO_4^- + 8H^+ + 7e \rightleftharpoons 1/2\,Cl_2 + 4H_2O$	1.39
$ClO_4^- + 8H^+ + 8e \rightleftharpoons Cl^- + 4H_2O$	1.389
$ClO_4^- + H_2O + 2e \rightleftharpoons ClO_3^- + 2OH^-$	0.36
$Cm^{4+} + e \rightleftharpoons Cm^{3+}$	3.0
$Cm^{3+} + 3e \rightleftharpoons Cm$	–2.04
$Co^{2+} + 2e \rightleftharpoons Co$	–0.28
$Co^{3+} + e \rightleftharpoons Co^{2+}$	1.92
$[Co(NH_3)_6]^{3+} + e \rightleftharpoons [Co(NH_3)_6]^{2+}$	0.108
$Co(OH)_2 + 2e \rightleftharpoons Co + 2OH^-$	–0.73
$Co(OH)_3 + e \rightleftharpoons Co(OH)_2 + OH^-$	0.17
$Cr^{2+} + 2e \rightleftharpoons Cr$	–0.913
$Cr^{3+} + e \rightleftharpoons Cr^{2+}$	–0.407
$Cr^{3+} + 3e \rightleftharpoons Cr$	–0.744
$Cr_2O_7^{2-} + 14H^+ + 6e \rightleftharpoons 2Cr^{3+} + 7H_2O$	1.232
$CrO_2^- + 2H_2O + 3e \rightleftharpoons Cr + 4OH^-$	–1.2
$HCrO_4^- + 7H^+ + 3e \rightleftharpoons Cr^{3+} + 4H_2O$	1.350
$CrO_2 + 4H^+ + e \rightleftharpoons Cr^{3+} \cdot 2H_2O$	1.48
$Cr_2O_5 + 10H^+ + e \rightleftharpoons CrO_2 \cdot 5H_2O$	1.3
$CrO_4^{2-} + 4H_2O + 3e \rightleftharpoons Cr(OH)_3 + 5OH^-$	–0.13
$Cr(OH)_3 + 3e \rightleftharpoons Cr + 3OH^-$	–1.48
$Cs^+ + e \rightleftharpoons Cs$	–3.026
$Cu^+ + e \rightleftharpoons Cu$	0.521
$Cu^{2+} + e \rightleftharpoons Cu^+$	0.153
$Cu^{2+} + 2e \rightleftharpoons Cu$	0.3419
$Cu^{2+} + 2e \rightleftharpoons Cu(Hg)$	0.345
$Cu^{3+} + e \rightleftharpoons Cu^{2+}$	2.4
$Cu_2O_3 + 6H^+ + e \rightleftharpoons Cu^{2+} + 3H_2O$	2.0
$Cu^{2+} + 2CN^- + e \rightleftharpoons [Cu(CN)_2]^-$	1.103
$CuI_2^- + e \rightleftharpoons Cu + 2I^-$	0.00
$Cu_2O + H_2O + 2e \rightleftharpoons 2Cu + 2OH^-$	–0.360
$Cu(OH)_2 + 2e \rightleftharpoons Cu + 2OH^-$	–0.222
$2Cu(OH)_2 + 2e \rightleftharpoons Cu_2O + 2OH^- + H_2O$	–0.080
$2D^+ + 2e \rightleftharpoons D_2$	–0.013
$Dy^{2+} + 2e \rightleftharpoons Dy$	–2.2
$Dy^{3+} + 3e \rightleftharpoons Dy$	–2.295
$Dy^{3+} + e \rightleftharpoons Dy^{2+}$	–2.6
$Er^{2+} + 2e \rightleftharpoons Er$	–2.0
$Er^{3+} + 3e \rightleftharpoons Er$	–2.331
$Er^{3+} + e \rightleftharpoons Er^{2+}$	–3.0
$Es^{3+} + e \rightleftharpoons Es^{2+}$	–1.3
$Es^{3+} + 3e \rightleftharpoons Es$	–1.91
$Es^{2+} + 2e \rightleftharpoons Es$	–2.23
$Eu^{2+} + 2e \rightleftharpoons Eu$	–2.812
$Eu^{3+} + 3e \rightleftharpoons Eu$	–1.991

TABLE 1
Alphabetical Listing (continued)

Reaction	E°/V
$Eu^{3+} + e \rightleftharpoons Eu^{2+}$	–0.36
$F_2 + 2H^+ + 2e \rightleftharpoons 2HF$	3.053
$F_2 + 2e \rightleftharpoons 2F^-$	2.866
$F_2O + 2H^+ + 4e \rightleftharpoons H_2O + 2F^-$	2.153
$Fe^{2+} + 2e \rightleftharpoons Fe$	–0.447
$Fe^{3+} + 3e \rightleftharpoons Fe$	–0.037
$Fe^{3+} + e \rightleftharpoons Fe^{2+}$	0.771
$HFeO_4^- + 2H^+ + 3e \rightleftharpoons Fe_2O_3 + H_2O$	2.09
$HFeO_4^- + 4H^+ + 3e \rightleftharpoons FeOOH + 2H_2O$	2.08
$HFeO_4^- + 8H^+ + 3e \rightleftharpoons Fe^{3+} + 4H_2O$	2.07
$Fe_2O_3 + 4H^+ + 2e \rightleftharpoons 2FeOH^+ + H_2O$	0.16
$[Fe(CN)_6]^{3-} + e \rightleftharpoons [Fe(CN)_6]^{4-}$	0.358
$FeO_4^{2-} + 8H^+ + 3e \rightleftharpoons Fe^{3+} + 4H_2O$	2.20
$[Fe(bipy)_2]^{3+} + e \rightleftharpoons Fe(bipy)_2]^{2+}$	0.78
$[Fe(bipy)_3]^{3+} + e \rightleftharpoons Fe(bipy)_3]^{2+}$	1.03
$Fe(OH)_3 + e \rightleftharpoons Fe(OH)_2 + OH^-$	–0.56
$[Fe(phen)_3]^{3+} + e \rightleftharpoons [Fe(phen)_3]^{2+}$	1.147
$[Fe(phen)_3]^{3+} + e \rightleftharpoons [Fe(phen)_3]^{2+}$ (1 molar H_2SO_4)	1.06
$[Ferricinium]^+ + e \rightleftharpoons$ ferrocene	0.400
$Fm^{3+} + e \rightleftharpoons Fm^{2+}$	–1.1
$Fm^{3+} + 3e \rightleftharpoons Fm$	–1.89
$Fm^{2+} + 2e \rightleftharpoons Fm$	–2.30
$Fr^+ + e \rightleftharpoons Fr$	–2.9
$Ga^{3+} + 3e \rightleftharpoons Ga$	–0.549
$Ga^+ + e \rightleftharpoons Ga$	–0.2
$GaOH^{2+} + H^+ + 3e \rightleftharpoons Ga + H_2O$	–0.498
$H_2GaO^-_3 + H_2O + 3e \rightleftharpoons Ga + 4OH^-$	–1.219
$Gd^{3+} + 3e \rightleftharpoons Gd$	–2.279
$Ge^{2+} + 2e \rightleftharpoons Ge$	0.24
$Ge^{4+} + 4e \rightleftharpoons Ge$	0.124
$Ge^{4+} + 2e \rightleftharpoons Ge^{2+}$	0.00
$GeO_2 + 2H^+ + 2e \rightleftharpoons GeO + H_2O$	–0.118
$H_2GeO_3 + 4H^+ + 4e \rightleftharpoons Ge + 3H_2O$	–0.182
$2H^+ + 2e \rightleftharpoons H_2$	0.00000
$H_2 + 2e \rightleftharpoons 2H^-$	–2.23
$HO_2 + H^+ + e \rightleftharpoons H_2O_2$	1.495
$2H_2O + 2e \rightleftharpoons H_2 + 2OH^-$	–0.8277
$H_2O_2 + 2H^+ + 2e \rightleftharpoons 2H_2O$	1.776
$Hf^{4+} + 4e \rightleftharpoons Hf$	–1.55
$HfO^{2+} + 2H^+ + 4e \rightleftharpoons Hf + H_2O$	–1.724
$HfO_2 + 4H^+ + 4e \rightleftharpoons Hf + 2H_2O$	–1.505
$HfO(OH)_2 + H_2O + 4e \rightleftharpoons Hf + 4OH^-$	–2.50
$Hg^{2+} + 2e \rightleftharpoons Hg$	0.851
$2Hg^{2+} + 2e \rightleftharpoons Hg_2^{2+}$	0.920
$Hg_2^{2+} + 2e \rightleftharpoons 2Hg$	0.7973
$Hg_2(ac)_2 + 2e \rightleftharpoons 2Hg + 2(ac)^-$	0.51163
$Hg_2Br_2 + 2e \rightleftharpoons 2Hg + 2Br^-$	0.13923
$Hg_2Cl_2 + 2e \rightleftharpoons 2Hg + 2Cl^-$	0.26808
$Hg_2HPO_4 + 2e \rightleftharpoons 2Hg + HPO_4^{2-}$	0.6359
$Hg_2I_2 + 2e \rightleftharpoons 2Hg + 2I^-$	–0.0405
$Hg_2O + H_2O + 2e \rightleftharpoons 2Hg + 2OH^-$	0.123
$HgO + H_2O + 2e \rightleftharpoons Hg + 2OH^-$	0.0977
$Hg(OH)_2 + 2H^+ + 2e \rightleftharpoons Hg + 2H_2O$	1.034
$Hg_2SO_4 + 2e \rightleftharpoons 2Hg + SO_4^{2-}$	0.6125
$Ho^{2+} + 2e \rightleftharpoons Ho$	–2.1
$Ho^{3+} + 3e \rightleftharpoons Ho$	–2.33
$Ho^{3+} + e \rightleftharpoons Ho^{2+}$	–2.8
$I_2 + 2e \rightleftharpoons 2I^-$	0.5355
$I_3^- + 2e \rightleftharpoons 3I^-$	0.536
$H_3IO_6 + 2e \rightleftharpoons IO_3 + 3OH^-$	0.7
$H_5IO_6 + H^+ + 2e \rightleftharpoons IO_3^- + 3H_2O$	1.601
$2HIO + 2H^+ + 2e \rightleftharpoons I_2 + 2H_2O$	1.439
$HIO + H^+ + 2e \rightleftharpoons I^- + H_2O$	0.987
$IO^- + H_2O + 2e \rightleftharpoons I^- + 2OH^-$	0.485
$2IO_3^- + 12H^+ + 10e \rightleftharpoons I_2 + 6H_2O$	1.195
$IO_3^- + 6H^+ + 6e \rightleftharpoons I^- + 3H_2O$	1.085
$IO_3^- + 2H_2O + 4e \rightleftharpoons IO^- + 4OH^-$	0.15
$IO_3^- + 3H_2O + 6e \rightleftharpoons IO^- + 6OH^-$	0.26
$In^+ + e \rightleftharpoons In$	–0.14
$In^{2+} + e \rightleftharpoons In^+$	–0.40
$In^{3+} + e \rightleftharpoons In^{2+}$	–0.49
$In^{3+} + 2e \rightleftharpoons In^+$	–0.443
$In^{3+} + 3e \rightleftharpoons In$	–0.3382
$In(OH)_3 + 3e \rightleftharpoons In + 3OH^-$	–0.99
$In(OH)_4^- + 3e \rightleftharpoons In + 4OH^-$	–1.007
$In_2O_3 + 3H_2O + 6e \rightleftharpoons 2In + 6OH^-$	–1.034
$Ir^{3+} + 3e \rightleftharpoons Ir$	1.156
$[IrCl_6]^{2-} + e \rightleftharpoons [IrCl_6]^{3-}$	0.8665
$[IrCl_6]^{3-} + 3e \rightleftharpoons Ir + 6Cl^-$	0.77
$Ir_2O_3 + 3H_2O + 6e \rightleftharpoons 2Ir + 6OH^-$	0.098
$K^+ + e \rightleftharpoons K$	–2.931
$La^{3+} + 3e \rightleftharpoons La$	–2.379
$La(OH)_3 + 3e \rightleftharpoons La + 3OH^-$	–2.90
$Li^+ + e \rightleftharpoons Li$	–3.0401
$Lr^{3+} + 3e \rightleftharpoons Lr$	–1.96
$Lu^{3+} + 3e \rightleftharpoons Lu$	–2.28
$Md^{3+} + e \rightleftharpoons Md^{2+}$	–0.1
$Md^{3+} + 3e \rightleftharpoons Md$	–1.65
$Md^{2+} + 2e \rightleftharpoons Md$	–2.40
$Mg^+ + e \rightleftharpoons Mg$	–2.70
$Mg^{2+} + 2e \rightleftharpoons Mg$	–2.372
$Mg(OH)_2 + 2e \rightleftharpoons Mg + 2OH^-$	–2.690
$Mn^{2+} + 2e \rightleftharpoons Mn$	–1.185
$Mn^{3+} + 3 \rightleftharpoons Mn^{2+}$	1.5415
$MnO_2 + 4H^+ + 2e \rightleftharpoons Mn^{2+} + 2H_2O$	1.224
$MnO_4^- + e \rightleftharpoons MnO_4^{2-}$	0.558
$MnO_4^- + 4H^+ + 3e \rightleftharpoons MnO_2 + 2H_2O$	1.679
$MnO_4^- + 8H^+ + 5e \rightleftharpoons Mn^{2+} + 4H_2O$	1.507
$MnO_4^- + 2H_2O + 3e \rightleftharpoons MnO_2 + 4OH^-$	0.595
$MnO_4^{2-} + 2H_2O + 2e \rightleftharpoons MnO_2 + 4OH^-$	0.60
$Mn(OH)_2 + 2e \rightleftharpoons Mn + 2OH^-$	–1.56
$Mn(OH)_3 + e \rightleftharpoons Mn(OH)_2 + OH^-$	0.15
$Mn_2O_3 + 6H + e \rightleftharpoons Mn^{2+} + 3H_2O$	1.485
$Mo^{3+} + 3e \rightleftharpoons Mo$	–0.200
$MoO_2 + 4H^+ + e \rightleftharpoons Mo + 4H_2O$	–0.152
$H_3Mo_7O_{24}^{3-} + 45H^+ + 42e \rightleftharpoons 7Mo + 24H_2O$	0.082
$MoO_3 + 6H^+ + 6e \rightleftharpoons Mo + 3H_2O$	0.075
$N_2 + 2H_2O + 6H^+ + 6e \rightleftharpoons 2NH_4OH$	0.092
$3N_2 + 2H^+ + 2e \rightleftharpoons 2HN_3$	–3.09
$N_5^+ + 3H^+ + 2e \rightleftharpoons 2NH_4^+$	1.275

ELECTROCHEMICAL SERIES (continued)

TABLE 1
Alphabetical Listing (continued)

Reaction	$E°$/V
$N_2O + 2H^+ + 2e \rightleftharpoons N_2 + H_2O$	1.766
$H_2N_2O_2 + 2H^+ + 2e \rightleftharpoons N_2 + 2H_2O$	2.65
$N_2O_4 + 2e \rightleftharpoons 2NO_2^-$	0.867
$N_2O_4 + 2H^+ + 2e \rightleftharpoons 2NHO_2$	1.065
$N_2O_4 + 4H^+ + 4e \rightleftharpoons 2NO + 2H_2O$	1.035
$2NH_3OH^+ + H^+ + 2e \rightleftharpoons N_2H_5^+ + 2H_2O$	1.42
$2NO + 2H^+ + 2e \rightleftharpoons N_2O + H_2O$	1.591
$2NO + H_2O + 2e \rightleftharpoons N_2O + 2OH^-$	0.76
$HNO_2 + H^+ + e \rightleftharpoons NO + H_2O$	0.983
$2HNO_2 + 4H^+ + 4e \rightleftharpoons H_2N_2O_2 + 2H_2O$	0.86
$2HNO_2 + 4H^+ + 4e \rightleftharpoons N_2O + 3H_2O$	1.297
$NO_2^- + H_2O + 3e \rightleftharpoons NO + 2OH^-$	–0.46
$2NO_2^- + 2H_2O + 4e \rightleftharpoons N_2^{2-} + 4OH^-$	–0.18
$2NO_2^- + 3H_2O + 4e \rightleftharpoons N_2O + 6OH^-$	0.15
$NO_3^- + 3H^+ + 2e \rightleftharpoons HNO_2 + H_2O$	0.934
$NO_3^- + 4H^+ + 3e \rightleftharpoons NO + 2H_2O$	0.957
$2NO_3^- + 4H^+ + 2e \rightleftharpoons N_2O_4 + 2H_2O$	0.803
$NO_3^- + H_2O + 2e \rightleftharpoons NO_2^- + 2OH^-$	0.01
$2NO_3^- + 2H_2O + 2e \rightleftharpoons N_2O_4 + 4OH^-$	–0.85
$Na^+ + e \rightleftharpoons Na$	–2.71
$Nb^{3+} + 3e \rightleftharpoons Nb$	–1.099
$NbO_2 + 2H^+ + 2e \rightleftharpoons NbO + H_2O$	–0.646
$NbO_2 + 4H^+ + 4e \rightleftharpoons Nb + 2H_2O$	–0.690
$NbO + 2H^+ + 2e \rightleftharpoons Nb + H_2O$	–0.733
$Nb_2O_5 + 10H^+ + 10e \rightleftharpoons 2Nb + 5H_2O$	–0.644
$Nd^{3+} + 3e \rightleftharpoons Nd$	–2.323
$Nd^{2+} + 2e \rightleftharpoons Nd$	–2.1
$Nd^{3+} + e \rightleftharpoons Nd^{2+}$	–2.7
$Ni^{2+} + 2e \rightleftharpoons Ni$	–0.257
$Ni(OH)_2 + 2e \rightleftharpoons Ni + 2OH^-$	–0.72
$NiO_2 + 4H^+ + 2e \rightleftharpoons Ni^{2+} + 2H_2O$	1.678
$NiO_2 + 2H_2O + 2e \rightleftharpoons Ni(OH)_2 + 2OH^-$	–0.490
$No^{3+} + e \rightleftharpoons No^{2+}$	1.4
$No^{3+} + 3e \rightleftharpoons No$	–1.20
$No^{2+} + 2e \rightleftharpoons No$	–2.50
$Np^{3+} + 3e \rightleftharpoons Np$	–1.856
$Np^{4+} + e \rightleftharpoons Np^{3+}$	0.147
$NpO_2 + H_2O + H^+ + e \rightleftharpoons Np(OH)_3$	–0.962
$O_2 + 2H^+ + 2e \rightleftharpoons H_2O_2$	0.695
$O_2 + 4H^+ + 4e \rightleftharpoons 2H_2O$	1.229
$O_2 + H_2O + 2e \rightleftharpoons HO_2^- + OH^-$	–0.076
$O_2 + 2H_2O + 2e \rightleftharpoons H_2O_2 + 2OH^-$	–0.146
$O_2 + 2H_2O + 4e \rightleftharpoons 4OH^-$	0.401
$O_3 + 2H^+ + 2e \rightleftharpoons O_2 + H_2O$	2.076
$O_3 + H_2O + 2e \rightleftharpoons O_2 + 2OH^-$	1.24
$O(g) + 2H^+ + 2e \rightleftharpoons H_2O$	2.421
$OH + e \rightleftharpoons OH^-$	2.02
$HO_2^- + H_2O + 2e \rightleftharpoons 3OH^-$	0.878
$OsO_4 + 8H^+ + 8e \rightleftharpoons Os + 4H_2O$	0.838
$OsO_4 + 4H^+ + 4e \rightleftharpoons OsO_2 + 2H_2O$	1.02
$[Os(bipy)_2]^{3+} + e \rightleftharpoons [Os(bipy)_2]^{2+}$	0.81
$[Os(bipy)_3]^{3+} + e \rightleftharpoons [Os(bipy)_3]^{2+}$	0.80
$P(red) + 3H^+ + 3e \rightleftharpoons PH_3(g)$	–0.111
$P(white) + 3H^+ + 3e \rightleftharpoons PH_3(g)$	–0.063
$P + 3H_2O + 3e \rightleftharpoons PH_3(g) + 3OH^-$	–0.87
$H_2P_2^- + e \rightleftharpoons P + 2OH^-$	–1.82
$H_3PO_2 + H^+ + 3e \rightleftharpoons P + 2H_2O$	–0.508
$H_3PO_3 + 2H^+ + 2e \rightleftharpoons H_3PO_2 + H_2O$	–0.499
$H_3PO_3 + 3H^+ + 3e \rightleftharpoons P + 3H_2O$	–0.454
$HPO_3^{2-} + 2H_2O + 2e \rightleftharpoons H_2PO_2^- + 3OH^-$	–1.65
$HPO_3^{2-} + 2H_2O + 3e \rightleftharpoons P + 5OH^-$	–1.71
$H_3PO_4 + 2H^+ + 2e \rightleftharpoons H_3PO_3 + H_2O$	–0.276
$PO_4^{3-} + 2H_2O + 3e \rightleftharpoons HPO_3^{2-} + 3OH^-$	–1.05
$Pa^{3+} + 3e \rightleftharpoons Pa$	–1.34
$Pa^{4+} + 4e \rightleftharpoons Pa$	–1.49
$Pa^{4+} + e \rightleftharpoons Pa^{3+}$	–1.9
$Pb^{2+} + 2e \rightleftharpoons Pb$	–0.1262
$Pb^{2+} + 2e \rightleftharpoons Pb(Hg)$	–0.1205
$PbBr_2 + 2e \rightleftharpoons Pb + 2Br^-$	–0.284
$PbCl_2 + 2e \rightleftharpoons Pb + 2Cl^-$	–0.2675
$PbF_2 + 2e \rightleftharpoons Pb + 2F^-$	–0.3444
$PbHPO_4 + 2e \rightleftharpoons Pb + HPO_4^{2-}$	–0.465
$PbI_2 + 2e \rightleftharpoons Pb + 2I^-$	–0.365
$PbO + H_2O + 2e \rightleftharpoons Pb + 2OH^-$	–0.580
$PbO_2 + 4H^+ + 2e \rightleftharpoons Pb^{2+} + 2H_2O$	1.455
$HPbO_2^- + H_2O + 2e \rightleftharpoons Pb + 3OH^-$	–0.537
$PbO_2 + H_2O + 2e \rightleftharpoons PbO + 2OH^-$	0.247
$PbO_2 + SO_4^{2-} + 4H^+ + 2e \rightleftharpoons PbSO_4 + 2H_2O$	1.6913
$PbSO_4 + 2e \rightleftharpoons Pb + SO_4^{2-}$	–0.3588
$PbSO_4 + 2e \rightleftharpoons Pb(Hg) + SO_4^{2-}$	–0.3505
$Pd^{2+} + 2e \rightleftharpoons Pd$	0.951
$[PdCl_4]^{2-} + 2e \rightleftharpoons Pd + 4Cl^-$	0.591
$[PdCl_6]^{2-} + 2e \rightleftharpoons [PdCl_4]^{2-} + 2Cl^-$	1.288
$Pd(OH)_2 + 2e \rightleftharpoons Pd + 2OH^-$	0.07
$Pm^{2+} + 2e \rightleftharpoons Pm$	–2.2
$Pm^{3+} + 3e \rightleftharpoons Pm$	–2.30
$Pm^{3+} + e \rightleftharpoons Pm^{2+}$	–2.6
$Po^{4+} + 2e \rightleftharpoons Po^{2+}$	0.9
$Po^{4+} + 4e \rightleftharpoons Po$	0.76
$Pr^{4+} + e \rightleftharpoons Pr^{3+}$	3.2
$Pr^{2+} + 2e \rightleftharpoons Pr$	–2.0
$Pr^{3+} + 3e \rightleftharpoons Pr$	–2.353
$Pr^{3+} + e \rightleftharpoons Pr^{2+}$	–3.1
$Pt^{2+} + 2e \rightleftharpoons Pt$	1.18
$[PtCl_4]^{2-} + 2e \rightleftharpoons Pt + 4Cl^-$	0.755
$[PtCl_6]^{2-} + 2e \rightleftharpoons [PtCl_4]^{2-} + 2Cl^-$	0.68
$Pt(OH)_2 + 2e \rightleftharpoons Pt + 2OH^-$	0.14
$PtO_3 + 2H^+ + 2e \rightleftharpoons PtO_2 + H_2O$	1.7
$PtO_3 + 4H^+ + 2e \rightleftharpoons Pt(OH)_2^{2+} + H_2O$	1.5
$PtOH^+ + H^+ + 2e \rightleftharpoons Pt + H_2O$	1.2
$PtO_2 + 4H^+ + 2e \rightleftharpoons PtO + 2H_2O$	1.01
$PtO_2 + 4H^+ + 4e \rightleftharpoons Pt + 2H_2O$	1.00
$Pu^{3+} + 3e \rightleftharpoons Pu$	–2.031
$Pu^{4+} + e \rightleftharpoons Pu^{3+}$	1.006
$Pu^{5+} + e \rightleftharpoons Pu^{4+}$	1.099
$PuO_2(OH)_2 + 2H^+ + 2e \rightleftharpoons Pu(OH)_4$	1.325
$PuO_2(OH)_2 + H^+ + e \rightleftharpoons PuO_2OH + H_2O$	1.062
$Ra^{2+} + 2e \rightleftharpoons Ra$	–2.8
$Rb^+ + e \rightleftharpoons Rb$	–2.98
$Re^{3+} + 3e \rightleftharpoons Re$	0.300

TABLE 1
Alphabetical Listing (continued)

Reaction	*E*°/V
$ReO_4^- + 4\,H^- + 3\,e \rightleftharpoons ReO_2 + 2\,H_2O$	0.510
$ReO_2 + 4\,H^+ + 4\,e \rightleftharpoons Re + 2\,H_2O$	0.2513
$ReO_4^- + 2\,H^+ + e \rightleftharpoons ReO_3 + H_2O$	0.768
$ReO_4^- + 4\,H_2O + 7\,e \rightleftharpoons Re + 8\,OH^-$	–0.584
$ReO_4^- + 8\,H^+ + 7\,e \rightleftharpoons Re + 4\,H_2O$	0.368
$Rh^+ + e \rightleftharpoons Rh$	0.600
$Rh^+ + 2e \rightleftharpoons Rh$	0.600
$Rh^{3+} + 3\,e \rightleftharpoons Rh$	0.758
$[RhCl_6]^{3-} + 3\,e \rightleftharpoons Rh + 6\,Cl^-$	0.431
$RhOH^{2+} + H^+ + 3\,e \rightleftharpoons Rh + H_2O$	0.83
$Ru^{2+} + 2\,e \rightleftharpoons Ru$	0.455
$Ru^{3+} + e \rightleftharpoons Ru^{2+}$	0.2487
$RuO_2 + 4\,H^+ + 2\,e \rightleftharpoons Ru^{2+} + 2\,H_2O$	1.120
$RuO_4^- + e \rightleftharpoons RuO_4^{2-}$	0.59
$RuO_4 + e \rightleftharpoons RuO_4^-$	1.00
$RuO_4 + 6\,H^+ + 4\,e \rightleftharpoons Ru(OH)_2^{2+} + 2\,H_2O$	1.40
$RuO_4 + 8\,H^+ + 8\,e \rightleftharpoons Ru + 4\,H_2O$	1.038
$[Ru(bipy)_3]^{3+} + e^- \rightleftharpoons [Ru(bipy)_3]^{2+}$	1.24
$[Ru(H_2O)_6]^{3+} + e^- \rightleftharpoons [Ru(H_2O)_6]^{2+}$	0.23
$[Ru(NH_3)_6]^{3+} + e^- \rightleftharpoons [Ru(NH_3)_6]^{2+}$	0.10
$[Ru(en)_3]^{3+} + e^- \rightleftharpoons Ru(en)_3^{2+}$	0.210
$[Ru(CN)_6]^{3-} + e^- \rightleftharpoons [Ru(CN)_6]^{4-}$	0.86
$S + 2\,e \rightleftharpoons S^{2-}$	–0.47627
$S + 2H^+ + 2\,e \rightleftharpoons H_2S(aq)$	0.142
$S + H_2O + 2\,e \rightleftharpoons SH^- OH^-$	–0.478
$2\,S + 2\,e \rightleftharpoons S_2^{2-}$	–0.42836
$S_2O_6^{2-} + 4\,H^+ + 2\,e \rightleftharpoons 2\,H_2SO_3$	0.564
$S_2O_8^{2-} + 2\,e \rightleftharpoons 2\,SO_4^{2-}$	2.010
$S_2O_8^{2-} + 2\,H^+ + 2\,e \rightleftharpoons 2\,HSO_4^-$	2.123
$S_4O_6^{2-} + 2\,e \rightleftharpoons 2\,S_2O_3^{2-}$	0.08
$2\,H_2SO_3 + H^+ + 2\,e \rightleftharpoons HS_2O_4^- + 2\,H_2O$	–0.056
$H_2SO_3 + 4\,H^+ + 4\,e \rightleftharpoons S + 3\,H_2O$	0.449
$2\,SO_3^{2-} + 2\,H_2O + 2\,e \rightleftharpoons S_2O_4^{2-} + 4\,OH^-$	–1.12
$2\,SO_3^{2-} + 3\,H_2O + 4\,e \rightleftharpoons S_2O_3^{2-} + 6\,OH^-$	–0.571
$SO_4^{2-} + 4\,H^+ + 2\,e \rightleftharpoons H_2SO_3 + H_2O$	0.172
$2\,SO_4^{2-} + 4\,H^+ + 2\,e \rightleftharpoons S_2O_6^{2-} + H_2O$	–0.22
$SO_4^{2-} + H_2O + 2\,e \rightleftharpoons SO_3^{2-} + 2\,OH^-$	–0.93
$Sb + 3\,H^+ + 3\,e \rightleftharpoons SbH_3$	–0.510
$Sb_2O_3 + 6\,H^+ + 6\,e \rightleftharpoons 2\,Sb + 3\,H_2O$	0.152
Sb_2O_5 (senarmontite) $+ 4\,H^+ + 4\,e \rightleftharpoons Sb_2O_3 + 2\,H_2O$	0.671
Sb_2O_5 (valentinite) $+ 4\,H^+ + 4\,e \rightleftharpoons Sb_2O_3 + 2\,H_2O$	0.649
$Sb_2O_5 + 6\,H^+ + 4\,e \rightleftharpoons 2\,SbO^+ + 3\,H_2O$	0.581
$SbO^+ + 2\,H^+ + 3\,e \rightleftharpoons Sb + 2\,H_2O$	0.212
$SbO_2^- + 2\,H_2O + 3\,e \rightleftharpoons Sb + 4\,OH^-$	–0.66
$SbO_3^- + H_2O + 2\,e \rightleftharpoons SbO_2^- + 2\,OH^-$	–0.59
$Sc^{3+} + 3\,e \rightleftharpoons Sc$	–2.077
$Se + 2\,e \rightleftharpoons Se^{2-}$	–0.924
$Se + 2\,H^+ + 2\,e \rightleftharpoons H_2Se(aq)$	–0.399
$H_2SeO_3 + 4\,H^+ + 4\,e \rightleftharpoons Se + 3\,H_2O$	0.74
$Se + 2\,H^+ + 2\,e \rightleftharpoons H_2Se$	–0.082
$SeO_3^{2-} + 3\,H_2O + 4\,e \rightleftharpoons Se + 6\,OH^-$	–0.366
$SeO_4^{2-} + 4\,H^+ + 2\,e \rightleftharpoons H_2SeO_3 + H_2O$	1.151
$SeO_4^{2-} + H_2O + 2\,e \rightleftharpoons SeO_3^{2-} + 2\,OH^-$	0.05
$SiF_6^{2-} + 4\,e \rightleftharpoons Si + 6\,F^-$	–1.24
$SiO + 2\,H^+ + 2\,e \rightleftharpoons Si + H_2O$	–0.8
SiO_2 (quartz) $+ 4\,H^+ + 4\,e \rightleftharpoons Si + 2\,H_2O$	0.857
$SiO_3^{2-} + 3\,H_2O + 4\,e \rightleftharpoons Si + 6\,OH^-$	–1.697
$Sm^{3+} + e \rightleftharpoons Sm^{2+}$	–1.55
$Sm^{3+} + 3\,e \rightleftharpoons Sm$	–2.304
$Sm^{2+} + 2\,e \rightleftharpoons Sm$	–2.68
$Sn^{2+} + 2\,e \rightleftharpoons Sn$	–0.1375
$Sn^{4+} + 2\,e \rightleftharpoons Sn^{2+}$	0.151
$Sn(OH)_3^+ + 3\,H^+ + 2\,e \rightleftharpoons Sn^{2+} + 3\,H_2O$	0.142
$SnO_2 + 4\,H^+ + 2\,e^- \rightleftharpoons Sn^{2+} + 2\,H_2O$	–0.094
$SnO_2 + 4\,H^+ + 4\,e \rightleftharpoons Sn + 2\,H_2O$	–0.117
$SnO_2 + 3\,H^+ + 2\,e \rightleftharpoons SnOH^+ + H_2O$	–0.194
$SnO_2 + 2\,H_2O + 4\,e \rightleftharpoons Sn + 4\,OH^-$	–0.945
$HSnO_2^- + H_2O + 2\,e \rightleftharpoons Sn + 3\,OH^-$	–0.909
$Sn(OH)_6^{2-} + 2\,e \rightleftharpoons HSnO_2^- + 3\,OH^- + H_2O$	–0.93
$Sr^+ + e \rightleftharpoons Sr$	–4.10
$Sr^{2+} + 2\,e \rightleftharpoons Sr$	–2.899
$Sr^{2+} + 2\,e \rightleftharpoons Sr(Hg)$	–1.793
$Sr(OH)_2 + 2\,e \rightleftharpoons Sr + 2\,OH^-$	–2.88
$Ta_2O_5 + 10\,H^+ + 10\,e \rightleftharpoons 2\,Ta + 5\,H_2O$	–0.750
$Ta^{3+} + 3\,e \rightleftharpoons Ta$	–0.6
$Tc^{2+} + 2\,e \rightleftharpoons Tc$	0.400
$TcO_4^- + 4\,H^+ + 3\,e \rightleftharpoons TcO_2 + 2\,H_2O$	0.782
$Tc^{3+} + e \rightleftharpoons Tc^{2+}$	0.3
$TcO_4^- + 8\,H^+ + 7\,e \rightleftharpoons Tc + 4\,H_2O$	0.472
$Tb^{4+} + e \rightleftharpoons Tb^{3+}$	3.1
$Tb^{3+} + 3\,e \rightleftharpoons Tb$	–2.28
$Te + 2\,e \rightleftharpoons Te^{2-}$	–1.143
$Te + 2\,H^+ + 2\,e \rightleftharpoons H_2Te$	–0.793
$Te^{4+} + 4\,e \rightleftharpoons Te$	0.568
$TeO_2 + 4\,H^+ + 4\,e \rightleftharpoons Te + 2\,H_2O$	0.593
$TeO_3^{2-} + 3\,H_2O + 4\,e \rightleftharpoons Te + 6\,OH^-$	–0.57
$TeO_4^- + 8\,H^+ + 7\,e \rightleftharpoons Te + 4\,H_2O$	0.472
$H_6TeO_6 + 2\,H^+ + 2\,e \rightleftharpoons TeO_2 + 4\,H_2O$	1.02
$Th^{4+} + 4\,e \rightleftharpoons Th$	–1.899
$ThO_2 + 4\,H^+ + 4\,e \rightleftharpoons Th + 2\,H_2O$	–1.789
$Th(OH)_4 + 4\,e \rightleftharpoons Th + 4\,OH^-$	–2.48
$Ti^{2+} + 2\,e \rightleftharpoons Ti$	–1.630
$Ti^{3+} + e \rightleftharpoons Ti^{2+}$	–0.9
$TiO_2 + 4\,H^+ + 2\,e \rightleftharpoons Ti^{2+} + 2\,H_2O$	–0.502
$Ti^{3+} + 3\,e \rightleftharpoons Ti$	–1.37
$TiOH^{3+} + H^+ + e \rightleftharpoons Ti^{3+} + H_2O$	–0.055
$Tl^+ + e \rightleftharpoons Tl$	–0.336
$Tl^+ + e \rightleftharpoons Tl(Hg)$	–0.3338
$Tl^{3+} + 2\,e \rightleftharpoons Tl^+$	1.252
$Tl^{3+} + 3\,e \rightleftharpoons Tl$	0.741
$TlBr + e \rightleftharpoons Tl + Br^-$	–0.658
$TlCl + e \rightleftharpoons Tl + Cl^-$	–0.5568
$TlI + e \rightleftharpoons Tl + I^-$	–0.752
$Tl_2O_3 + 3\,H_2O + 4\,e \rightleftharpoons 2\,Tl^+ + 6\,OH^-$	0.02
$TlOH + e \rightleftharpoons Tl + OH^-$	–0.34
$Tl(OH)_3 + 2\,e \rightleftharpoons TlOH + 2\,OH^-$	–0.05
$Tl_2SO_4 + 2\,e \rightleftharpoons Tl + SO_4^{2-}$	–0.4360
$Tm^{3+} + e \rightleftharpoons Tm^{2+}$	–2.2
$Tm^{3+} + 3\,e \rightleftharpoons Tm$	–2.319
$Tm^{2+} + 2\,e \rightleftharpoons Tm$	–2.4

TABLE 1
Alphabetical Listing (continued)

Reaction	$E°$/V
$U^{3+} + 3e \rightleftharpoons U$	–1.798
$U^{4+} + e \rightleftharpoons U^{3+}$	–0.607
$UO_2^+ + 4H^+ + e \rightleftharpoons U^{4+} + 2H_2O$	0.612
$UO_2^{2+} + e \rightleftharpoons UO^{+}_2$	0.062
$UO_2^{2+} + 4H^+ + 2e \rightleftharpoons U^{4+} + 2H_2O$	0.327
$UO_2^{2+} + 4H^+ + 6e \rightleftharpoons U + 2H_2O$	–1.444
$V^{2+} + 2e \rightleftharpoons V$	–1.175
$V^{3+} + e \rightleftharpoons V^{2+}$	–0.255
$VO^{2+} + 2H^+ + e \rightleftharpoons V^{3+} + H_2O$	0.337
$VO_2^+ + 2H^+ + e \rightleftharpoons VO^{2+} + H_2O$	0.991
$V_2O_5 + 6H^+ + 2e \rightleftharpoons 2VO^{2+} + 3H_2O$	0.957
$V_2O_5 + 10H^+ + 10e \rightleftharpoons 2V + 5H_2O$	–0.242
$V(OH)_4^+ + 2H^+ + e \rightleftharpoons VO^{2+} + 3H_2O$	1.00
$V(OH)_4^+ + 4H^+ + 5e \rightleftharpoons V + 4H_2O$	–0.254
$[V(phen)_3]^{3+} + e \rightleftharpoons [V(phen)_3]^{2+}$	0.14
$W^{3+} + 3e \rightleftharpoons W$	0.1
$W_2O_5 + 2H^+ + 2e \rightleftharpoons 2WO_2 + H_2O$	–0.031
$WO_2 + 4H^+ + 4e \rightleftharpoons W + 2H_2O$	–0.119
$WO_3 + 6H^+ + 6e \rightleftharpoons W + 3H_2O$	–0.090
$WO_3 + 2H^+ + 2e \rightleftharpoons WO_2 + H_2O$	0.036
$2WO_3 + 2H^+ + 2e \rightleftharpoons W_2O_5 + H_2O$	–0.029
$H_4XeO_6 + 2H^+ + 2e \rightleftharpoons XeO_3 + 3H_2O$	2.42
$XeO_3 + 6H^+ + 6e \rightleftharpoons Xe + 3H_2O$	2.10
$XeF + e \rightleftharpoons Xe + F^-$	3.4
$Y^{3+} + 3e \rightleftharpoons Y$	–2.372
$Yb^{3+} + e \rightleftharpoons Yb^{2+}$	–1.05
$Yb^{3+} + 3e \rightleftharpoons Yb$	–2.19
$Yb^{2+} + 2e \rightleftharpoons Yb$	–2.76
$Zn^{2+} + 2e \rightleftharpoons Zn$	–0.7618
$Zn^{2+} + 2e \rightleftharpoons Zn(Hg)$	–0.7628
$ZnO_2^{2-} + 2H_2O + 2e \rightleftharpoons Zn + 4OH^-$	–1.215
$ZnSO_4 + 7H_2O + 2e = Zn(Hg) + SO_4^{2-}$	–0.7993
(Saturated $ZnSO_4$)	
$ZnOH^+ + H^+ + 2e \rightleftharpoons Zn + H_2O$	–0.497
$Zn(OH)_4^{2-} + 2e \rightleftharpoons Zn + 4OH^-$	–1.199
$Zn(OH)_2 + 2e \rightleftharpoons Zn + 2OH^-$	–1.249
$ZnO + H_2O + 2e \rightleftharpoons Zn + 2OH^-$	–1.260
$ZrO_2 + 4H^+ + 4e \rightleftharpoons Zr + 2H_2O$	–1.553
$ZrO(OH)_2 + H_2O + 4e \rightleftharpoons Zr + 4OH^-$	–2.36
$Zr^{4+} + 4e \rightleftharpoons Zr$	–1.45

TABLE 2
Reduction Reactions Having $E°$ Values More Positive than that of the Standard Hydrogen Electrode

Reaction	$E°$/V
$2H^+ + 2e \rightleftharpoons H_2$	0.00000
$CuI_2^- + e \rightleftharpoons Cu + 2I^-$	0.00
$Ge^{4+} + 2e \rightleftharpoons Ge^{2+}$	0.00
$NO_3^- + H_2O + 2e \rightleftharpoons NO_2^- + 2OH^-$	0.01
$Tl_2O_3 + 3H_2O + 4e \rightleftharpoons 2Tl^+ + 6OH^-$	0.02
$SeO_4^{2-} + H_2O + 2e \rightleftharpoons SeO_3^{2-} + 2OH^-$	0.05
$WO_3 + 2H^+ + 2e \rightleftharpoons WO_2 + H_2O$	0.036
$UO_2^{2+} + e = UO_2^-$	0.062
$Pd(OH)_2 + 2e \rightleftharpoons Pd + 2OH^-$	0.07
$AgBr + e \rightleftharpoons Ag + Br^-$	0.07133
$MoO_3 + 6H^+ + 6e \rightleftharpoons Mo + 3H_2O$	0.075
$S_4O_6^{2-} + 2e \rightleftharpoons 2S_2O_3^{2-}$	0.08
$H_3Mo_7O_{24}^{3-} + 45H^+ + 42e \rightleftharpoons 7Mo + 24H_2O$	0.082
$AgSCN + e \rightleftharpoons Ag + SCN^-$	0.8951
$N_2 + 2H_2O + 6H^+ + 6e \rightleftharpoons 2NH_4OH$	0.092
$HgO + H_2O + 2e \rightleftharpoons Hg + 2OH^-$	0.0977
$Ir_2O_3 + 3H_2O + 6e \rightleftharpoons 2Ir + 6OH^-$	0.098
$2NO + 2e \rightleftharpoons N_2O_2^{2-}$	0.10
$[Ru(NH_3)_6]^{3+} + e \rightleftharpoons [Ru(NH_3)_6]^{2+}$	0.10
$W^{3+} + 3e \rightleftharpoons W$	0.1
$[Co(NH_3)_6]^{3+} + e \rightleftharpoons [Co(NH_3)_6]^{2+}$	0.108
$Hg_2O + H_2O + 2e \rightleftharpoons 2Hg + 2OH^-$	0.123
$Ge^{4+} + 4e \rightleftharpoons Ge$	0.124
$Hg_2Br_2 + 2e \rightleftharpoons 2Hg + 2Br^-$	0.13923
$Pt(OH)_2 + 2e \rightleftharpoons Pt + 2OH^-$	0.14
$[V(phen)_3]^{3+} + e \rightleftharpoons [V(phen)_3]^{2+}$	0.14
$S + 2H^+ + 2e \rightleftharpoons H_2S(aq)$	0.142
$Sn(OH)_3^+ + 3H^+ + 2e \rightleftharpoons Sn^{2+} + 3H_2O$	0.142
$Np^{4+} + e \rightleftharpoons Np^{3+}$	0.147
$Ag_4[Fe(CN)_6] + 4e \rightleftharpoons 4Ag + [Fe(CN)_6]^{4-}$	0.1478
$IO_3^- + 2H_2O + 4e \rightleftharpoons IO^- + 4OH^-$	0.15
$Mn(OH)_3 + e \rightleftharpoons Mn(OH)_2 + OH^-$	0.15
$2NO_2^- + 3H_2O + 4e \rightleftharpoons N_2O + 6OH^-$	0.15
$Sn^{4+} + 2e \rightleftharpoons Sn^{2+}$	0.151
$Sb_2O_3 + 6H^+ + 6e \rightleftharpoons 2Sb + 3H_2O$	0.152
$Cu^{2+} + e \rightleftharpoons Cu^+$	0.153
$BiOCl + 2H^+ + 3e \rightleftharpoons Bi + Cl^- + H_2O$	0.1583
$BiCl_4^- + 3e \rightleftharpoons Bi + 4Cl^-$	0.16
$Fe_2O_3 + 4H^+ + 2e \rightleftharpoons 2FeOH^+ + H_2O$	0.16
$Co(OH)_3 + e \rightleftharpoons Co(OH)_2 + OH^-$	0.17
$SO_4^{2-} + 4H^+ + 2e \rightleftharpoons H_2SO_3 + H_2O$	0.172
$Bi^{3+} + 2e \rightleftharpoons Bi^+$	0.2
$[Ru(en)_3]^{3+} + e \rightleftharpoons [Ru(en)_3]^{2+}$	0.210
$SbO^+ + 2H^+ + 3e \rightleftharpoons Sb + 2H_2O$	0.212
$AgCl + e \rightleftharpoons Ag + Cl^-$	0.22233
$[Ru(H_2O)_6]^{3+} + e \rightleftharpoons [Ru(H_2O)_6]^{2+}$	0.23
$As_2O_3 + 6H^+ + 6e \rightleftharpoons 2As + 3H_2O$	0.234
Calomel electrode, saturated NaCl (SSCE)	0.2360
$Ge^{2+} + 2e \rightleftharpoons Ge$	0.24
$Ru^{3+} + e \rightleftharpoons Ru^{2+}$	0.24
Calomel electrode, saturated KCl	0.2412
$PbO_2 + H_2O + 2e \rightleftharpoons PbO + 2OH^-$	0.247
$HAsO_2 + 3H^+ + 3e \rightleftharpoons As + 2H_2O$	0.248
$Ru^{3+} + e \rightleftharpoons Ru^{2+}$	0.2487

TABLE 2
Reduction Reactions Having $E°$ Values More Positive than that of the Standard Hydrogen Electrode (continued)

Reaction	$E°$/V
$ReO_2 + 4H^+ + 4e \rightleftharpoons Re + 2H_2O$	0.2513
$IO_3^- + 3H_2O + 6e \rightleftharpoons I^- + OH^-$	0.26
$Hg_2Cl_2 + 2e \rightleftharpoons 2Hg + 2Cl^-$	0.26808
Calomel electrode, 1 molal KCl	0.2800
Calomel electrode, 1 molar KCl (NCE)	0.2801
$At_2 + 2e \rightleftharpoons 2At^-$	0.3
$Re^{3+} + 3e \rightleftharpoons Re$	0.300
$Tc^{3+} + e \rightleftharpoons Tc^{2+}$	0.3
$Bi^{3+} + 3e \rightleftharpoons Bi$	0.308
$BiO^+ + 2H^+ + 3e \rightleftharpoons Bi + H_2O$	0.320
$UO_2^{2+} + 4H^+ + 2e \rightleftharpoons U^{4+} + 2H_2O$	0.327
$ClO_3^- + H_2O + 2e \rightleftharpoons ClO_2^- + 2OH^-$	0.33
$2HCNO + 2H^+ + 2e \rightleftharpoons (CN)_2 + 2H_2O$	0.330
Calomel electrode, 0.1 molar KCl	0.3337
$VO^{2+} + 2H^+ + e \rightleftharpoons V^{3+} + H_2O$	0.337
$Cu^{2+} + 2e \rightleftharpoons Cu$	0.3419
$Ag_2O + H_2O + 2e \rightleftharpoons 2Ag + 2OH^-$	0.342
$Cu^{2+} + 2e \rightleftharpoons Cu(Hg)$	0.345
$AgIO_3 + e \rightleftharpoons Ag + IO_3^-$	0.354
$[Fe(CN)_6]^{3-} + e \rightleftharpoons [Fe(CN)_6]^{4-}$	0.358
$ClO_4^- + H_2O + 2e \rightleftharpoons ClO^-_3 + 2OH^-$	0.36
$Ag_2SeO_3 + 2e \rightleftharpoons 2Ag + SeO_3^{2-}$	0.3629
$ReO_4^- + 8H^+ + 7e \rightleftharpoons Re + 4H_2O$	0.368
$(CN)_2 + 2H^+ + 2e \rightleftharpoons 2HCN$	0.373
$[Ferricinium]^+ + e \rightleftharpoons$ ferrocene	0.400
$Tc^{2+} + 2e \rightleftharpoons Tc$	0.400
$O_2 + 2H_2O + 4e \rightleftharpoons 4OH^-$	0.401
$AgOCN + e \rightleftharpoons Ag + OCN^-$	0.41
$[RhCl_6]^{3-} + 3e \rightleftharpoons Rh + 6Cl^-$	0.431
$Ag_2CrO_4 + 2e \rightleftharpoons 2Ag + CrO_4^{2-}$	0.4470
$H_2SO_3 + 4H^+ + 4e \rightleftharpoons S + 3H_2O$	0.449
$Ru^{2+} + 2e \rightleftharpoons Ru$	0.455
$Ag_2MoO_4 + 2e \rightleftharpoons 2Ag + MoO_4^{2-}$	0.4573
$Ag_2C_2O_4 + 2e \rightleftharpoons 2Ag + C_2O_4^{2-}$	0.4647
$Ag_2WO_4 + 2e \rightleftharpoons 2Ag + WO_4^{2-}$	0.4660
$Ag_2CO_3 + 2e \rightleftharpoons 2Ag + CO_3^{2-}$	0.47
$TcO_4^- + 8H^+ + 7e \rightleftharpoons Tc + 4H_2O$	0.472
$TeO_4^- + 8H^+ + 7e \rightleftharpoons Te + 4H_2O$	0.472
$IO^- + H_2O + 2e \rightleftharpoons I^- + 2OH^-$	0.485
$NiO_2 + 2H_2O + 2e \rightleftharpoons Ni(OH)_2 + 2OH^-$	0.490
$Bi^+ + e \rightleftharpoons Bi$	0.5
$ReO_4^- + 4H^+ + 3e \rightleftharpoons ReO_2 + 2H_2O$	0.510
$Hg_2(ac)_2 + 2e \rightleftharpoons 2Hg + 2(ac)^-$	0.51163
$Cu^+ + e \rightleftharpoons Cu$	0.521
$I_2 + 2e \rightleftharpoons 2I^-$	0.5355
$I_3^- + 2e \rightleftharpoons 3I^-$	0.536
$AgBrO_3 + e \rightleftharpoons Ag + BrO_3^-$	0.546
$MnO_4^- + e \rightleftharpoons MnO_2^-$	0.558
$H_3AsO_4 + 2H^+ + 2e \rightleftharpoons HAsO_2 + 2H_2O$	0.560
$S_2O_6^{2-} + 4H^+ + 2e \rightleftharpoons 2H_2SO_3$	0.564
$AgNO_2 + e \rightleftharpoons Ag + NO_2^-$	0.564
$Te^{4+} + 4e \rightleftharpoons Te$	0.568
$Sb_2O_5 + 6H^+ + 4e \rightleftharpoons 2SbO^+ + 3H_2O$	0.581
$RuO_4^- + e \rightleftharpoons RuO_4^{2-}$	0.59

Reaction	$E°$/V
$[PdCl_4]^{2-} + 2e \rightleftharpoons Pd + 4Cl^-$	0.591
$TeO_2 + 4H^+ + 4e \rightleftharpoons Te + 2H_2O$	0.593
$MnO_4^- + 2H_2O + 3e \rightleftharpoons MnO_2 + 4OH^-$	0.595
$Rh^{2+} + 2e \rightleftharpoons Rh$	0.600
$Rh^+ + e \rightleftharpoons Rh$	0.600
$MnO_4^{2-} + 2H_2O + 2e \rightleftharpoons MnO_2 + 4OH^-$	0.60
$2AgO + H_2O + 2e \rightleftharpoons Ag_2O + 2OH^-$	0.607
$BrO_3^- + 3H_2O + 6e \rightleftharpoons Br^- + 6OH^-$	0.61
$UO_2^+ + 4H^+ + e \rightleftharpoons U^{4+} + 2H_2O$	0.612
$Hg_2SO_4 + 2e \rightleftharpoons 2Hg + SO_4^{2-}$	0.6125
$ClO_3^- + 3H_2O + 6e \rightleftharpoons Cl^- + 6OH^-$	0.62
$Hg_2HPO_4 + 2e \rightleftharpoons 2Hg + HPO_4^{2-}$	0.6359
$Ag(ac) + e \rightleftharpoons Ag + (ac)^-$	0.643
Sb_2O_5(valentinite) $+ 4H^+ + 4e \rightleftharpoons Sb_2O_3 + 2H_2O$	0.649
$Ag_2SO_4 + 2e \rightleftharpoons 2Ag + SO_4^{2-}$	0.654
$ClO_2^- + H_2O + 2e \rightleftharpoons ClO^- + 2OH^-$	0.66
Sb_2O_5(senarmontite) $+ 4H^+ + 4e \rightleftharpoons Sb_2O_5 + 2H_2O$	0.671
$[PtCl_6]^{2-} + 2e \rightleftharpoons [PtCl_4]^{2-} + 2Cl^-$	0.68
$O_2 + 2H^+ + 2e \rightleftharpoons H_2O_2$	0.695
p–benzoquinone $+ 2H^+ + 2e \rightleftharpoons$ hydroquinone	0.6992
$H_3IO_6 + 2e \rightleftharpoons IO_3^- + 3OH^-$	0.7
$Ag_2O_3 + H_2O + 2e \rightleftharpoons 2AgO + 2OH^-$	0.739
$Tl^{3+} + 3e \rightleftharpoons Tl$	0.741
$[PtCl_4]^{2-} + 2e \rightleftharpoons Pt + 4Cl^-$	0.755
$Rh^{3+} + 3e \rightleftharpoons Rh$	0.758
$ClO_2^- + 2H_2O + 4e \rightleftharpoons Cl^- + 4OH^-$	0.76
$2NO + H_2O + 2e \rightleftharpoons N_2O + 2OH^-$	0.76
$Po^{4+} + 4e \rightleftharpoons Po$	0.76
$BrO^- + H_2O + 2e \rightleftharpoons Br^- + 2OH^-$	0.761
$ReO_4^- + 2H^+ + e \rightleftharpoons ReO_3 + H_2O$	0.768
$(CNS)_2 + 2e \rightleftharpoons 2CNS^-$	0.77
$[IrCl_6]^{3-} + 3e \rightleftharpoons Ir + 6Cl^-$	0.77
$Fe^{3+} + e \rightleftharpoons Fe^{2+}$	0.771
$AgF + e \rightleftharpoons Ag + F^-$	0.779
$[Fe(bipy)_2]^{3+} + e \rightleftharpoons [Fe(bipy)_2]^{2+}$	0.78
$TcO_4^- + 4H^+ + 3e \rightleftharpoons TcO_2 + 2H_2O$	0.782
$Hg_2^{2+} + 2e \rightleftharpoons 2Hg$	0.7973
$Ag^+ + e \rightleftharpoons Ag$	0.7996
$[Os(bipy)_3]^{3+} + e \rightleftharpoons [Os(bipy)_3]^{2+}$	0.80
$2NO_3^- + 4H^+ + 2e \rightleftharpoons N_2O_4 + 2H_2O$	0.803
$[Os(bipy)_2]^{3+} + e \rightleftharpoons [Os(bipy)_2]^{2+}$	0.81
$RhOH^{2+} + H + 3e \rightleftharpoons Rh + H_2O$	0.83
$OsO_4 + 8H^+ + 8e \rightleftharpoons Os + 4H_2O$	0.838
$ClO^- + H_2O + 2e \rightleftharpoons Cl^- + 2OH^-$	0.841
$Hg^{2+} + 2e \rightleftharpoons Hg$	0.851
$AuBr_4^- + 3e \rightleftharpoons Au + 4Br^-$	0.854
SiO_2(quartz) $+ 4H^+ + 4e \rightleftharpoons Si + 2H_2O$	0.857
$2HNO_2 + 4H^+ + 4e \rightleftharpoons H_2N_2O_2 + H_2O$	0.86
$[Ru(CN)_6]^{3-} + e^- \rightleftharpoons [Ru(CN)_6]^{4-}$	0.86
$[IrCl_6]^{2-} + e \rightleftharpoons [IrCl_6]^{3-}$	0.8665
$N_2O_4 + 2e \rightleftharpoons 2NO_2^-$	0.867
$HO_2^- + H_2O + 2e \rightleftharpoons 3OH^-$	0.878
$Po^{4+} + 2e \rightleftharpoons Po^{2+}$	0.9
$2Hg^+ + 2e \rightleftharpoons Hg_2^{2+}$	0.920

TABLE 2
Reduction Reactions Having $E°$ Values More Positive than that of the Standard Hydrogen Electrode (continued)

Reaction	$E°$/V
$NO_3^- + 3H^+ + 2e \rightleftharpoons HNO_2 + H_2O$	0.934
$Pd^{2+} + 2e \rightleftharpoons Pd$	0.951
$ClO_2(aq) + e \rightleftharpoons ClO_2^-$	0.954
$NO_3^- + 4H^+ + 3e \rightleftharpoons NO + 2H_2O$	0.957
$V_2O_5 + 6H^+ + 2e \rightleftharpoons 2VO^{2+} + 3H_2O$	0.957
$AuBr_2^- + e \rightleftharpoons Au + 2Br^-$	0.959
$HNO_2 + H^+ + e \rightleftharpoons NO + H_2O$	0.983
$HIO + H^+ + 2e \rightleftharpoons I^- + H_2O$	0.987
$VO_2^+ + 2H^+ + e \rightleftharpoons VO^{2+} + H_2O$	0.991
$PtO_2 + 4H^+ + 4e \rightleftharpoons Pt + 2H_2O$	1.00
$RuO_4 + e \rightleftharpoons RuO_4^-$	1.00
$V(OH)_4^+ + 2H^+ + e \rightleftharpoons VO^{2+} + 3H_2O$	1.00
$AuCl_4^- + 3e \rightleftharpoons Au + 4Cl^-$	1.002
$Pu^{4+} + e \rightleftharpoons Pu^{3+}$	1.006
$PtO_2 + 4H^+ + 2e \rightleftharpoons PtO + 2H_2O$	1.01
$OsO_4 + 4H + 4e \rightleftharpoons OsO_2 + 2H_2O$	1.02
$H_6TeO_6 + 2H^+ + 2e \rightleftharpoons TeO_2 + 4H_2O$	1.02
$[Fe(bipy)_3]^{3+} + e \rightleftharpoons [Fe(bipy)_3]^{2+}$	1.03
$Hg(OH)_2 + 2H^+ + 2e \rightleftharpoons Hg + 2H_2O$	1.034
$N_2O_4 + 4H^+ + 4e \rightleftharpoons 2NO + 2H_2O$	1.035
$RuO_4 + 8H^+ + 8e \rightleftharpoons Ru + 4H_2O$	1.038
$[Fe(phen)_3]^{3+} + e \rightleftharpoons [Fe(phen)_3]^{2+}$ (1 molar H_2SO_4)	1.06
$PuO_2(OH)_2 + H^+ + e \rightleftharpoons PuO_2OH + H_2O$	1.062
$N_2O_4 + 2H^+ + 2e \rightleftharpoons 2HNO_2$	1.065
$Br_2(\ell) + 2e \rightleftharpoons 2Br^-$	1.066
$IO_3^- + 6H^+ + 6e \rightleftharpoons I^- + 3H_2O$	1.085
$Br_2(aq) + 2e \rightleftharpoons 2Br^-$	1.0873
$Pu^{5+} + e \rightleftharpoons Pu^{4+}$	1.099
$Cu^{2+} + 2CN^- + e \rightleftharpoons [Cu(CN)_2]^-$	1.103
$RuO_2 + 4H^+ + 2e \rightleftharpoons Ru^{2+} + 2H_2O$	1.120
$[Fe(phen)_3]^{3+} + e \rightleftharpoons [Fe(phen)_3]^{2+}$	1.147
$SeO_4^{2-} + 4H^+ + 2e \rightleftharpoons H_2SeO_3 + H_2O$	1.151
$ClO_3^- + 2H^+ + e \rightleftharpoons ClO_2 + H_2O$	1.152
$Ir^{3+} + 3e \rightleftharpoons Ir$	1.156
$Pt^{2+} + 2e \rightleftharpoons Pt$	1.18
$ClO_4^- + 2H^+ + 2e \rightleftharpoons ClO_3^- + H_2O$	1.189
$2IO_3^- + 12H^+ + 10e \rightleftharpoons I_2 + 6H_2O$	1.195
$PtOH^+ + H^+ + 2e \rightleftharpoons Pt + H_2O$	1.2
$ClO_3^- + 3H^+ + 2e \rightleftharpoons HClO_2 + H_2O$	1.214
$MnO_2 + 4H^+ + 2e \rightleftharpoons Mn^{2+} + 2H_2O$	1.224
$O_2 + 4H^+ + 4e \rightleftharpoons 2H_2O$	1.229
$Cr_2O_7^{2-} + 14H^+ + 6e \rightleftharpoons 2Cr^{3+} + 7H_2O$	1.232
$O_3 + H_2O + 2e \rightleftharpoons O_2 + 2OH^-$	1.24
$[Ru(bipy)_3]^{3+} + e \rightleftharpoons [Ru(bipy)_3]^{2+}$	1.24
$Tl^{3+} + 2e \rightleftharpoons Tl^+$	1.252
$N_2H_5^+ + 3H^+ + 2e \rightleftharpoons 2NH_4^+$	1.275
$ClO_2 + H^+ + e \rightleftharpoons HClO_2$	1.277
$[PdCl_6]^{2-} + 2e \rightleftharpoons [PdCl_4]^{2-} + 2Cl^-$	1.288
$2HNO_2 + 4H^+ + 4e \rightleftharpoons N_2O + 3H_2O$	1.297
$Cr_2O_5 + 10H^+ + e \rightleftharpoons CrO_2 + 5H_2O$	1.3
$AuOH^{2+} + H^+ + 2e \rightleftharpoons Au^+ + H_2O$	1.32
$PuO_2(OH)_2 + 2H^- + 2e \rightleftharpoons Pu(OH)_4$	1.325
$HBrO + H^+ + 2e \rightleftharpoons Br^- + H_2O$	1.331
$HCrO_4^- + 7H^+ + 3e \rightleftharpoons Cr^{3+} + 4H_2O$	1.350

Reaction	$E°$/V
$Cl_2(g) + 2e \rightleftharpoons 2Cl^-$	1.35827
$ClO_4^- + 8H^+ + 8e \rightleftharpoons Cl^- + 4H_2O$	1.389
$ClO_4^- + 8H^+ + 7e \rightleftharpoons 1/2\,Cl_2 + 4H_2O$	1.39
$No^{3+} + e \rightleftharpoons No^{2+}$	1.4
$RuO_4 + 6H^+ + 4e \rightleftharpoons Ru(OH)_2^{2+} + 2H_2O$	1.40
$Au^{3+} + 2e \rightleftharpoons Au^+$	1.401
$2NH_3OH^+ + H^+ + 2e \rightleftharpoons N_2H_5^+ + 2H_2O$	1.42
$BrO_3^- + 6H^+ + 6e \rightleftharpoons Br^- + 3H_2O$	1.423
$2HIO + 2H^+ + 2e \rightleftharpoons I_2 + 2H_2O$	1.439
$Au(OH)_3 + 3H^+ + 3e \rightleftharpoons Au^- + 3H_2O$	1.45
$3IO_3^- + 6H^+ + 6e \rightleftharpoons Cl^- + 3H_2O$	1.451
$PbO_2 + 4H^+ + 2e \rightleftharpoons Pb^{2+} + 2H_2O$	1.455
$ClO_3^- + 6H^+ + 5e \rightleftharpoons 1/2\,Cl_2 + 3H_2O$	1.47
$CrO_2 + 4H^+ + e \rightleftharpoons Cr^{3+} + 2H_2O$	1.48
$BrO_3^- + 6H^+ + 5e \rightleftharpoons 1/2\,Br_2 + 3H_2O$	1.482
$HClO + H^+ + 2e \rightleftharpoons Cl^- + H_2O$	1.482
$Mn_2O_3 + 6H + e \rightleftharpoons Mn^{2+} + 3H_2O$	1.485
$HO_2 + H^+ + e \rightleftharpoons H_2O_2$	1.495
$Au^{3+} + 3e \rightleftharpoons Au$	1.498
$PtO_3 + 4H^+ + 2e \rightleftharpoons Pt(OH)_2^{2+} + H_2O$	1.5
$MnO_4^- + 8H^+ + 5e \rightleftharpoons Mn^{2+} + 4H_2O$	1.507
$Mn^{3+} + e \rightleftharpoons Mn^{2-}$	1.5415
$HClO_2 + 3H^+ + 4e \rightleftharpoons Cl^- + 2H_2O$	1.570
$HBrO + H^+ + e \rightleftharpoons 1/2\,Br_2(aq) + H_2O$	1.574
$2NO + 2H^+ + 2e \rightleftharpoons N_2O + H_2O$	1.591
$Bi_2O_4 + 4H^+ + 2e \rightleftharpoons 2BiO^+ + 2H_2O$	1.593
$HBrO + H^+ + e \rightleftharpoons 1/2\,Br_2(l) + H_2O$	1.596
$H_5IO_6 + H^+ + 2e \rightleftharpoons IO_3^- + 3H_2O$	1.601
$HClO + H^+ + e \rightleftharpoons 1/2\,Cl_2 + H_2O$	1.611
$HClO_2 + 3H^+ + 3e \rightleftharpoons 1/2\,Cl_2 + 2H_2O$	1.628
$HClO_2 + 2H^+ + 2e \rightleftharpoons HClO + H_2O$	1.645
$Bk^{4+} + e \rightleftharpoons Bk^{3+}$	1.67
$NiO_2 + 4H^+ + 2e \rightleftharpoons Ni^{2+} + 2H_2O$	1.678
$MnO_4^- + 4H^+ + 3e \rightleftharpoons MnO_2 + 2H_2O$	1.679
$PbO_2 + SO_4^{2-} + 4H^+ + 2e \rightleftharpoons PbSO_4 + 2H_2O$	1.6913
$Au^+ + e \rightleftharpoons Au$	1.692
$PtO_3 + 2H^+ + 2e \rightleftharpoons PtO_2 + H_2O$	1.7
$CeOH^{3+} + H^+ + e \rightleftharpoons Ce^{3+} + H_2O$	1.715
$Ce^{4+} + e \rightleftharpoons Ce^{3+}$	1.72
$N_2O + 2H^+ + 2e \rightleftharpoons N_2 + H_2O$	1.766
$H_2O_2 + 2H^+ + 2e \rightleftharpoons 2H_2O$	1.776
$Ag^{3+} + e \rightleftharpoons Ag^{2+}$	1.8
$Ag^{2+} + e^- \rightleftharpoons Au^+$	1.8
$Ag_2O_2 + 4H^+ + e \rightleftharpoons 2Ag + 2H_2O$	1.802
$Co^{3+} + e \rightleftharpoons Co^{2-}$ (2 molar H_2SO_4)	1.83
$Ag^{2+} + 2e \rightleftharpoons Ag^+$	1.9
$Co^{3+} + e \rightleftharpoons Co^{2+}$	1.92
$Ag^{2+} + e \rightleftharpoons Ag^-$	1.980
$Cu_2O_3 + 6H^+ + e \rightleftharpoons Cu^{2+} + 3H_2O$	2.0
$S_2O_8^{2-} + 2e \rightleftharpoons 2SO_4^{2-}$	2.010
$OH + e \rightleftharpoons OH^-$	2.02
$HFeO_4^- + 8H^+ + 3e \rightleftharpoons Fe^{3+} + 4H_2O$	2.07
$O_3 + 2H^+ + 2e \rightleftharpoons O_2 + H_2O$	2.076
$HFeO_4^- + 4H^+ + 3e \rightleftharpoons FeOOH + 2H_2O$	2.08

TABLE 2
Reduction Reactions Having $E°$ Values More Positive than that of the Standard Hydrogen Electrode (continued)

Reaction	$E°$/V	Reaction	$E°$/V
$HFeO_4^- + 2H^+ + 3e \rightleftharpoons Fe_2O_3 + H_2O$	2.09	$H_2N_2O_2 + 2H^+ + 2e \rightleftharpoons N_2 + 2H_2O$	2.65
$XeO_3 + 6H^+ + 6e \rightleftharpoons Xe + 3H_2O$	2.10	$F_2 + 2e \rightleftharpoons 2F^-$	2.866
$S_2O_8^{2-} + 2H^+ + 2e \rightleftharpoons 2HSO_4^-$	2.123	$Cm^{4+} + e \rightleftharpoons Cm^{3+}$	3.0
$F_2O + 2H^+ + 4e \rightleftharpoons H_2O + 2F^-$	2.153	$F_2 + 2H^+ + 2e \rightleftharpoons 2HF$	3.053
$FeO_4^{2-} + 8H^+ + 3e \rightleftharpoons Fe^{3+} + 4H_2O$	2.20	$Tb^{4+} + e \rightleftharpoons Tb^{3+}$	3.1
$Cu^{3+} + e \rightleftharpoons Cu^{2+}$	2.4	$Pr^{4+} + e \rightleftharpoons Pr^{3+}$	3.2
$H_4XeO_6 + 2H^+ + 2e \rightleftharpoons XeO_3 + 3H_2O$	2.42	$Cf^{4+} + e \rightleftharpoons Cf^{3+}$	3.3
$O(g) + 2H^+ + 2e \rightleftharpoons H_2O$	2.421	$XeF + e \rightleftharpoons Xe + F^-$	3.4
$Am^{4+} + e \rightleftharpoons Am^{3+}$	2.60		

TABLE 3
Reduction Reactions Having $E°$ Values More Negative than that of the Standard Hydrogen Electrode

Reaction	$E°$/V	Reaction	$E°$/V
$2H^+ + 2e \rightleftharpoons H_2$	0.00000	$Cu(OH)_2 + 2e \rightleftharpoons Cu + 2OH^-$	–0.222
$2D^+ + 2e = D_2$	0.013	$V_2O_5 + 10H^+ + 10e \rightleftharpoons 2V + 5H_2O$	–0.242
$AgCN + e = Ag + CN^-$	–0.017	$CdSO_4 + 2e \rightleftharpoons Cd + SO_4^{2-}$	–0.246
$2WO_3 + 2H^+ + 2e = W_2O_5 + H_2O$	–0.029	$V(OH)_4^+ + 4H^+ + 5e \rightleftharpoons V + 4H_2O$	–0.254
$W_2O_5 + 2H^+ + 2e = 2WO_2 + H_2O$	–0.031	$V^{3+} + e \rightleftharpoons V^{2+}$	–0.255
$Ag_2S + 2H^+ + 2e = 2Ag + H_2S$	–0.0366	$Ni^{2+} + 2e \rightleftharpoons Ni$	–0.257
$Fe^{3+} + 3e \rightleftharpoons Fe$	–0.037	$PbCl_2 + 2e \rightleftharpoons Pb + 2Cl^-$	–0.2675
$Hg_2I_2 + 2e \rightleftharpoons 2Hg + 2I^-$	–0.0405	$H_3PO_4 + 2H^+ + 2e \rightleftharpoons H_3PO_3 + H_2O$	–0.276
$Tl(OH)_3 + 2e \rightleftharpoons TlOH + 2OH^-$	–0.05	$Co^{2+} + 2e \rightleftharpoons Co$	–0.28
$TiOH^{3+} + H^+ + e \rightleftharpoons Ti^{3+} + H_2O$	–0.055	$PbBr_2 + 2e \rightleftharpoons Pb + 2Br^-$	–0.284
$2H_2SO_3 + H^+ + 2e \rightleftharpoons HS_2O_4^- + 2H_2O$	–0.056	$Tl^+ + e \rightleftharpoons Tl(Hg)$	–0.3338
$P(white) + 3H^+ + 3e \rightleftharpoons PH_3(g)$	–0.063	$Tl^+ + e \rightleftharpoons Tl$	–0.336
$O_2 + H_2O + 2e \rightleftharpoons HO_2^- + OH^-$	–0.076	$In^{3+} + 3e \rightleftharpoons In$	–0.3382
$2Cu(OH)_2 + 2e \rightleftharpoons Cu_2O + 2OH^- + H_2O$	–0.080	$TlOH + e \rightleftharpoons Tl + OH^-$	–0.34
$Se + 2H^+ + 2e \rightleftharpoons H_2Se$	–0.082	$PbF_2 + 2e \rightleftharpoons Pb + 2F^-$	–0.3444
$WO_3 + 6H^+ + 6e \rightleftharpoons W + 3H_2O$	–0.090	$PbSO_4 + 2e \rightleftharpoons Pb(Hg) + SO_4^{2-}$	–0.3505
$SnO_2 + 4H^+ + 2e \rightleftharpoons Sn^{2+} + 2H_2O$	–0.094	$Cd^{2+} + 2e \rightleftharpoons Cd(Hg)$	–0.3521
$Md^{3+} + e \rightleftharpoons Md^{2+}$	–0.1	$PbSO_4 + 2e \rightleftharpoons Pb + SO_4^{2-}$	–0.3588
$P(red) + 3H^+ + 3e \rightleftharpoons PH_3(g)$	–0.111	$Cu_2O + H_2O + 2e \rightleftharpoons 2Cu + 2OH^-$	–0.360
$SnO_2 + 4H^+ + 4e \rightleftharpoons Sn + 2H_2O$	–0.117	$Eu^{3+} + e \rightleftharpoons Eu^{2+}$	–0.36
$GeO_2 + 2H^+ + 2e \rightleftharpoons GeO + H_2O$	–0.118	$PbI_2 + 2e \rightleftharpoons Pb + 2I^-$	–0.365
$WO_2 + 4H^+ + 4e \rightleftharpoons W + 2H_2O$	–0.119	$SeO_3^{2-} + 3H_2O + 4e \rightleftharpoons Se + 6OH^-$	–0.366
$Pb^{2+} + 2e \rightleftharpoons Pb(Hg)$	–0.1205	$Se + 2H^+ + 2e \rightleftharpoons H_2Se(aq)$	–0.399
$Pb^{2+} + 2e \rightleftharpoons Pb$	–0.1262	$In^{2+} + e \rightleftharpoons In^+$	–0.40
$CrO_4^{2-} + 4H_2O + 3e \rightleftharpoons Cr(OH)_3 + 5OH^-$	–0.13	$Cd^{2+} + 2e \rightleftharpoons Cd$	–0.4030
$Sn^{2-} + 2e \rightleftharpoons Sn$	–0.1375	$Cr^{3+} + e \rightleftharpoons Cr^{2+}$	–0.407
$In^+ + e \rightleftharpoons In$	–0.14	$2S + 2e \rightleftharpoons S_2^{2-}$	–0.42836
$O_2 + 2H_2O + 2e \rightleftharpoons H_2O_2 + 2OH^-$	–0.146	$Tl_2SO_4 + 2e \rightleftharpoons Tl + SO_4^{2-}$	–0.4360
$MoO_2 + 4H^+ + e \rightleftharpoons Mo + 4H_2O$	–0.152	$In^{3+} + 2e \rightleftharpoons In^+$	–0.443
$AgI + e \rightleftharpoons Ag + I^-$	–0.15224	$Fe^{2+} + 2e \rightleftharpoons Fe$	–0.447
$2NO_2^- + 2H_2O + 4e \rightleftharpoons N_2O_2^{2} + 4OH^-$	–0.18	$H_3PO_3 + 3H^+ + 3e \rightleftharpoons P + 3H_2O$	–0.454
$H_2GeO_3 + 4H^+ + 4e \rightleftharpoons Ge + 3H_2O$	–0.182	$Bi_2O_3 + 3H_2O + 6e \rightleftharpoons 2Bi + 6OH^-$	–0.46
$SnO_2 + 3H^+ + 2e \rightleftharpoons SnOH^+ + H_2O$	–0.194	$NO_2^- + H_2O + e \rightleftharpoons NO + 2OH$	–0.46
$CO_2 + 2H^+ + 2e \rightleftharpoons HCOOH$	–0.199	$PbHPO_4 + 2e \rightleftharpoons Pb + HPO_4^{2-}$	–0.465
$Mo^{3+} + 3e \rightleftharpoons Mo$	–0.200	$S + 2e \rightleftharpoons S^{2-}$	–0.47627
$Ga^+ + e \rightleftharpoons Ga$	–0.2	$S + H_2O + 2e \rightleftharpoons HS^- + OH^-$	–0.478
$2SO_3^{2-} + 4H^+ + 2e \rightleftharpoons S_2O_6^{2-} + H_2O$	–0.22	$B(OH)_3 + 7H^+ + 8e \rightleftharpoons BH_4^- + 3H_2O$	–0.481

TABLE 3
Reduction Reactions Having $E°$ Values More Negative than that of the Standard Hydrogen Electrode (continued)

Reaction	$E°$/V
$In^{3+} + e \rightleftharpoons In^{2+}$	–0.49
$ZnOH^+ + H^+ + 2e \rightleftharpoons Zn + H_2O$	–0.497
$GaOH^{2+} + H^+ + 3e \rightleftharpoons Ga + H_2O$	–0.498
$H_3PO_3 + 2H^+ + 2e \rightleftharpoons H_3PO_2 + H_2O$	–0.499
$TiO_2 + 4H^+ + 2e \rightleftharpoons Ti^{2+} + 2H_2O$	–0.502
$H_3PO_2 + H^+ + e \rightleftharpoons P + 2H_2O$	–0.508
$Sb + 3H^+ + 3e \rightleftharpoons SbH_3$	–0.510
$HPbO_2^- + H_2O + 2e \rightleftharpoons Pb + 3OH^-$	–0.537
$Ga^{3+} + 3e \rightleftharpoons Ga$	–0.549
$TlCl + e \rightleftharpoons Tl + Cl^-$	–0.5568
$Fe(OH)_3 + e \rightleftharpoons Fe(OH)_2 + OH^-$	–0.56
$TeO_3^{2-} + 3H_2O + 4e \rightleftharpoons Te + 6OH^-$	–0.57
$2SO_3^{2-} + 3H_2O + 4e \rightleftharpoons S_2O_3^{2-} + 6OH^-$	–0.571
$PbO + H_2O + 2e \rightleftharpoons Pb + 2OH^-$	–0.580
$ReO_2^- + 4H_2O + 7e \rightleftharpoons Re + 8OH^-$	–0.584
$SbO_3^- + H_2O + 2e \rightleftharpoons SbO_2^- + 2OH^-$	–0.59
$Ta^{3+} + 3e \rightleftharpoons Ta$	–0.6
$U^{4+} + e \rightleftharpoons U^{3+}$	–0.607
$As + 3H^+ + 3e \rightleftharpoons AsH_3$	–0.608
$Nb_2O_5 + 10H^+ + 10e \rightleftharpoons 2Nb + 5H_2O$	–0.644
$NbO_2 + 2H^+ + 2e \rightleftharpoons NbO + H_2O$	–0.646
$Cd(OH)_4^{2-} + 2e \rightleftharpoons Cd + 4OH^-$	–0.658
$TlBr + e \rightleftharpoons Tl + Br^-$	–0.658
$SbO_2^- + 2H_2O + 3e \rightleftharpoons Sb + 4OH^-$	–0.66
$AsO_2^- + 2H_2O + 3e \rightleftharpoons As + 4OH^-$	–0.68
$NbO_2 + 4H^+ + 4e \rightleftharpoons Nb + 2H_2O$	–0.690
$Ag_2S + 2e \rightleftharpoons 2Ag + S^{2-}$	–0.691
$AsO_4^{3-} + 2H_2O + 2e \rightleftharpoons AsO_2^- + 4OH^-$	–0.71
$Ni(OH)_2 + 2e \rightleftharpoons Ni + 2OH^-$	–0.72
$Co(OH)_2 + 2e \rightleftharpoons Co + 2OH^-$	–0.73
$NbO + 2H^+ + 2e \rightleftharpoons Nb + H_2O$	–0.733
$H_2SeO_3 + 4H^+ + 4e \rightleftharpoons Se + 3H_2O$	–0.74
$Cr^{3+} + 3e \rightleftharpoons Cr$	–0.744
$Ta_2O_5 + 10H^+ + 10e \rightleftharpoons 2Ta + 5H_2O$	–0.750
$TlI + e \rightleftharpoons Tl + I^-$	–0.752
$Zn^{2+} + 2e \rightleftharpoons Zn$	–0.7618
$Zn^{2+} + 2e \rightleftharpoons Zn(Hg)$	–0.7628
$CdO + H_2O + 2e \rightleftharpoons Cd + 2OH^-$	–0.783
$Te + 2H^+ + 2e \rightleftharpoons H_2Te$	–0.793
$ZnSO_4 \cdot 7H_2O + 2e \rightleftharpoons Zn(Hg) + SO_4^{2-}$ (Saturated $ZnSO_4$)	–0.7993
$Bi + 3H^+ + 3e \rightleftharpoons BiH_3$	–0.8
$SiO + 2H^+ + 2e \rightleftharpoons Si + H_2O$	–0.8
$Cd(OH)_2 + 2e \rightleftharpoons Cd(Hg) + 2OH^-$	–0.809
$2H_2O + 2e \rightleftharpoons H_2 + 2OH^-$	–0.8277
$2NO^-_3 + 2H_2O + 2e \rightleftharpoons N_2O_4 + 4OH^-$	–0.85
$H_3BO_3 + 3H^+ + 3e \rightleftharpoons B + 3H_2O$	–0.8698
$P + 3H_2O + 3e \rightleftharpoons PH_3(g) + 3OH^-$	–0.87
$Ti^{3+} + e \rightleftharpoons Ti^{2+}$	–0.9
$HSnO_2^- + H_2O + 2e \rightleftharpoons Sn + 3OH^-$	–0.909
$Cr^{2+} + 2e \rightleftharpoons Cr$	–0.913
$Se + 2e \rightleftharpoons Se^{2-}$	–0.924
$SO_4^{2-} + H_2O + 2e \rightleftharpoons SO_3^{2-} + 2OH^-$	–0.93
$Sn(OH)_6^{2-} + 2e \rightleftharpoons HSnO_2^- + 3OH^- + H_2O$	–0.93
$SnO_2 + 2H_2O + 4e \rightleftharpoons Sn + 4OH^-$	–0.945
$In(OH)_3 + 3e \rightleftharpoons In + 3OH^-$	–0.99
$NpO_2 + H_2O + H^+ + e \rightleftharpoons Np(OH)_3$	–0.962
$In(OH)_4^- + 3e \rightleftharpoons In + 4OH^-$	–1.007
$In_2O_3 + 3H_2O + 6e \rightleftharpoons 2In + 6OH^-$	–1.034
$PO_4^{3-} + 2H_2O + 2e \rightleftharpoons HPO_3^{2-} + 3OH^-$	–1.05
$Yb^{3+} + e \rightleftharpoons Yb^{2+}$	–1.05
$Nb^{3+} + 3e \rightleftharpoons Nb$	–1.099
$Fm^{3+} + e \rightleftharpoons Fm^{2+}$	–1.1
$2SO_3^{2-} + 2H_2O + 2e \rightleftharpoons S_2O_4^{2-} + 2OH^-$	–1.12
$Te + 2e \rightleftharpoons Te^{2-}$	–1.143
$V^{2+} + 2e \rightleftharpoons V$	–1.175
$Mn^{2+} + 2e \rightleftharpoons Mn$	–1.185
$Zn(OH)_4^{2-} + 2e \rightleftharpoons Zn + 40H^-$	–1.199
$CrO_2 + 2H_2O + 3e \rightleftharpoons Cr + 4OH^-$	–1.2
$No^{3+} + 3e \rightleftharpoons No$	–1.20
$ZnO_2^- + 2H_2O + 2e \rightleftharpoons Zn + 4OH^-$	–1.215
$H_2GaO_3^- + H_2O + 3e \rightleftharpoons Ga + 4OH^-$	–1.219
$H_2BO_3^- + 5H_2O + 8e \rightleftharpoons BH_4^- + 8OH^-$	–1.24
$SiF_6^{2-} + 4e \rightleftharpoons Si + 6F^-$	–1.24
$Zn(OH)_2 + 2e \rightleftharpoons Zn + 2OH^-$	–1.249
$ZnO + H_2O + 2e \rightleftharpoons Zn + 2OH^-$	–1.260
$Es^{3+} + e \rightleftharpoons Es^{2+}$	–1.3
$Pa^{3+} + 3e \rightleftharpoons Pa$	–1.34
$Ti^{3+} + 3e \rightleftharpoons Ti$	–1.37
$Ce^{3+} + 3e \rightleftharpoons Ce(Hg)$	–1.4373
$UO_2^{2+} + 4H^+ + 6e \rightleftharpoons U + 2H_2O$	–1.444
$Zr^{4+} + 4e \rightleftharpoons Zr$	–1.45
$Cr(OH)_3 + 3e \rightleftharpoons Cr + 3OH^-$	–1.48
$Pa^{4+} + 4e \rightleftharpoons Pa$	–1.49
$HfO_2 + 4H^+ + 4e \rightleftharpoons Hf + 2H_2O$	–1.505
$Hf^{4+} + 4e \rightleftharpoons Hf$	–1.55
$Sm^{3+} + e \rightleftharpoons Sm^{2+}$	–1.55
$ZrO_2 + 4H^+ + 4e \rightleftharpoons Zr + 2H_2O$	–1.553
$Mn(OH)_2 + 2e \rightleftharpoons Mn + 2OH^-$	–1.56
$Ba^{2+} + 2e \rightleftharpoons Ba(Hg)$	–1.570
$Bk^{2+} + 2e \rightleftharpoons Bk$	–1.6
$Cf^{3+} + e \rightleftharpoons Cf^{2+}$	–1.6
$Ti^{2+} + 2e \rightleftharpoons Ti$	–1.630
$Md^{3+} + 3e \rightleftharpoons Md$	–1.65
$HPO_3^{2-} + 2H_2O + 2e \rightleftharpoons H_2PO_2^- + 3OH^-$	–1.65
$Al^{3+} + 3e \rightleftharpoons Al$	–1.662
$SiO_3^{2-} + H_2O + 4e \rightleftharpoons Si + 6OH^-$	–1.697
$HPO_3^{2-} + 2H_2O + 3e \rightleftharpoons P + 5OH^-$	–1.71
$HfO^{2+} + 2H^+ + 4e \rightleftharpoons Hf + H_2O$	–1.724
$ThO_2 + 4H^+ + 4e \rightleftharpoons Th + 2H_2O$	–1.789
$H_2BO_3^- + H_2O + 3e \rightleftharpoons B + 4OH^-$	–1.79
$Sr^{2+} + 2e \rightleftharpoons Sr(Hg)$	–1.793
$U^{3+} + 3e \rightleftharpoons U$	–1.798
$H_2PO^-_2 + e \rightleftharpoons P + 2OH^-$	–1.82
$Be^{2+} + 2e \rightleftharpoons Be$	–1.847
$Np^{3+} + 3e \rightleftharpoons Np$	–1.856
$Fm^{3+} + 3e \rightleftharpoons Fm$	–1.89
$Th^{4+} + 4e \rightleftharpoons Th$	–1.899

TABLE 3
Reduction Reactions Having $E°$ Values More Negative than that of the Standard Hydrogen Electrode (continued)

Reaction	$E°$/V
$Am^{2+} + 2e \rightleftharpoons Am$	–1.9
$Pa^{4+} + e \rightleftharpoons Pa^{3+}$	–1.9
$Es^{3+} + 3e \rightleftharpoons Es$	–1.91
$Cf^{3+} + 3e \rightleftharpoons Cf$	–1.94
$Lr^{3+} + 3e \rightleftharpoons Lr$	–1.96
$Eu^{3+} + 3e \rightleftharpoons Eu$	–1.991
$Er^{2+} + 2e \rightleftharpoons Er$	–2.0
$Pr^{2+} + 2e \rightleftharpoons Pr$	–2.0
$Pu^{3+} + e \rightleftharpoons Pu$	–2.031
$Cm^{3+} + 3e \rightleftharpoons Cm$	–2.04
$Am^{3+} + 3e \rightleftharpoons Am$	–2.048
$AlF_6^{3-} + 3e \rightleftharpoons Al + 6F^-$	–2.069
$Sc^{3+} + 3e \rightleftharpoons Sc$	–2.077
$Ho^{2+} + 2e \rightleftharpoons Ho$	–2.1
$Nd^{2+} + 2e \rightleftharpoons Nd$	–2.1
$Cf^{2+} + 2e \rightleftharpoons Cf$	–2.12
$Yb^{3+} + 3e \rightleftharpoons Yb$	–2.19
$Ac^{3+} + 3e \rightleftharpoons Ac$	–2.20
$Dy^{2+} + 2e \rightleftharpoons Dy$	–2.2
$Tm^{3+} + e \rightleftharpoons Tm^{2+}$	–2.2
$Pm^{2+} + 2e \rightleftharpoons Pm$	–2.2
$Es^{2+} + 2e \rightleftharpoons Es$	–2.23
$H_2 + 2e \rightleftharpoons 2H^-$	–2.23
$Gd^{3+} + 3e \rightleftharpoons Gd$	–2.279
$Tb^{3+} + 3e \rightleftharpoons Tb$	–2.28
$Lu^{3+} + 3e \rightleftharpoons Lu$	–2.28
$Dy^{3+} + 3e \rightleftharpoons Dy$	–2.295
$Am^{3+} + e \rightleftharpoons Am^{2+}$	–2.3
$Fm^{2+} + 2e \rightleftharpoons Fm$	–2.30
$Pm^{3+} + 3e \rightleftharpoons Pm$	–2.30
$Sm^{3+} + 3e \rightleftharpoons Sm$	–2.304
$Al(OH)_3 + 3e \rightleftharpoons Al + 3OH^-$	–2.31
$Tm^{3+} + 3e \rightleftharpoons Tm$	–2.319
$Nd^{3+} + 3e \rightleftharpoons Nd$	–2.323
$Al(OH)^- + 3e \rightleftharpoons Al + 4OH^-$	–2.328
$H_2AlO_3^- + H_2O + 3e \rightleftharpoons Al + 4OH^-$	–2.33
$Ho^{3+} + 3e \rightleftharpoons Ho$	–2.33
$Er^{3+} + 3e \rightleftharpoons Er$	–2.331
$Ce^{3+} + 3e \rightleftharpoons Ce$	–2.336
$Pr^{3+} + 3e \rightleftharpoons Pr$	–2.353

Reaction	$E°$/V
$ZrO(OH)_2 + H_2O + 4e \rightleftharpoons Zr + 4OH^-$	–2.36
$Mg^{2+} + 2e \rightleftharpoons Mg$	–2.372
$Y^{3+} + 3e \rightleftharpoons Y$	–2.372
$La^{3+} + 3e \rightleftharpoons La$	–2.379
$Tm^{2+} + 2e \rightleftharpoons Tm$	–2.4
$Md^{2+} + 2e \rightleftharpoons Md$	–2.40
$Th(OH)_4 + 4e \rightleftharpoons Th + 4OH^-$	–2.48
$HfO(OH)_2 + H_2O + 4e \rightleftharpoons Hf + 4OH^-$	–2.50
$No^{2+} + 2e \rightleftharpoons No$	2.50
$Dy^{3+} + e \rightleftharpoons Dy^{2+}$	–2.6
$Pm^{3+} + e \rightleftharpoons Pm^{2+}$	–2.6
$Be_2O_3^{2-} + 3H_2O + 4e \rightleftharpoons 2Be + 6OH^-$	–2.63
$Sm^{2+} + 2e \rightleftharpoons Sm$	–2.68
$Mg(OH)_2 + 2e \rightleftharpoons Mg + 2OH^-$	–2.690
$Nd^{3+} + e \rightleftharpoons Nd^{2+}$	–2.7
$Mg^+ + e \rightleftharpoons Mg$	–2.70
$Na^+ + e \rightleftharpoons Na$	–2.71
$Yb^{2+} + 2e \rightleftharpoons Yb$	–2.76
$Bk^{3+} + e \rightleftharpoons Bk^{2+}$	–2.8
$Ho^{3+} + e \rightleftharpoons Ho^{2+}$	–2.8
$Ra^{2+} + 2e \rightleftharpoons Ra$	–2.8
$Eu^{2+} + 2e \rightleftharpoons Eu$	–2.812
$Ca^{2+} + 2e \rightleftharpoons Ca$	–2.868
$Sr(OH)_2 + 2e \rightleftharpoons Sr + 2OH^-$	–2.88
$Sr^{2+} + 2e \rightleftharpoons Sr$	–2.89
$Fr^+ + e \rightleftharpoons Fr$	–2.9
$La(OH)_3 + 3e \rightleftharpoons La + 3OH^-$	–2.90
$Ba^{2+} + 2e \rightleftharpoons Ba$	–2.912
$K^+ + e \rightleftharpoons K$	–2.931
$Rb^+ + e \rightleftharpoons Rb$	–2.98
$Ba(OH)_2 + 2e \rightleftharpoons Ba + 2OH^-$	–2.99
$Er^{3+} + e \rightleftharpoons Er^{2+}$	–3.0
$Ca(OH)_2 + 2e \rightleftharpoons Ca + 2OH^-$	–3.02
$Cs^+ + e \rightleftharpoons Cs$	–3.026
$Li^+ + e \rightleftharpoons Li$	–3.0401
$3N_2 + 2H^+ + 2e \rightleftharpoons 2HN_3$	–3.09
$Pr^{3+} + e \rightleftharpoons Pr^{2+}$	–3.1
$Ca^+ + e \rightleftharpoons Ca$	–3.80
$Sr^+ + e \rightleftharpoons Sr$	–4.10

REDUCTION AND OXIDATION POTENTIALS FOR CERTAIN ION RADICALS*

Petr Vanýsek

There are two tables for ion radicals. The first table lists reduction potentials for organic compounds which produce anion radicals during reduction, a process described as $A + e^- \leftrightharpoons A^{-\cdot}$. The second table lists oxidation potentials for organic compounds which produce cation radicals during oxidation, a process described as $A \leftrightharpoons A^{+\cdot} + e^-$. To obtain reduction potential for a reverse reaction, the sign for the potential is changed.

Unlike the table of the Electrochemical Series, which lists *standard* potentials, values for radicals are experimental values with experimental conditions given in the second column. Since the measurements leading to potentials for ion radicals are very dependent on conditions, an attempt to report standard potentials for radicals would serve no useful purpose. For the same reason, the potentials are also reported as experimental values, usually a half-wave potential ($E_{1/2}$ in polarography) or a peak potential (E_p in cyclic voltammetry). Unless otherwise stated, the values are reported vs. SCE (saturated calomel electrode). To obtain a value vs. normal hydrogen electrode, 0.241 V has to be added to the SCE values. All the ion radicals chosen for inclusion in the tables result from electrochemically reversible reactions. More detailed data on ion radicals can be found in the *Encyclopedia of Electrochemistry of Elements,* (A. J. Bard, Ed.). Vol. XI and XII in particular, Marcel Dekker, New York, 1978.

* Abbreviations are: CV — cyclic voltammetry; DMF — N,N-Dimethylformamide; E swp — potential sweep; E^o — standard potential; E_p — peak potential; $E_{p/2}$ — half-peak potential; $E_{1/2}$ — half wave potential; *M* — mol/l; MeCN — acetonitrile; pol — polarography; rot Pt dsk — rotated Pt disk; SCE — saturated calomel electrode; $TBABF_4$ — tetrabutylammonium tetrafluoroborate; TBAI — tetra-*n*-butylammonium iodide; TBAP — tetra-*n*-butylammonium perchlorate; TEABr — tetra -*n*-ethylammonium bromide; TEAP — tetra-*n*-ethylammonium perchlorate; THF — tetrahydrofurane; $TPACF_3SO_3$ — tetra-*n*-proplyammonium trifluoromethanesulferate; TPAP — tetra-*n*-propylammonium perchlorate; and wr — wire.

REDUCTION POTENTIALS (PRODUCTS ARE ANION RADICALS)

Substance	Conditions/electrode/technique	Potential ([V] vs. SCE)
Acetone	DMF, 0.1 *M* TEABr/Hg/pol	$E_{1/2} = -2.84$
1-Naphthyphenylacetylene	DMF, 0.03 *M* TBAI/Hg/pol	$E_{1/2} = -1.91$
1-Naphthalenecarboxaldehyde	– /Hg/pol	$E_{1/2} = -0.91$
2-Naphthalenecarboxaldehyde	– /Hg/pol	$E_{1/2} = -0.96$
2-Phenanthrenecarboxaldehyde	– /Hg/pol	$E_{1/2} = -1.00$
3-Phenanthrenecarboxaldehyde	– /Hg/pol	$E_{1/2} = -0.94$
9-Phenanthrenecarboxaldehyde	– /Hg/pol	$E_{1/2} = -0.83$
1-Anthracenecarboxaldehyde	– /Hg/pol	$E_{1/2} = -0.75$
1-Pyrenecarboxaldehyde	– /Hg/pol	$E_{1/2} = -0.76$
2-Pyrenecarboxaldehyde	– /Hg/pol	$E_{1/2} = -1.00$
Anthracene	DMF, 0.1 *M* TBAP/Pt dsk/CV	$E_p = -2.00$
	DMF, 0.5 *M* $TBABF_4$/Hg/CV	$E_{1/2} = -1.93$
	MeCN, 0.1 *M* TEAP/Hg/CV	$E_{1/2} = -2.07$
	DMF, 0.1 *M* TBAI/Hg/pol	$E_{1/2} = -1.92$
9,10-Dimethylanthracene	DMF, 0.1 *M* TBAP/Pt/CV	$E_p = -2.08$
	MeCN, 0.1 *M* TBAP/Pt/CV	$E_p = -2.10$
1-Phenylanthracene	DMF, 0.5 *M* $TBABF_2$/Hg/CV	$E_{1/2} = -1.91$
	DMF, 0.1 *M* TBAI/Hg/pol	$E_{1/2} = -1.878$
2-Phenylanthracene	DMF, 0.1 *M* TBAI/Hg/pol	$E_{1/2} = -1.875$
8-Phenylanthracene	DMF, 0.5 *M* $TBABF_4$/Hg/CV	$E_{1/2} = -1.91$
9-Phenylanthracene	DMF, 0.5 *M* $TBABF_4$/Hg/CV	$E_{1/2} = -1.93$
	DMF, 0.1 *M* TBAI/Hg/pol	$E_{1/2} = -1.863$
1,8-Diphenylanthracene	DMF, 0.5 *M* $TBABF_4$/Hg/CV	$E_{1/2} = -1.88$
1,9-Diphenylanthracene	DMF, 0.1 *M* TBAI/Hg/pol	$E_{1/2} = -1.846$
1,10-Diphenylanthracene	DMF, 0.1 *M* TBAI/Hg/pol	$E_{1/2} = -1.786$
8,9-Diphenylanthracene	DMF, 0.5 *M* $TBABF_4$/Hg/CV	$E_{1/2} = -1.90$
9,10-Diphenylanthracene	MeCN, 0.1 *M* TBAP/rot Pt/E swp	$E_{1/2} = -1.83$
	DMF, 0.1 *M* TBAI/Hg/pol	$E_{1/2} = -1.835$
1,8,9-Triphenylanthracene	DMF, 0.5 M $TBABF_4$/Hg/CV	$E_{1/2} = -1.85$
1,8,10 Triphenylanthracene	DMF, 0.5 *M* $TBABF_4$/Hg/CV	$E_{1/2} = -1.81$
9,10-Dibiphenylanthracene	MeCN, 0.1 *M* TBAP/rot Pt/E swp	$E_{1/2} = -1.94$
Benz(a)anthracene	MeCN, 0.1 *M* TEAP/Hg/CV	$E_{1/2} = -2.11$
	MeCN, 0.1 *M* TEAP/Hg/pol	$E_{1/2} = -2.40$[a]
Azulene	DMF, 0.1 *M* TBAI/Hg/pol	$E_{1/2} = -1.10$[c]
Annulene	DMF, 0.5, *M* TBAP 0°C/Hg/pol	$E_{1/2} = -1.23$
Benzaldehyde	DMF, 0.1 *M* TBAP/Hg/pol	$E_{1/2} = -1.67$
Benzil	DMSO, 0.1 *M* TBAP/Hg/pol	$E_{1/2} = -1.04$
Benzophenone	– /Hg/pol	$E_{1/2} = -1.80$
Benzophenone	DMF/Pt dsk/CV	$E^o = -1.72$
Chrysene	MeCN, 0.1 *M* TEAP/Hg/pol	$E_{1/2} = -2.73$[a]
Fluoranthrene	DMF, 0.1 *M* TBAP/Pt dsk/CV	$E_p = -1.76$
Cyclohexanone	DMF, 0.1 *M* TEABr/Hg/pol	$E_{1/2} = -2.79$
5,5-Dimethyl-3-phenyl-2-cyclohexen-1-one	DMF, 0.5 *M*/Hg/pol	$E_{1/2} = -1.71$
1,2,3-Indanetrione hydrate (ninhydrin)	DMF, 0.2 *M* $NaNO_3$/Hg/pol	$E_{1/2} = -0.039$
Naphthacene	DMF, 0.1 *M* TBAI/Hg/pol	$E_{1/2} = -1.53$
Naphthalene	DMF, 0.1 *M* TBAP/Pt dsk/CV	$E_p = -2.55$
	DMF, 0.5 *M* $TBABF_4$/Hg/CV	$E_{1/2} = -2.56$
	DMF, MeCN, 0.1 *M* TEAP/Hg/CV	$E_{1/2} = -2.63$
	DMF, 0.1 *M* TBAI/Hg/pol	$E_{1/2} = -2.50$

REDUCTION POTENTIALS (continued)
(PRODUCTS ARE ANION RADICALS)

Substance	Conditions/electrode/technique	Potential ([V] vs. SCE)
1-Phenylnaphthalene	DMF, 0.5 *M* $TBBF_4$/Hg/CV	$E_{1/2}$ = −2.36
1,2-Diphenylnaphthalene	DMF, 0.5 *M* $TBABF_4$/Hg/CV	$E_{1/2}$ = −2.25
Cyclopentanone	DMF, 0.1 *M* TEABr/Hg/pol	$E_{1/2}$ = −2.82
Phenanthrene	MeCN, 0.1 *M* TBAP/Pt wr/CV	$E_{1/2}$ = −2.47
	MeCN, 0.1 *M* TEAP/Hg/pol	$E_{1/2}$ = −2.88[a]
Pentacene	THF, 0.1 *M* TBAP/rot Pt dsk/E swp	$E_{1/2}$ = −1.40
Perylene	MeCN, 0.1 *M* TEAP/Hg/CV	$E_{1/2}$ = −1.73
1,3-Diphenyl-1,3-propanedione	DMSO, 0.2 *M* TBAP/Hg/CV	$E_{1/2}$ = −1.42
2,2-Dimethyl-1,3-diphenyl-1,3 propanedione	DMSO, TBAP/Hg/CV	$E_{1/2}$ = −1.80
Pyrene	DMF, 0.1 *M* TBAP/Pt/CV	E_p = −2.14
	MeCN, 0.1 *M* TEAP/Hg/pol	$E_{1/2}$ = −2.49[a]
Diphenylsulfone	DMF, TEABr	$E_{1/2}$ = −2.16
Triphenylene	MeCN, 0.1 *M* TEAP/Hg/pol	$E_{1/2}$ = −2.87[a]
9,10-Anthraquinone	DMF, 0.5 *M* TBAP, 20°/Pt dsk/CV	E^{1}_{2} = −1.01
1,4-Benzoquinone	MeCN, 0.1 *M* TEAP/Pt/CV	E_p = −0.54
1,4-Naphthohydroquinone, dipotassium salt	DMF, 0.5 *M* TBAP, 20°/Pt dsk/CV	$E_{1/2}$ = −1.55
Rubrene	DMF, 0.1 *M* TBAP/Pt dsk/CV	E_p = −1.48
	DMF, 0.1 *M* TBAI/Hg/pol	$E_{1/2}$ = −1.410
Benzocyclooctatetraene	THF, 0.1 *M* TBAP/Hg/pol	$E_{1/2}$ = −2.13
sym-dibenzocyclooctatetraene,	THF, 0.1 *M* TBAP/Hg/pol	$E_{1/2}$ = −2.29
Ubiquinone-6	MeCN, 0.1 *M* TEAP/Pt/CV	E_p = −1.05[c]
(9-Phenyl-fluorenyl)$^+$	10.2 *M* H_2SO_4≡Hg/CV	E_p = −0.01[b]
(Triphenylcyclopropenyl)$^+$	MeCN, 0.1 *M* TEAP/Hg/CV	E_p = −1.87
(Triphenylmethyl)$^+$	MeCN, 0.1 *M* TBAP/Hg/pol	$E_{1/2}$ = 0.27
	H_2SO_4, 10.2 *M*/Hg/CV	E_p = −0.58[b]
(Tribiphenylmethyl)$^+$	MeCN, 0.1 *M* TBAP/Hg/pol	$E_{1/2}$ = 0.19
(Tri-4-*t*-butyl-5-phenylmethyl)$^+$	MeCN, 0.1 *M* TBAP/Hg/pol	$E_{1/2}$ = 0.13
(Tri-4-isopropylphenylmethyl)$^+$	MeCN, 0.1 *M* TBAP/Hg/pol	$E_{1/2}$ = 0.07
(Tri-4-methylphenylmethyl)$^+$	MeCN, 0.1 *M* TBAP/Hg/pol	$E_{1/2}$ = 0.05
(Tri-4-cyclopropylphenylmethyl)$^+$	MeCN, 0.1 *M* TBAP/Hg/pol	$E_{1/2}$ = 0.01
(Tropylium)$^+$	MeCN, 0.1 *M* TBAP/Hg/pol	$E_{1/2}$ = −0.17
	DMF, 0.15 *M* TBAI/Hg/pol	$E_{1/2}$ = −1.55
	DMF, 0.15 *M* TBAI/Hg/pol	$E_{1/2}$ = −1.55
	DMF, 0.15 *M* TBAI/Hg/pol	$E_{1/2}$ = −1.57
	DMF, 0.15 *M* TBAI/Hg/pol	$E_{1/2}$ = −1.60
	DMF, 0.15 *M* TBAI/Hg/pol	$E_{1/2}$ = −1.87
	DMF, 0.15 *M* TBAI/Hg/pol	$E_{1/2}$ = −1.96
	DMF, 0.15 *M* TBAI/Hg/pol	$E_{1/2}$ = −2.05

OXIDATION POTENTIALS
(Products are cation radicals)

Substance	Conditions/electrode/technique	Potential [V] vs. SCE
Anthracene	CH_2Cl_2, 0.2 *M* $TBABF_4$, −70°C/Pt dsk/CV	E_p = + 0.73[d]
9,10-dimethylanthracene	MeCN, 0.1 *M* $LiClO_4$/Pt wr/CV	E_p = +1.0
9,10-di-*n*-propylanthracene	MeCN, 0.1 *M* TEAP/Pt/CV	E_p = +1.08
1,8-diphenylanthracene	CH_2Cl_2, 0.2 *M* $TPrACF_3SO_3$/rot Pt wr/E swp	$E_{1/2}$ = +1.34
8,9-diphenylanthracene	CH_2Cl_2, 0.2 *M* $TPrACF_3SO_3$/rot Pt wr/E swp	$E_{1/2}$ = +1.30
9,10-diphenylanthracene	MeCN/Pt/CV	E_p = +1.22
Perylene	MeCN, 0.1 *M* TBAP/Pt/CV	E_p = +1.34
Pyrene	DMF, 0.1 *M* TBAP/Pt dsk/CV	E_p = +1.25
Rubrene	DMF, 0.1 *M* TBAP/Pt dsk/CV	E_p = +1.10
Tetracene	CH_2Cl_2, 0.2 *M* $TBABF_4$, −70°C/Pt wr/CV	E_p = +0.35[d]
1,4-dithiabenzene	MeCN, 0.1 *M* TEAP/Pt dsk/rot	$E_{1/2}$ = +0.69
1,4-dithianaphtalene	MeCN, 0.1 *M* TEAP/Pt dsk/rot	$E_{1/2}$ = +0.80
Thianthrene	0.1 *M* TPAP/Pt/CV	$E_{1/2}$ = +1.28

[a] Vs. 0.01 mol/1 Ag/$AgClO_4$
[b] Vs. Hg/Hg_2SO_4, 17 *M* H_2SO_4
[c] Vs. Hg pool
[d] Vs. Ag/saturated $AgNO_3$
[e] Vs. Ag/0.01 *m* Ag^+

pH SCALE FOR AQUEOUS SOLUTIONS

A. K. Covington

The pH value is the negative decadic logarithm of the (relative) ion activity of the hydrogen ion in the solution.

$$pH = -\log a_H \tag{1}$$

This is only a notional definition since Equation 1 involves a single ion activity, which is immeasurable, and has to be attained through a nonthermodynamic assumption such as that described in Equation 5 below. In terms of substance concentration, molarity, Equation 1 may be rewritten

$$pH = -\log (c_H y_H / c^o) \tag{2}$$

where c^o is an arbitrary constant representing the standard state condition and equal to 1 mol dm^{-3}, c_H is the concentration of hydrogen ion and y_H is the single ion activity of the hydrogen ion. In terms of molality, Equation 1 may be rewritten

$$pH = -\log (m_H \gamma_H / m^o) \tag{3}$$

where m^o is an arbitrary constant representing the standard state condition and equal to 1 mol kg^{-1}, m_H is the concentration of hydrogen ion and γ_H is the single ion activity of the hydrogen ion. For most purposes the difference between these two scales can be ignored for dilute aqueous solutions; the difference is 0.001 at 25° C and 0.02 at 100° C. Arising from the nonexperimental determinability of single ion activities, the definition and determination of pH have an operational basis, and depend on the assignment of pH values to a standard solution (or solutions) together with the determination of pH difference by a cell with liquid junction called the operational cell.

The Operational Definition of pH Difference[1,2]

The electromotive force, EMF, E(X) of the cell with liquid junction:

Reference electrode | KCl (aq., concentrated) ‖ Solution X | H_2 | Pt (I)

is measured, and likewise that, E(S), of the cell:

Reference electrode | KCl (aq., concentrated) ‖ Solution S | H_2 | Pt (II)

The temperature of both cells (I and II) must be equal and uniform throughout, and the hydrogen gas pressures identical. The two bridge solutions may be any molality of KCl not less than 3.5 mol kg^{-1} provided they are the same.

The pH of the solution X, pH(X), is then related to the assigned pH of the solution S, pH(S) by the definition:

$$pH(X) = pH(S) + [E(S) - E(X)]/[(RT/F) \ln 10] \tag{4}$$

where R is the gas constant, T the thermodynamic temperature, F the Faraday constant. The quantity $k = (RT/F) \ln 10$ is called the slope factor whose values are given as a function of temperature in Table 1. As a consequence of this definition any difference in liquid junction potential between cells I and II is subsumed into the value of pH(X).

The pH Scale

The pH scale at a particular temperature is defined by Equation 4 as a straight line, on the plot of pH against E(X), having a slope of k drawn through the pH value assigned to the Reference Value Standard (RVS) solution (as given in Table 2) and the value of E(S) for cell II when it contains the Reference Value Standard solution. The solution chosen for the RVS is 0.05 mol kg^{-1} aqueous potassium hydrogen phthalate. The procedure by which pH(RVS) values have been assigned to the Reference Value Standard (RVS) is the cell III without transference:[1,3]

Pt(Pd) | H_2 (g, p=1 atm = 101 325 Pa) | RVS, Cl^- | AgCl | Ag (III)

The palladised-platinum hydrogen electrode is used to reduce the catalytised chemical reduction of the phthalate by hydrogen gas. The calculation involves a non-thermodynamic assumption, the Bates-Guggenheim Convention, for the single ion activity of the chloride ion[1,2] as

$$\log (\gamma_{Cl})^o = -A(I/m^o)^{1/2}/[1 + 1.5\,(I/m^o)^{1/2}] \tag{5}$$

where I is the ionic strength = $(1/2)\Sigma m_i z_i^2$ = 0.0534 mol kg^{-1} for the RVS solution and A is a known function of temperature (Table 1).

To prepare the RVS solution, dry the sample at 110° C for 2 h before use. The water should have a conductivity of less than 0.1 mS m^{-1}. The required solution contains 10.211 g kg^{-1} water. It can be prepared on a volume basis by dissolving 10.138 g potassium hydrogen phthalate in water and making up to 1 L at 20° C. This solution is 0.04964 mol/L with a density of 1.00300 g/L at 20° C.

Primary Standards[2]

pH values may be assigned by the cell without transference method (cell III) to six other buffer solutions which meet certain criteria of reproducibility of preparation and properties. These solutions are called primary pH standards (PS), and details and the pH(PS) values assigned to them are given in Table 3. When these PS solutions are used in the operational cell I, the experimental value of the slope will not be in accord with the slope factor values of Table 1, and, moreover, the experimental value could change if additional primary solutions were to be defined. Hence the pH value determined for an unknown solution can be slightly dependent (±0.02) on the choice of primary standard.[2,4,5] Some useful data for standard buffers are given in Table 5.

Operational Standards[2,6]

Operational standards (OS) are also defined which are traceable to the Reference Value Standard (RVS). Values are assigned by means of the operational cells I and II where the liquid junctions are the free diffusion type reproducibly formed in 1 mm vertical capillary tubes. These operational standards are not restricted in number provided certain preparation criteria are met, and pH(OS) values for 16 solutions are given in Table 4.[2] These OS represent an alternative procedure and are in no way to be regarded as inferior to the primary standards. As a consequence of their definition, all pH(OS) values fall on the line with slope given by the slope factor value for the appropriate temperature in Table 1. Any difference in liquid junction potential between the solutions of cells I and II and KCl is subsumed into the assigned value of pH(OS).

Measurement of pH. Choice of Standard Reference Solution

1a. If pH is not required to better than ±0.05 any standard reference solution may be selected.

1b. If pH is required to ±0.002 and interpretation in terms of hydrogen ion concentration or activity is desired, choose a standard reference solution, pH(PS) or pH(OS), to match X as closely as possible in terms of pH, composition and ionic strength.

2. Alternatively, a bracketting procedure may be adopted whereby two standard reference solutions are chosen whose pH values, pH(S1), pH(S2) are on either side of pH(X). Then if the corresponding potential difference measurements are E(S1), E(S2), E(X), then pH(X) is obtained from

$$\mathrm{pH(X)} = \mathrm{pH(S1)} + [E(\mathrm{X}) - E(\mathrm{S1})]/\ \%k$$

where $\%k = 100[E(\mathrm{S2}) - E(\mathrm{S1})]/[\mathrm{pH(S2)} - \mathrm{pH(S1)}]$ is the apparent percentage slope. This procedure is very easily done on some pH meters simply by adjusting downwards the slope factor control with the electrodes in S2. The purpose of the bracketting procedure is to compensate for deficiencies in the electrodes and measuring system.

Information to be Given about the Measurement of pH(X)

The standard solutions selected for calibration of the pH meter system should be reported with the measurement as follows,

1.System calibrated with pH(RVS) = ... at ...K.
2.System calibrated with two primary standards pH(PS1) = ... and pH(PS2) = ... at ... K.
3.System calibrated with two operational standards pH(OS1) =..... and pH(OS2) = ... at .... K.

Interpretation of pH(X) in Terms of Hydrogen Ion Concentration

The operationally defined pH has no simple interpretation in terms of hydrogen ion concentration but the mean ionic activity coefficient of a typical 1:1 electrolyte can be substituted into equation 2 or 3 to obtain hydrogen ion concentration subject to an uncertainty of 3.9% in concentration corresponding to 0.02 in pH.

REFERENCES

1. R. G. Bates, *Measurement of pH. Theory and Practice*, 2nd ed., John Wiley & Sons,, New York, 1973.
2. A. K. Covington, R. G. Bates, and R. A. Durst, *Pure Appl. Chem.*, 57, 531, 1985.
3. H. P. Butikofer and A. K. Covington, *Anal. Chim. Acta*, 108, 179, 1979.
4. R. G. Bates, *Crit. Rev. Anal. Chem.*, 10, 247, 1981.
5. A. K. Covington, *Anal. Chim. Acta*, 127, 1, 1981.
6. A. K. Covington and M. J. Rebelo, *Anal. Chim. Acta.*, 200, 245, 1987.
7. V. E. Bower and R. G. Bates, *J. Res. Natl. Bur. Stand.*, 39, 263, 1954.

TABLE 1
Standard EMF, Slope Factor and Debye-Huckel Constant *A* (Unit Weight of Solvent) as Functions of Temperature

Temperature/°C	$E°$/mV[7]	Slope Factor k/mV	A[1]
0	236.55	54.199	0.4918
5	234.13	55.191	0.4952
10	231.42	56.183	0.4988
15	228.57	57.175	0.5026
20	225.57	58.167	0.5066
25	222.34	59.159	0.5108
30	219.04	60.152	0.5150
35	215.65	61.144	0.5196
40	212.08	62.136	0.5242
45	208.35	63.128	0.5291
50	204.49	64.120	0.5341
55	200.56	65.112	0.5393
60	196.49	66.104	0.5448
70	187.82	68.088	0.5562
80	178.73	70.073	0.5685
90	169.52	72.057	0.5817
95	165.11	73.049	0.5886

TABLE 2
Values of pH(RVS) for the Reference Value Standard of 0.05 mol kg^{-1} Potassium Hydrogen Phthalate at Various Temperatures

t/°C	pH(RVS)	t/°C	pH(RVS)	t/°C	pH(RVS)
0	4.000	35	4.018	65	4.097
5	3.998	37	4.022	70	4.116
10	3.997	40	4.027	75	4.137
15	3.998	45	4.038	80	4.159
20	4.001	50	4.050	85	4.183
25	4.005	55	4.064	90	4.21
30	4.011	60	4.080	95	4.24

TABLE 3
Values of pH(PS) for Primary Standard Reference Solutions

Primary ref. standard	t/°C 0	5	10	15	20	25	30	35	37	40	50	60	70	80	90	95
Saturated (at 25° C) Potassium hydrogen tartrate	—	—	—	—	—	3.557	3.552	3.549	3.548	3.547	3.549	3.560	3.580	3.610	3.650	3.674
0.1 mol/kg Potassium dihydrogen citrate	3.863	3.840	3.820	3.802	3.788	3.776	3.766	3.759	3.756	3.754	3.749	—	—	—	—	—
0.025 mol/kg Disodium hydrogen phosphate +0.025 mol/kg Potassium dihydrogen phosphate	6.984	6.951	6.923	6.900	6.881	6.865	6.853	6.844	6.841	6.838	6.833	6.836	6.845	6.859	6.876	6.886
0.03043 mol/kg Disodium hydrogen phosphate +0.008695 mol/kg Potassium dihydrogen phosphate	7.534	7.500	7.472	7.448	7.429	7.413	7.400	7.389	7.386	7.380	7.367	—	—	—	—	—
0.01 mol/kg Disodium tetraborate	9.464	9.395	9.332	9.276	9.225	9.180	9.139	9.102	9.088	9.068	9.011	8.962	8.921	8.884	8.850	8.833
0.025 mol/kg Sodium hydrogen carbonate +0.025 mol/kg sodium carbonate	10.317	10.245	10.179	10.118	10.062	10.012	9.966	9.926	9.910	9.889	9.828	—	—	—	—	—

Note: Based on an uncertainty of ±0.2 mV in determined (E-E^{θ}), the uncertainty is ±0.003 in pH in the range 0—50° C.

pH SCALE FOR AQUEOUS SOLUTIONS (continued)

TABLE 4
pH (OS) Values for Operational Reference Solutions

Operational standard ref. solution	t/°C														
	0	5	10	15	20	25	30	37	40	50	60	70	80	90	95
0.1 mol/kg Potassium tetroxalate[a]	—	—	—	—	1.475	1.479	1.483	1.490	1.493	1.503	1.513	1.52	1.53	1.53	1.53
0.05 mol/kg potassium tetroxalate[a]	—	—	1.638	1.642	1.644	1.646	1.648	1.649	1.650	1.653	1.660	1.671	1.689	1.72	1.73
0.05 mol/kg sodium hydrogen diglycolate[b]	—	3.466	3.470	3.476	3.484	3.492	3.502	3.519	3.527	3.558	3.595	—	—	—	—
Saturated (at 25° C) potassium hydrogen tartrate	—	—	—	—	—	3.556	3.549	3.544	3.542	3.544	3.553	3.570	3.596	3.627	3.649
0.05 mol/kg Potassium hydrogen phthalate (RVS)	4.000	3.998	3.997	3.998	4.000	4.005	4.011	4.022	4.027	4.050	4.080	4.115	4.159	4.21	4.24
0.1 mol/dm³ Acetic acid + 0.1 mol/dm sodium acetate	4.664	4.657	4.652	4.647	4.645	4.644	4.643	4.647	4.650	4.663	4.684	4.713	4.75	4.80	4.83
0.01 mol/dm³ Acetic acid + 0.1 mol/dm³ sodium acetate	4.729	4.722	4.717	4.714	4.712	4.713	4.715	4.722	4.726	4.743	4.768	4.800	4.839	4.88	4.91
0.02 mol/kg Piperazine phosphate[c]	—	6.477	6.419	6.364	6.310	6.259	6.209	6.143	6.116	6.030	5.952	—	—	—	—
0.025 mol/kg Disodium hydrogen phosphate + 0.025 mol/kg potassium dihydrogen phosphate	6.961	6.935	6.912	6.891	6.873	6.857	6.843	6.828	6.823	6.814	6.817	6.830	6.85	6.90	6.92
0.03043 mol/kg Disodium hydrogen phosphate + 0.008695 mol/kg potassium disodium phosphate	7.506	7.482	7.460	7.441	7.423	7.406	7.390	7.369	—	—	—	—	—	—	—
0.04 mol/kg Disodium hydrogen phosphate + 0.01 mol/kg potassium dihydrogen phosphate	—	7.512	7.488	7.466	7.445	7.428	7.414	7.404	—	—	—	—	—	—	—
0.05 mol/kg Tris hydrochloride + 0.01667 mol/kg Tris[d]	8.399	8.238	8.083	7.933	7.788	7.648	7.513	7.332	7.257	7.018	6.794	—	—	—	—
0.05 mol/kg Disodium tetraborate $(Na_2B_4O_7$	9.475	9.409	9.347	9.288	9.233	9.182	9.134	9.074	9.051	8.983	8.932	8.898	8.88	8.84	8.89
0.01 mol/kg Disodium tetraborate $(Na_2B_4O_7)$	9.451	9.388	9.329	9.275	9.225	9.179	9.138	9.086	9.066	9.009	8.965	8.932	8.91	8.90	8.89

pH SCALE FOR AQUEOUS SOLUTIONS (continued)

TABLE 4
pH(OS) Values for Operational Standard Reference Solutions (continued)

Operational standard ref. solution	t/°C 0	5	10	15	20	25	30	37	40	50	60	70	80	90	95
0.025 mol/kg Sodium hydrogen carbonate + 0.025 mol/kg sodium carbonate	10.273	10.212	10.154	10.098	10.045	9.995	9.948	9.889	9.866	9.800	9.753	9.728	9.725	9.75	9.77
Saturated (at 20° C) calcium hydroxide	13.360	13.159	12.965	12.780	12.602	12.431	12.267	12.049	11.959	11.678	11.423	11.192	10.984	10.80	10.71

Note: Uncertainty is ±0.003 in pH between 0 and 60° C rising to ±0.01 above 70° C.

[a] Potassium trihydrogen dioxalate ($KH_3C_4O_8$).
[b] Sodium hydrogen 2,2′-oxydiethanoate.
[c] $C_4H_{10}N_2 \cdot H_3PO_4$.
[d] 2-Amino-2-(hydroxymethyl)-1,3 propanediol or tris(hydroxymethyl)aminomethane.

TABLE 5
Useful Data on Some Standard Buffer Solutions

	Molecular formula	Molality (mol/kg)	Relative molar mass	Density at 20° C (g/cm³)	Molarity at at 20° C (mol/L)	Mass of 1 L at 20° C (g)	Mass tolerance for ±0.001 pH[a] (g)	Mass tolerance expressed as a percentage (%)
Potassium tetraoxalate	$KH_3C_4O_8 \cdot 2H_2O$	0.1	254.1913	1.0091	0.09875	25.1017	0.07	0.27
Potassium tetraoxalate	$KH_3C_4O_8 \cdot 2H_2O$	0.05	254.1913	1.0038	0.04965	12.6202	0.034	0.26
Disodium hydrogen orthophosphate	Na_2HPO_4	0.025	141.9588 }	1.0038	0.02492	3.5379	0.02	0.56
Potassium dihydrogen orthophosphate	KH_2PO_4	0.025	136.0852 }			3.3912	0.02	0.58
Disodium tetraborate	$Na_2B_4O_7 \cdot 10H_2O$	0.05	381.367	1.0075	0.04985	19.0117	0.9	4.73
Disodium tetraborate	$Na_2B_4O_7 \cdot 10H_2O$	0.01	381.367	1.0001	0.009981	3.8064	0.19	0.49
Sodium carbonate	Na_2CO_3	0.025	105.9887 }	1.0021	0.02494	2.6428	0.017	0.064
Sodium hydrogen carbonate	$NaHCO_3$	0.025	84.0069 }			2.0947	0.013	0.62

a Calculated from known dilution value of solution.

PRACTICAL pH MEASUREMENTS ON NATURAL WATERS

A. K. Covington and W. Davison

(1) Dilute solutions and freshwater including 'acid-rain' samples ($I < 0.02$ mol kg^{-1})

Major problems could be encountered due to errors associated with the liquid junction. It is recommended that either a free diffusion junction is used or it is verified that the junction is working correctly using dilute solutions as follows. For commercial electrodes calibrated with IUPAC aqueous RVS or PS standards, the pH(X) of dilute solutions should be within ±0.02 of those given in Table 1. The difference in determined pH(X) between a stirred and unstirred dilute solution should be < 0.02. The characteristics of glass electrodes are such that below pH 5 the readings should be stable within 2 min, but for pH 5 to 8, 8 or so minutes may be necessary to attain stability. Interpretation of pH(X) measured in this way in terms of activity of hydrogen ion, a_{H+} is subject[1] to an uncertainty of ±0.02 in pH.

(2) Seawater

Measurements made by calibration of electrodes with IUPAC aqueous RVS or PS standards to obtain pH(X) are perfectly valid. However, the interpretation of pH(X) in terms of the activity of hydrogen ion is complicated by the non zero residual liquid junction potential as well as by systematic differences between electrode pairs, principally attributable to the reference electrode. For 35‰ salinity seawater ($S = 0.035$) a_{H+} calculated from pH(X) is typically 12% too low. Special seawater pH scales have been devised to overcome this problem:
(i) The total hydrogen ion scale, pH_T, is defined in terms of the sum of free and complexed (total) hydrogen ion concentrations, where

$$^{T}C_H = [H^+] + [HSO_4^-] + [HF].$$

$$\text{So, } pH_T = -\log {}^{T}C_H$$

Calibration of the electrodes with a buffer having a composition similar to that of seawater, to which pH_T has been assigned, results in values of pHT(X) (Tables 2, 3) which are accurately interpretable in terms of $^{T}C_H$.
(ii) The free hydrogen ion scale, pH_F, is defined, and fully interpretable, in terms of the concentration of free hydrogen ions.

$$pH_F = -\log [H^+]$$

Values of pH_F as a function of temperature have been assigned to the same set of pH_T seawater buffers,and so alternatively can be used for calibration (Tables 2, 3) [2,3]

(3) Estuarine water

Prescriptions for seawater scale buffers are available for a range of salinities. Reliable estuarine pH measurements can be made by calibrating with a buffer of the same salinity as the sample. However, these buffers are difficult to prepare and their use presumes prior knowledge of salinity of the sample. Interpretable measurements of estuarine pH can be made by calibration with IUPAC aqueous RVS or PS standards if the electrode pair is additionally calibrated using a 20‰ salinity seawater buffer.[4] The difference between the assigned pH_{SWS} of the seawater buffer and its measured pH(X) value using RVS or PS standards is

$$\Delta pH = pH_{SWS} - pH(X)$$

Values of ΔpH should be in the range of 0.08 to 0.18. It empirically corrects for differences between the two pH scales and for measurement errors associated with the electrode pair. The pH(X) of samples measured using IUPAC aqueous buffers, can be converted to pH_T or pH_F using the appropriate measured ΔpH:

$$pH_T = pH(X) - \Delta pH$$
$$\text{or } pH_F = pH(X) - \Delta pH$$

This simple procedure is appropriate to pH measurement at salinities from 2‰ to 35‰. For salinities lower than 2‰ the procedures for freshwaters should be adopted.

REFERENCES

1. Davison, W. and Harbinson,T. R., *Analyst*, 113, 709, 1988.
2. Culberson, C. H., in *Marine Electrochemistry*, Whitfield, M. and Jagner,D., Eds., Wiley, 1981.
3. Millero, F. J., *Limnol. Oceanogr.*, 31, 839, 1986.
4. Covington, A. K., Whalley, P. D., Davison, W., and Whitfield, M., in *The Determination of Trace Metals in Natural Waters*, West, T. S. and Nurnberg, H. W., Eds., Blackwell, Oxford, 1988.
5. Koch, W. F., Marinenko, G., and Paule, R. C., *J. Res. NBS*, 91, 33, 1986.

Table 1
pH of Dilute Solutions at 25°C, Degassed and Equilibrated with Air, Suitable as Quality Control Standards

	Ionic strength mmol kg^{-1}	Concentration(x) mmol kg^{-1}	pH p_{CO_2} = 0	pH p_{CO_2} = air
Potassium hydrogen phthalate	10.7	10	4.12	4.12
	1.1	1	4.33	4.33
$x KH_2PO_4 + x Na_2HPO_4$	9.9	2.5	7.07	7.05
$x KH_2PO_4 + 3.5x Na_2HPO_4$	10	0.87	7.61	7.58
$Na_2B_4O_7 \cdot 10H_2O$	10	5	9.20	—
HCl	0.1	0.1	4.03	4.03
SRM2694-I[a]	—	—	4.30	—
SRM2694-II[a]	—	—	3.59	—

Note: The pH of solutions near to pH 4 is virtually independent of temperature over the range of 5 to 30°C.

[a] Simulated rainwater samples are available (Reference 5) from NIST containing sulfate, nitrate, chloride, fluoride, sodium, potassium, calcium and magnesium

Table 2
Composition of Seawater Buffer of Salinity S = 35‰ at 25°C (Reference 3)

Solute	mol dm^{-3}	mol kg^{-1}	g kg^{-1}	g dm^{-3}
NaCl	0.3666	0.3493	20.416	20.946
Na_2SO_4	0.02926	0.02788	3.96	4.063
KCl	0.01058	0.01008	0.752	0.772
$CaCl_2$	0.01077	0.01026	1.139	1.169
$MgCl_2$	0.05518	0.05258	5.006	5.139
Tris	0.06	0.05717	6.926	7.106
Tris · HCl	0.06	0.05717	9.010	9.244

Tris = tris(hydroxymethyl)aminomethane $(HOCH_2)_3CNH_2$.
A 20‰ buffer is made by diluting the 35‰ in the ratio 20:35.

Table 3
Assigned Values of 20‰ and 35‰ Buffers on Free and Total Hydrogen Ion Scales. Calculated from Equations Provided by Millero (Reference 3)

Temp (°C)	pH_T S = 20‰	pH_T S = 35‰	pH_F S = 20‰	pH_F S = 35‰
5	8.683	8.718	8.759	8.81
10	8.513	8.542	8.597	8.647
15	8.351	8.374	8.442	8.491
20	8.195	8.212	8.292	8.341
25	8.045	8.057	8.149	8.197
30	7.901	7.908	8.011	8.059
35	7.762	7.764	7.879	7.926

BUFFER SOLUTIONS GIVING ROUND VALUES OF pH AT 25° C

A*		B*		C*		D*		E*	
pH	x	pH	x	pH	x	pH	x	pH	x
1.00	67.0	2.20	49.5	4.10	1.3	5.80	3.6	7.00	46.6
1.10	52.8	2.30	45.8	4.20	3.0	5.90	4.6	7.10	45.7
1.20	42.5	2.40	42.2	4.30	4.7	6.00	5.6	7.20	44.7
1.30	33.6	2.50	38.8	4.40	6.6	6.10	6.8	7.30	43.4
1.40	26.6	2.60	35.4	4.50	8.7	6.20	8.1	7.40	42.0
1.50	20.7	2.70	32.1	4.60	11.1	6.30	9.7	7.50	40.3
1.60	16.2	2.80	28.9	4.70	13.6	6.40	11.6	7.60	38.5
1.70	13.0	2.90	25.7	4.80	16.5	6.50	13.9	7.70	36.6
1.80	10.2	3.00	22.3	4.90	19.4	6.60	16.4	7.80	34.5
1.90	8.1	3.10	18.8	5.00	22.6	6.70	19.3	7.90	32.0
2.00	6.5	3.20	15.7	5.10	25.5	6.80	22.4	8.00	29.2
2.10	5.1	3.30	12.9	5.20	28.8	6.90	25.9	8.10	26.2
2.20	3.9	3.40	10.4	5.30	31.6	7.00	29.1	8.20	22.9
		3.50	8.2	5.40	34.1	7.10	32.1	8.30	19.9
		3.60	6.3	5.50	36.6	7.20	34.7	8.40	17.2
		3.70	4.5	5.60	38.8	7.30	37.0	8.50	14.7
		3.80	2.9	5.70	40.6	7.40	39.1	8.60	12.2
		3.90	1.4	5.80	42.3	7.50	40.9	8.70	10.3
		4.00	0.1	5.90	43.7	7.60	42.4	8.80	8.5
						7.70	43.5	8.90	7.0
						7.80	44.5	9.00	5.7
						7.90	45.3		
						8.00	46.1		

F*		G*		H*		I*		J*	
pH	x	pH	x	pH	x	pH	x	pH	x
8.00	20.5	9.20	0.9	9.60	5.0	10.90	3.3	12.00	6.0
8.10	19.7	9.30	3.6	9.70	6.2	11.00	4.1	12.10	8.0
8.20	18.8	9.40	6.2	9.80	7.6	11.10	5.1	12.20	10.2
8.30	17.7	9.50	8.8	9.90	9.1	11.20	6.3	12.30	12.8
8.40	16.6	9.60	11.1	10.00	10.7	11.30	7.6	12.40	16.2
8.50	15.2	9.70	13.1	10.10	12.2	11.40	9.1	12.50	20.4
8.60	13.5	9.80	15.0	10.20	13.8	11.50	11.1	12.60	25.6
8.70	11.6	9.90	16.7	10.30	15.2	11.60	13.5	12.70	32.2
8.80	9.6	10.00	18.3	10.40	16.5	11.70	16.2	12.80	41.2
8.90	7.1	10.10	19.5	10.50	17.8	11.80	19.4	12.90	53.0
9.00	4.6	10.20	20.5	10.60	19.1	11.90	23.0	13.00	66.0
9.10	2.0	10.30	21.3	10.70	20.2	12.00	26.9		
		10.40	22.1	10.80	21.2				
		10.50	22.7	10.90	22.0				
		10.60	23.3	11.00	22.7				
		10.70	23.8						
		10.80	24.25						

*A. 25 ml of 0.2 molar KCl + x ml of 0.2 molar HCl.
*B. 50 ml of 0.1 molar potassium hydrogen phthalate + x ml of 0.1 molar HCl.
*C. 50 ml of 0.1 molar potassium hydrogen phthalate + x ml of 0.1 molar NaOH.
*D. 50 ml of 0.1 molar potassium dihydrogen phosphate + x ml 0.1 molar NaOH.
*E. 50 ml of 0.1 molar tris(hydroxymethyl) aminomethane + x ml of 0.1 M HCl.
*F. 50 ml of 0.025 molar borax + x ml of 0.1 molar HCl.
*G. 50 ml of 0.025 molar borax + x ml of 0.1 molar NaOH.
*H. 50 ml of 0.05 molar sodium bicarbonate + x ml of 0.1 molar NaOH.
*I. 50 ml of 0.05 molar disodium hydrogen phosphate + x ml of 0.1 molar NaOH.
*J. 25 ml of 0.2 molar KCl + x ml of 0.2 molar NaOH.
Final Volume of Mixtures = 100 ml

REFERENCES

1. Bower, V. E., and Bates, R. G., *J. Res. Natl. Bur. Stand.*, 55, 197, 1955 (A—D).
2. Bates, R. G., and Bower, V. E., *Anal. Chem.*, 28, 1322, 1956 (E—J).

DISSOCIATION CONSTANTS OF INORGANIC ACIDS AND BASES

The data in this table are presented as values of pK_a, defined as the negative logarithm of the acid dissociation constant K_a for the reaction

$$BH \rightleftharpoons B^- + H^+$$

Thus $pK_a = -\log K_a$, and the hydrogen ion concentration $[H^+]$ can be calculated from

$$K_a = \frac{[H^+][B^-]}{[BH]}$$

In the case of bases, the entry in the table is for the conjugate acid; e.g., ammonium ion for ammonia. The OH^- concentration in the system

$$NH_3 + H_2O \rightleftharpoons NH_4^+ + OH^-$$

can be calculated from the equation

$$K_b = K_{water} / K_a = \frac{[OH^-][NH_4^+]}{[NH_3]}$$

where $K_{water} = 1.01 \times 10^{-14}$ at 25 °C. Note that $pK_a + pK_b = pK_{water}$.

All values refer to dilute aqueous solutions at the temperature indicated. The table is arranged alphabetically by compound name.

REFERENCE

1. *Perrin, D. B., Ionization Constants of Inorganic Acids and Bases in Aqueous Solution, Second Edition, Pergamon, Oxford, 1982.*

Name	Formula	Step	T/°C	pK_a
Aluminum(III) ion	Al^{+3}		25	5.0
Ammonia	NH_3		25	9.25
Arsenic acid	H_3AsO_4	1	25	2.26
		2	25	6.76
		3	25	11.29
Arsenious acid	H_2AsO_3		25	9.29
Barium(II) ion	Ba^{+2}		25	13.4
Boric acid	H_3BO_3	1	20	9.27
		2	20	>14
Calcium(II) ion	Ca^{+2}		25	12.6
Carbonic acid	H_2CO_3	1	25	6.35
		2	25	10.33
Chlorous acid	$HClO_2$		25	1.94
Chromic acid	H_2CrO_4	1	25	0.74
		2	25	6.49
Cyanic acid	HCNO		25	3.46
Germanic acid	H_2GeO_3	1	25	9.01
		2	25	12.3
Hydrazine	N_2H_4		25	8.1
Hydrazoic acid	HN_3		25	4.6
Hydrocyanic acid	HCN		25	9.21
Hydrofluoric acid	HF		25	3.20
Hydrogen peroxide	H_2O_2		25	11.62
Hydrogen selenide	H_2Se	1	25	3.89
		2	25	11.0
Hydrogen sulfide	H_2S	1	25	7.05
		2	25	19
Hydrogen telluride	H_2Te	1	18	2.6
		2	25	11
Hydroxylamine	NH_2OH		25	5.94
Hypobromous acid	HBrO		25	8.55

Name	Formula	Step	T/°C	pK_a
Hypochlorous acid	$HClO$		25	7.40
Hypoiodous acid	HIO		25	10.5
Iodic acid	HIO_3		25	0.78
Lithium ion	Li^+		25	13.8
Magnesium(II) ion	Mg^{+2}		25	11.4
Nitrous acid	HNO_2		25	3.25
Perchloric acid	$HClO_4$		20	-1.6
Periodic acid	HIO_4		25	1.64
Phosphoric acid	H_3PO_4	1	25	2.16
		2	25	7.21
		3	25	12.32
Phosphorous acid	H_3PO_3	1	20	1.3
		2	20	6.70
Pyrophosphoric acid	$H_4P_2O_7$	1	25	0.91
		2	25	2.10
		3	25	6.70
		4	25	9.32
Selenic acid	H_2SeO_4	2	25	1.7
Selenious acid	H_2SeO_3	1	25	2.62
		2	25	8.32
Silicic acid	H_4SiO_4	1	30	9.9
		2	30	11.8
		3	30	12
		4	30	12
Sodium ion	Na^+		25	14.8
Strontium(II) ion	Sr^{+2}		25	13.2
Sulfamic acid	NH_2SO_3H		25	1.05
Sulfuric acid	H_2SO_4	2	25	1.98
Sulfurous acid	H_2SO_3	1	25	1.85
		2	25	7.2
Telluric acid	H_2TeO_4	1	18	7.68
		2	18	11.0
Tellurous acid	H_2TeO_3	1	25	6.27
		2	25	8.43
Tetrafluoroboric acid	HBF_4		25	0.5
Thiocyanic acid	$HSCN$		25	-1.8
Water	H_2O		25	13.995

DISSOCIATION CONSTANTS OF ORGANIC ACIDS AND BASES

This table lists the acid-base dissociation constants of over 600 organic compounds, including many amino acids. All data apply to dilute aqueous solutions and are presented in the form of pK_a, which is the negative of the logarithm of the acid dissociation constant K_a. See the preceding table, "Dissociation Constants of Inorganic Acids and Bases", for further details on notation.

Compounds are listed by molecular formula in the Hill order.

REFERENCES

1. Perrin, D. D., *Dissociation Constants of Organic Bases in Aqueous Solution*, Butterworths, London, 1965; Supplement, 1972.
2. Serjeant, E. P., and Dempsey, B., *Ionization Constants of Organic Acids in Aqueous Solution*, Pergamon, Oxford, 1979.
3. Albert, A., "Ionization Constants of Heterocyclic Substances", in *Physical Methods in Heterocyclic Chemistry*, Katritzky, A. R., Ed., Academic Press, New York, 1963.
4. Sober, H. A., Ed., *CRC Handbook of Biochemistry*, CRC Press, Cleveland, Ohio, 1968.
5. Perrin, D. D., Dempsey, B., and Serjeant, E. P., *pK_a Prediction for Organic Acids and Bases*, Chapman & Hall, London, 1981.
6. Dawson, R. M. C., Elliot, D. C., Elliot, W. H., and Jones, K. M., *Data for Biochemical Research*, Oxford Science Publications, Oxford, 1986.

Molecular formula	Name	Step	T/°C	pK_a
CH_2O_2	Formic acid		20	3.75
CH_4N_2O	Urea		21	0.10
CH_5N	Methylamine		25	10.63
$C_2HCl_3O_2$	Trichloroacetic acid		25	0.70
$C_2H_2Cl_2O_2$	Dichloroacetic acid		25	1.48
$C_2H_2O_3$	Glyoxylic acid		25	3.18
$C_2H_2O_4$	Oxalic acid	1	25	1.23
		2	25	4.19
$C_2H_3BrO_2$	Bromoacetic acid		25	2.69
$C_2H_3ClO_2$	Chloroacetic acid		25	2.85
$C_2H_3IO_2$	Iodoacetic acid		25	3.12
C_2H_4OS	Thioacetic acid		25	3.33
$C_2H_4O_2$	Acetic acid		25	4.76
$C_2H_4O_3$	Glycolic acid		25	3.83
C_2H_5N	Ethyleneimine		25	8.01
C_2H_5NO	Acetamide		25	0.63
$C_2H_5NO_2$	Glycine	1	25	2.35
		2	25	9.78
$C_2H_6O_2$	Ethylene glycol		25	14.22
$C_2H_7AsO_2$	Cacodylic acid	1	25	1.57
		2	25	6.27
C_2H_7N	Dimethylamine		25	10.68
C_2H_7N	Ethylamine		25	10.70
C_2H_7NO	Ethanolamine		25	9.50
$C_2H_7NO_3S$	Taurine	1	25	1.5
		2	25	9.061
C_2H_7NS	Cysteamine	1	25	8.35
		2	25	10.81
$C_2H_8N_2$	1,2-Ethanediamine	1	0	10.712
		2	0	7.564
$C_3H_3NO_2$	Cyanoacetic acid		25	2.45
C_3H_3NS	Thiazole		20	2.44
$C_3H_4N_2$	1H-Imidazole		25	6.953
$C_3H_4N_2S$	2-Thiazolamine		20	5.36
$C_3H_4O_2$	Acrylic acid		25	4.25
$C_3H_4O_3$	Pyruvic acid		25	2.39
$C_3H_4O_4$	Malonic acid	1	25	2.83
		2	25	5.69
$C_3H_5ClO_2$	2-Chloropropanoic acid		25	2.83
$C_3H_5ClO_2$	3-Chloropropanoic acid		25	3.98
$C_3H_6N_6$	Melamine		25	5.00
$C_3H_6O_2$	Propanoic acid		25	4.86
$C_3H_6O_3$	3-Hydroxypropanoic acid		25	4.51
$C_3H_6O_3$	Lactic acid		100	3.08
$C_3H_6O_4$	Glyceric acid		25	3.52

Molecular formula	Name	Step	T/°C	pK_a
C_3H_7N	Azetidine		25	11.29
$C_3H_7NO_2$	*L*-Alanine	1	25	2.34
		2	25	9.87
$C_3H_7NO_2$	β-Alanine	1	25	3.55
		2	25	10.24
$C_3H_7NO_2$	Methylglycine	1	25	2.21
		2	25	10.12
$C_3H_7NO_2S$	Cysteine	1	25	1.92
		2	25	8.37
		3	25	10.70
$C_3H_7NO_3$	Serine	1	25	2.19
		2	25	9.21
$C_3H_7NO_5S$	*l*-Cysteic acid	1	25	1.3
		2	25	1.9
		3	25	8.70
$C_3H_7N_3O_2$	Glycocyamine		25	2.82
$C_3H_8O_3$	Glycerol		25	14.15
C_3H_9N	Propylamine		20	10.60
C_3H_9N	Trimethylamine		25	9.80
C_3H_9NO	1-Amino-2-methoxyethane		10	9.89
C_3H_9NO	Trimethylamine oxide		25	4.65
$C_3H_{10}N_2$	1,2-Propanediamine	1	25	9.82
		2	25	6.61
$C_3H_{10}N_2$	1,3-Propanediamine	1	10	10.94
		2	10	9.03
$C_3H_{11}N_3$	1,2,3-Triaminopropane	1	20	9.59
		2	20	7.95
$C_4H_4N_2$	Pyrazine		27	0.65
$C_4H_4N_2$	Pyridazine		20	2.24
$C_4H_4N_2O_3$	Barbituric acid		25	4.01
$C_4H_4N_2O_5$	Alloxanic acid		25	6.64
$C_4H_4N_4O_2$	5-Nitropyrimidinamine		20	0.35
$C_4H_4O_4$	*trans*-Fumaric acid	1	18	3.03
		2	18	4.44
$C_4H_4O_4$	Maleic acid	1	25	1.83
		2	25	6.07
$C_4H_4O_5$	Oxaloacetic acid	1	25	2.22
		2	25	3.89
		3	25	13.03
$C_4H_5N_3$	2-Pyrimidinamine		20	3.45
$C_4H_6N_2$	1-Methylimidazol		25	6.95
$C_4H_6N_4O_3$	Allantoin		25	8.96
$C_4H_6O_2$	3-Butenoic acid		25	4.34
$C_4H_6O_2$	*trans*-Crotonic acid		25	4.69
$C_4H_6O_3$	Acetoacetic acid		18	3.58
$C_4H_6O_3$	2-Oxobutanoic acid		25	2.50
$C_4H_6O_4$	Methylmalonic acid	1	25	3.07
		2	25	5.76
$C_4H_6O_4$	Succinic acid	1	25	4.16
		2	25	5.61
$C_4H_6O_5$	Malic acid	1	25	3.40
		2	25	5.11
$C_4H_6O_6$	α-Tartaric acid	1	25	2.98
		2	25	4.34
$C_4H_6O_6$	*meso*-Tartaric acid	1	25	3.22
		2	25	4.82
$C_4H_6O_8$	Dihydroxytartaric acid		25	1.92
$C_4H_7ClO_2$	2-Chlorobutanoic acid		RT	2.86
$C_4H_7ClO_2$	3-Chlorobutanoic acid		RT	4.05
$C_4H_7ClO_2$	4-Chlorobutanoic acid		RT	4.52

Molecular formula	Name	Step	T/°C	pK_a
$C_4H_7NO_2$	4-Cyanobutanoic acid		25	2.42
$C_4H_7NO_3$	*N*-Acetylglycine		25	3.669
$C_4H_7NO_4$	Aspartic acid	1	25	1.99
		2	25	3.90
		3	25	9.90
$C_4H_7N_3O$	Creatinine	1	25	4.83
		2		9.2
$C_4H_8N_2O_3$	Asparagine	1	20	2.17
		2	20	8.80
$C_4H_8N_2O_3$	*N*-Glycylglycine		25	3.139
$C_4H_8O_2$	Butanoic acid		25	4.83
$C_4H_8O_2$	2-Methylpropanoic acid		25	4.88
$C_4H_8O_3$	3-Hydroxybutanoic acid		25	4.70
$C_4H_8O_3$	4-Hydroxybutanoic acid		25	4.72
C_4H_9N	Pyrrolidine		25	11.27
C_4H_9NO	Morpholine		25	8.33
$C_4H_9NO_2$	2-Aminobutanoic acid	1	25	2.29
		2	25	9.83
$C_4H_9NO_2$	4-Aminobutanoic acid	1	25	4.031
		2	25	10.556
$C_4H_9NO_2$	*N,N*-Dimethylglycine		25	9.89
$C_4H_9NO_2$	2-Methylalanine	1	25	2.36
		2	25	10.21
$C_4H_9NO_2S$	Homocysteine	1	25	2.22
		2	25	8.87
		3	25	10.86
$C_4H_9NO_3$	Homoserine	1	25	2.71
		2	25	9.62
$C_4H_9NO_3$	*DL*-Methoxyalanine		25	2.037
$C_4H_9NO_3$	Threonine	1	25	2.09
		2	25	9.10
$C_4H_9N_3O_2$	Creatine	1	25	2.63
		2	25	14.3
$C_4H_{10}N_2$	Piperazine	1	23	9.83
		2	23	5.56
$C_4H_{10}N_2O$	2,4-Diaminobutanoic acid	1	25	1.85
		2	25	8.24
		3	25	10.44
$C_4H_{11}N$	Butylamine		20	10.77
$C_4H_{11}N$	*sec*-Butylamine		25	10.56
$C_4H_{11}N$	*tert*-Butylamine		25	10.68
$C_4H_{11}N$	Diethylamine		40	11.02
$C_4H_{12}N_2$	1,4-Butanediamine	1	20	10.80
		2	20	9.35
$C_4H_{12}O_2$	1,2-Dimethylaminoethane	1	25	10.40
		2	25	8.26
C_5H_4BrN	3-Bromopyridine		25	2.84
C_5H_4ClN	3-Chloropyridine		25	2.84
$C_5H_4N_4$	Purine	1	20	2.30
		2	20	8.96
$C_5H_4N_4O_3$	Uric acid		12	3.89
C_5H_5N	Pyridine		25	5.25
C_5H_5NO	4-Pyridinol	1	20	3.20
		2	20	11.12
C_5H_5NO	2(1H)-Pyridinone	1	20	0.75
		2	20	11.65
$C_5H_5N_5$	1H-Purin-6-amine	1	25	4.12
		2	25	9.83
$C_5H_6N_2$	2-Methylpyrazine		27	1.45
$C_5H_6N_2$	2-Pyridinamine		20	6.82

Molecular formula	Name	Step	T/°C	pK_a
$C_5H_6N_2$	4-Pyridinamine		25	9.114
$C_5H_6O_4$	1,1-Cyclopropanedicarboxylic acid	1	25	1.82
		2	25	7.43
$C_5H_6O_4$	Itaconic acid	1	25	3.85
		2	25	5.45
$C_5H_6O_4$	Mesaconic acid	1	25	3.09
		2	25	4.75
$C_5H_6O_5$	2-Oxoglutaric acid	1	25	2.47
		2	25	4.68
$C_5H_7NO_3$	*L*-2-Pyrrolidone-5-carboxylic aci		25	3.32
$C_5H_7N_3$	Methylaminopyrazine		25	3.39
$C_5H_7N_3$	2,5-Pyridinediamine		20	6.48
$C_5H_8N_2$	2,4-Dimethylimidazol		25	8.36
$C_5H_8O_4$	Dimethylmalonic acid		25	3.15
$C_5H_8O_4$	Glutaric acid	1	25	4.31
		2	25	5.41
$C_5H_8O_4$	Methylsuccinic acid	1	25	4.13
		2	25	5.64
$C_5H_9NO_2$	Proline	1	25	1.952
		2	25	10.64
$C_5H_9NO_3$	5-Aminolevulinic acid	1	25	4.05
		2	25	8.90
$C_5H_9NO_3$	*trans*-4-Hydroxyproline	1	25	1.818
		2	25	9.662
$C_5H_9NO_3$	8-Hydroxypurine	1	20	2.56
		2	20	8.26
$C_5H_9NO_4$	*L*-Glutamic acid	1	25	2.13
		2	25	4.31
$C_5H_9N_3$	Histamine	1	25	6.04
		2	25	9.75
$C_5H_{10}N_2O_2$	Diaminopimelic acid	1	25	1.8
		2	25	2.2
		3	25	8.8
		4	25	9.9
$C_5H_{10}N_2O_3$	*L*-Glutamine	1	25	2.17
		2	25	9.13
$C_5H_{10}N_2O_3$	Glycylalanine		25	3.15
$C_5H_{10}N_2O_4$	Glycylserine	1	25	2.98
		2	25	8.38
$C_5H_{10}O_2$	2-Methylbutanoic acid		25	4.80
$C_5H_{10}O_2$	3-Methylbutanoic acid		25	4.77
$C_5H_{10}O_2$	Pentanoic acid		25	4.84
$C_5H_{10}O_2$	Trimethylacetic acid		25	5.03
$C_5H_{11}N$	*N*-Methylpyrrolidine		25	10.32
$C_5H_{11}N$	Piperidine		25	11.123
$C_5H_{11}NO_2$	5-Aminopentanoic acid	1	25	4.27
		2	25	10.766
$C_5H_{11}NO_2$	Betaine		0	1.83
$C_5H_{11}NO_2$	Norvaline	1	25	2.32
		2	25	9.81
$C_5H_{11}NO_2$	*N*-Propylglycine	1	25	2.35
		2	25	10.19
$C_5H_{11}NO_2$	Valine	1	25	2.29
		2	25	9.74
$C_5H_{11}NO_2S$	Methionine	1	25	2.13
		2	25	9.27
$C_5H_{12}N_2O_2$	Ornithine	1	25	1.705
		2	25	8.69
		3	25	10.76

Molecular formula	Name	Step	T/°C	pK_a
$C_5H_{13}N$	1-Amino-2,2-dimethylpropane		25	10.15
$C_5H_{13}N$	Diethylmethylamine		25	10.35
$C_5H_{13}N$	3-Methyl-1-butanamine		25	10.60
$C_5H_{13}N$	2-Methyl-2-butanamine		19	10.85
$C_5H_{13}N$	3-Pentanamine		17	10.59
$C_5H_{13}N$	Pentylamine		25	10.63
$C_5H_{14}NO$	Choline		25	13.9
$C_5H_{14}N_2$	Cadaverine	1	25	10.05
		2	25	10.93
$C_6H_3ClN_4$	6-Chloropteridine		20	3.68
$C_6H_3N_3O_7$	Picric acid		25	0.38
$C_6H_4Cl_2O$	2,3-Dichlorophenol		25	7.44
$C_6H_4N_2O_3$	5-Hydroxylysine		25	2.13
$C_6H_4N_2O_5$	2,4-Dinitrophenol		15	3.96
$C_6H_4N_2O_5$	3,6-Dinitrophenol		15	5.15
$C_6H_4N_4$	Pteridine		20	4.05
$C_6H_5Br_2N$	3,5-Dibromoaniline		25	2.34
C_6H_5ClO	2-Chlorophenol		25	8.49
C_6H_5ClO	3-Chlorophenol		25	8.85
C_6H_5ClO	4-Chlorophenol		25	9.18
$C_6H_5Cl_2N$	2,4-Dichloroaniline		22	2.05
C_6H_5NO	2-Pyridinecarboxaldehyde		20	3.80
$C_6H_5NO_2$	Nitrobenzene		0	3.98
$C_6H_5NO_2$	Picolinic acid	1	25	1.07
		2	25	5.25
$C_6H_5NO_2$	3-Pyridinecarboxylic acid		25	4.85
$C_6H_5NO_2$	4-Pyridinecarboxylic acid		25	4.96
$C_6H_5NO_3$	2-Nitrophenol		25	7.17
$C_6H_5NO_3$	3-Nitrophenol		25	8.28
$C_6H_5NO_3$	4-Nitrophenol		25	7.15
$C_6H_5N_5O$	2-Amino-4-hydroxypteridine	1	20	2.27
		2	20	7.96
$C_6H_5N_5O_2$	Xanthopterin	2	20	6.59
		3	20	9.31
C_6H_6BrN	2-Bromoaniline		25	2.53
C_6H_6BrN	3-Bromoaniline		25	3.58
C_6H_6BrN	4-Bromoaniline		25	3.86
C_6H_6ClN	2-Chloroaniline		25	2.65
C_6H_6ClN	3-Chloroaniline		25	3.46
C_6H_6ClN	4-Chloroaniline		25	4.15
C_6H_6FN	2-Fluoroaniline		25	3.20
C_6H_6FN	3-Fluoroaniline		25	3.50
C_6H_6FN	4-Fluoroaniline		25	4.65
C_6H_6IN	2-Iodoaniline		25	2.60
$C_6H_6N_2O$	2-Pyridinecarboxaldehyde, oxime	1	20	3.59
		2	20	10.18
$C_6H_6N_2O_2$	2-Nitroaniline		25	-0.26
$C_6H_6N_2O_2$	3-Nitroaniline		25	2.466
$C_6H_6N_2O_2$	4-Nitroaniline		25	1.0
C_6H_6O	Phenol		20	9.89
$C_6H_6O_2$	Hydroquinone		20	10.35
$C_6H_6O_2$	Pyrocatechol		20	9.85
$C_6H_6O_2$	Resorcinol		25	9.81
$C_6H_6O_3S$	Benzenesulfonic acid		25	0.70
$C_6H_6O_6$	*cis*-Aconitic acid		25	1.95
$C_6H_6O_6$	*trans*-Aconitic acid	1	25	2.80
		2	25	4.46
C_6H_7N	Aniline		25	4.63
C_6H_7N	2-Methylpyridine		20	5.97

Molecular formula	Name	Step	T/°C	pK_a
C_6H_7N	3-Methylpyridine		20	5.68
C_6H_7N	4-Methylpyridine		20	6.02
C_6H_7NO	Methoxypyridine		25	6.47
$C_6H_7NO_3S$	*o*-Aminobenzenesulfonic acid		25	2.48
$C_6H_7NO_3S$	*m*-Aminobenzenesulfonic acid		25	3.73
$C_6H_7NO_3S$	*p*-Aminobenzenesulfonic acid		25	3.24
$C_6H_8N_2$	*N*-Methylpyridinamine		20	9.65
$C_6H_8O_6$	Ascorbic acid	1	24	4.10
		2	16	11.79
$C_6H_8O_7$	Citric acid	1	20	3.14
		2	20	4.77
		3	20	6.39
$C_6H_8O_7$	Isocitric acid	1	25	3.29
		2	25	4.71
		3	25	6.40
$C_6H_9NO_6$	γ-Carboxyglutamic acid	1	25	1.7
		2	25	3.2
		3	25	4.75
		4	25	9.9
$C_6H_9N_3$	4,6-Dimethylpyrimidinamine		20	4.82
$C_6H_9N_3O_2$	Histidine	1	25	1.80
		2	25	6.04
		3	25	9.33
$C_6H_{10}O_3$	2-Oxo-3-methylpentanoic acid		25	2.3
$C_6H_{10}O_4$	Adipic acid	1	25	4.43
		2	25	5.41
$C_6H_{10}O_4$	3-Methylglutaric acid		25	4.24
$C_6H_{11}NO_2$	*l*-Pipecolic acid	1	25	2.28
		2	25	10.72
$C_6H_{11}NO_3$	Adipamic acid		25	4.63
$C_6H_{11}NO_4$	2-Aminoadipic acid	1	25	2.14
		2	25	4.21
		3	25	9.77
$C_6H_{11}N_3O_4$	Glycylasparagine	1	25	2.942
		2	18	8.44
$C_6H_{11}N_3O_4$	*N*-(*N*-Glycylglycyl)glycine	1	25	3.225
		2	25	8.09
$C_6H_{12}N_2O_4S_2$	Cystine	1	35	2.1
		2	35	8.0
$C_6H_{12}O_2$	Hexanoic acid		25	4.85
$C_6H_{12}O_2$	4-Methylpentanoic acid		18	4.84
$C_6H_{13}N$	Cyclohexylamine		24	10.66
$C_6H_{13}N$	1,2-Dimethylpyrrolidine		26	10.20
$C_6H_{13}N$	1-Methylpiperidine		25	10.08
$C_6H_{13}NO_2$	6-Aminohexanoic acid	1	25	4.373
		2	25	10.804
$C_6H_{13}NO_2$	Isoleucine	1	25	2.32
		2	25	9.76
$C_6H_{13}NO_2$	*L*-Leucine	1	25	2.328
		2	25	9.744
$C_6H_{13}NO_2$	Norleucine	1	25	2.335
		2	25	9.83
$C_6H_{13}N_3O_3$	Citrulline	1	25	2.43
		2	25	9.69
$C_6H_{14}N_2$	*cis*-1,2-Cyclohexanediamine	1	20	9.93
		2	20	6.13
$C_6H_{14}N_2$	*trans*-1,2-Cyclohexanediamine	1	20	9.94
		2	20	6.47
$C_6H_{14}N_2$	2,5-Dimethylpiperazine	1	25	9.66
		2	25	5.20

Molecular formula	Name	Step	T/°C	pK_a
$C_6H_{14}N_2O_2$	Lysine	1	25	2.16
		2	25	9.06
		3	25	10.54
$C_6H_{14}N_4O_2$	Arginine	1	25	1.82
		2	25	8.99
		3	25	12.48
$C_{10}H_{22}O_2$	3-Amino-3-methylpentane		16	11.01
$C_{10}H_{22}O_2$	Diisopropylamine		25	11.05
$C_{10}H_{22}O_2$	Hexylamine		25	10.56
$C_{10}H_{22}O_2$	Triethylamine		25	10.75
$C_6H_{16}N_2$	Hexamethylenediamine	1	0	11.857
		2	0	10.762
$C_7H_5BrO_2$	2-Bromobenzoic acid		25	2.84
$C_7H_5BrO_2$	3-Bromobenzoic acid		25	3.86
$C_7H_5ClO_2$	2-Chlorobenzoic acid		25	2.92
$C_7H_5ClO_2$	3-Chlorobenzoic acid		25	3.82
$C_7H_5ClO_2$	4-Chlorobenzoic acid		25	3.98
$C_7H_5IO_2$	2-Iodobenzoic acid		25	2.85
$C_7H_5IO_2$	3-Iodobenzoic acid		25	3.80
$C_7H_5NO_3S$	Saccharin		18	11.68
$C_7H_5NO_4$	Dinicotinic acid		25	2.80
$C_7H_5NO_4$	Dipicolinic acid	1	25	2.16
		2	25	4.76
$C_7H_5NO_4$	Lutidinic acid		25	2.15
$C_7H_5NO_4$	2-Nitrobenzoic acid		18	2.16
$C_7H_5NO_4$	3-Nitrobenzoic acid		25	3.47
$C_7H_5NO_4$	4-Nitrobenzoic acid		25	3.41
$C_7H_5NO_4$	Quinolinic acid	1	25	2.43
		2	25	4.78
$C_7H_6N_2$	Benzimidazole		25	5.532
$C_7H_6N_4O$	6-Hydroxy-4-methylpteridine	1	20	4.08
		2	20	6.41
$C_7H_6O_2$	Benzoic acid		25	4.19
$C_7H_6O_3$	*o*-Hydroxybenzoic acid	1	19	2.97
		2	18	13.40
$C_7H_6O_3$	*m*-Hydroxybenzoic acid	1	19	4.06
		2	19	9.92
$C_7H_6O_3$	*p*-Hydroxybenzoic acid	1	19	4.48
		2	19	9.32
$C_7H_6O_4$	2,5-Dihydroxybenzoic acid	1	25	2.97
$C_7H_6O_4$	3,4-Dihydroxybenzoic acid	1	25	4.48
		2	25	8.83
		3	25	12.6
$C_7H_6O_4$	3,5-Dihydroxybenzoic acid	1	25	4.04
$C_7H_6O_4$	Dihydroxymalic acid		25	1.92
$C_7H_6O_5$	Gallic acid		25	4.41
$C_7H_6O_5$	2,4,6-Trihydroxybenzoic acid		25	1.68
$C_7H_7NO_2$	2-Aminobenzoic acid	1	25	2.108
		2	25	4.946
$C_7H_7NO_2$	3-Aminobenzoic acid	2	25	4.78
$C_7H_7NO_2$	4-Animobenzoic acid	1	25	2.501
		2	25	4.874
$C_7H_8N_4O_2$	Theobromine		18	7.89
C_7H_8O	*o*-Cresol		25	10.20
C_7H_8O	*m*-Cresol		25	10.01
C_7H_8O	*p*-Cresol		25	10.17
C_7H_9N	Benzylamine		25	9.33
C_7H_9N	2,3-Dimethylpyridine		25	6.57
C_7H_9N	2,4-Dimethylpyridine		25	6.99

Molecular formula	Name	Step	$T/°C$	pK_a
C_7H_9N	2,5-Dimethylpyridine		25	6.40
C_7H_9N	2,6-Dimethylpyridine		25	6.65
C_7H_9N	3,4-Dimethylpyridine		25	6.46
C_7H_9N	3,5-Dimethylpyridine		25	6.15
C_7H_9N	2-Ethylpyridine		25	5.89
C_7H_9N	*N*-Methylaniline		25	4.84
C_7H_9N	*o*-Methylaniline		25	4.44
C_7H_9N	*m*-Methylaniline		25	4.73
C_7H_9N	*p*-Methylaniline		25	5.08
C_7H_9NO	*o*-Anisidine		25	4.52
C_7H_9NO	*m*-Anisidine		25	4.23
C_7H_9NO	*p*-Anisidine		25	5.34
C_7H_9NS	4-Methylthioaniline		25	4.35
$C_7H_9N_5$	2-Dimethylaminopurine	1	20	4.00
		2	20	10.24
$C_7H_{11}N_3O_2$	*l*-1-Methylhistidine	1	25	1.69
		2	25	6.48
		3	25	8.85
$C_7H_{11}N_3O_2$	*l*-3-Methylhistidine	1	25	1.92
		2	25	6.56
		3	25	8.73
$C_7H_{12}O_2$	Cyclohexanecarboxylic acid		25	4.90
$C_7H_{12}O_4$	Heptanedioic acid	1	25	4.71
		2	25	5.58
$C_7H_{13}NO$	3-Acetylpiperidine		25	3.18
$C_7H_{13}NO_4$	α-Ethylglutamic acid	1	25	3.846
		2	25	7.838
$C_7H_{14}O_2$	Heptanoic acid		25	4.89
$C_7H_{15}N$	1,2-Dimethylpiperidine		25	10.22
$C_7H_{15}N$	1-Ethylpiperidine		23	10.45
$C_7H_{15}NO_3$	Carnitine		25	3.80
$C_7H_{17}N$	2-Heptanamine		19	10.7
$C_7H_{17}N$	Heptylamine		25	10.67
$C_8H_6N_2$	Cinnoline		20	2.37
$C_8H_6N_2$	Quinazoline		20	3.43
$C_8H_6N_2$	Quinoxaline		20	0.56
$C_8H_6N_2O$	5-Hydoxyquinazoline	1	20	3.62
		2	20	7.41
$C_8H_6O_4$	*o*-Phthalic acid	1	25	2.89
		2	25	5.51
$C_8H_6O_4$	*m*-Phthalic acid	1	25	3.54
		2	18	4.60
$C_8H_6O_4$	*p*-Phthalic acid	1	25	3.51
		2	16	4.82
$C_8H_6O_4$	Terephthalic acid		25	3.51
$C_8H_7ClO_2$	2-Chlorophenylacetic acid		25	4.07
$C_8H_7ClO_2$	3-Chlorophenylacetic acid		25	4.14
$C_8H_7ClO_2$	4-Chlorophenylacetic acid		25	4.19
$C_8H_7ClO_3$	2-Chlorophenoxyacetic acid		25	3.05
$C_8H_7ClO_3$	3-Chlorophenoxyacetic acid		25	3.10
$C_8H_7NO_4$	2-Nitrophenylacetic acid		25	4.00
$C_8H_7NO_4$	3-Nitrophenylacetic acid		25	3.97
$C_8H_7NO_4$	4-Nitrophenylacetic acid		25	3.85
$C_8H_8N_2$	2-Methylbenzimidazole		25	6.19
$C_8H_8O_2$	Phenylacetic acid		18	4.28
$C_8H_8O_2$	*o*-Toluic acid		25	3.91
$C_8H_8O_2$	*m*-Toluic acid		25	4.27
$C_8H_8O_2$	*p*-Toluic acid		25	4.36
$C_8H_8O_3$	*DL*-Mandelic acid		25	3.85

Molecular formula	Name	Step	T/°C	pKa
$C_8H_8O_4$	Homogentisic acid		25	4.40
$C_8H_9NO_2$	2-(Methylamino)benzoic acid		25	5.34
$C_8H_9NO_2$	3-(Methylamino)benzoic acid		25	5.10
$C_8H_9NO_2$	4-(Methylamino)benzoic acid		25	5.04
$C_8H_9NO_2$	Phenylglycine	1	25	1.83
		2	25	4.39
$C_8H_{10}BrN$	4-Bromo-*N,N*-dimethylaniline		25	4.23
$C_8H_{10}ClN$	3-Chloro-*N,N*-dimethylaniline		20	3.83
$C_8H_{10}ClN$	4-Chloro-*N,N*-dimethylaniline		20	4.39
$C_8H_{10}N_2O_2$	*N,N*-Dimethyl-3-nitroaniline		25	2.62
$C_8H_{11}N$	*N,N*-Dimethylaniline		25	5.15
$C_8H_{11}N$	*N*-Ethylaniline		24	5.12
$C_8H_{11}N$	Phenylethylamine		25	9.81
$C_8H_{11}N$	2,4,6-Trimethylpyridine		25	7.43
$C_8H_{11}NO$	*o*-Phenetidine		28	4.43
$C_8H_{11}NO$	*m*-Phenetidine		25	4.18
$C_8H_{11}NO$	*p*-Phenetidine		28	5.20
$C_8H_{11}NO$	Tyramine	1	25	9.74
		2	25	10.52
$C_8H_{11}NO_2$	Dopamine	1	25	8.9
		2	25	10.6
$C_8H_{11}NO_3$	Noradrenaline	1	25	8.64
		2	25	9.70
$C_8H_{12}N_2O_3$	Veronal		25	7.43
$C_8H_{14}O_4$	Octanedioic acid		25	4.52
$C_8H_{16}N_2O_3$	*N*-Glycylleucine		25	3.18
$C_8H_{16}N_2O_3$	*N*-Leucylglycine	1	25	3.25
		2	25	8.2
$C_8H_{16}N_2O_4S_2$	Homocystine	1	25	1.59
		2	25	2.54
		3	25	8.52
		4	25	9.44
$C_8H_{16}O_2$	Octanoic acid		25	4.89
$C_8H_{17}N$	2,2,4-Trimethylpiperidine		30	11.04
$C_8H_{19}N$	Dibutylamine		21	11.25
$C_8H_{19}N$	*N*-Methyl-2-heptanamine		17	10.99
$C_8H_{19}N$	Octylamine		25	10.65
C_9H_6BrN	3-Bromoquinoline		25	2.69
$C_9H_7ClO_2$	*o*-Chlorocinnamic acid		25	4.23
$C_9H_7ClO_2$	*m*-Chlorocinnamic acid		25	4.29
$C_9H_7ClO_2$	*p*-Chlorocinnamic acid		25	4.41
C_9H_7N	Isoquinoline		20	5.42
C_9H_7N	Quinoline		20	4.90
C_9H_7NO	7-Isoquinolinol	1	20	5.68
		2	20	8.90
C_9H_7NO	3-Quinolinol	1	20	4.28
		2	20	8.08
C_9H_7NO	8-Quinolinol	1	20	5.017
		2	25	9.812
$C_9H_7NO_3$	*o*-Cyanophenoxyacetic acid		25	2.98
$C_9H_7NO_3$	*m*-Cyanophenoxyacetic acid		25	3.03
$C_9H_7NO_3$	*p*-Cyanophenoxyacetic acid		25	2.93
$C_9H_8N_2$	1-Isoquinolinamine		20	7.59
$C_9H_8N_2$	3-Quinolinamine		20	4.91
$C_9H_8O_2$	*cis*-Cinnamic acid		25	3.89
$C_9H_8O_2$	*trans*-Cinnamic acid		25	4.44
$C_9H_9ClO_2$	3-(2-Chlorophenyl)propanoic acid		25	4.58
$C_9H_9ClO_2$	3-(3-Chlorophenyl)propanoic acid		25	4.59
$C_9H_9ClO_2$	3-(4-Chlorophenyl)propanoic acid		25	4.61

Molecular formula	Name	Step	T/°C	pK_a
$C_9H_9I_2NO_3$	3,5-Diiodotyrosine	1	25	2.12
		2	25	5.32
		3	25	9.48
$C_9H_9NO_3$	Hippuric acid		25	3.62
$C_9H_9NO_4$	3-(2-Nitrophenyl)propanoic acid		25	4.50
$C_9H_9NO_4$	3-(4-Nitrophenyl)propanoic acid		25	4.47
$C_9H_{10}INO_3$	3-Iodotyrosine	1	25	2.2
		2	25	8.7
		3	25	9.1
$C_9H_{10}N_2$	2-Ethylbenzimidazole		25	6.18
$C_9H_{10}O_2$	Mesitylenic acid		25	4.32
$C_9H_{10}O_2$	α-Phenylpropanoic acid		25	4.64
$C_9H_{10}O_2$	β-Phenylpropanoic acid		25	4.37
$C_9H_{11}N$	*N*-Allylaniline		25	4.17
$C_9H_{11}N$	1-Aminoindane		22	9.21
$C_9H_{11}NO_2$	Phenylalanine	1	25	2.20
		2	25	9.31
$C_9H_{11}NO_3$	Tyrosine	1	25	2.20
		2	25	9.21
		3	25	10.46
$C_9H_{11}NO_4$	*L*-3,4-Dihydroxyphenylalanine	1	25	2.32
		2	25	8.72
		3	25	9.96
		4	25	11.79
$C_9H_{12}N_2O_2$	Tyrosineamide		25	7.33
$C_9H_{13}NO_3$	*D*-Adrenaline	1	25	8.66
		2	25	9.95
$C_9H_{14}N_4O_3$	*N*-β-Alanylhistidine	1	20	2.73
		2	20	6.87
		3	20	9.73
$C_9H_{14}N_4O_3$	*L*-Carnosine	1	25	2.62
		2	25	6.66
		3	25	9.24
$C_9H_{18}O_2$	Nonanic acid		25	4.96
$C_9H_{19}N$	1-Butylpiperidine		23	10.47
$C_9H_{19}N$	2,2,6,6-Tetramethylpiperidine		25	11.07
$C_9H_{21}N$	Nonylamine		25	10.64
$C_{10}H_7NO_2$	8-Quinolinecarboxylic acid		25	1.82
$C_{10}H_8O$	1-Naphthol		25	9.34
$C_{10}H_8O$	2-Naphthol		25	9.51
$C_{10}H_9N$	2-Methylquinoline		20	5.83
$C_{10}H_9N$	4-Methylquinoline		20	5.67
$C_{10}H_9N$	5-Methylquinoline		20	5.20
$C_{10}H_9N$	α-Naphthylamine		25	3.92
$C_{10}H_9N$	β-Naphthylamine		25	4.16
$C_{10}H_9NO$	1-Amino-6-hydroxynaphthalene		25	3.97
$C_{10}H_9NO$	6-Methoxyquinoline		20	5.03
$C_{10}H_{10}O_2$	*o*-Methylcinnamic acid		25	4.50
$C_{10}H_{10}O_2$	*m*-Methylcinnamic acid		25	4.44
$C_{10}H_{10}O_2$	*p*-Methylcinnamic acid		25	4.56
$C_{10}H_{12}N_2$	Tryptamine		25	10.2
$C_{10}H_{12}N_2O$	5-Hydroxytryptamine	1	25	9.8
		2	25	11.1
$C_{10}H_{12}O_2$	4-Phenylbutanoic acid		25	4.76
$C_{10}H_{12}O_3$	2-(*m*-Anisyl)propanoic acid		25	4.65
$C_{10}H_{12}O_3$	2-(*p*-Anisyl)propanoic acid		25	4.69
$C_{10}H_{12}O_3$	2-(*o*-Anisyl)propanonic acid		25	4.80
$C_{10}H_{14}N_2$	Nicotine	1	25	8.02
		2	25	3.12

Molecular formula	Name	Step	$T/°C$	pK_a
$C_{10}H_{15}N$	*N,N*-Diethylaniline		22	6.61
$C_{10}H_{15}NO$	*d*-Ephedrine		10	10.139
$C_{10}H_{15}NO$	*l*-Ephedrine		10	9.958
$C_{10}H_{17}N_3O_6S$	*l*-Glutathione	1	25	2.12
		2	25	3.59
		3	25	8.75
		4	25	9.65
$C_{10}H_{17}N_5O_6$	Tetraglycylglycine	1	20	3.10
		2	20	8.02
$C_{10}H_{18}N_4O_5$	*L*-Argininosuccinic acid	1	25	1.62
		2	25	2.70
		3	25	4.26
		4	25	9.58
$C_{10}H_{19}N$	Bornylamine		25	10.17
$C_{10}H_{19}N$	Neobornylamine		25	10.01
$C_{10}H_{21}N$	Butylcyclohexylamine		25	11.23
$C_{10}H_{23}N$	Decylamine		25	10.64
$C_{11}H_8N_2$	Perimidine		20	6.35
$C_{11}H_8O_2$	1-Naphthoic acid		25	3.70
$C_{11}H_8O_2$	2-Naphthoic acid		25	4.17
$C_{11}H_{11}N$	Methyl-1-naphthylamine		27	3.67
$C_{11}H_{12}N_2O_2$	Tryptophan	1	25	2.46
		2	25	9.41
$C_{11}H_{16}N_2O_2$	Pilocarpine		30	6.87
$C_{11}H_{25}N$	Undecylamine		25	10.63
$C_{12}H_8N_2$	1,10-Phenanthroline		25	4.84
$C_{12}H_{11}N$	2-Aminobiphenyl		22	3.82
$C_{12}H_{11}N$	2-Benzylpyridine		25	5.13
$C_{12}H_{11}N$	Diphenylamine		25	0.79
$C_{12}H_{11}N_3$	4-Aminoazobenzene		25	2.82
$C_{12}H_{12}N_2$	*p*-Benzidine	1	30	4.66
		2	30	3.57
$C_{12}H_{13}N$	*N,N*-Dimethyl-1-naphthylamine		25	4.83
$C_{12}H_{13}N$	*N,N*-Dimethyl-2-naphthylamine		25	4.566
$C_{12}H_{27}N$	Dodecylamine		25	10.63
$C_{13}H_9N$	Acridine		20	5.58
$C_{13}H_9N$	Phenanthridine		20	5.58
$C_{13}H_{10}N_2$	2-Phenylbenzimidazole	1	25	5.23
		2	25	11.91
$C_{13}H_{10}O_2$	2-Phenylbenzoic acid		25	3.46
$C_{13}H_{12}N_2O$	4-(*p*-Aminobenzoyl)aniline	1	25	2.93
$C_{13}H_{13}N$	4-Benzylaniline		25	2.17
$C_{13}H_{29}N$	Tridecylamine		25	10.63
$C_{14}H_{12}O_2$	Diphenylacetic acid		25	3.94
$C_{14}H_{15}N_3$	4-Dimethylaminoazobenzene		25	3.226
$C_{14}H_{31}N$	Tetradecylamine		25	10.62
$C_{15}H_{11}I_4NO_4$	*L*-Thyroxine	1	25	2.2
		2	25	6.45
		3	25	10.1
$C_{15}H_{33}N$	Pentadecylamine		25	10.61
$C_{16}H_{35}N$	Hexadecylamine		25	10.63
$C_{17}H_{19}NO_3$	Morphine		25	8.21
$C_{18}H_{21}NO_3$	Codeine		25	8.21
$C_{18}H_{39}N$	Octadecylamine		25	10.60
$C_{20}H_{21}NO_4$	Papaverine		25	6.40
$C_{20}H_{24}N_2O_2$	Quinine	1	25	8.52
		2	25	4.13
$C_{21}H_{22}N_2O_2$	Strychnine		25	8.26
$C_{23}H_{26}N_2O_4$	Brucine	1	25	8.28

ION PRODUCT OF WATER SUBSTANCE

William L. Marshall and E. U. Franck

Pressure (bars)	Temperature (°C) 0	25	50	75	100	150	200	250	300
Saturated vapor	14.938	13.995	13.275	12.712	12.265	11.638	11.289	11.191	11.406
250	14.83	13.90	13.19	12.63	12.18	11.54	11.16	11.01	11.14
500	14.72	13.82	13.11	12.55	12.10	11.45	11.05	10.85	10.86
750	14.62	13.73	13.04	12.48	12.03	11.36	10.95	10.72	10.66
1,000	14.53	13.66	12.96	12.41	11.96	11.29	10.86	10.60	10.50
1,500	14.34	13.53	12.85	12.29	11.84	11.16	10.71	10.43	10.26
2,000	14.21	13.40	12.73	12.18	11.72	11.04	10.57	10.27	10.08
2,500	14.08	13.28	12.62	12.07	11.61	10.92	10.45	10.12	9.91
3,000	13.97	13.18	12.53	11.98	11.53	10.83	10.34	9.99	9.76
3,500	13.87	13.09	12.44	11.90	11.44	10.74	10.24	9.88	9.63
4,000	13.77	13.00	12.35	11.82	11.37	10.66	10.16	9.79	9.52
5,000	13.60	12.83	12.19	11.66	11.22	10.52	10.00	9.62	9.34
6,000	13.44	12.68	12.05	11.53	11.09	10.39	9.87	9.48	9.18
7,000	13.31	12.55	11.93	11.41	10.97	10.27	9.75	9.35	9.04
8,000	13.18	12.43	11.82	11.30	10.86	10.17	9.64	9.24	8.93
9,000	13.04	12.31	11.71	11.20	10.77	10.07	9.54	9.13	8.82
10,000	12.91	12.21	11.62	11.11	10.68	9.98	9.45	9.04	8.71

Pressure (bars)	Temperature (°C) 350	400	450	500	600	700	800	900	1000
Saturated vapor	12.30	—	—	—	—	—	—	—	—
250	11.77	19.43	21.59	22.40	23.27	23.81	24.23	24.59	24.93
500	11.14	11.88	13.74	16.13	18.30	19.29	19.92	20.39	20.80
750	10.79	11.17	11.89	13.01	15.25	16.55	17.35	17.93	18.39
1,000	10.54	10.77	11.19	11.81	13.40	14.70	15.58	16.22	16.72
1,500	10.22	10.29	10.48	10.77	11.59	12.50	13.30	13.97	14.50
2,000	9.98	9.98	10.07	10.23	10.73	11.36	11.98	12.54	12.97
2,500	9.79	9.74	9.77	9.86	10.18	10.63	11.11	11.59	12.02
3,000	9.61	9.54	9.53	9.57	9.78	10.11	10.49	10.89	11.24
3,500	9.47	9.37	9.33	9.34	9.48	9.71	10.02	10.35	10.62
4,000	9.34	9.22	9.16	9.15	9.23	9.41	9.65	9.93	10.13
5,000	9.13	8.99	8.90	8.85	8.85	8.95	9.11	9.30	9.42
6,000	8.96	8.80	8.69	8.62	8.57	8.61	8.72	8.86	8.97
7,000	8.81	8.64	8.51	8.42	8.34	8.34	8.40	8.51	8.64
8,000	8.68	8.50	8.36	8.25	8.13	8.10	8.13	8.21	8.38
9,000	8.57	8.37	8.22	8.10	7.95	7.89	7.89	7.95	8.12
10,000	8.46	8.25	8.09	7.96	7.78	7.70	7.68	7.70	7.85

Data in this table were calculated from the equation, $\log_{10} K_w^* = A + B/T + C/T^2 + D/T^3 + (E + F/T + G/T^2) \log_{10} \rho_w^*$, where $K_w^* = K_w/(\text{mol kg}^{-1})$, and $\rho_w^* = \rho_w/(\text{g cm}^{-3})$. The parameters are:

$A = -4.098$	$E = +13.957$
$B = -3245.2$ K	$F = 1262.3$ K
$C = +2.2362 \times 10^5$ K^2	$G = +8.5641 \times 10^5$ K^2
$D = -3.984 \times 10^7$ K^3	

IONIZATION CONSTANT OF NORMAL AND HEAVY WATER

This table gives the ionization constant in molality terms for H_2O and D_2O at temperatures from 0 to 100°C at the saturated vapor pressure. The quantity tabulated is $-\log K_W$, where K_W is defined by

$$K_W = m_+ \times m_-$$

and m_+ and m_- are the molalities, in mol/kg of water, for H^+ and OH^-, respectively.

	$-\log K_W$	
t/°C	H_2O	D_2O
0	14.938	15.972
5	14.727	15.743
10	14.528	15.527
15	14.340	15.324
20	14.163	15.132
25	13.995	14.951
30	13.836	14.779
35	13.685	14.616
40	13.542	14.462
45	13.405	14.316
50	13.275	14.176
55	13.152	14.044
60	13.034	13.918
65	12.921	13.798
70	12.814	13.683
75	12.712	13.574
80	12.613	13.470
85	12.520	13.371
90	12.428	13.276
95	12.345	13.186
100	12.265	13.099

REFERENCES

1. W.L. Marshall and E.U. Franck, *J. Phys. Chem. Ref. Data*, 10, 295, 1981.
2. R.E. Mesmer and D.L. Herting, *J. Solution Chem.*, 7, 901, 1978.

SOLUBILITY PRODUCT CONSTANTS

The following solubility product constants are calculated from the Gibbs energies of formation of the substances as solids and those of the aqueous ions at their standard states of unit molality. Thus, for the reaction

$$M_mX_n(s) \rightleftharpoons mM^{n+}(aq) + nX^{m-}(aq),$$

$$\Delta G° = m\Delta G°_f(m^{n+}, aq) + n\Delta G°_f(X^{m-}, aq) - \Delta G°_f(M_mX_n, s)$$

where M_mX_n is the slightly soluble substance and M^{n+} and X^{m-} are the two ions produced in solution by the dissociationof M_mX_n. Then the solubility product constant, K_{sp}, is calculated by using the equation

$$\ln K_{sp} = -\frac{\Delta G°}{RT}$$

The values in the following table are for K_{sp} at 25°C.

For the sulfides, which are indicated by an asterisk*, the solubility product refers to the reaction $MS(s) + 2H^+(aq) \rightleftharpoons M^{2+}(aq) + H_2S(aq)$; see Myers, R. J., *J. Chem. Educ.*, 63, 687, 1986.

Substance	Formula	Solubility Product
Aluminum phosphate	$AlPO_4$	9.83×10^{-21}
Barium carbonate	$BaCO_3$	2.58×10^{-9}
Barium chromate	$BaCrO_4$	1.17×10^{-10}
Barium fluoride	BaF_2	1.84×10^{-7}
Barium hydroxide 8-hydrate	$Ba(OH)_2 \cdot 8H_2O$	2.55×10^{-4}
Barium iodate	$Ba(IO_3)_2$	4.01×10^{-9}
Barium iodate 1-hydrate	$Ba(IO_3)_2 \cdot H_2O$	1.67×10^{-9}
Barium sulfate	$BaSO_4$	1.07×10^{-10}
Bismuth arsenate	$BiAsO_4$	4.43×10^{-10}
Bismuth sulfide*	Bi_2S_3	2×10^{-78}
Cadmium arsenate	$Cd_3(AsO_4)_2$	2.17×10^{-33}
Cadmium carbonate	$CdCO_3$	6.18×10^{-12}
Cadmium fluoride	CdF_2	6.44×10^{-3}
Cadmium hydroxide	$Cd(OH)_2$	5.27×10^{-15}
Cadmium iodate	$Cd(IO_3)_2$	2.49×10^{-8}
Cadmium oxalate 3-hydrate	$CdC_2O_4 \cdot 3H_2O$	1.42×10^{-8}
Cadmium phosphate	$Cd_3(PO_4)_2$	2.53×10^{-33}
Cadmium sulfide*	CdS	8×10^{-7}
Calcium carbonate	$CaCO_3$	4.96×10^{-9}
Calcium fluoride	CaF_2	1.46×10^{-10}
Calcium hydroxide	$Ca(OH)_2$	4.68×10^{-6}
Calcium iodate	$Ca(IO_3)_2$	6.47×10^{-6}
Calcium iodate 6-hydrate	$Ca(IO_3)_2 \cdot 6H_2O$	7.54×10^{-7}
Calcium oxalate 1-hydrate	$CaC_2O_4 \cdot H_2O$	2.34×10^{-9}
Calcium phosphate	$Ca_3(PO_4)_2$	2.07×10^{-33}
Calcium sulfate	$CaSO_4$	7.10×10^{-5}
Cobalt(II) arsenate	$Co_3(AsO_4)_2$	6.79×10^{-29}
Cobalt(II) hydroxide (pink)	$Co(OH)_2$	1.09×10^{-15}
Cobalt(II) hydroxide (blue)	$Co(OH)_2$	5.92×10^{-15}
Cobalt(II) iodate 2-hydrate	$Co(IO_3)_2 \cdot 2H_2O$	1.21×10^{-2}
Cobalt(II) phosphate	$Co_3(PO_4)_2$	2.05×10^{-35}
Copper(I) bromide	$CuBr$	6.27×10^{-9}
Copper(I) chloride	$CuCl$	1.72×10^{-7}
Copper(I) iodide	CuI	1.27×10^{-12}
Copper(I) sulfide*	Cu_2S	2×10^{-27}
Copper(I) thiocyanate	$CuSCN$	1.77×10^{-13}
Copper(II) arsenate	$Cu_3(AsO_4)_2$	7.93×10^{-36}
Copper(II) iodate 1-hydrate	$Cu(IO_3)_2 \cdot H_2O$	6.94×10^{-8}
Copper(II) oxalate	CuC_2O_4	4.43×10^{-10}
Copper(II) phosphate	$Cu_3(PO_4)_2$	1.39×10^{-37}
Copper(II) sulfide*	CuS	6×10^{-16}
Iron(II) carbonate	$FeCO_3$	3.07×10^{-11}
Iron(II) fluoride	FeF_2	2.36×10^{-6}
Iron(II) hydroxide	$Fe(OH)_2$	4.87×10^{-17}
Iron(II) sulfide*	FeS	6×10^{2}
Iron(III) hydroxide	$Fe(OH)_3$	2.64×10^{-39}
Iron(III) phosphate 2-hydrate	$FePO_4 \cdot 2H_2O$	9.92×10^{-29}
Lead bromide	$PbBr_2$	6.60×10^{-6}
Lead carbonate	$PbCO_3$	1.46×10^{-13}
Lead chloride	$PbCl_2$	1.17×10^{-5}
Lead fluoride	PbF_2	7.12×10^{-7}
Lead hydroxide	$Pb(OH)_2$	1.42×10^{-20}
Lead iodate	$Pb(IO_3)_2$	3.68×10^{-13}
Lead iodide	PbI_2	8.49×10^{-9}
Lead oxalate	PbC_2O_4	8.51×10^{-10}
Lead sulfate	$PbSO_4$	1.82×10^{-8}
Lead sulfide*	PbS	3×10^{-7}
Lead thiocyanate	$Pb(SCN)_2$	2.11×10^{-5}
Lithium carbonate	Li_2CO_3	8.15×10^{-4}
Magnesium carbonate	$MgCO_3$	6.82×10^{-6}
Magnesium carbonate 3-hydrate	$MgCO_3 \cdot 3H_2O$	2.38×10^{-6}
Magnesium carbonate 5-hydrate	$MgCO_3 \cdot 5H_2O$	3.79×10^{-6}
Magnesium fluoride	MgF_2	7.42×10^{-11}
Magnesium hydroxide	$Mg(OH)_2$	5.61×10^{-12}
Manganese(II) carbonate	$MnCO_3$	2.24×10^{-11}
Manganese(II) hydroxide	$Mn(OH)_2$	2.06×10^{-13}
Manganese(II) iodate	$Mn(IO_3)_2$	4.37×10^{-7}
Manganese(II) oxalate 2-hydrate	$MnC_2O_4 \cdot 2H_2O$	1.70×10^{-7}
Manganese(II) sulfide (green)*	MnS	3×10^{7}
Mercury(I) bromide	Hg_2Br_2	6.41×10^{-23}
Mercury(I) carbonate	Hg_2CO_3	3.67×10^{-17}
Mercury(I) chloride	Hg_2Cl_2	1.45×10^{-18}
Mercury(I) fluoride	Hg_2F_2	3.10×10^{-6}
Mercury(I) iodide	Hg_2I_2	5.33×10^{-29}
Mercury(I) oxalate	$Hg_2C_2O_4$	1.75×10^{-13}
Mercury(I) sulfate	Hg_2SO_4	7.99×10^{-7}
Mercury(I) thiocyanate	$Hg_2(SCN)_2$	3.12×10^{-20}
Mercury(II) hydroxide	$Hg(OH)_2$	3.13×10^{-26}
Mercury(II) iodide	HgI_2	2.82×10^{-29}
Mercury(II) sulfide (black)*	HgS	2×10^{-32}
Mercury(II) sulfide (red)*	HgS	4×10^{-33}
Nickel(II) carbonate	$NiCO_3$	1.42×10^{-7}
Nickel(II) hydroxide	$Ni(OH)_2$	5.47×10^{-16}
Nickel(II) iodate	$Ni(IO_3)_2$	4.71×10^{-5}
Nickel(II) phosphate	$Ni_3(PO_4)_2$	4.73×10^{-32}
Nickel(II) sulfide*	NiS	1.1
Palladium(II) sulfide*	PdS	2×10^{-37}
Palladium(II) thiocyanate	$Pd(SCN)_2$	4.38×10^{-23}
Platinum(II) sulfide*	PtS	1×10^{-52}
Potassium hexachloroplatinate(IV)	$K_2[PtCl_6]$	7.48×10^{-6}
Potassium perchlorate	$KClO_4$	1.05×10^{-2}
Silver(I) acetate	$AgC_2H_3O_2$	1.94×10^{-3}
Silver(I) arsenate	Ag_3AsO_4	1.03×10^{-22}
Silver(I) bromate	$AgBrO_3$	5.34×10^{-5}
Silver(I) bromide	$AgBr$	5.35×10^{-13}
Silver(I) carbonate	Ag_2CO_3	8.45×10^{-12}
Silver(I) chloride	$AgCl$	1.77×10^{-10}
Silver(I) chromate	Ag_2CrO_4	1.12×10^{-12}
Silver(I) cyanide	$AgCN$	5.97×10^{-17}
Silver(I) iodate	$AgIO_3$	3.17×10^{-8}
Silver(I) iodide	AgI	8.51×10^{-17}
Silver(I) oxalate	$Ag_2C_2O_4$	5.40×10^{-12}
Silver(I) phosphate	Ag_3PO_4	8.88×10^{-17}
Silver(I) sulfate	Ag_2SO_4	1.20×10^{-5}
Silver(I) sulfide (α-form)	Ag_2S	6×10^{-30}
Silver(I) sulfide (β-form)	Ag_2S	1×10^{-29}
Silver(I) sulfite	Ag_2SO_3	1.49×10^{-14}
Silver(I) thiocyanate	$AgSCN$	1.03×10^{-12}
Strontium arsenate	$Sr_3(AsO_4)_2$	4.29×10^{-19}
Strontium carbonate	$SrCO_3$	5.60×10^{-10}
Strontium fluoride	SrF_2	4.33×10^{-9}
Strontium iodate	$Sr(IO_3)_2$	1.14×10^{-7}
Strontium iodate 1-hydrate	$Sr(IO_3)_2 \cdot H_2O$	3.58×10^{-7}
Strontium iodate 6-hydrate	$Sr(IO_3)_2 \cdot 6H_2O$	4.65×10^{-7}
Strontium sulfate	$SrSO_4$	3.44×10^{-7}
Tin(II) hydroxide	$Sn(OH)_2$	5.45×10^{-27}
Tin(II) sulfide*	SnS	1×10^{-5}
Zinc arsenate	$Zn_3(AsO_4)_2$	3.12×10^{-28}
Zinc carbonate	$ZnCO_3$	1.19×10^{-10}
Zinc carbonate 1-hydrate	$ZnCO_3 \cdot H_2O$	5.41×10^{-11}
Zinc fluoride	ZnF_2	3.04×10^{-2}
Zinc hydroxide (γ-form)	$Zn(OH)_2$	6.86×10^{-17}
Zinc hydroxide (β-form)	$Zn(OH)_2$	7.71×10^{-17}
Zinc hydroxide (ϵ-form)	$Zn(OH)_2$	4.12×10^{-17}
Zinc iodate	$Zn(IO_3)_2$	4.29×10^{-6}
Zinc oxalate 2-hydrate	$ZnC_2O_4 \cdot 2H_2O$	1.37×10^{-9}
Zinc sulfide (α-form)*	ZnS	2×10^{-4}

FLAME AND BEAD TESTS

Flame Colorations

Violet

Potassium compounds. Purple red through blue glass. Easily obscured by sodium flame. Bluish green through green glass. Rubidium and Cesium compounds impart same flame as potassium compounds

Blues

Azure.—Copper chloride. Copper bromide gives azure blue followed by green. Other copper compounds give same coloration when moistened with hydrochloric acid.

Light Blue.—Lead, Arsenic, Selenium.

Greens

Emerald.—Copper compounds except the halides, and when not moistened with hydrochloric acid.

Pure Green.—Compounds of thallium and tellurium.

Yellowish.—Barium compounds. Some molybdenum compounds. Borates, especially when treated with sulphuric acid or when burned with alcohol.

Bluish.—Phosphates with sulphuric acid.

Feeble.—Antimony compounds. Ammonium compounds.

Whitish.—Zinc.

Reds

Carmine.—Lithium compounds. Violet through blue glass. Invisible through green glass. Masked by barium flame.

Scarlet.—Strontium compounds. Violet through blue glass. Yellowish through green glass. Masked by barium flame.

Yellowish.—Calcium compounds. Greenish through blue glass. Green through green glass. Masked by barium flame.

Yellow

Yellow.—All sodium compounds. Invisible with blue glass.

Borax Beads

Abbreviations employed: s., saturated; s.s., supersaturated; n.s., not saturated; h., hot; c., cold.

Substance	Oxidizing flame	Reducing flame
Aluminum	Colorless (h.c., n.s.); opaque (s.s.)	Colorless; opaque (s.)
Antimony	Colorless, yellow or brownish (h., s.s.)	Gray and opaque
Barium	Colorless (n.s.)	
Bismuth	Colorless; yellow or brownish (h., s.s.)	Gray and opaque
Cadmium	Colorless	Gray and opaque
Calcium	Colorless (n.s.)	
Cerium	Red (h.)	Colorless (h.c.)
Chromium	Green (c.)	Green
Cobalt	Blue (h.c.)	Blue (h.c.)
Copper	Green (h.); blue (c.)	Red (c.): opaque (s.s.): colorless (h.)
Iron	Yellow or brownish red (h., n.s.)	Green (s.s.)
Lead	Colorless; yellow or brownish (h., s.s.)	Gray and opaque
Magnesium	Colorless (n.s.)	
Manganese	Violet (h.c.)	Colorless (h.c.)
Molybdenum	Colorless	Yellow or brown (h.)
Nickel	Brown; red (c.)	Gray and opaque
Silicon	Colorless (h.c.); opaque (s.s.)	Colorless; opaque (s.)
Silver	Colorless (n.s.)	Gray and opaque
Strontium	Colorless (n.s.)	
Tin	Colorless (h.c.); opaque (s.s.)	Colorless; opaque (s.)
Titanium	Colorless	Yellow (h.); violet (c.)
Tungsten	Colorless	Brown
Uranium	Yellow or brownish (h., n.s.)	Green
Vanadium	Colorless	Green

Beads of Microcosmic Salt
$NaNH_4HPO_4$

Substance	Oxidizing flame	Reducing flame
Aluminum	Colorless; opaque (s.)	Colorless; not clear (s.s.)
Antimony	Colorless (n.s.)	Gray and opaque
Barium	Colorless; opaque (s.)	Colorless; not clear (s.s.)
Bismuth	Colorless (n.s.)	Gray and opaque
Cadmium	Colorless (n.s.)	Gray and opaque
Calcium	Colorless; opaque (s.)	Colorless; not clear (s.s.)
Cerium	Yellow or brownish red (h., s.)	Colorless
Chromium	Red (h., s.); green (c.)	Green (c.)
Cobalt	Blue (h.c.)	Blue (h.c.)
Copper	Blue (c.); green (h.)	Red and opaque (c.)
Iron	Yellow or brown (h., s.)	Colorless; yellow or brownish (h.)
Lead	Colorless (n.s.)	Gray and opaque
Magnesium	Colorless; opaque (s.)	Colorless; not clear (s.s.)
Manganese	Violet (h.c.)	Colorless
Molybdenum	Colorless; green (h.)	Green (h.)
Nickel	Yellow (c.); red (h., s.)	Yellow (c.); red (h.); gray and opaque
Silicon	(Swims undissolved)	(Swims undissolved)
Silver		Gray and opaque
Strontium	Colorless; opaque (s.)	Colorless; not clear (s.s.)
Tin	Colorless; opaque (s.)	Colorless
Titanium	Colorless (n.s.)	Violet (c.); yellow or brownish (h.)
Uranium	Green; yellow or brownish (h., s.)	Green (h.)
Vanadium	Yellow	Green
Zinc	Colorless (n.s.)	Gray and opaque

Sodium Carbonate Bead

Substance	Oxidizing flame	Reducing flame
Manganese	Green	Colorless

SOLUBILITY CHART

Abbreviations: **W**, soluble in water; **A**, insoluble in water but soluble in acids; **w**, sparingly soluble in water but soluble in acids; a, insoluble in water and only sparingly soluble in acids; **I,** insoluble in water and acids; **d**, decomposes in water. * Indicates two modifications of the salt

No.		Al	NH_4	Sb	Ba	Bi	Cd	Ca	Cr	Co	Cu	Au (I)	Au (II)	H	Fe (II)	Fe (III)
1	Acetate —($C_2H_3O_2$)	W $Al(—)_3$	W $NH_4(—)$		W $Ba(—)_2$	W $Bi(—)_3$	W $Cd(—)_2$	W $Ca(—)_2$	W $Cr(—)_3$	W $Co(—)_2$	W $Cu(—)_2$	W	W	W $C_2H_4O_2$	$Fe(—)_2$	$Fe_2(—)_6$
2	Arsenate —(AsO_4)	a $Al(—)$	W $(NH_4)_3(—)$	A $Sb(—)$	w $Ba_3(—)_2$	A $Bi(—)$	A $Cd_3(—)_2$	w $Ca_3(—)_2$		A $Co_3(—)_2$	A $Cu_3(—)_2$			W H_3AsO_4	A $Fe_3(—)_2$	A $Fe(—)$
3	Arsenite —(AsO_3)		W NH_4AsO_2	A $Sb(—)$				w $Ca_3(—)_2$		A $Co_3H_6(—)_4$	A $CuH(—)$					
4	Benzoate —($C_7H_5O_2$)		W $NH_4(—)$		W $Ba(—)_2$	A $Bi(—)_3$	W $Cd(—)_2$	W $Ca(—)_2$		W $Co(—)_2$	w $Cu(—)_2$			W $C_7H_6O_2$	W $Fe(—)_2$	A $Fe_2(—)_6$
5	Bromide	W $AlBr_2$	W NH_4Br	d $SbBr_3$	W $BaBr_2$	d $BiBr_3$	W $CdBr_2$	W $CaBr_2$	W(I)* $CrBr_3$	W $CoBr_2$	W $CuBr_2$	w AuBr	W $AuBr_3$	W HBr	W $FeBr_2$	W $FeBr_3$
6	Carbonate		W $(NH_4)_2CO_3$		w $BaCO_3$		A $CdCO_3$	w $CaCO_3$	W $CrCO_3$	A $CoCO_3$					w $FeCO_3$	
7	Chlorate —(ClO_3)	W $Al(—)_3$	W $NH_4(—)$		W $Ba(—)_2$	W $Bi(—)_3$	W $Cd(—)_2$	W $Ca(—)_2$		W $Co(—)_2$	W $Cu(—)_2$			W $HClO_3$	W $Fe(—)_2$	W $Fe(—)_3$
8	Chloride	W $AlCl_3$	W NH_4Cl	W $SbCl_3$	W $BaCl_2$	d $BiCl_3$	W $CdCl_2$	W $CaCl_2$	I $CrCl_3$	W $CoCl_2$	W $CuCl_2$	w AuCl	W $AuCl_3$	W HCl	W $FeCl_2$	W $FeCl_3$
9	Chromate —(CrO_4)		W $(NH_4)_2(—)$		A $Ba(—)$		A $Cd(—)$	W $Ca(—)$		A $Co(—)$						A $Fe_2(—)_3$
10	Citrate —($C_6H_5O_7$)	W $Al(—)$	W $(NH_4)_3(—)$		w $Ba_3(—)_2$	A $Bi(—)$	A $Cd_3(—)_2$	w $Ca_3(—)_2$		w $Co_3(—)_2$				W $C_6H_8O_7$		W $Fe(—)$
11	Cyanide		W NH_4CN		W $Ba(CN)_2$	w $Bi(CN)_3$	W $Cd(CN)_2$	W $Ca(CN)_2$	A $Cr(CN)_3$	A $Co(CN)_2$	A $Cu(CN)_2$	w AuCN	W $Au(CN)_3$	W HCN	a $Fe(CN)_2$	
12	Ferricy'de —($Fe(CN)_6$)		W $(NH_4)_3(—)$		w $Ba_3(—)_2$		A $Cd_3(—)_2$	W $Ca_3(—)_2$		I $Co_3(—)_2$	I $Cu_3(—)_2$			W $H_3(—)$	I $Fe_3(—)_2$	
13	Ferrocy'de —($Fe(CN)_6$)	w $Al_4(—)_3$	W $(NH_4)_4(—)$		W $Ba_2(—)$		A $Cd_2(—)$	W $Ca_2(—)$		I $Co_2(—)$	I $Cu_2(—)$			W $H_4(—)$	I $Fe_2(—)$	a $Fe_4(—)_3$
14	Fluoride	W AlF_3	W NH_4F	W SbF_3	w BaF_2	W BiF_3	W CdF_2	w CaF_2	W(a)* CrF_3	W CoF_2	w CuF_2			W HF	w FeF_2	w FeF_3
15	Formate —(CHO_2)	W $Al(—)_3$	W $NH_4(—)$		W $Ba(—)_2$	W $Bi(—)_3$	W $Cd(—)_2$	W $Ca(—)_2$		W $Co(—)_2$	W $Cu(—)_2$			W CH_2O_2	W $Fe(—)_2$	W $Fe(—)_3$
16	Hydroxide	A $Al(OH)_3$	W NH_4OH		W $Ba(OH)_2$	A $Bi(OH)_3$	A $Cd(OH)_2$	W $Ca(OH)_2$	A $Cr(OH)_3$	A $Co(OH)_2$	A $Cu(OH)_2$	W AuOH	A $Au(OH)_3$		A $Fe(OH)_2$	A $Fe(OH)_3$
17	Iodide	W AlI_3	W NH_4I	d SbI_3	W BaI_2	A BiI_3	W CdI_2	W CaI_2	W CrI_3	W CoI_2	a CuI	a AuI	a AuI_3	W HI	W FeI_2	W FeI_3
18	Nitrate	W $Al(NO_3)_3$	W NH_4NO_3		W $Ba(NO_3)_2$	d $Bi(NO_3)_3$	W $Cd(NO_3)_2$	W $Ca(NO_3)_2$	W $Cr(NO_3)_3$	W $Co(NO_3)_2$	W $Cu(NO_3)_2$			W HNO_3	W $Fe(NO_3)_2$	W $Fe(NO)_3$
19	Oxalate —(C_2O_4)	A $Al_2(—)_3$	W $(NH_4)_2(—)$		w $Ba(—)$	A $Bi_2(—)_3$	w $Cd(—)$	A $Ca(—)$	W $Cr(—)$	A $Co(—)$	A $Cu(—)$			W $C_2H_2O_4$	A $Fe(—)$	W $Fe_2(—)_3$
20	Oxide	a Al_2O_3		w Sb_2O_3	W BaO	A Bi_2O_3	A CdO	w CaO	a Cr_2O_3	A CoO	A CuO	Au_2O	A Au_2O_3	W H_2O_2	A FeO	A Fe_2O_3
21	Phosphate	A $AlPO_4$	W $NH_4H_2PO_4$		A $Ba_3(PO_4)_2$	A $BiPO_4$	A $Cd_3(PO_4)_2$	w $Ca_3(PO_4)_2$	w $Cr_2(PO_4)_2$	A $Co_3(PO_4)_2$	A $Cu_3(PO_4)_2$		H_3PO_4	W $Fe_3(PO_4)_2$	A $FePO_4$	w
22	Silicate, —(SiO_3)	I $Al_2(—)_3$			W $Ba(—)$		A $Cd(—)$	w $Ca(—)$		A Co_2SiO_4	A $Cu(—)$			I H_2SiO_3		
23	Sulfate	W $Al_2(SO_4)_3$	W $(NH_4)_2SO_4$	A $Sb_2(SO_4)_3$	a $BaSO_4$	d $Bi_2(SO_4)_3$	W $CdSO_4$	w $CaSO_4$	W(I)* $Cr_2(SO_4)_3$	W $CoSO_4$	W $CuSO_4$			W H_2SO_4	W $FeSO_4$	w $Fe(SO_4)_3$
24	Sulfide	d Al_2S_3	W $(NH_4)_2S$	A Sb_2S_3	d BaS	A Bi_2S_3	A CdS	w CaS	d Cr_2S_3	A CoS	A CuS	I Au_2S	I Au_2S_3	W H_2S	A FeS	d Fe_2S_3
25	Tartrate —($C_4H_4O_6$)	w $Al_2(—)_3$	W $(NH_4)_2(—)$	W $Sb_2(—)_3$	w $Ba(—)$	A $Bi_2(—)_3$	A $Cd(—)$	w $Ca(—)$	d	A $Co(—)$	A $Cu(—)$	I	I	W $C_4H_6O_6$	A $Fe(—)$	d $Fe_2(—)_3$

SOLUBILITY CHART (continued)

No.		Al	NH_4	Sb	Ba	Bi	Cd	Ca	Cr	Co	Cu	Au (I)	Au (II)	H	Fe (II)	Fe (III)
26	Thiocy'te		W NH_4CNS		W $Ba(CNS)_2$			W Ca(CNS)		W $Co(CNS)_2$	d CuCNS			W CNSH	W $Fe(CNS)_2$	W $Fe(CNS)_3$

No.		Pb	Mg	Mn	Hg (I)	Hg (II)	Ni	K
1	Acetate —($C_2H_3O_2$)	W $Pb(—)_2$	W $Mg(—)_2$	W $Mn(—)_2$	w Hg(—)	W $Hg(—)_2$	W $Ni(—)_2$	W K(—)
2	Arsenate —(AsO_4)	A PbH(—)	A $Mg_3(—)$	w MnH(—)	A $Hg_3(—)$	w $Hg_3(—)_2$	A $Ni_3(—)_2$	W $K_3(—)$
3	Arsenite —(AsO_3)		W $Mg_3(—)_2$	A $Mn_3H_6(—)_4$	A $Hg_3(—)$	A $Hg_3(—)$	A $Ni_3H_6(—)_4$	W K_3AsO_3
4	Benzoate —($C_7H_5O_2$)	w $Pb(—)_2$	W $Mg(—)_2$	W $Mn(—)_2$	A $Hg_2(—)_2$	w $Hg(—)_2$	w $Ni(—)_2$	W K(—)
5	Bromide	W $PbBr_2$	W $MgBr_2$	W $MnBr_2$	A HgBr	W $HgBr_2$	W $NiBr_2$	W KBr
6	Carbonate	A $PbCO_3$	w $MgCO_3$	w $MnCO_3$	A Hg_2CO_3		w $NiCO_3$	W K_2CO_3
7	Chlorate —(ClO_3)	W $Pb(—)_2$	W $Mg(—)_2$	W $Mn(—)_2$	W Hg(—)	W $Hg(—)_2$	W $Ni(—)_2$	W K(—)
8	Chloride	W $PbCl_2$	W $MgCl_2$	W $MnCl_2$	a HgCl	W $HgCl_2$	W $NiCl_2$	W KCl
9	Chromate —(CrO_4)	A Pb(—)	W Mg(—)		w $Hg_2(—)$	w Hg(—)	A Ni(—)	W $K_2(—)$
10	Citrate —($C_6H_5O_7$)	W $Pb_3(—)_2$	W $Mg_3(—)_2$	w MnH(—)	w $Hg_3(—)$		W $Ni_3(—)_2$	W $K_3(—)$
11	Cyanide	w $Pb(CN)_2$	W $Mg(CN)_2$		A HgCN	W $Hg(CN)_2$	a $Ni(CN)_2$	W KCN
12	Ferricy'de —$Fe(CN)_6$	w $Pb_3(—)_2$	W $Mg_3(—)_2$			A $Hg_3(—)_2$	I $Ni_3(—)_2$	W $K_3(—)$
13	Ferrocy'de —$Fe(CN)_6$	a $Pb_2(—)$	W $Mg_2(—)$	A $Mn_2(—)$		I $Hg_2(—)$	I $Ni_2(—)$	W $K_4(—)$
14	Fluoride	w PbF_2	w MgF_2	A MnF_2	d HgF	d HgF_2	w NiF_2	W KF
15	Formate —(CHO_2)	W $Pb(—)_2$	W $Mg(—)_2$	W $Mn(—)_2$	w Hg(—)	W $Hg(—)_2$	W $Ni(—)_2$	W K(—)
16	Hydroxide	w $Pb(OH)_2$	A $Mg(OH)_2$	A $Mn(OH)_2$		A $Hg(OH)_2$	w $Ni(OH)_2$	W KOH
17	Iodide	w PbI_2	W MgI_2	W MnI_2	A HgI	w HgI_2	W NiI_2	W KI
18	Nitrate	W $Pb(NO_3)_2$	W $Mg(NO_3)_2$	W $Mn(NO_3)_2$	W $HgNO_3$	W $Hg(NO_3)_2$	W $Ni(NO_3)_2$	W KNO_3
19	Oxalate —(C_2O_4)	A Pb(—)	w Mg(—)	w Mn(—)	a $Hg_2(—)$	A Hg(—)	A Ni(—)	W $K_2(—)$
20	Oxide	w PbO	A MgO	A MnO	A Hg_2O	w HgO	A NiO	W K_2O
21	Phosphate	A $Pb_3(PO_4)_2$	w $Mg_3(PO_4)_2$	w $Mn_3(PO_4)_2$	A Hg_3PO_4	A $Hg_3(PO_4)_2$	A $Ni_3(PO_4)_2$	W K_3PO_4
22	Silicate —(SiO_3)	A Pb(—)	A Mg(—)	I Mn(—)				W $K_2(—)$
23	Sulfate	w $PbSO_4$	W $MgSO_4$	W $MnSO_4$	w Hg_2SO_4	d $HgSO_4$	W $NiSO_4$	W K_2SO_4

No.		Pt	Ag	Na	Sn (IV)	Sn (II)	Sr	Zn
1	Acetate —($C_2H_3O_2$)		w Ag(—)	W Na(—)	W $Sn(—)_4$	d $Sn(—)_2$	W $Sr(—)_2$	W $Zn(—)_2$
2	Arsenate —(AsO_4)		A $Ag_3(—)$	W $Na_3(—)$			w SrH(—)	A $Zn_3(—)_2$
3	Arsenite —(AsO_3)		A $Ag_3(—)$	W $Na_2H(—)$		A $Sn_3(—)_2$	w $Sr_3(—)_2$	
4	Benzoate —($C_7H_5O_2$)		w Ag(—)	W Na(—)				W $Zn(—)_2$
5	Bromide	w $PtBr_4$	a AgBr	W NaBr	W $SnBr_4$	W $SnBr_2$	W $SrBr_2$	W $ZnBr_2$
6	Carbonate		A Ag_2CO_3	W Na_2CO_3			w $SrCO_3$	w $ZnCO_3$
7	Chlorate —(ClO_3)		W Ag(—)	W Na(—)		W $Sn(—)_2$	W $Sr(—)_2$	W $Zn(—)_2$
8	Chloride	W $PtCl_4$	a AgCl	W NaCl	W $SnCl_4$	W $SnCl_2$	W $SrCl_2$	W $ZnCl_2$
9	Chromate —(CrO_4)		w $Ag_2(—)$	W $Ma_2(—)$	W $Sn(—)_2$	A Sn(—)	w Sr(—)	w Zn(—)
10	Citrate —($C_6H_5O_7$)		w $Ag_3(—)$	W $Na_3(—)$			A SrH(—)	w $Zn_3(—)_2$
11	Cyanide	I $Pt(CN)_2$	a AgCN	W NaCN			W $Sr(CN)_2$	A $Zn(CN)_2$
12	Ferricy'de —$Fe(CN)_6$		I $Ag_3(—)$	W $Na_3(—)$		A $Sn_3(—)_2$	W $Sr_3(—)_2$	A $Zn_3(—)_2$
13	Ferrocy'de —$Fe(CN)_6$		I $Ag_4(—)$	W $Na_4(—)$		a $Sn_2(—)$	W $Sr_2(—)$	I $Zn_2(—)$
14	Fluoride	[illegible] PtF_4	W AgF	W NaF	W SnF_4	W SnF_2	w SrF_2	w ZnF_2
15	Formate —(CHO_2)		W Ag(—)	W Na(—)			W $Sr(—)_2$	W $Zn(—)_2$
16	Hydroxide	[illegible] $Pt(OH)_4$		W NaOH	w $Sn(OH)_4$	A $Sn(OH)_2$	W $Sr(OH)_2$	A $Zn(OH)_2$
17	Iodide	[illegible] PtI_2	I AgI	W NaI	d SnI_4	W SnI_2	W SrI_2	W ZnI_2
18	Nitrate	[illegible] $Pt(NO_3)_4$	W $AgNO_3$	W $NaNO_3$		d $Sn(NO_3)_2$	W $Sr(NO_3)_2$	W $Zn(NO_3)_2$
19	Oxalate —(C_2O_4)		a $Ag_2(—)$	W $Na_2(—)$		A Sn(—)	w Sr(—)	A Zn(—)
20	Oxide	A PtO	w Ag_2O	d Na_2O	A SnO_2	A SnO	W SrO	w ZnO
21	Phosphate		A Ag_3PO_4	W Na_3PO_4		A $Sn_3(PO_4)$	A $Sr_3(PO_4)_2$	A $Zn_3(PO_4)_2$
22	Silicate —(SiO_3)			W $Na_2(—)$			A Sr(—)	A Zn(—)
23	Sulfate	W $Pt(SO_4)_2$	w Ag_2SO_4	W Na_2SO_4	W $Sn(SO_4)_2$	W $SnSO_4$	w $SrSO_4$	W $ZnSO_4$

SOLUBILITY CHART (continued)

No.		Pb	Mg	Mn	Hg (I)	Hg (II)	Ni	K	Pt	Ag	Na	Sn (IV)	Sn (II)	Sr	Zn
24	Sulfide	A PbS	d MgS	A MnS	I Hg_2S	I HgS	A NiS	W K_2S	I PtS	A Ag_2S	W Na_2S	A SnS_2	A SnS	W SrS	A ZnS
25	Tartrate —($C_4H_4O_6$)	A Pb(—)	w Mg(—)	w Mn(—)	I Hg_2(—)	I	A Ni(—)	W K_2(—)	I	A Ag_2(—)	W Na_2(—)	A	A Sn(—)	W Sr(—)	A Zn(—)
26	Thiocy'te	w $Pb(CNS)_2$	W $Mg(CNS)_2$	W $Mn(CNS)_2$	A HgCNS	w $Hg(CNS)_2$		W KCNS		I AgCNS	W NaCNS			W $Sr(CNS)_2$	W $Zn(CNS)_2$

REDUCTIONS OF WEIGHINGS IN AIR TO VACUO

When the mass M of a body is determined in air, a correction is necessary for the buoyancy of the air. The following table is computed for an air density of 0.0012 g/cm^3. The corrected mass is equal to $M + kM/1000$. Values of k are given in the table.

Density of body weighted (g/cm^3)	Correction factor, k			Density of body weighted (g/cm^3)	Correction factor, k		
	Pt Ir weights	Brass weights	Quartz or Al weights		Pt Ir weights	Brass weights	Quartz or Al weights
0.5	+2.34	+2.26	+1.95	1.6	+0.69	+0.61	+0.30
0.6	+1.94	+1.86	+1.55	1.7	+0.65	+0.56	+0.25
0.7	+1.66	+1.57	+1.26	1.8	+0.62	+0.52	+0.21
0.75	+1.55	+1.46	+1.15	1.9	+0.58	+0.49	+0.18
0.80	+1.44	+1.36	+1.05	2.0	+0.54	+0.46	+0.15
0.85	+1.36	+1.27	+0.96	2.5	+0.43	+0.34	+0.03
0.90	+1.28	+1.19	+0.88	3.0	+0.34	+0.26	–0.05
0.95	+1.21	+1.12	+0.81	4.0	+0.24	+0.16	–0.15
1.00	+1.14	+1.06	+0.75	6.0	+0.14	+0.06	–0.25
1.1	+1.04	+0.95	+0.64	8.0	+0.09	+0.01	–0.30
1.2	+0.94	+0.86	+0.55	10.0	+0.06	–0.02	–0.33
1.3	+0.87	+0.78	+0.47	15.0	+0.03	–0.06	–0.37
1.4	+0.80	+0.71	+0.40	20.0	+0.004	–0.08	–0.39
1.5	+0.75	+0.66	+0.35	22.0	–0.001	–0.09	–0.40

ION EXCHANGE RESINS

ANION EXCHANGE RESINS

The following table is divided into two parts; the first lists properties of some anionic resins and the second, properties of some cationic resins.

Character S = strong W = weak	Trade name	Manu-facturer*	Active group	Matrix	Effective pH	Selectivity	Order of selectivity	Total exchange capacity; meq/ml	Total exchange capacity; meq/gm	Maximum thermal stability; °C	Physical form; s = sphere b = beads	Standard mesh range	Ionic form as shipped	Shipping density; lb./cu. ft.
S	Dowex 1	1	Trimethyl benzyl ammonium	Polystyrene	0–14	Cl/H approx. 25	I, NO_3, Br, Cl, Acetate, OH, F	1.33	3.5	OH^- 50 Cl^- 150	s	20–50 (wet)	Cl^-	44
S	Dowex 21 K	1	Trimethyl benzyl ammonium	Polystyrene	0–14	Cl/H approx. 15	I, NO_3, Br, Cl, Acetate, OH, F	1.25	4.5	OH^- 50 Cl^- 150	s	20–50 (wet)	Cl^-	43
S	Duolite A-101 D	2	Quaternary ammonium	Polystyrene	0–14	—	—	1.4	4.2	OH^- 60 Cl^- 100	b	16–50	Cl^-	—
S	Ionac A-540	3	Quaternary ammonium	Polystyrene	0–14	—	—	1.0	3.6	salt 100 OH^- 60	b	16–50	salt	43–66
S	Dowex 2	1	Dimethyl ethanol benzyl ammonium	Polystyrene	0–14	Cl/H approx. 1.5	I, NO_3, Br, Cl, Acetate, OH, F	1.33	3.5	OH^- 30 Cl^- 150	s	20–50 (wet)	Cl^-	44
S	Duolite A-102 D	2	Quaternary ammonium	Polystyrene	0–14		—	1.4	4.2	OH^- 40 Cl^- 100	b	16–50	Cl^-	—
S	Ionac A-550	3	Dimethyl ethanol benzyl ammonium	Polystyrene	0–14	—	—	1.3	3.5	salt 100 OH^- 40	b	16–50	salt	43–46
W	Duolite A-30 B	2	Tertiary amine; Quaternary ammonium	Epoxy polyamines	0–9	—	—	2.6	8.7	60	b	16–50	salt	
W	Ionac A-300	3	Tertiary amine; Quaternary ammonium	Epoxy amine	0–12			1.8	5.5	40	g	16–50	salt	19–21
W	Duolite A-6	2	Tertiary amine	Phenolic	0–5	—	—	2.4	7.6	60	g	16–50	salt	—
W	Duolite A-7	2	Secondary amine	Phenolic	0–4			2.4	9.1	40	g	16–50	salt	

CATION EXCHANGE RESINS

Character S = strong W = weak	Trade name	Manu-facturer*	Active group	Matrix	Effective pH	Selectivity	Order of selectivity	Total exchange capacity; meq/ml	Total exchange capacity; meq/mg	Maximum thermal stability; °C	Physical form; s = sphere b = beads	Standard mesh range	Ionic form as shipped	Shipping density; lb./cu. ft.
S	Dowex 50	1	Nuclear sulfonic acid	Polystyrene	0–14	Na/H approx. 1.2	Ag, Cs, Rb, K, NH_4, Na, H, Li, Ba, Sr, Ca, Mg, Be	Na^+ 1.9 H^+ 1.7	Na^+ 4.8 H^+ 5.0	150	s	20–50 (wet)	H^+ or Na^+	H^+ 50 Na^+ 53
S	Dowex MPC-1	4	Nuclear sulfonic acid	Polystyrene	0–14	—	—	1.6–1.8 H^+ form	4.5–4.9 H^+ form	150	b	20–40 (wet)	Na^+	50
S	Duolite C-20	2	Nuclear sulfonic acid	Polystyrene	0–14	—	—	2.2	5.1	150	b	16–50	Na^+	—
S	Ionac 240	3	Nuclear sulfonic acid	Polystyrene	0–14	—	—	1.9	4.6	140 (Na^+) 130 (H^+)	b	16–50	Na^+	50–55
S	Duolite C-3	2	Methylene sulfonic	Phenolic	0–9			1.1	2.9	60	g	16–50	H^+	
W	Dowex CCR-1	4	Carboxylic	Phenolic	0–9		—	—		38	g	20–50 (wet)	H^+ (dry)	21
W	Duolite ES-63	2	Phosphonic	Polystyrene	4–14	—	—	3.3	6.5	100	b	16–50	H^+	
W	Duolite ES-80	2	Aliphatic	Acrylic	6–14		—	3.5	10.2	100	b	16–50	H^+	

* 1. Dow
2. Diamond Shamrock
3. Ionac
4. Nalco

SOLVENTS FOR LIQUID CHROMATOGRAPHY

Solvent	Viscosity (mPa·s; 20 °C)	UV cutoff (nm)	Refractive index (20 °C)	Normal boiling point (°C)	Dielectric constant (20 °C)
Acetic acid	1.31(15)		1.372	117.9	6.15
Acetone	0.30(25)	330	1.359	56.1	20.7(25)
Acetonitrile	0.34(25)	190	1.344	81.6	37.5
Benzene	0.65	278	1.501	80.1	2.284
1-Butanol	2.95	215	1.399	117.7	17.8
2-Butanol	4.21	260	1.397	99.5	15.8(25)
n-Butyl acetate	0.73	254	1.394	126.1	
n-Butyl chloride	0.47(15)	220	1.402	78.4	
Carbon tetrachloride	0.97	263	1.460	76.8	2.238
Chlorobenzene	0.80	287	1.525	131.7	2.708
Chloroform	0.58	245	1.446	61.2	4.806
Cyclohexane	0.98	200	1.426	80.7	2.023
Cyclopentane	0.44	200	1.406	49.3	1.965
o-Dichlorobenzene	1.32(25)	295	1.551	180.0	9.93(25)
N,N-Dimethylacetamide	2.14	268	1.438	166.0	37.8
Dimethylformamide	0.92	268	1.430	153.0	36.7
Dimethyl sulfoxide	2.20	286	1.478	189.0	4.7
Dioxane	1.44(15)	215	1.422	101.4	2.209(25)
2-Ethoxyethanol	2.05	210	1.408	135.0	
Ethyl acetate	0.46	256	1.372	77.1	6.02(25)
Ethyl ether	0.24	218	1.352	34.4	4.335
Glyme (ethylene glycol dimethyl ether)	0.46(25)	220	1.380	93.0	
Heptane	0.42	200	1.388	98.4	1.92
Hexadecane	3.34	200	1.434	286.9	
Hexane	0.31	200	1.375	68.7	1.890
Isobutyl alcohol	4.70(15)	200	1.396	107.9	15.8(25)
Methanol	0.55	205	1.328	64.6	32.63(25)
2-Methoxyethanol	1.72	210	1.402	124.1	16.9
2-Methoxyethyl acetate		254	1.402	143.0	
Methylene chloride	0.45(15)	233	1.424	39.8	9.08
Methylethylketone	0.42(15)	329	1.379	79.6	18.5
Methylisoamylketone		330	1.406	144.0	
Methylisobutylketone	0.54(25)	334	1.396	116.5	
N-Methyl-2-pyrrolidone	1.67(25)	285	1.488	202.0	32.0
Nonane	0.72	200	1.405	150.8	1.972
Pentane	0.24	200	1.357	36.1	1.84
Petroleum ether	0.30	226		30—60	
β-Phenethylamine		285	1.529(25)	197—198	
1-Propanol	2.26	210	1.386	97.2	20.1(25)
2-Propanol	2.86(15)	205	1.377	82.2	18.3(25)
Propylene carbonate			1.419	240.0	
Pyridine	0.95	330	1.510	115.3	12.3(25)
Tetrachloroethylene	0.93(15)	295	1.506	121.2	
Tetrahydrofuran	0.55	212	1.407	64.8	7.6
Tetramethyl urea		265	1.449(25)	175.2	23.0
Toluene	0.59	284	1.497	110.6	2:379(25)
Trichloroethylene	0.57	273	1.477	87.2	3.4(16)
1,2,2-Trichloro-1,2,2-trifluoroethane	0.71	231	1.356(25)	47.6	
2,2,4-Trimethylpentane	0.50	215	1.391	99.2	1.94
Water	1.00	$<$190	1.333	100.0	78.54
o-Xylene	0.81	288	1.505	144.4	2.568
p-Xylene		290	1.496	138.4	2.270

Reprinted from T. J. Bruno and P. D. N. Svoronos, *CRC Handbook of Basic Tables for Chemical Analysis*, CRC Press, Boca Raton, FL, 1989, pg. 89.

PROPERTIES OF CARRIER GASES FOR GAS CHROMATOGRAPHY

Carrier gas	Density (kg/m^3)	Thermal conductivity $\times 10^{-2}$ (W/[m·K])	Thermal conductivity differences			Viscosity $\times 10^{-5}$ (Pa·s)	Heat capacity (J/[kg·K])	Relative molecular mass
			δλ (He)	δλ (N_2)	δλ (Ar)			
Hydrogen	0.08988	19.71	3.97	16.96	17.81	0.876 (20.7 °C) 1.086 (129.4 °C) 1.381 (299.0 °C)	14,112.7	2.016
Helium	0.17847	15.74	—	12.99	13.84	1.941 (20.0 °C) 2.281 (100.0 °C) 2.672 (200.0 °C)	5,330.6	4.003
Methane	0.71680	3.74	−12.00	0.99	1.84	1.087 (20.0 °C) 1.331 (100.0 °C) 1.605 (200.5 °C)	2,217.2	16.04
Oxygen	1.42904	2.85	−12.89	0.10	0.95	2.018 (19.1 °C) 2.568 (127.7 °C) 3.017 (227.0 °C)	915.3	32.00
Nitrogen	1.25055	2.75	−12.99	—	0.85	1.781 (27.4 °C) 2.191 (127.2 °C) 2.559 (226.7 °C)	1,030.5	28.016
Carbon monoxide	1.25040	2.67	−13.07	−0.08	0.77	1.753 (21.7 °C) 2.183 (126.7 °C) 2.548 (227.0 °C)	1,030.7	28.01
Ethane	1.35660	2.44	−13.30	−0.31	0.54	0.901 (17.2 °C) 1.143 (100.4 °C) 1.409 (200.3 °C)	1,614.0	30.07
Ethene	1.26040	2.30	−13.44	−0.45	0.40	1.008 (20.0 °C) 1.257 (100.0 °C) 1.541 (200.0 °C)	—	28.05
Propane	2.00960	2.03	−13.71	−0.72	0.13	0.795 (17.9 °C) 1.009 (100.4 °C) 1.253 (199.3 °C)	—	44.09
Argon	1.78370	1.90	−13.84	−0.85	—	2.217 (20.0 °C) 2.695 (100.0 °C) 3.223 (200.0 °C)	523.7	39.94
Carbon dioxide	1.97690	1.83	−13.91	−0.92	−0.07	1.480 (20.0 °C) 1.861 (99.1 °C) 2.221 (182.4 °C)	836.6	44.01
n-Butane	2.51900	1.82	−13.92	−0.93	−0.08	0.840 (14.7 °C)	—	58.12
Sulfur hexafluoride	6.50 (20°C)	1.63	−14.11	−1.12	−0.27	1.450 (21.1 °C)	674.0	146.05

Note: Values refer to a pressure of 101 kPa (760 torr). The reference temperatures for the quantities are: density, 0°C, thermal conductivity, 48.9°C (120°F); heat capacity, 15°C.

Reprinted from T. J. Bruno and P. D. N. Svoronos, *CRC Handbook of Basic Tables for Chemical Analysis*. CRC Press, Boca Raton, FL, 1989, pg 4.

SOLVENTS FOR ULTRAVIOLET SPECTROPHOTOMETRY

Solvent	Wavelength cutoff (nm)	Dielectric constant (20°C)	
Acetic acid	260	6.15	
Acetone	330	20.7	(25°C)
Acetonitrile	190	37.5	
Benzene	280	2.284	
2-Butanol	260	15.8	(25°C)
n-Butyl acetate	254		
Carbon disulfide	380	2.641	
Carbon tetrachloride	265	2.238	
1-Chlorobutane	220	7.39	(25°C)
Chloroform[a]	245	4.806	
Cyclohexane	210	2.023	
1,2-Dichloroethane	226	10.19	(25°C)
1,2-Dimethoxyethane	240		
N,N-Dimethylacetamide	268	59	(83°C)
N,N-Dimethylformamide	270	36.7	
Dimethylsulfoxide	265	4.7	
1,4-Dioxane	215	2.209	(25°C)
Diethyl ether	218	4.335	
Ethanol	210	24.30	(25°C)
2-Ethoxyethanol	210		
Ethyl acetate	255	6.02	(25°C)
Glycerol	207	42.5	(25°C)
n-Hexadecane	200	2.06	(25°C)
n-Hexane	210	1.890	
Methanol	210	32.63	(25°C)
2-Methoxyethanol	210	16.9	
Methyl cyclohexane	210	2.02	(25°C)
Methyl ethyl ketone	330	18.5	
Methyl isobutyl ketone	335		
2-Methyl-1-propanol	230	1	
N-Methyl-2-pyrrolidone	285	32.0	
Pentane	210	1.844	
n-Pentyl acetate	212		
1-Propanol	210	20.1	(25°C)
2-Propanol	210	18.3	(25°C)
Pyridine	330	12.3	(25°C)
Tetrachloroethylene[b]	290		
Tetrahydrofuran	220	7.6	
Toluene	286	2.379	(25°C)
1,1,2-Trichloro-1,2,2-trifluoroethane	231		
2,2,4-Trimethylpentane	215	1.936	(25°C)
o-Xylene	290	2.568	
m-Xylene	290	2.374	
p-Xylene	290	2.270	
Water		78.54	(25°C)

[a] Stabilized with ethanol to avoid phosgene formation.

[b] Stabilized with thymol (isopropyl meta-cresol).

Reprinted from T. J. Bruno and P. D. N. Svoronos, *CRC Handbook of Basic Tables for Chemical Analysis*, CRC Press, Boca Raton, FL, 1989, pg. 212.

^{13}C CHEMICAL SHIFTS OF USEFUL NMR SOLVENTS

The following table gives the expected carbon-13 chemical shift(s) for various useful NMR solvents in parts per million (ppm). In some solvents, slight changes can occur with change of concentration.[1,2]

REFERENCES

1. **Silverstein, R. M., Bassler, G. C., and Morrill, T. C.,** *Spectrometric Identification of Organic Compounds,* John Wiley & Sons, New York, 1981.
2. **Rahman, A-U.,** *Nuclear Magnetic Resonance. Basic Principles,* Springer-Verlag, New York, 1986.

^{13}C CHEMICAL SHIFTS OF USEFUL NMR SOLVENTS

Solvent	Formula	Chemical shift (ppm)
Acetone-d_6	$(CD_3)_2C{=}O$	29.2 (CD_3) 204.1 (>C=O)
Acetonitrile-d_3	$CD_3C{\equiv}N$	1.3 (CD_3) 117.1 (C≡N)
Benzene-d_6	C_6D_6	128.4
Carbon disulfide	CS_2	192.3
Carbon tetrachloride	CCl_4	96.0
Chloroform-d_3	$CDCl_3$	77.05
Cyclohexane-d_{12}	C_6D_{12}	27.5
Dichloromethane-d_2	CD_2Cl_2	53.6
Dimethylformamide-d_7	$(CD_3)_2NCOD$	31 (CD_3) 36 (CD_3) 162.4 (DC=O)
Dimethylsulfoxide-d_6	$(CD_3)_2S{=}O$	39.6
Dioxane-d_8	$C_4D_8O_2$	67.4
Methanol-d_4	CD_3OD	49.3
Nitromethane-d_3	$CD_3{-}NO_2$	57.3
Pyridine-d_5	C_5D_5N	123.9 (C–3) 135.9 (C–4) 150.2 (C–2)
1,1,2,2-tetrachloroethane-d_2	$CDCl_2\ CDCl_2$	75.5
Tetrahydrofuran-d_8	C_4D_8O	25.8 (C–2) 67.9 (C–1)
Trichlorofluoromethane	$CFCl_3$	117.6
Water (heavy)	D_2O	—

Reprinted from T. J. Bruno and P. D. N. Svoronos, *CRC Handbook of Basic Tables for Chemical Analysis*, CRC Press, Boca Raton, FL, 1989, pg. 330.

IMPORTANT PEAKS IN THE MASS SPECTRA OF COMMON SOLVENTS

The following table gives the most important peaks that appear in the mass spectra of the most common solvents which might occur as an impurity in organic samples. The solvents are classified in ascending order of their M^+ peaks. The highest intensity peaks are indicated with (100%).[1-3]

REFERENCES

1. **Clere, J. T., Pretsch, E., and Seibl, J.,** *Studies in Analytical Chemistry,* Vol. I. *Structural Analysis of Organic Compounds by Combined Application of Spectroscopic Methods,* Elsevier, Amsterdam, 1981.
2. **McLafferty, F. W.,** *Interpretation of Mass Spectra,* University Science Books, Mill Valley, CA, 1980.
3. **Pasto, D. J. and Johnson, C. R.,** *Organic Structure Determination,* Prentice-Hall, Englewood Cliffs, NJ, 1969.

IMPORTANT PEAKS IN THE MASS SPECTRA OF COMMON SOLVENTS

Solvents	Formula	M^+	Important peaks (m/e)
Water	H_2O	18 (100%)	17
Methanol	CH_3OH	32	31 (100%), 29, 15
Acetonitrile	CH_3CN	41 (100%)	40, 39, 38, 28, 15
Ethanol	CH_3CH_2OH	46	45, 31 (100%), 27, 15
Dimethyl ether	CH_3OCH_3	46 (100%)	45, 29, 15
Acetone	CH_3COCH_3	58	43 (100%), 42, 39, 27, 15
Acetic acid	CH_3CO_2H	60	45, 43, 18, 15
Ethylene glycol	$HOCH_2CH_2OH$	62	43, 33, 31 (100%), 29, 18, 15
Furan	C_4H_4O	68 (100%)	42, 39, 38, 37, 29, 18
Tetrahydrofuran	C_4H_8O	72	71, 43, 42 (100%), 41, 40, 39, 27, 18, 15
n-Pentane	C_5H_{12}	72	57, 43 (100%), 42, 41, 39, 29, 28, 27, 15
Dimethyformamide (DMF)	$HCON(CH_3)_2$	73 (100%)	58, 44, 42, 30, 29, 28, 18, 15
Diethylether	$(C_2H_5)_2O$	74	59, 45, 41, 31 (100%), 29, 27, 15
Methylacetate	$CH_3CO_2CH_3$	74	59, 43 (100%), 42, 32, 29, 28, 15
Carbon disulfide	CS_2	76 (100%)	64, 44, 38, 32
Benzene	C_6H_6	78 (100%)	77, 52, 51, 50, 39, 28
Pyridine	C_5H_5N	79 (100%)	80, 78, 53, 52, 51, 50, 39, 26
Dichloromethane	CH_2Cl_2	84	86, 51, 49 (100%), 48, 47, 35, 28
Cyclohexane	C_6H_{12}	84	69, 56, 55, 43, 42, 41, 39, 27
n-Hexane	C_6H_{14}	86	85, 71, 69, 57 (100%), 43, 42, 41, 39, 29, 28, 27
p-Dioxane	$C_4H_8O_2$	88 (100%)	87, 58, 57, 45, 43, 31, 30, 29, 28
Tetramethylsilane (TMS)	$(CH_3)_4Si$	88	74, 73, 55, 45, 43, 29
1,2-Dimethoxy ethane	$(CH_3OCH_2)_2$	90	60, 58, 45 (100%), 31, 29
Toluene	$C_6H_5CH_3$	92	91 (100%), 65, 51, 39, 28
Chloroform	$CHCl_3$	118	120, 83, 81 (100%), 47, 35, 28
Chloroform-d_1	$CDCl_3$	119	121, 84, 82 (100%), 48, 47, 35, 28
Carbon tetrachloride	CCl_4	152 (not seen)	121, 119, 117 (100%), 84, 82, 58.5, 47, 35, 28
Tetrachloroethene	$CCl_2{=}CCl_2$	164 (not seen)	168, 166 (100%), 165, 164, 131, 128, 129, 95, 94, 82, 69, 59, 47, 31, 24

Reprinted from T. J. Bruno and P. D. N. Svoronos, *CRC Handbook of Basic Tables for Chemical Analysis*, CRC Press, Boca Raton, FL, 1989, pg. 357.

Section 9
Molecular Structure and Spectroscopy

BOND LENGTHS IN CRYSTALLINE ORGANIC COMPOUNDS

The following table gives average interatomic distances for bonds between the elements H, B, C, N, O, F, Si, P, S, Cl, As, Se, Br, Te, and I as determined from X-ray and neutron diffraction measurements on organic crystals. The table has been derived from an analysis of high-precision structure data on about 10,000 crystals contained in the 1985 version of the Cambridge Structural Database, which is maintained by the Cambridge Crystallographic Data Center. The explanation of the columns is:

Column 1: Specification of elements in the bond, with coordination number given in parentheses, and bond type (single, double, etc.). For carbon, the hybridization state is given.
Column 2: Substructure in which the bond is found. The target bond is set in boldface. Where X is not specified, it denotes any element type. C# indicates any sp^3 carbon atom, and C* denotes an sp^3 carbon whose bonds, in addition to those specified in the linear formulation, are to C and H atoms only.
Column 3: d is the unweighted mean in Å units of all the values for that bond length found in the sample.
Column 4: m is the median in Å units of all values.
Column 5: σ is the standard deviation in the sample.
Column 6: q_l is the lower quartile for the sample (i.e., 25% of values are less than q_l and 75% exceed it).
Column 7: q_2 is the upper quartile for the sample.
Column 8: n is number of observations in the sample.
Column 9: Notes refer to the footnotes in Appendix 1.

References to special cases are given in a shorthand form and listed in Appendix 2. Further information on the method of analysis of the data may be found in the reference cited below.

The table is reprinted with permission of the authors, the Royal Society of Chemistry, and the International Union of Crystallography.

REFERENCE

Frank H. Allen, Olga Kennard, David G. Watson, Lee Brammer, A. Guy Orpen, and Robin Taylor, *J. Chem. Soc. Perkin Trans. II*, S1—S19, 1987.

Bond	Substructure	d	m	σ	q_1	q_u	n	Note
As(3)–As(3)	X_2–**As–As**–X_2	2.459	2.457	0.011	2.456	2.466	8	
As–B	see CUDLOC (2.065), CUDLUI (2.041)							
As–Br	see CODDEE, CODDII (2.346—3.203)							
As(4)–C	X_3–**As–C**H_3	1.903	1.907	0.016	1.893	1.916	12	
	$(X)_2$(C,O,S=)**As–C***sp*^{3}	1.927	1.929	0.017	1.921	1.937	16	
	As–C*ar* in Ph_4As^+	1.905	1.909	0.012	1.897	1.912	108	
	$(X)_2$(C,O,S=)**As–C***ar*	1.922	1.927	0.016	1.908	1.934	36	
As(3)–C	X_2–**As–C***sp*^{3}	1.963	1.965	0.017	1.948	1.978	6	
	X_2–**As–C***ar*	1.956	1.956	0.015	1.944	1.964	41	
As(3)–Cl	X_2–**As–Cl**	2.268	2.256	0.039	2.247	2.281	10	
As(6)–F	in **As**F_6^-	1.678	1.676	0.020	1.659	1.695	36	
As(3)–I	see OPIMAS (2.579, 2.590)							
As(3)–N(3)	X_2–**As–N**–X_2	1.858	1.858	0.029	1.839	1.873	19	
As(4)=N(2)	see TPASSN (1.837)							
As(4)–O	$(X)_2$(O=)**As–O**H	1.710	1.712	0.017	1.695	1.726	6	
As(3)–O	see ASAZOC, PHASOC01 (1.787—1.845)							
As(4)=O	X_3–**As=O**	1.661	1.661	0.016	1.652	1.667	9	
As(3)=P(3)	see BELNIP (2.350, 2.362)							†
As(3)–P(3)	see BUTHAZ10 (2.124)							†
As(3)–S	X_2–**As–S**	2.275	2.266	0.032	2.247	2.298	14	
As(4)=S	X_3–**As=S**	2.083	2.082	0.004	2.080	2.086	9	
As(3)–Se(2)	see COSDIX, ESEARS (2.355—2.401)							†
As(3)–Si(4)	see BICGEZ, MESIAD (2.351—2.365)							†
As(3)–Te(2)	see ETEARS (2.571, 2.576)							†
B(n)–B(n)	n = 5—7 in boron cages	1.775	1.773	0.031	1.763	1.786	688	
B(4)–B(4)	see CETTAW (2.041)							
B(4)–B(3)	see COFVOI (1.698)							
B(3)–B(3)	X_2–**B–B**–X_2	1.701	1.700	0.014	1.691	1.712	8	
B(6)–Br		1.967	1.971	0.014	1.954	1.979	7	†
B(4)–Br		2.017	2.008	0.031	1.990	2.044	15	†
B(n)–C	n = 5—7: **B–C** in cages	1.716	1.717	0.020	1.707	1.728	96	
	n = 3—4: **B–C***sp*^{3} not cages	1.597	1.599	0.022	1.585	1.611	29	1
	n = 4: **B–C***ar*	1.606	1.607	0.012	1.596	1.615	41	
	n = 4: **B–C***ar* in Ph_4B^-	1.643	1.643	0.006	1.641	1.645	16	
B(n)–C	n = 3: **B–C***ar*	1.556	1.552	0.015	1.546	1.566	24	
B(n)–Cl	**B(5)–Cl** and **B(3)–Cl**	1.751	1.751	0.011	1.743	1.761	14	
	B(4)–Cl	1.833	1.833	0.013	1.821	1.843	22	
B(4)–F	**B–F** (B neutral)	1.366	1.368	0.017	1.356	1.375	25	
	B^-–**F** in BF_4^-	1.365	1.372	0.029	1.352	1.390	84	
B(4)–I	see TMPBTI (2.220, 2.253)							
B(4)–N(3)	X_3–**B–N**(=C)(X)	1.611	1.617	0.013	1.601	1.625	8	
	in pyrazaboles	1.549	1.552	0.015	1.536	1.560	10	

Bond	Substructure	d	m	σ	q_1	q_u	n	Note
B(3)–N(3)	X_2–**B**–N–C_2: all coplanar	1.404	1.404	0.014	1.389	1.408	40	2
	for τ(BN) > 30° see BOGSUL, BUSHAY, CILRUK (1.434—1.530)							
	S_2–**B**–N–X_2	1.447	1.443	0.013	1.435	1.470	14	
B(4)–O	**B^-–O** in BO_4^-	1.468	1.468	0.022	1.453	1.479	24	
	for neutral B–O see Note 3							3
B(3)–O(2)	X_2–**B–O**–X	1.367	1.367	0.024	1.349	1.382	35	
B(*n*)–P	n = 4: **B–P**	1.922	1.927	0.027	1.900	1.954	10	
	n = 3: see BUPSIB10 (1.892, 1.893)							
B(4)–S	**B**(4)–**S**(3)	1.930	1.927	0.009	1.925	1.934	10	
	B(4)–**S**(2)	1.896	1.896	0.004	1.893	1.899	6	
B(3)–S	N–**B–S**$_2$	1.806	1.806	0.010	1.799	1.816	28	
	(=X–)(N–)**B–S**	1.851	1.854	0.013	1.842	1.859	10	
Br–Br	see BEPZEB, TPASTB	2.542	2.548	0.015	2.526	2.551	4	
Br–C	**Br–C***	1.966	1.967	0.029	1.951	1.983	100	4
	Br–C*sp*^{3} (cyclopropane)	1.910	1.910	0.010	1.900	1.914	8	
	Br–C*sp*^{2}	1.883	1.881	0.015	1.874	1.894	31	4
	Br–C*ar* (mono-Br + *m,p*-Br_2)	1.899	1.899	0.012	1.892	1.906	119	4
	Br–C*ar* (*o*-Br_2)	1.875	1.872	0.011	1.864	1.884	8	4
$^-$Br(2)–Cl	see TEACBR (2.362—2.402)							†
Br–I	see DTHIBR10 (2.646), TPHOSI (2.695)							
Br–N	see NBBZAM (1.843)							
Br–O	see CIYFOF	1.581	1.581	0.007	1.574	1.587	4	
Br–P	see CISTED (2.366)							
Br–S(2)	see BEMLIO (2.206)							†
Br–S(3)	see CIWYIQ (2.435, 2.453)							†
Br–S(3)$^+$	see THINBR (2.321)							†
Br–Se	see CIFZUM (2.508, 2.619)							
Br–Si	see BIZJAV (2.284)							
Br–Te	In **Br_6Te^{2-}** see CUGBAH (2.692—2.716)							
	Br–Te(4) see BETUTE10 (3.079, 3.015)							
	Br–Te(3) see BTUPTE (2.835)							
C*sp*^{3}–C*sp*^{3}	C#–**CH$_2$–CH$_3$**	1.513	1.514	0.014	1.507	1.523	192	
	(C#)$_2$–**CH–CH$_3$**	1.524	1.526	0.015	1.518	1.534	226	
	(C#)$_3$–**C–CH$_3$**	1.534	1.534	0.011	1.527	1.541	825	
	C#–**CH$_2$–CH$_2$**–C#	1.524	1.524	0.014	1.516	1.532	2 459	
	(C#)$_2$–**CH–CH$_2$**–C#	1.531	1.531	0.012	1.524	1.538	1 217	
	(C#)$_3$–**C–CH$_2$**–C#	1.538	1.539	0.010	1.533	1.544	330	
	(C#)$_2$–**CH–CH**–(C#)$_2$	1.542	1.542	0.011	1.536	1.549	321	
	(C#)$_3$–**C–CH**–(C#)$_2$	1.556	1.556	0.011	1.549	1.562	215	
	(C#)$_3$–**C–C**–(C#)$_3$	1.588	1.580	0.025	1.566	1.610	21	
	C*–C* (overall)	1.530	1.530	0.015	1.521	1.539	5 777	5,6
	in cyclopropane (any subst.)	1.510	1.509	0.026	1.497	1.523	888	7
	in cyclobutane (any subst.)	1.554	1.553	0.021	1.540	1.567	679	8
	in cyclopentane (C,H-subst.)	1.543	1.543	0.018	1.532	1.554	1 641	
	in cyclohexane (C,H-subst.)	1.535	1.535	0.016	1.525	1.545	2 814	
	cyclopropyl-C* (exocyclic)	1.518	1.518	0.019	1.505	1.531	366	7
	cyclobutyl-C* (exocyclic)	1.529	1.529	0.016	1.519	1.539	376	8
	cyclopentyl-C* (exocyclic)	1.540	1.541	0.017	1.527	1.549	956	
	cyclohexyl-C* (exocyclic)	1.539	1.538	0.016	1.529	1.549	2 682	
	in cyclobutene (any subst.)	1.573	1.574	0.017	1.566	1.586	25	8
	in cyclopentene (C,H-subst.)	1.541	1.539	0.015	1.532	1.549	208	
	in cyclohexene (C,H-subst.)	1.541	1.541	0.020	1.528	1.554	586	
	in oxirane (epoxide)	1.466	1.466	0.015	1.458	1.474	249	9
	in aziridine	1.480	1.481	0.021	1.465	1.496	67	9
	in oxetane	1.541	1.541	0.019	1.527	1.557	16	
	in azetidine	1.548	1.543	0.018	1.536	1.558	22	
	oxiranyl-C* (exocyclic)	1.509	1.507	0.018	1.497	1.519	333	9
	aziridinyl-C* (exocyclic)	1.512	1.512	0.018	1.496	1.526	13	9
C*sp*^{3}–C*sp*^{2}	**CH$_3$–C**=C	1.503	1.504	0.011	1.497	1.509	215	
	C#–**CH$_2$–C**=C	1.502	1.502	0.013	1.494	1.510	483	
	(C#)$_2$–**CH–C**=C	1.510	1.510	0.014	1.501	1.518	564	
	(C#)$_3$–**C–C**=C	1.522	1.522	0.016	1.511	1.533	193	
C*sp*^{3}–C*sp*^{2}	**C*–C**=C (overall)	1.507	1.507	0.015	1.499	1.517	1 456	5
	C*–C=C (endocyclic)							
	in cyclopropene	1.509	1.508	0.016	1.500	1.516	20	10
	in cyclobutene	1.513	1.512	0.018	1.500	1.525	50	8
	in cyclopentene	1.512	1.512	0.014	1.502	1.521	208	
	in cyclohexene	1.506	1.505	0.016	1.495	1.516	391	
	in cyclopentadiene	1.502	1.503	0.019	1.490	1.515	18	
	in cyclohexa-1,3-diene	1.504	1.504	0.017	1.491	1.517	56	
	C*–C=C (exocyclic):							
	cyclopropenyl-C*	1.478	1.475	0.012	1.470	1.485	7	10
	cyclobutenyl-C*	1.489	1.483	0.015	1.479	1.496	11	8

Bond	Substructure	*d*	*m*	σ	q_l	q_u	*n*	Note
	cyclopentenyl-C*	1.504	1.506	0.012	1.495	1.512	115	
	cyclohexenyl-C*	1.511	1.511	0.013	1.502	1.519	292	
	C*–**C**H=O in aldehydes	1.510	1.510	0.008	1.501	1.518	7	
	(**C***)$_2$–**C**=O							
	in ketones	1.511	1.511	0.015	1.501	1.521	952	11
	in cyclobutanone	1.529	1.530	0.016	1.514	1.545	18	
	in cyclopentanone	1.514	1.514	0.016	1.505	1.523	312	
	acyclic and 6+ rings	1.509	1.509	0.016	1.499	1.519	626	
	C*–**C**OOH in carboxylic acids	1.502	1.502	0.014	1.495	1.510	176	
	C*–**C**OO$^-$ in carboxylate anions	1.520	1.521	0.011	1.516	1.528	57	
	C*–**C**(=O)(–OC*)							
	in acyclic esters	1.497	1.496	0.018	1.484	1.509	553	12
	in β-lactones	1.519	1.519	0.020	1.500	1.538	4	13
	in γ-lactones	1.512	1.512	0.015	1.501	1.521	110	12
	in δ-lactones	1.504	1.502	0.013	1.495	1.517	27	12
	cyclopropyl (**C**)–**C**=O in ketones, acids							
	and esters	1.486	1.485	0.018	1.474	1.497	105	7
	C*–**C**(=O)(–NH$_2$) in acyclic amides	1.514	1.512	0.016	1.506	1.526	32	14
	C*–**C**(=O)(–NHC*) in acyclic amides	1.506	1.505	0.012	1.498	1.515	78	14
	C*–**C**(=O)[–N(C*)$_2$] in acyclic amides	1.505	1.505	0.011	1.496	1.517	15	14
Csp^3–C*ar*	**C**H$_3$–**C***ar*	1.506	1.507	0.011	1.501	1.513	454	
	C#–**C**H$_2$–**C***ar*	1.510	1.510	0.009	1.505	1.516	674	
	(C#)$_2$–**C**H–**C***ar*	1.515	1.515	0.011	1.508	1.522	363	
	(C#)$_3$–**C**–**C***ar*	1.527	1.530	0.016	1.517	1.539	308	
	C*–**C***ar* (overall)	1.513	1.513	0.014	1.505	1.521	1 813	
	cyclopropyl (**C**)–**C***ar*	1.490	1.490	0.015	1.479	1.503	90	7
Csp^3–Csp^1	**C***–**C**≡C	1.466	1.465	0.010	1.460	1.469	21	15
	C#–**C**≡C	1.472	1.472	0.012	1.464	1.481	88	15
	C*–**C**≡N	1.470	1.469	0.013	1.463	1.479	106	7b
	cyclopropyl (**C**)–**C**≡N	1.444	1.447	0.010	1.436	1.451	38	7
Csp^2–Csp^2	C=**C**–**C**=C							
	(conjugated)	1.455	1.455	0.011	1.447	1.463	30	16,18
	(unconjugated)	1.478	1.476	0.012	1.470	1.479	8	17,18
	(overall)	1.460	1.460	0.015	1.450	1.470	38	
	C=C–**C**=**C**–C=C	1.443	1.445	0.013	1.431	1.454	29	18
	C=**C**–**C**=C (endocyclic in TCNQ)	1.432	1.433	0.012	1.424	1.441	280	19
	C=**C**–**C**(=O)(–C*)							
	(conjugated)	1.464	1.462	0.018	1.453	1.476	211	16,18
	(unconjugated)	1.484	1.486	0.017	1.475	1.497	14	17,18
	(overall)	1.465	1.462	0.018	1.453	1.478	226	
	C=**C**–**C**(=O)–C=C							
	in benzoquinone (C,H-subst. only)	1.478	1.476	0.011	1.469	1.488	28	
	in benzoquinone (any subst.)	1.478	1.478	0.031	1.464	1.498	172	
	non-quinonoid	1.456	1.455	0.012	1.447	1.464	28	
	C=**C**–**C**OOH	1.475	1.476	0.015	1.461	1.488	22	
	C=**C**–**C**OOC*	1.488	1.489	0.014	1.478	1.497	113	
	C=**C**–**C**OO$^-$	1.502	1.499	0.017	1.488	1.510	11	
	HOO**C**–**C**OOH	1.538	1.537	0.007	1.535	1.541	9	
	HOO**C**–**C**OO$^-$	1.549	1.552	0.009	1.546	1.553	13	
	$^-$OO**C**–**C**OO$^-$	1.564	1.559	0.022	1.554	1.568	9	
	formal Csp^2–Csp^2 single bond in selected							
	non-fused heterocycles:							
	in 1*H*-pyrrole (C3–C4)	1.412	1.410	0.016	1.401	1.427	29	
	in furan (C3–C4)	1.423	1.423	0.016	1.412	1.433	62	
	in thiophene (C3–C4)	1.424	1.425	0.015	1.415	1.433	40	
	in pyrazole (C3–C4)	1.410	1.412	0.016	1.400	1.418	20	
	in isoxazole (C3–C4)	1.425	1.425	0.016	1.413	1.438	9	
	in furazan (C3–C4)	1.428	1.427	0.007	1.422	1.435	6	
	in furoxan (C3–C4)	1.417	1.417	0.006	1.412	1.422	14	
Csp^2–C*ar*	C=**C**–**C***ar*							
	(conjugated)	1.470	1.470	0.015	1.463	1.480	37	16,18
Csp^2–C*ar*		1.488	1.490	0.012	1.480	1.496	87	17,18
	(overall)	1.483	1.483	0.015	1.472	1.494	124	
	cyclopropenyl (C=**C**)–**C***ar*	1.447	1.448	0.006	1.441	1.452	8	10
	C*ar*–**C**(=O)–C*	1.488	1.489	0.016	1.478	1.500	84	
	C*ar*–**C**(=O)–C*ar*	1.480	1.481	0.017	1.468	1.494	58	
	C*ar*–**C**OOH	1.484	1.485	0.014	1.474	1.491	75	
	C*ar*–**C**(=O)(–OC*)	1.487	1.487	0.012	1.480	1.494	218	
	C*ar*–**C**OO$^-$	1.504	1.509	0.014	1.495	1.512	26	
	C*ar*–**C**(=O)–NH$_2$	1.500	1.503	0.020	1.498	1.510	19	
	C*ar*–**C**=N–C#							
	(conjugated)	1.476	1.478	0.014	1.466	1.486	27	16
	(unconjugated)	1.491	1.490	0.008	1.485	1.496	48	17
	(overall)	1.485	1.487	0.013	1.481	1.493	75	

BOND LENGTHS IN CRYSTALLINE ORGANIC COMPOUNDS (continued)

Bond	Substructure	d	m	σ	q_l	q_u	n	Note
	in indole (C3–C3a)	1.434	1.434	0.011	1.428	1.439	40	
Csp^2–Csp^1	C=C–C≡C	1.431	1.427	0.014	1.425	1.441	11	7b
	C=C–C≡N in TCNQ	1.427	1.427	0.010	1.420	1.433	280	19
Car–*Car*	in biphenyls (*ortho* subst. all H)	1.487	1.488	0.007	1.484	1.493	30	
	(⩾1 non-H *ortho*-subst.)	1.490	1.491	0.010	1.486	1.495	212	
Car–Csp^1	*Car*–C≡C	1.434	1.436	0.006	1.430	1.437	37	
	Car–C≡N	1.443	1.444	0.008	1.436	1.448	31	
Csp^1–Csp^1	C≡C–C≡C	1.377	1.378	0.012	1.374	1.384	21	
Csp^2=Csp^2	C*–CH=CH_2	1.299	1.300	0.027	1.280	1.311	42	
	$(C^*)_2$–C=CH_2	1.321	1.321	0.013	1.313	1.328	77	
	C*–CH=CH–C*							
	(*cis*)	1.317	1.318	0.013	1.310	1.323	106	
	(*trans*)	1.312	1.311	0.011	1.304	1.320	19	
	(overall)	1.316	1.317	0.015	1.309	1.323	127	
	$(C^*)_2$–C=CH–C*	1.326	1.328	0.011	1.319	1.334	168	
	$(C^*)_2$–C=C–$(C^*)_2$	1.331	1.330	0.009	1.326	1.334	89	
	$(C^*,H)_2$–C=C–$(C^*,H)_2$ (overall)	1.322	1.323	0.014	1.315	1.331	493	5
	in cyclopropene (any subst.)	1.294	1.288	0.017	1.284	1.302	10	10
	in cyclobutene (any subst.)	1.335	1.335	0.019	1.324	1.347	25	8
	in cyclopentene (C,H-subst.)	1.323	1.324	0.013	1.314	1.331	104	
	in cyclohexene (C,H-subst.)	1.326	1.325	0.012	1.318	1.334	196	
	C=C=C (allenes, any subst.)	1.307	1.307	0.005	1.303	1.310	18	
	C=C–C=C (C,H subst., conjugated)	1.330	1.330	0.014	1.322	1.338	76	16
	C=C–C=C–C=C (C,H subst., conjugated)	1.345	1.345	0.012	1.337	1.350	58	16
	C=C-*Car* (C,H subst., conjugated)	1.339	1.340	0.011	1.334	1.346	124	16
	C=C in cyclopenta-1,3-diene (any subst.)	1.341	1.341	0.017	1.328	1.356	18	
	C=C in cyclohexa-1,3-diene (any subst.)	1.332	1.332	0.013	1.323	1.341	56	
	in C=C–C=O							
	(C,H subst., conjugated)	1.340	1.340	0.013	1.332	1.348	211	16,18
	(C,H subst., unconjugated)	1.331	1.330	0.008	1.326	1.339	14	17,18
	(C,H subst., overall)	1.340	1.339	0.013	1.332	1.348	226	
	in cyclohexa-2,5-dien-1-ones	1.329	1.327	0.011	1.321	1.335	28	
	in *p*-benzoquinones							
	(C*,H subst.)	1.333	1.337	0.011	1.325	1.338	14	
	(any subst.)	1.349	1.339	0.030	1.330	1.364	86	
	in TCNQ							
	(endocyclic)	1.352	1.353	0.010	1.345	1.358	142	19
	(exocyclic)	1.392	1.391	0.017	1.379	1.405	139	19
	C=C–OH in enol tautomers	1.362	1.360	0.020	1.349	1.370	54	
	in heterocycles (any subst.):							
	1*H*-pyrrole (C2–C3, C4–C5)	1.375	1.377	0.018	1.361	1.388	58	
	furan (C2–C3, C4–C5)	1.341	1.342	0.021	1.329	1.351	125	
	thiophene (C2–C3, C4–C5)	1.362	1.359	0.025	1.346	1.377	60	
	pyrazole (C4–C5)	1.369	1.372	0.019	1.362	1.383	20	
	imidazole (C4–C5)	1.360	1.361	0.014	1.352	1.367	44	
	isoxazole (C4–C5)	1.341	1.336	0.012	1.331	1.355	9	
	indole (C2–C3)	1.364	1.363	0.012	1.355	1.371	40	
Car≃*Car*	in phenyl rings with C*,H subst. only							
	H–C≃C–H	1.380	1.381	0.013	1.372	1.388	2 191	
	C*–C≃C–H	1.387	1.388	0.010	1.382	1.393	891	
	C*–C≃C–C*	1.397	1.397	0.009	1.392	1.403	182	
	C≃C (overall)	1.384	1.384	0.013	1.375	1.391	3 264	
	F–C≃C–F	1.372	1.374	0.011	1.366	1.380	84	4
	Cl–C≃C–Cl	1.388	1.389	0.014	1.380	1.398	152	4
	in naphthalene (D_{2h}, any subst.)							
	C1–C2	1.364	1.364	0.014	1.356	1.373	440	
	C2–C3	1.406	1.406	0.014	1.397	1.415	218	
	C1–C8a	1.420	1.419	0.012	1.412	1.426	440	
	C4a–C8a	1.422	1.424	0.011	1.417	1.429	109	
Car≃*Car*	in anthracene (D_{2h}, any subst.)							
	C1–C2	1.356	1.356	0.009	1.350	1.360	56	
	C2–C3	1.410	1.410	0.010	1.401	1.416	34	
	C1–C9a	1.430	1.430	0.006	1.426	1.434	56	
	C4a–C9a	1.435	1.436	0.007	1.429	1.440	34	
	C9–C9a	1.400	1.402	0.009	1.395	1.406	68	
	in pyridine (C,H subst.)	1.379	1.381	0.012	1.371	1.387	276	20
	(any subst.)	1.380	1.380	0.015	1.371	1.389	537	20
	in pyridinium cation							
	(N^+–H; C,H subst. on C)							
	C2–C3	1.373	1.375	0.012	1.368	1.380	30	
	C3–C4	1.379	1.380	0.011	1.371	1.388	30	
	(N^+–X; C,H subst. on C)							
	C2–C3	1.373	1.372	0.019	1.362	1.382	151	
	C3–C4	1.383	1.385	0.019	1.372	1.394	151	

BOND LENGTHS IN CRYSTALLINE ORGANIC COMPOUNDS (continued)

Bond	Substructure	d	m	σ	q_1	q_u	n	Note
	in pyrazine (H subst. on C)	1.379	1.377	0.010	1.370	1.388	10	
	(any subst. on C)	1.405	1.405	0.024	1.388	1.420	60	
	in pyrimidine (C,H subst. on C)	1.387	1.389	0.018	1.379	1.400	28	
Csp^1≡Csp^1	X–C≡C–X	1.183	1.183	0.014	1.174	1.193	119	15
	C,H–C≡C–C,H	1.181	1.181	0.014	1.173	1.192	104	15
	in C≡C–C(sp^2,ar)	1.189	1.193	0.010	1.181	1.195	38	15
	in C≡C–C≡C	1.192	1.192	0.010	1.187	1.197	42	15
	in CH≡C–C#	1.174	1.174	0.011	1.167	1.180	42	15
Csp^3–Cl	Omitting 1,2-dichlorides:							
	C–CH_2–Cl	1.790	1.790	0.007	1.783	1.795	13	4
	C_2–CH–Cl	1.803	1.802	0.003	1.800	1.807	8	4
	C_3–C–Cl	1.849	1.856	0.011	1.837	1.858	5	4
	X–CH_2–Cl (X = C,H,N,O)	1.790	1.791	0.011	1.783	1.797	37	4
	X_2–CH–Cl (X = C,H,N,O)	1.805	1.803	0.014	1.800	1.812	26	4
	X_3–C–Cl (X = C,H,N,O)	1.843	1.838	0.014	1.835	1.858	7	4
	X_2–C–Cl_2 (X = C,H,N,O)	1.779	1.776	0.015	1.769	1.790	18	4
	X–C–Cl_3 (X = C,H,N,O)	1.768	1.765	0.011	1.761	1.776	33	4
	Cl–CH(–C)–CH(–C)–Cl	1.793	1.793	0.013	1.786	1.800	66	4
	Cl–C(–C_2)–C(–C_2)–Cl	1.762	1.760	0.010	1.757	1.765	54	4
	cyclopropyl–Cl	1.755	1.756	0.011	1.749	1.763	64	
Csp^2–Cl	C=C–Cl (C,H,N,O subst. on C)	1.734	1.729	0.019	1.719	1.748	63	4
	C=C–Cl_2 (C,H,N,O subst. on C)	1.720	1.716	0.013	1.708	1.729	20	4
	Cl–C=C–Cl	1.713	1.711	0.011	1.705	1.720	80	4
Car–Cl	Car–Cl (mono-Cl + m,p-Cl_2)	1.739	1.741	0.010	1.734	1.745	340	4
	Car–Cl (o-Cl_2)	1.720	1.720	0.010	1.713	1.717	364	4
Csp^1–Cl	see HCLENE10 (1.634, 1.646)							
Csp^3–F	Omitting 1,2-difluorides							
	C-CH_2–F and C_2-CH–F	1.399	1.399	0.017	1.389	1.408	25	4
	C_3–C–F	1.428	1.431	0.009	1.421	1.435	11	4
	(C*,H)$_2$-C–F_2	1.349	1.347	0.012	1.342	1.356	58	4
	C*–C–F_3	1.336	1.334	0.007	1.330	1.344	12	4
	F–C*–C*–F	1.371	1.374	0.007	1.362	1.375	26	4
	X_3–C–F (X = C,H,N,O)	1.386	1.389	0.033	1.373	1.408	70	4
	X_2–C–F_2 (X = C,H,N,O)	1.351	1.349	0.013	1.342	1.356	58	4
	X–C–F_3 (X = C,H,N,O)	1.322	1.323	0.015	1.314	1.332	309	4
	F–C(–X)$_2$–C(–X)$_2$–F (X = C,H,N,O)	1.373	1.374	0.009	1.362	1.377	30	4
	F–C(–X)$_2$–NO_2 (X = any subst.)	1.320	1.319	0.009	1.312	1.327	18	
Csp^2–F	C=C–F (C,H,N,O subst. on C)	1.340	1.340	0.013	1.334	1.346	34	4
Car–F	Car–F (mono-F + m,p-F_2)	1.363	1.362	0.008	1.357	1.368	38	4
	Car–F (o-F_2)	1.340	1.340	0.009	1.336	1.344	167	4
Csp^3–H	C–C–H_3 (methyl)	1.059	1.061	0.030	1.039	1.083	83	21
	C_2–C–H_2 (primary)	1.092	1.095	0.013	1.088	1.099	100	21
	C_3–C–H (secondary)	1.099	1.097	0.004	1.095	1.103	14	21
	$C_{2,3}$–C–H (primary and secondary)	1.093	1.095	0.012	1.089	1.100	118	21
	X–C–H_3 (methyl)	1.066	1.074	0.028	1.049	1.087	160	21
	X_2–C–H_2 (primary)	1.092	1.095	0.012	1.088	1.099	230	21
	X_3–C–H (secondary)	1.099	1.099	0.007	1.095	1.103	117	21
	$X_{2,3}$–C–H (primary and secondary)	1.094	1.096	0.011	1.091	1.100	348	21
Csp^2–H	C–C=C–H	1.077	1.079	0.012	1.074	1.085	14	21
Car–H	Car–H	1.083	1.083	0.011	1.080	1.087	218	21
Csp^3–I	C*–I	2.162	2.159	0.015	2.149	2.179	15	4
Car–I	Car–I	2.095	2.095	0.015	2.089	2.104	51	4
Csp^3–N(4)	C*–NH_3^+	1.488	1.488	0.013	1.482	1.495	298	
	(C*)$_2$–NH_2^+	1.494	1.493	0.016	1.484	1.503	249	
	(C*)$_3$–NH^+	1.502	1.502	0.015	1.491	1.512	509	
	(C*)$_4$–N^+	1.510	1.509	0.020	1.496	1.523	319	
	C*–N^+ (overall)	1.499	1.498	0.018	1.488	1.510	1 370	
Csp^3–N(3)	C*–N^+ in N-subst. pyridinium	1.485	1.484	0.009	1.477	1.490	32	
	C*–NH_2 (Nsp^3: pyramidal)	1.469	1.470	0.010	1.462	1.474	19	22
	(C*)$_2$–NH (Nsp^3: pyramidal)	1.469	1.467	0.012	1.461	1.477	152	5,22
	(C*)$_3$–N (Nsp^3: pyramidal)	1.469	1.468	0.014	1.460	1.476	1 042	5,22
	C*–Nsp^3 (overall)	1.469	1.468	0.014	1.460	1.476	1 201	
	Csp^3–Nsp^3							
	in aziridine	1.472	1.471	0.016	1.464	1.482	134	
	in azetidine	1.484	1.481	0.018	1.472	1.495	21	
	in tetrahydropyrrole	1.475	1.473	0.016	1.464	1.483	66	
	in piperidine	1.473	1.473	0.013	1.460	1.479	240	
	Csp^3–Nsp^2 (N planar) in:							23
	acyclic amides C*–NH–C=O	1.454	1.451	0.011	1.446	1.461	78	14
	β-lactams C*–N(–X)–C=O (endo)	1.464	1.465	0.012	1.458	1.475	23	13
	γ-lactams							
	C*–NH–C=O (endo)	1.457	1.458	0.011	1.449	1.465	20	13
	C*–N(–C*)–C=O (endo)	1.462	1.461	0.010	1.453	1.466	15	13
	C*–N(–C*)–C=O (exo)	1.458	1.456	0.014	1.448	1.465	15	13

Bond	Substructure	d	m	σ	q_1	q_u	n	Note
	δ-lactams							
	C*–NH–C=O (endo)	1.478	1.472	0.016	1.467	1.491	6	14
	C*–N(–C*)–C=O (endo)	1.479	1.476	0.007	1.475	1.482	15	14
	C*–N(–C*)–C=O (exo)	1.468	1.471	0.009	1.462	1.477	15	14
	nitro compounds (1,2-dinitro omitted):							
	C–CH_2–NO_2	1.485	1.483	0.020	1.478	1.502	8	
	C_2–CH–NO_2	1.509	1.509	0.011	1.502	1.511	12	
	C_3–C–NO_2	1.533	1.533	0.013	1.530	1.539	17	
	C_2–C–$(NO_2)_2$	1.537	1.536	0.016	1.525	1.550	19	
	1,2-dinitro: NO_2–C*–C*–NO_2	1.552	1.550	0.023	1.536	1.572	32	
Csp^3–N(2)	C#–N=N	1.493	1.493	0.020	1.477	1.506	54	
	C*–N=C–*Car*	1.465	1.468	0.011	1.461	1.472	75	
Csp^2–N(3)	C=C–NH_2 Nsp^2 planar	1.336	1.344	0.017	1.317	1.348	10	23
	C=C–NH–C# Nsp^2 planar	1.339	1.340	0.016	1.327	1.351	17	23
	C=C–N–$(C\#)_2$							
	Nsp^2 planar	1.355	1.358	0.014	1.341	1.363	22	23
	Nsp^3 pyramidal	1.416	1.418	0.018	1.397	1.432	18	22
	Csp^2–Nsp^2 (N planar) in:							23
	acyclic amides							
	NH_2–C=O	1.325	1.323	0.009	1.318	1.331	32	14
	C*–NH–C=O	1.334	1.333	0.011	1.326	1.343	78	14
	$(C^*)_2$–N–C=O	1.346	1.342	0.011	1.339	1.356	5	14
	β-lactams C*–NH–C=O	1,385	1.388	0.019	1.374	1.396	23	13
	γ-lactams							
	C*–NH–C=O	1.331	1.331	0.011	1.326	1.337	20	13
	C*–N(–C*)–C=O	1.347	1.344	0.014	1.335	1.359	15	13
	δ-lactams							
	C*–NH–C=O	1.334	1.334	0.006	1.330	1.339	6	14
	C*–N(–C*)–C=O	1.352	1.353	0.010	1.344	1.356	15	14
	peptides C#–N(–X)–C(–C#)(=O)	1.333	1.334	0.013	1.326	1.340	380	24
	ureas							
	$(NH_2)_2$–C=O	1.334	1.334	0.008	1.329	1.339	48	25,26
	$(C\#–NH)_2$–C=O	1.347	1.345	0.010	1.341	1.354	26	25
	$[(C\#)_n–N]_2$–C=O	1.363	1.359	0.014	1.354	1.370	40	25,27
	thioureas	1.346	1.343	0.023	1.328	1.361	192	
	$(X_2N)_2$–C=S							
	imides							
	$[C\#–C(=O)]_2$–NH	1.376	1.377	0.012	1.369	1.383	64	
	$[C\#–C(=O)]_2$–N–C#	1.389	1.383	0.017	1.376	1.404	38	
	$[Csp^2–C(=O)]_2$–N–C#	1.396	1.396	0.010	1.389	1.403	46	
	$[Csp^2–C(=O)]_2$–N–Csp^2	1.409	1.406	0.020	1.391	1.419	28	
	guanidinium $[C–(NH_2)_3]^+$ (unsubst.)	1.321	1.320	0.008	1.314	1.327	39	
	(any subst.)	1.328	1.325	0.015	1.317	1.333	140	
	in heterocyclic systems (any subst.)							
	1*H*-pyrrole (N1–C2, N1–C5)	1.372	1.374	0.016	1.363	1.384	58	
	indole (N1–C2)	1.370	1.370	0.012	1.364	1.377	40	
	pyrazole (N1–C5)	1.357	1.359	0.012	1.347	1.365	20	
	imidazole (N1–C2)	1.349	1.349	0.018	1.338	1.358	44	
	imidazole (N1–C5)	1.370	1.370	0.010	1.365	1.377	44	
Csp^2–N(2)	in imidazole (N3–C4)	1.376	1.377	0.011	1.369	1.384	44	
Car–N(4)	*Car*–N^+–$(C,H)_3$	1.465	1.466	0.007	1.461	1.470	23	
Car–N(3)	*Car*–NH_2							
	(Nsp^2: planar)	1.355	1.360	0.020	1.340	1.372	33	23
	(Nsp^3: pyramidal)	1.394	1.396	0.011	1.385	1.403	25	22
	(overall)	1.375	1.377	0.025	1.363	1.394	98	28
Car–N(3)	*Car*–NH–C#							
	(Nsp^2: planar)	1.353	1.353	0.007	1.347	1.359	16	23
	(Nsp^3: pyramidal)	1.419	1.423	0.017	1.412	1.432	8	22
	(overall)	1.380	1.364	0.032	1.353	1.412	31	28
	Car–N–$(C\#)_2$							
	(Nsp^2: planar)	1.371	1.370	0.016	1.363	1.382	41	23
	(Nsp^3: pyramidal)	1.426	1.425	0.011	1.421	1.431	22	22
	(overall)	1.390	1.385	0.030	1.366	1.420	69	28
	in indole (N1–C7a)	1.372	1.372	0.007	1.367	1.376	40	
	Car–NO_2	1.468	1.469	0.014	1.460	1.476	556	
Car–N(2)	*Car*–N=N	1.431	1.435	0.020	1.422	1.442	26	
Csp^2=N(3)	in furoxan ($^+$N2=C3)	1.316	1.316	0.009	1.311	1.324	14	
Csp^2=N(2)	*Car*–C=N–C#	1.279	1.279	0.008	1.275	1.285	75	
	$(C,H)_2$–C=N–OH in oximes	1.281	1.280	0.013	1.273	1.288	67	
	S–C=N–X	1.302	1.302	0.021	1.285	1.319	36	
	in pyrazole (N2=C3)	1.329	1.331	0.014	1.315	1.339	20	
	in imidazole (C2=N3)	1.313	1.314	0.011	1.307	1.319	44	
	in isoxazole (N2=C3)	1.314	1.315	0.009	1.305	1.320	9	

Bond	Substructure	d	m	σ	q_1	q_u	n	Note
	in furazan (N2=C3, C4=N5)	1.298	1.299	0.006	1.294	1.303	12	
	in furoxan (C4=N5)	1.304	1.306	0.008	1.300	1.308	14	
Car≃N(3)	C~N^+–H (pyrimidinium)	1.335	1.334	0.015	1.325	1.342	30	
	C≃N^+–C* (pyrimidinium)	1.346	1.346	0.010	1.340	1.352	64	
	C≃N^+–O^- (pyrimidinium)	1.362	1.359	0.013	1.353	1.369	56	
Car≃N(2)	C≃N (pyridine)	1.337	1.338	0.012	1.330	1.344	269	
	C≃N (pyrazine)	1.336	1.335	0.022	1.319	1.347	120	
	C≃N≃C (pyrimidine)	1.339	1.338	0.015	1.333	1.342	28	
	N≃C≃N (pyrimidine)	1.333	1.335	0.013	1.326	1.337	28	
	C≃N (pyrimidine) (overall)	1.336	1.337	0.014	1.331	1.339	56	
	in any 6-membered N-containing aromatic ring:							
	H–C≃N≃C–H	1.334	1.334	0.014	1.327	1.341	146	
	H–C≃N≃C–C*	1.339	1.341	0.013	1.336	1.345	38	
	C*–C≃N≃C–C*	1.345	1.345	0.008	1.342	1.348	24	
	C≃N≃C (overall)	1.336	1.337	0.014	1.329	1.344	204	
Csp^1≡N(2)	X–S–N≡C^- (isothiocyanide)	1.144	1.147	0.006	1.140	1.148	6	
Csp^1≡N(1)	C*–C≡N	1.136	1.137	0.010	1.131	1.142	140	
	C=C–C≡N in TCNQ	1.144	1.144	0.008	1.139	1.149	284	19
	Car–C≡N	1.138	1.138	0.007	1.133	1.143	31	
	X–C≡N	1.144	1.141	0.012	1.138	1.151	10	
	$(S–C≡N)^-$	1.155	1.156	0.012	1.147	1.165	14	
Csp^3–O(2)	in alcohols							
	CH_3–OH	1.413	1.414	0.018	1.395	1.425	17	
	C–CH_2–OH	1.426	1.426	0.011	1.420	1.431	75	
	C_2–CH–OH	1.432	1.431	0.011	1.425	1.439	266	
	C_3–C–OH	1.440	1.440	0.012	1.432	1.449	106	
	C*–OH (overall)	1.432	1.431	0.013	1.424	1.441	464	
	in dialkyl ethers							29
	CH_3–O–C*	1.416	1.418	0.016	1.405	1.426	110	
	C–CH_2–O–C*	1.426	1.424	0.011	1.418	1.435	34	
	C_2–CH–O–C*	1.429	1.430	0.010	1.420	1.437	53	
	C_3–C–O–C*	1.452	1.450	0.011	1.445	1.458	39	
	C*–O–C* (overall)	1.426	1.425	0.019	1.414	1.437	236	5
	in aryl alkyl ethers							29
	CH_3–O–*Car*	1.424	1.424	0.012	1.417	1.431	616	
	C–CH_2–O–*Car*	1.431	1.430	0.013	1.422	1.438	188	
	C_2–CH–O–*Car*	1.447	1.446	0.020	1.435	1.466	58	
	C_3–C–O–*Car*	1.470	1.469	0.018	1.456	1.483	55	
	C*–O–*Car* (overall)	1.429	1.427	0.018	1.419	1.436	917	
	in alkyl esters of carboxylic acids							12,29
	CH_3–O–C(=O)–C*	1.448	1.449	0.010	1.442	1.455	200	
	C–CH_2–O–C(=O)–C*	1.452	1.453	0.009	1.445	1.458	32	
	C_2–CH–O–C(=O)–C*	1.460	1.460	0.010	1.454	1.465	78	
	C_3–C–O–C(=O)–C*	1.477	1.475	0.008	1.472	1.484	6	
	C*–O–C(=O)–C* (overall)	1.450	1.451	0.014	1.442	1.459	314	
	in alkyl esters of α,β-unsaturated acids:							
	C*–O–C(=O)–C=C (overall)	1.453	1.452	0.013	1.444	1.459	112	
	in alkyl esters of benzoic acid							
	C*–O–C(=O)–C(phenyl) (overall)	1.454	1.454	0.012	1.446	1.463	219	
	in ring systems							
	oxirane (epoxides) (any subst.)	1.446	1.446	0.014	1.438	1.456	498	9
	oxetane (any subst.)	1.463	1.460	0.015	1.451	1.474	16	
	tetrahydrofuran (C,H subst.)	1.442	1.441	0.017	1.430	1.451	154	
Csp^3–O(2)	tetrahydropyran (C,H subst.)	1.441	1.442	0.015	1.431	1.451	22	
	β-lactones: C*–O–C(=O)	1.492	1.494	0.010	1.481	1.501	4	16
	γ-lactones: C*–O–C(=O)	1.464	1.464	0.012	1.455	1.473	110	12
	δ-lactones: C*–O–C(=O)	1.461	1.464	0.017	1.452	1.473	27	12
	O–C–O system in *gem*-diols, and pyranose and furanose sugars:							30,31
	HO–C*–OH	1.397	1.401	0.012	1.388	1.405	18	
	C_5–O_5–C_1–O_1H in pyranoses							
	O_1 axial (α):							
	C_5–O_5	1.439	1.440	0.008	1.432	1.445	29	
	O_5–C_1	1.427	1.426	0.012	1.421	1.432	29	
	C_1–O_1	1.403	1.400	0.012	1.391	1.412	29	
	O_1 equatorial (β):							
	C_5–O_5	1.435	1.436	0.008	1.429	1.440	17	
	O_5–C_1	1.430	1.431	0.010	1.424	1.436	17	
	C_1–O_1	1.393	1.393	0.007	1.386	1.399	17	
	α + β (overall):							
	C_5–O_5	1.439	1.440	0.008	1.432	1.446	60	
	O_5–C_1	1.430	1.429	0.012	1.421	1.436	60	
	C_1–O_1	1.401	1.399	0.011	1.392	1.407	60	

Bond	Substructure	*d*	*m*	σ	q_l	q_u	*n*	Note
	C_4–O_4–C_1–O_1H in furanoses (overall values)							
	C_4–O_4	1.442	1.446	0.012	1.436	1.449	18	
	O_4–C_1	1.432	1.432	0.012	1.421	1.443	18	
	C_1–O_1	1.404	1.405	0.013	1.397	1.409	18	
	C_5–O_5–C_1–O_1–C* in pyranoses							
	O_1 axial (α):							
	C_5–O_5	1.439	1.438	0.010	1.433	1.446	67	
	O_5–C_1	1.417	1.417	0.009	1.410	1.424	67	
	C_1–O_1	1.409	1.409	0.014	1.401	1.417	67	
	O_1–C*	1.435	1.435	0.013	1.427	1.443	67	
	O_1 equatorial (β):							
	C_5–O_5	1.434	1.435	0.006	1.429	1.439	39	
	O_5–C_1	1.424	1.424	0.008	1.418	1.431	39	
	C_1–O_1	1.390	1.390	0.011	1.381	1.400	39	
	O_1–C*	1.437	1.438	0.013	1.428	1.445	39	
	α + β (overall):							
	C_5–O_5	1.436	1.436	0.009	1.431	1.442	126	
	O_5–C_1	1.419	1.419	0.011	1.412	1.426	126	
	C_1–O_1	1.402	1.403	0.016	1.391	1.413	126	
	O_1–C*	1.436	1.436	0.013	1.428	1.445	126	
	C_4–O_4–C_1–O_1–C* in furanoses (overall values)							
	C_4–O_4	1.443	1.445	0.013	1.429	1.453	23	
	O_4–C_1	1.421	1.418	0.012	1.413	1.431	23	
	C_1–O_1	1.410	1.409	0.014	1.401	1.420	23	
	O_1–C*	1.439	1.437	0.014	1.429	1.449	23	
	Miscellaneous:							
	C#–O–SiX_3	1.416	1.416	0.017	1.405	1.428	29	
	C*–O–SO_2–C	1.465	1.461	0.014	1.454	1.475	33	
Csp^2–O(2)	in enols: C=**C–OH**	1.333	1.331	0.017	1.324	1.342	53	
	in enol esters: C=**C–O**–C*	1.354	1.353	0.016	1.341	1.363	40	
	in acids:							
	C*–**C**(=O)–**OH**	1.308	1.311	0.019	1.298	1.320	174	
	C=C–**C**(=O)–**OH**	1.293	1.295	0.019	1.279	1.307	22	
	*C*ar–**C**(=O)–**OH**	1.305	1.311	0.020	1.291	1.317	75	
	in esters:							
	C*–**C**(=O)–**O**–C*	1.336	1.337	0.014	1.328	1.346	551	12,29
	C=C–**C**(=O)–**O**–C*	1.332	1.331	0.011	1.324	1.339	112	
	*C*ar–**C**(=O)–**O**–C*	1.337	1.335	0.013	1.329	1.344	219	12
	C*–**C**(=O)–**O**–C=C	1.362	1.359	0.018	1.351	1.374	26	
	C*–C(=O)–**O**–**C**=C	1.407	1.405	0.017	1.394	1.420	26	
	C*–**C**(=O)–**O**–*C*ar	1.360	1.359	0.011	1.355	1.367	40	12
	in anhydrides: O=**C–O**–C=O	1.386	1.386	0.011	1.379	1.393	70	
	in ring systems:							
	furan (O1–C2, O1–C5)	1.368	1.369	0.015	1.359	1.377	125	
	isoxazole (O1–C5)	1.354	1.354	0.010	1.345	1.360	9	
	β-lactones: C*–C(=O)–O–C*	1.359	1.359	0.013	1.348	1.371	4	13
	γ-lactones: C*–C(=O)–O–C*	1.350	1.349	0.012	1.342	1.359	110	12
	δ-lactones: C*–C(=O)–O–C*	1.339	1.339	0.016	1.332	1.347	27	12
*C*ar–O(2)	in phenols: ***C*ar–OH**	1.362	1.364	0.015	1.353	1.373	551	
	in aryl alkyl ethers: ***C*ar–O**–C*	1.370	1.370	0.011	1.363	1.377	920	29,32
*C*ar–O(2)	in diaryl ethers: ***C*ar–O**–*C*ar	1.384	1.381	0.014	1.375	1.391	132	
	in esters: ***C*ar–O**–C(=O)–C*	1.401	1.401	0.010	1.394	1.408	40	12
Csp^2=O(1)	in aldehydes and ketones:							
	C*–CH=**O**	1.192	1.192	0.005	1.188	1.197	7	
	$(C^*)_2$–**C=O**	1.210	1.210	0.008	1.206	1.215	474	5
	$(C\#)_2$–**C=O**							
	in cyclobutanones	1.198	1.198	0.007	1.194	1.204	12	
	in cyclopentanones	1.208	1.208	0.007	1.203	1.212	155	
	in cyclohexanones	1.211	1.211	0.009	1.207	1.216	312	
	C=C–**C=O**	1.222	1.222	0.010	1.216	1.229	225	
	$(C{=}C)_2$–**C=O**	1.233	1.229	0.010	1.226	1.242	28	
	*C*ar–**C=O**	1.221	1.218	0.014	1.212	1.229	85	
	$(Car)_2$–**C=O**	1.230	1.226	0.015	1.220	1.238	66	
	C=O in benzoquinones	1.222	1.220	0.013	1.211	1.231	86	
	delocalized double bonds in carboxylate anions:							
	H–**C** ≃ O_2^- (formate)	1.242	1.243	0.012	1.234	1.252	24	
	C*–**C** ≃ O_2^-	1.254	1.253	0.010	1.247	1.261	114	
	C=C–**C** ≃ O_2^-	1.250	1.248	0.017	1.238	1.261	52	
	*C*ar–**C** ≃ O_2^-	1.255	1.253	0.010	1.249	1.262	22	
	HOOC–**C** ≃ O_2^- (hydrogen oxalate)	1.243	1.247	0.015	1.232	1.256	26	
	$^-O_2$ ≃ C–C ≃ O_2^- (oxalate)	1.251	1.251	0.007	1.248	1.254	18	

Bond	Substructure	*d*	*m*	σ	q_1	q_u	*n*	Note
	in carboxylic acids (X–COOH)							
	C*–C(=O)–OH	1.214	1.214	0.019	1.203	1.224	175	
	C–C–C(=O)–OH	1.229	1.226	0.017	1.218	1.237	22	
	Car–C(=O)–OH	1.226	1.223	0.020	1.211	1.241	75	
	in esters:							
	C*–C(=O)–O–C*	1.196	1.196	0.010	1.190	1.202	551	12
	C=C–C(=O)–O–C*	1.199	1.198	0.009	1.193	1.203	113	
	Car–C(=O)–O–C*	1.202	1.201	0.009	1.196	1.207	218	12
	C*–C(=O)–O–C=C	1.190	1.190	0.014	1.184	1.198	26	
	C*–C(=O)–O–*Car*	1.187	1.188	0.011	1.181	1.195	40	12
	in anhydrides: O=C–O–C=O	1.187	1.187	0.010	1.184	1.193	70	
	in β-lactones: C*–C(=O)–O–C*	1.193	1.193	0.006	1.187	1.198	4	13
	γ-lactones: C*–C(=O)–O–C*	1.201	1.202	0.009	1.196	1.206	109	12
	δ-lactones: C*–C(=O)–O–C*	1.205	1.207	0.008	1.201	1.209	27	12
	in amides:							
	NH_2-C(–C*)=O	1.234	1.233	0.012	1.225	1.243	32	14
	(C*–)(C*,H–)N–C(–C*)=O	1.231	1.231	0.012	1.224	1.238	378	14
	β-lactams: C*–NH–C=O	1.198	1.200	0.012	1.193	1.204	23	13
	γ-lactams:							
	C*–NH–C=O	1.235	1.235	0.008	1.232	1.240	20	13
	C*–N(–C*)–C=O	1.225	1.226	0.011	1.217	1.233	15	13
	δ-lactams:							
	C*–NH–C=O	1.240	1.241	0.003	1.237	1.243	6	14
	C*–N(–C*)–C=O	1.233	1.233	0.007	1.229	1.239	15	14
	in ureas:							
	$(NH)_2)_2$–C=O	1.256	1.256	0.007	1.249	1.261	24	25,26
	$(C\#–NH)_2$–C=O	1.241	1.237	0.011	1.235	1.245	13	25
	$[(C\#)_n–N]_2$–C=O	1.230	1.230	0.007	1.224	1.234	20	25,27
Csp^3–P(4)	C_3–P^+–C*	1.800	1.802	0.015	1.790	1.812	35	33
	C_2–P(=O)–CH_3	1.791	1.790	0.006	1.786	1.795	10	
	C_2–P(=O)–CH_2–C	1.806	1.806	0.009	1.801	1.813	45	
	C_2–P(=O)–CH–C_2	1.821	1.821	0.009	1.815	1.828	15	
	C_2–P(=O)–C–C_3	1.841	1.842	0.008	1.835	1.847	14	
	C_2–P(=O)–C* (overall)	1.813	1.811	0.017	1.800	1.822	84	
Csp^3–P(3)	C_2–P–C*	1.855	1.857	0.019	1.840	1.870	23	
Car–P(4)	C_3–P^+–*Car*	1.793	1.792	0.011	1.786	1.800	276	
	C_2–P(=O)–*Car*	1.801	1.802	0.011	1.796	1.807	98	
	Ph_3–P=N^+=P–Ph_3	1.795	1.795	0.008	1.789	1.800	197	
Car–P(3)	C_2–P–*Car*	1.836	1.837	0.010	1.830	1.844	102	
	$(N\simeq)_2$P-*Car* ($P \simeq N$ aromatic)	1.795	1.793	0.011	1.788	1.803	43	
Csp^3–S(4)	C*–SO_2–C (C* = CH_3 excluded)	1.786	1.782	0.018	1.774	1.797	75	
	C*–SO_2–C (overall)	1.779	1.778	0.020	1.764	1.790	94	
	C*–SO_2–O–X	1.745	1.744	0.009	1.738	1.754	7	34
	C*–SO_2–N–X_2	1.758	1.756	0.018	1.746	1.773	17	34
Csp^3–S(3)	C*–S(=O)–C (C* = CH_3 excluded)	1.818	1.814	0.024	1.802	1.829	69	
	C*–S(=O)–C (overall)	1.809	1.806	0.025	1.793	1.820	88	
	CH_3–S^+–X_2	1.786	1.787	0.007	1.779	1.792	21	
	C*–S^+–X_2 (C* = CH_3 excluded)	1.823	1.820	0.016	1.812	1.834	18	
	C*–S^+–X_2 (overall)	1.804	1.794	0.025	1.788	1.820	41	
Csp^3–S(2)	C*–SH	1.808	1.805	0.010	1.800	1.819	6	
	CH_3–S–C*	1.789	1.787	0.008	1.784	1.794	9	
Csp^3–S(2)	C-CH_2–S–C*	1.817	1.816	0.013	1.808	1.824	92	
	C_2–CH–S–C*	1.819	1.819	0.011	1.811	1.825	32	
	C_3–C–S–C*	1.856	1.860	0.011	1.854	1.863	26	
	C*–S–C* (overall)	1.819	1.817	0.019	1.809	1.827	242	
	in thiirane	1.834	1.835	0.025	1.810	1.858	4	9
	in thietane: see ZCMXSP (1.817, 1.844)							
	in tetrahydrothiophene	1.827	1.826	0.018	1.811	1.837	20	
	in tetrahydrothiopyran	1.823	1.821	0.014	1.812	1.832	24	
	C–CH_2–S–S–X	1.823	1.820	0.014	1.813	1.832	41	
	C_3–C–S–S–X	1.863	1.865	0.015	1.848	1.878	11	
	C*–S–S–X (overall)	1.833	1.828	0.022	1.818	1.848	59	
Csp^2–S(2)	C=C–S–C*	1.751	1.755	0.017	1.740	1.764	61	
	C=C–S–C=C (in tetrathiafulvalene)	1.741	1.741	0.011	1.733	1.750	88	
	C=C–S–C=C (in thiophene)	1.712	1.712	0.013	1.703	1.722	60	
	O=C–S–C#	1.762	1.759	0.018	1.747	1.778	20	
Car–S(4)	*Car*–SO_2–C	1.763	1.764	0.009	1.756	1.769	96	
	Car–SO_2–O–X	1.752	1.750	0.008	1.749	1.756	27	
	Car–SO_2–N–X_2	1.758	1.759	0.013	1.749	1.765	106	35
Car–S(3)	*Car*–S(=O)–C	1.790	1.790	0.010	1.783	1.798	41	
	Car–S^+–X_2	1.778	1.779	0.010	1.771	1.787	10	
Car–S(2)	*Car*–S–C*	1.773	1.774	0.009	1.765	1.779	44	
	Car–S–*Car*	1.768	1.767	0.010	1.762	1.774	158	
	Car–S–*Car* (in phenothiazine)	1.764	1.764	0.008	1.760	1.769	48	

Bond	Substructure	d	m	σ	q_1	q_u	n	Note
	Car–S–S–X	1.777	1.777	0.012	1.767	1.785	47	
Csp^1–S(2)	N≡C–S–X	1.679	1.683	0.026	1.645	1.698	10	
Csp^1–S(1)	$(N{\equiv}C{-}S)^-$	1.630	1.630	0.014	1.619	1.641	14	
Csp^2=S(1)	$(C^*)_2$–C=S: see IPMUDS (1.599)							
	$(Car)_2$–C=S: see CELDOM (1.611)							
	$(X)_2$–C=S (X = C,N,O,S)	1.671	1.675	0.024	1.656	1.689	245	
	X_2N–C(=S)–S–X	1.660	1.660	0.016	1.648	1.674	38	
	$(X_2N)_2$–C=S (thioureas)	1.681	1.684	0.020	1.669	1.693	96	
	N–C(≃S)$_2$	1.720	1.721	0.012	1.709	1.731	20	
Csp^3–Se	C#–Se	1.970	1.967	0.032	1.948	1.998	21	
Csp^2–Se(2)	C=C–Se–C=C (in tetraselenafulvalene)	1.893	1.895	0.013	1.882	1.902	32	
Car–Se(3)	Ph_3–Se^+	1.930	1.929	0.006	1.924	1.936	13	
Csp^3–Si(5)	C#–Si^-–X_4	1.874	1.876	0.015	1.859	1.884	9	
Csp^3–Si(4)	CH_3–Si–X_3	1.857	1.857	0.018	1.848	1.869	552	
	C*–Si–X_3 (C* = CH_3 excluded)	1.888	1.887	0.023	1.872	1.905	124	
	C*–Si–X_3 (overall)	1.863	1.861	0.024	1.850	1.875	681	
Car–Si(4)	Car–Si–X_3	1.868	1.868	0.014	1.857	1.878	178	
Csp^1–Si(4)	C≡C–Si–X_3	1.837	1.840	0.012	1.824	1.849	8	
Csp^3–Te	C#–Te	2.158	2.159	0.030	2.128	2.177	13	
Car–Te	Car–Te	2.116	2.115	0.020	2.104	2.130	72	
Csp^2=Te	see CEDCUJ (2.044)							
Cl–Cl	see PHASCL (2.306, 2.227)							
Cl–I	see CMBIDZ (2.563), HXPASC (2.541, 2.513), METAMM (2.552), BQUINI (2.416, 2.718)							
Cl–N	see BECTAE (1.743—1.757), BOGPOC (1.705)							
Cl–O(1)	in ClO_4^-	1.414	1.419	0.026	1.403	1.431	252	
Cl–P	(N≃)$_2$P–Cl (N≃P aromatic)	1.997	1.994	0.015	1.989	2.004	46	
	Cl–P (overall)	2.008	2.001	0.035	1.986	2.028	111	
Cl–S	Cl–S (overall)	2.072	2.079	0.023	2.047	2.091	6	
	see also longer bonds in CILSAR (2.283), BIHXIZ (2.357), CANLUY (2.749)							
Cl–Se	see BIRGUE10, BIRHAL10, CTCNSE (2.234—2.851)							
Cl–Si(4)	Cl–Si–X_3 (monochloro)	2.072	2.075	0.009	2.066	2.078	5	
	Cl_2–Si–X_2 and Cl_3–Si–X	2.020	2.012	0.015	2.007	2.036	5	
Cl–Te	Cl–Te in range 2.34—2.60	2.520	2.515	0.034	2.493	2.537	22	36
	see also longer bonds in BARRIV, BOJPUL, CETUTE, EPHTEA, OPNTEC10 (2.73—2.94)							
F–N(3)	F–N–C_2 and F_2–N–C	1.406	1.404	0.016	1.395	1.416	9	
F–P(6)	in hexafluorophosphate, PF_6^-	1.579	1.587	0.025	1.563	1.598	72	
F–P(3)	(N≃)$_2$P–F (N≃P aromatic)	1.495	1.497	0.016	1.481	1.510	10	
F–S	43 observations in range 1.409—1.770 in a wide variety of environments; F–S(6) in F_2–SO_2–C_2 (see FPSULF10, BETJOZ)	1.640	1.646	0.011	1.626	1.649	6	
	F–S(4) in F_2–S(=O)–N (see BUDTEZ)	1.527	1.528	0.004	1.524	1.530	24	37
F–Si(6)	in SiF_6^{2-}	1.694	1.701	0.013	1.677	1.703	6	
F–Si(5)	F–Si^-–X_4	1.636	1.639	0.035	1.602	1.657	10	
F–Si(4)	F–Si–X_3	1.588	1.587	0.014	1.581	1.599	24	
F–Te	see CUCPIZ (F–Te(6) = 1.942, 1.937), FPHTEL (F–Te(4) = 2.006)							
H–N(4)	X_3–N^+–H	1.033	1.036	0.022	1.026	1.045	87	21
H–N(3)	X_2–N–H	1.009	1.010	0.019	0.997	1.023	95	21
H–O(2)	in alcohols C*–O–H	0.967	0.969	0.010	0.959	0.974	63	21
	C#–O–H	0.967	0.970	0.010	0.959	0.974	73	21
	in acids O=C–O–H	1.015	1.017	0.017	1.001	1.031	16	21,38
I–I	in I_3^-	2.917	2.918	0.011	2.907	2.927	6	
I–N	see BZPRIB, CMBIDZ, HMTITI, HMTNTI, IFORAM, IODMAM (2.042—2.475)							
I–O	X–I–O (see BZPRIB, CAJMAB, IBZDAC11)	2.144	2.144	0.028	2.127	2.164	6	
	for IO_6^- see BOVMEE (1.829—1.912)							
I–P(3)	see CEHKAB (2.490—2.493)							†
I–S	see DTHIBR10 (2.687), ISUREA10 (2.629), BZTPPI (3.251)							
I–Te(4)	I–Te–X_3	2.926	2.928	0.026	2.902	2.944	8	
N(4)–N(3)	X_3–N^+–N^0–X_2 (N^0 planar)	1.414	1.414	0.005	1.412	1.418	13	
N(3)–N(3)	(C)(C,H)–N_a–N_b–(C)(C,H)							5,39
	N_a, N_b pyramidal	1.454	1.452	0.021	1.444	1.457	44	40
	N_a pyramidal, N_b planar	1.420	1.420	0.015	1.407	1.433	68	40
	N_a, N_b planar	1.401	1.401	0.018	1.384	1.418	40	40
	overall	1.425	1.425	0.027	1.407	1.443	139	
N(3)–N(2)	in pyrazole (N1–N2)	1.366	1.366	0.019	1.350	1.375	20	
	in pyridazinium ($N1^+$≃N2)	1.350	1.349	0.010	1.345	1.361	7	

Bond	Substructure	*d*	*m*	σ	q_1	q_u	*n*	Note
N(2)≃N(2)	N≃N (aromatic) in pyridazine							
	with C,H as *ortho* substituents	1.304	1.300	0.019	1.287	1.326	6	
	with N,Cl as *ortho* substituents	1.368	1.373	0.011	1.362	1.375	9	
N(2)=N(2)	C#–N=N–C#							
	cis	1.245	1.244	0.009	1.239	1.252	21	
	trans	1.222	1.222	0.006	1.218	1.227	6	
	(overall)	1.240	1.241	0.012	1.230	1.251	27	
	C*ar*–N=N–C*ar*	1.255	1.253	0.016	1.247	1.262	13	
	X–N=N=N (azides)	1.216	1.226	0.028	1.202	1.237	19	
N(2)=N(1)	X–N=N=**N** (azides)	1.124	1.128	0.015	1.114	1.137	19	
N(3)–O(2)	$(C,H)_2$–**N–OH** (Nsp^2: planar)	1.396	1.394	0.012	1.390	1.401	28	
	C_2–N–**O**–C							
	(Nsp^3: pyramidal)	1.463	1.465	0.012	1.457	1.468	22	
	(Nsp^2: planar)	1.397	1.394	0.011	1.388	1.409	12	
	in furoxan (N2–O1)	1.438	1.436	0.009	1.430	1.447	14	
N(3)–O(1)	$(C\simeq)_2N^+$–$\mathbf{O}^-$ in pyridine *N*-oxides	1.304	1.299	0.015	1.291	1.316	11	
	in furoxan ($^+$N2–O6$^-$)	1.234	1.234	0.008	1.228	1.240	14	
N(2)–O(2)	in oximes							
	$(C\#)_2$–C=**N–OH**	1.416	1.418	0.006	1.416	1.420	7	
	(H)(Csp^2)–C=**N–OH**	1.390	1.390	0.011	1.380	1.401	20	
	(C#)(Csp^2)–C=**N–OH**	1.402	1.403	0.010	1.393	1.410	18	
	$(Csp^2)_2$–C=**N–OH**	1.378	1.377	0.017	1.365	1.393	16	
	$(C,H)_2$–C=**N–OH** (overall)	1.394	1.395	0.018	1.379	1.408	67	
	in furazan (O1–N2, O1–N5)	1.385	1.383	0.013	1.378	1.392	12	
	in furoxan (O1–N5)	1.380	1.380	0.011	1.370	1.388	14	
	in isoxazole (O1–N2)	1.425	1.425	0.010	1.417	1.434	9	
N(3)=O(1)	in nitrate ions $\mathbf{NO_3}^-$	1.239	1.240	0.020	1.227	1.251	105	
	in nitro groups							
	C*–$\mathbf{NO_2}$	1.212	1.214	0.012	1.206	1.221	84	
	C#–$\mathbf{NO_2}$	1.210	1.210	0.011	1.203	1.218	251	
	C*ar*– $\mathbf{NO_2}$	1.217	1.218	0.011	1.211	1.215	1 116	
	C–$\mathbf{NO_2}$ (overall)	1.218	1.219	0.013	1.210	1.226	1 733	
N(3)–P(4)	X_2–**P**(=X)–**N**X_2							
	Nsp^2: planar	1.652	1.651	0.024	1.634	1.670	205	
	Nsp^3: pyramidal	1.683	1.683	0.005	1.680	1.686	6	
	(overall)	1.662	1.662	0.029	1.639	1.682	358	
	subsets of this group are:							
	O_2–**P**(=S)–**N**X_2	1.628	1.624	0.015	1.615	1.634	9	
	C–**P**(=S)–$(\mathbf{N}X_2)_2$	1.691	1.694	0.018	1.678	1.703	28	
	O–**P**(=S)–$(\mathbf{N}X_2)_2$	1.652	1.654	0.014	1.642	1.664	28	
	P(=O)–$(\mathbf{N}X_2)_3$	1.663	1.668	0.026	1.640	1.679	78	
N(3)–P(3)	–NX–P(–X)–NX–P(–X)– (P_2N_2 ring)	1.730	1.721	0.017	1.716	1.748	20	
	–NX–P(=S)–NX–P(=S)– (P_2N_2 ring)	1.697	1.697	0.015	1.690	1.703	44	
	in *P*-substituted phosphazenes:							
	$(N\simeq)_2$**P–N** (amino)	1.637	1.638	0.014	1.625	1.651	16	
	(aziridinyl)	1.672	1.674	0.010	1.665	1.676	15	
N(2)=P(4)	Ph_3–**P=N**$^+$=P–Ph_3	1.571	1.573	0.013	1.563	1.580	66	
N(2)=P(3)	Ph_3–**P=N**–C,S	1.599	1.597	0.018	1.580	1.615	7	
N(2)≃P(3)	**N≃P** aromatic							
	in phosphazenes	1.582	1.582	0.019	1.571	1.594	126	
	in **P≃N**≃S	1.604	1.606	0.009	1.594	1.612	36	
N(3)–S(4)	C–$\mathbf{SO_2}$–$\mathbf{NH_2}$	1.600	1.601	0.012	1.591	1.610	14	35
	C–$\mathbf{SO_2}$–**NH**–C#	1.633	1.633	0.019	1.615	1.652	47	35
	C–$\mathbf{SO_2}$–**N**–$C(\#)_2$	1.642	1.641	0.024	1.623	1.659	38	35
N(3)–S(2)	C–**S–N**X_2 Nsp^2: planar	1.710	1.707	0.019	1.698	1.722	22	23
	(for Nsp^3 pyramidal see MODIAZ: 1.765)							
	X–**S–N**X_2 Nsp^2: planar	1.707	1.705	0.012	1.699	1.715	30	23
N(2)–S(2)	C=**N–S**–X	1.656	1.663	0.027	1.632	1.677	36	
N(2)≃S(2)	**N≃S** aromatic in P≃**N≃S**	1.560	1.558	0.011	1.554	1.563	37	
N(2)=S(2)	**N=S** in **N=S**=N and **N=S**=S	1.541	1.546	0.022	1.521	1.558	37	
N(3)–Se	see COJCUZ (1.830), DSEMOR10 (1.846, 1.852), MORTRS10 (1.841)							
N(2)–Se	see SEBZQI (1.805), NAPSEZ10 (1.809, 1.820)							
N(2)=Se	see CISMUM (1.790, 1.791)							
N(3)–Si(5)	see DMESIP01, BOJLER, CASSAQ, CASYOK, CECXEN, CINTEY, CIPBUY, FMESIB, MNPSIL, PNPOSI (1.973—2.344)							
N(3)–Si(4)	X_3–**Si–N**X_2 (overall)	1.748	1.746	0.022	1.735	1.757	170	
	subsets of this group are:							
	X_3–**Si–N**HX	1.714	1.719	0.014	1.702	1.727	16	
	X_3–**Si–N**X–Si–X_3 acyclic	1.743	1.744	0.016	1.731	1.755	45	
	N–Si–N in 4-membered rings	1.742	1.742	0.009	1.735	1.748	53	
	N–Si–N in 5-membered rings	1.741	1.742	0.019	1.726	1.749	33	
N(2)–Si(4)	X_3–**Si–N**$^-$–Si–X_3	1.711	1.712	0.019	1.693	1.729	15	

Bond	Substructure	d	m	σ	q_l	q_u	n	Note
N–Te	see ACLTEP (2.402), BIBLAZ (1.980), CESSAU (2.023)							
O(2)–O(2)	C*–**O–O**–C*,H							
	τ(OO) = 70—85°	1.464	1.464	0.009	1.458	1.472	12	
	τ(OO) *ca.* 180°	1.482	1.480	0.005	1.478	1.486	5	
	overall	1.469	1.471	0.012	1.461	1.478	17	
	O=C–**O–O**–C=O see ACBZPO01 (1.446), CEYLUN (1.452), CIMHIP (1.454)							
	Si–**O–O**–Si	1.496	1.499	0.005	1.490	1.499	10	
O(2)–P(5)	X–**P**–(**O**X)$_4$							41
	trigonal bipyramidal:							
	axial	1.689	1.685	0.024	1.675	1.712	20	
	equatorial	1.619	1.622	0.024	1.604	1.628	20	
	square pyramidal	1.662	1.661	0.020	1.649	1.673	28	
O(2)–P(4)	C–**O–P**($\simeq$O)$_3^{2-}$	1.621	1.622	0.007	1.615	1.628	12	
	(H–**O**)$_2$–**P**($\simeq$O)$_2^-$	1.560	1.561	0.009	1.555	1.566	16	
	(C–**O**)$_2$–**P**($\simeq$O)$_2^-$	1.608	1.607	0.013	1.599	1.615	16	
	(C#–**O**)$_3$–**P**=O	1.558	1.554	0.011	1.550	1.564	30	
	(*Car*–**O**)$_3$–**P**=O	1.587	1.588	0.014	1.572	1.599	19	
	X–**O–P**(=O)–(C,N)$_2$	1.590	1.585	0.016	1.577	1.601	33	
	(X–**O**)$_2$–**P**(=O)–(C,N)	1.571	1.572	0.013	1.563	1.579	70	
O(2)–P(3)	(N$\simeq$)$_2$**P–O**–C (N$\simeq$P aromatic)	1.573	1.573	0.011	1.563	1.584	16	
O(1)=P(4)	C–O–**P**($\simeq$**O**)$_3^{2-}$ (declocalized)	1.513	1.512	0.008	1.508	1.518	42	
	(H–O)$_2$–**P**($\simeq$**O**)$_2^-$ (delocalized)	1.503	1.503	0.005	1.499	1.508	16	
	(C–O)$_2$–**P**($\simeq$**O**)$_2^-$ (delocalized)	1.483	1.485	0.008	1.474	1.490	16	
	(C–O)$_3$–**P=O**	1.449	1.448	0.007	1.446	1.452	18	
	C$_3$–**P=O**	1.489	1.486	0.010	1.481	1.496	72	
	N$_3$–**P=O**	1.461	1.462	0.014	1.449	1.470	26	
	(C)$_2$(N)–**P=O**	1.487	1.489	0.007	1.479	1.493	5	
	(C,N)$_2$(O)–**P=O**	1.467	1.465	0.007	1.462	1.472	33	
	(C,N)(O)$_2$–**P=O**	1.457	1.458	0.009	1.454	1.462	35	
O(2)–S(4)	C–**O–S**O$_2$–C	1.577	1.576	0.015	1.566	1.584	41	
	C–**O–S**O$_2$–CH$_3$	1.569	1.569	0.013	1.556	1.582	7	
	C–**O–S**O$_2$–*Car*	1.580	1.578	0.015	1.571	1.588	27	
O(1)=S(4)	C–**SO**$_2$–C	1.436	1.437	0.010	1.431	1.442	316	42
	X–**SO**$_2$–NX$_2$	1.428	1.428	0.010	1.422	1.434	326	
	C–**SO**$_2$–N–(C,H)$_2$	1.430	1.430	0.009	1.425	1.435	206	
	C–**SO**$_2$–O–C	1.423	1.423	0.008	1.418	1.428	82	
	in **SO**$_4^{2-}$	1.472	1.473	0.013	1.463	1.481	104	
O(1)=S(3)	C–**S**(=**O**)–C	1.497	1.498	0.013	1.489	1.505	90	5
O–Se	see BAPPAJ, BIRGUE10, BIRHAL10, CXMSEO, DGLYSE, SPSEBU (1.597 for **O=Se** to 1.974 for **O–Se**)							
O(2)–Si(5)	(X–**O**)$_3$–**Si**–(N)(C)	1.663	1.658	0.023	1.650	1.665	21	
O(2)–Si(4)	X$_3$–**Si–O**–X (overall)	1.631	1.630	0.022	1.617	1.646	191	
O(2)–Si(4)	subsets of this group are:							
	X$_3$–**Si–O**–C#	1.645	1.647	0.012	1.634	1.652	29	
	X$_3$–**Si–O**–Si–X$_3$	1.622	1.625	0.014	1.614	1.631	70	
	X$_3$–**Si–O**–O–Si–X$_3$	1.680	1.676	0.008	1.673	1.688	10	
O(2)–Te(6)	(X–**O**)$_6$–**Te**	1.927	1.927	0.020	1.908	1.942	16	
O(2)–Te(4)	(X–**O**)$_2$–**Te**–X$_2$	2.133	2.136	0.054	2.078	2.177	12	
P(4)–P(4)	X$_3$–**P–P**–X$_3$	2.256	2.259	0.025	2.243	2.277	6	
P(4)–P(3)	see CECHEX (2.197), COZPIQ (2.249)							
P(3)–P(3)	X$_2$–**P–P**–X$_2$	2.214	2.210	0.022	2.200	2.224	41	
P(4)=P(4)	see BUTSUE (2.054)							
P(3)=P(3)	see BALXOB (2.034)							
P(4)=S(1)	C$_3$–**P=S**	1.954	1.952	0.005	1.950	1.957	13	
	(N,O)$_2$(C)–**P=S**	1.922	1.924	0.014	1.913	1.927	26	
	(N,O)$_3$–**P=S**	1.913	1.914	0.014	1.906	1.921	50	
P(4)=Se(1)	X$_3$–**P=Se**	2.093	2.099	0.019	2.075	2.108	12	
P(3)–Si(4)	X$_2$–**P–Si**–X$_3$: 3- and 4-rings excluded (see BOPFER, BOPFIV, CASTOF10, COZVIW: 2.201—2.317)	2.264	2.260	0.019	2.249	2.283	22	
P(4)=Te(1)	see MOPHTE (2.356), TTEBPZ (2.327)							
S(2)–S(2)	C–**S–S**–C							
	τ(SS) = 75—105°	2.031	2.029	0.015	2.021	2.038	46	
	τ(SS) = 0—20°	2.070	2.068	0.022	2.057	2.077	28	
	(overall)	2.048	2.045	0.026	2.028	2.068	99	
	in polysulphide chain–**S–S**–	2.051	2.050	0.022	2.037	2.065	126	
S(2)–S(1)	X–N=**S–S**	1.897	1.896	0.012	1.887	1.908	5	
S–Se(4)	see BUWZUO (2.264, 2.269)							
S–Se(2)	X–**Se–S** (any)	2.193	2.195	0.015	2.174	2.207	9	
S(2)–Si(4)	X$_3$–**Si–S**–X	2.145	2.138	0.020	2.130	2.158	19	

Bond	Substructure	d	m	σ	q_1	q_u	n	Note
S(2)–Te	X–S–Te (any)	2.405	2.406	0.022	2.383	2.424	10	
	X=S–Te (any)	2.682	2.686	0.035	2.673	2.694	28	
Se(2)–Se(2)	X–Se–Se–X	2.340	2.340	0.024	2.315	2.361	15	
Se(2)–Te(2)	see BAWFUA, BAWGAH (2.524—2.561)							†
Si(4)–Si(4)	X_3–Si–Si–X_3 3-membered rings excluded: see CIHRAM (2.511)	2.359	2.359	0.012	2.349	2.366	42	
Te–Te	see CAHJOK (2.751, 2.704)							

Appendix 1. (Footnotes to Table)

1. Sample dominated by B–CH_3. For longer bonds in B^-–CH_3 see LITMEB10 [B(4)–CH_3 = 1.621—1.644Å].
2. $p(\pi)$–$p(\pi)$ Bonding with Bsp^2 and Nsp^2 coplanar (τBN = 0 ± 15°) predominates. See G. Schmidt, R. Boese, and D. Bläser, *Z. Naturforsch.*, 1982, **37b**, 1230.
3. 84 observations range from 1.38 to 1.61 Å and individual values depend on substituents on B and O. For a discussion of borinic acid adducts see S. J. Rettig and J. Trotter, *Can. J. Chem.*, 1982, **60**, 2957.
4. See M. Kaftory in 'The Chemistry of Functional Groups. Supplement D: The Chemistry of Halides, Pseudohalides, and Azides' eds. S. Patai and Z. Rappoport, Wiley: New York, 1983, Part 2, ch. 24.
5. Bonds which are endocyclic or exocyclic to any 3- or 4-membered rings have been omitted from all averages in this section.
6. The overall average given here is for Csp^3–Csp^3 bonds which carry only C or H substituents. The value cited reflects the relative abundance of each 'substitution' group. The 'mean of means' for the 9 subgroups is 1.538 (σ = 0.022) Å.
7. See F. H. Allen, (*a*) *Acta Crystallogr.*, 1980, **B36**, 81; (*b*) 1981, **B37**, 890.
8. See F. H. Allen, *Acta Crystallogr.*, 1984, **B40**, 64.
9. See F. H. Allen, *Tetrahedron*, 1982, **38**, 2843.
10. See F. H. Allen, *Tetrahedron*, 1982, **38**, 645.
11. Cyclopropanones and cyclobutanones excluded.
12. See W. B. Schweizer and J. D. Dunitz, *Helv. Chim. Acta*, 1982, **65**, 1547.
13. See L. Norskov-Lauritsen, H.-B. Bürgi, P. Hoffmann, and H. R. Schmidt, *Helv. Chim. Acta*, 1985, **68**, 76.
14. See P. Chakrabarti and J. D. Dunitz, *Helv. Chim. Acta*, 1982, **65**, 1555.
15. See J. L. Hencher in 'The Chemistry of the C≡C Triple Bond,' ed. S. Patai, Wiley, New York, 1978, ch. 2.
16. Conjugated: torsion angle about central C–C single bond is 0 ± 20° (*cis*) or 180 ± 20° (*trans*).
17. Unconjugated: torsion angle about central C–C single bond is 20—160°.
18. Other conjugative substituents excluded.
19. TCNQ is tetracyanoquinodimethane.
20. No difference detected between C2≃C3 and C3≃C4 bonds.
21. Derived from neutron diffraction results only.
22. Nsp^3: pyramidal; mean valence angle at N is in range 108—114°.
23. Nsp^2: planar; mean valence angle at N is ⩾117.5°.
24. Cyclic and acyclic peptides.
25. See R. H. Blessing, *J. Am. Chem. Soc.*, 1983, **105**, 2776.
26. See L. Lebioda, *Acta Crystallogr.*, 1980, **B36**, 271.
27. n = 3 or 4, *i.e.* tri- or tetra-substituted ureas.
28. Overall value also includes structures with mean valence angle at N in the range 115—118°.
29. See F. H. Allen and A. J. Kirby, *J. Am. Chem. Soc.*, 1984, **106**, 6197.
30. See A. J. Kirby, 'The Anomeric Effect and Related Stereoelectronic Effects at Oxygen,' Springer, Berlin, 1983.
31. See B. Fuchs, L. Schleifer, and E. Tartakovsky, *Nouv. J. Chim.*, 1984, **8**, 275.
32. See S. C. Nyburg and C. H. Faerman, *J. Mol. Struct.*, 1986, **140**, 347.
33. Sample dominated by P–CH_3 and P–CH_2–C.
34. Sample dominated by C* = methyl.
35. See A. Kalman, M. Czugler, and G. Argay, *Acta Crystallogr.*, 1981, **B37**, 868.
36. Bimodal distribution resolved into 22 'short' bonds and 5 longer outliers.
37. All 24 observations come from BUDTEZ.
38. 'Long' O–H bonds in centrosymmetric O - - - H - - - O H-bonded dimers are excluded.
39. N–N bond length also dependent on torsion angle about N–N bond and on nature of substituent C atoms; these effects are ignored here.
40. N pyramidal has average angle at N in range 100—113.5°; N planar has average angle of ⩾117.5°.
41. See R. R. Holmes and J. A. Deiters, *J. Amer. Chem. Soc.*, 1977, **99**, 3318.
42. No detectable variation in S=O bond length with type of C-substituent.

Appendix 2.

Short-form references to individual CSD entries cited by reference code in the Table. A full list of CSD bibliographic entries is given in SUP 56701.

ACBZPO01	*J. Am. Chem. Soc.*, 1975, **97**, 6729.
ACLTEP	*J. Organomet. Chem.*, 1980, **184**, 417.
ASAZOC	*Dokl. Akad. Nauk SSSR*, 1979, **249**, 120.
BALXOB	*J. Am. Chem. Soc.*, 1981, **103**, 4587.
BAPPAJ	*Inorg. Chem.*, 1981, **20**, 3071.
BARRIV	*Acta Chem. Scand., Ser. A*, 1981, **35**, 443.
BAWFUA	*Cryst. Struct. Commun.*, 1981, **10**, 1345.
BAWGAH	*Cryst. Struct. Commun.*, 1981, **10**, 1353.
BECTAE	*J. Org. Chem.*, 1981, **46**, 5048, 1981.
BELNIP	*Z. Naturforsch., Teil B*, 1982, **37**, 299.
BEMLIO	*Chem. Ber.*, 1982, **115**, 1126.
BEPZEB	*Cryst. Struct. Commun.*, 1982, **11**, 175.
BETJOZ	*J. Am. Chem. Soc.*, 1982, **104**, 1683.
BETUTE10	*Acta Chem. Scand., Ser. A*, 1976, **30**, 719.
BIBLAZ	*Zh. Strukt. Khim.*, 1981, **22**, 118.
BICGEZ	*Z. Anorg. Allg. Chem.*, 1982, **486**, 90.
BIHXIZ	*J. Chem. Soc., Chem. Commun.*, 1982, 982.
BIRGUE10	*Z. Naturforsch., Teil B*, 1983, **38**, 20.
BIRHAL10	*Z. Naturforsch., Teil B*, 1982, **37**, 1410.
BIZJAV	*J. Organomet. Chem.*, 1982, **238**, C1.
BOGPOC	*Z. Naturforsch., Teil B*, 1982, **37**, 1402.
BOGSUL	*Z. Naturforsch., Teil* B, 1982, **37**, 1230.
BOJLER	*Z. Anorg. Allg. Chem.*, 1982, **493**, 53.
BOJPUL	*Acta Chem. Scand., Ser. A*, 1982, **36**, 829.
BOPFER	*Chem. Ber.*, 1983, **116**, 146.
BOPFIV	*Chem. Ber.*, 1983, **116**, 146.
BOVMEE	*Acta Crystallogr., Sect. B*, 1982, **38**, 1048.
BQUINI	*Acta Crystallogr., Sect. B*, 1979, **35**, 1930.

BTUPTE *Acta Chem. Scand., Ser. A*, 1975, **29**, 738.
BUDTEZ *Z. Naturforsch., Teil B*, 1983, **38**, 454.
BUPSIB10 *Z. Anorg. Allg. Chem.*, 1981, **474**, 31.
BUSHAY *Z. Naturforsch., Teil. B*, 1983, **38**, 692.
BUTHAZ10 *Inorg. Chem.*, 1984, **23**, 2582.
BUTSUE *J. Chem. Soc., Chem. Commun.*, 1983, 862.
BUWZUO *Acta Chem. Scand., Ser A*, 1983, **37**, 219.
BZPRIB *Z. Naturforsch., Teil B*, 1981, **36**, 922.
BZTPPI *Inorg. Chem.*, 1978, **17**, 894.
CAHJOK *Inorg. Chem.*, 1983, **22**, 1809.
CAJMAB *Chem. Z*, 1983, **107**, 169.
CANLUY *Tetrahedron Lett.*, 1983, **24**, 4337.
CASSAQ *J. Struct. Chem.*, 1983, **2**, 101.
CASTOF10 *Acta Crystallogr., Sect. C*, 1984, **40**, 1879.
CASYOK *J. Struct. Chem.*, 1983, **2**, 107.
CECHEX *Z. Anorg. Allg. Chem.*, 1984, **508**, 61.
CECXEN *J. Struct. Chem.*, 1983, **2**, 207.
CEDCUJ *J. Org. Chem.*, 1983, **48**, 5149.
CEHKAB *Z. Naturforsch., Teil B*, 1984, **39**, 139.
CELDOM *Acta Crystallogr., Sect. C*, 1984, **40**, 556.
CESSAU *Acta Crystallogr., Sect. C*, 1984, **40**, 653.
CETTAW *Chem. Ber.*, 1984, **117**, 1089.
CETUTE *Acta Chem. Scand., Ser A*, 1975, **29**, 763.
CEYLUN *Izv. Akad. Nauk SSSR, Ser. Khim.*, 1983, 2744.
CIFZUM *Acta Chem. Scand., Ser A*, 1984, **38**, 289.
CIHRAM *Angew. Chem., Int. Ed. Engl.*, 1984, **23**, 302.
CILRUK *J. Chem. Soc., Chem. Commun.*, 1984, 1023.
CILSAR *J. Chem. Soc., Chem. Commun.*, 1984, 1021.
CIMHIP *Acta Crystallogr., C*, 1984, **40**, 1458.
CINTEY *Dokl. Akad. Nauk SSSR*, 1984, **274**, 615.
CIPBUY *J. Struct. Chem.*, 1983, **2**, 281.
CISMUM *Z. Naturforsch., Teil B*, 1984, **39**, 485.
CISTED *Z. Anorg. Allg. Chem.*, 1984, **511**, 95.
CIWYIQ *Inorg. Chem.*, 1984, **23**, 1946.
CIYFOF *Inorg. Chem.*, 1984, **23**, 1790.
CMBIDZ *J. Org. Chem.*, 1979, **44**, 1447.
CODDEE *Z. Naturforsch., Teil B*, 1984, **39**, 1257.
CODDII *Z. Naturforsch., Teil B*, 1984, **39**, 1257.
COFVOI *Z. Naturforsch., Teil B*, 1984, **39**, 1027.
COJCUZ *Chem. Ber.*, 1984, **117**, 2686.
COSDIX *Z. Naturforsch., Teil B*, 1984, **39**, 1344.
COZPIQ *Chem. Ber.*, 1984, **117**, 2063.
COZVIW *Z. Anorg. Allg. Chem.*, 1984, **515**, 7.
CTCNSE *J. Am. Chem. Soc.*, 1980, **102**, 5430.
CUCPIZ *J. Am. Chem. Soc.*, 1984, **106**, 7529.
CUDLOC *J. Cryst. Spectrosc.*, 1985, **15**, 53.
CUDLUI *J. Cryst. Spectrosc.*, 1985, **15**, 53.
CUGBAH *Acta Crystallogr., Sect. C*, 1985, **41**, 476.
CXMSEO *Acta Crystallogr., Sect. B*, 1973, **29**, 595.
DGLYSE *Acta Crystallogr., Sect. B*, 1975, **31**, 1785.
DMESIP01 *Acta Crystallogr., Sect. C*, 1984, **40**, 895.
DSEMOR10 *J. Chem. Soc., Dalton Trans.*, 1980, 628.
DTHIBR10 *Inorg. Chem.*, 1971, **10**, 697.
EPHTEA *Inorg. Chem.*, 1980, **19**, 2487.
ESEARS *J. Chem. Soc. C*, 1971, 1511.
ETEARS *J. Chem. Soc. C*, 1971, 1511.
FMESIB *J. Organomet. Chem.*, 1980, **197**, 275.
FPHTEL *J. Chem. Soc., Dalton Trans.*, 1980, 2306.
FPSULF10 *J. Am. Chem. Soc.*, 1982, **104**, 1683.
HCLENE10 *Acta Crystallogr., Sect. B*, 1982, **38**, 3139.
HMTITI *Acta Crystallogr., Sect. B*, 1975, **31**, 1505.
HMTNTI *Z. Anorg. Allg. Chem.*, 1974, **409**, 237.
HXPASC *J. Chem. Soc., Dalton Trans.*, 1975, 1381.
IBZDAC11 *J. Chem. Soc., Dalton Trans.*, 1979, 854.
IFORAM *Monatsh. Chem.*, 1974, **105**, 621.
IODMAM *Acta Crystallogr., Sect. B*, 1977, **33**, 3209.
IPMUDS *Acta Crystallogr., Sect. B*, 1973, **29**, 2128.
ISUREA10 *Acta Crystallogr., Sect. B*, 1972, **28**, 643.
LITMEB10 *J. Am. Chem. Soc.*, 1975, **97**, 6401.
MESIAD *Z. Naturforsch., Teil B*, 1980, **35**, 789.
METAMM *Acta Crystallogr.*, 1964, **17**, 1336.
MNPSIL *J. Am. Chem. Soc.*, 1969, **91**, 4134.
MODIAZ *J. Heterocycl. Chem.*, 1980, **17**, 1217.
MOPHTE *Acta Chem. Scand., Ser. A*, 1980, **34**, 333.
MORTRS10 *J. Chem. Soc., Dalton Trans.*, 1980, 628.
NAPSEZ10 *J. Am. Chem. Soc.*, 1980, **102**, 5070.
NBBZAM *Z. Naturforsch., Teil B*, 1977, **32**, 1416.
OPIMAS *Aust. J. Chem.*, 1977, **30**, 2417.
OPNTEC10 *J. Chem. Soc., Dalton Trans.*, 1982, 251.
PHASCL *Acta Crystallogr., Sect. B*, 1981, **37**, 1357.
PHASOC01 *Aust. J. Chem.*, 1975, **28**, 15.
PNPOSI *J. Am. Chem. Soc.*, 1968, **90**, 5102.
SEBZQI *J. Chem. Soc., Chem. Commun.*, 1977, 325.
SPSEBU *Acta Chem. Scand., Ser. A*, 1979, **33**, 403.
TEACBR *Cryst. Struct. Commun.*, 1974, **3**, 753.
THINBR *J. Am. Chem. Soc.*, 1970, **92**, 4002.
TMPBTI *Acta Crystallogr., Sect. B*, 1975, **31**, 1116.
TPASSN *J. Chem. Soc., Dalton Trans.*, 1977, 514.
TPASTB *Cryst. Struct. Commun.*, 1976, **5**, 39.
TPHOSI *Z. Naturforsch., Teil B*, 1979, **34**, 1064.
TTEBPZ *Z. Naturforsch., Teil B*, 1979, **34**, 256.
ZCMXSP *Cryst. Struct. Commun.*, 1977, **6**, 93.

BOND LENGTHS AND ANGLES IN GAS-PHASE MOLECULES

This table is reprinted from *Kagaku Benran, 3rd Edition*, Vol. II, pp. 649—661 (1984), with permission of the publisher, Maruzen Company, LTD. (Copyright 1984 by the Chemical Society of Japan). Translation was carried out by Kozo Kuchitsu.

Internuclear distances and bond angles are represented in units of Å (1 Å = 10^{-10} m) and degrees, respectively. The same but inequivalent atoms are discriminated by subscripts a, b, etc. In some molecules ax for axial and eq for equatorial are also used. All measurements were made in the gas phase. The methods used are abbreviated as follows. UV: ultraviolet (including visible) spectroscopy; IR: infrared spectroscopy; R: Raman spectroscopy; MW: microwave spectroscopy; ED: electron diffraction; NMR: nuclear magnetic resonance; LMR: laser magnetic resonance; EPR: electron paramagnetic resonance; MBE: molecular beam electric resonance. If two methods were used jointly for structure determination, they are listed together, as (ED, MW). If the numerical values listed refer to the equilibrium values, they are specified by r_e and θ_e. In other cases the listed values represent various average values in vibrational states; it is frequently the case that they represent the r_s structure derived from several isotopic species for MW or the r_g structure (i.e., the average internuclear distances at thermal equilibrium) for ED. These internuclear distances for the same atom pair with different definitions may sometimes differ as much as 0.01 Å. Appropriate comments are made on the symmetry and conformation in the equilibrium structure.

In general, the numerical values listed in the following tables contain uncertainties in the last digits. However, for certain molecules such as diatomic molecules, with experimental uncertainties of the order of 10^{-5} Å or smaller, numerical values are listed to four decimal places.

REFERENCES

1. L. E. Sutton, ed., *Tables of Interatomic Distances and Configuration in Molecules and Ions*, The Chemical Society Special Publication, No. 11, 18, The Chemical Society (London) (1958, 1965).
2. K.-H. Hellwege, ed., *Landolt-Börnstein Numerical Data and Functional Relations in Science and Technology*, New Series, II/7, J. H. Callomon, E. Hirota, K. Kuchitsu, W. J. Lafferty, A. G. Maki, C. S. Pote, with assistance of I. Buck and B. Starck, *Structure Data of Free Polyatomic Molecules*, Springer-Verlag (1976).
3. K. P. Huber and G. Herzberg, *Molecular Spectra and Molecular Structure IV. Constants of Diatomic Molecules*, Van Nostrand Reinhold Co., London (1979).
4. B. Starck, *Microwave Catalogue and Supplements*.
5. B. Starck, *Electron Diffraction Catalogue and Supplements*.

STRUCTURES OF ELEMENTS AND INORGANIC COMPOUNDS
Compounds are Arranged in Alphabetical Order by their Chemical Formulas
(Lengths in Å and Angles in Degrees)

Compound	Structure				Method
AgBr	Ag—Br (r_e)	2.3931			MW
AgCl	Ag—Cl (r_e)	2.2808			MW
AgF	Ag—F (r_e)	1.9832			MW
AgH	Ag—H (r_e)	1.617			UV
AgI	Ag—I (r_e)	2.5446			MW
AgO	Ag—O (r_e)	2.0030			UV
AlBr	Al—Br (r_e)	2.295			UV
AlCl	Al—Cl (r_e)	2.1301			MW
AlF	Al—F (r_e)	1.6544			MW
AlH	Al—H (r_e)	1.6482			UV
AlI	Al—I (r_e)	2.5371			MW
AlO	Al—O (r_e)	1.6176			UV
Al_2Br_6	Br_a Br_b Br_a / Al Al / Br_a Br_b Br_a		Al—Br_a Al—Br_b ∠Br_bAlBr_b ∠Br_aAlBr_a (D_{2h})	2.22 2.38 82 118	ED
Al_2Cl_6	Cl_a Cl_b Cl_a / Al Al / Cl_a Cl_b Cl_a		Al—Cl_a Al—Cl_b ∠Cl_bAlCl_b ∠Cl_aAlCl_a (D_{2h})	2.04 2.24 87 122	ED
$AsBr_3$	As—Br	2.324	∠BrAsBr	99.6	ED
$AsCl_3$	As—Cl	2.165	∠ClAsCl	98.6	ED, MW
AsF_3	As—F	1.710	∠FAsF	95.9	ED, MW

Compound	Structure						Method
AsF_5	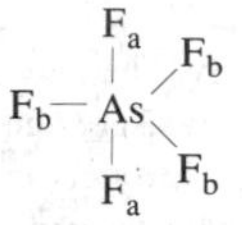 F_a, F_b, F_b—As, F_a, F_b	As—F_a	1.711	As—F_b	1.656	(D_{3h})	
AsH_3	As—H (r_e)	1.511		$\angle$HAsH (θ_e)	92.1		MW, IR
AsI_3	As—I	2.557		$\angle$IAsI	100.2		ED
AuH	Au—H (r_e)	1.5237					UV
BBr_3	B—Br	1.893		(D_{3h})			ED
BCl_3	B—Cl	1.742		(D_{3h})			ED
BF	B—F (r_e)	1.2626					UV
BF_2H	B—H	1.189	B—F	1.311	$\angle$FBF	118.3	MW
BF_2OH	B—F	1.32	B—O	1.34	O—H	0.941	MW
	$\angle$FBF	118	$\angle$FBO	123	$\angle$BOH	114.1	
BF_3	B—F	1.313		(D_{3h})			ED, IR
BH	B—H (r_e)	1.2325					UV
BH_3PH_3	B—P	1.937	B—H	1.212	P—H	1.399	MW
	$\angle$PBH	103.6	$\angle$BPH	116.9	$\angle$HBH	114.6	
	$\angle$HPH	101.3	staggered form				
BI_3	B—I	2.118		(D_{3h})			ED
BN	B—N (r_e)	1.281					UV
BO	B—O (r_e)	1.2045					EPR
BO_2	B—O	1.265		linear			UV
BS	B—S	1.6091					UV
B_2H_6	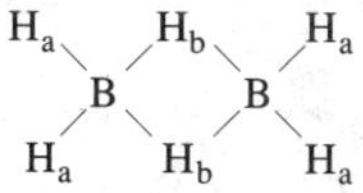 H_a, H_b, H_a, B, B, H_a, H_b, H_a			B—H_a	1.19		IR, ED
				B—H_b	1.33		
				B···B	1.77		
				$\angle H_aBH_a$	122		
				$\angle H_bBH_b$	97		
$B_3H_3O_3$	B—O	1.376		$\angle$BOB≅$\angle$OBO	120		ED
$B_3H_6N_3$	B—N	1.435	B—H	1.26	N—H	1.05	ED
	$\angle$NBN	118	$\angle$BNB	121	(C_2)		
BaH	Ba—H (r_e)	2.2318					UV
BaO	Ba—O (r_e)	1.9397					MW
BaS	Ba—S (r_e)	2.5074					MBE
BeF	Be—F (r_e)	1.3609					UV
BeH	Be—H (r_e)	1.3431					UV
BeO	Be—O (r_e)	1.3308					UV
BiBr	Bi—Br (r_e)	2.6095					MW
$BiBr_3$	Bi—Br	2.63		$\angle$BrBiBr	90	(C_{3v})	ED
BiCl	Bi—Cl (r_e)	2.4716					MW
$BiCl_3$	Bi—Cl	2.423		$\angle$ClBiCl	100	(C_{3v})	ED
BiF	Bi—F (r_e)	2.0516					MW
BiH	Bi—H (r_e)	1.805					UV
BiI	Bi—I (r_e)	2.8005					MW
BiO	Bi—O (r_e)	1.934					UV
BrCN	C—N (r_e)	1.157		C—Br (r_e)	1.790		IR
BrCl	Br—Cl (r_e)	2.1361					MW
BrF	Br—F (r_e)	1.7590					MW
BrF_3	F_a—Br—F_a, F_b	Br—F_a	1.810	Br—F_b	1.721		MW
		$\angle F_aBrF_b$	86.2	(C_{2v})			
BrF_5	Br—F (average)	1.753					ED, MW
	(Br—F_{eq}) – (Br—F_{ax}) = 0.069						
	$\angle F_{ax}BrF_{eq}$	85.1		(C_{4v})			
BrO	Br—O (r_e)	1.7172					MW
Br_2	Br—Br (r_e)	2.2811					R
CBr_4	C—Br	1.935		(T_d)			ED
CCl	C—Cl	1.6512					UV

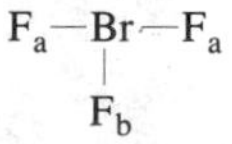

Compound	Structure						Method
$CClF_3$	C—Cl	1.752	C—F	1.325	∠FCF	108.6	ED, MW
CCl_3F	C—Cl	1.754	C—F	1.362	∠ClCCl	111	MW
					(C_{3v})		
CCl_4	C—Cl	1.767	(T_d)				ED
CF	C—F (r_e)	1.2718					EPR
CF_3I	C—I	2.138	C—F	1.330	∠FCF	108.1	ED, MW
CF_4	C—F	1.323	(T_d)				ED
CH	C—H (r_e)	1.1199					UV
CI_4	C—I	2.15	(T_d)				ED
CN	C—N (r_e)	1.1718					MW
CO	C—O (r_e)	1.1283					MW
$COBr_2$	C—O	1.178	C—Br	1.923			ED, MW
	∠BrCBr	112.3					
COClF	C—F	1.334	C—O	1.173	C—Cl	1.725	ED, MW
	∠FCCl	108.8	∠ClCO	127.5			
$COCl_2$	C—O	1.179	C—Cl	1.742			ED, MW
	∠ClCCl	111.8					
COF_2	C—F	1.3157	C—O	1.172			ED, MW
	∠FCF	107.71					
CO_2	C—O (r_e)	1.1600					IR
CP	C—P (r_e)	1.562					UV
CS	C—S (r_e)	1.5349					MW
CS_2	C—S (r_e)	1.5526					IR
C_2	C—C (r_e)	1.2425					UV
C_3O_2	C—O	1.163	C—C	1.289			ED
	linear (large-amplitude bending vibration)						
CaH	Ca—H (r_e)	2.002					UV
CaO	Ca—O (r_e)	1.8221					UV
CaS	Ca—S (r_e)	2.3178					UV
CdH	Cd—H (r_e)	1.781					EPR
$CdBr_2$	Cd—Br	2.35	linear				ED
$CdCl_2$	Cd—Cl	2.24	linear				ED
CdI_2	Cd—I	2.56	linear				ED
ClCN	C—Cl (r_e)	1.629	C—N (r_e)	1.160			MW
ClF	Cl—F (r_e)	1.6283					MW
ClF_3	F_a—Cl—F_a (with F_b on Cl)		Cl—F_a	1.698	Cl—F_b	1.598	MW
			∠F_aClF_b	87.5	(C_{2v})		
ClO	Cl—O (r_e)	1.5696					MW, UV
ClOH	O—Cl	1.690	O—H	0.975	∠HOCl	102.5	MW, IR
ClO_2	Cl—O	1.470	∠OClO	117.38			MW
$ClO_3(OH)$	O_a—Cl	1.407	O_b—Cl	1.639			ED
	H—O_b—Cl(O_a)$_3$		∠O_aClO_a	114.3	∠O_aClO_b	104.1	
Cl_2	Cl—Cl (r_e)	1.9878					UV
Cl_2O	Cl—O	1.6959	∠ClOCl	110.89			MW
CoH	Co—H (r_e)	1.542					UV
$Cr(CO)_6$	C—O	1.16	Cr—C	1.92			ED
	∠CrCO	180					
CrO	Cr—O (r_e)	1.615					UV
CsBr	Cs—Br (r_e)	3.0723					MW
CsCl	Cs—Cl (r_e)	2.9063					MW
CsF	Cs—F (r_e)	2.3454					MW
CsH	Cs—H (r_e)	2.4938					UV
CsI	Cs—I (r_e)	3.3152					MW

BOND LENGTHS AND ANGLES IN GAS-PHASE MOLECULES (continued)

Compound	Structure							Method
CsOH	Cs—O (r_e)	2.395	O—H (r_e)	0.97				MW
CuBr	Cu—Br (r_e)	2.1734						MW
CuCl	Cu—Cl (r_e)	2.0512						MW
CuF	Cu—F (r_e)	1.7449						MW
CuH	Cu—H (r_e)	1.4626						UV
CuI	Cu—I (r_e)	2.3383						MW
FCN	C—F	1.262	C—N	1.159				MW
FOH	O—H	0.96	O—F	1.442	∠HOF	97.2		MW
F_2	F—F (r_e)	1.4119						R
$Fe(CO)_5$	Fe—C (average)	1.821						ED
	$(Fe—C)_{eq} - (Fe—C)_{ax}$	0.020						
	C—O (average)	1.153	(D_{3h})					
GaBr	Ga—Br (r_e)	2.3525						MW
GaCl	Ga—Cl (r_e)	2.2017						MW
GaF	Ga—F (r_e)	1.7744						MW
GaF_3	Ga—F	1.88	(D_{3h})					ED
GaI	Ga—I (r_e)	2.5747						MW
GaI_3	Ga—I	2.458	(D_{3h})					ED
GdI_3	Gd—I	2.841	∠IGdI	108	(C_{3v})			ED
$GeBrH_3$	Ge—H	1.526	Ge—Br	2.299				MW, IR
	∠HGeH	106.2						
$GeBr_4$	Ge—Br	2.272	(T_d)					ED
$GeClH_3$	Ge—H	1.537	Ge—Cl	2.150				IR, MW
	∠HGeH	111.0						
$GeCl_2$	Ge—Cl	2.183	∠ClGeCl	100.3				ED
$GeCl_4$	Ge—Cl	2.113	(T_d)					ED
$GeFH_3$	Ge—H	1.522	Ge—F	1.732				MW, IR
	∠HGeH	113.0						
GeF_2	Ge—F (r_e)	1.7321	∠FGeF (θ_e)	97.17				MW
GeH	Ge—H (r_e)	1.5880						UV
GeH_4	Ge—H	1.5251	(T_d)					IR, R
GeO	Ge—O (r_e)	1.6246						MW
GeS	Ge—S (r_e)	2.0121						MW
GeSe	Ge—Se (r_e)	2.1346						MW
GeTe	Ge—Te (r_e)	2.3402						MW
Ge_2H_6	Ge—H	1.541	Ge—Ge	2.403				ED
	∠HGeH	106.4	∠GeGeH	112.5				
HBr	H—Br (r_e)	1.4145						MW
HCN	C—H (r_e)	1.0655	C—N (r_e)	1.1532				MW, IR
			linear					
HCNO	H—C	1.027	C—N	1.161	N—O	1.207		MW
					linear			
HCl	H-Cl (r_e)	1.2746						MW
HF	H—F (r_e)	0.9169						MW
HI	H—I (r_e)	1.6090						MW
HNCO	N—H	0.986	N—C	1.209	C—O	1.166		MW
	∠HNC	128.0						
HNCS	N—H	0.989	N—C	1.216	C—S	1.561		MW
	∠HNC	135.0	∠NCS	180				
HNO	N—H	1.063	N—O	1.212	∠HNO	108.6		UV
HNO_2	O_a=N—O_bH		*s-trans* conformer		*s-cis* conformer			MW
			O_b—H	0.958	0.98			
			N—O_b	1.432	1.39			
			N—O_a	1.170	1.19			
			∠O_aNO_b	110.7	114			
			∠NO_bH	102.1	104			
HNO_3	H—O_c—N(O_a)(O_b)		O_c—H	0.96	N—O_c	1.41		MW
			N—O_a	1.20	N—O_b	1.21		
			∠HO_cN	102.2	∠O_cNO_a	113.9		
			∠O_cNO_b	115.9	planar			

Compound	Structure						Method
HNSO	N—H	1.029	N—S	1.512	S—O	1.451	MW
	∠HNS	115.8	∠NSO	120.4			
			planar				
H_2	H—H (r_e)	0.7414					UV
H_2O	O—H (r_e)	0.9575		∠HOH (θ_e)	104.51		MW, IR
H_2O_2	O—O	1.475		∠OOH	94.8		IR
	dihedral angle of internal rotation			119.8	(C_2)		
H_2S	H—S (r_e)	1.3356		∠HSH (θ_e)	92.12		MW, IR
H_2SO_4	H_b O_b O_a H_a S O_d O_c		O—H	0.97	S—O_a	1.574	MW
			S—O_c	1.422	∠H_aO_aS	108.5	
			∠O_aSO_b	101.3	∠O_cSO_d	123.3	
			∠O_aSO_c	108.6	∠O_aSO_d	106.4	
	dihedral angle between the H_aO_aS and O_aSO_c planes				20.8		
	dihedral angle between the H_aO_aS and O_aSO_b planes				90.9		
	dihedral angle between the H_aSO_b and O_cSO_d planes				88.4	(C_2)	
H_2S_2	S—S	2.055	S—H	1.327	∠SSH	91.3	ED, MW
	dihedral angle of internal rotation		90.6	(C_2)			
$HfCl_4$	Hf—Cl	2.33		(T_d)			ED
$HgCl_2$	Hg—Cl	2.252		linear			ED
HgH	Hg—H (r_e)	1.7404					UV
HgI_2	Hg—I	2.553		linear			ED
IBr	I—Br (r_e)	2.4691					MW
ICN	C—I	1.995		C—N	1.159		MW
ICl	I—Cl (r_e)	2.3210					MW
IF_5	I—F (average)	1.860		$(I—F)_{eq} - (I—F)_{ax}$	0.03		ED, MW
	∠$F_{ax}IF_{eq}$	82.1		(C_{4v})			
IO	I—O (r_e)	1.8676					MW
I_2	I—I (r_e)	2.6663					R
InBr	In—Br (r_e)	2.5432					MW
InCl	In—Cl (r_e)	2.4012					MW
InF	In—F (r_e)	1.9854					MW
InH	In—H (r_e)	1.8376					UV
InI	In—I (r_e)	2.7537					MW
IrF_6	Ir—F	1.830		(O_h)			ED
KBr	K—Br (r_e)	2.8208					MW
KCl	K—Cl (r_e)	2.6667					MW
KF	K—F (r_e)	2.1716					MW
KH	K—H (r_e)	2.244					UV
KI	K—I (r_e)	3.0478					MW
KOH	O—H	0.91		K—O	2.212 linear		MW
K_2	K—K (r_e)	3.9051					UV
KrF_2	Kr—F	1.89		linear			ED
LiBr	Li—Br (r_e)	2.1704					MW
LiCl	Li—Cl (r_e)	2.0207					MW
LiF	Li—F (r_e)	1.5639					MW
LiH	Li—H (r_e)	1.5949					MW
LiI	Li—I (r_e)	2.3919					MW
Li_2	Li—Li (r_e)	2.6729					UV
Li_2Cl_2	Li Cl Cl Li			Li—Cl	2.23		ED
				Cl—Cl	3.61		
				∠ClLiCl	108		
$LuCl_3$	Lu—Cl	2.417		∠ClLuCl	112	(C_{3v})	ED
MgF	Mg—F (r_e)	1.7500					UV
MgH	Mg—H (r_e)	1.7297					UV
MgO	Mg—O (r_e)	1.749					UV
MnH	Mn—H (r_e)	1.7308					UV
$Mo(CO)_6$	Mo—C	2.063	C—O	1.145	(O_h)		ED
$MoCl_4O$	Mo—Cl	2.279		Mo—O	1.658		ED
	∠ClMoCl	87.2		(C_{4v})			

Compound	Structure	Method
MoF_6	Mo—F 1.820 (O_h)	ED
$NClH_2$	N—H 1.017; N—Cl 1.748	MW, IR
	∠HNCl 103.7; ∠HNH 107	
NCl_3	N—Cl 1.759; ∠ClNCl 107.1	ED
NF_2	N—F 1.3528; ∠FNF 103.18	MW
NH_2	N—H 1.024; ∠HNH 103.3	UV
NH_2CN	N—H 1.00; N_a—C 1.35	MW
	H, H–N_a—C≡N_b; C—N_b 1.160; ∠HNH 114	
	angle between the NH_2 plane and the N—C bond 142	
NH_2NO_2	N—N 1.427; N—H 1.005	MW
	∠HNH 115.2; ∠ONO 130.1	
	dihedral angle between the NH_2 and NNO_2 planes 128.2	
NH_3	N—H (r_e) 1.012; ∠HNH (θ_e) 106.7	IR
NH_4Cl	N—H 1.22; N—Cl 2.54; (C_{3v})	ED
NF_2CN	F_2N_b—C≡N_a; C—N_a 1.158; C—N_b 1.386	MW
	N_b—F 1.399; ∠N_aCN_b 174	
	∠CN_bF 105.4; ∠FN_bF 102.8	
NH	N—H (r_e) 1.0362	LMR
NH_2OH	N—H 1.02; N—O 1.453; O—H 0.962	MW
	∠HNH 107; ∠HNO 103.3; ∠NOH 101.4	
	The bisector of H—N—H angle is *trans* to the O—H bond	
NO	N—O (r_e) 1.1506	IR
NOCl	N—Cl 1.975; N—O 1.14; ∠ONCl 113	MW
NOF	O—N 1.136; N—F 1.512; ∠FNO 110.1	MW
NO_2	N—O 1.193; ∠ONO 134.1	MW
NO_2Cl	N—Cl 1.840; N—O 1.202	MW
	∠ONO 130.6; (C_{2v})	
NO_2F	N—O 1.1798; N—F 1.467	MW
	∠ONO 136; (C_{2v})	
NS	N—S (r_e) 1.4940	IR
N_2	N—N (r_e) 1.0977	UV
N_2H_4	N—H 1.021; N—N 1.449	ED, MW
	∠HNH 106.6 (assumed); ∠NNH_a 112	
	∠NNH_b 106; dihedral angle of internal rotation 91	
	H_a: the H atom closer to the C_2 axis, H_b: the H atom farther from the C_2 axis	
N_2O	N—N (r_e) 1.1284; N—O (r_e) 1.1841	MW, IR
N_2O_3	O_a–N_a–N_b(O_b)–O_c; N_a—N_b 1.864; N_a—O_a 1.142	MW
	N_b—O_b 1.202; N_b—O_c 1.217	
	∠$O_aN_aN_b$ 105.05	
	∠$N_aN_bO_b$ 112.72	
	∠$N_aN_bO_c$ 117.47	
N_2O_4	O_2N–NO_2; N—N 1.782; N—O 1.190	ED
	∠ONO 135.4; (D_{2h})	
NaBr	Na—Br (r_e) 2.5020	MW
NaCl	Na—Cl (r_e) 2.3609	MW
NaF	Na—F (r_e) 1.9260	MW
NaH	Na—H (r_e) 1.8873	UV
NaI	Na—I (r_e) 2.7115	MW
Na_2	Na—Na (r_e) 3.0789	UV
$NbCl_5$	Nb—Cl_{eq} 2.241; Nb—Cl_{ax} 2.338 (D_{3h})	ED
NbO	Nb—O (r_e) 1.691	UV
$Ni(CO)_4$	Ni—C 1.838; C—O 1.141; (T_d)	ED
NiH	Ni—H (r_e) 1.476	UV
NpF_6	Np—F 1.981; (O_h)	ED
OCS	C—O (r_e) 1.1578; C—S (r_e) 1.5601	MW

Compound	Structure						Method
OCSe	C—O	1.159	C—Se	1.709			MW
OF	O—F (r_e)	1.3579					LMR
OF_2	O—F (r_e)	1.4053	∠FOF (θ_e)	103.07	(C_{2v})		MW
$O(SiH_3)_2$	Si—H	1.486	Si—O	1.634			ED
	∠SiOSi	144.1					
O_2	O—O (r_e)	1.2074					MW
O_2F_2	O—O	1.217	F—O	1.575			MW
	∠OOF	109.5	dihedral angle of internal rotation	87.5	(C_2)		
O_3	O—O (r_e)	1.2716	∠OOO (θ_e)	117.47	(C_{2v})		MW
OsF_6	Os—F	1.831	(O_h)				ED
OsO_4	Os—O	1.712	(T_d)				ED
PBr_3	P—Br	2.220	∠BrPBr	101.0			ED
PCl_3	P—Cl	2.039	∠ClPCl	100.27			ED
PCl_5	Cl_a, Cl_b, Cl_b—P, Cl_a, Cl_b		P—Cl_a	2.124	P—Cl_b	2.020 (D_{3h})	ED
PF	P—F (r_e)	1.5896					UV
PF_3	P—F	1.570	∠FPF	97.8			ED, MW
PF_5	P—F_{ax}	1.577	P—F_{eq}	1.534	(D_{3h})		ED
PH	P—H (r_e)	1.4223					LMR
PH_2	P—H	1.418	∠HPH	91.70			UV
PH_3	P—H	1.4200	∠HPH	93.345			MW
PN	N—P (r_e)	1.4909					MW
PO	O—P (r_e)	1.4759					UV
$POCl_3$	P—O	1.449	P—Cl	1.993			ED
	∠ClPCl	103.3					
POF_3	P—O	1.436	P—F	1.524	∠FPF	101.3	ED, MW
P_2	P—P (r_e)	1.8931					UV
P_2F_4	P—F	1.587	P—P	2.281			ED
	∠PPF	95.4	∠FPF	99.1			
	The two PF_2 planes are *trans* to each other (the *gauche* conformer is less than 10%)						
P_4	P—P	2.21	(T_d)				ED
P_4O_6	P—O	1.638	∠POP	126.4	(T_d)		ED
PbH	Pb—H (r_e)	1.839					UV
PbO	Pb—O (r_e)	1.9218					MW
PbS	Pb—S (r_e)	2.2869					MW
PbSe	Pb—Se (r_e)	2.4022					MW
PbTe	Pb—Te (r_e)	2.5950					MW
PrI_3	Pr—I	2.904	∠IPrI	113	(C_{3v})		ED
PtO	Pt—O (r_e)	1.7273					UV
PuF_6	Pu—F	1.971	(O_h)				ED
RbBr	Rb—Br (r_e)	2.9447					MW
RbCl	Rb—Cl (r_e)	2.7869					MW
RbF	Rb—F (r_e)	2.2703					MW
RbH	Rb—H (r_e)	2.367					UV
RbI	Rb—I (r_e)	3.1768					MW
RbOH	Rb—O	2.301	O—H	0.957	linear		MW
$ReClO_3$	Re—O	1.702	Re—Cl	2.229			MW
	∠ClReO	109.4	(C_{3v})				
ReF_6	Re—F	1.832	(O_h)				ED
RuO_4	Ru—O	1.706	(T_d)				ED
SCSe	C—Se	1.693	C—S	1.553			MW
SCTe	C—S	1.557	C—Te	1.904			MW
SCl_2	S—Cl	2.006	∠ClSCl	103.0	(C_{2v})		ED
SF	S—F (r_e)	1.6006					MW
SF_2	S—F	1.5921	∠FSF	98.20			MW
SF_6	S—F	1.561	(O_h)				ED

Compound	Structure						Method
SO	S—O (r_e)	1.4811					MW
$SOCl_2$	S—O	1.44	S—Cl	2.072			MW
	∠ClSCl	97.2	∠OSCl	108.0			
SOF_2	S—O	1.420	S—F	1.583			ED
	∠OSF	106.2	∠FSF	92.2			
SOF_4	F_b F_b / F_a—S—F_a / O		S—O	1.403	S—F_a	1.575	ED
			S—F_b	1.552	∠OSF_a	90.7	
			∠OSF_b	124.9	∠F_aSF_b	89.6	
			∠F_bSF_b	110.2	(C_{2v})		
SO_2	S—O (r_e)	1.4308	∠OSO (θ_e)	119.329			MW
SO_2Cl_2	S—O	1.404	S—Cl	2.011	∠OSO	123.5	ED
	∠ClSCl	100.0	(C_{2v})				
SO_2F_2	S—O	1.397	S—F	1.530	∠OSO	123	ED
	∠FSF	97	(C_{2v})				
SO_3	S—O	1.4198	(D_{3h})				IR
$S(SiH_3)_2$	Si—H	1.494	Si—S	2.136	∠SiSSi	97.4	ED
S_2	S—S (r_e)	1.8892					R
S_2Br_2	S—Br	2.24	S—S	1.98			ED
	∠SSBr	105	dihedral angle of internal rotation	83.5			
S_2Cl_2	S—Cl	2.057	S—S	1.931			ED
	∠SSCl	108.2	dihedral angle of internal rotation	84.1	(C_2)		
S_2O_2	S—O	1.458	S—S	2.025	∠OSS	112.8	MW
					planar *cis* form		
S_8	S S S / S S / S / S S (ring)		S—S	2.07			ED
			∠SSS	105			
			(D_{4d})				
$SbCl_3$	Sb—Cl	2.333	∠ClSbCl	97.2			ED
SbH_3	Sb—H	1.704	∠HSbH	91.6			MW
SeF	Se—F	1.742					MW
SeF_6	Se—F	1.69	(O_h)				ED
SeO	Se—O (r_e)	1.6393					MW
$SeOF_2$	Se—O	1.576	Se—F	1.730			MW
	∠OSeF	104.82	∠FSeF	92.22			
SeO_2	Se—O (r_e)	1.6076	∠OSeO (θ_e)	113.83			MW
SeO_3	Se—O	1.69	(D_{3h})				ED
Se_2	Se—Se (r_e)	2.1660					UV
Se_6	Se—Se	2.34	∠SeSeSe	102			ED
		six-membered ring with chair conformation					
$SiBrF_3$	Si—F	1.560	Si—Br	2.153			MW
	∠FSiBr	108.5	(C_{3v})				
$SiBrH_3$	Si—H	1.485	Si—Br	2.210			MW
	∠HSiBr	107.8	(C_{3v})				
$SiClH_3$	Si—H	1.482	Si—Cl	2.048			MW
	∠HSiCl	107.9	(C_{3v})				
$SiCl_4$	Si—Cl	2.019	(T_4)				ED
SiF	Si—F	1.6008					UV
$SiFH_3$	Si—H	1.484	Si—F	1.593			MW, IR
	∠HSiH	110.63	(C_{3v})				
SiF_2	Si—F (r_e)	1.590	∠FSiF (θ_e)	100.8			MW
SiF_3H	Si—H (r_e)	1.4468	Si—F (r_e)	1.5624			MW
	∠HSiF (θ_e)	110.64					
SiF_4	Si—F	1.553	(T_d)				ED
SiH	Si—H (r_e)	1.5201					UV
SiH_3I	Si—H	1.485	Si—I	2.437			MW
	∠HSH	107.8					
SiH_4	Si—H	1.4798	(T_d)				IR
SiN	N—Si (r_e)	1.572					UV
SiO	Si—O (r_e)	1.5097					MW

Compound	Structure					Method
SiS	Si—S (r_e)	1.9293				MW
SiSe	Se—Si (r_e)	2.0583				MW
Si_2	Si—Si (r_e)	2.246				UV
Si_2Cl_6	Si—Si	2.32	Si—Cl	2.009		ED
	∠ClSiCl	109.7				
Si_2F_6	Si—Si	2.317	Si—F	1.564		ED
	∠FSiF	108.6				
Si_2H_6	Si—H	1.492	Si—Si	2.331		ED
	∠SiSiH	110.3	∠HSiH	108.6		
			staggered form (assumed)			
$SnCl_4$	Sn—Cl	2.280	(T_d)			ED
SnH	Sn—H (r_e)	1.7815				UV
SnH_4	Sn—H	1.711	(T_d)			R, IR
SnO	Sn—O	1.8325				MW
SnS	S—Sn (r_e)	2.2090				MW
SnSe	Se—Sn (r_e)	2.3256				MW
SnTe	Sn—Te (r_e)	2.5228				MW
SrH	Sr—H (r_e)	2.1455				UV
SrO	Sr—O (r_e)	1.9198				MW
SrS	S—Sr (r_e)	2.4405				UV
$TaCl_5$	Ta—Cl_{eq}	2.227	Ta—Cl_{ax}	2.369	(D_{3h})	ED
TaO	Ta—O (r_e)	1.6875				UV
TeF_6	Te—F	1.815	(O_h)			ED
Te_2	Te—Te (r_e)	2.5574				UV
$ThCl_4$	Th—Cl	2.58	(T_d)			ED
ThF_4	Th—F	2.14	(T_d)			ED
TlBr	Tl—Br (r_e)	2.6182				MW
TlCl	Tl—Cl (r_e)	2.4848				MW
TlF	Tl—F (r_e)	2.0844				MW
TlH	Tl—H (r_e)	1.870				UV
TlI	Tl—I (r_e)	2.8137				MW
$TiBr_4$	Ti—Br	2.339	(T_d)			ED
$TiCl_4$	Ti—Cl	2.170	(T_d)			ED
TiO	Ti—O (r_e)	1.620				UV
TiS	Ti—S (r_e)	2.0825				UV
UF_6	U—F	1.996	(O_h)			ED
$V(CO)_6$	V—C	2.015	C—O	1.138		ED
		(O_h, involving dynamic Jahn-Teller effect)				
VCl_3O	V—O	1.570	V—Cl	2.142		ED, MW
	∠ClVCl	111.3				
VCl_4	V—Cl	2.138	(T_d, involving dynamic Jahn-Teller effect)			ED
VF_5	V—F (average)	1.71				ED
VO	V—O (r_e)	1.5893				UV
$W(CO)_6$	W—C	2.059	C—O	1.149	(O_h)	ED
$WClF_5$	Cl, F_b / F_b—W—F_b / F_b, F_a		W—Cl	2.251		MW
			W—F (average)	1.836		
			∠F_aWF_b	88.7		
WF_4O	W—O	1.666	W—F	1.847		ED
	∠FWF	86.2	(C_{4v})			
WF_6	W—F	1.832	(O_h)			ED
XeF_2	Xe—F	1.977	linear			IR
XeF_4	Xe—F	1.94	(D_{4h})			ED
XeF_6	Xe—F	1.890	(large-amplitude bending vibration around the O_h structure)			ED
XeO_4	Xe—O	1.736	(T_d)			ED
ZnH	Zn—H (r_e)	1.5949				UV
$ZrCl_4$	Zr—Cl	2.32	(T_d)			ED
ZrF_4	Zr—F	1.902	(T_d)			ED
ZrO	Zr—O (r_e)	1.7116				UV

STRUCTURES OF ORGANIC MOLECULES

Compounds are Arranged in Alphabetical Order by Chemical Name; Cross References are Given for Common Synonyms (Lengths in Å and Angles in Degrees)

Compound	Structure				Method
Acetaldehyde	C_bH_3—C_a(=O)—H		C_a—O	1.210	ED, MW
			C_b—H	1.107	
			C_a—H	1.128	
	C_a—C_b	1.515	∠C_bC_aO	124.1	
	∠HC_bH	109.8	∠C_bC_aH	115.3	
Acetamide	C—O	1.220	C—N	1.380	ED
CH_3CONH_2	C—C	1.519	N—H	1.022	
	C—H	1.124	∠NCO	122.0	
	∠CCN	115.1			
Acetic acid	CH_3—C(=O_a)—O_b—H		C—C	1.520	ED
			C—O_a	1.214	
			C—O_b	1.364	
	C—H	1.10	∠CCO_a	126.6	
	∠CCO_b	110.6			
Acetone	C—C	1.520	C—O	1.213	ED, MW
$(CH_3)_2CO$	C—H	1.103	∠CCC	116.0	
	∠HCH	108.5	symmetry axis of each methyl group is tilted 2 from the C—C bond		
Acetonitrile	C—H	1.107	C—C	1.468	ED, MW
CH_3CN	C—N	1.159	∠CCH	109.7	
Acetonitrile oxide	C—C	1.442	C—N	1.169	MW
CH_3CNO	N—O	1.217	(C_{3v})		
Acetyl chloride	C—H	1.105	C—O	1.187	ED, MW
CH_3COCl	C—C	1.506	C—Cl	1.798	
	∠HCH	108.6	∠OCCl	121.2	
	∠CCCl	111.6			
Acetyl cyanide → Pyruvonitrile					
Acetylene	C—H (r_e)	1.060	C—C (r_e)	1.203	IR
HC≡CH					
Acrolein → Acrylaldehyde					
Acrylaldehyde	H_2C_a=C_bH—C_cH=O		C_b—C_c	1.484	ED, MW
			C_a—C_b	1.345	
			C_c—O	1.217	
			C_a—H	1.10	
			C_c—H	1.13	
	∠$C_aC_bC_c$	120.3	∠C_bC_cO	123.3	
	∠HC_cC_b	114	other CCH angles (average)	122	
	planar *s-trans* form				
Acrylonitrile	H_2C_a=C_bH—C_c≡N		C_a—C_b	1.343	ED, MW
			C_b—C_c	1.438	
			C_c—N	1.167	
			C_a—H	1.114	
			∠$C_aC_bC_c$	121.7	
	∠C_bC_cN	178	∠HCC	120	
Acryloyl chloride	C—H	1.086 (assumed)	C_b—C_c	1.48	MW
	C_c—Cl	1.82	C_a—C_b	1.35	
	H_2C_a=C_bH—C_c(=O)—Cl		C_c—O	1.19	
			∠C_bC_aH	120 (assumed)	
			∠C_aC_bH	121.5 (assumed)	
			∠$C_aC_bC_c$	123	
			∠C_bC_cCl	116	
			∠C_bC_cO	127	

Compound	Structure				Method
Allene	C—C	1.3084	C—H	1.087	IR
CH_2=C=CH_2	∠HCH	118.2			
Allyl chloride	Cl $H_2C_a{=}C_bH{-}C_cH_2$		*cis* conformer C—Cl	1.811	MW
			∠CCCl	115.2	
			skew conformer C—Cl	1.809	
	∠CCCl	109.6	CCCCl dihedral angle of internal rotation	122.4	
Aniline	C—C	1.392	C—N	1.431	MW
$C_6H_5NH_2$	N—H	0.998	∠HNH	113.9	
	dihedral angle between the NH_2 plane and the N—C bond			140.6	
Azetidine	CH_2—CH_2 | | CH_2—NH		C—N	1.482	ED
			C—C	1.553	
			C—H	1.107	
	N—H	1.03	∠CNC	92.2	
	∠CCC	86.9	∠CCN	85.8	
	dihedral angle between the CCC and CNC planes			147	
Aziridine	H_a | H_b N H_b C—C H_c H_c		N—H	1.016	MW
			N—C	1.475	
			C—C	1.481	
			C—H	1.084	
			∠CNC	60.3	
			∠H_aNC	109.3	
	∠H_bCH_c	115.7	∠H_bCC	117.8	
	∠H_bCN	118.3	∠H_cCC	119.3	
	∠H_cCN	114.3			
Azomethane	C—N	1.482	N—N	1.247	ED
CH_3N=NCH_3	∠CNN	112.3	*trans* conformer		
Benzene	C—C	1.399	C—H	1.101	ED, IR
C_6H_6					
p-Benzoquinone	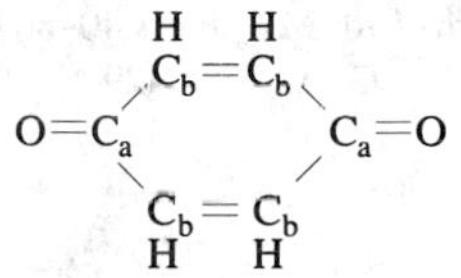 H H $C_b{=}C_b$ O=C_a C_a=O $C_b{=}C_b$ H H		C_a—O	1.225	ED
			C_b—C_b	1.344	
			C_a—C_b	1.481	
			∠$C_bC_aC_b$	118.1	
Biacetyl	C—O	1.215	C—C (average)	1.524	ED
$CH_3COCOCH_3$	C—H	1.108	∠CCO	119.5	
	∠CCC	116.2	*trans* conformer		
Bicyclo[1.1.0]butane	H_a | C_a H_b H_b C_a C_b C_b H_a H_c H_c		C_a—C_a	1.497	MW
			C_a—C_b	1.498	
			C_a—H_a	1.071	
			C_b—H_b, C_b—H_c	1.093	
			∠$H_bC_bH_c$	115.6	
	∠$C_bC_aH_a$	130.4	∠$C_aC_bC_a$	60.0	
	∠$C_aC_aH_a$	128.4	dihedral angle between the two $C_aC_aC_b$ planes	121.7	
Bicyclo[2.2.1]hepta-2,5-diene	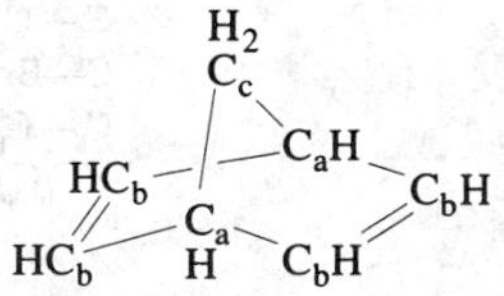 H_2 C_c C_aH HC_b C_bH C_a HC_b H C_bH		C_a—C_b	1.535	ED
			C_b—C_b	1.343	
			C_a—C_c	1.573	
			C—H	1.12	
			∠$C_aC_cC_a$	94	
	dihedral angle between the two $C_aC_bC_bC_a$ planes			115.6 (C_{2v})	
Bicyclo[2.2.1]heptane	See the preceding molecule for the labels of the C atoms				ED
C_7H_{12}	C_a—C_b	1.54	C_b—C_b	1.56	
	C_a—C_c	1.56	C—C (average)	1.549	
	∠$C_aC_cC_a$	93.1	dihedral angle between the two $C_aC_bC_bC_a$ planes	113.1	

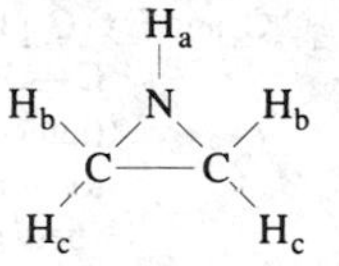

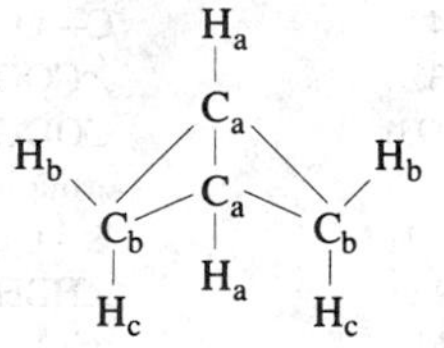

Compound	Structure	Method
Bicyclo[2.2.0]hexa-2,5-diene	C_b—C_b 1.345 C_a—C_a 1.574 C_a—C_b 1.524 dihedral angle between the two $C_aC_bC_bC_a$ planes 117.3	ED
Bicyclo[2.2.2]octane	$HC_a(C_bH_2C_bH_2)_3C_aH$ C_a—C_b 1.54 $\angle C_aC_bC_b$ 109.7 C_b—C_b 1.55 C—C (average) 1.542 large-amplitude torsional motion about the D_{3h} symmetry axis	ED
Bicyclo[1.1.1]pentane C_5H_8	C—C 1.557 ∠CCC 74.2	ED
Bicyclo[2.1.0]pentane	C_a—C_a 1.536 C_b—C_b 1.565 C_a—C_b 1.528 C_a—C_c 1.507 Dihedral angle between the $C_aC_aC_bC_b$ and $C_aC_aC_c$ planes 112.7	MW
Biphenyl	C—C (intra-ring) 1.396 (inter-ring) 1.49 torsional dihedral angle between the two rings ~40	ED
4,4′-Bipyridyl	C—C, C—N (intra-ring) 1.375 C—C (inter-ring) 1.465 torsional dihedral angle between the two rings ~37	ED
Bis (cyclopentadienyl) beryllium $(C_5H_5)_2Be$	Be—(cyclopentadienyl plane) 1.470, 1.92 C—C 1.423 (C_{5v}) (The Be atom has two equilibrium positions)	ED
Bis (cyclopentadienyl) iron → Ferrocene		
Bis (cyclopentadienyl) lead $(C_5H_5)_2Pb$	C—C 1.430 Pb—C 2.79 dihedral angle between the two C_5H_5 planes 40~50 (The two rings are not parallel.)	ED
Bis (cyclopentadienyl) manganese $(C_5H_5)_2Mn$	Mn—C 2.383 C—C 1.429 (D_{5h})	ED
Bis (cyclopentadienyl) nickel $(C_5H_5)_2Ni$	Ni—C 2.196 C—C 1.430 (D_{5h})	ED
Bis (cyclopentadienyl) ruthenium $(C_5H_5)_2Ru$	C—C 1.439 Ru—C 2.196	ED
Bis (cyclopentadienyl) tin $(C_5H_5)_2Sn$	C—C 1.431 Sn—C 2.71 C—H 1.14 (D_{5h})	ED
Bis (trifluoromethyl) peroxide CF_3OOCF_3	O—O 1.42 C—O 1.399 C—F 1.320 ∠COO 107 ∠FCF 109.0 COOC dihedral angle of internal rotation 123	ED
Borine carbonyl BH_3CO	B—H 1.194 B—C 1.540 C—O 1.131 ∠HBH 113.9 ∠BCO 180 (C_{3v})	MW
Bromobenzene	C—H 1.072 C_c—C_d 1.401 C_b—C_c 1.375 C—Br 1.85 C_a—C_b 1.42 $\angle C_bC_aC_b$ 117.4	MW
Bromoform $CHBr_3$	C—Br 1.924 C—H 1.11 ∠BrCBr 111.7 (C_{3v})	ED, MW
Bromoiodoacetylene IC≡CBr	C—I 1.972 C—C 1.206 C—Br 1.795	ED

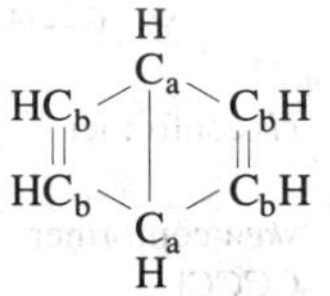

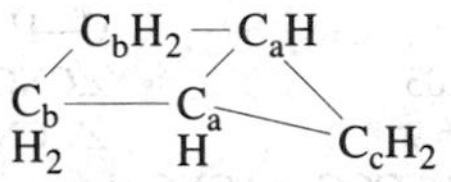

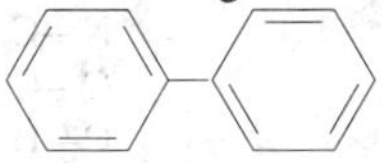

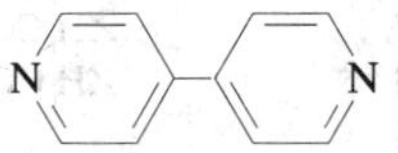

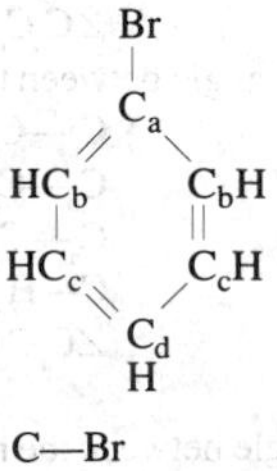

Compound	Structure				Method
1,3-Butadiene	C_aH_2		C_b—C_b	1.467	ED
			C_a—C_b	1.349	
	C_bH C_bH		C—H (average)	1.108	
	C_aH_2		∠CCC	124.4	
	∠C_bC_aH	120.9	*anti* conformer	(C_{2h})	
1,3-Butadiyne	$HC_a≡C_b$—$C_b≡C_aH$		C—H	1.09	ED
	C_a—C_b	1.218	C_b—C_b	1.384	
			linear		
Butane	C—C	1.531	C—H	1.117	ED
$CH_3CH_2CH_2CH_3$	∠CCC	113.8	∠CCH	111.0	
	trans conformer	54%	dihedral angle for the *gauche* conformer	65	
2-Butanone → Ethyl methyl ketone					
Butatriene	H_2C_a=C_b=C_b=C_aH_2		C—H	1.08	ED
	C_a—C_b	1.32	C_b—C_b	1.28 (D_{2h})	
2-Butene	C_aH_3—C_bH=C_bH—C_aH_3		ED		
	C_a—C_b	*cis* conformer	1.506	*trans* conformer 1.508	
	C_b—C_b	1.346	1.347		
	∠$C_aC_bC_b$	125.4	123.8		
3-Buten-1-yne → Vinylacetylene					
tert-Butyl chloride	C—H	1.102	C—C	1.528	ED, MW
$(CH_3)_3CCl$	C—Cl	1.828	∠CCCl	107.3	
	∠CCH	110.8	∠CCC	111.6	
tert-Butyl cyanide → Pivalonitrile					
2-Butyne	C_aH_3—$C_b≡C_b$—C_aH_3	C—H	1.116		ED
	C_b—C_b	1.214	C_a—C_b	1.468	
	∠C_bC_aH	110.7			
Carbon C_2	C—C (r_e)	1.3119			UV
Carbon C_3	C—C	1.277	linear		UV
Carbon suboxide → Tricarbon dioxide					
Carbon tetrabromide CBr_4	C—Br	1.935	(T_d)		ED
Carbon tetrachloride CCl_4	C—Cl	1.767	(T_d)		ED
Carbon tetrafluoride CF_4	C—F	1.323	(T_d)		ED
Carbon tetraiodide CI_4	C—I	2.15	(T_d)		ED
Carbonyl cyanide	C—O	1.209	C—C	1.466	ED, MW
$CO(CN)_2$	C—N	1.153	∠CCC	115	
	∠CCN	180			
Chloroacetylene	C—H	1.0550	C—C	1.2033	MW
HC≡CCl	C—Cl	1.6368			
Chlorobenzene	C—C	1.400	C—Cl	1.737	ED
C_6H_5Cl	C—H	1.083	∠CC(Cl)C	121.7	
	∠CC(H)C	120			
Chlorobromoacetylene	Cl—C	1.636	C—C	1.206	ED
ClC≡CBr	C—Br	1.784			
Chlorocyanoacetylene	C—Cl	1.624	C—C	1.205	ED
ClC≡CCN	C—CN	1.362	C—N	1.160	
Chloroethane → Ethyl chloride					
2-Chloroethanol	C—O	1.413	C—C	1.519	ED
$ClCH_2CH_2OH$	C—Cl	1.801	C—H	1.093	
	O—H 1.033	∠CCCl 110.7	∠CCO 113.8		
	fraction of the *gauche* conformer at 37C is 92 ~ 94%,				
	dihedral angle of internal rotation	62.4			
Chloroethylene → Vinyl chloride					
Chloroform	C—H	1.100	C—Cl	1.758	MW
$CHCl_3$	∠ClCCl	111.3	(C_{3v})		

Compound	Structure				Method
Chloroiodoacetylene	C—Cl	1.63	C—I	1.99	MW
ClC≡CI	C—C	1.209 (assumed)			
Chloromethane → Methyl chloride					
3-Chloropropene → Allyl chloride					
Cyanamide	N_a—C	1.346	C—N_b	1.160	MW
$H_2N_aCN_b$	N—H	1.00	∠HNH	114	
	dihedral angle between the NH_2 plane and the N—C bond			142	
Cyanoacetylene	C_b—H	1.058	C_a—C_b	1.205	MW
H—C_b≡C_a—C_c≡N	C_a—C_c	1.378	C_c—N	1.159	
Cyanocyclopropane	C—C (ring)	1.513	C—C_a	1.472	MW
$C_3H_5C_aN$	C—H	1.107	C_a—N	1.157	
	∠HCH	114.6	∠C_aCH	119.6	
Cyanogen	C—N	1.163	C—C	1.393	ED
$(CN)_2$				linear	
Cyclobutane	C—H	1.113	C—C	1.555	ED
$(CH_2)_4$	dihedral angle between the two CCC planes 145				
Cyclobutanone	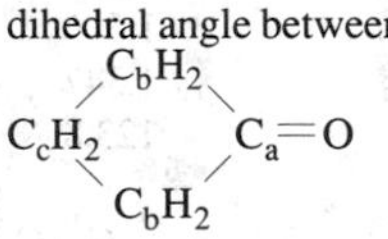 C_bH_2 / C_cH_2 C_a=O / C_bH_2		C_a—C_b	1.527	MW
			C_b—C_c	1.556	
			∠$C_bC_aC_b$	93.1	
			∠$C_aC_bC_c$	88.0	
Cyclobutene	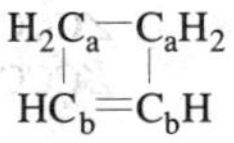 H_2C_a—C_aH_2 / HC_b=C_bH	C_b—C_b 1.342		C_a—C_a 1.566	MW
		C_a—C_b 1.517		C_a—H 1.094	
		C_b—H 1.083			
	∠$C_aC_bC_b$	94.2	∠C_bC_bH	133.5	
	∠C_aC_aH	114.5	∠$C_aC_aC_b$	85.8	
	∠HC_aH	109.2	dihedral angle between the CH_2 plane and the C_a—C_a bond	135.8	
Cyclohexane	C—C	1.536	C—H	1.119	ED
C_6H_{12}	∠CCC	111.3	chair form		
Cyclohexene	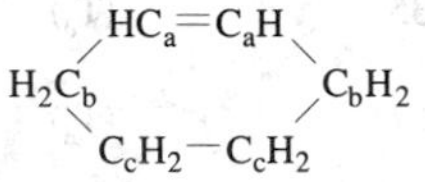 HC_a=C_aH / H_2C_b C_bH_2 / C_cH_2—C_cH_2		C_a—C_a	1.334	ED
			C_a—C_b	1.50	
			C_b—C_c	1.52	
			C_c—C_c	1.54	
	∠$C_aC_aC_b$	123.4	∠$C_aC_bC_c$	112.0	
	∠$C_bC_cC_c$	110.9	(C_2)	half-chair form	
Cyclooctatetraene	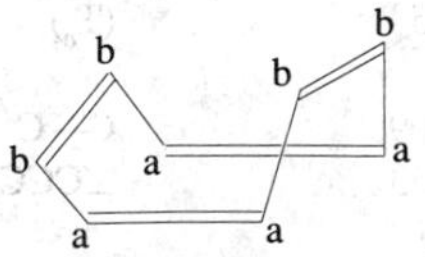		C_a—C_b	1.476	ED
			C—H	1.100	
			C_a—C_a, C_b—C_b	1.340	
			∠$C_bC_aC_a$, ∠$C_aC_bC_b$	126.1	
	dihedral angle between the $C_aC_aC_aC_a$ and $C_aC_bC_bC_a$ planes			136.9	
	tub form	(D_{2d})			
1,3-Cyclopentadiene	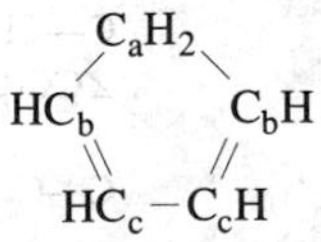 C_aH_2 / HC_b C_bH / HC_c—C_cH		C_a—C_b	1.509	MW
			C_b—C_c	1.342	
			C_c—C_c	1.469	
			∠$C_aC_bC_c$	109.3	
	∠$C_bC_cC_c$	109.4	∠$C_bC_aC_b$	102.8	
Cyclopentadienylindium	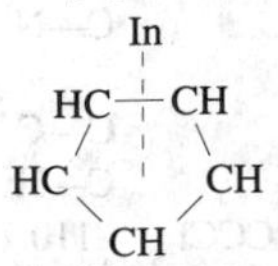 In / HC—CH / HC CH / CH		In—C	2.621	ED
			C—C	1.426	
			(C_{5v})		
Cyclopentane	C—H	1.114	C—C	1.546	ED
$(CH_2)_5$	∠CCH	111.7			
	(The out-of-plane vibration of the C atoms is essentially free pseudorotation; average value of the displacements of the C atoms from the molecular plane 0.43)				

Compound	Structure	Method
Cyclopentene	C_aH_2 ring: H_2C_b, C_bH_2, $C_cH=C_cH$ C_a—C_b 1.546 C_b—C_c 1.519 C_c—C_c 1.342 ∠$C_bC_aC_b$ 104.0 ∠$C_aC_bC_c$ 103.0 ∠$C_bC_cC_c$ 110.0 dihedral angle between the $C_bC_aC_b$ and $C_bC_cC_cC_b$ planes 151.2	ED
Cyclopropane $(CH_2)_3$	C—C 1.512 C—H 1.083 ∠HCH 114.0	R
Cyclopropanone	H_2C_b, H_2C_b, $C_a=O$ C—H 1.086 C_a—C_b 1.475 C_b—C_b 1.575 C_a—O 1.191 ∠$C_aC_bC_b$ 57.7 ∠HC_bH 114 dihedral angle between the CH_2 plane and the C_b—C_b bond 151	MW
Cyclopropene	C_aH_2, $HC_b=C_bH$ C_b—C_b 1.304 C_a—C_b 1.519 C_b—H 1.077 C_a—H 1.112 ∠C_bC_bH 133 ∠HC_aH 118	ED
Decalin $C_{10}H_{18}$	C—C (average) 1.530 C—H (average) 1.113 ∠CCC (average) 111.4	ED
Dewer benzene → Bicyclo[2.2.0] hexa-2,5-diene		
Diacetylene → 1,3-Butadiyne		
1,4-Diazabicyclo[2.2.2]octane	$N(CH_2CH_2)_3N$ C—N 1.472 C—C 1.562 ∠NCC 110.2 ∠CNC 108.7 large-amplitude torsional motion about the D_{3h} symmetry axis	ED
2,3-Diaza-1,3-butadiene → Formaldehyde azine		
Diazirine	CH_2 with N=N ring C—H 1.09 C—N 1.482 N—N 1.228 ∠HCH 117	MW
Diazoacetonitrile	H—C_b(=N_b=N_c)—C_a≡N_a C_b—N_b 1.280 N_b—N_c 1.132 C_a—N_a 1.165 C—H 1.082 C_a—C_b 1.424 ∠C_aC_bH 117 ∠$C_aC_bN_b$ 119.5	MW
Diazomethane CH_2N_2	C—H 1.075 C—N 1.32 N—N 1.12 ∠HCH 126.0 linear	MW, IR
1,2-Dibromoethane CH_2BrCH_2Br	C—C 1.506 C—Br 1.950 C—H 1.108 ∠CCBr 109.5 ∠CCH 110 fraction of the *trans* conformer at 25°C 95%	ED
Dibromomethane CH_2Br_2	C—H 1.08 C—Br 1.924 ∠HCBr 109 ∠BrCBr 113.2	ED
2,2′-Dichlorobiphenyl C_6H_4Cl—C_6H_4Cl	C—C 1.398 C—C inter-ring 1.495 C—Cl 1.732 C—H 1.10 ∠CCCl 121.4 ∠CCH 126 dihedral angle between the two aromatic rings 74 (defined to be 0 for that of the *cis* conformer)	ED
trans-1,4-Dichlorocyclohexane $C_6H_{10}Cl_2$	C—H 1.102 C—Cl 1.810 C—C 1.530 ∠CCC 111.5 ∠CCCl (*ee*) 108.6 ∠CCCl (*aa*) 110.6 ∠HCCl (*ee*) 111.5 ∠HCCl (*aa*) 107.6 *ee* 49% *aa* 51% e: equatorial, a: axial	ED

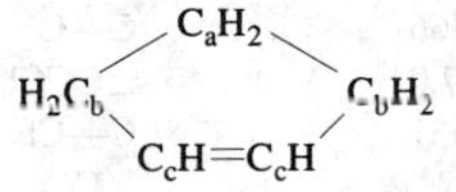

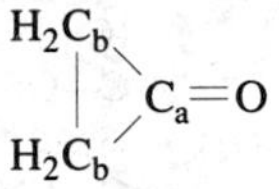

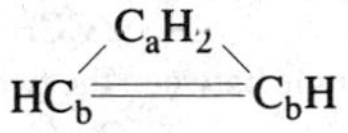

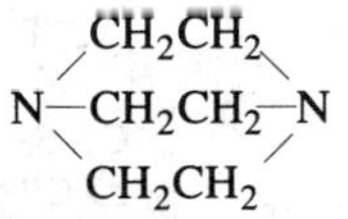

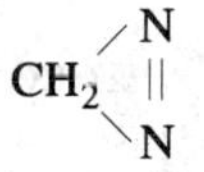

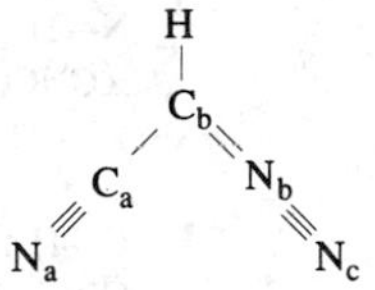

Compound	Structure				Method
1,1-Dichloroethane	C—Cl	1.766	C—C	1.540	MW
$CHCl_2CH_3$	∠ClCCl	112.0	∠CCCl	111.0	
1,2-Dichloroethane	C—C	1.531	C—Cl	1.790	ED
CH_2ClCH_2Cl	C—H	1.11	∠CCCl	109.0	
	∠CCH	113			
	fraction of the *trans* confórmer at room temperature 73%, that of the *gauche* conformer 27%				
1,1-Dichloroethylene	C—C	1.32 (assumed)	C—Cl	1.73	MW
$CH_2{=}CCl_2$	∠ClCC	123	(C_{2v})		
cis-1,2-Dichloroethylene	C—Cl	1.718	C—C	1.354	ED
CHCl=CHCl	∠ClCC	123.8			
Dichloromethane	C—H (r_e)	1.087	C—Cl (r_e)	1.765	MW, IR
CH_2Cl_2	∠HCH (θ_e)	111.5	∠ClCCl (θ_e)	112.0	
1,1-Difluoroethane	C—C	1.498	C—H (average)	1.081	ED
CH_3CHF_2	C—F	1.364	∠CCH (average)	111.0	
	∠CCF 110.7	dihedral angle between the two CCF planes		118.9	
1,2-Difluoroethane	C—F	1.389	C—C	1.503	ED
CH_2FCH_2F	C—H	1.103	∠CCF	110.3	
	∠CCH 111	dihedral angle of internal rotation		109	
	fraction of the *gauche* conformer at 22°C 94%				
1,1-Difluoroethane	C—C	1.340	C—F	1.315	ED, MW
$CH_2{=}CF_2$	C—H	1.091	∠CCF	124.7	
	∠CCH	119.0			
cis-1,2-Difluoroethylene	C—C	1.33	C—F	1.342	ED, MW
CHF=CHF	C—H	1.099	∠CCF	122.0	
	∠CCH	124.1			
Difluoromethane	C—H	1.093	C—F	1.357	MW
CH_2F_2	∠HCH	113.7	∠FCF	108.3	
Dimethoxymethane	H_3C_a—O—C_bH_2—O—C_aH_3		C_a—O	1.432	ED
			C_b—O	1.382	
			C—H (average)	1.108	
	∠COC	114.6	∠OCO	114.3	
	∠OCH	110.3			
Dimethylacetylene → 2-Butyne					
Dimethylamine	C—H	1.106	N—H	1.00	ED
$(CH)_2NH$	C—N	1.455	∠CNC	111.8	
	∠CNH	107	∠NCH	112	
	∠HCH	107			
Dimethylberyllium	Be—C	1.698	C—H	1.127	ED
$(CH_3)_2Be$	∠BeCH	113.9	CBeC linear		
Dimethylcadmium	C—Cd	2.112	∠HCH	108.4	R
$(CH_3)_2Cd$					
Dimethyl carbonate	$(C_aH_3O_a)_2C_b{=}O_b$		C_b—O_b	1.209	ED
			C_b—O_a	1.34	
			C_a—O_a	1.42	
	∠$O_aC_bO_a$	107	∠$C_bO_aC_a$	114.5	
Dimethylcyanamide	C_b—N_b	1.161	C_b—N_a	1.338	ED
$(C_aH_3)_2N_a$—$C_b{\equiv}N_b$	C_a—N_a	1.463	∠C_aNC_a	115.5	
	∠C_aNC_b	116.0			
1,2-Dimethyldiborane	$CH_3(H_t)B(H_b)_2B(H_t)CH_3$		B—B	1.799	ED
			B—C	1.580	
			B—H_b	1.358 (*cis*), 1.365 (*trans*)	
	B—H_t	1.24			
	∠BBC	122.6 (*cis*), 121.8 (*trans*)			
Dimethyl diselenide	C—H	1.13	C—Se	1.95	ED
$(CH_3)_2Se_2$	Se—Se	2.326	∠CSeSe	98.9	
	∠HCSe 108	dihedral angle between the CSeSe and SeSeC planes		88	

BOND LENGTHS AND ANGLES IN GAS-PHASE MOLECULES (continued)

Compound	Structure				Method
Dimethyl disulfide	C—S	1.816	S—S	2.029	ED
$(CH_3)_2S_2$	C—H	1.105	∠SSC	103.2	
	∠SCH 111.3		CSSC dihedral angle of internal rotation	85	
S,S′-Dimethyl dithiocarbonate	$C_aH_3S\underset{\underset{O}{\Vert}}{C_b}SC_aH_3$		C_b—O	1.206	ED
			C_b—S	1.777	
			C_a—S	1.802	
	∠OCS	124.9	∠CSC	99.3	
			syn-syn conformer		
Dimethyl ether	C—O	1.416	C—H	1.121	ED
$(CH_3)_2O$	∠COC	112	∠HCH	108	
Dimethylglyoxal → Biacetyl					
N,N′-Dimethylhydrazine	N—N	1.42	C—N	1.46	ED
$CH_3NH—NHCH_3$	N—H	1.03	C—H	1.12	
	∠NNC 112		CNNC dihedral angle of internal rotation	90	
Dimethylmercury	C—Hg	2.083	C—H	1.160 (assumed)	ED
$(CH_3)_2Hg$	Hg···H	2.71			
Dimethylphosphine	C—P	1.848	P—H	1.419	MW
$(CH_3)_2PH$	∠CPC	99.7	∠CPH	97.0	
Dimethyl selenide	C—H	1.093	Se—C	1.943	MW
$(CH_3)_2Se$	∠CSeC	96.2	∠SeCH	108.7	
	∠HCH	110.3			
Dimethyl sulfide	C—S	1.807	C—H	1.116	ED, MW
$(CH_3)_2S$	∠CSC	99.05	∠HCH	109.3	
Dimethyl sulfone	C—H	1.114	S—O	1.435	ED
$(CH_3)_2SO_2$	S—C	1.771	∠CSC	102	
	∠OSO	121			
Dimethyl sulfoxide	C—H	1.081	C—S	1.799	MW
$(CH_3)_2SO$	S—O	1.485	∠CSC	96.6	
	∠CSO	106.7	∠HCH	110.3	
	dihedral angle between the SCC plane and the S—O bond			115.5	
Dimethylzinc	Zn—C	1.929	∠HCH	107.7	R
$(CH_3)_2Zn$					
1,4-Dioxane	C—C	1.523	C—O	1.423	ED
CH_2CH_2	C—H	1.112	∠COC	112.45	
O O	∠CCO	109.2	chair form		
CH_2CH_2					
Ethanal → Acetaldehyde					
Ethane	C—C	1.5351	C—H	1.0940	MW
C_2H_6	∠CCH	111.17	staggered conformation		
Ethanethiol	$C_bH_3—C_aH_2—SH$		C_a—H	1.090	MW
	C_b—H	1.093	C_a—C_b	1.530	
	C_a—S	1.829	S—H	1.350	
	$∠C_bC_aH$	109.6	$∠C_aC_bH$	109.7	
	$∠C_bC_aS$	108.3	$∠C_aSH$	96.4	
Ethanol	$C_bH_3C_aH_2OH$		C—C	1.512	MW
	C—O	1.431	O—H	0.971	
	C_a—H	1.10	C_b—H	1.09	
	∠CCO	107.8	∠COH	105	
	$∠C_bC_aH$	111	$∠C_aC_bH$	110	
	staggered conformation				
Ethyl chloride	H_b Cl		C—C	1.528	ED, MW
	$H_b—C_b—C_a—H_a$		C—Cl	1.802	
	H_b H_a		C—H	1.103	
			$C_a—H_a$=$C_b—H_b$ (assumed)		
			∠CCCl	110.7	
	$∠H_bC_bH_b$	109.8	$∠H_aC_aH_a$	109.2	
	$∠C_bC_aH_a$	110.6			
Ethylene	C—H	1.087	C—C	1.339	MW
CH_2=CH_2	∠CCH	121.3			

Compound	Structure				Method
Ethylenediamine	C—N	1.469	C—C	1.545	ED
$H_2NCH_2CH_2NH_2$	C—H	1.11	∠CCN	110.2	
	gauche conformer	dihedral angle between the NCC and CCN planes		64	
Ethylene dibromide → 1,2-Dibromoethane					
Ethylene dichloride → 1,2-Dichloroethane					
Ethyleneimine → Aziridine					
Ethylene oxide	C—C	1.466	C—H	1.085	MW
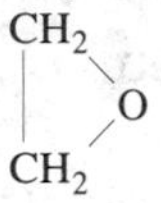 CH_2—O—CH_2 (ring)	C—O	1.431	∠HCH	116.6	
	dihedral angle between the NH_2 plane and the N—C bond			158.0	
Ethylene sulfide → Thiirane					
Ethyl methyl ether	C—O (average)	1.418	C—C	1.520	ED
$C_2H_5OCH_3$	C—H (average)	1.118	∠COC	111.9	
	∠OCC	109.4	∠HCH	109.0	
	fraction of the *trans* conformer at 20°C		80%		
Ethyl methyl ketone	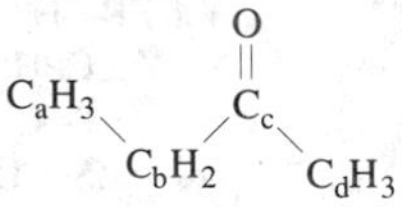 $C_aH_3C_bH_2C_c(=O)C_dH_3$		C—C (average)	1.518	ED
			C_c—O	1.219	
			C—H (average)	1.102	
			$\angle C_aC_bC_c$	113.5	
	$\angle C_bC_cO$, $\angle C_dC_cO$	121.9	*trans* conformer	95%	
Ethyl methyl sulfide	C—S (average)	1.813	C—C	1.536	ED
$C_2H_5SCH_3$	C—H	1.111	∠CSC	97	
	∠SCC	114.0	∠HCH	110	
	fraction of the *gauche* conformer at 20°C		75%		
Ferrocene	C—C	1.440	C—H	1.104	ED
$(C_5H_5)_2Fe$	Fe—C	2.064	(D_{5h})		
Fluoroform	C—H	1.098	C—F	1.332	MW
CHF_3	∠FCF	108.8	(C_{3v})		
Formaldehyde	C—H	1.116	C—O	1.208	MW
H_2CO	∠HCH	116.5			
Formaldehyde azine	H_2C=N—N=CH_2	N—N	1.418		ED
	C—N	1.277	C—H	1.094	
	∠CNN	111.4	∠HCN	120.7	
	fraction of the *trans* conformer at –30°C		91%		
Formaldehyde dimethylacetal → Dimethoxy-methane					
Formaldoxime	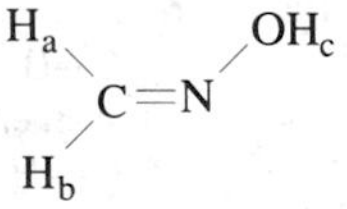 $(H_a)(H_b)C=N-OH_c$		$C—H_a$	1.085	MW
			$C—H_b$	1.086	
			C—N	1.276	
	N—O	1.408	$O—H_c$	0.956	
	$\angle H_bCN$	115.6	∠CNO	110.2	
	$\angle H_aCN$	121.8	$\angle NOH_c$	102.7	
Formamide	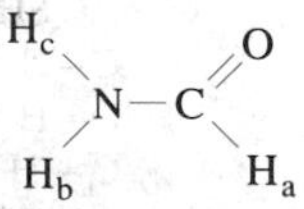 $(H_c)(H_b)N-C(=O)H_a$		$C—H_a$	1.125	ED, MW
			N—H	1.027	
			C—N	1.368	
	C—O	1.212	∠NCO	125.0	
	∠CNH (average)	119.2			
Formic acid	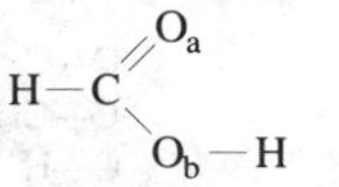 $H-C(=O_a)-O_b-H$		$C—O_a$	1.202	MW
			$C—O_b$	1.343	
	O_b—H	0.972	C—H	1.097	
	$\angle HCO_a$	124.1	$\angle O_aCO_b$	124.9	
	$\angle CO_bH$	106.3	planar		

Compound	Structure				Method
Formic acid dimer	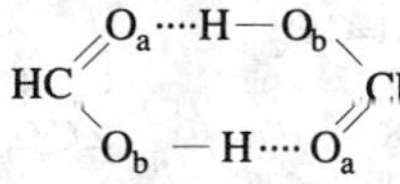		$O_a\cdots O_b$	2.703	ED
			$C—O_a$	1.220	
			$C—O_b$	1.323	
			$\angle O_aCO_b$	126.2	
			$\angle CO_aO_b$	108.5	
Formyl radical	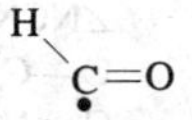 C—H	1.110	C—O	1.1712	MW
	∠HCO	127.43			
Fulvene	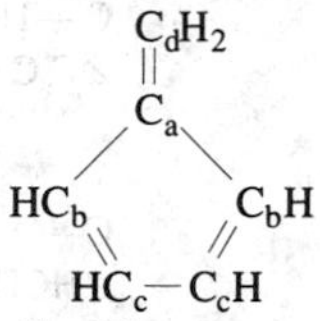		$C_a—C_d$	1.349	MW
			$C_a—C_b$	1.470	
			$C_b—C_c$	1.355	
			$C_c—C_c$	1.476	
			$C_b—H$	1.078	
			$C_c—H$	1.080	
	$C_d—H$	1.13	$\angle C_bC_aC_b$	106.6	
	$\angle C_aC_bC_c$	107.7	$\angle C_bC_cC_c$	109	
	$\angle C_aC_bH$	124.7	$\angle C_bC_cH$	126.4	
	$\angle HC_dH$	117			
2-Furaldehyde	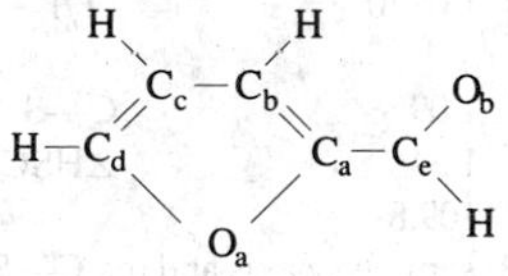		$C_a—C_e$	1.458	MW
			$C_e—O_b$	1.250	
			$C_e—H$	1.088	
			$\angle C_eC_aC_b$	133.9	
	$\angle C_aC_eH$	116.9	$\angle C_aC_eO$	121.6	
	trans conformer (with respect to the O_a and O_b atoms)				
Furan	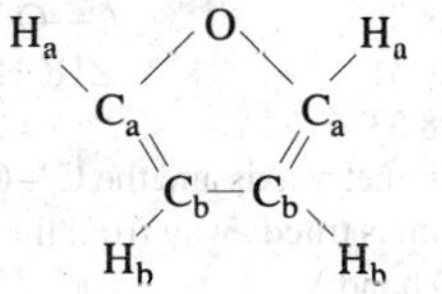		$C_b—C_b$	1.431	MW
			$C_a—C_b$	1.361	
			$C_a—O$	1.362	
			$C_a—H_a$	1.075	
			$C_b—H_b$	1.077	
	$\angle C_aC_bC_b$	106.1	$\angle C_aOC_a$	106.6	
	$\angle C_bC_aO$	110.7	$\angle OC_aH_a$	115.9	
	$\angle C_bC_bH_b$	128.0			
Furfural → 2-Furaldehyde					
Glycolaldehyde	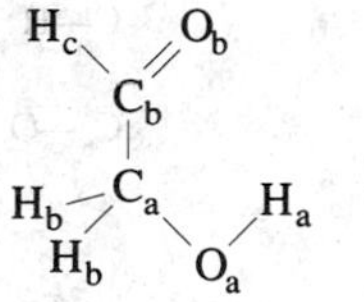		$C_b—O_b$	1.209	MW
			$C_a—O_a$	1.437	
			$C_a—C_b$	1.499	
			$O_a—H_a$	1.051	
			$C_b—H_c$	1.102	
			$C_a—H_b$	1.093	
	$\angle C_aC_bO_b$	122.7	$\angle C_aC_bH_c$	115.3	
	$\angle C_bC_aO_a$	111.5	$\angle C_aO_aH_a$	101.6	
	$\angle C_bC_aH_b$	109.2	$\angle H_bC_aH_b$	107.6	
	$\angle H_bC_aO_a$	109.7			
Glyoxal	C—C	1.526	C—O	1.212	ED, UV
CHOCHO	C—H	1.132	∠CCO	121.2	
	∠HCO	112	*trans* conformer	(C_{2h} (assumed))	
Hexachloroethane	C—C	1.56	C—Cl	1.769	ED
Cl_3CCCl_3	∠CCCl	110.0			
2,4-Hexadiyne	$C_aH_3—C_b\equiv C_c—C_c\equiv C_b—C_aH_3$				ED
	$C_a—C_b$	1.450	$C_b—C_c$	1.208	
	$C_c—C_c$	1.377	$C_a—H$	1.09	
Hexafluoroethane	C—C	1.545	C—F	1.326	ED
F_3CCF_3	∠CCF	109.8	staggered conformation		
Hexafluoropropene	average value of the C=C and C—F distances			1.329	ED
CF_2=CFCF$_3$	C—C	1.513	∠CCC	127.8	
	∠FCC (CF_2)	124	∠FCC (CF)	120	
	∠FCC (CF_3)	110			

Compound	Structure				Method
1,3,5-Hexatriene	H_2C_a=C_bH—C_cH=C_cH—C_bH=C_aH_2				ED
	C_a—C_b	1.337	C_b—C_c	1.458	
	C_c—C_c	1.368	∠$C_aC_bC_c$	121.7	
	∠$C_bC_cC_c$	124.4			
Iminocyanide radical	N—H	1.034	N···N	2.470	UV
HNCN	∠HNC	116.5	∠NCN	~180	
Iodocyanoacetylene	I—C_a	1.985	C_a—C_b	1.207	MW
I—C_a≡C_b—C_c≡N	C_b—C_c	1.370	C_c—N	1.160	
Isobutane	C_a—H	1.122	C_b—H	1.113	ED, MW
$(C_bH_3)_3C_aH$	C_a—C_b	1.535	∠$C_bC_aC_b$	110.8	
	∠C_aC_bH	111.4			
Isobutylene → 2-Methylpropene					
Ketene	C—C	1.317	C—O	1.161	MW
CH_2=C=O	C—H	1.080	∠HCH	123.0	
Malononitrile	C—H	1.091	C—C	1.480	MW
$C_aH_2(C_bN)_2$	C—N	1.147	∠CCC	110.4	
	∠HCH	108.4	∠CCN	176.6	
	(The two N atoms are bent away from each other in the plane of C_b—C_a—C_b)				
Methane	C—H (r_e)	1.0870	(T_d)		MW
CH_4					
Methanethiol	C—H	1.09	C—S	1.819	MW
CH_3SH	S—H	1.34	∠HSC	96.5	
	∠HCH	109.8			
	angle between the CH_3 symmetry axis and the C—S bond 2.2. (The axis of the CH_3 group is tilted away from the H atom with respect to the C—S bond.)				
Methanol	C—H	1.0936	C—O	1.4246	MW
CH_3OH	O—H	0.9451	∠HCH	108.63	
	∠COH	108.53			
	angle between the CH_3 symmetry axis and the C—O bond (The axis of the CH_3 group is tilted away from the H atom with respect to the C—O bond.)			3.27	
Methyl radical	C—H	1.08	planar		UV
$\cdot CH_3$					
N-Methylacetamide	H_3C_a \ C_b—N (H) / C_cH_3; C_b=O		C_a—C_b	1.520	ED
			N—C_c	1.469	
			C—H	1.107	
	C_b—N	1.386	C_b—O	1.225	
	∠C_bNC_c	119.7			
	∠NC_bO	121.8			
	∠C_aC_bN	114.1			
Methylacetylene → Propyne					
Methylal → Dimethoxymethane					
Methylamine	N—H	1.010	C—N	1.471	MW
CH_3NH_2	C—H	1.099	∠NHN	107.1	
	∠HNC	110.3	∠HCH	108.0	
	dihedral angle between the CH_3 symmetry axis and the C—N bond (The axis of the CH_3 group is tilted away from the NH_2 group with respect to the C—N bond.)			2.9	
Methyl azide	CH_3—N_a—N_b—N_c		C—H	1.09	ED
			C—N_a	1.468	
			N_a—N_b	1.216	
	N_b—N_c	1.113	∠CN_aN_b	116.8	
	NNN linear				
Methyl bromide	C—H (r_e)	1.086	C—Br (r_e)	1.933	MW, IR
CH_3Br	∠HCH (θ_e)	111.2	(C_{3v})		
Methyl chloride	C—H	1.090	C—Cl	1.785	MW, IR
CH_3Cl	∠HCH	110.8			

BOND LENGTHS AND ANGLES IN GAS-PHASE MOLECULES (continued)

Compound	Structure				Method
Methyldiazirine	$CH_3CH\langle N{=}N$ (ring)		C—N 1.481	C—C 1.501	MW
			N N 1.235	∠NCN 49.3	
	dihedral angle between the CNN plane and the C—C bond			122.3	
Methylene :CH_2	C—H	1.078	∠HCH	130	LMR
Methylenecyclopropane	C_cH_2–C_cH_2 ring, $C_b{=}C_aH_2$		C_a—C_b	1.332	MW
			C_b—C_c	1.457	
			C_c—C_c	1.542	
	C_c—H	1.09	$∠C_cC_bC_c$	63.9	
	$∠HC_aH$	114.3	$∠HC_cH$	113.5	
	dihedral angle between the C_cH_2 plane and the C_c—C_c bond			150.8	
3-Methyleneoxetane	O(C_cH_2)(C_cH_2) ring, $C_b{=}C_aH_2$		C_b—C_c	1.52	MW
			C_c—O	1.45	
			C_a—C_b	1.33	
	C—H	1.09 (assumed)	$∠C_cC_bC_c$	87	
	$∠HC_cH$	114 (assumed)	$∠HC_aH$	120 (assumed)	
Methyl fluoride CH_3F	C—H (r_e)	1.095			MW, IR
	C—F (r_e)	1.382	∠HCH (θ_e)	110.45 (C_{3v})	
Methyl formate	C_aH_3—O_a—C_b($=O_b$)—H_b		C_a—H	1.08	ED
			C_b—O_b	1.206	
			C—O (average)	1.393	
			C_b—H	1.101 (assumed)	
	$∠O_aC_bO_b$	127	∠COC	114	
	$∠O_aC_aH$	110			
Methylgermane CH_3GeH_3	C—H	1.083	Ge—H	1.529	MW
	C—Ge	1.945	∠HCH	108.4	
	∠HGeH	109.3			
Methyl hypochlorite CH_3OCl	C—H	1.103	O—Cl	1.674	MW
	O—C	1.389	∠HCH	109.6	
	∠COCl	112.8			
Methylidyne radical :ĊH	C—H (r_e)	1.1198			UV
Methylidyne phosphide HCP	H—C (r_e)	1.0692	C—P (r_e)	1.5398	MW
Methyl iodide CH_3I	C—H (r_e)	1.084	C—I (r_e)	2.132	MW, IR
	∠HCH (θ_e)	111.2	(C_{3v})		
Methyl isocyanide	C_aH_3—N≡C_b		C_a—H 1.102	C_a—N 1.424	MW
	N—C_b	1.166	$∠NC_aH$	109.12	
Methylketene	C_cH_3(C_bH)$=C_a=O$		O—C_a	1.171	MW
			C_b—C_c	1.518	
			C_c—H	1.10	
	C_a—C_b	1.306	C_b—H	1.083	
	$∠OC_aC_b$	180.5	$∠C_aC_bC_c$	122.6	
	$∠C_aC_bH$	113.7	$∠C_cC_bH$	123.7	
	∠HCH	109.2			
Methylmercury chloride CH_3HgCl	Hg—Cl	2.282	C—H	1.15	MW, NMR
	Hg—C	1.99	(C_{3v})		
Methyl nitrate	H_a,H_a,H_bC—O—N($=O_a$)$=O_b$		C—H_a	1.10	MW
			C—H_b	1.09	
			C—O	1.437	
			O—N	1.402	
	N—O_a	1.205	N—O_b	1.208	
	$∠OCH_a$	110	$∠OCH_b$	103	
	∠CON	112.7	$∠ONO_a$	118.1	
	$∠ONO_b$	112.4			

Compound	Structure						Method
Methylphosphine CH_3PH_2	C—P	1.858	C—H	1.094			ED
2-Methylpropane → Isobutane							
2-Methylpropene			C_a—H	1.119			ED, MW
			C_c—H_c	1.10			
			C_a—C_b	1.508			
			C_b—C_c	1.342			
	∠HC_aC_b (average)	111.4	∠$H_cC_cH_c$	118.5			
	∠$C_aC_bC_a$	115.6	∠$C_aC_bC_c$	122.2			
	∠HC_aH	107.9	∠C_bC_cH	121			
Methylsilane CH_3SiH_3	C—H	1.093	C—Si	1.867			MW
	Si—H	1.485	∠HCH	107.7			
	∠HSiH	108.3	(C_{3v})				
Methylstannane CH_3SnH_3	C—Sn	2.143	Sn—H	1.700			MW
			(C_{3v})				
Methyl thiocyanate			S—C_b	1.684	S—C_a	1.824	MW
			C_b—N	1.170	C—H	1.081	
	∠C_aSC_b	99.0	∠HCH	110.6	∠HCS	108.3	
Naphthalene			C_a—C_b	1.37			ED
			C_b—C_b	1.41			
			C_a—C_c	1.42			
			C_c—C_c	1.42			
			C—C (average)	1.40			
			∠$C_aC_cC_c$	119.4			
Neopentane $C(CH_3)_4$	C—C	1.537	C—H	1.114			ED
	∠CCH	112					
Nickelocene → Bis (cyclopentadienyl) nickel							
Nitromethane CH_3NO_2	C—H	1.088 (assumed)	C—N	1.489			MW
	N—O	1.224	∠NCH	107			
	∠ONO	125.3					
N-Nitrosodimethylamine $(CH_3)_2NNO$	N—O	1.235	N—N	1.344			ED
	C—N	1.461	∠ONN	113.6			
	∠CNC	123.2	∠CNN	116.4			
Nitrosomethane CH_3NO	C—N	1.49	N—O	1.22			MW
	C—H	1.084	∠CNO	112.6			
	∠NCH	109.0					
Norbornane → Bicyclo[2.2.1]heptane							
Norbornadiene → Bicyclo[2.2.1]hepta-2,5-diene							
1,2,5-Oxadiazole			O—N	1.380	∠NON	110.4	MW
			C—N	1.300	∠ONC	105.8	
			C—C	1.421	∠CCN	109.0	
			C—H	1.076	∠CCH	130.2	
			∠NCH	120.9	planar		
1,3,4-Oxadiazole			O—C	1.348	∠COC	102.0	MW
			C—N	1.297	∠OCN	113.4	
			N—N	1.399	∠CNN	105.6	
			C—H	1.075	∠OCH	118.1	
			∠NCH	128.5	planar		
Oxalic acid			C—C	1.544			ED
			C—O_a	1.205			
			C—O_b	1.336			
			O_b—H	1.05			
			∠CCO_a	123.1			
			∠O_aCO_b	125.0			
			∠CO_bH	104			

Compound	Structure				Method
Oxalyl dichloride	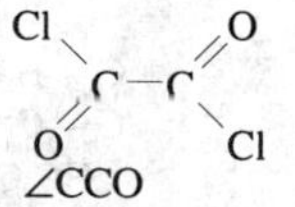 Cl(O=)C—C(=O)Cl		C—O	1.182	ED
			C—C	1.534	
			C—Cl	1.744	
	∠CCO	124.2	∠CCCl	111.7	
	fraction of the *trans* conformer at 0°C 68%, that of the gauche conformer 32%				
Oxetane	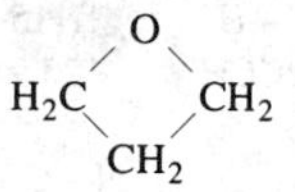 H_2C—O—CH_2, CH_2 (ring)		C—O	1.448	MW
			C—C	1.546	
			C—H (average)	1.090	
			∠COC	92	
	∠CCC	85	∠OCC	92	
	∠HCH (average)	109.9			
Oxirane → Ethylene oxide					
Phenol	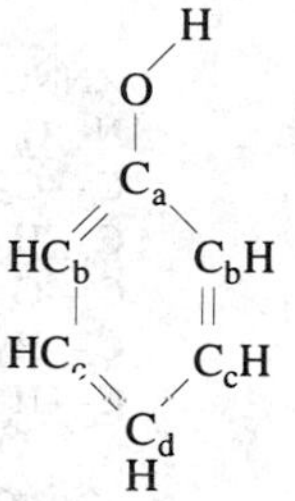 H—O—C_a; HC_b, C_bH, HC_c, C_cH, C_d—H (ring)		C—C (average)	1.397	MW
			C_b—H	1.084	
			C_c—H	1.076	
			C_d—H	1.082	
			C_a—O	1.364	
			O—H	0.956	
			∠COH	109.0	
Phosphirane	CH_2—CH_2—PH (ring)		C—P 1.867	P—H 1.43	MW
			C—C 1.502	C—H 1.09	
			∠CPC 47.4	∠HCH 114.4	
	∠HPC	95.2	∠CCH	118	
	dihedral angle between the PCC plane and the PH bond			95.7	
Piperazine	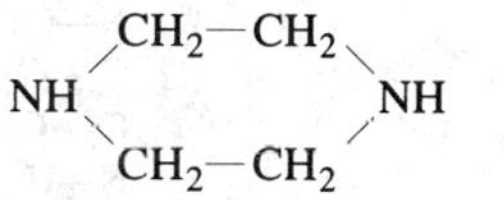 NH(CH_2—CH_2)$_2$NH (ring)		C—C	1.540	ED
			C—N	1.467	
			C—H	1.110	
	∠CNC	109.0	∠CCN (C_{2h})	110.4	
Pivalonitrile $(C_cH_3)_3C_b$—C_a≡N	C_a—C_b	1.495	C_a—N	1.159	MW
	C_b—C_c	1.536	∠$C_cC_bC_c$	110.5	
Propadiene → Allene					
Propane C_3H_8	C—C	1.532	C—H	1.107	ED
	∠CCC	112	∠HCH	107	
Propenal → Acrylaldehyde					
Propene	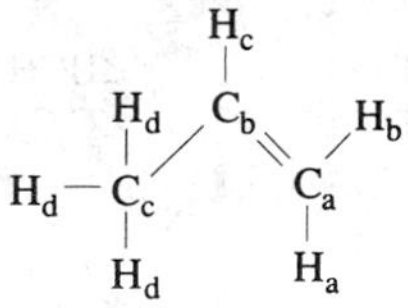 H_d—C_c(H_d)(H_d)—C_b(H_c)=C_a(H_a)—… H_b		C_a—H_a	1.104	ED, MW
			C_a—C_b	1.341	
			C_c—H_d	1.117	
			C_b—C_c	1.506	
			∠$C_bC_aH_{a,b,c}$	121.3	
	∠$C_bC_cH_d$	110.7	∠$C_aC_bC_c$	124.3	
1-Propenyl chloride	CH_3—C_bH=C_aH—Cl	C_a—Cl	1.728	MW	
	∠C_bC_aCl	121.9	*trans* conformer		
Propiolaldehyde	H_aC_a≡C_b—C_cH_cO		C_a—H_a	1.085	ED, MW
	C_a—C_b	1.211	C_b—C_c	1.453	
	C_c—H_c	1.130	C_c—O	1.214	
	∠C_bC_cO	124.2	∠$C_bC_cH_c$	113.7	
	∠$C_aC_bC_c$	178.6	planar		
Propylene → Propene					
Propylene oxide	$C_aH_3C_bH$—C_cH_2 (with O bridging C_b and C_c)		C_a—C_b	1.51	MW
			∠$C_aC_bC_c$	121.0	
			dihedral angle between the C_bC_cO plane and the C_aC_b bond	123.8	

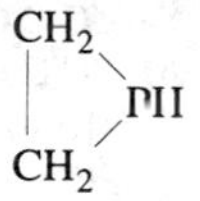

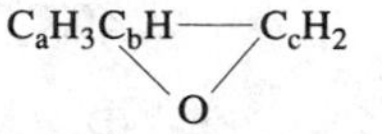

Compound	Structure					Method
Propynal → Propiolaldehyde						
Propyne	$H_3C_c-C_b\equiv C_aH$			C_c-H	1.105	MW
		C_c-C_b	1.459	C_b-C_a	1.206	
		C_a-H	1.056	$\angle HC_cC_b$	110.2	
Pyrazine		C—C	1.339	C—N	1.403	ED
		C—H	1.115	∠CCN	115.6	
		∠CCH	123.9			
Pyridazine				$N-C_a$	1.341	ED, MW
				C_a-C_b	1.393	
				N—N	1.330	
				C_b-C_b	1.375	
		∠NCC	123.7	∠NNC	119.3	
Pyridine				$N-C_a$	1.340	MW
				C_b-C_c	1.394	
				C_b-H_b	1.081	
				C_a-C_b	1.395	
				C_a-H_a	1.084	
				C_c-H_c	1.077	
		$\angle C_aNC_a$	116.8	$\angle NC_aC_b$	123.9	
		$\angle C_aC_bC_c$	118.5	$\angle C_bC_cC_b$	118.3	
		$\angle NC_aH_a$	115.9	$\angle C_cC_bH_b$	121.3	
Pyrimidine		N—C	1.340	C—C	1.393	ED
		∠NCN	127.6	∠CNC	115.5	
		(C_{2v} assumed)				
Pyrrole				N—Ca	1.370	MW
				C_b-C_b	1.417	
				C_a-C_b	1.382	
				N—H	0.996	
				C_a-H_a	1.076	
		C_b-H_b	1.077	$\angle C_aNC_a$	109.8	
		$\angle NC_aC_b$	107.7	$\angle C_aC_bC_b$	107.4	
		$\angle NC_aH_a$	121.5	$\angle C_bC_bH$	127.1	
Pyruvonitrile	$C_aH_3-C_b(=O)-C_c\equiv N$			C—H	1.12	ED, MW
				C—N	1.17	
				C—O	1.208	
				C_b-C_c	1.477	
		C_a-C_b	1.518	∠HCH	109.2	
		$\angle C_aC_bO$	124.5	$\angle C_aC_bC_c$	114.2	
		∠CCN	179			
Ruthenocene → Bis (cyclopentadienyl) ruthenium						
Silacyclobutane	CH_2-CH_2 / CH_2-SiH_2			Si—C	1.892	ED
				C—C	1.600	
				Si—H	1.47	
		C—H	1.14	∠CSiC	80.7	
		∠SiCC	84.8	∠CCC	99.8	
		dihedral angle between the CCC and CSiC planes			146	
Spiropentane				C_b-C_b	1.52	ED
				C_a-C_b	1.47	
				C—H	1.09	
				$\angle C_bC_aC_b$	62	
		∠HCH	118	(D_{2d})		

Compound	Structure	Method
Succinonitrile	CH_2CN | CH_2CN C—C 1.561; C—N 1.161; ∠CCC 110.4; C—C(N) 1.465; C—H 1.09 fraction of the *anti* conformer at 170 C 74%, dihedral angle of CCCC for the *gauche* conformer 75	ED
Tetrachloroethylene CCl_2=CCl_2	C—Cl 1.718; C—C 1.354 ∠ClCCl 115.7	ED
Tetracyanoethylene $(CN)_2C$=$C(CN)_2$	C—N 1.162; C—C 1.435 C=C 1.357; ∠CC=C 121.1	ED
Tetrafluoro-1,3-dithietane	F_2C(S)(S)CF_2 (four-membered ring) C—S 1.785; C—F 1.314; ∠CSC 83.2 ∠FCS 113.7; (D_{2h} assumed)	ED
Tetrafluoroethylene CF_2=CF_2	C—C 1.31; C—F 1.319 ∠CCF 123.8; (D_{2h} assumed)	ED
Tetrahydrofuran CH_2CH_2—O—CH_2CH_2 (ring)	C—H 1.115; C—O 1.428 C—C 1.536 The skeletal bending vibration of the molecular plane is essentially free pseudorotation	ED
Tetrahydropyran	H_2C, CH_2, H_2C, CH_2, H_2C, O (six-membered ring) C—O 1.420; C—C 1.531; C—H 1.116; ∠COC 111.5; ∠OCC 111.8 ∠CCC (C) 108; ∠CCC (O) 111 chair form	ED
Tetrahydrothiophene CH_2CH_2—S—CH_2CH_2 (ring)	C—S 1.839; C—H 1.120 C—C 1.536; ∠CSC 93.4 ∠SCC 106.1; ∠CCC 105.0	
Tetramethylgermane $(CH_3)_4Ge$	Ge—C 1.945; C—H 1.12 ∠GeCH 108; (T_d excluding the H atoms)	ED
Tetramethyllead $(CH_3)_4Pb$	Pb—C 2.238; (T_d excluding the H atoms)	ED
Tetramethylsilane $(CH_3)_4Si$	C—H 1.115; C—Si 1.875 ∠HCH 109.8; (T_d excluding the H atoms)	ED
Tetramethylstannane $(CH_3)_4Sn$	C—Sn 2.144 C—H 1.12; (T_d excluding the H atoms)	ED
1,2,5-Thiadiazole	S, N, N, HC—CH (ring) S—N 1.631; ∠NSN 99.6 C—N 1.328; ∠CCN 113.8 C—C 1.420; ∠CCH 126.2 C—H 1.079; planar	MW
1,3,4-Thiadiazole	S, HC, CH, N—N (ring) S—C 1.721; ∠CSC 86.4 N—N 1.371; ∠SCN 114.6 C—N 1.302; ∠CCN 112.2 C—H 1.08; ∠SCH 121.9 ∠NCH 123.5; planar	MW
Thietane	CH_2—CH_2 | | CH_2—S C—S 1.847; C—C 1.549; C—H (average) 1.100 ∠CSC 76.8; ∠HCH (average) 112 dihedral angle between the CCC and CSC planes 154	ED, MW
Thiirane	H_2C—S—H_2C (ring) C—C 1.484; ∠HCH 116 C—H 1.083; ∠CSC 48.3 C—S 1.815; ∠CCS 65.9 dihedral angle between the CH_2 plane and the C—C bond 152	MW

Compound	Structure				Method
Thioformaldehyde	C—S	1.611	C—H	1.093	MW
CH_2S	∠HCH	116.9			
Thioformamide	S=C(—H_c)—N(H_a)(H_b)		N—H_a	1.002	MW
			N—H_b	1.007	
			C—N	1.358	
	C—S	1.626	C—H_c	1.10	
	∠H_aNH_b	121.7	∠H_aNC	117.9	
	∠H_bNC	120.4	∠NCS	125.3	
	∠NCH_c	108	∠SCH_c	127	
Thiolane → Tetrahydrothiophene					
Thiophene	H_b, C_b—C_b, H_b; H_aC_a, C_aH_a; S		C_a—H_a	1.078	MW
			C_b—H_b	1.081	
			C_a—S	1.714	
			C_a—C_b	1.370	
			C_b—C_b	1.423	
			∠C_aSC_a	92.2	
	∠S$C_a$$C_b$	115.5	∠$C_a$$C_b$$C_b$	112.5	
	∠S$C_a$$H_a$	119.9	∠$C_b$$C_b$$H_b$	124.3	
Toluene	CH_3 (on benzene ring)	C—C (ring) 1.399	C—CH_3	1.524	ED
		C—H (average) 1.11			
		the difference between the C—H(CH_3) and C—H(ring): about 0.01			
1,1,1-Tribromoethane	C—Br	1.93	C—H	1.095 (assumed)	MW
CH_3CBr_3	C—C	1.51 (assumed)	∠CCBr	108	
	∠BrCBr	111	∠CCH	109.0 (assumed)	
Tribromomethane → Bromoform					
Tri-*tert*-butyl methane	C_a—C_b	1.611	C—H	1.111	ED
$HC_a[C_b(C_cH_3)_3]_3$	C_b—C_c	1.548	∠$C_a$$C_b$$C_c$	113.0	
Tricarbon dioxide	C—O	1.163	C—C	1.289	ED
OCCCO	linear (with a large-amplitude bending vibration)				
Trichloroacetonitrile	C—N	1.165	C—C	1.460	ED
CCl_3CN	C—Cl	1.763	∠ClCCl	110.0	
1,1,1-Trichloroethane	C—H	1.090	C—C	1.541	MW
CH_3CCl_3	C—Cl	1.771	∠HCH	110.0	
	∠CCH	108.9	∠ClCCl	109.4	
	∠CCCl	109.6			
Trichloro(methyl)germane	Ge—Cl	2.132	Ge—C	1.89	ED, MW
CH_3GeCl_3	C—H	1.103 (assumed)	∠ClGeCl	106.4	
	∠GeCH	110.5 (assumed)			
Trichloro(methyl)silane	C—Si	1.876	Si—Cl	2.021	MW
CH_3SiCl_3			(C_{3v})		
Trichloro(methyl)stannane	Sn—Cl	2.304	Sn—C	2.10	ED
CH_3SnCl_3	C—H	1.100	∠CSnCl	113.9	
	∠ClSnCl	104.7	∠SnCH	108	
Triethylenediamine → 1,4-Diazabicyclo [2.2.2]octane					
Trifluoroacetic acid	$CF_3C(=O_a)O_bH$		C—F	1.325	ED
			C—C	1.546	
			C—O_a	1.192	
	C—O_b	1.35	O—H	0.96 (assumed)	
	∠CCO_a	126.8	∠CCO_b	111.1	
	∠CCF	109.5			
1,1,1-Trifluoroethane	C—C	1.494	C—F	1.340	ED
CH_3CF_3	C—H	1.081	∠CCF	119.2	
	∠CCH	112			
Trifluoromethane → Fluoroform					

Compound	Structure				Method
1,1,1-Trifluoro-2,2,2-trichloroethane	C—C	1.54	C—F	1.33	MW
CF_3CCl_3	C—Cl	1.77	∠CCF	110	
	∠CCCl	109.6	staggered conformation		
Trimethylaluminium	C—H	1.113	Al—C	1.957	ED
$(CH_3)_3Al$	∠AlCH	111.7	∠CAlC	120	
Trimethylamine	C—N	1.458	C—H	1.100	ED
$(CH_3)_3N$	∠CNC	110.9	∠HCH	110	
Trimethylarsine	C—As	1.979	∠CAsC	98.8	ED
$(CH_3)_3As$	∠AsCH	111.4			
Trimethylbismuth	Bi—C	2.263	C—H	1.07	ED
$(CH_3)_3Bi$	∠CBiC	97.1			
Trimethylborane	C—B	1.578	C—H	1.114	ED
$(CH_3)_3B$	∠CBC	120.0	∠BCH	112.5	
Trimethyleneimine → Azetidine					
Trimethylphosphine	C—P	1.847	C—H	1.091	ED
$(CH_3)_3P$	∠CPC	98.6	∠PCH	110.7	
1,3,5-Trioxane	(ring: O, H_2C, CH_2, O, O, C H_2)		C—O	1.422	MW
			∠OCO	112.2	
			∠COC	110.3	
Triphenylamine	C—C	1.392	C—N	1.42	ED
$(C_6H_5)_3N$	∠CNC	116	(C_3)		
	torsional dihedral angle of the two phenyl rings 47 (defined to be 0 when the symmetry axis is contained in the phenyl planes)				
Tropone	(ring: O, C_a, HC_b, C_bH, HC_c, C_cH, C_d=C_d, H H)		C_a—O	1.23	ED
			C_a—C_b	1.45	
			C_b—C_c	1.36	
			C_c—C_d	1.46	
			C_d—C_d	1.34	
			$∠C_bC_aC_b$	122	
			$∠C_aC_bC_c$	133	
	$∠C_bC_cC_d$	126	$∠C_cC_dC_d$	130	
				(C_{2v})	
Vinylacetylene	(H_a, H_b, H_c, C_a=C_b, C_c≡C_d, H_d)		C_b—C_c	1.434	ED, MW
			C_a—C_b	1.344	
			C_c—C_d	1.215	
			C_a—H_a	1.11	
			C_d—H_d	1.09	
			$∠C_aC_bC_c$	123.1	
	$∠C_bC_cC_d$	178	$∠H_aC_aC_b$	119	
	$∠H_bC_aC_b$	122	$∠H_cC_bC_a$	122	
	$∠C_cC_dH_d$	182			
Vinyl chloride	(H_c, Cl, C=C, H_b, H_a)		C—C	1.342	ED, MW
			C—Cl	1.730	
			C—H	1.09	
	∠CCCl	122.5	$∠CCH_a$	124	
	$∠CCH_b$	120	$∠CCH_c$	121.1	

DIPOLE MOMENTS OF MOLECULES IN THE GAS PHASE

David R. Lide

This table gives values of the electric dipole moment for about 500 gas-phase molecules. When available, values determined by microwave spectroscopy, molecular beam electric resonance, and other spectroscopic techniques were selected; otherwise, the values come from measurements of the dielectric constant in the gas phase. Compounds are listed by molecular formula in the modified Hill order; compounds not containing carbon are listed first, followed by compounds containing carbon.

The dipole moment μ is given in debye units (D). The conversion factor to SI units is $1\ D = 3.33564 \times 10^{-30}$ C m. The column headed Acc. indicates the estimated accuracy of the value through the following codes:

A ±1% for μ > 1.0 D; ±0.01 D for μ < 1.0 D
B ±2% for μ > 1.0 D; ±0.02 D for μ < 1.0 D
C ±5% for μ > 1.0 D; ±0.05 D for μ < 1.0 D
D ±10% for μ > 1.0 D; ±0.10D for μ < 1.0D
Q Questionable value
i Indicates the value may be ambiguous because of the presence of rotational isomers

Dipole moments of individual rotational isomers are given when they have been measured. The isomers are designated as *gauche*, *trans-trans*, etc. The meaning of these designations can be found in the references. Other information, such as the components of the dipole moment in the molecular framework and the variation with vibrational state and isotopic species, may also be found in the references.

REFERENCES

1. R. D. Nelson, D. R. Lide, and A. A. Maryott, *Selected Values of Electric Dipole Moments for Molecules in the Gas Phase*, Natl. Stand. Ref. Data Ser. — Natl. Bur. Stnds. 10, 1967.
2. K. H. Hellwege, Ed., *Landolt-Börnstein, Numerical Data and Functional Relationships in Science and Technology*, Group II, Vol. 6, Molecular Constants, Springer-Verlag, Heidelberg, 1974.
3. K. H. Hellwege, Ed., *Landolt-Börnstein, Numerical Data and Functional Relationships in Science and Technology*, Group II, Vol. 14, Subvolume a, Springer-Verlag, Heidelberg, 1982.
4. K. H. Hellwege, Ed., *Landolt-Börnstein, Numerical Data and Functional Relationships in Science and Technology*, Group II, Vol. 14, Subvolume b, Springer-Verlag, Heidelberg, 1983.
5. R. Takashi, et. al., *J. Mol. Spec.*, 138, 450, 1989.
6. M. Hayashi and T. Inagusa, *J. Mol. Spec.*, 138, 135, 1989.
7. F. Scappini, A. C. Fantoni, and W. Caminati, *J. Mol. Spec.*, 120, 101, 1986.

Mol. form.	Name	μ/D	Acc.	Ref.
	Compounds not containing carbon			
AgCl	Silver chloride	5.70	C	2
AgF	Silver fluoride	6.22	C	2
AgI	Silver iodide	5.10	C	4
AlF	Aluminum fluoride (AlF)	1.53	D	1
$AsCl_3$	Arsenic trichloride	1.59	C	1
AsF_3	Arsenic trifluoride	2.59	B	1
AsH_3	Arsine	0.20	C	1
BF	Fluoroborane (BF)	0.5	Q	2
B_4H_{10}	Tetraborane	0.486	A	3
B_5H_9	Pentaborane	2.13	B	1
B_6H_{10}	Hexaborane	2.50	B	3
BaO	Barium oxide	7.954	A	2
BaS	Barium sulfide	10.86	A	2
BrCl	Bromine chloride	0.519	A	3
BrF	Bromine fluoride	1.42	A	3
BrF_5	Bromine pentafluoride	1.51	D	1
BrH	Hydrogen bromide	0.827	A	3
BrH_3Si	Bromosilane	1.319	A	3
BrI	Iodine bromide	0.726	A	4

DIPOLE MOMENTS OF MOLECULES IN THE GAS PHASE (continued)

Mol. form.	Name	μ/D	Acc.	Ref.
BrK	Potassium bromide	10.628	A	2
BrLi	Lithium bromide	7.268	A	2
BrNO	Nitrosyl bromide	1.8	Q	1
BrNa	Sodium bromide	9.118	A	2
BrO	Bromine oxide (BrO)	1.76	B	2
BrTl	Thallium bromide (TlBr)	4.49	A	2
CaCl	Calcium monochloride	3.6	Q	4
ClCs	Cesium chloride	10.387	A	2
ClF	Chlorine fluoride	0.888	A	2
$ClFO_3$	Perchloryl fluoride	0.023	A	3
ClF_3	Chlorine trifluoride	0.6	D	1
$ClGeH_3$	Chlorogermane	2.13	A	1
ClH	Hydrogen chloride	1.109	A	3
ClHO	Hypochlorous acid (HOCl)	1.3	Q	2
ClH_3Si	Chlorosilane	1.31	A	1
ClI	Iodine chloride	1.24	B	2
ClIn	Indium chloride	3.79	C	2
ClK	Potassium chloride	10.269	A	2
ClLi	Lithium chloride	7.129	A	2
$ClNO_2$	Nitryl chloride	0.53	A	1
ClNS	Thionitrosyl chloride (NSCl)	1.87	A	2
ClNa	Sodium chloride	9.001	A	2
ClO	Chlorine oxide (ClO)	1.239	A	2
ClRb	Rubidium chloride	10.510	A	2
ClTl	Thallium chloride (TlCl)	4.543	A	2
Cl_2H_2Si	Dichlorosilane	1.17	B	1
Cl_2OS	Thionyl chloride	1.45	B	1
Cl_2O_2S	Sulfuryl chloride	1.81	B	1
Cl_2S	Sulfur dichloride	0.36	A	3
Cl_3FSi	Trichlorofluorosilane	0.49	A	2
Cl_3HSi	Trichlorosilane	0.86	A	2
Cl_3N	Nitrogen trichloride	0.39	A	3
Cl_3OP	Phosphorous oxychloride	2.54	B	2
Cl_3P	Phosphorous trichloride	0.56	B	2
CsF	Cesium fluoride	7.884	A	2
CsNa	Cesium sodium (CsNa)	4.75	C	2
CuF	Copper fluoride (CuF)	5.77	C	2
FGa	Gallium fluoride (GaF)	2.45	B	2
$FGeH_3$	Fluorogermane	2.33	C	2
FH	Hydrogen fluoride	1.826	A	2
FHO	Hypofluorous acid (HOF)	2.23	C	3
FH_3Si	Fluorosilane	1.298	A	3
FI	Iodine fluoride	1.95	A	3
FIn	Indium fluoride (InF)	3.40	B	2
FK	Potassium fluoride	8.585	A	2
FLi	Lithium fluoride	6.326	A	3
FNO	Nitrosyl fluoride	1.730	A	3
FNO_2	Nitryl flouride	0.466	A	2
FNS	Thionitrosyl fluoride (NSF)	1.902	A	2
FNa	Sodium fluoride	8.156	A	2
FRb	Rubidium fluoride	8.546	A	2
FS	Sulfur fluoride (SF)	0.794	B	3
FTl	Thallium fuoride (TlF)	4.228	A	2
F_2Ge	Germanium fluoride (GeF_2)	2.61	A	2
F_2HN	Difluoramine	1.92	A	1
F_2H_2Si	Difluorosilane	1.55	A	1

Mol. form.	Name	μ/D	Acc.	Ref.
F_2N_2	*cis*-Difluorodiazine	0.16	A	1
F_2O	Oxygen difluoride	0.297	A	1
F_2OS	Thionyl fluoride	1.63	A	1
F_2O_2	Dioxygen difluoride (FOOF)	1.44	C	1
F_2O_2S	Sufuryl fluoride	1.12	B	1
F_2S	Sulfur difluoride	1.05	C	2
F_2Si	Difluorosilylene (SiF_2)	1.23	B	2
F_3HSi	Trifluorosilane	1.27	B	1
$F_3H_3Si_2$	1,1,1-Trifluorodisilane	2.03	C	3
F_3N	Nitrogen trifluoride	0.235	A	1
F_3NO	Trifluoramine oxide	0.039	A	3
F_3OP	Phosphorous oxyfluoride	1.868	A	3
F_3P	Phosphorous trifluoride	1.03	A	1
F_3PS	Thiophosphoryl trifluoride	0.64	B	1
F_4N_2	Tetrafluorohydrazine (*gauche*)	0.26	B	1
F_4S	Sulfur tetrafluoride	0.632	A	1
F_4Se	Selenium tetrafluoride	1.78	C	2
F_5I	Iodine pentafluoride	2.18	C	1
GeO	Germanium oxide (GeO)	3.282	A	2
GeS	Germanium sulfide (GeS)	2.00	C	2
HI	Hydrogen iodide	0.448	A	2
HLi	Lithium hydride	5.884	A	2
HLiO	Lithium hydroxide	4.754	A	3
HN	Imidogen (NH)	1.39	C	3
HNO	Nitrosyl hydride (HNO)	1.62	B	3
HNO_2	Nitrous acid (HONO) (*cis*)	1.423	A	2
	Nitrous acid (HONO) (*trans*)	1.855	A	2
HNO_3	Nitric acid	2.17	A	1
HN_3	Hydrazoic acid	1.70	C	3
HO	Hydroxyl (OH)	1.668	A	3
HS	Mercapto (SH)	0.758	A	3
H_2O	Water	1.854	A	3
H_2O_2	Hydrogen peroxide	1.573	A	3
H_2S	Hydrogen sulfide	0.97	A	1
H_3N	Ammonia	1.471	A	3
H_3NO	Hydroxylamine (NH_2OH)	0.59	C	2
H_3P	Phosphine	0.574	A	3
H_3Sb	Stibine	0.12	C	1
H_4N_2	Hydrazine	1.75	C	1
H_6OSi_2	Disiloxane	0.24	B	1
ILi	Lithium iodide	7.428	A	2
INa	Sodium iodide	9.236	A	2
IO	Iodine oxide (IO)	2.45	B	2
ITl	Thallium iodide (TlI)	4.61	B	2
KLi	Lithium potassium (LiK)	3.45	C	2
KNa	Potassium sodium (KNa)	2.73	C	3
LiNa	Lithium sodium (LiNa)	0.463	A	2
LiO	Lithium oxide (LiO)	6.84	A	2
LiRb	Lithium rubidium (LiRb)	4.00	C	2
NO	Nitric oxide	0.159	A	2
NO_2	Nitrogen dioxide	0.316	A	1
NS	Nitrogen sulfide (NS)	1.81	A	2
NP	Phosphorous nitride (PN)	2.747	A	2
N_2O	Nitrous oxide	0.161	A	3
N_2O_3	Nitrogen trioxide	2.122	A	2
NaRb	Rubidium sodium (RbNa)	3.1	D	2

Mol. form.	Name	μ/D	Acc.	Ref.
PbO	Lead oxide (PbO)	4.64	D	2
OS	Sulfur oxide (SO)	1.55	A	1
OS_2	Sulfur oxide (SSO)	1.47	B	1
OSi	Silicon monoxide (SiO)	3.098	A	2
OSn	Tin oxide (SnO)	4.32	C	2
OSr	Strontium oxide	8.900	A	2
O_2S	Sulfur dioxide	1.633	A	3
O_2Se	Selenium dioxide	2.62	B	2
O_3	Ozone	0.534	A	3
PbS	Lead sulfide (PbS)	3.59	C	2
SSi	Silicon monosulfide (SiS)	1.73	C	2
SSn	Tin sulfide (SnS)	3.18	C	2
	Compounds containing carbon			
$CBrF_3$	Bromotrifluoromethane	0.65	C	1
CBr_2F_2	Dibromodifluoromethane	0.66	C	1
$CClF_3$	Chlorotrifluoromethane	0.50	A	1
CClN	Cyanogen chloride	2.833	A	3
CCl_2F_2	Dichlorodifluoromethane	0.51	C	1
CCl_2O	Phosgene	1.17	A	1
CCl_3F	Trichlorofluoromethane	0.46	B	2
CF	Fluoromethylidene	0.645	A	3
CFN	Cyanogen fluoride	2.120	A	3
CF_2	Difluoromethylene	0.47	B	3
CF_2O	Carbonyl fluoride	0.95	A	1
CF_3I	Trifluoroiodomethane	1.048	A	3
CO	Carbon monoxide	0.110	A	3
COS	Carbon oxysulfide	0.715	A	3
COSe	Carbon oxyselenide	0.73	B	1
CS	Carbon sulfide	1.958	A	2
CSe	Carbon selenide	1.99	B	3
CH	Methylidyne	1.46	Q	2
$CHBr_3$	Tribromomethane (bromoform)	0.99	B	1
$CHClF_2$	Chlorodifluoromethane	1.42	B	1
$CHCl_2F$	Dichlorofluoromethane	1.29	B	1
$CHCl_3$	Trichloromethane (chloroform)	1.04	B	2
CHF_3	Trifluoromethane (fluoroform)	1.651	A	3
CHN	Hydrogen cyanide	2.984	A	3
CHN	Hydrogen isocyanide (HNC)	3.05	C	3
CHNO	Fulminic acid (HCNO)	3.099	A	5
CHNO	Isocyanic acid (HNCO)	1.6	Q	2
CH_2Br_2	Dibromomethane	1.43	B	1
CH_2ClF	Chlorofluoromethane	1.82	B	1
CH_2Cl_2	Dichloromethane	1.60	B	1
CH_2F_2	Difluoromethane	1.978	A	3
CH_2N_2	Cyanamide	4.27	C	1
CH_2N_2	Diazomethane	1.50	A	1
CH_2O	Formaldehyde	2.332	A	3
CH_2O_2	Formic acid	1.41	A	1
CH_3BO	Carbonyltrihydroboron (BH_3CO)	1.70	B	3
CH_3Br	Bromomethane (methyl bromide)	1.822	A	3
CH_3Cl	Chloromethane (methyl chloride)	1.892	A	2
CH_3F	Fluoromethane (methyl fluoride)	1.858	A	3
CH_3I	Iodomethane (methyl iodide)	1.62	B	1
CH_3NO	Formamide	3.73	B	1

Mol. form.	Name	μ/D	Acc.	Ref.
CH_3NO_2	Nitromethane	3.46	A	1
CH_3N_3	Methyl azide	2.17	B	2
CH_4O	Methanol	1.70	A	1
CH_4S	Methanethiol (methyl mercaptan)	1.52	C	1
CH_5N	Methylamine	1.31	B	1
CH_6OSi	Methoxysilane	1.15	B	2
CH_6Si	Methylsilane	0.735	A	1
C_2ClF_3	Chlorotrifluoroethylene	0.40	D	1
C_2ClF_5	Chloropentafluoroethane	0.52	C	1
$C_2Cl_2F_4$	1,2-Dichlorotetrafluoroethane	0.5	Q	1
C_2F_3N	Trifluoroacetonitrile	1.262	A	3
C_2HBr	Bromoacetylene	0.234	A	3
C_2HCl	Chloroacetylene	0.44	A	1
C_2HCl_5	Pentachloroethane	0.92	C	1
C_2HF	Fluoroacetylene	0.721	A	3
C_2HF_3	Trifluoroethylene	1.32	B	2
$C_2HF_3O_2$	Trifluoroacetic acid	2.28	D	1
$C_2H_2Cl_2$	1,1-Dichloroethylene	1.34	A	1
$C_2H_2Cl_2$	*cis*-1,2-Dichloroethylene	1.90	B	1
$C_2H_2Cl_2O$	Chloroacetyl chloride	2.23	Ci	1
$C_2H_2Cl_4$	1,1,2,2-Tetrachloroethane	1.32	Ci	1
$C_2H_2F_2$	1,1-Difluoroethylene	1.368	A	3
$C_2H_2F_2$	*cis*-1,2-Difluoroethylene	2.42	A	1
C_2H_2O	Ketene	1.422	A	3
$C_2H_2O_2$	Glyoxal (*cis*)	4.8	C	2
C_2H_3Br	Bromoethylene (vinyl bromide)	1.42	B	1
C_2H_3Cl	Chloroethylene (vinyl chloride)	1.45	B	1
$C_2H_3ClF_2$	1-Chloro-1,1-difluoroethane	2.14	B	1
C_2H_3ClO	Acetyl chloride	2.72	C	1
$C_2H_3Cl_3$	1,1,1-Trichloroethane	1.755	A	2
C_2H_3F	Fluoroethylene (vinyl fluoride)	1.468	A	6
C_2H_3FO	Acetyl fluoride	2.96	A	1
$C_2H_3F_3$	1,1,1-Trifluoroethane	2.347	A	3
C_2H_3N	Acetonitrile	3.924	A	3
C_2H_3NO	Methylisocyanate	2.8	Q	1
C_2H_4ClF	1-Chloro-1-fluoroethane	2.068	A	3
$C_2H_4Cl_2$	1,1-Dichloroethane	2.06	B	1
$C_2H_4F_2$	1,1-Difluoroethane	2.27	B	1
$C_2H_4F_2$	1,2-Difluoroethane (*gauche*)	2.67	C	2
C_2H_4O	Acetaldehyde	2.750	A	3
C_2H_4O	Ethylene oxide (oxirane)	1.89	A	1
$C_2H_4O_2$	Acetic acid	1.70	B	2
$C_2H_4O_2$	Methyl formate	1.77	B	1
$C_2H_4O_2$	Glycolaldehyde (hydroxyacetylene)	2.73	B	2
C_2H_5Br	Bromoethane (ethyl bromide)	2.03	A	1
C_2H_5Cl	Chloroethane (ethyl chloride)	2.05	A	1
C_2H_5ClO	2-Chloroethanol	1.78	Ci	1
C_2H_5F	Fluoroethane (ethyl fluoride)	1.94	B	1
C_2H_5I	Iodoethane (ethyl iodide)	1.91	B	1
C_2H_5N	Ethyleneimine	1.90	A	1
C_2H_5NO	Acetamide	3.76	Bi	1
C_2H_5NO	N-Methylformamide	3.83	Bi	1
$C_2H_5NO_2$	Nitroethane	3.23	A	2
C_2H_6O	Dimethyl ether	1.30	A	1
C_2H_6O	Ethanol (*average*)	1.69	Bi	1
	Ethanol (*trans*)	1.44	B	2

Mol. form.	Name	μ/D	Acc.	Ref.
	Ethanol (*gauche*)	1.68	B	3
C_2H_6OS	Dimethyl sulfoxide	3.96	A	1
$C_2H_6O_2$	Ethylene glycol	2.28	Ci	1
C_2H_6S	Dimethyl sulfide	1.554	A	3
C_2H_6S	Ethanethiol (*trans*)	1.58	C	3
	Ethanethiol (*gauche*)	1.61	C	3
C_2H_7N	Dimethylamine	1.01	B	2
C_2H_7N	Ethylamine	1.22	Ci	1
$C_2H_8N_2$	1,2-Ethanediamine	1.99	Ci	1
C_3HN	Cyanoacetylene	3.724	A	3
C_3H_2O	Propynal (propargyl aldehyde)	2.47	B	1
$C_3H_3Cl_2F$	1,1-Dichloro-2-fluoropropene	2.43	A	3
$C_3H_3F_3$	3,3,3-Trifluoropropene	2.45	B	1
C_3H_3N	Propenenitrile (acrylonitrile)	3.87	B	1
C_3H_3NO	Oxazole	1.503	B	3
C_3H_3NO	Isoxazole	2.90	B	2
C_3H_4	Propyne (methylacetylene)	0.784	A	3
C_3H_4	Cyclopropene	0.45	A	1
$C_3H_4F_2$	1,1-Difluoro-1-propene	0.889	A	2
$C_3H_4N_2$	Imidazole	3.8	D	2
C_3H_4O	Propenal (*trans*)	3.12	B	1
C_3H_4O	Propargyl alcohol	1.13	C	2
C_3H_4O	Cyclopropanone	2.67	C	2
$C_3H_4O_2$	Vinyl formate	1.49	A	1
$C_3H_4O_2$	2-Oxetanone (beta-propiolactone)	4.18	A	1
$C_3H_4O_2$	3-Oxetanone	0.887	A	2
C_3H_5Br	3-Bromopropene (allyl bromide)	1.9	Q	1
C_3H_5Cl	*cis*-1-Chloropropene	1.67	C	1
C_3H_5Cl	*trans*-1-Chloropropene	1.97	C	1
C_3H_5Cl	2-Chloropropene	1.647	A	3
C_3H_5Cl	3-Chloropropene (allyl chloride)	1.94	Ci	1
C_3H_5F	*cis*-1-Fluoropropene	1.46	B	1
C_3H_5F	*trans*-1-Fluoropropene	1.9	Q	1
C_3H_5F	2-Fluoropropene	1.61	B	1
C_3H_5F	3-Fluoropropene (*cis*)	1.76	A	1
	3-Fluoropropene (*gauche*)	1.94	A	1
C_3H_5N	Propanenitrile	4.05	A	3
C_3H_6	Propene (propylene)	0.366	A	1
$C_3H_6Cl_2$	1,3-Dichloropropane	2.08	Bi	1
C_3H_6O	Acetone	2.88	A	1
C_3H_6O	Allyl alcohol (*average*)	1.60	Ci	1
	Allyl alcohol (*gauche*)	1.55	C	3
C_3H_6O	Propanal (*cis*)	2.52	B	1
C_3H_6O	Methyloxirane	2.01	A	1
C_3H_6O	Oxetane (trimethylene oxide)	1.94	A	1
C_3H_6O	Methyl vinyl ether (*cis*)	0.96	B	2
$C_3H_6O_2$	Ethyl formate (*trans*)	1.98	A	2
	Ethyl formate (*gauche*)	1.81	A	2
$C_3H_6O_2$	Methyl acetate	1.72	Ci	1
$C_3H_6O_2$	Propanoic acid (*average*)	1.75	Ci	1
	Propanoic acid (*cis*)	1.46	C	2
$C_3H_6O_2$	1,3-Dioxolane	1.19	C	3
$C_3H_6O_3$	Trioxane	2.08	A	1
C_3H_6S	Thiacyclobutane (trimethylene sulfide)	1.85	C	1
C_3H_7Br	1-Bromopropane (propyl bromide)	2.18	Ci	1
C_3H_7Br	2-Bromopropane (isopropyl bromide)	2.21	C	1

Mol. form.	Name	μ/D	Acc.	Ref.
C_3H_7Cl	1-Chloropropane (propyl chloride)	2.05	Bi	1
C_3H_7Cl	2-Chloropropane (isopropyl chloride)	2.17	C	1
C_3H_7F	1-Fluoropropane (*trans*)	2.05	B	1
	1-Fluoropropane (*gauche*)	1.90	C	1
C_3H_7F	2-Fluoropropane	1.96	B	3
C_3H_7I	1-Iodopropane (propyl iodide)	2.04	Ci	1
C_3H_7N	Allylamine	1.2	Q	1
C_3H_7N	Cyclopropylamine	1.19	A	2
C_3H_7N	Propyleneimine (*cis*)	1.77	C	2
	Propyleneimine (*trans*)	1.57	B	2
C_3H_7NO	N,N-Dimethylformamide	3.82	Bi	1
$C_3H_7NO_2$	1-Nitropropane	3.66	Bi	1
$C_3H_7NO_2$	2-Nitropropane	3.73	B	1
C_3H_8	Propane	0.084	A	1
C_3H_8O	1-Propanol (*trans*)	1.55	B	2
	1-Propanol (*gauche*)	1.58	B	2
C_3H_8O	2-Propanol (*trans*)	1.58	B	2
C_3H_8O	Ethyl methyl ether (*trans*)	1.17	B	3
$C_3H_8O_2$	2-Methoxyethanol (*gauche*)	2.36	B	2
C_3H_8S	Ethyl methyl sulfide (*trans*)	1.56	B	3
C_3H_8S	1-Propanethiol (*trans*)	1.60	C	3
	1-Propanethiol (*gauche*)	1.683	A	3
C_3H_8S	2-Propanethiol (*trans*)	1.61	B	3
	2-Propanethiol (*gauche*)	1.53	B	3
C_3H_9N	Propylamine	1.17	Ci	1
C_3H_9N	Isopropylamine	1.19	C	3
C_3H_9N	Trimethylamine	0.612	A	1
C_4H_4	1-Buten-3-yne (vinylacetylene)	0.22	B	3
$C_4H_4N_2$	Pyridazine	4.22	A	2
$C_4H_4N_2$	Pyrimidine	2.334	A	2
C_4H_4O	Furan	0.66	A	1
$C_4H_4O_2$	Diketene	3.53	B	1
C_4H_4S	Thiophene	0.55	A	2
C_4H_5N	Pyrrole	1.74	B	2
C_4H_5N	Methylacrylonitrile	3.69	C	1
C_4H_6	1,2-Butadiene (methylallene)	0.403	A	1
C_4H_6	1-Butyne (ethylacetylene)	0.80	C	1
C_4H_6	Cyclobutene	0.132	A	1
C_4H_6O	*trans*-2-Butenal (*trans*-crotonaldehyde)	3.67	Bi	1
C_4H_6O	2-Methylpropenal (methacrolein)	2.68	Ci	1
C_4H_6O	Divinyl ether	0.78	C	2
C_4H_6O	Cyclobutanone	2.89	A	2
C_4H_6O	2,3-Dihydrofuran	1.32	B	2
$C_4H_6O_2$	*gamma*-Butyrolactone	4.27	A	3
$C_4H_6O_3$	Acetic anhydride	2.8	Q	1
C_4H_6S	2,5-Dihydrothiophene	1.75	A	3
C_4H_7N	Butanenitrile	4.07	Bi	1
C_4H_7N	2-Methylpropanenitrile (isobutyronitrile)	4.29	B	3
C_4H_8	1-Butene (*cis*)	0.438	A	2
	1-Butene (*skew*)	0.359	A	2
C_4H_8	*cis*-2-Butene	0.253	A	2
C_4H_8	Isobutene	0.503	A	1
C_4H_8	Methylcyclopropane	0.139	A	2
$C_4H_8Cl_2$	1,4-Dichlorobutane	2.22	Ci	1
C_4H_8O	Butanal (butyraldehyde)	2.72	Bi	1
C_4H_8O	2-Butanone (methyl ethyl ketone)	2.78	A	2

DIPOLE MOMENTS OF MOLECULES IN THE GAS PHASE (continued)

Mol. form.	Name	μ/D	Acc.	Ref.
C_4H_8O	Tetrahydrofuran	1.75	B	2
$C_4H_8O_2$	1,3-Dioxane	2.06	B	2
$C_4H_8O_2$	Ethyl acetate	1.78	Ci	1
C_4H_9Br	1-Bromobutane (butyl bromide)	2.08	Ci	1
C_4H_9Br	2-Bromobutane (*sec*-butyl bromide)	2.23	Ci	1
C_4H_9Cl	1-Chlorobutane (butyl chloride)	2.05	Bi	1
C_4H_9Cl	1-Chloro-2-methylpropane (isobutyl chloride)	2.00	Ci	1
C_4H_9Cl	2-Chlorobutane (*sec*-butyl chloride)	2.04	Ci	1
C_4H_9Cl	2-Chloro-2-methylpropane (*tert*-butyl chloride)	2.13	B	1
C_4H_9I	2-Iodobutane (*sec*-butyl iodide)	2.12	Ci	1
C_4H_9NO	Morpholine	1.55	B	3
C_4H_{10}	Isobutane	0.132	A	1
$C_4H_{10}O$	1-Butanol	1.66	Bi	1
$C_4H_{10}O$	2-Methyl-1-propanol (isobutyl alcohol)	1.64	C	1
$C_4H_{10}O$	Diethyl ether	1.15	Bi	1
$C_4H_{10}O$	Methyl propyl ether (*trans-trans*)	1.107	A	3
$C_4H_{10}S$	2-Methyl-2-propanethiol (*tert*-butyl mercaptan)	1.66	B	3
$C_4H_{10}S$	Diethyl sulfide	1.54	Ci	1
$C_4H_{11}N$	Butylamine	1.0	Q	1
$C_4H_{11}N$	Diethylamine	0.92	Ci	1
C_5H_5N	Pyridine	2.215	A	3
C_5H_6	1-Penten-3-yne	0.66	B	2
C_5H_6	*cis*-3-Penten-1-yne	0.78	B	2
C_5H_6	*trans*-3-Penten-1-yne	1.06	C	2
C_5H_6	2-Methyl-1-buten-3-yne	0.513	B	2
C_5H_6	1,3-Cyclopentadiene	0.419	A	1
C_5H_6O	2-Methylfuran	0.65	C	2
C_5H_6O	3-Methylfuran	1.03	B	2
C_5H_6S	2-Methylthiophene	0.674	A	2
C_5H_6S	3-Methylthiophene	0.95	C	2
C_5H_7N	Cyclobutanecarbonitrile	4.11	A	3
C_5H_8	2-Methyl-1,3-butadiene (isoprene)	0.25	A	1
C_5H_8	*cis*-1,3-Pentadiene	0.500	B	2
C_5H_8	*trans*-1,3-Pentadiene	0.585	A	2
C_5H_8	Cyclopentene	0.20	B	1
C_5H_8	1-Pentyne (*trans*)	0.842	A	2
	1-Pentyne (*gauche*)	0.77	C	2
C_5H_8O	Cyclopentanone	3.3	Q	1
C_5H_8O	Cyclopropyl methyl ketone	2.62	D	2
C_5H_9N	Pentanenitrile	4.12	Bi	1
C_5H_9N	2,2-Dimethylpropanenitrile (*tert*-butyl cyanide)	3.95	A	1
C_5H_{10}	1-Pentene	0.5	Q	1
C_5H_{10}	3-Methyl-1-butene (*trans*)	0.320	A	3
	3-Methyl-1-butene (*gauche*)	0.398	A	3
$C_5H_{10}O$	Tetrahydropyran (*chair form*)	1.74	B	2
$C_5H_{10}O$	2,2-Dimethylpropanal (pivaldehyde)	2.66	B	2
$C_5H_{10}O_3$	Diethyl carbonate	1.10	Ci	1
$C_5H_{10}S$	Thiacyclohexane	1.781	A	3
$C_5H_{11}Br$	1-Bromopentane (amyl bromide)	2.20	Ci	1
$C_5H_{11}Cl$	1-Chloropentane (amyl chloride)	2.16	Ci	1
$C_5H_{11}N$	Piperidine (*equatorial*)	0.82	B	3
	Piperidine (*axial*)	1.19	B	3
C_5H_{12}	Isopentane	0.13	C	1
$C_6H_4ClNO_2$	*o*-Chloronitrobenzene	4.64	B	1
$C_6H_4ClNO_2$	*m*-Chloronitrobenzene	3.73	B	1
$C_6H_4ClNO_2$	*p*-Chloronitrobenzene	2.83	B	1

Mol. form.	Name	μ/D	Acc.	Ref.
$C_6H_4Cl_2$	*o*-Dichlorobenzene	2.50	B	1
$C_6H_4Cl_2$	*m*-Dichlorobenzene	1.72	C	1
$C_6H_4FNO_2$	*p*-Fluoronitrobenzene	2.87	B	1
$C_6H_4F_2$	*o*-Difluorobenzene	2.46	B	2
$C_6H_4F_2$	*m*-Difluorobenzene	1.51	A	2
C_6H_5Br	Bromobenzene	1.70	B	1
C_6H_5Cl	Chlorobenzene	1.69	B	1
C_6H_5ClO	*p*-Chlorophenol	2.11	C	1
C_6H_5F	Fluorobenzene	1.60	C	1
C_6H_5I	Iodobenzene	1.70	C	1
$C_6H_5NO_2$	Nitrobenzene	4.22	B	1
C_6H_6	Fulvene	0.424	A	2
C_6H_6O	Phenol	1.224	A	3
C_6H_7N	Aniline	1.13	B	3
C_6H_7N	2-Methylpyridine	1.85	B	2
C_6H_7N	4-Methylpyridine	2.70	A	2
C_6H_{10}	Cyclohexene (*half-chair form*)	0.332	A	2
C_6H_{10}	1-Hexyne (butylacetylene)	0.83	Ci	1
C_6H_{10}	3,3-Dimethyl-1-butyne (*tert*-butylacetylene)	0.661	A	1
$C_6H_{10}O$	Cyclohexanone	2.87	B	2
$C_6H_{11}Cl$	Chlorocyclohexane (*equatorial*)	2.44	C	7
	Chlorocyclohexane (*axial*)	1.91	B	7
$C_6H_{11}F$	Fluorocyclohexane (*equatorial*)	2.11	B	2
	Fluorocyclohexane (*axial*)	1.81	B	2
$C_6H_{12}O_2$	Pentyl formate	1.90	Ci	1
$C_6H_{12}O_3$	Paraldehyde	1.43	C	1
$C_6H_{14}O$	Dipropyl ether	1.21	Ci	1
$C_6H_{14}O$	Diisopropyl ether	1.13	Di	1
$C_6H_{15}N$	Triethylamine	0.66	Ci	1
$C_7H_5F_3$	(Trifluoromethyl)benzene	2.86	B	1
C_7H_5N	Benzonitrile	4.18	B	1
C_7H_7Cl	*o*-Chlorotoluene	1.56	C	1
C_7H_7Cl	*p*-Chlorotoluene	2.21	B	1
C_7H_7F	*o*-Fluorotoluene	1.37	C	1
C_7H_7F	*m*-Fluorotoluene	1.82	B	2
C_7H_7F	*p*-Fluorotoluene	2.00	C	1
C_7H_8	Toluene	0.375	A	3
C_7H_8O	Benzyl alcohol	1.71	C	1
C_7H_8O	Anisole (methyl phenyl ether)	1.38	C	1
$C_7H_{14}O_2$	Pentyl acetate	1.75	Ci	1
$C_7H_{15}Br$	1-Bromoheptane	2.16	Ci	1
C_8H_8O	Acetophenone (methyl phenyl ketone)	3.02	B	1
C_8H_{10}	Ethylbenzene	0.59	C	1
C_8H_{10}	*o*-Xylene	0.640	A	2
$C_8H_{10}O$	Phenetole (ethyl phenyl ether)	1.45	Di	1
$C_8H_{11}N$	*N,N*-Dimethylaniline	1.68	D	1
$C_8H_{18}O$	Dibutyl ether	1.17	Ci	1
C_9H_7N	Quinoline	2.29	C	1
C_9H_7N	Isoquinoline	2.73	C	1
$C_9H_{10}O_2$	Ethyl benzoate	2.00	Ci	1
C_9H_{12}	Cumene (isopropylbenzene)	0.79	Q	1
$C_{10}H_8$	Azulene	0.80	B	1
$C_{10}H_{14}$	*tert*-Butylbenzene	0.83	Q	1
$C_{12}H_{10}$	Acenaphthene	0.85	Q	1
$C_{12}H_{10}O$	Diphenyl ether	1.3	Q	1

STRENGTHS OF CHEMICAL BONDS*

J. A. Kerr

The strength of a chemical bond, $D°(\text{R-X})$, often known as the bond dissociation energy, is defined as the standard enthalpy change of the reaction in which the bond is broken: $RX \rightarrow R + X$. It is given by the thermochemical equation, $D°(\text{R-X}) = \Delta_f H°(\text{R}) + \Delta_f H°(\text{X}) - \Delta_f H°(\text{RX})$. Some authors list bond strengths at a temperature of absolute zero but here the values at 298 K are given because more thermodynamic data are available for this temperature. Bond strengths or bond dissociation energies are not equal to, and may differ considerably from, mean bond energies solely from thermochemical data on atoms and molecules.

Table 1
BOND STRENGTHS IN DIATOMIC MOLECULES

These have usually been measured spectroscopically or by mass spectrometric analysis of hot gases effusing a Knudsen cell. Excellent accounts of these and other methods are given in (i) *Dissociation Energies and Spectra of Diatomic Molecules*, by A. G. Gaydon, 3rd. ed., Chapman & Hall, London, 1968 and (ii) "Mass Spectrometric Determination of Bond Energies of High-Temperature Molecules", K. A. Gingerich, *Chimia*, 26, 619, 1972. The errors quoted in the table are those given in the original paper or review article. The references have been chosen primarily as a key to the literature. It should not be assumed that the author referred to was responsible for the value quoted, as the reference may be to a review article.

Bond strength reported at a temperature of absolute zero, $D°_0$, have been converted to $D°_{298}$ by the use of enthalpy functions taken mainly from the JANAF Thermochemical Tables, Third Edition, *J. Phys. Chem. Ref. Data*, 14, Suppl. 1, 1985, wherever possible. For most bonds, however, where this data is not available, the conversion has been made by the approximate relation:

$$D°_{298} = D°_0 + (3/2)RT$$

The list below does not include the increasing number of bond strengths of diatomic molecules now being calculated by *ab initio* methods. The Table has been arranged in an alphabetical order of the atoms.

Molecule	$D°_{298}$/kJ mol^{-1}	Ref.
Ag-Ag	163 ± 8	69
Ag-Al	183.7 ± 9.2	68
Ag-Au	202.9 ± 9.2	3
Ag-Bi	193 ± 42	229
Ag-Br	293 ± 29	110
Ag-Cl	341.4	149
Ag-Cu	174.1 ± 9.2	126
Ag-D	226.8	189
Ag-Dy	130 ± 19	187
Ag-Eu	129.7 ± 12.6	55
Ag-F	354.4 ± 16.3	110
Ag-Ga	180 ± 15	38
Ag-Ge	174.5 ± 20.9	125
Ag-H	215.1 ± 8	200
Ag-Ho	123.4 ± 16.7	51
Ag-I	234 ± 29	110
Ag-In	166.5 ± 4.9	11
Ag-Li	177.4 ± 6.3	258
Ag-Mn	100 ± 21	229
Ag-Na	138.1 ± 8.4	272,280
Ag-Nd	<209	205
Ag-O	220.1 ± 20.9	268
Ag-S	217.1	326
Ag-Se	202.5	326
Ag-Si	177.8 ± 10.0	300
Ag-Sn	136.0 ± 20.9	2
Ag-Te	195.8	326
Al-Al	133 ± 6	108
Al-Ar	5.34 ± 0.78	37
Al-As	202.9 ± 7.1	276,284
Al-Au	325.9 ± 6.3	114
Al-Br	429 ± 6	174
Al-Cl	511.3 ± 0.8	292
Al-Cu	216.7 ± 10.5	274
Al-D	290.8	229
Al-F	663.6 ± 6.3	69
Al-H	284.9 ± 6.3	69
Al-I	369.9 ± 2.1	249
Al-Kr	5.89 ± 0.81	37
Al-Li	175.7 ± 14.6	150
Al-N	297 ± 96	110
Al-O	511 ± 3	63
Al-P	216.7 ± 12.6	69
Al-Pd	254.4 ± 12.1	53
Al-S	373.6 ± 7.9	349
Al-Sb	216.3 ± 5.9	275
Al-Se	337.6 ± 10.0	349
Al-Si	229.3 ± 30.1	42
Al-Te	267.8 ± 10.0	349
Al-U	326 ± 29	113
Al-Xe	7.43 ± 0.69	37
Ar-Ar	4.73 ± 0.04	170
Ar-He	3.89	229
Ar-Hg	6.15	229
Ar-I	10.0	34
Ar-K	4.2	189
As-As	382.0 ± 10.5	230
As-Cl	448	69
As-D	270.3	189
As-F	410	189
As-Ga	209.6 ± 1.2	72
As-H	274.0 ± 2.9	24
As-I	296.6 ± 28.0	305
As-In	201	279
As-N	489 ± 2	290
As-O	481 ± 8	244
As-P	433.5 ± 12.6	127
As-S	379.5 ± 6.3	244
As-Sb	330.5 ± 5.4	85
As-Se	96	269
As-Tl	198.3 ± 14.6	282
At-At	~80	79
Au-Au	224.7 ± 1.5	29,233
Au-B	367.8 ± 10.5	135
Au-Ba	254.8 ± 10.0	125
Au-Be	285 ± 8	110
Au-Bi	297 ± 8.4	125
Au-Ca	243	125
Au-Ce	339 ± 21	125
Au-Cl	343 ± 9.6	110
Au-Co	222 ± 17	125
Au-Cr	213 ± 17	125
Au-Cs	255 ± 3.3	35
Au-Cu	228.0 ± 5.0	29,125
Au-D	318.4	189
Au-Dy	259 ± 21	187
Au-Eu	241.0 ± 10.5	55
Au-Fe	187.0 ± 16.7	204
Au-Ga	234 ± 38	125
Au-Ge	274.1 ± 5.0	125

* Revised to December 1993.

Table 1
BOND STRENGTHS IN DIATOMIC MOLECULES (continued)

Molecule	$D°_{298}$/kJ mol^{-1}	Ref.
Au-H	292.0 ± 8	200
Au-Ho	267.4 ± 16.7	51,233
Au-In	286.0 ± 5.7	11
Au-La	336.4 ± 20.9	117
Au-Li	284.5 ± 6.7	258
Au-Lu	332.2 ± 16.7	122
Au-Mg	243 ± 42	229
Au-Mn	185.4 ± 12.6	319
Au-Na	215.1 ± 12.6	280
Au-Nd	299.2 ± 20.9	117
Au-Ni	255 ± 21	125
Au-O	221.8 ± 20.9	268
Au-Pb	130 ± 42	229
Au-Pd	155 ± 21	125
Au-Pr	310 ± 21	125
Au-Rb	243 ± 2.9	35
Au-Rh	231.0 ± 29	55
Au-S	418 ± 25	121
Au-Sc	280.3 ± 16.7	118
Au-Se	243.1	326
Au-Si	305.4 ± 5.9	129
Au-Sn	254.8 ± 7.1	216
Au-Sr	264 ± 42	229
Au-Tb	289.5 ± 33.5	142,233
Au-Te	317.6	326
Au-U	318 ± 29	113
Au-V	240.6 ± 12.1	163
Au-Y	307.1 ± 8.4	168
B-B	297 ± 21	69
B-Br	396	27
B-C	448 ± 29	229
B-Ce	305 ± 21	229
B-Cl	536	17
B-D	341.0 ± 6.3	229
B-F	757	239
B-H	340	285
B-I	220.5 ± 0.8	294
B-Ir	514.2 ± 17.2	354
B-La	339 ± 63	229
B-N	389 ± 21	69
B-O	808.8 ± 20.9	268
B-P	346.9 ± 16.7	137
B-Pd	329.3 ± 20.9	354
B-Pt	477.8 ± 16.7	250
B-Rh	475.7 ± 20.9	354
B-Ru	446.9 ± 20.9	354
B-S	580.7 ± 9.2	347
B-Sc	276 ± 63	229
B-Se	461.9 ± 14.6	347
B-Si	288.3	360
B-Te	354.4 ± 20.1	347
B-Th	297	130
B-Ti	276 ± 63	229
B-U	322 ± 33	229
B-Y	293 ± 63	229
Ba-Br	362.8 ± 8.4	95,184,213
Ba-Cl	436.0 ± 8.4	182,184
Ba-D	≤193.7	189
Ba-F	587.0 ± 6.7	92,181
Ba-H	176 ± 14.6	110
Ba-I	320.8 ± 6.3	176
Ba-O	561.9 ± 13.4	268
Ba-Pd	221.8 ± 5.0	115
Ba-Rh	259.4 ± 25.1	115
Ba-S	400.0 ± 18.8	59
Be-Be	59	30,77
Be-Br	381 ± 84	229
Be-Cl	388.3 ± 9.2	99,179,355
Be-D	203.05	189
Be-F	577 ± 42	69,99
Be-H	200.0 ± 1.3	56
Be-O	434.7 ± 13.4	268
Be-S	372 ± 59	110
Bi-Bi	200.4 ± 7.5	288,304
Bi-Br	267.4 ± 4.2	65
Bi-Cl	301 ± 4	67
Bi-D	283.7	248
Bi-F	>105	74
Bi-Ga	159 ± 17	278
Bi-H	≤283.3	248
Bi-I	218.0 ± 4.6	66
Bi-In	153.6 ± 1.7	301
Bi-Li	154.0 ± 5.0	260,286
Bi-O	337.2 ± 12.6	268
Bi-P	280 ± 13	127
Bi-Pb	141.8 ± 14.6	304
Bi-S	315.5 ± 4.6	348
Bi-Sb	251 ± 4	227
Bi-Se	280.3 ± 5.9	348
Bi-Sn	210.0 ± 8.4	125
Bi-Te	232.2 ± 11.3	348
Bi-Tl	121 ± 13	73
Br-Br	192.807	1
Br-C	280 ± 21	110
Br-Ca	310.9 ± 9.2	299
Br-Cd	159 ± 96	110
Br-Cl	217.53 ± 0.29	46
Br-Co	331 ± 42	229
Br-Cr	328.0 ± 24.3	110
Br-Cs	389.1 ± 4	266,338
Br-Cu	331 ± 25	110
Br-D	370.74	189
Br-F	280 ± 12	194
Br-Fe	247 ± 96	110
Br-Ga	444 ± 17	69
Br-Ge	255 ± 29	110
Br-H	366.35	189
Br-Hg	72.8 ± 4	69
Br-I	179.1 ± 0.4	110,289
Br-In	414 ± 21	69
Br-K	379.9 ± 0.8	338,351
Br-Li	418.8 ± 4	338
Br-Mg	≤327.2	189
Br-Mn	314.2 ± 9.6	110
Br-N	276 ± 21	110
Br-Na	367.4 ± 0.8	338,351
Br-Ni	360 ± 13	110
Br-O	235.1 ± 0.4	69
Br-Pb	247 ± 38	110
Br-Rb	380.7 ± 4	338
Br-Sb	314 ± 59	110
Br-Sc	444 ± 63	229
Br-Se	297 ± 84	229
Br-Si	367.8 ± 10.0	97
Br-Sn	≥552	267
Br-Sr	333.0 ± 9.2	184
Br-Th	364	175
Br-Ti	439	229
Br-Tl	333.9 ± 1.7	23
Br-U	377.4 ± 6.3	242
Br-V	439 ± 42	229
Br-W	329.3	207
Br-Y	485 ± 84	229
Br-Zn	142 ± 29	229
C-C	607 ± 21	69
C-Ce	444 ± 13	219
C-Cl	397 ± 29	264
C-D	341.4	189
C-F	552	180
C-Ge	460 ± 21	110
C-H	338.32	189
C-Hf	540 ± 25	333
C-I	209 ± 21	110
C-Ir	632 ± 4	162
C-La	462 ± 20	271
C-Mo	481 ± 15.9	158
C-N	754.3 ± 10	36
C-Nb	569 ± 13.0	158
C-O	1076.5 ± 0.4	69
C-Os	≥594	116
C-P	513.4 ± 8	327
C-Pt	598 ± 5.9	162,352
C-Rh	580.0 ± 3.8	309
C-Ru	616.2 ± 10.5	310
C-S	714.1 ± 1.2	60,330
C-Sc	≤444 ± 21	138
C-Se	590.4 ± 5.9	320
C-Si	451.5	81,360
C-Tc	565 ± 29	302
C-Th	453 ± 17	157,333
C-Ti	423 ± 29	153,333
C-U	454.8 ± 15.1	156,160
C-V	427 ± 23.8	158
C-Y	418 ± 14	315
C-Zr	561 ± 25	333
Ca-Ca	≤46	368,370
Ca-Cl	397 ± 13	182,355
Ca-D	≤169.9	189

Table 1
BOND STRENGTHS IN DIATOMIC MOLECULES (continued)

Molecule	$D°_{298}$/kJ mol^{-1}	Ref.
Ca-F	527 ± 21	92,178
Ca-H	167.8	110
Ca-I	284.7 ± 8.4	176
Ca-Li	84.9 ± 8.4	368
Ca-O	402.1 ± 16.7	268
Ca-S	337.6 ± 18.8	59,189
Cd-Cd	7.36	234
Cd-Cl	208.4	189
Cd-F	305 ± 21	26
Cd-H	69.0 ± 0.4	110
Cd-I	97.23	198
Cd-In	138	229
Cd-O	235.6 ± 83.7	147,268
Cd-S	208.4 ± 20.9	147
Cd-Se	127.6 ± 25.1	147
Cd-Te	100.0 ± 15.1	147
Ce-Ce	245.2	117
Ce-F	582 ± 42	229
Ce-Ir	586	139
Ce-N	519 ± 21	136
Ce-O	795 ± 8	86
Ce-Os	506 ± 33	116
Ce-Pd	322.2	52
Ce-Pt	556	139
Ce-Rh	548	139
Ce-Ru	531 ± 25	116
Ce-S	569	19
Ce-Se	494.5 ± 14.6	253
Ce-Te	189.4 ± 12.8	235
Cl-Cl	242.580 ± 0.004	189
Cl-Co	389	237
Cl-Cr	366.1 ± 24.3	110
Cl-Cs	448 ± 8	266,339
Cl-Cu	382.8 ± 4.6	152
Cl-D	436.47	189
Cl-Eu	~326	105
Cl-F	256.23	189,262
Cl-Fe	~351	110
Cl-Ga	481 ± 13	69
Cl-Ge	~431	189
Cl-H	431.62	189
Cl-Hg	100 ± 8	110
Cl-I	211.3 ± 0.4	110
Cl-In	439 ± 8	69
Cl-K	433.0 ± 8	339
Cl-Li	469 ± 13	69
Cl-Mg	327.6 ± 2.1	96,182,355
Cl-Mn	360.7 ± 9.6	110
Cl-N	333.9 ± 9.6	44
Cl-Na	412.1 ± 8	339
Cl-Ni	372 ± 21	110
Cl-O	269.1	189,298
Cl-P	289 ± 42	229
Cl-Pb	301 ± 29	110
Cl-Ra	343 ± 75	110
Cl-Rb	427.6 ± 8	339
Cl-S	277.0	208
Cl-Sb	360 ± 50	110
Cl-Sc	331	359
Cl-Se	322	229
Cl-Si	406	358
Cl-Sm	≥423 ± 13	371
Cl-Sn	414 ± 17	229
Cl-Sr	406 ± 13	182,184
Cl-Ta	544	18
Cl-Th	489	243
Cl-Ti	494	229
Cl-Tl	372.8 ± 2.1	23
Cl-U	452 ± 8	241
Cl-V	477 ± 63	229
Cl-W	423 ± 42	229
Cl-Xe	6.7	189
Cl-Y	527 ± 84	229
Cl-Yb	~322	105
Cl-Zn	228.9 ± 19.7	62
Cm-O	736	318
Co-Co	167 ± 25	201
Co-Cu	167 ± 17	125
Co-F	435 ± 63	229
Co-Ge	234 ± 21	125
Co-H	226 ± 42	344
Co-I	285 ± 21	229
Co-O	384.5 ± 13.4	268
Co-S	331	326
Co-Si	276 ± 17	353
Cr-Cr	155 ± 21	202
Cr-Cu	155 ± 21	206
Cr-F	444.8 ± 19.7	210
Cr-Ge	154 ± 7	186
Cr-H	280 ± 50	110
Cr-I	287.0 ± 24.3	110
Cr-N	377.8 ± 18.8	142,331
Cr-O	429.3 ± 29.3	268
Cr-Pb	105 ± 2	186
Cr-S	331	83
Cr-Sn	141 ± 3	186
Cs-Cs	43.919 ± 0.010	365
Cs-F	519 ± 8	266
Cs-H	175.364	372
Cs-Hg	8	189
Cs-I	337.2 ± 2.1	266,337
Cs-Na	63.2 ± 1.3	76
Cs-O	295.8 ± 62.8	268
Cs-Rb	49.57 ± 0.01	165
Cu-Cu	176.52 ± 2.38	125,303
Cu-D	270.3	189
Cu-Dy	142 ± 21	187
Cu-F	413.4 ± 13	90
Cu-Ga	215.9 ± 15.1	38
Cu-Ge	208.8 ± 21	255
Cu-H	277.8	200,297
Cu-Ho	142 ± 21	187
Cu-I	197 ± 21	110
Cu-In	187.4 ± 7.9	11
Cu-Li	192.9 ± 8.8	258
Cu-Mn	158.6 ± 17	206
Cu-Na	176.1 ± 16.7	281
Cu-Ni	205.9 ± 17	206
Cu-O	269.0 ± 20.9	268
Cu-S	276	326
Cu-Se	251	326
Cu-Si	221.3 ± 6.3	300
Cu-Sn	169.5 ± 6.7	2,220
Cu-Tb	193 ± 19	187
Cu-Te	278.7	1
D-D	443.533	189
D-F	576.6	189
D-Ga	<272.8	236
D-Ge	≤322	189
D-H	439.433	189
D-Hg	42.05	189
D-In	246.0	189
D-Li	240.1892 ± 0.0046	191,336
D-Mg	135.1	189
D-Ni	≤302.9	189
D-Pt	≤350.2	189
D-S	351	189
D-Si	302.5	189
D-Sr	≥275.7	189
D-Zn	88.7	189
Dy-F	531	377
Dy-O	607 ± 17	86
Dy-S	414 ± 42	229
Dy-Se	322 ± 42	229
Dy-Te	234 ± 42	229
Er-F	565 ± 17	377
Er-O	615 ± 13	86
Er-S	418 ± 42	229
Er-Se	326 ± 42	229
Er-Te	238 ± 42	229
Eu-Eu	33.5 ± 17	55
Eu-F	544	225
Eu-Li	66.9 ± 2.9	257
Eu-O	479 ± 10	86
Eu-Rh	233.9 ± 33	55
Eu-S	362.3 ± 13.0	253,324
Eu-Se	301 ± 14.6	20,169,253
Eu-Te	243 ± 14.6	20,253
F-F	158.78	189
F-Ga	577 ± 14.6	252
F-Gd	590.4 ± 27.2	374
F-Ge	485 ± 21	89
F-H	569.87 ± 0.06	373
F-Hf	650 ± 15	14
F-Hg	~180	189
F-Ho	540	377
F-I	≤271.5	6,28,47,64
F-In	506 ± 14.6	252

Table 1
BOND STRENGTHS IN DIATOMIC MOLECULES (continued)

Molecule	D°_{298}/kJ mol^{-1}	Ref.
F-K	497.5 ± 2.5	15
F-La	598 ± 42	229
F-Li	577 ± 21	69
F-Lu	333.5	192
F-Mg	461.9 ± 5.0	92,181
F-Mn	423.4 ± 14.6	211
F-Mo	464.8	183
F-N	343	189
F-Na	519	189
F-Nd	545.2 ± 12.6	375
F-Ni	430 ± 20	75
F-O	222 ± 17	48
F-P	439 ± 96	110
F-Pb	356 ± 8	379
F-Pm	540 ± 42	229
F-Pr	582 ± 46	229
F-Pu	538.5 ± 29	212
F-Rb	494 ± 21	69
F-Ru	402	173
F-S	342.7 ± 5.0	25,185,214
F-Sb	439 ± 96	110
F-Sc	589.1 ± 13	376
F-Se	339 ± 42	229
F-Si	552.7 ± 2.1	100
F-Sm	565	225
F-Sn	466.5 ± 13	379
F-Sr	541.8 ± 6.7	92,181
F-Ta	573 ± 13	238
F-Tb	561 ± 42	229
F-Th	652	245
F-Ti	569 ± 33	378
F-Tl	445.2 ± 19.2	23
F-Tm	510	225
F-U	659.0 ± 10.5	146,240
F-V	590 ± 63	229
F-W	548 ± 63	229
F-Xe	15.77	296,343
F-Y	605.0 ± 20.9	376
F-Yb	≥521.3 ± 9.6	16,105,371
F-Zn	368 ± 63	229
F-Zr	616 ± 15	14
Fe-Fe	75 ± 17	307
Fe-Ge	210.9 ± 29	203
Fe-H	180 ± 25	344
Fe-O	390.4 ± 17.2	268
Fe-S	322	83
Fe-Si	297 ± 25	353
Ga-Ga	112.1 ± 7	314
Ga-H	<274.1	236
Ga-I	339 ± 9.6	110
Ga-Li	133.1 ± 14.6	150
Ga-O	353.5 ± 41.8	268
Ga-P	229.7 ± 12.6	120
Ga-Sb	192.0 ± 12.6	277
Ga-Te	251 ± 25	350
Gd-O	719 ± 10	86
Gd-S	526.8 ± 10.5	107,322
Gd-Se	431 ± 14.6	20
Gd-Te	343 ± 14.6	20
Ge-Ge	263.6 ± 7.1	221
Ge-H	≤321.7	226
Ge-Ni	290.3 ± 10.9	312
Ge-O	659.4 ± 12.6	268
Ge-Pd	254.7 ± 10.5	311
Ge-S	551.0 ± 2.5	78,84
Ge-Sc	271.0 ± 11	217
Ge-Se	489 ± 4	263
Ge-Si	301 ± 21	110
Ge-Te	456 ± 13	104
Ge-Y	279.8 ± 11.4	218
H-H	435.990	189
H-Hg	39.844	189
H-I	298.407	189
H-In	243.1	189
H-K	174.576	190,372
H-Li	238.049 ± 0.004	364
H-Mg	126.4 ± 2.9	12,13,91
H-Mn	234 ± 29	110
H-N	≤339	189
H-Na	185.69 ± 0.25	256,295
H-Ni	252.3 ± 8	200
H-O	427.6	189
H-P	297	189
H-Pb	≤157	189
H-Pd	234 ± 25	344
H-Pt	≤335	189
H-Rb	167 ± 21	110
H-Rh	247 ± 21	344
H-Ru	234 ± 21	344
H-S	344.3 ± 12.1	195
H-Sc	~180	306
H-Se	314.47 ± 0.96	112
H-Si	≤299.2	189
H-Sn	264 ± 17	110
H-Sr	163 ± 8	110
H-Te	268 ± 2.1	109
H-Ti	204.6 ± 8.8	43
H-Tl	188 ± 8	110
H-Yb	159 ± 38	110
H-Zn	85.8 ± 2.1	110
He-He	3.8	189
He-Hg	6.61	229
Hf-C	548 ± 63	229
Hf-N	536 ± 29	142,228
Hf-O	801.7 ± 13.4	268
Hg-Hg	8 ± 2	187,188
Hg-I	34.69 ± 0.96	362
Hg-K	8.24 ± 0.21	229
Hg-Li	13.8	189
Hg-Na	9.2	189,382
Hg-O	220.9 ± 33.1	147
Hg-Rb	8.4	189
Hg-S	217.1 ± 22.2	147
Hg-Se	144.3 ± 30.1	147
Hg-Te	≤142	229
Hg-Tl	4	172
Ho-Ho	84 ± 17	51
Ho-O	611 ± 17	86
Ho-S	428.4 ± 14.6	322
Ho-Se	335 ± 17	20
Ho-Te	259 ± 17	20
I-I	151.088	189,345
I-In	331	357
I-K	325.1 ± 0.8	337,351
I-Li	345.2 ± 4.2	337
I-Mg	~285	21
I-Mn	282.8 ± 9.6	110
I-N	159 ± 17	229
I-Na	304.2 ± 2.1	337,351
I-Ni	293 ± 21	110
I-O	180	189,298
I-Pb	193 ± 4	317
I-Rb	318.8 ± 2.1	337
I-Si	293	189
I-Sn	234 ± 42	229
I-Sr	269.9 ± 5.9	222
I-Te	192 ± 42	229
I-Ti	310 ± 42	229
I-Tl	272 ± 8	21
I-Zn	108.29	198
I-Zr	305	224
In-In	100 ± 8	229
In-Li	92.5 ± 14.6	150
In-O	<320.1 ± 41.8	268
In-P	197.9 ± 8.4	276
In-S	289 ± 17	58
In-Sb	151.9 ± 10.5	70
In-Se	247 ± 17	58
In-Te	218 ± 17	58
Ir-La	577 ± 13	167
Ir-O	414.6 ± 42.3	268
Ir-Si	462.8 ± 20.9	354
Ir-Th	573	123
Ir-Ti	422 ± 13	270
Ir-Y	456.1 ± 16.7	168
K-K	54.63 ± 0.02	5,247
K-Kr	4.6	189
K-Li	82.0 ± 4.2	94,381
K-Na	65.994 ± 0.008	31,381
K-O	277.8 ± 20.9	268
K-Xe	5.0	189
Kr-Kr	5.23	40,189
Kr-O	<8	229
Kr-Xe	5.505 ± 0.002	7
La-La	247 ± 21	359
La-N	519 ± 42	229
La-O	799 ± 4	86
La-Pt	502 ± 21	254

Table 1
BOND STRENGTHS IN DIATOMIC MOLECULES (continued)

Molecule	$D°_{298}$/kJ mol^{-1}	Ref.
La-Rh	527 ± 17	54
La-S	573.2 ± 1.7	196,335
La-Se	477 ± 17	20,253
La-Te	381 ± 17	20,143
La-Y	202.1	359
Li-Li	110.21 ± 4	356,369
Li-Mg	67.4 ± 6.3	367
Li-Na	87.181 ± 0.001	93,102
Li-O	333.5 ± 8.4	268
Li-Pb	78.7 ± 7.9	260
Li-S	312.5 ± 7.5	215
Li-Sb	172.8 ± 10.0	261
Li-Sm	49.0 ± 4.2	257
Li-Tm	69.0 ± 3.3	257
Li-Yb	37.2 ± 2.9	257
Lu-Lu	142 ± 33	229
Lu-O	678 ± 8	86
Lu-Pt	402 ± 33	131
Lu-S	507.1 ± 14.6	106,322
Lu-Se	418 ± 17	20
Lu-Te	326 ± 17	20
Mg-Mg	8.552 ± 0.004	246,368
Mg-O	363.2 ± 12.6	268
Mg-S	234	59
Mn-Mn	25.9	205
Mn-O	402.9 ± 41.8	209,268
Mn-S	301 ± 17	366
Mn-Se	239.3 ± 9.2	328
Mo-Mo	406 ± 21	159
Mo-Nb	456 ± 25	154
Mo-O	560.2 ± 20.9	268
N-N	945.33 ± 0.59	189
N-O	630.57 ± 0.13	189
N-P	617.1 ± 20.9	61,141
N-Pu	473 ± 63	229
N-S	464 ± 21	229
N-Sb	301 ± 50	110
N-Sc	469 ± 84	229
N-Se	370 ± 11	330
N-Si	470 ± 15	291
N-Ta	611 ± 84	229
N-Th	577.4 ± 33.1	134,142
N-Ti	476.1 ± 33.1	142,332
N-U	531.4 ± 2.1	132
N-V	477.4 ± 17.2	98,142
N-Xe	23.0	171
N-Y	481 ± 63	229
N-Zr	564.8 ± 25.1	133,142
Na-Na	73.60 ± 0.25	361
Na-O	256.1 ± 16.7	268
Na-Rb	63.25	363
Nb-Nb	510 ± 10.0	155
Nb-O	771.5 ± 25.1	268
Nd-Nd	<163	229
Nd-O	703 ± 13	86
Nd-S	471.5	20
Nd-Se	385 ± 17	20,145,253
Nd-Te	305 ± 17	20
Ne-Ne	3.93	341
Ni-Ni	203.26 ± 0.96	251
Ni-O	382.0 ± 16.7	268
Ni-Pt	273.7 ± 0.3	342
Ni-S	344.3	83
Ni-Si	318 ± 17	353
Ni-V	206.3 ± 0.2	329
Np-O	718.4 ± 41.8	268
O-O	498.36 ± 0.17	33,189
O-Os	575	177
O-P	599.1 ± 12.6	268
O-Pa	788.3 ± 17.2	223
O-Pb	382.0 ± 12.6	268
O-Pd	380.7 ± 83.7	268
O-Pm	674 ± 63	229
O-Pr	753 ± 13	86
O-Pt	391.6 ± 41.8	268
O-Pu	715.9 ± 33.9	268
O-Rb	255 ± 84	32
O-Re	626.8 ± 83.7	268
O-Rh	405.0 ± 41.8	268
O-Ru	528.4 ± 41.8	268
O-S	521.7 ± 4.2	268
O-Sb	434.3 ± 41.8	268
O-Sc	681.6 ± 11.3	268
O-Se	464.8 ± 21.3	268,321
O-Si	799.6 ± 13.4	268
O-Sm	565 ± 13	86
O-Sn	531.8 ± 12.6	268
O-Sr	425.5 ± 16.7	268
O-Ta	799.1 ± 12.6	268
O-Tb	711 ± 13	86
O-Te	376.1 ± 20.9	268
O-Th	878.6 ± 12.1	268
O-Ti	672.4 ± 9.2	268
O-Tm	502 ± 13	86
O-U	759.4 ± 13.4	268
O-V	626.8 ± 18.8	10,268
O-W	672.0 ± 41.8	268
O-Xe	36.4	229
O-Y	719.6 ± 11.3	193,268
O-Yb	397 ± 17	86
O-Zn	159 ± 4	45
O-Zr	776.1 ± 13.4	268
P-P	489.5 ± 10.5	141
P-Pt	≤416.7 ± 17	325
P-Rh	353.1 ± 17	325
P-S	444 ± 8	82
P-Sb	356.9	232
P-Se	363.6 ± 10.0	82
P-Si	363.6	323
P-Te	297.9 ± 10.0	82
P-Th	550.2 ± 42	125
P-Tl	209 ± 13	275
P-U	297 ± 21	229
P-W	305 ± 4	140
Pb-Pb	86.6 ± 0.8	128,286
Pb-S	346.0 ± 1.7	348
Pb-Sb	161.5 ± 10.5	380
Pb-Se	302.9 ± 4	348
Pb-Te	251 ± 13	348
Pd-Pd	100 ± 15	308
Pd-Si	261 ± 12	313
Pd-Y	238 ± 17	293
Pm-S	423 ± 63	229
Pm-Se	339 ± 63	229
Pm-Te	255 ± 63	229
Po-Po	187.0	189
Pr-S	492.5 ± 4.6	103
Pr-Se	446.4 ± 23.0	144,253
Pr-Te	326 ± 42	229
Pt-Pt	357.3 ± 15.1	161
Pt-Si	501.2 ± 18.0	354
Pt-Th	552	123
Pt-Ti	397 ± 13	164
Pt-Y	474.0 ± 12.1	161
Rb-Rb	48.898 ± 0.005	4
Rh-Rh	285.3 ± 20.9	50,283
Rh-Sc	443.9 ± 10.5	166
Rh-Si	395.0 ± 18.0	354
Rh-Th	515 ± 21	119
Rh-Ti	390.8 ± 14.6	50
Rh-U	519 ± 17	119
Rh-V	364 ± 29	125
Rh-Y	445.2 ± 10.5	166
Ru-Si	397.1 ± 20.9	354
Ru-Th	591.6 ± 42	124
Ru-V	414 ± 29	125
S-S	425.30	189
S-Sb	378.7	101
S-Sc	477 ± 13	335,346
S-Se	371.1 ± 6.7	80
S-Si	623	189
S-Sm	389	103
S-Sn	464 ± 3.3	78
S-Sr	339	39
S-Tb	515 ± 42	229
S-Te	339 ± 21	78
S-Ti	418 ± 3	87,273
S-Tm	368 ± 42	229
S-U	522.6 ± 9.6	335
S-V	450	88,189
S-Y	528.4 ± 10.5	334
S-Yb	167	229
S-Zn	205 ± 13	71,147
S-Zr	575.3 ± 16.7	335
Sb-Sb	299.2 ± 6.3	70,231
Sb-Te	277.4 ± 3.8	287,340
Sb-Tl	126.8 ± 10.5	9,275
Sc-Sc	162.8 ± 21	126

Table 1
BOND STRENGTHS IN DIATOMIC MOLECULES (continued)

Molecule	$D°_{298}$/kJ mol^{-1}	Ref.
Sc-Se	385 ± 17	229
Sc-Si	228.7 ± 14	217
Sc-Te	289 ± 17	229
Se-Se	332.6 ± 0.4	80,348
Se-Si	548	189
Se-Sm	331.0 ± 14.6	253
Se-Sn	401.2 ± 5.9	57
Se-Sr	~285	22
Se-Tb	423 ± 42	229
Se-Te	291.6 ± 4	78,80,148
Se-Ti	381 ± 42	229
Se-Tm	276 ± 42	229
Se-V	347 ± 21	229
Se-Y	435 ± 13	229
Se-Zn	170.7 ± 25.9	71,147
Si-Si	326.8 ± 10.0	42
Si-Te	452	189
Si-Y	258.8 ± 17.3	218
Sm-Te	272.4 ± 14.6	253
Sn-Sn	187.1 ± 0.3	265
Sn-Te	359.8	189
Sr-Sr	15.5 ± 0.4	111
Tb-Tb	131.4 ± 25.1	233
Tb-Te	339 ± 42	229
Te-Te	259.8 ± 5.0	259
Te-Ti	289 ± 17	229
Te-Tm	276 ± 42	229
Te-Y	339 ± 13	229
Te-Zn	117.6 ± 18.0	147
Th-Th	≤289	130
Ti-Ti	141.4 ± 21	199
Tl-Tl	64.4 ± 17	8
U-U	222 ± 21	229
V-V	269.2 ± 0.2	329
Xe-Xe	6.53 ± 0.29	41
Y-Y	159 ± 21	229
Yb-Yb	20.5 ± 17	151
Zn-Zn	29	316

REFERENCES

1. Abbasov, A. S., Azizov, T. Kh., Alleva, N. A., Aliev, I. Ya, Mustafaev, F. M., and Mamedov, A. N., *Zh. Fiz. Khim.*, 50, 2172, 1976.
2. Ackerman, M., Drowart, J., Stafford, F. E., and Verhaegen, G., *J. Chem. Phys.*, 36, 1557, 1962.
3. Ackerman, M., Stafford, F. E., and Drowart, J., *J. Chem. Phys.*, 33, 1784, 1960.
4. Amiot, C., *J. Chem. Phys.*, 93, 8591, 1990.
5. Amiot, C., *J. Mol. Spectrosc.*, 147, 370, 1991.
6. Appelman, E. H. and Clyne, M. A. A., *J. Chem. Soc. Faraday Trans. 1*, 71, 2072, 1975.
7. Balakrishnan, A., Jones, W. J., Mahajan, C. G., and Stoicheff, B. P., *Chem. Phys. Lett.*, 155, 43, 1989.
8. Balducci, G. and Piacente, V., *J. Chem. Soc. Chem. Commun.*, 1287, 1980.
9. Balducci, G., Ferro, D., and Piacente, V., *High Temp. Sci.*, 14, 207, 1981.
10. Balducci, G., Gigli, G., and Guido, M., *J. Chem. Phys.*, 79, 5616, 1983.
11. Balducci, G., Nunzio, P. E., Gigli, G., and Guido, M., *J. Chem. Phys.*, 90, 406, 1989.
12. Balfour, W. J. and Cartwright, H. M., *Astron. Astrophys. Suppl. Ser.*, 26, 389, 1976.
13. Balfour, W. J. and Lingren, B., *Can. J. Chem.*, 56, 767, 1978.
14. Barkovskii, N. V., Tsirel'nikov, V. I., Emel'yanov, A. M., and Khodeev, Yu. S., *Teplofiz. Vys. Temp.*, 29, 474, 1991.
15. Barrow, R. F. and Caunt, A. D., *Proc. R. Soc. London Ser. A*, 219, 120, 1953.
16. Barrow, R. F. and Chojnicki, A. H., *J. Chem. Soc. Faraday Trans. 2*, 71, 728, 1975.
17. Barrow, R. F., *Trans. Faraday Soc.*, 56, 952, 1960.
18. Behrens, R. G. and Feber, R. C., *J. Less-Common Met.*, 75, 281, 1980.
19. Bergman, C. and Gingerich, K. A., *J. Phys. Chem.*, 76, 2332, 1972.
20. Bergman, C., Coppens P., Drowart, J., and Smoes, S., *Trans. Faraday Soc.*, 66, 800, 1970.
21. Berkowitz, J. and Chupka, W. A., *J. Chem. Phys.*, 45, 1287, 1966.
22. Berkowitz, J. and Chupka, W. A., *J. Chem. Phys.*, 45, 4289, 1966.
23. Berkowitz, J. and Walter, T., *J. Chem. Phys.*, 49, 1184, 1968.
24. Berkowitz, J., *J. Chem. Phys.*, 89, 7065, 1988.
25. Berneike ,W., Kreuttle, U., and Neuert, H., *Chem. Phys. Lett.*, 76, 525, 1980.
26. Besenbruch, G., Kana'an, A. S., and Margrave, J. L., *J. Phys. Chem.*, 69, 3174, 1965.
27. Bharate, N. S., Bhartiya, J. B., and Behere, S. H., *Proc. Indian Natl. Sci. Acad. Part A*, 57, 419, 1991.
28. Birks, J. W., Gabelnick, S. D., and Johnston, H. S., *J. Mol. Spectrosc.*, 57, 23, 1975.
29. Bishea, G. A. and Morse, M. D., *Chem. Phys. Lett.*, 171, 430, 1990.
30. Bondybey, V. E., *Chem. Phys. Lett.*, 109, 436, 1984.
31. Breford, E. J. and Engelke, F., *J. Chem. Phys.*, 71, 1994, 1979.
32. Brewer, L. and Rosenblatt, G. M., *Adv. High Temp. Sci.*, 2, 1, 1969.
33. Brix, P. and Herzberg, G., *J. Chem. Phys.*, 21, 2240, 1953.
34. Burns, G., LeRoy, L. J., Morris, D. J., and Blake, J. A., *Proc. R. Soc. London Ser. A*, 316, 81, 1970.
35. Busse, V. B. and Weil, K. G., *Ber. Bunsenges. Phys. Chem.*, 85, 309, 1981.
36. Calculated from $\Delta_f H°$ = 435.1 ± 10 kJ mol^{-1} (Table 4), but see also Costes, M., Naulin, C., and Dorthe, G., *Astron. Astronphys.*, 232, 270, 1990.
37. Callender, C. L., Mitchell, S. A., and Hackett, P. A., *J. Chem. Phys.*, 90, 5252, 1989.
38. Carbonel, M., Bergman, C., and Laffite, M., *Colloq. Int. Cent. Nat. Rech. Sci.*, 201, 311, 1972.
39. Cater, E. D. and Johnson, E. W., *J. Chem. Phys.*, 47, 5353, 1967.

Table 1
BOND STRENGTHS IN DIATOMIC MOLECULES (continued)

40. Chashchina, G. I. and Shreider, E. Ya., *Zh. Prikl. Spektrosk.*, 21, 696, 1974.
41. Chashchina, G. I. and Shreider, E. Ya., *Zh. Prikl. Spektrosk.*, 25, 163, 1976.
42. Chatillon, C., Allibert, M., and Pattoret, A., *C. R. Acad. Sci. Ser. C*, 280, 1505, 1975.
43. Chen, Y. M., Clemmer, D. E., and Armentrout, P. B., *J. Chem. Phys.*, 95, 1228, 1991.
44. Clarke, T. C. and Clyne, M. A. A., *Trans. Faraday Soc.*, 66, 877, 1970.
45. Clemmer, D. E., Daltaska, N. F., and Armentrout, P. B., *J. Chem. Phys.*, 95, 7263, 1991.
46. Clyne, M. A. A. and McDermid, I. S., *Faraday Discuss. Chem. Soc.*, 67, 316, 1979.
47. Clyne, M. A. A. and McDermid, I. S., *J. Chem. Soc. Faraday Trans. 2*, 72, 2252, 1976.
48. Clyne, M. A. A. and Watson, R. T., *Chem. Phys. Lett.*, 12, 344, 1971.
49. Clyne, M. A. A., Curran, A. H., and Coxon, J. A., *J. Mol. Spectrosc.*, 63, 43, 1976.
50. Cocke, D. L. and Gingerich, K. A., *J. Chem. Phys.*, 60, 1958, 1974.
51. Cocke, D. L. and Gingerich, K. A., *J. Phys. Chem.*, 75, 3264, 1971.
52. Cocke, D. L. and Gingerich, K. A., *J. Phys. Chem.*, 76, 2332, 1972.
53. Cocke, D. L., Gingerich, K. A., and Chang, C. A., *J. Chem. Soc. Faraday Trans. 1*, 72, 268, 1976.
54. Cocke, D. L., Gingerich, K. A., and Kordis, J., *High Temp. Sci.*, 5, 474, 1973.
55. Cocke, D. L., Gingerich, K. A., and Kordis, J., *High Temp. Sci.*, 7, 61, 1975.
56. Colin, R. and De Greef, D., *Can. J. Phys.*, 53, 2142, 1975.
57. Colin, R. and Drowart, J., *Trans. Faraday Soc.*, 60, 673, 1964.
58. Colin, R. and Drowart, J., *Trans. Faraday Soc.*, 64, 2611, 1968.
59. Colin, R., Goldfinger, P., and Jeunehomme, M., *Trans. Faraday Soc.*, 60, 306, 1964.
60. Coppens, P., Reynaert, J. C., and Drowart, J., *J. Chem. Soc. Faraday Trans. 2*, 75, 292, 1979.
61. Coquart, B. and Prudhomme, J. C., *J. Mol. Spectrosc.*, 87, 75, 1981.
62. Corbett, J. D. and Lynde, R. A., *Inorg. Chem.*, 6, 2199, 1967.
63. Costes, M., Naulin, C., Dorthe, G., Vaucamps, C., Nouchi, G., *Faraday Discuss. Chem. Soc.*, 84, 75, 1987.
64. Coxon, J. A., *Chem. Phys. Lett.*, 33, 136, 1975.
65. Cubicciotti, D., *Inorg. Chem.*, 7, 208, 1968.
66. Cubicciotti, D., *Inorg. Chem.*, 7, 211, 1968.
67. Cubicciotti, D., *J. Phys. Chem.*, 71, 3066, 1967.
68. Cuthill, A. M., Fabian, D. J., and Shu-Shou-Shen, S., *J. Phys. Chem.*, 77, 2008, 1973.
69. Darwent, B. de B., *Bond Dissociation Energies in Simple Molecules*, NSRDS-NBS-31, National Bureau of Standards, Washington, D. C., 1970.
70. De Maria, G., Drowart, J., and Inghram, M. G., *J. Chem. Phys.*, 31, 1076, 1959.
71. De Maria, G., Goldfinger, P., Malaspina, L., and Piacente, V., *Trans. Faraday Soc.*, 61, 2146, 1965.
72. De Maria, G., Malaspina, L., and Piacente, V., *J. Chem. Phys.*, 52, 1019, 1970.
73. De Maria, G., Malaspina, L., and Piacente, V., *J. Chem. Phys.*, 56, 1978, 1972.
74. Devore, T. C., Brock, L., Kohlscheuer, R., Dulaney, K., and Gole, J. L., *Chem. Phys.*, 155, 423, 1991
75. Devore, T. C., McQuaid, M., and Gole, J. L., *High Temp. Sci.*, 30, 83, 1990.
76. Diemer, U., Weickenmeier, H., Wahl, M., and Demtroeder, W., *Chem. Phys. Lett.*, 104, 489, 1984.
77. Drowart, J. and Goldfinger, P., *Angew. Chem.*, 6, 581, 1967.
78. Drowart, J. and Goldfinger, P., *Q. Rev. (London)*, 20, 545, 1966.
79. Drowart, J. and Honig, R. E., *J. Phys. Chem.*, 61, 980, 1957.
80. Drowart, J. and Smoes, S., *J. Chem. Soc. Faraday Trans. 2*, 73, 1755, 1977.
81. Drowart, J., De Maria, G., and Inghram, M. G., *J. Chem. Phys.*, 29, 1015, 1958.
82. Drowart, J., Myers, C. E., Szwarc, R.,Vander Auwera-Mahieu, A., and Uy, O. M., *High Temp. Sci.*, 5, 482, 1973.
83. Drowart, J., Pattoret, A., and Smoes, S., *Proc. Br. Ceramic Soc.*, No. 8, 67, 1967.
84. Drowart, J., Smets, J., Reynaert, J. C., and Coppens, P., *Adv. Mass Spectrom.*, 7A, 647, 1978.
85. Drowart, J., Smoes, S, and Vander Auwera-Mahieu, A., *J. Chem. Thermodyn.*, 10, 453, 1978.
86. Dulick, M., Murad, E., and Barrow, R. F., *J. Chem. Phys.*, 85, 385, 1986.
87. Edwards, J. G., Franklin, H. F., and Gilles, P. W., *J. Chem. Phys.*, 54, 545, 1971.
88. Edwards, J. G., *J. Chem. Phys.*, 96, 866, 1992
89. Ehlert, T. C. and Margrave, J. L., *J. Chem. Phys.*, 41, 1066, 1964.
90. Ehlert, T. C. and Wang, J. S., *J. Phys. Chem.*, 81, 2069, 1977.
91. Ehlert, T. C., Hilmer, R. M., and Beauchamp, E. A., *J. Inorg. Nucl. Chem.*, 30, 3112, 1968.
92. Engelke, F., *Chem. Phys.*, 39, 279, 1979.
93. Engelke, F., Ennen, G., and Meiwes, K. H., *Chem. Phys.*, 66, 391, 1982.
94. Engelke, F., Hage, H., and Sprick, U., *Chem. Phys.*, 88, 443, 1984.
95. Estler, C. and Zare, R. N., *Chem. Phys.*, 28, 253, 1978.
96. Farber, M. and Srivastava, R. D., *Chem. Phys. Lett.*, 42, 567, 1976.
97. Farber, M. and Srivastava, R. D., *High Temp. Sci.*, 12, 21, 1980.

Table 1
BOND STRENGTHS IN DIATOMIC MOLECULES (continued)

98. Farber, M. and Srivastava, R. D., *J. Chem. Soc. Faraday Trans. 1*, 69, 390, 1973.
99. Farber, M. and Srivastava, R. D., *J. Chem. Soc. Faraday Trans. 1*, 70, 1581, 1974.
100. Farber, M. and Srivastava, R. D., *J. Chem. Soc. Faraday Trans. 1*, 74, 1089, 1978.
101. Faure, F. M., Mitchell, M. J., and Bartlett, R. W., *High Temp. Sci.*, 4, 181, 1972.
102. Fellows, C. E., *J. Chem. Phys.*, 94, 5855, 1991.
103. Fenochka, B. V. and Gorkienko, S. P., *Zh. Fiz. Khim*, 47, 2445, 1973.
104. Ferro, D., Piacente, V., and Bardi, G., *High Temp. Sci.*, 21, 69, 1986.
105. Filippenko, N. V., Morozov, E. V., Giricheva, N. L., and Krasnev, K. S., *Izv. Vyssh. Ucheb. Zaved Khim.Technol.*, 15, 1416, 1972.
106. Franzen, H. and Hariharan, A. V., *J. Chem. Phys.*, 70, 4907, 1979.
107. Fries, J. A. and Cater, E. D., *J. Chem. Phys.*, 68, 3978, 1978.
108. Fu, Z., Lemire, G. W., Bishea, G. A., and Morse, M. D., *J. Chem. Phys.*, 93, 8420, 1990.
109. Gal, J. F., Maria, P. C., and Decouzon, M., *Int. J. Mass Spectrom. Ion Processes*, 93, 87, 1989.
110. Gaydon, A. G., *Dissociation Energies and Spectra of Diatomic Molecules*, 3rd ed., Chapman & Hall, London, 1968.
111. Gerber, G. and Moeller, R., *Contrib. Symp. At. Surf. Phys.*, 168, 1982.
112. Gibson, S. T., Greene, J. P., and Berkowitz, J., *J. Chem. Phys.*, 85, 4815, 1986.
113. Gingerich, K. A. and Blue, G. D., *J. Chem. Phys.*, 47, 5447, 1967.
114. Gingerich, K. A. and Blue, G. D., *J. Chem. Phys.*, 59, 186, 1973.
115. Gingerich, K. A. and Choudary, U. V., *J. Chem. Phys.*, 68, 3265, 1978.
116. Gingerich, K. A. and Cocke, D. L., *Inorg. Chim. Acta*, 28, L171, 1978.
117. Gingerich, K. A. and Finkbeiner, H. C., *J. Chem. Phys.*, 54, 2621, 1971.
118. Gingerich, K. A. and Finkbeiner, H. C., *Proc. 9th Rare Earth Res. Conf.*, 2, 795, 1971.
119. Gingerich, K. A. and Gupta, S. K., *J. Chem. Phys.*, 69, 505, 1978.
120. Gingerich, K. A. and Piacente, V., *J. Chem. Phys.*, 54, 2498, 1971.
121. Gingerich, K. A., *Chem. Commun.*, 580, 1970.
122. Gingerich, K. A., *Chem. Phys. Lett.*, 13, 262, 1972.
123. Gingerich, K. A., *Chem. Phys. Lett.*, 23, 270, 1973.
124. Gingerich, K. A., *Chem. Phys. Lett.*, 25, 523, 1974.
125. Gingerich, K. A., *Chem. Soc. Faraday, Symp.*, No.14, 109, 1980.
126. Gingerich, K. A., *Chimia*, 26, 619, 1972.
127. Gingerich, K. A., Cocke, D. L., and Kordis, J., *J. Phys. Chem.*, 78, 603, 1974.
128. Gingerich, K. A., Cocke, D. L., and Miller, F., *J. Chem. Phys.*, 64, 4027, 1976.
129. Gingerich, K. A., Haque, R., and Kingcade, J. E., *Thermochim. Acta*, 30, 61, 1979.
130. Gingerich, K. A., *High Temp. Sci.*, 1, 258, 1969.
131. Gingerich, K. A., *High Temp. Sci.*, 3, 415, 1971.
132. Gingerich, K. A., *J. Chem. Phys.*, 47, 2192, 1967.
133. Gingerich, K. A., *J. Chem. Phys.*, 49, 14, 1968.
134. Gingerich, K. A., *J. Chem. Phys.*, 49, 19, 1968.
135. Gingerich, K. A., *J. Chem. Phys.*, 54, 2646, 1971.
136. Gingerich, K. A., *J. Chem. Phys.*, 54, 3720, 1971.
137. Gingerich, K. A., *J. Chem. Phys.*, 56, 4239, 1972.
138. Gingerich, K. A., *J. Chem. Phys.*, 74, 6407, 1981.
139. Gingerich, K. A., *J. Chem. Soc. Faraday Trans. 2*, 70, 471, 1974.
140. Gingerich, K. A., *J. Phys. Chem.*, 68, 768, 1964.
141. Gingerich, K. A., *J. Phys. Chem.*, 73, 2734, 1969.
142. Gingerich, K. A., *NBS Spec. Publ. (U. S.)*, 561, 289, 1979.
143. Gordienko, S. P. and Fenochka, B. V., *Izv. Akad. Nauk. SSSR Neorg. Mater.*, 18, 1811, 1982.
144. Gordienko, S. P., Fenochka, B. V., Viksman, G. Sh., Klockkova, L. A., and Mikhlina, T. M., *Izv. Akad. Nauk. SSSR Neorg. Mater.*, 18, 18, 1982.
145. Gordienko, S. P., *Izv. Akad. Nauk. SSSR Neorg. Mater.*, 20, 1472, 1984.
146. Gorokhov, L. N., Smirnov, V. K., and Khodeev, Yu. S., *Zh. Fiz. Khim.*, 58, 1603, 1984.
147. Grade, M. and Hirschwald, W., *Ber. Bunsenges. Phys. Chem.*, 86, 899, 1982.
148. Grade, M., Wienecke, J., Rosinger, W., and Hirschwald, W., *Ber. Bunsenges. Phys. Chem.*, 87, 355, 1983.
149. Graeber, P. and Weil, K. G., *Ber. Bunsenges. Phys. Chem.*, 76, 417, 1972.
150. Guggi, D. J., Neubert, A., and Zmbov, K. F., *Conf. Int. Thermodyn. Chim. [C. R.] 4th*, 3, 124, 1975.
151. Guido, M. and Balducci, G., *J. Chem. Phys.*, 57, 5611, 1972.
152. Guido, M., Gigli, G., and Balducci, G., *J. Chem. Phys.*, 57, 3731, 1972.
153. Gupta, S. K. and Gingerich, K. A., *High Temp. - High Pressures*, 12, 273, 1980.
154. Gupta, S. K. and Gingerich, K. A., *J. Chem. Phys.*, 69, 4318, 1978.

Table 1
BOND STRENGTHS IN DIATOMIC MOLECULES (continued)

155. Gupta, S. K. and Gingerich, K. A., *J. Chem. Phys.*, 70, 5350, 1979..
156. Gupta, S. K. and Gingerich, K. A., *J. Chem. Phys.*, 71, 3072, 1979.
157. Gupta, S. K. and Gingerich, K. A., *J. Chem. Phys.*, 72, 2795, 1980.
158. Gupta, S. K. and Gingerich, K. A., *J. Chem. Phys.*, 74, 3584, 1981.
159. Gupta, S. K., Atkins, R. M., and Gingerich, K. A., *Inorg. Chem.*, 17, 3211, 1978.
160. Gupta, S. K., Kingcade, J. E., and Gingerich, K. A., *Adv. Mass Spectrom.*, 8A, 445, 1980.
161. Gupta, S. K., Nappi, B. M., and Gingerich, K. A., *Inorg. Chem.*, 20, 966, 1981.
162. Gupta, S. K., Nappi, B. M., and Gingerich, K. A., *J. Phys. Chem.*, 85, 971, 1981.
163. Gupta, S. K., Pelino, M., and Gingerich, K. A., *J. Chem. Phys.*, 70, 2044, 1979.
164. Gupta, S. K., Pelino, M., and Gingerich, K. A., *J. Phys. Chem.*, 83, 2335, 1979.
165. Gustavsson, T., Amiot, C., Verges, J., *Mol. Phys.*, 64, 279, 1988.
166. Haque, R. and Gingerich, K. A., *J. Chem. Thermodyn.*, 12, 439, 1980.
167. Haque, R., Pelino, M., and Gingerich, K. A., *J. Chem. Phys.*, 71, 2929, 1979.
168. Haque, R., Pelino, M., and Gingerich, K. A., *J. Chem. Phys.*, 73, 4045, 1980.
169. Hariharan, A. V. and Eick, H. A., *J. Chem. Thermodyn.*, 6, 373, 1974.
170. Herman, P. R., La Rocque, P. E., and Stoicheff, B., *J. Chem. Phys.*, 89, 4535, 1988.
171. Herman, R. and Herman, L., *J. Phys. Radium.*, 24, 73, 1963.
172. Herzberg, G., *Molecular Spectra and Molecular Structure. I. Spectra of Diatomic Molecules*, 2nd ed.,Van Nostrand, New York, 1950.
173. Hildenbrand, D. L. and Lau, K. H., *J. Chem. Phys.*, 89, 5825, 1988.
174. Hildenbrand, D. L. and Lau, K. H., *J. Chem. Phys.*, 91, 4909, 1989.
175. Hildenbrand, D. L. and Lau, K. H., *J. Chem. Phys.*, 93, 5983, 1990.
176. Hildenbrand, D. L. and Lau, K. H., *J. Chem. Phys.*, 96, 3830, 1992.
177. Hildenbrand, D. L. and Lau, K. H., *J. Phys. Chem.*, 96, 2325, 1992.
178. Hildenbrand, D. L. and Murad, E., *J. Chem. Phys.*, 44, 1524, 1966.
179. Hildenbrand, D. L. and Theard, L. P., *J. Chem. Phys.*, 50, 5350, 1969.
180. Hildenbrand, D. L., *Chem. Phys. Lett.*, 32, 523, 1975.
181. Hildenbrand, D. L., *J. Chem. Phys.*, 48, 3657, 1968.
182. Hildenbrand, D. L., *J. Chem. Phys.*, 52, 5751, 1970.
183. Hildenbrand, D. L., *J. Chem. Phys.*, 65, 614, 1976.
184. Hildenbrand, D. L., *J. Chem. Phys.*, 66, 3526, 1977.
185. Hildenbrand, D. L., *J. Phys. Chem.*, 77, 897, 1973.
186. Hilpert, K. and Ruthardt, K., *Ber. Bunsenges. Phys. Chem.*, 93, 1070, 1989.
187. Hilpert, K., *Ber. Kernforschungsanlage Juelich, JUEL-1744*, 272, 1981.
188. Hilpert, K., J. Chem. Phys., 77, 1425, 1982.
189. Huber, K. P. and Herzberg, G., *Molecular Spectra and Molecular Structure Constants of Diatomic Molecules*, Van Nostrand, New York, 1979.
190. Hussein, K., Effantin, C., D'Incan, J., Verges, J., and Barrow, R. F., *Chem. Phys. Lett.*, 124, 105, 1986.
191. Ihle, H. R. and Wu, C. H., *J. Chem. Phys.*, 63, 1605, 1975.
192. Ishwar, N. B., Varma, M. P., and Jha, B. L., *Acta Phys. Pol. A*, A61, 503, 1982.
193. Ishwar, N. B., Varma, M. P., and Jha, B. L., *Indian J. Pure Appl. Phys.*, 20, 992, 1982.
194. Jeyagopal, T., Rajavel, S. R. K., Ramakrishnan, M., and Rajamanickam, N., *Acta Phys. Hung.*, 68, 145, 1990.
195. Johns, J. W. C. and Ramsey, D. A., *Can. J. Phys.*, 39, 210, 1961.
196. Jones, R. W. and Gole, J. L., *Chem. Phys.*, 20, 311, 1977.
197. Jordan, K. J. and Lipson, R. H., *J. Phys. Chem.*, 95, 7204, 1991.
198. Jordan, K. J., Lipson, R. H., McDonald, N. A., and Le Roy, R. J., *J. Phys. Chem.*, 96, 4778, 1992.
199. Kant, A. and Lin, S.-S., *J. Chem. Phys.*, 51, 1644, 1969.
200. Kant, A. and Moon, K. A., *High Temp. Sci.*, 11, 55, 1979.
201. Kant, A. and Strauss, B. H., *J. Chem. Phys.*, 41, 3806, 1964.
202. Kant, A. and Strauss, B., *J. Chem. Phys.*, 45, 3161, 1966.
203. Kant, A. and Strauss, B., *J. Chem. Phys.*, 49, 3579, 1968.
204. Kant, A., *J. Chem. Phys.*, 49, 5144, 1968.
205. Kant, A., Lin, S.-S, and Strauss, B., *J. Chem. Phys.*, 49, 1983, 1968.
206. Kant, A., Strauss, B., and Lin, S.-S., *J. Chem. Phys.*, 52, 2384, 1970.
207. Kaposi, O., *Magy. Kem. Foly.*, 83, 356, 1977.
208. Kaufel, R., Vahl, G., Nunkwitz, R., and Baumgaertel, H., *Z. Anorg. Allg. Chem.*, 481, 207, 1981.
209. Kazenas ,E., Tagirov, V. K., and Zviadadze, G. N., *Izv. Akad. Nauk. SSSR Met.*, 58, 1984.
210. Kent, R. A. and Margrave, J. L., *J. Am. Chem. Soc.*, 87, 3582, 1965.
211. Kent, R. A., Ehlert, T. C., and Margrave, J. L., *J. Am. Chem. Soc.*, 86, 5090, 1964.
212. Kent, R. A., *J. Am. Chem. Soc.*, 90, 5657, 1968.

Table 1
BOND STRENGTHS IN DIATOMIC MOLECULES (continued)

213. Khitrov, A. N., Ryabova, V. G., and Gurvich, L. V., *Teplofiz. Vys. Tempo.*, 11, 1126, 1973.
214. Kiang, T. and Zare, R. N., *J. Am. Chem. Soc.*, 102, 4024, 1980.
215. Kimura, H., Asano, M., and Kubo, K., *J. Nucl. Mater.*, 97, 259, 1981.
216. Kingcade, J. E. Jr. and Gingerich, K. A., *J. Chem. Phys.*, 84, 3432, 1986.
217. Kingcade, J. E. Jr. and Gingerich, K. A., *J. Chem. Soc. Faraday Trans. 2*, 85, 195, 1989.
218. Kingcade, J. E. Jr., and Gingerich, K. A., *J. Chem. Phys.*, 84, 4574, 1986.
219. Kingcade, J. E., Cocke, D. L., and Gingerich, K. A., *High Temp. Sci.*, 16, 89, 1983,
220. Kingcade, J. E., Dufner, D. C., Gupta, S. K., and Gingerich, K. A., *High Temp. Sci.*, 10, 213, 1978.
221. Kingcade, J. E., Nagarathna, H. M., Shim, I., and Gingerich, K. A., *J. Phys. Chem.*, 90, 2830, 1986.
222. Kleinschmidt, P. D. and Hildenbrand, D. L., *J. Chem. Phys.*, 68, 2819, 1978.
223. Kleinschmidt, P. D. and Ward, J. W., *J. Less-Common. Met.*, 121, 61, 1986.
224. Kleinschmidt, P. D., Cubicciotti, D., and Hildenbrand, D. L., *J. Electrochem. Soc.*, 125, 1543, 1978; *Proc. Electrochem.Soc.*, 78, 217, 1978.
225. Kleinschmidt, P. D., Lau, K. H., and Hildenbrand, D. L., *J. Chem. Phys.*, 74, 653, 1981.
226. Klynning, L. and Lindgren, B., *Arkiv. Fysik.*, 32, 575, 1966.
227. Kohl, F. J. and Carlson, K. D., *J. Am. Chem. Soc.*, 90, 4814, 1968.
228. Kohl, F. J. and Stearns, C. A., *J. Phys. Chem.*, 78, 273, 1974.
229. Kondratiev, V. N., *Bond Dissociation Energies, Ionization Potentials and Electron Affinities*, Mauka Publishing House, Moscow, 1974.
230. Kordis, J. and Gingerich, K. A., *J. Chem. Eng. Data*, 18, 135, 1973.
231. Kordis, J. and Gingerich, K. A., *J. Chem. Phys.*, 58, 5141, 1973.
232. Kordis, J. and Gingerich, K. A., *J. Phys. Chem.*, 76, 2336, 1972.
233. Kordis, J., Gingerich, K. A., and Seyse, R. J., *J. Chem. Phys.*, 61, 5114, 1974.
234. Kowalski, A., Czaikowski, M., and Breckenridge, W. H., *Chem. Phys. Lett.*, 119, 368, 1985.
235. Koyama, T. and Yamawaki, M., *J. Nucl. Mater.*, 152, 30, 1988.
236. Kronekvist, M., Lagerqvist, A., and Neuhaus, H., *J. Mol. Spectrosc.*, 39, 516, 1971.
237. Kulkarni, M. P. and Dadape, V. V., *High Temp. Sci.*, 3, 277, 1971.
238. Lau, K. H. and Hildenbrand, D. L., *J. Chem. Phys.*, 71, 1572, 1979.
239. Lau, K. H. and Hildenbrand, D. L., *J. Chem. Phys.*, 72, 4928, 1980.
240. Lau, K. H. and Hildenbrand, D. L., *J. Chem. Phys.*, 76, 2646, 1982.
241. Lau, K. H. and Hildenbrand, D. L., *J. Chem. Phys.*, 80, 1312, 1984.
242. Lau, K. H. and Hildenbrand, D. L., *J. Chem. Phys.*, 86, 2949, 1987.
243. Lau, K. H. and Hildenbrand, D. L., *J. Chem. Phys.*, 92, 6124, 1990.
244. Lau, K. H., Brittain, R. D., and Hildenbrand, D. L., *Chem. Phys. Lett.*, 81, 227, 1981; *J. Phys. Chem.*, 86, 4429, 1982.
245. Lau, K. H., Brittain, R. D., and Hildenbrand, D. L., *J. Chem. Phys.*, 90, 1158, 1989.
246. Li, K. C. and Stwalley, W. C., *J. Chem. Phys.*, 59, 4423, 1973.
247. Li, L., Lyyra, A. M., Luh, W. T., and Stwalley, W. C., *J. Chem. Phys.*, 93, 8452, 1990.
248. Lindgren, B. and Nilsson, Ch., *J. Mol. Spectrosc.*, 55, 407, 1975.
249. Martin, E. and Barrow, R. F., *Phys. Scr.*, 17, 501, 1978.
250. McIntyre, N. S., Vander Auwera-Mahieu, A., and Drowart, J., *Trans. Faraday Soc.*, 64, 3006, 1968.
251. Morse, M. D., Hansen, G. P., Langridge-Smith, P. R. R., Zheng, L. S., Geusic, M. E., Michalopoulos, D. L., and Smalley, R. E., *J. Chem. Phys.*, 80, 5400, 1984.
252. Murad, E., Hildenbrand, D. L., and Main, R. P., *J. Chem. Phys.*, 45, 263, 1966.
253. Nagai, S., Shinmei, M., and Yokokawa, T., *J. Inorg. Nucl. Chem.*, 36, 1904, 1974.
254. Nappi, B. M. and Gingerich, K. A., *Inorg. Chem.*, 20, 522, 1981.
255. Neckel, A. and Sodeck, G., *Monatsch. Chem.*, 103, 367, 1972.
256. Nedelec, O. and Giroud, M., *J. Chem. Phys.*, 79, 2121, 1983.
257. Neubert, A. and Zmbov, K. F., *Chem. Phys.*, 76, 469, 1983.
258. Neubert, A. and Zmbov, K. F., *J. Chem. Soc. Faraday Trans. 1*, 70, 2219, 1974.
259. Neubert, A., *High Temp. Sci.*, 10, 213, 1978.
260. Neubert, A., Ihle, H. R., and Gingerich, K. A., *J. Chem. Phys.*, 73, 1406, 1980.
261. Neubert, A., Zmbov, K. F., Gingerich, K. A., and Ihle, H. R., *J. Chem. Phys.*, 77, 5218, 1982.
262. Nordine, P. C., *J. Chem. Phys.*, 61, 224, 1974.
263. O'Hare, P. A. G., Susman, S., and Volin, K.J., *J. Chem. Thermodyn.*, 21, 827, 1989.
264. Ovcharenko, I. E., Ya, Kuzyankow, Y., and Tatevaskii, V. M., *Opt. Spectrosk.*, 19, 528, 1965.
265. Pak, K., Cai, M. F., Dzugan, T. P., and Bondybey, V. E., *Faraday Discuss. Chem. Soc.*, 86, 153, 1988.
266. Parks, E. K. and Wexler, S., *J. Phys. Chem.*, 88, 4492, 1984.
267. Parr, T. P., Behrens, R., Freedman, A., and Heron, R. R., *Chem. Phys. Lett.*, 56, 71, 1978.
268. Pedley, J. B. and Marshall, E. M., *J. Phys. Chem. Ref. Data*, 12, 967, 1984.
269. Pelevin, O. V., Mil'vidskii, M. G., Belyaev, A. I., and Khotin, B. A., *Izv. Akad. Nauk. SSSR Neorg. Mater.*, 2, 924, 1966.

Table 1
BOND STRENGTHS IN DIATOMIC MOLECULES (continued)

270. Pelino, M. and Gingerich, K. A., *J. Chem. Phys.*, 90, 1286, 1989.
271. Pelino, M., and Gingerich, K. A., *J. Chem. Phys.*, 93, 1581, 1989.
272. Pelino, M., Piacente, V., and Ascenzo, G., *Thermochim. Acta*, 31, 383, 1979.
273. Pelino, M., Viswanadham, P., and Edwards, J. G., *J. Phys. Chem.*, 83, 2964, 1979.
274. Perakis, J., Chatillon, C., and Pattoret, A., *C. R. Acad. Sci. Ser. C*, 276, 1357, 1973.
275. Piacente, V. and Balducci, G., *Adv. Mass Spectrom.*, 7A, 626, 1978.
276. Piacente, V. and Balducci, G., *Dyn. Mass Spectrom.*, 4, 295, 1976.
277. Piacente, V. and Balducci, G., *High Temp. Sci.*, 6, 254, 1974.
278. Piacente, V. and Desideri, A., *J. Chem. Phys.*, 57, 2213, 1972.
279. Piacente, V. and Gigli, R., *J. Chem. Phys.*, 77, 4790, 1982.
280. Piacente, V. and Gingerich, K. A., *High Temp. Sci.*, 9, 189, 1977.
281. Piacente, V. and Gingerich, K. A., *Z. Naturforsch. Teil A*, 28, 316, 1973.
282. Piacente, V. and Malaspina, L., *J. Chem. Phys.*, 56, 1780, 1972.
283. Piacente, V., Balducci, G., and Bardi, G. , *J. Less-Commun. Metals.*, 37, 123, 1974.
284. Piacente, V., *J. Chem. Phys.*, 70, 5911, 1979.
285. Pianalto, F. S., O'Brien, L. C., Keller, P. C., and Bernath, P. F., *J. Mol. Spectrosc.*, 129, 348, 1988.
286. Pitzer, K. S., *J. Chem. Phys.*, 74, 3078, 1981.
287. Porter, R. F. and Spencer, C. W. J., *J. Chem. Phys.*, 32, 943, 1960.
288. Prasad, R., Venugopal, V., and Sood, D. D., *J. Chem. Thermodyn.*, 9, 593, 1977.
289. Rajamanickam, N., Palaniselvam, K., Rajavel, S. R. K., Rajesh, M., and Sureshkumar, G., *Acta Phys. Hung.*, 70, 141, 1991.
290. Rajamanickam, N., Senthilkumar, R. N., Ganesan, S., Gopalakrishnan, N., Rajkumer, J., Jegadesan, V., and Dandapani, C., *Acta Phys. Hung.*, 70, 71, 1991.
291. Rajamanickan, N., *Acta. Ciencia Indica Phys.*, 14, 18, 1988.
292. Ram, R. S., Rai, S. B., Ram, R. S., Upadhya, K. N., *J. Chim. Phys. Phys-Chim. Biol.*, 76, 560, 1979.
293. Ramakrishnan, E. S., Shim, I., and Gingerich, K. A., *J. Chem. Soc. Faraday Trans. 2*, 80, 395, 1984.
294. Rao, P. S. and Rao, T. V. R., *J. Quant. Spectrosc. Radiat. Transfer*, 27, 207, 1982.
295. Rao, S. P. and Rao, T. V. R., *Acta Ciencia Indica Phys.*, 7, 58, 1981.
296. Rao, T. V. R., Reddy, R. R., and Rao, P. S., *Indian J. Pure Appl. Phys.*, 19, 1219, 1981.
297. Rao, V. M., Rao, M. L. P., and Rao, P. T., *J. Quant. Spectrosc. Radiat. Transfer*, 25, 547, 1981.
298. Reddy, R. R., Rao, T. V. R., and Reddy, A. S. R., *Indian J. Pure Appl. Phys.*, 27, 243, 1989.
299. Reddy, R. R., Reddy, A. S. R., and Rao, T. V. R., *Acta Phys. Slovaca*, 36, 273, 1986.
300. Riekert, G., Lamparter, P., and Steeb, S., *Z. Metallkd.*, 72, 765, 1981.
301. Riekert, G., Rainer-Harbach, G., Lamparter, P., and Steeb, S., *Z. Metallkd.*, 76, 406, 1981.
302. Rinehart, G. H. and Behrens, R. G., *J. Phys. Chem.*, 83, 2052, 1979.
303. Rohlfing, E. A. and Valentine, J. J., *J. Chem. Phys.*, 84, 6560, 1986.
304. Rovner, L., Drowart, A., and Drowart, J., *Trans. Faraday Soc.*, 63, 2910, 1967.
305. Rusin, A. D., Zhukov, E., Agamirova, L. M., and Kalinnikov, V. T., *Zh. Neorg. Khim.*, 24, 1457, 1979.
306. Scott, P. R. and Richards, W. G., *J. Phys. B*, 7,1679, 1974.
307. Shim, I. and Gingerich, K. A., *J. Chem. Phys.*, 77, 2490, 1982.
308. Shim, I. and Gingerich, K. A., *J. Chem. Phys.*, 80, 5107, 1984.
309. Shim, I., and Gingerich, K. A., *J. Chem. Phys.*, 81, 5937, 1984.
310. Shim, I., Finkbeiner, H. C., and Gingerich, K. A., *J. Phys. Chem.*, 91, 3171, 1987.
311. Shim, I., Kingcade, J. E. Jr., and Gingerich, K. A., *J. Chem. Phys.*, 85, 6629, 1986.
312. Shim, I., Kingcade, J. E. Jr., and Gingerich, K. A., *J. Chem. Phys.*, 89, 3104, 1988.
313. Shim, I., Kingcade, J. E. Jr., and Gingerich, K. A., *Z. Phys. D - Atoms Molecules and Clusters*, 7, 261, 1987.
314. Shim, I., Mandix, K., and Gingerich, K. A., *J. Phys. Chem.*, 95, 5435, 1991.
315. Shim, I., Pelino, M., and Gingerich, K. A., *J. Chem. Phys.*, 97, 9240, 1992.
316. Siegel, B., *Q. Rev. (London)*, 19, 77, 1965.
317. Simons, J. W., Oldenberg, R. C., and Baughaim, S. L., *J. Phys. Chem.*, 91, 3840, 1987.
318. Smith, P. K. and Peterson, D. E., *J .Chem. Phys.*, 52, 4963, 1970.
319. Smoes, S. and Drowart, J., *Chem. Commun.*, p.534, 1968.
320. Smoes, S. and Drowart, J., *J. Chem. Soc. Faraday Trans. 2*, 73, 1746, 1977.
321. Smoes, S. and Drowart, J., *J. Chem. Soc. Faraday Trans. 2*, 80, 1171, 1984.
322. Smoes, S., Coppens, P., Bergman, C., and Drowart, J., *Trans. Faraday Soc.*, 65, 682, 1969.
323. Smoes, S., Depiere, D., and Drowart, J., *Rev. Int. Hautes Temp. Refractaires Paris*, 9, 171, 1972.
324. Smoes, S., Drowart, J., Welter, J. M., *J. Chem. Thermodyn.*, 9, 275, 1977; *Adv. Mass Spectrom.*, 7A, 622, 1978.
325. Smoes, S., Huguet, R., and Drowart, J., *Z. Naturforsch. Teil A*, 26, 1934, 1971.
326. Smoes, S., Mandy, F., Vander Auwera-Mahieu, A., and Drowart, J., *Bull. Soc. Chim. Belg.*, 81, 45, 1972.

Table 1
BOND STRENGTHS IN DIATOMIC MOLECULES (continued)

327. Smoes, S., Myers, C. E., and Drowart, J., *Chem. Phys. Lett.*, 8, 10, 1971.
328. Smoes, S., Pattje, W. R., and Drowart, J., *High Temp. Sci.*, 10, 109, 1978.
329. Spain, E. M. and Morse, M. D., *Int. J. Mass Spectrom. Ion Processes*, 102, 183, 1990.
330. Sreedhara Murthy, N., *Indian J. Phys.*, 62B, 92, 1988.
331. Srivastara, R. D. and Farber, M., *High Temp. Sci.*, 5, 489, 1973.
332. Stearns, C. A. and Kohl, F. J., *High Temp. Sci.*, 2, 146, 1970.
333. Stearns, C. A. and Kohl, F. J., *High Temp. Sci.*, 6, 284, 1974.
334. Steiger, R. A. and Cater, E. D., *High Temp. Sci.*, 7, 204, 1975.
335. Steiger, R. P. and Cater, E. D., *High Temp. Sci.*, 7, 288, 1975.
336. Stwalley, W. C., Way, K. R., and Velasco, R., *J. Chem. Phys.*, 60, 3611, 1974.
337. Su, T.-M. R. and Riley, S. J., *J. Chem. Phys.*, 71, 3194, 1979.
338. Su, T.-M. R. and Riley, S. J., *J. Chem. Phys.*, 72, 1614, 1980.
339. Su, T.-M. R. and Riley, S. J., *J. Chem. Phys.*, 72, 6632, 1980.
340. Sullivan, C. L., Zehe, M. J., and Carlson, K. D., *High Temp. Sci.*, 6, 80, 1974.
341. Tanaka, Y., Yushina, K., and Freeman, D. E., *J. Chem. Phys.*, 59, 564, 1973.
342. Taylor, S., Spain, E. M., and Morse, M. D., *J. Chem. Phys.*, 92, 2698, 1990.
343. Tellinghuisen, J., Tisone, G. C., Hoffmann, J. M., and Hays, A. K., *J. Chem. Phys.*, 64, 4796, 1976.
344. Tolbert, M. A. and Beauchamp, J. L., *J. Phys. Chem.*, 90, 5015, 1986.
345. Tromp, J. W., LeRoy, R. J., Gerstenkorn, S., and Luc, P., *J. Mol. Spectrosc.*, 100, 82, 1983.
346. Tuenge, R. T., Laabs, F., and Franzen, H. F., *J. Chem. Phys.*, 65, 2400, 1976.
347. Uy, O. M. and Drowart, J., *High Temp. Sci.*, 2, 293, 1970.
348. Uy, O. M. and Drowart, J., *Trans. Faraday Soc.*, 65, 3221, 1969.
349. Uy, O. M. and Drowart, J., *Trans. Faraday Soc.*, 67, 1293, 1971.
350. Uy, O. M., Muenow, D. W., Ficalora, P, J., and Margrave, J. L., *Trans. Faraday Soc.*, 64, 2998, 1968.
351. Van Veen, N. J. A., DeVries, M., and DeVries, A. E., *Chem. Phys. Lett.*, 64, 213, 1979.
352. Vander Auwera-Mahieu, A. and Drowart, J., *Chem. Phys. Lett.*, 1, 311, 1967.
353. Vander Auwera-Mahieu, A., McIntyre, N. S., and Drowart, J., *Chem. Phys. Lett.*, 4, 198, 1969.
354. Vander Auwera-Mahieu, A., Peeters, R., McIntyre, N. S., and Drowart, J., *Trans. Faraday Soc.*, 66, 809, 1970.
355. Varma, M. P., Ishwar, N. B., and Jha, B. L., *Indian J. Pure Appl. Phys.*, 20, 828, 1982.
356. Velasco, R., Ottinger, C., and Zare, R. N., *J. Chem. Phys.*, 51, 5522, 1969.
357. Vempati, S. N. and Jones, W. E., *J. Mol. Spectrosc.*, 127, 232, 1988.
358. Venkataramanaiah, M. and Lakshman, S. V. J., *J. Quant. Spectrosc. Radiat. Transfer*, 26, 11, 1981.
359. Verhaegen, G., Smoes, S., and Drowart, J., *J. Chem. Phys.*, 40, 239, 1964.
360. Verhaegen, G., Stafford, F. E., and Drowart, J., *J. Chem. Phys.*, 40, 1622, 1964.
361. Verma, K. K., Vu, T. H., and Stwalley, W. C., *J. Mol. Spectrosc.*, 85, 131, 1980.
362. Viswanathan, K. S. and Tellinghuisen, J., *J. Mol. Spectrosc.*, 98, 185, 1983.
363. Wang, Y. C., Kajitani, M., Kasahara, S., Bata, M., Ishikawa, K., and Kato, H., *J. Chem. Phys.*, 95, 6229, 1991.
364. Way, K. R. and Stwalley, W. C., *J. Chem. Phys.*, 59, 5298, 1973.
365. Weickenmeier, W., Diemer, U., Wahl, M., Raab, M., Demtroeder, W., and Mueller, W., *J. Phys.Chem.*, 82, 5354, 1985.
366. Wiedemeier, H. and Gilles, P. W., *J. Chem. Phys.*, 42, 2765, 1965.
367. Wu, C. H. and Ihle, H. R., *Adv. Mass Spectrom.*, 8A, 374, 1980.
368. Wu, C. H., Ihle, H. R., and Gingerich, K. A., *Int. J. Mass Spectrom. Ion Phys.*, 47, 235, 1983.
369. Wu, C. H., *J. Chem. Phys.*, 65, 3181, 1976; 65, 2040, 1976.
370. Wyss, J. C., *J. Chem. Phys.*, 71, 2949, 1979.
371. Yokozeki, A. and Menzinger, M., *Chem. Phys.*, 14, 427, 1976.
372. Zemke, W. T. and Stwalley, W. C., *Chem. Phys. Lett.*, 143, 84, 1988.
373. Zemke, W. T., Stwalley, W. C., Langhoff, S. R., Valderrama, G. L., and Berry, M. J., *J. Chem. Phys.*, 95, 7846, 1991.
374. Zmbov, K .F. and Margrave, J. L., *J. Inorg. Nucl. Chem.*, 29, 59, 1967.
375. Zmbov, K. F. and Margrave, J. L., *J. Chem. Phys.*, 45, 3167, 1966.
376. Zmbov, K. F. and Margrave, J. L., *J. Chem. Phys.*, 47, 3122, 1967.
377. Zmbov, K. F. and Margrave, J. L., *J. Phys. Chem.*, 70, 3379, 1966.
378. Zmbov, K. F. and Margrave, J. L., *J. Phys. Chem.*, 71, 2893, 1967.
379. Zmbov, K. F., Hastie, J. W., and Margrave, J. L., *Trans. Faraday Soc.*, 64, 861, 1968.
380. Zmbov, K. F., Neubert, A., and Ihle, H. R., *Z. Naturforsch., A*, 36A, 914, 1981.
381. Zmbov, K. F., Wu, C. H., and Ihle, H. R., *J. Chem. Phys.*, 67, 4603, 1977.
382. Zollweg, R. J., *Contrib. Pap. Int. Conf. Phenom. Ioniz. Gases. 11th*, 402, 1973.

Table 2
ENTHALPY OF FORMATION OF GASEOUS ATOMS FROM ELEMENTS IN THEIR STANDARD STATES

For elements that are diatomic gases in their standard states these are readily obtained from the bond strength. For elements that are crystalline in their standard states they are derived from vapor pressure data.

Atom	$\Delta_f H°_{298}$/kJ mol^{-1}	Ref.	Atom	$\Delta_f H°_{298}$/kJ mol^{-1}	Ref.	Atom	$\Delta_f H°_{298}$/kJ mol^{-1}	Ref.
Ag	284.9 ± 0.8	2	Hf	619 ± 4	1	Re	774 ± 6.3	1
Al	330.0 ± 4.0	2	Hg	61.38 ± 0.04	2	Rh	556 ± 4	1
As	302.5 ± 13	1	I	106.76 ± 0.04	2	Ru	650.6 ± 6.3	1
Au	368.2 ± 2.1	1	In	243 ± 4	1	S	277.17 ± 0.15	2
B	565 ± 5	2	Ir	669 ± 4	1	Sb	264.4 ± 2.5	1
Ba	177.8 ± 4	1	K	89.0 ± 0.8	2	Sc	377.8 ± 4	1
Be	324 ± 5	2	Li	159.3 ± 1.0	2	Se	227.2 ± 4	1
Bi	209.6 ± 2.1	1	Mg	147.1 ± 0.8	2	Si	450 ± 8	2
Br	111.87 ± 0.12	2	Mn	283.3 ± 4	1	Sn	301.2 ± 1.5	2
C	716.68 ± 0.45	2	Mo	658.1 ± 2.1	1	Sr	163.6 ± 2.1	1
Ca	177.8 ± 0.8	2	N	472.68 ± 0.40	2	Ta	782.0 ± 2.5	1
Cd	111.80 ± 0.20	2	Na	107.5 ± 0.7	2	Te	196.6 ± 2.1	1
Ce	423 ± 13	1	Nb	721.3 ± 4	1	Th	602 ± 6	2
Cl	121.301 ± 0.008	2	Ni	430.1 ± 2.1	1	Ti	473 ± 3	2
Co	428.4 ± 4	1	O	249.18 ± 0.10	2	Tl	182.21 ± 0.1	1
Cr	397 ± 4	1	Os	787 ± 6.3	1	U	533 ± 8	2
Cs	76.5 ± 1.0	2	P	316.5 ± 1.0	2	V	514.2 ± 1.3	1
Cu	337.4 ± 1.2	2	Pb	195.2 ± 0.8	2	W	849.8 ± 4	1
Er	317.1 ± 4	1	Pd	376.6 ± 2.1	1	Y	424.7 ± 2.1	1
F	79.38 ± 0.30	2	Pt	565.7 ± 1.3	1	Yb	152.09 ± 0.8	1
Ge	372 ± 3	2	Pu	364.4 ± 17	1	Zn	130.40 ± 0.40	2
H	217.998 ± 0.006	2	Rb	80.9 ± 0.8	2	Zr	608.8 ± 4	1

REFERENCES

1. Brewer, L. and Rosenblatt, G. M., *Adv. High Temp. Chem.*, 2, 1, 1969.
2. Cox, J. D., Wagman, D. D., and Medvedev, V. A., Eds., *CODATA Key Values for Thermodynamics*, Hemisphere Publishing Corporation, New York, 1989.

Table 3
BOND STRENGTHS IN POLYATOMIC MOLECULES

The values below refer to a temperature of 298 K and have mostly been determined by kinetic methods (see (i) S. W. Benson, *J. Chem. Educ.*, 42, 502, 1965, (ii) J. A. Kerr, *Chem. Rev.*, 66, 465, 1966 and (iii) D. F. McMillen and D. M. Golden, *Ann. Rev. Phys. Chem.*, 33, 493, 1982, for a full description of the methods). Bond strengths in polyatomic molecules are notoriously difficult to measure accurately since the mechanisms of the kinetic systems involved in the measurements are seldom straightforward. Thus much controversy has taken place in the literature over the past 15 years concerning C-H bond strengths in simple alkanes, although these now appear to have been resolved (P. W. Seakins, M. J. Pilling, J. T. Niiranen, D. Gutman, and L. N. Krasnoperov, *J. Phys. Chem.*, 96, 9847, 1992.). Other examples illustrating the difficulties involved are concerned with the C-H bond strengths in ethyne and ethene or the corresponding enthalpies of formation of the ethynyl and vinyl radicals, which are still in dispute. Here values from recent References have been selected as representing the presently accepted consensus of opinion, but future changes to the recommendations could well arise.

Some of the bond strengths have been calculated from the enthalpies of formation of the species involved according to the equations:

$$D°(\mathrm{R–X}) = \Delta_f H°(\mathrm{R}) + \Delta_f H°(\mathrm{X}) - \Delta_f H°(\mathrm{RX})$$
$$D°(\mathrm{R–R}) = 2\,\Delta_f H°(\mathrm{R}) - \Delta_f H°(\mathrm{RR})$$

The enthalpies of formation of the radicals are taken from Table 4 and for the molecules from the appropriate References following Table 3.

An attempt has been made to list all the important values obtained by methods that are considered to be valid. The references are intended to serve as a guide to the literature.

Table 3
BOND STRENGTHS IN POLYATOMIC MOLECULES (continued)

Bond	$D°_{298}$/kJ mol^{-1}	Ref.	Bond	$D°_{298}$/kJ mol^{-1}	Ref.
H-CH	421.7	1,19,92	H-C$(CH_3)_2$CN	361.9 ± 8.4	54
H-CH_2	464.8	1,78,92	H-CH_2NH_2	390.4 ± 8.4	54
H-CH_3	438.5 ± 1.5	1,67,78	H-CH_2NHCH_3	364 ± 8	54
H-CCH	556.1 ± 2.9	3,32	H-$CH_2N(CH_3)_2$	351 ± 8	54
H-$CHCH_2$	465.3 ± 3.4	32,75	H-CHO	364 ± 4	54
H-C_2H_5	422.8 ± 1.5	76	H-$COCH_3$	373.8 ± 1.5	59
H-Cycloprop-2-en-1-yl	379.1 ± 17	54	H-$COCHCH_2$	364.4 ± 4.2	54
H-CH_2CCH	374.0 ± 8	54	H-COC_2H_5	365.7 ± 4	54
H-CH_2CHCH_2	361.1 ± 6.3	54	H-COC_6H_5	363.6 ± 4	54
H-Cyclopropyl	444.8 ± 1.3	54	H-$COCF_3$	380.7 ± 8	54
H-n-C_3H_7	420.0 ± 2.5	18,67	H-CH_2CHO	396.6 ± 8.4	73
H-i-C_3H_7	412.7 ± 1.7	76	H-CH_2COCH_3	411.3 ± 7.5	54
H-CH_2CCCH_3	364.8 ± 8	54	H-$CH(CH_3)COCH_3$	386.2 ± 5.9	54
H-$CH(CH_3)CCH$	347.7 ± 9.2	54	H-CH_2OCH_3	389 ± 4	54
H-Cyclobutyl	403.8 ± 4	54	H-$CH(CH_3)OC_2H_5$	383.7 ± 1.7	46
H-Cyclopropylmethyl	407.5 ± 6.7	54	H-Tetrahydrofuran-2-yl	385 ± 4	54
H-$CH(CH_3)CHCH_2$	345.2 ± 5.4	54	H-2-Furylmethyl	361.9 ± 8	54
H-$CH_2CHCHCH_3$	358.2 ± 6.3	54	H-CH_2OH	410 ± 6	28
H-$CH_2C(CH_3)CH_2$	358.2 ± 4	83,87	H-$CH(CH_3)OH$	389 ± 4	54
H-s-C_4H_9	411.1 ± 2.2	76	H-$CH(OH)CHCH_2$	341.4 ± 7.5	54
H-t-C_4H_9	403.5 ± 1.8	76	H-$C(CH_3)_2OH$	381 ± 4	54
H-Cyclopenta-1,3-dien-5-yl	297.5 ± 6.3	54	H-$CH_2OCOC_6H_5$	419.2 ± 5.4	54
H-Spiropentyl	413.4 ± 4	54	H-$COOCH_3$	387.9 ± 4	54
H-Cyclopent-1-en-3-yl	344.3 ± 4	54	H-CH_2SH	402 ± 4	79
H-$CH_2CHCHCHCH_2$	347 ± 13	54	H-CH_2SCH_3	391.2 ± 5.0	57
H-$CH(C_2H_3)_2$	319.7	54,84	H-CH_2F	423.8 ± 4	70
H-$CH(CH_3)CCCH_3$	365.3 ± 11.3	54	H-CHF_2	431.8 ± 4	70
H-$C(CH_3)_2CCH$	338.9 ± 9.6	54	H-CF_3	446.4 ± 4	54
H-$C(CH_3)_2CHCH_2$	323.0 ± 6.3	54	H-CHFCl	421.7 ± 5.4	89
H-Cyclopentyl	403.5 ± 2.5	18,67	H-CF_2Cl	421.3 ± 8.3	55
H-$CH_2C(CH_3)_3$	418 ± 8	54	H-$CHFCl_2$	413.8 ± 5.0	89
H-$C(CH_3)_2C_2H_5$	404.0 ± 6.3	1,67,81	H-CH_2Cl	421.7 ± 4.2	89
H-C_6H_5	464.0 ± 8.4	54	H-$CHCl_2$	411.7 ± 5.0	89
H-Cyclohexa-1,3-dien-5-yl	305 ± 21	54	H-CCl_3	392.5 ± 2.5	42
H-Cyclohexa-1,4-dien-3-yl	305.4 ± 8.4	40	H-CH_2Br	425.1 ± 4.2	89
H-Cyclohexyl	399.6 ± 4	54	H-$CHBr_2$	417.2 ± 7.5	89
H-$C(CH_3)_2CCCH_3$	344.3 ± 11.3	54	H-CBr_3	401.7 ± 6.7	54
H-$CH_2C(CH_3)C(CH_3)_2$	326.4 ± 4.6	54	H-CH_2I	431 ± 8	54
H-$C(CH_3)_2C(CH_3)CH_2$	319.2 ± 4.6	54	H-CHI_2	431 ± 8	54
H-$CH_2C_6H_5$	368.2 ± 4	54	H-$CHCF_2$	448 ± 8	82
H-Cyclohepta-1,3,5-trien-7-yl	305.4 ± 8	54	H-CFCHF	448 ± 8	82
H-Norbornyl	404.6 ± 10.5	54	H-$CFCF_2$	452 ± 8	82
H-Cycloheptyl	387.0 ± 4	54	H-CH_2CF_3	446.4 ± 4.6	54
H-$CH(CH_3)C_6H_5$	357.3 ± 6.3	54	H-CF_2CH_3	416.3 ± 10.5	54
H-Inden-1-yl	351 ± 13	54	H-C_2F_5	429.7 ± 2.1	54
H-$C(CH_3)_2C_6H_5$	353.1 ± 6.3	54	H-CFCFCl	444 ± 8	82
H-1-Naphthylmethyl	356.1 ± 6.3	54	H-$CHClCF_3$	425.9 ± 6.3	54
H-$CH(C_6H_5)_2$	340.6	74	H-CClCFCl	439 ± 8	82
H-9,10-Dihydroanthracen-9-yl	315.1 ± 6.3	54	H-$CClCH_2$	>433.5	75
H-$C(CH_3)(C_6H_5)_2$	339 ± 8	54	H-CClCHCl	435 ± 8	82
H-9-Anthracenylmethyl	342.3 ± 6.3	54	H-CCl_2CHCl_2	393 ± 8	54
H-9-Phenanthrenylmethyl	356.1 ± 6.3	54	H-C_2Cl_5	397 ± 8	54
H-CN	518.0 ± 8	54	H-$CClBrCF_3$	404.2 ± 6.3	54
H-CH_2CN	389 ± 10.5	54	H-n-C_3F_7	435 ± 8	54
H-$CH(CH_3)CN$	376.1 ± 9.6	54	H-i-C_3F_7	433.5 ± 2.5	33

Table 3
BOND STRENGTHS IN POLYATOMIC MOLECULES (continued)

Bond	$D°_{298}$/kJ mol^{-1}	Ref.
$H-CHClCHCH_2$	370.7 ± 5.9	54
$H-C_6F_5$	476.6	54
$H-CH_2Si(CH_3)_3$	415.1 ± 4	91
$H-SiH$	351	54
$H-SiH_2$	268	54
$H-SiH_3$	384.1 ± 2.0	77
$H-SiH_2CH_3$	374.9	91
$H-SiH(CH_3)_2$	374.0	91
$H-Si(CH_3)_3$	377.8	54,91
$D-Si(CH_3)_3$	389 ± 7.1	31
$H-SiH_2C_6H_5$	369.0	54,91
$H-SiF_3$	418.8	54,91
$H-SiCl_3$	382.0	54,91
$H-Si_2H_5$	361.1	54
$H-Si(CH_3)_2Si(CH_3)_3$	356.9 ± 8.4	40
$H-Si(Si(CH_3)_3)_3$	330.5 ± 8.4	40
$H-GeH_3$	347 ± 8	60
$H-GeH_2I$	331 ± 8	61
$H-Ge(CH_3)_3$	339 ± 8	30
$H-Sn(n-C_4H_9)_3$	308.4 ± 8.4	13
$H-NH_2$	449.4 ± 4.6	54
$H-NHCH_3$	418.4 ± 10.5	54
$H-N(CH_3)_2$	382.8 ± 8	54
$H-NHC_6H_5$	368.2 ± 8	54
$H-N(CH_3)C_6H_5$	366.1 ± 8	54
$H-NO$	≤207.1	20
$H-NO_2$	327.6 ± 2.1	54
$H-NF_2$	316.7 ± 10.5	54
$H-NHNH_2$	366.1	39
$H-N_3$	385 ± 21	54
$H-OH$	498 ± 4	54
$H-OCH_3$	436.8 ± 4	54,56
$H-OC_2H_5$	436.0 ± 4	54
$H-OC(CH_3)_3$	439.7 ± 4	54
$H-OCH_2C(CH_3)_3$	428.0 ± 6.3	54
$H-OC_6H_5$	361.9 ± 8	54
$H-O_2H$	369.0 ± 4.2	80
$H-O_2CH_3$	370.3 ± 2.1	47
$H-O_2-t-C_4H_9$	374.0 ± 0.8	41
$H-OCOCH_3$	442.7 ± 8	54
$H-OCOC_2H_5$	445.2 ± 8	54
$H-OCO-n-C_3H_7$	443.1 ± 8	54
$H-ONO$	327.6 ± 2.1	13
$H-ONO_2$	423.4 ± 2.1	13
$H-SH$	381.6 ± 2.9	58
$H-SCH_3$	365.3 ± 2.5	58
$H-SC_6H_5$	348.5 ± 8	54
$H-SO$	172.8	93
$HC \equiv CH$	965 ± 8	1,19,67
$H_2C=CH_2$	733 ± 8	1,67,92
CH_3-CH_3	376.0 ± 2.1	1,67,78
CH_3-CH_2CCH	318.0 ± 8	54
$CH_3-CH_2CCCH_3$	308.4 ± 6.3	54
$CH_3-CH(CH_3)CCH$	305.4	54
$CH_3-C(CH_3)CCH_2$	320.1 ± 9.2	54
$CH_3-CH_2CHCHCH_3$	305.0 ± 3.3	54
$CH_3-CH_2C(CH_3)CH_2$	301.2 ± 3.3	83
$CH_3-t-C_4H_9$	425.9 ± 8	54
$CH_3-CH(CH_3)CCCH_3$	320.9 ± 6.3	54
$CH_3-C(CH_3)_2CCH$	295.8 ± 6.3	54
$n-C_3H_7-CH_2CCH$	306.3 ± 6.3	54
$CH_3-C(CH_3)_2CHCH_2$	284.9 ± 6.3	54
$n-C_3H_7-CH_2CHCH_2$	295.8	88
$CH_3-C_6H_5$	317.1 ± 6.3	54
$CH_3-C(CH_3)_2CCCH_3$	303.3 ± 6.3	54
$CHCCH_2-s-C_4H_9$	300.0 ± 6.3	54
$CH_3-CH_2C_6H_5$	332.2 ± 4	54
$CH_3-CH(CH_3)C_6H_5$	312.1 ± 6.3	54
$C_2H_5-CH_2C_6H_5$	294.1 ± 4	54
CH_3-1-Naphthylmethyl	305.0 ± 6.3	54
$CH_3-C(CH_3)_2C_6H_5$	308.4 ± 6.3	54
$CHCCH_2-CH_2C_6H_5$	256.9 ± 8	54
$n-C_3H_7-CH_2C_6H_5$	292.9 ± 4	54
CH_3-9-Anthracenylmethyl	282.8 ± 6.3	54
CH_3-9-Phenanthrenylmethyl	305.0 ± 6.3	54
$CH_3-CH(C_6H_5)_2$	301 ± 8	54
$CH_3-C(CH_3)(C_6H_5)_2$	289 ± 8	54
CH_3-CN	509.6 ± 8	54
C_2H-CN	602 ± 4	63
$C_2H_5-CH_2NH_2$	332.2 ± 8	54
CH_3-CH_2CN	336.4 ± 4	85
$C_2H_5-CH_2CN$	321.7 ± 7.1	54
$CH_3-CH(CH_3)CN$	329.7 ± 8	54
$C_2H_5-CH_2CN$	321.7 ± 7.1	54
$CH_3-C(CH_3)_2CN$	312.5 ± 6.7	54
$CH_3-C(CH_3)(CN)C_6H_5$	250.6	54
$C_6H_5CH_2-CH_2NH_2$	284.5 ± 8	54
$C_6H_5CH_2-C_5H_4N$	362.8	74
$CN-CN$	536 ± 4	26
CH_3-2-Furylmethyl	314 ± 8	54
$C_6H_5CH_2-COCH_2C_6H_5$	273.6 ± 8	54
$CH_3CO-COCH_3$	282.0 ± 9.6	54
$C_6H_5CH_2-COOH$	280	54
$C_6H_5CO-COC_6H_5$	277.8	54
$(C_6H_5)_2CH-COOH$	248.5 ± 13	54
$CF_3-COC_6H_5$	308.8 ± 8	54
$CF_2=CF_2$	319.2 ± 13	95
CH_2F-CH_2F	368 ± 8	43
CH_3-CF_3	423.4 ± 4.6	72
CF_3-CF_3	413.0 ± 10.5	54
$C_6F_5-C_6F_5$	487.9 ± 24.7	71
CH_3-BF_2	~473	54
$C_6H_5-BCl_2$	~510	54
$CH_2CHCH_2-Si(CH_3)_3$	293	54
$s-C_4H_9-Si(CH_3)_3$	414	54
$CH_3-NHC_6H_5$	298.7 ± 8	54
$C_6H_5CH_2-NH_2$	297.5 ± 4	54
$CH_3-N(CH_3)C_6H_5$	296.2 ± 8	54
$C_6H_5CH_2-NHCH_3$	287.4 ± 8	54
$C_6H_5CH_2-N(CH_3)_2$	259.8 ± 8	54

Table 3
BOND STRENGTHS IN POLYATOMIC MOLECULES (continued)

Bond	$D°_{298}$/kJ mol^{-1}	Ref.
CH_2=N_2	<175	49
CH_3-N_2CH_3	219.7	11
C_2H_5-$N_2C_2H_5$	209.2	11
i-C_3H_7-N_2i-C_3H_7	198.7	11
n-C_4H_9-N_2n-C_4H_9	209.2	11
i-C_4H_9-N_2i-C_4H_9	205.0	11
s-C_4H_9-N_2s-C_4H_9	195.4	11
t-C_4H_9-N_2t-C_4H_9	182.0	11
$C_6H_5CH_2$-$N_2CH_2C_6H_5$	157.3	11
CF_3-N_2CF_3	231.0	11
CH_3-NO	167.4 ± 3.3	54
i-C_3H_7-NO	152.7 ± 13	54
t-C_4H_9-NO	165.3 ± 6.3	54
C_6H_5-NO	212.5 ± 4	54
NC-NO	120.5 ± 10.5	38
CF_3-NO	179.1 ± 8	54
C_6F_5-NO	208.4 ± 4	54
CCl_3-NO	134 ± 13	54
t-C_4H_9-NOt-C_4H_9	121	17
CH_3-NO_2	254.4	54
$CH_2C(CH_3)$-NO_2	245.2	54
i-C_3H_7-NO_2	246.9	54
t-C_4H_9-NO_2	244.8	54
C_6H_5-NO_2	298.3 ± 4	54
$C(NO_2)_3$-NO_2	169.5 ± 4	54
CH_3-$OC(CH_3)CH_2$	277.4	94
CH_3-OC_6H_5	238 ± 8	66
CH_3-$OCH_2C_6H_5$	280.3	22
C_2H_5-OC_6H_5	264 ± 6.3	54
CH_2CHCH_2-OC_6H_5	208.4 ± 8	54
O=CO	532.2 ± 0.4	25
CH_3-O_2	135.6 ± 2.9	53
C_2H_5-O_2	147.2 ± 6.3	53
CH_2CHCH_2-O_2	76.2 ± 2.1	53
i-C_3H_7-O_2	157.9 ± 7.5	53
t-C_4H_9-O_2	153.6 ± 7.9	53
$C_6H_5CH_2$-O_2CCH_3	280 ± 8	54
$C_6H_5CH_2$-$O_2CC_6H_5$	289	11
CH_3-O_2SCH_3	279.5	54
CH_2CHCH_2-O_2SCH_3	207.5	54
$C_6H_5CH_2$-O_2SCH_3	221.3	54
CF_3-O_2CF_3	361.5	9
CH_2Cl-O_2	120.9	53
$CHCl_2$-O_2	105.9	53
CCl_3-O_2	82.4 ± 3.8	53
CH_3-SH	312.5 ± 4.2	58
t-C_4H_9-SH	286.2 ± 6.3	54
C_6H_5-SH	361.9 ± 8	54
CH_3-SCH_3	307.9 ± 3.3	58
CH_3-SC_6H_5	290.4 ± 8	54
$C_6H_5CH_2$-SCH_3	256.9 ± 8	54
S-CS	430.5 ± 13	54
F-CH_3	472	1,52
F-CN	469.9 ± 5.0	54
F-CHFCl	465.3 ± 9.6	89
F-CF_2Cl	490 ± 25	36
F-$CFCl_2$	462.3 ± 10.0	89
F-CF_2CH_3	522.2 ± 8	54
F-C_2F_5	530.5 ± 7.5	54
Cl-CN	421.7 ± 5.0	54
Cl-COC_6H_5	310 ± 13	54
Cl-CSCl	265.3 ± 2.1	64
Cl-CF_3	360.2 ± 3.3	23
Cl-CHFCl	354.4 ± 11.7	89
Cl-CF_2Cl	346.0 ± 13.4	89
Cl-$CFCl_2$	305 ± 8	35
Cl-CH_2Cl	350.2 ± 0.8	89
Cl-$CHCl_2$	338.5 ± 4.2	89
Cl-CCl_3	305.9 ± 7.5	54
Cl-C_2F_5	346.0 ± 7.1	23
Cl-CF_2CF_2Cl	326 ± 8	54
Cl-$SiCl_3$	464	91
Br-CH_3	292.9 ± 5.0	34
Br-C_6H_5	336.8 ± 8	54
Br-CN	367.4 ± 5.0	54
Br-CH_2COCH_3	261.5	94
Br-COC_6H_5	268.6	11
Br-CHF_2	289 ± 8	54
Br-CF_3	295.4 ± 13	54
Br-CF_2CH_3	287.0 ± 5.4	69
Br-C_2F_5	287.4 ± 6.3	54
Br-n-C_3F_7	278.2 ± 10.5	54
Br-i-C_3F_7	274.1 ± 4.6	54
Br-$CH_2C_6F_5$	225 ± 6	45
Br-$CHClCF_3$	274.9 ± 6.3	54
Br-CCl_3	231.4 ± 4	54
Br-$CClBrCF_3$	251.0 ± 6.3	54
Br-CH_2Br	296.7 ± 1.3	89
Br-$CHBr_2$	292.0 ± 8	89
Br-CBr_3	235.1 ± 7.5	54
Br-NO_2	82.0 ± 7.1	48
Br-NF_2	≤222	21
I-$CHCH_2$	259.0 ± 4.2	16
I-n-C_4H_9	205.0 ± 4	54
I-Norbornyl	261.5 ± 10.5	62
I-CN	305 ± 4	26
I-CF_3	223.8 ± 2.9	2
I-CF_2CH_3	218.0 ± 4.2	54
I-CH_2CF_3	235.6 ± 4	54
I-C_2F_5	218.8 ± 2.9	2
I-n-C_3F_7	208.4 ± 4.2	54
I-i-C_3F_7	215.1 ± 2.9	2
I-n-C_4F_9	205.0 ± 4.2	65
I-$C(CF_3)_3$	206	29
I-C_6H_5	273.6 ± 8	54
I-C_6F_5	~277.0	54
C_5H_5-FeC_5H_5	381 ± 13	50
CH_3-$ZnCH_3$	285 ± 17	54
C_2H_5-ZnC_2H_5	238 ± 17	54
CH_3-$Ga(CH_3)_2$	264 ± 17	54

Table 3
BOND STRENGTHS IN POLYATOMIC MOLECULES (continued)

Bond	$D°_{298}$/kJ mol^{-1}	Ref.	Bond	$D°_{298}$/kJ mol^{-1}	Ref.
$C_2H_5-Ga(C_2H_5)_2$	209 ± 17	54	I-NO	77.8 ± 0.4	37
$CH_3-Ge(CH_3)_3$	347 ± 17	54	$I-NO_2$	76.6 ± 4	90
$CH_3-As(CH_3)_2$	280 ± 17	54	HO-OH	213 ± 4	54
CH_3-CdCH_3	251 ± 17	54	$HO-OCH_2C(CH_3)_3$	193.7 ± 7.9	54
$CH_3-In(CH_3)_2$	205 ± 17	54	CH_3O-OCH_3	157.3 ± 8	54
$CH_3-Sn(CH_3)_3$	297 ± 17	54	$C_2H_5O-OC_2H_5$	158.6 ± 4	54
$C_2H_5-Sn(C_2H_5)_3$	264 ± 17	54	n-C_3H_7O-On-C_3H_7	155.2 ± 4	54
$CH_3-Sb(CH_3)_2$	255 ± 17	54	i-C_3H_7O-Oi-C_3H_7	157.7 ± 4	54
$C_2H_5-Sb(C_2H_5)_2$	243 ± 17	54	s-C_4H_9O-Os-C_4H_9	152.3 ± 4	54
CH_3-HgCH_3	255 ± 17	54	t-C_4H_9O-Ot-C_4H_9	159.0 ± 4	54
$C_2H_5-HgC_2H_5$	205 ± 17	54	$C_2H_5C(CH_3)_2O-OC(CH_3)_2C_2H_5$	164.4 ± 4	54
$CH_3-Tl(CH_3)_2$	167 ± 17	54	$(CH_3)_3CCH_2O-OCH_2C(CH_3)_3$	152.3 ± 4	54
$CH_3-Pb(CH_3)_3$	238 ± 17	54	CF_3O-OCF_3	193.3	54
$C_2H_5-Pb(C_2H_5)_3$	230 ± 17	54	$(CF_3)_3CO-OC(CF_3)_3$	148.5 ± 4.6	54
$CH_3-Bi(CH_3)_2$	218 ± 17	54	t-C_4H_9O-$OSi(CH_3)_3$	197	54
$CO-Cr(CO)_5$	155 ± 8	51	SF_5O-OSF_5	155.6	54
$CO-Fe(CO)_4$	172 ± 8	51	t-C_4H_9O-$OGe(C_2H_5)_3$	192	54
$CO-Mo(CO)_5$	167 ± 8	51	t-C_4H_9O-$OSn(C_2H_5)_3$	192	54
$CO-W(CO)_5$	192 ± 8	51	$FClO_2-O$	244.3	11
BH_3 BH_3	146	11	$CF_3O-O_2CF_3$	126.8 ± 8	54
NH_2-NH_2	275.3	54	$SF_5O-O_2SF_5$	126.8	54
NH_2-NHCH_3	268.2 ± 8	54	$CH_3CO_2-O_2CCH_3$	127.2 ± 8	54
$NH_2-N(CH_3)_2$	246.9 ± 8	54	$C_2H_5CO_2-O_2CC_2H_5$	127.2 ± 8	54
$NH_2-NHC_6H_5$	218.8 ± 8	54	n-$C_3H_7CO_2$-O_2Cn-C_3H_7	127.2 ± 8	54
$ON-NO_2$	40.6 ± 2.1	54	O-SO	552 ± 8	25
O_2N-NO_2	56.9	54	$F-OCF_3$	182.0 ± 2.1	24
NF_2-NF_2	88 ± 4	54	HO-Cl	251 ± 13	44
$O-N_2$	167	1,12	O-ClO	247 ± 13	25
O-NO	305	1,12	HO-Br	234 ± 13	44
$O-NO_2$	208.7 ± 1.1	27	HO-I	234 ± 13	44
HO-NO	206.3	54	$O=PF_3$	544 ± 21	44
$HO-NO_2$	206.7	54	$O=PCl_3$	510 ± 21	44
HO_2-NO_2	96 ± 8	54	$O=PBr_3$	498 ± 21	44
CH_3O-NO	174.9 ± 3.8	8,10	$HO-Si(CH_3)_3$	536	54
C_2H_5O-NO	175.7 ± 5.4	7,10	HS-SH	276 ± 8	54
$CH_3COO_2-NO_2$	118.8 ± 3.0	14	CH_3S-SCH_3	272.8 ± 3.8	58
n-C_3H_7O-NO	167.8 ± 7.5	10	$F-SF_5$	420 ± 10	86
i-C_3H_7O-NO	171.5 ± 5.4	6,10	I-SH	206.7 ± 8	54
n-C_4H_9O-NO	177.8 ± 6.3	10	I-SO	180	54
i-C_4H_9O-NO	175.7 ± 6.3	10	$I-SCH_3$	206.3 ± 7.1	79
s-C_4H_9O-NO	173.6 ± 3.3	4,10	$I-Si(CH_3)_3$	322	91
t-C_4H_9O-NO	171.1 ± 3.3	5,10	$H_3Si-SiH_3$	310	54,91
$HO-NCHCH_3$	207.9	11	$(CH_3)_3Si-Si(CH_3)_3$	336.8	54,91
$Cl-NF_2$	~134	1,68	$(C_6H_5)_3Si-Si(C_6H_5)_3$	368 ± 29	54,91

REFERENCES

1. A value calculated from one of the thermochemical equations above, taking enthalpy data from the references quoted.
2. Ahonkhai, S. I. and Whittle, E., *Int. J. Chem. Kinet.*, 16, 543, 1984.
3. Baldwin, D. P., Buntine, M. A., and Chandler, D.W., *J. Chem. Phys.*, 93, 6578, 1990.
4. Batt, L. and McCulloch, R. D., *Int. J. Chem. Kinet.*, 8, 911, 1976.
5. Batt, L. and Milne, R. T. , *Int. J. Chem. Kinet.*, 8, 59, 1976.
6. Batt, L. and Milne, R. T. , *Int. J. Chem. Kinet.*, 9, 141, 1977.
7. Batt, L. and Milne, R. T. , *Int. J. Chem. Kinet.*, 9, 549, 1977.
8. Batt, L. and Milne, R. T., and McCulloch, R. D., *Int. J. Chem. Kinet.*, 9, 567, 1977.

Table 3
BOND STRENGTHS IN POLYATOMIC MOLECULES (continued)

9. Batt, L. and Walsh, R., *Int. J. Chem. Kinet.*, 15, 605, 1983.
10. Batt, L., Christie, K., Milne, R. T., and Summers, A. J., *Int. J. Chem. Kinet.*, 6, 877, 1974.
11. Benson, S. W. and O'Neal, H. E., *Kinetic Data on Gas Phase Unimolecular Reactions*, National Bureau of Standards, NSRDS-NBS, Washington, D. C., 21, 1970.
12. Benson, S. W., *J. Chem. Educ.*, 42, 502, 1965.
13. Benson, S. W., *Thermochemical Kinetics, 2nd ed.*, John Wiley & Sons, New York, 1976.
14. Bridier, I., Caralp, F., Loirat, H., Lesclaux, R., Veyret, B., Becker, K. H., Reimer, A., and Zabel, F., *J. Phys. Chem.*, 95, 3594, 1991.
15. Burkey, T. J., Majewski, M., and Griller, D., *J. Am. Chem. Soc.*, 108, 2218, 1986.
16. Cao, J. R., Zhang, J. M., Zhong, X., Huang, Y. H., Fang, W. Q., Wu, X. J., and Zhu, Q. H., *Chem. Phys.*, 138, 377, 1989.
17. Carmichael, P. J., Gowenlock, B. G., and Johnson, C. A. F., *J. Chem. Soc. Perkin Trans. 2*, 1853, 1973.
18. Castelhano, A. L. and Griller, D., *J. Am. Chem. Soc.*, 104, 3655, 1982.
19. Chupka, W. A. and Liftshitz, C., *J. Chem. Phys.*, 48, 1109, 1968.
20. Clement, M. J. Y. and Ramsay, D. A., *Can. J. Phys.*, 39, 205, 1961.
21. Clyne, M. A. A. and Connor, J., J. *Chem. Soc. Faraday Trans. 1*, 68, 1220, 1972.
22. Colussi, A. J., Zabel, F., and Benson, S. W., *Int. J. Chem. Kinet.*, 9, 161, 1977.
23. Coomber, J. W. and Whittle, E., *Trans. Faraday Soc.*, 63, 2656, 1967.
24. Czarnarski, J., Castellano, E., and Schumacher, H. J., *Chem. Comm.*, p. 1255, 1968.
25. Darwent, D. deB., *Bond Dissociation Energies in Simple Molecules*, National Bureau of Standards, NSRDS-NBS, 31 Washington, D. C., 1970.
26. Davis, D. D. and Okabe, H., *J. Chem. Phys.*, 49, 5526, 1968.
27. Davis, H. F., Kim, B., Johnston, H. S., and Lee, Y. T., *J. Chem. Phys.*, 97, 2172, 1993.
28. Dóbé, S., Otting, M., Temps, F., Wagner, H. Gg., and Ziemer, H., *Ber. Bunsenges. Phys. Chem.*, 97, 881, 1993
29. Dobychin, S. L., Mashendzhinov, V. I., Mishin, V. I., Semenov, V. N., and Shpak, V. S., *Doklady Akademii Nauk SSSR 312*, 1166 1991.
30. Doncaster, A. M. and Walsh, R., *J. Phys. Chem.*, 83, 578, 1979.
31. Ellul, E., Potzinger, P., Reimann, B., and Camilleri, P., *Ber. Bunsenges. Phys. Chem.*, 85, 407, 1981.
32. Ervin, K. M., Gronert, S., Barlow, S. E., Gilles, M. K., Harrison, A. G., Bierbaum, V. M., DePuy, C. H., Lineberger, W. C., and Ellison, G. B., *J. Am. Chem. Soc.*, 112, 5750, 1990.
33. Evans, B. S., Weeks, I., and Whittle, E., *J. Chem. Soc. Faraday Trans. 1*, 79, 1471, 1983.
34. Ferguson, K. C., Okafo, E. N., and Whittle, E., *J. Chem. Soc. Faraday Trans. 1*, 69, 295, 1973.
35. Foon, R. and Tait, K. B., *J. Chem. Soc. Faraday Trans. 1*, 68, 104, 1972.
36. Foon, R. and Tait, K. B., *J. Chem. Soc. Faraday Trans. 1*, 68, 1121, 1972.
37. Forte, E., Hippler, H., and van den Bergh, H., *Int. J. Chem. Kinet.*, 13, 1227, 1981.
38. Gowenlock, B. G., Jonhson, C. A. F., Keary, C. M., and Pfaf, J., *J. Chem. Soc. Perkin Trans. 2*, 71, 351, 1975.
39. Grela, M. A. and Colussi, A. J., *Int. J. Chem. Kinet.*, 20, 713, 1988.
40. Griller, D. and Wayner, D. D. M., *Pure Appl. Chem.*, 61, 717, 1989.
41. Heneghan, S. P. and Benson, S. W., *Int. J. Chem. Kinet.*, 15, 815, 1983.
42. Hudgens, J. W., Johnson, R. D., Timonen, R. S., Seetula, J. A., and Gutman, D., J. *Phys. Chem.*, 95, 4400, 1991.
43. Kerr, J. A. and Timlin, D. M., *Int. J. Chem. Kinet.*, 3, 427, 1971.
44. Kerr, J. A., *Chem. Rev.*, 66, 465, 1966.
45. Kominar, R. J., Krech, M. J., and Price, S. J. W., *Can. J. Chem.*, 58, 1906, 1980.
46. Kondo, O. and Benson, S. W., *Int. J. Chem. Kinet.*, 16, 949, 1984.
47. Kondo, O. and Benson, S. W., *J. Phys. Chem.*, 88, 6675, 1984.
48. Kreutter, K. D., Nicovich, J. M., and Wine, P. H., *J. Phys. Chem.*, 95, 4020, 1991.
49. Laufer, A. H. and Okabe, H., *J. Am. Chem. Soc.*, 93, 4137,1971.
50. Lewis, K. E. and Smith, G. P., *J. Am. Chem. Soc.*, 106, 4650, 1984.
51. Lewis, K. E., Golden, D. M., and Smith, G. P., *J. Am. Chem. Soc.*, 106, 3905, 1984.
52. Lias, S. G., Bartmess, J. E., Liebman, J. F., Holmes, J. L., Levin, R. D., and Mallard, W. G., *J. Phys. Chem. Ref. Data 17*, Suppl. 1, 1988.
53. Lightfoot, P. D., Cox, R. A., Crowley, J. N., Destriau, M., Hayman G. D., Jenkin, M. E., Mootrgat, G. K., and Zabel, F., *Atmos. Environ.*, 26A, 1805, 1992.
54. McMillen, D. F. and Golden, D. M., *Ann. Rev. Phys. Chem.*, 33, 493, 1982.
55. Miyokawa, K. and Tschuikow-Roux, E., *J. Phys. Chem.*, 96, 7328, 1992.
56. Moylan, C. R. and Braumann, J. I., *J. Phys. Chem.*, 88, 3175, 1984.
57. Nicovich, J. M., Kreutter, K. D., Daykin, E. P., Chin, M., and Wine, P. H., Presented at 11th International Symposium on Gas Kinetics, Assisi, Italy, 1990.
58. Nicovich, J. M., Kreutter, K. D., van Dijk, C. A., and Wine, P. H., *J. Phys. Chem.*, 96, 2518, 1992.
59. Niiranen, J. T., Gutman, D., and Krasnoperov, L. N., *J. Phys. Chem.*, 96, 5881, 1992.
60. Noble, P. N. and Walsh, R., *Int. J. Chem. Kinet.*, 15, 547, 1983.
61. Noble, P. N. and Walsh, R., *Int. J. Chem. Kinet.*, 15, 561, 1983.
62. O'Neal, H. E., Bagg, J. W., and Richardson, W. H., *Int. J. Chem. Kinet.*, 2, 493, 1970.

Table 3
BOND STRENGTHS IN POLYATOMIC MOLECULES (continued)

63. Okabe, H. and Dibeler, V. H., *J. Chem. Phys.*, 59, 2430, 1973.
64. Okabe, H., *J. Chem. Phys.*, 66, 2058, 1977.
65. Okafo, E. N. and Whittle, E., *Int. J. Chem. Kinet.*, 7, 287, 1975.
66. Paul, S. and Back, M. H., *Can J. Chem.*, 53, 3330, 1975.
67. Pedley, J. B. and Rylance, J., "Sussex - N.P.L. Computer Analysed Thermochemical Data; Organic and Organometallic Compounds", University of Sussex, 1977.
68. Petry, R. C., *J. Am. Chem. Soc.*, 89, 4600, 1967.
69. Pickard, J. M. and Rodgers, A. S., *Int. J. Chem. Kinet.*, 9, 759, 1977.
70. Pickard, J. M. and Rodgers, A. S., *Int. J. Chem. Kinet.*, 15, 569, 1983.
71. Price, S. J. W. and Sapiano, H. J., *Can. J. Chem.*, 57, 1468, 1979.
72. Rogers, A. S. and Ford, W. G. F., *Int. J. Chem. Kinet.*, 5, 965, 1973.
73. Rossi, M. and Golden, D. M., *Int. J. Chem. Kinet.*, 11, 715, 1979.
74. Rossi, M., McMillen, D. F., and Golden, D. M., *J. Phys. Chem.*, 88, 5031, 1984.
75. Russell, J. J., Senkan, S. M., Seetula, J. A., and Gutman, D., *J. Phys. Chem.*, 93, 5184, 1989.
76. Seakins, P. W., Pilling, M. J., Niiranen, J. T., Gutman, D., and Krasnoperov, L. N., *J. Phys. Chem.*, 96, 9847, 1992.
77. Seetula, J. A., Feng, Y., Gutman, D., Seakins, P. W., and Pilling, M. J., *J. Phys. Chem.*, 95, 1658, 1991.
78. Seetula, J. A., Russell, J. J., and Gutman, D., *J. Am. Chem. Soc.*, 112, 1347, 1990.
79. Shum, L. G. S. and Benson, S. W., *Int. J. Chem.*, 15, 433, 1983.
80. Shum, L. G. S. and Benson, S. W., *J. Phys. Chem.*, 87, 3479, 1983.
81. Stein, S. E., SRD Thermochemical Database, 25. N.I.S.T. Structures and Properties Database and Estimation Program, U.S. Department of Commerce, 1992.
82. Steinkruger, F. J. and Rowland, F. S., *J. Phys. Chem.*, 85, 136, 1981.
83. Trenwith, A. B. and Wrigley, S. P., *J. Chem. Soc. Faraday Trans. 1*, 73, 817, 1977.
84. Trenwith, A. B., *J. Chem. Soc. Faraday Trans. 1*, 78, 3131, 1982.
85. Trenwith, A. B., *J. Chem. Soc. Faraday Trans. 1*, 79, 2755, 1983.
86. Tsang, W. and Herron, J. T., *J. Chem. Phys.*, 96, 4272, 1992.
87. Tsang, W., *Int. J. Chem. Kinet.*, 5, 929, 1973.
88. Tsang, W., *Int. J. Chem. Kinet.*, 10, 1119, 1978.
89. Tschuikow-Roux, E. and Paddison, S., *Int. J. Chem. Kinet.*, 19, 15, 1987.
90. van den Bergh, H. and Troe, J., *J. Chem. Phys.*, 64, 736, 1976.
91. Walsh, R., *Acc. Chem. Res.*, 14, 246, 1981.
92. Walsh, R., Private communication, calculated from enthalpy data listed in Table 4.
93. White, J. N. and Gardiner, W. C., *Chem. Phys. Lett.*, 58, 470, 1978.
94. Zabel, F., Benson, S. W., and Golden, D. M., *Int. J. Chem. Kinet.*, 10, 295, 1978.
95. Zmbov, K. F., Uy, O. M., and Margrave, J. L., *J. Am. Chem. Soc.*, 90, 5090, 1968.

Table 4
ENTHALPIES OF FORMATION OF FREE RADICALS

The enthalpies of formation of the free radicals are related to the corresponding bond strengths by the equations

$$D°(\text{R-X}) = \Delta_f H°(\text{R}) + \Delta_f H°(\text{X}) - \Delta_f H°(\text{RX})$$

or

$$D°(\text{R-R}) = 2\Delta_f H°(\text{R}) + \Delta_f H°(\text{RR})$$

For an excellent review of the methods of determining the enthalpies of formation of free radicals the reader is referred to "Thermochemistry of Free Radicals" by H. E. O'Neal and S. W. Benson in *Free Radicals*, Kochi, J. K., Ed., John Wiley & Sons, New York, 1973, 275.

Radical	$\Delta_f H°_{298}$/kJ mol^{-1}	Ref.
CH	596.35 ± 4	5,14
CH_2(triplet)	392.5 ± 2.1	11,29,60
CH_2(singlet)	430.1 ± 4.2	11,29,60
CH_3	146 ± 1.0	51
CH≡C	566.1 ± 2.9	4,18
CH_2=CH	300.0 ± 3.4	18,47

Table 4
ENTHALPIES OF FORMATION OF FREE RADICALS (continued)

Radical	$\Delta_f H°_{298}$/kJ mol^{-1}	Ref.
C_2H_5	121.0 ± 1.5	49,51
Cycloprop-2-en-1-yl	439.7 ± 17.2	30
$CH{\equiv}CCH_2$	340.6 ± 8.4	30
$CH_2{=}CHCH_2$	163.6 ± 6.3	30
Cyclopropyl	279.9 ± 1.1	30
n-C_3H_7	97.5 ± 2.5	12
i-C_3H_7	90.0 ± 1.7	49,51
$CH_3C{\equiv}CCH_2$	293.7	30
$CH_2{=}CHCHCH_3$	125.5 ± 6.3	30
$CH{\equiv}CCHCH_3$	295.0 ± 9.2	30
Cyclobutyl	214.2 ± 4.2	30
Cyclopropylmethyl	213.8 ± 6.7	30
$CH_2{=}C(CH_3)CH_2$	127.2 ± 5.4	30
$CH_3CH{=}CHCH_2$	125.5 ± 6.3	30
s-C_4H_9	67.5 ± 2.2	49,51
t-C_4H_9	51.3 ± 1.8	49,51
Cyclopenta-1,3-dien-5-yl	242.3 ± 6.3	30
Spiropentyl	380.7 ± 4	30
Cyclopent-1-en-3-yl	160.7 ± 4	30
$CH_2{=}CHCH{=}CHCH_2$	205 ± 13	30
$(C_2H_3)_2CH$	205 ± 13	30
$CH_3C{\equiv}CCHCH_3$	272.8 ± 9.6	30
$CH{\equiv}CC(CH_3)_2$	257.3 ± 8.4	30
$CH_2{=}CHC(CH_3)_2$	77.4 ± 6.3	30
Cyclopentyl	107.1 ± 2.5	12
$(CH_3)_3CCH_2$	36.4 ± 8	30
$C_2H_5C(CH_3)_2$	32.2 ± 6.3	54
C_6H_5	328.9 ± 8.4	30
Cyclohexa-1,3-dien-5-yl	197 ± 21	30
Cyclohexyl	58.2 ± 4	30
$CH_3C{\equiv}CC(CH_3)_2$	221.8 ± 9.6	30
$(CH_3)_2C{=}C(CH_3)CH_2$	39.8 ± 6.3	30
$CH_2{=}C(CH_3)C(CH_3)_2$	37.7 ± 6.3	30
$C_6H_5CH_2$	200.0 ± 6.3	30
Cyclohepta-1,3,5-trien-7-yl	271.1 ± 8	30
$CH_3CH_2CH_2C(CH_3)_2$	3.4 ± 8.4	52
Norbornyl	136.4 ± 10.5	30
Cycloheptyl	51.1 ± 4	30
$C_6H_5CHCH_3$	169.0	30
$C_6H_5C(CH_3)_2$	134.7	30
1-Naphthylmethyl	252.7	30
$(C_6H_5)_2CH$	289	46
9,10-Dihydroanthracen-9-yl	256.9 ± 6.3	30
9-Anthracenylmethyl	337.6	30
9-Phenanthrenylmethyl	311.3	30
CH_2CN	244.8 ± 10.5	30
CH_3CHCN	209.2 ± 9.6	30
$(CH_3)_2CCN$	166.5 ± 8.4	30
$C_6H_5C(CH_3)CN$	248.5	30
CH_2NH_2	149.4 ± 8	30
CH_3NHCH_2	126 ± 8	30
$(CH_3)_2NCH_2$	109 ± 8	30
CN	435 ± 8	30
CHO	37.2 ± 5.0	30
CH_3CO	-10.0 ± 1.2	36
$CH_2{=}CHCO$	72.4	30
C_2H_5CO	-42.7 ± 4.2	30
C_6H_5CO	109.2 ± 8	30
CH_2CHO	12.6 ± 8.4	45
CH_3COCH_2	-23.9 ± 10.9	30
$CH_3COCHCH_3$	-70.3 ± 7.1	30
CH_3OCH_2	-11.7 ± 5.0	30
$C_2H_5OCHCH_3$	-84.5	23
Tetrahydrofuran-2-yl	-18.0 ± 6.3	30
CH_2OH	-9 ± 6	17
CH_3CHOH	-63.6 ± 4.2	30
$CH_2{=}CHCHOH$	0.0	30
$(CH_3)_2COH$	-111.3 ± 4.6	30
COOH	-223.0	30
$COOCH_3$	-169.0 ± 4	30
$C_6H_5COOCH_2$	-69.9 ± 8.4	30
CHF	163 ± 13	43
CH_2F	-31.8 ± 8.4	40
FCO	-171.5	13
CHF_2	-238.9 ± 4	40
CF_2	-194.1 ± 9.2	38
CF_3	-467.4 ± 15.1	30
CH_2Cl	121.8 ± 4.2	56
CHFCl	-60.7 ± 10.0	56
CF_2Cl	-279.1 ± 8.3	31
$CHCl_2$	98.3 ± 5.0	56
$CFCl_2$	-89.1 ± 10.0	56
CCl_2	239	37
CCl_3	71.1 ± 2.5	22
CH_2Br	169.0 ± 4.2	56
$CHBr_2$	188.3 ± 9.2	56
CBr_3	207.1 ± 8	56
CH_2I	230.1 ± 6.7	30
CHI_2	333.9 ± 9.2	30
CH_3CF_2	-302.5 ± 8	30
CF_3CH_2	-517.1 ± 5.0	30
C_2F_5	-892.9 ± 4	30
$CCl{=}CH_2$	>251	47
$CHCl_2CCl_2$	23.4 ± 8	30
CF_2ClCF_2	-686 ± 17	30
C_2Cl_5	35.2 ± 8	30
C_6F_5	-547.7 ± 8	30
$(CH_3)_3SiCH_2$	-34.7	30
CS	278.5 ± 3.8	42
CH_3SCH_2	135.1 ± 5.0	33
NH	352.3 ± 9.6	41
NH_2	185.4 ± 4.6	30
NF_2	34 ± 4	30
N_2H_3	243.5	20
N_3	469 ± 21	30
CH_2NH	104.6 ± 13	20
CH_3NH	177.4 ± 8	30
$(CH_3)_3N$	145.2 ± 8	30
C_6H_5NH	237.2 ± 8	30
$C_6H_5NCH_3$	233.5 ± 8	30

Table 4
ENTHALPIES OF FORMATION OF FREE RADICALS (continued)

Radical	$\Delta_f H^\circ_{298}$/kJ mol^{-1}	Ref.
NCO	159	13
CH_3N_2	215.5 ± 7.5	2
$C_2H_5N_2$	187.4 ± 10.5	2
i-$C_3H_7N_2$	158.6 ± 9.2	2
OH	39.3	30
CH_3O	17.6	30
C_2H_5O	-17.2	30
n-C_3H_7O	-41.4	30
i-C_3H_7O	-52.3	30
n-C_4H_9O	-62.8	30
s-C_4H_9O	-69.5 ± 3.3	30
t-C_4H_9O	-90.8	30
C_6H_5O	47.7	30
CF_3O	-655.6	7
FO	109 ± 4	8
ClO	101.6 ± 0.1	1,44
BrO	125	58
IO	176	6,44
HO_2	14.6	53
CH_3O_2	10.4 ± 3.1	24
$C_2H_5O_2$	-28.7 ± 6.5	24
CH_2=$CHCH_2O_2$	87.9 ± 5.5	24
i-$C_3H_7O_2$	-68.9 ± 8.1	24
t-$C_4H_9O_2$	-105.0 ± 8.1	24
$HOCH_2O_2$	-162.1 ± 2.1	24
CH_2ClO_2	9.2	24
$CHCl_2O_2$	1.6	24
CF_3O_2	-614.0	24
$CFCl_2O_2$	-213.7	24
CF_2ClO_2	-406.5	24
CCl_3O_2	-11.3 ± 4.6	24
CH_3CO_2	-207.5 ± 4	30
CH_3COO_2	-172 ± 20	10
$C_2H_5CO_2$	-228.5 ± 4	30
n-$C_3H_7CO_2$	-249.4 ± 4	30
FO_2	26.1	26,39
ClO_2	97.5	3,34
NO_3	73.7 ± 1.4	16
sym-ClO_3	232.6	15
SiH	377	30,59
SiH_2	269.0 ± 1.3	19,28
SiH_3	200.5 ± 2.5	50
CH_3Si	310	59
CH_3SiH	213.0 ± 14.6	57
CH_3SiH_2	152.7	30,59
$(CH_3)_2Si$	109	59
$(CH_3)_2SiH$	59.8	30,59
$(CH_3)_3Si$	-3.3	30,59
$C_6H_5Si(CH_3)_2$	163	37
$(C_6H_5)_2SiCH_3$	326	37
$(C_6H_5)_3Si$	486.2	37
SiF	-19.3	30,59
SiF_2	-587.9	30,59
SiF_3	-1025	30,59
SiCl	195.8	30,59
$SiCl_2$	-163.6	30,59
$SiCl_3$	-318	30,59
SiH_3SiH	269.9 ± 14.6	57
Si_2H_5	220.0	30,59
GeH_3	239	37
PH_2	138 ± 8	37
HS	143.0 ± 2.8	35
CH_3S	124.6 ± 1.8	35
C_6H_5S	229.7 ± 8	30
SO	5.0	13
HSO	-4	25
HSO_2	-222	9
CH_3SO_2	-239.3	37
$HOSO_2$	-385	27
SO_3	-395.7	58
SF_4	-746 ± 12	55
SF_5	-879.9 ± 20	55
CH_3S_2	68.6 ± 8	21
$C_2H_5S_2$	43.5 ± 8	21
i-$C_3H_7S_2$	13.8 ± 8	21
t-$C_4H_9S_2$	-19.3 ± 8	21
$HOCS_2$	110.5	32

REFERENCES

1. Abramowitz, S. and Chase, M. W., *Pure Appl. Chem.*, 63, 1448, 1991.
2. Acs, G. and Peter, A., *Int. J. Chem. Kinet.*, 19, 929, 1987.
3. Baer, S., Hippler, H., Rabu, R., Siefke, M., Seitzinger, N., and Troe, J., *J. Chem. Phys.*, in press.
4. Baldwin, D. P., Buntine, M. A., and Chandler, D. W., *J. Chem. Phys.*, 93, 6578, 1990.
5. Based on D(C-H), see Table 1 and $\Delta_f H^\circ$(C) and $\Delta_f H^\circ$(H) see Table 2.
6. Based on D(I-O), see Table 1 and $\Delta_f H^\circ$(I) and $\Delta_f H^\circ$(O) see Table 2.
7. Batt, L. and Walsh, R., *Int. J. Chem. Kinet.*, 14, 933, 1982.
8. Benson, S. W., *Thermochemical Kinetics*, 2nd ed., John Wiley & Sons, New York, 1976.
9. Boyd, R. J., Gupta, A., Langler, R. F., Lownie, S. P., and Pincock, J. A., *Can. J. Chem.*, 58, 331, 1980.
10. Bridier, I., Caralp, F., Loirat, H., Lesclaux, R., Veyret, B., Becker, K. H., Reimer, A., and Zabel, F., *J. Phys. Chem.*, 95, 3594, 1991.
11. Bunker, P. R. and Sears, T. J., *J. Chem. Phys.*, 83, 4866, 1985.
12. Castelhano, A. L. and Griller, D., *J. Am. Chem. Soc.*, 104, 3655, 1982; value adjusted to $\Delta_f H^\circ(CH_3)$ = 146 kJ mol^{-1} and error limits assigned here.

Table 4
ENTHALPIES OF FORMATION OF FREE RADICALS (continued)

13. Chase, M. W. Jr., Davies, C. A., Downey, J.R. Jr., Frurip, D. J., McDonald, R. A., and Syverud, A. N., *J. Phys. Chem. Ref. Data*, 14, Suppl. 1, 1985.
14. Chupka, W. A. and Lifshitz, C., *J. Chem. Phys.*, 48, 1109, 1968.
15. Colussi, A. J., *J. Phys. Chem.*, 94, 8922, 1990.
16. Davis, H. F., Kim, B., Johnston, H. S., and Lee, Y. T., *J. Phys. Chem.*, 97, 2172, 1993.
17. Dóbé, S., Otting, M., Temps, F., Wagner, H. Gg., and Ziemer, H., *Ber. Bunsenges. Phys. Chem.*, 97, 881, 1993.
18. Ervin, K. M., Gronert, S., Barlow, S. E., Gilles, M. K., Harrison, A. G., Bierbaum, V. M., DePuy, C. H., Lineberger, W. C., and Ellison, G. B., *J. Am. Chem. Soc.*, 112, 5750, 1990.
19. Frey, H. M., Walsh, R., and Watts, I. M., *J. Chem. Soc. Chem. Comm.*, 1189, 1986.
20. Grela, M. A. and Colussi, A. J., *Int. J. Chem. Kinet.*, 20, 713, 1988.
21. Howari, J. A. Griller, D., and Lossing, F. P., *J. Am. Chem. Soc.*, 108, 3273, 1986.
22. Hudgens, J. W., Johnson, R. D., Timonen, R. S., Seetula, J. A., and Gutman, D., *J. Phys. Chem.*, 95, 4400, 1991.
23. Kondo, O. and Benson, S. W., *Int. J. Chem. Kinet.*, 16, 949, 1984.
24. Lightfoot, P. D., Cox, R. A., Crowley, J. N., Destriau, M., Hayman, G.D., Jenkin, M. E., Moortgat, G. K., and Zabel, F., *Atmos. Environ.*, 26A, 1805, 1992.
25. Lovejoy, E. R., Wang, N. S., and Howard, C. J., *J. Phys. Chem.*, 91, 5749, 1987.
26. Lyman, J. L. and Holland, R., *J. Phys. Chem.*, 92, 7232, 1988.
27. Margitan, J. J., *J. Phys. Chem.*, 88, 3314, 1984.
28. Martin, J. G., Ring, M. A., and O'Neal, H. E., *Int. J. Chem. Kinet.*, 19, 715, 1987.
29. McKellar, A. R. W., Bunker, P. R., Sears, T. J., Evenson, K. M., Saykally, R. J., and Langhoff, S. R., *J. Chem. Phys.*, 79, 5251, 1983.
30. McMillen, D. F. and Golden, D. M., *Ann. Rev. Phys. Chem.*, 33, 493, 1982.
31. Miyokawa, K. and Tschuikow-Roux, E., *J. Phys. Chem.*, 96, 7328, 1992.
32. Murrells, T. P., Lovejoy, E.R., and Ravishankara,A. R., *J. Phys. Chem.*, 80, 4065, 1984.
33. Nicovich, J. M., Kreutter, K. D., Daykin, E. P., Chin, M., and Wine, P.H., presented at 11th International Symposium on Gas Kinetics, Assisi, Italy, 1990.
34. Nicovich, J. M., Kreutter, K. D., Shockelford, C. J., and Wine, P. H., *Chem. Phys. Lett.*, 179, 367, 1991.
35. Nicovich, J. M., Kreutter, K. D., van Dijk, C. A., and Wine, P. H., *J. Phys. Chem.*, 96, 2518, 1992.
36. Niiranen, J. T., Gutman, D., and Krasnoperov, L. N., *J. Phys. Chem.*, 96, 5881, 1992.
37. O'Neal, H. E. and Benson, S. W., in "Free Radicals", Kochi, J. K., Ed., John Wiley & Sons, New York, 1973, 275.
38. Okafo, E. N. and Whittle, E., *J. Chem. Soc. Faraday Soc.*, 1, 70, 1366, 1974.
39. Pagsberg, P., Ratajczak, E., Sillesen, A., Jodkowski, J. T., *Chem. Phys. Lett.*, 141, 88, 1987.
40. Pickard, J. M. and Rodgers, A. S., *Int. J. Chem. Kinet.*, 15, 569, 1983.
41. Piper, L. G., *J. Chem. Phys.*, 70, 3417, 1979.
42. Prinslow, D. A. and Armentrout, P. B., *J. Chem. Phys.*, 94, 3563, 1991.
43. Pritchard, G. O., Nilson, W. B., and Kirtman, B., *Int. J. Chem. Kinet.*, 16, 1637, 1984.
44. Reddy, R. R., Rao, T. V. R., and Reddy, A. S. R., Indian, *J. Pure Appl. Phys.*, 27, 243, 1989.
45. Rossi, M. and Golden, D. M., *Int. J. Chem. Kinet.*, 11, 715, 1979.
46. Rossi, M., McMillen, D. F., and Golden, D. M., *J. Phys. Chem.*, 88, 5031, 1984.
47. Russell, J. R., Senkan, S. M., Seetula, J. A., and Gutman, D., *J. Phys. Chem.*, 93, 5184, 1989.
48. Seakins, P. W., and Pilling, M. J., *J. Phys. Chem.*, 95, 9874, 1991.
49. Seakins, P. W., Pilling, M. J., Niiranen, J. T., Gutman, D., and Krasnoperov, L. N., *J. Phys. Chem.*, 96, 9847, 1992.
50. Seetula, J. A., Feng, Y., Gutman, D., Seakins, P. W., and Pilling, M. J., *J. Phys. Chem.*, 95, 1658, 1991.
51. Seetula, J. A., Russell, J. A., and Gutman, D., *J. Am. Chem. Soc.*, 112, 1347, 1990.
52. Seres, L., Gorgenyi, M., and Farkas, J., *Int. J. Chem. Kinet.*, 15, 1133, 1983.
53. Shum, L. G. S. and Benson, S. W., *J. Phys. Chem.*, 87, 3479, 1983.
54. Stein, S. E., SRD Thermochemical Database, 25. N.I.S.T. Structures and Properties Database and Estimation Program, U.S. Department of Commerce, 1992.
55. Tsang, W. and Herron, J. T., *J. Chem. Phys.*, 96, 4272, 1992.
56. Tschuikow-Roux, E. and Paddison, S., *Int. J. Chem. Kinet.*, 19, 15, 1987.
57. Vanderwielen, A. J., Ring, M. A., and O'Neal, H. E., *J. Am. Chem. Soc.*, 97, 993, 1975.
58. Wagman, D. D., Evans, W. H., Parker, V. B., Schumm, R. H., Halow, I., Bailey, S. M., Churney, K. L., and Nuttall, R.L., *J. Phys. Chem. Ref. Data*, 11, Suppl. 2, 1978.
59. Walsh, R., *Acc. Chem. Res.*, 14, 246, 1981.
60. Walsh, R., private communication (listed values are based on data in References 11 and 29).

Table 5
BOND STRENGTHS OF SOME ORGANIC MOLECULES

Bond strengths at 298 K expressed in kJ/mol for some organic molecules of the general formula R-X are presented below. Some are experimental values taken from the preceding tables; the remainder are calculated from the enthalpies of formation of the radicals, listed in Table 4 above, and the enthalpies of formation of the parent compounds from sources indicated by the references below. The table also includes bond strengths for the inorganic molecules, hydrogen, the hydrogen halides, water and ammonia.

	H	F	Cl	Br	I	OH	NH_2	CH_3O	CH_3	CH_3CO	NO	CF_3	CCl_3
H	435.990	570.3	431.62	366.35	298.407	498	449	437	439	360	207	446	393
CH_3	439	472	349[c]	293	239[c]	387[i]	354[i]	348[i]	376[i]	339[i]	167	427	362[d]
C_2H_5	423	463[f]	353[d]	294[d]	235[d]	395[i]	354[i]	355[i]	372[i]	338[i]	—	—	—
i-C_3H_7	413	463[f]	356[i]	300[i]	238[i]	402[i]	359[i]	360[i]	371[i]	328[i]	153	—	—
t-C_4H_9	404	—	355[i]	296[i]	230[i]	403[i]	358[i]	353[i]	365[i]	317[i]	165	—	—
C_6H_5	464	524[i]	399[i]	337[i]	274[i]	465[i]	427[i]	415[i]	425[i]	391[i]	213	—	—
$C_6H_5CH_2$	368	—	303[i]	248[i]	207[i]	340[i]	—	—	332	274[i]	—	—	—
CCl_3	393	435[a]	288[c]	225[h]	167[e]	—	—	—	362[d]	—	134	335[d]	292[d]
CF_3	446	545[c]	362[c]	295[c]	224	—	—	—	427[d]	—	179	413	335[d]
C_2F_5	430	531[d]	347[j]	287[g]	219	—	—	—	—	—	—	—	—
CH_3CO	360	499[i]	340[i]	278[i]	209[i]	447[i]	399[i]	403	339[i]	282	—	—	—
CN	518	470	422	367	305	—	—	—	—	—	121	—	—
C_6F_5	477	477[b]	385[b]	—	276[b]	448[b]	—	—	444[b]	—	209[b]	—	—

REFERENCES

a. Chen, S. S., Wilhoit, R. C., and Zwolinski, B. J., *J. Phys. Chem. Ref. Data*, 5, 571, 1976.
b. Choo, K. Y., Mendenhall, G. D., Golden, D.M., and Benson, S. W., *Int. J. Chem. Kinet.*, 6, 813, 1974.
c. Kolesov, V. P., *Russ. Chem. Rev.*, 47, 599, 1978.
d. Kolesov, V. P. and Papina, T. S., *Russ. Chem. Rev.*, 52, 425, 1983.
e. Kudchadker, S. A. and Kudchadker, A. P., *J. Phys. Chem. Ref. Data*, 1, 1285, 1978.
f. Lias, S. G., Bartmess, J. E., Liebman, J. F., Holmes, J. L., Levin, R. D., and Mallard, W. G., *J. Phys. Chem. Ref. Data*, 17, 1, 1988.
g. McMillen, D. F. and Golden, D. M., *Ann. Rev. Phys. Chem.*, 33, 493, 1982.
h. Mendenhall, G. D., Golden, D. M., and Benson, S. W., *J. Phys. Chem.*, 77, 2707, 1973.
i. Pedley, J. B. and Rylance, J., "Sussex.N.P.L. Computer Analysed Thermochemical Data: Organic and Organometallic Compounds", University of Sussex, 1977.
j. Wu, E.-C. and Rodgers, A. S., *J. Am. Chem. Soc.*, 98, 6112, 1976.

FORCE CONSTANTS FOR BOND STRETCHING

Representative force constants (f) for stretching of chemical bonds are listed in this table. Except where noted, all force constants are derived from values of the harmonic vibrational frequencies ω_e. Values derived from the observed vibrational fundamentals ν, which are noted by a, are lower than the harmonic force constants, typically by 2 to 3% in the case of heavy atoms (often by 5 to 10% if one of the atoms is hydrogen). Values are given in the SI unit newton per centimeter (N/cm), which is identical to the commonly used cgs unit mdyn/Å.

REFERENCES

1. Huber, K. P., and Herzberg, G., *Molecular Spectra and Molecular Structure. IV. Constants of Diatomic Molecules*, Van Nostrand Reinhold, New York, 1979.
2. Shimanouchi, T., *The Molecular Force Field*, in Eyring, H., Henderson, D., and Yost, W., Eds., *Physical Chemistry: An Advanced Treatise*, Vol. IV, Academic Press, New York, 1970.
3. Tasumi, M., and Nakata, M., *Pure and Appl. Chem.*, 57, 121—147, 1985.

Bond	Molecule	f/(N/cm)	Note
H-H	H_2	5.75	
Be-H	BeH	2.27	
B-H	BH	3.05	
C-H	CH	4.48	
	CH_4	5.44	b
	C_2H_6	4.83	a,b,c
	CH_3CN	5.33	b
	CH_3Cl	5.02	a,b,c
	$CCl_2{=}CH_2$	5.57	b
	HCN	6.22	
N-H	NH	5.97	
O-H	OH	7.80	
	H_2O	8.45	
P-H	PH	3.22	
S-H	SH	4.23	
	H_2S	4.28	
F-H	HF	9.66	
Cl-H	HCl	5.16	
Br-H	HBr	4.12	
I-H	HI	3.14	
Li-H	LiH	1.03	
Na-H	NaH	0.78	
K-H	KH	0.56	
Rb-H	RbH	0.52	
Cs-H	CsH	0.47	
C-C	C_2	12.16	
	$CCl_2{=}CH_2$	8.43	
	C_2H_6	4.50	a,c
	CH_3CN	5.16	
C-F	CF	7.42	
	CH_3F	5.71	a,c
C-Cl	CCl	3.95	
	CH_3Cl	3.44	a,c
	$CCl_2{=}CH_2$	4.02	b
C-Br	CH_3Br	2.89	a,c
C-I	CH_3I	2.34	a,c
C-O	CO	19.02	
	CO_2	16.00	
	OCS	16.14	
	CH_3OH	5.42	a,c
C-S	CS	8.49	
	CS_2	7.88	
	OCS	7.44	
C-N	CN	16.29	
	HCN	18.78	
	CH_3CN	18.33	
	CH_3NH_2	5.12	a,c
C-P	CP	7.83	
Si-Si	Si_2	2.15	
Si-O	SiO	9.24	
Si-F	SiF	4.90	
Si-Cl	SiCl	2.63	
N-N	N_2	22.95	
	N_2O	18.72	
N-O	NO	15.95	
	N_2O	11.70	
P-P	P_2	5.56	
P-O	PO	9.45	
O-O	O_2	11.77	
	O_3	5.74	a
S-O	SO	8.30	
	SO_2	10.33	a
S-S	S_2	4.96	
F-F	F_2	4.70	
Cl-F	ClF	4.48	
Br-F	BrF	4.06	
Cl-Cl	Cl_2	3.23	
Br-Cl	BrCl	2.82	
Br-Br	Br_2	2.46	
I-I	I_2	1.72	
Li-Li	Li_2	0.26	
Li-Na	LiNa	0.21	
Na-Na	Na_2	0.17	
Li-F	LiF	2.50	
Li-Cl	LiCl	1.43	
Li-Br	LiBr	1.20	
Li-I	LiI	0.97	
Na-F	NaF	1.76	
Na-Cl	NaCl	1.09	
Na-Br	NaBr	0.94	
Na-I	NaI	0.76	
Be-O	BeO	7.51	
Mg-O	MgO	3.48	
Ca-O	CaO	3.61	

[a] Derived from fundamental frequency, without anharmonicity correction.
[b] Average of symmetric and antisymmetric (or degenerate) modes.
[c] Calculated from Local Symmetry Force Field (see Reference 2).

FUNDAMENTAL VIBRATIONAL FREQUENCIES OF SMALL MOLECULES

This table lists the fundamental vibrational frequencies of selected three-, four-, and five-atom molecules. Both stable molecules and transient free radicals are included. The data have been taken from evaluated sources. In general, the selected values are based on gas-phase infrared, Raman, or ultraviolet spectra; when these were not available, liquid-phase or matrix-isolation spectra were used.

Molecules are grouped by structural type. Within each group, related molecules appear together for convenient comparison.

The vibrational modes are described by their approximate character in terms of stretching, bending, deformation, etc. However, it should be emphasized that most such descriptions are only approximate, and that the true normal mode usually involves a mixture of motions. Abbreviations are:

sym.	symmetric
antisym.	antisymmetric
str.	stretch
deform.	deformation
scis.	scissors
rock.	rocking
deg.	degenerate

In the case of free radicals, strong interactions may exist between the electronic and bending vibrational motions. Details can be found in References 3 and 4. The references should be consulted for information on the accuracy of the data and for data on other molecules not listed here.

All fundamental frequencies (more precisely, wavenumbers) are given in units of cm^{-1}.

XY_2 Molecules
Point groups $D_{\infty h}$(linear) and C_{2v}(bent)

Molecule	Structure	Sym. str.	Bend	Antisym. str.
CO_2	Linear	1333	667	2349
CS_2	Linear	658	397	1535
C_3	Linear	1224	63	2040
CNC	Linear		321	1453
NCN	Linear	1197	423	1476
BO_2	Linear	1056	447	1278
BS_2	Linear	510	120	1015
KrF_2	Linear	449	233	590
XeF_2	Linear	515	213	555
$XeCl_2$	Linear	316		481
H_2O	Bent	3657	1595	3756
D_2O	Bent	2671	1178	2788
F_2O	Bent	928	461	831
Cl_2O	Bent	639	296	686
O_3	Bent	1103	701	1042
H_2S	Bent	2615	1183	2626
D_2S	Bent	1896	855	1999
SF_2	Bent	838	357	813
SCl_2	Bent	525	208	535
SO_2	Bent	1151	518	1362
H_2Se	Bent	2345	1034	2358
D_2Se	Bent	1630	745	1696
NH_2	Bent	3219	1497	3301
NO_2	Bent	1318	750	1618
NF_2	Bent	1075	573	942
ClO_2	Bent	945	445	1111
CH_2	Bent		963	
CD_2	Bent		752	
CF_2	Bent	1225	667	1114
CCl_2	Bent	721	333	748
CBr_2	Bent	595	196	641
SiH_2	Bent	2032	990	2022
SiD_2	Bent	1472	729	1468
SiF_2	Bent	855	345	870

FUNDAMENTAL VIBRATIONAL FREQUENCIES OF SMALL MOLECULES (continued)

Molecule	Structure	Sym. str.	Bend	Antisym. str.
$SiCl_2$	Bent	515		505
$SiBr_2$	Bent	403		400
GeH_2	Bent	1887	920	1864
$GeCl_2$	Bent	399	159	374
SnF_2	Bent	593	197	571
$SnCl_2$	Bent	352	120	334
$SnBr_2$	Bent	244	80	231
PbF_2	Bent	531	165	507
$PbCl_2$	Bent	314	99	299
ClF_2	Bent	500		576

XYZ Molecules
Point Groups $C_{\infty v}$ (linear) and C_s(bent)

Molecule	Structure	XY str.	Bend	YZ str.
HCN	Linear	3311	712	2097
DCN	Linear	2630	569	1925
FCN	Linear	1077	451	2323
ClCN	Linear	744	378	2216
BrCN	Linear	575	342	2198
ICN	Linear	486	305	2188
CCN	Linear	1060	230	1917
CCO	Linear	1063	379	1967
HCO	Bent	2485	1081	1868
HCC	Linear	3612		1848
OCS	Linear	2062	520	859
NCO	Linear	1270	535	1921
NNO	Linear	2224	589	1285
HNB	Linear	3675		2035
HNC	Linear	3653		2032
HNSi	Linear	3583	523	1198
HBO	Linear		754	1817
FBO	Linear		500	2075
ClBO	Linear	676	404	1958
BrBO	Linear	535	374	1937
FNO	Bent	766	520	1844
ClNO	Bent	596	332	1800
BrNO	Bent	542	266	1799
HNF	Bent		1419	1000
HNO	Bent	2684	1501	1565
HPO	Bent	2095	983	1179
HOF	Bent	3537	886	1393
HOCl	Bent	3609	1242	725
HOO	Bent	3436	1392	1098
FOO	Bent	579	376	1490
ClOO	Bent	407	373	1443
BrOO	Bent			1487
HSO	Bent		1063	1009
NSF	Bent	1372	366	640
NSCl	Bent	1325	273	414
HCF	Bent		1407	1181
HCCl	Bent		1201	815
HSiF	Bent	1913	860	834
HSiCl	Bent		808	522
HSiBr	Bent	1548	774	408

Symmetric XY_3 Molecules
Point Groups D_{3h} (planar) and C_{3v} (pyramidal)

Molecule	Structure	Sym. str.	Sym. deform.	Deg. str.	Deg. deform.
NH_3	Pyram.	3337	950	3444	1627
ND_3	Pyram.	2420	748	2564	1191
PH_3	Pyram.	2323	992	2328	1118
AsH_3	Pyram.	2116	906	2123	1003
SbH_3	Pyram.	1891	782	1894	831
NF_3	Pyram.	1032	647	907	492
PF_3	Pyram.	892	487	860	344
AsF_3	Pyram.	741	337	702	262
PCl_3	Pyram.	504	252	482	198
PI_3	Pyram.	303	111	325	79
AsI_3	Pyram.	219	94	224	71
$AlCl_3$	Pyram.	375	183	595	150
SO_3	Planar	1065	498	1391	530
BF_3	Planar	888	691	1449	480
BH_3	Planar		1125	2808	1640
CH_3	Planar		606	3161	1396
CD_3	Planar		453	2369	1029
CF_3	Pyram.	1090	701	1260	510
SiF_3	Pyram.	830	427	937	290

Linear XYYX Molecules
Point Group $D_{\infty h}$

Molecule	Sym. XY str.	Antisym. XY str.	YY str.	Bend	Bend
C_2H_2	3374	3289	1974	612	730
C_2D_2	2701	2439	1762	505	537
C_2N_2	2330	2158	851	507	233

Planar X_2YZ Molecules
Point Group C_{2v}

Molecule	Sym.XY str.	YZ str.	YX_2 scis.	Antisym. XY str.	YX_2 rock	YX_2 wag
H_2CO	2783	1746	1500	2843	1249	1167
D_2CO	2056	1700	1106	2160	990	938
F_2CO	965	1928	584	1249	626	774
Cl_2CO	567	1827	285	849	440	580
O_2NF	1310	822	568	1792	560	742
O_2NCl	1286	793	370	1685	408	652

Tetrahedral XY_4 Molecules
Point Group T_d

Molecule	Sym. str.	Deg. deform.(e)	Deg. str.(f)	Deg. deform.(f)
CH_4	2917	1534	3019	1306
CD_4	2109	1092	2259	996
CF_4	909	435	1281	632
CCl_4	459	217	776	314

FUNDAMENTAL VIBRATIONAL FREQUENCIES OF SMALL MOLECULES (continued)

Molecule	Sym. str.	Deg. deform.(e)	Deg. str.(f)	Deg. deform.(f)
CBr_4	267	122	672	182
CI_4	178	90	555	125
SiH_4	2187	975	2191	914
SiD_4	1558	700	1597	681
SiF_4	800	268	1032	389
$SiCl_4$	424	150	621	221
GeH_4	2106	931	2114	819
GeD_4	1504	665	1522	596
$GeCl_4$	396	134	453	172
$SnCl_4$	366	104	403	134
$TiCl_4$	389	114	498	136
$ZrCl_4$	377	98	418	113
$HfCl_4$	382	102	390	112
RuO_4	885	322	921	336
OsO_4	965	333	960	329

REFERENCES

1. T. Shimanouchi, Tables of Molecular Vibrational Frequencies, Consolidated Volume I, Natl. Stand. Ref. Data Ser. Natl. Bur. Stand. (U.S.), 39, 1972.
2. T. Shimanouchi, Tables of Molecular Vibrational Frequencies, Consolidated Volume II, *J. Phys. Chem. Ref. Data*, 6, 993, 1977.
3. G. Herzberg, *Electronic Spectra and Electronic Structure of Polyatomic Molecules*, D. Van Nostrand Co., Princeton, 1966.
4. M. E. Jacox, Ground state vibrational energy levels of polyatomic transient molecules, *J. Phys. Chem. Ref. Data*, 13, 945, 1984.

INFRARED CORRELATION CHARTS

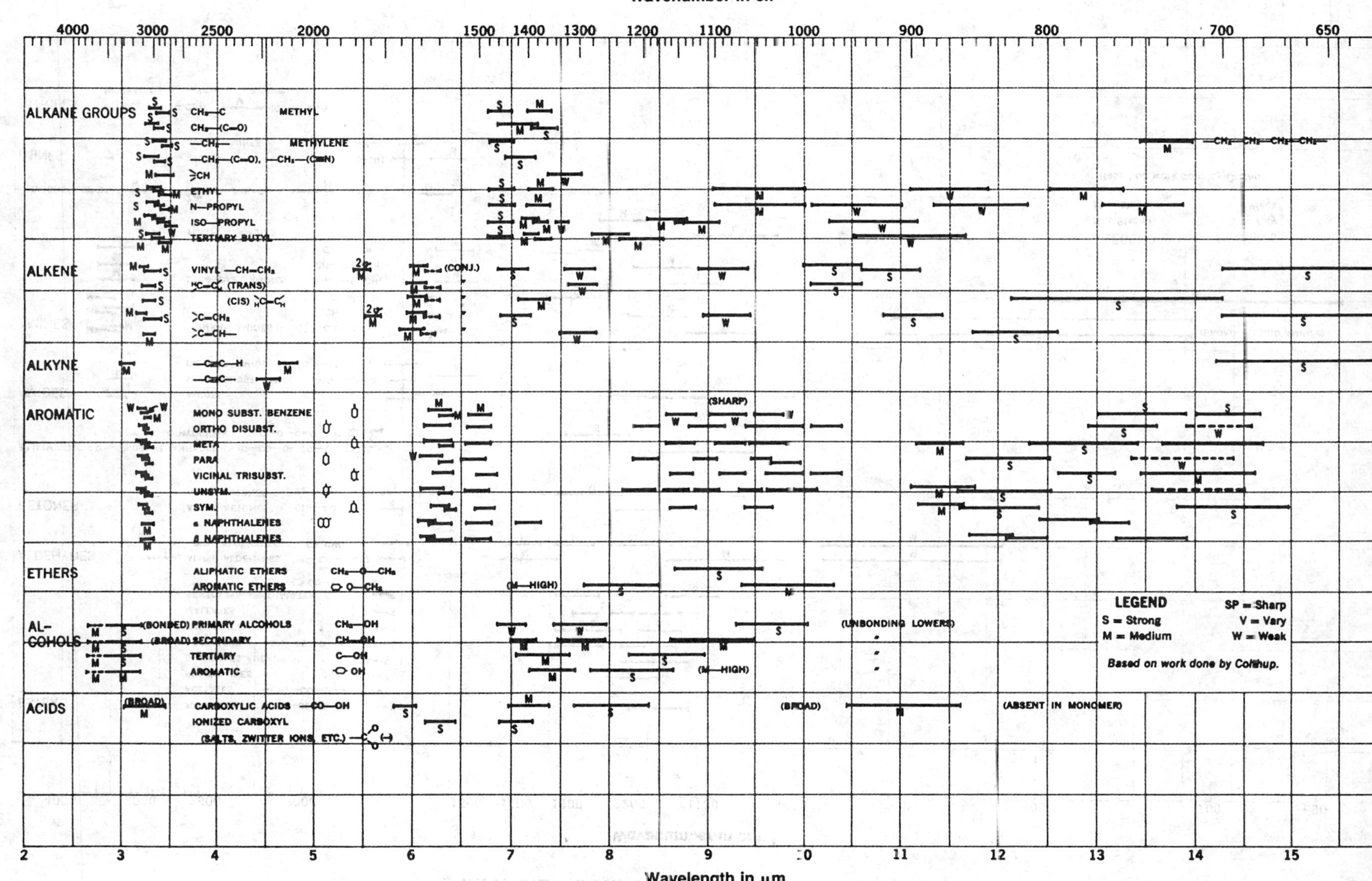

INFRARED CORRELATION CHARTS (continued)

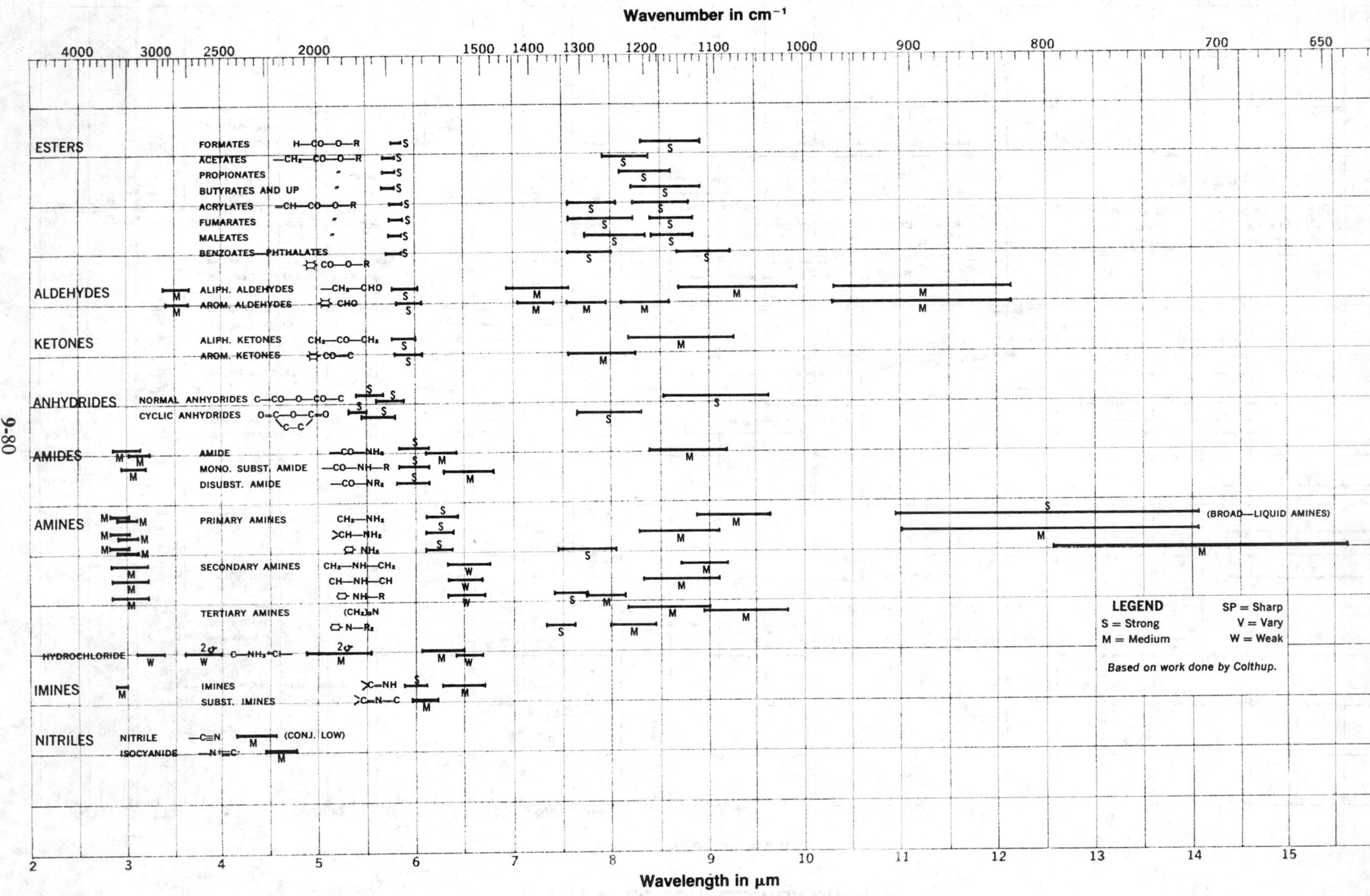

INFRARED CORRELATION CHARTS (continued)

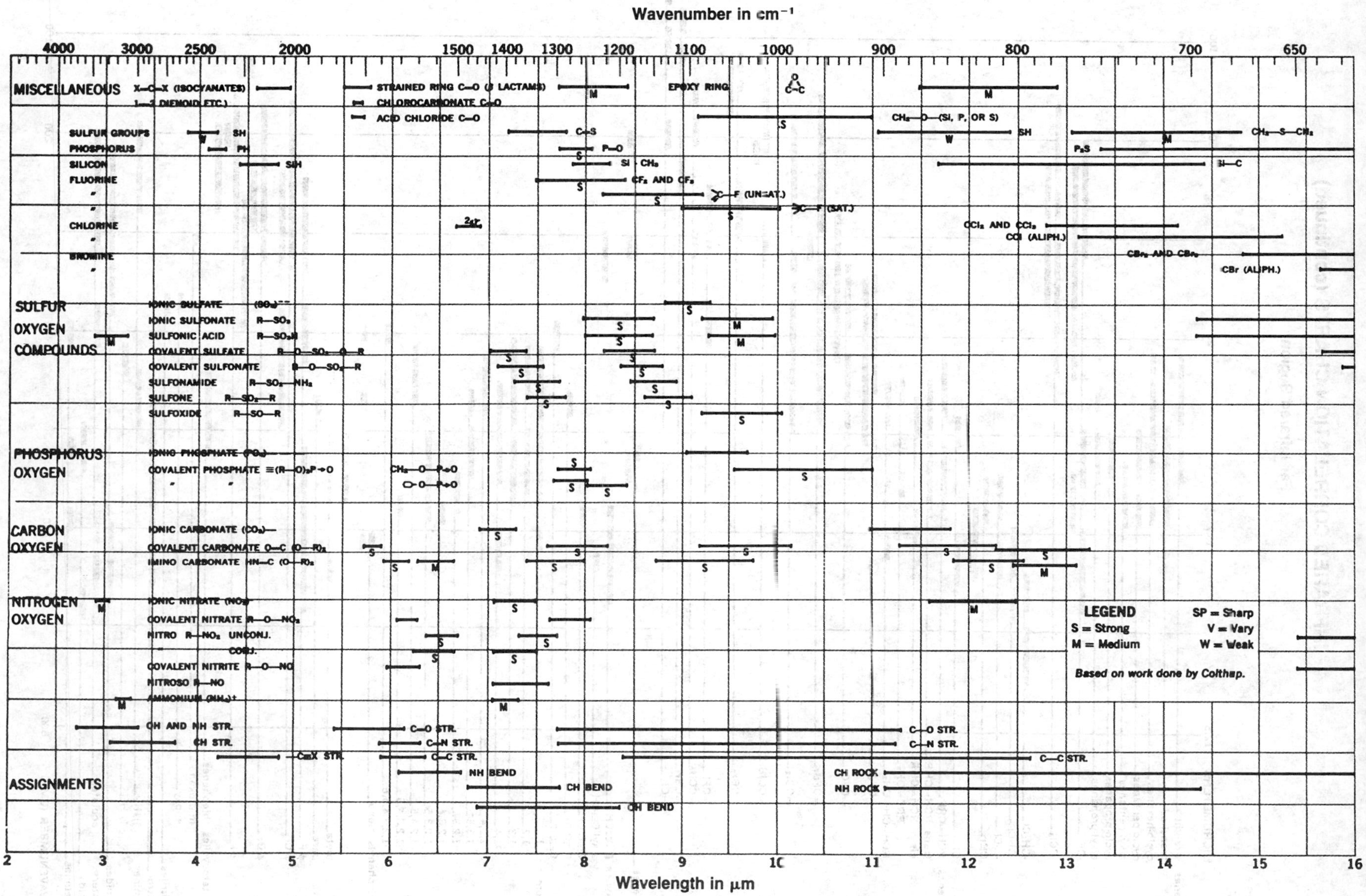

Wavenumber in cm^{-1}

4000 3000 2500 2000 1500 1400 1300 1200 1100 1000 900 800 700 650

MISCELLANEOUS

X=C=X (ISOCYANATES) — STRAINED RING C=O (β LACTAMS) — M — EPOXY RING — M

1—2 DIENOID ETC.) — CHLOROCARBONATE C=O — ACID CHLORIDE C=O

CH_2—O—(SI, P, OR S) — S

SULFUR GROUPS — W — SH — C—S — S — W — SH — M — CH_2—S—CH_2

PHOSPHORUS — PH — P—O — P_2S

SILICON — SiH — SI · CH_3 — Si—C

FLUORINE — S — CF_2 AND CF_3

" — S — C—F (UNSAT.)

" — S — C—F (SAT.)

CHLORINE — 2ν — CCl_2 AND CCl_3

" — CCl (ALIPH.)

BROMINE — CBr_2 AND CBr_3

" — CBr (ALIPH.)

SULFUR OXYGEN COMPOUNDS

IONIC SULFATE $(SO_4)^{--}$ — S

IONIC SULFONATE R—SO_3 — S — M

SULFONIC ACID R—SO_3H — M — S — M

COVALENT SULFATE R—O—SO_2—O—R — S — S

COVALENT SULFONATE R—O—SO_2—R — S — S

SULFONAMIDE R—SO_2—NH_2 — S — S

SULFONE R—SO_2—R — S — S

SULFOXIDE R—SO—R — S

PHOSPHORUS OXYGEN

IONIC PHOSPHATE (PO_4)

COVALENT PHOSPHATE ≡(R—O)$_3$P→O — CH_3—O—P→O — S — S

" " — ⬡—O—P→O — S — S

CARBON OXYGEN

IONIC CARBONATE (CO_3)— — S

COVALENT CARBONATE O=C (O—R)$_2$ — S — S — S — S — S

IMINO CARBONATE HN=C (O—R)$_2$ — S — M — S — S — S — M

NITROGEN OXYGEN

IONIC NITRATE (NO_3) — M — S — M

COVALENT NITRATE R—O—NO_2

NITRO R—NO_2 UNCONJ. — S — S

" CONJ. — S — S

COVALENT NITRITE R—O—NO

NITROSO R—NO

AMMONIUM $(NH_4)^+$ — M — M

ASSIGNMENTS

OH AND NH STR. — C=O STR. — C—O STR.

CH STR. — C=N STR. — C—N STR.

—C≡X STR. — C=C STR. — C—C STR.

NH BEND — CH ROCK

CH BEND — NH ROCK

OH BEND

2 3 4 5 6 7 8 9 10 11 12 13 14 15 16

Wavelength in μm

LEGEND

S = Strong

M = Medium

SP = Sharp

V = Vary

W = Weak

Based on work done by Colthup.

INFRARED CORRELATION CHARTS (continued)

Far Infrared Region

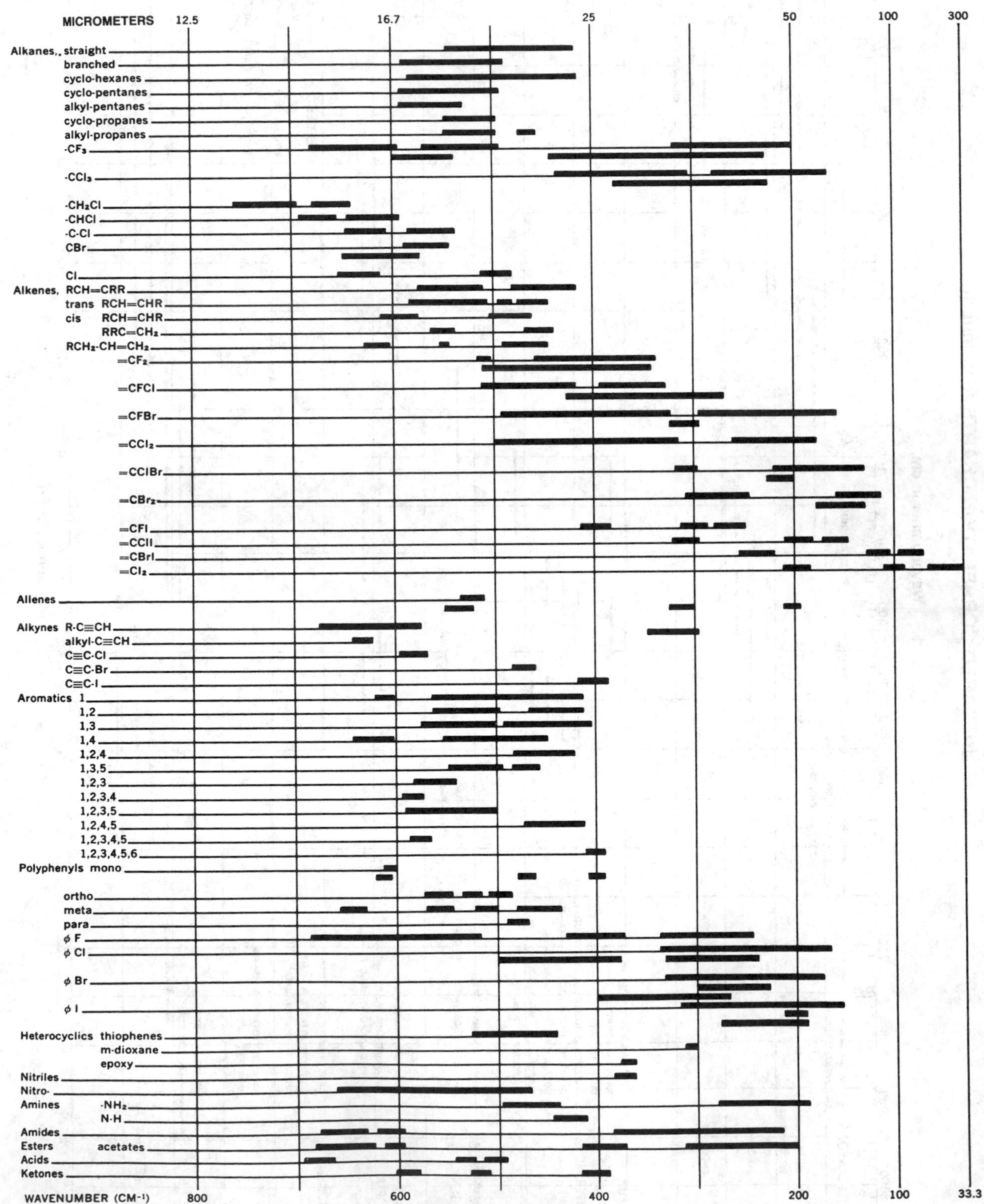

INFRARED CORRELATION CHARTS (continued)

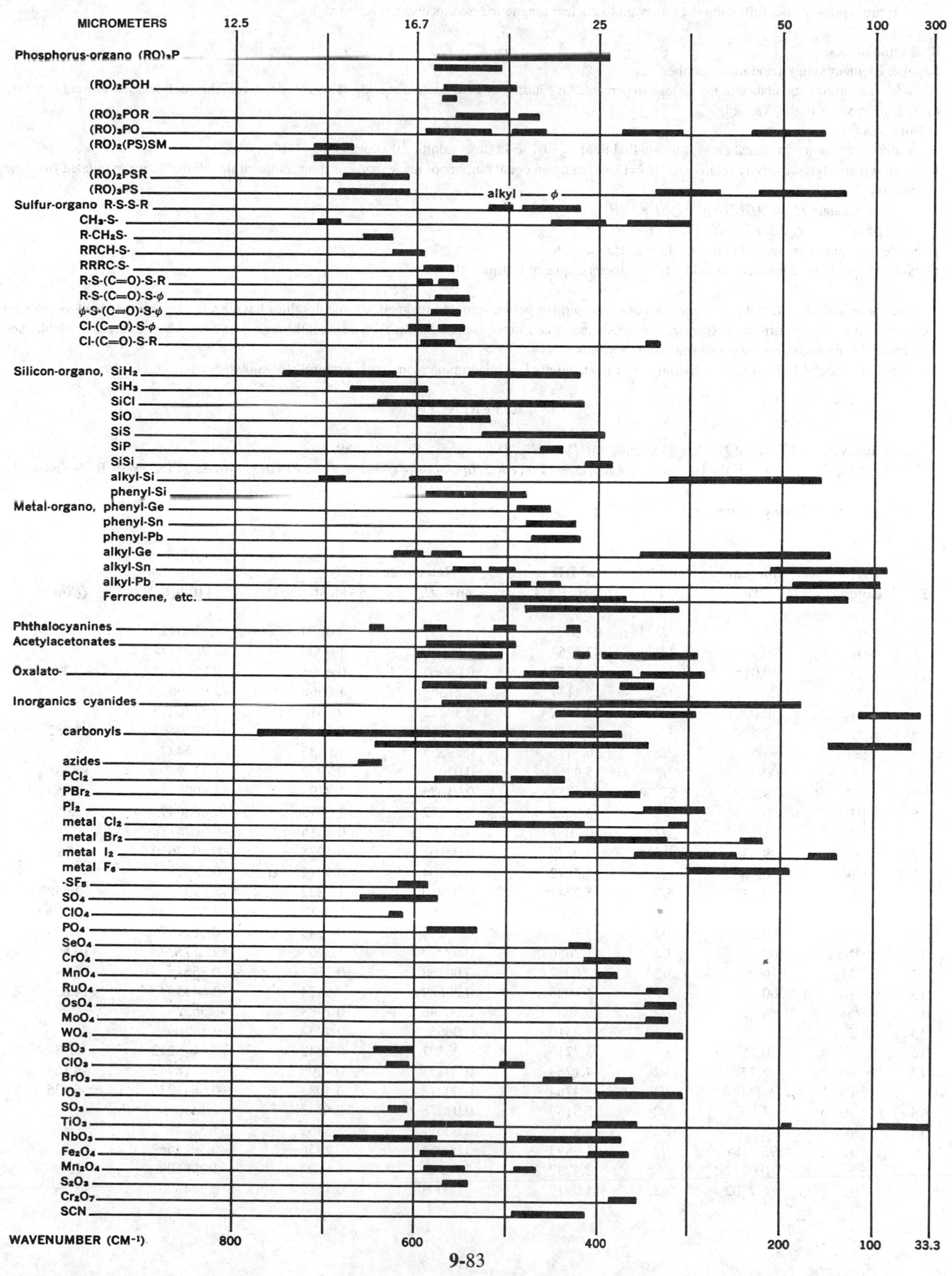

NUCLEAR SPINS, MOMENTS, AND OTHER DATA RELATED TO NMR SPECTROSCOPY

This table presents the following data relevant to nuclear magnetic resonance spectroscopy:

Z: Atomic number
Isotope: Element symbol and mass number
Abundance: Natural abundance of the isotope in percent. An * indicates a radioactive nuclide; if no value is given, the nuclide is not present in nature or its abundance is highly variable.
I: Nuclear spin
ν: Resonant frequency in megahertz for an applied field H_o of 1 tesla (in cgs units, 10 kilogauss)
Relative sensitivity: Sensitivity relative to ^{1}H (=1) assuming an equal number of nuclei and constant temperature. Values were calculated from the expressions:

For constant H_o: $0.0076508(\mu/\mu_N)^3(I + 1)/I^2$

For constant ν: $0.23871(\mu/\mu_N)\,(I + 1)$

μ: Nuclear magnetic moment in units of the nuclear magneton μ_N
Q: Nuclear quadrupole moment in units of femtometers squared (1 fm^2 = 10^{-2} barn)

The table includes all stable nuclides of non-zero spin for which spin and magnetic moment values have been measured, as well as selected radioactive nuclides of current or potential interest. At least one isotope is included for each element through $Z = 95$ for which data are available. See Reference 2 for a complete listing of spins and moments.

The assistance of P. Pyykkö in providing recent data on nuclear quadrupole moments is gratefully acknowledged.

REFERENCES

1. Raghavan, P., "Table of Nuclear Moments", *At. Data Nuc. Data Tables,* 42, 189, 1989.
2. Holden, N. E., "Table of the Isotopes", in Lide, D. R., Ed., *CRC Handbook of Chemistry and Physics,* 74th Ed., CRC Press, Boca Raton, FL, 1993.
3. Pyykkö, P., *Z. Naturforsch*., 47a, 189, 1992.

Z	Isotope	Abundance %	I	ν/MHz for H_o = 1 T	Relative sensitivity Const. H_o	Const. ν	μ/μ_N	Q/fm^2
1	^{1}n	*	1/2	29.1639	0.32139	0.6850	-1.91304275	
1	^{1}H	99.985	1/2	42.5764	1.00000	1.0000	+2.7928474	
1	^{2}H	0.015	1	6.53573	0.00965	0.4093	+0.8574382	+0.2860
1	^{3}H	*	1/2	45.4137	1.21354	1.0666	+2.9789625	
2	^{3}He	0.0001	1/2	32.4352	0.44212	0.7618	-2.1276248	
3	^{6}Li	7.5	1	6.2660	0.00850	0.3925	+0.8220467	-0.082
3	^{7}Li	92.5	3/2	16.5478	0.29355	1.9433	+3.256427	-4.01
4	^{9}Be	100	3/2	5.986	0.01389	0.7029	-1.1779	+5.288
5	^{10}B	19.9	3	4.5751	0.01985	1.7193	+1.800645	+8.459
5	^{11}B	80.1	3/2	13.6626	0.16522	1.6045	+2.688649	+4.059
6	^{13}C	1.10	1/2	10.7081	0.01591	0.2515	+0.7024118	
7	^{14}N	99.634	1	3.0776	0.00101	0.1928	+0.4037610	+2.02
7	^{15}N	0.366	1/2	4.3172	0.00104	0.1014	-0.2831888	
8	^{17}O	0.038	5/2	5.7741	0.02910	1.5822	-1.89379	-2.558
9	^{19}F	100	1/2	40.0765	0.83400	0.9413	+2.628868	
10	^{21}Ne	0.27	3/2	3.3630	0.00246	0.3949	-0.661797	+10.155
11	^{23}Na	100	3/2	11.2686	0.09270	1.3233	+2.217522	+10.89
12	^{25}Mg	10.00	5/2	2.6082	0.00268	0.7147	-0.85545	+19.94
13	^{27}Al	100	5/2	11.1028	0.20689	3.0424	+3.641507	+14.03
14	^{29}Si	4.67	1/2	8.4653	0.00786	0.1988	-0.55529	
15	^{31}P	100	1/2	17.2510	0.06652	0.4052	+1.13160	
16	^{33}S	0.75	3/2	3.2716	0.00227	0.3842	+0.6438212	-6.78
17	^{35}Cl	75.77	3/2	4.1764	0.00472	0.4905	+0.8218743	-8.165
17	^{37}Cl	24.23	3/2	3.4764	0.00272	0.4083	+0.6841236	-6.435
18	^{37}Ar	*	3/2	5.818	0.01276	0.6833	+1.145	
18	^{39}Ar	*	7/2	2.8	0.00617	1.3964	-1.3	
19	^{39}K	93.2581	3/2	1.9893	0.00051	0.2336	+0.3914662	+6.01
19	^{40}K	0.0117	4	2.4737	0.00523	1.5493	-1.298100	-7.49
19	^{41}K	6.7302	3/2	1.0919	0.00008	0.1282	+0.2148701	+7.33

Z	Isotope	Abundance %	I	ν/MHz for H_o = 1 T	Relative sensitivity Const. H_o	Const. ν	μ/μ$_N$	*Q*/fm^2
20	^{43}Ca	0.135	7/2	2.8688	0.00642	1.4150	-1.31726	-4.08
21	^{45}Sc	100	7/2	10.3588	0.30244	5.1093	+4.756487	-22
22	^{47}Ti	7.3	5/2	2.4040	0.00210	0.6587	-0.78848	+29
22	^{49}Ti	5.5	7/2	2.4047	0.00378	1.1861	-1.10417	+24
23	^{50}V	0.250	6	4.2504	0.05571	5.5904	+3.345689	+21
23	^{51}V	99.750	7/2	11.2130	0.38360	5.5306	+5.1487057	-5.2
24	^{53}Cr	9.501	3/2	2.4114	0.00091	0.2832	-0.47454	-15
25	^{55}Mn	100	5/2	10.5760	0.17881	2.8980	+3.46872	+33
26	^{57}Fe	2.1	1/2	1.3815	0.00003	0.0324	+0.0906230	
27	^{59}Co	100	7/2	10.077	0.27841	4.9702	+4.627	+42
28	^{61}Ni	1.140	3/2	3.8113	0.00359	0.4476	-0.75002	+16.2
29	^{63}Cu	69.17	3/2	11.2979	0.09342	1.3268	+2.22329	-22.0
29	^{65}Cu	30.83	3/2	12.1027	0.11484	1.4213	+2.38167	-20.4
30	^{67}Zn	4.1	5/2	2.6693	0.00287	0.7314	+0.875479	+15.0
31	^{69}Ga	60.108	3/2	10.2475	0.06971	1.2034	+2.01659	+17.0
31	^{71}Ga	39.892	3/2	13.0204	0.14300	1.5291	+2.56227	+10.0
32	^{73}Ge	7.73	9/2	1.4897	0.00141	1.1546	-0.8794677	-17.3
33	^{75}As	100	3/2	7.3148	0.02536	0.8590	+1.439475	+31.4
34	^{77}Se	7.63	1/2	8.1566	0.00703	0.1916	+0.535042	
35	^{79}Br	50.69	3/2	10.7039	0.07945	1.2570	+2.106400	+33.1
35	^{81}Br	49.31	3/2	11.5381	0.09951	1.3550	+2.270562	+27.6
36	^{83}Kr	11.5	9/2	1.6442	0.00190	1.2744	0.970669	+25.3
37	^{85}Rb	72.165	5/2	4.1253	0.01061	1.1304	+1.35303	+27.4
37	^{87}Rb	27.835	3/2	13.9807	0.17703	1.6418	+2.75124	+13.2
38	^{87}Sr	7.00	9/2	1.8524	0.00272	1.4358	-1.093603	+33.5
39	^{89}Y	100	1/2	2.0949	0.00012	0.0492	-0.1374154	
40	^{91}Zr	11.22	5/2	3.9747	0.00949	1.0891	-1.30362	-20.6
41	^{93}Nb	100	9/2	10.4520	0.48821	8.1011	+6.1705	-32
42	^{95}Mo	15.92	5/2	2.7874	0.00327	0.7638	-0.9142	-2.2
42	^{97}Mo	9.55	5/2	2.8462	0.00349	0.7799	-0.9335	+25.5
43	^{99}Tc	*	9/2	9.63	0.38174	7.4633	+5.6847	-12.9
44	^{99}Ru	12.7	5/2	1.9553	0.00113	0.5358	-0.6413	+7.9
44	^{101}Ru	17.0	5/2	2.192	0.00159	0.6005	-0.7188	+45.7
45	^{103}Rh	100	1/2	1.3476	0.00003	0.0317	-0.08840	
46	^{105}Pd	22.33	5/2	1.957	0.00113	0.5364	-0.642	+66.0
47	^{107}Ag	51.839	1/2	1.7330	0.00007	0.0407	-0.1136796	
47	^{109}Ag	48.161	1/2	1.9924	0.00010	0.0468	-0.1306906	
48	^{111}Cd	12.80	1/2	9.0689	0.00966	0.2130	-0.5948861	
48	^{113}Cd	12.22	1/2	9.4868	0.01106	0.2228	-0.6223009	
49	^{113}In	4.3	9/2	9.3652	0.35121	7.2588	+5.5289	+79.9
49	^{115}In	95.7	9/2	9.3854	0.35348	7.2744	+5.5408	+81
50	^{115}Sn	0.34	1/2	14.0074	0.03561	0.3290	-0.91883	
50	^{117}Sn	7.68	1/2	15.2606	0.04605	0.3584	-1.00104	
50	^{119}Sn	8.59	1/2	15.9656	0.05273	0.3750	-1.04728	
51	^{121}Sb	57.36	5/2	10.2549	0.16302	2.8100	+3.3634	-36
51	^{123}Sb	42.64	7/2	5.5530	0.04659	2.7389	+2.5498	-49
52	^{123}Te	0.908	1/2	11.2346	0.01837	0.2639	-0.7369478	
52	^{125}Te	7.139	1/2	13.5451	0.03220	0.3181	-0.8885051	
53	^{127}I	100	5/2	8.5776	0.09540	2.3504	+2.813273	-78.9
54	^{129}Xe	26.4	1/2	11.8601	0.02162	0.2786	-0.7779763	
54	^{131}Xe	21.2	3/2	3.5158	0.00282	0.4129	+0.6918619	-12.0
55	^{133}Cs	100	7/2	5.6232	0.04838	2.7735	+2.582025	-0.37
56	^{135}Ba	6.592	3/2	4.2581	0.00500	0.5001	+0.837943	+16.0
56	^{137}Ba	11.23	3/2	4.7633	0.00700	0.5594	+0.937365	+24.5
57	^{138}La	* 0.0902	5	5.6614	0.09404	5.3188	+3.713646	+45
57	^{139}La	99.9098	7/2	6.0610	0.06058	2.9895	+2.7830455	+20
58	^{137}Ce	*	3/2	4.62	0.00641	0.5431	0.91	

Z	Isotope	Abundance %	I	ν/MHz for H_o = 1 T	Relative sensitivity Const. H_o	Const. ν	μ/μ$_N$	Q/fm^2
58	^{139}Ce	*	3/2	4.62	0.00641	0.5431	0.91	
58	^{141}Ce	*	7/2	2.37	0.00364	1.1708	1.09	
59	^{141}Pr	100	5/2	13.0355	0.33483	3.5720	+4.2754	-5.9
60	^{143}Nd	12.18	7/2	2.319	0.00339	1.1440	-1.065	-63
60	^{145}Nd	8.30	7/2	1.429	0.00079	0.7047	-0.656	-33
61	^{143}Pm	*	5/2	11.6	0.23510	3.1748	3.8	
61	^{147}Pm	*	7/2	5.7	0.04940	2.7928	+2.6	+70
62	^{147}Sm	15.0	7/2	1.7747	0.00152	0.8753	-0.8149	-26
62	^{149}Sm	13.8	7/2	1.4631	0.00085	0.7216	-0.6718	+9.4
63	^{151}Eu	47.8	5/2	10.5854	0.17929	2.9006	+3.4718	+90.3
63	^{153}Eu	52.2	5/2	4.6744	0.01544	1.2809	+1.5331	+241
64	^{155}Gd	14.80	3/2	1.317	0.00015	0.1546	-0.2591	+127
64	^{157}Gd	15.65	3/2	1.727	0.00033	0.2028	-0.3399	+135
65	^{159}Tb	100	3/2	10.23	0.06945	1.2019	+2.014	+143.2
66	^{161}Dy	18.9	5/2	1.4653	0.00048	0.4015	-0.4806	+247
66	^{163}Dy	24.9	5/2	2.0507	0.00130	0.5619	+0.6726	+265
67	^{165}Ho	100	7/2	9.0881	0.20423	4.4825	+4.173	+358
68	^{167}Er	22.95	7/2	1.2281	0.00050	0.6057	-0.5639	+357
69	^{169}Tm	100	1/2	3.531	0.00057	0.0829	-0.2316	
70	^{171}Yb	14.3	1/2	7.5259	0.00552	0.1768	+0.49367	
70	^{173}Yb	16.12	5/2	2.0730	0.00135	0.5680	-0.67989	+280
71	^{175}Lu	97.41	7/2	4.8624	0.03128	2.3983	+2.2327	+349
71	^{176}Lu	* 2.59	7	3.451	0.03975	6.0516	+3.169	+497
72	^{177}Hf	18.606	7/2	1.7281	0.00140	0.8524	+0.7935	+336
72	^{179}Hf	13.629	9/2	1.0856	0.00055	0.8414	-0.6409	+379
73	^{180}Ta	0.012	9	4.04	0.10251	11.3862	4.77	
73	^{181}Ta	99.988	7/2	5.1625	0.03744	2.5463	+2.3705	+317
74	^{183}W	14.3	1/2	1.7956	0.00008	0.0422	+0.1177847	
75	^{185}Re	37.40	5/2	9.717	0.13870	2.6627	+3.1871	+218
75	^{187}Re	*62.60	5/2	9.817	0.14300	2.6900	+3.2197	+207
76	^{187}Os	1.6	1/2	0.9856	0.00001	0.0231	+0.06465189	
76	^{189}Os	16.1	3/2	3.3535	0.00244	0.3938	+0.659933	+85.6
77	^{191}Ir	37.3	3/2	0.766	0.00003	0.0899	+0.1507	+81.6
77	^{193}Ir	62.7	3/2	0.832	0.00004	0.0977	+0.1637	+75.1
78	^{195}Pt	33.8	1/2	9.2920	0.01039	0.2182	+0.60952	
79	^{197}Au	100	3/2	0.7406	0.00003	0.0870	+0.145746	+54.7
80	^{199}Hg	16.87	1/2	7.7121	0.00594	0.1811	+0.5058855	
80	^{201}Hg	13.18	3/2	2.8468	0.00149	0.3343	-0.5602257	+38.6
81	^{203}Tl	29.524	1/2	24.7310	0.19598	0.5809	+1.6222579	
81	^{205}Tl	70.476	1/2	24.9742	0.20182	0.5866	+1.6382146	
82	^{207}Pb	22.1	1/2	9.0338	0.00955	0.2122	+0.59258	
83	^{209}Bi	100	9/2	6.9628	0.14433	5.3967	+4.1106	-50
84	^{209}Po	*	1/2	11.7	0.02096	0.2757	+0.77	
86	^{211}Rn	*	1/2	9.16	0.00997	0.2152	+0.601	
87	^{223}Fr	*	3/2	5.94	0.01362	0.6982	+1.17	+117
88	^{223}Ra	*	3/2	1.3746	0.00017	0.1614	+0.2705	+119
88	^{225}Ra	*	1/2	11.187	0.01814	0.2627	-0.7338	
89	^{227}Ac	*	3/2	5.6	0.01131	0.6564	+1.1	+170
90	^{229}Th	*	5/2	1.40	0.00042	0.3843	+0.46	+430
91	^{231}Pa	*	3/2	10.2	0.06903	1.1995	2.01	-172
92	^{235}U	* 0.7200	7/2	0.83	0.00015	0.4082	-0.38	+493.6
93	^{237}Np	*	5/2	9.57	0.13264	2.6234	+3.14	+388.6
94	^{239}Pu	*	1/2	3.09	0.00038	0.0727	+0.203	
95	^{243}Am	*	5/2	4.91	0.01788	1.3451	+1.61	+421

CHARACTERISTIC NMR SPECTRAL POSITIONS FOR HYDROGEN IN ORGANIC STRUCTURES

By permission from Erno Mohacsi, J. of Chemical Education, 41, 38 (1964)

This table is useful for quick qualitative determination of proton spectrum lines by providing a tabulation of line positions obtained using tetramethylsilane as an internal reference. The listing has been kept as simple as possible for this purpose. The proton spectrum lines are arranged according to the chemical shift relative to tetramethylsilane and are given in values of τ and δ. The purpose of this table is to supplement tables available in standard references and to summarize information available in the literature.

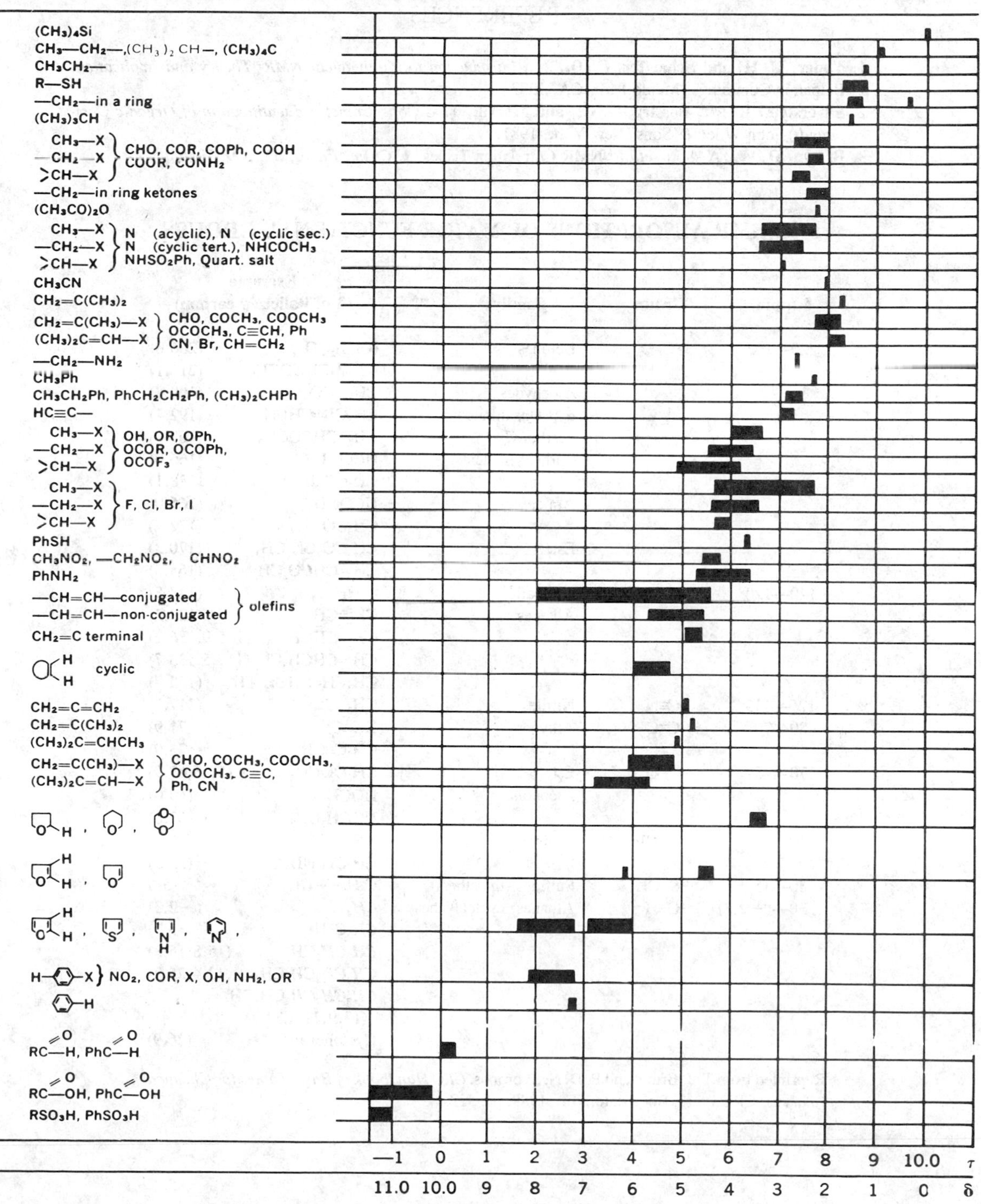

^{13}C-NMR ABSORPTIONS OF MAJOR FUNCTIONAL GROUPS

The table below lists the ^{13}C chemical shifts in parts per million (ppm), in descending order with the corresponding functional groups. A series of typical simple compounds for every family is added to illustrate the correlations above. The shifts for the carbons of interest are given in parentheses — either for each carbon as it appears from left to right in the formula, or by italics.[1-3]

REFERENCES

1. **Yoder, C. H. and Schaeffer, C. D., Jr.,** *Introduction to Multinuclear NMR: Theory and Application,* Benjamin/Cummings, Menlo Park, CA, 1987.
2. **Silverstein, R. M., Bassler, G. C., and Morrill, T. C.,** *Spectrometric Identification of Organic Compounds,* John Wiley & Sons, New York, 1981.
3. **Brown, D. W.,** A short set of ^{13}NMR Correlation Tables, *J. Chem. Educ.*, 62, 209, 1985.

^{13}C-NMR ABSORPTIONS OF MAJOR FUNCTIONAL GROUPS

δ (ppm)	Group	Family	Example (δ of italicized carbon)	
220—165	$>C=O$	Ketones	$(CH_3)_2CO$	(206.0)
			$(CH_3)_2CHCOCH_3$	(212.1)
		Aldehydes	CH_3CHO	(199.7)
		α,β-Unsaturated carbonyls	$CH_3CH{=}CHCHO$	(192.4)
			$CH_2{=}CHCOCH_3$	(169.9)
		Carboxylic acids	HCO_2H	(166.0)
			CH_3CO_2H	(178.1)
		Amides	$HCONH_2$	(165.0)
			CH_3CONH_2	(172.7)
		Esters	$CH_3CO_2CH_2CH_3$	(170.3)
			$CH_2{=}CHCO_2CH_3$	(165.5)
140—120	$>C=C<$	Aromatic	C_6H_6	(128.5)
		Alkenes	$CH_2{=}CH_2$	(123.2)
			$CH_2{=}CHCH_3$	(115.9,136.2)
			$CH_2{=}CHCH_2Cl$	(117.5.133.7)
			$CH_3CH{=}CHCH_2CH_3$	(132.7)
125—115	$-C\equiv N$	Nitriles	$CH_3-C\equiv N$	(117.7)
80—70	$-C\equiv C-$	Alkynes	$HC\equiv CH$	(71.9)
			$CH_3C\equiv CH_3$	(73.9)
70—45	$-C-O$	Esters	$CH_3OOCH_2CH_3$	(57.6,67.9)
		Alcohols	$HOCH_3$	(49.0)
			$HOCH_2CH_3$	(57.0)
40—20	$-C-NH_2$	Amines	CH_3NH_2	(26.9)
			$CH_3CH_2NH_2$	(35.9)
30—15	$-S-CH_3$	Sulfides (thioethers)	$C_6H_5-S-CH_3$	15.6
30—(−2.3)	$-C-H$	Alkanes, cycloalkanes	CH_4	(−2.3)
			CH_3CH_3	(5.7)
			$CH_3CH_2CH_3$	(15.8,16.3)
			$CH_3CH_2CH_2CH_3$	(13.4,25.2)
			$CH_3CH_2CH_2CH_2CH_3$ (13.9,22.8,34.7)	
			Cyclohexane	(26.9)

Reprinted from T. J. Bruno and P. D. N. Svoronos, *CRC Handbook of Basic Tables for Chemical Analysis*, CRC Press, Boca Raton, FL, 1989, pg. 329.

ULTRAVIOLET SPECTRA OF COMMON LIQUIDS

The following tables present the UV spectra of some common solvents and other liquids. The data were obtained, using a 1.00-cm path-length cell, against a water reference.[1]

REFERENCES

1. **Krieger, P. A.,** *High Purity Solvent Guide,* Burdick and Jackson, McGaw Park, IL, 1984.

ACETONE

Wavelength (nm)	Max absorbance
330	1.000
340	0.060
350	0.010
375	0.005
400	0.005

BENZENE

Wavelength (nm)	Max absorbance
278	1.000
300	0.020
325	0.010
350	0.005
400	0.005

ACETONITRILE

Wavelength (nm)	Max absorbance
190	1.000
200	0.050
225	0.010
250	0.005
350	0.005

1-BUTANOL

Wavelength (nm)	Max absorbance
215	1.000
225	0.500
250	0.040
275	0.010
300	0.005

2-BUTANOL

Wavelength (nm)	Max absorbance
260	1.000
275	0.300
300	0.010
350	0.005
400	0.005

CARBON TETRACHLORIDE

Wavelength (nm)	Max absorbance
263	1.000
275	0.100
300	0.005
350	0.005
400	0.005

n-BUTYL ACETATE

Wavelength (nm)	Max absorbance
254	1.000
275	0.050
300	0.010
350	0.005
400	0.005

CHLOROBENZENE

Wavelength (nm)	Max absorbance
287	1.000
300	0.050
325	0.040
350	0.020
400	0.005

Reprinted from T. J. Bruno and P. D. N. Svoronos, *CRC Handbook of Basic Tables for Chemical Analysis*, CRC Press, Boca Raton, FL, 1989, pg.213—218..

n-BUTYL CHLORIDE

Wavelength (nm)	Max absorbance
220	1.000
225	0.300
250	0.010
300	0.005
400	0.005

CHLOROFORM

Wavelength (nm)	Max absorbance
245	1.000
250	0.300
275	0.005
300	0.005
400	0.005

CYCLOHEXANE

Wavelength (nm)	Max absorbance
200	1.000
225	0.170
250	0.020
300	0.005
400	0.005

o-DICHLOROBENZENE

Wavelength (nm)	Max absorbance
295	1.000
300	0.300
325	0.100
350	0.050
400	0.005

CYCLOPENTANE

Wavelength (nm)	Max absorbance
200	1.000
215	0.300
225	0.020
300	0.005
400	0.005

DIETHYL CARBONATE

Wavelength (nm)	Max absorbance
256	1.000
265	0.150
275	0.050
300	0.040
400	0.010

DECAHYDRONAPHTHALENE

Wavelength (nm)	Max absorbance
200	1.000
225	0.500
250	0.050
300	0.005
400	0.005

DIMETHYL ACETAMIDE

Wavelength (nm)	Max absorbance
268	1.000
275	0.300
300	0.080
350	0.005
400	0.005

DIMETHYL FORMAMIDE

Wavelength (nm)	Max absorbance
268	1.000
275	0.300
300	0.050
350	0.005
400	0.005

2-ETHOXYETHANOL

Wavelength (nm)	Max absorbance
210	1.000
225	0.500
250	0.200
300	0.005
400	0.005

DIMETHYL SULFOXIDE

Wavelength (nm)	Max absorbance
268	1.000
275	0.500
300	0.200
350	0.020
400	0.005

ETHYL ACETATE

Wavelength (nm)	Max absorbance
256	1.000
275	0.050
300	0.030
325	0.005
350	0.005

1,4-DIOXANE

Wavelength (nm)	Max absorbance
215	1.000
250	0.300
300	0.020
350	0.005
400	0.005

DIETHYL ETHER

Wavelength (nm)	Max absorbance
215	1.000
250	0.080
275	0.010
300	0.005
400	0.005

ETHYLENE DICHLORIDE

Wavelength (nm)	Max absorbance
228	1.000
240	0.300
250	0.100
300	0.005
400	0.005

HEXADECANE

Wavelength (nm)	Max absorbance
190	1.000
200	0.500
250	0.020
300	0.005
400	0.005

ETHYLENE GLYCOL DIMETHYL ETHER (GLYME)

Wavelength (nm)	Max absorbance
220	1.000
250	0.250
300	0.050
350	0.010
400	0.005

HEXANE

Wavelength (nm)	Max absorbance
195	1.000
225	0.050
250	0.010
275	0.005
300	0.005

HEPTANE

Wavelength (nm)	Max absorbance
200	1.000
225	0.100
250	0.010
300	0.005
400	0.005

ISOBUTANOL

Wavelength (nm)	Max absorbance
220	1.000
250	0.050
275	0.030
300	0.020
400	0.010

METHANOL

Wavelength (nm)	Max absorbance
205	1.000
225	0.160
250	0.020
300	0.005
400	0.005

METHYL *t*-BUTYL ETHER

Wavelength (nm)	Max absorbance
210	1.000
225	0.500
250	0.100
300	0.005
400	0.005

2-METHOXYETHANOL

Wavelength (nm)	Max absorbance
210	1.000
250	0.130
275	0.030
300	0.005
400	0.005

METHYLENE CHLORIDE

Wavelength (nm)	Max absorbance
233	1.000
240	0.100
250	0.010
300	0.005
400	0.005

2-METHOXYETHYL ACETATE

Wavelength (nm)	Max absorbance
254	1.000
275	0.150
300	0.050
350	0.005
400	0.005

METHYL ETHYL KETONE

Wavelength (nm)	Max absorbance
329	1.000
340	0.100
350	0.020
375	0.010
400	0.005

METHYL ISOAMYL KETONE

Wavelength (nm)	Max absorbance
330	1.000
340	0.100
350	0.050
375	0.010
400	0.005

n-METHYLPYRROLIDONE

Wavelength (nm)	Max absorbance
285	1.000
300	0.500
325	0.100
350	0.030
400	0.010

METHYL ISOBUTYL KETONE

Wavelength (nm)	Max absorbance
334	1.000
340	0.500
350	0.250
375	0.050
400	0.005

PENTANE

Wavelength (nm)	Max absorbance
190	1.000
200	0.600
250	0.010
300	0.005
400	0.005

METHYL *n*-PROPYL KETONE

Wavelength (nm)	Max absorbance
331	1.000
340	0.150
350	0.020
375	0.005
400	0.005

β-PHENETHYLAMINE

Wavelength (nm)	Max absorbance
285	1.000
300	0.300
325	0.100
350	0.050
400	0.005

1-PROPANOL

Wavelength (nm)	Max absorbance
210	1.000
225	0.500
250	0.050
300	0.005
400	0.005

PYRIDINE

Wavelength (nm)	Max absorbance
330	1.000
340	0.100
350	0.010
375	0.010
400	0.005

2-PROPANOL

Wavelength (nm)	Max absorbance
205	1.000
225	0.160
250	0.020
300	0.005
400	0.010

TETRAHYDROFURAN

Wavelength (nm)	Max absorbance
212	1.000
250	0.180
300	0.020
350	0.005
400	0.005

PROPYLENE CARBONATE

Wavelength (nm)	Max absorbance
280	1.000
300	0.500
350	0.050
375	0.030
400	0.020

TOLUENE

Wavelength (nm)	Max absorbance
284	1.000
300	0.120
325	0.020
350	0.050
400	0.005

1,2,4-TRICHLOROBENZENE

Wavelength (nm)	Max absorbance
308	1.000
310	0.500
350	0.050
375	0.010
400	0.005

2,2,4-TRIMETHYLPENTANE

Wavelength (nm)	Max absorbance
215	1.000
225	0.100
250	0.020
300	0.005
400	0.005

TRICHLOROETHYLENE

Wavelength (nm)	Max absorbance
273	1.000
300	0.100
325	0.080
350	0.060
400	0.060

WATER

Wavelength (nm)	Max absorbance
190	0.010
200	0.010
250	0.005
300	0.005
400	0.005

1,1,2-TRICHLOROTRIFLUOROETHANE

Wavelength (nm)	Max absorbance
231	1.000
250	0.050
300	0.005
350	0.005
400	0.005

o-XYLENE

Wavelength (nm)	Max absorbance
288	1.000
300	0.200
325	0.050
350	0.010
400	0.005

Section 10
Atomic, Molecular, and Optical Physics

LINE SPECTRA OF THE ELEMENTS

Edited by Joseph Reader and Charles H. Corliss*
National Bureau of Standards

These tables were prepared under the auspices of the Committee on Line Spectra of the Elements of the National Academy of Sciences—National Research Council. They contain the outstanding spectral lines of neutral (I), singly ionized (II), doubly ionized (III), triply ionized (IV), and quadruply ionized (V) atoms. Listed are lines that appear in emission from the vacuum ultraviolet to the far infrared. For most atoms these lines were selected from much larger lists in such a way as to include the stronger observed lines in each spectral region. In a few cases prominent monoxide band heads are also given.

The data were compiled by the following contributors, whose initials are given in the headings of the tables that they prepared:

- J. G. Conway—Lawrence Berkeley Laboratory
- C. H. Corliss—National Bureau of Standards
- R. D. Cowan—Los Alamos Scientific Laboratory
- C. R. Cowley—University of Michigan
- Henry M. and Hannah Crosswhite—Argonne National Laboratory
- S. P. Davis—University of California, Berkeley
- V. Kaufman—National Bureau of Standards
- R. L. Kelly—Naval Postgraduate School
- J. F. Kielkopf—University of Louisville
- W. C. Martin—National Bureau of Standards
- T. K. McCubbin—Pennsylvania State University
- L. J. Radziemski—Los Alamos Scientific Laboratory
- J. Reader—National Bureau of Standards
- C. J. Sansonetti—National Bureau of Standards
- G. V. Shalimoff—Lawrence Berkeley Laboratory
- R. W. Stanley—Purdue University
- J. O. Stoner, Jr.—University of Arizona
- H. H. Stroke—New York University
- D. R. Wood—Wright State University
- E. F. Worden—Lawrence Livermore Laboratory
- J. J. Wynne—International Business Machines Corporation
- R. Zalubas—National Bureau of Standards

The literature references are collected at the end of the entire set of tables.

All wavelengths are given in Angstroms. Below 2000 Å the wavelengths are in vacuum; above 2000 Å the wavelengths are in air. Wavelengths given to three decimal places have an uncertainty of less than 0.001 Å and are therefore suitable for the calibration of most spectrographs. In the air region, the elements used most commonly for calibration purposes are Ne, Ar, Kr, Fe, Th, and Hg; in the vacuum region, the most common are C, N, O, Si, Cu.

A large number of the lines for neutral and singly ionized atoms were extracted from the National Bureau of Standards (NBS) Tables of Spectral-Line Intensities.[1] The intensities of these lines represent quantitative estimates of relative line strengths that take account of varying detection sensitivity at different wavelengths. They are on a linear scale. For nearly all of the other lines the intensities represent qualitative estimates of the relative strengths of lines not greatly separated in wavelength. Because different observers frequently use different scales for their intensity estimates, these intensities are useful only as a rough indication of the appearance of a spectrum. In some cases the intensity scale is not intended to be linear. In the tables of first and second spectra the intensities of the lines of the singly ionized atom relative to those of the neutral atom should be used with caution, inasmuch as the concentration of ions in a light source depends greatly on the excitation conditions.

Descriptive symbols used in the tables have the following meanings:

- c—complex
- d—line consists of two unresolved lines
- h—hazy
- l—shaded to longer wavelengths
- s—shaded to shorter wavelengths
- p—perturbed by a close line
- b—band head
- r—easily reversed
- w—wide

ACTINIUM (Ac)
Z = 89

Ac I and II
Ref. 193 — J.G.C.

Intensity		Wavelength	
		Air	
8	h	2100.00	II
20		2712.50	II
10		2726.23	II
10	h	2760.18	II
10	h	2781.56	II
20		2797.59	II
20		2806.76	II
8		2833.47	II
150	h	2847.16	II
8		2895.20	II
30		2896.82	II
30		2923.02	II
200		2994.17	II
500		3043.30	II
200		3069.36	II
100		3078.07	II
100		3086.04	II
100		3087.37	II
200		3112.83	II
100		3120.16	II
500	s	3153.09	II
600	s	3154.41	II
200	s	3164.81	II
300	s	3230.59	II
150	s	3237.70	II
500		3260.91	II
100	s	3318.01	II
200	s	3383.53	II
200		3413.84	II
500		3417.77	II
500	s	3481.16	II
200		3489.53	II
100		3529.24	II
100		3534.63	II
200	s	3554.99	II
1000	s	3565.59	II
100		3694.88	II
300	s	3756.67	II
200		3799.82	II
2000	s	3863.12	II
100		3914.47	II
400	s	4061.60	II
3000	s	4088.44	II
3000	s	4168.40	II
100		4179.98	I
20		4183.12	I
20		4194.40	I
300	s	4209.69	II
300		4359.13	II
20	l	4384.53	I
1000	l	4386.41	II
20		4396.71	I
20		4462.73	I
1000	l	4507.20	II
500		4605.45	II
10		4716.58	I
400	s	4720.16	II
300		4812.22	II
100		4945.18	II
100		4958.23	II
100		4960.87	II
150		5446.38	II
300	l	5732.05	II
400		5758.97	II
1000		5910.85	II
600	l	6164.75	II
200	l	6167.83	II
400		6242.83	II
20		6359.86	I
20	l	6691.27	I
6		7290.40	I
6	l	7866.10	I

Ac III
Ref. 193 — J.G.C.

Intensity		Wavelength	
		Air	
1000	h	2626.44	III
50	h	2682.90	III
2000	h	2952.55	III
2000	h	3392.78	III
3000		3487.59	III
2000	h	4413.09	III
3000	h	4569.87	III
8	h	5193.21	III

Ac IV
Ref. 193 — J.G.C.

Intensity		Wavelength	
		Air	
20	h	2062.00	IV
30	h	2502.12	IV
100	h	2558.08	IV
5	h	2790.83	IV
50	h	2793.90	IV
20	l	3224.7	IV

ALUMINUM (Al)
Z = 13

Al I and II
Ref. 81, 89, 144, 227, 228, 282 — E.F.W.

Intensity	Wavelength	
	Vacuum	
40	1177.43	II
50	1191.812	II
150	1350.18	II
800	1539.830	II
100	1569.385	II
125	1596.059	II
150	1625.627	II
100	1644.235	II
100	1644.809	II
1000	1670.787	II
100	1686.250	II
800	1719.440	II
500	1721.244	II
900	1721.271	II
500	1724.952	II
900	1724.984	II
350	1760.104	II
300	1761.975	II
290	1763.00	I
500	1763.869	II
700	1763.952	II
450	1765.64	I

* Charles H. Corliss is now retired.

Aluminum (Cont.)

Intensity		Wavelength	
300		1765.815	II
450		1766.38	I
400		1767.731	II
450		1769.14	I
600		1828.588	II
400		1832.837	II
250		1834.808	II
300		1855.929	II
700		1858.026	II
120		1859.980	II
1000		1862.311	II
200		1929.978	II
150		1931.048	II
200		1932.377	II
400		1934.503	II
150		1934.713	II
150		1936.907	II
220		1939.261	II
700		1990.531	II
		Air	
150		2016.052	II
150		2016.234	II
100		2016.368	II
200		2074.008	II
700		2094.264	II
150		2094.744	II
300		2094.791	II
100		2095.104	II
200		2095.141	II
60		2150.70	I
60		2181.00	I
400		2269.10	I
120		2269.22	I
60		2312.49	I
70		2313.53	I
90		2317.48	I
60		2319.06	I
140		2321.56	I
460		2367.05	I
110		2367.61	I
110		2368.11	I
180		2369.30	I
140		2370.22	I
70		2370.73	I
160		2372.07	I
850		2373.12	I
170		2373.35	I
110		2373.57	I
60		2378.40	I
60		2513.30	I
240		2567.98	I
480		2575.10	I
60		2575.40	I
80		2631.55	II
110		2637.70	II
150		2652.48	I
200		2660.39	I
160		2669.17	II
650		2816.19	II
90		2837.96	I
90		2840.10	I
150		3041.28	II
360		3050.07	I
60		3054.68	I
450		3057.14	I
90		3064.29	I
60		3066.14	I
150		3074.64	II
4500	rS	3082.153	I
7200	r	3092.710	I
1800	r	3092.839	I
150		3428.92	II
70		3439.35	I
150		3443.64	I
70		3444.86	I
70		3458.22	I
60		3479.81	I
60		3482.63	I
450		3586.56	II
360		3587.07	II
290		3587.45	II
220		3651.06	II
110		3651.10	II
150		3654.98	II
290		3655.00	II
450		3900.68	II
60		3932.00	I
4500	r	3944.006	I
9000	r	3961.520	I
110		3995.86	II
290		4226.81	II
150		4585.82	II
110		4588.19	II
550		4666.80	II
110		4898.76	II
110		4902.77	II
150		5280.21	II
70		5107.52	I
290		5283.77	II
150		5285.85	II
110		5312.32	II
220		5316.07	II
150		5371.84	II
180		5557.06	I
110		5557.95	I
450		5593.23	II
110		5853.62	II
220		5971.94	II
290		6001.76	II
220		6001.88	II
450		6006.42	II
150		6061.11	II
290		6068.43	II
110		6068.53	II
450		6073.23	II
110		6181.57	II
150		6181.68	II
290		6182.28	II
220		6182.45	II
450	h	6183.42	II
450		6201.52	II
360		6201.70	II
290		6226.18	II
360		6231.78	II
450		6243.36	II
450		6335.74	II
360		6696.02	I
230		6698.67	I
60		7083.97	I
70		7084.64	I
110		7361.57	I
140		7362.30	I
60		7606.16	I
90		7614.82	I
230		7835.31	I
290		7836.13	I
60		7993.05	I
90		8003.19	I
70		8065.97	I
110		8075.35	I
290		8640.70	II
360		8772.87	I
450		8773.90	I
110		8828.91	I
180		8841.28	I
90		8912.90	I
140		8923.56	I
60		9089.91	I
70		9139.95	I
150		9290.65	II
110		9290.75	II
150		10076.29	II
110		10768.36	I
140		10782.04	I
110		10872.98	I
230		10891.73	I
450		11253.19	I
570		11254.88	I
570		13123.41	I
450		13150.76	I
230		16718.96	I
300		16750.56	I
140		16763.36	I
300		21093.04	I
360		21163.75	I

Al III
Ref. 127 — E.F.W.

Intensity		Wavelength	
		Vacuum	
70		486.884	III
30		486.912	III
250		511.138	III
150		511.191	III
500		560.317	III
200		560.433	III
100		670.068	III
200		671.118	III
500		695.829	III
400		696.217	III
200		725.683	III
300		726.915	III
400		855.034	III
500		856.746	III
400		892.024	III
50		893.887	III
450		893.897	III
10		1162.59	III
5		1162.62	III
100		1352.81	III
5		1352.82	III
70		1352.86	III
600		1379.67	III
800		1384.13	III
700		1605.766	III
100		1611.814	III
800		1611.874	III
1000		1854.716	III
600		1862.790	III
300		1935.840	III
15		1935.863	III
200		1935.949	III
		Air	
110		2399.00	III
285		2762.77	III
220		2762.87	III
450		2906.93	III
360		3348.52	III
290		3350.88	III
870		3601.63	III
550		3601.93	III
750		3612.36	III
450		3702.11	III
550		3713.12	III
110	h	3980.14	III
110		4082.45	III
150		4088.61	III
110	h	4142.37	III
650		4149.92	III
650		4150.17	III
110	h	4364.64	III
650		4479.89	III
650		4479.97	III
760		4512.56	III
550		4528.94	III
870		4529.19	III
110		4701.15	III
150		4701.41	III
110	h	4904.10	III
110	h	5151.01	III
110	h	5163.89	III
1200		5696.60	III
1000		5722.73	III
110		6055.21	III
220	h	7635.37	III
150		7660.26	III
220		7681.97	III
360		7881.79	III
150		7882.52	III
290		7905.51	III
290	h	8243.59	III
360	h	8275.11	III
290		9571.52	III
360		9605.99	III

Al IV
Ref. 8, 146 — E.F.W.

Intensity	Wavelength	
	Vacuum	
400	124.03	IV
700	129.73	IV
800	160.07	IV
700	161.69	IV
500	1027.34	IV
800	1042.17	IV
700	1048.52	IV
500	1058.90	IV
500	1061.43	IV
600	1064.89	IV
500	1066.57	IV
600	1069.44	IV
400	1105.74	IV
600	1118.82	IV
500	1125.61	IV
400	1136.82	IV
400	1198.50	IV
400	1220.55	IV
900	1237.19	IV
600	1240.21	IV
700	1240.86	IV
700	1248.79	IV
900	1257.62	IV
800	1264.18	IV
1000	1272.76	IV
400	1337.90	IV
500	1376.62	IV
500	1388.79	IV
600	1431.94	IV
700	1441.82	IV
800	1447.51	IV
600	1457.96	IV
700	1486.89	IV
800	1494.79	IV
400	1519.07	IV
800	1537.54	IV
500	1550.19	IV
1000	1557.25	IV
500	1559.03	IV
700	1564.16	IV
900	1582.04	IV
800	1584.46	IV
400	1589.28	IV
400	1606.65	IV
400	1617.81	IV
600	1627.54	IV
500	1636.82	IV
800	1639.06	IV
1000	1818.56	IV
700	1881.16	IV
	Air	
400	2515.87	IV
500	3208.20	IV
500	3267.21	IV
600	3285.13	IV
400	3344.46	IV
500	3473.54	IV
900	3492.23	IV
800	3508.46	IV
500	3511.28	IV
700	3517.56	IV
400	3527.03	IV
500	3541.08	IV

Al V
Ref. 6 — E.F.W.

Intensity	Wavelength	
	Vacuum	
300	103.80	V
400	103.88	V
250	104.07	V
250	104.18	V
600	107.95	V
300	108.06	V
300	108.11	V
250	118.50	V
900	125.53	V
800	126.07	V
800	130.41	V
1000	130.85	V
900	131.00	V
900	131.44	V
500	132.63	V
1000	278.69	V
900	281.39	V
250	1068.26	V
500	1088.67	V
300	1090.14	V
300	1150.30	V
350	1165.42	V
250	1168.48	V
500	1287.70	V
400	1330.06	V
400	1350.52	V
400	1363.35	V
600	1369.20	V
300	1373.70	V
400	1445.87	V
300	1455.26	V
600	1475.64	V
300	1486.05	V
700	1508.37	V
1000	1526.14	V
500	1539.12	V
300	1577.90	V
350	1589.87	V

AMERICIUM (Am)
Z = 95

Am I and II
Ref. 92 — J.G.C.

Intensity		Wavelength	
		Air	
100	s	2706.35	II
100	s	2728.69	II
200	s	2756.55	II
100	1	2812.10	II
200	s	2812.92	II
1000	1	2815.28	II
100	s	2815.98	II
100	1	2831.24	II
5000	s	2832.26	II
100	1	2833.95	II
100	1	2861.92	II
100	s	2866.20	II
1000	1	2888.51	II
100	1	2893.29	II
200	1	2899.56	II
100	1	2909.86	II
200	1	2911.13	II
1000	s	2920.59	II
200	1	2927.53	II
200	1	2936.99	II
100	1	2939.08	II
500	1	2950.39	II
100	s	2957.05	II
100	1	2958.39	II
100	1	2963.02	II
1000	1	2966.71	II
1000	1	2969.29	II
1000	s	2987.24	II
500	1	2993.51	II
1000	s	3004.25	II

Americium (Cont.)

Intensity		Wavelength	Spectrum
500	s	3027.99	II
100	l	3028.86	II
500	l	3038.36	II
200	s	3053.69	II
2000	s	3120.49	II
200	s	3161.83	II
100	s	3167.06	II
100	l	3203.26	II
500	s	3282.32	II
200	l	3286.67	II
100	l	3343.87	I
500	s	3362.55	II
200	l	3395.01	I
200	s	3419.66	II
200		3446.19	I
1000	l	3452.10	II
5000	l	3483.31	II
5000		3510.13	I
1000	l	3530.95	I
200	s	3562.68	II
5000		3569.16	I
100	s	3596.07	II
500		3603.41	I
5000		3673.12	I
100	l	3684.57	II
1000	s	3696.42	II
100	l	3707.86	II
5000	l	3777.50	II
5000	l	3926.25	II
1000	l	3952.58	II
100	s	4020.25	I
100	s	4035.81	I
500	s	4036.37	II
5000	s	4089.29	II
100	l	4089.32	II
100	s	4140.96	I
1000	s	4188.12	II
1000		4265.55	I
5000		4289.26	I
200	s	4309.65	II
2000	s	4324.57	II
2000	s	4441.36	II
5000	l	4509.45	II
5000	l	4575.35	II
1000	s	4593.31	II
100	l	4649.12	I
100	l	4653.45	I
5000	l	4662.79	I
2000	l	4681.65	I
2000	l	4699.70	II
1000	l	4706.80	I
2000	l	4872.22	II
200	l	4990.79	I
100	s	5000.21	I
1000	l	5020.96	II
200	l	5215.99	II
1000	s	5402.62	I
1000	s	5424.70	I
1000	l	5584.21	II
1000	s	5598.13	I
10000	l	6054.64	I
1000	l	6405.11	I
500	l	6544.16	I
500	s	6955.58	I

ANTIMONY (Sb)
Z = 51

Sb I and II
Ref. 167, 194 — L.J.R. and J.R.

Intensity		Wavelength	
		Vacuum	
1		691.20	II
1		764.43	II
1		814.85	II
1		849.39	II
2		855.08	II
4		876.84	II
4		921.07	II
6		983.57	II
6		1001.13	II
6		1009.43	II
6		1052.21	II
8		1056.27	II
8		1057.32	II
6		1073.81	II
6		1230.30	II
8		1274.98	II
8		1327.40	II
6		1358.04	II
8		1384.70	II
6		1407.83	II
10		1430.76	I
8		1436.49	II
10	h	1464.19	I
20	r	1486.57	I
40	h	1491.36	I
50	r	1512.57	I
120	r	1532.74	I
80	r	1535.06	I
6		1565.51	II
8		1576.11	II
7		1581.36	II
80	r	1599.96	I
10		1606.98	II
200	w	1612.8	I
100	w	1623.3	I
50	h	1651.20	I
20		1657.04	II
100	w	1662.6	I
50		1698.85	I
80	r	1716.93	I
150	r	1717.45	I
150	r	1723.43	I
100	r	1736.19	I
8		1736.43	II
50		1757.79	I
100	h	1765.76	I
100	r	1780.87	I
100	r	1788.24	I
150		1800.18	I
50	r	1810.50	I
80	r	1814.20	I
100		1829.50	I
60		1858.89	I
50	r	1868.17	I
300	r	1871.15	I
150	r	1882.56	I
70		1891.28	I
70		1899.39	I
100		1927.08	I
200	r	1950.39	I
80	h	1964.3	I
60		1986.05	I
6		1990.60	II
		Air	
50		2024.00	I
60	r	2029.49	I
70	r	2039.77	I
150	r	2049.57	I
50		2063.43	I
1000	r	2068.33	I
100		2079.56	I
50	r	2098.41	I
80	r	2118.48	I
100	r	2127.39	I
50	r	2137.05	I
100	r	2139.69	I
10		2141.80	II
50	r	2141.83	I
100	r	2144.86	I
50		2158.91	I
1500	r	2175.81	I
250	r	2179.19	I
6		2179.25	II
200	r	2201.32	I
300	r	2208.45	I
6		2208.50	II
150	r	2220.73	I
100		2221.98	I
120	r	2224.93	I
6		2225.15	II
300	r	2262.51	I
120		2288.98	I
150	r	2293.44	I
300	r	2306.46	I
2500	r	2311.47	I
150		2315.89	I
400	h	2373.67	I
300	h	2383.64	I
100		2395.22	I
150		2422.13	I
250		2426.35	I
400	r	2445.51	I
400		2478.32	I
150		2480.44	I
8		2480.46	II
100		2510.54	I
2000	r	2528.52	I
15		2528.54	II
10		2567.75	II
150		2574.06	I
1500	r	2598.05	I
500	r	2598.09	I
300	r	2612.31	I
200	r	2652.60	I
12		2656.55	II
300	r	2670.64	I
200	r	2682.76	I
120		2692.25	I
150	r	2718.90	I
400	r	2769.95	I
12		2851.09	II
100		2851.11	I
1000	r	2877.92	I
12		2966.10	II
15		2980.96	II
500	r	3029.83	I
12		3034.01	II
12		3040.67	II
600	r	3232.52	I
20		3241.28	II
700	r	3267.51	I
12		3383.09	II
100		3383.15	I
15		3498.46	II
12		3520.47	II
25		3637.80	II
250		3637.83	I
20		3722.78	II
200	r	3722.79	I
20		3850.22	II
200		4033.55	I
20		4033.56	II
20		4133.63	II
15		4140.54	II
15		4195.17	II
20		4219.07	II
20		4314.32	II
12		4344.83	II
12		4411.42	II
12		4446.48	II
12		4506.92	II
15		4514.50	II
30		4596.90	II
20		4599.09	II
15		4604.77	II
30		4647.32	II
20		4675.74	II
40		4711.26	II
12		4735.44	II
20		4757.81	II
20		4765.36	II
12		4766.91	II
30		4784.03	II
30		4802.01	II
20		4832.82	II
20		4877.24	II
15		4947.40	II
15		5044.56	II
12		5166.32	II
15		5176.55	II
20		5238.94	II
20		5354.24	II
15		5464.08	II
30	h	5490.32	I
40	h	5556.10	I
15		5568.13	II
30	l	5602.19	I
100	l	5632.02	I
30		5639.75	II
60	h	5830.34	I
15		5895.09	II
100		6005.21	II
20		6053.41	II
30		6079.80	II
50		6130.04	II
20		6154.94	II
12		6302.76	II
20		6611.49	I
30		6647.44	II
15		6688.01	II
6		6806.67	II
30	h	7648.28	I
80		7844.44	I
200		7924.65	I
40	h	7969.55	I
60		8411.69	I
150		8572.64	I
100		8619.55	I
30	h	8682.7	I
30		9132.21	I
400		9518.68	I
30		9866.78	I
400		9949.14	I
200		10078.49	I
300		10261.01	I
50		10364.33	I
50	h	10488.3	I
200		10585.60	I
1000		10677.41	I
800		10741.94	I
80		10794.11	I
600		10839.73	I
200		10868.58	I
400		10879.55	I
300		11012.79	I
40	h	11079.95	I
30	h	11084.98	I
50	h	11104.84	I
50	h	11108.52	I
30		11189.61	I
150		11266.23	I
30		11863.37	I
1		11957.7	I
5		12116.06	I
2		12276.6	I
2		12466.75	I

Sb III
Ref. 164 — L.J.R. and J.R.

Intensity	Wavelength	
	Vacuum	
10	691.18	III
10	698.69	III
15	722.86	III
8	724.81	III
15	732.33	III
15	999.62	III
40	1011.94	III
10	1056.58	III
40	1065.90	III
20	1069.93	III
20	1070.43	III
5	1073.76	III
30	1075.82	III
5	1078.10	III
20	1084.06	III
10	1098.34	III
10	1135.43	III
30	1151.49	III
40	1157.74	III
12	1166.96	III
50	1205.20	III
50	1210.64	III
20	1306.69	III
8	1379.58	III
20	1404.18	III
10	1429.57	III
15	1673.89	III
3	1710.23	III
15	1711.84	III
15	1725.33	III
12	1762.30	III
12	1839.32	III
10	1946.13	III
	Air	
3	2054.10	III
2	2091.85	III
5	2127.00	III
5	2507.71	III
15	2590.13	III
1	2614.28	III
12	2617.17	III
1	2617.63	III
20	2669.39	III
5	2785.87	III
20	2790.27	III
20	3336.61	III
50	3504.07	III
15	3519.06	III
15	3533.45	III
40	3559.18	III
40	3566.25	III
30	3738.90	III
40	4265.09	III
50	4352.16	III
30	4591.89	III
30	4692.91	III
1	5247.71	III
1	5690.8	III
1	5717.3	III
3	5845.5	III
5	6246.7	III
3	6287.6	III

Sb IV
Ref. 386 — L.J.R.

Intensity	Wavelength			
	Vacuum			
	861.5	SB	IV	1
	873.5	SB	IV	1
	888.3	SB	IV	1
	891.2	SB	IV	1
	1087.6	SB	IV	1
	1099.3	SB	IV	1
	1115.1	SB	IV	1
	1120.4	SB	IV	1
	1145.9	SB	IV	1
	1151.5	SB	IV	1
	1171.4	SB	IV	1
	1192.9	SB	IV	1
	1199.1	SB	IV	1
	1499.2	SB	IV	1

Sb V
Ref. 406 — L.J.R.

Intensity	Wavelength	
	Vacuum	
3	699.22	V
1	746.06	V
6	831.00	V
1	898.02	V
8	1104.32	V
12	1226.00	V
12	1505.70	V
12	1524.47	V

ARGON (Ar)
Z = 18

Ar I and II
Ref. 190, 203, 204, 219, — E.F.W.

Intensity		Wavelength	
Vacuum			
30		487.227	II
50		490.650	II
30		490.701	II
30		519.327	II
30		542.912	II
200		543.203	II
70		547.461	II
70		556.817	II
70		573.362	II
30		576.736	II
70		580.263	II
30		583.437	II
70		597.700	II
30		602.858	II
30		612.372	II
500		661.867	II
30		664.562	II
200		666.011	II
1000		670.946	II
3000		671.851	II
70		676.242	II
30		677.952	II
30		679.218	II
200		679.401	II
200		718.090	II
3000		723.361	II
500		725.548	II
70		730.930	II
200		740.269	II
200		744.925	II
70		745.322	II
20		802.859	I
100		806.471	I
60		806.869	I
30		807.218	I
40		807.653	I
50		809.927	I
120		816.232	I
70		816.464	I
80		820.124	I
120		825.346	I
120		826.365	I
150		834.392	I
100		835.002	I
100		842.805	I
180		866.800	I
150		869.754	I
180	r	876.058	I
180	r	879.947	I
150		894.310	I
1000		919.781	II
1000		932.054	II
1000	r	1048.220	I
500	r	1066.660	I
Air			
5		2420.456	II
10		2516.789	II
10		2534.709	II
15		2562.087	II
25		2891.612	II
200		2942.893	II
100		2979.050	II
50		3033.508	II
50		3093.402	II
8		3200.37	I
20		3243.689	II
25		3293.640	II
20		3307.228	II
7		3319.34	I
25		3350.924	II
7		3373.47	I
25		3376.436	II
25		3388.531	II
7		3393.73	I
7		3461.07	I
70		3476.747	II
20		3478.232	II
50		3491.244	II
100		3491.536	II
70		3509.778	II
70		3514.388	II
70		3545.596	II
70		3545.845	II
7		3554.306	I
100		3559.508	II
100		3561.030	II
70		3576.616	II
25		3581.608	II
50		3582.355	II
70		3588.441	II
7		3606.522	I
25		3622.138	II
20		3639.833	II
35		3718.206	II
70		3729.309	II
50		3737.889	II
150		3765.270	II
50		3766.119	II
20		3770.369	I
20		3770.520	II
25		3780.840	II
25		3803.172	II
50		3809.456	II
7		3834.679	I
70		3850.581	II
35		3868.528	II
35		3925.719	II
50		3928.623	II
25		3932.547	II
70		3946.097	II
7		3947.505	I
35		3948.979	I
20		3979.356	II
35		3994.792	II
50		4013.857	II
50		4033.809	II
20		4035.460	II
150		4042.894	II
50		4044.418	I
100		4052.921	II
200		4072.005	II
70		4072.385	II
25		4076.628	II
35		4079.574	II
25		4082.387	II
150		4103.912	II
300		4131.724	II
35		4156.086	II
400		4158.590	I
50		4164.180	I
35		4179.297	II
50		4181.884	I
100		4190.713	I
50		4191.029	I
200		4198.317	I
400		4200.674	I
25		4218.665	II
25		4222.637	II
25		4226.988	II
100		4228.158	II
100		4237.220	II
25		4251.185	I
200		4259.362	I
100		4266.286	I
70		4266.527	II
150		4272.169	I
550		4277.528	II
20		4282.898	II
100		4300.101	I
25		4300.650	II
70		4309.239	II
200		4331.200	II
50		4332.030	II
100		4333.561	I
50		4335.338	I
25		4345.168	I
800		4348.064	II
50		4352.205	II
25		4362.066	II
50		4367.832	II
200		4370.753	II
70		4371.329	II
50		4375.954	II
150		4379.667	II
50		4385.057	II
70		4400.097	II
200		4400.986	II
400		4426.001	II
150		4430.189	II
50		4430.996	II
50		4433.838	II
20		4439.461	II
35		4448.879	II
100		4474.759	II
200		4481.811	II
100		4510.733	I
20		4522.323	I
20		4530.552	II
400		4545.052	II
20		4564.405	II
400		4579.350	II
400		4589.898	II
15		4596.097	I
550		4609.567	II
7		4628.441	I
35		4637.233	II
400		4657.901	II
15		4702.316	I
20		4721.591	II
550		4726.868	II
50		4732.053	II
300		4735.906	II
800		4764.865	II
550		4806.020	II
150		4847.810	II
50		4865.910	II
800		4879.864	II
70		4889.042	II
20		4904.752	II
35		4933.209	II
200		4965.080	II
50		5009.334	II
70		5017.163	II
70		5062.037	II
20		5090.495	II
100		5141.783	II
70		5145.308	II
5		5151.391	I
15		5162.285	I
25		5165.773	II
20		5187.746	I
20		5216.814	II
7		5221.271	I
5		5421.352	I
10		5451.652	I
25		5495.874	I
5		5506.113	I
25		5558.702	I
10		5572.541	I
35		5606.733	I
20		5650.704	I
10		5739.520	I
5		5834.263	I
10		5860.310	I
15		5882.624	I
25		5888.584	I
50		5912.085	I
15		5928.813	I
5		5942.669	I
7		5987.302	I
5		5998.999	I
5		6025.150	I
70		6032.127	I
35		6043.223	I
10		6052.723	I
20		6059.372	I
7		6098.803	I
10		6105.635	I
100		6114.923	II
10		6145.441	I
7		6170.174	I
150		6172.278	II
10		6173.096	I
10		6212.503	I
5		6215.938	I
25		6243.120	II
7		6296.872	I
15		6307.657	I
7		6369.575	I
20		6384.717	I
70		6416.307	I
25		6483.082	II
15		6538.112	I
15		6604.853	I
25		6638.221	II
20		6639.740	II
50		6643.698	II
5		6660.676	I
5		6664.051	I
25		6666.359	II
100		6677.282	I
35		6684.293	II
150		6752.834	I
5		6756.163	I
15		6766.612	I
20		6861.269	II
150		6871.289	I
5		6879.582	I
10		6888.174	I
50		6937.664	I
7		6951.478	I
7		6960.250	I
10000		6965.431	I
150		7030.251	I
10000		7067.218	I
100		7068.736	I
25		7107.478	I
25		7125.820	I
1000		7147.042	I
15		7158.839	I
70		7206.980	I
15		7265.172	I
7		7270.664	I
2000		7272.936	I
35		7311.716	I
25		7316.005	I
5		7350.814	I
70		7353.293	I
200		7372.118	I
20		7380.426	II
10000		7383.980	I
20		7392.980	I
15		7412.337	I
10		7425.294	I
25		7435.368	I
10		7436.297	I
20000		7503.869	I
15000		7514.652	I
25000		7635.106	I
15000		7723.761	I
10000		7724.207	I
10		7891.075	I
20000		7948.176	I
20000		8006.157	I
25000		8014.786	I
7		8053.308	I
20000		8103.693	I
35000		8115.311	I
10000		8264.522	I
20		8392.27	I
15000		8408.210	I
20000		8424.648	I
15000		8521.442	I
7		8605.776	I
4500		8667.944	I
20		8771.860	II
180		8849.91	I
20		9075.394	I
35000		9122.967	I
550		9194.638	I
15000		9224.499	I
400		9291.531	I
1600		9354.220	I
25000		9657.786	I
4500		9784.503	I
180		10052.06	I
30		10332.72	I
100		10467.177	II
1600		10470.054	I
13		10478.034	I
180		10506.50	I
200		10673.565	I
11		10681.773	I
7		10683.034	II
30		10733.87	I
30		10759.16	I
7		10812.896	II
11		11078.869	I
30		11106.46	I
12		11441.832	I
400		11488.109	I
200		11668.710	I
12		11719.488	I
200		12112.326	I
50		12139.738	I
50		12343.393	I
200		12402.827	I
200		12439.321	I
100		12456.12	I
200		12487.663	I
150		12702.281	I
30		12733.418	I
12		12746.232	I
200		12802.739	I
50		12933.195	I
500		12956.659	I
200		13008.264	I
200		13213.99	I
200		13228.107	I
100		13230.90	I
500		13272.64	I
1000		13313.210	I
1000		13367.111	I
30		13499.41	I
1000		13504.191	I
11		13573.617	I
30		13599.333	I
400		13622.659	I
200		13678.550	I
1000		13718.577	I
10		13825.715	I
10		13907.478	I
200		14093.640	I
100		15046.50	I
25		15172.69	I
10		15329.34	I
30		15989.49	I
30		16519.86	I
500		16940.58	I
12		18427.76	I
50		20616.23	I
30		20986.11	I
20		23133.20	I
20		23966.52	I

Ar III
Ref. 367, 372, 373, 375 — E.F.W.

Intensity	Wavelength	
Vacuum		
12	769.15	III

Argon (Cont.)

Intensity		Wavelength	
10		871.10	III
9		875.53	III
12		878.73	III
8		879.62	III
9		883.18	III
10		887.40	III
7		1669.67	III
7		1673.42	III
7		1675.48	III
9		1914.40	III
7		1915.56	III
		Air	
10		2125.16	III
15		2133.87	III
10		2138.59	III
10		2148.73	III
15		2166.19	III
10		2168.26	III
20		2170.23	III
25		2177.22	III
8		2184.06	III
10		2188.22	III
15		2192.06	III
7		2248.73	III
10		2279.10	III
7		2281.22	III
7		2282.21	III
12		2293.03	III
10		2300.85	III
15		2302.17	III
9		2317.00	III
15		2317.47	III
12		2318.04	III
10		2319.13	III
10		2319.37	III
9		2345.17	III
7		2351.67	III
9		2360.26	III
10		2395.63	III
12		2399.15	III
10		2413.20	III
7		2415.61	III
10		2418.82	III
12		2423.52	III
12		2423.93	III
7		2443.69	III
8		2472.95	III
7		2476.10	III
12		2488.86	III
7		2631.90	III
10		2654.63	III
8		2674.02	III
9		2678.38	III
10		2724.84	III
7		2762.23	III
7		2842.88	III
8		2855.29	III
9		2884.12	III
10		3010.02	III
12		3024.05	III
12		3054.82	III
10		3064.77	III
10		3078.15	III
7		3110.41	III
7		3127.90	III
25		3285.85	III
20		3301.88	III
15		3311.25	III
7		3323.59	III
25		3336.13	III
20		3344.72	III
15		3358.49	III
7		3361.28	III
15		3391.85	III
7		3417.49	III
9		3424.25	III
8		3438.04	III
9		3471.32	III
20		3480.55	III
12		3499.67	III
15		3503.58	III
8		3511.12	III
20		3795.37	III
10		3858.32	III
7		3907.84	III
8		3960.53	III
6		4023.60	III
5		4146.70	III

Ar IV
Ref. 367, 368, 374 — E.F.W.

Intensity		Wavelength	
		Vacuum	
4		396.87	IV
4		398.55	IV
6		623.77	IV
10		683.28	IV
7		688.39	IV
12	p	689.01	IV
6		699.41	IV
8		700.28	IV
4		754.20	IV
5		761.47	IV
5		800.57	IV
10		801.09	IV
10		801.41	IV
5		801.91	IV
15		840.03	IV
20		843.77	IV
25		850.60	IV
5		900.36	IV
9		901.17	IV
		Air	
4		2299.72	IV
8		2447.71	IV
12		2513.28	IV
6		2518.40	IV
9		2525.69	IV
12		2562.17	IV
10		2568.07	IV
7		2569.53	IV
12		2599.47	IV
10		2608.06	IV
7		2608.44	IV
12		2615.68	IV
6		2619.98	IV
12		2621.36	IV
12		2624.92	IV
15		2640.34	IV
9		2682.63	IV
14		2757.92	IV
10		2776.26	IV
12		2784.47	IV
14		2788.96	IV
7		2797.11	IV
16		2809.44	IV
10		2830.25	IV
8		2874.40	IV
12		2913.00	IV
11		2926.33	IV
6		3037.98	IV
8		3077.40	IV

Ar V
Ref. 414, 421 — E.F.W.

Intensity		Wavelength	
		Vacuum	
3		336.56	V
3		337.56	V
6		338.00	V
2		338.43	V
2		339.01	V
3		339.89	V
3		350.88	V
2		436.67	V
5		446.00	V
8		446.95	V
4		447.53	V
18		449.06	V
4		449.49	V
3		458.12	V
2		458.98	V
6	p	461.23	V
3		462.42	V
7		463.94	V
3		522.09	V
5		524.19	V
6		527.69	V
2		554.50	V
5		558.48	V
3		635.12	V
3		705.35	V
5		709.20	V
4		715.60	V
3		715.65	V
2		725.11	V
4		822.16	V
5		827.05	V
3		827.35	V
4	p	834.88	V
2		836.13	V

ARSENIC (As)
Z = 33

As I and II
Ref. 168, 197 — R.L.K.

Intensity		Wavelength	
		Vacuum	
165		761.24	II
165		802.83	II
340		1021.96	II
340		1082.35	II
500		1139.40	II
615		1149.31	II
555		1181.51	II
555		1189.87	II
615		1196.38	II
615		1196.56	II
340		1207.44	II
800		1211.17	II
800		1218.10	II
340		1223.15	II
760		1241.31	II
965		1243.08	II
870		1245.67	II
800		1258.58	II
965		1263.77	II
800		1266.34	II
800		1267.59	II
715		1280.99	II
715		1287.54	II
715		1305.70	II
340		1307.74	II
760		1333.15	II
965		1341.55	II
760		1355.93	II
965		1369.77	II
800		1373.65	II
1000		1375.07	II
760		1375.78	II
800		1394.64	II
800		1400.31	II
500		1448.59	II
500		1558.88	II
500		1570.99	II
100	r	1593.60	I
500		1660.55	II
100		1758.60	I
170		1806.15	I
340		1860.34	II
1000	r	1890.42	I
500		1912.94	II
800	r	1937.59	I
585	r	1972.62	I
170	r	1990.35	I
100	r	1991.13	I
100	r	1995.43	I
		Air	
230	r	2003.34	I
100	r	2009.19	I
100		2013.32	I
100		2112.99	I
100		2144.08	I
135		2165.52	I
350	r	2288.12	I
350	r	2349.84	I
100	r	2370.77	I
135	r	2381.18	I
170	r	2456.53	I
340		2602.00	II
170	r	2780.22	I
300		2830.359	II
300		2831.164	II
100	r	2860.44	I
300		2884.406	II
615		2959.572	II
300		3003.819	II
300		3116.516	II
340		3842.60	II
715		4190.082	II
615		4197.40	II
615		4242.982	II
500		4315.657	II
500		4323.867	II
500		4336.64	II
500		4352.145	II
425		4352.864	II
375		4371.17	II
615		4427.106	II
615		4431.562	II
715		4458.469	II
340		4461.075	II
715		4466.348	II
500		4474.46	II
800		4494.230	II
850		4507.659	II
615		4539.74	II
715		4543.483	II
615		4602.427	II
340		4629.787	II
340		4707.586	II
340		4730.67	II
340		4888.557	II
100		5068.98	I
340		5105.58	II
500		5107.55	II
100		5121.34	I
100		5141.63	I
425		5231.38	II
500		5331.23	II
100		5408.13	I
135		5451.32	I
340		5497.727	II
425		5558.09	II
425		5651.32	II
425		6110.07	II
500		6170.27	II
300		6511.74	II
300		7092.27	II
300		7102.72	II
340		7990.53	II
300		8174.51	II
100		8428.91	I
100		8564.71	I
100		8654.14	I
135		8821.73	I
100		8869.66	I
135		9267.28	I
200		9300.61	I
230		9597.95	I
290		9626.70	I
230		9833.76	I
100		9886.05	I
140		9900.55	I
170		9915.71	I
290		9923.05	I
100		10010.63	I
290		10024.04	I
100		10453.09	I
100		10575.02	I
170		10614.07	I

As III
Ref. 163 — R.L.K.

Intensity	Wavelength	
	Vacuum	
185	849.9	III
185	866.3	III
510	871.7	III
325	889.0	III
325	927.5	III
325	953.6	III
325	937.2	III
325	963.8	III
120	1172.2	III
185	1209.3	III
	Air	
80	2926.3	III
185	2982.0	III
325	3922.6	III
185	4037.2	III

As IV
Ref. 244 — R.L.K.

Intensity	Wavelength	
	Air	
150	2253.1	IV
200	2263.2	IV
200	2301.0	IV
250	2417.5	IV
150	2446.1	IV
250	2454.0	IV
200	2461.4	IV
150	3108.8	IV

As V
Ref. 280 — R.L.K.

Intensity	Wavelength	
	Vacuum	
25	600.7	V
40	616.0	V
120	715.5	V
150	734.8	V
60	737.2	V
250	987.7	V
250	1029.5	V
40	1051.6	V
60	1056.6	V

ASTATINE (At)
Z = 85

At I
Ref. 188 — E.F.W.

Astatine (Cont.)

Intensity		Wavelength	
		Air	
8		2162.25	I
10		2244.01	I

BARIUM (Ba)
Z = 56

Ba I and II
Ref. 1, 252, 277, 279 — J.J.W.

Intensity		Wavelength	
		Vacuum	
200		1486.72	II
400		1504.01	II
300		1554.38	II
200		1572.73	II
		1573.92	II
		1630.40	II
100		1674.51	II
400		1694.37	II
		1697.16	II
		1761.75	II
		1771.03	II
		1786.93	II
100		1904.15	II
500		1924.70	II
		1985.60	II
300		1999.54	II
		Air	
		2009.20	II
400		2023.95	II
		2052.68	II
		2054.57	II
500		2214.7	II
800		2245.61	II
1000		2254.73	II
1400		2304.24	II
2000		2335.27	II
190		2347.58	II
60		2528.51	II
8	h	2596.64	I
100		2634.78	II
8		2702.63	I
18		2771.36	II
15		2785.28	I
100	r	3071.58	I
10	h	3108.21	I
8		3132.60	I
8	h	3135.72	I
10		3137.70	I
10		3155.34	I
10		3155.67	I
12		3158.05	I
12	h	3158.54	I
25		3165.60	I
15	h	3173.69	I
30		3183.16	I
15		3183.96	I
10		3193.91	I
25	h	3203.70	I
30		3221.63	I
40		3222.19	I
50		3261.96	I
60	r	3262.34	I
40		3281.50	I
15		3281.77	I
50		3322.80	I
80	h	3356.80	I
60	r	3377.08	I
20		3377.39	I
70	r	3420.32	I
25		3421.01	I
30	h	3421.48	I
40		3463.74	I
200	r	3501.11	I
80	h	3524.97	I
30	h	3531.35	I
80	h	3544.66	I
20	h	3547.68	I
100		3552.45	II
200		3567.73	II
100		3576.28	II
30		3577.62	I
80	h	3579.67	I
200		3596.57	II
40		3630.64	I
40	h	3636.83	I
20	h	3688.47	I
400		3735.75	II
200		3816.69	II
200		3842.80	II
100		3854.76	II
20		3889.33	I
1400	l	3891.78	II
20		3892.65	I
40		3909.91	I
500		3914.73	II
50		3935.72	I
20		3937.87	I
200		3939.67	II
500		3949.51	II
80		3993.40	I
30		3995.66	I
300		4036.26	II
200		4083.77	II
30	h	4084.86	I
1500	h	4130.66	II
20		4132.43	I
200		4166.00	II
500		4216.04	II
800		4267.95	II
100		4283.10	I
300		4287.80	II
200		4297.60	II
800		4309.32	II
20	h	4323.00	I
600		4325.73	II
200		4326.74	II
300		4329.62	II
80		4350.33	I
60		4402.54	I
400		4405.23	II
60	h	4488.98	I
40		4431.89	I
50	h	4493.64	I
40		4505.92	I
200		4509.63	II
60	h	4523.17	I
130		4524.93	II
65000		4554.03	II
40		4573.85	I
80		4579.64	I
30		4599.75	I
20	h	4619.92	I
25	h	4628.33	I
300		4644.10	II
30		4673.62	I
35		4691.62	I
20		4700.43	I
800		4708.94	II
40		4726.44	I
800		4843.46	II
300		4847.14	II
200		4850.84	II
30	h	4877.65	I
400		4899.97	II
15		4902.90	I
20000		4934.09	II
8		4947.35	I
1000		4957.15	II
300		4997.81	II
1000		5013.00	II
20	h	5159.94	I
20		5267.03	I
800		5361.35	II
1000		5391.60	II
200		5421.05	II
100		5424.55	I
200		5428.79	II
300		5480.30	II
200		5519.05	I
1000	r	5535.48	I
20	h	5620.40	I
10		5680.18	I
400		5777.62	I
800		5784.18	II
100		5800.23	I
20		5805.69	I
150		5826.28	I
2800		5853.68	II
15		5907.64	I
100		5971.70	I
800		5981.25	II
100		5997.09	I
300		5999.85	II
100		6019.47	I
200		6063.12	I
300		6110.78	I
400		6135.83	II
20000		6141.72	II
150		6341.68	I
500		6378.91	II
90		6450.85	I
150		6482.91	I
12000		6496.90	II
300		6498.76	I
150		6527.31	I
3000		6595.33	I
150		6654.10	I
1500		6675.27	I
1800		6693.84	I
1000		6769.62	II
600		6865.69	I
300	h	6867.85	I
1000		6874.09	II
6000		7059.94	I
2400	hS	7120.33	I
600		7195.24	I
600	hL	7228.84	I
3000		7280.30	I
1200		7392.41	I
300		7417.53	I
900	hL	7459.78	I
600		7488.08	I
450	hL	7636.90	I
600	hL	7642.91	I
1800		7672.09	I
1200		7780.48	I
180	h	7839.57	I
1500		7905.75	I
600		7911.34	I
900	h	8210.24	I
1800	h	8559.97	I
100		8710.74	II
100		8737.71	II
300	h	8799.76	I
300		8860.98	I
450		8914.99	I
300		9219.69	I
300		9308.08	I
300	h	9324.58	I
1500		9370.06	I
300		9455.92	I
450		9589.37	I
900		9608.88	I
300	h	9645.72	I
1500	hL	9830.37	I
900		10001.08	I
600		10032.10	I
1200	h	10233.23	I
300		10471.26	I
120	hL	10791.25	I
180	hL	11012.69	I
150	h	11114.42	I
240		11303.04	I
120	h	11697.45	I
120		13207.30	I
120		13810.50	I
120		14077.90	I
120		15000.40	I
120		20712.00	I
150		25515.70	I
150		29223.90	I

Ba III
Ref. III — J.J.W.

Intensity		Wavelength	
		Vacuum	
5		403.82	III
2		407.12	III
7		420.12	III
4		423.84	III
9		448.95	III
8		456.96	III
14		555.48	III
14		587.57	III
18		647.27	III
9		653.36	III
15		743.12	III
12		1097.41	III
15		1113.67	III
11		1116.01	III
14		1133.05	III
12		1151.76	III
12		1170.62	III
13		1207.29	III
11		1218.92	III
12		1224.55	III
12		1288.53	III
11		1299.18	III
11		1307.40	III
12		1308.87	III
12		1315.72	III
12		1334.01	III
11		1354.71	III
11		1369.53	III
11		1416.61	III
12		1478.85	III
12		1510.68	III
12		1514.22	III
12		1565.61	III
12		1566.12	III
12		1574.55	III
12		1596.80	III
12		1610.95	III
12		1615.78	III
12		1711.53	III
12		1861.74	III
12		1883.92	III
11		1974.76	III
		Air	
10		2001.30	III
15		2008.40	III
13		2022.45	III
10		2038.84	III
12		2070.43	III
12		2071.68	III
10		2076.00	III
12		2081.35	III
10		2134.87	III
16		2156.37	III
10		2160.76	III
20		2230.33	III
30		2280.68	III
35		2323.51	III
60		2331.10	III
25		2476.73	III
25		2505.07	III
40		2512.28	III
40		2523.83	III
25		2530.92	III
50		2559.54	III
25		2570.48	III
40		2681.89	III
30		2745.78	III
25		2938.95	III
25		2960.05	III
30		2962.48	III
20		3014.22	III
30		3043.42	III
40		3079.14	III
30		3103.92	III
30		3119.22	III
30		3152.70	III
25		3195.17	III
25		3235.04	III
25		3281.65	III
20		3286.79	III
50		3368.18	III
30		3369.68	III
25		3649.18	III
25		3926.85	III
25		3993.06	III
18	p	4053.71	III
15		4697.44	III
10		5049.55	III
10		5097.54	III
12	p	5102.25	III
10		5134.54	III
10		5998.00	III
13		6101.99	III
10		6377.11	III
10		6383.76	III
8		6526.17	III
8		7095.49	III
8		8308.69	III
8		9521.76	III

Ba IV
Ref. 78 — J.J.W.

Intensity		Wavelength	
		Vacuum	
40000		794.89	IV
50000		923.74	IV

Ba V

Ref. 259—J.R.

Intensity		Wavelength	
		Vacuum	
15		612.55	V
5		658.11	V
5		681.09	V
300		719.86	V
150		721.85	V
100		760.45	V
1000		766.87	V
100		783.61	V
100		816.41	V
40		875.69	V
300		877.41	V
15		892.28	V
200		946.26	V

BERKELIUM (Bk)
Z = 97

Bk I and II
Ref.53, 339—J.G.C.

Intensity		Wavelength	
		Air	
10000	s	2748.02	II
10000	s	2827.57	II
10000	l	2872.11	II
10000	s	2878.57	II
10000	l	2884.77	II
10000	s	2889.80	II
10000	s	2893.66	II
10000	l	2910.65	II
10000	s	2926.49	
10000	s	2927.91	
10000	l	2941.71	II
10000	l	2951.76	II
10000	l	2969.13	II
10000	l	2987.76	II
10000	l	3178.47	II
10000		3239.72	I
10000	s	3247.26	II
10000		3252.19	I
10000	s	3263.47	II
10000	l	3288.75	I

Berkelium (Cont.)

10000		3289.35	I
10000	s	3302.35	II
10000		3335.26	I
10000	s	3387.45	II
10000		3408.28	I
10000	1	3412.13	II
10000		3426.95	I
10000	1	3432.62	I
10000	1	3437.47	I
10000		3442.66	I
10000	s	3461.24	II
10000	1	3453.90	I
10000	s	3464.13	II
10000	s	3472.02	II
10000	s	3477.62	II
10000		3528.72	I
10000	1	3531.40	I
10000	1	3535.73	I
10000	s	3542.19	II
10000		3553.60	I
10000	s	3555.88	I
10000		3556.52	I
10000		3565.41	
10000	1	3567.25	II
10000		3590.32	I
10000		3595.88	I
10000		3601.12	I
10000	s	3603.20	II
10000		3604.78	
10000	1	3608.49	
10000		3609.61	
10000		3611.03	
10000	1	3611.93	
10000		3613.91	
10000		3616.62	
10000		3619.37	
10000	s	3621.81	
10000		3627.61	I
10000		3633.28	
10000	1	3637.05	I
10000		3640.26	I
10000	s	3640.93	II
10000	1	3675.59	I
10000	s	3681.22	II
10000	1	3684.43	I
10000	s	3685.21	I
10000	1	3686.74	I
10000		3692.73	I
10000		3695.37	I
10000		3703.28	I
10000	s	3704.02	I
10000		3705.26	I
10000	s	3711.14	II
10000		3712.93	I
10000		3725.39	I
10000		3739.92	I
10000		3743.05	I
10000		3745.40	I
10000		3750.08	I
10000		3751.91	I
10000		3757.35	I
10000		3757.85	I
10000	s	3771.06	II
10000		3780.72	I
10000		3781.17	I
10000		3785.38	I
10000		3788.21	I
10000		3791.42	I
10000		3796.21	I
10000		3797.12	
10000		3798.63	I
10000		3801.08	II
10000	s	3802.35	I
10000		3802.47	I
10000		3815.29	I
10000	s	3823.10	II
10000	s	3824.08	II
10000		3825.19	I
10000	s	3825.84	II
10000		3827.61	I
10000		3830.55	I
10000	1	3831.57	II
10000	s	3833.48	
10000	s	3835.97	II
10000		3842.19	I
10000	1	3846.62	I
10000		3847.63	I
10000		3855.03	I
10000	1	3859.89	II
10000	1	3877.94	II
10000		3880.11	I
10000	1	3882.60	I
10000	1	3894.55	II
10000	s	3906.09	II
10000	1	3912.16	II
10000	1	3916.37	II
10000		3921.42	I
10000	1	3928.05	II
10000	1	4147.13	II
10000	s	4189.69	II
10000	s	4197.44	II
10000	1	4329.58	I
10000		4351.50	I
10000		4363.64	I
10000		4423.01	I
10000		4466.46	I
10000		4685.70	
10000	1	4765.40	
10000	s	5056.73	I
10000	1	5118.24	I
10000	s	5135.52	II
10000		5170.61	I
10000		5197.55	I
10000		5212.53	I
10000		5271.95	I
10000		5392.03	I
10000		5394.24	I
10000	1	5404.62	I
10000		5449.63	I
10000	1	5467.47	I
10000	s	5484.58	I
10000		5512.22	II
10000	1	5537.93	I
10000	1	5556.80	I
10000	s	5557.09	I
10000		5581.21	I
10000	1	5656.54	I
10000		5659.03	I
10000		5702.24	I
10000		5910.71	I
10000	1	7040.85	I
10000		7107.85	I
10000	1	7176.22	I
10000		7249.26	
10000		7252.50	I
10000	1	7257.21	I
10000		7306.94	I
10000		7394.26	
10000	1	7511.26	I
10000		7551.12	I
10000		7579.77	I
10000	1	7729.93	I
10000	s	7903.90	I
10000	s	9319.30	I
10000	s	9429.13	I
10000	1	9801.18	I
10000	1	9862.39	II
10000	1	9879.29	I
10000	1	9892.38	I
10000		10126.20	I
10000	1	10186.58	I
10000		10292.44	I
10000	1	10527.71	I
10000	1	10570.53	I
10000	1	11293.14	I
10000	1	11500.30	I
10000	s	11575.34	I
10000	s	11793.09	I
10000	s	12159.05	I
10000		13061.13	I
10000		13498.36	I
10000	s	14196.93	I
10000		15136.10	I
10000		18352.31	
10000	1	19273.87	I
10000	s	19653.22	I
10000	s	23902.85	I
10000	s	24192.62	I

BERYLLIUM (Be)
Z = 4

Be I and II
Ref. 15, 44, 115, 134, 135, 198, 335 — J.O.S.

Intensity		Wavelength	
		Vacuum	
		82.58	II
		83.66	II
		89.10	I
		89.80	II
		90.04	II
		90.21	I
		90.67	I
		91.06	II
		91.36	II
		91.74	II
		92.19	I
		92.61	II
		93.14	II
		93.42	II
		93.93	II
		94.78	II
		95.76	II
		96.29	I
		97.24	I
		97.44	I
		97.86	I
		97.97	I
		98.12	I
		98.37	I
		98.66	I
		98.94	I
		99.19	I
		100.86	I
		101.20	I
		102.13	I
		102.49	II
		104.40	II
		104.67	I
		105.80	I
		107.26	I
		107.38	I
		714.0	II
		725.71	II
		743.58	II
8		775.37	II
20		842.06	II
		865.3	II
2		925.25	II
10		943.56	II
10		973.27	II
		981.4	II
		1020.1	II
8		1026.93	II
5		1036.32	II
15		1048.23	II
20		1143.03	II
		1155.9	II
60		1197.19	II
		1426.12	I
		1491.76	I
20		1512.30	II
60		1512.43	II
100		1661.49	I
15		1776.12	II
20		1776.34	II
		1907.	I
		1909.0	II
		1912.	I
		1919.	I
5		1929.67	I
10		1943.68	I
		1956.	I
50		1964.59	I
5		1985.13	I
		1997.95	I
		1997.98	I
60		1998.01	I
		Air	
		2033.25	I
		2033.28	I
		2033.38	I
50		2055.90	I
100		2056.01	I
10		2125.57	I
20		2125.68	I
25		2145.	I
55		2174.99	I
55		2175.10	I
		2273.5	II
		2324.6	II
		2337.0	I
950		2348.61	I
20		2350.66	I
60		2350.71	I
200		2350.83	I
2		2413.34	II
16		2413.46	II
20		2453.84	II
		2480.6	I
35		2494.54	I
35		2494.58	I
100		2494.73	I
16		2507.43	II
5		2617.99	II
20		2618.13	II
100		2650.45	I
60		2650.55	I
200		2650.62	I
60		2650.[illegible]	I
100		2650.76	I
5		2697.46	II
20		2697.58	II
20		2728.88	II
30		2738.05	I
		2764.2	II
20		2898.13	I
10		2898.19	I
20		2898.25	I
30		2986.06	I
10		2986.42	I
60		3019.33	I
30		3019.49	I
30		3019.53	I
20		3019.60	I
10		3046.52	II
30		3046.69	II
		3090.3	I
10		3110.81	I
10		3110.92	I
20		3110.99	I
		3120.	I
480		3130.42	II
320		3131.07	II
		3136.	I
		3150.	I
		3160.6	I
		3163.	I
		3168.	I
		3180.7	II
		3187.	I
20		3193.81	I
20		3197.10	II
30		3197.15	II
20		3208.60	I
		3220.	I
60		3229.63	I
2		3233.52	II
10		3241.62	II
30		3241.83	II
15		3269.02	I
100		3274.58	II
30		3274.67	II
30		3282.91	I
30		3321.01	I
30		3321.09	I
220		3321.34	I
20		3345.43	I
60		3367.63	I
		3405.6	II
5		3451.37	I
300		3455.18	I
20		3476.56	I
300		3515.54	I
10		3555.	I
100		3736.30	I
700		3813.45	I
40		3865.13	I
80		3865.42	I
1		3865.51	I
6		3865.72	I
100		3866.03	I
100		4253.05	I
60		4253.76	I
300		4360.66	II
500		4360.99	II
400		4407.94	I
		4526.6	I
		4548.	I
12		4572.66	I
700		4673.33	II
1000		4673.42	II
6		4709.37	I
200		4828.16	II
40		4849.16	I
2	h	4858.22	II
80		5087.75	I
8		5218.12	II
20		5218.33	II
3		5255.86	II
64		5270.28	II
500		5270.81	II
20		5403.04	II
20		5410.21	II
		5558.	I
10		6229.11	I
16		6279.43	II
30		6279.73	II
30		6473.54	I
60		6547.89	II
60		6558.36	II
30		6564.52	I
2	h	6636.44	II
1		6756.72	II
2		6757.13	II
30		6786.56	I
1	h	6884.22	I
6	h	6884.44	I
100		6982.75	I
6	h	7154.40	I
40	h	7154.65	I
100		7209.13	I
5		7401.20	II
2		7401.43	II
10		7551.90	I
10	h	7618.68	I
20	h	7618.88	I
60		8090.06	I
5	h	8158.99	I
10	h	8159.24	I
4		8254.07	I
10	h	8287.07	I
30		8547.36	I
60		8547.67	I
300		8801.37	I
6		8882.18	I
40		9190.45	I
20	h	9243.92	I
1	h	9343.89	II
40		9392.74	I
2		9476.43	II
16		9477.03	II
20		9847.32	I
10	h	9895.63	I
20	h	9895.96	I
80		9939.78	I
16		10095.52	II

Beryllium (Cont.)

Intensity		Wavelength	
20		10095.73	II
60		10119.92	II
80		10331.03	I
30		11066.46	I
		11173.	II
1		11173.73	II
120		11496.39	I
2	h	11625.16	II
		11659.	II
2		11660.25	II
100		12095.36	II
30		12098.18	II
100		14643.92	I
60		14644.75	I
200		16157.72	I
80		17855.38	I
120		17856.63	I
100		18143.54	I
160		31775.05	I
200		31778.70	I

Be III
Ref. 73, 102, 175 — J.O.S.

Intensity		Wavelength	
		Vacuum	
1	h	76.10	III
2		76.48	III
3		78.53	III
4		78.66	III
1	h	78.92	III
5		81.89	III
10		82.38	III
20		83.20	III
30		84.76	III
50		88.31	III
100		100.25	III
3		509.99	III
2		549.31	III
6		582.08	III
4		661.32	III
8		675.59	III
4		725.59	III
7		746.23	III
2		767.75	III
1		1114.69	III
2		1213.12	III
1		1214.32	III
2		1362.25	III
1		1401.52	III
10		1421.26	III
5		1422.86	III
1		1435.17	III
2		1440.77	III
2	h	1754.69	III
3		1917.03	III
60	h	1954.97	III
		Air	
75	h	2076.94	III
60	h	2080.38	III
25		2118.56	III
15	h	2122.27	III
15	h	2127.20	III
5		2137.25	III
5		2191.57	III
100		3720.36	III
		3720.92	III
		3722.98	III
90	h	4249.14	III
2		4485.52	III
100	h	4487.30	III
1		4495.09	III
140	h	4497.8	III
140	h	6142.01	III

Be IV
Ref. 171 — J.O.S.

Intensity		Wavelength	
		Vacuum	
		58.13	IV
		58.57	IV
		59.32	IV
		60.74	IV
		64.06	IV
		75.93	IV

BISMUTH (Bi)
Z = 83

Bi I and II
Ref. 1, 357-359 — C.H.C.

Intensity		Wavelength	
		Vacuum	
15		1058.88	II
20		1085.47	II
10		1099.20	II
8		1163.19	II
8		1167.06	II
10		1225.43	II
15		1232.78	II
10		1241.05	II
10		1265.35	II
15		1283.73	II
10		1306.18	II
20		1325.46	II
20		1329.47	II
20		1350.07	II
25		1372.61	II
15		1376.02	II
20		1393.92	II
45		1436.83	II
25		1447.94	II
50		1455.11	II
25		1462.14	II
35		1486.93	II
20		1502.50	II
40		1520.57	II
40		1533.17	II
30		1536.77	II
35		1538.06	II
20		1563.67	II
40		1573.70	II
60		1591.79	II
25		1601.58	II
40		1609.70	II
40		1611.38	II
20		1652.81	II
20		1749.29	II
80		1777.11	II
60		1787.47	II
70		1791.93	II
70		1823.80	II
100		1902.41	II
9000		1954.53	I
7000		1960.13	I
25		1989.35	II
		Air	
7000		2021.21	I
9000		2061.70	I
45	h	2068.9	II
4600		2110.26	I
2500		2133.63	I
15		2143.40	II
15		2143.46	II
60		2186.9	II
40	h	2214.0	II
360		2228.25	I
1700		2230.61	I
340		2276.58	I
16		2368.12	II
12		2368.25	II
190		2400.88	I
10		2501.0	II
25		2515.69	I
70		2524.49	I
20	h	2544.5	II
700		2627.91	I
12		2693.0	II
280	c	2696.76	I
20		2713.3	II
140	d	2730.50	I
360		2780.52	I
15		2803.42	II
11		2803.70	II
12		2805.3	II
140	c	2809.62	I
4000		2897.98	I
15		2936.7	II
3200		2938.30	I
20		2950.4	II
12		2963.4	II
2800		2989.03	I
700		2993.34	I
2400		3024.64	I
60		3034.87	I
9000	c	3067.72	I
140		3076.66	I
550	c	3397.21	I
10		3430.83	II
12		3431.23	II
500	c	3510.85	I
380	c	3596.11	I
12		3654.2	II
70	h	3792.5	II
12		3811.1	II
20		3815.8	II
10		3845.8	II
30		3863.9	II
40	h	4079.1	II
10		4097.2	II
140		4121.53	I
140		4121.86	I
75	h	4259.4	II
25		4272.0	II
70	h	4301.7	II
12	h	4339.8	II
25	h	4340.5	II
12	h	4379.4	II
25	h	4476.8	II
60	h	4705.3	II
600	c	4722.52	I
30		4730.3	II
20		4749.7	II
12		4908.2	II
10		4916.6	II
12		4969.7	II
20		4993.6	II
10		5091.6	II
50	h	5124.3	II
60	h	5144.3	II
20		5201.5	II
75	h	5209.2	II
40	h	5270.3	II
10		5397.8	II
10	c	5552.35	I
3		5599.41	I
20		5655.2	II
40	h	5719.2	II
6		5742.55	I
12		5818.3	II
20		5860.2	II
20		5973.0	II
15		6059.1	II
15		6128.0	II
6		6134.82	I
3		6475.73	I
3		6476.24	I
15		6497.7	II
10		6577.2	II
40	h	6600.2	II
50	h	6808.6	II
4	h	6991.12	I
12		7033.	II
2		7036.15	I
10	h	7381.	II
2		7502.33	I
10	h	7637.	II
10		7750.	II
3		7838.70	I
2		7840.33	I
20		7965.	II
12	h	8050.	II
15		8328.	II
15		8388.	II
30		8532.	II
2		8544.54	I
1		8579.74	I
25		8653.	II
2		8754.88	I
3		8761.54	I
25		8863.	II
2		8907.81	I
2000	d	9657.04	I
40		9827.78	I
20		10104.5	I
15		10138.8	I
20		10300.6	I
20		10536.19	I
50		11072.44	I
15		11551.6	I
1500	d	11710.37	I
40		11999.49	I
200		12165.08	I
10		12374.64	I
200		12690.04	I
100		12817.8	I
200		14330.5	I
50		16001.5	I
60		22551.6	I

Bi III
Ref. 359 — C.H.C.

Intensity		Wavelength	
		Vacuum	
1		590.73	III
5		670.76	III
4		775.16	III
6		803.65	III
7		920.93	III
6		925.48	III
25		1039.99	III
50	h	1045.76	III
30		1051.81	III
20		1139.01	III
15		1145.91	III
50		1224.64	III
40		1326.84	III
60		1346.12	III
35		1423.33	III
35		1423.52	III
60	h	1461.00	III
60	h	1606.40	III
20	h	1691.5	III
20		1834.32	III
10		1863.9	III
10		1912.12	III
10		1988.26	III
		Air	
20		2020.75	III
20		2021.15	III
10		2073.22	III
14		2073.37	III
15		2103.42	III
30		2213.55	III
75	h	2414.6	III
10		2437.6	III
30	h	2847.4	III
80	h	2855.6	III
35		3115.0	III
40	h	3451.0	III
40		3473.8	III
35		3485.5	III
15		3540.8	III
45		3613.4	III
50		3695.32	III
50		3695.68	III
12		4224.6	III
25		4327.8	III
30		4560.84	III
30		4561.54	III
40	h	4797.4	III
45	h	5079.3	III
12		6623.4	III
10	h	7381.	III
12		7551.	III
25		7598.	III
10	h	7637.	III
40		8008.	III
50		8070.	III
20		8100.	III
15		8671.	III
20		8934.	III

Bi IV
Ref. 360 — C.H.C.

Intensity		Wavelength	
		Vacuum	
6		420.7	IV
6		431.2	IV
6		790.5	IV
6		790.6	IV
8		792.5	IV
10		820.3	IV
9		822.9	IV
12		824.9	IV
15		872.6	IV
8		876.8	IV
9		916.7	IV
12		923.9	IV
15		943.3	IV
9		967.6	IV
8		968.8	IV
8		989.8	IV
24		1103.4	IV
7		1128.8	IV
6		1138.6	IV
6		1139.8	IV
7		1149.7	IV
60		1317.0	IV
30		1910.0	IV
		Air	
30		2093.	IV
100		2311.	IV
100		2326.	IV
100		2376.	IV
100		2629.	IV
100		2677.	IV
100		2767.	IV
100		2772.	IV
100		2786.	IV
100		2842.	IV
100		2924.	IV
100		2933.	IV
100		2936.	IV
100		3012.	IV
100		3042.	IV
100		3239.	IV
100		3643.	IV
100		3682.	IV
100		3734.	IV
100		3868.	IV
30		4342.	IV

Bismuth (Cont.)

Intensity		Wavelength	
30		5347.	IV

Bi V
Ref. 361 — C.H.C.

Intensity		Wavelength	
		Vacuum	
1		355.77	V
1		369.52	V
1		429.78	V
1		435.63	V
2		488.39	V
1		492.72	V
3		563.62	V
2		678.87	V
6		686.88	V
1		706.54	V
5		730.71	V
10		738.17	V
6		849.86	V
5		855.68	V
15	d	864.45	V
6		880.17	V
6		929.81	V
15	d	1139.46	V

BORON (B)
Z = 5

B I and II
Ref. 66, 104, 171, 222 — R.L.K.

Intensity	Wavelength	
	Vacuum	
70	693.95	II
40	731.36	II
40	731.44	II
110	882.54	II
110	882.68	II
40	984.67	II
110	1081.88	II
110	1082.07	II
110	1230.16	II
220	1362.46	II
70	1600.46	I
120	1600.73	I
160	1623.58	II
110	1623.77	II
220	1624.02	II
70	1624.16	II
160	1624.34	II
100	1663.04	I
150	1666.87	I
200	1667.29	I
150	1817.86	I
200	1818.37	I
300	1825.91	I
300	1826.41	I
110	1842.81	II
	Air	
250	2066.38	I
250	2066.65	I
100	2066.93	I
300	2067.19	I
500	2088.91	I
500	2089.57	I
70	2220.30	II
40	2323.03	II
40	2328.67	II
40	2393.20	II
220	2393.05	II
40	2459.69	II
40	2459.90	II
1000	2496.77	I
1000	2497.73	I
160	2918.08	II
110	3032.26	II
70	3179.33	II
110	3323.18	II
110	3323.60	II
450	3451.29	II
285	4121.93	II
110	4194.79	II
110	4472.10	II
110	4472.85	II
70	4784.21	II
110	4940.38	II
110	6080.44	II
70	6285.47	II
70	7030.20	II
40	7031.90	II
70	8668.57	I
20	8667.22	I
800	11660.04	I
570	11662.47	I
125	15629.08	I
200	16240.38	I
250	16244.67	I
235	18994.33	I

B III
Ref. 60, 221 — R.L.K.

Intensity	Wavelength	
	Vacuum	
150	518.24	III
75	518.27	III
40	411.80	III
20	510.77	III
40	510.85	III
110	677.00	III
160	677.14	III
40	758.48	III
70	758.67	III
20	1953.83	III
	Air	
550	2065.78	III
450	2067.23	III
160	2077.09	III
40	2234.09	III
70	2234.59	III
40	4242.98	III
70	4243.61	III
220	4487.05	III
360	4497.73	III
110	7835.25	III
70	7841.41	III

B IV
Ref. 74 — R.L.K.

Intensity	Wavelength	
	Vacuum	
10	52.68	IV
30	60.31	IV
160	344.0	IV
450	385.0	IV
285	418.7	IV
70	1112.2	IV
450	1168.9	IV
70	1170.9	IV
	Air	
70	2524.7	IV
160	2530.3	IV
450	2821.68	IV
70	2824.57	IV
285	2825.85	IV

B V
Ref. 94 — R.L.K.

Intensity	Wavelength	
	Vacuum	
30	41.00	V
	48.59	V
	194.37	V
	262.37	V
	512.53	V
	749.74	V

BROMINE (Br)
Z = 35

Br I and II
Ref. 122, 124, 240, 248, 316
G.V.S.

Intensity		Wavelength	
		Vacuum	
300		711.68	II
250		815.48	II
350		856.19	II
1000		889.23	II
500		896.64	II
500		905.99	II
300		922.56	II
1000		948.97	II
500		984.93	II
500		1012.10	II
1000		1015.54	II
500		1037.02	II
1000		1049.00	II
450		1064.76	II
500		1071.87	II
250		1101.50	I
300		1134.59	I
250		1136.29	I
250		1177.23	I
400		1178.90	I
1000		1189.28	I
230		1189.38	I
1000		1189.50	I
500		1198.37	I
800		1209.76	I
1000		1210.73	I
750		1216.01	I
1000		1221.13	I
900		1221.87	I
1000		1223.24	I
1200		1224.41	I
1200		1226.90	I
750		1228.05	I
7500		1232.43	I
1200		1243.90	I
800		1249.59	I
1500		1251.66	I
1000		1255.80	I
1500		1259.20	I
1200		1261.66	I
1200		1266.20	I
1000		1279.48	I
1000		1286.26	I
3000		1309.91	I
3000		1316.74	I
1000		1317.37	I
2000		1317.70	I
12000		1384.60	I
3000		1449.90	I
50000		1488.45	I
30000		1531.74	I
25000		1540.65	I
30000		1574.84	I
20000		1576.39	I
25000		1582.31	I
75000		1633.40	I
		Air	
350		2285.17	II
350		2287.60	II
500	h	2317.30	II
400		2336.93	II
350		2386.45	II
500		2386.70	II
300		2388.69	II
450		2388.96	II
500		2389.69	II
350		2392.21	II
400		2392.42	II
300		2400.50	II
300	h	2495.22	II
450		2521.70	II
400		2541.48	II
400		2556.92	II
350		2690.17	II
400		2713.77	II
350	h	2746.52	II
300	h	2807.55	II
400	h	2893.40	II
400	h	2917.18	II
400	h	2967.21	II
500	h	2972.26	II
300		2981.86	II
300	h	2985.87	II
300	h	2986.53	II
300	h	3016.48	II
350		3423.82	II
300	h	3606.80	II
350		3714.30	II
1200		3815.65	I
350		3834.69	II
300		3871.21	II
400		3891.63	II
300		3901.24	II
300		3914.20	II
500		3914.38	II
350		3919.51	II
400		3924.09	II
300		3929.55	II
350		3939.69	II
350		3950.61	II
500		3980.38	II
1500		3992.36	I
300		4024.04	II
300		4135.66	II
300		4140.20	II
400		4179.63	II
300		4193.45	II
1000		4223.89	II
300		4236.89	II
300		4291.39	II
2000		4365.14	I
1000		4365.60	II
1500		4425.14	I
10000		4441.74	I
10000		4472.61	I
20000		4477.72	I
1000		4490.42	I
3000		4513.44	I
15000		4525.59	I
300		4529.60	II
500		4542.92	II
3000		4575.74	I
300		4601.36	II
2500		4614.58	I
350		4622.70	II
300		4642.02	II
300		4651.98	II
500		4678.70	II
400		4693.17	II
500		4704.85	II
350		4719.76	II
400		4720.36	II
300		4728.20	II
300		4735.41	II
400		4742.64	II
2500		4752.28	I
350		4766.00	II
400		4779.40	II
4000		4780.31	I
1600		4785.19	I
500		4785.50	II
300		4802.33	II
500		4816.70	II
300		4818.46	II
350		4844.81	II
350		4848.75	II
400		4921.12	II
400		4928.79	II
450		4930.66	II
300		4945.51	II
4000		4979.76	I
300		5038.74	II
300		5054.64	II
400		5164.38	II
300		5180.01	II
500		5182.35	II
300		5193.90	II
500		[illegible]	II
300		5272.68	II
350		5304.10	II
400		5330.57	II
500		5332.05	II
1200		5395.48	I
400		5422.78	II
350		5424.99	II
300		5435.07	II
1200		5466.22	I
350		5478.47	II
300		5488.79	II
300		5495.06	II
500		5506.69	II
350		5589.94	II
300		5718.71	II
300		5830.78	II
1800		5852.08	I
1600		5940.48	I
2400		6122.14	I
40000		6148.60	I
300		6161.74	II
2000		6177.39	I
1500		6335.48	I
60000		6350.73	I
400		6352.94	II
2500		6410.32	I
1800		6483.56	I
1000		6514.62	I
20000		6544.57	I
1500		6548.09	I
50000	c	6559.80	I
1000		6571.31	I
1800		6579.14	I
20000		6582.17	I
1500		6620.47	I
50000	c	6631.62	I
20000		6682.28	I
10000		6692.13	I
8000		6728.28	I
2000		6760.06	I
2000		6779.48	I
2200		6786.74	I
6500		6790.04	I
1600	c	6791.48	I
1800		6861.15	I
10000		7005.19	I
2000		7260.45	I
10000		7348.51	I
40000		7512.96	I
1600		7591.61	I
1800		7595.07	I
2000		7616.41	I
30000		7803.02	I
1200		7827.23	I
2500	s	7881.45	I
2500		7881.57	I
2500		7925.81	I
30000	c	7938.68	I
3000		7947.94	I
3000		7950.18	I

Bromine (Cont.)

Intensity		Wavelength	
8000		7978.44	I
10000		7978.57	I
30000		7989.94	I
2000		8026.35	I
2500		8026.54	I
30000		8131.52	I
1000	c	8152.65	I
10000		8153.75	I
25000		8154.00	I
5000		8246.86	I
15000		8264.96	I
75000	c	8272.44	I
20000		8334.70	I
10000		8343.70	I
1200		8384.04	I
40000		8446.55	I
4000		8477.45	I
1500		8513.38	I
1000		8557.73	I
1000		8566.28	I
20000		8638.66	I
4000		8698.53	I
10000	c	8793.47	I
15000		8819.96	I
25000		8825.22	I
4000		8888.98	I
30000		8897.62	I
6000		8932.40	I
1800		8949.39	I
9000		8964.00	I
350		9024.42	II
30000		9166.06	I
15000		9173.63	I
20000		9178.16	I
40000		9265.42	I
15000		9320.86	I
300		9434.04	II
6000		9793.48	I
10000		9896.40	I
3000		10140.08	I
6000		10237.74	I
1000		10299.62	I
1500		10377.65	I
30000		10457.96	I
1000		10742.14	I
3000		10755.92	I
1700		13217.17	I
1800		14354.57	I
1250		14888.70	I
1800		16731.19	I
1200		18568.31	I
3500		19733.62	I
1000		20281.73	I
1000		20624.67	I
1200		21787.24	I
4000		22865.65	I
1000		23513.15	I
500		28346.50	I
500		30380.85	I
600		31630.13	I
120		32693.90	I
150		34181.87	I
150		38345.75	I
120		39964.36	I

Br III
Ref. 246, 250 — G.V.S.

Intensity		Wavelength	
		Vacuum	
450		611.1	III
300		620.4	III
500		665.54	III
500		677.19	III
300		677.8	III
450		687.68	III
400		690.2	III
350		696.99	III
300		727.0	III
300		736.4	III
250		769.63	III
250		817.79	III
350		949.0	III
400		960.4	III
450		984.9	III
250		1313.5	III
250		1402.9	III
		Air	
400	h	2293.44	III
300		2313.29	III
300		2462.39	III
300		2482.60	III
350		2499.25	III
350		2529.49	III
350	h	2551.09	III
350	h	2570.83	III
300	h	2573.17	III
400	h	2584.99	III
500		2589.14	III
300	h	2594.48	III
400	h	2595.98	III
450	h	2606.20	III
350	h	2608.15	III
500	h	2613.13	III
350	h	2616.26	III
500	h	2626.52	III
350	h	2629.23	III
350	h	2639.60	III
350	h	2671.53	III
350	h	2735.83	III
300		2770.50	III
300	h	2785.28	III
300		2804.16	III
400		2926.96	III
300		2936.22	III
350		2969.00	III
400		2994.04	III
500		3020.76	III
300		3033.63	III
350		3036.45	III
500		3074.42	III
350		3091.94	III
350		3117.29	III
300		3147.81	III
400		3174.08	III
300		3321.08	III
450		3333.07	III
500		3349.64	III
300		3385.25	III
450		3447.36	III
400		3487.58	III
300		3506.47	III
450		3517.36	III
500		3540.16	III
300		3551.08	III
500		3562.43	III
450		3600.71	III
250		3693.53	III
450		3820.26	III
200		3903.95	III
350		4506.55	III
200		4519.74	III
150		5175.87	III
100		5446.80	III
100		7192.8	III
100		7673.1	III

Br IV
Ref. 139, 142, 243, 249
G.V.S.

Intensity		Wavelength	
		Vacuum	
700		379.73	IV
700		400.37	IV
1000		545.43	IV
1000		559.76	IV
1000		569.19	IV
1000		576.59	IV
1000		585.10	IV
1000		586.71	IV
1000		597.51	IV
1000		600.09	IV
1000		601.27	IV
1000		607.03	IV
1000		617.85	IV
1000		619.87	IV
1000		630.14	IV
1000		642.23	IV
1000		661.53	IV
1000		683.51	IV
1000		697.72	IV
1000		715.39	IV
1000		731.00	IV
1000		800.12	IV
1000		813.66	IV
900		1274.82	IV
1000		1703.51	IV
		Air	
1000		2133.79	IV
1000		2145.02	IV
1000		2257.21	IV
1000		2272.73	IV
1000		2307.40	IV
1000		2408.16	IV
1000		2411.58	IV
700		2491.14	IV
1000		2581.19	IV
600		2661.40	IV
700		2820.87	IV
1000		2842.88	IV
1100	h	2907.71	IV
500		3041.18	IV
500		3380.56	IV

Br V
Ref. 42 — G.V.S.

Intensity		Wavelength	
		Vacuum	
600		468.37	V
800		482.11	V
900		531.97	V
1000		547.90	V
700		549.77	V
800		621.03	V
800		632.22	V
700		645.44	V
400		652.64	V
800		657.54	V
800		679.62	V
700		812.95	V
1000		850.81	V
150		855.27	V
600		1041.60	V
1000		1069.15	V
500		1080.54	V
900		1112.13	V
1000		1143.56	V
150		1429.75	V
400		1442.60	V
150		1470.35	V

CADMIUM (Cd)
Z = 48

Cd I and II
Ref. 44, 285, 296 — R.D.C.

Intensity		Wavelength	
		Vacuum	
100		1256.00	II
150		1296.43	II
100		1326.50	II
150		1370.91	II
200		1514.26	II
200		1571.58	II
100		1668.60	II
50		1702.47	II
50		1724.41	II
100		1785.84	II
100		1827.70	II
300		1922.23	II
100		1943.54	II
40		1965.54	II
30		1986.89	II
200		1995.43	II
		Air	
100		2007.49	II
50		2032.45	II
75		2036.23	II
150		2096.00	II
1000	r	2144.41	II
50		2155.06	II
100		2187.79	II
1000		2194.56	II
1000		2265.02	II
1500	r	2288.022	I
1000		2312.77	II
200		2321.07	II
40		2376.82	II
50		2418.69	II
50		2469.73	II
40		2487.93	II
3		2491.00	I
40		2495.58	II
10		2508.91	I
50		2509.11	II
30		2516.22	II
15	h	2518.59	I
25	h	2525.196	I
50		2544.613	I
50		2551.98	II
25		2553.465	I
3		2565.789	I
500		2572.93	II
50		2580.106	I
3		2584.87	I
30		2592.026	I
25	h	2602.048	I
50		2628.979	I
40		2632.190	I
75		2639.420	I
40		2659.23	II
50	h	2660.325	I
25		2668.20	II
50		2672.62	II
100		2677.540	I
25		2677.748	I
50		2707.00	II
75		2712.505	I
50		2733.820	I
1000		2748.54	II
100	h	2763.894	I
50	h	2764.230	I
50		2774.958	I
30		2823.19	II
200		2836.900	I
25		2856.46	II
100		2868.180	I
200	r	2880.767	I
50	r	2881.224	I
200		2914.67	II
50		2927.87	II
200		2929.27	II
1000	r	2980.620	I
200	r	2981.362	I
50		2981.845	I
50		3030.60	II
150		3080.822	I
25		3081.48	II
30		3082.593	I
100		3092.34	II
200		3133.167	I
50		3146.79	II
150		3250.33	II
300		3252.524	I
300		3261.055	I
50		3343.21	II
50		3385.49	II
30		3388.88	II
800		3403.652	I
50		3417.49	II
50		3442.42	II
100		3464.43	II
1000		3466.200	I
800		3467.655	I
25		3483.08	II
150		3495.44	II
25		3499.952	I
100		3524.11	II
100		3535.69	II
1000		3610.508	I
800		3612.873	I
60		3614.453	I
20		3649.558	I
10		3981.926	I
100		4029.12	II
200		4134.77	II
50		4141.49	II
100		4285.08	II
8		4306.672	I
100		4412.41	II
3		4412.989	I
1000		4415.63	II
30		4440.45	II
8		4662.352	I
200		4678.149	I
30		4744.69	II
300		4799.912	I
50		4881.72	II
50		5025.50	II
1000	h	5085.822	I
6		5154.660	I
100		5268.01	II
100		5271.60	II
1000		5337.48	II
1000		5378.13	II
200		5381.89	II
40		5843.30	II
50		5880.22	II
300		6099.142	I
100		6111.49	I
100		6325.166	I
30		6330.013	I
400		6354.72	II
500		6359.98	II
2000		6438.470	I
400		6464.94	II
25		6567.65	II
500		6725.78	II
100		6759.19	II
30		6778.116	I
50		7237.01	II
100		7284.38	II
1000		7345.670	I
50		8066.99	II
5		8200.309	I
20		9292	I
15		11655	I
35		14491	I
80		15712	I
55	d	19125	I
25		24378	I
35		25455	I

Cd III
Ref. 296 — R.D.C.

Intensity		Wavelength	
		Vacuum	
8		677.39	III

Cadmium (Cont.)

Intensity	Wavelength	
15	684.58	III
10	720.70	III
5	1383.60	III
15	1392.10	III
10	1396.78	III
25	1416.28	III
5	1420.29	III
8	1420.54	III
5	1432.86	III
20	1446.08	III
25	1447.55	III
30	1455.74	III
8	1466.14	III
5	1471.97	III
15	1491.81	III
5	1511.01	III
10	1511.65	III
5	1513.13	III
10	1523.55	III
15	1528.40	III
30	1529.30	III
5	1532.10	III
50	1545.17	III
25	1547.57	III
10	1550.07	III
20	1550.45	III
15	1550.89	III
5	1552.18	III
5	1556.48	III
15	1560.66	III
5	1566.03	III
15	1568.98	III
10	1582.39	III
40	1601.59	III
20	1604.87	III
20	1606.64	III
10	1607.28	III
15	1608.91	III
10	1609.61	III
15	1612.51	III
15	1625.27	III
25	1628.54	III
20	1651.87	III
25	1655.63	III
30	1678.15	III
10	1699.70	III
40	1707.16	III
10	1721.93	III
40	1722.95	III
30	1725.66	III
25	1739.00	III
5	1745.69	III
40	1747.67	III
15	1748.15	III
30	1768.82	III
40	1773.06	III
30	1789.19	III
75	1793.40	III
15	1796.10	III
5	1800.57	III
40	1823.41	III
50	1844.66	III
40	1851.13	III
20	1851.37	III
40	1855.85	III
200	1856.67	III
150	1874.08	III
15	1886.49	III
15	1903.48	III
25	1909.98	III
15	1910.57	III
10	1939.59	III
15	1988.81	III
	Air	
20	2000.60	III
15	2004.07	III
15	2016.12	III
40	2039.83	III
50	2045.61	III
10	2061.25	III
75	2087.91	III
10	2097.45	III
5	2100.47	III
50	2111.60	III
5	2188.13	III
5	2218.43	III
5	2224.43	III
7	2418.24	III
10	2426.36	III
25	2499.81	III
15	2618.81	III
5	2630.56	III
20	2766.99	III
30	2805.59	III
10	3035.72	III

Cd IV
Ref. 353, 399 — R.D.C.

Intensity	Wavelength	
	Vacuum	
50	427.01	IV
20	437.88	IV
50	447.85	IV
60	480.90	IV
10	489.49	IV
70	493.00	IV
70	495.13	IV
70	498.14	IV
70	498.53	IV
80	504.09	IV
70	504.20	IV
70	504.50	IV
80	506.31	IV
60	508.01	IV
50	508.95	IV
70	509.55	IV
25	509.81	IV
70	511.40	IV
80	513.00	IV
70	514.50	IV
60	519.42	IV
15	520.97	IV
80	524.41	IV
70	524.47	IV
50	524.77	IV
70	525.10	IV
60	525.19	IV
70	527.07	IV
50	530.79	IV
80	531.09	IV
80	531.51	IV
70	534.29	IV
70	536.77	IV
50	537.24	IV
60	540.90	IV
70	541.74	IV
80	542.60	IV
80	546.55	IV
40	548.01	IV
20	548.33	IV
15	548.90	IV
25	551.27	IV
10	552.90	IV
60	553.06	IV
80	554.05	IV
25	560.26	IV
10	564.16	IV
60	567.01	IV
20	1062.23	IV
150	1118.16	IV
30	1126.00	IV
20	1134.08	IV
15	1139.04	IV
20	1154.64	IV
10	1155.73	IV
100	1164.65	IV
20	1165.78	IV
40	1167.30	IV
20	1179.73	IV
15	1183.07	IV
100	1183.40	IV
40	1194.13	IV
20	1195.63	IV
30	1196.47	IV
20	1198.93	IV
15	1215.38	IV
20	1223.52	IV
20	1246.06	IV
15	1246.56	IV
15	1249.94	IV
30	1266.47	IV
15	1274.41	IV
20	1285.63	IV
20	1287.58	IV
20	1299.46	IV
40	1304.36	IV
30	1306.07	IV
15	1316.89	IV
15	1321.85	IV
20	1325.55	IV
30	1340.97	IV
15	1346.15	IV
30	1354.78	IV
20	1358.11	IV
30	1362.55	IV
60	1370.48	IV
30	1380.98	IV
20	1397.65	IV
30	1403.68	IV
15	1406.58	IV
60	1418.89	IV
15	1429.83	IV
20	1447.54	IV
20	1452.63	IV
15	1465.97	IV
15	1466.67	IV
15	1482.95	IV
20	1491.79	IV
20	1570.20	IV
20	1598.73	IV
20	1600.42	IV
15	1622.87	IV

CALCIUM (Ca)
Z = 20

Ca I and II
Ref. 70, 150, 270 — J.J.W. and H.H.S.

Intensity	Wavelength	
	Vacuum	
24	1341.89	II
12	1342.54	II
20	1433.75	II
12	1432.50	II
20	1553.18	II
32	1554.64	II
4	1642.80	II
20	1643.77	II
36	1644.44	II
60	1649.86	II
32	1651.99	II
12	1673.86	II
20	1680.05	II
2	1680.13	II
8	1691.78	II
16	1698.18	II
20	1807.34	II
40	1814.50	II
4	1814.65	II
60	1840.06	II
40	1838.01	II
20	1843.09	II
40	1850.69	II
	Air	
	2103.24	II
	2112.76	II
	2113.15	II
	2128.75	II
	2131.51	II
	2132.30	II
2	2150.80	I
	2197.79	II
5	2200.73	I
	2208.61	II
6	2275.46	I
8	2398.56	I
7	2721.65	I
9	2994.96	I
8	2997.31	I
8	2999.64	I
9	3000.86	I
10	3006.86	I
9	3009.21	I
2	3024.94	I
2	3034.54	I
2	3045.74	I
3	3055.32	I
2	3071.57	I
2	3076.95	I
2	3080.79	I
2	3099.30	I
10	3125.18	II
5	3136.02	I
6	3140.79	I
7	3150.75	I
170	3158.87	II
180	3179.33	II
5	3180.52	I
150	3181.28	II
7	3209.96	I
8	3215.17	I
6	3215.34	I
9	3225.90	I
6	3226.15	I
5	3274.67	I
6	3286.07	I
10	3308.02	II
20	3316.51	II
10	3344.51	I
10	3347.04	II
11	3350.21	I
9	3350.36	I
12	3361.92	I
9	3362.14	I
10	3452.66	II
20	3461.87	II
9	3468.48	I
11	3474.76	I
10	3485.61	II
13	3487.60	I
10	3495.16	II
15	3624.11	I
17	3630.75	I
14	3630.97	I
20	3644.41	I
14	3644.77	I
8	3644.99	I
5	3675.29	I
6	3678.21	I
30	3683.70	II
40	3694.11	II
10	3694.36	II
170	3706.03	II
180	3736.90	II
10	3739.38	II
6	3748.35	I
8	3750.29	I
9	3753.34	I
20	3755.67	II
30	3758.39	II
9	3870.48	I
11	3872.54	I
11	3872.56	I
12	3875.78	I
12	3875.80	I
6	3889.10	I
6	3923.48	I
230	3933.66	II
9	3935.29	I
6	3946.04	I
15	3948.90	I
17	3957.05	I
220	3968.47	II
8	3972.57	I
18	3973.71	I
50	4097.10	II
15	4098.53	I
15	4098.57	I
60	4109.82	II
30	4110.28	II
40	4206.18	II
50	4220.07	II
50	4226.73	I
15	4240.46	I
24	4283.01	I
22	4289.36	I
22	4298.99	I
25	4302.53	I
23	4307.74	I
22	4318.65	I
20	[illegible]	I
25	4425.44	I
26	4434.96	I
25	4435.69	I
30	4454.78	I
28	4455.89	I
20	4456.61	I
20	4472.04	II
10	4479.23	II
20	4489.18	II
23	4526.94	I
22	4578.55	I
23	4581.40	I
23	4581.47	I
24	4585.87	I
24	4585.96	I
20	4685.27	I
30	4716.74	II
40	4721.03	II
40	4799.97	II
25	4878.13	I
70	5001.48	II
80	5019.97	II
40	5021.14	II
23	5041.62	I
25	5188.85	I
22	5261.71	I
23	5262.24	I
22	5264.24	I
24	5265.56	I
25	5270.27	I
60	5285.27	II
70	5307.22	II
50	5339.19	II
27	5349.47	I
23	5512.98	I
25	5581.97	I
27	5588.76	I
24	5590.12	I
26	5594.47	I
25	5598.49	I
24	5601.29	I
24	5602.85	I
30	5857.45	I
10	5922.72	II
10	5923.69	II
27	6102.72	I
29	6122.22	I
22	6161.29	I
30	6162.17	I
22	6163.76	I
24	6166.44	I
26	6169.06	I
28	6169.56	I
35	6439.07	I
30	6449.81	I
22	6455.60	I
80	6456.87	II
34	6462.57	I
29	6471.66	I

Calcium (Cont.)

Intensity	Wavelength	
32	6493.78	I
28	6499.65	I
23	6572.78	I
30	6717.69	I
33	7148.15	I
31	7202.19	I
	7291.47	II
	7323.89	II
33	7326.15	I
30	7575.81	II
60	7581.11	II
80	7601.30	II
20	7602.32	II
40	7820.78	II
60	7843.38	II
20	8017.50	II
20	8020.50	II
70	8133.05	II
100	8201.72	II
110	8248.80	II
70	8254.73	II
14	8256.67	I
10	8338.04	I
12	8339.12	I
10	8352.39	I
11	8357.17	I
130	8498.02	II
170	8542.09	II
10	8633.95	I
160	8662.14	II
12	8842.61	I
15	8909.18	I
100	8912.07	II
110	8927.36	II
12	8967.47	I
16	9099.10	I
13	9105.62	I
12	9108.82	I
10	9171.14	I
110	9213.90	II
90	9312.00	II
100	9319.56	II
110	9320.65	II
25	9416.97	I
10	9456.80	I
10	9534.88	I
11	9548.38	I
100	9567.97	II
110	9599.24	II
80	9601.82	II
10	9604.28	I
12	9663.65	I
10	9664.41	I
14	9676.30	I
14	9688.67	I
13	9701.94	I
80	9854.74	II
110	9890.63	II
90	9931.39	II
100	10223.04	II
20	10343.81	I
13	10838.97	I
13	10861.58	I
13	10863.87	I
14	10869.50	I
14	10879.87	I
20	11838.99	II
10	11949.72	II
25	12816.04	I
24	12823.86	I
25	12909.10	I
30	13033.57	I
21	13086.44	I
24	13134.95	I
20	16150.77	I
22	16157.36	I
21	16197.04	I
20	18925.47	I
24	18970.14	I
30	19046.14	I
48	19309.20	I
49	19452.99	I
47	19505.72	I
50	19776.79	I
35	19853.10	I
34	19862.22	I
23	19917.19	I
24	19933.70	I
	21389.00	II
	21428.90	II
20	22607.93	I
25	22624.93	I
30	22651.23	I

Ca III
Ref. 25, 16 — J.J.W. and H.H.S.

Intensity		Wavelength	
		Vacuum	
6		296.96	III
9		403.72	III
7		409.95	III
5		439.69	III
5		633.59	III
5		685.41	III
5		697.55	III
5		699.09	III
5		699.89	III
6		701.39	III
5		727.66	III
8		740.55	III
6		746.25	III
5		747.98	III
5		779.61	III
5		800.30	III
5		809.93	III
5		817.06	III
6		821.57	III
6		840.56	III
6		1020.07	III
5		1034.65	III
5		1187.30	III
8		1188.61	III
8		1188.61	III
5		1190.86	III
10		1262.65	III
11		1278.39	III
10		1281.55	III
12		1286.52	III
12		1298.04	III
11		1317.70	III
10		1328.95	III
11		1335.13	III
10		1360.01	III
11		1385.43	III
11		1397.69	III
13		1453.16	III
12		1459.79	III
11		1461.88	III
15		1463.34	III
16		1484.87	III
12		1496.88	III
11		1506.88	III
20		1545.29	III
15		1555.53	III
18		1562.47	III
13		1571.27	III
13		1586.13	III
10		1762.26	III
10		1783.93	III
10		1794.22	III
12		1800.21	III
13		1807.89	III
14		1812.15	III
11		1813.59	III
12		1830.06	III
10		1860.43	III
14		1870.26	III
14		1872.37	III
10		1894.12	III
11		1910.10	III
12		1935.72	III
10		1939.68	III
11		1943.01	III
12		1948.26	III
10		1953.55	III
10		1958.97	III
13		1964.61	III
13		1967.94	III
12		1972.82	III
10		1977.01	III
10		1978.55	III
11		1981.19	III
		Air	
12		2033.36	III
12		2041.53	III
13		2078.92	III
13		2098.49	III
15		2114.41	III
17		2123.03	III
14		2129.19	III
14		2133.96	III
13		2140.36	III
16		2152.43	III
12		2171.57	III
12		2276.52	III
15		2312.08	III
14		2497.74	III
15		2541.50	III
13		2587.15	III
12		2590.41	III
15		2620.82	III
15		2634.14	III
12		2686.72	III
16		2687.76	III
15		2704.86	III
14		2771.28	III
15		2791.59	III
16		2813.88	III
17		2866.54	III
18		2869.95	III
19		2881.78	III
21		2899.79	III
19		2924.33	III
20		2988.63	III
18		2989.27	III
15		3028.59	III
19		3119.67	III
15		3367.79	III
19		3372.67	III
18		3537.77	III
15		4081.77	III
15		4153.57	III
15		4164.31	III
15		4184.20	III
18		4207.24	III
17		4233.74	III
16		4240.74	III
15		4284.39	III
20		4302.81	III
15		4329.19	III
16		4333.57	III
15		4358.38	III
19		4399.59	III
17		4406.29	III
17		4431.30	III
19		4499.88	III
18		4516.59	III
18		4572.12	III
11		4708.83	III
11		4716.27	III
10		4859.17	III
10		5008.95	III
10		5050.07	III
10		5231.82	III
11		5247.37	III
13		5271.98	III
10		5301.32	III
11		5321.29	III
10		5328.06	III
11		5570.58	III
10		5579.06	III
13		6069.98	III
10		6173.22	III
12		6213.98	III
11		6294.89	III
11		6370.11	III
10		6387.55	III
12		6424.51	III
12		6485.35	III
10		6538.78	III
10		6542.24	III
10		7308.69	III
10		7843.06	III
12		7898.46	III
10	l	8217.20	III

Ca IV
Ref. 150 — J.J.W. and H.H.S.

Intensity		Wavelength	
		Vacuum	
150		249.41	IV
150		250.15	IV
150		251.35	IV
250		296.55	IV
200		299.32	IV
200		318.09	IV
50		318.39	IV
120		321.59	IV
250		329.12	IV
150		329.39	IV
200		331.44	IV
250		331.99	IV
235		332.53	IV
150		332.81	IV
200		338.83	IV
150		339.79	IV
200		340.29	IV
200		341.29	IV
200		341.46	IV
250	c	342.45	IV
100		343.19	IV
200		343.44	IV
250		343.93	IV
200		344.96	IV
215		345.13	IV
250		374.74	IV
600		434.57	IV
100		437.27	IV
250		437.77	IV
200		438.93	IV
750		443.82	IV
50		445.02	IV
500		450.57	IV
50		454.55	IV
250		456.98	IV
250		461.09	IV
150		565.46	IV
750		656.00	IV
500		669.70	IV

Ca V
Ref. 150 — J.J.W. and H.H.S.

Intensity		Wavelength	
		Vacuum	
200		190.36	V
250		190.46	V
250		196.97	V
300		199.55	V
250		200.51	V
265		257.98	V
165		260.45	V
400		267.77	V
300		270.31	V
200		271.14	V
250		272.27	V
200		272.98	V
400		280.99	V
300		284.98	V
450	c	286.96	V
500		322.17	V
250		322.76	V
300		323.22	V
250		324.48	V
250		325.28	V
300		330.94	V
200		333.44	V
200		333.57	V
300		334.55	V
250	c	335.34	V
200		336.55	V
200		337.54	V
250		338.06	V
200		343.64	V
450		352.92	V
250		356.25	V
250		377.18	V
200		387.08	V
750		425.00	V
500		558.60	V
400		637.93	V
300		643.12	V
400		646.57	V
250		647.88	V
250		651.55	V
300		656.76	V

CALIFORNIUM (Cf)
Z = 98

Cf I and II
Ref. 52, 331 — J.G.C.

Intensity		Wavelength	
		Air	
10000		2739.31	
10000	s	2759.10	
10000		2774.52	
10000	l	2852.03	
10000	s	2855.24	
10000		3298.14	
10000		3352.71	
10000	l	3367.79	
10000		3392.22	I
10000		3481.07	
10000	l	3513.47	
10000		3531.49	I
10000		3540.98	I
10000		3598.77	I
10000		3605.32	I
10000		3612.11	II
10000		3617.49	I
10000	s	3626.76	II
10000		3659.46	
10000		3662.70	I
10000	s	3699.49	
10000	l	3722.11	II
10000		3739.35	I
10000		3785.61	I
10000	l	3789.04	II
10000	s	3893.23	II
10000	l	3993.57	II
10000		4035.45	
10000		4099.12	I
10000		4242.38	I
10000		4329.03	I
10000		4335.22	I
10000		5173.96	I
10000		5179.08	I
10000		5219.24	I
10000	s	5279.01	
10000	s	5320.09	
10000	s	5339.13	
10000		5408.88	I
10000		5726.05	I
10000		6622.83	I
10000		6631.26	I

Californium (Cont.)

Intensity		Wavelength	
10000		6677.90	I
10000		6894.59	I
10000		6927.10	II
10000	l	7074.52	I
10000	s	7307.90	I
10000		8141.29	I
10000		8241.77	I
10000		8333.85	II
10000		8423.49	II
10000		8568.83	II
10000	l	9228.52	
10000		9337.70	
10000		9649.51	
10000	s	10308.41	
10000	l	10568.83	
10000	s	10614.84	
10000		11300.19	
10000		11681.85	
10000		11941.33	
10000	l	12183.05	
10000	s	12352.72	
10000	s	12437.48	
10000	l	12789.41	
10000	l	13329.98	
10000	s	13362.98	
10000	l	13376.89	
10000	l	13474.44	
10000	l	14772.49	
10000	s	15281.32	
10000		15587.12	
10000		15675.92	
10000		16759.06	
10000	s	17626.25	
10000	l	18718.69	
10000	h	19068.71	
10000	l	19336.96	
10000	l	19576.84	
10000	l	20393.38	
10000	s	20869.98	

CARBON (C)
Z = 6

C I and II
Ref. 211 — R.L.K.

Intensity		Wavelength	
		Vacuum	
9		595.022	II
30		687.053	II
50		687.345	II
10		858.092	II
20		858.559	II
30		903.624	II
60		903.962	II
150		904.142	II
30		904.480	II
9		1009.86	II
10		1010.08	II
10		1010.37	II
80		1036.337	II
150		1037.018	II
150		1157.910	I
150		1158.019	I
150		1158.035	I
150		1188.992	I
150		1189.447	I
200		1189.631	I
300		1193.009	I
300		1193.031	I
300		1193.240	I
300		1193.264	I
100		1193.393	I
150		1193.649	I
150		1193.679	I
100		1194.064	I
100		1194.488	I
100		1261.552	I
250		1277.245	I
250		1277.282	I
300		1277.513	I
300		1277.550	I
200		1280.333	I
100		1311.363	I
9		1323.951	II
120		1329.578	I
120		1329.600	I
150		1334.532	II
300		1335.708	II
100		1354.288	I
150		1355.84	I
120		1364.164	I
100		1459.032	I
200		1463.336	I
120		1467.402	I
150		1481.764	I
150		1560.310	I
400		1560.683	I
400		1560.708	I
100		1561.341	I
400		1561.438	I
150		1656.266	I
120		1656.928	I
300		1657.008	I
120		1657.380	I
120		1657.907	I
150		1658.122	I
500		1751.823	I
1000		1930.905	I
		Air	
800		2478.56	I
250		2509.12	II
350		2512.06	II
250	h	2574.83	II
350	l	2741.28	II
250		2746.49	II
1000		2836.71	II
800		2837.60	II
800	h	2992.62	II
350		3876.19	II
350		3876.41	II
350		3876.66	II
570		3918.98	II
800		3920.69	II
250		4074.52	II
350	l	4075.85	II
800		4267.00	II
1000		4267.26	II
200		4771.75	I
200		4932.05	I
200		5052.17	I
350		5132.94	II
350		5133.28	II
350		5143.49	II
570		5145.16	II
400		5151.09	II
300		5380.34	I
250		5648.07	II
350		5662.47	II
570		5889.77	II
350		5891.59	II
200		[illegible]	I
250		6006.03	I
110		6007.18	I
150		6010.68	I
300		6013.22	I
250		6014.84	I
800		6578.05	II
570		6582.88	II
200		6587.61	I
250		6783.90	II
250		7113.18	I
250		7115.19	I
250		7115.63	II
200		7116.99	I
350		7119.90	II
800		7231.32	II
1000		7236.42	II
200		7860.89	I
200		8058.62	I
520		8335.15	I
250		9061.43	I
200		9062.47	I
200		9078.28	I
250		9088.51	I
450		9094.83	I
300		9111.80	I
800		9405.73	I
150		9603.03	I
250		9620.80	I
300		9658.44	I
200		10683.08	I
300		10691.25	I
12		11619.29	I
23		11628.83	I
13		11658.85	I
47		11659.68	I
14		11669.63	I
85		11748.22	I
142		11753.32	I
114		11754.76	I
11		11777.54	I
17		11892.91	I
30		11895.75	I
26		12614.10	I
20		13502.27	I
38		14399.65	I
16		14403.25	I
61		14420.12	I
12		14429.03	I
13		14442.24	I
12		16559.66	I
50		16890.38	I
10		17338.56	I
11		17448.60	I
13		18139.80	I
23		19721.99	I

C III
Ref. 22, 211 — R.L.K.

Intensity		Wavelength	
		Vacuum	
250		371.69	III
250		371.75	III
150		371.78	III
500		386.203	III
200		450.734	III
400		459.46	III
500		459.52	III
570		459.63	III
250		511.522	III
250		535.288	III
300		538.080	III
350		538.149	III
400		538.312	III
350		574.281	III
800		977.03	III
370		1174.93	III
350		1175.26	III
330		1175.59	III
500		1175.71	III
350		1175.99	III
370		1176.37	III
		Air	
250		2162.94	III
800		2296.87	III
150		2697.75	III
110	l	2724.85	III
150	l	2725.30	III
150	l	2725.90	III
200		2982.11	III
150		4056.06	III
200		4067.94	III
250		4068.91	III
250		4070.26	III
150		4162.86	III
250	h	4186.90	III
200		4325.56	III
600		4647.42	III
520		4650.25	III
375		4651.47	III
200		4665.86	III
450		5695.92	III
150		5826.42	III
150		6744.38	III
150	h	7037.25	III
150		7612.65	III
300	h	8196.48	III
150		8332.99	III
300		8500.32	III

C IV
Ref. 66, 211 — R.L.K.

Intensity		Wavelength	
		Vacuum	
250		244.91	IV
200		289.14	IV
250		289.23	IV
570		312.42	IV
500		312.46	IV
650		384.03	IV
700		384.18	IV
400		419.52	IV
500		419.71	IV
1000		1548.202	IV
900		1550.774	IV
		Air	
200	l	2524.41	IV
300	s	2529.98	IV
200	w	4658.30	IV
250		5801.33	IV
200		5811.98	IV
90	w	7726.2	IV

C V
Ref. 211 — R.L.K.

Intensity		Wavelength	
		Vacuum	
110		34.973	V
450		40.268	V
110		227.19	V
160		248.66	V
160		248.74	V
		Air	
40		2270.91	V
5		2277.25	V
20		2277.92	V
5		4943.88	V
5		4944.56	V

CERIUM (Ce)
Z = 58

Ce I and II
Ref. 1 — C.H.C.

Intensity	Wavelength	
	Air	
130	2462.97	II
110	2518.51	II
200	2548.68	II
340	2651.01	II
120	2696.07	II
120	2706.88	II
120	2723.38	II
110	2741.96	II
100	2750.89	II
150	2761.42	II
120	2784.27	II
100	2785.35	II
100	2790.53	II
140	2791.42	II
270	2830.90	II
100	2833.31	II
250	2874.14	II
110	2908.42	II
100	2918.67	II
120	2955.94	II
110	2964.80	II
100	2972.58	II
400	2976.91	II
150	2977.46	II
120	2980.41	II
250	2990.87	II
110	[illegible]	II
320	2995.64	II
400	3008.79	II
370	3017.20	II
210	3037.73	II
200	3051.98	II
350	3055.24	II
320	3056.78	II
680	3063.01	II
320	3083.67	II
250	3084.44	II
200	3090.37	II
370	3103.38	II
200	3107.47	II
320	3110.28	II
300	3111.17	II
220	3127.53	II
200	3130.33	II
240	3130.87	II
200	3144.60	II
290	3145.28	II
290	3146.41	II
290	3164.15	II
290	3169.18	II
290	3171.61	II
480	3183.52	II
240	3186.13	II
200	3190.34	II
710	3194.83	II
200	3199.28	II
990	3201.71	II
200	3218.38	II
710	3218.94	II
880	3221.17	II
330	3225.67	II
710	3227.11	II
240	3229.36	II
480	3231.24	II
710	3234.16	II
330	3234.89	II
390	3236.74	II
390	3243.37	II
200	3246.67	II
200	3260.98	II
200	3263.88	II
990	3272.25	II
330	3274.86	II
200	3279.84	II
330	3285.22	II
240	3295.28	II
200	3296.88	II
220	3300.15	II
240	3304.84	II
200	3312.22	II
240	3314.72	II
200	3317.80	II
200	3334.46	II
240	3341.87	II
330	3343.86	II
440	3344.76	II
200	3355.02	II

Cerium (Cont.)

240	3357.22	II
200	3360.54	II
240	3366.55	II
200	3371.18	II
200	3373.46	II
200	3373.73	II
480	3377.13	II
200	3383.68	II
200	3404.91	II
240	3405.98	II
290	3417.45	II
600	3422.71	II
390	3426.21	II
290	3441.21	II
480	3476.84	II
240	3482.35	II
710	3485.05	II
210	3507.94	II
600	3517.38	II
210	3520.52	II
330	3521.88	II
210	3526.68	II
600	3534.05	II
770	3539.08	II
210	3545.60	II
290	3546.19	II
240	3552.73	II
420	3555.00	II
1200	3560.80	II
210	3576.23	II
1000	3577.45	II
330	3590.60	II
390	3607.63	II
550	3609.69	II
420	3613.70	II
440	3622.15	II
380	3623.74	II
440	3623.84	II
200	3631.19	II
350	3646.97	II
260	3647.75	II
260	3647.95	II
420	3653.11	II
660	3653.67	II
310	3654.97	II
1800	3655.85	II
440	3659.23	II
350	3659.97	II
880	3660.64	II
880	3667.98	II
220	3672.18	I
350	3672.79	II
220	3679.42	II
300	3694.91	II
220	3704.98	II
1000	3709.29	II
1000	3709.93	II
1400	3716.37	II
420	3718.19	II
420	3718.38	II
210	3719.80	II
420	3725.68	II
490	3728.02	II
800	3728.42	II
320	3748.06	II
250	3751.45	II
200	3755.43	II
300	3762.98	II
680	3764.12	II
200	3765.04	II
300	3768.76	II
210	3770.76	II
300	3771.60	II
250	3776.61	II
620	3781.62	II
440	3782.52	II
200	3783.58	II
860	3786.63	II
520	3788.75	II
300	3792.32	II
2500	3801.52	II
800	3803.09	II
1000	3808.11	II
490	3809.21	II
250	3812.20	II
490	3815.85	II
470	3817.46	II
300	3819.02	II
470	3823.90	II
470	3830.55	II
490	3831.08	II
490	3834.55	II
270	3836.10	II
1100	3838.54	II
200	3843.76	II
220	3846.52	II
250	3848.10	II
860	3848.59	II
860	3853.15	II
1200	3854.18	II
1200	3854.31	II
620	3855.29	II
390	3857.02	II
370	3857.64	II
200	3862.46	II
200	3868.13	II
270	3874.68	II
620	3876.97	II
1100	3878.36	II
1500	3882.45	II
1000	3889.98	II
210	3890.75	II
210	3890.98	II
620	3895.11	II
590	3896.80	II
490	3898.27	II
270	3898.94	II
200	3903.34	II
250	3904.34	II
200	3906.92	II
770	3907.29	II
560	3908.41	II
390	3908.54	II
270	3909.31	II
230	3912.19	II
980	3912.44	II
390	3915.52	II
390	3916.14	II
230	3917.64	II
770	3918.28	II
480	3919.81	II
590	3921.73	II
560	3923.11	II
450	3924.64	II
770	3931.09	II
310	3931.37	II
230	3931.83	II
310	3933.73	II
560	3938.09	II
770	3940.34	II
310	3940.97	II
2000	3942.15	II
2700	3942.75	II
770	3943.89	II
310	3947.97	II
3100	3952.54	II
340	3953.66	II
310	3955.36	II
230	3956.06	II
980	3956.28	II
230	3958.27	II
230	3958.87	II
770	3960.91	II
390	3964.50	II
770	3967.05	II
450	3971.68	II
270	3972.07	II
270	3975.07	II
770	3978.65	II
560	3980.88	II
560	3982.89	II
310	3983.29	II
770	3984.68	II
370	3989.44	II
700	3992.39	II
370	3992.91	II
910	3993.82	II
2800	3999.24	II
230	4001.56	II
910	4003.77	II
370	4005.64	II
210	4007.59	II
2700	4012.39	II
910	4014.90	II
250	4015.88	II
200	4019.04	II
240	4022.27	II
840	4024.49	II
240	4025.15	II
840	4028.41	II
250	4030.34	II
840	4031.34	II
340	4037.67	II
2100	4040.76	II
910	4042.58	II
230	4045.21	II
620	4046.34	II
210	4051.43	II
210	4051.99	II
700	4053.51	II
450	4054.99	II
280	4062.22	II
230	4062.94	II
280	4067.28	II
420	4068.84	II
1100	4071.81	II
270	4072.92	II
1800	4073.48	II
210	4073.74	II
1500	4075.71	II
1500	4075.85	II
210	4076.24	II
420	4077.47	II
530	4078.32	II
270	4078.52	II
270	4080.44	II
670	4081.22	II
910	4083.23	II
450	4085.23	II
250	4087.36	II
230	4088.85	II
450	4101.77	II
250	4105.00	II
510	4107.42	II
200	4110.38	II
250	4111.39	II
420	4115.37	II
250	4117.01	II
200	4117.29	II
200	4117.59	II
770	4118.14	II
250	4119.02	II
310	4119.79	II
310	4119.88	II
450	4120.83	II
510	4123.24	II
510	4123.49	II
980	4123.87	II
510	4124.79	II
980	4127.37	II
250	4127.74	II
200	4128.07	II
530	4130.71	II
480	4131.10	II
2700	4133.80	II
270	4135.44	II
270	4137.47	II
2000	4137.65	II
270	4138.10	II
210	4138.35	II
770	4142.40	II
390	4144.49	II
670	4145.00	II
480	4146.23	II
280	4148.90	II
420	4149.79	II
980	4149.94	II
420	4150.91	II
1400	4151.97	II
230	4153.13	II
450	4159.03	II
310	4163.52	II
1300	4165.61	II
620	4166.88	II
250	4167.80	II
320	4169.77	II
320	4169.88	II
340	4176.70	II
340	4181.08	II
340	4185.33	II
3500	4186.60	II
530	4187.32	II
560	4193.09	II
370	4193.28	II
370	4193.87	II
630	4196.34	II
280	4198.00	II
280	4198.67	II
840	4198.72	II
240	4201.24	II
910	4202.94	II
270	4209.41	II
370	4214.04	II
310	4217.59	II
1500	4222.60	II
770	4227.75	II
390	4231.74	II
240	4234.21	II
200	4236.02	II
980	4239.92	II
390	4242.72	II
310	4245.89	II
310	4245.98	II
390	4246.72	II
1100	4248.68	II
390	4253.37	II
620	4255.79	II
200	4263.43	II
620	4270.19	II
390	4270.72	II
200	4278.86	II
280	4285.37	II
200	4288.66	II
200	4289.44	II
2000	4289.94	II
200	4296.07	II
1500	4296.67	II
420	4296.78	II
590	4299.36	II
770	4300.33	II
420	4305.14	II
770	4306.72	II
390	4309.74	II
560	4320.72	II
310	4330.45	II
310	4332.71	II
240	4336.23	II
980	4337.77	II
340	4339.31	II
700	4349.79	II
560	4352.71	II
910	4364.66	II
350	4373.82	II
530	4375.92	II
910	4382.17	II
700	4386.84	II
310	4388.01	II
1700	4391.66	II
200	4398.79	II
510	4399.20	II
350	4410.64	II
350	4410.76	II
310	4416.90	II
980	4418.78	II
200	4423.68	II
310	4427.07	II
480	4427.92	II
310	4428.44	II
650	4429.27	II
480	4444.39	II
450	4444.70	II
770	4449.34	II
620	4450.73	II
2400	4460.21	II
450	4461.14	II
420	4463.41	II
280	4467.54	II
1400	4471.24	II
450	4472.72	II
700	4479.36	II
700	4483.90	II
840	4486.91	II
250	4497.85	II
100	4506.41	I
110	4515.86	II
100	4519.59	II
770	4523.08	II
840	4527.35	II
840	4528.47	II
110	4532.49	II
110	4539.07	II
840	4539.75	II
210	4544.96	II
250	4551.30	II
650	4560.28	II
310	4560.96	II
2100	4562.36	II
420	4565.84	II
1100	4572.28	II
420	4582.50	II
130	4591.12	II
840	4593.93	II
420	4606.40	II
420	4624.90	II
1700	4628.16	II
170	4632.32	I
110	4650.51	I
130	4654.29	II
110	4669.50	II
150	4680.13	II
270	4684.61	II
200	4714.00	II
100	4714.81	II
110	4725.09	II
100	4733.52	II
310	4737.28	II
100	4739.53	II
160	4747.17	II
110	4757.84	II
100	4768.77	II
230	4773.94	II
110	4822.55	I
140	4847.77	I
180	4882.46	II
110	4943.44	I
130	4971.50	II
130	4994.63	I
210	5009.10	I
100	5011.77	II
120	5022.87	II
120	5037.78	II
120	5040.85	I
180	5044.02	II
120	5071.78	I
240	5075.35	II
470	5079.68	II
130	5112.70	I
160	5117.17	II
170	5129.57	I
110	5147.57	II
100	5149.99	I
280	5159.69	I
280	5161.48	I
190	5174.55	I
370	5187.46	II
210	5191.66	II
190	5211.92	I
260	5223.46	I
180	5229.75	I
140	5232.92	II
260	5245.92	I
130	5265.71	II

Cerium (Cont.)

340		5274.23	II
130		5296.56	I
130		5328.08	I
190		5330.54	II
450		5353.53	II
300		5393.40	II
150		5397.64	I
280		5409.23	II
110		5420.38	I
140		5449.24	I
140		5468.37	II
140		5472.29	II
260		5512.08	II
110		5556.25	I
170		5564.97	I
130		5565.97	I
100		5595.88	I
240		5601.28	I
190		5655.14	I
240		5669.96	I
120		5677.75	I
120		5692.94	I
300		5696.99	I
370		5699.23	I
240		5719.03	I
140		5773.12	I
120		5788.15	I
120		5812.92	I
230		5940.86	I
11		6000.18	I
55		6001.90	I
55		6005.86	I
15		6006.20	I
55		6006.82	I
19		6007.37	I
75		6013.42	I
23		6016.59	I
110		6024.20	I
15		6027.16	I
11		6031.26	I
23		6033.58	II
35		6034.20	II
23		6034.41	I
35		6035.49	II
110		6043.39	II
28		6045.41	I
55		6047.40	I
19		6051.80	II
23		6057.50	I
35		6058.00	I
23		6066.75	I
19		6069.46	I
35		6069.48	I
35		6072.00	I
35		6076.61	I
17		6077.16	I
17		6080.37	I
17		6081.28	I
19		6088.86	I
19		6088.96	I
35		6093.19	I
45		6098.34	II
11		6099.80	I
28		6108.74	II
15		6118.56	I
17		6118.90	I
45		6123.67	I
19		6132.00	II
19		6132.18	I
11		6135.45	I
23		6139.03	I
15		6142.92	I
35		6143.36	II
23		6146.43	I
19		6147.84	I
23		6151.72	I
19		6159.82	I
19		6162.14	I
19		6165.45	I
19		6175.28	I
35		6186.17	I
15		6187.97	I
15		6195.23	I
19		6195.53	I
19		6198.05	I
35		6208.98	I
11		6216.82	I
35		6228.94	I
19		6229.13	I
23		6232.45	II
28		6237.45	I
13		6238.71	I
11		6241.87	I
13		6242.91	I
15		6253.65	I
13		6257.99	I
15		6264.27	I
45		6272.05	II
15		6276.47	I
35		6295.58	I
28		6299.51	II
23		6300.21	I
13		6306.64	I
35		6310.01	I
15		6335.40	I
11		6337.21	I
13		6340.70	I
35		6343.95	II
35		6371.11	II
28		6386.84	I
23		6393.02	II
11		6395.16	I
11		6425.29	II
35		6430.07	I
19		6434.39	I
23		6436.40	I
19		6446.12	I
35		6458.03	I
19		6466.88	II
28		6467.39	I
35		6473.72	I
17		6490.97	I
11		6503.27	II
11		6507.16	II
23		6513.59	II
19		6517.31	I
19		6551.70	I
45		6555.65	I
23		6579.10	I
15		6606.35	I
15		6606.86	II
22		6612.06	I
10		6623.00	I
30		6628.93	I
13		6650.89	I
22		6652.72	II
10		6661.41	I
13		6665.59	I
10		6675.54	II
15		6686.60	I
26		6700.66	I
35		6704.27	I
13		6704.52	II
10		6706.04	II
15		6728.71	I
15		6729.57	I
15		6744.70	II
10		6746.90	I
30		6774.28	II
35		6775.59	I
10		6778.28	I
18		6807.81	I
10		6808.82	I
15		6818.23	I
10		6829.73	II
13		6847.25	I
12		6856.55	I
10		6893.66	I
10		6898.45	II
30		6924.81	I
10		6939.45	I
19		6973.50	II
10		6983.82	II
30		6986.02	I
12		7054.51	I
11		7058.68	II
11		7060.00	I
35		7061.75	II
11		7064.49	I
35		7086.35	II
11		7105.04	II
11		7115.08	II
10		7124.73	I
16		7141.42	I
19		7150.23	II
10		7151.67	I
16		7155.25	I
16		7156.99	II
16		7189.40	II
10		7191.72	I
11		7201.56	II
16		7201.89	I
10		7203.55	I
12		7210.67	I
19		7217.36	I
16		7235.71	II
22		7238.36	II
12		7241.73	I
25		7252.75	I
12		7262.64	I
11	h	7277.90	I
11		7296.17	I
19		7301.42	II
19		7313.45	II
25		7329.91	I
16		7334.68	II
12		7343.44	I
25		7397.77	I
11		7401.27	I
12		7417.94	II
11		7424.70	C
12		7433.08	I
11		7438.56	I
12		7444.44	I
10		7472.41	I
16		7486.57	II
11		7527.46	I
11		7527.68	I
10		7533.73	I
10		7551.25	I
12		7562.44	I
10		7562.86	I
10	h	7563.60	I
10		7603.10	I
25		7616.11	II
12		7646.08	I
10		7647.88	I
12		7682.47	I
25		7689.17	II
10		7732.33	I
16		7748.35	I
10		7797.70	I
12		7842.59	I
22		7844.94	II
16		7850.02	II
16		7851.18	II
22		7857.54	II
12		7864.49	I
10		7866.04	I
16		7898.96	II
11		7913.52	I
10		7927.30	C
10		7927.72	I
10		7934.50	II
30		8025.56	II
16		8070.71	I
10		8094.43	I
16		8120.36	I
10		8241.55	II
12		8261.09	I
16		8418.23	II
11		8495.82	I
12		8539.08	II
10		8612.64	I
10	h	8647.66	I
11	h	8702.38	II
25		8772.14	II
12		8810.84	I
30		8891.20	II

Ce III
Ref. 136, 305 — J.R.

Intensity	Wavelength	
	Vacuum	
100	840.24	III
20	844.11	III
40	845.02	III
20	847.88	III
200	851.18	III
200	852.63	III
200	853.47	III
60	853.78	III
200	855.16	III
200	858.30	III
400	860.15	III
200	862.25	III
40	868.74	III
20	869.51	III
40	869.84	III
20	871.15	III
40	871.27	III
20	880.68	III
60	881.75	III
60	884.04	III
20	885.22	III
30	888.39	III
80	892.75	III
20	899.32	III
100	912.77	III
40	937.04	III
40	999.26	III
20	1025.25	III
20	1025.29	III
20	1026.28	III
40	1029.37	III
20	1034.55	III
20	1041.14	III
100	1042.74	III
50	1051.61	III
70	1057.40	III
100	1057.66	III
100	1058.46	III
50	1062.99	III
30	1063.26	III
50	1063.51	III
100	1067.76	III
100	1068.69	III
20	1070.54	III
200	1072.79	III
40	1073.69	III
30	1079.35	III
20	1080.82	III
30	1088.70	III
30	1090.03	III
20	1092.48	III
20	1099.25	III
40	1100.71	III
20	1107.09	III
20	1111.19	III
20	1116.30	III
20	1125.58	III
20	1129.73	III
20	1132.74	III
100	1142.55	III
20	1192.41	III
50	1201.87	III
20	1204.05	III
20	1719.43	III
20	1796.89	III
20	1836.66	III
20	1836.99	III
30	1862.32	III
30	1950.36	III
20	1990.54	III
	Air	
100	2033.34	III
100	2057.65	III
100	2077.87	III
200	2083.32	III
100	2089.96	III
400	2109.07	III
100	2122.55	III
500	2136.95	III
1000	2151.44	III
3000	2166.88	III
2000	2169.48	III
5000	2180.64	III
1000	2183.71	III
2000	2203.15	III
3000	2218.11	III
5000	2222.01	III
5000	2225.08	III
3000	2227.84	III
3000	2228.05	III
5000	2242.29	III
2000	2249.25	III
2000	2264.85	III
2000	2268.20	III
2000	2287.82	III
2000	2298.70	III
3000	2300.65	III
4000	2302.09	III
5000	2317.34	III
10000	2318.64	III
5000	2324.31	III
2000	2337.66	III
5000	2350.10	III
2000	2362.54	III
2000	2367.77	III
10000	2372.34	III
5000	2377.07	III
5000	2377.48	III
10000	2380.12	III
3000	2382.28	III
3000	2385.06	III
5000	2395.04	III
3000	2406.15	III
4000	2408.08	III
2000	2410.26	III
5000	2415.60	III
2000	2417.01	III
2000	2423.02	III
3000	2428.64	III
5000	2430.24	III
10000	2431.45	III
15000	2439.80	III
3000	2441.55	III
2000	2444.78	III
10000	2454.32	III
10000	2469.95	III
3000	2471.66	III
5000	2477.25	III
8000	2479.44	III
3000	2479.51	III
10000	2483.82	III
10000	2497.50	III
3000	2503.56	III
2000	2504.43	III
20000	2531.99	III
3000	2539.27	III
3000	2557.49	III
4000	2577.67	III
2000	2578.30	III
2000	2584.71	III
10000	2603.59	III
2000	2607.96	III
2000	2615.79	III
2000	2649.38	III
2000	2662.81	III
3000	2719.30	III
2000	2730.04	III
3000	2743.71	III
4000	2748.90	III
4000	2754.87	III
4000	2768.28	III
3000	2849.40	III

Cerium (Cont.)

Intensity	Wavelength	Spectrum
2000	2861.39	III
4000	2907.05	III
10000	2923.81	III
5000	2925.26	III
10000	2931.54	III
2000	2948.53	III
5000	2973.72	III
10000	3022.75	III
50000	3031.58	III
95000	3055.59	III
20000	3056.56	III
40000	3057.23	III
20000	3057.58	III
40000	3085.10	III
20000	106.98	III
30000	3110.53	III
30000	3121.56	III
20000	3141.29	III
20000	3143.96	III
20000	3147.06	III
20000	3228.57	III
3000	3234.20	III
4000	3267.76	III
3000	3267.94	III
20000	3353.29	III
10000	3395.77	III
4000	3398.91	III
30000	3427.36	III
40000	3443.63	III
30000	3454.39	III
40000	3459.39	III
60000	3470.92	III
50000	3497.81	III
60000	3504.64	III
500	3514.41	III
50000	3544.07	III
3000	3784.29	III
800	3936.80	III
300	3957.10	III
500	4169.42	III
300	4191.70	III
500	4194.83	III
300	4213.26	III
300	4217.13	III
400	4284.77	III
300	4304.71	III
600	4346.35	III
400	4389.97	III
600	4448.32	III
500	4485.27	III
1000	4521.92	III
1000	4535.73	III
300	4576.90	III
500	4627.60	III
300	4766.07	III
500	4976.45	III
500	5650.97	III
1000	5664.20	III
500	5691.08	III
300	5710.59	III
500	5749.47	III
500	5949.83	III
2000	5962.22	III
500	5962.71	III
400	5979.56	III
1000	5983.40	III
3000	6002.63	III
10000	6032.54	III
10000	6060.91	III
500	6061.79	III
500	6097.35	III
500	6098.87	III
500	6135.10	III
300	6287.79	III
500	6308.16	III
300	6341.75	III
1000	6944.94	III
700	7739.04	III
300	7758.27	III
500	7826.80	III
300	7948.64	III
500	7960.31	III
500	7991.01	III
400	8030.80	III
300	8084.12	III
400	8177.33	III
300	8186.03	III
300	8222.16	III
300	9056.53	III
300	9079.58	III
400	9328.20	III
300	9367.03	III
300	9567.37	III
400	10458.37	III
400	10494.42	III
300	10534.36	III
400	10684.46	III
15	12756.96	III
12	12821.62	III
80	15847.58	III
80	15956.79	III
12	15960.59	III
87	16128.75	III
42	18579.82	III
38	19141.29	III
27	19377.15	III
26	19466.14	III
20	19498.14	III
55	19524.18	III
30	20685.63	III
12	21380.23	III

Ce IV

Ref. 166 — J.R.

Intensity	Wavelength	
	Vacuum	
2	447.58	IV
1	443.11	IV
8	558.92	IV
8	571.59	IV
40	741.79	IV
30	754.60	IV
12	755.75	IV
6	975.20	IV
5	1009.31	IV
2	1022.12	IV
9	1057.67	IV
1	1059.64	IV
50	1289.41	IV
75	1332.16	IV
75	1372.72	IV
2	1577.60	IV
1	1572.62	IV
15	1641.58	IV
20	1775.30	IV
20	1779.03	IV
35	1914.75	IV
10	1937.21	IV
	Air	
100	2000.42	IV
35	2003.11	IV
100	2009.94	IV
3	2433.50	IV
5	2445.50	IV

Ce V

Ref. 261 — J.R.

Intensity	Wavelength	
	Vacuum	
100	365.66	V
300	399.36	V
150	404.21	V
200	482.96	V
100	552.13	V

CESIUM (Cs)
Z = 55

Cs I and II

Ref. 82, 154, 155, 200, 263, 325— C.J.S.

Intensity		Wavelength	
		Vacuum	
250		591.04	II
2000		639.36	II
500		668.39	II
15000		718.14	II
15000		808.76	II
15000		813.84	II
35000		901.27	II
40000		926.66	II
3		1656.15	II
7		1689.46	II
2		1691.83	II
7		1717.64	II
7		1718.97	II
3		1727.79	II
3		1736.77	II
8		1807.83	II
8		1813.75	II
7		1815.16	II
18		1840.50	II
13		1859.16	II
7		1864.83	II
7		1876.72	II
18		1883.93	II
5		1914.61	II
29		1935.19	II
		Air	
72		2024.97	II
46		2028.32	II
180		2080.05	II
160		2087.20	II
100		2102.22	II
160		2205.52	II
190		2220.53	II
120		2245.81	II
1600		2267.65	II
750		2273.84	II
220		2285.41	II
250		2286.15	II
330		2315.69	II
1400		2332.46	II
240		2375.86	II
220		2379.27	II
1000		2392.86	II
370		2425.17	II
220		2543.93	II
630		2596.99	II
450		2609.43	II
170		2628.86	II
300		2699.18	II
180		2789.78	II
150		2793.31	II
680	w	2816.92	II
150		2829.04	II
140	c	2866.36	II
1600		2931.08	II
290		3151.19	II
520	c	3265.91	II
650	w	3267.12	II
850	c	3271.63	II
570		3368.57	II
150		3514.05	II
110	w	3615.01	II
170	w	3680.11	II
130		3732.56	II
1100	w	3785.44	II
2700		3805.12	II
520	c	3861.50	II
2100	c	3876.15	I
600	c	3888.61	I
3400		3896.99	II
4200		3959.51	II
2500		3965.20	II
2100		3974.25	II
8000		4039.85	II
2000		4068.78	II
320	w	4151.27	II
2000		4213.14	II
1900		4232.20	II
980		4234.41	II
14000		4264.70	II
18000	w	4277.13	II
5100		4288.38	II
1900		4300.65	II
7600		4363.30	II
3700		4373.04	II
960		4384.44	II
3900	w	4405.26	II
12000		4501.55	II
20000		4526.74	II
4100		4538.97	II
1000	c	4555.28	I
460	c	4593.17	I
100000		4603.79	II
4200	c	4616.17	II
2800		4646.52	II
990		4670.29	II
1500		4732.99	II
7000		4763.64	II
1900		4786.38	II
25000		4830.19	II
19000		4870.04	II
4900	c	4880.05	II
37000		4952.85	II
8200		4972.60	II
27000		5043.80	II
2800	c	5059.87	II
2900		5096.60	II
6500	c	5209.58	II
75000		5227.04	II
29000		5249.38	II
11000		5274.05	II
1300		5306.60	II
10000	c	5349.13	II
22000		5370.99	II
1200		5402.78	II
2900		5419.67	II
60	c	5465.94	I
37		5502.88	I
39000		5563.02	II
100		5635.21	I
210	c	5664.02	I
27		5745.72	I
4500		5814.16	II
24000		5831.14	II
59	c	5838.83	I
300		5845.14	I
51000		5925.63	II
1400	c	5984.39	II
640	c	6010.49	I
86		6034.09	I
760		6076.72	II
9800		6128.61	II
1000		6213.10	I
170		6217.60	I
320	c	6354.55	I
2000		6419.52	II
8300		6495.53	II
10000	w	6536.44	II
490		6586.51	I
530	c	6628.01	II
97		6628.66	I
8800		6646.57	II
3300	c	6723.28	I
9600		6724.47	II
200		6824.65	I
300		6870.45	I
37000		6955.50	II
4800		6973.30	I
16000		6979.67	II
980		6983.49	I
2300		7130.54	II
13000	w	7149.54	II
1600		7160.90	II
630		7188.37	II
1100		7206.04	II
790		7228.53	I
960		7248.88	II
130		7279.90	I
1100		7279.96	I
780		7369.36	II
550		7437.78	II
440	w	7523.39	II
2600	c	7608.90	I
910	w	7651.95	II
310	h	7746.98	II
2400		7852.52	II
3300		7943.88	I
22000		7997.44	II
2100		8012.98	II
3500		8015.73	I
8200	c	8047.13	II
1300		8078.50	II
510		8078.94	I
4500		8079.04	I
59000	cr	8521.13	I
680		8521.62	II
1700		8608.31	II
420	w	8695.60	II
15000	c	8761.41	I
840		8775.42	II
340	w	8857.39	II
61000	cr	8943.47	I
18000		9172.32	I
5200		9208.53	I
540	c	9212.36	II
3300	c	9220.75	II
310	c	9718.11	II
630	c	9932.91	II
1400		9994.79	II
19000		10024.36	I
4800		10123.41	I
26000		10123.60	I
3000	c	10176.02	II
2700		10379.66	II
1300		10480.93	II
4700		10504.51	II
1900	w	10807.88	II
610	w	11324.34	II
530	c	11496.56	II
410	c	11704.18	II
330	c	11797.93	II
710	c	11840.84	II
1100		12604.29	II
2000	c	12735.52	II
400	c	12746.34	II
850		13406.83	II
2900		13424.31	I
38000	c	13588.29	I
8400		13602.56	I
4200	c	13692.91	II
5700		13758.81	I
1400	c	13868.82	II
880	c	14482.21	II
55000	c	14694.91	I
350		14906.91	II
1900	c	15293.80	II
1600	c	15356.61	II
620		15445.47	II
940	c	15735.48	II
1100	c	16426.14	II
820		16535.63	I
1500		17012.32	I
340		18160.84	II
430		18179.35	II
710		18221.36	II
390		18222.40	II
510		18404.77	II
310		18407.23	II
590		18509.52	II
610		18509.85	II
100	c	18742.18	II
530	c	18921.76	II
150	c	18986.48	II
180		19924.85	II
390		20110.77	II
760		20138.47	I
200		20301.54	II
220	c	20443.87	II
92	w	21103.42	II
79	c	22344.98	II
78	c	22448.98	II
880		22811.86	I
1100		23037.98	I
3900		23344.47	I

Cesium (Cont.)

Intensity		Wavelength	
320	c	23408.41	II
180		24132.52	II
4400		24251.21	I
850		24374.96	I
120	c	24528.78	II
140		24810.89	II
170		25189.83	II
900		25220.37	II
240		25733.29	II
890	d	25763.51	II
500		25764.73	I
120		26448.83	II
340		26503.26	II
190		26727.33	II
17		26954.56	II
22		26978.68	II
23		27187.98	II
9		28649.38	II
680	c	29310.06	I
7		29542.08	II
2800		30103.27	I
9		30487.16	II
610	c	30953.06	I
1100		34900.13	I
190		36131.00	I
2	c	39177.28	I
2	d	39421.25	I
1		39424.11	I

Cs III

Ref 78, 200, 201 C.J.S.

Intensity		Wavelength	
		Vacuum	
10000		614.01	III
2000		638.17	III
2500		666.25	III
5000		691.60	III
3500		703.89	III
20000		721.79	III
20000		722.20	III
5000		731.56	III
12000		740.29	III
7500		830.39	III
15000		920.35	III
25000	c	1054.79	III
17	c	1673.99	III
12	c	1705.25	III
10		1801.83	III
20	c	1822.40	III
11		1823.93	III
12		1824.70	III
12		1841.80	III
25		1915.50	III
25	c	1923.29	III
12		1961.33	III
17		1996.56	III
		Air	
710		2035.11	III
120		2056.43	III
330		2076.43	III
540		2077.30	III
410		2088.68	III
210		2101.63	III
200		2141.47	III
1000		2316.88	III
230		2325.95	III
390		2340.49	III
1600		2455.81	III
1600		2477.57	III
890		2485.45	III
410		2495.07	III
1400		2525.67	III
430		2573.05	III
16000		2596.86	III
390		2610.12	III
6200		2630.51	III
370		2700.32	III
710		2701.20	III
390		2776.44	III
370		[illegible]	III
630		2845.70	III
3100		2859.32	III
200		2893.85	III
180		2921.13	III
3200		2976.86	III
210		3001.28	III
1700		3066.59	III
1100	c	3149.36	III
1400		3152.36	III
8400		3268.32	III
1300		3315.51	III
550		3340.60	III
430		3344.02	III
1200		3349.46	III
400		3463.45	III
580		3476.83	III
480		3559.82	III
7200		3597.45	III
1300		3608.31	III
2300		3618.19	III
300	c	3641.34	III
520		3651.08	III
4800		3661.40	III
640		3699.50	III
430		3837.46	III
2900		3888.37	III
2700		3925.60	III
680	c	4001.70	III
3100		4006.55	III
420		4006.78	III
520		4043.42	III
370		4403.86	III
1200		4410.22	III
940		4425.68	III
530		4471.48	III
1200		4506.72	III
590		4522.86	III
420	h	4620.61	III
210		4665.52	III
140		4851.59	III
370		5035.72	III
230		5380.79	III
140		5950.14	III
110		5979.97	III
150		6043.99	III
870		6079.86	III
330		6150.42	III
450		6242.96	III
510		6456.33	III
400		6753.12	III
1900	c	7219.60	III

Cs IV

Ref. 259 — J.R.

Intensity		Wavelength	
		Vacuum	
35		707.20	IV
60		759.57	IV
5		778.21	IV
500		824.80	IV
350		828.86	IV
400		868.18	IV
1000		874.84	IV
400		896.92	IV
400		923.02	IV
600		986.14	IV
300		995.14	IV
[illegible]		1019.13	IV
550		1068.91	IV
200		1282.66	IV

CHLORINE (Cl)
Z = 17

Cl I and II

Ref 238, 239 — L.J.R.

Intensity		Wavelength	
		Vacuum	
350		559.305	II
400		571.904	II
800		574.406	II
500		586.24	II
700		618.057	II
600		619.982	II
800		620.298	II
700		626.735	II
800		635.881	II
1000		636.626	II
1000		650.894	II
1000		659.811	II
1300		661.841	II
2000		663.074	II
1500		682.053	II
1500		687.656	II
1500		693.594	II
2000		725.271	II
2500		728.951	II
2000		777.562	II
5000		787.580	II
5000		788.740	II
5000		793.342	II
6000		839.297	II
[illegible]		839.399	II
7000	p	841.41	II
5000		851.691	II
2000		888.026	II
2000		893.549	II
2000		961.499	II
30		969.92	I
40		978.284	I
25		998.372	I
25		998.432	I
75		1002.346	I
150		1013.664	I
90		1025.553	I
6000		1063.831	II
3000		1067.945	II
9000		1071.036	II
6000		1071.767	II
5000		1075.230	II
5000		1079.080	II
200		1084.667	I
200		1085.171	I
250		1085.304	I
400		1088.06	I
350		1090.271	I
250		1090.982	I
250		1092.437	I
400		1094.769	I
350		1095.148	I
350		1095.662	I
400		1095.797	I
250		1096.810	I
300		1097.369	I
200		1098.068	I
200		1099.523	I
500		1107.528	I
800		1139.214	II
800		1167.148	I
3000		1179.293	I
1200		1188.774	I
900		1201.353	I
3000		1335.726	I
10000		1347.240	I
5000		1351.657	I
12000		1363.447	I
2500		1373.116	I
20000		1379.528	I
25000		1389.693	I
20000		1389.957	I
12000		1396.527	I
500		1441.470	II
500		1528.569	II
500		1542.942	II
500		1558.144	II
500		1565.050	II
500		1857.488	II
450	h	1997.370	II
		Air	
450		2032.116	II
350	h	2088.583	II
350	h	2091.458	II
170		2427.79	II
360		2434.07	II
340		2498.53	II
470		2502.74	II
260		2546.96	II
500		2549.88	II
460		2564.84	II
320		2602.21	II
950		2658.72	II
750		2676.95	II
1200		2688.04	II
410		2912.05	II
950		2996.65	II
500		3006.06	II
950		3057.96	II
1300		3071.32	II
1400		3092.19	II
1200		3123.72	II
1900		3315.43	II
1200		3329.10	II
2500		3353.35	II
20		3726.54	I
1200		3749.96	II
1000		3781.17	II
1500		3798.76	II
1900		3805.18	II
1300		3809.46	II
1700		3820.20	II
2800		3827.59	II
4500		3833.35	II
2500		3843.20	II
3100		3845.37	II
3900		3845.65	II
1500		3845.80	II
10000		3850.99	II
7900		3851.37	II
1200		3851.65	II
25000		3860.83	II
4400		3860.99	II
1000		3861.37	II
1500		3913.87	II
1100		3916.63	II
20		3944.82	I
10		4104.79	I
10000	h	4132.50	II
65		4209.67	I
50		4226.42	I
60		4264.58	I
100		4363.27	I
100		4369.50	I
5000		4372.93	II
100		4379.90	I
100		4389.75	I
90		4390.40	I
90		4403.03	I
100		4438.49	I
90		4475.30	I
1500		4489.91	II
100		4526.19	I
80		4600.98	I
40		4623.938	I
50		4654.040	I
80		4661.208	I
45		4691.523	I
40		4721.255	I
45		4740.729	I
4300		4768.65	II
13000		4781.32	II
99000		4794.55	II
29000		4810.06	II
16000		4819.47	II
81000		4896.77	II
47000		4904.78	II
26000		4917.73	II
10000		4995.48	II
26000		5078.26	II
30		5099.789	I
56000		5217.94	II
23000		5221.36	II
15000		5392.12	II
99000		5423.23	II
10000		5423.51	II
19000		5443.37	II
10000		5444.21	II
5600		5457.02	II
40		5532.162	I
50	d	5796.305	I
45		5799.914	I
30		5856.742	I
100	d	5948.58	I
50		6019.812	I
35		6082.61	I
1900		6094.69	II
160		6114.43	I
200		6140.245	I
160		6194.757	I
160		6398.66	I
150		6434.833	I
150		6531.43	I
1400		6661.67	II
150		6678.43	I
1300		6686.02	II
1200		6713.41	II
150		6840.29	I
300		6932.903	I
300		6981.886	I
600		7086.814	I
7500		7256.62	I
5000		7414.11	I
550		7462.370	I
550		7489.47	I
700		7492.118	I
11000		7547.072	I
2300		7672.42	I
450		7702.828	I
7000		7717.581	I
10000		7744.97	I
2200		7769.16	I
650		7771.09	I
2200		7821.36	I
1700		7830.75	I
3000		7878.22	I
220		7893.34	I
2300		7899.31	I
1800		7915.08	I
3000		7924.645	I
2100		7933.89	I
1700		7935.012	I
650		7952.52	I
1500		7974.72	I
1300		7976.97	I
600		7980.60	I
2900		7997.85	I
2200		8015.61	I
1100		8023.33	I
400		8051.07	I
1700		8084.51	I
2200		8085.56	I
3000		8086.67	I
1300		8087.73	I
250		8094.67	I
2500		8194.42	I
2200		8199.13	I
2200		8200.21	I
800		8203.78	I
18000		8212.04	I
3000		8220.45	I
20000		8221.74	I
18000		8333.31	I
1000		8360.71	II
560		8361.84	II
99900		8375.94	I
180		8382.67	II
100		8392.02	II
400		8406.199	I
15000		8428.25	I
2200		8467.34	I
2200		8550.44	I
20000		8575.24	I
750		8578.02	I
75000		8585.97	I
450		8628.54	I
300		8641.71	I
3500		8686.26	I
2200		8912.92	I
3000		8948.06	I
2000		9038.982	I
2500		9045.43	I
1000		9069.656	I
2000		9073.17	I
7500		9121.15	I

Chlorine (Cont.)

Intensity	Wavelength	
3000	9191.731	I
500	9197.596	I
4000	9288.86	I
1500	9393.862	I
3500	9452.10	I
500	9486.964	I
1000	9584.801	I
3500	9592.22	I
250	9632.509	I
1000	9702.439	I
250	9744.426	I
200	9807.057	I
400	9875.970	I
331	10392.549	I
38	10432.83	II
10	10506.62	II
14	10509.12	II
19	10512.46	II
25	10514.17	II
9	10801.47	II
5	10885.42	II
1	10955.71	II
300	11123.05	I
231	11392.62	I
269	11409.69	I
1000	11436.33	I
180	11720.56	I
195	11866.76	I
172	12021.7	I
350	13243.8	I
310	13296.0	I
550	13346.8	I
525	13821.7	I
148	14369.7	I
294	14931.7	I
269	15108.0	I
381	15465.1	I
169	15467.6	I
1094	15520.3	I
1487	15730.1	I
193	15818.4	I
2780	15869.7	I
277	15883.3	I
342	15928.9	I
735	15960.0	I
283	15970.5	I
129	16077.6	I
259	16198.5	I
227	19370.3	I
717	19755.3	I
185	19766.8	I
227	20199.4	I
85	20370.1	I
100	24470.0	I
	39603.7	I
	39615.3	I
	39716.0	I
	39744.0	I
	39750.9	I
	39875.3	I
	39881.0	I
	39985.7	I
	40085.5	I
	40089.5	I
	40171.0	I
	40310.3	I
	40335.4	I
	40532.2	I

Cl III
Ref. 28, 30 — L.J.R.

Intensity	Wavelength	
	Vacuum	
100	406.27	III
400	411.37	III
400	411.81	III
600	556.23	III
700	556.61	III
700	557.12	III
600	2965.56	III
600	3104.46	III
800	3139.34	III
900	3191.45	III
700	3289.80	III
700	3320.57	III
800	3329.06	III
900	3340.42	III
800	3392.89	III
800	3393.45	III
900	3530.03	III
800	3560.68	III
900	3602.10	III
800	3612.85	III
700	3622.69	III
700	3656.95	III
700	3670.28	III
700	3682.05	III
600	3705.45	III
600	3707.34	III
800	3720.45	III
800	3748.81	III
500	3779.35	III
500	3925.87	III
700	3991.50	III
600	4018.50	III
600	4059.07	III
500	4104.23	III
500	4106.83	III
400	4370.91	III
500	4608.21	III
300	4703.14	III
100	4863.75	III
10	4971.64	III
700	561.53	III
700	561.68	III
700	561.74	III
500	606.35	III
400	621.28	III
300	670.38	III
300	673.13	III
100	936.28	III
500	1005.28	III
600	1008.78	III
700	1015.02	III
600	1822.50	III
500	1828.40	III
500	1901.61	III
500	1983.61	III
	Air	
400	2006.84	III
700	2253.07	III
500	2268.95	III
500	2278.34	III
700	2283.93	III
600	2323.50	III
500	2336.45	III
600	2340.64	III
600	2359.67	III
600	2370.37	III
700	2416.42	III
600	2447.14	III
600	2448.58	III
500	2486.91	III
500	2532.48	III
600	2580.67	III
500	2603.59	III
500	2632.67	III
500	2633.18	III
600	2665.54	III
700	2710.37	III

Cl IV
Ref. 11, 28, 30, 31 — L.J.R.

Intensity	Wavelength	
	Vacuum	
300	319.62	IV
200	331.84	IV
400	437.83	IV
400	464.86	IV
800	486.17	IV
800	534.73	IV
700	535.67	IV
600	536.15	IV
900	537.61	IV
600	538.12	IV
500	549.22	IV
400	550.02	IV
700	552.02	IV
600	553.30	IV
700	554.62	IV
500	601.50	IV
500	604.59	IV
400	608.90	IV
400	612.07	IV
400	653.70	IV
400	745.21	IV
400	831.43	IV
500	834.84	IV
500	834.97	IV
400	840.81	IV
600	840.93	IV
	865.3	IV
500	973.21	IV
600	977.56	IV
400	977.90	IV
700	984.95	IV
400	985.75	IV
300	1537.21	IV
200	1539.30	IV
200	1545.19	IV
200	1549.15	IV
200	1622.86	IV
	Air	
400	2701.36	IV
500	2724.03	IV
500	2751.23	IV
400	2770.64	IV
700	2782.47	IV
400	2835.4	IV
500	3063.13	IV
600	3076.68	IV
200	3167.87	IV

Cl V
Ref. 11, 28, 30, 85, 233 — L.J.R.

Intensity	Wavelength	
	Vacuum	
300	287.33	V
300	373.78	V
400	390.15	V
500	392.43	V
300	536.53	V
400	537.01	V
300	537.46	V
500	538.03	V
400	538.68	V
800	542.23	V
600	542.30	V
400	542.87	V
1000	545.11	V
600	546.33	V
1000	547.63	V
400	633.19	V
400	635.32	V
400	681.92	V
400	683.17	V
400	688.93	V
	715.55	V
	716.19	V
400	883.13	V
400	894.34	V
100	894.91	V
	914.5	V

CHROMIUM (Cr)
Z = 24

Cr I and II
Ref. 1 — C.H.C.

Intensity		Wavelength	
		Vacuum	
19000		2055.52	II
14000		2061.49	II
8900		2065.42	II
80	h	2364.71	I
130		2383.33	I
140		2408.62	I
170		2496.31	I
110		2502.53	I
190		2504.31	I
50		2508.11	I
60		2508.98	I
40		2513.62	I
110		2516.92	I
80		2518.71	I
390		2519.52	I
190		2527.12	I
40		2530.45	I
70		2534.34	II
50		2545.64	I
160		2549.54	I
40		2553.06	I
80		2557.15	I
130		2560.69	I
150		2571.74	I
100		2577.65	I
50		2588.20	I
380		2591.85	I
35		2603.57	I
35		2622.86	I
22		2625.32	I
18		2626.60	I
18		2629.82	I
35		2642.12	I
250		2653.59	II
250		2658.59	II
70		2661.73	II
320		2663.42	II
70		2663.68	II
440		2666.02	II
280		2668.71	II
350		2671.81	II
280		2672.83	II
1800		2677.16	II
35		2678.16	I
320		2678.79	II
18		2680.34	II
230		2687.09	II
60		2688.04	I
55		2688.29	II
26		2690.26	I
280		2691.04	II
35		2693.52	II
35		2697.91	II
180		2698.41	II
180		2698.69	II
18		2700.60	I
110		2701.99	I
18		2702.53	I
70		2703.48	I
		2703.55	II
35		2703.86	II
18		2705.43	I
60		2708.79	II
35		2709.31	II
140		2712.31	II
45		2716.18	I
55		2717.51	II
45		2718.43	II
170		2722.75	II
18		2724.04	II
420	h	2726.51	I
45		2727.26	II
280	h	2731.91	I
170	h	2736.47	I
70		2739.38	I
70		2740.10	II
95		2741.07	I
95		2742.03	II
95		2742.17	I
250		2743.64	II
35		2746.21	II
110	h	2748.29	I
330		2748.98	II
390		2750.73	II
45		2751.60	I
280		2751.87	II
110	h	2752.88	I
35		2754.28	II
22		2754.90	I
22		2755.27	I
22		2756.75	I
150		2757.10	I
350		2757.72	II
60		2758.98	II
80		2759.39	II
45		2759.73	II
90	h	2761.76	I
750		2762.59	II
22		2763.06	I
80	h	2764.35	I
750		2766.54	II
22		2767.54	I
250	h	2769.92	I
18		2771.45	I
45		2778.06	II
22		2779.14	I
80		2780.30	II
610		2780.70	I
70		2785.70	II
35		2787.63	II
35		2787.84	I
90		2792.16	II
55		2798.67	II
70		2800.77	II
80		2812.01	II
60		2818.36	II
45		2822.01	II
180		2822.37	II
22		2826.75	I
180		2830.47	II
70		2834.26	II
2500		2835.63	II
45		2836.48	II
55		2838.79	II
110		2840.02	II
1700		2843.25	II
22		2846.02	I
45		2849.29	I
1200		2849.84	II
120		2851.36	II
55		2853.22	II
55		2855.07	II
880		2855.68	II
90		2856.77	II
70		2857.40	II
610		2858.91	II
440		2860.93	II
790		2862.57	II
750		2865.11	II
55		2865.33	II
610		2866.74	II
90		2867.10	II
480		2867.65	II
210		2870.44	II
110		2871.63	I
160		2873.48	II
90		2873.82	II
320		2875.99	II
230		2876.24	II
180		2877.98	II
70		2878.45	II
120		2879.27	I
95		2880.87	II
30		2881.14	I
170		2887.00	I
55		2888.74	II
700		2889.29	I
55		2889.82	II
55		2891.42	I
370		2893.25	I
190		2894.17	I
55		2896.46	II

Chromium (Cont.)

210		2896.75	I
55	d	2897.67	II
		2897.73	II
90		2898.54	II
80		2899.21	I
55		2899.48	II
26		2903.97	II
55		2904.68	I
100		2905.45	I
260		2909.05	I
260		2910.90	I
250		2911.14	I
45		2911.68	II
60		2913.73	I
22		2915.23	II
22		2915.46	II
90		2921.24	II
60		2921.82	II
60		2927.08	II
80		2928.15	II
95		2928.30	II
26		2929.44	II
35		2930.85	II
26		2932.70	II
55		2933.97	II
90		2935.14	II
45		2940.22	II
60		2946.84	II
55		2953.36	II
45		2953.71	II
55		2961.73	II
45		2966.05	II
480		2967.64	I
480		2971.11	I
210		2971.91	II
480		2975.48	I
30		2976.72	II
190		2979.74	II
350		2980.79	I
110		2985.32	II
480		2985.85	I
1500		2986.00	I
2100		2986.47	I
660		2988.65	I
160		2989.19	II
680		2991.09	I
230		2994.07	I
300		2995.10	I
700		2996.58	I
210		2998.79	I
1100		3000.89	I
750		3005.06	I
140		3013.03	I
710		3013.71	I
710		3014.76	I
1400		3014.92	I
710		3015.19	I
2800		3017.57	I
430		3018.50	I
240		3018.82	I
430		3020.67	I
2800		3021.56	I
1100		3024.35	I
85		3026.65	II
170		3029.16	I
710		3030.24	I
140		3031.35	I
28		3032.93	II
390		3034.19	I
550		3037.04	I
80		3039.78	I
550		3040.85	I
		3040.91	II
55		3041.74	II
110		3050.14	II
710		3053.88	I
24		3059.52	II
85		3065.07	I
28		3067.16	II
85		3073.68	I
55		3077.83	I
28		3095.86	I
28		3109.34	I
28		3110.86	I
240		3118.65	II
45		3119.25	I
40		3119.71	I
430		3120.37	II
28		3122.60	II
470		3124.94	II
		3125.02	II
120		3128.70	II
590		3132.06	II
140		3136.68	II
140		3147.23	II
85		3148.44	I
100		3155.15	I
100		3163.76	I
240		3180.70	II
30		3181.43	II
65	h	3188.01	I
220		3197.08	II
24		3198.11	I
30		3208.59	II
170		3209.18	II
140		3217.40	II
30		3229.20	I
28		3234.06	II
65		3237.73	I
120		3245.54	I
130		3251.84	I
130		3257.82	I
95		3259.98	I
30		3295.43	II
24		3307.02	II
55		3324.06	II
28		3326.59	I
30		3328.35	II
30		3329.05	I
95		3336.33	II
130		3339.80	II
110		3342.59	II
30		3343.34	I
95		3346.02	I
95		3346.74	I
95		3347.84	II
65		3349.07	I
55		3349.32	I
30		3351.60	I
55		3351.97	I
55	h	3353.03	I
		3353.13	II
170		3358.50	II
160		3360.30	II
65		3361.77	II
55		3362.21	I
430		3368.05	II
30		3376.40	I
55		3378.34	II
30		3379.17	I
30		3379.37	II
95		3379.83	II
140		3382.68	II
95		3391.43	II
55		3392.99	II
70		3393.84	II
55		3394.30	II
30		3402.40	II
170		3403.32	II
360		3408.76	II
210		3421.21	II
270		3422.74	II
140		3433.31	II
270		3433.60	I
55		3434.11	I
160		3436.19	I
70		3441.12	I
140		3441.44	I
30		3443.79	I
170		3445.62	I
30		3447.02	I
170		3447.43	I
70		3447.76	I
190		3453.33	I
40		3453.74	I
130		3455.60	I
100		3460.43	I
65		3465.25	I
40		3467.02	I
70		3467.72	I
45		3469.59	I
16		3472.76	I
24		3472.91	I
40		3473.61	I
70		3481.30	I
55		3481.54	I
55		3494.97	I
40		3495.38	II
80		3510.54	I
40		3511.84	II
120		3550.64	I
80		3558.52	I
130		3566.16	I
130		3573.64	I
80		3574.04	I
330	h	3574.80	I
		3574.94	I
19000		3578.69	I
160	h	3584.33	I
130		3585.30	II
17000		3593.49	I
350		3601.67	I
40		3602.57	I
85		3603.74	I
		3603.78	II
13000		3605.33	I
40		3608.40	I
40		3609.48	I
40		3610.05	I
70		3612.61	I
85		3615.64	I
130		3632.84	I
350		3636.59	I
630		3639.80	I
85		3640.39	I
70		3641.47	I
220		3641.83	I
45		3646.16	I
85		3648.53	I
220		3649.00	I
170		3653.91	I
220		3656.26	I
45		3662.84	I
130		3663.21	I
45		3665.98	I
95		3666.64	I
55		3668.03	I
65		3676.32	I
40		3677.68	II
55		3677.89	II
40		3679.82	I
19		3681.69	I
120		3685.55	I
130		3686.80	I
130		3687.25	I
75		3687.54	I
19		3688.46	I
75		3712.95	II
40		3716.53	I
130		3730.81	I
150		3732.03	I
95		3742.97	I
480		3743.58	I
570		3743.88	I
85		3744.49	I
55		3748.61	I
340		3749.00	I
50		3757.17	I
230		3757.66	I
60		3758.04	I
24		3767.43	I
260		3768.24	I
95		3768.73	I
95		3788.86	I
95		3790.45	I
130		3791.38	I
130		3792.14	I
120		3793.29	I
130		3793.88	I
85		3794.61	I
160		3797.12	I
200		3797.72	I
530		3804.80	I
110		3806.83	I
110		3807.93	I
180		3815.43	I
70		3818.48	I
180		3819.56	I
70		3823.52	I
130		3826.42	I
130		3830.03	I
380		3841.28	I
190		3848.98	I
140		3849.36	I
290		3850.04	I
140		3852.22	I
190		3854.22	I
110		3855.29	I
140		3855.57	I
260		3857.63	I
70		3874.53	I
660		3883.29	I
50		3883.66	I
570		3885.22	I
380		3886.79	I
60		3891.93	I
260		3894.04	I
40		3897.65	I
35		3902.11	I
360		3902.92	I
60		3903.16	I
960		3908.76	I
120	Hd	3911.82	I
		3912.00	I
120		3915.84	I
190		3916.24	I
35		3917.60	I
1900		3919.16	I
600		3921.02	I
30		3926.65	I
600		3928.64	I
410		3941.49	I
30		3951.10	I
40		3952.40	I
35		3953.16	I
1900		3963.69	I
120		3969.06	I
1600		3969.75	I
85		3971.26	I
1600		3976.66	I
85		3978.68	I
40		3979.80	I
85		3981.23	I
960		3983.91	I
190		3984.34	I
160		3989.99	I
960		3991.12	I
160		3991.67	I
190		3992.84	I
40		3993.97	I
160		4001.44	I
120		4012.47	II
30		4014.67	I
85		4022.26	I
70		4025.01	I
120		4026.17	I
85		4027.10	I
85		4030.68	I
190		4039.10	I
160		4048.78	I
120		4058.77	I
40		4065.72	I
85		4066.94	I
35		4074.86	I
40		4076.06	I
40		4077.09	I
40		4077.68	I
40		4104.87	I
40		4109.58	I
40		4120.61	I
40		4121.82	I
35		4122.16	I
40		4123.39	I
140		4126.52	I
35		4127.30	I
40		4127.64	I
40		4131.36	I
30		4152.78	I
120		4153.82	I
85		4161.42	I
140		4163.62	I
70		4165.52	I
40		4169.84	I
35		4170.20	I
40		4172.77	I
170		4174.80	I
30		4175.94	I
170		4179.26	I
35		4184.90	I
30		4186.36	I
35		4190.13	I
85		4191.27	I
35		4192.10	I
85		4193.66	I
70		4194.95	I
40		4197.23	I
85		4198.52	I
60		4203.59	I
40		4204.47	I
35		4208.36	I
110		4209.37	I
40		4209.76	I
40		4211.35	I
40		4216.36	I
85		4217.63	I
40		4221.57	I
40		4222.73	I
40		4238.96	I
60		4240.70	I
20000		4254.35	I
70		4255.50	I
60		4261.35	I
110		4263.14	I
30		4271.06	I
40		4272.91	I
16000		4274.80	I
85		4280.40	I
10000		4289.72	I
40		4291.96	I
85		4295.76	I
70		4297.74	I
35		4300.51	I
50		4301.18	I
30		4305.45	I
35		4319.64	I
60		4325.08	I
780		4337.57	I
1100		4339.45	I
380		4339.72	I
60		4340.13	I
1900		4344.51	I
70		4346.83	I
380		4351.05	I
2300		4351.77	I
570		4359.63	I
70		4363.13	I
530		4371.28	I
70		4373.25	I
110		4374.16	I
70		4375.33	I
50		4381.11	I
530		4384.98	I
60		4387.50	I
70		4391.75	I
60		4403.50	I
24		4410.30	I
60		4411.09	I
35		4412.25	I
50		4413.87	I
60		4424.28	I
24		4428.50	I
50		4430.49	I

Chromium (Cont.)

Intensity		Wavelength		Spectrum
50		4432.18		I
110		4458.54		I
30		4459.74		I
30		4465.36		I
30		4482.88		I
40		4488.05		I
50		4489.47		I
60		4492.31		I
660		4496.86		I
50		4498.73		I
70		4500.30		I
50		4501.11		I
22		4501.79		I
24		4506.85		I
95		4511.90		I
12		4514.37		I
35		4514.53		I
24		4521.14		I
24		4526.11		I
380		4526.47		I
70	d	4527.34		I
		4527.47		I
24		4529.85		I
380		4530.74		I
50		4535.15		I
240		4535.72		I
40		4539.79		I
240		4540.50		I
240		4540.72		I
35		4541.07		I
19		4541.51		I
24		4542.62		I
140		4544.62		I
24		4545.34		I
600		4545.96		I
50		4556.17		I
22		4558.66		II
19		4564.17		I
120		4565.51		I
95		4569.64		I
120		4571.68		I
22		4575.12		I
360		4580.06		I
24		4586.14		I
360		4591.39		I
70		4595.59		I
50		4600.10		I
480		4600.75		I
50		4601.02		I
240		4613.37		I
600		4616.14		I
70		4619.55		I
85		4621.96		I
70		4622.49		I
24		4622.76		I
550		4626.19		I
24		4632.18		I
40		4637.18		I
50		4637.77		I
50	d	4639.52		I
		4639.70		I
1600		4646.17		I
24		4646.81		I
24		4648.13		I
24		4648.87		I
35		4649.46		I
570		4651.28		I
840		4652.16		I
35		4654.74		I
19		4656.19		I
40		4663.33		I
70		4663.83		I
95		4664.80		I
35		4665.90		I
22		4666.22		I
70		4666.51		I
50		4669.34		I
40		4680.54		I
19		4680.87		I
70		4689.37		I
60		4693.95		I
24		4695.15		I
60		4697.06		I
240	d	4698.46		I
		4698.62	CR	I
35		4700.61		I
190		4708.04		I
240		4718.43		I
50		4723.10		I
50		4724.42		I
50		4727.15		I
24		4729.72		I
120		4730.71		I
140		4737.35		I
19		4745.31		I
70		4752.08		I
340		4756.11		I
50		4764.29		I
22		4766.63		I
30		4767.86		I
190		4789.32		I
95		4792.51		I
120		4801.03		I
110		4829.38		I
14		4836.86		I
17		4861.20		I
70		4861.84		I
140		4870.80		I
35		4885.78		I
19		4885.96		I
130		4887.01		I
19		4888.53		I
35		4903.24		I
260		4922.27		I
110		4936.33		I
70		4942.50		I
110		4954.81		I
35		4964.93		I
60		5013.32		I
17		5051.90		I
17		5065.91		I
40		5067.71		I
40		5072.92		I
30		5110.75		I
17		5113.13		I
17		5123.46		I
50		5139.65		I
14		5144.67		I
70		5166.23		I
35		5177.43		I
70		5184.59		I
70		5192.00		I
12		5193.49		I
85		5196.44		I
35		5200.19		I
5300		5204.52		I
8400		5206.04		I
11000		5208.44		I
19		5214.13		I
30		5221.75		I
85		5224.94		I
12		5226.89		I
19		5238.97		I
30		5243.40		I
290		5247.56		I
60		5254.92		I
60		5255.13		I
19		5261.75		I
530		5264.15		I
30		5265.16		I
180		5265.72		I
35		5272.01		I
30		5273.44		I
95	h	5275.17		I
35	h	5275.69		I
70	h	5276.03		I
19		5280.29		I
10		5287.19		I
340		5296.69		I
70	h	5297.36		I
660		5298.27		I
85		5300.75		I
17		5304.21		I
24		5312.88		I
24		5318.78		I
340	h	5328.34		I
70	h	5329.17		I
17	h	5329.72		I
14		5340.44		I
10		5344.76		I
780		5345.81		I
380		5348.32		I
30		5386.98		I
22		5387.57		I
10		5390.39		I
40		5400.61		I
22		5405.00		I
1400		5409.79		I
12		5442.41		I
19		5463.97		I
19		5480.50		I
24		5628.64		I
7		5642.36		I
12	h	5649.37		I
24		5664.04		I
7	h	5681.20		I
7	h	5682.48		I
24		5694.73		I
40		5698.33		I
24		5702.31		I
12		5712.64		I
24		5712.78		I
7		5719.82		I
7		5746.43		I
7		5753.69		I
12	h	5781.20		I
6	h	5781.81		I
24	h	5783.11		I
30	h	5783.93		I
24	h	5785.00		I
19	h	5785.82		I
60	h	5787.99		I
180	h	5791.00		I
35		6330.10		I
22		6362.87		I
19		6661.08		I
11		6669.26		I
5	h	6881.62		I
10	h	6882.38		I
21	h	6883.03		I
27	h	6924.13		I
17	h	6925.20		I
30	h	6978.48		I
11	h	6979.82		I
7		7185.52		I
6	h	7236.20		I
85		7355.90		I
130		7400.21		I
150		7462.31		I
11	h	7942.04		I
5	h	8163.18		I
9		8348.28		I
6		8450.26		I
3		8455.24		I
6		8548.86		I
40		8947.15		I
19		8976.83		I

Cr III
Ref. 412 — C.H.C.

Intensity	Wavelength	
	Vacuum	
20	969.26	III
40	1000.86	III
40	1001.04	III
30	1002.96	III
50	1017.14	III
50	1017.31	III
50	1017.57	III
30	1028.33	III
60	1030.47	III
30	1030.89	III
50	1033.23	III
50	1033.45	III
100	1033.69	III
50	1035.93	III
100	1036.03	III
30	1040.17	III
40	1040.53	III
40	1045.06	III
40	1045.14	III
60	1059.13	III
60	1060.15	III
60	1061.04	III
50	1062.68	III
30	1064.32	III
30	1064.43	III
50	1066.23	III
80	1068.41	III
30	1100.61	III
30	1101.43	III
30	1102.88	III
30	1117.19	III
30	1132.75	III
50	1136.67	III
50	1161.43	III
30	1187.65	III
60	1206.38	III
80	1209.13	III
80	1211.12	III
40	1221.07	III
40	1221.90	III
30	1225.65	III
30	1228.65	III
30	1231.88	III
50	1232.96	III
40	1236.20	III
40	1238.51	III
50	1252.61	III
40	1259.02	III
40	1261.86	III
30	1262.34	III
35	1263.61	III
35	1264.21	III
40	1287.05	III
30	1455.27	III
40	1584.60	III
30	1603.19	III
30	1679.25	III
30	1690.28	III
60	1692.89	III
60	1696.64	III
60	1701.48	III
80	1707.43	III
40	1707.78	III
45	1762.81	III
30	1766.92	III
30	1769.17	III
30	1827.26	III
	AIR	
60	2036.39	III
50	2039.63	III
80	2047.23	III
100	2113.73	III
100	2113.83	III
50	2114.26	III
50	2114.53	III
100	2114.87	III
100	2117.53	III
80	2123.53	III
80	2139.11	III
100	2141.15	III
80	2144.15	III
50	2147.16	III
50	2147.56	III
50	2148.65	III
50	2149.48	III
50	2152.76	III
100	2157.17	III
50	2163.86	III
60	2166.25	III
100	2170.70	III
50	2183.71	III
100	2185.01	III
50	2190.09	III
100	2190.76	III
100	2191.58	III
100	2197.89	III
100	2198.62	III
100	2203.22	III
60	2208.70	III
200	2226.72	III
100	2231.81	III
100	2233.81	III
200	2235.91	III
150	2237.59	III
150	2244.10	III
80	2251.45	III
50	2257.92	III
100	2273.30	III
80	2275.43	III
100	2276.38	III
80	2277.47	III
150	2284.44	III
50	2289.23	III
80	2290.66	III
60	2295.55	III
50	2309.99	III
80	2314.63	III
100	2319.07	III
150	2324.88	III
60	2340.51	III
50	2456.83	III
100	2472.88	III
100	2479.77	III
100	2483.06	III
60	2488.26	III
80	2506.41	III
80	2530.99	III
80	2537.73	III
80	2544.37	III
50	2545.17	III
80	2564.76	III
80	2616.50	III
100	2626.08	III
100	2640.73	III
50	2647.50	III
40	2655.28	III
40	2916.57	III

Cr IV
Ref. 379, 412 — C.H.C.

Intensity	Wavelength	
	Vacuum	
50	575.05	IV
30	576.24	IV
30	576.62	IV
30	595.09	IV
50	612.64	IV
40	613.75	IV
40	614.03	IV
40	614.90	IV
30	615.34	IV
30	615.60	IV
50	616.82	IV
40	618.23	IV
40	619.13	IV
100	620.66	IV
60	621.36	IV
40	622.09	IV
30	623.54	IV
40	625.04	IV
40	625.99	IV
100	629.26	IV
50	629.74	IV
80	630.30	IV
30	632.62	IV
30	637.34	IV
50	637.55	IV
50	638.13	IV

Chromium (Cont.)

Intensity		Wavelength	
30		638.54	IV
100		666.55	IV
75		667.30	IV
40		677.55	IV
40		687.12	IV
50		688.46	IV
100		693.92	IV
50		695.21	IV
50		705.98	IV
30		712.90	IV
80		1055.89	IV
60		1057.85	IV
30		1367.39	IV
40		1375.05	IV
70		1401.82	IV
100		1417.42	IV
30		1485.05	IV
80		1595.04	IV
90	d	1595.59	IV
100		1658.08	IV
120		1672.66	IV
90		1686.07	IV
100		1690.88	IV
80		1725.26	IV
90		1727.07	IV
100		1732.04	IV
40		1733.98	IV
80		1734.16	IV
50		1739.19	IV
70		1746.88	IV
80		1747.13	IV
110		1755.64	IV
120		1758.51	IV
100		1769.64	IV
100		1777.82	IV
40		1791.09	IV
140		1802.72	IV
130		1812.41	IV
60		1819.23	IV
30		1826.21	IV
30		1826.86	IV
100		1840.14	IV
50		1851.89	IV
100		1863.11	IV
140		1873.89	IV
35		1883.16	IV
40		1937.63	IV
30		1946.59	IV
140		1967.18	IV
120		1972.07	IV
40		1990.25	IV

Air

Intensity		Wavelength	
50	d	2042.91	IV
40		2055.73	IV
70		2299.21	IV
90		2299.59	IV
100		2316.85	IV
40		2324.06	IV
50		2360.40	IV
70		2405.15	IV
60		2423.32	IV

Cr V
Ref. 380 — C.H.C.

Vacuum

Intensity		Wavelength	
100		438.62	V
100		464.02	V
50		469.64	V
50		825.60	V
50		968.70	V
50		1045.04	V
60		1060.65	V
50		1112.45	V
60		1114.35	V
100		1116.48	V
80		1117.56	V
50		1118.16	V
150		1121.07	V
150		1127.63	V
100		1193.95	V
80		1196.04	V
50		1210.50	V
50		1259.99	V
100		1263.50	V
150		1465.86	V
50		1481.65	V
50		1482.76	V
50		1484.67	V
100		1489.71	V
150		1497.97	V
170		1519.03	V
220		1579.70	V
170		1591.72	V
150		1603.19	V
60		1638.50	V
50		1639.40	V
200		1837.44	V

COBALT (Co)
Z = 27

Co I and II
Ref. 1, 125, 276 — C.R.C.

Vacuum

Intensity		Wavelength	
20		1265.93	II
40		1271.94	II
20		1276.90	II
30		1293.97	II
25		1295.53	II
30		1295.86	II
20		1297.10	II
80		1299.58	II
30		1302.39	II
20		1306.76	II
80		1306.95	II
40		1311.12	II
40		1311.86	II
30		1315.42	II
30		1316.09	II
20		1318.19	II
30		1318.60	II
20		1319.84	II
20		1409.33	II
20		1466.21	II
20		1471.87	II
30		1472.90	II
30		1475.81	II
20		1484.26	II
30		1486.50	II
20		1509.23	II
20		1590.54	II
20		1595.77	II
20		1599.30	II
20		1693.34	II
8		1706.05	II
10		1723.01	II
15	h	1740.55	II
15		1743.39	II
8		1754.21	II
20	d	1808.01	II
10		1837.56	II
15		1839.37	II
1500		1842.34	I
1800		1847.89	I
1800		1852.71	I
2400		1855.05	I
1500		1878.28	I
10		1917.62	II
1800		1936.58	I
1500		1946.79	I
30		1950.09	II
1500		1951.90	I
1800		1954.22	I
1800		1955.17	I
30		1957.42	II
1500		1958.55	I
1500		1961.59	I
1500	h	1968.69	I
1500	h	1968.93	I
3000		1970.71	I
1800	h	1971.16	I
1800	h	1972.52	I
1500		1973.85	I
1800		1976.97	I
2400	h	1980.89	I
1500		1989.80	I
1800		1990.34	I
1500	l	1998.49	I

Air

Intensity		Wavelength	
20		2000.79	II
1500		2002.32	I
900		2008.04	I
50		2011.51	II
1200	h	2014.58	I
900		2016.17	I
50		2022.35	II
40		2025.76	II
50		2027.04	II
900		2031.96	I
30		2036.58	II
1500		2039.95	I
1200		2041.11	I
20		2049.17	II
40		2058.82	II
40		2063.78	II
50		2065.54	II
1500	h	2077.76	I
900		2085.67	I
900		2087.55	I
900		2089.35	I
900		2093.40	I
900		2094.86	I
900		2095.77	I
1200		2097.51	I
1500		2104.73	I
1500		2106.80	I
900		2108.98	I
30		2111.44	II
900	s	2117.68	I
20		2128.79	II
900		2137.78	I
900		2138.97	I
900		2163.03	I
20		2164.44	II
30		2173.33	II
1100		2174.60	I
20		2181.99	II
30		2187.01	II
20		2190.68	II
20		2192.50	II
200		2193.60	II
20		2200.40	II
40	p	2202.95	II
200		2256.73	II
150		2260.00	II
200		2283.52	II
1000		2286.15	II
200		2291.98	II
300	d	2293.38	II
300		2301.40	II
800	d	2307.85	II
2600		2309.02	I
500		2311.60	II
500		2314.05	II
300		2314.96	II
200	p	2317.06	II
2400		2323.14	I
300	p	2324.31	II
200	d	2326.11	II
500		2326.47	II
1400		2335.99	I
1600		2338.67	I
200		2347.39	II
1600		2352.85	I
200	d	2353.41	II
2000		2353.42	I
500		2363.80	II
400		2378.62	II
1400		2380.48	I
200		2381.76	II
300	p	2383.45	II
1400		2384.86	I
200		2386.36	II
500		2388.92	II
200		2397.38	II
1100	d	2402.06	I
200	p	2404.16	II
5300		2407.25	I
5300		2411.62	I
1600		2412.76	I
4800		2414.46	I
4800		2415.30	I
300		2417.65	II
4100		2424.93	I
3300		2432.21	I
2900		2436.66	I
2400		2439.05	I
200		2442.63	II
200	d	2446.03	II
200	p	2447.69	II
200		2450.00	II
200		2464.20	II
200		2486.44	II
200		2498.82	II
570		2504.52	I
500		2506.46	II
360		2506.88	I
200		2511.16	II
880		2517.87	I
500		2519.82	II
4300		2521.36	I
200	h	2524.65	II
300		2524.97	II
500		2528.62	II
2900		2528.97	I
200	p	2530.09	II
720		2530.13	I
860		2532.18	I
200	d	2533.82	II
2900		2535.96	I
860		2536.49	I
300		2541.94	II
1700		2544.25	I
200		2546.74	II
340		2548.34	I
310		2553.37	I
310		2555.07	I
300		2559.41	II
200		2560.03	II
960		2562.15	I
500		2564.04	II
1100		2567.35	I
960		2574.35	I
800		2580.32	II
300	d	2582.22	II
500		2587.22	II
500		2587.52	II
200		2588.91	II
100	p	2605.71	II
100		2612.50	II
100		2614.36	II
100	p	2628.77	II
100		2632.26	II
100		2636.07	II
310		2646.42	I
770		2648.64	I
100		2653.72	II
100		2663.53	II
200		2666.73	II
100		2675.85	II
100		2684.42	II
100		2702.02	II
200		2706.62	II
200		2707.35	II
190		2715.99	I
100		2727.78	II
80		2734.54	II
190		2745.10	I
100		2753.22	II
190		2764.19	I
100		2766.70	II
100		2774.97	II
100		2791.00	II
100		2793.73	II
150		2815.56	I
80		2835.63	II
80		2847.35	II
80		2871.22	II
190		2886.44	I
100		2918.38	II
100		2930.24	II
100		2954.73	II
690		2987.16	I
690		2989.59	I
20		3008.86	II
60		3022.59	II
30		3035.13	II
3100		3044.00	I
1700		3061.82	I
20		3352.79	II
80		3387.70	II
1100		3388.17	I
2200		3395.38	I
11000		3405.12	I
4500		3409.18	I
6700		3412.34	I
2200		3412.63	I
30		3415.77	II
2700		3417.16	I
50		3423.84	II
2500		3431.58	I
4500		3433.04	I
1600		3442.93	I
8800		3443.64	I
50		3446.39	II
4100		3449.17	I
2100		3449.44	I
21000		3453.50	I
1000		3455.23	I
5100		3462.80	I
5100		3465.80	I
8000		3474.02	I
1900		3483.41	I
4800		3489.40	I
2400		3495.69	I
40		3497.33	II
50		3501.72	II
9600		3502.28	I
7000		3506.32	I
50		3507.77	II
2900		3509.84	I
1400		3510.43	I
4800		3512.64	I
3800		3513.48	I
10		3514.23	II
30		3517.50	II
4800		3518.35	I
1300		3520.08	I
2700		3521.57	I
3800		3523.43	I
60		3523.51	II
6400		3526.85	I
2700		3529.03	I
7300		3529.81	I
1900		3533.36	I
40		3535.92	II
50		3545.03	II
20		3555.93	II
1100		3560.89	I
80		3561.07	II
20		3566.98	II
8800		3569.38	I
50		3574.95	II

Cobalt (Cont.)

Intensity		Wavelength	
1600		3574.96	I
60		3575.32	II
2500		3575.36	I
60		3577.96	II
1000		3585.16	I
6700		3587.19	I
1900		3594.87	I
1600		3602.08	I
100		3621.21	II
1000		3627.81	I
80		3643.61	II
10		3656.75	II
60		3681.35	II
5		3695.32	II
5		3714.73	II
1100		3745.50	I
20		3754.69	II
1400		3842.05	I
6900		3845.47	I
5500		3873.12	I
2800		3873.96	I
7900		3894.08	I
20	h	3911.40	II
1500		3935.97	I
80	h	3963.10	II
40	h	3976.74	II
10		3983.02	II
6000		3995.31	I
970		3997.91	I
350		4020.90	I
10	h	4036.14	II
20	h	4037.37	II
4		4040.02	II
370		4045.39	I
5	h	4050.23	II
10	h	4052.40	II
20	h	4062.73	II
2	h	4064.50	II
350		4066.37	I
5	h	4074.34	II
830		4092.39	I
1		4096.57	II
550		4110.54	I
2800		4118.77	I
4400		4121.32	I
3		4130.88	II
3		4145.13	II
3	d	4160.67	II
1		4181.13	II
90		4190.71	I
1		4208.61	II
30	s	4244.25	II
8		4272.33	II
20	h	4288.25	II
3		4328.86	II
2		4384.26	II
2	h	4396.94	II
3		4413.91	II
90		4469.56	I
10		4482.50	II
2	h	4489.12	II
4	h	4497.44	II
10	d	4500.54	II
0		4516.65	II
690		4530.96	I
2	h	4533.22	II
0	h	4537.95	II
90		4549.66	I
1	h	4559.29	II
140		4565.59	I
1		4569.26	II
190		4581.60	I
5		4616.30	II
120		4629.38	I
25	h	4660.66	II
85		4663.41	I
110		4792.86	I
10	d	4831.16	II
100		4840.27	I
150		4867.88	I
80	h	4964.18	II
10	h	4970.05	II
10	h	4990.47	II
20	h	4995.98	II
35		5146.74	I
50		5212.71	I
50		5230.22	I
45		5235.21	I
50		5247.93	I
26		5266.30	I
45		5266.49	I
26		5268.52	I
45	h	5280.65	I
26		5301.06	I
50		5342.71	I
26		5343.39	I
50		5352.05	I
26		5353.48	I
35		5369.58	I
45		5483.34	I
17		5530.77	I
17		5647.22	I
17		5991.88	I
17		6082.44	I
17		6282.63	I
45		6450.24	I
21		6455.00	I
15		6563.42	I
15		6632.44	I
14		6814.94	I
14		6872.40	I
21		7052.89	I
45		7084.99	I
8		7417.38	I
8		7712.68	I
7		7908.71	I
9		7987.38	I
13		8007.27	I
9		8093.96	I
9		8372.84	I
4		8575.35	I
3		8819.15	I

Co III
Ref. 291 — C.R.C.

Intensity	Wavelength	
	Vacuum	
1000	1696.01	III
800	1697.99	III
500	1702.79	III
1000	1707.35	III
500	1707.95	III
500	1723.97	III
400	1745.67	III
500	1755.98	III
5000	1760.35	III
500	1769.96	III
500	1773.22	III
5000	1773.57	III
500	1774.42	III
1000	1777.14	III
2000	1780.05	III
3000	1782.97	III
500	1784.06	III
1000	1787.08	III
1000	1789.07	III
500	1790.26	III
500	1791.28	III
500	1798.06	III
500	1805.54	III
400	1811.47	III
400	1821.26	III
400	1821.69	III
400	1821.77	III
1000	1823.08	III
400	1825.36	III
750	1825.95	III
2000	1830.09	III
2000	1831.44	III
750	1831.92	III
400	1832.20	III
5000	1835.00	III
1000	1837.63	III
500	1846.16	III
500	1852.92	III
400	1854.39	III
400	1854.76	III
1000	1861.78	III
2000	1863.83	III
400	1864.19	III
500	1871.87	III
1000	1881.70	III
500	1895.37	III
500	1919.12	III
500	1928.57	III
500	1940.15	III
500	1953.94	III
500	1959.41	III
400	1989.60	III
	Air	
100	2001.09	III
200	2011.62	III
200	2013.88	III
100	2031.81	III
200	2053.11	III
100	2056.21	III
10	2062.17	III
10	2079.74	III
15	2088.58	III
10	2090.51	III
10	2097.63	III
10	2134.15	III
10	2452.16	III
20	2811.75	III
10	2888.31	III
10	2933.27	III
10	2978.01	III
20	2991.89	III
25	3010.92	III
10	3116.68	III
2	3151.40	III
2	3180.64	III
20	3232.11	III
2	3249.24	III
20	3259.68	III
2	3269.23	III
10	3287.68	III
15	3305.38	III
10	3451.25	III
2	3526.24	III
1	3634.21	III
1	3636.31	III
2	3667.52	III
2	3677.23	III
3	3680.74	III
1	3762.50	III
15	3782.27	III

Co IV
Ref. 236 — C.R.C.

Intensity	Wavelength	
	Vacuum	
81	606.79	IV
74	607.59	IV
55	608.24	IV
66	609.16	IV
70	609.21	IV
64	609.28	IV
43	610.04	IV
37	610.25	IV
24	610.79	IV

Co V
Ref. 100, 159 — C.R.C.

Intensity	Wavelength	
	Vacuum	
20	355.52	V
18	355.88	V
12	356.06	V
4	1006.86	V
10	1007.51	V
15	1009.02	V
10	1010.94	V
10	1013.80	V
1	1017.43	V
10	1018.36	V
10	1021.14	V
1	1028.08	V
3	1226.31	V
8	1228.19	V
15	1231.73	V
2	1234.55	V
20	1236.95	V
2	1239.85	V
8	1246.91	V
6	1258.61	V
5	1263.28	V
20	1270.70	V
20	1272.23	V
2	1275.52	V
50	1277.01	V
30	1281.63	V
15	1284.00	V
28	1286.95	V
25	1295.55	V
40	1295.87	V
35	1301.12	V
50	1345.67	V
6	1351.22	V
15	1353.42	V
40	1355.20	V
30	1357.67	V
20	1361.32	V
30	1362.46	V
30	1364.17	V
30	1368.24	V
4	1369.30	V
10	1371.01	V
30	1373.09	V
30	1375.20	V
25	1378.12	V
10	1379.05	V
10	1380.21	V
32	1389.11	V
15	1459.77	V
35	1468.98	V
30	1476.65	V
25	1482.62	V
20	1482.91	V
25	1486.02	V
20	1488.73	V

COPPER (Cu)
Z = 29

Cu I and II
Ref. 273, 290—V.K.

Intensity	Wavelength	
	Vacuum	
80	685.141	II
100	709.313	II
100	718.179	II
150	724.489	II
200	735.520	II
250	736.032	II
80	779.295	II
100	797.455	II
150	810.998	II
200	813.883	II
300	826.996	II
150	848.808	II
250	851.303	II
250	858.487	II
400	861.994	II
400	865.390	II
250	869.336	II
150	873.263	II
200	876.723	II
250	877.012	II
200	877.555	II
500	878.699	II
100	884.133	II
250	885.847	II
600	886.943	II
600	890.567	II
500	892.414	II
800	893.678	II
400	894.227	II
600	896.759	II
400	896.976	II
600	901.073	II
400	906.113	II
800	914.213	II
600	922.019	II
500	924.239	II
400	935.232	II
600	935.898	II
600	943.335	II
600	945.525	II
500	945.965	II
200	954.383	II
250	956.290	II
400	958.154	II
200	960.414	II
250	968.042	II
200	974.759	II
250	977.567	II
100	987.657	II
250	992.953	II
300	1004.055	II
300	1008.569	II
300	1008.728	II
300	1010.269	II
250	1012.597	II
500	1018.707	II
500	1027.831	II
250	1028.328	II
200	1030.263	II
600	1036.470	II
600	1039.348	II
600	1039.582	II
800	1044.519	II
800	1044.744	II
500	1049.755	II
600	1054.690	II
400	1055.797	II
600	1056.955	II
400	1058.799	II
600	1059.096	II
600	1060.634	II
600	1063.005	II
200	1065.782	II
200	1066.134	II
500	1069.195	II
300	1073.745	II
200	1088.395	II
300	1094.402	II
250	1097.053	II
150	1119.947	II
200	1142.640	II
300	1144.856	II
100	1250.048	II
150	1265.506	II
300	1275.572	II
150	1282.455	II
150	1287.468	II
150	1298.395	II
300	1308.297	II
300	1314.337	II
100	1320.686	II
100	1326.395	II
150	1350.594	II
250	1351.837	II
150	1355.305	II
300	1358.773	II
200	1359.009	II

Copper (Cont.)

200		1362.600	II
250		1367.951	II
200		1371.840	II
100		1393.128	II
100		1398.642	II
150		1402.777	II
150		1407.169	II
100		1414.898	II
250		1418.426	II
250		1421.759	II
200		1427.829	II
400		1430.243	II
250		1434.904	II
150		1436.236	II
150		1442.139	II
200		1445.984	II
200		1449.058	II
250		1450.304	II
200		1452.294	II
300		1458.002	II
250		1459.412	II
200		1463.752	II
400		1463.838	II
200		1466.070	II
400		1470.697	II
200		1472.395	II
250		1473.978	II
200		1474.935	II
150		1476.059	II
200		1481.544	II
200		1485.328	II
750		1488.831	II
300		1492.834	II
250		1493.366	II
250		1495.430	II
350		1496.687	II
150		1503.368	II
250		1504.757	II
200		1505.388	II
300		1508.632	II
350		1510.506	II
200		1512.465	II
200		1513.366	II
500		1514.492	II
200		1517.631	II
500		1519.492	II
600		1519.837	II
200		1520.540	II
200		1524.860	II
150		1525.764	II
500		1531.856	II
300		1532.131	II
250		1533.986	II
250		1535.002	II
500		1537.559	II
200		1540.239	II
300		1540.389	II
300		1540.588	II
750		1541.703	II
400		1544.677	II
100		1547.958	II
300		1550.653	II
300		1551.389	II
500		1552.646	II
250		1553.896	II
400		1555.134	II
500		1555.703	II
300		1558.345	II
400		1565.924	II
400		1566.415	II
100		1569.416	II
300		1579.492	II
300		1580.626	II
400		1581.995	II
500		1583.682	II
400		1590.165	II
600		1593.556	II
400		1598.402	II
400		1602.388	II
200		1604.040	II
300		1605.281	II
400		1606.834	II
250		1608.639	II
150		1610.296	II
200		1617.915	II
600		1621.426	II
400		1622.428	II
250		1630.268	II
100		1636.605	II
250		1649.458	II
30	r	1655.32	II
200		1656.322	II
200		1660.001	II
300		1663.002	II
100		1672.776	II
30		1688.09	I
30		1691.08	I
30	r	1703.84	I
50	r	1713.36	I
150		1717.721	II
50	r	1725.66	I
100		1736.551	II
50	r	1741.57	I
150		1753.281	II

200	r	1774.82	I
100	r	1825.35	I
250		1929.751	II
250		1944.597	II
100		1946.493	II
200		1957.518	II
150		1970.495	II
150		1977.027	II
500		1979.956	II
300		1989.855	II

Air

250		1999.698	II
270		2035.854	II
250		2037.127	II
350		2043.802	II
300		2054.980	II
100		2078.663	II
110		2098.398	II
320		2104.797	II
300		2112.100	II
320		2117.310	II
350		2122.980	II
350		2126.044	II
420		2134.341	II
900		2135.981	II
400		2148.984	II
150		2161.320	II
1300	r	2165.09	I
250		2174.982	II
1600	r	2178.94	I
700		2179.410	II
1700	r	2181.72	I
700		2189.630	II
900		2192.268	II
400		2195.683	II
1700	r	2199.58	I
1300	r	2199.75	I
100		2200.509	II
200		2209.806	II
750		2210.268	II
1600	r	2214.58	I
250		2215.106	II
1000	r	2215.65	I
750		2218.108	II
2100	r	2225.70	I
150		2226.780	II
1600	r	2227.78	I
350		2228.868	II
2500	r	2230.08	I
1100	r	2238.45	I
900		2242.618	II
2300	r	2244.26	I
1000		2247.002	II
1300	r	2260.53	I
2200	r	2263.08	I
150		2263.786	II
200		2276.258	II
100		2286.645	II
2500	r	2293.84	I
170		2294.368	II
1000		2303.12	I
150		2369.890	II
2500	r	2392.63	I
120		2403.337	II
1500		2406.66	I
1000	r	2441.64	I
100		2485.792	II
2000	r	2492.15	I
150		2506.273	II
120		2526.593	II
300		2544.805	II
100		2571.756	II
150		2590.529	II
200		2600.270	II
2500	r	2618.37	I
200		2666.291	II
750		2689.300	II
700		2700.962	II
650		2703.184	II
700		2713.508	II
650		2718.778	II
300		2721.677	II
120		2737.342	II
270		2745.271	II
2500	r	2766.37	I
800		2769.669	II
200		2791.795	II
170		2799.528	II
100		2810.804	II
1250	r	2824.37	I
350		2837.368	II
100		2857.748	II
600		2877.100	II
270		2884.196	II
2500	r	2961.16	I
100		2986.335	II
2000		2997.36	I
2000		3010.84	I
2500		3036.10	I
2500		3063.41	I
1400		3073.80	I
1500		3093.99	I

1250		3099.93	I
2000		3108.60	I
1400	h	3126.11	I
1500		3194.10	I
1400		3208.23	I
1500	h	3243.16	I
10000	r	3247.54	I
10000	r	3273.96	I
1400	h	3282.72	I
400		3290.418	II
1500	h	3290.54	I
110		3300.881	II
250		3301.229	II
2500	h	3307.95	I
200		3316.276	II
1500		3337.84	I
150		3338.648	II
200		3365.648	II
450		3370.454	II
300		3374.952	II
200		3380.712	II
100		3384.945	II
1250	h	3483.76	I
1250		3524.23	I
2000		3530.38	I
1400		3599.13	I
1400		3602.03	I
1000		3686.555	II
150		3786.270	II
170		3797.849	II
100		3818.879	II
140		3826.921	II
160		3864.137	II
280		3884.131	II
150		3892.924	II
170		3903.177	II
140		3920.654	II
120		3933.268	II
120		3987.024	II
150		3993.302	II
140		4003.476	II
1250		4022.63	I
100		4032.647	II
600		4043.484	II
500		4043.751	II
2000		4062.64	I
120		4068.106	II
500		4131.363	II
200		4143.017	II
300		4153.623	II
500		4161.140	II
370		4164.284	II
400		4171.851	II
500		4179.512	II
500		4211.866	II
320		4230.449	II
200		4255.635	II
950		4275.11	I
300		4279.962	II
500		4292.470	II
400		4365.370	II
100		4444.831	II
400		4506.002	II
150		4516.049	II
150		4541.032	II
500		4555.920	II
100		4596.906	II
120		4649.271	II
2000		4651.12	I
120		4661.363	II
320		4671.702	II
300		4673.577	II
450		4681.994	II
100		4758.433	II
400		4812.948	II
120		4851.262	II
300		4854.988	II
100		4873.304	II
150		[illegible]	II
1000		4909.734	II
500		4918.376	II
200		4926.424	II
900		4931.698	II
120		4943.026	II
700		4953.724	II
500		4985.506	II
400		5006.801	II
350		5009.851	II
400		5012.620	II
350		5021.279	II
200		5039.016	II
300		5047.348	II
900		5051.793	II
400		5058.910	II
500		5065.459	II
450		5067.094	II
350		5072.302	II
450		5088.277	II
420		5093.816	II
350		5100.067	II
1500		5105.54	I
250		5124.476	II
2000		5153.24	I
100		5158.093	II

100		5183.367	II
2500		5218.20	I
100		5269.991	II
100		5276.525	II
1650		5292.52	I
100		5368.383	II
1500		5700.24	I
1500		5782.13	I
150		5805.989	II
100		5833.515	II
200		5897.971	II
120		5937.577	II
400		5941.196	II
100		5993.260	II
650		6000.120	II
100		6023.264	II
250		6072.218	II
150		6080.343	II
150		6099.990	II
160		6107.412	II
300		6114.493	II
600		6150.384	II
750		6154.222	II
500		6172.037	II
550		6186.884	II
400		6188.676	II
300		6198.092	II
470		6204.261	II
450		6208.457	II
750		6216.939	II
700		6219.844	II
500		6261.848	II
1000		6273.349	II
350		6288.696	II
900		6301.009	II
550		6305.972	II
400		6312.492	II
120		6326.466	II
400		6373.268	II
750		6377.840	II
400		6403.384	II
850		6423.884	II
200		6442.965	II
750		6448.559	II
170		6466.246	II
950		6470.168	II
750		6481.437	II
400		6484.421	II
220		6517.317	II
400		6530.083	II
120		6551.286	II
200		6577.080	II
750		6624.292	II
800		6641.396	II
450		6660.962	II
100		6770.362	II
300		6806.216	II
400		6809.647	II
320		6823.202	II
250		6844.157	II
320		6868.791	II
270		6872.231	II
270		6879.404	II
220		6937.553	II
150		6952.871	II
150		6977.572	II
200		7022.860	II
300		7194.896	II
400		7326.008	II
300		7331.694	II
250		7382.277	II
1000		7404.354	II
270		7434.156	II
500		7562.015	II
700		7652.333	II
1000		7664.648	II
150		7681.788	II
450		7744.097	II
1000		7778.738	II
750		7805.184	II
1500		7807.659	II
1000		7825.654	II
350		7860.577	II
300		7890.567	II
700		7902.553	II
1500		7933.13	I
400		7944.438	II
400		7972.033	II
1200		7988.163	II
2000		8092.63	I
500		8277.560	II
800		8283.160	II
250		8503.396	II
750		8511.061	II
200		8609.134	II
500		9813.213	II
250		9827.978	II
200		9830.798	II
600		9861.280	II
600		9864.137	II
200		9883.969	II
550		9916.419	II
500		9917.954	II
550		9925.594	II

Copper (Cont.)

Intensity		Wavelength	
450		9938.998	II
500		9960.354	II
450		10006.588	II
550		10022.969	II
550		10038.093	II
650		10054.938	II
450		10080.354	II

Cu III
Ref. 295—V.K.

Vacuum

Intensity		Wavelength	
75		542.90	III
200		615.67	III
150		616.03	III
150		687.98	III
150		715.53	III
125		730.38	III
250		788.07	III
250		788.46	III
250		791.36	III
150		801.14	III
100		829.34	III
40		1048.88	III
50		1186.80	III
50		1200.96	III
300		1219.30	III
200		1244.38	III
100		1279.14	III
200		1312.39	III
300		1332.97	III
200		1339.48	III
150		1363.08	III
300	r	1376.79	III
200	r	1377.49	III
150		1423.48	III
300	r	1481.23	III
200		1543.46	III
500	r	1593.75	III
1000	r	1642.21	III
300		1679.14	III
400		1702.10	III
500		1722.37	III
600		1741.37	III
200		1768.86	III
200		1840.91	III
100		1971.95	III

Air

Intensity		Wavelength	
200		2013.22	III
150		2157.28	III
100		2299.47	III
500		2368.17	III
400		2391.74	III
800		2405.50	III
700		2412.34	III
2000		2444.44	III
500		2468.41	III
1000		2482.36	III
700		2486.46	III
500		2508.49	III
500		2522.38	III
500		2538.66	III
400		2566.37	III
400		2573.33	III
500		2609.32	III
200		2643.92	III
200		2696.38	III
20		2751.33	III
100		2812.94	III
100		2978.87	III
75		3548.87	III
100		3639.42	III
500		3702.92	III
800		3744.70	III
400		3748.27	III
600		3752.06	III
1000		3776.97	III
800		3790.80	III
600		3804.13	III
600		3809.18	III
300		3881.68	III
150		3953.81	III
100		4090.49	III
200		4283.40	III
500		4351.97	III
1000		4352.80	III
500		4355.24	III
500		4370.84	III
500		4371.40	III
500		4373.43	III
1000		4377.11	III
200		4386.42	III
150		4927.41	III
400		5094.28	III
200		5168.97	III
400		5208.34	III
600		5219.21	III
200		5268.59	III
400		5317.78	III
300		5369.79	III
350		5418.48	III
250	d	5494.94	III
50		5573.94	III
100		5609.00	III
75		5702.12	III
100		5768.56	III
100		5850.72	III
200		5965.25	III
30		6100.87	III
50		6369.27	III
20		6512.54	III
20		6644.13	III
50		6793.20	III

Cu IV
Ref. 199—V.K.

Vacuum

Intensity		Wavelength	
30		360.86	IV
20		374.40	IV
30		405.24	IV
80		406.45	IV
40		413.45	IV
70		443.68	IV
80		451.16	IV
80		463.72	IV
80		484.53	IV
90		497.00	IV
90		504.60	IV
70		509.38	IV
40		519.51	IV
40		540.65	IV
60		550.92	IV
20		584.85	IV
60		1056.13	IV
30		1074.72	IV
50		1091.65	IV
30		1105.50	IV
25		1119.43	IV
40	p	1152.18	IV
60		1227.44	IV
70		1228.87	IV
70		1258.69	IV
90		1274.84	IV
90		1293.46	IV
90		1309.41	IV
70		1321.17	IV
100	d	1340.08	IV
100		1350.42	IV
100		1362.05	IV
100		1372.14	IV
100		1377.82	IV
100		1388.80	IV
90		1405.49	IV
90		1415.27	IV
90		1434.34	IV
90		1449.69	IV
80		1466.18	IV
70		1482.77	IV
80		1499.81	IV
90		1515.28	IV
90		1535.12	IV
80		1551.12	IV
90		1567.35	IV
80		1583.47	IV
70		1595.12	IV
80	p	1608.14	IV
90		1639.75	IV
20		1650.16	IV
70		1704.37	IV
30		1797.99	IV
70		1817.56	IV
70		1819.23	IV
60		1837.04	IV
80		1849.62	IV
30		1867.24	IV
30		1918.71	IV
40		1966.31	IV

Cu V
Ref. 324—V.K.

Vacuum

Intensity		Wavelength	
9	h	258.95	V
49		271.33	V
49		283.97	V
22		293.41	V
56		299.64	V
65		305.83	V
51		312.51	V
66		321.05	V
74		326.57	V
82		333.56	V
81		339.88	V
81		346.00	V
86		355.41	V
77		363.96	V
65		370.63	V
74		377.76	V
70		387.40	V
51		396.06	V
25		406.94	V
13		1097.10	V
42		1106.24	V
77		1113.22	V
67		1121.20	V
76		1128.80	V
59		1133.86	V
63		1142.38	V
54		1149.06	V
72		1157.54	V
64		1167.35	V
77		1176.53	V
84		1183.63	V
76		1192.54	V
83		1201.22	V
77		1204.90	V
71		1214.36	V
70		1221.34	V
76		1230.11	V
80		1239.73	V
79		1246.99	V
65		1253.07	V
70		1260.24	V
77		1269.35	V
78		1278.20	V
68		1286.13	V
67		1292.08	V
73		1299.22	V
65		1309.72	V
55		1318.89	V
65		1323.28	V
64		1329.22	V

CURIUM (Cm)
Z = 96

Cm I and II
Ref. 51, 332—J.G.C.

Air

Intensity		Wavelength	
10000		2462.76	II
10000		2617.17	II
10000		2636.28	II
10000		2651.17	II
10000		2653.80	II
10000		2725.68	II
10000		2736.89	II
10000		2748.04	II
10000		2784.83	II
10000		2811.62	II
10000		2824.20	II
10000		2833.58	II
10000		2899.90	II
10000		2912.97	II
10000		2928.92	II
10000		2996.18	II
10000		2999.39	I
10000	b	3014.87	II
10000		3044.85	II
10000		3109.69	I
10000		3116.41	I
10000		3137.16	I
10000		3147.33	I
10000		3155.10	I
10000		3158.60	I
10000		3169.98	II
10000		3177.55	I
10000		3179.10	I
10000		3186.41	I
10000		3188.11	I
10000		3207.12	II
10000		3207.71	I
10000		3209.89	II
10000		3209.94	I
10000		3210.05	II
10000		3220.76	II
10000		3224.23	I
10000		3225.11	I
10000		3226.41	II
10000		3230.28	I
10000		3230.35	II
10000		3236.74	I
10000		3238.55	II
10000		3242.66	II
10000		3246.25	I
10000		3252.68	I
10000		3265.81	I
10000		3280.45	I
10000		3296.71	II
10000		3304.85	I
10000		3317.14	I
10000		3374.70	I
10000		3452.92	I
10000		3458.34	I
10000		3510.28	I
10000		3522.36	I
10000		3524.94	I
10000		3542.06	I
10000		3547.02	I
10000		3547.92	I
10000		3561.44	I
10000		3572.95	II
10000		3600.62	I
10000		3639.94	I
10000		3664.34	I
10000		3709.43	I
10000		3729.00	I
10000		3732.35	I
10000		3747.86	I
10000		3763.05	I
10000		3775.75	I
10000		3816.30	I
10000		3825.14	I
10000		3833.32	I
10000		3837.59	I
10000		3842.00	I
10000		3849.92	I
10000		3854.11	I
10000		3900.25	I
10000		3904.06	II
10000		3908.24	II
10000		3936.67	I
10000		3942.03	I
10000		3944.15	I
10000		3948.68	I
10000		3953.36	I
10000		3964.83	I
10000		3995.10	I
10000		4016.17	I
10000		4031.76	I
10000		4048.29	I
10000		4049.65	I
10000		4113.29	I
10000		4129.71	I
10000		4207.66	II
10000		4211.62	I
10000		4266.45	I
10000		4293.00	I
10000		4330.82	I
10000		4345.69	I
10000		4447.77	I
10000		4459.16	I
10000		4608.40	I
10000		5846.07	I
10000		5952.41	I
10000		6058.90	I
10000		6243.35	I
10000		6376.71	I
10000		6510.16	I
10000		6554.41	I
10000		6640.17	I
10000		6663.25	I
10000		6686.87	I
10000		6706.85	I
10000		6726.68	I
10000		6793.15	I
10000		7162.69	I
10000		7577.80	I
10000		7673.79	I
10000		7720.47	I
10000		8392.37	I
10000		9293.25	I
10000		9567.08	I
10000		9657.12	I
10000		10310.83	I
10000		10351.73	I
10000		10424.49	I
10000		10508.11	I
10000		10542.98	I
10000		10792.25	I
10000		10897.45	I
10000		11507.45	I
10000		11707.73	I
10000		11780.95	I
10000		11834.28	I
10000		12017.85	I
10000		12394.16	I
10000		12454.98	I
10000		12464.99	I
10000		13004.56	I
10000		13258.18	I
10000		13289.84	I
10000		13344.62	I
10000		13480.54	I
10000		13590.01	I
10000		13644.77	I
10000		13789.52	I
10000		13840.18	I
10000		13908.46	I
10000		13964.14	I
10000		14235.27	I
10000		14334.52	I
10000		14563.41	I
10000		14580.23	I
10000		15018.13	I
10000		15222.27	I
10000		15642.59	I

Curium (Cont.)

Intensity		Wavelength	
10000		15757.23	I
10000		15793.31	I
10000		16008.41	I
10000		17148.22	I
10000		17453.18	I
10000		17619.28	I
10000		18069.02	I
10000		19572.62	I
10000		19975.98	I
10000		20526.32	I
10000		20853.49	I
10000		20911.52	I
10000		20968.11	I
10000		21241.06	I
10000		21393.23	I

DYSPROSIUM (Dy)
Z = 66

Dy I and II
Ref. 1—C.H.C.

Intensity		Wavelength Air	
260		2356.91	II
65		2381.95	
130		2387.36	II
150		2392.15	
180		2402.29	II
240		2410.01	II
150		2422.75	II
260		2439.84	II
90		2455.15	II
110		2459.99	II
90		2471.40	II
110		2480.93	II
170		2490.61	II
90		2510.31	II
170		2513.55	II
170		2517.61	II
130		2543.81	II
90		2545.12	II
150		2552.29	II
180		2557.94	II
90		2560.21	II
90		2566.25	II
220		2585.30	I
90		2591.56	II
75		2592.54	II
130		2600.16	II
130		2600.76	II
75		2608.69	II
370		2623.69	I
440		2634.80	II
110		2642.15	I
110		2645.35	II
110		2667.94	I
55		2676.84	II
50		2677.34	II
85		2689.31	II
85		2692.83	II
55		2709.01	II
55		2727.17	II
85		2729.50	II
55		2735.79	I
40		2739.30	II
85		2740.70	II
220		2755.75	II
55		2757.08	II
70		2766.50	II
70		2772.42	II
110		2772.61	II
40		2779.58	II
55		2791.44	II
120		2800.33	II
110		2800.53	II
110		2801.41	II
300		2816.39	II
140		2825.42	II
140		2862.70	I
190		2877.88	II
110		2884.28	II
120		2885.53	I
120		2890.74	II
120		2900.82	II
110		2904.62	II
190		2906.39	II
390		2913.95	II
110		2934.31	II
250		2934.52	II
110		2941.05	II
140		2944.56	II
150		2947.06	II
150		2947.21	II
250		2948.31	II
170		2950.33	II
110		2952.12	II
140		2953.70	II
220		2964.60	I
110		2977.42	II
110		2985.97	II
220		3015.68	II
390		3026.16	II
210		3029.81	II
610		3038.28	II
280		3043.13	II
210		3047.56	II
280		3060.64	II
390		3062.62	II
220		3066.99	II
330		3071.91	II
280		3073.54	II
220		3078.68	II
280		3101.93	II
220		3103.24	II
410		3109.76	II
330		3128.41	II
830		3135.38	II
360		3140.64	II
500		3141.14	II
220		3143.83	II
250		3146.16	II
1200		3156.52	II
670		3162.83	II
1000		3169.99	II
400		3177.89	II
220		3178.37	II
200		3184.79	II
330		3186.38	II
240		3187.68	II
330		3193.30	II
240		3206.40	II
220		3207.12	II
290		3208.85	II
470		3215.19	II
830		3216.63	II
240		3221.49	II
290		3223.28	II
240		3225.08	II
330		3225.95	II
490		3235.89	II
290		3236.69	II
490		3245.12	II
200		3248.36	II
1200		3251.27	II
200		3252.19	II
290		3256.26	II
200		3266.21	II
240		3269.11	II
200		3272.73	II
890		3280.09	II
490		3282.77	II
200		3287.94	II
200		3293.88	II
200		3296.30	II
200		3305.40	II
200		3305.51	II
240		3306.19	II
440		3308.79	II
1100		3308.88	II
510		3312.72	II
780		3316.32	II
240		3317.12	II
1000		3319.88	II
270		3326.19	II
780		3341.00	II
270		3341.88	II
200		3347.83	II
270		3352.69	II
510		3353.58	II
240		3359.46	II
510		3368.11	II
5300		3385.02	II
210		3386.57	II
610		3388.85	II
210		3391.96	II
3800		3393.57	II
1300		3396.16	II
380		3407.16	II
5300		3407.80	II
420		3408.14	II
1300		3413.78	II
530		3414.82	II
780		3419.63	II
530		3425.06	II
420		3429.44	II
1900		3434.37	II
330		3438.94	II
560		3440.93	II
1300		3441.45	II
3800		3445.57	II
830		3446.99	II
440		3449.89	II
2700		3454.32	II
440		3454.51	II
1300		3456.56	II
4400		3460.97	II
720		3468.43	II
560		3471.14	II
560	d	3471.53	II
380		3473.70	II
1300		3477.07	II
4400		3494.49	II
560		3496.34	II
400		3497.81	II
830		3498.71	II
400		3501.50	II
830		3504.53	II
830		3505.45	II
1300		3506.81	II
560		3517.26	II
4400		3523.98	II
22000		3531.70	II
4400		3534.96	II
5500		3536.02	II
4400		3538.52	II
400		3539.37	II
1700		3542.33	II
400		3544.20	II
400		3544.35	II
1400		3546.83	II
330		3548.19	II
4400		3550.22	II
2200		3551.62	II
440	h	3558.23	II
440		3559.30	II
2200		3563.15	II
560		3563.69	II
780		3573.83	II
1400		3574.15	II
4400		3576.24	II
1700		3576.87	II
830		3577.98	II
440		3580.04	II
400		3584.42	II
3300		3585.06	II
1400		3585.78	II
560		3586.11	II
360		3590.07	II
1100		3591.41	II
560		3591.81	II
560		3592.11	II
1800		3595.04	II
400		3596.06	II
560		3600.38	II
360		3602.82	II
1800		3606.12	II
440		3618.51	II
560		3620.16	II
670		3624.27	II
1100		3629.42	II
4000		3630.24	II
440		3632.78	II
400		3635.27	II
360		3637.28	II
1100		3640.25	II
400		3643.92	II
11000		3645.40	II
360		3645.86	II
1000		3648.78	II
700		3664.62	II
400		3666.84	I
990		3672.30	II
420		3672.70	II
400		3673.14	II
1400		3674.08	II
2200		3676.59	II
640		3678.51	I
820		3684.85	I
1300		3685.78	I
4700		3694.81	II
370		3697.31	II
990		3698.21	II
540		3701.63	II
330		3707.40	II
440		3707.57	II
440		3708.22	II
420		3710.07	II
330		3711.66	II
1600		3724.45	II
300		3728.00	I
930		3739.34	I
1200		3747.82	II
1400		3753.51	II
1400		3753.75	II
1200		3757.05	I
4700		3757.37	II
640		3767.63	I
330		3771.11	I
640		3773.05	I
370		3774.71	I
420		3781.47	I
330		3785.41	II
3300		3786.18	II
1600		3788.44	II
700		3791.87	II
510		3804.14	II
580		3806.27	II
470		3812.27	I
470		3813.67	II
1400		3816.76	II
700		3825.68	II
2300		3836.50	II
370		3840.89	I
1400		3841.31	II
330		3842.00	II
330		3844.36	I
420		3846.34	II
420		3847.02	I
330		3849.39	II
1200		3853.03	II
420		3858.40	I
370		3866.58	II
560		3868.45	II
1600		3868.81	I
300		3869.42	II
820		3869.86	II
7000		3872.11	II
1200		3873.99	II
470		3879.11	II
300		3881.99	II
5800		3898.53	II
540		3914.87	II
540		3915.59	II
540	d	3917.29	I
320		3923.38	II
420		3927.86	I
540		3930.14	I
2100		3931.52	II
320		3932.22	II
370		3933.00	II
320		3934.21	II
420		3936.70	I
540		3942.53	II
10000		3944.68	II
420		3946.93	II
540		3950.39	II
420		3954.55	II
800		3957.79	II
370		3962.59	I
320		3967.51	I
14000		3968.39	II
2700		3978.57	II
1400		3981.92	II
1600		3983.65	II
800		3984.21	II
540		3991.32	II
1600		3996.69	II
8000		4000.45	II
420		4005.84	I
320		4006.07	I
540		4011.29	II
540		4013.82	I
540		4014.70	II
370		4023.71	I
420		4027.78	II
520	d	4028.32	II
520		4032.47	II
420		4033.65	II
420		4036.32	II
320		4041.98	II
12000		4045.97	I
1600		4050.56	II
520		4055.14	II
2500		4073.12	II
7400		4077.96	II
370		4085.34	I
390		4096.10	I
3900		4103.30	II
860		4103.87	I
1500		4111.34	II
490		4124.63	II
390		4128.24	II
350		4129.12	I
990		4129.42	II
350		4130.35	I
390		4133.85	I
470		4141.50	II
1200		4143.10	II
990		4146.06	I
5700		4167.97	I
370		4171.93	I
930		4183.72	I
12000		4186.82	I
320		4190.94	I
2200		4191.64	I
6800		4194.84	I
320		4195.19	II
800		4198.02	I
680		4201.30	I
680		4202.24	I
230		4205.06	I
370		4206.54	II
440		4211.24	I
16000		4211.72	I
1800		4213.18	I
3700		4215.16	I
4400		4218.09	I
4400		4221.11	I
540		4222.21	I
2700		4225.16	I
680		4232.02	I
680		4239.85	I
440		4245.91	I
440		4256.33	II
250		4276.69	I
370	d	4294.93	II
		4295.04	I
1000		4308.63	II
320		4325.86	I
200		4358.44	II

Dysprosium (Cont.)

Intensity		Wavelength	
320		4374.24	II
320		4374.76	II
540		4409.38	II
150		4444.58	I
740		4449.70	II
110		4455.60	II
250		4468.14	II
100	d	4527.58	I
		4527.76	II
100		4541.66	II
140		4565.09	I
420		4577.78	I
2100		4589.36	I
990		4612.26	I
50		4613.83	I
50		4614.82	I
60		4617.26	II
140		4620.03	II
50		4662.72	I
110		4664.66	II
85		4673.60	II
50		4682.03	II
50		4689.75	II
95		4698.68	II
85		4721.22	I
70		4727.13	II
170		4731.84	II
40		4745.73	II
60		4754.99	II
50		4760.04	II
60		4771.94	I
50		4774.80	I
120	h	4775.79	I
75		4786.92	II
95		4791.29	I
29		4800.64	I
50		4807.94	I
40		4810.28	I
50		4812.80	I
75		4819.04	I
85		4824.96	I
75		4828.88	I
50		4829.68	II
70		4832.38	I
35		4833.75	II
75		4841.75	I
40		4856.24	II
40		4868.05	II
40		4875.93	I
85		4880.16	I
40		4884.55	I
95		4888.08	I
40		4889.33	II
75		4890.10	II
50		4893.68	I
24		4899.24	I
55		4916.41	I
50		4922.22	II
65		4923.16	II
480		4957.34	II
24		4959.59	I
28		4973.57	I
40		4985.52	I
50		5003.87	I
55		5004.28	II
24		5010.60	I
24		5017.98	II
70		5022.12	I
30		5024.03	I
24		5024.54	I
40		5027.87	I
50		5033.00	I
160		5042.63	I
24		5047.25	I
50	h	5050.21	I
30		5053.35	I
24		5055.46	I
95		5070.68	I
120		5077.67	I
80		5090.38	II
80		5110.32	I
130	h	5120.04	I
30		5135.02	I
190		5139.60	II
40		5161.03	II
40		5164.12	II
50		5165.34	I
110		5169.69	II
20		5172.90	II
80		5185.30	I
40		5188.45	II
290		5192.86	II
95		5197.66	II
50		5246.94	II
70		5259.88	I
130		5260.56	I
55	Bl	5263.3	D
65		5267.11	I
50		5272.25	II
50		5275.29	II
50		5279.70	II
55		5282.07	I
28		5284.99	II
40		5297.82	II
160		5301.58	I
40		5309.02	II
50		5324.69	I
24		5337.43	II
65		5340.30	I
30		5352.11	I
30		5368.20	II
20		5385.63	II
85		5389.58	II
40		5395.57	I
20	h	5398.26	D
24		5399.93	II
50		5404.19	I
80		5419.13	I
70		5423.32	I
30		5424.27	I
40		5426.70	II
30		5443.34	II
95		5451.11	I
30		5455.47	II
24		5469.10	II
28		5496.83	I
24		5502.79	I
28		5506.52	I
24		5515.41	II
30		5528.01	I
65		5547.27	I
40	d	5600.65	II
24		5605.53	I
30		5613.23	I
20		5627.49	I
100		5639.50	I
55	h	5645.99	I
80		5652.01	I
24		5685.58	I
28	h	5693.67	D
24	h	5694.10	D
28	Cw	5694.54	D
28		5698.72	II
24		5702.91	I
70	h	5718.46	I
28	h	5725.84	D
55	h	5728.64	D
24	h	5738.73	D
50		5740.20	I
55		5745.53	I
24		5750.48	I
24		5758.79	I
80	h	5832.01	D
55	h	5833.85	D
40	h	5834.86	D
28	h	5844.41	D
24		5845.65	D
40	h	5848.05	D
40		5855.56	D
55	h	5868.11	II
40		5915.16	II
20		5924.56	II
70		5945.80	I
50	l	5964.46	I
120		5974.49	I
24		5984.86	I
140		5988.56	I
24	h	6005.75	D
24	h	6006.54	D
24	h	6006.97	D
30		6008.94	I
65		6010.82	I
24		6017.26	I
24		6030.98	I
24	Bl	6042.49	D
24		6058.18	I
30		6085.06	I
140		6088.26	I
24		6127.15	I
24		6133.64	I
24		6158.28	I
100		6168.43	I
20		6196.23	II
270		6259.09	I
30		6260.36	I
14		6343.32	I
40		6386.80	I
24		6396.60	II
50		6421.92	I
13	h	6436.55	I
8		6460.83	I
10		6468.58	II
11		6474.91	I
20		6483.59	II
28		6486.59	I
20		6558.02	I
160		6579.37	I
14		6594.14	II
15		6643.37	I
22		6658.36	I
29		6661.64	I
75		6667.86	I
10		6700.64	II
29		6747.93	I
10		6757.62	I
45		6765.89	I
12		6818.20	I
180		6835.42	I
80		6852.96	I
22		6856.46	I
22		6888.83	I
15		6897.97	II
65		6899.32	II
22		6906.53	II
15		6929.55	I
29		6950.28	II
11		6951.42	I
40		6958.08	I
13	h	6982.44	I
13		6991.30	I
45		6998.10	I
20		7017.42	I
35		7055.95	II
24		7075.14	II
17		7109.26	II
11		7120.81	II
13		7175.11	II
11		7213.27	I
17	h	7230.04	I
13		7250.01	I
17		7345.13	II
11		7370.23	II
20		7376.04	I
11		7407.59	I
24		7412.37	I
55		7426.86	II
20		7457.05	II
17		7516.61	II
55		7543.73	I
17	h	7553.00	I
27		7559.78	I
40		7562.96	II
20	h	7577.46	II
27	h	7591.30	I
13	h	7611.55	I
11	h	7617.70	I
35	h	7641.09	I
17		7645.86	I
13		7646.64	I
80		7662.36	I
11		7666.78	II
35		7715.33	I
45		7729.76	II
20		7751.62	II
35		7812.06	I
27		7909.38	I
11		7968.63	I
12		7982.85	II
13		8147.29	I
27		8198.77	II
100		8201.57	II
11		8218.62	II
20		8265.53	I
35		8326.10	I
35		8392.01	II
12		8405.85	II
20		8416.64	II
24		8438.58	II
11		8630.12	I
27		8655.94	II
17		8657.68	II
17		8678.49	II
11		8696.83	II
11	h	8715.95	II
20		8750.40	II
12		8780.83	I
45		8791.39	II
13		8833.08	II
24		8850.37	II

EINSTEINIUM (Es)
Z = 99

Es I and II
Ref. 333—J.G.C.

Intensity		Wavelength	
		Air	
300	l	2694.32	II
100	l	2703.84	
10000	s	2708.66	II
100	s	2716.02	II
1000	s	2724.57	II
1000	l	2765.76	
3000	l	2787.10	II
1000	l	2796.11	II
3000	l	2815.15	
100	s	2885.84	II
100	s	2886.44	
3000		2907.03	
100		3003.28	
1000	l	3065.40	
3000		3135.25	I
1000	l	3154.27	II
10000		3413.17	
300	l	3423.12	I
100	l	3424.28	
10000	s	3428.48	I
300		3437.31	
100		3437.34	
3000	s	3445.25	
300	l	3446.93	
3000		3452.36	
100		3453.16	
300	s	3470.77	
3000	s	3484.59	I
300		3494.30	
10000	s	3498.11	I
10000	l	3514.33	I
10000	s	3521.38	
10000	s	3523.49	I
300	l	3528.58	
10000	s	3536.01	
10000	s	3547.75	II
300		3549.97	
10000	l	3555.34	I
100	s	3555.53	
3000	s	3556.65	
3000	s	3560.92	
1000	s	3575.68	
100	s	3578.56	
300	s	3579.38	
1000	s	3582.95	
3000		3590.28	
1000		3595.47	
10000	s	3602.43	II
300		3605.58	
1000	s	3606.75	
100	l	3624.52	
3000	l	3631.09	
10000	s	3632.87	
100	l	3634.41	
1000	l	3651.94	
10000	l	3670.01	II
100	l	3672.32	
100	s	3713.56	
1000	l	3720.56	
300		3722.32	
10000	l	3728.55	II
100	h	3737.47	
300	s	3776.27	
10000	l	3792.99	
10000	l	3801.49	
300	s	3929.10	
3000		3930.77	
100		3957.19	
100		3995.35	
300		4077.71	
10000	s	4082.24	
3000	l	4107.59	
1000	l	4176.94	
100	s	4496.25	
100	l	4631.66	
300	l	4650.86	
1000		4789.93	
300	h	4802.17	
1000	h	4802.21	
3000	h	4958.29	
10000	s	5052.08	
1000	s	5102.93	
100	s	5155.82	
10000	s	5161.74	I
10000	l	5204.40	I
3000	L	5615.51	I
100	S	6539.71	ES

ERBIUM (Er)
Z = 68

Er I and II
Ref. 1—C.H.C.

Wavelength			Intensity
		Air	
110		2358.51	II
100		2386.58	II
120		2387.17	II
110		2396.38	III
140		2446.39	II
100		2537.02	II
110		2547.28	II
290		2586.73	II
110		2587.04	II
130		2592.57	II
120		2595.03	II
140		2624.18	II
490		2670.26	II
330		2672.25	II
100		2675.35	II
270		2739.31	II
310		2750.19	II
230		2755.01	II
610		2755.63	II
510		2770.02	II
230		2778.97	II
230		2802.53	II
310		2804.35	II
410		2820.19	II
270	d	2833.91	II

Erbium (Cont.)

Intensity		Wavelength	Spectrum
390		2838.71	II
270		2848.37	II
250		2855.41	II
310		2859.84	II
310		2896.96	II
390		2897.52	II
1000		2904.47	II
210		2909.58	II
1500		2910.36	II
270		2915.62	II
350		2929.27	II
270		2945.28	II
230		2946.62	II
1500		2964.52	II
410		2968.76	II
210		2974.47	II
230		2975.68	II
270		2983.80	II
1200		3002.41	II
310		3002.65	II
230		3012.47	II
230		3016.84	II
290		3025.95	II
270		3028.27	II
370		3031.31	II
310		3036.22	II
210		3054.42	II
230		3066.22	II
450		3070.74	II
560		3072.53	II
610		3073.34	II
210		3078.87	II
720		3082.08	II
610		3084.02	II
370		3099.19	II
230		3106.78	II
310	d	3113.43	II
		3113.54	II
770		3122.72	II
290		3132.52	II
470		3132.77	II
410		3141.10	II
		3141.15	II
250		3144.33	II
410		3154.29	II
870		3181.92	II
410		3183.42	II
250		3185.25	II
310		3200.58	II
230		3205.15	II
270		3214.44	II
870		3220.73	II
610		3223.31	II
210		3227.16	II
2300		3230.58	II
250		3232.03	II
330		3237.98	II
330		3249.34	II
560		3259.05	II
		3259.11	II
2700		3264.78	II
430		3267.10	II
		3267.18	II
330		3269.41	II
250		3278.22	II
720		3279.33	II
720		3280.22	II
470		3286.77	II
330	d	3303.88	II
		3303.95	II
370		3305.56	II
2300		3312.42	II
560		3316.39	II
770		3323.19	II
290		3329.66	II
770		3332.70	II
370		3337.25	II
290		3337.79	II
250		3340.03	II
290		3341.84	II
1300		3346.04	II
470		3350.06	II
350		3350.26	II
1400		3364.08	II
1400	d	3368.02	II
		3368.13	I
450		3370.55	II
7700		3372.71	II
970		3374.17	II
290		3381.32	II
230		3382.06	I
1700		3385.08	II
450		3389.74	II
2300		3392.00	II
350		3396.07	II
290		3396.84	II
390		3401.83	II
350		3417.63	II
490		3428.39	II
770		3441.13	II
390		3442.68	I
490		3469.51	I
970		3471.71	II
610		3479.41	II
970		3485.85	II
350		3486.82	II
350		3496.86	II
6700		3499.10	II
610		3502.78	I
390		3508.38	II
490		3514.89	II
390		3518.18	II
610		3524.91	II
410		3539.59	I
310		3548.26	II
820		3549.84	II
310		3553.20	II
1500		3558.02	I
510		3558.71	I
1000		3559.90	II
310		3565.17	I
920		3570.75	II
310		3578.24	I
1000		3580.52	II
370		3586.60	I
610		3590.76	I
410		3595.84	I
610		3599.50	II
1000		3599.83	II
510		3604.90	II
410		3607.42	I
3100		3616.56	II
510		3617.85	II
510		3618.92	II
720		3628.04	I
310		3629.37	I
1000		3633.54	II
510		3634.67	I
1600		3638.68	I
900		3645.94	II
520		3650.41	II
360		3652.58	II
500		3652.87	II
360		3664.45	I
470		3669.02	II
500		3682.70	II
320		3684.01	I
380		3684.28	II
7900		3692.65	II
430		3696.23	II
380		3697.68	I
540		3700.72	II
520		3707.64	II
520		3712.39	II
320		3719.35	I
1300		3729.52	II
450		3731.26	II
540		3738.16	II
340		3741.10	II
900		3742.64	II
900		3747.43	I
540		3756.05	I
410		3781.01	II
1800		3786.84	II
560		3787.86	II
560		3791.83	II
500		3792.79	I
560		3797.06	II
1600		3810.33	I
3600		3830.48	II
540		3849.91	I
320		3851.60	II
680		3855.90	I
540		3858.39	II
7500		3862.85	I
1500		3880.61	II
1200		3882.89	II
400		3890.61	II
4200		3892.68	I
5200		3896.23	II
810		3902.76	II
250		3903.98	I
250		3904.56	II
1200		3905.40	I
11000		3906.31	II
280		3918.05	I
210		3918.35	II
280		3921.88	II
810		3932.25	II
3200		3937.01	I
2100		3938.63	II
3200		3944.42	I
550		3948.06	I
250		3951.48	I
320		3956.42	I
280		3966.35	I
2700		3973.04	I
3200		3973.58	I
1400		3974.72	II
280		3976.73	I
810		3977.02	I
1100		3982.33	I
280		3987.53	I
810		3987.66	I
230		3991.15	I
230		4004.05	I
14000		4007.96	I
230		4008.18	II
280		4009.16	II
1100		4012.58	I
350		4015.57	II
3000		4020.51	I
450		4021.55	I
230		4043.01	II
1000		4046.96	I
280		4048.34	II
200		4049.49	II
940		4055.47	II
550		4059.51	I
690		4059.78	II
420		4077.88	I
550		4081.24	II
3500		4087.63	I
210		4092.90	I
1100		4098.10	I
350		4100.56	II
320		4116.36	I
320		4118.55	I
600		4131.50	I
550		4142.91	II
6900		4151.11	I
280		4189.98	II
1000		4190.70	I
130		4205.32	I
1400		4218.43	I
200		4220.99	I
320		4230.20	II
140		4234.78	II
200		4251.94	II
140		4276.48	II
690		4286.56	I
320		4298.91	I
320		4301.60	II
140		4303.81	II
110		4319.94	II
130		4328.81	I
110		4331.36	I
140		4340.92	I
190		4348.34	I
110		4369.39	II
160		4382.17	I
300		4384.70	II
800		4386.40	I
100		4403.17	II
810		4409.34	I
180		4418.70	I
570		4419.61	II
110		4422.51	II
320		4424.57	I
370		4426.77	I
110		4437.66	I
100		4459.24	II
100		4473.50	II
130		4496.39	I
200		4500.75	II
130		4522.74	I
160		4563.26	II
1000		4606.61	I
160		4630.88	II
110		4640.60	II
110		4665.44	II
310		4673.16	I
570		4675.62	II
150		4679.06	II
230		4722.69	I
150		4729.05	I
130		4751.52	II
170		4759.65	II
190		4820.35	II
140		4857.44	I
150		4872.09	II
210		4900.08	II
210		4934.11	II
130		4944.36	I
180		4951.74	II
130		4976.42	I
250		5007.25	I
160		5028.33	I
120		5028.91	II
200		5035.94	I
210		5042.05	II
130		5043.86	I
130		5044.89	I
130		5077.59	II
120		5124.56	I
130		5127.41	II
120		5131.53	I
130		5133.83	II
170		5164.77	II
130		5172.78	I
160		5188.90	II
150		5206.52	I
60		5212.91	II
30		5215.13	II
30		5218.26	II
45		5229.34	II
140		5255.93	II
22		5256.47	II
27		5257.02	II
35		5264.77	II
80		5272.91	I
55		5277.71	I
27		5279.34	II
45		5302.30	II
55		5333.06	I
27		5333.33	II
27		5334.23	II
22		5343.94	II
30		5344.50	II
90		5348.06	I
45		5350.47	I
35		5368.85	I
35		5395.87	II
60		5414.63	II
18		5422.81	II
18	h	5451.30	I
35		5454.27	II
180		5456.62	I
35		5462.43	II
90		5468.32	I
18		5477.47	II
80		5485.97	II
27		5497.44	II
27		5516.02	I
80		5593.46	I
45	d	5601.14	I
		5601.32	I
45	h	5609.94	I
60		5611.82	I
70		5622.01	I
80		5626.53	II
30		5636.20	I
90		5640.36	I
22		5641.42	I
22	h	5658.63	II
70		5664.95	I
45		5665.44	II
55		5675.48	I
14		5695.53	II
27		5710.87	II
55		5717.48	I
70		5719.55	I
55		5726.97	I
22		5733.43	II
22		5736.56	I
80		5737.04	I
100		5739.19	I
35		5740.61	I
60		5748.65	I
55		5752.53	I
70		5757.63	II
290		5762.80	I
70		5769.92	I
45		5782.82	I
70		5784.66	I
22		5791.15	II
70		5800.79	I
22		5806.10	I
430		5826.79	I
45		5835.84	I
100		5850.07	I
120		5855.31	I
140		5872.35	I
120		5881.14	I
27		5886.30	II
27		5902.08	II
55		5906.06	I
45		5909.24	I
35		5933.50	I
22		5946.37	I
55		5968.68	I
27		5975.49	I
35		6006.79	II
22		6008.75	II
55		6014.83	I
35		6015.74	II
70		6022.56	I
22		6032.12	II
22		6045.63	II
22		6048.14	II
45		6056.85	I
70		6061.25	I
60		6076.45	II
35		6116.01	I
35		6125.32	I
30		6170.06	II
27		6183.21	II
360		6221.02	I
35		6230.90	I
55		6262.56	I
45		6267.93	I
60		6268.87	I
35		6274.94	I
30		6286.86	I
45		6299.42	I
130		6308.77	I
55		6326.13	I
22		6347.16	II
45		6388.19	I
22		6432.53	I
27		6485.87	I
55		6492.35	I

Erbium (Cont.)

Intensity		Wavelength	
22		6541.57	I
60		6583.48	I
70		6601.11	I
27		6721.91	I
70		6759.87	I
22		6762.92	I
27		6773.37	I
35		6790.92	I
22		6825.44	I
22		6825.98	I
70		6848.10	I
55		6865.13	I
27		6879.98	I
22		7001.40	I
12		7058.55	I
12	h	7065.04	I
11		7070.99	II
18		7101.27	I
8		7109.67	I
11		7155.40	II
5		7161.91	I
14	h	7197.00	I
7	h	7264.82	II
7	h	7283.95	I
14		7329.73	II
18		7355.37	I
11		7356.34	I
18		7428.67	I
55		7459.55	I
9		7460.42	I
120		7469.51	I
22		7532.34	I
6	h	7539.18	I
27		7556.26	I
6	h	7574.21	I
5	h	7590.51	I
11		7597.33	I
6		7607.23	I
11		7613.52	I
6		7623.48	I
16	h	7645.67	I
8		7650.63	I
22		7654.45	II
12	h	7658.05	I
22		7659.25	I
4		7665.64	I
35		7680.01	I
9		7722.14	I
8		7726.19	II
11	h	7747.44	I
22		7754.63	I
4		7762.16	I
9		7796.69	I
35		7797.47	I
9		7838.80	I
11		7844.00	I
16		7847.55	I
5		7875.36	I
5		7879.36	I
18		7899.55	I
8		7913.08	I
35		7921.85	I
30		7937.84	I
8		7952.93	I
12		7964.51	I
8		7979.03	II
8		7980.87	I
5		8023.03	I
12		8035.91	I
12		8181.85	I
35		8312.82	I
18		8328.57	II
5		8367.58	II
55		8409.90	I
11		8466.18	II
35		8472.42	I
14		8517.71	II
18		8521.37	II
22		8768.64	I
11	h	8776.63	II
9		8866.84	II

Er III
Ref. 301—J.R.

Intensity		Wavelength	
		Air	
2		2165.26	III
3		2190.77	III
10		2198.15	III
1		2223.98	III
4		2232.35	III
60		2235.28	III
8		2245.60	III
2		2255.95	III
80		2269.36	III
600		2277.65	III
100		2309.19	III
40		2358.69	III
10		2358.79	III
50		2359.33	III
200		2367.64	III
20		2375.50	III
10		2377.07	III
80		2381.25	III
20		2381.40	III
40		2381.75	III
6		2391.96	III
60		2393.08	III
5		2393.60	III
250		2396.40	III
10		2398.91	III
2		2402.75	III
80		2404.58	III
100		2410.47	III
200		2419.81	III
200		2422.47	III
40		2431.51	III
60		2464.60	III
2		2492.04	III
100		2508.59	III
2		2531.03	III
100		2532.36	III
8		2536.76	III
80		2540.91	III
3		2543.31	III
10		2545.95	III
50		2557.22	III
80		2570.74	III
40		2580.02	III
2		2589.55	III
80		2590.72	III
20		2591.56	III
200		2591.83	III
20		2598.39	III
3		2599.18	III
100		2603.62	III
40		2604.91	III
25		2614.53	III
30		2617.64	III
2		2618.40	III
2		2618.94	III
8		2625.19	III
20		2626.37	III
4		2637.52	III
200		2637.77	III
5		2651.49	III
25		2683.10	III
400		2723.29	III
100		2738.53	III
500		2739.27	III
8		2741.41	III
80		2746.03	III
6		2752.20	III
80		2756.20	III
400		2759.23	III
150		2761.92	III
60		2762.66	III
15		2767.11	III
100		2768.72	III
60		2772.07	III
60		2774.80	III
20		2775.55	III
2		2780.60	III
80		2783.11	III
500		2792.54	III
10		2804.10	III
100		2805.87	III
6		2808.44	III
50		2824.75	III
150		2830.34	III
1		2831.95	III
8		2833.03	III
8		2845.29	III
60		2846.08	III
6		2849.63	III
1		2869.52	III
8		2878.24	III
1		2955.93	III
1		2958.63	III
1000		3055.10	III
1000		3070.40	III
500		3100.40	III
1500		3166.25	III
3		3172.47	III
1		3173.45	III
50		3175.74	III
400		3214.95	III
2000		3301.23	III
8		3341.00	III
200		3480.54	III
8		3592.96	III
600		3715.67	III
200		3739.43	III
4000		3816.78	III
600		3962.87	III
40		4009.70	III
2		4088.58	III
1000		4288.18	III
40000		4290.06	III
300		4338.24	III
20000		4386.86	III
30		4612.93	III
15000		4735.56	III
2000		4783.12	III
8		4876.07	III
8000		5903.30	III

EUROPIUM (Eu)
Z = 63

Eu I and II
Ref. 1—C.H.C.

Intensity		Wavelength	
		Air	
21		2499.39	II
26		2554.78	II
26		2559.18	II
160		2564.17	II
110		2568.17	II
26		2574.76	II
230		2577.14	II
26		2581.86	II
26		2604.61	I
30		2635.50	II
1000		2638.77	II
380		2641.27	II
40		2653.61	II
640		2668.34	II
110		2673.42	II
250		2678.29	II
250		2685.66	II
550		2692.03	II
700		2701.14	II
800		2701.90	II
240		2705.28	II
180		2709.99	I
700		2716.98	II
70		2723.96	I
4200		2727.78	II
190		2729.33	II
380		2729.44	II
50		2731.37	I
40		2732.61	I
80		2735.25	I
160		2740.62	II
70		2743.28	I
120		2744.26	II
40		2745.61	I
70		2747.29	II
80		2747.83	I
90		2752.17	II
480		2781.89	II
1900		2802.84	II
220		2811.75	II
30		2813.08	II
3400		2813.94	II
550		2816.18	II
2000		2820.78	II
400	Cw	2828.72	II
120		2829.30	II
140		2833.26	II
80		2843.96	II
60		2852.05	II
260		2859.67	II
280		2862.57	II
25		2864.42	II
60		2876.06	II
100		2878.87	I
80		2887.85	II
200		2892.54	I
140		2893.03	I
360		2893.83	I
3200		2906.68	II
160		2908.99	I
30		2917.44	II
850		2925.04	II
60		2947.29	II
200	Cw	2952.68	II
30		2958.91	I
35		2959.47	II
260		2960.21	II
300		2991.33	II
35		2995.22	II
40		3006.26	II
35		3022.15	I
30		3040.77	II
320	Cw	3054.94	II
120		3058.98	I
35		3069.11	II
35		3076.07	II
220		3077.36	II
35		3089.35	II
120		3097.45	II
320		3106.18	I
950		3111.43	I
120		3130.73	II
40		3132.16	I
45		3149.88	II
85		3173.61	II
40		3185.54	I
420		3210.57	I
1000		3212.81	I
420		3213.75	I
45		3235.13	I
95		3241.40	I
45		3246.03	I
45		3247.32	II
100		3247.55	I
100		3266.39	II
150		3272.77	II
210		3277.78	II
150		3301.95	II
45		3304.50	II
140		3308.02	II
140		3313.33	II
65		3319.89	II
95		3321.86	II
85		3322.26	I
950		3334.33	I
45		3338.75	II
110		3350.40	I
40		3351.56	II
40		3354.38	II
45		3367.64	II
140		3369.06	II
65		3380.25	II
75		3390.78	II
190		3391.99	II
280		3396.58	II
45		3419.84	II
65		3423.09	II
150		3425.02	II
45		3426.44	II
45		3435.05	II
65		3435.20	II
40		3435.72	II
45		3440.82	II
150		3441.00	II
45		3445.18	II
85		3457.05	I
45		3457.56	II
130		3461.38	II
85		3467.88	I
75		3477.07	I
75	h	3505.30	II
470	Cw	3521.09	II
75		3531.15	II
45		3532.23	II
65		3538.08	II
150		3542.15	II
85		3543.85	II
45		3549.71	II
180		3552.52	II
75		3589.27	I
45		3591.31	II
150		3603.20	II
75		3611.57	II
45		3616.15	II
95		3622.54	II
95		3632.18	II
45		3673.19	II
45		3674.63	II
45		3678.26	II
6400		3688.42	II
60		3710.87	II
95		3713.45	II
95		3714.90	II
35		3716.94	II
35		3717.69	II
40		3719.16	I
20000	Cw	3724.94	II
45		3729.68	II
45		3729.74	II
21		3732.20	I
45		3738.08	II
350		3741.31	II
100		3743.56	II
260		3761.12	II
95		3765.93	II
40		3774.10	I
60		3781.40	II
40		3788.76	II
45		3791.50	II
130		3799.01	II
70		3801.36	
95		3807.54	II
120		3811.33	I
120		3815.50	II
39000	Cw	3819.67	II
120		3826.68	II
140		3844.23	II
190		3865.57	I
45		3872.72	I
70		3877.27	II
150		3884.75	I
23		3896.78	I
23		3900.18	I
70		3900.51	I
28000	Cw	3907.10	II
45		3915.24	II
45		3916.00	I
230		3917.29	I
23		3917.70	II
40		3918.52	I
100		3919.09	II
40		3928.87	II
32000	Cw	3930.48	II
55		3941.56	II

Europium (Cont.)

Intensity		Wavelength	Spectrum
30	h	3942.21	II
60		3942.94	II
120		3943.08	II
30		3944.59	II
30		3945.67	II
30		3949.13	II
60		3949.60	I
45		3950.76	II
55		3951.33	II
60		3955.75	I
40		3957.92	II
30		3963.61	I
120		3964.90	II
150		3966.59	II
45		3967.18	I
30000	Cw	3971.96	II
60		3978.42	I
30		3979.63	II
55	h	3986.60	I
40		3988.24	II
30		3993.93	II
55		3995.98	II
60		4003.71	II
180		4011.69	II
150		4017.58	II
120		4039.19	I
45	h	4078.24	I
120		4085.38	II
75		4096.80	II
60		4106.88	I
90	h	4112.04	II
45		4119.30	II
75		4127.28	I
33000	Cw	4129.70	II
30		4136.59	II
40		4137.07	I
30		4141.02	II
60		4141.72	II
30		4151.52	II
45		4151.64	II
30		4157.72	I
110		4172.80	II
30		4175.16	II
110		4182.22	I
40		4196.18	II
60000	Cw	4205.05	II
45		4221.08	II
40		4223.88	II
90	h	4227.40	II
75		4229.33	II
75		4232.45	II
90		4237.51	II
45		4238.69	II
45		4244.74	I
45		4247.06	II
45		4253.80	II
30		4270.24	II
150		4298.73	I
90		4329.36	I
75		4329.97	I
60		4330.61	II
40		4331.18	I
90		4337.68	I
240		4355.09	II
27		4361.57	II
55		4369.47	II
45	h	4372.20	II
75		4383.17	II
90		4387.88	I
21	h	4405.27	II
55		4407.07	II
18		4419.66	II
120		4434.81	II
14000	Cw	4435.56	II
75	h	4464.97	II
24		4485.15	II
3000		4522.57	II
45	h	4535.59	I
11000		4594.03	I
21		4602.63	I
[illegible]		[illegible]	I
8300		4661.88	I
30		4713.59	I
27		4740.50	I
45		4792.59	I
40	h	4829.30	I
60		4830.33	I
40	h	4840.47	I
60	h	4849.64	I
110		4867.62	I
40	h	4884.05	I
90		4894.68	I
60		4900.86	I
150		4907.18	I
180		4911.40	I
55		4953.52	I
55		4960.21	I
55		4962.55	I
45		4975.76	I
180		5013.17	I
170		5022.91	I
110		5029.54	I
90		5033.55	I
75		5067.95	I
75	h	5092.69	I
90	h	5096.44	I
170		5114.37	I
90		5124.77	I
170		5129.10	I
90		5130.08	I
210		5133.52	I
270		5160.07	I
210		5166.70	I
60		5193.74	I
200		5199.85	I
110		5200.96	I
120		5206.44	I
750		5215.10	I
300		5223.49	I
120		5239.24	I
200		5266.40	I
390		5271.96	I
110		5272.48	I
150		5282.82	I
55		5287.25	I
60		5289.25	I
120		5291.26	I
60		5293.68	I
120		5294.64	I
90		5303.85	I
30	h	5350.41	I
75	h	5351.69	I
40		5352.84	I
90		5355.10	I
540		5357.61	I
60		5360.83	I
120		5361.61	I
110		5376.94	I
120		5392.94	I
450		5402.77	I
45		5405.33	I
45		5411.86	I
55		5421.07	I
90		5426.94	I
40		5443.56	I
380		5451.51	I
260		5452.94	I
40		5457.62	I
90		5472.32	I
100		[illegible]	I
45		5495.20	I
15		5500.83	I
120		5510.52	I
30		5526.63	I
30		5533.25	I
30		5542.54	I
200		5547.44	I
150		5570.33	I
200		5577.14	I
75		5579.63	I
120		5580.03	I
90		5586.24	I
75		5586.83	I
18		5592.25	I
18		5599.80	I
18		5605.86	I
40		5618.81	I
60		5622.44	I
75		5632.54	I
210		5645.80	I
15		5651.11	I
60		5673.85	I
27		5681.10	I
27		5684.24	I
60		5730.87	I
60		5739.00	I
330		5765.20	I
180		5783.69	I
15		5792.72	I
60		5800.27	I
170		5818.74	II
600	Cw	5830.98	I
27		5845.77	I
27		5860.97	I
15		5864.77	I
90		5872.98	II
15		5895.31	I
27		5902.97	I
12		5909.94	I
75		5915.74	I
12		5925.30	I
27		5926.52	I
45		5942.72	I
27		5953.49	I
27		5953.84	II
30		5954.28	I
90		5963.76	I
330		5966.07	II
480	Cw	5967.10	I
15	h	5968.43	I
30		5971.69	I
170		5972.75	I
15		5980.47	I
27		5983.14	I
27		5983.78	I
240		5992.83	I
60		6004.36	I
15	h	6005.61	I
60	h	6012.20	I
110		6012.56	I
60		6015.58	I
420		6018.15	I
60		6023.15	I
170		6029.00	I
60		6044.66	I
420		6049.51	II
140		6057.36	I
90		6075.58	I
30		6077.38	I
240		6083.84	I
240		6099.35	I
60		6108.15	I
120		6118.78	I
60		6124.67	I
330		6173.05	II
110		6178.76	I
260	Cw	6188.13	I
140		6195.07	I
15	h	6207.60	I
15		6230.51	I
90	h	6233.73	I
55		6250.47	I
240		6262.25	I
55		6266.95	I
15	h	6285.95	I
60		6291.34	I
170		6299.77	I
230		6303.41	II
24	h	6313.78	I
15		6318.58	I
75		6335.82	I
120	Cw	6350.04	I
60		6355.89	I
60		6369.25	I
55		6382.73	I
75		6383.86	I
120	Cw	6400.93	I
40		6406.11	I
180		6410.04	I
140		6411.32	I
55		6428.29	I
830		6437.64	II
18		6439.93	I
120		6457.96	I
12		6470.70	I
18		6483.02	I
45		6501.55	I
60		6519.59	I
15		6522.72	I
8	h	6549.12	I
75		6567.87	I
45		6593.79	I
18	h	6603.55	I
1400		6645.11	II
26		6685.21	I
95		6693.96	I
7	h	6701.06	I
12	h	6710.45	I
30		6744.88	I
30	h	6782.54	I
14	h	6787.48	I
140		6802.72	I
35		6816.06	I
11	h	6834.30	I
17		6840.93	I
17	h	6844.83	I
14	h	6847.04	I
360		6864.54	I
21		6898.21	I
60	h	6903.67	I
14	h	6910.17	I
30	h	6914.82	I
120		7040.20	I
12		7074.54	I
330		7077.10	II
100		7106.48	I
6		7164.66	I
30		7175.55	I
570		7194.81	II
570		7217.55	II
11	h	7224.68	I
15		7258.72	I
30		7262.77	I
11	h	7281.53	I
6	h	7297.56	I
540		7301.17	II
11		7310.46	I
12		7313.63	I
55	Cw	7336.18	I
4		7346.25	I
4		7356.65	I
11		7362.25	I
55	Cw	7369.60	I
720		7370.22	II
4		7387.36	I
12		7389.16	I
11	h	7404.41	I
300		7426.57	II
21	h	7436.59	I
8		7470.53	I
5	h	7491.00	I
50	Cw	7528.70	I
5	h	7533.02	I
6		7547.32	I
150		7583.91	I
60	Cw	7742.57	I
70		7746.19	I
8	h	7803.32	I
8		7818.21	I
35		7887.99	I
7		8015.47	I
24	Cw	8209.80	I
15	Cw	8226.81	I
6	h	8464.71	I
21	Cw	8642.67	I
7		8727.77	I
6		8782.46	I
12	Cw	8790.88	I
18		8870.30	I

Eu III
Ref. 312—J.R.

Intensity		Wavelength	
		Air	
10		2073.40	III
10		2093.50	III
30		2124.69	III
10		2167.12	III
10		2173.59	III
10		2184.68	III
10		2190.59	III
10		2194.81	III
20		2211.85	III
20		2212.63	III
20		2214.66	III
10		2215.34	III
30		2217.23	III
30		2219.33	III
20		2219.42	III
10		2223.13	III
10		2235.17	III
10		2240.14	III
10		2261.88	III
20		2265.74	III
10		2269.39	III
20		2276.85	III
40		2291.62	III
20		2304.37	III
10		2311.92	III
10		2327.69	III
10		2334.56	III
10		2336.96	III
10		2339.84	III
10		2343.10	III
10		2346.83	III
10		2347.64	III
10		2350.38	III
200		2350.51	III
10		2352.28	III
10		2357.87	III
10		2359.08	III
10		2360.65	III
20		2363.76	III
20		2368.04	III
20		2374.08	III
10		2375.20	III
4000		2375.46	III
10		2376.42	III
20		2377.23	III
10		2381.81	III
10		2383.62	III
20		2387.29	III
20		2389.11	III
10		2389.98	III
20		2391.11	III
20		2391.90	III
10		2392.59	III
10		2394.66	III
20		2395.62	III
20		2398.79	III
[illegible]		[illegible]	III
20		2402.34	III
20		2404.08	III
10		2406.14	III
20		2407.30	III
20		2408.32	III
10		2409.63	III
20		2410.08	III
40		2412.02	III
20		2412.96	III
20		2413.26	III
10		2413.41	III
10		2419.11	III
10		2419.25	III
10		2419.58	III
30		2422.00	III
10		2422.90	III
10		2425.33	III
50		2425.68	III
40		2427.67	III
40		2429.32	III
10	d	2429.66	III
10		2430.04	III

Europium (Cont.)

Intensity		Wavelength	
10		2431.49	III
10		2431.76	III
10		2432.55	III
10		2433.65	III
10		2434.19	III
100	d	2435.14	III
20		2436.39	III
10		2436.77	III
10		2438.83	III
10		2440.26	III
50		2440.67	III
1000		2444.38	III
4000		2445.99	III
30	h	2446.43	III
20		2448.57	III
20		2451.24	III
10		2451.73	III
30		2455.22	III
10		2461.79	III
30		2463.30	III
30		2464.47	III
40		2470.51	III
10		2474.94	III
10		2476.24	III
20		2476.45	III
10		2477.78	III
20		2480.02	III
20		2483.29	III
10		2486.92	III
10		2488.91	III
10		2490.50	III
10		2491.08	III
10		2492.48	III
10		2496.92	III
10		2499.17	III
2000		2513.76	III
20		2517.94	III
200		2522.14	III
10		2539.14	III
10		2548.30	III
20		2548.59	III
10		2554.50	III
10		2558.07	III
10		2560.36	III
10		2594.71	III
20		2594.76	III
10		2596.34	III
10		2604.44	III
30		2608.34	III
30		2610.09	III
50	c	2616.11	III
20		2616.26	III
10		2616.33	III
10		2616.35	III
10		2620.79	III
20		2623.33	III
10		2626.98	III
20	c	2628.46	III
10		2628.82	III
10		2631.98	III
30	c	2642.27	III
20		2645.22	III
20	c	2650.93	III
10		2653.19	III
10		2655.09	III
10		2662.24	III
20	c	2666.86	III
20	c	2668.21	III
40	c	2676.09	III
20	c	2683.21	III
20		2686.13	III
20	c	2687.74	III
40	c	2693.51	III
10		2694.80	III
20		2699.87	III
20	c	2700.78	III
10	c	2708.25	III
20	c	2708.84	III
10		2712.08	III
50	c	2720.67	III
20	c	2725.54	III
10		2743.94	III
10		2752.68	III
20		2755.12	III
10		2757.75	III
20	c	2760.21	III
10		2761.72	III
10	c	2766.26	III
20	c	2768.38	III
10	c	2768.54	III
10		2769.71	III
20	c	2780.48	III
20		2792.51	III
10		2808.09	III
10		2817.58	III
20	c	2839.56	III
20		2844.99	III
10	c	2848.44	III
20		2850.39	III
10	c	2892.60	III
40		2912.23	III
40	c	2912.64	III
10		2913.04	III
10		2928.91	III
20	c	2931.00	III
10	c	2950.20	III
20		2956.74	III
10		2956.90	III
10	c	2972.30	III
30	c	2982.29	III
20	c	3000.11	III
20		3006.37	III
20	c	3013.28	III
10	c	3018.43	III
20	c	3022.08	III
50	c	3022.69	III
20	c	3023.40	III
100	c	3023.93	III
10		3025.32	III
10	c	3026.09	III
200	c	3026.79	III
50	c	3029.92	III
20	c	3031.24	III
40	c	3032.84	III
20	c	3036.98	III
10		3038.64	III
10		3039.05	III
20	c	3039.98	III
10		3054.07	III
10	c	3054.97	III
20		3076.43	III
10	c	3089.09	III
10		3105.25	III
10		3109.67	III
10		3129.31	III
10		3142.54	III
50	c	3171.00	III
10		3178.08	III
20	h	3178.87	III
50	c	3183.78	III
10	h	3191.46	III
20	c	3194.34	III
10		3206.30	III
10		3208.95	III
10	h	3213.84	III
10		4837.98	III
50		6666.35	III
30		7221.84	III
20		7690.44	III
10		8079.07	III

FLUORINE (F)
Z = 9

F I and II
Ref. 169, 224—G.V.S.

Intensity		Wavelength	
		Vacuum	
30		375.30	II
30		380.90	II
40		407.04	II
50		430.91	II
40		431.55	II
40		435.64	II
70		457.18	II
40		471.95	II
60		472.00	II
50		472.71	II
40		473.02	II
90		484.60	II
50		513.64	II
70		514.94	II
70		546.85	II
60		547.87	II
50		548.32	II
40		548.52	II
90		605.67	II
80		606.29	II
100		606.80	II
70		606.92	II
80		607.47	II
90		608.06	II
15		780.39	I
10		780.52	I
10		782.38	I
12		791.88	I
10		792.54	I
10		794.42	I
150		806.96	I
125		809.60	I
500		951.87	I
1000		954.83	I
750		955.55	I
500		958.52	I
20		972.40	I
350		973.90	I
100		976.22	I
40		976.51	I
100		977.75	I
40		1129.76	II
40		1327.06	II
50		1328.11	II
40		1333.59	II
50		1343.60	II
40		1344.04	II
50		1400.61	II
40		1407.14	II
60		1493.09	II
50		1493.24	II
40		1493.31	II
40		1702.13	II
40		1744.75	II
50		1745.55	II
60		1747.39	II
		Air	
100		2556.11	II
100		2871.40	II
120		3059.99	II
140		3153.49	II
170		3202.76	II
140		3264.08	II
140		3414.65	II
150		3416.45	II
140		3416.80	II
160		3417.00	II
160		3472.96	II
150		3473.31	II
170		3474.78	II
190		3501.39	II
200		3501.45	II
200		3501.57	II
180		3502.84	II
200		3502.96	II
210		3503.11	II
170		3505.37	II
200		3505.52	II
220		3505.63	II
160		3522.89	II
150		3536.87	II
160		3541.77	II
160		3590.52	II
6		3594.10	I
170		3598.69	II
180		3601.39	II
190		3602.84	II
12		3668.17	I
180		3704.53	II
160		3710.35	II
160		3739.57	II
140		3805.83	II
270		3847.09	II
260		3849.99	II
250		3851.67	II
5		3898.48	I
190		3898.83	II
180		3901.93	II
170		3903.82	II
8		3930.69	I
5		3934.26	I
5		3948.56	I
150		3972.04	II
160		3972.67	II
170		3974.78	II
240		4024.73	II
220		4025.01	II
230		4025.49	II
160		4083.91	II
190		4103.07	II
170		4103.22	II
200		4103.51	II
180		4103.71	II
170		4103.87	II
170		4109.16	II
160		4116.54	II
150		4119.21	II
140		4207.15	II
170	h	4225.16	II
150	h	4244.12	II
200		4246.23	II
190		4246.39	II
180		4246.59	II
170		4246.77	II
160		4246.84	II
170	h	4275.36	II
160	h	4277.53	II
160	h	4278.93	II
200		4299.17	II
160		4446.53	II
170		4446.72	II
180		4447.19	II
140		4734.38	II
170		4859.39	II
160		4933.26	II
6		4960.65	I
140		5002.00	II
150		5173.25	II
15		5230.41	I
12		5279.01	I
18		5540.52	I
12		5552.43	I
10		5577.33	I
160		5589.27	II
20		5624.06	I
12		5626.93	I
15		5659.15	I
40		5667.53	I
90		5671.67	I
18		5689.14	I
25		5700.82	I
25		5707.31	I
12		5950.15	I
25		5959.19	I
70		5965.28	I
50		5994.43	I
150		6015.83	I
80		6038.04	I
900		6047.54	I
100		6080.11	I
800		6149.76	I
400		6210.87	I
13000		6239.65	I
140		6247.90	II
10000		6348.51	I
8000		6413.65	I
450		6569.69	I
300		6580.39	I
400		6650.41	I
1800		6690.48	I
400		6708.28	I
7000		6773.98	I
1500		6795.53	I
9000		6834.26	I
50000		6856.03	I
8000		6870.22	I
15000		6902.48	I
6000		6909.82	I
4000		6966.35	I
45000		7037.47	I
30000		7127.89	I
130		7179.90	II
15000		7202.36	I
130	h	7211.79	II
1000		7309.03	I
15000		7311.02	I
700		7314.30	I
5000		7331.96	I
10000		7398.69	I
4000		7425.65	I
2200		7482.72	I
2500		7489.16	I
900		7514.92	I
5000		7552.24	I
5000		7573.38	I
7000		7607.17	I
18000		7754.70	I
15000		7800.21	I
300		7879.18	I
500		7898.59	I
350		7936.31	I
300		7956.32	I
80		8016.01	II
1000		8040.93	I
900		8075.52	I
350		8077.52	I
350		8126.56	I
600		8129.26	I
300		8159.51	I
600		8179.34	I
300		8191.24	I
350		8208.63	I
2500		8214.73	I
3000		8230.77	I
500		8232.19	I
1500		8274.62	I
2000		8298.58	I
600		8302.40	I
900		8807.58	I
1000		8900.92	I
300		8912.78	I
350		9025.49	I
400		9042.10	I
350		9178.68	I
200		9433.67	I
25		9505.30	I
12		9662.04	I
25		9734.34	I
15		9822.11	I
12		9902.65	I
80	h	10047.98	II
15		10285.45	I
20		10862.31	I

F III
Ref. 225—G.V.S.

Intensity		Wavelength	
		Vacuum	
50	h	230.12	III
50		255.72	III
60		255.77	III
70		255.86	III
70		261.71	III
60		261.75	III
80		263.81	III

Fluorine (Cont.)

Intensity		Wavelength	Spectrum
70		279.69	III
80		315.22	III
70		315.54	III
60		315.75	III
100		429.51	III
110		430.15	III
80		430.22	III
90		464.29	III
100		465.11	III
120		508.39	III
120		567.69	III
110		567.75	III
80		630.14	III
90		630.20	III
120		656.12	III
130		656.87	III
140		658.33	III
80		1219.03	III
80		1266.87	III
90		1267.71	III
70		1297.54	III
70		1359.92	III
110		1498.93	III
120		1502.01	III
110		1504.18	III
140		1504.79	III
130		1506.30	III
110		1506.77	III
100		1553.02	III
110		1557.59	III
100		1563.73	III
100		1565.54	III
100		1623.40	III
100		1650.76	III
130		1670.39	III
140		1677.40	III
100		1716.99	III
120		1770.09	III
150		1770.67	III
110		1772.93	III
140		1773.36	III
160		1791.65	III
110		1803.03	III
100		1804.70	III
170		1805.90	III
110		1839.30	III
120		1839.97	III
110		1840.14	III
80		1900.76	III
		Air	
100		2027.44	III
120		2030.32	III
120		2217.17	III
120		2452.07	III
130		2464.85	III
130		2470.29	III
120		2478.73	III
150		2484.37	III
120		2542.77	III
120		2580.04	III
130		2583.81	III
120		2593.23	III
130		2595.53	III
140		2599.28	III
130		2625.01	III
140		2629.70	III
120		2656.44	III
130		2755.55	III
160		2759.63	III
120		2788.15	III
160		2811.45	III
140		2833.99	III
150		2835.63	III
150		2860.33	III
120		2862.86	III
140		2887.58	III
150		2889.45	III
120		2905.30	III
140		2913.29	III
160		2916.34	III
[illegible]		[illegible]	III
140		2994.28	III
120	h	2997.21	III
130		2997.53	III
120		2999.47	III
130		3039.25	III
120		3039.75	III
160		3042.80	III
150		3049.14	III
140		3113.62	III
160		3115.70	III
180		3121.54	III
140		3124.79	III
140		3134.23	III
140		3146.99	III
180		3174.17	III
170		3174.76	III
120		3214.00	III
140	h	4420.30	III
120	h	4427.35	III
120	h	4432.32	III
140	h	4479.99	III
150		5012.54	III
160		5110.99	III
140		5753.17	III
120		5761.20	III
150		6091.82	III
140		6125.50	III
130		6233.57	III
140		6363.05	III
120		7336.77	III
130		7354.94	III

F IV
Ref. 68, 226—G.V.S.

Intensity	Wavelength	
	Vacuum	
30	169.79	IV
30	169.84	IV
30	171.07	IV
40	176.37	IV
40	181.52	IV
40	181.57	IV
30	187.24	IV
50	196.39	IV
60	196.45	IV
50	199.76	IV
50	199.80	IV
50	199.85	IV
50	199.93	IV
50	200.00	IV
70	200.09	IV
60	201.01	IV
70	201.06	IV
60	201.10	IV
80	201.16	IV
60	201.22	IV
90	208.25	IV
70	213.85	IV
70	214.06	IV
70	220.77	IV
60	226.94	IV
50	227.10	IV
60	233.22	IV
[illegible]	[illegible]	IV
70	239.86	IV
70	240.02	IV
90	240.08	IV
70	240.15	IV
70	240.28	IV
70	240.37	IV
100	251.03	IV
60	270.23	IV
140	419.65	IV
150	420.05	IV
160	420.73	IV
150	430.76	IV
130	490.57	IV
160	491.00	IV
50	497.38	IV
60	497.83	IV
70	498.80	IV
140	570.64	IV
140	571.30	IV
150	571.39	IV
160	572.66	IV
140	676.12	IV
130	677.15	IV
150	677.22	IV
130	678.99	IV
160	679.21	IV
	Air	
40	2171.44	IV
50	2298.29	IV
40	2451.58	IV
50	2456.92	IV
40	2820.74	IV
50	2826.13	IV

F V
Ref. 68, 226—G.V.S.

Intensity	Wavelength	
	Vacuum	
40	134.54	V
40	147.95	V
50	148.00	V
40	152.51	V
40	158.54	V
40	162.27	V
40	163.50	V
50	163.56	V
90	165.98	V
100	166.18	V
40	174.70	V
50	178.43	V
40	178.59	V
40	182.98	V
40	186.72	V
40	186.79	V
50	186.84	V
40	186.97	V
40	187.01	V
60	190.57	V
70	190.84	V
40	191.97	V
40	205.55	V
100	464.37	V
110	465.37	V
120	465.98	V
100	466.99	V
90	506.16	V
100	508.08	V
70	513.97	V
60	514.08	V
80	524.59	V
90	525.29	V
100	526.30	V
70	647.67	V
100	647.77	V
110	647.87	V
70	647.97	V
130	654.03	V
110	657.23	V
140	657.33	V
60	757.04	V
60	1082.31	V
70	1088.39	V
	AIR	
10	2229.18	V
20	2252.72	V
20	2450.63	V
10	2461.33	V
10	2693.98	V
10	2702.30	V
10	2703.96	V
20	2707.17	V

FRANCIUM (Fr)
Z = 87

Fr I
Ref. 408 — C.H.C.

Intensity	Wavelength	
	Air	
	7177.	I

GADOLINIUM (Gd)
Z = 64

Gd I and II
Ref. 1—C.H.C.

Intensity	Wavelength	
	Air	
100	2468.22	II
55	2471.58	II
35	2485.67	II
70	2487.46	II
110	2488.72	II
55	2493.29	II
35	2496.35	II
45	2499.04	II
28	2543.68	II
28	2586.13	II
28	2661.50	II
70	2720.50	II
430	2750.22	II
[illegible]	[illegible]	II
40	2768.51	II
320	2769.81	II
230	2770.17	II
21	2770.98	II
45	2778.76	I
45	2779.14	II
440	2781.40	II
70	2787.68	I
390	2791.96	II
100	2794.66	II
930	2796.93	II
60	2808.38	II
750	2809.72	II
160	2810.93	II
45	2814.01	II
300	2833.75	II
35	2836.69	II
70	2837.00	II
560	2840.23	II
140	2841.33	II
40	2853.91	II
60	2856.52	II
19	2859.78	II
120	2862.48	II
60	2865.06	II
40	2866.33	II
40	2871.75	II
460	2881.33	II
40	2882.13	II
130	2885.60	II
35	2907.44	II
170	2910.53	II
60	2913.08	II
45	2918.52	II
95	2923.32	II
35	2924.25	II
35	2928.34	II
35	2947.80	II
70	2948.01	II
35	2952.43	I
35	2955.60	II
70	2960.93	II
130	2963.60	II
80	2965.43	II
29	2972.74	II
560	2980.15	II
35	2983.74	II
40	2991.52	II
95	2993.04	II
1200	2999.04	II
370	3002.86	II
100	3005.09	II
2100	3010.13	II
130	3012.19	II
1900	3027.60	II
120	3028.98	II
2100	3032.84	II
1600	3034.05	II
130	3043.01	I
160	3046.48	I
280	3053.57	II
100	3059.92	I
1000	3068.64	II
560	3072.56	II
640	3076.92	II
150	3077.08	II
2100	3081.99	II
140	3084.01	II
280	3089.95	II
140	3092.06	II
460	3098.64	II
190	3098.90	II
3500	3100.50	II
120	3101.18	II
230	3101.91	II
580	3102.55	II
130	3108.36	II
170	3111.19	I
160	3113.17	II
120	3118.60	II
120	3119.01	I
510	3119.94	II
100	3120.18	II
370	3123.99	II
120	3124.25	II
130	3128.56	II
130	3130.81	II
100	3133.09	II
460	3133.85	II
210	3135.03	II
190	3136.93	I
190	3137.30	I
120	3138.71	I
230	3143.13	II
930	3145.00	II
370	3145.52	II
230	3146.88	II
980	3156.53	II
200	3158.63	I
140	3160.69	II
980	3161.37	II
220	3190.28	I
220	3199.30	I
160	3199.58	I
110	3203.41	I
690	3223.74	II
	3223.78	I
110	3225.46	II
160	3226.32	II
220	3232.78	I
100	3250.19	II
110	3259.25	II
540	3266.73	I
250	3267.64	I
140	3268.34	II
110	3274.18	II
110	3279.53	II
100	3281.61	II
250	3282.25	I
	3282.30	II
430	3291.48	I
370	3292.21	II
430	3294.08	I
330	3313.73	II
200	3315.59	II

Gadolinium (Cont.)

Intensity		Wavelength	Spectrum
430		3330.34	II
1400		3331.38	II
830		3332.13	II
1100		3336.18	II
590		3345.98	II
200		3350.10	II
5400		3350.47	II
220		3357.61	I
270		3358.43	II
4300		3358.62	II
780		3360.71	II
5400		3362.23	II
270		3364.24	II
200		3365.59	II
220		3374.69	II
220		3379.76	II
220		3380.52	II
1100		3392.53	II
540		3395.12	II
220	d	3397.22	I
		3397.32	I
200		3399.41	II
540		3399.99	II
540		3402.07	II
200		3406.92	I
1100	d	3407.56	II
		3407.61	II
250		3409.30	II
220		3411.02	I
220		3413.27	II
1400		3416.95	II
1400		3418.73	II
6900		3422.47	II
390		3422.75	II
1100		3423.90	I
		3423.92	II
830		3424.59	II
390		3425.93	II
220		3428.47	II
690		3432.99	II
1700		3439.21	II
830		3439.78	II
2700		3439.99	II
390		3449.62	II
1400		3450.38	II
1100		3451.23	II
540		3454.14	II
880		3454.90	II
200		3455.27	I
200		3457.05	II
220		3461.95	II
220		3463.00	II
2700		3463.98	II
330		3466.95	II
1700		3467.27	II
1700		3468.99	II
1400		3473.22	II
2200		3481.28	II
1700		3481.80	II
490		3482.60	II
220		3486.20	I
980		3491.95	II
1700		3494.40	II
1400		3505.51	II
780		3512.22	II
1100		3512.50	II
830		3513.65	I
980		3524.20	II
430		3528.54	II
540		3542.77	II
4300		3545.80	II
3900		3549.36	II
1400		3557.05	II
540		3558.19	II
430		3558.47	II
200		3564.05	II
690		3571.93	II
330		3574.74	II
390		3578.36	II
980		3581.91	II
5400		3584.96	II
540		3590.47	II
1100		3592.71	II
200		3593.44	II
540		3600.96	II
1100		3604.87	I
270		3605.26	II
250		3605.66	II
830		3608.75	II
830		3610.76	II
220		3610.91	II
540		3613.39	II
270	d	3614.21	II
		3614.42	I
430		3617.16	II
390		3620.46	II
270		3624.89	II
250		3629.51	II
330		3634.76	II
220		3639.05	II
250		3640.18	II
330		3641.39	II
870		3645.62	II
6100		3646.19	II
310		3649.44	II
450		3650.95	II
620		3652.54	II
3900		3654.62	II
3100		3656.15	II
210		3658.19	I
1400		3662.26	II
2700		3664.60	II
2000		3671.20	II
1000		3674.05	I
350		3679.21	I
2000		3684.13	I
720		3686.33	II
3100		3687.74	II
210		3694.03	II
2000		3697.73	II
1300		3699.73	II
2700		3712.70	II
2000		3713.57	I
1400		3716.36	II
2000		3717.48	I
1800	d	3719.45	II
		3719.53	II
250		3722.07	II
430		3725.47	II
1500		3730.84	II
270		3732.32	I
230		3732.45	II
230		3732.67	I
510		3733.08	II
490		3739.76	I
330		3740.02	II
4500		3743.47	II
620		3744.83	I
230		3757.74	II
1000		3757.94	I
1400		3758.31	II
820		3759.00	II
620		3760.71	II
290		3760.92	II
870		3762.20	I
210		3763.33	II
370		3764.20	II
870		3767.04	II
8700		3768.39	II
620		3769.45	II
1400		3770.69	II
250		3771.26	I
210		3773.45	I
210		3776.83	I
1000		3782.34	II
2900		3783.05	I
1100		3787.56	II
200		3790.63	I
770		3791.17	II
490		3792.39	II
5100		3796.37	II
720		3801.29	II
210		3804.39	I
210		3805.09	II
560		3805.52	II
3700		3813.97	II
430		3814.74	II
770		3816.64	II
430		3818.75	II
350		3826.05	II
230		3827.33	II
230		3829.46	II
370		3831.80	II
210		3832.97	I
330		3834.99	II
970		3836.91	II
1000		3839.64	II
1200		3842.20	II
1400		3843.28	I
1400		3844.58	II
3300		3850.69	II
5100		3850.97	II
4300		3852.45	II
470		3855.56	II
250		3863.05	II
1600		3866.99	I
250		3873.57	I
220		3875.46	II
1500		3894.70	II
450		3895.79	II
750		3902.40	II
300		3902.71	I
240		3904.29	I
450		3905.65	I
2200		3916.51	II
450		3923.25	II
1200		3934.79	I
		3934.82	II
220		3935.38	I
450		3941.80	I
590		3942.63	I
270		3943.24	I
220		3943.62	I
1400		3945.54	I
300		3952.00	II
590		3953.37	I
1200		3957.67	II
750		3959.44	II
		3959.52 GD	II
220		3963.66	II
590		3966.28	I
590		3968.26	II
750		3969.00	I
270		3969.29	II
450		3971.75	II
390		3972.71	I
590		3973.98	II
300		3974.81	I
750		3979.33	I
450		3987.21	II
470		3987.84	I
320		3992.69	I
220		3993.21	II
650		3994.16	II
700		3996.32	II
320		3997.76	II
470		4001.26	II
260		4004.94	II
320		4008.33	I
300		4008.91	II
300		4013.80	II
200		4015.58	I
300		4017.25	I
430		4017.71	I
300		4019.73	I
300		4022.33	II
1100		4023.14	I
810		4023.35	I
220		4027.61	I
1100		4028.15	I
860		4030.88	I
700		4033.49	I
340		4035.40	I
260		4036.84	I
1400		4037.33	II
700		4037.90	II
410		4043.71	I
1600		4045.01	I
270		4046.84	II
270		4047.09	I
270		4049.20	I
1300		4049.43	II
2200		4049.86	II
270		4050.37	I
810		4053.29	II
2600		4053.64	I
810		4054.72	I
2600		4058.22	I
650		4059.88	I
270		4061.30	II
650		4062.59	II
1900		4063.39	II
540		4063.59	II
260		4066.04	I
520		4068.35	I
260		4068.74	I
750		4070.29	II
		4070.39	II
650		4073.20	II
300		4073.76	II
1300		4078.44	II
2800		4078.70	I
520		4083.70	I
1500		4085.56	II
260		4087.69	II
650		4090.41	I
1100		4092.71	I
260		4093.72	I
260		4094.48	II
2600		4098.61	II
520		4098.90	II
650		4100.26	I
390		4111.44	II
2200		4130.37	II
270		4131.48	II
1100		4132.28	II
750		4134.16	I
410		4137.10	II
280		4148.86	I
540		4162.73	II
280		4163.09	II
280		4167.16	II
		4167.27	I
2400		4175.54	I
2400		4184.25	II
2200		4190.78	I
750		4191.07	II
750		4191.63	I
450		4197.68	II
590		4204.86	II
1300		4212.00	II
970		4215.02	II
650		4217.20	II
320		4225.03	I
4800		4225.85	I
220		4227.14	II
220		4229.80	II
650		4238.78	II
200		4246.57	II
1700		4251.73	II
860		4253.37	II
650		4253.61	II
810		4260.12	I
1600		4262.09	I
650		4266.60	I
470		4267.00	I
300		4274.17	I
910		4280.49	II
430		4285.82	I
300		4286.12	I
540		4296.08	II
220		4297.17	II
430		4299.29	I
1100		4306.34	I
260		4309.29	I
1800		4313.84	I
520		4314.40	I
520		4316.05	II
370		4320.52	I
750		4321.11	II
		4321.20	I
2600	d	4325.57	II
		4325.69	I
1900		4327.12	I
370		4329.58	I
340		4330.61	II
240		4331.38	I
450		4341.28	II
910		4342.18	II
1000		4344.30	II
2200		4346.46	I
910		4346.62	I
220		4347.31	II
300		4369.77	II
970		4373.83	I
280		4392.06	I
1400		4401.86	I
520		4403.14	I
260		4406.67	II
260		4408.25	II
220		4409.25	I
520		4411.16	I
860		4414.16	I
700		4414.73	I
340		4419.03	II
1400		4422.41	I
1100		4430.63	I
240	d	4436.10	I
		4436.22	II
300		4464.74	I
300		4466.55	II
		4466.60	I
520		4467.08	I
700		4474.13	I
860		4476.12	I
220		4478.80	II
280		4481.06	II
220		4483.33	II
220		4484.70	I
280		4486.90	I
500		4497.13	I
220		4497.32	I
430		4506.21	I
140		4506.33	II
140		4514.50	II
1100		4519.66	I
300		4522.82	II
150		4524.12	I
910		4537.81	I
220		4540.02	II
300		4542.03	I
240		4548.00	I
120		4558.08	II
130		4573.81	I
260		4575.91	I
280		4579.59	I
410		4581.29	I
130		4582.53	II
410		4583.07	I
160		4586.99	I
220		4596.98	II
320		4597.91	II
410		4598.90	I
340		4601.05	II
240		4602.93	I
520		4614.50	I
140		4624.42	I
430		4636.64	I
110		4639.00	II
170		4640.04	I
170		4646.00	I
170		4647.64	I
170	d	4648.59	I
		4648.70	I
430		4653.54	I
140	h	4670.87	I
170		4679.18	I
260		4680.04	I
430		4683.33	I
140		4688.12	I
700		4694.33	I
170		4695.49	I

Gadolinium (Cont.)

430		4697.42	I
170		4703.13	I
200		4709.78	I
110		4721.46	I
150		4728.47	II
220		4732.60	II
260		4735.75	I
410		4743.65	I
110		4745.82	I
320		4758.70	I
110		4760.74	I
130		4763.82	I
470		4767.24	I
180		4781.92	I
300		4784.62	I
110		4786.75	I
140		4801.05	II
220		4807.45	I
320		4821.69	I
130		4835.26	I
110		4848.10	I
110		4862.59	I
170		4865.02	II
120		4871.50	I
280		4934.12	I
220		4938.61	I
110		4952.47	I
130		4958.79	I
65		5010.82	II
55		5011.74	I
750		5015.04	I
55		5023.13	II
65		5031.29	II
75		5039.09	I
65		5050.88	II
55		5073.74	I
55		5082.80	I
95		5092.25	II
65		5096.06	II
130		5098.38	II
55		5100.94	II
910		5103.45	I
180		5108.91	II
120		5125.56	II
65		5130.28	II
65		5135.59	I
75		5136.04	I
65		5140.04	II
75		5141.50	I
75		5142.68	I
860		5155.84	I
55		5156.76	II
75		5158.48	I
75		5163.70	I
55		5164.54	II
190		5176.28	II
55		5187.24	II
55		5187.88	I
55		5191.08	II
410		5197.77	I
55		5210.49	II
85		5217.48	I
280		5219.40	I
75		5220.30	II
130		5233.93	I
65		5246.87	I
320		5251.18	I
120		5252.14	II
85		5254.75	I
140		5255.80	I
65		5268.78	I
55		5272.91	I
55		5282.48	I
280		5283.08	I
280		5301.67	I
220		5302.76	I
55		5306.70	I
280		5307.30	I
130		5321.50	I
280		5321.78	I
110		5327.32	I
65		5328.30	I
170		5333.30	I
55		5337.53	I
300		5343.00	I
85		5345.13	I
75		5345.68	I
200		5348.67	I
300		5350.38	I
240		5353.26	I
55		5361.66	I
95		5365.38	I
95		5369.92	I
150		5370.63	I
85		5389.50	I
85		5413.20	I
85		5415.69	I
65		5453.46	I
55		5583.68	II
55	d	5591.85	I
190		5617.91	I
65		5629.55	I
110		5632.25	I
260		5643.24	I
55	Bl	5680.89	G
390		5696.22	I
95		5701.35	I
65		5709.42	I
120		5733.86	II
85		5746.36	I
85	d	5754.17	I
75		5776.02	I
240		5791.38	I
65	h	5796.80	I
55		5802.92	I
55	Hs	5807.72	I
55	h	5809.22	I
55		5815.85	II
65	Hs	5819.51	G
55		5840.47	II
220		5851.63	I
55		5855.24	II
280		5856.22	I
55		5860.73	II
65		5877.26	II
55		5886.46	I
55		5904.07	II
110		5904.56	I
170		5911.45	II
65		5913.55	II
55		5916.77	I
85		5930.29	I
85		5936.84	I
65		5937.71	I
55	h	5940.95	G
55	h	5942.78	G
55		5951.60	II
55		5956.48	II
85		5977.25	I
110	h	5988.02	I
85		5999.08	I
65		6000.96	G
75	h	6001.87	G
55		6004.57	II
55		6008.71	I
55		6021.13	I
55		6080.65	II
430		6114.07	I
55		6180.42	II
110	Bl	6182.68	G
110	Bl	6200.86	G
110	Bl	6211.71	G
110	Bl	6220.93	G
55	Bl	6231.62	G
75	Bl	6241.66	G
55	b	6252.12	G
55	Bl	6262.64	G
45	b	6273.00	G
85		6289.73	II
30		6292.87	I
75		6305.15	II
30		6309.11	II
27		6317.19	I
40		6331.35	I
17		6333.75	I
17		6336.34	I
27		6346.65	II
27	h	6351.72	I
17		6363.23	I
40		6380.95	II
17		6382.19	II
22		6408.55	I
22		6422.42	II
17		6424.52	I
19	h	6470.29	I
15		6480.11	II
40	h	6538.15	I
22		6549.25	I
55		6564.78	I
10		6568.00	II
10		6573.80	I
30		6591.60	I
15		6593.42	I
10		6610.04	II
50		6634.36	II
35		6640.08	I
10		[illegible]	I
30		6643.98	I
10		6646.85	I
10		6653.55	I
10		6679.56	II
35		6681.23	II
10		6692.86	I
10		6704.18	II
14		6718.14	II
17		6727.83	II
85		6730.73	I
50		6752.67	II
14		6753.91	II
14		6783.39	I
26		6786.33	II
10		6787.18	I
12		6814.56	I
26		6816.49	I
17		6820.90	I
100		6828.25	I
35		6846.60	II
30		6857.13	II
15		6864.25	I
21		6887.63	II
14		6900.73	II
100		6916.57	I
21		6920.62	II
15		6924.99	II
21		6926.49	I
17		6945.98	II
15		6957.74	II
15		6959.24	II
14		6964.33	I
15		6971.66	II
12		6976.35	II
10		6978.27	II
26	h	6980.86	I
50		6985.89	II
10		6988.75	II
75		6991.92	I
21		6993.18	I
60		6996.76	II
17		7000.75	II
45		7006.16	II
10		7016.60	I
21		7037.26	II
14		7051.00	II
13		7054.62	II
10		7058.02	II
10		7068.09	II
18		7071.00	I
18		7073.63	I
14		7098.11	I
14		7098.73	I
10	h	7116.77	II
21		7118.86	II
35		7122.57	I
13		7135.73	II
18		7147.31	II
13		7158.28	I
170		7168.37	I
21		7172.26	II
28		7189.57	II
13		7197.08	II
13		7201.41	II
10		7228.02	I
25		7233.45	I
14		7252.70	II
28		7262.66	I
14		7291.35	I
21		7313.28	I
18		7324.89	II
14		7373.81	I
14		7376.41	I
13		7377.27	II
13		7380.28	I
13		7394.90	II
13		7430.19	I
35		7441.85	I
40		7464.36	I
55		7562.97	I
10		7563.19	II
10		7588.20	I
10		7611.78	I
21		7621.96	I
21		7650.32	I
25		7672.56	I
10		7676.06	I
13		7694.45	I
80		7733.50	I
35		7749.30	I
10		7755.97	I
10		7766.48	II
11	h	7844.87	I
10		7845.80	I
35		7846.35	II
35		7856.93	I
14		7869.72	I
25		7930.25	II
13		8077.59	I
18		8146.15	I
11		8218.08	I
10		8275.42	I
[illegible]		[illegible]	I
11		8398.30	I
10		8445.47	I
13	h	8527.88	I
21		8668.63	I
11		8770.36	I
13		8784.85	I
10		8795.76	I
21	h	8832.06	II
14	h	8849.14	I
18	h	8867.31	I

Gd III

Ref. 46, 137, 151—J.F.K.

Intensity	Wavelength	
	Vacuum	
600	1813.47	III
900	1946.26	III
1100	1974.34	III
2200	1975.24	III
	Air	
900	2008.79	III
3400	2018.07	III
1800	2027.82	III
800	2046.02	III
500	2057.79	III
1500	2080.08	III
1800	2098.20	III
1300	2125.68	III
1400	2148.03	III
1700	2176.84	III
1700	2223.95	III
1700	2236.73	III
1200	2239.84	III
1500	2243.75	III
1700	2250.18	III
1300	2257.05	III
1300	2292.51	III
1000	2300.38	III
1200	2303.72	III
1500	2307.03	III
1100	2313.50	III
1700	2313.56	III
1000	2315.09	III
1400	2323.12	III
1700	2323.18	III
2200	2323.78	III
1900	2329.35	III
2100	2335.01	III
1600	2336.02	III
1900	2338.97	III
1600	2339.88	III
2100	2342.74	III
2500	2346.52	III
2800	2359.31	III
1900	2360.87	III
2300	2361.91	III
1400	2362.38	III
2100	2363.26	III
1600	2365.22	III
2000	2373.38	III
1300	2374.29	III
1300	2381.20	III
1600	2387.82	III
2000	2388.77	III
1200	2393.86	III
1200	2397.34	III
1200	2405.03	III
1300	2408.41	III
1200	2409.35	III
1500	2466.84	III
1100	2469.14	III
1300	2499.53	III
1600	2520.38	III
1600	2534.11	III
1300	2536.10	III
1600	2551.56	III
2200	2553.90	III
2100	2554.04	III
2500	2563.33	III
2100	2564.46	III
1000	2565.04	III
2400	2565.95	III
1800	2569.27	III
1800	2573.57	III
2000	2576.06	III
1300	2576.15	III
1400	2578.13	III
1600	2578.76	III
1700	2583.62	III
2000	2588.21	III
2000	2588.46	III
1300	2595.81	III
1800	2609.77	III
1200	2619.40	III
1200	2621.52	III
1400	2623.52	III
1400	2625.48	III
2000	2628.10	III
1300	2628.99	III
2400	2629.83	III
2100	2632.30	III
1800	2633.32	III
1400	2635.71	III
1600	2636.44	III
1700	2637.15	III
2100	2637.97	III
2100	2638.06	III
2200	2640.53	III
1600	2641.65	III
2100	2643.71	III
1600	2644.52	III
1800	2646.04	III
1800	2646.84	III
1600	2651.48	III
2000	2655.59	III
1900	2656.55	III
2200	2660.83	III

Gadolinium (Cont.)

Intensity	Wavelength	Spectrum
1800	2675.75	III
1800	2679.44	III
1700	2680.63	III
1800	2682.52	III
1500	2683.91	III
1600	2692.78	III
1900	2692.86	III
1500	2694.43	III
2800	2697.39	III
1500	2702.91	III
2800	2703.28	III
1600	2704.53	III
1800	2717.35	III
2700	2727.89	III
450	2751.24	III
1800	2833.83	III
9000	2904.73	III
1800	2918.40	III
9500	2955.53	III
1000	2975.42	III
1000	2984.10	III
1000	3116.59	III
2500	3118.04	III
4000	3176.66	III
400	3253.53	III
400	3330.34	III
400	3371.05	III
400	3402.97	III
450	3624.90	III
250	3700.47	III
300	3831.73	III
300	3910.24	III
300	4016.91	III
600	4177.26	III
400	4279.96	III
300	4314.28	III
300	4445.91	III
600	4684.25	III
600	4715.06	III
600	4782.79	III
250	4976.72	III
5000	5091.70	III
300	5124.06	III
1800	5347.95	III
3000	5365.96	III
1100	5412.62	III
4000	5553.30	III
3000	5587.88	III
3000	5658.98	III
1800	5786.96	III
1500	5862.09	III
1500	5987.85	III
5000	14332.88	III
2000	17474.78	III
800	19996.34	III
800	21259.44	III
600	22493.33	III

Gd IV
Ref. 152—J.F.K.

Intensity	Wavelength	
	Vacuum	
1000	967.92	IV
1000	983.42	IV
1000	987.10	IV
1000	987.91	IV
1000	995.04	IV
1000	995.80	IV
1000	996.49	IV
1000	999.24	IV
1000	1000.36	IV
1000	1002.73	IV
1000	1004.46	IV
1000	1005.66	IV
1000	1006.55	IV
1200	1007.24	IV
1200	1063.84	IV
500	1228.37	IV
500	1307.23	IV
500	1313.29	IV
500	1316.71	IV
600	1321.42	IV
500	1330.79	IV
1100	1393.24	IV
1600	1476.98	IV
1500	1705.03	IV
1600	1706.01	IV
2000	1736.24	IV
1500	1815.32	IV
400	1997.89	IV
	Air	
800	2049.28	IV
800	2061.30	IV
800	2070.40	IV
800	2076.66	IV
1000	2094.29	IV
1000	2296.89	IV
1000	2352.66	IV
800	2379.17	IV
900	2385.65	IV
500	2390.07	IV
700	2392.30	IV
700	2393.29	IV
700	2395.76	IV
500	2396.22	IV
600	2396.27	IV
1400	2397.87	IV
700	2402.70	IV
500	2412.21	IV
900	2419.26	IV
500	2439.84	IV
600	2440.38	IV
600	2468.60	IV

GALLIUM (Ga)
Z = 31

Ga I and II
Ref. 19, 132, 195, 281—L.J.R.

Intensity		Wavelength	
		Vacuum	
2		829.60	II
2		958.67	II
1		960.57	II
2		969.19	II
2		998.52	II
3		1002.95	II
5		1012.38	II
3		1019.10	II
5		1023.80	II
8		1033.69	II
1		1113.87	II
3		1119.25	II
5		1130.81	II
1		1167.62	II
2		1173.78	II
3		1186.81	II
1		1227.13	II
5		1286.38	II
5		1327.81	II
20		1414.44	II
5		1449.49	II
2		1463.65	II
3		1473.73	II
3		1483.52	II
3		1485.95	II
3		1495.21	II
3		1504.41	II
3		1505.01	II
5		1514.57	II
3		1515.19	II
8		1535.40	II
5		1536.37	II
1		1536.91	II
3		1669.83	II
5		1695.85	II
5		1799.42	II
10		1813.98	II
15		1845.30	II
		Air	
20		2091.34	II
1		2218.04	I
1		2255.03	I
1		2259.23	I
2		2294.19	I
1		2297.87	I
3		2338.24	I
1		2338.60	I
3		2371.29	I
2		2377.53	II
4		2418.69	I
5		2438.88	II
6		2450.08	I
7		2500.19	I
3		2500.71	I
5		2513.55	II
3		2514.15	II
2		2551.26	II
3		2552.87	II
4		2555.28	II
5		2607.47	I
8		2624.82	I
10		2632.66	I
3		2659.87	I
10		2665.05	I
8		2691.29	I
20		2700.47	II
3		2719.66	I
15		2780.15	II
6		2874.24	I
1		2886.45	II
		2893.65	GA II
2		2910.77	II
6		2943.64	I
6		2944.17	I
3		2969.41	II
1		2971.01	II
3		2971.60	II
5		2974.77	II
1		2992.84	II
2		3011.90	II
1		3158.18	II
4		3374.94	II
1		3375.95	II
2		3436.66	II
3		3446.46	II
2		3447.26	II
5		3470.34	II
1		3471.46	II
1		3472.52	II
2		3583.60	II
1		3693.93	II
2		3705.85	II
4		3734.85	II
9		3924.39	II
10		4032.99	I
10		4172.04	I
4		4251.11	II
15		4251.16	II
10	h	4254.04	II
4	h	4255.64	II
5		4255.70	II
10		4255.77	II
40		4262.00	II
3		5218.21	II
1		5338.3	II
2		5353.49	I
2		5360.6	II
1		5363.5	II
3		5416.8	II
1		5421.6	II
1		5425.6	II
10		6334.2	II
2000		6396.56	I
1000		6413.44	I
5		6419.4	II
3		6456.3	II
1		7000.0	II
3	h	7051.24	I
5	h	7106.82	I
1	h	7116.3	I
2	h	7172.9	I
5	h	7193.6	I
7		7198.7	II
10	h	7251.4	I
3	h	7289.6	I
5	h	7349.3	I
20	h	7403.0	I
30	h	7464.0	I
6	h	7556.6	I
10	h	7620.5	I
50	h	7734.77	I
2		7793.0	II
100	h	7800.01	I
4	h	7801.6	I
15	h	8002.55	I
20	h	8074.25	I
3	h	8167.5	I
5	h	8171.6	I
100	h	8311.86	I
200	h	8386.49	I
10	h	8389.30	I
7	h	8415.51	I
10	h	8419.91	I
20	h	8808.75	I
30	h	8813.56	I
20	h	8856.37	I
30	h	8944.33	I
200	h	9492.92	I
200	h	9493.12	I
300	h	9589.36	I
20	h	9594.25	I
60	h	10898.10	I
100	h	10905.95	I
10		10968.27	I
20		11103.51	I
400		11949.12	I
200		12109.78	I
40		12885.05	I
50		13057.50	I
50		14982.75	I
60		14996.64	I
20		17757.91	I
10		17868.96	I
60		22016.81	I
70		22568.71	I

Ga III
Ref. 141—L.J.R.

Intensity	Wavelength	
	Vacuum	
50	620.00	III
40	622.01	III
90	806.51	III
90	817.30	III
50	828.70	III
80	1085.00	III
60	1105.61	III
90	1150.27	III
90	1267.16	III
80	1293.46	III
60	1295.36	III
60	1323.15	III
70	1353.92	III
90	1495.07	III
50	1534.46	III
	Air	
90	2417.70	III
90	2423.98	III
15	2424.36	III
50	3521.77	III
80	3581.19	III
100	3589.34	III
10	3731.10	III
10	3806.60	III
100	4380.69	III
150	4381.76	III
100	4863.00	III
150	4993.78	III
10	5808.28	III
20	5848.25	III
15	5993.51	III

Ga IV
Ref. 141, 143—L.J.R.

Intensity	Wavelength	
	Vacuum	
14	294.53	IV
61	295.67	IV
41	304.99	IV
4	422.12	IV
25	423.18	IV
16	439.92	IV
67	1137.06	IV
70	1156.10	IV
70	1163.60	IV
75	1170.58	IV
48	1171.71	IV
68	1185.23	IV
40	1186.06	IV
73	1190.89	IV
73	1193.02	IV
75	1195.02	IV
69	1201.54	IV
72	1206.89	IV
63	1216.15	IV
50	1228.03	IV
60	1236.38	IV
60	1238.59	IV
45	1241.81	IV
75	1245.53	IV
83	1258.77	IV
81	1264.66	IV
82	1267.15	IV
81	1279.24	IV
80	1285.33	IV
82	1295.86	IV
83	1299.46	IV
82	1303.53	IV
80	1309.68	IV
80	1314.82	IV
85	1338.09	IV
77	1347.03	IV
76	1351.06	IV
74	1364.63	IV
60	1395.54	IV
77	1402.55	IV
70	1405.32	IV
73	1465.87	IV

Ga V
Ref. 2, 62, 140—L.J.R.

Intensity	Wavelength	
	Vacuum	
5	290.53	V
1	296.13	V
5	296.82	V
30	298.44	V
20	299.47	V
30	300.01	V
25	300.57	V
10	300.78	V
20	301.19	V
30	302.86	V
20	303.84	V
30	307.03	V
30	308.26	V
15	309.64	V
30	311.79	V
25	312.41	V
30	313.68	V

Gallium (Cont.)

Intensity	Wavelength	
15	315.95	V
20	316.48	V
40	319.41	V
12	320.53	V
40	322.31	V
50	322.99	V
30	323.10	V
40	324.25	V
40	324.95	V
40	326.14	V
30	326.77	V
30	328.65	V
5	336.61	V
20	878.17	V
40	973.21	V
10	977.89	V
15	979.60	V
20	984.95	V
40	989.75	V
90	1014.47	V
90	1019.71	V
20	1033.55	V
30	1038.76	V
30	1047.50	V
120	1050.48	V
80	1054.56	V
90	1058.12	V
80	1066.69	V
35	1068.59	V
30	1069.45	V
60	1069.60	V
55	1071.19	V
45	1071.41	V
80	1073.77	V
90	1078.83	V
110	1079.60	V
60	1080.99	V
250	1085.01	V
80	1087.37	V
40	1090.53	V
90	1091.71	V
100	1094.36	V
80	1095.10	V
70	1101.62	V
160	1102.83	V
110	1103.09	V
60	1104.93	V
75	1105.62	V
70	1106.17	V
40	1115.55	V
80	1118.34	V
55	1123.18	V
80	1123.66	V
120	1126.40	V
80	1127.75	V
130	1128.10	V
120	1128.53	V
100	1129.94	V
80	1131.43	V
40	1133.91	V
130	1136.07	V
65	1138.20	V
60	1144.30	V
50	1145.70	V
30	1148.42	V
45	1150.09	V
130	1150.23	V
120	1156.51	V
35	1157.74	V
25	1169.40	V
40	1178.95	V
80	1213.17	V
30	1265.45	V
30	1276.85	V
15	1283.64	V
10	1311.35	V

GERMANIUM (Ge)
Z = 32

Ge I and II
Ref. 5, 119, 293, 340—C.H.C.

Intensity		Wavelength	
		Vacuum	
1		822.97	II
3		835.08	II
10		850.50	II
10		862.234	II
15		875.493	II
15		905.977	II
20		920.554	II
50		999.101	II
100		1016.638	II
100		1075.072	II
300		1085.51	II
200		1098.71	II
500		1106.74	II
500		1120.46	II
200		1164.27	II
500		1181.19	II
500		1181.65	II
200		1188.73	II
100		1189.62	II
300		1191.26	II
50		1191.72	II
500		1237.059	II
500		1261.905	II
100		1264.710	II
100		1380.42	II
50		1392.26	II
200		1401.24	II
200		1538.091	II
500		1576.855	II
75		1581.070	II
100		1602.486	II
3	r	1615.57	I
2	r	1624.130	I
2	r	1630.173	I
3	r	1636.31	I
2		1638.96	I
4	r	1639.730	I
2		1647.531	I
200		1649.194	II
2		1651.528	I
4	r	1651.955	I
3		1661.345	I
4	r	1663.539	I
10	h	1665.275	I
4		1667.802	I
3	r	1670.608	I
100	r	1691.090	I
200	r	1716.784	I
100	h	1739.102	I
100		1742.195	I
50		1746.065	I
200		1750.043	I
100		1758.279	I
100	h	1764.185	I
100	h	1765.284	I
50	h	1766.433	I
200		1774.176	I
200		1785.046	I
100	h	1793.071	I
75	h	1801.432	I
200	h	1841.328	I
200	h	1842.410	I
100	h	1844.410	I
100	h	1845.872	I
100	h	1846.958	I
200		1853.134	I
500	r	1860.086	I
100		1865.052	I
300	r	1874.256	I
100		1895.197	I
500	r	1904.702	I
50	h	1908.434	I
30		1912.409	I
300	r	1917.592	I
100	h	1923.467	I
500	r	1929.826	I
10	h	1934.048	I
100	r	1937.483	I
500		1938.008	II
100	r	1938.300	I
500		1938.891	II
30	s	1944.116	I
200		1944.731	I
200		1955.115	I
500		1962.013	I
30	h	1963.373	I
30		1965.383	I
200		1970.880	I
200		1979.274	II
300	h	1987.849	I
300		1988.267	I
500	r	1998.887	I
		Air	
50		2007.04	II
200		2011.13	I
1700		2019.068	I
2400	r	2041.712	I
1600	r	2043.770	I
420		2054.461	I
220	h	2057.238	I
750	r	2065.215	I
2600	r	2068.656	I
420		2086.021	I
2000	r	2094.258	I
240		2105.824	I
95	h	2124.744	I
50	h	2186.451	I
100		2197.62	II
340	r	2198.714	I
100		2205.85	II
15		2220.375	I
18		2256.001	I
18		2314.201	I
24		2327.918	I
15		2359.233	I
20		2379.144	I
10		2389.472	I
15		2397.885	I
130		2417.367	I
30		2436.412	I
100		2478.66	II
90		2497.962	I
500		2500.54	II
70		2533.230	I
3		2556.298	I
28		2589.188	I
500		2592.534	I
8		2644.184	I
1200		2651.172	I
500		2651.568	I
500		2691.341	I
200		2704.03	II
850		2709.624	I
400		2729.78	II
40		2740.426	I
650		2754.588	I
50		2770.59	II
75		2772.35	II
70		2793.925	I
80		2829.008	I
1000		2831.843	II
50		2834.28	II
75		2839.68	II
1000		2845.527	II
75		2853.97	II
750		3039.067	I
600		3067.021	I
20		3124.816	I
50		3186.72	II
100		3221.64	II
110		3269.489	I
50		3312.56	II
75		3323.64	II
100		3455.72	II
300		3499.21	II
30		3845.11	II
70		4226.562	I
10		4685.829	I
75	h	4689.87	II
50	h	4690.02	II
1000		4741.806	II
1000		4814.608	II
50		4824.097	II
100		5131.752	II
200		5178.648	II
3		5194.583	I
6		5265.892	I
6		5513.263	I
8		5564.741	I
8		5607.010	I
6		5616.135	I
7		5621.426	I
8		5655.96	I
6		5664.226	I
5		5664.842	I
9		5691.954	I
6		5701.776	I
5		5717.877	I
6		5801.029	I
9		5802.093	I
1000		5893.389	II
500		6021.041	II
150		6078.39	II
50		6267.14	II
150		6268.07	II
100		6268.34	II
75		6283.452	II
100		6336.377	II
100		6484.181	II
6		6557.488	I
50		6780.51	II
50		7049.369	II
6		7130.12	I
30		7145.390	II
7		7330.38	I
5		7353.334	I
7		7384.208	I
6		7402.64	I
7		7511.57	I
5		7776.20	I
10		7833.575	I
7		7837.63	I
6		7853.77	I
7		7878.12	I
5		7962.26	I
5		7983.33	I
10		8031.039	I
6		8044.165	I
5		8095.29	I
5		8225.22	I
7		8226.09	I
10		8256.013	I
5		8264.15	I
5		8280.09	I
6		8281.04	I
8		8367.81	I
7		8391.70	I
5		8396.36	I
5		8429.42	I
10		8482.21	I
8		8506.70	I
5		8507.66	I
8		8564.89	I
6		8599.27	I
6		8652.42	I
5		8669.60	I
9		8700.60	I
5		8712.90	I
6		8734.78	I
6		8789.88	I
5		9068.785	I
5		9095.957	I
6		9398.868	I
20		9474.993	II
20		9475.645	II
4		9492.559	I
7		9625.664	I
5		10039.436	I
4		10200.952	I
10		10382.427	I
10		10404.913	I
8		10734.068	I
8		10947.416	I
10		11125.130	I
230		11252.83	I
24		11293.40	I
33		11318.13	I
55		11459.05	I
150		11483.77	I
175		11614.81	I
600		11714.76	I
10		11839.77	I
55		11917.01	I
10		12025.64	I
10		12055.49	I
30		12061.41	I
45		12065.76	I
1300		12069.20	I
30		12198.88	I
20		12207.73	I
60		12286.75	I
55		12338.76	I
1050		12391.58	I
48		12540.41	I
15		12636.80	I
150		12676.58	I
40		12681.28	I
115		12800.66	I
175		12836.38	I
12		12847.92	I
120		12955.73	I
15		13028.64	I
235		13107.61	I
20		13492.28	I
42		13534.85	I
28		13724.48	I
42		14116.70	I
42		14297.15	I
40		14569.84	I
12		14667.52	I
470		14822.38	I
16		14921.97	I
15		15001.75	I
13		15041.21	I
20		15504.34	I
14		16424.77	I
12		16626.64	I
70		16699.29	I
150		16759.79	I
135		17214.34	I
16		18428.30	I
35		18495.54	I
10		18764.11	I
70		18811.86	I
62		19279.24	I
28		20673.64	I
4		21518.30	I
9		22091.84	I
5		23921.92	I

Ge III
Ref. 341—C.H.C.

Intensity	Wavelength	
	Vacuum	
2	542.90	III
2	663.77	III
3	670.88	III
2	680.28	III
2	952.76	III
12	988.96	III
15	995.72	III
10	996.50	III
15	1011.21	III
10	1012.31	III
8	1032.62	III
12	1040.99	III
12	1058.91	III
40	1088.45	III
10	1137.92	III
12	1150.55	III

Germanium (Cont.)

Intensity		Wavelength	
8		1159.15	III
8		1159.62	III
8		1160.79	III
10		1173.78	III
8		1212.47	III
4		1323.24	III
10		1525.32	III
2		1527.15	III
9		1600.09	III
6		1883.26	III
2		1978.22	III
		Air	
2		2019.22	III
4		2022.25	III
3		2062.14	III
15		2100.05	III
15		2102.42	III
25		2104.45	III
3		2922.86	III
25		3197.56	III
35		3211.86	III
25		3214.95	III
40		3255.05	III
20		3259.90	III
5		3369.57	III
20		3414.27	III
40		3434.03	III
8		3464.59	III
40		3489.08	III
2		3724.51	III
15		3884.78	III
200		4178.96	III
12		4245.41	III
200		4260.85	III
150		4291.71	III
10		4674.36	III
10		5016.88	III
18		5134.75	III
5		5229.37	III
3		5256.61	III

Ge IV
Ref. 341—C.H.C.

Intensity		Wavelength	
		Vacuum	
1		440.11	IV
1		441.95	IV
3		847.80	IV
3		868.30	IV
8		915.00	IV
8		936.70	IV
4		938.90	IV
1		1073.44	IV
20		1188.99	IV
20		1229.81	IV
2		1494.89	IV
6		1500.61	IV
3		1648.14	IV
		Air	
2		2293.0	IV
2		2343.37	IV
15		2445.38	IV
15		2445.71	IV
30		2488.25	IV
20		2542.44	IV
5		2631.78	IV
3		2698.08	IV
15		2717.44	IV
30		2736.09	IV
30		2788.61	IV
5		3071.84	IV
60		3554.19	IV
50		3676.65	IV

Ge V
Ref. 342—C.H.C.

Intensity		Wavelength	
		Vacuum	
700		294.51	V
1000		295.64	V
200		304.98	V
20		621.52	V
35		716.26	V
50		724.21	V
35		733.54	V
35		735.35	V
35		741.52	V
60		746.88	V
40		750.26	V
35		755.84	V
60		760.05	V
60		958.51	V
300		971.35	V
150		984.92	V
200		988.13	V
300		990.66	V
300		1004.38	V
300		1016.66	V
250		1038.40	V
900		1045.71	V
400		1050.05	V
300		1054.59	V
300		1068.43	V
400		1069.13	V
700		1072.66	V
600		1086.65	V
500		1087.85	V
800		1089.49	V
300		1092.09	V
1000		1116.94	V
300		1122.01	V
700		1163.39	V
300		1165.26	V
200		1176.69	V
700		1222.30	V

GOLD (Au)
Z = 79

Au I and II
Ref. 38, 72, 234—C.H.C.

Intensity		Wavelength	
		Vacuum	
		925.72	II
		946.03	II
		950.39	II
20		957.78	II
3		967.94	II
3		974.47	II
2		982.24	II
		1062.67	II
8		1066.96	II
		1085.00	II
8		1090.78	II
5		1094.92	II
20		1103.31	II
3		1166.76	II
5		1210.86	II
20		1224.57	II
40	h	1305.34	I
25		1310.47	I
100	h	1328.37	I
3		1336.26	II
10	h	1338.37	I
10	h	1342.80	I
20		1350.09	I
20		1350.84	I
22		1351.74	I
25		1352.82	I
25		1354.14	I
30		1355.79	I
35		1357.86	I
40		1360.51	I
6		1362.33	II
20		1362.47	I
8		1363.15	II
50		1363.98	I
25		1364.15	I
10		1364.74	II
60		1368.62	I
35		1368.98	I
70		1374.82	I
50		1375.76	I
30		1378.87	I
8		1380.53	II
80		1382.75	I
50		1385.33	I
20		1389.14	I
60		1392.27	I
6		1393.80	II
50		1402.12	I
		1405.12	II
70		1407.38	I
100		1408.45	I
		1410.69	II
		1415.22	II
80		1429.19	I
50		1435.79	I
20		1436.61	II
10		1468.85	II
10		1469.17	II
10		1469.28	II
100		1481.76	I
25		1486.55	II
20		1532.82	I
20		1532.86	I
10		1562.04	II
200		1587.16	I
12		1593.41	II
70		1598.24	I
12		1611.11	II
		1616.65	AU II
2		1622.83	II
100		1624.34	I
2		1632.53	II
50		1639.90	I
150		1646.67	I
10		1656.99	II
100		1665.76	I
25		1673.59	II
7		1694.38	II
2		1698.65	II
200		1699.34	I
30		1700.69	II
10		1720.04	II
25		1725.75	II
45		1740.52	II
10		1749.80	II
35		1756.15	II
60		1783.22	II
35		1793.31	II
35		1800.58	II
25		1823.24	II
100		1879.83	I
20		1919.64	I
20		1921.64	II
45		1942.31	I
25		1951.93	I
30		1978.19	I
		Air	
25		2000.81	II
11000		2012.00	I
2600		2021.38	I
50		2044.54	II
150		2082.09	II
35		2095.13	II
20		2098.14	II
60		2110.68	II
30		2125.29	II
15		2126.63	I
20		2170.75	I
35		2188.81	II
25		2201.32	II
35		2215.63	II
45		2228.88	II
30		2231.18	II
25		2240.16	II
70		2248.56	II
80		2263.62	II
18		2263.88	II
25		2277.52	II
25		2283.30	II
25		2291.40	II
45		2304.69	II
25		2314.55	II
20		2315.75	II
25		2340.06	II
180		2352.65	I
20		2376.28	I
120		2387.75	I
2600		2427.95	I
60		2533.52	II
16		2544.19	I
45		2552.67	II
20		2589.25	I
30		2590.04	I
50		2616.40	II
20		2627.02	II
250		2641.48	I
3400		2675.95	I
20		2687.63	II
20		2688.16	II
30		2688.71	I
80		2700.89	I
1100		2748.25	I
20		2748.71	II
100		2780.82	I
30		2800.93	I
1000		2802.04	II
300		2819.79	II
100		2822.55	II
30		2823.13	II
100	h	2825.44	II
30		2833.03	II
300		2837.85	II
100		2846.92	II
100		2856.74	II
3		2872.36	I
300		2883.45	I
3		2886.96	I
10		2888.40	I
300		2891.96	I
100		2893.25	II
3		2905.74	I
30		2905.90	I
100		2907.04	II
300		2913.52	II
3		2914.82	I
300		2918.24	II
16		2932.19	
30		2940.67	I
100		2954.22	II
10		2973.33	I
100	h	2990.27	II
300		2994.80	II
10		3002.65	I
3		3005.85	I
10		3024.67	I
320		3029.20	I
30		3033.25	I
300		3065.42	I
10		3102.63	I
10	h	3117.01	I
100		3122.50	II
1600		3122.78	I
30		3126.86	II
30		3127.03	I
10		3164.88	I
10		3172.35	II
30		3191.76	I
100		3194.72	I
30		3200.37	I
30		3204.74	I
1		3221.86	I
30		3225.25	I
300		3230.63	I
10		3253.94	I
10	h	3265.10	I
30	h	3267.07	I
10		3271.63	I
10		3273.47	I
300		3308.30	I
300		3309.64	I
100		3320.12	I
100		3355.15	I
30	h	3368.44	I
10		3381.90	I
100	h	3391.31	I
100		3395.40	I
30	h	3440.36	I
100		3467.21	I
30		3471.61	I
30	h	3509.04	I
10		3510.82	I
3		3523.34	II
3		3545.61	I
30		3553.57	I
300		3557.36	I
10		3558.22	I
30		3565.97	I
30	h	3584.37	I
300		3586.73	I
30	h	3588.79	I
30		3598.06	I
100		3611.57	I
30	h	3614.00	I
30		3622.74	I
100	h	3631.31	I
10		3633.22	II
30		3634.53	I
10		3635.12	I
300		3637.90	I
3		3639.87	I
100	h	3645.02	I
10		3649.09	I
100		3650.74	I
10		3653.53	I
3		3654.69	I
10		3655.30	I
10		3656.90	I
30		3706.55	II
100		3709.62	I
10		3766.61	I
10		3770.76	I
100		3796.01	I
30		3801.92	I
30		3804.01	II
10		3821.85	I
10		3825.70	I
100		3874.73	I
30		3880.25	I
30		3889.48	I
100	h	3892.26	I
400		3897.86	I
30		3901.09	I
300		3909.38	I
100		3927.69	I
10	h	3959.10	I
30		3966.23	I
30		3976.65	I
30		3979.68	I
30		3991.37	I
3		4012.57	I
10		4016.07	II
400		4040.93	I
30		4052.79	II
700		4065.07	I
10		4076.35	II
3		4083.28	II
100		4084.10	I
30		4101.70	I
30		4128.59	I
30		4201.13	I
30	h	4227.88	I
100		4241.80	I
200		4315.11	I
30		4361.04	II
10		4420.61	II
120	h	4437.27	I

Gold (Cont.)

Intensity		Wavelength	
250		4488.25	I
900	h	4607.51	I
100	h	4620.56	I
1		4663.92	I
3		4663.97	I
10		4694.69	I
3		4760.17	II
500		4792.58	I
100		4811.60	I
10		4822.96	I
30	h	4950.82	I
30		5064.59	I
30	h	5108.84	I
100		5147.44	I
300		5230.26	I
100	h	5261.76	I
100		5655.77	I
100	h	5721.36	I
300		5837.37	I
100	h	5862.93	I
300	h	5956.96	I
30	h	5962.68	I
600		6278.17	I
100		6562.68	I
30		6652.89	I
600		7510.73	I
10		8145.06	I
10		9254.28	I

Au III
Ref. 72, 393, 395 — R.D.C.

Intensity		Wavelength	
		Vacuum	
30		779.73	III
30		788.78	III
50		799.93	III
40		811.83	III
50		817.95	III
40		820.06	III
80		833.16	III
100		843.44	III
100		845.14	III
80		855.49	III
80		859.90	III
90		883.42	III
80		901.03	III
80		910.45	III
80		924.02	III
200		945.10	III
100		1040.63	III
80		1044.49	III
80		1046.81	III
100	h	1239.96	III
100		1278.51	III
100		1314.84	III
200		1336.72	III
180		1341.68	III
100		1348.89	III
150		1350.32	III
150		1355.61	III
150		1356.13	III
80		1362.06	III
500		1365.40	III
200		1367.17	III
180		1377.73	III
150		1378.69	III
150		1379.98	III
125		1380.53	III
200		1381.36	III
300		1385.79	III
100		1389.41	III
180		1391.46	III
180		1396.00	III
100		1402.91	III
225		1409.50	III
250		1413.80	III
100		1414.27	III
80		1415.54	III
100		1417.09	III
125		1417.39	III
150		1427.42	III
300		1428.93	III
250		1430.06	III
275		1433.37	III
250		1435.81	III
80		1436.12	III
300		1439.12	III
200		1441.21	III
150		1446.37	III
80		1446.69	III
250		1448.42	III
250		1454.95	III
100		1464.72	III
150		1471.28	III
80		1473.32	III
100		1474.73	III
150		1481.10	III
300		1487.15	III
250		1487.91	III
200		1489.47	III
250		1500.37	III
200		1502.47	III
200		1503.74	III
80		1540.26	III
100		1542.00	III
80		1542.25	III
100		1548.50	III
80		1554.61	III
80		1562.33	III
80		1562.41	III
200		1567.54	III
80		1571.94	III
200		1574.85	III
200		1579.44	III
150		1584.10	III
200		1589.56	III
80		1589.68	III
150		1593.41	III
200		1600.51	III
250		1617.16	III
100		1617.78	III
500		1621.93	III
300	d	1629.13	III
250		1638.88	III
100		1644.17	III
250		1652.74	III
250		1664.77	III
100		1668.11	III
125		1673.93	III
1000		1693.94	III
150		1697.09	III
200		1698.98	III
200		1700.00	III
200		1702.25	III
100		1707.53	III
250		1710.16	III
200		1715.69	III
100		1716.71	III
300		1717.83	III
500		1727.31	III
100	d	1733.17	III
300		1738.48	III
150		1744.39	III
500		1746.10	III
500		1756.92	III
500		1761.95	III
300		1767.42	III
100		1774.42	III
800		1775.17	III
200		1776.40	III
100		1780.57	III
300		1786.11	III
150		1792.65	III
500		1793.76	III
200		1801.98	III
400		1805.24	III
100		1809.81	III
400		1821.17	III
400		1844.89	III
150		1848.83	III
80		1850.15	III
500		1861.80	III
150		1871.92	III
150		1918.28	III
100		1932.04	III
100		1935.42	III
200		1948.79	III
100		1958.47	III
400		1989.63	III
150		1996.85	III
		Air	
300		2083.09	III
80		2085.45	III
100		2159.08	III
80		2167.33	III
200		2172.20	III
100		2184.11	III
500		2188.97	III
300		2322.27	III
100		2382.40	III
150		2402.71	III
150		2405.12	III
100		3227.99	III
100		3309.86	III

HAFNIUM (Hf)
Z = 72

Hf I and II
Ref. 1—C.H.C.

Intensity	Wavelength	
	Air	
6200	2012.78	II
8500	2028.18	II
1200	2096.18	II
540	2210.82	II
320	2254.01	II
160	2255.15	II
250	2266.83	II
620	2277.16	II
230	2321.14	II
580	2322.47	II
300	2323.25	II
120	2324.50	II
300	2324.89	II
200	2332.97	II
200	2337.33	II
230	2343.32	II
320	2347.44	II
540	2351.22	II
110	2353.02	I
90	2365.98	II
250	2380.30	II
100	2381.00	II
170	2393.18	II
450	2393.36	II
670	2393.83	II
130	2400.78	II
70	2404.56	II
540	2405.42	II
130	2406.44	II
370	2410.14	II
90	2413.33	II
55	2415.96	II
320	2417.69	II
120	2425.98	II
45	2428.75	I
120	2428.99	II
130	2433.57	II
45	2434.74	II
35	2444.99	I
390	2447.25	II
140	2449.44	II
35	2452.30	II
110	2453.34	II
450	2460.49	II
70	2463.97	II
430	2464.19	II
90	2465.06	II
35	2465.67	I
140	2467.97	II
210	2469.18	II
100	2473.92	II
55	2481.66	II
55	2482.65	I
55	2487.16	I
290	2496.99	II
580	2512.69	II
580	2513.03	II
130	2515.48	II
890	2516.88	II
340	2531.19	II
200	2537.33	II
110	2548.20	II
320	2551.40	II
130	2559.19	II
250	2563.61	II
890	2571.67	II
320	2573.90	II
320	2576.82	II
300	2578.14	II
320	2582.54	II
130	2591.33	II
390	2606.37	II
450	2607.03	II
120	2608.45	I
230	2613.60	II
450	2622.74	II
160	2637.00	I
1100	2638.71	II
1100	2641.41	II
160	2642.75	I
670	2647.29	II
100	2651.16	II
160	2657.84	II
210	2661.88	II
290	2683.35	II
670	2705.61	I
110	2706.73	II
210	2712.42	II
140	2713.84	I
250	2718.59	I
120	2730.85	I
710	2738.76	II
200	2743.64	I
360	2751.81	II
450	2761.63	I
160	2766.96	I
170	2773.02	I
980	2773.36	II
180	2774.02	II
390	2779.37	I
100	2789.50	II
140	2789.73	II
230	2808.00	II
230	2813.86	II
170	2814.48	II
230	2817.68	I
140	2818.94	I
200	2819.74	I
1200	2820.22	II
490	2822.68	II
180	2833.28	I
110	2834.13	I
410	2845.83	I
270	2849.21	II
270	2850.96	I
180	2851.21	II
180	2860.56	I
760	2861.01	II
760	2861.70	II
2100	2866.37	I
130	2869.82	II
150	2876.33	II
210	2887.14	I
100	2887.54	I
800	2889.62	I
1800	2898.26	I
130	2898.71	II
1200	2904.41	I
890	2904.75	I
140	2909.91	II
2000	2916.48	I
580	2918.58	I
320	2919.59	II
180	2924.62	I
490	2929.63	II
450	2929.90	I
710	2937.80	II
2000	2940.77	I
160	2944.71	I
1200	2950.68	I
1100	2954.20	I
540	2958.02	I
120	2961.80	II
1400	2964.88	I
620	2966.93	I
140	2967.23	II
710	2968.81	II
110	2973.37	I
890	2975.88	II
150	2979.28	I
1100	2980.81	I
210	2982.72	I
170	3000.10	II
800	3005.56	I
1100	3012.90	II
540	3016.78	I
1100	3016.94	II
980	3018.31	I
1200	3020.53	I
140	3025.29	II
410	3031.16	II
110	3046.08	II
710	3050.76	I
1100	3057.02	I
130	3063.78	I
130	3064.68	II
850	3067.41	I
2100	3072.88	I
170	3074.10	I
250	3074.79	I
150	3080.66	II
430	3080.84	I
200	3096.76	I
340	3101.40	II
710	3109.12	II
130	3110.87	II
130	3119.98	I
710	3131.81	I
850	3134.72	II
130	3137.51	I
170	3139.65	II
120	3140.76	II
220	3145.32	II
220	3148.41	I
120	3151.63	I
450	3156.63	I
270	3159.82	I
710	3162.61	II
450	3168.39	I
890	3172.94	I
450	3176.86	II
220	3181.01	I
120	3181.15	I
130	3189.62	I
360	3193.53	II
670	3194.19	II
200	3196.93	I
130	3199.99	II
310	3206.11	I
180	3210.98	I
180	3217.30	II
180	3220.61	II
130	3230.06	I
130	3239.44	I
130	3243.35	I
360	3247.66	I
220	3249.53	I
890	3253.70	II
270	3255.28	II
120	3262.47	I
180	3273.66	II
270	3279.98	II

Hafnium (Cont.)

Intensity		Wavelength	Spectrum
160		3291.05	I
210		3306.12	I
120		3309.19	I
340		3310.27	I
670		3312.86	I
180		3317.99	II
130		3328.21	II
890		3332.73	I
370		3352.06	II
130		3356.78	I
230		3358.91	I
180		3360.06	I
140		3366.68	I
180		3378.93	I
140		3384.14	II
230		3384.70	II
170		3386.21	I
800		3389.83	II
230		3392.81	I
230		3394.59	II
140		3394.98	II
230		3397.26	I
230		3397.60	I
2300		3399.80	II
170		3400.21	I
180		3402.51	I
140		3407.76	II
230		3410.17	II
230		3417.34	I
410		3419.18	I
140		3427.44	I
200		3428.37	II
250		3438.24	II
140		3438.43	I
100		3441.84	I
100		3452.31	I
140		3462.64	II
140		3467.60	I
710		3472.40	I
200		3478.99	II
480		3479.28	II
250		3495.75	II
250		3497.16	I
980		3497.49	I
100		3498.98	I
1200		3505.23	II
150		3513.28	I
130		3518.75	II
980		3523.02	I
100		3530.87	I
100		3531.23	I
980		3535.54	II
760		3536.62	I
180		3548.81	I
540		3552.70	II
150		3554.00	I
1300		3561.66	II
150		3564.31	I
270		3567.36	I
1100		3569.04	II
150		3579.90	I
110		3583.28	I
210		3597.42	II
540		3599.87	I
110		3615.04	I
800		3616.89	I
110		3617.68	I
110		3624.00	II
320		3630.87	II
100		3635.43	I
800		3644.36	II
320		3649.10	I
200		3651.84	I
140		3661.05	II
220		3665.35	II
100		3668.21	I
200		3672.27	I
480		3675.74	I
2200		3682.24	I
280		3696.51	I
100		3698.40	II
240		3699.72	II
340		3701.15	II
100		3704.92	I
120		3705.40	II
1000		3717.80	I
650		3719.28	II
140		3726.49	I
160		3729.10	I
460		3733.79	I
160		3737.88	II
120		3739.04	I
100		3744.98	II
400		3746.80	I
140		3753.22	I
100		3765.05	I
100		3765.56	I
170		3766.92	II
200		3768.25	I
1400		3777.64	I
1400		3785.46	I
650		3793.37	II
100		3798.66	I
850	d	3800.38	I
140		3806.07	II
320		3811.78	I
100		3817.20	II
100		3819.38	I
1300		3820.73	I
140		3829.67	I
280		3830.02	I
800		3849.18	I
140		3849.52	II
600		3858.31	I
230		3860.91	I
200		3872.55	II
160		3877.10	II
380		3880.82	II
200		3882.52	I
150		3883.77	II
200		3889.23	I
200		3889.33	I
620		3899.94	I
620		3918.09	II
200		3923.90	II
120		3926.42	I
150		3927.57	I
110		3929.54	II
320		3931.38	I
120		3935.65	II
120		3939.04	I
410		3951.83	I
160		3968.01	I
150	Bl	3970.05	H
200		3973.48	I
180		4032.27	I
100		4047.96	II
230		4062.84	I
140		4066.21	I
180		4083.35	I
540		4093.16	II
110		4104.23	I
140		4106.58	I
110		4113.53	II
110		4118.60	I
150		4127.80	II
140		4145.76	I
150		4162.36	II
110		4162.69	I
1100		4174.34	I
120		4190.95	I
160		4206.58	II
190		4209.70	I
170		4228.08	I
170		4232.44	II
120	Bl	4252.08	H
170		4260.98	I
200		4263.39	I
170		4272.85	II
320		4294.79	I
120		4318.14	I
160		4330.27	I
180		4336.66	II
150		4350.51	II
250		4356.33	I
110		4367.90	II
180		4370.97	II
120		4417.35	II
160		4417.91	I
200		4438.04	I
140		4457.34	I
140		4461.18	I
140		4540.93	I
250		4565.94	I
500	d	4598.80	I
230		4620.86	I
210		4655.19	I
120		4699.01	I
160		4782.74	I
310		4800.50	I
130		4859.24	I
120		4975.25	I
95		5018.20	I
15		5021.75	I
55		5040.82	II
95		5047.45	I
55	b	5074.74	H
30		5079.65	II
55	b	5093.88	H
15		5112.13	I
19		5128.53	II
30		5157.96	I
55		5167.42	I
75		5170.18	I
230		5181.86	I
30		5186.84	I
30		5187.75	II
110		5243.99	I
55		5247.10	II
25		5260.44	II
30		5264.95	II
55		5275.04	I
22		5286.09	I
120		5294.87	I
45		5298.06	II
30		5307.82	I
45		5309.68	I
55		5311.60	II
12		5324.26	II
9		5346.30	II
110		5354.73	I
110		5373.86	I
40		5389.34	I
19		5391.36	II
19		5404.47	I
28		5424.02	I
12		5435.78	I
40		5438.74	I
14		5444.07	II
75		5452.92	I
30		5463.38	II
15		5497.30	I
15		5510.12	I
15		5510.45	I
19		5524.35	II
45		5538.02	I
28		5538.26	I
230		5550.60	I
230		5552.12	I
55		5575.86	I
14		5600.77	I
95		5613.27	I
25		5614.01	I
8		5628.27	I
19		5650.83	I
40	Bl	5698.03	H
25		5713.28	I
160		5719.18	I
25	Bl	5720.16	H
12		5748.72	I
14		5767.18	II
12		5809.50	II
19		5817.47	I
25		5842.23	II
25		5845.87	I
19		5847.77	I
22		5883.66	I
15		5926.47	I
60		5933.69	I
75		5974.28	I
25		5974.72	I
60		5978.66	I
25		5992.96	I
45	c	6016.79	I
28	b	6021.12	H
25	b	6043.19	H
25		6054.17	I
95		6098.67	I
95		6185.13	I
55		6210.70	I
28		6216.82	I
45		6238.58	I
60		6248.95	II
22	h	6299.54	I
25	h	6311.85	I
19		6318.33	I
30		6338.10	I
19	h	6380.19	I
60		6386.23	I
19	h	6409.52	I
15	h	6556.50	I
28		6587.23	I
45		6644.60	II
19		6647.06	II
11		6659.40	I
30		6713.48	I
17		6754.61	II
11		6769.95	I
85		6789.27	I
160		6818.94	I
15		6826.56	I
13		6850.07	I
35		6858.70	I
45		6911.40	I
10		6926.19	I
19		6979.59	I
21		6980.91	II
7		7019.25	I
7		7030.33	II
7		7035.13	I
11		7061.90	I
15		7062.87	I
160		7063.83	I
11	h	7094.40	I
15		7100.54	I
55		7119.52	I
570		7131.81	I
650		7237.10	I
410		7240.87	I
6		7262.62	I
75		7320.05	I
16		7321.76	I
6		7356.10	I
6		7365.28	I
20		7390.70	I
6		7423.69	I
25		7437.56	I
13		7463.86	I
7		7484.56	I
15		7556.37	I
75		7562.93	I
15		7564.22	I
11		7576.95	I
11		7592.96	I
13		7608.59	I
360		7624.40	I
20		7645.64	I
110		7740.17	I
8		7743.57	I
5		7757.89	II
40		7790.90	I
7		7796.81	I
35		7814.55	I
310		7845.35	I
7		7846.56	I
130		7920.71	I
29		7938.06	I
250		7994.73	I
7		8010.58	I
25		8056.52	I
25		8080.32	I
16		8173.89	I
130		8204.58	I
7		8248.81	I
55		8276.95	I
13		8305.91	II
25		8344.25	I
5		8380.06	I
5		8382.98	I
35		8460.01	I
150		8546.48	I
160		8640.06	I
40		8711.24	I
65		9004.73	I

Hf III
Ref. 404 — R.L.K.

Intensity		Wavelength	
		Vacuum	
20		1449.83	III
30		1507.82	III
50		1683.95	III
50		1756.91	III
60		1843.64	III
60		1870.58	III
50		1874.81	III
150		1885.15	III
100		1991.44	III
		Air	
100		2037.76	III
300		2070.94	III
150	h	2085.33	III
200	h	2099.30	III
200	h	2110.31	III
100	h	2119.69	III
200		2155.66	III
200		2183.50	III
200		2195.44	III
100		2213.54	III
200		2234.59	III
200	h	2313.44	III
300		2336.47	III
150	h	2355.48	III
100		2373.30	III
120		2377.57	III
250		2383.540	III
400		2461.74	III
2000		2495.16	III
1000		2515.16	III
100		2534.33	III
400	h	2560.74	III
300	h	2567.46	III
200		2687.22	III
500		2753.60	III
100	h	3060.08	III
200	h	3279.67	III
100	h	3741.94	III

Hf IV
Ref. 369, 425 — R.L.K.

Intensity	Wavelength	
	Vacuum	
40	520.04	IV
50	569.19	IV
100	596.56	IV
100	600.90	IV
100	603.16	IV
200	618.27	IV
100	620.19	IV
100	633.58	IV
100	643.05	IV
200	644.54	IV
400	647.39	IV
600	665.65	IV
100	671.36	IV
200	673.49	IV

Hafnium (Cont.)

15	1305.24	IV
12	1357.40	IV
40	1390.39	IV
50	1491.67	IV
35	1528.82	IV
15	1560.18	IV
25	1572.03	IV
100	1717.21	IV
20	1718.57	IV
	Air	
100	2054.46	IV
7	2014.06	IV
20	7751.29	IV
10	7267.58	IV

Hf V
Ref. 410 — R.L.K.

Intensity	Wavelength	
	Vacuum	
220	545.41	V
180	600.00	V
100	816.81	V
100	830.69	V
100	836.74	V
100	846.87	V
100	856.32	V
100	861.80	V
135	865.16	V
270	867.25	V
180	875.88	V
135	877.87	V
135	880.37	V
100	880.85	V
180	885.58	V
135	885.80	V
135	894.24	V
100	894.41	V
180	896.14	V
100	896.47	V
135	899.70	V
180	901.54	V
135	901.92	V
135	904.95	V
135	909.70	V
135	913.68	V
135	918.48	V
180	919.10	V
270	921.67	V
135	928.01	V
135	931.50	V
135	947.12	V
245	951.62	V
180	960.12	V
180	964.74	V
160	971.51	V
135	974.62	V
120	984.64	V
135	991.50	V
100	1078.42	V
100	1079.92	V
160	1092.76	V
135	1097.28	V
135	1137.49	V
160	1201.76	V
135	1208.88	V
135	1224.62	V
135	1227.98	V
135	1230.21	V
270	1232.03	V
200	1233.59	V
160	1237.42	V
100	1238.85	V
160	1239.53	V
160	1244.46	V
100	1259.25	V
440	1396.66	V
270	1400.09	V
160	1401.70	V
135	1405.77	V
370	1407.17	V
370	1408.38	V
270	1412.28	V
270	1413.51	V
160	1421.96	V
220	1422.53	V
370	1433.43	V
370	1437.27	V
500	1437.73	V
370	1445.40	V
270	1457.91	V
270	1719.32	V
550	1729.08	V
750	1731.83	V
750	1733.96	V
440	1741.74	V
1000	1749.11	V
1000	1750.19	V
500	1760.89	V
370	1765.62	V
270	1774.02	V
135	1792.39	V

HELIUM (He)
Z = 2

He I and II
Ref. 16, 94, 173, 183, 317
W.C.M.

Intensity	Wavelength	
	Vacuum	
15	231.454	II
20	232.584	II
30	234.347	II
50	237.331	II
100	243.027	II
300	256.317	II
1000	303.780	II
500	303.786	II
10	320.293	I
2	505.500	I
3	505.684	I
4	505.912	I
5	506.200	I
7	506.570	I
10	507.058	I
15	507.718	I
20	508.643	I
25	509.998	I
35	512.098	I
50	515.616	I
100	522.213	I
400	537.030	I
1000	584.334	I
50	591.412	I
5	958.70	II
6	972.11	II
8	992.36	II
15	1025.27	II
30	1084.94	II
35	1215.09	II
50	1215.17	II
120	1640.34	II
180	1640.47	II
	Air	
7	2385.40	II
9	2511.20	II
50	2577.6	I
1	2723.19	I
12	2733.30	II
2	2763.80	I
10	2818.2	I
4	2829.08	I
10	2945.11	I
40	3013.7	I
20	3187.74	I
3	3202.96	II
15	3203.10	II
1	3354.55	I
2	3447.59	I
1	3587.27	I
3	3613.64	I
2	3634.23	I
3	3705.00	I
1	3732.86	I
10	3819.607	I
1	3819.76	I
500	3888.65	I
20	3964.729	I
1	4009.27	I
50	4026.191	I
5	4026.36	I
12	4120.82	I
2	4120.99	I
[illegible]	[illegible]	[illegible]
10	4387.929	I
3	4437.55	I
200	4471.479	I
25	4471.68	I
6	4685.4	II
30	4685.7	II
30	4713.146	I
4	4713.38	I
20	4921.931	I
100	5015.678	I
10	5047.74	I
5	5411.52	II
500	5875.62	I
100	5875.97	I
8	6560.10	II
100	6678.15	I
3	6867.48	I
200	7065.19	I
30	7065.71	I
50	7281.35	I
1	7816.15	I
2	8361.69	I
2	9063.27	I
2	9210.34	I
10	9463.61	I
4	9516.60	I
3	9526.17	I
1	9529.27	I
1	9603.42	I
3	9702.60	I
6	10027.73	I
2	10031.16	I
15	10123.6	II
1	10138.50	I
10	10311.23	I
2	10311.54	I
3	10667.65	I
300	10829.09	I
1000	10830.25	I
2000	10830.34	I
9	10913.05	I
3	10917.10	I
4	11626.4	II
30	11969.12	I
20	12527.52	I
50	12784.99	I
20	12790.57	I
7	12845.96	I
10	12968.45	I
2	12984.89	I
12	15083.64	I
200	17002.47	I
1	18555.55	I
6	18636.8	II
500	18685.34	I
200	18697.23	I
100	19089.38	I
20	19543.08	I
1000	20581.30	I
80	21120.07	I
10	21121.43	I
20	21132.03	I
3	30908.5	II
4	40478.90	I

HOLMIUM (Ho)
Z = 67

Ho I and II
Ref. 1 — C.H.C.

Intensity		Wavelength	
		Air	
170		2502.91	II
80		2508.53	II
110		2513.55	II
95		2518.73	II
170		2533.80	I
130		2536.86	II
80		2556.84	I
80		2567.73	II
80		2586.52	I
60		2591.05	II
95		2592.99	II
190		2605.86	II
110		2610.51	II
95		2613.99	II
60		2625.20	II
80		2640.09	I
80		2640.30	II
60		2649.68	II
80		2666.24	II
70		2689.03	II
210		2713.65	II
230		2733.95	II
270		2750.35	II
110	c	2759.35	II
110		2766.85	II
270		2769.89	II
110		[illegible]	II
140		2777.10	II
140		2794.41	II
100		2799.99	II
100		2806.72	II
160	c	2809.99	II
220		2811.36	II
180		2812.00	II
190		2814.74	II
300		2824.20	II
140		2826.64	II
270	c	2831.69	II
210		2834.99	II
110		2835.85	II
110		2844.18	II
100		2844.68	II
270		2849.10	II
100		2861.23	II
250		2861.49	II
150		2862.72	II
210		2871.99	II
230		2874.06	II
160		2874.43	II
360		2880.26	II
460		2880.98	II
340		2894.99	II
160		2895.62	II
170		2900.84	II
570	c	2909.41	II
170		2915.82	II
300		2919.62	II
110		2925.35	II
160		2926.09	II
300	c	2928.30	II
220		2942.05	II
300		2944.49	II
250	c	2953.11	II
390		2973.00	II
410	c	2979.63	II
180		2981.46	II
140		2985.48	II
410		2987.64	II
250		2990.27	II
110		2995.86	II
320	:	3008.10	II
220		3014.60	II
270		3038.69	II
480	c	3049.38	II
410	c	3054.00	II
500	c	3057.45	II
230		3074.30	II
500	c	3082.34	II
910		3084.36	II
430	c	3086.54	II
200		3108.31	II
200	c	3109.91	II
760		3118.50	II
300	c	3130.99	II
200	c	3134.39	II
300	c	3144.36	II
200		3156.18	II
270		3156.97	II
200	c	3159.67	II
580	c	3166.62	II
390	D1	3171.72	II
810		3173.78	II
390		3174.84	II
270	c	3176.97	II
810	c	3181.50	II
390		3183.84	II
270	Cw	3184.48	II
200		3186.37	I
390	c	3197.83	II
390		3201.76	II
200		3206.86	II
270	c	3210.41	II
200	c	3221.42	II
320		3233.34	II
200		3236.90	II
200		3237.40	II
200	c	3257.45	II
390	c	3278.15	II
270		3279.25	II
980	c	3281.97	II
390		3288.46	II
270	c	3290.96	II
200	c	3305.16	II
200		3319.87	II
230		3320.25	II
200		3331.93	II
630	c	3337.23	II
390	c	3338.86	II
980	c	3343.58	II
200		3344.47	II
360		3350.49	II
320		3352.10	II
320	Cw	3353.55	II
320		3354.58	II
320		3357.91	II
320		3364.27	II
290		3370.87	II
230		3374.16	II
290	c	3390.75	II
[illegible]	[illegible]	[illegible]	II
8100	c	3398.98	II
810	c	3410.26	II
1200		3421.63	II
3200		3453.14	II
390	c	3410.65	II
1400	c	3414.90	II
5400		3416.46	II
2000	c	3425.34	II
2000	c	3428.13	II
630	c	3429.18	II
320	c	3432.10	II
390		3449.35	I
810	c	3455.70	II
16000	c	3456.00	II
1600		3461.97	II
360	c	3467.07	II
810	c	3473.91	II
5400	c	3474.26	II
6300		3484.84	II
490		3489.58	II
580	c	3493.09	II

Holmium (Cont.)

2500	c	3494.76	II
810	c	3498.88	II
410	c	3506.95	II
320		3509.37	II
810		3510.73	I
4100	c	3515.59	II
410	c	3519.94	II
630		3540.76	II
1600		3546.05	II
1100	c	3556.78	II
410		3560.15	II
410	c	3573.24	II
630	c	3574.80	II
810		3579.12	I
410		3580.75	II
410		3581.83	II
630	c	3592.23	II
1100	Cw	3598.77	II
340		3599.48	I
540	c	3600.95	II
340		3613.31	II
410		3618.43	I
430	c	3626.69	II
490		3627.25	II
430	c	3631.76	II
430	c	3638.30	II
1600	c	3662.29	I
430		3662.99	I
720		3666.65	I
1400		3667.97	I
320		3669.05	II
450		3669.52	I
450	c	3674.77	II
720		3679.19	I
670		3679.70	I
720		3682.65	I
430		3685.16	II
580		3690.65	I
340		3691.95	I
410		3700.04	I
490	c	3702.35	II
320		3709.76	I
430		3712.88	I
450		3720.72	I
1100		3731.40	I
360		3732.09	I
810		3736.35	I
3200	Cw	3748.17	II
320	c	3753.73	II
340		3769.09	I
320		3788.08	II
8900	c	3796.75	II
8900	c	3810.73	II
490		3811.86	I
900	c	3813.25	II
300		3821.73	II
390		3829.27	I
320	Cw	3831.9	II
410	c	3835.35	II
1300	Cw	3837.51	II
410	c	3842.05	II
1100		3843.86	II
490	c	3846.73	II
300		3849.88	I
320		3852.40	II
1800	c	3854.07	II
390	Cw	3856.94	II
720		3857.72	II
2700	c	3861.68	II
540		3862.62	I
360		3872.05	II
320	c	3874.09	II
630		3874.68	II
540		3881.61	II
3000	c	3888.96	II
490		3890.42	I
13000	c	3891.02	II
540		3896.76	II
290		3902.23	II
320		3904.44	I
1300	Cw	3905.68	II
320		3911.80	I
320		3919.45	I
320	c	3936.44	II
220		3938.85	I
320	Cw	3940.53	II
220		3950.56	I
580		3955.73	I
230	c	3959.51	II
490		3959.68	I
220		3975.88	I
390	c	3976.93	I
220	Cw	3985.71	II
220		3993.73	II
380		3999.58	I
160	Cw	4002.59	II
220		4003.39	I
110		4013.50	I
320		4014.20	II
160	c	4018.09	II
160	c	4022.76	II
160	c	4023.94	II
110		4025.39	I
320		4027.21	I
270		4028.86	I
180	c	4031.80	I
220		4037.62	I
220	c	4038.87	II
2700		4040.81	I
5400	c	4045.44	II
220	c	4047.52	I
8100		4053.93	I
540		4054.48	II
270		4057.55	I
220		4060.31	I
1700		4065.09	II
170		4067.57	I
720		4068.05	I
270		4071.83	I
270		4073.13	I
290		4073.51	I
120	c	4080.23	II
230		4083.67	I
140		4085.09	I
170		4087.35	I
200		4087.59	I
140		4091.64	I
120		4094.78	I
230		4100.22	I
8900		4103.84	I
120		4105.04	I
270		4106.50	I
100		4107.36	I
2900		4108.62	I
300		4112.00	I
100		4112.72	I
270		4116.73	I
1500		4120.20	I
1300		4125.65	I
4300		4127.16	I
300		4134.54	I
1500		4136.22	I
130		4139.34	I
230		4142.19	I
290		4148.97	I
980	Cw	4152.61	II
8100		4163.03	I
160		4172.23	I
2500		4173.23	I
540		4194.35	I
100		4198.08	I
130		4203.21	I
100		4211.30	II
290		4222.29	I
290		4223.47	I
2000		4227.04	I
390		4229.52	II
130	h	4231.24	I
290		4243.78	I
1300	Cw	4254.43	I
130	c	4258.61	II
490		4264.05	I
300		4266.04	I
100		4273.63	II
200		4311.04	I
250		4330.64	II
300		4337.13	II
100	Cw	4346.84	II
1300		4350.73	I
290		4356.73	II
140		4363.93	II
170		4379.14	II
180	c	4384.83	II
150		4400.55	II
120		4401.24	II
180		4403.27	I
200		4420.56	II
130		4444.63	I
100		4473.59	II
300		4477.64	II
120		4484.57	II
140		4510.82	I
100	c	4526.14	II
170		4530.08	II
170	c	4531.28	I
130	c	4531.65	II
170		4534.58	I
200		4562.52	I
120	Cw	4609.32	II
130		4613.37	I
100		4618.84	I
100	c	4628.22	I
290		4629.10	II
200	c	4649.77	II
130	c	4661.33	II
140	c	4674.62	II
70		4701.17	II
80		4701.69	II
130		4709.84	II
65		4711.39	I
130	c	4717.52	I
35	c	4728.72	II
35		4738.00	II
290		4742.04	II
35	c	4749.09	II
35		4751.40	I
100	c	4757.01	I
35		4762.39	II
35		4763.57	II
55		4777.48	II
30		4779.42	I
70	c	4781.19	I
65		4782.92	I
55	c	4786.29	I
35		4791.48	II
35		4795.92	II
45	h	4798.87	I
27		4812.92	II
55		4832.31	II
30		4833.32	I
30		4855.54	II
45		4860.39	I
27	c	4889.67	II
30		4892.35	I
35		4896.44	II
55		4906.99	II
45		4922.73	I
55	c	4934.89	I
290		4939.01	I
27	c	4946.80	I
45		4948.18	II
65	c	4959.42	II
35		4961.03	II
55	Cw	4966.73	II
250	c	4967.21	II
220		4979.97	I
35	c	4988.96	I
90		4995.05	I
35	c	5012.42	I
55		5013.28	II
65		5026.53	I
30		5028.17	I
55		5032.95	II
65	c	5037.60	I
130		5042.37	I
35		5044.73	I
30		5051.44	II
30		5054.92	II
35	c	5060.75	I
65		5074.34	I
80		5093.07	I
140		5127.81	I
55		5129.27	II
130		5142.59	II
110		5143.22	II
160		5149.59	II
90	c	5167.88	I
130	c	5182.11	I
55		5187.85	I
90		5190.11	II
18		5195.23	I
45		5221.54	I
35	c	5244.47	I
65		5251.82	I
55		5275.48	I
90		5301.25	I
35		5319.24	I
35		5319.65	I
80		5330.11	I
90		5359.99	I
55		5381.40	I
30		5384.56	I
30		5384.97	I
18	h	5393.85	I
70		5403.17	I
100		5407.08	I
14		5413.62	II
16		5434.39	II
18		5435.87	I
30		5445.39	I
18		5449.8	II
30	h	5451.90	I
14		5454.0	II
30	c	5498.57	I
30		5504.51	I
27		5515.56	II
18		5516.45	II
30		5534.33	I
27		5553.14	I
35	c	5560.94	I
35	b	5563.6	H
70		5566.52	I
18		5573.96	II
35	Bl	5584.7	H
55	b	5591.1	H
55	Bl	5592.3	H
30	b	5607.1	H
27		5613.64	I
45	b	5626.4	H
65		5627.60	I
30		5628.24	II
55		5640.62	I
70	Bs	5655.9	H
65	b	5658.9	H
140		5659.58	I
70	c	5671.84	I
65		5674.70	I
140	c	5691.47	I
70	Bs	5696.3	H
140	c	5696.57	I
27		5734.02	I
45		5736.4	H
55		5739.24	I
22		5749.58	I
30		5766.64	
27	b	5803.8	H
45	b	5819.2	H
27	h	5821.90	I
22		5839.47	I
45	b	5849.4	H
140	c	5860.28	I
27	h	5864.42	I
45		5870.85	I
27	b	5879.6	H
70	c	5882.99	I
35	c	5892.56	I
22		5904.29	I
70		5921.76	I
30	c	5933.71	I
70	Cw	5948.03	I
45		5955.98	I
70		5972.76	I
90		5973.52	I
22		5981.43	I
230	c	5982.90	I
55		6002.04	I
27		6005.33	I
35		6021.43	I
16		6038.97	I
27		6050.71	I
45		6060.31	I
120		6081.79	I
70	Cw	6133.60	I
35		6156.38	I
27		6156.58	I
55		6191.68	I
70		6208.65	I
18		6234.17	I
45	c	6255.75	I
70	c	6305.36	I
22		6306.68	I
30		6321.94	
30	c	6354.35	I
30	c	6372.59	I
14	h	6373.86	I
22	h	6413.41	I
27	c	6471.77	I
13		6479.17	I
11		6515.30	I
11	h	6538.99	I
70		6550.97	I
15		6560.08	I
35	d	6600.58	I
260		6604.94	I
55		6607.47	I
13		6628.35	I
120		6628.99	I
15		6632.24	I
9	h	6652.98	I
15		6662.52	I
19	c	6680.46	I
24	c	6681.62	I
15	h	6682.02	
55	Cw	6694.32	I
15	Cw	6722.34	I
40		6745.05	I
13		6766.74	I
28	c	6774.68	I
55	c	6785.43	I
13		6793.7	I
13	Cw	6811.04	I
15	Cw	6820.38	I
24		6821.64	I
17	c	6825.72	I
8	h	6826.62	
8	h	6852.97	
17	Cw	6865.85	I
9		6883.36	I
13		6888.50	I
15	c	6892.96	I
17		6897.95	I
15	h	6903.80	I
15	Cw	6913.47	I
9		6916.70	I
40	Cw	6939.49	I
45	Cw	6950.39	I
13	Hl	6955.3	I
19		6976.7	II
10		6985.11	
9		6994.38	
14	h	7000.71	
10		7079.07	
12		7098.58	
9		7242.08	I
9		7250.60	I
14		7308.55	
25		7341.43	
18		7389.40	
5	h	7496.20	I
10	h	7510.74	I
140		7555.09	I
18		7589.20	I
25		7591.87	I
9	h	7593.64	I
7	h	7594.35	I

Holmium (Cont.)

Intensity		Wavelength	
12		7602.31	II
16		7605.35	I
12		7617.05	I
14		7627.98	I
40	c	7628.42	I
9	c	7641.14	I
4		7648.16	I
14	c	7653.80	I
12	c	7667.30	I
20		7690.43	I
50	c	7693.15	I
40	Cw	7715.06	I
16		7719.05	I
16	h	7738.98	I
8	c	7752.01	I
60	Cw	7815.48	I
40	Cw	7823.63	I
8	h	7879.22	I
60		7894.64	I
10	h	8464.66	I
10	h	8482.67	I
50		8512.94	I
20		8545.61	II
18		8601.84	II
40		8670.19	I
8	h	8697.32	
16	h	8805.48	II
20	c	8834.49	I
90		8915.98	II

HYDROGEN (H)
Z = 1

H I
Ref. 214 — W.C.M.

Intensity	Wavelength	
	Vacuum	
15	926.226	I
20	930.748	I
30	937.803	I
50	949.743	I
100	[illegible]	I
300	1025.722	I
1000	1215.668	I
500	1215.674	I
	Air	
5	3835.384	I
6	3889.049	I
8	3970.072	I
15	4101.74	I
30	4340.47	I
80	4861.33	I
120	6562.72	I
180	6562.852	I
5	9545.97	I
7	10049.4	I
12	10938.1	I
20	12818.1	I
40	18751.0	I
5	21655.3	I
8	26251.5	I
15	40511.6	I
4	46525.1	I
6	74578	I
3	123685	I

INDIUM (In)
Z = 49

In I and II
Ref. 1, 132, 348—350 — C.H.C.

Intensity		Wavelength	
		Vacuum	
2		1648.00	I
1	h	1676.16	I
5	h	1711.54	I
2	h	1741.23	I
1	h	1758.49	I
		Air	
10		2103.89	II
10		2166.88	II
2		2179.90	I
2		2182.40	I
2		2187.40	I
2		2190.84	I
15		2195.67	II
2		2197.41	I
2		2202.24	I
50		2205.28	II
3		2211.14	I
5		2230.70	I
3		2241.66	I
30		2255.79	II
10		2259.99	I
5		2278.20	I
40		2281.64	II
2		2283.75	I
2		2298.33	I
2		2298.70	I
2		2302.49	I
100	c	2306.05	II
25		2306.86	I
3		2309.32	I
2		2309.75	I
90	d	2313.21	II
2		2315.09	I
30		2323.40	II
5		2324.41	I
3		2324.92	I
70	d	2327.95	II
3		2332.76	I
80	h	2334.57	II
10		2340.19	I
8		2345.90	I
5		2346.56	I
50	d	2350.75	II
5		2358.70	I
15		2378.14	I
10		2379.00	I
110	d	2382.63	II
40		2389.54	I
40		2393.18	II
10		2399.18	I
50	h	2406.47	II
50		2408.76	II
50		2419.06	II
50		2419.20	II
70	h	2427.20	II
20		2429.86	I
10		2430.99	I
50		2432.73	II
60		2442.63	II
100		2447.90	II
60		2453.23	II
60		2460.08	I
30	h	2468.02	I
70		2486.15	II
110	d	2488.62	II
90		2488.95	II
80		2498.59	II
100		2499.60	II
90	d	2500.99	II
60		2508.16	II
110	d	2512.31	II
100		2521.37	I
10		2522.98	I
70		2553.56	II
160	d	2554.44	II
1100		2560.15	I
70		2565.13	II
70	d	2598.75	II
200		2601.76	I
50		2604.04	II
90	d	2654.70	II
100	d	2662.63	II
140	d	2668.65	II
140	d	2674.56	II
80		2683.12	II
1600		2710.26	I
300		2713.94	I
130	d	2749.75	II
700		2753.88	I
40		2775.37	I
60		2798.76	II
90	d	2818.97	II
180	c	2836.92	I
30	c	2858.14	I
80		2865.68	II
120	d	2890.18	II
1100		2932.63	I
100		2941.05	II
20	c	2957.01	I
[illegible]	[illegible]	[illegible]	II
110	c	2999.40	II
8000		3039.36	I
8	d	3051.15	I
110	d	3099.80	II
180	c	3101.8	II
130	c	3138.60	II
80	c	3142.75	II
130	d	3146.70	II
150		3155.77	II
100	c	3158.40	II
90	c	3176.30	II
90	d	3198.11	II
13000		3256.09	I
3000		3258.56	I
90	c	3338.50	II
75	c	3376.59	II
100	c	3404.28	II
110	d	3438.40	II
180	c	3693.91	II
95	c	3708.13	II
380	w	3716.14	II
120	c	3718.30	II
160	c	3718.72	II
160	c	3723.40	II
170	w	3795.21	II
230	c	3799.21	II
250	c	3834.65	II
200	c	3842.18	II
100		3889.78	II
100	c	3902.07	II
60	d	3922.12	II
65	c	3934.40	II
250	w	3962.35	II
120	c	4004.66	II
140	d	4013.92	II
410	w	4056.94	II
17000		4101.76	I
140	c	4205.14	II
100	d	4213.04	II
110	d	4219.66	II
150	d	4372.87	II
150	c	4500.78	II
18000		4511.31	I
110	c	4549.01	II
140	c	4570.85	II
180	w	4578.02	II
180	w	4578.40	II
140	c	4616.08	II
170	c	4617.17	II
250	c	4620.14	II
150	w	4620.70	II
170	c	4627.30	II
140	c	4637.04	II
380	c	4638.16	II
220	c	4644.58	II
360	c	4655.62	II
320	w	4656.74	II
190	c	4681.11	II
450	w	4684.8	II
3		4878.37	I
90	d	4907.06	II
70	h	4924.93	II
150	c	4973.77	II
80	h	5109.36	II
100	w	5115.14	II
140	c	5117.40	II
270	c	5120.80	II
200	w	5121.75	II
80	d	5129.85	II
240	c	5175.42	II
140	c	5184.44	II
30		5254.32	I
12		5262.74	I
150	c	5309.45	II
80		5411.41	II
140	c	5418.45	II
220	w	5436.70	II
130	c	5497.50	II
140	c	5507.08	II
320	c	5513.00	II
250	w	5523.28	II
130	c	5536.50	II
190	w	5555.45	II
240	c	5576.90	II
200	w	5636.70	II
160	c	5708.50	II
50		5709.91	I
100	c	5721.80	II
50		5727.68	I
210	c	5853.15	II
490	w	5903.4	II
260	w	5915.4	II
120	c	5918.78	II
130	c	6062.9	II
250	c	6095.95	II
210	c	6108.66	II
180	w	6115.9	II
230	w	6128.7	II
240	w	6129.4	II
320	w	6132.1	II
150	c	6140.0	II
90		6143.23	II
[illegible]	[illegible]	6148.10	II
190	w	6149.5	II
80		6161.15	II
180	w	6162.45	II
100	c	6224.28	II
280	w	6228.3	II
140	w	6231.1	II
270	w	6304.8	II
290	w	6362.3	II
300	w	6469.0	II
210	c	6541.20	II
190	c	6751.88	II
180	c	6765.9	II
100	c	6783.72	II
8	h	6847.44	I
320	w	6891.5	II
4	h	6900.13	I
380	w	7182.9	II
180	c	7255.0	II
210	c	7276.5	II
180	c	7303.4	II
320	c	7350.6	II
100	c	7632.7	II
100	c	7682.9	II
210	c	7740.7	II
100	c	7776.96	II
180	c	7789.0	II
70	c	7806.8	II
70	c	7814.5	II
90	c	7840.9	II
20	h	8050.78	I
240	c	8227.0	II
30	h	8238.66	I
15	h	8314.92	I
50	c	8434.55	II
30		8678.95	I
20		8682.63	I
50		8700.25	I
100	w	8813.5	II
80	c	8832.6	II
40		8894.47	I
10		9170.08	I
120	c	9197.7	II
120	c	9202.0	II
220	w	9213.0	II
160	d	9241.1	II
40	h	9349.83	I
60	h	9370.27	I
20		9427.99	I
100		9977.86	I
200		10257.03	I
60	h	10717.42	I
100	h	10744.31	I
20		11334.72	I
20		11731.48	I
10		12912.59	I
9		13429.96	I
5		13824.48	I
6		14316.25	I
3		14419.20	I
6		14668.66	I
7		14719.08	I
2		16504.31	I
6		22291.06	I
7		23879.13	I

In III
Ref. 351 — C.H.C.

Intensity	Wavelength	
	Vacuum	
7	685.31	III
5	691.62	III
1	782.17	III
10	882.24	III
10	890.84	III
10	915.87	III
2	917.45	III
5	926.83	III
30	1403.08	III
30	1434.85	III
20	1487.70	III
20	1494.14	III
10	1524.78	III
20	1530.21	III
30	1532.95	III
100	1625.42	III
20	1642.28	III
20	1702.53	III
100	1748.83	III
2	1767.88	III
1	1810.71	III
30	1842.41	III
40	1850.30	III
15	1862.98	III
	Air	
30	2154.08	III
2	2154.42	III
10	2199.52	III
5	2232.18	III
20	2281.28	III
5	2266.26	III
5	2272.41	III
5	2272.84	III
10	2300.90	III
100	2527.41	III
50	2725.52	III
80	2726.15	III
100	2982.80	III
100	3008.08	III
30	3008.82	III
30	3293.55	III
8	3350.91	III
5	3562.32	III
100	3852.82	III
100	4023.77	III
150	4032.32	III
50	4062.30	III
100	4071.57	III
100	4072.93	III
100	4252.68	III
40	4509.58	III
200	5248.77	III

Indium (Cont.)

100	5645.15	III
40	5723.17	III
100	5819.50	III
200	6197.72	III

In IV
Refs. 352, 435, 436 — J.R.

Intensity	Wavelength	
	Vacuum	
622	472.71	IV
689	479.39	IV
709	498.62	IV
61	945.74	IV
85	954.67	IV
87	973.50	IV
86	991.60	IV
89	1024.68	IV
85	1024.79	IV
88	1031.45	IV
82	1031.98	IV
80	1054.43	IV
84	1063.03	IV
83	1068.25	IV
82	1069.82	IV
86	1077.64	IV
90	1082.10	IV
83	1086.33	IV
82	1096.81	IV
84	1097.18	IV
85	1116.10	IV
80	1124.06	IV
90	1131.46	IV
85	1144.43	IV
80	1145.41	IV
89	1146.62	IV
83	1154.11	IV
84	1154.60	IV
90	1157.71	IV
90	1157.82	IV
85	1159.78	IV
88	1176.50	IV
85	1191.58	IV
83	1204.87	IV
90	1206.55	IV
88	1221.50	IV
85	1221.90	IV
85	1233.58	IV
87	1235.84	IV
90	1373.20	IV
88	1398.77	IV
81	1412.09	IV

In V
Ref. 353 — C.H.C.

Intensity	Wavelength	
	Vacuum	
6	368.67	V
6	370.10	V
10	372.82	V
10	372.94	V
2	374.95	V
6	375.84	V
6	376.07	V
10	376.79	V
17	378.61	V
3	379.24	V
9	380.27	V
11	381.56	V
9	382.14	V
11	382.76	V
10	383.05	V
17	386.21	V
10	386.70	V
3	388.66	V
14	388.91	V
11	390.03	V
11	390.92	V
9	392.29	V
9	392.46	V
1	393.60	V
25	393.89	V
11	395.74	V
3	397.73	V
10	399.79	V
9	400.05	V
25	400.57	V
25	402.39	V
3	405.33	V
9	407.28	V
3	407.36	V
9	407.95	V
9	417.43	V
2	418.45	V
2	423.16	V

IODINE (I)

Z = 53

I I and II
Ref. 124, 153, 176, 184
L.J.R.

Intensity		Wavelength	
		Vacuum	
2		655.80	II
6		659.00	II
8		663.98	II
8		664.52	II
8		665.06	II
150		665.70	II
1000		719.55	II
1000		722.98	II
1000		798.16	II
1200		834.10	II
600		847.80	II
1500		873.49	II
1000		875.94	II
2000		879.84	II
1500		881.88	II
1000		891.00	II
1000		893.17	II
1200		1000.57	II
1000		1003.35	II
4000		1018.58	II
10000		1034.66	II
1500		1054.74	II
2000		1066.34	II
3000		1075.21	II
5000		1105.00	II
2500		1111.16	II
1500		1117.22	II
3500		1125.25	II
2000		1131.50	II
1200		1139.75	II
10000		1139.80	II
1500		1154.67	II
1000		1159.87	II
10000		1160.56	II
20000		1166.48	II
1500		1167.05	II
5000		1175.84	II
10000		1178.65	II
15000		1187.34	II
10000		1190.85	II
15		1195.29	I
5000		1198.88	II
7000		1200.22	II
200		1218.41	I
20000		1220.89	II
600		1224.05	I
600		1224.08	I
500		1228.89	I
20000		1234.06	II
600		1251.34	I
2500		1259.15	I
3000		1259.51	I
800		1261.27	I
600		1267.57	I
600		1267.60	I
1500		1275.26	I
3000		1289.40	I
10000		1300.34	I
3000		1302.98	I
3000		1313.95	I
3000		1317.54	I
2000		1330.19	I
20000		1336.52	II
5000		1355.10	I
3000		1357.97	I
5000		1360.97	I
3000		1361.11	I
2500		1367.71	I
2500		1368.22	I
4000		1383.23	I
3000		1390.75	I
2000		1392.90	I
2000		1400.01	I
8000		1425.49	I
5000		1446.26	I
5000		1453.18	I
5000		1457.39	I
5000		1457.47	I
10000		1457.98	I
2500		1458.79	I
4000		1459.15	I
2500		1465.83	I
1000		1485.92	I
5000		1492.89	I
5000		1507.04	I
5000		1514.68	I
15000		1518.05	I
2500		1526.45	I
5000		1593.58	I
5000		1617.60	I
2500		1640.78	I
15000		1702.07	I
12000		1782.76	I
5000		1799.09	I
75000		1830.38	I
15000		1844.45	I
		Air	
2000		2061.63	I
100		2408.01	II
100		2419.18	II
100		2494.74	II
100		2533.60	II
200		2534.27	II
1000		2566.24	II
2000		2582.79	II
300		2593.46	II
200	c	2688.98	II
500		2730.12	II
20		2765.15	II
200		2808.59	II
1500		2878.63	II
1000		2993.87	II
5000		3078.75	II
200		3161.03	II
1000		3175.07	II
300		3355.53	II
250	c	3424.99	II
300	c	3497.41	II
500		3526.90	II
200		3742.14	II
200		4102.23	I
200		4129.21	I
100	d	4134.15	I
500	d	4321.84	I
300		4452.86	II
200		4599.77	II
300	c	4632.45	II
500	d	4666.48	II
1000		4675.53	II
250		4763.31	I
1000		4862.32	I
200		4916.94	I
1000		4986.92	II
400	c	5065.37	II
10000		5119.29	I
200		5149.73	II
3000	c	5161.20	II
300		5176.19	II
600		5216.27	II
500	d	5228.97	II
1000		5234.57	I
3000	c	5245.71	II
500		5269.36	II
400		5299.78	II
400		5322.80	II
10000		5338.22	II
5000	c	5345.15	II
1000	c	5369.86	II
800	c	5405.42	II
800	c	5407.36	II
600	c	5427.06	I
3000		5435.83	II
1000		5438.00	II
2000	c	5464.62	II
800		5491.50	II
1000	c	5496.94	II
1000		5504.72	II
600	c	5522.06	II
600	c	5598.52	II
1000		5600.32	II
1500		5612.89	II
10000		5625.69	II
1000		5678.08	II
2000	c	5690.91	II
500		5702.05	II
4000	c	5710.53	II
1000		5738.27	II
1000		5760.72	II
1000	d	5764.33	I
500	c	5774.83	II
500		5787.02	II
2000		5894.03	I
5000		5950.25	II
300		5984.86	I
2000	d	6024.08	I
500		6068.93	II
2000	c	6074.98	II
1000		6082.43	I
2000	c	6127.49	II
800		6191.88	I
1000		6204.86	II
500		6213.10	I
800		6244.48	I
900	c	6257.49	II
1000		6293.98	I
500		6313.13	I
800		6330.37	I
400		6333.50	I
2000		6337.85	I
1000		6339.44	I
500		6359.16	I
1000		6566.49	I
2000		6583.75	I
1000		6585.27	I
5000		6619.66	I
500		6661.11	I
600		6665.96	II
500	c	6697.29	I
300		6718.83	II
400		6732.03	I
4000		6812.57	II
1000		6958.78	II
500		6989.78	I
200	c	7085.21	II
500		7120.05	I
1200		7122.05	I
2000		7142.06	I
1000		7164.79	I
400	d	7191.66	I
700		7227.30	I
1000		7236.78	I
500		7237.84	I
500		7351.35	II
5000		7402.06	I
1000		7410.50	I
500		7416.48	I
5000		7468.99	I
500	c	7490.52	I
2000		7554.18	I
500	d	7556.65	I
2000	c	7700.20	I
500		7798.98	II
600		7897.98	I
500		7969.48	I
1000		8003.63	I
99000		8043.74	I
300	d	8065.70	I
1000		8090.76	I
800	c	8169.38	I
500	d	8222.57	I
4000		8240.05	I
10000	c	8393.30	I
150		8414.60	II
1000		8486.11	I
1500	c	8664.95	I
500	c	8700.80	I
250	d	8748.22	I
1000		8853.24	I
2000		8853.80	I
3000		8857.50	I
1000	d	8898.50	I
400		8964.69	I
400		8993.13	I
5000		9022.40	I
15000		9058.33	I
1000		9098.86	I
12000		9113.91	I
600		9128.03	I
30		9195.30	II
600		9227.74	I
1000		9335.05	I
4000		9426.71	I
3000		9427.15	I
10	c	9480.33	II
2000		9598.22	I
2000		9649.61	I
3000	d	9653.06	I
5000		9731.73	I
500		10003.05	I
750		10131.16	I
1000		10238.82	I
400		10375.20	I
400		10391.74	I
6		10405.49	II
5000		10466.54	I
1		11084.68	II
400		11236.56	I
350		11558.46	I
320		11778.34	I
450		11996.86	I
300		12033.69	I
150		12304.58	I
60		13149.16	I
140		13958.27	I
200		14287.02	I
100		14460.00	I
225		15032.57	I
105		15528.65	I
150		16037.33	I
15		18275.71	I
20		18348.52	I
15		18982.41	I
35		19070.17	I
110		19105.12	I
50		19370.02	I
10		20648.69	I
220		22183.03	I
150		22226.53	I
30		22309.21	I
32		24420.82	I
12		27365.42	I
9		27573.05	I
10		30361.93	I
8		30383.88	I
10		34295.73	I
9		34513.11	I
3		40228.54	I
2		41633.80	I

Iodine (Cont.)

I III
Ref. 20, 21, 161 — L.J.R.

Intensity	Wavelength	
	Vacuum	
6	666.81	III
8	705.11	III
7	784.64	III
7	784.80	III
8	795.52	III
5	865.97	III
5	920.38	III
6	961.17	III
6	1078.58	III
8	1094.20	III
4	1244.66	III
8	1252.35	III
5	1306.93	III
	Air	
1	2224.43	III
1	2238.12	III
3	2249.31	III
2	2309.38	III
3	2340.85	III
3	2350.43	III
2	2353.46	III
4	2367.74	III
2	2371.45	III
3	2372.45	III
4	2376.47	III
4	2387.12	III
3	2392.01	III
2	2403.06	III
2	2403.63	III
2	2414.85	III
2	2418.49	III
2	2418.85	III
2	2423.91	III
5	2426.12	III
3	2434.88	III
2	2462.50	III
3	2466.69	III
3	2466.99	III
6	2475.36	III
4	2489.27	III
2	2493.21	III
2	2494.27	III
3	2495.16	III
2	2496.07	III
3	2501.41	III
2	2516.82	III
6	2519.75	III
4	2521.72	III
3	2531.99	III
2	2537.56	III
7	2545.71	III
4	2640.77	III
4	2642.11	III
6	2652.25	III
2	2818.48	III
2	2839.44	III
4	2864.67	III
4	2885.15	III
3	2910.98	III
3	2917.35	III
2	2931.11	III
2	3005.68	III
3	3069.23	III
3	3153.88	III
3	3170.14	III
3	3181.66	III
3	3210.14	III
4	3213.49	III
4	3224.93	III
2	3300.47	III
2	3479.53	III
3	3546.92	III
3	3613.81	III
[illegible]	3734.40	III
2	3754.55	III
3	3963.16	III
3	4077.14	III

I IV
Ref. 21, 58 — L.J.R.

Intensity	Wavelength	
	Vacuum	
5	601.86	IV
6	612.46	IV
4	615.17	IV
4	654.22	IV
4	654.56	IV
7	919.28	IV
	Air	
5	2249.30	IV
4	2340.84	IV
7	2361.13	IV
5	2367.75	IV
6	2372.45	IV
7	2376.46	IV
4	2385.28	IV
8	2387.11	IV
6	2392.00	IV
4	2403.05	IV
2	2418.45	IV
3	2423.89	IV
9	2426.10	IV
6	2434.85	IV
3	2466.68	IV
3	2466.96	IV
8	2475.35	IV
4	2485.51	IV
5	2489.24	IV
4	2493.20	IV
2	2501.38	IV
3	2513.74	IV
8	2519.74	IV
6	2521.72	IV
4	2531.98	IV
5	2537.54	IV
8	2545.67	IV
4	2640.77	IV
5	2642.11	IV
8	2652.23	IV
3	2818.45	IV
6	2864.68	IV
4	2910.97	IV
5	2917.33	IV
4	3069.17	IV
4	3170.11	IV
4	3181.64	IV
4	3210.12	IV
6	3213.48	IV
6	3224.90	IV
4	3546.90	IV

IV
Ref. 84 — L.J.R.

Intensity	Wavelength	
	Vacuum	
30	363.78	V
36	380.74	V
45	565.53	V
50	607.57	V

IRIDIUM (Ir)
Z = 77

Ir I and II
Ref. 1 — C.H.C.

Intensity	Wavelength	
	Air	
9900	2010.65	I
8700	2022.35	I
15000	2033.57	I
6200	2052.22	I
5000	2060.64	I
3700	2083.22	I
3100	2085.74	I
17000	2088.82	I
14000	2092.63	I
2700	2112.68	I
1800	2119.54	I
2000	2125.44	I
4500	2126.81	II
2000	2127.52	I
4500	2127.94	I
3700	2148.22	I
2500	2150.54	I
3500	2152.68	II
2900	2155.81	I
7900	2158.05	I
2100	2162.88	I
5800	2169.42	II
4500	2175.24	I
2700	2178.17	I
1600	2187.43	II
1100	2190.38	II
740	2191.64	I
910	2208.09	II
1300	2220.37	I
790	2221.07	II
2500	2242.68	II
620	2245.76	II
2100	2253.38	I
	2253.49	I
2100	2255.10	I
1400	2255.81	I
350	2258.51	I
1400	2258.86	I
830	2264.61	I
1100	2266.33	I
1000	2268.90	I
660	2280.00	I
950	2281.02	II
660	2281.91	I
330	2284.60	I
330	2295.08	I
790	2298.05	I
	2298.16	I
460	2299.53	I
910	2300.50	I
2700	2304.22	I
410	2305.47	I
210	2307.27	I
910	2308.93	I
460	2315.38	I
410	2321.45	I
410	2321.58	I
210	2327.98	I
540	2333.30	I
740	2333.84	I
580	2334.50	I
1600	2343.18	I
740	2343.61	I
100	2352.62	I
580	2355.00	I
230	2357.53	II
410	2358.16	I
500	2360.73	I
2500	2363.04	I
370	2368.04	II
3500	2372.77	I
290	2375.09	II
250	2377.28	I
250	2377.98	I
500	2379.38	I
540	2381.62	I
210	2383.17	I
120	2386.58	II
1300	2386.89	I
2500	2390.62	I
2700	2391.18	I
230	2407.59	I
290	2409.37	I
290	2410.17	I
290	2410.73	I
560	2413.31	I
370	2415.86	I
620	2418.11	I
120	2424.32	I
120	2424.66	I
210	2424.89	I
370	2424.99	I
290	2425.66	I
170	2426.53	II
540	2427.61	I
540	2431.24	I
1300	2431.94	I
170	2432.36	I
100	2432.58	I
270	2435.14	I
250	2445.34	I
250	2447.76	I
190	2448.23	I
910	2452.81	I
1300	2455.61	I
230	2455.87	I
210	2457.03	I
210	2457.23	I
120	2465.09	I
870	2467.30	I
3300	2475.12	I
210	2478.11	I
2100	2481.18	I
100	2485.38	I
620	2493.08	I
210	2496.27	I
250	2502.63	I
4100	2502.98	I
170	2504.37	I
120	2505.74	I
120	2507.63	[illegible]
170	2509.71	I
170	2511.94	I
170	2512.58	II
210	2513.71	I
120	2515.36	I
40	2524.88	II
170	2525.05	I
120	2532.52	I
990	2533.13	I
1100	2534.46	I
580	2537.22	I
170	2537.68	I
100	2541.48	I
580	2542.02	I
40	2542.80	II
7900	2543.97	I
150	2545.54	I
790	2546.03	I
120	2547.20	I
120	2547.69	I
210	2551.40	I
190	2554.40	I
210	2555.35	I
170	2555.88	I
150	2563.28	I
910	2564.18	I
210	2569.88	I
100	2570.62	I
230	2572.70	I
740	2577.26	I
100	2578.71	I
35	2579.49	II
740	2592.06	I
740	2599.04	I
150	2602.04	I
190	2604.55	I
190	2607.52	I
700	2608.25	I
1800	2611.30	I
210	2614.98	I
330	2617.78	I
210	2619.88	I
70	2623.64	II
250	2625.32	I
100	2626.76	I
700	2634.17	I
170	2635.27	I
250	2639.42	I
3500	2639.71	I
210	2644.19	I
170	2653.76	I
100	2656.81	I
1800	2661.98	I
350	2662.63	I
2700	2664.79	I
140	2668.99	I
520	2669.91	I
520	2671.84	I
330	2673.61	I
120	2676.83	I
110	2684.04	I
270	2692.34	I
3000	2694.23	I
110	2704.03	I
160	2712.74	I
140	2744.00	I
330	2772.46	I
250	2775.55	I
520	2781.29	I
330	2785.22	I
540	2797.35	I
1600	2797.70	I
380	2798.18	I
410	2800.82	I
680	2823.18	I
1200	2824.45	I
110	2833.24	II
110	2835.66	I
820	2836.40	I
160	2837.33	I
1100	2839.16	I
820	2840.22	I
160	2842.28	I
3800	2849.72	I
110	2863.84	I
380	2875.60	I
380	2875.98	I
270	2877.68	I
140	2879.41	I
820	2882.64	I
650	2897.15	I
260	2901.95	I
260	2904.80	I
200	2907.24	I
440	2916.36	I
230	2918.57	I
4400	2924.79	I
1200	2934.64	I
880	2936.68	I
250	2938.47	I
190	2939.27	I
140	2940.54	I
2700	[illegible]	I
230	2946.97	I
200	2949.76	I
1200	2951.22	I
150	2962.99	I
200	2974.95	I
440	2980.65	I
150	2985.80	I
190	2990.62	I
300	2996.08	I
180	2997.41	I
220	3002.25	I
600	3003.63	I
160	3011.69	I
120	3016.43	I
270	3017.31	I
140	3019.23	I
110	3025.82	I
380	3029.36	I
330	3039.26	I
35	3042.65	II
300	3047.16	I
300	3049.44	I
300	3057.28	I

Iridium (Cont.)

Intensity		Wavelength	
1600		3068.89	I
190		3069.09	I
190		3069.71	I
170		3076.69	I
320		3083.22	I
240		3086.44	I
390		3088.04	I
510		3100.29	I
510		3100.45	I
340		3120.76	I
200		3121.78	I
3400		3133.32	I
190		3150.61	I
190		3154.74	I
190		3159.15	I
140		3168.18	I
490		3168.88	I
370		3177.58	I
170		3180.35	I
370		3198.92	I
610		3212.12	I
370		3219.51	I
5100		3220.78	I
100		3221.28	I
300		3229.28	I
100		3230.76	I
470		3241.52	I
200		3262.01	I
390		3266.44	I
160		3277.28	I
100		3287.59	I
160		3310.52	I
200		3322.60	I
130		3334.16	I
560		3368.48	I
660		3437.02	I
100		3437.50	I
410		3448.97	I
3200		3513.64	I
220		3515.95	I
410		3522.03	I
160		3557.17	I
320		3558.99	I
1200		3573.72	I
320		3594.39	I
220		3609.77	I
190		3617.21	I
160		3626.29	I
660		3628.67	I
220		3636.20	I
300		3661.71	I
300		3664.62	I
320		3674.98	I
200		3687.08	I
140		3725.38	I
200		3731.36	II
130		3738.53	I
530		3747.20	I
120		3793.79	I
3100		3800.12	I
230		3817.24	I
170		3865.64	I
480		3902.51	I
480		3915.38	I
400		3934.84	I
120		3946.27	I
590		3976.31	I
460		3992.12	I
180		4020.03	I
350		4033.76	I
130		4040.08	I
370		4069.92	I
150		4070.68	I
100		4092.61	I
140		4115.78	I
23		4127.92	I
27		4155.70	I
15		4166.04	I
90		4172.56	I
35		4182.47	I
15	h	4183.21	I
18		4185.66	I
23		4197.54	I
27		4217.76	I
13		4220.80	I
75		4259.11	I
27		4265.30	I
260		4268.10	I
23		4286.62	I
75		4301.60	I
55		4310.59	I
220		4311.50	I
18		4351.30	I
18		4352.56	I
18		4392.59	I
160		4399.47	I
65		4403.78	I
110		4426.27	I
15		4450.18	I
55		4478.48	I
16		4495.35	I
11	h	4496.03	I
55		4545.68	I
30		4548.48	I
13		4550.78	I
35		4568.09	I
18		4570.02	I
18		4604.48	I
75		4616.39	I
26		4656.18	I
17	h	4668.99	I
21		4708.88	I
50		4728.86	I
21		4731.86	I
26		4756.46	I
13		4757.96	I
65		4778.16	I
30		4795.67	I
10		4807.14	I
21		4809.47	I
10		4840.77	I
17		4845.38	I
50		4938.09	I
26		4970.48	I
25		4999.74	I
25		5002.74	I
17		5009.17	I
30		5014.98	I
17		5046.06	I
30		5123.66	I
20		5177.95	I
22		5238.92	I
12		5340.74	I
35		5364.32	I
75		5449.50	I
30		5454.50	I
7		5469.40	I
10		5620.04	I
45		5625.55	I
10		5828.55	I
10		5882.30	I
7		5887.36	I
35		5894.06	I
7		6026.10	I
12		6067.83	I
20		6110.67	I
12		6288.28	I
7		6334.44	I
5		6624.73	I
10		6686.08	I
5		6830.01	I
5		6929.88	I
4		7183.71	I
6		7834.32	I

IRON (Fe)
Z = 26

Fe I and II
Ref. 56, 63, 105, 138, 174, 278
— H.M.C. and H.C.

Intensity	Wavelength	
	Vacuum	
12	1055.27	II
15	1068.36	II
15	1071.60	II
15	1096.89	II
12	1099.12	II
18	1112.09	II
12	1121.99	II
12	1122.86	II
12	1128.07	II
12	1130.43	II
15	1133.41	II
12	1133.68	II
12	1138.64	II
12	1142.33	II
12	1143.23	II
18	1144.95	II
12	1147.41	II
15	1148.29	II
12	1151.16	II
12	1267.44	II
12	1272.00	II
12	1371.02	II
12	1563.79	II
12	1580.62	II
18	1608.46	II
12	1618.47	II
15	1621.68	II
15	1629.15	II
15	1631.12	II
18	1635.40	II
15	1636.32	II
15	1639.40	II
12	1641.76	II
12	1647.16	II
12	1670.74	II
12	1702.04	II
12	1761.38	II
20	1785.26	II
20	1786.74	II
18	1788.07	II
30	1934.538	I
25	1937.269	I
50	1946.988	I
25	1951.571	I
30	1952.59	I
30	1953.005	I
60	1957.823	I
60	1960.144	I
30	1961.25	I
50	1962.111	I
12	1963.11	II
	Air	
100	2084.122	I
50	2157.794	I
15	2162.02	II
40	2166.773	I
300	2178.118	I
250	2186.486	I
60	2186.892	I
120	2187.195	I
250	2191.839	I
150	2196.043	I
80	2200.390	I
80	2200.724	I
15	2208.41	II
20	2213.65	II
12	2218.26	II
20	2220.38	II
25	2245.58	II
50	2250.790	I
60	2251.874	I
25	2255.77	II
300	2259.511	I
60	2264.389	I
80	2267.085	I
80	2267.469	I
50	2270.862	I
150	2272.070	I
150	2276.026	I
80	2279.937	I
150	2284.086	I
150	2287.250	I
300	2292.524	I
80	2294.41	I
200	2297.787	I
600	2298.169	I
80	2299.220	I
300	2300.142	I
50	2301.684	I
100	2303.424	I
150	2303.581	I
120	2308.999	I
150	2313.104	I
200	2320.358	I
100	2327.40	II
15	2327.88	II
100	2331.31	II
15	2331.97	II
300	2332.80	II
200	2338.01	II
600	2343.49	II
80	2343.96	II
150	2344.28	II
25	2344.98	II
50	2345.34	II
200	2348.11	II
250	2348.30	II
50	2351.20	II
15	2351.67	II
25	2352.31	II
30	2353.47	II
15	2353.68	II
50	2354.48	II
40	2354.89	II
200	2359.12	II
15	2359.59	II
150	2360.00	II
120	2360.29	II
30	2360.51	II
40	2362.02	II
60	2363.86	II
200	2364.83	II
80	2365.76	II
25	2366.59	II
80	2368.59	II
80	2369.456	I
80	2369.95	II
25	2370.50	II
120	2371.430	I
300	2373.624	I
150	2373.74	II
120	2374.518	I
60	2375.19	II
120	2376.43	II
20	2378.13	II
80	2379.27	II
20	2379.41	II
40	2380.20	II
120	2380.76	II
150	2381.835	I
1000	2382.04	II
20	2382.90	II
20	2383.06	II
60	2383.25	II
50	2384.39	II
40	2388.37	II
300	2388.63	II
200	2389.973	I
30	2390.10	II
20	2390.77	II
15	2391.48	II
20	2392.58	II
40	2395.42	II
1000	2395.62	II
15	2396.72	II
300	2399.24	II
20	2400.05	II
15	2401.29	II
50	2404.43	II
800	2404.88	II
250	2406.66	II
80	2406.97	II
300	2410.52	II
200	2411.07	II
50	2411.81	II
150	2413.31	II
20	2416.45	II
80	2417.87	II
15	2418.44	II
60	2420.396	I
60	2422.69	II
60	2423.089	I
40	2423.21	II
150	2424.14	II
15	2424.39	II
30	2424.59	II
30	2428.29	II
120	2428.36	II
25	2428.80	II
25	2429.03	II
20	2429.39	II
30	2429.86	II
120	2430.08	II
25	2431.02	II
80	2432.26	II
60	2432.87	II
25	2434.06	II
20	2434.24	II
20	2434.65	II
50	2434.73	II
50	2434.95	II
25	2436.62	II
60	2438.182	I
150	2439.30	II
150	2439.74	I
80	2440.11	I
40	2440.42	II
30	2442.37	II
100	2442.57	I
60	2443.71	II
250	2443.872	I
100	2444.51	II
50	2445.11	II
50	2445.212	I
100	2445.57	II
40	2445.80	II
50	2446.11	II
30	2446.47	II
40	2447.20	II
25	2447.33	II
60	2447.709	I
30	2447.75	II
25	2449.96	II
25	2450.20	II
100	2453.476	I
20	2453.98	II
30	2454.58	II
15	2455.71	II
15	2455.90	II
15	2457.09	II
1500	2457.598	I
150	2458.78	II
40	2458.97	II
60	2460.44	II
80	2461.28	II
100	2461.86	II
100	2462.181	I
1500	2462.647	I
50	2463.29	II
50	2463.730	I
40	2464.01	II
40	2464.90	II
800	2465.149	I
50	2465.91	II
15	2466.50	II
60	2466.67	II
60	2466.82	II
60	2467.732	I
15	2468.29	II
600	2468.879	I
60	2469.51	II
25	2470.41	II
80	2470.67	II

Iron (Cont.)

80	2470.965	I
800	2472.336	I
40	2472.43	II
40	2472.60	II
1000	2472.895	I
200	2473.16	I
50	2473.32	II
30	2474.05	II
600	2474.814	I
50	2475.12	II
40	2475.54	II
15	2476.26	II
60	2476.657	I
25	2477.34	II
60	2478.57	II
120	2479.480	I
1200	2479.776	I
100	2480.16	II
15	2481.05	II
80	2482.12	II
25	2482.32	II
100	2482.66	II
10000	2483.271	I
300	2483.533	I
1000	2484.185	I
60	2484.24	II
30	2484.44	II
50	2485.990	I
800	2486.373	I
100	2486.691	I
100	2487.066	I
120	2487.370	I
4000	2488.143	I
100	2488.945	I
80	2489.48	II
1000	2489.750	I
50	2489.83	II
50	2489.913	I
3000	2490.644	I
100	2490.71	II
60	2490.86	II
2000	2491.155	I
100	2491.40	II
25	2492.34	II
100	2493.18	II
500	2493.26	II
60	2494.000	I
50	2494.251	I
100	2495.87	I
600	2496.533	I
50	2497.82	II
150	2498.90	I
40	2500.92	II
1000	2501.132	I
40	2501.31	II
50	2501.693	I
60	2502.39	II
40	2503.33	II
60	2503.87	II
80	2506.09	II
40	2506.80	II
500	2507.900	I
30	2508.34	II
50	2508.753	I
1000	2510.835	I
120	2511.76	II
80	2512.275	I
400	2512.365	I
50	2514.38	II
80	2516.570	I
50	2517.13	II
300	2517.661	I
800	2518.102	I
60	2519.05	II
150	2519.629	I
40	2521.09	II
30	2521.82	II
50	2522.480	I
4000	2522.849	I
200	2523.66	I
500	2524.293	I
100	[illegible]	[illegible]
200	2525.39	II
25	2526.07	II
300	2526.29	II
2000	2527.435	I
30	2527.70	II
800	2529.135	I
25	2529.23	II
80	2529.31	I
250	2529.55	II
150	2529.836	I
40	2530.11	II
200	2530.687	I
120	2533.63	II
60	2533.80	I
100	2534.42	II
120	2535.49	II
400	2535.607	I
60	2536.67	II
200	2536.792	I
200	2536.80	II
50	2536.84	II
50	2537.14	II
50	2538.20	II
40	2538.50	II
100	2538.80	II
100	2538.91	II
150	2538.99	II
50	2539.357	I
200	2540.66	II
600	2540.972	I
80	2541.10	II
60	2541.84	II
300	2542.10	I
25	2542.78	II
60	2543.38	II
250	2543.92	I
150	2544.70	I
40	2544.97	II
40	2545.22	II
800	2545.978	I
40	2546.44	II
80	2546.67	II
80	2546.87	I
100	2548.74	II
80	2549.08	II
80	2549.39	II
60	2549.46	II
600	2549.613	I
40	2549.77	II
60	2550.03	II
25	2550.15	II
50	2550.68	II
40	2560.28	II
25	2562.09	II
400	2562.53	II
200	2563.48	II
60	2566.91	II
25	2570.52	II
30	2570.85	II
150	2574.36	II
50	2575.74	I
300	2576.691	I
25	2576.86	II
60	2577.92	II
50	2582.30	I
100	2582.58	II
1500	[illegible]	[illegible]
650	2585.88	II
90	2588.00	I
90	2591.54	II
30	2592.78	II
60	2593.51	I
90	2593.73	II
650	2598.37	II
2000	2599.40	II
300	2599.57	I
20	2605.34	II
20	2605.42	II
60	2605.657	I
300	2606.51	II
800	2606.827	I
650	2607.09	II
20	2611.07	II
600	2611.87	II
320	2613.82	II
320	2617.62	II
250	2618.018	I
20	2619.07	II
90	2620.41	II
20	2620.69	II
40	2621.67	II
400	2623.53	I
50	2625.49	II
200	2625.67	II
150	2628.29	II
20	2630.07	II
250	2631.05	II
250	2631.32	II
50	2631.61	II
100	2632.237	I
300	2635.809	I
50	[illegible]	I
200	2643.998	I
60	2664.66	II
30	2666.64	II
300	2666.812	I
60	2666.965	I
600	2679.062	I
500	2684.75	II
400	2689.212	I
60	2692.60	II
50	2696.28	I
200	2699.106	I
60	2703.99	II
80	2706.012	I
400	2706.582	I
60	2708.571	I
20	2709.05	II
200	2711.655	I
80	2714.41	II
50	2716.22	II
50	2716.257	I
50	2717.786	I
50	2717.87	II
250	2718.436	I
4000	2719.027	I
100	2719.420	I
50	2720.197	I
1500	2720.903	I
400	2723.578	I
30	2724.88	II
150	2724.953	I
80	2726.05	I
50	2726.235	I
25	2727.38	II
80	2727.54	II
200	2728.020	I
50	2728.820	I
80	2728.90	II
40	2730.73	II
1000	2733.581	I
60	2734.005	I
50	2734.268	I
500	2735.475	I
50	2735.612	I
500	2737.310	I
120	2737.83	I
400	2739.55	II
250	2742.254	I
800	2742.405	I
200	2743.20	II
150	2743.565	I
200	2744.068	I
80	2744.527	I
300	2746.48	II
100	2749.32	II
500	2749.48	II
1200	2750.140	I
20	2751.13	II
20	2752.15	II
80	2753.29	II
50	2753.69	I
150	2754.032	I
100	2754.426	I
30	2754.89	II
800	2755.73	II
250	2756.328	I
100	2757.316	I
50	2759.81	I
120	2761.780	I
150	2761.81	II
150	2762.026	I
120	2762.772	I
120	2763.109	I
20	2763.66	II
25	2765.13	II
80	2766.910	I
250	2767.522	I
50	2769.30	I
25	2769.35	II
300	2772.07	I
50	2773.23	I
20	2774.69	II
15	2776.91	II
60	2778.07	I
600	2778.220	I
40	2779.30	II
50	2783.69	II
30	2785.19	II
3000	2788.10	I
20	2793.89	II
200	2797.78	I
30	2799.29	II
400	2804.521	I
1500	2806.98	I
2500	2813.287	I
300	2823.276	I
600	2825.56	I
50	2825.687	I
120	2828.808	I
25	2831.56	II
1500	2832.436	I
120	2835.950	I
200	2838.119	I
30	2839.51	II
[illegible]	[illegible]	II
15	2840.65	II
200	2843.631	I
1000	2843.977	I
100	2845.594	I
15	2848.11	II
15	2848.32	II
800	2851.797	I
30	2856.91	II
25	2858.34	II
50	2869.307	I
50	2872.334	I
80	2874.172	I
50	2894.504	I
120	2912.157	I
120	2929.007	I
1200	2936.903	I
60	2941.343	I
12	2944.40	II
1000	2947.876	I
60	2950.24	I
600	2953.940	I
250	2957.364	I
80	2959.99	I
150	2965.254	I
1500	2966.898	I
120	2969.36	I
800	2970.099	I
15	2970.52	II
1200	2973.132	I
500	2973.235	I
600	2981.445	I
1000	2983.570	I
60	2984.77	I
50	2984.82	II
13	2985.54	II
1000	2994.427	I
250	2994.502	I
500	2999.512	I
120	3000.451	I
800	3000.948	I
60	3001.655	I
15	3002.64	II
200	3007.282	I
500	3008.14	I
120	3009.569	I
60	3017.627	I
60	3018.983	I
60	3020.01	II
500	3020.491	I
1500	3020.639	I
600	3021.073	I
500	3024.032	I
150	3025.638	I
500	3025.842	I
80	3030.148	I
60	3031.214	I
60	3034.484	I
40	3036.96	II
800	3037.389	I
80	3041.637	I
800	3047.604	I
600	3057.446	I
1000	3059.086	I
250	3067.244	I
120	3075.719	I
120	3091.577	I
80	3098.180	I
100	3099.895	I
100	3099.968	I
60	3100.303	I
100	3100.665	I
12	3154.20	II
80	3175.445	I
150	3184.895	I
250	3191.659	I
500	3193.226	I
800	3193.299	I
12	3196.08	II
200	3196.928	I
80	3199.500	I
60	3200.47	I
50	3205.398	I
50	3211.67	I
100	3211.88	I
13	3213.31	II
200	3214.011	I
200	3214.396	I
60	3215.938	I
50	3217.377	I
80	3219.583	I
60	3219.766	I
300	3222.045	I
600	3225.78	I
13	3227.73	II
80	3227.796	I
20	3230.42	II
80	3233.05	I
50	3233.967	I
120	3234.613	I
300	3236.222	I
100	3239.433	I
80	3244.187	I
80	3246.005	I
60	3254.36	I
80	3265.046	I
50	3265.617	I
50	3271.000	I
50	3280.26	I
150	3286.75	I
120	3305.97	I
200	3306.343	I
400	3355.227	I
80	3355.517	I
60	3369.546	I
120	3370.783	I
50	3378.678	I
50	3380.110	I
60	3383.978	I
12	3388.13	II
50	3392.304	I
150	3392.651	I
150	3399.333	I
80	3404.353	I
500	3407.458	I

Iron (Cont.)

250	3413.131	I
60	3424.284	I
500	3427.119	I
60	3428.748	I
6000	3440.606	I
2500	3440.989	I
1000	3443.876	I
200	3445.149	I
15	3453.61	II
1200	3465.860	I
2000	3475.450	I
500	3476.702	I
2500	3490.574	I
500	3497.840	I
250	3513.817	I
300	3521.261	I
400	3526.040	I
100	3526.166	I
60	3526.237	I
60	3526.381	I
60	3526.467	I
100	3533.199	I
200	3536.556	I
300	3541.083	I
250	3542.075	I
80	3553.739	I
400	3554.925	I
200	3556.878	I
400	3558.515	I
1000	3565.379	I
1200	3570.097	I
800	3570.25	I
120	3571.996	I
100	3573.393	I
60	3573.829	I
60	3573.888	I
4000	3581.19	I
150	3582.199	I
150	3584.660	I
120	3584.929	I
300	3585.319	I
150	3585.705	I
200	3586.103	I
400	3586.984	I
100	3594.633	I
150	3603.204	I
200	3605.454	I
500	3606.680	I
1500	3608.859	I
250	3610.16	I
60	3612.068	I
150	3617.788	I
1500	3618.768	I
200	3621.462	I
150	3622.004	I
150	3623.19	I
100	3631.096	I
1200	3631.463	I
60	3632.041	I
100	3638.298	I
200	3640.389	I
80	3643.717	I
1500	3647.842	I
250	3649.506	I
80	3650.279	I
200	3651.467	I
120	3670.024	I
150	3670.089	I
100	3676.311	I
150	3677.629	I
1500	3679.913	I
200	3682.242	I
120	3683.054	I
150	3684.107	I
120	3685.998	I
500	3687.456	I
120	3689.477	I
150	3694.008	I
120	3695.051	I
150	3701.086	I
80	3704.462	I
1200	3705.566	I
60	3707.041	I
150	3707.821	I
300	3707.919	I
600	3709.246	I
120	3716.442	I
8000	3719.935	I
1500	3722.563	I
120	3724.377	I
60	3725.491	I
60	3727.093	I
500	3727.619	I
150	3732.396	I
1200	3733.317	I
5000	3734.864	I
120	3735.324	I
6000	3737.131	I
100	3738.306	I
400	3743.362	I
80	3743.47	I
6000	3745.561	I
1200	3745.899	I
3000	3748.262	I
80	3748.964	I
3000	3749.485	I
1500	3758.232	I
400	3760.05	I
1500	3763.788	I
400	3765.54	I
600	3767.191	I
60	3776.452	I
250	3785.95	I
100	3786.68	I
250	3787.880	I
250	3790.092	I
150	3794.34	I
400	3795.002	I
120	3797.518	I
250	3798.511	I
400	3799.547	I
200	3805.345	I
80	3806.696	I
600	3812.964	I
60	3813.059	I
1500	3815.840	I
2500	3820.425	I
150	3821.179	I
80	3824.306	I
2500	3824.444	I
1500	3825.880	I
1200	3827.823	I
1000	3834.222	I
120	3839.257	I
500	3840.437	I
800	3841.047	I
120	3843.256	I
80	3846.800	I
200	3849.96	I
120	3850.817	I
2500	3856.372	I
150	3859.212	I
10000	3859.911	I
150	3865.523	I
60	3867.215	I
250	3872.501	I
150	3873.761	I
250	3878.018	I
2000	3878.573	I
4000	3886.282	I
200	3887.048	I
300	3888.513	I
800	3895.656	I
1200	3899.707	I
400	3902.945	I
250	3906.479	I
80	3916.731	I
600	3920.258	I
1200	3922.911	I
1200	3927.920	I
2000	3930.296	I
60	3948.774	I
60	3949.953	I
50	3951.164	I
50	3952.601	I
60	3956.454	I
250	3956.68	I
60	3966.614	I
100	3969.257	I
80	3977.741	I
40	3981.771	I
50	3983.956	I
60	3994.114	I
200	3997.392	I
40	3998.053	I
400	4005.241	I
60	4009.713	I
80	4014.53	I
100	4021.867	I
50	4040.638	I
4000	4045.813	I
1500	4063.594	I
50	4066.975	I
50	4067.977	I
1200	4071.737	I
40	4076.629	I
40	4100.737	I
40	4107.489	I
150	4118.544	I
40	4127.608	I
400	4132.058	I
80	4134.676	I
40	4136.997	I
200	4143.415	I
800	4143.869	I
40	4153.898	I
50	4154.500	I
60	4156.799	I
50	4172.744	I
60	4174.912	I
50	4175.635	I
50	4177.593	I
120	4181.754	I
50	4184.891	I
120	4187.038	I
120	4187.795	I
80	4191.430	I
40	4195.329	I
150	4198.304	I
40	4199.095	I
300	4202.029	I
40	4203.984	I
80	4206.696	I
80	4210.343	I
400	4216.183	I
100	4219.360	I
50	4222.212	I
50	4225.956	I
200	4227.423	I
11	4233.17	II
100	4233.602	I
250	4235.936	I
50	4238.809	I
50	4247.425	I
200	4250.118	I
300	4250.787	I
40	4258.315	I
800	4260.473	I
250	4271.153	I
1200	4271.759	I
1200	4282.402	I
80	4291.462	I
250	4299.234	I
1200	4307.901	I
150	4315.084	I
1500	4325.761	I
80	4352.734	I
80	4369.771	I
800	4375.929	I
3000	4383.544	I
1200	4404.750	I
300	4415.122	I
600	4427.299	I
400	4461.652	I
120	4466.551	I
80	4476.017	I
80	4482.169	I
200	4482.252	I
50	4489.739	I
50	4528.613	I
11	4583.83	II
30	4647.433	I
30	4736.771	I
50	4859.741	I
120	4871.317	I
60	4872.136	I
30	4878.208	I
100	4890.754	I
250	4891.492	I
30	4903.309	I
150	4918.992	I
500	4920.502	I
12	4923.92	II
1500	4957.597	I
11	4990.50	II
80	5001.862	I
18	5001.91	II
11	5004.20	II
30	5005.711	I
100	5006.117	I
60	5012.067	I
30	5014.941	I
12	5018.43	II
11	5030.64	II
25	5030.77	I
12	5035.71	II
150	5041.755	I
30	5049.819	I
30	5051.634	I
25	5074.748	I
18	5100.73	II
15	5100.95	II
150	5110.357	I
40	5133.69	I
40	5139.251	I
100	5139.462	I
11	5144.36	II
12	5149.46	II
25	5151.910	I
30	5162.27	I
80	5166.281	I
2500	5167.487	I
80	5168.897	I
12	5169.03	II
500	5171.595	I
50	5191.454	I
80	5192.343	I
200	5194.941	I
30	5204.582	I
25	5215.179	I
150	5216.274	I
18	5216.85	II
60	5226.862	I
1000	5227.150	I
13	5227.49	II
250	5232.939	I
13	5247.95	II
13	5251.23	II
18	5260.26	II
11	5264.18	II
100	5266.555	I
1200	5269.537	I
800	5270.357	I
30	5281.789	I
60	5283.621	I
25	5302.299	I
11	5306.18	II
13	5316.23	II
150	5324.178	I
800	5328.038	I
300	5328.531	I
100	5332.899	I
14	5339.59	II
80	5339.928	I
500	5341.023	I
25	5364.87	I
40	5367.47	I
50	5369.96	I
400	5371.489	I
60	5383.37	I
14	5387.06	II
40	5393.167	I
12	5395.86	II
300	5397.127	I
15	5402.06	II
60	5404.12	I
250	5405.774	I
30	5410.91	I
60	5415.20	I
60	5424.07	I
30	5427.83	II
250	5429.695	I
13	5429.99	II
100	5434.523	I
200	5446.871	I
25	5455.45	I
120	5455.609	I
16	5465.93	II
20	5466.94	II
16	5482.31	II
14	5493.83	II
25	5497.516	I
20	5501.464	I
18	5506.20	II
30	5506.778	I
12	5510.78	II
12	5529.06	II
13	5544.76	II
30	5569.618	I
60	5572.841	I
120	5586.755	I
200	5615.644	I
20	5624.541	I
12	5645.40	II
50	5662.515	I
20	5762.990	I
11	5783.63	II
30	5862.353	I
13	5885.02	II
16	5902.82	II
30	5914.114	I
14	5955.70	II
30	5986.956	I
18	5961.71	II
30	5962.4	II
13	5965.63	II
40	6065.482	I
30	6102.159	I
40	6136.614	I
40	6137.694	I
30	6147.73	II
20	6149.24	II
15	6175.16	II
40	6191.558	I
30	6213.429	I
30	6219.279	I
40	6230.726	I
20	6238.37	II
20	6246.317	I
80	6247.56	II
30	6252.554	I
15	6305.32	II
12	6331.97	II
15	6383.75	II
20	6393.602	I
30	6399.999	I
20	6411.647	I
20	6416.90	II
20	6421.349	I
30	6430.844	I
20	6446.43	II
200	6456.38	II
60	6494.981	I
20	6516.05	II
20	6546.239	I
20	6592.913	I
40	6677.989	I
15	6855.18	I
15	6945.21	I
20	7067.44	II
15	7130.94	I
25	7164.443	I
80	7187.313	I

Iron (Cont.)

Intensity	Wavelength	Spectrum
30	7207.381	I
12	7224.51	II
50	7307.97	II
40	7320.70	II
20	7376.46	II
30	7445.746	I
20	7462.38	II
40	7495.059	I
60	7511.045	I
15	7586.04	I
15	7711.71	II
30	7780.59	I
40	7832.22	I
80	7937.131	I
60	7945.984	I
80	7998.939	I
60	8046.047	I
50	8085.176	I
150	8220.41	I
120	8327.053	I
20	8331.908	I
120	8387.770	I
30	8468.404	I
15	8514.069	I
60	8661.898	I
150	8688.621	I
12	8793.38	I
12	8824.23	I
20	8866.96	I
15	8999.56	I
15	10216.32	I
13	10469.65	I
21	11119.80	I
14	11374.08	I
52	11422.32	I
87	11439.12	I
91	11593.59	I
255	11607.57	I
160	11638.26	I
230	11689.98	I
160	11783.26	I
580	11882.84	I
225	11884.08	I
1030	11973.05	I
15	12638.71	I
14	12879.76	I
17	13565.04	I
30	14236.25	I
14	14285.11	I
14	14292.38	I
16	14308.69	I
96	14400.56	I
20	14442.28	I
72	14512.23	I
50	14555.06	I
14	14565.95	I
40	14826.43	I
37	15051.77	I
28	15207.55	I
94	15294.58	I
16	15335.40	I
30	15621.67	I
25	15631.97	I
14	15723.59	I
41	15769.42	I
28	15813.13	I
13	16444.82	I
20	16486.69	I
105	18856.65	I
47	18987.01	I
25	19113.68	I
22	19791.88	I
14	22380.82	I
21	22619.85	I
38	26222.04	I
17	26659.22	I

Fe III
Ref. 71, 101 — J.R.

Vacuum

Intensity		Wavelength	Spectrum
6		728.81	III
5		730.00	III
5		737.71	III
5		739.26	III
9		807.55	III
8		807.86	III
8		808.84	III
8	p	811.28	III
10		813.38	III
8		838.05	III
10		844.28	III
9		845.41	III
8	w	847.42	III
8		859.72	III
8	p	861.76	III
10	p	861.83	III
8		873.46	III
9		890.76	III
10		891.17	III
8		891.44	III
8		899.42	III
10		950.33	III
10		981.37	III
10	w	983.88	III
8		985.82	III
9		991.23	III
9		1017.25	III
8		1017.74	III
8		1018.29	III
8		1032.12	III
8		1063.87	III
9		1122.53	III
9		1124.88	III
8		1128.02	III
10	h	1505.17	III
10	h	1538.63	III
12	h	1550.20	III
10	h	1601.21	III
10		1869.83	III
12		1877.99	III
10		1882.05	III
12		1886.76	III
13		1890.67	III
11		1893.98	III
20		1895.46	III
10	s	1907.58	III
19		1914.06	III
15		1915.08	III
15		1922.79	III
10	p	1926.01	III
10		1926.30	III
15		1930.39	III
14		1931.51	III
14		1937.34	III
10	l	1938.90	III
14	s	1943.48	III
12		1945.34	III
10		1950.33	III
12		1951.01	III
11		1952.65	III
13		1953.32	III
10		1953.49	III
10	w	1954.22	III
11		1958.58	III
13		1960.32	III
15		1987.50	III
14		1991.61	III
13		1994.07	III
12		1995.56	III
12		1996.42	III

Air

Intensity		Wavelength	Spectrum
10		2061.55	III
12		2068.24	III
14		2078.99	III
10		2084.35	III
12		2090.14	III
15		2097.48	III
12		2097.69	III
12		2103.80	III
10		2107.32	III
15		2151.78	III
12		2157.71	III
12		2158.47	III
10		2161.27	III
12		2166.95	III
12		2171.04	III
15		2174.66	III
12		2180.41	III
10	p	2208.85	III
10		2221.83	III
10		2229.27	III
10		2232.43	III
10		2232.69	III
10		2235.91	III
10		2238.16	III
12	p	2241.54	III
12		2261.59	III
10		2267.42	III
10		2293.06	III
15		2295.86	III
10	p	2317.70	III
10		2319.22	III
10	p	2321.71	III
10		2326.95	III
10	p	2336.77	III
10		2338.96	III
8		2389.53	III
8		2438.17	III
8	p	2582.37	III
8		2595.62	III
8		2617.15	III
9	p	2645.39	III
10	h	2695.13	III
9	h	2695.34	III
8	h	2700.02	III
8	h	2701.13	III
8		2773.31	III
10	p	2813.24	III
8	p	2895.08	III
9	p	2902.47	III
12		2904.43	III
8	p	2905.80	III
10		2907.50	III
12		2907.70	III
8		2923.90	III
8		2948.39	III
8		2963.23	III
12		3001.62	III
12	h	3007.28	III
15		3013.17	III
10	p	3136.43	III
10		3174.09	III
10		3175.99	III
10		3178.01	III
13		3266.88	III
11		3276.08	III
10		3288.81	III
9		3305.22	III
9		3339.39	III
9		3499.59	III
9		3500.28	III
10		3501.76	III
10		3586.04	III
11		3600.94	III
11		3603.88	III
16		3954.33	III
11		3968.72	III
9		3969.49	III
10	w	3979.42	III
10		4035.42	III
11		4053.11	III
12		4081.00	III
10		4120.90	III
11		4122.02	III
11		4122.78	III
15		4137.76	III
13		4139.35	III
9		4140.48	III
9		4154.96	III
18		4164.73	III
9		4164.92	III
13		4166.84	III
13		4174.26	III
9		4210.67	III
11		4222.27	III
13		4235.56	III
9		4238.62	III
12		4243.75	III
12	h	4273.40	III
12		4279.72	III
14	h	4286.16	III
16	h	4296.85	III
18	h	4304.78	III
20	h	4310.36	III
9		4323.68	III
9	h	4372.04	III
9	h	4372.14	III
11	h	4372.31	III
14	h	4372.53	III
18	h	4372.81	III
9		4395.76	III
12		4419.60	III
9		4431.02	III
9		5111.07	III
9		5127.35	III
12		5156.12	III
10		5199.08	III
10		5235.66	III
18		5243.31	III
13	l	5260.34	III
9		5272.37	III
14		5272.98	III
15		5276.48	III
16		5282.30	III
12		5284.83	III
11		5298.12	III
12		5299.93	III
14	w	5302.60	III
10		5306.76	III
9		5310.88	III
10		5322.74	III
11		5346.88	III
12		5353.77	III
12		5363.76	III
10		5368.06	III
11	l	5375.47	III
11		5719.88	III
9		5744.19	III
10		5756.38	III
18		5833.93	III
9		5848.76	III
10		5854.62	III
9		5876.26	III
15		5891.91	III
9		5898.68	III
9		5918.96	III
10	p	5920.13	III
13	p	5929.69	III
10		5952.31	III
14		5953.62	III
9		5968.48	III
12		5979.32	III
9	h	5981.01	III
12	h	5989.08	III
18		5999.54	III
9		6031.02	III
16		6032.59	III
13		6036.56	III
11		6048.72	III
11		6054.18	III
9		6056.36	III
9		6149.99	III
9		6169.74	III
9		6185.26	III
7		6186.56	III
7		6194.79	III
6		6195.43	III
6		6201.37	III
5	s	6203.04	III
5		6259.81	III
6	p	6294.50	III
5		6357.81	III
5	h	7317.63	III
6	h	7320.14	III
5	w	7921.17	III
5	w	8230.88	III
5	w	8231.79	III
9	w	8235.45	III
8	w	8236.75	III
6	w	8238.98	III
5		8563.49	III

Fe IV
Ref. 382 — J.R.

Vacuum

Intensity	Wavelength	Spectrum
10	502.42	IV
11	506.69	IV
11	505.35	IV
17	525.69	IV
15	526.29	IV
10	526.57	IV
13	526.63	IV
10	530.91	IV
10	531.78	IV
10	535.55	IV
14	536.61	IV
10	536.74	IV
15	537.10	IV
13	537.26	IV
14	537.79	IV
13	537.94	IV
10	538.44	IV
10	544.20	IV
10	546.22	IV
10	548.80	IV
11	550.32	IV
10	551.77	IV
13	552.14	IV
11	552.74	IV
10	554.26	IV
10	555.66	IV
10	572.88	IV
10	576.76	IV
10	579.76	IV
14	607.53	IV
13	608.80	IV
10	609.65	IV
12	1425.73	IV
13	1431.43	IV
12	1473.20	IV
12	1489.53	IV
12	1495.18	IV
13	1526.60	IV
13	1530.26	IV
14	1532.63	IV
13	1532.91	IV
15	1533.86	IV
13	1533.95	IV
14	1536.58	IV
12	1538.29	IV
13	1542.16	IV
14	1542.70	IV
12	1546.40	IV
12	1552.35	IV
12	1552.71	IV
12	1562.46	IV
13	1566.26	IV
14	1568.27	IV
12	1570.18	IV
12	1570.42	IV
12	1571.24	IV
12	1577.20	IV
12	1577.76	IV
12	1590.62	IV
13	1591.51	IV
13	1592.05	IV
12	1596.67	IV
13	1598.01	IV
12	1600.50	IV
13	1600.58	IV
13	1601.67	IV

Intensity		Wavelength	
12		1602.08	IV
13		1603.18	IV
13		1603.73	IV
13		1604.88	IV
13		1605.68	IV
15		1605.97	IV
13		1606.98	IV
17		1609.10	IV
14		1609.83	IV
13		1610.47	IV
13		1611.20	IV
13		1613.64	IV
15		1614.02	IV
13		1614.64	IV
13		1615.00	IV
12		1615.61	IV
16		1616.68	IV
14		1617.68	IV
14		1619.02	IV
12		1620.91	IV
13		1621.16	IV
14		1621.57	IV
13		1623.38	IV
13		1623.53	IV
15		1626.47	IV
14		1626.90	IV
13		1628.54	IV
13		1630.18	IV
17		1631.08	IV
12		1632.08	IV
14		1632.40	IV
13		1634.01	IV
12		1638.07	IV
12		1638.30	IV
14		1639.40	IV
16		1640.04	IV
14		1640.16	IV
15		1641.87	IV
12		1642.88	IV
15		1647.09	IV
15		1651.58	IV
15		1652.90	IV
13		1653.41	IV
13		1656.11	IV
15		1656.65	IV
12		1657.82	IV
12		1658.43	IV
14		1660.10	IV
12		1661.57	IV
13		1662.32	IV
13		1662.52	IV
13		1663.54	IV
13		1668.09	IV
12		1669.61	IV
14		1671.04	IV
12		1672.86	IV
13		1673.68	IV
14		1675.66	IV
12		1676.78	IV
12		1677.12	IV
13		1681.36	IV
12		1681.95	IV
15		1687.69	IV
15		1698.88	IV
12		1700.40	IV
12		1704.93	IV
13		1709.81	IV
15		1711.41	IV
14		1712.76	IV
12		1717.11	IV
14		1717.90	IV
14		1718.16	IV
12		1718.42	IV
14		1719.46	IV
14		1722.71	IV
14		1724.06	IV
12		1724.26	IV
16		1725.63	IV
13		1761.08	IV
12		1764.92	IV
12		1767.36	IV
13		1792.10	IV
13		1796.93	IV
12		1805.32	IV
12		1820.42	IV
13		1827.98	IV
12		1840.24	IV
12		1860.42	IV
12		1869.64	IV
12		1874.23	IV

Fe V
Ref. 381 — J.R.

Intensity		Wavelength	
		Vacuum	
300		361.28	V
300		365.43	V
300		365.86	V
300		374.24	V
300		374.87	V
300		375.98	V
300		379.59	V
300		380.31	V
300		381.27	V
300		384.96	V
300		384.97	V
300		385.03	V
300		385.11	V
300		385.25	V
300		385.26	V
300		385.30	V
300		385.75	V
300		385.88	V
350		386.16	V
300		386.74	V
300		386.78	V
300		386.85	V
300		386.88	V
350		386.88	V
400		387.20	V
400		387.50	V
300		387.62	V
400		387.76	V
400		387.78	V
300		387.98	V
300		388.61	V
300		388.82	V
300		390.11	V
300		390.19	V
300		390.78	V
300		391.94	V
300		392.06	V
300		392.38	V
300		392.50	V
300		392.51	V
300		392.70	V
300		392.91	V
300		393.27	V
300		393.72	V
300		393.73	V
300		393.91	V
300		393.97	V
300		394.04	V
300		394.64	V
300		395.15	V
300		395.79	V
400		395.90	V
300		399.84	V
300		400.11	V
300		400.51	V
300		400.52	V
300		400.63	V
300		401.04	V
300		401.64	V
300		401.86	V
300		402.87	V
300		403.06	V
400		404.62	V
400		405.50	V
800		407.42	V
600		407.44	V
400		407.49	V
500		407.75	V
400		409.71	V
400		410.20	V
600		411.55	V
300		415.01	V
300		416.66	V
300		416.84	V
700		417.39	V
700		418.04	V
500		418.47	V
300		420.56	V
700		421.06	V
500		421.78	V
300		422.28	V
500		422.31	V
300		423.23	V
500		426.06	V
500		426.11	V
300		426.83	V
350		426.97	V
300		434.42	V
300		439.22	V
300		444.70	V
300		445.44	V
300		446.04	V
300		458.16	V
300		486.17	V
400		1317.86	V
300		1318.35	V
300		1320.41	V
300		1321.34	V
300		1321.49	V
400		1323.27	V
400		1330.40	V
300		1345.61	V
400		1359.01	V
300		1361.28	V
300		1361.45	V
600		1361.82	V
300		1363.08	V
300		1363.64	V
300		1365.57	V
700		1373.59	V
600		1373.67	V
300		1374.12	V
500		1376.34	V
300		1376.46	V
500		1378.56	V
300		1385.68	V
800		1387.94	V
400		1397.97	V
600		1400.24	V
800		1402.39	V
400		1406.67	V
500		1406.82	V
400		1407.25	V
300		1409.03	V
300		1409.22	V
600		1409.45	V
400		1415.20	V
300		1418.12	V
600		1420.46	V
800		1430.57	V
800		1440.53	V
300		1440.79	V
400		1442.22	V
800		1446.62	V
700		1448.85	V
400		1449.93	V
300		1455.56	V
700		1456.16	V
500		1459.83	V
400		1460.73	V
500		1462.63	V
700		1464.68	V
500		1465.38	V
400		1466.65	V
500		1469.00	V
300		1475.60	V
500		1479.47	V
300		1554.22	V

KRYTPON (Kr)
Z = 36

Kr I and II
Ref. 61, 121, 123, 147, 208, 232
— E.F.W.

Intensity		Wavelength	
		Vacuum	
60		729.40	II
200		761.18	II
100		763.98	II
60		766.20	II
200		771.03	II
60	p	773.69	II
200		782.10	II
100		783.72	II
60		818.15	II
60		830.38	II
100		844.06	II
60		864.82	II
60		868.87	II
200		884.14	II
1000		886.30	II
400		891.01	II
200		911.39	II
2000		917.43	II
50		945.44	I
50		946.54	I
20		951.06	I
50		953.40	I
50		963.37	I
2000		964.97	II
100		1001.06	I
100		1003.55	I
100		1030.02	I
200		1164.87	I
650		1235.84	I
		Air	
100	h	2464.77	II
60		2492.48	II
80	h	2712.40	II
100		2833.00	II
100	h	3607.88	II
200		3631.889	II
250		3653.928	II
80		3665.324	I
150		3669.01	II
100		3679.559	I
80		3686.182	II
300	h	3718.02	II
200		3718.595	II
150		3721.350	II
200		3741.638	II
150		3744.80	II
80		3754.245	II
500		3778.089	II
500		3783.095	II
150	h	3875.44	II
150		3906.177	II
200		3920.081	II
100		3994.840	II
100	h	3997.793	II
300		4057.037	II
300		4065.128	II
500		4088.337	II
250		4098.729	II
100		4109.248	II
250		4145.122	II
150		4250.580	II
1000		4273.969	I
100		4282.967	I
600		4292.923	II
200		4300.49	II
500	h	4317.81	II
400		4318.551	I
1000		4319.579	I
150	h	4322.98	II
100		4351.359	I
3000		4355.477	II
500		4362.641	I
200		4369.69	II
800		4376.121	I
300	h	4386.54	II
200		4399.965	I
100		4425.189	I
500		4431.685	II
600		4436.812	II
600		4453.917	I
800		4463.689	I
800		4475.014	II
400	h	4489.88	II
600		4502.353	I
400	h	4523.14	II
200	h	4556.61	II
800		4577.209	II
300		4582.978	II
150	h	4592.80	II
500		4615.292	II
1000		4619.166	II
800		4633.885	II
2000		4658.876	II
500		4680.406	II
100		4691.301	II
200		4694.360	II
3000		4739.002	II
300		4762.435	II
1000		4765.744	II
300		4811.76	II
300		4825.18	II
800		4832.077	II
700		4846.612	II
150		4857.20	II
300		4945.59	II
200		5022.40	II
250		5086.52	II
400	h	5125.73	II
500		5208.32	II
200		5308.66	II
500		5333.41	II
200		5468.17	II
500		5562.224	I
2000		5570.288	I
80		5580.386	I
100		5649.561	I
400		5681.89	II
200	h	5690.35	II
100		5832.855	I
3000		5870.914	I
200		5992.22	II
60		5993.849	I
60		6056.125	I
300		6420.18	II
100		6421.026	I
200		6456.288	I
150		6570.07	II
60		6699.228	I
100		6904.678	I
250		7213.13	II
100		7224.104	I
80		7287.258	I
400		7289.78	II
400		7407.02	II
60		7425.541	I
200		7435.78	II
100		7486.862	I
300		7524.46	II
1000		7587.411	I
2000		7601.544	I
150		7641.16	II
1000		7685.244	I
1200		7694.538	I
250		7735.69	II
150		7746.827	I
800		7854.821	I
200		7913.423	I
180		7928.597	I
200		7933.22	II
120		7973.62	II
100		7982.401	I
1500		8059.503	I
4000		8104.364	I

Krypton (Cont.)

Intensity		Wavelength	
6000		8112.899	I
60		8132.967	I
3000		8190.054	I
200		8202.72	II
80		8218.365	I
3000		8263.240	I
100		8272.353	I
5000		8298.107	I
1500		8281.050	I
100		8412.430	I
3000		8508.870	I
150		8764.110	I
6000		8776.748	I
2000		8928.692	I
500		9238.48	II
500	hL	9293.82	II
200	h	9320.99	II
300		9361.95	II
100		9362.082	I
200	h	9402.82	II
200	h	9470.93	II
500		9577.52	II
500	h	9605.80	II
400	h	9619.61	II
200		9663.34	II
200	h	9711.60	II
2000		9751.758	I
500		9803.14	II
500		9856.314	I
1000		10221.46	II
100		11187.108	I
200		11257.711	I
150		11259.126	I
500		11457.481	I
150		11792.425	I
1500		11819.377	I
600		11997.105	I
160		12077.224	I
100		12861.892	I
1100		13177.412	I
1000		13622.415	I
2400		13634.220	I
800		13658.394	I
200		13711.036	I
600		13738.851	I
150		13974.027	I
550		14045.657	I
140		14104.298	I
100		14402.22	I
2000		14426.793	I
100		14517.84	I
1600		14734.436	I
550		14762.672	I
450		14765.472	I
400		14961.894	I
120		15005.307	I
140		15209.526	I
1700		15239.615	I
130		15326.480	I
1500		15334.958	I
700		15372.037	I
200		15474.026	I
180		15681.02	I
120		15820.09	I
200		16726.513	I
2000		16785.128	I
1000		16853.488	I
2400		16890.441	I
1600		16896.753	I
1800		16935.806	I
600		17098.771	I
700		17367.606	I
120		17404.443	I
150		17616.854	I
650		17842.737	I
700		18002.229	I
2600		18167.315	I
100		18399.786	I
150		18580.896	I
300		18696.294	I
170		18785.460	I
200		18797.703	I
140		20209.878	I
300		[illegible]	I
140		20446.971	I
600		21165.471	I
1800		21902.513	I
120		22485.775	I
180		23340.416	I
120		24260.506	I
180		24292.221	I
600		25233.820	I
180		28610.55	I
1000		28655.72	I
150		28769.71	I
140		28822.49	I
300		29236.69	I
300		30663.54	I
300		30979.16	I
500		39300.6	I
1100		39486.52	I
220		39557.25	I
100		39572.60	I
1400		39588.4	I
1100		39589.6	I
500		39954.8	I
300		39966.6	I
1300		40306.1	I
250		40685.16	I

Kr III
Ref. 208, 366, 390, 421 — E.F.W.

Intensity		Wavelength	
		Vacuum	
30		467.35	III
30		540.86	III
30		565.64	III
30		569.16	III
30		571.98	III
30		579.83	III
30		585.14	III
30		585.96	III
30		593.70	III
30		594.10	III
30		596.41	III
40		600.17	III
30		603.67	III
50		605.86	III
35		606.47	III
50		611.12	III
35		616.72	III
40		621.45	III
45		622.80	III
50		625.02	III
30		625.76	III
45		628.59	III
50		630.04	III
35		633.09	III
50		639.98	III
60		646.41	III
50		651.20	III
50		659.72	III
30		664.86	III
40		672.34	III
35		672.85	III
35		676.57	III
35		680.13	III
35		683.68	III
45		686.25	III
45		687.98	III
45		691.93	III
50		695.61	III
30		698.05	III
50		708.36	III
50		714.00	III
100	p	722.04	III
30		746.70	III
60		785.97	III
50		837.66	III
50		854.73	III
60		862.58	III
40		870.84	III
50		876.08	III
75		897.81	III
50		987.29	III
30		1158.74	III
6		1638.82	III
6		1914.09	III
		Air	
40		2393.94	III
40		2494.01	III
30		2563.25	III
60		2639.76	III
30		2680.32	III
40		2681.19	III
30		2841.00	III
30		2851.16	III
50		2870.61	III
100		2892.18	III
30		[illegible]	III
50		2952.56	III
60		2992.22	III
50		3022.30	III
80		3024.45	III
50		3046.93	III
30		3056.72	III
60		3063.13	III
40		3097.16	III
60		3112.25	III
30		3120.61	III
100		3124.39	III
60		3141.35	III
100		3189.11	III
80		3191.21	III
40		3239.52	III
40		3240.44	III
300		3245.69	III
150		3264.81	III
100		3268.48	III
30		3271.65	III
30		3285.89	III
30		3304.75	III
50		3311.47	III
200		3325.75	III
60		3330.76	III
50		3342.48	III
100		3351.93	III
40		3374.96	III
100		3439.46	III
70		3474.65	III
100		3488.59	III
200		3507.42	III
100		3564.23	III
30		3641.34	III
30		3690.65	III
40	h	3868.70	III
50		4067.37	III
40		4131.33	III
40		4154.46	III
20	h	5016.45	III
10		5501.43	III
10	h	6037.17	III
10	h	6078.38	III
10		6310.22	III

Kr IV
Ref. 366, 409, 417 — E.F.W.

Intensity	Wavelength	
	Vacuum	
	793.44	IV
	794.11	IV
7	805.76	IV
18	816.82	IV
22	842.04	IV
	Air	
3	2237.34	IV
6	2291.26	IV
3	2329.3	IV
4	2336.75	IV
4	2348.27	IV
3	2358.5	IV
3	2388.05	IV
4	2416.9	IV
3	2428.04	IV
5	2442.68	IV
4	2451.7	IV
6	2459.74	IV
5	2474.06	IV
4	2517.0	IV
5	2518.02	IV
6	2519.38	IV
5	2524.5	IV
5	2546.0	IV
6	2547.0	IV
4	2558.08	IV
3	2586.9	IV
5	2606.17	IV
10	2609.5	IV
8	2615.3	IV
7	2621.11	IV
3	2730.55	IV
8	2748.18	IV
6	2774.70	IV
3	2829.60	IV
3	2836.08	IV
5	2853.0	IV
3	2859.3	IV
3	3142.01	IV
6	3224.99	IV
3	3261.70	IV
3	3809.30	IV
5	3860.58	IV
5	3934.29	IV

Kr V
Ref. 409, 431 — E.F.W.

Intensity	Wavelength	
	Vacuum	
150	472.16	V
100	484.39	V
250	496.25	V
120	500.77	V
200	507.20	V
60	548.04	V
120	637.87	V
	690.86	V
	691.75	V
600	708.85	V
	810.70	V

LANTHANUM (La)
Z = 57

La I and II
Ref. 1 — C.H.C.

Intensity		Wavelength	
		Air	
240		2187.87	II
770		2256.76	II
200		2319.44	II
400		2610.34	II
420		2808.39	II
130		2885.14	II
160		2893.07	II
110		2950.50	II
180		3104.59	II
130		3142.76	II
510		3245.13	II
260		3249.35	II
550		3265.67	II
800		3303.11	II
1500		3337.49	II
870		3344.56	II
200		3376.33	II
1500		3380.91	II
130		3452.18	II
180		3453.17	II
200		3574.43	I
320		3628.83	II
120		3637.15	II
170	d	3641.53	I
		3641.66	II
1000		3645.42	II
390		3650.18	II
170		3662.08	II
120		3704.54	I
320		3705.82	II
550		3713.54	II
140		3714.87	II
270		3715.53	II
2400		3759.08	II
120		3780.67	II
3700		3790.83	II
3900		3794.78	II
190		3835.08	II
600		3840.72	II
130		3848.00	II
1600		3849.02	II
130		3854.91	II
3400		3871.64	II
1700		3886.37	II
1300		3916.05	II
1100		3921.54	II
160		3927.56	I
2200		3929.22	II
180		3936.22	II
9000		3949.10	II
4400		3988.52	II
3600		3995.75	II
180		4015.39	I
250		4025.88	II
2800		4031.69	II
140		4037.21	I
3000		4042.91	II
320		4050.08	II
220		4060.33	I
160		4064.79	I
850		4067.39	II
110		4076.71	II
2800		4077.35	II
120		4079.18	I
5500		4086.72	II
180		4089.61	I
280		4099.54	II
110		4104.87	I
4400		4123.23	II
110		4137.04	I
550		4141.74	II
1100		4151.97	II
220		4152.78	II
100		4160.26	I
280		4187.3[illegible]	[illegible]
100		4192.36	II
1500		4196.55	II
240		4204.04	II
300		4217.56	II
200		4230.95	II
1600		4238.38	II
140		4249.99	II
320		4263.59	II
480		4269.50	II
240		4275.64	II
300		4280.27	I
600		4286.97	II
600		4296.05	II
120		4300.44	II
440		4322.51	II
4600		4333.74	II
550		4354.40	II
110		4364.67	II
110	bl	4371.97	L
110	bl	4375.84	L
110		4378.10	II
280		4383.44	II
100		4385.20	II

Lanthanum (Cont.)

Intensity		Wavelength	
220	bl	4418.24	L
160	bl	4423.17	L
160		4423.90	I
260		4427.55	II
100	bl	4428.10	L
2000		4429.90	II
160	bl	4432.98	L
100	bl	4438.01	L
100		4452.15	I
100		4455.80	II
850		4522.37	II
170		4525.31	II
420		4526.12	II
400		4558.46	II
110		4559.29	II
160		4567.91	I
200		4570.02	I
400		4574.88	II
200		4580.06	II
160		4605.78	II
410		4613.39	II
410		4619.88	II
110		4645.28	II
540		4655.50	II
360		4662.51	II
230		4663.76	II
200		4668.91	II
160		4671.83	II
230		4692.50	II
140		4703.28	II
170		4716.44	II
140		4719.94	II
230		4728.42	II
500		4740.28	II
390		4743.09	II
320		4748.73	II
160		4766.89	I
160		4804.04	II
160		4809.01	II
200		4824.06	II
320		4860.91	II
850		4899.92	II
1000		4920.98	II
1000		4921.79	II
140		4934.83	II
110		4946.47	II
370		4949.77	I
340		4970.39	II
370		4986.83	II
140		4991.28	II
720		4999.47	II
140		5046.88	I
210		5050.57	I
170		5056.46	I
200		5106.23	I
470		5114.56	II
470		5122.99	II
450		5145.42	I
180		5156.74	II
180		5157.43	II
290		5158.69	I
120		5163.62	II
580		5177.31	I
850		5183.42	II
260		5188.22	II
170		5204.15	II
720		5211.86	I
520		5234.27	I
340		5253.46	I
110		5259.39	II
370		5271.19	I
140		5290.84	II
370		5301.98	II
140		5302.62	II
180		5303.55	II
110		5340.67	II
110		5357.86	I
130		5377.09	II
140		5380.99	II
500		5455.15	I
470		5501.34	I
110	bl	5602.50	L
160		5631.22	I
240		5648.25	I
130		5657.72	I
180		5740.66	I
160		5744.41	I
160		5761.84	I
160		5769.07	II
370		5769.34	I
320		5789.24	I
450		5791.34	I
220		5797.58	II
160		5805.78	II
140		5821.99	I
320		5930.62	I
720		6249.93	I
260	d	6262.30	II
180		6296.09	II
160		6320.39	II
110		6325.91	I
170		6390.48	II
450		6394.23	I
210		6410.99	I
250		6455.99	I
110		6526.99	II
130		6543.16	I
140		6578.51	I
180		6709.50	I
120		6774.26	II
13	b	7011.22	L
75		7023.67	I
26		7032.05	I
26	b	7040.84	L
110		7045.96	I
13	b	7054.80	L
160		7066.23	II
65		7068.37	I
21	bl	7070.79	L
13		7076.38	I
21	bl	7085.40	L
26	bl	7101.02	L
10	h	7116.8	II
19	bl	7131.58	L
10		7149.77	I
40	h	7158.08	I
50		7161.25	I
10		7162.60	L
21		7219.91	I
10	b	7257.16	L
26		7270.09	I
10		7270.30	I
110	cw	7282.34	II
10		7320.91	I
110	cw	7334.18	I
65		7345.34	I
50	bl	7379.71	L
85	bl	7380.08	L
35		7382.73	I
110	bl	7403.52	L
210	bl	7403.75	L
50	b	7411.34	L
65	bl	7434.28	L
110	bl	7434.36	L
30	b	7442.92	L
50	h	7463.08	I
50	bl	7465.25	L
95	bl	7465.48	L
75	cw	7483.50	II
40	bl	7496.50	L
95	bl	7496.78	L
50		7498.83	I
30	b	7506.79	L
19	bl	7528.21	L
50	bl	7528.39	L
30		7533.59	I
85		7539.23	I
35	bl	7560.09	L
35	bl	7592.26	L
19	h	7612.94	II
19	b	7624.99	L
21		7664.34	I
15	h	7841.80	I
21	b	7876.87	L
75	bl	7877.22	L
75	bl	7910.19	L
150	bl	7910.54	L
50	b	7944.61	L
110	bl	7944.95	L
40		7964.83	I
35	b	7979.34	L
75	bl	7979.70	L
35	h	8001.89	I
21	b	8014.43	L
65	bl	8014.79	L
30	b	8019.48	L
35	hc	8051.39	I
75		8086.05	I
15	b	8122.20	L
15	b	8159.02	L
7	h	8203.38	I
50		8247.44	I
13	h	8316.04	I
85		8324.69	I
95		8346.53	I
8	h	8379.80	I
8	b	8453.55	L
8	h	8467.62	I
26		8476.48	I
13	h	8507.37	I
13	h	8513.57	I
8	h	8514.65	II
17	b	8526.59	L
17	c	8543.46	I
65		8545.44	I
15	b	8563.54	L
9	h	8590.94	I
9	b	8600.81	L
7	h	8624.22	I
15		8638.47	I
19	hw	8672.11	I
40		8674.43	I
13	h	8720.41	I
35		8748.38	I
19		8818.93	I
35		8825.82	I
21		8839.63	I

La III
Ref. 220, 309 — J.R.

Intensity		Wavelength	
		Vacuum	
3		744.19	III
10		753.03	III
1		786.64	III
200		787.14	III
1		796.03	III
400		796.99	III
1		797.20	III
10		835.03	III
30		845.62	III
1		850.73	III
1		860.39	III
2		860.88	III
5		865.04	III
2000		870.40	III
30		872.43	III
1000		882.34	III
20		882.72	III
200		929.71	III
400		942.86	III
30		967.69	III
10		974.33	III
50		979.99	III
10		980.29	III
200		1058.63	III
1000		1072.59	III
5000		1076.91	III
50000		1081.61	III
95000		1099.73	III
5000		1100.70	III
30		1208.80	III
30		1212.29	III
200		1236.54	III
100		1253.99	III
2000		1255.63	III
100		1259.55	III
100		1322.42	III
5000		1330.04	III
10000		1349.18	III
5000		1459.49	III
2000		1466.44	III
10000		1523.79	III
500		1528.55	III
5000		1536.17	III
200		1923.34	III
500		1938.57	III
		Air	
60		2216.07	III
20		2238.36	III
25		2258.61	III
5		2260.30	III
250		2297.74	III
400		2379.37	III
10		2387.99	III
20		2392.49	III
100		2476.60	III
50		2478.65	III
2		2513.43	III
4		2588.87	III
2		2604.83	III
400		2651.50	III
100		2682.34	III
150		2684.76	III
110		2897.88	III
160		2904.58	III
7		2950.84	III
10		2953.77	III
40		2992.10	III
4		3006.19	III
15		3009.22	III
100		3075.17	III
4		3085.38	III
25		3093.03	III
15		3096.26	III
200		3111.97	III
50		3116.74	III
1000		3171.63	III
1500		3171.74	III
50		3172.69	III
20		3196.84	III
70		3289.11	III
15		3301.48	III
35		3327.66	III
500		3517.09	III
600		3517.22	III
3		4129.24	III
5		4137.43	III
200		4482.97	III
300		4499.05	III
5		5145.73	III
8		5158.41	III
6		5467.81	III
55		5491.90	III
2		5511.72	III
1		5518.19	III
45		5529.54	III
1		5744.09	III
200		5778.14	III
2		5813.45	III
3		5875.63	III
55		5888.62	III
3		5932.71	III
2		6017.11	III
20		6055.84	III
35		6119.25	III
120		6141.99	III
55		6220.00	III
60		6348.21	III
3		8114.42	III
2		8135.96	III
250		8252.60	III
100		8275.39	III
200		8287.75	III
250		8321.11	III
300		8583.45	III
120		9184.38	III
100		9212.63	III
80		9923.99	III
140		10284.79	III
20		10370.34	III
12		10937.90	III
		13894.47	III
		14096.18	III
		17898.09	III

La IV
Ref. 79 — J.R.

Intensity		Wavelength	
		Vacuum	
100		344.12	IV
7000		453.50	IV
10000		463.14	IV
15000		499.54	IV
40000		552.02	IV
30000		631.26	IV
25		724.92	IV
15		733.29	IV
10		797.03	IV
10	c	980.03	IV
50		1039.30	IV
60	p	1062.09	IV
75		1158.35	IV
50		1164.29	IV
400		1230.90	IV
75	p	1260.79	IV
300		1261.12	IV
150		1283.19	IV
2000		1302.31	IV
1200		1333.53	IV
3500		1334.96	IV
1000		1352.76	IV
25000		1368.04	IV
8000		1377.49	IV
3000		1394.32	IV
5000		1414.58	IV
7000		1432.55	IV
7000		1441.63	IV
7500		1462.15	IV
20000		1463.47	IV
7500		1467.54	IV
15000		1507.87	IV
5000		1527.19	IV
2500		1575.92	IV
1500		1583.61	IV
750		1585.11	IV
750		1637.42	IV
750	d	1645.21	IV
1000		1664.84	IV
750		1684.17	IV
2000	p	1767.65	IV
4000		1808.66	IV
1000		1851.81	IV
1500		1852.77	IV
750		1879.79	IV
1000		1881.57	IV
800		1889.22	IV
1000		1891.47	IV
5000		1902.97	IV
800		1907.44	IV
1500	c	1950.80	IV
1200	c	1957.57	IV
		Air	
3000		2012.42	IV
750		2037.43	IV
2000	c	2066.50	IV
3000	c	2073.18	IV
1500		2143.23	IV
4000	c	2197.45	IV
1000	w	2221.12	IV
900		2227.34	IV
3000		2244.95	IV
7500	w	2265.91	IV
2000		2315.89	IV
750		2348.36	IV
750		2355.31	IV
2000	c	2407.10	IV

Lanthanum (Cont.)

Intensity		Wavelength	
25000	w	2417.58	IV
1200	p	2443.92	IV
18000	c	2502.81	IV
15000		2515.02	IV
50000		2532.75	IV
900	d	2535.76	IV
45000		2582.05	IV
18000	c	2591.30	IV
95000	w	2597.50	IV
5000	c	2608.01	IV
70000	w	2662.75	IV
50000	w	2848.30	IV
12000	c	2863.30	IV
30000	c	2962.58	IV
70000	w	3009.51	IV
90000	c	3056.68	IV
3500		3522.28	IV
2000		3650.40	IV
2000	p	4270.76	IV
1500	w	4549.80	IV
500		4836.89	IV

La V
Ref. 78 — J.R.

Intensity		Wavelength	
		Vacuum	
2		389.03	V
400		390.72	V
1		398.53	V
30		399.34	V
350		405.10	V
50		416.13	V
3		421.55	V
50		423.07	V
400		424.78	V
1000		432.11	V
2500		435.28	V
700		436.14	V
700		436.84	V
300		437.11	V
700		437.55	V
20		444.01	V
10		444.07	V
1050		450.40	V
600		457.30	V
1000		463.85	V
150		476.67	V
5000		482.16	V
200		482.43	V
2000		483.30	V
7000		498.08	V
4000		499.03	V
10000		503.58	V
40		508.15	V
1500		525.71	V
12000		526.76	V
10000		531.07	V
15000		533.23	V
4000		540.20	V
6000		544.80	V
8000		547.44	V
3000		570.90	V
2500		593.18	V
750		597.70	V
2000		600.01	V
5000		600.24	V
700		611.70	V
500		617.60	V

LEAD (Pb)
Z = 82

Pb I and II
Ref. 64, 274, 283, 329, 330 — D.R.W.

Intensity		Wavelength	
		Vacuum	
2		846.04	II
2	h	849.88	II
3		855.57	II
3		863.00	II
6		873.71	II
2		877.96	II
8		889.68	II
3		896.30	II
5		926.44	II
2		958.76	II
2		960.21	II
3		965.36	II
10		967.23	II
9		972.56	II
8		982.17	II
10		986.71	II
10		995.89	II
6		1001.81	II
10		1016.61	II
10		1049.82	II
10		1050.77	II
10		1060.66	II
9		1065.58	II
10		1103.94	II
10		1108.43	II
10		1109.84	II
10		1119.57	II
10		1121.36	II
10		1133.14	II
4		1145.91	II
10		1203.63	II
10		1231.20	II
10		1331.65	II
10		1335.20	II
10		1348.37	II
10		1433.96	II
3		1449.35	II
10		1512.42	II
10		1671.53	II
10		1682.15	II
20		1726.75	II
2		1740.00	I
2		1766.64	I
2		1794.67	I
10		1796.670	II
5		1812.97	I
10		1822.050	II
4		1868.76	I
10		1904.77	I
7		1921.471	II
4		1972.44	I
2		1977.88	I
2		1991.60	I
2		1992.31	I
		Air	
5	r	2022.02	I
5		2050.88	I
8	r	2053.28	I
6		2111.758	I
10		2115.066	I
500	r	2170.00	I
7		2175.580	I
7		2187.888	I
8		[illegible]	I
10		2203.534	II
20		2237.425	I
20		2246.86	I
25		2246.89	I
150		2332.418	I
180		2388.797	I
550	r	2393.792	I
140		2399.597	I
320	r	2401.940	I
320	r	2411.734	I
150	r	2443.829	I
160	r	2446.181	I
130	r	2476.378	I
8	c	2526.69	II
8	c	2576.60	II
80	r	2577.260	I
2	c	2608.38	II
500	r	2613.655	I
900	r	2614.175	I
160		2628.262	I
4		2634.256	II
10		2657.094	I
700		2663.154	I
10		2697.541	I
25000	r	2801.995	I
100		2822.58	I
14000	r	2823.189	I
35000	r	2833.053	I
6		2840.557	II
14000	r	2873.311	I
3	c	2887.30	II
3		2914.442	II
2	c	2947.43	II
3	c	2948.53	II
15		2966.460	I
15		2972.991	I
15		2980.157	I
4		2986.876	II
10	c	3016.39	II
150		3118.894	I
600		3220.528	I
100		3229.613	I
400		3240.186	I
200		3262.355	I
35000		3572.729	I
50000	r	3639.568	I
20000		3671.491	I
70000	r	3683.462	I
10		3713.982	II
25000		3739.935	I
15000		4019.632	I
95000		4057.807	I
14000		4062.136	I
5		4110.76	II
4		4113.35	II
10		4152.82	II
10		4157.814	I
10000		4168.033	I
9	c	4242.14	II
20	c	4244.92	II
7		4293.82	II
6		4296.65	II
200		4340.413	I
10		4352.74	II
20	c	4386.46	II
10		4579.051	II
10		4582.27	II
1000		5005.416	I
100		5006.572	I
50		5042.58	II
10		5070.58	II
10		5074.53	II
10		5076.35	I
50		5089.484	I
20		5090.01	I
10		5107.242	I
10		5111.64	II
2000		5201.437	I
10		5367.64	II
10		5372.099	II
10	c	5544.25	II
20	c	5608.85	II
40		5692.346	I
200		5895.624	I
2000		6001.862	I
9	c	6009.58	II
500		6011.667	I
8	c	6041.17	II
500		6059.356	I
40		6075.74	II
40		6081.409	II
50		6110.520	I
10		6159.89	II
100		6235.266	I
50	c	6660.20	II
10		6892.11	I
5		7128.94	I
20		7193.60	II
20000		7228.965	I
5		7304.68	I
8		7330.15	I
10		7346.676	I
10		7558.97	II
10		7632.56	II
4		7732.96	II
20		7809.259	I
5		7817.97	I
6		7829.01	I
5		7896.737	I
2	d	8156.91	II
10		8168.001	I
6		8191.886	I
5		8217.711	I
8		8255.61	I
40		8272.690	I
6		8335.54	II
10		8395.68	II
20		8409.384	I
10		8478.492	I
8		8532.17	I
7	c	8544.95	II
7		8709.90	II
5		8719.39	II
5		8722.810	I
10		8857.457	I
10		9050.82	II
10		9063.43	II
2	d	9245.28	II
8		9293.476	I
5		9384.35	I
5		9385.89	I
15		9438.05	I
15		9604.297	I
6		9608.73	I
15		9674.351	I
200		10290.458	I
5		10434.32	I
100		10498.965	I
50		10649.249	I
5		10759.41	I
7		10759.74	I
15		10886.688	I
40		10969.53	I
6		11059.22	I
3		11333.08	I
2	d	11479.49	II
2		11488.76	I
5		11627.91	II
1		12561.37	I
		13495.3	I
		13498.2	I
		13512.6	I
		14722.8	I
		14742.1	I
		14743.0	I
		15314.8	I
		15327.6	I
		15331.0	I
		15349.6	I
		38831.1	I
		38950.1	I
		38958.6	I
		39039.4	I

Pb III
Ref. 54, 256, 297 — D.R.W.

Intensity		Wavelength	
		Vacuum	
1		961.01	III
3		1030.5	III
12		1048.9	III
4		1069.2	III
3		1074.7	III
4		1118.67	III
4		1167.0	III
4		1250.6	III
1		1266.9	III
20		1553.1	III
1		1610.1	III
4		1711.23	III
		Air	
10		3043.85	III
4		3089.08	III
4		3102.74	III
10		3137.81	III
10		3176.50	III
5		3242.84	III
1		3530.17	III
7		3589.87	III
7		3689.31	III
3		3706.02	III
5		3728.69	III
12		3854.08	III
8		3951.92	III
3		4031.16	III
3		4094.54	III
2		4128.11	III
8		4272.66	III
6		4499.34	III
7		4571.81	III
1		4596.45	III
6		4761.12	III
4		4798.59	III
1		4826.86	III
2		4855.06	III
3		5065.12	III
4		5191.56	III
5		5523.97	III
3		5779.41	III
6		5857.96	III

Pb IV
Ref. 106 — D.R.W.

Intensity		Wavelength	
		Vacuum	
8		475.36	IV
7		478.35	IV
10		496.38	IV
12		499.94	IV
9		515.07	IV
14		529.78	IV
20		570.16	IV
8		573.90	IV
8		584.52	IV
10		648.50	IV
9		656.10	IV
10		761.09	IV
18		802.07	IV
12		802.82	IV
10		812.59	IV
8		822.07	IV
10		827.41	IV
12		832.60	IV
8		840.99	IV
8		842.88	IV
12		845.94	IV
18		857.64	IV
8		859.02	IV
16		862.33	IV
14		870.44	IV
12		879.96	IV
7		880.35	IV
14		884.96	IV
14		884.99	IV
16		890.72	IV
12		908.51	IV
12		917.90	IV
10		922.12	IV
12		922.49	IV
7		924.52	IV
10		927.64	IV
14		932.20	IV
8		937.00	IV
7		952.85	IV

Lead (Cont.)

Intensity	Wavelength	
8	1012.44	IV
14	1028.61	IV
20	1032.05	IV
16	1041.24	IV
18	1044.14	IV
15	1056.53	IV
12	1072.09	IV
7	1079.88	IV
18	1080.81	IV
20	1084.17	IV
6	1089.94	IV
7	1099.47	IV
6	1115.30	IV
20	1116.08	IV
18	1137.84	IV
8	1142.77	IV
14	1144.93	IV
20	1189.95	IV
8	1267.55	IV
8	1290.82	IV
10	1291.10	IV
20	1313.05	IV
8	1323.92	IV
12	1343.06	IV
16	1388.94	IV
6	1397.02	IV
18	1400.26	IV
10	1404.34	IV
7	1510.76	IV
14	1535.71	IV
8	1798.39	IV
8	1893.19	IV
12	1959.34	IV
16	1973.16	IV
	Air	
10	2042.58	IV
12	2049.34	IV
12	2079.22	IV
8	2151.96	IV
15	2154.01	IV
12	2177.46	IV
16	2359.53	IV
4	2864.24	IV
4	2864.50	IV
16	2417.61	IV
4	2978.14	IV
4	3052.56	IV
4	3221.17	IV
4	3962.48	IV
4	4049.80	IV
10	4496.15	IV
16	4534.60	IV
8	4605.40	IV
2	5914.54	IV

Pb V
Ref. 106 — D.R.W.

Intensity	Wavelength	
	Vacuum	
2	367.40	V
2	372.53	V
2	387.87	V
2	394.38	V
5	424.64	V
3	431.03	V
3	436.60	V
2	438.47	V
6	438.91	V
4	453.45	V
3	461.70	V
3	496.20	V
4	694.42	V
8	696.20	V
20	703.73	V
4	706.29	V
6	707.66	V
5	730.85	V
12	749.46	V
10	752.52	V
10	755.80	V
6	762.76	V
10	765.87	V
18	767.45	V
18	769.49	V
14	771.42	V
14	782.79	V
10	787.05	V
15	797.02	V
5	799.80	V
18	809.63	V
5	812.32	V
8	814.10	V
8	820.09	V
5	825.52	V
8	829.32	V
5	851.98	V
20	863.97	V
10	867.10	V
6	880.50	V
18	883.90	V
14	888.37	V
14	894.40	V
12	896.08	V
8	915.09	V
14	915.71	V
12	918.09	V
12	920.28	V
12	920.66	V
6	940.74	V
8	946.20	V
6	950.93	V
12	954.35	V
4	954.95	V
10	955.28	V
4	964.38	V
6	989.14	V
8	1005.42	V
10	1051.26	V
4	1059.26	V
10	1088.86	V
9	1096.52	V
6	1104.79	V
4	1121.33	V
10	1137.50	V
4	1152.36	V
12	1157.88	V
14	1185.43	V
11	1233.50	V
10	1248.47	V
8	1635.75	V
2	1802.87	V
2	1843.00	V
2	1888.67	V
2	1897.02	V
2	1914.33	V
4	1919.74	V
5	1957.96	V
2	1998.58	V
10	1998.83	V
	Air	
8	2078.45	V
10	2142.55	V
10	2167.97	V
20	2259.01	V
10	2276.66	V
8	2301.49	V
15	2424.81	V
4	4809.36	V
5	6650.99	V
4	6753.20	V

LITHIUM (Li)
Z = 3

Li I and II
Ref. 3, 15, 17, 18, 37, 44, 112, 284, 321, 335 — J.D.S.

Intensity		Wavelength	
		Vacuum	
		125.5	II
		136.5	II
		140.5	II
		167.21	II
		168.74	II
		171.58	II
		178.02	II
		199.28	II
		207.5	II
		456.	II
		483.	II
		540.	II
		729.	II
		800.	II
		820.	II
		861.	II
		905.5	II
		917.5	II
		936.	II
		945.	II
		965.	II
		972.	II
		988.	II
		1018.	II
		1032.	II
		1036.	II
		1093.	II
		1103.	II
		1109.	II
		1116.	II
		1132.1	II
		1141.	II
		1166.4	II
		1198.09	II
		1215.	II
		1238.	II
		1253.8	II
		1420.89	II
		1424.	II
3		1492.93	II
5		1492.97	II
1		1493.04	II
		1555.	II
3		1653.08	II
5		1653.13	II
1		1653.21	II
		1681.66	II
		1755.33	II
		Air	
		2009.	II
		2039.	I
		2068.	II
		2131.	II
		2164.	II
		2173.4	I
		2183.	II
		2214.	II
		2222.	II
		2237.	II
	H	2249.21	II
		2286.82	II
		2302.57	II
		2303.33	II
		2304.59	I
		2304.92	I
		2305.36	I
		2305.83	I
		2306.29	I
		2306.82	I
		2307.44	I
		2308.97	I
		2309.88	I
		2310.94	I
		2312.11	I
		2313.49	I
		2315.08	I
		2316.95	I
		2319.18	I
		2321.88	I
		2325.11	I
		2329.02	I
		2329.84	II
		2333.94	I
3		2336.88	II
5		2336.91	II
2		2337.00	II
		2340.15	I
		2348.22	I
		2358.93	I
		2373.54	I
		2381.54	II
		2383.20	II
1		2394.39	I
		2402.33	II
		2410.84	II
3		2425.43	I
		2429.81	II
		2460.2	I
10		2475.06	I
		2506.94	II
		2508.78	II
		2518.	I
		2539.49	II
24		2551.7	II
		2559	II
15		2562.31	I
		2605.08	II
		2640.	
2		2657.29	II
3		2657.30	II
		2674.46	II
0		2728.24	II
5		2728.29	II
2		2728.32	II
3		2730.47	II
1		2730.55	II
5		2741.20	I
		2766.99	II
		2790.31	II
		2801.	I
		2846.	I
		2868.	I
		2895.	I
2		2934.02	II
2		2934.07	II
5		2934.12	II
1		2934.25	II
		2968.	I
3		3029.12	II
3		3029.14	II
		3144.	I
3		3155.31	II
4		3155.33	II
1		3196.26	II
9		3196.33	II
4		3196.36	II
5		3199.33	II
2		3199.43	II
17		3232.66	I
		3249.87	II
		3306.28	II
		3393.	
		3488.	I
		3579.8	I
		3618.	I
		3662.	I
		3684.32	II
1		3714.00	II
5		3714.16	II
6	d	3714.27	II
8		3714.29	II
7	d	3714.40	II
10		3714.41	II
1		3714.51	II
0		3714.58	II
3		3718.7	I
6		3794.72	I
20		3915.30	I
20		3915.35	I
10		3985.48	I
10		3985.54	I
40		4132.56	I
40		4132.62	I
		4196.	I
20		4273.07	I
20		4273.13	I
5		4325.42	II
5		4325.47	II
1		4325.54	II
		4516.45	II
		4590.	
13		4602.83	I
13		4602.89	I
		4607.34	II
0		4671.51	II
6		4671.65	II
2		4671.70	II
3		4678.06	II
1		4678.29	II
		4760.	I
		4763.	II
		4788.36	II
		4843.0	II
4		4881.32	II
4		4881.39	II
1		4881.49	II
8		4971.66	I
8		4971.75	I
		5037.92	II
		5095.	
		5114.	
		5190.	
		5271.	I
		5315.	I
		5395.	I
		5440.	I
600	c	5483.55	II
600	c	5485.65	II
320		6103.54	I
320		6103.65	I
3600		6707.76	I
3600		6707.91	I
48		8126.23	I
48		8126.45	I
		8517.37	II
		9581.42	II
		10120.	II
		12232.	I
		12782.	I
		13566.	I
		17552.	I
		18697.	I
		19290.	I
		24467.	I
		40475.	I

Li III
Ref. 335 — J.O.S.

Intensity	Wavelength	
	Vacuum	
	102.9	III
	103.4	III
	104.1	III
	105.5	III
	108.0	III
	113.9	III
	135.0	III
	540.0	III
	729.1	III

LUTETIUM (Lu)
Z = 71

Lu I and II
Ref. 1 — C.H.C.

Intensity	Wavelength
	Air

Lutetium (Cont.)

Intensity		Wavelength	
1700	h	2195.54	II
95		2276.94	II
190		2297.41	II
1300		2392.19	II
120		2399.14	II
80		2419.21	II
55		2430.26	II
130		2459.64	II
80		2469.27	II
21	h	2481.72	II
370		2536.95	II
40		2546.87	II
20		2549.44	I
20		2549.72	I
35		2561.80	II
930		2571.23	II
1700		2578.79	II
80	h	2582.13	II
1800		2613.40	II
18000		2615.42	II
1800		2619.26	II
90		2657.05	
2700		2657.80	II
90	h	2677.25	I
570	h	2685.08	I
90	h	2685.54	I
4200		2701.71	II
90	h	2715.91	I
180	d	2719.09	I
480	h	2728.95	I
75	c	2738.17	II
3600		2754.17	II
750	h	2765.74	I
2700		2796.63	II
35		2821.23	II
270	c	2834.35	II
330	h	2845.13	I
3000		2847.51	II
570	h	2885.14	I
6300		2894.84	II
4500		2900.30	II
300		2903.05	I
9000		2911.39	II
270	h	2949.73	I
1200		2951.69	II
60		1955.78	II
4200		2963.32	II
2400		2969.82	II
1800		2989.27	I
3000		3020.54	II
120		3027.29	II
2100		3056.72	II
7500		3077.60	II
390		3080.11	I
5100	h	3081.47	I
3000		3118.43	T
2400		3171.36	I
100		3183.73	II
260		3191.80	II
1400		3198.12	II
4800		3254.31	II
3800		3278.97	I
7600		3281.74	I
6200		3312.11	I
7600		3359.56	I
6200		3376.50	I
950		3385.50	I
160	h	3391.55	I
1400		3396.82	I
4100		3397.07	II
4800		3472.48	II
8300	c	3507.39	II
1600		3508.42	I
4800		3554.43	II
4800		3567.84	I
340		3596.34	I
800		3623.99	II
680		3636.25	I
2600		3647.77	I
60		3684.32	I
60		3710.95	I
110		3756.70	I
110		3756.79	I
30	h	3786.18	I
150		3800.67	I
75	h	3829.07	I
2700		3841.18	I
75		3843.61	I
95		3853.29	I
40		3874.61	I
530		3876.65	II
29		3911.77	I
50		3918.86	I
35	h	3926.62	I
480		3968.46	I
50		3991.38	I
670		4054.45	I
75	Bl	4096.13	L
35	h	4107.44	I
95	h	4112.67	I
310		4122.49	I
3100		4124.73	I
150	c	4131.79	I
460		4154.08	I
24		4158.98	I
1600		4184.25	II
150		4277.50	I
250		4281.03	I
330	d	4295.97	I
		4296.09	I
150		4309.57	I
75		4332.72	I
29		4341.98	II
65	h	4420.96	I
190	c	4430.48	I
35		4438.79	I
190		4450.81	I
50	h	4471.55	I
60	h	4498.85	I
3300		4518.57	I
24	b	4560.95	L
24	b	4575.31	L
85	c	4605.39	I
95	h	4645.47	I
100	h	4648.21	I
95	h	4648.85	I
65	b	4654.03	L
1000		4658.02	I
85	h	4659.03	I
630	Bl	4661.75	L
310	Bl	4672.31	L
420	Bl	4684.16	L
270	Bl	4695.46	L
190	Bl	4708.00	L
30		4716.70	I
65	b	4720.86	L
65	h	4726.20	I
100	Bl	4735.00	L
75	Bl	4749.11	L
40	Bl	4764.22	L
150		4785.42	II
85		4815.05	I
50	c	4839.62	II
18		4865.36	II
460		4904.88	I
180		4942.34	I
800		4994.13	II
800		5001.16	I
55	h	5057.60	I
140		5134.05	I
2700		5135.09	I
130	Bl	5170.11	L
170		5196.61	I
90		5206.47	I
40		5304.40	I
80		5349.12	I
500		5402.57	I
140	c	5421.90	I
100		5437.88	I
35		5453.57	I
2100		5476.69	II
9		5664.89	II
14		5713.49	II
550		5736.55	I
55		5775.40	I
80		5800.59	I
40	h	5860.79	I
9		5866.30	I
690	Cw	5983.9	II
140		5997.13	I
1400		6004.52	I
35	h	6041.66	I
440		6055.03	I
11		6140.71	I
150		6159.94	II
160		6199.66	II
2100		6221.87	II
35		6228.14	II
80		6235.36	II
160		6242.34	II
16	h	6248.80	I
70	h	6345.35	I
18	h	6354.85	I
9		6365.79	I
16		6366.00	I
22		6441.14	I
11		6444.89	II
1100		6463.12	II
29		6477.67	I
55	c	6523.18	I
35	Cw	6611.28	II
		6611.58	LU II
		6611.80	LU II
		6611.95	LU II
		6612.04	LU II
11		6619.15	II
23	c	6677.14	I
9	h	6735.76	I
30	c	6793.77	I
11		6826.59	II
45		6917.31	I
8		6943.96	II
23		7031.24	I
14	c	7096.34	I
45		7125.84	II
9		7142.79	I
7		7143.10	I
8		7165.94	II
14	Ch	7237.98	I
5		7409.70	II
11	c	7441.52	I
8		7456.96	II
7	Ch	7640.08	I
7	Cw	7758.30	I
7	h	7815.9	I
9	c	8178.16	I
17		8382.08	I
35		8459.19	II
10	d	8478.50	I
29	c	8508.08	I
35	c	8610.98	I

Lu III
Ref. 148 — J.R.

Vacuum

Intensity		Wavelength	
1		677.34	III
7		691.05	III
30		700.25	III
50		714.89	III
100		738.76	III
200		755.03	III
3		755.16	III
500		810.73	III
100		830.53	III
2000		832.28	III
10		972.66	III
2		991.26	III
100		996.44	III
400		1001.18	III
1		1022.40	III
100		1029.83	III
200		1030.33	III
100		1031.54	III
50		1056.53	III
20		1061.99	III
3		1092.86	III
200		1187.34	III
50		1228.7	III
200		1277.53	III
30		1283.41	III
100		1331.93	III
1000		1854.57	III

Air

Intensity		Wavelength	
40		2050.72	III
1500		2065.35	III
1500	c	2070.56	III
100		2083.34	III
200		2099.44	III
1000		2236.14	III
2000		2236.22	III
500	c	2381.59	III
300		2563.51	III
4500	c	2603.35	III
200		2721.65	III
2000		2772.55	III
20		2781.16	III
10		2788.37	III
500		2800.90	III
20	p	2993.21	III
1000		3057.86	III
200		4251.44	III
300		4271.91	III
200		4490.00	III
400		4956.43	III
10		5046.12	III
150	c	5145.86	III
70		5419.42	III
60		5519.88	III
5		5526.80	III
5		5748.71	III
70		5786.46	III
80		5869.71	III
60		5889.76	III
300		6197.96	III
600		6198.13	III
100		7309.95	III
200		7310.25	III
50		7534.27	III
70	l	7936.45	III
3		8008.59	III

Lu IV
Ref. 310 — J.R.

Vacuum

Intensity		Wavelength	
400		876.80	IV
100		902.06	IV
300		1015.18	IV
20		1136.17	IV
50		1189.27	IV
20	p	1194.59	IV
60		1213.08	IV
15		1220.74	IV
20		1223.75	IV
20		1240.07	IV
20		1248.10	IV
40		1266.27	IV
100		1272.42	IV
20		1273.02	IV
40		1274.77	IV
20		1276.54	IV
50		1289.38	IV
60		1310.08	IV
50		1323.02	IV
15		1331.04	IV
800		1333.79	IV
300		1334.94	IV
50		1338.20	IV
100		1339.49	IV
300		1342.58	IV
200		1351.68	IV
300		1353.74	IV
200		1355.85	IV
20		1359.67	IV
100		1363.24	IV
50		1363.37	IV
20		1367.34	IV
20		1373.54	IV
15		1375.36	IV
100		1376.02	IV
30		1379.56	IV
15		1383.18	IV
15		1389.85	IV
60		1390.07	IV
200		1390.30	IV
50		1390.69	IV
40		1392.38	IV
40		1397.18	IV
30		1401.32	IV
100		1401.46	IV
200		1406.64	IV
100		1407.00	IV
250		1407.04	IV
20		1420.32	IV
100		1421.59	IV
200		1429.08	IV
400		1429.38	IV
40		1430.80	IV
100		1440.62	IV
100		1448.14	IV
200		1452.33	IV
30		1462.65	IV
100		1483.79	IV
200		1493.24	IV
400		1511.26	IV
200		1521.06	IV
100		1522.21	IV
100		1537.77	IV
30		1549.35	IV
20		1551.59	IV
20		1562.06	IV
30		1592.55	IV
15		1594.92	IV
100	c	1607.72	IV
50	c	1631.65	IV
20		1684.50	IV
60	c	1693.67	IV
400	c	1721.42	IV
100		1725.14	IV
100	c	1735.79	IV
100		1736.78	IV
100		1741.74	IV
50	c	1743.84	IV
40	c	1752.60	IV
200		1759.61	IV
300		1772.08	IV
600		1772.57	IV
200		1782.45	IV
100		1797.52	IV
20	c	1901.63	IV
100		1983.92	IV
20		1990.52	IV
40		1996.18	IV

Air

Intensity		Wavelength	
100		2003.18	IV
100		2020.94	IV
20		2071.10	IV
100		2081.09	IV
400	c	2085.70	IV
600	c	2086.47	IV
400	c	2092.16	IV
100		2103.63	IV
1000	c	2104.41	IV
200	c	2107.85	IV
1000	c	2108.31	IV
100	c	2127.43	IV

Lu V
Ref. 401 — J.R.

Lutetium (Cont.)

Intensity		Wavelength	
		Vacuum	
40		555.44	V
100		563.72	V
50		601.54	V
60		614.23	V
40		628.79	V
50	p	637.44	V
40		637.53	V
50		663.29	V
60		850.06	V
100		861.92	V
70		866.93	V
40		870.84	V
50		875.89	V
50		876.45	V
100		880.32	V
40		884.21	V
70		886.16	V
50		886.32	V
50		886.44	V
100		891.81	V
60		895.01	V
40		895.15	V
40		898.42	V
100		914.72	V
40		918.26	V
50		920.92	V
40		921.32	V
60		921.90	V
40		922.73	V
80		925.79	V
50		927.22	V
40		947.80	V
50	w	1420.02	V
50		1432.50	V
100		1432.77	V
200		1441.76	V
40		1443.64	V
100		1448.14	V
40		1449.32	V
100		1450.36	V
100		1450.69	V
100		1452.64	V
200		1453.35	V
100		1454.38	V
100		1455.21	V
100		1460.11	V
40	c	1467.81	V
200		1468.99	V
50		1469.45	V
100		1471.20	V
400		1472.12	V
200		1473.71	V
40		1475.77	V
200		1485.58	V
50		1709.02	V
40		1728.90	V
50	c	1775.92	V
40	c	1777.68	V
40	c	1784.71	V
100	c	1786.25	V
60	c	1787.58	V
60	c	1793.85	V
60		1809.73	V
60		1814.24	V

MAGNESIUM (Mg)
Z = 12

Mg I and II
Ref. 49, 83, 103, 217, 269, 315, 335
— J.O.S.

Intensity	Wavelength	
	Vacuum	
	184.05	II
	184.31	II
	184.68	II
	184.81	II
	185.26	II
	185.59	II
	185.98	II
	186.47	II
	186.84	II
	187.19	II
	187.38	II
	188.54	II
	188.91	II
	189.01	II
	189.23	II
	189.37	II
	191.30	II
	191.56	II
	191.65	II
	192.40	II
	192.55	II
	192.84	II
	193.09	II
	193.31	II
	193.40	II
	193.64	II
	197.76	II
	199.31	II
	200.29	I
	202.00	II
	202.27	II
	202.51	II
	202.94	II
	203.15	I
	203.42	II
	203.53	II
	204.22	I
	209.09	II
	209.43	II
	209.84	I
	213.53	I
	215.12	I
	215.31	I
	215.45	I
	215.66	I
	215.79	I
	216.22	I
	216.36	I
	216.68	I
	217.21	I
	217.37	I
	218.19	I
	218.34	I
	218.42	I
	218.74	I
	219.04	I
	219.28	I
	220.03	I
	220.33	I
	222.03	I
	222.67	I
	223.45	I
	223.74	I
	225.18	I
	225.54	I
	226.26	I
	247.14	II
	248.47	II
	884.70	II
	884.72	II
	907.38	II
	907.41	II
8	946.70	II
9	946.77	II
14	1025.96	II
12	1026.11	II
25	1239.94	II
20	1240.40	II
6	1248.51	II
8	1249.93	II
8	1271.24	II
9	1271.94	II
8	1272.72	II
11	1273.43	II
11	1306.71	II
12	1307.88	II
12	1308.28	II
14	1309.44	II
14	1365.45	II
	1365.54	II
15	1367.26	II
15	1367.70	II
18	1369.42	II
20	1476.00	II
25	1478.01	II
20	1480.89	II
30	1482.90	II
	1625.22	I
	1625.50	I
	1625.81	I
	1626.16	I
	1626.36	I
	1626.56	I
	1626.79	I
	1627.02	I
	1627.27	I
	1627.53	I
	1627.82	I
	1628.12	I
	1628.46	I
	1628.80	I
	1629.21	I
	1629.59	I
	1630.52	I
	1631.62	I
	1632.93	I
	1634.52	I
	1636.48	I
	1638.90	I
	1641.97	I
1	1645.93	I
1	1651.16	I
2	1658.31	I
5	1668.43	I
10	1683.41	I
15	1707.06	I
40	1734.84	II
50	1737.62	II
20	1747.80	I
40	1750.65	II
50	1753.46	II
30	1827.93	I
	Air	
9	2025.82	I
3	2329.58	II
6	2449.57	II
1	2557.23	I
1	2560.94	I
1	2562.26	I
1	2564.94	I
1	2570.91	I
1	2572.25	I
2	2574.94	I
1	2577.89	I
1	2580.59	I
1	2584.22	I
2	2585.56	I
3	2588.28	I
1	2591.89	I
1	2593.23	I
2	2595.97	I
2	2602.50	I
4	2603.85	I
5	2606.62	I
1	2613.36	I
2	2614.73	I
3	2617.51	I
3	2628.66	I
6	2630.05	I
8	2632.87	I
2	2644.80	I
3	2646.21	I
4	2649.06	I
8	2660.76	II
8	2660.82	II
6	2668.12	I
8	2669.55	I
10	2672.46	I
3	2693.72	I
5	2695.18	I
6	2698.14	I
8	2731.99	I
10	2733.49	I
12	2736.53	I
5	2765.22	I
7	2768.34	I
38	2776.69	I
32	2778.27	I
90	2779.83	I
8	2781.29	I
32	2781.42	I
36	2782.97	I
13	2790.79	II
1000	2795.53	II
16	2798.06	II
600	2802.70	II
12	2846.75	I
14	2848.42	I
16	2851.65	I
6000	2852.13	I
3	2915.45	I
10	2936.74	I
12	2938.47	I
2	2942.00	I
3	2809.76	I
2	2811.11	I
1	2811.78	I
12	2846.72	I
14	2848.34	I
16	2851.66	I
2	2902.92	I
4	2906.36	I
3	2915.45	I
2	2928.75	II
3	2936.54	II
10	2936.74	I
12	2938.47	I
13	2942.00	I
1	2967.87	II
1	2971.70	II
20	3091.08	I
22	3092.99	I
14	3096.90	I
9	3104.71	II
8	3104.81	II
6	3168.98	II
6	3172.71	II
7	3175.78	II
2	3197.62	I
17	3329.93	I
6	3332.15	I
9	3336.68	I
7	3535.04	II
8	3538.86	II
7	3549.52	II
8	3553.37	II
140	3829.30	I
300	3832.30	I
500	3838.29	I
8	3848.24	II
1	3848.91	I
7	3850.40	II
2	3853.96	I
1	3854.96	I
2	3858.86	I
3	3878.31	I
2	3891.91	I
2	3893.30	I
3	3895.57	I
4	3903.86	I
6	3938.40	I
1	3984.21	I
8	3986.75	I
2	4054.69	I
10	4057.50	I
3	4075.06	I
2	4081.83	I
4	4165.10	I
15	4167.27	I
20	4351.91	I
6	4354.53	I
6	4380.38	I
9	4384.64	II
10	4390.59	II
8	4428.00	II
9	4433.99	II
5	4436.49	II
4	4436.60	II
14	4481.16	II
13	4481.33	II
6	4534.29	II
28	4571.10	I
3	4621.30	I
7	4702.99	I
10	4730.03	I
6	4739.59	II
5	4739.71	II
7	4851.10	II
75	5167.33	I
220	5172.68	I
400	5183.61	I
8	5264.21	II
7	5264.37	II
1	5345.98	I
9	5401.54	II
2	5509.60	I
6	5528.41	I
30	5711.09	I
5	5785.31	I
4	5785.56	I
7	5916.43	II
6	5918.16	II
10	6318.72	I
9	6319.24	I
7	6319.49	I
10	6346.74	II
9	6346.96	II
11	6545.97	II
5	6620.44	II
6	6620.57	II
2	6630.83	I
7	6781.45	II
8	6787.85	II
7	6812.86	II
8	6819.27	II
4	6894.90	I
6	6965.40	I
8	7060.41	I
10	7193.17	I
10	7291.06	I
5	7387.00	I
12	7387.69	I
4	7580.76	II
20	7657.60	I
19	7659.15	I
17	7659.90	I
8	7690.16	I
15	7691.55	I
1	7722.61	I
1	7746.34	I
1	7759.30	I
5	7786.50	II
4	7790.98	II
3	7811.14	I
12	7877.05	II
2	7881.67	I
13	7896.37	II
7	7930.81	I
3	8047.73	I
5	8049.85	I
7	8054.23	I
10	8098.72	I
9	8115.22	II
8	8120.43	II
1	8154.64	I
2	8159.13	I
10	8209.84	I
20	8213.03	I
10	8213.99	II
7	8222.92	II
7	8233.19	II
11	8234.64	II

Magnesium (Cont.)

Intensity		Wavelength	
7		8303.31	I
9		8305.60	I
10		8310.26	I
15		8346.12	I
2		8466.48	I
5		8468.84	I
7		8473.69	I
10		8710.18	I
12		8712.69	I
13		8717.83	I
10		8734.99	II
17		8736.02	I
11		8745.66	II
14		8806.76	I
10		8824.32	II
11		8835.08	II
20		8923.57	I
7		8989.03	I
9		8991.69	I
10		8997.16	I
14		9218.25	II
13		9244.27	II
12		9246.50	I
30		9255.78	I
10		9327.54	II
10		9340.54	II
25		9414.96	I
17		9429.81	I
19		9432.76	I
20		9438.78	I
8		9502.45	I
7		9503.11	I
5		9503.43	I
12		9631.89	II
11		9632.43	II
15		9953.20	I
15		9983.20	I
17		9986.47	I
18		9993.21	I
14		10092.16	II
5		10391.76	II
6		10392.23	II
35		10811.08	I
11		10914.23	II
7		10915.27	II
10		10951.78	II
[illegible]		[illegible]	I
27		10957.30	I
28		10965.45	I
15		11032.10	I
14		11033.66	I
5		11255.93	II
4		11256.35	II
45		11828.18	I
30		12083.66	I
28		14877.62	I
35		15024.99	I
30		15040.24	I
25		15047.70	I
6		15740.71	I
8		15748.99	I
10		15765.84	I
30		17108.66	I
5		26392.90	I

Mg III
Ref. 4, 83, 177 — J.O.S.

Intensity		Wavelength	
		Vacuum	
		106.30	III
		106.92	III
		108.08	III
		110.16	III
		114.32	III
		126.50	III
15		170.80	III
15		171.39	III
15		182.24	III
12		182.97	III
20		186.51	III
20		187.20	III
10		188.53	III
100		231.73	III
80		234.26	III
10		1274.83	III
11	h	1280.70	III
12		1391.27	III
15		1393.39	III
10		1431.14	III
16		1572.71	III
12		1586.24	III
13		1687.09	III
13		1697.28	III
10		1722.04	III
22		1738.84	III
12		1747.56	III
18		1748.93	III
15		1772.98	III
20		1783.25	III
14		1794.58	III
15		1800.66	III
13		1858.19	III
12		1879.49	III
10		1908.50	III
12		1923.90	III
13		1930.67	III
11		1937.84	III
		Air	
15		2039.55	III
15		2055.49	III
25		2064.90	III
15		2085.90	III
20		2091.96	III
13		2097.93	III
15		2112.77	III
16		2134.06	III
20		2177.70	III
20		2395.15	III
15		2467.75	III
10		2490.54	III
10		2529.19	III
12		3299.05	III
13		3306.39	III
12		3335.90	III
11		3342.58	III
12		3361.41	III
10		3381.24	III
11		3382.90	III
11		3387.37	III
10		3706.74	III
10		4916.00	III
10		5839.82	III
15		6256.75	III

Mg IV
Ref. 7, 128, 129 — J.O.S.

Intensity		Wavelength	
		Vacuum	
40		118.16	IV
80	p	118.81	IV
70		123.59	IV
240		124.65	IV
300		129.86	IV
300		132.81	IV
400		146.95	IV
300		147.41	IV
300		147.54	IV
350		180.07	IV
400		180.62	IV
400		180.80	IV
350		181.34	IV
4000		320.99	IV
3000		323.31	IV
40		800.41	IV
150		857.29	IV
30		866.74	IV
50		919.03	IV
30		929.78	IV
40		1008.76	IV
30		1026.41	IV
250		1037.41	IV
80		1044.37	IV
60		1055.76	IV
300		1210.99	IV
300		1342.19	IV
800		1346.57	IV
300		1346.68	IV
600		1352.05	IV
900		1384.46	IV
500		1385.77	IV
800		1387.53	IV
300		1404.68	IV
1000		1409.36	IV
500		1437.53	IV
1000		1437.64	IV
300		1447.42	IV
300		1459.54	IV
400		1459.62	IV
400		1481.51	IV
350		1490.45	IV
300		1495.50	IV
300		1607.11	IV
500		1683.02	IV
400		1698.81	IV
300		1844.17	IV
		Air	
12	p	2518.40	IV
4		2534.79	IV

Mg V
Ref. 128 — J.O.S.

Intensity		Wavelength	
		Vacuum	
5		251.58	V
35		276.58	V
10		312.30	V
20		351.09	V
18		352.20	V
30		353.09	V
15		353.30	V
18		354.22	V
20		355.33	V

MANGANESE (Mn)
Z = 25

Mn I and II
Ref. 1, 126 — C.H.C.

Intensity		Wavelength	
		Vacuum	
20		1726.47	II
30		1732.70	II
50		1733.55	II
40		1734.49	II
30		1737.93	II
20		1740.16	II
20		1742.00	II
30		1853.27	II
20		1857.92	II
30		1902.95	II
20		1907.84	II
30		1911.41	II
20	d	1914.68	II
100		1915.10	II
20		1918.64	II
30		1919.64	II
80		1921.25	II
20		1923.07	II
20		1923.34	II
30		1925.52	II
50		1926.59	II
30		1931.40	II
20		1945.15	II
20		1947.93	II
20		1950.14	II
30		1953.23	II
20	d	1954.81	II
30		1959.25	II
20		1969.24	II
30		1994.23	II
9700		1996.06	I
14000		1999.51	I
		Air	
18000		2003.85	I
50		2037.31	II
40		2037.64	II
40		2039.97	II
30		2076.21	II
1500		2092.16	I
20		2097.46	II
20		2102.50	II
1700		2109.58	I
30		2113.96	II
290		2208.81	I
540		2213.85	I
770		2221.84	I
20		2373.36	II
20		2427.38	II
50		2427.72	II
30		2427.94	II
30		2437.37	II
20		2437.84	II
30		2452.49	II
50		2499.00	II
30		2507.60	II
20		2516.60	II
30		2516.74	II
20		2521.66	II
20		2530.72	II
20		2531.80	II
50		2532.78	II
75		2533.06	I
50		2533.33	II
30		2534.10	II
80		2534.22	II
100		2535.66	II
30		2535.98	II
100		2537.92	II
50		2541.11	II
80		2542.92	II
50		2543.45	II
100		2548.75	II
50		2551.85	II
30		2553.27	II
75		2556.57	II
30		2556.89	II
50		2557.54	II
95		2558.59	II
30		2559.41	II
150		2563.65	II
30		2565.22	II
580		2572.76	I
480		2575.51	I
12000		2576.10	II
550		2584.31	I
30		2588.97	II
45		2589.71	II
250		2592.94	I
6200		2593.73	II
250		2595.76	I
95		2598.90	II
40		2602.14	I
30		2602.72	II
45		2603.72	II
4300		2605.69	II
190		2610.20	II
500		2618.14	II
140		2622.90	I
150		2624.04	I
40		2624.80	II
200		2625.58	II
95		2626.64	I
30		2630.26	I
60		2630.57	I
190		2632.35	II
130		2638.17	II
80		2639.84	II
27		2650.99	II
60		2655.91	II
30		2666.77	II
45		2667.00	I
30		2667.03	II
110		2672.59	II
55		2673.37	II
55		2674.43	II
30		2676.33	I
45		2680.34	II
30		2680.68	II
30		2681.25	II
40		2681.72	I
45		2683.02	I
23		2683.75	I
55		2684.55	II
55		2685.94	II
110		2688.25	II
[illegible]		[illegible]	I
27		2693.19	II
55		2695.36	II
27		2698.97	II
85		2701.00	II
50		2701.17	II
160		2701.70	II
100		2703.98	II
130		2705.74	II
80		2707.53	II
110		2708.45	II
45		2709.96	II
80		2710.33	II
110		2711.58	II
30		2716.80	II
30		2717.53	II
30		2719.01	II
50		2719.74	II
30		2722.10	II
30		2724.46	II
55		2728.61	II
30		2738.86	I
45	h	2760.93	I
30		2771.44	I
30	h	2776.23	I
30		2780.00	I
55		2789.20	I
60		2790.36	I
60		2791.08	I
6200		2794.82	I
5100		2798.27	I
220		2799.84	I
3700		2801.06	I
70		2804.10	I
60		2806.14	I
55		2808.02	I
110		2809.11	I
60		2812.84	I
70		2813.47	I
60		2815.02	II
30		2816.33	II
85		2817.97	I
40		2818.77	I
55		2821.45	I
55		2822.55	I
80		2830.79	I
27		2836.31	I
60		2870.08	II
30		2872.94	II
80		2879.49	II
40		2882.90	I
70		2886.68	II
160		2889.58	II
55		2892.39	II
50		2898.70	II
80		2900.16	II
40		2907.22	I
140	h	2914.60	I
190	h	2925.57	I

Manganese (Cont.)

Intensity		Wavelength	
27		2928.68	I
1100		2933.06	II
27		2934.02	I
1500		2939.30	II
250	h	2940.39	I
		2940.48	I
60		2941.04	I
1900		2949.20	II
40		3007.66	I
40		3011.16	I
40		3011.38	I
40		3014.67	I
60		3016.45	I
30		3019.92	II
70		3022.75	I
55		3031.06	II
30		3035.35	II
95		3040.60	I
27		3042.73	I
85		3043.36	I
330		3044.57	I
120		3045.59	I
200		3047.04	I
40		3048.86	I
30		3050.65	II
250		3054.36	I
140		3062.12	I
170		3066.02	I
170		3070.27	I
160		3073.13	I
90		3079.63	I
50		3081.33	I
23		3082.05	I
40		3097.06	I
40		3110.68	I
60	h	3148.18	I
90	h	3161.04	I
140	h	3178.50	I
220		3212.88	I
65		3216.95	I
1000		3228.09	I
300		3230.72	I
850		3236.78	I
330		3243.78	I
650		3248.52	I
100		3251.14	I
310		3252.95	I
65		3254.04	I
310		3256.14	I
220		3258.41	I
180		3260.23	I
180		3264.71	I
65		3296.88	I
65		3298.22	I
65		3320.69	I
70		3330.67	I
200		3330.78	II
100		3336.39	II
30		3365.02	II
30		3400.12	II
50		3438.97	II
720		3441.99	II
50		3460.03	II
360		3460.33	II
360	h	3474.04	II
		3474.13	II
290		3482.91	II
180		3488.68	II
140		3495.84	II
50		3496.81	II
100		3497.54	II
360		3531.85	I
		3532.00	I
1100		3532.12	I
1300		3547.80	I
1100		3548.03	I
390		3548.20	I
2200		3569.49	I
720		3569.80	I
		3570.04	I
1400		3577.88	I
720		3586.54	I
290		3595.12	I
420		3607.54	I
420		3608.49	I
360		3610.30	I
290		3619.28	I
220		3623.79	I
140		3629.74	I
100		3660.40	I
70		3670.52	I
70		3676.96	I
50		3682.09	I
280		3693.67	I
180		3696.57	I
70		3701.73	I
210		3706.08	I
130		3718.93	I
55		3728.89	I
130		3731.93	I
260		3790.22	I
55		3799.26	I
110		3800.55	I
55		3801.91	I
3200		3806.72	I
700		3809.59	I
55		3810.69	I
90		3816.75	I
2100		3823.51	I
390		3823.89	I
200		3829.68	I
480		3833.86	I
1300		3834.36	I
350		3839.78	I
670		3841.08	I
350		3843.98	I
65		3918.32	I
120		3926.47	I
65		3952.84	I
55		3975.89	I
65		3977.08	I
130		3982.58	I
150		3985.24	I
190		3986.83	I
150		3987.10	I
1500		4018.10	I
150		4026.44	I
27000		4030.76	I
19000		4033.07	I
11000		4034.49	I
1500		4035.73	I
55		4038.73	I
5600		4041.36	I
210	d	4045.13	I
		4045.21	I
1100		4048.76	I
80		4049.00	I
55		4051.73	I
65		4052.47	I
150		4055.21	I
1900		4055.54	I
210		4057.95	I
1100		4058.93	I
150		4059.39	I
730		4061.74	I
730		4063.53	I
80		4065.08	I
80		4068.00	I
290		4070.28	I
730		4079.24	I
730		4079.42	I
1100		4082.94	I
1100		4083.63	I
65		4089.94	I
55		4105.36	I
200		4110.90	I
150		4131.12	I
120		4135.04	I
80		4141.06	I
55		4147.53	I
80		4148.80	I
150		4176.60	I
120		4189.99	I
65		4201.76	I
65		4211.75	I
370		4235.14	I
510		4235.29	I
190		4239.72	I
290		4257.66	I
290		4265.92	I
270		4281.10	I
65		4284.08	I
65		4312.55	I
50		4323.63	II
45		4374.95	I
45		4381.70	I
55		4411.88	I
350		4414.88	I
55		4419.78	I
210		4436.35	I
800		4451.59	I
160		4453.00	I
130		4455.01	I
160		4455.32	I
110		4455.82	I
55		4457.04	I
210		4457.55	I
270		4458.26	I
55		4460.38	I
150		4461.08	I
510		4462.02	I
290		4464.68	I
200		4470.14	I
130		4472.79	I
40		4479.40	I
170		4490.08	I
240		4498.90	I
240		4502.22	I
80		4605.36	I
80		4626.54	I
35		4671.69	I
50		4701.16	I
160		4709.72	I
180		4727.48	I
130		4739.11	I
1000		4754.04	I
180		4761.53	I
750		4762.38	I
300		4765.86	I
500		4766.43	I
940		4783.42	I
1000		4823.52	I
25		4844.32	I
35		4965.88	I
19		5004.91	I
30		5074.79	I
60		5117.94	I
50		5150.89	I
50		5196.59	I
85		5255.32	I
160		5341.06	I
19		5349.88	I
95		5377.63	I
95		5394.67	I
50		5399.49	I
95		5407.42	I
35		5413.69	I
85		5420.36	I
35		5432.55	I
12		5457.47	I
60		5470.64	I
40		5481.40	I
30		5505.87	I
50		5516.77	I
40		5537.76	I
21		5551.98	I
8		5567.76	I
7		5573.01	I
8		5573.68	I
7		5738.29	I
7		5780.19	I
7		5816.84	I
140		6013.50	I
200		6016.64	I
290		6021.80	I
7		6384.67	I
17		6440.97	I
24		6491.71	I
14	h	6942.52	I
12		6989.96	I
14		7069.84	I
12		7184.25	I
10		7247.82	I
24	h	7283.82	I
35	h	7302.89	I
50		7326.51	I
12		7680.20	I
10		7712.42	I
10	h	7764.72	I
10	h	8670.92	I
12	h	8672.06	I
10	h	8673.97	I
12	h	8701.05	I
17	h	8703.76	I
30	h	8740.93	I

Mn III
Ref. 385 — C.H.C.

Intensity		Wavelength	
		Vacuum	
20		892.39	III
20		1108.16	III
30		1183.30	III
25	w	1183.86	III
30		1198.49	III
30		1219.80	III
100		1228.97	III
500		1283.58	III
400		1287.59	III
300		1291.62	III
1000		1360.72	III
800		1365.20	III
400		1369.43	III
300		1371.65	III
300		1596.95	III
500	h	1609.17	III
1000		1614.14	III
2000		1620.60	III
300		1623.91	III
400		1629.12	III
500		1633.80	III
250		1647.46	III
400		1653.57	III
400		1804.06	III
300		1806.47	III
300		1811.02	III
400		1877.62	III
300		1885.21	III
500		1941.28	III
250	w	1942.89	III
800		1943.21	III
500		1952.36	III
1000		1952.52	III
300		1956.61	III
250		1962.04	III
500		1978.95	III
400		1982.76	III
400		1989.59	III
		Air	
300		2022.19	III
1000	w	2027.83	III
500	w	2028.14	III
300		2044.57	III
400		2048.93	III
500		2049.68	III
300		2056.80	III
500		2066.38	III
1000		2069.02	III
900		2077.38	III
300		2078.13	III
800		2084.23	III
600		2090.05	III
300		2090.25	III
300		2094.14	III
500		2094.78	III
500		2097.93	III
500		2099.97	III
300		2123.25	III
1000		2169.78	III
700		2174.15	III
900		2176.87	III
800		2181.86	III
800	w	2184.87	III
600		2185.13	III
400		2211.95	III
600		2212.42	III
800		2215.21	III
900		2220.55	III
1000		2227.42	III
100		3287.49	III
100		3540.52	III
150		3601.72	III
100		3616.00	III
100		4246.17	III
200		5079.20	III
150		5100.03	III
100		5117.03	III
100		5252.23	III
100		5365.59	III
150		5454.07	III
200		5474.68	III
100		5671.12	III
200		5946.65	III
100	s	6213.11	III
200		6231.21	III
100		6238.64	III
100		6273.71	III

Mn IV
Ref. 433 — C.H.C.

Intensity		Wavelength	
		Vacuum	
60		579.79	IV
60		581.44	IV
60		581.65	IV
60		585.21	IV
90		1242.25	IV
90		1244.50	IV
85		1247.73	IV
95		1251.93	IV
95		1257.28	IV
90		1264.41	IV
70		1603.60	IV
70		1611.10	IV
75		1653.83	IV
70		1656.39	IV
70		1659.25	IV
75		1664.73	IV
80	b	1667.00	IV
70		1670.08	IV
75		1691.68	IV
75		1693.15	IV
80		1698.30	IV
75		1698.70	IV
70		1699.06	IV
75		1707.43	IV
65		1718.67	IV
75	b	1720.52	IV
75		1720.74	IV
75		1721.41	IV
65		1722.94	IV
75		1724.83	IV
85	b	1742.10	IV
85		1751.59	IV
75		1759.82	IV
70		1762.17	IV
75		1762.94	IV
85	d	1766.27	IV
75		1767.09	IV
65		1772.11	IV
75		1773.51	IV
75		1782.21	IV
75		1786.02	IV

Manganese (Cont.)

75	1787.04	IV
75	1787.38	IV
75	1788.64	IV
75	1790.44	IV
80	1795.65	IV
80	1795.79	IV
60	1907.03	IV
75	1910.25	IV
65	1997.54	IV

Mn V
Ref. 405 — C.H.C.

Intensity	Wavelength	
	Vacuum	
300	404.36	V
380	406.02	V
300	406.40	V
600	410.30	V
600	410.60	V
480	410.98	V
400	411.32	V
460	412.74	V
460	413.75	V
600	415.62	V
650	415.98	V
350	419.80	V
600	428.59	V
500	429.05	V
400	433.54	V
600	435.67	V
350	436.16	V
500	436.18	V
450	438.74	V
350	439.35	V
1000	441.72	V
850	442.49	V
400	467.32	V
300	474.82	V

MERCURY (198) (Hg)
Z = 80

Hg I and II (198)
Ref. 43, 50, 69, 145, 229, 242 — R.W.S.

Intensity	Wavelength	
	Vacuum	
80	1250.564	I
8	1259.242	I
100	1268.825	I
5	1307.751	I
20	1402.619	I
10	1435.503	I
1000	1849.492	I
	Air	
60	2262.210	II
20	2302.065	I
20	2345.440	I
100	2378.325	I
20	2380.004	I
40	2399.349	I
20	2399.729	I
20	2446.900	I
15	2464.064	I
40	2481.999	I
30	2482.713	I
40	2483.821	I
90	2534.769	I
15000	2536.506	I
25	2563.861	I
25	2576.290	I
250	2652.043	I
400	2653.683	I
100	2655.130	I
50	2698.831	I
80	2752.783	I
20	2759.710	I
40	2803.471	I
30	2804.438	I
750	2847.675	II
50	2856.939	I
150	2893.598	I
150	2916.227	II
60	2925.413	I
1200	2967.283	I
300	3021.500	I
120	3023.476	I
30	3025.608	I
50	3027.490	I
400	3125.670	I
320	3131.551	I
320	3131.842	I
80	3341.481	I
2800	3650.157	I
300	3654.839	I
80	3662.883	I
240	3663.281	I
30	3701.432	I
35	3704.170	I
30	3801.660	I
20	3901.867	I
60	3906.372	I
200	3983.839	II
1800	4046.572	I
150	4077.838	I
40	4108.057	I
250	4339.224	I
400	4347.496	I
4000	4358.337	I
80	4916.068	I
1100	5460.753	I
160	5675.922	I
240	5769.598	I
280	5790.663	I
20	6072.713	I
30	6234.402	I
160	6716.429	I
250	6907.461	I
240	11287.407	I

MERCURY (NATURAL) (Hg)
Z = 80

Hg I and II (nat.)
Ref. 34, 45, 90, 117, 133, 189, 235, 304, 327, 328 — R.W.S.

Intensity	Wavelength	
	Vacuum	
400	893.08	II
300	915.83	II
150	923.39	II
200	940.80	II
100	962.74	II
50	969.13	II
800	1099.26	II
80	1250.58	I
8	1259.24	I
100	1268.82	I
5	1307.75	I
300	1307.93	II
400	1321.71	II
400	1331.74	II
80	1350.07	II
200	1361.27	II
20	1402.62	I
200	1414.43	II
10	1435.51	I
15	1619.46	II
120	1623.95	II
20	1628.25	II
150	1649.94	II
50	1653.64	II
200	1672.41	II
100	1702.73	II
100	1707.40	II
120	1727.18	II
250	1732.14	II
20	1775.68	I
40	1783.70	II
30	1796.22	II
200	1796.90	II
60	1798.74	II
30	1803.89	II
40	1808.29	II
400	1820.34	II
5	1832.74	I
1000	1849.50	I
160	1869.23	II
300	1870.55	II
200	1875.54	II
20	1900.28	II
30	1927.60	II
300	1942.27	II
100	1972.94	II
200	1973.89	II
150	1987.98	II
	Air	
90	2026.97	II
90	2052.93	II
70	2148.00	II
5	2247.55	I
60	2262.23	II
20	2302.06	I
15	2323.20	I
5	2340.57	I
20	2345.43	I
20	2352.48	I
100	2378.32	I
20	2380.00	I
40	2399.38	I
20	2399.73	I
10	2400.49	I
60	2407.35	II
50	2414.13	II
5	2441.06	I
20	2446.90	I
15	2464.06	I
40	2482.00	I
30	2482.72	I
40	2483.82	I
90	2534.77	I
15000	2536.52	I
25	2563.86	I
25	2576.29	I
5	2578.91	I
15	2625.19	I
5	2639.78	I
250	2652.04	I
400	2653.69	I
100	2655.13	I
5	2674.91	I
50	2698.83	I
50	2699.38	I
80	2705.36	II
80	2752.78	I
20	2759.71	I
40	2803.46	I
30	2804.43	I
2	2805.34	I
2	2806.77	I
150	2814.93	II
750	2847.68	II
50	2856.94	I
150	2893.60	I
150	2916.27	II
60	2925.41	I
150	2935.94	II
400	2947.08	II
1200	2967.28	I
300	3021.50	I
120	3023.47	I
30	3025.61	I
50	3027.49	I
400	3125.67	I
320	3131.55	I
320	3131.84	I
400	[illegible]	II
400	3264.06	II
80	3341.48	I
100	3385.25	II
400	3451.69	II
200	3549.42	II
2800	3650.15	I
300	3654.84	I
80	3662.88	I
240	3663.28	I
30	3701.44	I
35	3704.17	I
30	3801.66	I
100	3806.38	II
20	3901.87	I
60	3906.37	I
100	3918.92	II
200	3983.96	II
1800	4046.56	I
150	4077.83	I
40	4108.05	I
250	4339.22	I
400	4347.49	I
4000	4358.33	I
100	4398.62	II
90	4660.28	II
80	4855.72	II
5	4883.00	I
5	4889.91	I
80	4916.07	I
5	4970.37	I
5	4980.64	I
20	5102.70	I
40	5120.64	I
100	5128.45	II
20	5137.94	I
20	5290.74	I
5	5316.78	I
60	5354.05	I
30	5384.63	I
1100	5460.74	I
30	5549.63	I
160	5675.86	I
240	5769.60	I
100	5789.66	I
280	5790.66	I
140	5803.78	I
60	5859.25	I
60	5871.73	II
20	5871.98	I
20	6072.72	I
1000	6149.50	II
30	6234.40	I
80	6521.13	II
160	6716.43	I
250	6907.52	I
250	7081.90	I
200	7091.86	I
40	7346.37	II
100	7485.87	II
20	7728.82	I
100	7944.66	II
2000	10139.75	I
240	11287.40	I
120	13209.95	I
140	13426.57	I
60	13468.38	I
80	13505.58	I
500	13570.21	I
450	13673.51	I
200	13950.55	I
500	15295.82	I
100	16881.48	I
400	16920.16	I
300	16942.00	I
500	17072.79	I
400	17109.93	I
20	17116.75	I
20	17198.67	I
20	17213.20	I
70	17329.41	I
30	17436.18	I
50	18130.38	I
40	19700.17	I
	22493.28	I
250	23253.07	I
	32148.06	I
	36303.03	I

Hg III
Ref. 343 — C.H.C.

Intensity	Wavelength	
	Vacuum	
3	621.44	III
2	679.68	III
2	878.59	III
1	886.48	III
1	988.89	III
2	1009.29	III
5	1068.03	III
2	1161.95	III
9	1681.40	III
15	1759.75	III
1	1894.77	III
	Air	
7	2314.15	III
4	2380.55	III
8	2431.65	III
5	2480.56	III
7	2484.50	III
2	2612.92	III
4	2617.97	III
3	2670.49	III
70	2724.43	III
6	2769.22	III
3	2844.76	III
15	3090.05	III
5	3283.02	III
12	3312.28	III
8	3389.01	III
5	3450.77	III
3	3500.35	III
4	3538.88	III
5	3557.24	III
15	3803.51	III
70	4122.07	III
10	4140.34	III
100	4216.74	III
15	4470.58	III
12	4552.84	III
50	4797.01	III
10	4869.85	III
80	4973.57	III
30	5210.82	III
6	5695.71	III
25	6220.35	III
35	6418.98	III
40	6501.38	III
10	6584.26	III
6	6610.12	III
30	6709.29	III
12	7517.46	III
7	7808.10	III
25	7946.75	III
50	7984.51	III
5	8151.64	III

MOLYBDENUM (Mo)
Z = 42

Mo I and II
Ref. 1 — C.H.C.

Intensity		Wavelength	
		AIR	
19000		2015.11	II
40000		2020.30	II
21000		2038.44	II
17000		2045.98	II
4800		2081.68	II
2400		2089.52	II
2200		2092.50	II
4000		2093.11	II
2700		2100.84	II
1500		2104.29	II
1400		2108.02	II
400		2269.69	II
160		2304.25	II
160		2306.97	II
130		2325.94	I
240		2330.46	I
110		2332.12	II
190		2340.47	I
190		2341.59	II
80		2352.61	I
80		2355.22	I
80		2355.42	II
70		2364.37	I
50		2366.09	II
140		2372.27	I
100		2380.41	I
150		2383.52	I
110		2389.20	II
140		2403.61	II
80		2404.66	II
140		2405.86	I
40		2408.39	I
40		2412.84	II
120		2413.01	II
70		2415.33	I
80		2417.96	II
65		2419.01	II
80		2420.18	II
70		2424.00	II
65		2430.43	I
65		2435.96	II
65		2440.28	II
40		2461.81	II
50		2466.68	II
50		2466.97	II
50		2468.78	II
30		2470.04	II
150	h	2471.97	I
70		2477.57	II
70	h	2481.81	I
65		2482.57	II
40		2484.75	II
40	h	2485.31	I
24		2496.24	II
85		2498.28	II
40		2500.44	II
65		2502.84	II
50		2511.80	II
65		2515.08	II
70		2527.14	II
50		2530.34	II
70		2532.31	II
440		2538.46	II
50		2539.44	II
110	h	2540.45	I
330		2542.67	II
40		2543.61	II
330		2548.22	I
110	h	2550.85	I
65		2555.42	II
40		2556.75	II
80		2558.88	II
65		2562.08	II
85		2564.34	II
40		2566.26	II
250		2567.05	I
20		2571.45	II
320		2572.34	I
50		2574.42	II
40		2576.56	II
40		2578.36	II
250		2582.16	I
30		2585.95	II
65		2588.78	II
40		2591.77	II
250		2593.70	II
100		2595.40	I
40		2597.38	II
250		2602.80	II
40		2605.08	II
40		2605.93	II
250		2607.37	I
190		2611.20	I
290		2613.08	I
130		2615.39	I
400		2616.78	I
70		2619.34	II
140		2621.07	I
320		2627.55	I
160		2628.74	I
440		2629.85	I
330		2636.67	II
250		2638.30	I
720		2638.76	II
410		2640.99	I
600		2644.35	II
370		2646.49	II
640		2649.46	I
480		2653.35	II
560	h	2655.03	I
290		2658.11	I
640		2660.58	II
110		2665.10	I
55		2671.83	II
720		2672.84	II
250		2673.27	II
1000		2679.85	I
95		2681.36	II
640		2683.23	II
880		2684.14	II
560		2687.99	II
30		2692.61	II
55		2695.22	II
30		2696.83	II
55		2699.41	II
140		2701.03	I
480		2701.42	II
30		2701.87	II
30		2704.93	II
40		2710.19	II
30		2711.49	II
50		2712.35	II
190		2713.51	II
290		2717.35	II
110		2724.41	I
180		2725.15	I
85		2726.97	II
140		2729.68	II
80		2730.20	II
330		2732.88	II
250		2733.39	I
160		2736.96	II
80	h	2737.88	II
50		2738.60	II
40		2741.32	II
55		2741.62	II
240		2743.07	I
290		2746.30	II
320		2751.47	I
110		2756.07	II
65	d	2758.63	II
20		2760.53	II
190		2761.53	I
220		2763.62	II
110		2766.26	I
240		2769.76	II
160		2773.78	II
190		2774.39	II
1700		2775.40	II
130		2777.74	I
65		2777.86	II
880		2780.04	II
400		2784.99	II
180		2787.83	I
40		2791.54	II
240	d	2797.93	I
220		2801.47	I
400		2807.76	II
28		2812.58	II
24		2814.67	II
1700		2816.15	II
220		2817.44	II
50		2822.03	II
240		2826.54	I
80		2827.74	II
40		2831.44	II
30		2832.07	II
80		2834.39	II
80		2835.33	II
160		2842.15	II
24		2843.73	II
220		2844.39	I
1700		2848.23	II
160		2849.38	I
370		2853.23	II
50		2856.00	II
24		2863.20	II
370		2863.81	II
160		2864.31	I
140		2864.66	I
40		2865.62	II
220		2866.69	II
40		2868.11	II
40		2868.32	II
1700		2871.51	II
85		2872.88	II
220		2879.05	II
65		2888.15	II
95		2891.28	II
1300		2890.99	II
190		2892.81	II
950		2894.45	II
140		2897.63	II
70		2900.80	II
290		2903.07	II
160		2905.27	I
80		2907.12	II
600		2909.12	II
1100		2911.92	II
55		2913.81	II
120		2918.83	II
1300		2923.39	II
140		2924.32	II
65		2927.54	II
50		2930.06	II
1100		2930.50	II
55		2930.77	II
800		2934.30	II
65		2935.20	II
120		2937.66	I
40		2938.30	II
95		2940.10	II
110		2941.22	II
140		2944.21	I
150		2944.82	II
140		2945.66	I
		2945.95	II
190		2946.01	I
140		2946.42	I
140		2946.69	II
95		2947.28	II
95		2955.84	II
240		2956.06	II
70		2956.90	II
95		2960.24	II
140		2962.89	I
250		2963.79	II
50		2964.96	II
210		2965.27	II
70		2971.91	II
250		2972.61	II
80		2975.40	II
180		2978.28	I
120		2981.52	I
110		2987.92	I
160		2988.68	I
190		2989.80	I
95		2992.84	II
50		2993.52	II
190		3002.21	I
40		3004.46	II
130		3013.39	I
140	h	3013.76	I
250		3025.00	I
95		3027.77	II
100		3036.31	I
300		3041.70	I
150		3046.80	I
210		3047.31	I
210		3055.32	I
100		3060.78	II
160		3061.59	I
800		3064.28	I
250		3065.04	II
100		3068.00	I
250		3070.90	I
800		3074.37	I
85		3077.66	II
150		3079.88	I
210		3080.41	I
800		3085.62	I
270		3087.62	II
100		3089.12	I
100		3089.71	I
190		3092.07	II
560		3094.66	I
110		3099.93	I
110		3100.88	I
560		3101.34	I
1400		3112.12	I
290		3122.00	II
14000		3132.59	I
110		3138.72	II
220		3147.35	I
220		3152.82	II
55		3155.64	II
6000		3158.16	I
120		3164.53	I
8700		3170.35	I
95		3172.03	II
160		3172.74	II
370		3183.03	I
120		3184.57	I
370		3185.10	I
180		3185.71	I
120	d	3187.59	II
7600		3193.97	I
290		3195.96	I
120		3198.85	I
40		3201.50	II
330		3205.22	I
880		3205.88	I
3000		3208.83	I
240		3210.97	I
560		3215.07	I
350		3221.74	I
880		3228.22	I
600		3229.79	I
1100		3233.14	I
950		3237.08	I
65		3240.71	II
950		3256.21	I
300		3262.63	I
480		3264.40	I
800		3270.90	I
240		3285.02	I
320		3285.36	I
1100		3289.02	I
950		3290.82	I
190		3292.31	II
320		3305.56	I
320		3307.12	I
100		3313.62	II
190		3320.90	II
640		3323.95	I
360		3325.67	I
360		3327.30	I
240		3340.17	I
1300		3344.75	I
95		3346.40	II
320		3347.02	I
1600		3358.12	I
250		3361.37	I
950		3363.78	I
950		3379.97	I
320		3382.48	I
1900		3384.62	I
130		3395.36	II
640		3404.34	I
1300		3405.94	I
240		3418.52	I
250		3420.04	I
250		3422.31	I
380		3434.79	I
320		3435.45	I
640		3437.22	I
250		3438.87	I
250		3441.44	I
250		3443.26	I
130		3446.08	II
3200		3447.12	I
640		3449.07	I
300		3451.75	I
250		3452.60	I
950		3456.39	I
640		3460.78	I
320		3466.83	I
250		3467.85	I
320		3469.22	I
240		3485.93	I
800		3504.41	I
240		3505.32	I
560		3508.12	I
480		3521.41	I
240		3524.98	I
640		3537.28	I
320		3542.17	I
520		3558.10	I
400		3563.14	I
300		3566.05	I
240		3570.65	I
320		3573.88	I
1400		3581.89	I
200		3590.74	I
210		3598.88	I
270		3602.94	I
210		3608.37	I
200		3623.23	I
1400		3624.46	I
330		3626.18	I
28		3635.14	II
1000		3635.43	I
400		3657.35	I
540		3664.81	I
290		3666.72	I
590		3672.82	I
1300		3680.60	I
45		3684.22	II
65		3688.31	II
240		3690.59	I
180		3692.64	II
1400		3694.94	I
220		3702.03	I
220		3715.65	I
500		3727.69	I
330	d	3732.71	I
240		3742.28	I
80		3744.37	II
360		3770.45	I
220		3779.77	I

Molybdenum (Cont.)

Intensity		Wavelength	
360		3781.59	I
250		3797.30	I
29000		3798.25	I
290		3801.84	I
520		3826.70	I
940		3828.87	I
1700		3833.75	I
380		3847.25	I
29000		3864.11	I
580		3869.08	I
580		3886.82	I
380		3901.77	I
19000		3902.96	I
65		3941.48	II
230		3943.04	I
270		4056.01	I
1400		4062.08	I
2300		4069.88	I
1300		4081.44	I
940		4084.38	I
250		4102.15	I
730		4107.47	I
630		4120.10	I
2900		4143.55	I
230		4148.94	I
250		4155.28	I
200		4157.40	I
200		4178.27	I
480		4185.82	I
2500		4188.32	I
250		4194.56	I
1500		4232.59	I
270		4269.28	I
890		4276.91	I
1200		4277.24	I
1400		4288.64	I
680		4292.13	I
890		4293.21	I
360		4293.88	I
840		4326.14	I
250		4326.74	I
230		4350.34	I
230		4369.04	I
1900		4381.64	I
2500		4411.57	I
210		4423.62	I
990		4434.95	I
200		4442.20	I
340		4449.74	I
480		4457.36	I
630		4474.56	I
230		4491.28	I
120		4504.90	I
140		4512.15	I
230		4517.13	I
230		4524.34	I
120		4529.40	I
400		4536.80	I
110		4558.11	I
210		4576.50	I
170		4595.16	I
360		4609.88	I
100		4621.38	I
460		4626.47	I
100		4627.48	I
220		4662.76	I
130		4671.90	I
130		4688.22	I
640		4707.26	I
150		4708.22	I
220		4717.92	I
100		4729.14	I
700		4731.44	I
100		4750.39	I
770		4760.19	I
150		4776.34	I
100		4796.52	I
410		4819.25	I
410		4830.51	I
360		4868.00	I
110		[illegible]	I
150		4957.54	I
210		4979.12	I
110		4999.91	I
20		5010.81	I
180		5014.60	I
26		5016.78	I
20		5019.85	I
80		5029.00	I
65		5030.78	I
23		5038.91	I
26		5046.52	I
100		5047.71	I
50		5055.00	I
35		5058.07	I
200		5059.88	I
35		5062.52	I
29		5064.64	I
35		5079.87	I
100		5080.02	I
35		5081.26	I
40		5090.97	I
35		5091.34	I
35		5092.16	I
40		5095.89	I
100		5096.65	I
130		5097.52	I
35		5098.03	I
130		5109.71	I
80		5114.97	I
35		5116.97	I
29		5123.83	I
150		5145.38	I
110		5147.39	I
80		5163.19	I
100		5167.76	I
160	d	5171.08	I
		5171.25	I
230	h	5172.94	I
160	h	5174.18	I
40		5191.44	I
110		5200.17	I
50		5200.74	I
26		5210.44	I
50		5211.86	I
80		5219.40	I
65		5231.06	I
26		5232.36	I
100		5234.26	I
460	h	5238.20	I
230	h	5240.88	I
110	h	5242.81	I
100		5245.51	I
150		5259.04	I
16		5260.17	I
65		5261.14	I
20		5268.95	I
35		5271.80	I
35		5276.28	I
65		5279.65	I
210		5280.86	I
20		5283.84	I
55		5292.08	I
35		5293.46	I
55		5295.47	I
20		5306.26	I
55		5313.89	I
35		5313.04	I
20		5319.89	I
20		5324.47	I
35		5327.06	I
20		5352.35	I
80		5354.88	I
35		5355.51	I
65		5356.48	I
560	H1	5360.56	I
110	H1	5364.28	I
35	H1	5367.11	I
35		5372.40	I
26		5388.69	I
65		5394.52	I
35		5397.38	I
50		5400.47	I
35		5405.79	I
35		5406.39	I
40		5417.38	I
23		5426.89	I
55		5435.68	I
65		5437.75	I
40		5450.51	I
35		5456.46	I
26		5460.53	I
23		5465.57	I
35		5475.90	I
35		5490.28	I
20		5492.17	I
26	h	5493.80	I
26		5498.49	I
50		5501.54	I
23		5501.87	I
26	h	5503.54	I
7800		5506.49	I
23		5520.04	I
20		5520.64	I
40		5526.52	I
40		5526.97	I
5200		5533.05	I
40		5539.41	I
50		5543.12	I
40		5544.49	I
55		5556.28	I
26		5556.72	I
20		5564.05	I
40		5568.62	I
26		5569.48	I
2500		5570.45	I
35		5575.19	I
20		5591.58	I
40		5602.76	I
23		5608.62	I
23		5609.23	I
100		5610.93	I
23		5613.07	I
20		5618.45	I
23		5619.38	I
330		5632.47	I
50		5634.86	I
230		5650.13	I
23		5673.63	I
55		5674.47	I
40		5677.89	I
35		5682.89	I
460		5689.14	I
23		5699.28	I
80		5705.72	I
23		5711.80	I
210		5722.74	I
23		5728.77	I
26	d	5729.45	I
		5729.59	I
620		5751.40	I
23		5774.55	I
40		5779.36	I
23	h	5783.33	I
520		5791.85	I
23	h	5795.77	I
26		5800.46	I
35		5802.67	I
23		5825.20	I
23		5835.59	I
20		5839.99	I
20	h	5848.86	I
55	h	5849.73	I
50	h	5851.52	I
520		5858.27	I
20		5861.38	I
50		5869.33	I
26		5876.59	I
820		5888.33	I
23		5892.29	I
50	h	5893.38	I
20		5898.78	I
		5898.82	MO I
40		5901.47	I
40	h	5926.36	I
160	h	5928.88	I
40		5988.17	I
35		6025.49	I
16		6027.27	I
1300		6030.66	I
20		6047.83	I
20		6054.81	I
20		6079.58	I
10		6081.27	I
40		6101.87	I
10		6130.63	I
10		6197.66	I
20		6217.89	I
10		6264.27	I
16		6265.88	I
15		6290.74	I
13		6301.75	I
11		6323.54	I
40		6357.22	I
16		6389.11	I
11		6391.12	I
35		6401.07	I
26		6409.11	I
10		6412.39	I
100		6424.37	I
20		6446.34	I
20		6471.20	I
20		6473.99	I
10		6493.13	I
23		6519.84	I
15	h	6611.20	I
230		6619.13	I
10		6624.57	I
50		6650.38	I
13		6659.68	I
18		6690.47	I
110		6733.98	I
21		6746.08	I
50		6746.27	I
35		6753.97	I
13		6763.50	I
10	h	6799.88	I
10		6802.62	I
10		6812.03	I
13		6825.63	I
18	d	6828.87	I
		6829.05	I
40		6838.88	I
16		6848.92	I
21		6886.28	I
16		6892.36	I
10		6898.01	I
10		6898.98	I
13		6908.20	I
35		6914.01	I
13		6934.10	I
10		6947.39	I
10		6960.64	I
16		6978.71	I
26		6988.94	I
12		6999.13	I
16		7001.60	I
22		7037.98	I
22		7060.21	I
13		7063.34	I
13		7081.22	I
110		7109.87	I
27		7134.08	I
150		7242.50	I
40		7245.85	I
22		7267.62	I
17		7300.19	I
13		7348.49	I
13		7361.65	I
10	h	7364.41	I
40		7391.36	I
10		7434.10	I
13		7447.34	I
13	h	7452.85	I
140		7485.74	I
13		7504.47	I
11		7572.64	I
11	h	7595.16	I
11		7601.84	I
17	h	7656.76	I
13		7679.49	I
27		7720.77	I
17		7829.65	I
15		7854.45	I
11		7923.15	I
15		7986.60	I
22	h	8245.06	I
40	h	8328.44	I
45	h	8389.32	I
45	h	8483.39	I

Mo III
Ref. 420 — C.H.C.

Intensity	Wavelength	
	Vacuum	
50	1166.07	III
100	1169.33	III
50	1173.67	III
50	1209.60	III
50	1225.46	III
50	1230.34	III
50	1234.63	III
30	1236.10	III
100	1254.93	III
100	1262.21	III
100	1263.74	III
100	1274.37	III
50	1274.94	III
100	1276.40	III
200	1277.40	III
200	1277.58	III
100	1278.06	III
200	1278.40	III
150	1281.90	III
150	1283.60	III
100	1258.52	III
100	1286.42	III
50	1288.07	III
50	1288.25	III
100	1290.49	III
100	1299.82	III
50	1305.58	III
50	1437.37	III
50	1452.38	III
50	1534.86	III
50	1751.22	III
50	1760.57	III
50	1807.60	III
100	1854.73	III
	Air	
75	2165.19	III
75	[illegible]	III
50	2172.46	III
75	2179.37	III
100	2184.37	III
100	2211.02	III
50	2223.19	III
100	2253.18	III
150	2269.71	III
50	2275.47	III
50	2275.64	III
200	2294.97	III
80	2304.26	III
50	2326.75	III
150	2330.93	III
100	2359.76	III
80	2386.96	III
50	2403.61	III
70	2412.71	III
50	2422.18	III
200	2506.19	III
90	2597.13	III
50	2756.06	III
100	2807.74	III
125	2947.32	III

Molybdenum (Cont.)

Intensity		Wavelength	
75		2983.94	III
80		3254.70	III
200		3271.69	III

Mo IV
Ref. 383 — C.H.C.

Intensity		Wavelength	
		Vacuum	
10		857.75	IV
10		859.72	IV
25		865.24	IV
20		865.53	IV
20		863.63	IV
50		867.92	IV
10		878.43	IV
100		884.19	IV
10		884.82	IV
60		886.05	IV
50		891.74	IV
40		894.80	IV
15		895.41	IV
20		1819.50	IV
30		1821.59	IV
30		1850.69	IV
20		1877.88	IV
80		1926.26	IV
100		1929.24	IV
20		1949.44	IV
80		1971.06	IV
20		1991.41	IV
		Air	
70		2010.92	IV
20		2023.78	IV
25		2055.64	IV
50		2060.38	IV
15		2091.89	IV
10		2113.78	IV
5		2140.33	IV

NEODYMIUM (Nd)
Z = 60

Nd I and II
Ref. 1 — C.H.C.

Intensity		Wavelength	
		Air	
75		2702.46	
75		2704.54	
75		2764.98	I
60		2785.79	I
50		2863.95	
50		2921.26	
55		2962.88	II
65		2963.58	II
80		2993.20	II
40		2994.73	
95		3007.97	II
95		3014.19	II
95		3018.35	II
80		3026.47	II
50		3038.98	II
50		3043.29	
50		3051.11	II
80		3052.15	II
140		3056.71	II
130		3069.73	II
65	d	3071.43	II
		3071.50	II
160		3075.38	II
95		3079.38	II
95		3080.94	II
95		3092.73	
240		3092.92	II
140		3098.48	II
55		3099.52	II
130		3105.43	II
95		3106.18	II
65		3108.01	II
260		3115.18	II
190		3116.15	II
50		3119.75	II
160		3123.06	II
190		3124.58	II
290		3133.60	II
220		3134.90	II
100		3137.24	II
170		3141.46	II
170		3142.44	II
100		3144.55	II
100		3144.82	II
100		3148.51	II
100		3149.29	II
100		3149.51	II
100		3162.62	II
100		3175.99	II
50		3181.54	II
50		3188.73	II
50		3200.62	II
150		3203.47	II
85		3211.00	II
100		3217.12	II
50		3222.62	I
50		3228.04	II
60		3234.62	
40		3237.91	II
100		3254.08	II
50		3256.91	II
220		3259.24	II
100		3260.66	II
220		3265.12	II
50		3265.38	II
170		3267.25	II
100		3273.18	II
320		3275.22	II
50		3281.49	II
50		3282.78	II
290		3285.10	II
100		3286.62	II
50		3289.52	
100		3290.65	II
70		3293.84	II
70		3294.68	II
70		3298.61	
300		3300.16	II
200		3312.75	II
200		3325.90	II
410		3328.28	II
250		3331.57	II
290		3334.48	II
290		3339.07	II
320		3353.59	II
200		3355.93	II
270		3364.96	II
290		3393.63	II
120	h	3484.88	I
200		3527.53	II
290		3543.35	II
200		3555.77	II
410		3560.75	II
340		3568.87	II
470		3587.51	II
300		3592.59	II
340		3598.02	II
300		3600.91	II
320		3609.79	II
370		3615.82	II
300		3618.96	II
300		3631.02	II
340		3634.30	II
240		3637.00	II
240		3637.23	II
240		3640.24	II
240		3645.78	II
340		3648.20	II
240		3649.46	
240		3650.42	II
410		3653.15	II
240		3654.16	
470		3662.26	II
540		3665.18	II
540		3672.36	II
580		3673.54	II
240		3678.18	II
1200		3685.80	II
440		3687.30	II
410		3689.69	II
300		3694.81	II
410		3697.56	II
240		3702.84	
240		3704.95	II
200		3712.81	
470		3713.70	II
370		3714.20	II
640	d	3714.73	II
250		3715.04	II
200		3715.39	II
470		3715.68	II
410		3718.54	II
410		3721.35	II
220		3722.42	II
780		3723.50	II
410		3724.87	II
250		3726.90	II
710		3728.13	II
470		3730.58	II
270		3732.78	II
1000	d	3735.54	II
		3735.60	II
440		3737.10	II
1000		3738.06	II
270		3741.42	II
200		3749.85	II
320		3750.31	II
580		3752.49	II
370		3752.67	II
250		3754.83	II
370		3755.60	II
510		3757.82	II
930		3758.95	II
300		3759.79	II
930		3763.47	II
300		3766.59	II
510		3769.65	II
1400		3775.50	II
250		3776.34	II
710		3779.47	II
580		3780.40	II
510		3781.32	II
300		3783.78	II
2400		3784.25	II
270		3784.73	II
340		3791.50	II
340		3795.45	II
240		3799.55	II
370		3801.12	II
200		3801.38	II
340		3802.30	II
1200		3803.47	II
200		3804.10	II
2500		3805.36	II
340		3805.55	II
470		3807.23	II
540		3808.77	II
440		3809.06	II
580		3810.49	II
240		3811.06	II
270		3811.77	II
200		3812.53	II
710		3814.73	II
240		3819.70	II
410		3822.47	II
1200		3826.42	II
240		3828.00	II
540		3828.85	II
440		3829.16	II
510		3830.47	II
740		3836.54	II
340		3837.91	II
1700		3838.98	II
340		3839.51	II
410	d	3841.82	II
		3841.88	II
1700	d	3848.24	II
		3848.31	II
1500		3848.52	II
470		3850.22	II
2400	d	3851.66	II
		3851.74	II
340		3858.55	II
270		3860.94	II
300		3862.52	II
3700	d	3863.33	II
		3863.40	II
240		3866.52	II
220		3866.81	II
850		3869.07	II
240		3875.74	II
470		3875.87	II
1100		3878.58	II
1000		3879.55	II
780		3880.38	II
1200		3880.78	II
200		3881.59	
540		3887.87	II
370	h	3889.66	II
1300		3889.93	II
1300		3890.58	II
1300		3890.94	II
580		3891.51	II
470		3892.06	II
810		3894.63	II
270		3896.13	II
440		3897.63	II
2000		3900.21	II
1300		3901.84	II
1700		3905.89	II
200		3907.70	II
510		3907.84	II
2000		3911.16	II
850		3912.23	II
340		3913.69	II
440		3915.13	II
610		3915.95	II
340		3917.65	II
220		3919.92	II
1100		3920.96	II
510		3927.10	II
200		3929.26	II
610		3934.82	II
410		3936.11	II
510		3938.86	II
2000		3941.51	II
2000		3951.16	II
810		3952.20	II
320		3952.87	II
320		3953.52	II
240		3957.45	II
590		3958.00	II
510		3962.21	II
1400		3963.12	II
270		3963.90	II
1100		3973.30	II
740		3973.69	II
740		3976.85	II
740		3979.49	II
320		3982.36	II
470		3986.25	II
1400		3990.10	II
1000		3991.74	II
1100		3994.68	II
410		4000.50	II
540		4004.02	II
410		4007.43	II
3700		4012.25	II
540		4012.70	II
370		4018.81	II
1000		4020.87	II
1000		4021.34	II
1000		4021.78	II
1200		4023.00	II
340		4024.78	II
410		4030.47	II
1200		4031.82	II
270		4038.12	II
3000		4040.80	II
200		4041.06	II
410		4043.59	II
410		4048.81	II
850		4051.15	II
850		4059.96	II
4700		4061.09	II
1100		4069.28	II
710		4075.12	II
470		4075.28	II
240		4077.62	II
470		4080.23	II
240		4085.82	II
270		4096.13	II
220		4098.18	II
200		4106.59	II
1400		4109.08	II
2500		4109.46	II
510		4110.48	II
300	h	4113.83	II
410		4123.88	II
470		4133.36	II
510		4135.33	II
3000		4156.08	II
510		4156.26	II
340		4160.57	II
410		4168.00	II
810		4175.61	II
2400		4177.32	II
200		4178.64	II
640		4179.59	II
250		4184.98	II
470		4205.60	II
470		4211.29	II
290		4220.25	II
440		4227.73	II
1300		4232.38	II
250		4234.19	II
290	h	4235.24	II
290		4239.84	II
2000		4247.38	II
850		4252.44	II
290		4254.29	
410		4261.84	II
340		4266.71	II
240		4270.56	II
340		4272.79	II
340		4275.09	II
470		4282.44	II
240		4282.57	II
710		4284.52	II
270		4297.80	II
5400		4303.58	II
340		4304.45	II
200		4307.78	II
470		4314.52	II
1100		4325.76	II
510		4327.93	II
540		4338.70	II
680		4351.29	II
850		4358.17	II
240		4366.38	II
340		4368.64	II
470	d	4374.93	II
		4375.04	II
710		4385.66	II
250		4390.66	II
540		4400.83	II
510		4411.06	II
580		4446.39	II
1400		4451.57	II
200		4451.99	II
300		4456.40	II
740		4462.99	II
410		4501.82	II
200		4506.59	II
170		4513.34	II
250		4516.36	II

Neodymium (Cont.)

Intensity		Wavelength	
120		4527.25	I
340		4541.27	II
340		4542.61	II
100		4556.14	II
170		4559.67	I
340		4563.22	II
200		4578.89	II
200		4579.32	II
100		4586.62	I
200		4597.02	II
100		4603.82	I
100		4609.87	I
300		4621.94	I
100		4627.98	I
510		4634.24	I
340		4641.10	I
250		4645.77	II
200		4646.40	I
300		4649.67	I
200		4654.73	I
130		4670.56	II
170		4680.74	II
310		4683.45	I
110		4684.04	I
110		4690.35	I
190		4696.44	I
130		4703.57	II
470		4706.54	II
140		4706.96	I
190		4709.71	II
190		4715.59	II
240		4719.02	I
190		4724.35	II
140		4731.77	I
120		4779.46	I
170		4789.41	II
120		4797.15	II
240		4811.34	II
140		4820.34	II
350		4825.48	II
130		4832.28	II
110		4849.06	II
280		4859.02	II
190		4866.74	I
350		4883.81	I
140		4885.16	II
220		4890.70	II
240		4891.07	I
280		4896.93	I
120		4901.53	I
210		4901.84	I
110		4902.03	II
190		4913.41	I
170		4914.37	II
330		4920.68	II
470		4924.53	I
260		4944.83	I
290		4954.78	I
290		4959.13	II
150		4961.39	II
250		4989.94	II
150		5033.52	II
110		5063.73	II
360		5076.59	II
150	h	5089.84	II
360		5092.80	II
180		5102.39	II
150	d	5105.21	II
		5105.35	I
360		5107.59	II
340		5123.79	II
680		5130.60	II
170		5132.33	II
170		5165.14	II
130		5181.17	II
120		5182.60	II
500		5191.45	II
630		5192.62	II
330		5200.12	II
310		5212.37	II
150		5213.23	I
130		5225.05	II
130		5228.43	II
450		5234.20	II
250		5239.79	II
720		5249.59	II
200		5250.82	II
360		5255.51	II
120		5269.48	II
590		5273.43	II
150		5276.88	II
110		5291.67	I
680		5293.17	II
160		5302.28	II
110		5306.47	II
220		5311.46	II
500		5319.82	II
180		5356.98	II
290		5361.47	II
150		5371.94	II
110		5385.90	II
160		5431.53	II
110		5451.12	II
170		5485.70	II
35		5501.47	I
45		5525.72	I
90		5533.82	I
55		5535.27	II
55		5543.24	I
55		5548.47	II
55		5561.17	I
27		5575.50	I
27		5576.70	I
27		5577.70	I
27		5587.61	I
240		5594.43	II
55		5601.43	I
45		5601.92	I
220		5620.54	I
65		5635.76	I
45		5639.54	I
35		5653.57	I
70		5668.87	II
65		5669.77	I
140	d	5675.97	I
55		5676.33	I
220		5688.53	II
23		5689.51	I
30		5701.57	I
130		5702.24	II
80		5706.21	II
160		5708.28	II
80		5718.12	II
65		5726.83	II
100		5729.29	I
23		5734.55	
70		5740.86	II
55		5749.19	I
27		5749.66	I
23		5767.33	I
45		5776.12	I
45		5784.96	I
45		5788.22	I
45		5800.09	I
160		5804.02	II
80		5811.57	II
45		5813.89	I
27		5820.37	I
70		5823.67	II
30		5826.74	I
80		5842.39	II
30		5844.66	I
23		5845.95	I
55		5858.91	I
35		5867.08	I
30		5868.90	I
27		5871.04	I
30		5883.29	I
23		5886.24	I
30		5887.91	I
27		5921.22	I
27		5955.87	I
30		5994.76	I
27		5996.47	I
45		6007.67	I
35		6031.27	II
27		6033.29	I
45		6034.24	II
55		6066.03	I
27		6071.70	I
30		6073.97	I
23	d	6133.47	II
27		6149.28	I
27		6155.06	I
35		6157.83	II
23		6166.67	II
35		6170.49	II
45		6178.59	I
27		6183.91	II
27		6208.24	I
45		6223.39	I
27		6226.50	I
23		6238.50	II
35		6244.08	I
23		6257.49	I
27		6258.73	II
23		6277.29	II
27		6285.79	I
23		6292.84	II
23		6297.07	I
55		6310.49	I
27		6341.51	II
23		6382.07	II
65		6385.20	II
35		6485.69	I
45		6630.14	I
35		6637.96	II
45		6650.57	II
30		6655.67	I
25		6737.79	II
40		6740.11	II
25		6742.54	I
30		6790.37	II
30		6804.00	II
25		6846.72	II
40		6900.43	II
24		6941.39	II
17	h	7010.80	II
8		7018.85	II
17		7020.92	II
17		7024.58	II
10		7033.21	
35		7037.30	II
7		7052.14	II
7		7054.74	II
7		7061.47	II
40		7066.89	II
8		7082.93	II
12	h	7089.71	II
12	h	7092.09	
12	h	7092.74	II
12	h	7092.94	
17	h	7093.98	I
20	h	7095.42	I
29		7129.35	II
12	h	7142.04	II
10		7143.72	
8		7151.03	II
6		7153.09	I
6	h	7185.01	
10		7189.09	II
24		7189.42	II
20		7192.01	II
10		7199.00	II
8	h	7227.01	I
15		7236.54	II
7	h	7261.64	II
9		7285.29	II
9		7288.56	II
6		7291.38	II
7		7298.72	II
12		7316.81	II
7	h	7321.43	I
7		7323.12	II
6		7334.54	I
6		7357.10	I
6		7374.04	II
7		7381.79	II
9		7401.31	I
10		7406.62	II
6		7411.20	II
10		7418.18	II
9		7427.41	II
9		7448.71	II
5		7481.28	II
12		7511.16	II
17		7513.73	II
7	h	7514.44	II
7		7516.02	II
9		7526.45	II
12		7528.99	II
10		7538.26	II
5		7540.97	II
7		7547.00	II
5		7577.54	II
7		7587.65	II
6		7590.75	II
6		7603.73	II
5		7605.92	II
5		7614.72	I
9		7639.79	II
8		7646.00	II
6		7663.52	II
12		7696.56	II
6		7718.20	II
4		7743.90	II
4		7748.92	II
10		7750.95	II
6		7773.06	II
7		7792.22	II
6		7796.40	II
8		7797.32	II
5		7798.32	II
10		7808.47	II
7		7818.83	II
5		7825.20	II
12		7863.04	II
5	h	7872.03	I
7		7886.60	II
4	h	7896.50	II
9		7900.40	II
5	h	7906.03	I
12		7917.01	II
10		7925.03	II
5		7947.93	II
10		7949.68	II
5	h	7955.38	II
12		7958.95	I
12		7965.73	II
15		7982.09	II
12		7982.68	II
12		8000.76	II
9		8007.70	I
4	h	8020.07	
8		8026.35	
10		8043.24	I
8		8051.33	II
5		8064.00	II
10		8099.17	I
10		8120.93	II
12		8122.07	II
12		8141.75	II
12		8143.27	II
7	h	8164.97	I
8		8172.56	II
9		8179.83	II
9		8182.41	II
4		8185.58	II
7	h	8205.38	II
10		8231.52	II
4		8248.76	II
5	h	8249.68	II
4	h	8262.80	II
7	h	8266.72	II
4		8272.79	II
4	h	8302.74	II
10		8307.72	II
6		8324.50	II
4		8332.01	II
12		8346.36	II
4		8375.16	II
4		8375.33	II
4		8394.71	II
7		8400.85	II
5	h	8456.87	II
4		8530.53	II
5		8582.03	II
5		8591.53	II
7		8594.87	II
8	c	8643.43	II
5		8667.07	II
5		8677.48	
6		8691.29	II
6		8695.07	II
6		8712.82	II
6		8715.03	I
17		8839.10	II

NEON (Ne)
Z = 10

Ne I and II
Ref. 56, 58, 118, 150, 230 —
S.P.D.

Intensity		Wavelength	
		Vacuum	
90		352.956	II
60		354.962	II
90		361.433	II
60		362.455	II
150		405.854	II
120		407.138	II
200		445.040	II
300		446.256	II
250		446.590	II
180		447.815	II
150		454.654	II
200		455.274	II
10		456.275	II
120		456.348	II
90		456.896	II
1000		460.728	II
500		462.391	II
35		587.213	I
35		589.179	I
35		589.911	I
70		591.830	I
100		595.920	I
75		598.706	I
35		598.891	I
70		600.036	I
170		602.726	I
170		615.628	I
170		618.672	I
120		619.102	I
200		626.823	I
200		629.739	I
1000		735.896	I
400		743.720	I
60		993.88	II
70		1068.65	II
90		1131.72	II
100		1131.85	II
90		1229.83	II
90		1418.38	II
90		1428.58	II
90		1436.09	II
120		1681.68	II
180		1688.36	II
100		1888.11	II
100		1889.71	II
200		1907.49	II
500		1916.08	II
300		1930.03	II
200		1938.83	II
100	c	1945.46	II
		AIR	

Neon (Cont.)

Intensity		Wavelength	Spectrum
80		2007.01	II
80		2025.56	II
150		2085.47	II
180		2096.11	II
120		2096.25	II
80	p	2562.12	II
90	w	2567.12	II
80		2623.11	II
80		2629.89	II
90	w	2636.07	II
80		2638.29	II
80		2644.10	II
80		2762.92	II
90		2792.02	II
80		2794.22	II
100		2809.48	II
80		2906.59	II
80		2906.82	II
90		2910.06	II
90		2910.41	II
80		2911.14	II
80		2915.12	II
80		2925.62	II
80	w	2932.10	II
80		2940.65	II
90		2946.04	II
150		2955.72	II
150		2963.24	II
150		2967.18	II
100		2973.10	II
15		2974.72	I
100		2979.46	II
12		2982.67	I
150		3001.67	II
120	p	3017.31	II
300		3027.02	II
300		3028.86	II
100		3030.79	II
120		3034.46	II
100		3035.92	II
100		3037.72	II
100		3039.59	II
100		3044.09	II
100		3045.56	II
120		3047.56	II
100		3054.34	II
100		3054.68	II
100		3059.11	II
100		3062.49	II
100		3063.30	II
100		3070.89	II
100		3071.53	II
100		3075.73	II
120		3088.17	II
100		3092.09	II
120		3092.90	II
100		3094.01	II
100		3095.10	II
100		3097.13	II
100		3117.98	II
120		3118.16	II
10		3126.199	I
300		3141.33	II
100		3143.72	II
100	p	3148.68	II
100		3164.43	II
100		3165.65	II
100		3188.74	II
120		3194.58	II
500		3198.59	II
60		3208.96	II
120		3209.36	II
120		3213.74	II
150		3214.33	II
150		3218.19	II
120		3224.82	II
120		3229.57	II
200		3230.07	II
120		3230.42	II
120		3232.02	II
150		3232.37	II
100		3243.40	II
100		3244.10	II
100		3248.34	II
100		3250.36	II
150		3297.73	II
150		3309.74	II
300		3319.72	II
1000		3323.74	II
150		3327.15	II
100		3329.16	II
200		3334.84	II
150		3344.40	II
300		3345.45	II
150		3345.83	II
200		3355.02	II
120		3357.82	II
200		3360.60	II
120		3362.16	II
100		3362.71	II
120		3367.22	II
12		3369.808	I
40		3369.908	I
100		3371.80	II
500		3378.22	II
150		3388.42	II
120		3388.94	II
300		3392.80	II
100		3404.82	II
120		3406.95	II
100		3413.15	II
120		3416.91	II
120		3417.69	II
50		3417.904	I
15		3418.006	I
120		3428.69	II
60		3447.703	I
50		3454.195	I
100		3456.61	II
100		3459.32	II
25		3460.524	I
30		3464.339	I
30		3466.579	I
60		3472.571	I
150		3479.52	II
200		3480.72	II
200		3481.93	II
25		3498.064	I
30		3501.216	I
25		3515.191	I
150		3520.472	I
120		3542.85	II
120		3557.80	II
100		3561.20	II
250		3568.50	II
100		3574.18	II
200		3574.61	II
50		3593.526	I
30		3593.640	I
15		3600.169	I
20		3633.665	I
150		3643.93	II
200		3664.07	II
20		3682.243	I
12		3685.736	I
200		3694.21	II
10		3701.225	I
150		3709.62	II
250		3713.08	II
250		3727.11	II
800		3766.26	II
1000		3777.13	II
100		3818.43	II
120		3829.75	II
150		4219.74	II
100		4233.85	II
120		4250.65	II
120		4369.86	II
70		4379.40	II
150		4379.55	II
100		4385.06	II
200		4391.99	II
150		4397.99	II
150		4409.30	II
100		4413.22	II
100		4421.39	II
100	p	4428.52	II
100	p	4428.63	II
150	p	4430.90	II
150	p	4430.94	II
120		4457.05	II
100		4522.72	II
10		4537.754	I
10		4540.380	I
100		4569.06	II
15		4704.395	I
12		4708.862	I
10		4710.067	I
10		4712.066	I
15		4715.347	I
10		4752.732	I
12		4788.927	I
10		4790.22	I
10		4827.344	I
10		4884.917	I
4		5005.159	I
10		5037.751	I
10		5144.938	I
25		5330.778	I
20		5341.094	I
8		5343.283	I
60		5400.562	I
5		5562.766	I
10		5656.659	I
5		5719.225	I
12		5748.298	I
80		5764.419	I
12		5804.450	I
40		5820.156	I
500		5852.488	I
100		5872.828	I
100		5881.895	I
60		5902.462	I
60		5906.429	I
100		5944.834	I
100		5965.471	I
100		5974.627	I
120		5975.534	I
80		5987.907	I
100		6029.997	I
100		6074.338	I
80		6096.163	I
60		6128.450	I
100		6143.063	I
120		6163.594	I
250		6182.146	I
150		6217.281	I
150		6266.495	I
60		6304.789	I
7		6328.165	I
100		6334.428	I
120		6382.992	I
200		6402.246	I
150		6506.528	I
60		6532.882	I
150		6598.953	I
70		6652.093	I
90		6678.276	I
20		6717.043	I
100		6929.467	I
90		7024.050	I
100		7032.413	I
50		7051.292	I
80		7059.107	I
100		7173.938	I
150		7213.20	II
150		7235.19	II
100		7245.167	I
150		7343.94	II
40		7472.439	I
90		7488.871	I
100		7492.10	II
150		7522.82	II
80		7535.774	I
60		7544.044	I
100		7724.628	I
120		7740.74	II
300		7839.055	I
120		7926.20	II
400		7927.118	I
700		7936.996	I
2000		7943.181	I
2000		8082.458	I
100		8084.34	II
1000		8118.549	I
600		8128.911	I
3000		8136.406	I
2500		8259.379	I
100		8264.81	II
2500		8266.077	I
800		8267.117	I
6000		8300.326	I
100		8315.00	II
1500		8365.749	I
100		8372.11	II
8000		8377.606	I
1000		8417.159	I
4000		8418.427	I
1500		8463.358	I
800		8484.444	I
5000		8495.360	I
600		8544.696	I
1000		8571.352	I
4000		8591.259	I
6000		8634.647	I
3000		8647.041	I
15000		8654.383	I
4000		8655.522	I
100		8668.26	II
5000		8679.492	I
5000		8681.921	I
2000		8704.112	I
4000		8771.656	I
12000		8780.621	I
10000		8783.753	I
500		8830.907	I
7000		8853.867	I
1000		8865.306	I
1000		8865.755	I
3000		8919.501	I
2000		8988.57	I
100		9079.46	II
6000		9148.67	I
6000		9201.76	I
4000		9220.06	I
2000		9221.58	I
2000		9226.69	I
1000		9275.52	I
200		9287.56	II
6000		9300.85	I
1500		9310.58	I
3000		9313.97	I
6000		9326.51	I
2000		9373.31	I
5000		9425.38	I
3000		9459.21	I
5000		9486.68	I
5000		9534.16	I
3000		9547.40	I
120		9577.01	II
1000		9665.42	I
100		9808.86	II
800		10295.42	I
2000		10562.41	I
1500		10798.07	I
2000		10844.48	I
3000		11143.020	I
3500		11177.528	I
1600		11390.434	I
1100		11409.134	I
3000		11522.746	I
1500		11525.020	I
950		11536.344	I
500		11601.537	I
1200		11614.081	I
300		11688.002	I
2000		11766.792	I
1500		11789.044	I
500		11789.889	I
1000		11984.912	I
3000		12066.334	I
800		12459.389	I
1000		12689.201	I
1100		12912.014	I
700		13219.241	I
800		15230.714	I
400		17161.930	I
400		18035.80	I
1000		18083.21	I
350		18221.11	I
250		18227.02	I
2500		18276.68	I
2000		18282.62	I
1200		18303.97	I
250		18359.12	I
1200		18384.85	I
2000		18389.95	I
1000		18402.84	I
1200		18422.39	I
300		18458.65	I
400		18475.79	I
900		18591.55	I
1600		18597.70	I
350		18618.96	I
550		18625.16	I
1200		21041.295	I
750		21708.145	I
300		22247.35	I
350		22428.13	I
2250		22530.40	I
400		22661.81	I
600		23100.51	I
1000		23260.30	I
1050		23373.00	I
850		23565.36	I
3500		23636.52	I
300		23701.64	I
1100		23709.2	I
1800		23951.42	I
600		23956.46	I
1000		23978.12	I
200		24098.54	I
500		24161.42	I
600		24249.64	I
1500		24365.05	I
800		24371.60	I
400		24447.85	I
700		24459.4	I
300		24776.46	I
550		24928.88	I
250		25161.69	I
650		25524.37	I
125		28386.21	I
150		30200.	I
250		33173.09	I
450		33352.35	I
1300		33901.	I
2200		33912.10	I
600		34131.31	I
100		34471.44	I
120		35834.78	I

Ne III

Ref. 365, 371, 402 — R.L.K.

Intensity	Wavelength	
	Vacuum	
20	251.14	III
20	251.56	III
20	251.73	III
40	267.06	III
40	267.52	III
20	267.71	III
40	283.18	III
160	283.21	III
110	283.69	III
40	283.89	III
220	301.12	III
220	313.05	III
220	313.68	III
40	313.95	III

Neon (Cont.)

Intensity	Wavelength	
220	379.31	III
285	488.10	III
220	488.87	III
450	489.50	III
70	489.64	III
220	490.31	III
360	491.05	III
20	1255.03	III
110	1255.68	III
160	1257.19	III
	Air	
200	2086.96	III
300	2089.43	III
240	2092.44	III
400	2095.54	III
200	2161.22	III
300	2163.77	III
200	2180.89	III
200	2209.35	III
200	2211.85	III
240	2213.76	III
300	2216.07	III
240	2263.21	III
200	2264.91	III
300	2412.73	III
240	2412.94	III
200	2413.78	III
200	2473.40	III
800	2590.04	III
600	2593.60	III
400	2595.68	III
300	2610.03	III
240	2613.41	III
200	2615.87	III
200	2638.70	III
200	2641.07	III
600	2677.90	III
500	2678.64	III

Ne IV
Ref. 69, 364, 388, 400, 413, 430, — R.L.K.

Intensity	Wavelength	
	Vacuum	
15	151.82	IV
15	152.23	IV
15	158.65	IV
15	158.82	IV
80	172.62	IV
80	177.16	IV
150	186.58	IV
100	194.28	IV
100	208.48	IV
100	208.73	IV
80	208.90	IV
150	212.56	IV
140	223.24	IV
120	223.60	IV
140	234.32	IV
120	234.70	IV
50	357.83	IV
200	358.72	IV
125	387.14	IV
100	388.22	IV
150	421.61	IV
140	469.77	IV
200	469.82	IV
180	469.87	IV
140	469.92	IV
120	521.74	IV
140	521.82	IV
80	541.13	IV
100	542.07	IV
150	543.89	IV
	Air	
65	2018.44	IV
110	2022.19	IV
30	2203.88	IV
10	2220.81	IV
250	2258.02	IV
175	2262.08	IV
110	2264.54	IV
550	2285.79	IV
30	2293.14	IV
250	2293.49	IV
250	2363.28	IV
110	2365.49	IV
250	2350.84	IV
450	2352.52	IV
700	2357.96	IV
250	2362.68	IV
350	2372.16	IV
65	2384.20	IV
350	2384.95	IV

Ne V
Ref. 69, 388, 389, 400, 413 — R.L.K.

Intensity	Wavelength	
	Vacuum	
66	119.01	V
200	122.52	V
66	125.12	V
45	131.99	V
50	132.04	V
150	140.76	V
150	140.79	V
100	142.44	V
100	142.50	V
150	142.72	V
100	143.27	V
150	143.34	V
150	147.13	V
66	151.23	V
120	151.42	V
45	154.50	V
100	164.02	V
100	164.14	V
500	173.93	V
400	357.96	V
500	358.47	V
500	359.38	V
1000	365.59	V
800	416.20	V
250	480.41	V
150	481.28	V
250	481.36	V
500	482.99	V
400	568.42	V
250	569.76	V
500	569.83	V
250	572.11	V
800	572.34	V
	Air	
75	[illegible]	V
110	2232.41	V
65	2245.48	V
65	2259.57	V
65	2263.39	V
250	2265.71	V

NEPTUNIUM (Np)
Z = 93

Np I and II
Ref. 93 = J.G.C.

Intensity		Wavelength	
		Air	
300		3481.93	I
300	h	3501.50	I
300	l	3986.89	I
300	s	5044.66	I
300	l	5601.70	I
300	l	5652.75	I
300	l	5784.39	I
300	l	5878.04	I
300	s	6011.22	I
300		6056.09	I
300	s	6073.90	I
300	s	6080.05	I
300	l	6120.49	I
300		6188.59	I
300	l	6200.00	I
300	s	6215.90	I
300	s	6317.84	I
300	l	6341.38	I
300	l	6566.11	I
300	l	6720.68	I
300	s	6751.32	I
300	s	6795.21	I
300	l	6802.62	I
300	l	6805.81	I
300	s	6816.44	I
300	l	6865.45	I
300	s	6907.13	I
300	h	6912.91	I
1000	s	6930.31	I
300	l	6963.63	I
3000	s	6972.09	I
300		7014.02	I
300	l	7018.91	I
300	s	7039.14	I
300	s	7080.01	I
300	l	7174.83	I
300	l	7184.93	I
300	l	7284.28	I
300	l	7292.29	I
300	l	7332.52	I
300	s	7370.60	I
300	l	7381.03	I
300	l	7381.65	I
300	l	7402.70	I
300	s	7512.22	I
300	l	7515.15	I
300	l	7546.05	I
300	l	7624.83	I
300		7626.85	I
300	s	7681.01	I
300	s	7685.25	I
1000	l	7735.14	I
300	l	7761.61	I
1000	l	7765.75	I
300	s	7776.07	I
300		7787.46	I
1000	l	7791.38	I
300	l	7851.44	I
300	l	7887.88	I
300	l	7901.71	I
300	l	7975.98	I
300	h	8080.32	I
300	s	8124.59	I
300		8155.11	I
300	l	8167.42	I
300	l	8183.06	I
300	l	8188.61	I
300	l	8247.82	I
300	l	8287.11	I
300	s	8287.75	I
300	l	8306.22	I
300	s	8313.66	I
1000	l	8339.12	I
300		8356.79	I
300	l	8367.11	I
3000		8372.88	I
3000		8529.96	I
1000	s	8696.23	I
1000	s	8906.02	I
1000		8942.70	I
1000	s	9004.75	I
1000	l	9006.31	I
10000	l	9016.18	I
3000	l	9141.30	I
3000	s	9379.33	I
[illegible]	l	[illegible]	I
3000	s	9679.13	I
3000	l	9930.55	I
10000	l	10091.99	I
10000	s	10817.45	I
10000	l	11695.15	I
10000	l	11776.64	I
10000	s	12148.18	I
10000	s	12377.42	I
10000	l	12407.99	I
10000	l	13834.33	I

NICKEL (Ni)
Z = 28

Ni I and II
Ref. 1,294 = C.H.C.

Intensity	Wavelength	
	Vacuum	
500	1317.22	II
400	1335.20	II
500	1370.14	II
1000	1741.55	II
500	1748.28	II
	Air	
1000	2165.55	II
2000	2169.10	II
2000	2174.67	II
1500	2175.15	II
500	2177.09	II
400	2177.36	II
400	2179.35	II
800	2180.47	II
800	2184.60	II
2500	2185.50	II
3000	2192.09	II
600	2201.41	II
5000	2205.55	II
4000	2206.72	II
6000	2216.48	II
800	2220.40	II
500	2221.06	II
900	2222.96	II
500	2242.68	II
500	2253.85	II
1000	2264.46	II
2000	2270.21	II
800	2277.28	II
400	2278.32	II
800	2278.77	II
500	2287.65	II
1600	2289.98	I
400	2296.55	II
400	2297.14	II
630	2300.78	I
1000	2303.00	II
2000	2310.96	I
1700	2312.34	I
1400	2313.66	I
1400	2313.98	I
1000	2316.04	II
1400	2317.16	I
500	2319.75	II
2600	2320.03	I
1900	2321.38	I
240	2322.68	I
1400	2325.79	I
940	2329.96	I
500	2334.58	II
460	2337.49	I
160	2337.82	I
500	2341.20	II
1200	2345.54	I
190	2346.63	I
400	2347.52	I
160	2360.63	I
200	2362.06	I
1000	2375.42	II
240	2386.58	I
1000	2394.52	II
2000	2416.13	II
240	2419.31	I
85	2421.23	I
70	2423.33	I
70	2423.66	I
70	2424.03	I
500	2437.89	II
85	2453.99	I
160	2472.06	I
85	2476.87	I
500	2510.87	II
500	2565.92	II
500	2606.26	II
500	2609.94	II
500	2615.06	II
45	2696.49	I
150	2798.65	I
230	2821.29	I
500	2864.02	II
50	2865.50	I
60	2907.46	I
25	2914.01	I
500	2943.91	I
570	2981.65	I
250	2984.13	I
500	2992.60	I
1000	2994.46	I
4000	3002.49	I
2200	3003.63	I
3700	3012.00	I
350	3019.14	I
120	3031.87	I
1700	3037.94	I
150	3045.01	I
3500	3050.82	I
1500	3054.32	I
1900	3057.64	I
500	3064.62	I
420	3080.76	I
260	3097.12	I
210	3099.12	I
2600	3101.55	I
1300	3101.88	I
220	3105.47	I
270	3114.12	I
2900	3134.11	I
55	3145.72	I
55	3181.74	I
100	3184.37	I
55	3195.57	I
150	3197.11	I
55	3202.14	I
180	3214.06	I
180	3217.83	I
100	3221.27	I
150	3221.65	I
210	3225.02	I
1100	3232.96	I
290	3234.65	I
600	3243.06	I
100	3248.46	I
120	3250.74	I
100	3271.12	I
120	3282.70	I
400	3292.87	II
500	3297.60	II
400	3305.71	II
660	3315.66	I
330	3320.26	I
310	3322.31	I
2000	3331.88	II
400	3335.64	II
500	3338.09	II
500	3348.84	II
500	3349.24	II

Nickel (Cont.)

Intensity		Wavelength	
600		3358.68	II
330		3361.56	I
500		3363.45	II
330		3365.77	I
330		3366.17	I
65		3366.81	I
65		3367.89	I
2900		3369.57	I
400		3371.99	I
260		3374.22	I
130		3374.64	I
500		3378.97	II
3300		3380.57	I
240		3380.85	I
1300		3391.05	I
3300		3392.99	I
500		3401.05	II
130		3409.58	I
330		3413.48	I
330		3413.94	I
8200		3414.76	I
1600		3423.71	I
2600		3433.56	I
990		3437.28	I
4800		3446.26	I
1300		3452.89	I
5000		3458.47	I
5000		3461.65	I
200		3467.50	I
240		3469.49	I
1600		3472.54	I
550		3483.77	I
130		3485.89	I
5500		3492.96	I
660		3500.85	I
65		3502.60	I
55		3507.69	I
2600		3510.34	I
260		3513.93	I
6600		3515.05	I
660		3519.77	I
8200		3524.54	I
110		3527.98	I
330		3548.18	I
55		3551.53	I
65		3561.75	I
5000		3566.37	I
990		3571.87	I
130		3587.93	I
1300		3597.70	I
1300		3610.46	I
530		3612.74	I
6600		3619.39	I
130		3624.73	I
200		3664.10	I
130		3669.24	I
180		3670.43	I
260		3674.15	I
160		3688.42	I
80		3693.93	I
120		3722.48	I
150		3736.81	I
60		3739.23	I
600		3775.57	I
700		3783.53	I
700		3807.14	I
110		3831.69	I
1200		3858.30	I
30		3889.67	I
35		3972.17	I
110		3973.56	I
110		4401.55	I
85		4459.04	I
18		4462.46	I
55		4470.48	I
35		4592.53	I
18		4600.37	I
65		4605.00	I
18		4606.23	I
75		4648.66	I
23		4686.22	I
110		4714.42	I
22		4715.78	I
30		4756.52	I
15		4763.95	I
45		4786.54	I
22		4807.00	I
22	h	4829.03	I
19		4831.18	I
45		4855.41	I
30		4866.27	I
17		4873.44	I
40		4904.41	I
22		4918.36	I
16		4935.83	I
45		4980.16	I
45		4984.13	I
500		4992.02	II
16	h	5000.34	I
18		5012.46	I
50		5017.59	I
100		5035.37	I
16		5048.85	I
100		5080.52	I
65		5081.11	I
26	h	5084.08	I
18		5099.32	I
26	h	5099.95	I
21		5115.40	I
18	h	5129.38	I
23		5137.08	I
23	h	5142.77	I
40	h	5146.48	I
40	h	5155.76	I
16		5168.66	I
13		5176.56	I
8		5435.87	I
180		5476.91	I
6		5510.00	I
6		5578.73	I
9		5587.86	I
13		5592.28	I
9		5614.79	I
5	h	5625.33	I
4		5649.70	I
5		5664.02	I
12		5682.20	I
8		5695.00	I
23		5709.56	I
10		5711.90	I
10		5715.09	I
16		5754.68	I
8		5760.85	I
10		5857.76	I
10		5892.88	I
10		6108.12	I
10		6176.81	I
10		6191.18	I
13		6256.36	I
10		6314.66	I
16		6643.64	I
22		6767.77	I
9		6772.32	I
10		6914.56	I
5		7110.90	I
26		7122.20	I
6		7182.00	I
5		7197.02	I
5		7261.93	I
5		7291.45	I
4		7385.24	I
16		7393.60	I
16		7409.35	I
5		7414.51	I
23		7422.28	I
13		7522.76	I
9		7525.12	I
19		7555.60	I
8		7574.05	I
23		7617.00	I
9		7619.21	I
16		7714.32	I
5	h	7715.58	I
19		7727.61	I
19		7748.89	I
10		7788.94	I
13		7797.59	I
2		7917.44	I
1000		8096.75	II
500		8114.21	II
700		8121.48	II
2		8809.42	I
9		8862.55	I
500	w	9900.92	II

Ni III
Ref. 422 — C.H.C.

Intensity		Wavelength	
		Vacuum	
100		625.68	III
500		630.71	III
200		637.54	III
200		662.37	III
150		663.57	III
500		676.94	III
200		700.17	III
300		713.33	III
300		713.38	III
500		718.48	III
200		721.26	III
300		722.09	III
250		725.20	III
500		729.82	III
250		730.11	III
400		731.70	III
300		732.16	III
200		738.26	III
300		747.99	III
200		749.68	III
300		750.05	III
200		752.02	III
300		757.80	III
250		758.73	III
250		758.27	III
400		770.22	III
200		772.04	III
500		778.81	III
200	d	785.02	III
300		788.04	III
200		788.30	III
200		805.01	III
500		811.57	III
500		826.14	III
200		826.50	III
500		842.14	III
400		845.24	III
300		847.43	III
200		857.09	III
300		860.64	III
300		862.88	III
300		863.22	III
300		867.51	III
200		869.70	III
200		870.84	III
300		973.79	III
400		979.59	III
200		1428.87	III
200		1434.31	III
200		1451.50	III
300		1604.54	III
300		1652.87	III
200		1653.12	III
250		1656.13	III
200		1661.79	III
400		1687.90	III
1000		1692.51	III
200		1707.35	III
200		1707.43	III
800		1709.90	III
650		1715.30	III
500		1719.46	III
200		1721.26	III
400		1722.28	III
250		1733.13	III
500		1738.25	III
300		1739.78	III
300		1741.96	III
550		1747.01	III
300		1752.43	III
400		1753.01	III
800		1764.69	III
500		1767.94	III
2000		1769.64	III
400		1776.07	III
200		1788.30	III
250		1790.40	III
200		1790.93	III
200		1791.64	III
200		1794.90	III
300		1807.24	III
200		1811.69	III
300		1819.28	III
800		1823.06	III
400		1830.01	III
200		1830.08	III
650		1847.28	III
800		1854.15	III
300		1858.75	III
200		1930.43	III
200		1952.54	III

Ni IV
Ref. 415 — C.H.C.

Intensity		Wavelength	
		Vacuum	
33		392.68	IV
32		393.24	IV
49		424.40	IV
57		444.21	IV
67		469.67	IV
65		471.24	IV
65		485.42	IV
66		536.28	IV
67		537.96	IV
58		1345.72	IV
69		1357.07	IV
76		1398.19	IV
74		1411.45	IV
69		1419.58	IV
74		1421.22	IV
70		1427.45	IV
67		1428.93	IV
67		1435.24	IV
70		1438.82	IV
73		1449.01	IV
76		1452.22	IV
70		1455.42	IV
69		1472.63	IV
68		1476.82	IV
73		1482.25	IV
67		1489.53	IV
72		1489.83	IV
69		1493.01	IV
74		1493.67	IV
68		1498.71	IV
71		1498.77	IV
72		1498.90	IV
67		1499.97	IV
70		1512.74	IV
70		1516.66	IV
73		1520.63	IV
75		1525.31	IV
74		1527.68	IV
74		1527.80	IV
76		1534.71	IV
73		1537.25	IV
69		1538.93	IV
75		1543.41	IV
74		1546.23	IV
68		1548.04	IV
69		1557.28	IV
67		1560.18	IV

Ni V
Ref. 416 — C.H.C.

Intensity		Wavelength	
		Vacuum	
29		304.02	V
55		315.24	V
56		315.71	V
63		336.79	V
68		343.93	V
78		347.34	V
70		347.46	V
67		347.72	V
71		348.10	V
69		350.77	V
69		353.59	V
72		354.18	V
76		354.42	V
68		354.49	V
68		355.61	V
70		355.78	V
65		357.37	V
69		358.57	V
68		358.58	V
66		359.47	V
69		365.62	V
70		370.62	V
67		371.31	V
68		371.76	V
67		373.60	V
72		377.68	V
70		393.91	V
66		394.31	V
66		395.24	V
41		400.59	V

NIOBIUM (Nb)
Z = 41

Nb I and II
Ref. I = C.H.C.

Intensity		Wavelength	
		Air	
3300		2029.32	II
3000		2032.99	II
2000		2109.42	II
1700		2125.21	II
1100		2126.54	II
1500		2131.18	II
370		2295.68	II
280		2302.08	II
170		2376.40	II
110		2387.09	II
140		2387.52	II
45		2388.27	II
160		2398.48	II
55		2405.34	II
55		2405.85	II
140		2412.46	II
160		2416.99	II
140		2418.69	II
75		2433.80	II
40		2435.95	II
35		2436.33	I
45		2437.42	II
40		2442.14	II
28		2442.68	II
65		2451.87	II
65		2453.95	II
55		2458.09	II
65		2462.89	I
35		2466.73	I
55		2469.08	I
110		2477.38	II
65		2478.29	II

Niobium (Cont.)

65		2479.94	II
35		2483.88	II
110		2504.65	I
110		2511.00	II
110		2521.40	II
390		2544.80	II
110		2551.38	II
130		2556.94	II
130		2562.41	II
130		2565.41	I
100		2569.03	I
110		2571.33	II
200		2578.74	I
390		2583.99	II
390		2590.94	II
270		2592.20	I
130		2616.48	I
130		2623.51	I
130		2627.44	I
130		2628.49	I
200		2642.24	II
320		2646.26	II
330		2647.50	I
240		2649.52	I
330		2654.45	I
310		2656.08	II
160		2657.62	I
110		2665.25	II
110		2666.59	II
110		2667.30	II
130		2668.29	I
400		2671.93	II
200		2673.57	II
200		2675.94	II
130		2687.15	I
160		2691.77	II
1000		2697.06	II
320		2698.86	II
320		2702.20	II
150		2702.52	II
470		2716.62	II
470		2721.98	II
310		2733.26	II
110		2737.09	II
100		2740.91	I
200		2748.85	I
190		2753.01	I
280		2758.61	I
240		2768.13	II
310		2773.20	I
270		2780.24	II
130		2782.36	I
110		2793.05	II
190		2827.08	II
150		2836.24	I
110		2840.94	I
250		2841.15	II
280		2842.65	II
160		2846.28	II
110		2851.45	I
240		2861.09	II
100		2864.32	I
100		2865.61	II
500		2868.52	II
800		2875.39	II
270		2876.95	II
530		2877.03	II
100		2880.72	II
570		2883.18	II
280		2888.83	II
470		2897.81	II
400		2899.24	II
470		2908.24	II
670		2910.59	II
470		2911.74	II
1100		2927.81	II
110		2931.47	II
870		2941.54	II
110	h	2945.88	II
110		2946.12	II
110		2946.90	II
1100		2950.88	II
400		2972.57	II
320		2974.10	II
210		2977.68	II
200		2982.11	II
330		2990.26	II
470		2994.73	II
140		3024.74	II
350		3028.44	II
300		3032.77	II
100		3044.76	II
150		3048.10	I
110		3053.09	I
100		3055.52	II
220		3064.53	II
110		3069.68	II
100		3070.90	II
110		3071.56	II
100		3073.24	II
400		3076.87	II
110		3080.35	II
1800		3094.18	II
140		3099.19	II
150		3111.45	I
270		3127.53	II
1500		3130.79	II
390		3145.40	II
140		3151.87	I
1200		3163.40	II
150		3175.78	II
390		3180.29	II
200		3187.49	I
300		3191.10	II
150		3191.43	II
1000		3194.98	II
120		3203.35	II
300		3206.34	II
390		3215.60	II
800		3225.48	II
140		3229.56	II
400		3236.40	II
200		3247.47	II
120		3248.94	II
160		3249.52	I
320		3254.07	II
230		3260.56	II
160		3263.37	II
160		3264.59	I
120		3270.47	I
100		3270.76	I
200		3272.07	I
160		3277.67	I
200		3283.46	II
230		3285.66	I
200		3287.59	I
160		3287.92	I
160		3292.02	II
320		3296.01	I
160		3299.61	I
120		3304.83	I
120		3308.05	I
120		3310.47	I
400		3312.60	I
200		3315.22	I
200		3318.98	I
120		3319.26	I
110		3319.98	II
240		3326.62	I
170		3329.36	I
110		3332.16	I
130		3341.60	II
1300		3341.97	I
1300		3343.71	I
130		3346.93	I
1700		3349.06	I
420		3349.52	I
340		3354.74	I
130		3357.04	I
1700		3358.42	I
130		3365.58	II
340		3366.96	I
130		3369.16	II
170		3371.33	I
350		3374.92	I
270		3380.41	I
130		3380.86	I
170		3386.24	II
350		3392.34	I
170		3395.93	I
120		3399.40	I
230		3405.41	I
130		3406.13	I
270		3408.38	I
230		3408.68	II
180		3409.19	II
230		3412.94	II
180		3415.97	I
180		3423.76	I
230		3425.42	II
130		3425.85	I
230		3426.57	II
230		3427.45	I
130		3429.04	I
180		3432.70	II
180		3440.59	II
170		3463.81	I
180		3465.86	I
130		3469.44	I
100		3471.19	I
140		3473.02	I
290		3478.69	I
200		3479.56	II
100		3484.05	II
230		3491.03	I
200		3497.81	I
500		3498.63	I
460		3507.96	I
200		3510.26	II
200		3515.42	II
200		3517.67	II
200		3520.06	I
2000		3535.30	I
1300		3537.48	I
250		3540.96	II
500		3544.02	I
250		3544.65	I
300		3550.45	I
250		3554.52	I
1000		3554.66	I
630		3563.50	I
630		3563.62	I
1500		3575.85	I
200		3577.72	I
5000		3580.27	I
500		3584.97	I
750		3589.11	I
500		3589.36	I
500		3593.97	I
500		3602.56	I
300		3619.51	II
200		3621.03	I
200		3639.33	I
420		3649.85	I
250		3650.81	I
400		3651.19	II
200		3659.61	II
630		3660.37	I
900		3664.70	I
220		3669.01	I
270		3674.78	I
1500		3697.85	I
330		3711.34	I
3300		3713.01	I
400		3716.99	I
2700		3726.24	I
270		3738.42	I
2700		3739.80	I
670		3740.73	II
270		3741.78	I
1700		3742.39	I
250		3753.18	I
210		3755.77	I
530		3763.49	I
350		3765.08	I
250		3766.13	I
530		3771.85	I
870		3781.01	I
1700		3787.06	I
1300		3790.15	I
3500		3791.21	I
2700		3798.12	I
270		3801.30	I
2700		3802.92	I
670		3803.88	I
530		3804.74	I
670		3810.49	I
530		3811.03	I
530		3815.51	I
210		3818.86	II
210		3819.15	I
670		3824.88	I
350		3835.18	I
250		3836.45	I
210		3845.90	I
290		3858.95	I
350		3863.38	I
270		3867.92	I
530		3877.56	I
870		3878.82	I
670		3883.14	I
1100		3885.44	I
670		3885.68	I
210		3886.07	I
580		3891.30	I
210		3908.97	I
670		3914.70	I
530		3920.20	I
670		3937.44	I
520		3943.67	I
250		3965.69	I
910	d	3966.09	I
210		3971.85	I
1100		4032.52	I
250		4039.53	I
16000	c	4058.94	I
210		4059.51	I
350		4060.79	I
210		4068.26	I
12000		4079.73	I
270		4084.86	I
440		4100.40	I
6700		4100.92	I
310		4116.90	I
5300		4123.81	I
670		4129.43	I
770		4129.93	I
2300		4137.10	I
440		4139.44	I
2700		4139.71	I
350		4143.21	I
870		4150.12	I
4400		4152.58	I
870		4163.47	I
4400		4163.66	I
4000		4164.66	I
3500		4168.13	I
310		4184.44	I
1200		4190.88	I
870		4192.07	I
870		4195.09	I
1300		4195.66	I
310		4198.51	I
350		4201.52	I
870		4205.31	I
350		4214.73	I
420		4217.94	I
420		4229.15	I
250		4255.44	I
770		4262.05	I
420		4266.02	I
290		4270.69	I
400		4286.99	I
580		4299.60	I
580		4300.99	I
120		4309.56	I
390		4311.27	I
120		4312.45	I
350		4326.33	I
120		4327.38	I
390		4331.37	I
140		4342.82	I
140		4348.65	I
110		4349.03	I
290		4351.57	I
210		4368.43	I
140		4377.96	I
130		4388.36	I
160		4392.69	I
330		4410.21	I
190		4419.44	I
230	c	4437.22	I
290		4447.18	I
140		4456.80	I
140		4457.42	I
140		4469.71	I
140		4471.29	I
140		4472.53	I
150		4503.04	I
530		4523.41	I
480		4546.82	I
370		4564.53	I
720		4573.08	I
480		4581.62	I
1200		4606.77	I
170		4616.17	I
450		4630.11	I
450		4648.95	I
110		4649.27	I
450		4663.83	I
340		4666.24	I
240		4667.22	I
580		4672.09	I
530		4675.37	I
110		4678.48	I
320		4685.14	I
130	c	4706.14	I
260		4708.29	I
150		4713.50	I
110	c	4733.89	I
220	c	4749.70	I
110		4816.38	I
110	c	4848.37	I
130	c	4967.78	I
110		4973.14	I
190		4988.97	I
85		5000.95	I
65		5002.25	I
40		5013.27	I
230		5017.75	I
40		5019.51	I
150		5026.36	I
40		5030.13	I
210		5039.04	I
40		5047.96	I
170		5058.01	I
65		5059.35	I
130		5065.25	I
40		5077.40	I
750		5078.96	I
40	c	5094.41	I
420		5095.30	I
170		5100.16	I
170		5120.30	I
85		5121.80	I
85		5127.66	I
40		5133.34	I
210		5134.75	I
75		5140.58	I
75		5147.54	I
40		5150.64	I
75		5152.63	I
250		5160.33	I
250		5164.38	I
230		5180.31	I
110		5186.98	I
190		5189.20	I
170		5193.08	I
150		5195.84	I
65		5203.22	I
35		5205.13	I
85	c	5219.10	I

Niobium (Cont.)

Intensity		Wavelength	Spectrum
65		5225.16	I
150		5232.81	I
85	c	5237.43	I
29		5240.39	I
150	d	5251.62	I
		5251.81	I
75		5253.03	I
85		5253.93	I
50		5269.92	I
270		5271.53	I
25		5272.48	I
130	c	5276.20	I
29	c	5279.43	I
50		5285.26	I
35		5296.34	I
50		5315.55	I
17		5317.01	I
250		5318.60	I
50		5319.49	I
75		5334.87	I
25		5336.81	I
50		5340.80	I
25		5343.58	I
460		5344.17	I
340		5350.74	I
40		5353.28	I
25		5355.31	I
40		5355.70	I
29		5359.19	I
17		5362.01	I
40		5375.27	I
40		5381.34	I
17		5388.30	I
21		5395.86	I
29		5396.33	I
29		5411.24	I
21		5416.30	I
65		5422.44	I
21		5431.26	I
110		5437.27	I
19		5448.31	I
19		5456.19	I
40		5458.04	I
19	h	5468.10	I
40		5481.00	I
13		5483.09	I
19		5483.49	I
13		5491.06	I
17		5499.53	I
40		5504.58	I
17		5509.12	I
35	c	5512.82	I
17		5517.39	I
50		5523.57	I
25		5541.47	I
85		5551.35	I
29		5563.00	I
17	c	5571.44	I
35	c	5576.16	I
35		5578.29	I
50		5586.97	I
17	c	5590.95	I
13		5594.89	I
17	c	5599.59	I
40		5603.52	I
13		5603.93	I
25		5628.26	I
65		5629.17	I
35	c	5635.42	I
170		5642.11	I
35		5645.30	I
17		5654.14	I
130		5664.71	I
170		5665.63	I
17		5666.86	I
65	Cw	5671.02	I
85		5671.91	I
25		5677.47	I
25		5693.09	I
35	d	5697.90	I
		5698.03	I
40		5706.16	I
85		5706.48	I
29		5709.33	I
17		5715.59	I
65		5716.35	I
25		5725.66	I
130		5729.19	I
21		5737.36	I
13		5738.20	I
85		5751.44	I
110		5760.34	I
65		5764.99	I
29		5771.08	I
50	c	5776.07	I
17		5780.34	I
85		5787.54	I
17		5789.79	I
50		5794.24	I
50		5804.03	I
29	h	5815.33	I
110		5819.43	I
35		5820.62	I
75		5834.90	I
25		5838.15	I
130	d	5838.64	I
50		5842.47	I
17		5846.09	I
65		5866.47	I
35		5874.70	I
17		5877.79	I
40		5893.44	I
190	Cw	5900.62	I
40	c	5903.80	I
29		5927.41	I
40	c	5934.16	I
40		5957.70	I
150		5983.22	I
65		5986.08	I
85	Cw	5997.93	I
50		6029.75	I
50		6031.84	I
50		6045.50	I
25		6048.72	I
29		6056.65	I
29		6107.71	I
40		6142.51	I
50		6148.13	I
50		6164.32	I
29		6213.06	I
75		6221.96	I
40	c	6251.76	I
21		6260.77	I
85	c	6430.46	I
50	c	6433.22	I
17		6497.84	I
65		6544.61	I
15		6574.73	I
19	Cw	6591.00	I
19		6606.16	I
19		6607.28	I
35		6614.15	I
19		6626.98	I
210	Cw	6660.84	I
150	Cw	6677.33	I
65		6701.20	I
130	c	6723.62	I
75		6739.88	I
25		6795.31	I
85		6828.11	I
25	c	6849.35	I
19		6870.92	I
40		6876.36	I
25	c	6902.89	I
35		6908.07	I
40		6918.32	I
17		6946.07	I
17		6972.49	I
25		6986.09	I
85		6990.32	I
17	c	6996.11	I
21		7023.48	I
17		7038.04	I
190	c	7046.81	I
8		7066.41	I
8		7075.23	I
40	c	7098.94	I
17	Cw	7102.01	I
19		7119.31	I
15		7122.95	I
35		7126.17	I
17		7130.06	I
130		7159.43	I
17		7191.37	I
19	c	7208.94	I
50		7252.35	I
15		7274.81	I
13		7317.03	I
17	c	7323.92	I
29	Cw	7328.38	I
65	c	7353.16	I
190	Cw	7372.50	I
13		7419.83	I
15		7436.02	I
19		7478.20	I
65		7515.93	I
29	c	7519.77	I
170	c	7574.58	I
17	c	7583.21	I
13		7639.81	I
13		7647.71	I
25		7703.33	I
75	c	7726.68	I
25		7757.31	I
6		7787.11	I
13	Cw	7873.41	I
35		7885.31	I
25		7938.89	I
8		7954.76	I
40		8135.20	I
13	Cw	8240.00	I
29	Cw	8320.93	I
29		8346.08	I
10		8350.04	I
17		8439.77	I
17	Cw	8475.98	I
25		8526.99	I
13	c	8547.25	I
17	c	8560.54	I
17		8575.87	I
21	c	8697.55	I
21		8740.96	I
21		8767.97	I
29	Cw	8815.56	I
35		8905.78	I

Nb III
Ref. 392 — C.H.C.

Intensity		Wavelength	
		Vacuum	
60		1314.56	III
50		1319.15	III
60		1431.92	III
60		1433.39	III
50		1435.26	III
80		1445.43	III
80		1445.98	III
80		1447.09	III
50		1448.50	III
60		1451.63	III
100		1456.68	III
80		1484.73	III
50		1486.79	III
100		1495.94	III
80		1498.02	III
80		1499.45	III
50		1501.53	III
100		1501.99	III
60		1505.03	III
50		1509.71	III
50		1512.34	III
50		1513.25	III
80		1513.81	III
50		1517.38	III
100		1524.91	III
60		1532.98	III
60		1537.50	III
50		1566.92	III
50		1570.19	III
50		1586.82	III
100		1590.21	III
80		1598.86	III
80		1604.72	III
80		1639.51	III
100		1682.77	III
60		1684.40	III
100		1705.44	III
100		1707.14	III
50		1739.30	III
50		1758.63	III
60		1763.72	III
50		1808.70	III
50		1863.13	III
100		1892.92	III
100		1938.84	III
60	h	1979.07	III
50		1985.15	III
50		1997.11	III
		Air	
50	h	2007.28	III
50		2032.47	III
50		2060.29	III
80	h	2130.24	III
60		2206.01	III
60		2240.31	III
60		2244.19	III
60		2265.63	III
80		2273.92	III
100		2275.23	III
80		2279.36	III
100		2281.51	III
80		2284.40	III
100		2290.36	III
60		2304.78	III
50		2309.92	III
100		2313.30	III
50		2330.22	III
100		2338.09	III
80		2344.12	III
90		2349.21	III
80		2355.54	III
100		2362.06	III
80		2362.50	III
80		2365.70	III
100		2372.73	III
100		2387.41	III
80		2388.23	III
50		2404.23	III
80		2404.89	III
100		2413.94	III
60		2414.50	III
100		2421.91	III
50		2437.74	III
60		2446.45	III
100		2456.99	III
50		2460.34	III
50		2460.45	III
50		2463.72	III
80		2468.72	III
60		2469.39	III
80		2475.87	III
50		2486.02	III
60		2488.74	III
60		2493.02	III
100		2499.73	III
60		2508.53	III
50		2511.95	III
100		2545.64	III
80		2557.94	III
50		2567.44	III
60		2593.75	III
80		2598.86	III
50		2628.67	III
80		2633.17	III
80		2657.99	III
50		2937.71	III
80		3001.84	III
80		3142.26	III
60		3266.11	III

Nb IV
Ref. 407 — C.H.C.

Intensity	Wavelength	
	Vacuum	
12	542.38	IV
12	543.09	IV
12	545.21	IV
10	559.94	IV
10	566.22	IV
18	981.27	IV
60	993.54	IV
18	996.16	IV
50	1002.76	IV
400	1005.72	IV
500	1007.05	IV
500	1010.19	IV
45	1030.27	IV
100	1116.08	IV
150	1120.02	IV
50	1447.48	IV
40	1473.43	IV
40	1487.23	IV
60	1502.30	IV
60	1524.36	IV
50	1534.06	IV
40	1635.68	IV
40	1910.70	IV
60	1922.41	IV
60	1978.22	IV
	Air	
40	2027.50	IV
65	2032.53	IV
40	2034.67	IV
40	2068.62	IV
55	2084.07	IV
35	2093.12	IV
50	2093.30	IV
20	2122.68	IV
25	2130.23	IV
45	2146.36	IV
18	2249.98	IV

Nb V
Ref. 431 — C.H.C.

Intensity	Wavelength	
	Vacuum	
80	464.55	V
80	468.32	V
60	753.01	V
80	763.77	V
80	774.02	V
40	1007.02	V
50	1044.90	V
70	1212.21	V
100	1258.87	V
40	1267.60	V
100	1758.33	V
100	1877.34	V

NITROGEN (N)
Z = 7

N I and II
Ref. 213 = R.L.K.

Intensity	Wavelength

Nitrogen (Cont.)

Intensity		Wavelength	
		Vacuum	
285		644.634	II
360		644.837	II
450		645.178	II
140		647.50	I
360		660.286	II
170		671.016	II
285		671.386	II
150		671.630	II
160		671.773	II
170		672.001	II
350		692.70	I
285		746.984	II
650		775.965	II
90		885.67	I
90		909.697	I
80		910.278	I
40		910.645	I
450		915.612	II
450		915.962	II
550		916.012	II
650		916.701	II
90		953.415	I
100		953.655	I
130		953.970	I
130		963.990	I
115		964.626	I
70		965.041	I
90		1067.614	I
60		1068.612	I
450		1083.990	II
600		1084.580	II
430		1085.546	II
650		1085.701	II
175		1097.237	I
115		1098.095	I
115		1098.260	I
105		1100.360	I
40		1100.465	I
90		1101.291	I
360		1134.165	I
385		1134.415	I
410		1134.980	I
105		1143.65	I
130		1163.884	I
60		1164.206	I
105		1164.325	I
270		1167.448	I
105		1168.334	I
60		1168.417	I
195		1168.536	I
230		1176.510	I
105		1176.630	I
195		1177.695	I
410		1199.550	I
385		1200.223	I
360		1200.710	I
175		1225.026	I
160		1225.37	I
130		1228.41	I
160		1228.79	I
360		1243.179	I
315		1243.306	I
290		1310.540	I
250		1310.95	I
230		1319.00	I
315		1319.68	I
115		1326.57	I
115		1327.92	I
360		1411.94	I
700		1492.625	I
490		1492.820	I
640		1494.675	I
775		1742.729	I
700		1745.252	I
		Air	
160		2095.53	II
70		2096.20	II
110		2096.86	II
110		2130.18	II
160		2142.78	II
160		2206.09	II
160		2286.69	II
110		2288.44	II
220		2316.49	II
160		2316.69	II
285		2317.05	II
160		2461.27	II
110		2496.83	II
70		2496.97	II
110		2520.22	II
160		2520.79	II
220		2522.23	II
110		2590.94	II
160		2709.84	II
110		2799.22	II
110		2823.64	II
160		2885.27	II
220		3006.83	II
360		3437.15	II
285		3838.37	II
360		3919.00	II
450		3955.85	II
1000		3995.00	II
360		4035.08	II
550		4041.31	II
360		4043.53	II
140		4099.94	I
185		4109.95	I
285		4176.16	II
285		4227.74	II
285		4236.91	II
220		4237.05	II
450		4241.78	II
285		4432.74	II
650		4447.03	II
360		4530.41	II
550		4601.48	II
450		4607.16	II
360		4613.87	II
450		4621.39	II
870		4630.54	II
550		4643.08	II
285		4788.13	II
450		4803.29	II
180		4847.38	I
285		4895.11	II
160		4914.94	I
210		4935.12	I
160		4950.23	I
350		4963.98	I
285		4987.37	II
450		4994.36	II
650		5001.48	II
360		5002.70	II
870		5005.15	II
550		5007.32	II
450		5010.62	II
360		5016.39	II
360		5025.66	II
550		5045.10	II
185		5281.20	I
140		5292.68	I
450		5495.67	II
285		5535.36	II
630		5668.63	II
550		5676.02	II
870		5679.56	II
450		5686.21	II
450		5710.77	II
285		5747.30	II
700		5752.50	I
240		5764.75	I
265		5829.54	I
235		5854.04	I
360		5927.81	II
550		5931.78	II
285		5940.24	II
650		5941.65	II
285		5952.39	II
160		5999.43	I
210		6008.47	I
285		6167.76	II
360		6379.62	II
185		6411.65	I
210		6420.64	I
210		6423.02	I
210		6428.32	I
185		6437.68	I
235		6440.94	I
185		6457.90	I
300		6468.44	I
750		6482.05	II
360		6482.70	I
300		6483.75	I
265		6481.71	I
325		6484.80	I
160		6491.22	I
210		6499.54	I
185		6506.31	I
750		6610.56	II
185		6622.54	I
185		6636.94	I
235		6644.96	I
185		6646.50	I
235		6653.46	I
210		6656.51	I
185		6722.62	I
210		7398.64	I
160		7406.12	I
265		7406.24	I
685		7423.64	I
785		7442.29	I
900		7468.31	I
185		7608.80	I
450		7762.24	II
400		8184.87	I
400		8188.02	I
250		8200.36	I
300		8210.72	I
570		8216.34	I
400		8223.14	I
400		8242.39	I
550		8438.74	II
500		8567.74	I
570		8594.00	I
650		8629.24	I
500		8655.89	I
220		8676.08	II
700		8680.28	I
650		8683.40	I
500		8686.15	I
110		8687.43	II
110	h	8699.00	II
500		8703.25	I
160	h	8710.54	II
570		8711.70	I
500		8718.83	I
250		8728.89	I
200		8747.36	I
500		9386.80	I
570		9392.79	I
250		9460.68	I
200		9863.33	I
160	h	9865.41	II
110	h	9868.21	II
160	h	9887.39	II
220	h	9891.09	II
160	h	9961.86	II
220	h	9969.34	II
285	h	10023.27	II
220	h	10035.45	II
220	h	10065.15	II
160	h	10070.12	II
250		10105.13	I
300		10108.89	I
350		10112.48	I
400		10114.64	I
110	h	10126.27	II
250		10539.57	I
200		12074.51	I
380		12186.82	I
225		12288.97	I
290		12328.76	I
310		12381.65	I
180		12438.40	I
510		12461.25	I
920		12469.62	I
300		13429.01	I
840		13581.33	I
180		13587.73	I
180		13602.27	I
290		13624.18	I
250		14757.07	I
100		14868.87	I
160		14966.60	I
180		15582.27	I
120	s	17516.58	I
100	l	17584.86	I
100		17878.26	I

N III
Ref. 66, 213 = R.L.K.

Intensity		Wavelength	
		Vacuum	
500		257.95	III
650		258.50	III
700		259.19	III
800		260.09	III
800		261.28	III
500		262.91	III
500		265.23	III
500		265.27	III
500		268.70	III
150		314.715	III
200		314.850	III
90		314.877	III
600		323.26	III
500		338.35	III
500		340.20	III
500		351.98	III
120		362.833	III
150		362.881	III
150		362.946	III
90		362.985	III
300		374.204	III
350		374.441	III
500		387.48	III
250		451.869	III
300		452.226	III
500		684.996	III
570		685.513	III
650		685.816	III
500		686.335	III
500		763.336	III
570		764.359	III
250		771.544	III
300		771.901	III
350		772.385	III
200		772.891	III
150		772.975	III
650		979.842	III
700		979.919	III
900		989.790	III
700		991.514	III
1000		991.579	III
500		1183.031	III
570		1184.550	III
150		1387.371	III
250		1729.945	III
570		1747.848	III
350		1751.218	III
650		1751.657	III
150		1804.486	III
200		1805.669	III
150		1846.42	III
350		1885.06	III
400		1885.22	III
200		1907.99	III
150		1919.55	III
150		1919.77	III
300		1920.65	III
150		1920.84	III
200		1921.30	III
		Air	
200		2064.01	III
250		2064.42	III
120		2068.68	III
90		2071.09	III
90		2117.59	III
90		2121.50	III
90		2147.31	III
200		2188.20	III
150		2188.38	III
250	w	2682.18	III
90		2689.20	III
120		3367.34	III
90		3754.67	III
120		3771.05	III
90		3938.52	III
150		3998.63	III
200		4003.58	III
250		4097.33	III
200		4103.43	III
120		4195.76	III
150		4200.10	III
90		4332.91	III
120		4345.68	III
300		4379.11	III
90		4510.91	III
120		4514.86	III
90		4634.14	III
120		4640.64	III
90		4858.82	III
150		4867.15	III
90		5314.35	III
200		5320.82	III
150		5327.18	III
90		6454.11	III
120		6467.02	III

N IV
Ref. 108, 212 = R.L.K.

Intensity		Wavelength	
		Vacuum	
400		181.75	IV
400		191.7	IV
400		192.9	IV
500		196.87	IV
500		197.23	IV
500		202.60	IV
500		205.94	IV
500		205.97	IV
500		206.03	IV
500		217.20	IV
500	d	217.90	IV
500	d	223.4	IV
800	w	225.12	IV
800		225.21	IV
600	w	234.12	IV
600	w	234.20	IV
600	w	234.25	IV
550		236.07	IV
500		237.99	IV
500	w	238.7	IV
600		238.80	IV
500	w	239.62	IV
900		247.20	IV
500	w	248.43	IV
500	w	248.46	IV
500	w	248.48	IV
600		260.45	IV
650		270.99	IV
250		283.42	IV
300		283.48	IV
350		283.58	IV
600		285.56	IV
600	w	297.7	IV
700		297.82	IV
650		300.32	IV
90		303.123	IV

Nitrogen (Cont.)

Intensity		Wavelength	
500		303.28	IV
150		315.053	IV
120		322.503	IV
150		322.570	IV
200		322.724	IV
120		323.175	IV
300		335.050	IV
500	w	351.93	IV
700		353.06	IV
500		420.77	IV
650		463.74	IV
570		765.148	IV
520		921.992	IV
500		922.519	IV
480		923.057	IV
520		921.992	IV
500		922.519	IV
520		924.283	IV
1000		955.335	IV
150	w	1036.16	IV
90		1078.71	IV
90		1188.01	IV
1000		1718.55	IV

Air

Intensity		Wavelength	
90		2080.34	IV
90	w	2318.09	IV
150		2477.69	IV
250		2645.65	IV
300		2646.18	IV
350		2646.96	IV
90		3078.25	IV
90		3463.37	IV
570		3478.71	IV
500		3482.99	IV
400		3484.96	IV
90		3747.54	IV
150		4057.76	IV
90		4606.33	IV
150		6380.77	IV

N V
Ref. 66, 107, 318 = R.L.K.

Vacuum

Intensity		Wavelength	
52		166.947	V
52		186.069	V
62		186.153	V
90		209.303	V
90		247.561	V
120		247.706	V
150		266.196	V
200		266.379	V
90		713.518	V
150		713.860	V
150		748.195	V
200		748.291	V
1000		1238.821	V
900		1242.804	V
90		1549.336	V
200	l	1616.33	V
350	l	1619.69	V
90	w	1860.37	V

Air

Intensity		Wavelength	
60	l	2859.16	V
90	l	2974.52	V
150	w	2980.78	V
250	w	2981.31	V
60	w	2998.43	V
350		4603.73	V
250		4619.98	V
200	w	4944.56	V
60	w	7618.46	V

OSMIUM (Os)
Z = 76

Os I and II
Ref. 1 = C.H.C.

Air

Intensity		Wavelength	
9600		2001.45	I
13000		2003.73	I
9000		2004.78	
17000		2010.15	I
29000		2018.14	I
29000		2020.26	
14000		2022.76	I
14000		2028.23	I
18000		2034.44	I
26000		2045.36	I
8600		2058.69	I
		2058.78	I
13000		2061.69	I
7800		2067.21	II
4200		2070.67	II
7200		2076.95	I
7200		2078.09	
14000		2079.97	I
2900		2082.54	I
2900		2089.03	I
2900		2089.21	I
6000		2097.60	I
5500		2100.63	I
2100		2117.66	I
4800		2117.96	I
6600		2119.79	
1900		2123.84	I
5300		2137.11	I
2400		2149.97	
2600		2154.59	I
1300		2157.84	I
1200		2158.53	I
2400		2161.00	
3100		2166.90	I
1100		2167.75	I
2100		2171.65	I
960		2184.68	I
840		2194.39	II
760		2202.49	I
600		2227.98	I
1100		2234.61	I
1300		2252.15	I
2000		2255.85	II
1400		2264.60	I
360		2268.28	I
960		2270.17	I
1400		2282.26	II
840		2283.67	I
570		2289.32	I
380		2297.31	I
660		2308.31	I
190		2313.75	II
550		2320.18	I
310		2323.98	I
660		2324.24	I
330		2326.99	I
310		2334.56	I
720		2336.80	II
430		2338.63	I
290		2340.69	I
430		2343.74	I
260		2345.75	I
430		2347.38	I
230		2350.23	II
360		2352.99	I
120		2355.28	II
240		2356.92	I
240		2357.25	I
310		2362.41	I
900		2362.77	I
500		2367.35	II
290		2369.24	I
500		2370.70	I
480		2371.18	I
95		2375.06	II
2600		2377.03	I
260		2377.61	I
900		2379.39	I
240		2384.62	I
1700		2387.29	I
330		2394.29	I
290		2395.39	I
1100		2395.88	I
220		2396.78	I
960		2401.13	I
260		2402.23	I
200		2403.54	I
330		2403.85	I
95		2405.08	II
290		2405.45	I
200		2405.96	I
360		2408.67	I
240		2410.98	I
290		2414.52	I
530		2417.99	I
530		2418.53	I
95		2420.02	II
200		2423.07	II
70		2424.02	II
500		2424.56	I
1400		2424.97	I
240		2426.81	I
70		2427.90	II
380		2431.19	I
380		2431.61	I
360		2446.02	I
900		2450.74	I
530		2451.73	I
530		2453.90	I
110		2454.91	II
530		2456.46	I
1800		2461.42	I
110		2468.90	II
290		2472.28	I
290		2474.78	I
900		2476.84	I
360		2482.43	I
530		2486.24	II
4500		2488.55	I
290		2491.02	I
290		2491.69	I
360		2492.42	I
2600		2498.41	I
330		2499.92	I
330		2502.29	I
500		2504.39	I
260		2504.51	I
35		2507.18	II
70		2509.71	II
660		2512.87	I
2400		2513.25	I
660		2515.04	I
500		2517.92	I
660		2518.44	I
200		2519.29	I
330		2519.79	I
200		2532.44	I
780		2538.00	II
240		2538.10	I
1000		2542.51	I
30		2548.83	II
310		2554.46	I
190		2563.16	II
600		2566.49	I
290		2566.88	I
480		2568.83	I
340		2571.78	I
150		2578.32	II
130		2580.03	II
360		2581.05	I
740		2581.96	I
1000		2590.76	I
200		2591.98	I
170		2596.00	II
210		2609.20	I
380		2609.56	I
400		2610.78	I
470		2612.63	I
1800		2613.06	I
800		2619.94	I
230		2620.62	I
530		2621.82	I
380		2628.48	I
27		2631.22	II
3800		2637.13	I
1900		2644.11	I
340		2646.89	I
380		2647.73	I
380		2649.34	I
490		2656.68	I
1900		2658.60	I
640		2659.83	I
380		2661.18	I
40		2664.29	II
580		2674.57	I
400		2674.88	I
2100		2689.82	I
510		2699.59	I
580		2706.70	I
3000		2714.64	I
580		2715.36	I
1300		2720.04	I
850		2721.86	I
580		2730.61	I
40		2731.36	II
580		2732.80	I
690		2761.42	I
470		2763.27	I
340		2765.04	I
960		2770.71	I
300		2776.91	I
740		2782.55	I
40		2783.88	II
640		2786.31	I
230		2793.99	I
230		2794.19	I
530		2796.73	I
320		2804.07	I
2800		2806.91	I
470		2808.94	I
420		2813.84	I
740		2814.20	I
300		2815.78	I
420		2829.27	I
230		2837.42	I
470		2838.17	I
5100		2838.63	I
740		2841.60	I
2300		2844.40	I
420		2846.39	I
420		2848.25	I
1500		2850.76	I
1500		2860.96	I
35		2863.37	II
360		2874.96	I
300		2878.40	I
35		2879.39	II
30		2880.20	II
260		2896.06	I
9600		2909.06	I
2100		2912.33	I
530		2917.26	I
2100		2919.79	I
300		2925.57	I
360		2929.51	I
510		2931.28	I
260		2934.64	I
200		2942.85	I
1100	h	2948.23	I
1400		2949.53	I
210	d	2949.81	I
300		2961.01	I
530		2962.15	I
450		2964.06	I
740		2970.97	I
450		2977.64	I
510		2982.90	I
340		2983.49	
260		2997.65	I
330		3013.07	I
570		3017.25	I
4400		3018.04	I
480		3019.38	I
1100		3030.70	I
2900		3040.90	I
210		3043.50	I
120		3042.74	II
230		3049.46	I
210		3050.39	I
8600		3058.66	I
290		3060.30	I
570		3062.19	I
210		3069.94	I
360		3074.08	I
290		3074.96	I
290		3077.44	I
1100		3077.72	I
360		3078.11	I
230		3078.38	I
230		3090.08	I
270		3093.59	I
310		3101.53	I
360		3105.99	I
310		3108.98	I
620		3109.38	I
250		3111.09	I
310		3118.33	I
480		3131.12	I
250		3152.67	I
290		3153.61	I
3100		3156.25	I
250		3156.78	I
310		3166.51	I
180		3173.93	II
420		3178.06	I
230		3181.88	I
230		3185.33	I
310		3186.98	I
310		3189.46	I
310		3194.23	I
150		3213.31	II
1900		3232.06	I
290		3238.63	I
190		3241.04	I
120		3248.00	I
190		3254.91	I
190		3256.92	I
190		3260.30	I
3100		3262.29	I
380		3262.75	I
3100		3267.94	I
620		3269.21	I
190		3272.16	I
530		3275.20	I
330		3277.97	I
190		3288.84	I
1200		3290.26	I
7600		3301.56	I
250		3306.23	I
620		3310.91	I
120		3315.42	I
250		3324.33	I
310		3327.42	I
960		3336.15	I
110		3351.74	I
120		3353.91	
230		3357.97	I
250		3361.15	I
190		3364.12	I
120		3370.20	I
960		3370.59	I
160		3372.08	I
120		3378.68	I
310		3384.00	I
190		3385.94	I
620		3387.84	I
120		3401.17	I
620		3401.86	I
250		3402.51	I
120		3408.76	I
120		3412.74	I

Osmium (Cont.)

Intensity		Wavelength	
120		3421.69	I
150		3427.67	I
250		3440.60	I
120		3444.46	I
160		3445.55	I
310		3449.20	I
120		3458.38	I
120		3465.44	I
120		3478.53	I
120		3482.11	I
120		3487.46	I
120		3490.33	
160		3498.54	I
250		3501.16	I
620		3504.66	I
440		3512.99	I
310		3518.72	I
120		3520.00	I
480		3523.64	I
120		3526.04	I
1200		3528.60	I
230		3530.06	I
230		3532.80	I
120		3533.41	I
230		3542.71	I
960		3559.79	I
1200		3560.86	I
120		3562.34	I
310		3569.78	I
120		3574.08	I
120		3587.32	I
620		3598.11	I
190		3601.83	I
95		3604.48	II
250		3616.57	I
120		3619.43	I
450		3640.33	I
230		3654.49	I
330		3656.90	I
120		3666.31	I
480		3670.89	I
120		3675.45	I
250		3689.06	I
190		3703.25	I
120		3706.56	I
120		3709.14	I
230		3713.73	I
210		3719.52	I
230		3720.13	I
180		3746.47	I
3700		3752.52	I
100		3757.12	I
130		3766.30	I
120		3768.14	I
120		3774.40	I
110		3774.62	I
120		3776.25	I
290		3776.99	I
2100		3782.20	I
620		3790.14	I
180		3790.73	I
370		3793.91	I
250		3836.06	I
150		3840.30	I
150		3841.29	I
190		3849.94	I
230		3857.09	I
230		3865.47	I
730		3876.77	I
250		3881.86	I
140		3900.39	I
190		3901.71	I
100		3930.00	I
250		3938.59	I
100		3949.78	I
200		3961.02	I
1000		3963.63	I
100		3964.96	I
150		3969.67	I
110	h	3975.44	I
730		3977.23	I
100		3988.18	I
150		4003.48	I
100		4004.02	I
150		4005.16	I
160		4018.26	I
100		4037.84	I
280		4041.92	I
160		4048.05	I
960		4066.69	I
250		4070.86	I
190		4071.56	I
230		4074.68	I
490		4091.82	I
120		4100.30	I
1200		4112.02	I
180		4124.60	I
180		4128.96	I
2500		4135.78	I
150		4137.84	I
180		4172.57	I
1200		4173.23	I
620		4175.63	I
120		4184.13	I
320		4189.91	I
180		4201.45	I
250		4202.06	I
1200		4211.86	I
120		4213.86	I
100		4215.16	I
170		4233.46	I
4900		4260.85	I
100		4264.75	I
120		4269.61	I
100		4285.90	I
560		4293.95	I
560		4311.40	I
110		4326.25	I
340		4328.68	I
100		4338.75	I
100		4351.53	I
210		4365.67	I
110		4370.66	I
520		4394.86	I
160		4397.26	I
160		4402.74	I
4900		4420.47	I
100		4432.41	I
290		4436.32	I
100		4439.64	I
230		4447.35	I
120		4484.76	I
110		4548.66	I
540		4550.41	I
140		4551.30	I
170		4616.78	I
170		4631.83	I
140		4663.82	I
670		4793.99	I
110		4865.60	I
55		5031.83	I
45		5039.12	I
35		5072.88	I
35		5074.77	I
35		5079.09	I
90		5103.50	I
55		5110.81	I
22		5122.23	I
22		5145.54	I
140		5149.74	I
28		5152.01	I
28		5168.98	I
40		5193.52	I
270		5202.63	I
35		5203.23	I
20		5250.46	
45		5255.82	I
55		5265.15	I
20		5283.89	I
20		5295.65	I
40		5298.78	I
13		5302.58	I
18		5336.23	I
11		5346.03	I
13		5352.25	I
110		5376.79	I
16		5403.43	I
13		5412.14	I
120		5416.34	I
45		5416.69	I
28		5417.51	I
16		5441.82	I
55		5443.31	I
22		5446.93	I
11		5447.76	I
20		5449.37	I
20		5453.40	I
22		5457.30	I
28		5470.00	I
13		5474.58	I
13		5475.13	I
9		5477.27	I
16		5481.85	I
22		5509.33	I
8		5516.01	I
270		5523.53	I
22		5546.82	I
9		5549.79	I
13		5552.88	I
11		5560.62	I
16		5580.66	I
80		5584.44	I
8		5600.50	I
35		5620.08	I
9		5637.41	I
22		5642.56	I
28		5645.25	I
7		5648.98	I
9		5660.21	I
7		5674.38	I
28		5680.88	I
11		5709.37	I
170		5721.93	I
8		5737.89	I
8		5739.72	I
22		5765.05	I
170		5780.82	I
40		5800.60	I
8		5842.49	I
110		5857.76	I
28		5860.64	I
11		5882.92	I
11		5903.98	I
11		5906.84	I
7		5908.95	I
7		5981.36	I
11		5983.22	I
65		5996.00	I
20		6015.79	I
7		6054.63	I
20		6144.53	I
11		6158.03	I
35		6227.70	I
7		6241.70	I
22		6269.41	I
11		6274.94	I
11		6286.83	I
9		6398.86	I
22		6403.15	I
9		6448.13	I
6		6520.85	I
7		6528.87	I
7		6533.14	I
11		6538.30	I
11		6576.83	I
8		6614.56	I
4		6615.43	I
7		6661.81	I
27		6729.56	I
18		6791.53	I
14		6806.61	I
5		6878.70	I
4		6901.58	I
11		6956.02	I
6		6984.95	I
15		7060.67	I
22		7145.54	I
10		7149.89	I
4		7184.10	I
10		7206.33	I
5		7209.96	I
9		7251.16	I
8		7255.43	I
6		7375.07	I
9		7407.95	I
26		7602.95	I
4		7701.46	I
7		7789.96	I
7		7852.17	I
6		7981.20	I
7		8041.29	I

OXYGEN (O)
Z = 8

OI and II
Ref. 66, 69, 209, 210, 215 —
R.L.K.

Intensity		Wavelength	
		Vacuum	
250		537.83	II
300		538.26	II
220		539.09	II
200		539.55	II
150		539.85	II
150		644.148	II
200		672.95	II
150		673.77	II
70		685.544	I
900		718.484	II
600		718.562	II
70		744.794	I
70		770.793	I
90		771.056	I
70		775.321	I
70		791.973	I
300		796.66	II
90		804.287	I
70		804.848	I
70		805.295	I
80		805.810	I
240		832.762	II
450		833.332	II
600		834.467	II
40		877.879	I
80		922.008	I
90		935.193	I
40		948.686	I
90		971.738	I
40		976.448	I
160		988.773	I
40		990.204	I
250		1025.762	I
90		1027.431	I
160		1039.230	I
60		1040.942	I
40		1152.152	I
900		1302.168	I
600		1304.858	I
300		1306.029	I
		Air	
30	d	2283.42	II
30	d	2284.89	II
110		2293.32	II
200		2300.35	II
30	d	2313.05	II
30	d	2316.12	II
30	d	2316.79	II
50	d	2319.68	II
30	d	2322.15	II
30	d	2339.31	II
110		2411.60	II
80		2425.55	II
250		2433.56	II
80	d	2436.06	II
80		2444.26	II
300		2445.55	II
300		2733.34	II
110		2747.46	II
265		2972.29	I
160		3122.62	II
220		3129.44	II
450		3134.82	II
285		3138.44	II
220		3270.98	II
220		3273.52	II
220		3277.69	II
360		3287.59	II
160		3305.15	II
160		3306.60	II
220		3377.20	II
285		3390.25	II
220		3407.38	II
160		3409.84	II
285		3470.81	II
220		3712.75	II
285		3727.33	II
160		3739.92	II
360		3749.49	II
160		3803.14	II
120		3823.41	I
450		3911.96	II
160		3919.29	II
185		3947.29	I
160		3947.48	I
140		3947.59	I
220		3954.37	II
100		3954.61	I
450		3973.26	II
220		3982.20	II
160		4069.90	II
285		4072.16	II
450		4075.87	II
80	d	4083.91	II
50	d	4087.14	II
150	d	4089.27	II
110		4097.24	II
220		4105.00	II
285		4119.22	II
160		4132.81	II
50		4146.06	II
220		4153.30	II
285		4185.46	II
450		4189.79	II
80		4233.27	I
50	d	4253.74	II
50	d	4253.98	II
50	d	4275.47	II
50	d	4303.78	II
285		4317.14	II
160		4336.86	II
220		4345.56	II
285		4349.43	II
220		4366.90	II
100		4368.25	I
220		4395.95	II
450		4414.91	II
285		4416.98	II
160		4448.21	II
160		4452.38	II
30		4463.43	II
50	d	4466.28	II
50		4467.83	II
50		4469.41	II
360		4590.97	II
285		4596.17	II
80	d	4609.39	II
160		4638.85	II
360		4641.81	II
450		4649.14	II
160		4650.84	II
360		4661.64	II
285		4676.23	II
220		4699.21	II
285		4705.36	II
160		4924.60	II
220		4943.06	II
135		5329.10	I
160		5329.68	I
190		5330.74	I
90		5435.18	I
110		5435.78	I

Oxygen (Cont.)

Intensity		Wavelength	Spectrum
135		5436.86	I
120		5577.34	I
160		5958.39	I
190		5958.58	I
80		5995.28	I
160		6046.23	I
190		6046.44	I
110		6046.49	I
100		6106.27	I
400		6155.98	I
450		6156.77	I
490		6158.18	I
80		6256.83	I
100		6261.55	I
100		6366.34	I
100		6374.32	I
320		6453.60	I
360		6454.44	I
400		6455.98	I
80		6604.91	I
100		6653.83	I
360		7001.92	I
450		7002.23	I
210		7156.70	I
400		7254.15	I
450		7254.45	I
320		7254.53	I
210		7476.44	I
100		7477.24	I
120		7479.08	I
120		7480.67	I
100		7706.75	I
870		7771.94	I
810		7774.17	I
750		7775.39	I
80		7886.27	I
100		7943.15	I
100		7947.17	I
235		7947.55	I
210		7950.80	I
185		7952.16	I
110		7981.94	I
135		7982.40	I
190		7986.98	I
135		7987.33	I
250		7995.07	I
400		8221.82	I
265		8227.65	I
265		8230.02	I
325		8233.00	I
120		8235.35	I
120		8426.16	I
810		8446.25	I
1000		8446.36	I
935		8446.76	I
325		8820.43	I
160	d	9057.01	I
120		9118.29	I
80		9134.71	I
80		9150.14	I
80		9151.48	I
235		9156.01	I
450		9260.81	I
490		9260.84	I
450		9260.94	I
400		9262.58	I
540		9262.67	I
590		9262.77	I
490		9265.94	I
640		9266.01	I
185		9399.19	I
120		9481.16	I
120	d	9482.88	I
235		9487.43	I
140		9492.71	I
265		9497.97	I
160		9499.30	I
235		9505.59	I
210		9521.96	I
120		9523.36	I
120		9523.96	I
100		9528.28	I
100		9622.13	I
120		9625.29	I
160		9677.38	I
80		9694.66	I
65		9694.91	I
235		9741.50	I
235		9760.65	I
120		9909.05	I
140		9936.98	I
120		9940.41	I
160		9995.31	I
120	d	10421.18	I
590		11286.34	I
640		11286.91	I
490		11287.02	I
490		11287.32	I
490		11295.10	I
540		11297.68	I
590		11302.38	I
265		11358.69	I
490		12464.02	I
450		12570.04	I
120		12990.77	I
160		13076.91	I
700		13163.89	I
750		13164.85	I
640		13165.11	I
160		16212.06	I
120		17966.70	I
590		18021.21	I
120		18041.48	I
120		18042.19	I
120		18046.23	I
140		18229.23	I
540		18243.63	I
140		26173.56	I

O III

Ref. 23, 66, 210 — R.L.K.

Intensity		Wavelength	
		Vacuum	
80	d	264.34	III
110		264.48	III
110		266.97	III
150		266.98	III
150		267.03	III
150		277.38	III
80		295.62	III
110		295.66	III
120		295.72	III
150		303.41	III
150		303.46	III
140		303.52	III
160		303.62	III
160		303.69	III
250		303.80	III
200		305.60	III
250		305.66	III
190		305.70	III
300		305.77	III
190		305.84	III
450		320.979	III
300		328.45	III
250		328.74	III
300		345.31	III
110		355.14	III
90		355.33	III
80		355.47	III
200		359.02	III
190		359.22	III
150		359.38	III
210		373.80	III
200		374.00	III
300		374.08	III
190		374.16	III
200		374.33	III
210		374.44	III
450		395.558	III
300		434.98	III
800		507.391	III
900		507.683	III
1000		508.182	III
1000		525.795	III
700		597.818	III
1000		599.598	III
110		609.70	III
160		610.04	III
200		610.75	III
100		610.85	III
800		702.332	III
800		702.822	III
900		702.899	III
1000		703.850	III
600		832.927	III
780		833.742	III
600		835.096	III
800		835.292	III
160		1476.89	III
285		1590.01	III
160		1591.33	III
220		1760.12	III
110		1760.42	III
220		1763.22	III
220		1764.48	III
750		1767.78	III
550		1768.24	III
360		1771.67	III
110		1773.00	III
110		1773.85	III
220		1779.16	III
160		1781.03	III
160		1784.85	III
220		1789.66	III
110		1848.26	III
110		1856.62	III
285		1872.78	III
285		1872.87	III
285		1874.94	III
160		1920.04	III
110		1920.75	III
110		1921.52	III
220		1923.49	III
110		1923.82	III
110		1926.94	III
		Air	
360		2013.27	III
160		2026.96	III
220		2045.67	III
160		2052.74	III
200	d	2390.44	III
80		2394.33	III
80		2422.84	III
80	d	2438.83	III
200		2454.99	III
200		2558.06	III
80		2687.53	III
110		2695.49	III
80		2959.68	III
250		2983.78	III
80		3017.63	III
80		3023.45	III
80		3043.02	III
200		3047.13	III
110		3059.30	III
80		3121.71	III
110		3132.86	III
80		3238.57	III
200		3260.98	III
300		3265.46	III
80		3267.31	III
80		3312.30	III
110		3340.74	III
80		3444.10	III
80		3455.12	III
80		3698.70	III
80		3702.75	III
80		3703.37	III
110		3707.24	III
110		3715.08	III
110		3744.00	III
150		3754.67	III
80		3757.21	III
250		3759.87	III
110		3791.26	III
200		3961.59	III
110		5592.37	III

O IV

Ref. 36, 66 = R.L.K.

Intensity		Wavelength	
		Vacuum	
150		195.86	IV
200		196.01	IV
110		207.18	IV
150		207.24	IV
140		233.46	IV
150		233.50	IV
110		233.52	IV
200		233.56	IV
110		233.60	IV
90		238.36	IV
180		238.57	IV
110		252.56	IV
110		252.95	IV
150		253.08	IV
300		260.39	IV
250		260.56	IV
300		279.63	IV
375		279.94	IV
110		285.71	IV
150		285.84	IV
200		306.62	IV
150		306.88	IV
700		553.330	IV
775		554.075	IV
850		554.514	IV
700		555.261	IV
580		608.398	IV
640		609.829	IV
270		616.952	IV
150		617.005	IV
200		617.036	IV
520		624.617	IV
580		625.130	IV
640		625.852	IV
200		779.734	IV
315		779.821	IV
360		779.912	IV
200		779.997	IV
640		787.711	IV
520		790.109	IV
700		790.199	IV
200		802.200	IV
160		802.255	IV
130		921.296	IV
160		921.366	IV
200		923.367	IV
130		923.433	IV
200		1338.612	IV
130		1342.992	IV
230		1343.512	IV
		Air	
200		2449.372	IV
200		2450.040	IV
200		2493.44	IV
200		2493.77	IV
200		2507.73	IV
230		2509.19	IV
200		2517.2	IV
160		2836.26	IV
160		2921.45	IV
460		3063.42	IV
410		3071.61	IV
160		3209.66	IV
230		3348.08	IV
270		3349.11	IV
160		3354.27	IV
200		3375.40	IV
130		3378.06	IV
360		3381.20	IV
360		3385.52	IV
270		3396.79	IV
360		3403.52	IV
230		3409.66	IV
410		3411.69	IV
230		3413.64	IV
200		3489.83	IV
160		3492.24	IV
230		3560.39	IV
270		3563.33	IV
315	w	3725.93	IV
360		3729.03	IV
410		3736.85	IV
230		3744.89	IV

O V

Ref. 24, 66 = R.L.K.

Intensity		Wavelength	
		Vacuum	
80		124.616	V
110		135.523	V
80		138.109	V
110		139.029	V
80		151.447	V
110		151.477	V
150		151.546	V
80		164.574	V
110		164.657	V
80		164.709	V
80		166.235	V
150		167.99	V
110		170.219	V
450		172.169	V
250		185.745	V
375		192.751	V
450		192.799	V
520		192.906	V
80		193.003	V
200		194.593	V
80		202.161	V
80		202.224	V
80		202.283	V
80		202.334	V
150		202.393	V
110		203.78	V
150		203.82	V
100		203.85	V
200		203.89	V
100		203.94	V
300		207.794	V
150		215.040	V
200		215.103	V
250		215.245	V
250		216.018	V
520		220.352	V
80		227.372	V
80		227.469	V
150		227.511	V
80		227.549	V
80		227.634	V
80		227.689	V
150		231.823	V
110		248.459	V
110		286.448	V
1000		629.730	V
230		681.272	V
700		758.678	V
640		759.441	V
580		760.228	V
775		760.445	V
640		761.128	V
700		762.003	V
520		774.518	V
640		1371.292	V
160	w	1506.72	V
315	w	1643.68	V
160		1707.996	V
		Air	
1000		2781.01	V
920		2786.99	V
775		2789.85	V
200		2941.33	V
210		2941.65	V
160		3144.66	V
100		4123.99	V
230	w	4930.27	V
130		5597.91	V
130		6500.24	V

PALLADIUM (Pd)

Z = 46

Pd I and II
Ref. 1, 287 — C.H.C.

Air

Intensity		Wavelength	
50		2162.27	II
50		2182.35	II
50		2212.15	II
100	r	2231.59	II
200	r	2296.53	II
50		2351.32	II
50		2362.31	II
75		2367.92	II
60		2372.16	II
50		2388.29	II
60		2414.73	II
75		2418.72	II
75		2424.49	II
100		2426.87	II
100		2430.94	II
100		2433.11	II
100		2435.32	II
150		2446.17	II
75		2446.72	II
1100		2447.91	I
80		2448.15	II
100		2457.29	II
60		2457.76	II
150		2469.29	II
80		2470.06	II
100		2471.18	II
50		2472.55	II
1700		2476.42	I
250		2486.52	II
300		2488.92	II
75		2489.61	II
200		2498.81	II
150		2505.73	II
50		2514.47	II
80		2534.57	II
50	h	2539.44	II
150		2551.84	II
150		2565.51	II
100		2569.56	II
60		2593.24	II
50		2628.24	II
70		2635.92	II
150		2658.75	II
1900		2763.09	I
150	h	2776.85	II
100	h	2787.92	II
50	h	2800.64	II
50	h	2807.59	II
200		2854.59	II
100	h	2871.37	II
100	h	2878.01	II
520		2922.49	I
50		2980.63	II
650		3002.65	I
45		3009.78	I
1500		3027.91	I
1100		3065.31	I
2600		3114.04	I
270		3142.81	I
11000		3242.70	I
2700		3251.64	I
3500		3258.78	I
460		3287.25	I
3600		3302.13	I
5000		3373.00	I
24000		3404.58	I
13000		3421.24	I
5000		3433.45	I
6400		3441.40	I
7700		3460.77	I
10000		3481.15	I
2000		3489.77	I
12000		3516.94	I
10000		[illegible]	I
4500		3571.16	I
20000		3609.55	I
20000		3634.70	I
5500		3690.34	I
1400		3718.91	I
1500		3799.19	I
1500		3832.29	I
2200		3894.20	I
1500		3958.64	I
290		4087.34	I
90		4169.84	I
2500		4212.95	I
180		4473.59	I
55	h	4788.18	I
45	h	4817.51	I
35		4875.43	I
55		5110.81	I
75		5117.02	I
160		5163.84	I
55		5234.86	I
120		5295.63	I
18		5312.57	I
15		5345.10	I
35		5395.24	I
55		5542.80	I
35		5547.02	I
27		5619.44	I
15		5642.69	I
14		5655.42	I
75		5670.07	I
11		5690.14	I
55	h	5695.09	I
18		5736.61	I
23		6774.54	I
65		6784.52	I
4	h	6833.42	I
11		7016.44	I
13	h	7310.06	I
75		7368.12	I
27		7391.92	I
16		7486.90	I
120		7764.03	I
27		7786.67	I
45		7915.80	I
18		7961.08	I
55		8132.82	I
45		8300.83	I
9	h	8353.58	I
18	h	8532.74	I
16	h	8599.10	I
65		8761.35	I

Pd III
Ref. 424 — L.J.R.

Vacuum

Intensity		Wavelength	
10		688.74	III
20		689.46	III
50		689.54	III
50		695.91	III
200		705.49	III
150		707.80	III
100		709.89	III
100		717.90	III
100		719.47	III
200		727.72	III
150		738.79	III
100		[illegible]	III
100		757.41	III
500		763.06	III
500		766.42	III
200		772.11	III
200		776.51	III
2000		781.02	III
200		784.99	III
200		787.31	III
200		787.95	III
200		789.58	III
500		794.08	III
500		797.52	III
500		800.03	III
500		800.10	III
500		803.67	III
500		825.35	III
500		840.58	III
500		856.47	III
500		864.04	III
500		880.59	III
500		888.84	III
1000		889.29	III
300		947.78	III
300		965.52	III
100		1505.40	III
200		1517.18	III
200	h	1526.88	III
100		1542.63	III
200	h	1545.95	III
300		1596.89	III
200	h	1606.10	III
150		1630.84	III
50		1679.73	III
100		1704.33	III
50		1706.40	III
200		1719.88	III
500		1741.62	III
400		1758.19	III
4000		1782.55	III
400		1804.91	III
400		1843.49	III
1500		1851.59	III
2000		1852.27	III
1000		1859.21	III
1500		1874.63	III
2000		1885.83	III
1000		1887.40	III
1500		1891.34	III
4000		1914.62	III
1000		1930.33	III
2000		1941.64	III
400		1951.56	III
300		1972.29	III
300		1977.53	III

Air

Intensity		Wavelength	
800		2002.16	III
1000		2004.47	III
500		2055.11	III
500		2149.82	III
500		2177.55	III
500		2177.63	III
100		2291.45	III
100		2452.42	III
100		2633.22	III

PHOSPHORUS (P)

Z = 15

P I and II
Ref. 182 = R.L.K.

Vacuum

Intensity	Wavelength	
10	810.24	II
10	865.44	II
20	1249.82	II
20	1301.87	II
20	1304.47	II
15	1304.68	II
35	1305.48	II
25	1309.87	II
60	1310.70	II
15	1372.033	I
15	1373.500	I
10	1374.732	I
15	1377.080	I
15	1377.937	I
25	1379.429	I
25	1381.469	I
15	1381.637	I
30	1452.89	II
80	1532.51	II
120	1535.90	II
80	1536.39	II
120	1542.29	II
140	1671.070	I
100	1671.510	I
180	1671.680	I
140	1672.035	I
140	1672.474	I
600	1674.591	I
600	1679.695	I
140	1685.976	I
100	1696.028	I
100	1694.486	I
100	1706.376	I
100	1707.553	I
600	1774.951	I
500	1782.838	I
400	1787.656	I
140	1834.801	I
140	1847.165	I
100	1849.820	I
140	1851.194	I
100	1852.069	I
500	1858.886	I
400	1859.393	I
140	1864.348	I
180	1905.481	I
140	1906.403	I
280	1907.665	I

Air

Intensity	Wavelength	
280	2023.489	I
180	2024.516	I
400	2032.432	I
400	2033.477	I
400	2135.465	I
400	2136.182	I
400	2149.145	I
280	2152.940	I
500	2154.080	I
180	2235.732	I
100	2484.19	II
750	2533.976	I
950	2535.603	I
750	2553.262	I
500	2554.915	I
150	2606.06	II
100	2626.18	II
90	2636.76	II
150	3308.92	II
125	3419.34	II
100	3425.00	II
100	4178.48	II
200	4288.60	II
200	4385.35	II
400	4420.71	II
100	4452.46	II
150	4463.00	II
120	4467.98	II
200	4475.26	II
200	4499.24	II
120	4530.81	II
120	4554.83	II
120	4558.07	II
120	4581.71	II
500	4588.04	II
500	4589.86	II
600	4602.08	II
300	4626.70	II
300	4658.31	II
200	4864.42	II
150	4927.20	II
500	4943.53	II
300	4954.39	II
300	4969.71	II
100	5079.381	I
100	5098.221	I
100	5100.974	I
140	5109.628	I
140	5154.844	I
180	5162.290	I
150	5191.41	II
300	5253.52	II
140	5293.539	I
400	5296.13	II
250	5316.07	II
300	5344.75	II
180	5345.851	I
100	5364.631	I
250	5378.20	II
300	5386.88	II
200	5409.72	II
400	5425.91	II
100	5428.094	I
400	5450.74	II
140	5458.305	I
125	5461.20	II
180	5477.672	I
140	5477.860	I
140	5478.267	I
200	5483.55	II
200	5499.73	II
200	5507.19	II
100	5514.774	I
100	5516.997	I
200	5541.14	II
200	5583.27	II
250	5588.34	II
100	5727.71	II
500	6024.18	II
400	6034.04	II
500	6043.12	II
250	6055.50	II
100	6057.86	II
350	6087.82	II
180	6097.690	I
350	6165.59	II
500	6199.024	I
180	6210.499	I
100	6232.29	II
200	6367.27	II
140	6375.681	I
100	6388.579	I
250	6435.32	II
130	6436.31	II
600	6459.99	II
600	6503.46	II
600	6507.97	II
150	6713.28	II
100	6717.411	I
100	7102.200	I
100	7158.367	I
180	7165.465	I
180	7175.102	I
180	7176.660	I
250	7845.63	II
100	8046.801	I
140	8278.058	I
100	8367.856	I
140	8531.475	I
140	8613.835	I
180	8637.578	I
400	8741.529	I
100	8872.174	I
140	9153.34	I
180	9175.819	I
950	9193.85	I
600	9278.88	I
1250	9304.94	I
500	9323.50	I
100	9327.13	I
110	9371.109	I
950	9435.069	I
950	9441.86	I
600	9452.83	I
100	9481.84	I
100	9492.12	I
1250	9493.56	I
180	9521.78	I
1700	9525.73	I
1500	9545.18	I
280	9556.81	I
1700	9563.439	I
280	9593.50	I
750	9609.04	I
180	9625.80	I
180	9628.42	I
400	9638.939	I
140	9675.41	I
500	9676.24	I
180	9706.533	I
1500	9734.750	I
280	9736.680	I
1500	9750.77	I

Phosphorus (Cont.)

Intensity	Wavelength	
100	9760.77	I
100	9776.85	I
100	9779.11	I
600	9790.21	I
1700	9796.85	I
280	9834.80	I
400	9903.68	I
280	9976.67	I
229	10084.27	I
174	10432.66	I
132	10455.87	I
458	10511.58	I
962	10529.52	I
1235	10581.57	I
415	10596.90	I
435	10681.40	I
265	10813.13	I
134	10932.72	I
103	10967.37	I
180	11160.05	I
764	11183.23	I
402	11186.75	I
76	13438.43	I
86	13485.19	I
479	14241.64	I
150	14272.75	I
256	14307.83	I
173	14430.50	I
135	14470.62	I
98	14646.42	I
714	15711.52	I
228	15962.53	I
296	16254.77	I
203	16292.97	I
1627	16482.92	I
588	16590.07	I
225	16613.05	I
221	16738.68	I
419	16803.39	I
471	17112.48	I
104	17223.28	I
289	17286.91	I
145	17359.00	I
299	17423.67	I
95	17665.68	I
186	18007.63	I
106	18518.90	I
92	18881.16	I
125	20841.62	I
124	23038.83	I
287	23844.97	I
98	26134.44	I
118	27959.52	I
188	28049.42	I
127	28154.52	I
132	28284.16	I
98	28288.69	I
311	29097.16	I
92	31483.21	I
91	32270.90	I
146	35551.93	I
90	35582.27	I
192	35802.53	I
146	36417.43	I

P III

Ref. 180 = R.L.K.

Intensity	Wavelength	
	Vacuum	
90	471.146	III
90	484.278	III
120	498.180	III
200	569.853	III
200	581.831	III
200	844.646	III
150	845.038	III
250	845.664	III
300	847.669	III
200	848.016	III
120	848.465	III
150	848.639	III
250	852.686	III
350	855.624	III
200	859.406	III
500	859.652	III
250	859.729	III
300	913.971	III
300	917.120	III
350	918.665	III
250	921.849	III
200	997.999	III
250	1003.598	III
500	1334.808	III
650	1344.327	III
300	1344.845	III
250	1380.463	III
150	1381.089	III
350	1502.228	III
250	1504.663	III
150	1618.632	III
200	1618.907	III
	Air	
200	2611.147	III
300	2632.713	III
200	2680.133	III
250	2895.241	III
250	3186.186	III
300	3219.307	III
150	3233.536	III
400	3233.602	III
200	3556.546	III
200	3577.526	III
200	3904.812	III
250	3914.314	III
300	3957.641	III
350	3978.307	III
200	4057.440	III
400	4059.312	III
300	4080.084	III
500	4222.195	III
350	4246.720	III
200	4428.171	III
200	4463.668	III
250	4479.776	III
150	6083.409	III
150	6409.204	III
150	6484.440	III
150	6486.381	III
150	6992.690	III
150	8113.528	III

P IV

Ref. 336 – R.L.K.

Intensity	Wavelength	
	Vacuum	
90	282.301	IV
90	304.996	IV
120	359.293	IV
150	359.899	IV
120	361.514	IV
150	361.629	IV
120	371.299	IV
150	371.504	IV
200	372.001	IV
500	388.318	IV
120	414.604	IV
200	414.999	IV
250	415.805	IV
250	444.245	IV
300	445.158	IV
250	568.038	IV
350	629.008	IV
400	629.914	IV
500	631.779	IV
350	648.482	IV
300	756.510	IV
300	776.353	IV
650	823.179	IV
700	824.730	IV
800	827.932	IV
250	847.019	IV
350	849.799	IV
200	850.392	IV
700	877.476	IV
1000	950.655	IV
570	1025.563	IV
500	1028.096	IV
570	1030.517	IV
500	1033.111	IV
500	1035.517	IV
570	1118.551	IV
200	1206.422	IV
200	1335.705	IV
500	1366.695	IV
400	1372.674	IV
350	1377.282	IV
500	1484.507	IV
400	1487.788	IV
300	1489.098	IV
250	1862.762	IV
120	1862.893	IV
200	1863.580	IV
650	1888.523	IV
200	1910.183	IV
120	1985.682	IV
150	1985.851	IV
200	1986.114	IV
150	1987.022	IV
	Air	
200	2477.823	IV
150	2478.070	IV
250	2478.256	IV
250	2605.506	IV
400	2644.295	IV
300	2724.764	IV
400	2728.770	IV
200	2729.120	IV
500	2739.309	IV
250	2739.872	IV
200	2740.223	IV
200	2961.242	IV
650	3347.736	IV
570	3364.467	IV
400	3371.122	IV
200	3413.543	IV
200	3733.393	IV
300	4249.656	IV
250	4540.288	IV
250	4541.112	IV
150	4548.056	IV
200	4548.449	IV
150	5235.499	IV
150	5989.774	IV
150	6142.605	IV
150	6713.939	IV
120	6715.906	IV
200	7443.657	IV

P V

Ref. 179 = R.L.K.

Intensity	Wavelength	
	Vacuum	
80	255.59	V
50	255.67	V
110	310.58	V
150	311.34	V
300	328.47	V
250	328.78	V
150	347.23	V
200	348.20	V
110	378.56	V
250	389.50	V
300	390.70	V
150	410.03	V
375	475.60	V
110	534.63	V
80	534.99	V
520	542.57	V
600	544.92	V
450	673.90	V
450	865.45	V
600	871.39	V
250	997.62	V
150	1000.38	V
900	1117.98	V
700	1128.01	V
150	1379.62	V
250	1385.05	V
375	1447.83	V
450	1610.50	V
	Air	
200	2180.29	V
150	2186.42	V
375	2424.40	V
450	2440.93	V
200	2441.24	V
300	2961.00	V
450	2978.55	V
700	3175.09	V
520	3204.04	V
150	4083.18	V
110	4094.95	V
110	5156.72	V

PLATINUM (Pt)

Z = 78

Pt I and II

Ref. 1, 288 — C.H.C.

Intensity		Wavelength	
		Vacuum	
30		1621.66	II
30		1723.13	II
30		1751.70	II
50	r	1777.09	II
30		1781.86	II
30		1879.09	II
40		1883.05	II
50		1889.52	II
50		1911.70	II
30		1929.25	II
30		1929.68	II
30		1939.80	II
30		1949.90	II
30		1983.74	II
		Air	
40		2014.93	II
3200		2030.63	I
4400		2032.41	I
100		2036.46	II
40		2041.57	II
5500		2049.37	I
1500		2067.50	I
3000		2084.59	I
1000		2103.33	I
30		2115.57	II
950		2128.61	I
30		2130.69	II
1900		2144.23	I
100		2144.24	II
600		2165.17	I
1500		2174.67	I
30		2190.32	II
400		2202.22	I
50	h	2202.58	II
320		2222.61	I
50	h	2233.11	II
150		2249.30	I
30	h	2240.99	II
100		2245.52	II
30		2251.52	II
30	h	2251.92	II
190		2268.84	I
30	h	2271.72	II
280		2274.38	I
50	h	2287.50	II
30		2288.20	II
150		2289.27	I
150		2292.40	I
240		2308.04	I
50		2310.96	II
90		2315.50	I
220		2318.29	I
100		2326.10	I
170		2340.18	I
280		2357.10	I
180		2368.28	I
50		2377.28	II
130		2383.64	I
40		2386.81	I
120		2389.53	I
35		2396.17	I
70		2401.87	I
200		2403.09	I
100		2418.06	I
50		2424.87	II
80		2428.04	I
50		2428.20	I
25		2429.10	I
180		2436.69	I
650		2440.06	I
60		2450.97	I
440		2467.44	I
35		2471.01	I
1000		2487.17	I
25		2488.74	II
200		2490.12	I
160		2495.82	I
240		2498.50	I
50		2505.93	I
120		2508.50	I
50		2514.07	I
60		2515.03	I
240		2515.58	I
140		2524.30	I
40		2529.41	I
50		2536.49	I
160		2539.20	I
18		2549.46	I
50		2552.25	I
50		2596.00	I
70		2603.14	I
30		2616.76	II
50		2619.57	I
30		2625.34	II
1100		2628.03	I
130		2639.35	I
1000		2646.89	I
500		2650.86	I
20		2658.17	I
2800		2659.45	I
40		2674.57	I
440		2677.15	I
200		2698.43	I
2000		2702.40	I
1600		2705.89	I
60		2713.13	I
1300		2719.04	I
130		2729.92	I
1800		2733.96	I
70		2738.48	I
70		2747.61	I
80		2753.86	I
200		2754.92	I
30		2769.84	I
500		2771.67	I
40		2773.24	I
20		2774.00	I
50		2774.77	II
50		2793.27	I
100		2794.21	II
40	h	2799.98	II
140		2803.24	I
10		2808.51	I
50		2818.25	I
30	h	2822.27	II
1400		2830.30	I
70		2834.71	I
16		2853.11	I
80	h	2860.68	II
40	h	2865.05	II
40	h	2875.85	II
100	h	2877.52	II
25		2888.20	I
25		2893.22	I
600		2893.86	I
300		2897.87	I
60		2905.90	I
120		2912.26	I
120		2913.54	I
70		2919.34	I
30		2921.38	I

Platinum (Cont.)

Intensity		Wavelength	
1700		2929.79	I
30		2942.76	I
30		2944.75	I
25		2959.10	I
60		2960.75	I
1800		2997.97	I
35		3001.17	II
220		3002.27	I
30		3017.88	I
30	h	3031.22	II
130		3036.45	I
800		3042.64	I
3200		3064.71	I
30		3071.94	I
130		3100.04	I
320		3139.39	I
140		3156.56	I
120		3200.71	I
320		3204.04	I
30		3230.29	I
20		3233.42	I
20		3250.36	I
40		3251.98	I
160		3255.92	I
25		3268.42	I
25		3281.97	I
120		3290.22	I
500		3301.86	I
60		3315.05	I
35		3323.80	I
340		3408.13	I
35		3427.93	I
60		3483.43	I
160		3485.27	I
120		3628.11	I
70		3638.79	I
70		3643.17	I
50		3663.10	I
80		3671.99	I
80		3674.04	I
35		3699.91	I
18		3706.53	I
20		3801.05	
80		3818.69	I
40		3900.73	I
110		3922.96	I
35		3948.40	I
100		3966.36	I
20		3996.57	I
110		4118.69	I
80		4164.56	I
40		4192.43	I
18		4327.06	I
18		4391.83	I
80		4442.55	I
14		4445.55	I
25		4498.76	I
12		4520.90	I
35		4552.42	I
12		4879.53	I
14		5044.04	I
30		5059.48	I
35		5227.66	I
40		5301.02	I
12		5368.99	I
12		5390.79	I
14		5475.77	I
14		5478.50	I
6		5763.57	I
20		5840.12	I
8		5844.84	I
6		6026.04	I
7		6318.37	I
8		6326.58	I
9		6523.45	I
10		6710.42	I
20		6760.02	I
60		6842.60	I
20		7113.73	I
10		8224.74	I

PLUTONIUM (Pu)
Z = 94

Pu I and II
Ref. 91 — J.G.C.

Intensity	Wavelength Air	
10000	2781.40	II
10000	2784.48	II
10000	2806.11	II
10000	2815.77	II
10000	2897.97	II
10000	2898.94	II
10000	2904.25	II
10000	2904.94	II
10000	2910.40	II
10000	2918.00	II
10000	2918.80	II
10000	2926.08	II
10000	2928.25	II
10000	2929.71	II
10000	2930.98	II
10000	2932.32	II
10000	2933.30	II
10000	2938.54	II
10000	2938.95	II
10000	2941.39	II
10000	2945.26	II
10000	2946.00	II
10000	2950.06	II
10000	2951.62	
10000	2954.46	II
10000	2963.47	II
10000	2966.84	II
10000	2967.54	II
10000	2972.50	II
10000	2977.81	II
10000	2978.37	II
10000	2980.23	II
10000	2981.23	II
10000	2986.95	II
10000	2988.21	II
10000	2991.31	II
10000	2996.40	II
10000	3000.31	II
10000	3009.57	II
10000	3028.85	II
10000	3042.61	II
10000	3043.12	II
10000	3060.32	II
10000	3069.32	II
10000	3091.33	II
10000	3091.94	II
10000	3092.59	II
10000	3104.12	II
10000	3105.04	II
10000	3106.03	
10000	3123.87	II
10000	3159.21	II
10000	3161.73	II
10000	3163.18	II
10000	3174.49	II
10000	3179.41	II
10000	3185.12	II
10000	3187.60	II
10000	3189.23	II
10000	3193.54	
10000	3193.55	II
10000	3194.56	II
10000	3198.47	II
10000	3200.23	II
10000	3201.00	II
10000	3201.66	II
10000	3204.48	II
10000	3206.80	II
10000	3207.97	II
10000	3215.08	I
10000	3216.15	II
10000	3220.94	II
10000	3224.87	II
10000	3231.86	II
10000	3232.24	
10000	3232.63	II
10000	3241.39	II
10000	3242.96	II
10000	3243.40	II
10000	3244.16	I
10000	3245.25	II
10000	3245.71	
10000	3246.35	II
10000	3247.50	
10000	3247.56	II
10000	3252.08	I
10000	3260.54	II
10000	3265.17	
10000	3273.11	II
10000	3274.71	II
10000	3275.24	I
10000	3292.56	I
10000	3293.61	I
10000	3296.91	I
10000	3297.87	I
10000	3298.47	II
10000	3301.76	I
10000	3306.59	I
10000	3306.66	I
10000	3307.66	II
10000	3308.75	I
10000	3312.65	II
10000	3315.34	II
10000	3316.96	II
10000	3320.61	I
10000	3320.84	I
10000	3323.48	
10000	3327.19	I
10000	3330.11	
10000	3331.52	II
10000	3332.34	I
10000	3333.03	
10000	3337.71	II
10000	3338.40	II
10000	3338.94	I
10000	3347.87	II
10000	3349.63	I
10000	3351.82	II
10000	3356.61	II
10000	3358.41	II
10000	3358.84	II
10000	3362.26	II
10000	3365.20	I
10000	3365.66	
10000	3368.86	I
10000	3370.64	II
10000	3371.19	I
10000	3375.80	I
10000	3376.76	II
10000	3376.94	II
10000	3377.37	II
10000	3379.51	I
10000	3381.82	I
10000	3381.97	II
10000	3382.70	I
10000	3390.33	II
10000	3391.41	II
10000	3393.67	I
10000	3394.32	I
10000	3418.88	II
10000	3465.10	II
10000	3473.64	II
10000	3483.20	I
10000	3585.87	II
10000	3632.21	II
10000	3699.19	II
10000	3720.59	I
10000	3725.98	I
10000	3726.11	II
10000	3726.79	II
10000	3732.03	II
10000	3744.78	I
10000	3753.63	I
10000	3755.94	I
10000	3757.82	I
10000	3758.34	I
10000	3774.38	I
10000	3776.71	I
10000	3792.22	I
10000	3799.37	I
10000	3805.93	I
10000	3811.40	I
10000	3812.30	II
10000	3827.57	I
10000	3835.52	I
10000	3836.96	I
10000	3838.92	I
10000	3842.10	I
10000	3851.01	I
10000	3851.85	I
10000	3878.54	I
10000	3895.89	I
10000	3928.53	I
10000	3975.43	II
10000	4097.12	I
10000	4101.96	I
10000	4105.95	II
10000	4111.07	I
10000	4114.91	I
10000	4128.12	I
10000	4129.93	II
10000	4133.01	I
10000	4135.97	I
10000	4140.04	I
10000	4141.20	II
10000	4151.09	I
10000	4151.45	I
10000	4155.46	I
10000	4159.39	II
10000	4167.77	I
10000	4170.95	I
10000	4178.28	II
10000	4189.90	II
10000	4190.06	II
10000	4196.20	II
10000	4206.48	I
10000	4208.23	I
10000	4221.87	I
10000	4224.20	I
10000	4229.77	II
10000	4254.76	II
10000	4261.88	I
10000	4269.77	I
10000	4273.34	II
10000	4281.17	I
10000	4289.08	II
10000	4337.18	II
10000	4352.71	II
10000	4367.41	I
10000	4379.91	II
10000	4385.35	II
10000	4393.93	II
10000	4404.90	I
10000	4441.65	II
10000	4468.54	II
10000	4472.79	II
10000	4493.78	II
10000	4504.91	II
10000	4536.15	II
10000	4735.40	I
10000	4989.34	I
10000	5269.86	I
10000	5381.02	I
10000	5498.50	I
10000	5510.72	I
10000	5537.59	I
10000	5549.62	I
10000	5590.54	I
10000	5592.33	I
10000	5712.39	I
10000	5770.26	I
10000	5839.05	I
10000	5983.35	I
10000	6012.78	I
10000	6192.80	I
10000	6304.66	I
10000	6449.75	I
10000	6486.71	I
10000	6488.86	I
10000	6488.89	
10000	6535.27	I
10000	6544.21	I
10000	6608.95	I
10000	6672.72	I
10000	6784.66	I
10000	6880.16	I
10000	6891.38	I
10000	7059.23	I
10000	7068.90	I
10000	7092.46	I
10000	7116.88	I
10000	7141.66	I
10000	7177.14	I
10000	7231.09	I
10000	7258.06	I
10000	7322.23	I
10000	7325.97	I
10000	7331.81	I
10000	7431.18	I
10000	7447.99	I
10000	7507.80	I
10000	7526.93	I
10000	7547.45	I
10000	7564.50	I
10000	7571.87	I
10000	7572.93	I
10000	7609.77	I
10000	7689.40	I
10000	7758.20	I
10000	7798.54	I
10000	7933.17	I
10000	8102.54	I
10000	8130.86	I
10000	8309.61	I
10000	8435.47	I
10000	8476.13	I
10000	8495.75	
10000	8597.26	I
10000	8665.02	I
10000	8691.94	I
3000	8729.82	I
3000	8836.16	I
3000	9533.07	I
3000	10046.75	I
3000	11114.82	I
3000	12144.46	I
3000	12231.22	I
3000	15377.31	I
3000	16397.38	I

POLONIUM (Po)
Z = 84

Po I and II
Ref. 47, 48 — E.F.W.

Intensity		Wavelength Air	
250	w	2139.02	I
300	h	2203.80	
300		2220.67	I
200		2222.13	I
200		2284.22	
250		2344.61	I
250		2421.72	I
300		2426.09	I
1500	w	2450.08	I
700		2483.94	I
700		2490.53	I
200	h	2502.18	
300		2534.95	I
300		2557.33	I
1500	w	2558.01	I
400		2562.31	
300		2578.80	
400		2587.64	I
200		2637.01	
300		2645.36	I
700	h	2663.33	
200		2671.67	I
600		2761.92	I
400		2800.26	I
250		2824.11	I
300		2866.01	I
400		2919.31	I
600		2958.92	I
2500	w	3003.21	I
450		3069.31	I
200		3115.95	

Potassium (Cont.)

Intensity		Wavelength	Spectrum
400		3189.02	I
600		3240.24	I
250		3286.38	I
600		3328.60	I
300		3489.79	
200		3493.65	
400		3588.33	
200		3671.36	
500		3861.93	I
200		4051.98	
1200		4170.52	I
250		4236.13	
200	h	4415.58	
800		4493.21	I
350		4611.44	I
200		4867.12	
400		4876.24	I
450		4946.81	
350		5323.23	I
300		5744.85	I
600		7962.62	I
300		8433.87	I
500		8618.26	I
250		9227.87	

POTASSIUM (K)
Z = 19

K I and II
Ref. 59, 76, 172, 268 — L.J.R.

Intensity		Wavelength	
		Vacuum	
5		261.20	II
25		441.81	II
5		465.08	II
		469.50	II
10		476.03	II
30		495.14	II
30		600.77	II
25		607.93	II
30		612.62	II
3		1725.0	II
		Air	
6		2190.00	II
4		2210.53	II
5		2265.04	II
6		2550.02	II
4		2743.55	II
		2992.12	I
		2992.42	I
		3034.76	I
		3034.92	I
5		3062.18	II
4		3101.79	I
3		3102.04	I
6		3105.00	II
5		3190.07	II
7		3217.16	I
6		3217.62	I
4		3220.60	II
5		3290.65	II
6		3345.32	II
6		3373.60	II
6		3380.62	II
6		3384.86	II
6		3404.24	II
7		3440.05	II
11		3446.37	I
10		3447.38	I
6		3481.11	II
7		3530.75	II
5		3608.88	II
6		3618.49	II
4		3626.42	II
3		3648.84	I
4		3648.98	I
6		3681.54	II
5		3716.60	II
5		3721.34	II
5		3739.13	II
5		3744.42	II
6		3767.36	II
6		3783.19	II
6		3800.14	II
6		3816.56	II
7		3817.50	II
5		3873.74	II
4		3878.62	II
8		3897.92	II
5		3923.00	II
5		3926.36	II
6		3942.53	II
6		3955.21	II
6		3966.72	II
6		3972.58	II
6		3995.10	II
7		4001.24	II
5		4012.10	II
6		4042.59	II
18		4044.14	I
17		4047.21	I
5		4093.69	II
6		4114.99	II
7		4134.72	II
7		4149.19	II
8		4186.24	II
7		4222.97	II
7		4225.67	II
7		4263.40	II
7		4305.00	II
7		4309.10	II
5		4340.03	II
7		4388.16	II
5		4466.65	II
6		4505.33	II
5		4595.65	II
8		4608.45	II
10		4641.88	I
11		4642.37	I
5		4659.38	II
4		4740.91	I
6		4744.35	I
5		4753.93	I
7		4757.39	I
5		4786.49	I
7		4791.05	I
6		4799.75	I
8		4804.35	I
9		4829.23	II
7		4849.86	I
8		4856.09	I
8		4863.48	I
9		4869.76	I
8		4942.02	I
6		4943.29	II
9		4950.82	I
9		4956.15	I
10		4965.03	I
8		5005.60	II
7		5056.27	II
10		5084.23	I
11		5097.17	I
11		5099.20	I
12		5112.25	I
5		5310.24	II
12		5323.28	I
13		5339.69	I
12		5342.97	I
14		5359.57	I
6		5470.13	II
5		5642.73	II
4		5772.32	II
16		5782.38	I
17		5801.75	I
15		5812.15	I
17		5831.89	I
2		5969.64	II
8		6120.27	II
6		6246.59	II
7		6307.29	II
5		6427.96	II
2		6595.00	II
19		6911.08	I
12		6936.28	I
20		6938.77	I
7		6964.18	I
12		6964.67	I
25		7664.90	I
24		7698.96	I
5		7955.37	I
4		7956.83	I
7		8078.11	I
6		8079.62	I
9		8250.18	I
8		8251.74	I
3		8390.22	I
		8391.44	I
2		8417.54	I
1		8420.00	I
11		8503.45	I
10		8505.11	I
4		8763.96	I
3		8767.05	I
13		8902.19	I
12		8904.02	I
5		8923.31	I
4		8925.44	I
7		9347.24	I
3		9349.25	I
6		9351.59	I
15		9595.70	I
14		9597.83	I
6		9949.67	I
5		9954.14	I
9		10479.63	I
5		10482.15	I
8		10487.11	I
17		11019.87	I
16		11022.67	I
17		11690.21	I
16		11769.62	I
17		11772.83	I
		12432.24	I
		12522.11	I
		13377.86	I
		13397.09	I
		15163.08	I
		15168.40	I
		40158.37	I

K III
Ref. 60, 76 — L.J.R.

Intensity	Wavelength	
	Vacuum	
2	325.28	III
5	327.60	III
25	330.68	III
30	341.92	III
15	348.00	III
30	379.12	III
25	380.48	III
30	382.23	III
15	398.63	III
20	402.10	III
30	406.48	III
40	408.96	III
50	413.79	III
30	414.87	III
30	416.00	III
30	417.54	III
30	418.62	III
75	434.72	III
50	435.68	III
75	444.34	III
75	448.60	III
75	466.79	III
100	470.09	III
75	471.57	III
45	474.92	III
40	479.18	III
10	482.11	III
10	482.41	III
75	497.10	III
10	514.94	III
50	520.61	III
25	523.79	III
40	529.80	III
15	539.71	III
15	546.12	III
20	708.84	III
20	765.31	III
30	765.64	III
35	778.53	III
20	872.31	III
10	873.86	III
15	874.04	III
	Air	
6	2550.02	III
5	2635.11	III
1	2736.96	III
5	2689.90	III
1	2898.90	III
5	2938.45	III
1	2948.94	III
5	2986.20	III
6	2992.24	III
6	3052.07	III
5	3056.84	III
6	3201.95	III
6	3209.34	III
6	3278.79	III
6	3289.06	III
6	3322.40	III
6	3364.22	III
6	3420.82	III
4	3421.83	III
6	3468.32	III
6	3481.11	III
5	3513.88	III
1	3885.50	III

K IV
Ref. 32, 76, 86, 150, 160, 314, 322 — L.J.R.

Intensity	Wavelength	
	Vacuum	
150	271.82	IV
100	273.06	IV
	279.88	IV
300	340.46	IV
150	340.74	IV
300	354.93	IV
150	356.26	IV
300	359.73	IV
200	359.91	IV
250	362.08	IV
150	362.15	IV
150	363.02	IV
300	375.96	IV
300	379.88	IV
250	380.48	IV
200	381.70	IV
300	382.23	IV
150	382.49	IV
200	382.65	IV
300	382.91	IV
250	384.10	IV
200	386.61	IV
250	388.92	IV
250	389.07	IV
250	390.42	IV
300	390.57	IV
200	391.46	IV
200	392.47	IV
500	393.14	IV
400	400.21	IV
300	402.91	IV
250	403.97	IV
150	404.41	IV
250	408.08	IV
150	417.28	IV
200	442.30	IV
300	443.57	IV
200	445.61	IV
250	446.83	IV
750	448.60	IV
400	456.33	IV
250	523.00	IV
200	526.45	IV
150	527.62	IV
750	646.19	IV
500	737.14	IV
500	741.95	IV
500	745.26	IV
400	746.35	IV
300	749.99	IV
150	754.19	IV
400	754.67	IV

K V
Ref. 32, 75, 76, 150, 322 — L.J.R.

Intensity	Wavelength	
	Vacuum	
100	214.35	V
150	282.35	V
150	293.33	V
300	294.84	V
200	296.17	V
200	297.06	V
200	300.25	V
200	300.50	V
200	311.24	V
250	312.77	V
200	315.18	V
250	327.38	V
250	389.07	V
200	349.50	V
500	372.15	V
200	372.46	V
200	372.77	V
300	375.96	V
250	377.76	V
300	379.12	V
300	387.80	V
250	390.11	V
250	395.40	V
200	398.36	V
200	398.88	V
200	399.75	V
250	415.05	V
200	415.79	V
400	422.18	V
300	425.16	V
500	425.59	V
250	438.02	V
200	449.71	V
200	452.90	V
250	455.67	V
400	456.33	V
200	482.71	V
200	483.75	V
750	580.32	V
250	585.51	V
500	586.32	V
250	602.27	V
400	603.43	V
250	638.67	V
300	687.50	V
300	720.43	V
400	724.42	V
600	731.86	V
150	770.29	V
150	771.46	V
	1035.60	V

PRASEODYMIUM (Pr)
Z = 59

Pr I and II
Ref. 1 — C.H.C.

Intensity		Wavelength	
		Air	
25		2558.58	II
25		2578.27	I
30		2579.31	I
40	h	2598.04	II
25		2608.92	II
25		2615.75	II
25		2648.48	II
30		2654.75	II
25		2666.70	II
20		2672.52	II
30		2685.19	II
45		2685.70	II

Praseodymium (Cont.)

Intensity		Wavelength	Spectrum
50		2698.92	II
60		2700.38	II
30		2702.25	II
100	h	2707.37	II
20		2714.16	II
60		2720.17	II
30		2721.90	II
50		2726.50	II
12		2731.78	II
25		2733.12	II
50		2734.30	II
25		2737.90	II
40		2742.12	II
25		2744.66	II
20		2746.28	II
60		2760.35	II
50		2769.60	II
50	d	2775.94	II
		2776.03	II
40		2778.80	II
50		2783.31	II
30		2789.05	II
35		2792.51	II
50		2802.05	II
20		2823.17	II
20		2824.14	II
20		2828.29	II
20		2842.98	I
20		2844.01	II
20		2850.62	I
25		2853.99	II
30		2865.64	II
50		2881.60	I
30		2882.31	II
30		2884.89	II
30		2943.97	II
30		2967.58	II
30		2971.13	II
40	d	2971.40	II
		2971.46	II
50		2984.98	II
30		2986.18	II
30		2990.22	II
110		3082.11	II
100		3111.34	II
140		3121.58	II
140		3163.73	II
270		3168.24	II
160		3172.31	II
110		3191.42	II
200	d	3195.99	II
110		3199.04	II
100		3207.89	II
190		3219.48	II
100		3234.27	II
100		3245.48	II
140		3355.67	II
110		3394.62	II
110		3465.74	II
200		3584.21	II
130		3611.94	II
170		3630.96	II
100		3645.55	II
250		3645.66	II
250		3646.30	II
100		3648.30	II
150	c	3660.36	II
100		3661.62	II
370		3668.83	II
250		3687.03	II
150		3687.19	II
100		3689.71	II
150		3698.06	II
230		3706.75	II
170	c	3711.10	II
290		3714.05	II
120	c	3733.03	II
210	c	3734.41	II
250		3735.76	II
190		3736.49	II
410		3739.18	II
150		3740.99	II
120		3743.98	II
190		3750.98	II
140		3759.60	II
120		3760.08	II
680		3761.87	II
230		3764.77	II
230		3768.94	II
170	c	3772.82	II
170		3774.06	II
140		3777.62	II
170		3780.66	II
150		3785.46	II
150		3786.86	II
210		3792.51	II
190		3794.93	II
680		3800.30	II
290		3804.84	II
140		3809.18	II
390		3811.84	II
1300	h	3816.02	II
120		3817.66	II
680		3818.28	II
120		3819.14	II
310		3821.80	II
150	c	3823.18	II
120		3826.67	II
960		3830.72	II
140		3834.93	II
480		3840.99	II
270		3842.34	II
150	c	3844.54	II
580		3846.59	II
1200		3850.79	II
720	c	3851.55	II
960		3852.80	II
120		3858.25	II
110		3859.14	II
480	c	3865.45	II
210		3867.52	II
210		3870.72	II
480		3876.19	II
1700	c	3877.18	II
270		3879.20	II
680		3880.47	II
440	c	3885.19	II
440	c	3889.34	II
120	c	3891.71	II
190		3897.25	II
210		3898.84	II
250		3902.45	II
770	c	3908.05	II
630		3912.90	II
310		3913.55	II
210		3914.76	II
1300	c	3918.85	II
420		3919.63	II
250		3920.53	II
960		3925.47	II
480		3927.46	II
370		3929.29	II
370		3935.82	II
250		3938.30	II
730	c	3947.63	II
900	c	3949.43	II
900	c	3953.51	II
380		3956.75	II
190		3959.44	I
470		3962.45	II
560		3964.26	II
1000	c	3964.81	II
560	c	3966.57	II
500		3971.16	II
320		3971.67	II
620	c	3972.14	II
320		3974.85	II
1300	c	3989.68	II
230		3991.91	II
340		3992.16	II
1600		3994.79	II
270		3995.83	II
560	c	3997.04	II
230		3997.96	II
320		3999.12	II
620	c	4000.17	II
730		4004.70	II
1900		4008.69	II
620		4010.60	II
730		4015.39	II
620		4020.96	II
470		4022.71	II
360		4025.54	II
230		4026.83	II
230		4029.00	II
360	c	4029.72	II
730	c	4031.75	II
230		4032.47	II
960		4033.83	II
230		4034.33	II
230		4038.22	II
730		4038.45	II
470		4039.34	II
1300		4044.81	II
230		4045.70	II
230		4046.63	II
340		4047.08	II
450		4051.13	II
2200		4054.88	II
2200		4056.54	II
450		4058.80	II
230		4062.22	II
3400		4062.81	II
210		4068.80	II
500	c	4079.77	II
500	c	4080.98	II
790		4081.85	II
500		4083.34	II
200	c	4087.21	II
560		4096.82	II
380		4098.40	II
2900	c	4100.72	II
270	c	4113.89	II
1700	c	4118.46	II
250		4129.15	II
340		4130.77	II
200		4133.61	II
1500	c	4141.22	II
2700		4143.11	II
270	c	4146.50	II
270		4148.44	II
200		4156.50	II
1700	c	4164.16	II
270		4168.04	II
230		4169.45	II
620		4171.82	II
730		4172.25	II
250		4175.32	II
250		4175.62	II
200		4178.63	II
5200		4179.39	II
2500		4189.48	II
560	c	4191.60	II
290		4201.17	II
2500	c	4206.72	II
500		4208.32	II
320		4211.86	II
320		4217.81	II
3800		4222.93	II
3800		4225.35	II
320		4233.11	II
320	c	4236.15	II
270		4240.02	II
960		4241.01	II
340		4243.51	II
840	c	4247.63	II
500		4254.40	II
270	c	4263.78	II
320		4269.09	II
790	c	4272.27	II
470	c	4280.07	II
790	c	4282.42	II
450	c	4298.98	II
290		4303.61	II
1500		4305.76	II
210		4323.55	II
270		4329.41	II
1300		4333.97	II
200		4335.74	II
360		4338.70	II
620	Cw	4344.30	II
470	c	4347.49	II
340		4350.40	II
450		4354.91	II
410	c	4359.79	II
1200		4368.33	II
320		4371.62	II
270		4396.08	II
170		4403.60	II
100		4405.12	II
430		4405.83	II
1700		4408.82	II
410		4413.77	II
160		4419.04	II
190		4419.65	II
160	c	4421.22	II
160		4424.58	II
1200	c	4429.13	II
110		4432.28	II
730		4449.83	II
140		4451.90	II
140		4454.68	II
100		4465.97	II
960		4468.66	II
140	c	4477.26	II
1100		4496.46	II
790		4510.15	II
200	c	4517.58	II
340	c	4534.15	II
340		4535.92	II
200		4563.12	II
140		4612.08	II
270	c	4628.74	II
140		4632.28	I
140		4635.68	I
200		4639.55	I
110	c	4643.49	II
140		4646.05	II
200	c	4651.50	II
140		4664.65	II
270	c	4672.09	II
180		4687.80	I
290		4695.77	I
140	Cw	4708.07	II
140		4709.52	I
180		4730.67	I
250		4736.69	I
100		4744.16	I
150		4746.92	II
100		4762.72	II
110		4783.35	II
110		4906.99	I
140		4914.02	I
200		4924.60	I
140		4936.00	I
320		4939.74	I
160		4940.30	I
380		4951.37	I
110		4975.75	I
120		5018.59	I
200		5019.76	I
200		5026.96	I
100		5033.38	I
270		5034.41	II
110		5043.83	I
320		5045.52	I
160		5053.40	I
180		5087.12	I
360		5110.38	II
560		5110.76	II
410		5129.52	II
270		5133.44	I
270		5135.14	II
100		5139.81	I
100	c	5152.30	II
200		5161.74	II
620		5173.90	II
200		5191.32	II
120		5194.43	I
150		5195.11	II
200		5195.31	II
360		5206.55	II
150		5207.90	II
360		5219.05	II
560		5220.11	II
110		5227.97	I
680		5259.73	II
180		5263.88	II
340	c	5292.02	II
340		5292.62	II
230		5298.09	II
430		5322.76	II
200		5352.40	II
16		5501.50	I
40		5508.79	II
65		5509.15	II
16	c	5511.63	II
55		5513.58	II
28		5515.12	II
13		5519.38	II
20	c	5520.31	II
45	c	5522.79	II
28	c	5524.15	I
28	c	5525.91	II
16	c	5527.93	I
13		5530.21	I
45		5531.16	I
150		5535.17	II
28		5538.37	I
20		5538.78	II
55		5545.01	II
20		5548.33	II
11		5553.42	II
22		5561.46	II
45	c	5562.06	I
13		5565.52	I
13		5566.91	II
45		5571.83	II
11		5574.61	II
11		5578.81	I
13		5582.35	II
11		5584.02	II
22		5594.92	I
22		5597.29	II
13		5601.30	II
90		5605.65	II
13		5606.68	I
28		5608.93	II
55		5610.22	II
11		5620.06	II
20		5620.26	I
45	c	5621.89	II
110		5623.05	II
90		5624.45	II
11	h	5633.03	I
22		5636.46	II
55	c	5638.79	II
16		5640.37	II
16	Cw	5643.16	I
22		5645.41	II
35		5654.23	II
55		5659.84	II
35	h	5661.57	I
16		5662.19	II
65	c	5668.46	I
45		5669.55	II
35		5669.99	II
16		5674.14	II
16		5677.03	II
55		5681.89	II
13		5685.60	II
16		5686.52	I
22	h	5687.17	II
65		5688.44	II
22		5689.21	II
55	h	5690.97	II
22		5695.90	II
22		5704.38	I
65		5707.61	I
40		5711.63	II
22		5713.83	II
16		5716.08	II
45		5719.08	II
45	d	5719.63	II
		5719.80	II
11		5728.38	I

Praseodymium (Cont.)

Intensity		Wavelength	
40		5731.88	II
20		5747.13	II
11		5747.74	I
11		5747.95	II
22		5753.02	II
90		5756.17	II
16		5759.40	II
22		5760.20	I
22		5769.16	II
16		5769.79	II
45		5773.16	II
11		5775.91	II
16		5777.29	II
90		5779.28	I
65	c	5785.28	II
65		5786.17	II
16	h	5788.29	II
16		5788.92	II
16		5790.86	II
45		5791.36	II
22		5792.95	I
40		5810.58	II
16		5813.55	II
160	d	5815.17	II
		5815.33	II
55		5818.57	II
40		5820.62	II
16	h	5821.36	I
55		5822.59	II
90		5823.72	II
45		5830.94	II
40		5835.13	I
35	c	5844.65	II
40		5844.98	II
65		5847.13	II
65	c	5850.64	II
45		5852.63	II
11	c	5854.44	I
45		5856.07	II
55		5856.90	II
90		5859.68	II
80		5868.83	II
22		5873.83	II
35		5874.72	I
35		5878.10	I
35		5879.04	I
80		5879.25	II
35	c	5884.72	I
55		5892.23	II
22		5894.22	II
40		5903.11	II
45		5904.45	II
40		5908.67	II
11		5915.31	I
11		5915.97	I
40		5920.76	I
40		5930.66	II
16		5936.33	II
160		5939.90	II
65		5940.72	II
22		5941.65	I
35		5947.16	II
22	c	5949.76	I
55		5951.27	II
20		5951.76	II
90		5956.60	II
		5956.70	I
13		5959.25	I
20		5962.18	I
28		5963.00	I
110		5967.82	II
13	c	5976.95	I
13		5978.88	I
65		5981.19	II
40		5986.14	I
45	c	5987.14	I
		5987.29	II
13		5991.27	I
13	c	5994.89	I
11		5996.06	I
29		6002.44	II
90		6006.33	II
13		6008.54	I
55		6016.48	II
150		6017.80	II
28	c	6019.85	I
150		6025.72	II
35		6042.87	II
55		6046.66	II
35		6049.26	I
28		6050.04	II
11		6050.88	I
140		6055.13	I
13		6067.27	II
13		6085.81	I
28		6086.16	II
65		6087.52	II
20		6090.38	II
28		6093.09	II
18		6096.28	I
22		6106.72	II
18		6109.08	I
65		6114.38	II
22	c	6118.02	I
22	c	6122.15	I
35		6141.51	II
65		6148.23	I
		6148.24	II
22		6157.82	II
13		6159.10	II
190		6161.18	II
18		6165.38	I
270		6165.94	II
55		6182.34	II
13		6187.96	I
35		6197.45	II
35		6200.81	II
13		6205.63	II
13		6210.59	I
22		6212.73	I
18		6218.06	I
20	h	6236.80	I
20	h	6241.05	I
45		6244.35	II
35		6255.10	II
40		6262.55	II
18		6264.54	II
22	c	6274.66	II
		6274.81	II
40		6278.68	II
110		6281.28	II
18	c	6289.02	I
11	c	6298.01	I
11		6302.05	I
35		6302.35	II
16		6304.05	I
35		6305.23	II
11	h	6318.13	II
45	c	6322.36	I
22	h	6343.88	I
28		6347.11	II
18	c	6350.98	I
22	c	6357.20	I
55	c	6359.03	I
11	h	6363.62	II
16		6377.61	I
16		6378.59	I
11		6389.57	I
18	c	6391.99	I
40		6393.18	I
45		6397.96	II
10	h	6410.69	I
55		6411.23	I
40		6413.68	II
10		6415.43	I
45		6429.63	II
45		6431.84	II
7	h	6442.78	II
9	h	6443.91	II
16	c	6453.44	I
9		6454.84	II
9		6456.18	I
9	h	6460.19	I
18		6467.72	II
9	h	6475.26	II
35	Cw	6478.02	II
45		6486.55	I
9	h	6486.97	II
40	h	6491.75	I
9		6493.49	I
11		6494.89	I
22	c	6497.11	I
18		6498.94	II
22		6500.72	I
9		6504.09	I
8		6517.14	I
16		6518.79	II
8		6534.52	I
16		6540.47	I
7	h	6553.30	I
22		6564.62	II
45		6566.77	II
7		6571.03	I
6		6578.00	I
6		6584.56	II
9	h	6593.74	II
11		6595.48	I
15		6609.86	I
55		6616.67	I
11		6618.34	II
7	h	6631.00	I
13	h	6632.06	I
14		6647.12	I
75		6656.83	II
55		6673.41	II
75		6673.78	II
5	h	6687.51	II
4	h	6699.25	I
13		6736.79	I
35	c	6747.09	I
19	c	6749.19	I
7	c	6784.99	II
55	Cw	6798.60	I
11		6811.76	II
17	Cw	6812.87	II
13		6814.04	II
9		6817.61	I
35	Cw	6827.60	II
19		6830.50	II
9		6844.39	I
9	h	6845.47	II
9		6846.59	II
17	c	6850.46	II
11		6852.77	I
11	c	6870.44	I
7		6884.66	I
8		6892.71	I
8	h	6970.38	
8	c	6980.12	I
40		7021.51	II
10		7024.53	I
13		7042.40	I
8		7044.45	II
7		7051.07	I
10		7079.99	I
11	c	7095.18	I
20		7114.55	I
10	h	7116.90	I
11		7118.24	II
7		7137.33	II
10	h	7159.88	I
7		7167.77	II
7	h	7189.95	I
10	c	7208.85	II
24		7227.70	II
13		7231.53	I
7	c	7243.26	I
7	c	7259.21	I
7	c	7287.61	I
7	h	7289.19	I
7	Cw	7324.42	I
7		7328.47	I
7		7344.86	
16		7407.56	II
20	c	7451.74	II
11	h	7495.59	I
6	h	7499.42	I
14		7541.02	II
6		7574.86	I
20		7645.66	II
7		7704.98	II
16		7721.84	I
6	h	7786.16	II
6	Cw	7841.27	I
14		7871.67	I
6		7881.09	I
6	Cw	7888.56	II
6		7915.19	II
6		8031.92	I
6		8055.43	I
14		8067.44	I
10	Cw	8122.78	II
11		8141.10	I
5		8181.34	II
5	c	8211.93	I
6		8289.93	I
6		8379.84	I
6	h	8427.82	I
6	h	8605.27	II
10		8714.59	II

Pr III
Ref. 306, 308 — J.R.

Intensity		Wavelength	
		Vacuum	
25		1008.61	III
50		1021.35	III
25		1026.18	III
100		1029.03	III
50		1038.29	III
50		1042.96	III
25		1043.80	III
25		1044.03	III
25		1046.20	III
150		1047.24	III
100		1049.09	III
50		1052.63	III
25		1061.60	III
25		1066.03	III
25	p	1068.85	III
25		1069.88	III
25		1084.42	III
25		1088.66	III
100		1104.84	III
25		1108.82	III
30		1352.70	III
25		1881.22	III
		Air	
50		2031.46	III
100		2033.30	III
25		2043.12	III
100		2052.30	III
50		2052.87	III
50		2053.85	III
200		2058.59	III
50		2064.08	III
25		2075.08	III
200		2090.75	III
200		2093.49	III
25		2096.85	III
50		2096.94	III
10	w	2148.14	III
10	w	2194.24	III
10	w	2197.25	III
10	w	2205.48	III
10	w	2206.26	III
10	w	2214.45	III
10	w	2215.25	III
10	w	2217.12	III
10	w	2223.23	III
10	w	2230.35	III
10	w	2237.26	III
10		2239.06	III
10		2239.42	III
10	w	2242.15	III
10	w	2284.62	III
10		2307.59	III
10		2307.77	III
10		2308.41	III
10		2311.29	III
10		2311.44	III
10		2314.18	III
10	w	2315.46	III
10		2318.15	III
10		2318.36	III
10		2318.64	III
10		2318.82	III
10		2318.97	III
10	w	2319.40	III
10		2320.41	III
10	w	2328.56	III
10	w	2336.13	III
10		2365.52	III
10		2368.78	III
10		2369.08	III
10	w	2378.06	III
10	w	2378.97	III
10		2395.44	III
10		2399.70	III
10	w	2405.56	III
10		2408.19	III
10		2409.80	III
10		2412.40	III
10		2417.69	III
10		2418.95	III
10		2426.14	III
10		2426.85	III
10	w	2430.32	III
10	w	2434.18	III
10		2434.39	III
10		2435.91	III
10		2436.89	III
10		2438.63	III
10	w	2444.93	III
10		2445.49	III
10		2446.77	III
10		2448.16	III
10		2452.02	III
10	w	2452.81	III
10	w	2452.85	III
10		2454.60	III
10		2454.82	III
10	w	2459.77	III
10	w	2460.72	III
10		2462.18	III
10		2462.90	III
10		2468.20	III
10		2468.97	III
10	w	2473.42	III
10	w	2478.32	III
10		2479.98	III
10		2481.02	III
10		2483.30	III
10		2483.99	III
10	w	2484.60	III
10	w	2485.16	III
10	w	2488.72	III
10		2491.97	III
10	w	2494.20	III
10	w	2495.37	III
10	w	2495.51	III
10		2499.97	III
20	w	2587.71	III
40	w	2624.91	III
20	w	2644.62	III
20	w	2656.88	III
20	w	2667.51	III
70	w	2679.47	III
40	w	2710.30	III
20	w	2718.65	III
100	w	2724.03	III
20	l	2841.94	III
70	s	2910.61	III
50	s	2911.77	III
100	l	2914.49	III
50	s	2930.19	III
70	s	2942.43	III
70	s	2953.58	III
90	w	2954.40	III
90	s	2964.85	III
150	s	2968.83	III
80	s	2969.41	III
150	l	2976.86	III
150	s	2977.06	III
500	s	2980.54	III

Praseodymium (Cont.)

Intensity		Wavelength	
100	l	2981.65	III
150	c	2982.42	III
500	s	2985.82	III
150	s	2997.12	III
70	l	2998.79	III
150	l	3000.46	III
150	s	3003.20	III
60	l	3006.47	III
150	s	3008.04	III
150	l	3010.61	III
90	s	3014.60	III
100	l	3015.13	III
90	s	3016.26	III
70	s	3021.77	III
70	s	3025.26	III
100	l	3029.38	III
100	l	3033.31	III
90	l	3034.25	III
70		3040.02	III
60	c	3040.94	III
60	l	3041.78	III
100	s	3042.35	III
100	l	3045.81	III
70	l	3046.98	III
120	l	3050.30	III
70	s	3055.30	III
150	l	3058.90	III
150	l	3066.71	III
70	s	3078.68	III
150	l	3080.20	III
50	w	3248.39	III
90	s	3280.92	III
90	l	3292.58	III
90	s	3296.10	III
100	l	3306.14	III
60	s	3333.26	III
90	s	3340.58	III
500	s	3341.43	III
70	s	3341.68	III
70	s	3345.38	III
70	l	3345.44	III
50	w	3351.07	III
50	s	3353.87	III
250	s	3354.91	III
500	l	3357.56	III
500	l	3359.41	III
100	l	3364.52	III
50		3364.88	III
50	s	3365.80	III
200	s	3367.35	III
500	s	3367.58	III
100	l	3371.92	III
100	s	3377.14	III
50	s	3379.13	III
150	s	3380.21	III
150	s	3381.26	III
300	s	3381.84	III
100	s	3391.08	III
1000	w	3394.22	III
600	d	3396.07	III
300	s	3396.62	III
300	l	3397.46	III
50	l	3402.97	III
500	c	3413.21	III
300	s	3415.15	III
150	s	3420.07	III
300	l	3422.22	III
50	c	3426.27	III
500	l	3427.02	III
300	l	3436.36	III
150	l	3440.62	III
50	l	3445.29	III
70	s	3454.05	III
180		3653.58	III
60		3817.25	III
60	l	3861.80	III
150		3980.51	III
200		4000.20	III
90		4018.36	III
180		4029.60	III
90		4142.46	III
120		4144.48	III
90		4147.85	III
90		4172.15	III
150		4179.77	III
180		4184.18	III
240		4197.01	III
120		4219.45	III
180		4231.45	III
180		4275.07	III
120		4286.32	III
90		4298.27	III
90		4301.73	III
90		4316.34	III
60		4354.28	III
90		4379.82	III
90		4381.47	III
120		4404.71	III
120		4421.10	III
120		4431.85	III
150	w	4447.93	III
180	w	4450.14	III
120	w	4451.00	III
120		4461.02	III
200		4461.81	III
300	w	4500.31	III
450	w	4612.02	III
600	w	4625.18	III
120		4654.16	III
600	w	4713.70	III
300	w	4725.55	III
270		4728.21	III
300		4747.11	III
300	w	4771.83	III
450	w	4775.30	III
600		4857.39	III
150		5208.51	III
150		5261.68	III
1000		5264.44	III
1500		5284.70	III
1500		5299.99	III
1500		5340.02	III
100		5427.70	III
100		5581.74	III
150		5646.80	III
600		5765.27	III
1500		5844.41	III
200		5947.98	III
7000	w	5956.05	III
900	w	5998.94	III
1500	w	6053.01	III
900		6071.09	III
9000	w	6090.02	III
5000		6160.24	III
1500	w	6161.22	III
100		6195.05	III
2000		6195.63	III
200		6310.36	III
100		6361.65	III
300		6429.26	III
300		6444.74	III
600		6500.04	III
300		6501.49	III
200		6578.90	III
100		6616.46	III
600		6706.70	III
100		6727.63	III
200		6827.96	III
100		6854.63	III
200		6857.30	III
1000		6866.80	III
1000		6899.06	III
500		6903.52	III
7000		6910.14	III
150	w	6934.55	III
500		6970.96	III
100		6979.83	III
5000		7030.39	III
100		7075.21	III
4500		7076.62	III
100		7083.99	III
500		7112.53	III
100	w	7165.64	III
250		7231.62	III
100	w	7238.26	III
250		7240.21	III
100	w	7262.32	III
150	w	7340.69	III
350		7343.70	III
200		7349.75	III
300	w	7350.61	III
100	w	7355.52	III
2000		7426.48	III
4000		7429.05	III
100		7463.96	III
250		7487.40	III
200		7493.20	III
100		7511.17	III
500		7529.11	III
100	w	7549.20	III
150	w	7588.64	III
300		7596.41	III
100	w	7625.63	III
100	w	7648.34	III
100	w	7670.65	III
200	w	7674.65	III
500		7742.34	III
250	w	7745.59	III
500		7754.31	III
100	w	7755.48	III
3000		7781.98	III
200	w	7814.74	III
1500	w	7866.14	III
1000		7888.12	III
400		7897.09	III
1000		7914.00	III
100		7923.16	III
1000	w	7972.75	III
250		8001.14	III
3000		8102.90	III
250		8119.54	III
400		8132.23	III
100		8138.34	III
150	w	8235.33	III
250	w	8244.89	III
100		8409.10	III
100	w	8494.99	III
200	w	8567.63	III
5000	w	8602.74	III
500		8691.58	III
500		8771.38	III
1000		8854.05	III
125		8886.17	III
100		8908.70	III
250		9099.98	III
250		9131.90	III
200		9222.32	III
250	w	9265.56	III
125		9320.54	III
250		9334.33	III
175		9377.44	III
175		9388.56	III
175		9549.77	III
100		9579.74	III
150		9802.98	III
175		9806.37	III
500	w	9991.16	III
500		10031.10	III
500		10160.33	III
500		10238.63	III
500		10301.58	III
500		10324.59	III
500		10716.58	III

Pr IV
Ref. 337, 338 — J.R.

Intensity	Wavelength	
	Vacuum	
20	718.23	IV
30	721.34	IV
60	722.41	IV
30	722.58	IV
20	726.04	IV
50	730.37	IV
30	731.77	IV
30	734.86	IV
20	735.04	IV
20	736.19	IV
50	736.32	IV
100	737.17	IV
100	741.45	IV
20	743.15	IV
20	743.89	IV
20	746.14	IV
40	763.16	IV
20	764.00	IV
300	1226.40	IV
2000	1228.59	IV
500	1230.69	IV
400	1238.19	IV
200	1249.35	IV
200	1255.64	IV
200	1261.27	IV
400	1268.32	IV
300	1270.58	IV
1000	1275.10	IV
200	1275.40	IV
1000	1278.65	IV
200	1279.34	IV
1000	1287.44	IV
300	1290.93	IV
1000	1292.30	IV
5000	1293.22	IV
5000	1295.28	IV
400	1296.50	IV
200	1298.26	IV
300	1298.54	IV
200	1304.71	IV
300	1306.86	IV
200	1308.08	IV
200	1310.71	IV
500	1314.96	IV
300	1315.28	IV
300	1316.96	IV
500	1320.10	IV
1000	1320.70	IV
5000	1321.36	IV
500	1322.51	IV
500	1326.38	IV
5000	1333.57	IV
300	1335.96	IV
500	1339.29	IV
1000	1340.74	IV
200	1341.32	IV
300	1344.23	IV
1000	1347.07	IV
1000	1352.81	IV
500	1354.35	IV
5000	1354.66	IV
2000	1360.64	IV
1000	1364.81	IV
2000	1365.77	IV
400	1368.90	IV
5000	1374.41	IV
1000	1382.62	IV
200	1384.23	IV
200	1385.91	IV
300	1394.11	IV
500	1397.11	IV
1000	1399.31	IV
1000	1400.96	IV
400	1410.90	IV
1000	1424.36	IV
500	1426.59	IV
5000	1435.56	IV
200	1459.95	IV
200	1461.76	IV
1000	1474.91	IV
200	1477.32	IV
500	1485.88	IV
500	1503.35	IV
400	1516.86	IV
200	1520.71	IV
2000	1520.98	IV
400	1525.46	IV
500	1553.62	IV
500	1559.49	IV
500	1570.13	IV
200	1572.80	IV
5000	1574.55	IV
5000	1575.10	IV
3000	1578.38	IV
500	1585.10	IV
300	1613.00	IV
400	1613.65	IV
1000	1618.03	IV
2000	1622.30	IV
300	1634.77	IV
400	1676.08	IV
200	1688.49	IV
200	1713.53	IV
500	1732.86	IV
300	1762.86	IV
1000	1766.88	IV
1000	1771.14	IV
500	1841.08	IV
10000	1884.87	IV
400	1951.23	IV
200	1954.61	IV

Intensity		Wavelength	
		Air	
200		2025.06	IV
1000		2039.15	IV
200		2047.05	IV
200		2050.73	IV
200		2058.48	IV
2000		2083.23	IV
500		2100.[illegible]	IV
1000		2154.31	IV
300		2193.37	IV
1000		2205.13	IV
300		2265.70	IV
200	c	2334.46	IV
200	c	2339.08	IV
500	c	2376.09	IV
2000	c	2378.98	IV
1000	c	2379.66	IV
500	c	2427.07	IV
500	c	2428.13	IV
500	c	2438.57	IV
500	c	2455.64	IV
500	c	2705.19	IV
200	c	2708.01	IV
200	c	2753.47	IV
200	c	2767.60	IV

Pr V
Ref. 149 — J.R.

Intensity	Wavelength	
	Vacuum	
200	843.78	V
7000	865.90	V
5000	869.17	V
80	869.66	V
1000	896.65	V
750	922.29	V
250	1234.07	V
250	1342.78	V
200	1958.09	V
400	1958.20	V
	AIR	
300	2246.06	V
300	2246.20	V

PROMETHIUM (Pm)
Z = 61

Pm I and II
Ref. 196, 260 — C.H.C.

Intensity		Wavelength	
		Air	
40	w	2502.12	II
40		2608.24	II
150		2632.00	II
70		2638.46	II
100		2671.05	II
50	w	2787.72	II
40		2808.05	II
100	h	2820.10	II
100	w	2840.82	II
150	w	2841.86	II
200	c	2857.46	II
100		3004.59	II
100		3008.85	II
300		3072.41	II
150		3086.02	II

Promethium (Cont.)

120		3090.19	II
150		3091.86	II
150		3108.11	II
100		3115.36	II
100		3117.22	II
100		3118.76	II
35		3162.23	I
35		3168.82	I
100		3172.77	II
35		3222.04	I
35		3238.55	I
60		3239.62	II
75		3296.63	I
60		3311.76	II
50		3313.38	I
50		3329.22	I
75		3331.57	I
100		3354.45	I
100		3358.14	II
80		3360.21	II
80		3364.44	II
300		3366.03	I
90		3377.68	II
100		3391.28	II
100		3408.06	II
500		3427.40	II
120		3441.15	II
400		3449.80	II
250		3460.25	II
200		3462.91	II
200		3480.61	II
150		3497.13	II
100		3514.85	II
200		3546.81	I
100		3559.43	II
200		3565.31	II
150		3580.10	II
200		3610.76	II
200		3629.84	II
300		3634.20	II
300		3659.39	II
300	r	3669.22	I
200		3674.85	I
200		3678.51	II
300	r	3679.85	I
200		3687.65	II
400		3689.79	II
300		3692.50	II
300		3697.50	II
300	r	3697.63	I
400		3702.63	II
800		3711.72	II
200		3715.75	II
200		3721.72	II
500		3726.01	I
200		3738.43	I
300		3740.68	I
300		3742.52	II
300	r	3742.97	I
500		3745.86	II
300		3747.09	II
500		3750.09	II
200		3761.68	I
300		3765.75	I
300		3775.42	I
300		3780.77	I
200		3781.43	I
400		3795.66	II
250		3806.06	II
300	r	3809.20	I
400		3810.93	I
200		3819.26	II
200		3820.53	II
300		3839.52	I
200		3842.88	II
300		3842.98	II
250		3845.38	II
300		3874.03	I
800		3877.62	II
300	r	3885.79	I
250		3890.97	I
1000		3892.15	II
300		3898.73	I
400		3899.78	II
250		3909.50	II
1000		3910.26	II
1000		3919.10	II
800		3936.48	II
300		3944.21	II
300	r	3954.76	I
1000		3957.74	II
500		3980.74	II
300		3995.05	II
1000	r	3998.96	II
500		4009.96	II
200		4012.72	II
250		4014.20	II
200		4019.34	II
250		4028.20	II
200		4045.36	II
300		4051.54	II
600	r	4055.20	II
200	r	4056.56	I
600		4075.84	II
200	r	4085.31	I
500		4086.10	II
250		4140.46	II
200		4185.74	II
300		4192.92	II
200		4194.70	II
200		4222.15	II
300	r	4264.32	I
300	r	4284.37	I
600		4297.78	II
200		4303.89	II
200	r	4305.64	I
400		4318.80	I
250		4325.92	II
200		4332.05	II
300		4336.54	II
200		4337.48	II
300		4342.12	II
200		4347.72	I
350	r	4363.92	I
300	r	4369.64	I
200		4381.88	II
400	r	4388.49	I
200		4388.76	II
400	r	4409.42	I
500	r	4412.47	I
1000		4417.96	II
400		4432.51	II
250	r	4435.86	I
300	r	4436.55	I
300	r	4438.68	I
500		4445.41	II
600		4446.90	II
800		4453.95	II
200		4459.97	II
250	r	4468.16	I
200		4471.48	II
300		4473.23	II
200		4477.46	II
350	r	4478.58	I
300	r	4481.60	I
300	r	4485.05	I
300	r	4490.50	I
250		4492.05	II
600		4500.15	II
350	r	4500.33	I
250		4506.84	I
100		4509.38	II
100		4513.56	II
200		4517.31	I
200		4523.32	II
600		4525.20	II
250	r	4526.12	I
250	r	4526.76	I
400	r	4527.70	I
800		4529.21	II
300		4540.06	I
300	r	4541.42	I
450	r	4541.75	I
500	r	4544.08	I
200		4545.17	I
400	r	4549.78	I
300	r	4554.03	I
200		4554.63	I
500	r	4555.34	I
200		4556.06	I
300		4557.03	I
300		4559.21	I
100		4564.83	II
300	r	4568.14	I
200		4570.37	I
300	r	4572.15	I
400	r	4575.27	I
300		4578.28	I
200	r	4578.41	I
300	r	4579.48	I
300		4581.14	I
300	r	4585.49	I
200		4593.82	I
400	r	4595.82	I
800		4597.55	I
500	r	4600.25	I
400	r	4602.96	I
400		4604.59	I
600	r	4605.66	I
500	r	4609.85	I
100		4615.87	II
600	r	4617.02	I
200		4618.40	I
400	r	4618.49	I
500	r	4619.75	I
500		4621.57	I
500		4623.31	I
700	r	4623.68	I
900		4624.41	I
500	r	4625.29	I
400	r	4627.60	I
200		4630.93	I
600	r	4633.45	I
400	r	4640.96	I
700	r	4643.36	I
700	r	4643.76	I
400		4645.94	I
600	r	4647.03	I
600	r	4650.42	I
500		4650.52	I
400	r	4653.41	I
400		4654.50	I
500	r	4655.05	I
300		4659.38	I
500	r	4660.79	I
300	r	4663.26	I
600	r	4663.46	I
400	r	4665.19	I
500	r	4671.23	I
400		4671.76	I
500	r	4674.42	I
200		4677.46	I
500	r	4677.92	I
400	r	4678.09	I
700	r	4682.92	I
500	r	4696.80	I
200		4699.51	I
250		4722.06	II
300		4727.06	I
900	r	4728.36	I
400	r	4728.68	I
800	r	4734.27	I
200	r	4737.99	I
100		4739.08	II
200		4739.78	I
350	r	4745.13	I
500	r	4757.73	I
800	r	4759.00	I
700	r	4762.57	I
700	r	4773.46	I
900	r	4781.29	I
250		4794.59	I
200		4795.43	I
700	r	4798.98	I
900	r	4801.36	I
700	r	4809.54	I
900	r	4811.96	I
400	r	4817.12	I
400	r	4827.72	I
800	r	4837.66	I
400	r	4838.92	I
300		4839.62	I
200		4844.01	I
350	r	4852.73	I
400	r	4860.62	I
700	r	4860.74	I
300	r	4865.30	I
500	r	4865.72	I
400	r	4869.80	I
700	r	4872.42	I
500	r	4887.02	I
700	r	4892.52	I
400	r	4900.30	I
300	r	4904.28	I
400	r	4918.28	I
600	r	4932.99	I
700	r	4959.46	I
100		4971.40	II
500	r	4997.10	I
200	r	5030.80	I
300	r	5058.31	I
100		5067.35	II
150		5080.52	II
150		5089.35	II
200		5092.42	I
400	r	5094.83	I
200		5096.18	I
150		5097.30	II
400	r	5100.77	I
250		5121.47	II
400	r	5127.34	I
200		5129.75	I
400	r	5145.13	I
500	r	5146.30	I
400		5153.86	II
300		5169.71	II
500		5171.58	II
300		5194.05	II
500		5208.09	II
150		5215.96	II
250		5225.12	II
500		5236.26	II
300		5236.66	II
400		5246.33	II
150		5262.42	II
500		5270.64	II
200		5293.92	II
100		5308.86	II
150		5318.58	II
200		5410.45	II
200		5424.54	II
180		5424.79	II
150		5429.04	II
100		5467.64	II
150		5495.45	II
100		5516.42	II
180		5534.96	II
200		5537.38	I
800		5546.08	II
120		5556.88	II
150		5558.39	II
200		5561.73	II
800		5576.02	II
200		5641.29	II
200		5730.81	I
200		5768.16	II
200		5776.99	I
500		5823.93	II
300	c	5868.79	II
200	c	5875.31	II
100	c	5878.76	II
150		5899.76	II
250		5904.71	I
100		5905.90	I
125		5914.96	I
250	c	5927.17	II
150		5939.66	I
400	c	5946.49	II
800		5956.42	I
200		5956.69	I
100	c	5960.08	II
150		5963.00	II
400		5967.89	I
200		5979.73	I
200		5984.82	I
100	c	5987.13	II
400		5997.12	I
200		6027.11	II
300		6030.06	I
400		6031.32	I
500		6043.39	I
150	c	6052.57	II
100	c	6067.00	II
500		6069.06	I
100		6076.40	II
200		6085.41	II
900		6100.21	I
400		6106.40	I
100		6114.90	II
400	h	6151.76	I
100		6159.53	II
400		6163.16	I
100		6184.52	II
200		6208.91	II
500		6229.64	I
400		6237.79	I
100		6263.25	II
400		6272.69	I
400		6286.06	I
500		6308.29	I
100		6314.20	II
700		6323.84	I
500		6390.31	I
100		6429.64	II
500	h	6431.93	I
100		6436.57	II
400		6487.61	I
400		6510.34	I
500		6517.25	I
200		6519.43	II
1000	d	6520.45	I
500		6542.20	I
100	h	6558.48	II
100		6586.39	II
100		6592.29	II
900		6598.15	I
800		6598.66	I
700		6606.37	I
800	w	6625.23	I
100	h	6625.54	II
700		6649.81	I
400		6659.05	II
100		6661.25	II
500		6661.68	I
400		6663.76	I
800	c	6667.51	I
700	h	6677.47	I
200		6680.89	II
500		6685.55	I
500		6685.68	I
150		6690.09	II
600		6700.33	I
100		6706.27	II
700		6714.67	I
500		6717.26	I
500		6720.71	I
700		6727.50	I
600		6743.71	I
900		6749.91	I
900		6750.48	I
200		6756.45	II
300		6772.29	II
400		6778.78	I
100		6783.09	II
100		6796.87	II
200		6811.68	II
800		6833.30	I
400		6848.37	I
50		6858.58	II

PROTACTINIUM (Pa)
Z = 91

Pa I and II
Ref. 96 — J.G.C.

Intensity		Wavelength	
		Air	
3000	h	2466.85	
3000	h	2492.85	
3000		2599.16	II
3000		2699.22	II
3000		2822.79	II
3000	l	2832.14	
3000		2870.01	
3000	h	2871.42	II
3000	h	2891.14	II
3000	h	2906.93	
3000	l	3011.10	II
3000	s	3033.59	II
3000	l	3071.24	II
3000	h	3083.19	
3000	l	3093.23	II
3000	l	3126.23	II
3000	l	3146.28	II
3000	l	3170.89	II
3000	l	3171.54	II
3000		3204.16	
3000	l	3240.58	II
3000		3274.46	II
3000	l	3332.69	II
3000	s	3346.66	II
3000	l	3394.49	
3000	l	3452.82	II
3000		3504.97	I
3000	s	3530.65	II
3000		3570.56	I
3000		3571.82	I
3000		3618.07	I
10000		3636.52	I
3000		3702.74	I
3000		3752.67	I
3000		3873.35	I
3000		3931.83	I
3000	s	3952.62	II
10000	l	3957.85	II
3000	s	3970.07	II
3000		3981.82	I
10000		3982.23	I
3000	l	4012.96	II
3000	s	4018.21	II
3000		4030.16	II
3000	s	4046.93	II
10000	s	4056.20	II
10000	s	4070.40	II
3000		4117.62	
3000	l	4176.18	II
10000	l	4217.23	II
10000	s	4248.08	II
3000	s	4291.34	II
3000		4400.77	
3000		4436.13	
3000	s	4601.43	II
3000		4628.19	
3000		4820.34	
3000	s	4861.49	
3000	l	6035.78	I
3000		6162.56	I
3000		6216.35	
3000	l	6358.61	I
3000		6379.25	I
3000	l	6438.97	I
3000	h	6792.75	I
10000		6945.72	I
3000		6960.09	I
3000	h	6961.78	I
3000	s	6992.73	I
3000		7076.27	I
3000	h	7100.94	I
10000	s	7114.89	I
3000	h	7171.55	I
3000		7227.13	I
3000		7318.79	I
10000	l	7368.25	I
3000	h	7471.89	I
10000	h	7493.15	I
3000	h	7558.26	I
10000	h	7608.20	I
10000		7626.79	I
10000	s	7635.18	I
10000		7669.34	I
3000		7679.20	I
10000	h	7749.19	I
3000		7872.95	I
3000	l	7945.56	I
10000		8039.34	I
10000	h	8099.84	I
10000		8199.04	I
10000		8271.87	I
3000	s	8358.98	I
3000	s	8369.60	I
3000	h	8441.04	I
10000	h	8532.66	I
10000	s	8572.96	I
3000	h	8639.91	I
3000	h	8653.51	I
10000		8735.27	I
3000		10594.38	
3000		10923.32	I
3000		11646.78	
10000		11791.73	I
3000		12279.01	
3000		13234.09	
10000		13522.40	
10000		14344.76	I
3000		18478.61	I

RADIUM (Ra)
Z = 88

Ra I and II
Ref. 253, 254 — E.F.W.

Intensity		Wavelength	
		Air	
8		2369.73	II
8		2460.55	II
10		2475.50	II
8		2586.61	II
10		2643.73	II
20		2708.96	II
10		2795.21	II
30		2813.76	II
10		3033.44	II
5		3101.80	I
100		3649.55	II
200		3814.42	II
8		4194.09	II
8		4244.72	II
100		4340.64	II
20		4436.27	II
30		4533.11	II
8		4641.29	I
100		4682.28	II
8		4699.28	I
100		4825.91	I
10		4856.07	I
10		4859.41	II
10		[illegible]	II
10		5097.56	I
10		5205.93	I
10		5283.28	I
10		5320.29	I
10		5399.80	I
20		5400.23	I
20		5406.81	I
8		5482.13	I
10		5501.98	I
10		5553.57	I
20		5555.85	I
10		5616.66	I
50		5660.81	I
20		5813.63	II
30		6200.30	I
10	p	6336.90	I
20		6446.20	I
20		6487.32	I
10		6593.34	II
10		6719.32	II
20		6980.22	I
20		7118.50	I
50		7141.21	I
20		7225.16	I
10		7310.27	I
20		7838.12	I
50		8019.70	II
6		8177.31	I
5		8335.07	I
5		9932.21	I

RADON (Rn)
Z = 86

Rn I
Ref. 251 — E.F.W.

Intensity	Wavelength	
	Air	
5	3514.60	I
10	3739.89	I
20	3753.65	I
10	3917.20	I
10	3941.72	I
10	3952.36	I
10	4226.06	I
80	4307.76	I
7	4335.78	I
100	4349.60	I
40	4435.05	I
50	4459.25	I
50	4508.48	I
50	4577.72	I
50	4609.38	I
30	4721.76	I
6	5722.58	I
10	6061.92	I
6	6200.75	I
6	6380.45	I
10	6557.49	I
10	6606.43	I
15	6627.23	I
6	6669.60	I
8	6704.28	I
20	6751.81	I
6	6806.79	I
8	6836.95	I
8	6837.57	I
10	6891.16	I
10	6998.90	I
200	7055.42	I
100	7268.11	I
20	7291.00	I
6	7320.98	I
10	7419.04	I
300	7450.00	I
8	7470.89	I
8	7483.13	I
8	7514.13	I
8	7516.92	I
6	7523.93	I
6	7597.55	I
8	7601.28	I
10	7657.48	I
10	7738.43	I
20	7746.64	I
100	7809.82	I
20	8049.00	I
100	8099.51	I
6	8173.84	I
100	8270.96	I
8	8314.51	I
6	8349.74	I
10	8381.05	I
10	8487.48	I
10	8494.89	I
20	8520.95	I
100	8600.07	I
10	[illegible]	I
15	8675.83	I
10	8807.75	I
50	9327.02	I
6	9948.57	I
5	10106.13	I

RHENIUM (Re)
Z = 75

Re I and II
Ref. 1 — C.H.C.

Intensity		Wavelength	
		Air	
25000		2003.53	I
16000		2017.87	I
27000		2049.08	I
4200		2074.70	I
3700		2083.92	I
10000		2085.59	I
4700		2092.41	II
9800		2097.12	I
2700		2109.22	I
3400		2139.04	II
1600		2142.74	II
		2142.97	I
3700		2156.67	I
4900		2167.94	I
3400		2176.21	I
4200	c	2214.26	II
2200		2214.58	I
1700		2226.42	I
920		2235.44	I
440		2255.73	I
860		2256.19	I
2000		2264.39	I
2100		2274.62	I
5200	c	2275.25	II
1600		2281.62	I
2900		2287.51	I
2700		2294.49	I
390		2298.09	II
390		2299.77	I
610		2302.99	I
680		2306.54	I
230		2312.97	I
220		2313.34	I
220		2319.19	I
370		2320.16	I
800		2322.49	I
300		2328.66	I
270		2334.33	I
270		2335.73	I
220		2336.10	I
270		2337.95	I
860		2344.78	I
230		2349.39	I
220	d	2350.46	I
680		2352.07	I
210	d	2353.95	I
250		2356.50	I
200		2365.32	I
1200		2365.90	I
570		2367.68	I
180		2368.53	II
520		2369.27	I
220		2370.76	II
210		2371.52	I
150		2373.48	II
320		2375.07	I
75		2378.53	II
370		2379.77	I
180		2386.90	II
340		2388.57	I
230		2393.65	I
320		2394.37	I
320		2396.79	I
200		2397.31	I
210	d	2400.72	I
		2400.89	I
210		2401.68	I
75		2403.04	II
1500		2405.06	I
740		2405.60	I
320		2406.70	I
270		2410.37	I
60		2418.20	II
1200		2419.81	I
300		2421.73	I
300		2421.88	I
60		2423.84	II
2500		2428.58	I
490		2431.54	I
420		2432.18	I
340	c	2441.47	I
230		2442.51	I
250		2444.94	I
610		2446.98	I
85		[illegible]	II
85		2449.52	II
610		2449.71	I
200		2455.83	II
390		2461.20	I
800	c	2461.84	II
200		2467.57	II
120		2467.85	II
150	c	2469.36	II
120		2470.61	II
75		2471.05	II
150		2473.72	II
160		2475.17	II
75		2477.43	II
200		2479.02	I
1200		2483.92	I
390		2485.81	I
980		2487.33	I
75		2490.16	II
200		2492.84	I
370		2496.04	I
200		2498.22	I
370		2501.72	I
570		2502.35	II
230		2504.60	II
270		2505.94	I
1800	c	2508.99	I
570		2520.01	I
540		2521.50	I
150		2534.10	II
370		2534.80	I
570		2540.51	I
740	d	2544.74	I
370		2545.48	I
160		2550.09	II
300		2552.02	I
150	c	2553.59	II
370		2554.63	II
1000		2556.51	I
250		2559.08	I
340		2564.19	I
540		2568.64	II
370		2571.81	II
380		2586.79	I
290		2599.86	I
290		2603.89	I
660		2608.50	II
610	d	2611.54	I
160	c	2616.72	II
200		2622.76	I
310		2635.83	II
550		2636.64	I
190		2637.01	II
90		2641.02	II
270		2642.75	I
65		2648.46	II
270		2649.05	I
660		2651.90	I

Rhenium (Cont.)

Intensity		Wavelength	Spectrum
400		2654.12	I
220		2663.63	I
940		2674.34	I
220		2688.53	I
1300		2715.47	I
200		2731.56	II
220		2732.21	I
610		2733.04	II
110	h	2753.64	II
220		2758.00	I
210		2763.79	I
200		2766.39	I
310		2767.74	I
220		2768.85	I
220		2769.32	I
350		2770.42	I
550		2783.57	I
220		2791.29	I
120		2803.28	II
220		2814.68	I
75		2819.78	II
880		2819.95	I
310		2834.08	I
200		2837.55	I
200		2840.35	I
220		2843.00	I
270		2850.98	I
240		2867.19	I
200		2875.28	I
200		2883.44	I
2900		2887.68	I
130	c	2888.06	II
490		2896.01	I
830	c	2902.48	I
210		2905.58	I
550		2909.82	I
65	h	2916.73	II
830	c	2927.42	I
270		2930.61	I
440		2943.14	I
130	h	2957.91	II
270		2962.27	I
720		2965.11	I
1500		2965.76	I
90		2968.98	II
310		2976.29	I
210		2978.15	I
220		2980.82	I
220		2982.19	I
220		2988.47	I
1800		2992.36	I
5500		2999.60	I
350		3001.14	I
220		3004.14	I
200		3006.42	I
500		3016.02	I
300		3016.49	I
380		3030.45	I
240		3047.25	I
200		3058.78	I
1600		3067.40	I
320		3069.94	I
260		3071.16	I
200		3072.96	I
550		3082.43	I
340		3088.76	I
200		3093.64	I
200		3095.06	I
700		3100.67	I
140		3103.06	II
700		3108.81	I
340		3110.86	I
340	c	3118.19	I
340		3121.36	I
420		3128.94	I
260		3134.02	I
250		3141.38	I
440		3151.64	I
330		3153.79	I
360	c	3158.31	I
220		3164.52	I
700		3168.37	I
220		3174.61	I
440		3177.71	I
260		3178.61	I
600		3182.87	I
1100		3184.76	I
1100		3185.57	I
260		3190.78	I
260		3192.36	I
200		3194.50	I
220		3198.58	I
1100	c	3204.25	I
380		3235.94	I
600		3258.85	I
600		3259.55	I
200		3261.56	I
300		3268.89	I
200		3294.83	I
280		3296.70	I
280		3296.99	I
280		3301.60	I
240		3302.23	I
320		3303.21	II
280		3303.75	I
740		3313.95	I
600		3322.48	I
200		3331.52	I
2000		3338.18	I
1600		3342.24	I
810		3344.32	I
320		3346.20	I
240	d	3356.33	I
200		3358.02	I
200		3362.74	I
240		3377.74	I
320		3379.06	II
320		3379.70	I
200		3385.76	I
240		3389.43	I
200		3390.25	I
4000		3399.30	I
650		3404.72	I
650		3405.89	I
240		3408.67	I
320		3409.83	I
320		3417.77	I
810		3419.41	I
8000		3424.62	I
400		3426.19	I
300		3427.61	I
320		3437.71	I
400		3449.37	I
16000	c	3451.88	I
240		3453.50	I
55000	c	3460.46	I
40000	c	3464.73	I
400		3467.96	I
240		3476.44	I
400		3480.38	I
320		3480.85	I
240		3482.23	I
560		3503.06	I
100	c	3512.28	I
320		3516.65	I
320		3517.33	I
120		3534.82	I
320		3537.46	I
160		3539.33	I
240		3549.89	I
160		3551.29	I
160		3553.65	I
160		3558.94	I
160		3568.23	I
240		3570.26	I
360		3579.12	I
810	c	3580.15	II
650		3580.97	I
810		3583.02	I
160		3596.39	I
160		3610.49	I
320		3617.08	I
160		3621.46	I
160		3625.91	I
140		3637.06	I
810		3637.84	I
440		3651.97	I
120		3669.78	I
320		3670.53	I
860	c	3689.50	I
1500	c	3691.48	I
100		3697.71	I
520		3703.24	I
100		3705.02	I
240		3709.93	I
360	c	3717.28	I
4000		3725.76	I
140		3731.87	I
140		3732.28	I
240	c	3735.01	I
810		3735.31	I
910		3740.10	I
140		3740.41	I
130		3742.26	II
300	Cw	3745.44	I
140		3766.48	I
120		3768.26	I
140		3777.66	I
700		3787.52	I
160		3796.59	I
160		3797.59	I
190		3807.74	I
120		3815.66	I
120		3836.30	I
240		3869.94	I
240		3875.26	I
240		3876.86	I
100		3908.21	I
130		3913.92	I
380	c	3917.27	I
550		3929.85	I
140		3936.90	I
110		3944.72	I
180		3945.91	I
280		3961.04	I
350	c	3962.48	I
100		4004.93	I
140		4022.96	I
100		4023.31	I
110	c	4029.63	I
220		4033.31	I
110		4037.49	I
200		4048.99	I
240		4081.43	I
140		4104.42	I
240	c	4110.89	I
190		4121.64	I
240	Cw	4133.42	I
1800		4136.45	I
700		4144.36	I
140		4149.96	I
160		4170.40	I
220		4182.90	I
220		4183.06	I
650		4221.08	I
3600	c	4227.46	I
150		4241.39	I
260	c	4257.60	I
120	c	4291.17	I
200		4304.40	I
200		4332.25	I
40		4357.98	II
380		4358.69	I
190		4367.58	I
140		4391.34	I
360	Cw	4394.38	I
110	Cw	4406.40	I
180		4415.82	I
150		4475.08	I
120		4478.39	I
120	c	4507.04	I
2600		4513.31	I
260		4516.64	I
500		4522.73	I
120		4523.88	I
120		4529.95	I
100		4545.17	I
120		4580.68	I
120		4605.73	I
100		4621.38	I
190	c	4791.42	I
2200	Cw	4889.14	I
220		4923.90	I
40		5058.56	I
70		5096.50	I
20		5120.32	I
25		5161.65	I
40	c	5178.89	I
20		5181.74	I
35		5234.31	I
50		5248.86	I
1300		5270.95	I
1600	Cw	5275.56	I
100		5278.24	I
30		5305.56	I
20		5317.28	I
35		5321.28	I
50		5327.46	I
20		5331.90	I
20		5332.76	I
20		5333.85	I
35		5369.48	I
50	c	5369.80	I
100	c	5377.10	I
25		5431.90	I
14		5437.03	I
14		5447.92	I
25		5460.64	I
14	h	5520.05	I
25		5521.10	I
50	c	5532.68	I
50	c	5563.24	I
25		5573.47	I
25		5584.72	I
10	h	5607.21	I
12	h	5612.27	I
100		5667.88	I
25		5711.43	I
18		5716.95	I
110	c	5752.93	I
110	Cw	5776.83	I
18		5791.60	I
10	c	5815.92	I
550		5834.31	I
10		5919.86	I
60		5943.24	I
10		5950.21	I
18	h	5969.77	I
10		5989.99	I
18	h	5995.73	I
30		6114.22	I
35	c	6145.81	I
50		6146.82	I
18		6203.24	I
25		6217.97	I
30	Cw	6229.42	I
35	Cw	6243.24	I
35	d	6260.02	I
		6260.24	I
18	c	6271.37	I
18		6278.76	I
10		6286.41	I
10		6303.42	I
200		6307.70	I
200		6321.90	I
80	d	6350.75	I
16	h	6382.94	I
14		6411.47	I
50		6511.47	I
14		6515.25	I
12		6544.91	I
35	c	6577.11	I
40	Cw	6592.52	I
100	Cw	6605.19	I
30	c	6623.91	I
10	c	6637.25	I
27	Cw	6652.39	I
15	h	6683.28	I
9	c	6711.30	I
30		6751.22	I
5	c	6761.19	I
180	c	6813.41	I
260		6829.90	I
85		6971.53	I
35	Cw	7006.63	I
65	Cw	7024.15	I
65	Cw	7246.67	I
13	Cw	7292.72	I
40	Cw	7578.73	I
13		7611.89	I
7	Cw	7620.25	I
50	Cw	7640.94	I
65	Cw	7912.94	I
35	Cw	7980.77	I
40		8417.13	I
29	Cw	8527.73	I

RHODIUM (Rh)
Z = 45

Rh I and II
Ref. 1 — C.H.C.

Intensity		Wavelength	
		Air	
150		2276.21	II
140		2288.57	I
110		2309.82	I
55		2318.36	I
95		2319.10	I
95		2321.73	I
350		2322.58	I
140		2326.47	I
80		2328.64	I
190		2334.77	II
55		2345.41	I
55		2352.47	I
55		2359.18	I
300		2361.92	I
110		2368.34	I
270		2382.89	I
230		2383.40	I
40		2384.65	I
270		2386.14	II
80		2407.88	I
27		2408.19	I
27		2410.25	I
80		2415.84	II
55		2418.64	
45		2419.75	I
45		2420.18	II
65		2420.98	II
75		2423.94	
65		2427.11	II
130		2427.68	I
230		2429.52	I
40		2431.85	II
40		2432.66	I
18		2437.08	I
110		2437.90	I
330		2440.34	I
50	h	2444.27	I
65		2448.84	I
50		2449.04	I
75		2450.56	I
30		2455.70	II
65		2458.90	II
90		2461.04	II
30		2463.61	I
75		2470.39	I
90		2471.47	I
30		2472.51	I
130		2473.09	I
15		2475.64	II
15		2477.54	II
25		2482.04	I
50		2483.33	I
150		2487.47	I
100		2490.77	II
30		2492.30	I
75	h	2494.51	
15		2499.02	I
40		2500.58	
130		2502.46	I

Rhodium (Cont.)

Intensity		Wavelength	Spectrum
15		2503.84	II
300		2504.29	II
40		2505.10	II
150		2505.67	I
350		2509.70	I
50		2510.66	II
300		2511.03	II
75		2513.36	II
200		2515.75	I
130		2520.53	II
13		2525.99	I
13		2531.74	I
50		2532.66	
13		2533.59	
50		2534.07	
110		2536.71	
110		2537.04	II
30		2539.72	
40		2544.22	
350		2545.70	I
13		2548.60	
550		2555.36	I
25		2558.62	I
50		2565.79	I
45		2566.04	I
25		2566.92	II
50		2567.28	
25		2574.66	
25		2575.75	I
13		2576.23	
40		2587.29	II
30		2598.07	
30		2603.32	II
75		2606.44	II
75		2613.60	
150		2622.58	I
230		2625.88	I
100		2630.42	I
40		2634.99	I
30		2638.74	II
75		2643.00	I
110		2647.28	I
400		2652.66	I
30		2659.01	I
30		2671.06	I
65		2676.11	I
75		2678.10	I
100		2680.63	I
30		2681.78	I
30	h	2686.50	
30	h	2686.91	
50		2694.31	I
400		2703.73	I
40		2705.63	II
40		2707.23	I
75		2714.41	I
100		2715.31	II
75		2717.51	I
180		2718.54	I
65		2720.14	I
30		2720.52	I
160		2728.94	I
40		2736.76	I
75		2741.75	I
50		2767.73	I
100		2771.51	I
50		2778.06	I
75		2779.54	I
130		2783.03	I
25		2791.16	I
75		2796.63	I
150		2826.43	I
180		2826.68	I
30		2827.31	I
75		2834.12	I
45		2835.44	I
75		2836.69	I
50		2856.16	I
50	d	2860.68	I
		2860.76	I
280		2862.94	I
65		2864.40	I
50		2871.35	I
30		2873.62	I
110		2878.66	I
75		2880.76	I
140		2882.37	I
75		2885.97	I
75		2889.11	I
75		2889.84	I
65		2899.96	I
25		2904.81	I
160		2907.21	I
65		2910.17	II
75		2912.62	I
90		2915.42	I
30		2923.10	I
180		2924.02	I
130		2929.11	I
130		2931.94	I
30		2955.41	I
230		2968.66	I
25		2974.03	I
160		2977.68	I
450		2986.20	I
90		2986.99	I
50		2987.45	I
110		3004.46	I
50		3019.54	I
130		3023.91	I
50		3028.43	I
30		3045.77	I
30		3046.76	I
25		3057.89	I
65		3067.30	I
180		3083.96	I
29		3087.42	I
70		3114.91	I
140		3121.76	I
240		3123.70	I
35		3130.79	I
95		3137.71	I
45		3151.36	I
45		3152.60	I
130		3155.78	I
70		3179.73	I
80		3185.59	I
140		3189.05	I
470		3191.19	I
190		3197.13	I
70		3214.32	I
80		3237.66	I
520		3263.14	I
520		3271.61	I
2300		3280.55	I
110		3281.70	I
2300		3283.57	I
280		3289.14	I
45		3289.64	I
210		3294.28	I
45		3296.72	I
260		3300.46	I
4200		3323.09	I
60		3331.09	I
45		3331.24	I
330		3338.54	I
70		3342.90	I
80		3344.20	I
60		3359.90	I
280		3360.80	I
60		3362.18	I
420		3368.38	I
45		3369.68	I
1100		3372.25	I
110		3377.14	I
80		3377.71	I
110		3385.78	I
5600		3396.82	I
820		3399.70	I
160		3406.55	I
820		3412.27	I
60		3420.16	I
330		3421.22	I
120	d	3424.38	I
8200		3434.89	I
1400		3440.53	I
35		3442.63	I
120		3447.74	I
60		3448.58	I
120		3450.29	I
60		3451.15	I
400		3455.22	I
60		3455.42	I
180		3457.07	I
220		3457.93	I
5900		3462.04	I
180		3469.62	I
4700		3470.66	I
120		3472.25	I
4700		3474.78	I
2100		3478.91	I
95		3484.04	I
80		3491.07	I
110		3494.44	I
1200		3498.73	I
5900		3502.52	I
60		3505.41	I
2800		3507.32	I
60		3511.78	I
60		3513.10	I
60		3519.54	I
8800		3528.02	I
880	d	3538.14	I
		3538.26	I
280		3541.91	I
1200		3543.95	I
1800		3549.54	I
240		3564.13	I
1200		3570.18	I
4700		3583.10	I
120		3583.53	I
4700		3596.19	I
5900		3597.15	I
310		3605.86	I
3100		3612.47	I
240		3614.78	I
200		3620.46	I
1800		3626.59	I
95		3627.80	I
310		3639.51	I
350		3654.87	I
8200		3657.99	I
280		3661.86	I
1300		3666.22	I
180		3666.91	I
140		3674.76	I
560		3681.04	I
1900		3690.70	I
9400		3692.36	I
60		3694.95	I
940		3695.52	I
280		3698.26	I
380		3698.60	I
7600		3700.91	I
940		3713.02	I
60		3713.43	I
45		3714.83	I
16		3724.94	I
650		3735.28	I
420		3737.27	I
420		3744.17	I
1200		3748.22	I
240		3754.12	I
380		3754.27	I
490		3755.58	I
1000		3760.40	I
2300		3765.08	I
490		3769.97	I
70		3775.72	I
380		3778.13	I
1000		3788.47	I
1300		3792.18	I
3800		3793.22	I
4900		3799.31	I
760		3805.92	I
1300		3806.76	I
45		3809.50	I
95		3812.45	I
470		3815.01	I
760		3816.47	I
1300		3818.19	I
3800		3822.26	I
2300		3828.48	I
2000		3833.89	I
45		3834.75	I
5900		3856.52	I
490		3870.01	I
70		3872.39	I
380		3877.34	I
70		3888.34	I
29		3904.22	I
23		3912.83	I
120		3913.51	I
240		3922.19	I
2000		3934.23	I
45		3934.98	I
50		3935.84	I
590		3942.72	I
95		3958.24	I
3800		3958.86	I
45		3964.54	II
380		3975.31	I
240		3984.40	I
240		3995.61	I
380		3996.15	I
120		4023.14	I
60		4048.41	I
23		4049.04	I
40		4053.44	I
23		4056.34	I
70		4077.57	I
560		4082.78	I
19		4084.28	I
45		4087.79	I
60		4088.50	I
140		4097.52	I
45		4107.49	I
70		4116.33	I
120		4119.68	I
1100		4121.68	I
1500		4128.87	I
2100		4135.27	I
240		4154.37	I
330		4196.50	I
70		4206.62	I
3300		4211.14	I
29		4230.20	I
40		4244.44	I
60		4273.43	I
60		4278.60	I
820		4288.71	I
70		4296.77	I
23		4342.44	I
45		4373.04	I
4200		4374.80	I
95		4379.92	I
23		4433.32	I
35		4492.47	I
29		4503.78	I
23		4528.72	I
16		4544.27	I
35		4548.73	I
40		4551.64	I
19		4560.89	I
16		4565.19	I
130		4569.00	I
14		4571.31	I
29		4608.12	I
14		4619.91	I
23		4643.18	I
150		4675.03	I
19		4721.00	I
70		4745.11	I
12		4755.58	I
23		4810.49	I
21		4842.43	I
45		4843.99	I
60		4851.63	I
60		4963.71	I
60		4977.75	I
40		4979.18	I
14		5085.52	I
70		5090.63	I
23		5120.69	I
19		5130.76	I
60		5155.54	I
14		5157.09	I
40		5158.69	I
60		5175.97	I
12		5177.27	I
35		5184.19	I
95		5193.14	I
16		5206.95	I
16		5211.52	I
19		5212.73	I
16		5214.79	I
19		5222.66	I
19		5230.62	I
45		5237.16	I
9		5237.80	I
14		5269.27	I
11	h	5280.12	I
14		5292.14	I
14		5314.79	I
40	h	5329.74	I
14	h	5331.08	I
9		5349.31	I
130		5354.40	I
23		5356.47	I
45		5379.10	I
95		5390.44	I
23	h	5404.73	I
60	h	5424.07	I
19		5424.72	I
19	h	5425.45	I
12		5439.58	I
12	h	5441.36	I
9	h	5444.32	I
35	h	5445.23	I
23	h	5468.11	I
35	h	5470.85	I
12		5476.12	I
12		5481.42	I
16		5484.23	I
9		5504.65	I
29		5535.04	I
21	l	5544.58	I
160		5599.42	I
7		5607.71	I
16		5608.35	I
5		5632.77	I
9		5659.62	I
40		5686.38	I
9	h	5702.47	I
6		5727.30	I
29		5792.66	I
9		5795.79	I
9		5803.34	I
40		5806.91	I
6		5821.84	I
35		5831.58	I
7		5907.31	I
9		5918.54	I
7		5941.46	I
130		5983.60	I
9		5991.19	I
35		6102.72	I
6		6116.15	I
8		6128.06	I
8		6186.89	I
14		6199.99	I
16		6253.72	I
5		6276.66	I
8		6277.46	I
6		6293.38	I
29		6319.53	I
12		6414.72	I
16		6510.41	I
19		6519.70	I
9		6627.80	I
19		6630.16	I
40		6752.35	I
9		6796.65	I
13		6827.33	I
11		6857.68	I
20		6879.94	I

Rhodium (Cont.)

Intensity		Wavelength	
65		6965.67	I
8		6972.91	I
16		6979.15	I
16		7001.58	I
11		7038.76	I
18		7101.64	I
15		7104.45	I
6		7142.55	I
9		7219.06	I
18		7268.18	I
35		7270.82	I
12		7271.94	I
5		7273.03	I
9		7375.57	I
5	h	7386.64	I
9		7430.80	I
18	h	7442.39	I
7		7446.77	I
12		7475.74	I
12		7495.24	I
8		7542.02	I
11		7557.67	I
8		7577.22	I
11		7690.05	I
18		7772.90	I
29		7791.61	I
55		7824.91	I
15		7830.05	I
15		7846.50	I
21		8029.91	I
11	h	8036.09	I
29		8045.36	I
7		8063.50	I
15		8136.20	I
7	h	8193.67	I
5	h	8369.67	I
8		8425.59	I

Rh III
Ref. 396 — L.J.R.

Intensity		Wavelength	
		Vacuum	
10		746.28	III
30		759.54	III
50		813.44	III
30		826.01	III
30		843.63	III
30		849.08	III
40		852.70	III
40		854.77	III
40		859.89	III
40		861.34	III
50		863.78	III
50		865.22	III
50		870.40	III
80		882.51	III
100		925.75	III
150		937.28	III
100		976.12	III
500		991.62	III
400		992.48	III
500	d	1009.60	III
200		1012.22	III
200		1015.17	III
100		1050.00	III
100		1058.97	III
200		1073.87	III
100		1100.58	III
100		1113.79	III
100		1768.43	III
150		1784.24	III
200		1784.94	III
150		1796.50	III
200		1816.03	III
1000		1832.05	III
500		1859.85	III
100		1874.70	III
800		1880.66	III
500		1884.91	III
500		1887.36	III
700		1888.62	III
800		1901.32	III
500		1910.16	III
600		1919.37	III
500		1927.07	III
700		1931.79	III
500		1954.25	III
400		1965.16	III
500		1994.26	III
		Air	
400		2005.14	III
800		2013.71	III
500		2017.47	III
500		2028.53	III
800		2036.72	III
600		2037.61	III
1000		2040.18	III
3000		2048.67	III
2000		2064.11	III
800		2076.84	III
1000		2118.53	III
1000		2118.63	III
1000		2139.44	III
1000		2152.23	III
3000		2158.17	III
3000		2163.19	III
3000		2167.33	III
100		2207.00	III
100		2230.66	III
50		2250.84	III
30		2374.84	III
20		2470.65	III
50		3006.43	III
50		3052.44	III
50		3310.69	III
1		3852.98	III

RUBIDIUM (Rb)
Z = 37

Rb I and II
Ref. 12, 130, 241, 257, 264 — J.R.

Intensity		Wavelength	
		Vacuum	
10		474.88	II
40		481.118	II
90		497.430	II
20		508.434	II
150		513.266	II
300		530.173	II
75		533.801	II
40		542.887	II
200		555.036	II
2500		589.419	II
1500		643.878	II
3000		697.049	II
6000		711.187	II
10000		741.456	II
1000		1604.12	II
200		1644.96	II
200		1707.52	II
600		1716.85	II
5000		1760.50	II
200		1803.47	II
500		1809.68	II
500		1865.33	II
500		1889.42	II
500		1954.24	II
300		1956.54	II
200		1971.42	II
500		1983.19	II
		Air	
300		2042.23	II
300		2052.21	II
500		2052.80	II
2000		2068.92	II
1000		2071.50	II
10000		2075.95	II
1000		2090.29	II
200		2108.06	II
300		2116.50	II
1000		2125.25	II
400		2129.82	II
200		2143.10	II
30000		2143.83	II
200		2190.36	II
600		2197.99	II
600		2198.26	II
300		2207.86	II
10000		2217.08	II
200		2223.79	II
400		2237.72	II
500		2250.65	II
200		2251.43	II
800		2254.19	II
200		2254.55	II
200		2263.54	II
500		2263.94	II
500		2286.82	II
5000		2291.71	II
300		2298.80	II
250		2333.01	II
2000		2333.39	II
350		2353.11	II
300		2353.96	II
400		2356.97	II
300		2358.04	II
300		2364.27	II
200		2364.32	II
200		2365.15	II
300		2367.51	II
200		2373.21	II
2000		2385.34	II
250		2405.94	II
400		2434.17	II
800		2459.14	II
50000		2472.20	II
300		2484.56	II
700		2484.70	II
2000		2496.38	II
200		2502.67	II
250		2514.18	II
1000		2524.24	II
200		2594.56	II
400		2623.76	II
400		2645.58	II
1000		2684.10	II
1000		2711.76	II
250		2741.01	II
500		2812.15	II
350		2838.51	II
750		2873.88	II
1000		3051.36	II
2		3082.02	I
250		3088.58	II
10		3112.57	I
3		3113.06	I
5000	c	3148.90	II
25		3157.54	I
5		3158.26	I
1200		3161.00	II
50		3227.98	I
6		3229.16	I
2000		3270.99	II
1500		3321.49	II
1200		3340.55	II
60		3348.72	I
75		3350.82	I
750		3353.89	II
1200		3393.03	II
750		3415.58	II
1000		3434.18	II
1500		3461.50	II
3000		3521.39	II
3000	l	3531.55	II
1000		3541.15	II
100		3587.05	I
40		3591.57	I
5000		3600.60	II
10000		3600.64	II
600	c	3639.80	II
400	c	3646.26	II
350	c	3647.56	II
1000	c	3662.74	II
900	c	3663.81	II
350		3666.72	II
300		3675.66	II
2500	c	3699.58	II
350		3746.33	II
3500		3796.81	II
2500		3801.90	II
1000		3826.66	II
450		3860.74	II
250		3907.29	II
500		3922.20	II
2500	l	3926.44	II
25000		3940.51	II
1000	c	3978.15	II
1700		4029.49	II
2500	c	4083.88	II
2000	c	4104.28	II
1700	c	4136.11	II
3500		4193.08	II
1000		4201.80	I
500		4215.53	I
90000		4244.40	II
500		4266.58	II
250	c	4270.25	II
15000		4273.14	II
2500	c	4287.97	II
1500		4293.97	II
500	c	4306.26	II
1000		4346.96	II
2500		4377.12	II
300		4440.10	II
1000		4469.47	II
400	c	4493.92	II
700		4519.04	II
3000		4530.34	II
500	l	4533.79	II
400		4540.74	II
20000		4571.77	II
3000	c	4622.42	II
350	c	4631.89	II
10000		4648.57	II
500		4659.28	II
1000		4730.45	II
1000		4755.30	II
400	c	4757.82	II
30000		4775.95	II
5000	c	4782.83	II
300	c	4855.34	II
1500	c	4885.59	II
2		5087.987	I
2		5132.471	I
10		5150.134	I
10000		5152.08	II
300		5164.58	II
1		5165.023	I
2		5165.142	I
1		5169.65	I
15		5195.278	I
2		5233.968	I
20		5260.034	I
1		5260.228	I
200		5270.51	II
3		5322.380	I
40		5362.601	I
4		5390.568	I
75		5431.532	I
3		5431.830	I
500		5512.55	II
5000		5522.78	II
6		5578.788	I
5000	c	5635.99	II
40		5647.774	I
20		5653.750	I
3000	d	5699.15	II
60		5724.121	I
3		5724.614	I
200		5739.64	II
75		6070.755	I
200		6135.27	II
30	c	6159.626	I
1000	c	6199.08	II
75	c	6206.309	I
300		6269.40	II
120	c	6298.325	I
5		6299.224	I
10000		6458.33	II
1000		6555.62	II
5000		6560.81	II
3000	l	6775.07	II
100	l	7279.997	I
300	c	7316.52	II
150		7408.173	I
200	l	7618.933	I
300		7757.651	I
60		7759.436	I
90000	c	7800.27	I
5	l	7925.26	I
4		7925.54	I
45000	c	7947.60	I
40	l	8271.41	I
30		8271.71	I
2000		8603.96	II
40	l	8868.512	I
30		8868.852	I
300		8978.88	II
300		9021.77	II
3		9224.64	I
2		9234.25	I
500	c	9246.41	II
300		9338.87	II
200	w	9373.50	II
300		9391.36	II
1000		9479.32	II
700	l	9493.72	II
30	l	9522.65	I
5		9523.05	I
20	l	9540.18	I
300		9612.99	II
300		9671.54	II
2000	c	9689.05	II
200		9776.06	II
200		9934.76	II
35	l	10075.282	I
30	l	10075.708	I
100		13235.17	I
20		13442.81	I
30		13443.57	I
75		13665.01	I
1000		14752.41	I
800		15288.43	I
150		15289.48	I
20		22529.65	I
10		22932.47	I
4		27314.31	I
2		27905.37	I

Rb III
Ref. 258, 262 — J.R.

Intensity		Wavelength	
		Vacuum	
30		465.85	III
35	p	482.43	III
30	p	482.47	III
500		482.83	III
300		484.84	III
500		489.66	III
100		489.96	III
600		493.48	III
50		497.82	III
100		500.28	III
30		508.33	III
400		516.79	III
800		533.64	III
1200		535.86	III
1200		556.19	III
500		558.36	III

Rubidium (Cont.)

Intensity		Wavelength	
700		564.77	III
1500		566.71	III
1000		572.82	III
1500		576.65	III
2500		579.63	III
1500		581.26	III
500		582.34	III
800		586.77	III
100		591.42	III
900		593.65	III
1000		594.94	III
1300		595.88	III
1200		598.49	III
450		602.09	III
50		605.51	III
500		607.28	III
400		613.31	III
500		619.67	III
20		620.83	III
100		622.24	III
250		630.06	III
500		645.67	III
20		674.81	III
5000		769.04	III
2500		815.28	III
		Air	
100		2153.21	III
250		2164.59	III
100		2268.00	III
150		2300.12	III
500		2304.14	III
150		2304.45	III
250		2312.46	III
200		2337.07	III
100		2341.90	III
200		2345.37	III
100		2349.81	III
150		2380.44	III
100		2381.29	III
150		2418.46	III
300		2561.86	III
100		2573.71	III
100		2577.07	III
200		2606.83	III
1000		2631.75	III
350		2636.83	III
100		2656.68	III
100		2713.86	III
500		2798.86	III
150		2800.27	III
500		2807.58	III
100		2845.44	III
150		2869.77	III
500		2903.69	III
150		2949.62	III
100		2951.01	III
2000		2956.07	III
500	l	2967.45	III
150		2968.13	III
500		2970.74	III
250		2987.40	III
350		3023.61	III
200		3039.62	III
200		3041.48	III
250		3070.70	III
500		3086.84	III
100		3098.49	III
500		3111.36	III
250	s	3114.82	III
120		3118.92	III
100		3169.34	III
200		3222.60	III
500		3286.41	III
100		3330.16	III
200		3346.92	III
250		3439.26	III
100		3492.68	III

Rb IV

Ref. 109 — J.R.

Intensity	Wavelength	
	Vacuum	
10	595.18	IV
25	663.76	IV
25	716.24	IV
20	733.41	IV
50	740.85	IV
20	749.86	IV
20	753.75	IV
10	771.54	IV
25	776.89	IV
9	817.92	IV
15	850.18	IV
10	988.00	IV

RUTHENIUM (Ru)
Z = 44

Ru I and II

Ref. 1 — C.H.C.

Intensity		Wavelength	
		Air	
2400		2076.43	I
2600		2083.77	I
2400		2090.89	I
690		2255.52	I
290		2259.53	I
780		2272.09	I
240		2278.19	I
780		2279.57	I
170		2285.38	I
290		2302.54	I
480		2317.80	I
150		2322.01	I
120		2334.96	II
240		2340.69	I
190	h	2342.85	II
190		2349.34	I
310		2351.33	I
170		2357.91	II
140		2360.56	I
170		2370.17	I
240		2375.27	I
80		2375.63	II
160		2392.42	I
95		2396.71	II
780		2402.72	II
150		2407.92	II
55		2410.89	I
55		2414.82	II
130		2420.82	I
55		2422.92	I
45		2429.60	I
65		2432.93	I
30		2447.45	I
30		2450.58	I
65		2454.92	I
180		2455.53	II
150		2456.44	II
370		2456.57	II
65	h	2458.62	I
55		2462.94	I
85		2464.70	I
30		2474.04	I
110		2475.41	I
100		2476.88	I
280		2478.93	II
28		2481.11	II
30		2489.91	I
18		2491.78	I
65		2493.69	II
85		2494.02	I
45		2494.48	II
85		2495.69	II
65		2496.56	I
140		2498.42	II
140		2498.57	II
85		2499.78	I
260		2507.01	II
130		2508.27	I
110		2509.07	I
110		2512.81	I
110		2513.32	II
110		2517.32	II
150		2535.59	II
65		2543.25	II
280		2544.22	I
120		2546.67	I
280		2549.48	I
550		2549.58	I
130		2560.26	I
120		2560.83	I
110		2563.15	I
160		2568.77	I
100		2570.97	I
100		2578.57	I
100		2579.53	I
100		2589.57	I
170		2591.12	I
120		2592.02	I
100		2593.70	I
110		2594.85	I
370		2609.06	I
830		2612.07	I
100		2615.09	I
220		2631.30	I
220		2635.86	I
170		2636.67	I
110		2640.33	I
460		2642.96	I
110		2647.32	I
110		2651.29	I
330		2651.84	I
28		2656.25	II
400		2659.62	I
23		2661.17	II
330		2661.61	II
200		2664.76	I
30		2667.40	II
690		2678.76	II
220		2686.29	I
28		2687.50	II
		2688.16	II
330		2692.06	II
110		2701.34	I
110		2702.83	I
170		2709.20	I
200		2712.41	II
690		2719.52	I
130		2722.65	I
140		2725.47	II
310		2734.35	II
1800		2735.72	I
170		2739.22	I
130		2744.45	I
35		2747.97	II
75		2752.45	II
75		2752.77	II
260		2763.42	I
35		2765.44	II
90		2768.93	II
100		2778.38	II
110		2787.83	II
140		2802.81	I
35		2806.74	II
350		2810.03	I
1700		2810.55	I
350		2818.36	I
110		2822.03	I
200		2827.87	I
400		2829.16	I
130		2834.00	I
150		2840.54	I
35		2841.68	II
640		2854.07	I
180		2860.02	I
420		2861.41	I
550		2866.64	I
110		2868.31	I
1800		2874.98	I
220		2879.76	I
55		2881.12	II
130		2883.60	I
740		2886.54	I
180		2892.56	I
110		2901.94	I
140		2905.65	I
370		2908.88	I
1100		2916.26	I
150		2919.61	I
35		2927.54	II
180		2945.67	II
180		2946.99	I
370		2949.50	I
150		2954.49	I
18		2963.40	II
550		2965.16	I
170		2965.55	II
140		2976.59	II
550		2976.92	I
45		2977.23	II
75		2979.96	II
1400		2988.95	I
35		2991.62	II
110		2993.27	I
460		2994.96	I
440		3006.59	I
330		3017.24	I
310		3020.88	I
240		3033.45	I
200		3040.31	I
220		3042.48	I
110		3045.71	I
110		3048.78	I
150		3054.94	I
390		3064.84	I
170		3089.14	I
120		3089.80	I
330		3096.57	I
120		3097.60	I
830		3099.28	I
740		3100.84	I
120		3125.96	I
120		3153.82	I
290		3159.92	I
200		3168.52	I
60		3177.05	II
180		3186.04	I
240		3188.34	I
240		3189.98	I
180		3196.59	I
180		3223.27	I
110		3226.37	I
100		3227.88	I
220		3228.53	I
220		3238.53	I
120		3241.24	I
120		3243.50	I
280		3260.35	I
120	d	3264.55	I
120		3266.44	I
200		3268.21	I
200		3273.08	I
200		3274.71	I
100		3277.57	I
490		3294.11	I
370		3301.59	I
220		3306.17	I
290		3315.23	I
290		3316.39	I
100		3325.00	I
120		3335.69	I
930		3339.55	I
240		3341.66	I
200		3361.15	I
370		3368.45	I
100		3371.86	I
130		3374.65	I
120		3378.02	I
100		3379.60	I
130		3380.18	I
130		3385.14	I
130		3388.71	I
100		3389.50	I
370		3392.54	I
310		3401.74	I
310		3409.28	I
3100		3417.35	I
4900		3428.31	I
490		3430.77	I
310		3432.74	I
6400		3436.74	I
260		3438.37	I
220		3440.20	I
260		3473.75	I
240		3481.30	I
8300		3498.94	I
640		3514.49	I
330		3519.64	I
200		3528.68	I
240		3532.81	I
390		3537.95	I
790		3539.37	I
200		3541.63	I
690		3570.59	I
200		3574.58	I
390		3587.20	I
6400		3589.22	I
6900		3593.02	I
6400		3596.18	I
1300		3599.76	I
350		3625.20	I
370		3626.74	I
3100		3634.93	I
210		3637.47	I
200		3640.64	I
290		3650.32	I
310		3654.40	I
6200		3661.35	I
830		3663.37	I
650		3669.49	I
240		3678.32	I
260		3696.59	I
410		3717.00	I
260		3719.33	I
550		3726.10	I
8700		3726.93	I
11000		3728.03	I
7100		3730.43	I
280		3737.40	I
410		3739.46	I
3500		3742.28	I
870		3742.78	I
280		3744.22	I
410		3744.40	I
2800		3745.59	I
760		3753.54	I
310		3755.09	I
870		3755.93	I
1200		[illegible]	I
370		3760.03	I
600		3761.51	I
600		3767.35	I
1500		3777.59	I
460		3781.18	I
600		3782.74	I
3900		3786.06	I
6000		3790.51	I
240		3794.92	I
760		3798.05	I
7600		3798.90	I
7600		3799.35	I
310		3800.26	I
310		3808.68	I
600		3812.72	I
760		3817.27	I
760		3819.03	I
650		3822.09	I
550		3824.93	I
760		3831.80	I
220		3835.05	I

Ruthenium (Cont.)

Intensity		Wavelength	
310		3838.07	I
930		3839.70	I
480		3846.68	I
760		3850.43	I
480		3856.46	I
1300		3857.55	I
220		3860.72	I
650		3862.69	I
1300		3867.84	I
260		3873.52	I
650		3892.21	I
760		3909.08	I
260		3920.92	I
1500		3923.47	I
3300		3925.92	I
600		3931.76	I
310		3933.55	I
760		3945.57	I
460		3950.21	I
310		3952.68	I
460		3964.90	I
600		3978.44	I
600		3979.42	I
870		3984.86	I
280		3995.98	I
1500		4022.16	I
600		4023.83	I
310		4039.21	I
1400		4051.40	I
710		4054.05	I
370		4064.46	I
200		4067.61	I
760		4068.37	I
200		4073.00	I
980		4076.73	I
6000		4080.60	I
310		4085.43	I
930		4097.79	I
350		4101.74	I
1900		4112.74	I
2000		4144.16	I
650		4145.74	I
260		4146.77	I
870		4167.51	I
550		4197.58	I
550		4198.88	I
7600		4199.90	I
1500		4206.02	I
5400		4212.06	I
760		4214.44	I
930		4217.27	I
370		4220.68	I
550		4230.31	I
760		4241.05	I
760		4243.06	I
370		4246.73	I
310		4258.99	I
760		4284.33	I
220		4293.28	I
260		4294.79	I
550		4295.93	I
3700		4297.71	I
930		4307.60	I
370		4318.43	I
550		4319.87	I
550		4342.07	I
350		4349.70	I
710		4354.13	I
870		4361.21	I
2400		4372.21	I
870		4385.39	I
1300		4385.65	I
1700		4390.44	I
1600		4410.03	I
160		4421.46	I
330		4428.46	I
460		4439.76	I
440		4449.34	I
1100		4460.04	I
190		4473.93	I
150		4480.45	I
350		4498.14	I
120		4510.10	I
220		4516.89	I
220		4517.82	I
110		4520.95	I
170		4547.33	I
110		4547.85	I
5400		4554.51	I
110		4559.98	I
1700		4584.44	I
110		4591.10	I
150		4592.52	I
330		4599.08	I
170		4635.69	I
200		4645.09	I
720		4647.61	I
290		4654.32	I
290		4681.79	I
190		4684.02	I
290		4690.11	I
1400		4709.48	I
140		4731.33	I
120		4733.52	I
500		4757.84	I
260		4815.52	I
120		4844.56	I
550		4869.15	I
160		4895.60	I
470		4903.05	I
120		4907.89	I
260		4921.07	I
180		4938.43	I
160		4968.90	I
160		4980.35	I
120		4992.74	I
160		5011.23	I
90		5014.95	I
90		5026.18	I
65		5028.16	I
35		5040.35	I
35		5040.74	I
65		5047.31	I
450		5057.33	I
21		5062.64	I
90		5072.97	I
120		5076.32	I
200		5093.83	I
80		5107.07	I
24		5123.73	I
55		5127.26	I
65		5133.89	I
530		5136.55	I
170		5142.76	I
250		5147.24	I
110		5151.07	I
55		5153.20	I
500		5155.14	I
55		5160.00	I
920		5171.03	I
180		5195.02	I
80		5199.87	I
45		5202.12	I
45		5213.43	I
65		5223.55	I
40		5242.38	I
55		5251.67	I
40		5257.07	I
40		5266.47	I
40		5266.83	I
40		5280.82	I
130		5284.08	I
40		5291.16	I
80		5304.86	I
260		5309.27	I
13		5315.33	I
40		5332.93	I
45	h	5334.70	I
110		5335.93	I
130		5361.77	I
65		5377.84	I
65		5385.88	I
110	h	5401.04	I
40		5401.39	I
40		5418.86	I
55		5427.59	I
26	l	5439.21	I
13		5452.71	I
80	h	5454.82	I
90		5456.13	I
13	h	5475.18	I
55		5479.40	I
26		5480.30	I
80		5484.32	I
18		5484.64	I
26		5496.69	I
13		5501.02	I
130		5510.71	I
20		5512.37	I
8		5517.86	I
12		5521.78	I
12		5530.99	I
24		5540.66	I
12		5556.52	I
90		5559.75	I
11		5569.03	I
21		5578.40	I
21		5603.14	I
8		5603.55	I
13		5606.73	I
11		5629.79	I
290		5636.24	I
11		5641.66	I
7		5649.56	I
7		5653.30	I
11		5665.20	I
16		5679.63	I
180		5699.05	I
13		5724.82	I
13		5725.73	I
16		5745.99	I
16		5747.47	I
11		5752.02	I
11		5756.83	I
11		5767.92	I
16		5804.39	I
65		5814.98	I
8		5828.06	I
16	h	5833.21	I
55		5919.34	I
80		5921.45	I
21		5926.87	I
26		5932.38	I
8		5936.65	I
8		5951.15	I
21	h	5973.38	I
8		5974.17	I
16		5988.67	I
35		5993.65	I
18		6116.77	I
26		6199.42	I
26		6225.20	I
9		6284.49	I
18		6295.22	I
13		6330.62	I
9		6336.12	I
9	h	6363.41	I
9		6376.45	I
16		6390.23	I
8		6417.57	I
26	h	6444.84	I
8		6496.44	I
11		6528.74	I
4		6560.45	I
4		6593.74	I
9		6618.20	I
21		6663.14	I
55		6690.00	I
11		6707.52	I
15		6718.30	I
15		6730.45	I
7		6756.54	I
21		6766.95	I
30		6775.02	I
13		6787.23	I
8		6813.51	I
15		6823.88	I
21		6824.17	I
7		6831.52	I
26		6911.48	I
110		6923.23	I
26		6982.01	I
26		7027.98	I
9		7086.06	I
12		7087.35	I
4		7141.72	I
6		7219.26	I
35		7238.92	I
7		7266.96	I
8		7323.56	I
16		7393.93	I
18		7468.91	I
12		7475.40	I
26		7485.79	I
70		7499.75	I
7		7532.07	I
26		7559.61	I
5		7612.94	I
18		7621.50	I
18		7722.87	I
5		7729.91	I
22		7791.86	I
4		7797.89	I
4		7806.82	I
3		7813.43	I
4		7829.81	I
5	h	7833.39	I
6	h	7841.90	I
30		7847.80	I
80		7881.49	I
16		7890.37	I
16		7924.43	I
5		7948.15	I
9		7967.84	I
9		8112.47	I
18		8264.96	I
11		8348.98	I
6		8352.94	I
4		8435.77	I
11		8473.64	I
11		8483.56	I
22		8710.84	I
14		8724.98	I
9		8777.36	I

Ru III
Ref. 423 — C.H.C.

Intensity		Wavelength	
		Vacuum	
250		850.09	III
200		850.30	III
50		851.22	III
50		852.49	III
150		856.32	III
50		867.48	III
250		919.74	III
50		921.78	III
250		928.08	III
150		937.16	III
500		940.09	III
250		940.68	III
50		941.85	III
50		942.63	III
50		943.06	III
150		945.68	III
100		946.05	III
100		947.14	III
100		949.83	III
50		950.35	III
100		950.45	III
50		952.59	III
50		957.06	III
50		957.18	III
50		961.58	III
250		961.68	III
100		962.56	III
500		966.54	III
250		967.09	III
150		967.85	III
150		967.92	III
150		971.83	III
250		972.40	III
100		973.54	III
150		973.78	III
750		974.14	III
250		974.46	III
250		977.51	III
100		978.18	III
900		979.43	III
500		981.35	III
250		983.81	III
250		983.91	III
250		985.55	III
900		986.84	III
200		987.87	III
250		991.67	III
250		992.75	III
900		994.56	III
250		995.30	III
200		1000.78	III
300		1001.65	III
250		1004.29	III
500		1009.13	III
900		1009.87	III
500		1014.68	III
100		1018.72	III
100		1019.33	III
100		1020.77	III
200		1080.00	III
100		1184.37	III
800		1190.51	III
500		1200.07	III
100		1204.57	III
200		1204.88	III
500		1207.17	III
500		1209.77	III
300		1211.31	III
200		1232.57	III
100	h	1653.77	III
200		1699.84	III
100	h	1715.97	III
200		1759.49	III
200	h	1880.95	III
100	h	1883.56	III
200	h	1899.04	III
100	h	1899.42	III
100		1908.31	III
500		1941.35	III
100		1981.82	III
100		1982.10	III
200		1989.22	III
200		1993.32	III
100		1997.55	III
		Air	
200		2005.71	III
100		2006.46	III
500		2009.28	III
100		2011.17	III
50		2011.56	III
50		2011.66	III
100		2015.20	III
50		2018.58	III
100		2044.59	III

SAMARIUM (Sm)
Z = 62

Sm I and II
Ref. 1 — C.H.C.

Intensity		Wavelength	
		Air	
45		2610.07	
90		2640.27	

Samarium (Cont.)

35		2649.17	
45		2657.68	
70		2662.42	
120		2675.15	
100		2688.60	
45		2690.90	II
130		2693.34	
45		2693.74	
60		2696.08	
85		2707.96	
50		2732.42	
35		2739.87	
29		2762.28	
35		2764.18	
85		2767.85	II
60		2774.77	
85		2776.11	
85		2779.23	II
85		2786.64	
150		2789.38	II
130	h	2796.70	
85		2807.36	
150		2809.50	
120		2810.86	II
85		2817.20	II
29		2820.96	II
220		2830.94	
60		2840.30	
60		2847.49	II
60		2851.35	
120		2866.09	II
70		2868.40	II
70		2881.34	
85		2881.68	
60		2883.09	
45		2889.06	
60		2891.34	
85	d	2907.88	
		2907.99	II
130		2910.28	II
85		2937.48	II
70		2943.49	II
150		2953.19	II
85		2962.74	II
160		2969.02	II
100		2983.43	II
60		2991.57	II
100		3021.01	
150		3034.84	II
100		3039.13	II
120		3046.93	II
150		3067.54	
120		3071.29	II
100		3086.45	II
120		3096.88	II
100	h	3102.30	II
250		3106.52	II
220		3110.20	II
200		3117.72	II
270		3136.30	II
150		3139.97	II
150		3147.19	II
180		3152.10	II
410		3152.52	II
150		3162.30	II
360		3169.88	II
180		3178.12	II
720		3183.92	II
310		3187.01	II
430		3187.22	II
360		3187.79	II
360		3193.01	II
360		3196.18	II
150		3201.80	II
150		3204.90	II
360		3207.18	II
180		3208.17	II
600		3211.73	II
150		3214.12	II
270		3215.26	II
530		3216.85	II
600		3218.61	II
150		3219.43	II
270		3226.84	II
180		3228.50	II
270		3228.78	II
720		3230.56	II
360		3231.53	II
150		3231.95	II
430		3233.68	II
720		3236.64	II
150		3237.89	II
720		3239.66	II
530		3241.16	II
180		3241.59	II
180		3242.04	II
150		3244.69	II
240		3249.75	II
720		3250.37	II
360		3253.40	II
270		3253.94	II
850		3254.38	II
110		3255.63	II
360		3262.28	II
430		3264.94	II
180		3270.49	II
180		3270.68	II
430	d	3272.48	II
		3272.60	II
430		3272.81	II
430		3273.48	II
430		3276.75	II
270		3280.84	II
180		3285.66	II
430		3286.23	II
720	d	3290.28	II
		3290.39	II
180		3290.65	II
240		3293.37	II
360		3295.44	II
430		3295.81	II
720		3298.10	II
170		3300.98	II
340		3301.68	II
340		3304.52	II
340		3305.18	II
1700		3306.39	II
170		3306.61	II
850		3307.02	II
340		3309.52	II
850		3310.66	II
600		3312.42	II
410		3316.58	II
430		3320.16	II
110		3320.59	II
1200		3321.18	II
340		3323.77	II
340		3325.26	II
170		3325.48	II
340		3327.88	II
170		3333.64	II
170		3336.12	II
850		3340.58	II
240		3343.49	II
110		3343.64	II
240		3344.35	II
170		3347.30	II
240		3348.68	II
220		3350.88	II
410	d	3354.18	II
		3354.30	II
170		3354.72	II
1200		3365.86	II
150		3367.27	II
340		3368.57	II
340		3369.46	II
170		3370.59	II
340		3371.21	II
150		3376.48	II
1200		3382.40	II
510		3384.66	II
150		3384.86	
150		3387.66	II
410		3389.32	II
150		3391.11	II
410		3396.19	II
150		3397.76	II
150		3399.84	II
600		3402.46	II
210		3403.09	II
850		3408.68	II
270		3418.15	II
430		3418.51	II
170		3419.77	II
120		3424.78	II
170		3426.20	II
170		3433.68	II
150		3437.10	II
170		3438.06	II
240		3440.50	II
170		3453.56	II
170		3459.20	II
120		3459.42	II
240		3461.13	II
120		3464.07	II
170		3467.87	II
130		3473.96	II
130		3479.53	II
130		3480.26	II
170		3480.56	II
170		3487.41	II
170		3493.61	II
220		3499.84	II
340		3511.23	II
310		3530.60	II
220		3532.57	II
270		3535.65	II
240		3554.15	II
510		3559.10	II
220		3566.84	II
4200		3568.27	II
270		3577.79	II
390		3580.94	II
310		3583.39	II
4200		3592.60	II
340		3601.69	II
1700		3604.28	II
3400		3609.49	II
240		3620.58	II
1700		3621.23	II
240		3623.32	II
850		3627.01	II
850		3631.13	II
3400		3634.29	II
240		3634.93	II
410		3638.77	II
360		3645.29	II
300		3645.39	II
660		3649.53	II
340		3650.19	II
340		3656.22	II
2200		3661.36	II
220		3662.69	II
340		3667.93	II
340		3670.66	II
2200		3670.84	II
340		3677.79	II
270		3681.73	II
270		3688.42	II
270		3692.22	II
1100		3693.99	II
480		3706.75	II
480		3706.98	II
480		3708.41	II
930		3708.65	II
480		3711.54	II
350		3712.76	II
930		3718.88	II
930		3721.85	II
420		3724.90	II
1600		3728.47	II
2100		3731.26	II
1600		3735.98	II
800		3737.14	II
320		3737.48	II
2900		3739.12	II
		3739.20	II
800		3741.29	II
1200		3743.87	II
930		3745.46	I
		3745.60	II
480		3747.62	II
800		3755.28	II
800		3756.41	I
1200		3757.53	II
450		3758.45	II
660		3758.97	II
350		3760.04	II
1900		3760.69	II
660		3762.59	II
1100		3764.37	II
480		3767.36	II
480		3767.76	II
370	d	3773.33	I
		3773.42	II
1100		3778.14	II
660		3780.76	II
420		3780.93	II
320		3787.20	II
1500		3788.12	II
1600		3793.97	II
420		3797.28	II
1600		3797.73	II
500		3799.54	II
800		3800.89	II
320		3805.63	II
420		3808.46	II
320		3809.75	II
320		3809.88	II
420		3810.43	II
500		3812.07	II
480		3813.63	II
420		3814.63	II
930	d	3820.82	
530		3824.18	II
1600		3826.20	II
530		3830.29	II
1100		3831.50	II
530		3833.83	II
560		3834.48	I
560		3834.60	II
370		3835.72	II
500		3838.94	II
400		3840.45	II
1600		3843.50	II
530		3847.51	II
640		3848.78	II
420		3851.88	II
530		3853.30	I
2700		3854.21	II
480		3854.56	I
800		3855.90	II
480		3857.91	II
400		3858.74	I
660		3862.05	II
350		3862.23	II
320		3865.24	II
800		3871.78	II
400		3875.19	II
560		3875.54	II
800		3880.77	II
450		3881.38	II
450		3881.79	II
320		3882.50	II
3700		3885.29	II
660		3889.16	II
		3889.22	II
610		3890.08	II
320		3891.21	II
400		3894.05	II
1600		3896.98	II
1300		3903.42	II
620		3917.44	II
2500		3922.40	II
1900		3928.28	II
470		3935.76	II
1300		3941.87	II
620		3943.24	II
500		3946.51	II
740		3948.11	II
470		3951.89	I
370		3959.53	II
1500		3963.00	II
620		3966.04	II
470		3967.68	II
740		3970.53	II
1500		3971.40	II
620		3974.66	I
960		3976.27	II
1000		3976.43	II
960		3979.20	II
740		3983.14	II
740		3986.68	II
370		3987.43	II
1500		3990.00	II
		3990.02	I
740		3993.31	II
280		4003.46	II
470		4007.48	II
280		4019.98	II
880		4023.23	II
740		4035.11	II
590		4041.68	II
740		4042.72	II
880		4042.90	II
240		4044.11	II
560		4045.05	II
440		4046.16	II
740		4047.16	II
210		4048.62	II
590		4049.81	II
440		4058.87	II
560		4063.54	II
280		4064.32	II
1400		4064.58	II
810		4066.74	II
710		4068.33	II
810		4075.84	II
280		4076.65	II
240		4080.56	II
410		4082.60	II
280		4083.58	II
220		4084.40	II
1000		4092.27	II
290		4094.05	II
240		4104.13	II
810		4107.28	II
		4107.39	II
410		4109.40	II
280		4110.19	II
410		4113.90	II
1900		4118.55	II
410		4121.36	II
280		4122.51	II
710		4123.96	II
280		4129.23	II
250		4135.14	II
320		4147.71	II
810		4149.83	II
1200		4152.21	II
530		4153.33	II
560		4155.22	II
810		4169.48	II
410		4171.57	II
440		4178.02	II
530		4181.10	II
210		4183.33	I
530		4183.76	II
1000		4188.13	II
410		4191.93	II
270		4199.45	II
650		4202.92	II
1100		4203.05	II
660		4206.13	II
270		4206.62	II
660		4210.35	II
740		4220.66	II
1000		4225.33	II
740		4229.70	II
620		4234.57	II
1200		4236.74	II
500		4237.66	II
620		4244.70	II
210		4249.55	II
250		4251.78	II

Samarium (Cont.)

2100		4256.39	II
210		4258.58	II
1300		4262.68	II
500		4265.08	II
1200		4279.68	II
		4279.75	II
240		4279.94	II
2200		4280.79	II
710		4282.21	I
470		4282.83	I
240		4283.50	I
350		4286.64	II
350		4292.18	II
1600		4296.74	I
320		4304.94	II
880		4309.01	II
240		4312.85	I
1900		4318.94	II
470		4319.53	I
590		4323.28	II
240		4324.46	I
1800		4329.02	II
440		4330.02	I
1300		4334.15	II
880		4336.14	I
560		4345.86	II
1100		4347.80	II
560		4350.46	II
560		4352.10	II
560		4360.72	II
220		4361.07	II
810		4362.04	II
440		4362.91	I
220		4363.45	II
500		4368.03	II
210		4369.92	II
440		4373.46	II
320		4374.98	II
880		4378.24	II
530		4380.42	I
290		4384.29	II
1600		4390.86	II
210		4393.35	I
290		4397.34	I
410		4401.17	I
810	d	4403.06	II
		4403.13	I
410		4403.36	II
520		4409.33	II
290		4411.58	I
380		4417.58	II
470		4419.33	I
1500		4420.53	II
960		4421.14	II
2900		4424.34	II
470		4429.66	I
1600		4433.88	II
1800		4434.32	II
530		4441.81	I
440		4442.28	I
710		4444.26	II
710		4445.15	I
1300		4452.73	II
250		4452.95	I
1200		4454.63	II
1000		4458.52	II
250		4459.29	I
2200		4467.34	II
810		4470.89	I
470		4472.43	II
620		4473.02	II
740		4478.66	II
370		4499.11	I
370		4499.48	II
240		4503.38	I
180		4505.05	II
120		4511.33	I
560		4511.83	II
440		4515.09	II
880		4519.63	II
440		4523.04	II
		4523.18	I
650		4523.91	II
290		4533.80	I
270		4536.51	II
710		4537.95	II
150		4538.53	II
290		4540.19	II
380		4542.06	II
810		4543.95	II
100		4544.83	II
410		4552.66	II
270		4554.45	II
240		4560.43	II
470		4566.21	II
590		4577.69	II
290		4581.58	I
440		4581.73	I
560		4584.83	II
290		4591.82	II
380		4593.54	II
560		4595.29	II
240		4596.74	I
220		4604.18	II
290		4606.51	II
290		4615.44	II
470		4615.69	II
150		4630.21	II
880		4642.24	II
290		4645.40	I
290		4646.68	II
240		4648.16	II
380		4649.49	I
150		4655.13	II
290		4663.56	I
740		4669.40	II
620		4669.65	II
470	d	4670.75	I
		4670.83	I
1100		4674.60	II
680		4676.91	II
210		4681.55	I
370		4687.18	II
370		4688.73	I
130		4693.63	II
120		4699.34	II
530		4704.40	II
270		4713.06	II
130		4715.26	II
730		4716.10	I
270		4717.07	I
210		4717.72	II
190		4718.33	II
270		4719.84	II
130		4726.02	II
770		4728.42	I
470		4745.68	II
150		4750.72	I
730		4760.27	I
110		4770.20	I
110		4774.15	II
190		4777.85	II
580		4783.10	I
350		4785.86	I
160		4789.96	I
230		4791.58	II
430		4815.81	II
130		4829.57	II
970		4841.70	I
310		4844.21	II
140		4847.76	II
270		4848.32	I
120		4854.36	II
210		4883.77	I
730		4883.97	I
170		4904.97	I
630		4910.40	I
350		4913.25	II
430		4918.99	I
110		4924.04	I
120		4938.10	II
170		4948.63	II
120		4952.37	II
170		4961.94	II
170		4975.98	I
140		5028.44	II
400		5044.28	I
200		5052.76	II
170		5069.46	II
540		5071.20	I
170		5100.22	II
		5100.39	I
260		5103.09	II
140		5104.48	II
140		5116.70	II
510		5117.16	I
350		5122.14	I
360		5155.03	II
250		5172.74	I
470		5175.42	I
250		5200.59	I
260		5251.92	I
400		5271.40	I
250		5282.91	I
190		5320.60	I
110		5341.29	I
140		5368.36	I
130		5405.23	I
220		5453.00	I
140		5466.72	I
230		5493.72	I
80		5512.10	I
230		5516.09	I
50		5548.95	I
140		5550.40	I
45		5573.42	I
35		5588.20	I
50		5600.86	II
50		5621.79	I
70		5626.01	I
85		5644.10	I
140		5659.86	I
120		5696.73	I
85		5706.20	I
35		5710.93	I
50		5732.95	I
50		5743.35	II
45		5759.52	II
70		5773.77	I
60		5778.33	I
45		5779.24	I
45		5781.93	II
70	d	5786.98	II
60		5788.38	I
60		5800.52	I
65		5802.84	I
45		5814.89	I
45		5831.02	II
45		5836.37	II
35		5860.78	I
65		5867.79	I
45		5868.61	I
35		5871.06	I
50		5874.21	I
45		5897.39	II
50		5898.96	I
35		5938.90	II
65		5965.71	II
35	h	5968.82	II
35		5984.29	I
50		6045.00	I
45		6045.39	I
50		6070.06	I
45		6084.12	I
35	h	6091.40	I
45		6110.66	II
45	h	6159.56	I
45		6246.76	II
45		6256.54	I
45		6256.66	II
100		6267.28	II
50		6291.82	II
35		6307.06	II
70		6327.47	II
45		6426.64	II
45		6472.34	II
35		6484.52	II
35		6498.67	II
50		6542.76	II
140		6569.31	II
35	h	6570.67	II
40	h	6585.21	II
110		6589.72	II
40		6601.83	II
95		6604.56	II
40	h	6632.28	II
50		6671.51	I
70		6679.21	II
70		6693.55	II
40	d	6723.07	I
120	d	6731.84	II
70	d	6734.06	II
40	d	6734.81	II
55		6741.47	II
40	h	6778.61	II
60		6790.00	II
95		6794.20	II
55		6844.71	II
75		6856.03	II
120		6860.93	I
40		6862.82	II
30		6950.51	II
120		6955.29	II
90		7020.44	II
13		7036.73	II
90		7039.22	II
90		7042.24	II
13		7049.15	II
90		7051.52	II
16		7054.97	II
19		7074.67	I
90		7082.37	II
40	d	7085.52	II
26		7088.30	I
16		7091.16	I
30		7095.50	I
16		7096.33	I
30		7104.54	I
19		7106.23	I
26		7115.96	I
23		7117.51	II
26	h	7119.81	II
12*		7122.40	II
23	h	7125.11	II
13		7131.80	I
10		7136.01	I
12		7139.39	II
40	d	7143.98	II
85	d	7149.60	II
10		7172.67	I
10		7189.57	II
9		7210.95	I
23		7213.82	I
26	d	7218.09	II
13		7220.07	I
13		7237.02	II
60		7240.90	II
9		7257.11	II
9	d	7261.52	II
13		7279.25	I
26		7281.47	II
8		7282.21	I
19		7283.33	II
16		7288.92	II
13		7290.23	I
26	h	7300.72	II
13		7327.08	II
13		7332.65	I
8		7338.04	I
26		7347.30	I
26		7376.69	II
13		7393.98	II
30		7444.56	I
26		7445.41	I
26	d	7453.03	II
13		7470.76	I
26		7481.99	II
23	h	7502.39	II
10	h	7517.00	II
23	h	7541.42	II
9		7544.74	I
10		7546.57	I
12	h	7560.03	II
19		7562.94	II
23		7570.95	II
23		7572.29	II
19		7578.09	II
30		7585.85	II
23		7588.31	II
10		7598.01	I
23	d	7607.48	II
		7607.74	I
12		7613.94	II
10	h	7631.77	II
23		7637.94	II
45		7645.09	II
12		7645.82	I
19		7648.02	II
10		7655.78	II
19		7667.20	II
8		7672.49	II
10	h	7678.79	II
10	h	7695.78	I
23		7712.04	II
30		7728.56	II
30		7736.26	II
30		7749.30	II
23		7755.20	II
10		7794.50	I
10		7801.54	I
8	h	7812.75	II
16		7820.15	II
10		7831.40	II
40	w	7835.08	II
26		7837.27	II
10		7844.82	II
6		7859.53	I
19		7863.65	II
10	h	7880.01	II
16		7895.96	I
26		7914.96	II
90		7928.14	II
9		7931.92	I
19		7937.09	II
16		7948.12	II
19	w	8001.61	II
19	w	8014.92	II
23		8025.12	II
23	w	8026.32	II
16		8032.03	II
40		8048.70	II
16		8065.16	I
45		8068.46	II
9	w	8117.16	II
9		8125.12	II
26		8161.82	II
19	w	8195.50	II
6		8206.30	II
26	w	8218.76	II
9		8230.33	I
16		8240.98	II
19	w	8289.26	II
10		8300.88	II
40	w	8305.79	II
10		8315.45	I
19	w	8348.68	II
19		8383.71	I
19		8387.77	II
30	w	8432.64	II
19	w	8473.54	II
45	w	8485.99	II
30	w	8510.90	II
23		8543.22	II
23	w	8617.03	II
23	w	8632.82	II
12	w	8677.81	II
13		8706.32	II
45	w	8708.43	II
30	w	8717.89	II
30	w	8758.28	II
16	w	8780.59	II
23	w	8788.83	II
26		8859.76	II
95		8913.66	II

SCANDIUM (Sc)
Z = 21

Sc I and II
Ref. 1, 88 — C.H.C.

Intensity		Wavelength Air	
65		2429.16	I
110		2438.62	I
560		2545.22	II
2900		2552.37	II
560		2555.82	II
2300		2560.25	II
1100		2563.21	II
40		2611.22	II
19		2684.23	II
120		2692.78	I
360		2706.77	I
210		2707.95	I
580		2711.35	I
30		2819.54	II
35		2822.15	II
60		2826.68	II
340		2965.86	I
1200		2974.01	I
1400		2980.75	I
340		2988.95	I
2200		3015.36	I
2700		3019.34	I
360		3030.76	I
30		3039.93	II
70		3045.72	II
85		3052.93	II
120	h	3056.31	I
130		3065.11	II
45		3139.75	II
990		3251.32	II
1500		3255.69	I
4400		3269.91	I
5500		3273.63	I
110	d	3343.28	II
270		3352.05	II
9900		3353.73	II
65	d	3357.30	II
2000		3359.68	II
1700		3361.27	II
1700		3361.94	II
4000		3368.95	II
6600		3372.15	II
90		3416.68	I
130		3418.51	I
65		3419.36	I
200		3429.21	I
200		3429.48	I
270		3431.36	I
530		3435.56	I
90		3439.41	I
65		3440.18	I
65		3448.49	I
270		3457.45	I
180		3462.19	I
130	d	3469.65	I
110		3471.13	I
200		3498.91	I
2700		3535.73	II
6600		3558.55	II
6100		3567.70	II
13000		3572.53	II
9900		3576.35	II
7700		3580.94	II
4000		3589.64	II
4000		3590.48	II
28000		3613.84	II
110		3617.43	I
20000		3630.75	II
13000		3642.79	II
6600		3645.31	II
110		3646.90	I
5300		3651.80	II
110		3664.25	II
290		3666.54	II
55		3675.26	II
40		3678.35	II
75	h	3717.10	I
270		3833.07	II
610		3843.03	II
90		3894.97	I
20000		3907.49	I
23000		3911.81	I
45		3923.51	II
4400		3933.38	I
45		3952.27	I
45		3989.06	II
5500		3996.61	I
530		4014.49	II
20000		4020.40	I
20000		4023.69	I
220		4030.67	I
140		4031.39	I
100		4034.23	I
220		4043.80	I
200		4046.48	I
2700		4047.79	I
120		4049.95	I
5500		4054.55	I
220		4056.59	I
160	h	4074.97	I
160		4078.57	I
6100		4082.40	I
200		4086.67	I
400		4087.16	I
40	h	4093.13	I
65		4094.85	I
55	h	4098.35	I
65		4100.33	I
440	h	4133.00	I
530	h	4140.30	I
65	h	4147.40	I
720		4152.36	I
55	h	4154.72	
90	Hd	4161.88	I
1100	h	4165.19	I
65	h	4171.56	I
45	h	4186.45	I
65	h	4187.62	I
75		4205.20	I
65		4212.34	I
45		4212.49	I
75	h	4216.10	I
110	h	4218.26	I
110	h	4219.73	I
40		4221.88	I
90	d	4225.59	I
180		4231.93	I
200		4233.61	I
100		4237.82	I
400		4238.05	I
90		4239.57	I
100		4246.12	I
15000		4246.83	II
55		4283.56	
290		4294.77	II
350		4305.71	II
4200		4314.09	II
3300		4320.74	II
8100		4325.01	II
28		4348.53	I
180		4354.61	II
110		4358.64	I
55		4359.08	I
28		4364.92	I
2000		4374.46	II
130		4384.81	II
45	h	4389.60	I
1100		4400.37	II
880		4415.56	II
28		4420.66	II
45		4431.36	II
65		4542.55	I
90		4544.68	I
120	h	4557.24	I
160	h	4573.99	I
65	h	4592.94	I
65	h	4598.45	I
55	h	4604.72	I
45		4609.53	I
45		4609.95	I
350		4670.40	II
40	h	4680.49	I
50		4698.29	II
120		4706.97	I
120		4709.34	I
200		4728.77	I
490		4729.23	I
40	h	4732.30	I
590		4734.10	I
60		4735.08	
690		4737.65	I
790		4741.02	I
1200		4743.81	I
200		4753.16	I
220		4779.35	I
90		4791.50	I
100		4827.28	I
100		4833.67	I
170		4839.44	I
40		4840.47	I
80		4847.68	I
80		4852.68	I
140	BLd	4857.79	S
		4858.09	SCO
80		4906.67	I
90		4909.76	I
90		4922.84	I
90		4934.25	I
45		4935.74	I
70		4941.33	I
170		4954.06	I
120		4973.66	I
150		4980.37	I
80		4983.45	I
140		4991.92	I
80		5018.39	I
70		5020.14	I
80		5021.51	I
530		5031.02	II
55		5032.74	I
250		5064.32	I
80		5068.86	I
530		5070.23	I
250		5075.81	I
2100		5081.56	I
1200		5083.72	I
1100		5085.55	I
750		5086.95	I
390		5087.14	I
270		5089.89	I
45		5092.46	I
390		5096.73	I
620		5099.23	I
370		5101.12	I
180		5109.06	I
150		5112.86	I
320		5116.69	I
70	b	5133.68	S
45	b	5171.06	S
390		5210.52	I
45		5211.28	I
280		5219.67	I
350		5239.82	II
280		5258.33	I
35		5284.97	I
210		5285.76	I
35		5301.94	I
22		5318.35	II
70		5331.77	I
14		5334.23	II
95		5339.41	I
120		5341.05	I
95		5342.96	I
350		5349.30	I
120		5349.71	I
60		5350.30	I
210		5355.75	I
530		5356.10	I
14		5357.19	II
270		5375.35	I
370		5392.08	I
45		5416.12	I
45		5425.57	I
45		5429.41	I
35		5432.94	I
55		5433.23	I
45		5438.22	I
55		5439.03	I
55	h	5442.60	I
270		5446.20	I
18		5447.39	I
120		5451.34	I
30		5455.21	I
18		5465.20	I
55		5468.40	I
60		5472.19	I
18		5474.64	I
750		5481.99	I
530		5484.62	I
570		5514.22	I
16		5515.39	I
660		5520.50	I
45		5526.06	I
660		5526.82	II
55		5541.04	I
30		5546.40	I
18		5550.40	I
5		5552.25	II
35		5553.59	I
16		5561.10	I
70		5564.86	I
18		5571.24	I
14		5579.76	I
110		5591.33	I
35	h	5593.38	I
22		5604.19	I
22		5631.02	I
80		5640.98	II
45		5646.36	I
16		5647.60	I
55		5649.56	I
250		5657.88	II
60		5658.34	II
55		5667.16	II
70		5669.04	II
1500		5671.81	I
95		5684.20	II
1200		5686.84	I
1100		5700.21	I
190		5708.61	I
880		5711.75	I
230		5717.28	I
180		5724.08	I
55	Bl	5736.85	S
55	Bl	5764.45	S
95	Bl	5772.74	S
55	Bl	5775.32	S
70	Bl	5809.84	S
70	Bl	5811.60	S
95	Bl	5847.73	S
70	Bl	5849.07	S
70	b	5887.38	S
35	Bl	5918.04	S
30		5919.11	I
60	Bl	5928.10	S
35		5961.49	I
60	Bl	5968.25	S
35		5969.19	I
90		5988.42	I
160	Bl	6017.07	S
60		6026.18	I
620	Bl	6036.17	S
490	Bl	6064.31	S
440	Bl	6072.65	S
620	Bl	6079.30	S
320	Bl	6101.87	S
370	Bl	6109.93	S
370	Bl	6115.97	S
180	b	6148.70	S
150	b	6153.93	S
150	b	6188.09	S
150	b	6192.90	S
620		6210.68	I
90		6239.41	I
320		6239.78	I
120		6245.63	II
110		6249.96	I
250		6258.96	I
60		6262.25	I
55		6276.31	I
45		6279.76	II
18		6300.70	II
750		6305.67	I
26		6309.90	II
16		6320.85	II
26		6344.83	I
60		6378.82	I
55	Bl	6408.41	S
90		6413.35	I
26	b	6437.08	S
55	Bl	6446.24	S
26	b	6457.78	S
35	b	6485.40	S
26	b	6495.90	S
55	b	6525.62	S
22	b	6535.30	S
45	b	6557.84	S
35	b	6566.88	S
18	b	6575.85	S
60		6604.60	II
26	Bl	6609.99	S
18	Bl	6617.94	S
18	Bl	6645.08	S
22	Bl	6654.42	S
26	Bl	6661.01	S
18	b	6700.48	S
18	b	6705.93	S
65		6737.87	I
35		6739.40	I
35		6817.08	I
50		6819.52	I
29		6829.54	I
50		6835.03	I
5	b	6963.12	S
5	Bl	6990.68	S
5	Bl	7025.72	S
8	b	7035.77	S
5	b	7072.37	S
5	b	7094.38	S
12	h	7138.14	I
14		7169.13	I
12		7257.57	I
8		7275.57	I
3	h	7300.62	I
12	h	7524.13	I
14	h	7553.96	I
15	h	7574.44	I
11		7617.45	I
14	h	7665.72	I
30		7697.73	I
18		7729.72	I
55	h	7741.17	I
5	h	7750.37	I
5		7752.72	I
6	h	7771.06	I
15		7785.17	I
8		7794.68	I
30		7800.44	I
11		7821.64	I
11	h	8196.98	I
15		8241.13	I
19	h	8761.40	I
11	h	8774.8	
15	h	8794.72	I
15	h	8823.8	I
30	h	8834.35	I
70		20616.32	
30		20985.81	
400		22051.86	I
150		22065.05	I

Scandium (Cont.)

Sc III
Ref. 323 — C.H.C.

Vacuum

Intensity	Wavelength	
10	730.60	III
15	731.65	III
15	1148.24	III
20	1154.52	III
20	1162.44	III
25	1168.61	III
10	1168.88	III
80	1598.00	III
180	1603.06	III
150	1610.19	III
40	1895.44	III
60	1912.62	III
90	1993.89	III

Air

Intensity	Wavelength	
160	2010.42	III
50	2012.26	III
350	2699.07	III
230	2734.05	III
10	2831.75	III
80	4061.21	III
100	4068.66	III
40	4309.47	III
10	4740.95	III
15	4780.87	III
50	4992.89	III
60	5032.09	III
80	6256.01	III
60	6307.60	III
90	7449.16	III
70	7548.15	III
70	7868.65	III
35	8814.29	III
50	8829.78	III
30	8865.89	III
15	8881.58	III

Sc IV
Ref. 298 — C.H.C.

Vacuum

Intensity		Wavelength	
8		220.28	IV
15		289.85	IV
15		296.31	IV
15		299.04	IV
10		371.16	IV
9		438.80	IV
8		557.50	IV
8		584.83	IV
9		617.08	IV
8		761.43	IV
8		769.70	IV
10		785.12	IV
8		789.00	IV
8		791.71	IV
8		861.24	IV
8		861.30	IV
8		890.87	IV
8		1219.40	IV
9		1228.20	IV
9		1424.66	IV
9		1444.10	IV
8		1489.64	IV
8		1514.96	IV
8		1535.76	IV
9		1543.86	IV
9		1549.55	IV
15		1550.80	IV
8		1555.72	IV
9		1563.81	IV
10		1574.92	IV
9		1583.41	IV
8		1584.64	IV
8		1592.23	IV
8		1660.71	IV
10		1665.92	IV
8		1746.23	IV

Air

Intensity		Wavelength	
10		2056.06	IV
8		2078.93	IV
12		2118.97	IV
9		2164.43	IV
11		2185.43	IV
11		2205.46	IV
14		2222.22	IV
11		2271.33	IV
9		2464.45	IV
8		2520.93	IV
11		2586.93	IV
9		2595.17	IV
8		2678.01	IV
8		2723.52	IV
8		2773.04	IV
8	d	4594.42	IV
8	d	4639.96	IV
8		5501.74	IV
9		5620.72	IV
10		5706.82	IV
14		5771.63	IV
9		6548.03	IV

Sc V
Ref. 150 — C.H.C.

Vacuum

Intensity		Wavelength	
150		179.42	V
350		180.14	V
200		180.82	V
200		180.96	V
50		181.55	V
200		182.39	V
300		228.56	V
100		230.85	V
40		243.82	V
500		243.87	V
400		246.42	V
400		250.98	V
500		252.85	V
500		253.73	V
50		255.38	V
300		255.64	V
200		257.16	V
150		258.24	V
40		258.81	V
50		260.05	V
400		281.00	V
900		283.91	V
800		284.45	V
600		288.29	V
900		289.59	V
1000	d	291.93	V
800		293.25	V
400		296.17	V
700		300.00	V
400		375.05	V
100		378.68	V
200		388.68	V
400		395.32	V
200		399.50	V
1000		573.36	V
600		587.94	V

SELENIUM (Se)
Z = 34

Se I and II
Ref. 80, 181, 216, 275 — R.L.K.

Vacuum

Intensity	Wavelength	
285	828.5	II
360	832.7	II
285	906.6	II
360	912.9	II
360	1013.4	II
360	1014.0	II
450	1033.6	II
450	1049.6	II
360	1057.4	II
285	1097.8	II
360	1141.9	II
220	1156.0	II
285	1156.9	II
285	1168.5	II
450	1192.3	II
220	1205.7	II
220	1234.9	II
285	1291.0	II
285	1308.9	II
100	1405.4	I
100	1406.4	I
100	1406.6	I
120	1435.3	I
120	1435.8	I
100	1444.8	I
100	1446.8	I
100	1447.0	I
150	1449.2	I
120	1456.3	I
150	1500.9	I
250	1530.4	I
150	1531.3	I
200	1531.8	I
120	1547.1	I
120	1560.3	I
150	1575.3	I
150	1577.6	I
150	1577.9	I
150	1579.5	I
200	1580.0	I
150	1587.5	I
150	1593.2	I
250	1606.5	I
100	1610.7	I
100	1611.3	I
200	1617.4	I
150	1621.2	I
100	1622.7	I
120	1626.2	I
150	1643.4	I
250	1671.2	I
250	1675.3	I
250	1690.7	I
250	1793.3	I
300	1795.3	I
300	1855.2	I
250	1858.8	I
400	1898.6	I
350	1913.8	I
300	1919.2	I
500	1960.9	I
150	1995.1	I

Air

Intensity	Wavelength	
500	2039.8	I
500	2074.8	I
500	2164.2	I
150	2332.8	I
600	2413.5	I
300	2548.0	I
220	3038.7	II
220	3041.3	II
285	4070.2	II
360	4175.3	II
450	4180.9	II
120	4328.7	I
100	4330.3	I
285	4382.9	II
285	4446.0	II
220	4449.2	II
285	4467.6	II
500	4730.8	I
400	4739.0	I
300	4742.2	I
285	4840.6	II
360	4845.0	II
450	5227.5	II
360	5305.4	II
100	5365.5	I
120	5369.9	I
110	5374.1	I
285	5522.4	II
285	5566.9	II
285	5866.3	II
450	6056.0	II
200	6325.6	I
360	6444.2	II
285	6490.5	II
285	6535.0	II
150	6831.3	I
120	6990.690	I
100	6991.792	I
200	7010.809	I
150	7013.875	I
300	7062.065	I
200	7575.1	I
250	7583.4	I
150	7592.2	I
120	7606.8	I
300	8001.0	I
200	8036.4	I
120	8060.9	I
120	8065.3	I
120	8081.1	I
150	8093.2	I
150	8094.7	I
180	8149.3	I
150	8152.0	I
200	8157.7	I
180	8163.1	I
150	8182.9	I
100	8185.0	I
120	8194.6	I
150	8440.47	I
150	8450.38	I
150	8742.33	I
300	8918.86	I
100	8969.69	I
200	9001.97	I
200	9038.61	I
80	9083.14	I
120	9088.79	I
80	9140.83	I
60	9181.88	I
60	9271.12	I
100	9432.50	I
60	9825.58	I
200	10217.25	I
377	10307.45	I
900	10327.26	I
640	10386.36	I
124	10650.30	I
125	11934.56	I
275	11946.87	I
100	11947.92	I
105	11952.27	I
170	11952.64	I
100	11966.04	I
205	11972.93	I
115	11973.07	I
315	14817.93	I
410	14917.47	I
500	15151.44	I
115	15469.06	I
320	15471.00	I
265	15520.97	I
395	15618.40	I
115	15620.38	I
360	16659.44	I
505	16813.78	I
165	16817.76	I
205	16866.54	I
115	16972.71	I
235	21374.24	I
680	21442.56	I
415	21473.48	I
270	21716.36	I
240	21730.60	I
105	23133.66	I
150	23388.85	I
110	23628.17	I
265	24148.18	I
170	24159.23	I
185	24204.44	I
375	24385.99	I
160	24413.67	I
225	24471.17	I
255	25017.51	I
510	25127.43	I

Se III
Ref. 9, 247 — R.L.K.

Vacuum

Intensity	Wavelength	
220	709.2	III
220	709.4	III
220	720.6	III
360	724.3	III
285	726.4	III
220	737.2	III
220	741.9	III
285	777.3	III
220	790.8	III
360	843.0	III
220	879.2	III
285	953.7	III
220	954.4	III
220	954.7	III
160	974.1	III
360	974.8	III
285	1079.8	III
360	1099.1	III
450	1119.2	III

Air

Intensity	Wavelength	
285	2057.5	III
285	2767.2	III
220	2773.8	III
285	3379.8	III
450	3387.2	III
450	3413.9	III
285	3428.4	III
450	3457.8	III
360	3543.6	III
285	3570.2	III
450	3637.6	III
360	3711.7	III
450	3738.7	III
285	3743.0	III
450	3800.9	III
360	4046.7	III
220	4083.2	III
450	4169.1	III
220	4637.9	III
285	6303.8	III

Se IV
Ref. 245 — R.L.K.

Vacuum

Intensity	Wavelength	
285	636.0	IV
285	654.2	IV
360	652.7	IV
450	670.1	IV
285	671.9	IV
220	722.8	IV
285	734.6	IV
450	746.4	IV
285	759.0	IV
285	776.5	IV
285	803.8	IV

Selenium (Cont.)

Intensity		Wavelength	
360		959.6	IV
450		996.7	IV
220		1307.2	IV
285		1314.4	IV
		Air	
220		2090.0	IV
285		2136.6	IV
160		2165.2	IV
160		2166.6	IV
360		2665.5	IV
285		2724.3	IV
160		2951.6	IV

Se V
Ref. 245 — R.L.K.

Intensity		Wavelength	
		Vacuum	
285		596.0	V
285		601.0	V
220		608.7	V
360		613.0	V
285		614.3	V
450		759.1	V
285		785.8	V
285		804.3	V
360		808.7	V
220		814.8	V
220		820.7	V
360		830.3	V
450		839.5	V
360		845.8	V
360		1094.7	V
220		1151.0	V
450		1227.6	V

SILICON (Si)
Z = 14

Si I and II
Ref. 170, 237, 292 — L.J.R.

Intensity		Wavelength	
		Vacuum	
10	h	805.10	II
20	h	820.52	II
20	h	843.72	II
40	h	845.77	II
10		850.14	II
100		889.72	II
200		892.00	II
10		899.41	II
20		901.74	II
10		913.01	II
20		913.85	II
20		929.81	II
100		989.87	II
200		992.68	II
25		1020.70	II
50		1023.69	II
30		1057.05	II
15		1057.50	II
20	h	1127.44	II
40	h	1127.91	II
100		1190.42	II
200		1193.28	II
250		1194.50	II
100		1197.39	II
10	h	1216.12	II
20		1223.91	II
20		1224.25	II
10		1224.97	II
50		1226.81	II
20		1226.89	II
40		1226.99	II
100		1227.60	II
10		1228.44	II
25		1228.62	II
150		1228.75	II
200		1229.39	II
10		1235.92	II
100		1246.74	II
150		1248.43	II
100		1250.09	II
150		1250.43	II
200		1251.16	II
10		1255.28	I
40		1256.49	I
50		1258.80	I
1000		1260.42	II
2000		1264.73	II
200		1265.02	II
100		1304.37	II
50	h	1305.59	II
200		1309.27	II
20	h	1309.46	II
100		1346.87	II
100		1348.54	II
150		1350.06	II
20		1350.52	II
20		1350.66	II
100		1352.64	II
100		1353.72	II
10	h	1409.07	II
20	h	1410.22	II
10	h	1416.97	II
15	h	1474.65	II
15		1484.87	II
90	h	1485.02	II
30		1485.22	II
100	h	1485.51	II
100	h	1509.10	II
50	h	1512.07	II
30	p	1513.57	II
60	p	1516.91	II
500		1526.72	II
1000		1533.45	II
10		1562.45	II
15		1562.85	II
10		1563.77	II
50		1573.87	I
50		1574.82	I
50		1592.41	I
150		1594.55	I
50		1594.93	I
30		1597.95	I
100		1622.87	I
30		1625.71	I
300		1629.43	I
200		1629.92	I
75		1631.13	I
50		1633.98	I
30	h	1653.35	I
30		1664.52	I
50		1666.37	I
100		1667.62	I
100		1668.52	I
100		1672.59	I
200		1675.20	I
30		1682.68	I
30		1686.82	I
50	h	1689.29	I
30	h	1690.79	I
50		1693.29	I
50		1695.51	I
200		1696.20	I
200		1697.94	I
50		1700.42	I
30		1700.63	I
30		1702.86	I
50		1704.43	I
10	h	1710.83	II
20	h	1711.30	II
30	h	1743.88	I
50		1747.40	I
30	h	1753.11	I
50		1763.66	I
40		1765.03	I
30	h	1765.60	I
30		1766.06	I
30		1770.63	I
100	h	1770.92	I
100	h	1776.83	I
50	h	1783.23	I
100	h	1799.12	I
150		1808.00	II
50	h	1809.09	I
500	h	1814.07	I
200		1816.92	II
10		1817.45	II
50		1822.45	I
200		1836.51	I
30	h	1838.01	I
100	h	1841.15	I
200		1841.44	I
200		1843.77	I
300		1845.51	I
100		1846.10	I
400		1847.47	I
200		1848.14	I
100		1848.74	I
500		1850.67	I
30	h	1851.79	I
200		1852.46	I
50		1853.15	I
20		1869.32	II
15		1870.23	II
100		1873.10	I
500	h	1874.84	I
100		1875.81	I
200		1881.85	I
200		1887.70	I
200	h	1893.25	I
1000	h	1901.33	I
100	h	1902.46	II
50	h	1904.66	I
50	h	1910.62	II
50		1941.67	II
15		1944.59	II
10		1949.33	II
100		1949.56	II
100		1954.97	I
30		1984.43	I
50		1991.85	I
		Air	
30		2010.97	I
50		2054.83	I
50		2058.65	II
50		2059.01	II
40		2061.19	I
30		2065.52	I
200		2072.02	II
200		2072.70	II
30	h	2103.21	I
30		2114.63	I
100		2124.12	I
10	h	2133.99	II
30	h	2136.40	II
50	h	2136.56	II
50	h	2147.91	I
110		2207.98	I
115		2210.89	I
110		2211.74	I
120		2216.67	I
120		2218.06	I
50		2218.91	I
35		2291.03	I
55		2303.06	I
30		2334.40	II
30		2334.61	II
10		2344.20	II
10	h	2349.54	II
20		2350.17	II
20	h	2353.09	II
100	h	2356.30	II
30	h	2357.18	II
50	h	2357.97	II
10	h	2360.20	II
30		2366.97	II
20		2374.26	II
10	h	2428.45	II
300		2435.15	I
65		2438.77	I
65		2443.36	I
70		2452.12	I
425		2506.90	I
375		2514.32	I
500		2516.113	I
350		2519.202	I
425		2524.108	I
450		2528.509	I
110		2532.381	I
30		2563.679	I
85		2568.641	I
45		2577.151	I
190		2631.282	I
10	h	2682.21	II
1000		2881.579	I
10	h	2887.51	II
300		2904.28	II
500		2905.69	II
55		2970.355	I
150		2987.645	I
50		3006.739	I
100	h	3030.00	II
75		3020.004	I
20	h	3021.55	II
20	h	3041.57	II
30	h	3042.19	II
100	h	3043.69	II
10	h	3043.85	II
10	h	3045.77	II
50	h	3048.30	II
150	h	3053.18	II
150		3188.97	II
50		3192.25	II
150		3193.09	II
50		3194.21	II
50		3194.69	II
100		3195.41	II
200		3199.51	II
20		3202.49	II
100	h	3203.87	II
200	h	3210.03	II
75		3214.66	II
15	h	3217.99	II
10		3220.44	II
20		3223.01	II
300		3333.14	II
500		3339.82	II
100	h	3853.66	II
500	h	3856.02	II
200	h	3862.60	II
300		3905.523	I
10	h	3955.74	II
10	h	3977.46	II
15	h	3991.77	II
10	h	3998.01	II
20	h	4075.45	II
15	h	4076.78	II
70		4102.936	I
300	h	4128.07	II
500	h	4130.89	II
10	h	4183.35	II
100	h	4190.72	II
50		4198.13	II
100		4621.42	II
150		4621.72	II
50		4782.991	I
35		4792.212	I
80		4792.324	I
15	h	4883.20	II
20	h	4906.99	II
20	h	4932.80	II
30		4947.607	I
40		5006.061	I
1000		5041.03	II
1000		5055.98	II
100		5181.90	II
100	h	5185.25	II
200	h	5192.86	II
500	h	5202.41	II
30	h	5295.19	II
100	h	5405.34	II
15	h	5417.24	II
15	h	5428.92	II
15	h	5432.89	II
100	h	5438.62	II
20	h	5447.26	II
15	h	5454.49	II
100	h	5456.45	II
500	h	5466.43	II
500	h	5466.87	II
100	h	5469.21	II
40		5493.23	I
200	h	5496.45	II
35		5517.535	I
100	h	5540.74	II
150	h	5576.66	II
30		5622.221	I
100	h	5632.97	II
200	h	5639.48	II
90		5645.611	I
150	h	5660.66	II
80		5665.554	I
1000	h	5669.56	II
30	h	5681.44	II
120		5684.484	I
300	h	5688.81	II
100		5690.425	I
90		5701.105	I
200	h	5701.37	II
100	h	5706.37	II
160		5708.397	I
45		5747.667	I
45		5753.625	I
45		5754.220	I
45		5762.977	I
70		5772.145	I
70		5780.384	I
30	h	5785.73	II
90		5793.071	I
30	h	5794.90	II
100		5797.859	I
150	h	5800.47	II
200		5806.74	II
30		5827.80	II
50		5846.13	II
10		5867.48	II
300	h	5868.40	II
40		5873.764	I
150		5915.22	II
200		5948.545	I
500		5957.56	II
500		5978.93	II
10	h	6067.45	II
20	h	6080.06	II
10	h	6086.67	II
90		6125.021	I
85		6131.574	I
90		6131.850	I
100		6142.487	I
100		6145.015	I
160		6155.134	I
160		6237.320	I
40		6238.287	I
125		6243.813	I
125		6244.468	I
180		6254.188	I
45		6331.954	I
1000		6347.10	II
1000		6371.36	II
45		6526.609	I
45		6527.199	I
45		6555.462	I
50	h	6660.52	II
15		6665.00	II
100		6671.88	II
20		6699.38	II
50	h	6717.04	II
100		6721.853	I
30		6741.64	I
20	h	6750.28	II
30		6818.45	II
50		6829.82	II
30		6848.568	I
80		6976.523	I

Silicon (Cont.)

Intensity		Wavelength	
180		7003.567	I
180		7005.883	I
30		7017.28	I
90		7017.646	I
250		7034.903	I
70		7164.69	I
200		7165.545	I
70		7184.89	I
65		7193.58	I
30		7193.90	I
100		7226.206	I
100		7235.326	I
60		7235.82	I
180		7250.625	I
160		7275.294	I
40		7282.81	I
400		7289.173	I
55		7290.26	I
35		7373.00	I
375		7405.774	I
200		7409.082	I
40		7415.35	I
275		7415.946	I
425		7423.497	I
85		7424.60	I
100		7680.267	I
40		7742.71	I
30		7800.008	I
400		7848.80	II
500		7849.72	II
30		7849.967	I
90		7918.386	I
120		7932.349	I
140		7944.001	I
35		7970.306	I
35		8035.619	I
70		8093.241	I
35		8230.642	I
40		8443.982	I
40		8501.547	I
60		8502.221	I
40		8536.165	I
120		8556.780	I
50		8648.462	I
40		8728.011	I
75		8742.451	I
100		8752.009	I
35		8790.389	I
100		9412.72	II
100		9413.506	I
30		10371.269	I
120		10585.141	I
120		10603.431	I
120		10660.975	I
30		10694.251	I
30		10727.408	I
60		10749.384	I
30		10784.550	I
80		10786.856	I
140		10827.091	I
60		10843.854	I
30		10868.79	I
130		10869.541	I
30		10882.802	I
30		10885.336	I
80		10979.308	I
30		10982.061	I
80		11017.965	I
13		11187.60	I
12		11289.84	I
12		11611.09	I
370		11984.19	I
220		11991.57	I
440		12031.51	I
150		12103.53	I
120		12270.68	I
11		13176.90	I
190		15888.39	I
40		15960.04	I
95		16060.03	I
20		16094.80	I
60		16163.71	I
11		16215.68	I
16		16381.55	I
29		16680.77	I
28		17327.29	I
26		18722.90	I
15		19385.94	I
48		19432.97	I
13		19493.38	I
110		19722.50	I
31		19928.88	I
12		20917.13	I
21		21354.24	I
12		22062.71	I

Si III

Ref. 320 — L.J.R.

Vacuum

Intensity		Wavelength	
8		566.61	III
6		652.22	III
8		653.33	III
5		673.48	III
5		800.07	III
9		823.41	III
5		883.40	III
7		939.09	III
9		967.95	III
10		993.52	III
13		994.79	III
16		997.39	III
7		1005.37	III
7		1031.16	III
8		1033.92	III
7		1037.05	III
6		1083.22	III
14		1108.37	III
16		1109.97	III
18		1113.23	III
6		1140.55	III
7		1141.58	III
6		1142.28	III
8		1144.31	III
6		1144.96	III
8		1145.11	III
7		1145.18	III
6		1155.00	III
6		1155.96	III
7		1158.10	III
6		1160.26	III
8		1161.58	III
5		1174.37	III
6		1174.43	III
8		1178.00	III
30		1206.51	III
30		1206.53	III
9		1207.52	III
10		1210.46	III
7		1235.43	III
6		1280.35	III
17		1294.54	III
14		1296.73	III
15		1298.89	III
18		1298.96	III
14		1301.15	III
16		1303.32	III
13		1312.59	III
8		1341.47	III
7		1342.39	III
6		1343.39	III
8		1361.60	III
5		1362.37	III
7		1363.47	III
8		1365.26	III
7		1367.05	III
5		1369.44	III
5		1373.03	III
5		1387.99	III
13		1417.24	III
6		1433.69	III
8		1435.77	III
7		1436.17	III
5		1441.73	III
6		1447.20	III
5		1457.25	III
12		1500.24	III
10		1501.19	III
9		1501.87	III
6		1506.06	III
7		1673.32	III
9		1842.55	III

Air

Intensity		Wavelength	
5		2176.89	III
6		2295.48	III
10		2296.87	III
8		2300.93	III
10		2308.19	III
11		2449.48	III
6		2483.20	III
25		2541.82	III
10		2546.09	III
14		2559.21	III
11		2640.79	III
14		2655.51	III
9		2817.11	III
7		2831.49	III
5		2839.62	III
5		2959.15	III
5		2980.52	III
5		3013.09	III
6		3034.73	III
8		3037.29	III
9		3040.93	III
7		3043.93	III
5		3045.08	III
7		3068.24	III
25		3086.24	III
6		3086.46	III
20		3093.42	III
5		3093.65	III
16		3096.83	III
6		3126.27	III
7		3147.37	III
8		3161.61	III
16		3185.13	III
13		3186.02	III
14		3196.50	III
15		3210.55	III
7		3216.25	III
12		3230.50	III
14		3233.95	III
15		3241.62	III
7		3253.40	III
5		3253.74	III
7		3254.80	III
12		3258.66	III
6		3270.46	III
10		3276.26	III
7		3279.26	III
15		3486.91	III
9		3525.94	III
8		3569.67	III
20		3590.47	III
8	h	3622.54	III
5	h	3639.45	III
6	h	3645.12	III
7	h	3681.40	III
5	h	3682.15	III
20	c	3791.41	III
25		3796.11	III
30		3806.54	III
7		3842.46	III
20		3924.47	III
6	h	3947.49	III
6		3963.84	III
5		3981.24	III
5	h	4101.86	III
8		4102.42	III
5	h	4115.50	III
9		4338.50	III
8		4341.40	III
8	h	4377.63	III
6	h	4405.90	III
8	h	4406.72	III
6		4494.05	III
30		4552.62	III
8		4554.00	III
25		4567.82	III
20		4574.76	III
7		4619.66	III
7		4638.28	III
8		4665.87	III
9		4683.02	III
7		4683.80	III
16		4716.65	III
7		4730.52	III
8		4800.43	III
15		4813.33	III
16		4819.72	III
18		4828.97	III
10	h	5091.42	III
7	h	5113.76	III
8	h	5114.12	III
5		5197.26	III
6		5451.46	III
7		5473.05	III
7		5704.60	III
8		5716.29	III
20		5739.73	III
10	h	5898.79	III
7		6314.46	III
6	h	6524.36	III
6	h	6831.56	III
7	h	6851.65	III
5	h	7461.89	III
8	h	7462.62	III
9	h	7466.32	III
12	h	7612.36	III
9	h	8102.86	III
11	h	8103.45	III
7	h	8190.43	III
6	h	8191.16	III
8	h	8191.68	III
9	h	8262.57	III
5	h	8265.64	III
8	h	8269.32	III
5	h	8271.38	III
6	h	8271.94	III

Si IV

Ref. 319 — L.J.R.

Vacuum

Intensity		Wavelength	
4		457.82	IV
3		458.16	IV
2		515.12	IV
3		516.35	IV
2		645.76	IV
5		749.94	IV
7		815.05	IV
8		818.13	IV
8		1066.63	IV
8		1122.49	IV
10		1128.34	IV
15		1393.76	IV
12		1402.77	IV
1		1634.61	IV
6		1722.53	IV
5		1727.38	IV

Air

Intensity		Wavelength	
3		2120.18	IV
4		2127.47	IV
5	h	2287.04	IV
2	h	2328.56	IV
2		2366.76	IV
3		2370.99	IV
2		2482.82	IV
1		2485.38	IV
7		2517.51	IV
3		2672.19	IV
4		2675.12	IV
4		2675.25	IV
1		2677.57	IV
3	h	2723.81	IV
3	h	2895.13	IV
2	h	2904.47	IV
1	h	2971.52	IV
7		3149.56	IV
9		3165.71	IV
1	h	3244.19	IV
8		3762.44	IV
6		3773.15	IV
1	h	4031.39	IV
2	h	4038.06	IV
10		4088.85	IV
9		4116.10	IV
7	h	4212.41	IV
3		4314.10	IV
5		4328.18	IV
2	h	4403.73	IV
1	h	4411.65	IV
1	h	4611.27	IV
3	h	4628.62	IV
9	h	4631.24	IV
10	h	4654.32	IV
3	h	4656.92	IV
1	h	4667.14	IV
2	h	4673.30	IV
1	h	4947.45	IV
3		4950.11	IV
2	h	5304.97	IV
1	h	5309.49	IV
5		6667.56	IV
7		6701.21	IV
3	h	6998.36	IV
6	h	7047.94	IV
4	h	7068.41	IV
2	h	7630.50	IV
4	h	7654.56	IV
4	h	7678.75	IV
5	h	7718.79	IV
6	h	7723.82	IV
2	h	7725.64	IV
1	h	7730.47	IV
1	h	7752.91	IV
1	h	8240.61	IV
2	h	8957.25	IV
1	h	9018.16	IV

Si V

Ref. 87 — L.J.R.

Vacuum

Intensity		Wavelength	
1		78.61	V
1		78.90	V
2		80.81	V
2		81.11	V
10		85.18	V
6		85.58	V
4		90.45	V
4		90.85	V
15		96.44	V
10		97.14	V
2		98.21	V
20		117.86	V
20		118.97	V

SILVER (Ag)
Z = 47

Ag I and II

Ref. 13, 99, 255, 286, 289 — C.H.C.

Vacuum

Intensity		Wavelength	
25		730.83	II

Silver (Cont.)

Intensity		Wavelength	
30		752.80	II
15		1005.32	II
10		1065.49	II
12		1072.23	II
250		1074.22	II
150		1107.03	II
150		1112.46	II
60		1195.83	II
50		1223.33	II
50		1240.80	II
50		1246.87	II
55		1256.81	II
55		1257.55	II
50		1266.63	II
70		1273.67	II
65		1297.51	II
85		1311.20	II
55		1313.81	II
50		1314.61	II
60		1323.84	II
60		1342.09	II
50		1342.57	II
70		1346.62	II
50		1353.54	II
150		1364.50	II
100		1396.00	II
100		1410.93	II
90		1419.72	II
95		1432.60	II
100		1464.72	II
50		1466.23	II
50	r	1507.37	I
100	r	1515.63	I
50	r	1548.58	I
100		1555.16	II
100		1644.50	II
60		1651.52	I
50		1652.10	I
120		1682.82	II
10		1708.11	I
50		1709.27	I
125		1736.44	II
10	h	1766.14	I
75		1790.37	II
20		1847.71	I
100		1967.38	II

Air

Intensity		Wavelength	
150		2015.96	II
150		2033.98	II
200		2061.17	I
100		2069.85	I
80	r	2113.82	II
60		2145.60	II
15		2170.00	I
50		2186.76	II
60		2229.53	II
100	r	2246.43	II
75	r	2248.74	II
75		2280.03	II
30	h	2309.56	I
10	h	2312.60	I
70	r	2317.05	II
80	r	2320.29	II
70	r	2324.68	II
80	r	2331.40	II
70		2357.92	II
50	h	2375.02	I
75		2411.41	II
90	r	2413.23	II
100	r	2437.81	II
80		2447.93	II
80		2473.84	II
60		2506.63	II
50	h	2575.63	I
60		2660.49	II
60		2721.77	I
75		2767.54	II
100	h	2824.39	I
10	h	2926.77	I
20	h	2938.42	I
20		3099.10	I
30	h	3130.02	I
10	h	3170.58	I
90		3180.70	II
15	h	3215.67	I
10		3225.15	I
15		3233.18	I
100		3267.35	II
55000	r	3280.68	I
10	h	3305.67	I
28000	r	3382.89	I
10	h	3403.78	I
30		3469.16	I
70		3475.82	II
80		3495.28	II
20	h	3501.92	I
20		3508.03	I
15	h	3513.38	I
10		3521.12	I
50		3542.61	I
10	h	3547.16	I
10	h	3557.01	I
20	h	3586.67	I
10	h	3623.49	I
50	h	3624.68	I
75		3682.46	II
30		3682.50	I
80		3683.34	II
50	h	3709.20	I
10	h	3727.42	I
20	h	3753.14	I
200		3810.94	I
50		3811.78	I
100	h	3840.74	I
15		3847.85	I
50	h	3907.41	I
50		3909.31	II
50	h	3914.40	I
70		3920.10	II
10	h	3928.01	I
10	h	3940.43	I
10	h	3942.97	I
60		3949.43	II
100	h	3981.58	I
70		3985.19	II
10	h	3992.15	I
100	h	4055.48	I
10	h	4083.43	I
80		4085.91	II
100		4185.48	II
90	h	4210.96	I
100		4212.82	I
50		4311.07	I
20		4396.23	I
50	h	4476.04	I
20	h	4556.0	I
30	h	4615.69	I
80		4620.04	II
50		4620.46	II
60	h	4668.48	I
30	h	4677.60	I
100		4788.40	II
20	h	4796.2	I
30	h	4847.82	I
100		4874.10	I
20		4888.21	I
10	h	4917.5	I
10		4935.75	I
20	h	4992.89	I
80		5027.35	II
15	h	5123.50	I
1000		5209.08	I
10	h	5333.62	I
1000		5465.50	I
100		5471.55	I
20		5475.38	I
20	h	5545.67	I
10	h	5559.58	I
100		5667.34	I
10	h	6083.78	I
10	h	6268.50	I
20		6621.08	I
20		7359.96	I
320		7687.78	I
25		8005.4	II
15		8254.7	II
500		8273.52	I
20		8324.4	II
15		8379.5	II
25		8403.8	II
15		8492.5	II
30	h	8645.70	I
10	h	8704.85	I
12		8747.6	II
15		9000.9	II
10		12551.0	I
60		16819.5	I
20		17416.7	I
15		18307.9	I
15		18382.3	I

Ag III
Ref. 363, 387, 398 — R.D.C.

Vacuum

Intensity	Wavelength	
200	709.80	III
200	713.85	III
100	717.73	III
200	718.53	III
300	726.96	III
350	730.04	III
150	730.28	III
150	730.94	III
200	736.57	III
100	738.13	III
200	740.98	III
200	742.29	III
200	748.30	III
150	755.73	III
100	758.27	III
200	767.19	III
250	768.33	III
150	769.61	III
350	776.38	III
200	782.91	III
150	785.76	III
200	789.08	III
250	792.35	III
200	796.54	III
250	797.91	III
400	799.41	III
300	808.88	III
150	816.12	III
180	822.39	III
200	838.11	III
120	1373.22	III
120	1374.76	III
110	1404.93	III
120	1413.90	III
120	1414.29	III
120	1428.61	III
120	1452.74	III
200	1456.41	III
100	1471.44	III
300	1489.01	III
150	1515.08	III
100	1524.23	III
120	1527.04	III
150	1541.14	III
150	1550.89	III
130	1553.04	III
130	1587.41	III
120	1589.28	III
100	1613.79	III
130	1619.14	III
100	1634.46	III
100	1652.24	III
130	1653.60	III
300	1654.43	III
700	1656.18	III
150	1657.10	III
100	1661.54	III
130	1670.75	III
150	1676.14	III
100	1681.07	III
500	1693.51	III
200	1705.06	III
150	1708.86	III
130	1717.68	III
200	1722.27	III
150	1726.76	III
250	1728.14	III
200	1747.34	III
120	1749.64	III
150	1750.89	III
750	1751.03	III
100	1760.57	III
150	1762.62	III
150	1768.70	III
100	1771.81	III
100	1783.85	III
100	1791.70	III
100	1792.69	III
150	1793.90	III
150	1802.24	III
150	1802.26	III
100	1802.77	III
300	1808.23	III
250	1816.83	III
150	1822.45	III
350	1828.83	III
250	1832.33	III
120	1832.50	III
100	1834.31	III
250	1836.10	III
150	1838.64	III
400	1840.14	III
100	1846.96	III
120	1849.93	III
150	1856.33	III
120	1858.91	III
100	1860.39	III
100	1860.64	III
350	1867.12	III
150	1868.10	III
100	1872.55	III
400	1873.45	III
250	1880.36	III
300	1889.57	III
400	1916.92	III
600	1917.08	III
200	1925.30	III
150	1946.32	III
100	1948.44	III
700	1957.62	III
120	1959.27	III
100	1960.86	III
400	1966.89	III
600	1975.92	III
500	1977.03	III
150	1981.87	III
200	1987.02	III
130	1995.16	III
600	2000.24	III

Air

Intensity	Wavelength	
300	2007.30	III
200	2011.49	III
150	2013.65	III
150	2041.33	III
150	2053.17	III
150	2053.83	III
200	2056.99	III
200	2081.04	III
300	2146.47	III
150	2149.19	III
600	2161.89	III
150	2166.21	III
150	2211.23	III
100	2238.40	III
500	2246.51	III
100	2286.50	III
700	2310.04	III
100	2386.85	III
300	2395.69	III
100	2469.62	III
100	2562.87	III

SODIUM (Na)
Z = 11

Na I and II
Ref. 268, 334 — T.K.M.

Vacuum

Intensity	Wavelength	
160	300.15	II
160	300.20	II
90	301.32	II
100	301.44	II
60	302.45	II
300	372.08	II
350	376.38	II
60	1293.97	II
50	1327.74	II
45	1347.54	II
90	1374.69	II
90	1404.08	II
45	1495.21	II
45	1497.73	II
90	1506.41	II
60	1506.91	II
70	1513.10	II
60	1519.63	II
60	1657.92	II
90	1776.57	II
60	1783.04	II
80	1787.19	II
45	1788.85	II
80	1798.41	II
45	1801.26	II
90	1807.09	II
60	1808.38	II
50	1821.70	II
45	1833.87	II
80	1835.22	II
45	1837.89	II
60	1841.82	II
70	1845.02	II
45	1850.15	II
70	1851.19	II
80	1853.17	II
45	1866.45	II
45	1873.37	II
60	1875.08	II
160	1881.91	II
50	1885.09	II
45	1885.74	II

Air

Intensity	Wavelength	
80	2228.53	II
80	2303.58	II
300	2315.65	II
130	2393.28	II
100	2401.01	II
300	2420.99	II
300	2424.73	II
200	2439.14	II
250	2441.50	II
200	2448.72	II
200	2452.18	II
1000	2493.15	II
300	2502.84	II
450	2506.30	II
600	2515.46	II
600	2531.54	II
20	2543.84	I
10	2543.87	I
550	2586.31	II
70	2593.87	I
35	2593.92	I
600	2594.96	II
850	2611.81	II
300	2627.41	II

Sodium (Cont.)

850	2661.00	II
350	2666.46	II
1000	2671.83	II
850	2678.09	II
200	2680.34	I
100	2680.43	I
650	2808.71	II
850	2809.52	II
600	2829.87	II
800	2839.56	II
1000	2841.72	II
400	2852.81	I
200	2853.01	I
650	2856.51	II
800	2859.49	II
750	2871.28	II
650	2872.95	II
900	2881.15	II
850	2886.26	II
2	2893.62	I
700	2893.95	II
900	2901.14	II
800	2904.72	II
1100	2904.92	II
1100	2917.52	II
1100	2919.05	II
1200	2919.85	II
1300	2920.95	II
1000	2923.49	II
750	2930.88	II
850	2934.08	II
950	2937.74	II
800	2945.70	II
950	2947.50	II
1200	2951.24	II
1100	2952.40	II
850	2960.12	II
500	2970.73	II
600	2974.24	II
750	2974.99	II
1000	2977.13	II
1100	2979.66	II
1100	2980.63	II
1300	2984.19	II
550	3004.15	II
750	3007.44	II
750	3009.14	II
600	3015.40	II
550	3053.67	II
550	3055.35	II
550	3056.16	II
550	3057.38	II
550	3057.95	II
550	3058.72	II
700	3060.25	II
800	3061.35	II
500	3064.38	II
500	3066.22	II
500	3066.54	II
550	3074.33	II
550	3078.32	II
550	3080.25	II
550	3087.06	II
550	3092.04	II
550	3092.73	II
650	3094.45	II
650	3095.55	II
500	3103.58	II
500	3104.40	II
500	3113.69	II
1700	3124.42	II
600	3125.21	II
600	3129.38	II
2500	3135.48	II
1700	3137.86	II
950	3145.71	II
2000	3149.28	II
2000	3163.74	II
700	3175.09	II
1000	3179.06	II
1700	3189.79	II
1600	3212.19	II
700	3234.93	II
1500	3257.96	II
650	3260.21	II
950	3274.22	II
1700	3285.60	II
1700	3301.35	II
1200	3302.37	I
600	3302.98	I
1500	3304.96	II
1000	3318.04	II
950	3327.69	II
50	3426.86	I
1500	3533.05	II
1200	3631.27	II
850	3711.07	II
300	4113.70	II
250	4123.08	II
250	4233.26	II
6	4238.99	I
250	4240.90	II
10	4242.08	I
1	4249.41	I
2	4252.52	I
15	4273.64	I
20	4276.79	I
2	4287.84	I
3	4291.01	I
250	4292.48	II
250	4292.86	II
250	4308.81	II
250	4309.04	II
250	4320.91	II
30	4321.40	I
40	4324.62	I
250	4337.29	II
3	4341.49	I
250	4344.11	II
5	4344.74	I
200	4368.60	II
200	4375.22	II
200	4387.49	II
40	4390.03	I
250	4392.81	II
60	4393.34	I
200	4405.12	II
5	4419.88	I
8	4423.25	I
200	4446.70	II
200	4447.41	II
200	4454.74	II
200	4455.23	II
200	4457.21	II
200	4474.63	II
200	4478.80	II
200	4481.67	II
200	4490.15	II
200	4490.87	II
60	4494.18	I
100	4497.66	I
200	4499.62	II
200	4506.97	II
200	4519.21	II
200	4524.98	II
200	4533.32	II
10	4541.63	I
15	4545.19	I
200	4551.53	II
160	4590.92	II
120	4664.811	I
200	4668.560	I
160	4722.23	II
160	4731.10	II
160	4741.67	II
20	4747.941	I
30	4751.822	I
160	4768.79	II
100	4788.79	II
200	4978.541	I
400	4982.813	I
40	5148.838	I
80	5153.402	I
100	5191.65	II
80	5208.55	II
70	5400.46	II
90	5414.55	II
280	5682.633	I
70	5688.193	I
560	5688.205	I
80000	5889.950	I
40000	5895.924	I
120	6154.225	I
240	6160.747	I
60	6175.25	II
70	6199.26	II
70	6234.68	II
80	6260.01	II
80	6274.74	II
70	6361.15	II
70	6366.41	II
90	6514.21	II
80	6524.68	II
130	6530.70	II
130	6544.04	II
130	6545.75	II
80	6552.43	II
20	7373.23	I
10	7373.49	I
50	7809.78	I
25	7810.24	I
4400	8183.256	I
800	8194.790	I
8800	8194.824	I
100	8649.92	I
60	8650.89	I
25	8942.96	I
40	9153.88	I
60	9465.94	I
80	9961.28	I
20	10566.00	I
60	10572.28	I
200	10746.44	I
80	10749.29	I
120	10834.87	I
35	11190.19	I
50	11197.21	I
400	11381.45	I
1000	11403.78	I
400	12679.17	I
60	14767.48	I
100	14779.73	I
60	16373.85	I
100	16388.85	I
400	18465.25	I
50	22056.44	I
25	22083.67	I
60	23348.41	I
100	23379.13	I

Na III
Ref. 178, 205, 207 — T.K.M.

Intensity		Wavelength	
		Vacuum	
5		183.95	III
5	h	189.35	III
5		193.80	III
5	h	194.04	III
5	h	194.17	III
5		194.29	III
6		194.68	III
6		195.53	III
6		202.15	III
6		202.19	III
8		202.49	III
5	d	202.71	III
7	d	202.72	III
8		202.76	III
8	p	203.06	III
8		203.28	III
8		203.33	III
10		207.30	III
10	c	215.34	III
12		215.86	III
12		216.12	III
15		229.87	III
12		230.59	III
50	c	250.52	III
30		251.37	III
25		266.90	III
70		267.65	III
50		267.87	III
50		268.63	III
20	p	272.08	III
20		272.45	III
100		378.14	III
70		380.10	III
7		1336.76	III
7		1337.36	III
8		1340.67	III
9	d	1342.39	III
10		1342.73	III
11		1355.28	III
12		1361.90	III
11		1372.34	III
10		1420.89	III
10		1444.19	III
12		1449.31	III
11		1562.87	III
10		1565.29	III
10		1598.18	III
11		1688.94	III
10		1699.29	III
10		1711.12	III
11		1728.27	III
10		1731.11	III
10		1755.48	III
15		1807.07	III
10		1810.77	III
11		1811.67	III
10		1816.81	III
10	d	1835.22	III
10		1838.94	III
11		1844.36	III
12	d	1847.53	III
10	d	1847.59	III
15		1849.56	III
12		1850.38	III
10		1855.92	III
10		1856.71	III
10		1861.21	III
10		1880.66	III
10	d	1887.39	III
20	d	1887.47	III
15	d	1890.75	III
15		1900.16	III
10		1918.45	III
11		1923.96	III
14		1926.26	III
12		1927.24	III
12		1932.74	III
13		1933.89	III
10		1943.52	III
12		1946.43	III
12		1950.91	III
14		1951.24	III
10		1977.16	III
13		1985.57	III
10		1995.68	III
		Air	
10		2004.21	III
11		2005.22	III
11		2008.47	III
15		2011.87	III
11		2014.17	III
12		2017.03	III
12		2028.56	III
12		2031.13	III
11		2035.90	III
12		2041.66	III
12		2043.29	III
10		2044.82	III
10		2045.44	III
11		2051.48	III
10		2060.36	III
15		2066.60	III
13		2082.91	III
15		2140.72	III
14		2144.54	III
15		2202.83	III
15		2225.93	III
30		2230.33	III
16		2232.19	III
20	h	2246.70	III
14		2251.47	III
15		2278.42	III
13		2285.66	III
15		2309.99	III
18		2386.99	III
17		2394.03	III
15		2406.59	III
25		2459.31	III
18		2468.85	III
20		2474.73	III
25		2497.03	III
17		2510.26	III
15		2530.25	III
14		2542.80	III

Na IV
Ref. 206 — T.K.M.

Intensity		Wavelength	
		Vacuum	
4		136.551	IV
4		136.854	IV
4		139.961	IV
7		142.232	IV
6		142.359	IV
8		146.064	IV
7		146.302	IV
9		150.298	IV
7		150.543	IV
7	c	150.64	IV
8		150.687	IV
7		151.299	IV
7		155.083	IV
7		155.240	IV
7		155.448	IV
8		155.510	IV
8		156.537	IV
12		162.448	IV
10		163.190	IV
12		168.411	IV
10		168.546	IV
10		190.445	IV
10		199.772	IV
10	c	205.49	IV
10		319.644	IV
10		360.76	IV
12		408.684	IV
10		409.614	IV
15		410.372	IV
10		411.334	IV
13		412.242	IV
10		1580.50	IV
11		1582.18	IV
10		1582.33	IV
11	d	1583.98	IV
12		1584.14	IV
10	d	1586.99	IV
12	d	1587.05	IV
10		1613.95	IV
11		1615.92	IV
12		1618.57	IV
11		1655.47	IV
15	c	1701.97	IV
10		1702.41	IV
12		1960.76	IV
11		1965.08	IV
10		1967.60	IV
		Air	
10		2018.39	IV
12	d	2106.33	IV
10		2114.53	IV

Sodium (Cont.)

Intensity		Wavelength	
10		2155.76	IV

Na V
Ref. 299 — T.K.M.

Intensity		Wavelength	
		Vacuum	
100		106.28	V
100		106.30	V
100		106.40	V
100		106.49	V
200	c	107.93	V
200		108.02	V
200	c	110.82	V
200		110.88	V
100		111.51	V
300	c	112.01	V
100	h	114.70	V
100		114.74	V
400		117.99	V
100		120.04	V
400		125.18	V
400		125.22	V
500		125.29	V
300		125.43	V
300		125.46	V
200		125.90	V
100		126.21	V
200		126.56	V
100		126.61	V
400		127.44	V
400		127.47	V
400		128.03	V
400		128.05	V
200		130.68	V
300		131.35	V
200		131.41	V
300	h	131.64	V
500		133.16	V
400		133.39	V
200		134.27	V
300		135.79	V
300		135.85	V
200		138.81	V
300		138.92	V
400		148.64	V
300		148.86	V
400		151.13	V
300		157.21	V
300		163.62	V
800		307.15	V
1000		308.26	V
800		332.55	V
900		333.91	V
800		360.32	V
800		360.37	V
1000		400.72	V
500		445.05	V
600		445.19	V
600		459.90	V
850		461.05	V
1000		463.26	V

STRONTIUM (Sr)
Z = 38

Sr I and II
Ref. 1, 218, 279, 313 — J.J.W.

Intensity		Wavelength	
		Air	
1400		2152.84	II
1400		2165.96	II
160		2428.10	I
120		2569.47	I
200		2931.83	I
300		3301.73	I
300		3329.99	I
400		3351.25	I
300		3366.33	I
650		3380.71	II
950		3464.46	II
120		3474.89	II
300	h	3940.80	I
600		3969.26	I
300		3970.04	I
1300		4030.38	I
300		4032.38	I
46000		4077.71	II
200		4161.80	II
32000		4215.52	II
340		4305.45	II
350	h	4438.04	I
		4526.10	II
		4585.91	II
65000		4607.33	I
3200		4722.28	I
2200		4741.92	I
1400		4784.32	I
4800		4811.88	I
3600		4832.08	I
500		4855.04	I
600		4868.70	I
3000		4872.49	I
600		4876.06	I
2000		4876.32	I
1000		4891.98	I
8000		4962.26	I
1300		4967.94	I
800	h	5156.07	I
1400		5222.20	I
2000		5225.11	I
2000		5229.27	I
2800		5238.55	I
4800		5256.90	I
		5303.13	II
350	h	5329.82	I
		5379.13	II
		5385.45	II
1500		5450.84	I
7000		5480.84	I
1100		5486.12	I
3500		5504.17	I
2600		5521.83	I
2000		5534.81	I
2000		5540.05	I
250	h	5543.36	I
		5622.94	II
		5650.54	II
		5723.70	II
		5819.00	II
200	h	5970.10	I
250	h	6345.75	I
250	h	6363.94	I
350	h	6369.96	I
1000		6380.75	I
900	h	6386.50	I
600	h	6388.24	I
9000		6408.47	I
250		6446.68	I
250	h	6465.79	I
		6483.17	II
3300		6504.00	I
		6509.20	II
1000		6546.79	I
1700		6550.26	I
3000		6617.26	I
800		6643.54	I
1800		6791.05	I
4800		6878.38	I
1200		6892.59	I
5500		7070.10	I
60		7153.09	I
250	h	7167.24	I
200		7232.27	I
2500		7309.41	I
500		7621.50	I
400	h	7673.06	I
50	hL	7850.00	I
30	h	7866.90	I
20	hL	7874.00	I
200	h	8422.80	I
120		8505.69	II
200		8688.91	II
30		8719.56	II
40		9170.00	I
30		9204.50	I
20		9283.90	I
100		9294.10	I
15		9306.60	I
30		9319.20	I
60		9380.45	I
40	h	9411.25	I
400	h	9448.95	I
600		9596.00	I
300		9624.70	I
100		9638.10	I
100	h	9647.70	II
300		10036.66	II
1000		10327.31	II
7		10872.70	II
200		10914.88	II
10		10984.00	I
13		11224.57	II
700		11241.25	I
100		12014.76	II
20		12236.20	I
60		12445.90	II
20		12479.60	I
40		12495.00	I
15		12652.20	I
75		12974.70	II
100		13123.80	II
15		13522.80	I
15		17140.90	I
30		17170.50	I
50		17447.40	I
4		17626.00	II
30		17743.00	I
15		19759.60	I
230		20261.40	I
120		20700.70	I
40		20764.50	I
15		20778.70	I
30		26023.60	I

Sr III
Ref. 231, 265 — J.J.W.

Intensity		Wavelength	
		Vacuum	
25		307.18	III
50		316.11	III
50		321.61	III
125		330.67	III
500		351.62	III
75		358.80	III
250		363.49	III
150		371.21	III
1000		437.24	III
1875		491.79	III
1250		507.04	III
3750		514.38	III
2500		562.75	III
20		968.37	III
20		975.78	III
25		992.98	III
50		1025.23	III
35		1044.91	III
20		1057.74	III
25		1060.20	III
20		1098.77	III
35		1125.49	III
20		1140.24	III
20		1168.27	III
20		1182.09	III
50		1236.23	III
20		1940.58	III
30	p	1958.44	III
30		1966.92	III
		Air	
25		2068.63	III
50		2099.59	III
25		2114.31	III
30		2118.48	III
50		2119.52	III
50		2133.12	III
30		2142.80	III
20		2145.74	III
30		2178.91	III
30		2180.14	III
30		2190.88	III
50		2203.86	III
50		2219.50	III
50		2220.05	III
50		2267.03	III
100		2273.71	III
50		2277.87	III
30		2310.33	III
50		2314.95	III
50		2334.79	III
100		2340.13	III
50		2404.17	III
30		2410.52	III
50		2454.03	III
100		2486.52	III
50		2503.59	III
30		2599.10	III
35		2622.69	III
30		2642.96	III
30		2648.51	III
35		2654.66	III
40		2722.47	III
50		2786.00	III
50		2821.42	III
30		2874.86	III
30		2929.34	III
30		2983.00	III
100		3002.61	III
200		3012.32	III
100		3021.73	III
30		3059.83	III
50		3061.43	III
30		3104.25	III
50		3182.61	III
100		3235.39	III
30		3302.72	III
50		3430.76	III
30		3874.26	III
30		3936.40	III
30		3936.72	III
30		3958.75	III
30		4094.03	III
30		4097.02	III
30		4105.63	III
35		4335.80	III
30		5071.09	III
30		5130.34	III
35		5158.26	III
40		5257.71	III
30		5262.21	III
30		5288.32	III
30		5391.03	III
40		5443.48	III
30		5463.90	III
30		5664.66	III
30		5689.72	III

Sr IV
Ref. 110 — J.J.W.

Intensity		Wavelength	
		Vacuum	
12		284.31	IV
12		291.09	IV
12		291.19	IV
12		293.22	IV
15		298.12	IV
15		300.12	IV
12		300.27	IV
12		301.67	IV
20		378.53	IV
75		392.44	IV
50		393.00	IV
45		394.90	IV
50		396.22	IV
40		399.92	IV
35		403.85	IV
35		406.94	IV
30		412.93	IV
40		413.07	IV
40		415.32	IV
30		419.78	IV
25		430.21	IV
30		430.65	IV
25		442.73	IV
25		471.76	IV
25		484.20	IV
25	p	508.14	IV
25		534.19	IV
200		664.43	IV
100		710.35	IV
20		1190.01	IV
30		1244.14	IV
20	p	1244.75	IV
20	p	1244.87	IV
20	p	1257.78	IV
20		1268.62	IV
20		1331.13	IV
30		1347.90	IV
20		1361.15	IV
25		1408.67	IV
20		1592.74	IV
25		1677.03	IV
20		1705.16	IV
20		1724.23	IV
25		1729.53	IV
20		1732.12	IV
20		1777.25	IV
20		1994.61	IV
		Air	
20		2104.38	IV
20		2117.90	IV
20		2217.99	IV
20		2230.41	IV
20		2240.49	IV
20		2253.38	IV
50		2346.97	IV
20		2357.34	IV
20		2438.93	IV
25		2441.41	IV
30		2482.79	IV
25		2483.57	IV
18		2500.57	IV
20		2508.02	IV
20		2534.03	IV
18		2548.02	IV
40		2555.60	IV
40		2571.04	IV
25		2571.58	IV
15		2589.34	IV
25		2620.35	IV
20		2621.16	IV
20		2642.16	IV
15		2830.53	IV
9		2934.60	IV
10		3019.29	IV
9		3266.52	IV
9		3566.43	IV
9		3741.05	IV
9		4298.57	IV
9		4685.08	IV

Sr V
Ref. 109 — J.J.W.

Intensity		Wavelength	
		Vacuum	

Strontium (Cont.)

Intensity	Wavelength	
10	517.28	V
6	540.51	V
25	578.01	V
30	624.93	V
25	642.23	V
50	649.21	V
20	659.15	V
25	660.94	V
9	669.93	V
35	686.23	V
6	715.79	V
12	747.82	V
9	862.32	V

SULFUR (S)
Z = 16

S I and II
Ref. 144, 209, 210, 266—R.L.K.

Intensity	Wavelength	
	Vacuum	
40	906.9	II
40	910.5	II
40	912.7	II
40	937.4	II
40	937.7	II
20	996.0	II
20	1000.5	II
20	1014.4	II
20	1019.5	II
20	1096.6	II
40	1102.3	II
20	1131.0	II
20	1131.6	II
40	1234.1	II
40	1250.5	II
110	1253.8	II
110	1259.5	II
275	1270.782	I
250	1277.216	I
280	1295.653	I
275	1302.337	I
235	1302.863	I
235	1303.110	I
245	1303.430	I
260	1305.883	I
265	1310.194	I
355	1316.542	I
290	1316.618	I
375	1323.515	I
355	1326.643	I
775	1381.552	I
710	1385.510	I
960	1388.435	I
640	1389.154	I
775	1392.588	I
1000	1396.112	I
300	1409.337	I
510	1425.030	I
425	1433.280	I
300	1436.968	I
300	1448.229	I
425	1472.972	I
550	1473.995	I
300	1474.380	I
355	1481.665	I
485	1483.039	I
300	1483.233	I
330	1485.622	I
390	1487.150	I
680	1666.688	I
640	1687.530	I
710	1807.311	I
680	1820.343	I
640	1826.245	I
710	1900.286	I
550	1914.698	I
	Air	
20	2629.1	II
40	2670.0	II
40	2847.7	II
285	3867.6	I
285	3902.0	I
360	3933.3	II
450	4120.8	I
280	4142.3	II
360	4145.1	II
450	4153.1	II
450	4162.7	II
450	4694.1	I
285	4695.4	I
160	4696.2	I
280	4716.2	II
450	4815.5	II
360	4924.1	II
450	4925.3	II
285	4993.5	I
360	5428.6	II
650	5432.8	II
1000	5453.8	II
1000	5473.6	II
1000	5509.7	II
280	5564.9	II
1000	5606.1	II
450	5640.0	II
450	5640.3	II
280	5647.0	II
650	5659.9	II
450	5664.7	II
160	5706.1	I
450	5819.2	II
450	6052.7	I
280	6286.4	II
450	6287.1	II
450	6305.5	II
450	6312.7	II
280	6384.9	II
280	6397.3	II
280	6398.0	II
360	6413.7	II
160	6743.6	I
285	6748.8	I
450	6757.2	I
450	7579.0	I
450	7629.8	I
285	7686.1	I
450	7696.7	I
1000	7924.0	I
160	7928.8	I
285	7930.3	I
450	7931.7	I
450	7967.4	I
450	7967.4	II
450	8314.7	I
450	8314.7	II
450	8585.6	I
285	8680.5	I
450	8694.7	I
360	8874.5	I
110	8882.5	I
220	8884.2	I
160	9035.9	I
450	9212.9	I
450	9228.1	I
450	9237.5	I
285	9413.5	I
285	9421.9	I
285	9437.1	I
650	9649.9	I
450	9672.3	I
450	9680.8	I
450	9693.7	I
285	9697.3	I
285	9739.7	I
110	9741.9	I
285	9932.3	I
285	9949.8	I
285	9958.9	I
285	10455.5	I
70	10456.8	I
285	10459.5	I

S III
Ref. 209, 210 — R.L.K.

Intensity	Wavelength	
	Vacuum	
70	729.5	III
110	732.42	III
70	735.2	III
70	738.5	III
70	789.0	III
70	796.7	III
70	824.9	III
70	836.3	III
285	1077.1	III
70	1194.0	III
70	1201.0	III
	Air	
110	2460.5	III
110	2489.6	III
160	2496.2	III
160	2499.1	III
220	2508.2	III
70	2636.9	III
220	2665.4	III
70	2680.5	III
110	2691.8	III
110	2702.8	III
220	2718.9	III
110	2721.4	III
220	2726.8	III
220	2731.1	III
110	2741.0	III
285	2756.9	III
110	2775.2	III
160	2785.5	III
70	2797.4	III
70	2856.0	III
110	2863.5	III
160	2904.3	III
70	2964.8	III
160	2986.0	III
70	3234.2	III
70	3324.9	III
110	3497.3	III
160	3632.0	III
70	3662.0	III
110	3709.4	III
160	3717.8	III
160	3838.3	III
160	3928.6	III
360	4253.6	III
110	4285.0	III
70	4332.7	III

S IV
Ref. 29, 202, 209 — R.L.K.

Intensity	Wavelength	
	Vacuum	
20	519.3	IV
20	520.1	IV
40	520.8	IV
20	522.0	IV
20	522.5	IV
20	551.2	IV
40	652.5	IV
40	653.0	IV
70	653.6	IV
40	654.0	IV
70	655.6	IV
20	655.9	IV
110	657.3	IV
40	660.9	IV
160	661.4	IV
40	663.7	IV
40	664.8	IV
70	666.1	IV
110	744.9	IV
110	748.4	IV
110	750.2	IV
110	753.8	IV
40	798.3	IV
70	800.5	IV
70	804.0	IV
70	809.7	IV
110	816.0	IV
160	1062.7	IV
160	1073.0	IV
70	1073.5	IV
20	1108.4	IV
20	1110.9	IV
20	1624.0	IV
20	1629.2	IV
	Air	
20	2387.0	IV
40	2398.9	IV
110	3097.5	IV
40	3117.7	IV

S V
Ref. 29 — R.L.K.

Intensity	Wavelength	
	Vacuum	
5	437.4	V
5	438.2	V
5	439.6	V
40	658.3	V
70	659.8	V
110	663.2	V
5	676.2	V
5	677.3	V
20	678.1	V
40	680.3	V
110	680.9	V
40	681.6	V
5	686.2	V
5	686.9	V
5	689.8	V
5	691.7	V
20	693.5	V
285	786.5	V
160	849.2	V
110	852.2	V
220	854.8	V
110	857.9	V
110	860.5	V
20	883.6	V
20	884.5	V
5	885.8	V
20	900.9	V
5	902.8	V
20	905.9	V

TANTALUM (Ta)
Z = 73

Ta I and II
Ref. 1 — C.H.C.

Intensity		Wavelength	
		Air	
1100		2140.13	II
1500		2146.87	II
740		2150.62	II
600		2165.01	II
740		2178.03	II
1200		2182.71	II
540		2193.20	II
1100		2193.88	II
1500		2196.03	II
1500		2199.67	II
500		2207.14	II
1400	d	2210.03	II
		2210.19	II
420		2215.60	II
1400		2239.48	II
240		2248.48	II
480		2249.79	II
1200		2250.76	II
260		2254.86	II
440		2255.77	II
360		2256.51	II
500		2258.71	II
840		2261.42	II
260		2261.62	II
990		2262.30	II
220		2269.56	II
740		2271.85	II
990		2272.59	II
200		2279.85	I
320		2282.19	II
130		2285.02	II
790		2285.25	II
600		2286.59	II
240		2287.27	II
990		2289.16	II
180		2292.54	II
160		2295.18	
160		2301.47	II
440		2302.24	II
440		2302.93	II
300		2303.49	II
100		2308.46	II
440		2312.60	II
420		2315.46	II
260		2319.16	II
100		2331.29	II
690		2331.98	II
550		2332.19	II
110		2334.13	II
180		2334.88	II
140		2335.75	II
300		2338.28	II
200		2340.94	II
200		2341.61	II
130		2343.64	II
100		2346.42	II
90		2351.99	II
170		2353.86	II
120		2355.22	II
170		2356.05	II
140		2356.90	II
250		2357.30	I
170		2359.16	II
260		2361.09	I
160		2362.78	II
130		2363.32	II
600		2364.24	II
50		2367.24	II
150		2369.32	II
300		2370.76	II
320		2371.58	I
100		2373.94	II
70		2375.91	I
150		2378.31	II
440		2381.13	II
240		2381.52	II
170		2383.72	II
240		2384.28	II
130		2385.73	I
1400		2387.06	II
80		2388.37	II
160		2389.11	II
70		2396.30	I
110		2399.15	I
50		2399.92	II
2400		2400.63	II
140		2402.13	II
100		2403.68	II
130		2406.55	I

Tantalum (Cont.)

130		2408.26	II
120		2414.32	I
240		2415.21	II
320		2416.89	II
220		2417.86	II
150		2418.77	II
140		2421.03	I
150		2421.85	II
170		2423.48	II
130		2425.91	II
360		2427.64	I
360		2429.71	II
170		2431.06	II
480		2432.70	II
130		2433.59	II
130		2436.51	II
110		2437.07	I
110		2438.64	II
200		2439.91	I
130		2442.39	I
100		2444.13	II
100		2447.17	I
100		2454.48	I
100		2458.68	I
100		2460.55	I
160		2463.82	II
130		2466.99	II
130		2467.37	II
380		2470.90	II
120		2471.38	I
120		2472.13	I
150		2473.13	I
120		2473.31	II
600		2474.62	I
120		2475.33	I
200		2476.67	II
150		2478.22	I
120		2481.86	II
100		2482.10	I
100		2482.58	II
100		2484.04	II
500		2484.95	I
120		2486.70	I
600		2488.70	II
500		2490.46	I
600		2504.45	I
600		2507.45	I
240		2512.65	I
1200	d	[illegible]	I
600		2532.12	II
240		2545.49	II
240		2546.80	I
460	d	2551.07	I
460		2554.62	II
240		2555.05	I
1200		2559.43	I
460		2562.10	I
340		2571.51	II
430		2573.54	I
390		2573.79	I
600		2577.37	II
340		2577.78	I
210		2580.16	I
340		2584.03	II
430		2593.08	I
410		2593.66	II
310		2594.25	II
560		2595.26	I
310	l	2596.45	II
220		2600.14	I
600		2603.49	II
1400		2608.63	I
210		2609.00	I
310	d	2611.34	I
340		2615.46	I
310		2615.66	I
1200		2635.58	II
470		2636.67	I
860		2636.90	I
510		2646.22	I
600		2646.37	I
2400		2647.47	I
270		2651.22	II
2600		2653.27	I
1900		2656.61	I
1500		2661.34	I
220		2665.60	II
220		2668.07	I
600		2668.62	I
770		2675.90	II
270		2680.06	II
220		2680.66	II
600		2684.28	I
1500		2685.17	II
340		2691.31	I
260		2692.40	I
470		2694.52	II
240		2696.81	I
1000		2698.30	I
470		2706.69	I
310		2709.27	II
1200		2710.13	I
2600		2714.67	I
240		2717.18	I
470		2720.76	I
470		2727.44	II
410		2727.78	I
310		2736.25	II
210		2739.26	II
210		2743.59	I
510		2746.68	I
1200		2748.78	I
860		2749.83	I
410		2752.49	II
1000		2758.31	I
430		2761.68	II
770		2775.88	I
390		2787.69	I
680		2796.34	I
680		2797.76	II
380		2802.07	I
430		2806.30	I
510		2806.58	I
260		2817.10	II
260		2842.82	I
640		2844.25	I
290		2844.46	II
290	c	2845.35	I
560		2848.52	I
1500		2850.49	I
1900		2850.98	I
220		2858.44	II
360		2861.98	I
310		2868.65	I
470		2871.42	I
270		2873.36	I
260		2873.56	I
210		2874.17	I
380		2880.02	I
770		2891.84	I
260		2899.04	I
560		2902.05	I
210		2914.12	I
310		2915.49	I
410		2925.19	I
310		2932.70	I
1700		2933.55	I
470		2940.06	I
1200		2940.22	I
240		2942.14	I
510		2951.92	I
340		2953.56	I
1500		2963.32	I
770		2965.13	II
770		2965.54	I
340		2969.47	I
430		2975.56	I
210		3011.88	I
1800		3012.54	II
290	d	3027.48	I
290		3042.06	II
530		3049.56	I
530		3069.24	I
360		3077.24	I
560		3103.25	I
380		3124.97	I
380		3130.58	I
270		3132.64	I
320		3170.29	I
270		3173.59	I
200		3176.29	I
600		3180.95	I
240		3184.55	I
200		3198.67	I
200		3213.91	II
300		3223.83	I
230		3229.24	I
200		3242.05	I
200		3242.83	I
210		3274.95	II
1100		3311.16	I
210		3317.93	I
680		3318.84	I
330	d	3330.99	II
230		3358.47	I
640		3371.54	I
360		3385.05	I
230		3398.33	I
450		3406.94	I
230		3463.77	I
490		3480.52	I
380		3497.85	I
240		3503.87	I
490		3511.04	I
200		3513.61	I
750		3607.41	I
980		3626.62	I
500		3642.06	I
100		3686.18	I
100		3689.73	I
130		3731.02	I
140		3736.76	I
130		3746.36	I
110		3754.52	I
110		3777.10	I
110		3792.02	I
210		3833.74	II
100		3848.05	I
100		3885.20	I
210		3918.51	I
140		3922.78	I
140		3922.92	I
210		3970.10	I
210		3996.17	I
100		3999.28	I
190		4006.84	I
190		4026.94	I
140		4029.94	I
120		4040.87	I
410		4061.40	I
210		4064.63	I
100		4067.24	I
310		4067.91	I
120		4105.02	I
210		4129.38	I
230		4136.20	I
230		4147.89	I
210		4175.21	I
100		4177.92	I
130		4181.15	I
300		4205.88	I
120		4206.40	I
130		4245.35	I
130		4268.26	I
160	c	4302.98	I
110		4355.14	I
100		4378.82	I
150		4386.07	I
110		4398.45	I
180		4402.50	I
130		4415.74	I
360	c	4510.98	I
190		4530.85	I
130		4551.95	I
170		4565.85	I
340		4574.31	I
260		4619.51	I
130		4669.14	I
450		4681.88	I
130		4691.90	I
150		4740.16	I
220		4756.51	I
120		4768.98	I
220		4812.75	I
110		4920.11	I
100		4921.27	I
110		4926.00	I
150		4936.42	I
200		5037.37	I
100		5067.87	I
110		5115.84	I
100		5141.62	I
100		5143.69	I
330		5156.56	I
110		5212.74	I
110	d	5218.45	I
		5218.66	I
140		5341.05	I
200		5402.51	I
130		5419.19	I
18		5500.68	I
20		5505.66	I
15	c	5516.27	I
90		5518.91	I
9		5521.15	I
10		5523.98	I
13		5528.36	I
10	l	5545.20	I
20		5548.32	I
30		5584.02	I
15		5598.75	I
30		5599.52	I
9	c	5605.50	I
9	c	5617.71	I
40		5620.68	I
13		5628.20	I
20		5635.71	I
40		5640.18	I
150		5645.91	I
130		5664.90	I
30		5688.25	I
40		5699.24	I
15		5704.31	I
25		5706.28	I
30		5715.24	I
8		5716.53	I
23		5746.71	I
30		5755.81	I
15	h	5761.61	I
25		5766.56	I
30	c	5767.91	I
10		5771.93	I
130		5776.77	I
25		5780.02	I
90		5780.71	I
130		5811.10	I
25		5816.51	I
45	c	5843.94	I
13		5849.68	I
15		5866.61	I
240		5877.36	I
130		5882.30	I
90		5901.91	I
30		5916.51	I
90		5918.95	I
15		5925.90	I
15		5930.62	I
23		5931.05	I
20		5931.68	I
18		5935.54	I
130		5939.76	I
240		5944.02	I
25		5951.78	I
18		5960.13	I
190	c	5997.23	I
25	h	6009.89	I
25		6015.90	I
100		6020.72	I
250		6045.39	I
100		6047.25	I
25		6053.70	I
30		6090.82	I
18		6092.06	I
100		6101.58	I
25		6140.07	I
65		6144.56	I
30		6152.54	I
130		6154.50	I
40		6158.84	I
15		6170.46	I
15		6189.66	I
15		6193.11	I
25		6208.37	I
40		6249.79	I
150		6256.68	I
150		6268.70	I
50		6278.34	I
65		6281.33	I
15		6287.36	I
40		6287.91	I
40		6289.34	I
50		6309.06	I
150		6309.58	I
25	c	6312.22	I
75		6325.08	I
50		6332.91	I
65		6341.17	I
30		6346.02	I
75		[illegible]	I
65		6360.84	I
40		6373.06	I
15		6379.07	I
90		6389.45	I
23	h	6392.21	I
65		6428.60	I
250		6430.79	I
13		6437.36	I
40		6444.61	I
30		6445.87	I
200		6450.36	I
20		6455.83	I
30		6459.92	I
380		6485.37	I
18	h	6502.43	I
65		6505.52	I
100		6514.39	I
100		6516.10	I
25	h	6561.60	I
25	Cw	6564.26	I
100		6574.84	I
10		6585.13	I
15	c	6587.16	I
110		6611.95	I
75		6621.30	I
15	Cw	6662.24	I
100		6673.73	I
180		6675.53	I
30		6684.00	I
15		6693.61	I
15		6706.46	I
25		6709.39	I
10		6714.44	I
15	h	6723.61	I
75	c	6740.73	I
40		6754.91	I
13	c	6755.85	I
13		6770.37	I
75		6771.74	I
40	Cw	6774.25	I
40	Cw	6788.99	I
13	c	6790.06	I
13		6799.27	I
40	c	6810.46	I
160	c	6813.25	I
20		6819.36	I
18	c	6824.96	I
13		6832.00	I
15		6850.83	I
15		6865.13	I
210		6866.23	I
180		6875.27	I
40		6877.49	I
15		6896.77	I
40		6900.55	I
150		6902.10	I
140		6927.38	I

Tantalum (Cont.)

Intensity		Wavelength	
140		6928.54	I
8	h	6939.33	I
20	c	6946.87	I
65		6951.26	I
45		6953.88	I
180		6966.13	I
8		6969.49	I
8	c	6971.31	I
9		6971.53	I
23		6983.52	I
110	d	6995.22	I
		6995.49	I
20		7000.21	I
40		7005.07	I
75		7006.96	I
50		7025.03	I
13		7031.51	I
40		7039.07	I
15	h	7081.30	I
20		7085.40	I
23		7093.02	I
8		7108.05	I
15	c	7117.52	I
20		7121.27	I
40		7125.72	I
150		7148.63	I
110		7172.90	I
13	c	7174.91	I
13		7191.35	I
8		7233.45	I
30	h	7250.27	I
11	h	7264.82	I
6		7272.29	I
30		7276.96	I
5		7277.54	I
9		7286.36	I
13	c	7296.32	I
140		7301.74	I
20		7319.84	I
11	c	7322.72	I
13		7325.95	I
11	Cw	7340.19	I
160		7346.41	I
140	c	7352.86	I
100		7356.96	I
90	Cw	7369.09	I
160		7407.89	I
11	c	7435.19	I
23		7440.17	I
30		7467.75	I
23		7486.01	I
30		7520.56	I
6		7569.23	I
9		7590.22	I
6		7649.62	I
11	h	7722.02	I
11		7763.11	I
9	c	7779.67	I
20	c	7842.76	I
100		7882.37	I
30		7950.19	I
5		7952.07	I
6		7998.75	I
6		8022.09	I
75		8026.50	I
5		8029.04	I
15		8039.08	I
8		8053.93	I
15		8068.98	I
5		8100.11	I
13		8128.76	I
9		8158.54	I
5	c	8180.74	I
13	c	8248.95	I
20	d	8264.85	I
75		8281.62	I
11		8389.06	I
5		8415.73	I
25	Cw	8447.62	I
11	h	8550.49	I
15		8575.92	I
10	Cw	8595.84	I

Ta IV
Ref. 411 — R.L.K.

Intensity	Wavelength	
	Vacuum	
10	763.14	IV
32	934.41	IV
67	999.34	IV
65	1063.53	IV
68	1067.17	IV
71	1074.47	IV
71	1086.39	IV
72	1094.60	IV
68	1100.13	IV
79	1116.10	IV
71	1118.83	IV
78	1136.17	IV
66	1138.26	IV
75	1149.72	IV
75	1150.42	IV
75	1151.92	IV
76	1172.51	IV
85	1175.51	IV
80	1189.28	IV
78	1192.52	IV
80	1192.67	IV
78	1211.94	IV
80	1212.68	IV
85	1213.09	IV
70	1214.66	IV
85	1215.53	IV
80	1220.73	IV
80	1220.96	IV
90	1223.73	IV
88	1238.12	IV
95	1240.06	IV
87	1258.34	IV
94	1264.91	IV
98	1272.42	IV
94	1275.48	IV
86	1275.94	IV
92	1308.51	IV
85	1311.35	IV
87	1315.58	IV
81	1325.19	IV
92	1332.38	IV
86	1343.30	IV
75	1350.46	IV
92	1365.88	IV
79	1376.62	IV
78	1388.80	IV
91	1398.78	IV
93	1413.40	IV
79	1430.11	IV
83	1441.54	IV
91	1454.32	IV
92	1464.41	IV
93	1469.82	IV
90	1495.25	IV
95	1514.19	IV
70	1525.69	IV
82	1565.97	IV
82	1584.64	IV
84	1594.91	IV
85	1607.70	IV
84	1631.65	IV
79	1639.82	IV
84	1668.76	IV
84	1676.45	IV
85	1712.16	IV
85	1716.13	IV
82	1753.90	IV
82	1759.04	IV
79	1763.03	IV
83	1865.92	IV
84	1901.63	IV
84	1907.66	IV
84	1924.75	IV
77	1940.25	IV
79	1985.68	IV
82	1989.44	IV
	Air	
85	2055.75	IV
83	2079.01	IV
75	2111.53	IV
90	2199.58	IV
90	2207.64	IV
68	2697.42	IV
10	3076.06	IV

Ta V
Ref. 426 — R.L.K.

Intensity	Wavelength	
	Vacuum	
20	478.29	V
60	493.07	V
200	841.31	V
1000	890.87	V
500	947.30	V
100	990.29	V
200	1066.64	V
200	1140.49	V
500	1213.42	V
100	1242.98	V
5000	1392.56	V
7000	1709.10	V

TECHNETIUM (Tc)
Z = 43

Tc I and II
Ref. 35 — C.H.C.

Intensity		Wavelength	
		Air	
15		2106.23	II
20		2116.44	II
15		2119.41	I
30		2156.27	I
30		2185.39	I
30		2189.06	I
40		2193.35	I
10		2266.22	II
10		2282.12	I
10		2282.71	I
50		2285.45	I
100		2298.08	I
30		2416.22	I
50		2423.23	I
20		2424.54	I
20		2435.83	I
10		2436.99	I
80	w	2463.69	I
20		2465.09	I
30		2466.87	I
20		2475.11	I
50		2480.70	I
50		2483.22	I
20		2486.50	I
25		2492.72	I
20		2493.43	I
100		2496.77	II
30		2510.17	I
80		2529.34	II
500		2543.23	II
60		2544.81	II
50		2547.92	II
50		2558.61	II
50		2567.01	II
30		2575.06	II
80		2576.28	II
40		2577.86	II
300	Hl	2578.79	I
200		2589.86	I
20	w	2590.19	I
100		2592.82	I
20		2597.19	II
500		2608.86	I
1000	c	2609.99	II
1500		2614.23	I
1000		2615.87	I
30		2618.28	I
200		2634.91	II
80		2636.36	I
30		2641.26	I
100		2642.37	I
40		2644.50	II
1000	c	2647.01	II
300	c	2649.21	I
100		2652.35	II
30		2653.57	I
100		2654.31	I
120		2660.88	I
100		2662.30	I
80		2681.19	II
60		2683.14	I
80		2683.89	I
80		2693.11	I
50		2696.64	I
70		2702.27	I
40		2702.96	II
100		2707.90	II
1000		2708.78	I
30		2715.20	I
30		2723.55	I
1000		2726.69	I
30	c	2728.47	I
500		2730.53	I
300		2732.87	I
150		2736.23	I
60	c	2736.83	II
100		2737.97	I
20	c	2738.83	II
100		2755.76	I
100		2762.13	I
200		2762.34	I
60		2765.95	I
500		2766.89	I
20		2777.31	II
150		2778.91	I
25		2781.22	I
1000		2782.05	I
500		2785.59	I
40		2788.89	I
500		2789.25	I
100		2794.53	I
80		2795.65	I
200		2795.78	II
1000		2802.81	I
150		2803.02	I
500		2808.36	I
50	c	2809.65	II
500		2811.61	II
30		2814.86	I
40		2819.46	I
100		2821.35	II
200		2828.04	I
60		2831.18	II
50		2840.38	II
60		2845.04	I
10		2846.39	II
60		2849.20	I
150		2850.96	I
500	h	2857.13	I
2000	c	2859.11	I
500		2864.49	I
100		2868.09	I
1000		2887.73	I
100		2888.46	I
30		2889.20	II
200		2893.16	I
150		2893.45	I
200		2894.32	I
1000		2896.34	I
40		2903.81	I
1000		2913.15	I
500		2921.91	I
20	c	2923.34	II
1000		2928.20	I
80		2933.89	I
200		2955.93	I
200		2973.65	I
100		2979.34	I
150		2985.36	I
100		3010.83	I
300		3017.23	I
150		3021.56	I
100		3022.66	I
200		3023.68	I
80		3025.26	I
300	w	3026.89	I
50		3033.16	I
80		3034.57	I
40		3036.88	I
20		3037.90	II
100		3038.23	I
40		3042.64	I
100	h	3051.55	I
40		3052.47	I
80		3062.11	I
200		3062.36	I
300		3064.67	I
100	c	3066.60	I
120	c	3068.34	I
80		3076.24	I
150		3089.34	I
1000		3099.10	I
200		3099.52	I
60		3108.25	I
40		3109.15	I
60		3115.98	I
80		3119.17	I
40		3119.66	I
700		3122.64	I
1500		3131.23	I
40	c	3150.26	I
300		3161.67	I
3000		3173.30	I
200		3180.30	I
2000		3182.37	I
2000		3183.11	I
800	c	3195.20	II
40	w	3197.53	I
300	c	3202.83	I
1000		3212.02	II
40		3220.74	I
60		3230.02	I
1000		3237.02	II
100		3241.84	I
500		3244.19	I
300		3252.05	I
40		3256.10	I
40		3261.94	I
100		3287.14	I
30		3298.84	II
100		3300.77	I
80		3305.89	I
200		3310.65	I
150		3313.65	I
200		3325.55	I
150	c	3327.10	I
100		3330.77	I
50		3332.47	I
60		3350.56	I
50	c	3350.83	I
400		3366.75	I
40		3386.67	I
50		3392.23	I
300		3394.18	I
60		3396.90	I
40		3397.83	I
300		3398.33	I
200		3402.10	I
200	c	3403.93	I
80		3405.33	I
80		3407.28	I
50		3408.33	I
40		3411.80	I
40	c	3418.20	I
100		3419.10	I

Technetium (Cont.)

Intensity		Wavelength	Spectrum
60		3427.85	I
40		3431.75	I
200		3434.70	I
40		3435.68	I
150		3437.44	I
80		3438.73	I
200	c	3443.47	I
200		3451.05	I
200		3456.85	I
400		3457.24	I
40		3457.60	I
5000	c	3466.28	I
150		3470.51	I
80		3475.18	I
1000		3475.59	I
60		3484.62	I
1000	c	3486.23	I
100		3490.30	I
400		3493.39	I
500		3494.62	I
40		3499.14	I
1000		3500.70	I
200		3501.24	I
800	c	3502.70	I
100	c	3507.19	I
100		3508.27	I
100		3510.91	I
800		3525.83	I
300		3526.18	I
100		3529.83	I
150		3534.88	I
500		3535.51	I
300		3538.12	I
800		3538.68	I
2000	c	3541.77	I
6000	c	3549.72	I
4000	c	3550.64	I
300		3559.75	I
800		3560.32	I
100		3565.22	I
800		3568.85	I
100		3570.65	I
100		3575.42	I
1000		3580.06	I
600		3581.26	I
800		3582.08	I
2000		3582.63	I
4000		3587.94	I
[illegible]		3593.47	I
300		3594.57	I
1000	c	3595.66	I
1000	c	3607.32	I
200		3607.62	I
2000	c	3608.27	I
200		3618.94	I
1000	c	3627.36	I
200		3630.39	I
3000	c	3635.15	I
10000	c	3636.07	I
1000		3638.22	I
200		3638.85	I
900		3639.38	I
400		3640.23	I
1000	c	3648.04	I
600		3651.47	I
1000	c	3658.59	I
400	c	3661.45	I
200		3664.92	I
1000		3679.15	I
300		3680.32	I
5000		3684.74	I
300		3692.76	I
800		3703.83	I
300		3704.80	I
200		3706.70	I
200		3707.63	I
200		3708.26	I
1000		3712.26	I
300		3712.82	I
500		3715.94	I
10000		3718.86	I
1500		3723.67	I
2000		3724.40	I
5000		3726.35	I
200		3727.36	I
400	c	3729.18	I
500		3731.74	I
300		3737.42	I
400		3745.01	I
1000		3746.15	I
5000		3746.84	I
1000		3752.13	I
4000		3754.37	I
1000		3758.54	I
2000		3761.81	I
5000		3768.77	I
3000		3771.03	I
500		3777.27	I
2000		3779.37	I
3000	c	3780.68	I
500		3784.06	I
200		3786.06	I
500		3791.28	I
300		3791.73	I
200		3797.44	I
1000		3797.77	I
200		3814.67	I
300		3816.89	I
300		3824.47	I
500		3828.54	I
200		3830.35	I
200		3832.45	I
600		3832.82	I
1500		3837.56	I
800		3841.31	I
800		3845.97	I
500		3847.60	I
300		3851.22	I
500	c	3856.73	I
200		3863.07	I
400		3864.11	I
1000		3868.24	I
200		3875.66	I
500	c	3879.16	I
600	c	3880.72	I
300	w	3892.12	I
200		3893.22	I
600		3899.83	I
300		3919.38	I
300	c	3923.66	I
200		3927.57	I
200		3933.70	I
4000	c	3946.57	I
2000		3947.09	I
200		3955.73	I
300		3979.64	I
500		3980.35	I
10000	c	3984.97	I
400		3987.78	I
300		3994.04	I
2000		3994.51	I
200		3996.97	I
300		4004.69	I
500		4007.14	I
1000		4012.00	I
400		4016.68	I
600		4017.22	I
2000		4020.76	I
20000	c	4031.63	I
1000		4039.25	I
200		4041.78	I
10000	c	4049.11	I
500		4051.95	I
200		4053.18	I
200	c	4056.08	I
400		4083.54	I
10000		4088.71	I
200		4093.69	I
15000		4095.67	I
1000		4110.22	I
10000		4115.08	I
600		4119.27	I
8000		4124.22	I
1000		4128.27	I
300		4134.81	I
300		4139.12	I
800		4139.85	I
400		4141.27	I
6000		4144.95	I
3000		4145.08	I
200		4147.62	I
10000		4165.61	I
500		4167.42	I
1000		4169.68	I
4000		4170.27	I
5000		4172.53	I
1000		4176.28	I
800		4186.51	I
300		4218.61	I
10000	c	4238.19	I
20000		4262.27	I
1000		4262.69	I
800		4274.97	II
800		4278.90	I
30000		4297.06	I
400	c	4336.86	I
400		4358.49	I
200		4359.26	I
1000		4429.59	I
1000		4481.53	I
3000		4487.06	I
400		4495.03	I
1000		4515.98	I
10000		4522.84	I
2000		4539.53	I
400		4542.09	I
400		4552.20	I
800		4552.85	I
1000		4557.05	I
2000		4564.54	I
1000		4578.45	I
1000		4593.35	I
300	c	4609.16	I
1000		4616.86	I
200	c	4622.69	I
300		4624.96	I
1000		4630.57	I
200		4633.15	I
3000		4637.50	I
500		4643.28	I
2000		4648.33	I
2000	c	4660.21	I
2000		4669.30	I
400		4672.17	I
200		4678.90	I
400		4689.36	I
300		4694.28	I
1000		4706.92	I
200		4714.22	I
2000		4717.77	I
500	c	4719.02	I
4000	c	4719.28	I
200	c	4736.51	I
10000		4740.61	I
500		4749.61	I
1000		4752.72	I
200		4762.36	I
4000		4771.54	I
200		4773.89	I
200		4783.92	I
500		4785.60	I
200	c	4790.48	I
250		4791.62	I
300		4799.98	I
100		4805.69	I
100		4809.42	I
500		4816.79	I
10000		4820.74	I
300		4831.35	I
1000		4834.37	I
1000		4835.39	I
100		4841.36	I
20000		4853.59	I
100		4857.21	I
100	c	4862.19	I
10000		4866.73	I
200		4870.77	I
100		4888.70	I
150	c	4890.88	I
8000		4891.92	I
150		4892.49	I
1000		4908.51	I
2000		4909.57	I
500		4913.02	I
150	c	4914.70	I
200	c	4920.67	I
300		4923.60	I
400		4948.06	I
5000		4976.34	I
400		4995.00	I
200		5002.67	I
100		5005.74	I
200	c	5014.52	I
500		5026.24	I
300		5026.79	I
150		5027.89	I
80		5032.45	I
300		5055.27	I
60		5058.33	I
500		5060.69	I
80		5090.74	I
5000		5096.28	I
200	c	5103.24	I
500		5104.32	I
200		5109.81	I
100		5120.60	I
500		5139.26	I
500		5150.63	I
2000		5161.81	I
2000		5174.81	I
100		5206.56	I
200		5225.55	I
200	c	5260.22	I
200	c	5261.44	I
1000		5275.51	I
800		5285.07	I
100		5305.31	I
400		5314.96	I
600		5320.20	I
200		5334.79	I
500	c	5353.48	I
200		5356.63	I
300		5358.65	I
200		5360.14	I
500	h	5375.20	I
150	c	5423.05	I
200		5447.40	I
500	c	5451.90	I
100		5455.95	I
300		5471.96	I
70		5483.01	I
60		5485.37	I
80	c	5506.89	I
150		5524.11	I
100		5528.23	I
200		5541.94	I
80		5543.63	I
100		5550.53	I
3000	c	5589.02	I
200		5602.23	I
2000	c	5620.45	I
300		5629.94	I
1500		5642.13	I
800		5644.94	I
100		5656.00	I
60		5672.15	I
200		5687.30	I
200		5689.05	I
700		5725.31	I
500	c	5771.47	I
100		5794.65	I
80	c	5799.85	I
100	c	5814.24	I
200		5831.48	I
150	c	5836.33	I
150		5923.36	I
1000	c	5924.47	I
200		5926.29	I
600	c	5931.93	I
60		6032.36	I
60		6047.99	I
200		6065.09	I
800		6085.23	I
300		6099.39	I
500	c	6102.96	I
1000		6120.68	I
1000		6130.80	I
150	c	6132.23	I
100		6184.70	I
800		6192.66	I
600	c	6244.18	I
100		6312.18	I
100		6354.86	I
100		6356.73	I
80		6389.87	I
100		6408.83	I
1000		6455.90	I
600	c	6461.93	I
100		6470.27	I
200	c	6491.68	I
200		6526.82	I
150		6579.24	I
500	c	6625.57	I
300	c	6673.66	I
100		6687.10	I
80		6786.00	I
70		6798.63	I
60		6838.90	I
150		7002.37	I
100		7016.57	I
500	c	7086.18	I
60		7093.12	I
200		7141.28	I
200	c	7157.62	I
70		7256.08	I
100		7322.38	I
80		7329.14	I
100		7396.80	I
100	c	7402.61	I
200		7405.36	I
60		7427.15	I
150		7434.12	I
600		7452.49	I
60		7461.59	I
80		7534.95	I
800		7540.26	I
80		7543.39	I
200		7574.02	I
500		7579.26	I
90		7624.53	I
100		7684.45	I
500		7697.37	I
80	c	7698.19	I
800	c	7793.04	I
60		7798.28	I
60		7816.74	I
800		7817.72	I
100		7856.38	I
200		7861.44	I
400	d	7871.25	I
60	c	7874.76	I
70		7965.45	I
500		7999.73	I
200		8126.55	I
200		8170.55	I
150		8205.27	I
100		8206.49	I
150		8211.31	I
500	c	8237.08	I
200		8308.15	I
200		8309.16	I
60		8315.50	I
100		8531.06	I
100		8543.61	I
100	c	8707.21	I
100	c	8737.93	I
200	c	8829.82	I

TELLURIUM (Te)
Z = 52

Te I and II
Ref. 1, 344—347—C.H.C.

Tellurium (Cont.)

Intensity		Wavelength	
		Vacuum	
6		799.60	II
8		802.28	II
6		942.62	II
6		1003.73	II
6		1007.80	II
5		1014.27	II
5		1022.79	II
6		1057.00	II
8		1059.51	II
6		1068.86	II
8		1077.66	II
6		1090.11	II
6		1144.04	II
5		1153.10	II
10		1161.42	II
10		1174.34	II
12		1175.79	II
9		1208.54	II
5		1213.00	II
9		1220.98	II
9		1253.62	II
9		1270.52	II
7		1274.76	II
8		1306.53	II
10		1324.92	II
7		1336.42	II
7		1345.20	II
9		1363.24	II
8		1366.73	II
10		1374.80	II
6		1395.22	II
6		1439.52	II
6		1465.25	II
7		1489.56	II
8		1607.99	II
10		1608.41	II
10		1613.15	II
6		1638.91	II
5		1655.4	I
5		1688.5	I
6		1700.0	I
6		1701.58	II
5		1708.0	I
6		1751.0	I
5		1759.4	I
5		1775.0	I
6		1795.7	I
6		1796.3	I
10		1822.4	I
6		1825.5	I
6		1850.6	I
6		1852.1	I
6		1853.8	I
8		1857.2	I
6		1860.4	I
3		1962.88	II
7		1994.83	I
		Air	
6		2000.2	I
26000		2002.02	I
8		2070.9	I
6500		2081.16	I
18000		2142.81	I
3200		2147.25	I
360		2159.85	I
9		2208.74	I
10		2255.49	I
500		2259.02	I
10		2265.52	I
20		2373.06	II
1200		2383.26	I
1500		2385.78	I
20		2387.82	II
10		2401.63	II
10		2436.47	II
50		2438.69	II
120		2530.72	I
20		2567.82	II
10		2574.96	II
5		2576.10	II
7		2579.24	II
10		2591.12	II
10		2592.85	II
10		2605.72	II
5		2621.92	II
10		2624.86	II
20		2627.96	II
20		2641.89	II
20		2648.48	II
100		2649.66	II
40		2657.70	II
80		2661.10	II
110		2677.13	I
20		2711.58	II
6		2769.65	I
10		2841.17	II
10		2846.15	II
100		2858.29	II
20		2861.00	II
40		2868.82	II
150		2895.41	II
30		2919.89	II
50		2942.11	II
50		2946.68	II
70		2967.29	II
20		2973.67	II
50		2975.90	II
15		2997.04	II
15		3012.02	II
50		3017.58	II
20		3023.31	II
70		3047.00	II
20		3052.46	II
10		3063.16	II
15		3073.56	II
8		3104.44	II
10		3132.58	II
20		3160.66	II
100		3175.14	I
10		3189.83	II
5		3211.21	II
60		3256.80	II
30		3268.77	II
30		3282.63	II
40		3321.92	II
40		3323.11	II
60		3329.22	II
60		3352.10	II
60		3362.79	II
25		3374.10	II
150		3406.79	II
20		3419.63	II
50		3442.25	II
40		3455.12	II
20		3456.88	II
20		3480.32	II
40		3483.67	II
20		3486.11	II
50		3521.11	II
50		3552.19	II
100		3611.78	II
50		3617.57	II
40		3644.46	II
20		3679.26	II
30		3725.66	II
40		3797.22	II
20		3800.92	II
20		3905.67	II
20		3918.54	II
30		3931.49	II
20		3947.98	II
40		3969.22	II
25		3975.94	II
20		3981.77	II
50		4006.52	II
20		4011.69	II
30		4029.73	II
40		4047.17	II
30		4048.88	II
15		4073.48	II
30		4101.04	II
70		4127.32	II
30		4163.55	II
100		4169.77	II
30		4179.29	II
25		4211.31	II
80		4225.73	II
30		4246.47	II
20		4251.15	II
100		4261.11	II
30		4264.36	II
60		4273.43	II
80		4285.85	II
40		4320.90	II
30		4361.28	II
150		4364.00	II
30		4377.12	II
75		4385.10	II
60		4396.00	II
170		4478.63	II
80		4537.07	II
100		4557.78	II
70		4630.62	II
100		4641.12	II
180		4654.37	II
200		4686.91	II
100		4696.38	II
100		4706.53	II
100		4766.05	II
70		4771.56	II
100		4784.87	II
100		4827.14	II
150		4831.28	II
150		4842.90	II
130		4865.12	II
200		4866.24	II
80		4885.22	II
80		4904.44	II
60		4961.88	II
60		5000.82	II
8		5083.0	I
7		5148.7	I
50		5449.84	II
50		5487.95	II
150		5576.35	II
150		5649.26	II
100		5666.20	II
200		5708.12	II
7		5733.5	I
150		5755.85	II
8		5789.1	I
50		5936.15	II
100		5974.68	II
8	d	6273.5	I
8	h	6349.7	I
50		6367.13	II
8		6405.9	I
7	h	6456.7	I
8	h	6613.4	I
10		6648.58	II
8		6660.2	I
8	h	6690.0	I
10	h	6790.0	I
20	h	6837.6	I
20	h	6854.7	I
10		7016.06	II
10		7039.13	II
15	h	7191.1	I
10		7236.62	II
20	h	7263.5	I
8		7280.9	I
10		7289.26	II
10		7445.39	II
12		7460.98	II
15		7468.75	II
10		7481.26	II
10		7556.8	I
6		7688.61	II
15		7759.1	I
8		7818.79	II
8		7861.61	II
15		7921.69	II
15		7943.14	II
10		7950.34	II
20		7972.9	I
6		8056.15	II
30	h	8061.4	I
10	h	8082.5	I
10		8122.44	II
8		8130.39	II
8		8154.47	II
20		8186.44	II
6		8190.94	II
10		8251.5	I
15		8273.53	II
10		8276.6	I
10		8291.1	I
15		8355.8	I
10		8372.12	II
7		8469.8	I
8		8492.2	I
8		8500.8	I
12		8521.4	I
10		8535.68	II
12		8575.78	II
10		8604.63	II
8		8621.68	II
7	h	8632.1	I
15		8672.95	II
12		8701.09	I
10		8733.81	II
205		8758.18	I
12		8831.52	I
18		8851.15	I
6		8897.92	II
81		9004.37	I
18		9043.39	I
12		9196.80	I
15		9206.78	I
17		9207.64	I
12		9469.00	I
5660		9722.74	I
185		9785.54	I
109		9842.30	I
532		9868.92	I
118		9902.61	I
689		9956.30	I
37		9959.93	I
325		9977.13	I
136		9979.31	I
45		9985.85	I
5950		10051.41	I
4097		10091.01	I
104		10099.57	I
279		10106.05	I
381		10118.08	I
296		10151.06	I
397		10300.56	I
205		10323.05	I
745		10493.57	I
197		10509.86	I
1880		10918.34	I
298		11007.80	I
10200		11089.56	I
508		11163.74	I
6620		11487.23	I
280		11978.96	I
188		12566.24	I
389		12589.19	I
161		12805.50	I
400		13104.18	I
1580		13247.75	I
483		13316.63	I
217		14037.09	I
144		14072.53	I
434		14335.74	I
220		14417.46	I
1050		14513.51	I
129		14554.68	I
1480		15452.45	I
2430		15546.23	I
3760		16403.90	I
1960		17303.54	I
2780		18291.59	I
394		18777.30	I
269		19623.52	I
239		20147.54	I
1020		21043.73	I
464		21602.50	I
37		21799.64	I
74		22555.29	I
48		22755.66	I
27		23294.94	I
17		23978.70	I
25		24059.04	I
13		26428.62	I
38		26539.17	I
15		26553.74	I
7		27179.26	I

TERBIUM (Tb)
Z = 65

Tb I and II
Ref. 1 — C.H.C.

Intensity	Wavelength	
	Air	
29	2577.73	II
110	2584.61	II
29	2590.31	II
29	2591.42	II
24	2592.64	II
55	2597.71	II
40	2602.93	II
110	2608.57	II
40	2616.90	
130	2628.69	II
55	2655.96	II
50	2661.40	II
24	2661.64	II
55	2667.64	II
50	2668.86	II
140	2669.29	II
40	2674.13	II
40	2674.69	II
29	2678.15	II
40	2683.97	II
35	2687.82	II
50	2691.90	II
35	2693.05	II
55	2693.41	II
35	2695.46	II
50	2696.83	II
190	2704.07	II
130	2736.24	II
160	2759.47	II
270	2769.53	II
130	2784.49	II
180	2800.51	II
250	2802.75	II
250	2809.30	II
180	2812.64	II
190	2852.14	II
110	2857.68	II
230	2886.29	II
160	2894.45	II
320	2897.44	II
160	2898.86	II
110	2901.54	II
110	2910.30	II
160	2914.75	II
160	2915.30	II
190	2915.60	II
120	2916.24	II
120	2918.89	II
120	2924.16	II
120	2924.53	II
160	2932.89	II
150	2940.05	II
250	2956.21	II
170	2968.87	II
170	2977.78	II
110	2987.03	II
110	2988.57	II

Terbium (Cont.)

130		2996.00	II
110		2999.03	II
130		3005.52	II
170		3009.30	II
230		3010.59	II
230		3016.18	II
130		3019.17	II
170		3020.29	II
110		3023.43	II
170		3027.33	II
230		3031.60	II
230		3044.96	II
190		3051.13	II
130		3053.24	II
460		3053.55	II
130		3062.78	II
230		3064.09	II
110		3065.69	II
230		3067.20	II
270		3069.03	II
460		3070.05	II
270		3072.60	II
670		3078.86	II
480		3082.36	II
120		3086.78	II
250		3088.43	II
480		3089.58	II
230		3102.54	II
480		3102.96	II
290		3117.89	II
290		3119.62	II
230		3121.94	II
230		3123.05	II
160		3124.54	II
110		3131.35	II
250		3134.26	II
440		3139.64	II
190		3140.06	II
230		3145.22	II
150		3146.67	II
310		3147.04	II
310		3147.15	II
310		3148.71	II
120		3155.62	II
130		3162.42	II
290		3162.93	II
190		3165.74	II
380		3167.31	II
140		3168.32	II
230		3169.84	II
190		3173.76	II
380		3174.66	II
380		3180.54	II
140		3183.88	II
480		3187.26	II
290		3188.03	II
190		3194.69	II
380		3195.60	II
480		3199.56	II
1100		3218.93	II
1200		3219.98	II
250		3230.03	II
250		3231.06	II
210	d	3239.60	II
250		3240.00	II
480		3252.32	II
250		3262.97	II
230	d	3263.87	II
230		3264.90	II
400		3266.40	II
250		3274.14	II
250		3274.33	II
210		3277.32	II
760		3280.31	II
760		3281.40	II
520		3283.10	II
1000		3285.04	II
310		3287.55	II
310		3291.56	II
1500		3293.07	II
210		3295.33	II
310		3298.66	II
210		3304.95	II
420	d	3307.44	II
210		3308.51	II
210		3314.38	II
340	d	3321.15	II
420		3322.28	II
210		3323.38	II
210		3323.89	II
3800		3324.40	II
520		3329.08	II
210		3334.48	II
250		3336.70	II
310		3338.03	II
250		3339.00	II
210		3347.27	II
210		3348.07	II
760		3349.42	II
320		3362.25	II
760		3364.93	II
230	d	3370.61	II
320		3371.50	II
520		3372.36	II
460	d	3372.72	II
520		3375.03	II
320		3378.73	II
520		3378.86	II
320		3382.80	II
210		3390.60	II
380		3391.28	II
270		3398.35	II
320		3399.10	II
270		3400.53	II
210	d	3400.86	II
420		3402.33	II
210		3410.40	II
210		3410.68	II
520		3413.76	II
270		3416.24	II
400	d	3420.34	II
210		3430.61	II
320		3433.26	II
270		3439.72	II
520		3440.37	II
320		3444.58	II
210		3446.40	II
270		3449.46	II
810		3454.06	II
380		3460.38	II
230	d	3462.97	II
620		3468.03	II
270		3471.73	II
270		3472.37	II
810	d	3472.79	II
210		3473.00	II
380		3480.17	II
230		3483.04	II
230		3483.69	II
290	d	3489.51	II
210	d	3492.00	II
270		3494.21	II
270		3495.36	II
810		3500.84	II
570		3507.45	II
5700		3509.17	II
380		3510.10	II
320		3513.10	II
570		3519.76	II
1300		3523.66	II
380		3525.14	II
440		3525.61	II
440		3536.32	II
570		3537.94	II
1100		3540.24	II
810		3543.89	II
310	d	3551.03	
320		3551.96	II
460	d	3558.77	
3200		3561.74	II
480		3562.90	II
570		3565.74	
810		3567.35	II
4200		3568.52	II
1600		3568.98	II
320		3572.07	II
1100		3579.20	II
710		3585.03	II
570		3587.44	II
810		3596.38	II
440		3598.06	II
1600		3600.44	II
320		3604.90	II
320		3611.33	II
320		3614.63	II
320	d	3615.66	II
320		3616.58	II
380		3617.86	II
380		3619.73	II
810		3625.54	II
570		3626.50	II
380		3629.44	II
670		3633.29	II
670		3638.46	II
670		3641.66	II
440		3647.06	II
570		3647.75	II
2300		3650.40	II
810		3654.88	II
2000		3658.88	II
450		3663.12	II
3800		3676.35	II
300		3677.89	II
810		3682.26	II
320		3688.15	II
610		3691.15	II
300		3692.95	II
450		3693.58	I
320		3696.85	II
450		3700.12	I
4700		3702.86	II
300		3703.12	I
2400		3703.92	II
370		3709.30	II
1000	d	3711.76	II
300		3719.45	II
650		3729.91	II
430		3732.39	II
430		3743.09	
650		3745.04	I
870		3747.17	II
870		3747.34	II
1100		3755.24	II
430		3757.44	II
430	d	3757.90	II
650		3759.35	I
350		3761.14	I
1700		3765.14	I
2100		3776.49	II
330		3779.22	
600		3783.53	I
410	d	3787.22	II
410		3789.92	I
390		3792.20	I
600		3793.55	
330		3801.80	II
760	d	3806.85	II
1500		3830.26	I
540		3833.42	I
920	d	3842.50	II
370	d	3845.61	II
3700		3848.73	II
450	d	3869.75	II
3500	w	3874.17	II
330		3883.34	I
480		3888.22	I
490		3894.64	I
330		3895.99	I
330		3896.58	II
330		3897.89	I
2400		3899.20	II
1600		3901.33	I
480		3908.06	I
380		3909.14	I
330		3909.55	I
650		3915.43	I
480		3919.52	II
300		3922.10	II
480		3922.74	II
760		3925.45	II
650		3935.24	II
810	d	3939.52	II
650		3946.89	II
350	d	3958.36	II
3000	d	3978.84	II
1800		3981.87	II
300		3983.85	II
350		3999.40	II
350	d	4002.19	II
970		4002.59	II
1900		4005.47	II
300		4010.04	I
760		4012.75	II
330		4013.26	I
370		4019.14	II
540		4020.47	II
220		4022.88	I
370		4024.77	I
520		4031.66	II
870		4032.28	I
2100		4033.03	II
350		4036.22	I
210		4038.86	I
300		4051.86	II
300		4052.87	II
430		4054.12	I
410		4060.37	I
220		4060.87	II
1300		4061.58	I
220		4063.89	II
390		4066.22	II
260		4075.22	I
390		4081.24	I
210		4086.60	I
210		4092.19	I
260		4094.37	II
260		4094.49	I
260		4103.90	II
650		4105.37	I
300		4112.50	I
260		4119.92	I
280		4143.51	I
1100		4144.41	II
350		4158.53	I
240		4169.09	I
240		4169.32	I
240		4171.05	I
240		4172.60	I
240		4172.82	I
260		4173.47	I
240		4186.21	I
300		4187.16	I
390		4196.74	I
450		4201.00	II
650		4203.74	I
600		4206.49	I
300	Cw	4213.50	I
300		4214.42	II
480		4215.09	I
300		4217.56	I
260		4219.16	I
260		4224.28	I
480	Cw	4226.45	II
260		4231.89	I
480		4232.82	I
300		4235.35	I
370		4255.24	I
480		4258.23	II
260		4263.66	I
650		4266.34	I
330		4269.69	I
220		4275.21	I
760	Cw	4278.52	II
300		4285.13	II
300		4289.70	I
370		4298.36	I
300		4299.90	I
240		4302.95	I
240		4307.18	I
450		4310.42	I
300		4311.56	I
370		4313.25	I
2200		4318.83	I
600		4322.23	I
600		4325.83	II
3000		4326.43	I
240		4328.90	I
600		4332.12	I
870		4336.43	I
600		4337.64	I
1700		4338.41	I
700		4340.62	I
430	Cw	4342.53	I
430	d	4353.20	II
280		4356.09	I
870		4356.81	I
280		4360.16	I
220		4367.30	II
220		4372.02	I
330		4382.45	I
300		4388.23	I
260		4390.91	I
200		4416.27	II
140		4420.19	I
350		4423.10	I
110		4432.72	I
240		4436.12	I
110		[illegible]	I
240		4448.04	I
110		4467.69	I
430		4493.07	I
45	d	4509.04	II
150	h	4511.52	I
45		4512.96	II
75		4514.31	II
45		4519.72	II
45		4525.01	II
45		4529.76	
45	h	4531.83	II
45		4534.13	I
45		4537.14	I
45		4537.23	I
110		4549.07	I
45		4549.72	II
110		4550.45	I
110		4556.46	I
55		4562.24	II
110		4563.69	II
30		4564.85	II
55		4573.19	II
210		4578.69	II
65		4584.84	II
65		4591.56	II
45		4592.38	I
45	h	4604.10	II
30		4611.96	I
45	h	4615.92	II
27		4617.49	I
30		4619.36	II
75	d	4626.32	II
95		4626.94	II
65		4632.07	I
65	h	4636.59	I
30		4636.99	II
85		4641.00	II
210		4641.98	II
260	Cw	4645.31	II
80		4647.23	I
60		4658.38	I
20		4658.73	
80		4662.79	I
50	c	4665.45	I
40		4669.40	I
80		4676.90	I
70	c	4681.87	I
50		4682.52	I
25	c	4682.79	II
80		4688.63	II
80		4693.11	II
30	h	4693.39	II
200		4702.41	II
110		4707.94	II
40	w	4716.07	II
40		4728.16	II
60	Cw	4734.20	I
80		4739.93	I

Terbium (Cont.)

Intensity		Wavelength	
70		4747.80	I
410	Cw	4752.53	II
40		4758.44	II
40		4760.19	II
30		4762.37	II
25		4764.47	II
35		4778.36	II
35		4778.80	II
180		4786.78	I
40	Cw	4789.91	II
30		4801.87	II
100		4813.77	I
60		4837.59	II
25		4840.39	I
30	c	4842.69	II
30		4844.89	II
30		4854.81	I
20		4856.54	II
30		4858.87	II
80		4875.57	II
25		4876.12	II
80		4881.15	II
29		4894.33	
95		4915.90	I
35		4924.09	I
35		4926.83	I
50		4928.93	I
65		4931.79	I
29		4970.99	II
29		4971.42	I
29		4973.04	I
29		4980.16	II
29		4980.56	I
85		4993.82	II
50		4995.84	II
55		4997.95	I
29		5006.10	II
50		5022.16	I
29		5024.24	II
29		5024.65	I
50		5033.12	I
50	w	5042.06	II
55		5054.30	I
55		5065.79	I
110		5078.25	I
24		5080.05	II
24		5081.11	I
75		5089.12	II
24		5089.66	I
24		5101.09	I
24		5108.56	I
35		5118.39	I
24		5120.18	I
50	w	5131.69	I
50	w	5141.08	II
50		5147.58	I
24		5164.27	I
29		5170.13	I
24		5170.61	I
50		5176.51	I
50		5179.97	I
50		5184.59	I
85		5186.13	I
50		5188.48	I
50		5198.86	I
35	w	5202.77	I
40		5204.55	I
40		5207.97	I
40		5214.28	I
40		5221.99	I
120		5228.12	I
40		5235.11	I
75		5248.71	I
75	w	5262.11	II
24		5275.03	I
75		5281.05	I
65		5304.72	I
29		5308.19	I
29		5309.46	I
110		5319.23	I
35		5331.04	I
65	w	5337.90	I
35	d	5338.59	I
24		5347.83	II
160		5354.88	I
75		5369.72	I
75		5375.98	I
29	d	5402.06	II
29		5413.65	I
29		5416.20	I
50		5424.10	II
29	c	5426.43	I
35		5443.38	I
29		5457.00	I
55		5459.81	I
29	w	5470.34	II
24		5481.45	I
55		5509.61	I
50		5514.54	I
65		5524.12	I
24	c	5525.62	I
35		5565.93	I
29	c	5638.80	I
29	c	5685.74	II
40	c	5686.48	I
85	c	5747.58	I
24		5762.66	I
24		5785.18	II
75		5795.64	I
75		5803.13	II
65		5815.36	I
29		5842.97	I
65		5851.07	I
65		5870.62	I
35		5898.84	I
24		5902.40	I
35		5904.71	I
65	c	5920.78	I
50	c	5939.38	I
35		5940.17	I
24		5951.17	I
75		5967.34	II
29		6038.97	I
29		6039.38	I
24		6104.29	II
24	c	6292.43	I
35		6331.68	II
24		6334.91	II
24		6446.87	II
35	Cw	6518.68	I
24	c	6574.04	II
35		6581.82	I
30		6607.17	II
90		6677.94	II
40	Cw	6702.61	I
20	c	6706.79	II
30		6785.12	II
130		6794.58	II
40		6874.18	II
55		6896.37	II
45	h	6899.95	I
40		6901.98	I
9		7005.99	II
17		7082.85	II
11		7089.22	II
11		7112.69	I
10		7187.48	I
10	h	7195.89	II
65		7204.28	I
19	h	7234.98	I
40		7257.73	I
17		7311.57	I
45		7348.88	II
10		7398.27	II
15	h	7424.24	II
10	h	7429.62	II
9		7472.15	I
22		7484.54	I
9		7495.45	I
45		7496.12	I
17		7499.69	II
27		7511.40	I
9		7519.77	II
6		7557.59	II
27	h	7582.03	II
27		7587.49	I
45		7590.24	I
65		7596.44	I
17	h	7601.18	II
17		7616.01	II
22	h	7624.05	I
30		7627.81	I
9	h	7639.05	II
8		7672.72	II
8	h	7694.74	II
22	h	7706.16	II
22	h	7726.97	II
30		7737.63	I
22		7793.20	I
8		7807.33	II
16		7832.91	II
30		7855.79	II
15		7864.99	I
6	h	7885.70	I
6		7913.11	I
27		7927.90	II
13		7955.31	I
11		7998.03	I
17		8001.04	I
13	h	8010.16	II
30		8025.42	II
6		8053.80	I
19		8067.35	II
30		8085.06	II
27		8164.17	I
13		8171.70	I
65		8194.82	II
95		8212.57	I
11		8214.33	I
8		8259.08	I
40		8450.06	II
8	h	8465.80	II
13		8502.70	I
30	h	8511.80	I
45		8583.45	II
30		8603.40	I
9		8678.25	I
65		8765.74	II

Tb IV
Ref. 302 — J.R.

Intensity	Wavelength	
	Vacuum	
30	1176.58	IV
30	1192.01	IV
70	1200.58	IV
50	1213.94	IV
80	1221.22	IV
500	1235.04	IV
1000	1259.40	IV
300	1301.48	IV
300	1308.30	IV
600	1311.70	IV
700	1315.12	IV
500	1325.56	IV
1000	1327.67	IV
100	1367.56	IV
400	1367.71	IV
700	1369.64	IV
1000	1373.86	IV
400	1376.46	IV
200	1378.23	IV
300	1381.00	IV
100	1382.83	IV
20	1389.92	IV
200	1516.17	IV
50	1530.10	IV
5000	1595.39	IV
2000	1633.19	IV
300	1649.38	IV
400	1654.75	IV
400	1667.58	IV
200	1672.55	IV
400	1681.98	IV
5	1684.46	IV
100	1685.37	IV
400	1691.95	IV
10	1695.23	IV
300	1698.36	IV
30	1701.60	IV
50	1704.79	IV
20	1705.05	IV
3	1943.94	IV
50	1970.90	IV
	Air	
2000	2027.79	IV
200	2029.22	IV
400	2048.88	IV
200	2078.83	IV
1000	2089.98	IV
1000	2332.54	IV
100	2436.01	IV

THALLIUM (Tl) Z = 81

Tl I and II
Ref. 1, 195, 348, 354 — C.H.C.

Intensity		Wavelength	
		Vacuum	
3		650.90	II
5	r	670.87	II
4		674.10	II
15	r	696.30	II
5	r	709.23	II
10	r	817.18	II
5	r	836.34	II
8	r	1018.85	II
10	r	1049.73	II
8	r	1050.30	II
5	r	1074.97	II
10	r	1130.17	II
15	r	1162.55	II
10	r	1167.43	II
10	r	1183.41	II
12	r	1194.84	II
8		1231.81	II
5	r	1246.00	II
15	r	1307.50	II
8	r	1310.20	II
25	r	1321.71	II
8	r	1330.40	II
10	r	1373.52	II
1		1423.2	I
8	r	1489.65	I
5		1490.50	II
10	r	1499.30	II
10	r	1507.82	II
15	r	1561.58	II
10	r	1568.57	II
7	r	1593.26	II
5	h	1616.	I
1		1650.2	I
5		1685.40	I
1		1728.	I
10	r	1792.76	II
12	r	1814.85	II
3	h	1847.	I
8		1892.72	II
25	r	1908.64	II
		Air	
100	r	2007.56	I
2		2209.75	I
100	r	2210.71	I
3		2287.6	I
30		2298.04	II
140		2315.98	I
900	h	2379.69	I
8		2451.83	II
6		2469.03	II
1		2508.2	I
20		2530.86	II
700		2580.14	I
60		2608.99	I
80		2665.57	I
420		2709.23	I
50	h	2710.67	I
4400	d	2767.87	I
280		2826.16	I
10		2849.80	II
2800		2918.32	I
440		2921.52	I
5		3029.01	II
20		3091.56	II
15		3185.51	II
15		3186.56	II
15		3187.74	II
1200		3229.75	I
15		3291.01	II
12		3319.91	II
12		3321.04	II
8		3322.25	II
15		3369.15	II
8		3381.00	II
6		3381.80	II
6	d	3460.48	II
20000		3519.24	I
5000		3529.43	I
8		3540.08	II
9		3560.68	II
5		3567.67	II
12000	Cw	3775.72	I
8		3793.95	II
10		3832.30	II
6		3869.15	II
10		3887.15	II
8		4223.05	II
20		4274.98	II
40		4306.80	II
2		4359.9	I
8		4490.77	II
20		4737.05	II
15		4981.35	II
25		5078.54	II
25		5152.14	II
6		5181.95	II
6		5183.10	II
18000		5350.46	I
15	d	5384.85	II
7		5409.92	II
10		5410.97	II
25		5949.48	II
10		6179.98	II
8	d	6239.03	II
10		6378.32	II
16	h	6549.84	I
6	h	6713.80	I
10		6966.5	II
3		7493.6	I
2		7678.93	I
10		7815.80	I
8		8130.0	I
20		8373.6	I
8		8445.8	II
10		8474.27	I
8		8632.9	II
10		8664.1	II
4		8850.4	I
5		8976.75	I
3		9038.4	I
20		9130.	II
20		9130.5	I
2	h	9183.1	I
4		9225.	II
2	h	9252.6	I
3		9254.	II
40		9509.4	I
10		9863.4	I
20		9930.4	I
2		9937.4	I
30		10011.9	I
40		10488.80	I
5		11101.61	I
4		11483.7	I
1000		11512.82	I

Thallium (Cont.)

Intensity	Wavelength	
5	11592.9	I
15	12491.8	I
150	12736.4	I
700	13013.2	I

Tl III
Ref 355 — C.H.C.

Intensity	Wavelength	
	Vacuum	
7	1231.57	III
10	1266.33	III
4	1332.36	III
10	1477.14	III
4	1506.37	III
8	1558.67	III
8	1660.05	III
	Air	
6	3163.53	III
3	3300.80	III
9	3456.34	III
4	3507.41	III
6	3933.05	III
2	3946.02	III
7	4109.85	III
4	4155.75	III
6	4269.81	III
2	4380.57	III
4	5086.99	III
4	5362.40	III
2	5499.4	III
5	5927.8	III
4	8001.	III

Tl IV
Ref. 356 — C.H.C.

Intensity	Wavelength	
	Vacuum	
7	531.26	IV
10	570.49	IV
4	597.01	IV
1	868.99	IV
3	912.74	IV
8	917.31	IV
30	1028.69	IV
20	1034.73	IV
20	1036.61	IV
10	1049.48	IV
10	1057.56	IV
20	1068.04	IV
20	1070.47	IV
30	1079.68	IV
5	1079.70	IV
2	1092.90	IV
4	1094.41	IV
6	1099.60	IV
4	1125.52	IV
5	1139.30	IV
3	1144.07	IV
3	1225.45	IV
6	1273.03	IV
3	1304.55	IV
6	1323.66	IV
7	1337.10	IV
7	1358.56	IV
5	1374.62	IV
7	1377.75	IV
8	1404.60	IV
5	1412.93	IV
6	1434.72	IV
5	1449.37	IV
5	1883.2	IV
3	1974.6	IV

THORIUM (Th)
Z = 90

Th I and II
Ref. 1, 97, 98, 434 — J.G.C. and R.Z.

Intensity	Wavelength	
	Air	
100	2326.926	II
190	2377.84	II
90	2404.504	II
100	2413.409	II
30	2439.433	I
5	2532.894	I
150	2547.901	II
500	2565.593	II
270	2566.588	II
200	2576.688	II
230	2589.059	II
230	2597.047	II
230	2600.882	II
100	2609.855	II
230	2618.91	II
270	2623.448	II
270	2625.737	II
55	2628.812	II
270	2641.488	II
170	2650.583	II
150	2658.663	II
50	2680.692	I
360	2684.288	II
480	2692.415	II
100	2695.553	II
270	2703.958	II
170	2708.176	II
230	2721.691	II
170	2722.380	II
250	2729.327	II
250	2732.808	II
28	2735.834	II
520	2747.156	II
100	2749.530	II
410	2752.166	II
130	2760.391	II
100	2765.123	II
270	2768.841	II
200	2770.816	II
70	2773.951	II
50	2774.066	II
70	2778.706	II
35	2791.496	I
90	2794.255	II
70	2797.737	II
55	2799.114	II
110	2807.827	II
180	2808.998	II
70	2814.319	II
45	2816.071	II
100	2819.322	II
100	2820.336	II
100	2822.025	II
170	2826.855	II
70	2830.442	II
800	[illegible]	II
1200	2837.295	II
320	2842.812	II
100	2848.084	I
270	2851.260	II
50	2854.342	I
30	2860.490	I
220	2861.42	II
35	2868.461	I
70	2869.916	II
550	2870.406	II
30	2878.657	I
40	2882.511	II
320	2884.289	II
360	2885.049	II
360	2887.817	II
40	2892.172	II
250	2899.720	II
45	2903.167	II
50	2908.506	I
200	2910.594	II
90	2911.320	II
90	2912.009	II
140	2919.840	II
250	2925.050	II
250	2928.254	II
50	2931.281	I
55	2934.135	II
100	2936.086	I
35	2940.589	II
340	2942.860	II
100	2943.729	I
150	2949.068	II
80	2950.438	II
35	2955.849	II
170	2957.580	II
28	2959.853	I
28	2963.607	I
270	2968.686	II
110	2971.481	II
220	2974.011	II
50	2976.104	I
55	2980.334	II
160	2985.243	II
360	2988.232	II
150	2991.062	II
110	2996.986	II
50	3002.686	I
30	3004.248	I
180	3008.497	II
50	3010.736	I
20	3018.644	I
50	3021.056	I
150	3026.575	II
40	3030.487	I
370	3034.065	II
170	3035.110	II
85	3038.598	II
130	3045.564	II
420	3049.092	II
30	3056.692	I
220	3061.699	II
450	3067.729	II
370	3072.114	II
670	3078.828	II
480	3080.217	II
240	3088.470	II
130	3090.093	II
50	3093.711	I
140	3097.266	II
200	3102.664	II
510	3108.296	II
50	3115.538	I
100	3116.263	I
510	3119.526	II
510	3122.963	II
370	3124.387	II
480	3125.507	II
150	3131.070	II
100	3136.216	I
420	3139.306	II
420	3142.835	II
310	3146.044	II
150	3150.455	II
310	3154.300	II
50	3157.221	I
30	3161.364	I
140	3166.099	II
110	3169.328	II
420	3175.726	II
270	3179.048	II
1100	3180.193	II
310	3184.948	II
770	3188.233	II
85	3191.221	II
55	3192.585	I
55	3195.689	I
30	3202.520	I
170	3210.308	II
55	3214.380	I
560	3221.292	II
30	3223.168	I
560	3229.009	II
110	3230.868	II
400	3235.84	II
590	3238.116	II
240	3241.108	II
110	3244.448	I
280	3251.915	II
910	3256.274	II
180	3257.366	I
910	3262.668	II
180	3267.003	II
110	3272.027	I
30	3278.733	I
50	3281.048	I
130	3285.752	I
620	3287.789	II
910	3291.739	II
620	3292.520	II
240	3297.832	II
240	3301.650	I
480	3304.238	I
130	3309.365	I
50	3314.790	I
30	3318.390	I
510	3321.450	II
390	3324.752	II
840	3325.120	II
55	3328.255	II
250	3330.476	I
620	3334.604	II
620	3337.870	II
30	3342.073	II
180	3346.557	II
310	3348.768	I
980	3351.228	II
310	3354.179	II
620	3358.602	II
75	3361.738	II
390	3367.819	II
250	3374.974	I
390	3378.573	II
130	3380.859	I
310	3385.531	II
310	3386.501	II
110	3387.920	I
1300	3392.035	II
200	3396.727	I
250	3398.544	I
200	3405.558	I
250	3413.012	I
50	3417.497	I
390	3421.210	I
270	3423.989	I
50	3428.622	I
980	3433.998	II
770	3435.976	II
340	3438.949	II
110	3442.578	I
50	3446.547	I
130	3451.702	I
50	3457.068	I
340	3462.850	II
130	3465.924	II
390	3468.219	II
1300	3469.920	II
170	3471.218	I
250	3479.173	II
70	3480.052	I
200	3486.552	I
100	3489.184	I
270	3493.518	II
70	3496.810	I
130	3498.621	I
70	3503.786	I
50	3506.645	I
110	3511.157	I
140	3518.404	I
70	3521.059	I
70	3526.633	I
140	3531.450	I
670	3539.587	II
180	3544.018	I
170	3549.595	I
140	3551.401	I
200	3555.013	I
530	3559.451	II
110	3563.375	I
70	3569.820	I
200	3576.557	I
100	3583.101	I
170	3589.750	I
270	3592.780	I
270	3598.120	I
390	3601.034	II
170	3608.377	I
980	3609.445	II
200	3612.427	I
480	3615.133	II
400	3617.118	II
270	3623.970	II
390	3625.627	II
140	3632.831	I
270	3635.943	I
70	3638.644	I
210	3642.248	I
170	[illegible]	I
50	3658.808	I
100	3659.629	I
220	3663.202	I
140	3668.140	I
280	3669.968	I
700	3675.567	II
150	3682.486	I
50	3688.658	I
100	3690.624	I
170	3692.566	I
180	3698.105	I
50	3703.229	I
340	3706.767	I
280	3711.305	II
590	3719.435	I
50	3719.836	I
770	3721.825	II
110	3727.902	I
50	3733.672	I
1300	3741.183	II
310	3747.539	I
650	3752.569	II
140	3757.694	I
110	3765.240	I
180	3770.056	I
50	3772.649	I
85	3776.271	I
50	3780.966	I
340	3785.600	II
100	3789.167	I
85	3795.386	I
50	3800.197	I
590	3803.075	I
50	3807.273	I
340	3813.068	II
50	3818.685	I
75	3825.133	I
450	3828.384	I
70	3836.584	I
840	3839.746	II
280	3841.960	II
100	3846.887	I
85	3852.135	I
390	3854.511	II
140	3859.840	II
450	3863.405	II
100	3869.663	I
210	3875.374	I
140	3879.644	I
100	3886.915	I
340	3895.419	I
50	3901.661	I
110	3903.102	I
170	3905.186	II
50	3908.750	I
85	3911.909	I
50	3916.417	I

Thorium (Cont.)

Intensity	Wavelength	Spectrum
110	3919.023	I
140	3925.093	I
590	3929.669	II
200	3932.911	I
140	3937.040	II
50	3942.072	I
50	3948.030	I
200	3948.964	II
50	3952.760	I
110	3959.300	I
390	3967.392	I
200	3972.155	I
150	3980.089	I
110	3991.730	I
530	3994.549	II
50	3998.061	I
240	4003.309	II
250	4007.021	II
220	4008.210	I
220	4009.056	I
280	4012.495	I
4200	4019.129	II
210	4025.656	II
140	4027.009	I
250	4030.842	I
250	4036.047	I
240	4036.565	II
240	4041.204	II
55	4048.287	I
110	4050.887	I
140	4059.253	I
250	4063.407	I
300	4069.201	II
100	4069.461	I
55	4075.503	I
110	4081.368	I
85	4085.434	I
700	4086.520	II
70	4088.726	I
700	4094.747	II
150	4100.341	I
270	4105.330	II
840	4108.421	II
240	4112.754	I
280	4115.758	I
1100	4116.713	II
30	4123.600	I
200	4127.411	I
110	4131.002	I
340	4132.753	II
200	4134.067	I
220	4140.235	II
250	4142.701	II
200	4148.182	II
450	4149.986	II
110	4158.535	I
140	4165.766	I
620	4178.060	II
250	4179.714	II
30	4184.138	I
130	4193.017	I
620	4208.890	II
130	4210.923	I
28	4214.828	I
55	4220.065	I
55	4227.387	I
30	4230.824	I
85	4235.463	I
20	4241.094	I
30	4247.989	II
110	4253.538	I
70	4256.254	I
110	4260.333	I
28	4269.942	I
280	4273.357	II
480	4277.313	II
700	4282.042	II
28	4288.669	I
55	4297.306	I
85	4299.839	I
100	4307.176	I
200	4309.991	II
55	4315.254	I
110	4318.416	I
30	4325.274	I
28	4330.844	I
130	4337.277	I
85	4342.256	II
130	4344.326	II
55	4349.072	I
55	4354.484	I
85	4359.372	I
85	4365.930	I
85	4374.123	I
1300	4381.860	II
1100	4391.110	II
55	4392.974	I
55	4401.580	I
85	4408.882	I
210	4412.741	II
28	4416.845	I
50	4422.048	I
250	4432.963	II
140	4440.866	II
25	4452.565	I
85	4458.002	I
220	4465.341	II
30	4475.221	I
75	4482.169	I
50	4489.664	I
110	4498.940	I
55	4505.216	I
280	4510.527	II
70	4521.194	I
22	4530.319	I
40	4535.255	I
30	4545.915	II
70	4555.812	I
40	4563.660	I
65	4570.972	I
50	4588.426	I
75	4595.421	I
26	4612.554	II
30	4621.163	I
140	4631.761	II
140	4631.761	II
30	4641.254	I
140	4651.558	II
23	4663.202	I
50	4669.984	I
65	4676.056	I
50	4686.195	I
140	4694.091	II
50	4703.990	I
20	4712.841	I
90	4723.438	I
30	4729.128	I
190	4740.529	II
140	4752.414	II
13	4766.600	I
50	4778.294	I
20	4786.531	I
40	4789.387	I
45	4808.134	I
20	4819.193	I
26	4822.855	I
40	4826.700	I
45	4831.121	I
50	4840.843	I
30	4848.362	I
15	4852.868	I
40	4858.333	II
280	4863.163	II
40	4872.917	I
26	4878.733	I
45	4894.955	I
20	4907.209	I
240	4919.816	II
18	4927.780	I
40	4939.642	I
60	4947.575	II
50	4954.659	II
30	4965.731	I
35	4975.950	II
24	4985.372	I
50	5002.097	I
50	5002.097	I
50	5015.889	II
260	5017.255	II
20	5029.892	I
24	5039.230	I
50	5044.719	I
240	5049.796	II
85	5055.347	II
70	5058.562	II
110	5067.974	I
30	5081.446	I
50	5090.051	I
50	5098.043	II
40	5101.130	I
50	5110.867	II
30	5115.044	I
10	5125.950	I
20	5134.746	I
95	5143.267	II
120	5148.211	II
50	5151.612	I
50	5154.243	I
85	5158.604	I
70	5160.730	I
20	5168.922	I
50	5176.961	I
35	5183.990	I
50	5190.871	II
50 h	5195.814	I
50	5198.800	I
95	5199.164	I
50	5211.230	I
95	5216.596	II
50	5218.528	II
35	5219.110	I
110	5231.160	I
85	5233.225	II
85	5233.229	II
95	5247.654	I
10	5255.573	I
35	5258.360	I
12	5266.710	I
70	5277.501	II
15	5281.069	I
10	5294.397	I
30	5297.743	I
30	5307.466	II
35	5312.002	I
20	5317.494	I
60	5325.145	II
50	5326.976	I
20	5330.080	I
60	5343.581	I
14	5351.126	I
30	5358.707	I
20	5369.281	I
30	5378.836	I
70	5390.466	II
50	5392.572	I
20	5399.175	I
24	5417.486	I
60	5425.678	II
50	5435.893	II
40	5449.479	II
30	5462.615	II
15	5470.759	I
24	5484.147	II
10	5496.137	I
19	5504.302	I
35	5509.994	I
12	5524.584	I
50	5539.262	I
70	5539.911	II
35	5548.176	I
50	5558.342	I
60	5564.203	II
40	5573.354	I
60	5587.026	I
24	5595.064	I
50	5604.515	II
35	5615.320	I
7	5630.297	I
70	5639.746	II
7	5648.991	I
12	5657.925	I
15	5667.128	I
20	5677.053	I
10	5685.192	I
65	5700.918	II
95	5707.103	II
50	5720.183	I
30	5732.975	II
24	5742.084	II
30	5749.388	II
70	5760.551	I
15	5773.946	I
20	5789.645	I
35	5804.141	I
19	5815.422	II
10	5832.370	I
10	5845.919	I
15	5854.121	I
15	5868.373	I
10	5878.933	I
8	5891.451	I
15	5899.844	I
20	5914.387	II
19	5925.893	II
10	5937.162	I
10	5944.648	I
8	5957.587	I
30	5973.665	I
85	5989.044	II
24	5994.129	I
21	6007.072	I
30	6015.426	II
17	6021.036	I
17	6037.698	I
24	6044.431	II
10	6053.381	I
5	6061.536	I
30	6077.106	I
30	6087.262	II
24	6099.083	I
30	6104.580	I
40	6112.837	I
30	6120.557	II
10	6124.480	I
14	6151.993	I
10	6161.354	I
60	6169.822	I
50	6182.622	I
12	6191.906	I
24	6193.858	II
12	6203.493	I
12	6224.528	I
24 h	6234.856	I
8	6240.954	I
10	6257.424	I
21	6261.063	I
21	6261.418	I
50	6274.116	II
50	6274.117	II
30	6279.172	II
8	6291.192	I
10	6303.251	II
8	6317.185	I
21	6327.278	I
35	6342.860	I
50	6355.911	II
14	6369.140	I
40	6376.931	I
30	6411.899	I
24	6413.615	I
15	6437.762	I
15	6450.005	I
60	6457.283	I
50	6462.614	I
5	6466.717	I
14	6490.738	I
5	6501.992	I
20	6512.364	I
5	6522.044	I
50 h	6531.342	I
6	6554.160	I
5	6558.876	I
3	6565.070	I
5	6577.215	I
24	6583.907	I
24	6588.540	I
24	6593.940	I
24	6605.416	II
24	6619.946	II
24	6619.947	II
21	6644.650	II
6	6658.678	II
30	6662.269	I
6	6674.697	I
3	6683.367	I
8	6692.724	II
5	6697.712	I
16	6727.459	I
3	6735.126	I
5	6742.884	I
20	6756.453	I
6	6765.677	I
15	6778.313	I
15	6780.413	I
6	6791.236	I
3	6798.747	I
3	6809.511	I
5	6823.509	I
11	6829.036	I
14	6834.925	I
4	6862.873	I
5	6866.367	I
8	6874.754	I
20	6889.303	II
3	6908.988	I
24	6911.227	I
5	6916.129	I
5	6936.652	I
35	6943.611	I
5	6954.657	I
15	6965.947	I
3	6981.086	I
55	6989.656	I
24	6993.038	II
18	7000.806	I
18	7000.806	I
3	7015.319	I
10	7018.569	I
3	7026.462	I
7	7036.281	I
30	7045.795	II
15	7053.619	II
6	7060.654	I
24	7075.333	II
30	7084.171	I
24	7089.339	II
10	7100.512	II
3	7109.861	I
11	7124.562	I
3	7132.613	I
5	7148.560	I
10	7154.954	I
30	7168.896	I
15	7173.373	I
40	7191.132	II
7	7200.046	I
35	7208.006	I
11	7212.69	I
10	7217.755	II
11	7218.054	I
3	7242.355	I
5	7255.354	I
3	7270.558	I
7	7284.904	I
5	7298.143	I
11	7305.405	II
5	7315.067	I
7	7324.808	I
5	7339.606	I
8	7341.152	I
5	7361.349	I
5	7384.175	I
18	7385.501	I
3	7393.431	II
5	7402.252	I
3	7418.550	I

Thorium (Cont.)

Intensity	Wavelength	Spectrum
21	7428.940	I
10	7430.254	I
3	7444.749	I
2	7462.993	I
10	7481.355	I
2	7493.427	I
50	7525.508	II
7	7549.314	I
18	7567.740	I
12	7585.69	I
12	7585.792	I
4	7598.204	I
2	7607.824	I
5	7627.176	I
3	7636.176	I
30	7647.380	I
4	7658.324	I
7	7676.219	I
21	7685.305	I
4	7710.269	I
4	7728.951	I
10	7731.72	II
4	7771.948	I
15	7787.79	II
15	7788.937	I
5	7798.360	I
4	7810.625	I
21	7817.771	I
8	7834.459	II
15	7847.540	I
4	7864.023	I
12	7865.95	I
6	7886.284	I
11	7900.31	I
4	7937.732	I
11	7941.72	I
5	7954.594	I
7	7972.598	I
24	7978.974	I
11	7987.97	I
5	7993.680	I
2	8014.502	I
5	8024.253	II
11	8032.433	I
11	8062.64	I
5	8085.220	I
5	8093.626	I
5	8129.407	I
11	8138.477	I
18	8143.139	I
12	8159.729	I
10	8162.103	II
7	8169.788	I
15	8186.914	I
12	8203.199	II
2	8231.408	I
5	8252.395	I
3	8259.512	I
18	8275.629	I
3	8292.529	I
15	8320.857	I
30	8330.451	I
4	8358.726	I
6	8387.104	II
18	8403.767	II
15	8416.729	I
12	8421.227	I
21	8446.509	I
5	8464.230	I
18	8478.360	I
5	8510.621	I
5	8516.557	I
5	8539.795	I
6	8554.946	I
11	8573.122	I
12	8591.838	I
3	8621.325	I
3	8638.363	I
10	8665.487	I
4	8668.116	I
2	8701.127	I
5	8709.236	I
8	8732.401	II
18	8748.033	I
15	8758.244	I
8	8775.573	I
4	8792.058	I
3	8804.590	I
5	8841.185	I
18	8842.073	II
15	8868.834	I
4	8875.233	I
2	8907.038	I
5	8955.848	I
15	8957.97	II
40	8967.641	I
3	8987.408	I
5	9031.819	I
25	9048.252	I
5	9063.953	I
6	9094.831	I
3	9107.225	I
2	9118.140	I
3	9170.825	I
10	9203.963	I
10	9266.208	I
10	9276.276	I
10	9289.563	I
4	9317.722	I
7	9340.706	I
15	9399.085	I
10	9431.603	I
15	9461.030	I
15	9474.882	I
15	9495.501	I
15	9497.191	I
10	9505.392	I
10	9561.24	I
8	9613.689	II
12	9632.647	I
10	9664.700	I
4	9676.106	I
15	9700.564	I
15	9746.46	I
10	9812.70	I
10	9826.45	I
20	9833.42	I
15	10039.364	I
15	10089.138	I
15	10133.56	II
15	10419.57	II
15	10556.45	I
15	10723.92	II
20	10726.93	I
20	10942.24	II
15	11051.90	I
30	11230.259	I
20	11354.719	I
15	11703.46	I
15	11864.25	I
15	11940.64	I
20	11984.67	II
15	12018.72	I
20	12127.30	I
20	12194.16	II
15	12206.89	I
20	12231.94	I
15	12338.00	I
15	12477.30	I
20	12646.54	I
15	12866.64	I
15	12940.65	II
15	12959.82	I
15	13145.90	II
15	13565.67	I
15	14090.25	I
15	14168.67	I
15	14424.54	I
15	14618.98	I
15	14654.91	I
15	14940.49	II
15	15240.24	II
15	15429.78	I
15	15831.75	I
20	17208.22	II
15	17307.66	I
15	17381.91	I
15	17481.04	I
15	17584.52	I
15	17936.43	II
15	18811.88	I
15	19145.60	II
15	19338.98	II
10	19774.30	II
10	20634.36	I
10	20692.06	II
10	22264.35	II

Th III
Ref. 157 — J.G.C.

Intensity	Wavelength	
	Vacuum	
100	1888.12	III
	Air	
50	2149.18	III
50	2162.82	III
50	2199.74	III
50	2206.62	III
50	2291.59	III
100	2301.18	III
100	2319.52	III
80	2324.68	III
150	2335.50	III
100	2340.58	III
100	2363.06	III
50	2368.91	III
100	2371.42	III
80	2381.47	III
100	2391.48	III
200	2413.50	III
50	2424.54	III
200	2427.94	III
200	2431.68	III
200	2441.24	III
100	2463.66	III
50	2473.93	III
100	2501.08	III
60	2512.69	III
50	2514.31	III
100	4555.73	III
50	4589.28	III
100	5376.13	III
50	5447.18	III
50	6242.95	III
50	6599.39	III
50	7461.59	III
50	8105.14	III

Th IV
Ref. 156, 165 — J.G.C.

Intensity	Wavelength	
	Vacuum	
4	797.53	IV
1	835.55	IV
30	846.91	IV
1	854.02	IV
30	882.39	IV
12	886.66	IV
100	1565.85	IV
70	1682.22	IV
30	1684.01	IV
150	1707.37	IV
200	1959.02	IV
	Air	
200	2002.34	IV
100	2066.70	IV
20	2143.91	IV
30	2146.81	IV
1	2242.11	IV
2	2261.26	IV
5	2296.81	IV
100	2693.99	IV
2	4937.09	IV
4	4952.52	IV
3	5420.38	IV
2	6711.87	IV
3	6740.37	IV
50	6901.16	IV
	9839.25	IV
	10875.05	IV

THULIUM (Tm)
Z = 69

Tm I and II
Ref. 1 — C.H.C.

Intensity		Wavelength	
		Air	
360		2284.79	II
120		2329.77	II
70		2340.92	II
120		2363.91	II
45		2365.96	II
160		2367.11	II
150		2383.68	II
110		2388.95	II
450		2409.02	II
110		2412.44	II
120		2421.65	II
450		2426.17	II
140		2445.47	II
770		2480.13	II
150		2481.15	II
130		2487.52	II
250		2491.60	II
100		2499.54	II
130		2507.15	II
1300		2509.08	II
200		2520.87	II
250		2522.17	II
180		2524.11	II
130		2527.02	I
110		2527.42	II
120		2542.66	II
360		2552.76	I
540		2561.65	II
150		2563.86	II
430		2588.27	II
170	h	2596.49	I
110		2601.09	I
220		2606.02	II
810		2607.06	II
730		2624.33	II
210		2640.76	II
130		2646.45	II
160		2650.27	II
190		2658.48	II
250		2660.09	II
140		2668.20	II
310		2679.57	II
170		2697.50	II
540		2721.19	II
200		2744.08	II
270		2779.55	II
350		2785.07	II
680		2794.60	II
730		2797.27	II
250		2818.47	II
250		2827.02	II
580		2827.92	II
200		2831.55	II
310		2844.67	II
200		2854.17	I
200		2860.12	II
200		2861.74	II
1600		2869.23	II
630		2890.94	II
210		2918.27	II
270		2925.65	II
680		2926.74	II
630		2935.99	II
350		2951.26	II
430		2965.86	II
490		2973.22	I
540		2981.48	II
350		2986.52	II
630		2990.54	II
200		2993.26	II
230		3013.71	II
430		3014.65	II
1500		3015.30	II
270		3017.09	II
330		3026.07	II
280		3042.35	II
340	d	3046.76	II
320		3050.73	II
340		3056.07	II
580		3073.08	II
360		3081.12	I
740		3098.60	II
7400		3131.26	II
2300		3133.89	II
230		3144.90	II
230		3146.16	II
1900		3151.04	II
1500		3157.34	II
450		3172.65	I
2300		3172.83	II
380		3173.58	II
230		3195.33	II
320		3210.56	II
320		3210.82	II
320		3212.01	II
230		3231.51	II
470		3235.44	II
1200		3236.81	II
1600		3240.23	II
2300		3241.54	II
320		3246.96	I
420		3247.46	II
1900		3258.05	II
400		3261.65	II
320		3264.10	II
1600		3266.64	II
1200		3267.40	II
790		3268.99	II
1100		3276.81	II
1200		3283.40	II
1200		3285.61	II
2300		3291.00	II
2000		3302.46	II
210		3306.01	II
210		3306.91	II
210		3308.01	II
1200		3309.80	II
640		3310.59	II
400		3316.88	II
210		3318.65	II
230		3349.99	I
230		3354.86	II
4000		3362.61	II
490		3374.50	II
420	d	3384.99	II
1700		3397.50	II
420		3399.95	II
850		3410.05	I
340		3412.59	I
340		3416.59	I
6400		3425.08	II
950		3425.63	II
340		3429.33	I
850		3429.96	II
420		3431.19	II
4900		3441.50	II
4900		3453.66	II
8500		3462.20	II
210		3467.51	I

Thulium (Cont.)

Intensity		Wavelength	Spectrum
340		3476.69	I
340		3480.98	I
340		3481.75	II
420		3487.28	I
210		3492.58	II
340		3499.95	I
250		3513.02	II
250		3517.60	I
250		3534.85	II
1700		3535.52	II
490		3536.21	II
850		3536.58	II
420		3537.91	I
210		3555.82	I
420		3557.79	II
340		3560.92	I
420		3563.88	I
490		3565.91	II
1300		3566.47	II
420		3567.36	I
280		3574.06	II
280		3586.07	I
2100		3608.77	II
250		3609.53	II
380		3638.41	I
950		3643.65	II
240		3647.72	II
600		3653.61	II
500		3665.81	II
1100		3668.09	II
410		3677.98	II
450	d	3678.85	II
410		3694.74	II
4800		3700.26	II
3800		3701.36	II
330		3704.85	II
7700		3717.91	I
890		3725.06	II
2400		3734.12	II
5000		3744.06	I
1700		3751.81	I
310		3756.86	II
6000		3761.33	II
4800		3761.91	II
260		3783.55	II
380		3795.16	II
7100		3795.75	II
770		3798.54	I
240		3798.75	II
600		3807.72	I
380		3810.72	II
550		3817.39	II
290		3826.39	I
1300		3838.20	II
290		3840.87	I
8900		3848.02	II
140		3857.84	II
6800		3883.13	I
1800		3883.44	II
5400		3887.35	I
440		3890.53	II
440		3896.62	I
680		3900.79	II
3500		3916.48	I
120		3928.66	II
570		3929.58	II
1500		3949.27	I
1500		3958.10	II
440		3995.58	II
1800		3996.52	II
220		4024.23	I
380		4044.47	I
10000		4094.19	I
9500		4105.84	I
120		4132.69	II
1100		4138.33	I
120		4149.14	I
120		4158.60	I
8800		4187.62	I
520		4199.92	II
6000		4203.73	I
220		4206.00	II
380		4222.67	I
3000		4242.15	II
270		4271.71	I
150		4298.36	I
2700		4359.93	I
1400		4386.43	I
200		4394.42	I
120		4395.96	I
140		4396.50	I
55		4437.40	II
80		4442.74	I
50		4447.58	I
120		4454.03	I
80		4459.99	I
50		4467.98	I
540		4481.26	II
80		4489.70	II
150		4519.60	I
260		4522.57	II
180		4529.38	II
80		4532.15	I
110		4548.60	I
40		4556.68	II
40		4561.86	II
80		4564.68	I
40		4567.11	II
95		4596.63	I
270		4599.02	I
35		4601.29	II
55		4603.43	II
40		4604.85	I
50		4613.97	I
40		4614.47	II
300		4615.94	II
35		4619.06	II
40		4621.72	I
80		4626.33	II
95		4626.56	II
40		4626.97	I
110		4634.26	II
40		4642.96	II
95		4643.12	I
35		4644.58	I
120		4655.09	I
35		4666.70	II
35		4671.99	II
35		4675.10	I
80		4675.31	I
40		4677.86	II
160		4681.92	I
70		4685.11	I
120		4691.11	I
110		4724.26	I
680		4733.34	I
35		4750.75	II
70		4759.90	I
27		4789.92	II
27		4807.48	I
35		4808.68	I
35		4813.50	I
27		4826.99	II
27		4828.97	I
80		4831.20	II
35		4835.75	I
27	d	4851.76	I
		4851.90	II
19		4872.28	II
27		4879.19	I
27		4891.64	I
24		4909.74	I
55		4923.83	I
140		4957.18	I
40		4970.87	II
27		4971.26	I
40		4975.12	II
50		4978.90	I
40		4980.68	II
55		4989.32	II
27		4993.79	II
19		4994.72	II
35		5001.02	I
27		5001.59	I
160		5009.77	II
35		5014.56	II
27		5017.87	II
160		5034.22	II
27	h	5041.00	II
22		5043.50	I
35		5045.41	I
27		5060.42	II
150		5060.90	I
27		5062.25	I
27		5065.88	I
80		5066.67	I
27		5072.42	I
27		5076.36	I
27		5077.18	I
35		5085.09	I
40		5107.53	I
95		5113.97	I
50		5114.55	II
22		5120.67	I
22		5140.28	II
40		5149.40	II
19		5182.68	I
40		5185.25	I
14		5204.51	II
80		5213.38	I
22		5228.23	II
14		5260.93	II
24		5267.34	II
40		5291.14	I
40		5294.32	I
35		5300.21	I
35		5302.69	I
55		5305.87	II
650		5307.12	I
16		5322.99	II
35		5338.90	I
		5339.03	I
80		5346.49	II
27		5372.98	II
14		5391.96	II
27		5400.46	II
27		5402.23	I
14		5405.98	II
14		5461.95	II
14		5464.14	I
14		5465.54	II
16		5500.30	II
14		5526.82	II
24		5528.34	I
14		5539.03	II
27		5566.00	I
22		5581.37	I
14		5586.65	II
14		5589.94	II
14		5606.64	I
270		5631.41	I
40		5642.60	I
27		5645.40	I
70		5658.30	I
520		5675.84	I
14		5683.59	I
40		5684.76	II
14		5696.42	II
35		5709.97	II
22		5715.79	I
14		5733.81	II
11	d	5737.20	II
		5737.25	II
14	h	5738.92	II
27		5758.02	I
55		5760.20	I
190		5764.29	I
5		5778.82	II
19		5782.36	II
22		5784.46	II
11		5799.97	II
14		5811.19	II
14	h	5816.46	I
35		5838.76	II
240		5895.63	I
35		5899.47	I
24		5901.57	I
8		5912.58	I
11		5931.70	I
27		5935.90	I
140		5971.26	I
27		5975.02	I
11		5984.87	I
19		6025.44	I
11		6067.78	II
16		6131.53	I
14		6175.29	I
14		6181.41	II
14		6299.46	II
27		6352.66	I
22		6401.44	I
8		6430.94	II
14		6440.54	I
200		6460.26	I
14		6490.70	I
14		6519.78	I
8		6575.54	I
95		6604.96	I
8		6627.25	I
35		6657.72	I
11		6658.64	I
11		6692.93	I
30		6721.36	I
9		6726.34	I
9		6727.94	II
18		6739.22	I
9	h	6767.48	I
9		6777.93	I
110		6779.77	I
14	h	6782.00	I
18		6788.52	I
13	h	6820.27	I
14		6826.95	I
14		6829.12	II
23		6831.09	I
120		6844.26	I
80		6845.76	I
18		6854.12	I
6		6898.56	I
6		6915.86	I
10		6937.37	I
5		6949.54	I
5	h	6976.69	II
5		7010.79	I
6	h	7014.31	II
10		7017.90	I
6	h	7029.40	I
12		7034.34	I
10		7056.43	II
5		7060.97	I
6		7079.78	II
10		7106.14	I
5	h	7231.33	I
5		7233.74	II
4		7257.72	I
17		7272.62	I
8		7284.30	I
11	h	7286.16	I
14		7310.51	I
11		7336.63	II
14		7432.18	I
5		7434.51	II
5		7439.95	II
75		7481.08	I
75		7490.20	I
10	h	7507.28	I
14		7545.78	I
140		7558.33	I
17	h	7580.61	I
20	h	7593.74	I
17		7595.07	II
5		7629.85	I
5	h	7648.76	II
17		7655.00	I
4		7660.32	I
7	h	7666.24	I
8		7676.04	II
8	h	7701.46	I
80		7731.53	I
4	h	7778.27	I
12	h	7782.35	I
8	h	7785.51	I
		7785.90	I
17		7803.93	I
4		7829.22	I
40		7856.08	I
3		7861.67	I
5		7918.10	I
55		7927.51	I
110		7930.84	I
6		7971.56	I
11	h	7985.93	I
14	h	8014.77	I
95		8017.90	I
3	h	8021.33	I
14		8194.19	I
5		8294.52	I
7		8365.75	I
7		8460.79	II
27		8472.01	II
7	h	8546.07	II
11		8565.73	II

Tm III
Ref. 307 — J.R.

Intensity	Wavelength	
	Air	
500	2099.11	III
500	2107.10	III
200	2136.67	III
200	2156.29	III
800	2182.98	III
300	2183.91	III
5000	2185.94	III
100	2212.25	III
300	2230.86	III
400	2231.25	III
200	2243.34	III
400	2243.98	III
200	2246.68	III
200	2269.39	III
1000	2276.91	III
100	2280.08	III
100	2281.27	III
100	2282.86	III
200	2282.98	III
200	2286.57	III
400	2287.21	III
500	2294.73	III
20000	2296.21	III
200	2297.43	III
100	2304.64	III
400	2304.82	III
5000	2305.03	III
20000	2311.16	III
5000	2312.72	III
200	2314.88	III
400	2317.35	III
500	2320.96	III
200	2322.83	III
100	2323.71	III
100	2323.77	III
100	2324.43	III
500	2324.62	III
5000	2326.19	III
100	2327.02	III
300	2327.25	III
6000	2328.50	III
6000	2329.29	III
200	2330.87	III
3000	2331.80	III
400	2335.01	III
1000	2338.36	III
500	2341.74	III
300	2342.04	III
100	2344.59	III
500	2345.61	III
300	2347.43	III
400	2353.10	III
100	2355.65	III
3000	2357.05	III

Thulium (Cont.)

Intensity	Wavelength	Spectrum
1000	2361.23	III
500	2363.97	III
1000	2375.32	III
700	2375.83	III
400	2389.52	III
4000	2406.63	III
500	2435.31	III
500	2457.86	III
500	2471.23	III
30000	2489.44	III
200	2496.25	III
2000	2504.71	III
3000	2519.78	III
10000	2552.46	III
500	2557.90	III
1000	2574.52	III
500	2574.98	III
100	2581.84	III
100	2585.48	III
300	2589.20	III
500	2608.96	III
300	2609.66	III
500	2617.22	III
500	2618.78	III
1000	2621.12	III
400	2621.35	III
400	2622.31	III
100	2627.09	III
100	2628.83	III
300	2634.66	III
200	2636.68	III
200	2637.30	III
100	2640.32	III
500	2643.58	III
100	2645.05	III
200	2649.27	III
100	2650.82	III
100	2654.05	III
100	2656.30	III
500	2661.51	III
1000	2663.00	III
500	2664.76	III
500	2664.88	III
200	2665.05	III
1000	2666.93	III
200	2668.59	III
100	2668.66	III
200	2669.18	III
100	2671.42	III
100	2675.30	III
1000	2676.64	III
500	2676.91	III
100	2678.28	III
100	2680.49	III
5000	2682.32	III
300	2682.64	III
300	2687.14	III
300	2695.69	III
400	2698.21	III
1000	2699.49	III
1000	2699.80	III
100	2703.63	III
200	2703.68	III
100	2704.93	III
2000	2707.03	III
300	2707.19	III
200	2707.44	III
500	2707.60	III
1000	2709.74	III
200	2710.79	III
1000	2713.38	III
200	2715.81	III
300	2717.56	III
100	2718.02	III
3000	2719.47	III
3000	2724.44	III
4000	2727.56	III
200	2728.13	III
1000	2731.38	III
300	2732.11	III
400	2737.98	III
800	2744.74	III
400	2745.99	III
500	2752.46	III
400	2753.20	III
400	2756.15	III
800	2765.98	III
700	2769.92	III
200	2772.64	III
100	2777.43	III
400	2781.12	III
2000	2806.77	III
300	2821.12	III
200	2849.52	III
700	2882.02	III
100	2899.29	III
100	2912.33	III
400	2921.08	III
200	2947.02	III
1000	2947.72	III
500	2953.18	III
100	2966.15	III
500	2966.85	III
400	2972.61	III
100	2974.85	III
1000	2998.28	III
100	3048.11	III
700	3078.87	III
200	3120.15	III
200	3277.26	III
200	3407.73	III
100	3415.40	III
100	3415.96	III
100	3436.93	III
400	3467.93	III
200	3529.29	III
200	3533.28	III
100	3537.47	III
300	3562.41	III
100	3563.42	III
200	3587.74	III
600	3617.96	III
1000	3629.09	III
100	3706.11	III
100	3799.41	III
300	3998.84	III
200	4021.92	III
200	4026.03	III
700	4032.13	III
100	4076.15	III
200	4335.47	III
500	4385.41	III

TIN (Sn)
Z = 50

Sn I and II
Ref. 187, 191 — C.H.C.

Intensity		Wavelength	
Vacuum			
1		899.92	II
2		917.40	II
1		935.63	II
3		945.83	II
4		954.50	II
7		985.13	II
4		997.21	II
2		1016.26	II
4		1040.78	II
1		1041.32	II
3		1062.10	II
8		1108.19	II
4		1159.05	II
10		1161.43	II
3		1162.94	II
4		1180.51	II
9		1219.07	II
13		1223.70	II
11		1243.00	II
20		1290.86	II
20		1316.59	II
25		1400.52	II
20		1475.15	II
9		1489.22	II
7		16 9.47	II
10	r	1737.21	I
15	r	1751.46	I
10	h	1753.3	I
7		1758.00	II
20	r	1764.98	I
20		1773.40	I
30	r	1790.75	I
80	r	1804.60	I
15		1811.34	II
30		1813.04	I
40	r	1815.74	I
25		1819.31	I
120	r	1823.00	I
9		1831.89	II
50	r	1848.75	I
30		1852.00	I
200	r	1860.32	I
20		1861.42	I
20		1865.52	I
30		1865.96	I
15		1873.29	I
30		1882.64	I
80		1886.05	I
100		1891.40	I
20		1897.29	I
12		1899.91	II
50		1909.30	I
40		1911.61	I
20		1913.52	I
80		1925.31	I
20		1926.77	I
15		1927.95	I
40	h	1928.9	I
25		1933.17	I
20		1942.69	I
150		1952.15	I
15		1960.21	I
30		1971.46	I
50	h	1977.6	I
80		1984.20	I
15		1991.88	I
20		1994.98	I
Air			
25		2008.05	I
30		2015.76	I
30		2026.98	I
50		2040.66	I
20		2040.90	I
50		2054.03	I
70		2058.31	I
20		2064.00	I
80		2068.58	I
100		2072.89	I
100		2073.08	I
25		2080.62	I
30		2091.58	I
40		2094.35	I
200		2096.39	I
100		2100.93	I
100	r	2113.93	I
50		2121.26	I
25		2140.73	I
20		2141.43	I
15		2148.46	I
1		2148.63	II
40	r	2148.73	I
20	r	2151.43	I
30		2151.54	II
80		2171.32	I
150	r	2194.49	I
300	r	2199.34	I
400	r	2209.65	I
4		2209.67	II
40		2211.05	I
80	r	2231.72	I
400	r	2246.05	I
6		2246.07	II
60		2251.17	I
30		2267.19	I
400	r	2268.91	I
20		2282.76	I
200	r	2286.68	I
600	r	2317.23	I
300	r	2334.80	I
1000	r	2354.84	I
20		2357.90	I
3		2360.34	II
22		2368.33	II
60		2380.72	I
4		2384.54	II
100		2408.15	I
800	r	2421.70	I
1000	r	2429.49	I
1		2433.52	II
15		2448.98	II
60		2455.24	I
20		2476.40	I
300		2483.39	I
13		2483.48	II
10		2486.99	II
200		2495.70	I
5		2522.61	II
90		2523.92	I
80	h	2531.17	I
400		2546.55	I
40	h	2558.01	I
500	r	2571.58	I
200		2594.42	I
50	h	2636.94	I
200	r	2661.24	I
2		2664.93	II
700	r	2706.51	I
2		2727.82	II
20		2761.78	I
150		2779.81	I
80		2785.03	I
60		2787.96	I
60		2812.59	I
80		2813.58	I
2		2825.52	II
1400	r	2839.99	I
1		2846.42	II
200		2850.62	I
1000	r	2863.32	I
1		2912.80	II
200		2913.54	I
6		2919.82	II
3		2991.00	II
7		2994.44	II
700	r	3009.14	I
1		3012.18	II
8		3023.94	II
200		3032.80	I
850	r	3034.12	I
12		3047.50	II
6		3094.69	II
60		3141.84	I
550	r	3175.05	I
40		3218.71	I
550	r	3262.34	I
50		3283.21	II
110		3330.62	I
60		3351.97	II
2		3407.48	II
10		3472.46	II
7		3537.57	II
11		3575.45	II
3		3582.39	II
2		3620.08	II
6		3620.54	II
40		3655.78	I
6		3715.23	II
280	r	3801.02	I
4		3841.44	II
1		4294.65	II
40		4524.74	I
1		4579.13	II
1		4580.29	II
2		4877.22	II
3		4944.31	II
20		4979.73	I
2		5071.14	II
2		5072.67	II
20		5174.54	I
10		5332.36	II
20		5561.95	II
25		5588.92	II
2		5596.20	II
500		5631.71	I
15		5753.59	I
1		5797.20	II
15		5799.18	II
50		5925.44	I
100		5970.30	I
150		6037.70	I
200		6054.86	+
250		6069.00	I
100		6073.46	I
6		6077.48	II
5		6079.70	II
400		6149.71	I
200		6154.60	I
150		6171.50	I
100		6310.78	I
40		6354.35	I
70		6453.50	II
8		6761.45	II
25		6844.05	II
20		7191.40	II
10		7387.79	II
20	h	7398.6	I
1		7408.62	II
30		7685.30	I
13		7741.80	II
100		7754.97	I
3		7904.00	II
100	h	8030.5	I
30	h	8039.3	I
200		8114.09	I
30	h	8121.0	I
30		8349.35	I
80		8357.04	I
300		8422.72	I
400		8552.60	I
50	h	8681.7	I
30	h	9018.95	I
50	h	9410.86	I
80	h	9415.37	I
150		9616.40	I
50		9741.1	I
100	h	9742.8	I
300	h	9805.38	I
500		9850.52	I
25		10456.47	I
11		10807.58	I
54		10894.00	I
70		11191.85	I
56		11277.66	I
17		11336.97	I
200		11454.59	I
200		11616.26	I
76		11670.77	I
25		11694.45	I
258		11739.78	I
96		11825.18	I
106		11835.82	I
254		11932.99	I
48		12009.50	I
111		12313.24	I
33		12335.6	I
42		12530.87	I
42		12536.5	I
37		12788.2	I
89		12888.5	I
187		12981.7	I
20		13000.3	I
187		13018.5	I
68		13081.5	I
378		13460.2	I
144		13608.2	I
13		15018.2	I
30		15464.2	I
20		17000.5	I

Tin (Cont.)

Intensity		Wavelength	
10		17807.5	I
20		20622.2	I
40		20861.7	I
8		21686.2	I
4		22131.7	I
3		22997.2	I
4		24327.2	I
4		24738.2	I

Sn III
Ref. 423 — C.H.C.

Intensity		Wavelength	
		Vacuum	
100		753.01	III
50		760.62	III
75		775.79	III
50		784.68	III
200		910.92	III
50		1010.92	III
50		1048.84	III
1000		1139.29	III
1000		1158.33	III
200		1161.09	III
100		1161.58	III
100		1180.62	III
1000		1184.25	III
200		1189.99	III
200		1204.06	III
2000		1210.52	III
100		1215.10	III
100		1218.14	III
100		1230.17	III
100		1231.38	III
500		1243.63	III
1000		1259.92	III
40		1276.31	III
1000		1305.97	III
1000		1327.34	III
200		1334.70	III
200		1346.05	III
1000		1347.65	III
200		1369.71	III
1000		1386.74	III
500		1410.61	III
200		1449.77	III
1000		1570.36	III
50		1674.29	III
500		1811.71	III
500		1941.86	III
50		1955.52	III

Sn IV
Ref. 423 — C.H.C.

Intensity		Wavelength	
		Vacuum	
50		605.23	IV
50		619.04	IV
50		628.73	IV
50		908.22	IV
500		956.25	IV
500		1019.72	IV
1000		1044.49	IV
100		1058.37	IV
50		1058.59	IV
1000		1073.41	IV
200		1087.50	IV
300		1096.92	IV
50		1103.24	IV
000		1119.34	IV
200		1120.68	IV
1000		1314.55	IV
1000		1437.52	IV
100		1532.90	IV

Sn V
Ref. 399, 423 — C.H.C.

Intensity		Wavelength	
		Vacuum	
120		355.14	V
150		361.01	V
100		372.55	V
200		1089.35	V
100		1132.79	V
200		1160.74	V
100		1176.26	V
100		1189.92	V
100		1205.72	V
2000		1251.38	V
100		1283.81	V
200		1294.36	V
100		1302.20	V

TITANIUM (Ti)
Z = 22

Ti I and II
Ref. 1 — C.H.C.

Intensity		Wavelength	
		Air	
140		2272.61	I
180		2273.28	I
130		2276.70	I
190		2279.96	I
150		2299.85	I
140		2302.73	I
190		2305.67	I
65		2380.81	I
35		2384.52	I
55		2418.36	I
75		2421.30	I
95		2424.24	I
40		2428.23	I
35		2433.22	I
19		2434.10	I
35		2440.21	II
65		2440.98	I
24		2450.44	II
24		2504.54	I
75		2517.43	II
40		2519.04	I
140		2520.54	I
75		2524.64	II
360		2525.60	II
29		2527.98	I
210		2529.85	I
190		2531.25	II
190		2534.62	II
130		2535.87	II
190		2541.92	I
65		2555.99	II
110		2571.03	II
50		2572.65	II
50		2580.82	I
35		2590.26	I
190		2593.64	I
65		2596.58	I
270		2599.92	I
340		2605.15	I
510		2611.28	I
75		2611.48	I
300		2619.94	I
170		2631.54	I
170		2632.42	I
640		2641.10	I
800		2644.26	I
950		2646.64	I
30		2649.30	I
15		2654.93	I
35		2657.19	I
85		2661.97	I
95		2669.60	I
130		2679.93	I
26		2684.80	I
30		2685.14	I
65		2688.82	I
26		2716.25	II
85		2725.07	I
75		2727.42	I
21		2731.13	I
40		2731.58	I
170		2733.26	I
55		2735.29	I
40		2735.61	I
85		2739.81	I
250		2742.32	I
40		2749.06	I
65		2757.40	I
95		2758.08	I
15		2761.29	II
250		2802.50	I
55		2805.70	I
30		2806.50	II
40		2809.17	I
75		2810.30	II
30		2812.98	I
30		2817.40	I
65		2817.84	I
		2817.87	II
65		2828.07	I
		2828.15	II
130		2832.16	II
190		2841.94	II
110		2851.10	II
40		2853.93	II
95		2862.32	II
55		2868.74	II
180		2877.44	II
280		2884.11	II
65		2888.93	II
55		2891.07	II
55		2905.66	I
30		2909.92	II
450		2912.08	I
340		2928.34	I
15		2931.03	
180		2933.55	I
26		2935.96	
150		2937.32	I
1100		2942.00	I
1300		2948.26	I
30		2954.58	
1600		2956.13	I
170		2956.80	I
30		2958.77	
26		2959.71	I
35		2959.99	I
170		2965.71	I
190		2967.22	I
26		2968.23	I
75		2970.38	I
30		2974.93	I
170		2983.31	I
35		3000.87	I
120		3017.19	II
140		3029.73	II
110		3046.68	II
130		3056.74	II
130		3057.40	II
170		3058.09	II
85		3059.74	II
1300	d	3066.22	II
		3066.35	II
70		3071.24	II
600		3072.11	II
1100		3072.97	II
1600		3075.22	II
2300		3078.64	II
3600		3088.02	II
180		3089.40	II
180		3097.19	II
180		3100.67	I
230		3103.80	II
230		3105.08	II
260		3106.23	II
70		3106.81	I
50		3110.67	II
50		3112.48	I
140		3117.67	II
720		3119.72	I
		3119.80	II
190		3123.07	I
240		3130.80	II
140		3141.54	I
95		3141.67	I
220		3143.76	II
240		3148.04	II
240		3152.25	II
240		3154.20	II
240		3155.67	II
500		3161.20	II
780		3161.77	II
1000		3162.57	II
1600		3168.52	II
2400		3186.45	I
1000		3190.87	II
3100		3191.99	I
50		3197.52	II
3800		3199.92	I
780		3202.54	II
50		3203.44	II
240		3203.83	I
50		3204.87	I
110		3213.14	II
260		3214.24	I
190		3214.75	II
1100		3217.06	II
110		3217.94	I
260		3218.27	II
110		3219.21	I
110		3221.38	I
1300		3222.84	II
220		3223.52	I
240		3224.24	II
140		3226.13	I
530		3228.60	II
780		3229.19	II
530		3229.42	II
110		3231.32	II
240		3232.28	II
6600		3234.52	II
220		3236.12	II
5200		3236.57	II
4100		3239.04	II
220		3239.66	II
2600		3241.99	II
1200		3248.60	II
950		3251.91	II
1200		3252.91	II
1200		3254.25	II
1200		3261.60	II
310		3271.65	II
310		3272.08	II
200		3278.29	II
260		3278.92	II
220		3282.33	II
530		3287.66	II
290		3292.08	I
170		3299.41	I
170		3306.88	I
220		3308.39	I
220		3308.81	II
260		3309.50	I
60		3309.73	I
110		3312.69	I
840		3314.42	I
		3314.52	I
290		3315.32	II
330		3318.02	II
550		3321.70	II
2900		3322.94	II
380		3326.76	II
2100		3329.46	II
550		3332.11	II
1800		3335.20	II
1100		3340.34	II
5700		3341.88	I
120		3342.15	I
260		3343.77	II
330		3346.73	II
4300		3349.04	II
12000		3349.41	II
120		3352.94	I
4100		3354.64	I
290		3358.28	I
290		3360.99	I
7200		3361.21	II
		3361.26	I
120		3361.84	I
1100		3370.44	I
4300		3371.45	I
140		3372.21	II
5700		3372.80	II
60		3374.35	II
2900	d	3377.48	I
		3377.58	I
290		3379.22	I
1400		3380.28	II
170		3382.31	I
5700		3383.76	II
170		3385.66	I
1400		3385.95	I
1400		3387.84	II
60		3388.76	II
140		3390.68	I
140		3392.71	I
1100		3394.58	II
60		3398.63	I
60		3402.42	II
60		3407.20	II
95		3409.81	II
60		3439.30	I
890		3444.31	II
60		3452.47	II
180		3456.39	II
600		3461.50	II
95		3467.26	I
600		3477.18	II
60		3478.92	I
240		3480.53	I
60		3485.69	I
60		3489.74	II
480		3491.05	II
60		3495.75	I
95		3499.10	I
890		3504.89	II
120		3506.64	I
600		3510.84	II
60		3520.25	II
310		3535.41	II
190		3547.03	I
120		3573.74	II
60		3574.24	I
60		3587.13	II
240		3596.05	II
190		3598.72	I
600		3610.16	I
190		3624.82	II
95		3635.20	I
4800		3635.46	I
120		3637.97	I
190		3641.33	II
6600		3642.68	I
180		3646.20	I
7200		3653.50	I
290		3654.59	I
660		3658.10	I
120		3659.76	II
380		3660.63	I
190		3662.24	II
380		3668.97	I
600		3671.67	I
3100		3685.20	II
120		3685.96	I
95		3687.35	I
600		3689.91	I

Titanium (Cont.)

Intensity		Wavelength	Spectrum
140		3694.45	I
30		3698.18	I
60		3698.43	I
60		3700.08	I
120		3702.29	I
190		3704.30	I
140		3706.23	II
50		3707.53	I
290		3709.96	I
30		3715.40	I
450		3717.40	I
140		3721.64	II
330		3722.57	I
600		3724.57	I
380		3725.16	I
2900		3729.82	I
50		3735.67	I
60		3738.90	I
3300		3741.06	I
330		3741.64	II
160		3748.10	I
5200		3752.86	I
600		3753.64	I
140		3757.69	II
3300		3759.30	II
2900		3761.32	II
50		3761.89	II
60		3766.45	I
600		3771.66	I
30		3776.06	II
840		3786.04	I
120		3789.30	I
70		3795.90	I
60		3798.31	I
70		3818.22	I
60		3822.03	I
240		3828.19	I
95		3833.68	I
95		3836.78	I
60		3846.45	I
130		3853.05	I
130		3853.73	I
170		3858.14	I
240		3866.44	I
170		3868.40	I
120		3873.21	I
260		3875.26	I
170		3882.15	I
170		3882.33	I
500		3882.89	I
60	h	3888.02	I
70		3889.95	I
200	h	3895.25	I
85		3900.49	I
330		3900.54	II
180		3900.96	I
2600		3904.78	I
110	h	3911.19	I
500		3913.46	II
500		3914.34	I
24		3914.74	I
35		3919.82	I
290		3921.42	I
1100		3924.53	I
110		3926.32	I
890		3929.88	I
35		3932.02	II
70		3934.24	I
1100		3947.78	I
4500		3948.67	I
4500		3956.34	I
5200		3958.21	I
950		3962.85	I
950		3964.27	I
4800		3981.76	I
570		3982.48	I
60		3984.33	I
35		3985.25	I
60		3985.59	I
5700		3989.76	I
35		3994.70	I
7800		3998.64	I
70		3999.36	I
70		4002.49	I
70		4003.81	I
35		4005.97	I
70		4008.06	I
950		4008.93	I
190		4009.66	I
70		4012.39	II
180		4013.58	I
70		4015.38	I
35		4016.28	I
120	h	4017.77	I
140		4021.83	I
1200		4024.57	I
40		4025.14	II
190	h	4026.54	I
40		4027.48	I
40		4028.34	II
190	h	4030.51	I
40		4033.91	I
30		4034.91	I
110		4035.83	I
35	h	4040.32	I
290		4055.02	I
85		4057.62	I
85		4058.14	I
410		4060.26	I
200		4064.22	I
200		4065.10	I
840		4078.47	I
40		4079.72	I
290		4082.46	I
85		4099.17	I
220		4112.71	I
85		4122.17	I
40		4123.31	I
85		4123.57	I
130		4127.54	I
40		4129.17	I
40		4131.25	I
140		4137.29	I
85		4143.05	I
170		4150.96	I
85		4159.64	I
70		4163.65	II
35		4164.14	I
40		4166.32	I
85		4169.35	I
120		4171.03	I
40		4171.90	II
35		4183.30	I
360		4186.12	I
40		4188.69	I
70		4200.75	I
85		4203.46	I
35		4211.73	I
40		4224.79	I
40		4227.65	I
130		4237.89	I
85		4249.12	I
130		4256.04	I
70		4258.54	I
70		4261.60	I
330		4263.13	I
35		4265.71	I
40		4266.22	I
70		4270.14	I
85		4272.43	I
240		4274.58	I
120		4276.43	I
120		4278.23	I
30		4278.81	I
110		4281.38	I
220		4282.71	I
160		4284.99	I
890		4286.01	I
840		4287.40	I
30		4288.16	I
950		4289.07	I
120		4290.23	II
840		4290.94	I
120		4291.14	I
140		4294.12	II
840		4295.76	I
2000		4298.66	I
200		4299.23	I
200		4299.64	I
200		4300.05	II
2900		4300.56	I
4100		4301.09	I
85		4301.93	II
6000		4305.92	I
180		4307.90	II
35		4308.50	I
40		4311.65	I
85		4312.87	II
85		4314.35	I
1200		4314.80	I
360		4318.64	I
180		4321.66	I
190		4325.13	I
160		4326.36	I
30		4334.84	I
160		4337.92	II
24		4344.29	II
70		4346.11	I
35		4354.06	I
95		4360.49	I
24		4368.96	I
95		4369.68	I
60		4372.38	I
30		4388.08	I
170		4393.92	I
330		4395.04	II
60		4399.77	II
240		4404.28	I
60		4404.90	I
30		4405.68	I
60		4416.54	I
220		4417.28	I
60		4417.72	II
120		4421.76	I
120		4422.82	I
24		4424.39	I
30		4425.83	I
120		4426.06	I
890		4427.10	I
21		4430.02	I
85		4430.37	I
50		4431.28	I
30		4432.60	I
24		4433.58	I
170		4434.00	I
70		4436.59	I
30		4438.23	I
130		4440.35	I
50		4441.27	I
230		4443.80	II
24		4444.27	I
840		4449.15	I
30		4450.49	II
550		4450.90	I
840		4453.32	I
290		4453.71	I
950		4455.33	I
1100		4457.43	I
21		4462.09	I
70		4463.38	I
95		4463.54	I
290		4465.81	I
240		4468.50	II
240		4471.24	I
95		4474.85	I
95		4479.70	I
50		4480.59	I
530		4481.26	I
95		4482.69	I
19		4488.32	II
260		4489.09	I
24		4492.55	I
40		4495.01	I
240		4496.15	I
24		4497.73	I
200		4501.27	II
40		4503.78	I
21		4506.36	I
50		4511.17	I
780		4512.74	I
19		4515.62	I
1000		4518.03	I
95		4518.70	I
1000		4522.80	I
780		4527.31	I
6000		4533.24	I
240		4533.97	II
3600		4534.78	I
2400		4535.58	I
1200		4535.92	I
1200		4536.05	I
24		4537.23	I
24		4539.10	I
720		4544.69	I
950		4548.77	I
240		4549.63	II
950		4552.46	I
24		4555.08	I
720		4555.49	I
19		4557.86	I
19		4558.11	I
60		4559.92	I
50		4562.63	I
35		4563.43	I
110		4563.77	II
35		4570.91	I
240		4571.98	II
19		4585.84	I
24		4589.95	II
60		4599.23	I
21		4609.37	I
950		4617.27	I
24		4619.52	I
480		4623.09	I
190		4629.34	I
50	d	4634.87	
60		4637.88	I
240		4639.37	I
220		4639.67	I
190		4639.95	I
140		4645.19	I
120		4650.02	I
24		4656.04	I
720		4656.47	I
840		4667.59	I
70		4675.12	I
950		4681.92	I
21		4686.92	I
24		4690.80	I
190		4691.34	I
40		4693.68	I
24		4696.94	I
190		4698.76	I
120		4710.19	I
24		4715.30	I
65		4722.62	I
65		4723.17	I
55		4731.17	I
45		4733.43	I
18		4734.68	I
22		4742.11	I
170		4742.79	I
22		4747.68	I
310		4758.12	I
310		4759.28	I
45		4766.33	I
28		4769.77	I
65		4778.26	I
45		4781.72	I
110		4792.49	I
45		4796.22	I
35		4797.98	I
110		4799.80	I
28		4805.10	II
110		4805.43	I
45		4808.53	I
22		4811.08	I
40		4812.25	I
200		4820.42	I
22		4825.46	I
40		4836.13	I
470		4840.87	I
65		4848.47	I
290		4856.01	I
35		4864.18	I
200		4868.26	I
250		4870.14	I
28		4880.91	I
45		4882.35	I
400		4885.08	I
380		4899.91	I
320		4913.62	I
55		4915.24	I
130		4919.87	I
180		4921.77	I
55		4925.41	I
30		4926.16	I
150		4928.34	I
30		4937.74	I
95		4938.29	I
30		4941.58	I
21		4948.19	I
21		4958.25	I
55		4964.75	I
21		4966.04	I
65		4968.58	I
75		4973.05	I
120		4975.35	I
65		4977.74	I
120		4978.20	I
5800		4981.73	I
150		4989.15	I
4600		4991.07	I
30		4995.08	I
140		4997.10	I
4000		4999.51	I
230		5001.01	I
3600		5007.21	I
120		5009.65	I
230		5013.30	I
3200	d	5014.19	I
		5014.24	I
580		5016.17	I
840		5020.03	I
840		5022.87	I
580		5024.84	I
300		5025.58	I
1200		5035.91	I
840		5036.47	I
740		5038.40	I
1200		5039.95	I
75		5040.62	I
85		5043.59	I
35		5044.27	I
55		5045.41	I
26		5048.21	I
110		5052.87	I
21		5054.08	I
110		5062.11	I
35		5064.07	I
1400		5064.66	I
95		5065.99	I
35	h	5068.33	I
65		5069.35	I
130		5071.48	I
60		5085.34	I
130		5087.07	I
21		5103.15	I
55		5109.44	I
190		5113.44	I
270		5120.42	I
30		5129.15	II
270		5145.47	I
230		5147.48	I
210		5152.20	I
21	Bl	5166.86	I
1100		5173.75	I
40		5186.34	I
85		5188.70	II
30		5189.58	I
1300		5192.98	I
85	h	5194.04	I
65		5201.10	I
120		5206.08	I
75		5207.87	I
65		5208.42	

Titanium (Cont.)

Intensity		Wavelength	
1400		5210.39	I
65		5212.29	I
150		5219.71	I
95		5222.69	I
85		5223.64	I
250		5224.32	I
95		5224.56	I
190		5224.95	I
65		5226.56	II
120		5238.58	I
21		5246.15	I
55		5246.57	I
75		5247.31	I
21		5250.95	I
110		5252.11	I
75		5255.83	I
55		5259.99	I
55		5263.50	I
150		5265.98	I
40		5282.39	I
140		5283.45	I
35		5284.39	I
26		5288.81	
65		5295.79	I
120		5297.26	I
65		5298.44	I
26		5336.81	II
17		5341.50	I
75		5351.08	I
26		5366.65	I
55		5369.64	I
40		5389.18	I
55		5389.99	I
17		5396.60	I
85		5397.09	I
35		5404.02	I
110		5409.61	I
40		5426.26	I
75		5429.15	I
26		5436.73	I
17		5438.32	I
40		5446.64	I
11	Bl	5448.34	T
30		5448.90	I
21		5449.16	I
35		5453.65	I
55		5460.51	I
75		5471.21	I
35		5472.70	I
40	h	5473.55	I
85		5474.23	I
30		5474.46	I
120	h	5477.71	I
110		5481.43	I
75		5481.87	I
85	h	5488.20	I
150		5490.15	I
26		5490.84	I
110		5503.90	I
40		5511.78	I
340		5512.53	I
270		5514.35	I
320		5514.54	I
26		5530.49	
110		5565.49	I
13		5579.16	
21	h	5582.98	
30	h	5585.68	
65	Bl	5597.85	T
55	Bl	5629.28	T
17		5635.84	
250		5644.14	I
75		5648.58	I
26	Bl	5661.55	T
190		5662.16	I
75		5662.91	I
21		5673.42	I
130		5675.44	I
30	h	5679.94	I
95		5689.47	I
75		5702.68	I
35		5708.23	I
65		5711.88	I
40	h	5713.92	I
95		5715.13	I
55		5716.48	I
35		5720.48	I
85		5739.51	I
40		5740.02	I
19		5741.22	I
21		5752.84	I
19		5756.86	I
40	h	5762.27	I
55	h	5766.35	I
75	h	5774.05	I
30		5780.78	I
75	h	5785.98	I
65	Hl	5804.26	I
21	Bl	5814.96	T
40		5823.71	I
21	h	5841.18	
21		5852.34	
400		5866.46	I
65		5880.31	I
21	h	5888.68	
230		5899.32	I
55		5903.33	I
120		5918.55	I
150		5922.12	I
75		5937.82	I
120		5941.76	I
300		5953.17	I
200		5965.84	I
270		5978.56	I
340		5999.04	I
65		5999.68	I
21		6012.73	
110		6064.63	I
120		6085.23	I
120		6091.17	I
40		6092.81	I
40	h	6098.67	I
35	h	6121.01	I
120		6126.22	I
19		6138.38	I
30		6146.22	I
21		6149.74	I
30	Bl	6162.23	T
35		6186.15	I
95	h	6215.28	I
75	h	6220.49	I
65	h	6221.41	I
380		6258.10	I
380		6258.70	I
300		6261.10	I
65		6303.75	I
55		6312.24	I
26		6318.03	I
30		6336.10	I
35		6366.35	I
11		6419.10	I
17		6497.69	I
19		6508.14	I
55		6546.28	I
65		6554.23	I
11	h	6554.83	
75		6556.07	I
19	h	6565.62	I
14	h	6575.18	I
35		6599.11	I
18	Bl	6651.46	T
18	h	6666.55	I
22	h	6667.74	
9		6668.39	
18		6677.18	I
22	b	6691.21	T
26		6716.68	I
16	Bl	6723.95	T
80		6743.12	I
22		6745.52	I
18		6844.64	
18		6860.39	
35		6861.47	I
9		6873.92	I
12		6913.19	I
14	h	6933.15	I
14	h	6943.70	I
23		6996.63	I
15		7004.66	I
14		7008.35	I
14		7010.94	I
14	h	7035.86	I
40		7038.80	I
14		7050.65	I
40	Bl	7054.51	T
23		7069.11	I
23		7072.05	
45	Bl	7087.89	T
30	b	7124.9	T
40	Bl	7125.61	T
26		7138.91	I
26		7167.13	
23		7171.53	
55		7189.89	I
26	b	7203.64	T
260		7209.44	I
60		7216.20	I
130		7244.86	I
130		7251.72	I
19		7263.40	
19		7266.29	I
19	b	7269.05	T
15		7315.56	I
26		7318.39	I
120		7344.72	I
11		7352.16	I
90		7357.74	I
60		7364.11	I
26		7440.60	I
9		7474.94	I
26		7489.61	I
19		7496.12	I
12		7580.55	I
9	Bl	7589.62	T
15		7614.50	I
23		7654.44	I
11	Bl	7705.21	T
30		7949.17	I
26	h	7961.58	I
60		7978.88	I
9		7979.07	I
30		7996.53	I
7	h	8003.55	
55		8024.84	I
30		8068.24	I
8		8267.62	
14	h	8306.31	I
9	h	8307.41	I
9	h	8311.76	I
8	h	8312.85	I
12		8334.37	I
14		8353.15	I
75		8364.24	I
100		8377.85	I
100		8382.54	I
55		8382.82	I
75		8396.87	I
120		8412.36	I
19		8416.98	I
15		8424.41	I
170		8426.52	I
490		8434.94	I
240		8435.70	I
40		8438.93	I
40		8450.89	I
9	h	8457.10	I
19	h	8467.15	I
45		8468.50	I
15		8496.04	I
19	h	8518.05	I
40		8518.32	I
14		8539.38	I
40		8548.12	I
9		8569.77	I
9	h	8598.18	I
90		8675.39	I
45		8682.99	I
23		8692.33	I
19		8734.69	I
23		8766.64	I
15	h	8778.71	I

Ti III
Ref. 378 — C.H.C.

Intensity		Wavelength	
		Vacuum	
6		1282.48	III
6		1286.23	III
15		1286.36	III
10		1289.30	III
10		1291.62	III
10		1293.23	III
15		1294.70	III
10		1295.88	III
20		1298.66	III
20		1298.97	III
12		1327.59	III
10		1420.04	III
10		1420.44	III
10		1421.63	III
10		1421.77	III
12		1422.40	III
10		1424.14	III
23		1455.19	III
10		1498.70	III
		Air	
10		2199.22	III
12		2237.77	III
10		2327.02	III
15		2331.35	III
15		2331.66	III
15		2334.34	III
17		2339.00	III
18		2346.79	III
18		2374.99	III
22		2413.99	III
25		2516.05	III
24		2527.84	III
23		2540.06	III
24		2563.44	III
23		2565.42	III
22		2567.56	III
15		2576.47	III
15		2580.46	III
10		2692.16	III
12		2701.96	III
22		2984.75	III
12	d	3354.71	III
12		3872.50	III
12		3881.21	III
12		3893.63	III
10		3896.33	III
15		3915.47	III
12		3921.38	III
10		3921.61	III
12		3922.95	III
10		3924.86	III
10		4060.21	III
10		4119.14	III
11		4215.52	III
11		4269.84	III
11		4296.70	III
10		4348.04	III
11		4433.91	III
10		4540.22	III
15		4549.84	III
10	d	4555.46	III
15	d	4572.20	III
10		4649.45	III
12		4652.86	III
10		4874.00	III
10		4950.10	III
10		4971.19	III
10		5083.80	III
14		5147.31	III
12	d	5226.28	III
11		5247.49	III
17		5278.12	III
10		5278.70	III
12		5298.43	III
16		5301.20	III
15		5306.88	III
10		5395.69	III
12		5533.01	III
10		5817.44	III
12		6611.38	III
18		6621.58	III
10		6629.37	III
14		6647.47	III
18		6667.99	III
15		6674.19	III
14		6707.76	III
12		6724.80	III
16		6734.10	III
15		6862.26	III
12		6874.35	III
10		6896.12	III
12		7015.38	III
10		7071.93	III
20		7072.64	III
18		7084.57	III
15		7124.13	III
11		7171.79	III
10		7175.92	III
10		7217.50	III
9		7225.55	III
12		7270.67	III
14		7316.30	III
10		7316.68	III
12		7379.96	III
10		7408.13	III
10		7457.85	III
15		7506.87	III
17		7507.68	III
10		7523.85	III
12		7544.29	III
9		7566.25	III
10	h	8172.21	III
9	h	8173.37	III
9	h	8178.00	III
10		8182.42	III
9	h	8192.68	III
9	h	8194.75	III
9	h	8263.67	III
15	h	8267.32	III
10		8338.54	III
12		8394.20	III
20		8466.87	III
5		8699.85	III
3		9017.10	III

Ti IV
Ref. 428 — C.H.C.

Intensity		Wavelength	
		Vacuum	
10		776.76	IV
18		779.07	IV
16		781.73	IV
8		1183.64	IV
10		1195.21	IV
18		1451.74	IV
20		1467.34	IV
12		1469.19	IV
		Air	
20		2067.56	IV
18		2103.16	IV
10		2359.14	IV
10		2359.50	IV

Titanium (Cont.)

Intensity		Wavelength	
8		2541.79	IV
10		2546.88	IV
5		2862.60	IV
6		2929.96	IV
14		2937.33	IV
12		2957.31	IV
15		3541.36	IV
17		3576.44	IV
10		3581.39	IV
13		4131.22	IV
14		4133.78	IV
10		4397.33	IV
9		4403.45	IV
15		4618.11	IV
20		5398.93	IV
8		5470.98	IV
18		5492.51	IV
10		5517.72	IV
14		5877.79	IV
15		5885.96	IV
7		5891.15	IV
6		6231.62	IV
17		6246.65	IV
11		6247.74	IV
15		6292.41	IV
12		6913.85	IV
15		6978.51	IV
9		7491.37	IV
8		7494.77	IV
5	h	7652.12	IV
8	h	7706.85	IV

Ti V
Ref. 427 — C.H.C.

Intensity		Wavelength	
		Vacuum	
12		225.35	V
10		228.91	V
17		252.96	V
7		323.36	V
7		461.41	V
8		474.69	V
8		483.99	V
10	d	488.58	V
15		498.26	V
14		502.08	V
7		502.71	V
12		504.66	V
7		506.47	V
8		513.37	V
7		523.05	V
12		524.58	V
13		526.57	V
8		529.32	V
10		535.84	V
10		535.89	V
8	d	540.14	V
8		541.46	V
9		541.71	V
7		543.10	V
7		543.34	V
7		1128.55	V
8		1192.35	V
9		1198.66	V
9		1222.36	V
10		1230.36	V
11		1239.96	V
10		1241.67	V
7		1246.13	V
8		1268.49	V
8		1306.11	V
8		1411.31	V
9		1675.15	V
8		1687.16	V
11		1717.40	V
8		1759.76	V
7		1771.45	V
10		1841.49	V
7		1864.45	V
7		1881.89	V
7		1920.16	V
7		1988.75	V

TUNGSTEN (W)
Z = 74

W I and II
Ref. 1 — C.H.C.

Intensity		Wavelength	
		Air	
5800		2001.71	II
13000		2008.07	II
5100		2009.98	II
4100		2010.23	II
4100		2014.23	II
7300		2026.08	II
15000		2029.98	II
2700		2035.03	II
5300		2049.63	II
2300		2065.57	II
3400		2071.21	II
2200		2075.59	II
9700		2079.11	II
3600		2088.19	II
2200		2089.14	II
1700		2090.48	I
6100		2094.75	II
2400		2098.60	II
2200		2100.67	II
1500		2101.54	I
1500		2106.18	II
1300		2110.34	II
2100		2118.87	II
2400		2121.59	II
850		2153.56	II
850		2157.80	II
1500		2166.32	II
480		2182.90	I
440		2194.52	II
1300		2204.48	II
460		2248.75	II
460		2249.80	I
180		2270.24	II
95		2271.37	I
510		2277.58	I
160		2284.91	I
320		2285.17	I
530	d	2294.49	I
		2294.54	II
270		2298.33	I
240		2303.83	II
240		2306.59	I
340		2309.02	I
440		2313.17	I
220		2314.17	I
190		2315.02	II
460		2321.63	I
290		2326.09	II
390	d	2326.56	I
		2326.70	I
75		2328.31	II
130		2333.77	II
210		2341.37	I
75		2349.26	II
120	d	2350.37	II
320		2354.61	I
60		2358.81	II
580		2360.44	I
850		2363.07	I
60		2364.22	II
510		2374.47	I
210		2382.99	I
670		2384.82	I
240		2389.08	I
120		2390.37	II
120		2392.93	II
730		2397.09	II
560		2397.73	I
560		2397.98	I
75		2404.24	II
1700	d	2405.58	I
		2405.69	I
75		2411.54	II
320		2414.04	I
610		2415.68	I
50		2419.34	II
50		2421.01	II
870		2424.21	I
190		2427.49	II
170		2429.39	II
580		2431.08	I
630		2433.98	I
60		2435.01	II
1800		2435.96	I
250		2436.62	I
580		2444.06	I
160		2446.39	II
270		2448.39	I
270		2451.35	I
780		2451.48	II
870		2452.00	I
430		2454.72	I
630		2454.98	I
780		2455.51	I
780		2456.53	I
1100		2459.30	I
270		2460.16	I
480		2462.79	I
270		2464.30	I
230		2466.52	II
1400		2466.85	I
75		2470.80	II
480		2472.51	I
1200		2474.15	I
290		2477.80	II
870		2480.13	I
390		2480.96	I
1500		2481.44	I
480	d	2482.10	I
		2482.21	I
29		2484.40	II
580		2484.74	I
390		2487.50	I
270		2488.77	II
390		2489.23	II
75		2492.93	II
630		2495.26	I
230		2496.64	II
95		2497.48	II
140		2499.69	II
40		2500.11	II
		2501.90	II
680		2504.70	I
270		2506.02	I
24		2508.00	II
250		2510.17	I
75		2510.47	II
60		2518.14	II
310		2520.46	I
780		2521.32	I
270		2522.04	II
780		2523.41	I
430		2527.76	I
780		2533.64	I
50		2534.82	II
580		2545.34	I
1200		2547.14	I
50		2549.09	II
40		2550.10	II
780		2550.38	I
2700		2551.35	I
450		2553.82	I
410		2554.86	II
580		2555.09	II
		2555.21	I
310		2556.75	I
290		2560.12	I
730		2561.97	I
230		2563.16	II
110		2563.91	II
530		2571.44	II
170	d	2572.24	II
		2572.35	II
75		2573.95	II
190		2579.26	II
290		2580.34	I
870		2580.49	I
40		2581.20	II
390		2584.39	I
390		2589.17	II
170		2591.49	II
110		2598.74	II
370		2601.96	I
75		2602.51	II
75		2603.02	II
270		2603.54	I
680		2606.39	I
320		2607.38	I
370		2608.32	I
970		2613.08	I
480		2613.82	I
230		2615.12	I
70		2615.44	II
210		2619.18	I
400		2620.25	I
400		2622.21	I
400		2625.22	I
210		2628.26	I
400		2632.48	I
400		2632.70	I
810		2633.13	I
290		2636.54	I
400	d	2638.62	I
		2638.75	I
210		2645.69	I
650		2646.18	I
400		2646.73	I
75		2647.74	II
40		2653.42	II
80		2653.57	II
1600		2656.54	I
400		2657.38	I
400	d	2658.04	II
		2658.18	I
810		2662.84	I
260		2664.97	I
75		2666.49	II
210		2669.30	I
810		2671.47	I
80		2673.59	II
650		2677.28	I
160	d	2677.79	II
		2677.91	I
400		2678.88	I
2100		2681.42	I
290		2683.35	I
210		2691.09	I
650		2695.67	I
210		2697.71	II
650		2699.59	I
400		2700.01	I
40		2701.48	II
160		2702.11	II
210		2702.52	I
400		2706.58	I
400		2708.59	I
400	d	2708.80	I
		2708.93	I
80		2709.58	II
40		2710.78	II
400		2715.50	I
80		2716.32	II
80		2718.04	II
2100		2718.91	I
320		2719.33	I
210		2719.86	I
2600		2724.35	I
210		2724.62	I
400		2725.03	I
		2725.06	I
80		2729.62	II
75		2740.79	II
650		2748.84	I
40	d	2760.74	II
80		2761.59	II
400		2762.34	I
400		2764.27	II
210		2768.98	I
400		2769.74	I
810		2770.88	I
210		2773.70	I
810		2774.00	I
810		2774.48	I
160		2776.50	II
40		2778.69	II
210		2787.98	I
340		2791.96	I
810		2792.70	I
80		2799.03	II
400		2799.93	I
160	d	2801.05	II
		2801.17	I
130		2805.92	II
40		2812.25	II
810		2818.06	I
160		2822.57	II
260		2829.82	I
1600		2831.38	I
810		2833.63	I
210		2835.64	I
400		2841.57	I
810		2848.02	I
650		2856.03	I
650		2866.06	I
230		2878.72	I
610		2879.11	I
610		2879.40	I
440		2896.01	I
1500		2896.44	I
230		2910.48	I
270		2911.00	I
360		2918.25	I
50		2918.63	II
360		2923.10	I
230		2923.54	I
230		2925.13	I
690		2935.00	I
2400		2944.40	I
2400		2946.99	I
480		2947.39	I
210		2952.29	II
440		2964.52	I
480		2977.11	I
		2977.21	I
730	d	2979.71	I
		2979.86	I
400		2993.61	I
240		2995.26	I
190		3009.09	I
360		3013.79	I
520		3016.47	I
770		3017.44	I
110		3024.50	II
210		3024.93	I
310	d	3026.67	I
		3026.79	I
160		3033.56	I
160		3034.19	I
160		3039.31	I
440	d	3041.73	I
		3041.86	I
270		3043.80	I
440		3046.44	I
110		3048.66	I
810		3049.69	I
110		3064.93	I
180		3073.28	I
110		3077.52	II
180	d	3084.83	I
		3084.91	I
370		3093.50	I
240		3107.23	I
240		3108.02	I

Tungsten (Cont.)

Intensity		Wavelength	Spectrum
230		3117.57	I
260		3120.18	I
160		3133.88	I
130		3141.42	I
65		3149.85	II
290		3163.42	I
130		3164.44	I
130		3165.38	I
320		3176.60	I
130		3179.06	I
190		3181.82	I
130		3184.05	I
130		3184.42	I
65		3189.24	II
390		3191.57	I
390		3198.84	I
520		3207.25	I
140		3208.28	I
1000		3215.56	I
140		3221.21	I
140		3221.91	I
190		3232.49	I
140		3237.09	I
140		3242.03	I
140		3252.29	I
210		3254.36	I
140		3259.43	I
210		3259.66	I
210	d	3266.62	I
		3266.77	I
150		3281.94	I
150		3293.71	I
730		3300.82	I
440		3311.38	I
440		3326.20	I
440		3331.69	I
150		3354.45	I
150		3371.04	I
390		3373.75	I
150		3412.96	I
150		3413.53	I
150		3422.42	I
150		3427.71	I
230		3429.59	I
240		3443.00	I
160		3477.94	I
400		3495.24	I
160		3508.73	I
160		3510.02	I
160		3526.85	I
160		3535.54	I
160		3537.45	I
650		3545.22	I
160		3568.04	I
240		3570.65	I
80		3572.48	II
160		3575.22	I
80		3592.42	II
240		3606.06	I
80		3613.79	II
1900		3617.52	I
160		3622.34	I
130		3627.24	I
320		3631.94	I
240		3641.41	II
80		3646.52	II
80		3657.59	II
160		3675.55	I
650		3682.08	I
400		3683.30	I
		3683.39	I
160		3683.93	I
570		3688.06	I
810		3707.92	I
60		3716.08	II
100		3719.39	I
50		3736.22	II
120		3741.71	I
510		3757.92	I
680		3760.13	I
1000		3768.45	I
120		3769.21	I
120		3769.86	I
340		3773.71	I
1000		3780.77	I
170		3792.76	I
290		3809.22	I
190		3810.38	I
260		3810.79	I
1400		3817.48	I
110		3829.13	I
1100		3835.06	I
290		3838.51	I
730		3846.22	I
250		3847.49	I
27		3851.57	II
150		3855.55	I
150		3859.30	I
180		3864.34	I
1800		3867.99	I
250		3872.84	I
110		3874.41	I
730		3881.41	I
110		3892.72	I
140	h	3897.91	I
150		3935.03	I
120		3936.97	I
120		3947.98	I
120		3952.52	I
120		3952.90	I
160		3953.15	I
200		3955.30	I
160		3965.14	I
130		3968.59	I
150	h	3970.80	I
130		3979.29	I
130		3980.64	I
250		3983.29	I
8600		4008.75	I
540		4015.22	I
170	h	4016.52	I
220		4019.23	I
130		4022.12	I
180		4028.79	I
180		4036.86	I
140		4039.85	I
140	h	4044.28	I
910		4045.59	I
180		4064.79	I
150		4069.79	I
730		4069.95	I
340		4070.61	I
100		4071.93	I
5000		4074.36	I
150		4082.96	I
130		4088.33	I
100		4095.69	I
1000		4102.70	I
150		4109.75	I
100		4111.82	I
150		4118.05	I
100		4118.19	I
100		4120.85	I
100		4125.16	I
150		4126.80	I
100		4133.48	I
540		4137.46	I
150		4138.02	I
110		4142.25	I
140		4145.16	I
110		4145.95	I
160		4154.66	I
160		4170.53	I
450		4171.17	I
160		4204.40	I
220		4207.05	I
110		4215.38	I
250		4219.37	I
110		4222.04	I
150		4234.34	I
290		4241.44	I
540		4244.36	I
290		4259.35	I
200		4260.29	I
200		4263.30	I
1400		4269.38	I
110		4269.77	I
220		4274.55	I
160		4275.49	I
160		4276.74	I
110		4282.34	I
110		4286.01	I
110		4294.10	I
4100		4294.61	I
2200		4302.11	I
160		4306.87	I
110		4307.64	I
110		4332.13	I
100		4347.00	I
150		4355.17	I
100		4361.81	I
150		4364.78	I
100	d	4365.95	I
		4366.07	I
150		4372.52	I
200		4378.48	I
180		4384.85	I
100		4403.95	I
200		4408.28	I
130		4412.19	I
160		4436.90	I
140		4460.49	I
140		4466.34	I
140		4466.74	I
640		4484.19	I
160		4504.84	I
130		4512.88	I
120		4513.25	I
150		4543.54	I
150		4546.47	I
150		4551.82	I
140		4570.64	I
170		4588.73	I
140		4599.94	I
140		4609.89	I
160		4613.30	I
100		4642.53	I
130		4657.42	I
640		4659.87	I
640		4680.51	I
100		4693.72	I
140		4757.54	I
790		4843.81	I
380		4886.90	I
220		4982.59	I
330		5006.15	I
220		5015.30	I
820		5053.28	I
210		5054.60	I
210		5069.12	I
120		5071.74	I
770		5224.66	I
27		5500.49	I
27		5503.44	I
10		5508.61	I
220		5514.68	I
15		5531.38	I
15		5537.72	I
13		5568.09	I
13		5604.31	I
11		5631.27	I
27		5631.94	I
65		5648.37	I
35		5660.72	I
27		5674.39	I
13		5676.60	I
15		5676.90	I
15		5697.79	I
55		5735.09	I
13		5749.24	I
11		5756.10	I
13		5793.06	I
13		5796.49	I
45		5804.85	I
13	d	5806.05	I
		5806.24	I
13		5833.61	I
13		5838.97	I
17		5845.27	I
28		5851.58	I
11		5856.61	I
22		5864.63	I
11		5874.22	I
13		5880.21	I
13		5891.61	I
13		5901.20	I
40		5902.64	I
13		5928.58	I
55		5947.57	I
13		5953.96	I
13		5956.19	I
27		5960.83	I
55		5965.86	I
27		5972.51	I
20		5978.86	I
20		5983.82	I
13		6009.01	I
55		6012.78	I
40		6021.52	I
20		6028.32	I
20		6043.31	I
13		6049.92	I
13		6065.08	I
22		6081.44	I
13		6111.66	I
13		6115.52	I
22		6128.25	I
13		6143.94	I
20		6153.72	I
20		6154.87	I
20		6203.51	I
20		6254.28	I
27		6285.88	I
45		6292.02	I
20		6303.21	I
13		6386.47	I
35		6404.21	I
40		6445.12	I
11		6508.05	I
15		6532.39	I
13		6538.11	I
13		6563.20	I
20		6573.93	I
11		6607.13	I
11		6609.05	I
17		6611.62	I
11		6621.74	I
13		6678.42	I
15		6693.08	I
5		6746.56	I
5		6764.45	I
7		6805.31	I
9		6814.92	I
9		6820.27	I
8		6828.43	I
4		6853.74	I
4		6876.01	I
5		6908.29	I
9		6934.23	I
8		6964.12	I
13		6984.27	I
8		6993.27	I
4		6994.06	I
8		7017.88	I
3		7028.68	I
3		7098.22	I
4	h	7111.18	I
15		7140.52	I
9		7162.64	I
5	h	7191.33	I
5		7198.62	I
11		7200.16	I
5		7216.35	I
4		7226.06	I
8		7237.12	I
5		7274.47	I
10		7278.24	I
15		7285.81	I
15		7296.55	I
3		7298.25	I
7		7385.08	I
4		7451.39	I
3		7456.37	I
8		7483.35	I
7		7504.13	I
10		7509.00	I
3		7520.66	I
9		7537.45	I
9		7550.48	I
17		7569.92	I
5		7582.88	I
3		7612.18	I
17		7614.15	I
3		7631.29	I
3		7654.81	I
13		7688.97	I
4		7701.01	I
5		7761.16	I
3		7776.73	I
11		7784.15	I
7		7808.96	I
2		7823.82	I
4		7863.47	I
2		7867.04	I
4		7880.40	I
5		7886.48	I
9		7940.92	I
3		7957.06	I
22		8017.19	I
7		8054.89	I
22		8055.64	I
5		8060.38	I
13		8123.82	I
5		8143.19	I
3		8165.72	I
5		8210.22	I
4		8322.05	I
10		8338.08	I
4		8348.81	I
7		8358.72	I
3		8382.94	I
4		8402.60	I
4		8475.14	I
27		8585.11	I
10		8594.42	I
8		8613.27	I
3		8614.50	I
13		8865.53	I

URANIUM (U)
Z = 92

U I and II
Ref. 1, 303 — J.G.C.

Intensity	Wavelength	
	Air	
440	2565.41	II
340	2569.71	II
340	2591.25	II
610	2635.53	II
470	2645.47	II
340	2669.17	II
470	2683.28	II
320	2691.04	II
370	2706.95	II
370	2733.97	II
470	2754.16	II
340	2762.85	II
390	2770.04	II
410	2784.45	II
830	2793.94	II
870	2802.56	II
630	2807.05	II
440	2808.98	II
630	2817.96	II
870	2821.12	II
390	2824.37	II
680	2828.90	II
920	2832.06	II
360	2837.19	II
460	2839.89	II

Uranium (Cont.)

360		2842.09	II
360		2849.48	II
390		2860.47	II
970		2865.68	II
340		2870.97	II
490		2882.74	II
460		2887.25	II
410		2888.26	II
1200		2889.62	II
320		2894.14	II
410		2894.51	II
780		2906.80	II
780		2908.28	II
320		2914.25	II
360		2914.63	II
440	p	2921.68	II
320		2927.38	II
490		2928.60	II
580		2931.41	II
440		2932.61	II
340		2933.86	II
530	p	2940.37	II
1300		2941.92	II
830		2943.90	II
340		2948.09	II
390		2954.77	II
580		2956.06	II
460		2965.03	II
580		2967.94	II
580		2971.06	II
410		2976.35	II
320		2982.74	II
530		2984.61	II
410		2992.72	II
360		3007.91	II
320		3021.22	II
630		3022.21	II
320		3024.51	II
320		3028.19	II
630		3031.99	II
490		3033.18	II
490		3044.16	II
580		3050.20	II
630		3057.91	II
460		3061.62	II
630		3062.54	II
580		3072.78	II
580		3093.01	II
320		3095.75	II
[illegible]		3098.01	II
580		3102.39	II
460		3104.15	II
970		3111.62	II
530		3119.35	II
680		3124.95	II
530		3139.61	II
410		3144.97	II
490		3145.56	II
680		3149.24	II
530		3153.11	II
340		3176.21	II
340		3177.33	II
340		3206.05	II
730		3229.50	II
680		3232.16	II
440		3244.22	II
340		3265.79	II
440		3270.12	II
440		3288.21	II
730		3291.33	II
1100		3305.89	II
390		3337.79	II
440		3341.66	II
390		3357.84	I
730		3390.38	I
340		3394.77	II
580		3424.56	II
580		3435.49	I
360		3453.55	II
320		3454.23	II
320		3457.05	II
320		3457.71	II
360		3459.92	I
320		3462.22	I
460		3463.55	I
630		3466.30	I
390		3472.52	II
320		3473.43	I
360		3480.36	I
680		3482.49	II
1600		3489.37	I
390		3493.33	II
340		3494.00	I
320		3494.84	II
530		3496.41	II
630		3500.08	I
320		3504.01	I
320		3505.07	II
780		3507.34	I
320		3508.84	II
1600		3514.61	I
390		3509.66	II
320		3513.67	I
390		3519.96	II
390		3531.11	II
630		3533.57	II
320		3534.33	I
530		3540.47	II
320		3542.57	I
390		3547.19	II
320		3549.20	I
1200		3550.82	II
320		3552.17	II
680		3555.32	I
320		3561.41	I
1200		3561.80	I
390		3563.66	I
2300		3566.59	I
530		3569.08	I
320		3574.76	I
360		3577.92	I
630		3578.72	II
360		3581.84	II
3200		3584.88	I
320		3590.50	II
390		3591.74	I
460		3593.52	II
460		3605.27	I
360		3606.32	II
320		3616.33	I
320		3616.76	II
320		3620.08	I
320		3622.70	I
390		3623.06	II
460		3630.73	II
840		3638.20	I
310		3640.76	II
420		3644.24	I
310		3645.03	II
660		3651.54	I
490		3652.06	I
960		3659.15	I
2800		3670.07	II
380		3678.75	II
540		3691.92	II
330		3693.70	II
540		3700.57	II
1100		3701.52	II
350		3713.55	I
300		3717.42	II
350		3718.11	II
350		3729.82	II
[illegible]		3732.62	II
350		3733.07	II
600		3738.04	II
300		3744.25	II
680		3746.42	II
350		3747.14	II
950		3748.68	II
600		3751.17	I
350		3752.66	II
350		3755.48	II
490		3758.35	I
350		3759.24	II
330		3763.26	I
490		3764.57	II
430		3766.89	I
330		3769.53	II
540		3773.43	I
300		3776.48	I
380		3780.71	II
1900		3782.84	II
430		3783.84	II
570		3793.10	II
380		3793.26	I
380		3793.57	II
380		3808.92	I
380		3809.22	II
1900		3811.99	I
380		3813.79	II
380		3814.06	II
750		3826.51	II
2000		3831.46	II
1200		3839.63	I
490	p	3848.60	II
620		3854.22	I
2400		3854.64	II
4900		3859.57	II
490		3861.17	II
1900		3865.92	II
380		3866.80	II
1500		3871.03	I
620		3874.04	II
620		3878.08	II
1000		3881.45	II
490		3882.36	II
380		3883.28	II
2200		3890.36	II
620		3892.68	II
490		3894.12	I
490		3896.77	II
620		3899.78	II
410		3902.55	II
460		3904.30	II
380		3906.45	I
330		3911.67	II
380		3915.88	II
330		3926.21	I
330		3926.72	I
430		3930.98	II
2000		3932.02	II
490		3935.38	II
330		3940.48	II
1200		3943.82	I
300		3948.44	I
300		3953.58	II
360		3954.67	II
350		3964.21	I
600		3966.52	II
1200		3985.79	II
460		3990.42	II
380		3992.53	II
350		3998.24	II
350		4004.06	II
430		4005.21	I
570		4017.72	II
300		4018.99	II
1000		4042.75	I
520		4044.41	II
410		4047.61	I
1600		4050.04	II
540		4051.91	II
300		4054.30	II
430		4058.19	II
880		4062.54	II
520		4067.75	II
410		4071.12	II
300		4074.48	II
330		4076.69	II
330		4080.60	II
2200		4090.13	II
460		4093.03	II
380		4106.38	II
810		4116.10	II
410		4124.73	II
410		4128.34	II
460		4141.22	II
880		4153.97	I
380		4156.65	I
350		4163.68	II
1400		4171.59	II
300		4189.27	II
350		4222.37	I
1000		4241.67	II
520		4244.37	II
680		4341.69	II
430	h	4355.74	I
430		4362.05	I
330		4393.59	I
600		4472.33	II
240		4515.28	II
620		4543.63	II
300		4620.21	I
240		4627.07	II
210		4631.62	I
220		4646.60	II
140		4666.85	II
100		4671.40	II
170		4689.07	II
100		4702.51	II
160		4722.72	II
120		4731.59	II
100		4755.74	II
150		4756.81	I
100		4772.70	II
100		4860.99	II
110		5008.21	II
170		5027.38	I
70		5117.24	II
80		5160.32	II
55		5164.14	I
55		5184.57	II
45		5204.31	II
45		5247.75	II
45		5257.04	II
70		5280.38	I
55		5386.19	II
80		5475.70	II
70		5480.26	II
70		5481.20	II
45		5482.53	II
160		5492.95	II
70		5527.82	II
70		5564.17	I
45		5581.59	II
55		5620.78	I
70		5780.59	I
70		5798.53	II
45		5836.02	I
55		5837.68	II
230		5915.39	I
55		5971.50	I
100		5976.32	I
45		5997.31	I
28		6017.38	II
55		6051.74	II
45		6067.22	II
90		6077.29	I
28		6087.34	II
40		6171.86	I
35		6175.39	I
28		6280.18	II
28		6359.29	I
55		6372.46	I
28		6378.52	II
28		6392.77	I
90		6395.42	I
110		6449.16	I
35		6464.98	I
90		6826.92	I
35		6876.74	II
23		7074.79	I
27		7101.61	I
30		7128.90	I
16		7147.89	I
16		7254.45	I
23		7425.50	I
45		7533.93	I
16		7619.35	I
50		7881.94	I
18		7970.46	I
16		8174.66	I
18		8262.06	I
16		8318.35	I
16		8337.50	II
18		8381.87	I
16		8441.21	I
35		8445.39	I
18		8450.03	I
16		8570.52	I
75		8607.95	I
23		8691.28	I
18		8710.76	I
18		8753.69	I
30		8757.76	I
16		8951.96	I
16		8989.92	I
10		9093.67	I
10		9139.56	I
10		9201.51	I
10		9265.34	I
10		9276.44	I
10		9385.90	I
10		9653.26	I
10		9819.00	I
10		9819.05	I
10		9868.36	I
10		[illegible]	I
10		9964.11	I
50		10157.91	I
50		10259.55	I
100		10554.93	I
50		10799.78	I
25		10823.93	I
25		11095.77	I
75		11167.84	I
50		11294.13	I
100		11384.13	I
25		11410.43	I
50		11503.38	I
25		11568.81	I
20		11784.72	II
100		11859.42	I
100		11908.83	I
100		12250.46	I
25		13088.28	I
100		13185.16	I
75		13306.23	I
100		13961.58	I
50		16906.00	I
50		17451.11	I
50		18136.65	I
50		18366.96	I
75		18634.43	I
25		19029.39	I
10		20201.13	I
10		20271.41	I
10		20374.13	I
10		20517.29	I
10		20690.64	I
10		20772.19	I
10		21008.38	I
10		21099.98	I
10		21112.14	I
20		21144.90	II
10		21674.51	I
10		21693.38	I
75		21910.22	I
10		22110.73	I
10		23156.76	I
10		23948.19	I
10		29557.07	I

VANADIUM (V)
Z = 23

V I and II
Ref. 1 — C.H.C.

Intensity	Wavelength	
	Air	
2100	2092.44	I

Vanadium (Cont.)

40		2384.00	II
40		2384.28	I
60		2386.96	I
60		2388.92	I
75		2390.87	I
75		2391.26	I
85		2392.90	I
70		2397.78	I
70		2398.27	I
70		2399.96	I
120		2406.75	I
110		2407.90	I
120		2415.33	I
120		2416.75	I
100		2420.12	I
100		2421.06	I
100		2421.98	I
110		2428.28	I
110		2435.52	I
140		2501.61	I
150		2506.90	I
240		2507.78	I
180		2511.65	I
180		2511.95	I
180		2517.14	I
240		2519.62	I
410		2526.22	I
210		2527.90	II
120		2528.47	II
150		2528.84	II
240		2530.18	I
110		2549.28	II
120		2552.65	I
210		2562.13	I
110		2564.82	I
230		2574.02	I
140		2630.67	II
130		2642.21	II
150		2645.26	I
140		2651.90	I
150		2656.22	I
180		2661.42	I
290		2672.00	II
380		2677.80	II
270		2678.57	II
380		2679.32	II
180		2682.87	II
180		2683.09	II
1100		2687.96	II
170		2688.72	II
150		2689.88	II
230		2690.24	II
240		2690.79	II
120		2696.99	I
120		2697.74	I
680		2700.94	II
380		2702.19	II
530		2706.17	II
150		2706.70	II
110		2707.86	II
170		2711.74	II
120		2714.20	II
640		2715.69	II
150		2722.56	I
240		2728.64	II
180		2731.35	I
100		2739.71	II
140		2753.40	II
140		2765.67	II
140		2777.73	II
120		2803.47	II
120		2846.57	I
110		2847.57	II
140		2852.87	I
140		2854.34	II
200		2855.22	I
180		2859.97	I
240		2864.36	I
170		2866.59	I
210		2868.10	I
140		2869.13	II
210		2870.55	I
110		2877.69	II
110		2879.16	II
350		2880.03	II
380		2882.50	II
380		2884.78	II
140		2888.25	II
380		2889.62	II
900		2891.64	II
530		2892.44	II
900		2892.66	II
1400		2893.32	II
360		2896.21	II
110		2899.60	I
360		2903.08	II
150		2906.13	I
900		2906.46	II
490		2907.47	II
2400		2908.82	II
710		2910.02	II
530		2910.39	II
560		2911.06	II
380		2914.93	I
120		2917.37	II
210		2919.99	II
380		2920.38	II
710		2923.62	I
2400		2924.02	II
1700		2924.64	II
710		2930.81	II
210		2934.40	II
110		2935.87	I
900		2941.37	II
450		2941.49	II
230	d	2942.33	I
230		2943.20	I
1100		2944.57	II
110		2946.53	I
230		2949.63	I
300		2950.35	II
640		2952.08	II
120		2954.33	I
260		2957.52	II
410		2962.77	I
600		2968.38	II
120		2972.25	II
120		2976.20	II
380		2976.52	II
240		2977.54	I
260		3001.20	II
140		3014.82	II
180		3016.78	II
270		3033.45	II
290		3033.82	II
230		3043.12	I
230		3043.56	I
230		3044.94	I
230		3048.22	II
170		3050.89	I
180		3053.39	II
450		3053.65	I
1200		3056.33	I
1400		3060.46	I
140		3063.25	II
2400		3066.38	I
200		3067.12	II
140		3069.64	I
170		3073.82	I
100		3075.27	I
150		3082.11	I
3800		3093.11	II
200		3094.20	II
180		3100.94	II
3000		3102.30	II
2600		3110.71	II
2000		3118.38	II
380		3121.14	II
150		3122.90	II
1500		3125.28	II
260		3126.22	II
530		3130.27	II
410		3133.33	II
210		3134.93	II
150		3136.51	II
150		3139.74	II
200		3142.48	II
150		3145.34	II
3200		3183.41	I
5300		3183.98	I
3800		3185.40	I
410		3187.71	II
530		3188.51	II
750		3190.68	II
530		3198.01	I
750		3202.38	I
450		3205.58	I
450		3207.41	I
410		3212.43	I
210		3217.11	II
150		3237.87	II
140		3254.77	II
140		3263.24	I
1100		3267.70	II
900		3271.12	II
750		3276.12	II
110		3279.84	II
140		3298.14	I
110		3329.86	I
110		3365.55	I
110		3377.62	I
170		3400.40	I
110		3425.07	I
110		3485.92	II
210		3504.44	II
560		3517.30	II
150		3520.02	II
110		3524.72	II
230		3529.74	I
230		3530.77	II
560		3533.68	I
110		3543.50	I
560		3545.20	II
110		3553.27	I
560		3556.80	II
110		3566.18	I
560		3589.76	II
490		3592.02	II
560		3592.53	I
270		3593.33	II
110		3606.69	I
110		3639.02	I
110		3644.71	I
250		3663.59	I
250		3667.74	I
110		3669.41	II
170		3671.20	I
280		3673.40	I
280		3675.70	I
170		3676.68	I
300		3680.11	I
570		3683.13	I
190		3686.26	I
470		3687.47	I
1300		3688.07	I
1000		3690.28	I
1500		3692.22	I
450		3695.34	I
1000		3695.86	I
3800		3703.58	I
1800		3704.70	I
570		3705.04	I
130		3708.72	I
320		3715.47	II
250		3727.34	II
280		3732.76	II
150		3734.43	I
230		3745.80	II
210		3750.87	II
210		3770.97	II
270		3778.68	I
520		3790.32	I
1100		3794.96	I
570		3799.91	I
570		3803.47	I
190		3806.80	I
300		3807.50	I
520		3808.52	I
230		3809.60	I
1000		3813.49	I
140	d	3817.84	I
1300		3818.24	I
230		3819.96	I
230		3821.49	I
570		3822.01	I
450		3822.89	I
300		3823.21	I
1700		3828.56	I
280		3834.22	I
160		3839.00	I
110		3839.38	I
570		3840.44	I
2600		3840.75	I
110		3841.89	I
380		3844.44	I
320		3847.33	I
110		3849.32	I
1200		3855.37	I
3000		3855.84	I
150		3862.22	I
130		3863.87	I
1300		3864.86	I
230		3867.60	I
170		3871.08	I
1500		3875.08	I
420		3875.90	I
570		3876.09	I
130		3878.71	II
700		3890.18	I
460		3892.86	I
280	h	3898.02	I
140		3899.13	II
140	h	3900.18	I
140	h	3901.15	I
2400		3902.25	I
100		3906.75	I
700		3909.89	I
100		3910.79	I
220		3912.21	I
140		3914.33	II
100		3916.41	II
100		3920.49	I
100		3921.90	I
230		3922.43	I
240		3924.66	I
150		3925.24	I
200		3927.93	I
260		3930.02	I
150		3931.34	I
260		3934.01	I
150		3935.14	I
100		3936.28	I
150		3943.66	I
100		3950.23	I
140		3951.97	II
100		3973.64	II
540		3990.57	I
260		3992.80	I
430		3998.73	I
170		4005.71	II
120		4023.39	II
120		4031.83	I
150		4035.63	II
120		4042.64	I
360		4050.96	I
360		4051.35	I
280		4057.07	I
130		4057.82	I
230		4063.93	I
230		4071.54	I
1100		4090.58	I
180		4092.41	I
1800		4092.69	I
120		4093.50	I
890		4095.49	I
2800		4099.80	I
590		4102.16	I
230		4104.40	I
260		4104.78	I
2800		4105.17	I
120		4108.22	I
2300		4109.79	I
8900		4111.78	I
120		4112.33	I
230		4113.52	I
4300		4115.18	I
1800		4116.47	I
180		4118.18	I
180		4118.64	I
230		4119.46	I
180		4120.54	I
180		4123.19	I
2000		4123.57	I
120		4124.07	I
3100		4128.07	I
120		4128.86	I
3100		4132.02	I
2300		4134.49	I
150		4159.69	I
100		4174.01	I
230		4179.42	I
150		4182.59	I
180		4189.84	I
180		4191.56	I
230		4209.86	I
120		4226.62	I
360		4232.46	I
180		4232.95	I
180		4234.00	I
120		4235.76	I
100		4257.37	I
120		4259.31	I
120		4262.16	I
560		4268.64	I
460		4271.55	I
460		4276.96	I
430		4284.06	I
330		4291.82	I
220		4296.11	I
170		4297.68	I
170		4298.03	I
170		4306.21	I
140		4307.18	I
170		4309.80	I
460		4330.02	I
510		4332.82	I
760		4341.01	I
1000		4352.87	I
130		4354.98	I
150		4355.94	I
150		4368.04	I
140	d	4373.23	I
100		4375.30	I
12000		4379.24	I
100		4380.55	I
7000		4384.72	I
4800		4389.97	I
3600		4395.23	I
1400		4400.58	I
2300		4406.64	I
2800		4407.64	I
3600		4408.20	I
4600		4408.51	I
140		4412.14	I
640		4416.47	I
120		4419.94	I
640		4421.57	I
460		4426.00	I
120		4427.31	I
310		4428.52	I
230		4429.80	I
430		4436.14	I
640		4437.84	I
830		4441.68	I
640		4444.21	I
610		4452.01	I
410		4457.48	I
120		4457.76	I
1000		4459.76	I
2000		4460.29	I
610		4462.36	I
120		4468.01	I
380		4469.71	I
120		4474.04	I
200		4474.71	I
380		4488.89	I

Vanadium (Cont.)

100		4496.06	I
120		4501.95	I
140		4524.22	I
360		4545.39	I
100		4549.65	I
280		4560.71	I
200		4571.78	I
510		4577.17	I
140		4578.73	I
640		4580.40	I
830		4586.36	I
170		4591.22	I
1300		4594.11	I
100		4606.15	I
30		4609.65	I
25		4611.74	I
230		4619.77	I
65		4624.41	I
50		4626.48	I
100		4635.18	I
65		4640.07	I
65		4640.74	I
130		4646.40	I
30		4648.89	I
30		4666.14	I
160		4670.49	I
24		4684.45	I
35		4686.92	I
55		4706.16	I
80		4706.57	I
80		4710.56	I
65		4714.12	I
35		4715.89	I
55		4717.69	I
40		4721.51	I
40		4722.86	I
40		4729.53	I
27		4730.38	I
27		4742.63	I
24		4746.63	I
40		4748.52	I
45		4750.98	I
35		4751.56	I
40		4753.93	I
65		4757.48	I
55		4766.63	I
130		4776.36	I
		4776.52	I
110		4786.51	I
130		4796.92	I
19		4799.77	I
130		4807.53	I
130		4827.45	I
150		4831.64	I
120		4832.43	I
19		4833.02	I
19		4848.81	I
320		4851.48	I
35		4862.61	I
480		4864.74	I
21		4871.26	I
620		4875.48	I
55		4880.56	I
740		4881.56	I
27		4891.60	I
21		4894.21	I
55		4900.62	I
95	d	4904.29	I
		4904.34	I
85		4925.65	I
35		4932.03	I
23		4966.12	I
70		5002.33	I
85		5014.62	I
28		5051.63	I
35		5064.12	I
35		5105.14	I
110		5128.53	I
110		5138.42	I
25		5139.53	I
70		5148.72	I
40		5159.35	I
23		5169.94	I
70		5176.77	I
20		5192.01	I
110		5192.99	I
23		5193.62	I
110		5194.83	I
55		5195.36	I
20		5206.61	I
40		5216.59	I
35		5225.77	I
35		5233.75	I
110		5234.07	I
20		5240.20	I
110		5240.87	I
17		5260.98	I
40		5353.41	I
35		5383.43	I
40		5385.14	I
14		5388.30	I
11		5397.87	I
100		5401.93	I
140		5415.26	I
28		5418.09	I
50		5424.08	I
40		5434.18	I
11		5437.66	I
17		5458.12	I
13		5471.33	
25		5487.22	I
85		5487.92	I
25		5489.94	I
28		5504.87	I
70		5507.75	I
14		5511.18	I
23		5545.93	I
70		5547.07	I
35		5558.75	I
28		5561.66	I
140		5584.50	I
23		5586.00	I
100		5592.42	I
28		5601.38	I
70		5604.94	I
13		5624.20	I
200		5624.60	I
70		5624.89	I
55		5626.01	I
400		5627.64	I
13		5632.46	I
10		5633.90	I
13		5635.51	I
85		5646.11	I
110		5657.44	I
110		5668.36	I
310		5670.85	I
20		5683.22	I
1200		5698.52	I
920		5703.56	I
570		5706.98	I
11		5708.95	I
11	h	5716.21	I
70		5725.64	I
850		5727.03	I
170		5727.66	I
230		5731.25	I
40		5734.01	I
230		5737.06	I
110		5743.45	I
17		5747.70	I
40		5768.87	I
17		5752.74	I
17		5761.41	I
70		5772.42	I
35		5776.64	I
11		5782.61	I
11		5783.50	I
40	h	5784.38	I
55	h	5786.16	I
23		5788.56	I
35	h	5807.14	I
23		5817.06	I
35	h	5817.53	I
55	h	5830.72	I
85	h	5846.30	I
11		5850.32	I
40		5924.57	I
28		5978.91	I
20		5980.78	I
28		6002.31	I
55		6002.63	I
28		6016.12	I
20		6025.41	I
450		6039.73	I
100		6058.14	I
20		6067.26	I
480		6081.44	I
1300		6090.22	I
28		6106.98	I
280		6111.67	I
600		6119.52	I
20		6128.34	I
280		6135.38	I
180		6150.15	I
85		6170.36	I
23		6189.35	I
450		6199.19	I
130		6213.87	I
450		6216.37	I
28	h	6218.31	I
130		6224.50	I
430		6230.74	I
100		6233.20	I
55		6240.13	I
170		6242.81	I
710		6243.10	I
280		6251.82	I
85		6256.90	I
85		6258.57	I
55		6261.22	I
85		6266.32	I
130		6268.82	I
170		6274.65	I
17	h	6282.33	I
200		6285.16	I
200		6292.83	I
170		6296.49	I
28	h	6311.50	I
14		6324.66	I
70		6326.84	I
55		6339.09	I
50		6349.48	I
14		6355.58	I
50		6357.30	I
25		6358.82	I
35		6361.27	I
23		6379.36	I
14		6393.28	I
35		6430.47	I
23		6431.63	I
14		6433.18	I
11		6435.16	I
70		6452.34	I
11		6488.05	I
55		6504.17	I
110		6531.43	I
28		6543.51	I
17		6558.02	I
11		6565.88	I
50		6605.97	I
15		6607.83	I
10		6623.54	I
50		6624.85	I
13		6633.26	I
13		6643.79	I
8		6693.66	I
8		6708.07	I
65	c	6753.00	I
10		6760.12	I
50	c	6766.49	I
40		6784.98	I
15		6786.32	I
26		6812.40	I
9	c	6829.94	I
15		6832.44	I
12		6839.58	I
12		6841.90	I
10	c	6870.88	I
8		6871.56	I
7		6894.00	I
12		6974.50	I
21		7026.07	I
7		7063.69	I
11	h	7092.08	I
6		7102.58	I
24		7148.15	I
7		7151.36	I
7		7182.08	I
14		7264.29	I
8		7321.44	I
40		7338.92	I
35		7356.54	I
11		7358.66	I
24		7361.39	I
12		7362.49	I
24		7363.16	I
9		7385.95	I
6	h	7393.49	
12	h	7485.90	I
12	h	7488.08	I
12	h	7492.44	
12	h	7578.75	I
9	h	7591.24	I
14	h	7596.92	I
12	h	7598.28	I
24		7624.81	I
5		7701.37	I
8		7704.81	I
8	h	7851.18	
14	Bl	7865.51	VO
12		7896.40	
14	h	7898.81	
24		7937.92	I
29	c	8027.39	I
14		8028.13	I
14	h	8035.38	
14	h	8045.71	
12		8051.89	
14		8093.48	I
8		8102.44	I
12		8108.59	I
9	h	8109.07	I
120	Cw	8116.80	I
11	h	8136.79	I
29		8144.59	I
9		8154.55	I
70	c	8161.07	I
14		8171.35	I
7		8180.21	I
35		8186.71	I
24		8187.33	I
29		8198.87	I
35		8203.07	I
24		8241.61	I
29	c	8253.51	I
29		8255.88	I
5	h	8280.39	I
19		8282.37	I
8		8324.42	I
14	h	8331.23	I
14		8342.03	I
7		8402.81	I
12		8499.52	I
6		8534.49	I
6	Bl	8624.86	VO
60	c	8919.85	I
29	c	8932.93	I
12		8971.62	I

V III

Ref. 394 — C.H.C.

Intensity		Wavelength	
		Vacuum	
25		616.09	III
50		633.94	III
40		635.41	III
100		864.27	III
75		948.84	III
500		1006.46	III
500		1149.94	III
400		1157.18	III
300		1160.77	III
500		1252.11	III
400		1254.01	III
500		1287.87	III
400		1289.42	III
300		1290.77	III
400		1313.35	III
300		1313.27	III
500		1331.99	III
500		1335.12	III
1000		1643.03	III
1000		1650.14	III
300		1668.03	III
300		1670.66	III
300		1679.19	III
1000		1694.78	III
400		1721.98	III
300		1724.63	III
500	d	1751.68	III
500		1757.73	III
1000		1760.07	III
300		1773.43	III
400		1778.02	III
500		1779.72	III
400		1784.44	III
1000		1788.26	III
500		1793.82	III
1000		1794.60	III
300		1796.77	III
500		1798.15	III
300		1802.55	III
500		1804.13	III
1000		1812.19	III
400		1831.15	III
400		1831.64	III
300		1845.07	III
300		1850.69	III
400		1852.01	III
500		1854.42	III
300		1855.06	III
500		1856.64	III
300		1864.51	III
300		1878.68	III
400		1880.41	III
300		1895.01	III
400		1899.81	III
500		1902.23	III
300		1934.00	III
		Air	
500		2232.91	III
400		2241.53	III
1000		2292.86	III
400		2314.18	III
500		2318.06	III
400		2319.00	III
500		2323.82	III
2500		2330.42	III
500		2331.75	III
400		2334.20	III
500		2343.10	III
500		2358.73	III
500		2366.31	III
2500		2371.06	III
1000		2382.46	III
500		2393.58	III
500		2404.18	III
500		2516.14	III
250		2521.16	III
250		2521.55	III
250		2548.21	III
150		2554.22	III
150		2563.32	III
250		2593.05	III
250		2595.10	III
100		3679.86	III
50	h	3705.35	III

Vanadium (Cont.)

Intensity		Wavelength	
40		4714.89	III
50		6597.20	III

V IV
Ref. 397 — C.H.C.

Intensity		Wavelength	
		Vacuum	
200		677.34	IV
60		678.74	IV
50		679.65	IV
500		684.37	IV
100		684.45	IV
100		691.53	IV
50		693.13	IV
400		737.85	IV
150		750.11	IV
60		1226.52	IV
50		1308.06	IV
80		1355.13	IV
60		1395.00	IV
50		1414.41	IV
80		1419.58	IV
100		1426.65	IV
60		1520.14	IV
80		1601.92	IV
80		1611.88	IV
80		1806.18	IV
60		1809.85	IV
100		1817.68	IV
200		1825.84	IV
300		1861.56	IV
500		1939.06	IV
400		1951.43	IV
300		1963.10	IV
500		1997.72	IV
200		1999.32	IV
		Air	
100		2002.48	IV
50	h	2088.74	IV
50	h	2146.83	IV
100	h	2155.34	IV
500		2268.30	IV
50	h	2421.32	IV
50	h	2433.53	IV
50	h	2446.80	IV
50	h	2450.87	IV
50	h	2556.92	IV
80	h	2570.72	IV
50	h	2624.21	IV
80	h	2645.54	IV
50	h	2655.41	IV
50	h	2656.87	IV
50		3284.56	IV
60		3334.79	IV
50		3448.41	IV
50		3496.42	IV
80	h	3514.25	IV
50	h	4985.65	IV
50	h	5130.78	IV
50	h	5262.16	IV
60	h	5352.32	IV
40	h	5940.12	IV

V V
Ref. 432 — C.H.C.

Intensity		Wavelength	
		Vacuum	
18		224.91	V
20		225.46	V
17		227.88	V
16		239.41	V
19		239.48	V
20		251.66	V
18		252.44	V
18		285.98	V
20		286.84	V
17		312.39	V
35		483.01	V
25		484.51	V
20		820.86	V
30		829.48	V
15		962.03	V
15		1142.74	V
25		1157.58	V
20		1490.11	V
100		1680.20	V
50		1716.72	V
15		1724.99	V
25		1792.99	V
30		1811.42	V
		Air	
20	d	2319.66	V
20		2577.90	V
15		2775.82	V
15		3617.97	V
12	d	3746.36	V
20		4200.32	V
15	w	4930.53	V
8		5356.07	V
7		6628.80	V
3		7595.51	V

XENON (Xe)
Z = 54

Xe I and II
Ref. 33, 116, 118, 120, 232 — S.P.D.

Intensity		Wavelength	
		Vacuum	
350		740.41	II
350		803.07	II
600		880.80	II
350		885.54	II
600		925.87	II
250		935.40	II
800		972.77	II
700		976.68	II
500		1032.44	II
700		1037.68	II
1100		1041.31	II
1000		1048.27	II
1200		1051.92	II
2000		1074.48	II
600		1083.86	II
1200		1100.43	II
600		1158.47	II
250		1169.63	II
800	p	1183.05	II
250		1192.04	I
600		1244.76	II
250		1250.20	I
1000		1295.59	I
600		1469.61	I
		Air	
200		2864.73	II
150	h	2895.22	II
400		2979.32	II
100	h	3017.43	II
300		3128.87	II
200	h	3366.72	II
2		3400.07	I
2		3418.37	I
2		3420.00	I
3		3442.66	I
100	h	3461.26	II
4		3469.81	I
4		3472.36	I
5		3506.74	I
10		3549.86	I
10		3554.04	I
15		3610.32	I
8		3613.06	I
6		3633.06	I
10		3669.91	I
40		3685.90	I
40		3693.49	I
100	l	3907.91	II
100		4037.59	II
200	l	4057.46	II
100	h	4098.89	II
200	l	4158.04	II
1000	h	4180.10	II
500	h	4193.15	II
300	h	4208.48	II
100	h	4209.47	II
300	h	4213.72	II
100		4215.60	II
300	h	4223.00	II
400	h	4238.25	II
500	h	4245.38	II
100	l	4251.57	II
500	h	4296.40	II
500	h	4310.51	II
1000	l	4330.52	II
200	h	4369.20	II
100	l	4373.78	II
500	h	4393.20	II
500	l	4395.77	II
200	l	4406.88	II
150	l	4416.07	II
500	h	4448.13	II
1000	h	4462.19	II
500	l	4480.86	II
100	l	4521.86	II
600		4734.152	I
150		4792.619	I
500		4807.02	I
400		4829.71	I
300		4843.29	I
500		4916.51	I
500		4923.152	I
200	l	4971.71	II
400		4972.71	II
300		4988.77	II
100	l	4991.17	II
200		5028.280	I
200		5044.92	II
1000		5080.62	II
300		5122.42	II
100		5125.70	II
100		5178.82	II
300		5188.04	II
400		5191.37	II
100		5192.10	II
500		5260.44	II
500		5261.95	II
2000		5292.22	II
300		5309.27	II
1000		5313.87	II
2000		5339.33	II
200		5363.20	II
200		5368.07	II
500		5372.39	II
100		5392.80	I
3000		5419.15	II
800		5438.96	II
300		5445.45	II
200		5450.45	II
400		5460.39	II
1000		5472.61	II
100	l	5494.86	II
200		5525.53	II
600		5531.07	II
100		5566.62	I
300		5616.67	II
300		5659.38	II
600		5667.56	II
150		5670.91	II
100		5695.75	I
200		5699.61	II
200		5716.10	II
500		5726.91	II
500		5751.03	II
300		5758.65	II
300		5776.39	II
100		5815.96	II
300		5823.89	I
150		5824.80	I
100		5875.02	I
300		5893.29	II
100		5894.99	I
200		5905.13	II
100		5934.17	I
500		5945.53	II
300		5971.13	II
2000		5976.46	II
200		6008.92	II
1000		6036.20	II
2000		6051.15	II
600		6093.50	II
1500		6097.59	II
400		6101.43	II
100		6115.08	II
100		6146.45	II
150		6178.30	I
120		6179.66	I
300		6182.42	I
500		6194.07	II
100		6198.26	I
100		6220.02	II
500		6270.82	II
400		6277.54	II
100		6284.41	II
100		6286.01	I
250		6300.86	II
500		6318.06	I
400		6343.96	II
600		6356.35	II
200		6375.28	II
100		6397.99	II
300		6469.70	I
150		6472.84	I
120		6487.76	I
100		6498.72	I
200	h	6504.18	I
300		6512.83	II
200		6528.65	II
100		6533.16	I
1000		6595.01	II
100		6595.56	I
400		6597.25	II
100		6598.84	II
150		6668.92	I
300		6694.32	II
200		6728.01	I
150		6788.71	II
100		6790.37	II
1000		6805.74	II
200		6827.32	I
100		6872.11	I
300		6882.16	I
80		6910.22	II
100		6925.53	I
800	h	6942.11	II
100		6976.18	I
2000		6990.88	II
150		7082.15	II
500		7119.60	I
50	s	7147.50	II
200		7149.03	II
500		7164.83	II
100		7284.34	II
200		7301.80	II
200		7339.30	II
100		7386.00	I
150		7393.79	I
300		7548.45	II
200		7584.68	I
80		7618.57	II
500		7642.02	I
100		7643.91	I
200		7670.66	II
60		7787.04	II
100		7802.65	I
100		7881.32	I
300		7887.40	I
500		7967.34	I
100		8029.67	I
200		8057.26	I
150		8061.34	I
100		8101.98	I
150	h	8151.80	II
100		8171.02	I
700		8206.34	I
10000		8231.635	I
500		8266.52	I
7000		8280.116	I
2000		8346.82	I
100		8347.24	II
2000		8409.19	I
50	h	8515.19	II
200		8576.01	I
50	h	8604.23	II
250		8648.54	I
100		8692.20	I
200		8696.86	I
50	h	8716.19	II
300		8739.39	I
100		8758.20	I
5000		8819.41	I
300		8862.32	I
200		8908.73	I
200		8930.83	I
1000		8952.25	I
100		8981.05	I
200		8987.57	I
400		9045.45	I
500		9162.65	I
100		9167.52	I
100		9374.76	I
200		9513.38	I
50	h	9591.35	II
150		9685.32	I
50	l	9698.68	II
100		9718.16	I
2000		9799.70	I
3000		9923.19	I
100		10838.37	I
90		11742.01	I
375		12235.24	I
100		12257.76	I
300		12590.20	I
2500		12623.391	I
250		13544.15	I
2000		13657.055	I
1250		14142.444	I
800		14240.96	I
375		14364.99	I
140		14660.81	I
3000		14732.806	I
100		15099.72	I
2500		15418.394	I
150		15557.13	I
250		15979.54	I
100		16039.90	I
1000		16053.28	I
125		16554.49	I
1500		16728.15	I
1500		17325.77	I
350		18788.13	I
150		20187.19	I
3000		20262.242	I
250		21470.09	I
1250		23193.33	I
110		23279.54	I
1800		24824.71	I
175		25145.84	I
2000		26269.08	I
2500		26510.86	I
250		28381.54	I
750		28582.25	I
300		29384.41	I
150		29448.06	I

Xenon (Cont.)

Intensity		Wavelength	
100		29649.58	I
100		29813.62	I
600		30253.14	I
1500		30475.46	I
100		30504.12	I
500		30794.18	I
6000		31069.23	I
125		31336.01	I
550		31607.91	I
100		32293.08	I
1800		32739.26	I
3500		33666.69	I
150		34014.67	I
450		34335.27	I
170		34744.00	I
5000		35070.25	I
110		35246.92	I
250		36209.21	I
150		36231.74	I
450		36508.36	I
850		36788.83	I
140		38685.98	I
175		38737.82	I
270		38939.60	I
120		39955.14	I

Xe III
Ref. 33, 384, 391, 429 — R.L.K.

Intensity		Wavelength	
		Vacuum	
8		657.8	III
8		660.1	III
9		673.8	III
9		674.0	III
9		676.6	III
10		694.0	III
20		698.5	III
12		705.1	III
10		721.2	III
15		731.0	III
10		733.3	III
15		742.6	III
10		756.0	III
10		761.3	III
10		769.1	III
25		779.1	III
15		792.9	III
12		796.1	III
15		802.0	III
25		823.2	III
30		824.9	III
25		853.0	III
15		889.3	III
20		894.0	III
20		896.0	III
10		965.5	III
35		1003.4	III
35		1017.7	III
10		1047.8	III
12		1066.4	III
30		1130.3	III
25		1232.1	III
		Air	
80		2668.98	III
100		2717.33	III
30		2814.45	III
40		2815.91	III
30		2827.45	III
40		2847.65	III
30		2862.40	III
80	w	2871.10	III
60	w	2871.24	III
30		2871.7	III
30		2[illegible]6.62	III
50		2906.6	III
40		2911.89	III
80	w	2912.36	III
40		2940.2	III
60		2945.2	III
40		2947.5	III
40		2948.1	III
80	w	2970.47	III
40		2992.87	III
30		3004.25	III
100		3023.81	III
40		3083.5	III
50		3091.1	III
30		3106.46	III
100	w	3138.3	III
80	c	3150.82	III
40		3185.2	III
100		3242.86	III
80		3268.98	III
30		3287.82	III
80	w	3301.55	III
40		3331.6	III
30		3358.0	III
80		3384.12	III
60		3444.2	III
70		3454.2	III
100	w	3458.7	III
40		3468.22	III
80		3522.83	III
50		3542.3	III
50		3552.1	III
40		3561.4	III
100		3579.7	III
80		3583.6	III
100	w	3595.4	III
100		3606.06	III
40		3607.0	III
100	w	3615.9	III
40		3623.1	III
600		3624.08	III
50		3676.67	III
40		3776.3	III
300		3781.02	III
100		3841.5	III
200		3877.8	III
60		3880.5	III
500		3922.55	III
300		3950.59	III
200		4050.07	III
60		4060.4	III
100		4109.1	III
100		4145.7	III
30		4285.9	III
50		4434.2	III
100	w	4462.1	III
100	w	4569.1	III
100	w	4570.1	III
100	w	4641.4	III
30		4673.7	III
60		4683.57	III
30		4723.60	III
100	w	4757.3	III
40		4869.5	III
60		5239.0	III
30		5367.1	III
50		5401.0	III
40		5524.4	III
60		6205.97	III
25		6221.7	III
60		6238.2	III
60		[illegible]	III

YTTERBIUM (Yb)
Z = 70

Yb I and II
Ref. I = C.H.C.

Intensity		Wavelength	
		Air	
2500		2116.67	II
3000		2126.74	II
370		2161.60	II
850		2185.71	II
640		2224.46	II
140		2320.81	II
50		2362.89	II
170		2390.74	II
18		2398.02	II
28		2421.35	II
25		2447.26	II
28		2460.25	II
460		2464.50	I
14		2484.89	II
70		2502.02	II
28		2505.48	II
11		2508.07	II
140		2512.06	II
18		2516.35	II
50		2522.44	II
65		2537.65	II
270		2538.67	II
14		2550.06	II
70		2552.15	II
55		2552.70	II
21		2565.57	II
28		2571.36	II
13		2573.15	II
18		2596.16	II
28		2596.32	II
21		2615.26	II
100		2617.01	II
55		2634.31	II
45		2639.45	II
85		2641.89	II
110		2644.31	II
28		2646.44	II
28		2647.46	II
28		2648.80	II
50		2649.79	II
28		2650.73	II
990		2653.75	II
35		2656.12	II
21		2659.27	II
200		2665.04	II
55		2668.75	II
390		2671.96	I
390		2672.66	II
21		2680.40	II
14		2683.42	II
70		2684.75	II
25		2687.98	II
28		2695.43	II
14		2696.62	II
18		2700.80	II
21		2708.84	II
65		2710.54	II
25		2711.78	II
55		2712.66	II
170		2718.35	II
21		2722.20	II
110		2732.74	II
21		2734.09	II
55		2741.71	II
55		2747.58	II
18		2748.04	II
230		2748.66	II
1300		2750.48	II
85		2751.45	II
21		2759.00	II
65		2760.78	II
65		2761.37	II
35		2764.41	II
85		2771.32	II
170		2776.28	II
100		2784.66	II
18		2787.96	II
45		2793.28	II
25		2794.44	II
21		2795.07	II
18		2795.29	II
35		2797.80	II
100		2798.21	II
45		2799.38	II
50		2800.00	II
35		2800.06	II
14		2810.72	II
65		2814.53	II
28		2816.[illegible]	II
140		2821.15	II
100		2824.97	II
190		2830.99	II
18		2832.20	II
28		2834.97	II
14		2842.59	II
230	h	2847.18	II
100		2848.44	II
21		2849.34	II
360		2851.13	II
55		2851.86	II
21		2853.41	II
18		2853.68	II
55		2854.14	II
45		2854.49	II
45		2858.33	II
45		2858.46	II
100		2859.39	II
430		2859.80	II
55		2860.39	II
140		2861.21	II
100		2861.34	II
200		2867.06	II
25		2870.06	II
45		2873.49	I
28		2885.97	II
70		2886.26	II
200		2888.04	II
3600		2891.38	II
45		2893.62	II
28		2896.90	II
85		2899.70	II
18		2902.41	II
21		2902.92	II
21		2906.88	II
28		2908.33	II
35		2909.19	II
55		2909.48	II
85		2911.52	II
18		2912.86	II
170		2914.21	II
140		2915.28	II
18		2916.43	II
280		2919.35	II
55		2921.12	II
45		2924.24	II
25		2927.85	II
35		2934.36	I
55		2935.11	II
21		2937.19	II
45		2939.53	II
45		2940.52	II
28		2942.04	II
140		2945.91	II
45		2946.30	II
18		2946.76	II
28		2950.33	II
45		2955.32	II
18		2957.63	II
65		2962.52	II
21		2963.26	II
45		2963.46	II
130		2964.76	II
2000		2970.56	II
45		2982.49	II
21		2982.66	II
28		2983.70	II
200		2983.99	II
90		2985.08	II
35		2985.88	II
45		2990.37	II
65		2991.87	II
28		2993.94	II
170		2994.80	II
28		2995.86	II
70		3000.46	II
25		3002.61	II
310		3005.77	II
100		3009.39	II
65		3010.62	II
55		3014.43	II
160		3017.56	II
160		3026.67	II
920		3031.11	II
55		3034.64	II
25		3037.99	II
55		3039.67	II
80		3042.65	II
21		3044.00	II
45		3046.48	II
35		3047.05	II
45		3063.12	II
21		3063.67	II
110		3065.04	II
18		3076.01	II
100		3089.10	II
70		3093.87	II
28		3100.74	I
45		3101.36	II
28		3102.07	II
55		3107.76	II
170		3107.90	II
85		3115.34	II
[illegible]		3116.70	II
190		3117.81	II
50		3136.76	II
230		3140.94	II
80		3141.73	II
80		3145.06	II
28		3145.54	II
28		3153.18	II
90		3153.88	II
50		3155.18	II
28		3162.29	I
70		3163.80	II
50		3165.21	II
120		3169.06	II
120		3180.92	II
390		3192.88	II
70		3198.65	II
240		3201.16	II
80		3217.18	II
50		3218.32	II
50		3225.88	II
45		3239.20	II
35		3239.58	I
35		3246.06	II
130		3261.51	II
18000		3289.37	II
130		3305.25	I
140		3305.73	II
50		3315.10	II
80		3319.41	I
50		3333.06	II
240		3337.17	II
280	d	3342.93	II
		3343.07	II
80		3346.50	
50		3347.54	II
35		3351.09	II
50		3351.26	
100		3352.49	II
100		3362.44	II
50		3363.64	II
240		3375.48	II
50		3376.62	II
28		3382.54	
140		3387.50	I
50		3390.25	II
28		3390.42	II
50		3391.10	II
50	h	3394.44	II
50		3401.01	II
35		3404.10	II
50		3412.45	I
140		3418.39	I
360		3426.04	I
80		3428.46	II
240		3431.11	I
45		3434.61	II

Ytterbium (Cont.)

Intensity		Wavelength	Spectrum
50		3438.71	II
100		3438.85	II
35		3443.59	
35		3446.89	II
85		3452.40	I
500		3454.08	II
190	d	3458.29	II
		3458.39	I
360		3460.27	I
35		3462.34	II
2400		3464.37	I
500		3476.30	II
500		3478.84	II
50		3482.56	II
85		3485.76	II
85		3488.43	II
100	Hw	3495.90	II
85		3507.83	II
50		3517.00	I
230		3520.29	II
50		3545.72	II
100		3549.82	II
35		3559.03	I
200		3560.33	II
170		3560.70	II
50	h	3563.94	II
85		3570.57	II
50		3572.50	II
50		3574.58	II
360		3585.47	II
130		3606.48	II
50		3610.23	II
70		3611.30	II
200		3619.80	II
110		3634.52	
240		3637.76	II
70		3648.15	I
90		3655.73	I
240		3669.69	II
50		3670.69	II
140		3675.08	II
50		3690.56	II
32000		3694.19	II
70		3698.60	II
70		3700.58	I
50		3710.34	II
60		3724.21	II
180		3734.69	I
550		3770.10	I
80		3774.32	I
60	h	3791.74	I
170		3839.91	I
340		3872.85	I
340		3900.85	I
50		3904.81	II
140		3911.27	I
32000		3987.99	I
930		3990.88	I
50		4007.36	I
70		4052.28	I
85		4077.28	II
440		4089.68	I
120	h	4119.25	II
70		4135.09	II
470		4149.07	I
120		4174.56	I
340		4180.81	II
150	d	4218.56	II
		4218.69	I
120		4231.97	I
70		4277.74	I
120		4305.97	I
70		4316.95	II
60	h	4393.69	I
60	h	4430.21	I
440		4439.19	I
85	h	4482.42	I
85		4515.16	II
35		4553.58	II
85	h	4563.95	I
640		4576.21	I
200		4582.36	I
70		4589.21	I
140		4590.83	I
40		4598.36	II
35		4683.81	II
40		4684.27	I
190		4726.08	II
170	h	4781.87	I
170		4786.61	II
35		4816.43	I
40		4820.24	II
35		4836.96	II
40		4837.46	I
17		4851.15	II
40	h	4894.60	I
27		4912.36	I
710		4935.50	I
24		4937.22	II
140		4966.90	I
24		5009.52	II
17		5067.30	II
30		5067.80	I
70		5069.14	I
220		5074.34	I
50		5076.74	I
20		5135.98	II
14		5147.02	II
20		5184.15	II
60		5196.08	I
85		5211.60	I
35		5240.51	II
100		5244.11	I
40		5257.49	II
150	h	5277.04	I
35		5279.53	II
17		5300.94	II
170		5335.15	II
30	d	5345.66	II
		5345.83	II
60		5347.22	II
30	h	5351.29	I
150		5352.95	II
30		5358.64	II
30		5363.66	I
17		5389.84	II
14		5432.71	II
40		5449.27	II
14		5478.50	II
60		5481.92	I
40		5505.49	I
17		5524.54	I
85	h	5539.05	I
2400		5556.47	I
35		5562.09	I
20		5568.11	I
20		5586.36	I
40		5588.45	II
60		5651.98	II
7		5686.53	II
220		5719.99	I
10		5749.91	II
10	h	5755.89	I
27		5771.66	II
10		5803.44	I
10		5819.41	II
35		5833.99	II
35		5837.14	II
27		5854.51	I
8		5897.21	II
20		5908.36	II
17		5989.33	I
40		5991.51	II
10		6052.88	II
10		6054.57	I
60		6152.57	II
30		6246.97	II
60		6274.78	II
14		6308.15	II
35	h	6400.35	I
35	h	6417.91	I
20		6432.73	II
17	h	6463.15	II
340		6489.06	I
20		6643.55	I
180		6667.82	I
15		6678.17	I
25		6727.61	II
25		6768.70	I
690		6799.60	I
18		6934.05	II
20		6999.88	II
10		7043.78	II
9	h	7244.41	I
8	h	7305.22	I
10	h	7313.05	I
16	h	7350.04	I
25		7448.28	I
30	h	7527.46	I
750		7699.48	I
7		7895.08	I
70	h	8922.56	II

Yb III
Ref. 40, 192 — J.R.

Intensity		Wavelength	
Vacuum			
5		968.46	III
20		973.16	III
10		994.56	III
10		1560.66	III
80		1561.42	III
30		1669.60	III
50		1670.78	III
50		1719.82	III
60		1739.18	III
70		1762.80	III
80	h	1765.21	III
65		1775.29	III
70		1779.74	III
70		1781.31	III
20		1793.70	III
60		1798.85	III
65		1810.88	III
60		1826.41	III
20		1826.77	III
30		1838.01	III
30		1847.30	III
30		1849.24	III
10		1849.42	III
75		1852.36	III
75		1852.94	III
90		1854.80	III
80		1857.16	III
100		1863.32	III
5		1864.85	III
10		1867.23	III
10		1867.63	III
10		1868.19	III
5		1868.92	III
10		1870.07	III
15		1870.83	III
10		1871.15	III
200		1872.03	III
800		1873.91	III
100		1875.41	III
75		1875.92	III
70		1880.30	III
80		1884.22	III
70		1885.07	III
70		1887.22	III
10		1890.34	III
10		1890.87	III
15		1892.42	III
5		1895.50	III
100		1896.18	III
10		1897.57	III
500		1898.25	III
7		1906.74	III
10		1908.50	III
100		1909.66	III
70		1910.86	III
20		1920.53	III
10		1926.76	III
70		1928.09	III
15		1930.63	III
55		1942.59	III
40		1950.34	III
15		1962.80	III
80		1967.13	III
20		1969.47	III
30		1969.73	III
10		1973.96	III
10		1974.18	III
25		1976.46	III
2		1981.74	III
5		1983.88	III
25		1984.62	III
7		1985.74	III
80		1986.43	III
80	h	1989.82	III
50		1991.14	III
45		1995.05	III
55		1997.28	III
55		1997.66	III
500		1998.82	III
Air			
20		2054.80	III
10		2066.49	III
10		2073.64	III
30		2078.05	III
10		2087.37	III
50		2087.98	III
20		2091.23	III
20		2092.26	III
10		2094.77	III
80		2095.31	III
15		2096.79	III
30		2098.36	III
10		2106.71	III
50		2109.54	III
20		2119.18	III
20		2198.14	III
80		2202.27	III
300		2240.11	III
100		2244.28	III
200		2257.03	III
100		2262.26	III
200		2265.67	III
150		2282.99	III
100		2283.99	III
300		2305.32	III
100		2309.27	III
200		2314.49	III
200		2337.97	III
40		2361.08	III
200		2365.43	III
50		2367.46	III
30		2369.99	III
20		2377.22	III
50		2403.95	III
20		2410.04	III
60		2412.33	III
10		2429.18	III
20		2433.43	III
100		2438.27	III
20		2439.31	III
20		2440.43	III
10		2458.64	III
10		2464.59	III
200		2490.42	III
20		2491.69	III
40		2506.25	III
300		2516.82	III
15		2522.07	III
20		2529.14	III
40		2550.39	III
300		2555.29	III
100		2560.56	III
10		2561.66	III
100		2566.78	III
2000		2567.61	III
1000		2579.57	III
100		2588.62	III
20		2592.69	III
500		2597.23	III
800		2599.14	III
30		2609.14	III
600		2621.11	III
300		2627.07	III
30		2635.37	III
500		2638.06	III
300		2640.48	III
1000		2642.56	III
100		2643.62	III
1000		2651.74	III
700		2652.25	III
100		2659.98	III
70		2664.89	III
2000		2666.13	III
2000		2666.99	III
30		2673.33	III
500		2677.39	III
500		2691.01	III
30		2708.04	III
400		2712.32	III
500		2749.91	III
200		2755.94	III
200		2756.76	III
100		2765.50	III
300		2788.24	III
600		2795.60	III
400		2803.32	III
1000		2803.43	III
10		2807.22	III
50		2808.51	III
600		2816.92	III
1000		2818.72	III
15		2826.01	III
300		2842.96	III
400		2875.86	III
600		2898.30	III
1000		2906.31	III
300		2928.97	III
50		2977.84	III
800		2998.00	III
2000		3029.49	III
100		3031.62	III
30		3040.65	III
3000		3092.50	III
20		3102.18	III
4000		3126.01	III
1000		3138.58	III
100		3151.44	III
70		3179.34	III
800		3191.35	III
50		3216.27	III
2000		3228.58	III
2000		3325.51	III
50		3358.25	III
20		3364.30	III
2000		3384.01	III
150		3392.56	III
100		3397.66	III
80		3432.94	III
40		3456.18	III
150		3463.51	III
20		3469.98	III
300		3550.87	III
200		3613.89	III
30		3659.84	III
30		3663.74	III
200		3664.74	III
20		3675.78	III
400		3711.91	III
20		3879.98	III
10		3882.58	III
20		3887.17	III
150		3896.55	III
15		3912.75	III
20		3913.23	III
500		3931.23	III
100		3985.56	III
10		3991.74	III
10		3997.67	III
2000		4028.14	III
10		4033.03	III
20		4074.53	III
20		4090.67	III

Ytterbium (Cont.)

Intensity		Wavelength	
20		4098.23	III
15		4121.06	III
10		4150.04	III
15		4153.11	III
100		4162.72	III
60		4172.95	III
30		4194.34	III
100		4194.95	III
10		4198.74	III
300		4213.64	III
10		4220.83	III
15		4231.07	III
20		4289.64	III
40		4301.14	III
20		4304.01	III
15		4350.80	III
10		4380.07	III
100		4517.58	III
40		4639.14	III
10		4834.93	III
15		5054.94	III
20		5256.85	III
20		5331.54	III
15		5740.83	III
10	d	5949.02	III
20		5973.05	III
40		6055.85	III
100		6214.22	III
200		6328.52	III
10		6365.88	III
150		6378.33	III
25		6466.33	III
20		6985.15	III
10		7037.04	III
15		7157.72	III
10		7311.02	III
10		7399.98	III
80		7410.01	III
15		7456.86	III
70		7664.41	III
80		7892.39	III
20		7893.10	III
100		7971.46	III
20		8056.02	III
10		8117.44	III
10		8326.86	III
30		8327.88	III
20		8400.01	III
30		8489.90	III
200		10110.60	III
100		10830.36	III

Yb IV
Ref. 40, 311 — J.R.

Intensity		Wavelength	
		Vacuum	
200		828.96	IV
200		870.35	IV
300		902.46	IV
300		927.01	IV
300		936.22	IV
400		943.04	IV
400		946.20	IV
200		975.21	IV
1000		1050.24	IV
1000		1054.46	IV
400		1092.51	IV
200		1109.96	IV
200		1110.55	IV
5000		1134.43	IV
300		1136.24	IV
500		1166.01	IV
600		1185.58	IV
200		1290.24	IV
600		1305.58	IV
900		1316.04	IV
200		1326.32	IV
800		1326.36	IV
200		1340.06	IV
300		1345.36	IV
900		1350.26	IV
200		1353.43	IV
400		1356.15	IV
200		1361.75	IV
300		1365.88	IV
300		1369.72	IV
400		1375.42	IV
300		1376.66	IV
200		1384.41	IV
250		1393.93	IV
350		1398.77	IV
400		1407.05	IV
300		1413.14	IV
400		1416.15	IV
400		1417.72	IV
200		1423.99	IV
300		1430.29	IV
200		1440.61	IV
400		1477.92	IV
300		1491.57	IV
400		1765.03	IV
200		1776.18	IV
200		1778.20	IV
200		1779.34	IV
300		1789.71	IV
800		1791.06	IV
200		1801.67	IV
250		1809.63	IV
600		1813.84	IV
400		1816.07	IV
250		1817.58	IV
300		1819.02	IV
200		1824.22	IV
		Air	
300		2106.48	IV
900		2116.65	IV
500		2121.29	IV
250		2122.84	IV
800		2123.32	IV
600		2125.72	IV
200		2129.65	IV
500		2135.21	IV
300		2137.58	IV
500		2138.35	IV
200		2138.53	IV
800		2139.99	IV
400		2141.04	IV
200		2142.20	IV
300		2143.42	IV
300		2143.89	IV
20000		2144.77	IV
400		2148.10	IV
300		2148.52	IV
15000		2154.18	IV
250		2165.55	IV
300		2169.12	IV
200		2172.16	IV
300		2177.53	IV
400		2183.32	IV
270	h	2186.13	IV
270		2187.17	IV
90	h	2189.90	IV
120		2193.36	IV
150		2198.27	IV
90		2224.64	IV
150		2231.28	IV
90		2233.30	IV
90		2244.20	IV
140		2331.36	IV

YTTRIUM (Y)
Z = 39

Y I and II
Ref. 1 — C.H.C.

Intensity		Wavelength	
		Air	
350		2243.06	II
50		2354.20	I
30		2373.83	
50		2385.24	
25		2413.93	II
560		2422.20	II
60		2460.61	II
25		2490.42	I
12		2540.28	
14		2547.57	
10		2550.17	
20		2681.65	I
60		2694.21	I
26		2695.39	I
95		2723.00	I
22		2730.08	I
22		2734.85	II
70		2742.53	I
140		2760.10	I
30		2785.21	II
12		2785.59	II
12		2791.20	I
30		2800.11	II
26		2813.64	I
18		2818.86	I
45		2822.56	I
22		2825.37	II
45		2826.38	II
70		2854.43	II
26		2856.30	II
11		2857.87	II
95		2886.48	I
18		2897.69	II
14		2898.82	II
160		2919.05	I
18	h	2930.03	II
390		2948.40	I
350		2964.96	I
18		2973.91	II
480		2974.59	I
30		2980.55	II
750		2984.26	I
70		2995.26	I
140		2996.94	I
70		3005.26	I
55		3018.95	I
130		3021.73	I
90		3022.28	I
26		3026.49	II
30		3036.59	II
45		3044.84	I
190		3045.37	I
22		3047.11	I
60		3055.22	II
60		3086.85	II
55	h	3091.70	I
22		3093.76	II
95		3095.88	II
45		3111.81	I
55		3112.04	II
22		3114.28	I
60		3128.77	II
80		3129.93	II
95		3135.17	II
110		3173.06	II
220		3179.41	II
70		3191.31	I
2300		3195.62	II
2200		3200.27	II
2200		3203.32	II
3900		3216.69	II
6200		3242.28	II
310		3280.91	II
19		3308.47	II
4700		3327.89	II
55		3340.38	I
160		3362.00	II
85		3388.59	I
45		3397.04	I
85		3412.47	I
200		3448.82	II
70		3450.95	I
110		3467.88	II
170		3485.73	I
1700		3606.00	II
80		3521.53	I
45		3546.01	II
3900		3549.01	II
130		3551.80	I
540		3552.69	I
170		3558.76	I
190		3571.43	I
260		3576.05	I
3300		3584.52	II
300		3587.75	I
100		3589.69	I
2800		3592.92	I
10000		3600.73	II
6200		3601.92	II
7800		3611.05	II
4300		3620.94	I
1900		3628.71	II
7800		3633.12	II
3000		3664.61	II
45		3668.49	II
170		3692.53	I
13000		3710.30	II
60		3718.12	I
60		3738.61	I
1200		3747.55	II
50		3749.89	I
10000		3774.33	II
1400		3776.56	II
50		3782.30	II
7400		3788.70	II
1300		3818.35	II
4000		3832.88	II
70		3847.87	II
80		3876.82	I
480		3878.28	II
30		3887.77	I
60	h	3904.59	I
50	h	3918.25	I
60	h	3930.11	I
240		3930.66	II
4400		3950.36	II
150		3951.60	II
60	h	3955.09	I
3600		3982.60	II
40		3987.50	I
940		4039.83	I
2400		4047.64	I
9400		4077.38	I
90	h	4081.22	I
2000		4083.71	I
9900		4102.38	I
60	h	4106.39	I
80		4110.81	I
320		4124.92	II
8900		4128.31	I
7500		4142.85	I
100	h	4157.63	I
2400		4167.52	I
2000		4174.14	I
8000		4177.54	II
120		4199.28	II
380		4204.70	II
80		4213.02	I
40		4213.54	I
160		4217.80	I
280	h	4220.63	I
80		4224.25	I
600		4235.73	II
2200		4235.94	I
300		4251.20	I
360	h	4302.30	I
2800		4309.63	II
50		4316.30	I
110		4330.78	I
30		4337.29	I
60		4344.65	I
440	h	4348.79	I
60		4352.33	I
60		4352.70	I
120		4357.73	I
800		4358.73	II
120		4366.03	I
12000		4374.94	II
150	h	4375.61	I
80		4379.33	I
30		4385.48	I
100		4387.74	I
30		4394.01	I
30		4394.67	I
1800		4398.02	II
890		4422.59	II
80		4437.34	I
100		4443.66	I
130		4446.63	I
20		4465.27	I
40		4473.89	I
170		4475.72	I
180		4476.96	I
160		4477.45	I
110		4487.28	I
300		4487.47	I
30		4491.75	I
25		4492.42	I
[illegible]		4503.33	I
50		4513.58	I
80		4514.01	I
40	h	4522.05	I
890		4527.25	I
440		4527.80	I
100		4544.32	I
100		4559.37	I
30		4564.39	I
60	h	4573.56	I
35		4581.32	I
30		4581.77	I
130		4596.55	I
95		4604.80	I
40		4613.00	I
2000		4643.70	I
200	h	4658.32	I
70		4658.89	I
85		4667.47	I
60		4670.82	I
2000		4674.84	I
60		4678.35	I
260		4682.32	II
85		4689.77	I
180		4696.81	I
35		4708.85	I
60		4725.85	I
170		4728.53	I
60	h	4732.37	I
85		4741.40	I
160		4752.79	I
410		4760.98	I
17		4780.18	I
120		4781.04	I
160		4786.58	II
170		4786.89	I
180		4799.30	I
50		4804.31	I
70		4804.81	I
85	Bl	4817.38	YO
140	Bl	4818.20	YO
140		4819.64	I
120		4822.13	I
190		4823.31	II
60		4839.15	I
770		4839.87	I
550		4845.68	I
410		4852.69	I
120		4854.25	I
890		4854.87	II
50		4856.70	I
330		4859.84	I
50		4879.65	I
1900		4883.69	II
50		4886.28	I
40		4886.65	I
95		4893.44	I
1100		4900.12	II
100		4906.11	I

Intensity		Wavelength	Spectrum
45		4909.00	I
150		4921.87	I
35		4930.93	I
45		4950.66	I
120		4974.30	I
120		4982.13	II
100		5006.97	I
75		5070.21	I
75		5072.19	I
1100		5087.42	II
30		5088.18	I
210		5119.11	II
450		5123.21	II
180		5135.20	I
120		5196.43	II
960		5200.41	II
1500		5205.72	II
180		5240.81	I
60		5289.82	II
45		5320.78	II
75		5380.62	I
220		5402.78	II
24		5417.03	I
90		5424.37	I
190		5438.24	I
710		5466.46	I
100		5468.47	I
90		5473.39	II
90		5480.74	II
60		5493.17	I
35		5495.59	I
240		5497.41	II
300		5503.45	I
250		5509.90	II
60		5513.64	I
120		5521.63	I
		5521.70	II
24		5526.76	I
740		5527.54	I
35		5541.63	I
120		5544.50	I
		5544.61	II
90		5546.02	II
75		5556.43	I
60		5567.75	I
180		5577.42	I
24		5581.08	I
620		5581.87	I
21		5590.96	I
21		5594.12	I
120		5606.33	I
15		5623.91	I
560		5630.13	I
24		5632.25	I
21		5632.89	I
120		5644.69	I
120		5648.47	I
740		5662.94	II
90		5675.27	I
18		5693.63	I
160		5706.73	I
24		5720.61	I
75		5728.89	II
150	Bl	5730.12	YO
21		5732.09	I
90		5743.85	I
18	Bl	5746.93	YO
24	Bl	5764.22	YO
75		5765.64	I
35		5773.95	I
100		5781.69	II
15	Bl	5800.00	YO
15	Bl	5818.58	YO
30		5821.87	I
21		5832.27	I
9	b	5838.07	YO
15		5858.83	YO
15		5871.83	I
24		5876.14	YO
24		5879.96	I
24	b	5893.94	YO
35		5902.96	I
24	b	5912.19	YO
24	b	5931.10	YO
90	Bl	5939.08	YO
45		5945.72	I
24		5950.02	I
75	b	5956.41	YO
1300	Bl	5972.04	YO
50		5981.86	I
1000	Bl	5987.64	YO
740	Bl	6003.60	YO
120		6004.65	
120		6009.19	I
620	Bl	6019.87	YO
120		6023.41	I
500	Bl	6036.60	YO
420	Bl	6053.81	YO
130	Bl	6072.78	YO
50		6088.00	I
210	Bl	6089.35	YO
160	Bl	6096.78	YO
130	Bl	6107.82	YO
130	Bl	6114.73	YO
75	Bl	6127.38	YO
1400	Bl	6132.06	YO
120		6135.04	I
150		6138.43	I
1100	Bl	6148.36	YO
120		6151.72	YO
820	Bl	6165.08	YO
560	Bl	6182.23	YO
1200		6191.73	I
590	Bl	6199.82	YO
450	Bl	6217.96	YO
300		6222.59	I
270	Bl	6236.72	YO
45		6251.05	I
120	Bl	6275.01	YO
60	b	6295.46	YO
24	b	6316.20	YO
24	b	6338.10	YO
15	b	6359.48	YO
15		6369.87	YO
75		6402.01	I
1000		6435.00	I
24		6437.18	I
18	h	6501.23	YO
18	h	6518.33	YO
18	h	6535.84	YO
90		6538.60	I
12	h	6553.84	YO
70		6557.39	I
12	h	6572.58	
35		6576.85	I
23		6584.87	I
95		6613.75	II
14		6622.49	I
19	h	6636.49	I
40		6650.61	I
21		6664.40	I
150		6687.58	I
14	h	6691.83	I
7		6694.75	I
16	h	6699.26	I
70		6700.71	I
35		6713.20	I
40		6735.99	I
190		6793.71	I
70	h	6795.41	II
12	h	6803.15	I
21		6815.16	I
14		6832.49	II
45		6845.24	I
14		6858.24	II
29		6887.22	I
21		6896.00	II
9		6908.26	I
14		6933.52	I
24	h	6950.31	I
10		6951.68	II
10		6958.04	I
24		6979.88	I
13	h	7008.97	I
10		7009.93	I
19	h	7035.18	I
29		7052.94	I
13	h	7054.28	I
9		7075.13	I
11		7127.92	I
35		7191.66	I
10	h	7195.93	I
35		7264.17	II
9	h	7293.08	I
9	h	7330.62	I
5		7332.96	II
50		7346.46	I
11	h	7398.77	I
29		7450.30	II
17		7494.88	I
7	h	7536.71	I
35		7563.13	I
8	h	7617.72	I
19	h	7622.94	I
7		7652.89	I
5		7689.49	I
8	h	7698.00	I
19		7719.89	I
19		7724.08	I
13		7788.42	I
13		7796.32	I
6		7802.52	I
17		7812.16	I
29		7855.52	I
110		7881.90	II
10	h	7999.33	I
9		8329.61	I
24		8344.43	I
8	h	8365.64	I
17		8450.36	I
8	h	8528.94	I
95		8800.62	I
19	h	8835.85	II

Y III
Ref. 77 — J.R.

Intensity		Wavelength	
		Vacuum	
1		643.68	III
4		646.69	III
6		653.87	III
10		656.98	III
25		668.74	III
40		671.98	III
100		691.72	III
4		693.85	III
200		695.20	III
9		727.91	III
4		728.47	III
2		728.83	III
20		729.73	III
600		730.49	III
15		732.70	III
800		734.36	III
15		770.78	III
10		771.79	III
20		804.26	III
5000		805.20	III
75		806.18	III
150		808.97	III
7000		809.92	III
100		855.64	III
60		857.82	III
25		984.23	III
15		987.96	III
15000		989.21	III
25000		996.37	III
20		999.19	III
150		1000.56	III
25		1003.35	III
1000		1006.58	III
1200		1007.86	III
120		1077.52	III
500		1081.35	III
75	p	1084.63	III
350	p	1088.39	III
250		1095.25	III
25		1095.87	III
150		1103.21	III
3000		1289.74	III
2500		1306.96	III
5000		1314.51	III
1500		1316.10	III
4000		1334.04	III
8		1549.08	III
15	p	1553.81	III
30		1635.14	III
75		1640.43	III
200		1779.80	III
600		1786.05	III
		Air	
10		2041.93	III
5		2042.07	III
1500		2060.58	III
4000		2068.98	III
10000		2127.98	III
16000		2191.16	III
8000		2200.76	III
8000		2206.03	III
150		2261.41	III
80		2261.57	III
10000		2284.34	III
3		2319.92	III
10000		2327.31	III
50000		2367.23	III
40000		2414.64	III
100	p	2710.30	III
90	h	2710.54	III
5		2780.11	III
70		2791.44	III
20		2803.27	III
100		2807.00	III
90000		2817.04	III
6000		2867.67	III
6000		2913.41	III
1500		2917.74	III
1600		2918.56	III
15		2940.53	III
99000		2946.01	III
20	p	2948.48	III
6000		2970.42	III
1400		3013.93	III
1500		3018.85	III
3		3267.10	III
25	l	3276.80	III
500		3866.96	III
3000		3900.74	III
4000		3914.58	III
3800		4039.60	III
3000		4040.11	III
120	h	4121.61	III
2000	c	4737.62	III
7500		5102.88	III
1300		5120.40	III
10000		5238.10	III
3000		5263.58	III
4000		5383.64	III
6000		5562.81	III
600		5567.27	III
4000		5572.24	III
400		5595.48	III
3000		5602.08	III
2000	h	7254.58	III
9000		7558.71	III
6000		7864.53	III
8000		7916.71	III
400		7989.41	III
10000		7991.43	III
8000		8171.41	III
4000		8645.09	III
10000		8796.21	III
8000		9116.59	III

Y IV
Ref. 265 — J.R.

Intensity		Wavelength	
		Vacuum	
3		211.80	IV
3		214.51	IV
6		215.97	IV
6		217.39	IV
12		221.71	IV
3		222.18	IV
6		222.98	IV
2		228.84	IV
20		228.94	IV
3		229.78	IV
3		235.17	IV
25		235.77	IV
1		242.12	IV
30		242.30	IV
3		244.14	IV
3		263.72	IV
150		264.64	IV
30		272.40	IV
150		273.03	IV
10		278.60	IV
900		355.86	IV
300		370.42	IV
500		386.82	IV
600		425.03	IV
300		473.10	IV

Y V
Ref. 419 — J.R.

Intensity		Wavelength	
		Vacuum	
5		289.18	V
50		299.99	V
3		312.89	V
200		313.35	V
40		320.47	V
40		321.69	V
150		325.58	V
200		326.57	V
175		328.34	V
50		330.40	V
900		333.09	V
500		333.80	V
100		335.12	V
400		335.14	V
500		336.62	V
500		339.02	V
200		340.02	V
5		340.42	V
500		344.59	V
100		349.65	V
2		349.75	V
10		351.36	V
100		353.98	V
100		355.56	V
300		372.05	V
400		379.96	V
200		397.77	V
300		403.45	V
1		408.81	V
200		409.31	V
200		415.03	V
100		418.18	V
150		418.59	V
50		419.79	V
300		420.74	V
1		427.87	V
50		430.75	V
100		437.66	V
3		441.62	V
30		442.96	V

Yttrium (Cont.)

2		451.97	V
15		455.84	V
85		457.84	V
4000		584.98	V
2000		630.97	V

ZINC (Zn)
Z = 30

Zn I and II
Ref. 39, 55, 113, 131, 185, 186 — R.D.C.

Intensity		Wavelength	
		Vacuum	
60		1193.23	II
60		1277.31	II
60	d	1366.68	II
		1404.12	I
60		1410.44	II
60		1439.09	II
60		1445.04	II
50		1456.91	II
		1457.57	I
60		1477.02	II
50		1514.76	II
90		1572.99	II
		1589.57	I
60		1617.68	II
60		1658.25	II
50		1713.25	II
60		1715.76	II
80	d	1735.61	II
60		1736.89	II
50		1737.90	II
75		1747.12	II
80	c	1762.19	II
75		1774.04	II
80		1790.76	II
100		1797.64	II
100	d	1811.05	II
80		1816.48	II
80		1831.38	II
100	d	1833.57	II
70		1836.01	II
75		1836.65	II
75		1847.56	II
100		1864.12	II
100		1866.08	II
100		1872.13	II
75		1894.26	II
60		1901.52	II
60		1914.81	II
100	d	1918.96	II
70		1920.27	II
100	d	1929.67	II
60		1945.58	II
60		1951.91	II
80		1953.00	II
75		1954.87	II
80		1964.54	II
100		1969.40	II
100		1982.11	II
70		1985.61	II
100		1986.99	II
50		1993.37	II
50		1996.92	II
		Air	
100		2011.94	II
500		2025.48	II
60		2039.31	II
500		2062.00	II
200		2064.23	II
120		2079.08	I
50		2079.93	II
60		2087.33	I
80		2096.93	I
300		2099.94	II
200		2102.18	II
150		2104.42	I
75		2122.74	II
800	r	2138.56	I
75		2147.42	II
60		2210.18	II
50		2273.15	II
1000		2501.99	II
150		2515.81	I
50		2527.96	II
1000		2557.95	II
50		2567.80	II
50		2567.98	II
100	h	2569.87	I
100		2582.44	I
300		2582.49	I
200		2608.56	I
300		2608.64	I
200		2670.53	I
300		2684.16	I
300		2712.49	I
200		2756.45	I
300		2770.86	I
300		2770.98	I
400		2800.87	I
100		2801.06	I
5		2801.17	I
100		2801.96	II
100		2902.30	II
125		3018.36	I
200		3035.78	I
200		3072.06	I
150		3075.90	I
100		3171.45	II
100		3172.23	II
300		3196.31	II
100		3197.10	II
500	r	3282.33	I
50		3299.42	II
800		3302.58	I
700	r	3302.94	I
75		3306.01	II
800		3345.02	I
500		3345.57	I
150		3345.94	I
5		3799.00	I
50		3806.34	II
100		3840.29	II
50		3883.34	I
15		3965.43	I
10		4113.21	I
25		4292.88	I
25		4298.33	I
35		4629.81	I
300		4680.14	I
400		4722.15	I
400		4810.53	I
800		4911.62	II
500		4924.03	II
7		5068.66	I
15		5069.58	I
200		5181.98	I
8		5308.65	I
7		5310.24	I
7		5311.02	I
4		5772.10	I
4		5775.50	I
10		5777.11	I
500		5894.33	II
500		6021.18	II
500		6102.49	II
100		6111.53	II
500		6214.61	II
8		6237.90	I
8		6239.17	I
1000	h	6362.34	I
10		6479.18	I
15		6928.32	I
8		6938.47	I
3		6943.20	I
200		7478.8	II
300		7588.5	II
100		7612.9	II
300		7732.5	II
200		7757.9	II
10		7799.36	I
100		11054.25	I
100		13053.63	I
100		13150.59	I
20		13196.61	I
100		14038.70	I
20		15680.29	I
20		16483.45	I
20		16491.98	I
20		16505.23	I
5		24044.16	I
10		24375.02	I

Zn III
Ref. 376, 377 — R.D.C.

Intensity	Wavelength	
	Vacuum	
1000	677.63	III
750	677.96	III
200	713.90	III
100	1432.15	III
200	1456.72	III
100	1464.20	III
100	1465.75	III
300	1473.41	III
100	1489.26	III
100	1490.96	III
200	1498.79	III
300	1499.42	III
300	1500.42	III
300	1505.92	III
300	1515.85	III
30	1533.09	III
300	1552.30	III
200	1552.94	III
80	1553.11	III
200	1560.79	III
150	1562.55	III
200	1581.53	III
100	1582.06	III
100	1598.52	III
100	1600.87	III
100	1619.61	III
150	1622.51	III
200	1629.19	III
200	1639.33	III
150	1644.82	III
100	1651.74	III
200	1673.05	III
100	1688.59	III
80	1695.40	III
80	1706.65	III
80	1749.63	III
50	1753.84	III
100	1767.69	III
80	1839.32	III

Zn IV
Ref. 370, 377 — R.D.C.

Intensity	Wavelength	
	Vacuum	
30	412.67	IV
30	423.42	IV
50	423.54	IV
200	425.90	IV
200	428.54	IV
80	428.79	IV
150	429.30	IV
200	430.59	IV
150	431.54	IV
120	431.62	IV
10	434.41	IV
150	435.02	IV
150	435.76	IV
80	436.25	IV
30	436.38	IV
50	436.82	IV
50	441.15	IV
20	441.52	IV
20	441.56	IV
100	441.70	IV
200	442.39	IV
150	444.39	IV
100	444.46	IV
100	446.58	IV
30	447.85	IV
10	449.13	IV
200	449.98	IV
200	450.99	IV
30	451.62	IV
20	452.80	IV
80	456.67	IV
150	457.32	IV
200	466.93	IV
150	468.43	IV
200	472.09	IV
200	472.66	IV
200	473.02	IV
200	473.51	IV
200	474.56	IV
50	475.78	IV
200	476.42	IV
200	478.65	IV
200	478.90	IV
200	482.10	IV
10	482.68	IV
10	485.48	IV
10	489.19	IV
10	490.96	IV
10	493.37	IV
10	496.70	IV
100	497.70	IV
15	1193.29	IV
15	1203.44	IV
8	1212.71	IV
5	1214.14	IV
25	1224.35	IV
20	1227.62	IV
40	1228.65	IV
30	1231.46	IV
30	1237.26	IV
50	1239.12	IV
3	1246.26	IV
15	1247.01	IV
50	1249.69	IV
30	1253.67	IV
30	1257.31	IV
8	1259.68	IV
500	1265.74	IV
5	1269.15	IV
100	1272.21	IV
200	1272.98	IV
30	1275.78	IV
25	1278.51	IV
100	1280.47	IV
100	1291.83	IV
100	1292.49	IV
10	1294.32	IV
100	1296.62	IV
100	1296.73	IV
20	1301.88	IV
500	1306.66	IV
200	1318.00	IV
200	1320.74	IV
200	1321.22	IV
200	1322.33	IV
200	1322.43	IV
200	1326.74	IV
200	1329.11	IV
100	1333.32	IV
150	1344.08	IV
200	1347.98	IV
200	1349.90	IV
50	1352.27	IV
100	1356.20	IV
200	1357.82	IV
100	1363.43	IV
200	1363.95	IV
40	1368.17	IV
200	1369.53	IV
50	1370.42	IV
100	1375.33	IV
50	1375.98	IV
200	1377.65	IV
50	1391.24	IV
80	1393.07	IV
50	1394.54	IV
30	1400.14	IV
15	1403.98	IV
30	1409.40	IV
5	1410.33	IV
30	1419.60	IV
50	1427.79	IV
20	1438.58	IV
80	1455.65	IV
200	1459.98	IV
100	1476.43	IV
100	1481.25	IV
100	1529.84	IV
50	1533.68	IV

ZIRCONIUM (Zr)
Z = 40

Zr I and II
Ref. 1 — C.H.C.

Intensity	Wavelength	
	Air	
60	2374.42	I
60	2384.17	I
50	2388.01	I
50	2389.21	I
45	2405.52	I
60	2419.41	II
150	2449.85	II
21	2457.44	II
75	2487.29	II
45	2496.48	II
180	2532.46	II
90	2539.65	I
220	2542.10	II
45	2550.51	I
220	2550.74	II
45	2556.43	I
60	2567.45	I
570	2567.64	II
1600	2568.87	II
2100	2571.39	II
75	2583.40	II
130	2589.07	II
22	2589.65	I
45	2609.43	I
150	2630.91	II
80	2635.42	I
210	2639.09	II
70	2643.40	II
55	2647.78	I
110	2650.38	II
70	2658.69	I
180	2667.80	II
55	2669.49	II
120	2670.96	II
1800	2678.63	II
35	2681.76	II
90	2687.75	I
90	2692.60	II
22	2692.92	I
160	2693.53	II
180	2694.06	II
70	2695.43	II

Zirconium (Cont.)

95		2699.60	II
750		2700.13	II
280		2711.51	II
140		2712.42	II
140		2714.26	II
1300		2722.61	II
140		2725.47	I
800		2726.49	II
490		2732.72	II
1400		2734.86	II
110		2740.51	II
140		2741.55	II
1100		2742.56	II
660		2745.86	II
660		2752.21	II
530		2758.81	II
200	d	2768.73	II
		2768.85	II
170	d	2774.04	I
		2774.16	II
200		2790.14	I
120		2792.04	I
160		2796.90	II
110		2799.15	II
180		2810.91	II
620		2814.90	I
390		2818.74	II
530		2825.56	II
110		2833.91	II
710		2837.23	I
120		2839.34	II
130		2843.52	II
660		2844.58	II
210		2848.19	II
350		2848.52	I
350		2851.97	II
340		2869.81	II
490		2875.98	I
120		2892.26	I
160		2905.23	II
300		2915.99	II
110		2916.64	II
270		2918.24	II
320		2926.99	II
160		2934.61	II
160		2936.31	II
320		2948.94	II
210		2951.48	II
320		2955.78	II
320		2960.87	I
320		2962.68	II
320		2968.96	II
120		2969.19	I
230		2969.63	II
130		2976.61	II
320		2978.05	II
230		2979.18	II
160		2981.02	II
820		2985.39	I
320		3003.74	II
100		3005.37	I
160		3005.50	I
820		3011.75	I
100		3013.32	II
160		3019.84	II
350		3020.47	II
500		3028.04	II
880		3029.52	I
180		3030.92	II
350	d	3036.39	II
100		3045.83	I
690		3054.84	II
100		3060.11	II
100		3064.63	II
110		3085.34	I
110		3094.80	I
250		3095.07	II
110		3095.82	I
280		3099.23	II
690		3106.58	II
110		3108.37	I
210		3110.88	II
350		3120.74	I
320		3125.92	II
500		3129.18	II
500		3129.76	II
140		3131.11	I
350		3132.07	I
110		3133.23	I
350		3133.48	II
180		3136.96	I
690		3138.68	II
140		3139.80	I
180		3148.82	I
290		3155.67	II
150		3157.00	II
320		3157.82	I
540		3164.31	II
150		3165.45	II
880		3165.97	II
150		3166.26	II
190		3178.09	II
190		3181.58	II
150		3181.92	II
880		3182.86	II
540		3191.21	I
210		3191.90	II
540		3212.01	I
760		3214.19	II
110		3222.47	II
200		3228.81	II
630		3231.69	II
630		3234.12	I
110		3236.58	II
760		3241.05	II
320		3250.39	I
200		3254.28	I
200		3260.11	I
190		3269.66	I
150		3271.13	II
540		3272.22	II
1000		3273.05	II
1300		3279.26	II
320	d	3282.73	I
880		3284.71	II
140		3285.88	II
150		3288.80	II
540		3305.15	II
880		3306.28	II
150		3313.70	II
210		3314.50	II
150		3319.02	II
380		3322.99	II
380		3326.80	II
380		3334.25	II
210		3334.62	II
190		3338.41	II
760		3340.56	II
380		3344.79	II
130		3353.66	I
180		3354.39	II
760		3356.09	II
540		3357.26	II
180		3359.96	II
150		3360.46	I
150		3363.82	II
150		3367.82	II
150		3370.59	I
180		3373.42	II
380		3374.73	II
110		3376.27	II
150		3377.46	II
570		3387.87	II
760		3388.30	II
5700		3391.98	II
570		3393.12	II
160		3396.33	II
380		3399.35	II
570		3404.83	II
760		3410.25	II
380		3414.66	I
1000		3430.53	II
380		3437.14	II
4700		3438.23	II
600		3447.36	I
200		3455.91	I
410		3457.56	II
200		3458.93	II
820		3463.02	II
600		3471.19	I
200		3478.79	I
1200		3479.39	II
1300		3481.15	II
760		3483.54	II
4100		3496.21	II
350		3505.48	II
820		3505.67	II
1000		3509.32	I
200		3510.46	II
2000		3519.60	I
440		3525.81	II
440		3533.22	I
210		3535.16	I
630		3542.62	II
1800		3547.68	I
210		3549.74	I
630		3550.46	I
1800		3551.95	II
2100		3556.60	II
1100		3566.10	I
210		3568.88	I
2100		3572.47	II
210		3573.08	II
1100		3575.79	I
1300		3576.85	II
880		3586.29	I
440		3587.98	II
3500		3601.19	I
690		3611.89	II
1100		3613.10	II
1100		3614.77	II
1100		3623.86	I
320		3634.15	I
260		3661.20	I
1100		3663.65	I
390		3671.27	II
800		3674.72	II
390		3697.46	II
960		3698.17	II
720		3709.26	II
270		3731.26	II
560		3745.98	II
880		3751.60	II
480		3764.39	I
480		3766.72	I
340		3766.82	II
720		3780.54	I
560		3791.40	I
210		3817.58	II
560		3822.41	I
2200		3835.96	I
1300		3836.76	II
550		3843.02	II
550		3847.01	I
550		3849.25	I
2900		3863.87	I
770		3864.34	I
990		3877.60	I
200		3879.05	I
1500		3885.42	I
2900		3890.32	I
2000		3891.38	I
400		3900.52	I
310		3915.94	II
610		3921.79	I
1200		3929.53	I
200		3934.12	II
200		3934.79	II
940		3958.22	II
490		3966.66	I
990		3968.26	I
660		3973.50	I
200		3975.29	I
200	h	3981.60	I
770		3991.13	II
770		3998.97	II
200		4007.60	I
200		4012.25	I
400		4023.98	I
770		4024.92	I
990		4027.20	I
240		4028.95	I
400		4029.68	II
490		4030.04	I
400		4035.89	I
240		4042.22	I
610		4043.58	I
490		4044.56	I
400		4045.61	II
610		4048.67	II
200		4050.33	II
200		4050.48	I
770		4055.03	I
600		4055.71	I
330		4061.53	I
1500		4064.16	I
2000		4072.70	I
310		4074.93	I
200		4076.53	I
240		4078.31	I
2000		4081.22	I
200		4108.40	I
400		4121.46	I
1200		4149.20	II
200		4152.64	I
290		4156.24	II
400		4161.21	II
400		4166.36	I
200		4183.32	I
660		4187.56	I
400		4194.76	I
610		4199.09	I
610		4201.46	I
610		4208.98	II
200		4211.88	II
400		4213.86	I
2000		4227.76	I
200		4236.06	I
2000		4239.31	I
770		4240.34	I
770		4241.20	I
1200		4241.69	I
310		4268.02	I
550		4282.20	I
550		4294.79	I
310		4302.89	I
550		4341.13	I
1000		4347.89	I
290		4359.74	II
310		4360.81	I
350		4366.45	I
240		4379.78	II
190		4413.04	I
240		4420.46	I
120		4427.24	I
160		4431.49	I
140		4443.00	II
110		4457.43	I
110		4466.91	I
110		4470.31	I
190		4470.56	I
200		4496.97	II
550		4507.12	I
610		4535.75	I
490		4542.22	I
200		4553.01	I
200		4555.13	I
140		4555.52	I
490		4575.52	I
100		4582.29	I
140		4590.55	I
350		4602.57	I
140		4604.42	I
210		4626.41	I
700		4633.98	I
210		4644.83	I
260		4683.42	I
2300		4687.80	I
510		4688.45	I
110		4707.79	I
1900		4710.08	I
160		4711.92	I
120		4717.62	I
210		4719.12	I
300		4732.33	I
1400		4739.48	I
190		4762.78	I
870		4772.31	I
210		4784.92	I
160		4788.67	I
260		4805.87	I
140		4809.47	I
190		4815.04	I
700		4815.63	I
280		4824.29	I
190		4828.04	I
110		4838.78	I
210		4851.36	I
160		4866.06	I
110		4881.24	I
110		4883.60	I
100		4994.76	I
30		5011.46	I
250		5046.58	I
85		5060.39	I
360		5064.91	I
110		5065.22	I
100		5070.26	I
75		5073.98	I
470		5078.25	I
85		5085.26	I
50		5112.27	II
140		5115.24	I
50		5120.42	I
85		5133.40	I
300		5155.45	I
200		5158.00	I
35		5158.67	I
75		5160.99	I
85		5165.96	I
17		5178.99	I
100		5183.70	I
30		5187.03	I
100		5191.60	II
100		5201.15	I
85		5209.30	I
85		5224.93	I
30		5243.47	I
120		5277.41	I
75		5280.05	I
60		5294.82	I
120		5296.79	I
60		5301.97	I
110		5311.40	I
25		5321.26	I
22		5330.84	I
12		5338.43	I
30		5350.09	II
30		5350.35	II
25		5350.90	I
25		5351.92	I
75		5362.56	I
12		5363.35	I
17		5369.39	I
20		5382.37	I
270		5385.14	I
30		5386.65	I
17		5391.18	I
17		5395.88	I
25		5405.13	I
85		5407.62	I
17		5413.93	I
20		5421.86	I
15		5426.36	I
25		5428.42	I
25		5437.76	I
15		5440.41	I
35		5448.57	I
10		5474.92	I
10		5477.40	I
35		5478.33	I
35		5480.83	I
10		5481.16	I
30		5486.09	I
140		5502.12	I

Zirconium (Cont.)

Intensity		Wavelength	
25		5507.87	I
30		5517.11	I
10		5518.05	I
75		5528.41	I
20		5532.30	I
45		5537.46	I
50		5545.32	I
22	Bl	5551.75	Z
25	Bl	5553.17	Z
12		5612.11	I
120		5620.14	I
35		5623.53	I
25	Bl	5629.02	Z
25	Bl	5629.58	Z
160		5664.51	I
20		5666.28	I
120		5680.90	I
15		5685.42	I
30		5708.89	I
75	Bl	5718.21	Z
120		5735.70	I
35	Bl	5748.17	Z
17	Bl	5778.57	Z
160		5797.74	I
30		5847.32	I
50		5868.27	I
110		5869.50	I
340		5879.80	I
85		5885.62	I
50		5901.09	I
30	Bl	5908.61	Z
140		5925.13	I
100		5935.20	I
110		5955.35	I
30	Bl	5977.80	Z
100		5984.23	I
17		5995.37	I
50		6001.05	I
30		6025.36	I
85		6032.61	I
170		6045.85	I
100		6049.24	I
140		6062.84	I
50		6120.83	I
170		6121.91	I
85		6124.84	I
680		6127.44	I
340		6134.55	I
100		6140.46	I
440		6143.20	I
30		6155.61	I
75		6157.71	I
25		6160.20	I
35		6189.40	I
60		6192.96	I
85		6213.05	I
100		6214.69	I
170	Bl	6226.51	Z
100		6257.26	I
50	b	6261.05	Z
35		6267.06	I
45	Bl	6292.84	Z
120		6299.66	I
15		6304.34	I
300		6313.02	I
30		6314.71	I
50		6321.35	I
22		6340.36	I
50	Bl	6345.10	Z
75		6345.22	I
75	Bl	6378.56	Z
35		6407.00	I
50	b	6412.39	Z
12		6426.17	I
35		6434.33	I
60		6445.74	I
20		6451.62	I
20		6457.63	I
110		6470.21	I
60	Bl	6473.79	Z
11		6484.35	I
110		6489.64	I
22		6493.10	I
50		6503.26	I
50		6506.36	I
50	Bl	6508.15	Z
30	Bl	6542.90	Z
35		6550.54	I
30		6569.43	I
20		6576.56	I
30	b	6578.06	Z
50		6591.99	I
10		6596.71	I
10		6598.84	I
50		6603.27	I
15		6620.56	I
11		6678.01	II
22		6688.18	I
11		6702.12	I
17		6709.61	I
27		6717.88	I
40		6752.73	I
75		6762.38	I

Intensity		Wavelength	
85		6769.16	I
27		6772.89	I
15		6787.15	II
35		6790.85	I
45		6828.78	I
45		6832.89	I
13		6845.33	I
17		6846.34	I
100		6846.97	I
27		6849.26	I
13		6852.56	I
120		6888.29	I
29		6900.59	I
20		6904.36	I
29		6907.37	I
20		6916.87	I
16		6932.38	I
29		6948.46	I
150		6953.84	I
60		6966.44	I
10		6975.91	I
150		6990.84	I
80		6994.32	I
10		7005.46	I
100		7027.40	I
25		7057.36	I
14		7057.96	I
140		7087.30	I
25		7089.43	I
35		7094.46	I
50		7095.59	I
540		7097.70	I
280		7102.91	I
170		7103.72	I
140		7111.68	I
40		7112.82	I
18		7113.52	I
12		7132.95	I
16		7140.74	I
12		7144.47	I
590		7169.09	I
50		7201.62	I
12		7258.17	I
35		7264.76	I
20		7306.21	I
25		7311.62	I
35		7313.72	I
90		7318.08	I
10		7327.82	I
50		7335.97	I
50		7343.96	I
20		7373.50	I
25		7383.63	I
14		7400.90	I
10		7411.39	I
10		7422.75	I
10		7433.10	I
110		7439.86	I
18		7467.57	I
16		7479.58	I
14	h	7515.70	I
12		7517.95	I
20	h	7540.62	I
20		7544.59	I
29		7551.46	I
40		7554.70	I
25		7558.45	I
12		7560.09	I
12		7562.12	I
80		7607.15	I
14		7612.08	I
20		7621.17	I
29		7658.60	I
18		7690.83	I
14		7704.27	I
10		7708.42	I
10		7766.55	I
12		7816.32	I
110		7819.35	I
35		7822.94	I
40		7826.72	I
90		7849.35	I
35		7869.99	I
14		7876.25	I
16		7882.18	I
10		7897.98	I
16		7908.46	I
20		7940.47	I
160		7944.61	I
80		7956.66	I
80		7959.98	I
20		7963.63	I
160		8005.27	I
25		8046.05	I
16		8053.06	I
20		8055.29	I
20		8055.76	I
60		8058.08	I
150		8063.09	I
790		8070.08	I
10		8114.28	I
20		8120.17	I
390		8132.99	I
20		8152.58	I

Intensity	Wavelength	
12	8188.77	I
40	8194.73	I
60	8201.73	I
280	8212.53	I
20	8240.37	I
40	8283.81	I
140	8305.90	I
14	8332.44	I
50	8370.23	I
120	8389.41	I
70	8414.00	I
50	8453.17	I
50	8464.65	I
40	8498.44	I
18	8584.21	I
10	8734.86	I
12	8749.48	I
10	8786.23	I
16	8804.98	I
70	8836.09	I
60	8899.52	I

Zr III
Ref. 403 — J.R.

Intensity	Wavelength	
	Vacuum	
25	687.64	III
50	690.39	III
25	819.59	III
30	820.21	III
35	823.69	III
25	829.50	III
30	850.61	III
25	859.56	III
30	864.86	III
25	868.64	III
25	868.99	III
25	919.59	III
30	1320.81	III
25	1375.13	III
30	1378.93	III
40	1403.48	III
35	1420.12	III
25	1420.87	III
25	1465.44	III
25	1563.24	III
40	1593.59	III
100	1612.38	III
50	1620.62	III
75	1631.31	III
50	1638.33	III
35	1675.06	III
35	1675.75	III
40	1703.36	III
40	1725.03	III
40	1754.38	III
35	1759.12	III
30	1764.75	III
30	1771.96	III
40	1773.90	III
100	1779.51	III
40	1783.35	III
200	1790.19	III
150	1793.56	III
125	1798.13	III
75	1800.03	III
25	1801.67	III
100	1805.26	III
35	1831.89	III
40	1850.06	III
40	1853.38	III
30	1859.12	III
25	1860.47	III
25	1861.77	III
75	1864.06	III
30	1865.45	III
35	1877.00	III
50	1892.07	III
65	1914.25	III
75	1921.96	III
75	1932.54	III
50	1934.32	III
40	1935.20	III
75	1936.48	III
75	1936.67	III
80	1937.27	III
200	1940.25	III
100	1941.08	III
25	1946.12	III
80	1946.61	III
50	1946.99	III
100	1953.95	III
50	1961.32	III
100	1962.01	III
40	1962.92	III
85	1966.22	III
25	1967.81	III
60	1974.99	III
40	1983.14	III
50	1989.83	III
30	1990.95	III
30	1994.46	III
	Air	
45	2000.23	III
100	2006.82	III
30	2013.30	III
35	2016.63	III
30	2021.52	III
60	2026.78	III
100	2035.42	III
50	2036.92	III
75	2056.13	III
40	2058.73	III
75	2060.83	III
50	2061.47	III
125	2070.43	III
50	2074.12	III
100	2077.92	III
100	2080.99	III
30	2081.81	III
25	2085.35	III
200	2086.78	III
40	2089.50	III
40	2097.03	III
40	2102.30	III
50	2103.16	III
25	2104.23	III
40	2113.98	III
35	2114.10	III
40	2125.06	III
35	2137.90	III
35	2138.45	III
25	2139.85	III
40	2159.24	III
40	2162.20	III
100	2175.80	III
100	2191.15	III
35	2192.05	III
60	2206.33	III
40	2206.97	III
30	2231.00	III
50	2245.36	III
25	2251.14	III
40	2257.83	III
30	2281.43	III
100	2301.60	III
75	2308.12	III
35	2405.81	III
40	2406.21	III
75	2420.65	III
25	2438.70	III
50	2444.58	III
100	2448.86	III
100	2593.64	III
250	2620.56	III
50	2621.28	III
60	2628.26	III
200	2643.79	III
100	2656.46	III
150	2664.26	III
100	2682.16	III
75	2686.28	III
70	2690.49	III
60	2698.31	III
50	2709.05	III
45	2715.76	III
40	2720.07	III
75	2735.76	III
25	2775.23	III
40	2836.18	III
40	3278.86	III

Zr IV
Ref. 362 — J.R.

Intensity	Wavelength	
	Vacuum	
15	478.97	IV
60	480.66	IV
60	497.23	IV
60	500.22	IV
4	500.34	IV
4	584.65	IV
15	585.42	IV
2	586.42	IV
60	588.89	IV
100	589.74	IV
600	628.66	IV
500	633.56	IV
200	633.63	IV
8	712.49	IV
90	754.39	IV
90	760.15	IV
150	846.40	IV
300	863.65	IV
500	864.59	IV
200	881.30	IV

Zirconium (Cont.)

Intensity		Wavelength	
200		882.59	IV
100		1099.76	IV
150		1100.00	IV
9000		1183.97	IV
9000		1201.77	IV
200		1212.71	IV
100		1213.01	IV
10000		1219.86	IV
500		1285.89	IV
500		1290.56	IV
70		1291.70	IV
500		1417.70	IV
500		1440.65	IV
500		1441.06	IV
1000		1469.47	IV
10000		1546.17	IV
10		1596.29	IV
10000		1598.95	IV
150		1605.26	IV
5000		1607.95	IV
90		1609.50	IV
35		1818.06	IV
500		1836.15	IV
500		1846.37	IV
200		1848.03	IV
70		1851.91	IV
		Air	
20		2045.12	IV
40		2047.15	IV
10000		2091.49	IV
10000		2092.36	IV
10000		2163.68	IV
10000		2286.67	IV
200		2473.75	IV
300		2476.71	IV
5		2572.32	IV
400		2573.66	IV
400		2583.32	IV

Zr V
Ref. 418 — J.R.

Intensity		Wavelength	
		Vacuum	
300		292.19	V
500		304.01	V
300		305.24	V
400	p	368.18	V
200		519.25	V
200		536.50	V
200	p	674.13	V
200		675.53	V
300		679.39	V
200		688.27	V
200	p	703.03	V
300		717.24	V
200		740.33	V
2000		740.61	V
500		742.74	V
300		752.48	V
300		764.58	V
500		766.20	V
200		773.80	V
200		775.58	V
200		779.21	V
400		784.50	V
400		797.23	V
10000		800.00	V
10000		806.89	V
200		809.75	V
10000		812.05	V
500		822.06	V
300	p	823.46	V
200		823.78	V
200		825.13	V
200		830.59	V
400		836.57	V
3000		841.40	V
300	p	852.87	V
700		853.68	V
200		873.11	V
300		885.68	V
300		900.48	V
500		906.66	V
500		915.30	V
400		923.10	V
400		940.41	V
200		949.70	V
200		978.06	V
200		980.70	V
200		984.18	V
300		995.59	V
400		1002.48	V
400		1038.69	V
200		1044.41	V
300		1047.77	V
400		1068.55	V
300		1072.25	V
300		1083.45	V
200		1087.05	V
200		1093.54	V
200		1108.79	V
300		1194.24	V
300		1200.76	V
200		1233.91	V
300		1238.93	V
300		1253.61	V
200		1259.70	V
400		1260.91	V
500		1265.38	V
300		1295.81	V
400		1302.80	V
500		1303.93	V
400		1306.76	V
200		1315.14	V
400		1320.74	V
500	p	1323.81	V
300		1332.06	V
300		1337.34	V
300		1355.21	V
500		1355.98	V
300		1361.39	V
500		1376.54	V
200		1396.79	V
300		1410.03	V
200		1413.40	V
300		1460.05	V
300		1486.90	V
300		1491.33	V
200		1520.47	V
200		1550.12	V
500		1633.03	V
500		1654.46	V
700		1725.02	V
500		1786.20	V
200	p	1790.81	V
500		1806.09	V
500		1860.48	V
600		1860.86	V
500		1878.33	V
400		1926.24	V
500		1927.43	V
400		1934.88	V
		Air	
500		2009.29	V
600		2028.54	V
600		2132.42	V
200		2150.18	V
200		2336.94	V

REFERENCES

1. **Meggers, W. F., Corliss, C. H., and Scribner, B. F.,** *Natl. Bur. Stand. (U.S.) Monogr.*, 145, Washington, D.C., 1975.
2. **Aksenov, V. P. and Ryabtsev, A. N.,** *Opt. Spectrosc.*, 37, 860, 1970.
3. **Andersen, N., Bickel, W. S., Carriveau, G. W., Jensen, K., and Veje, E.,** *Phys. Scr.*, 4, 113, 1971.
4. **Andersson, E. and Johannesson, G. A.,** *Phys. Scr.*, 3, 203, 1971.
5. **Andrew, K. L. and Meissner, K. W.,** *J. Opt. Soc. Am.*, 49, 146, 1959.
6. **Artru, M. C. and Brillet, W. U. L.,** *J. Opt. Soc. Am.*, 64, 1063, 1974.
7. **Artru, M. C. and Kaufman, V.,** *J. Opt. Soc. Am.*, 62, 949, 1972.
8. **Artru, M. C. and Kaufman, V.,** *J. Opt. Soc. Am.*, 65, 594, 1975.
9. **Badami, J. S. and Rao, K. R.,** *Proc. R. Soc. London,* 140(A), 387, 1933.
10. **Baird, K. M. and Smith, D. S.,** *J. Opt. Soc. Am.*, 48, 300, 1958.
11. **Bashkin, S. and Martinson, I.,** *J. Opt. Soc. Am.*, 61, 1686, 1971.
12. **Beacham, J. R., Ph.D. thesis, Purdue University, 1970.**
13. **Benschop, H., Joshi, Y. N., and van Kleef, T. A. M.,** *Can. J. Phys.*, 53, 700, 1975.
14. **Berry, H. G., Bromander, J., and Buchta, R.,** *Phys. Scr.*, 1, 181, 1970.
15. **Berry, H. G., Bromander, J., Martinson, I., and Buchta, R.,** *Phys. Scr.*, 3, 63, 1971.
16. **Berry, H. G., Desesquelles, J., and Dufay, M.,** *Phys. Rev. Sect. A.*, 6, 600, 1972.
17. **Berry, H. G., Desesquelles, J., and Dufay, M.,** *Nucl. Instrum. Methods,* 110, 43, 1973.
18. **Berry, H. G., Pinnington, E. H., and Subtil, J. L.,** *J. Opt. Soc. Am.*, 62, 767, 1972.
19. **Bidelman, W. P. and Corliss, C. H.,** *Astrophys. J.*, 135, 968, 1962.
20. **Bloch, L. and Bloch, E.,** *Ann. Phys.* (Paris), 10(11), 141, 1929.
21. **Bloch, L., Bloch, E., and Felici, N.,** *J. Phys. Radium,* 8, 355, 1937.
22. **Bockasten, K.,** *Ark. Fys.*, 9, 457, 1955.
23. **Bockasten, K., Hallin, R., Johansson, K. B., and Tsui, P.,** *Phys. Lett.* (Netherlands), 8, 181, 1964.
24. **Bockasten, K. and Johansson, K. B.,** *Ark. Fys.*, 38, 563, 1969.
25. **Borgstrom, A.,** *Ark. Fys.*, 38, 243, 1968.
26. **Borgstrom, A.,** *Phys. Scr.*, 3, 157, 1971.
27. **Bowen, I. S.,** *Phys. Rev.*, 29, 231, 1927.
28. **Bowen, I. S.,** *Phys. Rev.*, 31, 34, 1928.
29. **Bowen, I. S.,** *Phys. Rev.*, 39, 8, 1932.
30. **Bowen, I. S.,** *Phys. Rev.*, 45, 401, 1934.
31. **Bowen, I. S.,** *Phys. Rev.*, 46, 377, 1934.
32. **Bowen, I. S.,** *Phys. Rev.*, 46, 791, 1934.
33. **Boyce, J. C.,** *Phys. Rev.*, 49, 730, 1936.
34. **Boyce, J. C. and Robinson, H. A.,** *J. Opt. Soc. Am.*, 26, 133, 1936.
35. **Bozman, W. R., Meggers, W. F., and Corliss, C. H.,** *J. Res. Natl. Bur. Stand. Sect. A*, 71, 547, 1967.
36. **Bromander, J.,** *Ark. Fys.*, 40, 257, 1969.
37. **Bromander, J. and Buchta, R.,** *Phys. Scr.*, 1, 184, 1970.
38. **Brown, C. M. and Ginter, M. L.,** *J. Opt. Soc. Am.*, 68, 243, 1978.
39. **Brown, C. M., Tilford, S. G., and Ginter, M. L.,** *J. Opt. Soc. Am.*, 65, 1404, 1975.
40. **Bryant, B. W.,** *Johns Hopkins Spectroscopic Report* No. 21, 1961.
41. **Buchet, J. P., Buchet-Poulizac, M. C., Berry, H. G., and Drake, G. W. F.,** *Phys. Rev. Sect. A*, 7, 922, 1973.
42. **Budhiraja, C. J. and Joshi, Y. N.,** *Can. J. Phys.*, 49, 391, 1971.

43. **Burns, K. and Adams, K. B.,** *J. Opt. Soc. Am.*, 42, 56, 1952.
44. **Burns, K. and Adams, K. B.,** *J. Opt. Soc. Am.*, 46, 94, 1956.
45. **Burns, K., Adams, K. B., and Longwell, J.,** *J. Opt. Soc. Am.*, 40, 339, 1950.
46. **Callahan, W. R.,** Ph.D. thesis, Johns Hopkins University, 1962.
47. **Charles, G. W.,** *J. Opt. Soc. Am.*, 56, 1292, 1966.
48. **Charles, G. W., Hunt, D. J., Pish, G., and Timma, D. L.,** *J. Opt. Soc. Am.*, 45, 869, 1955.
49. **Codling, K.,** *Proc. Phys. Soc.*, 77, 797, 1961.
50. **Comite Consulatif Pour La Definition du Metre,** *J. Phys. Chem. Ref. Data*, 3, 852, 1974.
51. **Conway, J. G., Blaise, J., and Verges, J.,** *Spectrochim. Acta Part B*, 31, 31, 1976.
52. **Conway, J. G., Worden, E. F., Blaise, J., and Verges, J.,** *Spectrochim. Acta Part B*, 32, 97, 1977.
53. **Conway, J. G., Worden, E. F., Blaise, J., Camus, P., and Verges, J.,** *Spectrochim. Acta Part B*, 32, 101, 1977.
54. **Crooker, A. M.,** *Can. J. Res. Sect. A*, 14, 115, 1936.
55. **Crooker, A. M. and Dick, K. A.,** *Can. J. Phys.*, 46, 1241, 1968.
56. **Crosswhite, H. M.,** *J. Res. Natl. Bur. Stand. Sect. A*, 79, 17, 1975.
58. **Crosswhite, H. M. and Dieke, G. H.,** *American Institute of Physics Handbook*, Section 7, 1972.
59. **de Bruin, T. L.,** *Z. Phys.*, 38, 94, 1926.
60. **de Bruin, T. L.,** *Z. Phys.*, 53, 658, 1929.
61. **de Bruin, T. L., Humphreys, C. J., and Meggers, W. F.,** *J. Res. Natl. Bur. Stand.*, 11, 409, 1933.
62. **Dick, K. A.,** *J. Opt. Soc. Am.*, 64, 702, 1973.
63. **Dobbie, J. C.,** *Ann. Solar Phys. Observ.* (Cambridge), 5, 1, 1938.
64. **Earls, L. T. and Sawyer, R. A.,** *Phys. Rev.*, 47, 115, 1935.
65. **Edlen, B.,** *Z. Phys.*, 85, 85, 1933.
66. **Edlen, B.,** *Nova Acta Reglae Soc. Sci. Ups.*, (IV) 9, No. 6, 1934.
67. **Edlen, B.,** *Z. Phys.*, 93, 726, 1935.
68. **Edlen, B.,** *Z. Phys.*, 94, 47, 1935.
69. **Edlen, B.,** *Rep. Prog. Phys.*, 26, 181, 1963.
70. **Edlen, B. and Risberg, P.,** *Ark. Fys.*, 10, 553, 1956.
71. **Edlen, B. and Swings, P.,** *Astrophys. J.*, 95, 532, 1942.
72. **Ehrhardt, J. C. and Davis, S. P.,** *J. Opt. Soc. Am.*, 61, 1342, 1971.
73. **Eidelsberg, M.,** *J. Phys. B*, 5, 1031, 1972.
74. **Eidelsberg, M.,** *J. Phys. B*, 7, 1476, 1974.
75. **Ekberg, J. O. and Svensson, L. A.,** *Phys. Scr.*, 2, 283, 1970.
76. **Ekefors, E.,** *Z. Phys.*, 71, 53, 1931.
77. **Epstein, G. L. and Reader, J.,** *J. Opt. Soc. Am.*, 65, 310, 1975.
78. **Epstein, G. L. and Reader, J.,** *J. Opt. Soc. Am.*, 66, 590, 1976.
79. **Epstein, G. L. and Reader, J.,** unpublished.
80. **Eriksson, K. B. S.,** *Phys. Lett. A.*, 41, 97, 1972.
81. **Eriksson, K. B. S. and Isberg, H. B. S.,** [illegible]
82. **Eriksson, K. B. S. and Wenaker, I.,** *Phys. Scr.*, 1, 21, 1970.
83. **Esteva, J. M. and Mehlman, G.,** *Astrophys. J.*, 193, 747, 1974.
84. **Even-Zohar, M. and Fraenkel, B. S.,** *J. Phys. B*, 5, 1596, 1972.
85. **Fawcett, B. C.,** *J. Phys. B*, 3, 1732, 1970.
86. **Fawcett, B. C.,** Culham Laboratory Report ARU-R4, 1971.
87. **Ferner, E.,** *Ark. Mat. Astron. Fys.*, 28(A), 4, 1941.
88. **Fischer, R. A., Knopf, W. C., and Kinney, F. E.,** *Astrophys. J.*, 130, 683, 1959.
89. **Fowler, A.,** *Report on Series in Line Spectra*, Fleetway Press, London, 1922.
90. **Fowles, G. R.,** *J. Opt. Soc. Am.*, 44, 760, 1954.
91. **Fred, M.,** *Argonne Natl. Lab.*, unpublished, 1977.
92. **Fred, M. and Tomkins, F. S.,** *J. Opt. Soc. Am.*, 47, 1076, 1957.
93. **Fred, M., Tomkins, F. S., Blaise, J. E., Camus, P., and Verges, J.,** Argonne National Laboratory Report No. 76-68, 1976.
94. **Garcia, J. D. and Mack, J. E.,** *J. Opt. Soc. Am.*, 55, 654, 1965.
96. **Giacchetti, A.,** *Argonne Natl. Lab.*, unpublished, 1975.
97. **Giacchetti, A., Blaise, J., Corliss, C. H., and Zalubas, R.,** *J. Res. Natl. Bur. Stand. Sect. A*, 78, 247, 1974.
98. **Giacchetti, A., Stanley, R. W., and Zalubas, R.,** *J. Opt. Soc. Am.*, 69, 474, 1970.
99. **Gilbert, W. P.,** *Phys. Rev.*, 47, 847, 1935.
100. **Gilroy, H. T.,** *Phys. Rev.*, 38, 2217, 1931.
101. **Glad, S.,** *Ark. Fys.*, 10, 291, 1956.
102. **Goldsmith, S.,** *J. Phys. B*, 2, 1075, 1969.
103. **Goorvitch, D., Mehlman-Balloffet, G., and Valero, F. P. J.,** *J. Opt. Soc. Am.*, 60, 1458, 1970.
104. **Goorvitch, D. and Valero, F. P. J.,** *Astrophys. J.*, 171, 643, 1972.
105. **Green, L. C.,** *Phys. Rev.*, 55, 1209, 1939.
106. **Gutman, F.,** *Diss. Abstr. Int. B*, 31, 363, 1970.
107. **Hallin, R.,** *Ark. Fys.*, 31, 511, 1966.
108. **Hallin, R.,** *Ark. Fys.*, 32, 201, 1966.
109. **Hansen, J. E. and Persson, W.,** *J. Opt. Soc. Am.*, 64, 696, 1974.
110. **Hansen, J. E. and Persson, W.,** *Phys. Scr.*, 13, 166, 1976.
111. **Hellintin, P.,** *Phys. Scr.*, 13, 155, 1976.
112. **Herzberg, G. and Moore, H. R.,** *Can. J. Phys.*, 37, 1293, 1959.
113. **Hetzler, C. W., Boreman, R, W., and Burns, K.,** *Phys. Rev.*, 48, 656, 1935.
114. **Holmstrom, J. E. and Johansson, L.,** *Ark. Fys.*, 40, 133, 1969.
115. **Hontzeas, S., Martinson, I., Erman, P., and Buchta, R.,** *Nucl. Instrum. Methods*, 110, 51, 1973.
116. **Humphreys, C. J.,** *J. Res. Natl. Bur. Stand.*, 22, 19, 1939.
117. **Humphreys, C. J.,** *J. Opt. Soc. Am.*, 43, 1027, 1953.
118. **Humphreys, C. J.,** *J. Phys. Chem. Ref. Data*, 2, 519, 1973.
119. **Humphreys, C. J. and Andrew, K. L.,** *J. Opt. Soc. Am.*, 54, 1134, 1964.
120. **Humphreys, C. J. and Meggers, W. F.,** *J. Res. Natl. Bur. Stand.* , 10, 139, 1933.
121. **Humphreys, C. J. and Paul, E., Jr.,** *J. Opt. Soc. Am.*, 60, 200, 1970.
122. **Humphreys, C. J. and Paul, E., Jr.,** *J. Opt. Soc. Am.*, 62, 432, 1972.
123. **Humphreys, C. J., Paul, E., Jr., Cowan, R. D., and Andrew, K. L.,** *J. Opt. Soc. Am.*, 57, 855, 1967.
124. **Humphreys, C. J., Paul, E., Jr., and Minnhagen, L.,** *J. Opt. Soc. Am.*, 61, 110, 1971.
125. **Iglesias, L.,** Inst. of Optics, Madrid, unpublished, 1977.

126. Iglesias, L. and Velasco, R., *Publ. Inst. Opt. Madrid*, No. 23, 1964.
127. Isberg, B., *Ark. Fys.*, 35, 551, 1967.
128. Johannesson, G. A., Lundstrom, T., and Minnhagen, L., *Phys. Scr.*, 6, 129, 1972.
129. Johannesson, G. A. and Lundstrom, T., *Phys. Scr.*, 8, 53, 1973.
130. Johansson, I., *Ark. Fys.*, 20, 135, 1961.
131. Johansson, I. and Contreras, R., *Ark. Fys.*, 37, 513, 1968.
132. Johansson, I. and Litzen, U., *Ark. Fys.*, 34, 573, 1967.
133. Johansson, I. and Svensson, K. F., *Ark. Fys.*, 16, 353, 1960.
134. Johansson, L., *Ark. Fys.*, 20, 489, 1961.
135. Johansson, L., *Ark. Fys.*, 23, 119, 1963.
136. Johansson, S. and Litzen, U., *Phys. Scr.*, 6, 139, 1972.
137. Johansson, S. and Litzen, U., *Phys. Scr.*, 8, 43, 1973.
138. Johansson, S. and Litzen, U., *Phys. Scr.*, 10, 121, 1974.
139. Joshi, Y. N., St. Francis Xavier Univ., Nova Scotia, unpublished.
140. Joshi, Y. N., Bhatia, K. S., and Jones, W. E., *Sci. Light Tokyo*, 21, 113, 1972.
141. Joshi, Y. N., Bhatia, K. S., and Jones, W. E., *Spectrochim. Acta Part B*, 28, 149, 1973.
142. Joshi, Y. N. and Budhiraja, C. J., *Can. J. Phys.*, 49, 670, 1971.
143. Joshi, Y. N. and van Kleef, T. A. M., *Can. J. Phys.*, 52, 1891, 1974.
144. Kaufman, V., *Natl. Bur. Stand.*, unpublished.
145. Kaufman, V., *J. Opt. Soc. Am.*, 52, 866, 1962.
146. Kaufman, V., Artru, M. C., and Brillet, W. U. L., *J. Opt. Soc. Am.*, 64, 197, 1974.
147. Kaufman, V. and Humphreys, C. J., *J. Opt. Soc. Am.*, 59, 1614, 1969.
148. Kaufman, V. and Sugar, J., *J. Opt. Soc. Am.*, 61, 1693, 1971.
149. Kaufman, V. and Sugar, J., *J. Res. Natl. Bur. Stand. Sect. A*, 71, 583, 1967.
150. Kelly, R. L. and Palumbo, L. J., *Naval Research Laboratory Report 7599, Washington, D. C.*, 1973.
151. Kielkopf, J. F., *Univ. of Louisville*, unpublished, 1975.
152. Kielkopf, J. F., *Univ. of Louisville*, unpublished, 1976.
153. Kiess, C. C. and Corliss, C. H., *J. Res. Natl. Bur. Stand. Sect. A*, 63, 1, 1959.
154. Kleiman, H., *J. Opt. Soc. Am.*, 52, 441, 1962.
155. Eriksson, K. B., Johansson, I., and Norlen, G., *Ark. Fys.*, 28, 233, 1964.
156. Klinkenberg, P. F. A., *Physica*, 15, 774, 1949.
157. Klinkenberg, P. F. A., *Physica*, 16, 618, 1950.
158. Krishnamurty, S. G., *Proc. Phys. Soc. London*, 48, 277, 1936.
159. Kruger, P. G. and Gilroy, H. T., *Phys. Rev.*, 48, 720, 1935.
160. Kruger, P. G. and Pattin, H. S., *Phys. Rev.*, 52, 621, 1937.
161. Lacroute, P., *Ann. Phys.* (Paris), 3, 5, 1935.
162. Lang, R. J., *Phys. Rev.*, 30, 762, 1927.
163. Lang, R. J., *Phys. Rev.*, 32, 737, 1928.
164. Lang, R. J., *Phys. Rev.*, 35, 445, 1930.
165. Lang, R. J., *Can. J. Res. Sect. A*, 14, 43, 1936.
166. Lang, R. J., *Can. J. Res. Sect. A*, 14, 127, 1936.
167. Lang, R. J. and Vestine, E. H., *Phys. Rev.*, 42, 233, 1932.
168. Li, H. and Andrew, K. L., *J. Opt. Soc. Am.*, 61, 96, 1971.
169. Liden, K., *Ark. Fys.*, 1, 229, 1949.
170. Litzen, U., *Ark. Fys.*, 28, 239, 1965.
171. Litzen, U., *Phys. Scr.*, 1, 251, 1970.
172. Litzen, U., *Phys. Scr.*, 1, 253, 1970.
173. Litzen, U., *Phys. Scr.*, 2, 103, 1970.
174. Litzen, U. and Verges, J., *Phys. Scr.*, 13, 240, 1976.
175. Lofstrand, B., *Phys. Scr.*, 8, 57, 1973.
176. Luc-Koenig, E., Morillon, C., and Verges, J., *Phys. Scr.*, 12, 199, 1975.
177. Lundstrom, T., *Phys. Scr.*, 7, 62, 1973.
178. Lundstrom, T. and Minnhagen, L., *Phys. Scr.*, 5, 243, 1972.
179. Magnusson, C. E. and Zetterberg, P. O., *Phys. Scr.*, 10, 177, 1974.
180. Magnusson, C. E., and Zetterberg, P. O., *Phys. Scr.*, 15, 237, 1977.
181. Martin, D. C., *Phys. Rev.*, 48, 938, 1935.
182. Svendenius, N., *Phys. Scr.*, 22, 240, 1980.
183. Martin, W. C., *J. Res. Natl. Bur. Stand. Sect. A*, 64, 19, 1960.
184. Martin, W. C. and Corliss, C. H., *J. Res. Natl. Bur. Stand. Sect. A*, 64, 443, 1960.
185. Martin, W. C. and Kaufman, V., *J. Res. Natl. Bur. Stand. Sect. A*, 74, 11, 1970.
186. Martin, W. C. and Kaufman, V., *J. Opt. Soc. Am.*, 60, 1096, 1970.
187. McCormick, W. W. and Sawyer, R. A., *Phys. Rev.*, 54, 71, 1938.
188. McLaughlin, R., *J. Opt. Soc. Am.*, 54, 965, 1964.
189. McLennan, J. C., McLay, A. B., and Crawford, M. F., *Proc. R. Soc. London Ser. A*, 134, 41, 1931.
190. Meissner, K. W., *Z. Phys.*, 39, 172, 1926.
191. Meggers, W. F., *J. Res. Natl. Bur. Stand.*, 24, 153, 1940.
192. Meggers, W. F. and Corliss, C. H., *J. Res. Natl. Bur. Stand. Sect. A*, 70, 63, 1966.
193. Meggers, W. F., Fred, M., and Tomkins, F. S., *J. Res. Natl. Bur. Stand.*, 58, 297, 1957.
194. Meggers, W. F. and Humphreys, C. J., *J. Res. Natl. Bur. Stand.*, 28, 463, 1942.
195. Meggers, W. F. and Murphy, R. J., *J. Res. Natl. Bur. Stand.*, 48, 334, 1952.
196. Meggers, W. F., Scribner, B. F., and Bozman, W. R., *J. Res. Natl. Bur. Stand.*, 46, 85, 1951.
197. Meggers, W. F., Shenstone, A. G., and Moore, C. E., *J. Res. Natl. Bur. Stand.*, 45, 346, 1950.
198. Mehlman, G. and Esteva, J. M., *Astrophys. J.*, 188, 191, 1974.
199. Meinders, E., *Physica*, 84(C), 117, 1976.
200. Sansonetti, C. J., Dissertation, Purdue University, 1981.
201. Sansonetti, C. J., *Natl. Bur. Stand. (U.S.)*, unpublished.
202. Millikan, R. A. and Bowen, I. S. *Phys. Rev.*, 25, 600, 1925.
203. Minnhagen, L., *J. Opt. Soc. Am.*, 61, 1257, 1925.
204. Minnhagen, L., *J. Opt. Soc. Am.*, 63, 1185, 1973.
205. Minnhagen, L., *Phys. Scr.*, 11, 38, 1975.
206. Minnhagen, L., *J. Opt. Soc. Am.*, 66, 659, 1976.
207. Minnhagen, L. and Nietsche, H., *Phys. Scr.*, 5, 237, 1972.
208. Minnhagen, L., Strihed, H., and Petersson, B., *Ark. Fys.*, 39, 471, 1969.
209. Moore, C. E., *Natl. Bur. Stand. (U.S.) Circ.*, 488, 1950.
210. Moore, C. E., *Revised Multiplet Table*, Princeton University Observatory No. 20, 1945.

211. **Moore, C. E.**, National Standard Reference Data Series — National Bureau of Standards 3, Sect. 3, 1970.
212. **Moore, C. E.**, National Standard Reference Data Series — National Bureau of Standards 3, Sect. 4, 1971.
213. **Moore, C. E.**, National Standard Reference Data Series — National Bureau of Standards 3, Sect. 5, 1975.
214. **Moore, C. E.**, National Standard Reference Data Series — National Bureau of Standards 3, Sect. 6, 1972.
215. **Moore, C. E.**, *National Standard Reference Data Series — National Bureau of Standards 3, Sect. 7*, 1975.
216. **Morillon, C. and Verges, J.**, *Phys. Scr.*, 10, 227, 1974.
217. **Newsom, G. H.**, *Astrophys. J.*, 166, 243, 1971.
218. **Newsom, G. H., O'Connor, S., and Learner, R. C. M.**, *J. Phys. B*, 6, 2162, 1973.
219. **Norlen, G.**, *Phys. Scr.*, 8, 249, 1973.
220. **Odabasi, H.**, *J. Opt. Soc. Am.*, 57, 1459, 1967.
221. **Olme, A.**, *Ark. Fys.*, 40, 35, 1969.
222. **Olme, A.**, *Phys. Scr.*, 1, 256, 1970.
223. Johansson, S., and Litzen, U., *J. Opt. Soc. Am.*, 61, 1427, 1971.
224. **Palenius, H. P.**, *Ark. Fys.*, 39, 15, 1969.
225. **Palenius, H. P.**, *Phys. Scr.*, 1, 113, 1970.
226. **Palenius, H. P.**, *Univ. of Lund, Sweden*, unpublished.
227. **Paschen, F.**, *Ann. Phys.*, Series 5, 12, 509, 1932.
228. **Paschen, F. and Ritschl, R.**, *Ann. Phys.*, Series 5, 18, 867, 1933.
229. **Peck, E. R., Khanna, B. N., and Anderholm, N. C.** *J. Opt. Soc. Am.*, 52, 53, 1962.
230. **Persson, W.**, *Phys. Scr.*, 3, 133, 1971.
231. **Persson, W. and Valind S.**, *Phys. Scr.*, 5, 187, 1972.
232. **Petersson, B.**, *Ark. Fys.*, 27, 317, 1964.
233. **Phillips, L. W. and Parker, W. L.**, *Phys. Rev.*, 60, 301, 1941.
234. **Platt, J. R. and Sawyer, R. A.**, *Phys. Rev.*, 60, 866, 1941.
235. **Plyer, E. K., Blaine, L. R., and Tidwell, E.**, *J. Res. Natl. Bur. Stand.*, 55, 279, 1955.
236. **Poppe, R., van Kleef, T. A. M., and Raassen, A. J. J.**, *Physica*, 77, 165, 1974.
237. **Radziemski, L. J., Jr. and Andrew, K. L.**, *J. Opt. Soc. Am.*, 55, 474, 1965.
238. **Radziemski, L. J., Jr. and Kaufman, V.**, *J. Opt. Soc. Am.*, 59, 424, 1969.
239. **Radziemski, L. J., Jr. and Kaufman, V.**, *J. Opt. Soc. Am.*, 64, 366, 1974.
240. **Ramanadham, R. and Rao, K. R.**, *Indian J. Phys.*, 18, 317, 1944.
241. **Ramb, R.**, *Ann. Phys.*, 10, 311, 1931.
242. **Rank, D. H., Bennett, J. M., and Bennett, H. E.**, *J. Opt. Soc. Am.*, 40, 477, 1950.
243. **Rao, A. S. and Krishnamurty, S. G.**, *Proc. Phys. Soc. London*, 46, 531, 1943.
244. **Rao, K. R.**, *Proc. R. Soc. London, Ser. A*, 134, 604, 1932.
245. **Rao, K. R. and Badami, J. S.**, *Proc. R. Soc. London Ser. A*, 131, 154, 1931.
246. **Rao, K. R. and Krishnam[illegible]** *Proc. R. Soc. London Ser. A*, 161, 38, 1937.
247. **Rao, K. R. and Murti, S. G. K.**, *Proc. R. Soc. London Ser. A*, 145, 681, 1934.
248. **Rao, Y. B.**, *Indian J. Phys.*, 32, 497, 1958.
249. **Rao, Y. B.**, *Indian J. Phys.*, 33, 546, 1959.
250. **Rao, Y. B.**, *Indian J. Phys.*, 35, 386, 1961.
251. **Rasmussen, E.**, *Z. Phys.*, 80, 726, 1933.
252. **Rasmussen, E.**, *Z. Phys.*, 83, 404, 1933.
253. **Rasmussen, E.**, *Z. Phys.*, 86, 24, 1934.
254. **Rasmussen, E.**, *Z. Phys.*, 87, 607, 1934.
255. **Rasmussen, E.**, *Phys. Rev.*, 57, 840, 1940.
256. **Rau, A. S. and Narayan, A. L.**, *Z. Phys.*, 59, 687, 1930.
257. **Reader, J.**, *J. Opt. Soc. Am.*, 65, 286, 1975.
258. **Reader, J.**, *J. Opt. Soc. Am.*, 65, 988, 1975.
259. Reader, J., *J. Opt Soc. Am.*, 73, 349, 1983.
260. **Reader, J. and Davis, S.**, *J. Res. Natl. Bur. Stand. Sect. A*, 71, 587, 1967, and unpublished.
261. **Reader, J. and Ekberg, J. O.**, *J. Opt. Soc. Am.*, 62, 464, 1972.
262. **Reader, J. and Epstein, G. L.**, *J. Opt. Soc. Am.*, 62, 1467, 1972.
263. **Reader, J. and Epstein, G. L.**, *J. Opt. Soc. Am.*, 65, 638, 1975.
264. **Reader, J. and Epstein, G. L.**, *Natl. Bur. Stand.*, unpublished.
265. **Reader, J., Epstein, G. L., and Ekberg, J. O.**, *J. Opt. Soc. Am.*, 62, 273, 1972.
266. Kaufman, V., *Phys. Scr.*, 26, 439, 1982.
267. **Ricard, R., Givord, M., and George, F.**, *C. R. Acad. Sci. Paris*, 205, 1229, 1937.
268. **Risberg, P.**, *Ark. Fys.*, 10, 583, 1956.
269. **Risberg, G.**, *Ark. Fys.*, 28, 381, 1965.
270. **Risberg, G.**, *Ark. Fys.*, 37, 231, 1968.
271. **Robinson, H. A.**, *Phys. Rev.*, 49, 297, 1936.
272. **Robinson, H. A.**, *Phys. Rev.*, 50, 99, 1936.
273. **Ross, C. B., Jr.**, Doctoral dissertation, Purdue University, 1969.
274. **Ross, C. B., Wood, D. R., and Scholl, P. S.**, *J. Opt. Soc. Am.*, 66, 36, 1976.
275. **Ruedy, J. E. and Gibbs, R. C.**, *Phys. Rev.*, 46, 880, 1934.
276. **Russell, H. N., King, R. B., and Moore, C. E.**, *Phys. Rev.*, 58, 407, 1940.
277. **Russell, H. N. and Moore, C. E.**, *J. Res. Natl. Bur. Stand.*, 55, 299, 1955.
278. **Russell, H. N., Moore, C. E., and Weeks, D. W.**, *Trans. Am. Philos. Soc.*, 34(2), 111, 1944.
279. **Saunders, F., Schneider, E., and Buckingham, E.**, *Proc. Natl. Acad. Sci.*, 20, 291, 1934.
280. **Sawyer, R. A. and Humphreys, C. J.**, *Phys. Rev.*, 32, 583, 1928.
281. **Sawyer, R. A. and Lang, R. J.**, *Phys. Rev.*, 34, 712, 1929.
282. **Sawyer, R. A. and Paschen, F.** *Ann. Phys.*, 84(4), 1, 1927.
283. **Scholl, P. S.**, M.S. thesis, Wright State Univ., 1975.
284. **Schurmann, D.**, *Z. Phys.*, 17, 4, 1975.
285. **Seguier, J.**, *C. R. Acad. Sci. Paris*, 256, 1703, 1963.
286. **Shenstone, A. G.**, *Phys. Rev.*, 31, 317, 1928.
287. **Shenstone, A. G.**, *Phys. Rev.*, 32, 30, 1928.
288. **Shenstone, A. G.**, *Trans. R. Soc. London*, 237(A), 57, 1938.
289. **Shenstone, A. G.**, *Phys. Rev.*, 57, 894, 1940.
290. **Shenstone, A. G** *Philos. Trans. R. Soc. London Ser. A*, 241, 297, 1948.
291. **Shenstone, A. G.**, *Can. J. Phys.*, 38, 677, 1960.
292. **Shenstone, A. G.**, *Proc. R. Soc. London*, 261(A), 153, 1961.

293. **Shenstone, A. G.,** *Proc. R. Soc. London,* 276(A), 293, 1963.
294. **Shenstone, A. G.,** *J. Res. Natl. Bur. Stand. Sect. A,* 74, 801, 1970.
295. **Shenstone, A. G.,** *J. Res. Natl. Bur. Stand. Sect. A,* 79, 497, 1975.
296. **Shenstone, A. G. and Pittenger, J. T.,** *J. Opt. Soc. Am.,* 39, 219, 1949.
297. **Smith, S.,** *Phys. Rev.,* 36, 1, 1930.
298. **Smitt, R.,** *Phys. Scr.,* 8, 292, 1973.
299. **Soderqvist, J.,** *Ark. Mat. Astronom. Fys.,* 32(A), 1, 1946.
300. **Sommer, L. A.,** *Ann. Phys.,* 75, 163, 1924.
301. **Spector, N.,** *J. Opt. Soc. Am.,* 63, 358, 1973.
302. **Spector, N. and Sugar, J.,** *J. Opt. Soc. Am.,* 66, 436, 1976.
303. **Steinhaus, D. W., Radziemski, L. J., Jr., and Blaise, J.,** *Los Alamos Sci. Lab.,* unpublished, 1975.
304. **Subbaraya, T. S.,** *Z. Phys.,* 78, 541, 1932.
305. **Sugar, J.,** *J. Opt. Soc. Am.,* 55, 33, 1965.
306. **Sugar, J.,** *J. Res. Natl. Bur. Stand. Sect. A,* 73, 333, 1969.
307. **Sugar, J.,** *J. Opt. Soc. Am.,* 60, 454, 1970.
308. **Sugar, J.,** *J. Res. Natl. Bur. Stand. Sect. A,* 78, 555, 1974.
309. **Sugar, J. and Kaufman, V.,** *J. Opt. Soc. Am.,* 55, 1283, 1965.
310. **Sugar, J. and Kaufman, V.,** *J. Opt. Soc. Am.,* 62, 562, 1972.
311. **Sugar, J., Kaufman, V., and Spector, N.,** *J. Res. Natl. Bur. Stand.,* Sect. A, 83, 233, 1978.
312. **Sugar, J. and Spector, N.,** *J. Opt. Soc. Am.,* 64, 1484, 1974.
313. **Sullivan, F. J.** *Univ. Pittsburgh Bull.,* 35, 1, 1938.
314. **Svensson, L. A. and Ekberg, J. O.,** *Ark. Fys.,* 37, 65, 1968.
315. **Swensson, J. W. and Risberg, G.,** *Ark. Fys.,* 31, 237, 1966.
316. **Tech. J. L.,** *J. Res. Natl. Bur. Stand. Sect. A,* 67, 505, 1963.
317. **Tech, J. L. and Ward, J. F.,** *Phys. Rev. Lett.,* 27, 367, 1971.
318. **Tilford, S. G.,** *J. Opt. Soc. Am.,* 53, 1051, 1963.
319. **Toresson, Y. G.,** *Ark. Fys.,* 17, 179, 1960.
320. **Toresson, Y. G.,** *Ark. Fys.,* 18, 389, 1960.
321. **Toresson, Y. G. and Edlen, B.,** *Ark. Fys.,* 23, 117, 1963.
322. **Tsien, W. Z.,** *Chin. J. Phys.,* Peiping, 3, 117, 1939.
323. **van Deurzen, C. H. H., Conway, J., and Davis, S. P.,** *J. Opt. Soc. Am.,* 63, 158, 1973.
324. **van Kleef, T. A. M., Raassen, A. J. J., and Joshi, Y. N.,** *Physica,* 84(C), 401, 1976.
325. Sansonetti, C. J., Andrew, K. L., and Verges, J., *J. Opt. Soc. Am.,* 71, 423, 1981.
326. **Wheatley, M. A. and Sawyer, R. A.,** *Phys. Rev.,* 61, 591, 1942.
327. **Wilkinson, P. G.,** *J. Opt. Soc. Am.,* 45, 862, 1955.
328. **Wilkinson, P. G. and Andrew, K. L.,** *J. Opt. Soc. Am.,* 53, 710, 1963.
329. **Wood, D. and Andrew, K. L.,** *J. Opt. Soc. Am.,* 58, 818, 1968.
330. **Wood, D. R., Ross, C. B., Scholl, P. S., and Hoke, M.,** *J. Opt. Soc. Am.,* 64, 1159, 1974.
331. **Worden, E. F. and Conway, J. G.,** *Lawrence Livermore Lab.,* unpublished, 1977.
332. **Worden, E. F., Hulet, E. K., Gutmacher, R. G., Conway, J. G.,** *At. Data Nucl. Data Tables,* 18, 459, 1976.
333. **Worden, E. F., Lougheed, R. W., Gutmacher, R. G., and Conway, J. G.,** *J. Opt. Soc. Am.,* 64, 77, 1974.
334. **Wu, C. M.,** Ph.D. thesis, University of British Columbia, 1971.
335. **Zaidel, A. N., Prokofev, V. K., Raiskii, S. M., Slavnyi, V. A., and Schreider, E. Y.,** *Tables of Spectral Lines,* 3rd ed., Plenum, New York, 1970.
336. **Zetterberg, P. O. and Magnusson, C. E.,** *Phys. Scr.,* 15, 189, 1977.
337. **Sugar, J.,** *J. Opt. Soc. Am.,* 55, 1058, 1965.
338. **Sugar, J.,** *J. Opt. Soc. Am.,* 61, 727, 1971.
339. Worden, E. F., and Conway, J. G., *At. Data Nucl. Data Tables,* 22, 329, 1978.
340. **Kaufman, V. and Edlen, B.,** *J. Phys. Chem. Ref. Data,* 3, 825, 1974.
341. **Lang, R. J.,** *Phys. Rev.,* 34, 697, 1929.
342. **Ryabtsev, A. N.,** *Opt. Spectros.,* 39, 455, 1975.
343. **Foster, E. W.,** *Proc. R. Soc. London,* 200(A), 429, 1950.
344. **Morillon, C. and Verges, J.,** *Phys. Scr.,* 12, 129, 1975.
345. **Ruedy, J. E.,** *Phys. Rev.,* 41, 588, 1932.
346. **McLennan, J. C., McLay, A. B., and McLeod, J. H.,** *Philos. Mag.,* 4, 486, 1927.
347. **Handrup, M. B. and Mack, J. E.,** *Physica,* 30, 1245, 1964.
348. **Clearman, H. E.,** *J. Opt. Soc. Am.,* 42, 373, 1952.
349. **Paschen, F.,** *Ann. Physik,* 424, 148, 1938.
350. **Paschen, F. and Campbell, J. S.,** *Ann. Phys.,* 31(5), 29, 1938.
351. **Nodwell, R.,** *Univ. of British Columbia, Vancouver,* unpublished, 1955.
352. **Gibbs, R. C. and White, H. E.,** *Phys. Rev.,* 31, 776, 1928.
353. **Green, M.,** *Phys. Rev.,* 60, 117, 1941.
354. **Ellis, C. B. and Sawyer, R. A.,** *Phys. Rev.,* 49, 145, 1936.
355. **McLennan, J. C., McLay, A. B., and Crawford, M. F.,** *Proc. R. Soc. London Ser. A,* 125, 50, 1929.
356. **Mack, J. E. and Fromer, M.,** *Phys. Rev.,* 48, 346, 1935.
357. **Humphreys, C. J. and Paul, E.,** U.S. Nav. Ord. Lab., Navord Rep. 4589, 25, 1956.
358. **Walters, F. M.,** *Sci. Pap. Bur. Stand.,* 17, 161, 1921.
359. **Crawford, M. F. and McLay, A. B. ,** *Proc. R. Soc. London Ser. A,* 143, 540, 1934.
360. **McLay, A. B. and Crawford, M. F.,** *Phys. Rev.,* 44, 986, 1933.
361. **Schoepfle, G. K.,** *Phys. Rev.,* 47, 232, 1935.
362. Acquista, N., and Reader, J., *J. Opt. Soc. Am.,* 70, 789, 1980.
363. Benschop, H., Joshi, Y. N., and Van Kleef, T. A. M., *Can. J. Phys.,* 53, 498, 1975.
364. Bockasten, K., Hallin, R., and Hughes, T. P., *Proc. Phys. Soc.,* 81, 522, 1963.
365. Boyce, J. C., *Phys. Rev.,* 46, 378, 1934.
366. Boyce, J. C., *Phys. Rev.,* 47, 718, 1935.
367. Boyce, J. C., *Phys. Rev.,* 48, 396, 1935.
368. Boyce, J. C., *Phys. Rev.,* 49, 351, 1936.
369. Corliss, C. H. and Meggers, W. F., *J. Res. Natl. Bur. Stand.,* 61, 269, 1958.
370. Crooker, A. M. and Dick, K. A., *Can. J. Phys.,* 42, 766, 1964.
371. De Bruin, T. L., *Z. Physik,* 77, 505, 1932.
372. De Bruin, T. L., *Proc. Roy. Acad. Amsterdam,* 36, 727, 1933.
373. De Bruin, T. L., *Zeeman Verhandelingen,* (The Hague), 1935, p. 415.
374. De Bruin, T. L., *Physica,* 3, 809, 1936.
375. De Bruin, T. L., *Proc. Roy. Acad. Amsterdam,* 40, 339, 1937.

376. Dick, K. A., *Can. J. Phys.*, 46, 1291, 1968.
377. Dick, K. A., unpublished, 1978.
378. Edlen, B. and Swensson, J. W., *Phys. Scr.*, 12, 21, 1975.
379. Ekberg, J. O., *Phys. Scr.*, 7, 55, 1973.
380. Ekberg, J. O., *Phys. Scr.*, 7, 59, 1973.
381. Ekberg, J. O., *Phys. Scr.*, 12, 42, 1975.
382. Ekberg, J. O. and Edlen, B., *Phys. Scr.*, 18, 107, 1978.
383. Eliason, A. Y., *Phys. Rev.*, 43, 745, 1933.
384. Gallardo, M., Massone, C. A., Tagliaferri, A. A., Garavaglia, M., and Persson, W., *Phys. Scr.*, to be published, 1979.
385. Garcia-Riquelme, O., *Optica Pura Y Aplicada*, 1, 53, 1968.
386. Gibbs, R. C., Vieweg, A. M., and Gartlein, C. W., *Phys. Rev.*, 34, 406, 1929.
387. Gilbert, W. P., *Phys. Rev.*, 48, 338, 1935.
388. Goldsmith, S. and Kaufman, A. S., *Proc. Phys. Soc.*, 81, 544, 1963.
389. Hermansdorfer, H., *J. Opt. Soc. Am.*, 62, 1149, 1972.
390. Humphreys, C. J., *Phys. Rev.*, 47, 712, 1935.
391. Humphreys, C. J., *J. Res. Natl. Bur. Stand.*, 16, 639, 1936.
392. Iglesias, L., *J. Opt. Soc. Am.*, 45, 856, 1955.
393. Iglesias, L., *J. Res. Natl. Bur. Stand.*, 64A, 481, 1960.
394. Iglesias, L., *Anales Fisica Y Quimica*, 58A, 191, 1962.
395. Iglesias, L., *J. Res. Natl. Bur. Stand.*, 70A, 465, 1966.
396. Iglesias, L., *Can. J. Phys.*, 44, 895, 1966.
397. Iglesias, L., *J. Res. Natl. Bur. Stand.*, 72A, 295, 1968.
398. Joshi, Y. N., *Can. Spectrosc.*, 15, 96, 1970.
399. Joshi, Y. N. and Van Kleef, T. A. M., *Can. J. Phys.*, 55, 714, 1977.
400. Kaufman, A. S., Hughes, T. P., and Williams, R. V., *Proc. Phys. Soc.*, 76, 17, 1960.
401. Kaufman, V. and Sugar, J., *J. Opt. Soc. Am.*, 68, 1529, 1978.
402. Keussler, V., *Z. Physik*, 85, 1, 1933.
403. Kiess, C. C., *J. Res. Natl. Bur. Stand.*, 56, 167, 1956.
404. Klinkenberg, P. F. A., Van Kleef, T. A. M., and Noorman, P. E., *Physica*, 27, 1177, 1961.
405. Kovalev, V. I., Romanos, A. A., and Ryabtsev, A. N., *Opt. Spectrosc.*, 43, 10, 1977.
406. Lang, R. J., *Proc. Natl. Acad. Sci.*, 13, 341, 1927.
407. Lang, R. J., *Zeeman Verhandelingen*, (The Hague), 44, 1935.
408. Liberman, S., et al., *C. R. Acad. Sci. (Paris)*, 286, 253, 1978.
409. Livingston, A. E., *J. Phys.*, B9, L215, 1976.
410. Meijer, F. G., *Physica*, 72, 431, 1974.
411. Meijer, F. G. and Metsch, B. C., *Physica*, 94C, 259, 1978.
412. Moore, F. L., thesis, Princeton, 1949.
413. Paul, F. W. and Polster, H. D., *Phys. Rev.*, 59, 424, 1941.
414. Phillips, L. W. and Parker, W. L., *Phys. Rev.*, 60, 301, 1941.
415. Poppe, R., *Physica*, 81C, 351, 1976.
416. Raassen, A. J. J., Van Kleef, T. A. M., and Metsch, B. C., *Physica*, 84C, 133, 1976.
417. Rao, A. B. and Krishnamurty, S. G., *Proc. Phys. Soc. (London)*, 51, 772, 1939.
418. Reader, J. and Acquista, N., *J. Opt. Soc. Am.*, 69, 239, 1979.
419. Reader, J. and Epstein, G. L., *J. Opt. Soc. Am.*, 62, 619, 1972.
420. Rico, F. R., *Anales, Real Soc. Esp. Fis. Quim.*, 61, 103, 1965.
421. Schonheit, E., *Optik*, 23, 409, 1966.
422. Shenstone, A. G., *J. Opt. Soc. Am.*, 44, 749, 1954.
423. Shenstone, A. G., unpublished, 1958.
424. Shenstone, A. G., *J. Res. Natl. Bur. Stand.*, 67A, 87, 1963.
425. Sugar, J. and Kaufman, V., *J. Opt. Soc. Am.*, 64, 1656, 1974.
426. Sugar, J. and Kaufman, V., *Phys. Rev.*, C12, 1336, 1975.
427. Svensson, L. A., *Phys. Scr.*, 13, 235, 1976.
428. Swensson, J. W. and Edlen, B., *Phys. Scr.*, 9, 335, 1974.
429. Tagliaferri, A. A., Gallego Lluesma, E., Garavaglia, M., Gallardo, M., and Massone, C. A., *Optica Pura Y Aplica*, 7, 89.
430. Tilford, S. G. and Giddings, L. E., *Astrophys. J.*, 141, 1222, 1965.
431. Trawick, M. W., *Phys. Rev.*, 46, 63, 1934.
432. Van Deurzen, C. H. H., *J. Opt. Soc. Am.*, 67, 476, 1977.
433. Yarosewick, S. L. and Moore, F. L., *J. Opt. Soc. Am.*, 57, 1381, 1967.
434. Zalubas, R., unpublished, 1979.
435. Bhatia, K. S., Jones, W. E., and Crooker, A. M., *Can. J. Phys.*, 50, 2421, 1972.
436. van Kleef, A. M. and Joshi, Y. N., *Phys. Scr.*, 24, 557, 1981.

ATOMIC TRANSITION PROBABILITIES

J. R. Fuhr and W. L. Wiese

These tables are a substantial update and enlargement of the earlier tables by W. L. Wiese and G. A. Martin in this Handbook. The new tables contain critically evaluated atomic transition probabilities for about 8300 selected lines of all elements for which reliable data are available on an absolute scale. The material is largely for neutral and singly ionized spectra, but also includes a number of prominent lines of more highly charged ions of important elements.

Many of the data are obtained from comprehensive compilations of the Data Center on Atomic Transition Probabilities at the National Institute of Standards and Technology (formerly the National Bureau of Standards). Specifically, data have been taken from two recent comprehensive critical compilations on Sc through Mn[1] and Fe through Ni.[2] Material from earlier compilations for the elements H through Ne[3] and Na through Ca[4] was supplemented by more recent material taken directly from the original literature. For the highly charged ions, some of the data were derived from studies of the systematic behavior of transition probabilities.[5—7] Most of the original literature is cited in the above tables and in recent bibliographies;[8,9] for lack of space, individual literature references are not cited here.

The wavelength range for the neutral species is normally the visible spectrum or shorter wavelengths; only the very prominent near infrared lines are included. For the higher ions, most of the strong lines are located in the far UV. The tabulation is limited to electric dipole — including intercombination — lines and comprises essentially the fairly strong transitions with estimated uncertainties of 50% or less. With the exception of hydrogen, helium, and the alkalis, most transitions are between states with low principal quantum numbers.

The transition probability, A, is given in units of 10^8 s^{-1} and is listed to as many digits as is consistent with the indicated accuracy. The power of 10 is indicated by the E notation (i.e., E-02 means 10^{-2}). Generally, the estimated uncertainties of the A-values are ±25 to 50% for two-digit numbers and ±10 to 25% for three-digit numbers.

Each transition is identified by the wavelength, λ, in angstroms; and the statistical weights, g_i and g_k, of the lower (i) and upper (k) states [the product $g_k A$ (or $g_i f$) is needed for many applications]. Whenever the wavelengths of individual lines within a multiplet are extremely close, only an average wavelength for the multiplet as well as the multiplet A-value are given, and this is indicated by an asterisk (*) to the left of the wavelength. This also has been done when the transition probability for an entire multiplet has been taken from the literature and values for individual lines cannot be determined because of insufficient knowledge of the coupling of electrons. The wavelength data have been taken either from recent compilations or from the original literature cited in bibliographies published by the Atomic Energy Levels Data Center[10,11] at the National Institute of Standards and Technology (formerly the National Bureau of Standards). Wavelength values are consistent with those given in the table "Line Spectra of the Elements", which appears elsewhere in this Handbook.

The transition probabilities for hydrogen and hydrogen-like ions are known very precisely. Because of the hydrogen degeneracy, a "transition" is actually the sum of all fine-structure transitions between the principal quantum numbers listed in the transition column; therefore, the special hydrogen table which appears below gives weighted average A-values.

In addition to the transition probability A, the atomic oscillator strength f and the line strength S are often used in the literature. The conversion factors between these quantities are (for electric-dipole transitions):

$$g_i f = 1.499 \times 10^{-8} \lambda^2 g_k A = 303.8\, \lambda^{-1} S$$

where λ is in angstroms, A is in 10^8 s^{-1}, and S is in atomic units, which are $a_0^2 e^2 = 7.188 \times 10^{-59}$ m^2 C^2.

After the special table for hydrogen, the tables for other elements appear in alphabetical sequence by element name (not symbol). Within each element, the tables are ordered by increasing ionization stage (e.g., Al I, Al II, etc.).

We acknowledge the valuable preparatory work by Theresa Deters and David Worthington for this compilation.

REFERENCES

1. G. A. Martin, J. R. Fuhr, and W. L. Wiese, Atomic transition probabilities — scandium through manganese, *J. Phys. Chem. Ref. Data*, 17, *Suppl.* 3, 1988.
2. J. R. Fuhr, G. A. Martin, and W. L. Wiese, Atomic transition probabilities — iron through nickel, *J. Phys. Chem. Ref. Data*, 17, *Suppl.* 4, 1988.
3. W. L. Wiese, M. W. Smith, and B. M. Glennon, Atomic Transition Probabilities (H through Ne — A Critical Data Compilation), National Standard Reference Data Series, National Bureau of Standards 4, Vol. I, U.S. Government Printing Office, Washington, D. C., 1966.
4. W. L. Wiese, M. W. Smith, and B. M. Miles, Atomic Transition Probabilities (Na through Ca — A Critical Data Compilation), National Standard Reference Data Series, National Bureau of Standards 22, Vol. II, U. S. Government Printing Office, Washington, D. C., 1969.
5. W. L. Wiese and A. W. Weiss, *Phys. Rev.*, 175, 50, 1968.
6. M. W. Smith and W. L. Wiese, *Astrophys. J., Suppl. Ser.* 23, No. 196, 103, 1971.
7. G. A. Martin and W. L. Wiese, *J. Phys. Chem. Ref. Data*, 5, 537, 1976.
8. J. R. Fuhr, B. J. Miller, and G. A. Martin, Bibliography on Atomic Transition Probabilities (1914 through October 1977), National Bureau of Standards Special Publication 505, 1978; B. J. Miller, J. R. Fuhr, and G. A. Martin, Bibliography on Atomic Transition Probabilities (November 1977 through February 1980), National Bureau of Standards Special Publication 505, Supplement 1, 1980.
9. W. L. Wiese, Reports on astronomy, *Trans. Int. Astron. Union*, 18A, 116—123, 1982; 19A, 122—138, 1985; 20A, 117—123, 1988, D. Reidel, Kluwer, Dordrecht, Holland.
10. C. E., Moore, Bibliography on the Analyses of Optical Atomic Spectra, National Bureau of Standards Special Publication 306 — Section 1, 1968; Sections 2—4, 1969.

11. L. Hagan, and W. C. Martin, Bibliography on Atomic Energy Levels and Spectra (July 1968 through June 1971), National Bureau of Standards Special Publication 363, 1972; L. Hagan, Bibliography on Atomic Energy Levels and Spectra (July 1971 through June 1975), National Bureau of Standards Special Publication 363, Supplement 1, 1977; R. Zalubas and A. Albright, Bibliography on Atomic Energy Levels and Spectra (July 1975 through June 1979), National Bureau of Standards Special Publication 363, Supplement 2, 1980; A. Musgrove and R. Zalubas, Bibliography on Atomic Energy Levels and Spectra (July 1979 through December 1983), National Bureau of Standards Special Publication 363, Supplement 3, 1985.
12. S. M. Younger and A. W. Weiss, *J. Res. Natl. Bur. Stand.*, 79A, 629, 1975.

Transition Probabilities for Allowed Lines of Hydrogen

λ Å	Weights g_i	g_k	A 10^8 s^{-1}	λ Å	Weights g_i	g_k	A 10^8 s^{-1}	λ Å	Weights g_i	g_k	A 10^8 s^{-1}
Hydrogen				**Hydrogen**				**Hydrogen**			
H I				**H I**				**H I**			
912.768	2	1800	5.167E-06	3666.10	8	1458	4.826E-06	8665.02	18	338	1.343E-04
912.839	2	1682	6.122E-06	3667.68	8	1352	5.830E-06	8750.48	18	288	2.021E-04
912.918	2	1568	7.297E-06	3669.46	8	1250	7.096E-06	8862.79	18	242	3.156E-04
913.006	2	1458	8.753E-06	3671.48	8	1152	8.707E-06	9014.91	18	200	5.156E-04
913.104	2	1352	1.057E-05	3673.76	8	1058	1.078E-05	9229.02	18	162	8.905E-04
913.215	2	1250	1.286E-05	3676.36	8	968	1.347E-05	9545.97	18	128	1.651E-03
913.339	2	1152	1.578E-05	3679.35	8	882	1.700E-05	10049.4	18	98	3.358E-03
913.480	2	1058	1.952E-05	3682.81	8	800	2.172E-05	10938.1	18	72	7.783E-03
913.641	2	968	2.438E-05	3686.83	8	722	2.809E-05	12818.1	18	50	2.201E-02
913.826	2	882	3.077E-05	3691.55	8	648	3.685E-05	16407.2	32	288	1.620E-04
914.039	2	800	3.928E-05	3697.15	8	578	4.910E-05	16806.5	32	242	2.556E-04
914.286	2	722	5.077E-05	3703.85	8	512	6.658E-05	17362.1	32	200	4.235E-04
914.576	2	648	6.654E-05	3711.97	8	450	9.210E-05	18174.1	32	162	7.459E-04
914.919	2	578	8.858E-05	3721.94	8	392	1.303E-04	18751.0	18	32	8.986E-02
915.329	2	512	1.200E-04	3734.37	8	338	1.893E-04	19445.6	32	128	1.424E-03
915.824	2	450	1.657E-04	3750.15	8	288	2.834E-04	21655.3	32	98	3.041E-03
916.429	2	392	2.341E-04	3770.63	8	242	4.397E-04	26251.5	32	72	7.711E-03
917.181	2	338	3.393E-04	3797.90	8	200	7.122E-04	27575	50	288	1.402E-04
918.129	2	288	5.066E-04	3835.38	8	162	1.216E-03	28722	50	242	2.246E-04
919.351	2	242	7.834E-04	3889.05	8	128	2.215E-03	30384	50	200	3.800E-04
920.963	2	200	1.263E-03	3970.07	8	98	4.389E-03	32961	50	162	6.908E-04
923.150	2	162	2.143E-03	4101.73	8	72	9.732E-03	37395	50	128	1.388E-03
926.226	2	128	3.869E-03	4340.46	8	50	2.530E-02	40511.5	32	50	2.699E-02
930.748	2	98	7.568E-03	4861.32	8	32	8.419E-02	43753	72	288	1.288E-04
937.803	2	72	1.644E-02	6562.80	8	18	4.410E-01	46525	50	98	3.253E-03
949.743	2	50	4.125E-02	8392.40	18	800	1.517E-05	46712	72	242	2.110E-04
972.537	2	32	1.278E-01	8413.32	18	722	1.964E-05	51273	72	200	3.688E-04
1025.72	2	18	5.575E-01	8437.96	18	648	2.580E-05	59066	72	162	7.065E-04
1215.67	2	8	4.699E+00	8467.26	18	578	3.444E-05	74578	50	72	1.025E-02
3662.26	8	1800	2.847E-06	8502.49	18	512	4.680E-05	75004	72	128	1.561E-03
3663.40	8	1682	3.374E-06	8545.39	18	450	6.490E-05	123680	72	98	4.561E-03
3664.68	8	1568	4.022E-06	8598.40	18	392	9.211E-05				

For hydrogen-like ions of nuclear charge Z, the following scaling laws hold:

$$A_Z = Z^4 A_{\text{Hydrogen}};\ f_Z = f_H;\ S_Z = Z^{-2} S_{\text{H}}$$
$$\text{(For wavelengths, } \lambda_Z = Z^{-2}\lambda_{\text{H}})$$

For very highly charged hydrogen-like ions, starting at about $Z > 25$, relativistic corrections[12] must be applied.

Transition Probabilities for Other Elements*

λ Å	Weights g_i	Weights g_k	A 10^8 s^{-1}
Aluminum			
Al I			
2263.5	2	4	6.6E-01
2269.1	4	6	7.9E-01
2269.2	4	4	1.3E-01
2367.1	2	4	7.2E-01
2373.1	4	6	8.6E-01
2373.4	4	4	1.4E-01
2568.0	2	4	2.3E-01
2575.1	4	6	2.8E-01
2575.4	4	4	4.4E-02
2652.5	2	2	1.33E-01
2660.4	4	2	2.64E-01
3082.2	2	4	6.3E-01
3092.7	4	6	7.4E-01
3092.8	4	4	1.2E-01
3944.0	2	2	4.93E-01
3961.5	4	2	9.8E-01
6696.0	2	4	1.69E-02
6698.7	2	2	1.69E-02
7835.3	4	6	5.7E-02
7836.1	6	8	6.2E-02
Al II			
1047.9	1	3	3.6E-01
1048.6	3	5	4.8E-01
1539.8	3	5	8.8E+00
1670.8	1	3	1.46E+01
1719.4	1	3	6.79E+00
1764.0	5	5	9.8E+00
1772.8	1	3	9.5E+00
1777.0	5	7	1.7E+01
*1819.0	15	15	5.6E+00
1855.9	1	3	8.32E-01
1858.0	3	3	2.48E+00
1862.3	5	3	4.12E+00
1931.0	3	1	1.08E+01
1990.5	3	5	1.47E+01
2816.2	3	1	3.83E+00
4663.1	5	3	5.3E-01
6226.2	1	3	6.2E-01
6231.8	3	5	8.4E-01
6243.4	5	7	1.1E+00
6335.7	5	3	1.4E-01
6823.4	3	3	3.4E-01
6837.1	5	3	5.7E-01
6920.3	3	1	9.6E-01
7042.1	3	5	5.9E-01
7056.7	3	3	5.8E-01
7471.4	5	7	9.4E-01
Al III			
*560.36	2	6	4.0E-01
695.83	2	4	7.4E-01
696.22	2	2	7.2E-01
*1352.8	10	14	4.40E+00
1379.7	2	2	4.59E+00
1384.1	4	2	9.1E+00
1605.8	2	4	1.22E+01
1611.8	4	4	2.42E+00
1611.9	4	6	1.45E+01
1854.7	2	4	5.40E+00
1862.8	2	2	5.33E+00
*1935.9	10	14	1.22E+01
3601.6	6	4	1.34E+00
3601.9	4	4	1.49E-01
3612.4	4	2	1.5E+00
Al X			
39.925	1	3	2.22E+03
51.979	1	3	4.8E+03
55.227	1	3	5.2E+03
55.272	3	5	7.2E+03
55.376	5	7	9.5E+03
59.107	3	5	4.6E+03
332.78	1	3	5.6E+01
394.83	3	1	8.3E+01
395.36	3	5	1.2E+01
397.76	1	3	1.7E+01
400.43	3	3	1.3E+01
401.12	5	5	3.6E+01
403.55	3	1	4.9E+01
406.31	5	3	1.9E+01
670.06	3	5	9.8E+00
2535	1	3	3.8E-01
Al XI			
*36.675	2	6	1.5E+03
39.091	2	4	2.6E+03
39.180	4	6	3.1E+03
39.530	2	2	1.8E+02
39.623	4	2	3.7E+02
48.298	2	4	3.09E+03
48.338	2	2	3.08E+03
52.299	2	4	8.1E+03
52.446	4	6	9.6E+03
52.458	4	4	1.6E+03
54.217	2	2	4.8E+02
54.388	4	2	9.6E+02
*99.083	2	6	2.2E+02
103.6	2	4	4.2E+02
103.8	4	6	5.0E+02
*141.6	2	6	4.07E+02
150.31	2	4	8.5E+02
150.61	4	6	9.9E+02
157.0	2	2	1.3E+02
157.4	4	2	2.6E+02
*205.0	2	6	6.3E+01
*308.6	2	6	9.9E+01
*341.3	6	2	1.3E+02
550.05	2	4	8.55E+00
568.12	2	2	7.73E+00
1997	2	4	1.07E+00
2069	2	2	9.7E-01
*4761	2	6	2.55E-01
5172	2	4	3.95E-02
5551	4	6	3.85E-02
5687	4	4	6.0E-03
Argon			
Ar I			
3406.18	3	1	3.9E-03
3461.08	3	5	6.7E-04
3554.30	5	5	2.7E-03
3563.29	1	3	1.2E-03
3567.66	5	7	1.1E-03
3572.30	3	1	5.1E-03
3606.52	3	1	7.6E-03
3632.68	3	5	6.6E-04
3634.46	3	3	1.3E-03
3643.12	3	5	2.4E-04
3649.83	3	1	8.0E-03
3659.53	3	3	4.4E-04
3670.67	3	5	3.1E-04
3675.23	3	3	4.9E-04
3770.37	1	3	7.0E-04
3834.68	3	1	7.5E-03
3894.66	3	3	5.7E-04
3947.50	5	5	5.6E-04
3948.98	5	3	4.55E-03
4044.42	3	5	3.33E-03
4045.96	3	3	4.1E-04
4054.53	3	3	2.7E-04
4158.59	5	5	1.40E-02
4164.18	5	3	2.88E-03
4181.88	1	3	5.61E-03
4190.71	5	5	2.80E-03
4191.03	1	3	5.39E-03
4198.32	3	1	2.57E-02
4200.67	5	7	9.67E-03
4251.18	5	3	1.11E-03
4259.36	3	1	3.98E-02
4266.29	3	5	3.12E-03
4272.17	3	3	7.97E-03
4300.10	3	5	3.77E-03
4333.56	3	5	5.68E-03
4335.34	3	3	3.87E-03
4345.17	3	3	2.97E-03
4363.79	3	3	1.2E-04
4424.00	1	3	7.3E-05
4510.73	3	1	1.18E-02
4522.32	1	3	8.98E-04
4544.75	3	3	8.3E-04
4554.32	3	5	3.8E-04
4584.96	3	5	1.6E-03
4586.61	3	3	2.3E-03
4587.21	3	1	4.9E-03
4589.29	3	5	6.2E-05
4596.10	3	3	9.47E-04
4628.44	3	5	3.83E-04
4642.15	3	5	9.6E-04
4647.49	3	3	1.2E-03
4702.32	3	3	1.09E-03
4746.82	3	1	3.6E-03

λ Å	Weights g_i	g_k	A 10^8 s^{-1}	λ Å	Weights g_i	g_k	A 10^8 s^{-1}	λ Å	Weights g_i	g_k	A 10^8 s^{-1}
4752.94	3	3	4.5E-03	5524.96	7	7	1.7E-03	6085.86	3	3	9.0E-05
4768.68	3	5	8.6E-03	5528.97	1	3	1.2E-03	6090.79	1	3	3.0E-03
4798.74	7	9	8.8E-04	5534.49	5	3	2.7E-03	6098.81	3	3	5.2E-03
4835.97	7	9	9.3E-04	5540.87	7	5	4.1E-04	6101.16	3	3	3.3E-03
4836.70	3	5	1.02E-03	5552.77	3	3	7.9E-04	6104.58	3	1	3.4E-03
4876.26	3	5	7.8E-03	5558.70	3	5	1.42E-02	6105.64	3	5	1.21E-02
4886.29	7	9	1.2E-03	5559.66	3	5	2.2E-03	6113.46	3	5	4.7E-04
4887.95	3	3	1.3E-02	5572.54	5	7	6.6E-03	6119.66	3	3	5.1E-04
4894.69	3	1	1.8E-02	5574.22	3	5	4.6E-04	6121.86	3	5	1.3E-04
4921.04	5	7	5.9E-04	5581.87	7	5	5.6E-04	6127.42	5	3	1.1E-03
4937.72	7	5	3.6E-04	5588.72	5	5	1.5E-03	6128.73	3	5	8.6E-04
4956.75	7	9	1.8E-03	5597.48	5	7	4.2E-03	6145.44	5	7	7.6E-03
4989.95	5	7	1.1E-03	5606.73	3	3	2.20E-02	6155.24	5	3	5.1E-03
5032.03	7	5	8.2E-04	5618.01	3	3	2.1E-03	6165.12	5	5	9.89E-04
5048.81	3	5	4.6E-03	5620.92	3	1	3.6E-03	6170.17	5	5	5.0E-03
5054.18	3	3	4.5E-03	5623.78	5	5	1.4E-03	6173.10	3	5	6.7E-03
5056.53	3	1	5.7E-03	5635.58	3	5	9.6E-04	6179.41	5	3	6.6E-04
5060.08	7	9	3.7E-03	5637.33	1	3	9.1E-04	6212.50	5	7	3.9E-03
5070.99	5	3	2.6E-03	5639.12	1	3	2.1E-03	6215.94	5	5	5.7E-03
5073.08	3	5	5.9E-04	5641.39	3	5	8.7E-04	6230.93	5	5	1.2E-04
5078.03	7	7	4.7E-04	5648.69	5	3	1.2E-03	6243.40	3	1	1.3E-03
5087.09	5	7	1.6E-03	5650.70	3	1	3.20E-02	6244.73	3	5	2.0E-04
5104.74	3	5	8.7E-04	5659.13	5	5	2.6E-03	6248.41	3	5	6.8E-04
5118.21	5	7	2.7E-03	5681.90	5	7	2.0E-03	6278.63	5	7	2.0E-04
5127.80	5	5	3.3E-04	5683.73	5	5	2.0E-03	6296.87	3	5	9.0E-03
5151.39	3	1	2.39E-02	5700.87	5	7	5.9E-03	6307.66	5	5	6.0E-03
5152.30	3	5	1.1E-03	5712.51	1	3	8.7E-04	6309.14	3	3	7.6E-04
5162.29	3	3	1.90E-02	5739.52	3	5	8.7E-03	6364.89	3	1	5.6E-03
5177.54	7	5	2.4E-03	5772.11	5	7	2.0E-03	6369.58	5	3	4.2E-03
5192.72	7	7	1.2E-04	5773.99	5	5	1.1E-03	6384.72	3	3	4.21E-03
5194.02	3	1	7.8E 03	5783.54	3	5	8.1E-04	6416.31	3	5	1.16E-02
5210.49	7	7	1.1E-03	5789.48	5	5	4.6E-04	6431.56	5	3	5.1E-04
5214.77	5	3	2.1E-03	5790.40	5	3	3.4E-04	6466.55	1	3	1.5E-03
5216.28	5	3	1.3E-03	5802.08	5	3	4.2E-03	6481.14	1	3	9.4E-04
5221.27	7	9	8.8E-03	5843.77	3	5	3.3E-04	6513.85	3	3	5.4E-04
5241.09	5	5	1.3E-03	5882.62	3	1	1.23E-02	6538.11	7	7	1.1E-03
5246.24	5	7	1.2E-03	5888.58	7	5	1.29E-02	6596.12	7	5	2.3E-04
5249.20	5	5	7.9E-04	5916.58	5	3	5.9E-04	6598.68	5	5	3.6E-04
5252.79	5	7	5.4E-03	5927.11	7	7	3.7E-04	6604.02	7	5	2.8E-03
5254.47	3	5	3.6E-03	5928.81	5	3	1.1E-02	6604.85	5	7	1.3E-04
5286.07	5	7	9.6E-04	5940.86	1	3	1.2E-03	6632.09	3	3	5.3E-04
5290.00	5	3	9.0E-04	5942.67	5	5	1.8E-03	6656.88	3	3	3.1E-04
5309.52	5	5	1.2E-03	5943.89	7	5	3.6E-04	6660.68	3	1	7.8E-03
5317.73	5	7	2.6E-03	5949.26	3	3	1.5E-03	6664.05	5	5	1.5E-03
5373.50	3	5	2.7E-03	5964.48	1	3	7.7E-04	6677.28	3	1	2.36E-03
5393.27	5	5	9.6E-04	5968.32	3	3	1.8E-03	6684.73	3	5	3.9E-04
5410.48	5	7	2.0E-03	5971.60	3	1	1.1E-02	6698.47	3	3	2.5E-04
5421.35	7	5	6.0E-03	5981.90	5	7	1.2E-04	6698.88	5	3	1.6E-03
5439.99	3	3	1.9E-03	5987.30	7	7	1.2E-03	6719.22	1	3	2.4E-03
5442.24	7	7	9.3E-04	5988.13	3	5	6.1E-04	6722.88	5	7	3.2E-04
5451.65	3	5	4.7E-03	5994.66	3	5	2.6E-04	6752.84	3	5	1.93E-02
5457.42	5	3	3.6E-03	5999.00	5	5	1.4E-03	6754.37	3	3	2.1E-03
5459.65	7	7	3.8E-04	6005.73	5	3	1.4E-03	6756.10	5	5	3.6E-03
5467.16	5	5	7.6E-04	6013.68	7	5	1.4E-03	6766.61	5	3	4.0E-03
5473.46	5	3	2.0E-03	6025.15	5	3	9.0E-03	6779.93	1	3	1.21E-03
5490.12	5	5	8.5E-04	6043.22	5	7	1.47E-02	6818.29	3	1	2.0E-03
5492.09	3	1	5.6E-03	6052.73	3	5	1.9E-03	6827.25	5	3	2.4E-03
5495.87	7	9	1.69E-02	6064.76	5	7	5.8E-04	6851.88	3	5	6.7E-04
5506.11	5	7	3.6E-03	6081.25	3	3	7.5E-04	6871.29	3	3	2.78E-02

λ Å	Weights g_i	Weights g_k	A 10^8 s^{-1}
6879.59	3	5	1.8E-03
6887.10	5	7	1.3E-03
6888.17	3	5	2.5E-03
6925.01	3	3	1.2E-03
6937.67	3	1	3.08E-02
6951.46	5	5	2.2E-03
6960.23	5	5	2.4E-03
6965.43	5	3	6.39E-02
6992.17	3	1	7.5E-03
7030.25	7	5	2.67E-02
7067.22	5	5	3.80E-02
7068.73	5	3	2.0E-02
7086.70	1	3	1.5E-03
7107.48	5	5	4.5E-03
7125.83	3	3	6.0E-03
7147.04	5	3	6.25E-03
7158.83	3	1	2.1E-02
7162.57	1	3	5.8E-04
7206.98	5	3	2.48E-02
7229.93	5	5	6.6E-04
7265.17	3	3	1.7E-03
7270.66	7	7	1.1E-03
7272.93	3	3	1.83E-02
7285.44	5	3	1.2E-03
7311.72	3	3	1.7E-02
7316.01	3	3	9.6E-03
7350.78	3	1	1.2E-02
7353.32	5	7	9.6E-03
7372.12	7	9	1.9E-02
7383.98	3	5	8.47E-02
7392.97	5	3	7.2E-03
7412.33	3	5	3.9E-03
7422.26	3	5	6.6E-04
7425.29	5	7	3.1E-03
7435.33	5	5	9.0E-03
7436.25	7	5	2.7E-03
7471.17	3	3	2.2E-04
7484.24	3	5	3.4E-03
7503.84	3	1	4.45E-01
7510.42	5	5	4.5E-03
7514.65	3	1	4.02E-01
7618.33	3	5	2.9E-03
7628.86	3	5	2.9E-03
7635.11	5	5	2.45E-01
7670.04	5	3	2.8E-03
7704.81	5	7	6.3E-04
7723.76	5	3	5.18E-02
7724.21	1	3	1.17E-01
7798.55	3	5	8.7E-04
7868.20	1	3	3.50E-03
7891.08	5	5	9.5E-03
7916.45	3	3	1.2E-03
7948.18	1	3	1.86E-01
8006.16	3	5	4.90E-02
8014.79	5	5	9.28E-02
8037.23	1	3	3.59E-03
8046.13	3	1	1.12E-02
8053.31	5	3	8.6E-03
8066.60	5	5	1.4E-03
8103.69	3	3	2.5E-01
8115.31	5	7	3.31E-01
8264.52	3	3	1.53E-01
8384.73	5	7	2.4E-03
8408.21	3	5	2.23E-01
8424.65	3	5	2.15E-01
8490.30	3	5	9.6E-04
8521.44	3	3	1.39E-01
8605.78	5	5	1.04E-02
8620.46	1	3	9.2E-03
8667.94	1	3	2.43E-02
8761.69	3	5	9.5E-03
8784.61	3	1	2.4E-03
8799.08	5	3	4.6E-03
8962.19	3	3	1.6E-03
9075.42	3	1	1.2E-02
9122.97	5	3	1.89E-01
9194.64	3	3	1.76E-02
9224.50	3	5	5.03E-02
9291.53	3	1	3.26E-02
9354.22	3	3	1.06E-02
9657.78	3	3	5.43E-02
9784.50	3	5	1.47E-02
10470.05	1	3	9.8E-03
10478.0	3	3	2.44E-02
10950.7	5	3	3.96E-03
11078.9	5	5	8.3E-03
11393.7	3	1	2.22E-02
11441.8	5	3	1.39E-02
11467.5	3	5	3.69E-03
11488.11	3	3	1.9E-03
11668.7	5	5	3.76E-02
11719.5	5	3	9.52E-03
12026.6	1	3	4.2E-03
12112.2	7	7	3.1E-02
12139.8	3	3	4.5E-02
12343.7	5	7	2.0E-02
12402.9	3	3	1.1E-01
12439.2	3	5	4.9E-02
12456.1	5	3	8.9E-02
12487.6	7	5	1.1E-01
12554.4	7	5	1.2E-03
12702.4	3	3	7.1E-02
12733.6	5	5	1.1E-02
12746.3	3	3	2.0E-02
12802.7	5	5	5.7E-02
12933.3	3	1	1.0E-01
12956.6	3	3	7.4E-02
13008.5	5	3	8.9E-02
13214.7	3	1	8.1E-02
13273.1	5	7	1.5E-01
13313.4	3	5	1.3E-01
13504.0	5	7	1.1E-01
13599.2	5	5	2.2E-02
13622.4	3	5	7.3E-02
13678.5	3	5	6.2E-02
14093.6	1	3	4.3E-02
14739.1	5	7	8.8E-04
15046.4	1	3	5.2E-02
15172.3	1	3	1.3E-02
15329.6	5	5	1.2E-03
15555.5	5	7	9.8E-05
15734.9	5	3	2.9E-04
15816.8	5	3	8.7E-04
15989.3	1	3	1.9E-02
16122.7	5	3	3.9E-04
16180.0	5	5	1.2E-03
16264.1	3	3	3.0E-04
16520.1	3	5	2.6E-03
16739.8	3	5	3.1E-03
16940.4	5	5	2.5E-02
20317.0	1	3	1.6E-03
20616.5	5	5	3.9E-03
20812.0	5	7	7.6E-04
21332.2	3	3	3.2E-04
21534.9	3	5	1.1E-03
22039.2	3	1	1.2E-03
22077.4	5	3	1.4E-03
23133.4	3	3	1.7E-03
23844.8	9	7	1.1E-02
23967.5	3	1	3.6E-03
Ar II			
3000.4	4	4	1.5E+00
3028.9	2	4	2.3E+00
3093.4	4	6	4.4E+00
3139.0	6	6	1.0E+00
3161.4	2	4	1.8E+00
3169.7	4	6	8.2E-01
3181.0	6	4	6.3E-01
3236.8	2	4	5.2E-01
3243.7	4	2	2.0E+00
3249.8	2	4	1.0E+00
3293.6	4	4	1.7E+00
3307.2	2	2	3.4E+00
3350.9	6	6	1.5E+00
3376.4	8	8	1.5E+00
3388.5	2	4	1.9E+00
3454.1	6	4	4.5E-01
3464.1	6	6	3.7E-01
3476.7	6	6	1.34E+00
3491.2	4	4	2.2E+00
3509.8	2	2	2.5E+00
3514.4	4	6	1.23E+00
3520.0	6	6	8.0E-01
3521.3	8	8	2.3E-01
3535.3	2	4	8.2E-01
3545.6	4	6	3.4E+00
3545.8	6	8	3.9E+00
3548.5	4	4	1.1E+00
3559.5	6	8	3.9E+00
3565.0	2	4	1.1E+00
3576.6	6	8	2.77E+00
3581.6	2	4	1.8E+00
3582.4	4	6	3.72E+00
3588.4	8	10	3.39E+00
3600.2	4	4	2.2E+00
3622.1	4	2	6.4E-01

λ Å	Weights g_i	g_k	A 10^8 s^{-1}
3639.8	4	6	1.4E+00
3655.3	4	6	2.3E-01
3671.0	4	2	7.1E-01
3680.1	2	4	1.2E+00
3718.2	4	6	2.0E+00
3724.5	6	6	3.4E-01
3729.3	6	4	6.0E-01
3737.9	6	8	2.3E+00
3765.3	6	6	9.8E-01
3770.5	2	4	4.1E-01
3780.8	8	8	9.4E-01
3796.6	4	6	2.5E-01
3803.2	6	6	1.5E+00
3809.5	4	6	4.4E-01
3825.7	6	4	7.6E-01
3850.6	4	4	4.7E-01
3868.5	4	6	1.9E+00
3925.7	6	4	1.4E+00
3928.6	2	4	3.0E-01
3932.5	4	4	1.1E+00
3946.1	8	6	1.4E+00
3952.7	4	4	3.5E-01
3979.4	4	2	1.3E+00
4033.8	4	2	9.8E-01
4042.9	4	4	1.4E+00
4072.0	6	6	5.7E-01
4076.6	2	2	8.0E-01
4076.9	4	2	9.9E-01
4079.6	6	4	2.6E-01
4131.7	4	2	1.4E+00
4156.1	4	4	3.9E-01
4179.3	6	6	1.3E-01
4218.7	4	4	3.6E-01
4222.6	4	2	6.9E-01
4227.0	4	6	4.1E-01
4266.5	6	6	1.56E-01
4275.2	2	4	2.6E-01
4277.5	6	4	1.0E+00
4331.2	4	4	5.6E-01
4337.1	2	4	3.4E-01
4348.1	6	8	1.24E+00
4370.8	4	4	6.5E-01
4371.3	6	4	2.33E-01
4379.7	2	2	1.04E+00
4401.0	8	6	3.22E-01
4426.0	4	6	8.3E-01
4430.2	2	4	5.3E-01
4448.9	6	6	6.5E-01
4481.8	6	6	4.94E-01
4545.1	4	4	4.13E-01
4547.8	4	4	7.7E-02
4589.9	4	6	8.2E-01
4609.6	6	8	9.1E-01
4637.2	6	6	9.0E-02
4657.9	4	2	8.1E-01
4726.9	4	4	5.0E-01
4735.9	6	4	5.8E-01
4764.9	2	4	5.75E-01
4806.0	6	6	7.9E-01
4847.8	4	2	8.5E-01
4865.9	4	6	1.5E-01
4879.9	4	6	7.8E-01
4965.1	2	4	3.47E-01
5009.3	4	6	1.47E-01
5017.2	4	6	2.31E-01
5062.0	2	4	2.21E-01
5141.8	6	8	9.5E-02
6638.2	6	4	1.29E-01
6643.7	10	8	1.67E-01
6684.3	8	6	1.13E-01
Ar III			
769.15	5	3	6.0E+00
871.10	5	3	1.59E+00
875.53	3	1	3.74E+00
878.73	5	5	2.79E+00
879.62	3	3	9.2E-01
883.18	1	3	1.22E+00
887.40	3	5	9.0E-01
3024.1	5	7	2.6E+00
3027.2	5	5	6.4E-01
3054.8	3	5	1.9E+00
3064.8	3	3	1.0E+00
3078.2	1	3	1.4E+00
3285.9	5	7	2.0E+00
3301.9	5	5	2.0E+00
3311.3	5	3	2.0E+00
3336.1	7	9	2.0E+00
3344.7	5	7	1.8E+00
3352.1	7	7	2.2E-01
3358.5	3	5	1.6E+00
3361.3	5	5	3.0E-01
3472.6	5	7	2.0E-01
3480.6	7	7	1.6E+00
3499.7	3	3	1.3E+00
3500.6	3	5	2.6E-01
3502.7	5	3	4.3E-01
3503.6	5	5	1.2E+00
3511.7	7	5	2.6E-01
Ar IV			
840.03	4	2	2.73E+00
843.77	4	4	2.70E+00
850.60	4	6	2.63E+00
Ar VI			
292.15	2	2	6.9E+01
294.05	4	2	1.36E+02
Ar VII			
*250.41	9	3	2.78E+02
*477.54	9	15	9.92E+01
585.75	1	3	7.83E+01
*637.30	9	9	6.7E+01
Ar VIII			
158.92	2	4	1.1E+02
159.18	2	2	1.11E+02
229.44	2	2	1.12E+02
230.88	4	2	2.21E+02
337.09	4	4	1.2E+01
337.26	6	4	1.0E+02
338.22	4	2	1.1E+02
519.43	2	4	6.3E+01
526.46	4	6	7.2E+01
526.87	4	4	1.2E+01
700.24	2	4	2.55E+01
713.81	2	2	2.4E+01
Ar IX			
48.739	1	3	1.69E+03
Ar XIII			
162.96	5	3	3.4E+02
*163.08	9	3	5.3E+02
184.90	5	5	1.66E+02
186.38	1	3	8.8E+01
*207.89	9	9	9.5E+01
*245.10	9	15	3.7E+01
Ar XIV			
180.29	2	4	4.5E+01
183.41	2	2	1.69E+02
187.95	4	4	1.97E+02
191.35	4	2	7.5E+01
194.39	2	2	4.6E+01
203.35	4	2	7.8E+01
Ar XV			
25.05	1	3	1.7E+04
221.10	1	3	9.55E+01
*265.3	9	9	8.1E+01
Ar XVI			
*23.52	2	6	1.43E+04
*24.96	6	10	4.4E+04
353.88	2	4	1.5E+01
389.11	2	2	1.1E+01
1268	2	4	1.9E+00
1401	2	2	1.4E+00
2975	2	4	9.0E-02
3514	4	6	6.5E-02
Arsenic			
As I			
1890.4	4	6	2.0E+00
1937.6	4	4	2.0E+00
1972.6	4	2	2.0E+00
2288.1	6	4	2.8E+00
2344.0	2	4	3.5E-01
2349.8	4	2	3.1E+00
2369.7	4	4	6.0E-01
2370.8	4	6	4.2E-01
2456.5	6	4	7.2E-02
2492.9	4	2	1.2E-01
2745.0	2	4	2.6E-01
2780.2	4	4	7.8E-01
2860.4	2	2	5.5E-01

λ Å	Weights g_i	g_k	A $10^8\ s^{-1}$	λ Å	Weights g_i	g_k	A $10^8\ s^{-1}$	λ Å	Weights g_i	g_k	A $10^8\ s^{-1}$
2898.7	4	2	9.9E-02	5800.2	5	5	9.9E-02	1674.5	4	6	2.2E-01
Barium				5805.7	7	7	1.1E-02	1694.4	6	8	2.1E-01
Ba I				5826.3	5	3	5.6E-01	1697.2	6	6	1.7E-02
2409.2	1	3	8.6E-04	5907.6	3	5	1.5E-02	1761.8	4	4	3.9E-03
2414.1	1	3	1.5E-03	5971.7	5	5	1.8E-01	1771.0	4	2	3.4E-02
2420.1	1	3	2.3E-03	5997.1	3	3	2.7E-01	1786.9	6	4	4.4E-02
2427.4	1	3	5.6E-03	6019.5	3	1	1.4E+00	1892.7	2	4	9.0E-02
2432.5	1	3	7.2E-03	6063.1	5	3	5.7E-01	1904.2	4	6	1.1E-02
2438.8	1	3	1.4E-03	6083.4	3	1	1.1E-01	1906.8	2	2	5.1E-02
2444.6	1	3	4.5E-03	6110.8	7	5	5.5E-01	1924.7	6	8	3.1E-02
2452.4	1	3	8.1E-04	6129.2	3	1	6.0E-02	1954.2	4	6	1.3E-01
2473.2	1	3	4.6E-03	6341.7	5	7	1.9E-01	1955.1	4	4	1.8E-02
2500.2	1	3	1.5E-02	6450.9	3	5	1.1E-01	1970.2	4	2	6.7E-02
2543.2	1	3	4.1E-02	6482.9	5	7	4.4E-01	1985.6	2	4	2.5E-01
2596.6	1	3	1.2E-01	6498.8	7	7	8.6E-01	1999.5	2	4	1.0E-01
2646.5	1	3	1.1E-02	6527.3	5	5	5.9E-01	2009.2	2	2	8.6E-02
2702.6	1	3	2.5E-02	6595.3	3	3	3.9E-01	2052.7	4	6	2.0E-01
2739.2	1	3	9.1E-03	6675.3	5	3	1.9E-01	2054.6	4	4	2.9E-02
2785.3	1	3	2.8E-02	6693.8	7	5	2.8E-01	2080.0	4	2	1.0E-01
3071.6	1	3	4.1E-01	6865.7	5	5	2.3E-02	2153.9	2	4	5.3E-01
3501.1	1	3	1.9E-01	7059.9	7	9	7.1E-01	2200.9	2	2	2.0E-01
3889.3	1	3	8.8E-03	7120.3	3	5	2.1E-01	2232.8	4	6	2.9E-01
3909.9	3	5	4.9E-01	7195.2	1	3	2.4E-01	2235.4	4	4	4.4E-02
3935.7	5	7	4.7E-01	7280.3	5	7	5.3E-01	2286.0	4	2	1.3E-01
3937.9	5	5	1.1E-01	7392.4	3	3	5.0E-01	2528.5	2	4	7.1E-01
3993.4	7	9	5.5E-01	7417.5	7	5	2.5E-02	2634.8	4	6	7.6E-01
3995.7	7	7	8.8E-02	7488.1	7	7	1.0E-01	2641.4	4	4	1.2E-01
4132.4	1	3	7.1E-03	7528.2	5	5	2.7E-02	2647.3	2	2	2.0E-01
4239.6	5	3	2.4E-01	7672.1	3	5	3.1E-01	2771.4	4	2	4.0E-01
4242.6	3	5	5.6E-02	7780.5	5	5	1.3E-01	3816.7	4	6	2.3E-03
4264.4	1	3	1.5E-01	7905.8	5	3	6.3E-01	3842.8	6	8	2.2E-03
4283.1	5	7	6.4E-01	7911.3	1	3	2.98E-03	3891.8	2	4	1.67E+00
4323.0	3	5	1.5E-01	8147.7	5	5	6.3E-02	4024.1	6	4	5.3E-03
4325.2	5	7	7.1E-02	9645.6	7	5	1.1E-01	4057.5	8	6	1.2E-02
4332.9	3	3	1.5E-01	9704.3	3	1	1.6E-01	4130.7	4	6	1.80E+00
4350.3	3	5	6.0E-01	9821.5	3	1	5.5E-02	4166.0	4	4	3.7E-01
4402.5	3	5	2.7E-01	10370.3	3	5	1.3E-02	4216.0	2	4	5.8E-02
4406.8	5	5	1.0E-01	10649.1	5	5	2.7E-02	4287.8	2	2	2.4E-02
4431.9	1	3	1.2E+00	11075.7	3	3	3.6E-05	4325.7	4	6	5.9E-02
4467.1	5	7	6.6E-02	11303.1	5	3	1.2E-03	4329.6	4	4	8.8E-03
4489.0	5	7	4.2E-01	11373.8	3	1	1.3E-01	4405.2	4	2	3.9E-02
4493.6	5	5	3.6E-01	14158.4	9	7	2.0E-03	4470.7	6	4	1.4E-02
4505.9	3	3	1.1E+00	14723.2	3	5	8.6E-03	4509.6	8	6	1.2E-02
4523.2	5	5	9.6E-01	14999.9	5	3	2.8E-03	4524.9	2	2	7.2E-01
4573.9	3	1	1.21E+00	17123.7	7	7	3.3E-03	4554.0	2	4	1.17E+00
4579.6	5	5	7.0E-01	17187.1	3	1	2.7E-02	4708.9	2	4	9.7E-02
4591.8	5	5	1.6E-02	20563.9	5	7	2.6E-03	4843.5	4	6	9.3E-02
4599.7	3	1	4.07E-01					4847.1	2	2	4.1E-02
4605.0	3	1	7.7E-02	**Ba II**				4850.8	4	4	1.4E-02
4619.9	1	3	9.3E-02	1413.4	6	8	1.7E-02	4900.0	4	2	7.75E-01
4628.3	5	3	6.0E-02	1417.1	4	6	3.8E-02	4934.1	2	2	9.55E-01
4673.6	7	5	6.5E-02	1444.9	4	6	8.1E-02	4997.8	4	2	6.1E-02
4691.6	5	3	1.6E+00	1461.5	6	8	8.7E-02	5185.0	2	4	1.8E-02
4700.4	3	3	2.4E-01	1487.0	4	6	1.4E-01	5361.4	4	6	4.8E-02
4726.4	5	3	4.6E-01	1503.9	6	8	1.5E-01	5391.6	6	8	5.2E-02
5519.1	3	5	5.0E-01	1554.4	4	6	2.6E-01	5413.6	6	6	8.4E-04
5535.5	1	3	1.19E+00	1572.7	6	8	2.4E-01	5421.1	6	6	1.9E-03
5777.6	5	7	6.5E-01	1573.9	6	6	1.6E-02	5428.8	6	4	2.3E-02
5784.0	3	5	2.1E-01	1630.4	2	2	1.7E-02	5480.3	8	6	1.8E-02

λ Å	Weights g_i	Weights g_k	A 10^8 s^{-1}
5784.2	2	4	2.0E-01
5853.7	4	4	4.8E-02
5981.3	4	6	1.6E-01
5999.9	4	4	2.6E-02
6135.8	2	2	8.5E-02
6141.7	6	4	3.7E-01
6363.2	6	4	2.9E-03
6372.9	4	4	6.7E-04
6378.9	4	2	9.9E-02
6457.7	6	4	3.0E-03
6496.9	4	2	3.32E-01
7556.8	6	4	1.6E-03
7678.2	8	6	6.6E-04
8710.7	6	8	8.0E-01
8737.7	4	6	9.3E-01

Beryllium

Be I

λ Å	g_i	g_k	A 10^8 s^{-1}
1491.8	1	3	1.3E-02
1661.5	1	3	2.0E-01
2348.6	1	3	5.56E+00
*2494.7	9	15	1.6E+00
*2650.6	9	9	4.31E+00
4572.7	3	5	7.9E-01

Be II

λ Å	g_i	g_k	A 10^8 s^{-1}
1197.1	2	2	4.7E-01
1197.2	4	2	9.4E-01
1512.3	2	4	9.2E+00
1512.4	4	6	1.1E+01
1776.1	2	2	1.4E+00
1776.3	4	2	2.9E+00
*2453.8	2	6	1.42E-01
3046.5	2	4	4.8E-01
3046.7	4	6	5.9E-01
3130.4	2	4	1.14E+00
3131.1	2	2	1.15E+00
3241.6	2	2	1.41E-01
3241.8	4	2	2.8E-01
3274.6	2	4	1.9E-01
3274.7	2	2	1.9E-01
4360.7	2	4	9.2E-01
4361.0	4	6	1.1E+00
*5255.9	2	6	2.56E-02
5270.3	2	2	3.30E-01
5270.8	4	2	6.6E-01
6279.4	2	4	1.2E-01
6279.7	4	6	1.43E-01
6756.7	2	2	5.1E-02
6757.1	4	2	1.02E-01
7401.2	2	4	3.0E-02
7401.4	2	2	3.0E-02

Bismuth

Bi I

λ Å	g_i	g_k	A 10^8 s^{-1}
1954.5	4	6	1.2E+00
2021.2	4	4	6.0E-02
2061.7	4	6	9.9E-01
2110.3	4	2	9.1E-01
2177.3	4	2	2.6E-02
2228.3	4	4	8.9E-01
2230.6	4	6	2.6E+00
2276.6	4	4	2.5E-01
2515.7	4	6	4.3E-02
2627.9	4	4	4.7E-01
2696.8	4	6	6.4E-02
2780.5	4	2	3.09E-01
2798.7	6	6	3.6E-02
2898.0	4	2	1.53E+00
2938.3	6	4	1.23E+00
2989.0	4	4	5.5E-01
2993.3	4	6	1.6E-01
3024.6	6	6	8.8E-01
3067.7	4	2	2.07E+00
3076.7	4	4	3.5E-02
3397.2	6	4	1.81E-01
3402.9	6	6	1.6E-02
3510.9	6	4	6.8E-02
3596.1	2	4	1.98E-01
3888.2	2	2	6.9E-02
4121.5	2	2	1.64E-01
4308.5	2	4	1.6E-02
4493.0	2	4	1.5E-02
4722.3	4	2	1.17E-01
6134.8	4	4	1.8E-02

Boron

B I

λ Å	g_i	g_k	A 10^8 s^{-1}
1378.6	2	4	3.50E+00
1378.9	2	2	1.40E+01
1378.9	4	4	1.75E+01
1379.2	4	2	7.0E+00
1465.5	2	4	3.34E+00
1465.7	4	4	6.7E+00
1465.8	6	4	1.00E+01
1825.9	2	4	2.0E+00
1826.4	4	6	2.4E+00
2088.9	2	4	2.8E-01
2089.6	4	6	3.3E-01
2496.8	2	2	8.5E-01
2497.7	4	2	1.69E+00

Bromine

Br I

λ Å	g_i	g_k	A 10^8 s^{-1}
1488.5	4	4	1.2E+00
1540.7	4	4	1.4E+00
1574.8	2	4	2.0E-01
1576.4	4	6	2.1E-02
1633.4	2	4	8.1E-02
4365.1	2	4	7.5E-03
4425.1	4	2	4.2E-03
4441.7	6	4	7.5E-03
4472.6	4	4	9.3E-03
4477.7	6	8	1.3E-02
4513.4	6	4	2.8E-03
4525.6	6	6	7.2E-03
4575.7	4	4	1.6E-02
4614.6	4	6	5.4E-03
4979.8	4	4	2.6E-03
5245.1	2	4	3.1E-03
5345.4	2	4	7.6E-03
7348.5	4	6	1.2E-01
7513.0	6	4	1.2E-01
7803.0	2	4	5.3E-02
7938.7	6	6	1.9E-01
8131.5	2	4	3.8E-02
8343.7	2	2	2.2E-01
8446.6	4	4	1.2E-01
8638.7	6	4	9.7E-02

Br II

λ Å	g_i	g_k	A 10^8 s^{-1}
4704.9	5	7	1.1E+00
4785.5	5	5	9.4E-01
4816.7	5	3	1.1E+00

Cadmium

Cd I

λ Å	g_i	g_k	A 10^8 s^{-1}
2288.0	1	3	5.3E+00
2836.9	1	3	2.8E-01
2880.8	3	5	4.2E-01
2881.2	3	3	2.4E-01
2980.6	5	7	5.9E-01
2981.4	5	5	1.5E-01
3261.1	1	3	4.06E-03
3403.7	1	3	7.7E-01
3466.2	3	5	1.2E+00
3467.7	3	3	6.7E-01
3610.5	5	7	1.3E+00
3612.9	5	5	3.5E-01
4140.5	3	5	4.7E-02
4662.4	3	5	5.5E-02
4678.1	1	3	1.3E-01
4799.9	3	3	4.1E-01
5085.8	5	3	5.6E-01
6438.5	3	5	5.9E-01

Cd II

λ Å	g_i	g_k	A 10^8 s^{-1}
2144.4	2	4	2.8E+00
2265.0	2	2	3.0E+00
2572.9	2	2	1.7E+00
2748.5	4	2	2.8E+00
4415.6	4	6	1.4E-02

Calcium

Ca I

λ Å	g_i	g_k	A 10^8 s^{-1}
2275.5	1	3	3.01E-01
2995.0	1	3	3.67E-01
2997.3	3	5	2.41E-01
2999.6	3	3	2.79E-01
3000.9	3	1	1.58E+00
3006.9	5	5	7.5E-01
3009.2	5	3	4.30E-01
3344.5	1	3	1.51E-01
3350.2	3	5	1.78E-01
3361.9	5	7	2.23E-01
3624.1	1	3	2.12E-01
3630.8	3	5	2.97E-01

λ Å	Weights g_i	g_k	A 10^8 s^{-1}
3631.0	3	3	1.53E-01
3644.4	5	7	3.55E-01
3644.8	5	5	9.4E-02
3870.5	3	5	7.2E-02
3957.1	3	3	9.8E-02
3973.7	5	3	1.75E-01
4092.6	3	5	1.1E-01
4094.9	5	7	1.2E-01
4098.5	7	9	1.3E-01
4108.5	5	7	9.0E-01
4226.7	1	3	2.18E+00
4283.0	3	5	4.34E-01
4289.4	1	3	6.0E-01
4299.0	3	3	4.66E-01
4302.5	5	5	1.36E+00
4307.7	3	1	1.99E+00
4318.7	5	3	7.4E-01
4355.1	5	7	1.9E-01
4425.4	1	3	4.98E-01
4435.0	3	5	6.7E-01
4435.7	3	3	3.42E-01
4454.8	5	7	8.7E-01
4455.9	5	5	2.0E-01
4526.9	5	3	4.1E-01
4578.6	3	5	1.76E-01
4581.4	5	7	2.09E-01
4585.9	7	9	2.29E-01
4685.3	3	5	8.0E-02
4878.1	5	7	1.88E-01
5041.6	5	3	3.3E-01
5188.9	3	5	4.0E-01
5261.7	3	3	1.5E-01
5262.2	3	1	6.0E-01
5264.2	5	5	9.1E-02
5265.6	5	3	4.4E-01
5270.3	7	5	5.0E-01
5582.0	5	7	6.0E-02
5588.8	7	7	4.9E-01
5590.1	3	5	8.3E-02
5594.5	5	5	3.8E-01
5598.5	3	3	4.3E-01
5601.3	7	5	8.6E-02
5602.9	5	3	1.4E-01
5857.5	3	5	6.6E-01
6102.7	1	3	9.6E-02
6122.2	3	3	2.87E-01
6161.3	5	5	3.3E-02
6162.2	5	3	4.77E-01
6163.8	3	3	5.6E-02
6166.4	3	1	2.2E-01
6169.1	5	3	1.7E-01
6169.6	7	5	1.9E-01
6439.1	7	9	5.3E-01
6449.8	3	5	9.0E-02
6462.6	5	7	4.7E-01
6471.7	7	7	5.9E-02
6493.8	3	5	4.4E-01
6499.7	5	5	8.1E-02

λ Å	Weights g_i	g_k	A 10^8 s^{-1}
Ca II			
1341.9	2	4	1.5E-02
1342.5	2	2	1.5E-02
1649.9	2	4	3.2E-03
1652.0	2	2	3.1E-03
1673.9	2	4	2.24E-01
1680.1	4	6	2.65E-01
1680.1	4	4	4.41E-02
1807.3	2	4	3.54E-01
1814.5	4	6	4.2E-01
1814.7	4	4	7.0E-02
1843.1	2	2	1.6E-01
1850.7	4	2	3.08E-01
2103.2	2	4	8.2E-01
2112.8	4	6	9.7E-01
2113.2	4	4	1.6E-01
2197.8	2	2	3.1E-01
2208.6	4	2	6.2E-01
3158.9	2	4	3.1E+00
3179.3	4	6	3.6E+00
3181.3	4	4	5.8E-01
3706.0	2	2	8.8E-01
3736.9	4	2	1.7E+00
3933.7	2	4	1.47E+00
3968.5	2	2	1.4E+00
Ca III			
357.97	1	3	8.8E+02
439.69	1	3	1.9E-01
490.55	1	3	1.6E-02
Ca V			
558.60	5	3	2.2E+01
637.93	5	3	3.9E+00
643.12	3	1	9.1E+00
646.57	5	5	6.9E+00
647.88	3	3	2.3E+00
651.55	1	3	2.9E+00
656.76	3	5	2.1E+00
Ca VII			
550.20	5	5	1.8E+01
624.39	1	3	3.3E+00
630.54	3	5	4.5E+00
630.79	3	3	2.2E+00
639.15	5	7	5.7E+00
640.41	5	5	1.3E+00
Ca VIII			
182.71	2	2	1.6E+02
184.16	4	2	3.2E+02
Ca IX			
163.23	5	3	3.76E+02
371.89	1	3	8.8E+01
373.81	3	5	1.16E+02
378.08	5	7	1.5E+02
395.03	3	5	2.2E+02

λ Å	Weights g_i	g_k	A 10^8 s^{-1}
466.24	1	3	1.12E+02
498.01	3	5	2.49E+01
506.18	5	5	7.2E+01
515.57	5	3	3.75E+01
Ca X			
110.96	2	4	2.9E+02
111.20	2	2	2.92E+02
151.84	2	2	2.3E+02
153.02	4	2	4.5E+02
206.57	4	4	2.9E+01
206.75	6	4	2.6E+02
207.39	4	2	2.8E+02
411.70	2	4	8.3E+01
419.75	4	6	9.5E+01
420.47	4	4	1.6E+01
557.76	2	4	3.50E+01
574.01	2	2	3.2E+01
Ca XI			
30.448	1	3	6.2E+03
30.867	1	3	4.9E+04
35.212	1	3	2.0E+03
Ca XII			
140.05	4	2	3.7E+02
147.27	2	2	1.6E+02
Ca XV			
141.69	5	3	4.08E+02
*142.23	9	3	6.3E+02
161.00	5	5	1.9E+02
Ca XVII			
19.558	1	3	3.8E+04
21.198	3	5	4.9E+04
192.82	1	3	1.21E+02
218.82	3	5	2.76E+01
223.02	1	3	3.44E+01
228.72	3	3	2.37E+01
232.83	5	5	6.5E+01
244.06	5	3	3.28E+01
Ca XVIII			
*18.71	2	6	2.31E+04
*19.74	6	10	7.0E+04
302.19	2	4	2.0E+01
344.76	2	2	1.3E+01
Carbon			
C I			
945.19	1	3	6.2E+00
945.34	3	3	1.8E+01
945.58	5	3	3.1E+01
1260.7	1	3	4.0E-01
1260.9	3	1	1.2E+00
1261.0	3	3	3.1E-01
1261.1	3	5	3.0E-01

λ Å	Weights g_i	g_k	A 10^8 s^{-1}
1261.4	5	3	5.0E-01
1261.6	5	5	9.3E-01
1274.1	5	7	6.8E-03
1277.2	1	3	8.8E-01
1277.3	3	5	1.2E+00
1277.5	3	3	6.5E-01
1277.6	5	7	1.5E+00
1277.7	5	5	3.9E-01
1278.0	5	3	4.2E-02
1279.2	5	7	1.1E-01
1279.9	3	5	2.1E-01
1280.1	1	3	2.7E-01
1280.3	5	5	6.2E-01
1280.4	3	3	2.0E-01
1280.6	3	1	8.1E-01
1280.8	5	3	3.5E-01
1328.8	1	3	4.9E-01
1364.2	5	5	4.7E-02
1431.6	5	7	1.5E+00
1432.1	5	5	1.4E+00
1432.5	5	3	1.3E+00
1459.0	5	3	3.7E-01
1463.3	5	7	2.1E+00
1467.4	5	3	4.6E-01
1468.4	5	3	1.9E-02
1470.1	5	7	8.8E-03
1472.2	5	3	5.1E-03
1481.8	5	5	3.3E-01
1560.3	1	3	8.2E-01
1561.3	5	5	3.6E-01
1561.4	5	7	1.4E+00
1656.3	3	5	8.0E-01
1656.9	1	3	1.1E+00
1657.0	5	5	2.4E+00
1657.4	3	3	8.0E-01
1657.9	3	1	3.2E+00
1658.1	5	3	1.3E+00
1751.8	1	3	5.7E-01
1763.9	1	3	2.2E-02
1765.4	1	3	7.1E-03
1930.9	5	3	3.7E+00
2478.6	1	3	1.8E-01
2902.3	1	3	6.6E-03
2903.3	3	3	1.7E-02
2905.0	5	3	2.2E-02
4269.0	3	5	3.2E-03
4371.4	3	3	9.7E-03
4762.3	1	3	5.2E-03
4762.5	3	5	3.8E-03
4766.7	3	3	3.9E-03
4770.0	3	1	1.5E-02
4771.8	5	5	1.2E-02
4775.9	5	3	6.2E-03
4812.9	1	3	9.7E-04
4817.4	3	3	2.8E-03
4826.8	5	3	4.7E-03
4932.1	3	1	4.6E-02
5052.2	3	5	1.7E-02
5380.3	3	3	1.6E-02
5793.1	7	5	3.3E-03
5794.5	5	5	5.8E-04
5800.2	3	3	9.7E-04
5800.6	5	3	2.9E-03
5805.2	3	1	3.9E-03
6587.6	3	3	2.4E-02

C II

λ Å	g_i	g_k	A 10^8 s^{-1}
687.35	4	6	2.70E+01
858.09	2	2	3.69E-01
858.56	4	2	1.11E+00
903.62	2	4	6.6E+00
903.96	2	2	2.63E+01
904.14	4	4	3.30E+01
904.48	4	2	1.33E+01
1009.9	2	4	5.8E+00
1010.1	4	4	1.15E+01
1010.4	6	4	1.73E+01
1036.3	2	2	8.0E+00
1037.0	4	2	1.59E+01
1323.9	4	4	4.53E+00
1324.0	6	6	4.71E+00
1334.5	2	4	2.41E+00
1335.7	4	6	2.89E+00
2509.1	2	4	5.4E-01
2511.7	4	4	1.06E-01
2512.1	4	6	6.4E-01
6578.1	2	4	3.6E-01
6582.9	2	2	3.6E-01
7231.3	2	4	3.6E-01
7236.4	4	6	4.4E-01
7237.2	4	4	7.2E-02

C III

λ Å	g_i	g_k	A 10^8 s^{-1}
310.17	1	3	1.8E+01
386.20	1	3	3.22E+01
459.46	1	3	5.5E+01
459.52	3	5	7.5E+01
459.63	5	7	9.8E+01
574.28	3	5	6.3E+01
977.03	1	3	1.75E+01
1174.9	3	5	3.42E+00
1175.3	1	3	4.55E+00
1175.6	3	3	3.41E+00
1175.7	5	5	1.02E+01
1176.0	3	1	1.36E+01
1176.4	5	3	5.7E+00
1247.4	3	1	1.86E+01
2296.9	3	5	1.46E+00
4647.4	3	5	7.3E-01
4650.3	3	3	7.4E-01
4651.5	3	1	7.4E-01

C IV

λ Å	g_i	g_k	A 10^8 s^{-1}
*312.43	2	6	4.49E+01
*384.13	6	10	1.8E+02
1548.2	2	4	2.66E+00
1550.8	2	2	2.64E+00
5801.3	2	4	3.19E-01
5812.0	2	2	3.16E-01

C V

λ Å	g_i	g_k	A 10^8 s^{-1}
34.973	1	3	2.554E+03
40.268	1	3	8.873E+03
*227.19	3	9	1.363E+02
247.31	1	3	1.279E+02
*248.70	9	15	4.25E+02
*260.19	9	3	6.683E+01
267.27	3	5	3.96E+02
*2273.9	3	9	5.650E-01
3526.7	1	3	1.663E-01
*8432.2	3	9	6.870E-02

Cesium

Cs I

λ Å	g_i	g_k	A 10^8 s^{-1}
3203.5	2	4	7.6E-06
3205.3	2	4	7.9E-06
3207.5	2	4	8.5E-06
3210.0	2	4	9.4E-06
3212.8	2	4	1.19E-05
3216.2	2	4	1.49E-05
3220.1	2	4	1.7E-05
3220.2	2	2	1.07E-07
3224.8	2	4	2.0E-05
3225.0	2	2	1.43E-07
3230.5	2	4	2.5E-05
3230.7	2	2	1.97E-07
3237.4	2	4	2.8E-05
3237.6	2	2	2.63E-07
3245.9	2	4	3.45E-05
3246.2	2	2	3.7E-07
3256.7	2	4	4.25E-05
3257.1	2	2	7.0E-07
3270.5	2	4	5.6E-05
3271.0	2	2	9.8E-07
3288.6	2	4	1.0E-04
3289.3	2	2	2.7E-06
3313.1	2	4	1.6E-04
3314.0	2	2	5.2E-06
3347.5	2	4	2.2E-04
3348.8	2	2	1.1E-05
3397.9	2	4	4.0E-04
3400.0	2	2	2.4E-05
3476.8	2	4	6.6E-04
3480.0	2	2	6.6E-05
3611.4	2	4	1.5E-03
3617.3	2	2	2.5E-04
3876.1	2	4	3.8E-03
3888.6	2	2	9.7E-04
4555.3	2	4	1.88E-02
4593.2	2	2	8.0E-03

Chlorine

Cl I

λ Å	g_i	g_k	A 10^8 s^{-1}
1188.8	4	6	2.33E+00
1188.8	4	4	2.71E-01
1201.4	2	4	2.39E+00
1335.7	4	2	1.74E+00

λ Å	Weights g_i	g_k	A 10^8 s^{-1}
1347.2	4	4	4.19E+00
1351.7	2	2	3.23E+00
1363.4	2	4	7.5E-01
4323.3	4	4	1.1E-02
4363.3	4	6	6.8E-03
4379.9	4	4	1.4E-02
4389.8	6	8	1.4E-02
4526.2	4	4	5.1E-02
4601.0	2	2	4.2E-02
4661.2	2	4	1.2E-02
7256.6	6	4	1.5E-01
7414.1	6	4	4.7E-02
7547.1	4	4	1.2E-01
7717.6	4	4	3.0E-02
7745.0	2	4	6.3E-02
7769.2	6	6	6.0E-02
7821.4	6	8	9.8E-02
7830.8	4	4	9.7E-02
7878.2	6	6	1.8E-02
7899.3	4	6	5.1E-02
7924.6	2	4	2.1E-02
7935.0	6	8	3.9E-02
7997.9	4	4	2.1E-02
Cl II			
3329.1	5	7	1.5E+00
3522.1	7	7	1.4E+00
3798.8	5	7	1.6E+00
3805.2	7	9	1.8E+00
3809.5	3	5	1.5E+00
3851.0	5	7	1.8E+00
3851.4	5	5	1.6E+00
3854.7	3	5	2.2E+00
3861.9	5	7	2.4E+00
3868.6	7	9	2.7E+00
3913.9	9	9	8.2E-01
3990.2	5	7	8.4E-01
4132.5	5	5	1.6E+00
4276.5	9	7	7.6E-01
4768.7	3	5	7.7E-01
4781.3	5	7	1.0E+00
4794.6	5	7	1.04E+00
4810.1	5	5	9.9E-01
4819.5	5	3	1.00E+00
4904.8	5	7	8.1E-01
4917.7	3	5	7.5E-01
5078.3	7	7	7.7E-01
5219.1	3	9	8.6E-01
5392.1	5	7	1.0E+00
Cl III			
2298.5	4	4	4.2E+00
2340.6	6	6	4.2E+00
2370.4	8	6	2.8E+00
2531.8	2	4	4.4E+00
2532.5	4	6	5.3E+00
2577.1	4	6	4.3E+00
2580.7	6	8	4.7E+00
2601.2	2	4	4.6E+00
2603.6	4	6	5.0E+00
2609.5	6	8	5.7E+00
2617.0	8	10	6.6E+00
2661.6	4	6	3.4E+00
2665.5	6	8	4.8E+00
2691.5	4	4	3.5E+00
2710.4	4	6	3.5E+00
3340.4	6	6	1.5E+00
3392.9	4	4	1.9E+00
3393.5	6	6	1.9E+00
3530.0	6	8	1.8E+00
3560.7	4	6	1.7E+00
3602.1	6	8	1.7E+00
3612.9	4	6	1.2E+00
3720.5	4	6	1.7E+00
Chromium			
Cr I			
1999.95	9	9	1.4E+00
2383.30	9	11	4.1E-01
2389.21	3	5	2.3E-01
2408.60	9	7	6.7E-01
2408.72	7	5	2.9E-01
2492.57	3	5	4.5E-01
2495.08	3	3	2.7E-01
2496.30	5	7	5.6E-01
2502.55	7	9	2.2E-01
2504.31	7	9	4.5E-01
2508.11	5	5	2.1E-01
2508.97	5	3	3.8E-01
2527.11	9	9	5.3E-01
2549.55	3	3	4.8E-01
2560.70	5	5	4.3E-01
2571.74	7	5	6.4E-01
2577.66	7	7	2.6E-01
2591.84	9	7	6.5E-01
2620.48	5	3	1.9E-01
2673.64	3	3	1.8E-01
2701.99	9	11	2.1E-01
2726.50	5	7	7.5E-01
2731.90	5	5	7.8E-01
2736.46	5	3	7.5E-01
2752.85	3	3	8.7E-01
2757.09	5	5	6.8E-01
2761.74	5	3	6.8E-01
2764.36	7	7	3.7E-01
2769.90	7	5	1.1E+00
2780.70	9	7	1.4E+00
2879.27	5	7	2.1E-01
2887.00	3	5	2.7E-01
2889.22	9	9	6.6E-01
2893.25	7	7	5.2E-01
2894.17	1	3	3.3E-01
2896.76	5	5	3.0E-01
2905.48	3	1	1.3E+00
2909.05	5	3	6.8E-01
2910.89	7	5	3.4E-01
2911.15	9	7	2.6E-01
2967.64	7	9	3.9E-01
2971.10	5	7	7.1E-01
2975.48	3	5	8.9E-01
2980.78	1	3	5.10E-01
2988.64	5	7	5.2E-01
2991.88	3	1	3.0E+00
2994.06	5	5	2.5E-01
2995.09	5	5	4.3E-01
2996.57	5	3	2.0E+00
2998.78	5	3	4.07E-01
3000.88	7	5	1.6E+00
3005.06	9	7	9.2E-01
3013.72	3	5	8.3E-01
3015.20	1	3	1.63E+00
3020.67	3	3	1.5E+00
3021.58	9	11	2.91E+00
3024.36	5	5	1.27E+00
3029.17	5	3	3.8E-01
3030.25	7	7	1.1E+00
3031.35	5	3	3.1E-01
3034.19	7	7	3.5E-01
3037.05	9	9	5.4E-01
3040.84	7	5	7.4E-01
3053.87	9	7	7.97E-01
3148.44	9	11	5.6E-01
3155.16	11	13	5.7E-01
3163.76	13	15	6.0E-01
3237.73	9	9	1.3E+00
3238.09	11	11	2.0E-01
3578.68	7	9	1.48E+00
3593.48	7	7	1.50E+00
3605.32	7	5	1.62E+00
3639.80	13	11	1.8E+00
3743.89	13	13	7.61E-01
3757.66	7	7	4.13E-01
3768.24	5	5	5.10E-01
3804.80	9	9	6.9E-01
3963.69	13	15	1.3E+00
3969.75	11	13	1.2E+00
3983.90	7	9	1.05E+00
3991.12	5	7	1.07E+00
4001.44	9	11	6.8E-01
4039.10	15	15	6.7E-01
4048.78	13	13	6.4E-01
4058.78	11	11	6.7E-01
4065.71	9	11	3.5E-01
4165.52	11	13	7.5E-01
4204.48	13	11	3.1E-01
4254.33	7	9	3.15E-01
4263.15	15	17	6.4E-01
4274.81	7	7	3.07E-01
4275.98	11	11	2.2E-01
4280.42	13	15	4.7E-01
4289.73	7	5	3.16E-01
4291.97	7	5	2.4E-01
4297.75	11	13	4.9E-01
4298.05	9	9	2.6E-01
4300.52	9	7	1.9E-01
4301.19	11	9	2.6E-01
4302.78	11	11	2.5E-01

λ Å	Weights g_i	g_k	A 10^8 s^{-1}
4319.66	5	3	1.8E-01
4337.25	5	7	2.0E-01
4373.65	9	9	2.8E-01
4376.80	13	13	3.2E-01
4413.86	7	5	2.7E-01
4422.70	5	5	2.7E-01
4424.29	9	7	2.1E-01
4429.93	3	3	2.4E-01
4432.16	1	3	1.8E-01
4432.77	15	15	4.9E-01
4443.72	3	1	4.5E-01
4482.88	3	3	3.0E-01
4490.55	9	7	3.9E-01
4492.31	5	3	4.47E-01
4495.28	9	7	2.0E-01
4500.29	7	7	2.1E-01
4506.84	13	11	2.7E-01
4540.72	11	11	3.14E-01
4564.17	11	13	5.1E-01
4595.60	13	13	4.7E-01
4622.47	7	7	4.1E-01
4663.33	3	3	2.0E-01
4665.90	3	3	3.0E-01
4689.38	7	5	2.3E-01
4698.46	9	7	2.2E-01
4708.02	11	9	4.31E-01
4718.43	13	11	3.4E-01
4730.69	7	5	3.83E-01
4737.33	9	7	3.38E-01
4741.09	3	5	2.2E-01
4752.07	13	13	6.2E-01
4756.09	11	9	4.0E-01
4792.49	7	5	2.6E-01
4801.02	9	7	3.06E-01
4816.13	9	9	1.8E-01
4870.79	7	9	3.5E-01
4887.01	9	11	3.2E-01
4922.28	11	13	4.0E-01
4966.80	3	1	3.0E-01
5204.51	5	3	5.09E-01
5206.02	5	5	5.14E-01
5208.42	5	7	5.06E-01
5243.38	5	3	2.19E-01
5297.37	7	9	3.88E-01
5297.99	7	7	3.0E-01
5328.36	9	11	6.2E-01
5329.17	9	9	2.25E-01
5783.11	3	3	2.1E-01
5783.89	5	5	2.02E-01
5787.97	5	7	2.35E-01
Cr II			
2653.57	4	6	3.5E-01
2658.59	2	4	5.8E-01
2666.02	6	8	5.9E-01
2668.71	4	2	1.4E+00
2671.80	6	4	1.0E+00
2672.83	8	6	5.5E-01
2744.97	4	6	8.5E-01
2787.61	6	6	1.5E+00
2822.38	14	16	2.3E+00
2835.63	10	12	2.0E+00
2840.01	10	12	2.7E+00
2843.24	8	10	6.4E-01
2849.83	6	8	9.2E-01
2851.35	8	10	2.2E+00
2856.77	4	6	4.3E-01
2857.40	6	8	2.8E-01
2860.92	2	4	6.9E-01
2862.57	8	8	6.3E-01
2866.72	4	4	1.2E+00
2867.09	4	4	1.1E+00
2867.65	2	2	1.1E+00
2870.43	6	6	1.3E+00
2873.81	4	2	8.8E-01
2880.86	6	4	7.9E-01
2898.53	10	12	1.2E+00
2921.81	8	10	9.0E-01
2930.83	2	4	1.1E+00
2935.12	6	8	1.8E+00
2953.34	2	2	1.8E+00
2966.03	10	8	5.4E-01
2971.90	14	14	2.0E+00
2979.73	12	12	1.8E+00
2985.32	10	10	2.2E+00
2989.18	8	8	2.2E+00
3118.64	2	4	1.7E+00
3120.36	4	6	1.5E+00
3122.59	12	12	4.4E-01
3128.69	4	4	8.1E-01
3136.68	6	6	6.4E-01
4588.22	8	6	1.2E-01
Cr V			
434.306	9	9	1.5E+01
436.351	9	7	2.4E+01
436.601	7	5	2.1E+01
437.420	7	7	1.4E+01
437.655	5	5	1.3E+01
441.056	5	3	2.3E+01
456.357	1	3	9.5E+00
456.637	3	1	3.3E+01
456.743	3	3	9.1E+00
457.028	5	5	2.7E+01
457.504	5	3	1.2E+01
464.015	9	7	3.6E+01
469.634	5	5	2.3E+01
1106.25	7	9	1.2E+01
1121.07	7	9	2.1E+01
1127.63	9	11	3.5E+01
1465.86	5	3	1.1E+01
1481.65	3	1	1.0E+01
1519.03	5	7	9.5E+00
1579.70	7	9	8.6E+00
Cr VI			
161.687	6	6	1.7E+02
168.088	4	6	2.0E+02
201.007	4	4	2.5E+03
201.224	4	6	1.8E+02
201.388	6	4	2.7E+02
201.606	6	6	2.6E+03
202.442	6	4	1.0E+03
202.739	4	2	1.2E+03
226.241	6	8	7.2E+02
227.202	4	6	6.6E+02
Cr X			
216.72	6	8	9.0E+02
223.86	4	2	7.7E+02
224.74	4	4	7.6E+02
226.24	4	6	7.3E+02
227.42	4	4	5.2E+02
227.50	4	6	1.8E+01
228.63	6	4	8.1E+01
228.71	6	6	4.5E+02
231.21	2	4	1.2E+02
232.96	4	4	4.4E+02
242.20	2	4	5.0E+01
244.19	4	6	5.8E+01
395.984	4	4	2.4E+01
398.150	6	6	2.1E+01
Cr XI			
214.31	5	7	1.4E+01
226.45	5	7	6.0E+02
232	3	1	4.1E+02
235.53	5	7	5.5E+02
240.76	1	3	4.8E+02
250.28	5	7	1.0E+01
366.491	3	3	1.2E+01
366.942	3	1	3.0E+01
374.927	5	5	2.3E+01
422.083	3	5	1.0E+01
Cr XII			
216	4	6	2.4E+02
218	6	8	2.4E+02
239	2	2	1.6E+02
244.70	2	4	3.0E+02
247	4	2	2.4E+02
247	2	2	3.3E+02
248	6	8	1.4E+02
250	6	8	3.5E+02
250	6	6	2.2E+02
251.52	4	6	3.4E+02
252	4	6	2.0E+02
256	2	2	1.5E+02
259	2	4	3.2E+02
269	2	2	2.1E+02
300.32	2	2	1.4E+02
305.81	4	4	2.76E+02
309	4	2	2.7E+02
309	6	6	1.6E+02
311.55	4	2	1.6E+02
324	4	6	2.2E+02
327	6	8	2.2E+02

λ Å	Weights g_i	g_k	A 10^8 s^{-1}
332.06	6	4	1.4E+02
Cr XIII			
49.59	1	3	9.9E+02
67.01	1	3	1.67E+03
228	5	7	1.8E+02
267.73	5	7	1.9E+02
270	3	1	1.7E+02
276.4	5	7	2.2E+02
277	1	3	2.1E+02
279.32	3	5	3.5E+02
286	3	1	4.6E+02
328.29	1	3	1.86E+02
345	7	9	1.74E+02
Cr XIV			
*38.036	2	6	2.47E+02
39.796	2	4	3.05E+02
40.018	4	6	3.6E+02
40.782	2	4	3.9E+02
40.800	2	2	3.9E+02
41.556	2	4	4.5E+02
41.788	4	6	5.3E+02
44.597	2	4	7.1E+02
44.869	4	6	8.3E+02
46.125	4	2	3.1E+02
46.468	2	4	6.6E+02
46.527	2	2	6.7E+02
48.300	4	6	5.9E+02
48.338	6	8	6.3E+02
50.821	2	4	1.2E+03
51.172	4	6	1.4E+03
51.180	4	4	2.3E+02
52.321	4	6	1.0E+03
52.363	6	8	1.1E+03
53.760	2	2	3.0E+02
54.164	4	2	5.9E+02
60.699	4	6	2.05E+03
60.756	6	8	2.19E+03
63.324	2	4	1.07E+03
63.539	2	2	1.13E+03
68.594	2	4	1.98E+03
69.213	4	6	2.31E+03
69.247	4	4	3.8E+02
86.060	4	6	5.3E+03
86.169	6	8	5.9E+03
86.185	6	6	3.9E+02
101.05	6	4	4.4E+02
101.42	4	2	4.83E+02
104.4	4	6	3.0E+02
104.5	6	8	3.1E+02
109.8	2	4	2.3E+02
110.4	4	6	2.8E+02
118.3	4	2	2.1E+02
125.2	4	6	5.0E+02
125.3	6	8	5.4E+02
148.5	2	4	2.18E+02
149.1	2	2	2.1E+02
157.1	2	4	3.3E+02
158.4	4	6	3.7E+02
187.02	4	6	9.3E+02
187.30	6	8	9.6E+02
189.1	2	2	2.13E+02
191.0	4	2	4.11E+02
222.9	4	2	2.2E+02
346.3	4	6	2.4E+02
346.5	6	8	2.5E+02
Cr XV			
18.497	1	3	1.62E+05
18.782	1	3	2.8E+04
19.015	1	3	6.3E+02
20.863	1	3	6.0E+03
21.153	1	3	5.6E+03
102	3	3	1.6E+02
102.18	5	3	7.0E+02
103	3	1	3.8E+02
105	7	5	5.3E+02
111.27	3	3	1.7E+02
Cr XVI			
17.073	4	6	1.2E+04
17.242	2	4	8.6E+04
17.299	4	4	2.5E+04
17.372	4	4	1.4E+05
17.438	4	2	1.1E+05
17.514	2	4	1.1E+05
17.587	2	4	2.0E+04
17.656	2	2	2.0E+04
19.442	4	2	9.9E+03
19.714	2	2	1.1E+04
Cr XVII			
16.31	5	3	9.6E+03
16.32	5	7	3.2E+04
16.37	3	1	9.7E+04
16.44	5	7	1.3E+05
16.59	3	1	5.7E+04
16.65	5	5	1.1E+04
16.66	1	3	1.8E+05
16.68	5	7	6.8E+04
16.80	5	7	4.4E+04
16.97	1	3	2.63E+04
16.97	3	3	1.5E+04
17.968	5	3	8.6E+03
18.336	5	3	1.7E+04
18.336	5	5	1.6E+04
18.389	1	3	9.2E+03
Cr XVIII			
95.77	4	2	3.08E+02
102.32	4	4	1.54E+02
104.98	6	4	8.7E+02
106.84	4	2	3.4E+02
110.41	4	2	7.9E+02
112.27	4	2	4.24E+02
119.62	2	2	3.2E+02
123.87	6	4	3.9E+02
125.51	4	4	3.4E+02
128.10	6	6	2.8E+02
136.52	4	2	1.66E+02
139.87	4	4	1.49E+02
140.82	4	2	2.66E+02
155.46	2	2	2.84E+02
157.40	4	4	2.83E+02
Cr XIX			
14.73	3	3	7.1E+04
14.80	1	3	1.3E+05
14.81	5	3	3.4E+04
14.84	5	7	1.3E+05
109.64	3	3	2.46E+02
110.37	5	3	6.0E+02
113.97	5	3	5.5E+02
118.31	3	1	3.29E+02
118.67	5	3	2.1E+02
118.83	3	3	1.35E+02
126.30	1	3	1.56E+02
126.33	5	5	4.35E+02
130.99	7	5	2.9E+02
134.89	3	1	1.98E+02
138.15	3	1	1.75E+02
138.45	5	5	1.71E+02
140.92	5	3	1.38E+02
143.57	3	1	7.2E+02
163.94	5	5	3.1E+02
179.18	3	1	1.45E+02
Cr XX			
14.13	2	4	1.1E+05
14.26	4	6	1.3E+05
128.42	4	4	3.8E+02
131.31	6	4	1.27E+02
133.82	2	4	8.3E+01
135.26	4	2	2.41E+02
140.75	4	4	1.35E+02
148.99	6	4	1.75E+02
156.00	2	4	8.4E+01
167.97	6	6	1.12E+02
180.85	4	4	1.6E+02
Cr XXI			
12.97	3	1	4.8E+04
12.98	5	5	3.9E+04
13.02	3	5	3.8E+04
13.02	5	7	3.9E+04
13.08	1	3	5.2E+04
13.22	3	1	4.6E+04
13.34	3	5	5.2E+04
13.49	1	3	9.0E+04
13.53	3	3	6.6E+04
13.55	3	5	1.2E+05
13.65	5	7	1.5E+05
13.66	3	1	1.2E+05
13.67	5	5	3.9E+04
13.68	3	3	8.2E+04
13.75	5	3	4.5E+04

λ Å	Weights g_i	g_k	A 10^8 s^{-1}
13.75	5	5	9.5E+04
13.76	1	3	1.51E+05
13.78	5	7	1.7E+05
13.84	5	7	2.59E+05
13.87	3	5	8.5E+04
13.92	3	5	8.5E+04
13.93	5	7	4.2E+04
13.95	5	5	3.8E+04
14.04	3	5	1.2E+05
14.24	1	3	1.41E+05
Cr XXII			
2.190	4	2	1.7E+06
2.191	2	2	2.5E+06
2.198	4	4	4.5E+06
2.199	2	4	2.3E+06
2.202	4	6	1.6E+06
2.203	4	2	1.3E+06
13.149	2	4	1.29E+05
13.292	4	6	1.54E+05
Cr XXIII			
1.7632	1	3	3.68E+05
1.8557	1	3	8.97E+05
2.095	3	1	3.5E+06
2.101	1	3	2.0E+06
2.101	5	5	7.9E+05
2.102	3	5	2.1E+06
2.103	3	5	1.2E+06
2.104	1	3	1.4E+06
2.105	3	3	9.6E+05
2.106	3	3	2.0E+06
2.107	5	5	2.3E+06
2.107	3	5	3.3E+06
2.109	5	3	1.7E+06
2.113	3	5	5.9E+05
2.119	3	1	2.7E+05
2.129	3	1	5.1E+05
2.1818	1	3	3.37E+06
2.1923	1	3	2.34E+05
Cobalt			
Co I			
2287.80	8	8	8.6E-01
2295.22	10	8	2.2E-01
2309.03	10	10	5.6E-01
2323.13	8	8	5.0E-01
2325.53	6	8	1.1E-01
2335.98	6	6	5.1E-01
2338.66	4	4	7.7E-01
2353.36	8	10	1.5E-01
2355.48	6	8	1.3E-01
2358.18	4	6	1.4E-01
2365.06	10	10	1.3E-01
2371.85	6	8	7.3E-02
2384.86	10	8	2.4E-01
2392.03	6	6	4.0E-01
2402.06	8	6	5.1E-01
2407.25	10	12	3.6E+00
2412.76	4	6	6.5E-01
2414.46	6	8	3.4E+00
2415.29	4	6	3.6E+00
2424.93	10	10	3.2E+00
2432.21	8	8	2.6E+00
2436.66	6	6	2.6E+00
2439.04	4	4	2.7E+00
2460.80	4	6	1.2E-01
2467.69	6	8	7.0E-02
2470.27	10	12	1.5E-01
2476.64	10	8	2.2E-01
2504.52	10	8	1.8E-01
2511.02	10	10	9.2E-01
2521.36	10	8	3.0E+00
2528.97	8	6	2.8E+00
2530.13	6	6	7.1E-02
2535.96	6	4	1.9E+00
2536.50	8	8	3.0E-01
2544.25	4	2	3.0E+00
2562.12	4	4	3.9E-01
2567.34	6	6	3.0E-01
2574.35	8	8	1.7E-01
2685.34	6	8	5.5E-02
3017.55	8	6	6.9E-02
3044.00	10	10	1.0E-01
3048.89	6	4	7.5E-02
3061.82	8	8	1.6E-01
3072.34	6	6	1.5E-01
3086.78	4	4	1.9E-01
3354.37	8	6	1.1E-01
3367.11	10	8	6.0E-02
3385.22	8	6	1.1E-01
3388.16	6	4	2.4E-01
3395.37	6	8	2.9E-01
3405.12	10	10	1.0E+00
3409.17	8	8	4.2E-01
3412.34	8	10	6.1E-01
3412.63	10	8	1.2E-01
3414.74	4	4	8.8E-02
3417.15	6	6	3.2E-01
3431.58	8	6	1.1E-01
3433.05	4	4	1.0E+00
3442.92	6	4	1.2E-01
3443.64	8	8	6.9E-01
3449.17	6	6	7.6E-01
3449.44	10	10	1.8E-01
3453.51	10	12	1.1E+00
3455.24	4	2	1.9E-01
3462.80	4	6	7.9E-01
3465.79	10	12	9.2E-02
3474.02	6	8	5.6E-01
3483.41	8	10	5.5E-02
3489.40	8	6	1.3E+00
3491.32	4	4	5.0E-02
3495.68	4	6	4.9E-01
3502.28	10	8	8.0E-01
3502.63	6	6	5.2E-02
3506.32	8	6	8.2E-01
3509.84	6	8	3.2E-01
3512.64	6	4	1.0E+00
3513.48	8	10	7.8E-02
3518.34	6	4	1.6E+00
3521.58	10	8	1.8E-01
3523.42	4	2	9.8E-01
3526.85	10	10	1.3E-01
3529.03	6	8	8.8E-02
3529.82	8	10	4.6E-01
3533.36	4	6	9.1E-02
3560.89	4	4	2.3E-01
3564.95	6	8	7.0E-02
3569.37	8	8	1.5E+00
3574.97	6	6	1.5E-01
3575.36	8	8	9.6E-02
3585.15	8	8	7.1E-02
3587.19	6	6	1.4E+00
3594.87	6	6	9.2E-02
3602.08	4	4	1.0E-01
3704.06	6	8	1.2E-01
3745.49	8	8	7.5E-02
3842.05	8	6	1.3E-01
3845.47	8	10	4.6E-01
3861.16	6	4	1.4E-01
3873.12	10	8	1.2E-01
3873.95	8	6	1.0E-01
3881.87	6	4	8.2E-02
3894.07	6	8	6.9E-01
3894.98	4	2	8.8E-02
3935.96	8	10	6.2E-02
3995.31	8	10	2.5E-01
3997.90	6	8	7.0E-02
4092.39	8	8	5.7E-02
4110.53	6	6	5.5E-02
4118.77	6	8	1.6E-01
4121.32	8	10	1.9E-01
5146.75	8	8	1.5E-01
5212.70	10	10	1.9E-01
5265.79	6	8	5.0E-02
5280.63	10	8	2.8E-01
5352.05	12	10	2.7E-01
5477.09	6	8	6.8E-02
5483.96	8	10	7.3E-02
6082.43	10	10	5.4E-02
6455.00	8	10	9.0E-02
7838.12	8	10	5.4E-02
8093.93	12	10	2.0E-01
8372.79	10	10	8.7E-02
Co II			
2286.15	11	13	3.3E+00
2307.85	9	11	2.6E+00
2311.61	7	9	2.8E+00
2314.05	5	7	2.8E+00
2314.97	3	5	2.7E+00
2330.36	5	3	1.32E+00
2344.28	3	3	1.5E+00
2353.41	7	7	1.9E+00
2363.80	9	9	2.1E+00
2378.62	11	9	1.9E+00
2383.45	9	7	1.8E+00

λ Å	Weights g_i	g_k	A 10^8 s^{-1}
2388.92	11	11	2.8E+00
2389.54	5	3	1.5E+00
2404.17	3	3	1.5E+00
2417.66	9	9	8.5E-01
Copper			
Cu I			
*2024.3	2	6	9.8E-02
2165.1	2	4	5.1E-01
2178.9	2	4	9.13E-01
2181.7	2	2	1.0E+00
2225.7	2	2	4.6E-01
2244.3	2	4	1.19E-02
2441.6	2	2	2.0E-02
2492.2	2	4	3.11E-02
2618.4	6	4	3.07E-01
2766.4	4	4	9.6E-02
2824.4	6	6	7.8E-02
2961.2	6	8	3.76E-02
3063.4	4	4	1.55E-02
3194.1	4	4	1.55E-02
3247.5	2	4	1.39E+00
3274.0	2	2	1.37E+00
3337.8	6	8	3.8E-03
4022.6	2	4	1.90E-01
4062.6	4	6	2.10E-01
4249.0	2	2	1.95E-01
4275.1	6	8	3.45E-01
4480.4	2	2	3.0E-02
4509.4	4	2	2.75E-01
4530.8	4	2	8.4E-02
4539.7	6	4	2.12E-01
4587.0	8	6	3.20E-01
4651.1	10	8	3.80E-01
4704.6	8	8	5.5E-02
5105.5	6	4	2.0E-02
5153.2	2	4	6.0E-01
5218.2	4	6	7.5E-01
5220.1	4	4	1.50E-01
5292.5	8	8	1.09E-01
5700.2	4	4	2.4E-03
5782.1	4	2	1.65E-02
Cu II			
2489.7	5	5	1.5E-02
2544.8	9	7	1.1E+00
2689.3	7	7	4.1E-01
2701.0	5	5	6.7E-01
2703.2	3	3	1.2E+00
2713.5	5	5	6.8E-01
2769.7	7	7	6.1E-01
Dysprosium			
Dy I			
2862.7	17	15	6.5E-02
2964.6	17	17	6.5E-02
3147.7	15	17	1.1E-01
3263.2	15	13	1.4E-01
3511.0	15	13	3.1E-01
3571.4	15	13	2.0E-01
3757.1	17	19	3.0E+00
3868.8	17	17	3.1E+00
3967.5	17	19	8.7E-01
4046.0	17	15	1.5E+00
4103.9	13	11	1.7E+00
4186.8	17	17	1.32E+00
4194.8	17	17	7.2E-01
4211.7	17	19	2.08E+00
4218.1	15	15	1.85E+00
4221.1	15	17	1.52E+00
4225.2	13	15	4.5E+00
4268.3	15	15	3.6E-02
4276.7	13	13	7.3E-01
4292.0	15	15	5.8E-02
4577.8	17	19	2.2E-02
4589.4	17	15	1.3E-01
4612.3	17	15	8.2E-02
5077.7	17	17	5.7E-03
5301.6	17	15	1.1E-02
5547.3	17	17	2.7E-03
5639.5	17	19	4.7E-03
5974.5	17	17	4.0E-03
5988.6	17	15	5.3E-03
6010.8	15	15	2.6E-02
6088.3	15	13	3.5E-02
6168.4	15	17	2.5E-02
6259.1	17	19	8.5E-03
6579.4	17	15	7.5E-03
Erbium			
Er I			
3862.9	13	13	2.5E+00
4008.0	13	15	2.6E+00
4151.1	13	11	1.8E+00
Europium			
Eu I			
2372.9	8	6	1.9E-01
2375.3	8	8	2.0E-01
2379.7	8	10	2.0E-01
2619.3	8	10	7.0E-03
2643.8	8	8	6.6E-03
2659.4	8	10	1.2E-02
2682.6	8	6	1.2E-02
2710.0	8	10	1.4E-01
2724.0	8	8	1.2E-01
2731.4	8	8	3.1E-02
2732.6	8	6	3.7E-02
2735.3	8	10	4.7E-02
2738.6	8	10	1.3E-02
2743.3	8	6	1.1E-01
2745.6	8	6	5.0E-02
2747.8	8	8	5.2E-02
2772.9	8	6	1.0E-02
2878.9	8	10	2.8E-02
2892.5	8	8	1.0E-01
2893.0	8	6	1.0E-01
2909.0	8	10	6.9E-02
2958.9	8	6	1.6E-02
3059.0	8	8	3.8E-02
3067.0	8	10	9.1E-03
3106.2	8	10	5.5E-02
3111.4	8	10	3.0E-01
3168.3	8	10	6.9E-02
3185.5	8	10	5.8E-03
3210.6	8	8	1.1E-01
3212.8	8	8	2.9E-01
3213.8	8	6	1.8E-01
3235.1	8	10	1.0E-02
3241.4	8	8	2.3E-02
3246.0	8	6	1.4E-02
3247.6	8	8	2.3E-02
3322.3	8	6	3.5E-02
3334.3	8	6	3.4E-01
3350.4	8	10	1.5E-02
3353.7	8	8	5.8E-03
3457.1	8	8	8.4E-03
3467.9	8	8	1.0E-02
3589.3	8	6	6.9E-03
4594.0	8	10	1.4E+00
4627.2	8	8	1.3E+00
4661.9	8	6	1.3E+00
5645.8	8	6	5.4E-03
5765.2	8	8	1.1E-02
6018.2	8	10	8.5E-03
6291.3	8	6	1.8E-03
6864.5	8	10	5.8E-03
7106.5	8	8	2.6E-03
Fluorine			
F I			
806.96	4	6	3.3E+00
809.60	2	4	2.8E+00
951.87	4	2	2.6E+00
954.83	4	4	6.4E+00
955.55	2	2	5.1E+00
958.52	2	4	1.3E+00
6239.7	6	4	2.5E-01
6348.5	4	4	1.8E-01
6413.7	2	4	1.1E-01
6708.3	6	4	1.4E-02
6774.0	6	6	1.0E-01
6795.5	4	2	5.2E-02
6834.3	4	4	2.1E-01
6856.0	6	8	4.2E-01
6870.2	2	2	3.8E-01
6902.5	4	6	3.2E-01
6909.8	2	4	2.2E-01
6966.4	4	2	1.1E-01
7037.5	4	4	3.0E-01
7127.9	2	2	3.8E-01
7309.0	6	8	4.7E-01
7311.0	4	2	3.9E-01
7314.3	4	6	4.8E-01
7332.0	6	4	3.1E-01
7398.7	6	6	3.1E-01
7425.7	4	2	3.4E-01
7482.7	4	4	5.6E-02
7489.2	2	2	1.1E-01

λ Å	Weights g_i	Weights g_k	A $10^8\ s^{-1}$
7514.9	2	2	5.2E-02
7552.2	4	6	7.8E-02
7573.4	2	4	1.0E-01
7607.2	4	4	7.0E-02
7754.7	4	6	3.0E-01
7800.2	2	4	2.1E-01

Gallium

Ga I

λ Å	Weights g_i	Weights g_k	A $10^8\ s^{-1}$
2195.4	2	2	1.9E-02
2199.7	4	2	3.3E-02
2214.4	4	6	1.2E-02
2235.9	4	2	4.3E-02
2255.0	2	2	3.1E-02
2259.2	4	6	3.1E-02
2294.2	2	4	7.0E-02
2297.9	4	2	5.8E-02
2338.2	4	6	9.8E-02
2371.3	2	2	5.7E-02
2418.7	4	2	1.0E-01
2450.1	2	4	2.8E-01
2500.2	4	6	3.4E-01
2659.9	2	2	1.2E-01
2719.7	4	2	2.3E-01
2874.2	2	4	1.2E+00
2943.6	4	6	1.4E+00
2944.2	4	4	2.7E-01
4033.0	2	2	4.9E-01
4172.0	4	2	9.2E-01

Ga II

λ Å	Weights g_i	Weights g_k	A $10^8\ s^{-1}$
829.60	1	3	2.2E-01
1414.4	1	3	1.88E+01

Germanium

Ge I

λ Å	Weights g_i	Weights g_k	A $10^8\ s^{-1}$
1944.7	3	1	7.0E-01
1955.1	3	3	2.8E-01
1988.3	5	3	2.5E-01
1998.9	5	5	5.5E-01
2041.7	1	3	1.1E+00
2065.2	3	3	8.5E-01
2068.7	3	5	1.2E+00
2086.0	3	5	4.0E-01
2094.3	5	7	9.7E-01
2105.8	5	5	1.7E-01
2256.0	5	5	3.2E-02
2417.4	5	5	9.6E-01
2498.0	1	3	1.3E-01
2533.2	3	3	1.0E-01
2589.2	5	3	5.1E-02
2592.5	3	5	7.1E-01
2651.2	5	5	2.0E+00
2651.6	1	3	8.5E-01
2691.3	3	3	6.1E-01
2709.6	3	1	2.8E+00
2754.6	5	3	1.1E+00
3039.1	5	3	2.8E+00
3124.8	5	5	3.1E-02
3269.5	5	3	2.9E-01
4226.6	1	3	2.1E-01
4685.8	1	3	9.5E-02

Ge II

λ Å	Weights g_i	Weights g_k	A $10^8\ s^{-1}$
999.10	2	4	1.9E+00
1016.6	4	6	2.1E+00
1017.1	4	4	3.5E-01
1055.0	2	2	6.9E-01
1075.1	4	2	1.3E+00
1237.1	2	4	1.9E+01
1261.9	4	6	2.2E+01
1264.7	4	4	3.5E+00
1602.5	2	2	3.4E+00
1649.2	4	2	6.5E+00
4741.8	2	4	4.6E-01
4814.6	4	6	5.1E-01
4824.1	4	4	8.6E-02
5131.8	4	6	1.9E+00
5178.5	6	6	1.3E-01
5178.6	6	8	2.0E+00
5893.4	2	4	9.2E-01
6021.0	2	2	8.4E-01
6336.4	2	2	4.4E-01
6484.2	4	2	8.5E-01

Gold

Au I

λ Å	Weights g_i	Weights g_k	A $10^8\ s^{-1}$
2427.95	2	4	1.99E+00
2675.95	2	2	1.64E+00
3122.78	6	4	1.90E-01
6278.30	4	2	3.4E-02

Helium

He I

λ Å	Weights g_i	Weights g_k	A $10^8\ s^{-1}$
510.00	1	3	4.62E-01
512.10	1	3	7.17E-01
515.62	1	3	1.3E+00
522.21	1	3	2.46E+00
537.03	1	3	5.66E+00
584.33	1	3	1.799E+01
*2696.1	3	9	5.50E-03
*2723.2	3	9	7.80E-03
*2763.8	3	9	1.11E-02
*2829.1	3	9	1.7E-02
*2945.1	3	9	3.20E-02
*3187.7	3	9	5.639E-02
3354.6	1	3	1.30E-02
3447.6	1	3	2.32E-02
*3554.4	9	15	1.31E-02
*3587.3	9	15	2.05E-02
3613.6	1	3	3.90E-02
*3634.2	9	15	2.61E-02
*3705.0	9	15	4.44E-02
*3819.6	9	15	6.36E-02
3833.6	3	5	9.71E-03
*3867.5	9	3	2.5E-02
3871.8	3	5	1.26E-02
*3888.7	3	9	9.478E-02
3926.5	3	5	1.95E-02
3964.7	1	3	7.19E-02
4009.3	3	5	2.79E-02
*4026.2	9	15	1.16E-01
*4120.8	9	3	4.44E-02
4143.8	3	5	4.85E-02
4387.9	3	5	8.94E-02
4437.6	3	1	3.3E-02
*4471.5	9	15	2.46E-01
*4713.2	9	3	9.55E-02
4921.9	3	5	1.98E-01
5015.7	1	3	1.338E-01
5047.7	3	1	6.75E-02
*5875.7	9	15	7.053E-01
6678.2	3	5	6.339E-01
*7065.2	9	3	2.786E-01
7281.4	3	1	1.829E-01
*8361.7	3	9	3.34E-03
*9463.6	3	9	5.01E-03
9603.4	1	3	6.10E-03
*9702.6	9	3	8.58E-03
*10311	9	15	2.01E-02
*10668	9	3	1.52E-02
*10830	3	9	1.022E-01
*10913	15	21	2.12E-02
10917	3	7	2.12E-02
*10997	15	9	1.3E-03
11013	1	3	1.00E-02
11045	3	5	1.85E-02
11226	3	1	1.08E-02
*11969	9	15	3.58E-02
*12528	3	9	7.10E-03
12756	5	3	1.2E-03
*12785	15	21	4.62E-02
12791	5	7	4.61E-02
*12846	9	3	2.89E-02
12968	3	5	3.43E-02
*12985	15	9	2.5E-03

Indium

In I

λ Å	Weights g_i	Weights g_k	A $10^8\ s^{-1}$
2560.2	2	4	4.0E-01
2710.3	4	6	4.0E-01
3039.4	2	4	1.3E+00
3256.1	4	6	1.3E+00
4101.8	2	2	5.6E-01
4511.3	4	2	1.02E+00

In II

λ Å	Weights g_i	Weights g_k	A $10^8\ s^{-1}$
2941.1	3	1	1.4E+00

Iodine

I I

λ Å	Weights g_i	Weights g_k	A $10^8\ s^{-1}$
1782.8	4	4	2.71E+00
1830.4	4	6	1.6E-01

Iridium

Ir I

λ Å	Weights g_i	Weights g_k	A $10^8\ s^{-1}$
2475.12	10	10	2.1E-01
2502.98	10	12	3.2E-01
2639.71	10	10	4.7E-01

λ Å	Weights g_i	g_k	A 10^8 s^{-1}
2661.98	10	10	2.5E-01
2664.79	10	8	4.0E-01
2694.23	10	12	4.8E-01
2849.72	10	10	2.2E-01
2853.31	10	10	2.0E-03
2882.64	10	8	7.2E-02
2924.79	10	12	1.42E-01
2934.64	8	10	2.0E-01
2951.22	10	8	2.8E-02
3003.63	8	10	5.9E-02
3168.88	8	10	5.47E-02
3220.78	10	8	2.4E-01
3558.99	6	8	1.5E-02
3573.72	8	10	5.4E-02
3617.21	6	8	2.0E-02
3628.67	8	8	2.8E-02
3661.71	8	10	4.0E-02
3734.77	8	8	2.7E-02
4033.76	8	10	2.7E-02
4069.92	6	8	3.6E-02
4913.35	12	12	3.3E-02
4939.24	10	12	2.5E-03

Iron

Fe I

λ Å	Weights g_i	g_k	A 10^8 s^{-1}
1934.54	9	7	2.5E-01
1937.27	9	7	2.2E-01
1940.66	7	5	2.6E-01
2084.12	9	7	3.7E-01
2102.35	7	7	8.8E-02
2112.97	1	3	1.9E-01
2132.02	9	9	7.6E-02
2145.19	7	7	5.7E-02
2153.01	5	5	6.9E-02
2161.58	3	5	5.0E-02
2166.77	9	7	2.7E+00
2171.30	5	7	5.1E-02
2173.21	3	5	8.3E-02
2176.84	1	3	1.0E-01
2191.20	1	3	7.3E-02
2191.84	5	5	1.2E+00
2196.04	3	3	1.2E+00
2200.72	3	5	2.8E-01
2259.51	9	11	7.0E-02
2267.08	7	5	7.1E-02
2272.07	7	9	3.8E-02
2276.03	9	7	1.7E-01
2277.11	7	5	3.7E+01
2287.25	5	3	3.4E-01
2292.52	7	9	4.3E-02
2294.41	3	1	6.1E-01
2300.14	5	7	8.0E-02
2301.68	1	3	1.3E-01
2303.42	1	3	9.4E-02
2303.58	3	5	7.6E-02
2309.00	3	5	1.5E-01
2313.10	5	7	1.4E-01
2320.36	7	9	1.2E-01
2371.43	5	5	5.2E-02
2373.62	7	7	6.7E-02
2374.52	1	3	2.9E-01
2381.83	3	5	5.4E-02
2389.97	5	7	5.0E-02
2462.18	7	5	1.5E-01
2462.65	9	9	5.8E-01
2479.78	5	5	1.8E+00
2483.27	9	11	4.9E+00
2488.14	7	9	4.7E+00
2490.64	5	7	3.8E+00
2491.15	3	5	3.0E+00
2501.13	9	7	6.8E-01
2510.83	7	5	1.3E+00
2518.10	5	3	1.9E+00
2522.85	9	9	2.9E+00
2524.29	3	1	3.4E+00
2527.43	7	7	1.9E+00
2529.13	5	5	9.8E-01
2535.61	1	3	9.7E-01
2540.97	3	5	9.2E-01
2545.98	5	7	6.7E-01
2549.61	7	9	3.6E-01
2584.54	11	13	4.6E-01
2606.83	9	11	4.2E-01
2618.02	7	7	4.0E-01
2623.53	7	9	3.3E-01
2656.15	13	15	2.8E-01
2669.49	11	13	1.7E-01
2679.06	11	11	1.9E-01
2719.03	9	7	1.4E+00
2720.90	7	5	1.1E+00
2723.58	5	3	6.4E-01
2733.58	11	9	8.6E-01
2735.48	9	7	6.2E-01
2737.31	3	3	8.5E-01
2742.41	5	5	6.3E-01
2744.07	1	3	3.5E-01
2750.14	7	7	3.9E-01
2756.33	3	5	2.0E-01
2788.10	11	13	6.3E-01
2894.50	5	5	6.2E-01
2899.42	5	3	5.9E-01
2920.69	5	5	5.2E-02
2923.29	11	11	1.6E+00
2925.36	7	9	1.8E-01
2929.01	7	5	7.3E-02
2936.90	9	9	1.3E-01
2941.34	5	3	5.6E-02
2947.88	7	7	2.0E-01
2953.94	5	5	1.89E-01
2954.65	5	7	1.0E-01
2957.36	3	3	1.77E-01
2965.25	1	3	1.16E-01
2966.90	9	11	2.72E-01
2969.36	3	1	3.66E-02
2973.13	5	7	1.35E-01
2973.24	7	9	1.83E-01
2980.53	7	7	2.2E-01
2981.45	7	5	6.54E-02
2983.57	9	7	2.80E-01
2987.29	9	7	6.6E-02
2990.39	9	11	3.9E-01
2994.43	7	5	4.4E-01
2996.39	3	5	1.6E-01
2999.51	11	11	2.3E-01
3000.95	5	3	6.42E-01
3008.14	3	1	1.07E+00
3009.09	13	11	6.7E-02
3009.57	9	9	1.7E-01
3011.48	7	9	4.7E-01
3015.92	11	9	5.9E-02
3016.18	5	3	8.5E-02
3017.63	3	3	6.82E-02
3018.98	7	7	1.3E-01
3021.07	7	7	4.56E-01
3024.03	3	5	4.88E-02
3025.84	1	3	3.48E-01
3026.46	5	5	1.1E-01
3031.63	3	3	1.5E-01
3037.39	3	5	3.2E-01
3042.02	3	5	4.9E-02
3042.66	5	7	5.7E-02
3047.60	5	7	2.84E-01
3053.07	3	5	1.5E-01
3057.45	11	9	4.4E-01
3059.09	7	9	1.7E-01
3067.24	9	7	3.4E-01
3068.17	5	3	9.8E-02
3075.72	7	5	2.9E-01
3083.74	5	3	3.0E-01
3091.58	3	1	5.4E-01
3098.19	11	11	1.1E-01
3100.67	7	7	1.4E-01
3119.49	11	9	8.2E-02
3120.43	9	7	8.9E-02
3156.27	7	7	5.4E-01
3160.66	9	9	1.9E-01
3161.95	11	13	1.2E-01
3166.44	9	7	1.14E-01
3168.85	5	7	5.7E-02
3175.45	11	11	1.3E-01
3176.36	5	3	9.2E-02
3196.93	9	11	9.0E-01
3199.53	9	9	2.6E-01
3205.40	3	3	1.2E+00
3215.94	5	5	8.0E-01
3217.38	11	9	2.2E-01
3219.58	7	9	6.2E-01
3222.07	11	11	3.3E-01
3225.79	11	13	8.8E-01
3227.80	9	7	1.4E+00
3228.25	5	3	4.5E-01
3229.99	9	11	4.5E-01
3230.21	5	5	1.9E-01
3230.96	7	5	3.9E-01
3233.05	13	15	5.4E-01
3233.97	9	9	2.0E-01
3246.96	5	3	9.9E-02

λ Å	Weights g_i	g_k	A 10^8 s^{-1}	λ Å	Weights g_i	g_k	A 10^8 s^{-1}	λ Å	Weights g_i	g_k	A 10^8 s^{-1}
3248.20	7	7	2.2E-01	3510.44	1	3	4.4E-02	3630.35	9	7	7.6E-02
3253.60	7	9	1.8E-01	3516.56	7	5	3.7E-02	3631.46	7	9	5.17E-01
3254.36	11	13	5.1E-01	3521.84	3	5	9.6E-02	3632.04	3	5	4.8E-01
3257.59	7	5	1.4E-01	3523.31	5	3	7.6E-02	3632.55	11	9	5.2E-02
3265.62	7	5	3.8E-01	3524.08	7	5	7.5E-02	3635.19	5	3	1.4E-01
3268.23	3	3	5.9E-02	3524.24	5	7	4.2E-02	3637.86	9	9	5.5E-02
3271.00	5	3	6.6E-01	3527.79	9	9	2.0E-01	3638.30	7	9	2.6E-01
3280.26	9	11	5.4E-01	3529.82	3	3	7.6E-01	3640.39	9	11	3.8E-01
3282.89	3	5	3.0E-01	3536.56	5	7	7.8E-01	3644.80	7	5	7.8E-02
3284.59	5	5	5.4E-02	3537.73	5	3	1.1E-01	3645.82	1	3	5.7E-01
3290.99	3	5	6.0E-02	3537.90	11	11	8.4E-02	3647.84	9	11	2.92E-01
3292.02	7	9	6.1E-01	3540.12	7	9	1.2E-01	3649.51	11	9	4.2E-01
3292.59	3	3	2.6E-01	3541.08	9	11	6.2E-01	3650.03	7	7	9.9E-02
3298.13	3	5	8.1E-02	3542.08	7	9	7.4E-01	3651.47	7	9	6.2E-01
3305.97	5	7	4.7E-01	3543.67	3	5	1.8E-01	3655.46	5	5	1.0E-01
3306.36	3	5	6.1E-01	3548.02	5	3	9.7E-02	3659.52	9	9	5.8E-02
3307.23	13	13	2.0E-01	3552.11	3	5	4.5E-02	3667.25	9	7	1.4E-01
3314.74	5	7	6.9E-01	3552.83	5	5	1.5E-01	3669.15	9	7	7.4E-02
3322.47	9	11	6.2E-02	3553.74	11	9	8.1E-01	3669.52	9	7	3.0E-01
3323.74	5	5	3.0E-01	3556.88	9	11	4.4E-01	3670.09	11	13	7.6E-02
3328.87	11	11	2.7E-01	3559.50	3	3	1.9E-01	3674.77	5	3	6.7E-02
3337.66	11	9	5.7E-02	3560.70	7	9	6.5E-02	3676.31	9	11	4.63E-02
3347.93	5	5	4.0E-02	3565.38	7	9	3.8E-01	3677.31	5	7	3.1E-01
3354.06	1	3	7.7E-02	3567.03	5	7	6.5E-02	3677.63	7	5	8.0E-01
3355.23	9	9	3.2E-01	3568.42	5	3	5.3E-02	3678.86	3	5	4.1E-02
3369.55	9	9	2.4E-01	3568.82	7	9	5.6E-02	3682.24	5	5	1.7E+00
3370.78	11	11	3.3E-01	3570.10	9	11	6.77E-01	3684.11	9	7	3.4E-01
3380.11	7	7	2.4E-01	3572.00	11	11	2.4E-01	3686.00	9	11	2.6E-01
3383.98	7	7	9.3E-02	3573.39	5	7	7.5E-02	3686.26	3	1	1.2E-01
3392.65	7	7	2.6E-01	3576.76	11	9	9.6E-02	3687.46	11	9	8.01E-02
3394.58	5	3	9.9E-02	3578.38	1	3	6.3E-02	3688.48	7	9	6.9E-02
3399.33	5	5	3.8E-01	3581.19	11	13	1.02E+00	3690.73	11	11	2.7E-01
3402.26	13	13	2.8E-01	3582.20	13	11	2.5E-01	3694.01	5	7	6.8E-01
3406.44	3	5	3.0E-01	3583.33	1	3	2.3E-01	3697.43	7	7	2.1E-01
3407.46	7	9	5.8E-01	3585.32	7	7	1.3E-01	3698.60	5	7	3.8E-02
3410.17	3	5	4.7E-01	3585.71	9	9	3.75E-02	3699.15	5	7	4.5E-02
3411.35	9	9	5.5E-02	3586.98	5	5	1.6E-01	3701.09	7	9	4.8E-01
3413.13	5	7	3.6E-01	3591.48	1	3	6.0E-02	3702.03	3	1	3.5E-01
3417.84	3	3	5.1E-01	3592.67	7	5	4.0E-02	3703.69	9	11	5.3E-02
3418.51	3	1	1.3E+00	3594.63	9	9	2.7E-01	3703.82	1	3	1.2E-01
3424.28	7	7	2.0E-01	3595.30	5	5	5.4E-02	3704.46	11	9	1.3E-01
3425.01	9	7	2.8E-01	3597.02	5	3	1.7E-01	3709.25	9	7	1.56E-01
3427.12	7	9	5.5E-01	3599.62	11	9	1.8E-01	3711.41	3	5	7.3E-02
3428.19	5	5	2.1E-01	3603.20	11	11	2.6E-01	3718.41	7	7	5.3E-02
3428.75	7	5	2.7E-01	3603.82	3	3	1.7E-01	3719.93	9	11	1.62E-01
3440.99	7	5	8.4E-02	3605.45	9	9	6.4E-01	3722.56	5	5	4.97E-02
3442.36	5	5	4.55E-02	3606.68	11	13	8.2E-01	3724.38	5	7	1.3E-01
3443.88	5	3	6.2E-02	3608.86	3	5	8.14E-01	3726.93	5	5	4.6E-01
3445.15	5	7	2.8E-01	3610.16	13	13	4.8E-01	3727.09	9	7	2.0E-01
3447.28	5	5	9.1E-02	3610.70	5	3	7.1E-02	3727.62	7	5	2.25E-01
3450.33	3	3	2.0E-01	3612.07	11	13	7.5E-02	3730.39	9	11	1.3E-01
3476.70	1	3	5.4E-02	3613.45	7	7	6.7E-02	3730.95	5	7	3.8E-02
3477.85	3	1	4.2E-02	3615.19	3	3	5.8E-02	3732.40	5	5	2.8E-01
3485.34	5	3	1.4E-01	3617.79	5	7	6.5E-01	3733.32	3	3	6.2E-02
3495.29	9	7	9.46E-02	3618.77	5	7	7.3E-01	3734.86	11	11	9.02E-01
3497.10	7	7	1.4E-01	3621.46	9	11	5.1E-01	3735.32	9	9	2.4E-01
3505.07	5	3	9.9E-02	3622.00	7	7	5.1E-01	3737.13	7	9	1.42E-01
3506.50	5	5	7.1E-02	3623.19	13	13	7.4E-02	3738.31	11	13	3.8E-01
3508.49	9	11	5.7E-02	3624.06	5	3	5.4E-02	3740.24	7	9	1.4E-01

λ Å	Weights g_i	g_k	A 10^8 s^{-1}	λ Å	Weights g_i	g_k	A 10^8 s^{-1}	λ Å	Weights g_i	g_k	A 10^8 s^{-1}
3742.62	9	9	1.0E-01	3845.17	3	3	6.8E-02	3976.61	3	5	1.8E-01
3743.36	5	3	2.60E-01	3845.69	5	7	4.9E-02	3977.74	5	5	7.0E-02
3744.10	5	3	3.6E-01	3846.00	9	7	4.3E-02	3981.77	9	9	3.9E-02
3745.56	5	7	1.15E-01	3846.41	11	9	1.9E-01	3983.96	9	7	7.6E-02
3745.90	1	3	7.33E-02	3846.80	7	7	6.6E-01	3985.39	5	5	6.7E-02
3746.93	7	7	2.2E-01	3849.96	3	1	6.06E-01	3989.86	5	7	5.0E-02
3748.26	3	5	9.15E-02	3856.37	7	5	4.64E-02	3996.97	9	9	6.7E-02
3749.48	9	9	7.64E-01	3859.21	13	11	8.5E-02	3997.39	9	11	1.5E-01
3753.61	7	5	9.3E-02	3859.91	9	9	9.70E-02	3998.05	11	9	6.6E-02
3756.94	11	11	2.4E-01	3865.52	3	3	1.55E-01	4003.76	3	3	7.1E-02
3757.45	5	3	1.2E-01	3867.22	5	5	3.4E-01	4005.24	7	5	2.04E-01
3758.23	7	7	6.34E-01	3871.75	11	11	6.7E-02	4006.31	11	9	4.7E-02
3760.05	13	15	4.47E-02	3872.50	5	5	1.05E-01	4007.27	7	5	4.2E-02
3760.53	3	5	4.8E-02	3873.76	11	9	8.0E-02	4009.71	3	5	5.2E-02
3763.79	5	5	5.44E-01	3878.02	7	7	7.72E-02	4014.53	11	11	2.4E-01
3765.54	13	15	9.8E-01	3878.57	5	3	6.6E-02	4017.15	9	11	4.5E-02
3766.67	5	3	9.7E-02	3883.28	7	7	1.6E-01	4021.87	7	9	1.0E-01
3767.19	3	3	6.40E-01	3884.36	11	9	3.5E-02	4024.72	7	9	8.9E-02
3768.03	3	1	8.4E-02	3885.51	3	5	5.8E-02	4031.96	3	5	7.1E-02
3774.82	3	3	4.7E-02	3886.28	7	7	5.30E-02	4040.64	5	7	4.4E-02
3778.51	7	5	1.2E-01	3887.05	9	9	3.52E-02	4044.61	5	3	1.1E-01
3781.94	5	7	3.7E-02	3888.51	5	5	2.6E-01	4045.81	9	9	8.63E-01
3785.95	11	13	4.2E-02	3888.82	5	3	2.7E-01	4054.87	5	3	1.6E-01
3786.19	5	5	1.2E-01	3891.93	3	3	4.0E-01	4058.22	9	7	4.9E-02
3787.16	5	5	1.0E-01	3893.39	11	11	1.3E-01	4059.73	5	3	8.1E-02
3787.88	3	5	1.29E-01	3895.66	3	1	9.40E-02	4062.44	3	3	2.2E-01
3789.82	9	7	3.9E-02	3900.52	7	7	7.5E-02	4063.59	7	7	6.8E-01
3791.73	5	3	6.3E-02	3902.95	7	7	2.14E-01	4065.40	3	1	1.9E-01
3793.87	3	3	7.4E-02	3903.90	9	9	9.6E-02	4067.98	9	9	1.7E-01
3794.34	9	11	3.8E-02	3906.75	5	7	6.7E-02	4070.77	7	5	1.3E-01
3795.00	5	7	1.15E-01	3907.93	7	5	6.7E-02	4071.74	5	5	7.65E-01
3799.55	7	9	7.32E-02	3909.66	3	5	5.3E-02	4073.76	5	3	1.6E-01
3801.68	5	7	6.6E-02	3909.83	3	3	6.5E-02	4074.79	9	9	4.8E-02
3802.00	11	13	3.5E-02	3914.27	3	3	5.4E-02	4076.63	9	9	1.9E-01
3802.28	5	5	5.0E-02	3916.73	13	11	1.2E-01	4078.35	5	3	4.2E-02
3804.01	11	9	4.7E-02	3919.07	9	9	3.9E-02	4079.18	5	5	5.1E-02
3805.35	9	11	9.8E-01	3925.20	1	3	5.7E-02	4079.84	1	3	6.3E-02
3806.22	3	3	2.3E-01	3931.12	5	7	4.5E-02	4080.21	3	1	2.4E-01
3806.70	11	11	5.4E-01	3941.28	5	5	8.4E-02	4082.44	3	3	3.8E-02
3807.54	3	5	8.0E-02	3942.44	3	5	9.0E-02	4084.49	11	9	1.1E-01
3808.73	9	9	3.54E-02	3946.99	9	11	4.4E-02	4085.00	3	5	4.2E-02
3810.76	5	3	2.0E-01	3948.77	11	9	2.2E-01	4085.30	7	7	1.1E-01
3813.88	13	11	8.7E-02	3949.14	3	3	3.9E-02	4085.98	7	5	5.0E-02
3815.84	9	7	1.3E+00	3949.95	7	5	5.9E-02	4088.57	5	3	3.9E-02
3817.64	11	11	8.3E-02	3951.16	3	5	3.6E-01	4098.18	7	7	6.8E-02
3819.50	7	5	4.6E-02	3952.60	11	11	4.1E-02	4107.49	5	3	2.5E-01
3820.43	11	9	6.68E-01	3953.15	7	9	3.7E-02	4109.07	1	3	4.5E-02
3821.18	11	13	7.0E-01	3955.34	3	3	1.4E-01	4109.80	3	3	1.6E-01
3821.83	5	5	7.8E-02	3955.96	3	3	5.7E-02	4112.96	11	13	1.4E-01
3825.88	9	7	5.98E-01	3956.45	13	11	2.1E-01	4114.45	5	5	4.7E-02
3827.82	7	5	1.05E+00	3957.02	5	7	1.6E-01	4118.54	11	13	5.8E-01
3833.31	9	9	4.69E-02	3960.28	5	7	4.2E-02	4126.18	11	11	3.9E-02
3834.22	7	5	4.53E-01	3963.10	3	5	1.7E-01	4127.61	1	3	1.3E-01
3836.33	5	5	3.7E-01	3967.42	9	7	2.3E-01	4132.06	5	7	1.2E-01
3839.26	9	9	2.8E-01	3967.96	7	9	6.3E-02	4132.90	3	5	9.4E-02
3839.61	3	5	3.9E-01	3969.26	9	7	2.3E-01	4134.68	5	7	1.8E-01
3840.44	5	3	4.70E-01	3970.39	3	1	3.5E-01	4137.00	3	5	2.2E-01
3841.05	5	3	1.3E+00	3971.32	11	9	5.7E-02	4137.42	5	7	6.1E-02
3843.26	9	7	4.7E-01	3973.65	5	7	6.6E-02	4142.63	3	5	7.4E-02

ATOMIC TRANSITION PROBABILITIES (continued)

λ Å	Weights g_i	g_k	A 10^8 s^{-1}	λ Å	Weights g_i	g_k	A 10^8 s^{-1}	λ Å	Weights g_i	g_k	A 10^8 s^{-1}
4143.87	7	9	1.5E-01	4352.73	3	5	3.9E-02	5125.11	9	7	2.6E-01
4149.37	11	13	3.6E-02	4369.77	9	9	7.2E-02	5133.69	11	13	2.7E-01
4150.25	3	3	7.1E-02	4383.54	9	11	5.00E-01	5137.38	11	9	1.1E-01
4153.90	7	9	2.3E-01	4387.89	3	3	3.9E-02	5159.06	5	3	1.3E-01
4154.80	9	11	1.5E-01	4388.41	7	7	1.3E-01	5162.27	11	11	2.4E-01
4156.80	5	5	1.9E-01	4401.29	7	7	5.9E-02	5184.26	5	7	3.5E-02
4158.79	3	5	1.6E-01	4404.75	7	9	2.75E-01	5208.59	7	5	5.2E-02
4170.90	5	5	6.1E-02	4415.12	5	7	1.19E-01	5232.94	9	11	1.4E-01
4172.12	7	5	9.7E-02	4422.57	3	3	8.8E-02	5263.30	5	5	5.2E-02
4175.64	3	5	1.6E-01	4430.61	3	1	7.45E-02	5266.55	7	9	8.6E-02
4181.75	5	7	3.6E-01	4433.22	5	3	2.3E-01	5283.62	7	7	8.0E-02
4182.38	5	5	4.9E-02	4438.34	3	1	7.9E-02	5302.30	3	5	6.3E-02
4184.89	5	5	1.1E-01	4442.34	5	5	3.76E-02	5324.18	9	9	1.5E-01
4187.04	7	5	2.15E-01	4443.19	1	3	1.1E-01	5339.93	5	7	7.0E-02
4187.79	9	7	1.52E-01	4446.83	3	3	5.3E-02	5353.39	9	7	4.8E-02
4191.68	1	3	4.8E-02	4447.72	3	3	5.11E-02	5364.87	5	7	5.5E-01
4196.21	7	7	9.8E-02	4454.38	5	5	3.8E-02	5367.47	7	9	5.8E-01
4198.30	11	9	8.03E-02	4455.03	9	7	3.9E-02	5369.96	9	11	4.7E-01
4198.64	5	5	1.3E-01	4466.55	5	7	1.2E-01	5373.71	7	9	3.5E-02
4199.09	9	11	6.1E-01	4469.37	5	7	2.6E-01	5383.37	11	13	5.6E-01
4200.09	7	7	4.0E-02	4481.61	3	3	4.2E-02	5389.48	7	7	1.3E-01
4200.92	7	9	4.2E-02	4484.22	7	9	7.0E-02	5398.29	5	5	9.8E-02
4202.03	9	9	8.22E-02	4485.67	3	3	1.1E-01	5400.50	9	9	1.8E-01
4203.67	7	9	8.6E-02	4528.61	7	9	5.44E-02	5410.91	7	9	4.8E-01
4203.94	13	13	1.3E-01	4533.13	3	1	3.7E-02	5415.20	11	13	5.6E-01
4205.54	5	5	3.6E-02	4547.85	5	7	7.6E-02	5424.07	13	15	5.0E-01
4207.13	5	3	4.3E-02	4619.29	7	5	4.7E-02	5432.95	5	5	4.1E-02
4210.34	3	3	1.7E-01	4669.17	5	3	4.0E-02	5445.04	11	11	2.0E-01
4213.65	3	1	1.9E-01	4673.16	5	7	4.6E-02	5463.27	9	9	3.2E-01
4217.55	3	5	2.3E-01	4678.85	7	9	7.4E-02	5466.39	9	7	7.5E-02
4219.36	11	13	3.8E-01	4704.95	3	1	8.1E-02	5473.90	7	7	5.5E-02
4220.34	3	1	1.9E-01	4736.77	9	11	4.9E-02	5480.87	3	1	1.2E-01
4222.21	7	7	5.77E-02	4789.65	5	5	7.2E-02	5487.74	7	5	8.6E-02
4224.17	9	11	1.3E-01	4859.74	5	3	1.3E-01	5554.89	9	9	8.7E-02
4224.51	3	5	7.1E-02	4871.32	7	5	2.2E-01	5569.62	5	3	2.1E-01
4225.45	5	7	1.7E-01	4872.14	3	3	2.4E-01	5572.84	7	5	2.1E-01
4226.42	3	3	3.7E-02	4878.21	1	3	9.1E-02	5576.09	3	1	2.1E-01
4233.60	3	5	1.85E-01	4890.75	5	5	2.1E-01	5586.76	9	7	1.9E-01
4235.94	9	9	1.88E-01	4891.49	9	7	2.9E-01	5598.30	5	5	1.8E-01
4238.81	7	9	2.2E-01	4892.87	3	3	4.8E-02	5615.64	11	9	1.7E-01
4240.37	5	3	5.7E-02	4903.31	3	5	4.7E-02	5624.54	5	5	5.3E-02
4245.26	1	3	8.3E-02	4917.23	5	3	6.1E-02	5633.97	11	13	8.7E-02
4246.08	7	5	5.7E-02	4918.01	1	3	4.0E-02	5638.27	9	7	4.0E-02
4247.43	9	11	2.0E-01	4918.99	7	7	1.7E-01	5650.01	3	5	5.0E-02
4248.22	3	5	3.5E-02	4920.50	11	9	3.5E-01	5655.18	7	9	5.3E-02
4250.12	5	7	2.08E-01	4930.31	3	3	4.1E-02	5658.82	7	7	3.6E-02
4250.79	7	7	1.0E-01	4969.92	3	3	1.8E-01	5679.02	5	7	3.6E-02
4260.47	11	11	3.2E-01	4973.10	3	3	1.0E-01	5686.53	9	11	4.4E-02
4267.83	1	3	9.4E-02	4978.60	5	3	1.1E-01	5691.51	3	1	6.2E-02
4268.75	5	3	4.2E-02	4988.95	7	7	4.9E-02	5705.99	7	9	6.7E-02
4271.15	7	9	1.82E-01	4991.27	5	7	8.2E-02	5717.85	1	3	5.0E-02
4271.76	9	11	2.28E-01	5001.86	9	7	3.9E-01	5753.12	3	5	7.0E-02
4282.40	7	5	1.1E-01	5004.04	5	3	3.5E-02	5762.99	5	7	1.0E-01
4300.83	5	5	4.7E-02	5014.94	7	5	3.0E-01	5816.36	9	11	3.7E-02
4305.45	5	3	6.0E-02	5022.24	5	3	2.6E-01	5905.67	5	3	1.2E-01
4307.90	7	9	3.4E-01	5074.75	9	11	1.5E-01	5927.80	5	3	5.1E-02
4315.08	5	5	7.7E-02	5090.78	7	5	2.0E-01	5930.17	5	7	1.6E-01
4325.76	5	7	5.0E-01	5109.65	3	5	5.4E-02	6020.17	7	9	1.1E-01
4327.09	5	5	7.8E-02	5121.64	5	5	7.9E-02	6024.07	9	11	1.3E-01

λ Å	Weights g_i	g_k	A 10^8 s^{-1}	λ Å	Weights g_i	g_k	A 10^8 s^{-1}	λ Å	Weights g_i	g_k	A 10^8 s^{-1}
6055.99	7	9	7.0E-02	2395.42	6	4	3.3E-01	2543.43	6	4	7.1E-01
6170.49	5	5	1.3E-01	2395.62	8	10	2.5E+00	2544.97	4	6	4.0E-01
6336.84	3	3	4.9E-02	2399.24	6	6	1.4E+00	2545.22	8	10	3.3E-01
6338.90	5	3	4.8E-02	2400.06	12	14	5.2E+00	2545.44	8	10	1.4E-01
6400.00	7	9	5.5E-02	2401.29	6	8	2.5E+00	2546.67	8	8	6.2E-01
6411.65	5	7	3.5E-02	2404.43	4	2	7.1E-01	2547.34	8	8	2.0E-01
6419.98	7	7	1.3E-01	2404.89	6	8	1.7E+00	2548.33	4	6	2.0E-01
6469.21	3	3	9.0E-02	2406.66	4	4	1.6E+00	2548.59	10	10	1.9E-01
6495.78	3	3	6.0E-02	2410.52	4	6	1.5E+00	2548.74	4	2	1.7E+00
6496.46	5	5	8.5E-02	2411.07	2	2	2.4E+00	2548.92	12	10	4.8E-01
6569.23	7	9	6.5E-02	2413.31	2	4	1.1E+00	2549.08	10	8	1.5E+00
6633.76	7	7	3.6E-02	2416.45	8	10	1.6E+00	2549.40	4	4	1.3E+00
6733.16	3	1	3.9E-02	2418.44	6	8	1.6E+00	2549.46	6	6	8.0E-01
6841.35	5	7	3.6E-02	2423.21	4	6	1.4E+00	2549.77	8	6	2.5E-01
7130.94	3	5	4.3E-02	2428.36	8	10	2.7E+00	2550.03	10	10	1.2E+00
				2432.87	14	14	3.2E+00	2550.15	8	10	4.0E-01
Fe II				2434.06	8	6	7.0E-01	2550.68	12	12	8.9E-01
1144.94	10	12	4.8E+00	2434.24	8	10	2.0E+00	2551.21	10	8	3.2E-01
1635.40	8	6	2.4E+00	2434.73	12	12	3.2E+00	2555.07	6	8	1.8E-01
1641.76	6	4	1.8E+00	2439.30	12	14	2.8E+00	2555.45	4	6	2.5E-01
1647.16	6	6	5.2E-01	2445.11	12	12	1.9E+00	2557.51	10	8	1.3E-01
2208.41	10	10	1.8E+00	2445.80	4	6	1.5E+00	2559.77	6	8	2.4E-01
2213.66	14	14	4.4E-01	2446.47	12	14	2.9E-01	2559.92	6	8	2.4E-01
2218.27	8	10	1.9E+00	2447.20	6	6	1.2E+00	2560.28	4	4	1.5E+00
2327.40	6	4	5.9E-01	2453.98	8	10	7.3E-01	2562.09	4	2	1.5E+00
2331.31	10	8	2.9E-01	2455.71	8	8	1.0E+00	2562.54	8	6	1.5E+00
2332.80	8	6	1.5E+00	2458.78	10	12	2.7E+00	2563.48	6	4	1.3E+00
2338.01	4	4	1.1E+00	2458.97	6	4	2.0E+00	2566.22	8	10	2.5E+00
2343.49	10	8	1.7E+00	2460.44	10	12	5.3E+00	2566.40	8	6	2.1E+00
2343.96	8	6	2.9E-01	2461.28	6	8	2.6E+00	2566.91	4	2	1.1E+00
2344.28	2	4	8.2E-01	2461.86	8	10	2.6E+00	2568.41	2	4	4.4E-01
2348.11	10	8	5.1E-01	2466.52	2	4	2.1E+00	2569.78	2	4	1.2E+00
2348.30	6	6	1.2E+00	2469.51	8	6	2.8E+00	2570.53	6	8	1.2E+00
2351.67	6	6	1.7E+00	2472.61	8	10	3.7E+00	2570.85	8	6	1.7E+00
2352.31	2	4	4.2E+00	2475.12	4	6	3.9E+00	2573.21	8	10	1.4E-01
2353.68	8	8	1.3E+00	2475.54	6	8	3.5E+00	2574.36	6	4	1.6E+00
2354.89	6	4	2.4E-01	2481.05	12	12	1.9E-01	2576.86	10	12	1.1E+00
2360.00	10	10	2.4E-01	2484.44	8	8	2.3E+00	2577.92	2	2	1.3E+00
2360.29	8	6	5.9E-01	2492.34	10	12	1.6E-01	2582.41	6	8	2.4E-01
2362.02	8	8	1.3E-01	2493.26	14	16	3.4E+00	2582.58	4	4	7.7E-01
2363.86	8	10	5.1E+00	2501.31	2	2	1.4E+00	2585.63	10	10	3.6E-01
2364.83	8	8	6.1E-01	2503.87	10	10	2.4E+00	2585.88	10	8	8.1E-01
2365.77	6	6	2.1E+00	2508.34	8	10	2.7E+00	2587.95	8	10	1.4E+00
2366.59	6	6	9.9E-02	2533.63	12	12	1.3E+00	2588.18	2	2	1.6E-01
2368.60	6	4	5.9E-01	2534.42	8	8	1.2E+00	2590.55	4	6	9.1E-02
2369.95	10	12	5.7E+00	2535.36	6	4	3.3E+00	2591.54	6	6	5.1E-01
2370.50	4	4	1.4E-01	2535.49	10	8	5.4E-01	2592.78	14	16	2.1E+00
2373.74	10	10	3.3E-01	2536.67	12	12	4.0E-01	2593.72	2	4	1.3E-01
2375.19	4	2	9.8E-01	2537.14	10	10	1.4E+00	2594.96	8	8	1.0E-01
2379.27	8	8	1.5E-01	2538.20	14	12	1.2E+00	2598.37	8	6	1.3E+00
2380.76	6	8	3.1E-01	2538.50	8	6	3.3E-01	2599.40	10	10	2.2E+00
2382.04	10	12	3.8E+00	2538.80	12	10	8.2E-01	2604.05	8	8	1.1E-01
2382.90	12	14	2.2E-01	2538.91	10	8	7.8E-01	2605.04	6	8	2.1E+00
2383.25	6	6	3.4E-01	2538.99	14	12	1.2E+00	2605.34	4	4	1.6E+00
2384.39	4	4	2.3E-01	2540.52	2	2	1.5E+00	2605.42	6	6	2.6E-01
2388.37	10	12	2.2E-01	2541.10	8	6	7.3E-01	2605.90	4	2	1.2E+00
2388.63	8	8	1.0E+00	2541.84	8	6	7.7E-01	2606.51	6	6	1.8E+00
2390.10	14	16	5.5E+00	2542.73	2	2	1.9E+00	2607.09	6	4	1.7E+00
2390.77	6	6	9.3E-01	2543.38	10	12	4.4E-01	2609.13	8	10	3.0E-01

λ Å	Weights g_i	g_k	A 10^8 s^{-1}
2609.87	8	8	1.8E-01
2611.87	8	8	1.1E+00
2613.82	4	2	2.0E+00
2617.62	6	6	4.4E-01
2619.07	10	10	2.7E-01
2620.17	6	6	1.3E-01
2620.70	8	8	3.3E-01
2621.67	2	2	4.9E-01
2623.11	14	14	1.1E-01
2623.73	6	6	2.2E-01
2625.49	12	14	2.2E+00
2625.67	8	10	3.4E-01
2626.50	4	6	3.4E-01
2628.29	2	4	8.6E-01
2629.59	6	8	6.2E-01
2630.07	4	6	5.7E-01
2631.05	4	6	7.7E-01
2631.32	6	8	6.0E-01
2631.61	10	12	5.3E-01
2633.20	6	4	1.7E+00
2636.69	4	4	1.2E-01
2637.50	6	6	5.2E-01
2637.64	2	4	8.3E-01
2639.56	2	2	1.1E+00
2642.01	6	6	3.6E-01
2649.47	6	8	1.8E+00
2650.48	6	8	1.6E+00
2654.63	4	4	7.7E-01
2658.23	8	8	3.2E-01
2662.56	2	2	9.6E-01
2664.66	8	10	1.5E+00
2666.64	6	8	1.7E+00
2667.22	4	6	9.2E-01
2669.93	2	4	4.7E-01
2671.40	2	4	5.6E-01
2682.51	8	10	7.0E-01
2683.00	4	6	6.4E-01
2684.75	8	10	1.4E+00
2692.60	10	12	1.2E+00
2697.33	4	4	2.7E-01
2697.46	4	2	1.8E+00
2699.20	4	4	6.6E-01
2703.99	8	8	1.2E+00
2707.13	4	6	8.5E-01
2709.05	4	6	3.7E-01
2711.84	12	14	3.8E-01
2712.39	10	12	1.3E-01
2714.41	8	6	5.5E-01
2716.22	6	6	1.1E+00
2716.56	14	12	1.6E+00
2717.87	16	14	1.4E+00
2718.64	10	8	1.3E+00
2719.30	6	8	3.7E-01
2722.06	8	8	1.1E-01
2722.74	6	8	7.8E-01
2724.88	6	6	9.7E-01
2727.38	12	10	3.2E-01
2727.54	6	4	8.5E-01
2728.91	8	10	8.8E-02
2730.73	4	4	2.5E-01
2732.94	8	6	7.8E-01
2739.55	8	8	1.9E+00
2741.40	6	6	1.7E-01
2743.20	2	4	1.8E+00
2746.48	4	6	1.9E+00
2746.98	6	6	1.6E+00
2749.18	4	4	1.1E+00
2749.32	6	8	2.1E+00
2749.49	2	2	1.1E+00
2753.29	10	12	1.2E+00
2754.91	8	6	8.4E-01
2755.73	8	10	2.1E+00
2761.81	2	4	1.1E-01
2762.34	6	6	3.7E-01
2763.66	14	12	1.3E+00
2765.13	10	8	1.2E+00
2767.50	12	14	1.9E+00
2769.36	12	14	1.6E-01
2774.69	2	4	2.4E-01
2776.91	8	8	3.0E-01
2779.30	10	8	7.6E-01
2779.91	2	4	2.3E-01
2780.04	2	2	2.9E-01
2783.69	12	10	7.0E-01
2785.19	12	10	1.0E+00
2787.24	8	6	1.3E-01
2793.89	10	12	9.6E-02
2796.63	10	10	1.0E-01
2799.29	10	8	1.1E-01
2809.78	8	8	1.6E-01
2817.09	6	4	2.1E-01
2831.56	4	6	5.8E-01
2833.09	6	6	2.7E-01
2835.71	4	6	3.1E-01
2838.22	4	2	4.2E-01
2839.51	10	8	9.9E-01
2839.80	8	10	4.1E-01
2840.65	2	4	5.3E-01
2840.76	10	12	1.1E-01
2844.96	2	2	4.5E-01
2847.77	4	4	3.3E-01
2848.11	6	6	7.0E-01
2848.32	6	4	1.1E+00
2855.69	8	10	1.0E-01
2856.38	6	8	2.7E-01
2856.91	8	8	8.7E-01
2857.17	6	8	9.5E-02
2872.39	10	8	1.5E-01
2873.40	8	10	3.4E-01
2875.35	8	10	9.5E-02
2883.71	12	14	1.0E-01
2884.77	6	8	1.4E-01
2895.22	8	10	8.0E-02
2897.27	6	4	1.4E-01
2944.40	4	2	4.6E-01
2947.66	6	4	2.0E-01
2949.18	10	8	2.0E-01
2959.84	8	6	1.6E-01
2964.63	2	2	9.3E-02
2969.93	8	6	1.8E-01
2982.06	4	6	2.1E-01
2984.82	6	6	3.6E-01
2985.55	2	4	1.8E-01
2997.30	6	8	8.3E-02
3002.65	4	6	1.4E-01
3036.96	6	6	1.6E-01
3048.99	4	4	2.8E-01
3062.23	12	10	1.2E-01
3071.12	2	4	1.9E-01
3076.44	4	6	2.8E-01
3077.17	14	12	1.1E-01
3078.68	6	8	4.2E-01
3135.36	6	6	8.4E-02
3154.20	10	10	1.5E-01
3167.86	8	8	1.3E-01
3177.54	8	8	8.1E-02
3179.50	6	8	9.9E-02
5247.95	4	6	1.7E+00
5506.20	12	14	1.4E+00
5961.71	10	12	7.7E-01

Fe III

λ Å	Weights g_i	g_k	A 10^8 s^{-1}
1843.4	9	7	4.8E+00
1844.3	7	5	4.9E+00
1846.9	5	3	5.5E+00
1854.38	3	1	5.7E+00
1865.20	7	7	6.1E+00
1893.98	11	9	5.5E+00
1896.80	13	11	5.0E+00
1904.3	5	5	5.7E+00
1907.58	15	13	5.3E+00
1915.08	13	15	6.0E+00
1922.79	11	13	5.5E+00
1930.39	9	11	5.1E+00
1931.51	9	11	5.3E+00
1937.35	7	9	5.1E+00
1943.48	5	7	5.0E+00
1950.33	13	15	5.5E+00
1951.01	11	11	5.3E+00
1952.65	9	9	4.9E+00
1953.32	7	7	5.1E+00
1987.50	13	13	4.9E+00

Fe VII

λ Å	Weights g_i	g_k	A 10^8 s^{-1}
150.807	5	7	1.3E+03
150.852	7	9	1.3E+03
151.023	9	11	1.6E+03
151.046	7	7	2.2E+02
151.145	9	9	2.1E+02
151.432	5	7	2.2E+02
151.512	5	5	5.3E+02
151.675	7	7	3.9E+02
151.782	9	9	2.4E+02
154.307	3	1	8.9E+02
154.335	5	7	1.2E+03
154.363	3	3	4.2E+02
154.565	5	3	3.5E+02

λ Å	Weights g_i	g_k	A $10^8\ s^{-1}$
154.650	5	5	8.8E+02
154.848	1	3	7.7E+02
154.921	3	5	9.7E+02
154.941	3	3	2.4E+02
154.949	5	7	1.0E+03
155.994	9	11	1.8E+03
158.481	9	9	2.3E+02
165.087	1	3	6.9E+02
165.919	7	5	2.8E+03
166.365	9	7	2.9E+03
173.441	9	9	3.6E+03
176.744	9	9	2.7E+03
176.928	7	7	2.4E+03
177.172	5	5	1.5E+03
235.221	5	3	1.7E+02
240.053	3	1	1.3E+02
243.379	9	7	2.1E+02
Fe VIII			
112.472	4	4	3.6E+02
112.486	6	6	4.3E+02
116.196	4	6	4.5E+02
117.197	6	8	3.8E+02
167.486	4	4	3.0E+03
168.172	6	6	3.1E+03
168.545	6	4	2.0E+03
168.929	4	2	2.1E+03
185.213	6	8	1.0E+03
186.601	4	6	9.4E+02
Fe X			
76.822	2	2	1.8E+03
77.865	4	6	1.6E+03
100.026	8	10	2.6E+03
101.733	6	8	1.8E+03
101.846	4	6	1.7E+03
102.095	10	12	2.9E+03
102.192	10	12	2.9E+03
102.829	4	6	2.1E+03
103.319	6	8	2.6E+03
103.724	6	8	1.7E+03
104.638	8	10	2.1E+03
174.534	4	6	1.8E+03
175.266	2	4	1.72E+03
Fe XI			
72.166	5	7	2.9E+03
72.310	5	5	1.5E+03
72.635	5	7	1.6E+03
91.394	5	7	2.6E+03
91.472	7	9	2.5E+03
91.63	3	5	2.3E+03
91.63	7	9	3.4E+03
91.63	5	7	2.8E+03
91.733	9	11	4.1E+03
92.81	9	11	3.7E+03
92.87	11	13	3.9E+03
93.433	9	11	3.2E+03
179.762	5	7	1.67E+03
Fe XII			
65.905	4	4	2.0E+03
66.526	6	8	1.7E+03
66.960	4	6	1.6E+03
67.164	4	2	1.1E+03
67.821	4	6	1.4E+03
68.382	2	4	1.7E+03
80.541	6	6	8.7E+02
81.943	6	4	1.4E+03
82.226	4	2	1.9E+03
84.48	4	6	4.5E+03
84.48	8	10	4.9E+03
84.52	10	12	5.2E+03
84.52	6	8	4.0E+03
84.85	6	8	2.3E+03
85.14	8	10	3.4E+03
85.477	10	12	4.6E+03
186.880	6	8	1.0E+03
192.394	4	2	9.0E+02
193.509	4	4	9.1E+02
195.119	4	6	8.6E+02
Fe XIII			
62.353	1	3	2.0E+03
62.46	5	7	1.2E+03
62.699	3	5	2.3E+03
63.188	5	7	3.9E+03
64.139	1	3	2.1E+03
74.845	5	5	1.0E+03
75.892	5	3	7.7E+02
76.117	5	3	2.1E+03
78.452	9	11	6.3E+03
84.270	7	9	5.5E+03
107.384	7	5	1.8E+03
Fe XIV			
58.963	2	4	2.7E+03
59.579	4	6	3.1E+03
69.176	4	6	5.6E+02
69.386	2	4	7.6E+02
69.66	2	2	8.9E+02
69.66	6	6	1.3E+03
70.251	6	4	8.1E+02
70.613	4	2	1.7E+03
72.80	10	12	7.9E+03
76.022	4	6	6.6E+03
76.152	6	8	7.0E+03
91.009	6	4	5.1E+02
91.273	4	2	5.6E+02
188	4	6	2.7E+02
190	6	8	2.8E+02
207	2	2	2.1E+02
211.316	2	4	3.6E+02
213	4	2	2.8E+02
214	2	2	4.0E+02
216	6	8	1.7E+02
217	6	8	4.0E+02
217	6	6	2.6E+02
219	2	4	4.8E+02
219	4	6	2.4E+02
219.123	4	6	3.9E+02
220	4	4	3.2E+02
221	4	6	5.9E+02
226	2	4	3.9E+02
234	2	2	2.8E+02
264.787	4	4	3.38E+02
265	4	4	1.5E+02
266	6	4	1.7E+02
268	6	6	2.1E+02
268	4	2	3.3E+02
270.524	4	2	2.1E+02
274.203	2	2	1.8E+02
280	4	6	2.8E+02
283	6	8	2.7E+02
288.45	6	4	1.6E+02
Fe XV			
38.95	1	3	1.69E+03
52.911	1	3	2.94E+03
59.404	3	5	3.4E+03
63.959	5	7	1.6E+03
65.370	1	3	3.2E+02
65.612	3	3	9.8E+02
66.238	5	3	1.6E+03
68.860	9	11	9.2E+03
69.7	3	1	1.9E+03
69.942	3	5	7.4E+03
69.989	5	7	7.9E+03
70.052	7	9	8.8E+03
70.224	1	3	4.13E+03
70.53	7	5	2.6E+02
70.59	7	7	1.7E+03
73.199	7	9	8.8E+03
73.473	5	7	6.2E+03
233.857	5	7	2.2E+02
235	1	3	2.5E+02
243	1	3	2.4E+02
243	5	7	2.3E+02
243.790	3	5	4.2E+02
248	3	1	5.4E+02
284.160	1	3	2.28E+02
Fe XVI			
31.041	2	4	5.2E+02
31.242	4	6	6.1E+02
32.166	2	4	6.8E+02
32.192	2	2	6.7E+02
32.433	2	4	7.7E+02
32.652	4	6	9.1E+02
34.857	2	4	1.23E+03
35.106	4	6	1.44E+03
35.333	4	6	6.4E+02
35.368	6	8	6.8E+02
36.01	4	2	5.0E+02
36.749	2	4	1.1E+03
36.803	2	2	1.2E+03
37.096	4	6	1.0E+03
37.138	6	8	1.07E+03

λ Å	Weights g_i	g_k	A 10^8 s^{-1}
39.827	2	4	2.1E+03
40.153	4	6	2.5E+03
40.161	4	4	4.1E+02
40.199	4	6	1.7E+03
40.245	6	8	1.8E+03
41.91	2	2	4.72E+02
42.30	4	2	9.2E+02
46.661	4	6	3.46E+03
46.718	6	8	3.7E+03
50.350	2	4	1.86E+03
50.555	2	2	1.98E+03
54.142	2	4	3.41E+03
54.728	4	6	4.16E+03
54.769	4	4	6.97E+02
62.879	2	2	1.05E+03
63.719	4	2	2.18E+03
66.263	4	6	9.39E+03
66.368	6	8	1.00E+04
66.392	6	6	6.69E+02
76.502	6	4	6.7E+02
76.796	4	2	7.72E+02
80.192	4	6	5.2E+02
80.270	6	8	5.4E+02
85.587	2	4	4.0E+02
86.133	4	6	4.8E+02
96.256	4	6	8.7E+02
96.348	6	8	9.3E+02
117.2	2	4	3.93E+02
117.7	2	2	3.9E+02
123.4	2	4	5.9E+02
124.5	4	6	7.0E+02
144.06	4	6	1.6E+03
144.25	6	8	1.6E+03
148	4	2	6.5E+02
266.7	4	6	3.9E+02
267.0	6	8	4.3E+02

FeXVII

λ Å	Weights g_i	g_k	A 10^8 s^{-1}
11.023	1	3	2.1E+04
12.123	1	3	8.0E+04
12.264	1	3	5.9E+04
12.526	1	3	3.0E+03
12.681	1	3	3.5E+03
13.823	1	3	3.3E+04
13.891	1	3	3.4E+03
15.015	1	3	2.28E+05
15.262	1	3	6.0E+04
16.777	1	3	8.29E+03
17.054	1	3	9.33E+03
41.37	9	11	4.8E+03
49.427	3	3	4.0E+03
50.26	7	9	6.0E+03
58.76	9	11	1.2E+04

FeXIX

λ Å	Weights g_i	g_k	A 10^8 s^{-1}
13.413	5	3	1.3E+04
13.426	5	7	4.8E+04
13.47	3	1	1.5E+05
13.520	5	7	2.0E+05
13.56	3	5	1.0E+04
13.68	3	1	8.0E+04
13.69	5	7	2.3E+04
13.700	1	3	2.7E+05
13.71	5	5	2.2E+04
13.738	5	7	1.0E+04
13.796	5	7	7.0E+04
13.83	5	5	1.4E+04
13.934	1	3	4.51E+04
13.961	3	3	2.0E+04
14.668	5	7	1.1E+04
14.671	5	3	1.1E+04
14.929	3	3	1.2E+04
14.966	5	3	2.5E+04
14.995	5	5	2.2E+04
15.015	1	3	1.4E+04
16.668	3	1	1.1E+04

FeXX

λ Å	Weights g_i	g_k	A 10^8 s^{-1}
12.67	6	6	1.0E+04
12.69	4	6	1.2E+04
12.73	4	2	4.0E+04
12.77	4	4	2.1E+05
12.78	4	2	6.9E+04
12.78	2	4	1.4E+05
12.79	6	4	1.7E+04
12.82	4	4	1.1E+05
12.88	6	4	2.7E+04
12.89	4	4	4.4E+04
12.90	4	2	6.2E+03
12.90	4	6	1.4E+05
12.92	2	4	1.7E+04
12.93	4	6	1.6E+05
12.93	2	2	1.2E+04
12.98	2	2	6.7E+04
12.99	6	6	5.1E+04
13.00	6	4	1.1E+04
13.01	2	4	3.0E+04
13.03	4	2	8.6E+04
13.07	6	4	8.2E+03
13.13	2	4	8.9E+04
13.24	4	4	1.2E+04
13.28	4	4	6.1E+03
13.70	4	6	1.1E+04
13.71	2	2	9.9E+03
13.78	4	4	1.0E+04
13.79	6	6	1.2E+04
13.83	4	2	9.8E+03
13.90	4	2	1.2E+04
13.98	6	4	1.6E+04
13.99	4	2	2.2E+04
14.05	4	4	1.7E+04
14.23	2	2	6.3E+03

FeXXI

λ Å	Weights g_i	g_k	A 10^8 s^{-1}
8.53	3	1	1.8E+04
8.53	3	5	6.1E+03
8.53	3	3	1.5E+04
8.56	5	7	2.0E+04
8.56	1	3	2.1E+04
8.56	5	3	6.5E+03
8.64	5	7	1.5E+04
8.65	5	7	3.9E+04
8.66	5	5	4.4E+03
8.74	1	3	2.5E+04
9.42	3	1	4.3E+04
9.42	3	3	3.3E+04
9.44	3	5	1.7E+04
9.45	1	3	5.2E+04
9.46	5	3	1.5E+04
9.47	5	7	4.9E+04
9.47	5	5	6.1E+03
9.52	3	3	8.1E+03
9.58	5	5	5.2E+03
9.59	5	5	1.0E+04
9.67	1	3	5.7E+04
9.68	5	7	4.0E+03
9.74	5	3	5.3E+03
12.02	1	3	1.3E+04
12.13	3	3	1.8E+04
12.18	5	7	2.2E+04
12.19	5	3	6.4E+03
12.21	3	1	1.5E+05
12.21	3	3	1.2E+05
12.25	1	3	2.1E+05
12.28	5	3	5.2E+04
12.30	5	7	2.1E+05
12.36	3	3	3.6E+04
12.37	5	7	3.1E+05
12.38	5	3	6.9E+03
12.47	5	7	5.8E+04
12.47	5	3	1.3E+04
12.49	5	7	1.3E+04
12.53	5	5	1.5E+04
12.57	1	3	7.2E+04
12.73	5	5	8.2E+03
12.95	3	5	6.2E+03
13.00	1	3	7.2E+03
13.03	5	5	1.3E+04
13.14	3	1	2.0E+04
13.41	1	3	7.3E+03

FeXXII

λ Å	Weights g_i	g_k	A 10^8 s^{-1}
9.002	4	6	5.5E+04
9.006	6	8	5.7E+04
9.006	6	6	5.3E+04
9.163	4	6	6.9E+04
9.183	6	8	8.3E+04
9.241	4	6	5.1E+04
11.748	4	4	1.2E+05
11.748	4	6	1.6E+05
11.748	4	2	1.8E+05
11.763	2	4	1.6E+05
11.789	2	2	2.6E+05
11.789	6	8	1.2E+05
11.797	2	4	1.7E+05
11.823	6	4	7.9E+04
11.837	6	8	2.3E+05

λ Å	Weights g_i	g_k	A 10^8 s^{-1}
11.837	6	6	1.7E+05
11.886	4	6	1.3E+05
11.898	2	4	8.2E+04
11.922	4	6	1.8E+05
11.976	6	8	5.9E+04
12.027	2	4	6.9E+04
12.045	6	8	2.4E+05
12.045	4	4	9.7E+04
12.053	4	6	6.1E+04
12.077	2	4	1.0E+05
12.077	4	6	2.4E+05
12.095	6	6	7.8E+04
12.193	2	4	7.2E+04
12.193	4	6	9.9E+04
12.325	2	2	1.5E+05
FeXXIII			
7.733	5	7	3.0E+04
7.849	5	7	4.9E+04
8.307	1	3	4.8E+04
8.529	1	3	4.3E+04
8.550	3	5	6.0E+04
8.552	3	3	3.2E+04
8.614	5	7	7.7E+04
8.664	3	3	4.4E+04
8.669	5	7	6.1E+04
8.672	1	3	6.8E+04
8.752	5	7	1.2E+05
8.764	5	7	4.6E+04
8.814	3	5	6.2E+04
10.902	5	5	5.3E+04
10.910	3	1	6.7E+04
10.927	5	7	6.0E+04
10.934	3	5	5.4E+04
10.979	1	3	7.9E+04
11.018	1	3	4.9E+04
11.086	3	1	6.5E+04
11.165	3	5	6.7E+04
11.255	3	3	3.7E+04
11.298	1	3	1.3E+05
11.325	3	5	1.7E+05
11.338	3	3	9.3E+04
11.429	3	1	1.7E+05
11.433	3	3	1.2E+05
11.441	5	7	2.2E+05
11.445	5	5	5.6E+04
11.485	3	5	1.40E+05
11.491	5	3	5.9E+04
11.519	5	5	1.16E+05
11.520	1	3	2.16E+05
11.524	5	7	2.3E+05
11.593	5	7	3.58E+05
11.613	3	5	1.0E+05
11.615	3	3	4.4E+04
11.691	5	7	7.7E+04
11.698	5	5	7.3E+04
11.737	3	5	1.8E+05
11.898	1	3	2.03E+05

λ Å	Weights g_i	g_k	A 10^8 s^{-1}
FeXXIV			
1.8523	2	2	1.0E+05
1.8552	2	4	4.82E+06
1.8563	4	2	2.43E+06
1.8572	2	2	3.06E+06
1.858	2	4	1.2E+05
1.8614	4	4	6.24E+06
1.8626	2	4	3.16E+06
1.8627	2	2	5.47E+06
1.8637	2	2	1.91E+06
1.8655	4	6	2.14E+06
1.8672	4	2	1.63E+06
1.8678	4	4	3.5E+05
1.8721	4	6	3.2E+05
1.8721	2	2	2.0E+05
1.8730	2	4	1.5E+05
1.8739	4	4	8.3E+04
1.891	2	2	9.7E+04
1.897	4	2	9.8E+04
8.231	2	4	6.10E+04
8.316	4	6	7.07E+04
10.619	2	4	7.28E+04
10.663	2	2	7.51E+04
11.030	2	4	1.84E+05
11.171	4	6	2.18E+05
FeXXV			
1.4607	1	3	2.54E+05
1.4945	1	3	5.05E+05
1.5730	1	3	1.24E+06
1.5749	1	3	1.5E+05
1.778	3	3	8.7E+04
1.782	3	1	4.69E+06
1.787	1	3	2.57E+06
1.787	5	5	1.19E+06
1.788	3	5	2.68E+06
1.788	3	5	1.63E+06
1.789	1	3	1.78E+06
1.790	3	3	1.23E+06
1.791	3	5	4.10E+06
1.791	3	3	2.59E+06
1.792	3	1	4.92E+06
1.792	5	5	2.81E+06
1.793	3	1	2.67E+06
1.794	5	3	2.22E+06
1.797	3	5	8.8E+05
1.798	3	3	1.0E+05
1.800	1	3	8.6E+04
1.802	3	1	4.1E+05
1.810	3	1	5.9E+05
1.8502	1	3	4.57E+06
1.8593	1	3	4.42E+05
10.038	3	3	8.08E+04
Krypton			
Kr I			
4274.0	5	5	2.6E-02
4351.4	3	1	3.2E-02

λ Å	Weights g_i	g_k	A 10^8 s^{-1}
4362.6	5	3	8.4E-03
4376.1	3	1	5.6E-02
4400.0	3	5	2.0E-02
4410.4	3	3	4.4E-03
4425.2	3	3	9.7E-03
4453.9	3	5	7.8E-03
4463.7	3	3	2.3E-02
4502.4	3	5	9.2E-03
5562.2	5	5	2.8E-03
5570.3	5	3	2.1E-02
5649.6	1	3	3.7E-03
5870.9	3	5	1.8E-02
6904.7	3	5	1.3E-02
7224.1	3	5	1.4E-02
7587.4	3	1	5.1E-01
7601.5	5	5	3.1E-01
7685.2	3	1	4.9E-01
7694.5	5	3	5.6E-02
7854.8	1	3	2.3E-01
8059.5	1	3	1.9E-01
8104.4	5	5	1.3E-01
8112.9	5	7	3.6E-01
8190.1	3	5	1.1E-01
8263.2	3	5	3.5E-01
8281.1	3	3	1.9E-01
8298.1	3	3	3.2E-01
8508.9	3	3	2.4E-01
8776.7	3	5	2.7E-01
8928.7	5	3	3.7E-01
Kr II			
4250.6	4	4	1.2E-01
4292.9	4	4	9.6E-01
4355.5	6	8	1.0E+00
4431.7	2	2	1.8E+00
4436.8	2	4	6.6E-01
4577.2	6	8	9.6E-01
4583.0	6	4	7.6E-01
4615.3	4	4	5.4E-01
4619.2	4	6	8.1E-01
4633.9	4	6	7.1E-01
4658.9	6	4	6.5E-01
4739.0	6	6	7.6E-01
4762.4	2	4	4.2E-01
4765.7	4	6	6.7E-01
4811.8	2	4	1.7E-01
4825.2	2	4	1.9E-01
4832.1	4	2	7.3E-01
5208.3	4	4	1.4E-01
5308.7	4	6	2.4E-02
7407.0	6	6	7.0E-02
Lead			
Pb I			
2022.0	1	3	5.2E-02
2053.3	1	3	1.2E-01
2170.0	1	3	1.5E+00
2401.9	3	3	1.9E-01

λ Å	Weights g_i	g_k	A 10^8 s^{-1}
2446.2	3	3	2.5E-01
2476.4	3	5	2.8E-01
2577.3	5	3	5.0E-01
2613.7	3	3	2.7E-01
2614.2	3	5	1.9E+00
2628.3	5	3	3.1E-02
2657.1	3	5	9.8E-04
2663.2	5	5	7.1E-01
2802.0	5	7	1.6E+00
2823.2	5	5	2.6E-01
2833.1	1	3	5.8E-01
2873.3	5	5	3.7E-01
3572.7	5	3	9.9E-01
3639.6	3	3	3.4E-01
3671.5	5	3	4.4E-01
3683.5	3	1	1.5E+00
3739.9	5	5	7.3E-01
4019.6	5	7	3.5E-02
4057.8	5	3	8.9E-01
4062.1	5	3	9.2E-01
4168.0	5	5	1.2E-02
5005.4	1	3	2.7E-01
5201.4	1	3	1.9E-01
7229.0	5	3	8.9E-03
Lithium			
Li I			
*2741.2	2	6	1.3E-02
*3232.7	2	6	1.2E-02
4602.8	2	4	1.97E-01
4602.9	4	6	2.4E-01
6103.5	2	4	6.0E-01
6103.7	4	6	7.1E-01
6103.7	4	4	1.2E-01
6707.8	2	4	3.72E-01
6707.9	2	2	3.72E-01
Lutetium			
Lu I			
3376.5	4	4	2.23E+00
3567.8	4	6	5.9E-01
3620.3	6	4	1.1E-02
3841.2	6	6	2.5E-01
4518.6	4	4	2.1E-01
Magnesium			
Mg I			
2025.8	1	3	8.4E-01
*2779.8	9	9	5.2E+00
*2850.0	9	15	2.3E-01
2852.1	1	3	4.95E+00
*3094.9	9	15	5.2E-01
3329.9	1	3	3.3E-02
3332.2	3	3	9.7E-02
3336.7	5	3	1.6E-01
*3835.3	9	15	1.68E+00
4703.0	3	5	2.55E-01
5167.3	1	3	1.16E-01
5172.7	3	3	3.46E-01
5183.6	5	3	5.75E-01
5528.4	3	5	1.99E-01
Mg II			
1239.9	2	4	1.4E-02
1240.4	2	2	1.4E-02
*2660.8	10	14	3.8E-01
2790.8	2	4	4.0E+00
2795.5	2	4	2.6E+00
2797.9	4	4	7.9E-01
2798.1	4	6	4.8E+00
2802.7	2	2	2.6E+00
2928.8	2	2	1.2E+00
2936.5	4	2	2.3E+00
*3104.8	10	14	8.1E-01
3848.2	6	4	2.8E-02
3848.3	4	4	3.0E-03
3850.4	4	2	3.0E-02
*4481.2	10	14	2.23E+00
9218.3	2	4	3.6E-01
9244.3	2	2	3.6E-01
Mg IV			
320.99	4	2	1.2E+02
323.31	2	2	5.9E+01
1219.0	6	6	5.9E+00
1375.5	4	4	4.5E+00
1439.6	6	4	4.6E+00
1495.5	4	6	6.4E+00
1510.7	4	4	6.7E+00
1683.0	6	8	5.8E+00
1698.8	4	6	3.9E+00
1893.9	6	6	2.8E+00
Mg VI			
*269.92	10	6	3.1E+02
*292.53	6	6	9.0E+01
*314.64	6	2	1.8E+02
*349.15	10	10	6.1E+01
*387.94	6	10	1.3E+01
399.29	4	2	2.8E+01
400.68	4	4	2.8E+01
403.32	4	6	2.7E+01
Mg VII			
277.01	3	3	9.5E+01
278.41	5	3	1.5E+02
280.74	5	3	2.0E+02
319.02	5	5	8.9E+01
*366.42	9	9	4.4E+01
*433.04	9	15	1.6E+01
1334.3	5	5	5.3E+00
1410.0	5	5	2.57E+00
1487.0	3	5	3.02E+00
1487.9	5	7	3.66E+00
Mg VIII			
*74.976	6	10	4.3E+03
315.02	4	4	1.2E+02
*342.29	10	6	6.3E+01
353.86	4	4	3.89E+01
356.00	6	4	5.7E+01
*428.52	10	10	3.24E+01
*434.62	6	10	1.6E+01
*489.33	6	6	3.9E+01
*686.92	6	10	9.4E+00
Mg IX			
62.751	1	3	2.87E+03
*67.189	9	15	6.20E+03
*71.965	9	3	1.22E+03
72.312	3	5	4.43E+03
77.737	3	1	3.92E+02
368.07	1	3	5.27E+01
438.69	3	1	7.9E+01
*443.74	9	9	4.19E+01
749.55	3	5	8.2E+00
1639.8	3	5	2.1E+00
2814.2	1	3	3.35E-01
Mg X			
57.876	2	4	2.09E+03
57.920	2	2	2.09E+03
63.152	2	4	5.6E+03
63.295	4	6	6.7E+03
609.79	2	4	7.53E+00
624.94	2	2	7.01E+00
2212.5	2	4	9.64E-01
2278.7	2	2	8.82E-01
5918.7	2	4	3.20E-02
6229.6	4	6	3.30E-02
Mg XI			
7.310	1	3	1.15E+04
7.473	1	3	2.27E+04
7.850	1	3	5.50E+04
9.169	1	3	1.97E+05
Manganese			
Mn I			
2794.82	6	8	3.7E+00
2798.27	6	6	3.6E+00
2801.08	6	4	3.7E+00
3007.65	6	8	1.8E-01
3011.38	8	10	3.1E-01
3016.45	10	12	2.9E-01
3043.36	8	8	5.9E-01
3044.57	10	8	5.7E-01
3045.59	10	10	6.7E-01
3045.80	8	10	1.7E-01
3047.03	12	12	6.1E-01
3054.36	8	6	4.6E-01
3070.27	6	6	1.9E-01
3073.18	4	4	3.7E-01
3082.71	14	14	2.9E-01
3110.68	6	8	2.7E-01
3113.80	12	10	2.6E-01
3118.10	4	6	1.7E-01
3122.88	10	10	1.9E-01

λ Å	Weights g_i	g_k	A 10^8 s^{-1}	λ Å	Weights g_i	g_k	A 10^8 s^{-1}	λ Å	Weights g_i	g_k	A 10^8 s^{-1}
3126.85	8	6	2.3E-01	3952.84	6	6	4.1E-01	4739.11	4	4	2.40E-01
3132.28	10	10	2.1E-01	3975.88	2	4	1.8E-01	4754.05	6	8	3.03E-01
3132.79	8	8	2.7E-01	3982.16	4	2	3.5E-01	4761.53	2	4	5.35E-01
3175.58	8	10	1.8E-01	3982.58	6	4	2.3E-01	4762.38	8	10	7.83E-01
3201.11	4	6	2.2E-01	3982.90	6	4	5.5E-01	4765.86	4	6	4.1E-01
3228.09	10	12	6.4E-01	3991.60	2	2	2.1E-01	4766.43	6	8	4.6E-01
3230.23	10	12	1.9E-01	4011.91	8	8	2.3E-01	4783.43	8	8	4.01E-01
3230.72	8	8	3.5E-01	4018.11	10	8	2.54E-01	4823.53	10	8	4.99E-01
3240.88	6	4	2.2E-01	4030.76	6	8	1.7E-01	6013.48	4	6	1.72E-01
3243.78	6	6	5.3E-01	4033.07	6	6	1.65E-01	6021.79	8	6	3.32E-01
3251.13	4	2	2.3E-01	4041.36	10	10	7.87E-01				
3252.95	4	4	1.8E-01	4048.75	6	4	7.5E-01	**Mn II**			
3256.14	4	6	5.0E-01	4052.48	6	8	3.8E-01	2593.72	7	7	2.6E+00
3258.41	2	2	9.7E-01	4055.55	8	8	4.31E-01	2605.68	7	5	2.7E+00
3260.24	2	4	3.8E-01	4058.94	4	2	7.25E-01	2933.05	5	3	2.0E+00
3267.79	14	14	3.5E-01	4061.74	8	6	1.9E-01	2939.31	5	5	1.9E+00
3268.72	6	8	3.3E-01	4063.53	6	6	1.69E-01	2949.20	5	7	1.9E+00
3270.35	12	12	2.6E-01	4065.08	12	14	2.5E-01	3441.99	9	7	4.3E-01
3273.02	10	10	2.7E-01	4066.24	10	8	2.2E-01	3460.32	7	5	3.2E-01
3298.23	6	4	2.8E-01	4070.28	2	2	2.3E-01	3474.13	5	3	1.5E-01
3303.28	4	4	1.9E-01	4079.42	2	4	3.8E-01	3482.90	5	5	2.0E-01
3463.66	8	8	3.2E-01	4082.95	4	6	2.95E-01	3488.68	3	3	2.5E-01
3470.01	6	8	2.4E-01	4083.63	6	8	2.8E-01				
3511.83	12	12	2.7E-01	4089.94	8	10	1.7E-01	**Mn VI**			
3535.30	10	10	1.7E-01	4105.37	10	8	1.7E-01	307.999	9	9	3.7E+01
3559.81	6	6	2.1E-01	4135.03	12	12	3.0E-01	309.440	9	7	5.7E+01
3577.87	10	8	9.4E-01	4141.06	10	10	2.6E-01	309.579	7	5	4.4E+01
3595.11	6	4	1.8E-01	4148.80	8	8	2.3E-01	310.058	7	7	3.4E+01
3601.27	12	10	2.3E-01	4176.61	14	12	2.4E-01	310.182	5	5	2.8E+01
3607.53	8	8	2.3E-01	4189.99	12	10	2.0E-01	311.748	5	3	5.7E+01
3608.49	6	6	3.6E-01	4201.78	10	8	2.3E-01	320.598	3	5	1.5E+01
3610.30	4	4	4.2E-01	4235.30	8	6	9.17E-01	320.681	1	3	2.2E+01
3635.70	10	8	2.1E-01	4239.74	4	2	3.9E-01	320.874	3	1	7.8E+01
3660.40	12	14	9.1E-01	4257.67	2	2	3.7E-01	320.979	3	3	2.2E+01
3675.67	6	8	2.2E-01	4265.93	4	4	4.92E-01	321.176	5	5	6.0E+01
3676.96	10	12	7.3E-01	4281.10	6	6	2.3E-01	321.541	5	3	2.7E+01
3680.15	12	10	1.9E-01	4411.87	12	10	2.6E-01	325.146	9	7	1.3E+02
3682.09	8	10	7.6E-01	4414.89	8	6	2.93E-01	328.431	5	5	4.4E+01
3684.87	6	8	2.6E-01	4419.77	10	8	2.1E-01	328.558	3	5	1.2E+01
3706.08	12	14	1.4E+00	4436.36	6	4	4.37E-01	329.043	1	3	1.1E+01
3718.92	10	12	9.6E-01	4451.58	8	8	7.98E-01	1236.23	5	3	1.3E+01
3731.94	8	10	1.0E+00	4453.01	4	2	5.44E-01	1255.77	3	1	1.2E+01
3771.44	14	14	1.9E-01	4455.82	4	6	1.7E-01	1285.10	5	7	1.1E+01
3773.86	12	12	2.5E-01	4457.04	6	4	2.34E-01	1333.87	7	9	1.0E+01
3800.55	6	8	2.7E-01	4457.55	6	6	4.27E-01				
3806.72	10	12	5.9E-01	4458.26	6	8	4.62E-01	**Mercury**			
3823.51	8	10	5.21E-01	4461.09	8	8	1.7E-01	**Hg I**			
3823.89	6	6	2.31E-01	4462.03	8	10	7.00E-01	2536.52	1	3	8.00E-02
3833.87	4	4	3.14E-01	4464.68	6	6	4.39E-01	2652.04	3	5	3.88E-01
3834.37	6	8	4.29E-01	4470.14	4	4	3.00E-01	2655.13	3	5	1.1E-01
3839.78	2	2	4.64E-01	4472.79	2	2	4.35E-01	2752.78	1	3	6.10E-02
3841.07	4	6	3.3E-01	4479.40	8	10	3.4E-01	2856.94	3	1	1.1E-02
3843.99	2	4	2.11E-01	4490.08	2	4	2.49E-01	2893.60	3	3	1.6E-01
3889.46	12	14	3.1E-01	4498.90	4	6	2.49E-01	2925.4	5	3	7.7E-02
3898.37	6	8	1.7E-01	4502.22	6	8	1.86E-01	2967.3	1	3	4.5E-01
3899.34	4	6	2.4E-01	4605.37	10	12	3.6E-01	3021.50	5	7	5.09E-01
3924.08	2	4	9.4E-01	4626.54	12	14	3.6E-01	3023.48	5	5	9.4E-02
3926.48	6	8	5.4E-01	4709.71	8	8	1.72E-01	3027.49	5	5	2.0E-02
3951.98	2	2	3.1E-01	4727.46	6	6	1.7E-01	3125.66	3	5	6.56E-01

λ Å	Weights g_i	g_k	A $10^8 s^{-1}$
3341.48	5	3	1.68E-01
3650.15	5	7	1.3E+00
3654.83	5	5	1.8E-01
4046.56	1	3	2.1E-01
4077.81	3	1	4.0E-02
4108.1	3	1	3.0E-02
4339.22	3	5	2.88E-02
4347.50	3	5	8.4E-02
4358.34	3	3	5.57E-01
4916.07	3	1	5.8E-02
5025.64	3	3	2.7E-04
5460.75	5	3	4.87E-01
5769.59	3	5	2.36E-01
6234.4	1	3	5.3E-03
6716.4	1	3	4.3E-03
6907.5	3	5	2.8E-02
7728.8	1	3	9.7E-03
10139.79	3	1	2.71E 01

Molybdenum
Mo I

λ Å	Weights g_i	g_k	A $10^8 s^{-1}$
2616.79	3	5	7.34E-01
2621.06	7	7	1.16E-01
2628.96	3	3	2.81E-01
2629.85	5	7	7.75E-01
2631.50	1	3	2.54E-01
2638.30	5	5	7.57E-01
2640.98	7	5	1.20E+00
2644.36	5	7	1.96E-01
2649.46	7	9	9.84E-01
2655.02	9	7	4.08E-01
2658.11	7	7	6.43E-01
2665.09	7	9	1.32E-01
2679.85	9	11	1.31E+00
2684.16	9	9	4.18E-01
2706.11	3	5	2.03E-01
2710.74	3	3	1.57E-01
2725.15	3	5	2.79E-01
2728.71	3	3	1.26E-01
2733.39	5	7	2.95E-01
2743.71	1	3	2.47E-01
2745.38	13	11	1.29E-01
2751.47	7	9	2.54E-01
2756.26	5	3	1.18E-01
2761.53	9	11	2.06E-01
2763.02	3	1	4.44E-01
2766.25	3	5	1.17E-01
2787.83	9	7	2.85E-01
2792.96	5	3	1.53E-01
2798.02	7	5	1.22E-01
2801.47	5	7	1.24E-01
2825.68	5	7	2.53E-01
2826.75	7	7	4.23E-01
2876.54	9	9	2.84E-01
2886.60	11	11	4.74E-01
2906.06	3	3	8.04E-01
2913.52	5	3	1.38E-01
2915.38	5	3	7.31E-01
2918.84	5	3	3.79E-01
2930.39	1	3	1.91E-01
2936.50	11	11	2.33E-01
2945.43	7	7	3.66E-01
2945.66	3	3	4.08E-01
2946.01	5	5	1.68E-01
2951.45	9	9	1.43E-01
2959.48	9	11	1.75E-01
2972.96	5	3	2.69E-01
2977.27	9	7	3.28E-01
2978.28	7	5	1.50E-01
2983.04	1	3	2.82E-01
2987.92	3	5	8.43E-01
2988.23	5	7	4.28E-01
2988.68	7	9	1.61E-01
2989.80	9	7	9.27E-01
3000.24	9	9	1.40E-01
3000.44	5	5	1.25E-01
3000.85	5	7	2.58E-01
3001.43	5	5	2.31E-01
3007.71	7	5	1.90E-01
3013.39	7	5	6.06E-01
3016.78	9	9	2.75E 01
3025.00	5	5	8.49E-01
3036.31	3	5	5.81E-01
3041.70	13	11	5.94E-01
3046.80	13	11	1.63E-01
3047.31	11	9	5.01E-01
3055.32	9	7	4.29E-01
3057.56	7	5	2.64E-01
3061.59	7	5	4.41E-01
3064.27	13	13	8.46E-01
3065.04	13	13	3.08E-01
3069.51	5	5	1.52E-01
3069.96	11	11	2.72E-01
3070.89	9	11	1.87E-01
3074.37	11	11	1.42E+00
3079.88	9	11	9.55E-01
3080.40	7	9	3.61E-01
3081.16	3	5	2.35E-01
3085.62	9	9	1.63E+00
3089.13	11	9	1.53E-01
3089.71	5	7	2.34E-01
3094.66	7	7	1.63E+00
3099.92	9	7	1.45E-01
3100.88	7	9	1.20E+00
3101.34	5	5	1.92E+00
3106.34	7	5	2.21E-01
3117.54	13	13	1.89E-01
3123.03	3	3	2.81E-01
3125.03	5	3	1.98E-01
3132.59	7	9	1.79E+00
3135.90	9	11	3.68E-01
3136.75	9	11	1.57E-01
3142.75	3	5	4.10E-01
3147.35	13	11	2.41E-01
3155.19	7	7	2.75E-01
3158.17	7	7	4.63E-01
3170.34	7	7	1.37E+00
3171.38	5	7	2.03E-01
3175.59	13	11	8.40E-01
3179.78	11	13	2.33E-01
3183.03	11	9	3.98E-01
3184.58	7	5	2.77E-01
3185.10	7	7	2.54E-01
3185.71	5	3	6.10E-01
3188.10	7	9	3.45E-01
3188.41	5	7	4.40E-01
3192.79	9	11	1.88E-01
3193.98	7	5	1.53E+00
3194.88	9	11	1.75E-01
3195.96	9	7	4.10E-01
3197.18	1	3	1.47E-01
3198.85	15	13	7.22E-01
3200.89	3	5	1.82E-01
3205.22	1	3	4.27E-01
3205.43	9	11	2.55E-01
3205.89	9	9	5.35E-01
3208.84	7	5	2.77E-01
3210.97	7	5	6.94E-01
3214.44	9	7	2.01E-01
3215.07	3	5	4.20E-01
3216.78	15	13	2.10E-01
3221.73	3	1	1.41E+00
3228.21	5	7	3.85E-01
3229.79	9	11	1.44E-01
3233.14	13	13	6.33E-01
3237.06	7	9	2.95E-01
3244.47	5	3	2.80E-01
3247.61	5	5	1.71E-01
3249.93	5	3	1.87E-01
3251.65	3	5	3.05E-01
3256.21	5	3	6.89E-01
3256.72	3	3	1.31E-01
3259.16	11	13	1.62E-01
3262.63	7	9	3.62E-01
3264.40	11	9	5.42E-01
3265.14	5	7	2.60E-01
3266.16	9	11	1.95E-01
3270.90	7	7	3.59E-01
3276.07	11	9	1.18E-01
3285.03	1	3	1.41E-01
3285.35	9	7	4.49E-01
3287.38	5	5	1.38E-01
3289.01	9	9	5.08E-01
3290.82	7	5	5.44E-01
3305.56	5	3	1.74E-01
3305.91	7	9	3.06E-01
3307.13	7	9	1.25E-01
3312.33	7	5	1.62E-01
3323.95	9	7	2.82E-01
3325.13	5	3	2.26E-01
3325.67	5	5	1.72E-01
3327.30	1	3	2.88E-01
3336.56	9	9	1.64E-01
3340.16	5	3	1.20E-01
3344.73	3	5	6.04E-01
3346.83	11	11	1.13E-01
3347.00	3	3	2.72E-01

λ Å	Weights g_i	g_k	A 10^8 s^{-1}
3358.12	5	7	7.59E-01
3361.37	9	9	1.38E-01
3363.78	5	7	2.74E-01
3363.87	5	7	1.39E-01
3373.81	3	3	2.03E-01
3375.22	7	7	1.38E-01
3375.65	7	9	1.56E-01
3378.19	3	1	1.88E-01
3378.46	13	13	3.75E-01
3379.96	5	5	4.11E-01
3382.48	3	3	2.66E-01
3384.61	7	9	7.32E-01
3385.87	9	11	3.30E-01
3389.79	5	7	1.85E-01
3392.17	9	9	1.97E-01
3393.65	11	11	2.08E-01
3404.33	7	7	2.10E-01
3413.37	11	11	1.25E-01
3415.27	9	9	1.83E-01
3415.61	7	9	1.29E-01
3416.14	9	11	2.45E-01
3418.52	5	3	1.41E-01
3419.69	7	7	1.15E-01
3420.04	5	5	3.28E-01
3422.31	9	9	2.52E-01
3425.13	11	11	2.29E-01
3427.90	11	13	4.09E-01
3434.79	7	7	1.75E-01
3435.45	15	15	1.50E+00
3437.21	11	9	8.06E-01
3438.87	1	3	2.34E-01
3441.87	5	3	1.34E-01
3442.66	3	3	2.94E-01
3445.03	7	9	1.53E-01
3445.26	7	5	2.96E-01
3445.80	9	9	1.14E-01
3447.12	9	11	8.75E-01
3447.29	5	3	1.79E-01
3449.07	7	9	1.52E-01
3449.85	5	7	1.65E-01
3452.60	7	7	2.48E-01
3456.15	5	5	3.60E-01
3456.52	3	3	2.96E-01
3460.22	5	3	2.77E-01
3460.78	9	7	6.03E-01
3465.84	3	1	9.99E-01
3466.19	9	7	2.11E-01
3466.96	7	7	1.52E-01
3467.85	5	7	2.63E-01
3469.22	5	3	6.96E-01
3469.63	13	15	1.51E-01
3470.92	3	5	2.91E-01
3475.03	3	3	4.68E-01
3479.42	7	5	2.26E-01
3483.67	7	7	1.13E-01
3483.83	7	5	1.41E-01
3489.43	7	7	3.27E-01
3504.41	7	9	8.06E-01
3505.31	7	9	2.25E-01
3508.11	9	9	1.59E-01
3510.77	13	13	4.75E-01
3517.55	11	11	5.41E-01
3518.21	3	3	3.64E-01
3521.38	9	9	1.39E-01
3521.41	9	11	6.06E-01
3524.65	5	3	3.10E-01
3524.98	7	9	2.25E-01
3538.92	11	11	2.24E-01
3540.57	5	3	4.46E-01
3542.17	7	5	4.93E-01
3552.71	9	7	3.64E-01
3555.64	3	3	3.46E-01
3558.09	5	7	5.43E-01
3563.75	1	3	1.53E-01
3566.05	9	9	2.67E-01
3566.74	7	7	1.43E-01
3570.64	15	15	7.18E-01
3573.88	3	5	3.58E-01
3580.54	13	11	5.49E-01
3581.88	11	13	3.81E-01
3584.25	3	3	1.73E-01
3585.57	7	5	3.95E-01
3588.95	7	7	1.18E-01
3590.74	7	9	2.23E-01
3595.55	5	5	2.32E-01
3598.88	13	11	5.67E-01
3600.73	9	9	2.07E-01
3601.88	7	9	1.15E-01
3602.94	5	7	2.96E-01
3604.07	9	7	3.25E-01
3610.61	5	3	1.78E-01
3611.99	7	7	1.16E-01
3615.16	7	9	1.96E-01
3623.22	11	9	5.58E-01
3624.46	9	11	5.27E-01
3624.62	5	7	1.37E-01
3638.20	5	3	3.51E-01
3638.21	5	3	3.33E-01
3640.62	7	5	1.94E-01
3647.84	7	7	2.11E-01
3648.70	7	5	1.15E-01
3654.58	3	3	1.80E-01
3657.36	5	7	2.03E-01
3658.13	9	9	1.86E-01
3659.36	7	9	6.70E-01
3660.92	3	5	1.34E-01
3662.15	7	9	1.45E-01
3662.99	11	11	3.48E-01
3663.27	7	5	2.30E-01
3664.81	11	13	9.54E-01
3664.88	1	3	1.92E-01
3669.34	9	7	2.16E-01
3672.81	9	11	1.95E-01
3672.82	9	9	1.13E-01
3676.23	3	1	5.22E-01
3680.68	11	11	2.96E-01
3681.72	9	7	1.68E-01
3683.01	3	5	1.20E-01
3687.96	5	7	2.12E-01
3688.97	11	9	3.26E-01
3690.59	11	9	2.07E-01
3694.94	5	7	6.36E-01
3696.04	11	11	3.59E-01
3698.07	7	5	1.48E-01
3708.55	7	9	1.28E-01
3715.75	9	7	2.38E-01
3718.48	5	7	1.34E-01
3720.25	7	9	2.86E-01
3725.55	7	7	1.60E-01
3727.68	9	11	1.51E-01
3728.30	7	5	1.55E-01
3728.50	7	9	2.20E-01
3733.02	7	7	1.45E-01
3733.41	13	13	2.80E-01
3735.62	11	11	1.66E-01
3742.28	7	7	1.56E-01
3747.19	5	7	3.07E-01
3748.48	9	11	3.95E-01
3755.10	3	5	1.41E-01
3755.16	9	9	2.48E-01
3758.52	9	9	1.22E-01
3759.60	9	7	1.82E-01
3760.88	9	9	2.16E-01
3768.73	9	9	2.88E-01
3769.99	7	9	2.46E-01
3777.72	13	11	1.66E-01
3788.25	7	9	2.87E-01
3794.43	9	9	1.22E-01
3797.47	7	5	1.48E-01
3798.25	7	9	6.90E-01
3801.84	9	7	3.16E-01
3805.99	5	5	2.44E-01
3819.78	9	11	1.47E-01
3824.78	5	7	1.40E-01
3827.15	7	7	1.94E-01
3828.88	7	7	1.35E-01
3830.81	5	5	1.83E-01
3831.07	7	9	1.20E-01
3832.11	9	9	3.05E-01
3833.75	9	9	1.70E-01
3834.64	3	5	1.20E-01
3846.18	7	7	1.26E-01
3847.25	3	1	2.41E-01
3848.30	9	9	1.26E-01
3851.99	11	9	1.78E-01
3864.10	7	7	6.24E-01
3866.69	3	5	1.74E-01
3867.67	5	3	2.22E-01
3869.08	5	3	1.35E-01
3874.15	7	5	1.67E-01
3902.95	7	5	6.17E-01
3909.54	9	7	1.13E-01
3911.94	5	5	1.15E-01
3915.43	5	5	1.40E-01
3916.43	5	3	1.78E-01
3919.55	11	13	2.24E-01
3955.48	13	11	1.71E-01

λ Å	Weights g_i	g_k	A 10^8 s^{-1}
3973.76	11	13	4.39E-01
3977.90	9	7	1.35E-01
3980.20	5	3	2.70E-01
3991.85	11	9	1.29E-01
4010.13	5	3	4.38E-01
4021.01	9	11	2.65E-01
4051.18	13	11	1.36E-01
4062.08	11	9	1.96E-01
4069.88	13	11	3.25E-01
4076.19	9	9	1.16E-01
4084.37	9	7	1.94E-01
4102.15	5	3	1.22E-01
4107.46	7	5	2.02E-01
4120.09	13	15	6.05E-01
4131.92	9	11	1.56E-01
4148.98	9	11	1.56E-01
4157.40	13	11	2.17E-01
4157.90	9	11	1.60E-01
4185.82	11	13	3.82E-01
4188.32	11	13	3.32E-01
4194.56	11	11	2.70E-01
4232.59	9	11	3.17E-01
4240.83	5	5	1.68E-01
4246.02	11	13	2.00E-01
4251.88	13	11	1.76E-01
4254.95	7	9	2.01E-01
4269.28	11	11	1.36E-01
4276.91	7	9	2.85E-01
4277.24	9	11	1.35E-01
4317.92	15	15	1.28E-01
4325.80	3	3	1.84E-01
4326.14	5	7	2.56E-01
4340.74	5	7	1.23E-01
4381.63	13	13	2.93E-01
4382.41	11	13	3.83E-01
4409.94	13	13	1.38E-01
4411.69	11	11	2.63E-01
4434.95	9	9	2.51E-01
4446.42	11	11	1.90E-01
4457.35	7	7	1.28E-01
4474.57	5	5	2.10E-01
4491.65	11	11	2.09E-01
4536.80	13	15	5.03E-01
4598.23	1	3	1.47E-01
4624.23	9	9	1.32E-01
4633.08	3	5	2.35E-01
4649.06	3	1	1.25E-01
4652.24	5	7	1.55E-01
4686.08	3	3	1.72E-01
4688.21	13	15	1.54E-01
4707.25	7	9	3.63E-01
4718.86	5	5	2.17E-01
4723.05	9	9	1.23E-01
4731.44	9	11	4.49E-01
4758.50	11	9	3.01E-01
4760.18	11	13	4.67E-01
4764.11	9	7	2.16E-01
4811.05	13	11	4.36E-01
4819.25	11	9	2.71E-01

λ Å	Weights g_i	g_k	A 10^8 s^{-1}
4830.51	9	7	4.07E-01
4858.39	13	11	1.24E-01
4868.02	7	5	3.11E-01
5037.18	9	7	1.14E-01
5044.36	7	5	1.31E-01
5047.70	3	1	2.61E-01
5163.18	9	11	2.03E-01
5171.06	5	7	1.84E-01
5172.94	5	5	4.11E-01
5174.18	5	3	5.83E-01
5191.45	7	9	1.62E-01
5238.21	7	9	3.74E-01
5240.87	7	7	3.89E-01
5242.80	7	5	2.01E-01
5261.53	5	7	1.13E-01
5280.85	5	5	1.28E-01
5355.52	9	9	1.21E-01
5356.46	11	11	2.11E-01
5360.51	9	11	6.19E-01
5364.28	9	9	2.26E-01
5460.50	5	3	3.46E-01
5493.76	7	5	2.13E-01
5506.49	5	7	3.61E-01
5533.03	5	5	3.72E-01
5570.44	5	3	3.30E-01
5849.71	3	3	3.02E-01
5851.50	3	5	1.55E-01
5893.36	5	5	2.60E-01
5895.93	5	7	3.12E-01
5926.37	7	7	1.63E-01
5928.88	7	9	5.32E-01
7154.11	9	9	3.45E-01

Neodymium
Nd II

λ Å	Weights g_i	g_k	A 10^8 s^{-1}
3780.4	16	18	1.4E-01
3805.4	14	16	6.9E-01
3807.2	10	12	4.9E-02
3863.3	8	10	1.5E-01
3941.5	10	10	6.1E-01
3951.2	12	12	6.0E-01
3973.3	18	18	6.3E-01
3979.5	10	12	2.7E-01
3990.1	16	16	5.2E-01
4012.3	18	20	5.5E-01
4061.1	16	18	4.4E-01
4106.6	14	16	6.8E-02
4109.5	14	16	3.7E-01
4133.4	14	12	1.5E-01
4156.1	12	14	3.4E-01
4205.6	18	16	1.8E-01
4284.5	18	18	8.5E-02
4303.6	8	10	4.7E-01
4325.8	16	16	1.6E-01
4358.2	14	14	1.5E-01
4382.7	12	10	4.0E-02
4400.8	10	10	6.8E-02
4451.6	12	14	2.5E-01
4456.4	16	18	6.4E-02
4463.0	14	16	1.8E-01
4958.1	12	10	1.2E-02
5130.6	22	20	1.6E-01
5192.6	20	18	1.7E-01
5249.6	18	16	1.8E-01
5276.9	12	10	1.2E-01
5293.2	16	14	1.2E-01
5302.3	20	18	1.1E-01
5311.5	14	12	1.1E-01
5319.8	12	10	1.6E-01
5357.0	18	16	1.8E-01
5371.9	20	20	5.1E-02
5485.7	18	18	5.7E-02
5594.4	16	16	7.0E-02
5620.6	18	18	1.3E-01
5688.5	14	14	5.9E-02
5718.1	16	16	8.7E-02
5726.8	10	10	5.6E-02
5740.9	12	12	7.2E-02
5804.0	10	10	4.6E-02
5865.1	16	18	1.3E-02
6051.9	12	10	1.1E-02

Neon
Ne I

λ Å	Weights g_i	g_k	A 10^8 s^{-1}
615.63	1	3	3.8E-01
618.67	1	3	9.3E-01
619.10	1	3	3.3E-01
626.82	1	3	7.4E-01
629.74	1	3	4.8E-01
735.90	1	3	6.11E+00
743.72	1	3	4.86E-01
3369.8	5	5	1.0E-03
3369.9	5	3	7.6E-03
3375.6	5	3	2.2E-03
3417.9	3	5	9.2E-03
3418.0	3	3	2.2E-03
3423.9	3	3	1.0E-03
3447.7	5	5	2.1E-02
3450.8	5	3	4.9E-03
3454.2	3	1	3.7E-02
3460.5	1	3	7.0E-03
3464.3	5	5	6.7E-03
3466.6	1	3	1.3E-02
3472.6	5	7	1.7E-02
3498.1	3	5	5.1E-03
3501.2	3	3	1.2E-02
3510.7	5	3	2.2E-03
3515.2	3	5	6.9E-03
3520.5	3	1	9.3E-02
3593.5	3	5	9.9E-03
3593.6	3	3	6.6E-03
3600.2	3	3	4.3E-03
3633.7	3	1	1.1E-02
3682.2	3	5	1.6E-03
3685.7	3	3	3.9E-03
3701.2	3	5	2.2E-03
4536.3	3	3	5.0E-03
4702.5	3	3	2.1E-03

λ	Weights		A	λ	Weights		A	λ	Weights		A
Å	g_i	g_k	10^8 s^{-1}	Å	g_i	g_k	10^8 s^{-1}	Å	g_i	g_k	10^8 s^{-1}
4708.9	3	3	4.2E-02	7438.9	1	3	2.31E-02	3037.7	4	4	2.1E+00
4955.4	3	3	3.3E-03	7472.4	3	3	4.0E-02	3045.6	2	2	2.5E+00
5113.7	3	3	1.0E-02	7535.8	3	3	4.3E-01	3047.6	4	6	1.8E+00
5120.5	3	3	5.6E-03	7937.0	5	5	7.8E-03	3054.7	2	4	9.4E-01
5154.4	3	3	1.9E-02	8082.5	3	3	1.2E-03	3092.9	6	6	1.3E+00
5191.3	3	3	1.3E-02	8118.5	3	3	4.9E-02	3097.1	8	8	1.3E+00
5326.4	3	3	6.8E-03	8128.9	3	5	7.2E-03	3118.0	8	6	4.2E-02
5333.3	3	3	5.3E-03	8259.4	5	5	2.03E-02	3134.1	6	4	2.6E-01
5341.1	3	3	1.1E-01	8571.4	3	3	5.5E-02	3140.4	8	6	2.4E-01
5400.6	3	1	9.0E-03	8582.9	3	5	1.00E-02	3151.1	6	6	4.8E-02
5418.6	3	3	5.2E-03	8647.0	5	5	3.91E-02	3154.8	8	6	1.8E-02
5433.7	3	3	2.83E-03	8681.9	3	3	2.1E-01	3164.4	8	8	1.6E-01
5652.6	3	3	8.9E-03	8767.5	3	3	1.1E-03	3165.7	6	6	1.2E-01
5662.5	3	3	6.9E-03	8771.7	3	3	1.6E-01	3173.6	6	4	4.5E-02
5852.5	3	1	6.82E-01	8783.8	3	5	3.13E-01	3176.1	4	6	6.0E-02
5868.4	3	3	1.4E-02	8865.3	3	3	9.4E-03	3187.6	4	6	1.4E-02
5881.9	5	3	1.15E-01	9201.8	3	3	9.1E-02	3188.7	6	6	3.9E-01
5913.6	3	3	4.8E-02	9433.0	3	3	1.1E-03	3190.9	4	6	1.5E-01
5939.3	5	3	2.00E-03	9486.7	3	3	2.5E-02	3194.6	4	4	5.2E-01
5944.8	5	5	1.13E-01	9534.2	3	3	6.3E-02	3198.6	6	8	1.7E+00
5961.6	3	3	3.3E-02	10621	3	3	2.4E-03	3198.9	4	4	2.3E-01
5975.5	5	3	3.51E-02	11409	3	3	4.2E-02	3209.0	8	8	1.6E-01
6030.0	3	3	5.61E-02	11525	3	3	8.4E-02	3209.4	2	4	6.0E-01
6046.1	3	3	2.26E-03	11767	3	3	6.9E-02	3213.7	2	4	1.7E+00
6074.3	3	1	6.03E-01	12459	3	3	1.5E-02	3214.3	4	6	2.2E+00
6096.2	3	5	1.81E-01					3218.2	8	10	3.6E+00
6118.0	5	3	6.09E-03	**Ne II**				3224.8	6	8	3.5E+00
6128.5	3	3	6.7E-03	*357.03	6	10	3.8E+01	3229.5	8	8	1.3E-01
6143.1	5	5	2.82E-01	*361.77	6	2	1.6E+01	3229.6	8	10	3.6E+00
6150.3	3	3	1.5E-02	*406.28	6	10	1.8E+01	3230.1	6	6	1.8E+00
6163.6	1	3	1.46E-01	*446.37	6	6	4.07E+01	3230.4	4	6	1.4E-01
6217.3	5	3	6.37E-02	460.73	4	2	4.7E+01	3232.0	6	4	2.7E-01
6266.5	1	3	2.49E-01	462.39	2	2	2.3E+01	3232.4	4	4	1.6E+00
6273.0	3	3	9.7E-03	1907.5	4	2	2.8E-01	3243.4	6	6	2.3E-01
6293.7	3	3	6.39E-03	1916.1	4	4	6.9E-01	3244.1	6	8	1.5E+00
6304.8	3	5	4.16E-02	1930.0	2	2	5.7E-01	3248.1	4	4	2.4E-01
6328.2	5	3	3.39E-02	1938.8	2	4	1.3E-01	3255.4	6	4	3.8E-02
6330.9	3	3	2.3E-02	2858.0	6	6	7.9E-01	3263.4	2	4	3.9E-01
6334.4	5	5	1.61E-01	2870.0	6	6	1.7E-01	3269.9	4	6	5.1E-01
6351.9	1	3	3.45E-03	2873.0	6	4	3.8E-01	3270.8	6	4	5.7E-02
6383.0	3	3	3.21E-01	2876.3	4	6	7.8E-01	3297.7	6	6	4.3E-01
6401.1	3	3	1.39E-02	2876.5	6	4	3.3E-01	3309.7	4	2	3.1E-01
6402.2	5	7	5.14E-02	2878.1	2	2	6.9E-02	3310.5	4	4	6.9E-02
6506.5	3	5	3.00E-01	2888.4	4	6	7.0E-02	3311.3	4	2	2.6E-01
6532.9	1	3	1.08E-01	2891.5	4	4	6.1E-02	3314.7	6	6	4.4E-02
6599.0	3	3	2.32E-01	2897.0	6	8	5.2E-02	3319.7	4	2	1.6E+00
6602.9	3	3	5.9E-03	2906.8	2	4	5.5E-01	3320.2	8	6	2.1E-01
6652.1	3	1	2.9E-03	2910.1	4	2	1.7E+00	3323.7	4	4	1.6E+00
6678.3	3	5	2.33E-01	2910.4	2	4	5.9E-01	3327.2	4	4	9.1E-01
6717.0	3	3	2.17E-01	2916.2	6	4	9.6E-02	3329.2	8	8	8.8E-01
6721.1	3	3	4.9E-04	2925.6	2	2	5.6E-01	3330.7	6	6	3.9E-02
6929.5	3	5	1.74E-01	2933.7	6	6	6.9E-02	3334.8	6	8	1.8E+00
7024.1	3	3	1.89E-02	2955.7	6	4	1.2E+00	3336.1	4	6	1.1E+00
7032.4	5	3	2.53E-01	3001.7	4	4	8.7E-01	3344.4	2	2	1.5E+00
7051.3	3	3	3.0E-02	3017.3	6	4	3.5E-01	3345.5	6	4	1.4E+00
7059.1	3	5	6.8E-02	3027.0	6	6	1.4E+00	3345.8	4	4	2.2E-01
7173.9	3	5	2.87E-02	3028.7	4	2	8.5E-01	3353.6	4	2	1.2E-01
7245.2	3	3	9.35E-02	3028.9	2	4	4.7E-01	3355.0	4	6	1.3E+00
7304.8	1	3	2.55E-03	3034.5	6	8	3.1E+00	3356.3	6	6	2.0E-01

λ Å	Weights g_i	g_k	A 10^8 s^{-1}
3357.8	6	6	5.0E-01
3360.3	2	4	8.6E-01
3360.6	2	4	8.2E-01
3362.9	4	2	3.5E-01
3371.8	4	6	2.2E-01
3374.1	4	4	3.0E-01
3378.2	2	2	1.7E+00
3379.3	2	2	3.0E-01
3386.2	4	6	5.5E-02
3388.4	4	6	2.2E+00
3390.6	2	4	7.7E-02
3392.8	2	4	4.4E-01
3404.8	4	6	1.9E+00
3407.0	6	8	2.3E+00
3411.4	4	2	6.1E-01
3413.2	4	4	1.8E+00
3414.9	4	6	1.8E-02
3416.9	6	6	6.4E-01
3417.7	6	8	1.6E+00
3438.9	2	2	1.4E+00
3440.7	2	4	3.5E-01
3453.1	4	4	4.6E-01
3454.8	4	4	1.6E+00
3456.6	2	4	9.6E-01
3457.1	4	6	9.9E-02
3459.3	6	6	1.6E+00
3475.2	4	4	1.2E-02
3477.6	4	6	4.3E-01
3481.9	4	2	1.4E+00
3503.6	2	2	2.0E+00
3522.7	4	2	2.3E-02
3538.0	4	2	7.6E-01
3539.9	4	4	3.6E-02
3542.2	6	4	6.0E-01
3542.9	4	6	1.2E+00
3546.2	2	4	6.3E-02
3551.6	2	4	3.7E-02
3557.8	2	2	1.9E-01
3561.2	4	6	2.1E-01
3565.8	4	4	6.2E-01
3568.5	6	8	1.4E+00
3571.2	4	4	6.3E-01
3574.2	6	6	1.0E-01
3574.6	4	6	1.3E+00
3590.4	4	6	3.6E-02
3594.2	4	2	1.3E+00
3612.3	2	4	2.6E-01
3628.0	4	4	6.0E-01
3632.7	4	4	1.3E-01
3643.9	4	4	3.2E-01
3644.9	2	4	9.9E-01
3659.9	4	6	6.7E-02
3664.1	6	4	7.0E-01
3679.8	4	2	3.2E-01
3694.2	6	6	1.0E+00
3697.1	2	2	2.8E-01
3701.8	4	6	2.7E-01
3709.6	4	2	1.1E+00
3713.1	4	6	1.3E+00
3721.8	4	6	2.0E-01
3726.9	4	4	1.2E-01
3727.1	2	4	9.8E-01
3734.9	4	4	1.9E-01
3744.6	2	4	2.6E-01
3751.2	2	2	1.8E-01
3753.8	4	6	4.5E-01
3766.3	4	6	2.9E-01
3777.1	2	4	4.2E-01
3800.0	4	4	3.7E-01
3818.4	2	4	6.1E-01
3829.8	4	6	8.4E-01
3942.3	4	6	1.0E-02
Ne V			
*142.61	9	9	6.7E+02
*143.32	9	15	1.2E+03
147.13	5	7	1.5E+03
151.23	5	5	3.38E+02
154.50	1	3	7.0E+02
*167.69	9	9	1.5E+02
*358.93	9	3	2.1E+02
365.59	5	3	1.35E+02
*482.15	9	9	3.01E+01
*571.04	9	15	1.0E+01
2259.6	3	5	1.9E+00
2265.7	5	7	2.4E+00
Ne VII			
97.502	1	3	1.07E+03
*115.46	9	3	4.8E+02
116.69	3	5	1.6E+03
127.66	3	1	1.9E+02
465.22	1	3	4.09E+01
558.61	3	5	8.11E+00
559.95	1	3	1.07E+01
561.38	3	3	7.99E+00
561.73	5	5	2.39E+01
562.99	3	1	3.17E+01
564.53	5	3	1.31E+01
Ne VIII			
*88.09	2	6	8.4E+02
*98.208	6	10	2.77E+03
770.41	2	4	5.90E+00
780.32	2	2	5.69E+00
2820.7	2	4	7.20E-01
2860.1	2	2	6.88E-01
Nickel			
Ni I			
1963.85	7	7	1.1E-01
1976.87	7	9	1.1E+00
1981.61	5	5	1.3E-01
1990.25	5	7	8.3E-01
2007.01	5	5	1.7E-01
2007.69	7	7	9.0E-02
2014.25	3	5	9.3E-01
2025.40	7	5	2.3E-01
2026.62	9	7	2.4E-01
2047.35	7	5	1.8E-01
2052.04	9	9	9.7E-02
2055.50	5	3	3.3E-01
2059.92	7	5	2.1E-01
2060.20	5	3	2.3E-01
2064.39	3	1	4.0E-01
2069.52	5	5	1.1E-01
2085.57	5	5	2.6E+00
2089.09	7	5	9.7E-02
2095.13	5	7	1.1E-01
2114.43	5	5	9.7E-02
2121.40	7	5	2.8E-01
2124.80	5	3	3.8E-01
2147.80	5	3	4.7E-01
2157.83	5	3	4.1E-01
2158.31	7	5	6.9E-01
2161.04	5	5	1.3E-01
2173.54	5	3	1.5E-01
2174.48	3	1	8.9E-01
2182.38	7	5	1.3E-01
2183.91	5	5	1.2E-01
2190.22	5	5	3.0E-01
2197.35	3	3	7.8E-01
2201.59	5	3	7.3E-01
2221.94	5	3	2.2E-01
2244.46	5	5	3.8E-01
2253.57	7	7	1.9E-01
2254.81	9	9	9.6E-02
2258.15	7	5	1.7E-01
2259.56	5	3	2.0E-01
2261.42	9	7	9.1E-02
2287.32	3	5	1.8E-01
2289.98	9	7	2.1E+00
2293.11	5	5	3.8E-01
2300.77	7	7	7.5E-01
2302.97	3	3	4.5E-01
2307.35	5	7	1.6E-01
2312.34	7	7	5.5E+00
2313.98	5	5	5.0E+00
2317.16	7	5	3.8E+00
2320.03	9	11	6.9E+00
2321.38	5	7	5.6E+00
2324.65	7	9	1.8E-01
2325.79	7	9	3.5E+00
2329.96	5	3	5.3E+00
2345.54	9	7	2.2E+00
2346.63	7	5	5.5E-01
2347.51	9	9	2.2E-01
2348.73	7	7	2.2E-01
2419.31	7	5	2.0E-01
2943.91	7	5	1.1E-01
2981.65	5	3	2.8E-01
3002.48	7	7	8.0E-01
3003.62	5	5	6.9E-01
3012.00	5	5	1.3E+00
3037.93	7	7	2.8E-01
3050.82	7	9	6.0E-01
3054.31	5	5	4.0E-01

λ Å	Weights g_i	g_k	A 10^8 s^{-1}
3057.64	3	3	1.0E+00
3064.62	5	7	1.1E-01
3101.56	5	7	6.3E-01
3101.88	5	7	4.9E-01
3134.11	3	5	7.3E-01
3225.02	5	3	9.3E-02
3369.56	9	7	1.8E-01
3380.57	5	3	1.3E+00
3392.98	7	7	2.4E-01
3414.76	7	9	5.5E-01
3423.71	3	3	3.3E-01
3433.56	7	7	1.7E-01
3446.26	5	5	4.4E-01
3452.88	5	7	9.8E-02
3458.46	3	5	6.1E-01
3461.66	7	9	2.7E-01
3472.55	5	7	1.2E-01
3483.77	5	3	1.4E-01
3492.96	5	3	9.8E-01
3510.33	3	1	1.2E+00
3515.05	5	7	4.2E-01
3524.54	7	5	1.0E+00
3566.37	5	5	5.6E-01
3597.70	3	3	1.4E-01
3619.39	5	7	6.6E-01
4027.67	5	7	1.3E-01
4295.88	9	7	1.7E-01
4401.54	9	11	3.8E-01
4462.46	3	5	1.7E-01
4470.48	5	7	1.9E-01
4600.37	5	3	2.6E-01
4604.99	9	7	2.3E-01
4606.23	5	3	1.0E-01
4648.66	11	9	2.4E-01
4686.22	5	5	1.4E-01
4701.54	9	9	1.4E-01
4714.42	13	11	4.6E-01
4715.78	7	7	2.0E-01
4732.47	7	9	9.3E-02
4752.43	3	3	2.0E-01
4756.52	9	9	1.5E-01
4786.54	11	11	1.8E-01
4812.00	3	1	9.5E-02
4829.03	5	7	1.9E-01
4831.18	9	7	1.6E-01
4838.64	9	7	2.2E-01
4855.41	5	5	5.7E-01
4904.41	5	3	6.2E-01
4912.03	3	3	1.5E-01
4913.97	1	3	2.2E-01
4918.36	9	7	2.3E-01
4935.83	7	5	2.4E-01
4937.34	9	9	1.2E-01
4953.20	5	5	1.2E-01
4980.17	9	11	1.9E-01
5000.34	7	7	1.4E-01
5012.46	7	7	1.1E-01
5017.58	11	11	2.0E-01
5035.37	7	9	5.7E-01
5042.20	3	5	1.4E-01
5048.85	7	7	1.6E-01
5080.53	9	11	3.2E-01
5081.11	7	9	5.7E-01
5082.35	3	3	2.5E-01
5084.08	7	9	3.1E-01
5099.95	7	7	2.9E-01
5115.40	11	9	2.2E-01
5129.37	7	5	1.2E-01
5155.14	5	5	1.1E-01
5155.76	5	7	2.9E-01
5176.57	5	5	1.8E-01
5371.33	7	7	1.6E-01
5476.91	1	3	9.5E-02
5637.12	3	3	1.1E-01
5664.02	5	7	1.1E-01
5695.00	3	3	1.7E-01
6086.29	3	5	1.1E-01
6175.42	3	3	1.7E-01
7122.24	5	7	2.1E-01
7381.94	9	11	9.7E-02
7422.30	7	5	1.8E-01
7727.66	7	7	1.1E-01
Ni II			
2165.55	10	10	2.4E+00
2169.10	8	8	1.58E+00
2174.67	8	10	1.43E+00
2175.15	6	6	1.77E+00
2184.61	4	4	2.90E+00
2201.41	4	6	1.3E+00
2206.72	6	8	1.66E+00
2216.48	10	12	3.4E+00
2220.40	6	8	2.3E+00
2222.96	10	10	9.8E-01
2224.86	8	8	1.55E+00
2226.33	6	6	1.3E+00
2253.85	4	6	1.98E+00
2264.46	6	8	1.43E+00
2270.21	8	10	1.56E+00
2278.77	8	6	2.8E+00
2287.09	6	4	2.8E+00
2296.55	8	8	1.98E+00
2297.14	6	4	2.70E+00
2297.49	4	2	3.0E+00
2298.27	6	6	2.8E+00
2303.00	8	6	2.9E+00
2316.04	10	8	2.88E+00
2334.58	8	8	8.0E-01
2375.42	6	8	6.6E-01
2394.52	8	10	1.70E+00
2416.13	6	8	2.1E+00
2437.89	8	10	5.4E-01
2510.87	8	10	5.8E-01
Ni III			
1692.51	11	13	7.9E+00
1709.90	9	11	6.3E+00
1719.46	5	7	6.0E+00
1722.28	3	5	5.9E+00
1724.52	3	1	6.7E+00
1741.96	9	7	5.7E+00
1752.43	7	5	5.5E+00
1760.56	5	3	6.5E+00
1769.64	11	11	6.2E+00
1823.06	9	9	5.6E+00
Ni XIV			
164.13	6	8	1.2E+03
168	2	4	2.4E+02
168.12	4	2	8.5E+02
169.69	4	4	9.8E+02
170.50	4	4	7.1E+02
171.37	4	6	9.4E+02
172.16	6	6	4.7E+02
172.80	6	4	1.4E+02
177.28	4	4	5.6E+02
178	2	4	8.9E+01
181	4	6	7.4E+01
182.14	4	2	1.5E+02
196	4	2	3.8E+01
288.894	4	4	4.6E+01
292.399	6	6	3.6E+01
Ni XV			
50.249	5	7	6.8E+03
60.890	9	11	1.0E+04
64.635	7	9	9.6E+03
163.64	5	7	5.6E+01
173.73	5	7	7.6E+02
175	3	1	5.7E+02
179.28	5	7	7.5E+02
181	1	3	6.8E+02
269	3	1	5.3E+01
278.386	5	5	4.3E+01
Ni XVI			
166	4	6	3.1E+02
168	6	8	3.2E+02
182	2	2	2.5E+02
185.23	2	4	4.2E+02
187	4	6	1.2E+02
187	4	2	3.3E+02
188	2	2	4.7E+02
190	6	8	2.0E+02
192	6	8	4.54E+02
192	6	6	3.1E+02
194	4	6	2.8E+02
194	2	4	5.5E+02
194	2	2	1.1E+02
194	4	4	3.5E+02
194.04	4	6	4.6E+02
195.27	4	4	9.5E+01
196	4	6	6.7E+02
197	4	6	1.5E+02
197	4	2	1.2E+02
199	2	4	4.9E+02
206	2	2	3.7E+02

λ Å	Weights g_i	Weights g_k	A 10^8 s^{-1}
217	4	4	1.1E+02
218.391	2	4	9.5E+01
223.119	2	2	1.3E+02
231	4	4	1.6E+02
232.475	4	4	4.07E+02
233	6	4	2.4E+02
235	4	2	3.8E+02
235	6	6	2.5E+02
236	4	4	1.2E+02
237.875	4	2	2.6E+02
238	6	4	1.3E+02
239.550	2	2	2.6E+02
245	4	4	1.4E+02
245	4	6	3.2E+02
249	6	8	3.3E+02
249	6	4	1.2E+02
250	4	2	1.6E+02
254	6	4	1.8E+02
Ni XVII			
30.919	1	3	2.77E+03
42.855	1	3	4.75E+03
54.451	9	11	1.5E+04
55.361	1	3	6.7E+03
57.348	7	9	1.4E+04
197.39	1	3	1.6E+02
199.87	3	5	2.1E+02
204	3	3	1.8E+02
205	3	1	2.4E+02
206	1	3	3.0E+02
207.50	5	7	2.5E+02
215.89	3	5	4.8E+02
216	1	3	2.7E+02
217	5	7	2.4E+02
227	5	5	1.6E+02
249.180	1	3	2.75E+02
281.50	3	1	2.1E+02
282	3	1	2.4E+02
284	5	3	1.5E+02
292	5	7	2.2E+02
Ni XVIII			
24.881	2	4	8.6E+02
25.070	4	6	9.9E+02
26.02	2	4	1.26E+03
26.020	2	4	1.1E+03
26.046	2	2	1.1E+03
26.218	4	6	1.5E+03
27.98	4	6	1.0E+03
27.982	2	4	2.0E+03
28.018	6	8	1.1E+03
28.220	4	6	2.33E+03
29.383	4	6	1.58E+03
29.422	6	8	1.69E+03
29.779	2	4	1.9E+03
29.829	2	2	1.9E+03
31.845	4	6	2.7E+03
31.890	6	8	3.0E+03
32.034	2	4	3.4E+03
32.340	4	6	4.0E+03
36.990	4	6	5.5E+03
37.049	6	8	5.9E+03
41.015	2	4	2.97E+03
41.218	2	2	3.2E+03
43.814	2	4	5.5E+03
44.365	4	6	6.8E+03
44.405	4	4	1.14E+03
52.615	4	6	1.5E+04
52.720	6	8	1.6E+04
52.745	6	6	1.06E+03
59.950	6	4	9.6E+02
60.212	4	2	1.1E+03
63.512	4	6	7.9E+02
63.589	6	8	8.5E+02
69.075	4	6	8.0E+02
76.254	4	6	1.38E+03
76.359	6	8	1.47E+03
99.275	2	4	1.0E+03
100.4	4	6	1.2E+03
114.46	4	6	2.5E+03
114.74	6	8	2.7E+03
Ni XIX			
9.140	1	3	5.1E+04
9.153	1	3	5.2E+03
9.977	1	3	1.1E+05
10.110	1	3	9.4E+04
10.283	1	3	4.7E+03
10.433	1	3	5.1E+03
11.539	1	3	4.8E+04
11.599	1	3	6.3E+03
12.435	1	3	3.66E+05
12.656	1	3	1.0E+05
13.779	1	3	1.23E+04
14.043	1	3	1.31E+04
40.7	3	3	6.4E+03
40.7	3	1	8.4E+03
41.132	7	9	9.4E+03
Ni XXI			
11.13	3	3	1.7E+04
11.23	5	3	1.7E+04
11.239	5	7	5.7E+04
11.28	3	1	2.2E+05
11.318	5	7	2.8E+05
11.48	3	1	1.1E+05
11.48	1	3	4.0E+05
11.517	5	7	1.4E+05
11.539	5	7	1.2E+05
11.67	1	3	8.0E+04
11.72	3	3	2.3E+04
12.454	5	3	3.3E+04
12.472	3	3	1.8E+04
12.502	5	5	2.8E+04
Ni XXII			
72.52	4	2	2.84E+02
84.06	6	4	1.2E+03
84.24	4	2	5.6E+02
85.86	4	2	4.9E+02
88.00	4	2	1.2E+03
95.95	2	2	4.4E+02
98.16	4	4	5.2E+02
98.58	4	4	2.45E+02
100.60	6	6	3.9E+02
101.31	6	4	4.83E+02
103.31	4	2	2.66E+02
106.04	4	4	2.36E+02
106.16	4	2	5.1E+02
124.31	2	2	3.7E+02
126.32	4	4	3.3E+02
Ni XXIII			
87.66	3	3	2.8E+02
88.11	5	3	8.3E+02
90.49	3	3	1.77E+02
90.96	5	3	2.5E+02
91.83	5	3	7.5E+02
92.32	3	1	4.39E+02
100.42	1	3	2.1E+02
102.08	5	5	5.3E+02
103.23	3	3	2.4E+02
103.67	5	5	1.78E+02
104.70	3	1	2.94E+02
106.02	5	5	2.87E+02
108.27	7	5	3.32E+02
111.23	3	1	2.26E+02
111.78	5	3	2.19E+02
111.86	1	3	1.7E+02
112.55	3	1	1.0E+03
128.87	5	5	4.02E+02
133.54	3	3	1.86E+02
137.55	3	1	2.53E+02
Ni XXIV			
101.13	6	4	1.63E+02
102.11	4	4	5.4E+02
103.43	2	4	1.3E+02
103.53	4	2	4.17E+02
104.64	2	2	4.7E+02
106.68	4	2	3.67E+02
113.14	4	4	1.65E+02
118.52	2	4	1.5E+02
122.72	6	4	2.17E+02
134.73	6	6	1.44E+02
135.47	4	4	8.0E+01
137.01	4	4	2.6E+02
138.80	4	6	7.2E+01
153.47	2	2	1.27E+02
159.69	2	4	8.9E+01
Ni XXV			
9.30	3	1	9.3E+04
9.31	5	7	8.2E+04
9.32	3	5	7.8E+04
9.34	1	3	1.1E+05
9.42	3	1	9.0E+04

λ	Weights		A
Å	g_i	g_k	10^8 s^{-1}
9.49	3	5	8.9E+04
9.60	1	3	1.8E+05
9.63	3	5	2.4E+05
9.64	3	3	1.3E+05
9.71	3	1	2.3E+05
9.71	3	3	1.8E+05
9.74	5	7	3.0E+05
9.75	3	5	1.3E+05
9.76	1	3	3.03E+05
9.76	5	3	7.5E+04
9.78	5	7	2.9E+05
9.86	5	7	4.8E+05
9.87	3	5	2.03E+05
9.92	5	5	1.3E+05
9.94	5	7	1.29E+05
9.97	3	5	2.5E+05
10.08	1	3	2.80E+05

Ni XXVI

λ Å	g_i	g_k	A 10^8 s^{-1}
1.5930	4	2	3.4E+06
1.5935	2	2	4.0E+06
1.5973	4	4	8.1E+06
1.5977	2	4	4.4E+06
1.5982	2	2	7.3E+06
1.5996	2	2	2.7E+06
1.6005	4	6	2.7E+06
1.6036	4	2	2.1E+06
9.390	2	4	2.59E+05
9.535	4	6	2.96E+05

Ni XXVII

λ Å	g_i	g_k	A 10^8 s^{-1}
1.2534	1	3	3.35E+05
1.2824	1	3	6.38E+05
1.3500	1	3	1.63E+06
1.3516	1	3	2.4E+05
1.531	3	3	2.0E+05
1.534	3	1	6.9E+06
1.537	5	5	2.3E+06
1.537	1	3	3.7E+06
1.538	3	5	3.9E+06
1.539	1	3	2.6E+06
1.539	3	5	2.6E+06
1.540	3	3	1.7E+06
1.541	3	5	5.5E+06
1.542	3	3	3.6E+06
1.542	5	5	3.5E+06
1.544	5	3	3.2E+06
1.546	3	5	1.6E+06
1.547	3	3	2.1E+05
1.549	1	3	2.0E+05
1.551	3	1	8.2E+05
1.558	3	1	6.5E+05
1.5883	1	3	6.02E+06
1.5963	1	3	7.70E+05

Nitrogen

N I

λ Å	g_i	g_k	A 10^8 s^{-1}
1163.9	6	6	4.3E-01
1164.0	4	6	3.2E-02
1164.2	6	4	4.8E-02
1164.3	4	4	4.3E-01
1167.4	6	8	1.1E+00
1168.4	6	6	9.5E-02
1168.5	4	6	1.3E+00
1169.7	6	8	3.0E-02
1176.5	6	4	9.5E-01
1176.6	4	4	1.1E-01
1177.7	4	2	1.3E+00
1199.6	4	6	5.5E+00
1200.2	4	4	5.3E+00
1200.7	4	2	5.5E+00
1310.5	4	6	1.3E+00
1316.3	4	6	2.5E-02
1492.6	6	4	5.3E+00
1492.8	4	4	5.8E-01
1494.7	4	2	5.0E+00
4099.9	2	4	3.4E-02
4110.0	4	6	4.0E-02
4114.0	4	4	6.8E-03
4137.6	2	4	3.9E-03
4143.4	4	4	7.8E-03
4151.5	6	4	1.3E-02
4214.8	4	6	2.2E-02
4216.1	2	4	3.1E-02
4218.9	2	2	1.2E-02
4222.1	4	4	9.8E-03
4223.1	6	6	5.1E-02
4224.9	4	2	6.1E-02
4230.5	6	4	3.3E-02
4385.5	2	2	5.2E-03
4392.4	4	2	1.02E-02
4914.9	2	2	7.59E-03
4935.1	4	2	1.58E-02
5169.6	6	4	2.09E-03
5181.4	4	4	1.44E-03
5186.6	2	4	7.3E-04
5199.8	2	2	2.3E-02
5201.6	2	4	2.3E-02
5281.2	6	6	2.82E-03
5292.7	6	4	1.67E-03
5293.5	4	6	1.13E-03
5309.4	4	2	2.73E-03
5310.5	2	4	1.37E-03
5344.0	6	6	6.2E-04
5356.6	4	6	1.89E-03
5367.0	4	4	1.18E-03
5372.6	2	4	1.07E-03
5378.3	2	2	2.10E-03
5816.5	4	6	2.78E-03
5829.5	6	6	6.4E-03
5834.6	2	4	3.83E-03
5840.9	4	4	1.22E-03
5849.7	2	2	1.52E-03
5854.0	6	4	4.09E-03
5856.0	4	2	7.6E-03
6606.2	4	6	7.9E-04
6622.5	6	6	7.1E-03
6627.0	2	4	1.97E-03
6636.9	4	4	1.25E-02
6645.0	8	6	3.11E-02
6646.5	2	2	1.94E-02
6653.5	6	4	2.44E-02
6656.5	4	2	1.93E-02
6926.7	4	6	6.4E-03
6945.2	6	6	1.49E-02
6951.6	2	4	8.8E-03
6960.5	4	4	2.81E-03
6973.1	2	2	3.50E-03
6979.2	6	4	9.4E-03
6982.0	4	2	1.74E-02
7423.6	2	4	5.2E-02
7442.3	4	4	1.06E-01
7468.3	6	4	1.61E-01

N II

λ Å	g_i	g_k	A 10^8 s^{-1}
474.89	5	5	4.5E+00
475.65	1	3	1.4E+01
475.70	3	5	2.0E+01
475.76	3	3	1.1E+01
475.80	5	7	2.6E+01
475.88	5	5	6.8E+00
508.70	5	5	2.8E+00
510.76	5	7	3.1E+01
513.85	5	5	6.8E+00
529.36	1	3	6.5E+00
529.41	3	1	2.0E+01
529.49	3	3	4.9E+00
529.64	3	5	4.9E+00
529.72	5	3	8.1E+00
529.87	5	5	1.5E+01
533.51	1	3	2.0E+01
533.58	3	5	2.7E+01
533.65	3	3	1.5E+01
533.73	5	7	3.6E+01
533.82	5	5	9.1E+00
547.82	5	3	5.2E+00
559.76	1	3	1.2E+01
574.65	5	7	3.5E+01
582.16	5	5	1.3E+01
635.20	1	3	1.8E+01
644.63	1	3	1.2E+01
644.84	3	3	3.5E+01
645.18	5	3	5.8E+01
660.29	5	3	4.0E+01
671.02	3	5	2.9E+00
671.39	5	5	8.9E+00
671.41	1	3	3.6E+00
671.63	3	3	2.7E+00
671.77	3	1	1.2E+01
672.00	5	3	4.4E+00
745.84	1	3	1.0E+01
746.98	5	3	4.0E+01
748.37	5	3	2.0E+00
775.97	5	5	3.5E+01
915.61	1	3	3.6E+00
915.96	3	1	1.1E+01
1084.0	1	3	2.0E+00

λ Å	Weights g_i	g_k	A $10^8\ s^{-1}$
1085.5	5	5	9.0E-01
1085.7	5	7	3.6E+00
2139.5	3	3	2.9E-01
3593.6	3	5	2.4E-01
3609.1	3	3	2.4E-01
3615.9	3	1	2.4E-01
3829.8	3	5	1.5E-01
3838.4	5	5	4.5E-01
3842.2	1	3	2.0E-01
3847.4	3	3	1.5E-01
3855.1	3	1	6.0E-01
3856.1	5	3	2.5E-01
3919.0	3	3	1.00E+00
3995.0	3	5	1.3E+00
4114.4	3	3	1.9E-03
4447.0	3	5	1.30E+00
4477.7	5	3	3.5E-02
4507.6	7	5	3.8E-02
4601.5	3	5	2.70E-01
4607.2	1	3	3.40E-01
4613.9	3	3	1.96E-01
4621.4	3	1	9.0E-01
4630.5	5	5	8.4E-01
4643.1	5	3	4.66E-01
4774.2	3	5	5.4E-02
4779.7	3	3	2.69E-01
4781.2	5	7	4.0E-02
4788.1	5	5	2.48E-01
4793.7	5	3	8.9E-02
4803.3	7	7	3.13E-01
4810.3	7	5	5.5E-02
4987.4	3	1	6.3E-01
4994.4	3	3	7.4E-01
5001.1	3	5	1.02E+00
5001.5	5	7	1.08E+00
5002.7	1	3	8.5E-02
5005.2	7	9	1.22E+00
5007.3	3	5	7.7E-01
5010.6	3	3	2.68E-01
5025.7	7	7	1.34E-01
5040.7	7	5	5.3E-03
5045.1	5	3	4.10E-01
5452.1	1	3	1.4E-01
5454.2	3	1	4.1E-01
5462.6	3	3	1.0E-01
5478.1	3	5	1.0E-01
5480.1	5	3	1.7E-01
5495.7	5	5	3.0E-01
5666.6	3	5	4.23E-01
5676.0	1	3	3.10E-01
5679.6	5	7	5.6E-01
5686.2	3	3	2.31E-01
5710.8	5	5	1.37E-01
5927.8	1	3	3.15E-01
5931.8	3	5	4.25E-01
5940.2	3	3	2.35E-01
5941.7	5	7	5.6E-01
5952.4	5	5	1.40E-01
6482.1	3	3	3.7E-01
6610.6	5	7	5.9E-01

N III

λ Å	Weights g_i	g_k	A $10^8\ s^{-1}$
374.20	2	4	1.01E+02
451.87	2	2	8.9E+00
452.23	4	2	1.78E+01
685.00	2	4	9.3E+00
685.51	2	2	3.71E+01
685.82	4	4	4.68E+01
686.34	4	2	1.90E+01
763.34	2	2	9.6E+00
764.36	4	2	1.87E+01
771.54	2	4	8.2E+00
771.90	4	4	1.65E+01
772.39	6	4	2.47E+01
772.89	6	4	2.03E+01
772.98	4	2	2.27E+01
979.84	4	4	8.9E+00
979.92	6	6	9.3E+00
989.79	2	4	4.15E+00
991.51	4	4	8.2E-01
991.58	4	6	4.96E+00
1747.8	2	4	1.31E+00
1751.2	4	4	2.58E-01
1751.7	4	6	1.57E+00
4097.3	2	4	8.2E-01
4103.4	2	2	8.2E-01
4634.1	2	4	6.5E-01
4640.6	4	6	7.8E-01
4641.9	4	4	1.30E-01

N IV

λ Å	Weights g_i	g_k	A $10^8\ s^{-1}$
247.20	1	3	1.14E+02
*283.52	9	15	2.9E+02
*322.64	9	3	8.4E+01
335.05	3	5	1.85E+02
387.35	3	1	2.8E+01
765.15	1	3	2.40E+01
*923.16	9	9	1.8E+01
955.34	3	1	3.0E+01
1718.6	3	5	2.37E+00
*3480.8	3	9	1.1E+00
4057.8	3	5	6.8E-01
6380.8	1	3	1.4E-01
*7116.7	9	15	1.2E-01

N V

λ Å	Weights g_i	g_k	A $10^8\ s^{-1}$
*209.29	2	6	1.18E+02
*247.66	6	10	4.3E+02
1238.8	2	4	3.41E+00
1242.8	2	2	3.38E+00
4603.7	2	4	4.12E-01
4620.0	2	2	4.08E-01

N VI

λ Å	Weights g_i	g_k	A $10^8\ s^{-1}$
24.898	1	3	5.158E+03
28.787	1	3	1.809E+04
*161.22	3	9	2.859E+02
173.29	1	3	2.697E+02
*173.92	9	15	8.76E+02
185.19	3	5	8.24E+02
*1901.5	3	9	6.777E-01
2896.4	1	3	2.080E-01

Oxygen

O I

λ Å	Weights g_i	g_k	A $10^8\ s^{-1}$
1028.2	1	3	2.0E-01
1152.2	5	5	5.5E+00
1217.6	1	3	1.8E+00
1302.2	5	3	3.3E+00
1304.9	3	3	2.0E+00
1306.0	1	3	6.6E-01
5435.2	3	5	6.1E-03
5435.8	5	5	1.02E-02
5436.9	7	5	1.42E-02
6453.6	3	5	1.42E-02
6454.4	5	5	2.37E-02
6456.0	7	5	3.31E-02
6653.8	3	1	6.00E-01
7156.7	5	5	4.73E-01
7471.4	5	3	1.14E-02
7473.2	5	5	1.02E-01
7476.4	5	7	4.08E-01
7477.2	3	3	1.70E-01
7479.1	3	5	3.06E-01
7480.7	1	3	2.26E-01
7771.9	5	7	3.40E-01
7774.2	5	5	3.40E-01
7775.4	5	3	3.40E-01
7886.3	3	5	3.70E-01
7939.5	7	5	1.65E-03
7943.2	7	7	4.17E-02
7947.2	5	5	5.8E-02
7947.6	7	9	3.73E-01
7950.8	5	7	3.31E-01
7952.2	3	5	3.13E-01
7981.9	3	3	1.2E-01
7982.4	1	3	1.6E-01
7987.0	3	5	2.1E-01
7987.3	5	5	7.2E-02
7995.1	5	7	2.9E-01

O II

λ Å	Weights g_i	g_k	A $10^8\ s^{-1}$
429.92	4	2	3.9E+01
430.04	4	4	3.9E+01
430.18	4	6	3.9E+01
483.75	4	2	8.4E-01
483.98	6	4	7.6E-01
484.03	4	4	8.4E-02
485.09	6	8	2.5E+01
485.47	6	6	1.6E+00
485.52	4	6	2.3E+01
3007.1	8	10	8.4E-01
3007.7	6	8	7.2E-01
3013.4	6	8	7.4E-01
3032.1	8	10	8.5E-01
3032.5	6	8	8.2E-01
3134.8	8	6	1.23E+00

λ Å	Weights g_i	g_k	A 10^8 s^{-1}
3273.5	8	6	1.14E+00
3377.2	2	2	1.88E+00
3390.3	2	4	1.86E+00
3407.4	6	6	7.5E-01
3749.5	6	4	9.0E-01
3882.2	8	8	4.93E-01
3912.0	6	4	1.27E+00
3919.3	4	2	1.40E+00
3973.3	4	4	1.27E+00
4069.6	2	4	1.39E+00
4069.9	4	6	1.49E+00
4072.2	6	8	1.70E+00
4075.9	8	10	1.98E+00
4085.1	6	6	4.78E-01
4087.2	4	6	2.24E+00
4089.3	10	12	2.62E+00
4095.6	6	8	2.23E+00
4097.2	8	10	2.37E+00
4104.7	4	6	1.04E+00
4105.0	4	4	8.0E-01
4108.8	8	8	3.49E-01
4119.2	6	8	1.48E+00
4120.3	6	6	4.43E-01
4132.8	2	4	8.4E-01
4153.3	4	6	7.7E-01
4276.7	6	8	1.82E+00
4277.4	2	4	1.49E+00
4277.9	8	8	3.02E-01
4281.4	6	6	6.0E-01
4282.8	4	4	1.06E+00
4283.0	4	6	1.58E+00
4283.1	6	6	5.1E-01
4283.8	4	4	5.9E-01
4294.8	4	6	1.39E+00
4303.8	6	8	1.97E+00
4328.6	4	2	1.21E+00
4340.4	6	8	2.23E+00
4347.4	4	4	9.4E-01
4349.4	6	6	7.4E-01
4351.3	6	6	9.7E-01
4396.0	6	6	3.98E-01
4414.9	4	6	1.15E+00
4417.0	2	4	9.5E-01
4443.1	6	6	5.7E-01
4448.2	8	8	5.7E-01
4489.5	2	4	1.51E+00
4491.3	4	6	1.81E+00
4596.2	4	6	1.03E+00
4602.1	4	6	1.70E+00
4609.4	6	8	1.82E+00
4641.8	4	6	7.9E-01
4661.6	4	4	5.2E-01
4701.2	4	4	8.7E-01
4703.2	4	6	8.2E-01
4705.4	6	8	1.38E+00
4871.6	4	6	4.35E-01
4906.9	4	4	6.8E-01
4924.6	4	6	6.7E-01
4941.1	2	4	8.3E-01

λ Å	Weights g_i	g_k	A 10^8 s^{-1}
4943.1	4	6	1.06E+00
5206.7	4	4	3.91E-01
6627.6	4	4	8.9E-02
6666.9	4	2	3.49E-02
6678.2	2	4	1.73E-02
6718.1	2	2	6.8E-02
6721.4	4	2	1.89E-01
6810.6	6	8	1.80E-03
6844.1	4	6	3.25E-03
6847.0	8	8	3.47E-02
6869.7	6	6	5.9E-02
6885.1	4	4	6.7E-02
6895.3	10	8	2.98E-01
6906.5	8	6	2.72E-01
6908.1	4	2	3.32E-01
6910.8	6	4	2.67E-01
O III			
262.88	5	5	4.0E+01
263.69	1	3	5.2E+01
263.73	3	5	7.3E+01
263.77	3	3	4.0E+01
263.82	5	7	9.6E+01
263.86	5	5	2.4E+01
277.38	5	7	1.1E+02
279.79	5	5	2.6E+01
295.94	1	3	4.8E+01
303.41	1	3	3.4E+01
303.46	3	1	1.0E+02
303.52	3	3	2.6E+01
303.62	3	5	2.5E+01
303.69	5	3	4.2E+01
303.80	5	5	7.6E+01
305.60	1	3	1.0E+02
305.66	3	5	1.4E+02
305.70	3	3	7.6E+01
305.77	5	7	1.8E+02
305.84	5	5	4.6E+01
320.98	5	7	1.9E+02
328.45	5	5	6.1E+01
345.31	1	3	9.9E+01
374.08	5	5	2.6E+01
395.56	5	3	4.9E+01
507.39	1	3	1.6E+01
507.68	3	3	4.7E+01
508.18	5	3	7.9E+01
525.80	5	3	8.8E+01
597.82	1	3	1.8E+01
599.60	5	5	5.5E+01
702.33	1	3	5.7E+00
702.82	3	1	1.7E+01
832.93	1	3	3.2E+00
835.10	5	5	1.4E+00
835.29	5	7	5.7E+00
1109.5	3	3	2.8E+00
1679.1	3	5	9.4E-01
1686.9	3	3	9.4E-01
1760.4	3	5	6.0E-01
1764.5	5	5	1.8E+00

λ Å	Weights g_i	g_k	A 10^8 s^{-1}
1766.3	1	3	7.8E-01
1772.3	3	1	2.3E+00
1773.0	5	3	9.8E-01
2390.4	3	3	2.2E+00
2959.7	3	5	2.1E+00
2996.5	3	3	5.1E-01
3004.4	5	5	4.7E-01
3017.6	7	7	5.9E-01
3115.7	3	1	1.5E+00
3121.7	3	3	1.5E+00
3132.9	3	5	1.4E+00
3261.0	5	7	1.8E+00
3265.5	7	9	2.1E+00
3267.3	3	5	1.7E+00
3281.9	5	5	3.2E-01
3284.6	7	7	2.3E-01
3405.7	1	3	2.7E-01
3408.1	3	1	8.1E-01
3415.3	3	3	2.0E-01
3428.7	3	5	2.0E-01
3430.6	5	3	3.3E-01
3444.1	5	5	5.9E-01
3702.8	1	3	6.2E-01
3707.2	3	5	8.3E-01
3714.0	3	3	4.6E-01
3715.1	5	7	1.1E+00
3725.3	5	5	2.7E-01
3961.6	5	7	1.3E+00
5268.1	1	3	3.1E-01
5508.1	5	5	1.1E-01
5592.4	3	3	3.6E-01
O IV			
238.36	2	4	2.88E+02
238.57	4	6	3.46E+02
238.58	4	4	5.8E+01
279.63	2	2	2.37E+01
279.94	4	2	4.77E+01
553.33	2	4	1.20E+01
554.08	2	2	4.73E+01
554.51	4	4	6.0E+01
555.26	4	2	2.50E+01
608.40	2	2	1.25E+01
609.83	4	2	2.36E+01
616.95	6	4	2.55E+01
617.01	4	4	2.93E+00
617.04	4	2	2.86E+01
624.62	2	4	1.07E+01
625.13	4	4	2.15E+01
625.85	6	4	3.22E+01
779.73	6	4	1.53E+00
779.82	4	4	1.31E+01
779.91	6	6	1.37E+01
780.00	4	6	1.0E+00
787.71	2	4	5.9E+00
790.11	4	4	1.15E+00
790.20	4	6	7.1E+00
921.30	2	4	2.11E+00
921.37	2	2	9.2E+00

λ Å	Weights g_i	g_k	A 10^8 s^{-1}
923.37	4	4	1.14E+01
923.43	4	2	4.48E+00
1338.6	2	4	2.22E+00
1343	4	4	4.28E-01
1343.5	4	6	2.64E+00
O V			
172.17	1	3	2.96E+02
*192.85	9	15	6.9E+02
*215.17	9	3	1.7E+02
220.35	3	5	4.4E+02
248.46	3	1	6.5E+01
629.73	1	3	2.80E+01
758.68	3	5	5.68E+00
759.44	1	3	7.55E+00
760.23	3	3	5.64E+00
760.45	5	5	1.69E+01
761.13	3	1	2.25E+01
762.00	5	3	9.34E+00
774.52	3	1	3.54E+01
1371.3	3	5	3.29E+00
*2784.0	3	9	1.6E+00
3144.7	3	5	9.3E-01
5114.1	1	3	1.7E-01
*5589.9	9	15	1.5E-01
O VI			
*130.10	2	6	2.54E+02
*173.03	6	10	8.85E+02
1031.9	2	4	4.15E+00
1037.6	2	2	4.08E+00
3811.4	2	4	5.13E-01
3834.2	2	2	5.05E-01
O VII			
18.627	1	3	9.362E+03
21.602	1	3	3.309E+04
*120.33	3	9	5.335E+02
128.20	1	3	5.053E+02
*128.46	9	15	1.62E+03
135.82	3	5	1.53E+03
*1630.2	3	9	7.935E-01
2450.0	1	3	2.514E-01
Phosphorus			
P I			
1671.7	4	2	3.9E-01
1674.6	4	4	4.0E-01
1679.7	4	6	3.9E-01
1775.0	4	6	2.17E+00
1782.9	4	4	2.14E+00
1787.7	4	2	2.13E+00
2135.5	4	4	2.11E-01
2136.2	6	4	2.83E+00
2149.1	4	2	3.18E+00
2152.9	2	4	4.85E-01
2154.1	4	4	1.73E-01
2154.1	4	6	5.8E-01
2534.0	2	4	2.00E-01
2535.6	4	4	9.5E-01
2553.3	2	2	7.1E-01
2554.9	4	2	3.00E-01
P II			
1301.9	1	3	5.0E-01
1304.5	3	1	1.5E+00
1304.7	3	3	3.7E-01
1305.5	3	5	3.8E-01
1309.9	5	3	6.2E-01
1310.7	5	5	1.1E+00
4475.3	5	7	1.3E+00
4499.2	5	7	1.4E+00
4530.8	3	5	1.0E+00
4554.8	3	5	9.6E-01
4588.0	5	7	1.7E+00
4589.9	3	5	1.6E+00
4602.1	7	9	1.9E+00
4943.5	7	5	6.3E-01
5253.5	3	5	1.0E+00
5425.9	5	5	6.9E-01
6024.2	3	5	5.1E-01
6043.1	5	7	6.8E-01
P III			
1334.8	2	4	5.5E-01
1344.3	4	6	6.4E-01
1344.8	4	4	1.1E-01
4057.4	4	4	1.0E-01
4059.3	6	4	9.0E-01
4080.1	4	2	9.9E-01
Potassium			
K I			
4044.1	2	4	1.24E-02
4047.2	2	2	1.24E-02
5084.2	2	2	3.50E-03
5099.2	4	2	7.0E-03
5323.3	2	2	6.3E-03
5339.7	4	2	1.26E-02
5343.0	2	4	4.0E-03
5359.6	4	6	4.6E-03
5782.4	2	2	1.23E-02
5801.8	4	2	2.46E-02
5812.2	2	4	2.8E-03
5831.9	4	6	3.2E-03
6911.1	2	2	2.72E-02
6938.8	4	2	5.4E-02
7664.9	2	4	3.87E-01
7699.0	2	2	3.82E-01
K II			
607.93	1	3	1.3E-02
K III			
2550.0	6	4	2.0E+00
2635.1	4	4	1.2E+00
2992.4	6	8	2.5E+00
3052.1	4	6	1.7E+00
3202.0	4	4	1.8E+00
3289.1	4	6	2.0E+00
3322.4	6	6	1.3E+00
3421.8	2	4	1.5E+00
K XVI			
206.27	1	3	9.4E+01
K XVII			
22.020	2	4	4.7E+04
22.163	4	6	5.6E+04
22.18	4	4	9.3E+03
22.60	2	2	2.5E+03
22.76	4	2	4.7E+03
Praseodymium			
Pr II			
3997.0	15	15	1.87E-01
4062.8	13	15	1.00E+00
4100.7	17	19	8.4E-01
4143.1	15	17	5.8E-01
4179.4	13	15	5.2E-01
4222.9	11	13	3.91E-01
4241.0	17	15	2.30E-01
4359.8	15	15	1.1E-01
4405.8	17	17	9.0E-02
4429.3	15	15	2.28E-01
4449.8	13	13	1.24E-01
4468.7	11	13	1.54E-01
4510.2	13	15	1.16E-01
4534.2	15	17	4.9E-02
4734.2	15	13	2.5E-02
4879.1	15	15	1.8E-02
4886.0	15	15	1.3E-02
4912.6	17	15	5.7E-02
5034.4	19	19	1.1E-01
5110.8	21	19	2.78E-01
5135.1	17	17	1.25E-01
5173.9	19	17	3.18E-01
5219.1	15	15	9.5E-02
5220.1	17	15	2.35E-01
5251.7	15	13	1.1E-02
5259.7	15	13	2.24E-01
5292.6	13	13	9.3E-02
5810.6	17	19	2.3E-02
5879.3	15	15	7.6E-02
6200.8	15	17	1.8E-02
6278.7	13	15	2.6E-02
6398.0	11	13	1.9E-02
Rhodium			
Rh I			
3083.96	8	6	4.8E-02
3114.91	6	4	4.45E-02
3121.76	6	6	1.1E-01
3123.70	10	8	4.6E-02
3137.71	4	6	3.3E-02
3189.05	6	6	3.03E-01
3197.13	6	4	4.35E-02

λ Å	Weights g_i	g_k	A 10^8 s^{-1}
3263.14	6	6	1.3E-01
3271.61	6	4	2.0E-01
3280.55	8	8	2.36E-01
3283.57	6	8	4.4E-01
3289.14	4	4	1.0E-01
3323.09	8	10	6.3E-01
3331.09	4	2	5.40E-02
3338.54	8	6	3.5E-02
3360.80	4	4	1.2E-01
3368.38	6	4	1.1E-01
3396.82	10	10	6.5E-01
3399.70	6	8	1.2E-01
3462.04	6	6	6.2E-01
3470.66	4	4	8.5E-01
3478.91	6	6	3.32E-01
3484.04	6	8	9.3E-03
3498.73	4	6	2.12E-01
3502.52	10	10	4.3E-01
3507.32	6	8	3.4E-01
3528.02	8	8	8.5E-01
3543.95	4	4	4.65E-01
3549.54	6	6	2.22E-01
3570.18	4	6	1.82E-01
3583.10	8	10	2.6E-01
3596.19	6	4	5.5E-01
3597.15	6	8	5.9E-01
3612.47	4	2	8.90E-01
3620.46	6	4	8.5E-02
3654.87	8	8	6.0E-02
3657.99	8	6	8.8E-01
3666.22	6	8	8.4E-02
3690.70	6	4	3.23E-01
3692.36	10	8	9.1E-01
3700.91	8	10	3.9E-01
3713.02	4	4	8.3E-02
3788.47	4	6	1.4E-01
3793.22	8	6	4.2E-01
3799.31	8	8	5.5E-01
3806.76	6	6	6.2E-02
3818.19	6	4	5.8E-01
3822.26	6	6	8.5E-01
3828.48	6	6	6.2E-01
3833.89	6	4	5.8E-01
3856.52	8	10	5.9E-01
3872.39	4	6	6.7E-03
3877.34	8	6	3.7E-02
3913.51	8	8	2.5E-03
3922.19	4	2	6.25E-02
3934.23	8	8	1.58E-01
3942.72	4	2	7.15E-01
3958.86	6	8	5.5E-01
3984.40	4	4	1.1E-01
3995.61	4	6	4.7E-02
4053.44	2	2	2.8E-02
4056.34	6	4	9.5E-03
4082.78	6	4	1.4E-01
4097.52	2	4	7.0E-02
4121.68	6	6	9.8E-02
4128.87	6	8	1.73E-01
4135.27	8	8	1.0E-01
4196.50	6	8	3.9E-02
4211.14	8	10	1.62E-01
4244.44	4	4	6.5E-03
4278.60	4	6	9.2E-03
4288.71	6	8	6.1E-02
4373.04	2	4	1.8E-02
4374.80	8	10	1.64E-01
4379.92	6	6	2.48E-02
4492.47	6	6	4.5E-03
4528.72	6	8	1.35E-02
4548.73	4	6	5.5E-03
4551.64	4	4	4.00E-02
4565.19	4	4	1.1E-02
4569.00	6	8	1.0E-02
4608.12	2	2	2.1E-02
4675.03	8	8	6.4E-03
4721.00	6	4	3.43E-03
4745.11	6	6	5.2E-03
4755.58	4	4	6.0E-03
4842.43	6	8	1.6E-03
4963.71	2	2	3.0E-02
4977.75	4	4	9.8E-03
4979.18	4	6	1.0E-02
5090.63	6	6	5.0E-03
5120.69	6	8	3.1E-03
5130.76	4	4	4.35E-03
5155.54	2	4	9.8E-03
5184.19	6	8	1.6E-03
5212.73	4	2	5.95E-03
5292.14	10	10	3.7E-03
5390.44	4	6	9.5E-03
5424.72	4	4	5.0E-03
5599.42	6	8	1.3E-02
5983.60	10	10	2.1E-02

Rubidium

Rb I

λ Å	Weights g_i	g_k	A 10^8 s^{-1}
3022.5	2	4	4.13E-05
3032.0	2	4	4.93E-05
3044.2	2	4	8.2E-05
3060.2	2	4	1.05E-04
3082.0	2	4	1.49E-04
3112.6	2	4	2.5E-04
3113.1	2	2	1.3E-04
3157.5	2	4	3.38E-04
3158.3	2	2	2.0E-04
3228.0	2	4	6.4E-04
3229.2	2	2	3.8E-04
3348.7	2	4	1.37E-03
3350.8	2	2	8.9E-04
3587.1	2	4	3.97E-03
3591.6	2	2	2.9E-03
4201.8	2	4	1.8E-02
4215.5	2	2	1.5E-02
7800.3	2	4	3.70E-01
7947.6	2	2	3.40E-01

Scandium

Sc I

λ Å	Weights g_i	g_k	A 10^8 s^{-1}
2116.7	4	4	2.0E-01
2120.4	6	6	2.0E-01
2262.3	4	4	5.8E-02
2266.6	4	2	4.8E-01
2270.9	6	4	4.6E-01
2280.8	4	6	2.8E-01
2289.6	6	6	4.1E-02
2311.29	4	6	4.1E-02
2315.69	4	4	2.5E-01
2320.32	6	6	2.4E-01
2324.75	6	4	4.1E-02
2328.19	4	6	4.6E-02
2334.67	4	2	1.7E-01
2346.03	6	4	1.3E-01
2429.19	4	4	2.8E-01
2438.63	6	6	2.1E-01
2468.40	4	2	4.9E-02
2692.78	4	2	1.61E-01
2699.02	4	6	2.4E-02
2706.74	4	4	3.1E-01
2707.93	6	4	1.49E-01
2711.34	6	6	3.2E-01
2965.88	4	6	7.5E-02
2974.01	4	4	5.5E-01
2980.76	6	6	5.4E-01
2988.97	6	4	6.9E-02
3015.37	4	6	7.8E-01
3019.35	6	8	8.7E-01
3030.76	6	6	1.00E-01
3255.68	4	4	3.2E-01
3269.90	4	2	3.13E+00
3273.63	6	4	2.81E+00
3907.48	4	6	1.66E+00
3911.81	6	8	1.79E+00
3933.38	6	6	1.62E-01
3996.60	4	6	1.65E-01
4020.39	4	4	1.63E+00
4023.22	4	4	3.0E-01
4023.68	6	6	1.65E+00
4031.38	6	6	2.9E-01
4036.86	6	4	7.9E-02
4043.80	8	8	3.11E-01
4047.80	6	4	1.54E-01
4051.83	8	6	7.7E-02
4054.54	4	2	1.67E-01
4067.00	6	8	1.91E-01
4067.63	10	8	4.1E-02
4074.96	4	6	3.7E-01
4078.56	2	4	4.3E-01
4080.57	4	4	6.6E-02
4082.39	6	4	2.73E-01
4086.66	6	8	3.7E-01
4087.47	4	6	1.12E-01
4093.12	4	4	1.23E-01
4094.86	6	6	1.44E-01
4098.36	8	8	8.7E-02
4132.98	4	6	1.19E+00

λ Å	Weights g_i	g_k	A 10^8 s^{-1}
4140.27	6	8	1.17E+00
4147.38	6	6	1.74E-01
4161.85	8	8	1.77E-01
4171.53	6	4	1.36E-01
4186.42	6	8	8.4E-02
4187.61	8	6	1.28E-01
4193.53	4	6	6.1E-02
4204.52	6	8	3.5E-02
4205.20	10	8	1.12E-01
4212.32	4	6	1.58E-01
4212.48	6	6	8.6E-02
4216.08	2	4	2.36E-01
4218.23	4	4	2.26E-01
4225.54	6	8	9.5E-02
4225.69	4	6	7.6E-02
4231.64	4	4	1.31E-01
4233.59	6	6	4.0E-01
4238.05	8	8	7.1E-01
4239.55	6	4	2.27E-01
4246.14	8	6	1.15E-01
4542.55	6	4	1.28E-01
4544.67	8	6	1.33E-01
4706.94	4	6	2.81E-01
4709.31	6	8	4.0E-01
4711.72	2	4	1.81E-01
4714.30	4	4	2.14E-01
4719.31	6	6	1.04E-01
4728.77	8	8	1.16E-01
4729.20	4	4	2.20E-01
4729.24	6	6	1.93E-01
4734.11	4	2	1.10E+00
4737.65	6	4	8.8E-01
4741.02	8	6	9.1E-01
4743.82	10	8	9.8E-01
4973.67	4	2	8.4E-01
4980.36	6	4	5.6E-01
4983.43	4	4	2.58E-01
4991.91	6	6	3.8E-01
4995.00	4	6	5.9E-02
5018.41	6	4	2.09E-01
5021.52	4	4	2.30E-01
5064.31	8	10	7.3E-02
5066.38	6	6	3.6E-02
5070.17	6	8	1.16E-01
5072.71	2	4	2.0E-02
5075.82	4	6	1.15E-01
5080.22	4	4	4.1E-02
5081.56	10	10	7.0E-01
5083.72	8	8	6.2E-01
5085.55	6	6	5.7E-01
5086.94	4	4	6.6E-01
5096.72	6	4	1.69E-01
5099.27	4	6	1.50E-01
5101.12	10	8	8.8E-02
5331.79	4	4	1.11E-01
5339.43	6	6	1.06E-01
5341.07	4	2	3.8E-01
5349.34	6	4	5.9E-01
5350.28	8	8	6.8E-02
5355.79	6	4	3.0E-01
5356.10	8	6	5.7E-01
5375.37	8	6	3.4E-01
5392.06	10	8	4.2E-01
5416.16	4	6	4.4E-02
5416.41	6	6	2.0E-02
5425.55	6	8	4.5E-02
5429.42	2	4	9.0E-02
5432.98	4	4	5.4E-02
5433.25	6	4	9.7E-02
5438.28	4	6	3.4E-02
5439.04	2	2	1.74E-01
5442.62	4	2	2.15E-01
5446.20	8	8	2.8E-01
5451.37	6	6	1.50E-01
5455.24	4	4	6.6E-02
5464.95	4	2	3.2E-02
5468.40	6	4	9.7E-02
5472.19	8	6	9.7E-02
5482.01	8	8	5.2E-01
5484.63	6	6	5.2E-01
5514.23	6	8	4.1E-01
5520.52	8	10	4.3E-01
5526.10	4	4	7.1E-02
5541.07	6	6	5.5E-02
5631.04	2	4	3.0E-02
5671.83	10	12	5.4E-01
5686.86	8	10	4.9E-01
5700.19	6	8	4.6E-01
5708.64	10	10	4.7E-02
5711.79	4	6	4.5E-01
5717.31	8	8	7.5E-02
5724.13	6	6	7.4E-02
5988.43	6	6	6.6E-02
6026.16	4	4	7.2E-02
6146.20	6	8	4.2E-02
6198.43	4	6	3.5E-02
6249.96	6	8	3.2E-02
6262.22	4	6	8.4E-02
6280.16	2	4	4.0E-02
6284.16	6	6	3.9E-02
6284.73	4	4	7.1E-02
6293.02	2	2	1.04E-01
7741.16	10	10	3.8E-02
7800.42	8	8	5.1E-02
Sc II			
1880.6	5	3	5.0E+00
2064.3	7	5	2.2E+00
2068.0	5	3	2.0E+00
2273.1	1	3	7.7E+00
2545.20	5	5	4.0E-01
2552.35	7	5	2.21E+00
2555.79	3	3	6.9E-01
2560.23	5	3	2.01E+00
2563.19	3	1	2.70E+00
2611.19	5	5	2.2E+00
2667.70	3	5	1.5E+00
2746.36	3	1	3.9E+00
2782.31	5	5	1.3E+00
2789.15	7	7	1.3E+00
2801.31	9	9	1.3E+00
2819.49	3	5	2.3E+00
2822.12	5	7	2.5E+00
2826.64	7	9	2.8E+00
2870.85	5	3	1.1E+00
2912.98	5	3	1.1E+00
2979.68	3	5	1.2E+00
2988.92	5	7	2.9E+00
3039.92	7	9	3.5E+00
3045.73	5	7	3.68E+00
3052.92	7	9	3.92E+00
3060.54	7	7	3.0E-01
3065.12	9	11	4.00E+00
3075.36	9	9	2.5E-01
3128.27	3	3	1.9E+00
3133.07	5	5	1.8E+00
3139.72	7	7	2.1E+00
3190.98	3	3	1.1E+00
3199.33	5	3	1.9E+00
3312.72	5	7	1.2E+00
3320.40	5	3	1.2E+00
3343.23	9	7	1.1E+00
3353.72	5	7	1.51E+00
3359.67	5	5	2.16E-01
3361.26	3	3	3.4E-01
3361.93	3	1	1.17E+00
3368.94	5	3	8.3E-01
3372.15	7	5	9.9E-01
3379.16	3	3	2.5E+00
3535.71	5	3	6.1E-01
3558.53	5	7	3.0E-01
3567.70	3	5	3.5E-01
3572.53	7	7	1.38E+00
3576.34	5	5	1.06E+00
3580.93	3	3	1.23E+00
3589.63	5	3	4.6E-01
3590.47	7	5	2.9E-01
3613.83	7	9	1.48E+00
3630.74	5	7	1.20E+00
3642.78	3	5	1.13E+00
3645.31	7	7	3.71E-01
3651.80	5	5	3.0E-01
3859.59	7	5	1.1E+00
4246.82	5	5	1.29E+00
4314.08	9	7	4.1E-01
4320.75	7	5	4.0E-01
4325.00	5	3	4.3E-01
4374.46	9	9	1.48E-01
4400.39	7	7	1.43E-01
4415.54	5	5	1.47E-01
4670.41	5	7	1.16E-01
5031.01	5	3	3.5E-01
5239.81	1	3	1.39E-01
5526.79	9	7	3.3E-01
5657.91	5	5	1.04E-01
5669.06	3	1	1.31E-01

λ Å	Weights g_i	g_k	A $10^8\ s^{-1}$
Silicon			
Si I			
1977.6	1	3	1.8E-01
1979.2	3	1	5.1E-01
1980.6	3	3	1.3E-01
1983.2	3	5	1.4E-01
1986.4	5	3	2.1E-01
1989.0	5	5	4.1E-01
2208.0	1	3	3.11E-01
2210.9	3	5	4.16E-01
2211.7	3	3	2.32E-01
2216.7	5	7	5.5E-01
2218.1	5	5	1.38E-01
2506.9	3	5	4.66E-01
2514.3	1	3	6.1E-01
2516.1	5	5	1.21E+00
2519.2	3	3	4.56E-01
2524.1	3	1	1.81E+00
2528.5	5	3	7.7E-01
2532.4	1	3	2.6E-01
2631.3	1	3	9.7E-01
2881.6	5	3	1.89E+00
3905.5	1	3	1.18E-01
4738.8	3	3	1.0E-02
4783.0	5	3	1.7E-02
4792.3	5	5	1.7E-02
4818.1	5	7	1.1E-02
4821.2	3	5	8.0E-03
4947.6	3	1	4.2E-02
5006.1	3	5	2.8E-02
5622.2	3	3	1.6E-02
5690.4	3	3	1.2E-02
5708.4	5	5	1.4E-02
5754.2	5	3	1.5E-02
5772.1	3	1	3.6E-02
5948.5	3	5	2.2E-02
7226.2	3	5	7.9E-03
7405.8	3	5	3.7E-02
7409.1	5	7	2.3E-02
7680.3	3	5	4.6E-02
7918.4	3	5	5.2E-02
7932.3	5	7	5.1E-02
7944.0	7	9	5.8E-02
7970.3	5	5	7.1E-03
Si II			
989.87	2	4	6.7E+00
992.68	4	6	8.0E+00
1020.7	2	2	1.3E+00
1190.4	2	4	6.9E+00
1193.3	2	2	2.8E+01
1194.5	4	4	3.6E+01
1197.4	4	2	1.4E+01
1248.4	4	4	1.3E+01
1251.2	6	4	1.9E+01
1260.4	2	4	2.0E+01
1264.7	4	6	2.3E+01
1304.4	2	2	3.6E+00
1309.3	4	2	7.0E+00
1526.7	2	2	3.73E+00
1533.5	4	2	7.4E+00
1808.0	2	4	3.7E-02
2904.3	4	6	6.7E-01
2905.7	6	8	7.1E-01
3210.0	4	6	4.6E-01
4128.1	4	6	1.32E+00
4130.9	6	8	1.42E+00
5041.0	2	4	9.8E-01
5056.0	4	6	1.2E+00
5957.6	2	2	4.2E-01
5978.9	4	2	8.1E-01
6347.1	2	4	7.0E-01
6371.4	2	2	6.9E-01
7848.8	4	6	3.9E-01
7849.7	6	8	4.2E-01
Si III			
883.40	5	7	6.3E+01
994.79	3	3	7.89E+00
997.39	5	3	1.31E+01
1141.6	3	5	3.0E+01
1144.3	5	7	3.9E+01
1161.6	5	5	1.6E+01
1206.5	1	3	2.59E+01
1206.5	3	5	4.89E+01
1207.5	5	5	1.9E+01
1294.5	3	5	5.42E+00
1296.7	1	3	7.19E+00
1298.9	3	3	5.36E+00
1299.0	5	5	1.61E+01
1301.2	3	1	2.13E+01
1303.3	5	3	8.85E+00
1328.8	1	3	2.7E+01
1417.2	3	1	2.60E+01
1435.8	5	7	2.1E+01
1589.0	5	3	1.1E+01
1778.7	7	9	4.4E+00
1783.1	5	7	3.8E+00
3241.6	5	3	2.3E+00
*3486.9	15	21	1.8E+00
3590.5	3	5	3.9E+00
4552.6	3	5	1.26E+00
4554.0	5	3	7.6E-01
4567.8	3	3	1.25E+00
4683.0	5	5	9.5E-01
4716.7	5	7	2.8E+00
5451.5	3	5	6.0E-01
5473.1	5	7	7.9E-01
5716.3	9	7	1.9E-01
5739.7	1	3	4.7E-01
7462.6	5	3	4.9E-01
7466.3	7	5	5.4E-01
7612.4	3	5	1.1E+00
Si IV			
457.82	2	4	3.6E+00
458.16	2	2	3.6E+00
515.12	2	2	4.1E+00
516.35	4	2	8.2E+00
*560.50	6	10	1.0E+00
*749.94	10	14	1.45E+01
815.05	2	2	1.23E+01
818.13	4	2	2.44E+01
*860.74	10	6	1.8E+00
*1066.6	10	14	3.91E+01
1122.5	2	4	2.05E+01
1128.3	4	4	4.03E+00
1128.3	4	6	2.42E+01
1393.8	2	4	7.73E+00
1402.8	2	2	7.58E+00
*1724.1	10	6	5.5E+00
Si V			
96.439	1	3	4.8E+02
97.143	1	3	2.0E+03
117.86	1	3	3.0E+02
Si VI			
246.00	4	2	1.7E+02
249.12	2	2	8.5E+01
Si VII			
217.83	5	3	4.3E+02
272.64	5	3	5.1E+01
274.18	3	1	1.2E+02
275.35	5	5	8.9E+01
275.67	3	3	3.0E+01
276.84	1	3	3.9E+01
278.45	3	5	2.9E+01
Si VIII			
214.76	4	2	4.1E+02
216.92	6	4	3.6E+02
232.86	2	2	8.0E+01
235.56	4	4	9.7E+01
250.45	2	2	7.7E+01
250.79	4	2	1.6E+02
314.31	4	2	5.2E+01
316.20	4	4	5.0E+01
319.83	4	6	4.9E+01
Si IX			
223.73	1	3	4.2E+01
225.03	3	3	1.2E+02
227.01	5	3	2.0E+02
227.30	5	3	2.3E+02
258.10	5	5	1.04E+02
*294.37	9	9	5.9E+01
*347.36	9	15	2.2E+01
Si X			
253.77	2	4	2.9E+01
256.57	2	2	1.1E+02
258.35	4	4	1.4E+02
261.05	4	2	5.4E+01
272.00	2	2	3.0E+01
277.26	4	2	5.7E+01

λ Å	Weights g_i	g_k	A $10^8\ s^{-1}$
287.08	2	4	2.6E+01
289.19	4	4	5.0E+01
292.22	6	4	7.3E+01
*347.73	10	10	4.3E+01
*353.09	6	10	2.1E+01
Si XI			
43.763	1	3	6.11E+03
*49.116	9	3	2.45E+03
49.222	3	5	8.9E+03
52.296	3	1	7.6E+02
303.30	1	3	6.42E+01
358.29	3	1	1.03E+02
358.63	3	5	1.38E+01
361.41	1	3	1.80E+01
364.50	3	3	1.32E+01
365.42	5	5	3.90E+01
368.28	3	1	5.1E+01
371.48	5	3	2.07E+01
604.14	3	5	1.12E+01
2300.8	1	3	4.34E-01
Si XII			
*40.924	2	6	4.42E+03
*44.118	6	10	1.4E+04
499.43	2	4	9.56E+00
520.77	2	2	8.47E+00
1862	2	4	1.15E+00
1949	2	2	1.0E+00
4620	2	4	4.6E-02
4942	4	6	4.5E-02
Silver			
Ag I			
2061.2	2	4	3.1E-02
2069.9	2	2	1.5E-02
3280.7	2	4	1.4E+00
3382.9	2	2	1.3E+00
5209.1	2	4	7.5E-01
5465.5	4	6	8.6E-01
5471.6	4	4	1.4E-01
Sodium			
Na I			
3302.4	2	4	2.81E-02
3303.0	2	2	2.81E-02
4390.0	2	4	7.7E-03
4393.3	4	4	1.6E-03
4393.3	4	6	9.2E-03
4494.2	2	4	1.2E-02
4497.7	4	6	1.4E-02
4497.7	4	4	2.4E-03
4664.8	2	4	2.33E-02
4668.6	4	4	4.1E-03
4668.6	4	6	2.5E-02
4747.9	2	2	6.3E-03
4751.8	4	2	1.27E-02
4978.5	2	4	4.1E-02
4982.8	4	4	8.2E-03
4982.8	4	6	4.89E-02
5148.8	2	2	1.17E-02
5153.4	4	2	2.33E-02
5682.6	2	4	1.03E-01
5688.2	4	6	1.2E-01
5688.2	4	4	2.1E-02
5890.0	2	4	6.22E-01
5895.9	2	2	6.18E-01
6154.2	2	2	2.6E-02
6160.8	4	2	5.2E-02
8183.3	2	4	4.53E-01
8194.8	4	6	5.4E-01
8194.8	4	4	9.0E-02
11381	2	2	8.9E-02
11404	4	2	1.76E-01
Na II			
300.15	1	3	3.0E+01
301.44	1	3	4.9E+01
372.08	1	3	3.4E+01
Na III			
378.14	4	2	7.7E+01
380.10	2	2	3.7E+01
1991.0	4	6	8.3E+00
2004.2	2	4	4.6E+00
2011.9	6	8	8.4E+00
2151.5	2	4	4.4E+00
2174.5	4	6	5.3E+00
2230.3	6	8	3.7E+00
2232.2	4	4	3.3E+00
2246.7	4	6	2.4E+00
2459.3	4	6	3.0E+00
2468.9	2	4	2.4E+00
2497.0	6	6	1.7E+00
Na V			
*307.89	10	6	2.0E+02
*333.46	6	6	5.6E+01
*369.01	10	6	1.2E+02
*400.72	10	10	5.0E+01
*445.14	6	10	7.1E+00
459.90	4	2	2.3E+01
461.05	4	4	2.3E+01
463.26	4	6	2.2E+01
510.10	2	2	5.6E+01
511.19	4	4	6.8E+01
Na VI			
313.75	5	3	1.3E+02
361.25	5	5	7.7E+01
*416.53	9	9	3.7E+01
*492.80	9	15	1.3E+01
1550.6	5	5	4.35E+00
1567.8	5	3	2.68E+00
1608.5	3	1	2.6E+00
1649.4	5	5	2.05E+00
1741.5	3	5	2.59E+00
1747.5	5	7	3.1E+00
Na VII			
*94.409	6	10	2.7E+03
*105.27	6	2	4.5E+02
353.29	4	4	1.0E+02
381.30	4	2	4.0E+01
397.49	4	4	3.5E+01
399.18	6	4	5.2E+01
*483.28	10	10	2.9E+01
486.74	2	4	1.1E+01
491.95	4	6	1.3E+01
555.80	4	4	2.3E+01
777.83	4	6	6.8E+00
Na VIII			
*83.34	9	15	3.94E+03
*89.88	9	3	8.09E+02
90.536	3	5	2.86E+03
411.15	1	3	4.42E+01
1239.4	3	3	3.02E+00
1802.7	3	1	2.70E+00
1867.7	3	5	2.01E+00
2059.1	3	5	1.80E+00
2558.2	5	3	2.26E-02
2772.0	3	5	4.19E-01
3021.0	5	7	4.90E-01
3108.9	1	3	2.58E-01
3182.3	1	3	2.92E-01
Na IX			
70.615	2	4	1.35E+03
70.653	2	2	1.35E+03
77.764	2	4	3.6E+03
77.911	4	6	4.3E+03
681.72	2	4	6.63E+00
694.17	2	2	6.30E+00
2487.7	2	4	8.32E-01
2535.8	2	2	7.89E-01
6841.8	2	4	2.59E-02
7103.4	4	6	2.78E-02
Strontium			
Sr I			
2206.2	1	3	6.6E-03
2211.3	1	3	8.5E-03
2217.8	1	3	1.2E-02
2226.3	1	3	1.6E-02
2237.7	1	3	2.3E-02
2253.3	1	3	3.7E-02
2275.3	1	3	6.7E-02
2307.3	1	3	1.2E-01
2354.3	1	3	1.8E-01
2428.1	1	3	1.7E-01
2569.5	1	3	5.3E-02

λ Å	Weights g_i	g_k	A 10^8 s^{-1}
2931.8	1	3	1.9E-02
4607.3	1	3	2.01E+00
Sr II			
2018.7	2	2	1.2E-01
2051.9	4	2	2.4E-01
2282.0	2	4	8.3E-01
2322.4	4	6	9.1E-01
2324.5	4	4	1.5E-01
2423.5	2	2	2.4E-01
2471.6	4	2	4.8E-01
3464.5	4	6	3.1E+00
3474.9	4	4	5.1E-01
4077.7	2	4	1.42E+00
4161.8	2	2	6.5E-01
4215.5	2	2	1.27E+00
4305.5	4	2	1.4E+00
4414.8	4	6	1.1E-01
4417.5	4	4	1.8E-02
4585.9	4	2	7.0E-02
5303.1	2	4	1.9E-01
5379.1	4	6	2.2E-01
5385.5	4	4	3.7E-02
5723.7	2	2	7.1E-02
5819.0	4	2	1.4E-01
8688.9	4	6	5.5E-01
8719.6	4	4	9.7E-02
Sulfur			
S I			
1295.7	5	5	4.9E+00
1296.2	5	3	2.7E+00
1302.3	3	5	1.8E+00
1302.9	3	3	1.6E+00
1303.1	3	1	6.6E+00
1303.4	5	3	1.9E+00
1305.9	1	3	2.4E+00
1401.5	5	3	9.1E-01
1409.3	3	3	5.0E-01
1412.9	1	3	1.6E-01
1425.0	5	7	4.5E+00
1425.2	5	5	1.2E+00
1433.3	3	5	3.3E+00
1433.3	3	3	1.9E+00
1437.0	1	3	2.4E+00
1448.2	5	3	7.3E+00
1473.0	5	7	4.2E-01
1474.0	5	7	1.6E+00
1474.4	5	5	5.0E-01
1474.6	5	3	6.2E-02
1481.7	3	5	1.7E-01
1483.0	3	5	1.2E+00
1483.2	3	3	7.5E-01
1487.2	1	3	8.7E-01
1666.7	5	5	6.3E+00
1687.5	1	3	9.4E-01
1782.3	1	3	1.9E+00
1807.3	5	3	3.8E+00
1820.3	3	3	2.2E+00

λ Å	Weights g_i	g_k	A 10^8 s^{-1}
1826.2	1	3	7.2E-01
4694.1	5	7	6.7E-03
4695.4	5	5	6.7E-03
4696.2	5	3	6.5E-03
6403.6	3	5	5.7E-03
6408.1	5	5	9.5E-03
6415.5	7	5	1.3E-02
*6751.2	15	25	7.9E-02
7679.6	3	5	1.2E-02
7686.1	5	5	2.0E-02
7696.7	7	5	2.8E-02
S II			
1124.4	2	4	1.0E+00
1125.0	4	4	4.6E+00
1131.0	2	2	3.5E+00
1131.6	4	2	1.4E+00
1250.5	4	2	4.6E-01
1253.8	4	4	4.2E-01
1259.5	4	6	3.4E-01
4463.6	8	6	5.3E-01
4483.4	6	4	3.1E-01
4486.7	4	2	6.6E-01
4524.7	4	4	9.3E-02
4525.0	6	4	1.2E+00
4552.4	4	2	1.2E+00
4656.7	2	4	9.0E-02
4716.2	4	4	2.9E-01
4815.5	6	4	8.8E-01
4885.6	2	4	1.7E-01
4917.2	2	2	6.6E-01
4924.1	4	6	2.2E-01
4925.3	2	4	2.4E-01
4942.5	2	2	1.5E-01
4991.9	4	4	1.5E-01
5009.5	4	2	7.0E-01
5014.0	4	4	8.4E-01
5027.2	4	2	2.6E-01
5032.4	6	6	8.1E-01
5047.3	4	2	3.6E-01
5103.3	6	4	5.0E-01
5142.3	2	2	1.9E-01
5201.0	4	4	7.5E-01
5201.3	6	4	6.5E-02
5212.6	4	6	9.8E-02
5212.6	6	6	8.5E-01
5320.7	6	8	9.2E-01
5345.7	4	6	8.8E-01
5345.7	6	6	1.1E-01
5428.6	2	4	4.2E-01
5432.8	4	6	6.8E-01
5453.8	6	8	8.5E-01
5473.6	2	2	7.3E-01
5509.7	4	4	4.0E-01
5526.2	8	8	8.1E-02
5536.8	4	6	6.6E-02
5556.0	4	2	1.1E-01
5564.9	6	6	1.7E-01
5578.8	6	6	1.1E-01

λ Å	Weights g_i	g_k	A 10^8 s^{-1}
5606.1	10	8	5.4E-01
5616.6	4	4	1.2E-01
5640.0	4	6	6.6E-01
5645.6	6	4	1.8E-02
5647.0	2	4	5.7E-01
5659.9	6	4	4.6E-01
5664.7	4	2	5.8E-01
5819.2	4	4	8.5E-02
6305.5	8	6	1.8E-01
6312.7	6	4	3.0E-01
S III			
2496.2	7	5	2.5E+00
2508.2	5	3	2.3E+00
2636.9	3	5	4.5E-01
2665.4	5	5	1.4E+00
2680.5	1	3	6.2E-01
2691.8	3	3	4.6E-01
2702.8	3	1	1.9E+00
2718.9	3	3	1.2E+00
2721.4	5	3	7.7E-01
2726.8	3	5	6.0E-01
2731.1	5	5	1.1E+00
2756.9	7	7	1.4E+00
2785.5	3	3	6.1E-01
2856.0	5	7	5.1E+00
2863.5	7	9	5.7E+00
2872.0	3	5	4.7E+00
2950.2	3	5	3.0E+00
2964.8	5	7	4.0E+00
3662.0	3	3	6.4E-01
3717.8	5	3	1.0E+00
3778.9	3	5	4.4E-01
3831.8	1	3	5.6E-01
3837.8	3	3	4.2E-01
3838.3	5	5	1.3E+00
3860.6	3	1	1.6E+00
3899.1	5	3	6.7E-01
4253.6	5	7	1.2E+00
4285.0	3	5	9.0E-01
S IV			
551.17	2	2	2.06E+01
554.07	4	2	4.08E+01
3097.5	2	4	2.6E+00
3117.7	2	2	2.5E+00
S V			
437.37	1	3	1.12E+01
438.19	3	3	3.33E+01
439.65	5	3	5.5E+01
*661.52	9	15	6.44E+01
*679.01	9	15	8.6E+01
*690.75	9	9	5.0E+01
786.48	1	3	5.25E+01
*854.85	9	9	4.18E+01
S VI			
248.99	2	4	3.1E+01

λ Å	Weights g_i	g_k	A 10^8 s^{-1}
249.27	2	2	3.1E+01
388.94	2	2	4.5E+01
390.86	4	2	8.8E+01
706.48	2	4	4.17E+01
712.68	4	6	4.85E+01
712.84	4	4	8.1E+00
933.38	2	4	1.7E+01
944.52	2	2	1.6E+01
S VII			
60.161	1	3	9.46E+03
60.804	1	3	5.1E+02
72.029	1	3	8.61E+02
S VIII			
198.55	4	2	2.5E+02
202.61	2	2	1.2E+02
S XI			
*189.90	9	3	4.3E+02
190.37	5	3	2.8E+02
215.95	5	5	1.4E+02
217.63	1	3	7.2E+01
239.81	1	3	2.6E+01
242.57	3	5	1.9E+01
243.82	3	3	1.9E+01
246.90	5	5	5.4E+01
247.12	5	3	3.0E+01
*288.49	9	15	2.9E+01
S XII			
212.14	2	4	3.7E+01
215.18	2	2	1.4E+02
218.20	4	4	1.7E+02
221.44	4	2	6.4E+01
227.50	2	2	3.7E+01
234.48	4	2	6.8E+01
S XIII			
32.236	1	3	1.09E+04
37.600	3	1	1.3E+03
256.66	1	3	8.7E+01
299.89	3	5	1.78E+01
303.37	1	3	2.28E+01
307.36	3	3	1.64E+01
308.91	5	5	4.82E+01
312.68	3	1	6.3E+01
316.84	5	3	2.50E+01
500.42	3	5	1.43E+01
S XIV			
*30.434	2	6	8.28E+03
*32.517	6	10	2.6E+04
417.67	2	4	1.2E+01
445.71	2	2	1.0E+01
1550	2	4	1.4E+00
1663	2	2	1.2E+00
3967	2	4	5.4E-02
4153	4	6	5.7E-02

λ Å	Weights g_i	g_k	A 10^8 s^{-1}
Thallium			
Tl I			
2104.6	2	4	4.0E-02
2118.9	2	2	2.0E-02
2129.3	2	4	5.8E-02
2151.9	2	2	3.1E-02
2168.6	2	4	9.8E-02
2237.8	2	4	1.9E-01
2316.0	2	2	7.8E-02
2379.7	2	4	4.4E-01
2507.9	4	2	1.1E-02
2538.2	4	2	1.6E-02
2580.1	2	2	1.8E-01
2609.0	4	6	1.0E-01
2609.8	4	4	1.9E-02
2665.6	4	2	5.7E-02
2709.2	4	6	1.7E-01
2710.7	4	4	3.7E-02
2767.9	2	4	1.26E+00
2826.2	4	2	8.0E-02
2918.3	4	6	4.2E-01
2921.5	4	4	7.6E-02
3229.8	4	2	1.73E-01
3519.2	4	6	1.24E+00
3529.4	4	4	2.20E-01
3775.7	2	2	6.25E-01
5350.5	4	2	7.05E-01
Thulium			
Tm I			
2513.8	8	10	6.9E-02
2527.0	8	8	1.7E-01
2596.5	8	10	1.6E-01
2601.1	8	6	1.7E-01
2622.5	8	10	6.1E-02
2841.1	6	6	2.0E-01
2854.2	8	6	2.7E-01
2914.8	8	8	7.7E-02
2933.0	8	6	1.0E-01
2973.2	8	8	2.3E-01
3046.9	8	8	1.8E-01
3081.1	8	8	1.9E-01
3122.5	6	6	5.2E-01
3142.4	6	6	8.8E-02
3172.7	8	8	1.8E-01
3233.7	8	10	5.1E-02
3247.0	6	8	3.0E-01
3251.8	6	4	5.2E-01
3380.7	6	8	2.0E-01
3406.0	6	8	1.5E-01
3410.1	8	10	1.0E-01
3416.6	8	8	5.7E-02
3418.6	6	6	1.1E-01
3563.9	8	6	9.8E-02
3567.4	8	10	4.2E-02
3744.1	8	8	9.5E-01
3751.8	8	10	1.9E-01
3798.5	6	4	1.2E+00
3807.7	6	6	3.9E-01

λ Å	Weights g_i	g_k	A 10^8 s^{-1}
3883.1	8	6	1.0E+00
3887.4	8	8	3.8E-01
3916.5	6	8	1.5E+00
3949.3	6	6	1.0E+00
4022.6	6	8	4.0E-02
4044.5	6	4	2.9E-01
4094.2	8	6	9.0E-01
4105.8	8	10	6.0E-01
4138.3	6	4	7.0E-01
4158.6	6	8	5.5E-02
4187.6	8	8	6.1E-01
4203.7	8	10	2.5E-01
4222.7	6	8	1.5E-01
4271.7	6	6	1.1E-01
4359.9	8	6	1.3E-01
4386.4	8	8	4.2E-02
4394.4	6	4	1.1E-01
4643.1	6	6	3.4E-02
4681.9	6	8	3.9E-02
4691.1	6	6	3.9E-02
5307.1	8	10	2.3E-02
5658.3	6	8	1.0E-02
5675.8	8	10	1.3E-02
5760.2	6	6	1.3E-02
Tin			
Sn I			
2073.1	1	3	3.6E-02
2199.3	3	5	2.9E-01
2209.7	5	5	5.6E-01
2246.1	1	3	1.6E+00
2268.9	5	7	1.2E+00
2286.7	5	5	3.1E-01
2317.2	5	7	2.0E+00
2334.8	3	3	6.6E-01
2354.8	3	5	1.7E+00
2380.7	3	5	3.1E-02
2408.2	5	3	1.8E-01
2421.7	5	7	2.5E+00
2429.5	5	7	1.5E+00
2433.5	5	3	8.0E-03
2455.2	5	5	1.1E-02
2476.4	5	3	1.1E-02
2483.4	5	5	2.1E-01
2491.8	1	3	1.7E-01
2495.7	5	5	6.2E-01
2523.9	5	3	7.4E-02
2546.6	1	3	2.1E-01
2558.0	1	3	3.4E-01
2571.6	5	7	4.5E-01
2594.4	5	5	3.0E-01
2636.9	1	3	1.1E-01
2661.2	3	3	1.1E-01
2706.5	3	5	6.6E-01
2761.8	5	5	3.7E-03
2779.8	5	7	1.8E-01
2785.0	5	3	1.4E-01
2788.0	1	3	1.4E-01
2812.6	1	3	2.3E-01

λ Å	Weights g_i	g_k	A 10^8 s^{-1}
2813.6	5	5	1.2E-01
2840.0	5	5	1.7E+00
2850.6	5	5	3.3E-01
2863.3	1	3	5.4E-01
2913.5	1	3	8.3E-01
3009.1	3	3	3.8E-01
3032.8	1	3	6.2E-01
3034.1	3	1	2.0E+00
3141.8	1	3	1.9E-01
3175.1	5	3	1.0E+00
3218.7	1	3	4.7E-02
3223.6	5	5	1.2E-03
3262.3	5	3	2.7E+00
3330.6	5	5	2.0E-01
3655.8	1	3	4.1E-02
3801.0	5	3	2.8E-01
4524.7	1	3	2.6E-01
5631.7	1	3	2.4E-02
5970.3	5	3	9.6E-02
6037.7	5	5	5.0E-02
6069.0	1	3	4.6E-02
6073.5	3	1	6.3E-02
6171.5	3	3	4.9E-02

Sn II

λ Å	Weights g_i	g_k	A 10^8 s^{-1}
2368.3	4	2	4.4E-03
2449.0	4	6	3.7E-01
2487.0	6	8	5.5E-01
3283.2	4	6	1.0E+00
3352.0	6	8	1.0E+00
3472.5	2	4	1.6E-01
3575.5	4	6	1.3E-01
5332.4	2	4	8.6E-01
5562.0	4	6	1.2E+00
5588.9	4	6	8.5E-01
5596.2	4	4	1.5E-01
5797.2	6	6	2.8E-01
5799.2	6	8	8.1E-01
6453.5	2	4	1.2E+00
6761.5	2	2	3.2E-01
6844.1	2	2	6.6E-01

Titanium

Ti I

λ Å	Weights g_i	g_k	A 10^8 s^{-1}
2276.75	7	5	1.3E+00
2280.00	9	7	9.4E-01
2299.86	5	5	6.9E-01
2302.75	7	7	5.7E-01
2305.69	9	9	5.2E-01
2424.26	9	9	1.7E-01
2520.54	5	3	3.8E-01
2529.87	7	5	3.8E-01
2541.92	9	7	4.3E-01
2599.91	5	5	6.7E-01
2605.16	7	7	6.4E-01
2611.29	9	9	6.4E-01
2611.47	7	5	3.3E-01
2619.94	9	7	2.1E-01
2631.55	7	7	1.7E-01
2632.42	5	5	2.7E-01
2641.12	5	3	1.8E+00
2644.28	7	5	1.4E+00
2646.65	9	7	1.5E+00
2733.27	5	5	1.9E+00
2735.30	3	1	4.1E+00
2912.07	5	7	1.3E+00
2942.00	5	5	1.0E+00
2948.26	7	7	9.3E-01
2956.13	9	9	9.7E-01
2956.80	7	5	1.8E-01
3186.45	5	7	8.0E-01
3191.99	7	9	8.5E-01
3199.92	9	11	9.4E-01
3341.88	5	7	6.5E-01
3354.63	7	9	6.9E-01
3370.44	5	3	7.6E-01
3371.45	9	11	7.2E-01
3377.58	7	5	6.9E-01
3385.94	9	7	5.0E-01
3635.46	5	7	8.04E-01
3642.68	7	9	7.74E-01
3653.50	9	11	7.54E-01
3724.57	9	9	9.1E-01
3725.16	5	3	7.3E-01
3729.81	5	5	4.27E-01
3741.06	7	7	4.17E-01
3752.86	9	9	5.04E-01
3786.04	5	3	1.4E+00
3948.67	5	3	4.85E-01
3956.34	7	5	3.00E-01
3958.21	9	7	4.05E-01
3981.76	5	5	3.76E-01
3989.76	7	7	3.79E-01
3998.64	9	9	4.08E-01
4013.24	7	5	2.0E-01
4055.01	1	3	2.8E-01
4060.26	3	5	2.4E-01
4064.20	3	3	2.4E-01
4065.09	3	1	7.0E-01
4186.12	9	9	2.10E-01
4266.23	5	5	3.1E-01
4284.99	5	5	3.2E-01
4289.07	5	5	3.0E-01
4290.93	3	3	4.5E-01
4295.75	3	1	1.3E+00
4393.93	9	11	3.3E-01
4417.27	11	9	3.6E-01
4449.14	11	11	9.7E-01
4450.90	9	9	9.6E-01
4453.31	5	5	5.98E-01
4453.71	7	7	4.7E-01
4455.32	7	7	4.8E-01
4457.43	9	9	5.6E-01
4465.81	5	7	3.28E-01
4481.26	7	7	5.7E-01
4496.15	7	5	4.4E-01
4518.02	7	9	1.72E-01
4522.80	5	7	1.9E-01
4527.31	3	5	2.2E-01
4533.24	11	11	8.83E-01
4534.78	9	9	6.87E-01
4544.69	5	3	3.3E-01
4548.76	7	5	2.85E-01
4552.45	9	7	2.1E-01
4563.43	9	11	2.1E-01
4617.27	7	9	8.51E-01
4623.10	5	7	5.74E-01
4639.94	3	3	6.64E-01
4640.43	3	1	5.0E-01
4645.19	3	1	8.57E-01
4650.02	5	3	2.6E-01
4742.79	9	9	5.3E-01
4758.12	11	11	7.13E-01
4759.27	13	13	7.40E-01
4778.26	9	9	2.0E-01
4805.42	5	7	5.8E-01
4840.87	5	5	1.76E-01
4856.01	13	15	5.2E-01
4885.08	11	13	4.90E-01
4913.62	7	9	4.44E-01
4928.34	3	5	6.2E-01
4981.73	11	13	6.60E-01
4989.14	7	5	3.25E-01
4991.07	9	11	5.84E-01
4999.50	7	9	5.27E-01
5000.99	9	7	3.52E-01
5007.21	5	7	4.92E-01
5014.28	3	5	6.8E-01
5036.47	7	9	3.94E-01
5038.40	5	7	3.87E-01
5062.11	5	3	2.98E-01
5222.69	3	3	1.95E-01
5224.30	11	11	3.6E-01
5259.98	5	7	2.3E-01
5351.07	7	7	3.4E-01
5503.90	11	9	2.6E-01
5774.04	9	11	5.5E-01
5785.98	11	13	6.1E-01
5804.27	13	15	6.8E-01
6098.66	9	7	2.5E-01
6220.46	9	7	1.8E-01

Ti II

λ Å	Weights g_i	g_k	A 10^8 s^{-1}
2440.91	4	4	5.1E-01
2451.18	6	6	4.5E-01
2525.59	10	8	5.6E-01
2531.28	8	6	4.9E-01
2534.63	6	4	5.4E-01
2535.89	4	2	6.8E-01
2555.99	6	8	3.2E-01
2635.44	4	4	1.9E+00
2638.56	6	6	1.7E+00
2642.02	8	8	1.9E+00
2645.86	10	10	2.7E+00
2746.54	6	8	2.6E+00
2751.59	8	10	3.7E+00
2752.68	8	10	1.1E+00

λ Å	Weights g_i	g_k	A 10^8 s^{-1}
2757.62	6	8	7.2E-01
2758.35	4	6	9.9E-01
2758.79	2	4	4.4E-01
2764.28	4	4	7.4E-01
2804.82	6	8	4.6E+00
2810.30	8	10	5.1E+00
2817.83	10	12	3.8E+00
2819.87	8	8	6.5E-01
2821.26	6	8	7.9E-01
2827.12	8	10	1.0E+00
2828.06	12	14	4.4E+00
2828.64	6	6	1.2E+00
2828.83	10	10	9.1E-01
2834.02	10	12	7.9E-01
2836.47	8	8	1.2E+00
2839.64	12	12	8.3E-01
2845.93	10	10	1.2E+00
2851.11	2	4	4.1E-01
2856.10	12	12	1.5E+00
2862.33	4	6	4.0E-01
2877.47	8	8	5.7E-01
2884.13	10	10	5.2E-01
2910.65	8	8	4.6E-01
2926.64	10	8	8.9E-01
2931.10	6	6	3.2E+00
2936.02	4	6	2.7E+00
2938.57	6	8	2.4E+00
2941.90	8	10	1.8E+00
2942.97	8	8	1.1E+00
2945.30	10	12	2.7E+00
2952.00	8	8	3.0E-01
2954.59	10	12	4.0E+00
2958.80	8	10	4.0E+00
2979.06	4	6	1.2E+00
2990.06	6	8	5.6E-01
3017.17	12	12	3.6E-01
3022.64	10	10	1.2E+00
3023.67	8	8	1.0E+00
3029.76	10	10	3.5E-01
3056.75	2	4	3.2E-01
3058.08	6	6	5.0E-01
3066.34	4	4	3.3E-01
3071.25	6	4	3.6E-01
3072.99	4	2	1.6E+00
3075.23	6	4	1.13E+00
3078.65	8	6	1.09E+00
3081.52	10	8	1.1E+00
3088.04	10	8	1.25E+00
3089.44	8	6	1.3E+00
3097.20	4	6	4.4E-01
3103.81	10	8	1.1E+00
3105.10	2	4	6.3E-01
3106.26	6	6	7.8E-01
3117.67	4	2	1.1E+00
3119.83	6	4	5.9E-01
3127.86	6	6	1.6E+00
3128.50	8	8	1.1E+00
3161.23	4	2	5.9E-01
3161.80	6	4	4.6E-01

λ Å	Weights g_i	g_k	A 10^8 s^{-1}
3162.59	8	6	3.9E-01
3168.55	10	8	4.1E-01
3181.73	6	8	4.6E-01
3182.54	4	6	4.3E-01
3189.49	4	4	9.2E-01
3190.91	6	8	1.3E+00
3202.56	4	6	1.1E+00
3224.25	12	10	7.0E-01
3228.62	4	2	2.0E+00
3232.29	8	6	6.0E-01
3234.51	10	10	1.38E+00
3236.13	4	4	7.0E-01
3236.58	8	8	1.11E+00
3239.04	6	6	9.87E-01
3239.66	6	4	9.4E-01
3241.99	4	4	1.16E+00
3251.91	6	4	3.38E-01
3252.92	8	6	3.9E-01
3272.07	2	4	3.2E-01
3278.28	4	4	9.6E-01
3278.91	6	4	1.0E+00
3282.32	2	2	1.6E+00
3287.66	8	10	1.4E+00
3315.32	2	4	3.8E-01
3321.70	4	4	7.2E-01
3322.94	10	10	3.96E-01
3329.46	8	8	3.25E-01
3332.11	6	4	1.1E+00
3340.34	4	4	3.6E-01
3361.23	8	10	1.1E+00
3372.80	6	8	1.11E+00
3383.77	4	6	1.09E+00
3452.49	2	2	7.7E-01
3456.40	4	4	8.2E-01
3465.56	4	2	4.1E-01
3483.63	10	8	9.7E-01
3492.37	8	6	9.8E-01
3504.90	10	10	8.2E-01
3510.86	8	8	9.3E-01
3520.27	2	4	4.8E-01
3535.41	4	6	5.5E-01
3641.33	4	2	4.9E-01
3706.23	4	4	3.1E-01
3741.64	6	6	6.2E-01
3757.70	4	4	4.1E-01
3759.30	8	8	9.4E-01
3761.33	6	6	9.9E-01
4911.18	6	4	3.2E-01

Ti III

λ Å	Weights g_i	g_k	A 10^8 s^{-1}
865.79	5	3	6.6E+01
1002.37	5	5	7.6E+00
1004.67	7	5	4.3E+01
1005.80	3	3	1.3E+01
1007.16	5	3	3.8E+01
1008.12	3	1	5.1E+01
1286.37	9	9	2.0E+00
1289.30	7	7	2.2E+00
1291.62	5	5	2.4E+00
1293.23	9	7	1.0E+00
1298.97	7	5	4.9E+00
1327.59	5	3	3.2E+00
1420.44	1	3	1.2E+00
1421.63	3	1	4.0E+00
1422.41	5	5	3.0E+00
1424.14	5	3	1.6E+00
1455.19	9	7	6.4E+00
1498.70	5	5	2.8E+00
2007.36	3	3	3.4E+00
2007.60	1	3	1.2E+00
2010.80	5	3	5.4E+00
2097.30	5	7	3.3E+00
2099.86	3	5	2.5E+00
2104.86	3	3	1.1E+00
2105.09	1	3	1.7E+00
2199.22	3	3	5.7E+00
2237.77	7	7	2.4E+00
2331.35	3	1	4.3E+00
2331.66	3	3	1.2E+00
2339.00	5	3	3.0E+00
2346.79	7	5	3.3E+00
2374.99	5	3	4.0E+00
2413.99	5	7	3.8E+00
2516.05	7	9	3.4E+00
2567.56	3	3	2.3E+00
2984.75	5	5	1.9E+00
3066.51	3	3	2.5E+00
3228.89	3	3	1.5E+00
3278.31	7	9	3.4E+00
3320.94	3	5	2.8E+00
3340.20	7	9	3.7E+00
3346.18	9	11	3.7E+00
3354.71	11	13	4.4E+00
3397.24	3	1	1.8E+00
3404.46	3	3	1.8E+00
3417.62	3	5	1.9E+00
3915.47	9	11	2.1E+00
4119.14	5	5	9.9E-01
4213.26	9	11	2.2E+00
4215.53	9	11	2.2E+00
4247.62	11	13	1.1E+00
4248.54	5	7	2.3E+00
4250.09	3	5	9.5E-01
4259.01	11	13	9.4E-01
4269.84	9	11	1.7E+00
4285.61	13	15	3.0E+00
4288.66	11	13	1.1E+00
4296.70	11	13	1.6E+00
4319.56	9	11	1.1E+00
4343.25	3	1	1.0E+00
4378.94	3	5	1.6E+00
4433.91	11	13	1.8E+00
4440.66	1	3	1.2E+00
4533.26	3	5	1.5E+00
4576.53	9	7	1.3E+00
4628.07	3	1	1.5E+00
4652.86	7	9	2.6E+00
4874.00	5	7	1.5E+00

λ Å	Weights g_i	g_k	A 10^8 s^{-1}
4914.32	3	3	1.1E+00
4971.19	9	11	2.1E+00
5083.80	5	3	9.7E-01
5278.33	3	3	9.4E-01
7506.87	11	13	1.1E+00
Ti IV			
423.49	4	6	4.9E+01
424.16	6	8	5.3E+01
433.63	4	2	5.5E+00
433.76	6	4	5.0E+00
729.36	4	2	5.7E+00
1183.64	2	2	6.9E+00
1195.21	4	2	1.4E+01
1451.74	2	4	1.8E+01
1467.34	4	6	2.1E+01
2067.56	2	4	5.1E+00
2103.16	2	2	5.0E+00
2541.79	4	6	6.9E+00
2546.88	6	8	7.4E+00
2862.60	4	2	4.1E+00
3576.44	4	6	4.6E+00
Ti VIII			
249	6	4	1.0E+01
258.610	6	8	7.5E+02
269.533	4	6	6.0E+02
272.037	4	4	4.3E+02
272.843	6	4	6.2E+01
276.701	2	4	9.3E+01
277.813	4	4	3.8E+02
289.375	2	4	3.6E+01
478.971	4	4	1.7E+01
480.376	6	6	1.5E+01
Ti IX			
267.941	5	7	5.1E+02
278.713	5	7	4.7E+02
281.446	3	1	3.2E+02
285.128	1	3	4.1E+02
433.567	1	3	6.9E+00
439.513	3	3	7.5E+00
439.745	3	1	2.1E+01
447.484	5	5	1.6E+01
447.701	5	3	6.5E+00
507.174	3	5	6.5E+00
516.215	5	7	6.9E+00
Ti X			
253	4	6	2.1E+02
254	6	8	2.3E+02
281	2	2	1.1E+02
289.579	2	4	2.5E+02
290.294	4	6	1.1E+02
291	4	2	1.8E+02
291	2	2	2.3E+02
292	6	8	1.1E+02
293.684	6	8	2.97E+02
293.798	6	6	1.7E+02

λ Å	Weights g_i	g_k	A 10^8 s^{-1}
295.584	4	6	2.9E+02
296	4	6	1.4E+02
297	4	6	9.9E+01
298	4	6	4.3E+02
302	2	2	1.6E+02
305	2	4	2.5E+02
317	2	2	1.5E+02
355.815	2	2	1.3E+02
360.133	4	4	2.19E+02
363	4	2	2.1E+02
363	6	6	1.3E+02
365.628	4	2	1.2E+02
382	4	6	1.8E+02
385	6	8	1.8E+02
389.99	6	4	1.1E+02
Ti XI			
65.403	1	3	5.1E+02
87.725	1	3	8.5E+02
266	5	7	1.8E+02
308.250	3	5	1.3E+02
313.229	5	7	1.6E+02
318	3	1	1.4E+02
322.75	5	7	1.99E+02
323	1	3	1.8E+02
327.192	3	5	2.9E+02
332	3	1	3.25E+02
386.140	1	3	1.48E+02
408	7	9	1.37E+02
425.74	3	1	1.2E+02
446.69	3	1	1.2E+02
453	5	7	1.3E+02
Ti XII			
52.896	2	4	1.61E+02
53.140	4	6	1.9E+02
53.433	2	4	2.1E+02
53.457	2	2	2.1E+02
55.181	2	4	2.4E+02
55.443	4	6	2.81E+02
59.133	2	4	3.72E+02
59.435	4	6	4.41E+02
60.701	2	4	3.4E+02
60.762	2	2	3.5E+02
61.286	4	2	1.8E+02
62.433	4	6	2.08E+02
62.470	6	8	2.22E+02
65.540	4	6	3.2E+02
65.577	6	8	3.5E+02
67.171	2	4	6.2E+02
67.555	4	6	7.2E+02
70.986	4	6	5.7E+02
71.031	6	8	6.1E+02
71.545	2	2	1.8E+02
71.987	4	2	3.48E+02
82.121	2	4	5.9E+02
82.307	4	6	1.13E+03
82.344	2	2	5.8E+02
82.368	6	8	1.2E+03

λ Å	Weights g_i	g_k	A 10^8 s^{-1}
89.844	2	4	9.9E+02
90.512	4	6	1.16E+03
90.547	4	4	1.9E+02
116.497	4	6	3.0E+03
116.597	6	8	3.2E+03
116.62	6	6	2.1E+02
139.884	6	4	2.6E+02
140.361	4	2	2.9E+02
141.6	4	6	1.7E+02
141.7	6	8	1.7E+02
169.7	4	6	2.8E+02
169.8	6	8	2.9E+02
207.2	2	4	1.5E+02
208.5	4	6	1.8E+02
252.8	4	6	4.8E+02
253.1	6	8	5.2E+02
257.5	4	2	2.4E+02
Ti XIII			
23.356	1	3	1.02E+05
23.698	1	3	1.2E+04
23.991	1	3	3.4E+02
26.641	1	3	4.06E+03
26.960	1	3	3.06E+03
117.1	3	3	1.3E+02
117.3	3	1	2.8E+02
120.2	5	3	5.4E+02
120.2	7	5	4.4E+02
128.7	3	3	1.2E+02
Ti XIV			
21.341	4	6	9.8E+03
21.522	2	4	4.5E+04
21.657	4	4	1.3E+04
21.733	4	4	8.8E+04
21.82	4	2	6.4E+04
21.883	2	4	7.0E+04
21.958	2	4	1.2E+04
22.05	2	2	1.4E+04
24.592	4	2	6.1E+03
24.891	2	2	7.5E+03
Ti XV			
20.19	5	7	6.9E+03
20.234	5	7	1.9E+04
20.234	3	3	4.9E+04
20.246	1	3	4.2E+04
20.250	5	3	6.5E+03
20.29	3	3	1.1E+04
20.30	1	3	3.4E+04
20.30	1	1	5.8E+04
20.313	5	3	7.5E+04
20.418	5	7	8.0E+04
20.538	3	3	3.8E+04
20.54	3	1	4.1E+04
20.551	1	3	1.3E+04
20.689	5	7	4.3E+04
20.698	1	3	1.1E+05
20.771	5	3	1.1E+04

λ Å	Weights g_i	g_k	A 10^8 s^{-1}
20.897	5	7	2.85E+04
20.928	5	5	8.4E+03
21.065	3	3	1.1E+04
21.079	1	3	1.58E+04
21.102	3	5	1.3E+04
22.482	5	3	6.4E+03
22.936	5	5	1.1E+04
22.966	5	3	1.1E+04
23.034	1	3	6.3E+03
Ti XVI			
110.561	4	2	3.36E+02
116.198	4	4	1.45E+02
118.215	6	4	7.4E+02
121.382	4	2	2.4E+02
124.805	4	2	6.1E+02
129.075	4	2	3.81E+02
134.724	2	2	2.6E+02
138.800	6	4	3.5E+02
143.459	4	4	2.8E+02
145.665	6	6	2.3E+02
157.812	4	2	1.32E+02
161.168	4	4	1.2E+02
163.610	4	2	1.92E+02
169.740	4	6	1.0E+02
176.267	2	2	2.43E+02
178.240	4	4	2.52E+02
Ti XVII			
18.05	3	3	4.5E+04
18.13	5	3	2.4E+04
18.13	1	3	8.1E+04
18.176	5	7	9.2E+04
123.654	3	3	2.3E+02
124.553	5	3	5.2E+02
127.782	5	3	4.6E+02
135.202	3	1	2.93E+02
136.160	5	3	1.95E+02
136.393	3	3	1.14E+02
141.948	5	5	3.87E+02
142.589	1	3	1.35E+02
144.405	5	5	9.4E+01
146.067	7	5	2.6E+02
154.133	3	1	1.63E+02
156.54	3	1	1.44E+02
158.469	5	5	1.4E+02
159.62	5	3	1.03E+02
163.049	3	1	6.2E+02
186.863	5	5	2.66E+02
207.73	3	1	1.07E+02
Ti XVIII			
17.22	2	4	7.3E+04
17.365	4	6	8.6E+04
17.39	4	4	1.4E+04
133.852	2	4	5.2E+01
144.759	4	4	3.2E+02
150.15	6	4	1.15E+02
153.15	4	2	1.97E+02
153.23	2	4	6.7E+01
159.00	4	4	1.16E+02
166.225	6	4	1.54E+02
179.902	2	4	6.3E+01
189.663	6	6	9.6E+01
191.23	4	4	6.6E+01
197.838	4	6	4.56E+01
208.07	4	4	1.2E+02
Ti XIX			
15.67	3	1	3.3E+04
15.68	5	5	2.7E+04
15.74	5	7	2.7E+04
15.75	3	5	2.4E+04
15.83	1	3	3.2E+04
15.86	1	3	2.9E+04
16.02	3	1	3.1E+04
16.18	3	5	3.8E+04
16.41	1	3	6.1E+04
16.43	3	5	8.2E+04
16.46	3	3	4.4E+04
16.51	5	7	1.0E+05
16.55	5	5	2.7E+04
16.61	3	1	8.0E+04
16.64	3	3	5.3E+04
16.69	1	3	1.02E+05
16.71	3	5	7.3E+04
16.72	5	3	3.3E+04
16.72	5	5	7.3E+04
16.74	5	7	1.2E+05
16.77	3	3	2.6E+04
16.80	5	7	1.81E+05
16.85	3	5	4.4E+04
17.08	3	5	8.3E+04
17.36	1	3	9.5E+04
Ti XX			
2.629	2	4	4.9E+04
2.6295	4	4	3.2E+06
2.631	2	2	6.1E+05
2.6319	2	4	1.5E+06
2.632	2	2	2.7E+06
2.6355	4	6	1.2E+06
8.621	4	2	1.1E+06
9.788	4	6	5.26E+03
10.046	2	4	7.29E+03
10.109	4	6	8.6E+03
*10.278	2	6	8.4E+03
10.620	2	4	1.34E+04
10.690	4	6	1.58E+04
*11.452	2	6	1.7E+04
11.872	2	4	2.8E+04
11.958	4	6	3.4E+04
11.958	4	4	5.6E+03
15.211	2	4	3.50E+04
15.253	2	2	3.58E+04
15.907	2	4	8.84E+04
16.049	4	6	1.05E+05
16.067	4	4	1.8E+04
31.586	4	6	5.49E+03
45.650	2	4	9.6E+03
45.996	4	6	1.1E+04
Ti XXI			
2.0633	1	3	1.32E+05
2.1108	1	3	2.60E+05
2.2211	1	3	6.35E+05
2.497	3	1	2.4E+06
2.505	5	5	3.5E+05
2.505	1	3	1.4E+06
2.507	3	5	1.4E+06
2.508	3	5	7.9E+05
2.510	3	3	6.9E+05
2.510	1	3	9.6E+05
2.511	3	3	1.4E+06
2.512	5	5	1.8E+06
2.512	3	1	1.4E+06
2.513	3	1	2.7E+06
2.513	3	5	2.4E+06
2.514	5	3	1.2E+06
2.520	3	5	2.6E+05
2.527	3	1	1.2E+05
2.530	3	1	4.1E+05
2.6102	1	3	2.40E+06
2.6227	1	3	1.12E+05
Uranium			
U I			
3553.0	13	13	2.0E-02
3553.0	9	7	1.4E-02
3553.4	15	13	2.2E-02
3554.5	11	9	8.4E-03
3554.9	15	17	7.9E-03
3555.3	13	15	2.7E-02
3555.8	13	11	4.1E-03
3556.9	13	11	7.5E-03
3557.8	13	13	2.9E-02
3558.0	11	13	1.6E-02
3558.6	9	7	3.9E-02
3559.4	7	9	1.5E-02
3560.3	9	7	6.4E-02
3561.4	15	13	5.5E-02
3561.5	9	9	2.5E-02
3561.8	13	11	5.7E-02
3563.7	13	13	2.9E-02
3563.8	7	7	1.1E-02
3565.0	13	11	2.9E-02
3566.0	13	15	1.7E-02
3566.6	11	11	2.4E-01
3568.8	13	13	3.8E-02
3569.1	17	15	1.1E-01
3569.4	9	9	1.5E-02
3570.1	13	11	1.3E-02
3570.2	11	9	5.3E-03
3570.6	13	15	2.7E-02
3570.7	15	15	1.2E-02
3571.2	11	11	6.3E-03
3571.6	17	15	1.3E-01

λ Å	Weights g_i	g_k	A 10^8 s^{-1}
3572.9	13	15	1.5E-02
3573.9	13	11	4.0E-02
3574.1	13	15	3.5E-02
3574.8	13	15	1.9E-02
3577.1	17	15	4.3E-02
3577.5	15	13	7.8E-03
3577.8	11	11	8.3E-03
3577.9	13	13	2.3E-02
3578.3	13	11	2.0E-02
3580.0	9	9	1.2E-02
3580.2	11	9	2.9E-02
3580.4	11	13	7.5E-03
3580.9	13	13	2.1E-02
3582.6	13	13	2.9E-02
3584.6	7	5	2.4E-02
3584.9	13	15	1.8E-01
3585.4	11	11	1.9E-02
3585.8	11	9	2.8E-02
3587.8	9	11	1.3E-02
3588.3	7	9	1.8E-02
3589.7	11	13	2.1E-02
3589.8	15	13	5.9E-02
3590.7	9	7	2.2E-02
3591.7	11	9	5.3E-02
3593.0	11	11	1.4E-02
3593.2	13	15	4.2E-02
3593.7	11	11	7.2E-02

Vanadium

V I

λ Å	Weights g_i	g_k	A 10^8 s^{-1}
3043.12	6	8	2.3E-01
3050.39	10	8	5.3E-01
3053.65	4	4	1.3E+00
3056.33	6	6	1.3E+00
3060.46	8	8	1.4E+00
3066.37	10	10	2.1E+00
3066.53	6	4	3.2E-01
3075.93	4	6	2.8E-01
3080.33	2	4	2.7E-01
3083.54	6	8	2.5E-01
3087.06	2	2	9.2E-01
3088.11	4	6	4.9E-01
3089.13	4	4	5.3E-01
3093.79	6	6	4.1E-01
3094.69	2	4	4.3E-01
3112.92	4	2	5.0E-01
3183.41	6	8	2.4E+00
3183.96	8	10	2.5E+00
3183.98	4	6	2.4E+00
3185.38	10	12	2.7E+00
3198.01	6	6	3.9E-01
3202.39	8	8	4.0E-01
3205.58	8	10	1.3E+00
3207.41	10	10	2.6E-01
3212.43	10	12	1.4E+00
3218.87	8	6	3.5E-01
3233.19	10	8	3.2E-01
3273.03	8	8	2.7E-01
3284.36	10	10	2.8E-01
3309.18	4	4	3.2E-01
3329.85	6	4	7.7E-01
3356.35	4	6	3.1E-01
3365.55	2	4	4.8E-01
3376.05	4	4	3.2E-01
3377.39	4	2	9.0E-01
3377.62	6	6	6.0E-01
3397.58	6	4	2.3E-01
3400.39	8	8	2.5E-01
3529.73	4	6	4.1E-01
3533.68	6	8	5.2E-01
3533.76	2	4	3.7E-01
3543.49	2	2	6.7E-01
3545.33	4	4	3.7E-01
3553.27	6	6	2.2E-01
3555.14	4	2	2.6E-01
3663.60	4	6	3.1E+00
3667.74	6	8	2.7E+00
3672.41	12	12	9.2E-01
3673.41	8	10	2.7E+00
3676.70	14	14	1.3E+00
3680.12	10	12	2.2E+00
3686.26	10	12	2.3E-01
3687.50	12	14	2.9E+00
3688.07	8	8	3.5E-01
3690.28	2	4	4.5E-01
3692.22	6	6	5.4E-01
3695.34	14	16	2.8E+00
3695.86	4	4	6.6E-01
3703.57	10	8	9.2E-01
3704.70	8	6	6.6E-01
3705.04	6	4	3.6E-01
3706.03	10	10	5.2E-01
3708.71	12	12	4.4E-01
3790.46	10	8	2.3E-01
3794.96	10	10	2.3E-01
3806.79	10	10	2.5E-01
3818.24	4	2	6.73E-01
3828.56	6	4	5.33E-01
3840.75	8	6	5.48E-01
3855.36	4	4	3.30E-01
3855.85	10	8	5.78E-01
3863.86	8	6	3.1E-01
3864.86	6	6	2.70E-01
3871.07	10	8	2.8E-01
3875.07	8	8	2.36E-01
3902.26	10	10	2.68E-01
3921.86	4	2	2.7E-01
3922.43	6	6	2.6E-01
3930.02	10	10	3.3E-01
3934.01	8	8	6.2E-01
3992.80	12	10	1.2E+00
3998.73	14	12	1.0E+00
4050.96	10	10	1.4E+00
4051.35	12	12	1.3E+00
4090.57	8	10	8.5E-01
4092.68	8	10	2.30E-01
4095.48	6	8	7.2E-01
4099.78	6	8	4.10E-01
4102.15	4	6	7.1E-01
4104.77	10	8	2.1E+00
4105.16	4	6	4.9E-01
4109.78	2	4	5.00E-01
4111.78	10	10	1.01E+00
4115.18	8	8	5.80E-01
4116.47	6	6	3.2E-01
4116.59	2	2	2.90E-01
4123.50	4	2	1.00E+00
4128.06	6	4	7.70E-01
4131.99	8	6	5.5E-01
4134.49	10	8	2.90E-01
4232.46	10	10	9.8E-01
4232.95	8	8	7.7E-01
4268.64	14	14	1.2E+00
4271.55	12	12	9.6E-01
4276.95	10	10	9.4E-01
4284.05	8	8	1.2E+00
4291.82	12	14	8.8E-01
4296.10	10	12	7.7E-01
4297.67	8	10	7.0E-01
4298.03	6	8	7.8E-01
4379.23	10	12	1.1E+00
4384.71	8	10	1.1E+00
4389.98	6	8	6.9E-01
4395.22	4	6	5.5E-01
4400.57	2	4	3.4E-01
4406.64	10	10	2.2E-01
4407.63	8	8	4.4E-01
4408.20	6	6	6.0E-01
4416.47	4	2	2.6E-01
4452.01	14	16	9.2E-01
4457.75	10	12	2.7E-01
4460.33	10	8	3.0E-01
4462.36	12	14	7.6E-01
4468.00	8	10	2.3E-01
4469.71	10	12	6.2E-01
4474.04	10	8	4.7E-01
4496.06	8	6	4.0E-01
4514.18	6	4	3.3E-01
4524.21	12	10	3.0E-01
4525.17	4	2	4.1E-01
4529.58	10	8	2.4E-01
4545.40	10	12	7.6E-01
4560.72	8	10	7.0E-01
4571.79	6	8	6.0E-01
4578.73	4	6	6.8E-01
4706.16	6	4	2.4E-01
4757.47	4	2	7.6E-01
4766.62	6	4	5.6E-01
4776.36	8	6	5.1E-01
4786.50	10	8	4.7E-01
4796.92	12	10	4.8E-01
4807.52	14	12	5.8E-01
5193.00	12	12	4.0E-01
5195.39	8	8	2.3E-01
5234.08	10	10	4.9E-01
5240.87	12	12	4.3E-01
5415.25	12	14	3.1E-01

λ Å	Weights g_i	g_k	A 10^8 s^{-1}
5487.91	12	10	2.9E-01
5507.75	10	8	3.5E-01
6090.21	8	6	2.60E-01
V II			
2527.90	13	13	6.1E-01
2528.47	9	9	5.2E-01
2528.83	11	11	5.3E-01
2554.04	9	9	5.4E-01
2589.10	9	9	7.7E-01
2640.86	5	7	1.2E+00
2677.80	3	5	3.4E-01
2679.33	7	7	3.4E-01
2683.09	1	3	3.4E-01
2687.96	9	9	7.6E-01
2689.88	3	1	9.2E-01
2690.25	7	5	3.4E-01
2690.79	5	3	5.2E-01
2700.94	9	11	3.5E-01
2706.17	7	9	3.4E-01
2734.22	9	7	6.2E-01
2753.41	13	11	4.2E-01
2784.20	9	9	1.3E+00
2787.91	7	9	5.0E-01
2825.86	9	7	1.2E+00
2843.82	7	5	9.9E-01
2847.57	9	7	4.6E-01
2854.34	11	9	5.0E-01
2862.31	11	11	3.6E-01
2868.11	5	3	2.1E+00
2869.13	13	11	4.8E-01
2882.49	5	5	4.2E-01
2884.78	3	3	5.6E-01
2889.61	3	1	1.9E+00
2891.64	5	3	1.4E+00
2892.43	9	9	3.6E-01
2892.65	7	5	1.3E+00
2893.31	9	7	1.2E+00
2903.07	3	5	3.4E-01
2906.45	7	7	7.8E-01
2908.81	11	9	1.6E+00
2910.01	5	5	1.1E+00
2910.38	3	3	1.2E+00
2911.05	7	9	3.7E-01
2912.46	11	9	5.0E-01
2915.88	9	7	4.9E-01
2924.02	11	11	1.7E+00
2924.63	9	9	1.2E+00
2930.80	7	7	5.8E-01
2941.37	11	9	3.5E-01
2944.57	9	7	7.6E-01
2948.08	9	11	4.0E-01
2952.07	7	5	7.2E-01
2955.58	7	9	3.3E-01
2968.37	7	9	7.0E-01
2972.26	5	7	5.2E-01
2973.98	9	11	3.5E-01
2985.18	7	9	4.4E-01
3001.20	7	7	7.5E-01
3014.82	5	3	8.9E-01
3016.78	7	5	5.0E-01
3020.21	9	7	5.0E-01
3048.21	11	13	7.0E-01
3063.25	9	11	1.0E+00
3100.94	7	7	5.8E-01
3113.56	11	11	5.0E-01
3122.89	11	13	7.6E-01
3134.93	13	13	5.9E-01
3136.50	11	11	5.3E-01
3139.73	9	9	5.2E-01
3151.32	3	5	4.4E-01
3190.69	9	9	3.3E-01
3250.78	11	9	5.2E-01
3251.87	5	7	3.5E-01
3271.12	7	9	6.9E-01
3276.12	9	11	5.2E-01
3279.84	9	11	5.8E-01
3287.71	5	7	7.5E-01
3337.85	5	7	5.3E-01
3517.30	9	7	3.8E-01
3530.77	5	3	4.5E-01
3545.19	7	5	4.3E-01
3556.80	9	7	5.1E-01
3592.01	7	5	4.4E-01
3618.92	3	5	3.3E-01
V III			
2318.06	8	10	4.6E+00
2323.82	6	8	3.8E+00
2330.42	10	10	3.2E+00
2331.75	8	8	2.5E+00
2334.21	6	6	2.2E+00
2337.13	4	4	2.7E+00
2343.10	6	8	3.6E+00
2358.73	6	8	4.2E+00
2366.31	8	10	4.2E+00
2371.06	10	12	5.2E+00
2373.06	4	6	2.9E+00
2382.46	8	10	5.0E+00
2393.58	6	8	4.3E+00
2404.18	4	6	2.5E+00
2516.14	10	10	3.7E+00
2521.55	8	8	3.5E+00
2548.21	6	4	2.0E+00
2554.22	8	6	1.2E+00
2593.05	6	6	2.8E+00
2595.10	8	8	2.8E+00
V IV			
677.345	9	9	6.7E+00
680.632	9	7	1.2E+01
681.145	7	5	1.1E+01
682.455	7	7	6.5E+00
682.923	5	5	6.9E+00
684.450	7	5	7.7E+00
691.530	5	3	1.1E+01
723.537	3	1	1.5E+01
724.068	5	5	1.1E+01
724.809	5	3	5.6E+00
737.854	9	7	2.4E+01
750.110	5	5	1.0E+01
884.146	1	3	4.7E+00
1071.05	5	5	6.1E+00
1110.72	3	3	5.0E+00
1112.20	7	7	6.3E+00
1112.44	5	5	5.0E+00
1127.84	7	5	8.9E+00
1131.26	9	7	9.4E+00
1194.46	7	5	1.0E+01
1226.52	5	5	1.5E+01
1243.72	3	1	9.4E+00
1247.07	5	3	4.7E+00
1272.97	3	1	2.7E+01
1304.17	3	5	1.5E+01
1305.42	5	7	7.0E+00
1308.06	7	9	7.9E+00
1309.50	5	5	8.7E+00
1312.72	7	7	8.6E+00
1317.57	5	7	8.7E+00
1321.92	7	9	9.9E+00
1326.81	3	5	4.0E+00
1329.29	5	5	1.5E+01
1329.97	3	3	4.8E+00
1330.36	1	3	6.0E+00
1331.67	3	1	1.7E+01
1332.46	5	3	7.5E+00
1334.49	9	9	8.3E+00
1355.13	7	9	2.5E+01
1356.53	5	3	4.9E+00
1395.00	5	7	1.4E+01
1400.42	5	7	7.5E+00
1403.62	7	9	8.4E+00
1412.69	3	3	1.1E+01
1414.41	5	7	1.2E+01
1414.84	5	5	4.6E+00
1418.53	7	7	5.2E+00
1419.58	7	9	1.3E+01
1423.72	3	5	7.1E+00
1426.65	9	11	2.2E+01
1429.11	5	5	5.0E+00
1434.84	7	7	5.4E+00
1451.04	3	3	7.0E+00
1454.00	5	3	1.1E+01
1520.14	5	7	7.2E+00
1522.49	3	5	5.5E+00
1601.92	3	3	1.2E+01
1611.88	7	7	5.2E+00
1806.18	5	3	7.3E+00
1809.85	3	1	7.2E+00
1817.68	5	3	4.8E+00
1825.84	7	5	5.3E+00
1861.56	5	7	6.6E+00
1939.07	7	9	5.8E+00
1951.43	5	7	5.0E+00
1963.10	3	5	4.8E+00
1997.72	7	7	4.7E+00
2084.43	5	5	4.0E+00

λ Å	Weights g_i	g_k	A $10^8\ s^{-1}$
2120.05	7	9	8.1E+00
2141.20	3	5	7.0E+00
2146.83	7	9	6.6E+00
2149.85	5	7	5.1E+00
2151.09	7	9	4.3E+00
2155.34	11	13	1.2E+01
2446.80	9	11	5.3E+00
2570.72	9	11	7.6E+00
3284.56	7	9	5.3E+00
3496.42	7	9	4.4E+00
3514.25	9	11	4.7E+00
Xenon			
Xe I			
1043.8	1	3	5.9E-01
1047.1	1	3	1.3E+00
1050.1	1	3	8.5E-02
1056.1	1	3	2.45E+00
1061.2	1	3	1.9E-01
1068.2	1	3	3.99E+00
1085.4	1	3	4.10E-01
1099.7	1	3	4.34E-01
1110.7	1	3	1.5E+00
1129.3	1	3	4.4E-02
1170.4	1	3	1.6E+00
1192.0	1	3	6.2E+00
1250.2	1	3	1.4E-01
1295.6	1	3	2.5E+00
1469.6	1	3	2.8E+00
4501.0	5	3	6.2E-03
4524.7	5	5	2.1E-03
4624.3	5	5	7.2E-03
4671.2	5	7	1.0E-02
4807.0	3	1	2.4E-02
7119.6	7	9	6.6E-02
7967.3	1	3	3.0E-03
8409.2	5	3	1.0E-02
Xe II			
4180.1	4	4	2.2E+00
4330.5	6	8	1.4E+00
4414.8	6	6	1.0E+00
4603.0	4	4	8.2E-01
4844.3	6	8	1.1E+00
4876.5	6	8	6.3E-01
5260.4	2	4	2.2E-01
5262.0	4	4	8.5E-01
5292.2	6	6	8.9E-01
5372.4	4	2	7.1E-01
5419.2	4	6	6.2E-01
5439.0	4	2	7.4E-01
5472.6	8	8	9.9E-02
5531.1	8	6	8.8E-02
5719.6	4	6	6.1E-02
5976.5	4	4	2.8E-01
6036.2	6	6	7.5E-02
6051.2	8	6	1.7E-01
6097.6	6	4	2.6E-01
6270.8	4	6	1.8E-01
6277.5	4	6	3.6E-02
6805.7	8	6	6.1E-02
6990.9	10	8	2.7E-01
Ytterbium			
Yb I			
2464.5	1	3	9.1E-01
2672.0	1	3	1.18E-01
3464.4	1	3	6.2E-01
3988.0	1	3	1.76E+00
5556.5	1	3	1.14E-02
Yb II			
3289.4	2	4	1.8E+00
3694.2	2	2	1.4E+00
Yttrium			
Y I			
2948.41	4	4	3.5E-01
2974.59	4	6	3.5E-01
2984.25	6	8	4.8E-01
2995.26	6	4	5.1E-02
2996.94	4	6	8.4E-02
3005.26	4	4	4.8E-02
3022.28	6	6	6.6E-02
3045.36	6	6	1.07E-01
3053.95	6	4	1.9E-03
3155.65	4	6	2.7E-03
3172.84	4	4	9.9E-03
3185.96	6	8	1.2E-03
3209.38	6	6	3.0E-03
3227.16	6	4	1.10E-03
3484.05	4	6	1.2E-02
3549.66	6	6	1.0E-03
3552.69	4	4	2.3E-01
4077.36	4	6	1.1E+00
4083.71	4	4	2.5E-01
4102.36	6	8	1.3E+00
4128.30	6	6	1.6E+00
4142.84	4	4	1.6E+00
4167.51	6	6	2.38E-01
4235.93	6	4	3.0E-01
4352.40	4	4	6.7E-03
4379.33	6	4	7.83E-01
4385.47	4	4	6.9E-02
4394.01	8	8	1.9E-02
4409.70	4	6	2.7E-03
4417.43	10	8	3.2E-02
4437.34	6	6	8.64E-02
4443.65	10	8	1.1E-01
4459.01	4	6	1.8E-02
4476.95	8	6	2.8E-01
4491.74	10	10	2.3E-02
4514.01	4	6	3.34E-01
4527.78	8	6	8.33E-01
4534.09	6	8	4.4E-02
4544.31	6	6	4.10E-01
4559.36	2	4	4.0E-01
4581.33	6	4	1.5E-01
4613.00	6	4	1.8E-01
4643.70	4	6	1.8E-01
4653.78	4	6	1.6E-01
4674.85	6	8	1.3E-01
4725.84	4	4	1.5E-01
4762.96	6	4	4.2E-02
4780.16	2	4	8.9E-02
4781.03	8	10	1.0E-01
4799.30	6	8	1.6E-01
4804.31	6	4	2.6E-01
4804.80	4	4	3.84E-01
4821.63	6	6	1.0E-01
4845.67	8	8	6.8E-01
4852.68	6	6	6.2E-01
4856.71	6	6	2.0E-01
4859.84	4	4	7.26E-01
4893.44	6	4	2.2E-01
4900.08	8	6	2.0E-01
4906.11	10	8	1.2E-01
4950.01	8	6	2.0E-02
4963.49	4	4	1.4E-02
4981.97	4	6	4.7E-03
5004.44	6	4	1.2E-02
5205.01	4	4	8.4E-03
5258.47	6	6	2.9E-03
5271.82	8	6	1.1E-02
5380.63	6	4	3.2E-01
5381.24	4	4	9.9E-03
5388.39	6	8	1.1E-02
5390.81	8	6	2.9E-02
5401.88	6	8	6.0E-03
5424.36	6	4	3.47E-01
5466.24	4	4	1.0E-01
5466.47	10	12	6.3E-01
5469.10	4	6	3.6E-03
5513.65	6	6	2.39E-01
5519.88	4	6	1.2E-02
5526.43	6	4	3.9E-03
5527.56	8	10	5.4E-01
5541.63	8	8	5.2E-02
5551.00	4	4	6.9E-02
5573.03	6	4	1.8E-02
5594.12	6	8	5.0E-02
5606.34	10	10	5.84E-02
5619.96	6	4	2.0E-02
5630.14	4	6	4.9E-01
5641.78	2	4	1.9E-02
5675.27	6	6	9.3E-02
5675.64	4	6	4.3E-02
5693.63	4	4	1.1E-01
5714.94	8	6	2.0E-02
5729.25	6	6	2.2E-03
5732.09	6	6	7.5E-02
5740.22	8	6	4.0E-02
5757.59	4	6	7.6E-03
5788.36	4	4	9.4E-03
5844.13	6	4	5.6E-03
5879.93	4	2	8.5E-02
5902.91	6	8	4.0E-02

λ Å	Weights g_i	g_k	A $10^8\ s^{-1}$
6087.94	6	4	1.1E-01
6191.72	4	4	4.7E-02
6222.58	4	6	5.9E-03
6402.01	6	4	2.7E-03
6435.02	6	6	4.0E-02
6437.17	10	8	4.8E-02
6538.57	10	10	1.5E-01
6622.48	8	6	4.5E-03
6815.15	2	4	7.18E-02
7009.89	2	4	4.4E-02
7035.15	4	4	6.3E-02
Y II			
3112.03	1	3	1.3E-02
3179.42	3	5	3.8E-02
3195.62	3	3	8.23E-01
3200.27	5	5	4.8E-01
3203.32	3	1	2.77E+00
3216.69	5	3	2.0E+00
3242.28	7	5	2.0E+00
3448.81	5	5	4.1E-02
3467.88	5	3	2.7E-02
3496.08	1	3	3.49E-01
3549.01	5	7	3.97E-01
3584.51	3	5	4.02E-01
3600.74	7	7	1.4E+00
3601.91	3	3	1.13E+00
3611.04	5	5	1.04E+00
3628.70	5	3	3.3E-01
3664.62	7	5	3.7E-01
3710.29	7	9	1.5E+00
3747.55	3	3	1.9E-01
3774.34	5	7	1.1E+00
3776.56	5	3	2.42E-01

λ Å	Weights g_i	g_k	A $10^8\ s^{-1}$
3788.70	3	5	8.1E-01
3818.34	5	5	9.70E-02
3832.90	7	7	3.0E-01
3878.29	7	5	2.9E-02
3930.66	5	5	2.1E-02
3950.36	3	5	2.80E-01
3951.59	5	3	1.5E-02
3982.60	5	5	2.7E-01
4124.91	5	7	1.8E-02
4177.54	5	5	5.27E-01
4199.27	3	5	5.36E-03
4204.69	1	3	2.20E-02
4235.73	5	5	2.3E-02
4309.62	7	5	1.29E-01
4358.73	3	3	5.55E-02
4374.95	5	5	9.97E-01
4398.01	5	3	1.16E-01
4422.59	3	1	1.83E-01
4682.33	5	5	1.9E-02
4786.58	7	7	2.1E-02
4823.31	5	5	4.3E-02
4854.87	5	3	3.9E-01
4881.44	5	3	1.5E-03
4883.69	9	7	4.7E-01
4900.11	7	5	4.51E-01
4982.13	7	9	1.5E-02
5087.42	9	9	2.0E-01
5119.11	5	7	1.6E-02
5200.41	5	5	1.3E-01
5205.73	7	7	1.6E-01
5289.82	7	5	6.7E-03
5320.78	9	7	3.9E-03
5473.39	3	5	4.3E-02
5480.73	1	3	7.62E-02

λ Å	Weights g_i	g_k	A $10^8\ s^{-1}$
5497.41	5	5	1.2E-01
5509.90	5	5	4.24E-02
5544.61	3	1	1.8E-01
5546.01	5	3	5.8E-02
5728.89	5	5	3.0E-02
6613.74	5	7	1.7E-02
6832.48	5	5	3.3E-03
7264.16	5	3	1.3E-02
Zinc			
Zn I			
748.29	1	3	6.0E-02
765.60	1	3	7.6E-02
792.05	1	3	5.7E-02
793.85	1	3	1.8E-01
809.92	1	3	2.6E-01
1109.1	1	3	3.05E-01
2138.6	1	3	7.09E+00
3075.9	1	3	3.29E-04
3282.3	1	3	9.0E-01
3302.6	3	5	1.2E+00
3302.9	3	3	6.7E-01
3345.0	5	7	1.7E+00
3345.6	5	5	4.0E-01
3345.9	5	3	4.5E-02
6362.3	3	5	4.74E-01
11054	3	1	2.43E-01
Zn II			
2025.5	2	4	3.3E+00
2064.2	2	4	4.6E+00
2099.9	4	6	5.6E+00
2102.2	4	4	9.3E-01
4911.6	4	6	1.6E+00

ELECTRON AFFINITIES

Thomas M. Miller

Electron affinity is defined as the energy difference between the lowest (ground) state of the neutral and the lowest state of the corresponding negative ion. The accuracy of electron affinity measurements has been greatly improved since the advent of laser photodetachment experiments with negative ions. Electron affinities can be determined with optical precision, though a detailed understanding of atomic and molecular states and splittings is required to specify the photodetachment threshold corresponding to the electron affinity.

Atomic and molecular electron affinities are discussed in two excellent articles reviewing photodetachment studies which appear in *Gas Phase Ion Chemistry*, Vol. 3, Bowers, M. T., Ed., Academic Press, Orlando, 1984: Chapter 21 by Drzaic, P. S., Marks, J., and Brauman, J. I., "Electron Photodetachment from Gas Phase Negative Ions", p. 167, and Chapter 22 by Mead, R. D., Stevens, A. E., and Lineberger, W. C., "Photodetachment in Negative Ion Beams", p. 213. Persons interested in photodetachment details should consult these articles and the critical review of Hotop, H., and Lineberger, W. C., *J. Phys. Chem. Ref. Data*, 14, 731, 1985. For simplicity in the tables below, any electron affinity which was discussed in the articles by Drzaic *et al.* or Hotop and Lineberger is referenced to these sources, where original references are given. A great many additional electron affinities have been provided here by G. B. Ellison, W. C. Lineberger, H. Hotop, D. G. Leopold, and K. H. Bowen. Electron affinities for the lanthanides and actinides have not been measured, but theoretical estimates have been made by Bratch, S. G., *Chem. Phys. Lett.*, 98, 113, 1983, and Bratch, S. G., and Lagowski, J. J., *Chem. Phys. Lett.*, 107, 136, 1984. The development of cluster-ion photodetachment apparatuses has brought an explosion of electron affinity estimates for atomic and molecular clusters. [See Arnold, S. T., Eaton, J. G., Patel-Mistra, D., Sarkas, H. W., and Bowen, K. H., in *Ion and Cluster Ion Spectroscopy and Structure*, Maier, J. P., Ed., Elsevier Science, New York, 1989, p. 417.] The policy in this tabulation is to list the electron affinities for the atoms, diatoms, and triatoms, if adiabatic electron affinities have been determined, but to refer the reader to original sources for higher-order clusters. Additional data on molecular electron affinities may be found in Lias, S. G., Bartmess, J. E., Liebman, J. F., Holmes, J. L., Levin, R. D., and Mallard, W. G., *Gas Phase Ion and Neutral Thermochemistry, J. Phys. Chem. Ref. Data*, 17, (Supplement No. 1), 1988.

For the present tabulation the value $e/hc = 8065.5410 \pm 0.0024$ cm^{-1} eV^{-1} (Cohen, E. R., and Taylor, B. N., *Rev. Mod. Phys.*, 59, 1121, 1987) has been used to convert electron affinities from the units used in spectroscopic work, cm^{-1}, into eV for these tables. The uncertainties in the electron affinities given in the tables do not include the 0.3 ppm uncertainty in e/hc; only in a few cases, *e.g.*, atomic oxygen, is this additional uncertainty significant.

Abbreviations used in the tables: calc = calculated value; PT = photodetachment threshold using a lamp as a light source; LPT = laser photodetachment threshold; LPES = laser photoelectron spectroscopy; DA = dissociative attachment; e-scat = electron scattering; CT = charge transfer; and CD = collisional detachment.

Table 1
Atomic Electron Affinities

Atomic number	Atom	Electron affinity in eV	Uncertainty in eV	Method	Ref.
1	D	0.754593	0.000074	LPT	89
	H	0.754195	0.000019	LPT	89
		0.754209	0.000003	calc	1
2	He	not stable	—	calc	1
3	Li	0.6180	0.0005	LPT	1
4	Be	not stable	—	calc	1
5	B	0.277	0.010	LPES	1
6	C	1.2629	0.0003	LPT	1
7	N	not stable	—	DA	1
8	O	1.4611103	0.0000007	LPT	4
9	F	3.401190	0.000004	LPT	74
10	Ne	not stable	—	calc	1
11	Na	0.547926	0.000025	LPT	1
12	Mg	not stable	—	e-scat	1
13	Al	0.441	0.010	LPES	1
14	Si	1.385	0.005	LPES	1
15	P	0.7465	0.0003	LPT	1
16	S	2.077104	0.000001	LPT	1
17	Cl	3.61269	0.00006	LPT	52
18	Ar	not stable	—	calc	1
19	K	0.50147	0.00010	LPT	1
20	Ca	0.0184	0.0025	LPT	44
21	Sc	0.188	0.020	LPES	1
22	Ti	0.079	0.014	LPES	1
23	V	0.525	0.012	LPES	1
24	Cr	0.666	0.012	LPES	1
25	Mn	not stable	—	calc	1

Table 1
Atomic Electron Affinities (continued)

Atomic number	Atom	Electron affinity in eV	Uncertainty in eV	Method	Ref.
26	Fe	0.151	0.003	LPES	27
27	Co	0.662	0.003	LPES	27
28	Ni	1.156	0.010	LPES	1
29	Cu	1.235	0.005	LPES	37
30	Zn	not stable	—	e-scat	1
31	Ga	0.3	0.15	PT	1
32	Ge	1.233	0.003	LPES	28
33	As	0.81	0.03	PT	1
34	Se	2.020670	0.000025	LPT	1
35	Br	3.363590	0.000003	LPT	74
36	Kr	not stable	—	calc	1
37	Rb	0.48592	0.00002	LPT	1
38	Sr	0.11	—	calc	57
39	Y	0.307	0.012	LPES	1
40	Zr	0.426	0.014	LPES	1
41	Nb	0.893	0.025	LPES	1
42	Mo	0.746	0.010	LPES	1
43	Tc	0.55	0.20	calc	1
44	Ru	1.05	0.15	calc	1
45	Rh	1.137	0.008	LPES	1
46	Pd	0.562	0.005	LPES	116
47	Ag	1.302	0.007	LPES	1
48	Cd	not stable	—	e-scat	1
49	In	0.3	0.2	PT	1
50	Sn	1.112	0.004	LPES	28
51	Sb	1.046	0.005	LPES	108
52	Te	1.9708	0.0003	LPT	1
53	I	3.059038	0.000010	LPT	92
54	Xe	not stable	—	calc	1
55	Cs	0.471626	0.000025	LPT	1
56	Ba	0.15	—	calc	57
57	La	0.5	0.3	calc	1
72	Hf	≈0	—	calc	1
73	Ta	0.322	0.012	LPES	1
74	W	0.815	0.002	LPES	37
75	Re	0.15	0.15	calc	1
76	Os	1.1	0.2	calc	1
77	Ir	1.565	0.008	LPES	1
78	Pt	2.128	0.002	LPT	1
79	Au	2.30863	0.00003	LPT	1
80	Hg	not stable	—	e-scat	1
81	Tl	0.2	0.2	PT	1
82	Pb	0.364	0.008	LPES	1
83	Bi	0.946	0.010	LPES	1
84	Po	1.9	0.3	calc	1
85	At	2.8	0.2	calc	1
86	Rn	not stable	—	calc	1
87	Fr	0.46	—	calc	82

Table 2
Electron Affinities for Diatomic Molecules

Molecule	Electron affinity in eV	Uncertainty in eV	Method	Ref.
Ag_2	1.023	0.007	LPES	37
Al_2	1.10	0.15	LPES	68
As_2	0	—	PT	2
AsH	1.0	0.1	PT	2
Au_2	1.938	0.007	LPES	37
BO	3.12	0.09	PT	6
BeH	0.7	0.1	PT	2
Bi_2	1.271	0.008	LPES	119
Br_2	2.55	0.10	CT	2
BrO	2.353	0.006	LPES	88
C_2	3.269	0.006	LPES	87
CH	1.238	0.008	LPES	2
CN	3.862	0.004	LPES	111
CS	0.205	0.021	LPES	2
CaH	0.93	0.05	PT	2
Cl_2	2.38	0.10	CT	2
ClO	2.275	0.006	LPES	88
Co_2	1.110	0.008	LPES	27
CoD	0.680	0.010	LPES	29
CoH	0.671	0.010	LPES	29
Cr_2	0.505	0.005	LPES	114
CrD	0.568	0.010	LPES	29
CrH	0.563	0.010	LPES	29
CrO	1.222	0.010	LPES	5
Cs_2	0.469	0.015	LPES	104
CsCl	0.455	0.010	LPES	30
Cu_2	0.836	0.006	LPES	37
CuO	1.777	0.006	LPES	118
F_2	3.08	0.10	CT	2
FO	2.272	0.006	LPES	88
Fe_2	0.902	0.008	LPES	27
FeD	0.932	0.015	LPES	9
FeH	0.934	0.011	LPES	9
FeO	1.493	0.005	LPES	45
I_2	2.55	0.05	CT	2
IBr	2.55	0.10	CT	2
IO	2.378	0.006	LPES	88
K_2	0.497	0.012	LPES	104
KBr	0.642	0.010	LPES	30
KCl	0.582	0.010	LPES	30
KCs	0.471	0.020	LPES	104
KI	0.728	0.010	LPES	30
KRb	0.486	0.020	LPES	104
LiCl	0.593	0.010	LPES	30
LiD	0.337	0.012	LPES	102
LiH	0.342	0.012	LPES	102
MgCl	1.589	0.011	LPES	31
MgH	1.05	0.06	PT	2
MgI	1.899	0.018	LPES	31
MnD	0.866	0.010	LPES	9
MnH	0.869	0.010	LPES	9
NH	0.370	0.004	LPT	32
NO	0.026	0.005	LPES	73
NS	1.194	0.011	LPES	2
Na_2	0.430	0.015	LPES	104
NaBr	0.788	0.010	LPES	30
NaCl	0.727	0.010	LPES	30
NaF	0.520	0.010	LPES	30
NaI	0.865	0.010	LPES	30
NaK	0.465	0.030	LPES	104
Ni_2	0.926	0.010	LPES	112
NiD	0.477	0.007	LPES	29
NiH	0.481	0.007	LPES	29
O_2	0.451	0.007	LPES	73
OD	1.825548	0.000037	LPT	3
OH	1.827670	0.000021	LPT	3
P_2	0.589	0.025	LPES	42
PH	1.028	0.010	LPES	2
PO	1.092	0.010	LPES	2
Pb_2	1.366	0.010	LPES	117
PbO	0.722	0.006	LPES	105
Pd_2	1.685	0.008	LPES	112
Pt_2	1.898	0.008	LPES	112
PtN	1.240	0.010	LPES	46
Rb_2	0.498	0.015	LPES	104
RbCl	0.544	0.010	LPES	30
RbCs	0.478	0.020	LPES	104
Re_2	1.571	0.008	LPES	33
S_2	1.670	0.015	LPES	53
SD	2.315	0.002	LPES	10
SF	2.285	0.006	LPES	93
SH	2.314344	0.000004	LPT	47
SO	1.125	0.005	LPES	84
Sb_2	1.282	0.008	LPES	108
Se_2	1.94	0.07	LPES	38
SeH	2.212520	0.000025	LPT	48
SeO	1.456	0.020	LPES	41
Si_2	2.201	0.010	LPES	100
SiH	1.277	0.009	LPES	2
Sn_2	1.962	0.010	LPES	117
SnPb	1.569	0.008	LPES	117
Te_2	1.92	0.07	LPES	38
TeH	2.102	0.015	LPES	39
TeO	1.697	0.022	LPES	40
ZnH	$\leq$0.95	—	PT	2

Table 3
Electron Affinities for Triatomic Molecules

Molecule	Electron affinity in eV	Uncertainty in eV	Method	Ref.
Ag_3	2.32	0.05	LPES	37
Al_3	1.4	0.15	LPES	68
AsH_2	1.27	0.03	PT	2
Au_3	3.7	0.3	LPES	37
BO_2	3.57	0.13	CT	6
BO_2	4.3	0.2	CT	98
Bi_3	1.60	0.03	LPES	119
C_3	1.981	0.020	LPES	11
CCl_2	1.591	0.010	LPES	95
CD_2	0.645	0.006	LPES	12
CDF	0.535	0.005	LPES	95
CF_2	0.165	0.010	LPES	95
CH_2	0.652	0.006	LPES	12
CHBr	1.454	0.005	LPES	95
CHCl	1.210	0.005	LPES	95
CHF	0.542	0.005	LPES	95
CHI	1.42	0.17	LPES	95
C_2H	2.969	0.006	LPES	87
C_2O	1.848	0.027	LPES	13
COS	0.46	0.20	CD	2
CS_2	0.895	0.020	LPES	11
CoD_2	1.465	0.013	LPES	34
CoH_2	1.450	0.014	LPES	34
CrH_2	>2.5	—	LPES	34
Cr_2D	1.464	0.005	LPES	107
Cr_2H	1.474	0.005	LPES	107
Cs_3	0.864	0.030	LPES	18
Cu_3	2.11	0.05	LPES	37
DCO	0.301	0.005	LPES	35
DNO	0.330	0.015	LPES	14
DO_2	1.089	0.017	LPES	15
DS_2	1.912	0.015	LPES	53
HS_2	1.907	0.015	LPES	53
FeCO	1.157	0.005	LPES	103
FeD_2	1.038	0.013	LPES	34
FeH_2	1.049	0.014	LPES	34
GeH_2	1.097	0.015	LPES	28
HCO	0.313	0.005	LPES	35
HCl_2	4.896	0.005	LPES	69
HNO	0.338	0.015	LPES	14
HO_2	1.078	0.017	LPES	15
K_3	0.956	0.050	LPES	18
MnD_2	0.465	0.014	LPES	34
MnH_2	0.444	0.016	LPES	34
N_3	2.70	0.12	PT	2
NCO	3.609	0.005	LPES	111
NCS	3.537	0.005	LPES	111
NH_2	0.771	0.005	LPES	58
N_2O	0.22	0.10	CT	59
NO_2	2.273	0.005	LPES	63
(NO)R	R=Ar,Kr,Xe	—	LPES	90
Na_3	1.019	0.060	LPES	18
Ni_3	1.41	0.05	LPES	55
NiCO	0.804	0.012	LPES	2
NiD_2	1.926	0.007	LPES	34
NiH_2	1.934	0.008	LPES	34
O_3	2.1028	0.0025	LPT	2
O_2Ar	0.52	0.02	LPES	75
OClO	2.140	0.008	LPES	88
OIO	2.577	0.008	LPES	88
PH_2	1.271	0.010	LPES	2
PO_2	3.8	0.2	CT	98
Pt_3	1.87	0.02	LPES	55
Pd_3	<1.5	0.1	LPES	55
Rb_3	0.920	0.030	LPES	18
S_3	2.093	0.025	LPES	16
SO_2	1.107	0.008	LPES	16
S_2O	1.877	0.008	LPES	16
Sb_3	1.85	0.03	LPES	108
SeO_2	1.823	0.050	LPES	38
SiH_2	1.124	0.020	LPES	2
VO_2	2.3	0.2	CT	101

Table 4
Electron Affinities for Larger Polyatomic Molecules

Molecule	Electron Affinity in eV	Uncertainty in eV	Method	Ref.	Name
Ag_n	$n = 1$-10	—	LPES	37	
Al_n	$n = 3$-32	—	LPES	68	
Al_3O	1.00	0.15	LPES	68	
$Ar(H_2O)_n$	$n = 2,6,7$	—	LPES	77	
Au_n	$n = 2$-5	—	LPES	37	
AuF_6	7.5	estimate	CT	98	
BD_3	0.027	0.014	LPES	62	
BH_3	0.038	0.015	LPES	62	
Bi_4	1.05	0.010	LPES	119	

Table 4
Electron Affinities for Larger Polyatomic Molecules (continued)

Molecule	Electron Affinity in eV	Uncertainty in eV	Method	Ref.	Name
C_n	n = 2-84	—	LPES	70	
$(CO_2)_n$	n = 1,2	—	LPES	75	
$(CS)_n$	n = 2	—	LPES	75	
$(CS_2)_n$	n = 1,2	—	LPES	75	
CF_3Br	0.91	0.2	CD	2	
CF_3I	1.57	0.2	CD	2	
$CO_3(H_2O)$	2.1	0.2	PT	2	
CHO_2	3.498	0.005	LPES	109	
CH_2S	0.465	0.023	LPES	53	
CH_3	0.08	0.03	LPES	2	
CH_3I	0.2	0.1	CT	2	
CH_3NO_2	0.48	0.10	CT	61	
CH_3Si	0.852	0.010	LPES	97	CH_3-Si
CH_3Si	2.010	0.010	LPES	97	$CH_2 = SiH$
CD_3O	1.552	0.022	LPES	2	
CH_3O	1.570	0.022	LPES	2	
CD_3S	1.856	0.006	LPT	2	
CH_3S	1.861	0.004	LPT	2	
CD_3S_2	1.748	0.022	LPES	53	
CH_3S_2	1.757	0.022	LPES	53	
CH_3SiH_2	1.19	0.04	LPT	65	
CO_3	2.69	0.14	LPES	2	
C_2F_2	2.255	0.006	LPES	106	difluorovinylidene
C_2DO	2.350	0.020	LPES	13	
C_2HO	2.350	0.020	LPES	13	
C_2D_2	0.492	0.006	LPES	83	vinylidene-d_2
C_2HD	0.489	0.006	LPES	83	vinylidene-d_1
C_2HF	1.718	0.006	LPES	106	monofluorovinylidene
C_2H_2	0.490	0.006	LPES	83	vinylidene
C_2H_2FO	2.22	0.09	PT	2	acetyl fluoride enolate
C_2D_2N	1.538	0.012	LPES	21	cyanomethyl-d_2 radical
C_2D_2N	1.070	0.024	LPES	21	isocyanomethyl-d_2 radical
C_2H_2N	1.543	0.014	LPES	21	cyanomethyl radical
C_2H_2N	1.059	0.024	LPES	21	isocyanomethyl radical
C_2H_3	0.667	0.024	LPES	90	vinyl
C_2D_3O	1.81897	0.00012	LPT	22	acetaldehyde-d_3 enolate
C_2H_3O	1.82476	0.00012	LPT	22	acetaldehyde enolate
C_2H_5N	0.56	0.01	PT	2	ethyl nitrine
C_2D_5O	1.702	0.033	LPES	23	ethoxide-d_3
C_2H_5O	1.726	0.033	LPES	23	ethoxide
C_2H_5S	1.953	0.006	LPT	2	ethyl sulfide
C_2H_5S	0.868	0.051	LPES	53	CH_3SCH_2
$C_2H_7O_2$	2.26	0.08	PT	50	MeOHOMe
C_3H	1.858	0.023	LPES	11	
C_3H_2	1.794	0.025	LPES	11	
$C_3H_2F_3O$	2.625	0.010	LPT	113	1,1,1-trifluoroacetone enolate
C_3H_3	0.893	0.025	LPES	24	propargyl radical
C_3H_2D	0.88	0.15	LPES	24	propargyl-d_1 radical
C_3D_2H	0.907	0.023	LPES	24	propargyl-d_2 radical
C_3H_3N	1.247	0.012	LPES	21	CH_3CH-CN
C_3D_5	0.381	0.025	LPES	25	allyl-d_5
C_3H_5	0.362	0.019	LPES	25	allyl
C_3H_4D	0.373	0.019	LPES	25	allyl-d_1
C_3H_5O	1.758	0.019	LPT	113	acetone enolate
C_3H_5O	1.621	0.006	LPT	113	propionaldehyde enolate

Table 4
Electron Affinities for Larger Polyatomic Molecules (continued)

Molecule	Electron Affinity in eV	Uncertainty in eV	Method	Ref.	Name
$C_3H_5O_2$	1.80	0.06	PT	2	methyl acetate enolate
C_3H_7O	1.789	0.033	LPES	23	*n*-propyl oxide
C_3H_7O	1.839	0.029	LPES	23	isopropyl oxide
C_3H_7S	2.00	0.02	PT	2	*n*-propyl sulfide
C_3H_7S	2.02	0.02	PT	2	isopropyl sulfide
C_3O	1.34	0.15	LPES	11	
C_3O_2	0.85	0.15	LPES	11	
$C_4F_4O_3$	0.5	0.2	CD	2	tetrafluorosuccinic anhydride
$C_4H_2O_3$	1.44	0.10	CT	61	maleic anhydride
C_4H_4N	2.39	0.13	PT	2	pyrrolate
C_4H_5O	1.801	0.008	LPT	113	cyclobutanone enolate
$C_4H_6O_2$	0.69	0.10	CT	61	2,3-butanedione
C_4H_7O	1.67	0.05	PT	2	butyraldehyde enolate
C_4H_5DO	1.67	0.05	PT	2	2-butanone-3-d_1 enolate
$C_4H_5D_2O$	1.75	0.06	PT	2	2-butanone-3,3-d_2 enolate
C_4H_9O	1.912	0.054	LPES	23	*t*-butoxyl
C_4H_9S	2.03	0.02	PT	2	*n*-butyl sulfide
C_4H_9S	2.07	0.02	PT	2	*t*-butyl sulfide
C_4O	2.05	0.15	LPES	11	
C_4O_2	2.0	0.2	LPES	11	
C_5	2.853	0.001	LPT	99	
C_5F_5N	0.68	0.11	CT	67	pentafluoropyridine
$C_5F_6O_3$	1.5	0.2	CD	2	hexafluoroglutaric anhydride
C_5D_5	1.790	0.008	LPES	11	cyclopentadienyl-d_5
C_5H_5	1.804	0.007	LPES	11	cyclopentadienyl
C_5H_7	0.91	0.03	PT	2	pentadienyl
C_5H_7O	1.598	0.007	LPT	113	cyclopentanone enolate
C_5H_9O	1.69	0.05	PT	2	3-penanone enolate
$C_5H_{11}O$	1.93	0.05	LPT	2	neopentoxyl
$C_5H_{11}S$	2.09	0.02	PT	2	*n*-pentyl sulfide
C_5O_2	1.2	0.2	LPES	11	
C_6	4.180	0.001	LPT	8	
$C_6Br_4O_2$	2.44	0.20	CT	2	tetrabromo-BQ
$C_6Cl_4O_2$	2.78	0.10	CT	61	tetrachloro-BQ
$C_6F_4O_2$	2.70	0.10	CT	61	tetrafluoro-BQ
C_6F_5Br	1.15	0.11	CT	67	pentafluorobromobenzene
C_6F_5Cl	0.82	0.11	CT	67	pentafluorochlorobenzene
C_6F_5I	1.41	0.11	CT	67	pentafluoroiodobenzene
$C_6F_5NO_2$	1.52	0.11	CT	67	pentafluoro-NB
C_6F_6	0.52	0.10	CT	51	hexafluorobenzene
C_6F_{10}	>1.4	0.3	CT	2	perfluorocyclohexane
$C_6H_2Cl_2O_2$	2.48	0.10	CT	61	2,6-dichloro-BQ
$C_6H_3F_2NO_2$	1.17	0.10	CT	61	2,4-difluoro-NB
C_6D_4	0.551	0.010	LPES	36	*o*-benzyne-d_4
C_6H_4	0.560	0.010	LPES	36	*o*-benzyne
$C_6H_4BrNO_2$	1.16	0.10	CT	61	*o*-bromo-NB
$C_6H_4BrNO_2$	1.32	0.10	CT	61	*m*-bromo-NB
$C_6H_4BrNO_2$	1.29	0.10	CT	61	*p*-bromo-NB
$C_6H_4ClNO_2$	1.14	0.10	CT	61	*o*-chloro-NB
$C_6H_4ClNO_2$	1.28	0.10	CT	61	*m*-chloro-NB
$C_6H_4ClNO_2$	1.26	0.10	CT	61	*p*-chloro-NB
C_6H_4ClO	≤2.58	0.08	PT	2	*o*-chloroperoxide
$C_6H_4FNO_2$	1.07	0.10	CT	61	*o*-fluoro-NB
$C_6H_4FNO_2$	1.23	0.10	CT	61	*m*-fluoro-NB
$C_6H_4FNO_2$	1.12	0.10	CT	61	*p*-fluoro-NB

Table 4
Electron Affinities for Larger Polyatomic Molecules (continued)

Molecule	Electron Affinity in eV	Uncertainty in eV	Method	Ref.	Name
$C_6H_4N_2O_4$	1.65	0.10	CT	61	*o*-diNB
$C_6H_4N_2O_4$	1.65	0.10	CT	61	*m*-diNB
$C_6H_4N_2O_4$	2.00	0.10	CT	61	*p*-diNB
$C_6H_4O_2$	1.91	0.10	CT	61	1,4-benzoquinone (BQ)
C_6D_5	1.092	0.020	LPES	26	phenyl-d_5
C_6D_5N	1.44	0.02	LPES	96	phenylnitrene-d_5
C_6H_5	1.096	0.006	LPES	26	phenyl
C_6H_5N	1.429	0.011	LPT	115	phenylnitrene
$C_6H_5NO_2$	1.01	0.10	CT	61	nitrobenzene (NB)
C_6H_5O	2.253	0.006	LPES	26	phenoxyl
C_6H_5S	≤2.47	0.06	PT	2	thiophenoxide
C_6H_5NH	1.70	0.03	PT	2	anilide
C_6H_7	<1.67	0.04	PT	2	methylchylopentadienyl
C_6H_8Si	1.435	0.004	LPT	65	$C_6H_5SiH_3$
C_6H_9O	1.526	0.010	LPT	113	cyclohexanone enolate
$C_6H_{11}O$	1.755	+0.05/-0.005	LPT	113	pinacolone enolate
$C_6H_{11}O$	1.82	0.06	PT	2	3,3-dimethylbutananl enolate
C_6N_4	2.3	0.3	PT	2	TCNE
C_7F_5N	1.11	0.11	CT	67	pentafluorobenzonitrile
C_7F_8	0.86	0.11	CT	67	octafluorotoluene
C_7F_{14}	1.08	0.10	CT	61	perfluoromethylcyclohexane
C_7HF_5O	1.10	0.11	CT	67	pentafluorobenzaldehyde
$C_7H_3N_3O_4$	2.16	0.10	CT	61	3,5-$(NO_2)_2$-benzonitrile
$C_7H_4F_3NO_2$	1.41	0.10	CT	61	*m*-trifluoromethyl-NB
$C_7H_4N_2O_2$	1.61	0.10	CT	61	*o*-cyano-NB
$C_7H_4N_2O_2$	1.56	0.10	CT	61	*m*-cyno-NB
$C_7H_4N_2O_2$	1.72	0.10	CT	61	*p*-cyano-NB
C_7H_6FO	2.218	0.010	LPT	2	*m*-fluoroacetophenone enolate
C_7H_6FO	2.176	0.010	LPT	2	*p*-fluoroacetophenone enolate
$C_7H_6FeO_3$	0.990	0.10	CT	120	η^4-1,3-butadiene-$Fe(CO)_3$
$C_7H_6N_2O_4$	1.77	0.05	PT	60	3,4-dintrotoluene
$C_7H_6N_2O_4$	1.77	0.05	PT	60	2,3-dinitrotoluene
$C_7H_6N_2O_4$	1.60	0.05	PT	60	2,4-dinitrotoluene
$C_7H_6N_2O_4$	1.55	0.05	PT	60	2,6-dinitrotoluene
C_7H_7	0.912	0.006	LPES	26	benzyl
$C_7H_7NO_2$	0.92	0.10	CT	61	*o*-methyl-NB
$C_7H_7NO_2$	0.99	0.10	CT	61	*m*-methyl-NB
$C_7H_7NO_2$	0.95	0.10	CT	61	*p*-methyl-NB
$C_7H_7NO_3$	1.04	0.10	CT	61	*m*-OCH_3-NB
$C_7H_7NO_3$	0.91	0.10	CT	61	*p*-OCH_3-NB
C_7H_7O	≤2.36	0.06	PT	2	*o*-methyl phenoxide
C_7H_7O	2.14	0.02	PT	50	benzyloxide
$C_7H_7O_2$	1.85	0.10	CT	61	*o*-CH_3-BQ
C_7H_8FO	<3.05	0.06	PT	50	$PhCH_2OHF$
C_7H_9	1.27	0.03	PT	2	heptatrienyl
C_7H_9O	1.61	0.05	PT	2	2-norbornanone enolate
C_7H_9Si	1.33	0.04	LPT	65	$C_6H_5(CH_3)SiH$
$C_7H_{11}O$	1.598	0.007	LPT	113	cycloheptanone enolate
$C_7H_{11}O$	1.49	0.04	PT	2	2,5-dimethylcyclopentanone enolate
$C_7H_{13}O$	1.72	0.06	PT	2	4-heptanone enolate
$C_7H_{13}O$	1.46	0.04	PT	2	di-isopropyl ketone enolate
$C_8F_{14}N_2$	1.89	0.10	CT	51	1,4-$(CN)_2C_6F_4$
$C_8H_3F_5O$	0.88	0.11	CT	67	pentafluoroacetophenone
$C_8H_3F_6NO_2$	1.79	0.10	CT	61	3,5-$(CF_3)_2$-NB
$C_8H_4O_3$	1.21	0.10	CT	61	phthalic anhydride

Table 4
Electron Affinities for Larger Polyatomic Molecules (continued)

Molecule	Electron Affinity in eV	Uncertainty in eV	Method	Ref.	Name
C_8H_7O	2.057	0.010	PT	2	acetophenone enolate
C_8H_7O	2.10	0.08	LPT	2	phenylacetaldehyde enolate
$C_8H_9NO_2$	1.21	0.05	PT	60	3,5-dimethyl-NB
$C_8H_9NO_2$	2.61	0.05	PT	60	2,6-dimethyl-NB
$C_8H_9NO_2$	0.86	0.10	CT	61	2,3-dimethyl-NB
$C_8H_{13}O$	1.63	0.06	PT	2	cyclooctanone enolate
$C_9H_8FeO_3$	0.76	0.10	CT	120	η^4-1,3-cyclohexadiene-Fe(CO)$_3$
C_9H_9O	2.030	0.010	LPT	2	*m*-methylacetophenone enolate
C_9H_9SiN	1.43	0.10	PT	2	trimethylsilylnitrene
$C_9H_{11}NO_2$	0.70	0.10	CT	61	2,4,6-trimethyl NB
$C_9H_{15}O$	1.69	0.06	PT	2	cyclononanone enolate
$C_{10}H_4Cl_2O_2$	2.19	0.10	CT	61	2,3-dichloro-1,4-naphthoquinone
$C_{10}H_6N_2O_4$	1.78	0.10	CT	61	1,3-dinitronaphthalene
$C_{10}H_6N_2O_4$	1.77	0.10	CT	61	1,5-dinitronaphthalene
$C_{10}H_6O_2$	1.81	0.10	CT	61	1,4 naphthoquinone
$C_{10}H_7NO_2$	1.23	0.10	CT	61	1-nitronaphthalene
$C_{10}H_7NO_2$	1.18	0.10	CT	61	2-nitronaphthalene
$C_{10}H_8$	0.69	0.10	CT	61	azulene
$C_{10}H_8CrO_3$	0.93	0.10	CT	120	η^4-1,3,5-cycloheptatriene Cr(CO)$_3$
$C_{10}H_8FeO_3$	0.98	0.10	CT	120	η^4-1,3,5-cycloheptatriene-Fe(CO)$_3$
$C_{10}H_{17}O$	1.83	0.06	PT	2	cyclodecanone enolate
$C_{11}H_8FeO_3$	1.29	0.10	CT	120	η^4-1,3-butadiene-Fe(CO)$_3$
$C_{12}F_{10}$	0.82	0.11	CT	67	decafluorobiphenyl
$C_{12}H_4N_4$	2.8	0.3	CD	2	TCNQ
$C_{12}H_9$	1.07	0.10	PT	2	perinaphthenyl
$C_{12}H_{15}O$	2.032	0.010	LPT	2	*t*-butylacetophenone enolate
$C_{12}H_{21}O$	1.90	0.07	PT	2	cyclododecanone enolate
$C_{13}F_{10}O$	1.52	0.11	CT	67	decafluorobenzophenone
$C_{13}H_9F$	0.64	0.10	CT	61	4-fluorobenzophenone
$C_{13}H_{10}O$	0.62	0.10	CT	61	benzophenone
$C_{14}H_9NO_2$	1.43	0.10	CT	61	9-nitroanthracene
$C_{14}H_{10}$	0.57	0.10	CT	66	anthracene
$C_{18}H_{12}$	1.04	0.10	CT	66	tetracene
$C_{20}H_{12}$	0.79	0.10	CT	66	benz[*a*]pyrene
$C_{20}H_{12}$	0.97	0.10	CT	66	perylene
$C_{22}H_{14}$	1.35	0.10	CT	66	pentacene
CeF_4	3.8	0.4	CT	98	
CoF_4	6.4	0.3	CT	98	
$Cr(CO)_3$	1.349	0.006	LPES	94	
CrO_3	3.6	0.2	CT	98	
Cu_n	n = 1-41	—	LPES	37	
$Fe(CO)_2$	1.22	0.02	LPES	2	
$Fe(CO)_3$	1.8	0.2	LPES	2	
$Fe(CO)_4$	2.4	0.3	LPES	2	
FeF_3	3.6	0.1	CT	98	
FeF_4	6.0	estimate	CT	98	
Ge_n	n = 3-15	—	LPES	71	
Ge_xAs_y	n = 5-30	$n = x + y$	LPES	72	
GeH_3	≤1.74	0.04	PT	2	
$(NH_3)_n$	n = 41-1100	—	LPES	77	
$H(NH_3)_n$	n = 1,2	—	LPES	76	
HNO_3	0.57	0.15	CD	2	
$(H_2O)_n$	n = 2-19	—	LPES	77	
IrF_4	4.7	0.3	CT	98	
IrF_6	6.5	0.4	CT	98	

Table 4
Electron Affinities for Larger Polyatomic Molecules (continued)

Molecule	Electron Affinity in eV	Uncertainty in eV	Method	Ref.	Name
K_n	n = 2-7	—	LPES	18	
MnF_4	5.5	0.2	CT	98	
$Mo(CO)_3$	1.337	0.006	LPES	94	
MoF_5	3.5	0.2	CT	98	
MoF_6	3.8	0.2	CT	98	
MoO_3	2.9	0.2	CT	98	
$NH_2(NH_3)_n$	n = 1,2	—	LPES	78	
$NO(H_2O)_n$	n = 1,2	—	LPES	75	
NO_3	3.937	0.014	LPES	85	
$NO(N_2O)_n$	n = 1,2	—	LPES	79	
$(NO)_2$	≥2.1	—	LPES	75	
$(N_2O)_n$	n = 1,2	—	LPES	81	
Na_n	n = 2-5	—	LPES	18	
$(NaF)_n$	n = 1-7,12	—	LPES	64	
$Na(NaF)_n$	n = 5,7-12	—	LPES	64	
$Ni(CO)_2$	0.643	0.014	LPES	2	
$Ni(CO)_3$	1.077	0.013	LPES	2	
$OH(H_2O)$	<2.95	0.15	PT	2	
OsF_4	3.9	0.3	CT	98	
OsF_6	6.0	0.3	CT	98	
PBr_3	1.59	0.15	CD	2	
PBr_2Cl	1.63	0.20	CD	2	
PCl_2Br	1.52	0.20	CD	2	
PCl_3	0.82	0.10	CD	2	
PO_3	4.5	0.5	CT	98	
$POCl_2$	3.83	0.25	CD	2	
$POCl_3$	1.41	0.20	CD	2	
PtF_4	5.5	0.3	CT	98	
PtF_6	7.0	0.4	CT	98	
ReF_6	4.7	estimate	CT	98	
RhF_4	5.4	0.3	CT	98	
RuF_4	4.8	0.3	CT	98	
RuF_5	5.2	0.4	CT	98	
RuF_6	7.5	0.3	CT	98	
SF_4	1.5	0.2	CT	91	
SF_6	1.05	0.10	CT	56	
SO_3	≥1.70	0.15	CD	2	
$(SO_2)_2$	0.6	0.2	LPES	80	
SeF_6	2.9	0.2	CD	2	
Si_4	2.17	0.01	LPES	110	
Si_n	n = 3-20	—	LPES	71	
SiD_3	1.386	0.022	LPES	43	
SiF_3	≤2.95	0.10	PT	17	
SiH_3	1.406	0.014	LPES	43	
TeF_6	3.34	0.17	CD	2	
UF_5	3.7	0.2	CT	98	
UF_6	5.1	0.2	CT	98	
UO_3	<2.1	—	CT	98	
VF_4	3.5	0.2	CT	98	
V_4O_{10}	4.2	0.6	CT	101	
$W(CO)_3$	1.859	0.006	LPES	94	
WF_5	1.25	0.3	CD	18	
WF_6	3.36	+0.04/-0.20	CT	19	
WO_3	3.33	+0.04/-0.15	LPT	86	
WO_3	3.9	0.2	CT	98	

REFERENCES

1. Hotop, H., and Lineberger, W. C., *J. Phys. Chem. Ref. Data*, 14, 731, 1985.
2. Drzaic, P. S., Marks, J., and Brauman, J. I., in *Gas Phase Ion Chemistry*, Vol. 3, Bowers, M. T., Ed., Academic Press, Orlando, 1984, p. 167.
3. Schulz, P. A., Mead, R. D., Jones, P. L., and Lineberger, W. C., *J. Chem. Phys.*, 77, 1153, 1982.
4. Neumark, D. M., Lykke, K. R., Anderson, T., and Lineberger, W. C., *Phys. Rev. A*, 32, 1890, 1985.
5. Leopold, D. G., Murray, K. K., Miller, T. M., and Lineberger, W. C., unpublished data.
6. Srivastava, R. D., Uy, O. M., and Farber, M., *Trans. Faraday Soc.*, 67, 2941, 1971.
7. Klein, R., McGinnis, R. P., and Leone, S. R., *Chem. Phys. Lett.*, 100, 475, 1983.
8. Arnold, C. C., Zhao, Y., Kitsopoulos, T. N., and Neumark, D. M., *J. Chem. Phys.*, 97, 6121, 1992; linear C_6^- to linear C_6.
9. Stevens, A. E., Fiegerle, C. S., and Lineberger, W. C., *J. Chem. Phys.*, 78, 5420, 1983.
10. Breyer, F., Frey, P., and Hotop, H., *Z. Phys.*, A 300, 7, 1981.
11. Oakes, J. M., and Ellison, G. B., *Tetrahedron*, 42, 6263, 1986.
12. Leopold, D. G., Murray, K. K., Miller, A. E. S., and Lineberger, W. C., *J. Chem. Phys.*, 83, 4849, 1985.
13. Oakes, J. M., Jones, M.E., Bierbaum, V. M., and Ellison, G. B., *J. Chem. Phys.*, 87, 4810, 1983.
14. Ellis, H. B., Jr. and Ellison, G. B., *J. Chem. Phys.*, 78, 6541, 1983.
15. Oakes, J. M., Harding, L. B., and Ellison, G. B., *J. Chem. Phys.*, 83, 5400, 1985.
16. Nimlos, M. E., and Ellison, G. B., *J. Chem. Phys.*, 90, 2574, 1986.
17. Richardson, L. M., Stephenson, L. M., and Brauman, J. I., *Chem. Phys. Lett.*, 30, 17, 1975.
18. McHugh, K. M., Eaton, J. G., Lee, G. H., Sarkas, H. W., Kidder, L. H., Snodgrass, J. T., Manaa, M. R., and Bowen, K. H., *J. Chem. Phys.*, 91, 3792, 1989. See also Ref. 104.
19. Viggiano, A. A., Paulson, J. F., Dale, F., Henchman, M., Adams, N. G., and Smith, D., *J. Phys. Chem.*, 89, 2264, 1985.
20. Burnett, S. M., Stevens, A. E., Fiegerle, C. S., and Lineberger, W. C., *Chem. Phys. Lett.*, 100, 124, 1983.
21. Moran, S., Ellis, H. B., DeFrees, D. J., McLean, A. D., and Ellison, G. B., *J. Am. Chem. Soc.*, 109, 5996, 1987; Moran, S., Ellis, H. B., DeFrees, D. J., McLean, A. D., Paulson, S. E., and Ellison, G. B., *J. Am. Chem. Soc.*, 109, 6004, 1987; see also Lykke, K. R., Neumark, D. M., Andersen, T., Trapa, V. J., and Lineberger, W. C., *J. Chem. Phys.*, 87, 6842, 1987.
22. Mead, R. D., Lykke, K. R., Lineberger, W. C., Marks, J., and Brauman, J. I., *J. Chem. Phys.*, 81, 4883, 1984; Lykke, K. R., Mead, R. D., and Lineberger, W. C., *Phys. Rev. Lett.*, 52, 2221, 1984. The EAs are 14717.7 ± 1.0 cm^{-1} for actaldehyde enolate and 14671.0 ± 1.0 cm^{-1} for acetaldehyde-d_3 enolate.
23. Ellison, G. B., Engelking, P. C., and Lineberger, W. C., *J. Chem. Phys.*, 86, 4873, 1982.
24. Oakes, J. M., and Ellison, G. B., *J. Am. Chem. Soc.*, 105, 2969, 1983.
25. Ellison, G. B., and Oakes, J. M., *J. Am. Chem. Soc.*, 106, 7734, 1984.
26. Gunion, R. F., Gilles, M. K., Polak, M. L., and Lineberger, W. C., *Int. J. Mass Spectrom. Ion Processes*, 117, 601, 1992; also, Miller, A. E. S., and Lineberger, W. C., unpublished.
27. Leopold, D. G., and Lineberger, W. C., *J. Chem. Phys.*, 85, 51, 1986.
28. Miller, T. M., Miller, A. E. S., and Lineberger, W. C., *Phys. Rev. A*, 33, 3558, 1986.
29. Miller, A. E. S., Fiegerle, C. S., and Lineberger, W. C., *J. Chem. Phys.*, 87, 1549, 1987.
30. Miller, T. M., Leopold, D. G., Murray, K. K., and Lineberger, W. C., *J. Chem. Phys.*,85, 2368, 1986.
31. Miller, T. M., and Lineberger, W. C., *Chem. Phys. Lett.*, 146, 364, 1988.
32. Neumark, D. M., Lykke, K. R., Andersen, T., and Lineberger, W. C., *J. Chem. Phys.*, 83, 4364, 1985.
33. Leopold, D. G., Miller, T. M., and Lineberger, W. C., *J. Am. Chem. Soc.*, 108, 178, 1986.
34. Miller, A. E. S., Fiegerle, C. S., and Lineberger, W. C., *J. Chem. Phys.*, 84, 4127, 1986.
35. Murray, K. K., Miller, T. M., Leopold, D. G., and Lineberger, W. C., *J. Chem. Phys.*, 84, 2520, 1986.
36. Leopold, D. G., Miller, A. E. S., and Lineberger, W. C., *J. Am. Chem. Soc.*, 108, 1379, 1986.
37. Ho, J., Ervin, K. M., and Lineberger, W. C., *J. Chem. Phys.*, 93, 6987, 1990; Leopold, D. G., Ho, J., and Lineberger, W. C., *J. Chem. Phys.*, 86, 1715, 1987; Pettiette, C. L., Yang, S. H., Craycraft, M. J., Conceicao, J., Laaksonen, R. T., Cheshnovsky, O., and Smalley, R. E., *J. Chem. Phys.*, 88, 5377, 1988.
38. Snodgrass, J. T., Coe, J. V., McHugh, K. M., Friedhoff, C. B., and Bowen, K. H., *J. Phys. Chem.*, 93, 1249, 1989
39. Friedhoff, C. B., Snodgrass, J. T., Coe, J. V., McHugh, K. M., and Bowen, K. H., *J. Chem. Phys.*,84, 1051, 1986.
40. Friedhoff, C. B., Coe, J. V., Snodgrass, J. T., McHugh, K. M., and Bowen, K. H., *Chem. Phys. Lett.*, 124, 268, 1986.
41. Coe, J. V., Snodgrass, J. T., Friedhoff, C. B., McHugh, K. M., and Bowen, K. H., *J. Chem. Phys.*, 84, 619, 1986.
42. Snodgrass, J. T., Coe, J. V., Friedhoff, C. B., McHugh, K. M., and Bowen, K. H., *Chem. Phys. Lett.*, 122, 352, 1985.
43. Nimlos, M. R., and Ellison, G. B., *J. Am. Chem. Soc.*, 108, 6522, 1986.
44. Peterson, J. R., *Aust. J. Phys.*, 45, 293, 1992. See also Nadeau, M.-J., Zhao, X.-L., Garwan, M. A., and Litherland, A. E., *Phys. Rev. A*, 46, R3588, 1992; also, Fabrikant, I. I., *J. Phys. B: At. Mol. Opt. Phys.*, 26, 2533, 1993.
45. Andersen, T., Lykke, K. R., Neumark, D. M., and Lineberger, W. C., *J. Chem. Phys.*, 86, 1858, 1987.
46. Murray, K. K., Lykke, K. R., and Lineberger, W. C., *Phys. Rev. A*, 36, 699, 1987.
47. Mansour, N. B., and Larson, D. J., *Abstracts of the XV Int. Conf. on the Phys. of Electronic and Atomic Collisions*, p. 70, 1987.
48. Stonemann, R. C., and Larson, D. J., *Phys. Rev. A*, 35, 2928, 1987.
49. Nimlos, M. R., Harding, L. B., and Ellison, G. B., *J. Chem. Phys.*, 87, 5116, 1987.
50. Moylan, C. R., Dodd, J. A., Han, C.-C., and Braumann, J. I., *J. Chem. Phys.*, 86, 5350, 1987.
51. Chowdhury, S., Grimsrud, E. P., Heinis, T., and Kebarle, P., *J. Am. Chem. Soc.*, 108, 3630, 1986.
52. Trainham, R., Fletcher, G. D., and Larson, D. J., *J. Phys. B: At. Mol. Phys.*, 20, L777, 1987.

53. Moran, S., and Ellison, G. B., *J. Phys. Chem.*, 92, 1794, 1988.
54. Murray, K. K., Leopold, D. G., Miller, T. M., and Lineberger, W. C., *J. Chem. Phys.*, 89, 5442, 1988.
55. Ervin, K. M., Ho, J., and Lineberger, W. C., *J. Chem. Phys.*, 89, 4514, 1988.
56. Grimsrud, E. P., Chowdhury, S., and Kebarle, P., *J. Chem. Phys.*, 85, 4989, 1985.
57. Fischer, C. F., *Phys. Rev. A*, 39, 963, 1989. The calculated EA for Ca is larger than the experimental value, so the theoretical values for Sr and Ba may also be too large.
58. Wickham-Jones, C. T., Ervin, K. M., Ellision, G. B., and Lineberger, W. C., *J. Chem. Phys.*, 91, 2762, 1989.
59. Hopper, D. G., Wahl, A. C., Wu, R. L. C., and Tiernan, T. O., *J. Chem. Phys.*, 65, 5474, 1976.
60. Mock, R. S., and Grimsrud, E. P., *J. Am. Chem. Soc.*, 111, 2861, 1989.
61. Chowdhury, S., Heinis, T., Grimsrud, E. P., and Kebarle, P., *J. Phys. Chem.*, 90, 2747, 1986. The uncertainty and other results are quoted in Ref. 60.
62. Wickham-Jones, C. T., Moran, S., and Ellison, G. B., *J. Chem. Phys.*, 90, 795, 1989.
63. Ervin, K. M., Ho, J., and Lineberger, W. C., *J. Phys. Chem.*, 92, 5405, 1988.
64. Miller, T. M., and Lineberger, W. C., *Int. J. Mass Spectrom. Ion Processes*, 102, 239, 1990.
65. Wetzel, D. M., Salomon, K. E., Berger, S., and Brauman, J. I., *J. Am. Chem. Soc.*, 111, 3835, 1989.
66. Crocker, L., Wang, T., and Kebarle, P., *J. Am. Chem. Soc.*, 115, 7818, 1993.
67. Dillow, G. W., and Kebarle, P., *J. Am. Chem. Soc.*, 111, 5592, 1989.
68. Gantefor, G., Gausa, M., Meiwes-Broer, K. H., and Lutz, H. O., *Z. Phys. D*, 9, 253, 1988; Taylor, K. J., Petteitte, C. L., Craycraft, M. J., Chesnovsky, O., and Smalley, R. E., *Chem. Phys. Lett.*, 152, 347, 1988.
69. Metz, R. B., Kitsopoulos, T., Weaver, A., and Neumark, D. M., *J. Chem. Phys.*, 88, 1463, 1988.
70. Yang, S., Pettiette, C. L., Conceicao, J., Cheshnovsky, O., and Smalley, R. E., *Chem. Phys. Lett.*, 139, 233, 1987; Yang, S., Taylor, K. J., Craycraft, M. J., Conceicao, J., Pettiette, C. L., Cheshnovsky, O., and Smalley, R. E., *Chem. Phys. Lett.*, 144, 431, 1988; Arnold, D. W., Bradforth, S. E., Kitsopoulos, T. N., and Neumark, D. M., *J. Chem. Phys.*, 95, 5479, 1991.
71. Cheshnovsky, O., Yang, S., Pettiette, C. L., Craycraft, M. J., Liu, Y., and Smalley, R. E., *Chem. Phys. Lett.*, 138, 119, 1987.
73. Travers, M. J., Cowles, D. C., and Ellison, G. B., *Chem. Phys. Lett.*, 164, 449, 1989.
74. Blondel, C., Cacciani, P., Delsart, C., and Trainham, R., *Phys. Rev. A*, 40, 3698, 1989.
75. Bowen, K. H., and Eaton, J. G., in *The Structure of Small Molecules and Ions*, Naaman, R., and Vager, Z., Eds., Plenum, New York, 1988, pp. 147-169; Arnold, S. T., Eaton, J. G., Patel-Mistra, D., Sarkas, H. W., and Bowen, K. H., in *Ion and Cluster Ion Spectroscopy and Structure*, Maier, J. P., Ed., Elsevier Science, New York, 1989, p. 417.
76. Snodgrass, J. T., Coe, J. V., Friedhoff, C. B., McHugh, K. M., and Bowen, K. H., *Faraday Disc. Chem. Soc.*, 88, 1988.
77. Lee, G. H., Arnold, S. T., Eaton, J. G., Sarkas, H. W., Bowen, K. H., Ludewigt, C., and Haberland, H., *Z. Phys. D - Atoms, Mol. and Clusters*, 20, 9, 1991; Coe, J. V., Lee, G. H., Eaton, J. G., Arnold, S. T., Sarkas, H. W., Bowen, K. H., Ludewigt, C., Haberland, H., and Worsnop, D. R., *J. Chem. Phys.*, 92, 3980, 1990.
78. Snodgrass, J. T., Coe, J. V., Freidhoff, C. B., McHugh, K. M., and Bowen, K. H., to be published, quoted in Ref. 75.
79. Coe, J. V., Snodgrass, J. T., Friedhoff, C. B., McHugh, K. M., and Bowen, K. H., *J. Chem. Phys.*, 87, 4302, 1987.
80. Friedhoff, C. B., Snodgrass, J. T., and Bowen, K. H., to be published, quoted in Ref. 75.
81. Coe, J. V., Snodgrass, J. T., Friedhoff, C. B., McHugh, K. M., and Bowen, K. H., *Chem. Phys. Lett.*, 124, 274, 1986.
82. Greene, C. H., *Phys. Rev. A*, 42, 1405, 1990.
83. Ervin, K. M., Ho, J., and Lineberger, W. C., *J. Chem. Phys.*, 91, 5974, 1989.
84. Polak, M. L., Fiala, B. L., Ervin, K. M., and Lineberger, W. C., *J. Chem. Phys.*, 94, 6924, 1991.
85. Weaver, A., Arnold, D. W., Bradforth, S. E., Neumark, D. M., *J. Chem. Phys.*, 94, 1740, 1991.
86. Walter, C. W., Devynck, P., Hertzler, C. F., Bae, Y. K., Smith, G. P., and Peterson, J. R., *Bull. Am. Phys. Soc.*, 35, 1163, 1990.
87. Ervin, K. M., and Lineberger, W. C., *J. Phys. Chem.*, 95, 1167, 1991.
88. Gilles, M. K., Polak, M. L., and Lineberger, W. C., *J. Chem. Phys.*, 96, 8012, 1992.
89. Lykke, K. R., Murray, K. K., and Lineberger, W. C., *Phys. Rev. A*, 43, 6104, 1991.
90. Ervin, K. M., Gronert, S., Barlow, S. E., Gilles, M. K., Harrison, A. G., Bierbaum, V. M., DePuy, C. H., Lineberger, W. C., and Ellison, G. B., *J. Am. Chem. Soc.*, 112, 5750, 1990.
91. Viggiano, A. A., Miller, T. M., Miller, A. E. S., Morris, R. A., Van Doren, J. M., and Paulson, J. F., *Int. J. Mass Spectrom. Ion Processes*, 109, 327, 1991.
92. Hanstorp, D., and Gustafsson, M., *J. Phys. B: At. Mol. Opt. Phys.*, 25, 1773, 1992.
93. Polak, M. L., Gilles, M. K., and Lineberger, W. C., *J. Chem. Phys.*, 96, 7191, 1992.
94. Bengali, A. A., Casey, S. M., Cheng, C.-L., Dick, J. P., Fenn, P. T., Villalta, P. W., and Leopold, D. G., *J. Am. Chem. Soc.*, 114, 5257, 1992.
95. Gilles, M. K., Ervin, K. M., Ho, J., and Lineberger, W. C., *J. Phys. Chem.*, 96, 1130, 1992.
96. Travers, M. J., Cowles, D. C., Clifford, E. P., and Ellison, G. B., *J. Am. Chem. Soc.*, 114, 8699, 1992.
97. Bengali, A. A., and Leopold, D. G., *J. Am. Chem. Soc.*, 114, 9192, 1992.
98. Rudnyi, E. B., Kaibicheva, E. A., and Sidorov, L. N., *Rapid Comm. in Mass Spectrom.*, 6, 356, 1992; Sidorov, L. N., *High Temp. Sci.*, 29, 153, 1990.
99. Kitsopoulos, T. N., Chick, C. J., Zhao, Y., and Neumark, D. M., *J. Chem. Phys.*, 95, 5479, 1991.
100. Arnold, C. C., Kitsopoulos, T. N., and Neumark, D. M., *J. Chem. Phys.*, 99, 766, 1993.
101. Rudnyi, E. B., Kaibicheva, E. A., and Sidorov, L. N., *J. Chem. Thermodynamics*, 25, 929, 1993.
102. Sarkas, H. W., Hendricks, J. H., Arnold, S. T., and Bowen, K. H., *J. Chem. Phys.* 100, 1884, 1994.
103. Villalta, P. W., and Leopold, D. G., *J. Chem. Phys.* 98, 7730, 1993.
104. Eaton, J. G., Sarkas, H. W., Arnold, S. T., McHugh, K. M., and Bowen, K. H., *Chem. Phys. Lett.*, 193, 141, 1992. See also Ref. 18.

105. Polak, M. L., Gilles, M. K., Gunion, R. F., and Lineberger, W. C., *Chem. Phys. Lett.*, 210, 55, 1993.
106. Gilles, M. K., Lineberger, W. C., and Ervin, K. M., *J. Am. Chem. Soc.*, 115, 1031, 1993.
107. Casey, S. M., and Leopold, D. G., *Chem. Phys. Lett.*, 201, 205, 1993.
108. Polak, M. L., Gerber, G., Ho, J., and Lineberger, W. C., *J. Chem. Phys.*, 97, 8990, 1992.
109. Neumark, D. M., private communication.
110. Arnold, C. C., and Neumark, D. M., *J. Chem. Phys.*, 99, 3353, 1993.
111. Bradforth, S. E., Kim, E. H., Arnold, D. W., and Neumark, D. M., *J. Chem. Phys.*, 98, 800, 1993.
112. Ho, J., Polak, M. L., Ervin, K. M., and Lineberger, W. C., *J. Chem. Phys.*, 99, 8542, 1993.
113. Brinkman, E. A., Berger, S., Marks, J., and Brauman, J. I., *J. Chem. Phys.*, 99, 7586, 1993.
114. Casey, S. M., and Leopold, D. G., *J. Phys. Chem.*, 97, 816, 1993.
115. McDonald, R. N., and Davidson, S. J., *J. Am. Chem. Soc.*, 115, 10857, 1993.
116. Ho, J., Ervin, K. M., Polak, M. L., Gilles, M. K., and Lineberger, W. C., *J. Chem. Phys.*, 95, 4845, 1991.
117. Ho, J., Polak, M. L., and Lineberger, W. C., *J. Chem. Phys.*, 96, 144, 1992.
118. Polak, M. L., Gilles, M. K., Ho, J., and Lineberger, W. C., *J. Phys. Chem.*, 95, 3460, 1991.
119. Polak, M. L., Ho, J., Gerber, G., and Lineberger, W. C., *J. Chem. Phys.*, 95, 3053, 1991.
120. Sharpe, P., and Kebarle, P., *J. Am. Chem. Soc.*, 115, 782, 1993.

ATOMIC AND MOLECULAR POLARIZABILITIES

Thomas M. Miller

The *polarizability* of an atom or molecule describes the response of the electron cloud to an external field. The atomic or molecular energy shift ΔW due to an external electric field E is proportional to E^2 for external fields which are weak compared to the internal electric fields between the nucleus and electron cloud. The *electric dipole polarizability* α is the constant of proportionality defined by $\Delta W = -\alpha E^2/2$. The induced electric dipole moment is αE. *Hyperpolarizabilities*, coefficients of higher powers of E, are less often required. Technically, the polarizability is a tensor quantity but for spherically symmetric charge distributions reduces to a single number. In any case, an *average polarizability* is usually adequate in calculations. Frequency-dependent or *dynamic polarizabilities* are needed for electric fields which vary in time, except for frequencies which are much lower than electron orbital frequencies, where *static polarizabilities* suffice.

Polarizabilities for atoms and molecules in excited states are found to be larger than for ground states and may be positive or negative. Molecular polarizabilities are very slightly temperature dependent since the size of the molecule depends on its rovibrational state. Only in the case of dihydrogen has this effect been studied enough to warrant consideration in Table 3.

Polarizabilities are normally expressed in cgs units of cm^3. Ground state polarizabilities are in the range of 10^{-24} cm^3 = 1 $Å^3$ and hence are often given in $Å^3$ units. Theorists tend to use atomic units of a_o^3 where a_o is the Bohr radius. The conversion is $\alpha(cm^3) = 0.148184 \times 10^{-24} \times \alpha(a_o^3)$. Polarizabilities are only recently encountered in SI units, $C{\cdot}m^2/V = J/(V/m)^2$. The conversion from cgs units to SI units is $\alpha(C{\cdot}m^2/V) = 4\pi\varepsilon_o \times 10^{-6} \times \alpha(cm^3)$, where ε_o is the permittivity of free space in SI units and the factor 10^{-6} simply converts cm^3 into m^3. Thus, $\alpha(C{\cdot}m^2/V) = 1.11265 \times 10^{-16} \times \alpha(cm^3)$. Persons measuring excited state polarizabilities by optical methods tend to use units of $MHz/(V/cm)^2$, where the energy shift, ΔW, is expressed in frequency units with a factor of h understood. The polarizability is $-2\,\Delta W/E^2$. The conversion into cgs units is $\alpha(cm^3) = 5.95531 \times 10^{-16} \times \alpha[MHz/(V/cm)^2]$.

The polarizability appears in many formulas for low-energy processes involving the valence electrons of atoms or molecules. These formulas are given below in cgs units: the polarizability α is in cm^3; masses m or μ are in grams; energies are in ergs; and electric charges are in esu, where $e = 4.8032 \times 10^{-10}$ esu. The symbol $\alpha(\nu)$ denotes a frequency (ν) dependent polarizability, where $\alpha(\nu)$ reduces to the static polarizability α for $\nu = 0$. For further information and references, see Miller, T. M., and Bederson, B., *Advances in Atomic and Molecular Physics*, 13, 1, 1977. Details on polarizability-related interactions, especially in regard to hyperpolarizabilities and nonlinear optical phenomena, are given by Bogaard, M. P., and Orr, B. J., in *Physical Chemistry*, Series Two, Vol. 2, *Molecular Structure and Properties*, Buckingham, A. D., Ed., Butterworths, London, 1975, pp. 149-194. A tabulation of tensor and hyperpolarizabilities is included. The gas number density, n, in Table 1 is usually taken to be that of 1 atm at 0°C in reporting experimental data.

Table 1
Formulas Involving Polarizability

Description	Formula	Remarks
Lorentz-Lorenz relation	$\alpha(\nu) = \frac{3}{4\pi n}\left[\frac{\eta^2(\nu)-1}{\eta^2(\nu)+2}\right]$	For a gas of atoms or nonpolar molecules; the index of refraction is $\eta(\nu)$
Refraction by polar molecules	$\alpha(\nu) + \frac{d^2}{3kT} = \frac{3}{4\pi n}\left[\frac{\eta^2(\nu)-1}{\eta^2(\nu)+2}\right]$	The dipole moment is d, in esu·cm (= 10^{-18} D)
Dielectric constant (dimensionless)	$\kappa(\nu) = 1 + 4\pi n\ \alpha(\nu)$	From the Lorentz-Lorenz relation for the usual case of $\kappa(\upsilon) \approx 1$
Index of refraction (dimensionless)	$\eta(\nu) = 1 + 2\pi n\ \alpha(\nu)$	From $\eta^2(\nu) = \kappa(\nu)$
Diamagnetic susceptibility	$\chi_m = e^2(a_o N\alpha)^{1/2}/4m_e c^2$	From the approximation that the static polarizability is given by the variational formula $\alpha = (4/9a_o)\Sigma(N_i r_i^2)^2$; N is the number of electrons, m_e is the electron mass; a crude approximation is $\chi_m = (E_i/4m_e c^2)\alpha$, where E_i is the ionization energy
Long-range electron- or ion-molecule interaction energy	$V(r) = -e^2\alpha/2r^4$	The target molecule polarizability is α
Ion mobility in a gas	$\kappa = 13.87/(\alpha\mu)^{1/2}\ \mathrm{cm^2/V{\cdot}s}$	This one formula is not in cgs units. Enter α in $Å^3$ or 10^{-24} cm^3 units and the reduced mass μ of the ion-molecule pair in amu. Classical limit; pure polarization potential

Table 1
Formulas Involving Polarizability (continued)

Description	Formula	Remarks
Langevin capture cross section	$\sigma(v_o) = (2\pi e/v_o)(\alpha/\mu)^{1/2}$	The relative velocity of approach for an ion-molecule pair is v_o; the target molecular polarizability is α and the reduced mass of the ion-molecule pair is μ
Langevin reaction rate coefficient	$k = 2\pi e(\alpha/\mu)^{1/2}$	Collisional rate coefficient for an ion-molecule reaction
Rate coefficient for polar molecules	$k_d = 2\pi e\left[(\alpha/\mu)^{1/2} + cd(2/\mu\pi kT)^{1/2}\right]$	The dipole moment of the neutral is *d* in esu·cm; the number c is a "locking factor" that depends on α and *d*, and is between 0 and 1
Modified effective range cross section for electron-neutral scattering	$\sigma(k) = 4\pi A^2$ $+32\pi^4\mu e^2\alpha Ak/3h^2$ $+\cdots$	Here, *k* is the electron momentum divided by *h*/2π, where *h* is Planck's constant; *A* is called the "scattering length"; the reduced mass is μ
van der Waals constant between two systems A, B	$C_6 = \frac{3}{2}\left[\frac{\alpha^A\alpha^B E^A E^B}{E^A + E^B}\right]$	For the interaction potential term $V_6(r) = -C_6 r^{-6}$; $E^{A,B}$ represents average dipole transition energies and $\alpha^{A,B}$ the respective polarizabilities of A, B
Dipole-quadrupole constant between two systems A, B	$C_8 = \frac{15}{4}\left[\frac{\alpha^A\alpha_q{}^B E^A E_q{}^B}{E^A + E_q{}^B}\right] + \frac{15}{4}\left[\frac{\alpha_q{}^A\alpha^B E_q{}^A E^B}{E_q{}^A + E^B}\right]$	For the interaction potential term $V_8(r) = -C_8 r^{-8}$; $E_q{}^{A,B}$ represents average quadrupole transition energies and $\alpha_q{}^{A,B}$ are the respective quadrupole polarizabilities of A, B
van der Waals constant between an atom and a surface	$C_3 = \frac{\alpha g E^A E^S}{8(E^A + E^S)}$	For an interaction potential $V_3(r) = C_3 r^{-3}$; $E^{A,S}$ are characteristic energies of the atom and surface; $g = 1$ for a free-electron metal and $g = (\varepsilon_\infty - 1)/(\varepsilon_\infty + 1)$ for an ionic crystal
Relation between α(ν) and oscillator strengths	$\alpha(\nu) = \frac{e^2h^2}{4\pi^2 m_e}\Sigma\frac{f_k}{E_k^2 - (h\nu)^2}$	Here, f_k is the oscillator strength from the ground state to an excited state *k*, with excitation energy E_k. This formula is often used to estimate static polarizabilities (ν = 0)
Dynamic polarizability	$\alpha(\nu) = \frac{\alpha E_r{}^2}{E_r{}^2 - (h\nu)^2}$	Approximate variation of the frequency-dependent polarizability α(ν) from ν = 0 up to the first dipole-allowed electronic transition, of energy E_r; the static dipole polarizability is α(0); infrared contributions ignored
Rayleigh scattering cross section	$\alpha(\nu) = \frac{8\pi}{9c^4}(2\pi\nu)^4 \times \left[3\alpha^2(\nu) + 2\gamma^2(\nu)/3\right]$	The photon frequency is ν; the polarizability anisotropy (the difference between polarizabilities parallel and perpendicular to the molecular axis) is γ(ν)
Verdet constant	$V(\nu) = \frac{\nu n}{2m_e c^2}\left[\frac{d\alpha(\nu)}{d\nu}\right]$	Defined from θ = *V*(ν)*B*, where θ is the angle of rotation of linearly polarized light through a medium of number density *n*, per unit length, for a longitudinal magnetic field strength *B* (Faraday effect)

Table 2
Static Average Electric Dipole Polarizabiilities for Ground State Atoms (in Units of 10^{-24} cm^3)

Atomic number	Atom	Polarizability	Estimated accuracy (%)	Method	Ref.
1	H	0.666793	"exact"	calc	MB77
2	He	0.204956	"exact"	calc	MB77
		0.2050	0.1	index/diel	NB65/OC67
3	Li	24.3	2	beam	MB77
4	Be	5.60	2	calc	MB77
5	B	3.03	2	calc	MB77
6	C	1.76	2	calc	MB77
7	N	1.10	2	calc/index	MB77
8	O	0.802	2	calc/index	MB77
9	F	0.557	2	calc	MB77
10	Ne	0.3956	0.1	diel	OC67
11	Na	24.08	0.4	interferom	ESCHP94
12	Mg	10.6	2	calc	MB77
13	Al	6.8	4.4	beam	MMD90
14	Si	5.38	2	calc	MB77
15	P	3.63	2	calc	MB77
16	S	2.90	2	calc	MB77
17	Cl	2.18	2	calc	MB77
18	Ar	1.6411	0.05	index/diel	NB65/OC67
19	K	43.4	2	beam	MB77
20	Ca	22.8	2	calc	MB77
		25.0	8	beam	MB77
21	Sc	17.8	25	calc	D84
22	Ti	14.6	25	calc	D84
23	V	12.4	25	calc	D84
24	Cr	11.6	25	calc	D84
25	Mn	9.4	25	calc	D84
26	Fe	8.4	25	calc	D84
27	Co	7.5	25	calc	D84
28	Ni	6.8	25	calc	D84
29	Cu	6.1	25	calc	D84
		7.31	25	calc	G84
30	Zn	7.1	25	calc	MB77
		5.6	25	calc	D84
31	Ga	8.12	2	calc	MB77
32	Ge	6.07	2	calc	MB77
33	As	4.31	2	calc	MB77
34	Se	3.77	2	calc	MB77
35	Br	3.05	2	calc	MB77
36	Kr	2.4844	0.05	diel	OC67
37	Rb	47.3	2	beam	MB77
38	Sr	27.6	8	beam	MB77
39	Y	22.7	25	calc	D84
40	Zr	17.9	25	calc	D84
41	Nb	15.7	25	calc	D84
42	Mo	12.8	25	calc	D84
43	Tc	11.4	25	calc	D84
44	Ru	9.6	25	calc	D84
45	Rh	8.6	25	calc	D84
46	Pd	4.8	25	calc	D84
47	Ag	7.2	25	calc	D84
		8.56	25	calc	G84
48	Cd	7.2	25	calc	D84
49	In	10.2	12	beam	GMBSJ84
		9.1	25	calc	D84

Table 2
Static Average Electric Dipole Polarizabiilities for Ground State Atoms (in Units of 10^{-24} cm^3) (continued)

Atomic number	Atom	Polarizability	Estimated accuracy (%)	Method	Ref.
50	Sn	7.7	25	calc	D84
51	Sb	6.6	25	calc	D84
52	Te	5.5	25	calc	D84
53	I	5.35	25	index	A56
		4.7	25	calc	D84
54	Xe	4.044	0.5	diel	MB77
55	Cs	59.6	2	beam	MB77
56	Ba	39.7	8	beam	MB77
57	La	31.1	25	calc	D84
58	Ce	29.6	25	calc	D84
59	Pr	28.2	25	calc	D84
60	Nd	31.4	25	calc	D84
61	Pm	30.1	25	calc	D84
62	Sm	28.8	25	calc	D84
63	Eu	27.7	25	calc	D84
64	Gd	23.5	25	calc	D84
65	Tb	25.5	25	calc	D84
66	Dy	24.5	25	calc	D84
67	Ho	23.6	25	calc	D84
68	Er	22.7	25	calc	D84
69	Tm	21.8	25	calc	D84
70	Yb	21.0	25	calc	D84
71	Lu	21.9	25	calc	D84
72	Hf	16.2	25	calc	D84
73	Ta	13.1	25	calc	D84
74	W	11.1	25	calc	D84
75	Re	9.7	25	calc	D84
76	Os	8.5	25	calc	D84
77	Ir	7.6	25	calc	D84
78	Pt	6.5	25	calc	D84
79	Au	5.8	25	calc	D84
		6.48	25	calc	G84
80	Hg	5.7	25	calc	D84
		5.1	15	diel	MB77
81	Tl	7.6	15	beam	NYU84
		7.5	25	calc	D84
82	Pb	6.8	25	calc	D84
83	Bi	7.4	25	calc	D84
84	Po	6.8	25	calc	D84
85	At	6.0	25	calc	D84
86	Rn	5.3	25	calc	D84
87	Fr	48.7	25	calc	D84
88	Ra	38.3	25	calc	D84
89	Ac	32.1	25	calc	D84
90	Th	32.1	25	calc	D84
91	Pa	25.4	25	calc	D84
92	U	24.9	6	beam	KB94
93	Np	24.8	25	calc	D84
94	Pu	24.5	25	calc	D84
95	Am	23.3	25	calc	D84
96	Cm	23.0	25	calc	D84
97	Bk	22.7	25	calc	D84
98	Cf	20.5	25	calc	D84
99	Es	19.7	25	calc	D84

Table 2
Static Average Electric Dipole Polarizabiilities for Ground State Atoms (in Units of 10^{-24} cm^3) (continued)

Atomic number	Atom	Polarizability	Estimated accuracy (%)	Method	Ref.
100	Fm	23.8	25	calc	D84
101	Md	18.2	25	calc	D84
102	No	17.5	25	calc	D84

Note: calc = calculated value; beam = atomic beam deflection technique; interferom = atomic beam interference; index = determination based on the measured index of refraction; diel = determination based on the measured dielectric constant.

REFERENCES

A56. Atoji, M., *J. Chem. Phys.*, 25, 174, 1956. Semiempirical method based on molecular polarizabilities and atomic radii.

D84. Doolen, G. D., Los Alamos National Laboratory, unpublished. A relativistic linear response method was used. The method is that described by Zangwill, A., and Soven, P., *Phys. Rev. A*, 21, 1561, 1980. Adjustments of less than 10% across the periodic table have been made to these results to bring them into agreement with accurate experimental values where avaialble, for the purpose of presenting "recommended" polarizabilities in Table 2.

ESCHP94. Ekstrom, C. R., Schmiedmayer, J., Chapman, M. S., Hammond, T. D., and Pritchard, D. E., *Phys. Rev. Lett.*, in press, 1994.

G84. Gollisch, H., *J. Phys. B: Atom. Mol. Phys.*, 17, 1463, 1984. Other results and useful references are contained in this paper.

GMBSJ84. Guella, T. P., Miller, T. M., Bederson, B., Stockdale, J. A. D., and Jaduszliwer, B., *Phys. Rev. A*, 29, 2977, 1984.

KB94. Kadar-Kallen, M. A., and Bonin, K. D., *Phys. Rev. Lett.*, 72, 828, 1994.

MB77. Miller, T. M., and Bederson, B., *Adv. At. Mol. Phys.*, 13, 1, 1977. For simplicity, any value in Table 2 which has not changed since this 1977 review is referenced as MB77. Persons interested in original references and further details should consult MB77.

MMD90. Milani, P., Moullet, I., and de Heer, W. A., *Phys. Rev. A*, 42, 5150, 1990.

NB65. Newell, A. C., and Baird, R. D., *J. Appl. Phys.*, 36, 3751, 1965.

NYU84. Preliminary value from the New York University group. See GMBSJ84.

OC67. Orcutt, R. H., and Cole, R. H., *J. Chem. Phys.*, 46, 697, 1967; see also the later references from this group, given following the tables.

Table 3
Average Electric Dipole Polarizabilities for Ground State Diatomic Molecules (in Units of 10^{-24} cm^3)

Molecule	Polarizability	Ref.
Al_2	19	23
BH	3.32*	1
Br_2	7.02	2
CO	1.95	3
Cl_2	4.61	3
Cs_2	104	22
CsK	89	22
D_2 (v = 0,J = 0)	0.7921*	5
D_2 (293 K)	0.7954	6
DCl	2.84	2
F_2	1.38*	7
H_2 (v = 0,J = 0)	0.8023*	5
H_2 (293 K)	0.8045*	5
H_2 (293 K)	0.8042	6
H_2 (322 K)	0.8059	8
HBr	3.61	3
HCl	2.63	3
	2.77	2
HD (v = 0,J = 0)	0.7976*	5
HF	0.80	27
	2.46	3
HI	5.44	3
	5.35	2
HgCl	7.4*	9
ICl	12.3	2
K_2	77	22
	72	21
Li_2	34	22
LiCl	3.46*	10
LiF	10.8*	11
LiH	3.84*	12
	3.68*	13
	3.88*	14
N_2	1.7403	6,8
NO	1.70	2
Na_2	40	22
	38	21
NaK	51	22
NaLi	40	4
O_2	1.5812	6
Rb_2	79	22

Table 4
Average Electric Dipole Polarizabilities for Ground State Triatomic Molecules (in Units of 10^{-24} cm^3)

Molecule	Polarizability	Ref.	Molecule	Polarizability	Ref.	Molecule	Polarizability	Ref.
BeH_2	4.34*	14	HCN	2.59	3	O_3	3.21	2
CO_2	2.911	8		2.46	2	OCS	5.71	2
CS_2	8.74	3	$HgBr_2$	14.5	2		5.2	15
	8.86	2	$HgCl_2$	11.6	2	SO_2	3.72	3
D_2O	1.26	2	HgI_2	19.1	2		4.28	2
H_2O	1.45	2	N_2O	3.03	8			
H_2S	3.78	3	NO_2	3.02	2†			
	3.95	2	Na_3	70	21			

Table 5
Average Electric Dipole Polarizabilities for Ground State Inorganic Polyatomic Molecules (Larger than Triatomic) (in Units of 10^{-24} cm^3)

Molecule	Polarizability	Ref.	Molecule	Polarizability	Ref.
$AsCl_3$	14.9	2	$(NaBr)_2$	26.8	16
AsN_3	5.75	2	$(NaCl)_2$	23.4	16
BCl_3	9.38	20	$(NaF)_2$	20.7	16
BF_3	3.31	2	$(NaI)_2$	26.9	16
$(BN_3)_2$	5.73	2	OsO_4	8.17	2
$(BH_2N)_3$	8.0	2†	PCl_3	12.8	2
ClF_3	6.32	2	PF_5	6.10	2
$(CsBr)_2$	54.5	16	PH_3	4.84	2
$(CsCl)_2$	42.4	16	$(RbBr)_2$	48.2	16
$(CsF)_2$	28.4	16	$(RbCl)_2$	43.2	16
$(CsI)_2$	51.8	16	$(RbF)_2$	40.7	16
$GeCl_4$	15.1	2	$(RbI)_2$	46.3	16
GeH_3Cl	6.7	2†	SF_6	6.54	8
$(HgCl)_2$	14.7	9	$(SF_5)_2$	13.2	2
K_n	n = 2,5,7-9,11,20	21	SO_3	4.84	2
$(KBr)_2$	42.0	16	SO_2Cl_2	10.5	2
$(KCl)_2$	32.1	16	SeF_6	7.33	2
$(KF)_2$	21.0	16	SiF_4	5.45	2
$(KI)_2$	36.3	16	SiH_4	5.44	2
$(LiBr)_2$	18.9	16	$(SiH_3)_2$	11.1	2
$(LiCl)_2$	13.1	16	$SiHCl_3$	10.7	2
$(LiF)_2$	6.9	16	SiH_2Cl_2	8.92	2
$(LiI)_2$	23.4	16	SiH_3Cl	7.02	2
ND_3	1.70	2	$SnBr_4$	22.0	2
NF_3	3.62	2	$SnCl_4$	18.0	2
NH_3	2.81	20		13.8	15
	2.10	2	SnI_4	32.3	2
	2.26	3	TeF_6	9.00	2
$(NO_2)_2$	6.69	2	$TiCl_4$	16.4	2
Na_n	n = 1-40	21	UF_6	12.5	2

Table 6
Average Electric Dipole Polarizabilities for Ground State Hydrocarbon Molecules (in Units of 10^{-24} cm^3)

Molecule	Name	Polarizability	Ref.
CH_4	methane	2.593	8
C_2H_2	acetylene	3.33	3
		3.93	2
C_2H_4	ethylene	4.252	8
C_2H_6	ethane	4.47	3
		4.43	2
C_3H_4	propyne	6.18	2
C_3H_6	propene	6.26	2
	cyclopropane	5.66	2
C_3H_8	propane	6.29	3
		6.37	2
C_4H_6	1-butyne	7.41	2†
	1,3-butadiene	8.64	2
C_4H_8	1-butene	7.97	2
		8.52	2
	trans-2-butene	8.49	2
	2-methylpropene	8.29	2
C_4H_{10}	*n*-butane	8.20	2
	isobutane	8.14	27
C_5H_6	1,3-cyclopentadiene	8.64	2
C_5H_8	1-pentyne	9.12	2
	trans-1,3-pentadiene	10.0	2
	isoprene	9.99	2
C_5H_{10}	cyclopentane	9.15	18
	1-pentene	9.65	27
	2-pentene	9.84	27
C_5H_{12}	pentane	9.99	2
	neopentane	10.20	18
C_6H_6	benzene	10.0	25
		10.32	3
		10.74	2
C_6H_{10}	1-hexyne	10.9	2†
	2-ethyl-1,3-butadiene	11.8	2†
	3-methyl-1,3-pentadiene	11.8	2†
	2-methyl-1,3-pentadiene	12.1	2†
	2,3-dimethyl-1,3-butadiene	11.8	2†
	cyclohexene	10.7	2†
C_6H_{12}	cyclohexane	11.0	18
		10.87	15
	1-hexene	11.7	27
C_6H_{14}	*n*-hexane	11.9	2
C_7H_8	toluene	11.8	25
		12.26	15
		12.3	2
C_7H_{12}	1-heptyne	12.8	2†
C_7H_{14}	methylcyclohexane	13.1	2
	1-heptene	13.5	27
C_7H_{16}	*n*-heptane	13.7	2
C_8H_8	styrene	15.0	2
		14.4	27
C_8H_{10}	ethylbenzene	14.2	2
	o-xylene	14.9	2
		14.1	15
	p-xylene	13.7	25
		14.2	15
		14.9	2
	m-xylene	14.2	15
C_8H_{16}	ethylcyclohexane	15.9	2
C_8H_{18}	*n*-octane	15.9	2
	3-methylheptane	15.4	27
	2,2,4-trimethylpentane	15.4	27
C_9H_{10}	α-methylstyrene	16.05	27
C_9H_{12}	isopropylbenzene	16.0	2-
	mesitylene	15.5	25
		16.1	27
C_9H_{18}	isopropylcyclohexane	17.2	2
C_9H_{20}	nonane	17.4	27
$C_{10}H_8$	napthalene	16.5	17
		17.5	27
$C_{10}H_{14}$	durene	17.3	25
	t-butylbenzene	17.2	25
		17.8	2†
$C_{10}H_{20}$	*t*-butylcyclohexane	19.8	2
$C_{10}H_{22}$	decane	19.1	27
$C_{11}H_{10}$	α-methylnaphthalene	19.35	27
	β-methylnaphthalene	19.52	27
$C_{11}H_{14}$	α,β,β-trimethylstyrene	19.64	27
$C_{11}H_{16}$	pentamethylbenzene	19.1	25
$C_{11}H_{24}$	undecane	21.0	27
$C_{12}H_{10}$	acenaphthene	20.6	27
$C_{12}H_{12}$	α-ethylnaphthalene	21.19	27
	β-ethylnaphthalene	21.36	27
$C_{12}H_{18}$	hexamethylbenzene	20.9	25
$C_{12}H_{26}$	dodecane	22.8	27
$C_{13}H_{10}$	fluorene	21.7	27
$C_{14}H_{10}$	anthracene	25.4	17
		25.9	27
	phenanthrene	36.8*	17
		24.7	27
$C_{14}H_{22}$	*p*-di-*t*-butylbenzene	24.5	25
$C_{16}H_{10}$	pyrene	28.2	27
$C_{17}H_{12}$	2,3-benzfluorene	30.2	27
$C_{18}H_{12}$	napthacene	32.3	27
	1,2-benzanthracene	32.9	27
	chrysene	33.1	27
	triphenylene	31.1	27
$C_{18}H_{30}$	1,3,5-tri-*t*-butylbenzene	31.8	25
$C_{24}H_{12}$	coronene		

Table 7
Average Electric Dipole Polarizabilities for Ground State Organic Halides (in Units of 10^{-24} cm^3)

Molecule	Name	Polarizability	Ref.
CBr_2F_2	dibromodifluoromethane	9.0	2†
$CClF_3$	chlorotrifluoromethane	5.72	20
		5.59	2
CCl_2F_2	dichlorodifluoromethane	7.93	20
		7.81	2
CCl_2O	phosgene	7.29	2
CCl_2S	thiophosgene	10.2	2
CCl_3F	trichlorofluoromethane	9.47	2
CCl_3NO_2	trichloronitromethane	10.8	2†
CCl_4	carbon tetrachloride	11.2	2
		10.5	3
CF_4	carbon tetrafluoride	3.838	8
CF_2O	carbonylfluoride	1.88*	17
$CHBr_3$	bromoform	11.8	17
$CHBrF_2$	bromodifluoromethane	5.7	2†
$CHClF_2$	chlorodifluoromethane	6.38	20
		5.91	2
$CHCl_2F$	dichlorofluoromethane	6.82	2
$CHCl_3$	chloroform	9.5	8
		8.23	27
CHF_3	fluoroform	3.52	20
		3.57	8
CHFO	fluoroformaldehyde	1.76*	17
CHI_3	iodoform	18.0	17
CH_2Br_2	dibromomethane	9.32	2
		8.62	27
CH_2ClNO_2	chloronitromethane	6.9	2†
CH_2Cl_2	dichloromethane	6.48	3
		7.93	2
CH_2I_2	diiodomethane	12.90	27
CH_3Br	bromomethane	5.87	20
		6.03	2
		5.55	15
CH_3Cl	chloromethane	5.35	20
		4.72	8
CH_3F	fluoromethane	2.97	8
CH_3I	iodomethane	7.97	2
C_2ClF_5	chloropentafluoroethane	6.3	2†
$C_2Cl_2F_4$	1,2-dichlorotetrafluoroethane	8.5	2†
C_2Cl_3N	trichloroacetonitrile	6.10	18
C_2F_6	hexafluoroethane	6.82	2
C_2HBr	bromoacetylene	7.39	2
C_2HCl	chloroacetylene	6.07	2
C_2HCl_5	pentachloroethane	14.0	2
$C_2H_2Cl_2$	dichloroethylene	10.03	27
	trans-dichloroethylene	9.28	27
	cis-dichloroethylene	9.19	27
$C_2H_2Cl_2F_2$	1,1-dichloro-2,2-difluoroethane	8.4	2†
$C_2H_2Cl_2O$	chloroacetyl chloride	8.92	2
$C_2H_2Cl_3F$	1,2,2-trichloro-1-fluoroethane	10.2	2†
$C_2H_2Cl_4$	1,1,2,2-tetrachloroethane	12.1	2†
C_2H_2ClN	chloroacetonitrile	6.10	18
$C_2H_2F_2$	1,1-difluoroethylene	5.01	20
C_2H_3Br	bromoethylene	7.59	2
C_2H_3Cl	chloroethylene	6.41	2

Table 7
Average Electric Dipole Polarizabilities for Ground State Organic Halides (in Units of 10^{-24} cm³) (continued)

Molecule	Name	Polarizability	Ref.
$C_2H_3ClF_2$	1-chloro-1,1-difluoroethane	8.05	2
C_2H_3ClO	acetyl chloride	6.62	2
$C_2H_3ClO_2$	ethyl chloroformate	7.1	2†
$C_2H_3Cl_3$	1,1,1-trichloroethane	10.7	2
$C_2H_3F_3$	1,1,1-trifluoroethane	4.4	2†
C_2H_3I	iodoethylene	9.3	2†
C_2H_4BrCl	1-bromo-2-chloroethane	9.5	2†
$C_2H_4Br_2$	1,2-dibromomethane	10.7	2†
C_2H_4ClF	1-chloro-2-fluoroethane	6.5	2†
$C_2H_4ClNO_2$	1-chloro-1-nitroethane	10.9	2
$C_2H_4Cl_2$	1,1-dichloroethane	8.64	2
	1,2-dichloroethane	8.0	2†
C_2H_5Br	bromoethane	8.05	2
		7.28	27
C_2H_5Cl	chloroethane	7.27	20
		8.29	2
		6.4	15
C_2H_5ClO	2-chloroethanol	7.1	2†
		6.88	27
	chloromethyl methyl ether	7.1	2†
C_2H_5F	fluoroethane	4.96	2
C_2H_5I	iodoethane	10.0	2
$C_3H_4Cl_2$	dichloropropene	10.1	2†
C_3H_5Cl	chloropropene	8.3	2
C_3H_5ClO	chloroacetone	8.4	2†
$C_3H_3ClO_2$	ethyl chloroformate	9.0	2†
$C_3H_6ClNO_2$	1-chloro-1-nitropropane	10.4	2†
$C_3H_6Cl_2$	dichloropropane	10.9	2†
C_3H_7Br	1-bromopropane	9.4	2†
		9.07	27
	2-bromopropane	9.6	2†
C_3H_7Cl	chloropropane	10.0	2
C_3H_7ClO	β-chloroethyl methyl ether	8.71	27
	2-chloro-1-propanol	8.89	27
	3-chloro-1-propanol	8.84	27
C_3H_7I	1-iodopropane	11.5	2†
C_4H_5Cl	4-chloro-1,2-butadiene	10.0	2†
C_4H_7Cl	1-chloro-2-methylpropene	10.8	2
$C_4H_7ClO_2$	2-chlorobutyric acid	10.7	27
	3-chlorobutyric acid	10.7	27
	4-chlorobutyric acid	10.6	27
$C_4H_8Cl_2$	1,4-dichlorobutane	12.0	2†
C_4H_9Br	bromobutane	13.9	2
		10.86	27
C_4H_9Cl	1-chlorobutane	11.3	2
	1-chloro-2-methylpropane	11.1	2
	2-chloro-2-methylpropane	12.5	2†
	2-chlorobutane	12.4	2
C_4H_9ClO	β-chloroethyl ethyl ether	10.56	27
	2-chloro-1-butanol	10.70	27
	3-chloro-1-butanol	10.38	27
C_4H_9I	1-iodobutane	13.3	2†
		12.65	27
$C_5H_9ClO_2$	2-chlorobutanoate acid	12.33	27
	3-chlorobutanoate acid	12.31	27

Table 7
Average Electric Dipole Polarizabilities for Ground State Organic Halides (in Units of 10^{-24} cm^3) (continued)

Molecule	Name	Polarizability	Ref.
	4-chlorobutanoate acid	12.27	27
	2-chloropentanoic acid	12.69	27
	3-chloropentanoic acid	12.57	27
	4-chloropentanoic acid	12.53	27
$C_5H_{11}Br$	1-bromopentane	13.1	2†
$C_5H_{11}Cl$	1-chloropentane	12.0	2†
$C_5H_{11}F$	fluoropentane	9.95	27
C_6F_6	hexafluorobenzene	9.58	27
C_6HF_5	pentafluorobenzene	9.63	27
$C_6H_2Cl_2O_2$	2,5-dichloro-1,4-benzoquinone	18.4	2
$C_6H_2F_4$	1,2,3,4-tetrafluorobenzene	9.69	27
	1,2,4,5-tetrafluorobenzene	9.69	27
$C_6H_3F_3$	1,3,5-trifluorobenzene	9.74	27
C_6H_4BrF	*p*-bromofluorobenzene	13.4	2†
$C_6H_4ClNO_2$	chloronitrobenzene	14.6	2†
$C_6H_4Cl_2$	*o*-dichlorobenzene	14.17	27
	m-dichlorobenzene	14.23	27
	p-dichlorobenzene	14.20	27
C_6H_4FI	*p*-fluoroiodobenzene	15.5	2†
$C_6H_4FNO_2$	*p*-fluoronitrobenzene	12.8	2†
$C_6H_4F_2$	*o*-difluorobenzene	9.80	27
	m-difluorobenzene	10.3	2†
	p-difluorobenzene	9.80	27
C_6H_5Br	bromobenzene	14.7	2
		13.62	27
C_6H_5Cl	chlorobenzene	14.1	2
		12.3	15
C_6H_5ClO	chlorophenol	13.0	2†
C_6H_5F	fluorobenzene	10.3	2
C_6H_5I	iodobenzene	15.5	2†
$C_6H_{11}ClO_2$	2-chlorobutanoate acid	14.16	27
	3-chlorobutanoate acid	14.13	27
	4-chlorobutanoate acid	14.11	27
$C_6H_{13}Br$	bromohexane	14.44	27
$C_6H_{13}F$	fluorohexane	11.80	27
C_7H_7Br	*p*-bromotoluene	14.80	27
C_7H_7Cl	*p*-chlorotoluene	13.70	27
C_7H_7F	*p*-fluorotoluene	11.70	27
C_7H_7I	*p*-iodotoluene	17.10	27
$C_7H_{15}Br$	1-bromoheptane	16.8	2†
		16.23	27
$C_7H_{15}F$	fluoroheptane	13.66	27
$C_8H_{17}Br$	bromooctane	18.02	27
$C_8H_{17}F$	fluorooctane	15.46	27
$C_9H_{19}Br$	bromononane	19.81	27
$C_9H_{19}F$	fluorononane	17.34	27
$C_{10}F_8$	octafluoronaphthalene	17.64	27
$C_{10}H_7Br$	α-bromonaphthalene	20.34	27
$C_{10}H_7Cl$	α-chloronaphthalene	19.30	27
	β-chloronaphthalene	19.58	27
$C_{10}H_7I$	α-iodonaphthalene	22.41	27
	β-iodonaphthalene	22.95	27
$C_{10}H_{21}Br$	bromodecane	21.60	27
$C_{10}H_{21}F$	fluorodecane	19.18	27

Table 7
Average Electric Dipole Polarizabilities for Ground State Organic Halides (in Units of 10^{-24} cm^3) (continued)

Molecule	Name	Polarizability	Ref.
$C_{11}H_{23}F$	fluoroundecane	21.00	27
$C_{12}H_{25}Br$	bromododecane	25.18	27
$C_{12}H_{25}F$	fluorododecane	22.83	27
$C_{12}H_8Br_2O$	4,4′–dibromodiphenyl ether	27.8	2†
$C_{12}H_9BrO$	4-bromodiphenyl ether	24.2	2†
$C_{13}H_{11}BrO$	*p*-bromophenyl-*p*-tolyl ether	26.6	2†
$C_{14}H_9Br$	9-bromoanthracene	28.32	27
$C_{14}H_9Cl$	9-chloroanthracene	27.35	27
$C_{14}H_9F$	fluoranthracene	28.34	27
$C_{14}H_{29}F$	fluorotetradecane	26.57	27
$C_{16}H_{33}Br$	bromohexadecane	32.34	27
$C_{18}H_{37}Br$	bromooctadecane	35.92	27

Table 8
Static Average Electric Dipole Polarizabilities for Other Ground State Organic Molecules (in Units of 10^{-24} cm^3)

Molecule	Name	Polarizability	Ref.
CN_4O_8	tetranitromethane	15.3	2
CH_2O	formaldehyde	2.8	2†
		2.45	18
CH_2O_2	formic acid	3.4	2†
CH_3NO	formamide	4.2	2†
		4.08	18
CH_3NO_2	nitromethane	7.37	2
CH_4O	methanol	3.29	2
		3.23	15
		3.32	18
CH_5N	methyl amine	4.7	2
		4.01	19
C_2N_2	cyanogen	7.99	2
C_2H_2O	ketene	4.4	2†
C_2H_3N	acetonitrile	4.40	2†
		4.48	18
C_2H_4O	acetaldehyde	4.6	2†
		4.59	18
	ethylene oxide	4.43	18
$C_2H_4O_2$	acetic acid	5.1	2†
	methyl formate	5.05	27
$C_2H_4O_4$	formic acid dimer	12.7	2
C_2H_5NO	acetamide	5.67	18
	N-methyl formamide	5.91	18
$C_2H_5NO_2$	nitroethane	9.63	2
	ethyl nitrite	7.0	15
C_2H_6O	ethanol	5.41	2
		5.11	18
	methyl ether	5.29	20
		5.84	2
		5.16	15
$C_2H_6O_2$	ethylene glycol	5.7	2†
		5.61	27
$C_2H_6O_2S$	dimethyl sulfone	7.3	2†

Table 8
Static Average Electric Dipole Polarizabilities for Other Ground State Organic Molecules (in Units of 10^{-24} cm^3) (continued)

Molecule	Name	Polarizability	Ref.
C_2H_6S	ethanethiol	7.41	2
C_2H_7N	ethyl amine	7.10	2
	dimethyl amine	6.37	2
$C_2H_8N_2$	ethylene diamine	7.2	2†
$C_3H_2N_2$	malononitrile	5.79	18
C_3H_3N	acrylonitrile	8.05	2
$C_3H_4N_2$	pyrazole	7.23	27
C_3H_4O	propenal	6.38	2†
C_3H_5N	propionitrile	6.70	2
		6.24	18
C_3H_6O	acetone	6.33	15
		6.4	2†
		6.39	18
	allyl alcohol	7.65	2
	propionaldehyde	6.50	2
$C_3H_6O_2$	propionic acid	6.9	2†
	ethyl formate	8.01	2
		6.88	27
	methyl acetate	6.94	2
		6.81	27
$C_3H_6O_3$	dimethyl carbonate	7.7	2†
C_3H_7NO	*N*-methyl acetamide	7.82	18
	N,N-dimethyl formamide	7.81	18
$C_3H_7NO_2$	nitropropane	8.5	2†
C_3H_8O	2-propanol	7.61	2
		6.97	18
	1-propanol	6.74	2
	ethyl methyl ether	7.93	2
$C_3H_8O_2$	dimethoxymethane	7.7	2†
	monomethylether ethylene glycol	7.44	27
C_3H_9N	*n*-propylamine	7.70	27
		9.20	2
	isopropylamine	7.77	27
	trimethylamine	8.15	2
$C_4H_2N_2$	fumaronitrile	11.8	2
$C_4H_4N_2$	succinonitrile	8.1	2†
	pyrimidene	8.53*	17
	pyridazine	9.27*	17
$C_4H_4O_2$	diketene	8.0	2†
C_4H_4S	thiophene	9.67	2
C_4H_5N	methacrylonitrile	8.0	2†
	trans-crotononitrile	8.2	2†
$C_4H_6N_2$	*N*-methylpyrazole	8.99	27
C_4H_6O	crotonaldehyde	8.5	2†
	methacrylaldehyde	8.3	2†
$C_4H_6O_2$	biacetyl	8.2	2†
$C_4H_6O_3$	acetic anhydride	8.9	2†
C_4H_6S	divinyl sulfide	10.9	2†
C_4H_7N	butyronitrile	8.4	2†
	isobutyronitrile	8.05	18
C_4H_8O	butanal	8.2	2†
	methyl ethyl ketone	8.13	15
	trans-2,3-epoxy butane	8.22*	17
$C_4H_8O_2$	ethyl acetate	9.7	2
		8.62	27

Table 8
Static Average Electric Dipole Polarizabilities for Other Ground State Organic Molecules (in Units of 10^{-24} cm^3) (continued)

Molecule	Name	Polarizability	Ref.
	1,4-dioxane	10.0	2
	p-dioxane	8.60	18
	2-methyl-1,3-dioxolane	9.44	15
	butyric acid	8.38	27
	methyl propionate	8.97	27
$C_4H_9NO_2$	1-nitrobutane	10.4	2†
	2-methyl-2-nitropropane	10.3	2†
$C_4H_{10}O$	ethyl ether	10.2	2
		8.73	15
	1-butanol	8.88	2
	2-methylpropanol	8.92	2
	n-propyl methyl ether	8.86	27
$C_4H_{10}O_2$	monoethylether ethylene glycol	9.28	27
$C_4H_{10}S$	ethyl sulfide	10.8	2
$C_4H_{11}N$	*n*-butylamine	13.5	2
	diethylamine	10.2	2
		9.61	27
C_5H_5N	pyridine	9.5	15
		9.18	27
	4-cyano-1,2-butadiene	10.5	2†
$C_5H_8N_2$	1,5-dimethylpyrazole	10.72	27
$C_5H_8O_2$	acetyl acetone	10.5	2†
C_5H_9N	valeronitrile	10.4	2
	22-DMPN	9.59	18
$C_5H_{10}O$	diethyl ketone	9.93	15
	methyl propyl ketone	9.93	15
$C_5H_{10}O_2$	ethyl propionate	10.41	27
	methyl butanoate	10.41	27
$C_5H_{10}O_3$	diethyl carbonate	11.3	2
$C_5H_{12}O$	ethyl propyl ether	10.68	27
$C_5H_{12}O_4$	tetramethyl orthocarbonate	13.0	2†
$C_6H_4N_2O_4$	*p*-dinitrobenzene	18.4	2
$C_6H_4O_2$	*p*-benzoquinone	14.5	2
$C_6H_5NO_2$	nitrobenzene	14.7	2
		12.92	15
C_6H_6O	phenol	11.1	2†
		9.94*	17
C_6H_7N	aniline	12.1	2†
$C_6H_8N_2$	phenylenediamine	13.8	2†
	phenylhydrazine	12.91	27
$C_6H_{10}N_2$	1-ethyl-5-methylpyrazole	12.50	27
$C_6H_{10}O_3$	ethyl acetoacetate	12.9	2†
$C_6H_{12}N_2$	dimethylketazine	15.6	2
$C_6H_{12}O$	cyclohexanol	11.56	18
$C_6H_{12}O_2$	amyl formate	14.2	2
$C_6H_{12}O_3$	paraldehyde	17.9	2
$C_6H_{14}O$	propyl ether	12.8	2
		12.5	15
$C_6H_{14}O_2$	1,1-diethoxyethane	13.2	2†
	1,2-diethoxyethane	11.3	2†
$C_6H_{15}N$	triethylamine	13.1	2
		13.38	27
	di-*n*-propylamine	13.29	27
$C_7H_4N_2O_2$	*p*-cyanonitrobenzene	19.0	2
C_7H_5N	benzonitrile	12.5	2†

Table 8
Static Average Electric Dipole Polarizabilities for Other Ground State Organic Molecules (in Units of 10^{-24} cm^3) (continued)

Molecule	Name	Polarizability	Ref.
$C_7H_7NO_3$	nitroanisole	15.7	2†
C_7H_7O	anisole	13.1	2†
C_7H_9NO	*o*-anisidine	14.2	2†
$C_7H_{10}N_2$	1,1-methylphenylhydrazine	14.81	27
$C_7H_{14}O$	cyclohexyl methyl ether	13.4	2†
	2,4-dimethyl-3-pentanone	13.5	15
$C_7H_{14}O_2$	pentyl acetate	14.9	2
$C_8H_4N_2$	*p*-dicyanobenzene	19.2	2
$C_8H_6O_2$	quinoxaline	15.13	27
C_8H_8O	acetophenone	15.0	2
$C_8H_8O_2$	2,5-dimethyl-1,4-benzoquinone	18.8	2
$C_8H_{10}O$	phenetole	14.9	2
$C_8H_{11}N$	*N*-dimethylaniline	16.2	2†
$C_8H_{12}N_2$	1,1-ethylphenylhydrazine	16.62	27
$C_8H_{12}O_2$	ethyl sorbate	17.2	2†
	tetramethylcyclobutane-1,3-dione	18.6	2
$C_8H_{14}O_4$	diethyl succinate	16.8	2†
$C_8H_{18}O$	*n*-butyl ether	17.2	2
C_9H_7N	quinoline	15.70	27
	isoquinoline	16.43	27
$C_9H_{10}O_2$	ethyl benzoate	16.9	2†
$C_9H_{21}N$	tri-*n*-propylamine	18.87	27
$C_{10}H_9N$	α-naphthylamine	19.50	27
	β-naphthylamine	19.73	27
	1-methylquinoline	18.65	27
	1-methylisoquinoline	18.28	27
$C_{10}H_{10}Fe$	ferrocene	17.1	26
$C_{10}H_{10}N_2$	2,3-dimethylquinoxaline	18.70	27
$C_{10}H_{14}BeO_4$	beryllium acetylacetonate	34.1	2
$C_{11}H_8O$	1-naphthaldehyde	19.75	27
	2-naphthaldehyde	20.06	27
$C_{14}H_8O_2$	anthraquinone	24.46	27
$C_{12}H_8N_2$	phenazine	23.43	27
$C_{12}H_9NO_3$	4-nitrodiphenyl ether	24.7	2†
$C_{14}H_{14}O$	di-*p*-tolyl ether	24.9	2†
$C_{15}H_{21}AlO_6$	aluminum acetylacetonate	51.9	2
$C_{15}H_{21}CrO_6$	chromium acetylacetonate	53.7	2
$C_{15}H_{21}FeO_6$	ferric acetylacetonate	58.1	2
$C_{20}H_{28}O_8Th$	thorium acetylacetonate	79.0	2
C_{60}	buckminsterfullerene	~80	24

Note: All polarizabilities in the tables are experimental values except those values marked by an asterisk (*), which indicates a calculated result. The experimental polarizabilities are mostly determined by measurements of a dielectric constant or refractive index which are quite accurate (0.5% or better). However, one should treat many of the results with several percent of caution because of the age of the data and because some of the results refer to optical frequencies rather than static. Comments given with the references are intended to allow one to judge the degree of caution required. Interested persons should consult these references. In many cases, the reference given is to a theoretical paper in which the experimental results are quoted. These papers, noted in the References, contain valuable information on polarizability calculations and experimental data which often includes the tensor components of the polarizability.

REFERENCES

1. McCullough, E. A., Jr., *J. Chem. Phys.*, 63, 5050, 1975. This calculation is for the parallel component, not the average polarizability.
2. Maryott, A. A., and Buckley, F., U. S. National Bureau of Standards Circular No. 537, 1953. A tabulation of dipole moments, dielectric constants, and molar refractions measured between 1910 and 1952, and used here to determine polarizabilities if no more recent result exists. The polarizability is $3/(4\pi N_o)$ times the molar polarization or molar refraction, where N_o is Avogadro's number. The value $3/(4\pi N_o) = 0.3964308 \times 10^{-24}$ cm^3 was used for this conversion. A dagger (†) following the reference number in the tables indicates that the polarizability was derived from the molar refraction and hence may not include some low-frequency contributions to the static polarizability; these "static" polarizabilities are therefore low by 1 to 30%.
3. Hirschfelder, J. O., Curtis, C. F., and Bird, R. B., *Molecular Theory of Gases and Liquids*, Wiley, New York, 1954, p. 950. Fundamental information on molecular polarizabilities.
4. Miller, T. M., and Bederson, B., *Adv. At. Mol. Phys.*, 13, 1, 1977. Review emphasizing atomic polarizabilities and measurement techniques. The data quoted in Table 3 are accurate to 8 to 12%.
5. Kolos, W., and Wolniewicz, L., *J. Chem. Phys.*, 46, 1426, 1967. Highly accurate molecular hydrogen calculations.
6. Newell, A. C., and Baird, R. C., *J. Appl. Phys.*, 36, 3751, 1965. Highly accurate refractive index measurements at 47.7 GHz (essentially static).
7. Jao, T. C., Beebe, N. H. F., Person, W. B., and Sabin, J. R., *Chem. Phys. Lett.*, 26, 47, 1974. Tensor polarizabilities, derivatives, and other results are reported.
8. Orcutt, R. H., and Cole, R. H., *J. Chem. Phys.*, 46, 697, 1967 (He, Ne, Ar, Kr, H_2, N_2); Sutter, H., and Cole, R. H., *J. Chem. Phys.*, 52, 132, 1970 (CF_3H, CFH_3, $CClF_3$, $CClH_3$); Bose, T. K., and Cole, R. H., *J. Chem. Phys.*, 52, 140, 1970 (CO_2), and 54, 3829, 1971 (C_2H_2, C_2H_4); Nelson, R. D., and Cole, R. H., *J. Chem. Phys.*, 54, 4033, 1971 (SF_6, $CClF_3$); Bose, T. K., Sochanski, J. S., and Cole, R. H., *J. Chem. Phys.*, 57, 3592, 1972 (CH_4, CF_4); Kirouac, S., and Bose, T. K., *J. Chem. Phys.*, 59, 3043, 1973 (N_2O), and 64, 1580, 1976 (He). Highly accurate dielectric constant measurements. These modern data give the most accurate polarizabilities available. A criticism of the interpretation of these data in the case of polar molecules is given in Ref. 20, p. 2905.
9. Huestis, D. L., Technical Report #MP 78-25, SRI International (project PYU 6158), Menlo Park, CA 94025. Molar refractions for mercury-chlorine compounds are analyzed.
10. Bounds, D. G., Clarke, J. H. R., and Hinchliffe, A., *Chem. Phys. Lett.*, 45, 367, 1977. Theoretical tensor polarizability for LiCl.
11. Kolker, H. J., and Karplus, M., *J. Chem. Phys.*, 39, 2011, 1963. Theoretical.
12. Cutschick, V. P., and McKoy, V., *J. Chem. Phys.*, 58, 2397, 1973. Theoretical tensor polarizabilities.
13. Gready, J. E., Bacskay, G. B., and Hush, N. S., *Chem. Phys.*, 22, 141, 1977, and 23, 9, 1977. Theoretical.
14. Amos, A. T., and Yoffe, J. A., *J. Chem. Phys.*, 63, 4723, 1975. Theoretical.
15. Stuart, H. A., *Landolt-Börnstein Zahlenwerte and Funktionen*, Vol. 1, Part 3, Eucken, A., and Hellwege, K. H., Eds., Springer-Verlag, Berlin, 1951, p. 511. Tabulation of molecular polarizabilities. Two misprints in the chemical symbols have been corrected.
16. Guella, T., Miller, T. M., Stockdale, J. A. D., Bederson, B., and Vuskovic, L., *J. Chem. Phys.*, 94, 6857, 1991. Beam measurements with accuracies between 12-24%.
17. Marchese, F. T., and Jaffé, *Theoret. Chim. Acta (Berlin)*, 45, 241, 1977. Theoretical and experimental tensor polarizabilities are tabulated in this paper.
18. Applequist, J., Carl, J. R., and Fung, K.-K., *J. Am. Chem. Soc.*, 94, 2952, 1972. Excellent reference on the calculation of molecular polarizabilities, including extensive tables of tensor polarizabilities, both theoretical and experimental, at 589.3 nm wavelength.
19. Bridge, N. J., and Buckingham, A. D., *Proc. Roy. Soc. (London)*, A295, 334, 1966. Measured tensor polarizabilities at 633 nm wavelength.
20. Barnes, A. N. M., Turner, D. J., and Sutton, L. E., *Trans. Faraday Soc.*, 67, 2902, 1971. Dielectric constants yielding polarizabilities accurate from 0.3-8%.
21. Knight, W. D., Clemenger, K., de Heer, W. A., and Saunders, W. A., *Phys. Rev. B*, 31, 2539, 1985. These data probably correspond to a very low internal temperature.
22. Tarnovsky, V., Bunimovicz, M., Vuskovic, L., Stumpf, B., and Bederson, B., *J. Chem. Phys.*, 98, 3894, 1993. These data correspond to internal temperatures 480-948 K.
23. Milani, P., Moullet, I., and de Heer, W. A., *Phys. Rev. A*, 42, 5150, 1990. Beam measurements accurate to 11%.
24. Bonin, K. D., and Kadar-Kallen, M. A., *Int. J. Mod. Phys.*, 8, 3313, 1994. Review article.
25. Aroney, M. J., and Pratten, S. J., *J. Chem. Soc., Faraday Trans. 1*, 80, 1201, 1984. Uncertainties in the range 1-3%.
26. Le Fevre, R. J. W., Murthy, D. S. N., and Saxby, J. D., *Aust. J. Chem.*, 24, 1057, 1971. Kerr effect.
27. No, K. T., Cho, K. H., Jhon, M. S., and Scheraga, H. A., *J. Am. Chem. Soc.*, 115, 2005, 1993. Theoretical; these results are quoted in numerous valuable papers on calculated polarizabilities, e.g., Miller, K. J., and Savchik, J. A., *J. Am. Chem. Soc.*, 101, 7206, 1979.

IONIZATION POTENTIALS OF ATOMS AND ATOMIC IONS

The ionization potentials of neutral and partially ionized atoms are listed in this table. Data were obtained from the compilations cited below, supplemented by results from the recent research literature. All values have been corrected to the currently recommended value of the conversion factor from wave number to energy, namely 1 eV = 8065.541 cm^{-1} (Reference 5). Values are given in eV.

Following the traditional spectroscopic notation, columns are headed I, II, III, etc. up to XXX, where I indicates the neutral atom, II the singly ionized atom, III the doubly ionized atom, etc. The first section of the table includes spectra I to VIII of all the elements; subsequent sections cover higher spectra (ionization stages) for those elements for which data are available.

REFERENCES

1. Moore, C. E., *Ionization Potentials and Ionization Limits Derived from the Analysis of Optical Spectra*, Natl. Stand. Ref. Data Ser. — Natl. Bur. Stand. (U.S.) No. 34, 1970.
2. Martin, W. C., Zalubas, R., and Hagan, L., *Atomic Energy Levels — The Rare Earth Elements*, Natl. Stand. Ref. Data Ser. — Natl. Bur. Stand. (U.S.), No. 60, 1978.
3. Sugar, J. and Corliss, C., *Atomic Energy Levels of the Iron Period Elements: Potassium through Nickel, J. Phys. Chem. Ref. Data*, Vol.14, Suppl. 2, 1985.
4. References to papers in *J. Phys. Chem. Ref. Data*, in the period 1973—91 covering other elements may be found in the cumulative index to that journal.
5. Cohen, E. R. and Taylor, B. N., *J. Phys. Chem. Ref. Data*, 17, 1795, 1988.

Neutral Atoms to +7 Ions

Z	Element	I	II	III	IV	V	VI	VII	VIII
1	H	13.59844							
2	He	24.58741	54.41778						
3	Li	5.39172	75.64018	122.45429					
4	Be	9.32263	18.21116	153.89661	217.71865				
5	B	8.29803	25.15484	37.93064	259.37521	340.22580			
6	C	11.26030	24.38332	47.8878	64.4939	392.087	489.99334		
7	N	14.53414	29.6013	47.44924	77.4735	97.8902	552.0718	667.046	
8	O	13.61806	35.11730	54.9355	77.41353	113.8990	138.1197	739.29	871.4101
9	F	17.42282	34.97082	62.7084	87.1398	114.2428	157.1651	185.186	953.9112
10	Ne	21.56454	40.96328	63.45	97.12	126.21	157.93	207.2759	239.0989
11	Na	5.13908	47.2864	71.6200	98.91	138.40	172.18	208.50	264.25
12	Mg	7.64624	15.03528	80.1437	109.2655	141.27	186.76	225.02	265.96
13	Al	5.98577	18.82856	28.44765	119.992	153.825	190.49	241.76	284.66
14	Si	8.15169	16.34585	33.49302	45.14181	166.767	205.27	246.5	303.54
15	P	10.48669	19.7694	30.2027	51.4439	65.0251	220.421	263.57	309.60
16	S	10.36001	23.3379	34.79	47.222	72.5945	88.0530	280.948	328.75
17	Cl	12.96764	23.814	39.61	53.4652	67.8	97.03	114.1958	348.28
18	Ar	15.75962	27.62967	40.74	59.81	75.02	91.009	124.323	143.460
19	K	4.34066	31.63	45.806	60.91	82.66	99.4	117.56	154.88
20	Ca	6.11316	11.87172	50.9131	67.27	84.50	108.78	127.2	147.24
21	Sc	6.56144	12.79967	24.75666	73.4894	91.65	110.68	138.0	158.1
22	Ti	6.8282	13.5755	27.4917	43.2672	99.30	119.53	140.8	170.4
23	V	6.7463	14.66	29.311	46.709	65.2817	128.13	150.6	173.4
24	Cr	6.76664	16.4857	30.96	49.16	69.46	90.6349	160.18	184.7
25	Mn	7.43402	15.63999	33.668	51.2	72.4	95.6	119.203	194.5
26	Fe	7.9024	16.1878	30.652	54.8	75.0	99.1	124.98	151.06
27	Co	7.8810	17.083	33.50	51.3	79.5	102.0	128.9	157.8
28	Ni	7.6398	18.16884	35.19	54.9	76.06	108	133	162
29	Cu	7.72638	20.29240	36.841	57.38	79.8	103	139	166
30	Zn	9.39405	17.96440	39.723	59.4	82.6	108	134	174
31	Ga	5.99930	20.5142	30.71	64				
32	Ge	7.900	15.93462	34.2241	45.7131	93.5			
33	As	9.8152	18.633	28.351	50.13	62.63	127.6		
34	Se	9.75238	21.19	30.8204	42.9450	68.3	81.7	155.4	
35	Br	11.81381	21.8	36	47.3	59.7	88.6	103.0	192.8
36	Kr	13.99961	24.35985	36.950	52.5	64.7	78.5	111.0	125.802
37	Rb	4.17713	27.285	40	52.6	71.0	84.4	99.2	136
38	Sr	5.69484	11.03013	42.89	57	71.6	90.8	106	122.3
39	Y	6.217	12.24	20.52	60.597	77.0	93.0	116	129
40	Zr	6.63390	13.13	22.99	34.34	80.348			
41	Nb	6.75885	14.32	25.04	38.3	50.55	102.057	125	

IONIZATION POTENTIALS OF ATOMS AND ATOMIC IONS (continued)

Z	Element	I	II	III	IV	V	VI	VII	VIII
42	Mo	7.09243	16.16	27.13	46.4	54.49	68.8276	125.664	143.6
43	Tc	7.28	15.26	29.54					
44	Ru	7.36050	16.76	28.47					
45	Rh	7.45890	18.08	31.06					
46	Pd	8.3369	19.43	32.93					
47	Ag	7.57624	21.49	34.83					
48	Cd	8.99367	16.90832	37.48					
49	In	5.78636	18.8698	28.03	54				
50	Sn	7.34381	14.63225	30.50260	40.73502	72.28			
51	Sb	8.64	16.53051	25.3	44.2	56	108		
52	Te	9.0096	18.6	27.96	37.41	58.75	70.7	137	
53	I	10.45126	19.1313	33					
54	Xe	12.12987	21.20979	32.1230					
55	Cs	3.89390	23.15745						
56	Ba	5.21170	10.00390						
57	La	5.5770	11.060	19.1773	49.95	61.6			
58	Ce	5.5387	10.85	20.198	36.758	65.55	77.6		
59	Pr	5.464	10.55	21.624	38.98	57.53			
60	Nd	5.5250	10.73	22.1	40.4				
61	Pm	5.55	10.90	22.3	41.1				
62	Sm	5.6437	11.07	23.4	41.4				
63	Eu	5.6704	11.241	24.92	42.7				
64	Gd	6.1500	12.09	20.63	44.0				
65	Tb	5.8639	11.52	21.91	39.79				
66	Dy	5.9389	11.67	22.8	41.4				
67	Ho	6.0216	11.80	22.84	42.5				
68	Er	6.1078	11.93	22.74	42.7				
69	Tm	6.18431	12.05	23.68	42.7				
70	Yb	6.25416	12.1761	25.05	43.56				
71	Lu	5.42585	13.9	20.9594	45.25	66.8			
72	Hf	6.82507	14.9	23.3	33.33				
73	Ta	7.89							
74	W	7.98							
75	Re	7.88							
76	Os	8.7							
77	Ir	9.1							
78	Pt	9.0	18.563						
79	Au	9.22567	20.5						
80	Hg	10.43750	18.756	34.2					
81	Tl	6.10829	20.428	29.83					
82	Pb	7.41666	15.0322	31.9373	42.32	68.8			
83	Bi	7.289	16.69	25.56	45.3	56.0	88.3		
84	Po	8.41671							
85	At								
86	Rn	10.74850							
87	Fr								
88	Ra	5.27892	10.14716						
89	Ac	5.17	12.1						
90	Th	6.08	11.5	20.0	28.8				
91	Pa	5.89							
92	U	6.19405							
93	Np	6.2657							
94	Pu	6.06							
95	Am	5.993							
96	Cm	6.02							
97	Bk	6.23							
98	Cf	6.30							
99	Es	6.42							
100	Fm	6.50							
101	Md	6.58							
102	No	6.65							

IONIZATION POTENTIALS OF ATOMS AND ATOMIC IONS (continued)

+8 Ions to +15 Ions

Z	Element	IX	X	XI	XII	XIII	XIV	XV	XVI
9	F	1103.1176							
10	Ne	1195.8286	1362.1995						
11	Na	299.864	1465.121	1648.702					
12	Mg	328.06	367.50	1761.805	1962.6650				
13	Al	330.13	398.75	442.00	2085.98	2304.1410			
14	Si	351.12	401.37	476.36	523.42	2437.63	2673.182		
15	P	372.13	424.4	479.46	560.8	611.74	2816.91	3069.842	
16	S	379.55	447.5	504.8	564.44	652.2	707.01	3223.78	3494.1892
17	Cl	400.06	455.63	529.28	591.99	656.71	749.76	809.40	3658.521
18	Ar	422.45	478.69	538.96	618.26	686.10	755.74	854.77	918.03
19	K	175.8174	503.8	564.7	629.4	714.6	786.6	861.1	968
20	Ca	188.54	211.275	591.9	657.2	726.6	817.6	894.5	974
21	Sc	180.03	225.18	249.798	687.36	756.7	830.8	927.5	1009
22	Ti	192.1	215.92	265.07	291.500	787.84	863.1	941.9	1044
23	V	205.8	230.5	255.7	308.1	336.277	896.0	976	1060
24	Cr	209.3	244.4	270.8	298.0	354.8	384.168	1010.6	1097
25	Mn	221.8	248.3	286.0	314.4	343.6	403.0	435.163	1134.7
26	Fe	233.6	262.1	290.2	330.8	361.0	392.2	457	489.256
27	Co	186.13	275.4	305	336	379	411	444	511.96
28	Ni	193	224.6	321.0	352	384	430	464	499
29	Cu	199	232	265.3	369	401	435	484	520
30	Zn	203	238	274	310.8	419.7	454	490	542
36	Kr	230.85	268.2	308	350	391	447	492	541
37	Rb	150	277.1						
38	Sr	162	177	324.1					
39	Y	146.2	191	206	374.0				
42	Mo	164.12	186.4	209.3	230.28	279.1	302.60	544.0	570

+16 Ions to +23 Ions

Z	Element	XVII	XVIII	XIX	XX	XXI	XXII	XXIII	XXIV
17	Cl	3946.2960							
18	Ar	4120.8857	4426.2296						
19	K	1033.4	4610.8	4934.046					
20	Ca	1087	1157.8	5128.8	5469.864				
21	Sc	1094	1213	1287.97	5674.8	6033.712			
22	Ti	1131	1221	1346	1425.4	6249.0	6625.82		
23	V	1168	1260	1355	1486	1569.6	6851.3	7246.12	
24	Cr	1185	1299	1396	1496	1634	1721.4	7481.7	7894.81
25	Mn	1224	1317	1437	1539	1644	1788	1879.9	8140.6
26	Fe	1266	1358	1456	1582	1689	1799	1950	2023
27	Co	546.58	1397.2	1504.6	1603	1735	1846	1962	2119
28	Ni	571.08	607.06	1541	1648	1756	1894	2011	2131
29	Cu	557	633	670.588	1697	1804	1916	2060	2182
30	Zn	579	619	698	738	1856			
36	Kr	592	641	786	833	884	937	998	1051
42	Mo	636	702	767	833	902	968	1020	1082

+24 Ions to +29 Ions

Z	Element	XXV	XXVI	XXVII	XXVIII	XXIX	XXX
25	Mn	8571.94					
26	Fe	8828	9277.69				
27	Co	2219.0	9544.1	10012.12			
28	Ni	2295	2399.2	10288.8	10775.40		
29	Cu	2308	2478	2587.5	11062.38	11567.617	
36	Kr	1151	1205.3	2928	3070	3227	3381
42	Mo	1263	1323	1387	1449	1535	1601

IONIZATION POTENTIALS OF GAS-PHASE MOLECULES

Sharon G. Lias

This table gives the first ionization potential *IP* in electronvolts for about 1000 molecules and atoms. Substances are listed by molecular formula in the modified Hill order (see introduction). Values enclosed in parentheses are considered to be not well-established. The enthalpy of formation of the ion at 298 K is also listed. For details on the criteria for selecting the data and the method of calculating the enthalpy of formation, see the reference. The enthalpy of formation is given here according to the ion convention; to convert these values to the electron convention, add 6 kJ/mol.

See Appendix B for an index to molecular formulas by compound name.

REFERENCE

Lias, Sharon G., et al., *Gas-Phase Ion and Neutral Thermochemistry, J. Phys. Chem. Ref. Data*, Vol. 17, Suppl. No. 1, 1988.

Molecular formula	Name	*IP*/eV	$\Delta_f H$ (ion)/ kJ/mol
	Compounds not containing carbon		
Ac	Actinium	5.17	905
Ag	Silver	7.57624	1016
AgCl	Silver chloride	(≤10.08)	≤1065
AgF	Silver fluoride	(11.0 ± 0.3)	1071
Al	Aluminum	5.98577	905
AlBr	Aluminum bromide (AlBr)	(9.3)	913
$AlBr_3$	Aluminum tribromide	(10.4)	593
AlCl	Aluminum chloride (AlCl)	9.4	855
$AlCl_3$	Aluminum trichloride	(12.01)	573
AlF	Aluminum fluoride (AlF)	9.73 ± 0.01	673
AlF3	Aluminum trifluoride	≤15.45	≤282
AlI	Aluminum iodide (AlI)	9.3 ± 0.3	965
AlI_3	Aluminum triiodide	(9.1)	673
Am	Americium	5.993	862
Ar	Argon	15.75962	1521
As	Arsenic	9.8152	1250
$AsCl_3$	Arsenic trichloride	(10.55 ± 0.025)	754
AsF_3	Arsenic trifluoride	(12.84 ± 0.05)	452
AsH_3	Arsine	9.89	1021
Au	Gold	9.22567	1254
B	Boron	8.29803	1363
BBr_3	Boron tribromide	(10.51 ± 0.02)	809
BCl_3	Boron trichloride	11.60 ± 0.02	718
BF	Fluoroborane (BF)	11.12 ± 0.01	957
BF_2	Difluoroborane (BF_2)	(9.4)	317
BF_3	Boron trifluoride	15.56 ± 0.03	365
BH	Borane (BH)	9.77 ± 0.05	1385
BH_3	Borane (BH_3)	12.3 ± 0.1	1287
BI_3	Boron triiodide	(9.25 ± 0.03)	964
BO_2	Boron oxide (BO_2)	(13.5 ± 0.3)	1001
B_2H_6	Diborane	11.38 ± 0.03	1134
B_2O_3	Boron oxide (B_2O_3)	13.5 ± 0.15	460
B_4H_{10}	Tetraborane	10.76 ± 0.04	1105
B_5H_9	Pentaborane	9.90 ± 0.04	1028
B_6H_{10}	Hexaborane	(9.0)	965
Ba	Barium	5.21170	683
BaO	Barium oxide	6.91 ± 0.06	543
Be	Beryllium	9.32263	1224
BeO	Beryllium oxide	(10.1 ± 0.4)	1111
Bi	Bismuth	7.289	908
$BiCl_3$	Bismuth trichloride	(10.4)	736
Bk	Berkelium	6.23	911
Br	Bromine	11.81381	1252

Molecular formula	Name	*IP*/eV	$\Delta_f H$ (ion)/ kJ/mol
BrCl	Bromine chloride	11.01	1079
BrF	Bromine fluoride	11.77 ± 0.01	1077
BrF_5	Bromine pentafluoride	(13.17 ± 0.01)	840
BrH	Hydrogen bromide	11.66 ± 0.03	1087
BrH_3Si	Bromosilane	10.6	943
BrI	Iodine bromide	9.790 ± 0.004	986
BrK	Potassium bromide	7.85 ± 0.1	578
BrLi	Lithium bromide	(8.7)	685
BrNO	Nitrosyl bromide	10.17 ± 0.03	1065
BrNa	Sodium bromide	8.31 ± 0.1	660
BrO	Bromine oxide (BrO)	(10.2)	1110
BrRb	Rubidium bromide	7.94 ± 0.03	583
BrTl	Thallium bromide (TlBr)	9.14 ± 0.02	844
Br_2	Bromine (Br_2)	10.515 ± 0.005	1046
Br_2Hg	Mercury bromide ($HgBr_2$)	10.560 ± 0.003	935
Br_2Sn	Tin bromide ($SnBr_2$)	9.0	839
Br_3Ga	Gallium bromide ($GaBr_3$)	10.40	711
Br_3P	Phosphorous tribromide	9.7	798
Br_4Hf	Hafnium bromide ($HfBr_4$)	(10.9)	366
Br_4Sn	Tin bromide ($SnBr_4$)	10.6	709
Br_4Ti	Titanium bromide ($TiBr_4$)	10.3	375
Br_4Zr	Zirconium bromide ($ZrBr_4$)	(10.7)	388
C	Carbon	11.26030	1803
Ca	Calcium	6.11316	768
CaCl	Calcium chloride (CaCl)	5.61 ± 0.13	438
CaO	Calcium oxide	(6.9)	691
Cd	Cadmium	8.99367	980
Ce	Cerium	5.5387	957
Cf	Californium	6.30	805
Cl	Chlorine	12.96764	1373
ClCs	Cesium chloride	(7.84 ± 0.05)	510
ClF	Chlorine fluoride	12.65 ± 0.01	1170
$ClFO_3$	Perchloryl fluoride	(12.945 ± 0.005)	1224
ClF_2	Chlorine difluoride	(12.77 ± 0.05)	1128
ClF_3	Chlorine trifluoride	(12.65 ± 0.05)	1057
ClF_5S	Sulfur chloride pentafluoride	(12.335 ± 0.005)	144
ClH	Hydrogen chloride	12.747	1137
ClHO	Hypochlorous acid (HOCl)	(11.12 ± 0.01)	993
ClH_3Si	Chlorosilane	11.4	899
ClI	Iodine chloride	10.088 ± 0.01	991
ClIn	Indium chloride (InCl)	(9.51)	842
ClK	Potassium chloride	(8.0 ± 0.4)	557
ClLi	Lithium chloride	9.57	727
ClNO	Nitrosyl chloride	10.87 ± 0.01	1099
$ClNO_2$	Nitryl chloride	(11.84)	1155
ClNa	Sodium chloride	8.92 ± 0.06	681
ClO	Chlorine oxide (ClO)	10.95	1159
ClO_2	Chlorine dioxide (ClO_2)	10.36 ± 0.02	1096
ClRb	Rubidium chloride	(8.50 ± 0.03)	590
ClTl	Thallium chloride (TlCl)	9.70 ± 0.03	869
Cl_2	Chlorine (Cl_2)	11.480 ± 0.005	1108
Cl_2CrO_2	Chromyl chloride (CrO_2Cl_2)	11.6	580
Cl_2Ge	Germanium dichloride	(10.20 ± 0.05)	813
Cl_2H_2Si	Dichlorosilane	11.4	765
Cl_2Hg	Mercury chloride ($HgCl_2$)	11.380 ± 0.003	952
Cl_2O	Oxygen dichloride	10.94	1135
Cl_2OS	Thionyl chloride	10.96	844

Molecular formula	Name	*IP*/eV	$\Delta_f H$ (ion)/ kJ/mol
Cl_2O_2S	Sulfuryl chloride	12.05	807
Cl_2Pb	Lead chloride ($PbCl_2$)	(10.0)	789
Cl_2S	Sulfur chloride (SCl_2)	9.45 ± 0.03	895
Cl_2Si	Dichlorosilylene ($SiCl_2$)	(10.93 ± 0.10)	887
Cl_2Sn	Tin chloride ($SnCl_2$)	(10.0)	760
Cl_3Ga	Gallium chloride ($GaCl_3$)	11.52	664
Cl_3HSi	Trichlorosilane	(11.7)	648
Cl_3N	Nitrogen trichloride	(10.12 ± 0.1)	1244
Cl_3OP	Phosphoryl chloride	11.36 ± 0.02	540
Cl_3OV	Vanadium oxytrichloride	(11.6)	425
Cl_3P	Phosphorous trichloride	9.91	668
Cl_3PS	Phosphorous thiochloride	9.71 ± 0.03	573
Cl_3Sb	Antimony trichloride	(10.1 ± 0.1)	661
Cl_4Ge	Germanium tetrachloride	11.68 ± 0.05	629
Cl_4Hf	Hafnium chloride ($HfCl_4$)	(11.7)	246
Cl_4Si	Silicon tetrachloride	11.79 ± 0.01	527
Cl_4Sn	Tin chloride ($SnCl_4$)	(11.88 ± 0.05)	673
Cl_4Ti	Titanium chloride ($TiCl_4$)	11.65 ± 0.15	363
Cl_4V	Vanadium chloride (VCl_4)	(9.2)	361
Cl_4Zr	Zirconium chloride ($ZrCl_4$)	(11.2)	210
Cl_5Mo	Molybdenum pentachloride	(8.7)	392
Cl_5Nb	Niobium chloride ($NbCl_5$)	(10.97)	356
Cl_5P	Phosphorous pentachloride	10.7	656
Cl_5Ta	Tantalum chloride ($TaCl_5$)	11.08	303
Cl_6W	Tungsten chloride (WCl_6)	(9.5)	348
Cm	Curium	6.02	966
Co	Cobalt	7.8810	1186
Cr	Chromium	6.76664	1050
Cs	Cesium	3.89390	452
CsF	Cesium fluoride	(8.80 ± 0.10)	489
CsNa	Cesium sodium (CsNa)	(4.05 ± 0.04)	535
Cu	Copper	7.72638	1084
CuF	Copper fluoride (CuF)	10.15 ± 0.02	984
Dy	Dysprosium	5.9389	862
Er	Erbium	6.1078	907
Es	Einsteinium	6.42	753
Eu	Europium	5.6704	723
F	Fluorine	17.42282	1760
FGa	Gallium fluoride (GaF)	(9.6 ± 0.5)	700
FH	Hydrogen fluoride	16.044 ± 0.003	1276
FHO	Hypofluorous acid (HOF)	12.71 ± 0.01	1130
FH_3Si	Fluorosilane	11.7	752
FI	Iodine fluoride	10.62	930
FIn	Indium fluoride (InF)	(9.6 ± 0.5)	740
FNO	Nitrosyl fluoride	12.63 ± 0.03	1152
FNO_2	Nitryl fluoride	(13.09)	1154
FNS	Thionitrosyl fluoride (NSF)	11.51 ± 0.04	1090
FO	Fluorine oxide (FO)	12.77	1341
FO_2	Fluorine superoxide (FOO)	(12.6 ± 0.2)	1228
FS	Sulfur fluoride (SF)	10.09	986
FTl	Thallium fluoride (TlF)	10.52	835
F_2	Fluorine (F_2)	15.697 ± 0.003	1515
F_2Ge	Germanium fluoride (GeF_2)	(11.65)	551
F_2HN	Difluoramine	(11.53 ± 0.08)	1046
F_2H_2Si	Difluorosilane	12.2	386
F_2Mg	Magnesium fluoride	(13.40.4)	569
F_2N	Difluoroamidogen (NF_2)	11.628 ± 0.01	1155

Molecular formula	Name	*IP*/eV	$\Delta_f H$ (ion)/ kJ/mol
F_2N_2	*trans*-Difluorodiazine	(12.8)	1315
F_2O	Oxygen difluoride	13.11 ± 0.01	1290
F_2OS	Thionyl fluoride	12.25	688
F_2O_2S	Sufuryl fluoride	13.04 ± 0.01	501
F_2Pb	Lead fluoride (PbF_2)	(11.5)	679
F_2S	Sulfur fluoride (SF_2)	(10.08)	676
F_2Si	Difluorosilylene (SiF_2)	10.78 ± 0.05	450
F_2Sn	Tin fluoride (SnF_2)	(11.1)	586
F_2Xe	Xenon difluoride	12.35 ± 0.01	1083
F_3HSi	Trifluorosilane	(14.0)	150
F_3N	Nitrogen trifluoride	13.00 ± 0.02	1125
F_3NO	Trifluoramine oxide	13.26 ± 0.01	1116
F_3OP	Phosphoryl fluoride	12.76 ± 0.01	–24
F_3P	Phosphorous trifluoride	11.44	146
F_3PS	Thiophosphoryl trifluoride	≤11.05 ± 0.035	≤58
F_3Si	Trifluorosilyl (SiF_3)	(9.3)	–99
F_4Ge	Germanium tetrafluoride	(15.5)	307
F_4N_2	Tetrafluorohydrazine (gauche)	11.94 ± 0.03	1119
F_4S	Sulfur fluoride (SF_4)	12.03 ± 0.05	399
F_4Si	Silicon tetrafluoride	(15.7)	–100
F_4Xe	Xenon tetrafluoride	12.65 ± 0.1	1016
F_5I	Iodine pentafluoride	12.943 ± 0.005	408
F_5P	Phosphorous pentafluoride	(15.1)	–137
F_5S	Sulfur pentafluoride	10.5 ± 0.1	97
F_6Mo	Molybdenum fluoride (MoF_6)	(14.5 ± 0.1)	–159
F_6S	Sulfur fluoride (SF_6)	15.33 ± 0.03	259
F_6U	Uranium fluoride (UF_6)	14.00 ± 0.10	–796
Fe	Iron	7.9024	1177
Fm	Fermium	6.50	
Ga	Gallium	5.99930	851
GaI_3	Gallium iodide (GaI_3)	9.40	765
Gd	Gadolinium	6.1500	991
Ge	Germanium	7.900	1139
GeH_4	Germane	11.33	1185
GeI_4	Germanium tetraiodide	(9.42)	850
GeO	Germanium oxide (GeO)	11.25 ± 0.01	1044
GeS	Germanium sulfide (GeS)	9.98 ± 0.02	1055
H	Hydrogen	13.59844	1530
HI	Hydrogen iodide	10.386 ± 0.001	1028
HLi	Lithium hydride	7.7	882
HN	Imidogen (NH)	13.49 ± 0.01	1678
HNO	Nitrosyl hydride (HNO)	(10.1)	1075
HNO_2	Nitrous acid (HONO)	≤11.3	977
HNO_3	Nitric acid	11.95 ± 0.01	1019
HN_3	Hydrazoic acid	10.72 ± 0.025	1328
HO	Hydroxyl (OH)	13.00	1293
HO_2	Hydroperoxy (HOO)	11.35 ± 0.01	1106
HS	Mercapto (SH)	10.37 ± 0.01	1140
H_2	Hydrogen (H_2)	15.42589 ± 0.00005	1488
H_2N	Amidogen (NH_2)	11.14 ± 0.01	1264
H_2O	Water	12.612 ± 0.010	975
H_2O_2	Hydrogen peroxide	10.54	881
H_2S	Hydrogen sulfide	10.453 ± 0.008	988
H_2Se	Hydrogen selenide	9.882 ± 0.001	983
H_2Si	Silylene (SiH_2)	8.92 ± 0.07	1149
H_3N	Ammonia	10.16 ± 0.01	934
H_3NO	Hydroxylamine (NH_2OH)	10.00	923

Molecular formula	Name	*IP*/eV	$\Delta_f H$ (ion)/ kJ/mol
H_3P	Phosphine	9.869 ± 0.002	958
H_3Sb	Stibine	9.54 ± 0.03	1067
H_4N_2	Hydrazine	8.1 ± 0.15	877
H_4Si	Silane	11.65	1158
H_4Sn	Stannane	(10.75)	1200
H_6Si_2	Disilane	(9.7)	1015
H_8Si_3	Trisilane	(9.2)	1009
He	Helium	24.58741	2372
Hf	Hafnium	6.82507	1278
Hg	Mercury	10.43750	1069
HgI_2	Mercury iodide (HgI_2)	9.5088 ± 0.0022	900
Ho	Holmium	6.0216 ± 0.0006	882
I	Iodine	10.45126	1115
IK	Potassium iodide	(7.21 ± 0.3)	570
ILi	Lithium iodide	(7.5)	633
INa	Sodium iodide	7.64 ± 0.02	659
ITl	Thallium iodide (TlI)	8.47 ± 0.02	826
I_2	Iodine (I_2)	9.3995 ± 0.0012	969
I_4Ti	Titanium iodide (TiI_4)	(9.1)	602
I_4Zr	Zirconium iodide (ZrI_4)	(9.3)	534
In	Indium	5.78636	802
Ir	Iridium	9.1	1543
K	Potassium	4.34066	508
KLi	Lithium potassium (LiK)	4.57 ± 0.04	512
KNa	Potassium sodium (KNa)	4.41636 ± 0.00017	561
K_2	Potassium (K_2)	4.0637 ± 0.0002	519
Kr	Krypton	13.99961	1351
La	Lanthanum	5.5770	969
Li	Lithium	5.39172	680
LiNa	Lithium sodium (LiNa)	5.05 ± 0.04	571
LiO	Lithium oxide (LiO)	(8.45 ± 0.20)	895
LiRb	Lithium rubidium (LiRb)	4.3 ± 0.1	486
Li_2	Lithium (Li_2)	5.1127 ± 0.0003	709
Lu	Lutetium	5.42585	950
Md	Mendelevium	6.58	
Mg	Magnesium	7.64624	885
MgO	Magnesium oxide	9.7	992
Mn	Manganese	7.43402	998
Mo	Molybdenum	7.09243	1343
N	Nitrogen	14.53414	1875
NO	Nitric oxide	9.26436 ± 0.00006	985
NO_2	Nitrogen dioxide	9.75 ± 0.01	974
NS	Nitrogen sulfide (NS)	8.87 ± 0.01	1119
NP	Phosphorous nitride (PN)	11.85	1248
N_2	Nitrogen (N_2)	15.5808	1503
N_2O	Nitrous oxide	12.886	1325
N_2O_4	Nitrogen tetroxide	10.8 ± 0.2	1050
N_2O_5	Nitrogen pentoxide	(11.9)	1161
Na	Sodium	5.13908	603
NaRb	Rubidium sodium (RbNa)	4.32 ± 0.04	480
Na_2	Sodium (Na_2)	4.88898 ± 0.00016	614
Nb	Niobium	6.75885	1384
Nd	Neodymium	5.5250	859
Ne	Neon	21.56454	2081
Ni	Nickel	7.6398	1167
No	Nobelium	6.65	
Np	Neptunium	6.2657 ± 0.0005	1069

IONIZATION POTENTIALS OF GAS-PHASE MOLECULES (continued)

Molecular formula	Name	*IP*/eV	$\Delta_f H$ (ion)/ kJ/mol
O	Oxygen	13.61806	1563
OPb	Lead oxide (PbO)	9.08 ± 0.10	939
OS	Sulfur oxide (SO)	10.32 ± 0.02	1001
OS_2	Sulfur oxide (SSO)	10.54 ± 0.04	967
OSi	Silicon oxide (SiO)	11.43	1002
OSn	Tin oxide (SnO)	9.60 ± 0.02	944
OSr	Strontium oxide	7.0 ± 0.15	662
O_2	Oxygen (O_2)	12.071 ± 0.001	1165
O_2S	Sulfur dioxide	12.32 ± 0.02	892
O_2Th	Thorium oxide (ThO_2)	(8.7 ± 0.15)	342
O2Ti	Titanium oxide (TiO_2)	(9.54 ± 0.1)	623
O_2U	Uranium oxide (UO_2)	(5.4 ± 0.1)	57
O_3	Ozone	12.43	1342
O_3S	Sulfur trioxide	12.80 ± 0.04	839
O_3U	Uranium oxide (UO_3)	(10.5 ± 0.5)	214
O_4Os	Osmium oxide (OsO_4)	12.320	850
O_4Ru	Ruthenium oxide (RuO_4)	12.15 ± 0.03	988
O_7Re_2	Rhenium oxide (Re_2O_7)	(12.7 ± 0.2)	125
Os	Osmium	8.7	1630
P	Phosphorous	10.48669	1328
P_2	Phosphorous (P_2)	10.53	1160
Pa	Protactinium	5.89	1133
Pb	Lead	7.41666	911
PbS	Lead sulfide (PbS)	(8.5 ± 0.5)	954
Pd	Palladium	8.3369	1181
Pm	Promethium	5.55	
Pr	Praseodymium	5.464 ± 0.006	883
Pt	Platinum	9.0	1433
Pu	Plutonium	6.06	930
Ra	Radium	5.27892	668
Rb	Rubidium	4.17713	484
Re	Rhenium	7.88	1530
Rh	Rhodium	7.45890	1276
Rn	Radon	10.74850	1037
Ru	Ruthenium	7.36050	1355
S	Sulfur	10.36001	1277
SSn	Tin sulfide (SnS)	(8.8)	966
S_2	Sulfur (S_2)	9.356 ± 0.002	1031
Sb	Antimony	8.64	1096
Sc	Scandium	6.56144	1010
Se	Selenium	9.75238	1168
Si	Silicon	8.15169	1238
Sm	Samarium	5.6437 ± 0.0006	751
Sn	Tin	7.34381	1011
Sr	Strontium	5.69484	713
Ta	Tantalum	7.89	1544
Tb	Terbium	5.8639	955
Tc	Technetium	7.28	1380
Te	Tellurium	9.0096	1066
Th	Thorium	6.08	1185
Ti	Titanium	6.8282	1127
Tl	Thallium	6.10829	771
Tm	Thulium	6.18431	826
U	Uranium	6.19405	1129
V	Vanadium	6.7463	1165
W	Tungsten	7.98	1621
Xe	Xenon	12.12987	1170

Molecular formula	Name	*IP*/eV	$\Delta_f H$ (ion)/ kJ/mol
Y	Yttrium	6.217	1022
Yb	Ytterbium	6.25416	754
Zn	Zinc	9.39405	1037
Zr	Zirconium	6.63390	1251
	Compounds containing carbon		
C	Carbon	11.26030	1803
$CBrClF_2$	Bromochlorodifluoromethane	(≤11.83)	≤702
$CBrCl_3$	Bromotrichloromethane	(10.6)	980
$CBrF_3$	Bromotrifluoromethane	11.4	451
CBr_2F_2	Dibromodifluoromethane	11.07 ± 0.03	687
CBr_4	Tetrabromomethane	(10.31 ± 0.02)	1079
CCl	Chloromethylidene (CCl)	(8.9 ± 0.2)	1244
$CClF_3$	Chlorotrifluoromethane	12.39	485
CClN	Cyanogen chloride	12.34 ± 0.01	1329
CCl_2	Dichloromethylene	10.36	1163
CCl_2F_2	Dichlorodifluoromethane	11.75 ± 0.04	656
CCl_2O	Carbonyl chloride	(11.4)	878
CCl_3F	Trichlorofluoromethane	11.77 ± 0.02	868
CCl_4	Tetrachloromethane	11.47 ± 0.01	1010
CF	Fluoromethylidene	9.11 ± 0.01	1134
CFN	Cyanogen fluoride	13.32 ± 0.01	1323
CF_2	Difluoromethylene	11.42 ± 0.01	897
CF_2O	Carbonyl fluoride	13.03	617
CF_3	Trifluoromethyl	(≤8.9)	≤399
CF_3I	Trifluoroiodomethane	10.23	397
CN	Cyanide (CN)	(14.09)	1795
CNO	Cyanate (NCO)	(11.76 ± 0.01)	1290
CO	Carbon monoxide	14.0139	1242
COS	Carbon oxysulfide	11.1736 ± 0.0015	936
COSe	Carbon oxyselenide	10.36 ± 0.01	929
CO_2	Carbon dioxide	13.773 ± 0.002	935
CS	Carbon sulfide	11.33 ± 0.01	1368
CS_2	Carbon disulfide	10.0685 ± 0.0020	1089
CH	Methylidyne	10.64 ± 0.01	1622
$CHBrCl_2$	Bromodichloromethane	10.6	973
$CHBr_2Cl$	Chlorodibromomethane	10.59 ± 0.01	1030
$CHBr_3$	Tribromomethane	10.48 ± 0.02	1035
CHCl	Chloromethylene (HCCl)	9.84	1247
$CHClF_2$	Chlorodifluoromethane	(12.2)	693
$CHCl_2F$	Dichlorofluoromethane	(11.5)	829
$CHCl_3$	Trichloromethane	11.37 ± 0.02	992
CHF	Fluoromethylene (HCF)	(10.49)	1121
CHF_3	Trifluoromethane	13.86	643
CHI_3	Triiodomethane	9.25 ± 0.02	1010
CHN	Hydrogen cyanide	13.60 ± 0.01	1447
CHN	Hydrogen isocyanide (HNC)	(12.5 ± 0.1)	1407
CHNO	Fulminic acid (HCNO)	(10.83)	1263
CHNO	Isocyanic acid (HNCO)	11.61 ± 0.03	1016
CHO	Oxomethyl (HCO)	8.10 ± 0.05	826
CH_2	Methylene	10.396 ± 0.003	1386
CH_2BrCl	Bromochloromethane	10.77 ± 0.01	1085
CH_2Br_2	Dibromomethane	10.50 ± 0.02	1013
CH_2ClF	Chlorofluoromethane	11.71 ± 0.01	870
CH_2Cl_2	Dichloromethane	11.32 ± 0.01	996
CH_2F_2	Difluoromethane	12.71	774

Molecular formula	Name	IP/eV	$\Delta_f H$ (ion)/ kJ/mol
CH_2I_2	Diiodomethane	9.46 ± 0.02	1030
CH_2N_2	Cyanamide	(10.4)	1137
CH_2N_2	Diazomethane	8.999 ± 0.001	1098
CH_2O	Formaldehyde	10.874 ± 0.002	940
CH_2O_2	Formic acid	11.33 ± 0.01	715
CH_3	Methyl	9.84 ± 0.01	1095
CH_3BO	Carbonyltrihydroboron (BH_3CO)	11.14 ± 0.02	962
CH_3Br	Bromomethane	10.541 ± 0.003	979
CH_3Cl	Chloromethane	11.22 ± 0.01	1001
CH_3Cl_3Si	Methyltrichlorosilane	(11.36 ± 0.03)	548
CH_3F	Fluoromethane	12.47 ± 0.02	956
CH_3I	Iodomethane	9.538	936
CH_3NO	Formamide	10.16 ± 0.06	796
CH_3NO_2	Nitromethane	11.02 ± 0.04	988
CH_3N_3	Methyl azide	9.81 ± 0.02	1227
CH_3O	Methoxy	(8.6)	845
CH_4	Methane	12.51	1133
CH_4N_2O	Urea	9.7	690
CH_4O	Methanol	10.85 ± 0.01	845
CH_4S	Methanethiol	9.44 ± 0.005	888
CH_5N	Methylamine	8.97 ± 0.02	843
CH_6N_2	Methylhydrazine	7.67 ± 0.02	835
CH_6Si	Methylsilane	10.7	1003
C_2	Carbon (C_2)	12.11	2000
$C_2Br_2F_4$	1,2-Dibromotetrafluoroethane	(11.1)	280
C_2ClF_3	Chlorotrifluoroethylene	9.81 ± 0.03	373
C_2ClF_5	Chloropentafluoroethane	(12.6)	99
C_2Cl_2	Dichloroacetylene	10.09	1183
$C_2Cl_2F_4$	1,2-Dichlorotetrafluoroethane	12.2	252
$C_2Cl_3F_3$	1,1,1-Trichlorotrifluoroethane	11.5	386
$C_2Cl_3F_3$	1,1,2-Trichlorotrifluoroethane	11.99 ± 0.02	429
C_2Cl_4	Tetrachloroethylene	9.32	887
$C_2Cl_4F_2$	Tetrachloro-1,2-difluoroethane	11.3	563
C_2Cl_4O	Trichloroacetyl chloride	(11.0)	827
C_2Cl_6	Hexachloroethane	11.1	920
C_2F_3N	Trifluoroacetonitrile	13.86	838
C_2F_4	Tetrafluoroethylene	10.12 ± 0.02	315
C_2F_6	Hexafluoroethane	(13.4)	–50
C_2N_2	Cyanogen	13.37 ± 0.01	1597
C_2H	Ethynyl (HCC)	(11.7)	1694
C_2HBr	Bromoacetylene	10.31 ± 0.02	1242
$C_2HBrClF_3$	1,1,1-Trifluoro-2-bromo-2-chloroethane	11.0	363
C_2HCl	Chloroacetylene	10.58 ± 0.02	1276
C_2HClF_2	1-Chloro-2,2-difluoroethylene	9.80 ± 0.04	628
C_2HCl_3	Trichloroethylene	9.47 ± 0.01	895
C_2HCl_3O	Dichloroacetyl chloride	(11.0)	819
C_2HCl_5	Pentachloroethane	(11.0)	919
C_2HF	Fluoroacetylene	11.26	1195
C_2HF_3	Trifluoroethylene	10.14	489
$C_2HF_3O_2$	Trifluoroacetic acid	11.46	75
C_2H_2	Acetylene	11.400 ± 0.002	1328
$C_2H_2Cl_2$	1,1-Dichloroethylene	9.79 ± 0.04	947
$C_2H_2Cl_2$	*cis*-1,2-Dichloroethylene	9.66 ± 0.01	936
$C_2H_2Cl_2$	*trans*-1,2-Dichloroethylene	9.65 ± 0.02	935
$C_2H_2Cl_2O$	Chloroacetyl chloride	(11.0)	815
$C_2H_2Cl_4$	1,1,1,2-Tetrachloroethane	(11.1)	920

Molecular formula	Name	*IP*/eV	$\Delta_f H$ (ion)/ kJ/mol
$C_2H_2Cl_4$	1,1,2,2-Tetrachloroethane	(≤11.62)	≤971
$C_2H_2F_2$	1,1-Difluoroethylene	10.29 ± 0.01	650
$C_2H_2F_2$	*cis*-1,2-Difluoroethylene	10.23	690
C_2H_2O	Ketene	9.61 ± 0.02	880
$C_2H_2O_2$	Glyoxal	10.1	763
C_2H_3Br	Bromoethylene	9.80 ± 0.02	1025
C_2H_3Cl	Chloroethylene	9.99 ± 0.02	985
$C_2H_3ClF_2$	1-Chloro-1,1-difluoroethane	11.98 ± 0.01	626
C_2H_3ClO	Acetyl chloride	10.85 ± 0.05	804
C_2H_3ClO	Chloroacetaldehyde	10.48 ± 0.03	815
$C_2H_3ClO_2$	Chloroacetic acid	(10.7)	597
$C_2H_3Cl_3$	1,1,1-Trichloroethane	(11.0)	917
$C_2H_3Cl_3$	1,1,2-Trichloroethane	11.0	911
C_2H_3F	Fluoroethylene	10.363 ± 0.015	861
C_2H_3FO	Acetyl fluoride	11.51 ± 0.02	667
$C_2H_3F_3$	1,1,1-Trifluoroethane	12.9 ± 0.1	496
C_2H_3N	Acetonitrile	12.194 ± 0.005	1252
C_2H_3NO	Methylisocyanate	(10.67 ± 0.02)	900
C_2H_4	Ethylene	10.507 ± 0.004	1066
$C_2H_4Br_2$	1,2-Dibromoethane	10.37	963
$C_2H_4Cl_2$	1,1-Dichloroethane	11.06	937
$C_2H_4Cl_2$	1,2-Dichloroethane	11.04	931
$C_2H_4F_2$	1,1-Difluoroethane	11.87 ± 0.03	643
C_2H_4O	Acetaldehyde	10.229 ± 0.0007	821
C_2H_4O	Ethylene oxide	10.566 ± 0.01	967
$C_2H_4O_2$	Acetic acid	10.66 ± 0.02	596
$C_2H_4O_2$	Methyl formate	10.815 ± 0.005	688
C_2H_5Br	Bromoethane	10.28	930
C_2H_5Cl	Chloroethane	10.97 ± 0.02	946
C_2H_5ClO	2-Chloroethanol	(10.52)	756
C_2H_5F	Fluoroethane	(11.6)	856
C_2H_5I	Iodoethane	9.346	893
C_2H_5N	Ethyleneimine	9.2 ± 0.1	1014
C_2H_5NO	Acetamide	9.65 ± 0.03	693
C_2H_5NO	*N*-Methylformamide	9.79	756
$C_2H_5NO_2$	Nitroethane	10.88 ± 0.05	948
C_2H_6	Ethane	11.52 ± 0.01	1027
$C_2H_6Cl_2Si$	Dichlorodimethylsilane	(10.7)	576
C_2H_6O	Dimethyl ether	10.025 ± 0.025	783
C_2H_6O	Ethanol	10.47 ± 0.02	776
C_2H_6OS	Dimethyl sulfoxide	(9.01)	718
$C_2H_6O_2$	Ethylene glycol	10.16	593
C_2H_6S	Dimethyl sulfide	8.69 ± 0.01	801
C_2H_6S	Ethanethiol	9.285 ± 0.005	849
$C_2H_6S_2$	Dimethyl disulfide	(7.4 ± 0.3)	690
C_2H_7N	Dimethylamine	8.23 ± 0.08	776
C_2H_7N	Ethylamine	8.86 ± 0.02	808
C_2H_7NO	Ethanolamine	8.96	664
$C_2H_8N_2$	1,1-Dimethylhydrazine	7.28 ± 0.04	786
$C_2H_8N_2$	1,2-Ethanediamine	(8.6)	812
C_3F_6	Perfluoropropene	10.60 ± 0.03	–103
C_3F_6O	Perfluoroacetone	(11.44)	–294
C_3F_8	Perfluoropropane	13.38	–491
C_3HN	Cyanoacetylene	11.64 ± 0.01	1475
C_3H_2O	Propynal	(10.8)	1155
$C_3H_3F_3$	3,3,3-Trifluoropropene	(10.9)	437
C_3H_3N	Propenenitrile	10.91 ± 0.01	1237

Molecular formula	Name	IP/eV	$\Delta_f H$ (ion)/ kJ/mol
C_3H_3NO	Oxazole	(9.6)	910
C_3H_3NO	Isoxazole	9.93 ± 0.05	1038
C_3H_4	Propyne	10.36 ± 0.01	1186
C_3H_4	Allene	9.69 ± 0.01	1126
C_3H_4	Cyclopropene	9.67 ± 0.01	1209
$C_3H_4N_2$	Imidazole	8.81 ± 0.01	997
C_3H_4O	Propenal	10.103 ± 0.006	900
C_3H_4O	Propargyl alcohol	10.51	1060
C_3H_4O	Cyclopropanone	(9.1 ± 0.1)	895
$C_3H_4O_2$	Propenoic acid	10.60	701
$C_3H_4O_2$	2-Oxetanone	(9.70 ± 0.01)	653
C_3H_5Br	3-Bromopropene	10.06	1018
C_3H_5Cl	3-Chloropropene	9.9	950
C_3H_5ClO	Epichlorohydrin	(10.2)	875
$C_3H_5ClO_2$	Methyl chloroacetate	(10.3)	575
C_3H_5F	3-Fluoropropene	10.11	821
C_3H_5N	Propanenitrile	11.84 ± 0.02	1194
C_3H_5NO	Propenamide	9.5	720
C_3H_6	Propene	9.73 ± 0.02	959
C_3H_6	Cyclopropane	9.86	1005
$C_3H_6Br_2$	1,2-Dibromopropane	10.1	903
$C_3H_6Br_2$	1,3-Dibromopropane	≤10.26	≤919
$C_3H_6Cl_2$	1,2-Dichloropropane	(10.87 ± 0.05)	886
$C_3H_6Cl_2$	1,3-Dichloropropane	10.85 ± 0.05	888
C_3H_6O	Acetone	9.705	719
C_3H_6O	Allyl alcohol	9.67 ± 0.05	808
C_3H_6O	Propanal	9.953 ± 0.005	773
C_3H_6O	Methyloxirane	10.22 ± 0.02	892
C_3H_6O	Oxetane	9.668 ± 0.005	853
C_3H_6O	Methyl vinyl ether	(8.93 ± 0.02)	761
$C_3H_6O_2$	Ethyl formate	10.61 ± 0.01	639
$C_3H_6O_2$	Methyl acetate	10.27 ± 0.02	581
$C_3H_6O_2$	Propanoic acid	10.525 ± 0.003	568
$C_3H_6O_2$	1,3-Dioxolane	(9.9)	658
$C_3H_6O_3$	Trioxane	(10.3)	528
C_3H_6S	Thiacyclobutane	8.69	899
C_3H_7Br	1-Bromopropane	10.18 ± 0.01	898
C_3H_7Br	2-Bromopropane	10.07 ± 0.01	874
C_3H_7Cl	1-Chloropropane	10.82 ± 0.03	912
C_3H_7Cl	2-Chloropropane	10.78 ± 0.02	895
C_3H_7F	1-Fluoropropane	(11.3)	806
C_3H_7F	2-Fluoropropane	(11.08 ± 0.02)	776
C_3H_7I	1-Iodopropane	9.269	862
C_3H_7I	2-Iodopropane	9.175	844
C_3H_7N	Allylamine	8.76	891
C_3H_7N	Cyclopropylamine	(8.7)	916
C_3H_7N	Propyleneimine	(9.0)	960
C_3H_7NO	*N,N*-Dimethylformamide	9.13 ± 0.02	689
$C_3H_7NO_2$	1-Nitropropane	10.81 ± 0.03	919
$C_3H_7NO_2$	2-Nitropropane	10.71 ± 0.05	894
C_3H_8	Propane	10.95 ± 0.05	952
C_3H_8O	1-Propanol	10.22 ± 0.03	731
C_3H_8O	2-Propanol	10.12 ± 0.08	704
C_3H_8O	Ethyl methyl ether	9.72	722
$C_3H_8O_2$	2-Methoxyethanol	9.6	562
$C_3H_8O_2$	Dimethoxymethane	9.5	569
C_3H_8S	Ethyl methyl sulfide	8.54 ± 0.1	765

Molecular formula	Name	*IP*/eV	$\Delta_f H$ (ion)/ kJ/mol
C_3H_8S	1-Propanethiol	9.195 ± 0.005	819
C_3H_8S	2-Propanethiol	9.14	806
$C_3H_9BO_3$	Trimethylborate	(10.0)	65
C_3H_9ClSi	Trimethylchlorosilane	(10.15)	624
C_3H_9N	Propylamine	8.78 ± 0.02	777
C_3H_9N	Isopropylamine	8.72 ± 0.03	758
C_3H_9N	Trimethylamine	7.82 ± 0.06	731
C_3H_9NO	3-Amino-1-propanol	(9.0)	651
C_4NiO_4	Nickel carbonyl	8.27 ± 0.04	200
$C_4H_2O_3$	Maleic anhydride	(10.8)	645
C_4H_4	1-Buten-3-yne	9.58 ± 0.02	1230
$C_4H_4N_2$	Pyridazine	(8.64)	1112
$C_4H_4N_2$	Pyrimidine	9.23	1087
$C_4H_4N_2$	Succinonitrile	12.1 ± 0.25	1377
C_4H_4O	Furan	8.883 ± 0.003	822
$C_4H_4O_2$	Diketene	(9.6 ± 0.02)	736
$C_4H_4O_3$	Succinic anhydride	(10.6)	500
$C_4H_4O_4$	Fumaric acid	(10.7)	355
C_4H_4S	Thiophene	8.87 ± 0.04	971
C_4H_5N	Cyclopropanecarbonitrile	10.25	1173
C_4H_5N	Pyrrole	8.208 ± 0.005	900
C_4H_5N	Methylacrylonitrile	10.34	1127
C_4H_6	1,2-Butadiene	(9.03)	1034
C_4H_6	1,3-Butadiene	9.069	985
C_4H_6	1-Butyne	10.178 ± 0.005	1147
C_4H_6	2-Butyne	9.562 ± 0.005	1068
C_4H_6	Cyclobutene	9.43	1067
C_4H_6O	*trans*-2-Butenal	9.73 ± 0.01	835
C_4H_6O	2-Methylpropenal	(9.86)	834
C_4H_6O	Divinyl ether	(8.7)	827
C_4H_6O	Cyclobutanone	9.354	815
$C_4H_6O_2$	*cis*-Crotonic acid	(10.08)	625
$C_4H_6O_2$	*trans*-Crotonic acid	(9.9)	604
$C_4H_6O_2$	Methacrylic acid	(10.15)	611
$C_4H_6O_2$	Methyl acrylate	(9.9)	641
$C_4H_6O_2$	Vinyl acetate	9.19	572
$C_4H_6O_3$	Acetic anhydride	(10.0)	398
$C_4H_6O_4$	Dimethyl oxalate	(10.0)	287
C_4H_6S	2,5-Dihydrothiophene	(8.4)	898
C_4H_7N	Butanenitrile	(11.2)	1110
C_4H_7N	2-Methylpropanenitrile	(11.3)	1115
C_4H_7NO	2-Pyrrolidone	(9.2)	674
C_4H_8	1-Butene	9.58 ± 0.02	924
C_4H_8	*cis*-2-Butene	9.108 ± 0.008	871
C_4H_8	*trans*-2-Butene	9.100 ± 0.008	866
C_4H_8	Isobutene	9.239 ± 0.003	875
C_4H_8	Cyclobutane	(9.92 ± 0.05)	986
C_4H_8	Methylcyclopropane	(9.46)	936
$C_4H_8Br_2$	1,4-Dibromobutane	(10.15)	879
C_4H_8O	1,2-Epoxybutane	(10.15)	862
C_4H_8O	Butanal	9.84 ± 0.02	742
C_4H_8O	Isobutanal	9.705 ± 0.005	721
C_4H_8O	2-Butanone	9.51 ± 0.04	677
C_4H_8O	Tetrahydrofuran	9.41 ± 0.02	724
C_4H_8O	Ethyl vinyl ether	(8.8)	707
$C_4H_8O_2$	1,3-Dioxane	9.8	607
$C_4H_8O_2$	1,4-Dioxane	9.19 ± 0.01	571

Molecular formula	Name	*IP*/eV	$\Delta_f H$ (ion)/ kJ/mol
$C_4H_8O_2$	Ethyl acetate	10.01 ± 0.05	522
$C_4H_8O_2$	Methyl propionate	10.15 ± 0.03	548
$C_4H_8O_2$	Propyl formate	10.52 ± 0.02	555
$C_4H_8O_2$	Butanoic acid	10.17 ± 0.05	509
$C_4H_8O_2$	2-Methylpropanoic acid	10.33 ± 0.03	516
$C_4H_8O_2S$	Sulfolane	(9.8)	577
C_4H_8S	Tetrahydrothiophene	8.47	783
C_4H_9Br	1-Bromobutane	10.13	870
C_4H_9Br	1-Bromo-2-methylpropane	10.09 ± 0.02	861
C_4H_9Br	2-Bromobutane	9.98 ± 0.01	842
C_4H_9Br	2-Bromo-2-methylpropane	9.92 ± 0.03	823
C_4H_9Cl	1-Chlorobutane	10.67 ± 0.03	875
C_4H_9Cl	1-Chloro-2-methylpropane	10.66 ± 0.03	870
C_4H_9Cl	2-Chlorobutane	10.53	857
C_4H_9Cl	2-Chloro-2-methylpropane	10.61 ± 0.03	842
C_4H_9I	1-Iodobutane	9.229	840
C_4H_9I	1-Iodo-2-methylpropane	9.202	825
C_4H_9I	2-Iodobutane	9.09 ± 0.02	814
C_4H_9I	2-Iodo-2-methylpropane	9.02 ± 0.03	798
C_4H_9N	Pyrrolidine	(8.0)	769
C_4H_9NO	*N,N*-Dimethylacetamide	8.81	616
C_4H_9NO	Morpholine	(8.2)	841
C_4H_{10}	Butane	10.53 ± 0.10	890
C_4H_{10}	Isobutane	10.57	886
$C_4H_{10}O$	1-Butanol	10.06 ± 0.03	696
$C_4H_{10}O$	2-Butanol	9.88	658
$C_4H_{10}O$	2-Methyl-2-propanol	9.97 ± 0.02	649
$C_4H_{10}O$	2-Methyl-1-propanol	10.12 ± 0.04	693
$C_4H_{10}O$	Diethyl ether	9.51 ± 0.03	666
$C_4H_{10}O$	Methyl propyl ether	(9.42)	671
$C_4H_{10}O$	Isopropyl methyl ether	9.42	657
$C_4H_{10}O_2$	2-Ethoxyethanol	(9.6)	529
$C_4H_{10}O_2$	1,2-Dimethoxyethane	(9.3)	558
$C_4H_{10}S$	1-Butanethiol	9.14 ± 0.02	794
$C_4H_{10}S$	2-Butanethiol	(9.10)	781
$C_4H_{10}S$	2-Methyl-1-propanethiol	(9.12)	783
$C_4H_{10}S$	2-Methyl-2-propanethiol	(9.03)	762
$C_4H_{10}S$	Diethyl sulfide	8.43 ± 0.01	730
$C_4H_{10}S$	Methyl propyl sulfide	(8.8 ± 0.2)	767
$C_4H_{10}S$	Isopropyl methyl sulfide	(8.7 ± 0.2)	749
$C_4H_{10}S_2$	Diethyl disulfide	≤8.27 ± 0.03	≤724
$C_4H_{11}N$	Butylamine	8.71 ± 0.03	748
$C_4H_{11}N$	Isobutylamine	(8.70)	741
$C_4H_{11}N$	*sec*-Butylamine	(8.70)	735
$C_4H_{11}N$	*tert*-Butylamine	(8.64)	713
$C_4H_{11}N$	Diethylamine	8.01 ± 0.01	700
$C_4H_{12}Si$	Tetramethylsilane	9.80 ± 0.04	713
$C_4H_{12}Sn$	Tetramethylstannane	8.89 ± 0.05	837
$C_5H_4O_2$	Furfural	9.21 ± 0.01	738
C_5H_5N	Pyridine	9.25	1031
C_5H_6	1-Penten-3-yne	9.00 ± 0.01	1119
C_5H_6	*cis*-3-Penten-1-yne	9.14 ± 0.04	1137
C_5H_6	*trans*-3-Penten-1-yne	(9.05)	1128
C_5H_6	2-Methyl-1-buten-3-yne	9.23 ± 0.01	1150
C_5H_6	1,3-Cyclopentadiene	8.56 ± 0.01	956
C_5H_6O	2-Methylfuran	8.39 ± 0.01	730
C_5H_6O	3-Methylfuran	(8.64)	763

Molecular formula	Name	*IP*/eV	$\Delta_f H$ (ion)/ kJ/mol
C_5H_6S	2-Methylthiophene	8.61 ± 0.02	914
C_5H_6S	3-Methylthiophene	(8.40)	893
C_5H_8	2-Methyl-1,3-butadiene	8.84 ± 0.01	928
C_5H_8	*cis*-1,3-Pentadiene	8.63 ± 0.03	914
C_5H_8	*trans*-1,3-Pentadiene	8.59 ± 0.02	905
C_5H_8	1,4-Pentadiene	(9.62 ± 0.02)	1034
C_5H_8	Cyclopentene	9.01 ± 0.02	905
C_5H_8	Spiropentane	9.26	1078
C_5H_8	1-Pentyne	10.05	1114
C_5H_8O	Cyclopentanone	9.25 ± 0.01	700
C_5H_8O	Cyclopropyl methyl ketone	9.46	796
C_5H_8O	Dihydropyran	8.34 ± 0.01	680
$C_5H_8O_2$	2,4-Pentanedione	8.85 ± 0.02	469
$C_5H_8O_2$	Ethyl acrylate	(<10.3)	617
$C_5H_8O_2$	Methyl methacrylate	(9.7)	589
C_5H_9NO	*N*-Methyl-2-pyrrolidone	≤9.17	≤676
C_5H_{10}	1-Pentene	9.52 ± 0.02	897
C_5H_{10}	*cis*-2-Pentene	9.036 ± 0.005	846
C_5H_{10}	*trans*-2-Pentene	9.036 ± 0.005	841
C_5H_{10}	2-Methyl-1-butene	9.13 ± 0.03	845
C_5H_{10}	2-Methyl-2-butene	8.68 ± 0.01	795
C_5H_{10}	3-Methyl-1-butene	9.52 ± 0.02	891
C_5H_{10}	Cyclopentane	10.51 ± 0.05	936
$C_5H_{10}O$	Cyclopentanol	9.72	695
$C_5H_{10}O$	2-Pentanone	9.38 ± 0.01	646
$C_5H_{10}O$	3-Pentanone	9.31 ± 0.01	640
$C_5H_{10}O$	3-Methyl-2-butanone	9.30 ± 0.01	635
$C_5H_{10}O$	Tetrahydropyran	9.25 ± 0.01	670
$C_5H_{10}O$	Pentanal	9.74 ± 0.04	709
$C_5H_{10}O$	2,2-Dimethylpropanal	9.50	674
$C_5H_{10}O_2$	Butyl formate	10.50 ± 0.02	582
$C_5H_{10}O_2$	Propyl acetate	10.04 ± 0.03	513
$C_5H_{10}O_2$	Isopropyl acetate	9.99 ± 0.03	482
$C_5H_{10}O_2$	Ethyl propanoate	(10.00 ± 0.02)	500
$C_5H_{10}O_2$	Methyl butanoate	10.07 ± 0.03	520
$C_5H_{10}O_2$	Pentanoic acid	(≤10.53)	≤527
$C_5H_{10}O_2$	3-Methylbutanoic acid	(≤10.51)	≤499
$C_5H_{10}S$	Thiacyclohexane	(8.2)	728
$C_5H_{11}Br$	1-Bromopentane	10.09 ± 0.02	845
$C_5H_{11}I$	1-Iodopentane	9.201	817
$C_5H_{11}N$	*N*-Methylpyrrolidine	(≤8.41 ± 0.02)	≤809
$C_5H_{11}N$	Piperidine	8.05 ± 0.05	728
C_5H_{12}	Pentane	10.35 ± 0.01	852
C_5H_{12}	Isopentane	≤10.22	≤833
C_5H_{12}	Neopentane	≤10.21 ± 0.04	≤818
$C_5H_{12}O$	Butyl methyl ether	(9.54)	662
$C_5H_{12}O$	*tert*-Butyl methyl ether	(9.24)	608
$C_5H_{12}O$	Ethyl propyl ether	(9.45 ± 0.1)	640
$C_5H_{12}O$	1-Pentanol	10.00 ± 0.03	668
$C_5H_{12}O$	2-Pentanol	(9.78 ± 0.03)	630
$C_5H_{12}O$	3-Pentanol	9.78	628
$C_5H_{12}O$	2-Methyl-1-butanol	(9.86)	649
$C_5H_{12}O$	2-Methyl-2-butanol	9.80	615
$C_5H_{12}O$	3-Methyl-2-butanol	(10.01)	650
$C_5H_{12}S$	*tert*-Butyl methyl sulfide	(8.38 ± 0.05)	687
$C_5H_{12}S$	Ethyl propyl sulfide	(8.50 ± 0.05)	716
$C_5H_{12}S$	Ethyl isopropyl sulfide	(8.35 ± 0.01)	689

Molecular formula	Name	*IP*/eV	$\Delta_f H$ (ion)/ kJ/mol
$C_5H_{13}N$	Pentylamine	(8.67)	728
C_6BrF_5	Bromopentafluorobenzene	9.57 ± 0.02	212
C_6ClF_5	Chloropentafluorobenzene	(9.72 ± 0.02)	126
C_6Cl_6	Hexachlorobenzene	8.98	822
C_6F_6	Hexafluorobenzene	9.906	10
C_6F_{12}	Perfluorocyclohexane	(13.2)	–1095
C_6HF_5	Pentafluorobenzene	9.63	122
C_6HF_5O	Pentafluorophenol	9.20 ± 0.02	–71
$C_6H_2F_4$	1,2,3,4-Tetrafluorobenzene	9.53 ± 0.01	284
$C_6H_2F_4$	1,2,3,5-Tetrafluorobenzene	9.53 ± 0.01	263
$C_6H_2F_4$	1,2,4,5-Tetrafluorobenzene	9.35 ± 0.01	254
$C_6H_3Cl_3$	1,2,4-Trichlorobenzene	9.04	880
$C_6H_3Cl_3$	1,3,5-Trichlorobenzene	9.32 ± 0.02	899
$C_6H_4ClNO_2$	*m*-Chloronitrobenzene	(9.92 ± 0.1)	995
$C_6H_4ClNO_2$	*p*-Chloronitrobenzene	9.96 ± 0.1	999
$C_6H_4Cl_2$	*o*-Dichlorobenzene	9.08 ± 0.01	909
$C_6H_4Cl_2$	*m*-Dichlorobenzene	9.11 ± 0.01	907
$C_6H_4Cl_2$	*p*-Dichlorobenzene	8.89 ± 0.01	882
$C_6H_4FNO_2$	*p*-Fluoronitrobenzene	9.90	826
$C_6H_4F_2$	*o*-Difluorobenzene	(9.28 ± 0.01)	602
$C_6H_4F_2$	*m*-Difluorobenzene	9.33 ± 0.01	591
$C_6H_4F_2$	*p*-Difluorobenzene	9.14 ± 0.01	575
$C_6H_4O_2$	*p*-Benzoquinone	10.04 ± 0.18	847
C_6H_5Br	Bromobenzene	8.98 ± 0.02	971
C_6H_5Cl	Chlorobenzene	9.06 ± 0.02	929
C_6H_5ClO	*m*-Chlorophenol	(8.65)	680
C_6H_5ClO	*p*-Chlorophenol	(≤8.69)	≤692
C_6H_5F	Fluorobenzene	9.200 ± 0.005	772
C_6H_5I	Iodobenzene	8.685	1003
$C_6H_5NO_2$	Nitrobenzene	9.86 ± 0.02	1019
$C_6H_5NO_3$	*o*-Nitrophenol	(9.1)	782
$C_6H_5NO_3$	*m*-Nitrophenol	(9.0)	755
$C_6H_5NO_3$	*p*-Nitrophenol	(9.1)	761
C_6H_6	Benzene	9.2459 ± 0.0002	975
C_6H_6	Fulvene	(8.36)	1031
C_6H_6ClN	*o*-Chloroaniline	(8.50)	883
C_6H_6ClN	*m*-Chloroaniline	(8.09 ± 0.1)	835
C_6H_6ClN	*p*-Chloroaniline	(≤8.18)	≤844
$C_6H_6N_2O_2$	*o*-Nitroaniline	8.27 ± 0.01	861
$C_6H_6N_2O_2$	*m*-Nitroaniline	8.31 ± 0.02	865
$C_6H_6N_2O_2$	*p*-Nitroaniline	8.34 ± 0.01	859
C_6H_6O	Phenol	8.47	721
$C_6H_6O_2$	*p*-Hydroquinone	7.95 ± 0.03	504
C_6H_6S	Benzenethiol	8.30 ± 0.02	913
C_6H_7N	Aniline	7.720 ± 0.002	832
C_6H_7N	2-Methylpyridine	9.02 ± 0.03	970
C_6H_7N	3-Methylpyridine	9.04 ± 0.03	979
C_6H_7N	4-Methylpyridine	9.04 ± 0.03	976
$C_6H_8N_2$	*o*-Phenylenediamine	7.2	787
$C_6H_8N_2$	*m*-Phenylenediamine	7.14	777
$C_6H_8N_2$	*p*-Phenylenediamine	6.87 ± 0.05	759
C_6H_{10}	Cyclohexene	8.945 ± 0.01	859
C_6H_{10}	1,5-Hexadiene	9.29 ± 0.05	980
C_6H_{10}	1-Hexyne	(9.95 ± 0.05)	1081
C_6H_{10}	3,3-Dimethyl-1-butyne	(9.80 ± 0.05)	1050
$C_6H_{10}O$	Cyclohexanone	9.14 ± 0.01	656
$C_6H_{10}O$	Mesityl oxide	9.08 ± 0.03	692

Molecular formula	Name	*IP*/eV	$\Delta_f H$ (ion)/ kJ/mol
$C_6H_{10}O_4$	Diethyl oxalate	(9.8)	205
$C_6H_{11}NO$	Caprolactam	(9.07 ± 0.02)	629
C_6H_{12}	Cyclohexane	9.86 ± 0.03	828
C_6H_{12}	Methylcyclopentane	9.85 ± 0.03	845
C_6H_{12}	1-Hexene	9.44 ± 0.04	869
C_6H_{12}	*cis*-2-Hexene	(8.97 ± 0.01)	818
C_6H_{12}	*trans*-2-Hexene	(8.97 ± 0.01)	814
C_6H_{12}	2-Methyl-1-pentene	(9.08 ± 0.01)	817
C_6H_{12}	2-Methyl-2-pentene	(8.58)	761
C_6H_{12}	4-Methyl-1-pentene	(9.45 ± 0.01)	862
C_6H_{12}	4-Methyl-*cis*-2-pentene	(8.98 ± 0.01)	809
C_6H_{12}	4-Methyl-*trans*-2-pentene	(8.97 ± 0.01)	804
C_6H_{12}	2,3-Dimethyl-1-butene	(9.07 ± 0.01)	812
C_6H_{12}	2,3-Dimethyl-2-butene	8.27 ± 0.01	729
C_6H_{12}	2-Ethyl-1-butene	(9.06 ± 0.02)	818
$C_6H_{12}O$	Cyclohexanol	(9.75)	651
$C_6H_{12}O$	2-Hexanone	9.35 ± 0.02	626
$C_6H_{12}O$	3-Hexanone	9.12 ± 0.02	600
$C_6H_{12}O$	2-Methyl-3-pentanone	9.10 ± 0.01	592
$C_6H_{12}O$	3-Methyl-2-pentanone	9.21 ± 0.01	600
$C_6H_{12}O$	4-Methyl-2-pentanone	9.30 ± 0.01	609
$C_6H_{12}O$	3,3-Dimethyl-2-butanone	9.11 ± 0.02	589
$C_6H_{12}O$	1-Hexanal	9.67 ± 0.05	686
$C_6H_{12}O_2$	Butyl acetate	10.0	479
$C_6H_{12}O_2$	*sec*-Butyl acetate	9.90	453
$C_6H_{12}O_2$	Methyl pentanoate	(10.4 ± 0.2)	532
$C_6H_{12}O_2$	Methyl 2,2-dimethylpropanoate	(9.90 ± 0.04)	466
$C_6H_{12}O_2$	Hexanoic acid	≤10.12	≤463
$C_6H_{13}I$	1-Iodohexane	9.179	794
$C_6H_{13}N$	Cyclohexylamine	(8.62 ± 0.24)	727
C_6H_{14}	Hexane	10.13	810
C_6H_{14}	2-Methylpentane	(10.12)	802
C_6H_{14}	3-Methylpentane	(10.08)	801
C_6H_{14}	2,2-Dimethylbutane	(10.06)	787
C_6H_{14}	2,3-Dimethylbutane	(10.02)	791
$C_6H_{14}O$	Methyl pentyl ether	(≤9.67)	≤657
$C_6H_{14}O$	Butyl ethyl ether	9.36	610
$C_6H_{14}O$	Dipropyl ether	9.27 ± 0.05	602
$C_6H_{14}O$	Diisopropyl ether	9.20 ± 0.05	569
$C_6H_{14}O$	1-Hexanol	(9.89 ± 0.03)	639
$C_6H_{14}O$	2-Hexanol	(9.80 ± 0.03)	611
$C_6H_{14}O$	3-Hexanol	(9.63 ± 0.03)	599
$C_6H_{14}O_2$	1,1-Diethoxyethane	≤9.78	≤490
$C_6H_{14}O_3$	Diethylene glycol dimethyl ether	≤9.8	≤448
$C_6H_{14}S$	Dipropyl sulfide	8.30 ± 0.02	676
$C_6H_{14}S$	Diisopropyl sulfide	8.0	630
$C_6H_{15}N$	Hexylamine	(8.63 ± 0.05)	699
$C_6H_{15}N$	Triethylamine	7.50	631
$C_6H_{15}N$	Dipropylamine	7.84 ± 0.02	641
$C_6H_{15}N$	Diisopropylamine	(7.73 ± 0.03)	602
$C_6H_{15}NO_3$	Triethanolamine	(7.9)	206
C_7F_8	Perfluorotoluene	(9.9)	–233
$C_7H_3F_5$	2,3,4,5,6-Pentafluorotoluene	(9.4)	64
C_7H_5Cl0	Benzoyl chloride	9.54	816
$C_7H_5Cl_3$	Trichloromethylbenzene	≤9.60	≤914
$C_7H_5F_3$	(Trifluoromethyl)benzene	9.685 ± 0.004	335
C_7H_5N	Benzonitrile	9.62	1146

Molecular formula	Name	IP/eV	$\Delta_f H$ (ion)/ kJ/mol
C_7H_6O	Benzaldehyde	9.49 ± 0.02	878
$C_7H_6O_2$	Benzoic acid	(9.47)	620
C_7H_7Br	*p*-Bromotoluene	8.67 ± 0.01	908
C_7H_7Cl	*o*-Chlorotoluene	(8.83 ± 0.02)	869
C_7H_7Cl	*m*-Chlorotoluene	(8.83 ± 0.02)	869
C_7H_7Cl	*p*-Chlorotoluene	8.69 ± 0.02	855
C_7H_7F	*o*-Fluorotoluene	8.91 ± 0.01	709
C_7H_7F	*m*-Fluorotoluene	8.91 ± 0.01	709
C_7H_7F	*p*-Fluorotoluene	8.79 ± 0.01	701
C_7H_7NO	Benzamide	9.45	811
$C_7H_7NO_2$	*o*-Nitrotoluene	9.45 ± 0.04	966
$C_7H_7NO_2$	*m*-Nitrotoluene	(9.48 ± 0.02)	944
$C_7H_7NO_2$	*p*-Nitrotoluene	(9.4)	936
C_7H_8	Toluene	8.82 ± 0.01	901
C_7H_8O	*o*-Cresol	8.14	660
C_7H_8O	*m*-Cresol	8.29	668
C_7H_8O	*p*-Cresol	8.13	659
C_7H_8O	Benzyl alcohol	(8.5)	720
C_7H_8O	Anisole	8.21 ± 0.02	724
C_7H_9N	*N*-Methylaniline	7.33 ± 0.02	791
C_7H_9N	*o*-Methyl aniline	7.44 ± 0.02	772
C_7H_9N	*m*-Methyl aniline	7.50 ± 0.02	778
C_7H_9N	*p*-Methyl aniline	(7.24 ± 0.02)	753
C_7H_9N	2,3-Dimethylpyridine	(8.85 ± 0.02)	922
C_7H_9N	2,4-Dimethylpyridine	(8.85 ± 0.03)	918
C_7H_9N	2,5-Dimethylpyridine	(≤8.80 ± 0.05)	≤916
C_7H_9N	2,6-Dimethylpyridine	8.86 ± 0.03	913
C_7H_9N	3,4-Dimethylpyridine	(≤9.15)	≤953
C_7H_9N	3,5-Dimethylpyridine	(≤9.25)	≤965
C_7H_9N	Benzylamine	8.64 ± 0.05	917
$C_7H_{10}O$	Dicyclopropyl ketone	(9.1)	1041
C_7H_{14}	Cycloheptane	9.97	844
C_7H_{14}	Methylcyclohexane	9.64	775
C_7H_{14}	Ethylcyclopentane	(10.12 ± 0.02)	850
C_7H_{14}	*cis*-1,2-Dimethylcyclopentane	(9.92 ± 0.05)	828
C_7H_{14}	*trans*-1,2-Dimethylcyclopentane	(9.95 ± 0.05)	823
C_7H_{14}	1-Heptene	(9.44)	849
C_7H_{14}	2-Heptene	(8.84 ± 0.02)	782
C_7H_{14}	3-Heptene	(8.92)	790
$C_7H_{14}O$	2-Heptanone	9.30 ± 0.01	596
$C_7H_{14}O$	5-Methyl-2-hexanone	(9.28 ± 0.01)	586
$C_7H_{14}O$	2,4-Dimethyl-3-pentanone	8.95 ± 0.01	552
$C_7H_{14}O$	1-Heptanal	(9.65 ± 0.02)	668
$C_7H_{14}O$	1-Methylcyclohexanol	(9.8 ± 0.2)	586
C_7H_{16}	Heptane	9.92 ± 0.05	770
$C_7H_{16}O$	Ethyl pentyl ether	(≤9.49)	≤602
$C_7H_{16}O$	1-Heptanol	(9.84 ± 0.03)	614
$C_7H_{16}O$	2-Heptanol	(9.70 ± 0.03)	580
$C_7H_{16}O$	3-Heptanol	9.68 ± 0.03	578
$C_7H_{16}O$	4-Heptanol	(9.61 ± 0.03)	572
$C_8H_4O_3$	Phthalic anhydride	(10.0)	593
$C_8H_6O_4$	Isophthalic acid	(9.98 ± 0.2)	268
$C_8H_6O_4$	Terephthalic acid	(9.86 ± 0.2)	232
C_8H_7N	*o*-Tolunitrile	9.38	1085
C_8H_7N	*m*-Tolunitrile	9.34	1085
C_8H_7N	*p*-Tolunitrile	9.32	1083
C_8H_8	Styrene	8.43 ± 0.06	961

Molecular formula	Name	*IP*/eV	$\Delta_f H$ (ion)/ kJ/mol
C_8H_8O	Acetophenone	9.29 ± 0.03	810
C_8H_8O	*p*-Tolualdehyde	9.33 ± 0.05	825
$C_8H_8O_2$	*o*-Toluic acid	(9.1)	558
$C_8H_8O_2$	*m*-Toluic acid	(9.43 ± 0.2)	579
$C_8H_8O_2$	*p*-Toluic acid	(9.23 ± 0.2)	560
$C_8H_8O_2$	Phenylacetic acid	(8.26)	479
$C_8H_8O_2$	Methyl benzoate	9.32 ± 0.03	611
C_8H_{10}	Ethylbenzene	8.77 ± 0.01	876
C_8H_{10}	*o*-Xylene	8.56 ± 0.01	844
C_8H_{10}	*m*-Xylene	8.56 ± 0.01	843
C_8H_{10}	*p*-Xylene	8.44 ± 0.01	832
$C_8H_{10}O$	*p*-Ethylphenol	(7.84)	613
$C_8H_{10}O$	2,3-Xylenol	(8.26)	640
$C_8H_{10}O$	2,4-Xylenol	(8.0)	609
$C_8H_{10}O$	2,6-Xylenol	8.05 ± 0.02	615
$C_8H_{10}O$	3,4-Xylenol	(8.09)	624
$C_8H_{10}O$	Phenetole	8.13 ± 0.02	683
$C_8H_{11}N$	*N,N*-Dimethylaniline	7.12 ± 0.02	787
$C_8H_{11}N$	*N*-Ethylaniline	(≤7.67)	≤794
$C_8H_{11}N$	2,4,6-Trimethylpyridine	(≤8.9 ± 0.1)	≤880
C_8H_{14}	1-Octyne	(9.95 ± 0.02)	1040
C_8H_{14}	2-Octyne	9.31 ± 0.01	961
C_8H_{14}	3-Octyne	9.22 ± 0.01	952
C_8H_{14}	4-Octyne	9.20 ± 0.01	946
C_8H_{16}	Cyclooctane	9.76	817
C_8H_{16}	Propylcyclopentane	(10.00 ± 0.04)	817
C_8H_{16}	Ethylcyclohexane	9.54	748
C_8H_{16}	1,1-Dimethylcyclohexane	9.42	728
C_8H_{16}	*cis*-1,2-Dimethylcyclohexane	(<9.78)	772
C_8H_{16}	*trans*-1,2-Dimethylcyclohexane	9.41	728
C_8H_{16}	*cis*-1,3-Dimethylcyclohexane	(<9.98)	778
C_8H_{16}	*trans*-1,3-Dimethylcyclohexane	9.53	743
C_8H_{16}	*cis*-1,4-Dimethylcyclohexane	(<9.93)	782
C_8H_{16}	*trans*-1,4-Dimethylcyclohexane	9.56	738
C_8H_{16}	1-Octene	9.43 ± 0.01	829
$C_8H_{16}O$	2,2,4-Trimethyl-3-pentanone	(8.80 ± 0.01)	511
C_8H_{18}	Octane	(9.82)	739
C_8H_{18}	2-Methylheptane	9.84	734
C_8H_{18}	2,2,4-Trimethylpentane	9.86	713
C_8H_{18}	2,2,3,3-Tetramethylbutane	9.8	720
$C_8H_{18}O$	Dibutyl ether	≤9.43	≤575
$C_8H_{18}O$	Di-*sec*-butyl ether	(9.11)	511
$C_8H_{18}O$	Di-*tert*-butyl ether	8.81	486
$C_8H_{18}S$	Dibutyl sulfide	(8.2)	624
$C_8H_{18}S$	Diisobutyl sulfide	8.36 ± 0.05	627
$C_8H_{18}S$	Di-*tert*-butyl sulfide	(8.0)	583
$C_8H_{19}N$	Dibutylamine	(7.69 ± 0.03)	586
$C_8H_{19}N$	Diisobutylamine	(7.81)	574
$C_8H_{20}Si$	Tetraethylsilane	(8.9)	595
C_9H_7N	Quinoline	8.62 ± 0.01	1041
C_9H_7N	Isoquinoline	8.53 ± 0.03	1032
C_9H_8	Indene	8.14 ± 0.01	949
C_9H_{10}	Cyclopropylbenzene	8.35	956
C_9H_{10}	*o*-Methylstyrene	8.20 ± 0.02	908
C_9H_{10}	*m*-Methylstyrene	8.15 ± 0.02	899
C_9H_{10}	*p*-Methylstyrene	8.1 ± 0.1	895
C_9H_{10}	Indane	(8.3)	864

Molecular formula	Name	*IP*/eV	$\Delta_f H$ (ion)/ kJ/mol
$C_9H_{10}O_2$	Ethyl benzoate	(8.9)	537
C_9H_{12}	Cumene	8.73 ± 0.01	847
C_9H_{12}	Propylbenzene	8.72 ± 0.01	849
C_9H_{12}	1,2,3-Trimethylbenzene	8.42 ± 0.02	803
C_9H_{12}	1,2,4-Trimethylbenzene	8.27 ± 0.01	784
C_9H_{12}	Mesitylene	8.41 ± 0.01	796
$C_9H_{13}N$	*N,N*-Dimethyl-*o*-toluidine	7.40 ± 0.02	814
$C_9H_{14}O$	Isophorone	(≤9.07)	≤670
C_9H_{18}	Butylcyclopentane	(9.95 ± 0.03)	793
C_9H_{18}	Propylcyclohexane	(9.46)	720
C_9H_{18}	Isopropylcyclohexane	(9.33)	704
$C_9H_{18}O$	2-Nonanone	(9.16)	545
$C_9H_{18}O$	5-Nonanone	(9.07)	530
$C_9H_{18}O$	2,6-Dimethyl-4-heptanone	9.04 ± 0.03	515
C_9H_{20}	Nonane	(9.72)	710
$C_{10}F_8$	Perfluoronaphthalene	8.85	–368
$C_{10}H_7Br$	1-Bromonaphthalene	(8.09)	956
$C_{10}H_7Cl$	1-Chloronaphthalene	(8.13)	906
$C_{10}H_8$	Azulene	7.41 ± 0.02	1004
$C_{10}H_8$	Naphthalene	8.14 ± 0.01	936
$C_{10}H_8O$	1-Naphthol	7.76 ± 0.03	719
$C_{10}H_8O$	2-Naphthol	7.85 ± 0.05	727
$C_{10}H_{10}O_4$	Dimethyl phthalate	(9.64 ± 0.07)	277
$C_{10}H_{12}$	1,2,3,4-Tetrahydronaphthalene	8.47	[illegible]
$C_{10}H_{14}$	Butylbenzene	8.69 ± 0.01	826
$C_{10}H_{14}$	Isobutylbenzene	8.68 ± 0.01	816
$C_{10}H_{14}$	*sec*-Butylbenzene	8.68 ± 0.01	820
$C_{10}H_{14}$	*tert*-Butylbenzene	8.64 ± 0.02	812
$C_{10}H_{14}$	*p*-Cymene	(8.29)	771
$C_{10}H_{14}$	*o*-Diethylbenzene	≤8.51	≤804
$C_{10}H_{14}$	*m*-Diethylbenzene	(8.49 ± 0.01)	798
$C_{10}H_{14}$	*p*-Diethylbenzene	8.40	790
$C_{10}H_{14}$	1,2,4,5-Tetramethylbenzene	8.04 ± 0.01	730
$C_{10}H_{14}O$	*p-tert*-Butylphenol	(7.8)	552
$C_{10}H_{16}$	α-Pinene	(8.07)	808
$C_{10}H_{16}O$	Camphor	(8.76 ± 0.03)	577
$C_{10}H_{18}$	*cis*-Decahydronaphthalene	9.26	724
$C_{10}H_{18}$	*trans*-Decahydronaphthalene	9.24	710
$C_{10}H_{20}$	Butylcyclohexane	9.41	695
$C_{10}H_{20}$	1-Decene	9.42 ± 0.01	786
$C_{10}H_{22}$	Decane	9.65	682
$C_{11}H_{10}$	1-Methylnaphthalene	7.85	870
$C_{11}H_{10}$	2-Methylnaphthalene	(7.8)	866
$C_{11}H_{16}$	*p-tert*-Butyltoluene	8.28	745
$C_{11}H_{24}$	Undecane	(9.56)	650
$C_{11}H_{24}$	2-Methyldecane	(9.68)	658
$C_{12}H_8$	Acenaphthylene	(8.22 ± 0.04)	1053
$C_{12}H_9N$	Carbazole	7.57 ± 0.03	961
$C_{12}H_{10}$	Acenaphthene	(7.68)	896
$C_{12}H_{10}$	Biphenyl	7.95 ± 0.02	950
$C_{12}H_{10}N_2O$	Azoxybenzene	(8.1)	1123
$C_{12}H_{10}O$	Diphenyl ether	8.09 ± 0.03	766
$C_{12}H_{11}N$	Diphenylamine	7.16 ± 0.04	908
$C_{12}H_{18}$	5,7-Dodecadiyne	(8.67)	1079
$C_{12}H_{18}$	Hexamethylbenzene	7.85	670
$C_{12}H_{22}$	Cyclohexylcyclohexane	(9.41)	690
$C_{12}H_{27}N$	Tributylamine	(7.4)	492

Molecular formula	Name	IP/eV	$\Delta_f H$ (ion)/ kJ/mol
$C_{13}H_{10}$	Fluorene	7.89 ± 0.03	950
$C_{13}H_{10}O$	Benzophenone	9.05 ± 0.05	923
$C_{13}H_{12}$	Diphenylmethane	8.55 ± 0.03	963
$C_{14}H_{10}$	Anthracene	7.45 ± 0.03	949
$C_{14}H_{10}$	Phenanthrene	7.86 ± 0.02	963
$C_{14}H_{10}$	Diphenylacetylene	7.90 ± 0.02	1164
$C_{14}H_{12}$	*cis*-Stilbene	(7.80 ± 0.02)	1005
$C_{14}H_{12}$	*trans*-Stilbene	7.70 ± 0.03	977
$C_{14}H_{14}$	1,2-Diphenylethane	8.7 ± 0.1	983
$C_{16}H_{10}$	Fluoranthene	(7.95 ± 0.04)	1057
$C_{16}H_{10}$	Pyrene	7.41	933
$C_{18}H_{12}$	Chrysene	7.59 ± 0.02	1016
$C_{18}H_{14}$	*o*-Terphenyl	8.0	1056
$C_{18}H_{14}$	*m*-Terphenyl	8.01 ± 0.01	1057
$C_{18}H_{14}$	*p*-Terphenyl	7.78 ± 0.01	1035
$C_{20}H_{12}$	Perylene	6.90 ± 0.01	975
$C_{24}H_{12}$	Coronene	7.29	1026

X-RAY WAVELENGTHS

J. A. Bearden

These tables were originally published as the final report to the U.S. Atomic Energy Commission as Report NYO-10586 in partial fulfillment of Contract AT(30-1)-2543. The tables were later reproduced in *Review of Modern Physics*. The data may also be obtained from the Superintendent of Documents, U.S. Government Printing Office, Washington, D. C. 20402 in the publication NSRDS-NBS 14. Persons seeking discussion of the experimental work, conventions, secondary standards, etc. will find these in *Review of Modern Physics*, Vol. 39, No. 1, 78-124, January 1967.

THE W $K\alpha_1$ WAVELENGTH STANDARD

A wavelength standard should possess characteristics which permit its ready redetermination in other laboratories by different techniques. Considering all of the factors involved in the selection of a wavelength standard, the W $K\alpha_1$ line is superior to any other x-ray or γ-ray wavelength. Its advantages as the x-ray wavelength standard are discussed in *Review of Modern Physics* Vol. 39, page 82 (1967).

$$\lambda \text{W } K\alpha_1 = 0.2090100 \text{ Å} \pm 5 \text{ ppm.}$$

This numerical value of the wavelength of the W $K\alpha_1$ line is used to define the *x-ray wavelength standard* by the relation

$$\lambda(\text{W } K\alpha_1) = 0.2090100 \text{ Å*}.$$

This is a new unit of length which may differ from the angstrom by ±5 ppm (probable error), *but as a wavelength standard it has no error*. In order to clearly indicate that this unit is not exactly an angstrom, it has been designated Å*.

Wavelengths tabulated normally refer to the pure element in its solid form. However, there are many instances in which such data are not available. For example, rare gases are of necessity almost always used in the gaseous form, while the rare-earth elements were customarily used in the form of salts.

In high precision work there is some ambiguity as to exactly what feature of a line profile should be taken to be the "true wavelength." In double-crystal work the line peak is usually employed. In crystallography the centroid is widely used; in photographic work with visual observation of the plates, there is involved some subjective criterion of the observer which it is difficult to define precisely. In this survey the peak of the line profile has been adopted as the standard criterion.

In the study of the X-ray literature, the wavelengths of a number of lines were noted which appeared inconsistent with the remaining data. A Moseley-type diagram was constructed, and if the value was clearly outside estimated probable error, it was assumed that an experimental or typographical error had occurred, and the interpolated value was listed in the table. Such cases are marked with a dagger (†) as a superscript to the wavelength. For elements of atomic number 85 through 89 and 91, there are no measured lines of the *K* series and very few of other series except for 88 radium and 91 protactinium. Likewise there are very few measurements for 43 technetium and 54 xenon. In these cases, interpolated values are listed for the more prominent lines and marked with a dagger (†).

X-ray wavelengths in Å* units and in keV. The probable error (p.e.) is the error in the last digit of wavelength. Designation indicates both conventional Siegbahn notation (if applicable) and transition, e.g., $\beta_1\ L_{II}M_{IV}$ denotes a transition between the L_{II} and M_{IV} levels, which is the $L\beta_1$ line in Siegbahn notation.

Designation	Å*	p.e.	keV	Å*	p.e.	keV
	3 Lithium			**4 Beryllium**		
$\alpha\ KL$	228.	1	0.0543	114.	1	0.1085
	5 Boron			**6 Carbon**		
$\alpha\ KL$	67.6	3	0.1833	44.7	3	0.277
	7 Nitrogen			**8 Oxygen**		
$\alpha\ KL$	31.6	4	0.3924	23.62	3	0.5249
	9 Fluorine			**10 Neon**		
$\alpha_{1,2}\ KL_{II,III}$	18.32	2	0.6768	14.610	3	0.8486
$\beta\ KM$				14.452	5	0.8579
	11 Sodium			**12 Magnesium**		
$\alpha_{1,2}\ KL_{II,III}$	11.9101	9	1.0410	9.8900	2	1.25360
$\beta\ KM$	11.575	2	1.0711	9.521	2	1.3022
$L_{II,III}M$	407.1	5	0.03045	251.5	5	0.0493
$L_IL_{II,III}$	376	1	0.0330	317	1	0.0392
	13 Aluminum			**14 Silicon**		
$\alpha_2\ KL_{II}$	8.34173	9	1.48627	7.12791	9	1.73938
$\alpha_1\ KL_{III}$	8.33934	9	1.48670	7.12542	9	1.73998
$\beta\ KM$	7.960	2	1.5574	6.753	1	1.8359
$L_{II,III}$	171.4	5	0.0724	135.5	4	0.0915
$L_IL_{II,III}$	290.	1	0.0428			
	15 Phosphorus			**16 Sulfur**		
$\alpha_2\ KL_{II}$	6.160†	1	2.0127	5.37496	8	2.30664
$\alpha_1\ KL_{III}$	6.157†	1	2.0137	5.37216	7	2.30784
$\beta\ KM$	5.796	2	2.1390			
$\beta_1\ KM$				5.0316	2	2.4640
$\beta_2\ KM$				5.0233	3	2.4681
$L_{II,III}M$	103.8	4	0.1194			
$l,\eta\ L_{II,III}M_I$				83.4	3	0.1487
	17 Chlorine			**18 Argon**		
$\alpha_2\ KL_{II}$	4.7307	1	2.62078	4.19474	5	2.95563
$\alpha_1\ KL_{III}$	4.7278	1	2.62239	4.19180	5	2.95770
$\beta\ KM$	4.4034	3	2.8156			
$\beta_{1,3}\ KM_{II,III}$				3.8860	2	3.1905
$\eta\ L_{II}M_I$	67.33	9	0.1841	55.9†	1	0.2217
$l\ L_{III}M_I$	67.90	9	0.1826	56.3†	1	0.2201
	19 Potassium			**20 Calcium**		
$\alpha_2\ KL_{II}$	3.7445	2	3.3111	3.36166	3	3.68809
$\alpha_1\ KL_{III}$	3.7414	2	3.3138	3.35839	3	3.69168
$\beta_{1,3}\ KM_{II,III}$	3.4539	2	3.5896	3.0897	2	4.0127
$\beta_5\ KM_{IV,V}$	3.4113	4	3.6027	3.0746	3	4.0325
	19 Potassium (*Cont.*)			**20 Calcium (*Cont.*)**		
$\eta\ L_{II}M_I$	47.24	2	0.2625	40.46	2	0.3064
β_1				35.94	2	0.3449
$l\ L_{III}M_I$	47.74	1	0.25971	40.96	2	0.3027
$\alpha_{1,2}\ L_{III}M_{IV,V}$				36.33	2	0.3413
$M_{II,III}N_I$	692	9	0.0179	525.	9	0.0236
	21 Scandium			**22 Titanium**		
$\alpha_2\ KL_{II}$	3.0342	1	4.0861	2.75216	2	4.50486
$\alpha_1\ KL_{III}$	3.0309†	1	4.0906	2.74851	2	4.51084
$\beta_{1,3}\ KM_{II,III}$	2.7796	2	4.4605	2.51391	2	4.93181
$\beta_5\ KM_{IV,V}$	2.7634	3	4.4865	2.4985	2	4.9623
$\eta\ L_{II}M_I$	35.13	2	0.3529	30.89	3	0.4013
$\beta_1\ L_{II}M_{IV}$	31.02	2	0.3996	27.05	2	0.4584
$l\ L_{III}M_I$	35.59	3	0.3483	31.36	2	0.3953
$\alpha_{1,2}\ L_{III}M_{IV,V}$	31.35	3	0.3954	27.42	2	0.4522
	23 Vanadium			**24 Chromium**		
$\alpha_2\ KL_{II}$	2.50738	2	4.94464	2.293606	3	5.40551
$\alpha_1\ KL_{III}$	2.50356	2	4.95220	2.28970	2	5.41472
$\beta_{1,3}\ KM_{II,III}$	2.28440	2	5.42729	2.08487	2	5.94671
$\beta_5\ KM_{IV,V}$	2.26951	6	5.4629	2.07087	6	5.9869
$\beta_{3,4}\ L_IM_{II,III}$	21.19†	9	0.585	18.96	2	0.654
$\eta\ L_{II}M_I$	27.34	3	0.4535	24.30	3	0.5102
$\beta_1\ L_{II}M_{IV}$	23.88	4	0.5192	21.27	1	0.5828
$l\ L_{III}M_I$	27.77	1	0.4465	24.78	1	0.5003
$\alpha_{1,2}\ L_{III}M_{IV,V}$	24.25	3	0.5113	21.64	3	0.5728
$M_{II,III}M_{IV,V}$	337.	9	0.037	309.	9	0.040
	25 Manganese			**26 Iron**		
$\alpha_2\ KL_{II}$	2.10578	2	5.88765	1.939980	9	6.39084
$\alpha_1\ KL_{III}$	2.101820	9	5.89875	1.936042	9	6.40384
$\beta_{1,3}\ KM_{II,III}$	1.91021	2	6.49045	1.75661	2	7.05798
$\beta_5\ KM_{IV,V}$	1.8971	1	6.5352	1.7442	1	7.1081
$\beta_{3,4}\ L_IM_{II,III}$	17.19	2	0.721	15.65	2	0.792
$\eta\ L_{II}M_I$	21.85	2	0.5675	19.75	4	0.628
$\beta_1\ L_{II}M_{IV}$	19.11	2	0.6488	17.26	1	0.7185
$l\ L_{III}M_I$	22.29	1	0.5563	20.15	1	0.6152
$\alpha_{1,2}\ L_{III}M_{IV,V}$	19.45	1	0.6374	17.59	2	0.7050
$M_{II,III}M_{IV,V}$	273.	6	0.045	243.	5	0.051
	27 Cobalt			**28 Nickel**		
$\alpha_2\ KL_{II}$	1.792850	9	6.91530	1.661747	8	7.46089
$\alpha_1\ KL_{III}$	1.788965	9	6.93032	1.657910	8	7.47815
$\beta_{1,3}\ KM_{II,III}$	1.62079	2	7.64943	1.500135	8	8.26466
$\beta_5\ KM_{IV,V}$	1.60891	3	7.7059	1.48862	4	8.3286
$\beta_{3,4}\ L_IM_{II,III}$	14.31	3	0.870	13.18	1	0.941
$\eta\ L_{II}M_I$	17.87	3	0.694	16.27	3	0.762
$\beta_1\ L_{II}M_{IV}$	15.666	8	0.7914	14.271	6	0.8688
$l\ L_{III}M_I$	18.292	8	0.6778	16.693	9	0.7427
$\alpha_{1,2}\ L_{III}M_{IV,V}$	15.972	6	0.7762	14.561	3	0.8515
$M_{II,III}M_{IV,V}$	214.	6	0.058	190.	2	0.0651

TABLE A (*Continued*)

Designation	Å*	p.e.	keV	Å*	p.e.	keV
	29 Copper			**30 Zinc**		
$\alpha_2\ KL_{II}$	1.544390	2	8.02783	1.439000	8	8.61578
$\alpha_1\ KL_{III}$	1.540562	2	8.04778	1.435155	7	8.63886
$\beta_3\ KM_{II}$	1.3926	1	8.9029			
$\beta_{1,3}\ KM_{II,III}$	1.392218	9	8.90529	1.29525	2	9.5720
$\beta_2\ KN_{II,III}$				1.28372	2	9.6580
$\beta_5\ KM_{IV,V}$	1.38109	3	8.9770	1.2848	1	9.6501
$\beta_{3,4}\ L_IM_{II,III}$	12.122	8	1.0228	11.200	7	1.1070
$\eta\ L_{II}M_I$	14.90	2	0.832	13.68	2	0.906
$\beta_1\ L_{II}M_{IV}$	13.053	3	0.9498	11.983	3	1.0347
$l\ L_{III}M_I$	15.286	9	0.8111	14.02	2	0.884
$\alpha_{1,2}\ L_{III}M_{IV,V}$	13.336	3	0.9297	12.254	3	1.0117
$M_{II,III}M_{IV,V}$	173.	3	0.072	157.	3	0.079
	31 Gallium			**32 Germanium**		
$\alpha_2\ KL_{II}$	1.34399	1	9.22482	1.258011	9	9.85532
$\alpha_1\ KL_{III}$	1.340083	9	9.25174	1.254054	9	9.88642
$\beta_3\ KM_{II}$	1.20835	5	10.2603	1.12936	9	10.9780
$\beta_1\ KM_{III}$	1.20789	2	10.2642	1.12894	2	10.9821
$\beta_2\ KN_{II,III}$	1.19600	2	10.3663	1.11686	2	11.1008
$\beta_5\ KM_{IV,V}$	1.1981	2	10.348	1.1195	1	11.0745
$\beta_4\ L_IM_{II}$				9.640	2	1.2861
$\beta_3\ L_IM_{III}$				9.581	2	1.2941
$\beta_{3,4}\ L_IM_{II,III}$	10.359†	8	1.197			
$\eta\ L_{II}M_I$	12.597	2	0.9842	11.609	2	1.0680
$\beta_1\ L_{II}M_{IV}$	11.023	2	1.1248	10.175	1	1.2185
$l\ L_{III}M_I$	12.953	2	0.9572	11.965	4	1.0362
$\alpha_{1,2}\ L_{III}M_{IV,V}$	11.292	1	1.09792	10.4361	8	1.18800
	33 Arsenic			**34 Selenium**		
$\alpha_2\ KL_{II}$	1.17987	1	10.50799	1.10882	2	11.1814
$\alpha_1\ KL_{III}$	1.17588	1	10.54372	1.10477	2	11.2224
$\beta_3\ KM_{II}$	1.05783	5	11.7203	0.99268	5	12.4896
$\beta_1\ KM_{III}$	1.05730	2	11.7262	0.99218	3	12.4959
$\beta_2\ KN_{II,III}$	1.04500	3	11.8642	0.97992	5	12.6522
$\beta_5\ KM_{IV,V}$	1.0488	1	11.822	0.9843	1	12.595
$\beta_{3,4}\ L_IM_{II,III}$	8.929	1	1.3884	8.321†	9	1.490
$\eta\ L_{II}M_I$	10.734	1	1.1550	9.962	1	1.2446
$\beta_1\ L_{II}M_{IV}$	9.4141	8	1.3170	8.7358	5	1.41923
$l\ L_{III}M_I$	11.072	1	1.1198	10.294	1	1.2044
$\alpha_{1,2}\ L_{III}M_{IV,V}$	9.6709	8	1.2820	8.9900	5	1.37910
M_VN_{III}				230.	2	0.0538
	35 Bromine			**36 Krypton**		
$\alpha_2\ KL_{II}$	1.04382	2	11.8776	0.9841	1	12.598
$\alpha_1\ KL_{III}$	1.03974	2	11.9242	0.9801	1	12.649
$\beta_3\ KM_{II}$	0.93327	5	13.2845	0.8790	1	14.104
$\beta_1\ KM_{III}$	0.93279	2	13.2914	0.8785	1	14.112
$\beta_2\ KN_{II,III}$	0.92046	2	13.4695	0.8661	1	14.315
$\beta_5\ KM_{IV,V}$	0.9255	1	13.396	0.8708	2	14.238
$\beta_4\ KN_{IV,V}$				0.8653	2	14.328
$\beta_4\ L_IM_{II}$				7.304	5	1.697
$\beta_3\ L_IM_{III}$				7.264	5	1.707

Designation	Å*	p.e.	keV	Å*	p.e.	keV
	35 Bromine (*Cont.*)			**36 Krypton** (*Cont.*)		
$\beta_{3,4}\ L_IM_{II,III}$	7.767†	9	1.596			
$\eta\ L_{II}M_I$	9.255	1	1.3396			
$\beta_1\ L_{II}M_{IV}$	8.1251	5	1.52590	7.576†	3	1.6366
γ_5				7.279	5	1.703
$l\ L_{III}M_I$	9.585	1	1.2935			
$\alpha_{1,2}\ L_{III}M_{IV,V}$	8.3746	5	1.48043	7.817†	3	1.5860
β_6				7.510	4	1.6510
$L_{II}N_{III}$				7.250	5	1.710
M_IM_{II}	184.6	3	0.0672			
M_IM_{III}	164.7	3	0.0753			
$M_{II}M_{IV}$	109.4	3	0.1133			
$M_{II}N_I$	76.9	2	0.1613			
$M_{III}M_{IV,V}$	113.8	3	0.1089			
	79.8	3	0.1554			
$\zeta_2\ M_{IV}N_{II}$	191.1	2	0.06488			
$M_{IV}N_{III}$	189.5	3	0.0654			
$\zeta_1\ M_VN_{III}$	192.6	2	0.06437			
	37 Rubidium			**38 Strontium**		
$\alpha_2\ KL_{II}$	0.92969	1	13.3358	0.87943	1	14.0979
$\alpha_1\ KL_{III}$	0.925553	9	13.3953	0.87526	1	14.1650
$\beta_3\ KM_{II}$	0.82921	3	[illegible]	0.78345	3	15.8249
$\beta_1\ KM_{III}$	0.82868	2	14.9613	0.78292	2	15.8357
$\beta_2\ KN_{II,III}$	0.81645	3	15.1854	0.77081	3	16.0846
$\beta_5\ KM_{IV,V}$	0.8219	1	15.085	0.7764	1	15.969
$\beta_4\ KN_{IV,V}$	0.8154	2	15.205	0.76989	5	16.104
$\beta_4\ L_IM_{II}$	6.8207	3	1.81771	6.4026	3	1.93643
$\beta_3\ L_IM_{III}$	6.7876	3	1.82659	6.3672	3	1.94719
$\gamma_{2,3}\ L_IN_{II,III}$	6.0458	3	2.0507	5.6445	3	2.1965
$\eta\ L_{II}M_I$	8.0415	4	1.54177	7.5171	3	1.64933
$\beta_1\ L_{II}M_{IV}$	7.0759	3	1.75217	6.6239	3	1.87172
$\gamma_5\ L_{II}N_{IV}$	6.7553	3	1.83532	6.2961	3	1.96916
$l\ L_{III}M_I$	8.3636	4	1.48238	7.8362	3	1.58215
$\alpha_2\ L_{III}M_{IV}$	7.3251	3	1.69256	6.8697	3	1.80474
$\alpha_1\ L_{III}M_V$	7.3183	2	1.69413	6.8628	2	1.80656
$\beta_6\ L_{III}N_I$	6.9842	3	1.77517	6.5191	3	1.90181
M_IM_{III}	144.4	3	0.0859			
$M_{II}M_{IV}$	91.5	2	0.1355	85.7	2	0.1447
$M_{II}N_I$	57.0	2	0.2174	51.3	1	0.2416
$M_{III}M_{IV,V}$	96.7	2	0.1282	91.4	2	0.1357
$M_{III}N_I$	59.5	2	0.2083	53.6	1	0.2313
$\zeta_2\ M_{IV}N_{II}$	127.8	2	0.0970			
$M_{IV}N_{III}$	126.8	2	0.0978			
$\zeta_2\ M_{IV}N_{II,III}$				108.0	2	0.1148
$\zeta_1\ M_VN_{III}$	128.7	2	0.0964	108.7	1	0.1140
	39 Yttrium			**40 Zirconium**		
$\alpha_2\ KL_{II}$	0.83305	1	14.8829	0.79015	1	15.6909
$\alpha_1\ KL_{III}$	0.82884	1	14.9584	0.78593	1	15.7751
$\beta_3\ KM_{II}$	0.74126	3	16.7258	0.70228	4	17.654
$\beta_1\ KM_{III}$	0.74072	2	16.7378	0.70173	3	17.6678
$\beta_2\ KN_{II,III}$	0.72864	4	17.0154	0.68993	4	17.970
$\beta_5\ KM_{IV,V}$	0.7345	1	16.879	0.6959	1	17.815

TABLE A (*Continued*)

Designation	Å*	p.e.	keV	Å*	p.e.	keV
	39 Yttrium (*Cont.*)			**40 Zirconium** (*Cont.*)		
$\beta_4\ KN_{IV,V}$	0.72776	5	17.036	0.68901	5	17.994
$\beta_4\ L_IM_{II}$	6.0186	3	2.0600	5.6681	3	2.1873
$\beta_3\ L_IM_{III}$	5.9832	3	2.0722	5.6330	3	2.2010
$\gamma_{2,3}\ L_IN_{II,III}$	5.2830	3	2.3468	4.9536	3	2.5029
$\eta\ L_IM_I$	7.0406	3	1.76095	6.6069	3	1.87654
$\beta_1\ L_{III}M_{IV}$	6.2120	3	1.99584	5.8360	3	2.1244
$\gamma_5\ L_{II}N_I$	5.8754	3	2.1102	5.4977	3	2.2551
$\gamma_1\ L_{II}N_{IV}$				5.3843	3	2.3027
$l\ L_{III}M_I$	7.3563	3	1.68536	6.9185	3	1.79201
$\alpha_2\ L_{III}M_{IV}$	6.4558	3	1.92047	6.0778	3	2.0399
$\alpha_1\ L_{III}M_V$	6.4488	2	1.92256	6.0705	2	2.04236
$\beta_6\ L_{III}N_I$	6.0942	3	2.0344	5.7101	3	2.1712
$\beta_{2,15}$				5.5863	3	2.2194
$M_{II}M_{IV}$	81.5	2	0.1522	76.7	2	0.1617
$M_{II}N_I$	46.48	9	0.267			
$M_{III}M_V$				80.9	3	0.1533
$M_{III}N_I$	48.5	2	0.256			
$M_{III}M_{IV,V}$	86.5	2	0.1434			
$\zeta\ M_{IV,V}N_{II,III}$	93.4	2	0.1328	82.1	2	0.1511
$M_{IV,V}O_{II,III}$				70.0	4	0.177
	41 Niobium			**42 Molybdenum**		
$\alpha_2\ KL_{II}$	0.75044	1	16.5210	0.713590	6	17.3743
$\alpha_1\ KL_{III}$	0.74620	1	16.6151	0.709300	1	17.47934
$\beta_3\ KM_{II}$	0.66634	3	18.6063	0.632872	9	19.5903
$\beta_1\ KM_{III}$	0.66576	2	18.6225	0.632288	9	19.6083
β_2^{II}				0.62107	5	19.963
$\beta_2\ KN_{II,III}$	0.65416	4	18.953	0.62099	2	19.9652
$\beta_4\ KN_{IV,V}$	0.65318	5	18.981			
$\beta_5^{II}\ KM_{IV}$				0.62708	5	19.771
$\beta_5^{I}\ KM_V$				0.62692	5	19.776
$\beta_4\ KN_{IV,V}$				0.62001	9	19.996
$\beta_4\ L_IM_{II}$	5.3455	3	2.3194	5.0488	3	2.4557
$\beta_3\ L_IM_{III}$	5.3102	3	2.3348	5.0133	3	2.4730
$\gamma_{2,3}\ L_IN_{II,III}$	4.6542	2	2.6638	4.3800	2	2.8306
$\eta\ L_{II}M_I$	6.2109	3	1.99620	5.8475	3	2.1202
$\beta_1\ L_{II}M_{IV}$	5.4923	3	2.2574	5.17708	8	2.39481
$\gamma_5\ L_{II}N_I$	5.1517	3	2.4066	4.8369	2	2.5632
$\gamma_1\ L_{II}N_{IV}$	5.0361	3	2.4618	4.7258	2	2.6235
$l\ L_{III}M_I$	6.5176	3	1.90225	6.1508	3	2.01568
$\alpha_2\ L_{III}M_{IV}$	5.7319	3	2.1630	5.41437	8	2.28985
$\alpha_1\ L_{III}M_V$	5.7243	2	2.16589	5.40655	8	2.29316
$\beta_6\ L_{III}N_I$	5.3613	3	2.3125	5.0488	5	2.4557
$\beta_{2,15}\ L_{III}N_{IV,V}$	5.2379	3	2.3670	4.9232	2	2.5183
$M_{II}M_{IV}$	72.1	3	0.1718	68.9	2	0.1798
$M_{II}N_I$	38.4	3	0.323	35.3	3	0.351
$M_{II}N_{IV}$	33.1	2	0.375			
$M_{III}M_V$	78.4	2	0.1582	74.9	1	0.1656
$M_{III}N_I$	40.7	2	0.305	37.5	2	0.331
$\gamma\ M_{III}N_{IV,V}$	34.9	2	0.356			
$\zeta\ M_{IV,V}N_{II,III}$	72.19	9	0.1717	64.38	7	0.1926
$M_{IV,V}O_{II,III}$	61.9	2	0.2002	54.8	2	0.2262

Designation	Å*	p.e.	keV	Å*	p.e.	keV
	43 Technetium			**44 Ruthenium**		
$\alpha_2\ KL_{II}$	0.67932†	3	18.2508	0.647408	5	19.1504
$\alpha_1\ KL_{III}$	0.67502†	3	18.3671	0.643083	4	19.2792
$\beta_3\ KM_{II}$	0.60188†	4	20.599	0.573067	4	21.6346
$\beta_1\ KM_{III}$	0.60130†	4	20.619	0.572482	4	21.6568
$\beta_2\ KN_{II,III}$	0.59024†	5	21.005	0.56166	3	22.074
$\beta_5^{II}\ KM_{IV}$				0.5680	2	21.829
$\beta_5^{I}\ KM_V$				0.56785	9	21.834
β_4				0.56089	9	22.104
$\beta_4\ L_{II}M_{IV}$				4.5230	2	2.7411
$\beta_3\ L_IM_{III}$				4.4866	3	2.7634
$\gamma_{2,3}\ L_IN_{II,III}$				3.8977	2	3.1809
$\eta\ L_{II}M_I$				5.2050	2	2.38197
$\beta_1\ L_{II}M_{IV}$	4.8873†	8	2.5368	4.62058	3	2.68323
$\gamma_5\ L_{II}N_I$				4.2873	2	2.8918
$\gamma_1\ L_{II}N_{IV}$				4.1822	2	2.9645
$l\ L_{III}M_I$				5.5035	3	2.2528
$\alpha_2\ L_{III}M_{IV}$				4.85381	7	2.55431
$\alpha_1\ L_{III}M_V$	5.1148†	3	2.4240	4.84575	5	2.55855
$\beta_6\ L_{III}N_I$				4.4866	3	2.7634
$\beta_{2,15}\ L_{III}N_{IV,V}$				4.3718	2	2.8360
$M_{II}M_{IV}$				62.2	1	0.1992
$M_{II}N_I$				32.3	2	0.384
$M_{II}N_{IV}$				25.50	9	0.486
$M_{III}M_V$				68.3	1	0.1814
$\gamma\ M_{III}N_{IV,V}$				26.9	1	0.462
$\zeta\ M_{IV,V}N_{II,III}$				52.34	7	0.2369
$M_{IV,V}O_{II,III}$				44.8	1	0.2768
	45 Rhodium			**46 Palladium**		
$\alpha_2\ KL_{II}$	0.617630	4	20.0737	0.589821	3	21.0201
$\alpha_1\ KL_{III}$	0.613279	4	20.2161	0.585448	3	21.1771
$\beta_3\ KM_{II}$	0.546200	4	22.6989	0.521123	4	23.7911
$\beta_1\ KM_{III}$	0.545605	4	22.7236	0.520520	4	23.8187
$\beta_2^{II}\ KN_{II}$	0.53513	5	23.168			
$\beta_2\ KN_{II,III}$	0.53503	2	23.1728	0.510228	4	24.2991
$\beta_5^{II}\ KM_{IV}$	0.54118	9	22.909			
$\beta_5^{I}\ KM_V$	0.54101	9	22.917			
$\beta_4\ KN_{IV,V}$	0.53401	9	23.217	0.5093	2	24.346
$\beta_5\ KM_{IV,V}$				0.51670	9	23.995
$\beta_4\ L_IM_{II}$	4.2888	2	2.8908	4.0711	2	3.0454
$\beta_3\ L_IM_{III}$	4.2522	2	2.9157	4.0346	2	3.0730
$\gamma_{2,3}\ L_IN_{II,III}$	3.6855	2	3.3640	3.4892	2	3.5533
$\eta\ L_{II}M_I$	4.9217	2	2.5191	4.6605	2	2.6603
$\beta_1\ L_{II}M_{IV}$	4.37414	4	2.83441	4.14622	5	2.99022
$\gamma_5\ L_{II}N_I$	4.0451	2	3.0650	3.8222	2	3.2437
$\gamma_1\ L_{II}N_{IV}$	3.9437	2	3.1438	3.7246	2	3.3287
$l\ L_{III}M_I$	5.2169	3	2.3765	4.9525	3	2.5034
$\alpha_2\ L_{III}M_{IV}$	4.60545	9	2.69205	4.37588	7	2.83329
$\alpha_1\ L_{III}M_V$	4.59743	9	2.69674	4.36767	5	2.83861
$\beta_6\ L_{III}N_I$	4.2417	2	2.9229	4.0162	2	3.0870
$\beta_{2,15}\ L_{III}N_{IV,V}$	4.1310	2	3.0013	3.90887	4	3.17179
$\beta_{10}\ L_IM_{IV}$				3.7988	2	3.2637

Designation	Å*	p.e.	keV	Å*	p.e.	keV
	45 Rhodium (*Cont.*)			**46 Palladium** (*Cont.*)		
$\beta_9\ L_IM_V$				3.7920	2	3.2696
$M_IN_{II,III}$				20.1	2	0.616
$M_{II}M_{IV}$	59.3	1	0.2090	56.5	1	0.2194
$M_{II}N_I$	28.1	2	0.442	26.2	2	0.474
$M_{II}N_{IV}$				22.1	1	0.560
$M_{II}\ M_V$	65.5	1	0.1892	62.9	1	0.1970
$M_{II}{}^{I}N_I$	29.8	1	0.417	27.9	1	0.445
$\gamma\ M_{III}N_{IV,V}$	25.01	9	0.496	23.3†	1	0.531
$\zeta\ M_{IV,V}N_{II,III}$	47.67	9	0.2601	43.6	1	0.2844
$M_{IV,V}O_{II,III}$	40.9	2	0.303	37.4	2	0.332
	47 Silver			**48 Cadmium**		
$\alpha_2\ KL_{II}$	0.563798	4	21.9903	0.539422	3	22.9841
$\alpha_1\ KL_{III}$	0.5594075	6	22.16292	0.535010	3	23.1736
$\beta_3\ KM_{II}$	0.497685	4	24.9115	0.475730	5	26.0612
$\beta_1\ KM_{III}$	0.497069	4	24.9424	0.475105	6	26.0955
$\beta_2\ KN_{II,III}$	0.487032	4	25.4564	0.465328	7	26.6438
$\beta_5\ KM_{IV,V}$	0.49306	2	25.145			
$\beta_4\ KN_{IV,V}$	0.48598	3	25.512			
$\beta_4\ L_IM_{II}$	3.87023	5	3.20346	3.68203	9	3.36719
$\beta_3\ L_IM_{III}$	3.83313	9	3.23446	3.64495	9	3.40145
$\gamma_2\ L_IN_{II}$	3.31216	9	3.7432	3.1377	2	3.9513
$\gamma_3\ L_IN_{III}$	3.30635	9	3.7498			
$\eta\ L_{II}M_I$	4.4183	2	2.8061	4.19315	9	2.95675
$\beta_1\ L_{II}M_{IV}$	3.93473	3	3.15094	3.73823	4	3.31657
$\gamma_5\ L_{II}N_I$	3.61638	9	3.42832	3.42551	9	3.61935
$\gamma_1\ L_{II}N_{IV}$	3.52260	4	3.51959	3.33564	6	3.71686
$l\ L_{III}M_I$	4.7076	2	2.6337	4.48014	9	2.76735
$\alpha_2\ L_{III}M_{IV}$	4.16294	5	2.97821	3.96496	6	3.12691
$\alpha_1\ L_{III}M_V$	4.15443	3	2.98431	3.95635	4	3.13373
$\beta_6\ L_{III}N_I$	3.80774	9	3.25603	3.61467	9	3.42994
$\beta_{2,15}\ L_{III}N_{IV,V}$	3.70335	3	3.34781	3.51408	4	3.52812
$\beta_{10}\ L_IM_{IV}$	3.61158	9	3.43287	3.4367	2	3.6075
$\beta_9\ L_IM_V$	3.60497	9	3.43917	3.43015	9	3.61445
$M_IN_{II,III}$	18.8	2	0.658			
$M_{II}M_{IV}$	54.0	1	0.2295	52.0	2	0.2384
$M_{II}N_I$				22.9	2	0.540
$M_{II}N_{IV}$	20.66	7	0.600	19.40	7	0.639
$M_{III}M_V$	60.5	1	0.2048	58.7	2	0.2111
$M_{III}N_I$	26.0	1	0.478	24.5	1	0.507
$\gamma\ M_{III}N_{IV,V}$	21.82	7	0.568	20.47	7	0.606
$M_{IV}O_{II,III}$				30.4	1	0.408
$\zeta\ M_{IV,V}N_{II,III}$	39.77	7	0.3117	36.8	1	0.3371
M_VN_I	24.4	2	0.509			
M_VO_{III}				30.8	1	0.403
$M_{IV,V}O_{II,III}$	33.5	3	0.370			

Designation	Å*	p.e.	keV	Å*	p.e.	keV
	49 Indium			**50 Tin**		
$\alpha_2\ KL_{II}$	0.516544	3	24.0020	0.495053	3	25.0440
$\alpha_1\ KL_{III}$	0.512113	3	24.2097	0.490599	3	25.2713
$\beta_3\ KM_{II}$	0.455181	4	27.2377	0.435877	5	28.4440
	49 Indium (*Cont.*)			**50 Tin** (*Cont.*)		
$\beta_1\ KM_{III}$	0.454545	4	27.2759	0.435236	5	28.4860
$\beta_2\ KN_{II,III}$	0.44500	1	27.8608	0.425915	8	29.1093
$KO_{II,III}$	0.44374	3	27.940	0.42467	3	29.195
$\beta_5{}^{II}\ KM_{IV}$	0.45098	2	27.491	0.43184	3	28.710
$\beta_5{}^{I}\ KM_V$	0.45086	2	27.499	0.43175	3	28.716
$\beta_4\ KN_{IV,V}$	0.44393	4	27.928	0.42495	3	29.175
$\beta_4\ L_IM_{II}$	3.50697	9	3.5353	3.34335	9	3.7083
$\beta_3\ L_IM_{III}$	3.46984	9	3.5731	3.30585	3	3.7500
$\gamma_{2,3}\ L_IN_{II,III}$	2.9800	2	4.1605	2.8327	2	4.3768
$\gamma_4\ L_IO_{II,III}$	2.9264	2	4.2367	2.7775	2	4.4638
$\eta\ L_{II}M_I$	3.98327	9	3.11254	3.78876	9	3.27234
$\beta_1\ L_{II}M_{IV}$	3.55531	4	3.48721	3.38487	3	3.66280
$\gamma_5\ L_{II}N_I$	3.24907	9	3.8159	3.08475	9	4.0192
$\gamma_1\ L_{II}N_{IV}$	3.16213	4	3.92081	3.00115	3	4.13112
$l\ L_{III}M_I$	4.26873	9	2.90440	4.07165	9	3.04499
$\alpha_2\ L_{III}M_{IV}$	3.78073	6	3.27929	3.60891	4	3.43542
$\alpha_1\ L_{III}M_V$	3.77192	4	3.28694	3.59994	3	3.44398
$\beta_6\ L_{III}N_I$	3.43606	9	3.60823	3.26901	9	3.7926
$\beta_{2,15}\ L_{III}N_{IV,V}$	3.33838	3	3.71381	3.17505	3	3.90486
$\beta_7\ L_{III}O_I$	3.324	4	3.730	3.1564	3	3.9279
$\beta_{10}\ L_IM_{IV}$	3.27404	9	3.7868	3.12170	9	3.9716
$\beta_9\ L_IM_V$	3.26763	9	3.7942	3.11513	9	3.9800
$M_{II}M_{IV}$				47.3	1	0.2621
$M_{II}N_I$				20.0	1	0.619
$M_{II}N_{IV}$				16.93	5	0.733
$M_{III}M_V$				54.2	1	0.2287
$M_{III}N_I$				21.5	1	0.575
$\gamma\ M_{III}N_{IV,V}$				17.94	5	0.691
$M_{IV}O_{II,III}$				25.3	1	0.491
$\zeta\ M_{IV,V}N_{II,III}$				31.24	9	0.397
M_VO_{III}				25.7	1	0.483

Designation	Å*	p.e.	keV	Å*	p.e.	keV
	51 Antimony			**52 Tellurium**		
$\alpha_2\ KL_{II}$	0.474827	3	26.1108	0.455784	3	27.2017
$\alpha_1\ KL_{III}$	0.470354	3	26.3591	0.451295	3	27.4723
$\beta_3\ KM_{II}$	0.417737	4	29.6792	0.400659	4	30.9443
$\beta_1\ KM_{III}$	0.417085	3	29.7256	0.399995	5	30.9957
$\beta_2\ KN_{II,III}$	0.407973	5	30.3895	0.391102	6	31.7004
$KO_{II,III}$	0.40666	1	30.4875	0.38974	1	31.8114
$\beta_5{}^{II}\ KM_{IV}$	0.41388	1	29.9560			
$\beta_5{}^{I}\ KM_V$	0.41378	1	29.9632			
$\beta_4\ KN_{IV,V}$	0.40702	1	30.4604			
$\beta_4\ L_IM_{II}$	3.19014	9	3.8864	3.04661	9	4.0695
$\beta_3\ L_IM_{III}$	3.15258	9	3.9327	3.00893	9	4.1204
$\gamma_{2,3}\ L_IN_{II,III}$	2.6953	2	4.5999	2.5674	2	4.8290
$\gamma_4\ L_IO_{II,III}$	2.6398	2	4.6967	2.5113	2	4.9369
$\eta\ L_{II}M_I$	3.60765	9	3.43661	3.43832	9	3.60586
$\beta_1\ L_{II}M_{IV}$	3.22567	4	3.84357	3.07677	6	4.02958
$\gamma_5\ L_{II}N_I$	2.93187	9	4.2287	2.79007	9	4.4437
$\gamma_1\ L_{II}N_{IV}$	2.85159	3	4.34779	2.71241	6	4.5709
$l\ L_{III}M_I$	3.88826	9	3.18860	3.71696	9	3.33555
$\alpha_2\ L_{III}M_{IV}$	3.44840	6	3.59532	3.29846	9	3.7588

Designation	Å*	p.e.	keV	Å*	p.e.	keV
	51 Antimony (*Cont.*)			**52 Tellurium** (*Cont.*)		
$\alpha_1\ L_{III}M_V$	3.43941	4	3.60472	3.28920	6	3.76933
$\beta_6\ L_{III}N_I$	3.11513	9	3.9800	2.97088	9	4.1732
$\beta_{2,15}\ L_{III}N_{IV,V}$	3.02335	3	4.10078	2.88217	8	4.3017
$\beta_7\ L_{III}O_I$	3.0052	3	4.1255	2.8634	3	4.3298
$\beta_{10}\ L_IM_{IV}$	2.97917	9	4.1616	2.84679	9	4.3551
$\beta_9\ L_IM_V$	2.97261	9	4.1708	2.83897	9	4.3671
$M_{II}M_{IV}$	45.2	1	0.2743			
$M_{II}N_I$	18.8	1	0.658	17.6	1	0.703
$M_{II}N_{IV}$	15.98	5	0.776			
$M_{III}M_V$	52.2	1	0.2375	50.3	1	0.2465
$M_{III}N_I$	20.2	1	0.612	19.1	1	0.648
$\gamma\ M_{III}N_{IV,V}$	16.92	4	0.733	15.93	4	0.778
$M_{IV}O_{II,III}$				21.34	5	0.581
$\zeta\ M_{IV,V}N_{II,III}$	28.88	8	0.429	26.72	9	0.464
M_VO_{III}				21.78	5	0.569
	53 Iodine			**54 Xenon**		
$\alpha_2\ KL_{II}$	0.437829	7	28.3172	0.42087†	2	29.458
$\alpha_1\ KL_{III}$	0.433318	5	28.6120	0.41634†	2	29.779
$\beta_3\ KM_{II}$	0.384564	4	32.2394	0.36941†	2	33.562
$\beta_1\ KM_{III}$	0.383905	4	32.2947	0.36872†	2	33.624
$\beta_2\ KN_{II,III}$	0.37523†	2	33.042	0.36026†	3	34.415
$\beta_4\ L_IM_{II}$	2.91207	9	4.2575			
$\beta_3\ L_IM_{III}$	2.87429	9	4.3134			
$\gamma_{2,3}\ L_IN_{II,III}$	2.4475	2	5.0657			
$\gamma_4\ L_IO_{II,III}$	2.3913	2	5.1848			
$\eta\ L_{II}M_I$	3.27979	9	3.7801			
$\beta_1\ L_{II}M_{IV}$	2.93744	6	4.22072			
$\gamma_5\ L_{II}N_I$	2.65710	9	4.6660			
$\gamma_1\ L_{II}N_{IV}$	2.58244	8	4.8009			
$l\ L_{III}M_I$	3.55754	9	3.48502			
$\alpha_2\ L_{III}M_{IV}$	3.15791	6	3.92604			
$\alpha_1\ L_{III}M_V$	3.14860	6	3.93765	3.0166†	2	4.1099
$\beta_6\ L_{III}N_I$	2.83672	9	4.3706			
$\beta_{2,15}\ L_{III}N_{IV,V}$	2.75053	8	4.5075			
$\beta_7\ L_{III}O_I$	2.7288	3	4.5435			
$\beta_{10}\ L_IM_{IV}$	2.72104	9	4.5564			
$\beta_9\ L_IM_V$	2.71352	9	4.5690			
	55 Cesium			**56 Barium**		
$\alpha_2\ KL_{II}$	0.404835	4	30.6251	0.389668	5	31.8171
$\alpha_1\ KL_{III}$	0.400290	4	30.9728	0.385111	4	32.1936
$\beta_3\ KM_{II}$	0.355050	4	34.9194	0.341507	4	36.3040
$\beta_1\ KM_{III}$	0.354364	7	34.9869	0.340811	3	36.3782
$\beta_2\ KN_{II,III}$	0.34611	2	35.822	0.33277	1	37.257
$KO_{II,III}$				0.33127	2	37.426
$\beta_5^{II}\ KM_{IV}$				0.33835	2	36.643
$\beta_5^{I}\ KM_V$				0.33814	2	36.666
$\beta_4\ KN_{IV,V}$				0.33229	2	37.311
$\beta_4\ L_IM_{II}$	2.6666	2	4.6494	2.5553	2	4.8519
$\beta_3\ L_IM_{III}$	2.6285	2	4.7167	2.5164	2	4.9269
$\gamma_2\ L_IN_{II}$	2.2371	2	5.5420	2.1387	2	5.7969
$\gamma_3\ L_IN_{III}$	2.2328	2	5.5527	2.1342	2	5.8092

Designation	Å*	p.e.	keV	Å*	p.e.	keV
	55 Cesium (*Cont.*)			**56 Barium** (*Cont.*)		
$\gamma_4\ L_IO_{II,III}$	2.1741	2	5.7026	2.0756	3	5.9733
$\eta\ L_{II}M_I$	2.9932	2	4.1421	2.8627	3	4.3309
$\beta_1\ L_{II}M_{IV}$	2.6837	2	4.6198	2.56821	5	4.82753
$\gamma_5\ L_{II}N_I$	2.4174	2	5.1287	2.3085	3	5.3707
$\gamma_1\ L_{II}N_{IV}$	2.3480	2	5.2804	2.2415	2	5.5311
$l\ L_{III}M_I$	3.2670	2	3.7950	3.1355	2	3.9541
$\alpha_2\ L_{III}M_{IV}$	2.9020	2	4.2722	2.78553	5	4.45090
$\alpha_1\ L_{III}M_V$	2.8924	2	4.2865	2.77595	5	4.46626
$\beta_6\ L_{III}N_I$	2.5932	2	4.7811	2.4826	2	4.9939
$\beta_{2,15}\ L_{III}N_{IV,V}$	2.5118	2	4.9359	2.40435	6	5.1565
$\beta_7\ L_{III}O_I$	2.4849	2	4.9893	2.3806	2	5.2079
$\beta_{10}\ L_IM_{IV}$	2.4920	2	4.9752	2.3869	2	5.1941
$\beta_9\ L_IM_V$	2.4783	2	5.0026	2.3764	2	5.2171
$\gamma\ M_{III}N_{IV,V}$				12.75	3	0.973
$M_{IV}O_{II}$				15.91	5	0.779
$M_{IV}O_{III}$				15.72	9	0.789
$\zeta\ M_VN_{III}$				20.64	4	0.601
M_VO_{III}				16.20	5	0.765
$N_{IV}O_{II}$	188.6	1	0.06574	163.3	2	0.07590
$N_{IV}O_{III}$	183.8	1	0.06746	159.0	2	0.07796
N_VO_{III}	190.3	1	0.06515	164.6	2	0.07530
	57 Lanthanum			**58 Cerium**		
$\alpha_2\ KL_{II}$	0.375313	2	33.0341	0.361683	2	34.2789
$\alpha_1\ KL_{III}$	0.370737	2	33.4418	0.357092	2	34.7197
$\beta_3\ KM_{II}$	0.328686	4	37.7202	0.316520	4	39.1701
$\beta_1\ KM_{III}$	0.327983	3	37.8010	0.315816	2	39.2573
$\beta_2\ KN_{II,III}$	0.320117	7	38.7299	0.30816	1	40.233
$KO_{II,III}$	0.31864	2	38.909	0.30668	2	40.427
$\beta_5^{II}\ KM_{IV}$	0.32563	2	38.074	0.31357	2	39.539
$\beta_5^{I}\ KM_V$	0.32546	2	38.094	0.31342	2	39.558
$\beta_4\ KN_{IV,V}$	0.31931	2	38.828	0.30737	2	40.337
$\beta_4\ L_IM_{II}$	2.4493	3	5.0620	2.3497	4	5.2765
$\beta_3\ L_IM_{III}$	2.4105	3	5.1434	2.3109	3	5.3651
$\gamma_2\ L_IN_{II}$	2.0460	4	6.060	1.9602	3	6.3250
$\gamma_3\ L_IN_{III}$	2.0410	4	6.074	1.9553	3	6.3409
$\gamma_4\ L_IO_{II,III}$	1.9830	4	6.252	1.8991	4	6.528
$\eta\ L_{II}M_I$	2.740	3	4.525	2.6203	4	4.7315
$\beta_1\ L_{II}M_{IV}$	2.45891	5	5.0421	2.3561	3	5.2622
$\gamma_5\ L_{II}N_I$	2.2056	4	5.621	2.1103	3	5.8751
$\gamma_1\ L_{II}N_{IV}$	2.1418	3	5.7885	2.0487	4	6.052
$\gamma_8\ L_{II}O_I$				2.0237	4	6.126
$l\ L_{III}M_I$	3.006	3	4.124	2.8917	4	4.2875
$\alpha_2\ L_{III}M_{IV}$	2.67533	5	4.63423	2.5706	3	4.8230
$\alpha_1\ L_{III}M_V$	2.66570	5	4.65097	2.5615	2	4.8402
$\beta_6\ L_{III}N_I$	2.3790	4	5.2114	2.2818	3	5.4334
$\beta_{2,15}\ L_{III}N_{IV,V}$	2.3030	3	5.3835	2.2087	2	5.6134
$\beta_7\ L_{III}O_I$	2.275	3	5.450	2.1701	2	5.7132
$\beta_{10}\ L_IM_{IV}$	2.290	3	5.415	2.1958	5	5.646
$\beta_9\ L_IM_V$	2.282	3	5.434	2.1885	3	5.6650
$\gamma\ M_{III}N_{IV,V}$	12.08	4	1.027	11.53	1	1.0749
$\beta\ M_{IV}N_{VI}$	14.51	5	0.854	13.75	4	0.902
$\zeta\ M_VN_{III}$	19.44	5	0.638	18.35	4	0.676
$\alpha\ M_VN_{VI,VII}$	14.88	5	0.833	14.04	2	0.883

Designation	Å*	p.e.	keV	Å*	p.e.	keV
	57 Lanthanum (*Cont.*)			**58 Cerium** (*Cont.*)		
$M_VO_{II,III}$				14.39	5	0.862
$N_{IV,V}O_{II,III}$	152.6	6	0.0812	144.4	6	0.0859
	59 Praseodymium			**60 Neodymium**		
$\alpha_2\ KL_{II}$	0.348749	2	35.5502	0.336472	2	36.8474
$\alpha_1\ KL_{III}$	0.344140	2	36.0263	0.331846	2	37.3610
$\beta_3\ KM_{II}$	0.304975	5	40.6529	0.294027	3	42.1665
$\beta_1\ KM_{III}$	0.304261	4	40.7482	0.293299	2	42.2713
$\beta_2\ KN_{II,III}$	0.29679	2	41.773	0.2861†	1	43.33
$\beta_4\ L_IM_{II}$	2.2550	4	5.4981	2.1669	3	5.7216
$\beta_3\ L_IM_{III}$	2.2172	3	5.5918	2.1268	2	5.8294
$\gamma_2\ L_IN_{II}$	1.8791	4	6.598	1.8013	4	6.883
$\gamma_3\ L_IN_{III}$	1.8740	4	6.616	1.7964	4	6.902
$\gamma_4\ L_IO_{II,III}$	1.8193	4	6.815	1.7445	4	7.107
$\eta\ L_{II}M_I$	2.512	3	4.935	2.4094	4	5.1457
$\beta_1\ L_{II}M_{IV}$	2.2588	3	5.4889	2.1669	2	5.7216
$\gamma_5\ L_{II}N_I$	2.0205	4	6.136	1.9355	4	6.406
$\gamma_1\ L_{II}N_{IV}$	1.9611	3	6.3221	1.8779	2	6.6021
$\gamma_8\ L_{II}O_I$	1.9362	4	6.403	1.8552	5	6.683
$l\ L_{III}M_I$	2.7841	4	4.4532	2.6760	4	4.6330
$\alpha_2\ L_{III}M_{IV}$	2.4729	3	5.0135	2.3807	3	5.2077
$\alpha_1\ L_{III}M_V$	2.4630	2	5.0337	2.3704	2	5.2304
$\beta_6\ L_{III}N_I$	2.1906	4	5.660	2.1039	3	5.8930
$\beta_{2,15}\ L_{III}N_{IV,V}$	2.1194	4	5.850	2.0360	3	6.0894
$\beta_7\ L_{III}O_I$	2.0919	4	5.927	2.0092	3	6.1708
$\beta_{10}\ L_IM_{IV}$	2.1071	4	5.884	2.0237	3	6.1265
$\beta_9\ L_IM_V$	2.1004	4	5.903	2.0165	3	6.1484
$\gamma\ M_{III}N_{IV,V}$	10.998	9	1.1273	10.505	9	1.180
$\beta\ M_{IV}N_{VI}$	13.06	2	0.950	12.44	2	0.997
$\zeta\ M_VN_{III}$	17.38	4	0.714	16.46	4	0.753
$\alpha\ M_VN_{VI,VII}$	13.343	5	0.9292	12.68	2	0.978
$N_{IV,V}N_{VI,VII}$	113.	1	0.1095	107.	1	0.116
$N_{IV,V}O_{II,III}$	136.5	4	0.0908	128.9	7	0.0962
	61 Promethium			**62 Samarium**		
$\alpha_2\ KL_{II}$	0.324803	4	38.1712	0.313698	2	39.5224
$\alpha_1\ KL_{III}$	0.320160	4	38.7247	0.309040	2	40.1181
$\beta_3\ KM_{II}$	0.28363†	4	43.713	0.27376	2	45.289
$\beta_1\ KM_{III}$	0.28290†	3	43.826	0.27301	2	45.413
$\beta_2\ KN_{II,III}$	0.2759†	1	44.94	0.2662	1	46.58
$KO_{II,III}$				0.26491	3	46.801
$\beta_5\ KM_{IV,V}$				0.27111	3	45.731
$\beta_4\ L_IM_{II}$				2.00095	6	6.1963
$\beta_3\ L_IM_{III}$	2.0421	4	6.071	1.96241	3	6.3180
$\gamma_2\ L_IN_{II}$				1.66044	6	7.4668
$\gamma_3\ L_IN_{III}$				1.65601	3	7.4867
$\gamma_4\ L_IO_{II,III}$				1.60728	3	7.7137
$\eta\ L_{II}M_I$				2.21824	3	5.5892
$\beta_1\ L_{II}M_{IV}$	2.0797	4	5.961	1.99806	3	6.2051
$\gamma_5\ L_{II}N_I$				1.77934•	3	6.9678
$\gamma_1\ L_{II}N_{IV}$	1.7989	9	6.892	1.72724	3	7.1780
$\gamma_6\ L_{II}O_{IV}$				1.6966	9	7.3076
$l\ L_{III}M_I$				2.4823	4	4.9945

Designation	Å*	p.e.	keV	Å*	p.e.	keV
	61 Promethium (*Cont.*)			**62 Samarium** (*Cont.*)		
$\alpha_2\ L_{III}M_{IV}$	2.2926	4	5.4078	2.21062	3	5.6084
$\alpha_1\ L_{III}M_V$	2.2822	3	5.4325	2.1998	2	5.6361
$\beta_6\ L_{III}N_I$				1.94643	3	6.3697
$\beta_{2,15}\ L_{III}N_{IV,V}$	1.9559	6	6.339	1.88221	3	6.5870
$\beta_7\ L_{III}O_I$				1.85626	3	6.6791
$\beta_5\ L_{III}O_{IV,V}$				1.84700	9	6.7126
$\beta_{10}\ L_IM_{IV}$				1.86990	3	6.6304
$\beta_9\ L_IM_V$				1.86166	3	6.6597
$\gamma\ M_{III}N_{IV,V}$				9.600	9	1.291
$\beta\ M_{IV}N_{VI}$				11.27	1	1.0998
$\zeta\ M_VN_{III}$				14.91	4	0.831
$\alpha\ M_VN_{VI,VII}$				11.47	3	1.081
$N_{IV,V}N_{VI,VII}$				98.	1	0.126
$N_{IV,V}O_{II,III}$				117.4	4	0.1056
	63 Europium			**64 Gadolinium**		
$\alpha_2\ KL_{II}$	0.303118	2	40.9019	0.293038	2	42.3089
$\alpha_1\ KL_{III}$	0.298446	2	41.5422	0.288353	2	42.9962
$\beta_3\ KM_{II}$	0.264332	5	46.9036	0.25534	2	48.555
$\beta_1\ KM_{III}$	0.263577	5	47.0379	0.25460	2	48.697
$\beta_2\ KN_{II,III}$	0.256923	8	48.256	0.24816	3	49.959
$KO_{II,III}$	0.255645	7	48.497	0.24687	8	50.221
$\beta_5\ KM_{IV,V}$				0.25275	3	49.052
$\beta_4\ L_IM_{II}$	1.9255	2	6.4389	1.8540	2	6.6871
$\beta_3\ L_IM_{III}$	1.8867	2	6.5713	1.8150	2	6.8311
$\gamma_2\ L_IN_{II}$	1.5961	2	7.7677	1.5331	2	8.087
$\gamma_3\ L_IN_{III}$	1.5903	2	7.7961	1.5297	2	8.105
$\gamma_4\ L_IO_{II,III}$	1.5439	1	8.0304	1.4839	2	8.355
$\eta\ L_{II}M_I$	2.1315	2	5.8166	2.0494	1	6.0495
$\beta_1\ L_{II}M_{IV}$	1.9203	2	6.4564	1.8468	2	6.7132
$\gamma_5\ L_{II}N_I$	1.7085	2	7.2566	1.6412	2	7.5543
$\gamma_1\ L_{II}N_{IV}$	1.6574	2	7.4803	1.5924	2	7.7858
$\gamma_8\ L_{II}O_I$	1.6346	2	7.5849	1.5707	2	7.894
$\gamma_6\ L_{II}O_{IV}$	1.6282	2	7.6147	1.5644	2	7.925
$l\ L_{III}M_I$	2.3948	2	5.1772	2.3122	2	5.3621
$\alpha_2\ L_{III}M_{IV}$	2.1315	2	5.8166	2.0578	2	6.0250
$\alpha_1\ L_{III}M_V$	2.1209	2	5.8457	2.0468	2	6.0572
$\beta_6\ L_{III}N_I$	1.8737	2	6.6170	1.8054	2	6.8671
$\beta_{2,15}\ L_{III}N_{IV,V}$	1.8118	2	6.8432	1.7455	2	7.1028
$\beta_7\ L_{III}O_I$	1.7851	2	6.9453	1.7203	2	7.2071
$\beta_5\ L_{III}O_{IV,V}$	1.7772	2	6.9763	1.7130	2	7.2374
$\beta_{10}\ L_IM_{IV}$	1.7993	3	6.890	1.7315	3	7.160
$\beta_9\ L_IM_V$	1.7916	3	6.920	1.7240	3	7.192
$L_IO_{IV,V}$				1.4807	3	8.373
$\gamma\ M_{III}N_{IV,V}$	9.211	9	1.346	8.844	9	1.402
$\beta\ M_{IV}N_{VI}$	10.750	7	1.1533	10.254	6	1.2091
$\zeta\ M_VN_{III}$	14.22	2	0.872	13.57	2	0.914
$\alpha\ M_VN_{VI,VII}$	10.96	3	1.131	10.46	3	1.185
$N_{IV,V}O_{II,III}$	112.0	6	0.1107			
	65 Terbium			**66 Dysprosium**		
$\alpha_2\ KL_{II}$	0.283423	2	43.7441	0.274247	2	45.2078
$\alpha_1\ KL_{III}$	0.278724	2	44.4816	0.269533	2	45.9984
$\beta\ KM_{II}$	0.24683	2	50.229	0.23862	2	51.957

Designation	Å*	p.e.	keV	Å*	p.e.	keV
	65 Terbium (*Cont.*)			**66 Dysprosium** (*Cont.*)		
$\beta_1\ KM_{III}$	0.24608	2	50.382	0.23788	2	52.119
$\beta_2\ KN_{II,III}$	0.2397†	2	51.72	0.2317†	2	53.51
$KO_{II,III}$	0.23858	3	51.965	0.23056	3	53.774
$\beta_5\ KM_{IV,V}$				0.23618	3	52.494
$\beta_4\ L_IM_{II}$	1.7864	2	6.9403	1.72103	7	7.2039
$\beta_3\ L_IM_{III}$	1.7472	2	7.0959	1.6822	2	7.3702
$\gamma_2\ L_IN_{II}$	1.4764	2	8.398	1.42278	7	8.7140
$\gamma_3\ L_IN_{III}$	1.4718	2	8.423	1.41640	7	8.7532
$\gamma_4\ L_IO_{II,III}$	1.4276	2	8.685	1.37459	7	9.0195
$\eta\ L_{II}M_I$	1.9730	2	6.2839	1.89743	7	6.5342
$\beta_1\ L_{II}M_{IV}$	1.7768	3	6.978	1.71062	7	7.2477
$\gamma_5\ L_{II}N_I$	1.5787	2	7.8535	1.51824	7	8.1661
$\gamma_1\ L_{II}N_{IV}$	1.5303	2	8.102	1.47266	7	8.4188
$\gamma_8\ L_{II}O_I$	1.5097	2	8.212			
$\gamma_6\ L_{II}O_{IV}$	1.5035	2	8.246	1.44579	7	8.5753
$l\ L_{III}M_I$	2.2352	2	5.5467	2.15877	7	5.7431
$\alpha_2\ L_{III}M_{IV}$	1.9875	2	6.2380	1.91991	3	6.4577
$\alpha_1\ L_{III}M_V$	1.9765	2	6.2728	1.90881	3	6.4952
$\beta_6\ L_{III}N_I$	1.7422	2	7.1163	1.68213	7	7.3705
$\beta_{2,15}\ L_{III}N_{IV,V}$	1.6830	2	7.3667	1.62369	7	7.6357
$\beta_7\ L_{III}O_I$	1.6585	2	7.4753	1.60447	7	7.7272
$\beta_5\ L_{III}O_{IV,V}$	1.6510	2	7.5094	1.58837	7	7.8055
$\beta_{10}\ L_IM_{IV}$	1.6673	3	7.436	1.60743	9	7.7130
$\beta_9\ L_IM_V$				1.59973	9	7.7501
$L_IO_{IV,V}$	1.4228	3	8.714			
$\gamma\ M_{III}N_{IV,V}$	8.486	9	1.461	8.144	9	1.522
$\beta\ M_{IV}N_{VI}$	9.792	6	1.2661	9.357	6	1.3250
$\zeta\ M_VN_{III}$	12.98	2	0.955	12.43	2	0.998
$\alpha\ M_VN_{VI,VII}$	10.00	2	1.240	9.59	2	1.293
$N_{IV,V}N_{VI,VII}$	86.	1	0.144	83.	1	0.149
$N_{IV,V}O_{II,III}$	102.2	4	0.1213	97.2	8	0.128
	67 Holmium			**68 Erbium**		
$\alpha_2\ KL_{II}$	0.265486	2	46.6997	0.257110	2	48.2211
$\alpha_1\ KL_{III}$	0.260756	2	47.5467	0.252365	2	49.1277
$\beta_3\ KM_{II}$	0.23083	2	53.711	0.22341	2	55.494
$\beta_1\ KM_{III}$	0.23012	2	53.877	0.22266	2	55.681
$\beta_2\ KN_{II,III}$	0.2241†	2	55.32	0.2167†	2	57.21
$KO_{II,III}$	0.22305	3	55.584	0.21581	3	57.450
$\beta_5\ KM_{IV,V}$	0.22855	3	54.246	0.22124	3	56.040
$\beta_4\ L_IM_{II}$	1.6595	2	7.4708	1.6007	1	7.7453
$\beta_3\ L_IM_{III}$	1.6203	2	7.6519	1.5616	1	7.9392
$\gamma_2\ L_IN_{II}$	1.3698	2	9.051	1.3210	2	9.385
$\gamma_3\ L_IN_{III}$	1.3643	2	9.087	1.3146	1	9.4309
$\gamma_4\ L_IO_{II,III}$	1.3225	2	9.374	1.2752	2	9.722
$\eta\ L_{II}M_I$	1.8264	2	6.7883	1.7566	1	7.0579
$\beta_1\ L_{II}M_{IV}$	1.6475	2	7.5253	1.5873	1	7.8109
$\gamma_5\ L_{II}N_I$	1.4618	2	8.481	1.4067	3	8.814
$\gamma_1\ L_{II}N_{IV}$	1.4174	2	8.747	1.3641	2	9.089
$\gamma_8\ L_{II}O_I$	1.3983	2	8.867			
$\gamma_6\ L_{II}O_{IV}$	1.3923	2	8.905	1.3397	3	9.255
$l\ L_{III}M_I$	2.0860	2	5.9434	2.015	1	6.152
$\alpha_2\ L_{III}M_{IV}$	1.8561	2	6.6795	1.7955	2	6.9050
$\alpha_1\ L_{III}M_V$	1.8450	2	6.7198	1.78425	9	6.9487
$\beta_6\ L_{III}N_I$	1.6237	2	7.6359	1.5675	2	7.909
$\beta_{2,15}\ L_{III}N_{IV,V}$	1.5671	2	7.911	1.51399	9	8.1890
$\beta_7\ L_{III}O_I$				1.4941	3	8.298
$\beta_5\ L_{III}O_{IV,V}$	1.5378	2	8.062	1.4848	3	8.350

Designation	Å*	p.e.	keV	Å*	p.e.	keV
	67 Holmium (*Cont.*)			**68 Erbium** (*Cont.*)		
$\beta_{10}\ L_IM_{IV}$	1.5486	3	8.006	1.4941	3	8.298
$L_IO_{IV,V}$	1.3208	3	9.387			
$\beta_9\ L_IM_V$				1.4855	5	8.346
$M_{II}N_{IV}$				7.60	1	1.632
$\gamma\ M_{III}N_{IV,V}$	7.865	9	1.576			
$\gamma\ M_{III}\ _V$				7.546	8	1.643
$\beta\ M_{IV}N_{VI}$	8.965	4	1.3830	8.592	3	1.4430
$\zeta\ M_VN_{III}$	11.86	1	1.0450	11.37	1	1.0901
$\alpha\ M_VN_{VI,VII}$	9.20	2	1.348	8.82	1	1.406
$N_{IV}N_{VI}$				72.7	9	0.171
$N_VN_{VI,VII}$				76.3	7	0.163
	69 Thulium			**70 Ytterbium**		
$\alpha_2\ KL_{II}$	0.249095	2	49.7726	0.241424	2	51.3540
$\alpha_1\ KL_{III}$	0.244338	2	50.7416	0.236655	2	52.3889
$\beta_3\ KM_{II}$	0.21636	2	57.304	0.2096†	1	59.14
$\beta_1\ KM_{III}$	0.21556	2	57.517	0.20884	8	59.37
$\beta_2\ KN_{II,III}$	0.2098†	2	59.09	0.2033†	2	60.98
$KO_{II,III}$	0.20891	2	59.346	0.20226	2	61.298
$\beta_5\ KM_{IV,V}$	0.21404	2	57.923	0.20739	2	59.782
$\beta_4\ L_IM_{II}$	1.5448	2	8.026	1.49138	3	8.3132
$\beta_3\ L_IM_{III}$	1.5063	2	8.231	1.45233	5	8.5367
$\gamma_2\ L_IN_{II}$	1.2742	2	9.730	1.22879	7	10.0897
$\gamma_3\ L_IN_{III}$	1.2678	2	9.779	1.22232	5	10.1431
$\gamma_4\ L_IO_{II,III}$	1.2294	2	10.084	1.1853	1	10.4603
$\eta\ L_{II}M_I$	1.6963	2	7.3088	1.63560	5	7.5802
$\beta_1\ L_{II}M_{IV}$	1.5304	2	8.101	1.47565	5	8.4018
$\gamma_5\ L_{II}N_I$	1.3558	2	9.144	1.3063	1	9.4910
$\gamma_1\ L_{II}N_{IV}$	1.3153	2	9.426	1.26769	5	9.8701
$\gamma_8\ L_{II}O_I$				1.24923	5	9.9246
$\gamma_6\ L_{II}O_{IV}$	1.2905	2	9.607	1.24271	3	9.9766
$l\ L_{III}M_I$	1.9550	2	6.3419	1.89415	5	6.5455
$\alpha_2\ L_{III}M_{IV}$	1.7381	2	7.1331	1.68285	5	7.3673
$\alpha_1\ L_{III}M_V$	1.7268†	2	7.1799	1.67189	4	7.4156
$\beta_6\ L_{III}N_I$	1.5162	2	8.177	1.4661	1	8.4563
$\beta_{2,15}\ L_{III}N_{IV,V}$	1.4640	2	8.468	1.41550	5	8.7588
$\beta_7\ L_{III}O_I$				1.3948	1	8.8889
$\beta_5\ L_{III}O_{IV,V}$	1.4349	2	8.641	1.38696	7	8.9390
$\beta_{10}\ L_IM_{IV}$	1.4410	3	8.604	1.3915	1	8.9100
$\beta_9\ L_IM_V$	1.4336	3	8.648	1.3838	1	8.9597
L_IO_I				1.1886	1	10.4312
$L_IO_{IV,V}$	1.2263	3	10.110	1.1827	1	10.4833
$L_{II}M_{II}$				1.58844	9	7.8052
$L_{II}O_{II,III}$				1.2453	1	9.9561
$t\ L_{III}M_{II}$				1.83091	9	6.7715
$L_{III}O_{II,III}$				1.3898	1	8.9209
$M_{III}N_I$				8.470	9	1.464
$\gamma\ M_{III}N_V$				7.024	8	1.765
$\beta\ M_{IV}N_{VI}$	8.249	7	1.503	7.909	2	1.5675
$\zeta\ M_VN_{III}$				10.48	1	1.183
$\alpha\ M_VN_{VI,VII}$	8.48	1	1.462	8.149	5	1.5214
$N_{IV}N_{VI}$				65.1	7	0.190
$N_VN_{VI,VII}$				69.3	5	0.179
	71 Lutetium			**72 Hafnium**		
$\alpha_2\ KL_{II}$	0.234081	2	52.9650	0.227024	3	54.6114
$\alpha_1\ KL_{III}$	0.229298	2	54.0698	0.222227	3	55.7902
$\beta_3\ KM_{II}$	0.20309†	4	61.05	0.19686†	4	62.98

Designation	Å*	p.e.	keV	Å*	p.e.	keV
	71 Lutetium (*Cont.*)			**72 Hafnium** (*Cont.*)		
$\beta_1\ KM_{III}$	0.20231†	3	61.283	0.19607†	3	63.234
$\beta_2\ KN_{II,III}$	0.1969†	2	62.97	0.1908†	2	64.98
$KO_{II,III}$	0.19589	2	63.293			
$\beta_5\ KM_{IV,V}$	0.20084	2	61.732			
$\beta_4\ L_IM_{II}$	1.44056	5	8.6064	1.39220	5	8.9054
$\beta_3\ L_IM_{III}$	1.40140	5	8.8469	1.35300	5	9.1634
$\gamma_2\ L_IN_{II}$	1.1853	2	10.460	1.14442	5	10.8335
$\gamma_3\ L_IN_{III}$	1.17953	4	10.5110	1.13841	5	10.8907
$\gamma'_4\ L_IO_{II}$				1.10376	5	11.2326
$\gamma_4\ L_IO_{II,III}$	1.1435	1	10.8425	1.10303	5	11.2401
$\eta\ L_{II}M_I$	1.5779	1	7.8575	1.52325	5	8.1393
$\beta_1\ L_{II}M_{IV}$	1.42359	3	8.7090	1.37410	5	9.0227
$\gamma_5\ L_{II}N_I$	1.2596	1	9.8428	1.21537	5	10.2011
$\gamma_1\ L_{II}N_{IV}$	1.22228	4	10.1434	1.17900	5	10.5158
$\gamma_8\ L_{II}O_I$	1.2047	1	10.2915	1.16138	5	10.6754
$\gamma_6\ L_{II}O_{IV}$	1.1987	1	10.3431	1.15519	5	10.7325
$l\ L_{III}M_I$	1.8360	1	6.7528	1.78145	5	6.9596
$\alpha_2\ L_{III}M_{IV}$	1.63029	5	7.6049	1.58046	5	7.8446
$\alpha_1\ L_{III}M_{IV}$	1.61951	3	7.6555	1.56958	5	7.8990
$\beta_6\ L_{III}N_I$	1.4189	1	8.7376	1.37410	5	9.0227
$\beta_{15}\ L_{III}N_{IV}$	1.3715	1	9.0395	1.32783	5	9.3371
$\beta_2\ L_{III}N_V$	1.37012	3	9.0489	1.32639	5	9.3473
$\beta_7\ L_{III}O_I$	1.34949	5	9.1873	1.30564	5	9.4958
$\beta_5\ L_{III}O_{IV,V}$	1.34183	7	9.2397	1.29761	5	9.5546
L_IM_I				1.43025	9	8.6685
$\beta_{10}\ L_IM_{IV}$	1.3430	2	9.232	1.29819	9	9.5503
$\beta_9 L_IM_V$	1.3358	1	9.2816	1.29025	9	9.6090
L_IN_{IV}	1.16227	9	10.6672	1.12250	9	11.0451
$\gamma_{11}\ L_IN_V$	1.16107	9	10.6782	1.12146	9	11.0553
L_IO_I				1.10664	9	11.2034
L_IO_{IV}				1.10086	9	11.2622
$L_{II}M_{II}$	1.53333	9	8.0858	1.48064	9	8.3735
$\beta_{17}\ L_{II}M_{III}$				1.43643	9	8.6312
$L_{II}N_V$				1.17788	9	10.5258
$v\ L_{II}N_{VI}$				1.15830	9	10.7037
$L_{II}O_{II,III}$	1.2014	1	10.3198			
$t\ L_{III}M_{II}$	1.7760	1	6.9810	1.72305	9	7.1954
$s\ L_{III}M_{III}$				1.66346	9	7.4532
$L_{III}N_{II}$				1.35887	9	9.1239
$L_{III}N_{III}$				1.35053	9	9.1802
$u\ L_{III}N_{VI,VII}$				1.30165	9	9.5249
$L_{III}O_{II,III}$	1.34524	9	9.2163			
$M_{III}N_I$				7.887	9	1.572
$\gamma\ M_{III}N_V$	6.768	6	1.832	6.544	4	1.895
ζ_2				9.686	7	1.2800
$\beta\ M_{IV}N_{VI}$	7.601	2	1.6312	7.303	1	1.6976
ζ_1				9.686	7	1.2800
$\alpha\ M_VN_{VI,VII}$	7.840	2	1.5813	7.539	1	1.6446
$N_{IV}N_{VI}$	63.0	5	0.197			
$N_VN_{VI,VII}$	65.7	2	0.1886			
	73 Tantalum			**74 Tungsten**		
$\alpha_2\ KL_{II}$	0.220305	8	56.277	0.213828	2	57.9817
$\alpha_1\ KL_{III}$	0.215497	4	57.532	0.2090100	Std	59.31824
$\beta_3\ KM_{II}$	0.190890	2	64.9488	0.185181	2	66.9514
$\beta_1\ KM_{III}$	0.190089	4	65.223	0.184374	2	67.2443
$\beta_2^{II}\ KN_{II}$	0.185188	9	66.949	0.17960	1	69.031
$\beta_2^{I}\ KN_{III}$	0.185011	8	67.013	0.179421	7	69.101

Designation	Å*	p.e.	keV	Å*	p.e.	keV
	73 Tantalum (*Cont.*)			**74 Tungsten** (*Cont.*)		
$KO_{II,III}$	0.184031	7	67.370	0.178444	5	69.479
KL_I				0.21592	4	57.42
$\beta_5^{II}\ KM_{IV}$	0.188920	6	65.626	0.183264	5	67.652
$\beta_5^{I}\ KM_V$	0.188757	6	65.683	0.183092	7	67.715
$\beta_4\ KN_{IV,V}$	0.18451	1	67.194	0.17892	2	69.294
$\beta_4\ L_IM_{II}$	1.34581	3	9.2124	1.30162	5	9.5252
$\beta_3\ L_IM_{III}$	1.30678	3	9.4875	1.26269	5	9.8188
$\gamma_2\ L_IN_{II}$	1.1053	1	11.217	1.06806	3	11.6080
$\gamma_3\ L_IN_{III}$	1.09936	4	11.2776	1.06200	6	11.6743
$\gamma'_4\ L_IO_{II}$	1.06544	3	11.6366	1.02863	3	12.0530
$\gamma_4\ L_IO_{III}$	1.06467	3	11.6451	1.02775	3	12.0634
$\eta\ L_{II}M_I$	1.47106	5	8.4280	1.42110	3	8.7243
$\beta_1\ L_{II}M_{IV}$	1.32698	3	9.3431	1.281809	9	9.67235
$\gamma_5\ L_{II}N_I$	1.1729	1	10.5702	1.13235	3	10.9490
$\gamma_1\ L_{II}N_{IV}$	1.13794	3	10.8952	1.09855	3	11.2859
$\gamma_8\ L_{II}O_I$	1.1205	1	11.0646	1.08113	4	11.4677
$\gamma_6\ L_{II}O_{IV}$	1.11388	3	11.1306	1.07448	5	11.5387
$l\ L_{III}M_I$	1.72841	5	7.1731	1.6782	1	7.3878
$\alpha_2\ L_{III}M_{IV}$	1.53293	2	8.0879	1.48743	2	8.3352
$\alpha_1\ L_{III}M_V$	1.52197	2	8.1461	1.47639	2	8.3976
$\beta_6\ L_{III}N_I$	1.33094	8	9.3153	1.28989	7	9.6117
$\beta_{15}\ L_{III}N_{IV}$	1.28619	5	9.6394	1.24631	3	9.9478
$\beta_2\ L_{III}N_V$	1.28454	2	9.6518	1.24460	3	9.9615
$\beta_7\ L_{III}O_I$	1.26385	5	9.8098	1.22400	4	10.1292
$\beta_5\ L_{III}O_{IV,V}$	1.2555	1	9.8750	1.21545	3	10.2004
L_IM_I				1.3365	3	9.277
$\beta_{10}\ L_IM_{IV}$	1.2537	2	9.889	1.21218	3	10.2279
$\beta_9\ L_IM_V$	1.2466	2	9.946	1.20479	7	10.2907
L_IN_I	1.11521	9	11.1173			
L_IN_{IV}	1.08377	7	11.4398	1.0468	2	11.844
$\gamma_{11}\ L_IN_V$	1.08205	7	11.4580	1.0458	1	11.856
$L_IN_{VI,VII}$	1.06357	9	11.6570			
L_IO_I	1.06771	9	11.6118	1.0317	3	12.017
$L_IO_{IV,V}$	1.06192	9	11.6752	1.0250	2	12.095
$L_{II}M_{II}$	1.43048	9	8.6671			
$\beta_{17}\ L_{II}M_{III}$	1.3864	1	8.9428	1.3387	2	9.261
$L_{II}M_V$	1.31897	9	9.3998	1.2728	2	9.741
$L_{II}N_{II}$	1.1600	2	10.688	1.1218	3	11.052
$L_{II}N_{III}$	1.1553	1	10.7316	1.1149	2	11.120
$L_{II}N_V$	1.13687	9	10.9055			
$v\ L_{II}N_{VI}$	1.1158	1	11.1113	1.0771	1	11.510
$L_{II}O_{II}$	1.11789	9	11.0907			
$L_{II}O_{III}$	1.11693	9	11.1001	1.0792	2	11.488
$t\ L_{III}M_{II}$	1.67265	9	7.4123	1.6244	3	7.632
$s\ L_{III}M_{III}$	1.61264	9	7.6881	1.5642	3	7.926
$L_{III}N_{II}$	1.3167	1	9.4158	1.2765	2	9.712
$L_{III}N_{III}$	1.3086	1	9.4742	1.2672	2	9.784
$u\ L_{III}N_{VI,VII}$	1.25778	4	9.8572	1.21868	5	10.1733
$L_{III}O_{II,III}$	1.2601	3	9.839	1.2211	2	10.153
M_IN_{III}	5.40	2	2.295	5.172	9	2.397
$M_IO_{II,III}$				4.44	2	2.79
$M_{II}N_I$				6.28	2	1.973
$M_{II}N_{IV}$	5.570	4	2.226	5.357	4	2.314
$M_{III}N_I$	7.612	9	1.629	7.360	8	1.684
$M_{III}N_{IV}$	6.353	5	1.951	6.134	4	2.021
$\gamma\ M_{III}N_V$	6.312	4	1.964	6.092	3	2.035
$M_{III}O_I$	5.83	2	2.126	5.628	8	2.203
$M_{III}O_{IV,V}$	5.67	3	2.19			

Designation	Å*	p.e.	keV	Å*	p.e.	keV
	73 Tantalum (*Cont.*)			**74 Tungsten** (*Cont.*)		
$\zeta_2\ M_{IV}N_{II}$	9.330	5	1.3288	8.993	5	1.3787
$M_{IV}N_{III}$	8.90	2	1.393	8.573	8	1.446
$\beta\ M_{IV}N_{VI}$	7.023	1	1.7655	6.757	1	1.8349
$M_{IV}O_{II}$	7.09	2	1.748	6.806	9	1.822
$\zeta_1\ M_VN_{III}$	9.316	4	1.3308	8.962	4	1.3835
$\alpha\ M_VN_{VI,VII}$	7.252	1	1.7096			
$\alpha_2\ M_VN_{VI}$				6.992	2	1.7731
$\alpha_1\ M_VN_{VII}$				6.983	1	1.7754
M_VO_{III}	7.30	2	1.700	7.005	9	1.770
$N_{II}N_{IV}$				54.0	2	0.2295
$N_{IV}N_{VI}$	58.2	1	0.2130	55.8	1	0.2221
$N_VN_{VI,VII}$	61.1	2	0.2028			
N_VN_{VI}				59.5	3	0.208
N_VN_{VII}				58.4	1	0.2122
	75 Rhenium			**76 Osmium**		
$\alpha_2\ KL_{II}$	0.207611	1	59.7179	0.201639	2	61.4867
$\alpha_1\ KL_{III}$	0.202781	2	61.1403	0.196794	2	63.0005
$\beta_3\ KM_{II}$	0.179697	3	68.994	0.174431	3	71.077
$\beta_1\ KM_{III}$	0.178880	3	69.310	0.173611	3	71.413
$\beta_2^{II}\ KN_{II}$	0.17425	1	71.151	0.16910	1	73.318
$\beta_2^{I}\ KN_{III}$	0.174054	6	71.232	0.168906	6	73.402
$KO_{II,III}$	0.17308	1	71.633	0.16798	1	73.808
$\beta_5^{II}\ KM_{IV}$	0.17783	1	69.719	0.17262	1	71.824
$\beta_5^{I}\ KM_V$	0.17766	1	69.786	0.17245	1	71.895
$\beta_4\ KN_{IV,V}$	0.17362	2	71.410	0.16842	2	73.615
$\beta_4\ L_IM_{II}$	1.25917	5	9.8463	1.21844	5	10.1754
$\beta_3\ L_IM_{III}$	1.22031	5	10.1598	1.17955	7	10.5108
$\gamma_2\ L_IN_{II}$	1.03233	5	12.0098	0.99805	5	12.4224
$\gamma_3\ L_IN_{III}$	1.02613	7	12.0824	0.99186	5	12.4998
$\gamma'_4\ L_IO_{II}$	0.99334	5	12.4813	0.96033	8	12.910
$\gamma_4\ L_IO_{III}$	0.99249	5	12.4920	0.95938	8	12.923
$\eta\ L_{II}M_I$	1.37342	5	9.0272	1.32785	7	9.3370
$\beta_1\ L_{II}M_{IV}$	1.23858	2	10.0100	1.19727	7	10.3553
$\gamma_5\ L_{II}N_I$	1.09388	5	11.3341	1.05693	5	11.7303
$\gamma_1\ L_{II}N_{IV}$	1.06099	5	11.6854	1.02503	5	12.0953
$\gamma_8\ L_{II}O_I$	1.04398	5	11.8758	1.00788	5	12.3012
$\gamma_6\ L_{II}O_{IV}$	1.03699	9	11.956	1.00107	5	12.3848
$l\ L_{III}M_I$	1.63056	5	7.6036	1.58498	7	7.8222
$\alpha_2\ L_{III}M_{IV}$	1.44396	5	8.5862	1.40234	5	8.8410
$\alpha_1\ L_{III}M_V$	1.43290	4	8.6525	1.39121	5	8.9117
$\beta_6\ L_{III}N_I$	1.25100	5	9.9105	1.21349	5	10.2169
$\beta_{15}\ L_{III}N_{IV}$	1.20819	5	10.2617	1.17167	5	10.5816
$\beta_2\ L_{III}N_V$	1.20660	4	10.2752	1.16979	8	10.5985
$\beta_7\ L_{III}O_I$	1.18610	5	10.4529	1.14933	8	10.7872
$\beta_5\ L_{III}O_{IV,V}$	1.17721	5	10.5318	1.1405	1	10.8711
$\beta_{10}\ L_IM_{IV}$	1.17218	5	10.5770	1.13353	5	10.9376
$\beta_9\ L_IM_V$	1.16487	4	10.6433	1.12637	6	11.0071
L_IN_I	1.0420	1	11.899			
L_IN_{IV}	1.0119	1	12.252	0.9772	3	12.687
$\gamma_{11}\ L_IN_V$	1.0108	1	12.266	0.9765	3	12.696
L_IO_I	0.9965	1	12.442	0.96318	7	12.8721
$L_IO_{IV,V}$	0.9900	1	12.524	0.95603	5	12.9683
$L_{II}M_{II}$	1.3366	1	9.2761	1.2934	2	9.586
$\beta_{17}\ L_{II}M_{III}$	1.2927	1	9.5910	1.2480	2	9.934

Designation	Å*	p.e.	keV	Å*	p.e.	keV
	75 Rhenium (*Cont.*)			**76 Osmium** (*Cont.*)		
$L_{II}M_V$	1.2305	1	10.0753	1.18977	7	10.4205
$L_{II}N_{II}$	1.0839	1	11.438			
$L_{II}N_{III}$	1.0767	1	11.515	1.03973	5	11.9243
$v\ L_{II}N_{VI}$	1.0404	1	11.917	1.0050	2	12.337
$L_{II}O_{III}$	1.0397	1	11.925	1.0047	2	12.340
$t\ L_{III}M_{II}$	1.5789	1	7.8525	1.5347	2	8.079
$s\ L_{III}M_{III}$	1.5178	1	8.1682	1.4735	2	8.414
$L_{III}N_I$				1.20086	7	10.3244
$L_{III}N_{III}$	1.2283	1	10.0933			
$u\ L_{III}N_{VI,VII}$	1.1815	1	10.4931	1.14537	7	10.8245
M_IN_{III}				4.79	2	2.59
$M_{II}N_I$				5.81	2	2.133
$M_{II}N_{IV}$				4.955	4	2.502
$M_{III}N_I$				6.89	2	1.798
$M_{III}N_{IV}$	5.931	5	2.090	5.724	5	2.166
$\gamma\ M_{III}N_V$	5.885	2	2.1067	5.682	4	2.182
$\zeta_2\ M_{IV}N_{II}$	8.664	5	1.4310	8.359	5	1.4831
$M_{IV}N_{III}$	8.239	8	1.505			
$\beta\ M_{IV}N_{VI}$	6.504	1	1.9061	6.267	1	1.9783
$\zeta_1\ M_VN_{III}$	8.629	4	1.4368	8.310	4	1.4919
$\alpha\ M_VN_{VI,VII}$	6.729	1	1.8425	6.490	1	1.9102
$N_{IV}N_{VI}$				51.9	1	0.2388
$N_VN_{VI,VII}$				54.7	2	0.2266
	77 Iridium			**78 Platinum**		
$\alpha_2\ KL_{II}$	0.195904	2	63.2867	0.190381	4	65.122
$\alpha_1\ KL_{III}$	0.191047	2	64.8956	0.185511	4	66.832
$\beta_3\ KM_{II}$	0.169367	2	73.2027	0.164501	3	75.368
$\beta_1\ KM_{III}$	0.168542	2	73.5608	0.163675	3	75.748
$\beta_2^{II}\ KN_{II}$	0.16415	1	75.529	0.15939	1	77.785
$\beta_2^{I}\ KN_{III}$	0.163956	7	75.619	0.15920	1	77.878
$KO_{II,III}$	0.163019	5	76.053	0.15826	1	78.341
$\beta_5^{II}\ KM_{IV}$	0.16759	2	73.980	0.16271	2	76.199
$\beta_5^{I}\ KM_V$	0.167373	9	74.075	0.16255	3	76.27
$\beta_4\ KN_{IV,V}$	0.16352	2	75.821	0.15881	2	78.069
$\beta_4\ L_IM_{II}$	1.17958	3	10.5106	1.14223	5	10.8543
$\beta_3\ L_IM_{III}$	1.14085	3	10.8674	1.10394	5	11.2308
$\gamma_2\ L_IN_{II}$	0.96545	3	12.8418	0.93427	5	13.2704
$\gamma_3\ L_IN_{III}$	0.95931	5	12.9240	0.92791	5	13.3613
$\gamma'_4\ L_IO_{II}$	0.92831	3	13.3555	0.89747	4	13.8145
$\gamma_4\ L_IO_{III}$	0.92744	3	13.3681	0.89659	4	13.8281
$\eta\ L_{II}M_I$	1.28448	3	9.6522	1.2429	2	9.975
$\beta_1\ L_{II}M_{IV}$	1.15781	3	10.7083	1.11990	2	11.0707
$\gamma_5\ L_{II}N_I$	1.02175	5	12.1342	0.9877	2	12.552
$\gamma_1\ L_{II}N_{IV}$	0.99085	3	12.5126	0.95797	3	12.9420
$\gamma_8\ L_{II}O_I$	0.97409	3	12.7279	0.9411	1	13.173
$\gamma_6\ L_{II}O_{IV}$	0.96708	4	12.8201	0.9342	2	13.271
$l\ L_{III}M_I$	1.54094	3	8.0458	1.4995	2	8.268
$\alpha_2\ L_{III}M_{IV}$	1.36250	5	9.0995	1.32432	2	9.3618
$\alpha_1\ L_{III}M_V$	1.35128	3	9.1751	1.31304	3	9.4423
$\beta_6\ L_{III}N_I$	1.17796	3	10.5251	1.14355	5	10.8418
$\beta_{15}\ L_{III}N_{IV}$	1.13707	3	10.9036			
$\beta_2\ L_{III}N_V$	1.13532	3	10.9203	1.10200	3	11.2505
$\beta_7\ L_{III}O_I$	1.11489	3	11.1205	1.08168	3	11.4619

Designation	Å*	p.e.	keV	Å*	p.e.	keV
	77 Iridium (*Cont.*)			**78 Platinum** (*Cont.*)		
$\beta_6\ L_{III}O_{IV,V}$	1.10585	3	11.2114	1.0724	2	11.561
L_IM_I	1.2102	2	10.245	1.16962	9	10.6001
$\beta_{10}\ L_IM_{IV}$	1.09702	4	11.3016	1.06183	7	11.6762
$\beta_9\ L_IM_V$	1.08975	5	11.3770	1.05446	5	11.7577
L_IN_I	0.9766	2	12.695	0.9455	2	13.113
L_IN_{IV}	0.9459	2	13.108			
$\gamma_{11}\ L_IN_V$	0.9446	2	13.126	0.9143	2	13.560
$L_IO_{IV,V}$	0.9243	3	13.413			
L_IO_I				0.8995	2	13.784
L_IO_{IV}				0.8943	1	13.864
L_IO_V				0.8934	1	13.878
$L_{II}M_{II}$	1.2502	3	9.917	1.213	1	10.225
$\beta_{17}\ L_{II}M_{III}$	1.2069	2	10.273	1.1667	1	10.6265
$L_{II}M_V$	1.1489	2	10.791	1.1129	2	11.140
$L_{II}N_{II}$	1.0120	2	12.251	0.9792	2	12.661
$L_{II}N_{III}$	1.0054	3	12.332	0.97173	4	12.7588
$v\ L_{II}N_{VI}$	0.97161	6	12.7603	0.93931	5	13.1992
$L_{II}O_{III}$	0.96979	5	12.7843			
$t\ L_{III}M_{II}$	1.4930	3	8.304	1.4530	2	8.533
$s\ L_{III}M_{III}$	1.4318	2	8.659	1.3895	2	8.923
$L_{III}N_{II}$	1.16545	5	10.6380	1.1310	2	10.962
$L_{III}N_{III}$	1.1560	3	10.725	1.1226	2	11.044
$u\ L_{III}N_{VI,VII}$	1.11145	4	11.1549	1.07896	5	11.4908
$L_{III}O_{II,III}$	1.10923	6	11.177[illegible]	1.0761	3	11.521
M_IN_{III}	4.631	9	2.677	4.460	9	2.780
$M_{II}N_{IV}$	4.780	4	2.594	4.601	4	2.695
$M_{III}N_I$	6.669	9	1.859	6.455	9	1.921
$M_{III}N_{IV}$	5.540	5	2.238	5.357	5	2.314
$\gamma\ M_{III}N_V$	5.500	4	2.254	5.319	4	2.331
$M_{III}O_I$				4.876	9	2.543
$M_{III}O_{IV,V}$	4.869	9	2.546	4.694	8	2.641
$\zeta_2\ M_{IV}N_{II}$	8.065	5	1.5373	7.790	5	1.592
$M_{IV}N_{III}$	7.645	8	1.622	7.371	8	1.682
$\beta\ M_{IV}N_{VI}$	6.038	1	2.0535	5.828	1	2.1273
$\zeta_1\ M_VN_{III}$	8.021	4	1.5458	7.738	4	1.6022
$\alpha_2\ M_VN_{VI}$	6.275	3	1.9758	6.058	3	2.047
$\alpha_1\ M_VN_{VII}$	6.262	1	1.9799	6.047	1	2.0505
M_VO_{III}				5.987	9	2.071
$N_{IV}N_{VI}$	50.2	1	0.2470	48.1	2	0.258
$N_VN_{VI,VII}$	52.8	1	0.2348	50.9	1	0.2436
	79 Gold			**80 Mercury**		
$\alpha_2\ KL_{II}$	0.185075	2	66.9895	0.179958	3	68.895
$\alpha_1\ KL_{III}$	0.180195	2	68.8037	0.175068	3	70.819
$\beta_3\ KM_{II}$	0.159810	2	77.580	0.155321	3	79.822
$\beta_1\ KM_{III}$	0.158982	3	77.984	0.154487	3	80.253
$\beta_2^{II}\ KN_{II}$	0.15483	2	80.08	0.15040	2	82.43
$\beta_2^{I}\ KN_{III}$	0.154018	9	80.185	0.15020	2	82.54
$KO_{II,III}$	0.153694	7	80.667	0.14931	2	83.04
KL_I	0.18672	4	66.40			
$\beta_5^{II}\ KM_{IV}$	0.158062	7	78.438			
$\beta_5^{I}\ KM_V$	0.157880	5	78.529			
$\beta_5\ KM_{IV,V}$				0.15353	2	80.75
$\beta_4\ KN_{IV,V}$	0.154224	5	80.391	0.14978	2	82.78
$\beta_4\ L_IM_{II}$	1.10651	3	11.2047	1.07222	7	11.5630
$\beta_3\ L_IM_{III}$	1.06785	9	11.6103	1.03358	7	11.9953

Designation	Å*	p.e.	keV	Å*	p.e.	keV
	79 Gold (*Cont.*)			**80 Mercury** (*Cont.*)		
$\gamma_2\ L_IN_{II}$	0.90434	3	13.7095	0.87544	7	14.162
$\gamma_3\ L_IN_{III}$	0.89783	5	13.8090	0.86915	7	14.265
$\gamma'_4\ L_IO_{II}$	0.86816	4	14.2809	0.84013	7	14.757
$\gamma_4\ L_IO_{III}$	0.86703	4	14.2996	0.83894	7	14.778
$\eta\ L_{II}M_I$	1.20273	3	10.3083	1.1640	1	10.6512
$\beta_1\ L_{II}M_{IV}$	1.08353	3	11.4423	1.04868	5	11.8226
$\gamma_5\ L_{II}N_I$	0.95559	3	12.9743	0.92453	7	13.410
$\gamma_1\ L_{II}N_{IV}$	0.92650	3	13.3817	0.89646	5	13.8301
$\gamma_8\ L_{II}O_I$	0.90989	5	13.6260	0.87995	7	14.090
$\gamma_6\ L_{II}O_{IV}$	0.90297	3	13.7304	0.87319	7	14.199
$l\ L_{III}M_I$	1.45964	9	8.4939	1.4216	1	8.7210
$\alpha_2\ L_{III}M_{IV}$	1.28772	3	9.6280	1.25264	7	9.8976
$\alpha_1\ L_{III}M_V$	1.27640	3	9.7133	1.24120	5	9.9888
$\beta_6\ L_{III}N_I$	1.11092	3	11.1602	1.07975	7	11.4824
$\beta_{15}\ L_{III}N_{IV}$	1.07188	5	11.5667	1.04151	7	11.9040
$\beta_2\ L_{III}N_V$	1.07022	3	11.5847	1.03975	7	11.9241
$\beta_7\ L_{III}O_I$	1.04974	8	11.8106	1.01937	7	12.1625
$\beta_5\ L_{III}O_{IV,V}$	1.04044	3	11.9163	1.00987	7	12.2769
L_IM_I	1.13525	5	10.9210	1.0999	2	11.272
$\beta_{10}\ L_IM_{IV}$	1.02789	7	12.0617	0.9962	2	12.446
$\beta_9\ L_IM_V$	1.02063	7	12.1474	0.9871	2	12.560
L_IN_I	0.9131	1	13.578	0.8827	[illegible]	14.015
L_IN_{IV}	0.88560	[illegible]	13.999			
$\gamma_{11}\ L_IN_V$	0.88433	7	14.020	0.85657	7	14.474
L_IO_I	0.87074	5	14.2385	0.8452	2	14.670
$L_IO_{IV,V}$	0.86400	5	14.3497	0.8350	2	14.847
$L_{II}M_{II}$	1.1708	1	10.5892	1.1387	5	10.888
$\beta_{17}\ L_{II}M_{III}$	1.12798	5	10.9915	1.0916	5	11.358
$L_{II}M_V$	1.0756	2	11.526			
$L_{II}N_{III}$	0.9402	2	13.186	0.90894	7	13.640
$v\ L_{II}N_{VI}$	0.90837	5	13.6487	0.87885	7	14.107
$L_{II}O_{II}$	0.90746	7	13.662	0.8784	1	14.114
$L_{II}O_{III}$	0.90638	7	13.679	0.8758	1	14.156
$t\ L_{III}M_{II}$	1.41366	7	8.7702	1.3746	2	9.019
$s\ L_{III}M_{III}$	1.35131	7	9.1749	1.3112	2	9.455
$L_{III}N_{II}$	1.09968	7	11.2743	1.0649	2	11.642
$L_{III}N_{III}$	1.09026	7	11.3717	1.0585	1	11.713
$u\ L_{III}N_{VI,VII}$	1.04752	5	11.8357			
$u'\ L_{III}N_{VI}$				1.01769	7	12.1826
$u\ L_{III}N_{VII}$				1.01674	7	12.1940
$L_{III}O_{II,III}$	1.0450	2	11.865			
$L_{III}O_{II}$				1.01558	7	12.2079
$L_{III}O_{III}$				1.01404	7	12.2264
$L_{III}P_{II,III}$	1.03876	7	11.9355			
M_IN_{III}	4.300	9	2.883			
$M_{II}N_{IV}$	4.432	1	2.797			
$M_{III}N_I$	6.259	9	1.981	6.09	2	2.036
$M_{III}N_{IV}$	5.186	5	2.391			
$\gamma\ M_{III}N_V$	5.145	4	2.410	4.984†	2	2.4875
$M_{III}O_I$	4.703	9	2.636			
$M_{III}O_{IV,V}$	4.522	6	2.742			
$\zeta_2\ M_{IV}N_{II}$	7.523	5	1.648			
$M_{IV}N_{III}$	7.101	8	1.746	6.87	2	1.805
$\beta\ M_{IV}N_{VI}$	5.624	1	2.2046	5.4318†	9	2.2825
$\zeta_1\ M_VN_{III}$	7.466	4	1.6605			
$\alpha_2\ M_VN_{VI}$	5.854	3	2.118			

Designation	Å*	p.e.	keV	Å*	p.e.	keV
	79 Gold (*Cont.*)			**80 Mercury** (*Cont.*)		
$\alpha_1\ M_VN_{VII}$	5.840	1	2.1229	5.6476†	9	2.1953
M_VO_{III}	5.767	9	2.150			
$N_{IV}N_{VI}$	46.8	2	0.265	45.2†	3	0.274
$N_VN_{VI,VII}$	49.4	1	0.2510	47.9†	3	0.259
	81 Thallium			**82 Lead**		
$\alpha_2\ KL_{II}$	0.175036	2	70.8319	0.170294	2	72.8042
$\alpha_1\ KL_{III}$	0.170136	2	72.8715	0.165376	2	74.9694
$\beta_3\ KM_{II}$	0.150980	6	82.118	0.146810	4	84.450
$\beta_1\ KM_{III}$	0.150142	5	82.576	0.145970	6	84.936
$\beta_2^{II}\ KN_{II}$	0.14614	1	84.836	0.14212	2	87.23
$\beta_2^{I}\ KN_{III}$	0.14595	1	84.946	0.14191	1	87.364
$KO_{II,III}$	0.14509	1	85.451	0.141012	8	87.922
KP				0.1408	1	88.06
$\beta_5\ KM_{IV,V}$	0.14917	1	83.114			
$\beta_5^{II}\ KM_{IV}$				0.14512	2	85.43
$\beta_5^{I}\ KM_V$				0.14495	3	85.53
$\beta_4\ KN_{IV,V}$	0.14553	2	85.19	0.14155	3	87.59
$\beta_4\ L_IM_{II}$	1.03918	3	11.9306	1.0075	1	12.306
$\beta_3\ L_IM_{III}$	1.00062	3	12.3904	0.96911	7	12.7933
$\gamma_2\ L_IN_{II}$	0.84773	5	14.6251	0.8210	2	15.101
$\gamma_3\ L_IN_{III}$	0.84130	4	14.7368	0.8147	1	15.218
$\gamma'_4\ L_IO_{II}$	0.81308	5	15.2482	0.78706	7	15.752
$\gamma_4\ L_IO_{III}$	0.81184	5	15.2716	0.7858	1	15.777
$\eta\ L_{II}M_I$	1.12769	3	10.9943	1.09241	7	11.3493
$\beta_1\ L_{II}M_{IV}$	1.01513	4	12.2133	0.98291	3	12.6137
$\gamma_5\ L_{II}N_I$	0.89500	4	13.8526	0.86655	5	14.3075
$\gamma_1\ L_{II}N_{IV}$	0.86752	3	14.2915	0.83973	3	14.7644
$\gamma_8\ L_{II}O_I$	0.8513	2	14.564	0.82365	5	15.0527
$\gamma_6\ L_{II}O_{IV}$	0.8442	2	14.685	0.81683	5	15.1783
$L_{II}P_I$				0.81583	5	15.1969
$l\ L_{III}M_I$	1.38477	3	8.9532	1.34990	7	9.1845
$\alpha_2\ L_{III}M_{IV}$	1.21875	3	10.1728	1.18648	5	10.4495
$\alpha_1\ L_{III}M_V$	1.20739	4	10.2685	1.17501	2	10.5515
$\beta_6\ L_{III}N_I$	1.04963	5	11.8118	1.0210	1	12.143
$\beta_{15}\ L_{III}N_{IV}$	1.01201	3	12.2510	0.98389	7	12.6011
$\beta_2\ L_{III}N_V$	1.01031	3	12.2715	0.98221	7	12.6226
$\beta_7\ L_{III}O_I$	0.99017	5	12.5212	0.9620	1	12.888
$\beta_5\ L_{III}O_{IV,V}$	0.98058	3	12.6436	0.9526	1	13.015
L_IM_I	1.0644	2	11.648	1.0323	2	12.010
$\beta_{10}\ L_IM_{IV}$	0.96389	7	12.8626	0.9339	2	13.275
$\beta_9\ L_IM_V$	0.95675	7	12.9585	0.9268	1	13.377
L_IN_I	0.8549	1	14.503	0.82859	7	14.963
L_IN_{IV}	0.83001	7	14.937	0.80364	7	15.427
$\gamma_{11}\ L_IN_V$	0.82879	5	14.9593	0.80233	9	15.453
$L_IN_{VI,VII}$				0.7884	1	15.725
L_IO_I	0.8158	1	15.198	0.7897	1	15.699
$L_IO_{IV,V}$	0.80861	5	15.3327	0.78257	7	15.843
$L_{II}M_{II}$	1.0997	1	11.274	1.0644	2	11.648
$\beta_{17}\ L_{II}M_{III}$	1.05609	7	11.7397	1.0223	1	12.127
$L_{II}M_V$	1.00722	5	12.3093	0.9747	1	12.720
$L_{II}N_{II}$	0.882	2	14.057	0.8585	3	14.442

Designation	Å*	p.e.	keV	Å*	p.e.	keV
	81 Thallium (*Cont.*)			**82 Lead** (*Cont.*)		
$L_{II}N_{III}$	0.87996	5	14.0893	0.85192	7	14.553
$L_{II}N_V$				0.8382	2	14.791
$v\ L_{II}N_{VI}$	0.85048	5	14.5777	0.82327	7	15.060
$L_{II}O_{II}$	0.8490	1	14.604			
$L_{II}O_{III}$				0.8200	1	15.120
$t\ L_{III}M_{II}$	1.34154	5	9.2417	1.30767	7	9.4811
$s\ L_{III}M_{III}$	1.27807	5	9.7007	1.24385	7	9.9675
$L_{III}N_{II}$				1.01040	7	12.2705
$L_{III}N_{III}$	1.0286	1	12.053	1.0005	1	12.392
$u\ L_{III}N_{VI,VII}$	0.9888	1	12.538	0.96133	7	12.8968
$L_{III}O_{II}$	0.98738	5	12.5566	0.9586	1	12.934
$L_{III}O_{III}$	0.98538	5	12.5820	0.9578	1	12.945
$L_{III}P_{II,III}$	0.97926	5	12.6607	0.95118	7	13.0344
M_IN_{III}	4.013	9	3.089	3.872	9	3.202
$M_{II}N_I$				4.655	8	2.664
$M_{II}N_{IV}$	4.116	4	3.013	3.968	5	3.124
$M_{III}N_I$	5.884	8	2.107	5.704	8	2.174
$M_{III}N_{IV}$	4.865	5	2.548	4.715	3	2.630
$\gamma\ M_{III}N_V$	4.823	4	2.571	4.674	1	2.6527
$M_{III}O_I$				4.244	9	2.921
$M_{II}O_{IV,V}$	4.216	6	2.941	4.069	6	3.047
$\zeta_2\ M_{IV}N_{II}$	7.032	5	1.763	6.802	5	1.823
$M_{IV}N_{III}$				6.384	7	1.942
$\beta\ M_{IV}N_{VI}$	5.249	1	2.3621	5.076	1	2.4427
$M_{IV}O_{II}$	5.196	9	2.386	5.004	9	2.477
$\zeta_1\ M_VN_{III}$	6.974	4	1.778	6.740	3	1.8395
$\alpha_2\ M_VN_{VI}$	5.472	2	2.2656	5.299	2	2.3397
$\alpha_1\ M_VN_{VII}$	5.460	1	2.2706	5.286	1	2.3455
M_VO_{III}				5.168	9	2.399
$N_{IV}N_{VI}$				42.3	2	0.293
$N_VN_{VI,VII}$	46.5	2	0.267	45.0	1	0.2756
$N_{VI}O_{IV}$	115.3	2	0.1075	102.4	1	0.1211
$N_{VI}O_V$	113.0	1	0.10968	100.2	2	0.1237
$N_{VII}O_V$	117.7	1	0.10530	104.3	1	0.1189
	83 Bismuth			**84 Polonium**		
$\alpha_2\ KL_{II}$	0.165717	2	74.8148	0.16130†	1	76.862
$\alpha_1\ KL_{III}$	0.160789	2	77.1079	0.15636†	1	79.290
$\beta_3\ KM_{II}$	0.142779	7	86.834	0.13892†	2	89.25
$\beta_1\ KM_{III}$	0.141948	3	87.343	0.13807†	2	89.80
$\beta_2^{II}\ KN_{II}$	0.13817	1	89.733	0.13438†	2	92.26
$\beta_2^{I}\ KN_{III}$	0.13797	1	89.864	0.13418†	2	92.40
$KO_{II,III}$	0.13709	1	90.435			
$\beta_5\ KM_{IV,V}$	0.14111	1	87.860			
$\beta_4\ KN_{IV,V}$	0.13759	2	90.11			
$\beta_4\ L_IM_{II}$	0.97690	4	12.6912	0.9475	3	13.086
$\beta_3\ L_IM_{III}$	0.93855	3	13.2098	0.9091	3	13.638
$\gamma_2\ L_IN_{II}$	0.79565	3	15.5824	0.772	1	16.07
$\gamma_3\ L_IN_{III}$	0.78917	5	15.7102			
$\gamma'_4\ L_IO_{II}$	0.76198	3	16.2709			
$\gamma_4\ L_IO_{III}$	0.76087	3	16.2947			
$\gamma_{13}\ L_IP_{II,III}$	0.75690	3	16.3802			
$\eta\ L_{II}M_I$	1.05856	3	11.7122			
$\beta_1\ L_{II}M_{IV}$	0.951978	9	13.0235	0.9220	2	13.447
$\gamma_5\ L_{II}N_I$	0.83923	5	14.7732			

Designation	Å*	p.e.	keV	Å*	p.e.	keV
	83 Bismuth (*Cont.*)			84 Polonium (*Cont.*)		
$\gamma_1\ L_{II}N_{IV}$	0.81311	2	15.2477	0.78748	9	15.744
$\gamma_8\ L_{II}O_I$	0.7973	1	15.551			
$\gamma_6\ L_{II}O_{IV}$	0.79043	3	15.6853	0.7645	2	16.218
$l\ L_{III}M_I$	1.31610	7	9.4204	1.2829	5	9.664
$\alpha_2\ L_{III}M_{IV}$	1.15536	1	10.73091	1.12548†	5	11.0158
$\alpha_1\ L_{III}M_V$	1.14386	2	10.8388	1.11386	4	11.1308
$\beta_6\ L_{III}N_I$	0.99331	3	12.4816	0.9672	2	12.819
$\beta_{15}\ L_{III}N_{IV}$	0.95702	5	12.9549	0.9312	2	13.314
$\beta_2\ L_{III}N_V$	0.95518	4	12.9799	0.92937	5	13.3404
$\beta_7\ L_{III}O_I$	0.93505	5	13.2593			
$\beta_5\ L_{III}O_{IV,V}$	0.92556	3	13.3953	0.8996	2	13.782
L_IM_I	1.0005	9	12.39			
$\beta_{10}\ L_IM_{IV}$	0.90495	4	13.7002			
$\beta_9\ L_IM_V$	0.89791	3	13.8077			
L_IN_I	0.8022	1	15.456			
L_IN_{IV}	0.7795	5	15.904			
$\gamma_{11}\ L_IN_V$	0.77728	5	15.951			
$L_IN_{VI,VII}$	0.7641	5	16.23			
$L_IO_{IV,V}$	0.75791	5	16.358			
$L_{II}M_{II}$	1.0346	9	11.98			
$\beta_{17}\ L_{II}M_{III}$	0.98913	5	12.5344			
$L_{II}M_V$	0.94419	5	13.1310			
$L_{II}N_{II}$	0.8344	9	14.86			
$L_{II}N_{III}$	0.8248	1	15.031			
$v\ L_{II}N_{VI}$	0.79721	9	15.552			
$L_{II}O_{III}$	0.79384	5	15.6178			
$t\ L_{III}M_{II}$	1.2748	1	9.7252			
$s\ L_{III}M_{III}$	1.2105	1	10.2421			
$L_{III}N_{II}$	0.98280	5	12.6151			
$L_{III}N_{III}$	0.97321	5	12.7394			
$u\ L_{III}N_{VI,VII}$	0.93505	5	13.2593			
$L_{III}O_{II}$	0.9323	2	13.298			
$L_{III}O_{III}$	0.9302	2	13.328			
$L_{III}P_{II,III}$	0.92413	4	13.4159			
M_IN_{II}	3.892	9	3.185			
M_IN_{III}	3.740	9	3.315			
$M_{II}N_{IV}$	3.834	4	3.234			
$M_{III}N_I$	5.537	8	2.239			
$M_{III}N_{IV}$	4.571	5	2.712			
$\gamma\ M_{III}N_V$	4.532	2	2.735			
$M_{III}O_I$	4.105	9	3.021			
$M_{III}O_{IV,V}$	3.932	6	3.153			
$\zeta_2\ M_{IV}N_{II}$	6.585	5	1.883			
$M_{IV}N_{III}$	6.162	8	2.012			
$\beta\ M_{IV}N_{VI}$	4.909	1	2.5255			
$M_{IV}O_{II}$	4.823	3	2.571			
$M_{IV}P_{II,III}$	4.59	2	2.70			
$\zeta_1\ M_VN_{III}$	6.521	4	1.901			
$\alpha_2\ M_VN_{VI}$	5.130	2	2.4170			
$\alpha_1\ M_VN_{VII}$	5.118	1	2.4226			
$N_IP_{II,III}$	13.30	6	0.932			
$N_{VI}O_{IV}$	91.6	1	0.1354			
$N_{VII}O_V$	93.2	1	0.1330			

Designation	Å*	p.e.	keV	Å*	p.e.	keV
	85 Astatine			**86 Radon**		
$\alpha_2\ KL_{II}$	0.15705†	2	78.95	0.15294†	3	81.07
$\alpha_1\ KL_{III}$	0.15210†	2	81.52	0.14798†	3	83.78
$\beta_3\ KM_{II}$	0.13517†	4	91.72	0.13155†	5	94.24
$\beta_1\ KM_{III}$	0.13432†	4	92.30	0.13069†	5	94.87
$\beta_2^{II}\ KN_{II}$	0.13072†	4	94.84	0.12719†	5	97.47
$\beta_2^{I}\ KN_{III}$	0.13052†	4	94.99	0.12698†	5	97.64
$\beta_3\ L_IM_{III}$	0.88135†	9	14.067	0.85436†	9	14.512
$\beta_1\ L_{II}M_{IV}$	0.89349†	9	13.876	0.86605†	9	14.316
$\gamma_1\ L_{II}N_{IV}$	0.76289†	9	16.251	0.73928†	9	16.770
$\alpha_2\ L_{III}M_{IV}$	1.09671†	5	11.3048	1.06899†	5	11.5979
$\alpha_1\ L_{III}M_V$	1.08500†	5	11.4268	1.05723†	5	11.7270
	87 Francium			**88 Radium**		
$\alpha_2\ KL_{II}$	0.14896†	3	83.23	0.14512†	2	85.43
$\alpha_1\ KL_{III}$	0.14399†	3	86.10	0.14014†	2	88.47
$\beta_3\ KM_{II}$	0.12807†	5	96.81	0.12469†	3	99.43
$\beta_1\ KM_{III}$	0.12719†	5	97.47	0.12382†	3	100.13
$\beta_2^{II}\ KN_{II}$	0.12379†	5	100.16	0.12050†	3	102.89
$\beta_2^{I}\ KN_{III}$	0.12358†	5	100.33	0.12029†	3	103.07
$\beta_4\ L_IM_{II}$				0.84071	5	14.7472
$\beta_3\ L_IM_{III}$	0.82789†	9	14.976	0.80273	5	15.4449
$\gamma_2\ L_IN_{II}$				0.68199	5	18.179
$\gamma_3\ L_IN_{III}$				0.67538	5	18.357
$\gamma'_4\ L_IO_{II}$				0.65131	5	19.036
$\gamma_4\ L_IO_{III}$				0.64965	5	19.084
$\gamma_{13}\ L_IP_{II,III}$				0.64513	5	19.218
$\eta\ L_{II}M_I$				0.90742	5	13.6630
$\beta_1\ L_{II}M_{IV}$	0.83940†	9	14.770	0.81375	5	15.2358
$\gamma_5\ L_{II}N_I$				0.71774	5	17.274
$\gamma_1\ L_{II}N_{IV}$	0.71652†	9	17.303	0.69463	5	17.849
γ_8				0.6801	1	18.230
$\gamma_6\ L_{II}O_{IV}$				0.67328	5	18.414
$L_{II}P_I$				0.6724	1	18.439
$l\ L_{III}M_I$				1.16719	5	10.6222
$\alpha_2\ L_{III}M_{IV}$	1.04230	5	11.8950	1.01656	5	12.1962
$\alpha_1\ L_{III}M_V$	1.03049	5	12.0313	1.00473	5	12.3397
$\beta_6\ L_{III}N_I$				0.87088	5	14.2362
$\beta_{15}\ L_{III}N_{IV}$				0.83722	5	14.8086
$\beta_2\ L_{III}N_V$	0.858	2	14.45	0.83537	5	14.8414
$\beta_7\ L_{III}O_I$				0.8162	1	15.190
$\beta_5\ L_{III}O_{IV,V}$				0.80627	5	15.3771
$L_{III}P_I$				0.8050	1	15.402
$\beta_{10}\ L_IM_{IV}$				0.77546	5	15.988
$\beta_9\ L_IM_V$				0.76857	5	16.131
L_IN_I				0.6874	1	18.036
L_IN_{IV}				0.6666	1	18.600
$\gamma_{11}\ L_IN_V$				0.6654	1	18.633
$L_IO_{IV,V}$				0.6468	1	19.167
$\beta_{17}\ L_{II}M_{III}$				0.8438	1	14.692
$L_{II}N_{III}$				0.7043	1	17.604
$L_{II}N_V$				0.6932	1	17.884
$L_{II}O_{II}$				0.6780	1	18.286

Designation	Å*	p.e.	keV	Å*	p.e.	keV
	87 Francium (*Cont.*)			**88 Radium** (*Cont.*)		
$L_{II}O_{III}$				0.6764	1	18.330
$L_{II}P_{II,III}$				0.6714	1	18.466
$L_{III}N_{II}$				0.8618	1	14.387
$L_{III}N_{III}$				0.8512	1	14.566
$u\ L_{III}N_{VI,VII}$				0.8186	1	15.146
$L_{III}P_{II,III}$				0.8038	1	15.425
	89 Actinium			**90 Thorium**		
$\alpha_2\ KL_{II}$	0.14141†	2	87.67	0.137829	2	89.953
$\alpha_1\ KL_{III}$	0.136417†	8	90.884	0.132813	2	93.350
$\beta_3\ KM_{II}$	0.12143†	2	102.10	0.118268	3	104.831
$\beta_1\ KM_{III}$	0.12055†	2	102.85	0.117396	9	105.609
$\beta_2^{II}\ KN_{II}$	0.11732†	2	105.67	0.11426	1	108.511
$\beta_2^{I}\ KN_{III}$	0.11711†	2	105.86	0.114040	9	108.717
$KO_{II,III}$				0.11322	1	109.500
$\beta_5\ KM_{IV,V}$				0.116667	9	106.269
$\beta_4\ KN_{IV,V}$				0.11366	2	109.08
$\beta_4\ L_IM_{II}$				0.79257	4	15.6429
$\beta_3\ L_IM_{III}$	0.77822†	9	15.931	0.75479	3	16.4258
$\gamma_2\ L_{II}N_{II}$				0.64221	4	19.305
$\gamma_3\ L_IN_{III}$				0.63559	4	19.507
$\gamma'_4\ L_IO_{II}$				0.61251	4	20.242
$\gamma_4\ L_IO_{III}$				0.61098	4	20.292
$\gamma_{13}\ L_IP_{II,III}$				0.60705	8	20.424
$\eta\ L_{II}M_I$				0.85446	4	14.5099
$\beta_1\ L_{II}M_{IV}$	0.78903†	9	15.713	0.765210	9	16.2022
$\gamma_5\ L_{II}N_I$				0.67491	4	18.370
$\gamma_1\ L_{II}N_{IV}$	0.67351†	9	18.408	0.65313	3	18.9825
$\gamma_8\ L_{II}O_I$				0.63898	5	19.403
$\gamma_6\ L_{II}O_{IV}$				0.63258	4	19.599
$L_{II}P_I$				0.6316	1	19.629
$L_{II}P_{IV}$				0.62991	9	19.682
$l\ L_{III}M_I$				1.11508	4	11.1186
$\alpha_2\ L_{III}M_{IV}$	0.99178†	5	12.5008	0.96788	2	12.8096
$\alpha_1\ L_{III}M_V$	0.97993†	5	12.6520	0.95600	3	12.9687
$\beta_6\ L_{III}N_I$				0.82790	8	14.975
$\beta_{15}\ L_{III}N_{IV}$				0.79539	5	15.5875
$\beta_2\ L_{III}N_V$				0.79354	3	15.6237
$\beta_7\ L_{III}O_I$				0.77437	4	16.0105
$\beta_5\ L_{III}O_{IV,V}$				0.76468	5	16.213
$L_{III}P_I$				0.76338	5	16.241
$L_{III}P_{IV,V}$				0.76087	9	16.295
$\beta_{10}\ L_IM_{IV}$				0.7301	1	16.981
$\beta_9\ L_IM_V$				0.7234	1	17.139
L_IN_I				0.64755	5	19.146
L_IN_{IV}				0.6276	1	19.755
$\gamma_{11}\ L_IN_V$				0.62636	9	19.794
$L_IN_{VI,VII}$				0.6160	1	20.128
L_IO_I				0.6146	1	20.174
$L_IO_{IV,V}$				0.6083	1	20.383
$L_{II}M_{II}$				0.8338	1	14.869
$\beta_{17}\ L_{II}M_{III}$				0.79257	4	15.6429
$L_{II}M_V$				0.7579	1	16.359
$L_{II}N_{III}$				0.6620	1	18.729
$L_{II}N_V$				0.6521	1	19.014

Designation	Å*	p.e.	keV	Å*	p.e.	keV
	89 Actinium (*Cont.*)			**90 Thorium** (*Cont.*)		
$v\ L_{II}N_{VI}$				0.64064	9	19.353
$L_{II}O_{II}$				0.6369	1	19.466
$L_{II}O_{III}$				0.6356	1	19.506
$L_{II}P_{II,III}$				0.6312	1	19.642
$t\ L_{III}M_I$				1.08009	9	11.4788
$s\ L_{III}M_{II}$				1.0112	1	12.261
$L_{III}N_{II}$				0.8190	2	15.138
$L_{III}N_{III}$				0.8082	1	15.341
$u\ L_{III}N_{VI,VII}$				0.77661	5	15.964
$L_{III}O_{II}$				0.7713	1	16.074
$L_{III}O_{III}$				0.7690	1	16.123
$L_{III}P_{II,III}$				0.7625	2	16.260
M_IN_{III}				2.934	8	4.23
M_IO_{III}				2.442	9	5.08
$M_{II}N_I$				3.537	9	3.505
$M_{II}N_{IV}$				3.011	2	4.117
$M_{II}O_{IV}$				2.618	5	4.735
$M_{III}N_I$				4.568	5	2.714
$M_{III}N_{IV}$				3.718	3	3.335
$\gamma\ M_{II}N_V$				3.679	2	3.370
$M_{III}O_I$				3.283	9	3.78
$M_{III}O_{IV,V}$				3.131	3	3.959
$\zeta_2\ M_VN_{II}$				5.340	5	2.322
$M_{IV}N_{III}$				4.911	5	2.524
$\beta\ M_{IV}N_{VI}$				3.941	1	3.1458
$M_{IV}O_{II}$				3.808	4	3.256
$\zeta_1\ M_VN_{III}$				5.245	5	2.364
$\alpha_2\ M_VN_{VI}$				4.151	2	2.987
$\alpha_1\ M_VN_{VII}$				4.1381	9	2.9961
M_VP_{III}				3.760	9	3.298
$N\ P_{II}$				9.44	7	1.313
$N\ P_{III}$				9.40	7	1.1319
$N_{II}O_{IV}$				11.56	5	1.072
N_IP_I				11.07	7	1.120
$N_{III}O_V$				13.8	1	0.897
$N_{IV}N_{VI}$				33.57	9	0.3693
$N_VN_{VI,VII}$				36.32	9	0.3414
$N_{VI}O_{IV}$				49.5	1	0.2505
$N_{VI}O_V$				48.2	1	0.2572
$N_{VII}O_V$				50.0	1	0.2479
$O_{III}P_{IV,V}$				68.2	3	0.1817
$O_{IV,V}Q_{II,III}$				181.	5	0.068
	91 Protactinium			**92 Uranium**		
$\alpha_2\ KL_{II}$	0.134343†	9	92.287	0.130968	4	94.665
$\alpha_1\ KL_{III}$	0.129325†	3	95.868	0.125947	3	98.439
$\beta_3\ KM_{II}$	0.11523†	2	107.60	0.112296	4	110.406
$\beta_1\ KM_{III}$	0.114345†	8	108.427	0.111394	5	111.300
$\beta_2^{II}\ KN_{II}$	0.11129†	2	111.40	0.10837	1	114.40
$\beta_2^{I}\ KN_{III}$	0.11107†	2	111.62	0.10818	1	114.60
$KO_{II,III}$				0.10744	1	115.39
$\beta_5\ KM_{IV,V}$				0.11069	1	112.01
$\beta_4\ KN_{IV,V}$				0.10780	2	115.01
$\beta_4\ L_IM_I$	0.7699	1	16.104	0.747985	9	16.5753
$\beta_3\ L_IM_{III}$	0.73230	5	16.930	0.71029	2	17.4550

Designation	Å*	p.e.	keV	Å*	p.e.	keV
	91 Protactinium (*Cont.*)			**92 Uranium** (*Cont.*)		
$\gamma_2\ L_IN_{II}$	0.6239	1	19.872	0.605237	9	20.4847
$\gamma_3\ L_IN_{III}$	0.6169	1	20.098	0.598574	9	20.7127
$\gamma'_4\ L_IO_{II}$				0.576700	9	21.4984
$\gamma_4\ L_IO_{II,III}$	0.5937	1	20.882	0.57499	9	21.562
γ_{13}				0.5706	1	21.729
$\eta\ L_{II}M_I$	0.8295	1	14.946	0.80509	2	15.3997
$\beta_1\ L_{II}M_{IV}$	0.74232	5	16.702	0.719984	8	17.2200
$\gamma_5\ L_{II}N_I$	0.6550	1	18.930	0.63557	2	19.5072
$\gamma_1\ L_{II}N_{IV}$	0.63358†	9	19.568	0.614770	9	20.1671
$\gamma_8\ L_{II}O_I$				0.60125	5	20.621
$\gamma_6\ L_{II}O_{IV}$	0.6133	1	20.216	0.594845	9	20.8426
$L_{II}P_{IV}$				0.59203	5	20.942
$l\ L_{III}M_I$	1.0908	1	11.366	1.06712	2	11.6183
$\alpha_2\ L_{III}M_{IV}$	0.94482†	5	13.1222	0.922558	9	13.4388
$\alpha_1\ L_{III}M_V$	0.93284	5	13.2907	0.910639	9	13.6147
$\beta_6\ L_{III}N_I$	0.8079	1	15.347	0.78838	2	15.7260
$\beta_{15}\ L_{III}N_{IV}$				0.756642	9	16.3857
$\beta_2\ L_{III}N_V$	0.7737	1	16.024	0.754681	9	16.4283
$\beta_7\ L_{III}O_I$	0.7546	2	16.431	0.73602	6	16.845
$\beta_5\ L_{III}O_{IV,V}$	0.7452	2	16.636	0.726305	9	17.0701
$L_{III}P_I$				0.72521	5	17.096
$L_{III}P_{IV,V}$				0.72240	5	17.162
$\beta_{10}\ L_IM_V$	0.7088	2	17.492	0.68760	5	18.031
$\beta_9\ L_IM_V$	0.7018	1	17.667	0.681014	8	18.2054
L_IN_{IV}				0.59096	5	20.979
$\gamma_{11}\ L_IN_V$				0.58986	5	21.019
$L_IO_{IV,V}$				0.5725	1	21.657
$\beta_{17}\ L_{II}M_{III}$				0.74503	5	16.641
$L_{II}N_{III}$				0.6228	1	19.907
$v\ L_{II}N_{VI}$				0.6031	1	20.556
$L_{III}O_{III}$				0.59728	5	20.758
$L_{II}P_{II,III}$				0.5930	2	20.906
$t\ L_{III}M_{II}$				1.0347	1	11.982
$s\ L_{III}M_{III}$				0.9636	1	12.866
$L_{III}N_{II}$				0.78017	9	15.892
$L_{III}N_{III}$				0.7691	1	16.120
$u\ L_{III}N_{VI,VII}$				0.738603	9	16.7859
$L_{III}O_{II}$				0.7333	1	16.907
$L_{III}O_{III}$				0.7309	1	16.962
$L_{III}P_{II,III}$				0.72426	5	17.118
M_IN_{II}				2.92	2	4.25
M_IN_{III}				2.753	8	4.50
M_IO_{III}				2.304	7	5.38
M_IP_{III}				2.253	6	5.50
$M_{II}N_I$	3.441	5	3.603	3.329	4	3.724
$M_{II}N_{IV}$	2.910	2	4.260	2.817	2	4.401
$M_{II}O_{IV}$	2.527	4	4.906	2.443	4	5.075
$M_{III}N_I$	4.450	4	2.786	4.330	2	2.863
$M_{III}N_{IV}$	3.614	2	3.430	3.521	2	3.521
$\gamma\ M_{III}N_V$	3.577	1	3.4657	3.479	1	3.563
$M_{III}O_I$	3.245	9	3.82	3.115	7	3.980
$M_{III}O_{IV,V}$	3.038	2	4.081	2.948	2	4.205
$\zeta_2\ M_{IV}N_{II}$	5.193	2	2.3876	5.050	2	2.4548
$M_{IV}N_{III}$				4.625	5	2.681
$\beta\ M_{IV}N_{VI}$	3.827	1	3.2397	3.716	1	3.3367

Designation	Å*	p.e.	keV	Å*	p.e.	keV
	91 Protactinium (*Cont.*)			**92 Uranium** (*Cont.*)		
$M_{IV}O_{II}$	3.691	2	3.359	3.576	1	3.4666
$\zeta_1\ M_VN_{III}$	5.092	2	2.4350	4.946	2	2.507
$\alpha_2\ M_VN_{VI}$	4.035	3	3.072	3.924	1	3.1595
$\alpha_1\ M_VN_{VII}$	4.022	1	3.0823	3.910	1	3.1708
N_IO_{III}				10.09	7	1.229
N_IP_{II}				8.81	7	1.41
N_IP_{III}				8.76	7	1.42
$N_{II}P_I$				10.40	7	1.192
$N_{III}O_V$				12.90	9	0.961
$N_{IV}N_{VI}$				31.8	1	0.390
$N_VN_{VI,VII}$				34.8	1	0.357
$N_{IV}O_{IV}$				43.3	2	0.286
$N_{VI}O_V$				42.1	2	0.295
$N_IP_{IV,V}$				8.60	7	1.44
	93 Neptunium			**94 Plutonium**		
$\beta_4\ L_IM_{II}$	0.72671	2	17.0607	0.70620	2	17.5560
$\beta_3\ L_IM_{III}$	0.68920†	9	17.989	0.66871	2	18.5405
$\gamma_2\ L_IN_{II}$	0.5873	5	21.11	0.57068	2	21.7251
$\gamma_3\ L_IN_{III}$	0.5810	5	21.34	0.564001	9	21.9824
$\gamma'_4\ L_IO_{II}$				0.5432	1	22.823
$\gamma_4\ L_IO_{II,III}$	0.5585	5	22.20	0.5416	1	22.891
$\eta\ L_{II}M_I$	0.7809	2	15.876	0.7591	1	16.333
$\beta_1\ L_{II}M_{IV}$	0.698478	9	17.7502	0.67772	2	18.2937
$\gamma_5\ L_{II}N_I$	0.616	1	20.12	0.5988	1	20.704
$\gamma_1\ L_{II}N_{IV}$	0.596498	9	20.7848	0.578882	9	21.4173
γ_8				0.5658	1	21.914
$\gamma_6\ L_{II}O_{IV}$	0.57699	5	21.488	0.55973	2	22.1502
$l\ L_{III}M_I$	1.0428	6	11.890	1.0226	1	12.124
$\alpha_2\ L_{III}M_{IV}$	0.901045	9	13.7597	0.88028	2	14.0842
$\alpha_1\ L_{III}M_V$	0.889128	9	13.9441	0.86830	2	14.2786
$\beta_6\ L_{III}N_I$	0.769	1	16.13	0.75148	2	16.4983
$\beta_{15}\ L_{III}N_{IV}$				0.7205	1	17.208
$\beta_2\ L_{III}N_V$	0.736230	9	16.8400	0.71851	2	17.2553
$\beta_7\ L_{III}O_I$				0.7003	1	17.705
$\beta_5\ L_{III}O_{IV,V}$	0.70814	2	17.5081	0.69068	2	17.9506
$\beta_{10}\ L_IM_{IV}$				0.6482	1	19.126
$\beta_9\ L_IM_V$				0.6416	1	19.323
$u\ L_{III}N_{VI,VII}$				0.7031	1	17.635
	95 Americium					
$\beta_4\ L_IM_{II}$	0.68639	2	18.0627			
$\beta_3\ L_IM_{III}$	0.64891	2	19.1059			
$\gamma_2\ L_IN_{II}$	0.5544	2	22.361			
$\beta_1\ L_{II}M_{IV}$	0.657655	9	18.8520			
$\gamma_1\ L_{II}N_{IV}$	0.561886	9	22.0652			
$\gamma_6\ L_{II}O_{IV}$	0.54311	2	22.8282			
$l\ L_{III}M_I$	1.0012	6	12.384			
$\alpha_2\ L_{III}M_{IV}$	0.860266	9	14.4119			
$\alpha_1\ L_{III}M_V$	0.848187	9	14.6172			
$\beta_6\ L_{III}N_I$	0.73418	2	16.8870			
$\beta_{15}\ L_{III}N_{IV}$	0.70341	2	17.6258			
$\beta_2\ L_{III}N_V$	0.701390	9	17.6765			
$\beta_5\ L_{III}O_{IV,V}$	0.67383	2	18.3996			

TABLE B. Wavelengths in numerical order of the emission lines and absorption edges.

Wavelength Å*	p.e.	Element	Designation		keV
0.10723	1	92 U	K	Abs. Edge	115.62
0.10744	1	92 U		$KO_{II,III}$	115.39
0.10780	2	92 U	$K\beta_4$	$KN_{IV,V}$	115.01
0.10818	1	92 U	$K\beta_2^{I}$	KN_{III}	114.60
0.10837	1	92 U	$K\beta_2^{II}$	KN_{II}	114.40
0.11069	1	92 U	$K\beta_5$	$KM_{IV,V}$	112.01
0.11107	2	91 Pa	$K\beta_2^{I}$	KN_{III}	111.62
0.11129	2	91 Pa	$K\beta_2^{II}$	KN_{II}	111.40
0.111394	5	92 U	$K\beta_1$	KM_{III}	111.300
0.112296	4	92 U	$K\beta_3$	KM_{II}	110.406
0.11307	1	90 Th	K	Abs. Edge	109.646
0.11322	1	90 Th		$KO_{II,III}$	109.500
0.11366	2	90 Th	$K\beta_4$	$KN_{IV,V}$	109.08
0.114040	9	90 Th	$K\beta_2^{I}$	KN_{III}	108.717
0.11426	1	90 Th	$K\beta_2^{II}$	KN_{II}	108.511
0.114345	8	91 Pa	$K\beta_1$	KM_{III}	108.427
0.11523	2	91 Pa	$K\beta_3$	KM_{II}	107.60
0.116667	9	90 Th	$K\beta_5$	$KM_{IV,V}$	106.269
0.11711	2	89 Ac	$K\beta_2^{I}$	KN_{III}	105.86
0.11732	2	89 Ac	$K\beta_2^{II}$	KN_{II}	105.67
0.117396	9	90 Th	$K\beta_1$	KM_{III}	105.609
0.118268	3	90 Th	$K\beta_3$	KM_{II}	104.831
0.12029	3	88 Ra	$K\beta_2^{I}$	KN_{III}	103.07
0.12050	3	88 Ra	$K\beta_2^{II}$	KN_{II}	102.89
0.12055	2	89 Ac	$K\beta_1$	KM_{III}	102.85
0.12143	2	89 Ac	$K\beta_3$	KM_{II}	102.10
0.12358	5	87 Fr	$K\beta_2^{I}$	KN_{III}	100.33
0.12379	5	87 Fr	$K\beta_2^{II}$	KN_{II}	100.16
0.12382	3	88 Ra	$K\beta_1$	KM_{III}	100.13
0.12469	3	88 Ra	$K\beta_3$	KM_{II}	99.43
0.125947	3	92 U	$K\alpha_1$	KL_{III}	98.439
0.12698	5	86 Rn	$K\beta_2^{I}$	KN_{III}	97.64
0.12719	5	86 Rn	$K\beta_2^{II}$	KN_{II}	97.47
0.12719	5	87 Fr	$K\beta_1$	KM_{III}	97.47
0.12807	5	87 Fr	$K\beta_3$	KM_{II}	96.81
0.129325	3	91 Pa	$K\alpha_1$	KL_{III}	95.868
0.13052	4	85 At	$K\beta_2^{I}$	KN_{III}	94.99
0.13069	5	86 Rn	$K\beta_1$	KM_{III}	94.87
0.13072	4	85 At	$K\beta_2^{II}$	KN_{II}	94.84
0.130968	4	92 U	$K\alpha_2$	KL_{II}	94.665
0.13155	5	86 Rn	$K\beta_3$	KM_{II}	94.24
0.132813	2	90 Th	$K\alpha_1$	KL_{III}	93.350
0.13418	2	84 Po	$K\beta_2^{I}$	KN_{III}	92.40
0.13432	4	85 At	$K\beta_1$	KM_{III}	92.30
0.134343	9	91 Pa	$K\alpha_2$	KL_{II}	92.287
0.13438	2	84 Po	$K\beta_2^{II}$	KN_{II}	92.26
0.13517	4	85 At	$K\beta_3$	KM_{II}	91.72
0.136417	8	89 Ac	$K\alpha_1$	KL_{III}	90.884
0.13694	1	83 Bi	K	Abs. Edge	90.534
0.13709	1	83 Bi		$KO_{II,III}$	90.435
0.13759	2	83 Bi	$K\beta_4$	$KN_{IV,V}$	90.11
0.137829	2	90 Th	$K\alpha_2$	KL_{II}	89.953
0.13797	1	83 Bi	$K\beta_2^{I}$	KN_{III}	89.864
0.13807	2	84 Po	$K\beta_1$	KM_{III}	89.80
0.13817	1	83 Bi	$K\beta_2^{II}$	KN_{II}	89.733
0.13892	2	84 Po	$K\beta_3$	KM_{II}	89.25
0.14014	2	88 Ra	$K\alpha_1$	KL_{III}	88.47
0.1408	1	82 Pb		KP	88.06
0.140880	5	82 Pb	K	Abs. Edge	88.005
0.141012	8	82 Pb		$KO_{II,III}$	87.922
0.14111	1	83 Bi	$K\beta_5$	$KM_{IV,V}$	87.860
0.14141	2	89 Ac	$K\alpha_2$	KL_{II}	87.67
0.14155	3	82 Pb	$K\beta_4$	$KN_{IV,V}$	87.59
0.14191	1	82 Pb	$K\beta_2^{I}$	KN_{III}	87.364
0.141948	3	83 Bi	$K\beta_1$	KM_{III}	87.343
0.14212	2	82 Pb	$K\beta_2^{II}$	KN_{II}	87.23
0.142779	7	83 Bi	$K\beta_3$	KM_{II}	86.834
0.14399	3	87 Fr	$K\alpha_1$	KL_{III}	86.10
0.14495	1	81 Tl	K	Abs. Edge	85.533
0.14495	3	82 Pb	$K\beta_5^{I}$	KM_{V}	85.53
0.14509	1	81 Tl		$KO_{II,III}$	85.451
0.14512	2	82 Pb	$K\beta_5^{II}$	KM_{IV}	85.43
0.14512	2	88 Ra	$K\alpha_2$	KL_{II}	85.43
0.14553	2	81 Tl	$K\beta_4$	$KN_{IV,V}$	85.19
0.14595	1	81 Tl	$K\beta_2^{I}$	KN_{III}	84.946
0.145970	6	82 Pb	$K\beta_1$	KM_{III}	84.936
0.14614	1	81 Tl	$K\beta_2^{II}$	KN_{II}	84.836
0.146810	4	82 Pb	$K\beta_3$	KM_{II}	84.450
0.14798	3	86 Rn	$K\alpha_1$	KL_{III}	83.78
0.14896	3	87 Fr	$K\alpha_2$	KL_{II}	83.23
0.14917	1	81 Tl	$K\beta_5$	$KM_{IV,V}$	83.114
0.14918	1	80 Hg	K	Abs. Edge	83.109
0.14931	2	80 Hg		$KO_{II,III}$	83.04
0.14978	2	80 Hg	$K\beta_4$	$KN_{IV,V}$	82.78
0.150142	5	81 Tl	$K\beta_1$	KM_{III}	82.576
0.15020	2	80 Hg	$K\beta_2^{I}$	KN_{III}	82.54
0.15040	2	80 Hg	$K\beta_2^{II}$	KN_{II}	82.43
0.150980	6	81 Tl	$K\beta_3$	KM_{II}	82.118
0.15210	2	85 At	$K\alpha_1$	KL_{III}	81.52
0.15294	3	86 Rn	$K\alpha_2$	KL_{II}	81.07
0.15353	2	80 Hg	$K\beta_5$	$KM_{IV,V}$	80.75
0.153593	5	79 Au	K	Abs. Edge	80.720
0.153694	7	79 Au		$KO_{II,III}$	80.667
0.154224	5	79 Au	$K\beta_4$	$KN_{IV,V}$	80.391
0.154487	3	80 Hg	$K\beta_1$	KM_{III}	80.253
0.154618	9	79 Au	$K\beta_2^{I}$	KN_{III}	80.185
0.15483	2	79 Au	$K\beta_2^{II}$	KN_{II}	80.08
0.155321	3	80 Hg	$K\beta_3$	KM_{II}	79.822
0.15636	1	84 Po	$K\alpha_1$	KL_{III}	79.290
0.15705	2	85 At	$K\alpha_2$	KL_{II}	78.95
0.157880	5	79 Au	$K\beta_5^{I}$	KM_{V}	78.529
0.158062	7	79 Au	$K\beta_5^{II}$	KM_{IV}	78.438
0.15818	1	78 Pt	K	Abs. Edge	78.381
0.15826	1	78 Pt		$KO_{II,III}$	78.341
0.15881	2	78 Pt	$K\beta_4$	$KN_{IV,V}$	78.069
0.158982	3	79 Au	$K\beta_1$	KM_{III}	77.984
0.15920	1	78 Pt	$K\beta_2^{I}$	KN_{III}	77.878
0.15939	1	78 Pt	$K\beta_2^{II}$	KN_{II}	77.785
0.159810	2	79 Au	$K\beta_3$	KM_{II}	77.580
0.160789	2	83 Bi	$K\alpha_1$	KL_{III}	77.1079
0.16130	1	84 Po	$K\alpha_2$	KL_{II}	76.862
0.16255	3	78 Pt	$K\beta_5^{I}$	KM_{V}	76.27
0.16271	2	78 Pt	$K\beta_5^{II}$	KM_{IV}	76.199
0.16292	1	77 Ir	K	Abs. Edge	76.101

Wavelength Å*	p.e.	Element	Designation		keV
0.163019	5	77 Ir		$KO_{II,III}$	76.053
0.16352	2	77 Ir	$K\beta_4$	$KN_{IV,V}$	75.821
0.163675	3	78 Pt	$K\beta_1$	KM_{III}	75.748
0.163956	7	77 Ir	$K\beta_2^{I}$	KN_{III}	75.619
0.16415	1	77 Ir	$K\beta_2^{II}$	KN_{II}	75.529
0.164501	3	78 Pt	$K\beta_3$	KM_{II}	75.368
0.165376	2	82 Pb	$K\alpha_1$	KL_{III}	74.9694
0.165717	2	83 Bi	$K\alpha_2$	KL_{II}	74.8148
0.167373	9	77 Ir	$K\beta_5^{I}$	KM_{V}	74.075
0.16759	2	77 Ir	$K\beta_5^{II}$	KM_{IV}	73.980
0.16787	1	76 Os	K	Abs. Edge	73.856
0.16798	1	76 Os		$KO_{II,III}$	73.808
0.16842	2	76 Os	$K\beta_4$	$KN_{IV,V}$	73.615
0.168542	2	77 Ir	$K\beta_1$	KM_{III}	73.5608
0.168906	6	76 Os	$K\beta_2^{I}$	KN_{III}	73.402
0.16910	1	76 Os	$K\beta_2^{II}$	KN_{II}	73.318
0.169367	2	77 Ir	$K\beta_3$	KM_{II}	73.2027
0.170136	2	81 Tl	$K\alpha_1$	KL_{III}	72.8715
0.170294	2	82 Pb	$K\alpha_2$	KL_{II}	72.8042
0.17245	1	76 Os	$K\beta_5^{I}$	KM_{V}	71.895
0.17262	1	76 Os	$K\beta_5^{II}$	KM_{IV}	71.824
0.17302	1	75 Re	K	Abs. Edge	71.658
0.17308	1	75 Re		$KO_{II,III}$	71.633
0.173611	3	76 Os	$K\beta_1$	KM_{III}	71.413
0.17362	2	75 Re	$K\beta_4$	$KN_{IV,V}$	71.410
0.174054	6	75 Re	$K\beta_2^{I}$	KN_{III}	71.232
0.17425	1	75 Re	$K\beta_2^{II}$	KN_{II}	71.151
0.174431	3	76 Os	$K\beta_3$	KM_{II}	71.077
0.175036	2	81 Tl	$K\alpha_2$	KL_{II}	70.8319
0.175068	3	80 Hg	$K\alpha_1$	KL_{III}	70.819
0.17766	1	75 Re	$K\beta_5^{I}$	KM_{V}	69.786
0.17783	1	75 Re	$K\beta_5^{II}$	KM_{IV}	69.719
0.17837	1	74 W	K	Abs. Edge	69.508
0.178444	5	74 W		$KO_{II,III}$	69.479
0.178880	3	75 Re	$K\beta_1$	KM_{III}	69.310
0.17892	2	74 W	$K\beta_4$	$KN_{IV,V}$	69.294
0.179421	7	74 W	$K\beta_2^{I}$	KN_{III}	69.101
0.17960	1	74 W	$K\beta_2^{II}$	KN_{II}	69.031
0.179697	3	75 Re	$K\beta_3$	KM_{II}	68.994
0.179958	3	80 Hg	$K\alpha_2$	KL_{II}	68.895
0.180195	2	79 Au	$K\alpha_1$	KL_{III}	68.8037
0.183092	7	74 W	$K\beta_5^{I}$	KM_{V}	67.715
0.183264	5	74 W	$K\beta_5^{II}$	KM_{IV}	67.652
0.18394	1	73 Ta	K	Abs. Edge	67.403
0.184031	7	73 Ta		$KO_{II,III}$	67.370
0.184374	2	74 W	$K\beta_1$	KM_{III}	67.2443
0.18451	1	73 Ta	$K\beta_4$	$KN_{IV,V}$	67.194
0.185011	8	73 Ta	$K\beta_2^{I}$	KN_{III}	67.013
0.185075	2	79 Au	$K\alpha_2$	KL_{II}	66.9895
0.185181	2	74 W	$K\beta_3$	KM_{II}	66.9514
0.185188	9	73 Ta	$K\beta_2^{II}$	KN_{II}	66.949
0.185511	4	78 Pt	$K\alpha_1$	KL_{III}	66.832
0.18672	4	79 Au		KL_{I}	66.40
0.188757	6	73 Ta	$K\beta_5^{I}$	KM_{V}	65.683
0.188920	6	73 Ta	$K\beta_5^{II}$	KM_{IV}	65.626
0.18982	5	72 Hf	K	Abs. Edge	65.31
0.190089	4	73 Ta	$K\beta_1$	KM_{III}	65.223
0.190381	4	78 Pt	$K\alpha_2$	KL_{II}	65.122
0.1908	2	72 Hf	$K\beta_2$	$KN_{II,III}$	64.98
0.190890	2	73 Ta	$K\beta_3$	KM_{II}	64.9488
0.191047	2	77 Ir	$K\alpha_1$	KL_{III}	64.8956
0.19585	5	71 Lu	K	Abs. Edge	63.31
0.19589	2	71 Lu		$KO_{II,III}$	63.293
0.195904	2	77 Ir	$K\alpha_2$	KL_{II}	63.2867
0.19607	3	72 Hf	$K\beta_1$	KM_{III}	63.234
0.196794	2	76 Os	$K\alpha_1$	KL_{III}	63.0005
0.19686	4	72 Hf	$K\beta_3$	KM_{II}	62.98
0.1969	2	71 Lu	$K\beta_2$	$KN_{II,III}$	62.97
0.20084	2	71 Lu	$K\beta_5$	$KM_{IV,V}$	61.732
0.201639	2	76 Os	$K\alpha_2$	KL_{II}	61.4867
0.20224	5	70 Yb	K	Abs. Edge	61.30
0.20226	2	70 Yb		$KO_{II,III}$	61.298
0.20231	3	71 Lu	$K\beta_1$	KM_{III}	61.283
0.202781	2	75 Re	$K\alpha_1$	KL_{III}	61.1403
0.20309	4	71 Lu	$K\beta_3$	KM_{II}	61.05
0.2033	2	70 Yb	$K\beta_2$	$KN_{II,III}$	60.89
0.20739	2	70 Yb	$K\beta_5$	$KM_{IV,V}$	59.782
0.207611	1	75 Re	$K\alpha_2$	KL_{II}	59.7179
0.20880	5	69 Tm	K	Abs. Edge	59.38
0.20884	8	70 Yb	$K\beta_1$	KM_{III}	59.37
0.20891	2	69 Tm		$KO_{II,III}$	59.346
0.2090100	Std.	74 W	$K\alpha_1$	KL_{III}	59.31824
0.2096	1	70 Yb	$K\beta_3$	KM_{II}	59.14
0.2098	2	69 Tm	$K\beta_2$	$KN_{II,III}$	59.09
0.213828	2	74 W	$K\alpha_2$	KL_{II}	57.9817
0.21404	2	69 Tm	$K\beta_5$	$KM_{IV,V}$	57.923
0.215497	4	73 Ta	$K\alpha_1$	KL_{III}	57.532
0.21556	2	69 Tm	$K\beta_1$	KM_{III}	57.517
0.21567	1	68 Er	K	Abs. Edge	57.487
0.21581	3	68 Er		$KO_{II,III}$	57.450
0.21592	4	74 W		KL_{I}	57.42
0.21636	2	69 Tm	$K\beta_3$	KM_{II}	57.304
0.2167	2	68 Er	$K\beta_2$	$KN_{II,III}$	57.21
0.220305	8	73 Ta	$K\alpha_2$	KL_{II}	56.277
0.22124	3	68 Er	$K\beta_5$	$KM_{IV,V}$	56.040
0.222227	3	72 Hf	$K\alpha_1$	KL_{III}	55.7902
0.22266	2	68 Er	$K\beta_1$	KM_{III}	55.681
0.22291	1	67 Ho	K	Abs. Edge	55.619
0.22305	3	67 Ho		$KO_{II,III}$	55.584
0.22341	2	68 Er	$K\beta_3$	KM_{II}	55.494
0.2241	2	67 Ho	$K\beta_2$	$KN_{II,III}$	55.32
0.227024	3	72 Hf	$K\alpha_2$	KL_{II}	54.6114
0.22855	3	67 Ho	$K\beta_5$	$KM_{IV,V}$	54.246
0.229298	2	71 Lu	$K\alpha_1$	KL_{III}	54.0698
0.23012	2	67 Ho	$K\beta_1$	KM_{III}	53.877
0.23048	1	66 Dy	K	Abs. Edge	53.793
0.23056	3	66 Dy		$KO_{II,III}$	53.774
0.23083	2	67 Ho	$K\beta_3$	KM_{II}	53.711
0.2317	2	66 Dy	$K\beta_2$	$KN_{II,III}$	53.47
0.234081	2	71 Lu	$K\alpha_2$	KL_{II}	52.9650
0.23618	3	66 Dy	$K\beta_5$	$KM_{IV,V}$	52.494
0.236655	2	70 Yb	$K\alpha_1$	KL_{III}	52.3889
0.23788	2	66 Dy	$K\beta_1$	KM_{III}	52.119
0.23841	1	65 Tb	K	Abs. Edge	52.002

Wavelength Å*	p.e.	Element	Designation		keV
0.23858	3	65 Tb		$KO_{II,III}$	51.965
0.23862	2	66 Dy	$K\beta_3$	KM_{II}	51.957
0.2397	2	65 Tb	$K\beta_2$	$KN_{II,III}$	51.68
0.241424	2	70 Yb	$K\alpha_2$	KL_{II}	51.3540
0.244338	2	69 Tm	$K\alpha_1$	KL_{III}	50.7416
0.24608	2	65 Tb	$K\beta_1$	KM_{III}	50.382
0.24681	1	64 Gd	K	Abs. Edge	50.233
0.24683	2	65 Tb	$K\beta_3$	KM_{II}	50.229
0.24687	3	64 Gd		$KO_{II,III}$	50.221
0.24816	3	64 Gd	$K\beta_2$	$KN_{II,III}$	49.959
0.249095	2	69 Tm	$K\alpha_2$	KL_{II}	49.7726
0.252365	2	68 Er	$K\alpha_1$	KL_{III}	49.1277
0.25275	3	64 Gd	$K\beta_5$	$KM_{IV,V}$	49.052
0.25460	2	64 Gd	$K\beta_1$	KM_{III}	48.697
0.25534	2	64 Gd	$K\beta_3$	KM_{II}	48.555
0.25553	1	63 Eu	K	Abs. Edge	48.519
0.255645	7	63 Eu		$KO_{II,III}$	48.497
0.256923	8	63 Eu	$K\beta_2^{I}$	$KN_{II,III}$	48.256
0.257110	2	68 Er	$K\alpha_2$	KL_{II}	48.2211
0.260756	2	67 Ho	$K\alpha_1$	KL_{III}	47.5467
0.263577	5	63 Eu	$K\beta_1$	KM_{III}	47.0379
0.264332	5	63 Eu	$K\beta_3$	KM_{II}	46.9036
0.26464	5	62 Sm	K	Abs. Edge	46.849
0.26491	3	62 Sm		$KO_{II,III}$	46.801
0.265486	2	67 Ho	$K\alpha_2$	KL_{II}	46.6997
0.2662	1	62 Sm	$K\beta_2$	$KN_{II,III}$	46.57
0.269533	2	66 Dy	$K\alpha_1$	KL_{III}	45.9984
0.27111	3	62 Sm	$K\beta_5$	$KM_{IV,V}$	45.731
0.27301	2	62 Sm	$K\beta_1$	KM_{III}	45.413
0.27376	2	62 Sm	$K\beta_3$	KM_{II}	45.289
0.274247	2	66 Dy	$K\alpha_2$	KL_{II}	45.2078
0.27431	5	61 Pm	K	Abs. Edge	45.198
0.2759	1	61 Pm	$K\beta_2$	$KN_{II,III}$	44.93
0.278724	2	65 Tb	$K\alpha_1$	KL_{III}	44.4816
0.28290	3	61 Pm	$K\beta_1$	KM_{III}	43.826
0.283423	2	65 Tb	$K\alpha_2$	KL_{II}	43.7441
0.28363	4	61 Pm	$K\beta_3$	KM_{II}	43.713
0.28453	5	60 Nd	K	Abs. Edge	43.574
0.2861	1	60 Nd	$K\beta_2$	$KN_{II,III}$	43.32
0.288353	2	64 Gd	$K\alpha_1$	KL_{III}	42.9962
0.293038	2	64 Gd	$K\alpha_2$	KL_{II}	42.3089
0.293299	2	60 Nd	$K\beta_1$	KM_{III}	42.2713
0.294027	3	60 Nd	$K\beta_3$	KM_{II}	42.1665
0.29518	5	59 Pr	K	Abs. Edge	42.002
0.29679	2	59 Pr	$K\beta_2$	$KN_{II,III}$	41.773
0.298446	2	63 Eu	$K\alpha_1$	KL_{III}	41.5422
0.303118	2	63 Eu	$K\alpha_2$	KL_{II}	40.9019
0.304261	4	59 Pr	$K\beta_1$	KM_{III}	40.7482
0.304975	5	59 Pr	$K\beta_3$	KM_{II}	40.6529
0.30648	5	58 Ce	K	Abs. Edge	40.453
0.30668	2	58 Ce		$KO_{II,III}$	40.427
0.30737	2	58 Ce	$K\beta_4^{I}$	$KN_{IV,V}$	40.337
0.30816	1	58 Ce	$K\beta_2$	$KN_{II,III}$	40.233
0.309040	2	62 Sm	$K\alpha_1$	KL_{III}	40.1181
0.31342	2	58 Ce	$K\beta_5^{I}$	KM_{V}	39.558
0.31357	2	58 Ce	$K\beta_5^{II}$	KM_{IV}	39.539
0.313698	2	62 Sm	$K\alpha_2$	KL_{II}	39.5224
0.315816	2	58 Ce	$K\beta_1$	KM_{III}	39.2573
0.316520	4	58 Ce	$K\beta_3$	KM_{II}	39.1701
0.31844	5	57 La	K	Abs. Edge	38.934
0.31864	2	57 La		$KO_{II,III}$	38.909
0.31931	2	57 La	$K\beta_4^{I}$	$KN_{IV,V}$	38.828
0.320117	7	57 La	$K\beta_2$	$KN_{II,III}$	38.7299
0.320160	4	61 Pm	$K\alpha_1$	KL_{III}	38.7247
0.324803	4	61 Pm	$K\alpha_2$	KL_{II}	38.1712
0.32546	2	57 La	$K\beta_5^{I}$	KM_{V}	38.094
0.32563	2	57 La	$K\beta_5^{II}$	KM_{IV}	38.074
0.327983	3	57 La	$K\beta_1$	KM_{III}	37.8010
0.328686	4	57 La	$K\beta_3$	KM_{II}	37.7202
0.33104	1	56 Ba	K	Abs. Edge	37.452
0.33127	2	56 Ba		$KO_{II,III}$	37.426
0.331846	2	60 Nd	$K\alpha_1$	KL_{III}	37.3610
0.33229	2	56 Ba	$K\beta_4^{II}$	KN_{IV}	37.311
0.33277	1	56 Ba	$K\beta_2$	$KN_{II,III}$	37.257
0.336472	2	60 Nd	$K\alpha_2$	KL_{II}	36.8474
0.33814	2	56 Ba	$K\beta_5^{I}$	KM_{V}	36.666
0.33835	2	56 Ba	$K\beta_5^{II}$	KM_{IV}	36.643
0.340811	3	56 Ba	$K\beta_1$	KM_{III}	36.3782
0.341507	4	56 Ba	$K\beta_3$	KM_{II}	36.3040
0.344140	2	59 Pr	$K\alpha_1$	KL_{III}	36.0263
0.34451	1	55 Cs	K	Abs. Edge	35.987
0.34611	2	55 Cs	$K\beta_2$	$KN_{II,III}$	35.822
0.348749	2	59 Pr	$K\alpha_2$	KL_{II}	35.5502
0.354364	7	55 Cs	$K\beta_1$	KM_{III}	34.9869
0.355050	4	55 Cs	$K\beta_3$	KM_{II}	34.9194
0.357092	2	58 Ce	$K\alpha_1$	KL_{III}	34.7197
0.3584	5	54 Xe	K	Abs. Edge	34.59
0.36026	3	54 Xe	$K\beta_2$	$KN_{II,III}$	34.415
0.361683	2	58 Ce	$K\alpha_2$	KL_{II}	34.2789
0.36872	2	54 Xe	$K\beta_1$	KM_{III}	33.624
0.36941	2	54 Xe	$K\beta_3$	KM_{II}	33.562
0.370737	2	57 La	$K\alpha_1$	KL_{III}	33.4418
0.37381	1	53 I	K	Abs. Edge	33.1665
0.37523	2	53 I	$K\beta_2$	$KN_{II,III}$	33.042
0.375313	2	57 La	$K\alpha_2$	KL_{II}	33.0341
0.383905	4	53 I	$K\beta_1$	KM_{III}	32.2947
0.384564	4	53 I	$K\beta_3$	KM_{II}	32.2394
0.385111	4	56 Ba	$K\alpha_1$	KL_{III}	32.1936
0.389668	5	56 Ba	$K\alpha_2$	KL_{II}	31.8171
0.38974	1	52 Te		$KO_{II,III}$	31.8114
0.38974	1	52 Te	K	Abs. Edge	31.8114
0.391102	6	52 Te	$K\beta_2$	$KN_{II,III}$	31.7004
0.399995	5	52 Te	$K\beta_1$	KM_{III}	30.9957
0.400290	4	55 Cs	$K\alpha_1$	KL_{III}	30.9728
0.400659	4	52 Te	$K\beta_3$	KM_{II}	30.9443
0.404835	4	55 Cs	$K\alpha_2$	KL_{II}	30.6251
0.40666	1	51 Sb		$KO_{II,III}$	30.4875
0.40668	1	51 Sb	K	Abs. Edge	30.4860
0.40702	1	51 Sb	$K\beta_4^{I}$	$KN_{IV,V}$	30.4604
0.407973	5	51 Sb	$K\beta_2$	$KN_{II,III}$	30.3895
0.41378	1	51 Sb	$K\beta_5^{I}$	KM_{V}	29.9632
0.41388	1	51 Sb	$K\beta_5^{II}$	KM_{IV}	29.9560
0.41634	2	54 Xe	$K\alpha_1$	KL_{III}	29.779
0.417085	3	51 Sb	$K\beta_1$	KM_{III}	29.7256

Wavelength Å*	p.e.	Element		Designation	keV
0.417737	4	51 Sb	$K\beta_3$	KM_{II}	29.6792
0.42087	2	54 Xe	$K\alpha_2$	KL_{II}	29.458
0.42467	3	50 Sn		$KO_{II,III}$	29.195
0.42467	1	50 Sn	K	Abs. Edge	29.1947
0.42495	3	50 Sn	$K\beta_4^{I}$	$KN_{IV,V}$	29.175
0.425915	8	50 Sn	$K\beta_2$	$KN_{II,III}$	29.1093
0.43175	3	50 Sn	$K\beta_5^{I}$	KM_V	28.716
0.43184	3	50 Sn	$K\beta_5^{II}$	KM_{IV}	28.710
0.433318	5	53 I	$K\alpha_1$	KL_{III}	28.6120
0.435236	5	50 Sn	$K\beta_1$	KM_{III}	28.4860
0.435877	5	50 Sn	$K\beta_3$	KM_{II}	28.4440
0.437829	7	53 I	$K\alpha_2$	KL_{II}	28.3172
0.44371	1	49 In	K	Abs. Edge	27.9420
0.44374	3	49 In		$KO_{II,III}$	27.940
0.44393	4	49 In	$K\beta_4^{I}$	$KN_{IV,V}$	27.928
0.44500	1	49 In	$K\beta_2$	$KN_{II,III}$	27.8608
0.45086	2	49 In	$K\beta_5^{I}$	KM_V	27.499
0.45098	2	49 In	$K\beta_5^{II}$	KM_{IV}	27.491
0.451295	3	52 Te	$K\alpha_1$	KL_{III}	27.4723
0.454545	4	49 In	$K\beta_1$	KM_{III}	27.2759
0.455181	4	49 In	$K\beta_3$	KM_{II}	27.2377
0.455784	3	52 Te	$K\alpha_2$	KL_{II}	27.2017
0.46407	1	48 Cd	K	Abs. Edge	26.7159
0.465328	7	48 Cd	$K\beta_2$	$KN_{II,III}$	26.6438
0.470354	3	51 Sb	$K\alpha_1$	KL_{III}	26.3591
0.474827	3	51 Sb	$K\alpha_2$	KL_{II}	26.1108
0.475105	6	48 Cd	$K\beta_1$	KM_{III}	26.0955
0.475730	5	48 Cd	$K\beta_3$	KM_{II}	26.0612
0.48589	1	47 Ag	K	Abs. Edge	25.5165
0.4859	9	47 Ag	$K\beta_4$	$KN_{IV,V}$	25.512
0.487032	4	47 Ag	$K\beta_2$	$KN_{II,III}$	25.4564
0.490599	3	50 Sn	$K\alpha_1$	KL_{III}	25.2713
0.49306	2	47 Ag	$K\beta_5$	$KM_{IV,V}$	25.145
0.495053	3	50 Sn	$K\alpha_2$	KL_{II}	25.0440
0.497069	4	47 Ag	$K\beta_1$	KM_{III}	24.9424
0.497685	4	47 Ag	$K\beta_3$	KM_{II}	24.9115
0.5092	1	46 Pd	K	Abs. Edge	24.348
0.5093	2	46 Pd	$K\beta_4$	$KN_{IV,V}$	24.346
0.510228	4	46 Pd	$K\beta_2$	$KN_{II,III}$	24.2991
0.512113	3	49 In	$K\alpha_1$	KL_{III}	24.2097
0.516544	3	49 In	$K\alpha_2$	KL_{II}	24.0020
0.51670	9	46 Pd	$K\beta_5$	$KM_{IV,V}$	23.995
0.520520	4	46 Pd	$K\beta_1$	KM_{III}	23.8187
0.521123	4	46 Pd	$K\beta_3$	KM_{II}	23.7911
0.53395	1	45 Rh	K	Abs. Edge	23.2198
0.53401	9	45 Rh	$K\beta_4^{I}$	$KN_{IV,V}$	23.217
0.535010	3	48 Cd	$K\alpha_1$	KL_{III}	23.1736
0.53503	2	45 Rh	$K\beta_2$	$KN_{II,III}$	23.1728
0.53513	5	45 Rh	$K\beta_2^{II}$	KN_{II}	23.168
0.5365	1	94 Pu	L_I	Abs. Edge	23.109
0.539422	3	48 Cd	$K\alpha_2$	KL_{II}	22.9841
0.54101	9	45 Rh	$K\beta_5^{I}$	KM_V	22.917
0.54118	9	45 Rh	$K\beta_5^{II}$	KM_{IV}	22.909
0.5416	1	94 Pu	$L\gamma_4$	L_IO_{III}	22.891
0.54311	2	95 Am	$L\gamma_6$	$L_{II}O_{IV}$	22.8282
0.5432	1	94 Pu	$L\gamma_4'$	L_IO_{II}	22.823
0.545605	4	45 Rh	$K\beta_1$	KM_{III}	22.7236
0.546200	4	45 Rh	$K\beta_3$	KM_{II}	22.6989
0.5544	2	95 Am	$L\gamma_2$	L_IN_{II}	22.361
0.5572	1	94 Pu	L_{II}	Abs. Edge	22.253
0.5585	5	93 Np	$L\gamma_4$	$L_IO_{II,III}$	22.20
0.5594075	6	47 Ag	$K\alpha_1$	KL_{III}	22.16292
0.55973	2	94 Pu	$L\gamma_6$	$L_{II}O_{IV}$	22.1502
0.56051	1	44 Ru	K	Abs. Edge	22.1193
0.56089	9	44 Ru	$K\beta_4$	$KN_{IV,V}$	22.104
0.56166	3	44 Ru	$K\beta_2$	$KN_{II,III}$	22.074
0.561886	9	95 Am	$L\gamma_1$	$L_{II}N_{IV}$	22.0652
0.563798	4	47 Ag	$K\alpha_2$	KL_{II}	21.9903
0.564001	9	94 Pu	$L\gamma_3$	L_IN_{III}	21.9824
0.5658	1	94 Pu	$L\gamma_8$	$L_{II}O_I$	21.914
0.56785	9	44 Ru	$K\beta_5^{I}$	KM_V	21.834
0.5680	2	44 Ru	$K\beta_5^{II}$	KM_{IV}	21.829
0.5695	1	92 U	L_I	Abs. Edge	21.771
0.5706	1	92 U	$L\gamma_{13}$	$L_IP_{II,III}$	21.729
0.57068	2	94 Pu	$L\gamma_2$	L_IN_{II}	21.1251
0.572482	4	44 Ru	$K\beta_1$	KM_{III}	21.6568
0.5725	1	92 U		$L_IO_{IV,V}$	21.657
0.573067	4	44 Ru	$K\beta_3$	KM_{II}	21.6346
0.57499	9	92 U	$L\gamma_4$	L_IO_{III}	21.562
0.576700	9	92 U	$L\gamma_4'$	L_IO_{II}	21.4984
0.57699	5	93 Np	$L\gamma_6$	$L_{II}O_{IV}$	21.488
0.578882	9	94 Pu	$L\gamma_1$	$L_{II}N_{IV}$	21.4173
0.5810	5	93 Np	$L\gamma_3$	L_IN_{III}	21.34
0.585448	3	46 Pd	$K\alpha_1$	KL_{III}	21.1771
0.5873	5	93 Np	$L\gamma_2$	L_IN_{II}	21.11
0.58906	1	43 Tc	K	Abs. Edge	21.0473
0.589821	3	46 Pd	$K\alpha_2$	KL_{II}	21.0201
0.58986	5	92 U	$L\gamma_{11}$	L_IN_V	21.019
0.59024	5	43 Tc	$K\beta_2$	$KN_{II,III}$	21.005
0.59096	5	92 U		L_IN_{IV}	20.979
0.5919	1	92 U	L_{II}	Abs. Edge	20.945
0.59203	5	92 U		$L_{II}P_{IV}$	20.942
0.5930	2	92 U		$L_{II}P_{II,III}$	20.906
0.5937	1	91 Pa	$L\gamma_4$	$L_IO_{II,III}$	20.882
0.594845	9	92 U	$L\gamma_6$	$L_{II}O_{IV}$	20.8426
0.596498	9	93 Np	$L\gamma_1$	$L_{II}N_{IV}$	20.7848
0.59728	5	92 U		$L_{II}O_{III}$	20.758
0.598574	9	92 U	$L\gamma_3$	L_IN_{III}	20.7127
0.5988	1	94 Pu	$L\gamma_5$	$L_{II}N_I$	20.704
0.60125	5	92 U	$L\gamma_8$	$L_{II}O_I$	20.621
0.60130	4	43 Tc	$K\beta_1$	KM_{III}	20.619
0.60188	4	43 Tc	$K\beta_3$	KM_{II}	20.599
0.6031	1	92 U	Lv	$L_{II}N_{VI}$	20.556
0.605237	9	92 U	$L\gamma_2$	L_IN_{II}	20.4847
0.6059	1	90 Th	L_I	Abs. Edge	20.464
0.60705	8	90 Th	$L\gamma_{13}$	$L_IP_{II,III}$	20.424
0.6083	1	90 Th		$L_IO_{IV,V}$	20.383
0.61098	4	90 Th	$L\gamma_4$	L_IO_{III}	20.292
0.61251	4	90 Th	$L\gamma_4'$	L_IO_{II}	20.242
0.6133	1	91 Pa	$L\gamma_6$	$L_{II}O_{IV}$	20.216
0.613279	4	45 Rh	$K\alpha_1$	KL_{III}	20.2161
0.6146	1	90 Th		L_IO_I	20.174
0.614770	9	92 U	$L\gamma_1$	$L_{II}N_{IV}$	20.1671
0.6160	1	90 Th		$L_IN_{VI,VII}$	20.128

Wavelength Å*	p.e.	Element	Designation		keV
0.616	1	93 Np	$L\gamma_5$	$L_{II}N_I$	20.12
0.6169	1	91 Pa	$L\gamma_3$	L_IN_{III}	20.098
0.617630	4	45 Rh	$K\alpha_2$	KL_{II}	20.0737
0.61978	1	42 Mo	K	Abs. Edge	20.0039
0.62001	9	42 Mo	$K\beta_4^{I}$	$KN_{IV,V}$	19.996
0.62099	2	42 Mo	$K\beta_2$	$KN_{II,III}$	19.9652
0.62107	5	42 Mo	$K\beta_2^{II}$	KN_{II}	19.963
0.6228	1	92 U		$L_{II}N_{III}$	19.907
0.6239	1	91 Pa	$L\gamma_2$	L_IN_{II}	19.872
0.62636	9	90 Th	$L\gamma_{11}$	L_IN_V	19.794
[illegible]	5	42 Mo	$K\beta_5^{I}$	KM_V	19.776
[illegible]	[illegible]	42 Mo	$K\beta_5^{II}$	KM_{IV}	19.771
[illegible]	1	90 Th		L_IN_{IV}	19.755
0.6299	1	90 Th	[illegible]	Abs. Edge	19.683
0.62991	9	90 Th		[illegible]	19.682
0.6312	1	90 Th		[illegible]	19.642
[illegible]	1	90 Th		[illegible]	[illegible]
[illegible]	[illegible]	42 Mo	[illegible]	KM_{III}	[illegible]
[illegible]	[illegible]	90 Th	$L\gamma_6$	$L_{II}O_{IV}$	[illegible]
[illegible]	[illegible]	42 Mo	$K\beta_3$	KM_{II}	19.5903
[illegible]	9	[illegible]	$L\gamma_1$	[illegible]	[illegible]
0.63551	[illegible]	[illegible] U	[illegible]	[illegible]	[illegible]
[illegible]	4	[illegible] Th	[illegible]	[illegible]	[illegible]
0.6356	[illegible]	[illegible] Th		[illegible]	[illegible]
0.6369	1	90 Th		[illegible]	[illegible]
[illegible]	5	90 Th	[illegible]	$L_{II}O_I$	[illegible]
[illegible]	[illegible]	90 Th	Lv	[illegible]	[illegible]
0.6410	[illegible]	[illegible]	$L\beta_9$	L_IM_V	[illegible]
[illegible]	4	[illegible]	[illegible]	L_IN_{II}	[illegible]
[illegible]	[illegible]	44 Ru	[illegible]	[illegible]	19.2792
[illegible]	1	[illegible]	L_I	[illegible]	19.236
[illegible]	[illegible]	88 Ra	[illegible]	[illegible]	[illegible]
[illegible]	1	[illegible]		[illegible]	[illegible]
[illegible]	[illegible]	44 Ru	[illegible]	KL_{II}	[illegible]
[illegible]	[illegible]	[illegible]		[illegible]	[illegible]
[illegible]	[illegible]	[illegible]	[illegible]	L_IM_{IV}	[illegible]
[illegible]	[illegible]	[illegible]	[illegible]	[illegible]	[illegible]
[illegible]	[illegible]	88 Ra	[illegible]	[illegible]	19.084
0.65131	[illegible]	[illegible]	[illegible]	[illegible]	[illegible]
[illegible]	1	90 Th		[illegible]	[illegible]
[illegible]	1	41 Nb	K	Abs. Edge	[illegible]
[illegible]	[illegible]	90 Th	$L\gamma_1$	[illegible]	[illegible]
[illegible]	[illegible]	41 Nb	$K\beta_4$	$KN_{IV,V}$	[illegible]
[illegible]	4	[illegible]	[illegible]	[illegible]	[illegible]
0.6550	[illegible]	91 Pa	[illegible]	[illegible]	18.930
0.657655	9	[illegible]	[illegible]	[illegible]	[illegible]
0.6620	1	90 Th		[illegible]	[illegible]
0.6654	1	88 Ra	$L\gamma_{11}$	[illegible]	[illegible]
[illegible]	2	41 Nb	$K\beta_1$	KM_{III}	[illegible]
[illegible]	[illegible]	41 Nb	$K\beta_3$	KM_{II}	[illegible]
0.6666	[illegible]	[illegible]		L_IN_{IV}	18.600
[illegible]	2	94 Pu	[illegible]	L_IM_{III}	18.5405
[illegible]	[illegible]	88 Ra	[illegible]	[illegible]	18.486
[illegible]	[illegible]	88 Ra		[illegible]	18.466
0.6724	[illegible]	88 Ra		[illegible]	[illegible]
[illegible]	5	[illegible]	[illegible]	[illegible]	[illegible]
0.67351	[illegible]	[illegible] Ac	[illegible]	[illegible]	18.408
0.67383	2	95 Am	$L\beta_5$	$L_{III}O_{IV,V}$	18.3996
0.67491	4	90 Th	$L\gamma_5$	$L_{II}N_I$	18.370
0.67502	3	43 Tc	$K\alpha_1$	KL_{III}	18.3671
0.67538	5	88 Ra	$L\gamma_3$	L_IN_{III}	18.357
0.6764	1	88 Ra		$L_{II}O_{III}$	18.330
0.67772	2	94 Pu	$L\beta_1$	$L_{II}M_{IV}$	18.2937
0.6780	1	88 Ra		$L_{II}O_{II}$	18.286
0.67932	3	43 Tc	$K\alpha_2$	KL_{II}	18.2508
0.6801	1	88 Ra	$L\gamma_8$	$L_{II}O_I$	18.230
0.681014	8	92 U	$L\beta_9$	L_IM_V	18.2054
0.68199	5	88 Ra	$L\gamma_2$	L_IN_{II}	18.179
0.68639	2	95 Am	$L\beta_4$	L_IM_{II}	18.0627
0.6867	1	94 Pu	L_{III}	Abs. Edge	18.054
0.6874	1	88 Ra		L_IN_I	18.036
0.68760	5	92 U	$L\beta_{10}$	L_IM_{IV}	18.031
0.68883	1	40 Zr	K	Abs. Edge	17.9989
0.68901	5	40 Zr	$K\beta_4$	$KN_{IV,V}$	17.994
0.68920	9	93 Np	$L\beta_3$	L_IM_{III}	17.989
0.68993	4	40 Zr	$K\beta_2$	$KN_{II,III}$	17.970
0.69068	2	[illegible] Pu	$L\beta_5$	$L_{III}O_{IV,V}$	17.9506
0.6932	1	[illegible] Ra		$L_{II}N_V$	17.884
0.69463	5	[illegible] Ra	[illegible]	$L_{II}N_{IV}$	17.849
[illegible]	1	40 Zr	$K\beta_5$	$KM_{IV,V}$	17.815
0.698478	[illegible]	93 Np	$L\beta_1$	$L_{II}M_{IV}$	17.7502
0.7003	1	94 Pu	$L\beta_7$	$L_{III}O_I$	17.705
0.701390	[illegible]	[illegible] Am	[illegible]	$L_{III}N_V$	17.6765
[illegible]	3	[illegible] Zr	[illegible]	[illegible]	[illegible]
[illegible]	[illegible]	[illegible] Pa	[illegible]	[illegible]	17.667
[illegible]	4	40 [illegible]	$K\beta_3$	KM_{II}	[illegible]
[illegible]	[illegible]	[illegible]	[illegible]	[illegible]	17.635
[illegible]	[illegible]	93 [illegible]	[illegible]	[illegible]	[illegible]
0.7043	[illegible]	[illegible]		[illegible]	17.604
0.70620	2	94 Pu	[illegible]	[illegible]	[illegible]
[illegible]	2	93 Np	$L\beta_5$	[illegible]	[illegible]
[illegible]	[illegible]	[illegible] Pa	$L\beta_{10}$	[illegible]	[illegible]
0.709300	[illegible]	[illegible]	[illegible]	KL_{III}	[illegible]
[illegible]	2	[illegible]	[illegible]	[illegible]	[illegible]
[illegible]	[illegible]	[illegible] Mo	[illegible]	[illegible]	17.3743
[illegible]	[illegible]	[illegible]	[illegible]	[illegible]	17.303
0.71774	[illegible]	[illegible]	[illegible]	$L_{II}N_I$	[illegible]
[illegible]	2	94 [illegible]	[illegible]	[illegible]	[illegible]
[illegible]	8	[illegible] U	$L\beta_1$	[illegible]	17.2200
[illegible]	[illegible]	94 [illegible]	[illegible]	[illegible]	[illegible]
[illegible]	[illegible]	[illegible]	[illegible]	Abs. Edge	[illegible]
[illegible]	[illegible]	[illegible]		[illegible]	[illegible]
0.7234	[illegible]	[illegible] Th	[illegible]	[illegible]	17.139
0.72426	[illegible]	[illegible]		[illegible]	[illegible]
[illegible]	5	92 U		[illegible]	[illegible]
[illegible]	9	[illegible] U	[illegible]	[illegible]	[illegible]
[illegible]	[illegible]	[illegible]	[illegible]	L_IM_{II}	[illegible]
0.72766	[illegible]	[illegible] Y	[illegible]	Abs. Edge	[illegible]
[illegible]	5	[illegible]	[illegible]	[illegible]	[illegible]
0.72864	[illegible]	39 Y	[illegible]	[illegible]	[illegible]
0.7301	1	90 Th	$L\beta_{10}$	[illegible]	[illegible]
0.7309	1	92 U		$L_{III}O_{III}$	[illegible]
0.73230	5	91 Pa	$L\beta_3$	[illegible]	[illegible]
[illegible]	1	92 U		$L_{III}O_{II}$	[illegible]

Wavelength Å*	p.e.	Element		Designation	keV
0.417737	4	51 Sb	$K\beta_3$	KM_{II}	29.6792
0.42087	2	54 Xe	$K\alpha_2$	KL_{II}	29.458
0.42467	3	50 Sn		$KO_{II,III}$	29.195
0.42467	1	50 Sn	K	Abs. Edge	29.1947
0.42495	3	50 Sn	$K\beta_4^{I}$	$KN_{IV,V}$	29.175
0.425915	8	50 Sn	$K\beta_2$	$KN_{II,III}$	29.1093
0.43175	3	50 Sn	$K\beta_5^{I}$	KM_V	28.716
0.43184	3	50 Sn	$K\beta_5^{II}$	KM_{IV}	28.710
0.433318	5	53 I	$K\alpha_1$	KL_{III}	28.6120
0.435236	5	50 Sn	$K\beta_1$	KM_{III}	28.4860
0.435877	5	50 Sn	$K\beta_3$	KM_{II}	28.4440
0.437829	7	53 I	$K\alpha_2$	KL_{II}	28.3172
0.44371	1	49 In	K	Abs. Edge	27.9420
0.44374	3	49 In		$KO_{II,III}$	27.940
0.44393	4	49 In	$K\beta_4^{I}$	$KN_{IV,V}$	27.928
0.44500	1	49 In	$K\beta_2$	$KN_{II,III}$	27.8608
0.45086	2	49 In	$K\beta_5^{I}$	KM_V	27.499
0.45098	2	49 In	$K\beta_5^{II}$	KM_{IV}	27.491
0.451295	3	52 Te	$K\alpha_1$	KL_{III}	27.4723
0.454545	4	49 In	$K\beta_1$	KM_{III}	27.2759
0.455181	4	49 In	$K\beta_3$	KM_{II}	27.2377
0.455784	3	52 Te	$K\alpha_2$	KL_{II}	27.2017
0.46407	1	48 Cd	K	Abs. Edge	26.7159
0.465328	7	48 Cd	$K\beta_2$	$KN_{II,III}$	26.6438
0.470354	3	51 Sb	$K\alpha_1$	KL_{III}	26.3591
0.474827	3	51 Sb	$K\alpha_2$	KL_{II}	26.1108
0.475105	6	48 Cd	$K\beta_1$	KM_{III}	26.0955
0.475730	5	48 Cd	$K\beta_3$	KM_{II}	26.0612
0.48589	1	47 Ag	K	Abs. Edge	25.5165
0.4859	9	47 Ag	$K\beta_4$	$KN_{IV,V}$	25.512
0.487032	4	47 Ag	$K\beta_2$	$KN_{II,III}$	25.4564
0.490599	3	50 Sn	$K\alpha_1$	KL_{III}	25.2713
0.49306	2	47 Ag	$K\beta_5$	$KM_{IV,V}$	25.145
0.495053	3	50 Sn	$K\alpha_2$	KL_{II}	25.0440
0.497069	4	47 Ag	$K\beta_1$	KM_{III}	24.9424
0.497685	4	47 Ag	$K\beta_3$	KM_{II}	24.9115
0.5092	1	46 Pd	K	Abs. Edge	24.348
0.5093	2	46 Pd	$K\beta_4$	$KN_{IV,V}$	24.346
0.510228	4	46 Pd	$K\beta_2$	$KN_{II,III}$	24.2991
0.512113	3	49 In	$K\alpha_1$	KL_{III}	24.2097
0.516544	3	49 In	$K\alpha_2$	KL_{II}	24.0020
0.51670	9	46 Pd	$K\beta_5$	$KM_{IV,V}$	23.995
0.520520	4	46 Pd	$K\beta_1$	KM_{III}	23.8187
0.521123	4	46 Pd	$K\beta_3$	KM_{II}	23.7911
0.53395	1	45 Rh	K	Abs. Edge	23.2198
0.53401	9	45 Rh	$K\beta_4^{I}$	$KN_{IV,V}$	23.217
0.535010	3	48 Cd	$K\alpha_1$	KL_{III}	23.1736
0.53503	2	45 Rh	$K\beta_2$	$KN_{II,III}$	23.1728
0.53513	5	45 Rh	$K\beta_2^{II}$	KN_{II}	23.168
0.5365	1	94 Pu	L_I	Abs. Edge	23.109
0.539422	3	48 Cd	$K\alpha_2$	KL_{II}	[illegible]
0.54101	9	45 Rh	$K\beta_5^{I}$	KM_V	22.917
0.54118	9	45 Rh	$K\beta_5^{II}$	KM_{IV}	22.909
0.5416	1	94 Pu	$L\gamma_4$	L_IO_{III}	22.891
0.54311	2	95 Am	$L\gamma_6$	$L_{II}O_{IV}$	22.8282
0.5432	1	94 Pu	$L\gamma_4'$	L_IO_{II}	22.823
0.545605	4	45 Rh	$K\beta_1$	KM_{III}	22.7236
0.546200	4	45 Rh	$K\beta_3$	KM_{II}	22.6989
0.5544	2	95 Am	$L\gamma_2$	L_IN_{II}	22.361
0.5572	1	94 Pu	L_{II}	Abs. Edge	22.253
0.5585	5	93 Np	$L\gamma_4$	$L_IO_{II,III}$	22.20
0.5594075	6	47 Ag	$K\alpha_1$	KL_{III}	22.16292
0.55973	2	94 Pu	$L\gamma_6$	$L_{II}O_{IV}$	22.1502
0.56051	1	44 Ru	K	Abs. Edge	22.1193
0.56089	9	44 Ru	$K\beta_4$	$KN_{IV,V}$	22.104
0.56166	3	44 Ru	$K\beta_2$	$KN_{II,III}$	22.074
0.561886	9	95 Am	$L\gamma_1$	$L_{II}N_{IV}$	22.0652
0.563798	4	47 Ag	$K\alpha_2$	KL_{II}	21.9903
0.564001	9	94 Pu	$L\gamma_3$	L_IN_{III}	21.9824
0.5658	1	94 Pu	$L\gamma_8$	$L_{II}O_I$	21.914
0.56785	9	44 Ru	$K\beta_5^{I}$	KM_V	21.834
0.5680	2	44 Ru	$K\beta_5^{II}$	KM_{IV}	21.829
0.5695	1	92 U	L_I	Abs. Edge	21.771
0.5706	1	92 U	$L\gamma_{13}$	$L_IP_{II,III}$	21.729
0.57068	2	94 Pu	$L\gamma_2$	L_IN_{II}	21.1251
0.572482	4	44 Ru	$K\beta_1$	KM_{III}	21.6568
0.5725	1	92 U		$L_IO_{IV,V}$	21.657
0.573067	4	44 Ru	$K\beta_3$	KM_{II}	21.6346
0.57499	9	92 U	$L\gamma_4$	L_IO_{III}	21.562
0.576700	9	92 U	$L\gamma_4'$	L_IO_{II}	21.4984
0.57699	5	93 Np	$L\gamma_6$	$L_{II}O_{IV}$	21.488
0.578882	9	94 Pu	$L\gamma_1$	$L_{II}N_{IV}$	21.4173
0.5810	5	93 Np	$L\gamma_3$	L_IN_{III}	21.34
0.585448	3	46 Pd	$K\alpha_1$	KL_{III}	21.1771
0.5873	5	93 Np	$L\gamma_2$	L_IN_{II}	21.11
0.58906	1	43 Tc	K	Abs. Edge	21.0473
0.589821	3	46 Pd	$K\alpha_2$	KL_{II}	21.0201
0.58986	5	92 U	$L\gamma_{11}$	L_IN_V	21.019
0.59024	5	43 Tc	$K\beta_2$	$KN_{II,III}$	21.005
0.59096	5	92 U		L_IN_{IV}	20.979
0.5919	1	92 U	L_{II}	Abs. Edge	20.945
0.59203	5	92 U		$L_{II}P_{IV}$	20.942
0.5930	2	92 U		$L_{II}P_{II,III}$	20.906
0.5937	1	91 Pa	$L\gamma_4$	$L_IO_{II,III}$	20.882
0.594845	9	92 U	$L\gamma_6$	$L_{II}O_{IV}$	20.8426
0.596498	9	93 Np	$L\gamma_1$	$L_{II}N_{IV}$	20.7848
0.59728	5	92 U		$L_{II}O_{III}$	20.758
0.598574	9	92 U	$L\gamma_3$	L_IN_{III}	20.7127
0.5988	1	94 Pu	$L\gamma_5$	$L_{II}N_I$	20.704
0.60125	5	92 U	$L\gamma_8$	$L_{II}O_I$	20.621
0.60130	4	43 Tc	$K\beta_1$	KM_{III}	20.619
0.60188	4	43 Tc	$K\beta_3$	KM_{II}	20.599
0.6031	1	92 U	Lv	$L_{II}N_{VI}$	20.556
0.605237	9	92 U	$L\gamma_2$	L_IN_{II}	20.4847
0.6059	1	90 Th	L_I	Abs. Edge	20.464
0.60705	8	90 Th	$L\gamma_{13}$	$L_IP_{II,III}$	20.424
0.6083	1	90 Th		$L_IO_{IV,V}$	20.383
0.61090	1	90 Th	$L\gamma_4$	L_IO_{III}	20.292
0.61251	4	90 Th	$L\gamma_4'$	L_IO_{II}	20.242
0.6133	1	91 Pa	$L\gamma_6$	$L_{II}O_{IV}$	20.216
0.613279	4	45 Rh	$K\alpha_1$	KL_{III}	20.2161
0.6146	1	90 Th		L_IO_I	20.174
0.614770	9	92 U	$L\gamma_1$	$L_{II}N_{IV}$	20.1671
0.6160	1	90 Th		$L_IN_{VI,VII}$	20.128

Wavelength Å*	p.e.	Element	Designation		keV
0.616	1	93 Np	$L\gamma_5$	$L_{II}N_{I}$	20.12
0.6169	1	91 Pa	$L\gamma_3$	$L_{I}N_{III}$	20.098
0.617630	4	45 Rh	$K\alpha_2$	KL_{II}	20.0737
0.61978	1	42 Mo	K	Abs. Edge	20.0039
0.62001	9	42 Mo	$K\beta_4^{I}$	$KN_{IV,V}$	19.996
0.62099	2	42 Mo	$K\beta_2$	$KN_{II,III}$	19.9652
0.62107	5	42 Mo	$K\beta_2^{II}$	KN_{II}	19.963
0.6228	1	92 U		$L_{II}N_{III}$	19.907
0.6239	1	91 Pa	$L\gamma_2$	$L_{I}N_{II}$	19.872
0.62636	9	90 Th	$L\gamma_{11}$	$L_{I}N_{V}$	19.794
0.62692	5	42 No	$K\beta_5^{I}$	KM_{V}	19.776
0.62708	5	42 Mo	$K\beta_5^{II}$	KM_{IV}	19.771
0.6276	1	90 Th		$L_{I}N_{IV}$	19.755
0.6299	1	90 Th	L_{II}	Abs. Edge	19.683
0.62991	9	90 Th		$L_{II}P_{IV}$	19.682
0.6312	1	90 Th		$L_{II}P_{II,III}$	19.642
0.6316	1	90 Th		$L_{II}P_{I}$	19.629
0.632288	9	42 Mo	$K\beta_1$	KM_{III}	19.6083
0.63258	4	90 Th	$L\gamma_6$	$L_{II}O_{IV}$	19.599
0.632872	2	42 Mo	$K\beta_3$	KM_{II}	19.5903
0.63358	9	91 Pa	$L\gamma_1$	$L_{II}N_{IV}$	19.568
0.63557	2	92 U	$L\gamma_5$	$L_{II}N_{I}$	19.5072
0.63559	4	90 Th	$L\gamma_3$	$L_{I}N_{III}$	19.507
0.6356	1	90 Th		$L_{II}O_{III}$	19.506
0.6369	1	90 Th		$L_{II}O_{II}$	19.466
0.63898	5	90 Th	$L\gamma_8$	$L_{II}O_{I}$	19.403
0.64064	9	90 Th	Lv	$L_{II}N_{VI}$	19.353
0.6416	1	94 Pu	$L\beta_9$	$L_{I}M_{V}$	19.323
0.64221	4	90 Th	$L\gamma_2$	$L_{I}N_{II}$	19.305
0.643083	4	44 Ru	$K\alpha_1$	KL_{III}	19.2792
0.6445	1	88 Ra	L_{I}	Abs. Edge	19.236
0.64513	5	88 Ra	$L\gamma_{13}$	$L_{I}P_{II,III}$	19.218
0.6468	1	88 Ra		$L_{I}O_{IV,V}$	19.167
0.647408	5	44 Ru	$K\alpha_2$	KL_{II}	19.1504
0.64755	5	90 Th		$L_{I}N_{I}$	19.146
0.6482	1	94 Pu	$L\beta_{10}$	$L_{I}M_{IV}$	19.126
0.64891	2	95 Am	$L\beta_3$	$L_{I}M_{III}$	19.1059
0.64965	5	88 Ra	$L\gamma_4$	$L_{I}O_{III}$	19.084
0.65131	5	88 Ra	$L\gamma_4'$	$L_{I}O_{II}$	19.036
0.6521	1	90 Th		$L_{II}N_{V}$	19.014
0.65298	1	41 Nb	K	Abs. Edge	18.9869
0.65313	3	90 Th	$L\gamma_1$	$L_{II}N_{IV}$	18.9825
0.65318	5	41 Nb	$K\beta_4$	$KN_{IV,V}$	18.981
0.65416	4	41 Nb	$K\beta_2$	$KN_{II,III}$	18.953
0.6550	1	91 Pa	$L\gamma_5$	$L_{II}N_{I}$	18.930
0.657655	9	95 Am	$L\beta_1$	$L_{II}M_{IV}$	18.8520
0.6620	1	90 Th		$L_{II}N_{III}$	18.729
0.6654	1	88 Ra	$L\gamma_{11}$	$L_{I}N_{V}$	18.633
0.66576	2	41 Nb	$K\beta_1$	KM_{III}	18.6225
0.66634	3	41 Nb	$K\beta_3$	KM_{II}	18.6063
0.6666	1	88 Ra		$L_{I}N_{IV}$	18.600
0.66871	2	94 Pu	$L\beta_3$	$L_{I}M_{III}$	18.5405
0.6707	1	88 Ra	L_{II}	Abs. Edge	18.486
0.6714	1	88 Ra		$L_{II}P_{II,III}$	18.466
0.6724	1	88 Ra		$L_{II}P_{I}$	18.439
0.67328	5	88 Ra	$L\gamma_6$	$L_{II}O_{IV}$	18.414
0.67351	9	89 Ac	$L\gamma_1$	$L_{II}N_{IV}$	18.408
0.67383	2	95 Am	$L\beta_5$	$L_{III}O_{IV,V}$	18.3996
0.67491	4	90 Th	$L\gamma_5$	$L_{II}N_{I}$	18.370
0.67502	3	43 Tc	$K\alpha_1$	KL_{III}	18.3671
0.67538	5	88 Ra	$L\gamma_3$	$L_{I}N_{III}$	18.357
0.6764	1	88 Ra		$L_{II}O_{III}$	18.330
0.67772	2	94 Pu	$L\beta_1$	$L_{II}M_{IV}$	18.2937
0.6780	1	88 Ra		$L_{II}O_{II}$	18.286
0.67932	3	43 Tc	$K\alpha_2$	KL_{II}	18.2508
0.6801	1	88 Ra	$L\gamma_8$	$L_{II}O_{I}$	18.230
0.681014	8	92 U	$L\beta_9$	$L_{I}M_{V}$	18.2054
0.68199	5	88 Ra	$L\gamma_2$	$L_{I}N_{II}$	18.179
0.68639	2	95 Am	$L\beta_4$	$L_{I}M_{II}$	18.0627
0.6867	1	94 Pu	L_{III}	Abs. Edge	18.054
0.6874	1	88 Ra		$L_{I}N_{I}$	18.036
0.68760	5	92 U	$L\beta_{10}$	$L_{I}M_{IV}$	18.031
0.68883	1	40 Zr	K	Abs. Edge	17.9989
0.68901	5	40 Zr	$K\beta_4$	$KN_{IV,V}$	17.994
0.68920	9	93 Np	$L\beta_3$	$L_{I}M_{III}$	17.989
0.68993	4	40 Zr	$K\beta_2$	$KN_{II,III}$	17.970
0.69068	2	94 Pu	$L\beta_5$	$L_{III}O_{IV,V}$	17.9506
0.6932	1	88 Ra		$L_{II}N_{V}$	17.884
0.69463	5	88 Ra	$L\gamma_1$	$L_{II}N_{IV}$	17.849
0.6959	1	40 Zr	$K\beta_5$	$KM_{IV,V}$	17.815
0.698478	9	93 Np	$L\beta_1$	$L_{II}M_{IV}$	17.7502
0.7003	1	94 Pu	$L\beta_7$	$L_{III}O_{I}$	17.705
0.701390	9	95 Am	$L\beta_2$	$L_{III}N_{V}$	17.6765
0.70173	3	40 Zr	$K\beta_1$	KM_{III}	17.6678
0.7018	1	91 Pa	$L\beta_9$	$L_{I}M_{V}$	17.667
0.70228	4	40 Zr	$K\beta_3$	KM_{II}	17.654
0.7031	1	94 Pu	Lu	$L_{III}N_{VI,VII}$	17.635
0.70341	2	95 Am	$L\beta_{15}$	$L_{III}N_{IV}$	17.6258
0.7043	1	88 Ra		$L_{II}N_{III}$	17.604
0.70620	2	94 Pu	$L\beta_4$	$L_{I}M_{II}$	17.5560
0.70814	2	93 Np	$L\beta_5$	$L_{III}O_{IV,V}$	17.5081
0.7088	2	91 Pa	$L\beta_{10}$	$L_{I}M_{IV}$	17.492
0.709300	1	42 Mo	$K\alpha_1$	KL_{III}	17.47934
0.71029	2	92 U	$L\beta_3$	$L_{I}M_{III}$	17.4550
0.713590	6	42 Mo	$K\alpha_2$	KL_{II}	17.3743
0.71652	9	87 Fr	$L\gamma_1$	$L_{II}N_{IV}$	17.303
0.71774	5	88 Ra	$L\gamma_5$	$L_{II}N_{I}$	17.274
0.71851	2	94 Pu	$L\beta_2$	$L_{III}N_{V}$	17.2553
0.719984	8	92 U	$L\beta_1$	$L_{II}M_{IV}$	17.2200
0.7205	1	94 Pu	$L\beta_{15}$	$L_{III}N_{IV}$	17.208
0.7223	1	92 U	L_{III}	Abs. Edge	17.165
0.72240	5	92 U		$L_{III}P_{IV,V}$	17.162
0.7234	1	90 Th	$L\beta_9$	$L_{I}M_{V}$	17.139
0.72426	5	92 U		$L_{III}P_{II,III}$	17.118
0.72521	5	92 U		$L_{III}P_{I}$	17.096
0.726305	9	92 U	$L\beta_5$	$L_{III}O_{IV,V}$	17.0701
0.72671	2	93 Np	$L\beta_4$	$L_{I}M_{II}$	17.0607
0.72766	5	39 Y	K	Abs. Edge	17.038
0.72776	5	39 Y	$K\beta_4$	$KN_{IV,V}$	17.036
0.72864	4	39 Y	$K\beta_2$	$KN_{II,III}$	17.0154
0.7301	1	90 Th	$L\beta_{10}$	$L_{I}M_{IV}$	16.981
0.7309	1	92 U		$L_{III}O_{III}$	16.962
0.73230	5	91 Pa	$L\beta_3$	$L_{I}M_{III}$	16.930
0.7333	1	92 U		$L_{III}O_{II}$	16.907

Wavelength Å*	p.e.	Element		Designation	keV
0.73418	2	95 Am	$L\beta_6$	$L_{III}N_I$	16.8870
0.7345	1	39 Y	$K\beta_5$	$KM_{IV,V}$	16.879
0.73602	6	92 U	$L\beta_7$	$L_{III}O_I$	16.845
0.736230	9	93 Np	$L\beta_2$	$L_{III}N_V$	16.8400
0.738603	9	92 U	Lu	$L_{III}N_{VI,VII}$	16.7859
0.73928	9	86 Rn	$L\gamma_1$	$L_{II}N_{IV}$	16.770
0.74072	2	39 Y	$K\beta_1$	KM_{III}	16.7378
0.74126	3	39 Y	$K\beta_3$	KM_{II}	16.7258
0.74232	5	91 Pa	$L\beta_1$	$L_{II}M_{IV}$	16.702
0.74503	5	92 U	$L\beta_{17}$	$L_{II}M_{III}$	16.641
0.7452	2	91 Pa	$L\beta_5$	$L_{III}O_{IV,V}$	16.636
0.74620	1	41 Nb	$K\alpha_1$	KL_{III}	16.6151
0.747985	9	92 U	$L\beta_4$	L_IM_{II}	16.5753
0.75044	1	41 Nb	$K\alpha_2$	KL_{II}	16.5210
0.75148	2	94 Pu	$L\beta_6$	$L_{III}N_I$	16.4983
0.7546	2	91 Pa	$L\beta_7$	$L_{III}O_I$	16.431
0.754681	9	92 U	$L\beta_2$	$L_{III}N_V$	16.4283
0.75479	3	90 Th	$L\beta_3$	L_IM_{III}	16.4258
0.756642	9	92 U	$L\beta_{15}$	$L_{III}N_{IV}$	16.3857
0.75690	3	83 Bi	$L\gamma_{13}$	$L_IP_{II,III}$	16.3802
0.7571	1	83 Bi	L_I	Abs. Edge	16.376
0.7579	1	90 Th		$L_{II}M_V$	16.359
0.75791	5	83 Bi		$L_IO_{IV,V}$	16.358
0.7591	1	94 Pu	$L\eta$	$L_{II}M_I$	16.333
0.7607	1	90 Th	L_{III}	Abs. Edge	16.299
0.76087	9	90 Th		$L_{III}P_{IV,V}$	16.295
0.76087	3	83 Bi	$L\gamma_4$	L_IO_{III}	16.2947
0.76198	3	83 Bi	$L\gamma_4'$	L_IO_{II}	16.2709
0.7625	2	90 Th		$L_{III}P_{II,III}$	16.260
0.76289	9	85 At	$L\gamma_1$	$L_{II}N_{IV}$	16.251
0.76338	5	90 Th		$L_{III}P_I$	16.241
0.7641	5	83 Bi		$L_IN_{VI,VII}$	16.23
0.7645	2	84 Po	$L\gamma_6$	$L_{II}O_{IV}$	16.218
0.76468	5	90 Th	$L\beta_5$	$L_{III}O_{IV,V}$	16.213
0.765210	9	90 Th	$L\beta_1$	$L_{II}M_{IV}$	16.2022
0.76857	5	88 Ra	$L\beta_9$	L_IM_V	16.131
0.769	1	93 Np	$L\beta_6$	$L_{III}N_I$	16.13
0.7690	1	90 Th		$L_{III}O_{III}$	16.123
0.7691	1	92 U		$L_{III}N_{III}$	16.120
0.76973	5	38 Sr	K	Abs. Edge	16.107
0.7699	1	91 Pa	$L\beta_4$	L_IM_{II}	16.104
0.76989	5	38 Sr	$K\beta_4$	$KN_{IV,V}$	16.104
0.77081	3	38 Sr	$K\beta_2$	$KN_{II,III}$	16.0846
0.7713	1	90 Th		$L_{III}O_{II}$	16.074
0.772	1	84 Po	$L\gamma_2$	L_IN_{II}	16.07
0.7737	1	91 Pa	$L\beta_2$	$L_{III}N_V$	16.024
0.77437	4	90 Th	$L\beta_7$	$L_{III}O_I$	16.0105
0.77546	5	88 Ra	$L\beta_{10}$	L_IM_{IV}	15.988
0.7764	1	38 Sr	$K\beta_5$	$KM_{IV,V}$	15.969
0.77661	5	90 Th	Lu	$L_{III}N_{VI,VII}$	15.964
0.77728	5	83 Bi	$L\gamma_{11}$	L_IN_V	15.951
0.77822	9	89 Ac	$L\beta_3$	L_IM_{III}	15.931
0.77954	5	83 Bi		L_IN_{IV}	15.904
0.78017	9	92 U		$L_{III}N_{II}$	15.892
0.7809	2	93 Np	$L\eta$	$L_{II}M_I$	15.876
0.78196	5	82 Pb	L_I	Abs. Edge	15.855
0.78257	7	82 Pb		$L_IO_{IV,V}$	15.843
0.78292	2	38 Sr	$K\beta_1$	KM_{III}	15.8357
0.78345	3	38 Sr	$K\beta_3$	KM_{II}	15.8249
0.7858	1	82 Pb	$L\gamma_4$	L_IO_{III}	15.777
0.78593	1	40 Zr	$K\alpha_1$	KL_{III}	15.7751
0.78706	7	82 Pb	$L\gamma_4'$	L_IO_{II}	15.752
0.78748	9	84 Po	$L\gamma_1$	$L_{II}N_{IV}$	15.744
0.78838	2	92 U	$L\beta_6$	$L_{III}N_I$	15.7260
0.7884	1	82 Pb		$L_IN_{VI,VII}$	15.725
0.7887	1	83 Bi	L_{II}	Abs. Edge	15.719
0.78903	9	89 Ac	$L\beta_1$	$L_{II}M_{IV}$	15.713
0.78917	5	83 Bi	$L\gamma_3$	L_IN_{III}	15.7102
0.7897	1	82 Pb		L_IO_I	15.699
0.79015	1	40 Zr	$K\alpha_2$	KL_{II}	15.6909
0.79043	3	83 Bi	$L\gamma_6$	$L_{II}O_{IV}$	15.6853
0.79257	4	90 Th	$L\beta_4$	L_IM_{II}	15.6429
0.79257	4	90 Th	$L\beta_{17}$	$L_{II}M_{III}$	15.6429
0.79354	3	90 Th	$L\beta_2$	$L_{III}M_V$	15.6237
0.79384	5	83 Bi		$L_{II}O_{III}$	15.6178
0.79539	5	90 Th	$L\beta_{15}$	$L_{III}N_{IV}$	15.5875
0.79565	3	83 Bi	$L\gamma_2$	L_IN_{II}	15.5824
0.79721	9	83 Bi	Lv	$L_{II}N_{VI}$	15.552
0.7973	1	83 Bi	$L\gamma_8$	$L_{II}O_I$	15.551
0.8022	1	83 Bi		L_IN_I	15.456
0.80233	9	82 Pb	$L\gamma_{11}$	L_IN_V	15.453
0.80273	5	88 Ra	$L\beta_3$	L_IM_{III}	15.4449
0.8028	1	88 Ra	L_{III}	Abs. Edge	15.444
0.80364	7	82 Pb		L_IN_{IV}	15.427
0.8038	1	88 Ra		$L_{III}P_{II,III}$	15.425
0.8050	1	88 Ra		$L_{III}P_I$	15.402
0.80509	2	92 U	$L\eta$	$L_{II}M_I$	15.3997
0.80627	5	88 Ra	$L\beta_5$	$L_{III}O_{IV,V}$	15.3771
0.8079	1	91 Pa	$L\beta_6$	$L_{III}N_I$	15.347
0.8081	1	81 Tl	L_I	Abs. Edge	15.343
0.8082	1	90 Th		$L_{III}N_{III}$	15.341
0.80861	5	81 Tl		$L_IO_{IV,V}$	15.3327
0.81163	9	90 Th		L_IM_I	15.276
0.81184	5	81 Tl	$L\gamma_4$	L_IO_{III}	15.2716
0.81308	5	81 Tl	$L\gamma_4'$	L_IO_{II}	15.2482
0.81311	2	83 Bi	$L\gamma_1$	$L_{II}N_{IV}$	15.2477
0.81375	5	88 Ra	$L\beta_1$	$L_{II}M_{IV}$	15.2358
0.8147	1	82 Pb	$L\gamma_3$	L_IN_{III}	15.218
0.81538	5	82 Pb	L_{II}	Abs. Edge	15.2053
0.8154	2	37 Rb	$K\beta_4$	$KN_{IV,V}$	15.205
0.81554	5	37 Rb	K	Abs. Edge	15.2023
0.8158	1	81 Tl		L_IO_I	15.198
0.81583	5	82 Pb		$L_{II}P_I$	15.1969
0.8162	1	88 Ra	$L\beta_7$	$L_{III}O_I$	15.190
0.81645	3	37 Rb	$K\beta_2$	$KN_{II,III}$	15.1854
0.81683	5	82 Pb	$L\gamma_6$	$L_{II}O_{IV}$	15.1783
0.8186	1	88 Ra	Lu	$L_{III}N_{VI,VII}$	15.146
0.8190	2	90 Th		$L_{III}N_{II}$	15.138
0.8200	1	82 Pb		$L_{II}O_{III}$	15.120
0.8210	2	82 Pb	$L\gamma_2$	L_IN_{II}	15.101
0.8219	1	37 Rb	$K\beta_5$	$KM_{IV,V}$	15.085
0.82327	7	82 Pb	Lv	$L_{II}N_{VI}$	15.060
0.82365	5	82 Pb	$L\gamma_8$	$L_{II}O_I$	15.0527
0.8248	1	83 Bi		$L_{II}N_{III}$	15.031

Wavelength Å*	p.e.	Element		Designation	keV
0.82789	9	87 Fr	$L\beta_3$	L_IM_{III}	14.976
0.82790	8	90 Th	$L\beta_6$	$L_{III}N_I$	14.975
0.82859	7	82 Pb		L_IN_I	14.963
0.82868	2	37 Rb	$K\beta_1$	KM_{III}	14.9613
0.82879	5	81 Tl	$L\gamma_{11}$	L_IN_V	14.9593
0.82884	1	39 Y	$K\alpha_1$	KL_{III}	14.9584
0.82921	3	37 Rb	$K\beta_3$	KM_{II}	14.9517
0.8295	1	91 Pa	$L\eta$	$L_{II}M_I$	14.946
0.83001	7	81 Tl		L_IN_{IV}	14.937
0.83305	1	39 Y	$K\alpha_2$	KL_{II}	14.8829
0.8338	1	90 Th		$L_{II}M_{II}$	14.869
0.8344	9	83 Bi		$L_{II}N_{II}$	14.86
0.8350	2	80 Hg		$L_IO_{IV,V}$	14.847
0.8353	1	80 Hg	L_I	Abs. Edge	14.842
0.83537	5	88 Ra	$L\beta_2$	$L_{III}N_V$	14.8414
0.83722	5	88 Ra	$L\beta_{15}$	$L_{III}N_{IV}$	14.8086
0.8382	2	82 Pb		$L_{II}N_V$	14.791
0.83894	7	80 Hg	$L\gamma_4$	L_IO_{III}	14.778
0.83923	5	83 Bi	$L\gamma_5$	$L_{II}N_I$	14.7732
0.83940	9	87 Fr	$L\beta_1$	$L_{II}M_{IV}$	14.770
0.83973	3	82 Pb	$L\gamma_1$	$L_{II}N_{IV}$	14.7644
0.84013	7	80 Hg	$L\gamma_4'$	L_IO_{II}	14.757
0.84071	5	88 Ra	$L\beta_4$	L_IM_{II}	14.7472
0.84130	4	81 Tl	$L\gamma_3$	L_IN_{III}	14.7368
0.8434	1	81 Tl	L_{II}	Abs. Edge	14.699
0.8438	1	88 Ra	$L\beta_{17}$	$L_{II}M_{III}$	14.692
0.8442	2	81 Tl	$L\gamma_6$	$L_{II}O_{IV}$	14.685
0.8452	2	80 Hg		L_IO_I	14.670
0.84773	5	81 Tl	$L\gamma_2$	L_IN_{II}	14.6251
0.848187	9	95 Am	$L\alpha_1$	$L_{III}M_V$	14.6172
0.8490	1	81 Tl		$L_{II}O_{II}$	14.604
0.85048	5	81 Tl	Lv	$L_{II}N_{VI}$	14.5777
0.8512	1	88 Ra		$L_{III}N_{III}$	14.566
0.8513	2	81 Tl	$L\gamma_8$	$L_{II}O_I$	14.564
0.85192	7	82 Pb		$L_{II}N_{III}$	14.553
0.85436	9	86 Rn	$L\beta_3$	L_IM_{III}	14.512
0.85446	4	90 Th	$L\eta$	$L_{II}M_I$	14.5099
0.8549	1	81 Tl		L_IN_I	14.503
0.85657	7	80 Hg	$L\gamma_{11}$	L_IN_V	14.474
0.858	2	87 Fr	$L\beta_2$	$L_{III}N_V$	14.45
0.8585	3	82 Pb		$L_{II}N_{II}$	14.442
0.860266	9	95 Am	$L\alpha_2$	$L_{III}M_{IV}$	14.4119
0.8618	1	88 Ra		$L_{III}N_{II}$	14.387
0.86376	5	79 Au	L_I	Abs. Edge	14.3537
0.86400	5	79 Au		$L_IO_{IV,V}$	14.3497
0.8653	2	36 Kr	$K\beta_4$	$KN_{IV,V}$	14.328
0.86552	1	36 Kr	K	Abs. Edge	14.3244
0.86605	9	86 Rn	$L\beta_1$	$L_{II}M_{IV}$	14.316
0.8661	1	36 Kr	$K\beta_2$	$KN_{II,III}$	14.315
0.86655	5	82 Pb	$L\gamma_5$	$L_{II}N_I$	14.3075
0.86703	4	79 Au	$L\gamma_4$	L_IO_{III}	14.2996
0.86752	3	81 Tl	$L\gamma_1$	$L_{II}N_{IV}$	14.2915
0.86816	4	79 Au	$L\gamma_4'$	L_IO_{II}	14.2809
0.86830	2	94 Pu	$L\alpha_1$	$L_{III}M_V$	14.2786
0.86915	7	80 Hg	$L\gamma_3$	L_IN_{III}	14.265
0.87074	5	79 Au		L_IO_I	14.2385
0.8708	2	36 Kr	$K\beta_5$	$KM_{IV,V}$	14.238
0.87088	5	88 Ra	$L\beta_6$	$L_{III}N_I$	14.2362
0.8722	1	80 Hg	L_{II}	Abs. Edge	14.215
0.87319	7	80 Hg	$L\gamma_6$	$L_{II}O_{IV}$	14.199
0.87526	1	38 Sr	$K\alpha_1$	KL_{III}	14.1650
0.87544	7	80 Hg	$L\gamma_2$	L_IN_{II}	14.162
0.8758	1	80 Hg		$L_{II}O_{III}$	14.156
0.8784	1	80 Hg		$L_{II}O_{II}$	14.114
0.8785	1	36 Kr	$K\beta_1$	KM_{III}	14.112
0.87885	7	80 Hg	Lv	$L_{II}N_{VI}$	14.107
0.8790	1	36 Kr	$K\beta_3$	KM_{II}	14.104
0.87943	1	38 Sr	$K\alpha_2$	KL_{II}	14.0979
0.87995	7	80 Hg	$L\gamma_8$	$L_{II}O_I$	14.090
0.87996	5	81 Tl		$L_{II}N_{III}$	14.0893
0.88028	2	94 Pu	$L\alpha_2$	$L_{III}M_{IV}$	14.0842
0.88135	9	85 At	$L\beta_3$	L_IM_{III}	14.067
0.8827	2	80 Hg		L_IN_I	14.045
0.88433	7	79 Au	$L\gamma_{11}$	L_IN_V	14.020
0.88563	7	79 Au		L_IN_{IV}	13.999
0.8882	2	81 Tl		$L_{II}M_{II}$	13.959
0.889128	9	93 Np	$L\alpha_1$	$L_{III}M_V$	13.9441
0.8931	1	78 Pt	L_I	Abs. Edge	13.883
0.8934	1	78 Pt		L_IO_V	13.878
0.89349	9	85 At	$L\beta_1$	$L_{II}M_{IV}$	13.876
0.8943	1	78 Pt		L_IO_{IV}	13.864
0.89500	4	81 Tl	$L\gamma_5$	$L_{II}N_I$	13.8526
0.89646	5	80 Hg	$L\gamma_1$	$L_{II}N_{IV}$	13.8301
0.89659	4	78 Pt	$L\gamma_4$	L_IO_{III}	13.8281
0.89747	4	78 Pt	$L\gamma_4'$	L_IO_{II}	13.8145
0.89783	5	79 Au	$L\gamma_3$	L_IN_{III}	13.8090
0.89791	3	83 Bi	$L\beta_9$	L_IM_V	13.8077
0.8995	2	78 Pt		L_IO_I	13.784
0.8996	2	84 Po	$L\beta_5$	$L_{III}O_{IV,V}$	13.782
0.901045	9	93 Np	$L\alpha_2$	$L_{III}M_{IV}$	13.7597
0.90259	5	79 Au	L_{II}	Abs. Edge	13.7361
0.90297	3	79 Au	$L\gamma_6$	$L_{II}O_{IV}$	13.7304
0.90434	3	79 Au	$L\gamma_2$	L_IN_{II}	13.7095
0.90495	4	83 Bi	$L\beta_{10}$	L_IM_{IV}	13.7002
0.90638	7	79 Au		$L_{II}O_{III}$	13.679
0.90742	5	88 Ra	$L\eta$	$L_{II}M_I$	13.6630
0.90746	7	79 Au		$L_{II}O_{II}$	13.662
0.90837	5	79 Au	Lv	$L_{II}N_{VI}$	13.6487
0.90894	7	80 Hg		$L_{II}N_{III}$	13.640
0.9091	3	84 Po	$L\beta_3$	L_IM_{III}	13.638
0.90989	5	79 Au	$L\gamma_8$	$L_{II}O_I$	13.6260
0.910639	9	92 U	$L\alpha_1$	$L_{III}M_V$	13.6147
0.9131	1	79 Au		L_IN_I	13.578
0.9143	2	78 Pt	$L\gamma_{11}$	L_IN_V	13.560
0.9204	1	35 Br	K	Abs. Edge	13.470
0.92046	2	35 Br	$K\beta_2$	$KN_{II,III}$	13.4695
0.9220	2	84 Po	$L\beta_1$	$L_{II}M_{IV}$	13.447
0.922558	9	92 U	$L\alpha_2$	$L_{III}M_{IV}$	13.4388
0.9234	1	83 Bi	L_{III}	Abs. Edge	13.426
0.9236	1	77 Ir	L_I	Abs. Edge	13.423
0.92413	4	83 Bi		$L_{III}P_{II,III}$	13.4159
0.9243	3	77 Ir		$L_IO_{IV,V}$	13.413
0.92453	7	80 Hg	$L\gamma_5$	$L_{II}N_I$	13.410
0.9255	1	35 Br	$K\beta_5$	$KM_{IV,V}$	13.396

Wavelength Å*	p.e.	Element	Designation		keV
0.925553	9	37 Rb	$K\alpha_1$	KL_{III}	13.3953
0.92556	3	83 Bi	$L\beta_5$	$L_{III}O_{IV,V}$	13.3953
0.92650	3	79 Au	$L\gamma_1$	$L_{II}N_{IV}$	13.3817
0.9268	1	82 Pb	$L\beta_9$	L_IM_V	13.377
0.92744	3	77 Ir	$L\gamma_4$	L_IO_{III}	13.3681
0.92791	5	78 Pt	$L\gamma_3$	L_IN_{III}	13.3613
0.92831	3	77 Ir	$L\gamma_4'$	L_IO_{II}	13.3555
0.92937	5	84 Po	$L\beta_2$	$L_{III}N_V$	13.3404
0.92969	1	37 Rb	$K\alpha_2$	KL_{II}	13.3358
0.9302	2	83 Bi		$L_{III}O_{III}$	13.328
0.9312	2	84 Po	$L\beta_{15}$	$L_{III}N_{IV}$	13.314
0.9323	2	83 Bi		$L_{III}O_{II}$	13.298
0.93279	2	35 Br	$K\beta_1$	KM_{III}	13.2914
0.93284	5	91 Pa	$L\alpha_1$	$L_{III}M_V$	13.2907
0.93327	5	35 Br	$K\beta_3$	KM_{II}	13.2845
0.9339	2	82 Pb	$L\beta_{10}$	L_IM_{IV}	13.275
0.93414	5	78 Pt	L_{II}	Abs. Edge	13.2723
0.9342	2	78 Pt	$L\gamma_6$	$L_{II}O_{IV}$	13.271
0.93427	5	78 Pt	$L\gamma_2$	L_IN_{II}	13.2704
0.93505	5	83 Bi	$L\beta_7$	$L_{III}O_I$	13.2593
0.93505	5	83 Bi	Lu	$L_{III}N_{VI,VII}$	13.2593
0.93855	3	83 Bi	$L\beta_3$	L_IM_{III}	13.2098
0.93931	5	78 Pt	Lv	$L_{II}N_{VI}$	13.1992
0.9402	2	79 Au		$L_{II}N_{III}$	13.186
0.9411	1	78 Pt	$L\gamma_8$	$L_{II}O_I$	13.173
0.94419	5	83 Bi		$L_{II}M_V$	13.1310
0.9446	2	77 Ir	$L\gamma_{11}$	L_IN_V	13.126
0.94482	5	91 Pa	$L\alpha_2$	$L_{III}M_{IV}$	13.1222
0.9455	2	78 Pt		L_IN_I	13.113
0.9459	2	77 Ir		L_IN_{IV}	13.108
0.9475	3	84 Po	$L\beta_4$	L_IM_{II}	13.086
0.95073	5	82 Pb	L_{III}	Abs. Edge	13.0406
0.95118	7	82 Pb		$L_{III}P_{II,III}$	13.0344
0.951978	9	83 Bi	$L\beta_1$	$L_{II}M_{IV}$	13.0235
0.9526	1	82 Pb	$L\beta_5$	$L_{III}O_{IV,V}$	13.015
0.95518	4	83 Bi	$L\beta_2$	$L_{III}N_V$	12.9799
0.95559	3	79 Au	$L\gamma_5$	$L_{II}N_I$	12.9743
0.9558	1	76 Os	L_I	Abs. Edge	12.972
0.95600	3	90 Th	$L\alpha_1$	$L_{III}M_V$	12.9687
0.95603	5	76 Os		$L_IO_{IV,V}$	12.9683
0.95675	7	81 Tl	$L\beta_9$	L_IM_V	12.9585
0.95702	5	83 Bi	$L\beta_{15}$	$L_{III}N_{IV}$	12.9549
0.9578	1	82 Pb		$L_{III}O_{III}$	12.945
0.95797	3	78 Pt	$L\gamma_1$	$L_{II}N_{IV}$	12.9420
0.9586	1	82 Pb		$L_{III}O_{II}$	12.934
0.95931	5	77 Ir	$L\gamma_3$	L_IN_{III}	12.9240
0.95938	8	76 Os	$L\gamma_4$	L_IO_{III}	12.923
0.96033	8	76 Os	$L\gamma_4'$	L_IO_{II}	12.910
0.96133	7	82 Pb	Lu	$L_{III}N_{VI,VII}$	12.8968
0.9620	1	82 Pb	$L\beta_7$	$L_{III}O_I$	12.888
0.96318	7	76 Os		L_IO_I	12.8721
0.9636	1	92 U	Ls	$L_{III}M_{III}$	12.866
0.96389	7	81 Tl	$L\beta_{10}$	L_IM_{IV}	12.8626
0.96545	3	77 Ir	$L\gamma_2$	L_IN_{II}	12.8418
0.96708	4	77 Ir	$L\gamma_6$	$L_{II}O_{IV}$	12.8201
0.9671	1	77 Ir	L_{II}	Abs. Edge	12.820
0.9672	2	84 Po	$L\beta_6$	$L_{III}N_I$	12.819
0.96788	2	90 Th	$L\alpha_2$	$L_{III}M_{IV}$	12.8096
0.96911	7	82 Pb	$L\beta_3$	L_IM_{III}	12.7933
0.96979	5	77 Ir		$L_{II}O_{III}$	12.7843
0.97161	6	77 Ir	Lv	$L_{II}N_{VI}$	12.7603
0.97173	4	78 Pt		$L_{II}N_{III}$	12.7588
0.97321	5	83 Bi		$L_{III}N_{III}$	12.7394
0.97409	3	77 Ir	$L\gamma_8$	$L_{II}O_I$	12.7279
0.9747	1	82 Pb		$L_{II}M_V$	12.720
0.9765	3	76 Os	$L\gamma_{11}$	L_IN_V	12.696
0.9766	2	77 Ir		L_IN_I	12.695
0.97690	4	83 Bi	$L\beta_4$	L_IM_{II}	12.6912
0.9772	3	76 Os		L_IN_{IV}	12.687
0.9792	2	78 Pt		$L_{II}N_{II}$	12.661
0.97926	5	81 Tl		$L_{III}P_{II,III}$	12.6607
0.9793	1	81 Tl	L_{III}	Abs. Edge	12.660
0.97974	1	34 Se	K	Abs. Edge	12.6545
0.97992	5	34 Se	$K\beta_2$	$KN_{II,III}$	12.6522
0.97993	5	89 Ac	$L\alpha_1$	$L_{III}M_V$	12.6520
0.9801	1	36 Kr	$K\alpha_1$	KL_{III}	12.649
0.98058	3	81 Tl	$L\beta_5$	$L_{III}O_{IV,V}$	12.6436
0.98221	7	82 Pb	$L\beta_2$	$L_{III}N_V$	12.6226
0.98280	5	83 Bi		$L_{III}N_{II}$	12.6151
0.98291	3	82 Pb	$L\beta_1$	$L_{II}M_{IV}$	12.6137
0.98389	7	82 Pb	$L\beta_{15}$	$L_{III}N_{IV}$	12.6011
0.9841	1	36 Kr	$K\alpha_2$	KL_{II}	12.598
0.9843	1	34 Se	$K\beta_5$	$KM_{IV,V}$	12.595
0.98538	5	81 Tl		$L_{III}O_{III}$	12.5820
0.9871	2	80 Hg	$L\beta_9$	L_IM_V	12.560
0.98738	5	81 Tl		$L_{III}O_{II}$	12.5566
0.9877	2	78 Pt	$L\gamma_5$	$L_{II}N_I$	12.552
0.9888	1	81 Tl	Lu	$L_{III}N_{VI,VII}$	12.538
0.98913	5	83 Bi	$L\beta_{17}$	$L_{II}M_{III}$	12.5344
0.9894	1	75 Re	L_I	Abs. Edge	12.530
0.9900	1	75 Re		$L_IO_{IV,V}$	12.524
0.99017	5	81 Tl	$L\beta_7$	$L_{III}O_I$	12.5212
0.99085	3	77 Ir	$L\gamma_1$	$L_{II}N_{IV}$	12.5126
0.99178	5	89 Ac	$L\alpha_2$	$L_{III}M_{IV}$	12.5008
0.99186	5	76 Os	$L\gamma_3$	L_IN_{III}	12.4998
0.99218	3	34 Se	$K\beta_1$	KM_{III}	12.4959
0.99249	5	75 Re	$L\gamma_4$	L_IO_{III}	12.4920
0.99268	5	34 Se	$K\beta_3$	KM_{II}	12.4896
0.99331	3	83 Bi	$L\beta_6$	$L_{III}N_I$	12.4816
0.99334	5	75 Re	$L\gamma_4'$	L_IO_{II}	12.4813
0.9962	2	80 Hg	$L\beta_{10}$	L_IM_{IV}	12.446
0.9965	1	75 Re		L_IO_I	12.442
0.99805	5	76 Os	$L\gamma_2$	L_IN_{II}	12.4224
1.0005	1	82 Pb		$L_{III}N_{III}$	12.392
1.0005	9	83 Bi		L_IM_I	12.39
1.00062	3	81 Tl	$L\beta_3$	L_IM_{III}	12.3904
1.00107	5	76 Os	$L\gamma_6$	$L_{II}O_{IV}$	12.3848
1.0012	6	95 Am	Ll	$L_{III}M_I$	12.384
1.0014	1	76 Os	L_{II}	Abs. Edge	12.381
1.0047	2	76 Os		$L_{II}O_{III}$	12.340
1.00473	5	88 Ra	$L\alpha_1$	$L_{III}M_V$	12.3397
1.0050	2	76 Os	Lv	$L_{II}N_{VI}$	12.337
1.0054	3	77 Ir		$L_{II}N_{III}$	12.332
1.00722	5	81 Tl		$L_{II}M_V$	12.3093

Wavelength Å*	p.e.	Element	Designation		keV
1.0075	1	82 Pb	$L\beta_4$	L_IM_{II}	12.306
1.00788	5	76 Os	$L\gamma_8$	$L_{II}O_I$	12.3012
1.0091	1	80 Hg	L_{III}	Abs. Edge	12.286
1.00987	7	80 Hg	$L\beta_5$	$L_{III}O_{IV,V}$	12.2769
1.01031	3	81 Tl	$L\beta_2$	$L_{III}N_V$	12.2715
1.01040	7	82 Pb		$L_{III}N_{II}$	12.2705
1.0108	1	75 Re	$L\gamma_{11}$	L_IN_V	12.266
1.0112	1	90 Th	Ls	$L_{III}M_{III}$	12.261
1.0119	1	75 Re		L_IN_{IV}	12.252
1.0120	2	77 Ir		$L_{II}N_{II}$	12.251
1.01201	3	81 Tl	$L\beta_{15}$	$L_{III}N_{IV}$	12.2510
1.01404	7	80 Hg		$L_{III}O_{III}$	12.2264
1.01513	4	81 Tl	$L\beta_1$	$L_{II}M_{IV}$	12.2133
1.01558	7	80 Hg		$L_{III}O_{II}$	12.2079
1.01656	5	88 Ra	$L\alpha_2$	$L_{III}M_{IV}$	12.1962
1.01674	7	80 Hg	Lu	$L_{III}N_{VII}$	12.1940
1.01769	7	80 Hg	Lu'	$L_{III}N_{VI}$	12.1826
1.01937	7	80 Hg	$L\beta_7$	$L_{III}O_I$	12.1625
1.02063	7	79 Au	$L\beta_9$	L_IM_V	12.1474
1.0210	1	82 Pb	$L\beta_6$	$L_{III}N_I$	12.143
1.02175	5	77 Ir	$L\gamma_5$	$L_{II}N_I$	12.1342
1.0223	1	82 Pb	$L\beta_{17}$	$L_{II}M_{III}$	12.127
1.0226	1	94 Pu	Ll	$L_{III}M_I$	12.124
1.02467	5	74 W	L_I	Abs. Edge	12.0996
1.0250	2	74 W		$L_IO_{IV,V}$	12.095
1.02503	5	76 Os	$L\gamma_1$	$L_{II}N_{IV}$	12.0953
1.02613	7	75 Re	$L\gamma_3$	L_IN_{III}	12.0824
1.02775	3	74 W	$L\gamma_4$	L_IO_{III}	12.0634
1.02789	7	79 Au	$L\beta_{10}$	L_IM_{IV}	12.0617
1.0286	1	81 Tl		$L_{III}N_{III}$	12.053
1.02863	3	74 W	$L\gamma_4'$	L_IO_{II}	12.0530
1.03049	5	87 Fr	$L\alpha_1$	$L_{III}M_V$	12.0313
1.0317	3	74 W		L_IO_I	12.017
1.03233	5	75 Re	$L\gamma_2$	L_IN_{II}	12.0098
1.0323	2	82 Pb		L_IM_I	12.010
1.03358	7	80 Hg	$L\beta_3$	L_IM_{III}	11.9953
1.0346	9	83 Bi		L_IM_{II}	11.98
1.0347	1	92 U	Lt	$L_{III}M_{II}$	11.982
1.03699	9	75 Re	$L\gamma_6$	$L_{II}O_{IV}$	11.956
1.0371	1	75 Re	L_{II}	Abs. Edge	11.954
1.03876	7	79 Au		$L_{III}P_{II,III}$	11.9355
1.03918	3	81 Tl	$L\beta_4$	L_IM_{II}	11.9306
1.0397	1	75 Re		$L_{II}O_{III}$	11.925
1.03973	5	76 Os		$L_{II}N_{III}$	11.9243
1.03974	2	35 Br	$K\alpha_1$	KL_{III}	11.9242
1.03975	7	80 Hg	$L\beta_2$	$L_{III}N_V$	11.9241
1.04000	5	79 Au	L_{III}	Abs. Edge	11.9212
1.0404	1	75 Re	Lv	$L_{II}N_{VI}$	11.917
1.04044	3	79 Au	$L\beta_5$	$L_{III}O_{IV,V}$	11.9163
1.04151	7	80 Hg	$L\beta_{15}$	$L_{III}N_{IV}$	11.9040
1.0420	1	75 Re		L_IN_I	11.899
1.04230	5	87 Fr	$L\alpha_2$	$L_{III}M_{IV}$	11.8950
1.0428	6	93 Np	Ll	$L_{III}M_I$	11.890
1.04382	2	35 Br	$K\alpha_2$	KL_{II}	11.8776
1.04398	5	75 Re	$L\gamma_8$	$L_{II}O_I$	11.8758
1.0450	2	79 Au		$L_{III}O_{II,III}$	11.865
1.0450	1	33 As	K	Abs. Edge	11.865
1.04500	3	33 As	$K\beta_2$	$KN_{II,III}$	11.8642
1.0458	1	74 W	$L\gamma_{11}$	L_IN_V	11.856
1.0468	2	74 W		L_IN_{IV}	11.844
1.04752	5	79 Au	Lu	$L_{III}N_{VI,VII}$	11.8357
1.04868	5	80 Hg	$L\beta_1$	$L_{II}M_{IV}$	11.8226
1.0488	1	33 As	$K\beta_5$	$KM_{IV,V}$	11.822
1.04963	5	81 Tl	$L\beta_6$	$L_{III}N_I$	11.8118
1.04974	8	79 Au	$L\beta_7$	$L_{III}O_I$	11.8106
1.05446	5	78 Pt	$L\beta_9$	L_IM_V	11.7577
1.05609	7	81 Tl	$L\beta_{17}$	$L_{II}M_{III}$	11.7397
1.05693	5	76 Os	$L\gamma_5$	$L_{II}N_I$	11.7303
1.05723	5	86 Rn	$L\alpha_1$	$L_{III}M_V$	11.7270
1.05730	2	33 As	$K\beta_1$	KM_{III}	11.7262
1.05783	5	33 As	$K\beta_3$	KM_{II}	11.7203
1.0585	1	80 Hg		$L_{III}N_{III}$	11.713
1.05856	3	83 Bi	$L\eta$	$L_{II}M_I$	11.7122
1.06099	5	75 Re	$L\gamma_1$	$L_{II}N_{IV}$	11.6854
1.0613	1	73 Ta	L_I	Abs. Edge	11.682
1.06183	7	78 Pt	$L\beta_{10}$	L_IM_{IV}	11.6762
1.06192	9	73 Ta		$L_IO_{IV,V}$	11.6752
1.06200	6	74 W	$L\gamma_3$	L_IN_{III}	11.6743
1.06357	9	73 Ta		$L_IN_{VI,VII}$	11.6570
1.0644	2	82 Pb		$L_{II}M_{II}$	11.648
1.0644	2	81 Tl		L_IM_I	11.648
1.06467	3	73 Ta	$L\gamma_4$	L_IO_{III}	11.6451
1.0649	2	80 Hg		$L_{III}N_{II}$	11.642
1.06544	3	73 Ta	$L\gamma_4'$	L_IO_{II}	11.6366
1.06712	2	92 U	Ll	$L_{III}M_I$	11.6183
1.06771	9	73 Ta		L_IO_I	11.6118
1.06785	9	79 Au	$L\beta_3$	L_IM_{III}	11.6103
1.06806	3	74 W	$L\gamma_2$	L_IN_{II}	11.6080
1.06899	5	86 Rn	$L\alpha_2$	$L_{III}M_{IV}$	11.5979
1.07022	3	79 Au	$L\beta_2$	$L_{III}N_V$	11.5847
1.07188	5	79 Au	$L\beta_{15}$	$L_{III}N_{IV}$	11.5667
1.07222	7	80 Hg	$L\beta_4$	L_IM_{II}	11.5630
1.0723	1	78 Pt	L_{III}	Abs. Edge	11.562
1.0724	2	78 Pt	$L\beta_5$	$L_{III}O_{IV,V}$	11.561
1.07448	5	74 W	$L\gamma_6$	$L_{II}O_{IV}$	11.5387
1.0745	1	74 W	L_{II}	Abs. Edge	11.538
1.0756	2	79 Au		$L_{II}M_V$	11.526
1.0761	3	78 Pt		$L_{III}O_{II,III}$	11.521
1.0767	1	75 Re		$L_{II}N_{III}$	11.515
1.0771	1	74 W	Lv	$L_{II}N_{VI}$	11.510
1.07896	5	78 Pt	Lu	$L_{III}N_{VI,VII}$	11.4908
1.0792	2	74 W		$L_{II}O_{III}$	11.488
1.07975	7	80 Hg	$L\beta_6$	$L_{III}N_I$	11.4824
1.08009	9	90 Th	Lt	$L_{III}M_{II}$	11.4788
1.08113	4	74 W	$L\gamma_8$	$L_{II}O_I$	11.4677
1.08168	3	78 Pt	$L\beta_7$	$L_{III}O_I$	11.4619
1.08205	7	73 Ta	$L\gamma_{11}$	L_IN_V	11.4580
1.08353	3	79 Au	$L\beta_1$	$L_{II}M_{IV}$	11.4423
1.08377	7	73 Ta		L_IN_{IV}	11.4398
1.0839	1	75 Re		$L_{II}N_{II}$	11.438
1.08500	5	85 At	$L\alpha_1$	$L_{III}M_V$	11.4268
1.08975	5	77 Ir	$L\beta_9$	L_IM_V	11.3770
1.09026	7	79 Au		$L_{III}N_{III}$	11.3717
1.0908	1	91 Pa	Ll	$L_{III}M_I$	11.366

Table B (*Continued*)

Wavelength Å*	p.e.	Element	Designation		keV
1.0916	5	80 Hg	$L\beta_{17}$	$L_{II}M_{III}$	11.358
1.09241	7	82 Pb	$L\eta$	$L_{II}M_{I}$	11.3493
1.09388	5	75 Re	$L\gamma_5$	$L_{II}N_{I}$	11.3341
1.09671	5	85 At	$L\alpha_2$	$L_{III}M_{IV}$	11.3048
1.09702	4	77 Ir	$L\beta_{10}$	$L_{I}M_{IV}$	11.3016
1.09855	3	74 W	$L\gamma_1$	$L_{II}N_{IV}$	11.2859
1.09936	4	73 Ta	$L\gamma_3$	$L_{I}N_{III}$	11.2776
1.0997	1	81 Tl		$L_{II}M_{II}$	11.274
1.0997	1	72 Hf	L_I	Abs. Edge	11.274
1.09968	7	79 Au		$L_{III}N_{II}$	11.2743
1.0999	2	80 Hg		$L_{I}M_{I}$	11.272
1.10086	9	72 Hf		$L_{I}O_{IV}$	11.2622
1.10200	3	78 Pt	$L\beta_2$	$L_{III}N_{V}$	11.2505
1.10303	5	72 Hf	$L\gamma_4$	$L_{I}O_{III}$	11.2401
1.10376	5	72 Hf	$L\gamma_4'$	$L_{I}O_{II}$	11.2326
1.10394	5	78 Pt	$L\beta_3$	$L_{I}M_{III}$	11.2308
1.10477	2	34 Se	$K\alpha_1$	KL_{III}	11.2224
1.1053	1	73 Ta	$L\gamma_2$	$L_{I}N_{II}$	11.217
1.1058	1	77 Ir	L_{III}	Abs. Edge	11.212
1.10585	3	77 Ir	$L\beta_5$	$L_{III}O_{IV,V}$	11.2114
1.10651	3	79 Au	$L\beta_4$	$L_{I}M_{II}$	11.2047
1.10664	9	72 Hf		$L_{I}O_{I}$	11.2034
1.10882	2	34 Se	$K\alpha_2$	KL_{II}	11.1814
1.10923	6	77 Ir		$L_{III}O_{II,III}$	11.1772
1.11092	3	79 Au	$L\beta_6$	$L_{III}N_{I}$	11.1602
1.11145	4	77 Ir	Lu	$L_{III}N_{VI,VII}$	11.1549
1.1129	2	78 Pt		$L_{II}M_{V}$	11.140
1.1137	1	73 Ta	L_{II}	Abs. Edge	11.132
1.11386	4	84 Po	$L\alpha_1$	$L_{III}M_{V}$	11.1308
1.11388	3	73 Ta	$L\gamma_6$	$L_{II}O_{IV}$	11.1306
1.11489	3	77 Ir	$L\beta_7$	$L_{III}O_{I}$	11.1205
1.1149	2	74 W		$L_{II}N_{III}$	11.120
1.11508	4	90 Th	Ll	$L_{III}M_{I}$	11.1186
1.11521	9	73 Ta		$L_{I}N_{I}$	11.1173
1.1158	1	73 Ta	Lv	$L_{II}N_{VI}$	11.1113
1.11658	5	32 Ge	K	Abs. Edge	11.1036
1.11686	2	32 Ge	$K\beta_2$	$KN_{II,III}$	11.1008
1.11693	9	73 Ta		$L_{II}O_{III}$	11.1001
1.11789	9	73 Ta		$L_{II}O_{II}$	11.0907
1.1195	1	32 Ge	$K\beta_5$	$KM_{IV,V}$	11.0745
1.11990	2	78 Pt	$L\beta_1$	$L_{II}M_{IV}$	11.0707
1.1205	1	73 Ta	$L\gamma_8$	$L_{II}O_{I}$	11.0646
1.12146	9	72 Hf	$L\gamma_{11}$	$L_{I}N_{V}$	11.0553
1.1218	3	74 W		$L_{II}N_{II}$	11.052
1.12250	9	72 Hf		$L_{I}N_{IV}$	11.0451
1.1226	2	78 Pt		$L_{III}N_{III}$	11.044
1.12548	5	84 Po	$L\alpha_2$	$L_{III}M_{IV}$	11.0158
1.12637	6	76 Os	$L\beta_9$	$L_{I}M_{V}$	11.0071
1.12769	3	81 Tl	$L\eta$	$L_{II}M_{I}$	10.9943
1.12798	5	79 Au	$L\beta_{17}$	$L_{II}M_{III}$	10.9915
1.12894	2	32 Ge	$K\beta_1$	KM_{III}	[illegible]
1.1[illegible]	9	32 Ge	$K\beta_3$	KM_{II}	10.9780
1.1310	2	78 Pt		$L_{III}N_{II}$	10.962
1.13235	3	74 W	$L\gamma_5$	$L_{II}N_{I}$	10.9490
1.13353	5	76 Os	$L\beta_{10}$	$L_{I}M_{IV}$	10.9376
1.13525	5	79 Au		$L_{I}M_{I}$	10.9210
1.13532	3	77 Ir	$L\beta_2$	$L_{III}N_{V}$	10.9203
1.13687	9	73 Ta		$L_{II}N_{V}$	10.9055
1.13707	3	77 Ir	$L\beta_{15}$	$L_{III}N_{IV}$	10.9036
1.13794	3	73 Ta	$L\gamma_1$	$L_{II}N_{IV}$	10.8952
1.13841	5	72 Hf	$L\gamma_3$	$L_{I}N_{III}$	10.8907
1.1387	5	80 Hg		$L_{II}M_{II}$	10.888
1.1402	1	71 Lu	L_I	Abs. Edge	10.8740
1.1405	1	76 Os	$L\beta_5$	$L_{III}O_{IV,V}$	10.8711
1.1408	1	76 Os	L_{III}	Abs. Edge	10.8683
1.14085	3	77 Ir	$L\beta_3$	$L_{I}M_{III}$	10.8674
1.14223	5	78 Pt	$L\beta_4$	$L_{I}M_{II}$	10.8543
1.1435	1	71 Lu	$L\gamma_4$	$L_{I}O_{II,III}$	10.8425
1.14355	5	78 Pt	$L\beta_6$	$L_{III}N_{I}$	10.8418
1.14386	2	83 Bi	$L\alpha_1$	$L_{III}M_{V}$	10.8388
1.14442	5	72 Hf	$L\gamma_2$	$L_{I}N_{II}$	10.8335
1.14537	7	76 Os	Lu	$L_{III}N_{VI,VII}$	10.8245
1.1489	2	77 Ir		$L_{II}M_{V}$	10.791
1.14933	8	76 Os	$L\beta_7$	$L_{III}O_{I}$	10.7872
1.1548	1	72 Hf	L_{II}	Abs. Edge	10.7362
1.15519	5	72 Hf	$L\gamma_6$	$L_{II}O_{IV}$	10.7325
1.1553	1	73 Ta		$L_{II}N_{III}$	10.7316
1.15536	1	83 Bi	$L\alpha_2$	$L_{III}M_{IV}$	10.73091
1.1560	3	77 Ir		$L_{III}N_{III}$	10.725
1.15781	3	77 Ir	$L\beta_1$	$L_{II}M_{IV}$	10.7083
1.15830	9	72 Hf	Lv	$L_{II}N_{VI}$	10.7037
1.1600	2	73 Ta		$L_{II}N_{II}$	10.688
1.16107	9	71 Lu	$L\gamma_{11}$	$L_{I}N_{V}$	10.6782
1.16138	5	72 Hf	$L\gamma_8$	$L_{II}O_{I}$	10.6754
1.16227	9	71 Lu		$L_{I}N_{IV}$	10.6672
1.1640	1	80 Hg	$L\eta$	$L_{II}M_{I}$	10.6512
1.16487	4	75 Re	$L\beta_9$	$L_{I}M_{V}$	10.6433
1.16545	5	77 Ir		$L_{III}N_{II}$	10.6380
1.1667	1	78 Pt	$L\beta_{17}$	$L_{II}M_{III}$	10.6265
1.16719	5	88 Ra	Ll	$L_{III}M_{I}$	10.6222
1.16962	9	78 Pt		$L_{I}M_{I}$	10.6001
1.16979	8	76 Os	$L\beta_2$	$L_{III}N_{V}$	10.5985
1.1708	1	79 Au		$L_{II}M_{II}$	10.5892
1.17167	5	76 Os	$L\beta_{15}$	$L_{III}N_{IV}$	10.5816
1.17218	5	75 Re	$L\beta_{10}$	$L_{I}M_{IV}$	10.5770
1.1729	1	73 Ta	$L\gamma_5$	$L_{II}N_{I}$	10.5702
1.17501	2	82 Pb	$L\alpha_1$	$L_{III}M_{V}$	10.5515
1.17588	1	33 As	$K\alpha_1$	KL_{III}	10.54372
1.17721	5	75 Re	$L\beta_5$	$L_{III}O_{IV,V}$	10.5318
1.1773	1	75 Re	L_{III}	Abs. Edge	10.5306
1.17788	9	72 Hf		$L_{II}N_{V}$	10.5258
1.17796	3	77 Ir	$L\beta_6$	$L_{III}N_{I}$	10.5251
1.17900	5	72 Hf	$L\gamma_1$	$L_{II}N_{IV}$	10.5158
1.17953	4	71 Lu	$L\gamma_3$	$L_{I}N_{III}$	10.5110
1.17955	7	76 Os	$L\beta_3$	$L_{I}M_{III}$	10.5108
1.17958	3	77 Ir	$L\beta_4$	$L_{I}M_{II}$	10.5106
1.17987	1	33 As	$K\alpha_2$	KL_{II}	[illegible]
1.181[illegible]	1	75 Re	Lu	$L_{III}N_{VI,VII}$	10.4931
1.1818	1	70 Yb	L_I	Abs. Edge	10.4904
1.1827	1	70 Yb		$L_{I}O_{IV,V}$	10.4833
1.1853	1	70 Yb	$L\gamma_4$	$L_{I}O_{II,III}$	10.4603
1.1853	2	71 Lu	$L\gamma_2$	$L_{I}N_{II}$	10.460
1.18610	5	75 Re	$L\beta_7$	$L_{III}O_{I}$	10.4529
1.18648	5	82 Pb	$L\alpha_2$	$L_{III}M_{IV}$	10.4495

Wavelength Å*	p.e.	Element		Designation	keV
1.1886	1	70 Yb		L_IO_I	10.4312
1.18977	7	76 Os		$L_{II}M_V$	10.4205
1.1958	1	31 Ga	K	Abs. Edge	10.3682
1.19600	2	31 Ga	$K\beta_2$	$KN_{II,III}$	10.3663
1.19727	7	76 Os	$L\beta_1$	$L_{II}M_{IV}$	10.3553
1.1981	2	31 Ga	$K\beta_5$	$KM_{IV,V}$	10.348
1.1985	1	71 Lu	L_{II}	Abs. Edge	10.3448
1.1987	1	71 Lu	$L\gamma_6$	$L_{II}O_{IV}$	10.3431
1.20086	7	76 Os		$L_{III}N_{II}$	10.3244
1.2014	1	71 Lu		$L_{II}O_{II,III}$	10.3198
1.20273	3	79 Au	$L\eta$	$L_{II}M_I$	10.3083
1.2047	1	71 Lu	$L\gamma_8$	$L_{II}O_I$	10.2915
1.20479	7	74 W	$L\beta_9$	L_IM_V	10.2907
1.20660	4	75 Re	$L\beta_2$	$L_{III}N_V$	10.2752
1.2069	2	77 Ir	$L\beta_{17}$	$L_{II}M_{III}$	10.273
1.20739	4	81 Tl	$L\alpha_1$	$L_{III}M_V$	10.2685
1.20789	2	31 Ga	$K\beta_1$	KM_{III}	10.2642
1.20819	5	75 Re	$L\beta_{15}$	$L_{III}N_{IV}$	10.2617
1.20835	5	31 Ga	$K\beta_3$	KM_{II}	10.2603
1.2102	2	77 Ir		L_IM_I	10.245
1.2105	1	83 Bi	Ls	$L_{III}M_{III}$	10.2421
1.21218	3	74 W	$L\beta_{10}$	L_IM_{IV}	10.2279
1.213	1	78 Pt		$L_{II}M_{II}$	10.225
1.21349	5	76 Os	$L\beta_6$	$L_{III}N_I$	10.2169
1.21537	5	72 Hf	$L\gamma_5$	$L_{II}N_I$	10.2011
1.21545	3	74 W	$L\beta_5$	$L_{III}O_{IV,V}$	10.2004
1.2155	1	74 W	L_{III}	Abs. Edge	10.1999
1.21844	5	76 Os	$L\beta_4$	L_IM_{II}	10.1754
1.21868	5	74 W	Lu	$L_{III}N_{VI,VII}$	10.1733
1.21875	3	81 Tl	$L\alpha_2$	$L_{III}M_{IV}$	10.1728
1.22031	5	75 Re	$L\beta_3$	L_IM_{III}	10.1598
1.2211	2	74 W		$L_{III}O_{II,III}$	10.153
1.22228	4	71 Lu	$L\gamma_1$	$L_{II}N_{IV}$	10.1434
1.22232	5	70 Yb	$L\gamma_3$	L_IN_{III}	10.1431
1.22400	4	74 W	$L\beta_7$	$L_{III}O_I$	10.1292
1.2250	1	69 Tm	L_I	Abs. Edge	10.1206
1.2263	3	69 Tm		$L_IO_{IV,V}$	10.110
1.2283	1	75 Re		$L_{III}N_{III}$	10.0933
1.22879	7	70 Yb	$L\gamma_2$	L_IN_{II}	10.0897
1.2294	2	69 Tm	$L\gamma_4$	$L_IO_{II,III}$	10.084
1.2305	1	75 Re		$L_{II}M_V$	10.0753
1.23858	2	75 Re	$L\beta_1$	$L_{II}M_{IV}$	10.0100
1.24120	5	80 Hg	$L\alpha_1$	$L_{III}M_V$	9.9888
1.24271	3	70 Yb	$L\gamma_6$	$L_{II}O_{IV}$	9.9766
1.2428	1	70 Yb	L_{II}	Abs. Edge	9.9761
1.2429	2	78 Pt	$L\eta$	$L_{II}M_I$	9.975
1.24385	7	82 Pb	Ls	$L_{III}M_{III}$	9.9675
1.24460	3	74 W	$L\beta_2$	$L_{III}N_V$	9.9615
1.2453	1	70 Yb		$L_{II}O_{II,III}$	9.9561
1.24631	3	74 W	$L\beta_{15}$	$L_{III}N_{IV}$	9.9478
1.2466	2	73 Ta	$L\beta_9$	L_IM_V	9.946
1.2480	2	76 Os	$L\beta_{17}$	$L_{II}M_{III}$	9.934
1.24923	5	70 Yb	$L\gamma_8$	$L_{II}O_I$	9.9246
1.2502	3	77 Ir		$L_{II}M_{II}$	9.917
1.25100	5	75 Re	$L\beta_6$	$L_{III}N_I$	9.9105
1.25264	7	80 Hg	$L\alpha_2$	$L_{III}M_{IV}$	9.8976
1.2537	2	73 Ta	$L\beta_{10}$	L_IM_{IV}	9.889
1.254054	9	32 Ge	$K\alpha_1$	KL_{III}	9.88642
1.2553	1	73 Ta	L_{III}	Abs. Edge	9.8766
1.2555	1	73 Ta	$L\beta_5$	$L_{III}O_{IV,V}$	9.8750
1.25778	4	73 Ta	Lu	$L_{III}N_{VI,VII}$	9.8572
1.258011	9	32 Ge	$K\alpha_2$	KL_{II}	9.85532
1.25917	5	75 Re	$L\beta_4$	L_IM_{II}	9.8463
1.2596	1	71 Lu	$L\gamma_5$	$L_{II}N_I$	9.8428
1.2601	3	73 Ta		$L_{III}O_{II,III}$	9.839
1.26269	5	74 W	$L\beta_3$	L_IM_{III}	9.8188
1.26385	5	73 Ta	$L\beta_7$	$L_{III}O_I$	9.8098
1.2672	2	74 W		$L_{III}N_{III}$	9.784
1.26769	5	70 Yb	$L\gamma_1$	$L_{II}N_{IV}$	9.7801
1.2678	2	69 Tm	$L\gamma_3$	L_IN_{III}	9.779
1.2706	1	68 Er	L_I	Abs. Edge	9.7574
1.2728	2	74 W		$L_{II}M_V$	9.741
1.2742	2	69 Tm	$L\gamma_2$	L_IN_{II}	9.730
1.2748	1	83 Bi	Lt	$L_{III}M_{II}$	9.7252
1.2752	2	68 Er	$L\gamma_4$	$L_IO_{II,III}$	9.722
1.27640	3	79 Au	$L\alpha_1$	$L_{III}M_V$	9.7133
1.2765	2	74 W		$L_{III}N_{II}$	9.712
1.27807	5	81 Tl	Ls	$L_{III}M_{III}$	9.7007
1.281809	9	74 W	$L\beta_1$	$L_{II}M_{IV}$	9.67235
1.2829	5	84 Po	Ll	$L_{III}M_I$	9.664
1.2834	1	30 Zn	K	Abs. Edge	9.6607
1.28372	2	30 Zn	$K\beta_2$	$KN_{II,III}$	9.6580
1.28448	3	77 Ir	$L\eta$	$L_{II}M_I$	9.6522
1.28454	2	73 Ta	$L\beta_2$	$L_{III}N_V$	9.6518
1.2848	1	30 Zn	$K\beta_5$	$KM_{IV,V}$	9.6501
1.28619	5	73 Ta	$L\beta_{15}$	$L_{III}N_{IV}$	9.6394
1.28772	3	79 Au	$L\alpha_2$	$L_{III}M_{IV}$	9.6280
1.2892	1	69 Tm	L_{II}	Abs. Edge	9.6171
1.28989	7	74 W	$L\beta_6$	$L_{III}N_I$	9.6117
1.29025	9	72 Hf	$L\beta_9$	L_IM_V	9.6090
1.2905	2	69 Tm	$L\gamma_6$	$L_{II}O_{IV}$	9.607
1.2927	1	75 Re	$L\beta_{17}$	$L_{II}M_{III}$	9.5910
1.2934	2	76 Os		$L_{II}M_{II}$	9.586
1.29525	2	30 Zn	$K\beta_{1,3}$	$KM_{II,III}$	9.5720
1.2972	1	72 Hf	L_{III}	Abs. Edge	9.5577
1.29761	5	72 Hf	$L\beta_5$	$L_{III}O_{IV,V}$	9.5546
1.29819	9	72 Hf	$L\beta_{10}$	L_IM_{IV}	9.5503
1.30162	5	74 W	$L\beta_4$	L_IM_{II}	9.5252
1.30165	9	72 Hf	Lu	$L_{III}N_{VI,VII}$	9.5249
1.30564	5	72 Hf	$L\beta_7$	$L_{III}O_I$	9.4958
1.3063	1	70 Yb	$L\gamma_5$	$L_{II}N_I$	9.4910
1.30678	3	73 Ta	$L\beta_3$	L_IM_{III}	9.4875
1.30767	7	82 Pb	Lt	$L_{III}M_{II}$	9.4811
1.3086	1	73 Ta		$L_{III}N_{III}$	9.4742
1.3112	2	80 Hg	Ls	$L_{III}M_{III}$	9.455
1.31304	3	78 Pt	$L\alpha_1$	$L_{III}M_V$	9.4423
1.3146	1	68 Er	$L\gamma_3$	L_IN_{III}	9.4309
1.3153	2	69 Tm	$L\gamma_1$	$L_{II}N_{IV}$	9.426
1.31610	7	83 Bi	Ll	$L_{III}M_I$	9.4204
1.3167	1	73 Ta		$L_{III}N_{II}$	9.4158
1.31897	9	73 Ta		$L_{II}M_V$	9.3998
1.3190	1	67 Ho	L_I	Abs. Edge	9.3994
1.3208	3	67 Ho		$L_IO_{IV,V}$	9.387
1.3210	2	68 Er	$L\gamma_2$	L_IN_{II}	9.385

Wavelength Å*	p.e.	Element	Designation		keV
1.3225	2	67 Ho	$L\gamma_4$	$L_IO_{II,III}$	9.374
1.32432	2	78 Pt	$L\alpha_2$	$L_{III}M_{IV}$	9.3618
1.32639	5	72 Hf	$L\beta_2$	$L_{III}N_V$	9.3473
1.32698	3	73 Ta	$L\beta_1$	$L_{II}M_{IV}$	9.3431
1.32783	5	72 Hf	$L\beta_{15}$	$L_{III}N_{IV}$	9.3371
1.32785	7	76 Os	$L\eta$	$L_{II}M_I$	9.3370
1.33094	8	73 Ta	$L\beta_6$	$L_{III}N_I$	9.3153
1.3358	1	71 Lu	$L\beta_9$	L_IM_V	9.2816
1.3365	3	74 W		L_IM_I	9.277
1.3366	1	75 Re		$L_{II}M_{II}$	9.2761
1.3386	1	68 Er	L_{II}	Abs. Edge	9.2622
1.3387	2	74 W	$L\beta_{17}$	$L_{II}M_{III}$	9.261
1.3397	3	68 Er	$L\gamma_6$	$L_{II}O_{IV}$	9.255
1.340083	9	31 Ga	$K\alpha_1$	KL_{III}	9.25174
1.3405	1	71 Lu	L_{III}	Abs. Edge	9.2490
1.34154	5	81 Tl	Lt	$L_{III}M_{II}$	9.2417
1.34183	7	71 Lu	$L\beta_5$	$L_{III}O_{IV,V}$	9.2397
1.3430	2	71 Lu	$L\beta_{10}$	L_IM_{IV}	9.232
1.34399	1	31 Ga	$K\alpha_2$	KL_{II}	9.22482
1.34524	9	71 Lu		$L_{III}O_{II,III}$	9.2163
1.34581	3	73 Ta	$L\beta_4$	L_IM_{II}	9.2124
1.34949	5	71 Lu	$L\beta_7$	$L_{III}O_I$	9.1873
1.34990	7	82 Pb	Ll	$L_{III}M_I$	9.1845
1.35053	9	72 Hf		$L_{III}N_{III}$	9.1802
1.35128	3	77 Ir	$L\alpha_1$	$L_{III}M_V$	9.1751
1.35131	7	79 Au	Ls	$L_{III}M_{III}$	9.1749
1.35300	5	72 Hf	$L\beta_3$	L_IM_{III}	9.1634
1.3558	2	69 Tm	$L\gamma_5$	$L_{II}N_I$	9.144
1.35887	9	72 Hf		$L_{III}N_{II}$	9.1239
1.36250	5	77 Ir	$L\alpha_2$	$L_{III}M_{IV}$	9.0995
1.3641	2	68 Er	$L\gamma_1$	$L_{II}N_{IV}$	9.089
1.3643	2	67 Ho	$L\gamma_8$	L_IN_{III}	9.087
1.3692	1	66 Dy	L_I	Abs. Edge	9.0548
1.3698	2	67 Ho	$L\gamma_2$	L_IN_{II}	9.051
1.37012	3	71 Lu	$L\beta_2$	$L_{III}N_V$	9.0489
1.3715	1	71 Lu	$L\beta_{15}$	$L_{III}N_{IV}$	9.0395
1.37342	5	75 Re	$L\eta$	$L_{II}M_I$	9.0272
1.37410	5	72 Hf	$L\beta_1$	$L_{II}M_{IV}$	9.0227
1.37410	5	72 Hf	$L\beta_6$	$L_{III}N_I$	9.0227
1.37459	7	66 Dy	$L\gamma_4$	$L_IO_{II,III}$	9.0195
1.3746	2	80 Hg	Lt	$L_{III}M_{II}$	9.019
1.38059	5	29 Cu	K	Abs. Edge	8.9803
1.38109	3	29 Cu	$K\beta_2$	$KM_{IV,V}$	8.9770
1.3838	1	70 Yb	$L\beta_9$	L_IM_V	8.9597
1.38477	3	81 Tl	Ll	$L_{III}M_I$	8.9532
1.3862	1	70 Yb	L_{III}	Abs. Edge	8.9441
1.3864	1	73 Ta	$L\beta_{17}$	$L_{II}M_{III}$	8.9428
1.38696	7	70 Yb	$L\beta_5$	$L_{III}O_{IV,V}$	8.9390
1.3895	2	78 Pt	Ls	$L_{III}M_{III}$	8.923
1.3898	1	70 Yb		$L_{III}O_{II,III}$	8.9209
1.3905	1	67 Ho	L_{II}	Abs. Edge	8.9104
1.39121	5	76 Os	$L\alpha_1$	$L_{III}M_V$	8.9117
1.3915	1	70 Yb	$L\beta_{10}$	L_IM_{IV}	8.9100
1.39220	5	72 Hf	$L\beta_4$	L_IM_{II}	8.9054
1.392218	9	29 Cu	$K\beta_{1,3}$	$KM_{II,III}$	8.90529
1.3923	2	67 Ho	$L\gamma_6$	$L_{II}O_{IV}$	8.905
1.3926	1	29 Cu	$K\beta_3$	KM_{II}	8.9029
1.3948	1	70 Yb	$L\beta_7$	$L_{III}O_I$	8.8889
1.3983	2	67 Ho	$L\gamma_8$	$L_{II}O_I$	8.867
1.40140	5	71 Lu	$L\beta_3$	L_IM_{III}	8.8469
1.40234	5	76 Os	$L\alpha_2$	$L_{III}M_{IV}$	8.8410
1.4067	3	68 Er	$L\gamma_5$	$L_{II}N_I$	8.814
1.41366	7	79 Au	Lt	$L_{III}M_{II}$	8.7702
1.41550	5	70 Yb	$L\beta_{2,15}$	$L_{III}N_{IV,V}$	8.7588
1.41640	7	66 Dy	$L\gamma_3$	L_IN_{III}	8.7532
1.4174	2	67 Ho	$L\gamma_1$	$L_{II}N_{IV}$	8.747
1.4189	1	71 Lu	$L\beta_6$	$L_{III}N_I$	8.7376
1.42110	3	74 W	$L\eta$	$L_{II}M_I$	8.7243
1.4216	1	80 Hg	Ll	$L_{III}M_I$	8.7210
1.4223	1	65 Tb	L_I	Abs. Edge	8.7167
1.42278	7	66 Dy	$L\gamma_2$	L_IN_{II}	8.7140
1.4228	3	65 Tb		$L_IO_{IV,V}$	8.714
1.42359	3	71 Lu	$L\beta_1$	$L_{II}M_{IV}$	8.7090
1.4276	2	65 Tb	$L\gamma_4$	$L_IO_{II,III}$	8.685
1.43025	9	72 Hf		L_IM_I	8.6685
1.43048	9	73 Ta		$L_{II}M_{II}$	8.6671
1.4318	2	77 Ir	Ls	$L_{III}M_{III}$	8.659
1.43290	4	75 Re	$L\alpha_1$	$L_{III}M_V$	8.6525
1.4334	1	69 Tm	L_{III}	Abs. Edge	8.6496
1.4336	3	69 Tm	$L\beta_9$	L_IM_V	8.648
1.4349	2	69 Tm	$L\beta_5$	$L_{III}O_{IV,V}$	8.641
1.435155	7	30 Zn	$K\alpha_1$	KL_{III}	8.63886
1.43643	9	72 Hf	$L\beta_{17}$	$L_{II}M_{III}$	8.6312
1.439000	8	30 Zn	$K\alpha_2$	KL_{II}	8.61578
1.44056	5	71 Lu	$L\beta_4$	L_IM_{II}	8.6064
1.4410	3	69 Tm	$L\beta_{10}$	L_IM_{IV}	8.604
1.44396	5	75 Re	$L\alpha_2$	$L_{III}M_{IV}$	8.5862
1.4445	1	66 Dy	L_{II}	Abs. Edge	8.5830
1.44579	7	66 Dy	$L\gamma_6$	$L_{II}O_{IV}$	8.5753
1.45233	5	70 Yb	$L\beta_3$	L_IM_{III}	8.5367
1.4530	2	78 Pt	Lt	$L_{III}M_{II}$	8.533
1.45964	9	79 Au	Ll	$L_{III}M_I$	8.4939
1.4618	2	67 Ho	$L\gamma_5$	$L_{II}N_I$	8.481
1.4640	2	69 Tm	$L\beta_{2,15}$	$L_{III}N_{IV,V}$	8.468
1.4661	1	70 Yb	$L\beta_6$	$L_{III}N_I$	8.4563
1.47106	5	73 Ta	$L\eta$	$L_{II}M_I$	8.4280
1.4718	2	65 Tb	$L\gamma_3$	L_IN_{III}	8.423
1.47266	7	66 Dy	$L\gamma_1$	$L_{II}N_{IV}$	8.4188
1.4735	2	76 Os	Ls	$L_{III}M_{III}$	8.414
1.47565	5	70 Yb	$L\beta_1$	$L_{II}M_{IV}$	8.4018
1.4764	2	65 Tb	$L\gamma_2$	L_IN_{II}	8.398
1.47639	2	74 W	$L\alpha_1$	$L_{III}M_V$	8.3976
1.4784	1	64 Gd	L_I	Abs. Edge	8.3864
1.48064	9	72 Hf		$L_{II}M_{II}$	8.3735
1.4807	3	64 Gd		$L_IO_{IV,V}$	8.373
1.4835	1	68 Er	L_{III}	Abs. Edge	8.3575
1.4839	2	64 Gd	$L\gamma_4$	$L_IO_{II,III}$	8.355
1.4848	3	68 Er	$L\beta_5$	$L_{III}O_{IV,V}$	8.350
1.4855	5	68 Er	$L\beta_9$	L_IM_V	8.346
1.48743	2	74 W	$L\alpha_2$	$L_{III}M_{IV}$	8.3352
1.48807	1	28 Ni	K	Abs. Edge	8.33165
1.48862	4	28 Ni	$K\beta_5$	$KM_{IV,V}$	8.3286
1.49138	3	70 Yb	$L\beta_4$	L_IM_{II}	8.3132
1.4930	3	77 Ir	Lt	$L_{III}M_{II}$	8.304

Wavelength Å*	p.e.	Element	Designation		keV
1.4941	3	68 Er	$L\beta_7$	$L_{III}O_I$	8.298
1.4941	3	68 Er	$L\beta_{10}$	L_IM_{IV}	8.298
1.4995	2	78 Pt	Ll	$L_{III}M_I$	8.268
1.500135	8	28 Ni	$K\beta_{1,3}$	$KM_{II,III}$	8.26466
1.5023	1	65 Tb	L_{II}	Abs. Edge	8.2527
1.5035	2	65 Tb	$L\gamma_6$	$L_{II}O_{IV}$	8.246
1.5063	2	69 Tm	$L\beta_3$	L_IM_{III}	8.231
1.5097	2	65 Tb	$L\gamma_8$	$L_{II}O_I$	8.212
1.51399	9	68 Er	$L\beta_{2,15}$	$L_{III}N_{IV,V}$	8.1890
1.5162	2	69 Tm	$L\beta_6$	$L_{III}N_I$	8.177
1.5178	1	75 Re	Ls	$L_{III}M_{III}$	8.1682
1.51824	7	66 Dy	$L\gamma_5$	$L_{II}N_I$	8.1661
1.52197	2	73 Ta	$L\alpha_1$	$L_{III}M_V$	8.1461
1.52325	5	72 Hf	$L\eta$	$L_{II}M_I$	8.1393
1.5297	2	64 Gd	$L\gamma_3$	L_IN_{III}	8.105
1.5303	2	65 Tb	$L\gamma_1$	$L_{II}N_{IV}$	8.102
1.5304	2	69 Tm	$L\beta_1$	$L_{II}M_{IV}$	8.101
1.53293	2	73 Ta	$L\alpha_2$	$L_{III}M_{IV}$	8.0879
1.5331	2	64 Gd	$L\gamma_2$	L_IN_{II}	8.087
1.53333	9	71 Lu		$L_{II}M_{II}$	8.0858
1.5347	2	76 Os	Lt	$L_{III}M_{II}$	8.079
1.5368	1	67 Ho	L_{III}	Abs. Edge	8.0676
1.5378	2	67 Ho	$L\beta_5$	$L_{III}O_{IV,V}$	8.062
1.5381	1	63 Eu	L_I	Abs. Edge	8.0607
1.540562	2	29 Cu	$K\alpha_1$	KL_{III}	8.04778
1.54094	3	77 Ir	Ll	$L_{III}M_I$	8.0458
1.5439	1	63 Eu	$L\gamma_4$	$L_IO_{II,III}$	8.0304
1.544390	9	29 Cu	$K\alpha_2$	KL_{II}	8.02783
1.5448	2	69 Tm	$L\beta_4$	L_IM_{II}	8.026
1.5486	3	67 Ho	$L\beta_{10}$	L_IM_{IV}	8.006
1.5616	1	68 Er	$L\beta_3$	L_IM_{III}	7.9392
1.5632	1	64 Gd	L_{II}	Abs. Edge	7.9310
1.5642	3	74 W	Ls	$L_{III}M_{III}$	7.926
1.5644	2	64 Gd	$L\gamma_6$	$L_{II}O_{IV}$	7.925
1.5671	2	67 Ho	$L\beta_{2,15}$	$L_{III}N_{IV,V}$	7.911
1.5675	2	68 Er	$L\beta_6$	$L_{III}N_I$	7.909
1.56958	5	72 Hf	$L\alpha_1$	$L_{III}M_V$	7.8990
1.5707	2	64 Gd	$L\gamma_8$	$L_{II}O_I$	7.894
1.5779	1	71 Lu	$L\eta$	$L_{II}M_I$	7.8575
1.5787	2	65 Tb	$L\gamma_5$	$L_{II}N_I$	7.8535
1.5789	1	75 Re	Lt	$L_{III}M_{II}$	7.8525
1.58046	5	72 Hf	$L\alpha_2$	$L_{III}M_{IV}$	7.8446
1.58498	7	76 Os	Ll	$L_{III}M_I$	7.8222
1.5873	1	68 Er	$L\beta_1$	$L_{II}M_{IV}$	7.8109
1.58837	7	66 Dy	$L\beta_5$	$L_{III}O_{IV,V}$	7.8055
1.58844	9	70 Yb		$L_{II}M_{II}$	7.8052
1.5903	2	63 Eu	$L\gamma_3$	L_IN_{III}	7.7961
1.5916	1	66 Dy	L_{III}	Abs. Edge	7.7897
1.5924	2	64 Gd	$L\gamma_1$	$L_{II}N_{IV}$	7.7858
1.5961	2	63 Eu	$L\gamma_2$	L_IN_{II}	7.7677
1.59973	9	66 Dy	$L\beta_9$	L_IM_V	7.7501
1.6002	1	62 Sm	L_I	Abs. Edge	7.7478
1.6007	1	68 Er	$L\beta_4$	L_IM_{II}	7.7453
1.60447	7	66 Dy	$L\beta_7$	$L_{III}O_I$	7.7272
1.60728	3	62 Sm	$L\gamma_4$	$L_IO_{II,III}$	7.714
1.60743	9	66 Dy	$L\beta_{10}$	L_IM_{IV}	7.7130
1.60815	1	27 Co	K	Abs. Edge	7.70954

Wavelength Å*	p.e.	Element	Designation		keV
1.60891	3	27 Co	$K\beta_5$	$KM_{IV,V}$	7.7059
1.61264	9	73 Ta	Ls	$L_{III}M_{III}$	7.6881
1.61951	3	71 Lu	$L\alpha_1$	$L_{III}M_V$	7.6555
1.6203	2	67 Ho	$L\beta_3$	L_IM_{III}	7.6519
1.62079	2	27 Co	$K\beta_{1,3}$	$KM_{II,III}$	7.64943
1.6237	2	67 Ho	$L\beta_6$	$L_{III}N_I$	7.6359
1.62369	7	66 Dy	$L\beta_{2,15}$	$L_{III}N_{IV,V}$	7.6357
1.6244	3	74 W	Lt	$L_{III}M_{II}$	7.6324
1.6271	1	63 Eu	L_{II}	Abs. Edge	7.6199
1.6282	2	63 Eu	$L\gamma_6$	$L_{II}O_{IV}$	7.6147
1.63029	5	71 Lu	$L\alpha_2$	$L_{III}M_{IV}$	7.6049
1.63056	5	75 Re	Ll	$L_{III}M_I$	7.6036
1.6346	2	63 Eu	$L\gamma_8$	$L_{II}O_I$	7.5849
1.63560	5	70 Yb	$L\eta$	$L_{II}M_I$	7.5802
1.6412	2	64 Gd	$L\gamma_5$	$L_{II}N_I$	7.5543
1.6475	2	67 Ho	$L\beta_1$	$L_{II}M_{IV}$	7.5253
1.6497	1	65 Tb	L_{III}	Abs. Edge	7.5153
1.6510	2	65 Tb	$L\beta_5$	$L_{III}O_{IV,V}$	7.5094
1.65601	3	62 Sm	$L\gamma_3$	L_IN_{III}	7.487
1.6574	2	63 Eu	$L\gamma_1$	$L_{II}N_{IV}$	7.4803
1.657910	8	28 Ni	$K\alpha_1$	KL_{III}	7.47815
1.6585	2	65 Tb	$L\beta_7$	$L_{III}O_I$	7.4753
1.6595	2	67 Ho	$L\beta_4$	L_IM_{II}	7.4708
1.66044	6	62 Sm	$L\gamma_2$	L_IN_{II}	7.467
1.661747	8	28 Ni	$K\alpha_2$	KL_{II}	7.46089
1.66346	9	72 Hf	Ls	$L_{III}M_{III}$	7.4532
1.6673	3	65 Tb	$L\beta_{10}$	L_IM_{IV}	7.436
1.6674	5	61 Pm	L_I	Abs. Edge	7.436
1.67189	4	70 Yb	$L\alpha_1$	$L_{III}M_V$	7.4156
1.67265	9	73 Ta	Lt	$L_{III}M_{II}$	7.4123
1.6782	1	74 W	Ll	$L_{III}M_I$	7.3878
1.68213	7	66 Dy	$L\beta_6$	$L_{III}N_I$	7.3705
1.6822	2	66 Dy	$L\beta_3$	L_IM_{III}	7.3702
1.68285	5	70 Yb	$L\alpha_2$	$L_{III}M_{IV}$	7.3673
1.6830	2	65 Tb	$L\beta_{2,15}$	$L_{III}N_{IV,V}$	7.3667
1.6953	1	62 Sm	L_{II}	Abs. Edge	7.3132
1.6963	2	69 Tm	$L\eta$	$L_{II}M_I$	7.3088
1.6966	9	62 Sm	$L\gamma_6$	$L_{II}O_{IV}$	7.308
1.7085	2	63 Eu	$L\gamma_5$	$L_{II}N_I$	7.2566
1.71062	7	66 Dy	$L\beta_1$	$L_{II}M_{IV}$	7.2477
1.7117	1	64 Gd	L_{III}	Abs. Edge	7.2430
1.7130	2	64 Gd	$L\beta_5$	$L_{III}O_{IV,V}$	7.2374
1.7203	2	64 Gd	$L\beta_7$	$L_{III}O_I$	7.2071
1.72103	7	66 Dy	$L\beta_4$	L_IM_{II}	7.2039
1.72305	9	72 Hf	Lt	$L_{III}M_{II}$	7.1954
1.7240	3	64 Gd	$L\beta_9$	L_IM_V	7.192
1.72724	3	62 Sm	$L\gamma_1$	$L_{II}N_{IV}$	7.178
1.7268	2	69 Tm	$L\alpha_1$	$L_{III}M_V$	7.1799
1.72841	5	73 Ta	Ll	$L_{III}M_I$	7.1731
1.7315	3	64 Gd	$L\beta_{10}$	L_IM_{IV}	7.160
1.7381	2	69 Tm	$L\alpha_2$	$L_{III}M_{IV}$	7.1331
1.7390	1	60 Nd	L_I	Abs. Edge	7.1294
1.7422	2	65 Tb	$L\beta_6$	$L_{III}N_I$	7.1163
1.74346	1	26 Fe	K	Abs. Edge	7.11120
1.7442	1	26 Fe	$K\beta_5$	$KM_{IV,V}$	7.1081
1.7445	4	60 Nd	$L\gamma_4$	$L_IO_{II,III}$	7.107
1.7455	2	64 Gd	$L\beta_{2,15}$	$L_{III}N_{IV,V}$	7.1028

Wavelength Å*	p.e.	Element	Designation		keV
1.7472	2	65 Tb	$L\beta_3$	L_IM_{III}	7.0959
1.75661	2	26 Fe	$K\beta_{1,3}$	$KM_{II,III}$	7.05798
1.7566	1	68 Er	$L\eta$	$L_{II}M_I$	7.0579
1.7676	5	61 Pm	L_{II}	Abs. Edge	7.014
1.7760	1	71 Lu	Lt	$L_{III}M_{II}$	6.9810
1.7761	1	63 Eu	L_{III}	Abs. Edge	6.9806
1.7768	3	65 Tb	$L\beta_1$	$L_{II}M_{IV}$	6.978
1.7772	2	63 Eu	$L\beta_5$	$L_{III}O_{IV,V}$	6.9763
1.77934	3	62 Sm	$L\gamma_5$	$L_{II}N_I$	6.968
1.78145	5	72 Hf	Ll	$L_{III}M_I$	6.9596
1.78425	9	68 Er	$L\alpha_1$	$L_{III}M_V$	6.9487
1.7851	2	63 Eu	$L\beta_7$	$L_{III}O_I$	6.9453
1.7864	2	65 Tb	$L\beta_4$	L_IM_{II}	6.9403
1.788965	9	27 Co	$K\alpha_1$	KL_{III}	6.93032
1.7916	3	63 Eu	$L\beta_9$	L_IM_V	6.920
1.792850	9	27 Co	$K\alpha_2$	KL_{II}	6.91530
1.7955	2	68 Er	$L\alpha_2$	$L_{III}M_{IV}$	6.9050
1.7964	4	60 Nd	$L\gamma_3$	L_IN_{III}	6.902
1.7989	9	61 Pm	$L\gamma_1$	$L_{II}N_{IV}$	6.892
1.7993	3	63 Eu	$L\beta_{10}$	L_IM_{IV}	6.890
1.8013	4	60 Nd	$L\gamma_2$	L_IN_{II}	6.883
1.8054	2	64 Gd	$L\beta_6$	$L_{III}N_I$	6.8671
1.8118	2	63 Eu	$L\beta_{2,15}$	$L_{III}N_{IV,V}$	6.8432
1.8141	5	59 Pr	L_I	Abs. Edge	6.834
1.8150	2	64 Gd	$L\beta_3$	L_IM_{III}	6.8311
1.8193	4	59 Pr	$L\gamma_4$	$L_IO_{II,III}$	6.815
1.8264	2	67 Ho	$L\eta$	$L_{II}M_I$	6.7883
1.83091	9	70 Yb	Lt	$L_{III}M_{II}$	6.7715
1.8360	1	71 Lu	Ll	$L_{III}M_I$	6.7528
1.8440	1	60 Nd	L_{II}	Abs. Edge	6.7234
1.8450	2	67 Ho	$L\alpha_1$	$L_{III}M_V$	6.7198
1.8457	1	62 Sm	L_{III}	Abs. Edge	6.7172
1.8468	2	64 Gd	$L\beta_1$	$L_{II}M_{IV}$	6.7132
1.84700	9	62 Sm	$L\beta_5$	$L_{III}O_{IV,V}$	6.7126
1.8540	2	64 Gd	$L\beta_4$	L_IM_{II}	6.6871
1.8552	5	60 Nd	$L\gamma_8$	$L_{II}O_I$	6.683
1.8561	2	67 Ho	$L\alpha_2$	$L_{III}M_{IV}$	6.6795
1.85626	3	62 Sm	$L\beta_7$	$L_{III}O_I$	6.679
1.86166	3	62 Sm	$L\beta_9$	L_IM_V	6.660
1.86990	3	62 Sm	$L\beta_{10}$	L_IM_{IV}	6.634
1.8737	2	63 Eu	$L\beta_6$	$L_{III}N_I$	6.6170
1.8740	4	59 Pr	$L\gamma_3$	L_IN_{III}	6.616
1.8779	2	60 Nd	$L\gamma_1$	$L_{II}N_{IV}$	6.6021
1.8791	4	59 Pr	$L\gamma_2$	L_IN_{II}	6.598
1.8821	3	62 Sm	$L\beta_{2,15}$	$L_{III}N_{IV,V}$	6.586
1.8867	2	63 Eu	$L\beta_3$	L_IM_{III}	6.5713
1.8934	5	58 Ce	L_I	Abs. Edge	6.548
1.89415	5	70 Yb	Ll	$L_{III}M_I$	6.5455
1.89643	5	25 Mn	K	Abs. Edge	6.5376
1.8971	1	25 Mn	$K\beta_5$	$KM_{IV,V}$	6.5352
1.89743	7	66 Dy	$L\eta$	$L_{II}M_I$	6.5342
1.8991	4	58 Ce	$L\gamma_4$	$L_IO_{II,III}$	6.528
1.90881	3	66 Dy	$L\alpha_1$	$L_{III}M_V$	6.4952
1.91021	2	25 Mn	$K\beta_{1,3}$	$KM_{II,III}$	6.49045
1.9191	1	61 Pm	L_{III}	Abs. Edge	6.4605
1.91991	3	66 Dy	$L\alpha_2$	$L_{III}M_{IV}$	6.4577
1.9203	2	63 Eu	$L\beta_1$	$L_{II}M_{IV}$	6.4564
1.9255	2	63 Eu	$L\beta_4$	L_IM_{II}	6.4389
1.9255	5	59 Pr	L_{II}	Abs. Edge	6.439
1.9355	4	60 Nd	$L\gamma_5$	$L_{II}N_I$	6.406
1.936042	9	26 Fe	$K\alpha_1$	KL_{III}	6.40384
1.9362	4	59 Pr	$L\gamma_8$	$L_{II}O_I$	6.403
1.939980	9	26 Fe	$K\alpha_2$	KL_{II}	6.39084
1.94643	3	62 Sm	$L\beta_6$	$L_{III}N_I$	6.3693
1.9550	2	69 Tm	Ll	$L_{III}M_I$	6.3419
1.9553	3	58 Ce	$L\gamma_3$	L_IN_{III}	6.3409
1.9559	6	61 Pm	$L\beta_{2,15}$	$L_{III}N_{IV,V}$	6.339
1.9602	3	58 Ce	$L\gamma_2$	L_IN_{II}	6.3250
1.9611	3	59 Pr	$L\gamma_1$	$L_{II}N_{IV}$	6.3221
1.96241	3	62 Sm	$L\beta_3$	L_IM_{III}	6.318
1.9730	2	65 Tb	$L\eta$	$L_{II}M_I$	6.2839
1.9765	2	65 Tb	$L\alpha_1$	$L_{III}M_V$	6.2728
1.9780	5	57 La	L_I	Abs. Edge	6.268
1.9830	4	57 La	$L\gamma_4$	$L_IO_{II,III}$	6.252
1.9875	2	65 Tb	$L\alpha_2$	$L_{III}M_{IV}$	6.2380
1.9967	1	60 Nd	L_{III}	Abs. Edge	6.2092
1.99806	3	62 Sm	$L\beta_1$	$L_{II}M_{IV}$	6.2051
2.00095	6	62 Sm	$L\beta_4$	L_IM_{II}	6.196
2.0092	3	60 Nd	$L\beta_7$	$L_{III}O_I$	6.1708
2.0124	5	58 Ce	L_{II}	Abs. Edge	6.161
2.015	1	68 Er	Ll	$L_{III}M_I$	6.152
2.0165	3	60 Nd	$L\beta_9$	L_IM_V	6.1484
2.0205	4	59 Pr	$L\gamma_5$	$L_{II}N_I$	6.136
2.0237	4	58 Ce	$L\gamma_8$	$L_{II}O_I$	6.126
2.0237	3	60 Nd	$L\beta_{10}$	L_IM_{IV}	6.1265
2.0360	3	60 Nd	$L\beta_{2,15}$	$L_{III}N_{IV,V}$	6.0894
2.0410	4	57 La	$L\gamma_3$	L_IN_{III}	6.074
2.0421	4	61 Pm	$L\beta_3$	L_IM_{III}	6.071
2.0460	4	57 La	$L\gamma_2$	L_IN_{II}	6.060
2.0468	2	64 Gd	$L\alpha_1$	$L_{III}M_V$	6.0572
2.0487	4	58 Ce	$L\gamma_1$	$L_{II}N_{IV}$	6.052
2.0494	1	64 Gd	$L\eta$	$L_{II}M_I$	6.0495
2.0578	2	64 Gd	$L\alpha_2$	$L_{III}M_{IV}$	6.0250
2.0678	5	56 Ba	L_I	Abs. Edge	5.996
2.07020	5	24 Cr	K	Abs. Edge	5.9888
2.07087	6	24 Cr	$K\beta_5$	$KM_{IV,V}$	5.9869
2.0756	3	56 Ba	$L\gamma_4$	$L_IO_{II,III}$	5.9733
2.0791	5	59 Pr	L_{III}	Abs. Edge	5.963
2.0797	4	61 Pm	$L\beta_1$	$L_{II}M_{IV}$	5.961
2.08487	2	24 Cr	$K\beta_{1,3}$	$KM_{II,III}$	5.94671
2.0860	2	67 Ho	Ll	$L_{III}M_I$	5.9434
2.0919	4	59 Pr	$L\beta_7$	$L_{III}O_I$	5.927
2.1004	4	59 Pr	$L\beta_9$	L_IM_V	5.903
2.101820	9	25 Mn	$K\alpha_1$	KL_{III}	5.89875
2.1039	3	60 Nd	$L\beta_6$	$L_{III}N_I$	5.8930
2.1053	5	57 La	L_{II}	Abs. Edge	5.889
2.10578	2	25 Mn	$K\alpha_2$	KL_{II}	5.88765
2.1071	4	59 Pr	$L\beta_{10}$	L_IM_{IV}	5.884
2.1103	3	58 Ce	$L\gamma_5$	$L_{II}N_I$	5.8751
2.1194	4	59 Pr	$L\beta_{2,15}$	$L_{III}N_{IV,V}$	5.850
2.1209	2	63 Eu	$L\alpha_1$	$L_{III}M_V$	5.8457
2.1268	2	60 Nd	$L\beta_3$	L_IM_{III}	5.8294
2.1315	2	63 Eu	$L\eta$	$L_{II}M_I$	5.8166
2.1315	2	63 Eu	$L\alpha_2$	$L_{III}M_{IV}$	5.8166

Wavelength Å*	p.e.	Element	Designation		keV
2.1342	2	56 Ba	$L\gamma_3$	$L_I N_{III}$	5.8092
2.1387	2	56 Ba	$L\gamma_2$	$L_I N_{II}$	5.7969
2.1418	3	57 La	$L\gamma_1$	$L_{II} N_{IV}$	5.7885
2.15877	7	66 Dy	Ll	$L_{III} M_I$	5.7431
2.166	1	58 Ce	L_{III}	Abs. Edge	5.723
2.1669	3	60 Nd	$L\beta_4$	$L_I M_{II}$	5.7216
2.1669	2	60 Nd	$L\beta_1$	$L_{II} M_{IV}$	5.7216
2.1673	5	55 Cs	L_I	Abs. Edge	5.721
2.1701	2	58 Ce	$L\beta_7$	$L_{III} O_I$	5.7132
2.1741	2	55 Cs	$L\gamma_4$	$L_I O_{II,III}$	5.7026
2.1885	3	58 Ce	$L\beta_9$	$L_I M_V$	5.6650
2.1906	4	59 Pr	$L\beta_6$	$L_{III} N_I$	5.660
2.1958	5	58 Ce	$L\beta_{10}$	$L_I M_{IV}$	5.646
2.1998	2	62 Sm	$L\alpha_1$	$L_{III} M_V$	5.6361
2.2048	1	56 Ba	L_{II}	Abs. Edge	5.6233
2.2056	4	57 La	$L\gamma_5$	$L_{II} N_I$	5.621
2.2087	2	58 Ce	$L\beta_{2,15}$	$L_{III} N_{IV,V}$	5.6134
2.21062	3	62 Sm	$L\alpha_2$	$L_{III} M_{IV}$	5.6090
2.2172	3	59 Pr	$L\beta_3$	$L_I M_{III}$	5.5918
2.21824	3	62 Sm	$L\eta$	$L_{II} M_I$	5.589
2.2328	2	55 Cs	$L\gamma_3$	$L_I N_{III}$	5.5527
2.2352	2	65 Tb	Ll	$L_{III} M_I$	5.5467
2.2371	2	55 Cs	$L\gamma_2$	$L_I N_{II}$	5.5420
2.2415	2	56 Ba	$L\gamma_1$	$L_{II} N_{IV}$	5.5311
2.253	6	92 U		$M_I P_{III}$	5.50
2.2550	4	59 Pr	$L\beta_4$	$L_I M_{II}$	5.4981
2.2588	3	59 Pr	$L\beta_1$	$L_{II} M_{IV}$	5.4889
2.261	1	57 La	L_{III}	Abs. Edge	5.484
2.2691	1	23 V	K	Abs. Edge	5.4639
2.26951	6	23 V	$K\beta_5$	$K M_{IV,V}$	5.4629
2.2737	1	54 Xe	L_I	Abs. Edge	5.4528
2.275	3	57 La	$L\beta_7$	$L_{III} O_I$	5.450
2.282	3	57 La	$L\beta_9$	$L_I M_V$	5.434
2.2818	3	58 Ce	$L\beta_6$	$L_{III} N_I$	5.4334
2.2822	3	61 Pm	$L\alpha_1$	$L_{III} M_V$	5.4325
2.28440	2	23 V	$K\beta_{1,3}$	$K M_{II,III}$	5.42729
2.28970	2	24 Cr	$K\alpha_1$	$K L_{III}$	5.41472
2.290	3	57 La	$L\beta_{10}$	$L_I M_{IV}$	5.415
2.2926	4	61 Pm	$L\alpha_2$	$L_{III} M_{IV}$	5.4078
2.293606	3	24 Cr	$K\alpha_2$	$K L_{II}$	5.405509
2.3030	3	57 La	$L\beta_{2,15}$	$L_{III} N_{IV,V}$	5.3835
2.304	7	92 U		$M_I O_{III}$	5.38
2.3085	3	56 Ba	$L\gamma_5$	$L_{II} N_I$	5.3707
2.3109	3	58 Ce	$L\beta_3$	$L_I M_{III}$	5.3651
2.3122	2	64 Gd	Ll	$L_{III} M_I$	5.3621
2.3139	1	55 Cs	L_{II}	Abs. Edge	5.3581
2.3480	2	55 Cs	$L\gamma_1$	$L_{II} N_{IV}$	5.2804
2.3497	4	58 Ce	$L\beta_4$	$L_I M_{II}$	5.2765
2.3561	3	58 Ce	$L\beta_1$	$L_{II} M_{IV}$	5.2622
2.3629	1	56 Ba	L_{III}	Abs. Edge	5.2470
2.3704	2	60 Nd	$L\alpha_1$	$L_{III} M_V$	5.2304
2.3764	2	56 Ba	$L\beta_9$	$L_I M_V$	5.2171
2.3790	4	57 La	$L\beta_6$	$L_{III} N_I$	5.2114
2.3806	2	56 Ba	$L\beta_7$	$L_{III} O_I$	5.2079
2.3807	3	60 Nd	$L\alpha_2$	$L_{III} M_{IV}$	5.2077
2.3869	2	56 Ba	$L\beta_{10}$	$L_I M_{IV}$	5.1941
2.3880	5	53 I	L_I	Abs. Edge	5.192
2.3913	2	53 I	$L\gamma_4$	$L_I O_{II,III}$	5.1848
2.3948	2	63 Eu	Ll	$L_{III} M_I$	5.1772
2.40435	6	56 Ba	$L\beta_{2,15}$	$L_{III} N_{IV,V}$	5.1565
2.4094	4	60 Nd	$L\eta$	$L_{II} M_I$	5.1457
2.4105	3	57 La	$L\beta_3$	$L_I M_{III}$	5.1434
2.4174	2	55 Cs	$L\gamma_5$	$L_{II} N_I$	5.1287
2.4292	1	54 Xe	L_{II}	Abs. Edge	5.1037
2.442	9	90 Th		$M_I O_{III}$	5.08
2.443	4	92 U		$M_{II} O_{IV}$	5.075
2.4475	2	53 I	$L\gamma_{2,3}$	$L_I N_{II,III}$	5.0657
2.4493	3	57 La	$L\beta_4$	$L_I M_{II}$	5.0620
2.45891	5	57 La	$L\beta_1$	$L_{II} M_{IV}$	5.0421
2.4630	2	59 Pr	$L\alpha_1$	$L_{III} M_V$	5.0337
2.4729	3	59 Pr	$L\alpha_2$	$L_{III} M_{IV}$	5.0135
2.4740	1	55 Cs	L_{III}	Abs. Edge	5.0113
2.4783	2	55 Cs	$L\beta_9$	$L_I M_V$	5.0026
2.4823	4	62 Sm	Ll	$L_{III} M_I$	4.9945
2.4826	2	56 Ba	$L\beta_6$	$L_{III} N_I$	4.9939
2.4849	2	55 Cs	$L\beta_7$	$L_{III} O_I$	4.9893
2.4920	2	55 Cs	$L\beta_{10}$	$L_I M_{IV}$	4.9752
2.49734	5	22 Ti	K	Abs. Edge	4.96452
2.4985	2	22 Ti	$K\beta_5$	$K M_{IV,V}$	4.9623
2.50356	2	23 V	$K\alpha_1$	$K L_{III}$	4.95220
2.50738	2	23 V	$K\alpha_2$	$K L_{II}$	4.94464
2.5099	1	52 Te	L_I	Abs. Edge	4.9397
2.5113	2	52 Te	$L\gamma_4$	$L_I O_{II,III}$	4.9369
2.5118	2	55 Cs	$L\beta_{2,15}$	$L_{III} N_{IV,V}$	4.9359
2.512	3	59 Pr	$L\eta$	$L_{II} M_I$	4.935
2.51391	2	22 Ti	$K\beta_{1,3}$	$K M_{II,III}$	4.93181
2.5164	2	56 Ba	$L\beta_3$	$L_I M_{III}$	4.9269
2.527	4	91 Pa		$M_{II} O_{IV}$	4.906
2.5542	5	53 I	L_{II}	Abs. Edge	4.8540
2.5553	2	56 Ba	$L\beta_4$	$L_I M_{II}$	4.8519
2.5615	2	58 Ce	$L\alpha_1$	$L_{III} M_V$	4.8402
2.5674	2	52 Te	$L\gamma_{2,3}$	$L_I N_{II,III}$	4.8290
2.56821	5	56 Ba	$L\beta_1$	$L_{II} M_{IV}$	4.82753
2.5706	3	58 Ce	$L\alpha_2$	$L_{III} M_{IV}$	4.8230
2.58244	8	53 I	$L\gamma_1$	$L_{II} N_{IV}$	4.8009
2.5926	1	54 Xe	L_{III}	Abs. Edge	4.7822
2.5932	2	55 Cs	$L\beta_6$	$L_{III} N_I$	4.7811
2.618	5	90 Th		$M_{II} O_{IV}$	4.735
2.6203	4	58 Ce	$L\eta$	$L_{II} M_I$	4.7315
2.6285	2	55 Cs	$L\beta_3$	$L_I M_{III}$	4.7167
2.6388	1	51 Sb	L_I	Abs. Edge	4.6984
2.6398	2	51 Sb	$L\gamma_4$	$L_I O_{II,III}$	4.6967
2.65710	9	53 I	$L\gamma_5$	$L_{II} N_I$	4.6660
2.66570	5	57 La	$L\alpha_1$	$L_{III} M_V$	4.65097
2.6666	2	55 Cs	$L\beta_4$	$L_I M_{II}$	4.6494
2.67533	5	57 La	$L\alpha_2$	$L_{III} M_{IV}$	4.63423
2.6760	4	60 Nd	Ll	$L_{III} M_I$	4.6330
2.6837	2	55 Cs	$L\beta_1$	$L_{II} M_{IV}$	4.6198
2.6879	1	52 Te	L_{II}	Abs. Edge	4.6126
2.6953	2	51 Sb	$L\gamma_{2,3}$	$L_I N_{II,III}$	4.5999
2.71241	6	52 Te	$L\gamma_1$	$L_{II} N_{IV}$	4.5709
2.71352	9	53 I	$L\beta_9$	$L_I M_V$	4.5690
2.7196	5	53 I	L_{III}	Abs. Edge	4.5587
2.72104	9	53 I	$L\beta_{10}$	$L_I M_{IV}$	4.5564

Wavelength Å*	p.e.	Element	Designation		keV
2.7288	3	53 I	$L\beta_7$	$L_{III}O_I$	4.5435
2.740	3	57 La	$L\eta$	$L_{II}M_I$	4.525
2.74851	2	22 Ti	$K\alpha_1$	KL_{III}	4.51084
2.75053	8	53 I	$L\beta_{2,15}$	$L_{III}N_{IV,V}$	4.5075
2.75216	2	22 Ti	$K\alpha_2$	KL_{II}	4.50486
2.753	8	92 U		M_IN_{III}	4.50
2.762	1	21 Sc	K	Abs. Edge	4.489
2.7634	3	21 Sc	$K\beta_5$	$KM_{IV,V}$	4.4865
2.77595	5	56 Ba	$L\alpha_1$	$L_{III}M_V$	4.46626
2.7769	1	50 Sn	L_I	Abs. Edge	4.4648
2.7775	2	50 Sn	$L\gamma_4$	$L_IO_{II,III}$	4.4638
2.7796	2	21 Sc	$K\beta_{1,3}$	$KM_{II,III}$	4.4605
2.7841	4	59 Pr	Ll	$L_{III}M_I$	4.4532
2.78553	5	56 Ba	$L\alpha_2$	$L_{III}M_{IV}$	4.45090
2.79007	9	52 Te	$L\gamma_5$	$L_{II}N_I$	4.4437
2.817	2	92 U		$M_{II}N_{IV}$	4.401
2.8294	5	51 Sb	L_{II}	Abs. Edge	4.3819
2.8327	2	50 Sn	$L\gamma_{2,3}$	$L_IN_{II,III}$	4.3768
2.83672	9	53 I	$L\beta_6$	$L_{III}N_I$	4.3706
2.83897	9	52 Te	$L\beta_9$	L_IM_V	4.3671
2.84679	9	52 Te	$L\beta_{10}$	L_IM_{IV}	4.3551
2.85159	3	51 Sb	$L\gamma_1$	$L_{II}N_{IV}$	4.34779
2.8555	1	52 Te	L_{III}	Abs. Edge	4.3418
2.8627	3	56 Ba	$L\eta$	$L_{II}M_I$	4.3309
2.8634	3	52 Te	$L\beta_7$	$L_{III}O_I$	4.3298
2.87429	9	53 I	$L\beta_3$	L_IM_{III}	4.3134
2.88217	8	52 Te	$L\beta_{2,15}$	$L_{III}N_{IV,V}$	4.3017
[illegible].004	5	92 U	M_{III}	Abs. Edge	4.299
2.8917	4	58 Ce	Ll	$L_{III}M_I$	4.2875
2.8924	2	55 Cs	$L\alpha_1$	$L_{III}M_V$	4.2865
2.9020	2	55 Cs	$L\alpha_2$	$L_{III}M_{IV}$	4.2722
2.910	2	91 Pa		$M_{II}N_{IV}$	4.260
2.91207	9	53 I	$L\beta_4$	L_IM_{II}	4.2575
2.92	2	92 U		M_IN_{II}	4.25
2.9260	1	49 In	L_I	Abs. Edge	4.2373
2.9264	2	49 In	$L\gamma_4$	$L_IO_{II,III}$	4.2367
2.93187	9	51 Sb	$L\gamma_5$	$L_{II}N_I$	4.2287
2.934	8	90 Th		M_IN_{III}	4.23
2.93744	6	53 I	$L\beta_1$	$L_{II}M_{IV}$	4.22072
2.948	2	92 U		$M_{III}O_{IV,V}$	4.205
2.97088	9	52 Te	$L\beta_6$	$L_{III}N_I$	4.1732
2.97261	9	51 Sb	$L\beta_9$	L_IM_V	4.1708
2.97917	9	51 Sb	$L\beta_{10}$	L_IM_{IV}	4.1616
2.9800	2	49 In	$L\gamma_{2,3}$	$L_IN_{II,III}$	4.1605
2.9823	1	50 Sn	L_{II}	Abs. Edge	4.1573
2.9932	2	55 Cs	$L\eta$	$L_{II}M_I$	4.1421
3.0003	1	51 Sb	L_{III}	Abs. Edge	4.1323
3.00115	3	50 Sn	$L\gamma_1$	$L_{II}N_{IV}$	4.13112
3.0052	3	51 Sb	$L\beta_7$	$L_{III}O_I$	4.1255
3.006	3	57 La	Ll	$L_{III}M_I$	4.124
3.00893	[illegible]	[illegible]	[illegible]	L_IM_{III}	4.1204
3.011	2	90 Th		$M_{II}N_{IV}$	4.117
3.0166	2	54 Xe	$L\alpha_1$	$L_{III}M_V$	4.1099
3.02335	3	51 Sb	$L\beta_{2,15}$	$L_{III}N_{IV,V}$	4.10078
3.0309	1	21 Sc	$K\alpha_1$	KL_{III}	4.0906
3.0342	1	21 Sc	$K\alpha_2$	KL_{II}	4.0861
3.038	2	91 Pa		$M_{III}O_{IV,V}$	4.081
3.04661	9	52 Te	$L\beta_4$	L_IM_{II}	4.0695
3.068	5	90 Th	M_{III}	Abs. Edge	4.041
3.0703	1	20 Ca	K	Abs. Edge	4.0381
3.0746	3	20 Ca	$K\beta_5$	$KM_{IV,V}$	4.0325
3.07677	6	52 Te	$L\beta_1$	$L_{II}M_{IV}$	4.02958
3.08475	9	50 Sn	$L\gamma_5$	$L_{II}N_I$	4.0192
3.0849	1	48 Cd	L_I	Abs. Edge	4.0190
3.0897	2	20 Ca	$K\beta_{1,3}$	$KM_{II,III}$	4.0127
3.094	5	83 Bi	M_I	Abs. Edge	4.007
3.11513	9	50 Sn	$L\beta_9$	L_IM_V	3.9800
3.11513	9	51 Sb	$L\beta_6$	$L_{III}N_I$	3.9800
3.115	7	92 U		$M_{III}O_I$	3.980
3.12170	9	50 Sn	$L\beta_{10}$	L_IM_{IV}	3.9716
3.131	3	90 Th		$M_{III}O_{IV,V}$	3.959
3.1355	2	56 Ba	Ll	$L_{III}M_I$	3.9541
3.1377	2	48 Cd	$L\gamma_2$	L_IN_{II}	3.9513
3.1473	1	49 In	L_{II}	Abs. Edge	3.9393
3.14860	6	53 I	$L\alpha_1$	$L_{III}M_V$	3.93765
3.15258	9	51 Sb	$L\beta_3$	L_IM_{III}	3.9327
3.1557	1	50 Sn	L_{III}	Abs. Edge	3.9288
3.1564	3	50 Sn	$L\beta_7$	$L_{III}O_I$	3.9279
3.15791	6	53 I	$L\alpha_2$	$L_{III}M_{IV}$	3.92604
3.16213	4	49 In	$L\gamma_1$	$L_{II}N_{IV}$	3.92081
3.17505	3	50 Sn	$L\beta_{2,15}$	$L_{III}N_{IV,V}$	3.90486
3.19014	9	51 Sb	$L\beta_4$	L_IM_{II}	3.8364
3.217	5	82 Pb	M_I	Abs. Edge	3.854
3.22567	4	51 Sb	$L\beta_1$	$L_{II}M_{IV}$	3.84357
3.245	9	91 Pa		$M_{III}O_I$	3.82
3.24907	9	49 In	$L\gamma_5$	$L_{II}N_I$	3.8159
3.2564	1	47 Ag	L_I	Abs. Edge	3.8072
3.2670	2	55 Cs	Ll	$L_{III}M_I$	3.7950
3.26763	9	49 In	$L\beta_9$	L_IM_V	3.7942
3.26901	9	50 Sn	$L\beta_6$	$L_{III}N_I$	3.7926
3.27404	9	49 In	$L\beta_{10}$	L_IM_{IV}	3.7868
3.27979	9	53 I	$L\eta$	$L_{II}M_I$	3.7801
3.283	9	90 Th		$M_{III}O_I$	3.78
3.28920	6	52 Te	$L\alpha_1$	$L_{III}M_V$	3.76933
3.29846	9	52 Te	$L\alpha_2$	$L_{III}M_{IV}$	3.7588
3.30585	3	50 Sn	$L\beta_3$	L_IM_{III}	3.7500
3.30635	9	47 Ag	$L\gamma_3$	L_IN_{III}	3.7498
3.31216	9	47 Ag	$L\gamma_2$	L_IN_{II}	3.7432
3.3237	1	49 In	L_{III}	Abs. Edge	3.7302
3.324	4	49 In	$L\beta_7$	$L_{III}O_I$	3.730
3.3257	1	48 Cd	L_{II}	Abs. Edge	3.7280
3.329	4	92 U		$M_{II}N_I$	3.724
3.333	5	92 U	M_{IV}	Abs. Edge	3.720
3.33564	6	48 Cd	$L\gamma_1$	$L_{II}N_{IV}$	3.71686
3.33838	3	49 In	$L\beta_{2,15}$	$L_{III}N_{IV,V}$	3.71381
3.34335	9	50 Sn	$L\beta_4$	L_IM_{II}	3.7083
3.346	[illegible]	[illegible]	M_I	Abs. Edge	3.705
3.35839	3	20 Ca	$K\alpha_1$	KL_{III}	3.69168
3.359	5	83 Bi	M_{II}	Abs. Edge	3.691
3.36166	3	20 Ca	$K\alpha_2$	KL_{II}	3.68809
3.38487	3	50 Sn	$L\beta_1$	$L_{II}M_{IV}$	3.66280
3.42551	9	48 Cd	$L\gamma_5$	$L_{II}N_I$	3.61935
3.43015	9	48 Cd	$L\beta_9$	L_IM_V	3.61445
3.43606	9	49 In	$L\beta_6$	$L_{III}N_I$	3.60823

Wavelength Å*	p.e.	Element		Designation	keV
3.4365	1	19 K	K	Abs. Edge	3.6078
3.4367	2	48 Cd	$L\beta_{10}$	$L_{I}M_{IV}$	3.6075
3.437	1	46 Pd	L_{I}	Abs. Edge	3.607
3.43832	9	52 Te	$L\eta$	$L_{II}M_{I}$	3.60586
3.43941	4	51 Sb	$L\alpha_1$	$L_{III}M_{V}$	3.60472
3.441	5	91 Pa		$M_{II}N_{I}$	3.603
3.4413	4	19 K	$K\beta_5$	$KM_{IV,V}$	3.6027
3.44840	6	51 Sb	$L\alpha_2$	$L_{III}M_{IV}$	3.59532
3.4539	2	19 K	$K\beta_{1,3}$	$KM_{II,III}$	3.5896
3.46984	9	49 In	$L\beta_3$	$L_{I}M_{III}$	3.57311
3.478	5	80 Hg	M_{I}	Abs. Edge	3.565
3.479	1	92 U	$M\gamma$	$M_{III}N_{V}$	3.563
3.4892	2	46 Pd	$L\gamma_{2,3}$	$L_{I}N_{II,III}$	3.5533
3.492	5	82 Pb	M_{II}	Abs. Edge	3.550
3.497	5	92 U	M_{V}	Abs. Edge	3.545
3.5047	1	48 Cd	L_{III}	Abs. Edge	3.5376
3.50697	9	49 In	$L\beta_4$	$L_{I}M_{II}$	3.53528
3.51408	4	48 Cd	$L\beta_{2,15}$	$L_{III}N_{IV,V}$	3.52812
3.5164	1	47 Ag	L_{II}	Abs. Edge	3.5258
3.521	2	92 U		$M_{III}N_{IV}$	3.521
3.52260	4	47 Ag	$L\gamma_1$	$L_{II}N_{IV}$	3.51959
3.537	9	90 Th		$M_{II}N_{I}$	3.505
3.55531	4	49 In	$L\beta_1$	$L_{II}M_{IV}$	3.48721
3.557	5	90 Th	M_{IV}	Abs. Edge	3.485
3.55754	9	53 I	Ll	$L_{III}M_{I}$	3.48502
3.576	1	92 U		$M_{IV}O_{II}$	3.4666
3.577	1	91 Pa	$M\gamma$	$M_{III}N_{V}$	3.4657
3.59994	3	50 Sn	$L\alpha_1$	$L_{III}M_{V}$	3.44398
3.60497	9	47 Ag	$L\beta_9$	$L_{I}M_{V}$	3.43917
3.60765	9	51 Sb	$L\eta$	$L_{II}M_{I}$	3.43661
3.60891	4	50 Sn	$L\alpha_2$	$L_{III}M_{IV}$	3.43542
3.61158	9	47 Ag	$L\beta_{10}$	$L_{I}M_{IV}$	3.43287
3.614	2	91 Pa		$M_{III}N_{IV}$	3.430
3.61467	9	48 Cd	$L\beta_6$	$L_{III}N_{I}$	3.42994
3.61638	9	47 Ag	$L\gamma_5$	$L_{II}N_{I}$	3.42832
3.616	5	79 Au	M_{I}	Abs. Edge	3.428
3.629	5	45 Rh	L_{I}	Abs. Edge	3.417
3.634	5	81 Tl	M_{II}	Abs. Edge	3.412
3.64495	9	48 Cd	$L\beta_3$	$L_{I}M_{III}$	3.40145
3.679	2	90 Th	$M\gamma$	$M_{III}N_{V}$	3.370
3.68203	9	48 Cd	$L\beta_4$	$L_{I}M_{II}$	3.36719
3.6855	2	45 Rh	$L\gamma_{2,3}$	$L_{I}N_{II,III}$	3.3640
3.691	2	91 Pa		$M_{IV}O_{II}$	3.359
3.6999	1	47 Ag	L_{III}	Abs. Edge	3.35096
3.70335	3	47 Ag	$L\beta_{2,15}$	$L_{III}N_{IV,V}$	3.34781
3.716	1	92 U	$M\beta$	$M_{IV}N_{VI}$	3.3367
3.71696	9	52 Te	Ll	$L_{III}M_{I}$	3.33555
3.718	3	90 Th		$M_{III}N_{IV}$	3.335
3.7228	1	46 Pd	L_{II}	Abs. Edge	3.33031
3.7246	2	46 Pd	$L\gamma_1$	$L_{II}N_{IV}$	3.3287
3.729	5	90 Th	M_{V}	Abs. Edge	3.325
3.73823	4	48 Cd	$L\beta_1$	$L_{II}M_{IV}$	3.31657
3.740	9	83 Bi		$M_{I}N_{III}$	3.315
3.7414	2	19 K	$K\alpha_1$	KL_{III}	3.3138
3.7445	2	19 K	$K\alpha_2$	KL_{II}	3.3111
3.760	9	90 Th		$M_{V}P_{III}$	3.298
3.762	5	78 Pt	M_{I}	Abs. Edge	3.296
3.77192	4	49 In	$L\alpha_1$	$L_{III}M_{V}$	3.28694
3.78073	6	49 In	$L\alpha_2$	$L_{III}M_{IV}$	3.27929
3.783	5	80 Hg	M_{II}	Abs. Edge	3.277
3.78876	9	50 Sn	$L\eta$	$L_{II}M_{I}$	3.27234
3.7920	2	46 Pd	$L\beta_9$	$L_{I}M_{V}$	3.2696
3.7988	2	46 Pd	$L\beta_{10}$	$L_{I}M_{IV}$	3.2637
3.80774	9	47 Ag	$L\beta_6$	$L_{III}N_{I}$	3.25603
3.808	4	90 Th		$M_{IV}O_{II}$	3.256
3.8222	2	46 Pd	$L\gamma_5$	$L_{II}N_{I}$	3.2437
3.827	1	91 Pa	$M\beta$	$M_{IV}N_{VI}$	3.2397
3.83313	9	47 Ag	$L\beta_3$	$L_{I}M_{III}$	3.23446
3.834	4	83 Bi		$M_{II}N_{IV}$	3.234
3.835	5	44 Ru	L_{I}	Abs. Edge	3.233
3.87023	5	47 Ag	$L\beta_4$	$L_{I}M_{II}$	3.20346
3.87090	5	18 A	K	Abs. Edge	3.20290
3.872	9	82 Pb		$M_{I}N_{III}$	3.202
3.8860	2	18 A	$K\beta_{1,3}$	$KM_{II,III}$	3.1905
3.88826	9	51 Sb	Ll	$L_{III}M_{I}$	3.18860
3.892	9	83 Bi		$M_{I}N_{II}$	3.185
3.8977	2	44 Ru	$L\gamma_{2,3}$	$L_{I}N_{II,III}$	3.1809
3.904	5	83 Bi	M_{III}	Abs. Edge	3.176
3.9074	1	46 Pd	L_{III}	Abs. Edge	3.17298
3.90887	4	46 Pd	$L\beta_{2,15}$	$L_{III}N_{IV,V}$	3.17179
3.910	1	92 U	$M\alpha_1$	$M_{V}N_{VII}$	3.1708
3.915	5	77 Ir	M_{I}	Abs. Edge	3.167
3.924	1	92 U	$M\alpha_2$	$M_{V}N_{VI}$	3.1595
3.932	6	83 Bi		$M_{III}O_{IV,V}$	3.153
3.93473	3	47 Ag	$L\beta_1$	$L_{II}M_{IV}$	3.15094
3.936	5	79 Au	M_{II}	Abs. Edge	3.150
3.941	1	90 Th	$M\beta$	$M_{IV}N_{VI}$	3.1458
3.9425	5	45 Rh	L_{II}	Abs. Edge	3.1448
3.9437	2	45 Rh	$L\gamma_1$	$L_{II}N_{IV}$	3.1438
3.95635	4	48 Cd	$L\alpha_1$	$L_{III}M_{V}$	3.13373
3.96496	6	48 Cd	$L\alpha_2$	$L_{III}M_{IV}$	3.12691
3.968	5	82 Pb		$M_{II}N_{IV}$	3.124
3.98327	9	49 In	$L\eta$	$L_{II}M_{I}$	3.11254
4.013	9	81 Tl		$M_{I}N_{III}$	3.089
4.0162	2	46 Pd	$L\beta_6$	$L_{III}N_{I}$	3.0870
4.022	1	91 Pa	$M\alpha_1$	$M_{V}N_{VII}$	3.0823
4.0346	2	46 Pd	$L\beta_3$	$L_{I}M_{III}$	3.0730
4.035	3	91 Pa	$M\alpha_2$	$M_{V}N_{VI}$	3.072
4.0451	2	45 Rh	$L\gamma_5$	$L_{II}N_{I}$	3.0650
4.047	1	82 Pb	M_{III}	Abs. Edge	3.0632
4.058	5	43 Te	L_{I}	Abs. Edge	3.055
4.069	6	82 Pb		$M_{III}O_{IV,V}$	3.047
4.0711	2	46 Pd	$L\beta_4$	$L_{I}M_{II}$	3.0454
4.071	5	76 Os	M_{I}	Abs. Edge	3.045
4.07165	9	50 Sn	Ll	$L_{III}M_{I}$	3.04499
4.093	5	78 Pt	M_{II}	Abs. Edge	3.029
4.105	9	83 Bi		$M_{III}O_{I}$	3.021
4.116	4	81 Tl		$M_{II}N_{IV}$	3.013
4.1299	5	45 Rh	L_{III}	Abs. Edge	3.0021
4.1310	2	45 Rh	$L\beta_{2,15}$	$L_{III}N_{IV,V}$	3.0013
4.1381	9	90 Th	$M\alpha_1$	$M_{V}N_{VII}$	2.9961
4.14622	5	46 Pd	$L\beta_1$	$L_{II}M_{IV}$	2.99022
4.151	2	90 Th	$M\alpha_2$	$M_{V}N_{VI}$	2.987
4.15443	3	47 Ag	$L\alpha_1$	$L_{III}M_{V}$	2.98431

Wavelength Å*	p.e.	Element	Designation		keV
4.16294	5	47 Ag	$L\alpha_2$	$L_{III}M_{IV}$	2.97821
4.180	1	44 Ru	L_{II}	Abs. Edge	2.9663
4.1822	2	44 Ru	$L\gamma_1$	$L_{II}N_{IV}$	2.9645
4.19180	5	18 A	$K\alpha_1$	KL_{III}	2.95770
4.19315	9	48 Cd	$L\eta$	$L_{II}M_I$	2.95675
4.19474	5	18 A	$K\alpha_2$	KL_{II}	2.95563
4.198	1	81 Tl	M_{III}	Abs. Edge	2.9535
4.216	6	81 Tl		$M_{III}O_{IV,V}$	2.941
4.236	5	75 Re	M_I	Abs. Edge	2.927
4.2417	2	45 Rh	$L\beta_6$	$L_{III}N_I$	2.9229
4.244	9	82 Pb		$M_{III}O_I$	2.921
4.2522	2	45 Rh	$L\beta_3$	L_IM_{III}	2.9157
4.260	5	77 Ir	M_{II}	Abs. Edge	2.910
4.26873	9	49 In	Ll	$L_{III}M_I$	2.90440
4.2873	2	44 Ru	$L\gamma_5$	$L_{II}N_I$	2.8918
4.2888	2	45 Rh	$L\beta_4$	L_IM_{II}	2.8908
4.300	9	79 Au		M_IN_{III}	2.883
4.304	5	42 Mo	L_I	Abs. Edge	2.881
4.330	2	92 U		$M_{III}N_I$	2.863
4.355	1	80 Hg	M_{III}	Abs. Edge	2.8469
4.36767	5	46 Pd	$L\alpha_1$	$L_{III}M_V$	2.83861
4.369	1	44 Ru	L_{III}	Abs. Edge	2.8377
4.3718	2	44 Ru	$L\beta_{2,15}$	$L_{III}N_{IV,V}$	2.8360
4.37414	4	45 Rh	$L\beta_1$	$L_{II}M_{IV}$	2.83441
4.37588	7	46 Pd	$L\alpha_2$	$L_{III}M_{IV}$	2.83329
4.3800	2	42 Mo	$L\gamma_{2,3}$	$L_IN_{II,III}$	2.8306
4.3971	1	17 Cl	K	Abs. Edge	2.81960
4.4034	3	17 Cl	$K\beta$	KM	2.8156
4.407	5	74 W	M_I	Abs. Edge	2.813
4.4183	2	47 Ag	$L\eta$	$L_{II}M_I$	2.8061
4.432	4	79 Au		$M_{II}N_{IV}$	2.797
4.433	5	76 Os	M_{II}	Abs. Edge	2.797
4.436	1	43 Te	L_{II}	Abs. Edge	2.7948
4.44	2	74 W		$M_IO_{II,III}$	2.79
4.450	4	91 Pa		$M_{III}N_I$	2.786
4.460	9	78 Pt		M_IN_{III}	2.780
4.48014	9	48 Cd	Ll	$L_{III}M_I$	2.76735
4.4866	3	44 Ru	$L\beta_3$	L_IM_{III}	2.7634
4.4866	3	44 Ru	$L\beta_6$	$L_{III}N_I$	2.7634
4.518	1	79 Au	M_{III}	Abs. Edge	2.7439
4.522	6	79 Au		$M_{III}O_{IV,V}$	2.742
4.5230	2	44 Ru	$L\beta_4$	L_IM_{II}	2.7411
4.532	2	83 Bi	$M\gamma$	$M_{III}N_V$	2.735
4.568	5	90 Th		$M_{III}N_I$	2.714
4.571	5	83 Bi		$M_{III}N_{IV}$	2.712
4.572	5	83 Bi	M_{IV}	Abs. Edge	2.711
4.575	5	41 Nb	L_I	Abs. Edge	2.710
4.585	5	73 Ta	M_I	Abs. Edge	2.704
4.59	2	83 Bi		$M_{IV}P_{II,III}$	2.70
4.59743	9	45 Rh	$L\alpha_1$	$L_{III}M_V$	2.69674
4.601	4	78 Pt		$M_{II}N_{IV}$	2.695
4.60545	9	45 Rh	$L\alpha_2$	$L_{III}M_{IV}$	2.69205
4.620	5	75 Re	M_{II}	Abs. Edge	2.684
4.62058	3	44 Ru	$L\beta_1$	$L_{II}M_{IV}$	2.68323
4.625	5	92 U		$M_{IV}N_{III}$	2.681
4.630	1	43 Tc	L_{III}	Abs. Edge	2.6780
4.631	9	77 Ir		M_IN_{III}	2.677
4.6542	2	41 Nb	$L\gamma_{2,3}$	$L_IN_{II,III}$	2.6638
4.655	8	82 Pb		$M_{II}N_I$	2.664
4.6605	2	46 Pd	$L\eta$	$L_{II}M_I$	2.6603
4.674	1	82 Pb	$M\gamma$	$M_{III}N_V$	2.6527
4.686	1	78 Pt	M_{III}	Abs. Edge	2.6459
4.694	8	78 Pt		$M_{III}O_{IV,V}$	2.641
4.703	9	79 Au		$M_{III}O_I$	2.636
4.7076	2	47 Ag	Ll	$L_{III}M_I$	2.6337
4.715	3	82 Pb		$M_{III}N_{IV}$	2.630
4.719	1	42 Mo	L_{II}	Abs. Edge	2.6274
4.7258	2	42 Mo	$L\gamma_1$	$L_{II}N_{IV}$	2.6235
4.7278	1	17 Cl	$K\alpha_1$	KL_{III}	2.62239
4.7307	1	17 Cl	$K\alpha_2$	KL_{II}	2.62078
4.757	5	82 Pb	M_{IV}	Abs. Edge	2.606
4.764	5	83 Bi	M_V	Abs. Edge	2.603
4.780	4	77 Ir		$M_{II}N_{IV}$	2.594
4.79	2	76 Os		M_IN_{III}	2.59
4.815	5	74 W	M_{II}	Abs. Edge	2.575
4.823	3	83 Bi		$M_{IV}O_{II}$	2.571
4.823	4	81 Tl	$M\gamma$	$M_{III}N_V$	2.571
4.8369	2	42 Mo	$L\gamma_5$	$L_{II}N_I$	2.5632
4.84575	5	44 Ru	$L\alpha_1$	$L_{III}M_V$	2.55855
4.85381	7	44 Ru	$L\alpha_2$	$L_{III}M_{IV}$	2.55431
4.861	1	77 Ir	M_{III}	Abs. Edge	2.5505
4.865	5	81 Tl		$M_{III}N_{IV}$	2.548
4.869	9	77 Ir		$M_{III}O_{IV,V}$	2.546
4.876	9	78 Pt		$M_{III}O_I$	2.543
4.879	5	40 Zr	L_I	Abs. Edge	2.541
4.8873	8	43 Tc	$L\beta_1$	$L_{II}M_{IV}$	2.5368
4.909	1	83 Bi	$M\beta$	$M_{IV}N_{VI}$	2.5255
4.911	5	90 Th		$M_{IV}N_{III}$	2.524
4.913	1	42 Mo	L_{III}	Abs. Edge	2.5234
4.9217	2	45 Rh	$L\eta$	$L_{II}M_I$	2.5191
4.9232	2	42 Mo	$L\beta_{2,15}$	$L_{III}N_{IV,V}$	2.5183
4.946	2	92 U	$M\zeta_1$	M_VN_{III}	2.507
4.952	5	81 Tl	M_{IV}	Abs. Edge	2.504
4.9525	3	46 Pd	Ll	$L_{III}M_I$	2.5034
4.9536	3	40 Zr	$L\gamma_{2,3}$	$L_IN_{II,III}$	2.5029
4.955	4	76 Os		$M_{II}N_{IV}$	2.502
4.955	5	82 Pb	M_V	Abs. Edge	2.502
4.984	2	80 Hg	$M\gamma$	$M_{III}N_V$	2.4875
5.004	9	82 Pb		$M_{IV}O_{II}$	2.477
5.0133	3	42 Mo	$L\beta_3$	L_IM_{III}	2.4730
5.0185	1	16 S	K	Abs. Edge	2.47048
5.020	5	73 Ta	M_{II}	Abs. Edge	2.470
5.0233	3	16 S	$K\beta_x$	KM	2.4681
5.031	1	41 Nb	L_{II}	Abs. Edge	2.4641
5.0316	2	16 S	$K\beta_1$	KM	2.46404
5.0361	3	41 Nb	$L\gamma_1$	$L_{II}N_{IV}$	2.4618
5.043	5	76 Os	M_{III}	Abs. Edge	2.458
5.0488	3	42 Mo	$L\beta_4$	L_IM_{II}	2.4557
5.0488	5	42 Mo	$L\beta_6$	$L_{III}N_I$	2.4557
5.050	2	92 U	$M\zeta_2$	$M_{IV}N_{II}$	2.4548
5.076	1	82 Pb	$M\beta$	$M_{IV}N_{VI}$	2.4427
5.092	2	91 Pa	$M\zeta_1$	M_VN_{III}	2.4350
5.1148	3	43 Tc	$L\alpha_1$	$L_{III}M_V$	2.4240
5.118	1	83 Bi	$M\alpha_1$	M_VN_{VII}	2.4226

Wavelength Å*	p.e.	Element	Designation		keV
5.130	2	83 Bi	$M\alpha_2$	M_VN_{VI}	2.4170
5.145	4	79 Au	$M\gamma$	$M_{III}N_V$	2.410
5.1517	3	41 Nb	$L\gamma_5$	$L_{II}N_I$	2.4066
5.153	5	81 Tl	M_V	Abs. Edge	2.406
5.157	5	80 Hg	M_{IV}	Abs. Edge	2.404
5.168	9	82 Pb		M_VO_{III}	2.399
5.172	9	74 W		M_IN_{III}	2.397
5.17708	8	42 Mo	$L\beta_1$	$L_{II}M_{IV}$	2.39481
5.186	5	79 Au		$M_{III}N_{IV}$	2.391
5.193	2	91 Pa	$M\zeta_2$	$M_{IV}N_{II}$	2.3876
5.196	9	81 Tl		$M_{IV}O_{II}$	2.386
5.2050	2	44 Ru	$L\eta$	$L_{II}M_I$	2.38197
5.217	5	39 Y	L_I	Abs. Edge	2.377
5.2169	3	45 Rh	Ll	$L_{III}M_I$	2.3765
5.230	1	41 Nb	L_{III}	Abs. Edge	2.3706
5.234	5	75 Re	M_{III}	Abs. Edge	2.369
5.2379	3	41 Nb	$L\beta_{2,15}$	$L_{III}N_{IV,V}$	2.3670
5.245	5	90 Th	$M\zeta_1$	M_VN_{III}	2.364
5.249	1	81 Tl	$M\beta$	$M_{IV}N_{VI}$	2.3621
5.2830	3	39 Y	$L\gamma_{2,3}$	$L_IN_{II,III}$	2.3468
5.286	1	82 Pb	$M\alpha_1$	M_VN_{VII}	2.3455
5.299	2	82 Pb	$M\alpha_2$	M_VN_{VI}	2.3397
5.3102	3	41 Nb	$L\beta_3$	L_IM_{III}	2.3348
5.319	4	78 Pt	$M\gamma$	$M_{III}N_V$	2.331
5.340	5	90 Th	$M\zeta_2$	$M_{IV}N_{II}$	2.322
5.3455	3	41 Nb	$L\beta_4$	L_IM_{II}	2.3194
5.357	4	74 W		$M_{II}N_{IV}$	2.314
5.357	5	78 Pt		$M_{III}N_{IV}$	2.314
5.36	1	80 Hg	M_V	Abs. Edge	2.313
5.3613	3	41 Nb	$L\beta_6$	$L_{III}N_I$	2.3125
5.37216	7	16 S	$K\alpha_1$	KL_{III}	2.30784
5.374	5	79 Au	M_{IV}	Abs. Edge	2.307
5.37496	8	16 S	$K\alpha_2$	KL_{II}	2.30664
5.378	1	40 Zr	L_{II}	Abs. Edge	2.3053
5.3843	3	40 Zr	$L\gamma_1$	$L_{II}N_{IV}$	2.3027
5.40	2	73 Ta		M_IN_{III}	2.295
5.40655	8	42 Mo	$L\alpha_1$	$L_{III}M_V$	2.29316
5.41437	8	42 Mo	$L\alpha_2$	$L_{III}M_{IV}$	2.28985
5.4318	9	80 Hg	$M\beta$	$M_{IV}N_{VI}$	2.2825
5.435	1	74 W	M_{III}	Abs. Edge	2.2811
5.460	1	81 Tl	$M\alpha_1$	M_VN_{VII}	2.2706
5.472	2	81 Tl	$M\alpha_2$	M_VN_{VI}	2.2656
5.4923	3	41 Nb	$L\beta_1$	$L_{II}M_{IV}$	2.2574
5.4977	3	40 Zr	$L\gamma_5$	$L_{II}N_I$	2.2551
5.500	4	77 Ir	$M\gamma$	$M_{III}N_V$	2.254
5.5035	3	44 Ru	Ll	$L_{III}M_I$	2.2528
5.537	8	83 Bi		$M_{III}N_I$	2.239
5.540	5	77 Ir		$M_{III}N_{IV}$	2.238
5.570	4	73 Ta		$M_{II}N_{IV}$	2.226
5.579	1	40 Zr	L_{III}	Abs. Edge	2.2225
5.584	5	79 Au	M_V	Abs. Edge	2.220
5.5863	3	40 Zr	$L\beta_{2,15}$	$L_{III}N_{IV,V}$	2.2194
5.59	1	78 Pt	M_{IV}	Abs. Edge	2.217
5.592	5	38 Sr	L_I	Abs. Edge	2.217
5.624	1	79 Au	$M\beta$	$M_{IV}N_{VI}$	2.2046
5.628	8	74 W		$M_{III}O_I$	2.203
5.6330	3	40 Zr	$L\beta_3$	L_IM_{III}	2.2010
5.6445	3	38 Sr	$L\gamma_{2,3}$	$L_IN_{II,III}$	2.1965
5.6476	9	80 Hg	$M\alpha_1$	M_VN_{VII}	2.1953
5.650	5	73 Ta	M_{III}	Abs. Edge	2.194
5.6681	3	40 Zr	$L\beta_4$	L_IM_{II}	2.1873
5.67	3	73 Ta		$M_{III}O_{IV,V}$	2.19
5.682	4	76 Os	$M\gamma$	$M_{III}N_V$	2.182
5.704	8	82 Pb		$M_{III}N_I$	2.174
5.7101	3	40 Zr	$L\beta_6$	$L_{III}N_I$	2.1712
5.724	5	76 Os		$M_{III}N_{IV}$	2.166
5.7243	2	41 Nb	$L\alpha_1$	$L_{III}M_V$	2.16589
5.7319	3	41 Nb	$L\alpha_2$	$L_{III}M_{IV}$	2.1630
5.756	1	39 Y	L_{II}	Abs. Edge	2.1540
5.767	9	79 Au		M_VO_{III}	2.150
5.784	1	15 P	K	Abs. Edge	2.1435
5.796	2	15 P	$K\beta$	KM	2.1391
5.81	2	76 Os		$M_{II}N_I$	2.133
5.81	1	78 Pt	M_V	Abs. Edge	2.133
5.828	1	78 Pt	$M\beta$	$M_{IV}N_{VI}$	2.1273
5.83	2	73 Ta		$M_{III}O_I$	2.126
5.83	1	77 Ir	M_{IV}	Abs. Edge	2.126
5.8360	3	40 Zr	$L\beta_1$	$L_{II}M_{IV}$	2.1244
5.840	1	79 Au	$M\alpha_1$	M_VN_{VII}	2.1229
5.8475	3	42 Mo	$L\eta$	$L_{II}M_I$	2.1202
5.854	3	79 Au	$M\alpha_2$	M_VN_{VI}	2.118
5.8754	3	39 Y	$L\gamma_5$	$L_{II}N_I$	2.1102
5.884	8	81 Tl		$M_{III}N_I$	2.107
5.885	2	75 Re	$M\gamma$	$M_{III}N_V$	2.1067
5.931	5	75 Re		$M_{III}N_{IV}$	2.090
5.962	1	39 Y	L_{III}	Abs. Edge	2.0794
5.9832	3	39 Y	$L\beta_3$	L_IM_{III}	2.0722
5.987	9	78 Pt		M_VO_{III}	2.071
6.008	5	37 Rb	L_I	Abs. Edge	2.063
6.0186	3	39 Y	$L\beta_4$	L_IM_{II}	2.0600
6.038	1	77 Ir	$M\beta$	$M_{IV}N_{VI}$	2.0535
6.0458	3	37 Rb	$L\gamma_{2,3}$	$L_IN_{II,III}$	2.0507
6.047	1	78 Pt	$M\alpha_1$	M_VN_{VII}	2.0505
6.05	1	77 Ir	M_V	Abs. Edge	2.048
6.058	3	78 Pt	$M\alpha_2$	M_VN_{VI}	2.047
6.0705	2	40 Zr	$L\alpha_1$	$L_{III}M_V$	2.04236
6.073	5	76 Os	M_{IV}	Abs. Edge	2.042
6.0778	3	40 Zr	$L\alpha_2$	$L_{III}M_{IV}$	2.0399
6.09	2	80 Hg		$M_{III}N_I$	2.036
6.092	3	74 W	$M\gamma$	$M_{III}N_V$	2.035
6.0942	3	39 Y	$L\beta_6$	$L_{III}N_I$	2.0344
6.134	4	74 W		$M_{III}N_{IV}$	2.021
6.1508	3	42 Mo	Ll	$L_{III}M_I$	2.01568
6.157	1	15 P	$K\alpha_1$	KL_{III}	2.0137
6.160	1	15 P	$K\alpha_2$	KL_{II}	2.0127
6.162	8	83 Bi		$M_{IV}N_{III}$	2.012
6.173	1	38 Sr	L_{II}	Abs. Edge	2.0085
6.2109	3	41 Nb	$L\eta$	$L_{II}M_I$	1.99620
6.2120	3	39 Y	$L\beta_1$	$L_{II}M_{IV}$	1.99584
6.259	9	79 Au		$M_{III}N_I$	1.981
6.262	1	77 Ir	$M\alpha_1$	M_VN_{VII}	1.9799
6.267	1	76 Os	$M\beta$	$M_{IV}N_{VI}$	1.9783
6.275	3	77 Ir	$M\alpha_2$	M_VN_{VI}	1.9758
6.28	2	74 W		$M_{II}N_I$	1.973

TABLE B (*Continued*)

Wavelength Å*	p.e.	Element	Designation		keV
6.2961	3	38 Sr	$L\gamma_5$	$L_{II}N_I$	1.96916
6.30	1	76 Os	M_V	Abs. Edge	1.967
6.312	4	73 Ta	$M\gamma$	$M_{III}N_V$	1.964
6.33	1	75 Re	M_{IV}	Abs. Edge	1.958
6.353	5	73 Ta		$M_{III}N_{IV}$	1.951
6.3672	3	38 Sr	$L\beta_3$	L_IM_{III}	1.94719
6.384	7	82 Pb		$M_{IV}N_{III}$	1.942
6.387	1	38 Sr	L_{III}	Abs. Edge	1.9411
6.4026	3	38 Sr	$L\beta_4$	L_IM_{II}	1.93643
6.4488	2	39 Y	$L\alpha_1$	$L_{III}M_V$	1.92256
6.455	9	78 Pt		$M_{III}N_I$	1.921
6.4558	3	39 Y	$L\alpha_2$	$L_{III}M_{IV}$	1.92047
6.47	1	36 Kr	L_I	Abs. Edge	1.915
6.490	1	76 Os	$M\alpha$	$M_VN_{VI,VII}$	1.9102
6.504	1	75 Re	$M\beta$	$M_{IV}N_{VI}$	1.9061
6.5176	3	41 Nb	Ll	$L_{III}M_I$	1.90225
6.5191	3	38 Sr	$L\beta_6$	$L_{III}N_I$	1.90181
6.521	4	83 Bi	$M\zeta_1$	M_VN_{III}	1.901
6.544	4	72 Hf	$M\gamma$	$M_{III}N_V$	1.895
6.560	5	75 Re	M_V	Abs. Edge	1.890
6.585	5	83 Bi	$M\zeta_2$	$M_{IV}N_{II}$	1.883
6.59	1	74 W	M_{IV}	Abs. Edge	1.880
6.6069	3	40 Zr	$L\eta$	$L_{II}M_I$	1.87654
6.6239	3	38 Sr	$L\beta_1$	$L_{II}M_{IV}$	1.87172
6.644	1	37 Rb	L_{II}	Abs. Edge	1.8661
6.669	9	77 Ir		$M_{III}N_I$	1.859
6.729	1	75 Re	$M\alpha$	$M_VN_{VI,VII}$	1.8425
6.738	1	14 Si	K	Abs. Edge	1.8400
6.740	3	82 Pb	$M\zeta_1$	M_VN_{III}	1.8395
6.7530	1	14 Si	$K\beta$	KM	1.83594
6.755	3	37 Rb	$L\gamma_5$	$L_{II}N_{IV}$	1.83532
6.757	1	74 W	$M\beta$	$M_{IV}N_{VI}$	1.8349
6.768	6	71 Lu	$M\gamma$	$M_{III}N_V$	1.832
6.7876	3	37 Rb	$L\beta_3$	L_IM_{III}	1.82659
6.802	5	82 Pb	$M\zeta_2$	$M_{IV}N_{II}$	1.823
6.806	9	74 W		$M_{IV}O_{II}$	1.822
6.8207	3	37 Rb	$L\beta_4$	L_IM_{II}	1.81771
6.83	1	74 W	M_V	Abs. Edge	1.814
6.862	1	37 Rb	L_{III}	Abs. Edge	1.8067
6.8628	2	38 Sr	$L\alpha_1$	$L_{III}M_V$	1.80656
6.8697	3	38 Sr	$L\alpha_2$	$L_{III}M_{IV}$	1.80474
6.87	1	73 Ta	M_{IV}	Abs. Edge	1.804
6.87	2	80 Hg	δ	$M_{IV}N_{III}$	1.805
6.89	2	76 Os		$M_{III}N_I$	1.798
6.9185	3	40 Zr	Ll	$L_{III}M_I$	1.79201
6.959	5	35 Br	L_I	Abs. Edge	1.781
6.974	4	81 Tl	$M\zeta_1$	M_VN_{III}	1.778
6.983	1	74 W	$M\alpha_1$	M_VN_{VII}	1.7754
6.9842	3	37 Rb	$L\beta_6$	$L_{III}N_I$	1.77517
6.992	2	74 W	$M\alpha_2$	M_VN_{VI}	1.7731
7.00[illegible]	9	74 W		M_VO_{III}	1.770
7.023	1	73 Ta	$M\beta$	$M_{IV}N_{VI}$	1.7655
7.024	8	70 Yb	$M\gamma$	$M_{III}N_V$	1.765
7.032	5	81 Tl	$M\zeta_2$	$M_{IV}N_{II}$	1.763
7.0406	3	39 Y	$L\eta$	$L_{II}M_I$	1.76095
7.0759	3	37 Rb	$L\beta_1$	$L_{II}M_{IV}$	1.75217
7.09	2	73 Ta		$M_{IV}O_{II,III}$	1.748
7.101	8	79 Au		$M_{IV}N_{III}$	1.746
7.11	1	73 Ta	M_V	Abs. Edge	1.743
7.12542	9	14 Si	$K\alpha_1$	KL_{III}	1.73998
7.12791	9	14 Si	$K\alpha_2$	KL_{II}	1.73938
7.168	1	36 Kr	L_{II}	Abs. Edge	1.7297
7.250	5	36 Kr		$L_{II}N_{III}$	1.710
7.252	1	73 Ta	$M\alpha$	$M_VN_{VI,VII}$	1.7096
7.264	5	36 Kr	$L\beta_3$	L_IM_{III}	1.707
7.279	5	36 Kr	$L\gamma_5$	$L_{II}N_I$	1.703
7.30	2	73 Ta		M_VO_{III}	1.700
7.303	1	72 Hf	$M\beta$	$M_{IV}N_{VI}$	1.6976
7.304	5	36 Kr	$L\beta_4$	L_IM_{II}	1.697
7.3183	2	37 Rb	$L\alpha_1$	$L_{III}M_V$	1.69413
7.3251	3	37 Rb	$L\alpha_2$	$L_{III}M_{IV}$	1.69256
7.3563	3	39 Y	Ll	$L_{III}M_I$	1.68536
7.360	8	74 W		$M_{III}N_I$	1.684
7.371	8	78 Pt		$M_{IV}N_{III}$	1.682
7.392	1	36 Kr	L_{III}	Abs. Edge	1.6772
7.466	4	79 Au	$M\zeta_1$	M_VN_{III}	1.6605
7.503	1	34 Se	L_I	Abs. Edge	1.6525
7.510	4	36 Kr	$L\beta_6$	$L_{III}N_I$	1.6510
7.5171	3	38 Sr	$L\eta$	$L_{II}M_I$	1.64933
7.523	5	79 Au	$M\zeta_2$	$M_{IV}N_{II}$	1.648
7.539	1	72 Hf	$M\alpha$	$M_VN_{VI,VII}$	1.6446
7.546	8	68 Er	$M\gamma$	$M_{III}N_V$	1.643
7.576	3	36 Kr	$L\beta_1$	$L_{II}M_{IV}$	1.6366
7.60	1	68 Er		$M_{III}N_{IV}$	1.632
7.601	2	71 Lu	$M\beta$	$M_{IV}N_{VI}$	1.6312
7.612	9	73 Ta		$M_{III}N_I$	1.629
7.645	8	77 Ir		$M_{IV}N_{III}$	1.622
7.738	4	78 Pt	$M\zeta_1$	M_VN_{III}	1.6022
7.753	5	35 Br	L_{II}	Abs. Edge	1.599
7.767	9	35 Br	$L\beta_{3,4}$	$L_IM_{II,III}$	1.596
7.790	5	78 Pt	$M\zeta_2$	$M_{IV}N_{II}$	1.592
7.817	3	36 Kr	$L\alpha_{1,2}$	$L_{III}M_{IV,V}$	1.5860
7.8362	3	38 Sr	Ll	$L_{III}M_I$	1.58215
7.840	2	71 Lu	$M\alpha$	$M_VN_{VI,VII}$	1.5813
7.865	9	67 Ho	$M\gamma$	$M_{III}N_{IV,V}$	1.576
7.887	9	72 Hf		$M_{III}N_I$	1.572
7.909	2	70 Yb	$M\beta$	$M_{IV}N_{VI}$	1.5675
7.94813	5	13 Al	K	Abs. Edge	1.55988
7.960	2	13 Al	$K\beta$	KM	1.55745
7.984	5	35 Br	L_{III}	Abs. Edge	1.5530
8.021	4	77 Ir	$M\zeta_1$	M_VN_{III}	1.5458
8.0415	4	37 Rb	$L\eta$	$L_{II}M_I$	1.54177
8.065	5	77 Ir	$M\zeta_2$	$M_{IV}N_{II}$	1.5373
8.107	1	33 As	L_I	Abs. Edge	1.5293
8.1251	5	35 Br	$L\beta_1$	$L_{II}M_{IV}$	1.52590
8.144	9	66 Dy	$M\gamma$	$M_{III}N_{IV,V}$	1.522
8.14[illegible]	[illegible]	70 Yb	$M\alpha$	$M_VN_{VI,VII}$	1.5214
8.239	8	75 Re		$M_{IV}N_{III}$	1.505
8.249	7	69 Tm	$M\beta$	$M_{IV}N_{VI}$	1.503
8.310	4	76 Os	$M\zeta_1$	M_VN_{III}	1.4919
8.321	9	34 Se	$L\beta_{3,4}$	$L_IM_{II,III}$	1.490
8.33934	9	13 Al	$K\alpha_1$	KL_{III}	1.48670
8.34173	9	13 Al	$K\alpha_2$	KL_{II}	1.48627
8.359	5	76 Os	$M\zeta_2$	$M_{IV}N_{II}$	1.4831

Wavelength Å*	p.e.	Element	Designation		keV
8.3636	4	37 Rb	Ll	$L_{III}M_{I}$	1.48238
8.3746	5	35 Br	$L\alpha_{1,2}$	$L_{III}M_{IV,V}$	1.48043
8.407	1	34 Se	L_{II}	Abs. Edge	1.4747
8.470	9	70 Yb		$M_{III}N_{I}$	1.464
8.48	1	69 Tm	$M\alpha$	$M_{V}N_{VI,VII}$	1.462
8.486	9	65 Tb	$M\gamma$	$M_{III}N_{IV,V}$	1.461
8.487	5	69 Tm	M_{V}	Abs. Edge	1.4609
8.573	8	74 W		$M_{IV}N_{III}$	1.446
8.592	3	68 Er	$M\beta$	$M_{IV}N_{VI}$	1.4430
8.60	7	92 U		$N_{I}P_{IV,V}$	1.44
8.601	5	68 Er	M_{IV}	Abs. Edge	1.4415
8.629	4	75 Re	$M\zeta_{1}$	$M_{V}N_{III}$	1.4368
8.646	1	34 Se	L_{III}	Abs. Edge	1.4340
8.664	5	75 Re	$M\zeta_{2}$	$M_{IV}N_{II}$	1.4310
8.7358	5	34 Se	$L\beta_{1}$	$L_{II}M_{IV}$	1.41923
8.76	7	92 U		$N_{I}P_{III}$	1.42
8.773	1	32 Ge	L_{I}	Abs. Edge	1.4132
8.81	7	92 U		$N_{I}P_{II}$	1.41
8.82	1	68 Er	$M\alpha$	$M_{V}N_{VI,VII}$	1.406
8.844	9	64 Gd	$M\gamma$	$M_{III}N_{IV,V}$	1.402
8.847	5	68 Er	M_{V}	Abs. Edge	1.4013
8.90	2	73 Ta		$M_{IV}N_{III}$	1.393
8.929	1	33 As	$L\beta_{3,4}$	$L_{I}M_{II,III}$	1.3884
8.962	4	74 W	$M\zeta_{1}$	$M_{V}N_{III}$	1.3835
8.965	4	67 Ho	$M\beta$	$M_{IV}N_{VI}$	1.3830
8.9900	5	34 Se	$L\alpha_{1,2}$	$L_{III}M_{IV,V}$	1.37910
8.993	5	74 W	$M\zeta_{2}$	$M_{IV}N_{II}$	1.3787
9.125	1	33 As	L_{II}	Abs. Edge	1.3587
9.20	2	67 Ho	$M\alpha$	$M_{V}N_{VI,VII}$	1.348
9.211	9	63 Eu	$M\gamma$	$M_{III}N_{IV,V}$	1.346
9.255	1	35 Br	$L\eta$	$L_{II}M_{I}$	1.3396
9.316	4	73 Ta	$M\zeta_{1}$	$M_{V}N_{III}$	1.3308
9.330	5	73 Ta	$M\zeta_{2}$	$M_{IV}N_{II}$	1.3288
9.357	6	66 Dy	$M\beta$	$M_{IV}N_{VI}$	1.3250
9.367	1	33 As	L_{III}	Abs. Edge	1.3235
9.40	7	90 Th		$N_{I}P_{III}$	1.319
9.4141	8	33 As	$L\beta_{1}$	$L_{II}M_{IV}$	1.3170
9.44	7	90 Th		$N_{I}P_{II}$	1.313
9.5122	1	12 Mg	K	Abs. Edge	1.30339
9.517	5	31 Ga	L_{I}	Abs. Edge	1.3028
9.521	2	12 Mg	$K\beta$	KM	1.3022
9.581	2	32 Ge	$L\beta_{3}$	$L_{I}M_{III}$	1.2941
9.585	1	35 Br	Ll	$L_{III}M_{I}$	1.2935
9.59	2	66 Dy	$M\alpha$	$M_{V}N_{VI,VII}$	1.293
9.600	9	62 Sm	$M\gamma$	$M_{III}N_{IV,V}$	1.291
9.640	2	32 Ge	$L\beta_{4}$	$L_{I}M_{II}$	1.2861
9.6709	8	33 As	$L\alpha_{1,2}$	$L_{III}M_{IV,V}$	1.2820
9.686	7	72 Hf	$M\zeta_{2}$	$M_{IV}N_{II}$	1.2800
9.686	7	72 Hf	$M\zeta_{1}$	$M_{V}N_{III}$	1.2800
9.792	6	65 Tb	$M\beta$	$M_{IV}N_{VI}$	1.2661
9.8900	2	12 Mg	$K\alpha_{1,2}$	$KL_{II,III}$	1.25360
9.924	1	32 Ge	L_{II}	Abs. Edge	1.2494
9.962	1	34 Se	$L\eta$	$L_{II}M_{I}$	1.2446
10.00	2	65 Tb	$M\alpha$	$M_{V}N_{VI,VII}$	1.240
10.09	7	92 U		$N_{I}O_{III}$	1.229
10.175	1	32 Ge	$L\beta_{1}$	$L_{II}M_{IV}$	1.2185
10.187	1	32 Ge	L_{III}	Abs. Edge	1.2170
10.254	6	64 Gd	$M\beta$	$M_{IV}N_{VI}$	1.2091
10.294	1	34 Se	Ll	$L_{III}M_{I}$	1.2044
10.359	9	31 Ga	$L\beta_{3,4}$	$L_{I}M_{II,III}$	1.197
10.40	7	92 U		$N_{II}P_{I}$	1.192
10.4361	8	32 Ge	$L\alpha_{1,2}$	$L_{III}M_{IV,V}$	1.18800
10.46	3	64 Gd	$M\alpha$	$M_{V}N_{VI,VII}$	1.185
10.48	1	70 Yb	$M\zeta$	$M_{V}N_{III}$	1.183
10.505	9	60 Nd	$M\gamma$	$M_{III}N_{IV,V}$	1.180
10.711	5	63 Eu	M_{IV}	Abs. Edge	1.1575
10.734	1	33 As	$L\eta$	$L_{II}M_{I}$	1.1550
10.750	7	63 Eu	$M\beta$	$M_{IV}N_{VI}$	1.1533
10.828	5	31 Ga	L_{II}	Abs. Edge	1.1450
10.96	3	63 Eu	$M\alpha$	$M_{V}N_{VI,VII}$	1.131
10.998	9	59 Pr	$M\gamma$	$M_{III}N_{IV,V}$	1.1273
11.013	5	63 Eu	M_{V}	Abs. Edge	1.1258
11.023	2	31 Ga	$L\beta_{1}$	$L_{II}M_{IV}$	1.1248
11.072	1	33 As	Ll	$L_{III}M_{I}$	1.1198
11.07	7	90 Th		$N_{II}P_{I}$	1.120
11.100	1	31 Ga	L_{III}	Abs. Edge	1.1169
11.200	7	30 Zn	$L\beta_{3,4}$	$L_{I}M_{II,III}$	1.1070
11.27	1	62 Sm	$M\beta$	$M_{IV}N_{VI}$	1.0998
11.288	5	62 Sm	M_{IV}	Abs. Edge	1.0983
11.292	1	31 Ga	$L\alpha_{1,2}$	$L_{III}M_{IV,V}$	1.09792
11.37	1	68 Er	$M\zeta$	$M_{V}N_{III}$	1.0901
11.47	3	62 Sm	$M\alpha$	$M_{V}N_{VI,VII}$	1.081
11.53	1	58 Ce	$M\gamma$	$M_{III}N_{IV,V}$	1.0749
11.552	5	62 Sm	M_{V}	Abs. Edge	1.0732
11.56	5	90 Th		$N_{II}O_{IV}$	1.072
11.569	1	11 Na	K	Abs. Edge	1.07167
11.575	2	11 Na	$K\beta$	KM	1.0711
11.609	2	32 Ge	$L\eta$	$L_{II}M_{I}$	1.0680
11.862	1	30 Zn	L_{II}	Abs. Edge	1.04523
11.86	1	67 Ho	$M\zeta$	$M_{V}N_{III}$	1.0450
11.9101	9	11 Na	$K\alpha_{1,2}$	$KL_{II,III}$	1.04098
11.965	2	32 Ge	Ll	$L_{III}M_{I}$	1.0362
11.983	3	30 Zn	$L\beta_{1}$	$L_{II}M_{IV}$	1.0347
12.08	4	57 La	$M\gamma$	$M_{III}N_{IV,V}$	1.027
12.122	3	29 Cu	$L\beta_{3,4}$	$L_{I}M_{II,III}$	1.0228
12.131	1	30 Zn	L_{III}	Abs. Edge	1.02201
12.254	3	30 Zn	$L\alpha_{1,2}$	$L_{III}M_{IV,V}$	1.0117
12.43	2	66 Dy	$M\zeta$	$M_{V}N_{III}$	0.998
12.44	2	60 Nd	$M\beta$	$M_{IV}N_{VI}$	0.997
12.459	5	60 Nd	M_{IV}	Abs. Edge	0.9951
12.597	2	31 Ga	$L\eta$	$L_{II}M_{I}$	0.9842
12.68	2	60 Nd	$M\alpha$	$M_{V}N_{VI,VII}$	0.978
12.737	5	60 Nd	M_{V}	Abs. Edge	0.9734
12.75	3	56 Ba	$M\gamma$	$M_{III}N_{IV,V}$	0.973
12.90	9	92 U		$N_{III}O_{V}$	0.961
12.953	2	31 Ga	Ll	$L_{III}M_{I}$	0.9572
12.98	2	65 Tb	$M\zeta$	$M_{V}N_{III}$	0.955
13.014	1	29 Cu	L_{II}	Abs. Edge	0.95268
13.053	3	29 Cu	$L\beta_{1}$	$L_{II}M_{IV}$	0.9498
13.06	2	59 Pr	$M\beta$	$M_{IV}N_{VI}$	0.950
13.06	1	30 Zn	L_{I}	Abs. Edge	0.9495
13.122	5	59 Pr	M_{IV}	Abs. Edge	0.9448
13.18	2	28 Ni	$L\beta_{3,4}$	$L_{I}M_{II,III}$	0.941
13.288	1	29 Cu	L_{III}	Abs. Edge	0.93306

Wavelength Å*	p.e.	Element	Designation		keV
13.30	6	83 Bi		$N_IP_{II,III}$	0.932
13.336	3	29 Cu	$L\alpha_{1,2}$	$L_{III}M_{IV,V}$	0.9297
13.343	5	59 Pr	$M\alpha$	$M_VN_{VI,VII}$	0.9292
13.394	5	59 Pr	M_V	Abs. Edge	0.9257
13.57	2	64 Gd	$M\zeta$	M_VN_{III}	0.914
13.68	2	30 Zn	$L\eta$	$L_{II}M_I$	0.906
13.75	4	58 Ce	$M\beta$	$M_{IV}N_{VI}$	0.902
13.8	1	90 Th		$N_{III}O_V$	0.897
14.02	2	30 Zn	Ll	$L_{III}M_I$	0.884
14.04	2	58 Ce	$M\alpha$	$M_VN_{VI,VII}$	0.883
14.22	2	63 Eu	$M\zeta$	M_VN_{III}	0.872
14.242	5	28 Ni	L_{II}	Abs. Edge	0.8706
14.271	6	28 Ni	$L\beta_1$	$L_{II}M_{IV}$	0.8688
14.3018	1	10 Ne	K	Abs. Edge	0.866889
14.31	3	27 Co	$L\beta_{3,4}$	$L_IM_{II,III}$	0.870
14.39	5	58 Ce		$M_VO_{II,III}$	0.862
14.452	5	10 Ne	$K\beta$	KM	0.8579
14.51	5	57 La	$M\beta$	$M_{IV}N_{VI}$	0.854
14.525	5	28 Ni	L_{III}	Abs. Edge	0.8536
14.561	3	28 Ni	$L\alpha_{1,2}$	$L_{III}M_{IV,V}$	0.8515
14.610	3	10 Ne	$K\alpha_{1,2}$	$KL_{II,III}$	0.8486
14.88	5	57 La	$M\alpha$	$M_VN_{VI,VII}$	0.833
14.90	2	29 Cu	$L\eta$	$L_{II}M_I$	0.832
14.91	4	62 Sm	$M\zeta$	M_VN_{III}	0.831
15.286	9	29 Cu	Ll	$L_{III}M_I$	0.8111
15.56	1	56 Ba	M_{IV}	Abs. Edge	0.7967
15.618	5	27 Co	L_{II}	Abs. Edge	0.7938
15.65	4	26 Fe	$L\beta_{3,4}$	$L_IM_{II,III}$	0.792
15.666	8	27 Co	$L\beta_1$	$L_{II}M_{IV}$	0.7914
15.72	9	56 Ba		$M_{IV}O_{III}$	0.789
15.89	1	56 Ba	M_V	Abs. Edge	0.7801
15.91	5	56 Ba		$M_{IV}O_{II}$	0.779
15.915	5	27 Co	L_{III}	Abs. Edge	0.7790
15.93	4	52 Te	$M\gamma$	$M_{III}N_{IV,V}$	0.778
15.972	6	27 Co	$L\alpha_{1,2}$	$L_{III}M_{IV,V}$	0.7762
15.98	5	51 Sb		$M_{II}N_{IV}$	0.776
16.20	5	56 Ba		M_VO_{III}	0.765
16.27	3	28 Ni	$L\eta$	$L_{II}M_I$	0.762
16.46	4	60 Nd	$M\zeta$	M_VN_{III}	0.753
16.693	9	28 Ni	Ll	$L_{III}M_I$	0.7427
16.7	1	24 Cr	L_I	Abs. Edge	0.741
16.92	4	51 Sb	$M\gamma$	$M_{III}N_{IV,V}$	0.733
16.93	5	50 Sn		$M_{II}N_{IV}$	0.733
17.19	4	25 Mn	$L\beta_{3,4}$	$L_IM_{II,III}$	0.721
17.202	5	26 Fe	L_{II}	Abs. Edge	0.7208
17.26	1	26 Fe	$L\beta_1$	$L_{II}M_{IV}$	0.7185
17.38	4	59 Pr	$M\zeta$	M_VN_{III}	0.714
17.525	5	26 Fe	L_{III}	Abs. Edge	0.7074
17.59	2	26 Fe	$L\alpha_{1,2}$	$L_{III}M_{IV,V}$	0.7050
17.6	1	52 Te		$M_{II}N_I$	0.703
17.87	3	27 Co	$L\eta$	$L_{II}M_I$	0.694
17.94	5	50 Sn	$M\gamma$	$M_{III}N_{IV,V}$	0.691
17.9	1	24 Cr	L_{II}	Abs. Edge	0.691
18.292	8	27 Co	Ll	$L_{III}M_I$	0.6778
18.32	2	9 F	$K\alpha$	KL	0.6768
18.35	4	58 Ce	$M\zeta$	M_VN_{III}	0.676
18.8	1	51 Sb		$M_{II}N_I$	0.658
18.8	2	47 Ag		$M_IN_{II,III}$	0.658
18.96	4	24 Cr	$L\beta_{3,4}$	$L_IM_{II,III}$	0.654
19.11	2	25 Mn	$L\beta_1$	$L_{II}M_{IV}$	0.6488
19.1	1	52 Te		$M_{III}N_I$	0.648
19.40	7	48 Cd		$M_{II}N_{IV}$	0.639
19.44	5	57 La	$M\zeta$	M_VN_{III}	0.638
19.45	1	25 Mn	$L\alpha_{1,2}$	$L_{III}M_{IV,V}$	0.6374
19.66	5	53 I	$M_{IV,V}$	Abs. Edge	0.631
19.75	4	26 Fe	$L\eta$	$L_{II}M_I$	0.628
20.0	1	50 Sn		$M_{II}N_I$	0.619
20.1	2	46 Pd		$M_IN_{II,III}$	0.616
20.15	1	26 Fe	Ll	$L_{III}M_I$	0.6152
20.2	1	51 Sb		$M_{III}N_I$	0.612
20.47	7	48 Cd	$M\gamma$	$M_{III}N_{IV,V}$	0.606
20.64	4	56 Ba	$M\zeta$	M_VN_{III}	0.601
20.66	7	47 Ag		$M_{II}N_{IV}$	0.600
20.7	1	24 Cr	L_{III}	Abs. Edge	0.598
21.19	5	23 Va	$L\beta_{3,4}$	$L_IM_{II,III}$	0.585
21.27	1	24 Cr	$L\beta_1$	$L_{II}M_{IV}$	0.5828
21.34	5	52 Te		$M_{IV}O_{II,III}$	0.581
21.5	1	50 Sn		$M_{III}N_I$	0.575
21.64	3	24 Cr	$L\alpha_{1,2}$	$L_{III}M_{IV,V}$	0.5728
21.78	5	52 Te		M_VO_{III}	0.569
21.82	7	47 Ag	$M\gamma$	$M_{III}N_{IV,V}$	0.568
21.85	2	25 Mn	$L\eta$	$L_{II}M_I$	0.5675
22.1	1	46 Pd		$M_{II}N_{IV}$	0.560
22.29	1	25 Mn	Ll	$L_{III}M_I$	0.5563
22.9	2	48 Cd		$M_{II}N_I$	0.540
23.32	1	8 O	K	Abs. Edge	0.5317
23.3	1	46 Pd	$M\gamma$	$M_{III}N_{IV,V}$	0.531
23.62	3	8 O	$K\alpha$	KL	0.5249
23.88	4	23 Va	$L\beta_1$	$L_{II}M_{IV}$	0.5192
24.25	3	23 Va	$L\alpha_{1,2}$	$L_{III}M_{IV,V}$	0.5113
24.28	5	50 Sn	$M_{IV,V}$	Abs. Edge	0.511
24.30	3	24 Cr	$L\eta$	$L_{II}M_I$	0.5102
24.4	2	47 Ag		M_VN_I	0.509
24.5	1	48 Cd		$M_{III}N_I$	0.507
24.78	1	24 Cr	Ll	$L_{III}M_I$	0.5003
25.01	9	45 Rh	$M\gamma$	$M_{III}N_{IV,V}$	0.496
25.3	1	50 Sn		$M_{IV}O_{II,III}$	0.491
25.50	9	44 Ru		$M_{II}N_{IV}$	0.486
25.7	1	50 Sn		M_VO_{III}	0.483
26.0	1	47 Ag		$M_{III}N_I$	0.478
26.2	2	46 Pd		$M_{II}N_I$	0.474
26.72	9	52 Te	$M\zeta$	$M_{IV,V}N_{II,III}$	0.464
26.9	1	44 Ru	$M\gamma$	$M_{III}N_{IV,V}$	0.462
27.05	2	22 Ti	$L\beta_1$	$L_{II}M_{IV}$	0.4584
27.29	1	22 Ti	$L_{II,III}$	Abs. Edge	0.4544
27.34	3	23 Va	$L\eta$	$L_{II}M_I$	0.4535
27.42	2	22 Ti	$L\alpha_{1,2}$	$L_{III}M_{IV,V}$	0.4522
27.77	1	23 Va	Ll	$L_{III}M_I$	0.4465
27.9	1	46 Pd		$M_{III}N_I$	0.445
28.1	2	45 Rh		$M_{II}N_I$	0.442
28.13	5	48 Cd	$M_{IV,V}$	Abs. Edge	0.4408
28.88	8	51 Sb	$M\zeta$	$M_{IV,V}N_{II,III}$	0.429
29.8	1	45 Rh		$M_{III}N_I$	0.417
30.4	1	48 Cd		$M_{IV}O_{II,III}$	0.408

Wavelength Å*	p.e.	Element	Designation		keV
30.8	1	48 Cd		$M_{V}O_{III}$	0.403
30.82	5	47 Ag	M_{IV}	Abs. Edge	0.4022
30.89	3	22 Ti	$L\eta$	$L_{II}M_{I}$	0.4013
30.99	1	7 N	K	Abs. Edge	0.4000
31.02	2	21 Sc	$L\beta_1$	$L_{II}M_{IV}$	0.3996
31.14	5	47 Ag	M_{V}	Abs. Edge	0.3981
31.24	9	50 Sn	$M\zeta$	$M_{IV,V}N_{II,III}$	0.397
31.35	3	21 Sc	$L\alpha_{1,2}$	$L_{III}M_{IV,V}$	0.3954
31.36	2	22 Ti	Ll	$L_{III}M_{I}$	0.3953
31.60	4	7 N	$K\alpha$	KL	0.3924
31.8	1	92 U		$N_{IV}N_{VI}$	0.390
32.3	2	44 Ru		$M_{II}N_{I}$	0.384
33.1	2	41 Nb		$M_{II}N_{IV}$	0.375
33.5	3	47 Ag		$M_{IV,V}O_{II,III}$	0.370
33.57	9	90 Th		$N_{IV}N_{VI}$	0.3693
34.8	1	92 U		$N_{V}N_{VI,VII}$	0.357
34.9	2	41 Nb	$M\gamma$	$M_{III}N_{IV,V}$	0.356
35.13	2	21 Sc	$L\eta$	$L_{II}M_{I}$	0.3529
35.13	1	20 Ca	L_{II}	Abs. Edge	0.3529
35.3	3	42 Mo		$M_{II}N_{I}$	0.351
35.49	1	20 Ca	L_{III}	Abs. Edge	0.34931
35.59	3	21 Sc	Ll	$L_{III}M_{I}$	0.3483
35.63	1	20 Ca	$L_{II,III}$	Abs. Edge	0.34793
35.94	2	20 Ca	$L\beta_1$	$L_{II}M_{IV}$	0.3449
36.32	9	90 Th		$N_{V}N_{VI,VII}$	0.3414
36.33	2	20 Ca	$L\alpha_{1,2}$	$L_{III}M_{IV,V}$	0.3413
36.8	1	48 Cd	$M\zeta$	$M_{IV,V}N_{II,III}$	0.3371
37.4	2	46 Pd		$M_{IV,V}O_{II,III}$	0.332
37.5	2	42 Mo		$M_{III}N_{I}$	0.331
38.4	3	41 Nb		$M_{II}N_{I}$	0.323
39.77	7	47 Ag	$M\zeta$	$M_{IV,V}N_{II,III}$	0.3117
40.46	2	20 Ca	$L\eta$	$L_{II}M_{I}$	0.3064
40.7	2	41 Nb		$M_{III}N_{I}$	0.305
40.9	2	45 Rh		$M_{IV,V}O_{II,III}$	0.303
40.96	2	20 Ca	Ll	$L_{III}M_{I}$	0.3027
42.1	2	92 U		$N_{VI}O_{V}$	0.295
42.1	1	19 K	$L_{II,III}$	Abs. Edge	0.2946
42.3	2	82 Pb		$N_{IV}N_{VI}$	0.293
43.3	2	92 U		$N_{VI}O_{IV}$	0.286
43.6	1	46 Pd	$M\zeta$	$M_{IV,V}N_{II,III}$	0.2844
43.68	1	6 C	K	Abs. Edge	0.28384
44.7	3	6 C	$K\alpha$	KL	0.277
44.8	1	44 Ru		$M_{IV,V}O_{II,III}$	0.2768
45.0	1	82 Pb		$N_{V}N_{VI,VII}$	0.2756
45.2	3	80 Hg		$N_{IV}N_{VI}$	0.274
45.2	1	51 Sb		$M_{II}M_{IV}$	0.2743
46.48	9	39 Y		$M_{II}N_{I}$	0.267
46.5	2	81 Tl		$N_{V}N_{VI,VII}$	0.267
46.8	2	79 Au		$N_{IV}N_{VI}$	0.265
47.24	2	19 K	Ll	$L_{II}M_{I}$	0.2625
47.3	1	50 Sn		$M_{II}M_{IV}$	0.2621
47.67	9	45 Rh	$M\zeta$	$M_{IV,V}N_{II,III}$	0.2601
47.74	1	19 K	Ll	$L_{III}M_{I}$	0.25971
47.9	3	80 Hg		$N_{V}N_{VI,VII}$	0.259
48.1	2	78 Pt		$N_{IV}N_{VI}$	0.258
48.2	1	90 Th		$N_{VI}O_{V}$	0.2572
48.5	2	39 Y		$M_{III}N_{I}$	0.256

Wavelength Å*	p.e.	Element	Designation		keV
49.4	1	79 Au		$N_{V}N_{VI,VII}$	0.2510
49.5	1	90 Th		$N_{VI}O_{IV}$	0.2505
50.0	1	90 Th		$N_{VII}O_{V}$	0.2479
50.2	1	77 Ir		$N_{IV}N_{VI}$	0.2470
50.3	1	52 Te		$M_{III}M_{V}$	0.2465
50.9	1	78 Pt		$N_{V}N_{VI,VII}$	0.2436
51.3	1	38 Sr		$M_{II}N_{I}$	0.2416
51.9	1	76 Os		$N_{IV}N_{VI}$	0.2388
52.0	2	48 Cd		$M_{II}M_{IV}$	0.2384
52.2	1	51 Sb		$M_{III}M_{V}$	0.2375
52.34	7	44 Ru	$M\zeta$	$M_{IV,V}N_{II,III}$	0.2369
52.8	1	77 Ir		$N_{V}N_{VI,VII}$	0.2348
53.6	1	38 Sr		$M_{III}N_{I}$	0.2313
54.0	2	74 W		$N_{II}N_{IV}$	0.2295
54.0	1	47 Ag		$M_{II}M_{IV}$	0.2295
54.2	1	50 Sn		$M_{III}M_{V}$	0.2287
54.7	2	76 Os		$N_{V}N_{VI,VII}$	0.2266
54.8	2	42 Mo		$M_{IV,V}O_{II,III}$	0.2262
55.8	1	74 W		$N_{IV}N_{VI}$	0.2221
55.9	1	18 A	$L\eta$	$L_{II}M_{I}$	0.2217
56.3	1	18 A	Ll	$L_{III}M_{I}$	0.2201
56.5	1	46 Pd		$M_{II}M_{IV}$	0.2194
57.0	2	37 Rb		$M_{II}N_{I}$	0.2174
58.2	1	73 Ta		$N_{IV}N_{VI}$	0.2130
58.4	1	74 W		$N_{V}N_{VII}$	0.2122
58.7	2	48 Cd		$M_{III}M_{V}$	0.2111
59.3	1	45 Rh		$M_{II}M_{IV}$	0.2090
59.5	3	74 W		$N_{V}N_{VI}$	0.208
59.5	2	37 Rb		$M_{III}N_{I}$	0.2083
60.5	1	47 Ag		$M_{III}M_{V}$	0.2048
61.1	2	73 Ta		$N_{V}N_{VI,VII}$	0.2028
61.9	2	41 Nb		$M_{IV,V}O_{II,III}$	0.2002
62.2	1	44 Ru		$M_{II}M_{IV}$	0.1992
62.9	1	46 Pd		$M_{III}M_{V}$	0.1970
63.0	5	71 Lu		$N_{IV}N_{VI}$	0.197
64.38	7	42 Mo	$M\zeta$	$M_{IV,V}N_{II,III}$	0.1926
65.1	7	70 Yb		$N_{IV}N_{VI}$	0.190
65.5	1	45 Rh		$M_{III}M_{V}$	0.1892
65.7	2	71 Lu		$N_{V}N_{VI,VII}$	0.1886
67.33	9	17 Cl	$L\eta$	$L_{II}M_{I}$	0.1841
67.6	3	5 B	$K\alpha$	KL	0.1833
67.90	9	17 Cl	Ll	$L_{III}M_{I}$	0.1826
68.2	3	90 Th		$O_{III}P_{IV,V}$	0.1817
68.3	1	44 Ru		$M_{III}M_{V}$	0.1814
68.9	2	42 Mo		$M_{II}M_{IV}$	0.1798
69.3	5	70 Yb		$N_{V}N_{VI,VII}$	0.179
70.0	4	40 Zr		$M_{IV,V}O_{II,III}$	0.177
72.1	3	41 Nb		$M_{II}M_{IV}$	0.1718
72.19	9	41 Nb	$M\zeta$	$M_{IV,V}N_{II,III}$	0.1717
72.7	9	68 Er		$N_{IV}N_{VI}$	0.171
74.9	1	42 Mo		$M_{III}M_{V}$	0.1656
76.3	7	68 Er		$N_{V}N_{VI,VII}$	0.163
76.7	2	40 Zr		$M_{II}M_{IV}$	0.1617
76.9	2	35 Br		$M_{II}N_{I}$	0.1613
78.4	2	41 Nb		$M_{III}M_{V}$	0.1582
79.8	3	35 Br		$M_{III}N_{I}$	0.1554
80.9	3	40 Zr		$M_{III}M_{V}$	0.1533

TABLE B (*Continued*)

Wavelength Å*	p.e.	Element	Designation		keV
81.5	2	39 Y		$M_{II}M_{IV}$	0.1522
82.1	2	40 Zr	$M\zeta$	$M_{IV,V}N_{II,III}$	0.1511
83.	1	66 Dy		$N_{IV,V}N_{VI,VII}$	0.149
83.4	3	16 S	Ll, η	$L_{II,III}M_I$	0.1487
85.7	2	38 Sr		$M_{II}M_{IV}$	0.1447
86.	1	65 Tb		$N_{IV,V}N_{VI,VII}$	0.144
86.5	2	39 Y		$M_{III}M_{IV,V}$	0.1434
91.4	2	38 Sr		$M_{III}M_{IV,V}$	0.1357
91.5	2	37 Rb		$M_{II}M_{IV}$	0.1355
91.6	1	83 Bi		$N_{VI}O_{IV}$	0.1354
93.2	1	83 Bi		$N_{VII}O_V$	0.1330
93.4	2	39 Y	$M\zeta$	$M_{IV,V}N_{II,III}$	0.1328
94.	1	15 P	$L_{II,III}$	Abs. Edge	0.132
96.7	2	37 Rb		$M_{III}M_{IV,V}$	0.1282
97.2	8	66 Dy		$N_{IV,V}O_{II,III}$	0.128
98.	1	62 Sm		$N_{IV,V}N_{VI,VII}$	0.126
100.2	2	82 Pb		$N_{VI}O_V$	0.1237
102.2	4	65 Tb		$N_{IV,V}O_{II,III}$	0.1213
102.4	1	82 Pb		$N_{VI}O_{IV}$	0.1211
103.8	4	15 P		$L_{II,III}M$	0.1194
104.3	1	82 Pb		$N_{VII}O_V$	0.1189
107.	1	60 Nd		$N_{IV,V}N_{VI,VII}$	0.116
108.0	2	38 Sr	$M\zeta_2$	$M_{IV}N_{II,III}$	0.1148
108.7	1	38 Sr	$M\zeta_1$	M_VN_{III}	0.1140
109.4	3	35 Br		$M_{II}M_{IV}$	0.1133
110.6	5	29 Cu	M_I	Abs. Edge	0.1121
111.	1	4 Be	K	Abs. Edge	0.111
112.0	6	63 Eu		$N_{IV,V}O_{II,III}$	0.1107
113.0	1	81 Tl		$N_{VI}O_V$	0.10968
113.	1	59 Pr		$N_{IV,V}N_{VI,VII}$	0.1095
113.8	3	35 Br		$M_{III}M_{IV,V}$	0.1089
114.	1	4 Be	$K\alpha$	KL	0.1085
115.3	2	81 Tl		$N_{VI}O_{IV}$	0.1075
117.4	4	62 Sm		$N_{IV,V}O_{II,III}$	0.1056
117.7	1	81 Tl		$N_{VII}O_V$	0.10530
123.	1	14 Si	$L_{II,III}$	Abs. Edge	0.1006
126.8	2	37 Rb		$M_{IV}N_{III}$	0.0978
127.8	2	37 Rb	$M\zeta_2$	$M_{IV}N_{II}$	0.0970
128.7	2	37 Rb	$M\zeta_1$	M_VN_{III}	0.0964
128.9	7	60 Nd		$N_{IV,V}O_{II,III}$	0.0962
135.5	4	14 Si		$L_{II,III}M$	0.0915
136.5	4	59 Pr		$N_{IV,V}O_{II,III}$	0.0908
137.0	5	30 Zn	M_{II}	Abs. Edge	0.0905
142.5	1	13 Al	L_I	Abs. Edge	0.08701
143.9	5	30 Zn	M_{III}	Abs. Edge	0.0862
144.4	6	58 Ce		$N_{IV,V}O_{II,III}$	0.0859
144.4	3	37 Rb		M_IM_{III}	0.0859
152.6	6	57 La		$N_{IV,V}O_{II,III}$	0.0812
157.	3	30 Zn		$M_{II,III}M_{IV,V}$	0.079
159.0	2	56 Ba		$N_{IV}O_{III}$	0.07796
159.5	5	29 Cu	M_{II}	Abs. Edge	0.0777
163.3	2	56 Ba		$N_{IV}O_{II}$	0.07590
164.6	2	56 Ba		N_VO_{III}	0.07530
164.7	3	35 Br		M_IM_{III}	0.0753
166.0	5	29 Cu	M_{III}	Abs. Edge	0.0747
170.4	1	13 Al	$L_{II,III}$	Abs. Edge	0.07278
171.4	5	13 Al		$L_{II,III}M$	0.0724
173.	3	29 Cu		$M_{II,III}M_{IV,V}$	0.072
181.	5	90 Th		$O_{IV,V}Q_{II,III}$	0.068
183.8	1	55 Cs		$N_{IV}O_{III}$	0.06746
184.6	3	35 Br		M_IM_{II}	0.0672
188.4	1	28 Ni	M_{III}	Abs. Edge	0.06581
188.6	1	55 Cs		$N_{IV}O_{II}$	0.06574
189.5	3	35 Br		$M_{IV}N_{III}$	0.0654
190.3	1	55 Cs		N_VO_{III}	0.06515
190.	2	28 Ni		$M_{II,III}M_{IV,V}$	0.0651
191.1	2	35 Br	$M\zeta_2$	$M_{IV}N_{II}$	0.06488
192.6	2	35 Br	$M\zeta_1$	M_VN_{III}	0.06437
197.3	1	12 Mg	L_I	Abs. Edge	0.06284
202.	5	27 Co	$M_{II,III}$	Abs. Edge	0.061
203.	1	16 S		$L_IL_{II,III}$	0.061
214.	6	27 Co		$M_{II,III}M_{IV,V}$	0.058
224.	1	53 I	$N_{IV,V}$	Abs. Edge	0.0552
226.5	1	3 Li	K	Abs. Edge	0.05475
227.8	1	34 Se	M_V	Abs. Edge	0.05443
228.	1	3 Li	$K\alpha$	KL	0.0543
230.	2	34 Se		M_VN_{III}	0.0538
230.	1	26 Fe	$M_{II,III}$	Abs. Edge	0.0538
243.	5	26 Fe		$M_{II,III}M_{IV,V}$	0.051
249.3	1	12 Mg	L_{II}	Abs. Edge	0.04973
250.7	1	12 Mg	L_{III}	Abs. Edge	0.04945
251.5	5	12 Mg		$L_{II,III}M$	0.04929
273.	6	25 Mn		$M_{II,III}M_{IV,V}$	0.045
290.	1	13 Al		$L_IL_{II,III}$	0.0428
309.	9	24 Cr		$M_{II,III}M_{IV,V}$	0.040
317.	1	12 Mg		$L_IL_{II,III}$	0.0392
337.	9	23 V		$M_{II,III}M_{IV,V}$	0.0368
376.	1	11 Na		$L_IL_{II,III}$	0.03299
399.	5	35 Br	N_I	Abs. Edge	0.0311
405.	5	11 Na	$L_{II,III}$	Abs. Edge	0.0306
407.1	5	11 Na		$L_{II,III}M$	0.03045
417.	5	17 Cl	M_I	Abs. Edge	0.0297
444.	5	53 I	O_I	Abs. Edge	0.0279
525.	9	20 Ca		$M_{II,III}N_I$	0.0236
692.	9	19 K		$M_{II,III}N_I$	0.0179

X-RAY ATOMIC ENERGY LEVELS*

J. A. Bearden and A. F. Burr

These tables were originally published as the final report to the U.S. Atomic Energy Commission as Report NYO-2543-1 in partial fulfillment of Contract AT(30-1)-2543. The tables were later reproduced in *Review of Modern Physics*. The data may also be obtained from the Superintendent of Documents, U.S. Government Printing Office, Washington, D. C. 20402 in the publication NSRDS-NBS 14. Persons seeking discussion of the details of calculations, sources of energy level information and the problem of properly interpreting the experimental measurements should refer to the original publication or to *Review of Modern Physics*, Vol. 39, 125–142, January 1967.

All of the x-ray emission wavelengths have recently been reevaluated and placed on a consistent Å* scale. For most elements these data give a highly overdetermined set of equations for energy level differences, which have been solved by least-squares adjustment for each case. This procedure makes "best" use of all x-ray wavelength data, and also permits calculation of the probable error for each energy difference. Photoelectron measurements of absolute energy levels are more precise than x-ray absorption edge data. These have been used to establish the absolute scale for eighty-one elements and, in many cases, to provide additional energy level difference data. The x-ray absorption wavelengths were used for eight elements and ionization measurements for two; the remaining five were interpolated by a Moseley diagram involving the output values of energy levels from adjacent elements. Probable errors are listed on an absolute energy basis. In the original source of the present data, a table of energy levels in Rydberg units is given. Difference tables in volts, Rydbergs, and milli-Å* wavelength units, with the respective probable errors, are also included there.

Recommended values of the atomic energy levels, and probable errors in eV. Where available, photoelectron direct measurements are listed in brackets [ ] immediately under the recommended values. The measured values of the x-ray absorption energies are shown in parentheses ( ). Interpolated values are enclosed in angle brackets ⟨ ⟩.

Level	1 H	2 He	3 Li	4 Be	5 B	6 C	7 N	8 O
K	13.59811[a]	24.58678[b]	54.75±0.02 (54.75)	111.0±1.0 (111.0)	188.0±0.4 [188.0][c]	283.8±0.4 [283.8][c] (283.8)	401.6±0.4 [401.6][c]	532.0±0.4 [532.0][c]
L_I								23.7±0.4 [23.7][d]
$L_{II,III}$					4.7±0.9	6.4±1.9	9.2±0.6	7.1±0.8

Level	9 F	10 Ne	11 Na	12 Mg	13 Al	14 Si	15 P	16 S
K	685.4±0.4 [685.4][c]	866.9±0.3 (866.9)	1072.1±0.4 [1072.1][c] (1072.)	1305.0±0.4 [1305.0][c] (1303.)	1559.6±0.4 [1559.6][c] (1559.8)	1838.9±0.4 [1838.9][c]	2145.5±0.4 [2145.5][d]	2472.0±0.4 [2472.0][c] (2470.)
L_I	⟨31.⟩	⟨45.⟩	63.3±0.4 [63.3][d]	89.4±0.4 [89.4][d] (63.)	117.7±0.4 [117.7][d] (87.)	148.7±0.4 [148.7][d]	189.3±0.4 [189.3][d]	229.2±0.4 [229.2][d]
$L_{II,III}$	8.6±0.8	18.3±0.4	31.1±0.4 (31.)	51.4±0.5 (50.)	73.1±0.5 (72.8)	99.2±0.5 (100.6)	132.2±0.5 (132.)	164.8±0.7

Level	17 Cl	18 Ar	19 K	20 Ca	21 Sc	22 Ti	23 V	24 Cr
K	2822.4±0.3 [2822.4][c] (2020.)	3202.9±0.3 (3202.9)	3607.4±0.4 [3607.4][c] (3607.8)	4038.1±0.4 [4038.1][c] (4038.1)	4492.8±0.4 [4492.8][c]	4966.4±0.4 [4966.4][d] (4964.5)	5465.1±0.3 [5465.1][c] (5464.)	5989.2±0.3 [5989.2][c] (5989.)
L_I	270.2±0.4 [270.2][d]	320. ⟨320.⟩[d]	377.1±0.4 [377.1][d]	437.8±0.4 [437.8][d]	500.4±0.4 [500.4][d]	563.7±0.4 [563.7][d]	628.2±0.4 [628.2][d]	694.6±0.4 [694.6][d]
L_{II}	201.6±0.3	247.3±0.3	296.3±0.4	350.0±0.4	406.7±0.4	461.5±0.4	520.5±0.3	583.7±0.3
L_{III}	200.0±0.3	245.2±0.3	293.6±0.4	346.4±0.4	402.2±0.4	455.5±0.4	512.9±0.3	574.5±0.3
M_I	17.5±0.4	25.3±0.4	33.9±0.4	43.7±0.4	53.8±0.4	60.3±0.4	66.5±0.4	74.1±0.4
$M_{II,III}$	6.8±0.4	12.4±0.3	17.8±0.4	25.4±0.4	32.3±0.5	34.6±0.4	37.8±0.3	42.5±0.3
$M_{IV,V}$					6.6±0.5	3.7	2.2±0.3	2.3±0.4

Level	25 Mn	26 Fe	27 Co	28 Ni	29 Cu	30 Zn	31 Ga	32 Ge
K	6539.0±0.4 [6539.0][e] (6538.)	7112.0±0.9 [7111.3][e,f] (7111.2)	7708.9±0.3 [7708.9][e] (7709.5)	8332.8±0.4 [8332.8][e] (8331.6)	8978.9±0.4 [8978.9][e,g] (8980.3)	9658.6±0.6 [9658.6][g] (9660.7)	10367.1±0.5 [10367.1][g] (10368.2)	11103.1±0.7 [11103.8][g] (11103.6)
L_I	769.0±0.4 [769.0][d]	846.1±0.4 [846.1][d]	925.6±0.4 [925.6][d]	1008.1±0.4 [1008.1][d]	1096.1±0.4 [1096.0][d]	1193.6±0.9	1297.7±1.1	1414.3±0.7 [1413.6][g]
L_{II}	651.4±0.4	721.1±0.9 (720.8)	793.6±0.3 (793.8)	871.9±0.4 (870.6)	951.0±0.4 [950.0][h] (953.)	1042.8±0.6 (1045.)	1142.3±0.5	1247.8±0.7 (1249.)
L_{III}	640.3±0.4	708.1±0.9 (707.4)	778.6±0.3 (779.0)	854.7±0.4 (853.6)	931.1±0.4 [931.4][h] (933.)	1019.7±0.6 (1022.)	1115.4±0.5 (1117.)	1216.7±0.7 (1217.0)

* Wavelengths corresponding to these energy levels may be calculated from $\lambda\ (\text{nm}) = \dfrac{1.239852}{E(\text{keV})}$.

	25 Mn	26 Fe	27 Co	28 Ni	29 Cu	30 Zn	31 Ga	32 Ge
M_I	83.9±0.5	92.9±0.9	100.7±0.4	111.8±0.6	119.8±0.6	135.9±1.1	158.1±0.5	180.0±0.8
M_{II}							106.8±0.7	127.9±0.9
M_{II}, M_{III}	48.6±0.4	54.0±0.9 (54.)	59.5±0.3 (61.)	68.1±0.4 (66.)	73.6±0.4 (75.)	86.6±0.6 (86.)		
M_{III}							102.9±0.5	120.8±0.7
$M_{IV,V}$	3.3±0.5	3.6±0.9	2.9±0.3	3.6±0.4	1.6±0.4	8.1±0.6	17.4±0.5	28.7±0.7

	33 As	34 Se	35 Br	36 Kr	37 Rb	38 Sr	39 Y	40 Zr
K	11866.7±0.7 [11866.7]^{j} (11865.)	12657.8±0.7 [12657.8]^{k} (12654.5)	13473.7±0.4 (13470.)	14325.6±0.8 (14324.4)	15199.7±0.3 (15202.)	16104.6±0.3 (16107.)	17038.4±0.3 (17038.)	17997.6±0.4 (17999.)
L_I	1526.5±0.8 (1529.)	1653.9±3.5 (1652.5)	1782.0±0.4 [1782.0]^{j}	1921.0±0.6 [1921.2]^{k}	2065.1±0.3 [2065.4]^{j}	2216.3±0.3 [2216.2]^{l}	2372.5±0.3 [2372.7]^{l}	2531.6±0.3 [2531.6]^{l}
L_{II}	1358.6±0.7 (1358.7)	1476.2±0.7 (1474.7)	1596.0±0.4 [1596.2]^{j}	1727.2±0.5 [1727.2]^{k} (1730.)	1863.9±0.3 [1863.4]^{j}	2006.8±0.3 [2006.6]^{l} (2008.5)	2155.5±0.3 [2155.0]^{l} (2154.0)	2306.7±0.3 [2306.5]^{l} (2305.3)
L_{III}	1323.1±0.7 (1323.5)	1435.8±0.7 (1434.0)	1549.9±0.4 [1549.7]^{j}	1674.9±0.5 [1674.8]^{k} (1677.)	1804.4±0.3 [1804.6]^{j}	1939.6±0.3 [1939.9]^{l} (1941.)	2080.0±0.3 [2080.2]^{l} (2079.4)	2222.3±0.3 [2222.5]^{l} (2222.5)
M_I	203.5±0.7	231.5±0.7	256.5±0.4		322.1±0.3	357.5±0.3	393.6±0.3	430.3±0.3
M_{II}	146.4±1.2	168.2±1.3	189.3±0.4	222.7±1.1	247.4±0.3	279.8±0.3	312.4±0.4	344.2±0.4
M_{III}	140.5±0.8	161.9±1.0	181.5±0.4	213.8±1.1	238.5±0.3	269.1±0.3	300.3±0.4	330.5±0.4
M_{IV}			70.1±0.4		111.8±0.3	135.0±0.3	159.6±0.3	182.4±0.3
M_{IV}, M_V	41.2±0.7	56.7±0.8		88.9±0.8				
M_V			69.0±0.4		110.3±0.3	133.1±0.3	157.4±0.3	180.0±0.3
N_I			27.3±0.5	24.0±0.8	29.3±0.3	37.7±0.3	45.4±0.3	51.3±0.3
N_{II}			5.2±0.4		14.8±0.4			
N_{II}, N_{III}	2.5±1.0	5.6±1.3		10.6±1.9		19.9±0.3	25.6±0.4	28.7±0.4
N_{III}			4.6±0.4		14.0±0.3			

	41 Nb	42 Mo	43 Tc	44 Ru	45 Rh	46 Pd	47 Ag	48 Cd
K	[illegible] (18987.)	19999.5±0.3 (20004.)	21044.0±0.7	22117.2±0.3 (22119.)	23219.9±0.3 (23219.8)	24350.3±0.3 (24348.)	25514.0±0.3 (25516.)	26711.2±0.3 (26716.)
L_I	2697.7±0.3 [2697.7]^{l}	2865.5±0.3 [2866.0]^{l}	3042.5±0.4 [3042.5]^{l}	3224.0±0.3 [3224.3]^{l}	3411.9±0.3 [3412.0]^{l} (3417.)	3604.3±0.3 [3604.6]^{l} (3607.)	3805.8±0.3 [3806.2]^{m} (3807.)	4018.0±0.3 [4018.1]^{m} (4019.)
L_{II}	2464.7±0.3 [2464.7]^{l}	2625.1±0.3 [2624.5]^{l} (2627.)	2793.2±0.4 [2973.2]^{l}	2966.9±0.3 [2966.8]^{l} (2966.3)	3146.1±0.3 [3146.3]^{l} (3145.)	3330.3±0.3 [3330.3]^{l} (3330.3)	3523.7±0.3 [3523.6]^{k,m} (3526.)	3727.0±0.3 [3727.1]^{m} (3728.)
L_{III}	2370.5±0.3 [2370.6]^{l}	2520.2±0.3 [2520.2]^{l} (2523.2)	2676.9±0.4 [2676.9]^{l}	2837.9±0.3 [2837.7]^{l} (2837.7)	3003.8±0.3 [3003.5]^{k,l} (3002.)	3173.3±0.3 [3173.0]^{k,l} (3173.0)	3351.1±0.3 [3350.8]^{k} (3351.0)	3537.5±0.3 [3537.3]^{k} (3537.6)
M_I	468.4±0.3	504.6±0.3		585.0±0.3	627.1±0.3	669.9±0.3	717.5±0.3	770.2±0.3
M_{II}	378.4±0.4	409.7±0.4	444.9±1.5	482.8±0.3	521.0±0.3	559.1±0.3	602.4+0.3	650.7±0.3
M_{III}	363.0±0.4	392.3±0.3	425.0±1.5	460.6±0.3	496.2±0.3	531.5±0.3	571.4±0.3	616.5±0.3
M_{IV}	207.4±0.3	230.3±0.3	256.4±0.5	283.6±0.3	311.7±0.3	340.0±0.3	372.8±0.3	410.5±0.3
M_V	204.6±0.3	227.0±0.3	252.9±0.4	279.4±0.3	307.0±0.3	334.7±0.3	366.7±0.3	403.7±0.3
N_I	58.1±0.3	61.8±0.3		74.9±0.3	81.0±0.3	86.4±0.3	95.2±0.3	107.6±0.3
N_{II}							62.6±0.3	
N_{II}, N_{III}	33.9±0.4	34.8±0.4	38.9±1.9	43.1±0.4	47.9±0.4	51.1±0.4		66.9±0.4
N_{III}							55.9±0.3	
$N_{IV,V}$	3.2±0.3	1.8±0.3		2.0±0.3	2.5±0.4	1.5±0.3	3.3±0.3	9.3±0.3

	49 In	50 Sn	51 Sb	52 Te	53 I	54 Xe	55 Cs	56 Ba
K	27939.9±0.3	29200.1±0.4 (29195.)	30491.2±0.3 (30486.)	31813.8±0.3 (31811.)	33169.4±0.4 (33167.)	34561.4±1.1 (34590.)	[illegible] (35987.)	[illegible] (37452.)
L_I	4237.5±0.3 [4237.7]^{m} (4237.3)	4464.7±0.3 [4464.5]^{k} (4464.8)	4698.3±0.3 [4698.3]^{m} (4698.4)	4939.2±0.3 [4939.3]^{m} (4939.7)	5188.1±0.3 [5188.1]^{j}	5452.8±0.4 (5452.8)	5714.3±0.4 [5712.7]^{j} (5721.)	5988.8±0.4 [5986.8]^{j} (5996.)
L_{II}	3938.0±0.3 [3937.8]^{m} (3939.3)	4156.1±0.3 [4156.2]^{k} (4157.)	4380.4±0.3 [4380.6]^{m} (4382.)	4612.0±0.3 [4612.0]^{m} (4612.6)	4852.1±0.3 [4852.0]^{j}	5103.7±0.4 (5103.7)	5359.4±0.3 [5359.5]^{j} (5358.)	5623.6±0.3 [5623.6]^{j} (5623.3)
L_{III}	3730.1±0.3 [3730.0]^{k} (3730.2)	3928.8±0.3 [3928.8]^{k} (3928.8)	4132.2±0.3 [4132.2]^{k} (4132.3)	4341.4±0.3 [4341.2]^{k} (4341.8)	4557.1±0.3 [4557.1]^{j}	4782.2±0.4 (4782.2)	5011.9±0.3 [5012.0]^{j} (5011.3)	5247.0±0.3 [5247.3]^{j} (5247.0)
M_I	825.6±0.3	883.8±0.3	943.7±0.3	1006.0±0.3	1072.1±0.3		1217.1±0.4	1292.8±0.4
M_{II}	702.2±0.3	756.4±0.4	811.9±0.3	869.7±0.3	930.5±0.3	999.0±2.1	1065.0±0.5	1136.7±0.5

	49 In	50 Sn	51 Sb	52 Te	53 I	54 Xe	55 Cs	56 Ba
M_{III}	664.3±0.3	714.4±0.3	765.6±0.3	818.7±0.3	874.6±0.3	937.0±2.1	997.6±0.5	1062.2±0.5
M_{IV}	450.8±0.3	493.3±0.3	536.9±0.3	582.5±0.3	631.3±0.3		739.5±0.4	796.1±0.3
M_V	443.1±0.3	484.8±0.3	527.5±0.3	572.1±0.3	619.4±0.3	672.3±0.5	725.5±0.5	780.7±0.3
N_I	121.9±0.3	136.5±0.4	152.0±0.3	168.3±0.3	186.4±0.3		230.8±0.4	253.0±0.5
N_{II}	77.4±0.4	88.6±0.4	98.4±0.5	110.2±0.5	122.7±0.5	146.7±3.1	172.3±0.6	191.8±0.7
N_{III}							161.6±0.6	179.7±0.6
N_{IV}	16.2±0.3	23.9±0.3	31.4±0.3	39.8±0.3	49.6±0.3		78.8±0.5	92.5±0.5
N_V							76.5±0.5	89.9±0.5
O_I	0.1±4.5	0.9±0.5	6.7±0.5	11.6±0.6	13.6±0.6		22.7±0.5	39.1±0.6
O_{II}	0.8±0.4	1.1±0.5	2.1±0.4	2.3±0.5	3.3±0.5		13.1±0.5	16.6±0.5
O_{III}							11.4±0.5	14.6±0.5

	57 La	58 Ce	59 Pr	60 Nd	61 Pm	62 Sm	63 Eu	64 Gd
K	38924.6±0.4 (38934.)	40443.0±0.4 (40453.)	41990.6±0.5 (42002.)	43568.9±0.4 (43574.)	45184.0±0.7 (45198.)	46834.2±0.5 (46849.)	48519.0±0.4 (48519.)	50239.1±0.5 (50233.)
L_I	6266.3±0.5 [6266.3][n]	6548.8±0.5 [6548.5][n]	6834.8±0.5 [6834.9][n]	7126.0±0.4 [7125.8][n] (7129.)	7427.9±0.8 [7427.9][o]	7736.8±0.5 [7736.2][n] (7748.)	8052.0±0.4 [8051.7][n] (8061.)	8375.6±0.5 [8375.4][n] (8386.)
L_{II}	5890.6±0.4 [5890.7][n]	6164.2±0.4 [6164.3][n]	6440.4±0.5 [6440.2][n]	6721.5±0.4 [6721.8][n] (6723.)	7012.8±0.6 [7012.8][o]	7311.8±0.4 [7312.0][n] (7313.)	7617.1±0.4 [7617.6][n] (7620.)	7930.3±0.4 [7930.5][n] (7931.)
L_{III}	5482.7±0.4 [5482.6][n]	5723.4±0.4 [5723.6][n]	5964.3±0.4 [5964.3][n]	6207.9±0.4 [6208.0][n] (6209.)	6459.3±0.6 [6459.4][o]	6716.2±0.5 [6716.8][n] (6717.)	6976.9±0.4 [6976.7][n] (6981.)	7242.8±0.4 [7242.8][n] (7243.)
M_I	1361.3±0.3	1434.6±0.6	1511.0±0.8	1575.3±0.7		1722.8±0.8	1800.0±0.5	1880.8±0.5
M_{II}	1204.4±0.6	1272.8±0.6	1337.4±0.7	1402.8±0.6	1471.4±6.2	1540.7±1.2	1613.9±0.7	1688.3±0.7
M_{III}	1123.4±0.5	1185.4±0.5	1242.2±0.6	1297.4±0.5	1356.9±1.4	1419.8±1.1	1480.6±0.6	1544.0±0.8
M_{IV}	848.5±0.4	901.3±0.6	951.1±0.6	999.9±0.6	1051.5±0.9	1106.0±0.8	1160.6±0.6	1217.2±0.6
M_V	831.7±0.4	883.3±0.5	931.0±0.6	977.7±0.6	1026.9±1.0	1080.2±0.6	1130.9±0.6	1185.2±0.6
N_I	270.4±0.8	289.6±0.7	304.5±0.9	315.2±0.8		345.7±0.9	360.2±0.7	375.8±0.7
N_{II}	205.8±1.2	223.3±1.1	236.3±1.5	243.3±1.6	242.±16.	265.6±1.9	283.9±1.0	288.5±1.2
N_{III}	191.4±0.9	207.2±0.9	217.6±1.1	224.6±1.3		247.4±1.5	256.6±0.8	270.9±0.9
$N_{IV,V}$	98.9±0.8	110.0±0.6	113.2±0.7	117.5±0.7	120.4±2.0	129.0±1.2	133.2±0.6	140.5±0.8
$N_{VI,VII}$		0.1±1.2	2.0±0.6	1.5±0.9		5.5±1.1	0.0±3.2	0.1±3.5
O_I	32.3±7.2	37.8±1.3	37.4±1.0	37.5±0.9		37.4±1.5	31.8±0.7	36.1±0.8
$O_{II,III}$	14.4±1.2	19.8±1.2	22.3±0.7	21.1±0.8		21.3±1.5	22.0±0.6	20.3±1.2

	65 Tb	66 Dy	67 Ho	68 Er	69 Tm	70 Yb	71 Lu	72 Hf
K	51995.7±0.5 (52002.)	53788.5±0.5 (53793.)	55617.7±0.5 (55619.)	57485.5±0.5 (57487.)	59389.6±0.5	61332.3±0.5 (61300.)	63313.8±0.5 (63310.)	65350.8±0.6 (65310.)
L_I	8708.0±0.5 [8707.6][n] (8717.)	9045.8±0.5 [9046.5][n]	9394.2±0.4 [9394.3][n] (9399.)	9751.3±0.4 [9751.5][n] (9757.)	10115.7±0.4 [10115.6][n] (10121.)	10486.4±0.4 [10487.3][n] (10490.)	10870.4±0.4 [10870.1][n] (10874.)	11270.7±0.4 [11271.6][o] (11274.)
L_{II}	8251.6±0.4 [8251.8][n] (8253.)	8580.6±0.4 [8580.4][n] (8583.)	8917.8±0.4 [8918.2][n] (8916.)	9264.3±0.4 [9264.3][n] (9262.)	9616.9±0.4 [9617.1][n] (9617.1)	9978.2±0.4 [9977.9][n] (9976.)	10348.6±0.4 [10349.0][n] (10345.)	10739.4±0.4 [10738.9][o] (10736.)
L_{III}	7514.0±0.4 [7514.2][n] (7515.)	7790.1±0.4 [7789.6][n] (7789.7)	8071.1±0.4 [8070.6][n] (8068.)	8357.9±0.4 [8357.6][n] (8357.5)	8648.0±0.4 [8647.8][n] (8649.6)	8943.6±0.4 [8942.6][n] (8944.1)	9244.1±0.4 [9243.8][n]	9560.7±0.4 [9560.4][o] (9558.)
M_I	1967.5±0.6	2046.8±0.4	2128.3±0.6	2206.5±0.6	2306.8±0.7	2398.1±0.4	2491.2±0.5	2600.9±0.4
M_{II}	1767.7±0.9	1841.8±0.5	1922.8±1.0	2005.8±0.6	2089.8±1.1	2173.0±0.4	2263.5±0.4	2365.4±0.4
M_{III}	1611.3±0.8	1675.6±0.9	1741.2±0.9	1811.8±0.6	1884.5±1.1	1949.8±0.5	2023.6±0.5	2107.6±0.4
M_{IV}	1275.0±0.6	1332.5±0.4	1391.5±0.7	1453.3±0.5	1514.6±0.7	1576.3±0.4	1639.4±0.4	1716.4±0.4
M_V	1241.2±0.7	1294.9±0.4	1351.4±0.8	1409.3±0.5	1467.7±0.9	1527.8±0.4	1588.5±0.4	1661.7±0.4
N_I	397.9±0.8	416.3±0.5	435.7±0.8	449.1±1.0	471.7±0.9	487.2±0.6	506.2±0.6	538.1±0.4
N_{II}	310.2±1.2	331.8±0.6	343.5±1.4	366.2±1.5	385.9±1.6	396.7±0.7	410.1±1.8	437.0±0.5
N_{III}	385.0±1.0	292.9±0.6	306.6±0.9	320.0±0.7	336.6±1.6	343.5±0.5	359.3±0.5	380.4±0.5
N_{IV}	147.0±0.8	154.2±0.5	161.0±1.0	176.7±1.2	179.6±1.2	198.1±0.5	204.8±0.5	223.8±0.4
N_V				167.6±1.5		184.9±1.3	195.0±0.4	213.7±0.5

	65 Tb	66 Dy	67 Ho	68 Er	69 Tm	70 Yb	71 Lu	72 Hf
$N_{VI,VII}$	2.6±1.5	4.2±1.6	3.7±3.0	4.3±1.4	5.3±1.9	6.3±1.0	6.9±0.5	17.1±0.5
O_I	39.0±0.8	62.9±0.5	51.2±1.3	59.8±1.7	53.2±3.0	54.1±0.5	56.8±0.5	64.9±0.4
O_{II}	25.4±0.8	26.3±0.6	20.3±1.5	29.4±1.6	32.3±1.6	23.4±0.6	28.0±0.6	38.1±0.6
O_{III}								30.6±0.6

	73 Ta	74 W	75 Re	76 Os	77 Ir	78 Pt	79 Au	80 Hg
K	67416.4±0.6 (67403.)	69525.0±0.3 (69508.)	71676.4±0.4 (71658.)	73870.8±0.5	76111.0±0.5	78394.8±0.7 (78381.)	80724.9±0.5 (80720.)	83102.3±0.8
L_I	11681.5±0.3 [11680.2][p] (11682.)	12099.8±0.3 [12098.2][p] (12099.6)	12526.7±0.4 (12530.)	12968.0±0.4 (12972.)	13418.5±0.3 (13423.)	13879.9±0.4 (13883.)	14352.8±0.4 (14353.7)	14839.3±1.0 (14842.)
L_{II}	11136.1±0.3 [11136.1][p] (11132.)	11544.0±0.3 [11541.4][p] (11538.)	11958.7±0.3 [11956.9][p] (11954.)	12385.0±0.4 (12381.)	12824.1±0.3 [12824.0][e,p] (12820.)	13272.6±0.3 [13272.6][e,p] (13272.3)	13733.6±0.3 [13733.5][e,p] (13736.)	14208.7±0.7 (14215.)
L_{III}	9881.1±0.3 [9880.3][p] (9877.7)	10206.8±0.3 [10204.2][p] (10200.)	10535.3±0.3 [10534.2][p] (10531.)	10870.9±0.3 [10870.7][p] (10868.)	11215.2±0.3 [11215.1][e,p] (11212.)	11563.7±0.3 [11563.7][e,p] (11562.)	11918.7±0.3 [11918.2][e,p] (11921.)	12283.9±0.4 [12284.0][e,p] (12286.)
M_I	2708.0±0.4	2819.6±0.4	2931.7±0.4	3048.5±0.4	3173.7±1.7	3296.0±0.9	3424.9±0.3 [3424.8][p]	3561.6±1.1
M_{II}	2468.7±0.3 [2468.6][p]	2574.9±0.3 [2575.0][p]	2681.6±0.4	2792.2±0.3 [2791.9][p]	2908.7±0.3 [2909.1][p]	3026.5±0.4 [3026.5][p'] (3029.)	3147.8±0.4 [3149.5][p]	3278.5±1.3
M_{III}	2194.0±0.3 [2194.1][p]	2281.0±0.3 [2281.0][p]	2367.3±0.3 [2367.3][p]	2457.2±0.4 [2457.4][p]	2550.7±0.3 [2550.5][p] (2550.5)	2645.4±0.4 [2645.5][p] (2645.9)	2743.0±0.3 [2743.1][p] (2744.0)	2847.1±0.4 [2847.1][p]
M_{IV}	1793.2±0.3 [1793.1][p]	1871.6±0.3 [1871.4][p]	1948.9±0.3 [1948.9][p]	2030.8±0.3 [2031.0][p]	2116.1±0.3 [2116.1][p]	2201.9±0.3 [2201.9][p]	2291.1±0.3 [2291.2][p] (2307.)	2384.9±0.3 [2384.9][p]
M_V	1735.1±0.3 [1735.2][p]	1809.2±0.3 [1809.3][p]	1882.9±0.3 [1882.9][p]	1960.1±0.3 [1960.2][p]	2040.4±0.3 [2040.5][p]	2121.6±0.3 [2121.6][p]	2205.7±0.3 [2206.1][p] (2220.)	2294.9±0.3 [2294.9][p]
N_I	565.5±0.5	595.0±0.4	625.0±0.4	654.3±0.5	690.1±0.4	722.0±0.6	758.8±0.4	800.3±1.0
N_{II}	464.8±0.5	491.6±0.4	517.9±0.5	546.5±0.5	577.1±0.4	609.2±0.6	643.7±0.5	676.9±2.4
N_{III}	404.5±0.4	425.3±0.5	444.4±0.5	468.2±0.6	494.3±0.6	519.0±0.6	545.4±0.5	571.0±1.4
N_{IV}	241.3±0.4	258.8±0.4	273.7±0.5	289.4±0.5	311.4±0.4	330.8±0.5	352.0±0.4	378.3±1.0
N_V	229.3±0.3	245.4±0.4	260.2±0.4	272.8±0.6	294.9±0.4	313.3±0.4	333.9±0.4	359.8±1.2
N_{VI}	25.0±0.4	36.5±0.4	40.6±0.4	46.3±0.6	63.4±0.4	74.3±0.4	86.4±0.4	102.2±0.5
N_{VII}		33.6±0.4			60.5±0.4	71.1±0.5	82.8±0.5	98.5±0.5
O_I	71.1±0.5	77.1±0.4	82.8±0.5	83.7±0.6	95.2±0.4	101.7±0.4	107.8±0.7	120.3±1.3
O_{II}	44.9±0.4	46.8±0.5	45.6±0.7	58.0±1.1	63.0±0.6	65.3±0.7	71.7±0.7	80.5±1.3
O_{III}	36.4±0.4	35.6±0.5	34.6±0.6	45.4±1.0	50.5±0.6	51.7±0.7	53.7±0.7	57.6±1.3
$O_{IV,V}$	5.7±0.4	6.1±0.4	3.5±0.5		3.8±0.4	2.2±1.3	2.5±0.5	6.4±1.4

	81 Tl	82 Pb	83 Bi	84 Po	85 At	86 Rn	87 Fr	88 Ra
K	85530.4±0.6	88004.5±0.7 (88005.)	90525.9±0.7 (90534.)	93105.0±3.8	95729.9±7.7	98404.±12.	101137.±13.	103921.9±7.2
L_I	15346.7±0.4 (15343.)	15860.8±0.5 (15855.)	16387.5±0.4 (16376.)	16939.3±9.8	17493.±29.	18049.±38.	18639.±40.	19236.7±1.5 (19236.0)
L_{II}	14697.9±0.3 [14697.3][p] (14699.)	15200.0±0.4 (15205.)	15711.1±0.3 [15708.4][p] (15719.)	16244.3±2.4	16784.7±2.5	17337.1±3.4	17906.5±3.5	18484.3±1.5 (18486.0)
L_{III}	12657.5±0.3 [12656.3][e,p] (12660.)	13035.2±0.3 [13034.9][e,p] (13041.)	13418.6±0.3 [13418.3][e,p] (13426.)	13813.8±1.0 ⟨13813.8⟩	14213.5±2.0 ⟨14213.5⟩	14619.4±3.0 ⟨14619.4⟩	15031.2±3.0 ⟨15031.2⟩	15444.4±1.5 (15444.0)
M_I	3704.1±0.4	3850.7±0.5	3999.1±0.3 [3999.1][p]	4149.4±3.9	⟨4317.⟩	⟨4482.⟩	⟨4652.⟩	4822.0±1.5
M_{II}	3415.7±0.3 [3415.7][p]	3554.2±0.3 [3554.2][p]	3696.3±0.3 [3696.4][p]	3854.1±9.8	4008.±28.	4159.±38.	4327.±40.	4489.5±1.8
M_{III}	2956.6±0.3 [2956.5][p]	3066.4±0.4 [3066.3][p]	3176.9±0.3 [3176.8][p]	3301.9±9.9	3426.±29.	3538.±38.	3663.±40.	3791.8±1.7
M_{IV}	2485.1±0.3 [2485.2][p]	2585.6±0.3 [2585.5][p] (2606.)	2687.6±0.3 [2687.4][p]	2798.0±1.2	2908.7±2.1	3021.5±3.1	3136.2±3.1	3248.4±1.6

	81 Tl	82 Pb	83 Bi	84 Po	85 At	86 Rn	87 Fr	88 Ra
M_V	2389.3±0.3 [2389.4][p]	2484.0±0.3 [2484.2][p] (2502.)	2579.6±0.3 [2579.5][p]	2683.0±1.1	2786.7±2.1	2892.4±3.1	2999.9±3.1	3104.9±1.6
N_I	845.5±0.5	893.6±0.7	938.2±0.3 [938.7][p]	995.3±2.9	⟨1042.⟩	⟨1097.⟩	⟨1153.⟩	1208.4±1.6
N_{II}	721.3±0.8	763.9±0.8	805.3±0.3 [805.3][p]	851.±12.	886.±30.	929.±40.	980±42.	1057.6±1.8
N_{III}	609.0±0.5	644.5±0.6	678.9±0.3 [678.9][p]	705.±14.	740.±30.	768.±40.	810±43.	879.1±1.8
N_{IV}	406.6±0.4	435.2±0.5	463.6±0.3 [463.6][p]	500.2±2.4	533.2±3.2	566.6±4.0	603.3±4.1	635.9±1.6
N_V	386.2±0.5	412.9±0.6	440.0±0.3 [440.1][p]	473.4±1.3			577.±34.	602.7±1.7
N_{VI}	122.8±0.4	142.9±0.4	161.9±0.5	}				298.9±2.4 (N_{VI}, N_{VII})
N_{VII}	118.5±0.4	138.1±0.4	157.4±0.6	}				
O_I	136.3±0.7	147.3±0.8	159.3±0.7					254.4±2.1
O_{II}	99.6±0.6	104.8±1.0	116.8±0.7					200.4±2.0
O_{III}	75.4±0.6	86.0±1.0	92.8±0.6					152.8±2.0
O_{IV}	15.3±0.4	21.8±0.4	26.5±0.5	} 31.4±3.2 (O_{IV}, O_V)				67.2±1.7
O_V	13.1±0.4	19.2±0.4	24.4±0.6					
P_I		3.1±1.0						43.5±2.2
$P_{II,III}$		0.7±1.0	2.7±0.7					18.8±1.8

	89 Ac	90 Th	91 Pa	92 U	93 Np	94 Pu	95 Am	96 Cm
K	106755.3±5.3	109650.9±0.9	112601.4±2.4	115606.1±1.6	118678.±33.	121818.±44.	125027.±55.	128220
L_I	19840.±18.	20472.1±0.5 (20464.)	21104.6±1.8 (21128.)	21757.4±0.3 (21771.)	22426.8±0.9	23097.2±1.6 (23109.)	23772.9±2.0 ⟨23772.9⟩	24460
L_{II}	19083.2±2.8	19693.2±0.4 (19683.)	20313.7±1.5 (20319.)	20947.6±0.3 (20945.)	21600.5±0.4	22266.2±0.7 (22253.)	22944.0±1.0	23779
L_{III}	15871.0±2.0 ⟨15871.0⟩	16300.3±0.3 [16299.6][q] (16299.)	16733.1±1.4 (16733.)	17166.3±0.3 [17168.5][r] (17165.)	17610.0±0.4 (17606.2)	18056.8±0.6 (18053.1)	18504.1±0.9 (18504.1)	18930
M_I	⟨5002.⟩	5182.3±0.3 [5182.3][q]	5366.9±1.6	5548.0±0.4	5723.2±3.6	5932.9±1.4	6120.5±7.5	6288
M_{II}	4656.±18.	4830.4±0.4 [4830.6][q]	5000.9±2.3	5182.2±0.4 [5180.9][r]	5366.2±0.7 [5366.4][s]	5541.2±1.7	5710.2±2.1	5895
M_{III}	3909.±18.	4046.1±0.4 [4046.1][q] (4041.)	4173.8±1.8	4303.4±0.3 [4303.6][r] (4299.)	4434.7±0.5 [4434.6][s]	4556.6±1.5	4667.0±2.1	4797
M_{IV}	3370.2±2.1	3490.8±0.3 [3490.7][q] (3485.)	3611.2±1.4 (3608.)	3727.6±0.3 [3728.1][r] (3720.)	3850.3±0.4 [3849.8][s]	3972.6±0.6 [3972.7][t]	4092.1±1.0	4227
M_V	3219.0±2.1	3332.0±0.3 [3332.1][q] (3325.)	3441.8±1.4 (3436.)	3551.7±0.3 [3551.7][r] (3545.)	3665.8±0.4 [3664.2][s]	3778.1±0.6 [3778.0][t]	3886.9±1.0	3971
N_I	⟨1269.⟩	1329.5±0.4 [1329.8][q]	1387.1±1.9	1440.8±0.4 [1441.3][r]	1500.7±0.8 [1500.7][s]	1558.6±0.8	1617.1±1.1	1643
N_{II}	1080.±19.	1168.2±0.4 [1168.3][q]	1224.3±1.6	1272.6±0.3 [1272.5][r]	1327.7±0.8 [1327.7][s]	1372.1±1.8	1411.8±8.3	1440
N_{III}	890.±19.	967.3±0.4 [967.6][q]	1006.7±1.7	1044.9±0.3 [1044.9][r]	1086.8±0.7 [1086.8][s]	1114.8±1.6	⟨1135.7⟩	1154
N_{IV}	674.9±3.7	714.1±0.4 [714.4][q]	743.4±2.1	780.4±0.3 [779.7][r]	815.9±0.5 [817.1][s]	848.9±0.6 [848.9][t]	878.7±1.0	
N_V		676.4±0.4 [676.4][q]	708.2±1.8	737.7±0.3 [737.6][r]	770.3±0.4 [773.2][s]	801.4±0.6 [801.4][t]	827.6±1.0	
N_{VI}		344.4±0.3 [344.2][q]	371.2±1.6	391.3±0.6	415.0±0.8 [415.0][s]	445.8±1.7		
N_{VII}		335.2±0.4 [335.0][q]	359.5±1.6	380.9±0.9	404.4±0.5 [404.4][s]	432.4±2.1		
O_I		290.2±0.8	309.6±4.3	323.7±1.1		351.9±2.4		385
O_{II}		229.4±1.1 }	222.9±3.9 (O_{II}, O_{III})	{ 259.3±0.5	283.4±0.8 [283.4][s]	274.1±4.7		
O_{III}		181.8±0.4 [181.8][q] }		{ 195.1±1.3	206.1±0.7 [206.1][s]	206.5±4.7		

	89 Ac	90 Th	91 Pa	92 U	93 Np	94 Pu	95 Am	96 Cm
O_{IV}		94.3±0.4 [94.4][q]	94.1±2.8	105.0±0.5	109.3±0.7 [108.8][s]	116.0±1.2	115.8±1.3	
O_{V}		87.9±0.3 [88.1][q]		96.3±1.4	101.3±0.5 [101.4][s]	105.4±1.0	103.3±1.1	
P_{I}		59.5±1.1		70.7±1.2				
P_{II}		49.0±2.5		42.3±9.0				
P_{III}		43.0±2.5		32.3±9.0				

	97 Bk	98 Cf	99 Es	100 Fm	101 Md	102 No	103 Lw
K	[131590±40][u]	135960	139490	143090	146780	150540	154380
L_{I}	[25275±17][u]	26110	26900	27700	28530	29380	30240
L_{II}	[24385±17][u]	25250	26020	26810	27610	28440	29280
L_{III}	[19452±20][u]	19930	20410	20900	21390	21880	22360
M_{I}	[6556±21][u]	6754	6977	7205	7441	7675	7900
M_{II}	[6147±31][u]	6359	6574	6793	7019	7245	7460
M_{III}	[4977±31][u]	5109	5252	5397	5546	5688	5710
M_{IV}	4366	4497	4630	4766	4903	5037	5150
M_{V}	4132	4253	4374	4498	4622	4741	4860
N_{I}	[1755±22][u]	1799	1868	1937	2010	2078	2140
N_{II}	1554	1616	1680	1747	1814	1876	1930
N_{III}	1235	1279	1321	1366	1410	1448	1480
O_{I}	[398±22][u]	419	435	454	472	484	490

[a] J. E. Mack, 1949, as given in C. E. Moore, *Atomic Energy Levels* (U. S. National Bureau of Standards, Washington, D. C., 1949), Vol. 1, p. 1.
[b] G. Herzberg, 1957, as given in C. E. Moore, *Atomic Energy Levels* (U. S. National Bureau of Standards, Washington, D. C., 1958), Vol. 3, p. 238.
[c] See Ref. 18.
[d] A. Fahlman, D. Hamrin, R. Nordberg, C. Nordling, and K. Siegbahn, Phys. Rev. Letters **14**, 127 (1965). See also Ref. 26.
[e] See Ref. 15.
[f] See Ref. 11.
[g] C. Nordling, Arkiv Fysik **15**, 397 (1959).
[h] E. Sokolowski, C. Nordling, and K. Siegbahn, Arkiv Fysik **12**, 301 (1957).
[i] C. Nordling and S. Hagström, Arkiv Fysik **16**, 515 (1960).
[j] I. Andersson and S. Hagström, Arkiv Fysik **27**, 161 (1964).
[k] M. O. Krause, Phys. Rev. **140**, A1845 (1965).
[l] A. Fahlman, O. Hörnfeldt, and C. Nordling, Arkiv Fysik **23**, 75 (1962).
[m] P. Bergvall, O. Hörnfeldt, and C. Nordling, Arkiv Fysik **17**, 113 (1960).
[n] P. Bergvall and S. Hagström, Arkiv Fysik **17**, 61 (1960).
[o] S. Hagström, Z. Physik **178**, 82 (1964).
[p] A. Fahlman and S. Hagström, Arkiv Fysik **27**, 69 (1964).
[q] C. Nordling and S. Hagström, Z. Physik **178**, 418 (1964).
[r] C. Nordling and S. Hagström, Arkiv Fysik **15**, 431 (1959).
[s] S. Hagström, Bull. Am. Phys. Soc. **11**, 389 (1966).
[t] A. Fahlman, K. Hamrin, R. Nordberg, C. Nordling, K. Siegbahn, and L. W. Holm, Phys. Letters **19**, 643 (1966).
[u] J. M. Hollander, M. D. Holtz, T. Novakov, and R. L. Graham, Arkiv Fysik **28**, 375 (1965).

ELECTRON BINDING ENERGIES OF THE ELEMENTS

Gwyn P. Williams

This table gives the binding energies in electron volts (eV) for selected electronic levels of the elements. For metallic elements the binding energy is referred to the Fermi level; for semiconductors, to the valence band maximum; and for gases and insulators, to the vacuum level. The atomic number is listed after the element name.

REFERENCES

1. Fluggle and Martensson, J. *Elect. Spect.*, 21, 275, 1980.
2. Cardona, M. and Ley, L., *Photoemission from Solids*, Springer Verlag, Heidelberg, 1978.
3. Bearden, J. A. and Burr, A. F., *Rev. Mod. Phys.*, 39, 125, 1967.

Actinium (89)

K	1s	106755
L I	2s	19840
L II	$2p_{1/2}$	19083
L III	$2p_{3/2}$	15871
M I	3s	5002
M II	$3p_{1/2}$	4656
M III	$3p_{3/2}$	3909
M IV	$3d_{3/2}$	3370
M V	$3d_{5/2}$	3219
N I	4s	1269[a]
N II	$4p_{1/2}$	1080[a]
N III	$4p_{3/2}$	890[a]
N IV	$4d_{3/2}$	675[a]
N V	$4d_{5/2}$	639[a]
N VI	$4f_{5/2}$	319[a]
N VII	$4f_{7/2}$	319[a]
O I	5s	272[a]
O II	$5p_{1/2}$	215[a]
O III	$5p_{3/2}$	167[a]
O IV	$5d_{3/2}$	80[a]
O V	$5d_{5/2}$	80[a]
P I	6s	—
P II	$6p_{1/2}$	—
P III	$6p_{3/2}$	—

Aluminum (13)

K	1s	1559.0
L I	2s	117.8[a]
L II	$2p_{1/2}$	72.9[a]
L III	$2p_{3/2}$	72.5[a]

Antimony (51)

K	1s	30419
L I	2s	4698
L II	$2p_{1/2}$	4380
L III	$2p_{3/2}$	4132
M I	3s	946[b]
M II	$3p_{1/2}$	812.7[b]
M III	$3p_{3/2}$	766.4[b]
M IV	$3d_{3/2}$	537.5[b]
M V	$3d_{5/2}$	528.2[b]
N I	4s	153.2[b]
N II	$4p_{1/2}$	95.6[b,c]
N III	$4p_{3/2}$	95.6[b]
N IV	$4d_{3/2}$	33.3[b]
N V	$4d_{5/2}$	32.1[b]

Argon (18)

K	1s	3205.9[a]
L I	2s	326.3[a]
L II	$2p_{1/2}$	250.6[a]
L III	$2p_{3/2}$	248.4[a]
M I	3s	29.3[a]
M II	$3p_{1/2}$	15.9[a]
M III	$3p_{3/2}$	15.7[a]

Arsenic (33)

K	1s	11867
L I	2s	1527.0[a,d]
L II	$2p_{1/2}$	1359.1[a,d]
L III	$2p_{3/2}$	1323.6[a,d]
M I	3s	204.7[a]
M II	$3p_{1/2}$	146.2[a]
M III	$3p_{3/2}$	141.2[a]
M IV	$3d_{3/2}$	41.7[a]
M V	$3d_{5/2}$	41.7[a]

Astatine (85)

K	1s	95730
L I	2s	17493
L II	$2p_{1/2}$	16785
L III	$2p_{3/2}$	14214
M I	3s	4317
M II	$3p_{1/2}$	4008
M III	$3p_{3/2}$	3426
M IV	$3d_{3/2}$	2909
M V	$3d_{5/2}$	2787
N I	4s	1042[a]
N II	$4p_{1/2}$	886[a]
N III	$4p_{3/2}$	740[a]
N IV	$4d_{3/2}$	533[a]
N V	$4d_{5/2}$	507[a]
N VI	$4f_{5/2}$	210[a]
N VII	$4f_{7/2}$	210[a]
O I	5s	195[a]
O II	$5p_{1/2}$	148[a]
O III	$5p_{3/2}$	115[a]
O IV	$5d_{3/2}$	40[a]
O V	$5d_{5/2}$	40[a]

Barium (56)

K	1s	37441
L I	2s	5989
L II	$2p_{1/2}$	5624
L III	$2p_{3/2}$	5247
M I	3s	1293[a,d]
M II	$3p_{1/2}$	1137[a,d]
M III	$3p_{3/2}$	1063[a,d]
M IV	$3d_{3/2}$	795.7[a]
M V	$3d_{5/2}$	780.5[a]
N I	4s	253.5[b]
N II	$4p_{1/2}$	192
N III	$4p_{3/2}$	178.6[b]
N IV	$4d_{3/2}$	92.6[b]
N V	$4d_{5/2}$	89.9[b]
N VI	$4f_{5/2}$	—
N VII	$4f_{7/2}$	—
O I	5s	30.3[b]
O II	$5p_{1/2}$	17.0[b]
O III	$5p_{3/2}$	14.8[b]

Beryllium (4)

K	1s	111.5[a]

Bismuth (83)

K	1s	90526
L I	2s	16388
L II	$2p_{1/2}$	15711
L III	$2p_{3/2}$	13419
M I	3s	3999
M II	$3p_{1/2}$	3696
M III	$3p_{3/2}$	3177
M IV	$3d_{3/2}$	2688
M V	$3d_{5/2}$	2580
N I	4s	939[b]
N II	$4p_{1/2}$	805.2[b]
N III	$4p_{3/2}$	678.8[b]
N IV	$4d_{3/2}$	464.0[b]
N V	$4d_{5/2}$	440.1[b]
N VI	$4f_{5/2}$	162.3[b]
N VII	$4f_{7/2}$	157.0[b]
O I	5s	159.3[a,d]
O II	$5p_{1/2}$	119.0[b]
O III	$5p_{3/2}$	92.6[b]
O IV	$5d_{3/2}$	26.9[b]
O V	$5d_{5/2}$	23.8[b]

Boron (5)

K	1s	188[a]

Bromine (35)

K	1s	13474
L I	2s	1782[a]
L II	$2p_{1/2}$	1596[a]
L III	$2p_{3/2}$	1550[a]
M I	3s	257[a]
M II	$3p_{1/2}$	189[a]
M III	$3p_{3/2}$	182[a]
M IV	$3d_{3/2}$	70[a]
M V	$3d_{5/2}$	69[a]

Cadmium (48)

K	1s	26711
L I	2s	4018
L II	$2p_{1/2}$	3727
L III	$2p_{3/2}$	3538
M I	3s	772.0[b]
M II	$3p_{1/2}$	652.6[b]
M III	$3p_{3/2}$	618.4[b]
M IV	$3d_{3/2}$	411.9[b]
M V	$3d_{5/2}$	405.2[b]
N I	4s	109.8[b]
N II	$4p_{1/2}$	63.9[b,c]
N III	$4p_{3/2}$	63.9[b,c]
N IV	$4d_{3/2}$	11.7[b]
N V	$4d_{5/2}$	10.7[b]

Calcium (20)

K	1s	4038.5[a]
L I	2s	438.4[b]
L II	$2p_{1/2}$	349.7[b]
L III	$2p_{3/2}$	346.2[b]
M I	3s	44.3[b]
M II	$3p_{1/2}$	25.4[b]
M III	$3p_{3/2}$	25.4[b]

Carbon (6)

K	1s	284.2[a]

Cerium (58)

K	1s	40443
L I	2s	6548
L II	$2p_{1/2}$	6164
L III	$2p_{3/2}$	5723
M I	3s	1436[a,d]
M II	$3p_{1/2}$	1274[a,d]
M III	$3p_{3/2}$	1187[a,d]
M IV	$3d_{3/2}$	902.4[a]
M V	$3d_{5/2}$	883.8[a]

N I	4s	291.0[a]
N II	$4p_{1/2}$	223.3
N III	$4p_{3/2}$	206.5[a]
N IV	$4d_{3/2}$	109[a]
N V	$4d_{5/2}$	—
N VI	$4f_{5/2}$	0.1
N VII	$4f_{7/2}$	0.1
O I	5s	37.8
O II	$5p_{1/2}$	19.8[a]
O III	$5p_{3/2}$	17.0[a]

Cesium (55)

K	1s	35985
L I	2s	5714
L II	$2p_{1/2}$	5359
L III	$2p_{3/2}$	5012
M I	3s	1211[a,d]
M II	$3p_{1/2}$	1071[a]
M III	$3p_{3/2}$	1003[a]
M IV	$3d_{3/2}$	740.5[a]
M V	$3d_{5/2}$	726.6[a]
N I	4s	232.3[a]
N II	$4p_{1/2}$	172.4[a]
N III	$4p_{3/2}$	161.3[a]
N IV	$4d_{3/2}$	79.8[a]
N V	$4d_{5/2}$	77.5[a]
N VI	$4f_{5/2}$	—
N VII	$4f_{7/2}$	—
O I	5s	22.7
O II	$5p_{1/2}$	14.2[a]
O III	$5p_{3/2}$	12.1[a]

Chlorine (17)

K	1s	2822.0
L I	2s	270[a]
L II	$2p_{1/2}$	202[a]
L III	$2p_{3/2}$	200[a]

Chromium(24)

K	1s	5989
L I	2s	696.0[b]
L II	$2p_{1/2}$	583.8[b]
L III	$2p_{3/2}$	574.1[b]
M I	3s	74.1[b]
M II	$3p_{1/2}$	42.2[b]
M III	$3p_{3/2}$	42.2[b]

Cobalt (27)

K	1s	7709
L I	2s	925.1[b]
L II	$2p_{1/2}$	793.2[b]
L III	$2p_{3/2}$	778.1[b]
M I	3s	101.0[b]
M II	$3p_{1/2}$	58.9[b]
M III	$3p_{3/2}$	58.9[b]

Copper (29)

K	1s	8979
L I	2s	1096.7[b]
L II	$2p_{1/2}$	952.3[b]
L III	$2p_{3/2}$	932.5[b]
M I	3s	122.5[b]
M II	$3p_{1/2}$	77.3[b]
M III	$3p_{3/2}$	75.1[b]

Dysprosium (66)

K	1s	53789
L I	2s	9046
L II	$2p_{1/2}$	8581
L III	$2p_{3/2}$	7790
M I	3s	2047
M II	$3p_{1/2}$	1842
M III	$3p_{3/2}$	1676
M IV	$3d_{3/2}$	1333
M V	$3d_{5/2}$	1292[a]
N I	4s	414.2[a]
N II	$4p_{1/2}$	333.5[a]
N III	$4p_{3/2}$	293.2[a]
N IV	$4d_{3/2}$	153.6[a]
N V	$4d_{5/2}$	153.6[a]
N VI	$4f_{5/2}$	8.0[a]
N VII	$4f_{7/2}$	4.3[a]
O I	5s	49.9[a]
O II	$5p_{1/2}$	26.3
O III	$5p_{3/2}$	26.3

Erbium (68)

K	1s	57486
L I	2s	9751
L II	$2p_{1/2}$	9264
L III	$2p_{3/2}$	8358
M I	3s	2206
M II	$3p_{1/2}$	2006
M III	$3p_{3/2}$	1812
M IV	$3d_{3/2}$	1453
M V	$3d_{5/2}$	1409
N I	4s	449.8[a]
N II	$4p_{1/2}$	366.2
N III	$4p_{3/2}$	320.2[a]
N IV	$4d_{3/2}$	167.6[a]
N V	$4d_{5/2}$	167.6[a]
N VI	$4f_{5/2}$	—
N VII	$4f_{7/2}$	4.7[a]
O I	5s	50.6[a]
O II	$5p_{1/2}$	31.4[a]
O III	$5p_{3/2}$	24.7[a]

Europium (63)

K	1s	48519
L I	2s	8052
L II	$2p_{1/2}$	7617
L III	$2p_{3/2}$	6977
M I	3s	1800
M II	$3p_{1/2}$	1614
M III	$3p_{3/2}$	1481
M IV	$3d_{3/2}$	1158.6[a]
M V	$3d_{5/2}$	1127.5[a]
N I	4s	360
N II	$4p_{1/2}$	284
N III	$4p_{3/2}$	257
N IV	$4d_{3/2}$	133
N V	$4d_{5/2}$	1227[a]
N VI	$4f_{5/2}$	0
N VII	$4f_{7/2}$	0
O I	5s	32
O II	$5p_{1/2}$	22
O III	$5p_{3/2}$	22

Fluorine (9)

K	1s	696.7[a]

Francium (87)

K	1s	101137
L I	2s	18639
L II	$2p_{1/2}$	17907
L III	$2p_{3/2}$	15031
M I	3s	4652
M II	$3p_{1/2}$	4327
M III	$3p_{3/2}$	3663
M IV	$3d_{3/2}$	3136
M V	$3d_{5/2}$	3000
N I	4s	1153[a]
N II	$4p_{1/2}$	980[a]
N III	$4p_{3/2}$	810[a]
N IV	$4d_{3/2}$	603[a]
N V	$4d_{5/2}$	577[a]
N VI	$4f_{5/2}$	268[a]
N VII	$4f_{7/2}$	268[a]
O I	5s	234[a]
O II	$5p_{1/2}$	182[a]
O III	$5p_{3/2}$	140[a]
O IV	$5d_{3/2}$	58[a]
O V	$5d_{5/2}$	58[a]
P I	6s	34
P II	$6p_{1/2}$	15
P III	$6p_{3/2}$	15

Gadolinium (64)

K	1s	50239
L I	2s	8376
L II	$2p_{1/2}$	7930
L III	$2p_{3/2}$	7243
M I	3s	1881
M II	$3p_{1/2}$	1688
M III	$3p_{3/2}$	1544
M IV	$3d_{3/2}$	1221.9[a]
M V	$3d_{5/2}$	1189.6[a]
N I	4s	378.6[a]
N II	$4p_{1/2}$	286
N III	$4p_{3/2}$	271
N IV	$4d_{3/2}$	—
N V	$4d_{5/2}$	142.6[a]
N VI	$4f_{5/2}$	8.6[a]
N VII	$4f_{7/2}$	8.6[a]
O I	5s	36
O II	$5p_{1/2}$	20
O III	$5p_{3/2}$	20

Gallium (31)

K	1s	10367
L I	2s	1299.0[a,d]
L II	$2p_{1/2}$	1143.2[b]
L III	$2p_{3/2}$	1116.4[b]
M I	3s	159.5[b]
M II	$3p_{1/2}$	103.5[b]
M III	$3p_{3/2}$	100.0[b]
M IV	$3d_{3/2}$	18.7[b]
M V	$3d_{5/2}$	18.7[b]

Germanium (32)

K	1s	11103
L I	2s	1414.6[a,d]
L II	$2p_{1/2}$	1248.1[a,d]
L III	$2p_{3/2}$	1217.0[a,d]
M I	3s	180.1[a]
M II	$3p_{1/2}$	124.9[a]
M III	$3p_{3/2}$	120.8[a]
M IV	$3d_{3/2}$	29.8[a]
M V	$3d_{5/2}$	29.2[a]

Gold (79)

K	1s	80725
L I	2s	14353
L II	$2p_{1/2}$	13734
L III	$2p_{3/2}$	11919
M I	3s	3425
M II	$3p_{1/2}$	3148
M III	$3p_{3/2}$	2743
M IV	$3d_{3/2}$	2291
M V	$3d_{5/2}$	2206
N I	4s	762.1[b]
N II	$4p_{1/2}$	642.7[b]
N III	$4p_{3/2}$	546.3[b]
N IV	$4d_{3/2}$	353.2[b]
N V	$4d_{5/2}$	335.1[b]
N VI	$4f_{5/2}$	87.6[b]
N VII	$4f_{7/2}$	83.9[b]
O I	5s	107.2[a,d]
O II	$5p_{1/2}$	74.2[b]
O III	$5p_{3/2}$	57.2[b]

Hafnium (72)

K	1s	65351
L I	2s	11271
L II	$2p_{1/2}$	10739
L III	$2p_{3/2}$	9561
M I	3s	2601
M II	$3p_{1/2}$	2365
M III	$3p_{3/2}$	2107
M IV	$3d_{3/2}$	1176
M V	$3d_{5/2}$	1662
N I	4s	538[a]
N II	$4p_{1/2}$	438.2[b]
N III	$4p_{3/2}$	380.7[b]
N IV	$4d_{3/2}$	220.0[b]
N V	$4d_{5/2}$	211.5[b]
N VI	$4f_{5/2}$	15.9[b]

N VII	$4f_{7/2}$	14.2[b]
O I	5s	64.2[b]
O II	$5p_{1/2}$	38[a]
O III	$5p_{3/2}$	29.9[b]

Helium (2)

K	1s	24.6[a]

Holmium (67)

K	1s	55618
L I	2s	9394
L II	$2p_{1/2}$	8918
L III	$2p_{3/2}$	8071
M I	3s	2128
M II	$3p_{1/2}$	1923
M III	$3p_{3/2}$	1741
M IV	$3d_{3/2}$	1392
M V	$3d_{5/2}$	1351
N I	4s	432.4[a]
N II	$4p_{1/2}$	343.5
N III	$4p_{3/2}$	308.2[a]
N IV	$4d_{3/2}$	160[a]
N V	$4d_{5/2}$	160[a]
N VI	$4f_{5/2}$	8.6[a]
N VII	$4f_{7/2}$	5.2[a]
O I	5s	49.3[a]
O II	$5p_{1/2}$	30.8[a]
O III	$5p_{3/2}$	24.1[a]

Hydrogen (1)

K	1s	13.6

Indium (49)

K	1s	27940
L I	2s	4238
L II	$2p_{1/2}$	3938
L III	$2p_{3/2}$	3730
M I	3s	827.2[b]
M II	$3p_{1/2}$	703.2[b]
M III	$3p_{3/2}$	665.3[b]
M IV	$3d_{3/2}$	451.4[b]
M V	$3d_{5/2}$	443.9[b]
N I	4s	122.9[b]
N II	$4p_{1/2}$	73.5[b,c]
N III	$4p_{3/2}$	73.5[b,c]
N IV	$4d_{3/2}$	17.7[b]
N V	$4d_{5/2}$	16.9[b]

Iodine (53)

K	1s	33169
L I	2s	5188
L II	$2p_{1/2}$	4852
L III	$2p_{3/2}$	4557
M I	3s	1072[a]
M II	$3p_{1/2}$	931[a]
M III	$3p_{3/2}$	875[a]
M IV	$3d_{3/2}$	631[a]
M V	$3d_{5/2}$	620[a]
N I	4s	186[a]
N II	$4p_{1/2}$	123[a]
N III	$4p_{3/2}$	123[a]
N IV	$4d_{3/2}$	50[a]
N V	$4d_{5/2}$	50[a]

Iridium (77)

K	1s	76111
L I	2s	13419
L II	$2p_{1/2}$	12824
L III	$2p_{3/2}$	11215
M I	3s	3174
M II	$3p_{1/2}$	2909
M III	$3p_{3/2}$	2551
M IV	$3d_{3/2}$	2116
M V	$3d_{5/2}$	2040
N I	4s	691.1[b]
N II	$4p_{1/2}$	577.8[b]
N III	$4p_{3/2}$	495.8[b]
N IV	$4d_{3/2}$	311.9[b]
N V	$4d_{5/2}$	296.3[b]
N VI	$4f_{5/2}$	63.8[b]
N VII	$4f_{7/2}$	60.8[b]
O I	52	95.2[a,d]
O II	$5p_{1/2}$	63.0[a,d]
O III	$5p_{3/2}$	48.0[b]

Iron (26)

K	1s	7112
L I	2s	844.6[b]
L II	$2p_{1/2}$	719.9[b]
L III	$2p_{3/2}$	706.8[b]
M I	3s	91.3[b]
M II	$3p_{1/2}$	52.7[b]
M III	$3p_{3/2}$	52.7[b]

Krypton (36)

K	1s	14326
L I	2s	1921
L II	$2p_{1/2}$	1730.9[a]
L III	$2p_{3/2}$	1678.4[a]
M I	3s	292.8[a]
M II	$3p_{1/2}$	222.2[a]
M III	$3p_{3/2}$	214.4[a]
M IV	$3d_{3/2}$	95.0[a]
M V	$3d_{5/2}$	93.8[a]
N I	4s	27.5[a]
N II	$4p_{1/2}$	14.1[a]
N III	$4p_{3/2}$	14.1[a]

Lanthanum (57)

K	1s	38925
L I	2s	6266
L II	$2p_{1/2}$	5891
L III	$2p_{3/2}$	5483
M I	3s	1362[a,d]
M II	$3p_{1/2}$	1209[a,d]
M III	$3p_{3/2}$	1128[a,d]
M IV	$3d_{3/2}$	853[a]
M V	$3d_{5/2}$	836[a]
N I	4s	247.7[a]
N II	$4p_{1/2}$	205.8
N III	$4p_{3/2}$	196.0[a]
N IV	$4d_{3/2}$	105.3[a]
N V	$4d_{5/2}$	102.5[a]
N VI	$4f_{5/2}$	—
N VII	$4f_{7/2}$	—
O I	5s	34.3[a]
O II	$5p_{1/2}$	19.3[a]
O III	$5p_{3/2}$	16.8[a]

Lead (82)

K	1s	88005
L I	2s	15861
L II	$2p_{1/2}$	15200
L III	$2p_{3/2}$	13055
M I	3s	3851
M II	$3p_{1/2}$	3554
M III	$3p_{3/2}$	3066
M IV	$3d_{3/2}$	2586
M V	$3d_{5/2}$	2484
N I	4s	891.8[b]
N II	$4p_{1/2}$	761.9[b]
N III	$4p_{3/2}$	643.5[b]
N IV	$4d_{3/2}$	434.3[b]
N V	$4d_{5/2}$	412.2[b]
N VI	$4f_{5/2}$	141.7[b]
N VII	$4f_{7/2}$	136.9[b]
O I	5s	147[a,d]
O II	$5p_{1/2}$	106.4[b]
O III	$5p_{3/2}$	83.3[b]
O IV	$5d_{3/2}$	20.7[b]
O V	$5d_{5/2}$	18.1[b]

Lithium (3)

K	1s	54.7[a]

Lutetium

K	1s	63314
L I	2s	10870
L II	$2p_{1/2}$	10349
L III	$2p_{3/2}$	9244
M I	3s	2491
M II	$3p_{1/2}$	2264
M III	$3p_{3/2}$	2024
M IV	$3d_{3/2}$	1639
M V	$3d_{5/2}$	1589
N I	4s	506.8[a]
N II	$4p_{1/2}$	412.4[a]
N III	$4p_{3/2}$	359.2[a]
N IV	$4d_{3/2}$	206.1[a]
N V	$4d_{5/2}$	196.3[a]
N VI	$4f_{5/2}$	8.9[a]
N VII	$4f_{7/2}$	7.5[a]
O I	5s	57.3[a]
O II	$5p_{1/2}$	33.6[a]
O III	$5p_{3/2}$	26.7[a]

Magnesium (12)

K	1s	1303.0[b]
L I	2s	88.6[a]
L II	$2p_{1/2}$	49.6[b]
L III	$2p_{3/2}$	49.2[a]

Manganese (25)

K	1s	6539
L I	2s	769.1[b]
L II	$2p_{1/2}$	649.9[b]
L III	$2p_{3/2}$	638.7[b]
M I	3s	82.3[b]
M II	$3p_{1/2}$	47.2[b]
M III	$3p_{3/2}$	47.2[b]

Mercury (80)

K	1s	83102
L I	2s	14839
L II	$2p_{1/2}$	14209
L III	$2p_{3/2}$	12284
M I	3s	3562
M II	$3p_{1/2}$	3279
M III	$3p_{3/2}$	2847
M IV	$3d_{3/2}$	2385
M V	$3d_{5/2}$	2295
N I	4s	802.2[b]
N II	$4p_{1/2}$	680.2[b]
N III	$4p_{3/2}$	576.6[b]
N IV	$4d_{3/2}$	378.2[b]
N V	$4d_{5/2}$	358.8[b]
N VI	$4f_{5/2}$	104.0[b]
N VII	$4f_{7/2}$	99.9[b]
O I	5s	127[b]
O II	$5p_{3/2}$	83.1[b]
O III	$5p_{3/2}$	64.5[b]
O IV	$5d_{3/2}$	9.6[b]
O V	$5d_{5/2}$	7.8[b]

Molybdenum (42)

K	1s	20000
L I	2s	2866
L II	$2p_{1/2}$	2625
L III	$2p_{3/2}$	2520
M I	3s	506.3[b]
M II	$3p_{1/2}$	411.6[b]
M III	$3p_{3/2}$	394.0[b]
M IV	$3d_{3/2}$	231.1[b]
M V	$3d_{5/2}$	227.9[b]
N I	4s	63.2[b]
N II	$4p_{1/2}$	37.6[b]
N III	$4p_{3/2}$	35.5[b]

Neodymium (60)

K	1s	43569
L I	2s	7126
L II	$2p_{1/2}$	6722
L III	$2p_{3/2}$	6208
M I	3s	1575
M II	$3p_{1/2}$	1403
M III	$3p_{3/2}$	1297
M IV	$3d_{3/2}$	1003.3[a]
M V	$3d_{5/2}$	980.4[a]
N I	4s	319.2[a]
N II	$4p_{1/2}$	243.3
N III	$4p_{3/2}$	224.6
N IV	$4d_{3/2}$	120.5[a]

N V	$4d_{5/2}$	120.5[a]
N VI	$4f_{5/2}$	1.5
N VII	$4f_{7/2}$	1.5
O I	5s	37.5
O II	$5p_{1/2}$	21.1
O III	$5p_{3/2}$	21.1

Neon (10)

K	1s	870.2[a]
L I	2s	48.5[a]
L II	$2p_{1/2}$	21.7[a]
L III	$2p_{3/2}$	21.6[a]

Nickel (28)

K	1s	8333
L I	2s	1008.6[b]
L II	$2p_{1/2}$	870.0[b]
L III	$2p_{3/2}$	852.7[b]
M I	3s	110.8[b]
M II	$3p_{1/2}$	68.0[b]
M III	$3p_{3/2}$	66.2[b]

Niobium (41)

K	1s	18986
L I	2s	2698
L II	$2p_{1/2}$	2465
L III	$2p_{3/2}$	2371
M I	3s	466.6[b]
M II	$3p_{1/2}$	376.1[b]
M III	$3p_{3/2}$	360.6[b]
M IV	$3d_{3/2}$	205.0[b]
M V	$3d_{5/2}$	202.3[b]
N I	4s	56.4[b]
N II	$4p_{1/2}$	32.6[b]
N III	$4p_{3/2}$	30.8[b]

Nitrogen (7)

K	1s	409.9[a]
L I	2s	37.3[a]

Osmium (76)

K	1s	73871
L I	2s	12968
L II	$2p_{1/2}$	12385
L III	$2p_{3/2}$	10871
M I	3s	3049
M II	$3p_{1/2}$	2792
M III	$3p_{3/2}$	2457
M IV	$3d_{3/2}$	2031
M V	$3d_{5/2}$	1960
N I	4s	658.2[b]
N II	$4p_{1/2}$	549.1[b]
N III	$4p_{3/2}$	470.7[b]
N IV	$4d_{3/2}$	293.1[b]
N V	$4d_{5/2}$	278.5[b]
N VI	$4f_{5/2}$	53.4[b]
N VII	$4f_{7/2}$	50.7[b]
O I	5s	84[a]
O II	$5p_{1/2}$	58[a]
O III	$5p_{3/2}$	44.5[b]

Oxygen (8)

K	1s	543.1[a]
L I	2s	41.6[a]

Palladium (46)

K	1s	24350
L I	2s	3604
L II	$2p_{1/2}$	3330
L III	$2p_{3/2}$	3173
M I	3s	671.6[b]
M II	$3p_{1/2}$	559.9[b]
M III	$3p_{3/2}$	532.3[b]
M IV	$3d_{3/2}$	340.5[b]
M V	$3d_{5/2}$	335.2[b]
N I	4s	87.1[a,d]
N II	$4p_{1/2}$	55.7[b,c]
N III	$4p_{3/2}$	50.9[b,c]

Phosphorus (15)

K	1s	2145.5
L I	2s	189[a]
L II	$2p_{1/2}$	136[a]
L III	$2p_{3/2}$	135[a]

Platinum (78)

K	1s	78395
L I	2s	13880
L II	$2p_{1/2}$	13273
L III	$2p_{3/2}$	11564
M I	3s	3296
M II	$3p_{1/2}$	3027
M III	$3p_{3/2}$	2645
M IV	$3d_{3/2}$	2202
M V	$3d_{5/2}$	2122
N I	4s	725.4[b]
N II	$4p_{1/2}$	609.1[b]
N III	$4p_{3/2}$	519.4[b]
N IV	$4d_{3/2}$	331.6[b]
N V	$4d_{5/2}$	314.6[b]
N VI	$4f_{5/2}$	74.5[b]
N VII	$4f_{7/2}$	71.2[b]
O I	5s	101.7[a,d]
O II	$5p_{1/2}$	65.3[a,b]
O III	$5p_{3/2}$	51.7[b]

Polonium (84)

K	1s	93105
L I	2s	16939
L II	$2p_{1/2}$	16244
L III	$2p_{3/2}$	13814
M I	3s	4149
M II	$3p_{1/2}$	3854
M III	$3p_{3/2}$	3302
M IV	$3d_{3/2}$	2798
M V	$3d_{5/2}$	2683
N I	4s	995[a]
N II	$4p_{1/2}$	851[a]
N III	$4p_{3/2}$	705[a]
N IV	$4d_{3/2}$	500[a]
N V	$4d_{5/2}$	473[a]
N VI	$4f_{5/2}$	184[a]
N VII	$4f_{7/2}$	184[a]
O I	5s	177[a]
O II	$5p_{1/2}$	132[a]
O III	$5p_{3/2}$	104[a]
O IV	$5d_{3/2}$	31[a]
O V	$5d_{5/2}$	31[a]

Potassium (19)

K	1s	3608.4[a]
L I	2s	378.6[a]
L II	$2p_{1/2}$	297.3[a]
L III	$2p_{3/2}$	294.6[a]
M I	3s	34.8[a]
M II	$3p_{1/2}$	18.3[a]
M III	$3p_{3/2}$	18.3[a]

Praseodymium (59)

K	1s	41991
L I	2s	6835
L II	$2p_{1/2}$	6440
L III	$2p_{3/2}$	5964
M I	3s	1511
M II	$3p_{1/2}$	1337
M III	$3p_{3/2}$	1242
M IV	$3d_{3/2}$	948.3[a]
M V	$3d_{5/2}$	928.8[a]
N I	4s	304.5
N II	$4p_{1/2}$	236.3
N III	$4p_{3/2}$	217.6
N IV	$4d_{3/2}$	115.1[a]
N V	$4d_{5/2}$	115.1[a]
N VI	$4f_{5/2}$	2.0
N VII	$4f_{7/2}$	2.0
O I	5s	37.4
O II	$5p_{1/2}$	22.3
O III	$5p_{3/2}$	22.3

Promethium (61)

K	1s	45184
L I	2s	7428
L II	$2p_{1/2}$	7013
L III	$2p_{3/2}$	6459
M I	3s	—
M II	$3p_{1/2}$	1471.4
M III	$3p_{3/2}$	1357
M IV	$3d_{3/2}$	1052
M V	$3d_{5/2}$	1027
N I	4s	—
N II	$4p_{1/2}$	242
N III	$4p_{3/2}$	242
N IV	$4d_{3/2}$	120
N V	$4d_{5/2}$	120

Protactinium (91)

K	1s	112601
L I	2s	21105
L II	$2p_{1/2}$	20314
L III	$2p_{3/2}$	16733
M I	3s	5367
M II	$3p_{1/2}$	5001
M III	$3p_{3/2}$	4174
M IV	$3d_{3/2}$	3611
M V	$3d_{5/2}$	3442
N I	4s	1387[a]
N II	$4p_{1/2}$	1224[a]
N III	$4p_{3/2}$	1007[a]
N IV	$4d_{3/2}$	743[a]
N V	$4d_{5/2}$	708[a]
N VI	$4f_{5/2}$	371[a]
N VII	$4f_{7/2}$	360[a]
O I	5s	310[a]
O II	$5p_{1/2}$	232[a]
O III	$5p_{3/2}$	232[a]
O IV	$5d_{3/2}$	94[a]
O V	$5d_{5/2}$	94[a]
P I	6s	—
P II	$6p_{1/2}$	—
P III	$6p_{3/2}$	—

Radium (88)

K	1s	103922
L I	2s	19237
L II	$2p_{1/2}$	18484
L III	$2p_{3/2}$	15444
M I	3s	4822
M II	$3p_{1/2}$	4490
M III	$3p_{3/2}$	3792
M IV	$3d_{3/2}$	3248
M V	$3d_{5/2}$	3105
N I	4s	1208[a]
N II	$4p_{1/2}$	1058
N III	$4p_{3/2}$	879[a]
N IV	$4d_{3/2}$	636[a]
N V	$4d_{5/2}$	603[a]
N VI	$4f_{5/2}$	299[a]
N VII	$4f_{7/2}$	299[a]
O I	5s	254[a]
O II	$5p_{1/2}$	200[a]
O III	$5p_{3/2}$	153[a]
O IV	$5d_{3/2}$	68[a]
O V	$5d_{5/2}$	68[a]
P I	6s	44
P II	$6p_{1/2}$	19
P III	$6p_{3/2}$	19

Radon (86)

K	1s	98404
L I	2s	18049
L II	$2p_{1/2}$	17337
L III	$2p_{3/2}$	14619
M I	3s	4482
M II	$3p_{1/2}$	4159
M III	$3p_{3/2}$	3538
M IV	$3d_{3/2}$	3022
M V	$3d_{5/2}$	2892

Level	Orbital	Energy
N I	4s	1097[a]
N II	$4p_{1/2}$	929[a]
N III	$4p_{3/2}$	768[a]
N IV	$4d_{3/2}$	567[a]
N V	$4d_{5/2}$	541[a]
N VI	$4f_{5/2}$	238[a]
N VII	$4f_{7/2}$	238[a]
O I	5s	214[a]
O II	$5p_{1/2}$	164[a]
O III	$5p_{3/2}$	127[a]
O IV	$5d_{3/2}$	48[a]
O V	$5d_{5/2}$	48[a]
P I	6s	26

Rhenium (75)

Level	Orbital	Energy
K	1s	71676
L I	2s	12527
L II	$2p_{1/2}$	11959
L III	$2p_{3/2}$	10535
M I	3s	2932
M II	$3p_{1/2}$	2682
M III	$3p_{3/2}$	2367
M IV	$3d_{3/2}$	1949
M V	$3d_{5/2}$	1883
N I	4s	625.4[b]
N II	$4p_{1/2}$	518.7[b]
N III	$4p_{3/2}$	446.8[b]
N IV	$4d_{3/2}$	273.9[b]
N V	$4d_{5/2}$	260.5[b]
N VI	$4f_{5/2}$	42.9[a]
N VII	$4f_{7/2}$	40.5[a]
O I	5s	83[b]
O II	$5p_{1/2}$	45.6[b]
O III	$5p_{3/2}$	34.6[a,d]

Rhodium (45)

Level	Orbital	Energy
K	1s	23220
L I	2s	3412
L II	$2p_{1/2}$	3146
L III	$2p_{3/2}$	3004
M I	3s	628.1[b]
M II	$3p_{1/2}$	521.3[b]
M III	$3p_{3/2}$	496.5[b]
M IV	$3d_{3/2}$	311.9[b]
M V	$3d_{5/2}$	307.2[b]
N I	4s	81.4[a,d]
N II	$4p_{1/2}$	50.5[b]
N III	$4p_{3/2}$	47.3[b]

Rubidium (37)

Level	Orbital	Energy
K	1s	15200
L I	2s	2065
L II	$2p_{1/2}$	1864
L III	$2p_{3/2}$	1804
M I	3s	326.7[a]
M II	$3p_{1/2}$	248.7[a]
M III	$3p_{3/2}$	239.1[a]
M IV	$3d_{3/2}$	113.0[a]
M V	$3d_{5/2}$	112[a]
N I	4s	30.5[a]
N II	$4p_{1/2}$	16.3[a]
N III	$4p_{3/2}$	15.3[a]

Ruthenium (44)

Level	Orbital	Energy
K	1s	22117
L I	2s	3224
L II	$2p_{1/2}$	2967
L III	$2p_{3/2}$	2838
M I	3s	586.2[b]
M II	$3p_{1/2}$	483.3[b]
M III	$3p_{3/2}$	461.5[b]
M IV	$3d_{3/2}$	284.2[b]
M V	$3d_{5/2}$	280.0[b]
N I	4s	75.0[b]
N II	$4p_{1/2}$	46.5[b]
N III	$4p_{3/2}$	43.2[b]

Samarium (62)

Level	Orbital	Energy
K	1s	46834
L I	2s	7737
L II	$2p_{1/2}$	7312
L III	$2p_{3/2}$	6716
M I	3s	1723
M II	$3p_{1/2}$	1541
M III	$3p_{3/2}$	1419.8
M IV	$3d_{3/2}$	1110.9[a]
M V	$3d_{5/2}$	1083.4[a]
N I	4s	347.2[a]
N II	$4p_{1/2}$	265.6
N III	$4p_{3/2}$	247.4
N IV	$4d_{3/2}$	129.0
N V	$4d_{5/2}$	129.0
N VI	$4f_{5/2}$	5.2
N VII	$4f_{7/2}$	5.2
O I	5s	37.4
O II	$5p_{1/2}$	21.3
O III	$5p_{3/2}$	21.3

Scandium (21)

Level	Orbital	Energy
K	1s	4492
L I	2s	498.0[a]
L II	$2p_{1/2}$	403.6[a]
L III	$2p_{3/2}$	389.7[a]
M I	3s	51.1[a]
M II	$3p_{1/2}$	28.3[a]
M III	$3p_{3/2}$	28.3[a]

Selenium (34)

Level	Orbital	Energy
K	1s	12658
L I	2s	1652.0[a,d]
L II	$2p_{1/2}$	1474.3[a,d]
L III	$2p_{3/2}$	1433.9[a,d]
M I	3s	229.6[a]
M II	$3p_{1/2}$	166.5[a]
M III	$3p_{3/2}$	160.7[a]
M IV	$3d_{3/2}$	55.5[a]
M V	$3d_{5/2}$	54.6[a]

Silicon (14)

Level	Orbital	Energy
K	1s	1839
L I	2s	149.7[a,d]
L II	$2p_{1/2}$	99.8[a]
L III	$2p_{3/2}$	99.2[a]

Silver (47)

Level	Orbital	Energy
K	1s	25514
L I	2s	3806
L II	$2p_{1/2}$	3524
L III	$2p_{3/2}$	3351
M I	3s	719.0[b]
M II	$3p_{1/2}$	603.8[b]
M III	$3p_{3/2}$	573.0[b]
M IV	$3d_{3/2}$	374.0[b]
M V	$3d_{5/2}$	368.0[b]
N I	4s	97.0[b]
N II	$4p_{1/2}$	63.7[b]
N III	$4p_{3/2}$	58.3[b]

Sodium (11)

Level	Orbital	Energy
K	1s	1070.8[b]
L I	2s	63.5[b]
L II	$2p_{1/2}$	30.4[b]
L III	$2p_{3/2}$	30.5[a]

Strontium (38)

Level	Orbital	Energy
K	1s	16105
L I	2s	2216
L II	$2p_{1/2}$	2007
L III	$2p_{3/2}$	1940
M I	3s	358.7[b]
M II	$3p_{1/2}$	280.3[b]
M III	$3p_{3/2}$	270.0[b]
M IV	$3d_{3/2}$	136.0[b]
M V	$3d_{5/2}$	134.2[b]
N I	4s	38.9[b]
N II	$4p_{1/2}$	21.6[b]
N III	$4p_{3/2}$	20.1[b]

Sulfur (16)

Level	Orbital	Energy
K	1s	2472
L I	2s	230.9[a,d]
L II	$2p_{1/2}$	163.6[a]
L III	$2p_{3/2}$	162.5[a]

Tantalum (73)

Level	Orbital	Energy
K	1s	67416
L I	2s	11682
L II	$2p_{1/2}$	11136
L III	$2p_{3/2}$	9881
M I	3s	2708
M II	$3p_{1/2}$	2469
M III	$3p_{3/2}$	2194
M IV	$3d_{3/2}$	1793
M V	$3d_{5/2}$	1735
N I	4s	563.4[b]
N II	$4p_{1/2}$	463.4[b]
N III	$4p_{3/2}$	400.9[b]
N IV	$4d_{3/2}$	237.9[b]
N V	$4d_{5/2}$	226.4[b]
N VI	$4f_{5/2}$	23.5[b]
N VII	$4f_{7/2}$	21.6[b]
O I	5s	69.7[b]
O II	$5p_{1/2}$	42.2[a]
O III	$5p_{3/2}$	32.7[b]

Technetium (43)

Level	Orbital	Energy
K	1s	21044
L I	2s	3043
L II	$2p_{1/2}$	2793
L III	$2p_{3/2}$	2677
M I	3s	586.1[a]
M II	$3p_{1/2}$	447.6[a]
M III	$3p_{3/2}$	417.7[a]
M IV	$3d_{3/2}$	257.6[a]
M V	$3d_{5/2}$	253.9[a]
N I	4s	69.5[a]
N II	$4p_{1/2}$	42.3[a]
N III	$4p_{3/2}$	39.9[a]

Tellurium (52)

Level	Orbital	Energy
K	1s	31814
L I	2s	4939
L II	$2p_{1/2}$	4612
L III	$2p_{3/2}$	4341
M I	3s	1006[b]
M II	$3p_{1/2}$	870.8[b]
M III	$3p_{3/2}$	820.0[b]
M IV	$3d_{3/2}$	583.4[b]
M V	$3d_{5/2}$	573.0[b]
N I	4s	169.4[b]
N II	$4p_{1/2}$	103.3[b,c]
N III	$4p_{3/2}$	103.3[b,c]
N IV	$4d_{3/2}$	41.9[b]
N V	$4d_{5/2}$	40.4[b]

Terbium (65)

Level	Orbital	Energy
K	1s	51996
L I	2s	8708
L II	$2p_{1/2}$	8252
L III	$2p_{3/2}$	7514
M I	3s	1968
M II	$3p_{1/2}$	1768
M III	$3p_{3/2}$	1611
M IV	$3d_{3/2}$	1267.9[a]
M V	$3d_{5/2}$	1241.1[a]
N I	4s	396.0[a]
N II	$4p_{1/2}$	322.4[a]
N III	$4p_{3/2}$	284.1[a]
N IV	$4d_{3/2}$	150.5[a]
N V	$4d_{5/2}$	150.5[a]
N VI	$4f_{5/2}$	7.7[a]
N VII	$4f_{7/2}$	2.4[a]
O I	5s	45.6[a]
O II	$5p_{1/2}$	28.7[a]
O III	$5p_{3/2}$	22.6[a]

Thallium (81)

K	1s	85530
L I	2s	15347
L II	$2p_{1/2}$	14698
L III	$2p_{3/2}$	12658
M I	3s	3704
M II	$3p_{1/2}$	3416
M III	$3p_{3/2}$	2957
M IV	$3d_{3/2}$	2485
M V	$3d_{5/2}$	2389
N I	4s	846.2[b]
N II	$4p_{1/2}$	720.5[b]
N III	$4p_{3/2}$	609.5[b]
N IV	$4d_{3/2}$	405.7[b]
N V	$4d_{5/2}$	385.0[b]
N VI	$4f_{5/2}$	122.2[b]
N VII	$4f_{7/2}$	117.8[b]
O I	5s	136[a,d]
O II	$5p_{1/2}$	94.6[b]
O III	$5p_{3/2}$	73.5[b]
O IV	$5d_{3/2}$	14.7[b]
O V	$5d_{5/2}$	12.5[b]

Thorium (90)

K	1s	109651
L I	2s	20472
L II	$2p_{1/2}$	19693
L III	$2p_{3/2}$	16300
M I	3s	5182
M II	$3p_{1/2}$	4830
M III	$3p_{3/2}$	4046
M IV	$3d_{3/2}$	3491
M V	$3d_{5/2}$	3332
N I	4s	1330[a]
N II	$4p_{1/2}$	1168[a]
N III	$4p_{3/2}$	966.4[b]
N IV	$4d_{3/2}$	712.1[b]
N V	$4d_{5/2}$	675.2[b]
N VI	$4f_{5/2}$	342.4[b]
N VII	$4f_{7/2}$	333.1[b]
O I	5s	290[a,c]
O II	$5p_{1/2}$	229[a,c]
O III	$5p_{3/2}$	182[a,c]
O IV	$5d_{3/2}$	92.5[b]
O V	$5d_{5/2}$	85.4[b]
P I	6s	41.4[b]
P II	$6p_{1/2}$	24.5[b]
P III	$6p_{3/2}$	16.6[b]

Thulium (69)

K	1s	59390
L I	2s	10116
L II	$2p_{1/2}$	9617
L III	$2p_{3/2}$	8648
M I	3s	2307
M II	$3p_{1/2}$	2090
M III	$3p_{3/2}$	1885
M IV	$3d_{3/2}$	1515
M V	$3d_{5/2}$	1468
N I	4s	470.9[a]
N II	$4p_{1/2}$	385.9[a]
N III	$4p_{3/2}$	332.6[a]
N IV	$4d_{3/2}$	175.5[a]
N V	$4d_{5/2}$	175.5[a]
N VI	$4f_{5/2}$	—
N VII	$4f_{7/2}$	4.6
O I	5s	54.7[a]
O II	$5p_{1/2}$	31.8[a]
O III	$5p_{3/2}$	25.0[a]

Tin (50)

K	1s	29200
L I	2s	4465
L II	$2p_{1/2}$	4156
L III	$2p_{3/2}$	3929
M I	3s	884.7[b]
M II	$3p_{1/2}$	756.5[b]
M III	$3p_{3/2}$	714.6[b]
M IV	$3d_{3/2}$	493.2[b]
M V	$3d_{5/2}$	484.9[b]
N I	4s	137.1[b]
N II	$4p_{1/2}$	83.6[b,c]
N III	$4p_{3/2}$	83.6[b,c]
N IV	$4d_{3/2}$	24.9[b]
N V	$4d_{5/2}$	23.9[b]

Titanium (22)

K	1s	4966
L I	2s	560.9[b]
L II	$2p_{1/2}$	460.2[b]
L III	$2p_{3/2}$	453.8[b]
M I	3s	58.7[b]
M II	$3p_{1/2}$	32.6[b]
M III	$3p_{3/2}$	32.6[b]

Tungsten (74)

K	1s	69525
L I	2s	12100
L II	$2p_{1/2}$	11544
L III	$2p_{3/2}$	10207
M I	3s	2820
M II	$3p_{1/2}$	2575
M III	$3p_{3/2}$	2281
M IV	$3d_{3/2}$	1949
M V	$3d_{5/2}$	1809
N I	4s	594.1[b]
N II	$4p_{1/2}$	490.4[b]
N III	$4p_{3/2}$	423.6[b]
N IV	$4d_{3/2}$	255.9[b]
N V	$4d_{5/2}$	243.5[b]
N VI	$4f_{5/2}$	33.6[a]
N VII	$4f_{7/2}$	31.4[b]
O I	5s	75.6[b]
O II	$5p_{1/2}$	453[a,d]
O III	$5p_{3/2}$	36.8[b]

Uranium (92)

K	1s	115606
L I	2s	21757
L II	$2p_{1/2}$	20948
L III	$2p_{3/2}$	17166
M I	3s	5548
M II	$3p_{1/2}$	5182
M III	$3p_{3/2}$	4303
M IV	$3d_{3/2}$	3728
M V	$3d_{5/2}$	3552
N I	4s	1439[a,d]
N II	$4p_{1/2}$	1271[a,d]
N III	$4p_{3/2}$	1043[b]
N IV	$4d_{3/2}$	778.3[b]
N V	$4d_{5/2}$	736.2[b]
N VI	$4f_{5/2}$	388.2[a]
N VII	$4f_{7/2}$	377.4[b]
O I	5s	321[a,c,d]
O II	$5p_{1/2}$	257[a,c,d]
O III	$5p_{3/2}$	192[a,c,d]
O IV	$5d_{3/2}$	102.8[b]
O V	$5d_{5/2}$	94.2[b]
P I	6s	43.9[b]
P II	$6p_{1/2}$	26.8[b]
P III	$6p_{3/2}$	16.8[b]

Vanadium (23)

K	1s	5465
L I	2s	626.7[b]
L II	$2p_{1/2}$	519.8[b]
L III	$2p_{3/2}$	521.1[b]
M I	3s	66.3[b]
M II	$3p_{1/2}$	37.2[b]
M III	$3p_{3/2}$	37.2[b]

Xenon (54)

K	1s	34561
L I	2s	5453
L II	$2p_{1/2}$	5107
L III	$2p_{3/2}$	4786
M I	3s	1148.7[a]
M II	$3p_{1/2}$	1002.1[a]
M III	$3p_{3/2}$	940.6[a]
M IV	$3d_{3/2}$	689.0[a]
M V	$3d_{5/2}$	676.4[a]
N I	4s	213.2[a]
N II	$4p_{1/2}$	146.7
N III	$4p_{3/2}$	145.5[a]
N IV	$4d_{3/2}$	69.5[a]
N V	$4d_{5/2}$	67.5[a]
N VI	$4f_{5/2}$	—
N VII	$4f_{7/2}$	—
O I	5s	23.3[a]
O II	$5p_{1/2}$	13.4[a]
O III	$5p_{3/2}$	12.1[a]

Ytterbium (70)

K	1s	61332
L I	2s	10486
L II	$2p_{1/2}$	9978
L III	$2p_{3/2}$	8944
M I	3s	2398
M II	$3p_{1/2}$	2173
M III	$3p_{3/2}$	1950
M IV	$3d_{3/2}$	1576
M V	$3d_{5/2}$	1528
N I	4s	480.5[a]
N II	$4p_{1/2}$	388.7[a]
N III	$4p_{3/2}$	339.7[a]
N IV	$4d_{3/2}$	191.2[a]
N V	$4d_{5/2}$	182.4[a]
N VI	$4f_{5/2}$	2.5[a]
N VII	$4f_{7/2}$	1.3[a]
O I	5s	52.0[a]
O II	$5p_{1/2}$	30.3[a]
O III	$5p_{3/2}$	24.1[a]

Yttrium (39)

K	1s	17038
L I	2s	2373
L II	$2p_{1/2}$	2156
L III	$2p_{3/2}$	2080
M I	3s	392.0[a,d]
M II	$3p_{1/2}$	310.6[a]
M III	$3p_{3/2}$	298.8[a]
M IV	$3d_{3/2}$	157.7[b]
M V	$3d_{5/2}$	155.8[b]
N I	4s	43.8[a]
N II	$4p_{1/2}$	24.4[a]
N III	$4p_{3/2}$	23.1[a]

Zinc (30)

K	1s	9659
L I	2s	1196.2[a]
L II	$2p_{1/2}$	1044.9[a]
L III	$2p_{3/2}$	1021.8[a]
M I	3s	139.8[a]
M II	$3p_{1/2}$	91.4[a]
M III	$3p_{3/2}$	88.6[a]
M IV	$3d_{3/2}$	10.2[a]
M V	$3d_{5/2}$	10.1[a]

Zirconium (40)

K	1s	17998
L I	2s	2532
L II	$2p_{1/2}$	2307
L III	$2p_{3/2}$	2223
M I	3s	430.3[b]
M II	$3p_{1/2}$	343.5[b]
M III	$3p_{3/2}$	329.8[b]
M IV	$3d_{3/2}$	181.1[b]
M V	$3d_{5/2}$	178.8[b]
N I	4s	50.6[b]
N II	$4p_{1/2}$	28.5[b]
N III	$4p_{3/2}$	27.1[b]

[a] Reference 1.

[b] Reference 2 (remaining values from Reference 3).

[c] One-particle approximation not valid.

[d] Derived using energy differences from Reference 3.

RADIATIVE TRANSITION PROBABILITIES FOR X-RAY LINES

Ratios of Transition Probabilities for *K* X-Ray Lines

Element	$K\alpha_2/K\alpha_1$	$K\alpha_3/K\alpha_1$	$K\beta_1/K\alpha_1$	$K\beta'_2/K\alpha_1$	$K\beta_4/K\alpha_1$	$K\beta_5/K\alpha_1$	$K\beta_3/K\beta_1$	$K\beta/K\alpha$
$_{12}$Mg								0.013
$_{14}$Si								0.027
$_{16}$S								0.059
$_{18}$Ar								0.105
$_{20}$Ca	0.502							0.128
$_{22}$Ti	0.503							0.133
$_{24}$Cr	0.504							0.133
$_{26}$Fe	0.506							0.134
$_{28}$Ni	0.508							0.135
$_{30}$Zn	0.510							0.136
$_{32}$Ge	0.513							0.147
$_{34}$Se	0.515							0.157
$_{36}$Kr	0.517			0.019				0.172
$_{38}$Sr	0.520			0.030				0.180
$_{40}$Ze	0.523			0.037				0.190
$_{42}$Mo	0.525			0.041				0.197
$_{44}$Ru	0.527			0.045			0.513	0.204
$_{46}$Pd	0.529			0.048			0.513	0.210
$_{48}$Cd	0.532			0.053			0.514	0.213
$_{50}$Sn	0.534			0.055			0.514	0.220
$_{52}$Te	0.537		0.187	0.058			0.514	0.225
$_{54}$Xe	0.539		0.188	0.064			0.516	0.232
$_{56}$Ba	0.543		0.189	0.070			0.516	0.237
$_{58}$Ce	0.546		0.190	0.076			0.516	0.242
$_{60}$Nd	0.549	0.11×10^{-3}	0.191	0.083			0.516	0.247
$_{62}$Sm	0.552	0.14×10^{-3}	0.192	0.086			0.516	0.250
$_{64}$Gd	0.556	0.17×10^{-3}	0.194	0.089	0.85×10^{-3}	3.02×10^{-3}	0.517	0.255
$_{66}$Dy	0.560	0.21×10^{-3}	0.198	0.089	0.92×10^{-3}	3.43×10^{-3}	0.517	0.257
$_{68}$Er	0.564	0.26×10^{-3}	0.202	0.088	0.96×10^{-3}	3.85×10^{-3}	0.518	0.260
$_{70}$Yb	0.567	0.30×10^{-3}	0.207	0.087	1.04×10^{-3}	4.23×10^{-3}	0.518	0.264
$_{72}$Hf	0.572	0.36×10^{-3}	0.212	0.085	1.16×10^{-3}	4.62×10^{-3}	0.518	0.267
$_{74}$W	0.576	0.43×10^{-3}	0.216	0.086	1.28×10^{-3}	5.04×10^{-3}	0.518	0.269
$_{76}$Os	0.580	0.51×10^{-3}	0.222	0.087	1.43×10^{-3}	5.44×10^{-3}	0.519	0.273
$_{78}$Pt	0.583	0.63×10^{-3}	0.226	0.091	1.61×10^{-3}	5.84×10^{-3}	0.520	0.275
$_{80}$Hg	0.588	0.76×10^{-3}	0.228	0.096	1.80×10^{-3}	6.24×10^{-3}	0.520	0.278
$_{82}$Pb	0.593	0.91×10^{-3}	0.228	0.102	2.02×10^{-3}	6.64×10^{-3}	0.521	0.280
$_{84}$Po	0.597	1.12×10^{-3}	0.228	0.108	2.26×10^{-3}	7.05×10^{-3}	0.522	0.283
$_{86}$Rn	0.602	1.32×10^{-3}	0.228	0.113	2.52×10^{-3}	7.48×10^{-3}	0.523	0.286
$_{88}$Ra	0.608	1.58×10^{-3}	0.230	0.117	2.80×10^{-3}	7.80×10^{-3}	0.524	0.287
$_{90}$Th	0.613	1.85×10^{-3}	0.232	0.120	3.13×10^{-3}	8.25×10^{-3}	0.525	0.288
$_{92}$U	0.619	2.15×10^{-3}	0.234	0.123	3.47×10^{-3}	8.65×10^{-3}	0.527	0.289
$_{94}$Pu	0.625		0.234	0.125			0.528	0.291
$_{96}$Cm	0.632		0.234	0.128			0.529	0.293
$_{98}$Cf	0.642		0.238	0.132			0.531	0.295
$_{100}$Fm	0.648		0.240	0.135			0.533	0.297

L_1 X-Ray Lines Normalized to $L\beta_3$ = 100

Element	$L\beta_3$	$L\beta_4$	$L\gamma_2$	$L\gamma_3$
$_{32}$Ge	100			17.3
$_{34}$Se	100			18.0
$_{36}$Kr	100			18.2
$_{38}$Sr	100			18.8
$_{40}$Zr	100			19.0
$_{42}$Mo	100	70.6		19.6
$_{44}$Ru	100	67.8		20.2
$_{46}$Pd	100	65.5		20.6
$_{48}$Cd	100	63.5		21.3
$_{50}$Sn	100	62.1		22.0
$_{52}$Te	100	60.7		22.6
$_{54}$Xe	100	59.8		23.3
$_{56}$Ba	100	59.5		24.0
$_{58}$Ce	100	59.2		24.6
$_{60}$Nd	100	59.4		25.4
$_{62}$Sm	100	60.0		26.3
$_{64}$Gd	100	60.8	19.2	27.0
$_{66}$Dy	100	61.8	19.5	28.0
$_{68}$Er	100	63.5	19.8	29.0
$_{70}$Yb	100	65.5	20.7	29.8
$_{72}$Hf	100	67.8	21.2	30.7
$_{74}$W	100	70.5	21.8	31.8
$_{76}$Os	100	73.2	23.0	32.8
$_{78}$Pt	100	76.5	24.5	33.8
$_{80}$Hg	100	80.3	26.3	35.0
$_{82}$Pb	100	84.2	28.6	36.0
$_{84}$Po	100	88.5	31.3	37.2
$_{86}$Rn	100	93.4	34.2	38.2
$_{88}$Ra	100	98.9	37.5	39.6
$_{90}$Th	100	104.5	41.2	41.0
$_{92}$U	100	110.2	45.0	42.6
$_{94}$Pu	100	116.2	49.5	44.0
$_{96}$Cm	100	123.0	55.7	45.7

L_2 X-Ray Lines Normalized to $L\beta_1$ = 100

Element	$L\beta_1$	$L\eta$	$L\gamma_1$	$L\gamma_6$
$_{28}$Ni	100	7.60		
$_{30}$Zn	100	6.80		
$_{32}$Ge	100	6.28		
$_{34}$Se	100	5.80		
$_{36}$Kr	100	5.35		
$_{38}$Sr	100	4.93		
$_{40}$Zr	100	4.60	3.30	
$_{42}$Mo	100	4.30	5.50	
$_{44}$Ru	100	4.00	7.33	
$_{46}$Pd	100	3.75	10.67	
$_{48}$Cd	100	3.55	10.60	
$_{50}$Sn	100	3.35	11.80	
$_{52}$Te	100	3.20	12.70	
$_{54}$Xe	100	3.00	14.00	
$_{56}$Ba	100	2.85	14.50	
$_{58}$Ce	100	2.70	15.30	
$_{60}$Nd	100	2.60	16.00	
$_{62}$Sn	100	2.45	16.50	
$_{64}$Gd	100	2.35	17.00	
$_{66}$Dy	100	2.25	17.40	
$_{68}$Er	100	2.16	17.80	
$_{70}$Yb	100	2.10	18.17	
$_{72}$Hf	100	2.08	18.43	
$_{74}$W	100	2.10	18.80	0.72
$_{76}$Os	100	2.12	19.34	1.65
$_{78}$Pt	100	2.18	19.73	2.40
$_{80}$Hg	100	2.25	20.35	3.10
$_{82}$Pb	100	2.30	20.93	3.65
$_{84}$Po	100	2.40	21.54	4.15
$_{86}$Rn	100	2.46	22.20	4.55
$_{88}$Ra	100	2.50	22.87	4.87
$_{90}$Th	100	2.60	23.43	5.02
$_{92}$U	100	2.65	24.10	5.12
$_{94}$Pu	100	2.70	24.40	5.16
$_{96}$Cm	100	2.75	25.07	5.20

L_3 X-Ray Lines Normalized to $L\alpha_1$ = 100

Element	$L\alpha_1$	$L\beta_{2,15}$	$L\alpha_2$	$L\beta_5$	$L\beta_6$	L_ℓ
$_{26}$Fe	100					12.22
$_{28}$Ni	100					8.95
$_{30}$Zn	100					7.34
$_{32}$Ge	100					6.45
$_{34}$Se	100					7.76
$_{36}$Kr	100					5.28
$_{38}$Sr	100					4.92
$_{40}$Zr	100	0.70	11.10			4.67
$_{42}$Mo	100	5.17	11.10			4.45
$_{44}$Ru	100	9.30	11.12			4.28
$_{46}$Pd	100	11.80	11.12			4.11
$_{48}$Cd	100	14.33	11.12			4.07
$_{50}$Sn	100	16.00	11.13			4.00
$_{52}$Te	100	18.00	11.13			4.00
$_{54}$Xe	100	19.40	11.13			4.00
$_{56}$Ba	100	20.67	11.13			4.02
$_{58}$Ce	100	21.00	11.14			4.09
$_{60}$Nd	100	21.33	11.14		0.875	4.13
$_{62}$Sm	100	21.07	11.14		0.925	4.16
$_{64}$Gd	100	20.83	11.14		0.99	4.20
$_{66}$Dy	100	20.50	11.14		1.05	4.26
$_{68}$Er	100	20.04	11.15		1.12	4.33
$_{70}$Yb	100	19.40	11.15		1.17	4.47
$_{72}$Hf	100	21.33	11.15	0.30	1.21	4.59
$_{74}$W	100	22.74	11.16	0.50	1.25	4.76
$_{76}$Os	100	23.40	11.16	1.32	1.37	4.95
$_{78}$Pt	100	24.00	11.16	1.98	1.43	5.14
$_{80}$Hg	100	24.50	11.17	2.62	1.50	5.37
$_{82}$Pb	100	24.83	11.17	3.21	1.56	5.58
$_{84}$Po	100	25.13	11.17	3.73	1.62	5.80
$_{86}$Rn	100	25.60	11.18	4.25	1.68	6.00
$_{88}$Ra	100	25.92	11.18	4.73	1.76	6.26
$_{90}$Th	100	26.17	11.18	5.18	1.82	6.54
$_{92}$U	100	26.40	11.18	5.58	1.89	6.79
$_{94}$Pu	100	26.67	11.18	5.92	1.95	7.02
$_{96}$Cm	100	26.93	11.18	6.26	2.01	7.34

From Salem, S. I., Panossian, S. L., and Krause, R. A., *At. Data Nucl. Data Tables,* 14, 91, 1974. Reproduced by permission of the copyright owner, Academic Press.

NATURAL WIDTH OF X-RAY LINES

Natural widths of K X-ray lines in eV:

Element	$K\alpha_1$	$K\alpha_2$	$K\beta_1$	$K\beta_3$
Ca	1.00	0.98		
Ti	1.45	2.13		
Cr	2.05	2.64		
Fe	2.45	3.20		
Ni	3.00	3.70		
Zn	3.40	3.96		
Ge	3.75	4.18		
Se	4.10	4.43		
Kr	4.23	4.62		
Sr	5.17	4.97		
Zr	5.70	5.25		
Mo	6.82	6.80		
Ru	7.41	7.96		
Pd	8.80	9.20		
Cd	9.80	10.40		
Sn	11.20	12.40	11.80	11.00
Te	12.80	14.20	13.30	13.10
Xe	14.20	15.10	15.30	14.50
Ba	16.10	16.80	18.15	16.70

Element	$K\alpha_1$	$K\alpha_2$	$K\beta_1$	$K\beta_3$
Ce	18.60	19.50	20.60	18.60
Nd	21.50	21.50	23.25	21.33
Sm	26.00	24.70	25.65	24.65
Gd	29.50	28.00	29.37	28.00
Dy	33.90	32.20	32.73	32.00
Er	35.00	35.50	36.20	35.70
Yb	38.80	40.60	41.43	41.15
Hf	42.70	44.30	46.00	46.10
W	46.80	48.00	51.83	51.50
Os	49.00	49.40	55.90	55.95
Pt	54.10	54.30	59.98	62.13
Hg	64.75	68.20	65.75	68.95
Pb	67.10	72.30	72.20	73.80
Po	73.20	75.10	78.60	80.10
Rn	80.00	81.50	85.50	86.50
Ra	87.00	88.20	94.20	95.50
Th	94.70	95.00	99.70	101.00
U	103.00	104.30	105.00	107.30

From Salem, S. I. and Lee, P. L., *At. Data Nucl. Data Tables*, 18, 233, 1976.

Natural widths of L X-ray lines in eV:

Element	$L\alpha_1$	$L\alpha_2$	$L\beta_1$	$L\beta_2$	$L\beta_3$	$L\beta_4$	$L\gamma_1$
Zr	1.68	1.52	1.87	5.13	5.50	5.60	3.34
Mo	1.86	1.80	2.03	5.30	5.90	5.78	3.76
Ru	2.03	1.98	2.18	5.45	6.35	5.96	4.15
Pd	2.21	2.16	2.36	5.63	6.80	6.18	4.50
Gd	2.43	2.40	2.54	5.82	7.23	6.28	4.83
Sn	2.62	2.62	2.75	6.10	7.70	6.60	5.23
Tc	2.88	2.88	2.96	6.25	8.22	6.82	5.60
Xe	3.15	3.15	3.20	6.43	8.70	7.15	5.95
Ba	3.39	3.45	3.45	6.70	9.20	7.42	6.35
Ce	3.70	3.78	3.73	6.86	9.70	7.82	6.75
Nd	3.93	4.08	4.00	7.18	10.30	8.15	7.16
Sm	4.13	4.50	4.33	7.42	10.80	8.60	7.50
Cd	4.46	4.90	4.63	7.70	11.20	9.08	7.83
Dy	4.81	5.35	5.03	7.90	11.50	9.60	8.30
Er	5.17	5.73	5.45	8.28	11.85	10.03	8.75
Yb	5.40	6.22	5.90	8.58	12.20	11.00	9.20
Hf	5.83	6.70	6.36	8.92	12.40	12.80	9.63
W	6.50	7.20	6.90	9.06	13.10	14.60	10.20
Os	7.04	7.70	7.42	9.60	14.60	16.50	10.65
Pt	7.60	8.28	8.00	9.95	16.10	18.00	11.20
Hg	8.10	8.80	8.70	10.40	17.40	19.70	11.80
Pb	8.82	9.35	9.35	10.75	18.65	21.30	12.30
Po	9.50	9.95	10.10	11.25	19.90	22.70	13.05
Rn	10.03	10.50	10.65	11.65	21.00	24.00	13.55
Ra	11.00	11.20	11.60	12.20	22.00	25.20	14.30
Th	11.90	11.80	12.40	12.80	22.85	26.35	15.00
U	12.40	12.40	13.50	13.30	23.70	27.50	15.70
Pu	13.20	13.00	14.10	13.90	24.10	28.30	16.40
Cm	14.80	13.60	15.70	14.60	25.00	29.40	17.10

PHOTON ATTENUATION COEFFICIENTS

Martin J. Berger and John H. Hubbell

This table gives mass attenuation coefficients for photons for all elements at energies between 1 keV (soft x-rays) and 1 GeV (hard gamma rays). The mass attenuation coefficient μ describes the attenuation of radiation as it passes through matter by the relation

$$I(x)/I_o = e^{-\mu\rho x}$$

where I_o is the initial intensity, $I(x)$ the intensity after path length x, and ρ is the mass density of the element in question. To a high approximation the mass attenuation coefficient is additive for the elements present, independent of the way in which they are bound in chemical compounds.

The power of ten is indicated beside each number in the table; i.e., 7.41 + 03 means 7.41×10^3. A vertical line between two columns indicates that an absorption edge lies between those energy values. The various edges are labeled at the bottom of the table.

The attenuation coefficients were calculated with the computer program XCOM (Reference 1), which uses a cross-section database compiled at the Photon and Charged Particle Data Center at the National Institute of Standards and Technology. Their accuracy has been confirmed at all energies by extensive comparisons with experimental attenuation coefficients. Such comparisons for X-ray energies up to 100 keV can be found in Reference 2.

REFERENCES

1. Berger, M. J. and Hubbell, J. H., National Bureau of Standards Report NBSIR-87-3597, 1987.
2. Saloman, E. B., Hubbell, J. H., and Scofield, J. H., *Atomic Data and Nuclear Data Tables*, 38, 1, 1988.

PHOTON ATTENUATION COEFFICIENTS (continued)

Mass attenuation coefficient, cm^2/g

	Atomic no.	Photon energy, MeV 0.001	0.002	0.005	0.01	0.02	0.05	0.1	0.2	0.5
H	1	7.21 + 00	1.06 + 00	4.19-01	3.85-01	3.69-01	3.36-01	2.94-01	2.43-01	1.73-01
He	2	6.08 + 01	6.86 + 00	5.77-01	2.48-01	1.96-01	1.70-01	1.49-01	1.22-01	8.71-02
Li	3	2.34 + 02	2.71 + 01	1.62 + 00	3.40-01	1.86-01	1.49-01	1.29-01	1.06-01	7.53-02
Be	4	6.04 + 02	7.47 + 01	4.37 + 00	6.47-01	2.25-01	1.55-01	1.33-01	1.09-01	7.74-02
B	5	1.23 + 03	1.60 + 02	9.68 + 00	1.25 + 00	3.01-01	1.66-01	1.39-01	1.14-01	8.07-02
C	6	2.21 + 03	3.03 + 02	1.91 + 01	2.37 + 00	4.42-01	1.87-01	1.51-01	1.23-01	8.72-02
N	7	3.31 + 03	4.77 + 02	3.14 + 01	3.88 + 00	6.18-01	1.98-01	1.53-01	1.23-01	8.72-02
O	8	4.59 + 03	6.95 + 02	4.79 + 01	5.95 + 00	8.65-01	2.13-01	1.55-01	1.24-01	8.73-02
F	9	5.65 + 03	9.05 + 02	6.51 + 01	8.21 + 00	1.13 + 00	2.21-01	1.50-01	1.18-01	8.27-02
Ne	10	7.41 + 03	1.24 + 03	9.34 + 01	1.20 + 01	1.61 + 00	2.58-01	1.60-01	1.24-01	8.66-02
Na	11	6.54 + 02	1.52 + 03	1.19 + 02	1.56 + 01	2.06 + 00	2.80-01	1.59-01	1.20-01	8.37-02
Mg	12	9.22 + 02	1.93 + 03	1.58 + 02	2.11 + 01	2.76 + 00	3.29-01	1.69-01	1.24-01	8.65-02
Al	13	1.19 + 03	2.26 + 03	1.93 + 02	2.62 + 01	3.44 + 00	3.68-01	1.70-01	1.22-01	8.44-02
Si	14	1.57 + 03	2.78 + 03	2.45 + 02	3.39 + 01	4.46 + 00	4.38-01	1.84-01	1.28-01	8.75-02
P	15	1.91 + 03	3.02 + 02	2.86 + 02	4.04 + 01	5.35 + 00	4.92-01	1.87-01	1.25-01	8.51-02
S	16	2.43 + 03	3.85 + 02	3.49 + 02	5.01 + 01	6.71 + 00	5.85-01	2.02-01	1.30-01	8.78-02
Cl	17	2.83 + 03	4.52 + 02	3.90 + 02	5.73 + 01	7.74 + 00	6.48-01	2.05-01	1.27-01	8.45-02
Ar	18	3.18 + 03	5.12 + 02	4.23 + 02	6.32 + 01	8.63 + 00	7.01-01	2.04-01	1.20-01	7.96-02
K	19	4.06 + 03	6.59 + 02	5.19 + 02	7.91 + 01	1.09 + 01	8.68-01	2.34-01	1.32-01	8.60-02
Ca	20	4.87 + 03	8.00 + 02	6.03 + 02	9.34 + 01	1.31 + 01	1.02 + 00	2.57-01	1.38-01	8.85-02
Sc	21	5.24 + 03	8.70 + 02	6.31 + 02	9.95 + 01	1.41 + 01	1.09 + 00	2.58-01	1.31-01	8.31-02
Ti	22	5.87 + 03	9.86 + 02	6.84 + 02	1.11 + 02	1.59 + 01	1.21 + 00	2.72-01	1.31-01	8.19-02
V	23	6.50 + 03	1.11 + 03	9.29 + 01	1.22 + 02	1.77 + 01	1.35 + 00	2.88-01	1.32-01	8.07-02
Cr	24	7.40 + 03	1.28 + 03	1.08 + 02	1.39 + 02	2.04 + 01	1.55 + 00	3.17-01	1.38-01	8.28-02
Mn	25	8.09 + 03	1.42 + 03	1.21 + 02	1.51 + 02	2.25 + 01	1.71 + 00	3.37-01	1.39-01	8.19-02
Fe	26	9.09 + 03	1.63 + 03	1.40 + 02	1.71 + 02	2.57 + 01	1.96 + 00	3.72-01	1.46-01	8.41-02
Co	27	9.80 + 03	1.78 + 03	1.54 + 02	1.84 + 02	2.80 + 01	2.14 + 00	3.95-01	1.48-01	8.32-02
Ni	28	9.86 + 03	2.05 + 03	1.79 + 02	2.09 + 02	3.22 + 01	2.47 + 00	4.44-01	1.58-01	8.70-02
Cu	29	1.06 + 04	2.15 + 03	1.90 + 02	2.16 + 02	3.38 + 01	2.61 + 00	4.58-01	1.56-01	8.36-02
Zn	30	1.55 + 03	2.37 + 03	2.12 + 02	2.33 + 02	3.72 + 01	2.89 + 00	4.97-01	1.62-01	8.45-02
Ga	31	1.70 + 03	2.52 + 03	2.27 + 02	3.42 + 01	3.93 + 01	3.08 + 00	5.20-01	1.62-01	8.24-02
Ge	32	1.89 + 03	2.71 + 03	2.47 + 02	3.74 + 01	4.22 + 01	3.34 + 00	5.55-01	1.66-01	8.21-02
As	33	2.12 + 03	2.93 + 03	2.71 + 02	4.12 + 01	4.56 + 01	3.63 + 00	5.97-01	1.72-01	8.26-02
Se	34	2.32 + 03	3.10 + 03	2.90 + 02	4.41 + 01	4.82 + 01	3.86 + 00	6.28-01	1.74-01	8.13-02
Br	35	2.62 + 03	3.41 + 03	3.21 + 02	4.91 + 01	5.27 + 01	4.26 + 00	6.86-01	1.84-01	8.33-02
Kr	36	2.85 + 03	3.60 + 03	3.43 + 02	5.26 + 01	5.55 + 01	4.52 + 00	7.22-01	1.87-01	8.23-02
Rb	37	3.17 + 03	3.41 + 03	3.74 + 02	5.77 + 01	5.98 + 01	4.92 + 00	7.80-01	1.96-01	8.36-02
Sr	38	3.49 + 03	2.59 + 03	4.06 + 02	6.27 + 01	6.39 + 01	5.31 + 00	8.37-01	2.04-01	8.44-02
Y	39	3.86 + 03	7.42 + 02	4.42 + 02	6.87 + 01	6.86 + 01	5.76 + 00	9.05-01	2.15-01	8.61-02
Zr	40	4.21 + 03	8.12 + 02	4.76 + 02	7.42 + 01	7.24 + 01	6.17 + 00	9.66-01	2.24-01	8.69-02
Nb	41	4.60 + 03	8.89 + 02	5.13 + 02	8.04 + 01	7.71 + 01	6.64 + 00	1.04 + 00	2.34-01	8.83-02
Mo	42	4.94 + 03	9.60 + 02	5.45 + 02	8.58 + 01	1.31 + 01	7.04 + 00	1.10 + 00	2.42-01	8.85-02
Tc	43	5.36 + 03	1.04 + 03	5.84 + 02	9.23 + 01	1.41 + 01	7.52 + 00	1.17 + 00	2.53-01	8.97-02
Ru	44	5.72 + 03	1.12 + 03	6.17 + 02	9.80 + 01	1.50 + 01	7.92 + 00	1.23 + 00	2.62-01	8.99-02
Rh	45	6.17 + 03	1.21 + 03	[illegible]	1.05 + 02	1.61 + 01	8.45 + 00	1.31 + 00	2.74-01	9.13-02
Pd	46	6.54 + 03	1.29 + 03	6.91 + 02	1.11 + 02	1.70 + 01	8.85 + 00	1.38 + 00	2.8[illegible]-01	[illegible]
Ag	47	7.04 + 03	1.40 + 03	7.39 + 02	1.19 + 02	1.84 + 01	9.45 + 00	1.47 + 00	2.97-01	9.32-02
Cd	48	7.35 + 03	1.47 + 03	7.69 + 02	1.24 + 02	1.92 + 01	9.78 + 00	1.52 + 00	3.04-01	9.25-02
In	49	7.81 + 03	1.58 + 03	8.13 + 02	1.32 + 02	2.04 + 01	1.03 + 01	1.61 + 00	3.17-01	9.37-02
Sn	50	8.16 + 03	1.66 + 03	8.47 + 02	1.38 + 02	2.15 + 01	1.07 + 01	1.68 + 00	3.26-01	9.37-02

L_3 L_1 L_2 K EDGE

Mass attenuation coefficient, cm²/g

	Atomic no.	Photon energy, MeV 1.0	2.0	5.0	10.0	20.0	50.0	100.0	500.0	1000.0
H	1	1.26-01	8.77-02	5.05-02	3.25-02	2.15-02	1.42-02	1.19-02	1.14-02	1.16-02
He	2	6.36-02	4.42-02	2.58-02	1.70-02	1.18-02	8.61-03	7.78-03	7.79-03	7.95-03
Li	3	5.50-02	3.83-02	2.26-02	1.53-02	1.11-02	8.68-03	8.21-03	8.61-03	8.87-03
Be	4	5.65-02	3.94-02	2.35-02	1.63-02	1.23-02	1.02-02	9.94-03	1.08-02	1.12-02
B	5	5.89-02	4.11-02	2.48-02	1.76-02	1.37-02	1.19-02	1.19-02	1.32-02	1.37-02
C	6	6.36-02	4.44-02	2.71-02	1.96-02	1.58-02	1.43-02	1.46-02	1.64-02	1.70-02
N	7	6.36-02	4.45-02	2.74-02	2.02-02	1.67-02	1.57-02	1.63-02	1.85-02	1.92-02
O	8	6.37-02	4.46-02	2.78-02	2.09-02	1.77-02	1.71-02	1.79-02	2.06-02	2.13-02
F	9	6.04-02	4.23-02	2.66-02	2.04-02	1.77-02	1.75-02	1.86-02	2.14-02	2.21-02
Ne	10	6.32-02	4.43-02	2.82-02	2.20-02	1.95-02	1.96-02	2.11-02	2.43-02	2.51-02
Na	11	6.10-02	4.28-02	2.75-02	2.18-02	1.97-02	2.03-02	2.19-02	2.53-02	2.62-02
Mg	12	6.30-02	4.43-02	2.87-02	2.31-02	2.13-02	2.23-02	2.42-02	2.81-02	2.90-02
Al	13	6.15-02	4.32-02	2.84-02	2.32-02	2.17-02	2.31-02	2.52-02	2.93-02	3.03-02
Si	14	6.36-02	4.48-02	2.97-02	2.46-02	2.34-02	2.52-02	2.76-02	3.23-02	3.34-02
P	15	6.18-02	4.36-02	2.91-02	2.45-02	2.36-02	2.58-02	2.84-02	3.33-02	3.45-02
S	16	6.37-02	4.50-02	3.04-02	2.59-02	2.53-02	2.79-02	3.08-02	3.62-02	3.75-02
Cl	17	6.13-02	4.33-02	2.95-02	2.55-02	2.52-02	2.81-02	3.11-02	3.67-02	3.80-02
Ar	18	5.76-02	4.07-02	2.80-02	2.45-02	2.45-02	2.76-02	3.07-02	3.62-02	3.75-02
K	19	6.22-02	4.40-02	3.05-02	2.70-02	2.74-02	3.11-02	3.46-02	4.09-02	4.24-02
Ca	20	6.39-02	4.52-02	3.17-02	2.84-02	2.90-02	3.32-02	3.71-02	4.40-02	4.56-02
Sc	21	5.98-02	4.24-02	3.00-02	2.72-02	2.80-02	3.23-02	3.62-02	4.30-02	4.45-02
Ti	22	5.89-02	4.18-02	2.98-02	2.73-02	2.84-02	3.30-02	3.71-02	4.40-02	4.56-02
V	23	5.79-02	4.11-02	2.96-02	2.74-02	2.88-02	3.36-02	3.78-02	4.49-02	4.65-02
Cr	24	5.93-02	4.21-02	3.06-02	2.86-02	3.03-02	3.56-02	4.01-02	4.76-02	4.93-02
Mn	25	5.85-02	4.16-02	3.04-02	2.87-02	3.07-02	3.63-02	4.09-02	4.86-02	5.04-02
Fe	26	5.99-02	4.26-02	3.15-02	2.99-02	3.22-02	3.83-02	4.33-02	5.15-02	5.33-02
Co	27	5.91-02	4.20-02	3.13-02	3.00-02	3.26-02	3.88-02	4.40-02	5.23-02	5.41-02
Ni	28	6.16-02	4.39-02	3.29-02	3.18-02	3.48-02	4.17-02	4.73-02	5.61-02	5.81-02
Cu	29	5.90-02	4.20-02	3.18-02	3.10-02	3.41-02	4.10-02	4.66-02	5.53-02	5.72-02
Zn	30	5.94-02	4.24-02	3.22-02	3.18-02	3.51-02	4.24-02	4.82-02	5.72-02	5.91-02
Ga	31	5.77-02	4.11-02	3.16-02	3.13-02	3.48-02	4.22-02	4.80-02	5.70-02	5.89-02
Ge	32	5.73-02	4.09-02	3.16-02	3.16-02	3.53-02	4.30-02	4.89-02	5.80-02	6.00-02
As	33	5.73-02	4.09-02	3.19-02	3.21-02	3.60-02	4.40-02	5.01-02	5.95-02	6.15-02
Se	34	5.62-02	4.01-02	3.14-02	3.19-02	3.60-02	4.41-02	5.03-02	5.97-02	6.17-02
Br	35	5.73-02	4.09-02	3.23-02	3.29-02	3.74-02	4.60-02	5.24-02	6.22-02	6.43-02
Kr	36	5.63-02	4.02-02	3.20-02	3.28-02	3.74-02	4.61-02	5.26-02	6.25-02	6.46-02
Rb	37	5.69-02	4.06-02	3.25-02	3.36-02	3.85-02	4.75-02	5.43-02	6.45-02	6.67-02
Sr	38	5.71-02	4.08-02	3.29-02	3.41-02	3.93-02	4.87-02	5.56-02	6.61-02	6.83-02
Y	39	5.80-02	4.14-02	3.35-02	3.50-02	4.05-02	5.03-02	5.75-02	6.83-02	7.06-02
Zr	40	5.81-02	4.15-02	3.38-02	3.55-02	4.12-02	5.13-02	5.87-02	6.98-02	7.22-02
Nb	41	5.87-02	4.18-02	3.44-02	3.63-02	4.22-02	5.27-02	6.03-02	7.17-02	7.42-02
Mo	42	5.84-02	4.16-02	3.44-02	3.65-02	4.26-02	5.33-02	6.10-02	7.26-02	7.51-02
Tc	43	5.88-02	4.19-02	3.48-02	3.71-02	4.35-02	5.45-02	6.24-02	7.43-02	7.68-02
Ru	44	5.85-02	4.16-02	3.48-02	3.73-02	4.39-02	5.50-02	6.30-02	7.51-02	7.77-02
Rh	45	5.89-02	4.20-02	3.53-02	3.80-02	4.48-02	5.63-02	6.45-02	7.69-02	7.94-02
Pd	46	5.85-02	4.16-02	3.52-02	3.80-02	4.50-02	5.66-02	6.49-02	7.73-02	8.00-02
Ag	47	5.92-02	4.21-02	3.58-02	3.88-02	4.61-02	5.81-02	6.67-02	7.93-02	8.20-02
Cd	48	5.83-02	4.14-02	3.54-02	3.85-02	4.59-02	5.79-02	6.64-02	7.91-02	8.18-02
In	49	5.85-02	4.15-02	3.56-02	3.90-02	4.65-02	5.88-02	6.75-02	8.04-02	8.32-02
Sn	50	5.80-02	4.11-02	3.55-02	3.90-02	4.66-02	5.90-02	6.78-02	8.07-02	8.35-02

PHOTON ATTENUATION COEFFICIENTS (continued)

		Mass attenuation coefficients, cm^2/g								
		Photon energy, MeV								
	Atomic no.	0.001	0.002	0.005	0.01	0.02	0.05	0.1	0.2	0.5
Sb	51	8.58 + 03	1.77 + 03	8.85 + 02	1.46 + 02	2.27 + 01	1.12 + 01	1.76 + 00	3.38-01	9.45-02
Te	52	8.43 + 03	1.83 + 03	9.01 + 02	1.50 + 02	2.34 + 01	1.14 + 01	1.80 + 00	3.43-01	9.33-02
I	53	9.10 + 03	2.00 + 03	8.43 + 02	1.63 + 02	2.54 + 01	1.23 + 01	1.94 + 00	3.66-01	9.70-02
Xe	54	9.41 + 03	2.09 + 03	6.39 + 02	1.69 + 02	2.65 + 01	1.27 + 01	2.01 + 00	3.76-01	9.70-02
Cs	55	9.37 + 03	2.23 + 03	2.30 + 02	1.79 + 02	2.82 + 01	1.34 + 01	2.12 + 00	3.94-01	9.91-02
ZB	56	8.54 + 03	2.32 + 03	2.41 + 02	1.86 + 02	2.94 + 01	1.38 + 01	2.20 + 00	4.05-01	9.92-02
La	57	9.09 + 03	2.46 + 03	2.58 + 02	1.97 + 02	3.12 + 01	1.45 + 01	2.32 + 00	4.24-01	1.01-01
Ce	58	9.71 + 03	2.61 + 03	2.74 + 02	2.08 + 02	3.31 + 01	1.52 + 01	2.45 + 00	4.45-01	1.04-01
Pr	59	1.06 + 04	2.77 + 03	2.92 + 02	2.21 + 02	3.53 + 01	1.60 + 01	2.59 + 00	4.69-01	1.07-01
Nd	60	6.63 + 03	2.88 + 03	3.06 + 02	2.30 + 02	3.68 + 01	1.65 + 01	2.69 + 00	4.84-01	1.08-01
Pm	61	2.06 + 03	3.05 + 03	3.26 + 02	2.44 + 02	3.92 + 01	1.73 + 01	2.84 + 00	5.10-01	1.12-01
Sm	62	2.11 + 03	3.12 + 03	3.36 + 02	2.50 + 02	4.03 + 01	1.77 + 01	2.90 + 00	5.19-01	1.11-01
Eu	63	2.22 + 03	3.28 + 03	3.54 + 02	2.63 + 02	4.24 + 01	1.85 + 01	3.04 + 00	5.43-01	1.14-01
Gd	64	2.29 + 03	3.36 + 03	3.65 + 02	2.69 + 02	4.36 + 01	3.86 + 00	3.11 + 00	5.54-01	1.14-01
Tb	65	2.40 + 03	3.51 + 03	3.84 + 02	2.82 + 02	4.59 + 01	4.06 + 00	3.25 + 00	5.77-01	1.17-01
Dy	66	2.49 + 03	3.47 + 03	3.99 + 02	2.90 + 02	4.76 + 01	4.23 + 00	3.36 + 00	5.95-01	1.18-01
Ho	67	2.62 + 03	3.59 + 03	4.17 + 02	3.01 + 02	4.98 + 01	4.43 + 00	3.49 + 00	6.18-01	1.20-01
Er	68	2.75 + 03	3.52 + 03	4.36 + 02	3.13 + 02	5.20 + 01	4.63 + 00	3.63 + 00	6.41-01	1.23-01
Tm	69	2.90 + 03	3.69 + 03	4.57 + 02	2.83 + 02	5.45 + 01	4.87 + 00	3.78 + 00	6.68-01	1.26-01
Yb	70	3.02 + 03	3.80 + 03	4.72 + 02	2.94 + 02	5.63 + 01	5.04 + 00	3.88 + 00	6.86-01	1.27-01
Lu	71	3.19 + 03	3.45 + 03	4.94 + 02	2.21 + 02	5.88 + 01	5.28 + 00	4.03 + 00	7.13-01	1.30-01
Hf	72	3.34 + 03	3.60 + 03	5.11 + 02	2.30 + 02	6.09 + 01	5.48 + 00	4.15 + 00	7.34-01	1.32-01
Ta	73	3.51 + 03	3.77 + 03	5.33 + 02	2.38 + 02	6.33 + 01	5.72 + 00	4.30 + 00	7.60-01	1.35-01
W	74	3.68 + 03	3.92 + 03	5.53 + 02	9.69 + 01	6.57 + 01	5.95 + 00	4.44 + 00	7.84-01	1.38-01
Re	75	3.87 + 03	3.77 + 03	5.76 + 02	1.01 + 02	6.84 + 01	6.21 + 00	4.59 + 00	8.12-01	1.41-01
Os	76	4.03 + 03	2.22 + 03	5.93 + 02	1.04 + 02	7.04 + 01	6.41 + 00	4.70 + 00	8.33-01	1.43-01
Ir	77	4.24 + 03	1.03 + 03	6.18 + 02	1.09 + 02	7.32 + 01	6.69 + 00	4.86 + 00	8.63-01	1.46-01
Pt	78	4.43 + 03	1.08 + 03	6.40 + 02	1.13 + 02	7.57 + 01	6.95 + 00	4.99 + 00	8.90-01	1.49-01
Au	79	4.65 + 03	1.14 + 03	6.66 + 02	1.18 + 02	7.88 + 01	7.26 + 00	5.16 + 00	9.22-01	1.53-01
Hg	80	4.83 + 03	1.18 + 03	6.87 + 02	1.22 + 02	8.12 + 01	7.50 + 00	5.28 + 00	9.46-01	1.56-01
Tl	81	5.01 + 03	1.23 + 03	7.07 + 02	1.26 + 02	8.36 + 01	7.75 + 00	5.40 + 00	9.69-01	1.58-01
Pb	82	5.21 + 03	1.29 + 03	7.30 + 02	1.31 + 02	8.64 + 01	8.04 + 00	5.55 + 00	9.99-01	1.61-01
Bi	83	5.44 + 03	1.35 + 03	7.58 + 02	1.36 + 02	8.95 + 01	8.38 + 00	5.74 + 00	1.03 + 00	1.66-01
Po	84	5.72 + 03	1.42 + 03	7.93 + 02	1.43 + 02	9.35 + 01	8.80 + 00	5.99 + 00	1.08 + 00	1.71-01
At	85	5.87 + 03	1.49 + 03	8.25 + 02	1.49 + 02	9.70 + 01	9.19 + 00	6.17 + 00	1.12 + 00	1.77-01
Rn	86	5.83 + 03	1.49 + 03	8.16 + 02	1.48 + 02	9.56 + 01	9.12 + 00	6.09 + 00	1.10 + 00	1.73-01
Fr	87	6.08 + 03	1.56 + 03	8.49 + 02	1.54 + 02	9.93 + 01	9.52 + 00	1.66 + 00	1.14 + 00	1.78-01
Ra	88	6.20 + 03	1.62 + 03	8.74 + 02	1.59 + 02	1.02 + 02	9.85 + 00	1.71 + 00	1.17 + 00	1.82-01
Ac	89	6.47 + 03	1.70 + 03	8.69 + 02	1.65 + 02	1.06 + 02	1.03 + 01	1.79 + 00	1.21 + 00	1.87-01
Th	90	6.61 + 03	1.74 + 03	8.88 + 02	1.69 + 02	9.37 + 01	1.05 + 01	1.83 + 00	1.23 + 00	1.90-01
Pa	91	6.53 + 03	1.83 + 03	8.76 + 02	1.77 + 02	7.03 + 01	1.10 + 01	1.92 + 00	1.29 + 00	1.97-01
U	92	6.63 + 03	1.86 + 03	8.89 + 02	1.79 + 02	7.11 + 01	1.12 + 01	1.95 + 00	1.30 + 00	1.98-01
Np	93	6.95 + 03	1.96 + 03	9.32 + 02	1.87 + 02	7.45 + 01	1.18 + 01	2.05 + 00	1.35 + 00	2.05-01
Pu	94	7.19 + 03	2.04 + 03	9.65 + 02	1.94 + 02	7.71 + 01	1.22 + 01	2.13 + 00	1.39 + 00	2.10-01
Am	95	7.37 + 03	2.10 + 03	9.90 + 02	1.98 + 02	7.93 + 01	1.25 + 01	2.19 + 00	1.42 + 00	2.14-01
Cm	96	7.54 + 03	2.15 + 03	1.02 + 03	2.03 + 02	8.14 + 01	1.28 + 01	2.25 + 00	1.44 + 00	2.18-01
Bk	97	7.84 + 03	2.25 + 03	1.06 + 03	2.10 + 02	8.39 + 01	1.34 + 01	2.35 + 00	1.50 + 00	2.25-01
Cf	98	7.89 + 03	2.31 + 03	9.27 + 02	2.15 + 02	8.58 + 01	1.37 + 01	2.41 + 00	1.52 + 00	2.29-01
Es	99	7.79 + 03	2.40 + 03	9.59 + 02	2.22 + 02	4.01 + 01	1.42 + 01	2.51 + 00	1.57 + 00	2.36-01
Fm	100	7.13 + 03	2.46 + 03	9.77 + 02	2.26 + 02	4.09 + 01	1.45 + 01	2.57 + 00	1.59 + 00	2.39-01

N_5 N_4 N_3 N_2 N_1 M_5 M_4 M_3 M_2 M_1 L_3 L_2 L_1 K EDGE

Mass attenuation coefficients, cm^2/g

	Atomic no.	1.0	2.0	5.0	10.0	20.0	50.0	100.0	500.0	1000.0
		Photon Energy, MeV								
Sb	51	5.80-02	4.10-02	3.56-02	3.92-02	4.70-02	5.96-02	6.85-02	8.16-02	8.44-02
Te	52	5.67-02	4.01-02	3.49-02	3.86-02	4.64-02	5.89-02	6.77-02	8.07-02	8.35-02
I	53	5.84-02	4.12-02	3.61-02	4.00-02	4.82-02	6.13-02	7.04-02	8.40-02	8.69-02
Xe	54	5.78-02	4.08-02	3.58-02	3.99-02	4.82-02	6.12-02	7.04-02	8.40-02	8.69-02
Cs	55	5.85-02	4.12-02	3.64-02	4.06-02	4.91-02	6.25-02	7.19-02	8.58-02	8.88-02
ZB	56	5.80-02	4.08-02	3.61-02	4.04-02	4.90-02	6.25-02	7.19-02	8.58-02	8.88-02
La	57	5.88-02	4.12-02	3.66-02	4.11-02	5.00-02	6.37-02	7.34-02	8.76-02	9.06-02
Ce	58	5.96-02	4.18-02	3.73-02	4.19-02	5.10-02	6.52-02	7.50-02	8.96-02	9.27-02
Pr	59	6.07-02	4.24-02	3.80-02	4.29-02	5.23-02	6.68-02	7.69-02	9.19-02	9.50-02
Nd	60	6.07-02	4.24-02	3.81-02	4.30-02	5.26-02	6.72-02	7.74-02	9.25-02	9.56-02
Pm	61	6.19-02	4.31-02	3.88-02	4.40-02	5.38-02	6.89-02	7.94-02	9.48-02	9.81-02
Sm	62	6.11-02	4.24-02	3.83-02	4.35-02	5.34-02	6.84-02	7.88-02	9.41-02	9.73-02
Eu	63	6.19-02	4.28-02	3.88-02	4.42-02	5.42-02	6.96-02	8.02-02	9.57-02	9.90-02
Gd	64	6.12-02	4.23-02	3.84-02	4.38-02	5.38-02	6.91-02	7.97-02	9.51-02	9.83-02
Tb	65	6.20-02	4.27-02	3.89-02	4.45-02	5.47-02	7.03-02	8.11-02	9.67-02	1.00-01
Dy	66	6.20-02	4.26-02	3.90-02	4.46-02	5.49-02	7.06-02	8.15-02	9.72-02	1.00-01
Ho	67	6.26-02	4.29-02	3.93-02	4.50-02	5.55-02	7.14-02	8.24-02	9.83-02	1.02-01
Er	68	6.32-02	4.32-02	3.96-02	4.55-02	5.61-02	7.23-02	8.34-02	9.95-02	1.03-01
Tm	69	6.40-02	4.36-02	4.01-02	4.61-02	5.70-02	7.35-02	8.48-02	1.01-01	1.04-01
Yb	70	6.40-02	4.35-02	4.00-02	4.61-02	5.70-02	7.35-02	8.49-02	1.01-01	1.04-01
Lu	71	6.48-02	4.39-02	4.05-02	4.66-02	5.77-02	7.45-02	8.60-02	1.02-01	1.06-01
Hf	72	6.50-02	4.39-02	4.05-02	4.68-02	5.80-02	7.48-02	8.64-02	1.03-01	1.06-01
Ta	73	6.57-02	4.41-02	4.08-02	4.72-02	5.85-02	7.56-02	8.73-02	1.04-01	1.07-01
W	74	6.62-02	4.43-02	4.10-02	4.75-02	5.89-02	7.62-02	8.80-02	1.05-01	1.08-01
Re	75	6.69-02	4.46-02	4.14-02	4.79-02	5.95-02	7.70-02	8.89-02	1.06-01	1.09-01
Os	76	6.71-02	4.46-02	4.13-02	4.79-02	5.96-02	7.71-02	8.90-02	1.06-01	1.10-01
Ir	77	6.79-02	4.50-02	4.17-02	4.84-02	6.02-02	7.80-02	9.01-02	1.07-01	1.11-01
Pt	78	6.86-02	4.52-02	4.20-02	4.87-02	6.06-02	7.86-02	9.08-02	1.08-01	1.12-01
Au	79	6.95-02	4.57-02	4.24-02	4.93-02	6.14-02	7.95-02	9.19-02	1.09-01	1.13-01
Hg	80	6.99-02	4.57-02	4.25-02	4.94-02	6.15-02	7.98-02	9.22-02	1.10-01	1.13-01
Tl	81	7.03-02	4.58-02	4.25-02	4.94-02	6.16-02	8.00-02	9.24-02	1.10-01	1.14-01
Pb	82	7.10-02	4.61-02	4.27-02	4.97-02	6.21-02	8.06-02	9.31-02	1.11-01	1.15-01
Bi	83	7.21-02	4.66-02	4.32-02	5.03-02	6.28-02	8.15-02	9.42-02	1.12-01	1.16-01
Po	84	7.39-02	4.75-02	4.40-02	5.12-02	6.40-02	8.32-02	9.61-02	1.15-01	1.18-01
At	85	7.54-02	4.82-02	4.46-02	5.20-02	6.49-02	8.44-02	9.76-02	1.16-01	1.20-01
Rn	86	7.30-02	4.65-02	4.30-02	5.01-02	6.26-02	8.14-02	9.42-02	1.12-01	1.16-01
Fr	87	7.45-02	4.72-02	4.36-02	5.08-02	6.35-02	8.26-02	9.56-02	1.14-01	1.18-01
Ra	88	7.53-02	4.75-02	4.38-02	5.10-02	6.38-02	8.31-02	9.61-02	1.15-01	1.19-01
Ac	89	7.69-02	4.82-02	4.44-02	5.17-02	6.47-02	8.43-02	9.75-02	1.16-01	1.20-01
Th	90	7.71-02	4.81-02	4.42-02	5.15-02	6.45-02	8.40-02	9.72-02	1.16-01	1.20-01
Pa	91	7.94-02	4.93-02	4.52-02	5.26-02	6.59-02	8.60-02	9.95-02	1.19-01	1.23-01
U	92	7.90-02	4.88-02	4.46-02	5.19-02	6.51-02	8.49-02	9.83-02	1.17-01	1.21-01
Np	93	8.13-02	4.99-02	4.56-02	5.30-02	6.65-02	8.68-02	1.01-01	1.20-01	1.24-01
Pu	94	8.26-02	5.05-02	4.60-02	5.34-02	6.71-02	8.76-02	1.01-01	1.21-01	1.25-01
Am	95	8.33-02	5.06-02	4.60-02	5.34-02	6.70-02	8.77-02	1.02-01	1.21-01	1.25-01
Cm	96	8.41-02	5.08-02	4.60-02	5.34-02	6.70-02	8.77-02	1.02-01	1.21-01	1.26-01
Bk	97	8.62-02	5.18-02	4.68-02	5.42-02	6.81-02	8.92-02	1.03-01	1.24-01	1.28-01
Cf	98	8.70-02	5.20-02	4.68-02	5.42-02	6.81-02	8.92-02	1.04-01	1.24-01	1.28-01
Es	99	8.89-02	5.28-02	4.74-02	5.48-02	6.89-02	9.04-02	1.05-01	1.25-01	1.29-01
Fm	100	8.94-02	5.28-02	4.72-02	5.45-02	6.86-02	9.00-02	1.05-01	1.25-01	1.29-01

UNITS, SYMBOLS, AND EQUATIONS FOR RADIOMETRIC AND PHOTOMETRIC QUANTITIES

Submitted by Abraham Abramowitz
from Z-7.1-1967

Radiometric Quantities
(See Note at bottom of page)

Quantity*	Symbol*	Defining Equation**	Commonly Used Units	Symbol
Radiant energy	$Q, (Q_e)$		erg ‡joule kilowatt-hour	 J kWh
Radiant density	$w, (w_e)$	$w = dQ/dV$	‡joule per cubic meter erg per cubic centimeter	J/m^3 erg/cm^3
Radiant flux	$\Phi, (\Phi_e)$	$\Phi = dQ/dt$	erg per second †watt	erg/s W
Radiant flux density at a surface				
Radiant exitance (Radiant emittance)†	$M, (M_e)$	$M = d\Phi/dA$	watt per square centimeter	W/cm^2
Irradiance	$E, (E_e)$	$E = d\Phi/dA$	‡watt per square meter, etc.	W/m^2
Radiant intensity	$I, (I_e)$	$I = d\Phi/d\omega$ (ω = solid angle through which flux from point source is radiated)	‡watt per steradian	W/sr
Radiance	$L, (L_e)$	$L = d^2\Phi/d\omega\,(dA\cos\theta)$ $= dI/(dA\cos\theta)$ (θ = angle between line of sight and normal to surface considered)	watt per steradian and square centimeter ‡watt per steradian and square meter	$W \cdot sr^{-1}\,cm^{-2}$ $W \cdot sr^{-1}\,m^{-2}$
Emissivity	ε	$\varepsilon = M/M_{blackbody}$ (M and $M_{blackbody}$ are respectively the radiant exitance of the measured specimen and that of a blackbody at the same temperature as the specimen)	one (numeric)	—

Note: The symbols for photometric quantities (see following table) are the same as those for the corresponding radiometric quantities (see above). When it is necessary to differentiate them the subscripts v and e respectively should be used, e.g., Q_v and Q_e.

*Quantities may be restricted to a narrow wavelength band by adding the word spectral and indicating the wavelength. The corresponding symbols are changed by adding a subscript λ, e.g., Q_λ for a spectral concentration or a λ in parentheses, e.g., $K(\lambda)$, for a function of wavelength.

**The equations in this column are given merely for identification.

***Φ_i = incident flux
Φ_a = absorbed flux
Φ_r = reflected flux
Φ_t = transmitted flux

†to be deprecated.

‡International System (SI) unit.

UNITS, SYMBOLS, AND EQUATIONS FOR RADIOMETRIC AND PHOTOMETRIC QUANTITIES (continued)

Photometric Quantities

Quantity*	Symbol*	Defining Equation**	Commonly Used Units	Symbol
Absorptance	$\alpha, (\alpha_v, \alpha_e)$	$\alpha = \Phi_a/\Phi_i$***	one (numeric)	—
Reflectance	$\rho, (\rho_v, \rho_e)$	$\rho = \Phi_r/\Phi_i$***	one (numeric)	—
Transmittance	$\tau, (\tau_v, \tau_e)$	$\tau = \Phi_t/\Phi_i$***	one (numeric)	—
Luminous energy (quantity of light)	$Q, (Q_v)$	$Q_v = \int_{380}^{760} K(\lambda)\, Q_e\lambda\, d\lambda$	lumen-hour ‡lumen-second (talbot)	$lm \cdot h$ $lm \cdot s$
Luminous density	$w, (w_v)$	$w = dQ/dV$	‡lumen-second per cubic meter	$lm \cdot s \cdot m^{-3}$
Luminous flux	$\Phi, (\Phi_v)$	$\Phi = dQ/dt$	‡lumen	lm
Luminous flux density at a surface				
Luminous exitance (Luminous emittance)†	$M, (M_v)$	$M = d\Phi/dA$	lumen per square foot	lm/ft^2
Illumination (illuminance)	$E, (E_v)$	$E = d\Phi/dA$	footcandle (lumen per square foot) ‡lux (lm/m^2) phot (lm/cm^2)	fc lx ph
Luminous intensity (candlepower)	$I, (I_v)$	$I = d\Phi/d\omega$ (ω = solid angle through which flux from point source is radiated)	‡candela (lumen per steradian)	cd
Luminance (photometric brightness)	$L, (L_v)$	$L = d^2\Phi/d\omega\,(dA \cos\theta)$ $= dI/(dA \cos\theta)$ (θ = angle between line of sight and normal to surface considered)	candela per unit area stilb (cd/cm^2) nit (†cd/m^2) footlambert ($cd/\pi ft^2$) lambert ($cd/\pi cm^2$) apostilb ($cd/\pi m^2$)	cd/in^2, etc. sb nt fL L asb
Luminous efficacy	K	$K = \Phi_v/\Phi_e$	‡lumen per watt	lm/W
Luminous efficiency	V	$V = K/K_{maximum}$ ($K_{maximum}$ = maximum value of $K(\lambda)$ function)	one (numeric)	—

ILLUMINATION CONVERSION FACTORS

1 lumen = 1/680 lightwatt
1 lumen-hour = 60 lumen-minutes
1 footcandle = 1 lumen/ft^2
1 watt-second = 1 joule = 10^7 ergs
1 phot = 1 lumen/cm^2
1 lux = 1 lumen/m^2

Number of → Multiplied by ↘ Equals Number of ↓	Footcandles	*Lux	Phots	Milliphots
Footcandles	1	0.0929	929	0.929
*Lux	10.76	1	10,000	10
Phot	0.00108	0.0001	1	0.001
Milliphot	1.076	0.1	1,000	1

*The International System (SI) unit.

CLASSIFICATION OF ELECTROMAGNETIC RADIATION

Hans Dolezalek

Basic Conversions: $c = \lambda\nu = \nu/k;\ \nu = c/\lambda = ck;\ \lambda = c/\nu = 1/k;\ k = \nu/c = 1/\lambda$

c = speed of light = 2.99792458×10^{8} m/s

Frequency (ν)	Wavelength (λ)	Wave number (k)	Names of bands	Approximate photon energies
3×10^{0} — 3×10^{1} Hz 3 — 30 Hz	10^{8} — 10^{7} m 100 — 10 Mm	10^{-8} — 10^{-7} m^{-1} 10 — 100 Gm^{-1}	ELF-(ELF 1), ITU band no. 1	
3×10^{1} — 3×10^{2} Hz 30 — 300 Hz	10^{7} — 10^{6} m 10 — 1 Mm	10^{-7} — 10^{-6} m^{-1} 100 Gm^{-1} — 1 Mm^{-1}	SLF-(ELF 2), ITU band no. 2, megameter waves	
3×10^{2} — 3×10^{3} Hz 300 Hz — 3 kHz	10^{6} — 10^{5} m 1 Mm — 100 km	10^{-6} — 10^{-5} m^{-1} 1 — 10 Mm^{-1}	ULF-(ELF 3), ITU band no. 3	
3×10^{3} — 3×10^{4} Hz 3 — 30 kHz	10^{5} — 10^{4} m 100 — 10 km	10^{-5} — 10^{-4} m^{-1} 10 — 100 Mm^{-1}	VLF, ITU band no. 4, myriameter waves	
3×10^{4} — 3×10^{5} Hz 30 — 300 kHz	10^{4} — 10^{3} m 10 — 1 km	10^{-4} — 10^{-3} m^{-1} 100 Mm^{-1} — 1 km^{-1}	LF, ITU band no. 5, kilometer waves	
3×10^{5} — 3×10^{6} Hz 300 kHz — 3 MHz	10^{3} — 10^{2} m 1 km — 100 m	10^{-3} — 10^{-2} m^{-1} 1 — 10 km^{-1}	MF, ITU band no. 6, hectometer waves	
3×10^{6} — 3×10^{7} Hz 3 — 30 MHz	10^{2} — 10^{1} m 100 — 10 m	10^{-2} — 10^{-1} m^{-1} 10 — 100 km^{-1}	HF, ITU band no. 7, decameter waves	
3×10^{7} — 3×10^{8} Hz 30 — 300 MHz	10^{1} — 10^{0} m 10 — 1 m	10^{-1} — 10^{0} m^{-1} 100 km^{-1} — 1 m^{-1}	VHF, ITU band no. 8, meter waves	
3×10^{8} — 3×10^{9} Hz 300 MHz — 3 GHz	10^{0} — 10^{-1} m 1 m — 100 mm	10^{0} — 10^{1} m^{-1} 1 — 10 m^{-1}	UHF, ITU band no. 9, decimeter waves[a]	
3×10^{9} — 3×10^{10} Hz 3 — 30 GHz	10^{-1} — 10^{-2} m 100 — 10 mm	10^{1} — 10^{2} m^{-1} 10 — 100 m^{-1}	SHF, ITU band no. 10, centimeter waves[a]	
3×10^{10} — 3×10^{11} Hz 30 — 300 GHz	10^{-2} — 10^{-3} m 10 — 1 mm	10^{2} — 10^{3} m^{-1} 100 m^{-1} — 1 mm^{-1} (1 — 10 cm^{-1})	EHF, ITU band no. 11, millimeter waves	
3×10^{11} — 3×10^{12} Hz 300 GHz — 3 THz	10^{-3} — 10^{-4} m 1 mm — 100 μm	10^{3} — 10^{4} m^{-1} 1 — 10 mm^{-1} (10 — 100 cm^{-1})	Part of micrometer waves, includes part of far or thermal infrared; ITU band no. 12	
3×10^{12} — 3×10^{13} Hz 3 — 30 THz	10^{-4} — 10^{-5} m 100 — 10 μm	10^{4} — 10^{5} m^{-1} 10 — 100 mm^{-1} (100 — 1000 cm^{-1})	Part of micrometer waves includes part of far (thermal) infrared	
3×10^{13} — 3×10^{14} Hz 30 — 300 THz	10^{-5} — 10^{-6} m 10 — 1 μm (100,000 — 10,000 Å)	10^{5} — 10^{6} m^{-1} 100 mm^{-1} — 1 μm^{-1}	Part of μm waves, part of infrared	(1.6—16) × 10^{-20} joule {0.1 — 1 eV}
3×10^{14} — 3×10^{15} Hz 300 THz — 3 PHz	10^{-6} — 10^{-7} m 1 μm — 100 nm (10,000 — 1000 Å)	10^{6} — 10^{7} m^{-1} 1 — 10 μm^{-1}	Near infrared, visible, near ultraviolet	(1.6—16) × 10^{-19} joule {1 — 10 eV}
3×10^{15} — 3×10^{16} Hz 3 — 30 PHz	10^{-7} — 10^{-8} m 100 — 10 nm (1000 — 100 Å)	10^{7} — 10^{8} m^{-1} 10 — 100 μm^{-1}	Part of "vacuum" - ultraviolet	(1.6—16) × 10^{-18} joule {10 — 100 eV}
3×10^{16} — 3×10^{17} Hz 30 — 300 PHz	10^{-8} — 10^{-9} m 10 — 1 nm (100 — 10 Å)	10^{8} — 10^{9} m^{-1} 100 μm^{-1} — 1 nm^{-1}	Part of soft X-rays	(1.6—16) × 10^{-17} joule {100 — 1000 eV)}
3×10^{17} — 3×10^{18} Hz 300 PHz — 3 EHz	10^{-9} — 10^{-10} m 1 nm — 100 pm (10 — 1 Å)	10^{9} — 10^{10} m^{-1} 1 — 10 nm^{-1}	Part of soft X-rays	(1.6—16) × 10^{-16} joule {1 — 10 keV}
3×10^{18} — 3×10^{19} Hz 3 — 30 EHz	10^{-10} — 10^{-11} m 100 — 10 pm (1 — 0.1 Å)	10^{10} — 10^{11} m^{-1} 10 — 100 nm^{-1}	Hard X-rays and part of soft γ-rays	(1.6—16) × 10^{-15} joule {10 — 100 keV}
3×10^{19} — 3×10^{20} Hz 30 — 300 EHz	10^{-11} — 10^{-12} m 10 — 1 pm (0.1 — 0.01 Å)	10^{11} — 10^{12} m^{-1} 100 nm^{-1} — 1 pm^{-1}	Part of soft and part of hard γ-rays (limit at 510 keV)	(1.6—16) × 10^{-14} joule {100 keV — 1 MeV}
3×10^{20} — 3×10^{21} Hz 300 — 3000 EHz	10^{-12} — 10^{-13} m 1 pm — 100 fm (0.01 — 0.001 Å)	10^{12} — 10^{13} m^{-1} 1 — 10 pm^{-1}	Part of hard γ-rays and part of "cosmic" γ-rays	(1.6—16) × 10^{-13} joule {1 — 10 MeV}
3×10^{21} — 3×10^{22} Hz 3000 — 30,000 EHz	10^{-13} — 10^{-14} m 100 — 10 fm (0.001 — 0.0001 Å)	10^{13} — 10^{14} m^{-1} 10 — 100 pm^{-1}	γ-rays produced by cosmic rays	(1.6—16) × 10^{-12} joule {10 — 100 MeV}

Note: Abbreviations used in this table: Å—ångstrom (1 Å = 10^{-10} m); EHz—exahertz (10^{18} hertz); EHF—extremely high frequency; ELF—extremely low frequency; eV—electron volt (1 eV = 1.60219×10^{-19} joule); PHz—petahertz (10^{15} hertz); fm—femtometer (10^{-15} m); GHz—gigahertz (10^9 hertz); Gm—gigameter (10^9 m); HF—high frequency; Hz—hertz (s^{-1}); ITU—International Telecommunications Union; keV—kiloelectron volt (10^3 eV); km—kilometer (10^3 m); LF—low frequency; m—meter; MeV—megaelectron volt (10^6 eV); MF—medium frequency; MHz—megahertz (10^6 hertz); Mm—megameter (10^6 meter); mm—millimeter (10^{-3} meter); µm—micrometer (10^{-6} meter); nm—nanometer (10^{-9} meter); pm—picometer (10^{-12} meter); SHF—super high frequency; SLF—super low frequency; THz—terahertz; UHF—ultra high frequency; ULF—ultra low frequency; VHF—very high frequency; VLF—very low frequency.

[a] Also called "microwaves"; not to be confused with "micrometer waves".

LETTER DESIGNATIONS OF MICROWAVE BANDS

Frequency (GHz)	Wavelength (cm)	Wavenumber (cm^{-1})	Band
1—2	30—15	0.033—0.067	L-Band
1—4	15—7.5	0.067—0.133	S-Band
4—8	7.5—3.7	0.133—0.267	C-Band
8—12	3.7—2.5	0.267—0.4	X-Band
12—18	2.5—1.7	0.4—0.6	Ku-Band
18—27	1.7—1.1	0.6—0.9	K-Band
27—40	1.1—0.75	0.9—1.33	Ka-Band

BLACK BODY RADIATION

The total power radiated from an ideal black body and the wavelength corresponding to maximum power are given here as a function of absolute temperature. Constants used in the calculation are taken from the table "Fundamental Physical Constants" in Section 1. The radiated power in a band $\Delta\lambda$ at λ_{max} may be calculated from:

$$P_{max} = 0.657548\ (\Delta\lambda/\lambda_{max})\ P_{tot}$$

T/K	P_{tot}	λ_{max}/μm	T/K	P_{tot}	λ_{max}/μm	T/K	P_{tot}	λ_{max}/μm
50	0.354 W/m^2	57.955	740	17.004	3.916	1520	302.689	1.906
100	5.671	28.978	750	17.942	3.864	1540	318.937	1.882
150	28.707	19.318	760	18.918	3.813	1560	335.831	1.858
200	90.728	14.489	770	19.934	3.763	1580	353.387	1.834
250	221.504	11.591	780	20.989	3.715	1600	371.623	1.811
273	314.973	10.614	790	22.087	3.668	1620	390.555	1.789
280	348.541	10.349	800	23.226	3.622	1640	410.202	1.767
290	401.064	9.992	810	24.410	3.577	1660	430.581	1.746
300	459.311	9.659	820	25.638	3.534	1680	451.710	1.725
310	523.684	9.348	830	26.911	3.491	1700	473.607	1.705
320	594.596	9.055	840	28.232	3.450	1720	496.290	1.685
330	672.478	8.781	850	29.600	3.409	1740	519.779	1.665
340	757.771	8.523	860	31.018	3.369	1760	544.093	1.646
350	850.931	8.279	870	32.486	3.331	1780	569.249	1.628
360	952.428	8.049	880	34.006	3.293	1800	595.267	1.610
370	1.063 kW/m^2	7.832	890	35.578	3.256	1820	622.168	1.592
380	1.182	7.626	900	37.204	3.220	1840	649.970	1.575
390	1.312	7.430	910	38.886	3.184	1860	678.694	1.558
400	1.452	7.244	920	40.623	3.150	1880	708.359	1.541
410	1.602	7.068	930	42.418	3.116	1900	738.987	1.525
420	1.764	6.899	940	44.272	3.083	1920	770.597	1.509
430	1.939	6.739	950	46.187	3.050	1940	803.210	1.494
440	2.125	6.586	960	48.162	3.018	1960	836.848	1.478
450	2.325	6.439	970	50.201	2.987	1980	871.531	1.464
460	2.539	6.299	980	52.303	2.957	2000	907.282	1.449
470	2.767	6.165	990	54.471	2.927	2020	944.121	1.435
480	3.010	6.037	1000	56.705	2.898	2040	982.071	1.420
490	3.269	5.914	1020	61.379	2.841	2060	1.021 MW/m^2	1.407
500	3.544	5.796	1040	66.337	2.786	2080	1.061	1.393
510	3.836	5.682	1060	71.589	2.734	2100	1.103	1.380
520	4.146	5.573	1080	77.147	2.683	2120	1.145	1.367
530	4.474	5.467	1100	83.022	2.634	2140	1.189	1.354
540	4.822	5.366	1120	89.227	2.587	2160	1.234	1.342
550	5.189	5.269	1140	95.773	2.542	2180	1.281	1.329
560	5.577	5.175	1160	102.672	2.498	2200	1.328	1.317
570	5.986	5.084	1180	109.939	2.456	2220	1.377	1.305
580	6.417	4.996	1200	117.584	2.415	2240	1.428	1.294
590	6.871	4.911	1220	125.621	2.375	2260	1.479	1.282
600	7.349	4.830	1240	134.063	2.337	2280	1.532	1.271
610	7.851	4.750	1260	142.924	2.300	2300	1.587	1.260
620	8.379	4.674	1280	152.217	2.264	2320	1.643	1.249
630	8.933	4.600	1300	161.955	2.229	2340	1.700	1.238
640	9.514	4.528	1320	172.154	2.195	2360	1.759	1.228
650	10.122	4.458	1340	182.827	2.163	2380	1.819	1.218
660	10.760	4.391	1360	193.989	2.131	2400	1.881	1.207
670	11.427	4.325	1380	205.655	2.100	2420	1.945	1.197
680	12.124	4.261	1400	217.838	2.070	2440	2.010	1.188
690	12.853	4.200	1420	230.556	2.041	2460	2.077	1.178
700	13.615	4.140	1440	243.822	2.012	2480	2.145	1.168
710	14.410	4.081	1460	257.652	1.985	2500	2.215	1.159
720	15.239	4.025	1480	272.063	1.958	2550	2.398	1.136
730	16.103	3.970	1500	287.070	1.932	2600	2.591	1.115

T/K	P_{tot}	λ_{max}/μm	T/K	P_{tot}	λ_{max}/μm	T/K	P_{tot}	λ_{max}/μm
2650	2.796	1.093	3600	9.524	0.805	5100	38.362	0.568
2700	3.014	1.073	3650	10.065	0.794	5200	41.461	0.557
2750	3.243	1.054	3700	10.627	0.783	5300	44.743	0.547
2800	3.485	1.035	3750	11.214	0.773	5400	48.217	0.537
2850	3.741	1.017	3800	11.824	0.763	5500	51.889	0.527
2900	4.011	0.999	3850	12.458	0.753	5600	55.767	0.517
2950	4.294	0.982	3900	13.118	0.743	5700	59.858	0.508
3000	4.593	0.966	3950	13.804	0.734	5800	64.170	0.500
3050	4.907	0.950	4000	14.517	0.724	5900	68.712	0.491
3100	5.237	0.935	4100	16.024	0.707	6000	73.490	0.483
3150	5.583	0.920	4200	17.645	0.690	6500	101.222	0.446
3200	5.946	0.906	4300	19.386	0.674	7000	136.149	0.414
3250	6.326	0.892	4400	21.254	0.659	7500	179.418	0.386
3300	6.725	0.878	4500	23.253	0.644	8000	232.264	0.362
3350	7.142	0.865	4600	25.389	0.630	8500	296.004	0.341
3400	7.578	0.852	4700	27.670	0.617	9000	372.042	0.322
3450	8.033	0.840	4800	30.101	0.604	9500	461.867	0.305
3500	8.509	0.828	4900	32.689	0.591	10000	567.051	0.290
3550	9.006	0.816	5000	35.441	0.580			

NOMOGRAPH AND TABLE FOR DOPPLER LINEWIDTHS

Sidney O. Kastner

The Doppler width of a spectral line is given by the well-known relation $\Delta\lambda = (7.162 \times 10^{-7})\,\lambda T^{1/2} M^{-1/2}$, where wavelength units are in Angstroms, temperature is in degrees Kelvin and M is the atomic mass. This relation between four variables is amenable to representation by a nomograph, which does not appear to have been constructed but which would seem to be of practical value. Therefore such a nomograph is presented here. Its construction is briefly described so that the reader who has not made a plot of this type may follow the steps.

The Doppler relation is first rewritten as

$$\ell n\,\Delta\lambda = \ell n\,k\lambda + \tfrac{1}{2}\ell n\,T - \tfrac{1}{2}\ell n\,M$$

or equivalently $u_4 = u_1 + u_2 + u_3$, with $k = 7.162 \times 10^{-7}$. Putting $\xi - u_1 + u_2$, first, one has a linear relation $\xi - u_1 - u_2 = 0$ between three variables which can be represented (Menzel*) by the determinantal equation

$$\begin{vmatrix} \xi/2 & \tfrac{1}{2} & 1 \\ u_1 & 1 & 1 \\ u_2 & 0 & 1 \end{vmatrix} = 0$$

so that in the Cartesian (x,y) plane the function u_2 lies along the x axis, the function u_1 along the line $y = 1$ and the function $\xi/2$ along the line $y = \tfrac{1}{2}$. These lines form three parallel scales with which to obtain a value of ξ, given the pair (u_1, u_2).

The original equation $u_4 = \xi + u_3$ provides a second linear relation $u_4 - \xi - u_3 = 0$, so that a second set of scales similarly results with u_3 along the x axis, ξ along the line $y = 1$ and $u_4/2$ along the line $y = \tfrac{1}{2}$. The function $u_4 \equiv \log \Delta\lambda$ can then be obtained from the known pair (ξ, u_3).

In practice, for the nomograph scales to represent useful ranges of the physical variables, some shifting and magnification of the individual scales will be found to be necessary, after completion of the first diagram. The nomograph arrived at here is shown in Figure 1. The x axis runs horizontally through the values $x = 0$, $T = 10^5$, and the ordinates of the six scales from left to right are given by:

$$y_1 = 5\ell n(10^{5/2}\,k\lambda) \qquad y_2 = 10\ell n(k\lambda T^{1/2})$$

$$y_3 = \frac{5}{2}\ell n(k\lambda T^{1/2}) \qquad y_4 = 5\ell n\Delta\lambda + \frac{5}{2}\ell n 10$$

$$y_5 = 5\ell n(T^{1/2}/10^{5/2}) \qquad y_6 = 5\ell n(10/M)$$

The approximate range of temperatures covered by the figure is a slowly varying function of wavelength, being 1000°K-10^6 K for $\lambda = 10{,}000$Å; 4000 K-5×10^6 K for $\lambda = 5000$Å; 80,000 K-10^7 K for $\lambda = 1000$Å; and 300,000 K-10^7 K for $\lambda = 500$Å.

In use, the left-hand sides of the three lines are used to find the value of the intermediate variable x corresponding to given λ and T. The right-hand sides then give the Doppler width $\Delta\lambda$ in Å, for this value of x and the given M.

For example, suppose one wishes to find the Doppler width of the solar coronal forbidden line of Fe XIV (atomic mass $M = 56$) at 5303Å, emitted at a temperature of about $T = 2\times10^6$ K. A straight line drawn between $\lambda = 5300$Å and $T = 2\times10^6$ intersects the x scale at about 4.2. A second line then drawn between $x = 4.2$ and $M = 56$ intersects the middle line at $\Delta\lambda \approx 0.73$Å.

To use the nomograph, one must know the appropriate atomic weight M for a given element of interest. An alternative and more accurate procedure, if a calculator is available, for obtaining any required Doppler width is to use Table I, which gives the value of the constant $k_Z \equiv 7.162\,M_Z^{-1/2}$ for any given atomic number Z, in the equivalent Doppler relation $\Delta\lambda = k_Z(\lambda T^{1/2}) \times 10^{-7}$. This table thereby avoids the necessity of looking up the atomic weight M. For the forbidden line example above, where $Z = 26$, the Doppler width obtained from the table is $\Delta\lambda = (0.9584)(5303)(2{,}000{,}000)^{1/2}(10^{-7}) = 0.719$Å.

* D. H. Menzel, *Fundamental Formulas of Physics,* Dover Publications, New York, 1960, ch. 3.

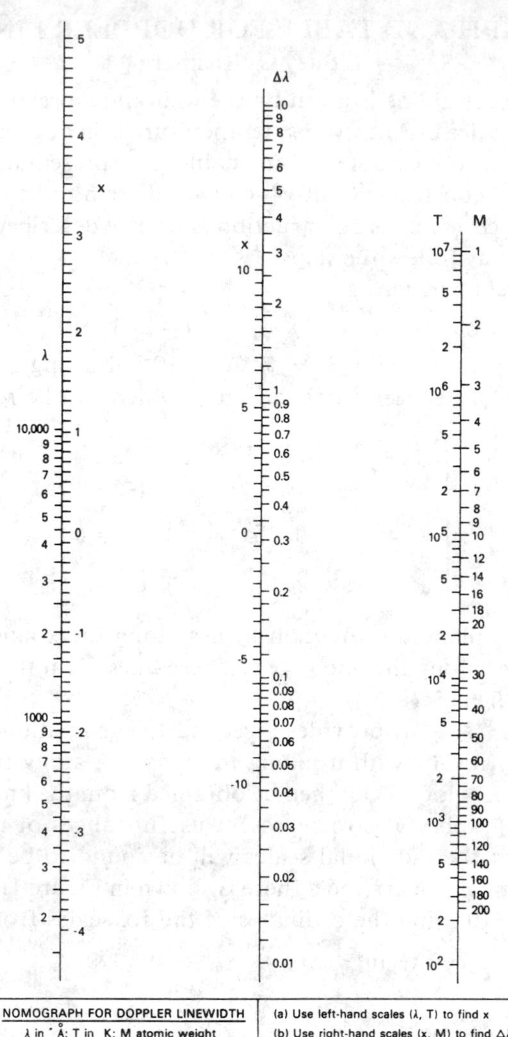

NOMOGRAPH FOR DOPPLER LINEWIDTH
λ in Å; T in K; M atomic weight

(a) Use left-hand scales (λ, T) to find x
(b) Use right-hand scales (x, M) to find Δλ

TABLE 1
Values of k_Z (Z Atomic Number) for Use in Doppler Linewidth Formula
$\Delta\lambda = k_Z(\lambda T^{1/2}) \times 10^{-7}$

Z	k_Z	Z	k_Z	Z	k_Z	Z	k_Z
1	7.1335	21	1.0682	41	0.7430	61	0.5927
2	3.5798	22	1.0348	42	0.7312	62	0.5840
3	2.7185	23	1.0035	43	0.7202	63	0.5810
4	2.3857	24	0.9932	44	0.7124	64	0.5711
5	2.1782	25	0.9663	45	0.7060	65	0.5681
6	2.0665	26	0.9584	46	0.6943	66	0.5618
7	1.9137	27	0.9329	47	0.6896	67	0.5577
8	1.7905	28	0.9347	48	0.6755	68	0.5538
9	1.6431	29	0.8984	49	0.6684	69	0.5510
10	1.5944	30	0.8858	50	0.6574	70	0.5492
11	1.4937	31	0.8577	51	0.6491	71	0.5414
12	1.4527	32	0.8406	52	0.6340	72	0.5361
13	1.3788	33	0.8274	53	0.6358	73	0.5324
14	1.3514	34	0.8060	54	0.6250	74	0.5282
15	1.2869	35	0.8012	55	0.6212	75	0.5249
16	1.2649	36	0.7824	56	0.6111	76	0.5193
17	1.2028	37	0.7747	57	0.6077	77	0.5166
18	1.1331	38	0.7651	58	0.6050	78	0.5128
19	1.1453	39	0.7596	59	0.6033	79	0.5103
20	1.1313	40	0.7499	60	0.5963	80	0.5057

EMISSIVITY OF TOTAL RADIATION FOR VARIOUS MATERIALS

Material	Temperature (°C)	ϵ_{tot}
Alloys		
Nickel-Chromium		
20 Ni—25 Cr—55 Fe, oxidized	200	0.90
	500	0.97
60 Ni—12 Cr—28 Fe, oxidized	270	0.89
	560	0.82
80 Ni—20 Cr	100	0.87
	600	0.87
	1300	0.89
Aluminum		
Polished	50—500	0.04—0.06
Rough surface	20—50	0.06—0.07
Strongly oxidized	55—500	0.2—0.3
	25	0.022
	100	0.028
	500	0.060
Oxidized	200	0.11
	600	0.19
Asbestos board	20	0.96
Bismuth		
Unoxidized	25	0.048
Brass		
Dull tarnished	200	0.61
Oxidized at 600° C	200	0.61
	600	0.59
Unoxidized	25	0.035
	100	0.035
Polished	200	0.03
Rolled sheet	20	0.06
Bronze		
Polished	50	0.1
Carbon		
Filament	1000—1400	0.53
Graphite	0—3600	0.7—0.8
Lamp black	20—400	0.96
Soot applied to solid	50—1000	0.96
Soot with water glass	20—200	0.96
Unoxidized	100	0.81
Chromium		
Polished	50	0.1
	500—1000	0.28—0.38
Colbalt		
Unoxidized	500	0.13
	1000	0.23
Columbium		
Unoxidized	1500	0.19
Copper		
Calorized	100	0.26
Calorized, oxidized	200	0.18
	600	0.19
Commercial, scoured to a shine	20	0.07
Oxidized	50	0.6—0.7
	500	0.88
Polished	50—100	0.02
Unoxidized	100	0.02
Unoxidized, liquid	—	0.15
Fire brick	1000	0.75
Glass	20—100	0.94—0.91
	250—1000	0.87—0.72
	1100—1500	0.7—0.67
Gold		
Carefully polished	200—600	0.02—0.03
Unoxidized	100	0.02
Enamel	100	0.37
Graphite	0—3600	0.7—0.8
Gypsum	20	0.93
[illegible]		
Cast		
Oxidized	200	0.64
	600	0.78
Strongly oxidized	40	0.95
	250	0.95
Unoxidized	100	0.21
Unoxidized, liquid	—	0.29
Oxidized	100	0.74
	500	0.84
	1200	0.89
Rusted	25	0.65
Wrought, dull	100	0.05
Lamp black	25	0.94
	20—400	0.96
Lead		
Oxidized	200	0.05
Unoxidized	200	0.63
Mercury		
Unoxidized	25	0.10
	100	0.12
Molybdenum	600—1000	0.08—0.13
	1500—2200	0.19—0.26
Monel metal		
Oxidized	200	0.43
	600	0.43
Nichrome		
Wire		
Clean	50	0.65
	500—1000	0.71—0.79
Oxidized	50—500	0.95—0.98
Nickel		
Industrial, polished	200—400	0.07—0.09
Oxidized	200	0.37
Oxidized at 600°C	200—600	0.37—0.48
Unoxidized	25	0.045
	100	0.06
	500	0.12
	1000	0.19
Platinum		
Clean, polished	200—600	0.05—0.1
Unoxidized	25	0.037
	100	0.047
	500	0.096
	1000	0.152
	1500	0.191
Wire	50—200	0.06—0.07
	500—1000	0.1—0.16
	1400	0.18
Porcelain		
Glazed	20	0.92
Rubber		
Hard	20	0.95
Soft, gray, rough	20	0.86
Silica brick	1000	0.80
	1100	0.85
Silver		
Clean, polished	200—600	0.02—0.03
Unoxidized	100	0.02
	500	0.035
Soot applied to a solid surface	50—1000	0.94—0.91
Soot with water glass	20—200	0.96
Steel		
Alloyed (8% Ni, 18% Cr)	500	0.35
Aluminized	50—500	0.79
Dull nickel plated	20	0.11
Flat, rough surface	50	0.95—0.98
Cast, polished	750—1050	0.52—0.56
Sheet, ground	50	0.56
	950—1100	0.55—0.61
Oxidized	200—600	0.8
Calorized, oxidized	200	0.52
	600	0.57
Sheet with shiny layer of oxide	20	0.82
Strongly oxidized	50	0.88
	500	0.98
Unoxidized	100	0.08
Unoxidized, liquid	—	0.28
Tantalum		
Unoxidized	[illegible]	0.21
	2000	0.26
Tungsten		
Unoxidized	25	0.024
	100	0.032
	500	0.071
	1000	0.15
	1500	0.23
	2000	0.28
Varnish	40—100	0.8—0.95
Dull black	40—100	0.96—0.98
Glossy black sprayed on iron	20	0.87
Zinc		
Polished	200—300	0.04—0.05
Unoxidized	300	0.05

SPECTRAL EMISSIVITY

Prepared by Roeser and Wensel, National Bureau of Standards
Spectral Emissivity of Materials, Surface Unoxidized for 0.65μm

Element	Solid	Liquid	Element	Solid	Liquid
Beryllium	0.61	0.61	Thorium	0.36	0.40
Carbon	0.80—0.93	—	Titanium	0.63	0.65
Chromium	0.34	0.39	Tungsten	0.43	—
Cobalt	0.36	0.37	Uranium	0.54	0.34
Columbium	0.37	0.40	Vanadium	0.35	0.32
Copper	0.10	0.15	Yttrium	0.35	0.35
Erbium	0.55	0.38	Zirconium	0.32	0.30
Gold	0.14	0.22	Steel	0.35	0.37
Iridium	0.30	—	Cast Iron	0.37	0.40
Iron	0.35	0.37	Constantan	0.35	—
Manganese	0.59	0.59	Monel	0.37	—
Molybdenum	0.37	0.40	Chromel P (90Ni-10Cr)	0.35	—
Nickel	0.36	0.37	80Ni-20Cr	0.35	—
Palladium	0.33	0.37	60Ni-24Fe-16Cr	0.36	—
Platinum	0.30	0.38	Alumel (95Ni; Bal. Al, Mn, Si)	0.37	—
Rhodium	0.24	0.30	90Pt-10Rh	0.27	—
Silver	0.07	0.07			
Tantalum	0.49	—			

SPECTRAL EMISSIVITY OF OXIDES

The emissivity of oxides and oxidized metals depends to a large extent upon the roughness of the surface. In general, higher values of emissivity are obtained on the rougher surfaces.

Material	Range of observed values	Probable value for oxide formed on smooth metal
Oxide		
Aluminum	0.22—0.40	0.30
Beryllim	0.07—0.37	0.35
Cerium	0.58—0.80	—
Chromium	0.60—0.80	0.70
Cobalt	—	0.75
Columbium	0.55—0.71	0.70
Copper	0.60—0.80	0.70
Iron	0.63—0.98	0.70
Magnesium	0.10—0.43	0.20
Nickel	0.85—0.96	0.90
Thorium	0.20—0.57	0.50
Tin	0.32—0.60	—
Titanium	—	0.50
Uranium	—	0.30
Vanadium	—	0.70
Yttrium	—	0.60
Zirconium	0.18—0.43	0.40
Oxidized		
Alumel	—	0.87
Cast Iron	—	0.70
Chromel P (90Ni-10Cr)	—	0.87
80Ni-20Cr	—	0.90
60Ni-24Fe-16Cr	—	0.83
55Fe-37.5Cr-7.5Al	—	0.78
70Fe-23Cr-5Al-2Co	—	0.75
Constantan (55Cu-45Ni)	—	0.84
Carbon Steel	—	0.80
Stainless Steel (18-8)	—	0.85
Porcelain	0.25—0.50	—

PROPERTIES OF TUNGSTEN

Jones and Langmuir, General Electric Review

Temp. K	Resistivity microhm cm	Electron emission amp./cm²	Evaporation g/cm²·s	Vapor pressure dynes/cm²	Thermal expansion per cent l_0 at 293°	Atomic heat cal/g atom/°C
300	5.65	—	—	—	.003	6.0
400	8.06	—	—	—	.044	6.0
500	10.56	—	—	—	.086	6.1
600	13.23	—	—	—	.130	6.1
700	16.09	—	—	—	.175	6.2
800	19.00	—	—	—	.222	6.2
900	21.94	—	—	—	.270	6.3
1000	24.93	1.07×10^{-15}	5.32×10^{-34}	1.98×10^{-29}	.320	6.4
1100	27.94	1.52×10^{-13}	2.17×10^{-30}	1.22×10^{-25}	.371	6.4
1200	30.98	9.73×10^{-12}	3.21×10^{-27}	1.87×10^{-22}	.424	6.5
1300	34.08	3.21×10^{-10}	1.35×10^{-24}	8.18×10^{-20}	.479	6.7
1400	37.19	6.62×10^{-9}	2.51×10^{-22}	1.62×10^{-17}	.535	6.8
1500	40.36	9.14×10^{-8}	2.37×10^{-20}	1.54×10^{-15}	.593	7.0
1600	43.55	9.27×10^{-7}	1.25×10^{-18}	8.43×10^{-14}	.652	7.1
1700	46.78	7.08×10^{-6}	4.17×10^{-17}	2.82×10^{-12}	.713	7.2
1800	50.05	4.47×10^{-5}	8.81×10^{-16}	6.31×10^{-11}	.775	7.4
1900	53.35	2.28×10^{-4}	1.41×10^{-14}	1.01×10^{-9}	.839	7.6
2000	56.67	1.00×10^{-3}	1.76×10^{-13}	1.33×10^{-8}	.904	7.7
2100	60.06	3.93×10^{-3}	1.66×10^{-12}	1.28×10^{-7}	.971	7.8
2200	63.48	1.33×10^{-2}	1.25×10^{-11}	9.88×10^{-7}	1.039	8.0
2300	66.91	4.07×10^{-2}	8.00×10^{-11}	6.47×10^{-6}	1.109	8.2
2400	70.39	1.16×10^{-1}	4.26×10^{-10}	3.52×10^{-5}	1.180	8.3
2500	73.91	2.98×10^{-1}	2.03×10^{-9}	1.71×10^{-4}	1.253	8.4
2600	77.49	7.16×10^{-1}	8.41×10^{-9}	7.24×10^{-4}	1.328	8.6
2700	81.04	1.63	3.19×10^{-8}	2.86×10^{-3}	1.404	8.7
2800	84.70	3.54	1.10×10^{-7}	9.84×10^{-3}	1.479	8.9
2900	88.33	7.31	3.30×10^{-7}	3.00×10^{-2}	1.561	9.0
3000	92.04	1.42×10	9.95×10^{-7}	9.20×10^{-2}	1.642	9.2
3100	95.76	2.64×10	2.60×10^{-6}	2.50×10^{-1}	1.724	9.4
3200	99.54	4.78×10	6.38×10^{-6}	6.13×10^{-1}	1.808	9.5
3300	103.3	8.44×10	1.56×10^{-5}	1.51	1.893	9.6
3400	107.2	1.42×10^{2}	3.47×10^{-5}	3.41	1.980	9.8
3500	111.1	2.33×10^{2}	7.54×10^{-5}	7.52	2.068	9.9
3600	115.0	3.73×10^{2}	1.51×10^{-4}	1.53×10	2.158	10.1
3655	117.1	4.79×10^{2}	2.28×10^{-4}	2.33×10	2.209	10.2

Roeser and Wensel, National Bureau of Standards

Temp. K	Normal brightness new candles per cm²	Spectral emissivity		Color emissivity	Total emissivity	Brightness temp. 0.65μm	Color temp
		0.65μm	0.467μm				
300	—	0.472	0.505	—	0.032	—	—
400	—	—	—	—	.042	—	—
500	—	—	—	—	.053	—	—
600	—	—	—	—	.064	—	—
700	—	—	—	—	.076	—	—
800	—	—	—	—	.088	—	—
900	—	—	—	—	.101	—	—
1000	0.0001	.458	.486	.395	.114	966	1007
1100	0.001	.456	.484	.392	.128	1059	1108
1200	0.006	.454	.482	.390	.143	1151	1210
1300	0.029	.452	.480	.387	.158	1242	1312
1400	0.11	.450	.478	.385	.175	1332	1414
1500	0.33	.448	.476	.382	.192	1422	1516
1600	0.92	.446	.475	.380	.207	1511	1619
1700	2.3	.444	.473	.377	.222	1599	1722
1800	5.1	.442	.472	.374	.236	1687	1825
1900	10.4	.440	.470	.371	.249	1774	1928
2000	20.0	.438	.469	.368	.260	1861	2032
2100	36	.436	.467	.365	.270	1946	2136
2200	61	.434	.466	.362	.279	2031	2241
2300	101	.432	.464	.359	.288	2115	2345
2400	157	.430	.463	.356	.296	2198	2451
2500	240	.428	.462	.353	.303	2280	2556
2600	350	.426	.460	.349	.311	2362	2662
2700	500	.424	.459	.346	.318	2443	2769
2800	690	.422	.458	.343	.323	2523	2876
2900	950	.420	.456	.340	.329	2602	2984
3000	1260	.418	.455	.336	.334	2681	3092
3100	1650	.416	.454	.333	.337	2759	3200
3200	2100	.414	.452	.330	.341	2837	3310
3300	2700	.412	.451	.326	.344	2913	3420
3400	3400	.410	.450	.323	.348	2989	3530
3500	4200	.408	.449	.320	.351	3063	3642
3600	5200	.406	.447	.317	.354	3137	3754

EMISSIVITY OF TUNGSTEN

Wavelengths in μm

Temperature K	0.25	0.30	0.35	0.40	0.50	0.60	0.70
1600	.448*	.482	.478	.481	.469	.455	.444
1800	.442*	.478*	.476	.477	.465	.452	.44
2000	.436*	.474	.473	.474	.462	.448	.436
2200	.429*	.470	.470	.471	.458	.445	.431
2400	.422	.465	.466	.468	.455	.441	.427
2600	.418	.461	.464	.464	.451	.437	.423
2800	.411	.456	.461	.461	.448	.434	.419

Temperature K	0.80	0.90	1.0	1.1	1.2	1.3	1.4
1600	.431	.413	.39	.366	.345	.322*	.300*
1800	.425	.407	.385	.364	.344	.323*	.302*
2000	.419	.401	.381	.361	.343	.323	.305
2200	.415	.396	.378	.359	.342	.324	.306

Temperature K	0.80	0.90	1.0	1.1	1.2	1.3	1.4
2400	.408	.391	.372	.355	.340	.324	.309
2600	.404	.386	.369	.352	.338	.325	.310
2800	.400	.383	.367	.352	.337	.325	.313

Temperature K	1.5	1.6	1.8	2.0	2.2	2.4	2.6
1600	.279*	.263*	.234*	.210*	.19*	.175*	.164*
1800	.282	.267*	.241*	.218*	.20*	.182*	.174*
2000	.288	.273	.247	.227	.209*	.197	.175
2200	.291	.278	.254	.235	.218	.205	.194
2400	.296	.283	.262	.244	.228	.215	.205
2600	.299	.288	.269	.251	.236	.224	.214
2800	.302	.292	.274	.259	.245	.233	.224

* Values by extrapolation.

EFFICACIES AND OTHER CHARACTERISTICS OF ILLUMINANTS

Gordon D. Rowe

A. Incandescent Lamps (120 volt; Inside Frost)

Nominal lamp watts	Bulb	Base	Approximate rated life (hr)	Approximate initial lumens	Approximate initial efficacy
25	A-19	Medium	2500	235	9.4
40	A-19	Medium	1000	480	12.0
60	A-19	Medium	1000	870	14.5
75	A-19	Medium	750	1,190	15.9
100	A-19	Medium	750	1,750	17.5
150	A-21	Medium	750	2,850	19.0
200	A-23	Medium	750	4,010	20.0
500	PS-35	Mogul	1000	10,850	21.7
750	PS-52	Mogul	1000	17,040	22.7
1000	PS-52	Mogul	1000	23,740	23.7
1500	Clear PS-52	Mogul	1000	34,400	22.9
5000	Clear G-64	Mogul bipost	150	145,000	29.0
10000	Clear G-96	Mogul bipost	75	335,000	33.6

Chromaticity (°K)

Note: The chromaticity of incandescent lamps is is generally in the order of 2900—3000 K; the color rendering index, 99+. These factors will vary with changes in the voltage.

B. Fluorescent Lamp (F40T12)

	Nominal lamp watts	Approximate rated life (h)[a]	Approximate initial lumens	Approximate initial efficacy (lm/lamp W)	Chromaticity (K)	Chromaticity coordinates X	Chromaticity coordinates Y	Color-rendering index
Cool white	40	20,000+	3150	78.75	4150	0.377	0.385	62
Deluxe cool white	40	20,000+	2250	56.25	4175	0.372	0.371	89
Warm white	40	20,000+	3200	80.00	3000	0.443	0.411	52
Deluxe warm white	40	20,000+	2200	55.00	3025	0.435	0.404	77
Daylight	40	20,000+	2600	65.00	6250	0.316	0.341	75
White	40	20,000+	3150	78.75	3450	0.410	0.400	60
Sign white	40	20,000+	2400	60.00	5200	0.340	0.354	82
Natural	40	20,000+	2100	52.50	3700	0.389	0.368	90
Chroma 50	40	20,000+	2210	55.25	5000	0.346	0.362	90
Chroma 75	40	20,000+	2000	50.00	7500	0.303	0.319	92
Regal white®	40	20,000+	2850	71.25	3000	0.434	0.403	81
Lite white	40	15,000[b]	3450	86.25	4200	0.377	0.393	49
SP30	40	15,000	3325	83.10	3000	0.438	0.403	70
SP35	40	15,000[b]	3325	83.10	3500	0.413	0.401	73
SP41	40	15,000	3265	81.60	4100	0.376	0.390	70
Green	40	20,000+	4350	108.75	6975	0.244	0.628	—
Cool green	40	20,000+	2850	71.25	6450	0.307	0.376	68
Blue	40	20,000+	1200	30.00	—	0.191	0.202	—
Gold	40	20,000+	2400	60.00	2500	0.516	0.476	38
Pink	40	20,000+	1100	27.50	—	0.538	0.343	—
Red	40	20,000+	200	5.00	—	0.691	0.303	—
Plant light	40	20,000+	850	21.25	6750	0.323	0.240	−2
Plant light (wide spectrum)	40	20,000+	1950	48.75	3050	0.412	0.360	90
Cool white (Watt-Miser® Plus)	32	15,000+	2750	86.00	4150	0.377	0.385	62
Warm white (Watt-Miser® Plus)	32	15,000+	2800	87.50	3000	0.443	0.411	52
SP30 (Watt-Miser® Plus)	32	15,000+	2900	90.60	3000	0.438	0.403	70
SP35 (Watt-Miser® Plus)	32	15,000+	2900	90.60	3500	0.413	0.401	73
SP41 (Watt-Miser® Plus)	32	15,000+	2850	89.00	4100	0.376	0.390	70
Lite white (Watt-Miser® Plus)	32	15,000+	2925	91.40	4200	0.377	0.393	49
Cool white (Watt-Miser®)	34	20,000	2750	80.90	4150	0.377	0.385	62
Deluxe cool white (Watt-Miser®)	32	20,000	1925	56.60	4175	0.372	0.371	89
Warm white (Watt-Miser®)	32	20,000	2800	82.35	3000	0.443	0.411	52
Deluxe warm white (Watt-Miser®)	32	20,000	1925	56.60	3025	0.435	0.404	77
Daylight (Watt-Miser®)	32	20,000	2300	67.65	6250	0.316	0.341	75
White (Watt-Miser®)	32	20,000	2800	82.35	3450	0.410	0.400	60

	Nominal lamp watts	Approximate rated life (h)[a]	Approximate initial lumens	Approximate initial efficacy (lm/lamp W)	Chromaticity (K)	Chromaticity coordinates X	Chromaticity coordinates Y	Color-rendering index
Regal White® (Watt-Miser®)	32	20,000	2450	72.00	3000	0.434	0.403	81
SP30 (Watt-Miser®)	32	20,000	2900	85.30	3000	0.438	0.403	70
SP35 (Watt-Miser®)	32	20,000	2900	85.30	3500	0.413	0.401	73
SP41 (Watt-Miser®)	32	20,000	2850	83.80	4100	0.376	0.390	70
Lite white (Watt-Miser® II)	32	20,000	2925	86.00	4200	0.377	0.393	49
Cool white	75	12,000[c]	6300	84.00				
Deluxe cool white	75	12,000[c]	4500	60.00				
Warm white	75	12,000[c]	6500	86.65				
Deluxe warm white	75	12,000[c]	4350	58.00				
Daylight	75	12,000[c]	5450	72.65				
White	75	12,000[c]	6400	85.30				
Cool white (Watt-Miser®)	60	12,000[c]	5600	93.30				
Deluxe cool white (Watt-Miser®)	60	12,000[c]	4000	66.65				
Warm white (Watt-Miser®)	60	12,000[c]	5800	96.65				
Daylight (Watt-Miser®)	60	12,000[c]	4840	80.65				
White (Watt-Miser®)	60	12,000[c]	5600	93.30				
Lite white (Watt-Miser® II)	60	12,000[c]	6000	100.00				
High-output (F96T12, 800ma)								
Cool white	110	12,000[c]	9200	83.60				
Deluxe cool white	110	12,000	6600	60.00				
Warm white	110	12,000[c]	9200	83.60				
Deluxe warm white	110	12,000[c]	6550	59.50				
Daylight	110	12,000[c]	7800	71.00				
White	110	12,000[c]	9200	83.60				
Cool white (Watt-Miser®)	95	12,000[c]	8300	87.35				
Deluxe cool white (Watt-Miser®)	95	12,000[c]	6100	64.20				
Warm white (Watt-Miser®)	95	12,000[c]	8500	89.50				
White (Watt-Miser®)	95	12,000[c]	8300	87.35				
Lite white (Watt-Miser® II)	95	12,000[c]	8800	92.60				
Cool white	215	12,000[c]	16,000	74.40				
Deluxe cool white	215	12,000[c]	11,000	51.15				
Warm white	215	12,000[c]	15,000	69.75				
Daylight	215	12,000[c]	13,300	61.85				
Cool white (Watt-Miser®)	185	12,000[c]	14,000	75.65				
Lite white (Watt-Miser® II)	185	12,000[c]	14,900	80.50				

Note: Chromaticity, Chromaticity Index, and Color Rendering Index (CRI) are the same as for the equivalent colors of the 40-W lamps listed in the F40T12 section of this table.

C. High-Intensity Discharge (HID)

	Bulb	ANSI Code	Nominal lamp watts	Approximate rated life (h)[a]	Approximate initial lumens[d]	Approximate initial efficacy (lm/lamp W)[d]
Mercury-vapor						
Clear	E-37	H33CD-R400	400	24,000+	21,000	52.50
Deluxe white	E-37	H33GL-R400/DX	400	24,000+	22,500	56.25
Warm deluxe white	E-37	H33GL-R400WDX	400	24,000+	19,500	48.75
Clear	BT-56	H36GV-R1000	1000	24,000+	57,000	57.00
Deluxe white	BT-56	H36GW-R1000/DX	1000	24,000+	63,000	63.00
Warm deluxe white	BT-56	H36GW-R1000/WDX	1000	24,000+	58,000	58.00
Metal halide (Multi-Vapor®)						
Clear; vertical, base up	E-23 1/2	M57PE-R175/XBU	175	10,000	16,600	94.85
Diffuse; vertical, base up	E-23 1/2	M57PF-R175/XBU	175	10,000	15,750	90.00
Clear; vertical, base down	E-23 1/2	M57PE-R175/XBD	175	10,000	16,600	94.85
Diffuse; vertical, base down	E-23 1/2	M57PF-R175/XBD	175	10,000	15,750	90.00
Clear; "any position"	E-28	M57PE-R175/U	175	10,000	14,000	80.00
Phosphor-coated; "any position"	E-28	M57PF-R175/U	175	10,000	14,000	80.00
Clear; "any position"	E-28	M58PG-R250/U	250	10,000	20,500	82.00
Phosphor-coated; "any position"	E-28	M58PH-R250/U	250	10,000	20,500	82.00
Clear; "any position"	E-37	M59PJ-R400/U	400	20,000	36,000	90.00
Phosphor-coated; "any position"	E-37	M59PK-R400/U	400	20,000	36,000	90.00
Clear; "any position"	BT-56	M47PA-R1000/U	1000	12,000	110,000	111.00
Phosphor-coated; "any position"	BT-56	M47PB-R1000/U	1000	12,000	105,000	105.00
High-pressure sodium (Lucalox®)						
Clear; medium base	E-17	S76HA-35	35	16,000	2,250	64.80
Diffuse; medium base	E-17	S76HB-35	35	16,000	2,150	61.40
Clear; mogul base	E-23 1/2	S68MS-50	50	24,000+	4,000	80.00
Diffuse; mogul base	E-23 1/2	S68MY-50	50	24,000+	3,800	76.00
Clear; medium base	E-17	S68XX-50	50	24,000+	4,000	80.00
Diffuse; medium base	E-17	S68YY-50	50	24,000+	3,800	76.00
Clear; mogul base	E-23 1/2	S62-ME70	70	24,000+	5,800	82.85
Diffuse; mogul base	E-23 1/2	S62-MF70	70	24,000+	5,400	77.00
Clear; medium base	E-17	S62-LG70	70	24,000+	5,800	82.85
Diffuse; medium base	E-17	S62LH-70	70	24,000+	5,400	77.00
Clear; mogul base	E-23 1/2	S54SB-100	100	24,000+	9,500	95.00
Diffuse; mogul base	E-23 1/2	S54MC-100	100	24,000+	8,800	88.00
Clear; medium base	E-17	S54SG-100	100	24,000+	9,500	95.00
Diffuse; medium base	E-17	S54SH-100	100	24,000+	8,800	88.00
Clear; mogul base	E-23 1/2	S55SC-150	150	24,000+	16,000	106.65
Diffuse; mogul base	E-23 1/2	S55MD-150	150	24,000+	15,000	100.00
Clear; mogul base	E-28	S56SD-150	150	24,000+	15,000	100.00
Clear; mogul base	E-18	S66MN-200	200	24,000+	22,000	110.00
Clear; mogul base	E-18	S50VA-250	250	24,000+	27,500	110.00
Clear; mogul base	E-18	S50VA-250/S	250	24,000+	30,000	120.00
Diffuse; mogul base	E-28	S50VC-250	250	24,000+	26,000	104.00

	Bulb	ANSI Code	Nominal lamp watts	Approximate rated life (hr.)	Approximate initial lumens[d]	Approximate initial efficacy (lm/lamp W)[d]
Clear; mogul base (deluxe color)	E-18	S50VA-250/DX	250	10,000	22,500	90.00
Clear; mogul base	E-18	S67MR-310	310	24,000+	37,000	119.35
Clear; mogul base	E-18	S51WA-400	400	24,000+	50,000	125.00
Diffuse; mogul base	E-37	S51WB-400	400	24,000+	47,500	118.75
Clear; mogul base	E-25	S52XB-1000	1000	24,000+	140,000	140.00

Note: ® = Registered trademark of the General Electric Company.

[a] Life rating at 3 or more hours per start on rapid-start circuits, 15,000 hr.
[b] Life with rapid-start circuits; on preheat circuits, 12,000 hr.
[c] Rated average life at 3 hr/start; at 12 hr/start, 18,000 hr.
[d] Initial lumen rating as measured when lamp is operated in vertical position.

APPROXIMATE LUMINANCE OF VARIOUS LIGHT SOURCES

Compiled by C.J. Allen and G.D. Rowe

Luminance of source is given in candelas per square centimeter

Source		Approx. avg. luminance* ($cd\ cm^{-2}$)
Natural sources		
Clear sky	Average luminance	0.8
Sun (as observed from earth's surface)	At meridian	160000
Sun (as observed from earth's surface)	Near horizon	600
Moon (as observed from earth's surface)	Bright spot	0.25
Combustion sources		
Candle flame (sperm)	Bright spot	1.0
Welsbach mantle	Bright spot	6.2
Acetylene flame	Mees burner	10.5
Photoflash lamps		16000-40000 peak
Incandescent electric lamps		
Carbon filament	3 Lumens per watt	52
Tungsten filament	Vacuum lamp–10 lumens per watt	200
Tungsten filament	Gas-filled lamp–20 lumens per watt	1200
Tungsten filament	750-watt projection lamp–26 lumens per watt	7500
Fluorescent lamps (cool white)		
Rapid start (40wT12)	430 mA	0.82
High-output (96T12)	800 mA	1.13
Grooved bulb (96T17)	1500 mA	1.50
Fluorescent lamps (other than cool white)		
Daylight (40wT12)	430 mA	0.62
Blue (40wT12)	430 mA	0.30
Green (40wT12)	430 mA	1.17
Red (40wT12)	430 mA	0.05
High-intensity discharge (HID)		
Mercury-vapor (E-37)		
Clear	400 watt	970
Color-improved	400 watt	11.0
Deluxe-white	400 watt	12.1
Mercury-vapor (BT-56)		
Clear	1000 watt	980
Color-improved	1000 watt	15.0
Deluxe-white	1000 watt	17.2
Metal halide (E-37)		
Clear	400 watt	810
Color-improved	1000 watt	930
Deluxe white	1500 watt	1620
High pressure sodium (Lucalox®)	250 watt	520
	400 watt	780
	1000 watt	810

®Registered trademark of the General Electric Company
*Luminance values perpendicular to lamp axis

INDEX OF REFRACTION OF WATER

This table gives the index of refraction of water at wavelengths corresponding to prominent emission lines of mercury, sodium, and helium.The data were generated from an equation for the index of refraction as a function of wavelength,temperature, and pressure.This table applies to a pressure of 1 bar (100 kPa).

T/°C	λ = 404.66 nm	λ = 589.32 nm	λ = 706.52 nm
0	1.34359	1.33346	1.33086
10	1.34351	1.33341	1.33073
20	1.34287	1.33283	1.33007
30	1.34180	1.33184	1.32903
40	1.34039	1.33052	1.32766
50	1.33867	1.32892	1.32603
60	1.33669	1.32707	1.32417
70	1.33447	1.32500	1.32209
80	1.33204	1.32274	1.31983
90	1.32942	1.32029	1.31739
100	1.32663	1.31766	1.31481

Reference

I. Thormahlen, J. Straub, and U. Grigul, Refractive index of water and its dependence on wavelength, temperature and density, *J. Phys. Chem. Ref. Data*, 14, 933, 1985.

INDEX OF REFRACTION OF ROCK SALT, SYLVINE, CALCITE, FLUORITE AND QUARTZ

(Compiled from data of Martens, Paschen, and others.)

Wave length	Rock salt	Silvine, KCl	Fluorite	Calcspar, ordinary ray	Calcspar, extraordinary ray	Quartz, ordinary ray	Quartz, extraordinary ray
0.185	1.893	1.827				1.676	1690
0.198			1.496		1.578	1.651	1.664
0.340				1.701	1.506	1.567	1.577
0.589	1.544	1.490	1.434	1.658	1.486	1.544	1.553
0.760			1.431	1.650	1.483	1.539	1.548
0.884	1.534	1.481	1.430				
1.179	1.530	1.478	1.428				
1.229				1.639	1.479		
2.324					1.474	1.516	
2.357	1.526	1.475	1.421				
3.536	1.523	1.473	1.414				
5.893	1.516	1.469	1.387				
8.840	1.502	1.461	1.331				

INDEX OF REFRACTION OF GLASS

Relative to Air

Variety	Wave length in microns							
	.361	.434	.486	.589 (Na)	.656	.768	1.20	2.00
Zinc crown	1.539	1.528	1.523	1.517	1.514	1.511	1.505	1.497
Higher dispersion crown	1.546	1.533	1.527	1.520	1.517	1.514	1.507	1.497
Light flint	1.614	1.594	1.585	1.575	1.571	1.567	1.559	1.549
Heavy flint	1.705	1.675	1.664	1.650	1.644	1.638	1.628	1.617
Heaviest flint		1.945	1.919	1.890	1.879	1.867	1.848	1.832

INDEX OF REFRACTION OF FUSED QUARTZ

λ mμ, 15°C	n, 18°C	λ mμ, 15°C	n, 18°C	λ mμ, 15°C	n, 18°C	λ mμ, 15°C	n, 18°C
185.467	1.57436	434.047	1.46690	250.329	1.50745	546.072	1.46013
193.583	1.55999	435.834	1.46675	257.304	1.50379	589.29	1.45845
202.55	1.54727	467.815	1.46435	274.867	1.49617	643.847	1.45674
214.439	1.53386	479.991	1.46355	303.412	1.48594	656.278	1.45640
219.462	1.52907	486.133	1.46318	340.365	1.47867	706.520	1.45517
226.503	1.52308	508.582	1.46191	396.848	1.47061	794.763	1.45340
231.288	1.51941	533.85	1.46067	404.656	1.46968		

INDEX OF REFRACTION OF LIQUIDS FOR CALIBRATION PURPOSES

This table gives the index of refraction of six liquids which are available in highly pure form and whose index of refraction has been accurately measured as a function of wavelength and temperature. They are therefore useful for calibration of refractometers. The estimated uncertainty in the values is:

2,2,4-Trimethylpentane	±0.00003
Hexadecane	±0.00008
trans-Bicyclo[4.0.0]decane	±0.00008
1-Methylnaphthalene	±0.00008
Toluene	±0.00003
Methylcyclohexane	±0.00003

Full details are given in the references. This table is reprinted from Reference 1 by permission of the International Union of Pure and Applied Chemistry.

REFERENCES

1. Marsh, K. N., Editor, *Recommended Reference Materials for the Realization of Physicochemical Properties*, Blackwell Scientific Publications, Oxford, 1987.
2. Tilton, L. W., J. *Opt. Soc. Am.*, 32, 71, 1941.

λ	2,2,4-Trimethylpentane			Hexadecane		
nm	20°C	25°C	30°C	20°C	25°C	30°C
667.81	1.38916	1.38670	1.38424	1.43204	1.43001	1.42798
656.28	1.38945	1.38698	1.38452	1.43235	1.43032	1.42829
589.26	1.39145	1.38898	1.38650	1.43453	1.43250	1.43047
546.07	1.39316	1.39068	1.38820	1.43640	1.43436	1.43232
501.57	1.39544	1.39294	1.39044	1.43888	1.43684	1.43480
486.13	1.39639	1.39389	1.39138	1.43993	1.43788	1.43583
435.83	1.40029	1.39776	1.39523	1.44419	1.44213	1.44007

λ	*trans*-Bicyclo[4.4.0]decane			1-Methylnaphthalene		
nm	20°C	25°C	30°C	20°C	25°C	30°C
667.81	1.46654	1.46438	1.46222	1.60828	1.60592	1.60360
656.28	1.46688	1.46472	1.46256	1.60940	1.60703	1.60471
589.26	1.46932	1.46715	1.46498	1.61755	1.61512	1.61278
546.07	1.47141	1.46923	1.46705	1.62488	1.62240	1.62005
501.57	1.47420	1.47200	1.46980	1.63513	1.63259	1.63022
486.13	1.47535	1.47315	1.47095	1.63958	1.63701	1.63463
435.83	1.48011	1.47789	1.47567		1.65627	1.65386

λ	Toluene			Methylcyclohexane		
nm	20°C	25°C	30°C	20°C	25°C	30°C
667.81	1.49180	1.48903	1.48619	1.42064	1.41812	1.41560
656.28	1.49243	1.48966	1.48682	1.42094	1.41842	1.41591
589.26	1.49693	1.49413	1.49126	1.42312	1.42058	1.41806
546.07	1.50086	1.49803	1.49514	1.42497	1.42243	1.41989
501.57	1.50620	1.50334	1.50041	1.42744	1.42488	1.42233
486.13	1.50847	1.50559	1.50265	1.42847	1.42590	1.42334
435.83	1.51800	1.51506	1.51206	1.43269	1.43010	1.42752

INDEX OF REFRACTION OF AIR

This is a table of the index of refraction n of dry air at 15°C and a pressure of 101.325 kPa and containing 0.03% by volume of carbon dioxide ("standard air"). The index of refraction is defined by $n = \lambda_{vac}/\lambda_{air}$, where λ is the wavelength of the radiation. The index is calculated from the expression:

$$(n-1)\times 10^8 = 8342.13 + 2406030(130-\sigma^2)^{-1} + 15997(38.9-\sigma^2)^{-1}$$

where $\sigma = 1/\lambda_{vac}$ and λ_{vac} has units of μm. The equation is valid for λ_{vac} from 200 nm to 2 μm.
The table also gives the correction $(n - 1)\lambda_{air}$ which must be added to the wavelength in air to obtain λ_{vac}.
If the air is at a temperature t in °C and a pressure p in pascals, a value of $(n - 1)$ from this table should be multiplied by

$$\frac{p\left[1+p(61.3-t)\times 10^{-10}\right]}{96095.4(1+0.003661t)}$$

REFERENCE

Edlen, B., *Metrologia*, 2, 71, 1966.

λ_{vac}	$(n - 1)\times 10^8$	$\lambda_{vac} - \lambda_{air}$
200 nm	32408	0.0648 nm
210	31746	0.0666
220	31224	0.0687
230	30799	0.0708
240	30445	0.0730
250	30146	0.0753
260	29890	0.0777
270	29669	0.0801
280	29475	0.0825
290	29306	0.0850
300	29155	0.0874
310	29022	0.0899
320	28902	0.0925
330	28795	0.0950
340	28698	0.0975
350	28611	0.1001
360	28531	0.1027
370	28458	0.1053
380	28392	0.1079
390	28331	0.1105
400	28275	0.1131
410	28223	0.1157
420	28175	0.1183
430	28131	0.1209
440	28090	0.1236
450	28052	0.1262
460	28016	0.1288
470	27983	0.1315
480	27952	0.1341
490	27923	0.1368
500	27896	0.1394
510	27870	0.1421
520	27846	0.1448
530	27824	0.1474
540	27803	0.1501
550	27782	0.1528
560	27763	0.1554
570	27745	0.1581
580	27728	0.1608
590	27712	0.1635
600	27697	0.1661
610	27682	0.1688
620	27669	0.1715
630	27656	0.1742
640	27643	0.1769
650	27631	0.1796
660	27620	0.1822
670	27609	0.1849
680	27599	0.1876
690	27589	0.1903
700	27579	0.1930
710	27570	0.1957
720	27562	0.1984
730	27553	0.2011
740	27545	0.2038
750	27538	0.2065
760	27530	0.2092
770	27523	0.2119
780	27516	0.2146
790	27510	0.2173
800	27504	0.2200
810	27498	0.2227
820	27492	0.2254
830	27486	0.2281
840	27481	0.2308
850	27476	0.2335
860	27471	0.2362
870	27466	0.2389
880	27461	0.2416
890	27457	0.2443
900	27452	0.2470
910	27448	0.2497
920	27444	0.2524
930	27440	0.2551
940	27436	0.2578
950	27432	0.2605
960	27429	0.2632
970	27425	0.2660
980	27422	0.2687
990 nm	27419	0.2714 nm
1.00 μm	27416	0.000274 μm
1.05	27401	0.000288
1.10	27389	0.000301
1.15	27378	0.000315
1.20	27368	0.000328
1.25	27360	0.000342
1.30	27352	0.000355
1.35	27346	0.000369
1.40	27340	0.000383
1.45	27334	0.000396
1.50	27330	0.000410
1.55	27325	0.000423
1.60	27321	0.000437
1.65	27318	0.000451
1.70	27314	0.000464
1.75	27311	0.000478
1.80	27309	0.000491
1.85	27306	0.000505
1.90	27304	0.000519
1.95	27302	0.000532
2.00 μm	27300	0.000546 μm

TRANSMISSION OF LIGHT BY COMMON OPTICAL MATERIALS

Ratio of the transmitted light to the incident light for a definite thickness of the substance, usually 1 cm.

Glass

Glass in general is opaque to the ultra-violet and infrared. Uviol glass is transparent to the longer radiations of the ultra-violet.

Coefficient of transparency of glass for visible and ultra-violet radiations.

Normal incidence, thickness 1 cm.

Wave length, μm	0.309	0.330	0.347	0.357	0.361	0.375	0.384	0.388	0.396
Crown, ordinary	—	—	—	—	—	.947	—	—	—
Crown, borosilicate	0.08	0.65	0.88	—	0.95	—	0.972	0.975	0.986
Flint, ordinary	—	—	—	0.72	—	—	—	0.904	—
Flint, heavy	—	—	0.01	—	0.16	—	0.58	—	—

Normal incidence, thickness 1 cm.

Wave length, μm	0.400	0.415	0.419	0.425	0.434	0.455	0.500	0.580	0.677
Crown, ordinary	0.964	—	0.952	—	0.960	0.981	—	0.986	0.990
Crown, borosilicate	—	0.985	—	0.993	—	—	0.993	—	—
Flint, ordinary	—	0.959	—	—	—	—	1.00	—	—
Flint, heavy	—	—	—	0.905	—	—	—	—	—

Quartz

Quartz is very transparent to the ultra-violet and to the visible spectrum, but opaque for the infrared beyond 7.0 μm.

(Pflüger.)

Wave length, μm	0.19	0.20	0.21	0.22
Transmission for 1 mm	.67	.84	.92	.94

Fluorite

Fluorite is very transparent to the ultra-violet, nearly to 0.10 μm. Coefficient of transparency at λ = 186 is found by Pflüger to be 0.80.

For the infrared the values are given in a table below.

Rock Salt and Sylvine and Fluorite
Transparency for the Infrared.
Thickness 1 cm.

Wave length μm	Rock salt	Sylvine KCl	Fluorite
8		—	.844
9.	0.995	1.000	.543
10.	.995	.988	.164
12.	.993	.995	.010
14.	.931	.975	.000
16.	.661	.936	—
18.	.275	.862	—
19.	.096	.758	—
20.7	.006	.585	—
23.7	.000	.155	—

CHARACTERISTICS OF LASER SOURCES

William F. Krupke

Light Amplification by Stimulated Emission of Radiation was first demonstrated by Maiman in 1960, the result of a population inversion produced between energy levels of chromium ions in a ruby crystal when irradiated with a xenon flashlamp. Since then population inversions and coherent emission have been generated in literally thousands of substances (neutral and ionized gases, liquids, and solids) using a variety of incoherent excitation techniques (optical pumping, electrical discharges, gas-dynamic flow, electron-beams, chemical reactions, nuclear decay).

The extrema of laser output parameters which have been demonstrated to date and the laser media used are summarized in Table 1. Note that the extreme power and energy parameters listed in this table were attained with laser *systems* rather than with simple laser oscillators.

Laser sources are commonly classified in terms of the state-of-matter of the active medium: gas, liquid, and solid. Each of these classes is further subdivided into one or more types as shown in Table 2. A well-known representative example of each type of laser is also given in Table 2 together with its nominal operation wavelength and the methods by which it is pumped.

The various lasers together cover a wide spectral range from the far ultraviolet to the far infrared. The particular wavelength of emission (usually a narrow line) is presented for some six dozen lasers in Figures 1A and 1B.

By suitably designing the excitation source and/or by controlling the laser resonator structure, laser systems can provide continuous or pulsed radiation as shown in Table 3.

Besides the method of excitation and the temporal behavior of a laser, there are many other parameters that characterize its operation and efficiency, as shown in Tables 4 and 5.

Although many lasers only emit in one or more narrow spectral "lines", an increasing number of lasers can be tuned by changing the composition or the pressure of the medium, or by varying the wavelength of the pump bands. The spectral regions in which these tunable lasers operate are presented in Figure 2.

REFERENCE

Krupke, W. F., in *Handbook of Laser Science and Technology,* Vol. I, Weber, M. J., Ed., CRC Press, Boca Raton, FL, 1986.

TABLE 1
Extrema of Output Parameters of Laser Devices or Systems

Parameter	Value	Laser medium
Peak power	1×10^{14} W (collimated)	Nd:glass
Peak power density	10^{18} W/cm^2 (focused)	Nd:glass
Pulse energy	$>10^5$ J	CO_2, Nd:glass
Average power	10^5 W	CO_2
Pulse duration	3×10^{-15} s continuous wave (cw)	Rh6G dye; various gases, liquids, solids
Wavelength	60 nm ↔ 385 μm	Many required
Efficiency (nonlaser pumped)	70%	CO
Beam quality	Diffraction limited	Various gases, liquids, solids
Spectral linewidth	20 Hz (for 10^{-1} s)	Neon-helium
Spatial coherence	10 m	Ruby

TABLE 2
Classes, Types, and Representative Examples of Laser Sources

Class	Type (characteristic)	Representative example	Nominal operating wavelength (nm)	Method(s) of excitation
Gas	Atom, neutral (electronic transition)	Neon-Helium (Ne-He)	633	Glow discharge
	Atom, ionic (electronic transition)	Argon (Ar^+)	488	Arc discharge
	Molecule, neutral (electronic transition)	Krypton fluoride (KrF)	248	Glow discharge; e-beam
	Molecule, neutral (vibrational transition)	Carbon dioxide (CO_2)	10600	Glow discharge; gas-dynamic flow
	Molecule, neutral (rotational transition)	Methyl fluoride (CH_3F)	496000	Laser pumping
	Molecule, ionic (electronic transition)	Nitrogen ion (N_2^+)	420	E-beam
Liquid	Organic solvent (dye-chromophore)	Rhodamine dye (Rh6G)	580–610	Flashlamp; laser pumping
	Organic solvent (rare earth chelate)	Europium:TTF	612	Flashlamp
	Inorganic solvent (trivalent rare earth ion)	Neodymium:$POCl_4$	1060	Flashlamp
Solid	Insulator, crystal (impurity)	Neodymium:YAG	1064	Flashlamp, arc lamp
	Insulator, crystal (stoichiometric)	Neodymium:UP(NdP_5O_{14})	1052	Flashlamp
	Insulator, crystal (color center)	F_2^-:LiF	1120	Laser pumping
	Insulator, amorphous (impurity)	Neodymium:glass	1061	Flashlamp
	Semiconductor (p-n junction)	GaAs	820	Injection current
	Semiconductor (electron-hole plasma)	GaAs	890	E-beam, laser pumping

Table 3
Temporal Characteristics of Lasers and Laser Systems

Form	Technique	Pulse width range (s)
Continuous wave	Excitation is continuous; resonator Q is held constant at some moderate value	∞
Pulsed	Excitation is pulsed; resonator Q is held constant at some moderate value	10^{-8}–10^{-3}
Q-Switched	Excitation is continuous or pulsed; resonator Q is switched from a very low value to a moderate value	10^{-8}–10^{-6}
Cavity dumped	Excitation is continuous or pulsed; resonator Q is switched from a very high value to a low value	10^{-7}–10^{-5}
Mode locked	Excitation is continuous or pulsed; phase or loss of the resonator modes is modulated at a rate related to the resonator transit time	10^{-12}–10^{-9}

Table 4
Properties and Performance of Some Continuous Wave (CW) Lasers

Parameter	Unit	Gas			Liquid	Solid	
		Neon helium	Argon ion	Carbon dioxide	Rhodamine 6G dye	Nd:YAG	GaAs
Excitation method		DC discharge	DC discharge	DC discharge	Ar⁺ laser pump	Krypton arc lamp	DC injection
Gain medium composition		Neon:helium	Argon	CO_2:N_2:He	Rh 6G:H_2O	Nd:YAG	p:n:GaAs
Gain medium density	Torr	0.1:1.0	0.4	0.4:0.8:5.0			
	ions/cm³				2(18):2(22)	1.5(20):2(22)	2(19):3(18):3(22)
Wavelength	nm	633	488	10600	590	1064	810
Laser cross-section	cm^{-2}	3(-13)	1.6(-12)	1.5(-16)	1.8(-16)	7(-19)	~6(-15)
Radiative lifetime (upper level)	s	~1(-7)	7.5(-9)	4(-3)	6.5(-9)	2.6(-4)	~1(-9)
Decay lifetime (upper level)	s	~1(-7)	~5.0(-9)	~4(-3)	6.0(-9)	2.3(-4)	~1(-9)
Gain bandwidth	nm	2(-3)	5(-3)	1.6(-2)	80	0.5	10
Type, gain saturation		Inhomogeneous	Inhomogeneous	Homogeneous	Homogeneous	Homogeneous	Homogeneous
Homogeneous saturation flux	W cm^{-2}			~20	3(5)	2.3(3)	~2(4)
Decay lifetime (lower level)	s	~1(-8)	~4(-10)	~5(-6)[b]	<1(-12)	<1(-7)	<1(-12)
Inversion density	cm^{-3}	~1(9)	2(10)	2(15)	2(16)	6(16)	1(16)
Small signal gain coefficient	cm^{-1}	~1(-3)	~3(-2)	1(-2)	4	5(-2)	40
Pump power density	W cm^{-3}	3	900	0.15	1(6)	150	7(7)
Output power density	W cm^{-3}	2.6(-3)	~1	2(-2)	3(5)	95	5(6)
Laser size (diameter:length)	cm:cm	0.5:100	0.3:100	5.0:600	1(-3):0.3	0.6:10	5(-4):7(-3);2(-2)[a]
Excitation current/voltage	A/V	3(-2):2(3)	30:300	0.1:1.5(4)		90:125	1.0/1.7
Excitation current density	A cm^{-2}	0.15	600	6(-3)		140	4.5(3)
Excitation power	W	60	9(3)	1.5(3)	4	1.1(4)	1.7
Output power	W	0.06	10	240	0.3	300	0.12
Efficiency	%	0.1	0.1	13	7	2.6	7

[a] Junction thickness:width:length.
[b] Pressure dependent.

Table 5
Properties and Performance of Some Pulsed Lasers

Parameter	Unit	Gas				Liquid	Solid	
		Carbon dixoxide		Krypton fluoride		Rhodamine 6G	Nd:YAG	Nd:glass
Excitation method		TEA-discharge	E-beam/sust.	Glow discharge	E-beam	Xenon flashlamp	Xenon flashlamp	Xenon flashlamp
Gain medium composition		CO_2:N_2:He	CO_2:N_2:He	He:Kr:F_2	Ar:Kr:F_2	Rh6G:alcohol	Nd:YAG	Nd:Glass
Gain medium density	orr	100:50:600	240:240:320	1070:70:3	1235:52:3			-
	ons/cm^3					1(18):1.5(22)	1.5(20):1(22)	3(20):2(22)
Wavelength	m	10600	10600	249	249	590	1064	1061
Laser cross-section	cm^{-2}	2(-18)	2(-18)	2(-16)	2(-16)	1.8(-16)	7(-19)	2.8(-20)
Radiative lifetime (upper level)		4(-3)	4(-3)	7(-9)	7(-9)	6.5(-9)	2.6(-4)	4.1(-4)
Decay lifetime (upper level)		~1(-4)	5(-5)	2(-9)	3(-9)	6.0(-9)	2.3(-4)	3.7(-4)
Gain bandwidth	m	1	1	2	2	80	0.5	26
Homogeneous saturation fluence	/cm^2	0.2	0.2	4(-3)	4(-3)	2(-3)	0.6	~5
Decay lifetime (lower level)		5(-8)[a]	1(-8)[a]	<1(-12)	<1(-12)	<1(-12)	<1(-7)	<1(-8)
Inversion density	cm^{-3}	3(17)	6(17)	4(14)	2(14)	2(16)	4(17)	3(18)
Small signal gain coefficient	cm^{-1}	2(-2)	4(-2)	8–92)	4(-2)	4	0.3	8(-2)
Medium excitation energy density	/cm^3	0.1	0.36	0.15	0.13	2.8	0.15	0.6
Output energy density	/cm^3	2(-2)	1.8(-2)	1.5(-3)	1.2(-2)	0.85	5(-2)	2(-2)
Laser dimensions	cm:cm:cm	4.5:4.5:87	10:10:100	1.5:4.5:100	8.5:10:100	1.2ϕ25	0.6ϕ7.5	0.6ϕ8.3
Excitation current/voltage	A/V	6(4)/3.3(3)	2.4(4)/4(4)	2.5(4)/1.5(5)	1.2(4)/2.5(5)	2(5)/2.5(4)		
Excitation current density	A cm^2	8.5	22	170	11.5	2.6(3)		
Excitation peak power	W	2(8)	9(8)	4(9)	3(9)	5.4(9)	4(4)	9(4)
Output pulse energy		35	180	1	102	32	0.1	1.0
Output pulse length	s	1(-6)	4(-6)	2.5(-8)	6(-7)	3.2(-6)	2(-8)	1(-4)
Output pulse power	W	3.5(7)	4(7)	4(7)	2(3)	1(7)	5(6)	1(4)
Efficiency	%	17	5	1	10[b]	0.2	1.5	3.7

[a] Pressure dependent.
[b] Intrinsic efficiency ≡ energy outpu⊏/energy deposited in gas.

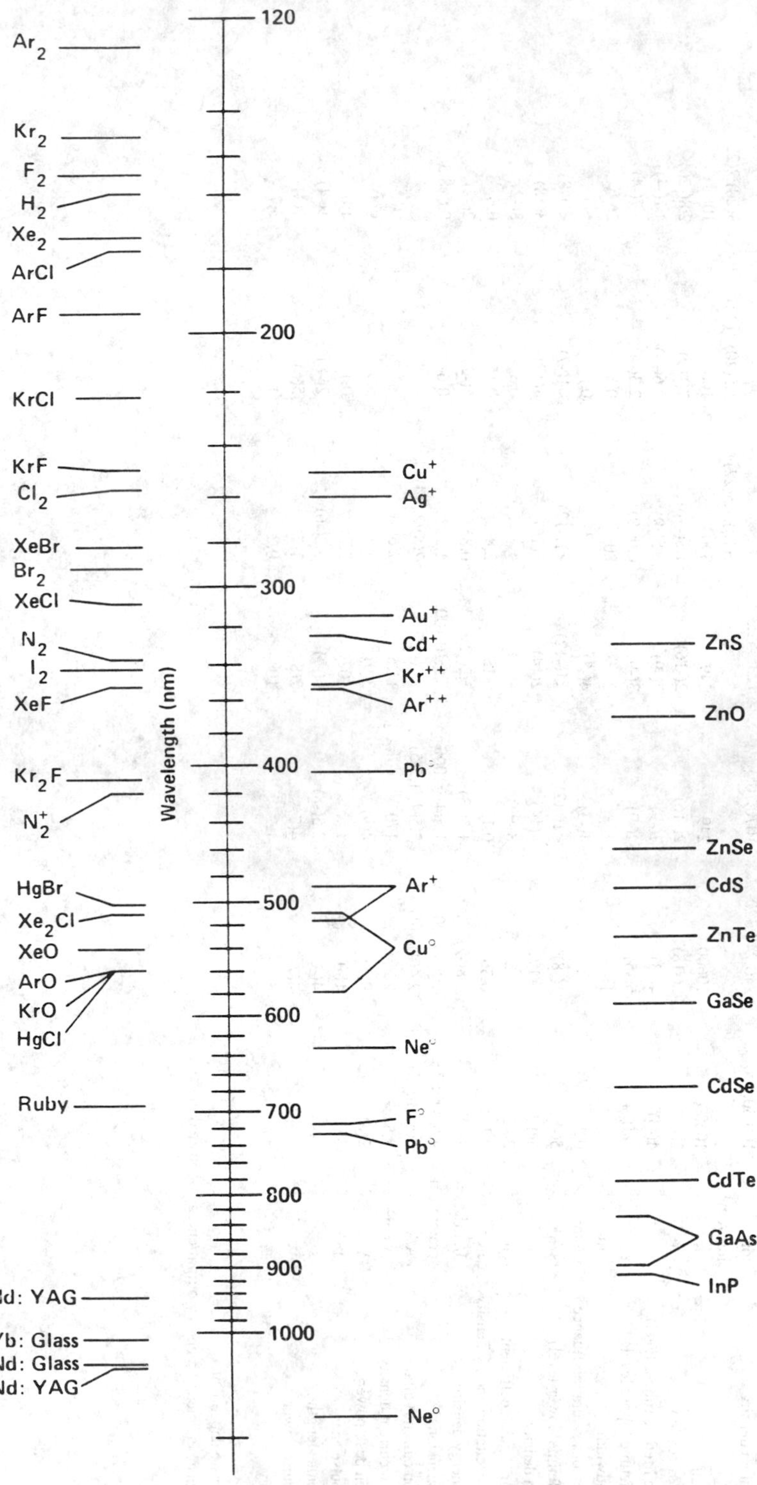

FIGURE 1A. Wavelengths of lasers operating in the 120 to 1200 nm spectral region.

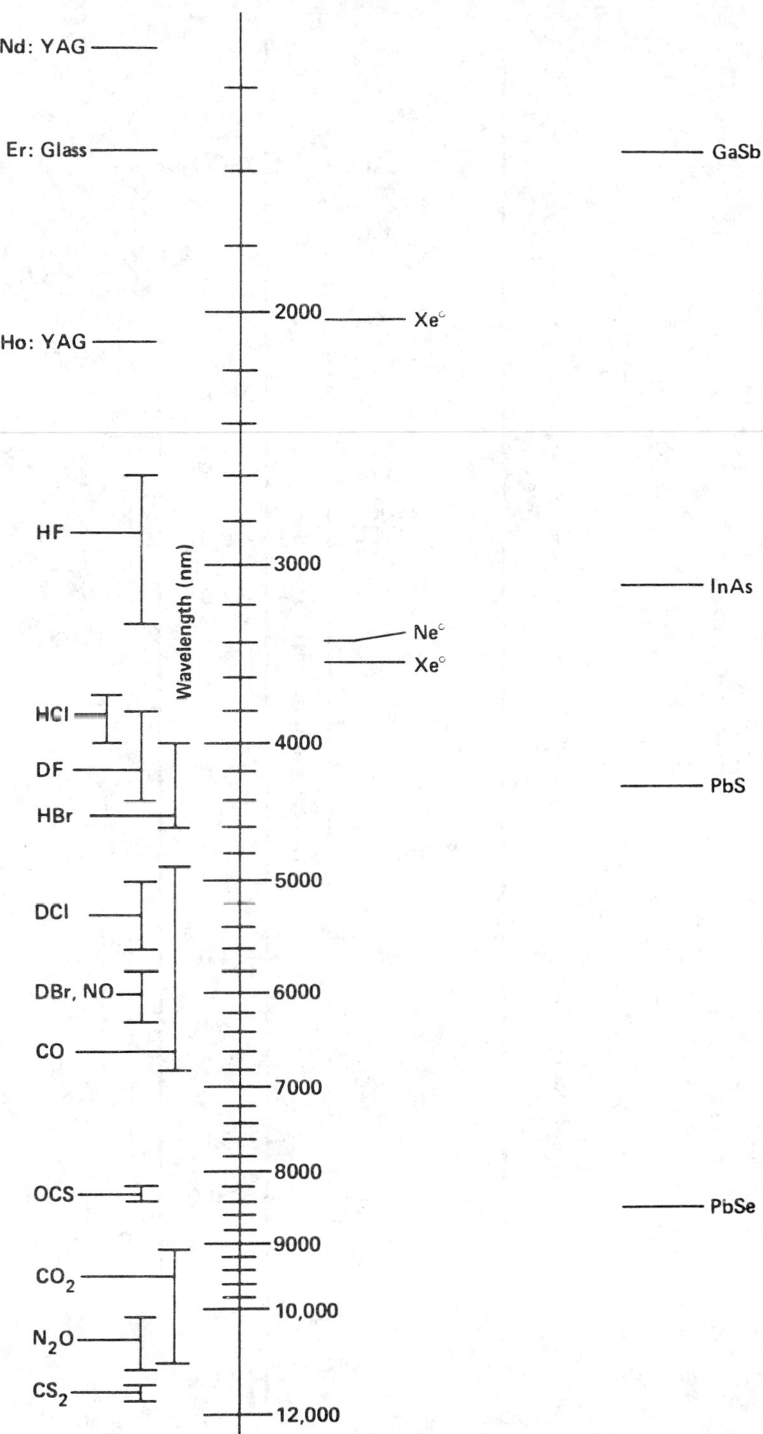

FIGURE 1B. Wavelength of lasers operating in the 1300 to 12,000 nm spectral region.

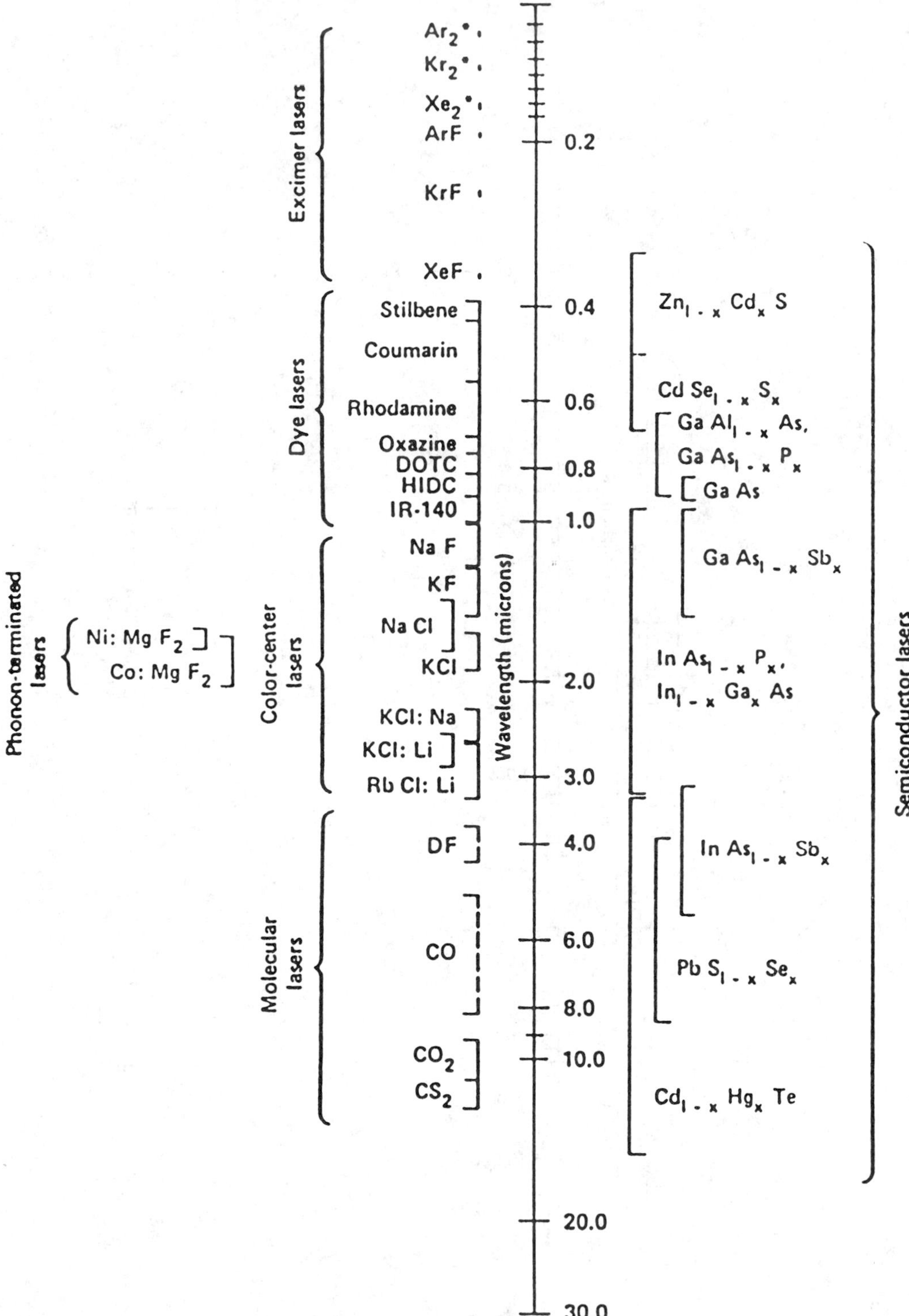

FIGURE 2. Spectral tuning ranges of various types of tunable lasers.

INFRARED LASER FREQUENCIES

Arthur Maki

The CO_2 laser has been the subject of a number of very accurate frequency measurements. Most of the earlier measurements are given by Bradley et al.[1] That analysis was based on a single absolute frequency measurement and many laser frequency differences. New measurements of the methane frequency[2-4] have made it necessary to slightly revise that single absolute frequency measurement. In addition, there have been several other absolute frequency measurements[5-7] that have been used here to improve the accuracy of the present tables. New frequency difference measurements have also been added to the database used for the present tables.[8]

REFERENCES

1. Bradley, L. C., Soohoo, K. L., and Freed, C., *IEEE J. Quantum Electron.*, QE-22, 234-267, 1986.
2. Clairon, A., Dahmani, B., Filimon, A., and Rutman, J., *IEEE Trans. Inst. Meas.*, IM-34, 265-268, 1985.
3. Weiss, C. O., Kramer, G., Lipphardt, B., and Garcia, E., *IEEE J. Quantum Electron.*, QE-24, 1970-1972, 1988.
4. Bagayev, S. N., Baklanov, A. E., Chebotayev, V. P., and Dychkov, A. S., *Appl. Phys.*, B-48, 31-35, 1989.
5. Blaney, T. G., Bradley, C. C., Edwards, G. J., Jolliffe, B. W., Knight, D. J. E., Rowley, W. R. C., Shotten, K. C., and Woods, P. T., *Proc. R. Soc. Lond.*, A-355, 61-88, 1977.
6. Chardonnet, Ch., Van Lerberghe, A., and Bordé, Ch. J., *Opt. Comm.*, 58, 333-337, 1986.
7. Clairon, A., Acef, O., Chardonnet, Ch., and Bordé, Ch. J., *Frequency Standards and Metrology*, De Marchi, A., Ed., Springer-Verlag, Berlin, Heidelberg, 1989, p. 212.
8. Evenson, K., private communication.

Frequencies for the 00°1-(10°0,02°0)$_I$ and 00°1-(10°0,02°0)$_{II}$ Bands of $^{12}C^{16}O_2$ with the Estimated 2-σ Uncertainties

Line	Band I Frequency (MHz)	Uncertainty (MHz)	Line	Band II Frequency (MHz)	Uncertainty (MHz)
P(70)	26721305.4647	0.1680	P(70)	29789856.3783	0.0308
P(68)	26794232.6712	0.1217	P(68)	29861850.7690	0.0192
P(66)	26866318.8073	0.0867	P(66)	29933216.1760	0.0122
P(64)	26937571.7234	0.0606	P(64)	30003944.2861	0.0086
P(62)	27007998.9216	0.0415	P(62)	30074026.9127	0.0072
P(60)	27077607.5643	0.0279	P(60)	30143456.0039	0.0066
P(58)	27146404.4834	0.0185	P(58)	30212223.6504	0.0061
P(56)	27214396.1873	0.0121	P(56)	30280322.0930	0.0055
P(54)	27281588.8696	0.0081	P(54)	30347743.7306	0.0049
P(52)	27347988.4161	0.0057	P(52)	30414481.1273	0.0044
P(50)	27413600.4119	0.0043	P(50)	30480527.0196	0.0041
P(48)	27478430.1487	0.0036	P(48)	30545874.3239	0.0039
P(46)	27542482.6310	0.0032	P(46)	30610516.1429	0.0039
P(44)	27605762.5826	0.0030	P(44)	30674445.7724	0.0039
P(42)	27668274.4525	0.0028	P(42)	30737656.7080	0.0039
P(40)	27730022.4206	0.0027	P(40)	30800142.6511	0.0039
P(38)	27791010.4036	0.0026	P(38)	30861897.5150	0.0038
P(36)	27851242.0594	0.0025	P(36)	30922915.4310	0.0037
P(34)	27910720.7927	0.0024	P(34)	30983190.7534	0.0037
P(32)	27969449.7593	0.0023	P(32)	31042718.0652	0.0037
P(30)	28027431.8708	0.0022	P(30)	31101492.1833	0.0036
P(28)	[illegible]	0.0021	P(28)	[illegible]	[illegible]
P(26)	28141165.9762	0.0020	P(26)	31216761.3029	0.0037
P(24)	28196922.6067	0.0019	P(24)	31273247.1487	0.0037
P(22)	28251941.6622	0.0017	P(22)	31328961.4978	0.0037
P(20)	28306224.8888	0.0016	P(20)	31383900.4028	0.0037
P(18)	28359773.8090	0.0014	P(18)	31438060.1749	0.0037
P(16)	28412589.7245	0.0012	P(16)	31491437.3872	0.0036
P(14)	28464673.7184	0.0011	P(14)	31544028.8776	0.0036
P(12)	28516026.6574	0.0009	P(12)	31595831.7516	0.0036
P(10)	28566649.1935	0.0008	P(10)	31646843.3843	0.0035
P(8)	28616541.7661	0.0008	P(8)	31697061.4225	0.0035
P(6)	28665704.6027	0.0008	P(6)	31746483.7868	0.0035

Line	Band I Frequency (MHz)	Uncertainty (MHz)	Line	Band II Frequency (MHz)	Uncertainty (MHz)
P(4)	28714137.7205	0.0008	P(4)	31795108.6724	0.0035
P(2)	28761840.9272	0.0008	P(2)	31842934.5511	0.0035
R(0)	28832026.2198	0.0008	R(0)	31913172.5691	0.0035
R(2)	28877902.4382	0.0007	R(2)	31958996.0621	0.0034
R(4)	28923046.4303	0.0006	R(4)	32004017.3822	0.0034
R(6)	28967457.0657	0.0005	R(6)	32048236.2498	0.0034
R(8)	29011133.0054	0.0003	R(8)	32091652.6619	0.0034
R(10)	29054072.7010	0.0001	R(10)	32134266.8917	0.0034
R(12)	29096274.3935	0.0003	R(12)	32176079.4878	0.0034
R(14)	29137736.1129	0.0005	R(14)	32217091.2721	0.0035
R(16)	29178455.6759	0.0007	R(16)	32257303.3386	0.0036
R(18)	29218430.6852	0.0009	R(18)	32296717.0510	0.0037
R(20)	29257658.5269	0.0010	R(20)	32335334.0408	0.0038
R(22)	29296136.3689	0.0011	R(22)	32373156.2044	0.0039
R(24)	29333861.1583	0.0012	R(24)	32410185.7003	0.0041
R(26)	29370829.6191	0.0011	R(26)	32446424.9459	0.0042
R(28)	29407038.2491	0.0011	R(28)	32481876.6140	0.0042
R(30)	29442483.3168	0.0011	R(30)	32516543.6293	0.0042
R(32)	29477160.8582	0.0012	R(32)	32550429.1641	0.0042
R(34)	29511066.6733	0.0013	R(34)	32583536.6340	0.0042
R(36)	29544196.3221	0.0015	R(36)	32615869.6937	0.0041
R(38)	29576545.1205	0.0017	R(38)	32647432.2320	0.0040
R(40)	29608108.1360	0.0019	R(40)	32678228.3665	0.0039
R(42)	29638880.1831	0.0022	R(42)	32708262.4386	0.0038
R(44)	29668855.8183	0.0024	R(44)	32737539.0081	0.0039
R(46)	29698029.3350	0.0027	R(46)	32766062.8469	0.0041
R(48)	29726394.7582	0.0032	R(48)	32793838.9334	0.0045
R(50)	29753945.8385	0.0037	R(50)	32820872.4463	0.0055
R(52)	29780676.0464	0.0042	R(52)	32847168.7576	0.0071
R(54)	29806578.5659	0.0047	R(54)	32872733.4269	0.0099
R(56)	29831646.2878	0.0052	R(56)	32897572.1935	0.0141
R(58)	29855871.8032	0.0058	R(58)	32921690.9701	0.0202
R(60)	29879247.3960	0.0074	R(60)	32945095.8355	0.0288
R(62)	29901765.0357	0.0113	R(62)	32967793.0268	0.0407
R(64)	29923416.3695	0.0186	R(64)	32989788.9322	0.0567
R(66)	29944192.7145	0.0302	R(66)	33011090.0831	0.0780
R(68)	29964085.0488	0.0475	R(68)	33031703.1467	0.1060
R(70)	29983084.0036	0.0720	R(70)	33051634.9172	0.1423

Frequencies for the 00°1-(10°0,02°0)$_I$ and 00°1-(10°0,02°0)$_{II}$ Bands of $^{13}C^{16}O_2$ with the Estimated 2-σ Uncertainties

Line	Band I Frequency (MHz)	Uncertainty (MHz)	Line	Band II Frequency (MHz)	Uncertainty (MHz)
P(66)	25523832.1808	0.7836	P(66)	28512082.5283	1.2894
P(64)	25590013.4703	0.5415	P(64)	28585121.9396	0.9194
P(62)	25655543.6502	0.3629	P(62)	28657449.4180	0.6420
P(60)	25720428.2487	0.2339	P(60)	28729056.6374	0.4375
P(58)	25784672.4840	0.1430	P(58)	28799935.4147	0.2897
P(56)	25848281.2771	0.0810	P(56)	28870077.7187	0.1853
P(54)	25911259.2627	0.0405	P(54)	28939475.6771	0.1135
P(52)	25973610.8005	0.0157	P(52)	29008121.5846	0.0659
P(50)	26035339.9857	0.0045	P(50)	29076007.9109	0.0357
P(48)	26096450.6582	0.0079	P(48)	29143127.3077	0.0180
P(46)	26156946.4123	0.0101	P(46)	29209472.6164	0.0090

INFRARED LASER FREQUENCIES (continued)

Band I Line	Band I Frequency (MHz)	Band I Uncertainty (MHz)	Band II Line	Band II Frequency (MHz)	Band II Uncertainty (MHz)
P(44)	26216830.6053	0.0101	P(44)	29275036.8754	0.0058
P(42)	26276106.3655	0.0090	P(42)	29339813.3270	0.0050
P(40)	26334776.6003	0.0077	P(40)	29403795.4243	0.0044
P(38)	26392844.0030	0.0068	P(38)	29466976.8383	0.0037
P(36)	26450311.0599	0.0063	P(36)	29529351.4635	0.0032
P(34)	26507180.0565	0.0061	P(34)	29590913.4252	0.0029
P(32)	26563453.0836	0.0060	P(32)	29651657.0844	0.0028
P(30)	26619132.0428	0.0058	P(30)	29711577.0447	0.0028
P(28)	26674218.6515	0.0055	P(28)	29770668.1566	0.0031
P(26)	26728714.4479	0.0054	P(26)	29828925.5239	0.0035
P(24)	26782620.7952	0.0054	P(24)	29886344.5074	0.0041
P(22)	26835938.8858	0.0054	P(22)	29942920.7308	0.0046
P(20)	26888669.7451	0.0055	P(20)	29998650.0838	0.0051
P(18)	26940814.2347	0.0055	P(18)	30053528.7271	0.0054
P(16)	26992373.0555	0.0055	P(16)	30107553.0955	0.0055
P(14)	27043346.7508	0.0054	P(14)	30160719.9016	0.0055
P(12)	27093735.7083	0.0052	P(12)	30213026.1388	0.0054
P(10)	27143540.1624	0.0051	P(10)	30264469.0839	0.0054
P(8)	27192760.1962	0.0049	P(8)	30315046.2994	0.0054
P(6)	27241395.7431	0.0048	P(6)	30364755.6359	0.0055
P(4)	27289446.5880	0.0047	P(4)	30413595.2335	0.0056
P(2)	27336912.3682	0.0046	P(2)	30461563.5231	0.0057
R(0)	27407012.8882	0.0045	P(0)	30531879.5415	0.0057
R(2)	27453013.4589	0.0043	P(2)	30577664.6138	0.0056
R(4)	27498426.5430	0.0040	P(4)	30622575.1885	0.0054
R(6)	27543251.1200	0.0037	P(6)	30666611.0128	0.0051
R(8)	27587486.0225	0.0034	P(8)	30709772.1257	0.0047
R(10)	27631129.9356	0.0031	P(10)	30752058.8571	0.0045
R(12)	27674181.3963	0.0029	P(12)	30793471.8269	0.0044
R(14)	27716638.7917	0.0029	P(14)	30834011.9425	0.0043
R(16)	27758500.3577	0.0029	P(16)	30873680.3976	0.0044
R(18)	27799764.1770	0.0029	P(18)	30912478.6694	0.0044
R(20)	27840428.1773	0.0030	P(20)	30950408.5159	0.0044
R(22)	27880490.1283	0.0029	P(22)	30987471.9732	0.0043
R(24)	27919947.6395	0.0029	P(24)	31023671.3517	0.0042
R(26)	27958798.1567	0.0028	P(26)	31059009.2327	0.0042
R(28)	27997038.9591	0.0028	P(28)	31093488.4642	0.0042
R(30)	28034667.1551	0.0027	P(30)	31127112.1569	0.0043
R(32)	28071679.6785	0.0027	P(32)	31159883.6793	0.0045
R(34)	28108073.2842	0.0026	P(34)	31191806.6529	0.0046
R(36)	28143844.5432	0.0026	P(36)	31222884.9469	0.0048
R(38)	28178989.8377	0.0026	P(38)	31253122.6730	0.0053
R(40)	28213505.3554	0.0028	P(40)	31282524.1795	0.0061
R(42)	28247387.0838	0.0033	P(42)	31311094.0452	0.0077
R(44)	28280630.8035	0.0046	P(44)	31338837.0736	0.0108
R(46)	28313232.0818	0.0083	P(46)	31365758.2858	0.0173
R(48)	28345186.2652	0.0161	P(48)	31391862.9147	0.0295
R(50)	28376488.4720	0.0301	P(50)	31417156.3972	0.0505
R(52)	28407133.5839	0.0531	P(52)	31441644.3679	0.0845
R(54)	28437116.2372	0.0887	P(54)	31465332.6516	0.1366
R(56)	28466430.8141	0.1419	P(56)	31488227.2557	0.2138
R(58)	28495071.4324	0.2188	P(58)	31510334.3631	0.3247
R(60)	28523031.9357	0.3271	P(60)	31531660.3243	0.4800
R(62)	28550305.8819	0.4763	P(62)	31552211.6497	0.6932
R(64)	28576886.5323	0.6781	P(64)	31571995.0017	0.9805
R(66)	28602766.8393	0.9467	P(66)	31591017.1868	1.3619

INFRARED LASER FREQUENCIES (continued)

Frequencies for the 00°1-(10°0,02°0)$_I$ and 00°1-(10°0,02°0)$_{II}$ Bands of $^{12}C^{18}O_2$ with the Estimated 2-σ Uncertainties

Line	Band I Frequency (MHz)	Uncertainty (MHz)	Line	Band II Frequency (MHz)	Uncertainty (MHz)
P(70)	27045326.3119	0.4540	P(70)	30695237.5856	0.0858
P(68)	27114914.0922	0.3324	P(68)	30755520.2231	0.0570
P(66)	27183635.7945	0.2392	P(66)	30815311.4928	0.0364
P(64)	27251496.4118	0.1688	P(64)	30874607.2084	0.0223
P(62)	27318500.7361	0.1165	P(62)	30933403.2309	0.0131
P(60)	27384653.3618	0.0783	P(60)	30991695.4724	0.0075
P(58)	27449958.6881	0.0510	P(58)	31049479.9009	0.0049
P(56)	27514420.9224	0.0319	P(56)	31106752.5446	0.0041
P(54)	27578044.0828	0.0191	P(54)	31163509.4964	0.0040
P(52)	27640832.0010	0.0108	P(52)	31219746.9183	0.0040
P(50)	27702788.3248	0.0059	P(50)	31275461.0455	0.0039
P(48)	27763916.5206	0.0035	P(48)	31330648.1908	0.0039
P(46)	27824219.8762	0.0028	P(46)	31385304.7490	0.0039
P(44)	27883701.5029	0.0026	P(44)	31439427.2006	0.0039
P(42)	27942364.3379	0.0025	P(42)	31493012.1163	0.0038
P(40)	28000211.1464	0.0024	P(40)	31546056.1605	0.0038
P(38)	28057244.5242	0.0022	P(38)	31598556.0954	0.0037
P(36)	28113466.8992	0.0021	P(36)	31650508.7847	0.0037
P(34)	28168880.5335	0.0020	P(34)	31701911.1970	0.0037
P(32)	28223487.5256	0.0019	P(32)	31752760.4093	0.0037
P(30)	28277289.8118	0.0017	P(30)	31803053.6105	0.0037
P(28)	28330289.1679	0.0016	P(28)	31852788.1043	0.0038
P(26)	28382487.2111	0.0015	P(26)	31901961.3125	0.0038
P(24)	28433885.4012	0.0013	P(24)	31950570.7773	0.0038
P(22)	28484485.0420	0.0012	P(22)	31998614.1649	0.0038
P(20)	28534287.2828	0.0011	P(20)	32046089.2669	0.0037
P(18)	28583293.1193	0.0010	P(18)	32092994.0036	0.0037
P(16)	28631503.3952	0.0010	P(16)	32139326.4254	0.0036
P(14)	28678918.8025	0.0009	P(14)	32185084.7154	0.0036
P(12)	28725539.8830	0.0010	P(12)	32230267.1907	0.0036
P(10)	28771367.0288	0.0010	P(10)	32274872.3041	0.0037
P(8)	28816400.4829	0.0010	P(8)	32318898.6455	0.0038
P(6)	28860640.3403	0.0011	P(6)	32362344.9434	0.0039
P(4)	28904086.5477	0.0011	P(4)	32405210.0652	0.0041
P(2)	28946738.9048	0.0011	P(2)	32447493.0185	0.0041
R(0)	29009228.1702	0.0010	P(0)	32509824.0580	0.0042
R(2)	29049894.0586	0.0010	P(2)	32550648.1723	0.0042
R(4)	29089764.2368	0.0009	P(4)	32590887.7542	0.0042
R(6)	29128837.8426	0.0008	P(6)	32630542.4457	0.0041
R(8)	29167113.8668	0.0008	P(8)	32669612.0295	0.0041
R(10)	29204591.1529	0.0009	P(10)	32708096.4282	0.0040
R(12)	29241268.3964	0.0010	P(12)	32745995.7040	0.0040
R(14)	29277144.1444	0.0011	P(14)	32783310.0573	0.0040
R(16)	29312216.7955	0.0012	P(16)	32820039.8258	0.0040
R(18)	29346484.5984	0.0012	P(18)	32856185.4827	0.0040
R(20)	29379945.6517	0.0013	P(20)	32891747.6358	0.0040
R(22)	29412597.9024	0.0013	P(22)	32926727.0254	0.0040
R(24)	29444439.1458	0.0013	P(24)	32961124.5220	0.0040
R(26)	29475467.0236	0.0014	P(26)	32994941.1249	0.0040
R(28)	29505679.0230	0.0015	P(28)	33028177.9594	0.0040
R(30)	29535072.4755	0.0016	P(30)	33060836.2743	0.0040
R(32)	29563644.5557	0.0018	P(32)	33092917.4394	0.0041
R(34)	29591392.2794	0.0020	P(34)	33124422.9429	0.0043
R(36)	29618312.5023	0.0023	P(36)	33155354.3878	0.0046
R(38)	29644401.9182	0.0028	P(38)	33185713.4894	0.0049

Line	Band I Frequency (MHz)	Uncertainty (MHz)	Line	Band II Frequency (MHz)	Uncertainty (MHz)
R(40)	29669657.0575	0.0036	P(40)	33215502.0716	0.0056
R(42)	29694074.2853	0.0053	P(42)	33244722.0637	0.0068
R(44)	29717649.7992	0.0082	P(44)	33273375.4969	0.0092
R(46)	29740379.6276	0.0128	P(46)	33301464.5003	0.0134
R(48)	29762259.6274	0.0200	P(48)	33328991.2976	0.0199
R(50)	29783285.4820	0.0307	P(50)	33355958.2027	0.0294
R(52)	29803452.6988	0.0461	P(52)	33382367.6161	0.0427
R(54)	29822756.6072	0.0681	P(54)	33408222.0209	0.0607
R(56)	29841192.3558	0.0985	P(56)	33433523.9780	0.0848
R(58)	29858754.9100	0.1401	P(58)	33458276.1228	0.1165
R(60)	29875439.0495	0.1960	P(60)	33482481.1601	0.1576
R(62)	29891239.3658	0.2702	P(62)	33506141.8605	0.2104
R(64)	29906150.2589	0.3673	P(64)	33529261.0556	0.2775
R(66)	29920165.9352	0.4930	P(66)	33551841.6335	0.3621
R(68)	29933280.4042	0.6540	P(68)	33573886.5352	0.4679
R(70)	29945487.4756	0.8581	P(70)	33595398.7493	0.5992

Frequencies for the 00°1-(10°0,02°0)$_I$ and 00°1-(10°0,02°0)$_{II}$ Bands of $^{13}C^{18}O_2$ with the Estimated 2-σ Uncertainties

Line	Band I Frequency (MHz)	Uncertainty (MHz)	Line	Band II Frequency (MHz)	Uncertainty (MHz)
P(70)	25967863.7652	1.1146	P(70)	28960176.2278	0.4069
P(68)	26033448.2798	0.8152	P(68)	29022326.9578	0.2861
P(66)	26098273.9159	0.5860	P(66)	29083661.3546	0.1961
P(64)	26162346.4813	0.4129	P(64)	29144473.5795	0.1303
P(62)	26225671.5466	0.2844	P(62)	29204757.8761	0.0833
P(60)	26288254.4494	0.1906	P(60)	29264508.5768	0.0507
P(58)	26350100.2984	0.1237	P(58)	29323720.1086	0.0290
P(56)	26411213.9778	0.0772	P(56)	29382386.9988	0.0152
P(54)	26471600.1504	0.0459	P(54)	29440503.8809	0.0073
P(52)	26531263.2618	0.0258	P(52)	29498065.4997	0.0038
P(50)	26590207.5442	0.0138	P(50)	29555066.7172	0.0032
P(48)	26648437.0195	0.0077	P(48)	29611502.5178	0.0031
P(46)	26705955.5026	0.0057	P(46)	29667368.0132	0.0031
P(44)	26762766.6051	0.0055	P(44)	29722658.4475	0.0034
P(42)	26818873.7378	0.0055	P(42)	29777369.2022	0.0039
P(40)	26874280.1143	0.0056	P(40)	29831495.8006	0.0044
P(38)	26928988.7531	0.0056	P(38)	29885033.9125	0.0049
P(36)	26983002.4809	0.0056	P(36)	29937979.3584	0.0053
P(34)	27036323.9351	0.0055	P(34)	29990328.1139	0.0054
P(32)	27088955.5657	0.0054	P(32)	30042076.3132	0.0055
P(30)	27140899.6384	0.0051	P(30)	30093220.2534	0.0055
P(28)	27192138.2303	0.0049	P(28)	30143756.3978	0.0054
P(26)	27242733.2620	0.0047	P(26)	30193681.3793	0.0053
P(24)	27292626.4396	0.0044	P(24)	30242992.0038	0.0052
P(22)	27341839.3165	0.0042	P(22)	30291685.2529	0.0051
P(20)	27390373.2651	0.0040	P(20)	30339758.2870	0.0049
P(18)	27438229.4843	0.0037	P(18)	30387208.4477	0.0048
P(16)	27485409.0008	0.0035	P(16)	30434033.2603	0.0046
P(14)	27531912.6704	0.0033	P(14)	30480230.4356	0.0045
P(12)	27577741.1795	0.0031	P(12)	30525797.8725	0.0044
P(10)	27622895.0455	0.0031	P(10)	30570733.6593	0.0043
P(8)	27667374.6182	0.0031	P(8)	30615036.0750	0.0043
P(6)	27711180.0803	0.0033	P(6)	30658703.5912	0.0044
P(4)	27754311.4480	0.0034	P(4)	30701734.8727	0.0045

	Band I			Band II	
Line	Frequency (MHz)	Uncertainty (MHz)	Line	Frequency (MHz)	Uncertainty (MHz)
P(2)	27796768.5718	0.0036	P(2)	30744128.7785	0.0045
R(0)	27859189.3155	0.0036	P(0)	30806522.5414	0.0045
R(2)	27899959.0889	0.0035	P(2)	30847319.2956	0.0044
R(4)	27940052.7921	0.0033	P(4)	30887476.2168	0.0043
R(6)	27979469.5315	0.0031	P(6)	30926993.0424	0.0042
R(8)	28018208.2478	0.0028	P(8)	30965869.7046	0.0041
R(10)	28056267.7161	0.0026	P(10)	31004106.3298	0.0040
R(12)	28093646.5448	0.0025	P(12)	31041703.2379	0.0040
R(14)	28130343.1757	0.0025	P(14)	31078660.9408	0.0040
R(16)	28166355.8825	0.0025	P(16)	31114980.1420	0.0040
R(18)	28201682.7706	0.0025	P(18)	31150661.7340	0.0041
R(20)	28236321.7757	0.0025	P(20)	31185706.7976	0.0042
R(22)	28270270.6628	0.0024	P(22)	31220116.5992	0.0043
R(24)	28303527.0249	0.0024	P(24)	31253892.5891	0.0043
R(26)	28336088.2817	0.0023	P(26)	31287036.3991	0.0044
R(28)	28367951.6781	0.0024	P(28)	31319549.8396	0.0043
R(30)	28399114.2823	0.0025	P(30)	31351434.8973	0.0043
R(32)	28429572.9843	0.0026	P(32)	31382693.7318	0.0042
R(34)	28459324.4940	0.0028	P(34)	31413328.6728	0.0042
R(36)	28488365.3390	0.0029	P(36)	31443342.2165	0.0041
R(38)	28516691.8625	0.0029	P(38)	31472737.0219	0.0040
R(40)	28544300.2211	0.0031	P(40)	31501515.9074	0.0039
R(42)	28571186.3823	0.0032	P(42)	31529681.8467	0.0040
R(44)	28597346.1222	0.0032	P(44)	31557237.9646	0.0042
R(46)	28622775.0223	0.0038	P(46)	31584187.5329	0.0046
R(48)	28647468.4672	0.0071	P(48)	31610533.9656	0.0057
R(50)	28671421.6417	0.0148	P(50)	31636280.8146	0.0088
R(52)	28694629.5272	0.0286	P(52)	31661431.7650	0.0151
R(54)	28717086.8993	0.0510	P(54)	31685990.6298	0.0261
R(56)	28738788.3239	0.0852	P(56)	31709961.3449	0.0434
R(58)	28759728.1540	0.1355	P(58)	31733347.9642	0.0693
R(60)	28779900.5263	0.2075	P(60)	31756154.6537	0.1068
R(62)	28799299.3572	0.3078	P(62)	31778385.6867	0.1594
R(64)	28817918.3393	0.4447	P(64)	31800045.4375	0.2317
R(66)	28835750.9374	0.6283	P(66)	31821138.3761	0.3291
R(68)	28852790.3843	0.8707	P(68)	31841669.0622	0.4581
R(70)	28869029.6768	1.1863	P(70)	31861642.1394	0.6268

Frequencies for the $01^{1e}1$-$(11^{1e}0,03^{1e}0)_I$ and $01^{1e}1$-$(11^{1e}0,03^{1e}0)_{II}$ Bands of $^{12}C^{16}O_2$ with the Estimated 2-σ Uncertainties

	Band I			Band II	
Line	Frequency (MHz)	Uncertainty (MHz)	Line	Frequency (MHz)	Uncertainty (MHz)
P(59)	26125213.2723	1.6633	P(59)	30427055.2899	0.1962
P(57)	26191576.6703	1.0880	P(57)	30494640.3229	0.1332
P(55)	26257240.7898	0.6844	P(55)	30561557.5929	0.0865
P(53)	26322208.2302	0.4094	P(53)	30627802.0344	0.0530
P(51)	26386481.4313	0.2286	P(51)	30693368.7014	0.0306
P(49)	26450062.6783	0.1155	P(49)	30758252.7710	0.0175
P(47)	26512954.1076	0.0498	P(47)	30822449.5469	0.0123
P(45)	26575157.7109	0.0191	P(45)	30885954.4624	0.0114
P(43)	26636675.3402	0.0160	P(43)	30948763.0834	0.0109
P(41)	26697508.7115	0.0182	P(41)	31010871.1119	0.0100
P(39)	26757659.4084	0.0177	P(39)	31072274.3882	0.0091
P(37)	26817128.8857	0.0160	P(37)	31132968.8940	0.0091
P(35)	26875918.4726	0.0144	P(35)	31192950.7549	0.0102

Line	Band I Frequency (MHz)	Uncertainty (MHz)	Line	Band II Frequency (MHz)	Uncertainty (MHz)
P(33)	26934029.3751	0.0131	P(33)	31252216.2430	0.0118
P(31)	26991462.6787	0.0119	P(31)	31310761.7788	0.0134
P(29)	27048219.3509	0.0106	P(29)	31368583.9339	0.0147
P(27)	27104300.2431	0.0096	P(27)	31425679.4328	0.0155
P(25)	27159706.0925	0.0093	P(25)	31482045.1550	0.0157
P(23)	27214437.5237	0.0097	P(23)	31537678.1367	0.0154
P(21)	27268495.0505	0.0104	P(21)	31592575.5725	0.0147
P(19)	27321879.0769	0.0108	P(19)	31646734.8172	0.0137
P(17)	27374589.8987	0.0108	P(17)	31700153.3868	0.0127
P(15)	27426627.7040	0.0104	P(15)	31752828.9602	0.0119
P(13)	27477992.5747	0.0098	P(13)	31804759.3803	0.0113
P(11)	27528684.4867	0.0096	P(11)	31855942.6551	0.0113
P(9)	27578703.3113	0.0101	P(9)	31906376.9582	0.0116
P(7)	27628048.8151	0.0113	P(7)	31956060.6304	0.0122
P(5)	27676720.6609	0.0127	P(5)	32004992.1796	0.0129
P(3)	27724718.4080	0.0141	P(3)	32053170.2819	0.0136
R(1)	27841759.7696	0.0152	P(1)	32170312.0391	0.0149
R(3)	27887393.2105	0.0146	P(3)	32215845.0845	0.0151
R(5)	27932349.2934	0.0135	P(5)	32260620.8121	0.0152
R(7)	27976627.0108	0.0124	P(7)	32304638.8261	0.0152
R(9)	28020225.2521	0.0115	P(9)	32347898.8990	0.0150
R(11)	28063142.8031	0.0110	P(11)	32390400.9714	0.0148
R(13)	28105378.3457	0.0109	P(13)	32432145.1513	0.0145
R(15)	28146930.4576	0.0109	P(15)	32473131.7137	0.0142
R(17)	28187797.6116	0.0107	P(17)	32513361.0997	0.0140
R(19)	28227978.1750	0.0103	P(19)	32552833.9153	0.0140
R(21)	28267470.4088	0.0099	P(21)	32591550.9309	0.0141
R(23)	28306272.4666	0.0099	P(23)	32629513.0796	0.0143
R(25)	28344382.3939	0.0107	P(25)	32666721.4564	0.0144
R(27)	28381798.1267	0.0122	P(27)	32703177.3164	0.0142
R(29)	28418517.4902	0.0141	P(29)	32738882.0732	0.0136
R(31)	28454538.1976	0.0165	P(31)	32773837.2976	0.0136
R(33)	28489857.8477	0.0213	P(33)	32808044.7156	0.0174
R(35)	28524473.9240	0.0312	P(35)	32841506.2063	0.0279
R(37)	28558383.7917	0.0486	P(37)	32874223.8000	0.0462
R(39)	28591584.6963	0.0754	P(39)	32906199.6761	0.0735
R(41)	28624073.7602	0.1131	P(41)	32937436.1606	0.1114
R(43)	28655847.9806	0.1644	P(43)	32967935.7238	0.1624
R(45)	28686904.2261	0.2328	P(45)	32997700.9775	0.2292
R(47)	28717239.2334	0.3239	P(47)	33026734.6728	0.3151
R(49)	28746849.6038	0.4465	P(49)	33055039.6965	0.4238
R(51)	28775731.7988	0.6142	P(51)	33082619.0689	0.5595
R(53)	28803882.1361	0.8465	P(53)	33109475.9403	0.7272

Frequencies for the $01^{1f}1$-$(11^{1f}0,03^{1f}0)_I$ and $01^{1e}1$-$(11^{1e}0,03^{1e}0)_{II}$ Bands of $^{14}C^{17}O_2$ with the Estimated 2σ Uncertainties

Line	Band I Frequency (MHz)	Uncertainty (MHz)	Line	Band II Frequency (MHz)	Uncertainty (MHz)
P(60)	26051570.0104	4.4521	P(60)	30355115.0204	0.2752
P(58)	26120964.4932	3.0629	P(58)	30425283.5969	0.1926
P(56)	26189552.8496	2.0516	P(56)	30494732.8293	0.1301
P(54)	26257339.6006	1.3305	P(54)	30563455.6325	0.0840
P(52)	26324329.0344	0.8289	P(52)	30631445.1076	0.0512
P(50)	26390525.2136	0.4901	P(50)	30698694.5456	0.0292
P(48)	26455931.9824	0.2698	P(48)	30765197.4310	0.0163

Line	Band I Frequency (MHz)	Uncertainty (MHz)	Line	Band II Frequency (MHz)	Uncertainty (MHz)
P(46)	26520552.9722	0.1334	P(46)	30830947.4444	0.0111
P(44)	26584391.6075	0.0551	P(44)	30895938.4662	0.0104
P(42)	26647451.1105	0.0181	P(42)	30960164.5794	0.0105
P(40)	26709734.5057	0.0151	P(40)	31023620.0723	0.0105
P(38)	26771244.6242	0.0174	P(38)	31086299.4415	0.0107
P(36)	26831984.1067	0.0157	P(36)	31148197.3941	0.0114
P(34)	26891955.4069	0.0126	P(34)	31209308.8510	0.0126
P(32)	26951160.7945	0.0105	P(32)	31269628.9481	0.0138
P(30)	27009602.3576	0.0096	P(30)	31329153.0395	0.0147
P(28)	27067282.0045	0.0092	P(28)	31387876.6994	0.0151
P(26)	27124201.4662	0.0090	P(26)	31445795.7236	0.0149
P(24)	27180362.2977	0.0089	P(24)	31502906.1318	0.0141
P(22)	27235765.8792	0.0090	P(22)	31559204.1695	0.0128
P(20)	27290413.4182	0.0093	P(20)	31614686.3091	0.0113
P(18)	27344305.9494	0.0096	P(18)	31669349.2515	0.0098
P(16)	27397444.3368	0.0097	P(16)	31723189.9280	0.0086
P(14)	27449829.2733	0.0096	P(14)	31776205.5007	0.0081
P(12)	27501461.2824	0.0096	P(12)	31828393.3642	0.0085
P(10)	27552340.7179	0.0101	P(10)	31879751.1463	0.0095
P(8)	27602467.7649	0.0111	P(8)	31930276.7092	0.0107
P(6)	27651842.4399	0.0125	P(6)	31979968.1497	0.0120
P(4)	27700464.5912	0.0139	P(4)	32028823.8002	0.0131
P(2)	27748333.8988	0.0148	P(2)	32076842.2290	0.0139
R(2)	27864709.8633	0.0146	P(2)	32193218.1935	0.0150
R(4)	27909939.2762	0.0135	P(4)	32238298.4853	0.0151
R(6)	27954412.3294	0.0122	P(6)	32282538.0393	0.0153
R(8)	27998127.7801	0.0112	P(8)	32325936.7244	0.0153
R(10)	28041084.2173	0.0107	P(10)	32368494.6458	0.0154
R(12)	28083280.0620	0.0108	P(12)	32410212.1438	0.0155
R(14)	28124713.5668	0.0110	P(14)	32451089.7941	0.0155
R(16)	28165382.8151	0.0112	P(16)	32491128.4063	0.0154
R(18)	28205285.7213	0.0111	P(18)	32530329.0234	0.0152
R(20)	28244420.0302	0.0110	P(20)	32568692.9211	0.0149
R(22)	28282783.3158	0.0114	P(22)	32606221.6061	0.0147
R(24)	28320372.9812	0.0129	P(24)	32642916.8154	0.0144
R(26)	28357186.2574	0.0149	P(26)	32678780.5147	0.0140
R(28)	28393220.2023	0.0168	P(28)	32713814.8971	0.0133
R(30)	28428471.6994	0.0175	P(30)	32748022.3813	0.0120
R(32)	28462937.4565	0.0165	P(32)	32781405.6101	0.0106
R(34)	28496614.0042	0.0142	P(34)	32813967.4482	0.0122
R(36)	28529497.6934	0.0163	P(36)	32845710.9809	0.0212
R(38)	28561584.6939	0.0309	P(38)	32876639.5111	0.0385
R(40)	28592870.9914	0.0593	P(40)	32906756.5580	0.0646
R(42)	28623352.3850	0.1054	P(42)	32936065.8540	0.1015
R(44)	28653024.4839	0.1779	P(44)	32964571.3426	0.1518
R(46)	28681882.7038	0.2907	P(46)	32992277.1760	0.2184
R(48)	28709922.2632	0.4635	P(48)	33019187.7118	0.3048
R(50)	28737138.1785	0.7231	P(50)	33045307.5105	0.4152

INFRARED AND FAR-INFRARED ABSORPTION FREQUENCY STANDARDS

Arthur Maki

Aside from the CO_2 laser transitions, the absorption spectrum of CO has been more accurately and thoroughly measured than any other spectrum. A bibliography of earlier measurements on CO is given by Maki and Wells,[1] and the present tables were calculated from the measurements referred to in that work. In addition, some new and very accurate frequency measurements[2,3] have been made and were incorporated in the present tables. The frequencies of the rotational transitions of HF and HCl were calculated from constants obtained from fitting the measurements of Evenson et al.[4,5] and Jennings and Wells.[6]

REFERENCES

1. Maki, A. G. and Wells, J. S., *Wavenumber Calibration Tables from Heterodyne Frequency Measurements*, NIST Special Publication 821, U.S. Dept. of Commerce, Washington, D.C., 1991.
2. Evenson, K. and Stroh, F., private communication.
3. George, T., Urban, W., and co-workers, private communication.
4. Jennings, D. A., Evenson, K. M., Zink, L. R., Demuynck, C., Destombes, J. L., Lemoine, B., and Johns, J. W. C., *J. Mol. Spectrosc.*, 122, 477-480, 1987.
5. Nolt, I. G., Radostitz, J. V., DiLonardo, G., Evenson, K. M., Jennings, D. A., Leopold, K. R., Vanek, M. D., Zink, L. R., Hinz, A., and Chance, K. V., *J. Mol. Spectrosc.*, 125, 274-287, 1987.
6. Jennings, D. A. and Wells, J. S., *J. Mol. Spectrosc.*, 130, 267-268, 1988.

Wavenumbers for the v = 1-0 Band of CO

Wavenumber (unc)* cm^{-1}	Transition	Wavenumber (unc) cm^{-1}	Transition
		2147.081132(01)	R(0)
2139.426071(01)	P(1)	2150.856006(01)	R(1)
2135.546178(01)	P(2)	2154.595581(01)	R(2)
2131.631574(01)	P(3)	2158.299710(01)	R(3)
2127.682404(01)	P(4)	2161.968245(01)	R(4)
2123.698816(01)	P(5)	2165.601041(01)	R(5)
2119.680957(01)	P(6)	2169.197949(01)	R(6)
2115.628973(01)	P(7)	2172.758824(01)	R(7)
2111.543012(01)	P(8)	2176.283519(01)	R(8)
2107.423221(01)	P(9)	2179.771887(01)	R(9)
2103.269746(01)	P(10)	2183.223782(01)	R(10)
2099.082734(01)	P(11)	2186.639057(01)	R(11)
2094.862333(01)	P(12)	2190.017565(01)	R(12)
2090.608688(01)	P(13)	2193.359161(01)	R(13)
2086.321947(01)	P(14)	2196.663698(01)	R(14)
2082.002256(01)	P(15)	2199.931030(01)	R(15)
2077.649762(01)	P(16)	2203.161010(01)	R(16)
2073.264612(01)	P(17)	2206.353492(01)	R(17)
2068.846952(01)	P(18)	2209.508331(02)	R(18)
2064.396929(01)	P(19)	2212.625379(02)	R(19)
2059.914688(02)	P(20)	2215.704492(02)	R(20)
2055.400377(02)	P(21)	2218.745522(02)	R(21)
2050.854140(02)	P(22)	2221.748326(03)	R(22)
2046.276126(03)	P(23)	2224.712755(03)	R(23)
2041.666479(03)	P(24)	2227.638666(03)	R(24)
2037.025345(03)	P(25)	2230.525912(04)	R(25)
2032.352870(04)	P(26)	2233.374349(04)	R(26)
2027.649200(04)	P(27)	2236.183829(04)	R(27)
2022.914480(04)	P(28)	2238.954210(05)	R(28)
2018.148857(05)	P(29)	2241.685344(05)	R(29)
2013.352474(05)	P(30)	2244.377088(06)	R(30)
2008.525477(06)	P(31)	2247.029296(07)	R(31)
2003.668012(06)	P(32)	2249.641824(08)	R(32)
1998.780224(07)	P(33)	2252.214527(10)	R(33)
1993.862257(09)	P(34)	2254.747262(14)	R(34)
1988.914257(11)	P(35)	2257.239883(18)	R(35)

Wavenumber (unc)* cm^{-1}	Transition	Wavenumber (unc) cm^{-1}	Transition
1983.936367(14)	P(36)	2259.692248(24)	R(36)
1978.928733(18)	P(37)	2262.104213(33)	R(37)
1973.891500(25)	P(38)	2264.475634(45)	R(38)
1968.824811(34)	P(39)	2266.806368(61)	R(39)
1963.728813(46)	P(40)	2269.096273(81)	R(40)
1958.603648(61)	P(41)	2271.345206(106)	R(41)
1953.449462(82)	P(42)	2273.553027(139)	R(42)

* The uncertainty in the last digits (twice the standard error) is given in parentheses.

Wavenumbers for the v = 2-0 Band of CO

Wavenumber (unc)* cm^{-1}	Transition	Wavenumber (unc) cm^{-1}	Transition
		4263.837198(02)	R(0)
4256.217140(02)	P(1)	4267.542066(02)	R(1)
4252.302244(02)	P(2)	4271.176630(02)	R(2)
4248.317633(02)	P(3)	4274.740746(02)	R(3)
4244.263453(02)	P(4)	4278.234264(02)	R(4)
4240.139852(02)	P(5)	4281.657039(02)	R(5)
4235.946975(02)	P(6)	4285.008924(02)	R(6)
4231.684972(02)	P(7)	4288.289772(02)	R(7)
4227.353987(02)	P(8)	4291.499437(02)	R(8)
4222.954169(02)	P(9)	4294.637773(02)	R(9)
4218.485665(02)	P(10)	4297.704631(02)	R(10)
4213.948620(02)	P(11)	4300.699868(02)	R(11)
4209.343182(02)	P(12)	4303.623334(02)	R(12)
4204.669499(02)	P(13)	4306.474886(02)	R(13)
4199.927716(02)	P(14)	4309.254375(02)	R(14)
4195.117980(02)	P(15)	4311.961657(02)	R(15)
4190.240439(02)	P(16)	4314.596584(02)	R(16)
4185.295239(02)	P(17)	4317.159011(02)	R(17)
4180.282526(02)	P(18)	4319.648791(02)	R(18)
4175.202447(02)	P(19)	4322.065779(03)	R(19)
4170.055149(03)	P(20)	4324.409829(03)	R(20)
4164.840777(03)	P(21)	4326.680794(03)	R(21)
4159.559478(03)	P(22)	4328.878530(03)	R(22)
4154.211398(03)	P(23)	4331.002889(04)	R(23)
4148.796683(04)	P(24)	4333.053728(04)	R(24)
4143.315479(04)	P(25)	4335.030899(05)	R(25)
4137.767932(04)	P(26)	4336.934259(06)	R(26)
4132.154187(05)	P(27)	4338.763661(07)	R(27)
4126.474391(06)	P(28)	4340.518961(09)	R(28)
4120.728689(07)	P(29)	4342.200014(11)	R(29)
4114.917226(09)	P(30)	4343.806675(16)	R(30)
4109.040148(12)	P(31)	4345.338799(21)	R(31)
4103.097600(16)	P(32)	4346.796243(29)	R(32)
4097.089728(21)	P(33)	4348.178862(40)	R(33)
4091.016676(29)	P(34)	4349.486513(54)	R(34)
4084.878591(40)	P(35)	4350.719052(73)	R(35)
4078.675618(54)	P(36)	4351.876336(96)	R(36)
4072.407901(73)	P(37)	4352.958224(127)	R(37)
4066.075588(97)	P(38)	4353.964572(166)	R(38)
4059.678822(127)	P(39)	4354.895240(214)	R(39)

* The uncertainty in the last digits (twice the standard error) is given in parentheses.

Wavenumbers for the v = 3-0 Band of CO

Wavenumber (unc)* cm^{-1}	Transition	Wavenumber (unc) cm^{-1}	Transition
		6354.179057(13)	R(0)
6346.594000(13)	P(1)	6357.813923(13)	R(1)
6342.644103(13)	P(2)	6361.343487(13)	R(2)
6338.589491(13)	P(3)	6364.767599(13)	R(3)
6334.430309(13)	P(4)	6368.086115(13)	R(4)
6330.166705(13)	P(5)	6371.298887(13)	R(5)
6325.798826(13)	P(6)	6374.405768(12)	R(6)
6321.326819(13)	P(7)	6377.406611(12)	R(7)
6316.750831(12)	P(8)	6380.301271(12)	R(8)
6312.071008(12)	P(9)	6383.089600(12)	R(9)
6307.287498(12)	P(10)	6385.771452(12)	R(10)
6302.400447(12)	P(11)	6388.346680(13)	R(11)
6297.410003(12)	P(12)	6390.815139(13)	R(12)
6292.316311(13)	P(13)	6393.176681(13)	R(13)
6287.119520(13)	P(14)	6395.431160(13)	R(14)
6281.819775(13)	P(15)	6397.578430(13)	R(15)
6276.417224(13)	P(16)	6399.618344(13)	R(16)
6270.912012(13)	P(17)	6401.550757(13)	R(17)
6265.304287(13)	P(18)	6403.375523(13)	R(18)
6259.594194(13)	P(19)	6405.092495(14)	R(19)
6253.781880(13)	P(20)	6406.701527(14)	R(20)
6247.067492(14)	P(21)	6408.202474(14)	R(21)
6241.851176(14)	P(22)	6409.595189(15)	R(22)
6235.733077(14)	P(23)	6410.879527(15)	R(23)
6229.513342(15)	P(24)	6412.055343(16)	R(24)
6223.192117(15)	P(25)	6413.122491(17)	R(25)
6216.769547(16)	P(26)	6414.080825(19)	R(26)
6210.245778(17)	P(27)	6414.930201(23)	R(27)
6203.620957(19)	P(28)	6415.670474(28)	R(28)
6196.895229(23)	P(29)	6416.301500(37)	R(29)
6190.068739(28)	P(30)	6416.823133(50)	R(30)
6183.141633(37)	P(31)	6417.235231(67)	R(31)
6176.114058(50)	P(32)	6417.537649(90)	R(32)
6168.986159(67)	P(33)		
6161.758082(90)	P(34)		

* The uncertainty in the last digits (twice the standard error) is given in parentheses.

Frequencies and Wavenumbers for the Rotational Lines of CO

Frequency MHz	Uncertainty* MHz	J'	J''	Wavenumber cm^{-1}	Uncertainty* cm^{-1}
115271.2029	0.0004	1	0	3.84503345	0.00000001
230538.0016	0.0008	2	1	7.68991999	0.00000003
345795.9923	0.0012	3	2	11.53451273	0.00000004
461040.7712	0.0016	4	3	15.37866477	0.00000005
576267.9350	0.0019	5	4	19.22222923	0.00000006
691473.0809	0.0021	6	5	23.06505926	0.00000007
806651.8065	0.0023	7	6	26.90700800	0.00000008
921799.7104	0.0025	8	7	30.74792863	0.00000008
1036912.3919	0.0027	9	8	34.58767438	0.00000009
1151985.4515	0.0029	10	9	38.42609848	0.00000010
1267014.4906	0.0031	11	10	42.26305422	0.00000010
1381995.1119	0.0034	12	11	46.09839491	0.00000011
1496922.9195	0.0038	13	12	49.93197392	0.00000013

Frequency MHz	Uncertainty* MHz	*J′*	*J″*	Wavenumber cm⁻¹	Uncertainty* cm⁻¹
1611793.5189	0.0042	14	13	53.76364468	0.00000014
1726602.5173	0.0047	15	14	57.59326065	0.00000016
1841345.5237	0.0052	16	15	61.42067535	0.00000017
1956018.1486	0.0057	17	16	65.24574239	0.00000019
2070616.0050	0.0061	18	17	69.06831542	0.00000020
2185134.7075	0.0065	19	18	72.88824816	0.00000022
2299569.8733	0.0069	20	19	76.70539441	0.00000023
2413917.1217	0.0071	21	20	80.51960806	0.00000024
2528172.0747	0.0073	22	21	84.33074306	0.00000024
2642330.3567	0.0074	23	22	88.13865346	0.00000025
2756387.5949	0.0075	24	23	91.94319341	0.00000025
2870339.4194	0.0077	25	24	95.74421713	0.00000026
2984181.4631	0.0080	26	25	99.54157896	0.00000027
3097909.3621	0.0085	27	26	103.33513334	0.00000028
3211518.7558	0.0090	28	27	107.12473480	0.00000030
3325005.2869	0.0096	29	28	110.91023800	0.00000032
3438364.6013	0.0102	30	29	114.69149772	0.00000034
3551592.3489	0.0107	31	30	118.46836884	0.00000036
3664684.1829	0.0111	32	31	122.24070637	0.00000037
3777635.7608	0.0118	33	32	126.00836545	0.00000039
3890442.7435	0.0137	34	33	129.77120137	0.00000046
4003100.7965	0.0179	35	34	133.52906952	0.00000060
4115605.5892	0.0254	36	35	137.28182546	0.00000085
4227952.7954	0.0370	37	36	141.02932487	0.00000123
4340138.0932	0.0531	38	37	144.77142361	0.00000177
4452157.1657	0.0746	39	38	148.50797766	0.00000249
4564005.7001	0.1025	40	39	152.23884318	0.00000342

* The uncertainty given is twice the standard error.

Frequencies and Wavenumbers for the Rotational Lines of HF

Frequency MHz	Uncertainty* MHz	*J′*	*J″*	Wavenumber cm⁻¹	Uncertainty* cm⁻¹
1232476.21	0.12	1	0	41.110981	0.000004
2463428.09	0.19	2	1	82.171116	0.000006
3691334.81	0.25	3	2	123.129676	0.000008
4914682.58	0.51	4	3	163.936165	0.000017
6131968.11	1.10	5	4	204.540439	0.000037
7341702.00	2.00	6	5	244.892818	0.000067
8542412.1	3.21	7	6	284.944197	0.000107
9732646.8	4.72	8	7	324.646153	0.000157
10910978.2	6.51	9	8	363.951056	0.000217
12076004.8	8.55	10	9	402.81216	0.000285
13226355.2	10.81	11	10	441.18372	0.000361
14360689.8	13.25	12	11	479.02105	0.00044
15477704.4	15.86	13	12	516.28065	0.00053
16576131.8	18.61	14	13	552.92024	0.00062
17654744.4	21.48	15	14	588.89888	0.00072
18712356.5	24.44	16	15	624.17703	0.00082
19747825.6	27.43	17	16	658.71656	0.00092
20760054.3	30.32	18	17	692.4809	0.00101
21747991.7	32.91	19	18	725.4349	0.00110
22710634.7	34.94	20	19	757.5452	0.00117
23647028.7	36.08	21	20	788.7800	0.00120
24556268.8	35.93	22	21	819.1090	0.00120
25437499.9	34.12	23	22	848.5037	0.00114

Frequency MHz	Uncertainty* MHz	*J′*	*J″*	Wavenumber cm⁻¹	Uncertainty* cm⁻¹
26289917.4	30.32	24	23	876.9373	0.00101
27112767.2	24.41	25	24	904.38457	0.00081
27905345.6	16.88	26	25	930.82214	0.00056
28666999.3	10.80	27	26	956.22817	0.00036
29397124.8	14.65	28	27	980.58253	0.00049
30095168.2	24.62	29	28	1003.86676	0.00082
30760624.2	33.36	30	29	1026.0640	0.00111
31393035.7	36.17	31	30	1047.1590	0.00121

* The uncertainty given is twice the standard error.

Frequencies and Wavenumbers for the Rotational Lines of $H^{35}Cl$

Frequency MHz	Uncertainty* MHz	*J′*	*J″*	Wavemumber cm⁻¹	Uncertainty* cm⁻¹
1876226.517	0.065	3	2	62.584180	0.000002
2499864.439	0.066	4	3	83.386502	0.000002
3121986.563	0.064	5	4	104.138262	0.000002
3742216.601	0.076	6	5	124.826909	0.000003
4360180.042	0.098	7	6	145.439951	0.000003
4975504.51	0.11	8	7	165.964966	0.000004
5587820.10	0.12	9	8	186.389615	0.000004
6196759.76	0.22	10	9	206.701656	0.000007
6801959.63	0.50	11	10	226.888951	0.000017
7403059.41	1.02	12	11	246.939481	0.000034
7999702.7	1.8	13	12	266.841359	0.000062
8591537.3	3.1	14	13	286.582837	0.000103
9178215.8	4.8	15	14	306.152324	0.000161

* The uncertainty given is twice the standard error.

Frequencies and Wavenumbers for the Rotational Lines of $H^{37}Cl$

Frequency MHz	Uncertainty* MHz	*J′*	*J″*	Wavenumber cm⁻¹	Uncertainty* cm⁻¹
1873410.72	0.05	3	2	62.490255	0.000002
2496115.33	0.05	4	3	83.261445	0.000002
3117308.69	0.05	5	4	103.982225	0.000002
3736615.64	0.06	6	5	124.640082	0.000002
4353662.84	0.08	7	6	145.222561	0.000003
4968079.04	0.09	8	7	165.717279	0.000003
5579495.53	0.10	9	8	186.111938	0.000003
6187546.42	0.19	10	9	206.394332	0.000006
6791062.04	0.15	11	10	226.552365	0.000015
7392104.3	0.9	12	11	246.574057	0.000030
7987896.9	1.6	13	12	266.447561	0.000054
8578896.1	2.7	14	13	286.161170	0.000089

* The uncertainty given is twice the standard error.

Section 11
Nuclear and Particle Physics

SUMMARY TABLES OF PARTICLE PROPERTIES

July 1994

Particle Data Group

M. Aguilar-Benitez, R.M. Barnett, C. Caso, G. Conforto, R.L. Crawford, S. Eidelman, C. Grab, D.E. Groom, A. Gurtu, K.G. Hayes, J.J. Hernandez, K. Hikasa, G. Höhler, S. Kawabata, D.M. Manley, A. Manohar, L. Montanet, R.J. Morrison, H. Murayama, K. Olive, F.C. Porter, M. Roos, R.H. Schindler, R.E. Shrock, J. Stone, N.A. Törnqvist, T.G. Trippe, C.G. Wohl, and R.L. Workman
Technical Associates: B. Armstrong, K. Gieselmann, P. Lantero, G.S. Wagman

Reprinted from Physical Review **D50**, (August 1994)
(Approximate closing date for data: January 1, 1994)

GAUGE AND HIGGS BOSONS

γ

$I(J^{PC}) = 0,1(1^{--})$

Mass $m < 3 \times 10^{-27}$ eV
Charge $q < 2 \times 10^{-32}$ e
Mean life τ = Stable

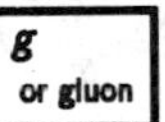

g or gluon

$I(J^P) = 0(1^-)$

Mass $m = 0$ [a]
SU(3) color octet

W

$J = 1$

Charge = ± 1 e
Mass $m = 80.22 \pm 0.26$ GeV
$m_Z - m_W = 10.96 \pm 0.26$ GeV
$m_{W^+} - m_{W^-} = -0.2 \pm 0.6$ GeV
Full width $\Gamma = 2.08 \pm 0.07$ GeV

W^- modes are charge conjugates of the modes below.

W^+ DECAY MODES		Fraction (Γ_i/Γ)	Confidence level	p (MeV/c)
$e^+\nu$		(10.8 ± 0.4) %		40100
$\mu^+\nu$		(10.6 ± 0.7) %		40100
$\tau^+\nu$		(10.8 ± 1.0) %		40100
$\ell^+\nu$	[b]	(10.7 ± 0.5) %		40100
hadrons		(67.8 ± 1.5) %		–
$\pi^+\gamma$		$< 5 \times 10^{-4}$	95%	40110

Z

$J = 1$

Charge = 0
Mass $m = 91.187 \pm 0.007$ GeV [c]
Full width $\Gamma = 2.490 \pm 0.007$ GeV
$\Gamma(\ell^+\ell^-) = 83.84 \pm 0.27$ MeV [b]
$\Gamma(\text{invisible}) = 498.2 \pm 4.2$ MeV [d]
$\Gamma(\text{hadrons}) = 1740.7 \pm 5.9$ MeV
$\Gamma(\mu^+\mu^-)/\Gamma(e^+e^-) = 1.000 \pm 0.005$
$\Gamma(\tau^+\tau^-)/\Gamma(e^+e^-) = 0.998 \pm 0.005$ [e]
$g_V^\ell = -0.0377 \pm 0.0016$
$g_A^\ell = -0.5008 \pm 0.0008$

Asymmetry parameters

$A_e = 0.161 \pm 0.012$ [f] (S = 1.7)
$A_\tau = 0.141 \pm 0.021$ [f] (S = 1.2)

Charge asymmetry at Z pole

$A_{FB}^{(0,\ell)} = (1.59 \pm 0.18) \times 10^{-2}$
$A_{FB}^{(0,c)} = (5.8 \pm 2.2) \times 10^{-2}$
$A_{FB}^{(0,b)} = (10.7 \pm 1.3) \times 10^{-2}$

Z DECAY MODES			Fraction (Γ_i/Γ)	Confidence level	p (MeV/c)
e^+e^-			(3.366 ± 0.008) %		45600
$\mu^+\mu^-$			(3.367 ± 0.013) %		45600
$\tau^+\tau^-$			(3.360 ± 0.015) %		45600
$\ell^+\ell^-$		[b]	(3.367 ± 0.006) %		45600
invisible			(20.01 ± 0.16) %		45600
hadrons			(69.90 ± 0.15) %		–
$(u\overline{u}+c\overline{c})/2$			(9.7 ± 1.8) %		–
$(d\overline{d}+s\overline{s}+b\overline{b})/3$			(16.8 ± 1.2) %		–
$c\overline{c}$			(11.9 ± 1.4) %		–
$b\overline{b}$			(15.45 ± 0.21) %		–
$\pi^0\gamma$			$< 5.5 \times 10^{-5}$	95%	45600
$\eta\gamma$			$< 5.1 \times 10^{-5}$	95%	45600
$\omega\gamma$			$< 6.5 \times 10^{-4}$	95%	45600
$\eta'(958)\gamma$			$< 4.2 \times 10^{-5}$	95%	45600
$\gamma\gamma$			$< 5.5 \times 10^{-5}$	95%	45600
$\gamma\gamma\gamma$			$< 1.7 \times 10^{-5}$	95%	45600
$\pi^\pm W^\mp$		[g]	$< 7 \times 10^{-5}$	95%	10300
$\rho^\pm W^\mp$		[g]	$< 8.3 \times 10^{-5}$	95%	10300
K^0X			(61.5 ± 0.6) %		–
$K^*(892)^+X$			(51 ± 5) %		–
ΛX			(20.9 ± 0.6) %		–
Ξ^-X			(1.42 ± 0.14) %		–
$\Sigma(1385)^\pm X$			(2.6 ± 0.4) %		–
$\Xi(1530)^0X$			$(4.4\pm1.0) \times 10^{-3}$		–
Ω^-X			$(3.5\pm1.0) \times 10^{-3}$		–
$J/\psi(1S)X$			$(3.8\pm0.5) \times 10^{-3}$		–
$\chi_{c1}(1P)X$			$(7.5\pm3.0) \times 10^{-3}$		–
$(D^0/\overline{D}^0)X$			(28 ± 4) %		–
$D^\pm X$			(13.9 ± 2.1) %		–
$D^*(2010)^\pm X$		[g]	(12.5 ± 1.3) %		–
B_s^0X			seen		–
anomalous γ+ hadrons		[h]	$< 3.2 \times 10^{-3}$	95%	–
$e^+e^-\gamma$		[h]	$< 5.2 \times 10^{-4}$	95%	45600
$\mu^+\mu^-\gamma$		[h]	$< 5.6 \times 10^{-4}$	95%	45600
$\tau^+\tau^-\gamma$		[h]	$< 7.3 \times 10^{-4}$	95%	45600
$\ell^+\ell^-\gamma\gamma$		[i]	$< 6.8 \times 10^{-6}$	95%	45600
$q\overline{q}\gamma\gamma$		[i]	$< 5.5 \times 10^{-6}$	95%	–
$\nu\overline{\nu}\gamma\gamma$		[i]	$< 3.1 \times 10^{-6}$	95%	45600
$e^\pm\mu^\mp$	LF	[g]	$< 6 \times 10^{-6}$	95%	45600
$e^\pm\tau^\mp$	LF	[g]	$< 1.3 \times 10^{-5}$	95%	45600
$\mu^\pm\tau^\mp$	LF	[g]	$< 1.9 \times 10^{-5}$	95%	45600

Searches for Higgs Bosons — H^0 and $H^\pm$

H^0 Mass $m > 58.4$ GeV, CL = 95%

H_1^0 in Supersymmetric Models ($m_{H_1^0} < m_{H_2^0}$) [j]

Mass $m > 44$ GeV, CL = 95% for $\tan\beta > 1$

A^0 Pseudoscalar Higgs Boson in Supersymmetric Models [i]

Mass $m > 22$ GeV, CL = 95% for $50 > \tan\beta > 1$

$H^\pm$ Mass $m > 41.7$ GeV, CL = 95%

See the Full Listings for a Note giving details of Higgs Bosons.

Searches for Heavy Bosons Other Than Higgs Bosons

Additional W Bosons

W_R — right-handed W
Mass $m > 406$ GeV, CL = 90%
(assuming light right-handed neutrino)

W' with standard couplings decaying to $e\nu$, $\mu\nu$
Mass $m > 520$ GeV, CL = 95%

Additional Z Bosons

Z'_{SM} with standard couplings
Mass $m >$ 412 GeV, CL = 95% ($p\overline{p}$ direct search)
Mass $m >$ 779 GeV, CL = 95% (electroweak fit)

Z_{LR} of $SU(2)_L \times SU(2)_R \times U(1)$
(with $g_L = g_R$)
Mass $m >$ 310 GeV, CL = 95% ($p\overline{p}$ direct search)
Mass $m >$ 389 GeV, CL = 95% (electroweak fit)

Z_χ of $SO(10) \rightarrow SU(5) \times U(1)_\chi$
(coupling constant derived from G.U.T.)
Mass $m >$ 340 GeV, CL = 95% ($p\overline{p}$ direct search)
Mass $m >$ 321 GeV, CL = 95% (electroweak fit)

Z_ψ of $E_6 \rightarrow SO(10) \times U(1)_\psi$
(coupling constant derived from G.U.T.)
Mass $m >$ 320 GeV, CL = 95% ($p\overline{p}$ direct search)
Mass $m >$ 160 GeV, CL = 95% (electroweak fit)

Z_η of $E_6 \rightarrow SU(3) \times SU(2) \times U(1) \times U(1)_\eta$
(coupling constant derived from G.U.T.;
charges are $Q_\eta = \sqrt{3/8}Q_\chi - \sqrt{5/8}Q_\psi$)
Mass $m >$ 340 GeV, CL = 95% ($p\overline{p}$ direct search)
Mass $m >$ 182 GeV, CL = 95% (electroweak fit)

Scalar Leptoquarks

Mass $m >$ 120 GeV, CL = 95% (1st generation, pair prod.)
Mass $m >$ 181 GeV, CL = 95% (1st gener., single prod.)
Mass $m >$ 44.5 GeV, CL = 95% (2nd gener., pair prod.)
Mass $m >$ 73 GeV, CL = 95% (2nd gener., single prod.)
Mass $m >$ 45 GeV, CL = 95% (3rd gener., pair prod.)
(last four limits are for charge $-1/3$, weak isoscalar)

Searches for Axions (A^0) and Other Very Light Bosons

The standard Peccei-Quinn axion is ruled out. Variants with reduced couplings or much smaller masses are constrained by various data. The Full Listings in the full *Review* contain a Note discussing axion searches.

The best limit for the half-life of neutrinoless double beta decay with Majoron emission is $> 7.2 \times 10^{24}$ years (CL = 90%).

NOTES

In this Summary Table:

When a quantity has "(S = ...)" to its right, the error on the quantity has been enlarged by the "scale factor" S, defined as $S = \sqrt{\chi^2/(N-1)}$, where N is the number of measurements used in calculating the quantity. We do this when $S > 1$, which often indicates that the measurements are inconsistent. When $S > 1.25$, we also show in the Full Listings an ideogram of the measurements. For more about S, see the Introduction.

A decay momentum p is given for each decay mode. For a 2-body decay, p is the momentum of each decay product in the rest frame of the decaying particle. For a 3-or-more-body decay, p is the largest momentum any of the products can have in this frame.

[*a*] Theoretical value. A mass as large as a few MeV may not be precluded.

[*b*] ℓ indicates each type of lepton (e, μ, and τ), not sum over them.

[*c*] The Z-boson mass listed here corresponds to a Breit-Wigner resonance parameter. It lies approximately 34 MeV above the real part of the position of the pole (in the energy plane) in the Z-boson propagator.

[*d*] This partial width takes into account Z decays into $\nu\overline{\nu}$ and any other possible undetected modes.

[*e*] This ratio has not been corrected for the τ mass.

[*f*] Here $A \equiv 2g_V g_A/(g_V^2+g_A^2)$.

[*g*] The value is for the sum of the charge states indicated.

[*h*] See the Z Full Listings for the γ energy range used in this measurement.

[*i*] For $m_{\gamma\gamma} = (60 \pm 5)$ GeV.

[*j*] The limits assume no invisible decays.

LEPTONS

Neutrinos

See the Full Listings for a Note giving details of neutrinos, masses, mixing, and the status of experimental searches.

ν_e

$J = \frac{1}{2}$

Mass m: The formal upper limit, as obtained from the m^2 average (see the Full Listings), is 5.1 eV at the 95% CL. Caution is urged in interpreting this result, since the m^2 average is positive with only a 3.5% probability. If the weighted average m^2 were forced to zero, the limit would increase to 7.0 eV.

Mean life/mass, $\tau/m_{\nu_e} > 300$ s/eV, CL = 90%

Magnetic moment $\mu < 1.08 \times 10^{-9}\ \mu_B$, CL = 90%

ν_μ

$J = \frac{1}{2}$

Mass $m < 0.27$ MeV, CL = 90%

Mean life/mass, $\tau/m_{\nu_\mu} > 15.4$ s/eV, CL = 90%

Magnetic moment $\mu < 7.4 \times 10^{-10}\ \mu_B$, CL = 90%

ν_τ

$J = \frac{1}{2}$

Mass $m < 31$ MeV, CL = 95%

Magnetic moment $\mu < 5.4 \times 10^{-7}\ \mu_B$, CL = 90%

e

$J = \frac{1}{2}$

Mass $m = 0.51099906 \pm 0.00000015$ MeV [a]

$= (5.48579903 \pm 0.00000013) \times 10^{-4}$ u

$(m_{e^+} - m_{e^-})/m < 4 \times 10^{-8}$, CL = 90%

$|q_{e^+} + q_{e^-}|/e < 4 \times 10^{-8}$

Magnetic moment $\mu = 1.001159652193 \pm 0.000000000010\ \mu_B$

$(g_{e^+} - g_{e^-}) / g_{\text{average}} = (-0.5 \pm 2.1) \times 10^{-12}$

Electric dipole moment $d = (-0.3 \pm 0.8) \times 10^{-26}$ $e$cm

Mean life $\tau > 2.7 \times 10^{23}$ yr, CL = 68% [b]

μ

$J = \frac{1}{2}$

Mass $m = 105.658389 \pm 0.000034$ MeV [a]

$= 0.113428913 \pm 0.000000017$ u

Mean life $\tau = (2.19703 \pm 0.00004) \times 10^{-6}$ s

$\tau_{\mu^+}/\tau_{\mu^-} = 1.00002 \pm 0.00008$

$c\tau = 658.654$ m

Magnetic moment $\mu = 1.001165923 \pm 0.000000008\ e\hbar/2m_\mu$

$(g_{\mu^+} - g_{\mu^-}) / g_{\text{average}} = (-2.6 \pm 1.6) \times 10^{-8}$

Electric dipole moment $d = (3.7 \pm 3.4) \times 10^{-19}$ $e$cm

Decay parameters [c]

$\rho = 0.7518 \pm 0.0026$

$\eta = -0.007 \pm 0.013$

$\delta = 0.749 \pm 0.004$

$\xi P_\mu = 1.003 \pm 0.008$ [d]

$\xi P_\mu \delta/\rho > 0.99682$, CL = 90% [d]

$\xi' = 1.00 \pm 0.04$

$\xi'' = 0.7 \pm 0.4$

$\alpha/A = (0 \pm 4) \times 10^{-3}$

$\alpha'/A = (0 \pm 4) \times 10^{-3}$

$\beta/A = (4 \pm 6) \times 10^{-3}$

$\beta'/A = (2 \pm 6) \times 10^{-3}$

$\overline{\eta} = 0.02 \pm 0.08$

μ^+ modes are charge conjugates of the modes below.

μ^- DECAY MODES		Fraction (Γ_i/Γ)	Confidence level	p (MeV/c)
$e^-\overline{\nu}_e\nu_\mu$		$\approx 100\%$		53
$e^-\overline{\nu}_e\nu_\mu\gamma$		[e] (1.4 ± 0.4) %		53
$e^-\overline{\nu}_e\nu_\mu e^+e^-$		[f] $(3.4\pm0.4) \times 10^{-5}$		53
Lepton Family number (*LF*) violating modes				
$e^-\nu_e\overline{\nu}_\mu$	*LF*	[g] < 1.2 %	90%	53
$e^-\gamma$	*LF*	$< 4.9 \times 10^{-11}$	90%	53
$e^-e^+e^-$	*LF*	$< 1.0 \times 10^{-12}$	90%	53
$e^-2\gamma$	*LF*	$< 7.2 \times 10^{-11}$	90%	53

τ

$J = \frac{1}{2}$

Mass $m = 1777.1^{+0.4}_{-0.5}$ MeV

Mean life $\tau = (295.6 \pm 3.1) \times 10^{-15}$ s

$c\tau = 88.6\ \mu$m

Electric dipole moment $d < 5 \times 10^{-17}$ $e$cm, CL = 95%

Weak dipole moment $< 3.7 \times 10^{-17}$ $e$cm, CL = 95%

Decay parameters

See the τ Full Listings for a note concerning τ-decay parameters.

$\rho^\tau(e \text{ or } \mu) = 0.74 \pm 0.04$

$\rho^\tau(e) = 0.72 \pm 0.04$

$\rho^\tau(\mu) = 0.76 \pm 0.05$

$\xi^\tau(e \text{ or } \mu) = 0.90 \pm 0.18$

W-τ couplings $2g_A g_V/(g_A^2 + g_V^2) = 1.25^{+0.27}_{-0.24}$

τ^+ modes are charge conjugates of the modes below. "$h^\pm$" stands for $\pi^\pm$ or $K^\pm$. "ℓ" stands for e or μ. "Neutral" means neutral hadron whose decay products include γ's and/or π^0's.

τ^- DECAY MODES	Fraction (Γ_i/Γ)	Scale factor/ Confidence level	p (MeV/c)
Modes with one charged particle			
particle$^-$ ≥ 0 neutrals ν_τ ("1-prong")	(85.49 ± 0.24) %	S=1.6	–
$\mu^-\overline{\nu}_\mu\nu_\tau$	(17.65 ± 0.24) %	S=1.1	885
$\mu^-\overline{\nu}_\mu\nu_\tau\gamma$ ($E_\gamma > 37$ MeV)	$(2.3 \pm1.1) \times 10^{-3}$		–
$e^-\overline{\nu}_e\nu_\tau$	(18.01 ± 0.18) %	S=1.1	889
$h^- \geq 0$ neutrals ν_τ	(49.83 ± 0.35) %	S=1.3	–
$h^-\nu_\tau$	(12.88 ± 0.34) %	S=1.2	–
$\pi^-\nu_\tau$	(11.7 ± 0.4) %	S=1.3	883
$K^- \geq 0$ neutrals ν_τ	(1.60 ± 0.24) %		–
$K^-\nu_\tau$	$(6.7 \pm2.3) \times 10^{-3}$	S=1.3	820
$K^- \geq 1$ neutrals ν_τ	$(1.2\ ^{+0.5}_{-0.6})$ %		–
$h^- \geq 1$ neutrals ν_τ	(36.9 ± 0.4) %	S=1.3	–
$h^-\pi^0\nu_\tau$	(25.7 ± 0.4) %	S=1.7	–
$\pi^-\pi^0\nu_\tau$	(25.2 ± 0.4) %	S=1.7	878
$h^- \geq 2\pi^0\nu_\tau$	(11.2 ± 0.4) %	S=1.5	–
$h^-2\pi^0\nu_\tau$	(9.6 ± 0.4) %	S=1.5	–
$h^- \geq 3\pi^0\nu_\tau$	(1.48 ± 0.26) %	S=1.7	–
$h^-3\pi^0\nu_\tau$	(1.28 ± 0.24) %	S=1.7	–
$h^-4\pi^0\nu_\tau$	$(1.9\ ^{+1.1}_{-1.0}) \times 10^{-3}$	S=1.6	–
Modes with three charged particles			
$2h^-h^+ \geq 0$ neutrals ν_τ ("3-prong")	(14.38 ± 0.24) %	S=1.5	–
$h^-h^-h^+\nu_\tau$	(8.42 ± 0.31) %	S=1.3	–
$h^-h^-h^+ \geq 1$ neutrals ν_τ	(5.63 ± 0.30) %	S=1.2	–
$h^-h^-h^+2\pi^0\nu_\tau$	$(4.9 \pm0.5) \times 10^{-3}$		–
$\omega\pi^- \geq 0$ neutrals ν_τ	(1.6 ± 0.4) %		–
$\omega\pi^-\nu_\tau$	(1.6 ± 0.5) %		708
$h^-\omega\pi^0\nu_\tau$	$(4.0 \pm0.6) \times 10^{-3}$		–
$K^-h^+h^- \geq 0$ neutrals ν_τ	$< 6 \times 10^{-3}$	CL=90%	–
$K^-\pi^+\pi^- \geq 0$ neutrals ν_τ	$(2.2\ ^{+1.6}_{-1.3}) \times 10^{-3}$		–
$K^-K^+\pi^-\nu_\tau$	$(2.2\ ^{+1.7}_{-1.1}) \times 10^{-3}$		685
Modes with five charged particles			
$3h^-2h^+ \geq 0$ neutrals ν_τ ("5-prong")	$(1.25\pm0.24) \times 10^{-3}$		–
$3h^-2h^+\nu_\tau$	$(5.6 \pm1.6) \times 10^{-4}$		–
$3h^-2h^+\pi^0\nu_\tau$	$(5.1 \pm2.2) \times 10^{-4}$		–

Mode	Violation	Fraction	Confidence level	p
Miscellaneous other allowed modes				
$4h^- 3h^+ \geq 0$ neutrals ν_τ ("7-prong")		$< 1.9 \times 10^{-4}$	CL=90%	–
$K^*(892)^- \geq 0$ neutrals ν_τ		(1.43 ± 0.17) %		–
$K^*(892)^- \nu_\tau$		(1.45 ± 0.18) %		665
$K^*(892)^0 K^- \geq 0$ neutrals ν_τ		$(3.2 \pm 1.4) \times 10^{-3}$		–
$\overline{K}^*(892)^0 \pi^- \geq 0$ neutrals ν_τ		$(3.8 \pm 1.7) \times 10^{-3}$		–
$K^0 h^- \geq 0$ neutrals ν_τ		(1.30 ± 0.30) %		–
$K^- K^0 \geq 0$ neutrals ν_τ		$< 8 \times 10^{-3}$	CL=90%	–
$K^0 K^- \nu_\tau$		$< 2.6 \times 10^{-3}$	CL=95%	737
$K^0 K^- \geq 1$ neutrals ν_τ		$< 2.6 \times 10^{-3}$	CL=95%	–
$K^0 h^+ h^- h^- \geq 0$ neutrals ν_τ		$< 1.7 \times 10^{-3}$	CL=95%	–
$K_2^*(1430)^- \nu_\tau$		$< 3 \times 10^{-3}$	CL=95%	314
$\eta \pi^- \geq 0$ neutrals ν_τ		< 1.3 %	CL=95%	–
$\eta \pi^- \nu_\tau$		$< 3.4 \times 10^{-4}$	CL=95%	798
$\eta \pi^- \pi^0 \nu_\tau$		$(1.70 \pm 0.28) \times 10^{-3}$		778
$\eta \pi^- \pi^0 \pi^0 \nu_\tau$		$< 4.3 \times 10^{-4}$	CL=95%	746
$\eta K^- \nu_\tau$		$< 4.7 \times 10^{-4}$	CL=95%	720
$\eta \pi^+ \pi^- \pi^- \geq 0$ neutrals ν_τ		$< 3 \times 10^{-3}$	CL=90%	–
$\eta \eta \pi^- \geq 0$ neutrals ν_τ		$< 5 \times 10^{-3}$	CL=90%	–
$\eta \eta \pi^- \nu_\tau$		$< 1.1 \times 10^{-4}$	CL=95%	637
$\eta \eta \pi^- \pi^0 \nu_\tau$		$< 2.0 \times 10^{-4}$	CL=95%	559

Lepton Family number (*LF*), Lepton number (*L*), or Baryon number (*B*) violating modes
(In the modes below, ℓ means a sum over e and μ modes)

L means lepton number violation (*e.g.* $\tau^- \to e^+ \pi^- \pi^-$). Following common usage, *LF* means lepton family violation *and not* lepton number violation (*e.g.* $\tau^- \to e^- \pi^+ \pi^-$).

Mode	Violation	Fraction	Confidence level	p
$e^- \gamma$	*LF*	$< 1.2 \times 10^{-4}$	CL=90%	889
$\mu^- \gamma$	*LF*	$< 4.2 \times 10^{-6}$	CL=90%	885
$e^- \pi^0$	*LF*	$< 1.4 \times 10^{-4}$	CL=90%	883
$\mu^- \pi^0$	*LF*	$< 4.4 \times 10^{-5}$	CL=90%	880
$e^- K^0$	*LF*	$< 1.3 \times 10^{-3}$	CL=90%	819
$\mu^- K^0$	*LF*	$< 1.0 \times 10^{-3}$	CL=90%	815
$e^- \eta$	*LF*	$< 6.3 \times 10^{-5}$	CL=90%	804
$\mu^- \eta$	*LF*	$< 7.3 \times 10^{-5}$	CL=90%	800
$e^- \rho^0$	*LF*	$< 1.9 \times 10^{-5}$	CL=90%	723
$\mu^- \rho^0$	*LF*	$< 2.9 \times 10^{-5}$	CL=90%	718
$e^- K^*(892)^0$	*LF*	$< 3.8 \times 10^{-5}$	CL=90%	665
$\mu^- K^*(892)^0$	*LF*	$< 4.5 \times 10^{-5}$	CL=90%	660
$\pi^- \gamma$	*L*	$< 2.8 \times 10^{-4}$	CL=90%	883
$\pi^- \pi^0$	*L*	$< 3.7 \times 10^{-4}$	CL=90%	878
$\ell^- \ell^- \ell^+$	*LF* [h]	$< 3.4 \times 10^{-5}$	CL=90%	–
$e^- e^+ e^-$	*LF*	$< 1.3 \times 10^{-5}$	CL=90%	889
$(e \mu \mu)^-$	*LF*	$< 2.7 \times 10^{-5}$	CL=90%	882
$e^- \mu^+ \mu^-$	*LF*	$< 1.9 \times 10^{-5}$	CL=90%	882
$e^+ \mu^- \mu^-$	*LF*	$< 1.6 \times 10^{-5}$	CL=90%	882
$(\mu e e)^-$	*LF*	$< 2.7 \times 10^{-5}$	CL=90%	885
$\mu^- e^+ e^-$	*LF*	$< 1.4 \times 10^{-5}$	CL=90%	885
$\mu^+ e^- e^-$	*LF*	$< 1.4 \times 10^{-5}$	CL=90%	885
$\mu^- \mu^+ \mu^-$	*LF*	$< 1.7 \times 10^{-5}$	CL=90%	873
$\ell^\pm \pi^\mp \pi^-$	*LF,L* [h,i]	$< 6.3 \times 10^{-5}$	CL=90%	–
$e^\mp \pi^\pm \pi^-$	*LF,L* [i]	$< 6.0 \times 10^{-5}$	CL=90%	877
$e^- \pi^+ \pi^-$	*LF*	$< 2.7 \times 10^{-5}$	CL=90%	877
$e^+ \pi^- \pi^-$	*L*	$< 1.7 \times 10^{-5}$	CL=90%	877
$\mu^\mp \pi^\pm \pi^-$	*LF,L* [i]	$< 3.9 \times 10^{-5}$	CL=90%	866
$\mu^- \pi^+ \pi^-$	*LF*	$< 3.6 \times 10^{-5}$	CL=90%	866
$\mu^+ \pi^- \pi^-$	*L*	$< 3.9 \times 10^{-5}$	CL=90%	866
$\ell^\pm \pi^\mp K^-$	*LF,L* [h,i]	$< 1.2 \times 10^{-4}$	CL=90%	–
$(e \pi K)^-$, all charged	*LF,L*	$< 7.7 \times 10^{-5}$	CL=90%	814
$e^- \pi^\pm K^\mp$	*LF* [i]	$< 5.8 \times 10^{-5}$	CL=90%	814
$e^- \pi^+ K^-$	*LF*	$< 2.9 \times 10^{-5}$	CL=90%	814
$e^- \pi^- K^+$	*LF*	$< 5.8 \times 10^{-5}$	CL=90%	814
$e^+ \pi^- K^-$	*L*	$< 2.0 \times 10^{-5}$	CL=90%	814
$(\mu \pi K)^-$, all charged	*LF,L*	$< 7.7 \times 10^{-5}$	CL=90%	800
$\mu^- \pi^\pm K^\mp$	*LF* [i]	$< 7.7 \times 10^{-5}$	CL=90%	800
$\mu^- \pi^+ K^-$	*LF*	$< 7.7 \times 10^{-5}$	CL=90%	800
$\mu^- \pi^- K^+$	*LF*	$< 7.7 \times 10^{-5}$	CL=90%	800
$\mu^+ \pi^- K^-$	*L*	$< 4.0 \times 10^{-5}$	CL=90%	800
$\overline{p} \gamma$	*L,B*	$< 2.9 \times 10^{-4}$	CL=90%	641
$\overline{p} \pi^0$	*L,B*	$< 6.6 \times 10^{-4}$	CL=90%	632
$\overline{p} \eta$	*L,B*	$< 1.30 \times 10^{-3}$	CL=90%	476
e^- light spinless boson	*LF*	$< 3.2 \times 10^{-3}$	CL=95%	–
μ^- light spinless boson	*LF*	$< 6 \times 10^{-3}$	CL=95%	–

Number of Light Neutrino Types

(including ν_e, ν_μ, and ν_τ)

Number $N = 2.983 \pm 0.025$ (Standard Model fits to Z data)
Number $N = 2.97 \pm 0.17$ (Direct measurement of invisible Z width)

Heavy Lepton Searches

$L^\pm$ – charged lepton

Mass $m >$ 44.3 GeV, CL = 95% $m_\nu \approx 0$

$L^\pm$ – stable charged heavy lepton

Mass $m >$ 42.8 GeV, CL = 95%

L^0 – stable neutral heavy lepton

Mass $m >$ 45.0 GeV, CL = 95% (Dirac)
Mass $m >$ 39.5 GeV, CL = 95% (Majorana)

Neutral heavy lepton

Mass $m >$ 19.6 GeV, CL = 95% (all $|U_{\ell j}|^2$) (Dirac)
Mass $m >$ 45.7 GeV or $m < 25$, CL = 95% ($|U_{\ell j}|^2 > 10^{-13}$ (Dirac)

Searches for Massive Neutrinos and Lepton Mixing

For excited leptons, see Compositeness Limits below.

See the Full Listings for a Note giving details of neutrinos, masses, mixing, and the status of experimental searches.

No direct, uncontested evidence for massive neutrinos or lepton mixing has been obtained. Sample limits are:

ν oscillation: $\overline{\nu}_e \not\to \overline{\nu}_e$

$\Delta(m^2) < 0.0083$ eV^2, CL = 90% (if $\sin^2 2\theta = 1$)
$\sin^2 2\theta < 0.14$, CL = 68% (if $\Delta(m^2)$ is large)

ν oscillation: $\nu_\mu \to \nu_e$ (θ = mixing angle)

$\Delta(m^2) < 0.09$ eV^2, CL = 90% (if $\sin^2 2\theta = 1$)
$\sin^2 2\theta < 2.5 \times 10^{-3}$, CL = 90% (if $\Delta(m^2)$ is large)

QUARKS

The *u*-, *d*-, and *s*-quark masses are estimates of so-called "current-quark masses," in a mass-independent subtraction scheme such as $\overline{MS}$ at a scale $\mu \approx 1$ GeV. The *c*- and *b*-quark masses are estimated from charmonium, bottomonium, *D*, and *B* masses. They are the "running" masses in the $\overline{MS}$ scheme. These can be different from the heavy quark masses obtained in potential models.

u

$I(J^P) = \frac{1}{2}(\frac{1}{2}^+)$

Mass m = 2 to 8 MeV [j] Charge = $\frac{2}{3}$ e $I_z = +\frac{1}{2}$
m_u/m_d = 0.25 to 0.70

d

$I(J^P) = \frac{1}{2}(\frac{1}{2}^+)$

Mass m = 5 to 15 MeV [j] Charge = $-\frac{1}{3}$ e $I_z = -\frac{1}{2}$
m_s/m_d = 17 to 25

s

$I(J^P) = 0(\frac{1}{2}^+)$

Mass m = 100 to 300 MeV [j] Charge = $-\frac{1}{3}$ e Strangeness = −1
$(m_s - (m_u + m_d)/2)/(m_d - m_u)$ = 34 to 51

c

$I(J^P) = 0(\frac{1}{2}^+)$

Mass m = 1.0 to 1.6 GeV Charge = $\frac{2}{3}$ e Charm = +1

b

$I(J^P) = 0(\frac{1}{2}^+)$

Mass m = 4.1 to 4.5 GeV Charge = $-\frac{1}{3}$ e Bottom = −1

Searches for *t* Quark

$I(J^P) = 0(\frac{1}{2}^+)$

Charge = $\frac{2}{3}$ e Top = +1

Mass $m >$ 62 GeV, CL = 95% (all decays)
Mass $m >$ 131 GeV, CL = 95% (assumes $t \rightarrow W b$ decay)
Mass $m = 174 \pm 10^{+13}_{-12}$ GeV (top candidate events)
Mass $m = 169^{+16}_{-18}{}^{+17}_{-20}$ GeV (Standard Model electroweak fit)

The first result is from a CDF $\Gamma(W)$ measurement; the second is from a DØ direct search; the third is from a CDF observation of top candidate events. CDF observes a 2.8σ effect which is not sufficient to firmly establish the existence of top but which, if interpreted as top, yields the third result. The fourth result is from a Standard Model electroweak fit to Z, W, and νN data not including direct m_t measurements. The central value assumes m_H = 300 GeV while the second upper (lower) error corresponds to m_H = 1000 (60) GeV.

Searches for *b′* (4th Generation) Quark

Mass $m >$ 85 GeV, CL = 95% ($p\overline{p}$, charged current decays)
Mass $m >$ 46.0 GeV, CL = 95% (e^+e^-, all decays)

NOTES

In this Summary Table:

When a quantity has "(S = ...)" to its right, the error on the quantity has been enlarged by the "scale factor" S, defined as $S = \sqrt{\chi^2/(N-1)}$, where N is the number of measurements used in calculating the quantity. We do this when S > 1, which often indicates that the measurements are inconsistent. When S > 1.25, we also show in the Full Listings an ideogram of the measurements. For more about S, see the Introduction.

A decay momentum p is given for each decay mode. For a 2-body decay, p is the momentum of each decay product in the rest frame of the decaying particle. For a 3-or-more-body decay, p is the largest momentum any of the products can have in this frame.

[a] The masses of the e and μ are most precisely known in u (unified atomic mass units). The conversion factor to MeV, 1 u = 931.49432(28) MeV, is less well known than are the masses in u.

[b] This is the best "electron disappearance" limit. The best limit for the mode $e^- \rightarrow \nu\gamma$ is > 2.35×10^{25} yr (CL=68%).

[c] See the "Note on Muon Decay Parameters" in the μ Full Listings for definitions and details.

[d] P_μ is the longitudinal polarization of the muon from pion decay. In standard $V-A$ theory, $P_\mu = 1$ and $\rho = \delta = 3/4$.

[e] This only includes events with the γ energy > 10 MeV. Since the $e^-\overline{\nu}_e\nu_\mu$ and $e^-\overline{\nu}_e\nu_\mu\gamma$ modes cannot be clearly separated, we regard the latter mode as a subset of the former.

[f] See the μ Full Listings for the energy limits used in this measurement.

[g] A test of additive vs. multiplicative lepton family number conservation.

[h] ℓ means a sum over e and μ modes.

[i] The value is for the sum of the charge states indicated.

[j] The ratios m_u/m_d and m_s/m_d are extracted from pion and kaon masses using chiral symmetry. The estimates of u and d masses are not without controversy and remain under active investigation. Within the literature there are even suggestions that the u quark could be essentially massless. The s-quark mass is estimated from SU(3) splittings in hadron masses.

LIGHT UNFLAVORED MESONS
$(S = C = B = 0)$

For $I = 1$ (π, b, ρ, a): $u\overline{d}$, $(u\overline{u}-d\overline{d})/\sqrt{2}$, $d\overline{u}$;
for $I = 0$ (η, η', h, h', ω, ϕ, f, f'): $c_1(u\overline{u} + d\overline{d}) + c_2(s\overline{s})$

$\pi^\pm$

$I^G(J^P) = 1^-(0^-)$

Mass $m = 139.56995 \pm 0.00035$ MeV [a]
Mean life $\tau = (2.6030 \pm 0.0024) \times 10^{-8}$ s
$c\tau = 7.804$ m

$\pi^\pm \to \ell^\pm \nu\gamma$ form factors [b]

$F_V = 0.017 \pm 0.008$
$F_A = 0.0116 \pm 0.0016$ (S = 1.3)
$R = 0.059^{+0.009}_{-0.008}$

π^- modes are charge conjugates of the modes below.

π^+ DECAY MODES	Fraction (Γ_i/Γ)	Confidence level	p (MeV/c)
$\mu^+\nu_\mu$	[c] (99.98770±0.00004) %		30
$\mu^+\nu_\mu\gamma$	[d] (1.24 ±0.25) $\times 10^{-4}$		30
$e^+\nu_e$	[c] (1.230 ±0.004) $\times 10^{-4}$		70
$e^+\nu_e\gamma$	[d] (1.61 ±0.23) $\times 10^{-7}$		70
$e^+\nu_e\pi^0$	(1.025 ±0.034) $\times 10^{-8}$		4
$e^+\nu_e e^+ e^-$	(3.2 ±0.5) $\times 10^{-9}$		70
$e^+\nu_e\nu\overline{\nu}$	< 5 $\times 10^{-6}$	90%	70
Lepton Family number (*LF*) or Lepton number (*L*) violating modes			
$\mu^+\overline{\nu}_e$	*L* [e] < 1.5 $\times 10^{-3}$	90%	30
$\mu^+\nu_e$	*LF* [e] < 8.0 $\times 10^{-3}$	90%	30
$\mu^- e^+ e^+ \nu$	*LF* < 1.6 $\times 10^{-6}$	90%	30

π^0

$I^G(J^{PC}) = 1^-(0^{-+})$

Mass $m = 134.9764 \pm 0.0006$ MeV [a]
$m_{\pi^\pm} - m_{\pi^0} = 4.5936 \pm 0.0005$ MeV
Mean life $\tau = (8.4 \pm 0.6) \times 10^{-17}$ s (S = 3.0)
$c\tau = 25.1$ nm

π^0 DECAY MODES	Fraction (Γ_i/Γ)	Scale factor/ Confidence level	p (MeV/c)
2γ	(98.798±0.032) %	S=1.1	67
$e^+e^-\gamma$	(1.198±0.032) %	S=1.1	67
γ positronium	(1.82 ±0.29) $\times 10^{-9}$		67
$e^+e^+e^-e^-$	(3.14 ±0.30) $\times 10^{-5}$		67
e^+e^-	(7.5 ±2.0) $\times 10^{-8}$		67
4γ	< 2 $\times 10^{-8}$	CL=90%	67
$\nu\overline{\nu}$	[f] < 8.3 $\times 10^{-7}$	CL=90%	67
$\nu_e\overline{\nu}_e$	< 1.7 $\times 10^{-6}$	CL=90%	67
$\nu_\mu\overline{\nu}_\mu$	< 3.1 $\times 10^{-6}$	CL=90%	67
$\nu_\tau\overline{\nu}_\tau$	< 2.1 $\times 10^{-6}$	CL=90%	67
Charge conjugation (*C*) or Lepton Family number (*LF*) violating modes			
3γ	*C* < 3.1 $\times 10^{-8}$	CL=90%	67
$\mu^+e^- + e^-\mu^+$	*LF* < 1.72 $\times 10^{-8}$	CL=90%	26

η

$I^G(J^{PC}) = 0^+(0^{-+})$

Mass $m = 547.45 \pm 0.19$ MeV (S = 1.6)
Full width $\Gamma = 1.20 \pm 0.11$ keV [g] (S = 1.8)

C-nonconserving decay parameters [h]

$\pi^+\pi^-\pi^0$ Left-right asymmetry $= (0.09 \pm 0.17) \times 10^{-2}$
$\pi^+\pi^-\pi^0$ Sextant asymmetry $= (0.18 \pm 0.16) \times 10^{-2}$
$\pi^+\pi^-\pi^0$ Quadrant asymmetry $= (-0.17 \pm 0.17) \times 10^{-2}$
$\pi^+\pi^-\gamma$ Left-right asymmetry $= (0.9 \pm 0.4) \times 10^{-2}$
$\pi^+\pi^-\gamma$ β (*D*-wave) $= 0.05 \pm 0.06$ (S = 1.5)

η DECAY MODES	Fraction (Γ_i/Γ)	Scale factor/ Confidence level	p (MeV/c)
neutral modes	(70.8 ±0.8) %	S=1.2	–
2γ	[g] (38.8 ±0.5) %	S=1.2	274
$3\pi^0$	(31.9 ±0.4) %	S=1.2	180
$\pi^0 2\gamma$	(7.1 ±1.4) $\times 10^{-4}$		258
charged modes	(29.2 ±0.8) %	S=1.2	–
$\pi^+\pi^-\pi^0$	(23.6 ±0.6) %	S=1.2	175
$\pi^+\pi^-\gamma$	(4.88±0.15) %	S=1.2	236
$e^+e^-\gamma$	(5.0 ±1.2) $\times 10^{-3}$		274
$\mu^+\mu^-\gamma$	(3.1 ±0.4) $\times 10^{-4}$		253
e^+e^-	< 3 $\times 10^{-4}$	CL=90%	274
$\mu^+\mu^-$	(5.7 ±0.8) $\times 10^{-6}$		253
$\pi^+\pi^-e^+e^-$	(1.3 $^{+1.3}_{-0.8}$) $\times 10^{-3}$		236
$\pi^+\pi^-2\gamma$	< 2.1 $\times 10^{-3}$		236
$\pi^+\pi^-\pi^0\gamma$	< 6 $\times 10^{-4}$	CL=90%	175
$\pi^0\mu^+\mu^-\gamma$	< 3 $\times 10^{-6}$	CL=90%	211
Charge conjugation (*C*), Parity (*P*), or Charge conjugation × Parity (*CP*) violating modes			
$\pi^+\pi^-$	*P,CP* < 1.5 $\times 10^{-3}$		236
3γ	*C* < 5 $\times 10^{-4}$	CL=95%	274
$\pi^0 e^+e^-$	*C* [i] < 4 $\times 10^{-5}$	CL=90%	258
$\pi^0\mu^+\mu^-$	*C* [i] < 5 $\times 10^{-6}$	CL=90%	211

$\rho(770)$

$I^G(J^{PC}) = 1^+(1^{--})$

Mass $m = 769.9 \pm 0.8$ MeV (S = 1.8)
Full width $\Gamma = 151.2 \pm 1.2$ MeV
$\Gamma_{ee} = 6.77 \pm 0.32$ keV

$\rho(770)$ DECAY MODES	Fraction (Γ_i/Γ)	Scale factor/ Confidence level	p (MeV/c)
$\pi\pi$	~ 100 %		359
$\rho(770)^\pm$ decays			
$\pi^\pm\gamma$	(4.5 ±0.5) $\times 10^{-4}$	S=2.2	372
$\pi^\pm\eta$	< 6 $\times 10^{-3}$	CL=84%	147
$\pi^\pm\pi^+\pi^-\pi^0$	< 2.0 $\times 10^{-3}$	CL=84%	250
$\rho(770)^0$ decays			
$\pi^+\pi^-\gamma$	(9.9 ±1.6) $\times 10^{-3}$		359
$\pi^0\gamma$	(7.9 ±2.0) $\times 10^{-4}$		373
$\eta\gamma$	(3.8 ±0.7) $\times 10^{-4}$		190
$\mu^+\mu^-$	[j] (4.60±0.28) $\times 10^{-5}$		370
e^+e^-	[j] (4.46±0.21) $\times 10^{-5}$		385
$\pi^+\pi^-\pi^0$	< 1.2 $\times 10^{-4}$	CL=90%	320
$\pi^+\pi^-\pi^+\pi^-$	< 2 $\times 10^{-4}$	CL=90%	247
$\pi^+\pi^-\pi^0\pi^0$	< 4 $\times 10^{-5}$	CL=90%	253

$\omega(782)$

$I^G(J^{PC}) = 0^-(1^{--})$

Mass $m = 781.94 \pm 0.12$ MeV (S = 1.5)
Full width $\Gamma = 8.43 \pm 0.10$ MeV
$\Gamma_{ee} = 0.60 \pm 0.02$ keV

$\omega(782)$ DECAY MODES	Fraction (Γ_i/Γ)	Confidence level	p (MeV/c)
$\pi^+\pi^-\pi^0$	(88.8 ±0.7) %		327
$\pi^0\gamma$	(8.5 ±0.5) %		379
$\pi^+\pi^-$	(2.21±0.30) %		365
neutrals (excluding $\pi^0\gamma$)	(5.3 $^{+8.7}_{-3.5}$) $\times 10^{-3}$		–
$\eta\gamma$	(8.3 ±2.1) $\times 10^{-4}$		199
$\pi^0 e^+e^-$	(5.9 ±1.9) $\times 10^{-4}$		379
$\pi^0\mu^+\mu^-$	(9.6 ±2.3) $\times 10^{-5}$		349
e^+e^-	(7.15±0.19) $\times 10^{-5}$		391
$\pi^+\pi^-\pi^0\pi^0$	< 2 %	90%	261
$\pi^+\pi^-\gamma$	< 3.6 $\times 10^{-3}$	95%	365
$\pi^+\pi^-\pi^+\pi^-$	< 1 $\times 10^{-3}$	90%	256
$\pi^0\pi^0\gamma$	< 4 $\times 10^{-4}$	90%	367
$\mu^+\mu^-$	< 1.8 $\times 10^{-4}$	90%	376

$\eta'(958)$

$I^G(J^{PC}) = 0^+(0^{-+})$

Mass $m = 957.77 \pm 0.14$ MeV
Full width $\Gamma = 0.201 \pm 0.016$ MeV (S = 1.3)

$\eta'(958)$ DECAY MODES	Fraction (Γ_i/Γ)	Scale factor/ Confidence level	p (MeV/c)
$\pi^+\pi^-\eta$	(43.7 ±1.5) %	S=1.1	232
$\rho^0\gamma$	(30.2 ±1.3) %	S=1.1	169
$\pi^0\pi^0\eta$	(20.8 ±1.3) %	S=1.2	239
$\omega\gamma$	(3.02±0.30) %		160
$\gamma\gamma$	(2.12±0.13) %	S=1.2	479
$3\pi^0$	(1.55±0.26) × 10^{-3}		430
$\mu^+\mu^-\gamma$	(1.04±0.26) × 10^{-4}		467
$\pi^+\pi^-\pi^0$	< 5 %	CL=90%	427
$\pi^0\rho^0$	< 4 %	CL=90%	118
$\pi^+\pi^-$	< 2 %	CL=90%	458
$\pi^0 e^+ e^-$	< 1.3 %	CL=90%	469
$\eta e^+ e^-$	< 1.1 %	CL=90%	322
$\pi^+\pi^+\pi^-\pi^-$	< 1 %	CL=90%	372
$\pi^+\pi^+\pi^-\pi^-$ neutrals	< 1 %	CL=95%	–
$\pi^+\pi^+\pi^-\pi^-\pi^0$	< 1 %	CL=90%	298
6π	< 1 %	CL=90%	189
$\pi^+\pi^- e^+ e^-$	< 6 × 10^{-3}	CL=90%	458
$\pi^0\pi^0$	< 9 × 10^{-4}	CL=90%	459
$\pi^0\gamma\gamma$	< 8 × 10^{-4}	CL=90%	469
$4\pi^0$	< 5 × 10^{-4}	CL=90%	379
3γ	< 1.0 × 10^{-4}	CL=90%	479
$\mu^+\mu^-\pi^0$	< 6.0 × 10^{-5}	CL=90%	445
$\mu^+\mu^-\eta$	< 1.5 × 10^{-5}	CL=90%	274
$\pi^+\pi^-\gamma$ (including $\rho^0\gamma$)	(27.9 ±2.3) %		–
e^+e^-	< 2.1 × 10^{-7}	CL=90%	479

$f_0(980)$
was $S(975)$

$I^G(J^{PC}) = 0^+(0^{++})$

Mass $m = 980 \pm 10$ MeV
Full width $\Gamma = 40$ to 400 MeV

$f_0(980)$ DECAY MODES	Fraction (Γ_i/Γ)	Confidence level	p (MeV/c)
$\pi\pi$	(78.1 ±2.4) %		470
$K\overline{K}$	(21.9 ±2.4) %		–
$\gamma\gamma$	(1.19±0.33) × 10^{-5}		490
e^+e^-	< 3 × 10^{-7}	90%	490

$a_0(980)$
was $\delta(980)$

$I^G(J^{PC}) = 1^-(0^{++})$

Mass $m = 982.4 \pm 1.4$ MeV
Full width $\Gamma = 50$ to 300 MeV

$a_0(980)$ DECAY MODES	Fraction (Γ_i/Γ)	p (MeV/c)
$\eta\pi$	dominant	319
$K\overline{K}$	seen	–
$\gamma\gamma$	seen	491

$\phi(1020)$

$I^G(J^{PC}) = 0^-(1^{--})$

Mass $m = 1019.413 \pm 0.008$ MeV
Full width $\Gamma = 4.43 \pm 0.06$ MeV
$\Gamma_{ee} = 1.37 \pm 0.05$ keV

$\phi(1020)$ DECAY MODES	Fraction (Γ_i/Γ)	Scale factor/ Confidence level	p (MeV/c)
K^+K^-	(49.1 ±0.9) %	S=1.3	127
$K^0_L K^0_S$	(34.3 ±0.7) %	S=1.2	110
$\rho\pi$	(12.9 ±0.7) %		181
$\pi^+\pi^-\pi^0$	(2.5 ±0.9) %	S=1.1	462
$\eta\gamma$	(1.28±0.06) %	S=1.2	363
$\pi^0\gamma$	(1.31±0.13) × 10^{-3}		501
e^+e^-	(3.09±0.07) × 10^{-4}		510
$\mu^+\mu^-$	(2.48±0.34) × 10^{-4}		499
$\eta e^+ e^-$	(1.3 $^{+0.8}_{-0.6}$) × 10^{-4}		363
$\pi^+\pi^-$	(8 $^{+5}_{-4}$) × 10^{-5}	S=1.5	490
$\omega\gamma$	< 5 %	CL=84%	210
$\rho\gamma$	< 2 %	CL=84%	219
$\pi^+\pi^-\gamma$	< 7 × 10^{-3}	CL=90%	490
$f_0(980)\gamma$	< 2 × 10^{-3}	CL=90%	39
$\pi^0\pi^0\gamma$	< 1 × 10^{-3}	CL=90%	492
$\pi^+\pi^-\pi^+\pi^-$	< 8.7 × 10^{-4}	CL=90%	410
$\eta'(958)\gamma$	< 4.1 × 10^{-4}	CL=90%	60
$\pi^+\pi^+\pi^-\pi^-\pi^0$	< 1.5 × 10^{-4}	CL=95%	341
$\pi^0 e^+ e^-$	< 1.2 × 10^{-4}	CL=90%	501
$\pi^0\eta\gamma$	< 2.5 × 10^{-3}	CL=90%	346
$a_0(980)\gamma$	< 5 × 10^{-3}	CL=90%	36

$h_1(1170)$

$I^G(J^{PC}) = 0^-(1^{+-})$

Mass $m = 1170 \pm 20$ MeV
Full width $\Gamma = 360 \pm 40$ MeV

$h_1(1170)$ DECAY MODES	Fraction (Γ_i/Γ)	p (MeV/c)
$\rho\pi$	seen	310

$b_1(1235)$

$I^G(J^{PC}) = 1^+(1^{+-})$

Mass $m = 1231 \pm 10$ MeV [k]
Full width $\Gamma = 142 \pm 8$ MeV (S = 1.1)

$b_1(1235)$ DECAY MODES	Fraction (Γ_i/Γ)	Confidence level	p (MeV/c)
$\omega\pi$	dominant		348
[D/S amplitude ratio = 0.26 ± 0.04]			
$\pi^\pm\gamma$	(1.6±0.4) × 10^{-3}		608
$\eta\rho$	seen		–
$\pi^+\pi^+\pi^-\pi^0$	< 50 %	84%	536
$(K\overline{K})^\pm\pi^0$	< 8 %	90%	248
$K^0_S K^0_L \pi^\pm$	< 6 %	90%	238
$K^0_S K^0_S \pi^\pm$	< 2 %	90%	238
$\pi\phi$	< 1.5 %	84%	146

$a_1(1260)$

$I^G(J^{PC}) = 1^-(1^{++})$

Mass $m = 1230 \pm 40$ MeV [k]
Full width $\Gamma \sim 400$ MeV

$a_1(1260)$ DECAY MODES	Fraction (Γ_i/Γ)	Confidence level	p (MeV/c)
$\rho\pi$	dominant		356
$\pi\gamma$	seen		607
$\pi(\pi\pi)_{S\text{-wave}}$	[k] <0.7 %	90%	575

$f_2(1270)$

$I^G(J^{PC}) = 0^+(2^{++})$

Mass $m = 1275 \pm 5$ MeV [k]
Full width $\Gamma = 185 \pm 20$ MeV [k]

$f_2(1270)$ DECAY MODES	Fraction (Γ_i/Γ)	Scale factor/ Confidence level	p (MeV/c)
$\pi\pi$	(84.9 $^{+2.5}_{-1.3}$) %	S=1.3	622
$\pi^+\pi^- 2\pi^0$	(6.9 $^{+1.5}_{-2.7}$) %	S=1.4	562
$K\overline{K}$	(4.6 ±0.5) %	S=2.8	403
$2\pi^+ 2\pi^-$	(2.8 ±0.4) %	S=1.2	559
$\eta\eta$	(4.5 ±1.0) × 10^{-3}	S=2.4	327
$4\pi^0$	(3.0 ±1.0) × 10^{-3}		564
$\gamma\gamma$	(1.32 $^{+0.18}_{-0.16}$) × 10^{-5}	S=1.1	637
$\eta\pi\pi$	< 8 × 10^{-3}	CL=95%	475
$K^0 K^- \pi^+$ + c.c.	< 3.4 × 10^{-3}	CL=95%	293
e^+e^-	< 9 × 10^{-9}	CL=90%	637

$f_1(1285)$

$I^G(J^{PC}) = 0^+(1^{+\,+})$

Mass $m = 1282 \pm 5$ MeV [k]
Full width $\Gamma = 24 \pm 3$ MeV [k]

$f_1(1285)$ DECAY MODES	Fraction (Γ_i/Γ)	Scale factor/ Confidence level	p (MeV/c)
4π	(29 ± 6) %		563
$\pi^0\pi^0\pi^+\pi^-$	$(15\ {}^{+\ 9}_{-\ 8})$ %	S=1.1	–
$2\pi^+2\pi^-$	(15 ± 6) %		563
$\rho^0\pi^+\pi^-$	dominates $2\pi^+\ 2\pi^-$		340
$4\pi^0$	$< 7 \times 10^{-4}$	CL=90%	568
$\eta\pi\pi$	(54 ± 15) %		479
$a_0(980)\pi$ [ignoring $a_0(980) \to K\overline{K}$]	(44 ± 7) %	S=1.1	234
$\eta\pi\pi$ [excluding $a_0(980)\pi$]	$(10\ {}^{+\ 7}_{-\ 6})$ %	S=1.1	–
$K\overline{K}\pi$	(9.7 ± 1.6) %	S=1.2	308
$K\overline{K}^*(892)$	not seen		–
$\gamma\rho^0$	(6.6 ± 1.3) %	S=1.5	410
$\phi\gamma$	$(8.0 \pm 3.1) \times 10^{-4}$		236

$\eta(1295)$

$I^G(J^{PC}) = 0^+(0^{-\,+})$

Mass $m = 1295 \pm 4$ MeV
Full width $\Gamma = 53 \pm 6$ MeV

$\eta(1295)$ DECAY MODES	Fraction (Γ_i/Γ)	p (MeV/c)
$\eta\pi^+\pi^-$	seen	488
$a_0(980)\pi$	seen	245

$f_0(1300)$ was $f_0(1400)$ was $\epsilon(1200)$

$I^G(J^{PC}) = 0^+(0^{+\,+})$

Mass $m =$ 1000-1500 MeV
Full width $\Gamma =$ 150 to 400 MeV
$\Gamma_{\gamma\gamma} = 5.4 \pm 2.3$ keV
$\Gamma_{ee} < 20$ eV, CL = 90%

$f_0(1300)$ DECAY MODES	Fraction (Γ_i/Γ)	p (MeV/c)
$\pi\pi$	$(93.6^{+1.9}_{-1.5})$ %	–
$K\overline{K}$	(7.5 ± 0.9) %	–
$\eta\eta$	seen	–
$\gamma\gamma$	seen	–
e^+e^-	not seen	–

$\pi(1300)$

$I^G(J^{PC}) = 1^-(0^{-\,+})$

Mass $m = 1300 \pm 100$ MeV [k]
Full width $\Gamma =$ 200 to 600 MeV

$\pi(1300)$ DECAY MODES	Fraction (Γ_i/Γ)	p (MeV/c)
$\rho\pi$	seen	406
$\pi(\pi\pi)_{S\text{-wave}}$	seen	612

$a_2(1320)$

$I^G(J^{PC}) = 1^-(2^{+\,+})$

Mass $m = 1318.4 \pm 0.6$ MeV (S = 1.1) (3π and $K^\pm K^0_S$ modes)
Full width $\Gamma = 107 \pm 5$ MeV [k] ($K^\pm K^0_S$ and $\eta\pi$ modes)

$a_2(1320)$ DECAY MODES	Fraction (Γ_i/Γ)	Scale factor/ Confidence level	p (MeV/c)
$\rho\pi$	(70.1 ± 2.7) %	S=1.2	419
$\eta\pi$	(14.5 ± 1.2) %		535
$\omega\pi\pi$	(10.6 ± 3.2) %	S=1.3	362
$K\overline{K}$	(4.9 ± 0.8) %		437
$\eta'(958)\pi$	$(5.7 \pm 1.1) \times 10^{-3}$		287
$\pi^\pm\gamma$	$(2.8 \pm 0.6) \times 10^{-3}$		652
$\gamma\gamma$	$(9.7 \pm 1.0) \times 10^{-6}$		659
$\pi^+\pi^-\pi^-$	< 8 %	CL=90%	621
e^+e^-	$< 2.3 \times 10^{-7}$	CL=90%	659

$f_1(1420)$ [l]

$I^G(J^{PC}) = 0^+(1^{+\,+})$

Mass $m = 1426.8 \pm 2.3$ MeV (S = 1.3)
Full width $\Gamma = 52 \pm 4$ MeV

$f_1(1420)$ DECAY MODES	Fraction (Γ_i/Γ)	p (MeV/c)
$K\overline{K}\pi$	dominant	439
$\eta\pi\pi$	possibly seen	571

$\omega(1420)$ [m]

$I^G(J^{PC}) = 0^-(1^{-\,-})$

Mass $m = 1419 \pm 31$ MeV
Full width $\Gamma = 174 \pm 60$ MeV

$\omega(1420)$ DECAY MODES	Fraction (Γ_i/Γ)	p (MeV/c)
$\rho\pi$	dominant	488

$\eta(1440)$ [n] was $\iota(1440)$

$I^G(J^{PC}) = 0^+(0^{-\,+})$

Mass $m = 1420 \pm 20$ MeV [k]
Full width $\Gamma = 60 \pm 30$ MeV [k]

$\eta(1440)$ DECAY MODES	Fraction (Γ_i/Γ)	p (MeV/c)
$K\overline{K}\pi$	seen	433
$\eta\pi\pi$	seen	567
$a_0(980)\pi$	seen	350
4π	seen	640

$\rho(1450)$ [o]

$I^G(J^{PC}) = 1^+(1^{-\,-})$

Mass $m = 1465 \pm 25$ MeV [k]
Full width $\Gamma = 310 \pm 60$ MeV [k]

$\rho(1450)$ DECAY MODES	Fraction (Γ_i/Γ)	Confidence level	p (MeV/c)
$\pi\pi$	seen		719
4π	seen		665
e^+e^-	seen		732
$\eta\rho$	<4 %		317
$\omega\pi$	<2.0 %	95%	512
$\phi\pi$	<1 %		358
$K\overline{K}$	$<1.6 \times 10^{-3}$	95%	541

$f_1(1510)$

$I^G(J^{PC}) = 0^+(1^{+\,+})$

Mass $m = 1512 \pm 4$ MeV
Full width $\Gamma = 35 \pm 15$ MeV

$f_1(1510)$ DECAY MODES	Fraction (Γ_i/Γ)	p (MeV/c)
$K\overline{K}^*(892)+$ c.c.	seen	292

$f'_2(1525)$

$I^G(J^{PC}) = 0^+(2^{+\,+})$

Mass $m = 1525 \pm 5$ MeV [k]
Full width $\Gamma = 76 \pm 10$ MeV [k]

$f'_2(1525)$ DECAY MODES	Fraction (Γ_i/Γ)	p (MeV/c)
$K\overline{K}$	$(71.2\ {}^{+2.0}_{-2.5})$ %	581
$\eta\eta$	$(27.9\ {}^{+2.5}_{-2.0})$ %	531
$\pi\pi$	$(8.2 \pm 1.6) \times 10^{-3}$	750
$\gamma\gamma$	$(1.23 \pm 0.22) \times 10^{-6}$	763

$f_0(1590)$

$I^G(J^{PC}) = 0^+(0^{++})$

Seen by one group only.

Mass $m = 1581 \pm 10$ MeV
Full width $\Gamma = 180 \pm 17$ MeV (S = 1.2)

$f_0(1590)$ DECAY MODES	Fraction (Γ_i/Γ)	p (MeV/c)
$\eta\eta'(958)$	dominant	234
$\eta\eta$	large	570
$4\pi^0$	large	732

$\omega(1600)$ [p]

$I^G(J^{PC}) = 0^-(1^{--})$

Mass $m = 1662 \pm 13$ MeV
Full width $\Gamma = 280 \pm 24$ MeV

$\omega(1600)$ DECAY MODES	Fraction (Γ_i/Γ)	p (MeV/c)
$\rho\pi$	seen	644
$\omega\pi\pi$	seen	610
e^+e^-	seen	831

$\omega_3(1670)$

$I^G(J^{PC}) = 0^-(3^{--})$

Mass $m = 1668 \pm 5$ MeV
Full width $\Gamma = 173 \pm 11$ MeV [k]

$\omega_3(1670)$ DECAY MODES	Fraction (Γ_i/Γ)	p (MeV/c)
$\rho\pi$	seen	647
$\omega\pi\pi$	seen	614
$b_1(1235)\pi$	possibly seen	359

$\pi_2(1670)$

$I^G(J^{PC}) = 1^-(2^{-+})$

Mass $m = 1670 \pm 20$ MeV [k]
Full width $\Gamma = 240 \pm 15$ MeV [k] (S = 1.1)
$\Gamma_{ee} = 1.35 \pm 0.26$ keV

$\pi_2(1670)$ DECAY MODES	Fraction (Γ_i/Γ)	p (MeV/c)
$f_2(1270)\pi$	(56.2 ± 3.2) %	325
$\pi^\pm\pi^+\pi^-$	(53 ± 4) %	–
$\rho\pi$	(31 ± 4) %	649
$f_0(1300)\pi$	(8.7 ± 3.4) %	–
$K\overline{K}^*(892)+$ c.c.	(4.2 ± 1.4) %	453
$\gamma\gamma$	$(5.6 \pm 1.1) \times 10^{-6}$	835
$\eta\pi$	< 5 %	738
$\pi^\pm 2\pi^+ 2\pi^-$	< 5 %	734

$\phi(1680)$

$I^G(J^{PC}) = 0^-(1^{--})$

Not a well-established resonance.

Mass $m = 1680 \pm 50$ MeV [k]
Full width $\Gamma = 150 \pm 50$ MeV [k]

$\phi(1680)$ DECAY MODES	Fraction (Γ_i/Γ)	p (MeV/c)
$K\overline{K}^*(892)+$ c.c.	dominant	462
$K^0_S K\pi$	seen	619
$K\overline{K}$	seen	680
e^+e^-	seen	840
$\omega\pi\pi$	not seen	621

$\rho_3(1690)$

$I^G(J^{PC}) = 1^+(3^{--})$

J^P from the 2π and $K\overline{K}$ modes.

Mass $m = 1691 \pm 5$ MeV [k] (2π, $K\overline{K}$, and $K\overline{K}\pi$ modes)
Full width $\Gamma = 215 \pm 20$ MeV [k] (2π, $K\overline{K}$, and $K\overline{K}\pi$ modes)

$\rho_3(1690)$ DECAY MODES	Fraction (Γ_i/Γ)	Scale factor	p (MeV/c)
4π	(71.1 ± 1.9) %		788
$\pi^\pm\pi^+\pi^-\pi^0$	(67 ± 22) %		788
$\pi\pi$	(23.6 ± 1.3) %		834
$\omega\pi$	(16 ± 6) %		656
$K\overline{K}\pi$	(3.8 ± 1.2) %		628
$K\overline{K}$	(1.58 ± 0.26) %	1.2	686
$\eta\pi^+\pi^-$	seen		728

$\rho(1700)$ [o]

$I^G(J^{PC}) = 1^+(1^{--})$

Mass $m = 1700 \pm 20$ MeV [k] ($\eta\rho^0$ and mixed modes)
Full width $\Gamma = 235 \pm 50$ MeV [k] ($\eta\rho^0$, $\pi^+\pi^-$, and mixed modes)

$\rho(1700)$ DECAY MODES	Fraction (Γ_i/Γ)	p (MeV/c)
$\rho\pi\pi$	dominant	640
$\rho^0\pi^+\pi^-$	large	640
$\rho^\pm\pi^\mp\pi^0$	[q] large	642
$2(\pi^+\pi^-)$	large	792
$\pi^+\pi^-$	seen	838
$K\overline{K}^*(892)+$ c.c.	seen	479
$\eta\rho$	seen	533
$K\overline{K}$	seen	692
e^+e^-	seen	850

$f_J(1710)$
was $\theta(1690)$

$I^G(J^{PC}) = 0^+(\text{even}^{++})$

Mass $m = 1709 \pm 5$ MeV
Full width $\Gamma = 140 \pm 12$ MeV

$f_J(1710)$ DECAY MODES	Fraction (Γ_i/Γ)	p (MeV/c)
$K\overline{K}$	seen	697
$\pi\pi$	seen	843

$\phi_3(1850)$

$I^G(J^{PC}) = 0^-(3^{--})$

Mass $m = 1854 \pm 7$ MeV
Full width $\Gamma = 87^{+28}_{-23}$ MeV (S = 1.2)

$\phi_3(1850)$ DECAY MODES	Fraction (Γ_i/Γ)	p (MeV/c)
$K\overline{K}$	seen	785
$K\overline{K}^*(892)+$ c.c.	seen	602

$f_2(2010)$

$I^G(J^{PC}) = 0^+(2^{++})$

Seen by one group only.

Mass $m = 2011^{+60}_{-80}$ MeV
Full width $\Gamma = 202 \pm 60$ MeV

$f_2(2010)$ DECAY MODES	Fraction (Γ_i/Γ)	p (MeV/c)
$\phi\phi$	seen	–

$f_4(2050)$ $I^G(J^{PC}) = 0^+(4^{++})$

Mass $m = 2044 \pm 11$ MeV (S = 1.4)
Full width $\Gamma = 208 \pm 13$ MeV (S = 1.2)

$f_4(2050)$ DECAY MODES	Fraction (Γ_i/Γ)	p (MeV/c)
$\omega\omega$	(26 ±6) %	658
$\pi\pi$	(17.0±1.5) %	1012
$K\overline{K}$	($6.8^{+3.4}_{-1.8}$) × 10^{-3}	895
$\eta\eta$	(2.1±0.8) × 10^{-3}	863
$4\pi^0$	< 1.2 %	977

$f_2(2300)$ $I^G(J^{PC}) = 0^+(2^{++})$

Mass $m = 2297 \pm 28$ MeV
Full width $\Gamma = 149 \pm 40$ MeV

$f_2(2300)$ DECAY MODES	Fraction (Γ_i/Γ)	p (MeV/c)
$\phi\phi$	seen	529

$f_2(2340)$ $I^G(J^{PC}) = 0^+(2^{++})$

Mass $m = 2339 \pm 60$ MeV
Full width $\Gamma = 319^{+80}_{-70}$ MeV

$f_2(2340)$ DECAY MODES	Fraction (Γ_i/Γ)	p (MeV/c)
$\phi\phi$	seen	573

STRANGE MESONS
($S = \pm1$, $C = B = 0$)

$K^+ = u\overline{s}$, $K^0 = d\overline{s}$, $\overline{K}^0 = \overline{d}s$, $K^- = \overline{u}s$, similarly for K^*'s

$K^\pm$ $I(J^P) = \frac{1}{2}(0^-)$

Mass $m = 493.677 \pm 0.016$ MeV (S = 2.8)
Mean life $\tau = (1.2371 \pm 0.0029) \times 10^{-8}$ s (S = 2.2)
$c\tau = 3.709$ m

Slope parameter g [r]

(See Full Listings for quadratic coefficients)

$K^+ \to \pi^+\pi^+\pi^- = -0.2154 \pm 0.0035$ (S = 1.4)
$K^- \to \pi^-\pi^-\pi^+ = -0.217 \pm 0.007$ (S = 2.5)
$K^\pm \to \pi^\pm\pi^0\pi^0 = 0.594 \pm 0.019$ (S = 1.3)

$K^\pm$ decay form factors [b,s]

K^+_{e3} $\lambda_+ = 0.0286 \pm 0.0022$
$K^+_{\mu3}$ $\lambda_+ = 0.033 \pm 0.008$ (S = 1.6)
$K^+_{\mu3}$ $\lambda_0 = 0.004 \pm 0.007$ (S = 1.6)
K^+_{e3} $|f_S/f_+| = 0.084 \pm 0.023$ (S = 1.2)
K^+_{e3} $|f_T/f_+| = 0.38 \pm 0.11$ (S = 1.1)
$K^+_{\mu3}$ $|f_T/f_+| = 0.02 \pm 0.12$
$K^+ \to e^+\nu_e\gamma$ $|F_A + F_V| = 0.148 \pm 0.010$
$K^+ \to \mu^+\nu_\mu\gamma$ $|F_A + F_V| < 0.23$, CL = 90%
$K^+ \to e^+\nu_e\gamma$ $|F_A - F_V| < 0.49$
$K^+ \to \mu^+\nu_\mu\gamma$ $|F_A - F_V| = -2.2$ to 0.3

K^- modes are charge conjugates of the modes below.

K^+ DECAY MODES		Fraction (Γ_i/Γ)	Scale factor/ Confidence level	p (MeV/c)
$\mu^+\nu_\mu$		(63.51±0.18) %	S=1.3	236
$e^+\nu_e$		(1.55±0.07) × 10^{-5}		247
$\pi^+\pi^0$		(21.16±0.14) %	S=1.1	205
$\pi^+\pi^+\pi^-$		(5.59±0.05) %	S=1.9	125
$\pi^+\pi^0\pi^0$		(1.73±0.04) %	S=1.2	133
$\pi^0\mu^+\nu_\mu$ Called $K^+_{\mu3}$.		(3.18±0.08) %	S=1.5	215
$\pi^0e^+\nu_e$ Called K^+_{e3}.		(4.82±0.06) %	S=1.3	228
$\pi^0\pi^0e^+\nu_e$		(2.1 ±0.4) × 10^{-5}		206
$\pi^+\pi^-e^+\nu_e$		(3.91±0.17) × 10^{-5}		203
$\pi^+\pi^-\mu^+\nu_\mu$		(1.4 ±0.9) × 10^{-5}		151
$\pi^0\pi^0\pi^0e^+\nu_e$		< 3.5 × 10^{-6}	CL=90%	135
$\pi^+\gamma\gamma$	[t]	< 1 × 10^{-6}	CL=90%	227
$\pi^+3\gamma$	[t]	< 1.0 × 10^{-4}	CL=90%	227
$e^+\nu_e\nu\overline{\nu}$		< 6 × 10^{-5}	CL=90%	247
$\mu^+\nu_\mu\nu\overline{\nu}$		< 6.0 × 10^{-6}	CL=90%	236
$\mu^+\nu_\mu e^+e^-$		(1.06±0.32) × 10^{-6}		236
$e^+\nu_e e^+e^-$		($2.1^{+2.1}_{-1.1}$) × 10^{-7}		247
$\mu^+\nu_\mu\mu^+\mu^-$		< 4.1 × 10^{-7}	CL=90%	185
$\mu^+\nu_\mu\gamma$	[t,u]	(5.50±0.28) × 10^{-3}		236
$\pi^+\pi^0\gamma$	[t,u]	(2.75±0.15) × 10^{-4}		205
$\pi^+\pi^0\gamma$(DE)	[t,v]	(1.8 ±0.4) × 10^{-5}		205
$\pi^+\pi^+\pi^-\gamma$	[t,u]	(1.04±0.31) × 10^{-4}		125
$\pi^+\pi^0\pi^0\gamma$	[t,u]	($7.4^{+5.5}_{-2.9}$) × 10^{-6}		133
$\pi^0\mu^+\nu_\mu\gamma$	[t,u]	< 6.1 × 10^{-5}	CL=90%	215
$\pi^0e^+\nu_e\gamma$	[t,u]	(2.62±0.20) × 10^{-4}		228
$\pi^0e^+\nu_e\gamma$(SD)	[w]	< 5.3 × 10^{-5}	CL=90%	228
$\pi^0\pi^0e^+\nu_e\gamma$		< 5 × 10^{-6}	CL=90%	206

Lepton Family number (*LF*), Lepton number (*L*), $\Delta S = \Delta Q$ (*SQ*) violating modes, or $\Delta S = 1$ weak neutral current (*S1*) modes

Mode			Fraction	CL	p (MeV/c)
$\pi^+\pi^+e^-\overline{\nu}_e$	*SQ*		< 1.2 × 10^{-8}	CL=90%	203
$\pi^+\pi^+\mu^-\overline{\nu}_\mu$	*SQ*		< 3.0 × 10^{-6}	CL=95%	151
$\pi^+e^+e^-$	*S1*		(2.74±0.23) × 10^{-7}		227
$\pi^+\mu^+\mu^-$	*S1*		< 2.3 × 10^{-7}	CL=90%	172
$\pi^+\nu\overline{\nu}$	*S1*		< 5.2 × 10^{-9}	CL=90%	227
$\mu^-\nu e^+e^+$	*LF*		< 2.0 × 10^{-8}	CL=90%	236
$\mu^+\nu_e$	*LF*	[e]	< 4 × 10^{-3}	CL=90%	236
$\pi^+\mu^+e^-$	*LF*		< 2.1 × 10^{-10}	CL=90%	214
$\pi^+\mu^-e^+$	*LF*		< 7 × 10^{-9}	CL=90%	214
$\pi^-\mu^+e^+$	*L*		< 7 × 10^{-9}	CL=90%	214
$\pi^-e^+e^+$	*L*		< 1.0 × 10^{-8}	CL=90%	227
$\pi^-\mu^+\mu^+$	*L*		< 1.5 × 10^{-4}	CL=90%	172
$\mu^+\overline{\nu}_e$	*L*	[e]	< 3.3 × 10^{-3}	CL=90%	236
$\pi^0e^+\overline{\nu}_e$	*L*	[e]	< 3 × 10^{-3}	CL=90%	228

K^0 $I(J^P) = \frac{1}{2}(0^-)$

50% K_S, 50% K_L
Mass $m = 497.672 \pm 0.031$ MeV
$m_{K^0} - m_{K^\pm} = 3.995 \pm 0.034$ MeV (S = 1.1)

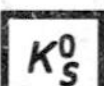

K_S^0

$I(J^P) = \frac{1}{2}(0^-)$

Mean life $\tau = (0.8926 \pm 0.0012) \times 10^{-10}$ s

$c\tau = 2.676$ cm

CP-violation parameters [x]

$\text{Im}(\eta_{+-0})^2 < 0.12$, CL = 90%

$\text{Im}(\eta_{000})^2 < 0.1$, CL = 90%

K_S^0 DECAY MODES		Fraction (Γ_i/Γ)	Scale factor/ Confidence level	p (MeV/c)
$\pi^+\pi^-$		(68.61±0.28) %	S=1.2	206
$\pi^0\pi^0$		(31.39±0.28) %	S=1.2	209
$\pi^+\pi^-\gamma$	[u,y]	(1.78±0.05) × 10^{-3}		206
$\gamma\gamma$		(2.4 ±1.2) × 10^{-6}		249
$\pi^+\pi^-\pi^0$		< 8.5 × 10^{-5}	CL=90%	133
$3\pi^0$		< 3.7 × 10^{-5}	CL=90%	139
$\pi^\pm e^\mp \nu$	[z]	(6.68±0.10) × 10^{-4}	S=1.3	229
$\pi^\pm \mu^\mp \nu$	[z]	(4.66±0.07) × 10^{-4}	S=1.2	216
$\Delta S = 1$ weak neutral current ($S1$) modes				
$\mu^+\mu^-$	$S1$	< 3.2 × 10^{-7}	CL=90%	225
e^+e^-	$S1$	< 1.0 × 10^{-5}	CL=90%	249
$\pi^0 e^+ e^-$	$S1$	< 1.1 × 10^{-6}	CL=90%	231

K_L^0

$I(J^P) = \frac{1}{2}(0^-)$

$m_{K_L} - m_{K_S} = (0.5333 \pm 0.0027) \times 10^{10}\ \hbar\ \text{s}^{-1}$ (S = 1.2)

$= (3.510 \pm 0.018) \times 10^{-12}$ MeV

Mean life $\tau = (5.17 \pm 0.04) \times 10^{-8}$ s

$c\tau = 15.49$ m

Slope parameter g [r]

(See Full Listings for quadratic coefficients)

$K_L^0 \to \pi^+\pi^-\pi^0 = 0.670 \pm 0.014$ (S = 1.6)

K_L decay form factors [s]

K_{e3}^0 $\lambda_+ = 0.0300 \pm 0.0016$ (S = 1.2)

$K_{\mu 3}^0$ $\lambda_+ = 0.034 \pm 0.005$ (S = 2.3)

$K_{\mu 3}^0$ $\lambda_0 = 0.025 \pm 0.006$ (S = 2.3)

K_{e3}^0 $|f_S/f_+| < 0.04$, CL = 68%

K_{e3}^0 $|f_T/f_+| < 0.23$, CL = 68%

$K_{\mu 3}^0$ $|f_T/f_+| = 0.12 \pm 0.12$

$K_L \to e^+e^-\gamma$: $\alpha_{K^*} = -0.28 \pm 0.08$

CP-violation parameters [x]

$\delta = (0.327 \pm 0.012)\%$

$|\eta_{00}| = (2.259 \pm 0.023) \times 10^{-3}$ (S = 1.1)

$|\eta_{+-}| = (2.269 \pm 0.023) \times 10^{-3}$ (S = 1.1)

$|\eta_{00}/\eta_{+-}| = 0.9955 \pm 0.0023$ [aa] (S = 1.8)

$\epsilon'/\epsilon = (1.5 \pm 0.8) \times 10^{-3}$ [aa] (S = 1.8)

$\phi_{+-} = (44.3 \pm 0.8)^\circ$

$\phi_{00} = (43.3 \pm 1.3)^\circ$

$\Delta S = -\Delta Q$ in $K_{\ell 3}^0$ decay

Re $x = 0.006 \pm 0.018$ (S = 1.3)

Im $x = -0.003 \pm 0.026$ (S = 1.2)

K_L^0 DECAY MODES		Fraction (Γ_i/Γ)	Scale factor/ Confidence level	p (MeV/c)
$3\pi^0$		(21.6 ±0.8) %	S=1.5	139
$\pi^+\pi^-\pi^0$		(12.38±0.21) %	S=1.5	133
$\pi^\pm\mu^\mp\nu$ Called $K_{\mu 3}^0$.	[q]	(27.0 ±0.4) %	S=1.3	216
$\pi^\pm e^\mp \nu$ Called K_{e3}^0.	[q]	(38.7 ±0.5) %	S=1.4	229
2γ		(5.73±0.27) × 10^{-4}	S=2.0	249
$\pi^0 2\gamma$	[bb]	(1.70±0.28) × 10^{-6}		231
$\pi^0\pi^\pm e^\mp \nu$	[q]	(5.18±0.29) × 10^{-5}		207
$(\pi\mu\text{atom})\nu$		(1.05±0.11) × 10^{-7}		216
$\pi^\pm e^\mp \nu_e \gamma$	[q,u,bb]	(1.3 ±0.8) %		229
$\pi^+\pi^-\gamma$	[u,bb]	(4.61±0.14) × 10^{-5}		206
$\pi^0\pi^0\gamma$		< 5.6 × 10^{-6}		–
Charge conjugation × Parity (CP) or Lepton Family number (LF) violating modes, or $\Delta S = 1$ weak neutral current ($S1$) modes				
$\pi^+\pi^-$	CPV	(2.03±0.04) × 10^{-3}	S=1.2	206
$\pi^0\pi^0$	CPV	(9.14±0.34) × 10^{-4}	S=1.8	209
$\mu^+\mu^-$	$S1$	(7.4 ±0.4) × 10^{-9}		225
$\mu^+\mu^-\gamma$	$S1$	(2.8 ±2.8) × 10^{-7}		225
e^+e^-	$S1$	< 4.1 × 10^{-11}	CL=90%	249
$e^+e^-\gamma$	$S1$	(9.1 ±0.5) × 10^{-6}		249
$e^+e^-\gamma\gamma$	$S1$ [bb]	(6.6 ±3.2) × 10^{-7}		249
$\pi^+\pi^-e^+e^-$	$S1$	< 2.5 × 10^{-6}	CL=90%	206
$\mu^+\mu^-e^+e^-$	$S1$	< 4.9 × 10^{-6}	CL=90%	225
$e^+e^-e^+e^-$	$S1$ [cc]	(3.9 ±0.7) × 10^{-8}		249
$\pi^0\mu^+\mu^-$	$CP,S1$ [dd]	< 5.1 × 10^{-9}	CL=90%	177
$\pi^0 e^+e^-$	$CP,S1$ [dd]	< 4.3 × 10^{-9}	CL=90%	231
$\pi^0\nu\overline{\nu}$	$CP,S1$ [ee]	< 2.2 × 10^{-4}	CL=90%	231
$e^\pm\mu^\mp$	LF [q]	< 3.3 × 10^{-11}	CL=90%	238

$K^*(892)$

$I(J^P) = \frac{1}{2}(1^-)$

$K^*(892)^\pm$ mass $m = 891.59 \pm 0.24$ MeV (S = 1.1)

$K^*(892)^0$ mass $m = 896.10 \pm 0.28$ MeV (S = 1.4)

$K^*(892)^\pm$ full width $\Gamma = 49.8 \pm 0.8$ MeV

$K^*(892)^0$ full width $\Gamma = 50.5 \pm 0.6$ MeV (S = 1.1)

$K^*(892)$ DECAY MODES	Fraction (Γ_i/Γ)	Confidence level	p (MeV/c)
$K\pi$	~ 100 %		291
$K^0\gamma$	(2.30±0.20) × 10^{-3}		310
$K^\pm\gamma$	(1.01±0.09) × 10^{-3}		309
$K\pi\pi$	< 7 × 10^{-4}	95%	224

$K_1(1270)$

$I(J^P) = \frac{1}{2}(1^+)$

Mass $m = 1273 \pm 7$ MeV [k]

Full width $\Gamma = 90 \pm 20$ MeV [k]

$K_1(1270)$ DECAY MODES	Fraction (Γ_i/Γ)	p (MeV/c)
$K\rho$	(42 ±6) %	76
$K_0^*(1430)\pi$	(28 ±4) %	–
$K^*(892)\pi$	(16 ±5) %	301
$K\omega$	(11.0±2.0) %	–
$K f_0(1300)$	(3.0±2.0) %	–

$K_1(1400)$

$I(J^P) = \frac{1}{2}(1^+)$

Mass $m = 1402 \pm 7$ MeV

Full width $\Gamma = 174 \pm 13$ MeV (S = 1.6)

$K_1(1400)$ DECAY MODES	Fraction (Γ_i/Γ)	p (MeV/c)
$K^*(892)\pi$	(94 ±6) %	401
$K\rho$	(3.0±3.0) %	298
$K f_0(1300)$	(2.0±2.0) %	–
$K\omega$	(1.0±1.0) %	285

K*(1410)

$I(J^P) = \frac{1}{2}(1^-)$

Mass $m = 1412 \pm 12$ MeV (S = 1.1)
Full width $\Gamma = 227 \pm 22$ MeV (S = 1.1)

K*(1410) DECAY MODES	Fraction (Γ_i/Γ)	Confidence level	p (MeV/c)
$K^*(892)\pi$	> 40 %	95%	408
$K\pi$	(6.6±1.3) %		611
$K\rho$	< 7 %	95%	309

$K_0^*(1430)$

$I(J^P) = \frac{1}{2}(0^+)$

Mass $m = 1429 \pm 6$ MeV
Full width $\Gamma = 287 \pm 23$ MeV

$K_0^*(1430)$ DECAY MODES	Fraction (Γ_i/Γ)	p (MeV/c)
$K\pi$	(93±10) %	621

$K_2^*(1430)$

$I(J^P) = \frac{1}{2}(2^+)$

$K_2^*(1430)^{\pm}$ mass $m = 1425.4 \pm 1.3$ MeV (S = 1.1)
$K_2^*(1430)^0$ mass $m = 1432.4 \pm 1.3$ MeV
$K_2^*(1430)^{\pm}$ full width $\Gamma = 98.4 \pm 2.3$ MeV
$K_2^*(1430)^0$ full width $\Gamma = 109 \pm 5$ MeV (S = 1.9)

$K_2^*(1430)$ DECAY MODES	Fraction (Γ_i/Γ)	Scale factor/ Confidence level	p (MeV/c)
$K\pi$	(49.7±1.2) %		622
$K^*(892)\pi$	(25.2±1.7) %		423
$K^*(892)\pi\pi$	(13.0±2.3) %		375
$K\rho$	(8.8±0.8) %	S=1.2	331
$K\omega$	(2.9±0.8) %		319
$K^+\gamma$	(2.4±0.5) × 10^{-3}		627
$K\eta$	($1.4^{+2.8}_{-0.9}$) × 10^{-3}	S=1.1	492
$K\omega\pi$	< 7.2 × 10^{-4}	CL=95%	110
$K^0\gamma$	< 9 × 10^{-4}	CL=90%	631

K*(1680)

$I(J^P) = \frac{1}{2}(1^-)$

Mass $m = 1714 \pm 20$ MeV (S = 1.1)
Full width $\Gamma = 323 \pm 110$ MeV (S = 4.2)

K*(1680) DECAY MODES	Fraction (Γ_i/Γ)	p (MeV/c)
$K\pi$	(38.7±2.5) %	779
$K\rho$	($31.4^{+4.7}_{-2.1}$) %	571
$K^*(892)\pi$	($29.9^{+2.2}_{-4.7}$) %	615

$K_2(1770)$ [ff]
was L(1770)

$I(J^P) = \frac{1}{2}(2^-)$

Mass $m = 1773 \pm 8$ MeV
Full width $\Gamma = 186 \pm 14$ MeV

$K_2(1770)$ DECAY MODES	Fraction (Γ_i/Γ)	p (MeV/c)
$K\pi\pi$		–
$K_2^*(1430)\pi$	dominant	287
$K^*(892)\pi$	seen	653
$K f_2(1270)$	seen	–
$K\phi$	seen	441
$K\omega$	seen	608

$K_3^*(1780)$

$I(J^P) = \frac{1}{2}(3^-)$

Mass $m = 1770 \pm 10$ MeV (S = 1.7)
Full width $\Gamma = 164 \pm 17$ MeV (S = 1.1)

$K_3^*(1780)$ DECAY MODES	Fraction (Γ_i/Γ)	Scale factor/ Confidence level	p (MeV/c)
$K\rho$	(45 ±4) %	S=1.4	612
$K^*(892)\pi$	(27.3±3.2) %	S=1.5	651
$K\pi$	(19.3±1.0) %		810
$K\eta$	(8.0±1.5) %	S=1.4	715
$K_2^*(1430)\pi$	< 21 %	CL=95%	284

$K_2(1820)$

$I(J^P) = \frac{1}{2}(2^-)$

Mass $m = 1816 \pm 13$ MeV
Full width $\Gamma = 276 \pm 35$ MeV

$K_2(1820)$ DECAY MODES	Fraction (Γ_i/Γ)	p (MeV/c)
$K\phi$	possibly seen	481
$K_2^*(1430)\pi$	seen	325
$K^*(892)\pi$	seen	680
$K f_2(1270)$	seen	186
$K\omega$	seen	638

$K_4^*(2045)$

$I(J^P) = \frac{1}{2}(4^+)$

Mass $m = 2045 \pm 9$ MeV (S = 1.1)
Full width $\Gamma = 198 \pm 30$ MeV

$K_4^*(2045)$ DECAY MODES	Fraction (Γ_i/Γ)	p (MeV/c)
$K\pi$	(9.9±1.2) %	958
$K^*(892)\pi\pi$	(9 ±5) %	800
$K^*(892)\pi\pi\pi$	(7 ±5) %	764
$\rho K\pi$	(5.7±3.2) %	742
$\omega K\pi$	(5.0±3.0) %	736
$\phi K\pi$	(2.8±1.4) %	591
$\phi K^*(892)$	(1.4±0.7) %	363

CHARMED MESONS
($C = \pm 1$)

$D^+ = c\bar{d}$, $D^0 = c\bar{u}$, $\overline{D}^0 = \bar{c}u$, $D^- = \bar{c}d$, similarly for D^*'s

$D^{\pm}$

$I(J^P) = \frac{1}{2}(0^-)$

Mass $m = 1869.4 \pm 0.4$ MeV
Mean life $\tau = (1.057 \pm 0.015) \times 10^{-12}$ s
$c\tau = 317$ μm

D^- modes are charge conjugates of the modes below.

D^+ DECAY MODES		Fraction (Γ_i/Γ)	Scale factor/ Confidence level	p (MeV/c)
Inclusive modes				
e^+ anything		(17.2 ±1.9) %		–
K^- anything		(24.2 ±2.8) %	S=1.4	–
$\overline{K}^0$ anything + K^0 anything		(59 ±7) %		–
K^+ anything		(5.8 ±1.4) %		–
η anything	[gg]	< 13 %	CL=90%	–
Leptonic and semileptonic modes				
$\mu^+\nu_\mu$		< 7.2 × 10^{-4}	CL=90%	932
$\overline{K}^0$ "e^+" ν_e	[hh]	(6.7 ±0.8) %		868
$\overline{K}^0 e^+\nu_e$		(6.6 ±0.9) %		868
$\overline{K}^0 \mu^+\nu_\mu$		(7.0 $^{+3.0}_{-2.0}$) %		865
$\overline{K}^0 \ell^+\nu_\ell$		(6.7 ±3.5) %		–
$K^-\pi^+e^+\nu_e$		(4.2 $^{+0.9}_{-0.7}$) %		863
$\overline{K}^*(892)^0 e^+\nu_e$ × B($\overline{K}^{*0} \to K^-\pi^+$)		(3.2 ±0.33) %		720
$K^-\pi^+e^+\nu_e$ nonresonant		< 7 × 10^{-3}	CL=90%	863
$K^-\pi^+\mu^+\nu_\mu$		(3.2 ±1.7) %		851
$\overline{K}^*(892)^0 \mu^+\nu_\mu$ × B($\overline{K}^{*0} \to K^-\pi^+$)		(3.0 ±0.4) %		715
$K^-\pi^+\mu^+\nu_\mu$ nonresonant		(2.7 ±1.1) × 10^{-3}		851
$(\overline{K}^*(892)\pi)^0 e^+\nu_e$		< 1.2 %	CL=90%	714
$(\overline{K}\pi\pi)^0 e^+\nu_e$ non-$\overline{K}^*(892)$		< 9 × 10^{-3}	CL=90%	846
$K^-\pi^+\pi^0\mu^+\nu_\mu$		< 1.4 × 10^{-3}	CL=90%	825
$\pi^0\ell^+\nu_\ell$	[ii]	(5.7 ±2.2) × 10^{-3}		–
Fractions of some of the following modes with resonances have already appeared above as submodes of particular charged-particle modes.				
$\overline{K}^*(892)^0$ "e^+" ν_e	[hh]	(4.8 ±0.4) %		720
$\overline{K}^*(892)^0 e^+\nu_e$		(4.8 ±0.5) %		720
$\overline{K}^*(892)^0 \mu^+\nu_\mu$		(4.5 ±0.6) %	S=1.1	715
$\rho^0 e^+\nu_e$		< 3.7 × 10^{-3}	CL=90%	776
$\rho^0 \mu^+\nu_\mu$		(2.0 $^{+1.5}_{-1.3}$) × 10^{-3}		772
$\phi e^+\nu_e$		< 2.09 %	CL=90%	657
$\phi \mu^+\nu_\mu$		< 3.72 %	CL=90%	651
$\eta'(958)\mu^+\nu_\mu$		< 9 × 10^{-3}	CL=90%	684
Hadronic modes with one or three K's				
$\overline{K}^0\pi^+$		(2.74±0.29) %		862
$K^-\pi^+\pi^+$	[jj]	(9.1 ±0.6) %		845
$\overline{K}^*(892)^0\pi^+$ × B($\overline{K}^{*0} \to K^-\pi^+$)		(1.5 ±0.3) %		712
$\overline{K}_0(1430)^0\pi^+$ × B($\overline{K}^*(1430)^0 \to K^-\pi^+$)		(2.3 ±0.3) %		368
$\overline{K}^*(1680)^0\pi^+$ × B($\overline{K}^*(1680)^0 \to K^-\pi^+$)		(2.6 ±1.3) × 10^{-3}		65
$K^-\pi^+\pi^+$ nonresonant		(7.3 ±1.4) %		845
$\overline{K}^0\pi^+\pi^0$	[jj]	(9.7 ±3.0) %	S=1.1	845
$\overline{K}^0\rho^+$		(6.6 ±2.5) %		680
$\overline{K}^*(892)^0\pi^+$ × B($\overline{K}^{*0} \to \overline{K}^0\pi^0$)		(0.7 ±0.2) %		712
$\overline{K}^0\pi^+\pi^0$ nonresonant		(1.3 ±1.1) %		845
$K^-\pi^+\pi^+\pi^0$	[jj]	(6.4 ±1.1) %		816
$\overline{K}^*(892)^0\rho^+$ total × B($\overline{K}^{*0} \to K^-\pi^+$)		(1.4 ±0.9) %		423
$\overline{K}_1(1400)^0\pi^+$ × B($\overline{K}_1(1400)^0 \to K^-\pi^+\pi^0$)		(2.2 ±0.6) %		390
$K^-\rho^+\pi^+$ total		(3.1 ±1.1) %		616
$\overline{K}^*(892)^0\pi^+\pi^0$ total × B($\overline{K}^{*0} \to K^-\pi^+$)		(4.5 ±0.9) %		687
$\overline{K}^*(892)^0\pi^+\pi^0$ 3-body × B($\overline{K}^{*0} \to K^-\pi^+$)		(2.8 ±0.9) %		687
$K^*(892)^-\pi^+\pi^+$ 3-body × B($K^{*-} \to K^-\pi^0$)		(1.4 ±0.6) %		688
$K^-\pi^+\pi^+\pi^0$ nonresonant	[kk]	(1.2 ±0.6) %		816
$\overline{K}^0\pi^+\pi^+\pi^-$	[jj]	(7.0 ±1.0) %		814
$\overline{K}^0 a_1(1260)^+$ × B($a_1(1260)^+ \to \pi^+\pi^+\pi^-$)		(4.0 ±0.8) %		328
$\overline{K}_1(1400)^0\pi^+$ × B($\overline{K}_1(1400)^0 \to \overline{K}^0\pi^+\pi^-$)		(2.2 ±0.6) %		390
$K^*(892)^-\pi^+\pi^+$ 3-body × B($K^{*-} \to \overline{K}^0\pi^-$)		(1.4 ±0.6) %		688
$\overline{K}^0\rho^0\pi^+$ total		(4.2 ±0.9) %		614
$\overline{K}^0\pi^+\pi^+\pi^-$ nonresonant		(8 ±4) × 10^{-3}		814
$K^-\pi^+\pi^+\pi^+\pi^-$		(8.2 ±1.4) × 10^{-3}		772
$\overline{K}^*(892)^0\pi^+\pi^+\pi^-$ × B($\overline{K}^{*0} \to K^-\pi^+$)		(6.8 ±1.8) × 10^{-3}		642
$\overline{K}^*(892)^0\rho^0\pi^+$ × B($\overline{K}^{*0} \to K^-\pi^+$)		(5.1 ±2.2) × 10^{-3}		242
$K^-\pi^+\pi^+\pi^0\pi^0$		(2.2 $^{+5.0}_{-0.9}$) %		775
$\overline{K}^0\pi^+\pi^+\pi^-\pi^0$		(5.4 $^{+3.0}_{-1.4}$) %		773
$\overline{K}^0\pi^+\pi^+\pi^+\pi^-\pi^-$		(8 ±7) × 10^{-4}		714
$K^-\pi^+\pi^+\pi^+\pi^-\pi^0$		(2.0 ±1.8) × 10^{-3}		719
$\overline{K}^0\overline{K}^0K^+$		(3.1 ±0.7) %		545
Fractions of some of the following modes with resonances have already appeared above as submodes of particular charged-particle modes.				
$\overline{K}^0\rho^+$		(6.6 ±2.5) %		680
$\overline{K}^0 a_1(1260)^+$		(8.1 ±1.7) %		328
$\overline{K}^0 a_2(1320)^+$		< 3 × 10^{-3}	CL=90%	199
$\overline{K}^*(892)^0\pi^+$		(2.2 ±0.4) %		712
$\overline{K}^*(892)^0\rho^+$ total		(2.1 ±1.4) %		423
$\overline{K}^*(892)^0\rho^+$ S-wave	[kk]	(1.7 ±1.6) %		423
$\overline{K}^*(892)^0\rho^+$ P-wave		< 1 × 10^{-3}	CL=90%	423
$\overline{K}^*(892)^0\rho^+$ D-wave		(10 ±7) × 10^{-3}		423
$\overline{K}^*(892)^0\rho^+$ D-wave longitudinal		< 7 × 10^{-3}	CL=90%	423
$\overline{K}_1(1270)^0\pi^+$		< 7 × 10^{-3}	CL=90%	487
$\overline{K}_1(1400)^0\pi^+$		(5.0 ±1.3) %		390
$\overline{K}^*(1410)^0\pi^+$		< 7 × 10^{-3}	CL=90%	382
$\overline{K}^*_0(1430)^0\pi^+$		(3.4 ±0.4) %		368
$\overline{K}^*(1680)^0\pi^+$		(1.0 ±0.5) %		65
$\overline{K}^*(892)^0\pi^+\pi^0$ total		(6.7 ±1.4) %		687
$\overline{K}^*(892)^0\pi^+\pi^0$ 3-body		(4.2 ±1.4) %		687
$K^*(892)^-\pi^+\pi^+$ 3-body		(2.1 ±0.9) %		688
$K^-\rho^+\pi^+$ total		(3.1 ±1.1) %		616
$K^-\rho^+\pi^+$ 3-body		(1.1 ±0.4) %		616
$\overline{K}^0\rho^0\pi^+$ total		(4.2 ±0.9) %	CL=90%	614
$\overline{K}^0\rho^0\pi^+$ 3-body		(5 ±5) × 10^{-3}		614
$\overline{K}^0 f_0(980)\pi^+$		< 5 × 10^{-3}	CL=90%	461
$\overline{K}^*(892)^0\pi^+\pi^+\pi^-$		(1.02±0.27) %		642
$\overline{K}^*(892)^0\rho^0\pi^+$		(7.7 ±3.3) × 10^{-3}		242
Pionic modes				
$\pi^+\pi^0$		(2.5 ±0.7) × 10^{-3}		925
$\pi^+\pi^+\pi^-$		(3.2 ±0.6) × 10^{-3}		908
$\rho^0\pi^+$		< 1.4 × 10^{-3}	CL=90%	769
$\pi^+\pi^+\pi^-$ nonresonant		(2.5 ±0.7) × 10^{-3}		908
$\pi^+\pi^+\pi^-\pi^0$		(1.9 $^{+1.5}_{-1.2}$) %		883
$\eta\pi^+$ × B($\eta \to \pi^+\pi^-\pi^0$)		(1.8 ±0.6) × 10^{-3}		848
$\omega\pi^+$ × B($\omega \to \pi^+\pi^-\pi^0$)		< 6 × 10^{-3}	CL=90%	764
$\pi^+\pi^+\pi^+\pi^-\pi^-$		(1.0 $^{+0.8}_{-0.7}$) × 10^{-3}		845
$\pi^+\pi^+\pi^+\pi^-\pi^-\pi^0$		(2.9 $^{+2.9}_{-2.0}$) × 10^{-3}		799

Fractions of some of the following modes with resonances have already appeared above as submodes of particular charged-particle modes.

Mode		Fraction	CL	p
$\rho^0\pi^+$		$< 1.4 \times 10^{-3}$	CL=90%	769
$\eta\pi^+$		$(7.5\ +2.5) \times 10^{-3}$		848
$\omega\pi^+$		$< 7 \times 10^{-3}$	CL=90%	764
$\eta\rho^+$		< 1.2 %	CL=90%	658
$\eta'(958)\pi^+$		$< 9 \times 10^{-3}$	CL=90%	680
$\eta'(958)\rho^+$		< 1.5 %	CL=90%	355
Hadronic modes with two K's				
$\overline{K}^0 K^+$		$(7.8 \pm 1.7) \times 10^{-3}$		792
$K^+K^-\pi^+$		(1.13 ± 0.13) %		744
$\phi\pi^+ \times B(\phi \to K^+K^-)$		$(3.3 \pm 0.4) \times 10^{-3}$		647
$\overline{K}^*(892)^0 K^+ \times B(\overline{K}^{*0} \to K^-\pi^+)$		$(3.4 \pm 0.7) \times 10^{-3}$		610
$K^+K^-\pi^+$ nonresonant		$(4.6 \pm 0.9) \times 10^{-3}$		744
$K^+K^-\pi^+\pi^0$				682
$\phi\pi^+\pi^0 \times B(\phi \to K^+K^-)$		(1.2 ± 0.5) %		619
$\phi\rho^+ \times B(\phi \to K^+K^-)$		$< 7 \times 10^{-3}$	CL=90%	268
$K^+K^-\pi^+\pi^0$ non-ϕ		$(1.5\ ^{+0.7}_{-0.6})$ %		682
$K^+\overline{K}^0\pi^+\pi^-$		< 2 %	CL=90%	678
$K^0K^-\pi^+\pi^+$		(1.0 ± 0.6) %		678
$K^*(892)^+\overline{K}^*(892)^0 \times B^2(K^* \to K\pi^+)$		(1.2 ± 0.5) %		273
$K^0K^-\pi^+\pi^+$ non-$K^{*+}\overline{K}^{*0}$		$< 7.9 \times 10^{-3}$	CL=90%	678
$K^+K^-\pi^+\pi^+\pi^-$				600
$\phi\pi^+\pi^+\pi^- \times B(\phi \to K^+K^-)$		$< 1 \times 10^{-3}$	CL=90%	566
$K^+K^-\pi^+\pi^+\pi^-$ nonresonant		< 3 %	CL=90%	600

Fractions of the following modes with resonances have already appeared above as submodes of particular charged-particle modes.

Mode		Fraction	CL	p
$\phi\pi^+$		$(6.7 \pm 0.8) \times 10^{-3}$		647
$\overline{K}^*(892)^0 K^+$		$(5.1 \pm 1.0) \times 10^{-3}$		610
$\phi\pi^+\pi^0$		(2.3 ± 1.0) %		619
$\phi\rho^+$		< 1.5 %	CL=90%	268
$K^*(892)^+\overline{K}^*(892)^0$		(2.6 ± 1.1) %		273
$\phi\pi^+\pi^+\pi^-$		$< 2 \times 10^{-3}$	CL=90%	566

Doubly Cabibbo suppressed (DC) modes, $\Delta C = 1$ weak neutral current ($C1$) modes, or Lepton Family number (LF) or Lepton number (L) violating modes

Mode			Fraction	CL	p
$K^+\pi^+\pi^-$	DC		$< 5 \times 10^{-3}$	CL=90%	845
$K^+K^+K^-$	DC		$(5.2 \pm 2.0) \times 10^{-3}$		550
ϕK^+	DC		$(3.9\ ^{+2.2}_{-1.9}) \times 10^{-4}$		527
$\pi^+e^+e^-$	C1		$< 2.5 \times 10^{-3}$	CL=90%	929
$\pi^+\mu^+\mu^-$	C1		$< 2.9 \times 10^{-3}$	CL=90%	917
$K^+e^+e^-$		[ll]	$< 4.8 \times 10^{-3}$	CL=90%	870
$K^+\mu^+\mu^-$		[ll]	$< 9.2 \times 10^{-3}$	CL=90%	856
$\pi^+e^\pm\mu^\mp$	LF	[q]	$< 3.8 \times 10^{-3}$	CL=90%	926
$\pi^+e^+\mu^-$	LF		$< 3.3 \times 10^{-3}$	CL=90%	926
$\pi^+e^-\mu^+$	LF		$< 3.3 \times 10^{-3}$	CL=90%	926
$K^+e^+\mu^-$	LF		$< 3.4 \times 10^{-3}$	CL=90%	866
$K^+e^-\mu^+$	LF		$< 3.4 \times 10^{-3}$	CL=90%	866
$\pi^-e^+e^+$	L		$< 4.8 \times 10^{-3}$	CL=90%	929
$\pi^-\mu^+\mu^+$	L		$< 6.8 \times 10^{-3}$	CL=90%	917
$\pi^-e^+\mu^+$	L		$< 3.7 \times 10^{-3}$	CL=90%	926
$K^-e^+e^+$	L		$< 9.1 \times 10^{-3}$	CL=90%	870
$K^-\mu^+\mu^+$	L		$< 4.3 \times 10^{-3}$	CL=90%	856
$K^-e^+\mu^+$	L		$< 4.0 \times 10^{-3}$	CL=90%	866

D^0

$I(J^P) = \frac{1}{2}(0^-)$

Mass $m = 1864.6 \pm 0.5$ MeV

$|m_{D^0_1} - m_{D^0_2}| < 20 \times 10^{10}\ \hbar\ s^{-1}$, CL = 90% [mm]

$m_{D^\pm} - m_{D^0} = 4.78 \pm 0.10$ MeV

Mean life $\tau = (0.415 \pm 0.004) \times 10^{-12}$ s

$c\tau = 124.4\ \mu m$

$|\tau_{D^0_1} - \tau_{D^0_2}|/\tau_{D^0} < 0.17$, CL = 90% [mm]

$\Gamma(K^+\pi^- (\text{via } \overline{D}^0))/\Gamma(K^-\pi^+) < 0.0037$, CL = 90%

$\Gamma(\mu^- X\ (\text{via } \overline{D}^0))/\Gamma(\mu^+ X) < 0.0056$, CL = 90%

$[\Gamma(D^0 \to K^+K^-) - \Gamma(\overline{D}^0 \to K^+K^-)]/\text{sum} < 0.45$, CL = 90%

$\overline{D}^0$ modes are charge conjugates of the modes below.

D^0 DECAY MODES		Fraction (Γ_i/Γ)	Scale factor/ Confidence level	p (MeV/c)
Inclusive modes				
e^+ anything		(7.7 ± 1.2) %	S=1.1	–
μ^+ anything		(10.0 ± 2.6) %		–
K^- anything		(53 ± 4) %	S=1.3	–
$\overline{K}^0$ anything + K^0 anything		(42 ± 5) %		–
K^+ anything		$(3.4\ ^{+0.6}_{-0.4})$ %		–
η anything	[gg]	< 13 %	CL=90%	–
Semileptonic modes				
K^- "e^+" ν_e	[hh]	(3.68 ± 0.21) %	S=1.1	867
$K^-e^+\nu_e$		(3.80 ± 0.22) %	S=1.1	867
$K^-\mu^+\nu_\mu$		(3.2 ± 0.4) %		864
$K^-\pi^0e^+\nu_e$	[nn]	$(1.6\ ^{+1.3}_{-0.5})$ %		861
$\overline{K}^0\pi^-e^+\nu_e$	[nn]	$(2.8\ ^{+1.7}_{-0.9})$ %		860
$\overline{K}^*(892)^-e^+\nu_e \times B(K^{*-} \to \overline{K}^0\pi^+)$		(1.3 ± 0.3) %		719
$\overline{K}^*(892)^0\pi^-e^+\nu_e$	[oo]	< 1.3 %	CL=90%	709
$K^-\pi^+\pi^-\mu^+\nu_\mu$		$< 1.2 \times 10^{-3}$	CL=90%	821
$(\overline{K}^*(892)\pi)^-\mu^+\nu_\mu$		$< 1.4 \times 10^{-3}$	CL=90%	694
$\pi^-e^+\nu_e$		$(3.9\ ^{+2.3}_{-1.2}) \times 10^{-3}$		927

A fraction of the following resonance mode has already appeared above as a submode of a particular charged-particle mode.

Mode		Fraction	Scale factor/CL	p
$K^*(892)^-e^+\nu_e$		(2.0 ± 0.4) %		719
Hadronic modes with one or three K's				
$K^-\pi^+$		(4.01 ± 0.14) %		861
$\overline{K}^0\pi^0$		(2.05 ± 0.26) %	S=1.1	860
$\overline{K}^0\pi^+\pi^-$	[ii]	(5.3 ± 0.6) %	S=1.2	842
$\overline{K}^0\rho^0$		(1.10 ± 0.18) %		676
$\overline{K}^0 f_0(980) \times B(f_0 \to \pi^+\pi^-)$		$(2.4 \pm 1.0) \times 10^{-3}$		549
$\overline{K}^0 f_2(1270) \times B(f_2 \to \pi^+\pi^-)$		$(2.6 \pm 1.2) \times 10^{-3}$		263
$\overline{K}^0 f_0(1300) \times B(f_0 \to \pi^+\pi^-)$		$(4.3 \pm 1.7) \times 10^{-3}$		223
$K^*(892)^-\pi^+ \times B(K^{*-} \to \overline{K}^0\pi^-)$		(3.3 ± 0.4) %		711
$K_0^*(1430)^-\pi^+ \times B(K_0^*(1430)^- \to \overline{K}^0\pi^-)$		$(7 \pm 3) \times 10^{-3}$		364
$\overline{K}^0\pi^+\pi^-$ nonresonant		(1.43 ± 0.26) %		842
$K^-\pi^+\pi^0$	[ii]	(13.8 ± 1.0) %	S=1.1	844
$K^-\rho^+$		(10.4 ± 1.3) %		678
$K^*(892)^-\pi^+ \times B(K^{*-} \to K^-\pi^0)$		(1.6 ± 0.2) %		711
$\overline{K}^*(892)^0\pi^0 \times B(\overline{K}^{*0} \to K^-\pi^+)$		(2.0 ± 0.3) %		709
$K^-\pi^+\pi^0$ nonresonant		$(6.0 \pm 2.7) \times 10^{-3}$		844
$\overline{K}^0\pi^0\pi^0$				843
$\overline{K}^*(892)^0\pi^0 \times B(\overline{K}^{*0} \to \overline{K}^0\pi^0)$		(1.0 ± 0.2) %		709
$\overline{K}^0\pi^0\pi^0$ nonresonant		$(7.6 \pm 2.1) \times 10^{-3}$		843

Mode		Fraction	Scale factor/CL	p
$K^-\pi^+\pi^+\pi^-$	[jj]	(8.1 ± 0.5) %		812
$K^-\pi^+\rho^0$ total		(6.8 ± 0.5) %		612
$K^-\pi^+\rho^0$ 3-body		$(5.1 \pm 2.3) \times 10^{-3}$		612
$\overline{K}^*(892)^0\rho^0$ $\times$ B($\overline{K}^{*0} \to K^-\pi^+$)		(1.1 ± 0.3) %		418
$K^-a_1(1260)^+$ $\times$ B($a_1(1260)^+ \to \pi^+\pi^+\pi^-$)		(3.9 ± 0.6) %		327
$\overline{K}^*(892)^0\pi^+\pi^-$ total $\times$ B($\overline{K}^{*0} \to K^-\pi^+$)		(1.6 ± 0.4) %		683
$\overline{K}^*(892)^0\pi^+\pi^-$ 3-body $\times$ B($\overline{K}^{*0} \to K^-\pi^+$)		(1.01 ± 0.22) %		683
$K_1(1270)^-\pi^+$ $\times$ B($K_1(1270)^- \to K^-\pi^+\pi^-$)		$(3.5 \pm 1.1) \times 10^{-3}$		483
$K^-\pi^+\pi^+\pi^-$ nonresonant		(1.89 ± 0.28) %		812
$\overline{K}^0\pi^+\pi^-\pi^0$	[jj]	(9.8 ± 1.4) %	S=1.1	812
$\overline{K}^0\eta$ $\times$ B($\eta \to \pi^+\pi^-\pi^0$)		$(1.61 \pm 0.26) \times 10^{-3}$		772
$\overline{K}^0\omega$ $\times$ B($\omega \to \pi^+\pi^-\pi^0$)		(1.8 ± 0.4) %		670
$K^*(892)^-\rho^+$ $\times$ B($K^{*-} \to \overline{K}^0\pi^-$)		(3.9 ± 1.6) %		422
$\overline{K}^*(892)^0\rho^0$ $\times$ B($\overline{K}^{*0} \to \overline{K}^0\pi^0$)		$(5.3 \pm 1.4) \times 10^{-3}$		418
$K_1(1270)^-\pi^+$ $\times$ B($K_1(1270)^- \to \overline{K}^0\pi^-\pi^0$)	[kk]	$(5.0 \pm 1.5) \times 10^{-3}$		483
$\overline{K}^*(892)^0\pi^+\pi^-$ 3-body $\times$ B($\overline{K}^{*0} \to \overline{K}^0\pi^0$)		$(5.1 \pm 1.1) \times 10^{-3}$		683
$\overline{K}^0\pi^+\pi^-\pi^0$ nonresonant		(2.1 ± 2.1) %		812
$K^-\pi^+\pi^0\pi^0$		(15 ± 5) %		815
$K^-\pi^+\pi^+\pi^-\pi^0$		(4.3 ± 0.4) %		771
$\overline{K}^*(892)^0\pi^+\pi^-\pi^0$ $\times$ B($\overline{K}^{*0} \to K^-\pi^+$)		(1.3 ± 0.6) %		641
$\overline{K}^*(892)^0\eta$ $\times$ B($\overline{K}^{*0} \to K^-\pi^+$) $\times$ B($\eta \to \pi^+\pi^-\pi^0$)		$(3.0 \pm 0.8) \times 10^{-3}$		580
$K^-\pi^+\omega$ $\times$ B($\omega \to \pi^+\pi^-\pi^0$)		(2.8 ± 0.5) %		605
$\overline{K}^*(892)^0\omega$ $\times$ B($\overline{K}^{*0} \to K^-\pi^+$) $\times$ B($\omega \to \pi^+\pi^-\pi^0$)		$(7 \pm 3) \times 10^{-3}$		406
$\overline{K}^0\pi^+\pi^+\pi^-\pi^-$		$(5.6 \pm 1.7) \times 10^{-3}$		768
$\overline{K}^0\pi^+\pi^-\pi^0\pi^0(\pi^0)$		$(10.6 {}^{+7.3}_{-3.0})$ %		771
$\overline{K}^0K^+K^-$		$(9.1 \pm 1.2) \times 10^{-3}$		544
$\overline{K}^0\phi$ $\times$ B($\phi \to K^+K^-$)		$(4.2 \pm 0.6) \times 10^{-3}$		520
$\overline{K}^0K^+K^-$ non-ϕ		$(4.9 \pm 0.9) \times 10^{-3}$		544
$K^0_S K^0_S K^0_S$		$(8.6 \pm 2.5) \times 10^{-4}$		538
$K^+K^-\overline{K}^0\pi^0$		$(7.2 {}^{+4.8}_{-3.5}) \times 10^{-3}$		435

Fractions of many of the following modes with resonances have already appeared above as submodes of particular charged-particle modes. (Modes for which there are only upper limits and $\overline{K}^*(892)\rho$ submodes only appear below.)

Mode		Fraction	Scale factor/CL	p
$\overline{K}^0\eta$		$(6.8 \pm 1.1) \times 10^{-3}$		772
$\overline{K}^0\rho^0$		(1.10 ± 0.18) %		676
$K^-\rho^+$		(10.4 ± 1.3) %	S=1.2	679
$\overline{K}^0\omega$		(2.0 ± 0.4) %		670
$\overline{K}^0\eta'(958)$		(1.66 ± 0.29) %		565
$\overline{K}^0 f_0(980)$		$(4.6 \pm 2.0) \times 10^{-3}$		549
$\overline{K}^0\phi$		$(8.3 \pm 1.2) \times 10^{-3}$	S=1.1	520
$K^-a_1(1260)^+$		(7.9 ± 1.2) %		327
$\overline{K}^0a_1(1260)^0$		< 1.9 %	CL=90%	322
$\overline{K}^0f_2(1270)$		$(4.6 \pm 2.1) \times 10^{-3}$		263
$\overline{K}^0f_0(1300)$		$(6.9 \pm 2.7) \times 10^{-3}$		223
$K^-a_2(1320)^+$		$< 2 \times 10^{-3}$	CL=90%	197
$K^*(892)^-\pi^+$		(4.9 ± 0.6) %	S=1.3	711
$\overline{K}^*(892)^0\pi^0$		(3.0 ± 0.4) %		709
$\overline{K}^*(892)^0\pi^+\pi^-$ total		(2.4 ± 0.6) %		683
$\overline{K}^*(892)^0\pi^+\pi^-$ 3-body		(1.52 ± 0.33) %		683
$K^-\pi^+\rho^0$ total		(6.8 ± 0.5) %		612
$K^-\pi^+\rho^0$ 3-body		$(5.1 \pm 2.3) \times 10^{-3}$		612
$\overline{K}^*(892)^0\rho^0$		(1.6 ± 0.4) %		418
$\overline{K}^*(892)^0\rho^0$ transverse		(1.6 ± 0.5) %		418
$\overline{K}^*(892)^0\rho^0$ S-wave		(3.0 ± 0.6) %		418
$\overline{K}^*(892)^0\rho^0$ S-wave long.		$< 3 \times 10^{-3}$	CL=90%	418
$\overline{K}^*(892)^0\rho^0$ P-wave		$< 3 \times 10^{-3}$	CL=90%	418
$\overline{K}^*(892)^0\rho^0$ D-wave		(2.1 ± 0.6) %		418
$K^*(892)^-\rho^+$		(5.9 ± 2.4) %		422
$K^*(892)^-\rho^+$ longitudinal		(2.8 ± 1.2) %		422
$K^*(892)^-\rho^+$ transverse		(3.1 ± 1.8) %		422
$K^*(892)^-\rho^+$ P-wave		< 1.5 %	CL=90%	422

Mode		Fraction	Scale factor/CL	p
$K^-\pi^+f_0(980)$		< 1.1 %	CL=90%	459
$\overline{K}^*(892)^0 f_0(980)$		$< 7 \times 10^{-3}$	CL=90%	–
$K_1(1270)^-\pi^+$	[kk]	(1.04 ± 0.31) %		483
$K_1(1400)^-\pi^+$		< 1.2 %	CL=90%	386
$\overline{K}_1(1400)^0\pi^0$		< 3.7 %	CL=90%	387
$K^*(1410)^-\pi^+$		< 1.2 %	CL=90%	378
$K^*_0(1430)^-\pi^+$		(1.1 ± 0.4) %		364
$K^*_2(1430)^-\pi^+$		$< 8 \times 10^{-3}$	CL=90%	367
$\overline{K}^*_2(1430)^0\pi^0$		$< 4 \times 10^{-3}$	CL=90%	363
$\overline{K}^*(892)^0\pi^+\pi^-\pi^0$		(1.9 ± 0.9) %		641
$\overline{K}^*(892)^0\eta$		(1.9 ± 0.5) %		580
$K^-\pi^+\omega$		(3.1 ± 0.6) %		605
$\overline{K}^*(892)^0\omega$		(1.1 ± 0.5) %		406
$K^-\pi^+\eta'(958)$		$(7.5 \pm 2.0) \times 10^{-3}$		479
$\overline{K}^*(892)^0\eta'(958)$		$< 1.1 \times 10^{-3}$	CL=90%	100

Pionic modes

Mode		Fraction	Scale factor/CL	p
$\pi^+\pi^-$		$(1.59 \pm 0.12) \times 10^{-3}$		922
$\pi^0\pi^0$		$(8.8 \pm 2.3) \times 10^{-4}$		922
$\pi^+\pi^-\pi^0$		(1.6 ± 1.1) %	S=2.7	907
$\pi^+\pi^+\pi^-\pi^-$		$(8.3 \pm 0.9) \times 10^{-3}$		880
$\pi^+\pi^+\pi^-\pi^-\pi^0$		(1.9 ± 0.4) %		844
$\pi^+\pi^+\pi^+\pi^-\pi^-\pi^-$		$(4.0 \pm 3.0) \times 10^{-4}$		795

Hadronic modes with two K's

Mode		Fraction	Scale factor/CL	p
K^+K^-		$(4.54 \pm 0.29) \times 10^{-3}$		791
$K^0\overline{K}^0$		$(1.1 \pm 0.4) \times 10^{-3}$		788
$K^0K^-\pi^+$		$(6.3 \pm 1.1) \times 10^{-3}$	S=1.2	739
$\overline{K}^*(892)^0K^0$ $\times$ B($\overline{K}^{*0} \to K^-\pi^+$)		$< 1.0 \times 10^{-3}$	CL=90%	605
$K^*(892)^+K^-$ $\times$ B($K^{*+} \to K^0\pi^+$)		$(2.3 \pm 0.5) \times 10^{-3}$		610
$K^0K^-\pi^+$ nonresonant		$(2.4 \pm 2.4) \times 10^{-3}$		739
$\overline{K}^0K^+\pi^-$		$(4.9 \pm 1.0) \times 10^{-3}$		739
$K^*(892)^0\overline{K}^0$ $\times$ B($K^{*0} \to K^+\pi^-$)		$< 5 \times 10^{-4}$	CL=90%	605
$K^*(892)^-K^+$ $\times$ B($K^{*-} \to \overline{K}^0\pi^-$)		$(1.2 \pm 0.7) \times 10^{-3}$		610
$\overline{K}^0K^+\pi^-$ nonresonant		$(4.0 {}^{+2.4}_{-2.0}) \times 10^{-3}$		739
$K^+K^-\pi^+\pi^-$		$(2.4 \pm 0.5) \times 10^{-3}$		677
$\phi\pi^+\pi^-$ $\times$ B($\phi \to K^+K^-$)		$(1.3 \pm 0.4) \times 10^{-3}$		614
$\phi\rho^0$ $\times$ B($\phi \to K^+K^-$)		$(1.0 \pm 0.25) \times 10^{-3}$		260
$K^*(892)^0K^-\pi^+$ + c.c. $\times$ B($K^{*0} \to K^+\pi^-$)		$(5 {}^{+9}_{-5}) \times 10^{-4}$		528
$K^*(892)^0\overline{K}^*(892)^0$ $\times$ B^2($K^{*0} \to K^+\pi^-$)		$(1.3 {}^{+0.7}_{-0.6}) \times 10^{-3}$		257
$K^+K^-\pi^+\pi^-$ non-ϕ		$(1.7 \pm 0.5) \times 10^{-3}$		677
$K^+K^-\pi^+\pi^-$ nonresonant		$(8 {}^{+90}_{-8}) \times 10^{-5}$		677
$K^+K^-\pi^+\pi^-\pi^0$		$(3.1 \pm 2.0) \times 10^{-3}$		600

Fractions of the following modes with resonances have already appeared above as submodes of particular charged-particle modes.

Mode		Fraction	Scale factor/CL	p
$\overline{K}^*(892)^0K^0$		$< 1.5 \times 10^{-3}$	CL=90%	605
$K^*(892)^+K^-$		$(3.4 \pm 0.8) \times 10^{-3}$		610
$K^*(892)^0\overline{K}^0$		$< 8 \times 10^{-4}$	CL=90%	605
$K^*(892)^-K^+$		$(1.8 \pm 1.0) \times 10^{-3}$		610
$\phi\pi^+\pi^-$		$(2.6 \pm 0.7) \times 10^{-3}$		614
$\phi\rho^0$		$(1.9 \pm 0.5) \times 10^{-3}$		260
$K^*(892)^0K^-\pi^+$ + c.c.		$(8 {}^{+13}_{-8}) \times 10^{-4}$		528
$K^*(892)^0\overline{K}^*(892)^0$		$(2.9 {}^{+1.6}_{-1.3}) \times 10^{-3}$		257

Doubly Cabibbo suppressed (*DC*) modes, $\Delta C = 2$ forbidden via mixing (*C2M*) modes, $\Delta C = 1$ weak neutral current (*C1*) modes, or Lepton Family number (*LF*) violating modes

Mode		Fraction	Scale factor/CL	p
$K^+\pi^-$	DC	$(3.1 \pm 1.4) \times 10^{-4}$		861
$K^+\pi^-$ (via $\overline{D}^0$)	C2M	$< 1.5 \times 10^{-4}$	CL=90%	861
$K^+\pi^+\pi^-\pi^-$	DC	$< 1.5 \times 10^{-3}$	CL=90%	812
μ^- anything (via $\overline{D}^0$)	C2M	$< 6 \times 10^{-4}$	CL=90%	–
e^+e^-	C1	$< 1.3 \times 10^{-4}$	CL=90%	932
$\mu^+\mu^-$	C1	$< 1.1 \times 10^{-5}$	CL=90%	926
$\overline{K}^0e^+e^-$		$< 1.7 \times 10^{-3}$	CL=90%	866
$\rho^0e^+e^-$	C1	$< 4.5 \times 10^{-4}$	CL=90%	773
$\rho^0\mu^+\mu^-$	C1	$< 8.1 \times 10^{-4}$	CL=90%	756
$\mu^{\pm}e^{\mp}$	LF [q]	$< 1.0 \times 10^{-4}$	CL=90%	929

$D^*(2007)^0$

$I(J^P) = \frac{1}{2}(1^-)$
I, J, P need confirmation.

Mass $m = 2006.7 \pm 0.5$ MeV
$m_{D^{*0}} - m_{D^0} = 142.12 \pm 0.07$ MeV
Full width $\Gamma < 2.1$ MeV, CL = 90%

$\overline{D}^*(2007)^0$ modes are charge conjugates of modes below.

$D^*(2007)^0$ DECAY MODES	Fraction (Γ_i/Γ)	p (MeV/c)
$D^0\pi^0$	(63.6±2.8) %	43
$D^0\gamma$	(36.4±2.8) %	137

$D^*(2010)^{\pm}$

$I(J^P) = \frac{1}{2}(1^-)$
I, J, P need confirmation.

Mass $m = 2010.0 \pm 0.5$ MeV
$m_{D^*(2010)^+} - m_{D^+} = 140.64 \pm 0.09$ MeV
$m_{D^*(2010)^+} - m_{D^0} = 145.42 \pm 0.05$ MeV
Full width $\Gamma < 0.131$ MeV, CL = 90%

$D^*(2010)^-$ modes are charge conjugates of the modes below.

$D^*(2010)^{\pm}$ DECAY MODES	Fraction (Γ_i/Γ)	p (MeV/c)
$D^0\pi^+$	(68.1±1.3) %	39
$D^+\pi^0$	(30.8±0.8) %	38
$D^+\gamma$	($1.1^{+1.4}_{-0.7}$) %	136

$D_1(2420)^0$

$I(J^P) = \frac{1}{2}(1^+)$
I, J, P need confirmation.

Mass $m = 2422.8 \pm 3.2$ MeV (S = 1.6)
Full width $\Gamma = 18^{+6}_{-4}$ MeV

$\overline{D}_1(2420)^0$ modes are charge conjugates of modes below.

$D_1(2420)^0$ DECAY MODES	Fraction (Γ_i/Γ)	p (MeV/c)
$D^*(2010)^+\pi^-$	seen	355
$D^+\pi^-$	not seen	474

$D_2^*(2460)$

$I(J^P) = \frac{1}{2}(2^+)$

$J^P = 2^+$ assignment strongly favored (ALBRECHT 89B).

Mass $m_{D_2^*(2460)^0} = 2457.7 \pm 1.9$ MeV
Mass $m_{D_2^*(2460)^{\pm}} = 2456 \pm 6$ MeV (S = 2.0)
$m_{D_2^*(2460)^{\pm}} - m_{D_2^*(2460)^0} = 2 \pm 5$ MeV (S = 1.4)
Full width $\Gamma_{D_2^*(2460)^0} = 21 \pm 5$ MeV
Full width $\Gamma_{D_2^*(2460)^{\pm}} = 23 \pm 10$ MeV

$\overline{D}_2^*(2460)$ modes are charge conjugates of modes below.

$D_2^*(2460)$ DECAY MODES	Fraction (Γ_i/Γ)	p (MeV/c)
$D_2^*(2460)^0 \to D^+\pi^-$	seen	503
$D_2^*(2460)^0 \to D^*(2010)^+\pi^-$	seen	387
$D_2^*(2460)^{\pm} \to D^0\pi^+$	seen	505

CHARMED, STRANGE MESONS ($C = S = \pm1$)

$D_s^+ = c\overline{s}$, $D_s^- = \overline{c}s$, similarly for D_s^*'s

$D_s^{\pm}$ was $F^{\pm}$

$I(J^P) = 0(0^-)$

Mass $m = 1968.5 \pm 0.7$ MeV (S = 1.2)
$m_{D_s^{\pm}} - m_{D^{\pm}} = 99.1 \pm 0.6$ MeV (S = 1.1)
Mean life $\tau = (0.467 \pm 0.017) \times 10^{-12}$ s
$c\tau = 140$ μm

Branching fractions for modes below with a resonance in the final state include all the decay modes of the resonance. D_s^- modes are charge conjugates of the modes below.

Nearly all other modes are measured relative to the $\phi\pi^+$ mode. However, none of the determinations of the $\phi\pi^+$ branching fraction are direct measurements: all rely on calculated relations between D^+ and D_s^+ decay widths, on estimates of D_s^+ cross sections, or on other model-dependent assumptions. Thus a better determination of the $\phi\pi^+$ branching fraction could cause the other branching fractions to slide up or down, all together.

D_s^+ DECAY MODES		Fraction (Γ_i/Γ)	Scale factor/ Confidence level	p (MeV/c)
Inclusive modes				
K^- anything		($13 \ ^{+14}_{-12}$) %		–
$\overline{K}^0$ anything + K^0 anything		(39 ±28) %		–
K^+ anything		($20 \ ^{+18}_{-14}$) %		–
non-$K\overline{K}$ anything		(64 ±17) %		–
e^+ anything		< 20 %	CL=90%	–
Leptonic and semileptonic modes				
$\mu^+\nu_\mu$		(5.9 ± 2.2) × 10^{-3}	S=1.1	981
$\phi\ell^+\nu_\ell$	[pp]	(1.88± 0.29) %		–
$\eta\mu^+\nu_\mu$ + $\eta'(958)\mu^+\nu_\mu$		(7.4 ± 3.2) %		–
$\eta\mu^+\nu_\mu$				905
$\eta'(958)\mu^+\nu_\mu$		< 3.0 %	CL=90%	747
Hadronic modes with two K's (including from ϕ's)				
$K^+\overline{K}^0$		(3.5 ± 0.7) %		850
$K^+K^-\pi^+$	[qq]	(4.8 ± 0.7) %		805
$\phi\pi^+$		(3.5 ± 0.4) %		712
$K^+\overline{K}^*(892)^0$		(3.3 ± 0.5) %		682
$K^+K^-\pi^+$ nonresonant		(8.7 ± 3.2) × 10^{-3}		805
$K^0\overline{K}^0\pi^+$				802
$K^*(892)^+\overline{K}^0$		(4.2 ± 1.0) %		683
$K^+K^-\pi^+\pi^0$				748
$\phi\pi^+\pi^0$		(8 ± 4) %		687
$\phi\rho^+$		($6.5 \ ^{+1.6}_{-1.8}$) %		407
$\phi\pi^+\pi^0$ 3-body		< 2.5 %	CL=90%	687
$K^+K^-\pi^+\pi^0$ non-ϕ		< 8 %	CL=90%	748
$K^+\overline{K}^0\pi^+\pi^-$		< 2.7 %	CL=90%	744
$K^0K^-\pi^+\pi^+$		(4.2 ± 1.1) %		744
$K^*(892)^+\overline{K}^*(892)^0$		(5.6 ± 2.1) %		412
$K^0K^-\pi^+\pi^+$ non-$K^{*+}\overline{K}^{*0}$		< 2.8 %	CL=90%	744
$K^+K^-\pi^+\pi^+\pi^-$				673
$\phi\pi^+\pi^+\pi^-$		(1.8 ± 0.5) %		640
$K^+K^-\pi^+\pi^+\pi^-$ non-ϕ		($3.0 \ ^{+3.0}_{-2.0}$) × 10^{-3}		673

Other hadronic modes			
$\pi^+\pi^+\pi^-$	(1.35± 0.31) %		959
$\rho^0\pi^+$	< 2.8 × 10^{-3}	CL=90%	827
$f_0(980)\pi^+$	(10 ± 4) × 10^{-3}		732
$\pi^+\pi^+\pi^-$ nonresonant	(1.01± 0.35) %		959
$\pi^+\pi^+\pi^-\pi^0$	< 12 %	CL=90%	935
$\eta\pi^+$	(1.9 ± 0.4) %		902
$\omega\pi^+$	< 1.7 %	CL=90%	822
$\pi^+\pi^+\pi^+\pi^-\pi^-$	(3.0 $^{+4.0}_{-3.0}$) × 10^{-3}		899
$\pi^+\pi^+\pi^-\pi^0\pi^0$			902
$\eta\rho^+$	(10.0 ± 2.2) %		727
$\eta\pi^+\pi^0$ 3-body	< 2.9 %	CL=90%	787
$\pi^+\pi^+\pi^+\pi^-\pi^-\pi^0$	(4.9 ± 3.2) %		856
$\eta'(958)\pi^+$	(4.7 ± 1.4) %		743
$\pi^+\pi^+\pi^+\pi^-\pi^-\pi^0\pi^0$			803
$\eta'(958)\rho^+$	(12.0 ± 3.0) %		470
$\eta'(958)\pi^+\pi^0$ 3-body	< 3.0 %	CL=90%	720
$K^0\pi^+$	< 7 × 10^{-3}	CL=90%	916
$K^+\pi^+\pi^-$	(3.0 $^{+4.0}_{-3.0}$) × 10^{-3}		900
$K^+K^-K^+$			628
ϕK^+	< 2.5 × 10^{-3}	CL=90%	607

$D_s^{*\pm}$

$I(J^P) = ?(?^?)$

Mass $m = 2110.0 \pm 1.9$ MeV (S = 1.2)
$m_{D_s^{*\pm}} - m_{D_s^\pm} = 141.6 \pm 1.8$ MeV (S = 1.2)
Full width Γ < 4.5 MeV, CL = 90%

D_s^{*-} modes are charge conjugates of the modes below.

D_s^{*+} DECAY MODES	Fraction (Γ_i/Γ)	p (MeV/c)
$D_s^+\gamma$	dominant	137

$D_{s1}(2536)^\pm$

$I(J^P) = 0(1^+)$
I, J, P need confirmation.

Mass $m = 2535.35 \pm 0.34$ MeV
Full width Γ < 2.3 MeV, CL = 90%

$D_{s1}(2536)^-$ modes are charge conjugates of the modes below.

$D_{s1}(2536)^+$ DECAY MODES	Fraction (Γ_i/Γ)	p (MeV/c)
$D^*(2010)^+K^0$	seen	150
$D^*(2007)^0K^+$	seen	169
D^+K^0	not seen	382
D^0K^+	not seen	392
$D_s^{*+}\gamma$	possibly seen	389

BOTTOM MESONS
$(B = \pm 1)$

$B^+ = u\overline{b}$, $B^0 = d\overline{b}$, $\overline{B}^0 = \overline{d}\,b$, $B^- = \overline{u}\,b$, similarly for B^*'s

$B^\pm$

$I(J^P) = \frac{1}{2}(0^-)$

I, J, P need confirmation. Quantum numbers shown are quark-model predictions. Measurements which do not identify the charge state of B also appear here.

Mass $m_{B^\pm} = 5278.7 \pm 2.0$ MeV
Mean life $\tau = (1.54 \pm 0.11) \times 10^{-12}$ s
Mean life τ (avg over B hadrons) $= (1.537 \pm 0.021) \times 10^{-12}$ s [a]
$c\tau = 388\ \mu$m

B^- modes are charge conjugates of the modes below.

Only data from $\Upsilon(4S)$ decays are used for branching fractions, with rare exceptions. The branching fractions listed below assume a 50:50 $B^0\overline{B}^0$:B^+B^- production ratio at the $\Upsilon(4S)$. We have attempted to bring older measurements up to date by rescaling their assumed $\Upsilon(4S)$ production ratio to 50:50 and their assumed D, D_s, D^*, and ψ branching ratios to current values whenever this would effect our averages and best limits significantly.

Indentation is used to indicate a subchannel of a previous reaction. All resonant subchannels have been corrected for resonance branching fractions to the final state so the sum of the subchannel branching fractions can exceed that of the final state.

B^+ DECAY MODES	Fraction (Γ_i/Γ)	Scale factor/ Confidence level	p (MeV/c)
Semileptonic modes			
$B^+ \to \overline{D}^0\ell^+\nu$	[b] (1.6 ± 0.7) %		–
$B^+ \to \overline{D}^*(2007)^0\ell^+\nu$	[b] (6.6 ± 2.2) %		–
$B^+ \to \pi^0e^+\nu_e$	< 2.2 × 10^{-3}	CL=90%	2638
$B^+ \to \omega\ell^+\nu_\ell$	[b] < 2.1 × 10^{-4}	CL=90%	–
$B^+ \to \omega\mu^+\nu_\mu$	seen		2580
$B^+ \to \rho^0\ell^+\nu_\ell$	[b] < 2.1 × 10^{-4}	CL=90%	–
D, D^*, or D_s modes			
$B^+ \to D^0\pi^+$	(5.3 ± 0.5) × 10^{-3}		2308
$B^+ \to \overline{D}^0\rho^+$	(1.34± 0.18) %		2237
$B^+ \to \overline{D}^0\pi^+\pi^+\pi^-$	(1.1 ± 0.4) %		2289
$B^+ \to \overline{D}^0\pi^+\pi^+\pi^-$ nonresonant	(5 ± 4) × 10^{-3}		2289
$B^+ \to \overline{D}^0\pi^+\rho^0$	(4.2 ± 3.0) × 10^{-3}		2208
$B^+ \to \overline{D}^0a_1(1260)^+$	(5 ± 4) × 10^{-3}		2123
$B^+ \to D^*(2010)^-\pi^+\pi^+$	(2.1 ± 0.6) × 10^{-3}		2247
$B^+ \to D^-\pi^+\pi^+$	< 1.4 × 10^{-3}	CL=90%	2299
$B^+ \to \overline{D}^*(2007)^0\pi^+$	(5.2 ± 0.8) × 10^{-3}		2255
$B^+ \to \overline{D}^*(2007)^0\rho^+$	(1.55± 0.31) %		2182
$B^+ \to \overline{D}^*(2007)^0\pi^+\pi^+\pi^-$	(9.4 ± 2.6) × 10^{-3}		2236
$B^+ \to D^*(2010)^-\pi^+\pi^+\pi^0$	(1.5 ± 0.7) %		2235
$B^+ \to D^*(2010)^-\pi^+\pi^+\pi^+\pi^-$	< 1 %	CL=90%	2217
$B^+ \to \overline{D}_1^*(2420)^0\pi^+$	(1.1 ± 0.5) × 10^{-3}		2081
$B^+ \to \overline{D}_1^*(2420)^0\rho^+$	< 1.4 × 10^{-3}	CL=90%	1996
$B^+ \to \overline{D}_2^*(2460)^0\pi^+$	[illegible] × 10^{-3}	[illegible]	[illegible]
$B^+ \to \overline{D}_2^*(2460)^0\rho^+$	< 4.7 × 10^{-3}	CL=90%	1979
$B^+ \to \overline{D}^0D_s^+$	(1.7 ± 0.6) %		1814
$B^+ \to \overline{D}^0D_s^{*+}$	(1.2 ± 1.0) %		1735
$B^+ \to \overline{D}^*(2007)^0D_s^+$	(1.0 ± 0.7) %		1737
$B^+ \to \overline{D}^*(2007)^0D_s^{*+}$	(2.4 ± 1.3) %		1652
$B^+ \to D_s^+\pi^0$	< 2.1 × 10^{-4}	CL=90%	2270
$B^+ \to D_s^{*+}\pi^0$	< 3.4 × 10^{-4}	CL=90%	2215
$B^+ \to D_s^+\eta$	< 5 × 10^{-4}	CL=90%	2235
$B^+ \to D_s^{*+}\eta$	< 8 × 10^{-4}	CL=90%	2178
$B^+ \to D_s^+\rho^0$	< 4 × 10^{-4}	CL=90%	2197
$B^+ \to D_s^{*+}\rho^0$	< 5 × 10^{-4}	CL=90%	2139
$B^+ \to D_s^+\omega$	< 5 × 10^{-4}	CL=90%	2195
$B^+ \to D_s^{*+}\omega$	< 7 × 10^{-4}	CL=90%	2137
$B^+ \to D_s^+a_1(1260)^0$	< 2.3 × 10^{-3}	CL=90%	2079

$B^+ \to D_s^{*+} a_1(1260)^0$	$< 1.7 \times 10^{-3}$	CL=90%	2015
$B^+ \to D_s^+ \phi$	$< 3.3 \times 10^{-4}$	CL=90%	2140
$B^+ \to D_s^{*+} \phi$	$< 4 \times 10^{-4}$	CL=90%	2080
$B^+ \to D_s^+ \overline{K}^0$	$< 1.1 \times 10^{-3}$	CL=90%	2241
$B^+ \to D_s^{*+} \overline{K}^0$	$< 1.2 \times 10^{-3}$	CL=90%	2185
$B^+ \to D_s^+ \overline{K}^*(892)^0$	$< 5 \times 10^{-4}$	CL=90%	2171
$B^+ \to D_s^{*+} \overline{K}^*(892)^0$	$< 5 \times 10^{-4}$	CL=90%	2111
$B^+ \to D_s^- \pi^+ K^+$	$< 9 \times 10^{-4}$	CL=90%	2222
$B^+ \to D_s^{*-} \pi^+ K^+$	$< 1.2 \times 10^{-3}$	CL=90%	2165
$B^+ \to D_s^- \pi^+ K^*(892)^+$	$< 7 \times 10^{-3}$	CL=90%	2137
$B^+ \to D_s^{*-} \pi^+ K^*(892)^+$	$< 9 \times 10^{-3}$	CL=90%	2076

Charmonium modes

$B^+ \to J/\psi(1S) K^+$	$(1.02 \pm 0.14) \times 10^{-3}$		1683
$B^+ \to J/\psi(1S) K^+ \pi^+ \pi^-$	$(1.4 \pm 0.6) \times 10^{-3}$		1612
$B^+ \to J/\psi(1S) K^*(892)^+$	$(1.7 \pm 0.5) \times 10^{-3}$		1571
$B^+ \to \psi(2S) K^+$	$(6.9 \pm 3.1) \times 10^{-4}$	S=1.3	1284
$B^+ \to \psi(2S) K^*(892)^+$	$< 3.0 \times 10^{-3}$	CL=90%	1115
$B^+ \to \psi(2S) K^*(892)^+ \pi^+ \pi^-$	$(1.9 \pm 1.2) \times 10^{-3}$		909
$B^+ \to \chi_{c1}(1P) K^+$	$(1.0 \pm 0.4) \times 10^{-3}$		1411
$B^+ \to \chi_{c1}(1P) K^*(892)^+$	$< 2.1 \times 10^{-3}$	CL=90%	1265

K or K^* modes

$B^+ \to K^0 \pi^+$	$< 1.0 \times 10^{-4}$	CL=90%	2614
$B^+ \to K^*(892)^0 \pi^+$	$< 1.5 \times 10^{-4}$	CL=90%	2561
$B^+ \to K^+ \pi^- \pi^+$ (no charm)	$< 1.9 \times 10^{-4}$	CL=90%	2609
$B^+ \to K_1(1400)^0 \pi^+$	$< 2.6 \times 10^{-3}$	CL=90%	2451
$B^+ \to K_2^*(1430)^0 \pi^+$	$< 6.8 \times 10^{-4}$	CL=90%	2443
$B^+ \to K^+ \rho^0$	$< 8 \times 10^{-5}$	CL=90%	2559
$B^+ \to K^*(892)^+ \pi^+ \pi^-$	$< 1.1 \times 10^{-3}$	CL=90%	2556
$B^+ \to K^*(892)^+ \rho^0$	$< 9.0 \times 10^{-4}$	CL=90%	2505
$B^+ \to K_1(1400)^+ \rho^0$	$< 7.8 \times 10^{-4}$	CL=90%	2388
$B^+ \to K_2^*(1430)^+ \rho^0$	$< 1.5 \times 10^{-3}$	CL=90%	2382
$B^+ \to K^+ K^- K^+$	$< 3.5 \times 10^{-4}$	CL=90%	2522
$B^+ \to K^+ \phi$	$< 9 \times 10^{-5}$	CL=90%	2516
$B^+ \to K^*(892)^+ K^+ K^-$	$< 1.6 \times 10^{-3}$	CL=90%	2466
$B^+ \to K^*(892)^+ \phi$	$< 1.3 \times 10^{-3}$	CL=90%	2460
$B^+ \to K_1(1400)^+ \phi$	$< 1.1 \times 10^{-3}$	CL=90%	2339
$B^+ \to K_2^*(1430)^+ \phi$	$< 3.4 \times 10^{-3}$	CL=90%	2332
$B^+ \to K^+ f_0(980)$	$< 8 \times 10^{-5}$	CL=90%	2524
$B^+ \to K^*(892)^+ \gamma$	$(5.7 \pm 3.3) \times 10^{-5}$		2564
$B^+ \to K_1(1270)^+ \gamma$	$< 7.3 \times 10^{-3}$	CL=90%	2486
$B^+ \to K_1(1400)^+ \gamma$	$< 2.2 \times 10^{-3}$	CL=90%	2453
$B^+ \to K_2^*(1430)^+ \gamma$	$< 1.4 \times 10^{-3}$	CL=90%	2447
$B^+ \to K^*(1680)^+ \gamma$	$< 1.9 \times 10^{-3}$	CL=90%	2361
$B^+ \to K_3^*(1780)^+ \gamma$	$< 5.5 \times 10^{-3}$	CL=90%	2343
$B^+ \to K_4^*(2045)^+ \gamma$	$< 9.9 \times 10^{-3}$	CL=90%	2243

Light unflavored meson modes

$B^+ \to \pi^+ \pi^0$	$< 2.4 \times 10^{-4}$	CL=90%	2636
$B^+ \to \pi^+ \pi^+ \pi^-$	$< 1.9 \times 10^{-4}$	CL=90%	2630
$B^+ \to \rho^0 \pi^+$	$< 1.5 \times 10^{-4}$	CL=90%	2581
$B^+ \to \pi^+ f_0(980)$	$< 1.4 \times 10^{-4}$	CL=90%	2546
$B^+ \to \pi^+ f_2(1270)$	$< 2.4 \times 10^{-4}$	CL=90%	2483
$B^+ \to \pi^+ \pi^0 \pi^0$	$< 8.9 \times 10^{-4}$	CL=90%	2631
$B^+ \to \rho^+ \pi^0$	$< 5.5 \times 10^{-4}$	CL=90%	2581
$B^+ \to \pi^+ \pi^- \pi^+ \pi^0$	$< 4.0 \times 10^{-3}$	CL=90%	2621
$B^+ \to \rho^+ \rho^0$	$< 1.0 \times 10^{-3}$	CL=90%	2525
$B^+ \to a_1(1260)^+ \pi^0$	$< 1.7 \times 10^{-3}$	CL=90%	2494
$B^+ \to a_1(1260)^0 \pi^+$	$< 9.0 \times 10^{-4}$	CL=90%	2494
$B^+ \to \omega \pi^+$	$< 4.0 \times 10^{-4}$	CL=90%	2580
$B^+ \to \eta \pi^+$	$< 7.0 \times 10^{-4}$	CL=90%	2609
$B^+ \to \pi^+ \pi^+ \pi^+ \pi^- \pi^-$	$< 8.6 \times 10^{-4}$	CL=90%	2608
$B^+ \to \rho^0 a_1(1260)^+$	$< 6.2 \times 10^{-4}$	CL=90%	2433
$B^+ \to \rho^0 a_2(1320)^+$	$< 7.2 \times 10^{-4}$	CL=90%	2411
$B^+ \to \pi^+ \pi^+ \pi^+ \pi^- \pi^- \pi^0$	$< 6.3 \times 10^{-3}$	CL=90%	2592
$B^+ \to a_1(1260)^+ a_1(1260)^0$	< 1.3 %	CL=90%	2335

Baryon modes

$B^+ \to p \overline{p} \pi^+$	$< 1.6 \times 10^{-4}$	CL=90%	2438
$B^+ \to p \overline{p} \pi^+ \pi^+ \pi^-$	$< 5.2 \times 10^{-4}$	CL=90%	2369
$B^+ \to p \overline{\Lambda}$	$< 6 \times 10^{-5}$	CL=90%	2430
$B^+ \to p \overline{\Lambda} \pi^+ \pi^-$	$< 2.0 \times 10^{-4}$	CL=90%	2367
$B^+ \to \overline{\Delta}^0 p$	$< 3.8 \times 10^{-4}$	CL=90%	2402
$B^+ \to \Delta^{++} \overline{p}$	$< 1.5 \times 10^{-4}$	CL=90%	2402

Lepton Family number (LF) or Lepton number (L) violating modes, or $\Delta B = 1$ weak neutral current ($B1$) modes

$B^+ \to \pi^+ e^+ e^-$	$B1$	$< 3.9 \times 10^{-3}$	CL=90%	2638
$B^+ \to \pi^+ \mu^+ \mu^-$	$B1$	$< 9.1 \times 10^{-3}$	CL=90%	2633
$B^+ \to K^+ e^+ e^-$	$B1$	$< 6 \times 10^{-5}$	CL=90%	2616
$B^+ \to K^+ \mu^+ \mu^-$	$B1$	$< 1.7 \times 10^{-4}$	CL=90%	2612
$B^+ \to K^*(892)^+ e^+ e^-$	$B1$	$< 6.9 \times 10^{-4}$	CL=90%	2564
$B^+ \to K^*(892)^+ \mu^+ \mu^-$	$B1$	$< 1.2 \times 10^{-3}$	CL=90%	2560
$B^+ \to \pi^+ e^+ \mu^-$	LF	$< 6.4 \times 10^{-3}$	CL=90%	2636
$B^+ \to \pi^+ e^- \mu^+$	LF	$< 6.4 \times 10^{-3}$	CL=90%	2636
$B^+ \to K^+ e^+ \mu^-$	LF	$< 6.4 \times 10^{-3}$	CL=90%	2615
$B^+ \to K^+ e^- \mu^+$	LF	$< 6.4 \times 10^{-3}$	CL=90%	2615
$B^+ \to \pi^- e^+ e^+$	L	$< 3.9 \times 10^{-3}$	CL=90%	2638
$B^+ \to \pi^- \mu^+ \mu^+$	L	$< 9.1 \times 10^{-3}$	CL=90%	2633
$B^+ \to \pi^- e^+ \mu^+$	L	$< 6.4 \times 10^{-3}$	CL=90%	2636
$B^+ \to K^- e^+ e^+$	L	$< 3.9 \times 10^{-3}$	CL=90%	2616
$B^+ \to K^- \mu^+ \mu^+$	L	$< 9.1 \times 10^{-3}$	CL=90%	2612
$B^+ \to K^- e^+ \mu^+$	L	$< 6.4 \times 10^{-3}$	CL=90%	2615

B DECAY MODES

$\overline{B}$ modes are charge conjugates of the modes below.

For the following modes, the charge of B was not determined. The measurements are for an admixture of B mesons at the $\Upsilon(4S)$ unless otherwise indicated by a footnote and a "$\overline{b}$" instead of "B" in the initial state.

Semileptonic and leptonic modes

$B \to e^+ \nu_e$ anything	[c]	(10.4 ± 0.4) %	S=1.3	–
$B \to \overline{D}^*(2010) e^+ \nu_e$		(7.0 ± 2.3) %		–
$B \to \overline{p} e^+ \nu_e$ anything		$< 1.6 \times 10^{-3}$	CL=90%	–
$B \to \mu^+ \nu_\mu$ anything	[c]	(10.3 ± 0.5) %		–
$B \to \ell^+ \nu_\ell$ anything	[b,c]	(10.43 ± 0.24) %		–
$B \to D^- \ell^+ \nu_\ell$ anything	[b]	(2.7 ± 0.8) %		–
$B \to \overline{D}^0 \ell^+ \nu_\ell$ anything	[b]	(7.0 ± 1.4) %		–
$B \to D^{**} \ell^+ \nu_\ell$	[b,d]	(2.7 ± 0.7) %		–
$B \to D_s^- \ell^+ \nu_\ell$ anything	[b]	$< 9 \times 10^{-3}$	CL=90%	–
$B \to D_s^- \ell^+ \nu_\ell K^+$ anything	[b]	$< 6 \times 10^{-3}$	CL=90%	–
$B \to D_s^- \ell^+ \nu_\ell K^0$ anything	[b]	$< 9 \times 10^{-3}$	CL=90%	–
$B \to K^+ \ell^+ \nu_\ell$ anything	[b]	(5.6 ± 1.0) %		–
$B \to K^- \ell^+ \nu_\ell$ anything	[b]	(1.0 ± 0.6) %		–
$B \to K^0 / \overline{K}^0 \ell^+ \nu_\ell$ anything	[b]	(4.1 ± 0.8) %		–
$\overline{b} \to \tau^+ \nu_\tau$ anything	[e]	(4.1 ± 1.0) %		–

D, D^*, or D_s modes

$B \to D^-$ anything		(26 ± 4) %		–
$B \to \overline{D}^0$ anything		(54 ± 6) %		–
$B \to D^*(2010)^-$ anything		(23 ± 4) %	S=1.4	–
$B \to D_s^\pm$ anything	[f]	(8.9 ± 1.1) %		–
$B \to D_s D$, $D_s^* D$, $D_s D^*$, or $D_s^* D^*$	[f]	(5.0 ± 0.9) %		–
$B \to D^*(2010)\gamma$		$< 1.1 \times 10^{-3}$	CL=90%	–
$B \to D_s^+ \pi^-$, $D_s^{*+} \pi^-$, $D_s^+ \rho^-$, $D_s^{*+} \rho^-$, $D_s^+ \pi^0$, $D_s^{*+} \pi^0$, $D_s^+ \eta$, $D_s^{*+} \eta$, $D_s^+ \rho^0$, $D_s^{*+} \rho^0$, $D_s^+ \omega$, $D_s^{*+} \omega$		$< 5 \times 10^{-4}$	CL=90%	–

Charmonium modes

$B \to J/\psi(1S)$ anything		(1.30 ± 0.17) %		–
$B \to \psi(2S)$ anything		$(4.6 \pm 2.0) \times 10^{-3}$		–
$B \to \chi_{c1}(1P)$ anything		(1.1 ± 0.4) %		–

K or K^* modes

$B \to K^\pm$ anything	[f]	(85 ± 11) %		–
$B \to K^0 / \overline{K}^0$ anything		(63 ± 8) %		–
$b \to s\gamma$	[g]	$< 1.2 \times 10^{-3}$	CL=90%	–
$B \to K^*(892)\gamma$		$< 2.4 \times 10^{-4}$	CL=90%	–
$B \to K_1(1400)\gamma$		$< 4.1 \times 10^{-4}$	CL=90%	–
$B \to K_2^*(1430)\gamma$		$< 8.3 \times 10^{-4}$	CL=90%	–
$B \to K_2(1770)\gamma$		$< 1.2 \times 10^{-3}$	CL=90%	–
$B \to K_3^*(1780)\gamma$		$< 3.0 \times 10^{-3}$	CL=90%	–
$B \to K_4^*(2045)\gamma$		$< 1.0 \times 10^{-3}$	CL=90%	–

Light unflavored meson modes

$B \to \phi$ anything		(2.3 ± 0.8) %		–

Mode	Fraction (Γ_i/Γ)	Confidence level	p (MeV/c)
Baryon modes			
$B \to$ charmed-baryon anything	(6.4 ± 1.1) %		–
$B \to \overline{\Sigma}_c^{--}$ anything	$(4.8 \pm 2.5) \times 10^{-3}$		–
$B \to \overline{\Sigma}_c^{-}$ anything	< 1.1 %	CL=90%	–
$B \to \overline{\Sigma}_c^{0}$ anything	$(5.3 \pm 2.5) \times 10^{-3}$		–
$B \to \overline{\Sigma}_c^{0} N (N = p$ or $n)$	$< 1.7 \times 10^{-3}$	CL=90%	–
$B \to p$anything + $\overline{p}$anything	(8.0 ± 0.5) %		–
$B \to p$(direct) anything + $\overline{p}$(direct) anything	(5.6 ± 0.7) %		–
$B \to \Lambda$anything + $\overline{\Lambda}$anything	(4.0 ± 0.5) %		–
$B \to \Xi^-$ anything + $\overline{\Xi}^+$ anything	$(2.7 \pm 0.6) \times 10^{-3}$		–
$B \to$ baryons anything	(6.8 ± 0.6) %		–
$B \to p\overline{p}$anything	(2.47 ± 0.23) %		–
$B \to \Lambda\overline{p}$anything + $\overline{\Lambda}p$anything	(2.5 ± 0.4) %		–
$B \to \Lambda\overline{\Lambda}$anything	$< 5 \times 10^{-3}$	CL=90%	–
$\Delta B = 1$ weak neutral current ($B1$) modes			
$\overline{b} \to e^+ e^-$ anything	$B1$ [g] $< 2.4 \times 10^{-3}$		–
$\overline{b} \to \mu^+ \mu^-$ anything	$B1$ [g] $< 5.0 \times 10^{-5}$	CL=90%	–

B^0

$I(J^P) = \frac{1}{2}(0^-)$

I, J, P need confirmation. Quantum numbers shown are quark-model predictions.

Mass $m_{B^0} = 5279.0 \pm 2.0$ MeV
$m_{B^0} - m_{B^\pm} = 0.34 \pm 0.29$ MeV $(S = 1.1)$
Mean life $\tau = (1.50 \pm 0.11) \times 10^{-12}$ s
$c\tau = 449\ \mu$m
$\tau_{B^+}/\tau_{B^0} = 0.98 \pm 0.09$

B^0-$\overline{B}^0$ mixing parameters

$\chi_d = 0.156 \pm 0.024$
$\Delta m_{B^0} = m_{B^0_H} - m_{B^0_L} = (0.51 \pm 0.06) \times 10^{12}\ \hbar\ s^{-1}$
$x_d = \Delta m_{B^0}/\Gamma_{B^0} = 0.71 \pm 0.06$ [a]

$\overline{B}^0$ modes are charge conjugates of the modes below. Reactions indicate the weak decay vertex and do not include mixing. Decays in which the charge of the B is not determined are in the $B^\pm$ section.

Only data from $\Upsilon(4S)$ decays are used for branching fractions, with rare exceptions. The branching fractions listed below assume a 50:50 $B^0\overline{B}^0$:B^+B^- production ratio at the $\Upsilon(4S)$. We have attempted to bring older measurements up to date by rescaling their assumed $\Upsilon(4S)$ production ratio to 50:50 and their assumed D, D_s, D^*, and ψ branching ratios to current values whenever this would effect our averages and best limits significantly.

Indentation is used to indicate a subchannel of a previous reaction. All resonant subchannels have been corrected for resonance branching fractions to the final state so the sum of the subchannel branching fractions can exceed that of the final state.

B^0 DECAY MODES	Fraction (Γ_i/Γ)	Confidence level	p (MeV/c)
Semileptonic and leptonic modes			
$\ell^+ \nu_\ell$anything	[b] (9.5 ± 1.6) %		–
$D^- \ell^+ \nu_\ell$	[b] (1.9 ± 0.5) %		–
$D^*(2010)^- \ell^+ \nu_\ell$	[b] (4.4 ± 0.4) %		–
$\rho^- \ell^+ \nu_\ell$	[b] $< 4.1 \times 10^{-4}$	90%	–
$\pi^- \mu^+ \nu_\mu$	seen		2636
D, D^*, or D_s modes			
$D^- \pi^+$	$(3.0 \pm 0.4) \times 10^{-3}$		2306
$D^- \rho^+$	$(7.8 \pm 1.4) \times 10^{-3}$		2236
$\overline{D}^0 \pi^+ \pi^-$	$< 1.6 \times 10^{-3}$	90%	2301
$D^*(2010)^- \pi^+$	$(2.6 \pm 0.4) \times 10^{-3}$		2254
$D^- \pi^+ \pi^+ \pi^-$	$(8.0 \pm 2.5) \times 10^{-3}$		2287
$(D^- \pi^+ \pi^+ \pi^-)$ nonresonant	$(3.9 \pm 1.9) \times 10^{-3}$		2287
$D^- \pi^+ \rho^0$	$(1.1 \pm 1.0) \times 10^{-3}$		2207
$D^- a_1(1260)^+$	$(6.0 \pm 3.3) \times 10^{-3}$		2121
$D^*(2010)^- \pi^+ \pi^0$	(1.5 ± 0.5) %		2247
$D^*(2010)^- \rho^+$	$(7.3 \pm 1.5) \times 10^{-3}$		2181
$D^*(2010)^- \pi^+ \pi^+ \pi^-$	(1.19 ± 0.27) %		2235
$(D^*(2010)^- \pi^+ \pi^+ \pi^-)$ non-resonant	$(0.0 \pm 2.5) \times 10^{-3}$		2235
$D^*(2010)^- \pi^+ \rho^0$	$(5.7 \pm 3.1) \times 10^{-3}$		2151
$D^*(2010)^- a_1(1260)^+$	(1.5 ± 0.7) %		2061
$D^*(2010)^- \pi^+ \pi^+ \pi^- \pi^0$	(3.4 ± 1.8) %		2218
$\overline{D}_2^*(2460)^- \pi^+$	$< 2.2 \times 10^{-3}$	90%	2065
$\overline{D}_2^*(2460)^- \rho^+$	$< 4.9 \times 10^{-3}$	90%	1980
$D^- D_s^+$	$(8 \pm 4) \times 10^{-3}$		1812
$D^*(2010)^- D_s^+$	(1.2 ± 0.6) %		1735
$D^- D_s^{*+}$	(2.1 ± 1.5) %		1733
$D^*(2010)^- D_s^{*+}$	(2.0 ± 1.2) %		1650
$D_s^+ \pi^-$	$< 2.9 \times 10^{-4}$	90%	2270
$D_s^{*+} \pi^-$	$< 5 \times 10^{-4}$	90%	2215
$D_s^+ \rho^-$	$< 7 \times 10^{-4}$	90%	2198
$D_s^{*+} \rho^-$	$< 8 \times 10^{-4}$	90%	2140
$D_s^+ a_1(1260)^-$	$< 2.7 \times 10^{-3}$	90%	2079
$D_s^{*+} a_1(1260)^-$	$< 2.2 \times 10^{-3}$	90%	2015
$D_s^- K^+$	$< 2.4 \times 10^{-4}$	90%	2242
$D_s^{*-} K^+$	$< 1.8 \times 10^{-4}$	90%	2186
$D_s^- K^*(892)^+$	$< 1.0 \times 10^{-3}$	90%	2172
$D_s^{*-} K^*(892)^+$	$< 1.2 \times 10^{-3}$	90%	2113
$D_s^- \pi^+ K^0$	$< 6 \times 10^{-3}$	90%	2221
$D_s^{*-} \pi^+ K^0$	$< 3.2 \times 10^{-3}$	90%	2164
$D_s^- \pi^+ K^*(892)^0$	$< 4 \times 10^{-3}$	90%	2136
$D_s^{*-} \pi^+ K^*(892)^0$	$< 2.1 \times 10^{-3}$	90%	2075
$\overline{D}^0 \pi^0$	$< 4.8 \times 10^{-4}$	90%	2308
$\overline{D}^0 \rho^0$	$< 5.5 \times 10^{-4}$	90%	2238
$\overline{D}^0 \eta$	$< 6.8 \times 10^{-4}$	90%	2274
$\overline{D}^0 \eta'$	$< 8.6 \times 10^{-4}$	90%	2197
$\overline{D}^0 \omega$	$< 6.3 \times 10^{-4}$	90%	2235
$\overline{D}^*(2007)^0 \pi^0$	$< 9.7 \times 10^{-4}$	90%	2256
$\overline{D}^*(2007)^0 \rho^0$	$< 1.17 \times 10^{-3}$	90%	2182
$\overline{D}^*(2007)^0 \eta$	$< 6.9 \times 10^{-4}$	90%	2220
$\overline{D}^*(2007)^0 \eta'$	$< 2.7 \times 10^{-3}$	90%	2140
$\overline{D}^*(2007)^0 \omega$	$< 2.1 \times 10^{-3}$	90%	2180
Charmonium modes			
$J/\psi(1S) K^0$	$(7.5 \pm 2.1) \times 10^{-4}$		1682
$J/\psi(1S) K^+ \pi^-$	$(1.2 \pm 0.6) \times 10^{-3}$		1652
$J/\psi(1S) K^*(892)^0$	$(1.58 \pm 0.28) \times 10^{-3}$		1569
$\psi(2S) K^0$	$< 8 \times 10^{-4}$	90%	1283
$\psi(2S) K^+ \pi^-$	$< 1 \times 10^{-3}$	90%	1238
$\psi(2S) K^*(892)^0$	$(1.4 \pm 0.9) \times 10^{-3}$		1113
$\chi_{c1}(1P) K^0$	$< 2.7 \times 10^{-3}$	90%	1410
$\chi_{c1}(1P) K^*(892)^0$	$< 2.1 \times 10^{-3}$	90%	1263
K or K^* modes			
$K^+ \pi^-$	$< 2.6 \times 10^{-5}$	90%	2615
$K^+ K^-$	$< 7 \times 10^{-6}$	90%	2593
$K^0 \pi^+ \pi^-$	$< 4.4 \times 10^{-4}$	90%	2609
$K^0 \rho^0$	$< 3.2 \times 10^{-4}$	90%	2559
$K^0 f_0(980)$	$< 3.6 \times 10^{-4}$	90%	2523
$K^*(892)^+ \pi^-$	$< 3.8 \times 10^{-4}$	90%	2562
$K_2^*(1430)^+ \pi^-$	$< 2.6 \times 10^{-3}$	90%	2445
$K^0 K^+ K^-$	$< 1.3 \times 10^{-3}$	90%	2522
$K^0 \phi$	$< 4.2 \times 10^{-4}$	90%	2516
$K^*(892)^0 \pi^+ \pi^-$	$< 1.4 \times 10^{-3}$	90%	2556
$K^*(892)^0 \rho^0$	$< 4.6 \times 10^{-4}$	90%	2504
$K^*(892)^0 f_0(980)$	$< 1.7 \times 10^{-4}$	90%	2467
$K_1(1400)^+ \pi^-$	$< 1.1 \times 10^{-3}$	90%	2451
$K^*(892)^0 K^+ K^-$	$< 6.1 \times 10^{-4}$	90%	2465
$K^*(892)^0 \phi$	$< 3.2 \times 10^{-4}$	90%	2459
$K_1(1400)^0 \rho^0$	$< 3.0 \times 10^{-3}$	90%	2388
$K_1(1400)^0 \phi$	$< 5.0 \times 10^{-3}$	90%	2339
$K_2^*(1430)^0 \rho^0$	$< 1.1 \times 10^{-3}$	90%	2380
$K_2^*(1430)^0 \phi$	$< 1.4 \times 10^{-3}$	90%	2330
$K^*(892)^0 \gamma$	$(4.0 \pm 1.9) \times 10^{-5}$		2563
$K_1(1270)^0 \gamma$	$< 7.0 \times 10^{-3}$	90%	2486
$K_1(1400)^0 \gamma$	$< 4.3 \times 10^{-3}$	90%	2453
$K_2^*(1430)^0 \gamma$	$< 4.0 \times 10^{-4}$	90%	2445
$K^*(1680)^0 \gamma$	$< 2.0 \times 10^{-3}$	90%	2361
$K_3^*(1780)^0 \gamma$	< 1.0 %	90%	2343
$K_4^*(2045)^0 \gamma$	$< 4.3 \times 10^{-3}$	90%	2243

Light unflavored meson modes					
$\pi^+\pi^-$		< 2.9	$\times 10^{-5}$	90%	2636
$\pi^+\pi^-\pi^0$		< 7.2	$\times 10^{-4}$	90%	2631
$\rho^0\pi^0$		< 4.0	$\times 10^{-4}$	90%	2582
$\rho^\mp\pi^\pm$		[c] < 5.2	$\times 10^{-4}$	90%	2581
$\pi^+\pi^-\pi^+\pi^-$		< 6.7	$\times 10^{-4}$	90%	2621
$\rho^0\rho^0$		< 2.8	$\times 10^{-4}$	90%	2525
$a_1(1260)^\mp\pi^\pm$		[c] < 4.9	$\times 10^{-4}$	90%	2494
$a_2(1320)^\mp\pi^\pm$		[c] < 3.0	$\times 10^{-4}$	90%	2473
$\pi^+\pi^-\pi^0\pi^0$		< 3.1	$\times 10^{-3}$	90%	2622
$\rho^+\rho^-$		< 2.2	$\times 10^{-3}$	90%	2525
$a_1(1260)^0\pi^0$		< 1.1	$\times 10^{-3}$	90%	2494
$\omega\pi^0$		< 4.6	$\times 10^{-4}$	90%	2580
$\eta\pi^0$		< 1.8	$\times 10^{-3}$	90%	2609
$\pi^+\pi^+\pi^-\pi^-\pi^0$		< 9.0	$\times 10^{-3}$	90%	2609
$a_1(1260)^+\rho^-$		< 3.4	$\times 10^{-3}$	90%	2433
$a_1(1260)^0\rho^0$		< 2.4	$\times 10^{-3}$	90%	2433
$\pi^+\pi^+\pi^+\pi^-\pi^-\pi^-$		< 3.0	$\times 10^{-3}$	90%	2591
$a_1(1260)^+a_1(1260)^-$		< 2.8	$\times 10^{-3}$	90%	2335
$\pi^+\pi^+\pi^+\pi^-\pi^-\pi^-\pi^0$		< 1.1	%	90%	2572
Baryon modes					
$p\overline{p}$		< 3.4	$\times 10^{-5}$	90%	2467
$p\overline{p}\pi^+\pi^-$		< 2.5	$\times 10^{-4}$	90%	2406
$p\overline{\Lambda}\pi^-$		< 1.8	$\times 10^{-4}$	90%	2401
$\Delta^0\overline{\Delta}^0$		< 1.5	$\times 10^{-3}$	90%	2334
$\Delta^{++}\Delta^{--}$		< 1.1	$\times 10^{-4}$	90%	2334
$\overline{\Sigma}_c^{--}\Delta^{++}$		< 1.2	$\times 10^{-3}$	90%	1839
Lepton Family number (*LF*) violating modes, $\Delta B = 2$ forbidden decay via mixing (*B2M*) modes, or $\Delta B = 1$ weak neutral current (*B1*) modes					
e^+e^-	B1	< 5.9	$\times 10^{-6}$	90%	2639
$\mu^+\mu^-$	B1	< 5.9	$\times 10^{-6}$	90%	2637
$K^0e^+e^-$	B1	< 3.0	$\times 10^{-4}$	90%	2616
$K^0\mu^+\mu^-$	B1	< 3.6	$\times 10^{-4}$	90%	2612
$K^*(892)^0e^+e^-$	B1	< 2.9	$\times 10^{-4}$	90%	2563
$K^*(892)^0\mu^+\mu^-$	B1	< 2.3	$\times 10^{-5}$	90%	2559
$e^\pm\mu^\mp$	LF	[c] < 5.9	$\times 10^{-6}$	90%	2638
$e^\pm\tau^\mp$	LF	[c] < 5.3	$\times 10^{-4}$	90%	2340
$\mu^\pm\tau^\mp$	LF	[c] < 8.3	$\times 10^{-4}$	90%	2339

B^*

$I(J^P) = \frac{1}{2}(1^-)$

I, *J*, *P* need confirmation. Quantum numbers shown are quark-model predictions.

Mass $m_{B^*} = 5324.8 \pm 2.1$ MeV

$m_{B^*} - m_B = 46.0 \pm 0.6$ MeV

BOTTOM, STRANGE MESONS $(B = \pm1, S = \mp1)$

$B_s^0 = s\overline{b}$, $\overline{B}_s^0 = \overline{s}b$, similarly for B_s^*'s

B_s^0

$I(J^P) = \frac{1}{2}(0^-)$

I, *J*, *P* need confirmation. Quantum numbers shown are quark-model predictions.

Mass $m_{B_s^0} = 5375 \pm 6$ MeV (S = 1.3)

Mean life $\tau = (1.34^{+0.32}_{-0.27}) \times 10^{-12}$ s (S = 1.4)

B_s^0-$\overline{B}_s^0$ mixing parameters

$\chi_s = 0.62 \pm 0.13$

$\Delta m_{B_s^0} = m_{B_{sH}^0} - m_{B_{sL}^0} > 1.8 \times 10^{12}\ \hbar\ s^{-1}$, CL = 95%

$x_s = \Delta m_{B_s^0}/\Gamma_{B_s^0} > 2.0$, CL = 95%

B_s^0 DECAY MODES	Fraction (Γ_i/Γ)	*p* (MeV/c)
D_s^- anything	seen	–
$D_s^-\ell^+\nu_\ell$ anything (ℓ means sum of *e* and μ)	seen	–
$D_s^-\pi^+$	seen	2325
$J/\psi(1S)\phi$	seen	1594
$\psi(2S)\phi$	seen	1128

HEAVY QUARK SEARCHES

Searches for Top and Fourth Generation Hadrons

See the sections "Searches for *t* Quark" and "Searches for *b*′ (4th Generation) Quark" at the end of the QUARKS section.

$c\overline{c}$ MESONS

$\eta_c(1S)$ or $\eta_c(2980)$

$I^G(J^{PC}) = 0^+(0^{-+})$

Mass $m = 2978.8 \pm 1.9$ MeV (S = 1.8)

Full width $\Gamma = 10.3^{+3.8}_{-3.4}$ MeV

$\eta_c(1S)$ DECAY MODES	Fraction (Γ_i/Γ)	Confidence level	*p* (MeV/c)
Decays involving hadronic resonances			
$\eta'(958)\pi\pi$	(4.1 ± 1.7) %		1319
$\rho\rho$	(2.6 ± 0.9) %		1275
$K^*(892)^0K^-\pi^+ +$ c.c.	(2.0 ± 0.7) %		1273
$K^*(892)\overline{K}^*(892)$	$(8.5 \pm 3.1) \times 10^{-3}$		1193
$\phi\phi$	$(7.1 \pm 2.8) \times 10^{-3}$		1086
$a_0(980)\pi$	< 2 %	90%	1323
$a_2(1320)\pi$	< 2 %	90%	1193
$K^*(892)\overline{K} +$ c.c.	< 1.28 %	90%	1307
$f_2(1270)\eta$	< 1.1 %	90%	1142
$\omega\omega$	$< 3.1 \times 10^{-3}$	90%	1268
Decays into stable hadrons			
$K\overline{K}\pi$	(6.6 ± 1.8) %		1378
$\eta\pi\pi$	(4.9 ± 1.8) %		1425
$\pi^+\pi^-K^+K^-$	$(2.0\ ^{+0.7}_{-0.6})$ %		1342
$2(\pi^+\pi^-)$	(1.2 ± 0.4) %		1457
$p\overline{p}$	$(1.2 \pm 0.4) \times 10^{-3}$		1157
$K\overline{K}\eta$	< 3.1 %	90%	1262
$\pi^+\pi^-p\overline{p}$	< 1.2 %	90%	1023
$\Lambda\overline{\Lambda}$	$< 2 \times 10^{-3}$	90%	987
Radiative decays			
$\gamma\gamma$	$(6\ ^{+6}_{-5}) \times 10^{-4}$		1489

$J/\psi(1S)$ or $J/\psi(3097)$

$I^G(J^{PC}) = 0^-(1^{--})$

Mass $m = 3096.88 \pm 0.04$ MeV
Full width $\Gamma = 88 \pm 5$ keV
$\Gamma_{ee} = 5.26 \pm 0.37$ keV (Assuming $\Gamma_{ee} = \Gamma_{\mu\mu}$)

$J/\psi(1S)$ DECAY MODES		Fraction (Γ_i/Γ)	Scale factor/ Confidence level	p (MeV/c)
hadrons		(86.0 ±2.0) %		–
virtual$\gamma \to$ hadrons		(17.0 ±2.0) %		–
e^+e^-		(5.99±0.25) %		1548
$\mu^+\mu^-$		(5.97±0.25) %	S=1.1	1545
Decays involving hadronic resonances				
$\rho\pi$		(1.28±0.10) %		1449
$\rho^0\pi^0$		(4.2 ±0.5) $\times 10^{-3}$		1449
$a_2(1320)\rho$		(1.09±0.22) %		1125
$\omega\pi^+\pi^+\pi^-\pi^-$		(8.5 ±3.4) $\times 10^{-3}$		1392
$\omega\pi^+\pi^-$		(7.2 ±1.0) $\times 10^{-3}$		1435
$K^*(892)^0\overline{K}{}_2^*(1430)^0$ + c.c.		(6.7 ±2.6) $\times 10^{-3}$		1005
$\omega K^*(892)\overline{K}$ + c.c.		(5.3 ±2.0) $\times 10^{-3}$		1098
$\omega f_2(1270)$		(4.3 ±0.6) $\times 10^{-3}$		1143
$K^+\overline{K}{}^*(892)^-$ + c.c.		(5.0 ±0.4) $\times 10^{-3}$		1373
$K^0\overline{K}{}^*(892)^0$ + c.c.		(4.2 ±0.4) $\times 10^{-3}$		1371
$\omega\pi^0\pi^0$		(3.4 ±0.8) $\times 10^{-3}$		1436
$b_1(1235)^\pm\pi^\mp$	[q]	(3.0 ±0.5) $\times 10^{-3}$		1299
$\omega K^\pm K_S^0\pi^\mp$	[q]	(3.0 ±0.7) $\times 10^{-3}$		1210
$b_1(1235)^0\pi^0$		(2.3 ±0.6) $\times 10^{-3}$		1299
$\phi K^*(892)\overline{K}$ + c.c.		(2.04±0.28) $\times 10^{-3}$		969
$\omega K\overline{K}$		(1.9 ±0.4) $\times 10^{-3}$		1268
$\omega f_J(1710) \to \omega K\overline{K}$		(4.8 ±1.1) $\times 10^{-4}$		878
$\phi 2(\pi^+\pi^-)$		(1.60±0.32) $\times 10^{-3}$		1318
$\Delta(1232)^{++}\overline{p}\pi^-$		(1.6 ±0.5) $\times 10^{-3}$		1030
$\omega\eta$		(1.58±0.16) $\times 10^{-3}$		1394
$\phi K\overline{K}$		(1.48±0.22) $\times 10^{-3}$		1179
$\phi f_J(1710) \to \phi K\overline{K}$		(3.6 ±0.6) $\times 10^{-4}$		875
$p\overline{p}\omega$		(1.30±0.25) $\times 10^{-3}$	S=1.3	769
$\Delta(1232)^{++}\overline{\Delta}(1232)^{--}$		(1.10±0.29) $\times 10^{-3}$		938
$\Sigma(1385)^-\overline{\Sigma}(1385)^+$ (or c.c.)	[q]	(1.03±0.13) $\times 10^{-3}$		692
$p\overline{p}\eta'(958)$		(9 ±4) $\times 10^{-4}$	S=1.7	596
$\phi f_2'(1525)$		(8 ±4) $\times 10^{-4}$	S=2.7	871
$\phi\pi^+\pi^-$		(8.0 ±1.2) $\times 10^{-4}$		1365
$\phi K^\pm K_S^0\pi^\mp$	[q]	(7.2 ±0.9) $\times 10^{-4}$		1114
$\omega f_1(1420)$		(6.8 ±2.4) $\times 10^{-4}$		1062
$\phi\eta$		(6.5 ±0.7) $\times 10^{-4}$		1320
$\Xi(1530)^-\overline{\Xi}^+$		(5.9 ±1.5) $\times 10^{-4}$		597
$pK^-\overline{\Sigma}(1385)^0$		(5.1 ±3.2) $\times 10^{-4}$		645
$\omega\pi^0$		(4.2 ±0.6) $\times 10^{-4}$	S=1.4	1447
$\phi\eta'(958)$		(3.3 ±0.4) $\times 10^{-4}$		1192
$\phi f_0(980)$		(3.2 ±0.9) $\times 10^{-4}$	S=1.9	1182
$\Xi(1530)^0\overline{\Xi}^0$		(3.2 ±1.4) $\times 10^{-4}$		608
$\Sigma(1385)^-\overline{\Sigma}^+$ (or c.c.)	[q]	(3.1 ±0.5) $\times 10^{-4}$		857
$\phi f_1(1285)$		(2.6 ±0.5) $\times 10^{-4}$	S=1.1	1032
$\rho\eta$		(1.93±0.23) $\times 10^{-4}$		1398
$\omega\eta'(958)$		(1.67±0.25) $\times 10^{-4}$		1279
$\omega f_0(980)$		(1.4 ±0.5) $\times 10^{-4}$		1271
$\rho\eta'(958)$		(1.05±0.18) $\times 10^{-4}$		1283
$p\overline{p}\phi$		(4.5 ±1.5) $\times 10^{-5}$		527
$a_2(1320)^\pm\pi^\mp$	[q]	< 4.3 $\times 10^{-3}$	CL=90%	1263
$K\overline{K}{}_2^*(1430)$ + c.c.		< 4.0 $\times 10^{-3}$	CL=90%	1159
$K_2^*(1430)^0\overline{K}{}_2^*(1430)^0$		< 2.9 $\times 10^{-3}$	CL=90%	588
$K^*(892)^0\overline{K}{}^*(892)^0$		< 5 $\times 10^{-4}$	CL=90%	1263
$\phi f_2(1270)$		< 3.7 $\times 10^{-4}$	CL=90%	1036
$p\overline{p}\rho$		< 3.1 $\times 10^{-4}$	CL=90%	779
$\phi\eta(1440) \to \phi\eta\pi\pi$		< 2.5 $\times 10^{-4}$	CL=90%	946
$\omega f_2'(1525)$		< 2.2 $\times 10^{-4}$	CL=90%	1003
$\Sigma(1385)^0\overline{\Lambda}$		< 2 $\times 10^{-4}$	CL=90%	911
$\Delta(1232)^+\overline{p}$		< 1 $\times 10^{-4}$	CL=90%	1100
$\Sigma^0\overline{\Lambda}$		< 9 $\times 10^{-5}$	CL=90%	1032
$\phi\pi^0$		< 6.8 $\times 10^{-6}$	CL=90%	1377
Decays into stable hadrons				
$2(\pi^+\pi^-)\pi^0$		(3.37±0.26) %		1496
$3(\pi^+\pi^-)\pi^0$		(2.9 ±0.6) %		1433
$\pi^+\pi^-\pi^0$		(1.50±0.20) %		1533
$\pi^+\pi^-\pi^0K^+K^-$		(1.20±0.30) %		1368
$4(\pi^+\pi^-)\pi^0$		(9.0 ±3.0) $\times 10^{-3}$		1345
$\pi^+\pi^-K^+K^-$		(7.2 ±2.3) $\times 10^{-3}$		1407
$K\overline{K}\pi$		(6.1 ±1.0) $\times 10^{-3}$		1440
$p\overline{p}\pi^+\pi^-$		(6.0 ±0.5) $\times 10^{-3}$	S=1.3	1107
$2(\pi^+\pi^-)$		(4.0 ±1.0) $\times 10^{-3}$		1517
$3(\pi^+\pi^-)$		(4.0 ±2.0) $\times 10^{-3}$		1466
$n\overline{n}\pi^+\pi^-$		(4 ±4) $\times 10^{-3}$		1106
$\Sigma\overline{\Sigma}$		(3.8 ±0.5) $\times 10^{-3}$		992
$2(\pi^+\pi^-)K^+K^-$		(3.1 ±1.3) $\times 10^{-3}$		1320
$p\overline{p}\pi^+\pi^-\pi^0$	[xx]	(2.3 ±0.9) $\times 10^{-3}$	S=1.9	1033
$p\overline{p}$		(2.14±0.10) $\times 10^{-3}$		1232
$p\overline{p}\eta$		(2.09±0.18) $\times 10^{-3}$		948
$p\overline{n}\pi^-$		(2.00±0.10) $\times 10^{-3}$		1174
$n\overline{n}$		(1.9 ±0.5) $\times 10^{-3}$		1231
$\Xi\overline{\Xi}$		(1.8 ±0.4) $\times 10^{-3}$	S=1.8	818
$\Lambda\overline{\Lambda}$		(1.35±0.14) $\times 10^{-3}$	S=1.2	1074
$p\overline{p}\pi^0$		(1.09±0.09) $\times 10^{-3}$		1176
$\Lambda\overline{\Sigma}^-\pi^+$ (or c.c.)	[q]	(1.06±0.12) $\times 10^{-3}$		945
$pK^-\overline{\Lambda}$		(8.9 ±1.6) $\times 10^{-4}$		876
$2(K^+K^-)$		(7.0 ±3.0) $\times 10^{-4}$		1131
$pK^-\overline{\Sigma}^0$		(2.9 ±0.8) $\times 10^{-4}$		820
K^+K^-		(2.37±0.31) $\times 10^{-4}$		1468
$\Lambda\overline{\Lambda}\pi^0$		(2.2 ±0.7) $\times 10^{-4}$		998
$\pi^+\pi^-$		(1.47±0.23) $\times 10^{-4}$		1542
$K_S^0K_L^0$		(1.08±0.14) $\times 10^{-4}$		1466
$\Lambda\overline{\Sigma}$ + c.c.		< 1.5 $\times 10^{-4}$	CL=90%	1032
$K_S^0K_S^0$		< 5.2 $\times 10^{-6}$	CL=90%	1466
Radiative decays				
$\gamma\eta_c(1S)$		(1.3 ±0.4) %		116
$\gamma\pi^+\pi^-2\pi^0$		(8.3 ±3.1) $\times 10^{-3}$		1518
$\gamma\eta\pi\pi$		(6.1 ±1.0) $\times 10^{-3}$		1487
$\gamma\eta(1440) \to \gamma K\overline{K}\pi$	[n]	(9.1 ±1.8) $\times 10^{-4}$		1223
$\gamma\eta(1440) \to \gamma\gamma\rho^0$		(6.4 ±1.4) $\times 10^{-5}$		1223
$\gamma\rho\rho$		(4.5 ±0.8) $\times 10^{-3}$		1343
$\gamma\eta'(958)$		(4.31±0.30) $\times 10^{-3}$		1400
$\gamma 2\pi^+2\pi^-$		(2.8 ±0.5) $\times 10^{-3}$	S=1.9	1517
$\gamma f_4(2050)$		(2.7 ±0.7) $\times 10^{-3}$		874
$\gamma\omega\omega$		(1.59±0.33) $\times 10^{-3}$		1337
$\gamma\eta(1440) \to \gamma\rho^0\rho^0$		(1.4 ±0.4) $\times 10^{-3}$		1223
$\gamma f_2(1270)$		(1.38±0.14) $\times 10^{-3}$		1286
$\gamma f_J(1710) \to \gamma K\overline{K}$		(9.7 ±1.2) $\times 10^{-4}$		1075
$\gamma\eta$		(8.6 ±0.8) $\times 10^{-4}$		1500
$\gamma f_1(1420) \to \gamma K\overline{K}\pi$		(8.3 ±1.5) $\times 10^{-4}$		1220
$\gamma f_1(1285)$		(6.5 ±1.0) $\times 10^{-4}$		1283
$\gamma f_2'(1525)$		(6.3 ±1.0) $\times 10^{-4}$		1173
$\gamma\phi\phi$		(4.0 ±1.2) $\times 10^{-4}$	S=2.1	1166
$\gamma p\overline{p}$		(3.8 ±1.0) $\times 10^{-4}$		1232
$\gamma\eta(2225)$		(2.9 ±0.6) $\times 10^{-4}$		834
$\gamma\eta(1760) \to \gamma\rho^0\rho^0$		(1.3 ±0.9) $\times 10^{-4}$		1048
$\gamma\pi^0$		(3.9 ±1.3) $\times 10^{-5}$		1546
$\gamma p\overline{p}\pi^+\pi^-$		< 7.9 $\times 10^{-4}$	CL=90%	1107
$\gamma\gamma$		< 5 $\times 10^{-4}$	CL=90%	1548
$\gamma\Lambda\overline{\Lambda}$		< 1.3 $\times 10^{-4}$	CL=90%	1074
3γ		< 5.5 $\times 10^{-5}$	CL=90%	1548

$\chi_{c0}(1P)$ or $\chi_{c0}(3415)$

$I^G(J^{PC}) = 0^+(0^{++})$

Mass $m = 3415.1 \pm 1.0$ MeV
Full width $\Gamma = 14 \pm 5$ MeV

$\chi_{c0}(1P)$ DECAY MODES	Fraction (Γ_i/Γ)	Confidence level	p (MeV/c)
Hadronic decays			
$2(\pi^+\pi^-)$	(3.7 ± 0.7) %		1679
$\pi^+\pi^-K^+K^-$	(3.0 ± 0.7) %		1580
$\rho^0\pi^+\pi^-$	(1.6 ± 0.5) %		1608
$3(\pi^+\pi^-)$	(1.5 ± 0.5) %		1633
$K^+\overline{K}^*(892)^0\pi^-$ + c.c.	(1.2 ± 0.4) %		1522
$\pi^+\pi^-$	$(7.5\pm2.1)\times10^{-3}$		1702
K^+K^-	$(7.1\pm2.4)\times10^{-3}$		1635
$\pi^+\pi^-p\overline{p}$	$(5.0\pm2.0)\times10^{-3}$		1320
$\pi^0\pi^0$	$(3.1\pm0.6)\times10^{-3}$		1702
$\eta\eta$	$(2.5\pm1.1)\times10^{-3}$		1617
$p\overline{p}$	$< 9.0 \times10^{-4}$	90%	1427
Radiative decays			
$\gamma J/\psi(1S)$	$(6.6\pm1.8)\times10^{-3}$		303
$\gamma\gamma$	$(4.0\pm2.3)\times10^{-4}$		1708

$\chi_{c1}(1P)$ or $\chi_{c1}(3510)$

$I^G(J^{PC}) = 0^+(1^{++})$

Mass $m = 3510.53 \pm 0.12$ MeV
Full width $\Gamma = 0.88 \pm 0.14$ MeV

$\chi_{c1}(1P)$ DECAY MODES	Fraction (Γ_i/Γ)	p (MeV/c)
Hadronic decays		
$3(\pi^+\pi^-)$	(2.2 ± 0.8) %	1683
$2(\pi^+\pi^-)$	(1.6 ± 0.5) %	1727
$\pi^+\pi^-K^+K^-$	$(9\ \pm4\ )\times10^{-3}$	1632
$\rho^0\pi^+\pi^-$	$(3.9\pm3.5)\times10^{-3}$	1659
$K^+\overline{K}^*(892)^0\pi^-$ + c.c.	$(3.2\pm2.1)\times10^{-3}$	1576
$\pi^+\pi^-p\overline{p}$	$(1.4\pm0.9)\times10^{-3}$	1381
$p\overline{p}$	$(8.6\pm1.2)\times10^{-5}$	1483
$\pi^+\pi^- + K^+K^-$	$< 2.1 \times10^{-3}$	–
Radiative decays		
$\gamma J/\psi(1S)$	(27.3 ± 1.6) %	389

$\chi_{c2}(1P)$ or $\chi_{c2}(3555)$

$I^G(J^{PC}) = 0^+(2^{++})$

Mass $m = 3556.17 \pm 0.13$ MeV
Full width $\Gamma = 2.00 \pm 0.18$ MeV

$\chi_{c2}(1P)$ DECAY MODES	Fraction (Γ_i/Γ)	Confidence level	p (MeV/c)
Hadronic decays			
$2(\pi^+\pi^-)$	$(2.2\ \pm0.5\ )$ %		1751
$\pi^+\pi^-K^+K^-$	$(1.9\ \pm0.5\ )$ %		1656
$3(\pi^+\pi^-)$	$(1.2\ \pm0.8\ )$ %		1707
$\rho^0\pi^+\pi^-$	$(7\ \pm4\ )\times10^{-3}$		1683
$K^+\overline{K}^*(892)^0\pi^-$ + c.c.	$(4.8\ \pm2.8\ )\times10^{-3}$		1601
$\pi^+\pi^-p\overline{p}$	$(3.3\ \pm1.3\ )\times10^{-3}$		1410
$\pi^+\pi^-$	$(1.9\ \pm1.0\ )\times10^{-3}$		1773
K^+K^-	$(1.5\ \pm1.1\ )\times10^{-3}$		1708
$p\overline{p}$	$(10.0\ \pm1.0\ )\times10^{-5}$		1510
$\pi^0\pi^0$	$(1.10\pm0.28)\times10^{-3}$		1773
$\eta\eta$	$(8\ \pm5\ )\times10^{-4}$		1692
$J/\psi(1S)\pi^+\pi^-\pi^0$	< 1.5 %	90%	185
Radiative decays			
$\gamma J/\psi(1S)$	$(13.5\ \pm1.1\ )$ %		430
$\gamma\gamma$	$(1.6\ \pm0.5\ )\times10^{-4}$		1778

$\psi(2S)$ or $\psi(3685)$

$I^G(J^{PC}) = 0^-(1^{--})$

Mass $m = 3686.00 \pm 0.09$ MeV
Full width $\Gamma = 277 \pm 31$ keV (S = 1.1)
$\Gamma_{ee} = 2.14 \pm 0.21$ keV (Assuming $\Gamma_{ee} = \Gamma_{\mu\mu}$)

$\psi(2S)$ DECAY MODES	Fraction (Γ_i/Γ)	Scale factor/ Confidence level	p (MeV/c)
hadrons	(98.10 ± 0.30) %		–
virtual$\gamma \to$ hadrons	$(2.9\ \pm0.4\ )$ %		–
e^+e^-	$(8.8\ \pm1.3\ )\times10^{-3}$		1843
$\mu^+\mu^-$	$(7.7\ \pm1.7\ )\times10^{-3}$		1840
Decays into $J/\psi(1S)$ and anything			
$J/\psi(1S)$anything	$(57\ \pm4\ )$ %		–
$J/\psi(1S)$neutrals	$(23.2\ \pm2.6\ )$ %		–
$J/\psi(1S)\pi^+\pi^-$	$(32.4\ \pm2.6\ )$ %		477
$J/\psi(1S)\pi^0\pi^0$	$(18.4\ \pm2.7\ )$ %		481
$J/\psi(1S)\eta$	$(2.7\ \pm0.4\ )$ %	S=1.7	200
$J/\psi(1S)\pi^0$	$(9.7\ \pm2.1\ )\times10^{-4}$		527
Hadronic decays			
$3(\pi^+\pi^-)\pi^0$	$(3.5\ \pm1.6\ )\times10^{-3}$		1746
$2(\pi^+\pi^-)\pi^0$	$(3.1\ \pm0.7\ )\times10^{-3}$		1799
$\pi^+\pi^-K^+K^-$	$(1.6\ \pm0.4\ )\times10^{-3}$		1726
$\pi^+\pi^-p\overline{p}$	$(8.0\ \pm2.0\ )\times10^{-4}$		1491
$K^+\overline{K}^*(892)^0\pi^-$ + c.c.	$(6.7\ \pm2.5\ )\times10^{-4}$		1673
$2(\pi^+\pi^-)$	$(4.5\ \pm1.0\ )\times10^{-4}$		1817
$\rho^0\pi^+\pi^-$	$(4.2\ \pm1.5\ )\times10^{-4}$		1751
$\overline{p}p$	$(1.9\ \pm0.5\ )\times10^{-4}$		1586
$3(\pi^+\pi^-)$	$(1.5\ \pm1.0\ )\times10^{-4}$		1774
$\overline{p}p\pi^0$	$(1.4\ \pm0.5\ )\times10^{-4}$		1543
K^+K^-	$(1.0\ \pm0.7\ )\times10^{-4}$		1776
$\pi^+\pi^-\pi^0$	$(9\ \pm5\ )\times10^{-5}$		1830
$\pi^+\pi^-$	$(8\ \pm5\ )\times10^{-5}$		1838
$\Lambda\overline{\Lambda}$	$< 4 \times10^{-4}$	CL=90%	1467
$\Xi^-\overline{\Xi}^+$	$< 2 \times10^{-4}$	CL=90%	1285
$\rho\pi$	$< 8.3 \times10^{-5}$	CL=90%	1760
$K^+K^-\pi^0$	$< 2.96 \times10^{-5}$	CL=90%	1754
$K^+\overline{K}^*(892)^-$ + c.c.	$< 1.79 \times10^{-5}$	CL=90%	1698
Radiative decays			
$\gamma\chi_{c0}(1P)$	$(9.3\ \pm0.8\ )$ %		261
$\gamma\chi_{c1}(1P)$	$(8.7\ \pm0.8\ )$ %		171
$\gamma\chi_{c2}(1P)$	$(7.8\ \pm0.8\ )$ %		127
$\gamma\eta_c(1S)$	$(2.8\ \pm0.6\ )\times10^{-3}$		639
$\gamma\pi^0$	$< 5.4 \times10^{-3}$	CL=95%	1841
$\gamma\eta'(958)$	$< 1.1 \times10^{-3}$	CL=90%	1719
$\gamma\gamma$	$< 1.6 \times10^{-4}$	CL=90%	1843
$\gamma\eta(1440) \to \gamma K\overline{K}\pi$	[n] $< 1.2 \times10^{-4}$	CL=90%	1569

$\psi(3770)$

$I^G(J^{PC}) = ?^?(1^{--})$

Mass $m = 3769.9 \pm 2.5$ MeV (S = 1.8)
Full width $\Gamma = 23.6 \pm 2.7$ MeV (S = 1.1)
$\Gamma_{ee} = 0.26 \pm 0.04$ keV (S = 1.2)

$\psi(3770)$ DECAY MODES	Fraction (Γ_i/Γ)	Scale factor	p (MeV/c)
$D\overline{D}$	dominant		242
e^+e^-	$(1.12\pm0.17)\times10^{-5}$	1.2	1885

$\psi(4040)$ [yy]

$I^G(J^{PC}) = ?^?(1^{--})$

Mass $m = 4040 \pm 10$ MeV
Full width $\Gamma = 52 \pm 10$ MeV
$\Gamma_{ee} = 0.75 \pm 0.15$ keV

$\psi(4040)$ DECAY MODES	Fraction (Γ_i/Γ)	p (MeV/c)
e^+e^-	$(1.4\pm0.4)\times10^{-5}$	2020
$D^0\overline{D}^0$	seen	777
$D^*(2007)^0\overline{D}^0$ + c.c.	seen	578
$D^*(2007)^0\overline{D}^*(2007)^0$	seen	232

$\psi(4160)$ [yy]

$I^G(J^{PC}) = ?^?(1^{--})$

Mass $m = 4159 \pm 20$ MeV
Full width $\Gamma = 78 \pm 20$ MeV
$\Gamma_{ee} = 0.77 \pm 0.23$ keV

$\psi(4160)$ DECAY MODES	Fraction (Γ_i/Γ)	p (MeV/c)
e^+e^-	$(10\pm4) \times 10^{-6}$	2079

$\psi(4415)$ [yy]

$I^G(J^{PC}) = ?^?(1^{--})$

Mass $m = 4415 \pm 6$ MeV
Full width $\Gamma = 43 \pm 15$ MeV (S = 1.8)
$\Gamma_{ee} = 0.47 \pm 0.10$ keV

$\psi(4415)$ DECAY MODES	Fraction (Γ_i/Γ)	p (MeV/c)
hadrons	dominant	–
e^+e^-	$(1.1\pm0.4) \times 10^{-5}$	2207

$b\overline{b}$ MESONS

$\Upsilon(1S)$ or $\Upsilon(9460)$

$I^G(J^{PC}) = ?^?(1^{--})$

Mass $m = 9460.37 \pm 0.21$ MeV (S = 2.7)
Full width $\Gamma = 52.5 \pm 1.8$ keV
$\Gamma_{ee} = 1.32 \pm 0.03$ keV

$\Upsilon(1S)$ DECAY MODES	Fraction (Γ_i/Γ)	Scale factor/ Confidence level	p (MeV/c)
$\tau^+\tau^-$	(2.97 ± 0.35) %		4384
e^+e^-	(2.52 ± 0.17) %		4730
$\mu^+\mu^-$	(2.48 ± 0.07) %	S=1.1	4729
Hadronic decays			
$J/\psi(1S)$anything	$(1.1 \pm0.4) \times 10^{-3}$		4223
$\rho\pi$	$< 2 \times 10^{-4}$	CL=90%	4698
$\pi^+\pi^-$	$< 5 \times 10^{-4}$	CL=90%	4728
K^+K^-	$< 5 \times 10^{-4}$	CL=90%	4704
$p\overline{p}$	$< 9 \times 10^{-4}$	CL=90%	4636
Radiative decays			
$\gamma 2h^+2h^-$	$(7.0 \pm1.5) \times 10^{-4}$		4720
$\gamma 3h^+3h^-$	$(5.4 \pm2.0) \times 10^{-4}$		4703
$\gamma 4h^+4h^-$	$(7.4 \pm3.5) \times 10^{-4}$		4679
$\gamma\pi^+\pi^-K^+K^-$	$(2.9 \pm0.9) \times 10^{-4}$		4686
$\gamma 2\pi^+2\pi^-$	$(2.5 \pm0.9) \times 10^{-4}$		4720
$\gamma 3\pi^+3\pi^-$	$(2.5 \pm1.2) \times 10^{-4}$		4703
$\gamma 2\pi^+2\pi^-K^+K^-$	$(2.4 \pm1.2) \times 10^{-4}$		4658
$\gamma\pi^+\pi^-p\overline{p}$	$(1.5 \pm0.6) \times 10^{-4}$		4604
$\gamma 2\pi^+2\pi^-p\overline{p}$	$(4 \pm6) \times 10^{-5}$		4563
$\gamma 2K^+2K^-$	$(2.0 \pm2.0) \times 10^{-5}$		4601
$\gamma\eta'(958)$	$< 1.3 \times 10^{-3}$	CL=90%	4682
$\gamma\eta$	$< 3.5 \times 10^{-4}$	CL=90%	4714
$\gamma f'_2(1525)$	$< 1.4 \times 10^{-4}$	CL=90%	4607
$\gamma f_2(1270)$	$< 1.3 \times 10^{-4}$	CL=90%	4644
$\gamma\eta(1440)$	$< 8.2 \times 10^{-5}$	CL=90%	4624
$\gamma f_J(1710) \to \gamma K\overline{K}$	$< 2.6 \times 10^{-4}$	CL=90%	4576
$\gamma f_4(2220) \to \gamma K^+K^-$	$< 1.5 \times 10^{-5}$	CL=90%	4469

$\chi_{b0}(1P)$ [zz] or $\chi_{b0}(9860)$

$I^G(J^{PC}) = ?^?(0 \text{ preferred}^{++})$
J needs confirmation.

Mass $m = 9859.8 \pm 1.3$ MeV

$\chi_{b0}(1P)$ DECAY MODES	Fraction (Γ_i/Γ)	Confidence level	p (MeV/c)
$\gamma\Upsilon(1S)$	<6 %	90%	391

$\chi_{b1}(1P)$ [zz] or $\chi_{b1}(9890)$

$I^G(J^{PC}) = ?^?(1^{++})$
J needs confirmation.

Mass $m = 9891.9 \pm 0.7$ MeV

$\chi_{b1}(1P)$ DECAY MODES	Fraction (Γ_i/Γ)	p (MeV/c)
$\gamma\Upsilon(1S)$	(35 ± 8) %	422

$\chi_{b2}(1P)$ [zz] or $\chi_{b2}(9915)$

$I^G(J^{PC}) = ?^?(2^{++})$
J needs confirmation.

Mass $m = 9913.2 \pm 0.6$ MeV

$\chi_{b2}(1P)$ DECAY MODES	Fraction (Γ_i/Γ)	p (MeV/c)
$\gamma\Upsilon(1S)$	(22 ± 4) %	443

$\Upsilon(2S)$ or $\Upsilon(10023)$

$I^G(J^{PC}) = ?^?(1^{--})$

Mass $m = 10.02330 \pm 0.00031$ GeV
Full width $\Gamma = 44 \pm 7$ keV

$\Upsilon(2S)$ DECAY MODES	Fraction (Γ_i/Γ)	Confidence level	p (MeV/c)
$\Upsilon(1S)\pi^+\pi^-$	(18.5 ± 0.8) %		475
$\Upsilon(1S)\pi^0\pi^0$	(8.8 ± 1.1) %		480
$\tau^+\tau^-$	(1.7 ± 1.6) %		4686
$\mu^+\mu^-$	(1.31 ± 0.21) %		5011
e^+e^-	seen		5012
$\Upsilon(1S)\pi^0$	$< 8 \times 10^{-3}$	90%	531
$\Upsilon(1S)\eta$	$< 2 \times 10^{-3}$	90%	127
$J/\psi(1S)$anything	$< 6 \times 10^{-3}$	90%	4533
Radiative decays			
$\gamma\chi_{b1}(1P)$	(6.7 ± 0.9) %		131
$\gamma\chi_{b2}(1P)$	(6.6 ± 0.9) %		110
$\gamma\chi_{b0}(1P)$	(4.3 ± 1.0) %		162
$\gamma f_J(1710)$	$< 5.9 \times 10^{-4}$	90%	4866
$\gamma f'_2(1525)$	$< 5.3 \times 10^{-4}$	90%	4896
$\gamma f_2(1270)$	$< 2.41 \times 10^{-4}$	90%	4931

$\chi_{b0}(2P)$ [zz] or $\chi_{b0}(10235)$

$I^G(J^{PC}) = ?^?(0 \text{ preferred}^{++})$
J needs confirmation.

Mass $m = 10.2321 \pm 0.0006$ GeV

$\chi_{b0}(2P)$ DECAY MODES	Fraction (Γ_i/Γ)	p (MeV/c)
$\gamma\Upsilon(2S)$	(4.6 ± 2.1) %	210
$\gamma\Upsilon(1S)$	$(9 \pm6) \times 10^{-3}$	746

$\chi_{b1}(2P)$ [zz] or $\chi_{b1}(10255)$

$I^G(J^{PC}) = ?^?(1 \text{ preferred}^{++})$
J needs confirmation.

Mass $m = 10.2552 \pm 0.0005$ GeV
$m_{\chi_{b1}(2P)} - m_{\chi_{b0}(2P)} = 23.5 \pm 1.0$ MeV

$\chi_{b1}(2P)$ DECAY MODES	Fraction (Γ_i/Γ)	Scale factor	p (MeV/c)
$\gamma\Upsilon(2S)$	(21 ± 4) %	1.5	229
$\gamma\Upsilon(1S)$	(8.5 ± 1.3) %	1.3	764

$\chi_{b2}(2P)$ [zz] or $\chi_{b2}(10270)$

$I^G(J^{PC}) = ?^?(2 \text{ preferred}^{++})$
J needs confirmation.

Mass $m = 10.2685 \pm 0.0004$ GeV
$m_{\chi_{b2}(2P)} - m_{\chi_{b1}(2P)} = 13.5 \pm 0.6$ MeV

$\chi_{b2}(2P)$ DECAY MODES	Fraction (Γ_i/Γ)	p (MeV/c)
$\gamma\Upsilon(2S)$	(16.2 ± 2.4) %	242
$\gamma\Upsilon(1S)$	(7.1 ± 1.0) %	776

$\Upsilon(3S)$ or $\Upsilon(10355)$

$I^G(J^{PC}) = ?^?(1^{--})$

Mass $m = 10.3553 \pm 0.0005$ GeV
Full width $\Gamma = 26.3 \pm 3.5$ keV

$\Upsilon(3S)$ DECAY MODES	Fraction (Γ_i/Γ)	Scale factor	p (MeV/c)
$\Upsilon(2S)$anything	(10.6 ± 0.8) %		296
$\Upsilon(2S)\pi^+\pi^-$	(2.8 ± 0.6) %	2.2	177
$\Upsilon(2S)\pi^0\pi^0$	(2.00 ± 0.32) %		190
$\Upsilon(2S)\gamma\gamma$	(5.0 ± 0.7) %		–
$\Upsilon(1S)\pi^+\pi^-$	(4.48 ± 0.21) %		814
$\Upsilon(1S)\pi^0\pi^0$	(2.06 ± 0.28) %		816
$\mu^+\mu^-$	(1.81 ± 0.17) %		5177
e^+e^-	seen		5177
Radiative decays			
$\gamma\chi_{b2}(2P)$	(11.4 ± 0.8) %	1.3	87
$\gamma\chi_{b1}(2P)$	(11.3 ± 0.6) %		100
$\gamma\chi_{b0}(2P)$	(5.4 ± 0.6) %	1.1	123

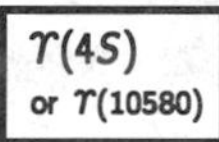

$\Upsilon(4S)$ or $\Upsilon(10580)$

$I^G(J^{PC}) = ?^?(1^{--})$

Mass $m = 10.5800 \pm 0.0035$ GeV
Full width $\Gamma = 23.8 \pm 2.2$ MeV
$\Gamma_{ee} = 0.24 \pm 0.05$ keV (S = 1.7)

$\Upsilon(4S)$ DECAY MODES	Fraction (Γ_i/Γ)	Confidence level	p (MeV/c)
e^+e^-	$(1.01 \pm 0.21) \times 10^{-5}$		5290
D^{*+}anything + c.c.	< 7.4 %	90%	5099
ϕanything	$< 2.3 \times 10^{-3}$	90%	5240
$\Upsilon(1S)$anything	$< 4 \times 10^{-3}$	90%	1053

$\Upsilon(10860)$

$I^G(J^{PC}) = ?^?(1^{--})$

Mass $m = 10.865 \pm 0.008$ GeV (S = 1.1)
Full width $\Gamma = 110 \pm 13$ MeV
$\Gamma_{ee} = 0.31 \pm 0.07$ keV (S = 1.3)

$\Upsilon(10860)$ DECAY MODES	Fraction (Γ_i/Γ)	p (MeV/c)
e^+e^-	$(2.8 \pm 0.7) \times 10^{-6}$	5432

$\Upsilon(11020)$

$I^G(J^{PC}) = ?^?(1^{--})$

Mass $m = 11.019 \pm 0.008$ GeV
Full width $\Gamma = 79 \pm 16$ MeV
$\Gamma_{ee} = 0.130 \pm 0.030$ keV

$\Upsilon(11020)$ DECAY MODES	Fraction (Γ_i/Γ)	p (MeV/c)
e^+e^-	$(1.6 \pm 0.5) \times 10^{-6}$	5509

Searches for Top and Fourth Generation Hadrons

See the sections "Searches for t Quark" and "Searches for b' (4th Generation) Quark" at the end of the QUARKS section.

NOTES

In this Summary Table:

When a quantity has "(S = ...)" to its right, the error on the quantity has been enlarged by the "scale factor" S, defined as $S = \sqrt{\chi^2/(N-1)}$, where N is the number of measurements used in calculating the quantity. We do this when S > 1, which often indicates that the measurements are inconsistent. When S > 1.25, we also show in the Full Listings an ideogram of the measurements. For more about S, see the Introduction.

A decay momentum p is given for each decay mode. For a 2-body decay, p is the momentum of each decay product in the rest frame of the decaying particle. For a 3-or-more-body decay, p is the largest momentum any of the products can have in this frame.

[a] The $\pi^\pm$ mass has increased by three (old) standard deviations since our 1992 edition, and the π^0 mass, which is determined using the mass difference $(m_{\pi^\pm} - m_{\pi^0})$, has increased accordingly. See the "Note on the Charged Pion Mass" in the $\pi^\pm$ Full Listings for a discussion.

[b] See the "Note on $\pi^\pm \to \ell^\pm\nu\gamma$ and $K^\pm \to \ell^\pm\nu\gamma$ Form Factors" in the $\pi^\pm$ Full Listings for definitions and details.

[c] Measurements of $\Gamma(e^+\nu_e)/\Gamma(\mu^+\nu_\mu)$ always include decays with γ's, and measurements of $\Gamma(e^+\nu_e\gamma)$ and $\Gamma(\mu^+\nu_\mu\gamma)$ never include low-energy γ's. Therefore, since no clean separation is possible, we consider the modes with γ's to be subreactions of the modes without them, and let $[\Gamma(e^+\nu_e) + \Gamma(\mu^+\nu_\mu)]/\Gamma_{\text{total}} = 100\%$.

[d] See the $\pi^\pm$ Full Listings for the energy limits used in this measurement; low-energy γ's are not included.

[e] Derived from an analysis of neutrino-oscillation experiments.

[f] Astrophysical and cosmological arguments give limits of order 10^{-13}; see the π^0 Full Listings.

[g] See the "Note on the Decay Width $\Gamma(\eta \to \gamma\gamma)$" in the η Full Listings.

[h] See the "Note on η Decay Parameters" in the η Full Listings.

[i] C parity forbids this to occur as a single-photon process.

[j] The e^+e^- branching fraction is from $e^+e^- \to \pi^+\pi^-$ experiments only. The $\omega\rho$ interference is then due to $\omega\rho$ mixing only, and is expected to be small. If $e\mu$ universality holds, $\Gamma(\rho^0 \to \mu^+\mu^-) = \Gamma(\rho^0 \to e^+e^-) \times 0.99785$.

[k] This is only an educated guess; the error given is larger than the error on the average of the published values. See the Full Listings for details.

[l] See the "Note on the $f_1(1420)$" in the $f_1(1420)$ Full Listings.

[m] See also the $\omega(1600)$ Full Listings.

[n] See the "Note on the $\eta(1440)$" in the $\eta(1440)$ Full Listings.

[o] See the "Note on the $\rho(1450)$ and the $\rho(1700)$" in the $\rho(1700)$ Full Listings.

[p] See also the $\omega(1420)$ Full Listings.

[q] The value is for the sum of the charge states indicated.

[r] The definition of the slope parameter g of the $K \to 3\pi$ Dalitz plot is as follows (see also "Note on Dalitz Plot Parameters for $K \to 3\pi$ Decays" in the $K^\pm$ Full Listings):
$$|M|^2 = 1 + g(s_3 - s_0)/m_{\pi^+}^2 + \cdots .$$

[s] For more details and definitions of parameters see the Full Listings.

[t] See the $K^\pm$ Full Listings for the energy limits used in this measurement.

[u] Most of this radiative mode, the low-momentum γ part, is also included in the parent mode listed without γ's.

[v] Direct-emission branching fraction.

[*w*] Structure-dependent part.

[*x*] The *CP*-violation parameters are defined as follows (see also "Note on *CP* Violation in $K_S \to 3\pi$" and "Note on *CP* Violation in K^0_L Decay" in the Full Listings):

$$\eta_{+-} = |\eta_{+-}|e^{i\phi_{+-}} = \frac{A(K^0_L \to \pi^+\pi^-)}{A(K^0_S \to \pi^+\pi^-)} = \epsilon + \epsilon'$$

$$\eta_{00} = |\eta_{00}|e^{i\phi_{00}} = \frac{A(K^0_L \to \pi^0\pi^0)}{A(K^0_S \to \pi^0\pi^0)} = \epsilon - 2\epsilon'$$

$$\delta = \frac{\Gamma(K^0_L \to \pi^-\ell^+\nu) - \Gamma(K^0_L \to \pi^+\ell^-\nu)}{\Gamma(K^0_L \to \pi^-\ell^+\nu) + \Gamma(K^0_L \to \pi^+\ell^-\nu)},$$

$$\text{Im}(\eta_{+-0})^2 = \frac{\Gamma(K^0_S \to \pi^+\pi^-\pi^0)^{CP\ \text{viol.}}}{\Gamma(K^0_L \to \pi^+\pi^-\pi^0)},$$

$$\text{Im}(\eta_{000})^2 = \frac{\Gamma(K^0_S \to \pi^0\pi^0\pi^0)}{\Gamma(K^0_L \to \pi^0\pi^0\pi^0)}.$$

where for the last two relations *CPT* is assumed valid, *i.e.*, $\text{Re}(\eta_{+-0}) \simeq 0$ and $\text{Re}(\eta_{000}) \simeq 0$.

[*y*] See the K^0_S Full Listings for the energy limits used in this measurement.

[*z*] Calculated from K^0_L semileptonic rates and the K^0_S lifetime assuming $\Delta S = \Delta Q$.

[*aa*] ϵ'/ϵ is derived from $|\eta_{00}/\eta_{+-}|$ measurements using theoretical input on phases.

[*bb*] See the K^0_L Full Listings for the energy limits used in this measurement.

[*cc*] $m_{e^+e^-} > 470$ MeV

[*dd*] Allowed by higher-order electroweak interactions.

[*ee*] Violates *CP* in leading order. Test of direct *CP* violation since the indirect *CP*-violating and *CP*-conserving contributions are expected to be suppressed.

[*ff*] See the note in the $L(1770)$ Full Listings in Reviews of Modern Physics **56** No. 2 Pt. II (1984), p. S200.

[*gg*] This is a weighted average of $D^{\pm}$ (44%) and D^0 (56%) branching fractions. See "D^+ and $D^0 \to$ (ηanything) / (total D^+ and D^0)" under "D^+ Branching Ratios" in the Full Listings.

[*hh*] This value combines the e^+ and μ^+ branching fractions, making a small phase-space adjustment to the μ^+ fraction to be able to use it as an e^+ fraction; hence the "e^+." In fact, some of the e^+ measurements already use μ^+ events in this way.

[*ii*] ℓ indicates e or μ mode, not sum over modes.

[*jj*] The branching fractions for this mode may differ from the sum of the submodes that contribute to it, due to interference effects. See the relevant papers in the Full Listings.

[*kk*] The two experiments determining this ratio are in serious disagreement. See the Full Listings.

[*ll*] This mode is not a useful test for a $\Delta C=1$ weak neutral current because both quarks must change flavor in this decay.

[*mm*] The D^0_1-D^0_2 limits are inferred from the limit on $D^0 \to \overline{D}{}^0 \to K^+\pi^-$.

[*nn*] See the "Note on Semileptonic Decays of D and B Mesons" in the D^+ Full Listings for a comparison of inclusive and summed-inclusive branching fractions.

[*oo*] The limit on $(\overline{K}{}^*(892)\pi)^- \mu^+\nu_\mu$ just below is much stronger.

[*pp*] For now, we average together measurements of the $\phi e^+\nu_e$ and $\phi\mu^+\nu_\mu$ branching fractions.

[*qq*] This branching fraction is calculated from appropriate fractions of the next three branching fractions.

[*rr*] For admixture of B hadrons at LEP and Tevatron energies.

[*ss*] These values are model dependent. See note on "Semileptonic Decays" in the B^+ Full Listings.

[*tt*] D^{**} stands for the sum of the $D(1\,^1P_1)$, $D(1\,^3P_0)$, $D(1\,^3P_1)$, $D(1\,^3P_2)$, $D(2\,^1S_0)$, and $D(2\,^1S_1)$ resonances.

[*uu*] B^0, B^+, B^0_s, and B baryon states not separated.

[*vv*] B^0, B^+, and B^0_s not separated.

[*ww*] Derived from measurements of χ_d and of Δm_{B^0} times B^0 mean life.

[*xx*] Includes $p\overline{p}\pi^+\pi^-\gamma$ and excludes $p\overline{p}\eta$, $p\overline{p}\omega$, $p\overline{p}\eta'$.

[*yy*] J^{PC} known by production in e^+e^- via single photon annihilation. I^G is not known; interpretation of this state as a single resonance is unclear because of the expectation of substantial threshold effects in this energy region.

[*zz*] Spectroscopic labeling for these states is theoretical, pending experimental information.

Meson Summary Table

See also the table of suggested $q\overline{q}$ quark-model assignments in the Quark Model section.

- • Indicates particles that appear in the preceding Meson Summary Table. We do not regard the other entries as being established.
- † Indicates that the value of J given is preferred, but needs confirmation.

LIGHT UNFLAVORED
($S = C = B = 0$)

	$I^G(J^{PC})$		$I^G(J^{PC})$
• $\pi^\pm$	$1^-(0^-)$	• $\omega_3(1670)$	$0^-(3^{--})$
• π^0	$1^-(0^{-+})$	• $\pi_2(1670)$	$1^-(2^{-+})$
• η	$0^+(0^{-+})$	• $\phi(1680)$	$0^-(1^{--})$
• $\rho(770)$	$1^+(1^{--})$	• $\rho_3(1690)$	$1^+(3^{--})$
• $\omega(782)$	$0^-(1^{--})$	• $\rho(1700)$	$1^+(1^{--})$
• $\eta'(958)$	$0^+(0^{-+})$	$X(1700)$	$\text{even}^+(?^{?+})$
• $f_0(980)$	$0^+(0^{++})$	• $f_J(1710)$	$0^+(\text{even}^{++})$
• $a_0(980)$	$1^-(0^{++})$	$X(1740)$	$0^+(\text{even}^{++})$
• $\phi(1020)$	$0^-(1^{--})$	$\eta(1760)$	$0^+(0^{-+})$
• $h_1(1170)$	$0^-(1^{+-})$	$\pi(1770)$	$1^-(0^{-+})$
• $b_1(1235)$	$1^+(1^{+-})$	$X(1775)$	$1^-(?^{-+})$
• $a_1(1260)$	$1^-(1^{++})$	$f_2(1810)$	$0^+(2^{++})$
• $f_2(1270)$	$0^+(2^{++})$	$X(1830)$	$1^-(?^{?+})$
• $f_1(1285)$	$0^+(1^{++})$	• $\phi_3(1850)$	$0^-(3^{--})$
• $\eta(1295)$	$0^+(0^{-+})$	$\eta_2(1870)$	$0^+(2^{-+})$
• $f_0(1300)$	$0^+(0^{++})$	$X(1910)$	$0^+(?^{?+})$
• $\pi(1300)$	$1^-(0^{-+})$	$X(1950)$	$0^+(\text{even}^{++})$
• $a_2(1320)$	$1^-(2^{++})$	• $f_2(2010)$	$0^+(2^{++})$
$f_0(1370)$	$0^+(0^{++})$	$a_4(2040)$	$1^-(4^{++})$
$h_1(1380)$	$?^-(1^{+?})$	$a_3(2050)$	$1^-(3^{++})$
$\hat{\rho}(1405)$	$1^-(1^{-+})$	• $f_4(2050)$	$0^+(4^{++})$
• $f_1(1420)$	$0^+(1^{++})$	$\pi_2(2100)$	$1^-(2^{-+})$
• $\omega(1420)$	$0^-(1^{--})$	$f_2(2150)$	$0^+(2^{++})$
$f_2(1430)$	$0^+(2^{++})$	$\rho(2150)$	$1^+(1^{--})$
• $\eta(1440)$	$0^+(0^{-+})$	$X(2200)$	$?^?(\text{even}^{++})$
• $\rho(1450)$	$1^+(1^{--})$	$\rho(2210)$	$1^+(1^{--})$
• $f_1(1510)$	$0^+(1^{++})$	$f_4(2220)$	$0^+(4^{++})$
$f_2(1520)$	$0^+(2^{++})$	$\eta(2225)$	$0^+(0^{-+})$
• $f_2'(1525)$	$0^+(2^{++})$	$\rho_3(2250)$	$1^+(3^{--})$
$f_0(1525)$	$0^+(0^{++})$	• $f_2(2300)$	$0^+(2^{++})$
• $f_0(1590)$	$0^+(0^{++})$	$f_4(2300)$	$0^+(4^{++})$
• $\omega(1600)$	$0^-(1^{--})$	• $f_2(2340)$	$0^+(2^{++})$
$X(1600)$	$2^+(2^{++})$	$\rho_5(2350)$	$1^+(5^{--})$
$f_2(1640)$	$0^+(2^{++})$	$a_6(2450)$	$1^-(6^{++})$
		$f_6(2510)$	$0^+(6^{++})$
		$X(3250)$	$?^?(?^{??})$

OTHER LIGHT UNFLAVORED
($S = C = B = 0$)

$e^+e^-(1100\text{–}2200)$	$?^?(1^{--})$
$\overline{N}N(1100\text{–}3600)$	
$X(1900\text{–}3600)$	

STRANGE
($S = \pm1$, $C = B = 0$)

	$I(J^P)$
• $K^\pm$	$1/2(0^-)$
• K^0	$1/2(0^-)$
• K_S^0	$1/2(0^-)$
• K_L^0	$1/2(0^-)$
• $K^*(892)$	$1/2(1^-)$
• $K_1(1270)$	$1/2(1^+)$
• $K_1(1400)$	$1/2(1^+)$
• $K^*(1410)$	$1/2(1^-)$
• $K_0^*(1430)$	$1/2(0^+)$
• $K_2^*(1430)$	$1/2(2^+)$
$K(1460)$	$1/2(0^-)$
$K_2(1580)$	$1/2(2^-)$
$K_1(1650)$	$1/2(1^+)$
• $K^*(1680)$	$1/2(1^-)$
• $K_2(1770)$	$1/2(2^-)$
• $K_3^*(1780)$	$1/2(3^-)$
• $K_2(1820)$	$1/2(2^-)$
$K(1830)$	$1/2(0^-)$
$K_0^*(1950)$	$1/2(0^+)$
$K_2^*(1980)$	$1/2(2^+)$
• $K_4^*(2045)$	$1/2(4^+)$
$K_2(2250)$	$1/2(2^-)$
$K_3(2320)$	$1/2(3^+)$
$K_5^*(2380)$	$1/2(5^-)$
$K_4(2500)$	$1/2(4^-)$
$K(3100)$	$?^?(?^{??})$

CHARMED
($C = \pm1$)

• $D^\pm$	$1/2(0^-)$
• D^0	$1/2(0^-)$
• $D^*(2007)^0$	$1/2(1^-)$
• $D^*(2010)^\pm$	$1/2(1^-)$
• $D_1(2420)^0$	$1/2(1^+)$
$D_J(2440)^\pm$	$1/2(?^?)$
• $D_2^*(2460)$	$1/2(2^+)$

CHARMED, STRANGE
($C = S = \pm1$)

• $D_s^\pm$	$0(0^-)$
• $D_s^{*\pm}$	$?(?^?)$
• $D_{s1}(2536)^\pm$	$0(1^+)$
$D_{sJ}(2573)^\pm$	$?(?^?)$

BOTTOM
($B = \pm1$)

• $B^\pm$	$1/2(0^-)$
• B^0	$1/2(0^-)$
• B^*	$1/2(1^-)$

BOTTOM, STRANGE
($S = \pm1$, $C = B = 0$)

	$I^G(J^{PC})$
• B_s^0	$1/2(0^-)$
B_s^*	$?(?^?)$

$c\overline{c}$

• $\eta_c(1S) =$ $\eta_c(2980)$	$0^+(0^{-+})$
• $J/\psi(1S) =$ $J/\psi(3097)$	$0^-(1^{--})$
• $\chi_{c0}(1P) =$ $\chi_{c0}(3415)$	$0^+(0^{++})$
• $\chi_{c1}(1P) =$ $\chi_{c1}(3510)$	$0^+(1^{++})$
$h_c(1P)$	$?^?(?^{??})$
• $\chi_{c2}(1P) =$ $\chi_{c2}(3555)$	$0^+(2^{++})$
$\eta_c(2S) =$ $\eta_c(3590)$	$?^?(?^{?+})$
• $\psi(2S) =$ $\psi(3685)$	$0^-(1^{--})$
• $\psi(3770)$	$?^?(1^{--})$
• $\psi(4040)$	$?^?(1^{--})$
• $\psi(4160)$	$?^?(1^{--})$
• $\psi(4415)$	$?^?(1^{--})$

$b\overline{b}$

• $\Upsilon(1S) =$ $\Upsilon(9460)$	$?^?(1^{--})$
• $\chi_{b0}(1P) =$ $\chi_{b0}(9860)$	$?^?(0^{++})$†
• $\chi_{b1}(1P) =$ $\chi_{b1}(9890)$	$?^?(1^{++})$
• $\chi_{b2}(1P) =$ $\chi_{b2}(9915)$	$?^?(2^{++})$
• $\Upsilon(2S) =$ $\Upsilon(10023)$	$?^?(1^{--})$
• $\chi_{b0}(2P) =$ $\chi_{b0}(10235)$	$?^?(0^{++})$†
• $\chi_{b1}(2P) =$ $\chi_{b1}(10255)$	$?^?(1^{++})$†
• $\chi_{b2}(2P) =$ $\chi_{b2}(10270)$	$?^?(2^{++})$†
• $\Upsilon(3S) =$ $\Upsilon(10355)$	$?^?(1^{--})$
• $\Upsilon(4S) =$ $\Upsilon(10580)$	$?^?(1^{--})$
• $\Upsilon(10860)$	$?^?(1^{--})$
• $\Upsilon(11020)$	$?^?(1^{--})$

NON-$q\overline{q}$ CANDIDATES

Non-$q\overline{q}$ Candidates

This short table gives the name, the quantum numbers (where known), and the status of baryons in the Review. Only the baryons with 3- or 4-star status are included in the main Baryon Summary Table. Due to insufficient data or uncertain interpretation, the other entries in the short table are not established as baryons. The names with masses are of baryons that decay strongly. See our 1986 edition (Physics Letters **170B**) for listings of evidence for Z baryons (KN resonances).

p	P_{11}	****	$\Delta(1232)$	P_{33}	****	Λ	P_{01}	****	Σ^+	P_{11}	****	Ξ^0	P_{11}	****
n	P_{11}	****	$\Delta(1600)$	P_{33}	***	$\Lambda(1405)$	S_{01}	****	Σ^0	P_{11}	****	Ξ^-	P_{11}	****
$N(1440)$	P_{11}	****	$\Delta(1620)$	S_{31}	****	$\Lambda(1520)$	D_{03}	****	Σ^-	P_{11}	****	$\Xi(1530)$	P_{13}	****
$N(1520)$	D_{13}	****	$\Delta(1700)$	D_{33}	****	$\Lambda(1600)$	P_{01}	***	$\Sigma(1385)$	P_{13}	****	$\Xi(1620)$		*
$N(1535)$	S_{11}	****	$\Delta(1750)$	P_{31}	*	$\Lambda(1670)$	S_{01}	****	$\Sigma(1480)$		*	$\Xi(1690)$		***
$N(1650)$	S_{11}	****	$\Delta(1900)$	S_{31}	***	$\Lambda(1690)$	D_{03}	****	$\Sigma(1560)$		**	$\Xi(1820)$	D_{13}	***
$N(1675)$	D_{15}	****	$\Delta(1905)$	F_{35}	****	$\Lambda(1800)$	S_{01}	***	$\Sigma(1580)$	D_{13}	**	$\Xi(1950)$		***
$N(1680)$	F_{15}	****	$\Delta(1910)$	P_{31}	****	$\Lambda(1810)$	P_{01}	***	$\Sigma(1620)$	S_{11}	**	$\Xi(2030)$		***
$N(1700)$	D_{13}	***	$\Delta(1920)$	P_{33}	***	$\Lambda(1820)$	F_{05}	****	$\Sigma(1660)$	P_{11}	***	$\Xi(2120)$		*
$N(1710)$	P_{11}	***	$\Delta(1930)$	D_{35}	***	$\Lambda(1830)$	D_{05}	****	$\Sigma(1670)$	D_{13}	****	$\Xi(2250)$		**
$N(1720)$	P_{13}	****	$\Delta(1940)$	D_{33}	*	$\Lambda(1890)$	P_{03}	****	$\Sigma(1690)$		**	$\Xi(2370)$		**
$N(1900)$	P_{13}	*	$\Delta(1950)$	F_{37}	****	$\Lambda(2000)$		*	$\Sigma(1750)$	S_{11}	***	$\Xi(2500)$		*
$N(1990)$	F_{17}	**	$\Delta(2000)$	F_{35}	*	$\Lambda(2020)$	F_{07}	*	$\Sigma(1770)$	P_{11}	*			
$N(2000)$	F_{15}	**	$\Delta(2150)$	S_{31}	*	$\Lambda(2100)$	G_{07}	****	$\Sigma(1775)$	D_{15}	****	Ω^-		****
$N(2080)$	D_{13}	**	$\Delta(2200)$	G_{37}	*	$\Lambda(2110)$	F_{05}	***	$\Sigma(1840)$	P_{13}	*	$\Omega(2250)^-$		***
$N(2090)$	S_{11}	*	$\Delta(2300)$	H_{39}	**	$\Lambda(2325)$	D_{03}	*	$\Sigma(1880)$	P_{11}	**	$\Omega(2380)^-$		**
$N(2100)$	P_{11}	*	$\Delta(2350)$	D_{35}	*	$\Lambda(2350)$	H_{09}	***	$\Sigma(1915)$	F_{15}	****	$\Omega(2470)^-$		**
$N(2190)$	G_{17}	****	$\Delta(2390)$	F_{37}	*	$\Lambda(2585)$		**	$\Sigma(1940)$	D_{13}	***			
$N(2200)$	D_{15}	**	$\Delta(2400)$	G_{39}	**				$\Sigma(2000)$	S_{11}	*	Λ_c^+		****
$N(2220)$	H_{19}	****	$\Delta(2420)$	$H_{3,11}$	****				$\Sigma(2030)$	F_{17}	****	$\Lambda_c(2625)^+$		***
$N(2250)$	G_{19}	****	$\Delta(2750)$	$I_{3,13}$	**				$\Sigma(2070)$	F_{15}	*	$\Sigma_c(2455)$		****
$N(2600)$	$I_{1,11}$	***	$\Delta(2950)$	$K_{3,15}$	**				$\Sigma(2080)$	P_{13}	**	$\Sigma_c(2530)$		*
$N(2700)$	$K_{1,13}$	**							$\Sigma(2100)$	G_{17}	*	Ξ_c^+		***
									$\Sigma(2250)$		***	Ξ_c^0		***
									$\Sigma(2455)$		**	Ω_c^0		**
									$\Sigma(2620)$		**			
									$\Sigma(3000)$		*	Λ_b^0		***
									$\Sigma(3170)$		*			

**** Existence is certain, and properties are at least fairly well explored.

*** Existence ranges from very likely to certain, but further confirmation is desirable and/or quantum numbers, branching fractions, etc. are not well determined.

** Evidence of existence is only fair.

* Evidence of existence is poor.

N BARYONS
$(S = 0, I = 1/2)$

$p, N^+ = uud; \quad n, N^0 = udd$

p

$I(J^P) = \frac{1}{2}(\frac{1}{2}^+)$

Mass $m = 938.27231 \pm 0.00028$ MeV [a]
$= 1.007276470 \pm 0.000000012$ u

$m_{\overline{p}}/m_p = 0.99999998 \pm 0.00000004$

$|q_p + q_{\overline{p}}|/e < 2 \times 10^{-5}$

$|q_p + q_e|/e < 1.0 \times 10^{-21}$ [b]

Magnetic moment $\mu = 2.79284739 \pm 0.00000006\ \mu_N$

Electric dipole moment $d = (-4 \pm 6) \times 10^{-23}$ $e$cm

Electric polarizability $\overline{\alpha} = (10.2 \pm 0.9) \times 10^{-4}$ fm^3

Magnetic polarizability $\overline{\beta} = (4.0 \pm 0.9) \times 10^{-4}$ fm^3

Mean life $\tau > 1.6 \times 10^{25}$ years (independent of mode)
$> 10^{31} - 5 \times 10^{32}$ years [c] (mode dependent)

For N decays, p and n distinguish proton and neutron partial lifetimes. See also the "Note on Proton Mean Life Limits" in the Full Listings.

The "partial mean life" limits tabulated here are the limits on τ/B_i, where τ is the total mean life and B_i is the branching fraction for the mode in question.

p DECAY MODES	Partial mean life (10^{30} years)	Confidence level	p (MeV/c)
Antilepton + meson			
$N \to e^+\pi$	> 130 (n), > 550 (p)	90%	459
$N \to \mu^+\pi$	> 100 (n), > 270 (p)	90%	453
$N \to \nu\pi$	> 100 (n), > 25 (p)	90%	459
$p \to e^+\eta$	> 140	90%	309
$p \to \mu^+\eta$	> 69	90%	296
$n \to \nu\eta$	> 54	90%	310
$N \to e^+\rho$	> 58 (n), > 75 (p)	90%	153
$N \to \mu^+\rho$	> 23 (n), > 110 (p)	90%	119
$N \to \nu\rho$	> 19 (n), > 27 (p)	90%	153
$p \to e^+\omega$	> 45	90%	142
$p \to \mu^+\omega$	> 57	90%	104
$n \to \nu\omega$	> 43	90%	144
$N \to e^+K$	> 1.3 (n), > 150 (p)	90%	337
$p \to e^+K^0_S$	> 76	90%	337
$p \to e^+K^0_L$	> 44	90%	337
$N \to \mu^+K$	> 1.1 (n), > 120 (p)	90%	326
$p \to \mu^+K^0_S$	> 64	90%	326
$p \to \mu^+K^0_L$	> 44	90%	326
$N \to \nu K$	> 86 (n), > 100 (p)	90%	339
$p \to e^+K^*(892)^0$	> 52	90%	45
$N \to \nu K^*(892)$	> 22 (n), > 20 (p)	90%	45
Antilepton + mesons			
$p \to e^+\pi^+\pi^-$	> 21	90%	448
$p \to e^+\pi^0\pi^0$	> 38	90%	449
$n \to e^+\pi^-\pi^0$	> 32	90%	449
$p \to \mu^+\pi^+\pi^-$	> 17	90%	425
$p \to \mu^+\pi^0\pi^0$	> 33	90%	427
$n \to \mu^+\pi^-\pi^0$	> 33	90%	427
$n \to e^+K^0\pi^-$	> 18	90%	319
Lepton + meson			
$n \to e^-\pi^+$	> 65	90%	459
$n \to \mu^-\pi^+$	> 49	90%	453
$n \to e^-\rho^+$	> 62	90%	154
$n \to \mu^-\rho^+$	> 7	90%	120
$n \to e^-K^+$	> 32	90%	340
$n \to \mu^-K^+$	> 57	90%	330
Lepton + mesons			
$p \to e^-\pi^+\pi^+$	> 30	90%	448
$n \to e^-\pi^+\pi^0$	> 29	90%	449
$p \to \mu^-\pi^+\pi^+$	> 17	90%	425
$n \to \mu^-\pi^+\pi^0$	> 34	90%	427
$p \to e^-\pi^+K^+$	> 20	90%	320
$p \to \mu^-\pi^+K^+$	> 5	90%	279
Antilepton + photon(s)			
$p \to e^+\gamma$	> 460	90%	469
$p \to \mu^+\gamma$	> 380	90%	463
$n \to \nu\gamma$	> 24	90%	470
$p \to e^+\gamma\gamma$	> 100	90%	469
Three leptons			
$p \to e^+e^+e^-$	> 510	90%	469
$p \to e^+\mu^+\mu^-$	> 81	90%	457
$p \to e^+\nu\nu$	> 11	90%	469
$n \to e^+e^-\nu$	> 74	90%	470
$n \to \mu^+e^-\nu$	> 47	90%	464
$n \to \mu^+\mu^-\nu$	> 42	90%	458
$p \to \mu^+e^+e^-$	> 91	90%	464
$p \to \mu^+\mu^+\mu^-$	> 190	90%	439
$p \to \mu^+\nu\nu$	> 21	90%	463
$p \to e^-\mu^+\mu^+$	> 6	90%	457
$n \to 3\nu$	> 0.0005	90%	470
Inclusive modes			
$N \to e^+$ anything	> 0.6 (n, p)	90%	–
$N \to \mu^+$ anything	> 12 (n, p)	90%	–
$N \to e^+\pi^0$ anything	> 0.6 (n, p)	90%	–
$\Delta B = 2$ dinucleon modes			
The following are lifetime limits per iron nucleus.			
$pp \to \pi^+\pi^+$	> 0.7	90%	–
$pn \to \pi^+\pi^0$	> 2	90%	–
$nn \to \pi^+\pi^-$	> 0.7	90%	–
$nn \to \pi^0\pi^0$	> 3.4	90%	–
$pp \to e^+e^+$	> 5.8	90%	–
$pp \to e^+\mu^+$	> 3.6	90%	–
$pp \to \mu^+\mu^+$	> 1.7	90%	–
$pn \to e^+\overline{\nu}$	> 2.8	90%	–
$pn \to \mu^+\overline{\nu}$	> 1.6	90%	–
$nn \to \nu_e\overline{\nu}_e$	> 0.000012	90%	–
$nn \to \nu_\mu\overline{\nu}_\mu$	> 0.000006	90%	–

$\overline{p}$ DECAY MODES

$\overline{p}$ DECAY MODES	Partial mean life (years)	Confidence level	p (MeV/c)
$\overline{p} \to e^-\gamma$	> 1848	95%	469
$\overline{p} \to e^-\pi^0$	> 554	95%	459
$\overline{p} \to e^-\eta$	> 171	95%	309
$\overline{p} \to e^-K^0_S$	> 29	95%	337
$\overline{p} \to e^-K^0_L$	> 9	95%	337

n

$I(J^P) = \frac{1}{2}(\frac{1}{2}^+)$

Mass $m = 939.56563 \pm 0.00028$ MeV [a]
$= 1.008664904 \pm 0.000000014$ u

$m_n - m_p = 1.293318 \pm 0.000009$ MeV
$= 0.001388434 \pm 0.000000009$ u

Mean life $\tau = 887.0 \pm 2.0$ s (S = 1.3)
$c\tau = 2.659 \times 10^8$ km

Magnetic moment $\mu = -1.9130428 \pm 0.0000005\ \mu_N$

Electric dipole moment $d < 11 \times 10^{-26}$ $e$cm, CL = 95%

Electric polarizability $\alpha = (1.16^{+0.19}_{-0.23}) \times 10^{-3}$ fm^3

Charge $q = (-0.4 \pm 1.1) \times 10^{-21}$ e

Mean time for $n\overline{n}$ oscillations $> 1.2 \times 10^8$ s, CL = 90% [d]

Decay parameters [e]

$pe^-\overline{\nu}_e$	$g_A/g_V = -1.2573 \pm 0.0028$
"	$A = -0.1127 \pm 0.0011$
"	$B = 0.997 \pm 0.028$
"	$a = -0.102 \pm 0.005$
"	$\phi_{AV} = (180.07 \pm 0.18)^\circ$ [f]
"	$D = (-0.5 \pm 1.4) \times 10^{-3}$

n DECAY MODES	Fraction (Γ_i/Γ)	Confidence level	p (MeV/c)
$pe^-\overline{\nu}_e$	100 %		1.19
Charge conservation (Q) violating mode			
$p\nu_e\overline{\nu}_e$	Q < 9×10^{-24}	90%	1.29

$N(1440)\ P_{11}$

$I(J^P) = \frac{1}{2}(\frac{1}{2}^+)$

Mass m = 1430 to 1470 ($\approx$ 1440) MeV
Full width Γ = 250 to 450 ($\approx$ 350) MeV
p_{beam} = 0.61 GeV/c $4\pi\lambda\!\!\!^{-2}$ = 31.0 mb

N(1440) DECAY MODES	Fraction (Γ_i/Γ)	p (MeV/c)
$N\pi$	60–70 %	397
$N\pi\pi$	30–40 %	342
$\Delta\pi$	20–30 %	143
$N\rho$	<8 %	†
$N(\pi\pi)^{I=0}_{S\text{-wave}}$	5–10 %	–
$p\gamma$	0.04–0.07 %	414
$n\gamma$	0.001–0.05 %	413

$N(1520)\ D_{13}$

$I(J^P) = \frac{1}{2}(\frac{3}{2}^-)$

Mass m = 1515 to 1530 ($\approx$ 1520) MeV
Full width Γ = 110 to 135 ($\approx$ 120) MeV
p_{beam} = 0.74 GeV/c $4\pi\lambda\!\!\!^{-2}$ = 23.5 mb

N(1520) DECAY MODES	Fraction (Γ_i/Γ)	p (MeV/c)
$N\pi$	50–60 %	456
$N\pi\pi$	40–50 %	410
$\Delta\pi$	15–25 %	228
$N\rho$	15–25 %	†
$N(\pi\pi)^{I=0}_{S\text{-wave}}$	<8 %	–
$p\gamma$	0.45–0.53 %	470
$n\gamma$	0.34–0.48 %	470

$N(1535)\ S_{11}$

$I(J^P) = \frac{1}{2}(\frac{1}{2}^-)$

Mass m = 1520 to 1555 ($\approx$ 1535) MeV
Full width Γ = 100 to 250 ($\approx$ 150) MeV
p_{beam} = 0.76 GeV/c $4\pi\lambda\!\!\!^{-2}$ = 22.5 mb

N(1535) DECAY MODES	Fraction (Γ_i/Γ)	p (MeV/c)
$N\pi$	35–55 %	467
$N\eta$	30–55 %	182
$N\pi\pi$	1–10 %	422
$\Delta\pi$	<1 %	242
$N\rho$	<4 %	†
$N(\pi\pi)^{I=0}_{S\text{-wave}}$	<3 %	–
$N(1440)\pi$	<7 %	†
$p\gamma$	0.45–0.53 %	481
$n\gamma$	0.34–0.48 %	480

$N(1650)\ S_{11}$

$I(J^P) = \frac{1}{2}(\frac{1}{2}^-)$

Mass m = 1640 to 1680 ($\approx$ 1650) MeV
Full width Γ = 145 to 190 ($\approx$ 150) MeV
p_{beam} = 0.96 GeV/c $4\pi\lambda\!\!\!^{-2}$ = 16.4 mb

N(1650) DECAY MODES	Fraction (Γ_i/Γ)	p (MeV/c)
$N\pi$	60–80 %	547
ΛK	3–11 %	161
$N\pi\pi$	5–20 %	511
$\Delta\pi$	3–7 %	344
$N\rho$	4–14 %	†
$N(\pi\pi)^{I=0}_{S\text{-wave}}$	<4 %	–
$N(1440)\pi$	<5 %	147
$p\gamma$	0.10–0.18 %	558
$n\gamma$	0.03–0.18 %	557

$N(1675)\ D_{15}$

$I(J^P) = \frac{1}{2}(\frac{5}{2}^-)$

Mass m = 1670 to 1685 ($\approx$ 1675) MeV
Full width Γ = 140 to 180 ($\approx$ 150) MeV
p_{beam} = 1.01 GeV/c $4\pi\lambda\!\!\!^{-2}$ = 15.4 mb

N(1675) DECAY MODES	Fraction (Γ_i/Γ)	p (MeV/c)
$N\pi$	40–50 %	563
ΛK	<1 %	209
$N\pi\pi$	50–60 %	529
$\Delta\pi$	50–60 %	364
$N\rho$	< 1–3 %	†
$p\gamma$	0.005–0.014 %	575
$n\gamma$	0.07–0.11 %	574

$N(1680)\ F_{15}$

$I(J^P) = \frac{1}{2}(\frac{5}{2}^+)$

Mass m = 1675 to 1690 ($\approx$ 1680) MeV
Full width Γ = 120 to 140 ($\approx$ 130) MeV
p_{beam} = 1.01 GeV/c $4\pi\lambda\!\!\!^{-2}$ = 15.2 mb

N(1680) DECAY MODES	Fraction (Γ_i/Γ)	p (MeV/c)
$N\pi$	60–70 %	567
$N\pi\pi$	30–40 %	532
$\Delta\pi$	5–15 %	369
$N\rho$	3–15 %	†
$N(\pi\pi)^{I=0}_{S\text{-wave}}$	5–20 %	–
$p\gamma$	0.21–0.35 %	578
$n\gamma$	0.02–0.04 %	577

$N(1700)\ D_{13}$

$I(J^P) = \frac{1}{2}(\frac{3}{2}^-)$

Mass m = 1650 to 1750 ($\approx$ 1700) MeV
Full width Γ = 50 to 150 ($\approx$ 100) MeV
p_{beam} = 1.05 GeV/c $4\pi\lambda\!\!\!^{-2}$ = 14.5 mb

N(1700) DECAY MODES	Fraction (Γ_i/Γ)	p (MeV/c)
$N\pi$	5–15 %	580
ΛK	<3 %	250
$N\pi\pi$	85–95 %	547
$N\rho$	<35 %	†
$p\gamma$	~ 0.01 %	591

$N(1710)\ P_{11}$

$I(J^P) = \frac{1}{2}(\frac{1}{2}^+)$

Mass m = 1680 to 1740 ($\approx$ 1710) MeV
Full width Γ = 50 to 250 ($\approx$ 100) MeV
p_{beam} = 1.07 GeV/c $4\pi\lambda\!\!\!^{-2}$ = 14.2 mb

N(1710) DECAY MODES	Fraction (Γ_i/Γ)	p (MeV/c)
$N\pi$	10–20 %	587
ΛK	5–25 %	264
$N\pi\pi$	40–90 %	554
[illegible]	15–40 %	393
$N\rho$	5–25 %	48
$N(\pi\pi)^{I=0}_{S\text{-wave}}$	10–40 %	–

$N(1720)\ P_{13}$

$I(J^P) = \frac{1}{2}(\frac{3}{2}^+)$

Mass m = 1650 to 1750 ($\approx$ 1720) MeV
Full width Γ = 100 to 200 ($\approx$ 150) MeV
p_{beam} = 1.09 GeV/c $4\pi\lambda\!\!\!^{-2}$ = 13.9 mb

N(1720) DECAY MODES	Fraction (Γ_i/Γ)	p (MeV/c)
$N\pi$	10–20 %	594
ΛK	1–15 %	278
$N\pi\pi$	>70 %	561
$N\rho$	70–85 %	104
$p\gamma$	0.01–0.06 %	–

$N(2190)$ G_{17}

$I(J^P) = \frac{1}{2}(\frac{7}{2}^-)$

Mass $m = 2100$ to 2200 (≈ 2190) MeV
Full width $\Gamma = 350$ to 550 (≈ 450) MeV
$p_{beam} = 2.07$ GeV/c $4\pi\lambda\!\!\!^{-2} = 6.21$ mb

N(2190) DECAY MODES	Fraction (Γ_i/Γ)	p (MeV/c)
$N\pi$	10–20 %	888

$N(2220)$ H_{19}

$I(J^P) = \frac{1}{2}(\frac{9}{2}^+)$

Mass $m = 2180$ to 2310 (≈ 2220) MeV
Full width $\Gamma = 320$ to 550 (≈ 400) MeV
$p_{beam} = 2.14$ GeV/c $4\pi\lambda\!\!\!^{-2} = 5.97$ mb

N(2220) DECAY MODES	Fraction (Γ_i/Γ)	p (MeV/c)
$N\pi$	10–20 %	905

$N(2250)$ G_{19}

$I(J^P) = \frac{1}{2}(\frac{9}{2}^-)$

Mass $m = 2170$ to 2310 (≈ 2250) MeV
Full width $\Gamma = 290$ to 470 (≈ 400) MeV
$p_{beam} = 2.21$ GeV/c $4\pi\lambda\!\!\!^{-2} = 5.74$ mb

N(2250) DECAY MODES	Fraction (Γ_i/Γ)	p (MeV/c)
$N\pi$	5–15 %	923

$N(2600)$ $I_{1,11}$

$I(J^P) = \frac{1}{2}(\frac{11}{2}^-)$

Mass $m = 2550$ to 2750 (≈ 2600) MeV
Full width $\Gamma = 500$ to 800 (≈ 650) MeV
$p_{beam} = 3.12$ GeV/c $4\pi\lambda\!\!\!^{-2} = 3.86$ mb

N(2600) DECAY MODES	Fraction (Γ_i/Γ)	p (MeV/c)
$N\pi$	5–10 %	1126

Δ BARYONS
($S = 0, I = 3/2$)

$\Delta^{++} = uuu$, $\Delta^{+} = uud$, $\Delta^{0} = udd$, $\Delta^{-} = ddd$

$\Delta(1232)$ P_{33}

$I(J^P) = \frac{3}{2}(\frac{3}{2}^+)$

Mass $m = 1230$ to 1234 (≈ 1232) MeV
Full width $\Gamma = 115$ to 125 (≈ 120) MeV
$p_{beam} = 0.30$ GeV/c $4\pi\lambda\!\!\!^{-2} = 94.8$ mb

Δ(1232) DECAY MODES	Fraction (Γ_i/Γ)	p (MeV/c)
$N\pi$	>99 %	227
$N\gamma$	0.55–0.61 %	259

$\Delta(1600)$ P_{33}

$I(J^P) = \frac{3}{2}(\frac{3}{2}^+)$

Mass $m = 1550$ to 1700 (≈ 1600) MeV
Full width $\Gamma = 250$ to 450 (≈ 350) MeV
$p_{beam} = 0.87$ GeV/c $4\pi\lambda\!\!\!^{-2} = 18.6$ mb

Δ(1600) DECAY MODES	Fraction (Γ_i/Γ)	p (MeV/c)
$N\pi$	10–25 %	512
$N\pi\pi$	75–90 %	473
$\Delta\pi$	40–70 %	301
$N\rho$	<25 %	†
$N(1440)\pi$	10–35 %	74
$p\gamma$	~0 %	–

$\Delta(1620)$ S_{31}

$I(J^P) = \frac{3}{2}(\frac{1}{2}^-)$

Mass $m = 1615$ to 1675 (≈ 1620) MeV
Full width $\Gamma = 120$ to 180 (≈ 150) MeV
$p_{beam} = 0.91$ GeV/c $4\pi\lambda\!\!\!^{-2} = 17.7$ mb

Δ(1620) DECAY MODES	Fraction (Γ_i/Γ)	p (MeV/c)
$N\pi$	20–30 %	526
$N\pi\pi$	70–80 %	488
$\Delta\pi$	30–60 %	318
$N\rho$	7–25 %	†
$N\gamma$	0.02–0.06 %	538

$\Delta(1700)$ D_{33}

$I(J^P) = \frac{3}{2}(\frac{3}{2}^-)$

Mass $m = 1670$ to 1770 (≈ 1700) MeV
Full width $\Gamma = 200$ to 400 (≈ 300) MeV
$p_{beam} = 1.05$ GeV/c $4\pi\lambda\!\!\!^{-2} = 14.5$ mb

Δ(1700) DECAY MODES	Fraction (Γ_i/Γ)	p (MeV/c)
$N\pi$	10–20 %	580
$N\pi\pi$	80–90 %	547
$\Delta\pi$	30–60 %	385
$N\rho$	30–55 %	†
$N\gamma$	0.16–0.28 %	591

$\Delta(1900)$ S_{31}

$I(J^P) = \frac{3}{2}(\frac{1}{2}^-)$

Mass $m = 1850$ to 1950 (≈ 1900) MeV
Full width $\Gamma = 140$ to 240 (≈ 200) MeV
$p_{beam} = 1.44$ GeV/c $4\pi\lambda\!\!\!^{-2} = 9.71$ mb

Δ(1900) DECAY MODES	Fraction (Γ_i/Γ)	p (MeV/c)
$N\pi$	10–30 %	710

$\Delta(1905)$ F_{35}

$I(J^P) = \frac{3}{2}(\frac{5}{2}^+)$

Mass $m = 1870$ to 1920 (≈ 1905) MeV
Full width $\Gamma = 280$ to 440 (≈ 350) MeV
$p_{beam} = 1.45$ GeV/c $4\pi\lambda\!\!\!^{-2} = 9.62$ mb

Δ(1905) DECAY MODES	Fraction (Γ_i/Γ)	p (MeV/c)
$N\pi$	5–15 %	713
$N\pi\pi$	85–95 %	687
$\Delta\pi$	<25 %	542
$N\rho$	>60 %	421
$N\gamma$	0.01–0.04 %	721

$\Delta(1910)$ P_{31}

$I(J^P) = \frac{3}{2}(\frac{1}{2}^+)$

Mass $m = 1870$ to 1920 (≈ 1910) MeV
Full width $\Gamma = 190$ to 270 (≈ 250) MeV
$p_{beam} = 1.46$ GeV/c $4\pi\lambda\!\!\!^{-2} = 9.54$ mb

Δ(1910) DECAY MODES	Fraction (Γ_i/Γ)	p (MeV/c)
$N\pi$	15–30 %	716

$\Delta(1920)$ P_{33}

$I(J^P) = \frac{3}{2}(\frac{3}{2}^+)$

Mass $m = 1900$ to 1970 (≈ 1920) MeV
Full width $\Gamma = 150$ to 300 (≈ 200) MeV
$p_{beam} = 1.48$ GeV/c $4\pi\lambda\!\!\!^{-2} = 9.37$ mb

Δ(1920) DECAY MODES	Fraction (Γ_i/Γ)	p (MeV/c)
$N\pi$	5–20 %	722

$\Delta(1930)\ D_{35}$

$I(J^P) = \frac{3}{2}(\frac{5}{2}^-)$

Mass m = 1920 to 1970 ($\approx$ 1930) MeV
Full width Γ = 250 to 450 ($\approx$ 350) MeV
p_{beam} = 1.50 GeV/c $\quad 4\pi\lambda\!\!\!^{-2}$ = 9.21 mb

Δ(1930) DECAY MODES	Fraction (Γ_i/Γ)	p (MeV/c)
$N\pi$	10–20 %	729

$\Delta(1950)\ F_{37}$

$I(J^P) = \frac{3}{2}(\frac{7}{2}^+)$

Mass m = 1940 to 1960 ($\approx$ 1950) MeV
Full width Γ = 290 to 350 ($\approx$ 300) MeV
p_{beam} = 1.54 GeV/c $\quad 4\pi\lambda\!\!\!^{-2}$ = 8.91 mb

Δ(1950) DECAY MODES	Fraction (Γ_i/Γ)	p (MeV/c)
$N\pi$	35–40 %	741
$N\pi\pi$		716
$\quad\Delta\pi$	20–30 %	574
$\quad N\rho$	<10 %	469
$N\gamma$	0.10–0.15 %	749

$\Delta(2420)\ H_{3,11}$

$I(J^P) = \frac{3}{2}(\frac{11}{2}^+)$

Mass m = 2300 to 2500 ($\approx$ 2420) MeV
Full width Γ = 300 to 500 ($\approx$ 400) MeV
p_{beam} = 2.64 GeV/c $\quad 4\pi\lambda\!\!\!^{-2}$ = 4.68 mb

Δ(2420) DECAY MODES	Fraction (Γ_i/Γ)	p (MeV/c)
$N\pi$	5–15 %	1023

Λ BARYONS
($S = -1, I = 0$)

$\Lambda^0 = uds$

Λ

$I(J^P) = 0(\frac{1}{2}^+)$

Mass m = 1115.684 $\pm$ 0.006 MeV
Mean life τ = (2.632 $\pm$ 0.020) $\times 10^{-10}$ s (S = 1.6)
$c\tau$ = 7.89 cm
Magnetic moment $\mu = -0.613 \pm 0.004\ \mu_N$
Electric dipole moment $d < 1.5 \times 10^{-16}$ $e$cm, CL = 95%

Decay parameters

$p\pi^-$	$\alpha_- = 0.642 \pm 0.013$
"	$\phi_- = (-6.5 \pm 3.5)^\circ$
"	$\gamma_- = 0.76$ [g]
"	$\Delta_- = (8 \pm 4)^\circ$ [g]
$n\pi^0$	$\alpha_0 = +0.65 \pm 0.05$
$pe^-\overline{\nu}_e$	$g_A/g_V = -0.718 \pm 0.015$ [e]

Λ DECAY MODES	Fraction (Γ_i/Γ)	p (MeV/c)
$p\pi^-$	(63.9 $\pm$0.5) %	101
$n\pi^0$	(35.8 $\pm$0.5) %	104
$n\gamma$	(1.75$\pm$0.15) $\times 10^{-3}$	162
$p\pi^-\gamma$	[h] (8.4 $\pm$1.4) $\times 10^{-4}$	101
$pe^-\overline{\nu}_e$	(8.32$\pm$0.14) $\times 10^{-4}$	163
$p\mu^-\overline{\nu}_\mu$	(1.57$\pm$0.35) $\times 10^{-4}$	131

$\Lambda(1405)\ S_{01}$

$I(J^P) = 0(\frac{1}{2}^-)$

Mass m = 1407 $\pm$ 4 MeV
Full width Γ = 50.0 $\pm$ 2.0 MeV
Below $\overline{K}N$ threshold

Λ(1405) DECAY MODES	Fraction (Γ_i/Γ)	p (MeV/c)
$\Sigma\pi$	100 %	152

$\Lambda(1520)\ D_{03}$

$I(J^P) = 0(\frac{3}{2}^-)$

Mass m = 1519.5 $\pm$ 1.0 MeV [i]
Full width Γ = 15.6 $\pm$ 1.0 MeV [i]
p_{beam} = 0.39 GeV/c $\quad 4\pi\lambda\!\!\!^{-2}$ = 82.8 mb

Λ(1520) DECAY MODES	Fraction (Γ_i/Γ)	p (MeV/c)
$N\overline{K}$	45 $\pm$ 1%	244
$\Sigma\pi$	42 $\pm$ 1%	267
$\Lambda\pi\pi$	10 $\pm$ 1%	252
$\Sigma\pi\pi$	0.9 $\pm$ 0.1%	152
$\Lambda\gamma$	0.8 $\pm$ 0.2%	351

$\Lambda(1600)\ P_{01}$

$I(J^P) = 0(\frac{1}{2}^+)$

Mass m = 1560 to 1700 ($\approx$ 1600) MeV
Full width Γ = 50 to 250 ($\approx$ 150) MeV
p_{beam} = 0.58 GeV/c $\quad 4\pi\lambda\!\!\!^{-2}$ = 41.6 mb

Λ(1600) DECAY MODES	Fraction (Γ_i/Γ)	p (MeV/c)
$N\overline{K}$	15–30 %	343
$\Sigma\pi$	10–60 %	336

$\Lambda(1670)\ S_{01}$

$I(J^P) = 0(\frac{1}{2}^-)$

Mass m = 1660 to 1680 ($\approx$ 1670) MeV
Full width Γ = 25 to 50 ($\approx$ 35) MeV
p_{beam} = 0.74 GeV/c $\quad 4\pi\lambda\!\!\!^{-2}$ = 28.5 mb

Λ(1670) DECAY MODES	Fraction (Γ_i/Γ)	p (MeV/c)
$N\overline{K}$	15–25 %	414
$\Sigma\pi$	20–60 %	393
$\Lambda\eta$	15–35 %	64

$\Lambda(1690)\ D_{03}$

$I(J^P) = 0(\frac{3}{2}^-)$

Mass m = 1685 to 1695 ($\approx$ 1690) MeV
Full width Γ = 50 to 70 ($\approx$ 60) MeV
p_{beam} = 0.78 GeV/c $\quad 4\pi\lambda\!\!\!^{-2}$ = 26.1 mb

Λ(1690) DECAY MODES	Fraction (Γ_i/Γ)	p (MeV/c)
$N\overline{K}$	20–30 %	433
$\Sigma\pi$	20–40 %	409
$\Lambda\pi\pi$	~ 25 %	415
$\Sigma\pi\pi$	~ 20 %	350

$\Lambda(1800)\ S_{01}$

$I(J^P) = 0(\frac{1}{2}^-)$

Mass m = 1720 to 1850 ($\approx$ 1800) MeV
Full width Γ = 200 to 400 ($\approx$ 300) MeV
p_{beam} = 1.01 GeV/c $\quad 4\pi\lambda\!\!\!^{-2}$ = 17.5 mb

Λ(1800) DECAY MODES	Fraction (Γ_i/Γ)	p (MeV/c)
$N\overline{K}$	25–40 %	528
$\Sigma\pi$	seen	493
$\Sigma(1385)\pi$	seen	345
$N\overline{K}^*(892)$	seen	†

$\Lambda(1810)\ P_{01}$

$I(J^P) = 0(\frac{1}{2}^+)$

Mass m = 1750 to 1850 ($\approx$ 1810) MeV
Full width Γ = 50 to 250 ($\approx$ 150) MeV
p_{beam} = 1.04 GeV/c $\quad 4\pi\lambda\!\!\!^{-2}$ = 17.0 mb

Λ(1810) DECAY MODES	Fraction (Γ_i/Γ)	p (MeV/c)
$N\overline{K}$	20–50 %	537
$\Sigma\pi$	10–40 %	501
$\Sigma(1385)\pi$	seen	356
$N\overline{K}^*(892)$	30–60 %	†

$\Lambda(1820)\ F_{05}$

$I(J^P) = 0(\frac{5}{2}^+)$

Mass m = 1815 to 1825 ($\approx$ 1820) MeV
Full width Γ = 70 to 90 ($\approx$ 80) MeV
p_{beam} = 1.06 GeV/c $4\pi\lambda\!\!\!^{-2}$ = 16.5 mb

Λ(1820) DECAY MODES	Fraction (Γ_i/Γ)	p (MeV/c)
$N\overline{K}$	55–65 %	545
$\Sigma\pi$	8–14 %	508
$\Sigma(1385)\pi$	5–10 %	362

$\Lambda(1830)\ D_{05}$

$I(J^P) = 0(\frac{5}{2}^-)$

Mass m = 1810 to 1830 ($\approx$ 1830) MeV
Full width Γ = 60 to 110 ($\approx$ 95) MeV
p_{beam} = 1.08 GeV/c $4\pi\lambda\!\!\!^{-2}$ = 16.0 mb

Λ(1830) DECAY MODES	Fraction (Γ_i/Γ)	p (MeV/c)
$N\overline{K}$	3–10 %	553
$\Sigma\pi$	35–75 %	515
$\Sigma(1385)\pi$	>15 %	371

$\Lambda(1890)\ P_{03}$

$I(J^P) = 0(\frac{3}{2}^+)$

Mass m = 1850 to 1910 ($\approx$ 1890) MeV
Full width Γ = 60 to 200 ($\approx$ 100) MeV
p_{beam} = 1.21 GeV/c $4\pi\lambda\!\!\!^{-2}$ = 13.6 mb

Λ(1890) DECAY MODES	Fraction (Γ_i/Γ)	p (MeV/c)
$N\overline{K}$	20–35 %	599
$\Sigma\pi$	3–10 %	559
$\Sigma(1385)\pi$	seen	420
$N\overline{K}^*(892)$	seen	233

$\Lambda(2100)\ G_{07}$

$I(J^P) = 0(\frac{7}{2}^-)$

Mass m = 2090 to 2110 ($\approx$ 2100) MeV
Full width Γ = 100 to 250 ($\approx$ 200) MeV
p_{beam} = 1.68 GeV/c $4\pi\lambda\!\!\!^{-2}$ = 8.68 mb

Λ(2100) DECAY MODES	Fraction (Γ_i/Γ)	p (MeV/c)
$N\overline{K}$	25–35 %	751
$\Sigma\pi$	~ 5 %	704
$\Lambda\eta$	<3 %	617
ΞK	<3 %	483
$\Lambda\omega$	<8 %	443
$N\overline{K}^*(892)$	10–20 %	514

$\Lambda(2110)\ F_{05}$

$I(J^P) = 0(\frac{5}{2}^+)$

Mass m = 2090 to 2140 ($\approx$ 2110) MeV
Full width Γ = 150 to 250 ($\approx$ 200) MeV
p_{beam} = 1.70 GeV/c $4\pi\lambda\!\!\!^{-2}$ = 8.53 mb

Λ(2110) DECAY MODES	Fraction (Γ_i/Γ)	p (MeV/c)
$N\overline{K}$	5–25 %	757
$\Sigma\pi$	10–40 %	711
$\Lambda\omega$	seen	455
$\Sigma(1385)\pi$	seen	589
$N\overline{K}^*(892)$	10–60 %	524

$\Lambda(2350)\ H_{09}$

$I(J^P) = 0(\frac{9}{2}^+)$

Mass m = 2340 to 2370 ($\approx$ 2350) MeV
Full width Γ = 100 to 250 ($\approx$ 150) MeV
p_{beam} = 2.29 GeV/c $4\pi\lambda\!\!\!^{-2}$ = 5.85 mb

Λ(2350) DECAY MODES	Fraction (Γ_i/Γ)	p (MeV/c)
$N\overline{K}$	~ 12 %	915
$\Sigma\pi$	~ 10 %	867

Σ BARYONS
$(S = -1, I = 1)$

$\Sigma^+ = uus,\quad \Sigma^0 = uds,\quad \Sigma^- = dds$

Σ^+

$I(J^P) = 1(\frac{1}{2}^+)$

Mass m = 1189.37 $\pm$ 0.07 MeV (S = 2.2)
Mean life τ = (0.799 $\pm$ 0.004) $\times 10^{-10}$ s
$c\tau$ = 2.396 cm
Magnetic moment μ = 2.458 $\pm$ 0.010 μ_N (S = 2.1)
$\Gamma(\Sigma^+ \to n\ell^+\nu)/\Gamma(\Sigma^- \to n\ell^-\overline{\nu})$ < 0.043

Decay parameters

$p\pi^0$	$\alpha_0 = -0.980^{+0.017}_{-0.015}$
"	$\phi_0 = (36 \pm 34)°$
"	$\gamma_0 = 0.16$ [g]
"	$\Delta_0 = (187 \pm 6)°$ [g]
$n\pi^+$	$\alpha_+ = 0.068 \pm 0.013$
"	$\phi_+ = (167 \pm 20)°$ (S = 1.1)
"	$\gamma_+ = -0.97$ [g]
"	$\Delta_+ = (-73^{+133}_{-10})°$ [g]
$p\gamma$	$\alpha_\gamma = -0.76 \pm 0.08$

Σ^+ DECAY MODES		Fraction (Γ_i/Γ)	Confidence level	p (MeV/c)
$p\pi^0$		(51.57$\pm$0.30) %		189
$n\pi^+$		(48.30$\pm$0.30) %		185
$p\gamma$		(1.25$\pm$0.07) $\times 10^{-3}$		225
$n\pi^+\gamma$	[h]	(4.5 $\pm$0.5) $\times 10^{-4}$		185
$\Lambda e^+\nu_e$		(2.0 $\pm$0.5) $\times 10^{-5}$		71
$\Delta S = \Delta Q$ (*SQ*) violating modes or ΔS = 1 weak neutral current (*S1*) modes				
$ne^+\nu_e$	*SQ*	< 5 $\times 10^{-6}$	90%	224
$n\mu^+\nu_\mu$	*SQ*	< 3.0 $\times 10^{-5}$	90%	202
pe^+e^-	*S1*	< 7 $\times 10^{-6}$		225

Σ^0

$I(J^P) = 1(\frac{1}{2}^+)$

J^P not measured; assumed to be the same as for the Σ^+ and Σ^-.

Mass m = 1192.55 $\pm$ 0.08 MeV (S = 1.2)
$m_{\Sigma^-} - m_{\Sigma^0}$ = 4.88 $\pm$ 0.08 MeV (S = 1.2)
$m_{\Sigma^0} - m_\Lambda$ = 76.87 $\pm$ 0.08 MeV (S = 1.2)
Mean life τ = (7.4 $\pm$ 0.7) $\times 10^{-20}$ s
$c\tau$ = 2.22 $\times 10^{-11}$ m
Transition magnetic moment $|\mu_{\Sigma\Lambda}|$ = 1.61 $\pm$ 0.08 μ_N

Σ^0 DECAY MODES		Fraction (Γ_i/Γ)	Confidence level	p (MeV/c)
$\Lambda\gamma$		100 %		74
$\Lambda\gamma\gamma$		< 3 %	90%	74
Λe^+e^-	[i]	5 $\times 10^{-3}$		74

Σ^-

$I(J^P) = 1(\frac{1}{2}^+)$

Mass m = 1197.436 $\pm$ 0.033 MeV (S = 1.2)
$m_{\Sigma^-} - m_{\Sigma^+}$ = 8.07 $\pm$ 0.08 MeV (S = 1.9)
$m_{\Sigma^-} - m_\Lambda$ = 81.752 $\pm$ 0.034 MeV (S = 1.2)
Mean life τ = (1.479 $\pm$ 0.011) $\times 10^{-10}$ s (S = 1.3)
$c\tau$ = 4.434 cm
Magnetic moment μ = $-$1.160 $\pm$ 0.025 μ_N (S = 1.7)

Decay parameters

$n\pi^-$	$\alpha_- = -0.068 \pm 0.008$
"	$\phi_- = (10 \pm 15)°$
"	$\gamma_- = 0.98$ [g]
"	$\Delta_- = (249^{+12}_{-120})°$ [g]
$ne^-\overline{\nu}_e$	$g_A/g_V = 0.340 \pm 0.017$ [e]
"	$f_2(0)/f_1(0) = 0.97 \pm 0.14$
"	$D = 0.11 \pm 0.10$
$\Lambda e^-\overline{\nu}_e$	$g_V/g_A = 0.01 \pm 0.10$ [e] (S = 1.5)
"	$g_{WM}/g_A = 2.4 \pm 1.7$ [e]

Σ^- DECAY MODES	Fraction (Γ_i/Γ)	p (MeV/c)
$n\pi^-$	(99.848±0.005) %	193
$n\pi^-\gamma$	[h] (4.6 ±0.6) × 10^{-4}	193
$ne^-\overline{\nu}_e$	(1.017±0.034) × 10^{-3}	230
$n\mu^-\overline{\nu}_\mu$	(4.5 ±0.4) × 10^{-4}	210
$\Lambda e^-\overline{\nu}_e$	(5.73 ±0.27) × 10^{-5}	79

$\Sigma(1385)$ P_{13}

$I(J^P) = 1(\frac{3}{2}^+)$

$\Sigma(1385)^+$ mass $m = 1382.8 \pm 0.4$ MeV (S = 2.0)
$\Sigma(1385)^0$ mass $m = 1383.7 \pm 1.0$ MeV (S = 1.4)
$\Sigma(1385)^-$ mass $m = 1387.2 \pm 0.5$ MeV (S = 2.2)
$\Sigma(1385)^+$ full width $\Gamma = 35.8 \pm 0.8$ MeV
$\Sigma(1385)^0$ full width $\Gamma = 36 \pm 5$ MeV
$\Sigma(1385)^-$ full width $\Gamma = 39.4 \pm 2.1$ MeV (S = 1.7)
Below $\overline{K}N$ threshold

$\Sigma(1385)$ DECAY MODES	Fraction (Γ_i/Γ)	p (MeV/c)
$\Lambda\pi$	88±2 %	208
$\Sigma\pi$	12±2 %	127

$\Sigma(1660)$ P_{11}

$I(J^P) = 1(\frac{1}{2}^+)$

Mass m = 1630 to 1690 (≈ 1660) MeV
Full width Γ = 40 to 200 (≈ 100) MeV
$p_{beam} = 0.72$ GeV/c $4\pi\lambda\!\!^{-2} = 29.9$ mb

$\Sigma(1660)$ DECAY MODES	Fraction (Γ_i/Γ)	p (MeV/c)
$N\overline{K}$	10–30 %	405
$\Lambda\pi$	seen	439
$\Sigma\pi$	seen	385

$\Sigma(1670)$ D_{13}

$I(J^P) = 1(\frac{3}{2}^-)$

Mass m = 1665 to 1685 (≈ 1670) MeV
Full width Γ = 40 to 80 (≈ 60) MeV
$p_{beam} = 0.74$ GeV/c $4\pi\lambda\!\!^{-2} = 28.5$ mb

$\Sigma(1670)$ DECAY MODES	Fraction (Γ_i/Γ)	p (MeV/c)
$N\overline{K}$	7–13 %	414
$\Lambda\pi$	5–15 %	447
$\Sigma\pi$	30–60 %	393

$\Sigma(1750)$ S_{11}

$I(J^P) = 1(\frac{1}{2}^-)$

Mass m = 1730 to 1800 (≈ 1750) MeV
Full width Γ = 60 to 160 (≈ 90) MeV
$p_{beam} = 0.91$ GeV/c $4\pi\lambda\!\!^{-2} = 20.7$ mb

$\Sigma(1750)$ DECAY MODES	Fraction (Γ_i/Γ)	p (MeV/c)
$N\overline{K}$	10–40 %	486
$\Lambda\pi$	seen	507
$\Sigma\pi$	<8 %	455
$\Sigma\eta$	15–55 %	81

$\Sigma(1775)$ D_{15}

$I(J^P) = 1(\frac{5}{2}^-)$

Mass m = 1770 to 1780 (≈ 1775) MeV
Full width Γ = 105 to 135 (≈ 120) MeV
$p_{beam} = 0.96$ GeV/c $4\pi\lambda\!\!^{-2} = 19.0$ mb

$\Sigma(1775)$ DECAY MODES	Fraction (Γ_i/Γ)	p (MeV/c)
$N\overline{K}$	37–43%	508
$\Lambda\pi$	14–20%	525
$\Sigma\pi$	2–5%	474
$\Sigma(1385)\pi$	8–12%	324
$\Lambda(1520)\pi$	17–23%	198

$\Sigma(1915)$ F_{15}

$I(J^P) = 1(\frac{5}{2}^+)$

Mass m = 1900 to 1935 (≈ 1915) MeV
Full width Γ = 80 to 160 (≈ 120) MeV
$p_{beam} = 1.26$ GeV/c $4\pi\lambda\!\!^{-2} = 12.8$ mb

$\Sigma(1915)$ DECAY MODES	Fraction (Γ_i/Γ)	p (MeV/c)
$N\overline{K}$	5–15 %	618
$\Lambda\pi$	seen	622
$\Sigma\pi$	seen	577
$\Sigma(1385)\pi$	<5 %	440

$\Sigma(1940)$ D_{13}

$I(J^P) = 1(\frac{3}{2}^-)$

Mass m = 1900 to 1950 (≈ 1940) MeV
Full width Γ = 150 to 300 (≈ 220) MeV
$p_{beam} = 1.32$ GeV/c $4\pi\lambda\!\!^{-2} = 12.1$ mb

$\Sigma(1940)$ DECAY MODES	Fraction (Γ_i/Γ)	p (MeV/c)
$N\overline{K}$	<20 %	637
$\Lambda\pi$	seen	639
$\Sigma\pi$	seen	594
$\Sigma(1385)\pi$	seen	460
$\Lambda(1520)\pi$	seen	354
$\Delta(1232)\overline{K}$	seen	410
$N\overline{K}^*(892)$	seen	320

$\Sigma(2030)$ F_{17}

$I(J^P) = 1(\frac{7}{2}^+)$

Mass m = 2025 to 2040 (≈ 2030) MeV
Full width Γ = 150 to 200 (≈ 180) MeV
$p_{beam} = 1.52$ GeV/c $4\pi\lambda\!\!^{-2} = 9.93$ mb

$\Sigma(2030)$ DECAY MODES	Fraction (Γ_i/Γ)	p (MeV/c)
$N\overline{K}$	17–23 %	702
$\Lambda\pi$	17–23 %	700
$\Sigma\pi$	5–10 %	657
ΞK	<2 %	412
$\Sigma(1385)\pi$	5–15 %	529
$\Lambda(1520)\pi$	10–20 %	430
$\Delta(1232)\overline{K}$	10–20 %	498
$N\overline{K}^*(892)$	<5 %	438

$\Sigma(2250)$

$I(J^P) = 1(?^?)$

Mass m = 2210 to 2280 (≈ 2250) MeV
Full width Γ = 60 to 150 (≈ 100) MeV
$p_{beam} = 2.04$ GeV/c $4\pi\lambda\!\!^{-2} = 6.76$ mb

$\Sigma(2250)$ DECAY MODES	Fraction (Γ_i/Γ)	p (MeV/c)
$N\overline{K}$	<10 %	851
$\Lambda\pi$	seen	842
$\Sigma\pi$	seen	803

Ξ BARYONS
($S=-2$, $I=1/2$)

$\Xi^0 = uss$, $\Xi^- = dss$

Ξ^0

$I(J^P) = \frac{1}{2}(\frac{1}{2}^+)$

P is not yet measured; + is the quark model prediction.

Mass $m = 1314.9 \pm 0.6$ MeV
$m_{\Xi^-} - m_{\Xi^0} = 6.4 \pm 0.6$ MeV
Mean life $\tau = (2.90 \pm 0.09) \times 10^{-10}$ s
$c\tau = 8.71$ cm
Magnetic moment $\mu = -1.250 \pm 0.014\ \mu_N$

Decay parameters

$\Lambda\pi^0$	$\alpha = -0.411 \pm 0.022$ (S = 2.1)
"	$\phi = (21 \pm 12)°$
"	$\gamma = 0.85$ [g]
"	$\Delta = (218^{+12}_{-19})°$ [g]
$\Lambda\gamma$	$\alpha = 0.4 \pm 0.4$
$\Sigma^0\gamma$	$\alpha = 0.20 \pm 0.32$

Ξ^0 DECAY MODES		Fraction (Γ_i/Γ)	Confidence level	p (MeV/c)
$\Lambda\pi^0$		(99.54 ± 0.05) %		135
$\Lambda\gamma$		$(1.06 \pm 0.16) \times 10^{-3}$		184
$\Sigma^0\gamma$		$(3.5 \pm 0.4) \times 10^{-3}$		117
$\Sigma^+ e^- \overline{\nu}_e$		$< 1.1 \times 10^{-3}$	90%	120
$\Sigma^+ \mu^- \overline{\nu}_\mu$		$< 1.1 \times 10^{-3}$	90%	64
$\Delta S = \Delta Q$ (*SQ*) violating modes or $\Delta S = 2$ forbidden (*S2*) modes				
$\Sigma^- e^+ \nu_e$	*SQ*	$< 9 \times 10^{-4}$	90%	112
$\Sigma^- \mu^+ \nu_\mu$	*SQ*	$< 9 \times 10^{-4}$	90%	49
$p\pi^-$	*S2*	$< 4 \times 10^{-5}$	90%	299
$pe^- \overline{\nu}_e$	*S2*	$< 1.3 \times 10^{-3}$		323
$p\mu^- \overline{\nu}_\mu$	*S2*	$< 1.3 \times 10^{-3}$		309

Ξ^-

$I(J^P) = \frac{1}{2}(\frac{1}{2}^+)$

P is not yet measured; + is the quark model prediction.

Mass $m = 1321.32 \pm 0.13$ MeV
Mean life $\tau = (1.639 \pm 0.015) \times 10^{-10}$ s
$c\tau = 4.91$ cm
Magnetic moment $\mu = -0.6507 \pm 0.0025\ \mu_N$

Decay parameters

$\Lambda\pi^-$	$\alpha = -0.456 \pm 0.014$ (S = 1.8)
"	$\phi = (4 \pm 4)°$
"	$\gamma = 0.89$ [g]
"	$\Delta = (188 \pm 8)°$ [g]
$\Lambda e^- \overline{\nu}_e$	$g_A/g_V = -0.25 \pm 0.05$ [e]

Ξ^- DECAY MODES		Fraction (Γ_i/Γ)	Confidence level	p (MeV/c)
$\Lambda\pi^-$		(99.887 ± 0.035) %		139
$\Sigma^-\gamma$		$(1.27 \pm 0.23) \times 10^{-4}$		118
$\Lambda e^- \overline{\nu}_e$		$(5.63 \pm 0.31) \times 10^{-4}$		190
$\Lambda \mu^- \overline{\nu}_\mu$		$(3.5\ ^{+3.5}_{-2.2}) \times 10^{-4}$		163
$\Sigma^0 e^- \overline{\nu}_e$		$(8.7 \pm 1.7) \times 10^{-5}$		122
$\Sigma^0 \mu^- \overline{\nu}_\mu$		$< 8 \times 10^{-4}$	90%	70
$\Xi^0 e^- \overline{\nu}_e$		$< 2.3 \times 10^{-3}$	90%	6
$\Delta S = 2$ forbidden (*S2*) modes				
$n\pi^-$	*S2*	$< 1.9 \times 10^{-5}$	90%	303
$ne^- \overline{\nu}_e$	*S2*	$< 3.2 \times 10^{-3}$	90%	327
$n\mu^- \overline{\nu}_\mu$	*S2*	< 1.5 %	90%	314
$p\pi^-\pi^-$	*S2*	$< 4 \times 10^{-4}$	90%	223
$p\pi^- e^- \overline{\nu}_e$	*S2*	$< 4 \times 10^{-4}$	90%	304
$p\pi^- \mu^- \overline{\nu}_\mu$	*S2*	$< 4 \times 10^{-4}$	90%	250
$p\mu^-\mu^-$	*L*	$< 4 \times 10^{-4}$	90%	–

$\Xi(1530)$ P_{13}

$I(J^P) = \frac{1}{2}(\frac{3}{2}^+)$

$\Xi(1530)^0$ mass $m = 1531.80 \pm 0.32$ MeV (S = 1.3)
$\Xi(1530)^-$ mass $m = 1535.0 \pm 0.6$ MeV
$\Xi(1530)^0$ full width $\Gamma = 9.1 \pm 0.5$ MeV
$\Xi(1530)^-$ full width $\Gamma = 9.9^{+1.7}_{-1.9}$ MeV

$\Xi(1530)$ DECAY MODES	Fraction (Γ_i/Γ)	Confidence level	p (MeV/c)
$\Xi\pi$	100 %		152
$\Xi\gamma$	<4 %	90%	200

$\Xi(1690)$

$I(J^P) = \frac{1}{2}(?^?)$

Mass $m = 1690 \pm 10$ MeV [l]
Full width $\Gamma < 50$ MeV

$\Xi(1690)$ DECAY MODES	Fraction (Γ_i/Γ)	p (MeV/c)
$\Lambda\overline{K}$	seen	240
$\Sigma\overline{K}$	seen	51
$\Xi^-\pi^+\pi^-$	possibly seen	214

$\Xi(1820)$ D_{13}

$I(J^P) = \frac{1}{2}(\frac{3}{2}^-)$

Mass $m = 1823 \pm 5$ MeV [l]
Full width $\Gamma = 24^{+15}_{-10}$ MeV [l]

$\Xi(1820)$ DECAY MODES	Fraction (Γ_i/Γ)	p (MeV/c)
$\Lambda\overline{K}$	large	400
$\Sigma\overline{K}$	small	320
$\Xi\pi$	small	413
$\Xi(1530)\pi$	small	234

$\Xi(1950)$

$I(J^P) = \frac{1}{2}(?^?)$

Mass $m = 1950 \pm 15$ MeV [l]
Full width $\Gamma = 60 \pm 20$ MeV [l]

$\Xi(1950)$ DECAY MODES	Fraction (Γ_i/Γ)	p (MeV/c)
$\Lambda\overline{K}$	seen	522
$\Sigma\overline{K}$	possibly seen	460
$\Xi\pi$	seen	518

$\Xi(2030)$

$I(J^P) = \frac{1}{2}(\geq \frac{5}{2}^?)$

Mass $m = 2025 \pm 5$ MeV [l]
Full width $\Gamma = 20^{+15}_{-5}$ MeV [l]

$\Xi(2030)$ DECAY MODES	Fraction (Γ_i/Γ)	p (MeV/c)
$\Lambda\overline{K}$	~ 20 %	589
$\Sigma\overline{K}$	~ 80 %	533
$\Xi\pi$	small	573
$\Xi(1530)\pi$	small	421
$\Lambda\overline{K}\pi$	small	501
$\Sigma\overline{K}\pi$	small	430

Ω BARYONS
$(S=-3, I=0)$

$\Omega^- = sss$

Ω^-

$I(J^P) = 0(\frac{3}{2}^+)$

J^P is not yet measured; $\frac{3}{2}^+$ is the quark model prediction.

Mass $m = 1672.45 \pm 0.29$ MeV
Mean life $\tau = (0.822 \pm 0.012) \times 10^{-10}$ s
$c\tau = 2.46$ cm
Magnetic moment $\mu = -1.94 \pm 0.22\ \mu_N$

Decay parameters

ΛK^-	$\alpha = -0.026 \pm 0.026$
$\Xi^0\pi^-$	$\alpha = 0.09 \pm 0.14$
$\Xi^-\pi^0$	$\alpha = 0.05 \pm 0.21$

Ω^- DECAY MODES		Fraction (Γ_i/Γ)	Confidence level	p (MeV/c)
ΛK^-		(67.8 ± 0.7) %		211
$\Xi^0\pi^-$		(23.6 ± 0.7) %		294
$\Xi^-\pi^0$		(8.6 ± 0.4) %		290
$\Xi^-\pi^+\pi^-$		$(4.3^{+3.4}_{-1.3}) \times 10^{-4}$		190
$\Xi(1530)^0\pi^-$		$(6.4^{+5.1}_{-2.0}) \times 10^{-4}$		17
$\Xi^0 e^- \overline{\nu}_e$		$(5.6 \pm 2.8) \times 10^{-3}$		319
$\Xi^-\gamma$		$< 2.2 \times 10^{-3}$	90%	314
$\Delta S = 2$ forbidden ($S2$) modes				
$\Lambda\pi^-$	$S2$	$< 1.9 \times 10^{-4}$	90%	449

$\Omega(2250)^-$

$I(J^P) = 0(?^?)$

Mass $m = 2252 \pm 9$ MeV
Full width $\Gamma = 55 \pm 18$ MeV

$\Omega(2250)^-$ DECAY MODES	Fraction (Γ_i/Γ)	p (MeV/c)
$\Xi^-\pi^+K^-$	seen	531
$\Xi(1530)^0 K^-$	seen	437

CHARMED BARYONS
$(C=+1)$

$\Lambda_c^+ = udc,\quad \Sigma_c^{++} = uuc,\quad \Sigma_c^+ = udc,\quad \Sigma_c^0 = ddc,$
$\Xi_c^+ = usc,\quad \Xi_c^0 = dsc,\quad \Omega_c^0 = ssc$

Λ_c^+

$I(J^P) = 0(\frac{1}{2}^+)$

J not confirmed; $\frac{1}{2}$ is the quark model prediction.

Mass $m = 2285.1 \pm 0.6$ MeV
Mean life $\tau = (0.200^{+0.011}_{-0.010}) \times 10^{-12}$ s
$c\tau = 60.0\ \mu$m

Decay asymmetry parameters

$\Lambda\pi^+$	$\alpha = -1.03 \pm 0.29$
$\Lambda e^+\nu_e$	$\alpha = -0.89^{+0.19}_{-0.12}$

Λ_c^+ DECAY MODES		Fraction (Γ_i/Γ)	Scale factor/ Confidence level	p (MeV/c)
Hadronic modes with a p and one K				
$p\overline{K}^0$		(2.1 ± 0.4) %		872
$pK^-\pi^+$		(4.4 ± 0.6) %		822
$p\overline{K}^*(892)^0$	[k]	(1.6 ± 0.4) %		681
$\Delta(1232)^{++}K^-$		$(7 \pm 4) \times 10^{-3}$		709
$\Lambda(1520)\pi^+$	[k]	$(3.9^{+2.0}_{-1.7}) \times 10^{-3}$		626
$pK^-\pi^+$ nonresonant		$(2.4^{+0.5}_{-0.6})$ %		822
$p\overline{K}^0\pi^+\pi^-$		(2.4 ± 0.8) %	S=1.3	753
$pK^-\pi^+\pi^0$		seen		758
$pK^*(892)^-\pi^+$		seen		579
$p(K^-\pi^+)_{\text{nonresonant}}\pi^0$		(3.2 ± 0.7) %		758
$\Delta(1232)\overline{K}^*(892)$		seen		417
$pK^-\pi^+\pi^+\pi^-$		$(10 \pm 7) \times 10^{-4}$		670
$pK^-\pi^+\pi^0\pi^0$		$(7.0 \pm 3.5) \times 10^{-3}$		676
$pK^-\pi^+\pi^0\pi^0\pi^0$		$(4.4 \pm 2.8) \times 10^{-3}$		573
Hadronic modes with a p and zero or two K's				
$p\pi^+\pi^-$		$(3.0 \pm 1.6) \times 10^{-3}$		926
$pf_0(980)$	[k]	$(2.4 \pm 1.6) \times 10^{-3}$		621
$p\pi^+\pi^+\pi^-\pi^-$		$(1.6 \pm 1.0) \times 10^{-3}$		851
pK^+K^-		$(3.0 \pm 1.1) \times 10^{-3}$		615
$p\phi$	[k]	$< 1.7 \times 10^{-3}$	CL=90%	589
Hadronic modes with a hyperon				
$\Lambda\pi^+$		$(7.9 \pm 1.8) \times 10^{-3}$		863
$\Lambda\pi^+\pi^0$		(3.2 ± 0.9) %		843
$\Lambda\rho^0$		< 4 %	CL=95%	639
$\Lambda\pi^+\pi^+\pi^-$		(2.7 ± 0.6) %		806
$\Sigma^0\pi^+$		$(8.7 \pm 2.0) \times 10^{-3}$		824
$\Sigma^0\pi^+\pi^0$		(1.6 ± 0.6) %		802
$\Sigma^0\pi^+\pi^+\pi^-$		$(9.2 \pm 3.3) \times 10^{-3}$		762
$\Sigma^+\pi^0$		$(8.7 \pm 2.2) \times 10^{-3}$		826
$\Sigma^+\pi^+\pi^-$		(3.0 ± 0.6) %		803
$\Sigma^+\rho^0$		< 1.2 %	CL=95%	579
$\Sigma^-\pi^+\pi^+$		(1.6 ± 0.6) %		798
$\Sigma^+\pi^+\pi^-\pi^0$				569
$\Sigma^+\omega$	[k]	(2.4 ± 0.7) %		569
$\Sigma^+\pi^+\pi^+\pi^-\pi^-$		$(2.6^{+3.5}_{-1.8}) \times 10^{-3}$		707
$\Sigma^+K^+K^-$		$(3.1 \pm 0.8) \times 10^{-3}$		346
$\Sigma^+\phi$	[k]	$(3.0 \pm 1.3) \times 10^{-3}$		292
$\Sigma^+K^+\pi^-$		$(5.7^{+5.3}_{-3.1}) \times 10^{-3}$		668
Ξ^0K^+		$(3.4 \pm 0.9) \times 10^{-3}$		652
$\Xi^-K^+\pi^+$		$(3.8 \pm 1.2) \times 10^{-3}$		564
$\Xi(1530)^0K^+$	[k]	$(2.3 \pm 0.9) \times 10^{-3}$		471
Inclusive modes				
p anything		(50 ± 16) %		–
p anything (no Λ)		(12 ± 19) %		–
n anything		(50 ± 16) %		–
n anything (no Λ)		(29 ± 17) %		–
Λ anything		(35 ± 11) %	S=1.4	–
Σ^+ anything	[l]	(10 ± 5) %		–
e^+ anything		(4.5 ± 1.7) %		–
pe^+ anything		(1.8 ± 0.9) %		–
Λe^+ anything		(1.4 ± 0.5) %		–
$\Lambda\mu^+$ anything		(1.5 ± 0.9) %		–

$\Lambda_c(2625)^+$

$I = 0$

Mass $m = 2625.6 \pm 0.8$ MeV
$m - m_{\Lambda_c^+} = 340.6 \pm 0.6$ MeV
Full width $\Gamma < 3.2$ MeV, CL = 90%

$\Lambda_c(2625)^+$ DECAY MODES	Fraction (Γ_i/Γ)	p (MeV/c)
$\Lambda_c^+\pi^+\pi^-$	seen	182
$\Sigma_c(2455)^{++}\pi^- + \Sigma_c(2455)^0\pi^+$	seen	99
$\Lambda_c^+\pi^+\pi^-$ nonresonant	seen	182

$\Sigma_c(2455)$

$I(J^P) = 1(\frac{1}{2}^+)$

J^P not confirmed; $\frac{1}{2}^+$ is the quark model prediction.

$\Sigma_c(2455)^{++}$ mass $m = 2453.1 \pm 0.6$ MeV
$\Sigma_c(2455)^+$ mass $m = 2453.8 \pm 0.9$ MeV
$\Sigma_c(2455)^0$ mass $m = 2452.4 \pm 0.7$ MeV (S = 1.1)

$\Sigma_c(2455)$ DECAY MODES	Fraction (Γ_i/Γ)	p (MeV/c)
$\Lambda_c^+\pi$	100 %	91

Ξ_c^+

$I(J^P) = \frac{1}{2}(\frac{1}{2}^+)$

$I(J^P)$ not confirmed; $\frac{1}{2}(\frac{1}{2}^+)$ is the quark model prediction.

Mass $m = 2465.1 \pm 1.6$ MeV
Mean life $\tau = (0.35^{+0.07}_{-0.04}) \times 10^{-12}$ s
$c\tau = 106\ \mu$m

Ξ_c^+ DECAY MODES	Fraction (Γ_i/Γ)	p (MeV/c)
$\Lambda K^- \pi^+ \pi^+$	seen	785
$\Sigma^+ K^- \pi^+$	seen	808
$\Sigma^0 K^- \pi^+ \pi^+$	seen	733
$\Xi^- \pi^+ \pi^+$	seen	850

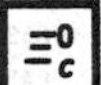

$I(J^P) = \frac{1}{2}(\frac{1}{2}^+)$

$I(J^P)$ not confirmed; $\frac{1}{2}(\frac{1}{2}^+)$ is the quark model prediction.

Mass $m = 2470.3 \pm 1.8$ MeV (S = 1.3)
$m_{\Xi_c^0} - m_{\Xi_c^+} = 5.2 \pm 2.2$ MeV (S = 1.1)
Mean life $\tau = (0.098^{+0.023}_{-0.015}) \times 10^{-12}$ s
$c\tau = 29\ \mu$m

A few branching *ratios* but no absolute branching *fractions* have been measured.

Ξ_c^0 DECAY MODES	Fraction (Γ_i/Γ)	p (MeV/c)
$\Xi^- \ell^+$ anything	[m] seen	–
$\Xi^- \pi^+$	seen	875
$\Xi^- \pi^+ \pi^+ \pi^-$	seen	816
$p K^- \overline{K}^*(892)^0$	seen	406
$\Omega^- K^+$	seen	522

BOTTOM (BEAUTY) BARYON
$(B = -1)$

$\Lambda_b^0 = u\,d\,b$

$I(J^P) = 0(\frac{1}{2}^+)$

$I(J^P)$ not yet measured; $0(\frac{1}{2}^+)$ is the quark model prediction.

Mass $m = 5641 \pm 50$ MeV
Mean life $\tau = (1.07^{+0.19}_{-0.16}) \times 10^{-12}$ s

Λ_b^0 DECAY MODES	Fraction (Γ_i/Γ)	p (MeV/c)
$J/\psi(1S)\Lambda$	seen	1756
$p D^0 \pi^-$	seen	2383
$\Lambda_c^+ \pi^+ \pi^- \pi^-$	seen	2336
$\Lambda \ell^- X$	seen	–
$\Lambda_c^+ \ell^- X$	seen	–

NOTES

This Summary Table only includes established baryons. The Full Listings include evidence for other baryons. The masses, widths, and branching fractions for the resonances in this Table are Breit-Wigner parameters. The Full Listings also give, where available, pole parameters. See, in particular, the *Note on N and Δ Resonances.*

For most of the resonances, the parameters come from various partial-wave analyses of more or less the same sets of data, and it is not appropriate to treat the results of the analyses as independent or to average them together. Furthermore, the systematic errors on the results are not well understood. Thus, we usually only give ranges for the parameters. We then also give a best guess for the mass (as part of the name of the resonance) and for the width. The *Note on N and Δ Resonances* and the *Note on Λ and Σ Resonances* in the Full Listings review the partial-wave analyses.

When a quantity has "(S = ...)" to its right, the error on the quantity has been enlarged by the "scale factor" S, defined as $S = \sqrt{\chi^2/(N-1)}$, where N is the number of measurements used in calculating the quantity. We do this when $S > 1$, which often indicates that the measurements are inconsistent. When $S > 1.25$, we also show in the Full Listings an ideogram of the measurements. For more about S, see the Introduction.

A decay momentum p is given for each decay mode. For a 2-body decay, p is the momentum of each decay product in the rest frame of the decaying particle. For a 3-or-more-body decay, p is the largest momentum any of the products can have in this frame. For any resonance, the *nominal* mass is used in calculating p. A dagger ("†") in this column indicates that the mode is forbidden when the nominal masses of resonances are used, but is in fact allowed due to the nonzero widths of the resonances.

[a] The masses of the p and n are most precisely known in u (unified atomic mass units). The conversion factor to MeV, 1 u = 931.49432 ± 0.00028 MeV, is less well known than are the masses in u.

[b] The limit is from neutrality-of-matter experiments; it assumes $q_n = q_p + q_e$. See also the charge of the neutron.

[c] The first limit is geochemical and independent of decay mode. The second limit assumes the dominant decay modes are among those investigated. For antiprotons the best limit, inferred from the observation of cosmic ray $\overline{p}$'s is $\tau_{\overline{p}} > 10^7$ yr, the cosmic-ray storage time, but this limit depends on a number of assumptions. The best direct observation of stored antiprotons gives $\tau_{\overline{p}}/B(\overline{p} \to e^-\gamma) > 1848$ yr.

[d] There is some controversy about whether nuclear physics and model dependence complicate the analysis for bound neutrons (from which the best limit comes). For reactor experiments with free neutrons, the best limit is $> 10^7$ s.

[e] The parameters g_A, g_V, and g_{WM} for semileptonic modes are defined by $\overline{B}_f[\gamma_\lambda(g_V + g_A\gamma_5) + i(g_{WM}/m_{B_i})\,\sigma_{\lambda\nu}\,q^\nu]B_i$, and ϕ_{AV} is defined by $g_A/g_V = |g_A/g_V|e^{i\phi_{AV}}$. See the "Note on Baryon Decay Parameters" in the neutron Full Listings.

[f] Time-reversal invariance requires this to be 0° or 180°.

[g] The decay parameters γ and Δ are calculated from α and ϕ using

$$\gamma = \sqrt{1-\alpha^2}\,\cos\phi\,, \qquad \tan\Delta = -\frac{1}{\alpha}\sqrt{1-\alpha^2}\,\sin\phi\,.$$

See the "Note on Baryon Decay Parameters" in the neutron Full Listings.

[h] See the Full Listings for the pion momentum range used in this measurement.

[i] The error given here is only an educated guess. It is larger than the error on the weighted average of the published values.

[j] A theoretical value using QED.

[k] The branching fraction includes all the decay modes of the final-state resonance.

[l] The value is for the sum of the charge states indicated.

[m] ℓ indicates e or μ mode, not sum over modes.

SEARCHES FOR FREE QUARKS, MONOPOLES, SUPERSYMMETRY, COMPOSITENESS, etc.

Free Quark Searches

All searches since 1977 have had negative results.

Magnetic Monopole Searches

Isolated candidate events have not been confirmed. Most experiments obtain negative results.

Supersymmetric Particle Searches

Limits are based on the Minimal Supersymmetric Standard Model.

Assumptions include: 1) $\widetilde{\chi}_1^0$ (or $\widetilde{\gamma}$) is lightest supersymmetric particle; 2) R-parity is conserved; 3) $m_{\widetilde{f}_L} = m_{\widetilde{f}_R}$, and all scalar quarks (except $\widetilde{t}_L$ and $\widetilde{t}_R$) are degenerate in mass.

See the Full Listings for a Note giving details of supersymmetry.

$\widetilde{\chi}_i^0$ — neutralinos (mixtures of $\widetilde{\gamma}$, $\widetilde{Z}^0$, and $\widetilde{H}_i^0$)

- Mass $m_{\widetilde{\gamma}} > 15$ GeV, CL = 90% [if $m_{\widetilde{f}} = 100$ GeV (from cosmology)]
- Mass $m_{\widetilde{\chi}_1^0} > 18$ GeV, CL = 90% [GUT relations assumed]
- Mass $m_{\widetilde{\chi}_2^0} > 45$ GeV, CL = 95% [GUT relations assumed]
- Mass $m_{\widetilde{\chi}_3^0} > 70$ GeV, CL = 95% [GUT relations assumed]
- Mass $m_{\widetilde{\chi}_4^0} > 108$ GeV, CL = 95% [GUT relations assumed]

$\widetilde{\chi}_i^\pm$ — charginos (mixtures of $\widetilde{W}^\pm$ and $\widetilde{H}_i^\pm$)

- Mass $m_{\widetilde{\chi}_1^\pm} > 45$ GeV, CL = 95% [all $m_{\widetilde{\chi}_1^0}$]
- Mass $m_{\widetilde{\chi}_2^\pm} > 99$ GeV, CL = 95% [GUT relations assumed]

$\widetilde{\nu}$ — scalar neutrino (sneutrino)

- Mass $m > 37.1$ GeV, CL = 95% [one flavor]
- Mass $m > 41.8$ GeV, CL = 95% [three degenerate flavors]

$\widetilde{e}$ — scalar electron (selectron)

- Mass $m > 65$ GeV, CL = 95% [if $m_{\widetilde{\gamma}} = 0$]
- Mass $m > 50$ GeV, CL = 95% [if $m_{\widetilde{\gamma}} < 5$ GeV]
- Mass $m > 45$ GeV, CL = 95% [if $m_{\widetilde{\chi}_1^0} < 41$ GeV]

$\widetilde{\mu}$ — scalar muon (smuon)

- Mass $m > 45$ GeV, CL = 95% [if $m_{\widetilde{\chi}_1^0} < 41$ GeV]

$\widetilde{\tau}$ — scalar tau (stau)

- Mass $m > 45$ GeV, CL = 95% [if $m_{\widetilde{\chi}_1^0} < 38$ GeV]

$\widetilde{q}$ — scalar quark (squark)

These limits include the effects of cascade decay, for a particular choice of parameters, $\mu = -250$ GeV, $\tan\beta = 2$. The limits are weakly sensitive to these parameters over much of parameter space. Limits assume GUT relations between gaugino masses and the gauge coupling; in particular that for $|\mu|$ not small, $m_{\widetilde{\chi}_1^0} \approx m_{\widetilde{g}}/6$.

- Mass $m > 90$ GeV, CL = 90% [any $m_{\widetilde{g}} < 410$ GeV]
- Mass $m > 218$ GeV, CL = 90% [if $m_{\widetilde{g}} = m_{\widetilde{q}}$]

$\widetilde{g}$ — gluino

There is some controversy about a low-mass window ($1 \lesssim m_{\widetilde{g}} \lesssim 4$ GeV). Several experiments cast doubt on the existence of this window.

These limits include the effects of cascade decay, for a particular choice of parameters, $\mu = -250$ GeV, $\tan\beta = 2$. The limits are weakly sensitive to these parameters over much of parameter space. Limits assume GUT relations between gaugino masses and the gauge coupling; in particular that for $|\mu|$ not small, $m_{\widetilde{\chi}_1^0} \approx m_{\widetilde{g}}/7$.

- Mass $m > 100$ GeV, CL = 90% [any $m_{\widetilde{q}}$]
- Mass $m > 218$ GeV, CL = 90% [if $m_{\widetilde{q}} \leq m_{\widetilde{g}}$]

Searches for Quark and Lepton Compositeness

Scale Limits Λ for Contact Interactions (the lowest dimensional interactions with four fermions)

If the Lagrangian has the form

$$\pm \frac{g^2}{2\Lambda^2} \overline{\psi}_L \gamma_\mu \psi_L \overline{\psi}_L \gamma^\mu \psi_L$$

(with $g^2/4\pi$ set equal to 1), then we define $\Lambda \equiv \Lambda_{LL}^\pm$. For the full definitions and for other forms, see the Note in the Listings on Searches for Quark and Lepton Compositeness in the full *Review* and the original literature.

- $\Lambda_{LL}^+(eeee)$ > 1.6 TeV, CL = 95%
- $\Lambda_{LL}^-(eeee)$ > 3.6 TeV, CL = 95%
- $\Lambda_{LL}^+(ee\mu\mu)$ > 2.6 TeV, CL = 95%
- $\Lambda_{LL}^-(ee\mu\mu)$ > 1.9 TeV, CL = 95%
- $\Lambda_{LL}^+(ee\tau\tau)$ > 1.9 TeV, CL = 95%
- $\Lambda_{LL}^-(ee\tau\tau)$ > 2.9 TeV, CL = 95%
- $\Lambda_{LL}^+(\ell\ell\ell\ell)$ > 3.5 TeV, CL = 95%
- $\Lambda_{LL}^-(\ell\ell\ell\ell)$ > 2.8 TeV, CL = 95%
- $\Lambda_{LL}^+(eeqq)$ > 1.7 TeV, CL = 95%
- $\Lambda_{LL}^-(eeqq)$ > 2.2 TeV, CL = 95%
- $\Lambda_{LL}^+(\mu\mu qq)$ > 1.4 TeV, CL = 95%
- $\Lambda_{LL}^-(\mu\mu qq)$ > 1.6 TeV, CL = 95%
- $\Lambda_{LR}^\pm(\nu_\mu \nu_e \mu e)$ > 3.1 TeV, CL = 90%
- $\Lambda_{LL}^\pm(qqqq)$ > 1.4 TeV, CL = 95%

Excited Leptons

The limits from $\ell^{*+}\ell^{*-}$ do not depend on λ (where λ is the $\ell\ell^*$ transition coupling). The λ-dependent limits assume chiral coupling, except for the third limit for e^* which is for nonchiral coupling. For chiral coupling, this limit corresponds to $\lambda_\gamma = \sqrt{2}$.

$e^{*\pm}$ — excited electron

- Mass $m > 46.1$ GeV, CL = 95% (from $e^{*+}e^{*-}$)
- Mass $m > 91$ GeV, CL = 95% (if $\lambda_Z > 1$)
- Mass $m > 127$ GeV, CL = 95% (if $\lambda_\gamma = 1$)

$\mu^{*\pm}$ — excited muon

- Mass $m > 46.1$ GeV, CL = 95% (from $\mu^{*+}\mu^{*-}$)
- Mass $m > 91$ GeV, CL = 95% (if $\lambda_Z > 1$)

$\tau^{*\pm}$ — excited tau

- Mass $m > 46.0$ GeV, CL = 95% (from $\tau^{*+}\tau^{*-}$)
- Mass $m > 90$ GeV, CL = 95% (if $\lambda_Z > 0.18$)

ν^* — excited neutrino

- Mass $m > 47$ GeV, CL = 95% (from $\nu^*\overline{\nu}^*$)
- Mass $m > 91$ GeV, CL = 95% (if $\lambda_Z > 1$)

q^* — excited quark

- Mass $m > 45.6$ GeV, CL = 95% (from $q^*\overline{q}^*$)
- Mass $m > 88$ GeV, CL = 95% (if $\lambda_Z > 1$)
- Mass $m > 510$ GeV, CL = 95% ($p\overline{p} \to q^* X$)

Color Sextet and Octet Particles

Color Sextet Quarks (q_6)

- Mass $m > 84$ GeV, CL = 95% (Stable q_6)

Color Octet Charged Leptons (ℓ_8)

- Mass $m > 86$ GeV, CL = 95% (Stable ℓ_8)

Color Octet Neutrinos (ν_8)

- Mass $m > 110$ GeV, CL = 90% ($\nu_8 \to \nu g$)

TABLE OF THE ISOTOPES
(Revised 1995)
Norman E. Holden

This table presents an evaluated set of values for the experimental quantities which characterize the decay of radioactive nuclides. A list of the major references used in this evaluation is given below. When uncertainties are not listed, they are assumed to be five or less in the last digit quoted. If they exceed five in the last digit, the value is prefaced by an approximate sign. The effective literature cutoff date for data in this edition of the Table is January, 1995. The assistance of Felicia Kramer in assembling this file is gratefully acknowledged.

TABLE LAYOUT

Column No.	Column Title	Description
1	Isotope or Element	For elements, the atomic number and chemical symbol are listed. For nuclides, the mass number and chemical symbol are listed. Isomers are indicated by the addition of m, m1, or m2.
2	Isotopic Abundance	in atom percent.
3	Atomic Mass or Atomic Weight	Atomic mass relative to ^{12}C = 12. Atomic weight is given on the same scale.
4	Half-life	Half-life in decimal notation. µs = microseconds; ms = milliseconds; s = seconds; m = minutes; h = hours; d = days; and y = years.
5	Decay Mode/Energy	Decay modes are α = alpha particle emission; β^- = negative beta emission; β^+ = positron emission; EC = orbital electron capture; IT = isomeric transition from upper to lower isomeric state; n = neutron emission; SF = spontaneous fission. Total disintegration energy in MeV units.
6	Particle Energy/Intensity	End point energies of beta transitions and discrete energies of alpha particles in MeV and their intensities in percent.
7	Spin and Parity	Nuclear spin or angular momentum of the nuclides in units of $h/2\pi$; parity is positive or negative.
8	Magnetic Dipole Moment	Magnetic dipole moments in nuclear magneton units.
9	Electric Quadrupole Moment	Electric quadrupole moments in barn units (10^{-24} cm^2).
10	Gamma Ray Energy/Intensity	Gamma ray energies in MeV and intensities in percent. Ann. rad. refers to the 511.006 keV photons emitted in the annihilation of positrons in matter.

GENERAL NUCLEAR DATA REFERENCES

The following references represent the major sources of the nuclear data presented, along with subsequent published journals and reports:

1. G. Audi, A.H. Wapstra, *The 1993 Atomic Mass Evaluation,* Nuclear Physics *A565*, 1 (1993).
2. International Commission on Atomic Weights, *Atomic Weights of the Elements - 1993,* Pure & Applied Chemistry *66*, 2423 (1994).
3. J.R. Parrington, S. Breneman, F. Feiner, H. Knox, *Chart of the Nuclides, 15th Edition,* Knolls Atomic Power Lab. (To be published).
4. N.E. Holden, *Total and Spontaneous Fission Half-lives for Uranium, Plutonium, Americium and Curium Nuclides,* Pure & Applied Chemistry *61*, 1483 (1989).
5. N.E. Holden, *Half-lives of Selected Nuclides,* Pure & Applied Chemistry *62*, 941 (1990).
6. N.E. Holden, *Review of Thermal Neutron Cross Sections and Isotopic Composition of the Elements,* BNL-NCS-42224 (March 1989).
7. P. Ragahavan, *Table of Nuclear Moments,* Atomic Data Nuclear Data Tables *42*, 189 (1989).
8. E. Brown, R. Firestone, *Radioactivity Handbook,* Wiley Interscience Press (1986).
9. J.K. Tuli, *Nuclear Wallet Cards,* Brookhaven National Laboratory (July 1990).

* This research was carried out under the auspices of the U.S. Department of Energy Contract No. DE-AC02–76CH00016.

TABLE OF THE ISOTOPES

Elem. or Isot.	Natural Abundance (%)	Atomic Mass or Weight	Half-Life	Decay Mode/Energy (/MeV)	Particle Energy /Intensity (MeV / %)	Spin ($h/2\pi$)	Nuclear Magnetic Mom.(nm)	Elect. Quadr. Mom. (b)	γ-ray Energy /Intensity (MeV / %)
$_0$n		**1.00866492**	**10.3 m**	**β^-/0.78235**	**0.782/100.**	**1/2+**	**-1.913043**		
$_1$H		**1.00794(7)**							
^{1}H	99.985(1)	1.007825032				1/2+	+2.79285		
^{2}H	0.015(1)	2.014101778				1+	+0.85744	+2.86 mb	
^{3}H		3.01604927	12.32 y	β^-/0.01859	0.01860/100.	1/2+	+2.97896		
$_2$He		**4.002602(2)**							
^{3}He	1.37x10^{-4}	3.01602931				1/2+	-2.12762		
^{4}He	≈100.	4.00260325				0+			
^{5}He		5.01222	7.6x10^{-22} s	n,α		3/2-			
^{6}He		6.018888	0.807 s	β^-/3.508,d	3.510/100.	0+			
^{7}He		7.02803	3.x10^{-21} s	n		(3/2)-			
^{8}He		8.03392	0.119 s	β^-/10.65, t	13/88.	0+			0.9807/84.
				n/	/12.				0.4776/5.
^{9}He		9.0438		n					
^{10}He				2n					
$_3$Li		**6.941(2)**							
^{5}Li		5.01254	≈3.x10^{-22} s			3/2-			
^{6}Li	7.5(2)	6.015122				1+	+0.82205	-0.8 mb	
^{7}Li	92.5(2)	7.016004				3/2-	+3.25644	-0.041	
^{8}Li		8.022486	0.84 s	β^-/16.004	12.5/100.	2+	+1.6536	+0.032	
				α/	α(1.6)				
^{9}Li		9.026789	0.178 s	β^-/13.606	13.5/75.	3/2-	3.439	0.028	
				β^-/	11/25.				
^{10}Li		10.03590	4. x 10^{-22} s	β^- /20.84					
^{11}Li		11.04379	8.7 ms	β^-/20.6		(1/2-)	3.668	0.031	3.367/35.
$_4$Be		**9.012182(3)**							
^{6}Be		6.01973	5.0x10^{-21} s	2p,α		0+			
^{7}Be		7.016929	53.28 d	EC/0.862		3/2-			0.4776/10.4
^{8}Be		8.0053051	≈ 7.x10^{-17} s	2α/0.046		0+			
^{9}Be	100.	9.012182				3/2-	-1.1776	+0.053	
^{10}Be		10.013534	1.52x10^6 y	β^-/0.5559	0.555/100.	0+			
^{11}Be		11.02166	13.8 s	β^-,$\beta^-\alpha$/11.51	11.48/61.	1/2+			2.125/35.5
									(0.478-7.97)
^{12}Be		12.02692	24. ms	β^-/11.71		0+			(0.95 - 4.4)
^{14}Be		14.0428	4. ms	β^- /16.2		0+			
$_5$B		**10.811(5)**							
^{7}B		7.0299	4.x10^{-22} s	p					
^{8}B		8.024607	0.770 s	β^+, 2α/17.979	13.7(β^+)/93.	2+	1.0355	0.068	ann.rad.
^{9}B		9.013329	8x10^{-19} s	p2α/		3/2-			
^{10}B	19.9(2)	10.012937				3+	+1.8006	+0.085	
^{11}B	80.1(2)	11.009306				3/2-	+2.6886	+0.0406	
^{12}B		12.014352	0.0202 s	β^-/13.369		1+	+1.0031	0.0132	4.438/1.3
				β^- α/1.6/					3.215/0.00065
^{13}B		13.017780	0.0174 s	β^- /13.437	13.4	3/2-	+3.17778	0.037	3.68/7.6
				β^- n/0.25/	2.43(n)/0.09				
					3.55(n)/0.16				
^{14}B		14.02540	14. ms	β^- /20.64		2-			6.094/90.
^{15}B		15.03110	10.4 ms	β^- ,(n)/19.09					
^{17}B		17.0469	5.1 ms	β^-/22.7					
^{19}B		19.0037		β^- /20.3					
$_6$C		**12.011(1)**							
^{8}C		8.03768	2.0x10^{-21} s	p					
^{9}C		9.031040	127. ms	β^+ ,p, 2α/16.498		(3/2-)			ann.rad.
^{10}C		10.016853	19.3 s	β^+ /3.648	1.865	0+			ann.rad.
									0.71829/100.
^{11}C		11.011433	20.3 m	β^+ ,EC/1.982	0.9608/99.	3/2-	-0.964	0.0333	ann.rad.
^{12}C	98.89(1)	12.000000				0+			
^{13}C	1.11(1)	13.00335484				1/2-	+0.70241		
^{14}C		14.00324199	5715. y	β^- /0.15648	0.1565/100.	0+			
^{15}C		15.010599	2.45 s	β^- /9.772	4.51/68.	1/2+	1.32		5.298/68.
					9.82/32.				(7.30-9.05)
^{16}C		16.014701	0.75 s	β^- ,n/8.012					
^{17}C		17.02258	0.19 s	β^- ,n/13.17					1.375
									1.849
									1.906

Elem. or Isot.	Natural Abundance (%)	Atomic Mass or Weight	Half-Life	Decay Mode/Energy (/MeV)	Particle Energy /Intensity (MeV / %)	Spin ($h/2\pi$)	Nuclear Magnetic Mom.(nm)	Elect. Quadr. Mom. (b)	γ-ray Energy /Intensity (MeV / %)
^{18}C		18.02676	0.09 s	β^- ,n/11.81					
^{19}C		19.0353	0.05 s	n					
^{20}C		20.0403	0.01 s						
$_7$N		**14.00674(7)**							
^{12}N		12.018613	11.00 ms	β^+ ,β^+ α/17.338	16.38/95.	1+	+0.457	+0.026	ann.rad.
									4.438/2.
^{13}N		13.0057386	9.97 m	β^+ /2.2204	1.190/100.	1/2-	0.3222		
^{14}N	99.634(9)	14.00307401				1+	+0.40376	+0.020	
^{15}N	0.366(9)	15.0001090				1/2-	-0.28319		
^{16}N		16.006100	7.13 s	β^- /10.419	4.27/68.	2-			6.129/68.8
					10.44/26.				7.115/4.7
				β^-, α	1.85/.0012				(0.99-8.87)
^{17}N		17.00845	4.17 s	β^- ,β^- n/8.68	3.7/100.	1/2-			0.871/3.
				0.4-1.7n/95.					2.1842/0.3
				β^- α/	8.0, 8.2				
^{18}N		18.01408	0.62 s	β^- /13.90	9.4/100.	1-			0.822/61.
									1.65/60.5
									1.982/98.
									(0.535-7.13)
^{19}N		19.01703	0.32 s	β^-/12.53					(0.096-3.14)
^{20}N		20.02337	0.14 s	β^- /17.97					
^{21}N		21.0271	0.08 s						
^{22}N		22.0344	0.02s						
^{23}N		23.0405							
$_8$O		**15.9994(3)**							
^{12}O		12.03442	$\approx$ 1.x10^{-21} s	2p					
^{13}O		13.02481	8.9 ms	β^+ ,p/17.77	1.56 (p)/	(3/2-)			ann.rad.
									4.438/0.56
^{14}O		14.0085953	70.60 s	β^+ /5.1430	1.81/99.	0+			ann.rad.
									2.312/99.4
^{15}O		15.003065	122.2 s	β^+ /2.754	1.723/100.	1/2-	0.719		ann.rad.
^{16}O	99.762(15)	15.99491462				0+			
^{17}O	0.038(3)	16.9991315				5/2+	-1.8938	-0.026	
^{18}O	0.200(12)	17.999160				0+			
^{19}O		19.003577	26.9 s	β^- /4.820	3.25/60.	5/2+			0.197/95.9
					4.60/40.				1.3569/50.4
									(0.11-4.18)
^{20}O		20.004076	13.5 s	β^- /3.814		0+			1.057/100.
^{21}O		21.00866	3.4 s	β^- /8.11					(0.28-4.6)
^{22}O		22.00997	2.2 s	β^- /6.5					(0.64-1.86)
^{23}O		23.0157	0.08 s						
^{24}O		24.0204	0.06 s						
$_9$F		**18.9984032(5)**							
^{15}F		15.0180	5.x10^{-22} s	p		(1/2+)			
^{16}F		16.01147	$\approx$ 1.x10^{-20} s	p		0-			
^{17}F		17.0020952	64.5 s	β^+ /2.761	1.75/	5/2+	+4.722	0.058	ann.rad.
^{18}F		18.000938	1.830 h	β^+ ,EC/1.656	0.635/97.	1+			
^{19}F	100.	18.9984032				1/2+	+2.62887	0.072	
^{20}F		19.9999813	11.00 s	β^- /7.0245	5.398/100.	2+	+2.094	0.042	1.634/100.
									3.33/0.009
^{21}F		20.999949	4.16 s	β^- /5.684	3.7/8.	5/2+			0.3507/90.
					5.0/63.				1.395/15.
					5.4/29.				(1.746-4.684)
^{22}F		22.00300	4.23 s	β^- /10.82	3.48/15.	4+			1.2746/100.
					4.67/7.				2.0826/82.
					5.50/62.				(0.82-4.37)
^{23}F		23.00357	2.2 s	β^- /8.5		5/2+			1.701/48.
									2.129/34.
									(0.493-3.83)
^{24}F		24.0081	0.3 s	β^- /13.5					1.9816/
^{25}F		25.0121	0.06 s						
^{26}F		26.0196	0.03 s						
^{27}F		27.0269							
$_{10}$Ne		**20.1797(6)**							
^{16}Ne		16.02575	4.x10^{-21} s	2p					
^{17}Ne		17.01770	109. ms	β^+ .p/14.53	1.4-10.6/	1/2-			ann.rad./
^{18}Ne		18.005710	1.67 s	β^+ /4.446	3.416/92.	0+			ann.rad./
									1.0413/7.8
									(0.658-1.70)

Elem. or Isot.	Natural Abundance (%)	Atomic Mass or Weight	Half-Life	Decay Mode/Energy (/MeV)	Particle Energy /Intensity (MeV / %)	Spin (h/2π)	Nuclear Magnetic Mom.(nm)	Elect. Quadr. Mom. (b)	γ-ray Energy /Intensity (MeV / %)
^{19}Ne		19.001880	17.22 s	β^+ /3.238	2.24/99.	1/2+	-1.885		ann.rad./
									(0.11-1.55)
^{20}Ne	90.48(3)	19.99244018				0+			
^{21}Ne	0.27(1)	20.9938467				3/2+	-0.66180	+0.103	
^{22}Ne	9.25(3)	21.9913855				0+		-0.19	
^{23}Ne		22.9944673	37.2 s	β^- /4.376	3.95/32.	5/2+	-1.08		0.440/33.
					4.39/67.				(1.64-2.98)
^{24}Ne		23.99362	3.38 m	β^- /2.47	1.10/8.	0+			0.4723/100.
					1.98/92.				0.874/7.9
^{25}Ne		24.99779	0.61 s	β^- /7.30	6.3/	1/2+			0.0895/96.
					7.3/				(0.98-3.69)
^{26}Ne		26.00046	197 ms	β^- /7.3					0.233/
^{27}Ne		27.0076	32 ms	β^-, n/12.7					
^{28}Ne		28.0121	17 ms	β^-, n/12.3					
^{29}Ne		29.0194	0.2 s	β^- /15.4					
$_{11}$Na		**22.989770(2)**							
^{19}Na		19.01388	0.03 s	β^+ ,p/11.18					
^{20}Na		20.00735	0.446 s	β^+ /13.89		2+	+0.3694		ann.rad./
				α	2.15/				1.634/79.
^{21}Na		20.997655	22.48 s	β^+ /3.547	2.50/95.	3/2+	+2.3863	+0.05	ann.rad./
									0.351/5.
^{22}Na		21.994437	2.605 y	β^+ /90/2.842	0.545/90.	3+	+1.746		ann.rad./
				EC/10/					1.2745/99.9
^{23}Na	100.	22.9897697				3/2+	+2.21752	+0.101	
^{24m}Na			20.2 ms	I.T.,β^-		1+			0.4723/100.
^{24}Na		23.9909633	14.96 h	β^- /5.5158	1.389/>99.	4+	+1.690		1.3686/100.
									2.754/100.
									(0.997-4.238)
^{25}Na		24.989954	59.3 s	β^- /3.835	2.6/7.	5/2+	+3.683	-0.10	0.3897/12.7
					3.15/25.				0.5850/13.
					4.0/65.				0.9747/14.9
									(0.836-2.80)
^{26}Na		25.99259	1.07 s	β^- /9.31		3+	+2.851	-0.08	1.809/98.9
^{27}Na		26.99401	0.290 s	β^- /9.01	7.95/	5/2+	+3.90	-0.06	0.9847/87.4
				β^- ,n/					1.698/11.9
^{28}Na		27.9989	31. ms	β^- /14.0	12.3/	1+	+2.43	-0.02	1.473/37.
				β^- ,n/					2.389/18.6
^{29}Na		29.0028	44. ms	β^- ,n/13.3	11.5/	3/2+	+2.45	-0.03	2.560/36.
									(1.04-3.99)
^{30}Na		30.0092	50. ms	β^- /17.5		2	+2.08		1.483/46.
^{31}Na		31.0136	17.2 ms	/15.9		(3/2-)	+2.31		1.483/14.
				β^- ,n					(0.05-3.54)
^{32}Na		32.0197	13.5 ms	β^- /19.1					0.886/60.
^{33}Na		33.027	8.1 ms	β^- /20.					0.886/16.
^{34}Na		34.035	5. ms	β^- /24.					0.886/60.
^{35}Na		35.044	1.5 ms	β^- /24					
$_{12}$Mg		**24.3050(6)**							
^{20}Mg		20.01886	0.1 s	β^+ ,p/10.73					
^{21}Mg		21.01171	122. ms	β^+ ,p/13.10		5/2+			0.332/51.
^{22}Mg		21.999574	3.86 s	β^+ /4.786	3.05/	0+			0.0729/60.
									0.5820/100.
									(1.28-1.93)
^{23}Mg		22.994125	11.32 s	β^+ /4.057	3.09/92.	3/2+	0.536		0.440/8.2
^{24}Mg	78.99(3)	23.9850419				0+			
^{25}Mg	10.00(1)	24.9858370				5/2+	-0.85545	+0.199	
^{26}Mg	11.01(2)	25.9825930				0+			
^{27}Mg		26.9843407	9.45 m	β^- /2.6103	1.59/41.	1/2+			0.17068/0.8
					1.75/58.				0.84376/73.
					2.65/0.3				1.01443/29.
^{28}Mg		27.983877	20.9 h	β^- /1.832	0.459/95.	0+			0.0306/95.
									0.4006/36.
									0.9418/36.
									1.342/54.
^{29}Mg		28.98855	1.3 s	β^- /7.55	5.4/	3/2+			0.960/15.
									1.398/16.
									2.224/36.
^{30}Mg		29.99046	0.32 s	β^- /7.0		0+			0.224/85.
^{31}Mg		30.99655	0.24 s	β^- /11.7		(3/2+)			1.61/26.
^{32}Mg		31.9992	0.12 s	β^- /10.3					2.765/25.
^{33}Mg		33.0056	0.09 s	β^- /13.7					1.848/
^{34}Mg		34.0091	0.02 s	β^- /11.3					

Elem. or Isot.	Natural Abundance (%)	Atomic Mass or Weight	Half-Life	Decay Mode/Energy (/MeV)	Particle Energy /Intensity (MeV / %)	Spin (h/2π)	Nuclear Magnetic Mom.(nm)	Elect. Quadr. Mom. (b)	γ-ray Energy /Intensity (MeV / %)
^{22}Al		22.0195	70. ms	β⁺ /18.6 β⁺ ,p/		4+			ann.rad./
^{23}Al		23.00727	0.47 s	β⁺ /12.24 β⁺ ,p/					ann.rad./
^{24m}Al			0.129 s	I.T./0.4259 β⁺ /	13.3/	1+			1.3686/5.3
^{24}Al		23.999941	2.07 s	β⁺ /13.878,p	3.40/48. 4.42/41. 6.80/3. 8.74/8.	4+			1.078(2)/16. 1.368(2)/96. 2.753(2)/43. 4.315(3)/15. 5.392(3)/20. 7.0662(2)/41.
^{25}Al		24.990429	7.17 s	β⁺ /4.277	3.27/	5/2+	3.646		ann.rad./ 1.6115(2)/100. 0.975(2)/5.
^{26m}Al			6.345 s	β⁺ /	3.2/	0+			ann.rad./
^{26}Al		25.9868917	7.1x10⁵ y	β⁺ /82/4.0042 EC/18/	1.16/	5+			ann.rad./ 1.80865(7)/99.8
^{27}Al	100.	26.9815384				5/2+	+3.64151	+0.15	
^{28}Al		27.9819102	2.25 m	β⁻ /4.6422	2.865/100.	3+	3.24	0.18	1.7778(6)/100.
^{29}Al		28.980445	6.5 m	β⁻ /3.680	1.4/30. 2.5/70.	5/2+			1.2732(8)/89. 2.0282(8)/4. 2.4262(8)/7.
^{30}Al		29.98296	3.68 s	β⁻ /8.56	5.05/	3+			1.26313(3)/35. 2.23525(5)/65.
^{31}Al		30.98395	0.64 s	β⁻ /8.00	6.25/				0.75223(3)/18. 1.69473(3)/59. 2.31664(4)/73.
^{32}Al		31.9881	33. ms	β⁻ /13.0		1+			
^{33}Al		32.9909							
^{34}Al		33.9969	0.05 s	β⁻ /17.1					
^{35}Al		34.9999							
^{36}Al		36.0064							
$_{14}$Si		**28.0855(3)**							
^{22}Si		22.0345							
^{24}Si		24.01155	0.10 s	β⁺ ,p/10.81					ann.rad./
^{25}Si		25.00411	221 ms	β⁺ ,p/12.74		5/2+			ann.rad./
^{26}Si		25.992330	2.23 s	β⁺ /5.066	3.282/	0+			ann.rad./ 0.8294(8)/22.
^{27}Si		26.9867041	4.14 s	β⁺ /4.8118	3.85/100.	5/2+	-0.8554		ann.rad./ 2.211(5)/0.2
^{28}Si	92.23(2)	27.9769265				0+			
^{29}Si	4.67(2)	28.9764947				1/2+	-0.5553		
^{30}Si	3.10(1)	29.9737702				0+			
^{31}Si		30.9753632	2.62 h	β⁻ /1.4920	1.471/99.9	3/2+			1.2662(5)/0.05
^{32}Si		31.974148	1.6x10² y	β⁻ /0.224	0.213/100.	0+			
^{33}Si		32.97800	6.1 s	β⁻ /5.85	3.92	(3/2+)			1.4313(5)/13. 1.84769(5)/100. 2.538(2)/10.
^{34}Si		33.97858	2.8 s	β⁻ /4.60	3.09/	0+			0.42907(5)/60. 1.17852(2)/64. 1.60756(5)/36.
^{35}Si		34.98458	0.9 s	β⁻ /10.50					
^{36}Si		35.9867	0.5 s	β⁻ /7.9					
^{37}Si		36.9930							
^{38}Si		37.9960							
^{39}Si									
$_{15}$P		**30.973761(2)**							
^{26}P		26.0118	≈20. ms	β⁺ ,p/18.1		3+			
^{27}P		26.99919	0.3 s	β⁻ /11.63		1/2+			
^{28}P		27.992312	270. ms	β⁺ /14.332	3.94/13. 5.25/13. 6.96/16. 8.8/7. 11.49/52.	3+			ann.rad./ 1.779(2)/98. 2.839(2)/2.8 3.040(2)/3.2 4.498(2)/12. 7.537(2)/9.
^{29}P		28.981801	4.14 s	β⁺ /4.9431	3.945/98.	1/2+	1.2349		ann.rad./ 1.273/1.32 2.426/0.39
^{30}P		29.978314	2.50 m	β⁺ /4.2323	3.245/99.9	1+			ann.rad./ 2.230(3)/0.07
^{31}P	100.	30.9737615				1/2+	+1.13160		
^{32}P		31.9739071	14.28 d	β⁻ /1.7106	1.710/100.	1+	-0.2524		
^{33}P		32.971725	25.3 d	β⁻ /0.249	0.249/100.	1/2+			

Elem. or Isot.	Natural Abundance (%)	Atomic Mass or Weight	Half-Life	Decay Mode/Energy (/MeV)	Particle Energy /Intensity (MeV / %)	Spin $(h/2\pi)$	Nuclear Magnetic Mom.(nm)	Elect. Quadr. Mom. (b)	γ-ray Energy /Intensity (MeV / %)
^{34}P		33.973636	12.4 s	β^- /5.374	3.2/15.	1+			1.78-4.1/
					5.1/85.				2.127(5)/15.
^{35}P		34.973314	47. s	β^- /3.989	2.34/100.	1/2+			1.572(1)/100.
^{36}P		35.97826	5.7 s	β^- /10.41					0.902/77.
									3.291/100.
^{37}P		36.97961	2.3 s	β^- /7.90					0.6462/
									1.5829/
^{38}P		37.9845	0.6 s	β^- /12.4					1.2923/
									2.224/
^{39}P		38.9864	$\approx$ 0.16 s						
^{40}P		39.9911							
^{41}P		40.9948	0.12 s						
^{42}P		42.0001	0.11 s						
$_{16}$**S**		**32.066(6)**							
^{27}S		27.0188	21. ms	β^+, 2p /18.3					
^{29}S		28.99661	0.188 s	β^+ /13.79		5/2+			ann.rad./
				β^+ ,p/					
^{30}S		29.984903	1.18 s	β^+ /6.138	4.42/78.	0+			ann.rad./
					5.08/20.				0.678/79.
^{31}S		30.979555	2.56 s	β^+ /5.396	4.39/99.	1/2+	0.48793		ann.rad./
									1.2662(5)/1.2
^{32}S	95.02(9)	31.9720707				0+			
^{33}S	0.75(1)	32.9714585				3/2+	+0.64382	-0.07	
^{34}S	4.21(8)	33.9678669				0+			
^{35}S		34.9690322	87.2 d	β^- /0.1672	0.1674/100.	3/2+	+1.00	+0.047	
^{36}S	0.02(1)	35.9670809				0+			
^{37}S		36.9711257	5.05 m	β^- /4.8653	1.64/94.				0.9083(4)/0.06
					4.75/5.6				3.1033(2)/94.2
^{38}S		37.97116	2.84 h	β^- /2.94	1.00/	0+			0.1962(4)/0.2
									1.9421(3)/84.
^{39}S		38.97514	11.5 s	β^- /6.64					1.301/52.
									1.697/44.
^{40}S		39.9755	9. s	β^- /4.7					
^{41}S		40.9800							
^{42}S		41.9815							
^{43}S		42.987							
^{44}S		43.988	0.12 s	β^-, n/9.					
$_{17}$**Cl**		**35.4527(9)**							
^{31}Cl		30.99242	0.15 s	β^+ ,p/11.98		3/2+			ann.rad./
^{32}Cl		31.98569	297. ms	β^+ /12.69	4.75/25.	1+			ann.rad./
					6.18/10.				1.548(2)/3.5
					7.48/14.				2.2305(1)/92.
					9.47/50.				2.4638(1)/4.
					11.6/1.				2.885(1)/1.
									4.770(1)/20.
^{33}Cl		32.977452	2.511 s	β^+ /5.583	4.51/98.	3/2+	+0.752		ann.rad./
									0.8409/0.52
									1.966/0.45
									2.866/0.44
^{34m}Cl			32.2 m	β^+ /	1.35/24.	3+			ann.rad./
					2.47/28.				
				I.T./					0.1457(8)/42.
									2.1276(5)/42.
^{34}Cl		33.9737630	1.528 s	β^+ /5.4922	4.50/100.	0+			ann.rad./
^{35}Cl	75.77(7)	34.96885271				3/2+	+0.82187	-0.08	
^{36}Cl		35.9683069	3.01×10^5 y	β^- /0.7086	0.7093/98.	0+	+1.28547	-0.018	
				β^+,EC/1.1421	0.115/0.002				
									ann.rad./
^{37}Cl	24.23(7)	36.9659026				3/2+	+0.68412	-0.065	
^{38m}Cl			0.715 s	I.T./		5-			0.67138(2)/100.
^{38}Cl		37.9680106	37.2 m	β^- /4.9168	1.11/31.	2-	2.05		1.64216(1)/31.
					2.77/11.				2.16760(2)/42.
					4.91/58.				
^{39}Cl		38.968009	55.6 m	β^- /3.442	1.91/85.	3/2+			0.25026(1)/47.
					2.18/8.				1.26720(5)/54.
					3.45/7.				0.986-1.517
^{40}Cl		39.97042	1.38 m	β^- /7.48		2-			0.6431(3)/6.
									1.4608(1)/77.
									2.8402(2)/17.
^{41}Cl		40.9707	34. s	β^- /5.7	3.8/				(0.167-1.359)
^{42}Cl		41.9732	6.8 s	β^- /9.4					
^{43}Cl		42.9742	3.3 s	β^- /8.0					
^{44}Cl		43.9785		β^- /12.3					
^{45}Cl		44.980	0.41	β^-, n/11.					
^{46}Cl		45.984	0.20	β^-, n/14.9					
^{47}Cl		46.988		β^-, n/15.					

Elem. or Isot.	Natural Abundance (%)	Atomic Mass or Weight	Half-Life	Decay Mode/Energy (/MeV)	Particle Energy /Intensity (MeV / %)	Spin (h/2π)	Nuclear Magnetic Mom.(nm)	Elect. Quadr. Mom. (b)	γ-ray Energy /Intensity (MeV / %)
$_{18}$Ar		**39.948(1)**							
^{31}Ar		31.0121	15. ms	β^+ /18.4					
				β^+, 2p					
				β^+, 3p					
^{32}Ar		31.99766	98. ms	β^+ ,p/11.2					ann.rad./
^{33}Ar		32.98993	174. ms	β^+ /11.62	3.12/	1/2+			ann.rad./
				β^+ ,p/					0.810(2)/48.
^{34}Ar		33.980270	0.844 s	β^+/6.061	5.0/95.	0+			ann.rad./
									0.6658(1)/2.5
									3.1290(1)/1.3
^{35}Ar		34.975257	1.77 s	β^+/5.965	4.94/93.	3/2+	+0.633		ann.rad./
									1.2185(5)/1.22
									1.763(1)/0.25
									2.964(1)/0.2
^{36}Ar	0.3365(30)	35.9675463				0+			
^{37}Ar		36.9667759	35.0 d	EC/.813		3/2+	+1.15		
^{38}Ar	0.0632(5)	37.9627322				0+			
^{39}Ar		38.964313	268. y	β^-/0.565	0.565/100.	7/2-	-1.3		
^{40}Ar	99.6003(30)	39.962383124				0+			
^{41}Ar		40.964501	1.82 h	β^-/2.492	1.198/	7/2-			1.29364(5)/99.
									1.6770(3)/0.05
^{42}Ar		41.96305	33. y	β^-/0.60	0.60/100.	0+			
^{43}Ar		42.9657	5.4 m	β^-/4.6					0.4791(2)/10.
									0.7380(1)/43.
									0.9752(1)/100.
									1.4400(3)/39.
^{44}Ar		43.96537	11.87 m	β^-/3.55		0+			0.182-1.866
^{45}Ar		44.96809	21.5 s	β^-/6.9		7/2-			0.0610/25.
									1.020/35.
									3.707/34.
^{46}Ar		45.96809	8.4 s	β^-/5.70		0+			1.944/
^{47}Ar		46.9722							
$_{19}$K		**39.0983(1)**							
^{35}K		34.98801	0.19 s	β^+ /11.88		3/2+			ann.rad./
				β^+ ,p/					1.751/14.
									2.5698/26.
									2.9827/51.
^{36}K		35.98129	0.342 s	β^+ /12.81	5.3/42.	2+	+0.548		ann.rad./
					9.9/44.				1.97044(5)/82.
									2.20783(5)/30.
									2.43343(2)/32.
^{37}K		36.9733769	1.23 s	β^+ /6.149	5.13/	3/2+	+0.2032		ann.rad./
									2.7944(8)/2.
									3.602(2)/0.05
^{38m}K			0.924 s	β^+ /6.742	5.02/100.	0+			ann.rad./
^{38}K		37.969080	7.63 m	β^+ /5.913	2.60/99.8	3+	+1.37		ann.rad./
									2.1675(3)/99.8
									3.9356(5)/0.2
^{39}K	93.2581(44)	38.9637069				3/2+	+0.39146	+0.049	
^{40}K	0.0117(1)	39.9639987	1.26x10^9 y	β^- /1.3111	1.312/89.	4-	-1.298	-0.061	ann.rad./
				β^+ ,EC/1.505	1.50/10.7				1.46081(5)/10.5
^{41}K	6.7302(44)	40.9618260				3/2+	+0.21487	+0.060	
^{42}K		41.9624031	12.36 h	β^- /3.525	1.97/19.	2-	-1.1425		0.31260(2)/0.3
					3.523/81.				1.5246(3)/18.
^{43}K		42.96072	22.3 h	β^- /1.82	0.465/8.	3/2+	+0.163		0.2211(2)/4.
					0.825/87.				0.3729(2)/88.
					1.24/3.5				0.3971(2)/11.
					1.814/1.3				0.6178(2)/81.
^{44}K		43.96156	22.1 m	β^- /5.66	5.66/34.	2-	-0.856		0.368207(1)/2.2
									1.15700(1)/58.
									2.15079(2)/22.
^{45}K		44.96070	17.8 m	β^- /4.20	1.1/23.	3/2+	+0.1734		0.1743(5)/80.
					2.1/69.				1.2607(8)/7.
					4.0/8.				1.7056(6)/69.
									2.3542(5)/14.
^{46}K		45.96198	1.8 m	β^- /7.72	6.3/	2-	-1.05		1.347(1)/91.
									3.700(5)/28.
^{47}K		46.96168	17.5 s	β^- /6.64	4.1/99.	1/2+	+1.93		0.56474(3)/15.
					6.0/1.				0.58575(3)/85.
									2.01313(3)/100.
^{48}K		47.96551	6.8 s	β^- /12.09	5.0/	(2-)			0.67122(1)/4.
									0.6723(5)/20.
									0.78016(1)/32.
									3.83153(7)/80.
^{49}K		48.96745	1.26 s	β^- /11.0					2.025/
									2.252/

Elem. or Isot.	Natural Abundance (%)	Atomic Mass or Weight	Half-Life	Decay Mode/Energy (/MeV)	Particle Energy /Intensity (MeV / %)	Spin ($h/2\pi$)	Nuclear Magnetic Mom.(nm)	Elect. Quadr. Mom. (b)	γ-ray Energy /Intensity (MeV / %)
^{50}K		49.9728	0.472 s	β^- /14.2					
^{51}K			0.365 s	β^- /					
^{52}K			0.105 s	β^-					
^{53}K			30. ms	β^-		3/2+			
^{54}K			10. ms	β^-					
$_{20}$Ca		**40.078(4)**							
^{35}Ca		35.0048	0.05 s	β^+ /15.6					
^{36}Ca		35.99309	0.1 s	β^+ /10.99 β^+ ,n/					ann.rad./
^{37}Ca		36.98587	0.173 s	β^+ /11.64 β^+ ,n/		3/2+			ann.rad./
^{38}Ca		37.976319	0.44 s	β^+ /6.74		0+			ann.rad./ 1.5677(5)/25. 3.210(2)/1.
^{39}Ca		38.970718	0.861 s	β^+ /6.531	5.49/100.	3/2+	1.02168		ann.rad./
^{40}Ca	96.941(18)	39.9625912				0+			
^{41}Ca		40.9622783	1.02x10^5 y	EC/0.4214		7/2-	-1.595	-0.08	
^{42}Ca	0.647(9)	41.9586183				0+			
^{43}Ca	0.135(6)	42.9587668				7/2-	-1.3173	-0.05	
^{44}Ca	2.086(12)	43.955481				0+			
^{45}Ca		44.956186	162.7 d	β^- /0.257	0.257/100.	7/2-	-1.327	+0.05	
^{46}Ca	0.004(3)	45.953693				0+			
^{47}Ca		46.954546	4.536 d	β^- /1.992	0.684/84. 1.98/16.	7/2-	-1.38	+0.02	1.297/75 (0.041-1.88)
^{48}Ca	0.187(4)	47.952533				0+			
^{49}Ca		48.955673	8.72 m	β^- /5.262	0.89/7. 1.95/92.	3/2-			3.0844(1)/92. 4.0719(1)/7.
^{50}Ca		49.95752	14. s	β^- /4.97	3.12/	0+			0.2569/98. (0.0715 - 1.59)
^{51}Ca		50.9615	10. s	β^- /7.3		(3/2-)			
^{52}Ca		51.9651	4.6 s	β^- /8.0					
^{53}Ca		52.9700	0.09 s	β^- /10.9					
$_{21}$Sc		**44.955910(8)**							
^{40}Sc		39.977964	0.182 s	β^+ /14.320	5.73/50. 7.53/15. 8.76/15. 9.58/20.	4-			ann.rad./ 0.752/41. 3.732/99.5 (1.12-3.92)
^{41}Sc		40.9692513	0.596 s	β^+ /6.4953	5.61/100.	7/2-	5.431	0.166	ann.rad./
^{42m}Sc			61.6 s	β^+ /	2.82/	7+			ann.rad./ 0.4375(5)/100. 1.2270(5)/100. 1.5245(5)/100.
^{42}Sc		41.9655168	0.682 s	β^+ /6.4259	5.32/100.	0+			ann.rad./
^{43}Sc		42.961151	3.89 h	β^+ ,EC/2.221	0.82/22. 1.22/78.	7/2-	+4.62	-0.26	ann.rad./ 0.3729(1)/22.
^{44m}Sc			2.44 d	I.T./0.27 EC/3.926		6+	+3.88		0.27124(1)/87. (1.00-1.16)
^{44}Sc		43.959403	3.93 h	β^+, EC/3.653	1.47/	2+	+2.56	+0.10	ann.rad./ 1.15700(1)/100.
^{45}Sc	100.	44.955910				7/2-	+4.75649	-0.22	
^{46m}Sc			18.7 s	I.T./0.14253		1-			0.14253(2)/62.
^{46}Sc		45.955170	83.81 d	β^- /2.367	0.357/100.	4+	+3.03	+0.12	0.88925(3)/100. 1.12051(1)/100.
^{47}Sc		46.952408	3.349 d	β^- /0.600	0.439/69. 0.601/31.	7/2-	+5.34	-0.22	0.15938(1)/68.
^{48}Sc		47.952235	43.7 h	β^- /3.99	0.655/	6+			0.98350(1)/100. 1.03750(1)/97. 1.31209(3)/100.
^{49}Sc		48.950024	57.3 m	β^- /2.006	2.00/99.9.	7/2-			1.7619(3)/0.05
^{50}Sc		49.95219	1.71 m	β^- /6.89	3.05/76. 3.60/24.	(5+)			0.5235(1)/88. 1.1210(1)/100. 1.5537(2)/100.
^{51}Sc		50.95360	12.4 s	β^- /6.51	4.4/ 5.0/	7/2-			1.4373(4)/52. 0.718-2.144
^{52}Sc		51.9566	8.2 s	β^- /9.0		(3+)			
^{53}Sc		52.9584		β^- /8.1					
^{54}Sc		53.9635		β^- /11.6					
^{55}Sc		54.969		β^- /13					
$_{22}$Ti		**47.867(1)**							
^{39}Ti		39.0013	28. ms	β^+ /15.4					
^{40}Ti		39.9905	0.06 s	β^+ /11.7					
^{41}Ti		40.98313	80. ms	β^+ ,p/12.93		3/2+			ann.rad./

Elem. or Isot.	Natural Abundance (%)	Atomic Mass or Weight	Half-Life	Decay Mode/Energy (/MeV)	Particle Energy /Intensity (MeV / %)	Spin (h/2π)	Nuclear Magnetic Mom.(nm)	Elect. Quadr. Mom. (b)	γ-ray Energy /Intensity (MeV / %)
^{42}Ti		41.97303	0.20 s	β^+ /7.000	6.0/				ann.rad./
									0.6107(5)/56.
^{43}Ti		42.96852	0.50 s	β^+ /6.87	5.80/	7/2-	0.85		ann.rad./
^{44}Ti		43.959690	67. y	EC/0.268		0+			0.067868(1)/91.
									0.078322(1)/97.
^{45}Ti		44.958124	3.078 h	β^+/86/2.062	1.04	7/2-	0.095	0.015	ann.rad./
				EC/14/					(0.36-1.66)/weak
^{46}Ti	8.25(3)	45.952630				0+			
^{47}Ti	7.44(2)	46.951764				5/2-	-0.78848	+0.30	
^{48}Ti	73.72(3)	47.947947				0+			
^{49}Ti	5.41(2)	48.947871				7/2-	-1.10417	+0.24	
^{50}Ti	5.18(2)	49.944792				0+			
^{51}Ti		50.946616	5.76 m	β^- /2.471	1.50/92.	3/2-			0.3197(2)/93.
					2.13/				0.6094-0.9291
^{52}Ti		51.94690	1.7 m	β^- /1.97	1.8/100.	0+			0.0170(5)/100.
									0.12445(5)/100.
^{53}Ti		52.9497	33. s	β^- /5.0	(2.2-3)/	3/2-			0.1008(1)/20.
									0.1276(1)/45.
									0.2284(1)/39.
									1.6755(5)/45.
									(1.72-2.8)/
^{54}Ti		53.9510		β^- /4.3					
^{55}Ti		54.9552		β^- /13					
^{56}Ti		55.9580		β^- /7.0					
^{57}Ti		56.964		β^- /11.					
^{58}Ti									
$_{23}$V		**50.9415(1)**							
^{43}V		42.9807	>0.8 s	β^+ /11.3					
^{44}V		43.9744	0.09 s	β^+ ,α/13.7					ann.rad./
^{45}V		44.96578	0.54 s	β^+ /7.13		7/2-			
^{46}V		45.96020	0.4223 s	β^+ /7.051	6.03/100.	0+			ann.rad./
^{47}V		46.954907	32.6 m	β^+ ,EC/2.928	1.90/99.+	3/2-			ann.rad./
									1.7949(8)/0.19
									(0.2-2.16)
^{48}V		47.952254	15.98 d	β^+ /4.012	0.698/50.	4+	2.01		ann.rad./
									0.98350(2)/100.
									(1.3-2.4)
^{49}V		48.948517	337. d	EC/0.602		7/2-	4.47		
^{50}V	0.250(2)	49.947163	>1.4x10^{17}y	EC, β^-		6+	+3.34569	+0.21	
^{51}V	99.750(2)	50.943964				7/2-	+5.148706	-0.04	
^{52}V		51.944780	3.76 m	β^- /3.976	2.47/	3+			1.4341(1)/100.
^{53}V		52.944342	1.61 m	β^- /3.436	2.52/	7/2-			1.0060(5)/90.
									1.2891(3)/10.
^{54}V		53.94644	49.8 s	β^- /7.04	1.00/5.	3+			0.8348/97.
					2.00/12.				0.9887/80.
					2.95/45.				2.259/46.
					5.20/11.				(0.56-3.38)
^{55}V		54.9472	6.5 s	β^- /6.0	6.0/	(7/2-)			0.5177/73.
									(0.224-1.21)
^{56}V		55.9505		β^- /9.1					
^{57}V		56.9525		β^- /8.1					
^{58}V		57.9567		β^- /11.6					
^{59}V		58.9593		β^- /9.9					
^{60}V		59.965		β^- /14.					
^{61}V									
$_{24}$Cr		**51.9961(6)**							
^{43}Cr		42.9977	21. ms						
^{44}Cr		43.9855	53. ms	β^+ ,(p) /10.3					
^{45}Cr		44.9792	0.05 s	β^+ ,p/12.5		7/2-			ann.rad./
^{46}Cr		45.96836	0.3 s	β^+ /7.60					ann.rad./
^{47}Cr		46.96291	0.51 s	β^+ /7.45		3/2-			ann.rad./
^{48}Cr		47.95404	21.6 h	EC/1.66					ann.rad./
									0.116(2)/95.
									0.305(10)/100.
^{49}Cr		48.951341	42.3 m	β^+,EC/2.631	1.39/	5/2-	0.476		ann.rad./
					1.45/				0.09064(1)/51.
					1.54/				0.15293(1)/27.
									(0.062-1.6)
^{50}Cr	4.345(13)	49.946050				0+			
^{51}Cr		50.944772	27.70 d	EC/0.7527		7/2-	-0.934		0.320076(1)/10.2
^{52}Cr	83.789(18)	51.940512				0+			
^{53}Cr	9.501(17)	52.940653				3/2-	-0.47454	-0.15	
^{54}Cr	2.365(7)	53.938885				0+		-0.2	

Elem. or Isot.	Natural Abundance (%)	Atomic Mass or Weight	Half-Life	Decay Mode/Energy (/MeV)	Particle Energy /Intensity (MeV / %)	Spin (h/2π)	Nuclear Magnetic Mom.(nm)	Elect. Quadr. Mom. (b)	γ-ray Energy /Intensity (MeV / %)
^{55}Cr		54.940844	3.497 m	β^- /2.603	2.5/	3/2-			1.5282(2)/0.04
									(0.13-2.37)
^{56}Cr		55.94065	5.9 m	β^- /1.62	1.50/100.	0+			0.026(2)/100.
									0.083(3)/100.
^{57}Cr		56.9438	21. s	β^- /5.1	3.3/	3/2-	0.0834		0.850/8.
					3.5/				(0.083-2.62)
^{58}Cr		57.9443	7.0 s	β^- /4.0					
^{59}Cr		58.9487	1.0 s	β^- /7.7					
^{60}Cr		59.9497	0.6 s	β^- /6.0					
^{61}Cr		60.9541		β^- /8.8					
^{62}Cr		61.9558		β^- /7.3					
^{63}Cr									
$_{25}$Mn		**54.938049(9)**							
^{46}Mn		45.9867	≈41. ms	β^+ /17.1					
^{47}Mn		46.9761		β^+ /12.3					
^{48}Mn		47.9686	0.15 s	β^+ /13.5	5.79/58.	4+			
					4.43/10.				
^{49}Mn		48.95962	0.38 s	β^+ /7.72	6.69/	5/2-			ann.rad./
^{50m}Mn			1.74 m	β^+ /7.887	3.54/	5+			ann.rad./
									1.0980/94.
									0.783/91.
									(0.66-3.11)
^{50}Mn		49.954244	0.283 s	β^+ /7.6330	6.61/	0+			ann.rad./
^{51}Mn		50.948215	46.2 m	β^+ ,EC/3.208	2.2/	5/2-	3.568	0.4	ann.rad./
									0.7491(1)/0.26
									(1.148-1.164)
^{52m}Mn			21.1 m	β^+ /98/5.09	2.631/	2+	0.0076		ann.rad./
				I.T./2/0.378					0.3778 (I.T.)
									1.43406(1)/98.
									(0.7-4.8)
^{52}Mn		51.945570	5.591 d	β^+ /4.712	0.575/	6+	+3.063	+0.5	ann.rad./
				EC/					0.74421(1)/90.
									1.43406(1)/100.
^{53}Mn		52.941294	3.7x10^6 y	EC/0.5970		7/2-	5.024		
^{54}Mn		53.940363	312.2 d	EC/1.377		3+	+3.282	+0.33	0.83403(5)/100.
^{55}Mn	100.	54.938049				5/2-	+3.4687	+0.32	
^{56}Mn		55.938909	2.579 h	β^- /3.6954	0.718/18.	3+	+3.2266		0.84675(2)/98.9
					1.028/34.				1.81072(4)/27.
									2.11305(5)/14.5
^{57}Mn		56.938287	1.45 m	β^- /2.691		5/2-			
^{58}Mn		57.93999	65 s	β^- /6.25	3.8/	3+			0.45916(2)/20.
					5.1/				0.81076(1)/82.
									1.32309(5)/53.
^{59}Mn		58.94045	4.6 s	β^- /5.19	4.5/				0.471/
									0.531-0.726
^{60m}Mn			1.77 s	β^- /IT	5.7/	3+			0.824/
^{60}Mn		59.9433	50. s	β^- /8.6		0+			1.969/
^{61}Mn		60.9446	0.71 s	β^- /7.4		(5/2)-			
^{62}Mn		61.9480	0.9 s	β^- /10.4		(3+)			0.877/
									0.942-1.299
^{63}Mn		62.9498		β^- /8.8					
^{64}Mn		63.9537		β^- /11.8					
^{65}Mn		64.956		β^- /10.					
^{66}Mn									
$_{26}$Fe		**55.845(2)**							
^{46}Fe		46.0008	≈0.02 s	β^+ /13.1					
^{47}Fe		46.9929	≈0.03 s	β^+ /15.6					
^{48}Fe		47.9806		β^+ /11.2					
^{49}Fe		48.9763	0.08 s	β^+ /13.0		(7/2-)			ann.rad./
^{50}Fe		49.9630	0.15 s	β^+ /8.2					0.651
^{51}Fe		50.95683	0.31 s	β^+ /8.02		(5/2-)			ann.rad./
^{52m}Fe			46. s	β^+ /4.4		(12+)			ann.rad./
									(0.622-2.286)/
^{52}Fe		51.94812	8.28 h	β^+ /57/2.37	0.804/	0+			ann.rad./
				EC/43/					0.16868(1)/99.
				I.T./					0.377 (I.T.)/
^{53m}Fe			2.6 m	I.T./3.0407		19/2-			0.7011(1)/99.
									1.0115(1)/87.
									1.3281(1)/87.
									2.3396(1)/13.
^{53}Fe		52.945312	8.51 m	β^+ /3.743	2.40/42.	7/2-			ann.rad./
					2.80/57.				0.3779(1)/42.
									(1.2 - 3.2)
^{54}Fe	5.845(35)	53.939615				0+			
^{55}Fe		54.938298	2.73 y	EC/0.2316		3/2-			

Elem. or Isot.	Natural Abundance (%)	Atomic Mass or Weight	Half-Life	Decay Mode/Energy (/MeV)	Particle Energy /Intensity (MeV / %)	Spin (h/2π)	Nuclear Magnetic Mom.(nm)	Elect. Quadr. Mom. (b)	γ-ray Energy /Intensity (MeV / %)
^{56}Fe	91.754(36)	55.934942				0+			
^{57}Fe	2.119(10)	56.935398				1/2-	+0.0906		
^{58}Fe	0.282(4)	57.933280				0+			
^{59}Fe		58.934880	44.51 d	β^- /1.565	0.273/48.	3/2-	0.29		1.099/57
					0.475/51.				1.292/43.
									(0.14-1.48)
^{60}Fe		59.934077	1.5x10^6 y	β^- /0.237	0.184/100.	0+			0.0586/100.(IT)
^{61}Fe		60.93675	6.0 m	β^- /3.98	2.5/13.				1.205/44.
					2.63/54.				1.028/43.
					2.80/31.				(0.12-3.37)
^{62}Fe		61.93677	68. s	β^- /2.53	2.5/100.	0+			0.5061(1)/100.
^{63}Fe		62.9404	6. s	β^- /6.3		5/2-			0.995/
									(1.365-1.427)
^{64}Fe		63.9411	2.0 s	β^- /4.9					
^{65}Fe		64.9449	≈0.45 s	β^- /7.9					
^{66}Fe		65.9460		β^- /5.7					
^{67}Fe		66.9500		β^- /8.8					
^{68}Fe		67.953	≈0.10 s	β^- /≈7.6					
^{69}Fe									
$_{27}$Co		**58.933200(9)**							
^{50}Co		49.9812		β^+ /17.0					
^{51}Co		50.9705		β^+ /12.8					
^{52}Co		51.9632		β^+ /14.0					
^{53m}Co			0.25 s	β^+ ,p/		19/2-			ann.rad./
^{53}Co		52.95423	0.26 s	β^+ /8.30		7/2-			ann.rad./
^{54m}Co			1.46 m	β^+ /8.44	4.25/100.	7+			ann.rad./
									0.411(1)/99.
									1.130(1)/100.
									1.408(1)/100.
^{54}Co		53.948464	0.1932 s	β^+ /8.2430	7.34/100.	0+			ann.rad./
^{55}Co		54.942003	17.53 h	β^+ /3.4513	0.53/	7/2-	+4.822		ann.rad./
				EC/	1.03/				0.9312/75.
					1.50/				0.4772/20.
									(0.092-3.11)
^{56}Co		55.939844	77.3 d	β^+ /4.566	1.459/18.	4+	3.85	+0.25	ann.rad./
				EC/					0.84678(6)/99.9
									1.2383/68.
									(0.26-3.61)
^{57}Co		56.936296	271.8 d	EC/0.8361		7/2-	+4.72	+0.5	0.12206(2)/85.6
									(0.014-0.706)
^{58m}Co			9.1 h	I.T./		5+			0.02489/0.035
^{58}Co		57.935757	70.88 d	β^+ /2.307		2+	+4.04	+0.22	ann.rad./
				EC/					0.810755(3)/99.
^{59}Co	100.	58.933200				7/2-	+4.63	+0.41	
^{60m}Co			10.47 m	I.T./99.8/0.059		2+	+4.40	+0.3	0.058603(1)/2.0
				β^- /0.2/1.56					
^{60}Co		59.933822	5.271 y	β^- /2.824	0.315/99.7	5+	+3.799	+0.44	1.173210(2)/100.
									1.332470(2)/100.
^{61}Co		60.932479	1.650 h	β^- /1.322	1.22/95.	7/2-			0.067415(1)/86.
									(0.8417-0.9092)
^{62m}Co			13.9 m	β^- /	0.88/25.	5+			1.1635(3)/70.
					2.88/75.				1.1730(3)/98.
									2.0039(3)/19.
^{62}Co		61.93405	1.50 m	β^- /5.32	1.03/10.	2+			1.1292(3)/13.
					1.76/5.				1.1730(3)/83.
					2.9/20.				1.9851(1)/3.
					4.05/60.				2.3020(1)/19.
^{63}Co		62.93362	27.5 s	β^- /3.67	3.6/	7/2-			0.08713(1)/49.
									0.9817(3)/2.6
									0.156-2.17
^{64}Co		63.93581	0.30 s	β^- /7.31	7.0/	1+			
^{65}Co		64.93648	1.14 s	β^- /5.96		(7/2)-			
^{66}Co		65.9398	≈ 0.23 s	β^- /10.0					
^{67}Co		66.9406		β^- /8.4					
^{68}Co		67.9444	≈ 0.18 s	β^- /11.7					
^{69}Co		68.9452	0.27 s	β^- /9.3					
^{70}Co		69.9498		β^- 13.					
^{71}Co									
$_{28}$Ni		**58.6934(2)**							
^{51}Ni		50.9877		β^+ /16.0					
^{52}Ni		51.9757		β^+ /11.7					
^{53}Ni		52.9685	0.05 s	β^+ ,p/13.3		7/2-			ann.rad./
^{54}Ni		53.95791		β^+ /8.80					
^{55}Ni		54.95134	0.20 s	β^+ /8.70	7.66/	7/2-			ann.rad./

Elem. or Isot.	Natural Abundance (%)	Atomic Mass or Weight	Half-Life	Decay Mode/Energy (/MeV)	Particle Energy /Intensity (MeV / %)	Spin (h/2π)	Nuclear Magnetic Mom.(nm)	Elect. Quadr. Mom. (b)	γ-ray Energy /Intensity (MeV / %)
^{56}Ni		55.94214	6.08 d	EC/2.14		0+			0.15838(3)/98.8
									0.81185(3)/87.
									0.2695-0.7500
^{57}Ni		56.939800	35.6 h	β^+ /3.264	0.712/10.	3/2-	0.88		ann.rad./
				EC/	0.849/76.				1.3776/78.
									(0.127-3.177)
^{58}Ni	68.077(9)	57.935348				0+			
^{59}Ni		58.934351	≈7.6x10^4 y	EC/		3/2-			
^{60}Ni	26.223(8)	59.930790				0+			
^{61}Ni	1.140(1)	60.931060				3/2-	-0.75002	+0.16	
^{62}Ni	3.634(2)	61.928348				0+			
^{63}Ni		62.929673	100. y	β^- /0.066945	0.065/	1/2-			
^{64}Ni	0.926(1)	63.927969				0+			
^{65}Ni		64.930088	2.517 h	β^- /2.137	0.65/30.	5/2-	0.69		0.36627(3)/5.
					1.020/11.				1.11553(4)/16.
					2.140/58.				1.48184(5)/23.
^{66}Ni		65.92912	54.6 h	β^- /0.23		0+			
^{67}Ni		66.93157	21. s	β^- /3.56	3.8/	1/2-			1.0722/100.
									1.6539/100.
									(0.10-1.98)
^{68}Ni		67.93185	19. s	β^- /2.06					
^{69}Ni		68.9352	10. s	β^- /5.4					0.6807(3)/100.
^{70}Ni		69.9361		β^- /3.5					
^{71}Ni		70.9400	1.9 s	β^- /6.9					
^{72}Ni		71.9413	2.1 s	β^- /5.2					
^{73}Ni		72.946	0.9 s	β^- /9.					
^{74}Ni		73.948	1.1 s	β^- /7.					
$_{29}$Cu		**63.546(3)**							
^{55}Cu		54.9655		β^+ /13.2					
^{56}Cu		55.9586		β^+ /15.3					
^{57}Cu		56.94922	0.23 s	β^+ /8.77		3/2-			
^{58}Cu		57.944541	3.21 s	β^+ /8.563	4.5/15.	1+			ann.rad./
				EC/	7.439/83.				0.0403(4)/5.
									1.4483(2)/11.
									1.4546(2)/16.
^{59}Cu		58.939504	1.36 m	β^+ /4.800	1.9/	3/2-			ann.rad./
					3.75/				0.3393(1)/8.
									0.8780(1)/12.
									1.3015(1)/15.
									(0.4 - 2.6)
^{60}Cu		59.937368	23.7 m	β^+ /6.127	2.00/69.	2+	+1.219		ann.rad./
				EC/	3.00/18.				1.3325/88.
					3.92/6.				1.7915/45.
									(0.12-5.048)
^{61}Cu		60.933462	3.35 h	β^+ /2.237	0.56/3.	3/2-	+2.14		ann.rad./
					0.94/5.				0.2830/13.
					1.15/2.				0.6560/11.
					1.220/51.				(0.067-2.123)
^{62}Cu		61.932587	9.74 m	β^+ /98/3.948	2.93/98.	1+	-0.380		ann.rad./
				EC/					1.17302(1)/0.6
									(0.87-3.37)
^{63}Cu	69.17(3)	62.929601				3/2-	+2.2233	-0.211	
^{64}Cu		63.929768	12.701 h	β^- /39/0.579	0.578/	1+	-0.217		ann.rad./
				β^+ /19/1.6751	0.65/				1.3459(3)/0.6
				EC/41/					
^{65}Cu	30.83(3)	64.927794				3/2-	+2.3817	-0.195	
^{66}Cu		65.928873	5.09 m	β^- /2.642	1.65/6.	1+	-0.282		0.8330(1)/0.22
					2.7/94.				1.0392(2)/9.2
^{67}Cu		66.92775	2.580 d	β^- /0.58	0.395/56.	3/2-			0.09125(1)/7.
					0.484/23.				0.09323(1)/17.
					0.577/20				0.18453(1)/47.
^{68m}Cu			3.70 m	I.T./86/		6-			0.0843(5)/70.
				β^- /14/1.8					0.1112(5)/18.
									0.5259(5)/74.
									(0.64-1.34)
^{68}Cu		67.92964	31. s	β^- /4.46	3.5/40.	1+			1.0774(5)/58.
					4.6/31.				1.2613(5)/17.
									(0.15-2.34)
^{69}Cu		68.92943	2.8 m	β^- /2.68	2.48/80.	3/2-			0.5307(3)/3.
									0.8340(5)/6.
									1.0065(8)/10.
^{70m}Cu			47. s	β^- /	2.52/10.	5-			0.8848(2)/100.
									0.9017(2)/90.
									1.2517(5)/60.
									(0.39-3.06)
^{70}Cu		69.93241	5. s	β^- /6.60	5.42/54.	1+			0.8848(2)/54.
					6.09/46.				

Elem. or Isot.	Natural Abundance (%)	Atomic Mass or Weight	Half-Life	Decay Mode/Energy (/MeV)	Particle Energy /Intensity (MeV / %)	Spin (h/2π)	Nuclear Magnetic Mom.(nm)	Elect. Quadr. Mom. (b)	γ-ray Energy /Intensity (MeV / %)
^{71}Cu		70.93262	20. s	β^- /4.56		3/2-			0.490/
^{72}Cu		71.9357	6.6 s	β^- /8.2		(1+)			0.652/
^{73}Cu		72.9365	3.9 s	β^- /6.3					0.450/
^{74}Cu		73.9401	1.6 s	β^- /9.9					
^{75}Cu		74.9414	1.2 s	β^- /7.9					
^{76}Cu		75.9455	0.64 s	β^- /11.					
^{77}Cu		76.947	0.47 s	β^- /≈10.					
^{78}Cu		77.952	0.34 s	β^- /12.					
^{79}Cu		78.954	0.19 s	β^- /11.					
$_{30}$Zn		**65.39(2)**							
^{57}Zn		56.9649	0.04 s	β^+,p/14.6		(7/2-)			ann.rad./
^{59}Zn		58.94927	183. ms	β^+,p/9.09	8.1/	3/2-			ann.rad./ (0.491-0.914)
^{60}Zn		59.94183	2.40 m	β^+ /97/4.16 EC/3/		0+			ann.rad./ 0.669/47. (0.062-0.947)
^{61}Zn		60.93951	1.485 m	β^+ /5.64	4.38/68.	3/2-			ann.rad./ 0.4748/17. (0.15-3.52)
^{62}Zn		61.93433	9.22 h	β^+ /3/1.63 EC/93/	0.66/7.	0+			ann.rad./ 0.0408/25 0.5967/26. (0.20-1.526)/
^{63}Zn		62.933215	38.5 m	β^+ /93/3.367 EC/7/	1.02/ 1.40/ 1.71/ 2.36/84.	3/2-	-0.28164	+0.29	ann.rad./ 0.66962(5)/8.4 0.96206(5)/6.6 (0.24-3.1)
^{64}Zn	48.6(3)	63.929146				0+			
^{65}Zn		64.929245	243.8 d	β^+ /98/1.3514 EC/1.5/	0.325/	5/2-	+0.7690	-0.023	ann.rad./ 1.11552(2)/50.8
^{66}Zn	27.9(2)	65.926036				0+			
^{67}Zn	4.1(1)	66.927131				5/2-	+0.8755	+0.15	
^{68}Zn	18.8(4)	67.924847				0+			
^{69m}Zn			13.76 h	I.T./99+/0.439		9/2+			0.4390(2)/95.
^{69}Zn		68.926553	56. m	β^- /0.906	0.905/99.9	1/2-			0.318/
^{70}Zn	0.6(1)	69.925325				0+			
^{71m}Zn			3.97 h	β^- /	1.45/	9/2+			0.3864/93. 0.4874/62. 0.6203/57. (0.099-2.489)
^{71}Zn		70.92773	2.4 m	β^- /2.81		1/2-			0.5116(1)/30. 0.9103(1)/7.5 (0.12-2.29)
^{72}Zn		71.92686	46.5 h	β^- /0.46	0.25/14. 0.30/86.	0+			0.0164(3)/8. 0.1447(1)/83. 0.1915(2)/9.4
^{73m}Zn			6. s		I.T./0.196	(7/2+)			0.042
^{73}Zn		72.92978	24. s	β^- /4.29	4.7/	(1/2-)			0.216(1)/100. 0.496-0.911
^{74}Zn		73.92946	1.60 m	β^- /2.3	2.1/				0.0565/ 0.1401/ (0.05-0.35)
^{75}Zn		74.9329	10.2 s	β^- /6.0					0.229/
^{76}Zn		75.9334	5.7 s	β^- /4.2	3.6/				0.119/
^{77m}Zn			1.0 s	β^- /		(1/2-)			0.772
^{77}Zn		76.9371	2.1 s	β^- /7.3	4.8/				0.189/
^{78}Zn		77.9386	1.5 s	β^- /6.4					0.225/
^{79}Zn		78.9421	1.0 s	β^- /8.6					0.702/
^{80}Zn		79.9444	0.54 s	β^- /7.3					0.713/ 0.2248/
^{81}Zn		80.9505	0.29 s	β^- /11.9					
$_{31}$Ga		**69.723(1)**							
^{61}Ga		60.9492	0.15 s	β^+ /9.0					
^{62}Ga		61.94418	0.116 s	β^+ /9.17 EC/	8.3/	0+			ann.rad./
^{63}Ga		62.9391	32. s	β^+ /5.5 EC/	4.5/				ann.rad./ 0.6271(2)/10. 0.6370(2)/11. 1.0652(4)/45.
^{64}Ga		63.936838	2.63 m	β^+ /7.165	2.79/ 6.05/	0+			ann.rad./ 0.80785(1)/14. 0.99152(1)/43. 1.38727(1)/12. 3.3659(1)/13.

Elem. or Isot.	Natural Abundance (%)	Atomic Mass or Weight	Half-Life	Decay Mode/Energy (/MeV)	Particle Energy /Intensity (MeV / %)	Spin (h/2π)	Nuclear Magnetic Mom.(nm)	Elect. Quadr. Mom. (b)	γ-ray Energy /Intensity (MeV / %)
^{65}Ga		64.9394	15.2 m	β^+ /86/3.255 EC/	0.82/10. 1.39/19. 2.113/56. 2.237/15.	3/2-			ann.rad./ 0.1151(2)/55. 0.1530(2)/96. 0.2069(2)/39. (0.06-2.4)
^{66}Ga		65.931592	9.5 h	β^+ /56/5.175 EC/43/	0.74/1. 1.84/54. 4.153/51.	0+			ann.rad./ 1.03935(8)/38. 2.7523(1)/23. (0.28-5.01)
^{67}Ga		66.928205	3.260 d	EC/1.001		3/2-	+1.8507	0.20	0.09332/37. 0.18459/20. 0.30024/17. (0.091-0.89)
^{68}Ga		67.927983	1.130 h	β^+ /90/2.921 EC/10/	1.83/	1+	0.01175	0.028	ann.rad./ 1.0774(1)/3. (0.57-2.33)/
^{69}Ga	60.108(6)	68.925581				3/2-	+2.01659	+0.17	
^{70}Ga		69.926027	21.1 m	EC/0.2/0.655 β^- /99.8/1.656	1.65/99.	1+			0.1755(5)/0.15 1.042(5)/0.48
^{71}Ga	39.892(6)	70.924707				3/2-	+2.56227	+0.11	
^{72}Ga		71.926372	14.10 h	β^- /4.001	0.64/40. 1.51/9. 2.52/8. 3.15/11.	3-	-0.13224	+0.5	0.62986(5)/24. 2.2016(2)/26. 2.5077(2)/12.8 (0.11-3.3)/
^{73}Ga		72.92517	74.87 h	β^- /1.59		3/2-			0.05344(5)/10. 0.29732(5)/47. (0.01-1.00)/
^{74m}Ga			10. s	I.T./		1+			0.0565(1)/75.
^{74}Ga		73.92694	8.1 m	β^- /5.4	2.6/	3-			0.5959/92. 2.354/45. (0.23-3.99)
^{75}Ga		74.92650	2.10 m	β^- /3.39	3.3/	3/2			0.2529/ 0.5746/ (0.12-2.10)
^{76}Ga		75.9289	29. s	β^- /7.0		3-			0.5629/66. 0.5455/26. (0.34-4.25)
^{77}Ga		76.9293	13.0 s	β^- /5.3	5.2/				0.469/ 0.459/
^{78}Ga		77.9317	5.09 s	β^- /8.2		3+			0.619/77. 1.187/20.
^{79}Ga		78.9329	2.85 s	β^- /7.0	4.6/				0.465/
^{80}Ga		79.9366	1.68 s	β^- /10.4	10./				0.659/
^{81}Ga		80.9377	1.22 s	β^- /8.3	5.1/				0.217/
^{82}Ga		81.9432	0.599 s	β^- /12.6					1.348/
^{83}Ga		82.9469	0.308 s	β^- /$\approx$ 11.5					
^{84}Ga		83.952	$\approx$0.085 s	β^- /14					
$_{32}$Ge		**72.61(2)**							
^{61}Ge		60.9638	0.04 s	β^+ /13.6					
^{63}Ge		62.9496	0.10 s	β^- /9.8					
^{64}Ge		63.9416	1.06 m	β^+ /4.4 EC/	3.0/	0+			ann.rad./ 0.1282(2)/11. 0.4270(3)/37. 0.6671(3)/17.
^{65}Ge		64.9394	31. s	β^+ /6.2 EC/	0.82/10. 1.39/19. 2.113/56. 2.237/15.				ann.rad./ 0.0620/27. 0.6497/33. 0.8091/21. (0.19-3.28)
^{66}Ge		65.93385	2.26 h	β^+ /27/2.10 EC/73/		0+			ann.rad./ 0.0438/29. 0.3819/28. (0.022-1.77)
^{67}Ge		66.932738	19.0 m	β^+ /96/4.225 EC/4/	1.6/ 2.3/ 3.15/	1/2-			ann.rad./ 0.1670/84. (0.25-3.73)
^{68}Ge		67.92810	270.8 d	EC/0.11		0+			Ga k x-ray/39.
^{69}Ge		68.927973	1.63 d	β^+ /36/2.2273 EC/64/	0.70/ 1.2/	5/2-	0.735	0.02	ann.rad./ 0.574/13. 1.1068/36. (0.2-2.04)
^{70}Ge	21.23(4)	69.924250				0+			
^{71m}Ge			20.4 ms		I.T./0.0234	9/2+			0.1749
^{71}Ge		70.924954	11.2 d	EC/0.229		1/2-	+0.547		
^{72}Ge	27.66(3)	71.922076				0+			
^{73}Ge	7.73(1)	72.923460				9/2+	-0.879467	-0.17	
^{74}Ge	35.94(2)	73.921178				0+			

Elem. or Isot.	Natural Abundance (%)	Atomic Mass or Weight	Half-Life	Decay Mode/Energy (/MeV)	Particle Energy /Intensity (MeV / %)	Spin (h/2π)	Nuclear Magnetic Mom.(nm)	Elect. Quadr. Mom. (b)	γ-ray Energy /Intensity (MeV / %)
^{75m}Ge			48. s	I.T./		7/2+			0.13968(3)/39.
^{75}Ge		74.922860	1.380 h	β^- /1.177	1.19/	1/2-	+0.510		0.26461(5)/11. 0.41931(5)/0.2
^{76}Ge	7.44(2)	75.921403				0+			
^{77m}Ge			53. s	I.T./20/ β^- /80/2.861	2.9/	1/2-			0.1597(1)/11. 0.21551/21. (0.19-0.61)
^{77}Ge		76.923549	11.30 h	β^- /2.702	0.71/23. 1.38/35. 2.19/42.	7/2+			0.2110/29. 0.2155/27. 0.2644/51. (0.15-2.35)
^{78}Ge		77.922853	1.45 h	β^- /0.95	0.70/	0+			0.2773(5)/96. 0.2939(5)/4.
^{79m}Ge			39. s	β^- /IT		7/2+			
^{79}Ge		78.9254	19.1 s	β^- /4.2	4.0/20. 4.3/80.	1/2-			0.1096/21. (0.10-2.59) 0.5427(4)/15.
^{80}Ge		79.92545	29.5 s	β^- /2.67	2.4/	0+			0.1104(4)/6. 0.2656(4)/25.
^{81m}Ge			≈7.6 s	β^- /	3.75/	1/2+			0.3362(4)/ 0.7935(4)/
^{81}Ge		80.9288	≈7.6 s	β^- /6.2	3.44/	9/2+			0.1976(4)/21. 0.3362(4)/100.
^{82}Ge		81.9296	4.6 s	β^- /4.7		0+			1.093/
^{83}Ge		82.9345	1.9 s	β^- /8.9					
^{84}Ge		83.9373	0.98 s	β^- /7.7					
^{85}Ge		84.943	0.54 s	β^- /10.					
$_{33}$As		**74.92160(2)**							
^{65}As		64.9495	≈0.19 s	β^+ /9.4					
^{66}As		65.94410	95.8 ms	β^+ /9.55					
^{67}As		66.9392	42. s	β^+ /6.0 EC/	5.0/	5/2-			0.121/ 0.123/ 0.244/
^{68}As		67.9368	2.53 m	β^+ /8.1		3+			ann.rad./ 0.652/32. 0.762/33. 1.016/77. (0.61-3.55)
^{69}As		68.93228	15.2 m	β^+ /98/4.01 EC/2/	2.95/	5/2-	1.2		ann.rad./ 0.0868(5)/1.5 0.1458(3)/2.4
^{70}As		69.93093	52.6 m	β^+ /84/6.22 EC/16/2.14 /2.89	1.44/	4+	+2.1061	+0.09	ann.rad./ 1.0395(7)/82. (0.17-4.4)/
^{71}As		70.927114	2.72 d	β^+ /32/2.013 EC/68/		5/2-	+1.6735	-0.02	ann.rad./ 0.1749(2)/84. 1.0957(2)/4.2
^{72}As		71.926753	26.0 h	β^+ /77/4.356	0.669/5. 1.884/12. 2.498/62. 3.339/19.	2-	-2.1566	-0.08	ann.rad./ 0.83395(5)/80. 1.0507(1)/9.6 (0.1-4.0)
^{73}As		72.923825	80.3 d	EC/0.341		3/2-			0.013263(1)/0.1 0.053437(1)/10.5 Se k x-ray/90.
^{74}As		73.923829	17.78 d	β^+ /31/2.562 EC/37/ β^- /1.353	0.94/26. 1.53/3. 0.71/16. 1.35/16.	2-	-1.597		ann.rad./ 0.59588(1)/60. 0.6084(1)/0.6 0.6348(1)/15.
^{75}As	100.	74.921597				3/2-	+1.43947	+0.31	
^{76}As		75.922394	26.3 h	β^- /2.962	0.54/3. 1.785/8. 2.410/36. 2.97/51.	2-	-0.906		0.5591(1)/45. 0.65703(5)/6.2 1.21602(1)/3.4 (0.3-2.67)
^{77}As		76.920648	38.8 h	β^- /0.683	0.70/98.	3/2-			0.2391(2)/1.6 0.2500(3)/0.4 0.52078(1)/0.43
^{78}As		77.92183	1.512 h	β^- /4.21	3.00/12. 3.70/17. 4.42/37.	2-			0.6136(3)/54. 0.6954(3)/18. 1.3088(3)/10.
^{79}As		78.92095	9.0 m	β^- /2.28	1.80/95.	3/2-			0.0955(5)/16. 0.3645(5)/1.9
^{80}As		79.92258	16. s	β^- /5.64	3.38/	1+			0.6662(2)/42. (2.5-3.0)
^{81}As		80.92213	33. s	β^- /3.856		3/2-			0.4676(2)/20. 0.4911(2)/8.

Elem. or Isot.	Natural Abundance (%)	Atomic Mass or Weight	Half-Life	Decay Mode/Energy (/MeV)	Particle Energy /Intensity (MeV / %)	Spin (h/2π)	Nuclear Magnetic Mom.(nm)	Elect. Quadr. Mom. (b)	γ-ray Energy /Intensity (MeV / %)
^{82m}As			13.7 s	β^- /	3.6/	5-			0.6544(1)/72. 0.8186(4)/27. 1.7313(2)/27. 1.8954(2)/38.
^{82}As		81.9246	19. s	β^- /7.4	7.2/80.	1+			0.6544(1)/15.
^{83}As		82.9250	13.4 s	β^- /5.5					0.7345/100. 1.1131/34. 2.0767/28.
^{84m}As			0.6 s	β^-					
^{84}As		83.9291	4. s	β^-, n/7.2		1-			0.6671(2)/21. 1.4439(5)/49. (0.325-5.150)
^{85}As		84.9318	2.03 s	β^-, n/8.9		3/2-			0.667(1)/42. 1.4551(2)/100.
^{86}As		85.9362	0.95 s	β^-, n/11.4					0.704/
^{87}As		86.9396	0.49 s	β^-, n/10.					0.704/
$_{34}$Se		**78.96(3)**							
^{65}Se		64.965	0.011 s	β^+ /60/14. β^+, p	 3.55/				
^{67}Se		66.9501	0.11 s	β^+ /10.2					ann.rad./ 0.352
^{68}Se		67.9419	36. s	β^+ /4.7					ann.rad./ (0.050-0.426)
^{69}Se		68.93956	27.4 s	β^+ /6.78 EC/	5.006/				ann.rad./ 0.0664(4)/27. 0.0982(4)/63.
^{70}Se		69.9335	41.1 m	β^+ /2.4		0+			ann.rad 0.04951(5)/35. 0.4262(2)/29.
^{71}Se		70.9319	4.7 m	β^+ /4.4 EC/	3.4/36.	5/2-			ann.rad 0.1472(3)/47. 0.8309(3)/13. 1.0960(3)/10.
^{72}Se		71.92711	8.5 d	EC/0.34		0+			0.0460(2)/57.
^{73m}Se			40. m	I.T./73/0.0257 β^+ /27/2.77	0.85 1.45/ 1.70/	3/2-			ann.rad. 0.0257(2)/27. 0.2538(1)/2.5
^{73}Se		72.92678	7.1 h	β^+ /65/2.74 EC/35/	0.80/ 1.32/95. 1.68/1.	9/2+			ann.rad 0.0670(1)/72. 0.3609(1)/97. (0.6-1.5)
^{74}Se	0.89(2)	73.922477				0+			
^{75}Se		74.922524	119.78 d	EC/0.864		5/2+	0.67	1.0	0.136000(1)/55. 0.264651(1)/58.
^{76}Se	9.36(11)	75.919214				0+			
^{77m}Se			17.4 s	I.T./		7/2+			0.1619(2)/52.
^{77}Se	7.63(6)	76.919915				1/2-	+0.53506		
^{78}Se	23.78(9)	77.917310				0+			
^{79m}Se			3.92 m	I.T./					0.09573(3)/9.5
^{79}Se		78.918500	6.5x10^4 y	β^-/0.151		7/2+	-1.018	+0.8	
^{80}Se	49.61(10)	79.916522				0+			
^{81m}Se			57.3 m	I.T./99/0.1031		7/2+			0.1031(3)/9.7 0.2602(2)/0.06 0.27599(1)/0.06
^{81}Se		80.917993	18.5 m	β^-/1.585	1.6/98.	1/2-			0.27594(5)/0.85 0.29008(5)/0.75 0.82827(5)/0.32
^{82}Se	8.73(6)	81.916700				0+			
^{83m}Se			1.17 m	β^- /3.96	2.88/ 3.92/	1/2-			0.35666(6)/17. 0.9879(1)/15. 1.0305(1)/21. 2.0514(2)/11. (0.19-3.1)
^{83}Se		82.919119	22.3 m	β^- /3.668	0.93/ 1.51/	9/2+			0.22516(6)/33. 0.35666(6)/69. 0.51004(8)/45. (0.21-2.42)
^{84}Se		83.91847	3.3 m	β^- /1.83	1.41/100.	0+			0.4088(5)/100.
^{85}Se		84.92225	32. s	β^- /6.18	5.9/	5/2+			0.3450(1)/22. 0.6094(1)/41.
^{86}Se		85.92428	15. s	β^- /5.10		5/2+			2.0124(1)/24. 2.4433(8)/100. 2.6619(1)/49.
^{87}Se		86.92853	5.4 s	β^- /7.28 n/					0.468(1)/100. 1.4979(1)/23.
^{88}Se		87.93143	1.5 s	β^- ,n/6.85					0.5346/
^{89}Se		88.9360	0.41s	β^- ,n/9.0					

Elem. or Isot.	Natural Abundance (%)	Atomic Mass or Weight	Half-Life	Decay Mode/Energy (/MeV)	Particle Energy /Intensity (MeV / %)	Spin (h/2π)	Nuclear Magnetic Mom.(nm)	Elect. Quadr. Mom. (b)	γ-ray Energy /Intensity (MeV / %)
^{91}Se		90.945	0.27 s	β^- ,n/8.					
$_{35}$**Br**		**79.904(1)**							
^{69}Br		68.9499		β^+ /9.6					
^{70}Br		69.9442	0.08 s	β^+ /10.0					
^{71}Br		70.9393	21. s	β^+ /6.9					
^{72}Br		71.9365	1.31 m	β^+ /8.7		3	≈0.55		0.4547-1.3167
^{73}Br		72.9318	3.4 m	β^+ /4.7	3.7/	3/2-			ann.rad (0.0649-0.6995)
^{74m}Br			46. m	β^+ /	4.5/	4-	1.8		ann.rad 0.6348 0.7285 (0.2 - 4.38)
^{74}Br		73.92989	25.4 m	β^+ /6.91					ann.rad 0.6341 0.6348 (0.2-4.7)
^{75}Br		74.92578	1.62	β^+ /76/3.03		3/2-	0.75(11)		ann.rad 0.28650 (0.1-1.56)
^{76m}Br			1.4 s	I.T./5.05		4+			0.104548 0.05711
^{76}Br		75.92454	16.0 h	β^+ /57 /4.96	1.9/ 3.68/	1-	0.54821	0.270	ann.rad 0.55911 1.85368 (0.4-4.6)
^{77m}Br			4.3 m	I.T./0.1059		9/2+	0.97		0.1059
^{77}Br		76.921380	2.376 d	EC/99 /1.365		3/2-			ann.rad. 0.23898 0.52069 (0.08-1.2)
^{78}Br		77.921146	6.45 m	β^+ /92/3.574 EC/8 /	1.2/ 2.5/	1+	0.1		ann.rad. 0.61363 (0.7-3.0)
^{79m}Br			4.86 s	I.T./0.207		9/2+	+2.106		0.2072
^{79}Br	50.69(7)	78.918338				3/2-	+2.106400	+0.331	
^{80m}Br			4.42 h	I.T./0.04885		5-	+1.3177		Br k x-ray 0.03705/39.1 0.04885/0.3
^{80}Br		79.918530	17.66 m	β^- /92 /2.004 EC/5.7/1.8706 β^+ /2.6/	1.38 β^- /7.6 1.99 β^- /82 0.85 β^+ /2.8	1+	0.5140	0.196	ann.rad. 0.6169/6.7 (0.64-1.45)
^{81}Br	49.31(7)	80.916291				3/2-	+2.270562	+0.276	
^{82m}Br			6.1 m	I.T./98/0.046 β^- /2 /3.139		2-			0.046/0.24 (0.62-2.66)
^{82}Br		81.916805	1.471 d	β^- /3.093	0.444/	5-	+1.6270	0.751	0.5544/71 0.61905/43 0.77649/84 (0.013-1.96)
^{83}Br		82.915181	2.40 h	β^- /0.972	0.395/1 0.925/99	3/2-			0.52964 (0.12-0.68)
^{84m}Br			6.0 m	β^- /4.97	2.2/100	(6-)			0.4240/100 0.8817/98 1.4637/101
^{84}Br		83.91651	31.8 m	β^- /4.65	2.70/11 3.81/20 4.63/34	2-	2.		0.8816/41 1.8976/13 (0.23-4.12)
^{85}Br		84.91561	2.87 m	β^- /2.87	2.57	3/2-			0.80241/2.56 0.92463/1.6 (0.09-2.4)
^{86}Br		85.91880	55.5 s	β^- /7.63	3.3 7.4	(2-)			1.56460/64 2.75106/21 (0.5-6.8)
^{87}Br		86.92072	55.6 s	β^- /6.85 n/	6.1/	3/2-			1.41983 1.4762 (0.2-6.1)
^{88}Br		87.92407	16.3 s	β^- /8.96 n/		1-			0.7649 0.7753 0.8021 (0.1-6.99)
^{89}Br		88.92640	4.35 s	β^- /8.16 n/		3/2-			0.7753 1.0978
^{90}Br		89.9306	1.91 s	β^- /10.4 n/	8.3/ 9.8/	2-			0.6555 0.7071 1.3626
^{91}Br		90.9339	0.54 s	β^- /90 /9.80 β^- n/10 /					0.263 0.803

Elem. or Isot.	Natural Abundance (%)	Atomic Mass or Weight	Half-Life	Decay Mode/Energy (/MeV)	Particle Energy /Intensity (MeV / %)	Spin (h/2π)	Nuclear Magnetic Mom.(nm)	Elect. Quadr. Mom. (b)	γ-ray Energy /Intensity (MeV / %)
^{92}Br		91.9392	0.31 s	β^- /12.20 β^- n/					0.740
^{93}Br		92.9431	0.10 s	β^- n/11.1					
^{94}Br			0.07 s	β^- n/					
$_{36}$Kr		**83.80(1)**							
^{71}Kr		70.9505	0.10 s	β^+ ,EC/10.5					
^{72}Kr		71.9419	17. s	β^+ /5.0 EC/		0+			ann.rad 0.3100/29 0.4150/36 (0.12-0.58)
^{73}Kr		72.9389	27. s	β^+ /6.7 EC/ β^+ ,p/					ann.rad. 0.1781/66 (0.06-0.86)
^{74}Kr		73.9333	11.5 m	β^+ /3.1 EC/		0+			ann.rad. 0.08970/31 0.2030/20 (0.010-1.06)
^{75}Kr		74.93104	4.3 m	β^+ /4.90 EC/	3.2/				ann.rad. 0.1325/68 0.1547/21 (0.02-1.7)
^{76}Kr		75.92595	14.8 h	EC/1.31		0+			Br k x-ray 0.270/21 0.3158/39 (0.03-1.07)
^{77}Kr		76.92467	1.24 h	β^+ /80 /3.06 EC/20 /	1.55/ 1.70/ 1.87/	5/2+			ann.rad. 0.1297/80 0.1465/38 (0.02-2.3)
^{78}Kr	0.35(2)	77.92039				0+			
^{79m}Kr			53. s	I.T./0.1299		7/2+			Kr x-ray
^{79}Kr		78.920083	1.455 d	β^+ /7 /1.626 EC/93 /		1/2-			ann.rad. 0.2613/13 0.39756/19 0.6061/8 (0.04-1.3)
^{80}Kr	2.25(2)	79.916379				0+			
^{81m}Kr			13.1 s	I.T./0.1904		1/2-			0.1904
^{81}Kr		80.916593	2.1x10^5 y	EC/0.2807		7/2+			Br k x-ray 0.2760
^{82}Kr	11.6(1)	81.913485				0+			
^{83m}Kr			1.86 h	I.T./0.0416		1/2-			Kr k x-ray 0.00940 0.03216
^{83}Kr	11.5(1)	82.914137				9/2+	-0.970699	+0.253	
^{84}Kr	57.0(3)	83.911508						0+	
^{85m}Kr			4.48 h	β^- /79 / I.T./21 /0.305	0.83/79	1/2-			0.30487 0.15118
^{85}Kr		84.912530	10.73 y	β^- /0.687	0.15/0.4	9/2+	1.005	+0.43	0.51399
^{86}Kr	17.3(2)	85.910615				0+			
^{87}Kr		86.913359	1.27 h	β^- /3.887	1.33/8 3.49/43 3.89/30	5/2+	-1.018		0.40258/49.6 2.5548/9.2 (0.13-3.31)
^{88}Kr		87.91445	2.84 h	β^- /2.91		0+			0.19632/26. 2.392/34.6 (0.03-2.8)
^{89}Kr		88.91764	3.15 m	β^- /4.99	3.8/ 4.6/ 4.9/	5/2+			0.19746 0.2209/19.9 0.5858/16.4 [illegible] (0.2-4.7)
^{90}Kr		89.91953	32.3 s	β^- /4.39	2.6/77 2.8/6	0+			0.12182/32.9 0.5395/28.6 1.1187/36.2 (0.1 - 4.2)
^{91}Kr		90.9234	8.6 s	β^- /6.4	4.33/ 4.59/	5/2+			0.10878/43.5 0.50658/19. (0.2-4.4)
^{92}Kr		91.92611	1.84 s	β^- /5.99 n/					0.1424/66. (0.14 - 3.7)
^{93}Kr		92.9312	1.29 s	β^- /8.6 n/	7.1/				0.1820 0.2534/42. 0.32309/24.6 (0.057-4.03)

Elem. or Isot.	Natural Abundance (%)	Atomic Mass or Weight	Half-Life	Decay Mode/Energy (/MeV)	Particle Energy /Intensity (MeV / %)	Spin (h/2π)	Nuclear Magnetic Mom.(nm)	Elect. Quadr. Mom. (b)	γ-ray Energy /Intensity (MeV / %)
^{94}Kr		93.9343	0.21 s	β^- /7.3					0.2196/67 0.6293/100. (0.098-0.985)
^{95}Kr		94.9397	0.78 s	β^- /9.7					
^{97}Kr			< 0.1 s	β^-					
$_{37}$Rb		**85.4678(3)**							
^{74}Rb		73.9445	65. ms	β^+ /10.4					
^{75}Rb		74.93857	19. s	β^+ /7.02	2.31/				ann. rad. 0.179
^{76}Rb		75.93508	39. s	β^+ /8.50	4.7/	1-	-0.372623	+0.4	ann.rad. 0.4240/92. (0.064-1.68)
^{77}Rb		76.93041	3.8 m	β^+ /5.34	3.86/	3/2-	+0.654468	+0.70	ann.rad. 0.0665/59 (0.04 - 2.82)
^{78m}Rb			5.7 m	I.T./0.1034 β^+ / EC/	 3.4	4-			ann.rad. 0.4553/81. (0.103-4.01)
^{78}Rb		77.92814	17.7 m	β^+ /7.22 EC/		0+			ann.rad. 0.4553/63. (0.42-5.57)
^{79}Rb		78.92400	23. m	β^+ /84 /3.65 EC/16 /		5/2+	+0.3358	-0.10	ann.rad. 0.68812/23. (0.017-3.02)
^{80}Rb		79.92252	34. s	β^+ /5.72	4.1/22 4.7/74	1+	-0.0836	+0.35	ann.rad. 0.6167/25.
^{81m}Rb			30.5 m	I.T./0.85 β^+,EC/	1.4	9/2+			ann.rad. (0.085-1.9)
^{81}Rb		80.91900	4.57 h	β^+/27 /2.24 EC/73	1.05/	3/2-	+2.060	+0.40	ann.rad./ 0.19030/64. (0.05 - 1.9)
^{82m}Rb			6.47 h	β^+/26 / EC/74 /	0.80/	5-	+1.5		ann.rad./ 0.5544/63. 0.7765/85. (0.092 - 2.3)
^{82}Rb		81.91821	1.258 m	β^+/96 /4.40 EC/4 /	3.3/	1+	+0.554508	+0.19	ann.rad./ 0.7665/13. (0.47 - 3.96)
^{83}Rb		82.91511	86.2 d	EC/0.91		5/2-	+1.425	+0.20	Kr x-ray 0.5205/46. (0.03-0.80)
^{84m}Rb			20.3 m	I.T./0.216		6-			0.2163/34. 0.2482/63. 0.4645/32.
^{84}Rb		83.914387	32.9 d	β^+/22 /2.681 EC/75 / β^-/3 /0.894	0.780/11 1.658/11 0.893/	2-	-1.32412	-0.015	ann.rad./ 0.8817/68. (1.02-1.9)
^{85}Rb	72.165(20)	84.911792				5/2-	+1.353	+0.23	
^{86m}Rb			1.018 m	I.T./0.5560		6-			0.556/98.
^{86}Rb		85.911170	18.65 d	β^-/1.775	1.774/8.8	2-	-1.6920		1.0768/8.8
^{87}Rb	27.835(20)	86.909186	4.88×10^{10}y	β^-/0.283	0.273/100	3/2-	+2.7512		
^{88}Rb		87.911323	17.7 m	β^-/5.316	5.31	2-	0.508		0.8980/14. 1.8360/21. (0.34-4.85)
^{89}Rb		88.91229	15.4 m	β^-/4.50	1.26/38 1.9/5 2.2/34 4.49/18	3/2-	+2.304	+0.14	1.032/58. 1.248/42. 2.1960/13 (0.12-4.09)
^{90m}Rb			4.3 m	β^-/4.50	1.7/ 6.5/	4-			0.1069(IT) 0.8317/94 (0.20-5.00)
^{90}Rb		89.91481	2.6 m	β^-/6.59	6.6	1-			0.8317/28. (0.31-5.60)
^{91}Rb		90.91649	58.0 s	β^-/5.861	5.9	3/2-	+2.182	+0.15	0.0936/34. (0.35-4.70)
^{92}Rb		91.91968	4.48 s	β^-/8.11	8.1/94	1-			0.8148/8. (0.1-6.1)
^{93}Rb		92.92195	5.85 s	β^-/7.46 n/1	7.4/	5/2	+1.410	+0.18	0.2134/4.8 0.4326/12.5 0.9861/4.9 (0.16-5.41)
^{94}Rb		93.92643	2.71 s	β^-/10.31 n/10	9.5/	3	+1.498	+0.16	0.8369/87. 1.5775/32. (0.12-6.35)
^{95}Rb		94.92929	0.377 s	β^-/9.30 n/8	8.6/	5/2	+1.334	+0.21	0.352/65. 0.680/22. (0.20-2.27)

Elem. or Isot.	Natural Abundance (%)	Atomic Mass or Weight	Half-Life	Decay Mode/Energy (/MeV)	Particle Energy /Intensity (MeV / %)	Spin (h/2π)	Nuclear Magnetic Mom.(nm)	Elect. Quadr. Mom. (b)	γ-ray Energy /Intensity (MeV / %)
^{96}Rb		95.93427	0.199 s	β^-/11.76	10.8/	2+	+1.466	+0.25	0.815/76.
				n/13/					(0.20-5.42)
^{97}Rb		96.93733	0.169 s	β^-/10.42	10.0	3/2	+1.841	+0.58	0.167/100.
				n/27/					0.585/79.
									0.599/56.
									1.258/52.
									(0.14-2.08)
^{98}Rb		97.94174	0.107 s	β^-/12.34	0.144/				
				n/13					(0.07-3.68)
^{99}Rb		98.9453	59. ms	β^-/11.3					
^{100}Rb		99.9499	53. ms	β^- /13.5					
^{101}Rb		100.9532	0.03 s	β^- /11.8					
^{102}Rb			0.09 s	β^-					
$_{36}$Sr		**87.62(1)**							
^{76}Sr		75.9416	8.9 s	β^+ /6.1					
^{77}Sr		76.9378	9.0 s	β^+ /6.9	5.6		-0.35	1.4	0.147
^{78}Sr		77.93218	2.7 m	β^+ /3.76					(0.047-0.793)
^{79}Sr		78.92971	2.1 m	β^+ /5.32	4.1	3/2-	-0.474	+0.74	ann.rad./
									0.039/28.
									0.105/22.
									(0.135-0.612)
^{80}Sr		79.92453	1.77 h	β^+ /1.87		0+			ann.rad./
									0.174/10.
									0.589/39.
									(0.24-0.55)
^{81}Sr		80.92322	22.3 m	β^+ /87 /3.93	2.43/	1/2-	+0.544		ann.rad./
				EC/13 /	2.68/				0.148/31.
									0.1534/35
									(0.06-1.7)
^{82}Sr		81.91840	25.36 d	EC/0.18					Rb x-ray
^{83m}Sr			5.0 s	I.T./0.2591		1/2-			0.2591/87.5
^{83}Sr		82.91756	1.350 d	β^+/24 /2.28	0.465/	7/2+	-0.898	+0.8	ann.rad./
				EC/76/	0.803/				0.3816/12.
					1.227/				0.3816
									0.7627/30.
									(0.094-2.15)
^{84}Sr	0.56(1)	83.913426				0+			
^{85m}Sr			1.127 h	I.T./87 /0.2387		1/2-			0.2318/84.
				EC/13					(0.15-0.24)
^{85}Sr		84.912936	64.85 d	EC/1.065		9/2+	-1.001	+0.32	0.51399/99.3
^{86}Sr	9.86(1)	85.909265				0+			
^{87m}Sr			2.81 h	I.T./0.3884		1/2-			0.3884(IT)
^{87}Sr	7.00(1)	86.908882				9/2+	-1.09360	+0.34	
^{88}Sr	82.58(1)	87.905617				0+			
^{89}Sr		88.907455	50.52 d	β^-/1.497	1.492/100	5/2+	-1.149	0.3	0.9092
^{90}Sr		89.907738	29.1 y	β^-/0.546	0.546/100	0+			
^{91}Sr		90.91020	9.5 h	β^-/2.70	0.61/7	5/2+	-0.887	+0.044	0.5556/61.
					1.09/33				0.7498/24.
					1.36/29				1.0243/33.
					2.66/26				(0.12-2.4)
^{92}Sr		91.91098	2.71 h	β^-/1.91	0.55/96	0+			1.3831/90.
					1.5/3				(0.24-1.1)
^{93}Sr		92.91394	7.4 m	β^-/4.08	2.2/10	5/2+	-0.794	+0.27	0.5903/
					2.6/25				0.7104
					3.2/65				0.87573
									0.8883/
									(0.17-3.97)
^{94}Sr		93.91537	1.25 m	β^-/3.511	2.1/	0+			0.6219
					3.3/				0.7043
									0.7241
									0.8064
									1.4283
^{95}Sr		94.91931	25.1 s	β^-/6.08		1/2+	-0.5379		0.6859
					6.1/50				0.8269
									2.7173
									2.9332
^{96}Sr		95.92165	1.06 s	β^-/5.37	4.2/	0+			0.1222
									0.5305
									0.8094
									0.9318
^{97}Sr		96.92615	0.42 s	β^-/7.47	5.3	(1/2+)	-0.500		0.2164
									0.3071
									0.6522
									0.9538
									1.2580
									1.9050

Elem. or Isot.	Natural Abundance (%)	Atomic Mass or Weight	Half-Life	Decay Mode/Energy (/MeV)	Particle Energy /Intensity (MeV / %)	Spin (h/2π)	Nuclear Magnetic Mom.(nm)	Elect. Quadr. Mom. (b)	γ-ray Energy /Intensity (MeV / %)
^{98}Sr		97.92845	0.65 s	β^-/5.83	5.1				0.0365 0.1190 0.4286 0.4447 0.5636
^{99}Sr		98.9333	0.27 s	β^-/8.0					
^{100}Sr		99.9354	0.201 s	β^-/7.1					
^{101}Sr		100.9405	0.115 s	β^-/9.5					
^{102}Sr		101.9430	68. ms	β^-/8.8					
$_{39}$Y		**88.90585(2)**							
^{78}Y		77.9435		β^+/10.5					
^{79}Y		78.9374	15. s	β^+/7.1					
^{80}Y		79.9320	36. s	β^+/7.0	5.5 5.0/	(4)			(0.152-1.106) ann.rad./ 0.3858 0.5951 1.1852
^{81}Y		80.9291	1.21 m	β^+/5.5	3.7/ 4.2/				ann.rad./ 0.428 0.469
^{82}Y		81.9268	9.5 s	β^+/7.8	6.3/	1+			ann.rad./ 0.5736 0.6017 0.7375
^{83m}Y			2.85 m	β^+/95/4.6 EC/5 /	2.9	1/2-			ann.rad./ 0.2591 0.4218 0.4945
^{83}Y		82.92235	7.1 m	β^+/4.47 EC/	3.3	9/2+			ann.rad./ 0.0355 0.4899 0.8821
^{84m}Y			4.6 s	β^+/ EC/		1+			(0.03 - 3.4) ann.rad./ 0.7930
^{84}Y		83.9203	40. m	β^+/6.4 EC/	1.64/47 2.24/25 2.64/21 3.15/7	5-			ann.rad./ 0.4628 0.6606 0.7931 0.9744 1.0398
^{85m}Y			4.9 h	β^+/70 / EC/30 /		9/2+			(0.2 - 3.3) ann.rad./ 0.2317 0.5356 0.7673 2.1238
^{85}Y		84.91643	2.6 h	β^+/55 /3.26 EC/45 /	1.54/	1/2-			(0.1 - 3.1) ann.rad./ 0.2317 0.5045 0.9140
^{86m}Y			48. m	I.T./99 / β^+/ EC/		8+			(0.07 - 1.4) ann.rad./ 0.0102(IT) 0.2080
^{86}Y		85.91489	14.74 h	β^+/5.24 EC/		4-	<0.6		(0.09 - 1.1) ann.rad./ 0.3070 0.6277 1.0766 1.1531 1.9207
^{87m}Y			13. h	I.T./98 / β^+/0.7 / EC/	 1.15/0.7	9/2+			(0.1 - 3.8) 0.3807
^{87}Y		86.910880	3.35 d	EC/99+/1.862	0.78/	1/2-			0.3880 0.4870
^{88}Y		87.909506	106.6 d	EC/99+ /3.623 β^+/0.2 /	0.76/	4-			ann.rad./ 0.89802 1.83601 2.73404 3.2190
^{89m}Y			15.7 s	I.T./0.909		9/2+			0.9092(IT)
^{89}Y	100.	88.905849				1/2-		-0.13742	

Elem. or Isot.	Natural Abundance (%)	Atomic Mass or Weight	Half-Life	Decay Mode/Energy (/MeV)	Particle Energy /Intensity (MeV / %)	Spin (h/2π)	Nuclear Magnetic Mom.(nm)	Elect. Quadr. Mom. (b)	γ-ray Energy /Intensity (MeV / %)
^{90m}Y			3.24 h	I.T./99+ /0.68204 β^-/0.002/		7+			0.2025 0.4794 0.6820
^{90}Y		89.907152	2.67 d	β^-/2.282	2.28/	2-	-1.630	-0.155	
^{91m}Y			49.7 m	I.T./0.555		9/2+	6.0		0.5556(IT)
^{91}Y		90.907301	58.5 d	β^-/1.544	1.545/	1/2-	0.1641		1.208
^{92}Y		91.90893	3.54 h	β^-/3.63	3.64/	2-			0.4485 0.5611 0.9345 1.4054 (0.4 - 3.3)
^{93m}Y			0.82 s	I.T./0.759		9/2+			0.1686(IT) 0.5902
^{93}Y		92.90956	10.2 h	β^-/2.87	2.88/90	1/2-			0.2669 0.9471 1.9178
^{94}Y		93.91160	18.7 m	β^-/4.919	4.92/	2-			0.3816 0.9188 1.1389 (0.3 - 4.1)
^{95}Y		94.91279	10.3 m	β^-/4.42		1/2-			0.4324 0.9542 2.1760 3.5770
^{96m}Y			9.6 s	β^-/		(3+)			0.1467 0.6174 0.9150 1.1071 1.7507
^{96}Y		95.91588	6.2 s	β^- /7.09	7.12/	0-			1.594
^{97m}Y			1.21 s	β^- /7.4	4.8/ 6.0/	9/2+			0.1614 0.9700 1.1030
^{97}Y		96.91813	3.76 s	β^- /6.69	6.7	1/2-			0.2969 1.9960 3.2876 3.4013
^{98m}Y			2.1 s	β^- /9.8	5.5/	(4-)			0.2415 0.6205 0.6473 1.2228 1.8016
^{98}Y		97.92224	0.59 s	β^- /8.83	8.7/	1+			0.2131 1.2228 1.5907 2.9413 4.4501
^{99}Y		98.92463	1.47 s	β^- /7.57		1/2-			0.1218/43.8 0.5362 0.7242 1.0130
^{100m}Y			0.94 s	β^- /		3+			
^{100}Y		99.9278	0.73 s	β^- /9.3		1+			
^{101}Y		100.9303	0.43 s	β^- /8.6		(5/2)			
^{102}Y		101.9336	0.36 s	β^- /9.9					
$_{40}$Zr		**91.224(2)**							
^{80}Zr		79.9406		β^+ /8.0					0.290 0.538
^{81}Zr		80.9368	15. s	β^+ /7.2	6.1				
^{82}Zr		81.9311	32. s	β^+ /4.0	3				ann.rad./
^{83m}Zr			7. s	β^+ /7.0		(7/2+)			ann.rad./
^{83}Zr		82.9287	44. s	β^+ /5.9 EC	4.8	(1/2-)			ann.rad./ 0.0556 0.1050 0.2560 0.474 1.525
^{84}Zr		83.9233	26. m	β^+ /2.7 EC/		0+			ann.rad./ 0.0449 0.1125 0.3729 0.667
^{85m}Zr			10.9 s	I.T./0.2922 β^+ ,EC/		1/2-			ann.rad./ 0.2922(IT) 0.4165

Elem. or Isot.	Natural Abundance (%)	Atomic Mass or Weight	Half-Life	Decay Mode/Energy (/MeV)	Particle Energy /Intensity (MeV / %)	Spin (h/2π)	Nuclear Magnetic Mom.(nm)	Elect. Quadr. Mom. (b)	γ-ray Energy /Intensity (MeV / %)
^{85}Zr		84.9215	7.9 m	β⁺ /4.7 EC/	3.1	7/2+			ann.rad./ 0.2663 0.4163 0.4543
^{86}Zr		85.91647	16.5 h	EC/1.47		0+			0.0280 0.243 0.612
^{87m}Zr			14.0 s	I.T./0.3362		1/2-			0.1352(IT) 0.2010
^{87}Zr		86.91482	1.73 h	β⁺ /3.67 EC/	2.26	9/2+			ann.rad./ 0.3811 1.228
^{88}Zr		87.91023	83.4 d	EC/0.67		0+			0.3929
^{89m}Zr			4.18 m	I.T./94 /0.5877 β⁺ /1.5 / EC/4.7 /		1/2-			ann.rad./ 0.5877(IT) 1.507
^{89}Zr		88.908889	3.27 d	β⁺ /23 /2.832 EC/77 /	0.9/	9/2+			ann.rad./ 0.9092
^{90m}Zr			0.809 s	I.T./		5-			0.1326 2.1862 2.3189(IT)
^{90}Zr	51.45(3)	89.904702				0+			
^{91}Zr	11.22(4)	90.905643				5/2+	-1.30362	-0.21	
^{92}Zr	17.15(2)	91.905039				0+			
^{93}Zr		92.906474	1.5x10^{6}y	β⁻ /0.091		5/2+			0.0304
^{94}Zr	17.38(4)	93.906314				0+			
^{95}Zr		94.908041	64.02 d	β⁻ /1.125	0.366/55 0.400/44	5/2+	1.10	+0.29	0.7242 0.7567
^{96}Zr	2.80(2)	95.908275				0+			
^{97}Zr		96.910950	16.8 h	β⁻ /2.658	1.91/	1/2-			0.7434
^{98}Zr		97.91276	30.7 s	β⁻ /2.26	2.2/100	0+			
^{99}Zr		98.91651	2.2 s	β⁻ /4.56	3.9/ 3.5/	1/2+			0.4692/55.2 0.5459 0.5940
^{100}Zr		99.91776	7.1 s	β⁻ /3.34		0+			0.4006 0.5043
^{101}Zr		100.92114	2.1 s	β⁻ /5.49	6.2/	3/2-			0.1194 0.2057 0.2089
^{102}Zr		101.92298	2.9 s	β⁻ /4.61					
^{103}Zr		102.9266	1.3 s	β⁻ /7.0					
^{104}Zr		103.9288	1.2 s	β⁻ /5.9					
^{105}Zr		104.9331	≈1. s	β⁻ /8.5					
$_{41}$Nb		**92.90638(2)**							
^{82}Nb		81.9431		β⁺ /11.					
^{83}Nb		82.9367		β⁺ /7.5					
^{84}Nb		83.9336	12. s	β⁺ ,EC/9.6		(3+)			
^{85}Nb		84.9279	2.3 m	β⁺ /6.0					
^{86m}Nb			56. s	β⁺					
^{86}Nb		85.9250	1.46 m	β⁺ /8.0					ann.rad./ 0.751 1.003
^{87m}Nb			3.7 m	β⁺ / EC/		1/2-			ann.rad./ 0.1352 0.2010
^{87}Nb		86.92036	2.6 m	β⁺ 5.2/ EC/		(9/2+)			ann.rad./ 0.2010 0.4706 0.6165 1.0665 1.8842
^{88m}Nb			7.7 m	β⁺ / EC/		4-			ann.rad./ 0.2625 0.3996 1.0569 1.0825
^{88}Nb		87.9183	14.3 m	β⁺ /7.6 EC/	3.2/	8+			ann.rad./ 1.0570 1.0828 (0.07 - 2.5)
^{89m}Nb			2.0 h	β⁺ / EC/	3.3/	9/2+			0.5880/10(D) (0.17 - 4.0)
^{89}Nb		88.91349	1.10 h	β⁺ /74 /4.29 EC/26 /	2.8/	1/2-			ann.rad./ 0.5074 0.5880 0.7696 1.2775

Elem. or Isot.	Natural Abundance (%)	Atomic Mass or Weight	Half-Life	Decay Mode/Energy (/MeV)	Particle Energy /Intensity (MeV / %)	Spin ($h/2\pi$)	Nuclear Magnetic Mom.(nm)	Elect. Quadr. Mom. (b)	γ-ray Energy /Intensity (MeV / %)
^{90m}Nb			18.8 s	I.T./0.1246		4-			0.002 0.1225
^{90}Nb		89.911263	14.6 h	β^+ /53 /6.111 EC/47 /	0.86/5 1.5/92	8+	4.961		ann.rad./ 0.1412 1.1292 2.1862 2.3189 (0.1 - 3.3)
^{91m}Nb			62. d	I.T./97 / EC/3 /		1/2-			0.1045(IT) 1.2050
^{91}Nb		90.906989	7x10^{2}y	EC/1.253		9/2+			Mo k x-ray
^{92m}Nb			10.13 d	EC/99+ /		2+	6.114		0.9126 0.9345 1.8475
^{92}Nb		91.907192	3.7x10^{7}y	EC/2.006		7+			0.5611 0.9345
^{93m}Nb			16.1 y	I.T./0.0304		1/2-			Nb x-ray 0.0304
^{93}Nb	100.	92.906376				9/2+	+6.1705	-0.32	
^{94m}Nb			6.26 m	I.T./99+ /2.086 β^- /0.5 /		3+			Nb k x-ray 0.0409 0.87109
^{94}Nb		93.907282	2.4x10^{4}y	β^- /2.045	0.47/	6+			0.70263 0.87109
^{95m}Nb			3.61 d	I.T./97.5 /0.2357 β^- /2.5 /		1/2-			0.2040 0.2356
^{95}Nb		94.906834	34.97 d	β^- /0.926	0.160/	9/2+	6.141		0.76578
^{96}Nb		95.908099	23.4 h	β^- /3.187	0.5/10 0.75/90	6+	4.976		0.7782 0.2191-1.498
^{97m}Nb			58.1 s	I.T./0.7434	0.734/98	1/2-			0.7434
^{97}Nb		96.908096	1.23 h	β^- /1.934	1.27/98	9/2+	6.15		0.4809 0.6579
^{98m}Nb			51. m	β^- /4.67		5+			0.7874 0.1726-1.89
^{98}Nb		97.91033	2.9 s	β^- /4.59	4.6/	1+			0.6451 0.7874 1.0243
^{99m}Nb			2.6 m	β^- /	3.2/	1/2-			0.0978/100 (0.138-3.010)
^{99}Nb		98.91162	15.0 s	β^- /3.64	3.5/100	9/2+			0.0977 0.1378/3.1
^{100m}Nb			3.0 s	β^- /6.74	5.8				Nb k x-ray 0.159 0.6364 1.0637
^{100}Nb		99.91418	1.5 s	β^- /6.25	6.2/ 5.3/				0.5354 0.6001-1.566
^{101}Nb		100.91525	7.1 s	β^- /4.57	4.3/				0.1105-0.810
^{102m}Nb			4.3 s	β^- /					
^{102}Nb		101.91804	1.3 s	β^- /7.21	7.2/				0.2960-2.184
^{103}Nb		102.91914	1.5 s	β^- /5.53	5.3/	5/2+			
^{104m}Nb			0.9 s	β^- /					
^{104}Nb		103.9225	4.8 s	β^- /8.1					
^{105}Nb		104.9239	3.0 s	β^- /6.5					
^{106}Nb		105.9281	1.0 s	β^- /9.3					
^{107}Nb			0.3 s	β^- /7.9					
$_{42}$Mo		**95.94(1)**							
^{84}Mo		83.9401		β^+ /6.					
^{85}Mo		84.9366		β^+/8.1					
^{86}Mo		85.9302	20. s	β^+ /4.8					
^{87}Mo		86.9273	14. s	EC, β^+/6.5					
^{88}Mo		87.92195	8.0 m	β^+ /3.4 EC		0+			ann.rad./ 0.0800 0.1399 0.1707
^{89m}Mo			0.19 s	I.T./0.118		1/2-			0.118(IT) 0.268
^{89}Mo		88.91948	2.2 m	β^+ /5.58 EC/		9/2+			ann.rad./ 0.659 0.803 1.155 1.272
^{90}Mo		89.91394	5.7 h	β^+ /25 /2.489 EC/75 /	1.085/	0+			ann.rad./ 0.04274 0.12237 0.25734

Elem. or Isot.	Natural Abundance (%)	Atomic Mass or Weight	Half-Life	Decay Mode/Energy (/MeV)	Particle Energy /Intensity (MeV / %)	Spin (h/2π)	Nuclear Magnetic Mom.(nm)	Elect. Quadr. Mom. (b)	γ-ray Energy /Intensity (MeV / %)
^{91m}Mo			1.08 m	I.T./50 /0.653		1/2-			ann.rad./
				β^+ ,EC/50 /	2.5/				0.6529
					2.8/				1.2081
					4.0/				1.5080
									2.2407
^{91}Mo		90.91175	15.5 m	β^+ /94 /4.43	3.44/94	9/2-			ann.rad./
				EC/6 /					1.6373
									2.6321
									3.0286
									(0.1 - 4.2)
^{92}Mo	14.84(4)	91.906810				0+			
^{93m}Mo			6.9 h	I.T./99+ /2.425	21/2+	+9.21			0.26306(IT)
									0.68461
									1.47711
^{93}Mo		92.906811	3.5x10^{3}y	EC/0.405		5/2+			0.0304
^{94}Mo	9.25(3)	93.905087				0+			
^{95}Mo	15.92(5)	94.905841				5/2+	-0.9142	-0.02	
^{96}Mo	16.68(5)	95.904678				0+			
^{97}Mo	9.55(3)	96.906020				5/2+	-0.9335	+0.26	
^{98}Mo	24.13(7)	97.905407				0+			
^{99}Mo		98.907711	2.7476 d	β^- /1.357	0.45/14	1/2+	0.375		0.144048
					0.84/2				0.18109
					1.21/84				0.36644
									0.73947
^{100}Mo	9.63(3)	99.90748				0+			
^{101}Mo		100.91035	14.6 m	β^- /2.82	2.23/	1/2+			0.0063
					0.7/				0.19193
									0.5909
									(0.0809-2.405)
^{102}Mo		101.91030	11.3 m	β^- /1.01	1.2/	0+			0.1493/89.
									0.2116/100.
									0.2243/32.
^{103}Mo		102.91320	1.13 m	β^- /3.8		3/2+			0.1028(2)/
									0.1440(2)
									0.2511(2)
^{104}Mo		103.91376	1.00 m	β^- /2.16		0+			0.0686(1)/100.
									0.4239(4)/21.
^{105}Mo		104.9170	36. s	β^- /4.95		3/2+			0.0642/
									0.0856/
									0.2495/
^{106}Mo		105.91814	8.4 s	β^- /3.52		0+			0.1894(2)/22.
									0.3644(2)/6.
									0.3723(2)/12.
^{107}Mo		106.9217	3.5 s	β^- /6.2					
^{108}Mo		107.9240	1.5 s	β^- /5.1					
^{109}Mo		108.9277	0.5 s	β^- /7.2					
^{110}Mo		109.9295	0.30 s	β^- /5.7					Tc k x-ray
									0.142
									(0.121-0.599)
$_{43}$Tc									
^{86}Tc		85.9430		β^+ /11.9					
^{87}Tc		86.9365		β^+ /8.6					
^{88}Tc		87.9328		β^+ /10.1					
^{89m}Tc			13. s						
^{89}Tc		88.9275	13. s	β^+ /7.5					
^{90m}Tc			49.2 s	β^+	5.3/	6+			ann.rad./
									0.9479/
									1.0542/
^{90}Tc		89.9235	8.3 s	β^+ /8.9	7.0/15	1+			ann.rad./
					7.9/95.				0.9479/
^{91m}Tc			3.3 m	β^+		1/2+			ann.rad./170.
				EC					0.3375(2)/1.1
									0.8110(5)/5.
									1.6052(1)/7.8
									1.6339(1)/9.1
									1.9023(1)/6.
									2.4509(1)/13.5
^{91}Tc		90.9184	3.14 m	β^+ /6.2	5.2	9/2+			ann.rad./200.
^{92}Tc		91.91526	4.4 m	β^+ /7.87	4.1	8+			ann.rad./200.
				EC					0.0850/
									0.1475
									0.3293
									0.7731
									1.5096
^{93m}Tc			43. m	I.T./13		1/2-			0.3924(IT)
				EC/20					0.9437
									2.6445

Elem. or Isot.	Natural Abundance (%)	Atomic Mass or Weight	Half-Life	Decay Mode/Energy (/MeV)	Particle Energy /Intensity (MeV / %)	Spin (h/2π)	Nuclear Magnetic Mom.(nm)	Elect. Quadr. Mom. (b)	γ-ray Energy /Intensity (MeV / %)
^{93}Tc		92.910248	2.73 h	β^+ /13/3.201 EC/87/	0.81	9/2+	6.26		ann.rad./ 1.3629 1.4771 1.5203 (0.1 - 3.0)
^{94m}Tc			52. m	β^+ /72/4.33 EC/28/		2+			ann.rad./ 0.8710 1.8686
^{94}Tc		93.909655	4.88 h	β^+ /11/4.256 EC/89/		7+	5.08		ann.rad./ 0.4491 0.7026 0.8496 0.8710
^{95m}Tc			61. d	I.T./4/ β^+ /0.3 EC/96	 0.5/ 0.7/	1/2-			ann.rad./ 0.0389(IT) 0.2041 0.5821 0.5821 0.8351
^{95}Tc		94.90766	20.0 h	EC/100/1.691		9/2+	5.89		0.7657 1.0738
^{96m}Tc			52. m	I.T./90/ β^+ ,EC/2/		4+			0.0342(IT) 0.7782 1.2002
^{96}Tc		95.90787	4.3 d	EC/2.973		7+	+5.04		Mo k x-ray 0.7782 0.8125 0.8498 1.12168
^{97m}Tc			90. d	I.T./0.0965		1/2-			Tc k x-ray 0.0965
^{97}Tc		96.906364	2.6x10^6 y	EC/100/0.320		9/2+			Mo k x-ray
^{98}Tc		97.907215	4.2x10^6 y	β^- /1.80	0.40/100	6+			0.65241 0.74535
^{99m}Tc			6.01 h	I.T./100/0.142		1/2-			Tc k x-ray 0.14049 0.14261
^{99}Tc		98.906254	2.13x10^5 y	β^- /0.294	0.293/100	9/2+	+5.6847	-0.129	
^{100}Tc		99.907657	15.8 s	β^- /3.202 EC /1.8(10)$^{-3}$/0.17	2.2/ 2.9/ 3.3	1+			0.5396 0.5908 1.5122 (0.3 - 2.6)
^{101}Tc		100.90731	14.2 m	β^- /1.61	1.32/	9/2+			0.1272 0.1841 0.3068 0.5451 (0.073-0.969)
^{102m}Tc			4.4 m	I.T./2/4.8 β^- /98/	1.8/				0.4184 0.4752 0.6281 0.6302 1.0464 1.1033 1.6163 2.2447
^{102}Tc		101.90921	5.3 s	β^- /4.53	3.4/ 4.2 2.2/	1+			0.4686 0.4751 1.1055
^{103}Tc		102.90918	54. s	β^- /2.66	2.0/ 2.2/	5/2+			0.1361 0.1743 0.2104 0.3464 0.5629 (0.13 - 1.0)
^{104}Tc		103.91144	18.2 m	β^- /5.60	5.3/	(3+)			0.3483 0.3580 0.5305 0.5351 0.8844 0.8931 1.6768 (0.3 - 3.7)
^{105}Tc		104.91166	7.6 m	β^- /3.6	3.4/	5/2+			0.1079 0.1432 0.3215

Elem. or Isot.	Natural Abundance (%)	Atomic Mass or Weight	Half-Life	Decay Mode/Energy (/MeV)	Particle Energy /Intensity (MeV / %)	Spin (h/2π)	Nuclear Magnetic Mom.(nm)	Elect. Quadr. Mom. (b)	γ-ray Energy /Intensity (MeV / %)
^{106}Tc		105.91436	36. s	β^- /6.55		2+			0.2703
									0.5222
									1.9694
									2.2393
									2.7893
^{107}Tc		106.9151	21.2 s	β^- /4.8					0.1027
									0.1063
									0.1770
									0.4587
^{108}Tc		107.9185	5.1 s	β^- /7.72		(3)			0.2422
									0.4656
									0.7078
									0.7326
									1.5835
^{109}Tc		108.9200	1.4 s	β^- /6.3					
^{110}Tc		109.9234	0.83 s	β^- /8.8					0.2407
^{111}Tc		110.9251	0.30 s	β^- /7.0					
^{113}Tc		112.931	0.13 s	β^- /8.					
$_{44}$Ru		**101.07(2)**							
^{88}Ru									
^{89}Ru		88.936		β^+ /8.					
^{90}Ru		89.9298	11. s	β^+ /5.9					ann.rad./
									0.155 - 1.551
^{91}Ru		90.9264	9. s	β^+ ,EC/7.4		9/2+			ann.rad./
^{92}Ru		91.9201	3.7 m	β^+ /53/4.5		0+			ann.rad./
				EC/47/					0.1346
									0.2138
									0.2593
^{93m}Ru			10.8 s	I.T./21/		1/2-			ann.rad./
				β^+ ,EC/79/	5.3/				0.7344
									1.1112
									1.3962
									2.0931
^{93}Ru		92.9171	1.0 m	β^+ /6.3		9/2+			ann.rad./
				EC/					0.6807
									1.4349
									(0.5- 4.2)weak
^{94}Ru		93.91137	52. m	EC/100/1.59					0.3672
									0.5247
									0.8922
^{95}Ru		94.91042	1.64 h	EC/85/2.57	1.20/	5/2+	0.86		ann.rad./
				β^+ /15/	0.91/				0.2904
									0.3364
									0.6268
^{96}Ru	5.52(6)	95.90760				0+			
^{97}Ru		96.90756	2.89 d	EC/1.12		5/2+	-0.78		Tc k x-ray
									0.2157
									0.3245
									0.4606
^{98}Ru	1.88(6)	97.90529				0+			
^{99}Ru	12.7(1)	98.905939				5/2+	-0.6413	+0.079	
^{100}Ru	12.6(1)	99.904219				0+			
^{101}Ru	17.0(1)	100.905582				5/2+	-0.7188	+0.46	
^{102}Ru	31.6(2)	101.904349				0+			
^{103}Ru		102.906323	39.27 d	β^- /0.763	0.223	3/2+	0.20	+0.62	0.05329
									0.29498
									0.4438
									0.49708
									0.55704
									0.61033
									(0.04 - 1.6)
^{104}Ru	18.7(2)	103.905430				0+			
^{105}Ru		104.907750	4.44 h	β^- /1.917	1.11/22	3/2+	-0.3		0.12968 / 0.0
					1.134/13				0.1491
					1.187/49				0.2629
									0.31664
									0.46943
									0.67634
									0.72420
									(0.1 - 1.8)
^{106}Ru		105.90733	1.020 y	β^- /0.0394	0.0394/100	0+			
^{107}Ru		106.9099	3.8 m	β^- /2.9	2.1/				0.1939
					3.2/				0.3741
									0.4625
									0.8488

Elem. or Isot.	Natural Abundance (%)	Atomic Mass or Weight	Half-Life	Decay Mode/Energy (/MeV)	Particle Energy /Intensity (MeV / %)	Spin (h/2π)	Nuclear Magnetic Mom.(nm)	Elect. Quadr. Mom. (b)	γ-ray Energy /Intensity (MeV / %)
^{108}Ru		107.9102	4.5 m	β^- /1.4	1.2/				0.0923 0.1651 0.4339 0.4975 0.6189
^{109}Ru		108.91320	34.5 s	β^- /4.2					0.1164 0.3584
^{110}Ru		109.9140	15. s	β^- /2.81					0.1121 0.3737 0.4397 0.7967
^{111}Ru		110.9176	1.5 s	β^- /5.5					
^{112}Ru		111.9188	4.5 s	β^- /4.5					
^{113}Ru		112.9225	2.7 s	β^- /7.					
^{114}Ru		113.9239	0.57 s	β^- /6.1					0.127/24 (0.053-0.180)
^{115}Ru		114.928	≈0.74 s	β^- /8.					
$_{45}$Rh		**102.90550(2)**							
^{90}Rh									
^{91}Rh									
^{92}Rh		91.9320		β^+ /11.1					
^{93}Rh		92.9257		β^+ /8.1					
^{94m}Rh			25.8 s	β^+ /		8+			ann.rad./ 0.1264 0.3117 0.7562 1.0752 1.4307
^{94}Rh		93.9217	1.18 m	β^+ /9.6	6.4/	3+			ann.rad./ 0.1461 0.3117 0.7562 1.4307
^{95m}Rh			1.96 m	I.T./88/ β^+ ,EC/12/		1/2+			ann.rad./ 0.5433(IT) 0.7837
^{95}Rh		94.9159	5.0 m	β^+ /5.1	3.2	9/2+			ann.rad./ 0.2293 0.4103 0.6610 0.9416 1.3520 (0.2 - 3.8)
^{96m}Rh			1.51 m	I.T./60 /0.052 β^+ ,EC/40/	 4.70/	2+			ann.rad./ Tc,Ru x-rays 0.8326 1.0985 1.6921 (0.4 - 3.3)
^{96}Rh		95.91452	9.6 m	β^+/6.45 EC/	3.3/	5+			ann.rad./ 0.4299 0.6315 0.6853 0.7418 0.8326 (0.2 - 3.4)
^{97m}Rh			46.m	I.T./5 / β^+,EC/95 /	2.6/	1/2-			ann.rad./ 0.1886 0.4215 [illegible]
^{97}Rh		96.91131	31.0 m	β^+ /3.52	2.1/	9/2+			ann.rad./ 0.1886 0.3892 0.4515 0.8398 0.8788 (0.2 - 3.5)
^{98m}Rh			3.5 m	β^+ /		5+			ann.rad./ 0.6154 0.6524 0.7452
^{98}Rh		97.91072	8.7 m	β^+ /90 /5.06	3.4/	2+			ann.rad./ 0.6524 0.7623

Elem. or Isot.	Natural Abundance (%)	Atomic Mass or Weight	Half-Life	Decay Mode/Energy (/MeV)	Particle Energy /Intensity (MeV / %)	Spin (h/2π)	Nuclear Magnetic Mom.(nm)	Elect. Quadr. Mom. (b)	γ-ray Energy /Intensity (MeV / %)
^{99m}Rh			4.7 h	β^+ /8 / EC/92 /	.74/	9/2+			ann.rad./ 0.2766/ 0.3408 0.6178 1.2612
^{99}Rh		98.90820	16. d	β^+/4 /2.10 EC/97 /	0.54/ 0.68/	1/2-			ann.rad./ 0.0894/ 0.3530 0.5277 (0.1 - 2.0)
^{100m}Rh			4.7 m	I.T./99 / β^+ /0.4 /		5+			ann.rad./ 0.0748/ 0.2647(IT)
^{100}Rh		99.90812	20.8 h	β^+ /3.63 EC/	2.62/ 2.07/	1-			0.4462 0.5396 0.5882 0.8225 1.5534 2.3761
^{101m}Rh			4.35 d	EC/92 / I.T./8 /0.1573		9/2+	+5.51		Rh k x-ray 0.1272/ 0.3069 0.5451
^{101}Rh		100.90616	3.3 y	EC/0.54		1/2-			Ru k x-ray 0.1272 0.1980 0.3252
^{102m}Rh			207. d	I.T./5 / β^- /19/ β^+ /14/ EC/62/					ann.rad./ 0.4686 0.4751 0.5566 0.6280 1.1032 (0.4 - 1.6)
^{102}Rh		101.906842	2.9 y	EC/2.323		6+	4.04		Ru k x-ray 0.4751 0.6313 0.6975 0.7668 1.0466 1.1032
^{103m}Rh			56.12 m	IT		7/2+			
^{103}Rh	100.	102.905504				1/2-	-0.0884		
^{104m}Rh			4.36 m	I.T./99+ / β^-	 1.3/	5+			Rh k x-ray 0.0514 0.0971 0.5558
^{104}Rh		103.906655	42.3 s	β^- /99+ /2.441 EC/0.4/1.141	1.88/2 2.44/98	1+			0.3581 0.5558 1.2370 (0.35 - 1.8)
^{105m}Rh			40. s	I.T./1.296		1/2-			Rh k x-ray 0.1296
^{105}Rh		104.905692	35.4 h	β^- /0.567	0.247/30 0.567/70	7/2+	+4.45		0.2801 0.3061 0.3189
^{106m}Rh			2.18 h	β^- /	0.92/	6+			0.2217 0.4510 0.5119 0.6162 0.7173 0.7484 1.0458 1.5277
^{106}Rh		105.90729	29.9 s	β^- /3.54	2.4/2 3.0/12 3.54/79	1+	2.58		0.51186/ 0.61612 0.62187 (0.05 - 3.04)
^{107}Rh		106.90675	21.7 m	β^- /1.51	1.20/65 1.5/17	7/2+			0.2776 0.3028 0.3925
^{108m}Rh			6.0 m	β^- /	1.57/				0.4339 0.4973 0.6189

Elem. or Isot.	Natural Abundance (%)	Atomic Mass or Weight	Half-Life	Decay Mode/Energy (/MeV)	Particle Energy /Intensity (MeV / %)	Spin (h/2π)	Nuclear Magnetic Mom.(nm)	Elect. Quadr. Mom. (b)	γ-ray Energy /Intensity (MeV / %)
^{108}Rh		107.9087	17. s	β^- /4.5		1+			0.4046 0.4339 0.4973 0.5811 0.6146 0.9014 0.9471
^{109}Rh		108.90874	1.34 m	β^- /2.59	2.25/	7/2+			0.1134 0.1780 0.2914 0.3254 0.3268 0.4261 (0.1 - 1.6)
^{110m}Rh			29. s	β^- /	[.6/				0.3737 0.4397 0.7967
^{110}Rh		109.9110	3.1 s	β^- /5.4	5.5/	1+			0.3737 0.4400 0.5463 0.6877 0.8381 0.9045
^{111}Rh		110.9117	11. s	β^- /3.7					0.275
^{112m}Rh			6.8 s	β^- /					
^{112}Rh		111.9140	4. s	β^- /6.2		1+			0.3489
^{113}Rh		112.9154	0.9 s	β^- /4.9					0.1285
^{114m}Rh			1.8 s	β^- /					
^{114}Rh		113.9173	1.8 s	β^- /6.5		1+			
^{115}Rh		114.9201	0.99 s	β^- /6.0					
^{116m}Rh			0.9 s	β^- /					0.3405
^{116}Rh		115.9228	0.7 s	β^- /8.0		1+			
^{117}Rh		116.925	0.44 s	β^- /7.					0.0346 0.1317
$_{46}$Pd		**106.42(1)**							
^{92}Pd									
^{93}Pd									
^{94}Pd		93.9288	9. s	EC,β^+ /≈ 6.6					0.5582 (0.0546-0.798)
^{95m}Pd		94.92684	13.4 s	EC,β^+ /10.2		21/2+			
^{95}Pd									
^{96}Pd		95.9182	2.03 m	EC,β^+ /3.5	1.15/				0.1248 0.4995
^{97}Pd		96.9165	3.1 m	β^+ ,EC/4.8	3.5/	5/2+			ann.rad./ 0.2653 0.4752 0.7927 (0.2 - 3.4)
^{98}Pd		97.91273	17.7 m	β^+ /1.87 EC/		0+			ann.rad./ 0.0677 0.1125 0.6630 0.8379
^{99}Pd		98.91181	21.4 m	β^+ /49 /3.37 EC/51 /	2.18/	5/2+			ann.rad./ 0.1360 0.2636 0.6734 (0.2 - 2.85)
^{100}Pd		99.90851	3.7 d	EC/0.36		0+			0.03271 0.0748 0.0840
^{101}Pd		100.90829	8.4 h	β^+ /5 /1.980 EC/95 /	0.776/	5/2+	-0.66		ann.rad./ 0.0244 0.2963 0.5904
^{102}Pd	1.02(1)	101.905607				0+			
^{103}Pd		102.906087	16.99 d	EC/0.543		5/2+			Rh k x-ray 0.03975 0.3575 0.4971
^{104}Pd	11.14(8)	103.904034				0+			
^{105}Pd	22.33(8)	104.905083				5/2+	-0.642	+0.66	
^{106}Pd	27.33(3)	105.903484				0+			
^{107m}Pd			20.9 s	I.T./0.2149		11/2-			Pd k x-ray 0.2149(IT)
^{107}Pd		106.90513	6.5x10^{6}y	β^- /0.033	0.03/	5/2+			
^{108}Pd	26.46(9)	107.903895				0+			

Elem. or Isot.	Natural Abundance (%)	Atomic Mass or Weight	Half-Life	Decay Mode/Energy (/MeV)	Particle Energy /Intensity (MeV / %)	Spin (h/2π)	Nuclear Magnetic Mom.(nm)	Elect. Quadr. Mom. (b)	γ-ray Energy /Intensity (MeV / %)
^{109m}Pd			4.75 m	I.T./0.1889		11/2-			Pd x-ray 0.1889(IT)
^{109}Pd		108.905954	13.5 h	β^- /1.116	1.028	5/2+			0.0880 (0.08 - 1.0)
^{110}Pd	11.72(9)	109.905153				0+			
^{111m}Pd			5.5 h	I.T./73 /0.172 β^- /27 /	 0.35 0.77	11/2-			0.0704 0.1722 0.3912 (0.1 - 1.97)
^{111}Pd		110.90764	23.4 m	β^- /2.19	2.2/95	5/2+			0.0598 0.2454 0.5800 0.6504 1.3885 1.4590
^{112}Pd		111.90731	21.04 h	β^- /0.29	0.28/	0+			0.018
^{113m}Pd			1.48 m	β^- /		5/2+			0.0959
^{113}Pd		112.91015	1.64 m	β^- /3.34					0.0958 0.4824 0.6436 0.7394
^{114}Pd		113.91037	2.48 m	β^- /1.45		0+			0.1266 0.2320 0.5582 0.5760
^{115}Pd		114.9137	47. s	β^- /4.58					0.1255 0.2554 0.3428
^{116}Pd		115.9142	12.7 s	β^- /2.61					0.1015 0.1147 0.1778
^{117}Pd		116.9178	4.4 s	β^- /5.7					0.2473 (0.0766-0.4025)
^{118}Pd		117.9189	2.4 s	β^- /4.1					0.1254 (0.0284-0.5957)
^{119}Pd		118.9227	0.9 s	β^- /6.5					0.2566 (0.0699-0.3261)
^{120}Pd		119.9240	0.5 s	β^- /5.0					0.1581 (0.0525-0.5952)
$_{47}$Ag		**107.8682(2)**							
^{94}Ag			0.42 s	β^+, p/					
^{95}Ag			2.0 s	β^+, p/					
^{96}Ag		95.9307	5.1 s	β^+ /11.6 EC/					ann.rad./ 0.1248 0.4995
^{97}Ag		96.9240	19. s	β^+ /7.0 EC/					ann.rad./ 0.6862 1.2941
^{98}Ag		97.9218	47. s	β^+ /8.4 EC/		5+			ann.rad./ 0.5711 0.6786 0.8631
^{99m}Ag			11. s	I.T./100/		1/2-			Ag k x-ray 0.1636(IT) 0.3426
^{99}Ag		98.9176	2.07 m	β^+ /87 /5.4 EC/13 /		9/2+			ann.rad./ 0.2199 0.2645 0.8056 0.8323 (0.2 - 3.5)
^{100m}Ag			2.3 m	β^+ / EC/		2+			ann.rad./ 0.6657 1.6941
^{100}Ag		99.9161	2.0m	β^+/7.1 EC/	4.7/	5+			ann.rad./ 0.2807 0.4503 0.6657 0.7508 0.7732
^{101m}Ag			3.1 s	I.T./0.23		1/2-			Ag k x-ray 0.0981 0.176(IT)

Elem. or Isot.	Natural Abundance (%)	Atomic Mass or Weight	Half-Life	Decay Mode/Energy (/MeV)	Particle Energy /Intensity (MeV / %)	Spin (h/2π)	Nuclear Magnetic Mom.(nm)	Elect. Quadr. Mom. (b)	γ-ray Energy /Intensity (MeV / %)
^{101}Ag		100.9128	11.1 m	β^+ /69 /4.2 EC/31 /	2.7/ 2.18/ 2.73/ 3.38/	9/2+	5.7		ann.rad./ 0.2610 0.2747 0.3269 0.4392 0.6673 1.1739 (0.2 - 3.1)
^{102m}Ag			7.8 m	β^+ /38 / EC/13 / I.T./49 /	3.4	2+	+4.14		ann.rad./ 0.5567 0.9777 1.8347 2.0545 2.1594 3.2386
^{102}Ag		101.91197	13.0 m	β^+ /78 /5.92 EC/22 /	2.26/	5+	4.6		ann.rad./ 0.5567 0.7194 0.8354 1.2571 1.5816 1.7446
^{103m}Ag			5.7 s	I.T./0.134		1/2-			Ag k x-ray 0.1344
^{103}Ag		102.90897	1.10 h	β^+ /28 /2.69 EC/72 /	1.7 1.3	7/2+	+4.47		ann.rad./ 0.1187 0.1482
^{104m}Ag			33. m	β^+ /64/ EC/36/ I.T./0.07/	2.71/	2+	+3.7		ann.rad./ 0.5558 0.7657 (0.5 3.4)
^{104}Ag		103.90863	69. m	β^+ /16 /4.28 EC/84 /	0.99/	5+	3.92		ann.rad./ 0.5558 0.9259 0.9416 (0.18 - 2.27)
^{105m}Ag			7.2 m	I.T./98 /0.0255 EC/2 /		7/2+			Ag x-ray 0.3063 0.3192 (0.1 - 1.0)
^{105}Ag		104.90653	41.3 d	EC/1.35		1/2-	0.1014		0.0640 0.2804 0.3445 0.4434
^{106m}Ag			8.4 d	EC/		6+	3.71		Pd k x-ray 0.4510 0.5118 0.7173 1.0458
^{106}Ag		105.90667	24.0 m	β^+ /59 /2.965 EC/41 /	/1.96	1+	+2.85		ann.rad./ 0.5119
^{107m}Ag			44.2 s	I.T./0.093		7/2+			Ag x-ray 0.0931
^{107}Ag	51.839(7)	106.905093				1/2-	-0.11357		
^{108m}Ag			1.3x10^{2}y	EC/92 / I.T./8 /0.079		6+	3.580		Ag k x-ray Pd k x-ray 0.43392 0.61427 0.72290
^{108}Ag		107.905954	2.39 m	β^- /97/1.65 EC/2/ β^- /1/1.92	1.02/1.7 1.65/96 0.88/0.3	1+	+2.6884		ann.rad./ 0.43392 0.61885 0.63298
^{109m}Ag			39.8 s	I.T./0.088		7/2+			Ag k x-ray 0.0880
^{109}Ag	48.161(7)	108.904756				1/2-	-0.13056		
^{110m}Ag			249.8 d	β^- /99 / I.T./1 /0.1164	0.087 0.530	6+	+3.607		0.65774 0.76393 0.88467 0.93748 1.38427 (0.447-1.56)
^{110}Ag		109.906111	24.6 s	β^- /2.892	2.22/5 2.89/95	1+	+2.7271		0.65774 0.8154 1.1257
^{111m}Ag			1.08 m	I.T./99 /0.0598 β^- /1/		7/2+			Ag k x-ray 0.0598 0.2454

Elem. or Isot.	Natural Abundance (%)	Atomic Mass or Weight	Half-Life	Decay Mode/Energy (/MeV)	Particle Energy /Intensity (MeV / %)	Spin ($h/2\pi$)	Nuclear Magnetic Mom.(nm)	Elect. Quadr. Mom. (b)	γ-ray Energy /Intensity (MeV / %)
^{111}Ag		110.905295	7.47 d	β^- /1.037	1.035/	1/2-	-0.146		0.2454 0.3421
^{112}Ag		111.90701	3.13 h	β^- /3.96	3.94/ 3.4	2-	0.0547		0.6067 0.6174 1.3877 (0.4 - 2.9)
^{113m}Ag			1.14 m	I.T./80 /0.043 β^- /20 /	 1.5	7/2+			0.1422 0.2983 0.3161 0.3923
^{113}Ag		112.90657	5.3 h	β^- /2.02	2.01/	1/2-	0.159		0.2588 0.2986
^{114}Ag		113.90881	4.6 s	β^- /5.08	4.9/	1+			0.5582 0.5760 1.9946
^{115m}Ag			18.7 s	β^- /		7/2+			0.1134 0.1315 0.2288 0.3887
^{115}Ag		114.90876	20. m	β^- /3.10		1/2-			0.1316 0.2128 0.2291 0.4727 (0.13 - 2.49)
^{116m}Ag			10.5 s	I.T./2 / β^- /98 /	3.2/ 2.9	5+			0.1027 0.2549 0.5134 0.7055 1.0289
^{116}Ag		115.91137	2.68 m	β^- /6.16	5.3	2-			0.5134 0.6993 2.4779
^{117m}Ag			5.3 s	β^- /	3.2/	7/2+			0.1354 0.2981 0.3868
^{117}Ag		116.91171	1.22 m	β^- /4.18	2.3	1/2-			0.1354 0.1571 0.3377
^{118m}Ag			2.8 s	β^- /59 / I.T./41 /0.1277					0.1277 0.4878 0.6771 0.7709 1.0586
^{118}Ag		117.9145	4.0 s	β^- /7.1					0.4878 0.6771 3.2259
^{119}Ag		118.9157	2.1 s	β^- /5.35		7/2+			0.0674 0.3662 0.3991 0.6264
^{120m}Ag			0.32 s	β^- / I.T./					0.2030 0.5059 0.6978 0.8300 0.9258
^{120}Ag		119.9187	1.23 s	β^- /8.2					0.5059 0.6978 0.8171 1.3231
^{121}Ag		120.9200	0.78 s	β^- /6.4					0.1150 0.3148 0.3537 0.3696 0.5007 1.5105 (0.11 - 2.5)
^{122m}Ag			1. s	β^- /					
^{122}Ag		121.9233	0.44 s	β^- /9.2					
^{123}Ag		122.9249	0.31 s	β^- /7.4					
^{124}Ag		123.9285	0.22 s	β^- /10.1					
$_{48}$Cd		**112.411(8)**							
^{97}Cd			3. s	β^+,(p)					
^{98}Cd		97.9276	9.2 s	β^+ /5.4					
^{99}Cd		98.9250	16. s	β^+ ,EC/6.9					ann.rad./
^{100}Cd		99.9203	1.1 m	β^+ ,EC/3.9					ann.rad./ (0.0935-0.935)

Elem. or Isot.	Natural Abundance (%)	Atomic Mass or Weight	Half-Life	Decay Mode/Energy (/MeV)	Particle Energy /Intensity (MeV / %)	Spin (h/2π)	Nuclear Magnetic Mom.(nm)	Elect. Quadr. Mom. (b)	γ-ray Energy /Intensity (MeV / %)
^{101}Cd		100.9187	1.2 m	β^+ /83 /5.5 EC/17 /	4.5	5/2+			In k x-ray 0.0985 1.7225 0.31 - 2.84)
^{102}Cd		101.91474	5.8 m	β^+ /27 /2.59 EC/73		0+			ann.rad./ 0.0974 0.4810 1.0366 1.3598
^{103}Cd		102.91342	7.5 m	β^+ /33 /4.14 EC/67 /		5/2+	-0.81	-0.8	ann.rad./ Ag k x-ray 1.0799 1.4487 1.4618 (0.1 - 2.8)
^{104}Cd		103.90985	58. m	EC/1.14		0+			Ag k x-ray 0.0835 0.7093
^{105}Cd		104.90947	55.5 m	β^+ /26 /2.739 EC/74 /	1.69/	5/2+	-0.7393	+0.43	Ag k x-ray 0.3469 0.6072 0.9618 1.3025 (0.25 - 2.4)
^{106}Cd	1.25(4)	105.90646				0+			
^{107}Cd		106.90661	6.52 h	EC/99+ /1.417 β^+ /		5/2+	-0.615055	+0.68	Ag k x-ray 0.0931 0.8289
^{108}Cd	0.89(2)	107.90418				0+			
^{109}Cd		108.904985	462.0 d	EC/0.214		5/2+	-0.827846	+0.69	Ag k x-ray 0.08804
^{110}Cd	12.49(12)	109.903006				0+			
^{111m}Cd			48.5 m	I.T./		11/2-			Cd k x-ray 0.1508(IT) 0.2454
^{111}Cd	12.80(8)	110.904182				1/2+	-0.594886		
^{112}Cd	24.13(14)	111.902758				0+			
^{113m}Cd			14.1 y	β^- /99.9 /0.59	0.59/99.9	11/2-	-1.087		0.2637
^{113}Cd	12.22(8)	112.904401				1/2+	-0.622301		
^{114}Cd	28.73(28)	113.903359				0+			
^{115m}Cd			44.6 d	β^- /1.629	0.68/1.6 1.62/97	11/2-	-1.042		0.48450 0.93381 1.29064
^{115}Cd		114.905431	2.228 d	β^- /1.446	0.593/42 1.11/58	1/2+	-0.648426		0.23141 0.26085 0.33624 0.49227 0.52780
^{116}Cd	7.49(12)	115.904756				0+			
^{117m}Cd			3.4 h	β^- /2.66	0.72/	11/2-			0.1586 0.5529 0.37 - 2.42
^{117}Cd		116.907219	2.49 h	β^- /2.52	0.67/51 2.2/10	1/2+			0.2209 0.2733 0.3445 1.3033
^{118}Cd		117.90692	50.3 m	β^- /0.52		0+			
^{119m}Cd			2.20 m	β^- /		11/2-			0.1056 0.7208 1.0250 2.0213
^{119}Cd		118.90992	2.69 m	β^- /3.8	≈ 3.5/	1/2+			0.1340 0.2929 0.3429
^{120}Cd		119.90985	50.8 s	β^- /1.76	1.5/	0+			
^{121m}Cd			8. s	β^- /		11/2-			0.1008 0.9878 1.0209 1.1815 2.0594
^{121}Cd		120.9131	13.5 s	β^- /4.9		(3/2+)			0.2102 0.3242 0.3492 1.0403
^{122}Cd		121.9135	5.3 s	β^- /3.0		0+			
^{123m}Cd			1.9 s	β^- /					
^{123}Cd		122.91770	2.09 s	β^- /6.12		3+			

Elem. or Isot.	Natural Abundance (%)	Atomic Mass or Weight	Half-Life	Decay Mode/Energy (/MeV)	Particle Energy /Intensity (MeV / %)	Spin (h/2π)	Nuclear Magnetic Mom.(nm)	Elect. Quadr. Mom. (b)	γ-ray Energy /Intensity (MeV / %)
^{124}Cd		123.9177	1.24 s	β^- /4.17		0+			0.0365 0.0628 0.1799
^{125m}Cd			0.66 s	β^- /					
^{125}Cd		124.92129	0.68 s	β^- /7.16		3/+			
^{126}Cd		125.9224	0.52 s	β^- /5.49		0+			0.2601
^{127}Cd		126.9264	0.4 s	β^- /8.5		3/+			
^{128}Cd		127.9278	0.28 s	β^- /7.1		0+			0.247
^{129}Cd			0.27 s	β^- /5.9					0.281
^{130}Cd			0.20 s	β^- /		0+			
$_{49}$In		**114.818(3)**							
^{99}In		98.9346		β^+ /8.9					
^{100}In		99.9316	5. s	β^+,(p)/10.5					
^{101}In		100.9266		β^+ /7.3					
^{102}In		101.9243	23. s	EC/8.9		(5)			0.1566 0.7767 0.8612
^{103}In		102.91991	1.1 m	β^+ ,EC/6.05	4.2	9/2+			ann.rad./ 0.1879 0.7200
^{104m}In			16. s	IT/0.0935					
^{104}In		103.9183	1.84 m	β^+ ,EC/7.9	4.8	5+	+4.44	+0.7	ann.rad./ 0.6580 0.8341 0.8781
^{105m}In			43. s	I.T.		1/2-			In k x-ray 0.6740
^{105}In		104.91467	5.1 m	β^+ ,EC/4.85	3.7	9/2+	+5.675	+0.83	0.1310 0.2600 0.6038
^{106m}In			5.3 m	β^+ /85 / EC/15 /	4.90	3+			ann.rad./ 0.6326 0.8611 1.7164
^{106}In		105.91346	6.2 m	β^+ /65 /6.52 EC/35 /	2.6	7+	+4.92	+0.97	ann.rad./ 0.2259 0.6327 0.8611 0.9978 1.0091
^{107m}In			51. s	I.T./0.6786		1/2-			In k x-ray 0.6785
^{107}In		106.91029	32.4 m	β^+ /35 /3.43 E.C/65 /	2.20/	9/2+	+5.59	+0.81	ann.rad./ Cd k x-ray 0.2050 0.3209 0.5055 (0.2 - 2.99)
^{108m}In			57. m	β^+ /53 / EC/47 /	1.3	6+			ann.rad./ Cd k x-ray 0.6329 1.9863 3.4522
^{108}In		107.90971	40. m	β^+ /33 /5.15 EC/67 /	3.49/	3+	+4.56	+1.01	ann.rad./ Cd k x-ray 0.2429 0.6331 0.8756
^{109m}In			1.3 m	I.T./0.650		1/2-			In k x-ray 0.6498
^{109}In		108.90715	4.2 h	β^+ /8 /2.02 EC/92 /	0.79/	9/2+	+5.54	+0.84	ann.rad./ Cd k x-ray 0.2035 0.6235
^{110m}In			4.9 h	EC/		7+	+4.72	+1.00	Cd k x-ray 0.6577 0.8847 0.9375 (0.1 - 1.98)
^{110}In		109.90717	1.15 h	β^+ /62 /3.88 EC/38 /	2.22/	2+	+4.37	+0.35	ann.rad./ Cd k x-ray 0.6577 (0.6 - 3.6)
^{111m}In			7.7 m	I.T./0.537		1/2-	+5.53		In k x-ray 0.537

Elem. or Isot.	Natural Abundance (%)	Atomic Mass or Weight	Half-Life	Decay Mode/Energy (/MeV)	Particle Energy /Intensity (MeV / %)	Spin (h/2π)	Nuclear Magnetic Mom.(nm)	Elect. Quadr. Mom. (b)	γ-ray Energy /Intensity (MeV / %)
^{111}In		110.90511	2.8049 d	EC/0.866		9/2+	+5.50	+0.80	Cd k x-ray 0.1712 0.2453
^{112m}In			20.8 m	I.T./0.155		4+			In k x-ray 0.1555
^{112}In		111.90553	14.4 m	β^+ /22 /2.586 EC/34 / β^- /0.663		1+	+2.82	+0.09	ann.rad./ Cd k x-ray 0.6171
^{113m}In			1.658 h	I.T./0.3917		1/2-	-0.210		In k x-ray 0.3917
^{113}In	4.29(2)	112.904062				9/2+	+5.529	+0.80	
^{114m}In			49.51 d	I.T./97 /0.190 EC/3 /		5+	+4.7		In k x-ray 0.19027
^{114}In		113.904918	1.198 m	β^- /97 /1.989 EC/3 /1.453	1.984/	1+	+2.82		Cd k x-ray 0.5584 0.5727 1.2998
^{115m}In			4.486 h	I.T./95 /0.336 β^- /5 /0.83		1/2-	-0.255		In k x-ray 0.3362 0.4974
^{115}In	95.71(2)	114.903879	$4.4x10^{14}$ y	β^- /0.495		9/2+	+5.541	+0.81	
^{116m2}In			2.16 s	I.T./0.162		8-			In k x-ray 0.1624
^{116m1}In			54.1 m	β^- /	1.0	5+	+4.3		0.13792 0.41688 1.09723 1.29349
^{116}In		115.905261	14.1 s	β^- /3.274	3.3/99	1+	2.788	0.11	0.46313 1.2526 1.29349
^{117m}In			1.94 h	β^- /53 /1.769 I.T./47 /	1.77/	1/2-	0.25		In k x-ray 0.15855 0.31531 0.55294
^{117}In		116.90452	44. m	β^- /1.455	0.74/	9/2+	+5.52	+0.83	0.15855 0.3966 0.55294
^{118m2}In			8.5 s	I.T./98 / β^- /2 /		(8-)			In k x-ray 0.1382
^{118m1}In			4.40 m	β^- /	1.3 2.0	5+			0.2086 0.6833 1.2295
^{118}In		117.90636	5.0 s	β^- /4.42	4.2/	1+			0.5282 1.1734 1.2295 2.0432
^{119m}In			17.9 m	β^- /97 / I.T./3 /0.311	2.7/	1/2-			0.3114 0.7631
^{119}In		118.90585	2.3 m	β^- /2.36	1.6/	9/2+	+5.52	+0.85	0.0239 0.6495 0.7631 1.2149
^{120m}In			47. s	β^- /	2.2/	8-			0.7042 1.1725 2.0398 2.3902
^{120}In		119.90796	3.1 s	β^- /5.37	5.6/ 3.1/	(1+)	+4.30	+0.81	0.4146 0.5924 0.8637 1.0232 1.1714 (0.4 - 2.7)
^{121m}In			3.8 m	β^- /99 / I.T./1 /0.313	3.7/	1/2-			0.0601 0.3136 0.9256 1.0412 1.1022 1.1204
^{121}In		120.90785	23. s	β^- /3.36	2.5	9/2+	+5.50	+0.81	0.2620 0.6573 0.9256
^{122m}In			10. s	β^- /	4.4/	8-			1.0014 1.1403
^{122}In		121.91028	1.5 s	β^- /6.37	5.3/	(1+)			0.2391 1.0014 1.1403 1.164 1.1903

Elem. or Isot.	Natural Abundance (%)	Atomic Mass or Weight	Half-Life	Decay Mode/Energy (/MeV)	Particle Energy /Intensity (MeV / %)	Spin (h/2π)	Nuclear Magnetic Mom.(nm)	Elect. Quadr. Mom. (b)	γ-ray Energy /Intensity (MeV / %)
^{123m}In			47. s	β^- /	4.6/	(1/2-)	-0.40		0.1258 1.170 3.234
^{123}In		122.91044	6.0 s	β^- /4.39	3.3/	(9/2+)	+5.49	+0.76	0.6188 1.0197 1.1305
^{124m}In			3.4 s	β^-		8-	+3.89	+0.66	0.1029 0.9699 1.0729 1.1316
^{124}In		123.91318	3.18 s	β^- /7.36	5/	3+	+4.04	+0.61	0.7070 0.9978 1.1316 3.2142 (0.3 - 4.6)
^{125m}In			12.2 s	β^- /	5.5/	1/2-	-0.43		0.1876
^{125}In		124.91360	2.33 s	β^- /5.42	4.1/	9/2+	+5.50	+0.71	0.4260 1.0318 1.3350
^{126m}In			1.53 s		4.9/	3+	+4.03	+0.49	0.9086 0.9696 1.1411
^{126}In		125.91646	1.63 s	β^- /8.21	4.2/	8-	+4.06		0.1118 0.9086 1.1411
^{127m}In			3.73 s	β^- /	6.4/	(1/2-)			0.2523 3.074
^{127}In		126.91734	1.14 s	β^- /6.51	4.9/	(9/2+)	+5.52	+0.59	0.4680 0.6461 0.8051
^{128m}In			0.7 s	β^- /	5.4/	(8-)			1.5977 1.8670 1.9739 (0.1205-2.12)
^{128}In		127.92017	0.80 s	β^- /8.98	5.0/	3+			0.9352 1.1688 3.5198 4.2970
^{129m}In			1.23 s	β^- /98 / n/2 /	≈ 7.5/	1/2-			0.3153 0.9067 1.2220
^{129}In		128.9217	0.63 s	β^- /7.66	5.5/	9/2+			0.2853 0.7693 1.8650 2.1180
^{130m2}In			0.53 s	β^- /	8.8/	5+			0.0892 0.7744 1.2212
^{130m1}In			0.51 s	β^- /	6.1/	10-			0.0892 0.1298 0.7744 1.2212 1.9052
^{130}In		129.92486	0.29 s	β^- /10.25	10.0/	1-			
^{131m2}In			0.3 s	β^- /		(21/2+)			
^{131m1}In			0.35 s	β^- /		(1/2-)			
^{131}In		130.9268	0.28 s	β^- /9.18	6.4/	(9/2+)			0.3328 2.433
^{132}In		131.9323	0.20 s	β^- /13.6	6.0/ 8.8/	(7-)			0.1320 0.2992 0.3747 4.0406
^{133}In			0.18 s	β^- ,(n)					
$_{50}$Sn		**118.710(7)**							
^{100}Sn		99.9394		β^+ /7.3					
^{101}Sn		100.9361		β^+ /9.					
^{102}Sn		101.9243		β^+ /5.8					
^{103}Sn		102.9281	7. s	β^+ /7.7					
^{104}Sn		103.9232	21. s	β^+, EC/4.5					
^{105}Sn		104.9214	28. s	β^+ /6.3					
^{106}Sn		105.91688	2.0 m	β^+ /20 /3.18 EC/80 /					ann.rad./ In k x-ray 0.3865 0.4772

Elem. or Isot.	Natural Abundance (%)	Atomic Mass or Weight	Half-Life	Decay Mode/Energy (/MeV)	Particle Energy /Intensity (MeV / %)	Spin (h/2π)	Nuclear Magnetic Mom.(nm)	Elect. Quadr. Mom. (b)	γ-ray Energy /Intensity (MeV / %)
^{107}Sn		106.9157	2.92 m	EC/5.0 β^+ /	1.2/				0.4218 0.6105 0.6785 1.0013 1.1290 1.542
^{108}Sn		107.91196	10.3 m	β^+ /1 /2.09 EC/99 /	0.36/	0+			In k x-ray 0.2724 0.3965 (0.105-1.68)
^{109}Sn		108.91129	18.0 m	β^+ /9 /3.85 EC/91 /	1.52/	7/2+	-1.08	+0.3	ann.rad./ In k x-ray 0.6498 1.0992
^{110}Sn		109.90785	4.1 h	EC/0.64		0+			In k x-ray 0.283
^{111}Sn		110.90774	35. m	β^+ /31 /2.45 EC/69 /	1.5/	7/2+	+0.61	+0.2	In k x-ray 0.7620 1.1530 1.9147
^{112}Sn	0.97(1)	111.904822				0+			
^{113m}Sn			21.4 m	I.T./92 /0.077 EC/8 /		7/2+			Sn k x-ray In x-ray 0.0774
^{113}Sn		112.905174	115.1 d	EC/1.036		1/2+	-0.879		In k x-ray 0.25511 0.39169
^{114}Sn	0.65(1)	113.902783				0+			
^{115}Sn	0.34(1)	114.903347				1/2+	-0.9188		
^{116}Sn	14.54(11)	115.901745				0+			
^{117m}Sn			13.60 d	I.T./0.3146		11/2-	-1.40	-0.4	Sn k x-ray 0.15856
^{117}Sn	7.68(7)	116.902955				1/2+	-1.0010		
^{118}Sn	24.22(11)	117.901608				0+			
^{119m}Sn			293. d	I.T./0.0896		11/2-	-1.4	0.21	Sn k x-ray 0.02387
^{119}Sn	8.59(4)	118.903311				1/2+	1.0473		
^{120}Sn	32.59(10)	119.902199				0+			
^{121m}Sn			≈ 55. y	I.T./78 /0.006 β^- /22 /	 0.354/	11/2-			Sn k x-ray 0.03715
^{121}Sn		120.904239	1.128 d	β^- /0.388	0.383/100	3/2+	0.698	-0.02	
^{122}Sn	4.63(3)	121.903441				0+			
^{123m}Sn			40.1 m	β^- /1.428	1.26/99	3/2+			0.1603 0.3814
^{123}Sn		122.905723	129.2 d	β^- /1.404	1.42/99.4	11/2-	-1.370	+0.03	0.1603 1.0302 1.0886
^{124}Sn	5.79(5)	123.905275				0+			
^{125m}Sn			9.51 m	β^- /2.387	2.03/98	3/2+			0.3321 1.4040
^{125}Sn		124.907785	9.63 d	β^- /2.364	2.35/82	11/2-	-1.35	+0.1	1.0671 (0.2-2.3)
^{126}Sn		125.90765	1.x10^{5}y	β^- /0.38	0.25/100	0+			0.0643 0.0876 0.4148 0.6663 0.6950
^{127m}Sn			4.15 m	β^- /3.21	2.72/	3/2+			0.4909 1.3480 1.5640
^{127}Sn		126.91035	2.12 h	β^- /3.20	2.42/ 3.2/	11/2-			0.8231 1.0956 (0.120-2.84)
^{128m}Sn			6.5 s	IT/0.091		(7-)			
^{128}Sn		127.91054	59.1 m	β^- /1.27	0.48/ 0.63/	0+			0.4823 0.5573 0.6805
^{129m}Sn			6.9 m	β^- /		11/2-			1.1611
^{129}Sn		128.9134	2.4 m	β^- /4.0		3/2+			0.6456
^{130m}Sn			1.7 m	β^- /		(7-)			0.1449 0.8992
^{130}Sn		129.91386	3.7 m	β^- /2.15	1.10/	0+			0.0700 0.1925 0.7798
^{131m}Sn			1.02m	β^-/	3.4/	11/2-			0.3043 0.4500 0.7985 1.2260 (0.08 - 3.21)

Elem. or Isot.	Natural Abundance (%)	Atomic Mass or Weight	Half-Life	Decay Mode/Energy (/MeV)	Particle Energy /Intensity (MeV / %)	Spin (h/2π)	Nuclear Magnetic Mom.(nm)	Elect. Quadr. Mom. (b)	γ-ray Energy /Intensity (MeV / %)
^{131}Sn		130.9169	39. s	β^- /4.64	3.8/	3/2+			see ^{131m} Sn
^{132}Sn		131.91775	40. s	β^- /3.3	1.8/				0.0855 0.2467 0.3402 0.8985
^{133}Sn		132.9236	1.44 s	β^- /7.8	7.5/	7/2-			
^{134}Sn		133.9278	1.04 s	β^- /6.8					
$_{51}Sb$		**121.760(1)**							
^{106}Sb		105.9282		β^+ /10.5					
^{107}Sb		106.9242		β^+ /7.9					
^{108}Sb		107.9222	7.0 s	β^+ /9.5					
^{109}Sb		108.91814	17. s	β^+ /6.38 EC/	4.42/ 4.67/ 4.33/	5/2+			ann.rad./ 0.2467 0.6645 0.9254 1.0617 1.4958
^{110}Sb		109.9175	24. s	β^+ /9.0 EC/	6.8/	3+			ann.rad./ 0.6365 0.9847 1.2117 1.2433
^{111}Sb		110.91254	1.25 m	β^+ /87 /4.47 EC/13 /	3.3/	5/2+			ann.rad./ 0.1002 0.1545 0.4891 1.0326
^{112}Sb		111.91240	51.4 s	β^+ /90 /7.06 EC/10 /	4.75/	3+			ann.rad./ 0.6700 0.9909 1.2571 (0.3 - 3.6)
^{113}Sb		112.90937	6.7 m	β^+ /65 /3.91 EC/35 /	2.42/	5/2+			ann.rad./ Sn k x-ray 0.3324 0.4980
^{114}Sb		113.9091	3.49 m	β^+ /78 /5.9 EC/22 /	3.4/	3+	1.7		ann.rad./ Sn k x-ray 0.8876 1.2999
^{115}Sb		114.90660	32.1 m	β^+ /67 /3.03 EC/33 /	1.51/	5/2+	+3.46	-0.4	ann.rad./ Sn k x-ray 0.4973
^{116m}Sb			1.00 h	β^+ /78 / EC/22 /	1.16/	8-			ann.rad./ Sn k x-ray 0.4073 0.5429 0.9725 1.2935 (0.0998-1.501)
^{116}Sb		115.90680	16. m	β^+ /50 /4.707 EC/50 /	1.3/ 2.3/	3+	2.72		ann.rad./ Sn k x-ray 0.93180 1.29354 (0.138-3.903)
^{117}Sb		116.90484	2.80 h	β^+ /2 /1.76 EC/98 /	0.57/	5/2+	+3.4		Sn k x-ray 0.1586
^{118m}Sb			5.00 h	EC/99 /		8-	2.3		Sn k x-ray 0.25368 1.05069 1.22964
^{118}Sb		117.905533	3.6 m	β^+ /74 /3.657 EC/26 /	2.65/	1+	2.5		ann.rad./ Sn k x-ray 1.22964
^{119}Sb		118.90395	38.1 h	EC/0.59		5/2+	+3.45	-0.4	Sn k x-ray 0.0239
^{120m}Sb			5.76 d	EC/		8-	2.34		Sn k x-ray 0.0898 0.19730 1.02301 1.17121
^{120}Sb		119.90508	15.89 m	β^+ /41 /2.68 EC/59 /	1.72/	1+	+2.3		ann.rad./ Sn k x-ray 0.7038 1.17121
^{121}Sb	57.21(5)	120.903822				5/2+	+3.363	-0.4	

Elem. or Isot.	Natural Abundance (%)	Atomic Mass or Weight	Half-Life	Decay Mode/Energy (/MeV)	Particle Energy /Intensity (MeV / %)	Spin (h/2π)	Nuclear Magnetic Mom.(nm)	Elect. Quadr. Mom. (b)	γ-ray Energy /Intensity (MeV / %)
^{122m}Sb			4.19 m	I.T./0.162		8-			Sb x-ray 0.0614 0.0761
^{122}Sb		121.90518	2.72 d	β^- /98 /1.979 β^+ /2 /1.620	1.414/65 1.980/26	2-	-1.90	+0.9	0.56409 0.69277 1.14050 1.2569
^{123}Sb	42.79(5)	122.904216				7/2+	+2.550	-0.5	
$^{124m2}Sb$			20.3 m	I.T./0.035		8-			
$^{124m1}Sb$			1.6 m	I.T./80 / β^- /20 /	1.2/ 1.7/	5+			0.4984 0.6027 0.6458 1.1010
^{124}Sb		123.905938	60.20 d	β^- /2.905	0.61/52 2.301/23	3-	1.2	+1.9	0.60271/97.8 0.64583/7.4 0.72277/10.5 1.69094/48.2 (0.0274-2.808)
^{125}Sb		124.905247	2.758 y	β^- /0.767	0.13/30 0.302/45 0.62/13	7/2+	+2.63		0.0355 0.17632 0.38044 0.42786 0.46336 0.60060 0.63595
$^{126m2}Sb$			11. s	I.T./		3-			L x-ray 0.0227
$^{126m1}Sb$			19.0 m	β^- /86 / I.T./14 /	1.9	5+			0.4148 0.6663 0.6950
^{126}Sb		125.90725	12.4 d	β /3.67	1.9	8-	1.3		0.2786 0.4148 0.6663 0.6950 0.7205
^{127}Sb		126.906914	3.84 d	β^- /1.581	0.89/ 1.10/ 1.50/	7/2+	2.70		0.2524 0.2908 0.4121 0.4370 0.6857 0.7837
^{128m}Sb			10.1 m	β^- /96 / I.T./4 /	2.6/	5+			0.3140 0.5941 0.7432 0.7539
^{128}Sb		127.90917	9.1 h	β^- /4.38	2.3/	8-	1.3		0.2148 0.3141 0.5265 0.7433 0.7540
^{129m}Sb			17.7 m	β^- /					0.4338 0.6578 0.7598
^{129}Sb		128.90915	4.40 h	β^- /2.38	0.65/	7/2-			0.0278 0.1808 0.3594 0.4596 0.5447 0.8128 0.9146 1.0301
^{130m}Sb			6.5 m	β^- /2.6	2.12/				0.1023 0.7934 0.8394
^{130}Sb		129.91155	38.4 m	β^- /4.96	2.9/	8-			0.1823 0.3309 0.4680 0.7394 0.8394
^{131}Sb		130.9120	23.0 m	β^- /3.2	1.31/ 3.0/	7/2+			0.6423 0.6579 0.9331 0.9434
^{132m}Sb			2.8 m	β^- /	3.9/	4+			0.1034 0.3538 0.6968 0.9739 0.9896

Elem. or Isot.	Natural Abundance (%)	Atomic Mass or Weight	Half-Life	Decay Mode/Energy (/MeV)	Particle Energy /Intensity (MeV / %)	Spin (h/2π)	Nuclear Magnetic Mom.(nm)	Elect. Quadr. Mom. (b)	γ-ray Energy /Intensity (MeV / %)
^{132}Sb		131.91420	4.2 m	β^- /5.29		8-			0.1034 0.1506 0.6968 0.9739
^{133}Sb		132.9152	2.5 m	β^- /4.00	1.20/	7/2+			0.4235 0.6318 0.8165 1.0764
^{134m}Sb			10.4 s	β^- /	6.1	7-			
^{134}Sb		133.9206	0.8 s	β^- /8.4	8.4	0-			0.1152 0.2970 0.7063 1.2791
^{135}Sb		134.9252	1.71 s	β^- /8.12		7/2+			1.127 1.279
^{136}Sb		135.9301	0.82 s	β^- /9.3					
$_{52}$Te		**127.60(3)**							
^{106}Te		105.9377	0.06 ms	α/4.32					
^{107}Te		106.9350	3.1 ms	α/ 70/ β^+ ,EC/10.1	3.86(1)/				
^{108}Te		107.9295	2.1 s	α/68 / β^+ ,EC/32 /6.8	3.314(4)/	0+			
^{109}Te		108.9275	4.6 s	β^+ EC/96 /8.7 α/4 /	 3.107(4)/				
^{110}Te		109.9224	19. s	β^+ ,EC/4.5		0+			ann.rad./ 0.2191 0.6059
^{111}Te		110.9211	19.3 s	β^+ ,EC/8.0		(7/2+)			ann.rad./ 0.267 0.322 0.341
^{112}Te		111.9171	2.0 m	β^+ ,EC/4.3		0+			ann.rad./ 0.2962 0.3727 0.4187
^{113}Te		112.9154	1.7 s	β^+ /85 /5.7 EC/15 /	4.5/	(7/2+)			ann.rad./ Sb k x-ray 0.8144 1.0181 1.1812
^{114}Te		113.9125	15. m	β^+ /40 /3.2 EC/60 /		0+			ann.rad./ Sb k x-ray 0.0838 0.0903
^{115m}Te			6.7 m	β^+ /45 / EC/55 /		(1/2+)			ann.rad./ Sb k x-ray 0.7236 0.7704
^{115}Te		114.9116	5.8 m	β^+ /45 /4.6 EC/55 /	2.7/	7/2+			ann.rad./ Sb k x-ray 0.7236 1.3268 1.3806 (0.22 - 2.7)
^{116}Te		115.9084	2.49 h	EC/1.5		0+			Sb k x-ray 0.0937
^{117}Te		116.90864	1.03 h	EC/75 /3.54 β^+ /25 /	1.78/	1/2+			ann.rad./ Sb k x-ray 0.9197 1.7164 2.3000
^{118}Te		117.90583	6.00 d	EC/0.28		0+			Sb k x-ray
^{119m}Te			4.69 d	EC/		11/2-	0.89		Sb k x-ray 0.15360 0.2705 1.21271
^{119}Te		118.90641	16.0 h	β^+ /2 /2.293 EC/98 /	0.627/	1/2+	0.25		ann.rad./ Sb k x-ray 0.6440 0.6998
^{120}Te	0.096(2)	119.90403				0+			
^{121m}Te			≈ 154. d	I.T.(89%) EC(11%)		11/2-			Te k x-ray 0.2122
^{121}Te		120.90494	16.8 d	EC/1.04		1/2+			Sb k x-ray 0.5076 0.5731

Elem. or Isot.	Natural Abundance (%)	Atomic Mass or Weight	Half-Life	Decay Mode/Energy (/MeV)	Particle Energy /Intensity (MeV / %)	Spin (h/2π)	Nuclear Magnetic Mom.(nm)	Elect. Quadr. Mom. (b)	γ-ray Energy /Intensity (MeV / %)
^{122}Te	2.603(4)	121.903056				0+			
^{123m}Te			119.7 d	I.T./0.247		11/2-	-0.93		Te k x-ray 0.1590/84.1
^{123}Te	0.908(2)	122.904271	1.3x10^{13} y	EC/0.051		1/2+	-0.73695		
^{124}Te	4.816(6)	123.902819				0+			
^{125m}Te			58. d	I.T./0.145		11/2-	-0.99	-0.06	Te k x-ray 0.0355
^{125}Te	7.139(6)	124.904424				1/2+	-0.8885		
^{126}Te	18.952(11)	125.903305				0+			
^{127m}Te			109. d	I.T./98 /0.088 β^- /2 /0.77		11/2-	-1.04		Te k x-ray 0.0883
^{127}Te		126.905217	9.4 h	β^- /0.698	0.696/	3/2+	0.64		0.3603
^{128}Te	31.687(11)	127.904462				0+			
^{129m}Te			33.6 d	I.T./63 /0.105 β^- /37 /	 1.60/	11/2-			Te k x-ray 0.45984 0.6959
^{129}Te		128.906596	1.16 h	β^- /1.498	0.99/9 1.45/89	3/2+	0.70	0.06	0.0278 0.45984 0.48728
^{130}Te	33.799(10)	129.906223	2.5x10^{21} y			0+			
^{131m}Te			1.35 d	β^- /78 /2.4 I.T./22 /0.18	0.42/	11/2-	-1.04		0.0811 0.1021 0.14973 0.77369 0.79375 0.85225
^{131}Te		130.908522	25.0 m	β^- /2.233	1.35/12 1.69/22 2.14/60	3/2+	0.70		0.14973 0.45327 0.49269
^{132}Te		131.90852	3.26 d	β^- /0.493	0.215	0+			0.049725 0.11198 0.22830
^{133m}Te			55.4 m	β^- /82 / I.T./18 /0.334	2.4/30	11/2-			Te k x-ray 0.0949 0.1689 0.3121 0.3341
^{133}Te		132.9109	12.4 m	β^- /2.9	2.25/25 2.65	3/2+			0.3121 0.4079 1.3334
^{134}Te		133.9116	42. m	β^- /1.6	0.6/ 0.7/	0+			0.7672/29 0.0794-0.9255
^{135}Te		134.9165	19.0 s	β^- /6.0	5.4/ 6.0				0.267 0.603 0.870
^{136}Te		135.92010	17.5 s	β^- /5.1	2.5/	0+			2.0779/25 0.0873-3.235
^{137}Te		136.9253	2.5 s	β^- /98 /6.9 n/2 /	6.8	7/2-			0.2436
^{138}Te		137.9292	1.4 s	β^- /6.4					
$_{53}$I		**126.90447(3)**							
^{108}I		107.9436	0.04 s	α/91 /4.	3.95				
^{109}I		108.9382	0.11 ms	p					
^{110}I		109.9346	0.65 s	β^+ ,EC/83 /11.4 α/17 /≈3.6 p /11 /	 3.457(10)/				ann.rad./
^{111}I		110.9303	2.5 s	β^+ ,E../8.5					ann.rad./ 0.2665 0.3215 0.3412
^{112}I		111.9280	3.4 s	β^+ ,EC/10.2					ann.rad./ 0.6889 0.7869
^{113}I		112.9237	5.9 s	β^+ ,EC/7.6					ann.rad./ 0.4625/100 0.6224/74 0.0550-1.422
^{114}I		113.9219	2.1 s	β^+ ,EC/8.7					ann.rad./ 0.6826 0.7088
^{115}I		114.9188	1.3 m	β^+ ,EC/6.7		5/2+			ann.rad./ 0.275 0.284 0.460 0.709

TABLE OF THE ISOTOPES

Elem. or Isot.	Natural Abundance (%)	Atomic Mass or Weight	Half-Life	Decay Mode/Energy (/MeV)	Particle Energy /Intensity (MeV / %)	Spin (h/2π)	Nuclear Magnetic Mom.(nm)	Elect. Quadr. Mom. (b)	γ-ray Energy /Intensity (MeV / %)
^{116}I		115.9167	2.9 s	β^+ /97 /7.8 EC/3 /	6.7/	1+			ann.rad./ 0.5402 0.6789
^{117}I		116.9136	2.22 m	β^+ ,EC/4.7	3.2/	(5/2+)	3.1		ann.rad./ 0.2744 0.3259
^{118m}I			8.5 m	β^+ ,EC/ I.T.	4.9/	7-	4.2		ann.rad./ 0.104 0.5998 0.6052 0.6138
^{118}I		117.9134	14. m	β^+ ,EC/7.0		2-	2.0		ann.rad./ 0.5448 0.6052 1.3384
^{119}I		118.9102	19. m	β^+ /54 /3.5 EC/46 /	2.4/	(5/2+)	+2.9		ann.rad./ Te k x-ray 0.2575
^{120m}I			53. m	β^+ /80 / EC/20 /	3.8		4.2		ann.rad. Te k x-ray 0.4257 0.5604 0.6147 1.3459
^{120}I		119.91005	1.35 h	β^+ /81 /5.62 EC/19 /	4.03 4.60	2-	1.23		ann.rad./ Tek x-ray 0.5604 0.6411 1.5230 (0.43 - 3.1)
^{121}I		120.90737	2.12 h	β^+ /13 /2.27 EC/87 /	1.2/	5/2+	2.3		ann.rad./ Te k x-ray 0.2122 (0.14 - 1.1)
^{122}I		121.90760	3.6 m	β^+ /4.234 EC/	3.1/	1+	0.94		ann.rad./ Te k x-ray 0.5641
^{123}I		122.905605	13.2 h	EC/1.242		5/2+	2.82		Te k x-ray 0.1590
^{124}I		123.906211	4.18 d	β^+ /23 /3.160 EC/77 /	1.54/ 2.14/ 0.75/	2-	1.44		ann.rad./ Te k x-ray 0.6027/62.9 0.7228/10.3 1.6910/11.2 (0.31-1.73)
^{125}I		124.904624	59.4 d	EC/0.1861		5/2+	2.82	-0.89	Te k x-ray 0.0355
^{126}I		125.905619	13.0 d	EC/ β^+ /2.155 β^- /1.258	 1.13/ 0.87/ 1.25/	2-	1.44		ann.rad./ Te k x-ray 0.3887 0.6622
^{127}I	100.	126.904468				5/2+	+2.8133	-0.79	
^{128}I		127.905805	25.00 m	β^- /2.118 EC/1.251	2.13/	1+			Te k x-ray 0.44287 0.52658
^{129}I		128.904988	1.7×10^7 y	β^- /0.194	0.15/	7/2+	+2.621	-0.55	Xe k x-ray 0.0396
^{130m}I			9.0 m	I.T./83 /0.048 β^- /17 /		2+			I k x-ray 0.5361
^{130}I		129.906674	12.36 h	β^- /2.949	1.04/ 0.62	5+	3.35		0.4180 0.5361 0.6685 0.7395
^{131}I		130.906125	8.040 d	β^- /0.971	0.606/	7/2+	+2.742	-0.40	0.08017 0.28431 0.36446 0.63699
^{132m}I			1.39 h	IT		8-			
^{132}I		131.90800	2.28 h	β^- /14 /3.58 I.T./86 /	0.80/ 1.03/ 1.2/ 1.6/ 2.16/	4+	3.09	0.09	I k x-ray 0.0980 0.5059 0.52264 0.63019 0.6506 0.66768 0.77260 0.95457

Elem. or Isot.	Natural Abundance (%)	Atomic Mass or Weight	Half-Life	Decay Mode/Energy (/MeV)	Particle Energy /Intensity (MeV / %)	Spin (h/2π)	Nuclear Magnetic Mom.(nm)	Elect. Quadr. Mom. (b)	γ-ray Energy /Intensity (MeV / %)
^{133m}I			9. s	I.T./1.63		19/2-			I kx-ray 0.0730 0.6474 0.9126
^{133}I		132.90781	20.8 h	β^- /1.77	1.24/85	7/2+	+2.86	-0.27	0.51056 0.52989 0.87537
^{134m}I			3.7 m	I.T./98 /0.316 β^- /2 /		8-			I k x-ray 0.0444 0.2719
^{134}I		133.9099	52.6 m	β^- /4.2	1.2/	4+			0.1354 0.84702 0.88409
^{135}I		134.91005	6.57 h	β^- /2.65	0.9/ 1.3/	7/2+			0.2884 0.41768 0.52658 1.13156 1.26046
^{136m}I			47. s	β^- /	4.7/ 5.2/	6-			0.1973 0.3468 0.3701 0.3814 1.3130 (0.16 - 2.36)
^{136}I		135.91466	1.39 m	β^- /6.93	4.3/ 5.6/	2-			0.3447 1.3130 1.3211 2.2896 (0.3 - 6.1)
^{137}I		136.91787	24.5 s	β^- /5.88	5.0/	(7/2+)			0.6010 1.2180 1.2201 1.3026 1.5343 (0.25 - 4.4)
^{138}I		137.9224	6.5 s	β^- /7.8	6.9/ 7.4/	2-			0.4836 0.5888 0.8752 (0.4 - 5.3)
^{139}I		138.92609	2.30 s	β^- /6.81 n/					0.192 0.198 0.273 0.382 0.386 0.468 0.683 1.313
^{140}I		139.9310	0.86 s	β^- /8.8 n/		(3)			0.372 0.377 0.457
^{141}I		140.9351	0.45 s	β^- /7.8					
^{142}I			$\approx$ 0.2 s	β^-					
$_{54}$Xe		**131.29(2)**							
^{110}Xe		109.9445	0.2 s	β^+ /9.2					
^{111m}Xe			0.9 s	EC,β^+					
^{111}Xe		110.9416	0.7 s	EC,β^+ /10.6 α/	 3.58(1)/				
^{112}Xe		111.9357	3. s	EC,β^+ /7.2 α/0.8 /					
^{113}Xe		112.9334	2.8 s	EC,β^+ /9.1					
^{114}Xe		113.9281	10.0 s	β^+ ,EC/5.9		0+			ann.rad./ 0.1031 0.1616 0.3085 0.6826 0.7088
^{115}Xe		114.9270	18. s	β^+ ,EC/7.6		(5/2+)			ann.rad./
^{116}Xe		115.9214	56. s	β^+ ,EC/4.3	3.3/	0+			ann.rad./ 0.1042 0.1916 0.2477 0.3107 0.4127
^{117}Xe		116.9206	1.02 m	β^+ ,EC/6.5		(5/2+)			ann.rad./ 0.2214 0.5190 0.6389 0.6613

Elem. or Isot.	Natural Abundance (%)	Atomic Mass or Weight	Half-Life	Decay Mode/Energy (/MeV)	Particle Energy /Intensity (MeV / %)	Spin (h/2π)	Nuclear Magnetic Mom.(nm)	Elect. Quadr. Mom. (b)	γ-ray Energy /Intensity (MeV / %)
^{118}Xe		117.917	$\approx$ 4. m	β^+ ,EC/3.	2.7/	0+			ann.rad./ 0.0535 0.0600 0.1199
^{119}Xe		118.9156	5.8 m	β^+ ,EC/5.0	3.5/	7/2+	0.6		0.0873 0.1000 0.2318 0.4615
^{120}Xe		119.91216	40. m	β^+ ,EC/97 /1.96 β^+ /3 /		0+			I k x-ray 0.0251 0.0726 0.1781 (0.1 - 1.03)
^{121}Xe		120.91138	39. m	β^+ /44 /3.73 EC/56 /	2.8/	5/2+	0.65		ann.rad./ I k x-ray 0.1328 0.2527 0.4452 (0.1 - 3.1)
^{122}Xe		121.9086	20.1 h	EC/0.9		0+			I k x-ray 0.3501
^{123}Xe		122.90848	2.00 h	β^+ /23 /2.68 EC/77 /	1.51/	1/2+			ann.rad./ I k x-ray 0.1489 0.1781 (0.1 - 2.1)
^{124}Xe	0.10(1)	123.905895							
^{125m}Xe			57. s	I.T./0.252		(9/2-)			Xe k x-ray 0.1111 0.141
^{125}Xe		124.906398	17.1 h	EC/1.653	0.47/	1/2+			I k x-ray 0.1884 0.2434
^{126}Xe	0.09(1)	125.90427				0+			
^{127m}Xe			1.15 m	I.T./0.297		(9/2-)			Xe k x-ray 0.1246 0.1725
^{127}Xe		126.905179	36.4 d	EC/0.662		1/2+	-0.504		I k x-ray 0.1721 0.2029 0.3750
^{128}Xe	1.91(3)	127.903531				0+			
^{129m}Xe			8.89 d	I.T./0.236		11/2-			Xe k x-ray 0.0396 0.1966
^{129}Xe	26.4(6)	128.904780				1/2+	-0.7780		
^{130}Xe	4.1(1)	129.903509				0+			
^{131m}Xe			11.9 d	I.T./0.164		11/2-	+0.6908		Xe k x-ray 0.16398
^{131}Xe	21.2(4)	130.905083				3/2+	+0.69186	-0.12	
^{132}Xe	26.9(5)	131.904155				0+			
^{133m}Xe			2.19 d	I.T./0.233		11/2-	-1.082		Xe k x-ray 0.23325
^{133}Xe		132.905906	5.243 d	β^- /0.427	0.346/99	3/2+	+0.813	+0.14	Cs k x-ray 0.080998 0.1606
^{134}Xe	10.4(2)	133.905395				0+			
^{135m}Xe			15.3 m	I.T./		11/2-	1.103		Xe k x-ray 0.52658
^{135}Xe		134.90721	9.10 h	β^- /1.15	0.91/	3/2+	0.903		0.24975 0.60807
^{136}Xe	8.9(1)	135.90722				0+			
^{137}Xe		136.91156	3.82 m	β^- /4.17	4.1/ 3.6/	7/2-			0.45549 0.8489 0.9822 1.2732 1.7834 2.8498
^{138}Xe		137.91399	14.1 m	β^- /2.77	0.8/ 2.4/	0+			0.1538 0.2426 0.2583 0.4345 1.76826 2.0158
^{139}Xe		138.91879	39.7 s	β^- /5.06	4.5/ 5.0/				0.1750 0.2186 0.2965 (0.1 - 3.37)

Elem. or Isot.	Natural Abundance (%)	Atomic Mass or Weight	Half-Life	Decay Mode/Energy (/MeV)	Particle Energy /Intensity (MeV / %)	Spin (h/2π)	Nuclear Magnetic Mom.(nm)	Elect. Quadr. Mom. (b)	γ-ray Energy /Intensity (MeV / %)
^{140}Xe		139.9216	13.6 s	β^- /4.1	2.6	0+			0.0801 0.6220 0.8055 1.4137 (0.04 - 2.3)
^{141}Xe		140.9267	1.72 s	β^- /6.2	6.2/	5/2+			0.1187 0.9095 (0.05 - 2.55)
^{142}Xe		141.9297	1.22 s	β^- /5.0	3.7/ 4.2/	0+			0.0338 0.0729 0.2038 0.3091 0.4145 0.5382 0.5718 0.6181 0.6448
^{143m}Xe			0.96 s	β^-					
^{143}Xe		142.9352	0.30 s	β^- /7.3					
^{144}Xe		143.9385	1.2 s	β^- /6.1					
^{145}Xe			0.9 s	β^- ,(n)					
$_{55}$Cs		**132.90545(2)**							
^{112}Cs		111.9503	0.5 ms	p	0.81				
^{113}Cs		112.9445	17. μs	p	0.96				
^{114}Cs		113.9408	0.58 s	β^+ ,EC/11.8		1+			ann.rad./ 0.6826 0.7088
^{115}Cs		114.9359	≈ 1.4 s	β^+ ,EC/8.4					ann.rad./
^{116m}Cs			0.7 s	β ,EC/					ann.rad./ 0.3935
^{116}Cs		115.9330	3.8 s	β^+ ,EC/10.8					ann.rad./ 0.3935 0.5243 0.6151 0.6223
^{117m}Cs			6.5 s	β^+ ,EC/					
^{117}Cs		116.9286	≈ 8.4 s	β^+ ,EC/7.5					ann.rad./
^{118m}Cs			17. s	β^+ ,EC/					
^{118}Cs		117.92654	14. s	β^+ ,EC/9.		2			ann.rad./ 0.3372 0.4727 0.5865 0.5906
^{119m}Cs			28. s			3/2			
^{119}Cs		118.92234	38. s	β^+ ,EC/6.3		9/2+	+5.5	+2.8	ann.rad./ 0.169 0.176 0.224 0.257
^{120m}Cs			60. s	β^+ ,EC/					
^{120}Cs		119.92066	64. s	β^+ ,EC/7.92		2+	+3.87	+1.45	ann.rad./ 0.3224 0.4735 0.5534 (0.3 - 3.28)
^{121m}Cs			2.0 m	I.T./60 / β^+ /40 /	 4.4	(9/2+)	+5.41	+2.7	ann.rad./ 0.1794 0.1961
^{121}Cs		120.91718	2.3 m	β^+ ,EC/5.40	4.38/	3/2+	+0.77	+0.84	ann.rad./ 0.1337 (0.08 - 0.56)
^{122m2}Cs			4.4 m	β^+ ,EC		8-	+4.77	+3.3	ann.rad./ 0.3311 0.4971 0.6385 (0.27 - 2.22)
^{122m1}Cs			0.36 s	IT					
^{122}Cs		121.91614	21. s	β^+ ,EC/7.1	5.8/	(1+)	-0.133	-0.19	ann.rad./ 0.3311 0.5120 0.8179
^{123m}Cs			1.6 s	I.T./		11/2-			Cs k x-ray 0.0946

Elem. or Isot.	Natural Abundance (%)	Atomic Mass or Weight	Half-Life	Decay Mode/Energy (/MeV)	Particle Energy /Intensity (MeV / %)	Spin (h/2π)	Nuclear Magnetic Mom.(nm)	Elect. Quadr. Mom. (b)	γ-ray Energy /Intensity (MeV / %)
^{123}Cs		122.91299	5.87 m	β^+ /75 /4.20 EC/25 /	3.0/	1/2+	+1.38		ann.rad./ Xe k x-ray 0.0974 0.5964
^{124m}Cs			6.3 s	IT		7+			
^{124}Cs		123.91225	30. s	β^+ /92 /5.92 EC/8 /	≈ 5.	1+	+0.673	-0.74	ann.rad./ Xe k x-ray 0.3539 0.4925 0.9418
^{125}Cs		124.90972	45. m	β^+ /40 /3.09 EC/60 /	2.06/	1/2+	+1.41		ann.rad./ Xe k x-ray 0.112 0.526
^{126}Cs		125.90945	1.64 m.	β^+ /81 /4.83 EC/19 /	3.4 3.7/	1+	+0.78	-0.68	ann.rad./ Xe k x-ray 0.3886 0.4912 0.9252
^{127}Cs		126.90741	6.2 h	β^+ /96 /2.08 EC/4 /	0.65/ 1.06	1/2+	+1.46		Xe k x-ray 0.1247 0.4119
^{128}Cs		127.90775	3.62 m	β^+ /68 /3.930 EC/32 /	2.44/ 2.88/	1+	+0.97	-0.57	ann.rad./ Xe k x-ray 0.4429
^{129}Cs		128.90606	1.336 d	EC/1.195		1/2+	+1.49		Xe k x-ray 0.3719 0.4115
^{130m}Cs			3.5 m	IT,β^+ ,EC		5-			
^{130}Cs		129.90671	29.21 m	β^+ /55 /2.98 EC/43 / β^- /1.6 /0.37	1.98/ 0.44/1.6	1+	+1.46	-0.06	ann.rad./ Xe k x-ray 0.5361
^{131}Cs		130.90546	9.69 d	EC/0.352		5/2+	+3.54	-0.58	Xe k x-ray
^{132}Cs		131.906430	6.48 d	EC/98 / β^+ /0.3 /2.120 β^- / /1.280		2-	+2.22	+0.51	Xe k x-ray 0.4646 0.6302 0.66769
^{133}Cs	100.	132.905447				7/2+	+2.582	-0.0037	
^{134m}Cs			2.91 h	I.T./0.139		8-	+1.098	+1.0	Cs k x-ray 0.12749
^{134}Cs		133.906714	2.065 y	β^- /2.059 EC/1.22	0.089/27 0.658/70	4+	+2.994	+0.39	0.56327 0.56935 0.60473 0.79584
^{135m}Cs			53. m	I.T./1.627		19/2-	+2.18	+0.9	0.7869 0.8402
^{135}Cs		134.905972	2.3x10^6 y	β^- /0.269	0.205/100	7/2+	+2.732	+0.05	
^{136m}Cs			19. s	I.T./		8			
^{136}Cs		135.907307	13.16 d	β^- /2.548	0.341/	5+	+3.71	+0.2	0.06691 0.34057 0.81850 1.04807
^{137}Cs		136.907085	30.2 y	β^- /1.176	0.514/95	7/2+	+2.84	+0.05	Ba k x-ray 0.66164
^{138m}Cs			2.9 m	I.T./75 /0.080 β^- /25 /	 3.3	6-	+1.71	-0.40	Cs k x-ray 0.0799 0.1917 0.4628 1.43579
^{138}Cs		137.91101	32.2 m	β^- /5.37	2.9/	3-	+0.700	+0.12	0.1381 0.46269 1.00969 1.43579 2.21788
^{139}Cs		138.913359	9.3 m	β^- /4.213	4.21	7/2+	+2.70	-0.07	0.6272 1.2832 (0.4 - 3.66)
^{140}Cs		139.91727	1.06 m	β^- /6.22	5.7/ 6.21/	1-	+0.13390	-0.11	0.5283 0.6023 0.9084 (0.41 - 3.94)

Elem. or Isot.	Natural Abundance (%)	Atomic Mass or Weight	Half-Life	Decay Mode/Energy (/MeV)	Particle Energy /Intensity (MeV / %)	Spin (h/2π)	Nuclear Magnetic Mom.(nm)	Elect. Quadr. Mom. (b)	γ-ray Energy /Intensity (MeV / %)
^{141}Cs		140.92005	24.9 s	β^- /5.26	5.20/	7/2+	+2.44	-0.4	Ba k x-ray
									0.0485
									0.5616
									0.5887
									1.1940
									(0.05 - 3.33)
^{142}Cs		141.92430	1.8 s	β^- /7.31	6.9/				0.3596
					7.28				0.9668
									1.1759
									1.3265
^{143}Cs		142.92732	1.78 s	β^- /6.24	6.1	(3/2+)	+0.87	+0.47	0.1955
									0.2324
									0.3064
									(0.17 - 1.98)
^{144}Cs		143.93203	1.01 s	β^- /8.47	8.46/	1	-0.546	+0.30	0.1993
					7.9/				0.5598
									0.6392
									0.7587
^{145}Cs		144.93541	0.59 s	β^- /7.89	7.4/	3/2+	+0.784	+0.6	0.1126
					7.9/				0.1755
									0.1990
^{146}Cs		145.94024	0.322 s	β^- ,(n)/9.38	≈ 9.0	2-			
^{147}Cs		146.9439	0.227 s	β^- ,(n)/9.3					
^{148}Cs		147.9490	0.15 s	β^- ,(n)/10.5					
$_{56}$Ba		**137.327(7)**							
^{117}Ba		116.9377	1.8 s	β^+ ,EC/8.4		(3/2-)			
^{119}Ba		118.931	5.4 s	β^+ ,EC/8.					
^{120}Ba		119.9260	24. s	β^+ ,EC/5.0		0+			ann.rad./
									0.140
									(0.075-0.146)
^{121}Ba		120.9245	30. s	β^+ ,EC/6.8		5/2	+0.660	+1.8	ann.rad./
^{122}Ba		121.9203	2.0 m	β^+ ,EC/3.8		0+			ann.rad./
^{123}Ba		122.9189	2.7 m	β^+ ,EC/5.5			-0.68	+1.5	ann.rad./
									0.0306
									0.0927
									0.1161
									0.1235
^{124}Ba		123.91509	12. m	β^+ ,EC/2.65					ann.rad./
									0.1695
									0.1888
									1.2160
^{125m}Ba			8. m	β^+ ,EC/	4.5				
^{125}Ba		124.9146	3.5 m	β^+ ,EC/4.6	3.4	1/2+	+0.18		ann.rad./
									0.0550
									0.0776
									0.0854
									0.1409
^{126}Ba		125.91124	1.65 h	β^+ /2 /1.67		0+			Cs k x-ray
				EC/98 /					0.2179
									0.2336
									0.2576
^{127}Ba		126.9111	12.9 m	β^+ /54 /3.5		1/2+	+0.09		ann.rad./
				EC/46 /					Cs k x-ray
									0.1148
									0.1808
									(0.07 - 2.5)
^{128}Ba		127.90831	2.43 d	EC/0.52		0+			Cs k x-ray
									0.27344
^{129m}Ba			2.17 h	EC/98 /		7/2+			Cs k x-ray
				β^+ /2 /					0.1769
									0.1823
									0.2023
									1.4593
^{129}Ba		128.90868	2.2 h	β^+ /20 /2.43	1.42/	1/2+	-0.40		ann.rad./
				EC/80 /					Cs k x-ray
									0.1291
									0.2143
									0.2208
^{130}Ba	0.106(2)	129.90631				0+			
^{131m}Ba			14.6 m	I.T./0.187		9/2-	-0.87	+1.5	Ba k x-ray
									0.1085

Elem. or Isot.	Natural Abundance (%)	Atomic Mass or Weight	Half-Life	Decay Mode/Energy (/MeV)	Particle Energy /Intensity (MeV / %)	Spin (h/2π)	Nuclear Magnetic Mom.(nm)	Elect. Quadr. Mom. (b)	γ-ray Energy /Intensity (MeV / %)
^{131}Ba		130.90693	11.7 d	EC/1.37		1/2+	0.7081		Cs k x-ray 0.12381 0.21608 0.49636
^{132}Ba	0.101(3)	131.905056				0+			
^{133m}Ba			1.621 d	I.T./0.288		11/2-	-0.91	+0.9	Ba k x-ray 0.2761
^{133}Ba		132.906003	10.53 y	EC/0.517		1/2+	0.7717		Cs k x-ray 0.08099 0.35600
^{134}Ba	2.417(27)	133.904504				0+			
^{135m}Ba			1.20 d	I.T./0.2682		11/2-	-1.00	+1.0	Ba k x-ray 0.2682
^{135}Ba	6.592(18)	134.905684				3/2+	+0.838	+0.16	
^{136m}Ba			0.308 s	I.T./2.0305		7-			Ba k x-ray 0.8185 1.0481
^{136}Ba	7.854(36)	135.904571				0+			
^{137m}Ba			2.552 m	I.T./0.6617		11/2-	-0.99	+0.8	Ba k x-ray 0.66164
^{137}Ba	11.23(4)	136.905822				3/2+	+0.9374	+0.245	
^{138}Ba	71.70(7)	137.905242				0+			
^{139}Ba		138.908836	1.396 h	β^- /2.317	2.14/27 2.27/72	7/2-	-0.97	-0.57	0.16585 1.2544 1.42033
^{140}Ba		139.91060	12.75 d	β^- /1.05	0.48 1.0/ 1.02/	0+			0.16268 0.30485 0.53727
^{141}Ba		140.91441	18.3 m	β^- /3.22	2.59/ 2.73/	3/2-	-0.34	+0.45	0.1903 0.2770 0.3042 (0.1 - 2.5)
^{142}Ba		141.91645	10.7 m	β^- /2.212	1.0/ 1.10/	0+			0.23152 0.25512 0.3090 1.2040
^{143}Ba		142.92061	14.3 s	β^- /4.24	4.2/	5/2+	+0.44	-0.88	0.1786 0.21148 0.7988 (0.17 - 2.4)
^{144}Ba		143.92294	11.4 s	β^- /3.1	2.4/ 2.9/	0+			La k x-ray 0.10386 0.1566 0.1728 0.3882 0.43048
^{145}Ba		144.9269	4.0 s	β^- /4.9	4.9/	(5/2-)	-0.28	+1.22	La k x-ray 0.0918 0.09709
^{146}Ba		145.9302	2.20 s	β^- /4.12	3.9/	0+			0.0644 0.2513 0.3270 0.3329 0.3622
^{147}Ba		146.9340	0.892 s	β^- /5.75	5.5/				
^{148}Ba		147.9377	0.64 s	β^- ,n/5.11					
^{149}Ba		148.9421	0.36 s	β^- ,(n)/7.3					
$_{57}$La		**138.9055(2)**							
^{120}La		119.938	2.8 s	EC,β^+ /11.					
^{122}La		121.931	9. s	EC,β^+ /≈ 9.7					
^{123}La		122.9262	17. s	EC /7.					
^{124}La		123.9245	30. s	EC/≈ 8.8		(7+)			
^{125}La		124.9207	1.2 m	β^+ ,EC/5.6		11/2-			ann.rad./ 0.0436 0.0676
^{126}La		125.9194	1.0 m	β^+ ,EC/7.6					ann.rad./ 0.2561 0.340 0.4555 0.6214
^{127}La		126.9162	3.8 m	β^+ ,EC/4.7		3/2+			ann.rad./ 0.025 0.0562

Elem. or Isot.	Natural Abundance (%)	Atomic Mass or Weight	Half-Life	Decay Mode/Energy (/MeV)	Particle Energy /Intensity (MeV / %)	Spin (h/2π)	Nuclear Magnetic Mom.(nm)	Elect. Quadr. Mom. (b)	γ-ray Energy /Intensity (MeV / %)
^{128}La		127.9155	5.0 m	β^+ /80 /6.7		(5-)			ann.rad./
				EC/20 /					Ba k x-ray
									0.2841
									0.4793
									0.6436
^{129m}La			0.56 s	IT		(11/2-)			
^{129}La		128.91267	11.6 m	β^+ /58 /3.72	2.42/	3/2+			ann.rad./
				EC/42 /					Ba k x-ray
									0.1105
									0.2786
									(0.1 - 1.8)
^{130}La		129.9123	8.7 m	β^+ /78 /5.6		3+			ann.rad./
				EC/22 /					Ba k x-ray
									0.3573
									0.5444
									0.5506
									0.9079
^{131}La		130.9101	59. m	β^+ /76 /3.0	1.42/	3/2+			ann.rad./
				EC/24 /	1.94/				Ba k x-ray
									0.1085
									0.3658
									0.5263
^{132m}La			24. m	I.T./76 /		6-			La k x-ray
				β^+ ,EC/24 /					0.1352
									0.4645
^{132}La		131.91011	4.8 h	β^+ /40 /4.71	2.6/	2-			ann.rad./
				EC/60 /	3.2				Ba k x-ray
					3.7/				0.4645
									0.5671
^{133}La		132.9084	3.91 h	β^+ /4 /2.2	1.2/	5/2+			Ba k x-ray
				EC/96 /					0.2788
									0.2901
									0.3024
^{134}La		133.90849	6.5 m	β^+ /63 /3.71	2.67/	1+			ann.rad./
				EC/37/					Ba k x-ray
									0.6047
									(0.5 - 1.9)
^{135}La		134.90697	19.5 h	EC/1.20		5/2+			Ba k x-ray
									0.4805
^{136}La		135.9077	9.87 m	β^+ /36 /2.9	1.8/	1+			ann.rad./
				EC/64 /					Ba k x-ray
									0.8185
^{137}La		136.90647	6x10^4 y	EC/0.60		7/2+	+2.70	+0.2	
^{138}La	0.0902(2)	137.907107	1.06x10^{11} y			5+			
^{139}La	99.9098(2)	138.906349				7/2+			
^{140}La		139.909473	1.678 d	β^- /3.762	1.35	3-			
					1.24/				
					1.67/				
^{141}La		140.910958	3.90 h	β^- /2.502	2.43/	7/2+			
^{142}La		141.91408	1.54 h	β^- /4.505	2.11/	2-			
					2.98/				
					4.52/				
^{143}La		142.91606	14.1 m	β^- /3.43	3.3/	7/2-			
^{144}La		143.9196	40.7 s	β^- /5.5	4.1/				
^{145}La		144.9217	24. s	β^- /4.1	4.1/	3/2+			
^{146m}La			10.0 s	β^- /6.7	5.5/	(6)			
^{146}La		145.9258	6.3 s	β^- /6.6	6.2/	(2-)			
^{147}La		146.9278	4.02 s	β^- /5.0	4.6/				
^{148}La		147.9322	1.1 s	β^- /7.26		2-			
^{149}La		148.9342	1.10 s	β^- /5.5					
$_{58}$Ce		**140.115(4)**							
^{123}Ce		122.936	3.8 s	β^+ ,EC/≈8.6					ann.rad./
^{124}Ce		123.931	6. s	EC /≈5.6					
^{125}Ce		124.929	10. s	β^+ ,EC/7.		(5/2+)			ann.rad./
^{126}Ce		125.9241	50. s	EC/4.					
^{127}Ce		126.9228	32. s	β^+ ,EC/6.1					ann.rad./
									0.06-0.25
^{128}Ce		127.9189	6. m	β^+ ,EC/3.2					ann.rad./
^{129}Ce		128.9187	3.5 m	β^+ ,EC/5.6					ann.rad./
									0.0675
^{130}Ce		129.9147	26. m	β^+ ,EC/2.2		0+			ann.rad./
									La k x-ray

Elem. or Isot.	Natural Abundance (%)	Atomic Mass or Weight	Half-Life	Decay Mode/Energy (/MeV)	Particle Energy /Intensity (MeV / %)	Spin (h/2π)	Nuclear Magnetic Mom.(nm)	Elect. Quadr. Mom. (b)	γ-ray Energy /Intensity (MeV / %)
^{131m}Ce			5. m	β^+ ,EC/					ann.rad./ 0.2304 0.3955 0.4213
^{131}Ce		130.9144	10. m	β^+ ,EC/4.0	2.8/				ann.rad./ 0.119 0.169 0.414
^{132}Ce		131.9115	3.5 h	EC/1.3		0+			La k x-ray 0.1554 0.1821
^{133m}Ce			1.6 h	β^+ ,EC/		1/2+			ann.rad./ 0.0769 0.0973 0.5577
^{133}Ce		132.9116	5.4 h	β^+ /8 /2.9 EC/92 /	1.3/	9/2-			ann.rad./ La k x-ray 0.0584 0.1308 0.4722 0.5104
^{134}Ce		133.9090	3.16 d	EC/0.5		0+			La k x-ray 0.1304 0.1623 0.6047
^{135m}Ce			20. s	I.T./0.446		11/2-			Ce k x-ray 0.0826 0.1497 0.2134
^{135}Ce		134.90915	17.7 h	β^+ /1 /2.026 EC/99 /	0.8/	1/2+			La k x-ray 0.0345 0.2656 0.3001 0.6068
^{136}Ce	0.19(1)	135.90714				0+			
^{137m}Ce			1.43 d	I.T./99 /0.254 EC/0.8 /		11/2-	0.70		Ce k x-ray 0.1693 0.2543
^{137}Ce		136.90778	9.0 h	β^+ /1.222		3/2+	0.9		La k x-ray 0.4472
^{138}Ce	0.25(1)	137.90599				0+			
^{139m}Ce			56.4 s	I.T./0.7542		11/2-			Ce k x-ray 0.7542
^{139}Ce		138.90665	137.6 d	EC/0.28		3/2+	0.9		La k x-ray 0.16585
^{140}Ce	88.48(10)	139.905435				0+			
^{141}Ce		140.908272	32.50 d	β^- /0.581	0.436/69 0.581/31	7/2-	1.1		Pr k x-ray 0.14544/48.0
^{142}Ce	11.08(10)	141.909241				0+			
^{143}Ce		142.912382	1.38 d	β^- /1.462	1.404/ 1.110/47	3/2-	≈ 1.		Pr k x-ray 0.0574 0.2933
^{144}Ce		143.913643	284.6 d	β^- /0.319	0.185/20 0.318/	0+			Pr k x-ray 0.0801 0.1335
^{145}Ce		144.91723	3.00 m	β^- /2.54	1.7/24 1.3	3/2-			Pr k x-ray 0.0627 0.7245
^{146}Ce		145.9187	13.5 m	β^- /1.04	0.7/90	0+			Pr k x-ray 0.0986 0.2182 0.3167
^{147}Ce		146.9225	56. s	β^- /3.29	3.3/				0.0930 0.2687
^{148}Ce		147.9244	56. s	β^- /2.1	1.66/	0+			0.0904 0.0985 0.1212 0.2918
^{149}Ce		148.9283	5.2 s	β^- /4.2					0.0577 0.0864 0.3800
^{150}Ce		149.9302	4.4 s	β^- /3.0					0.1099
^{151}Ce		150.9340	1.0 s	β^- /5.3					0.0526
^{152}Ce		151.9366	1.4 s	β^- /4.4					Pr k x-ray 0.098 0.115
$_{59}$Pr		**140.90765(2)**							

Elem. or Isot.	Natural Abundance (%)	Atomic Mass or Weight	Half-Life	Decay Mode/Energy (/MeV)	Particle Energy /Intensity (MeV / %)	Spin (h/2π)	Nuclear Magnetic Mom.(nm)	Elect. Quadr. Mom. (b)	γ-ray Energy /Intensity (MeV / %)
^{124}Pr		123.943	1.2 s	β⁺ ,EC/12.					ann.rad./
^{126}Pr		125.935	3. s	β⁺ ,EC/≈10.4					ann.rad./
^{127}Pr		126.931	≈6. s	β⁺ /≈7.5					ann.rad./ (0.12-0.16)
^{128}Pr		127.9288	3.2 s	β⁺ ,EC/≈9.3					ann.rad./
^{129}Pr		128.9249	0.5 m	β⁺ ,EC/5.8					ann.rad./ (0.04-0.20)
^{130}Pr		129.9234	40. s	β⁺ ,EC/8.1					ann.rad./
^{131m}Pr			5.7 s						(0.06 - 0.16)
^{131}Pr		130.9201	1.7 m	β⁺ ,EC/5.3				≈5.5	ann.rad./ (0.07-0.34)
^{132}Pr		131.9191	1.6 m	β⁺ ,EC/7.1					ann.rad./ 0.325 0.496 0.533
^{133}Pr		132.9162	6.5 m	β⁺ ,EC/4.3		5/2+			ann.rad./ 0.074 0.1343 0.2419 0.3156 0.3308 0.4650
^{134m}Pr			≈ 11. m	β⁺ ,EC/					ann.rad./ 0.294 0.460 0.495 0.632
^{134}Pr		133.9157	17. m	β⁺ ,EC/6.2		2+			ann.rad./ 0.294 0.495
^{135}Pr		134.9131	24. m	β⁺ ,EC/3.7	2.5/	3/2+			ann.rad./ 0.0826 0.2135 0.2961 0.5832
^{136}Pr		135.91265	13.1 m	β⁺ /57 /5.13 EC/43	2.98/	2+			ann.rad./ Ce k x-ray 0.5398 0.5522
^{137}Pr		136.91068	1.28 h	β⁺ /26 /2.70 EC/74 /	1.68/	5/2+			ann.rad./ Ce k x-ray 0.4339 0.5140 0.8367 (0.16 - 1.8)
^{138m}Pr			2.1 h	β⁺ /24 / EC/76 /	1.65/	7-			ann.rad./ Ce k x-ray 0.3027 0.7887 1.0378 (0.07 - 2.0)
^{138}Pr		137.91075	1.45 m	β⁺ /75 /4.44 EC/25 /	3.42/	1+			ann.rad./ Ce k x-ray 0.7887
^{139}Pr		138.90893	4.41 h	β⁺ /8 /2.129 EC/92 /	1.09/	5/2+			ann.rad./ Ce k x-ray 0.2551 1.3473 1.6307
^{140}Pr		139.90907	3.39 m	β⁺ /51 /3.39 EC/49 /	2.37/	1+			ann.rad./ Ce k x-ray 0.3069 1.5965
^{141}Pr	100.	140.907648				5/2+	+4.275	-0.059	
^{142m}Pr			14.6 m	I.T./0.004	c.e./	5-	2.2		
^{142}Pr		141.910041	19.12 h	β⁻ /2.162 EC/0.744	0.58/4 2.16/96	2-	+0.234	+0.030	0.5088 1.57580
^{143}Pr		142.910813	13.57 d	β⁻ /0.934	0.933/	7/2+	+2.7	+0.8	0.7420
^{144m}Pr			7.2 m	I.T./99+ /0.059 β⁻ /		3-			Pr k x-ray 0.0590 0.6965 0.8142
^{144}Pr		143.913301	17.28 m	β⁻ /2.998	0.807/1 2.30/ 2.996/98	0-			0.69649 1.48912 2.18562
^{145}Pr		144.91451	5.98 h	β⁻ /1.81	1.80/97	7/2+			0.0725 0.6758 0.7483

Elem. or Isot.	Natural Abundance (%)	Atomic Mass or Weight	Half-Life	Decay Mode/Energy (/MeV)	Particle Energy /Intensity (MeV / %)	Spin (h/2π)	Nuclear Magnetic Mom.(nm)	Elect. Quadr. Mom. (b)	γ-ray Energy /Intensity (MeV / %)
^{146}Pr		145.9176	24.2 m	β^- /4.2	2.2/30	2-			0.4539/48
					3.7/10				1.5247
					4.2/40				
^{147}Pr		146.91898	13.4 m	β^- /2.69	1.5/	3/2+			0.3146/24.
					2.1/				0.5779/16
									0.6413/19.
^{148m}Pr			2.0 m	β^- /	4.0/	(4)			0.3016
					3.8/				0.4506
									0.6975
^{148}Pr		147.9222	2.27 m	β^- /4.9	4.8/	1-			0.3017
					4.5/				
^{149}Pr		148.92379	2.3 m	β^- /3.40	3.0	(5/2+)			0.1085
									0.1385
									0.1651
^{150}Pr		149.9270	6.2 s	β^- /5.7		1-			0.1302
					≈ 5.5				0.8044
									0.8527
^{151}Pr		150.9283	22.4 s	β^- /4.2					
^{152}Pr		151.9319	3.2 s	β^- /6.7		(3)			0.0726
									0.164
									0.285
^{153}Pr		152.9339	4.3 s	β^- /5.5					
^{154}Pr		153.9381	2.3 s	β^- /7.9					
$_{60}$Nd		**144.24(3)**							
^{127}Nd		126.941	1.8 s	β^+ ,EC/9.		(5/2)			ann.rad./
^{128}Nd		127.935	4. s	β^+ ,EC/6.					ann.rad./
^{129}Nd		128.933	4.9 s	β^+ ,EC/8.		5/2(-)			ann.rad./
									(0.091-0.33)
^{130}Nd		129.929	28. s	β^+ ,EC/5.					ann.rad./
^{131}Nd		130.9271	0.5 m	β^+ ,EC/6.6					ann.rad./
									(0.09-0.36)
^{132}Nd		131.9231	1.5 m	β^+ ,EC/3.7					ann.rad./
									(0.099-0.567)
^{133}Nd		132.9222	1.2 m	β^+ ,EC/5.6					ann.rad./
									(0.06-0.37)
^{134}Nd		133.9187	≈ 8.5 m	β^+ /17 /2.8		0+			ann.rad./
				EC/83 /					Pr k x-ray
									0.1631/58
									(0.09-1.00)
^{135m}Nd			5.5 m	β^+ /					
^{135}Nd		134.9182	12. m	β^+ /65 /4.8		9/2-	0.78	+2.1	ann.rad./
				EC/35 /					Pr k x-ray
									0.0415/23.
									0.204/51.
									(0.11-1.8)
^{136}Nd		135.9150	50.6 m	EC/94 /2.21	1.04/	0+			Pr kx-ray
				β^+ /6 /					0.0401/21.
									0.1091/35.
									(0.10-0.97)
^{137m}Nd			1.6 s	I.T./0.5196		11/2-			Nd k x-ray
									0.1084
									0.1775
									0.2337
^{137}Nd		136.9146	38. m	β^+ /40 /3.69	1.7/20	1/2+	-0.63		ann.rad./
				EC/60 /	2.40/20				Pr k x-ray
									0.0755
									0.5806
^{138}Nd		137.9119	5.1 h	EC/1.1		0+			Pr k x-ray
									0.1995
									0.3258
^{139m}Nd			5.5 h	I.T./12 /0.231	1.17/	11/2-			Nd k x-ray
				β^+ /88 /					Pr k x-ray
									0.1139/34.
									0.7382/30.
^{139}Nd		138.91192	30. m	β^+ /25 /2.79	1.77/	3/2+	0.91	+0.3	ann.rad./
				EC/75 /					Pr k x-ray
									0.4050
^{140}Nd		139.90931	3.37 d	EC /0.22		0+			Pr k x-ray
^{141m}Nd			1.04 m	I.T./99+ /0.756		11/2-			Nd k x-ray
									0.7565
^{141}Nd		140.909605	2.49 h	EC/98 /1.823	0.802/	3/2+	1.01	+0.3	Pr k x-ray
				β^+ /2 /					(0.15-1.7)
^{142}Nd	27.13(12)	141.907719				0+			
^{143}Nd	12.18(6)	142.909810				7/2-	-1.07	-0.6	
^{144}Nd	23.80(12)	143.910083	2.1x10^{15} y			0+			
^{145}Nd	8.30(6)	144.912569				7/2-	-0.66	-0.33	
^{146}Nd	17.19(9)	145.913113				0+			

Elem. or Isot.	Natural Abundance (%)	Atomic Mass or Weight	Half-Life	Decay Mode/Energy (/MeV)	Particle Energy /Intensity (MeV / %)	Spin (h/2π)	Nuclear Magnetic Mom.(nm)	Elect. Quadr. Mom. (b)	γ-ray Energy /Intensity (MeV / %)
^{147}Nd		146.916096	10.98 d	β^- /0.896	0.805/	5/2-	0.58	0.9	Pr k x-ray 0.09111 0.53102
^{148}Nd	5.76(3)	147.916889				0+			
^{149}Nd		148.920145	1.73 h	β^- /1.691	1.03/25 1.13/26 1.42/	5/2-	0.35	1.3	Pr k x-ray 0.1143/19. 0.2113/27. (0.06 - 1.6)
^{150}Nd	5.64(3)	149.920887				0+			
^{151}Nd		150.923825	12.4 m	β^- /2.442	1.2/	(3/2+)			Pm k x-ray 0.1168 0.2557 1.1806 (0.10 - 1.9)m
^{152}Nd		151.92468	11.4 m	β^- /1.1		0+			0.2785/29. 0.2501/18. (0.016 - 0.66)
^{153}Nd		152.9280	28.9 s	β^- /3.6					0.418
^{154}Nd		153.9296	25.9 s	β^- /2.8					0.1519 0.7998
^{155}Nd		154.9334	8.9 s	β^- /5.0					0.1807
^{156}Nd		155.9355	5.5 s	β^- /4.1					0.0848
$_{61}$Pm									
^{130}Pm		129.940	2.2 s	β^+ ,EC/11.					ann.rad./
^{132}Pm		131.934	6. s	β^+ ,EC/10.					ann.rad./
^{133}Pm		132.930	12. s	β^+ ,EC/≈ 7.0					ann.rad./
^{134}Pm		133.9282	24. s	β^+ ,EC/≈ 8.9		(5+)			ann.rad./ 0.294 0.495
^{135}Pm		134.9247	0.8 m	β^+ ,EC/6.0		11/2-			(0.13-0.47)
^{136}Pm		135.9235	1.8 m	β^+ /89 /7.9 EC/11 /		(3+)			ann.rad./ Nd k x-ray 0.3735 0.6027
^{137}Pm		136.9206	2.4 m	β^+ ,EC/5.6		(11/2-)			ann.rad./ 0.1086 0.1775
^{138m}Pm			3.2 m	β^+ /50 /≈ 7.0 EC/50 /	3.9/	3+			ann.rad./ Nd k x-ray 0.5209 0.7290
^{138}Pm		137.9193	10. s	β^+ /6.9	6.1/	1+			ann.rad./
^{139m}Pm			0.18 s	IT/		(11/2-)			0.1887
^{139}Pm		138.91678	4.14 m	β^+ /68 /4.52 EC/32 /	3.52/	(5/2+)			ann.rad./ Nd k x-ray 0.4028 (0.27 - 2.4)
^{140m}Pm			5.87 m	β^+ /70 / EC/30 /	3.2	7/2-			ann.rad./ Nd k x-ray 0.4199 0.7738 1.0283
^{140}Pm		139.91585	9.2 s	β^+ /89 /6.09 EC/11 /	5.07/74	1+			ann.rad./ Nd k x-ray 0.7738 1.4898
^{141}Pm		140.91359	20.9 m	β^+ /52 /3.72 EC/48 /	2.71	5/2+			ann.rad./ Nd k x-ray 0.8862 1.2233
^{142}Pm		141.91295	40.5 s	β^+ /86 /4.87 EC/20 /	3.8/	1+			ann.rad./ Nd k x-ray 0.6414 1.5758
^{143}Pm		142.910928	265. d	EC/1.041 β^+ /$<6 \times 10^{-6}$/		5/2+	3.8		Nd k x-ray 0.7420
^{144}Pm		143.912586	360. d	EC/2.332 β^+ /$<8 \times 10^{-5}$/		5-	1.7		Nd k x-ray 0.6180 0.6965
^{145}Pm		144.912745	17.7 y	EC/0.163		5/2+			Nd k x-ray 0.0723
^{146}Pm		145.914693	5.53 y	EC/63 /1.472 β^- /37 /1.542	0.795/	3-			Nd k x-ray 0.4538 0.7362 0.7474

TABLE OF THE ISOTOPES

Elem. or Isot.	Natural Abundance (%)	Atomic Mass or Weight	Half-Life	Decay Mode/Energy (/MeV)	Particle Energy /Intensity (MeV / %)	Spin (h/2π)	Nuclear Magnetic Mom.(nm)	Elect. Quadr. Mom. (b)	γ-ray Energy /Intensity (MeV / %)
^{147}Pm		146.915134	2.6234 y	β^- /0.224	0.224/	7/2+	+2.6	+0.7	0.1213
^{148m}Pm			41.3 d	β^- /95 /2.6 I.T./5 /0.137	0.4/60 0.5/17 0.7/21	6-	1.8		0.1974 0.5503/94. 0.6300/89. 0.7257/33
^{148}Pm		147.91747	5.37 d	β^- /2.47	1.02/ 2.47/	1-	+2.0	+0.2	0.5503 0.9149 1.4651
^{149}Pm		148.918330	2.212 d	β^- /1.071	0.78/9 1.072/90	7/2+	3.3		0.2859 0.5909 0.8594
^{150}Pm		149.92098	2.68 h	β^- /3.45	1.6/ 2.3/ 1.8/	(1-)			0.3339/69. 1.1658/16. 1.3245/17. (0.25 - 2.9)
^{151}Pm		150.92120	1.183 d	β^- /1.187	0.84/	5/2+	+1.8	1.9	0.1677/8 0.2751/7 0.3401/22 (0.14-1.4)
^{152m2}Pm			15. m	β^- ,I.T./		(>6)			
^{152m1}Pm			7.5 m	β^- /		(4-)			0.1218 0.2447 0.3404 1.0971 1.4375
^{152}Pm		151.9235	4.1 m	β^- /3.5	3.5/20 3.50/60	1+			0.1218 (0.12 - 2.1)
^{153}Pm		152.92414	5.4 m	β^- /1.90	1.7/	(5/2-)			0.0910 0.1198 0.1273
^{154m}Pm			2.7 m	β^- /	2.0/				0.0820 0.1848 1.4403
^{154}Pm		153.9266	1.7 m	β^- /4.1	1.9/				0.0820 0.8396 1.3940 2.0589
^{155}Pm		154.9280	48. s	β^- /3.2		(5/2-)			(0.08 - 2.8)
^{156}Pm		155.93106	26.7 s	β^- /5.16					(0.05-0.78)
^{157}Pm		156.9332	10.9 s	β^- /4.6					
^{158}Pm		157.9367	5. s	β^- /6.3					
$_{62}$Sm		**150.36(3)**							
^{131}Sm			1.2 s	β^+ ,EC/					ann.rad./
^{133}Sm		132.939	2.9 s	β^+ ,EC/≈8.4		5/2+			ann.rad./
^{134}Sm		133.934	11. s	β^+ ,EC/5.		0+			ann.rad./
^{135}Sm		134.932	10. s	β^+ ,EC/7.		7/2+			ann.rad./
^{136}Sm		135.9283	42. s	β^+ ,EC/4.5		0+			ann.rad./
^{137}Sm		136.9271	45. s	β^+ ,EC/6.1					ann.rad./
^{138}Sm		137.9235	3.0 m	β^+ ,EC/3.9		0+			ann.rad./
^{139m}Sm			10. s	I.T./94 /0.457 β^+ /6 /	 4.7	(11/2-)			0.0536 0.0747 Sm k x-ray 0.1118 0.1553 0.1901
^{139}Sm		138.9226	2.6 m	β^+ /75 /5.5 EC/25 /	4.1/	1/2+	-0.53		0.2673 Pm k x-ray 0.3678 0.4028 (0.27 - 2.4)
^{140}Sm		139.9195	14.8 m	β^+ ,EC/3.4	1.9/	0+			ann.rad./ Pm k x-ray 0.1396 0.2255 (0.07 - 1.7)
^{141m}Sm			22.6 m	β^+ /32 / EC/68 / I.T./0.3 /0.1758	1.6/ 2.19/	11/2-	-0.83	+1.6	ann.rad./ Pm k x-ray 0.1966 0.4318 0.7774
^{141}Sm		140.91847	10.2 m	β^+ /52 /4.54 EC/48 /	3.2/	1/2+	-0.74		ann.rad./ Pm k x-ray 0.4382
^{142}Sm		141.91520	1.208 h	β^+ /6 /2.10 EC/94 /	1.0/	0+			ann.rad./ Pm k x-ray
^{143m}Sm			1.10 m	I.T./99 /0.7540		11/2-			Sm k x-ray 0.7540

Elem. or Isot.	Natural Abundance (%)	Atomic Mass or Weight	Half-Life	Decay Mode/Energy (/MeV)	Particle Energy /Intensity (MeV / %)	Spin (h/2π)	Nuclear Magnetic Mom.(nm)	Elect. Quadr. Mom. (b)	γ-ray Energy /Intensity (MeV / %)
^{143}Sm		142.914624	8.83 m	β^+ /46 /3.443 EC/54 /	2.47/	3/2+	+1.01	+0.4	ann.rad./ Pm k x-ray 1.0565
^{144}Sm	3.1(1)	143.911996				0+			
^{145}Sm		144.913407	340. d	EC/0.617		7/2-	-1.1	-0.6	Pm k x-ray 0.0613 0.4924
^{146}Sm		145.913038	1.03x10^{8}y	α/	2.50/	0+			
^{147}Sm	15.0(2)	146.914894	1.06x10^{11}y	α/	2.23/	7/2-	-0.815	-0.26	
^{148}Sm	11.3(1)	147.914818	7x10^{15}y	α/	1.96/	0+			
^{149}Sm	13.8(1)	148.917180	10^{16}y	α/		7/2-	-0.672	+0.075	
^{150}Sm	7.4(1)	149.917272				0+			
^{151}Sm		150.919929	90. y	β^- /0.0768	0.076/	5/2-	-0.363	+0.7	0.02154
^{152}Sm	26.7(2)	151.919729				0+			
^{153}Sm		152.922094	1.929 d	β^- /0.808	0.64/ 0.69/	3/2+	-0.0216	+1.3	Eu k x-ray 0.069676 0.10318
^{154}Sm	22.7(2)	153.922206				0+			
^{155}Sm		154.924636	22.2 m	β^- /1.627	1.52	3/2-		1.1	Eu k x-ray 0.1043/75.
^{156}Sm		155.92553	9.4 h	β^- /0.72	0.43/ 0.71/	0+			0.0872 0.1657 0.2038
^{157}Sm		156.9283	8.0 m	β^- /2.7	2.4/	3/2-			Eu k x-ray 0.1964 0.1978 0.3942
^{158}Sm		157.9299	5.5 m	β^- /2.0		0+			0.1894/100. 0.3636/82.
^{159}Sm		158.9332	11.3 s	β^- /3.8					0.1898
^{160}Sm		159.9353	9.6 s	β^- /3.6		0+			0.110
$_{63}$Eu		**151.965(9)**							
^{134}Eu			0.5 s	EC,β^+					ann.rad./
^{135}Eu		134.942	1.5 s	EC,β^+ /≈8.7					ann.rad./
^{136m}Eu			≈ 3.2 s			7+			0.255
^{136}Eu		135.940	≈ 3.9 s	EC,β^+ /10.		1+			ann.rad./
^{137}Eu		136.935	11. s	EC/≈7.5		11/2-			ann.rad./
^{138}Eu		137.9335	12. s	EC,β^+ /≈ 9.2		7+			ann.rad./
^{139}Eu		138.9298	18. s	EC,β^+ /6.7					ann.rad./
^{140m}Eu			0.125 s	EC,β^+					ann.rad./
^{140}Eu		139.9285	1.51 s	EC,β^+ /8.4		1-			ann.rad./
^{141m}Eu			3.0 s	β^+ /58 / EC/9 / I.T./33 /0.0964		11/2-			ann.rad./ Eu k x-ray (0.09 - 1.6)
^{141}Eu		140.9244	40. s	β^+/81 /5.6 EC/15 /		5/2+	+3.49	+0.85	ann.rad./ Sm k x-ray 0.3845 0.3940
^{142m}Eu			1.22 m	β^+/83 / EC/17 /	4.8/	8-	+2.98	+1.4	ann.rad./ Sm k x-ray 0.5566 0.7680 1.0233
^{142}Eu		141.9231	2.4 s	β^- /94/7.4 EC/6 /	7.0/	1+	+1.54	+0.12	ann.rad./ 0.7680
^{143}Eu		142.92017	2.62 m	β^+ /72/5.17 EC/28/	4.1/ 5.1/	5/2+	+3.67	+0.51	ann.rad./ Sm k x-ray 0.1107/7 1.5368/3. 1.9127/2.
^{144}Eu		143.91879	10.2 s	β^+ /86 /6.33 EC/13 /	5.31/	1+	+1.89	+0.10	ann.rad./ Sm k x-ray 1.6601
^{145}Eu		144.916263	5.93 d	β^+ /2 /2.660 EC/98 /1.71	0.79/	5/2+	+3.99	+0.29	ann.rad./ Sm k x-ray 0.6535 0.8937 1.6587
^{146}Eu		145.91720	4.57 d	β^+ /5 /3.88 EC/95 /	1.47/	4-	+1.43	-0.18	ann.rad./ Sm k x-ray 0.6336 0.6341 0.7470 (0.27 - 2.64)

Elem. or Isot.	Natural Abundance (%)	Atomic Mass or Weight	Half-Life	Decay Mode/Energy (/MeV)	Particle Energy /Intensity (MeV / %)	Spin ($h/2\pi$)	Nuclear Magnetic Mom.(nm)	Elect. Quadr. Mom. (b)	γ-ray Energy /Intensity (MeV / %)
^{147}Eu		146.916742	24.4 d	EC/99. /1.722 β^+ /0.4 /		5/2+	+3.72	+0.55	Sm k x-ray 0.12113 0.19725 0.6776
^{148}Eu		147.91815	54.5 d	EC/3.11	0.92	5-	+2.34	+0.35	Sm k x-ray 0.5503/99. 0.6299/71. (0.067-2.17)
^{149}Eu		148.91792	93.1 d	EC/0.692		5/2+	+3.57	+0.75	Sm k x-ray 0.2770 0.3275
^{150}Eu		149.91970	36. y	EC/2.26		5-	+2.71	+1.13	Sm k x-ray 0.3340 0.4394 0.5843 (0.25 - 1.8)
^{150m}Eu			12.8 h	β^- /92 / β^+ /0.4 / EC/8 /	1.013/ 1.24/	0-			Sm k x-ray 0.3339 0.4065
^{151}Eu	47.8(1.5)	150.919846				5/2+	+3.472	+0.90	
^{152m2}Eu			1.60 h	I.T./0.1478		8-			Eu k x-ray 0.0898
^{152m1}Eu			9.30 h	β^- /72 / EC/28 /	1.85/ 0.89/	0-			Sm k x-ray 0.12178 0.84153 0.96334
^{152}Eu		151.921741	13.5 y	EC/72 /1.874 β^- /28 /1.818	0.69/ 1.47/	3-	-1.941	+2.71	Sm k x-ray Gd k x-ray 0.12178 0.34427 1.40802 (0.252-1.528)
^{153}Eu	52.2(15)	152.921227				5/2+	+1.533	+2.41	
^{154m}Eu			46.1 m	I.T./≈ 0.16		8-			Eu k x-ray 0.0682 0.1009
^{154}Eu		153.922976	8.59 y	β^- /99.9 /1.969 EC./0.02 /0.717	0.27/29 0.58/38 0.84/17 0.98/4 1.87/11	3-	-2.01	+2.8	Gd k x-ray 0.12299/40. 0.72331/20. 1.2745/36 (0.059-1.90)
^{155}Eu		154.922890	4.76 y	β^- /0.252	0.15/	5/2+	+1.6	+2.3	Gd k x-ray 0.0865/30 0.1053/20
^{156}Eu		155.92475	15.2 d	β^- /2.451	0.30/11 0.49/30 1.2/12 2.45/31	1+	≈ 1.1		0.08899/9. 0.64623/7. 0.723441/6. 0.8118/10.
^{157}Eu		156.92542	15.13 h	β^- /1.36	0.98/ 1.30/41	(5/2+)			Gd k x-ray 0.0639/100. 0.3705/48. 0.4107/76.
^{158}Eu		157.9278	45.9 m	β^- /3.5	2.5/	(1-)			0.0795 0.8976 0.9442 0.9771
^{159}Eu		158.92909	18.1 m	β^- /2.51	2.4/ 2.57/	(5/2+)			0.0678 0.0786 0.0957
^{160}Eu		159.9315	38. s	β^- /4.1	2.7/ 4.1/	(0-)			0.0753 0.1735 0.4131 0.5155 0.8217 0.9110 0.9246
^{161}Eu		160.9337	27. s	β^- /3.7					0.0719
^{162}Eu		161.9370	11. s	β^- /5.6					
$_{64}$Gd		**157.25(3)**							
^{137}Gd		136.945	7. s	EC,β^+ /≈8.8					ann.rad./
^{139}Gd		138.9381	5. s	EC,β^+ /≈7.7				ann.rad./	
^{140}Gd		139.934	16. s	EC/4.8		0+			0.1748
^{141m}Gd			25. s	EC,β^+ /		11/2-			ann.rad./
^{141}Gd		140.9322	21. s	β^+ /7.3		0+			ann.rad./
^{142}Gd		141.9276	1.17 m	EC,β^+ /4.2		1/2+			ann.rad./

Elem. or Isot.	Natural Abundance (%)	Atomic Mass or Weight	Half-Life	Decay Mode/Energy (/MeV)	Particle Energy /Intensity (MeV / %)	Spin (h/2π)	Nuclear Magnetic Mom.(nm)	Elect. Quadr. Mom. (b)	γ-ray Energy /Intensity (MeV / %)
^{143m}Gd			1.84 m	β⁺ /67 / EC/33 / I.T./		11/2-			ann.rad./ Eu k x-ray 0.1176 0.2719 0.5880 0.6681 0.7999
^{143}Gd		142.9266	39. s	β⁺ /82 /6.0 EC/18 /		1/2+			ann.rad./ Eu k x-ray 0.2048 0.2588
^{144}Gd		143.9234	4.5 m	β⁺ /45 /4.3 EC/55 /	3.3/	0+			ann.rad./ Eu k x-ray 0.3332
^{145m}Gd			1.44 m	I.T./95 /0.749 β⁺ /4 /5.7		11/2-			0.0273 0.3295 0.3866 0.7214
^{145}Gd		144.92169	23.4 m	β⁺ /33 /5.05 EC/67 /	2.5/	1/2+			ann.rad./ Eu k x-ray 1.7579 1.8806 (0.32 - 3.69)
^{146}Gd		145.91831	48.3 d	EC/99.9 /1.03 β⁺ /0.2	0.35/	0+			Eu k x-ray 0.1147 0.1155 0.1546
^{147}Gd		146.919090	1.588 d	EC/99.8 /2.188 EC/0.2 /	0.93/	7/2-	1.0		Eu k x-ray 0.2293 0.3699 0.3960 0.9289 (0.1 - 1.8)
^{148}Gd		147.918111	75. y	α/3.27	3.1828/		0+		
^{149}Gd		148.919339	9.3 d	EC/1.32		7/2-	0.9		Eu k x-ray 0.1496 0.2985 0.3465
^{150}Gd		149.91866	1.8x10^{6} y	α/2.80	2.73/	0+			
^{151}Gd		150.920345	124. d	EC/0.464		7/2-	0.8		Eu k x-ray 0.1536 0.2432
^{152}Gd	0.20(1)	151.919789				0+			
^{153}Gd		152.921747	241.6 d	EC/0.485		3/2-	0.4		Eu k x-ray 0.09743 0.10318
^{154}Gd	2.18(3)	153.920862				0+			
^{155}Gd	14.80(5)	154.922619				3/2-	-2.59	+1.30	
^{156}Gd	20.47(4)	155.922120				0+			
^{157}Gd	15.65(3)	156.923957				3/2-	-3.40	+1.36	
^{158}Gd	24.84(12)	157.924101				0+			
^{159}Gd		158.926385	18.6 h	β⁻ 0.971	0.60/11 0.89/26 0.96/63	3/2-	-0.44		Tb k x-ray 0.05845 0.36351
^{160}Gd	21.86(4)	159.927051				0+			
^{161}Gd		160.929666	3.66 m	β⁻ /1.956	1.56/85	5/2-			Tb k x-ray 0.1023 0.3149 0.3609
^{162}Gd		161.930981	8.4 m	β⁻ /1.39	1.0/	0+			0.4030 0.4421
^{163}Gd		162.9340	1.13 m	β⁻ /3.1					0.2060 0.214 1.685
^{164}Gd		163.9359	45. s	β⁻ /2.3					
$_{65}$Tb		**158.92534(2)**							
^{140}Tb		139.946	2.4 s	β⁺ ,EC/11					
^{141}Tb		140.941	3.5 s	β⁺ ,EC/≈ 8.3					
^{142m}Tb			0.30 s	β⁺ ,EC/		4-			
^{142}Tb		141.939	0.60 s	β⁺ ,EC/10.		0+			
^{143}Tb		142.9346	12. s	β⁺ ,EC/7.4		11/2-			
^{144m}Tb			4.1 s	IT		5-			
^{144}Tb		143.9324	< 1.5 s	β⁺ ,EC/8.4		1+			
^{145m}Tb			30. s	β⁺ ,EC/≈ 6.6		11/2-			ann.rad./ 0.2577 0.5370 0.9876

TABLE OF THE ISOTOPES

Elem. or Isot.	Natural Abundance (%)	Atomic Mass or Weight	Half-Life	Decay Mode/Energy (/MeV)	Particle Energy /Intensity (MeV / %)	Spin (h/2π)	Nuclear Magnetic Mom.(nm)	Elect. Quadr. Mom. (b)	γ-ray Energy /Intensity (MeV / %)
^{145}Tb		144.9287		β^+ ,EC/6.5		1/2+			
^{146m}Tb			23. s	β^+ /76 / EC/24 /		(5-)			ann.rad./ Gd k x-ray 1.0789 1.5795
^{146}Tb		145.9270	≈ 8. s	β^+ /8.1		1+			
^{147m}Tb			1.8 m	β^+ /35 / EC/65 /		11/2-			ann.rad./ Gd k x-ray 1.3977 1.7978
^{147}Tb		146.92404	1.6 h	β^+ /42 /4.61 EC/58 /		5/2+			ann.rad./ Gd k x-ray 0.1398 0.6944 1.1522
^{148m}Tb			2.3 m	β^+ /25 / EC/75 /		9+			ann.rad./ Gd k x-ray 0.3945 0.6319 0.7845 0.8824
^{148}Tb		147.92422	1.00 h	β^+ ,EC/5.69		2-			ann.rad./ Gd k x-ray 0.4888 0.7845 (0.14 - 3.8)
^{149m}Tb			4.16 m	EC/88 / β^+ /12 /		11/2-			ann.rad./ Gd k x-ray 0.1650 0.7960
^{149}Tb		148.923243	4.13 h	β^+ /4 /3.636 α/16/	1.8/ 3.97/	1/2+	<0.9		Gd k x-ray 0.1650 0.3522 0.3886 (0.1 - 3.2)
^{150m}Tb			6.0 m	β^+ /17 / EC/83 /					ann.rad./ Gd k x-ray 0.4384 0.6380 0.6504 0.8275
^{150}Tb		149.92366	3.3 h	β^+ ,EC/4.66		2-			ann.rad./ 0.4963 0.6380 (0.3 - 4.29)
^{151m}Tb			25. s	I.T./95 / β^+ ,EC/7 /		11/2-			0.0229 0.0495 0.3797 0.8305
^{151}Tb		150.923099	17.61 h	β^+ /1 /2.565 EC/99 /	0.70/	1/2+			Gd k x-ray 0.1083 0.2517 0.2870 (0.1 - 1.8)
^{152m}Tb			4.3 m	I.T./79 /0.5018 EC/21 /4.35		(8+)			Tb k x-ray Gd k x-ray 0.2833 0.3443 0.4111
^{152}Tb		151.92407	17.5 h	β^+ /20 /3.99 EC/80 /	2.5/ 2.8/	2-	0.9		ann.rad./ Gd k x-ray 0.3443 (0.2 - 2.88)
^{153}Tb		152.923433	2.34 d	EC/1.570		5/2+	3.5		Gd k x-ray 0.2119 (0.05 - 1.1)
^{154m2}Tb			23.1 h	EC/98 / I.T./2 /		(7-)			Gd k x-ray 0.1231 0.2479 0.3467 1.4199
^{154m1}Tb			9. h	β^+ /78 / I.T./22 /		(3-)	1.8	+3.	Gd k x-ray 0.1231 0.2479 0.5401 (0.12 - 2.57)

Elem. or Isot.	Natural Abundance (%)	Atomic Mass or Weight	Half-Life	Decay Mode/Energy (/MeV)	Particle Energy /Intensity (MeV / %)	Spin (h/2π)	Nuclear Magnetic Mom.(nm)	Elect. Quadr. Mom. (b)	γ-ray Energy /Intensity (MeV / %)
^{154}Tb		153.92469	21.5 h	EC/99 /3.56 β^+ /1 /	1.86/ 2.45	0-	0.9		Gd k x-ray 0.1231 1.2744 2.1872 (0.12 - 3.14)
^{155}Tb		154.92350	5.3 d	EC/0.82		3/2+	2.0		Gd k x-ray 0.08654 0.10530
^{156m2}Tb			1.02 d	I.T./		(7-)			Tb k x-ray 0.0496
^{156m1}Tb			5.3 h	I.T./0.0884		(0+)			Tb k x-ray 0.0884
^{156}Tb		155.924744	5.3 d	EC/2.444		3-	1.4	+2.3	Gd k x-ray 0.08896 0.19921 0.53435 1.22245
^{157}Tb		156.924021	1.1x10^2 y	EC/0.0601		3/2+	2.0		Gd k x-ray 0.0545
^{158m}Tb			10.5 s	I.T./0.11		0-			Gd k x-ray 0.0110
^{158}Tb		157.925410	1.8x10^2 y	EC/80 /1.220 β^- /20 /0.937		3-	+1.76	+2.7	Gd k x-ray 0.0795 0.9442 0.9621
^{159}Tb	100.	158.925343				3/2+	+2.014	+1.43	
^{160}Tb		159.927164	72.3 d	β^- /1.835	0.57/47 0.86/27	3-	+1.79	3.8	Dy k x-ray 0.08678 0.29857 0.87936 0.96615
^{161}Tb		160.927566	6.91 d	β^- /0.593	0.46/23 0.52/66 0.6/10	3/2+	2.2	+1.2	Dy k x-ray 0.02565 0.04892 0.07458
^{162}Tb		161.92948	7.6 m	β^- /2.51	1.4	(1/2-)			Dy k x ray 0.2600 0.8075 0.8882
^{163}Tb		162.930644	19.5 m	β^- /1.785	0.80/	3/2+			Dy k x-ray 0.3511 0.3897 0.4945
^{164}Tb		163.9334	3.0 m	β^- /3.9	1.7/	(5+)			Dy k x-ray 0.1689 0.2157 0.6110 0.6885 0.7548
^{165}Tb		164.9349	2.1 m	β^- /3.0		3/2+			0.5389 1.1785 1.2920 1.6648
$_{66}$Dy		**162.50(3)**							
^{141}Dy		140.951	0.9 s	EC,β^+ /9.					
^{142}Dy		141.946	2.3 s	EC,β^+ /7.1					
^{143}Dy		142.9440	3.9 s	EC,β^+ /≈ 8.8					
^{144}Dy		143.9391	9.1 s	EC,β^+ /≈ 6.2					
^{145m}Dy		[illegible]	14. s	EC,β^+		11/2			
^{146m}Dy			0.15 s	I.T.		10+			
^{146}Dy		145.9325	30. s	EC,β^+ /5.2					
^{147m}Dy			56. s	I.T./40 / β^+ ,EC/60 /		(11/2-)	-0.66	+0.7	Dy k x-ray 0.072 0.6787
^{147}Dy		146.9309	75. s	EC,β^+ /6.37		1/2+	-0.92	0.1007	ann.rad./ 0.2534 0.3653
^{148}Dy		147.92710	3.1 m	β^+ /4 /2.68 EC/96 /	1.2/	0+			ann.rad./ Tb k x-ray 0.6202

Elem. or Isot.	Natural Abundance (%)	Atomic Mass or Weight	Half-Life	Decay Mode/Energy (/MeV)	Particle Energy /Intensity (MeV / %)	Spin (h/2π)	Nuclear Magnetic Mom.(nm)	Elect. Quadr. Mom. (b)	γ-ray Energy /Intensity (MeV / %)
^{149}Dy		148.92734	4.2 m	β^+ ,EC/3.81		(7/2-)	-0.12	-0.62	ann.rad./
									0.1008
									0.1063
									0.2534
									0.6536
									0.7894
									1.7765
									1.8062
^{150}Dy		149.92558	7.18 m	β^+ ,EC/67 /1.79		0+			Tb k x-ray
				α /33 /	4.233/				0.3967
^{151}Dy		150.926181	17. m	β^+ /5 /2.871		7/2-	-0.95	-0.30	Tb k x-ray
				EC/89 /					0.1764
				α /6 /	4.067/				0.3030
									0.3861
									0.5463
									(0.16 - 2.09)
^{152}Dy		151.92472	2.37 h	EC/0.60		0+			Tb k x-ray
				α /	3.63/				0.2569
^{153}Dy		152.925763	6.3 h	β^+ /1 /2.171	0.89/	(7/2-)	-0.78	-0.02	Tb k x-ray
				EC/99 /					0.0807
				α /0.01 /	3.46/				0.0997
									0.2137
									(0.08 - 1.66)
^{154}Dy		153.92442	3.x10^6 y	α/2.95	2.87/	0+			
^{155}Dy		154.92575	9.9 h	β^+ /2 /2.095	0.845/	3/2-	-0.385	+1.04	Tb k x-ray
				EC/98 /					0.0655
									0.2269
^{156}Dy	0.06(1)	155.92428				0+			
^{157}Dy		156.92546	8.1 h	EC/1.34		3/2-	-0.301	+1.30	Tb k x-ray
									0.3262
^{158}Dy	0.10(1)	157.924405				0+			
^{159}Dy		158.925736	144. d	EC/0.366		3/2-	-0.354	+1.37	Tb k x-ray
									0.3262
^{160}Dy	2.34(6)	159.925194				0+			
^{161}Dy	18.9(2)	160.926930				5/2+	-0.480	+2.51	
^{162}Dy	25.5(2)	161.926795				0+			
^{163}Dy	24.9(2)	162.928728				5/2-	+0.673	+2.65	
^{164}Dy	28.2(2)	163.929171				0+			
^{165m}Dy			1.26 m	I.T./98 /0.108		1/2-			Dy k x-ray
				β^- /2 /					0.1082
									0.5155
^{165}Dy		164.931700	2.33 h	β^- /1.286	1.29/	7/2+	-0.52	-3.5	Ho k x-ray
									0.09468
^{166}Dy		165.932803	3.400 d	β^- /0.486	0.40/	0+			Ho k x-ray
									0.0282
									0.0825
^{167}Dy		166.9357	6.2 m	β^- /$\approx$ 2.35	1.78	(1/2-)			Ho k x-ray
									0.2593
									0.3103
									0.5697
									(0.06 - 1.4)
^{168}Dy		167.9372	8.5 m	β^- /1.6		0+			Ho k x-ray
									0.1925
									0.4867
^{169}Dy		168.9403	$\approx$ 39. s	β^- /3.2					
$_{67}$Ho		**164.93032(2)**							
^{144}Ho		143.952	0.7 s	β^+ ,EC/12					
^{146}Ho		145.9440	3.3 s	β^+ ,EC/10.7		(10+)			ann.rad./
^{147}Ho		146.9396	5.8 s	β^+ ,EC/8.2		11/2-			ann.rad./
^{148m}Ho			9. s	β^+ ,EC/		4-			ann.rad./
^{148}Ho		147.9372	2. s	β^+ ,EC/9.4		1+			ann.rad./
									0.6615
									1.6883
^{149m}Ho			21. s	β^+ ,EC/		11/2-			ann.rad./
									1.0733
									1.0911
^{149}Ho		148.93379	> 30. s	β^+ ,EC/6.01		1/2+			
^{150m}Ho			25. s	β^+ ,EC/		(9+)			ann.rad./
									0.3939
									0.5511
									0.6534
									0.8034
^{150}Ho		149.9326	1.3 m	β^+ ,EC/6.6					ann.rad./
									0.5913
									0.6534
									0.8034

Elem. or Isot.	Natural Abundance (%)	Atomic Mass or Weight	Half-Life	Decay Mode/Energy (/MeV)	Particle Energy /Intensity (MeV / %)	Spin (h/2π)	Nuclear Magnetic Mom.(nm)	Elect. Quadr. Mom. (b)	γ-ray Energy /Intensity (MeV / %)
^{151m}Ho			47. s	β^+ ,EC/87 / α/13	 4.605/				ann.rad./ 0.2102 0.4889 0.6948 0.7762
^{151}Ho		150.93169	35.2 s	β^+ ,EC/80/5.13 α/20 /	 4.519/				ann.rad./ 0.3522 0.5274 0.9676 1.0471
^{152m}Ho			50. s	β^+ ,EC/90 / α/10 /	 4.453/	(9+)	+5.9	-1.	ann.rad./ 0.4929 0.6138 0.6474 0.6835
^{152}Ho		151.93166	2.4 m	β^+ ,EC/88 /6.47 α/12 /	 4.387/	(3+)	-1.02	+0.1	ann.rad./ 0.6140 0.6476
^{153m}Ho			9.3 m	β^+ ,EC/99+ /4.12 α/	 4.01/	5/2	+1.19		ann.rad./ 0.0905 0.1089 0.1618 0.2302 0.2707 0.3659 0.4565
^{153}Ho		152.93020	2.0 m	β^+ ,EC/99+ /4.13 α/	 3.91/	11/2-	+6.8	-1.	ann.rad./ 0.2958 0.3346 0.4381 0.6383
^{154m}Ho			3.3 m	β^+ ,EC/		(8+)	5.7	-.1.	ann.rad./ 0.3346 0.4124 0.4771
^{154}Ho		153.93060	12. m	β^+ ,EC/5.75		1-	-0.64	+0.2	ann.rad./ Dy k x-ray 0.3346 0.5700 0.8734
^{155}Ho		154.92908	48. m	β^+ /6 /3.10 EC/94 /		(5/2+)	+3.51	+1.5	ann.rad./ Dy k x-ray 0.0474 0.1363 0.3254 (0.06 - 2.24)
^{156m}Ho			5.8 m	I.T./0.0352 β^+ /25 / EC/75 /	 1.8/ 2.9/		+2.99	+2.3	ann.rad./ Dy k x-ray 0.1378 0.2666 (0.28 - 2.9)
^{156}Ho		155.9290	56. m	β^+ ,EC/4.4		(5+)			ann.rad./ 0.1378 0.2665
^{157}Ho		156.92819	12.6 m	β^+ /5 /2.54 EC/95 /	1.18/	7/2-	+4.35	+3.0	ann.rad./ Dy k x-ray 0.2800 0.3411
^{158m2}Ho			28. m	I.T./44 / EC/56 /		2-	+2.44	+1.6	ann.rad./ Dy k x-ray 0.0989 0.2182
[illegible]			21. m	β^+ ,EC/		(9+)			ann.rad./ 0.0981 0.1664 0.2182 0.3205 0.4062 0.9774 1.0532 0.4846
^{158}Ho		157.92895	11.3 m	β^+ /8 /4.24 EC/92 /	1.30/	5+	+3.77	+4.1	ann.rad./ Dy k x-ray 0.0989 0.2182 0.9488
^{159m}Ho			8.3 s	I.T./0.206		1/2+			Ho k x-ray 0.1660 0.2059

TABLE OF THE ISOTOPES

Elem. or Isot.	Natural Abundance (%)	Atomic Mass or Weight	Half-Life	Decay Mode/Energy (/MeV)	Particle Energy /Intensity (MeV / %)	Spin (h/2π)	Nuclear Magnetic Mom.(nm)	Elect. Quadr. Mom. (b)	γ-ray Energy /Intensity (MeV / %)
^{159}Ho		158.927708	33.0 m	EC/1.838		7/2-	+4.28	+3.2	Dy k x-ray 0.1210 0.1320 0.2529 0.3096 (0.06 - 1.2)
^{160m2}Ho			3. s			1+			
^{160m}Ho			5.0 h	I.T./67 /0.060 EC/33 /3.35		2-	+2.52	+1.8	0.0868 0.1970 0.6464 0.7281 0.8791 0.9619 0.9658
^{160}Ho		159.92873	25.6 m	β^+ ,EC/3.29	0.57/	5+	+3.71	+4.0	See Ho[166m] 0.7282 0.8794
^{161m}Ho			6.8 s	I.T./0.211					Ho k x-ray 0.2112
^{161}Ho		160.927852	2.48 h	EC/0.859		7/2-	+4.25	+3.2	Dy k x-ray 0.0256 0.0592 0.0774 0.1031
^{162m}Ho			1.12 h	I.T./61 / EC/39 /		6-	+3.60	+4.	Dy k x-ray Ho k x-ray 0.0807 0.1850 0.2828 0.9372 1.2200
^{162}Ho		161.929092	15. m	EC/96 /0.295 β^+ /4 /		1+			Dy k x-ray 0.0807 1.3196 1.3728
^{163m}Ho			1.09 s	I.T./0.298		(1/2+)			Ho k x-ray 0.2798
^{163}Ho		162.928730	4.57x10^3 y	EC/0.00258		7/2-	+4.23	+3.6	Dy M x-rays
^{164m}Ho			38. m	I.T./0.140		(6-)			Ho k x-ray 0.0373 0.0566 0.0940
^{164}Ho		163.930231	29. m	EC/58 /0.987 β^- /42 /0.963		1+			Dy k x-ray 0.0734 0.0914
^{165}Ho	100.	164.930319				7/2-	+4.17	+3.49	
^{166m}Ho			1.2x10^3 y	β^- /		7-	3.6	-3.	Er k x-ray 0.18407 0.71169 0.81031
^{166}Ho		165.932281	1.117 d	β^- /1.855	1.776/48 1.855/51	0-			Er k x-ray 0.08057 1.37943
^{167}Ho		166.933127	3.1 h	β^- /1.007	0.31/43 0.61/21 0.96/15 0.97/15	(7/2-)			Er k x-ray 0.0793 0.0835 0.2379 0.3213 0.3465
^{168m}Ho			2.2 m	I.T./					
^{168}Ho		167.93550	3.0 m	β^- /2.91	2.0/	3+			Er k x-ray 0.7413 0.8159 0.8211 (0.08 - 2.34)
^{169}Ho		168.93687	4.7 m	β^- /2.12	1.2/ 2.0/	(7/2-)			0.1496 0.7610 0.7784 0.7884 0.8529
^{170m}Ho			43. s	β^- /		1+			0.0787 0.8123 1.8940 1.9726

Elem. or Isot.	Natural Abundance (%)	Atomic Mass or Weight	Half-Life	Decay Mode/Energy (/MeV)	Particle Energy /Intensity (MeV / %)	Spin (h/2π)	Nuclear Magnetic Mom.(nm)	Elect. Quadr. Mom. (b)	γ-ray Energy /Intensity (MeV / %)
^{170}Ho		169.93962	2.8 m	β^- /3.87		6+			Er k x-ray 0.1816 0.2582 0.8902 0.9321 0.9414 1.1387
^{172}Ho			25. s	β^- /					Er k x-ray (0.077-1.186)
$_{68}$Er		**167.26(3)**							
^{147}Er		146.9494	2.5 s	E.C,β^+ /≈ 9.1					
^{148}Er		147.9444	4.5 s	β^+ ,EC/6.8					
^{149m}Er			10. s	IT		11/2-			
^{149}Er		148.9425	10.7 s	ECβ^+ /8.1		1/2+			
^{150}Er		149.9370	18. s	β^+ /36 /4.11 EC/64 /		0+			ann.rad./ Ho k x-ray 0.4758
^{151}Er		150.9373	23. s	β^+ ,EC/5.2		7/2-			ann.rad./
^{152}Er		151.93500	10.2 s	β^+ ,EC/10 /3.11 α/90 /	4.804/	0+			ann.rad./
^{153}Er		152.93509	37.1 s	α/ β^+ ,EC/47 /4.56	4.674 4.35/				0.351 (0.0945-1.700)
^{154}Er		153.93278	3.7 m	β^+ ,EC/99+ /2.03 α/0.5 /	4.166/	0+			ann.rad./
^{155}Er		154.93321	5.3 m	β^+ ,EC/47 /3.84 EC/53 /		(7/2-)	-0.669	-0.27	ann.rad./ Ho k x-ray 0.1101 0.2415
^{156}Er		155.9308	20. m	β^+ ,EC/1.7		0+			ann.rad./ 0.0298 0.0352 0.0522 0.1336
^{157}Er		156.9319	25. m	β^+ ,EC/3.5		3/2-	-0.412	+0.92	ann.rad./ 0.117 0.385 1.320 1.660 1.820 2.000
^{158}Er		157.93087	2.2 h	EC/99.5 /1.78 β^+ /0.5 /	0.74/	0+			Ho k x-ray 0.0719 0.2486 0.3868
^{159}Er		158.930681	36. m	β^+ /7 /2.769 EC/93 /		3/2-	-0.304	+1.17	ann.rad./ Ho k x-ray 0.6245 0.6493 (0.07 - 2.5)
^{160}Er		159.92908	1.191 d	EC/0.33		0+			Ho k x-ray (0.05 - 0.96)
^{161}Er		160.93000	3.21 h	EC/2.00		3/2-	-0.37	+1.36	Ho k x-ray 0.8265 (0.07 - 1.74)
^{162}Er	0.14(1)	161.928775				0+			
^{163}Er		162.93003	1.25 h	EC/1.210		5/2-	+0.557	+2.55	Ho k x-ray 0.4361 0.4399 1.1135
^{164}Er	1.61(2)	163.929197				0+			
^{165}Er		164.930723	10.36 h	EC/0.376		5/2-	+0.643	+2.71	Ho k x-ray
^{166}Er	33.6(2)	165.930290				0+			
^{167m}Er			2.27 s	I.T./0.208		1/2-			Er k x-ray 0.2078
^{167}Er	22.95(15)	166.932046				7/2+	-0.5639	+3.57	
^{168}Er	26.8(2)	167.932368				0+			
^{169}Er		168.934588	9.40 d	β^- /0.351	0.35/≈ 100	1/2-	+0.515		Tm k x-ray 0.1098 0.1182
^{170}Er	14.9(2)	169.935461				0+			
^{171}Er		170.938026	7.52 h	β^- /1.491		5/2-	0.66	2.9	Tm k x-ray 0.11160 0.29591 0.30832 (0.08 - 1.4)

Elem. or Isot.	Natural Abundance (%)	Atomic Mass or Weight	Half-Life	Decay Mode/Energy (/MeV)	Particle Energy /Intensity (MeV / %)	Spin (h/2π)	Nuclear Magnetic Mom.(nm)	Elect. Quadr. Mom. (b)	γ-ray Energy /Intensity (MeV / %)
^{172}Er		171.939352	2.05 d	β^- /0.891	0.28/48 0.36/46				Tm k x-ray 0.0597 0.4073 0.6101
^{173}Er		172.9424	1.4 m	β^- /2.6		(7/2-)			Tm k x-ray 0.1928 0.1992 0.8952
^{174}Er		173.9441	3.1 m	β^- /1.8					Tm k x-ray (0.100-0.152)
$_{69}$Tm		**168.93421(2)**							
^{146}Tm		145.967		β^+ /14. p	 1.119/ 1.189/				
^{147}Tm		146.961	0.6 s	EC, β^+ /85/$\approx$10.7 p/15/	 1.052/				
^{148m}Tm		147.9573	0.7 s	β^+ ,EC/12.					ann.rad./
^{148}Tm									
^{149}Tm		148.9524	0.9 s	β^+ ,EC/$\approx$9.2		11/2-			
^{150}Tm		149.9494	2.3 s	β^+ ,EC/$\approx$11.5		6-			
^{151}Tm		150.9454	4. s	β^+ ,EC/7.5					ann.rad./
^{152m}Tm			8. s	β^+ ,EC/		9+			
^{152}Tm		151.9443	5. s	β^+ ,EC/8.8					ann.rad./
^{153}Tm		152.94203	1.6 s	β^+ ,EC/10 /6.46 α/90 /	 5.109/				ann.rad./
^{154m}Tm			3.3 s	β^+ ,EC/15 / α /	 5.031/				ann.rad./
^{154}Tm		153.9407	8.1 s	β^+ ,EC/56 /7.4 α /44 /	 4.45/	9+			ann.rad./
^{155}Tm		154.93919	30. s	β^+ ,EC/5.58 α/	 4.46/				0.0315 0.0638 0.0881 0.2268 0.5320 0.6067
^{156m}Tm			19. s	α/	4.46/				
^{156}Tm		155.9389	1.40 m	β^+ ,EC/7.6 α/	 4.23/	2-	+0.40	-0.5	ann.rad./ 0.3446 0.4529 0.5860
^{157}Tm		156.9367	3.6 m	β^+ ,EC/4.5 α/	2.6 3.97/	1/2	+0.48		ann.rad./ 0.1104 0.3484 0.3855 0.4550 (0.1 - 1.58)
^{158}Tm		157.9379	4.0 m	β^+ ,EC/74 /6.5 EC/26 /		(2-)	+0.04	+0.7	ann.rad./ Er k x-ray 0.1921 0.3351 0.6280 1.1498 (0.18 - 2.81)
^{159}Tm		158.9348	9.1 m	β^+ /23 /3.9 EC/77 /		5/2+	+3.42	+1.9	ann.rad./ Er k x-ray 0.0591 0.0848 0.2713 (0.05 - 1.27)
^{160m}Tm			1.24 m	IT		(5)			
^{160}Tm		159.9354	9.4 m	β^+ /15 /5.9 EC/85 /		1-	+0.16	+0.58	ann.rad./ Er k x-ray 0.1264 0.2642 0.7285 0.8544 0.8614 1.3685
^{161}Tm		160.9334	31. m	β^+ ,EC/3.2		7/2+	+2.40	+2.9	ann.rad./ Er k x-ray 0.0595 0.0844 1.6481 (0.04 - 2.15)

Elem. or Isot.	Natural Abundance (%)	Atomic Mass or Weight	Half-Life	Decay Mode/Energy (/MeV)	Particle Energy /Intensity (MeV / %)	Spin (h/2π)	Nuclear Magnetic Mom.(nm)	Elect. Quadr. Mom. (b)	γ-ray Energy /Intensity (MeV / %)
^{162m}Tm			24. s	I.T./90 / β^+ ,EC/10 /		5+			Tm k x-ray Er k x-ray 0.0669 0.8115 0.9003
^{162}Tm		161.93394	21.7 m	β^+ /8 /4.81 EC/92 /		1-	+0.07	+0.69	ann.rad./ Er k x-ray 0.1020 0.7987 (0.1 - 3.75)m
^{163}Tm		162.93265	1.81 h	EC/98 /2.439 β^+ /1 /		1/2+	0.082		Er k x-ray 0.0692 0.1043 0.2414
^{164m}Tm			5.1 m	I.T./80 / β^+ ,EC/20 /		6-			0.0914 0.1394 0.2081 0.2405 0.3149
^{164}Tm		163.93345	2.0 m	β^+ /36 /3.96 EC/64 /	2.94/	1+	+2.38	+0.71	ann.rad./ Er k x-ray 0.0914
^{165}Tm		164.932433	1.253 d	EC/1.593		1/2+	0.139		Er k x-ray 0.0472 0.0544 0.29728 0.80636
^{166}Tm		165.93355	7.70 h	EC/98 /3.04 β^+ /2 /		2+	+0.092	+2.14	Er k x-ray 0.0806 0.1844 0.7789 1.2734 2.0524
^{167}Tm		166.932849	9.24 d	EC/0.748		1/2+	-0.197		Er k x-ray 0.0571 0.20778
^{168}Tm		167.934171	93.1 d	EC/1.679		3+	0.23	+3.2	Er k x-ray 0.19825 0.4475 0.81595
^{169}Tm	100	168.934211				1/2+	-0.2316	-1.2	
^{170}Tm		169.935798	128.6 d	β^- /99.8 /0.968 EC/0.2 /0.314	0.883/24 0.968/76	1-	0.2476	+0.74	Yb k x-ray 0.08425
^{171}Tm		170.936426	1.92 y	β^- /0.096	0.03/2 0.096/98	1/2+	-0.2303		0.06674
^{172}Tm		171.93840	2.65 d	β^- /1.88	1.79/36 1.88/29	2-			Yb k x-ray 0.07879 1.38722 1.46601 1.52982 1.60861
^{173}Tm		172.93960	8.2 h	β^- /1.298	0.80/21 0.86/71	1/2+			Yb k x-ray 0.3988 0.4613
^{174}Tm		173.94216	5.4 m	β^- /3.08	0.70/14 1.20/83	(4-)			Yb k x-ray 0.07664 0.17669 0.27332 0.3666 0.99205 (0.08 - 1.6)
^{175}Tm		174.94383	15.2 m	β^- /2.39	0.9/36 1.9/23	(1/2+)			Yb k x-ray 0.36396 0.51487 0.94125 0.98247
^{176}Tm		175.9471	1.9 m	β^- /4.2	2.0/ 1.2/	(4+)			Yb k x-ray 0.1898 0.3819 1.0691
$_{70}$Yb		**173.04(3)**							
^{151}Yb		150.9545	1.6 s	β^+ /8.5					
^{152}Yb		151.9502	3.2 s	β^+ EC/5.5					
^{153}Yb		152.9492	4. s	β^+ EC/6.7					
^{154}Yb		153.9455	0.40 s	β^+ EC/7 /4.49 α/93 /	 5.32/				ann.rad./

Elem. or Isot.	Natural Abundance (%)	Atomic Mass or Weight	Half-Life	Decay Mode/Energy (/MeV)	Particle Energy /Intensity (MeV / %)	Spin (h/2π)	Nuclear Magnetic Mom.(nm)	Elect. Quadr. Mom. (b)	γ-ray Energy /Intensity (MeV / %)
^{155}Yb		154.9456	1.7 s	β^+ ,EC/16 /6.0 α/84 /	 5.19/		-0.8		ann.rad./
^{156}Yb		155.94277	26. s	β^+ ,EC/21 /3.57 α/79 /	 4.69/	0+			ann.rad./
^{157}Yb		156.9427	39. s	β^+ ,EC/99+ /5.5 α/0.5 /	 4.69/				ann.rad./ 0.231 (0.035-0.670)
^{158}Yb		157.93986	1.5 m	β^+ ,EC/1.9		0+			ann.rad./ 0.0741 0.2526
^{159}Yb		158.9402	1.4 m	EC,β^+ /5.1					Tm k x-ray 0.1661 0.1772 0.3297 0.3903
^{160}Yb		159.9376	4.8 m	β^+ ,EC/2.0		0+			ann.rad./ 0.1404 0.1737 0.2158
^{161}Yb		160.9375	4.2 m	β^+ ,EC/3.9		3/2-	-0.33	+1.03	ann.rad./ Tm k x-ray 0.0782 0.5999 0.6315
^{162}Yb		161.9358	18.9 m	β^+ ,EC/1.7		0+			ann.rad./ Tm k x-ray 0.1188 0.1635
^{163}Yb		162.9363	11.1 m	β^+ /26 /3.4	1.4/	3/2-	-0.37	+1.2	ann.rad./ Tm k x-ray 0.0636 0.8603 (0.06 - 1.9)
^{164}Yb		163.9345	1.26 h	EC/1.0		0+			Tm k x-ray 0.0914 0.6752
^{165}Yb		164.93540	9.9 m	β^+ /10 /2.76 EC/90 /	1.58/	(5/2-)	+0.48	+2.48	ann.rad./ Tm k x-ray 0.0801 1.0903
^{166}Yb		165.93388	2.363 d	EC/0.30		0+			Tm k x-ray 0.0828 0.1844 0.7789 1.2734 2.0524
^{167}Yb		166.934947	17.5 m	β^+ /0.5 /1.954 EC/99.5 /	0.639/	5/2-	+0.62	+2.70	Tm k x-ray 0.06296 0.10616 0.11337 0.17633
^{168}Yb	0.13(1)	167.933895				0+			
^{169m}Yb			46. s	I.T./0.0242		1/2-			Yb L x-ray 0.0242
^{169}Yb		168.935187	32.03 d	EC/0.909		7/2+	-0.63	+3.5	Tm k x-ray 0.06306 0.10977 0.17718 0.19795
^{170}Yb	3.05(6)	169.934759				0+			
^{171}Yb	14.3(2)	170.936323				1/2-	+0.49367		
^{172}Yb	21.9(3)	171.936378				0+			
^{173}Yb	16.12(21)	172.938207				5/2-	-0.67989	+2.80	
^{174}Yb	31.8(4)	173.938858				0+			
^{175}Yb		174.941273	4.19 d	β^- /0.470	0.466/73 0.071/21 0.353/6.2	7/2-	0.6		Lu k x-ray 0.3963/13 (0.114 - 0.28)
^{176m}Yb			11.4 s	I.T./1.051		(8-)			Yb k x-ray 0.0961 0.1901 0.2929 0.3897
^{176}Yb	12.7(2)	175.942569				0+			
^{177m}Yb			6.41 s	I.T./0.3315		1/2-			Yb k x-ray 0.1131 0.2084
^{177}Yb		176.945257	1.9 h	β^- /1.399	1.40	9/2+			Lu k x-ray 0.1504

Elem. or Isot.	Natural Abundance (%)	Atomic Mass or Weight	Half-Life	Decay Mode/Energy (/MeV)	Particle Energy /Intensity (MeV / %)	Spin (h/2π)	Nuclear Magnetic Mom.(nm)	Elect. Quadr. Mom. (b)	γ-ray Energy /Intensity (MeV / %)
^{178}Yb		177.94664	1.23 h	β^- /0.65	0.25/	0+			0.1415
									0.3246
									0.3516
									0.3815
									0.6125
^{179}Yb		178.9499	8. m	β^- /2.4					
^{180}Yb			2. m	β^-					
$_{70}$Lu		**174.967(1)**							
^{150}Lu		149.973	≈35. ms	p					
^{151}Lu		150.967	0.08 s	p/1.231					
^{152}Lu		151.963	0.7 s						
^{153}Lu		152.959							
^{154}Lu		153.9571	1.0 s	β^+ ,EC/10.8					
^{155m}Lu			2.6 ms	α/7.41					
^{155}Lu		154.9542	0.07 s	EC/8.0					
				α/	5.66/				
^{156m}Lu			0.20 s	α/	5.57/				
^{156}Lu		155.9529	≈ 0.5 s	β^+ ,EC/9.5					ann.rad./
				α/	5.45/				
^{157m}Lu			≈9.6 s	α	4.925/				
^{157}Lu		156.95010	4.8 s	β^+ ,EC/94 /6.93					ann.rad./
				α/	5.00/				
^{158}Lu		157.94984	10.4 s	β^+ ,EC/99 /8.0					ann.rad./
				α/	4.67/				0.3682
									0.4770
^{159}Lu		158.9467	12.3 s	β^+ ,EC/6.0					ann.rad./
									0.1505
									0.1875
									0.3693
^{160}Lu		159.94654	36.1 s	β^+ ,EC/7.3					ann.rad./
									0.2434
									0.3957
									0.5773
^{161}Lu		160.9432	1.2 m	β^+ ,EC/5.3					ann.rad./
									0.0437
									0.0671
									0.1003
									0.1108
									0.1562
									0.2562
^{162m}Lu			≈ 1.5 m	EC/		4-			
^{162}Lu		161.9432	1.37 m	β^+ ,EC/6.9		1-			ann.rad./
									0.1666
									0.6314
^{163}Lu		162.9412	4.1 m	β^+ ,EC/4.6					ann.rad./
									0.0539
									0.0581
									0.1504
									0.1631
									0.3717
^{164}Lu		163.9412	3.14 m	β^+ ,EC/6.3	1.6/				0.1238
					3.8/				0.2621
									0.7404
									0.8639
									0.8804
^{165}Lu		164.9396	10.7 m	β^+ ,EC/3.9	2.06/	1/2+			ann.rad./
									0.1206
									0.1324
									0.1742
									0.2036
									(0.04 - 2.0)
^{166m2}Lu			2.1 m	β^+ /35 /		(0-)			ann.rad./
				EC/65 /					Yb k x-ray
									1.0673
									1.2566
									2.0986
^{166m1}Lu			1.4 m	β^+ ,EC/58 /		(3-)			ann.rad./
				I.T./42 /0.0344					0.1024
									0.2281
									0.2861
									0.8119
									0.8301

Elem. or Isot.	Natural Abundance (%)	Atomic Mass or Weight	Half-Life	Decay Mode/Energy (/MeV)	Particle Energy /Intensity (MeV / %)	Spin (h/2π)	Nuclear Magnetic Mom.(nm)	Elect. Quadr. Mom. (b)	γ-ray Energy /Intensity (MeV / %)
^{166}Lu		165.9398	2.8 m	β⁺ /25 /5.5 EC/75 /		(6-)			ann.rad./ Yb k x-ray 0.1024 0.2281 0.3375 0.3679
^{167}Lu		166.9383	52. m	β⁺ /2 /3.1 EC/98 /	2.1/	7/2+			Yb k x-ray 0.0297 0.2392 (0.03 - 2.0)
^{168m}Lu			6.7 m	β⁺ /12 / EC/88 /		3+			ann.rad./ Yb k x-ray 0.1988 0.8960 0.9792
^{168}Lu		167.9387	5.5 m	β⁺ /6 /4.5 EC/94 /	1.2/	(6-)			ann.rad./ Yb k x-ray 0.1114 0.1124 0.2286 0.3483 1.4836
^{169m}Lu			2.7 m	I.T./0.0290		1/2-			Lu L x-ray 0.0290
^{169}Lu		168.93765	1.419 d	EC/2.293	1.271/	7/2+			Yb k x-ray 0.19121 0.9606 (0.08 - 2.1)
^{170m}Lu			0.7 s	I.T./0.0929		4-			Lu L x-ray 0.04449 0.0484
^{170}Lu		169.93847	2.01 d	EC/3.46	2.44/	0+			Yb k x-ray 0.58711 0.5908 1.28029 (0.1 - 3.38)
^{171m}Lu			1.31 m	I.T./0.0711		1/2-			Lu k x-ray 0.07119
^{171}Lu		170.937910	8.24 d	EC/1.479	0.362/	7/2+	2.0		Yb k x-ray 0.01939 0.66744 (0.02 - 1.3)
^{172m}Lu			3.7 m	I.T./0.0419		1-			Lu L x-rays 0.04186
^{172}Lu		171.939082	6.70 d	EC/2.519		4-	2.25		Yb k x-ray 0.18156 1.09367 (0.07 - 2.2)
^{173}Lu		172.938927	1.37 y	EC/0.671		7/2+	2.3		Yb k x-ray 0.07860 0.27198
^{174m}Lu			142. d	I.T./99.3 /0.17086 EC/0.7 /		6-	1.5		Lu k x-ray 0.067055
^{174}Lu		173.940334	3.3 y	EC/1.374		1-	1.9		Yb k x-ray 0.07664 1.2419
^{175}Lu	97.41(2)	174.940768				7/2+	+2.2327	+3.49	
^{176m}Lu			3.66 h	β⁻ /1.315	1.229/ 1.317/	1-	+0.318	-1.47	Hf k x-ray 0.088372
^{176}Lu	2.59(2)	175.942683	3.8x10^{10} y	β⁻ /1.192		7-	+3.169	+4.92	Hf k x-ray 0.20187 0.30691
^{177m}Lu			160.7 d	I.T./22 /0.9702 β⁻ /78		23/2-	2.9	4.2	Lu k x-ray Hf k x-ray 0.11295 0.20836 0.37850 0.41853
^{177}Lu		176.943755	6.75 d	β⁻ /0.498	0.497/	7/2+	+2.239	+3.39	0.11295 0.20836
^{178m}Lu			23.1 m	β⁻ /		(9-)			0.2166 0.3317
^{178}Lu		177.945952	28.5 m	β⁻ /2.099	2.03/	1+			Hf k x-ray 0.0932 1.3099 1.3408 (0.09 - 1.7)
^{179}Lu		178.94732	4.6 h	β⁻ /1.405	1.35/	7/2+			0.2143 0.3377

Elem. or Isot.	Natural Abundance (%)	Atomic Mass or Weight	Half-Life	Decay Mode/Energy (/MeV)	Particle Energy /Intensity (MeV / %)	Spin (h/2π)	Nuclear Magnetic Mom.(nm)	Elect. Quadr. Mom. (b)	γ-ray Energy /Intensity (MeV / %)
^{180}Lu		179.9499	5.7 m	β^- /3.1	1.49/				0.40795/50.
									(0.07-1.9)
^{181}Lu		180.9518	3.5 m	β^- /2.5		(7/2+)			0.0458
									0.2059
									0.5749
^{182}Lu			2.0 m	β^- /≈ 4.1					0.0978
									0.7208
									0.8182
^{183}Lu			58. s	β^- /		7/2+			
$_{72}$Hf		**178.49(2)**							
^{154}Hf		153.964	2. s	EC,β^+ /≈ 6.7					
^{155}Hf		154.963	0.9 s	EC,β^+ /8.					
^{156}Hf		155.9593	25. ms	α/					
^{157}Hf		156.9581	0.11 s	α/					
^{158}Hf		157.9539	2.9 s	EC/54 /5.1		0+			
				α/46 /	5.27/				
^{159}Hf		158.9538	5.6 s	β^+ ,EC/88 /6.9					ann.rad./
				α/12 /	5.09/				
^{160}Hf		159.95063	≈ 12. s	β^+ ,EC/97 /4.9		0+			ann.rad./
				α /4.78					
^{161}Hf		160.9503	17. s	α/	4.60/				
^{162}Hf		161.94720	38. s	β^+ ,EC/3.7		0+			ann.rad./
									0.1739
									0.1963
									0.4101
^{163}Hf		162.9471	40. s	β^+ ,EC/5.5					ann.rad./
									0.0454
									0.0621
									0.0710
									0.6882
^{164}Hf		163.9536	2.8 m	EC,β^+ /3.0					
^{165}Hf		164.9445	1.7 m	EC/4.6		11/2-			
^{166}Hf		165.9423	6.8 m	EC/93 /2.3					ann.rad./
				β^+ /7 /					Lu k x-ray
									0.0788
^{167}Hf		166.9426	2.0 m	β^+ /40 /4.0		(5/2-)			ann.rad./
				EC/60 /					Lu k x-ray
									0.1754
									0.3152
^{168}Hf		167.9406	25.9 m	β^+ ,EC/1.8		0+			ann.rad./
									0.1572
									0.1838
									0.1988
^{169}Hf		168.9412	3.25 m	EC/85 /3.3		(5/2-)			ann.rad./
				β^+ /15 /					Lu k x-ray
									0.3695
									0.4929
^{170}Hf		169.9397	16.0 h	EC/1.1		0+			Lu k x-ray
									0.0985
									0.1202
									0.1647
									0.5729
									0.6207
^{171}Hf		170.9405	12.2 h	EC,β^+ /2.4		7/2+			ann.rad./
									Lu k x-ray
									0.1221
									0.6620
									1.0714
^{172}Hf		171.93946	1.87 y	EC/0.35		0+			Lu k x-ray
									0.02399
									0.12582
									(0.0818-0.123)
^{173}Hf		172.9407	23.6 h	EC/1.6		1/2-			Lu k x-ray
									0.12367
									0.13963
									0.29697
									0.31124
									(0.1 - 2.1)
^{174}Hf	0.162(3)	173.940042	2.0x10^{15} y			0+			
^{175}Hf		174.941504	70. d	EC/0.686		5/2-	0.54	+2.8	Lu k x-ray
									0.08936
									0.34340
^{176}Hf	5.206(5)	175.941403				0+			

Elem. or Isot.	Natural Abundance (%)	Atomic Mass or Weight	Half-Life	Decay Mode/Energy (/MeV)	Particle Energy /Intensity (MeV / %)	Spin (h/2π)	Nuclear Magnetic Mom.(nm)	Elect. Quadr. Mom. (b)	γ-ray Energy /Intensity (MeV / %)
^{177m2}Hf			51.4 m	I.T./2.740		37/2-			Hf k x-ray 0.2140 0.2951 0.3115 0.3267
^{177m1}Hf			1.1 s	I.T./		23/2+			Hf k x-ray 0.20836 0.22847 0.37851
^{177}Hf	18.606(4)	176.943220				7/2-	+0.7935	+0.337	
^{178m2}Hf			31. y	I.T./		16+	+8.2	+6.0	Hf k x-ray 0.21342 0.25761 0.32555 0.42635 0.57418
^{178m1}Hf			4.0 s	I.T./		8-			Hf k x-ray 0.21342 0.32555 0.42635
^{178}Hf	27.297(4)	177.943698				0+			
^{179m2}Hf			25.1 d	I.T./1.1057		25/2-	7.4		Hf k x-ray 0.1227 0.1461 0.3626 0.4537
^{179m1}Hf			18.7 s	I.T./0.375		1/2-			Hf k x-ray 0.1607 0.2141
^{179}Hf	13.629(6)	178.945815				9/2+	-0.6409	+3.79	
^{180m}Hf			5.52 h	I.T./1.1416		8-	+8.7	+4.6	Hf k x-ray 0.2152 0.3323 0.4432
^{180}Hf	35.100(7)	179.946549				0+			
^{181}Hf		180.949099	42.4 d	β^- /1.027	0.408/	1/2-			Ta k x-ray 0.13294 0.48200
^{182m}Hf			62. m	β^- /54 /1.60 I.T./46 /1.1729	0.49/43 0.95/10	8-			Hf k x-ray 0.0509 0.2244 0.3441 0.4558 0.5066 0.9428
^{182}Hf		181.95055	9.x10^6 y	β^- /0.37		0+			Ta k x-ray 0.2704
^{183}Hf		182.95353	1.07 h	β^- /2.01	1.18/68 1.54/25	3/2-			Ta k x-ray 0.0732 0.4591 0.7837
^{184}Hf		183.95545	4.1 h	β^- /1.34	0.74/38 0.85/16 1.10/46	0+			Ta k x-ray 0.0414 0.1391 0.3449
^{185}Hf			≈3.5m	β^- /					0.165
$_{73}$Ta		**180.9479(1)**							
^{156}Ta		155.972	> 0.01 s	β^+ /≈11.6 p/	 1.02/≈100				
^{157}Ta		156.968	5. ms	α/6.22					
^{158}Ta		157.9664	37. ms	α/6.02					
^{159}Ta		158.9629	0.6 s	β^+ ,EC/20 /8.5 α/80 /	 5.60				ann.rad./
^{160}Ta		159.9615	1.4 s	β^+ ,EC/10.1 α	 5.41/				ann.rad./
^{161}Ta		160.9584	2.9 s	β^+ ,EC/7.5 α/	 5.15				ann.rad./
^{162}Ta		161.9564	4. s	EC/8.6					
^{163}Ta		162.9544	10.6 s	EC/6.8					
^{164}Ta		163.9536	14.2 s	β^+ /8.5 α/	 4.62/	3+			ann.rad./ 0.2110 0.3768
^{165}Ta		164.9508	31. s	ECβ^+ /5.9					

Elem. or Isot.	Natural Abundance (%)	Atomic Mass or Weight	Half-Life	Decay Mode/Energy (/MeV)	Particle Energy /Intensity (MeV / %)	Spin (h/2π)	Nuclear Magnetic Mom.(nm)	Elect. Quadr. Mom. (b)	γ-ray Energy /Intensity (MeV / %)
^{166}Ta		165.9505	34. s	β^+ /82 /7.7 EC/18 /					ann.rad./ Hf k x-ray 0.1587 0.3117 0.8101
^{167}Ta		166.9486	1.4 m	β^+ ,EC/5.6					ann.rad./
^{168}Ta		167.9478	2.4 m	β^+ /77 /6.7 EC/23 /		3+			ann.rad./ Hf k x-ray 0.1239 0.2615 0.7502
^{169}Ta		168.9459	4.9 m	β^+ ,EC/4.4					ann.rad./ 0.0288 0.1535 0.1924
^{170}Ta		169.9461	6.8 m	β^+ /70 /6.0 EC/35 /		(3+)			ann.rad./ Hf k x-ray 0.1008 0.2212
^{171}Ta		170.9445	23.3 m	β^+ ,EC/3.7		(5/2-)			0.0496 0.5018 0.5064 (0.05 - 1.02)
^{172}Ta		171.9447	36.8 m	β^+ /25 /4.9 EC/75 /		(3-)			ann.rad./ Hf k x-ray 0.21396 1.10923 (0.09 - 3.8)
^{173}Ta		172.9446	3.6 h	β^+ /24 /3.7 EC/76 /		(5/2-)	1.7	-1.9	ann.rad./ Hf k x-ray 0.06972 0.17219 (0.06 - 2.7)
^{174}Ta		173.9442	1.12 h	β^+ /27 /3.8 EC/73 /		(3+)			ann.rad./ Hf k x-ray 0.09089 0.20638 (0.09 - 3.64)
^{175}Ta		174.9437	10.5 h	EC/2.0		7/2+	2.27	+3.7	Hf k x-ray 0.2077 0.2671 0.3487
^{176}Ta		175.9447	8.1 h	EC/3.1		1-			Hf k x-ray 0.08837 1.15735
^{177}Ta		176.944472	2.356 d	EC/1.166		7/2+	2.25		Hf k x-ray 0.11295 (0.07 - 1.06)
^{178m}Ta			2.4 h	EC/		(7-)			Hf k x-ray 0.08886 0.21342 0.32555 0.42635
^{178}Ta		177.9458	9.29 m	EC/99 /1.9 β^+ /1 /		1+	+2.74	+0.65	ann.rad./ Hf k x-ray 0.09316
^{179}Ta		178.94593	1.8 y	EC/0.110		7/2+			Hf k x-ray
^{180m}Ta	0.012(2)		> 1.2×10^{15} y			(9-)	4.77		
^{180}Ta		179.947466	8.15 h	EC/87 /0.854 β^- /13 /0.708	0.61/3 0.71/10	1+			Hf k x-ray W k x-ray 0.09333 0.10340
^{181}Ta	99.988(2)	180.947996				7/2+	+2.370	+3.3	
^{182m}Ta			15.8 m	I.T./0.5198		10-			Ta k x-ray 0.14678 0.17157
^{182}Ta		181.950152	114.43 d	β^- /1.814	0.25/30 0.44/20 0.52/40	3-	+3.02		W k x-ray 0.06775 0.10010 1.12127 1.22138
^{183}Ta		182.951373	5.1 d	β^- /1.070	0.45/5 0.62/91	7/2+	+2.36		W k x-ray 0.0847 0.0991 0.1079 0.2461 0.3540

Elem. or Isot.	Natural Abundance (%)	Atomic Mass or Weight	Half-Life	Decay Mode/Energy (/MeV)	Particle Energy /Intensity (MeV / %)	Spin (h/2π)	Nuclear Magnetic Mom.(nm)	Elect. Quadr. Mom. (b)	γ-ray Energy /Intensity (MeV / %)
^{184}Ta		183.95401	8.7 h	β^- /2.87	1.11/15 1.17/81	(5-)			W k x-ray 0.2528/44. 0.4140/74. (0.09-1.4)
^{185}Ta		184.95556	49. m	β^- /1.99	1.21/5 1.77/81	(7/2+)			W k x-ray 0.0697 0.1739 0.1776
^{186}Ta		185.9586	10.5 m	β^- /3.9	2.2/	(3-)			W k x-ray 0.1979 0.2149 0.5106 (0.09 - 1.5)
$_{74}$**W**		**183.84(1)**							
^{158m}W			<1. ms	α	8.28(3)/				
^{158}W		157.974	0.9 ms	α/	6.433				
^{159}W		158.972	7. ms	α/					
^{160}W		159.9684	0.08 s	α/	5.92/	0+			
^{161}W		160.9671	0.41 s	β^+ ,EC/18 /8.1 α/82 /	 5.78/				
^{162}W		161.9626	1.39 s	β^+ ,EC/54 /5.8 α/46 /	 5.54/	0+			
^{163}W		162.9624	2.8 s	β^+ ,EC/59 /7.5 α/41 /	 5.38/				
^{164}W		163.95890	6. s	β^+ ,EC/97 /5.0 α/3 /	 5.15/	0+			ann.rad./
^{165}W		164.9583	5.1 s	β^+ ,EC/99 /7.0 α/1 /	 4.91/				ann.rad./
^{166}W		165.95502	16. s	β^+ ,EC/99 /4.2 α/1 /	 4.74/	0+			ann.rad./
^{167}W		166.9547	20. s	EC/5.6					
^{168}W		167.9519	53. s	EC/3.8 α/10^{-5}/	 4.40(1)				ann.rad./ Ta k x-ray 0.1755 (0.037-0.573)
^{169}W		168.9518	1.3 m	EC/5.4					ann.rad./ Ta k x-ray 0.123 (0.097-0.699)
^{170}W		169.9485	2.4 m	EC/2.2					ann.rad./ Ta k x-ray 0.3162 (0.060-0.144)
^{171}W		170.9494	2.4 m	EC/4.6					ann.rad./ Ta k x-ray 0.1842 (0.052-0.479)
^{172}W		171.9474	6.6 m	β^+ ,EC/2.5					ann.rad./ Ta k x-ray 0.0389 (0.034-0.674)
^{173}W		172.9489	6.3 m	EC/4.0					ann.rad./ Ta k x-ray 0.4576 (0.035-0.623)
^{174}W		173.9462	35. m	EC/1.9		0+			ann.rad./ Ta k x-ray 0.3287 0.4288 (0.056-0.429)
^{175}W		174.9468	35. m	EC/2.9		1/2-			(0.015-0.27)
^{176}W		175.9456	2.5 h	β^+ ,EC/0.8		0+			0.03358 0.06129 0.09487 0.10020
^{177}W		176.9466	2.21 h	EC/2.0		(1/2-)			Ta k x-ray 0.15505 0.18569 0.42694
^{178}W		177.9459	21.6 d	EC/0.091		0+			Ta k x-ray
^{179m}W			6.4 m	I.T./99.7 /0.222 EC/0.3 /		(1/2-)			W k x-ray 0.2220
^{179}W		178.94707	38. m	EC/1.06		(7/2-)			Ta k x-ray 0.0307
^{180}W	0.120(1)	179.946706				0+			

Elem. or Isot.	Natural Abundance (%)	Atomic Mass or Weight	Half-Life	Decay Mode/Energy (/MeV)	Particle Energy /Intensity (MeV / %)	Spin (h/2π)	Nuclear Magnetic Mom.(nm)	Elect. Quadr. Mom. (b)	γ-ray Energy /Intensity (MeV / %)
^{181}W		180.94820	121.2 d	EC/0.188		9/2+			Ta k x-ray
									0.13617
									0.15221
^{182}W	26.498(29)	181.948205				0+			
^{183m}W			5.15 s	I.T./		(11/2+)			W k x-ray
									0.0465
									0.0526
									0.0991
									0.1605
^{183}W	14.314(4)	182.950224				1/2-	+0.1177848		
^{184}W	30.642(8)	183.950932				0+			
^{185m}W			1.6 m	I.T./0.1974		11/2+			W k x-ray
									0.0659
									0.1315
									0.1737
^{185}W		184.953420	74.8 d	β^- /0.433	0.433/99.9	3/2-			0.12536
^{186}W	28.426(37)	185.954362				0+			
^{187}W		186.957158	23.9 h	β^- /1.311	0.624/	3/2-	0.62		Re k x-ray
					1.315/				0.0725
									0.47951
									0.68572
^{188}W		187.958487	69.4 d	β^- /0.349	0.349/99	0+			0.0636
									0.2271
									0.2907
^{189}W		188.9619	11.5 m	β^- /2.5	1.4/	(3/2-)			0.258
					2.5/				0.417
									0.550
^{190}W		189.9632	30. m	β^- /1.3	0.95/	0+			Re k x-ray
									0.1576
									0.1621
$_{75}$Re		**186.207(1)**							
^{160}Re		159.981	0.8 ms	p/	1.261(6)/91				
				α/	6.54/				
^{161}Re		160.978	0.01 s	α/					
^{162}Re		161.9757	0.10 s	α/	6.12/				
^{163}Re		162.9721	0.26 s	β^+ ,EC/9.0					
				α/	5.92				
^{164}Re		163.9704	0.9 s	β^+ ,EC/10.7					
				α/	5.78/				
^{165}Re		164.9671	2. s	β^+ ,EC/87 /8.1					
				α/	5.51/				
^{166}Re		165.9651	2.5 s	β^+ ,EC/9.4					
				α/	5.50/				
^{167m}Re			6.2 s	α, EC/					
^{167}Re		166.9626	3.4 s	β^+ ,EC/7.4					
				α/	5.015/				
^{168}Re		167.9616	4.4 s	β^+ ,EC/9.1					
				α/	4.833/				0.1117
^{169m}Re			8.1 s	α	4.70/				
					4.87/				
^{169}Re		168.9588	16. s						
^{170}Re		169.9582	9.2 s	β^+, EC/9.0					0.1560
									0.3055
									0.4125
^{171}Re		170.9555	15.2 s	EC/≈ 5.7					
^{172m}Re			55. s	β^+ ,EC/		(2)			ann.rad./
									0.1234
									0.2537
									0.3504
^{172}Re		171.9553	15. s	β^+ ,EC/7.3					ann.rad./
									0.1234
									0.2537
^{173}Re		172.9531	2.0 m	EC/≈3.9					ann.rad./
^{174}Re		173.9521	2.4 m	β^+ ,EC/5.6					ann.rad./
									0.1119
									0.2430
^{175}Re		174.9514	5.8 m	β^+ ,EC/4.3					ann.rad./
^{176}Re		175.9516	5.3 m	β^+ ,EC/5.6		(3+)			ann.rad./
									0.1089
									0.2406
^{177}Re		176.9503	14. m	EC/78 /3.4		(5/2-)			ann.rad./
				β^+ /22 /					W k x-ray
									0.0797
									0.0843
									0.1968

Elem. or Isot.	Natural Abundance (%)	Atomic Mass or Weight	Half-Life	Decay Mode/Energy (/MeV)	Particle Energy /Intensity (MeV / %)	Spin (h/2π)	Nuclear Magnetic Mom.(nm)	Elect. Quadr. Mom. (b)	γ-ray Energy /Intensity (MeV / %)
^{178}Re		177.9509	13.2 m	β^+ /11 /4.7 EC/89 /	3.3/	(3)			ann.rad./ W k x-ray 0.1059 0.2373 0.9391
^{179}Re		178.9500	19.7 m	EC/99 /2.71 β^+ /1 /	0.95/	(5/2+)			W k x-ray 0.1199 0.2900 0.4154 0.4302 1.6803
^{180}Re		179.95079	2.45 m	EC/92 /3.80 β^+ /8 /	1.76/	1-			ann.rad./ W k x-ray 0.1036 0.9028 (0.07 - 2.2)
^{181}Re		180.95006	20. h	EC /1.74		5/2+	3.19		W k x-ray 0.3607 0.3655 0.6390
^{182m}Re			12.7 h	EC/	0.55/ 1.74/	2+	3.3	+1.8	W k x-ray 0.0677 1.1214 1.2215 (0.06 - 2.2)
^{182}Re		181.9512	2.67 d	EC/2.8		(7+)	2.8	+4.1	W k x-ray 0.0678 0.2293 1.1213 1.2214
^{183}Re		182.95082	70. d	EC/0.56		(5/2+)	+3.17	+2.3	W k x-ray 0.16232
^{184m}Re			165. d	I.T./75 /0.188 EC/25 /		8+	+2.9		Re k x-ray 0.1047 0.2165 0.92093 (0.10 - 1.1)
^{184}Re		183.95252	38. d	EC/1.48		3-	+2.53	+2.8	W k x-ray 0.79207 0.90328 (0.1 - 1.4)
^{185}Re	37.40(2)	184.952955				5/2+	+3.1871	+2.18	
^{186m}Re			2.0x10^5 y	I.T./0.150		8+			Re k x-ray 0.0590
^{186}Re		185.954986	3.718 d	β^- /92 /1.070 EC/8 /0.582	0.973/21 1.07/71	1-	+1.739	+0.62	W k x-ray 0.1227/0.6 0.1372/9.5 (0.63-0.77)
^{187}Re	62.60(2)	186.955751	4.4x10^{10} y	β^- /0.00266	0.0025/	5/2+	+3.2197	+2.07	
^{188m}Re			18.6 m	I.T./0.172		(6-)			Re k x-ray 0.0925 0.1059
^{188}Re		187.958112	16.94 h	β^- /2.120	1.962/20 2.118/79	1-	+1.788	+0.57	Os k x-ray 0.15502
^{189}Re		188.959228	24. h	β^- /1.01	1.01/	(5/2+)			0.1471 0.2167 0.2194 0.2451
^{190m}Re			3.0 h	β^- /51 / I.T./49 /		(6-)			Re k x-ray 0.1191 0.2238 0.6731 (0.1 - 1.79)
^{190}Re		189.9618	3.0 m	β^- /3.2	1.8/	(2-)			Os k x-ray 0.1867 0.5580 0.6051
^{191}Re		190.96312	9.7 m	β^- /2.05	1.8/				
^{192}Re		191.9660	16. s	β^- /4.2	$\approx$ 2.5/				(0.2-0.75)
$_{76}$Os		**190.23(3)**							
^{162}Os		161.984	$\approx$ 1.9 ms	α/					
^{163}Os		162.982		α					
^{164}Os		163.9779	0.04 s	α					
^{165}Os		164.9765	0.07 s	α					
^{166}Os		165.9718	0.18 s	β^+ ,EC/28 /6.3 α/72 /	6.27/ 5.98/	0+			ann. rad./

Elem. or Isot.	Natural Abundance (%)	Atomic Mass or Weight	Half-Life	Decay Mode/Energy (/MeV)	Particle Energy /Intensity (MeV / %)	Spin (h/2π)	Nuclear Magnetic Mom.(nm)	Elect. Quadr. Mom. (b)	γ-ray Energy /Intensity (MeV / %)
^{167}Os		166.9714	0.7 s	β^+ ,EC/76 /8.2					ann.rad./
				α/24 /	5.84/				
^{168}Os		167.96775	2.2 s	β^+ ,EC/51 /5.7		0+			ann. rad./
				α/49 /					
^{169}Os		168.9671	3.3 s	β^+ ,EC/89 /7.7					ann.rad./
				α/17 /	5.57/				
^{170}Os		169.96357	7.1 s	β^+ ,EC/5.0		0+			ann.rad./
				α/	5.40/				
^{171}Os		170.9630	8. s	β^+ ,EC/98 /7.1					ann.rad./
				α/2 /	5.24/				
^{172}Os		171.9601	19. s	β^+ ,EC/99 /4.5		0+			ann.rad./
				α/1 /	5.10/				0.177
									0.187
^{173}Os		172.9598	16. s	β^+ ,EC/6.3					ann.rad./
				α/0.02 /	4.94/				ann.rad./
^{174}Os		173.9563	44. s	β^+ ,EC/3.9		0+			0.118
				α/0.02 /	4.76/				0.138 / 0.001
									0.158
									0.325
^{175}Os		174.9570	1.4 m	β^+ ,EC/5.3					0.125
									0.181
									0.248
^{176}Os		175.9550	3.6 m	β^+ ,EC/3.2		0+			0.8155
									0.7758
									0.8573
									1.2093
									1.2909
^{177}Os		176.9551	2.8 m	β^+ ,EC/4.5		(1/2-)			0.0848
									0.1958
									0.3002
									1.2686
^{178}Os		177.9334	3.0 m	β^+ ,EC/2.3		0+			ann.rad./
									0.5946
									0.6850
									0.9687
									1.3311
^{179}Os		178.9539	7. m	β^+ ,EC/3.7					ann.rad./
									0.0654
									0.2186
									0.5938
^{180}Os		179.9524	21.5 m	β^+ ,EC/1.5		0+			Re k x-ray
									0.0202-0.7174
^{181m}Os			1.75 h	EC/		(1/2-)			ann.rad./
									0.14493
^{181}Os		180.9532	2.7 m	EC/2.9		(7/2-)			ann.rad./
									0.11794
									0.23868
									0.8267
									(0.07 - 2.64)
^{182}Os		181.95219	21.5 h	EC/0.9		0+			Re k x-ray
									0.1802
									0.5100
^{183m}Os			9.9 h	EC/84 /		1/2-			Os k x-ray
				I.T./16 /					Re k x-ray
									1.1020
									1.1080
^{183}Os		182.9531	13. h	EC/2.1		9/2+	-0.79	+3.1	Re k x-ray
									0.1144
									0.3818
^{184}Os	0.020(3)	183.952491				0+			
^{185}Os		184.954043	93.6 d	EC/1.013		1/2-			Re k x-ray
									0.6461
									0.8748
									0.8805
^{186}Os	1.58(10)	185.953838	$2.x10^{15}$ y	α/	≈ 2.75/	0+			
^{187}Os	1.6(1)	186.955748				1/2-	+0.0646519		
^{188}Os	13.3(2)	187.955836				0+			
^{189m}Os			5.8 h	I.T./0.0308		9/2-			Os L x-ray
									0.0308
^{189}Os	16.1(3)	188.958145				3/2+	+0.65993	+0.86	
^{190m}Os			9.9 m	I.T./1.705		10-	-0.6		Os k x-ray
									0.1867
									0.3611
									0.5026
									0.6161
^{190}Os	26.4(4)	189.958445				0+			
^{191m}Os			13.1 h	I.T./0.0744		3/2-			Os k x-ray
									0.0744

Elem. or Isot.	Natural Abundance (%)	Atomic Mass or Weight	Half-Life	Decay Mode/Energy (/MeV)	Particle Energy /Intensity (MeV / %)	Spin (h/2π)	Nuclear Magnetic Mom.(nm)	Elect. Quadr. Mom. (b)	γ-ray Energy /Intensity (MeV / %)
^{191}Os		190.960928	15.4 d	β^- /0.314	0.140/100	9/2-		+2.5	Ir k x-ray 0.1294
^{192m}Os			6.0 s	I.T./2.0154		(10-)			Os k x-ray 0.2058 0.3024 0.4531 0.5692
^{192}Os	41.0(3)	191.961479				0+			
^{193}Os		192.964148	30.5 h	β^- /1.141	1.04/20	3/2-	+0.730	+0.47	Ir k x-ray 0.1389 0.4605
^{194}Os		193.965179	6.0 y	β^- /0.097	0.054/33 0.096/67	0+			Ir L x-ray 0.0429
^{195}Os		194.9681	6.5 m	β^- /2.0	2.0/				
^{196}Os		195.96962	34.9 m	β^- /1.16	0.84/	0+			0.1262/5 0.4079/5.9
$_{77}$Ir		**192.217(3)**							
^{166}Ir		165.9855	≥ 5. ms	α/					
^{167}Ir		166.9817	≥ 5. ms	α/					
^{168}Ir		167.9799		α/					
^{169}Ir		168.9764	0.4 s	α/					
^{170}Ir		169.9743	1.0 s	α/	6.03/				
^{171}Ir		170.9718	1.5 s	α/	5.91/				
^{172}Ir		171.9706	2.1 s	α/	5.811/				0.228 (0.379-0.475)
^{173}Ir		172.9677	3.0 s	α/	5.665/				0.0493 (0.092-0.296)
^{174}Ir		173.9668	4. s	α/	5.478/				0.1587 (0.276-1.33)
^{175}Ir		174.9641	≈ 4.5 s	α/	5.393/				0.1056
^{176}Ir		175.9635	8. s	EC, β^+/80 α/3.2/	5.118/				0.260 (0.135-0.415)
^{177}Ir		176.9612	30. s	EC, β^+/5.7 α/0.06/	5.011/				0.184 (0.062-0.194)
^{178}Ir		177.9601	12. s	β^+ ,EC/6.3					0.1320 0.2667 0.3633
^{179}Ir		178.9592	4. m	EC/4.9					0.0975 (0.045-0.220)
^{180}Ir		179.9593	1.5 m	EC/6.4					0.2765 ((0.132-1.106)
^{181}Ir		180.9576	4.9 m	β^+ ,EC/4.1		(7/2+)			ann.rad./ 0.1076 0.2270 0.3189 1.6396
^{182}Ir		181.9582	15. m	β^+ /44 /5.6 EC/56 /					ann.rad./ Os k x-ray 0.1273 0.2370
^{183}Ir		182.9568	57. m	β^+ ,EC/3.5					ann.rad./ 0.0877 0.2285 0.2824
^{184}Ir		183.9574	3.0 h	β^+ /12 /4.6 EC/88 /	2.3/ 2.9/	5-	0.70	+2.0	ann.rad./ Os k x-ray 0.11968 0.2640 0.3904
^{185}Ir		184.9566	14. h	β^+ /3 /2.4 EC/97 /		(5/2-)	2.60	-2.1	ann.rad./ Os k x-ray 0.2543 1.8288
^{186m}Ir			1.7 h	EC /		(2-)			Os k x-ray 0.1371 0.7675
^{186}Ir		185.95795	15.7 h	EC/98 /3.83 β^+ /2 /		(5+)	3.9	-2.5	Os k x-ray 0.1372 0.2968 0.4348 (0.13 - 3.0)

Elem. or Isot.	Natural Abundance (%)	Atomic Mass or Weight	Half-Life	Decay Mode/Energy (/MeV)	Particle Energy /Intensity (MeV / %)	Spin (h/2π)	Nuclear Magnetic Mom.(nm)	Elect. Quadr. Mom. (b)	γ-ray Energy /Intensity (MeV / %)
^{187}Ir		186.95736	10.5 h	EC/1.50		3/2+			Os k x-ray 0.0743 0.4009 0.4271 0.6109 0.9128
^{188}Ir		187.95885	1.72 d	β^+ /2.81 EC/99+ /	1.13/ 1.64/	(2-)	0.30	+5.4	Os k x-ray 0.1550 0.4780 0.6330 2.2146
^{189}Ir		188.95872	13.2 d	EC/0.53		3/2+	0.13	+1.0	Os k x-ray 0.2449
^{190m2}Ir			3.2 h	β^+ ,EC/95 / I.T./5 /		(11-)			
^{190m1}Ir			1.2 h	I.T. /0.0263		7+			Ir L x-ray
^{190}Ir		189.9606	11.8 d	EC/2.0		(4+)	0.04	+2.8	Os k x-ray 0.1867 0.4072 0.5186 0.5580 0.6051 (0.2 - 1.4)
^{191m}Ir			4.93 s	I.T./0.1714		11/2-	0.603		Ir k x-ray 0.1294
^{191}Ir	37.3(5)	190.960591				3/2+	+0.151	+0.82	
^{192m2}Ir			241. y	I.T./0.161		(9+)			Ir k x-ray
^{192m1}Ir			1.44 m	I.T./0.0580		(1+)			Ir L x-ray 0.0580 0.3165
^{192}Ir		191.962602	73.83 d	β^- /1.460		(4-)	+1.92	+2.3	Pt k x-ray 0.31649/83. 0.46806/48.
^{193m}Ir			10.53 d	I.T./0.0802		11/2-			Ir L x-ray 0.0803
^{193}Ir	62.7(5)	192.962923				3/2+	+0.164	+0.75	
^{194m}Ir			170. d	β^- /		11			Pt k x-ray 0.3284 0.4829 0.5624
^{194}Ir		193.965075	19.3 h	β^- /2.247	1.92/9 2.25/86	1-	0.39	+0.34	0.2935 0.3284 0.6451 (0.1 - 2.2)
^{195m}Ir			3.9 h	β^- /	0.41/ 0.97/	(11/2-)			Pt k x-ray 0.3199/9.6 0.3649/9.5 0.4329/9.6 0.6849/9.6
^{195}Ir		194.965976	2.8 h	β^- /1.120	1.0/80 1.11/13	(3/2+)			Pt k x-ray 0.0989/9.7
^{196m}Ir			1.40 h	β^- /	1.16/				Pt k x-ray 0.3557 0.3935 0.4471 0.5214 0.6473
^{196}Ir		195.96838	52.s	β^- /3.21	2.1/15 3.2/80	0-			0.3329 0.3557 0.7796
^{197m}Ir			8.9 m	β^- / I.T./		(11/2-)			[illegible] See Ir[197]
^{197}Ir		196.96964	5.8 m	β^- /2.16	1.5/ 2.0/	(3/2+)			0.0531 0.1351 0.4306 0.4697
^{198}Ir		197.9723	8. s	β^- /4.1					0.4074 0.5070
$_{78}$Pt		**195.08(3)**							
^{168}Pt		167.9880		α					
^{169}Pt		168.9864	3. ms	α					
^{170}Pt		169.9816	6. ms	α	6.55				
^{171}Pt		170.9811	0.03 s	α					
^{172}Pt		171.97730	0.10 s	α /	6.31/94	0+			
^{173}Pt		172.9765	0.34 s	β^+ ,EC/8.2 α/	 6.20/				

Elem. or Isot.	Natural Abundance (%)	Atomic Mass or Weight	Half-Life	Decay Mode/Energy (/MeV)	Particle Energy /Intensity (MeV / %)	Spin (h/2π)	Nuclear Magnetic Mom.(nm)	Elect. Quadr. Mom. (b)	γ-ray Energy /Intensity (MeV / %)
^{174}Pt		173.97281	0.89 s	β^+ ,EC/17 /5.6		0+			
				α/83 /	6.040/				
^{175}Pt		174.9723	2.5 s	β^+ ,EC/65 /7.6					0.0774
				α/35 /	5.831/5				0.1354
					5.96/54				0.2128
					6.038/				
^{176}Pt		175.9690	6.3 s	β^+ ,EC/60 /5.1		0+			ann.rad./
				α/40 /	5.528/0.6				0.2277
					5.750/41				
^{177}Pt		176.9685	11. s	EC/91 /6.8	5.53/				0.0908
				α/9 /	5.485/3				
					5.525/6				
^{178}Pt		177.9649	21. s	EC/93 /4.5		0+			
				α/7 /	5.286/0.2				
					5.442/7				
^{179}Pt		178.9653	33. s	β^+ ,EC/5.7					
				α /	5.16/				
^{180}Pt		179.9632	52. s	β^+ ,EC/99.7 /3.7		0+			
				α/0.3 /	5.140/				
^{181}Pt		180.9632	51. s	β^+ ,EC/5.2					
^{182}Pt		181.9613	2.7 m	β^+ ,EC/2.9		0+			ann.rad./
									0.1360
									0.1460
									0.2100
^{183m}Pt			43. s	β^+ ,EC/		(7/2-)			ann.rad./
				I.T./					0.3132
									0.3164
									0.6296
^{183}Pt		182.9617	7. m	β^+ ,EC/4.6					ann.rad./
^{184}Pt		183.9599	17.3 m	β^+ ,EC/2.3					ann.rad./
									0.1549
									0.1919
									0.5484
^{185m}Pt			33. m	β^+ ,EC/		1/2-			
^{185}Pt		184.9607	1.18 h	β^+ ,EC/3.8		(9/2+)			ann.rad./
									0.1353
									0.1974
									0.2296
									0.2551
^{186}Pt		185.95943	2.0 h	β^+ ,EC/1.38		0+			ann.rad./
									0.6115
									0.6892
^{187}Pt		186.9607	2.35 h	β^+ ,EC/3.1		3/2			ann.rad./
									Ir k x-ray
									0.1064
									0.1100
									0.2015
									0.2849
									0.7092
^{188}Pt		187.95940	10.2 d	EC/0.51		0+			Ir k x-ray
									0.1876
									0.1951
^{189}Pt		188.96083	10.9 h	β^+ ,EC/1.97		3/2-	0.43	-0.6	Ir k x-ray
									0.0943
									0.6076
									0.7214
									(0.09 - 1.47)
^{190}Pt	0.01(1)	189.95993	6.5x10^{11} y			0+			
^{191}Pt		190.961684	2.96 d	EC/1.02		(3/2-)	0.50	-0.6	Ir k x-ray
									0.3599
									0.4094
									0.5389
^{192}Pt	0.79(6)	191.961035				0+			
^{193m}Pt			4.33 d	I.T./0.1498		13/2+	-0.75		Pt k x-ray
									0.1355
^{193}Pt		192.962984	60. y	EC/0.0566		(1/2-)			Ir k x-rays
^{194}Pt	32.9(6)	193.962663				0+			
^{195m}Pt			4.02 d	I.T./0.2952		13/2+	0.61	+1.4	Pt k x-ray
									0.0989
^{195}Pt	33.8(6)	194.964774				1/2-	+0.6095		
^{196}Pt	25.3(6)	195.964934				0+			
^{197m}Pt			1.590 h	I.T./97 /		13/2+			Pt k x-ray
				β^- /3 /					0.0530
									0.3465
^{197}Pt		196.967323	18.3 h	β^- /0.719		1/2-	0.51		Au k x-ray
									0.1914
									0.2688
^{198}Pt	7.2(2)	197.967875				0+			

Elem. or Isot.	Natural Abundance (%)	Atomic Mass or Weight	Half-Life	Decay Mode/Energy (/MeV)	Particle Energy /Intensity (MeV / %)	Spin (h/2π)	Nuclear Magnetic Mom.(nm)	Elect. Quadr. Mom. (b)	γ-ray Energy /Intensity (MeV / %)
^{199m}Pt			13.6 s	I.T./0.424		13/2+			Pt k x-ray 0.3919
^{199}Pt		198.970576	30.8 m	β^- /1.70	0.90/18 1.14/14	(5/2-)			0.0772 0.18579 0.31703 0.49375 0.54298
^{200}Pt		199.97142	12.5 h	β^- /$\approx$0.66		0+			Au k x-ray 0.13590 0.22747 0.24371
^{201}Pt		200.9745	2.5 m	β^- /2.66		(5/2-)			0.070 0.152 0.222 1.760
^{202}Pt			1.8 d						0.440
$_{79}$Au		**196.96655(2)**							
^{172}Au			4 ms	α/7.02	6.86				
^{173}Au		172.9864	0.06 s	α					
^{174}Au		173.9842	0.12 s	α					
^{175}Au		174.9817	0.20 s	α					
^{176}Au		175.9803	1.2 s	β^+ ,EC/10.5 α /	 6.260/80 6.290/20				
^{177}Au		176.9772	1.2 s	α /	6.115/ 6.150/				
^{178}Au		177.9760	2.6 s	α /	5.920/				
^{179}Au		178.9732	7.5 s	α /	5.85/				
^{180}Au		179.9724	8.1 s	EC/8.6 α/	5.65 5.61 5.50				0.1522 0.2564 0.5242 0.6765 0.8084 0.8597
^{181}Au		180.9700	11.4 s	EC/97.5/6.3 α/2.7/	5.482/				
^{182}Au		181.9686	21. s	β^+ ,EC/6.9 α/0.13/					ann.rad./ 0.1549 0.2649 (0.13 - 1.4)
^{183}Au		182.9676	42. s	EC/5.5 α/0.8/			+1.97		0.1630 0.2730 0.3625
^{184}Au		183.9675	53. s	EC,β^+ /7.1 α/0.013/		(5+)	+1.65		
^{185m}Au			6.8 m	β^+ ,EC/ I.T./0.145					
^{185}Au		184.9657	4.3 m	β^+ ,EC/4.71 α/0.26/		(5/2-)	+2.17	-1.1	ann.rad./
^{186m}Au			< 2. m	β^+ ,EC/					0.1915
^{186}Au		185.9659	10.7 m	β^+ ,EC/6.0 α/8(10)$^{-4}$/		3-	-1.26	+3.1	ann.rad./ 0.1915 0.2988
^{187m}Au			2.3 s	IT		9/2-			
^{187}Au		186.9646	8.3 m	β^+ ,EC/3.60		1/2+	+0.54		ann.rad./ 0.9152 1.2668 1.3321 1.4081
^{188}Au		187.9651	8.8 m	β^+ ,EC/5.3		(1-)	-0.07		ann.rad./ 0.2660 0.3404 0.6061
^{189m}Au			4.6 m	β^+ ,EC/		11/2-	+6.19		0.1667
^{189}Au		188.9642	28.7 m	EC/96 /3.2 β^+ /4 /		1/2+	+0.49		ann.rad./ Pt k x-ray 0.4478 0.7133 0.8128
^{190}Au		189.96470	43. m	β^+ /2 /4.44 EC/98 /		1-	-0.07		ann.rad./ Pt k x-ray 0.2958 0.3018 0.5977
^{191m}Au			0.9 s	I.T./0.2663		(11/2-	6.6		Au k x-ray 0.2414 0.2526

Elem. or Isot.	Natural Abundance (%)	Atomic Mass or Weight	Half-Life	Decay Mode/Energy (/MeV)	Particle Energy /Intensity (MeV / %)	Spin (h/2π)	Nuclear Magnetic Mom.(nm)	Elect. Quadr. Mom. (b)	γ-ray Energy /Intensity (MeV / %)
^{191}Au		190.96365	3.2 h	EC/1.83		3/2+	+0.137	+0.72	Pt k x-ray 0.2779 0.2839 0.5864 0.6742 (0.08 - 1.3)
^{192}Au		191.96481	4.9 h	β^+ /5 /3.52 EC/95 /	2.19/ 2.49/	1-	-0.011	-0.23	ann.rad./ Pt k x-ray 0.2959 0.3165
^{193m}Au			3.9 s	I.T./0.2901		11/2-	6.2		Au k x-ray 0.2580
^{193}Au		192.96413	17.6 h	EC/1.07		3/2+	+0.140	+0.66	Pt k x-ray 0.1862 0.2556
^{194}Au		193.96534	1.64 d	β^+ /3 /2.49 EC/97 /	1.49/	1-	+0.075	-0.24	ann.rad./ Pt k x-ray 0.2935 0.3284/61
^{195m}Au			30.5 s	I.T./0.3186		11/2-	6.2	+1.4	Au k x-ray 0.2617
^{195}Au		194.965017	186.12 d	EC/0.227		3/2+	+0.149	+0.61	Pt k x-ray
^{196m2}Au			9.7 h	I.T./0.5954		12-	5.7		Au k x-ray 0.1478 0.1883
^{196m1}Au			8.1 s	I.T./0.0846		8+			0.0847
^{196}Au		195.966551	6.18 d	EC/92 /1.506		2-	+0.591	0.81	Pt k x-ray
^{197m}Au			7.8 s	I.T./0.4094 β^- /8 /0.686		11/2-	+6.0	+1.4	Au k x-ray 0.1302 0.2790
^{197}Au	100.	196.966551				3/2+	+0.14575	+0.55	
^{198m}Au			2.30 d	I.T./0.812		(12-)			Au k x-ray 0.0972 0.1803 0.2419
^{198}Au		197.968225	2.694 d	β^- /1.372	0.290/1 0.961/99	2-	+0.5934	+0.68	Hg k x-ray 0.411794
^{199}Au		198.968748	3.14 d	β^- /0.453	0.25/22 0.292/72 0.462/6	3/2+	+0.2715	0.55	Hg k x-ray 0.15837 0.20820
^{200m}Au			18.7 h	β^- /84 /1.0 I.T./16 /	0.56/	12-	5.9		Au k x-ray 0.1111 0.2559 0.36797 0.4978 0.5793 0.7595
^{200}Au		199.97072	48.4 m	β^- /2.24	0.7/15 2.2/77	1-			0.3679 1.2254 (0.3 - 1.6)
^{201}Au		200.97165	26. m	β^- /1.28	1.27/82	3/2+			0.1674 0.5170 0.5426 0.6132
^{202}Au		201.9738	29. s	β^- /3.0		(1-)			0.4396
^{203}Au		202.97515	1.0 m	β^- /2.14	≈ 1.9/	3/2+			(0.04-0.37)
^{204}Au		203.9783	40. s	β^- /4.5		(2-)			0.4366 1.5113
^{205}Au			31. s	β^- /					(0.38 - 1.33)
$_{80}$Hg		**200.59(2)**							
^{175}Hg		174.9912	0.02 s	α					
^{176}Hg		175.98733	0.03 s	α	6.77				
^{177}Hg		176.9863	0.13 s	α					
^{178}Hg		177.98248	0.26 s	EC/50 /6.1 α/50 /	 6.43/	0+			
^{179}Hg		178.9818	1.09 s	EC/8.0 α /	 6.29/				
^{180}Hg		179.9783	2.6 s	EC/5.5 α /	 6.12/33 5.69/.03	0+			0.1250 0.3005 0.3812
^{181}Hg		180.9778	3.6 s	β^+ EC/74 /≈7.3 α/26 /		(1/2-)	+0.5071		0.0663 0.0811 0.0924 0.1474 0.1587 0.2142 0.2398

Elem. or Isot.	Natural Abundance (%)	Atomic Mass or Weight	Half-Life	Decay Mode/Energy (/MeV)	Particle Energy /Intensity (MeV / %)	Spin (h/2π)	Nuclear Magnetic Mom.(nm)	Elect. Quadr. Mom. (b)	γ-ray Energy /Intensity (MeV / %)
^{182}Hg		181.9739	10.8 s	β⁺ ,EC/85 /5.0		0+			0.1289
				α/15 /	5.87/8.6				0.2168
					5.45/0.03				0.4126
^{183}Hg		182.9744	9. s	β⁺ ,EC/77 /6.3		1/2-	+0.524		0.0714
				α /	5.83/				0.0874
					5.91/				0.1538
^{184}Hg		183.9719	30.9 s	β⁺ ,EC/99 /4.1		0+			0.0915
				α/1 /	5.54/1.3				0.1265
					5.07/2 x 10^{-3}				0.1560
									0.2362
^{185m}Hg			21. s	β⁺ ,EC,IT,α /	5.37/	13/2+	-1.02	+0.2	0.211
									0.292
^{185}Hg		184.9720	51. s	β⁺ ,EC/95 /5.8		1/2-	+0.509		(0.02 - 0.55)
^{186}Hg		185.9695	1.4 m	β⁺ ,EC/3.3		0+			0.1119
				α	5.09/0.02				0.2518
^{187m}Hg			1.7 m	β⁺ ,EC/		13/2+	-1.04	+0.5	See Hg[187]
^{187}Hg		186.9698	2.4 m	β⁺ ,EC/4.9		3/2-	-0.594		0.1034/32.
									0.2334/100.
									0.2403/33.
									0.27151/31.
									0.3763/38.
									0.5254/30.
									(0.10-2.18)
^{188}Hg		187.9676	3.2 m	β⁺ ,EC/2.3		0+			0.0988
				α	4.61				0.1148
									0.1424
									0.1900
^{189m}Hg			8.6 m	EC/		13/2+	-1.06	+0.7	0.0780
									0.3210
									0.4345
									0.5655
									(0.08 - 2.10)
^{189}Hg		188.9687	7.6 m	EC/4.2		3/2-	-0.6086	0.8	0.2005
									0.2038
									0.2386
									0.2485
^{190}Hg		189.9663	20.0 m	EC/1.5		0+			0.1296
									0.1426
^{191m}Hg			51. m	β⁺ /6 /		13/2+	-1.07	+0.6	ann.rad./
				EC/94 /					Au k x-ray
									0.2741
									0.4203
									0.5787
									(0.07 - 1.9)
^{191}Hg		190.9671	50. m	β⁺ ,EC/3.2		(3/2-)	-0.62	-0.8	0.1963
									0.2247
									0.2524
^{192}Hg		191.9653	5.0 h	EC/≈0.5		0+			Au k x-ray
									0.1572
									0.2748
									0.3065
^{193m}Hg			11.8 h	β⁺ ,EC/91 /		13/2+	-1.05843	+0.92	Hg k x-ray
				I.T./9 /0.2901					0.1866
									0.2580
									0.4076
									0.5733
									0.9324
									(0.1 - 1.96)
^{193}Hg		192.96664	3.8 h	EC,B⁺/2.34		3/2-	-0.6276		0.1866
									0.2580
									0.8611
^{194}Hg		193.96538	520. y	EC/0.04		0+			Au L x-rays
^{195m}Hg			1.67 d	I.T./(54)/0.3186		13/2+	-1.04465	+1.1	Hg k x-ray
				EC/(46)/					Au k x-ray
									0.2617
									0.5603
									0.7798
^{195}Hg		194.96664	9.5 h	EC/1.51		1/2-	+0.541475		Au k x-ray
									0.0614
									0.7798
^{196}Hg	0.15(1)	195.965814				0+			
^{197m}Hg			23.8 h	I.T./(93)/0.2989		13/2+	-1.02768	+1.2	Hg k x-ray
									Au k x-ray
									0.13398
^{197}Hg		196.967195	2.672 d	EC/0.600		1/2-	+0.527374		Au k x-ray
									0.07735
^{198}Hg	9.97(8)	197.966752				0+			
^{199m}Hg			42.6 m	I.T./0.532		13/2+	-1.014703	+1.2	Hg k x-ray
									0.15841

Elem. or Isot.	Natural Abundance (%)	Atomic Mass or Weight	Half-Life	Decay Mode/Energy (/MeV)	Particle Energy /Intensity (MeV / %)	Spin (h/2π)	Nuclear Magnetic Mom.(nm)	Elect. Quadr. Mom. (b)	γ-ray Energy /Intensity (MeV / %)
^{199}Hg	16.87(10)	198.968262				1/2-	+0.505885		
^{200}Hg	23.10(16)	199.968309				0+			
^{201}Hg	13.18(8)	200.970285				3/2-	-0.560226	+0.39	
^{202}Hg	29.86(20)	201.970625				0+			
^{203}Hg		202.972857	46.61 d	β^- /0.492	0.213/100	5/2-	+0.8489	+0.34	Tl k x-ray
									0.279188
^{204}Hg	6.87(4)	203.973475				0+			
^{205}Hg		204.976056	5.2 m	β^- /1.531	1.33/4	1/2-	+0.6010		0.20378
									(0.2 - 1.4)
^{206}Hg		205.97750	8.2 m	β^- /1.31	0.935/34	0+			Tl k x-ray
					1.3/63				0.3052
									0.6502
^{207}Hg		206.9825	2.9 m	β^- /4.8		(9/2+)			
^{208}Hg			0.7 h	β^-					0.474
$_{81}$Tl		**204.3833(2)**							
^{179}Tl		178.9917	0.2 s	α					
^{182}Tl		181.9856	3. s	β^+, EC/10.9					0.351
									(0.26 - 0.41)
^{183m}Tl			0.06 s	α		9/2-			
^{183}Tl		182.9826	5. s	β^+, EC/7.7		1/2+			0.208
^{184}Tl		183.9818	11. s	β^+, EC/(98)/9.2					0.2868
				α/(2)/	6.16/				0.3399
									0.3667
^{185m}Tl			1.8 s	I.T./0.453		(9/2-)			0.1688
				α/5.97					0.2840
					6.01/				
^{185}Tl		184.9791	20. s	EC/β^+/6.6					
^{186m}Tl			4. s	I.T./0.374					0.3738
^{186}Tl		185.9776	28. s	β^+ ,EC/7.5					0.3567
									0.4026
									0.4053
^{187m}Tl			15.6 s	I.T./$\approx$ 0.33		(9/2+)	3.8	-2.4	0.2995
^{187}Tl		186.9762	50. s	β^+ ,EC/6.0		1/2+	1.6		
^{188m}Tl			1.18 m	β^+ ,EC/		(7+)			Hg k x-ray
									0.4129
									0.5043
									0.5921
^{188}Tl		187.9759	1.2 m	β^+ ,EC/7.8		(2-)	0.48	0.13	See Tl[188m]
									0.4129
^{189m}Tl			1.4 m	β^+ ,EC/		(9/2-)	+3.878	-2.29	0.2156
									0.2284
									0.3175
									0.4452
^{189}Tl		188.9743	2.3 m	β^+ ,EC/5.2		(1/2+)			0.3337
									0.4510
									0.5223
									0.9422
^{190m}Tl			3.7 m	β^+ ,EC/	4.2/	(7+)	+0.495	0.29	0.1968
									0.4164
									0.7311
^{190}Tl		189.9738	2.6 m	β^+ ,EC/7.0	5.7/	(2-)	+0.25	-0.33	0.4164
									0.6254
									0.6838
									1.0999
^{191m}Tl			5.2 m	β^+ ,EC/(98)/		(9/2+)	+3.903	-2.3	0.2157
									0.2647
									0.3256
									0.3359
^{191}Tl		190.9723				(1/2)	1.59		
^{192m}Tl			10.8 m	β^+ ,EC/		(7+)	+0.518	0.46	0.1740
									0.4228
									0.6348
									0.7863
									0.7455
^{192}Tl		191.972	9.6 m	β^+ ,EC/6.4		(2-)	+0.20	-0.33	0.3975
									0.4228
									0.6908
^{193m}Tl			2.1 m	I.T./(75)/		(9/2-)	+3.948	-2.2	0.3650
^{193}Tl		192.9706	22. m	β^+ ,EC/3.6		(1/2+)	+1.591		0.2077
									0.3244
									0.3440
									0.6761
									1.0447
									1.5793

Elem. or Isot.	Natural Abundance (%)	Atomic Mass or Weight	Half-Life	Decay Mode/Energy (/MeV)	Particle Energy /Intensity (MeV / %)	Spin (h/2π)	Nuclear Magnetic Mom.(nm)	Elect. Quadr. Mom. (b)	γ-ray Energy /Intensity (MeV / %)
^{194m}Tl			32.8 m	β^+ /(20)/≈ 0.30 EC/(80)/		(7+)	+0.540	0.62	ann.rad./ Hg k x-ray 0.4282 0.6363 0.7490
^{194}Tl		193.9711	34. m	β^+ ,EC/5.3		2-	0.14	-0.28	0.3955 0.4282 0.6363
^{195m}Tl			3.6 s	I.T./0.483		9/2-			Tl k x-ray 0.0990 0.3836
^{195}Tl		194.9697	1.16 h	EC/(97)/2.8 β^+ /(3)/		1/2+	+1.58		ann.rad./ Hg k x-ray 0.2422 0.5635 0.8845 1.3639 (0.13 - 2.5)
^{196m}Tl			1.41 h	β^+ ,EC/(95)/4.9		(7+)	0.55	+0.76	0.0840 0.4261 0.6353 0.6954 (0.08 - 1.0)
^{196}Tl		195.9705	1.84 h	β^+ /(15)/4.4 EC/(85)/		2-	0.07	-0.18	ann.rad./ Hg k x-ray 0.4257 0.6105 (0.03 - 2.4)
^{197m}Tl			0.54 s	I.T./(53)/0.608 β^+ ,EC/(47)/		9/2-			Tl k x-ray 0.2262 0.4118 0.5872 0.6367
^{197}Tl		196.96954	2.83 h	β^+ /(1)/2.18 EC/(99)/		1/2+	+1.58		Hg k x-ray 0.1522/8.2 0.4258
^{198m}Tl			1.87 h	β^+ ,EC/(53)/ I.T./(47)/0.5347		7+	+0.64		Hg k x-ray Tl k x-ray 0.4118 0.5872 0.6367
^{198}Tl		197.9405	5.3 h	EC,β^+ /(1)/3.5	1.4/ 2.1/ 2.4/	2-	0.00(1)		Hg k x-ray 0.4118 0.6367 0.6759 (0.23 - 2.8)
^{199}Tl		198.9698	7.4 h	EC/1.4		1/2-	+1.60		Hg k x-ray 0.2082 0.2473 0.4555
^{200}Tl		199.97095	1.087 d	EC/2.46	1.07/ 1.44/	2-	0.04		Hg k x-ray 0.36799 1.2057 (0.11 - 2.3)
^{201}Tl		200.97080	3.040 d	EC/0.48		1/2+	+1.605		Hg k x-ray 0.13528 0.16740/10.0
^{202}Tl		201.97209	12.23 d	EC/1.36		2-	0.06		Hg k x-ray 0.43957
^{203}Tl	29.524(14)	202.972329				1/2+	+1.622258		
^{204}Tl		203.973848	3.78 y	β^- /(97)/0.7637 EC/(3)/0.347	0.763/97	2-	0.09		Hg k x-ray
^{205}Tl	70.476(14)	204.974412				1/2+	+1.638215		
^{206m}Tl			3.76 m	I.T./2.644		12-			Tl k x-ray 0.2166 0.2661 0.4534 0.6866 1.0219
^{206}Tl		205.976095	4.20 m	β^- /1.533	1.53/99.9	0-			Pb k x-ray 0.80313
^{207m}Tl			1.3 s	I.T./1.350		11/2-			Tl k x-ray 0.3501 1.0000
^{207}Tl		206.97741	4.77 m	β^- /1.423	1.43/99.8	1/2+	+1.88		0.89723

Elem. or Isot.	Natural Abundance (%)	Atomic Mass or Weight	Half-Life	Decay Mode/Energy (/MeV)	Particle Energy /Intensity (MeV / %)	Spin (h/2π)	Nuclear Magnetic Mom.(nm)	Elect. Quadr. Mom. (b)	γ-ray Energy /Intensity (MeV / %)
^{208}Tl		207.982004	3.053 m	β^- /5.001	1.28/23	(5+)	+0.29		Pb k x-ray
					1.52/22				0.27728
					1.796/51				0.51061
									0.58302
									2.61448
^{209}Tl		208.98535	2.16 m	β^- /3.98	1.8 /100	(1/2+)			Pb k x-ray
									1.5670/100
									0.4651/95
									(0.12 - 1.33)
^{210}Tl		209.99006	1.30 m	β^- /5.48	1.3/25	(5+)			Pb k x-ray
					1.9/56				0.081
									0.2981
									0.79788
$_{82}$Pb		**207.2(1)**							
^{182}Pb		181.99268	0.06 s	α	6.92				
^{183}Pb		182.9919	0.3 s	α/		1/2+			
^{184}Pb		183.9882	0.6 s	α/	6.63/	0+			
^{185}Pb		184.9876	4.1 s	α/	6.34/				
					6.40/				
					6.48/				
^{186}Pb		185.9835	5. s	β^+ ,EC/(95)/5.5		0+			
				α/(5)/	6.32/				
					6.34/<100				
					6.01/<0.2				
^{187m}Pb			15.2 s	β^+ ,EC/	5.99/	(1/2-)			0.0674
					6.19/				0.2080
									0.2755
									0.2995
									0.4487
									0.7477
^{187}Pb		186.9839	18.3 s	EC/7.2		13/2+			0.1930
				α/	6.08/				0.3314
									0.3435
									0.3934
^{188}Pb		187.9811	23. s	EC/(78)/4.8		0+			0.1850
				α/(22)/	5.98/<10				0.7582
					5.61/<0.1				
^{189}Pb		188.9809	51. s	EC/6.1					
				α/	5.58/				
^{190}Pb		189.9782	1.2 m	β^+ (13)/4.1		0+			ann.rad./
				EC/(86)/					Tl k x-ray
				α/(0.9)/	5.58/				0.1415
									0.1512
									0.9422
^{191m}Pb			2.2 m	β^+ ,EC/		13/2+			ann.rad./
									0.3871
									0.6135
									0.7122
^{191}Pb		190.9782	1.3 m	β^+ ,EC/5.5					ann.rad./
									0.9368
^{192}Pb		191.9758	3.5 m	β^+ ,EC/$\approx$3.4		0+			ann.rad./
				α/.006/	5.11				0.1675
									0.6082
									1.1954
^{193m}Pb			5.8 m	β^+ ,EC/		13/2+	-1.16		ann.rad./
									0.3650
									0.3922
^{193}Pb		192.9761	$\approx$ 2. m	EC/5.2		3/2			
^{194}Pb		193.9740	11. m	β^+ ,EC/2.7		0+			ann.rad./
				α	4.64				0.2036
^{195m}Pb			15. m	β^+ /(8)/		13/2+	-1.132	+0.29	ann.rad./
				EC/(92)/					Tl k x-ray
									0.3836
									0.3942
									0.8784
^{195}Pb		194.976	$\approx$ 15. m	β^+,EC/5.8					ann.rad./
									0.3836
									0.3937
									0.7776
^{196}Pb		195.9727	37. m	β^+ ,EC/2.1		0+			Tl k x-ray
									0.2531
									0.5021

Elem. or Isot.	Natural Abundance (%)	Atomic Mass or Weight	Half-Life	Decay Mode/Energy (/MeV)	Particle Energy /Intensity (MeV / %)	Spin (h/2π)	Nuclear Magnetic Mom.(nm)	Elect. Quadr. Mom. (b)	γ-ray Energy /Intensity (MeV / %)
^{197m}Pb			43. m	EC/(79)/ β⁺ /(2)/ I.T./(19)/0.3193		13/2+	-1.105	+0.5	Tl k x-ray 0.3079 0.3877 0.7743 (0.2 - 2.2)
^{197}Pb		196.9734	≈ 8. m	EC/(97)/3.6 β⁺ /(3)/		(3/2-)	-1.075	-0.08	Tl k x-ray 0.3755 0.3858 0.7611
^{198}Pb		197.9720	2.4 h	EC/1.4		0+			Tl k x-ray 0.1734 0.2903 0.3654
^{199m}Pb			12.2 m	I.T./(93)/0.4248 β⁺ ,EC/(7)/		13/2+	-1.074	+0.08	Pb k x-ray 0.4255
^{199}Pb		198.9729	1.5 h	EC/(99)/2.9 β⁺ /(1)/		5/2-			Tl k x-ray 0.3534 0.7202 1.1350 (0.22 - 2.4)
^{200}Pb		199.97182	21.5 h	EC/0.81		0+			Tl k x-ray 0.14763
^{201m}Pb			1.02 m	I.T./0.6291		13/2+			Pb k x-ray 0.6288
^{201}Pb		200.97285	9.33 h	EC/1.90		5/2-	+0.675	-0.009	Tl k x-ray 0.33120 0.36131 (0.11 - 1.8)
^{202m}Pb			3.53 h	I.T./(90)/2.170 β⁺ /(10)/		9-	-0.228	+0.58	Pb k x-ray Tl k x-ray 0.42219 0.78700 0.96271
^{202}Pb		201.97214	5.3x10⁴ y	EC/0.05		0+			Tl L x-ray
^{203m}Pb			6.2 s	I.T./0.8252		13/2+			Pb k x-ray 0.8203 0.8252
^{203}Pb		202.97338	2.1615 d	EC/0.98		5/2-	+0.686	+0.10	Tl k x-ray 0.279188
^{204m}Pb			1.12 h	I.T./2.185		9-			Pb k x-ray 0.37481 0.89922 0.91175
^{204}Pb	1.4(1)	203.973028				0+			
^{205}Pb		204.974467	1.51x10⁷ y	EC/0.0512		5/2-	+0.712	+0.23	Tl L x-ray
^{206}Pb	24.1(1)	205.974449				0+			
^{207m}Pb			0.80 s	I.T./1.632		13/2+			Pb k x-ray 0.56915 1.06310
^{207}Pb	22.1(1)	206.975880				1/2-	+0.59258		
^{208}Pb	52.4(1)	207.976636				0+			
^{209}Pb		208.981075	3.25 h	β⁻ /0.644	0.645/100	9/2+	-1.474	-0.3	
^{210}Pb		209.984174	22.6 y	β⁻ /0.0635 α	0.017/81 0.061/19 3.72	0+			
^{211}Pb		210.988732	36.1 m	β⁻ /1.37	0.57/5 1.36/92	(9/2+)	-1.404	+0.09	0.40486 0.42700 0.83186 (0.09 - 1.27)
^{212}Pb		211.991887	10.64 h	β⁻ /0.574	0.28/83 0.57/12	0+			Bi k x-ray [illegible]
^{213}Pb		212.9966	10.2 m	β⁻ /2.1					
^{214}Pb		[illegible]	[illegible]	β⁻ /1.0	0.67/48 0.73/42	0+			Bi k x-ray 0.24192 0.29509 0.35187
$_{83}$Bi		**208.98038(2)**							
^{187m}Bi			≈ 8. ms	α					
^{187}Bi		186.9935	35. ms	α					
^{188}Bi		187.9922		α					
^{189m}Bi			≈ 5. ms	α					
^{189}Bi		188.9895	0.68 s	α					
^{190}Bi		189.9875	5. s	β⁺ ,EC/(10)/8.7 α/(90)/	 6.45/				
^{191}Bi		190.9861	12. s	β⁺ ,EC/(60)/7.3 α/(40)/	 6.32/				

Elem. or Isot.	Natural Abundance (%)	Atomic Mass or Weight	Half-Life	Decay Mode/Energy (/MeV)	Particle Energy /Intensity (MeV / %)	Spin (h/2π)	Nuclear Magnetic Mom.(nm)	Elect. Quadr. Mom. (b)	γ-ray Energy /Intensity (MeV / %)
^{192}Bi		191.9854	40. s	β^+ ,EC/(80)/9.0					
				α/(20)/	6.06/				
^{193m}Bi			3.2 s	β^+ ,EC/		1/2+			
				α/	6.48/				
^{193}Bi		192.9837	1.11 m	β^+ ,EC/(40)/7.1		9/2+			
				α/(60)/	5.91/				
^{194}Bi		193.9828	1.8 m	β^+ ,EC/(99.9)/8.2		(10-)			0.1661
				α/0.1/					0.1740
									0.2802
									0.421
									0.5754
									0.9650
^{195m}Bi			1.45 m	β^+ ,EC/(94)/					
				α/(6)/	6.11/				
^{195}Bi		194.9811	2.9 m	β^+ ,EC/(99.8)/5.8		3/2-			
				α/(0.2)	5.45/				
^{196}Bi		195.9806	5. m	EC/$\approx$7.4					0.1376
									0.3720
									0.6880
									1.0486
^{197}Bi		196.9789	5. m	β^+ ,EC/5.2		1/2+			
^{198m}Bi			7.7 s	I.T./0.2485		(10-)			0.2485
^{198}Bi		197.9790	11.8 m	β^+ ,EC/6.6		(7+)			0.0900
									0.1976
									0.5624
									1.0635
^{199m}Bi			24.7 m	β^+ ,EC/					ann.rad./
^{199}Bi		198.9776	27. m	β^+ ,EC/4.3		9/2-			0.7203
									0.8374
									0.8417
									0.9460
									1.0528
									1.3056
									(0.12 - 3.2)
^{200m}Bi			31. m	β^+ ,EC/		(2+)			0.2453
									0.4198
									0.4624
									1.0265
^{200}Bi		199.9781	36. m	EC/(90)/5.9		7+			ann.rad./
				β^+ /(10)/					Pb k x-ray
									0.4198
									0.4623
									1.0265
^{201m}Bi			59.1 m	I.T./0.846		(1/2+)			Bi k x-ray
				β^+ ,EC/					0.8464
^{201}Bi		200.97697	1.8 h	EC/3.84		9/2-			Pb k x-ray
									0.6288
									0.9357
									1.0138
									(0.13 - 2.4)
^{202}Bi		201.97768	1.72 h	β^+ /(3)/5.16		5+			ann.rad./
				EC/(97)/					Pb k x-ray
									0.57860
									0.92734
									(0.08 - 3.5)
^{203}Bi		202.97687	11.8 h	EC/(99.8)/3.25		9/2-	+4.62	-0.7	Pb k x-ray
				β^+ /(0.2)/	1.35/				0.1865
									0.8203
									0.8969
									1.8475
									(0.1 - 2.9)
^{204}Bi		203.97779	11.2 h	EC/4.44		6+	+4.28	-0.43	Pb k x-ray
									0.37481
									0.89922
									0.98409
^{205}Bi		204.97737	15.31 d	EC/2.71		9/2-	+4.16		Pb k x-ray
									0.70347
									1.76435
^{206}Bi		205.97848	6.243 d	EC/3.76		6+	+4.60	-0.20	Pb k x-ray
									0.51619
									0.80313
									0.88100
^{207}Bi		206.978456	35. y	EC/2.399		9/2-	4.08	-0.6	Pb k x-ray
									0.56915
									1.06310
^{208}Bi		207.979727	3.68x10^5 y	EC/2.880		5+			Pb k x-ray
									2.61435
^{209}Bi	100.	208.980384				9/2-	+4.111	-0.37	

TABLE OF THE ISOTOPES

Elem. or Isot.	Natural Abundance (%)	Atomic Mass or Weight	Half-Life	Decay Mode/Energy (/MeV)	Particle Energy /Intensity (MeV / %)	Spin (h/2π)	Nuclear Magnetic Mom.(nm)	Elect. Quadr. Mom. (b)	γ-ray Energy /Intensity (MeV / %)
^{210m}Bi			3.0x10^{6} y	α/	4.420(3)/0.29 4.569(3)/3.9 4.584(3)/1.4 4.908(4)/39 4.946(3)/55	9-			Tl k x-ray 0.2661 0.3052 0.6502
^{210}Bi		209.984105	5.01 d	β^- /1.163	1.16/99	1-	-0.0445	+0.136	0.2661 0.3.52
^{211}Bi		210.98726	2.14 m	α/(99.7)/ β^- /(0.3)/0.58	6.279/16 6.623/84	9/2-			Tl k x-ray 0.3501
^{212m2}Bi			7. m	β^- /		(15-)			
^{212m1}Bi			25.0 m	α/(93)/ β^- /(7)/	6.300/40 6.340/53	(9-)			0.120 0.233 0.275 0.404 0.727
^{212}Bi		211.991271	1.009 h	β^- /(64)/2.254 α/(36)/	6.051/25 6.090/9.6	(1-)	0.41		Tl k x-ray Po k x-ray 0.2881 0.72725 0.78551 1.62066
^{213}Bi		212.99437	45.6 m	β^- /(98)/1.43 α/(2)/	1.02/31 1.42/66 5.549/0.16 5.869/2.0	9/2-	+3.9		Po k x-ray 0.4404 (0.15 - 1.328)
^{214}Bi		213.99870	19.7 m	β^- /3.27					1.10006 0.60931 1.12027 1.76449 (0.19 - 3.2)
^{215}Bi		215.0018	7.7 m	β^- /2.3					0.2937 (0.27 - 0.835)
^{216}Bi		216.0062	3.6 m	β^-/4.0					0.5498 0.4192
$_{84}$Po									
^{192}Po		191.9915	34. ms	α/	7.17				
^{193m}Po			0.24 s	α/	7.00				
^{193}Po		192.9911	0.45 s	α/	6.95				
^{194}Po		193.9883	0.39 s	α/	6.84/93 6.19/0.22	0+			
^{195m}Po			1.9 s	α/	6.70/				
^{195}Po		194.9881	4.6 s	α/	6.61/				
^{196}Po		195.9855	5.8 s	α/(95)/ β^+ ,EC/(5)/≈4.6	6.52/94 5.77/0.02	0+			
^{197m}Po			25.8 s	α/(84)/ β^+ ,EC/(16)/	6.385(3)/55	13/2+			
^{197}Po		196.9856	53. s	α/(44)/ β^+ ,EC/(56)/6.2	6.282(4)/76	(3/2-)			
^{198}Po		197.9834	1.76 m	α/(70)/ β^+ ,EC/(30)/4.0	6.18/57 5.27/7.6 x 10^{-4}	0+			
^{199m}Po			4.2 m	β^+ ,EC/(51)/ α/(39)/	6.059/24	13/2+	0.99		ann.rad./ 0.2745 0.4998 1.0020
^{199}Po		198.985	5.2 m	β^+ ,EC/(88)/7. α/(12)/	5.952/7.5	(3/2-)			Bi k x-ray 0.1877 0.3616 1.0214 1.0344
^{200}Po		199.9817	11.5 m	β^+ ,EC/(85)/3.4 α/(15)/	5.863/11.1	0+			0.14748 0.32792 0.6176 0.6709
^{201m}Po			8.9 m	β^+ ,EC/(57)/ I.T./(40)/0.418 α/(3)/	5.786/≈3.	13/2+	1.0		Bi k x-ray Po k x-ray 0.2726 0.4123 0.4179 0.9670
^{201}Po		200.9822	15.3 m	β^+ ,EC/(98)/4.9 α/(2)/	5.683(3)/1.1	3/2-	0.94		Bi k x-ray 0.2056 0.2250 0.8483 0.9048

Elem. or Isot.	Natural Abundance (%)	Atomic Mass or Weight	Half-Life	Decay Mode/Energy (/MeV)	Particle Energy /Intensity (MeV / %)	Spin ($h/2\pi$)	Nuclear Magnetic Mom.(nm)	Elect. Quadr. Mom. (b)	γ-ray Energy /Intensity (MeV / %)
^{202}Po		201.9807	45. m	β^+ ,EC/(98)/2.8		0+			0.0410
				α/(2)/	5.588/1.9				0.1656
									0.3158
									0.6884
^{203m}Po			1.2 m	I.T./(96)/0.6414		13/2+			Bi k x-ray
				β^- EC/(4)/					Po k x-ray
									0.6414
^{203}Po		202.9814	35. m	β^+ ,EC/4.2		5/2-	+0.74		0.17516
									0.21477
									0.89350
									0.90863
									1.09095
^{204}Po		203.98031	3.53 h	EC/2.34		0+			Bi k x-ray
				α	5.377/0.66				0.2702
									0.8844
									1.0162
									(0.11 - 1.9)
^{205}Po		204.98117	1.7 h	β^+ ,EC/3.53		5/2-	+0.76	+0.17	Bi k x-ray
									0.83681
									0.84983
									0.87241
									1.00124
									(0.12 - 2.7)
^{206}Po		205.98047	8.8 d	EC/(95)/1.85		0+			Bi k x-ray
				α/(5)/	5.223/5.5				0.28644
									0.31156
									0.51134
									0.80737
									1.03228
									(0.11 - 1.5)
^{207m}Po			2.8 s	I.T./1.383		19/2-			Po k x-ray
									0.2682
									0.30074
									0.81448
^{207}Po		206.98158	5.80 h	EC,β^+ /2.91		5/2-	+0.79	+0.28	Bi k x-ray
									0.74263
									0.91176
									0.99225
^{208}Po		207.981231	2.898 y	α/5.213	4.233/0.0002	0+			
					5.1158/100				
^{209}Po		208.982415	102. y	α/4.976	4.624/0.56	1/2-	+0.77		0.26049
					4.879/99.2				0.8964
^{210}Po		209.982857	138.38 d	α/5.407	4.516/0.001	0+			0.80313
					5.304/100				
^{211m}Po			25.2 s	α/	7.273/91	25/2+			Pb k x-ray
					7.994/1.7				0.32808
					8.316/0.25				0.56915
					8.875/7.0				0.89723
									1.06310
^{211}Po		210.986637	0.516 s	α/7.594	6.570/0.54	9/2+			0.56915
					6.892/0.55				0.89723
					7.450/98.9				
^{212m}Po			45. s	α/	8.514/2.0	16+			
					9.086/1.0				
					11.650/97				
^{212}Po		211.988852	0.298 μs	α/8.953	8.784/100	0+			
^{213}Po		212.992843	4. μs	α/8.537	7.614/0.003	9/2+			
					8.375/100				
^{214}Po		213.995186	163.7 μs	α/7.833	6.904/0.01	0+			
					7.686/99.99				
^{215}Po		214.999415	1.780 ms	α/7.526	6.950/0.02	(9/2+)			
					6.957/0.03				
					7.386/100				
^{216}Po		216.001905	0.145 s	α/6.906	5.895/0.002	0+			
					6.778/99.99				
^{217}Po		217.0064	< 10. s	α/6.662	6.539/				
^{218}Po		218.008965	3.04 m	α/6.114	5.181/1.00	0+			
$_{85}$At									
^{196}At		195.9957	0.3 s	α/	7.06/				
^{197m}At			4. s	α		(1/2+)			
^{197}At		196.9939	0.35 s	β^+ ,EC/7.8		(9/2-)			
				α/	6.96/				
^{198m}At			1.5 s	β^+ ,EC/(75)/					
				α/(25)/	6.85/				
^{198}At		197.9928	5. s	α/	6.75/				
^{199}At		198.9910	7.1 s	β^+ ,EC/(8)/5.6		9/2-			
				α/(92)/	6.64/				

Elem. or Isot.	Natural Abundance (%)	Atomic Mass or Weight	Half-Life	Decay Mode/Energy (/MeV)	Particle Energy /Intensity (MeV / %)	Spin (h/2π)	Nuclear Magnetic Mom.(nm)	Elect. Quadr. Mom. (b)	γ-ray Energy /Intensity (MeV / %)
^{200m}At			4.3 s	β$^+$,EC/(80)		10-			
				α/(20)/	6.536/				
^{200}At		199.990	43. s	β$^+$,EC/(65)/≈8.0		5+			
				α/(35)/	6.412/21				
					6.465/14				
^{201}At		200.9885	1.48 s	β$^+$,EC/(29)/5.9		9/2-			
				α/(71)/6.474	6.344/				
^{202m}At			≤ 1.5 s	I.T./0.391					
^{202}At		201.9885	3.02 m	β$^+$,EC/(88)/7.2		5+			ann.rad./
				α/(12)/	6.135/7.7				0.4413
					6.225/4.3				0.5697
									0.6753
^{203}At		202.9868	7.4 m	β$^+$,EC/(69)/5.1		9/2-			0.1458
				α/(31)/6.210	6.088/				0.2459
									0.6414
									1.0020
									1.0340
^{204}At		203.9873	9.1 m	β$^+$,EC/(95)/6.5		(5+)			Po k x-ray
				α/(5)/	5.951/				0.3271
									0.4254
									0.5156
									0.6837
^{205}At		204.98604	26. m	β$^+$,EC/(90)/4.54		(9/2-)			Po k x-ray
				α/(10)/6.020	5.902/				0.1543
									0.6696
									0.7194
^{206}At		205.98660	29.4 m	β$^+$,EC/(99)/5.72		5+			Po k x-ray
				α/(1)/5.881	5.703/				0.20186
									0.39561
									0.47716
									0.70071
^{207}At		206.98578	1.81 h	β$^+$,EC/(90)/3.91		9/2-			Po k x-ray
				α/(10)/5.873	5.758/				0.16801
									0.58842
									0.81448
^{208}At		207.98657	1.63 h	β$^+$,EC/(99)/4.97		(6+)			Po k x-ray
				α/(1)/5.752	5.626/0.01				0.1770
					5.641/0.53				0.2060
									0.6601
									0.6852
									0.8450
									1.0281
^{209}At		208.98616	5.4 h	β$^+$,EC/(96)/3.49		(6+)			Po k x-ray
				α/(4)/5.757	5.647/4.1				0.10422
									0.54503
									0.78189
									0.79020
									(0.1 - 2.6)
^{210}At		209.98713	8.1 h	EC/99.8/3.98		5+			Po k x-ray
				α/(0.2)/5.632	5.361/0.05				0.24535
					5.442/0.05				0.52758
									1.18143
									1.43678
									1.48335
									(0.04 - 2.4)
^{211}At		210.987481	7.21 h	EC/(58)/0.787		9/2-			Po k x-ray
				α/(42)/5.980	5.211/0.004				0.66956
					5.868/42				0.6870
									0.74263
^{212m}At			0.119 s	α/	7.837/65	(9-)			
					7.897/33				
^{212}At		211.990735	0.314 s	α/7.828	7.058/0.1	(1-)			
					7.088/0.6				
					7.618/15				
					7.681/84				
^{213}At		212.992922	0.11 μs	α/9.254	9.080/	9/2-			
^{214m}At			0.76 μs	α/8.762		(9-)			
^{214}At		213.996357	0.56 μs	α/8.987	8.819/100	(1-)			
^{215}At		214.99864	0.10 ms	α/8.178	7.626/0.045	(9/2-)			0.40486
					8.023/99.9				
^{216}At		216.002408	0.30 ms	α/7.947	7.595/0.2	(1-)			
					7.697/2.1				
					7.800/97				
^{217}At		217.00471	32. ms	α/7.202	6.812/0.06	(9/2-)			0.2595
					7.067/99.9				0.3345
									0.5940
^{218}At		218.00868	1.6 s	α/6.883	6.654/6				
					6.695/90				
					6.748/4				

Elem. or Isot.	Natural Abundance (%)	Atomic Mass or Weight	Half-Life	Decay Mode/Energy (/MeV)	Particle Energy /Intensity (MeV / %)	Spin (h/2π)	Nuclear Magnetic Mom.(nm)	Elect. Quadr. Mom. (b)	γ-ray Energy /Intensity (MeV / %)
^{219}At		219.0113	50. s	α/6.390	6.275/				
^{220}At		220.0153	3.71 m	β^- /3.7					(0.24-0.70)
^{221}At			2.3 m						
^{222}At			5.4 s						
^{223}At			50. s						
$_{86}$Rn									
^{198}Rn		197.9988	0.05 s	α					
^{199m}Rn			0.3 s	α		(13/2+)			
^{199}Rn		198.9983	0.62 s	α/		3/2-			
^{200}Rn		199.9957	1.06 s	α/(98)/	6.909/	0+			
				EC/(2)/5.					
^{201m}Rn			3.8 s	EC/(10)/		13/2+			
				α/(90)/	6.770/				
^{201}Rn		200.9955	7.0 s	α/(80)/6.636	6.721/	(3/2-)			
				EC/(20)/6.6					
^{202}Rn		201.9932	9.9 s	α/(12)/6.771	6.636(3)/	0+			
				EC/(88)/4.5					
^{203m}Rn			28. s	α/ 6.548(3)/		13/2+	-0.96	+1.3	
^{203}Rn		202.9948	45. s	α/(66)/6.629	6.498/	0			
				EC/(34)/≈7.4					
^{204}Rn		203.9914	1.24 m	α/(68)/6.546	6.417(3)/	0+			
				EC/(32)/3.8					
^{205}Rn		204.9917	2.8 m	α/(23)/6.390	6.123(3)/0.02	(5/2-)	+0.80	+0.06	0.2652
				EC/(77)/5.2	6.262(3)/23				0.3553
									0.4648
									0.6205
									0.6753
									0.7300
^{206}Rn		205.9902	5.7 m	α/(68)/6.384	6.258(3)/	0+			0.06170
				EC/(32)/3.3					0.0968
									0.3245
									0.3862
									0.4822
									0.4973
									0.7728
^{207}Rn		206.9907	9.3 m	β^+ ,EC/(77)/4.6		5/2-	+0.82	+0.22	At k x-ray
				α/(23)/6.252	5.995(4)/0.02				0.32947
					6.068(3)/0.15				0.34455
					6.126(3)/22.8				0.36767
									0.40267
									0.74723
									(0.18 - 1.4)
^{208}Rn		207.98963	24.3 m	α/(60)/6.260	5.469(2)/0.003	0+			
				EC/(40)/2.85	6.140(2)/60				
^{209}Rn		208.99038	29. m	β^+ /(83)/3.93	2.16/2.3	5/2-	+0.8388	+0.31	At k x-ray
				α/(17)/	5.887(3)/0.04				0.27933
					5.898(3)/0.02				0.33753
					6.039(2)/16.9				0.40841
									0.68942
									0.74594
									(0.18 - 3.2)
^{210}Rn		209.98968	2.4 h	α/(96)/6.157	5.351(2)/0.005	0+			At k x-ray
				EC/(4)/2.37	6.039(2)/96				0.19625
									0.45824
									0.57104
									0.64868
									(0.14 - 1.7)
^{211}Rn		210.99059	14.6 h	β^+ ,EC/(74)/2.89		1/2-	+0.60		At k x-ray
				α/(26)/5.964	5.619(1)/0.7				0.16877
					5.784(1)/16.4				0.25022
					5.851(1)/8.8				0.37049
									0.67412
									0.67839
									1.36298
									(0.11 - 2.7)
^{212}Rn		211.990689	24. m	α/6.385	5.587(4)/0.05	0+			
					6.260(4)/99.95				
^{213}Rn		212.99387	25.0 ms	α/8.243	7.552(8)/1.0	9/2+			
					8.087(8)/99				
^{214}Rn		213.99535	0.27 μs	α/9.209	9.037(9)/	0+			
^{215}Rn		214.99873	2.3 μs	α/8.840	8.674(8)/	(9/2+)			
^{216}Rn		216.00026	45. μs	α					
^{217}Rn		217.003915	0.6 ms	α/7.885	7.500/0.1	9/2+			
					7.742(4)/100				
^{218}Rn		218.005586	35. ms	α/7.267	6.534(1)/0.16	0+			
					7.133(1)/99.8				

Elem. or Isot.	Natural Abundance (%)	Atomic Mass or Weight	Half-Life	Decay Mode/Energy (/MeV)	Particle Energy /Intensity (MeV / %)	Spin (h/2π)	Nuclear Magnetic Mom.(nm)	Elect. Quadr. Mom. (b)	γ-ray Energy /Intensity (MeV / %)
^{219}Rn		219.009475	3.96 s	α/6.946(1)	6.3130(5)/0.05 6.425(3)/7.5 6.5309(4)/0.12 6.5531(3)/12.2 6.8193(3)/81	(5/2+)	-0.44	+0.93	Po k x-ray 0.13057 0.27113 0.40170 (0.1 - 1.05)
^{220}Rn		220.011384	55.6 s	α/6.404	5.7486(5)/0.07 6.2883(1)/99.9	0+			
^{221}Rn		221.0156	25. m	α/(22)/6.148 β^- /(78)/1.2	5.778(3)/1.8 5.788(3)/2.2 6.037(3)/18	7/2+	-0.020	-0.38	Fr L x-ray 0.07384 0.08323 0.0610 0.18639
^{222}Rn		222.017570	3.8235 d	α/5.590	4.987(1)/0.08 5.4897(3)/99.9	0+			0.510
^{223}Rn			23. m	β^- /					
^{224}Rn			1.8 h	β^- /		0+			0.1085 0.2601 0.2655
^{225}Rn			4.5 m	β^- /		7/2	-0.70	+0.84	
^{226}Rn			7.4 m	β^- /					
^{227}Rn			2. s	β^- /					
^{228}Rn			65. s	β^- /					
$_{87}$Fr									
^{201}Fr		201.0046	0.05 s	α/7.54	7.388(15)/	(9/2-)			
^{202}Fr		202.0033	0.34 s	α/7.590	7.237(8)/100				
^{203}Fr		203.0014	0.55 s	α/7.280	7.132(5)/	(9/2-)			
^{204}Fr		204.001	2.1 s	α/7.170	6.967(5)/30 7.027(5)/70				
^{205}Fr		204.9987	3.9 s	α/7.050	6.914(5)/	(9/2-)			
^{206m}Fr			0.7 s	α/	6.93				0.531(IT)
^{206}Fr		205.9985	16.0 s	α/7.416	6.792(5)/84				
^{207}Fr		206.9969	14.8 s	α/6.900	6.766(5)/	9/2-	+3.9	-0.16	
^{208}Fr		207.99713	59.1 s	α/(77)/6.770 EC/(23)/6.99	6.636(5)/	7+	-4.8	+0.004	
^{209}Fr		208.99592	50.0 s	α/(89)/5.1 EC/(11)/5.16	6.646(3)/	9/2-	+3.9	-0.24	
^{210}Fr		209.99640	3.2 m	α/6.670 EC/6.26	6.543(5)/	6+	+4.4	+0.19	0.2030 0.6438 0.8175 0.9008
^{211}Fr		210.99553	3.10 m	α/6.660 EC/4.61	6.534(5)/	9/2-	+4.0	-0.19	0.220 0.2799 0.5389 0.9169
^{212}Fr		211.99618	20. m	EC/(57)/5.12 α/(43)/6.529	6.261(1)/16 6.335(1)/4 6.335(1)/4 6.343(1)/1.3 6.383(1)/10 6.406(1)/9.5 6.08-6.18	(5+)	+4.6	-0.10	Rn x-ray 0.08107 0.08378 0.2277 1.1856 1.2748 0.014-1.178
^{213}Fr		212.99617	34.6 s	α/6.905	8.476(4)/51	9/2-	+4.0	-0.14	
^{214m}Fr			3.4 ms	α/	8.547(4)/46 6.775-8.046	9-			
^{214}Fr		213.99895	5.1 ms	α/8.587	7.409(3)/0.3 7.605(8)/1.0 7.940(3)/1.0 8.355(3)/4.7 8.427(3)/93	(1-)			
^{215}Fr		215.00033	0.12 μs	α/9.537	9.360(8)/	(9/2-)			
^{216}Fr		216.00319	0.70 μs	α/9.175	9.005(10)/				
^{217}Fr		217.00462	0.016 ms	α/8.471	8.315(8)/	(9/2-)			
^{218m}Fr			22. ms	α					
^{218}Fr		218.00756	1. ms	α/8.014	7.384(10)/0.5 7.542(15)/1.0 7.572(10)/5 7.732(10)/0.5 7.867(2)/93	(1-)			
^{219}Fr		219.00924	21. ms	α/8.132	6.802(2)/0.25 6.967(2)/0.6 7.146(2)/0.25 7.313(2)/99	(9/2-)			
^{220}Fr		220.012313	27.4 s	α/6.800	6.582(1)/10 6.630(2)/6 6.641(1)/12 6.686(1)/61 6.39-6.58	1+	-0.67	+0.47	0.0450 0.061 0.1060 0.1539 0.1617

TABLE OF THE ISOTOPES

Elem. or Isot.	Natural Abundance (%)	Atomic Mass or Weight	Half-Life	Decay Mode/Energy (/MeV)	Particle Energy /Intensity (MeV / %)	Spin ($h/2\pi$)	Nuclear Magnetic Mom.(nm)	Elect. Quadr. Mom. (b)	γ-ray Energy /Intensity (MeV / %)
^{221}Fr		221.01425	4.8 m	α/6.457	5.9393(7)/0.17	(5/2)	+1.58	-1.0	At k x-ray
					5.9797(7)/0.49				0.0995
					6.0751(7)/0.15				0.21798
					6.1270(7)/				0.4091
					6.2433(3)/1.3				
					6.3410(7)/83.4				
^{222}Fr		222.01754	14.3 m	β^- /2.03	1.78/	2-	+0.63	+0.51	
				α/5.850					
^{223}Fr		223.019731	22.0 m	β^- /1.149	1.17/65	(3/2+)	+1.17	+1.17	0.05014
									0.07972
									(0.13 - 0.9)
^{224}Fr		224.02323	3.0 m	β^- /2.82		1-	+0.40	+0.517	0.13150
									0.21575
									0.8367
									(0.1 - 2.21)
^{225}Fr		225.02561	3.9 m	β^- /1.87		3/2	+1.07	+1.3	
^{226}Fr		226.0293	49. s	β^- /3.6		1	+0.071	-1.35	0.18606
									0.25373
^{227}Fr		227.0318	2.48 m	β^- /2.5		1/2	+1.50		
^{228}Fr		228.035	39. s	β^- /$\approx$3.5		2-	-0.76	+2.4	
^{229}Fr			50. s	β^- /					
^{230}Fr			19. s	β^- /		(3)			
^{231}Fr			17. s	β^- /					
^{232}Fr			5. s	β^- /					
$_{88}$Ra									
^{206}Ra		206.0038	0.4 s	α/7.416	7.272(5)/	0+			
^{207}Ra		207.005	1.3 s	α/7.270	7.133(5)/				
^{208}Ra		208.0018	1.4 s	α/7.273	7.133(5)/	0+			
^{209}Ra		209.0019	4.6 s	α/7.150	7.008(5)/	5/2	+0.87	+0.38	
^{210}Ra		210.0005	3.7 s	α/7.610	7.020(5)/	0+			
^{211}Ra		211.0009	13. s	α/7.046	6.912(5)/	(5/2-)	+0.878	+0.46	
				EC/5.0					
^{212}Ra		211.99978	13.0 s	α/7.033	6.901(2)/	0+			
^{213m}Ra			2.1 ms	IT					
^{213}Ra		213.00034	2.7 m	EC/(20)/3.88		(1/2-)	+0.613		0.1024
				α/(80)/6.860	6.521(3)/4.8				0.11010
					6.622(3)/39				0.2125
					6.730(3)/36				
^{214}Ra		214.00009	2.46 s	α/7.272	7.136(4)/	0+			
^{215}Ra		215.00270	1.6 ms	α/8.864	7.883(6)/2.8	(9/2+)			
					8.171(3)/1.4				
					8.700(3)/95.9				
^{216}Ra		216.00352	0.18 μs	α/9.526	9.349(8)/	0+			
^{217}Ra		217.00631	1.6 μs	α/9.161	8.992(8)/	9/2-			
^{218}Ra		218.00712	26. μs	α/8.547	8.390(8)/	0+			
^{219}Ra		219.01006	0.010 s	α/8.132	7.680(10)/65				
					7.982(9)/35				
^{220}Ra		220.01101	25. ms	α/7.593	6.998(7)/1.0	0+			0.465
					7.455(7)/99				
^{221}Ra		221.01391	29. s	α/6.879	6.254(10)/0.7	5/2	-0.180	+1.9	
					6.578(5)/3				
					6.585(3)/8				
					6.608(3)/35				
					6.669(3)/21				
					6.758(3)/31				
^{222}Ra		222.015361	38. s	α/5.590	6.237(2)/3.0	0+			
					6.556(2)/97				
^{223}Ra		223.018497	11.435 d	α/5.979	5.287(1)/0.15	(3/2+)	+0.271	+1.2	Rn k x-ray
					5.338(1)/0.13				0.12231
					5.365(1)/0.13				0.14418
					5.433(5)/2.3				0.15418
					5.502(1)/1.0				0.15859
					5.540(1)/9.2				0.26939
					5.607(3)/24				0.32388
					5.716(3)/52				0.33328
					5.747(1)/9				0.44494
					5.857(1)/0.32				(0.10 - 0.7)
					5.872(1)/0.85				
^{224}Ra		224.020202	3.66 d	α/5.789	5.034(10)/0.003	0+			Rn k x-ray
					5.047(1)/0.007				0.2407
					5.164(5)/0.007				0.4093
					5.449(2)/4.9				0.6501
					5.685(2)/95				
^{225}Ra		225.023603	14.9 d	β^- /0.36	0.32/100	(3/2+)	-0.734		Ac k x-ray
									0.0434

Elem. or Isot.	Natural Abundance (%)	Atomic Mass or Weight	Half-Life	Decay Mode/Energy (/MeV)	Particle Energy /Intensity (MeV / %)	Spin (h/2π)	Nuclear Magnetic Mom.(nm)	Elect. Quadr. Mom. (b)	γ-ray Energy /Intensity (MeV / %)
^{226}Ra		226.025402	1599. y	α/4.870	4.194(1)/0.001	0+			Rn k x-ray
					4.343(1)/0.006				0.1861
					4.601(1)/5.5				0.2624
					4.784(1)/94				
^{227}Ra		227.029170	42. m	β⁻ /1.325	1.03/	(3/2+)	-0.404	+1.5	Ac L x-ray
					1.30/				Ac k x-ray
									0.02739
^{228}Ra		228.031063	5.76 y	β⁻ /0.046		0+			0.0135
									0.016
^{229}Ra		229.0348	4.0 m	β⁻ /1.76	1.76/	(3/2+)	+0.503	+3.0	
^{230}Ra		230.03708	1.5 h	β⁻ /1.0	0.7/	0+			0.0631
									0.0720
									0.2028
									0.4698
									0.4787
^{231}Ra			1.7 m	β⁻					
^{232}Ra			4. m	β⁻					
^{233}Ra			30. s	β⁻					
^{234}Ra			≈ 30. s	β⁻/					
$_{89}$Ac									
^{207}Ac			≈0.02 s	α/7.71					
^{208m}Ac			≈25. ms	α/7.76					
^{208}Ac			≈0.1 s	α/7.57					
^{209}Ac		209.0096	≈0.09 s	α/7.58					
^{210}Ac		210.0093	0.4 s	α/7.610	7.462(8)/				
^{211}Ac		211.0076	0.3 s	α/7.620	7.480(8)/				
^{212}Ac		212.0078	0.9 s	α/7.520	7.379(8)/				
^{213}Ac		213.0066	0.8 s	α/7.500	7.364(8)/	(9/2-)			
^{214}Ac		214.0069	8.2 s	α/(86)/7.350	7.007(8)/3	(5+)			
				EC/(14)/6.34	7.082(5)/38				
					7.214(5)/45				
^{215}Ac		215.0065	0.17 s	α/7.750	7.604(5)/	(9/2-)			
^{216m}Ac			0.33 ms	α/	8.198(8)/1.7	(9-)			
					8.283(8)/2.5				
					9.028(5)/49				
					9.106(5)/46				
^{216}Ac		216.00871	≈ 0.3 ms	α/9.241	8.990(2)/10	(1)			
					9.070(8)/90				
^{217m}Ac			0.7 μs	α/	10.540/100				
^{217}Ac		217.00933	0.07 μs	α/9.832	9.650(10)/100	9/2-			
^{218}Ac		218.01162	1.1 μs	α/9.380	9.205(15)/				
^{219}Ac		219.01241	0.012 ms	α/8.830	8.664(10)/	(9/2-)			
^{220}Ac		220.0148	26. ms	α/8.350	7.610(20)/23				
					4.680(20)/21				
					7.790(10)/13				
					7.850(10)/24				
					7.985(10)/4				
					8.005(10)/5				
					8.060(10)/6				
					8.195(10)/3				
^{221}Ac		221.01558	52. ms	α/7.790	7.170(10)/2				
					7.375(10)/10				
					7.440(15)/20				
					7.645(10)/70				
^{222m}Ac			63. s	α/(>89)/	6.710(20)/7				
				EC/(1)/	6.750(20)/13				
				I.T./(<10)/	6.810(20)/24				
					6.840(20)/9				
					6.890(20)/13				
					6.970(20)/7				
					7.000(20)/13				
^{222}Ac		222.01782	5. s	α/7.141	6.967(10)/6	1-			
					7.013(2)/94				
^{223}Ac		223.01913	2.1 m	α/(99)/6.783	6.131(2)/0.12	(5/2-)			0.0725
				EC/(1)/0.59	6.177(2)/0.94				0.0839
					6.293(1)/0.47				0.0927
					6.326(1)/0.3				0.0990
					6.332(2)/0.14				0.1917
					6.360(1)/0.22				0.2158
					6.397(1)/0.13				0.3588
					6.448(1)/0.2				0.4768
					6.473(1)/3.1				
					6.523(2)/0.6				
					6.528(1)/3.1				
					6.563(1)/13.6				
					6.582(3)/0.3				
					6.646(1)/44				
					6.661(1)/31				

Elem. or Isot.	Natural Abundance (%)	Atomic Mass or Weight	Half-Life	Decay Mode/Energy (/MeV)	Particle Energy /Intensity (MeV / %)	Spin ($h/2\pi$)	Nuclear Magnetic Mom.(nm)	Elect. Quadr. Mom. (b)	γ-ray Energy /Intensity (MeV / %)
^{224}Ac		224.021708	2.7 h	EC/(90)/1.403	5.841(1)/0.5	0-			Ra L kx-ray
				α/(10)/6.323	5.860(1)/0.75				Ra k x-ray
					5.875(1)/1.7				0.08426
					5.941(1)/4.4				0.13150
					6.000(1)/6.7				0.1571
					6.013(1)/1.4				0.21575
					6.056(1)/22				0.2619
					6.138(1)/26				(0.03 - 0.3)
					6.154(1)/1.0				
					6.204(1)/12				
					6.210(1)/20				
^{225}Ac		225.02322	10.0 d	α/5.935	5.286(1)/0.2	3/2			Fr k x-ray
					5.444(3)/0.1				0.9958
					5.554(1)/0.1				0.9982
					5.608(1)/1.1				0.1084
					5.636(1)/4.5				0.1116
					5.681(1)/1.4				0.1451
					5.722(1)/2.9				0.1539
					5.731(1)/10				0.15724
					5.791(1)/9				0.18799
					5.793(1)/18				0.19575
									0.2162
									0.21686
									(0.025 - 0.52)
^{226}Ac		226.026089	1.224 d	EC/(17)/0.640		(1-)			Ra k x-ray
				β^- /(83)/1.116					Th k x-ray
				α/(0.006)/5.510	5.399(5)/0.006				0.07218
									0.15816
									0.23034
^{227}Ac		227.027747	21.77 y	β^- /(98.6)/0.045	0.045/54	(3/2-)	+1.1	+1.7	0.0838/23.
				α/(1.4)/5.043	4.869(1)/0.09				0.0811/14.
					4.938(1)/0.52				0.2696/13.
					4.951(1)/0.65				(0.044 - 1.27)
^{228}Ac		228.031014	6.15 h	β^- /2.127	1.11/32	(3+)			Th L x-ray
					1.85/12				Th k x-ray
					2.18/11				0.12903
									0.33842
									0.91116
									0.96897
									(0.2 - 1.96)
^{229}Ac		229.03293	1.04 h	β^- /1.10	1.1/	(3/2+)			0.07450
									0.16451
									0.26188
									0.5085
									0.56916
^{230}Ac		230.0360	2.03 m	β^- /2.7	1.4/	1+			Th k x-ray
									0.45497
									0.50820
									(0.12 - 2.5)
^{231}Ac		231.0386	7.5 m	β^- /2.1	2.1/100	(1/2+)			0.14379
									0.18574
									0.22140
									0.28250
									0.3070
^{232}Ac		232.0420	2.0 m	β^- /3.7		(2-)			
^{233}Ac			2.4 m	β^- /		(1/2+)			
^{234}Ac			40. s	β^- /		(1+)			
$_{90}$Th		**232.03805(2)**							
^{212}Th		212.0129	≈ 30. ms	α/	7.80/	0+			
^{213}Th		213.0130	0.14 s	α/7.840	7.692(10)/				
^{214}Th		214.0115	0.09 s	α/7.825	7.677(10)/	0+			
^{215}Th		215.0117	1.2 s	α/7.660	7.33(10)/8	(1/2-)			
					7.395(8)/52				7.524(8)/40
^{216m}Th			0.18 ms	α					
^{216}Th		216.01105	28. ms	α/8.071	7.921(8)/	0+			
^{217}Th		217.01306	0.25 ms	α/9.424	9.250(10)/				
^{218}Th		218.01327	0.11 μs	α/9.847	9.665(10)/	0+			
^{219}Th		219.01552	1.05 μs	α/9.510	9.340(20)/				
^{220}Th		220.01573	10. μs	α/8.953	8.790(20)/	0+			
^{221}Th		221.01817	1.7 ms	α/8.628	7.743(8)/6				
					8.146(5)/56				
					8.4272(5)/39				
^{222}Th		222.01845	2.8 ms	α/8.129	7.982(8)/9.7	0+			
					7.600(15)/3				

Elem. or Isot.	Natural Abundance (%)	Atomic Mass or Weight	Half-Life	Decay Mode/Energy (/MeV)	Particle Energy /Intensity (MeV / %)	Spin (h/2π)	Nuclear Magnetic Mom.(nm)	Elect. Quadr. Mom. (b)	γ-ray Energy /Intensity (MeV / %)
^{223}Th		223.02079	0.65 s	α/7.454	7.29(1)/41(5) 7.32(1)/29(5) 7.350(15)/20(5) 7.390(15)/10(4)				
^{224}Th		224.02146	1.05 s	α/7.305	6.768(5)/1.2 6.997(5)/19 7.170(5)/79				
^{225}Th		225.02394	8.72 m	EC/(10)/0.68 α/(90)/6.920	6.441(2)/15 6.479(2)/43 6.501(3)/14 6.627(3)/3 6.650(5)/3 6.700(5)/2 6.743(3)/7 6.796(2)/9	(3/2+)			
^{226}Th		226.024891	30.6 m	α/6.454	6.026(1)/0.2 6.041(1)/0.19 6.098(1)/1.3 6.2283(4)/23 6.3375(4)/75	0+			Ra k x-ray 0.11110 0.13100 0.19028 0.20621 0.24210 (0.1 - 0.8)
^{227}Th		227.027699	18.72 d	α/6.146		(3/2+)			Ra L x-ray Ra k x-ray 0.05014 0.23597 0.25624 (0.02 - 1.0)
^{228}Th		228.028731	1.913 y	α/5.520	5.1770(2)/0.18 5.2114(1)/0.4 5.3405(1)/26.7 5.4233(1)/73	0+			
^{229}Th		229.031754	7.9x10^{3} y	α/5.168	4.814/9.3 4.845(5)/56 4.9008(5)/10.2 4.689-5.077	5/2+	+0.46	+4.	
^{230}Th		230.033126	7.54x10^{4} y	α/4.771	4.4383(6)/0.03 4.4798(6)/0.12 4.6211(6)/23.4 4.6876(6)/76.3	0+			0.0677/0.46 0.1439/0.078
^{231}Th		231.036296	1.063 d	β^- /0.390	0.138/22 0.218/20 0.305/52	5/2+			Pa L x-ray Pa k x-ray 0.02564 0.084203/ (0.02 - 0.3)
^{232}Th	100.	232.038050	1.4x10^{10} y	α/4.081	3.830(10)/0.2 3.952(5)/23 4.010(5)/77	0+			0.0590 0.124
^{233}Th		233.041576	22.3 m	β^- /1.245	1.245/	1/2+			Pa L x-ray Pa k x-ray 0.02938 0.08653 0.45930 (0.02 - 1.2)
^{234}Th		234.043596	24.10 d	β^- /0.273	0.102/20 0.198/72	0+			Pa L x-ray 0.06329/4.1 0.09235/2.4 0.09278/2.4
^{235}Th		235.04751	7.2 m	β^- /1.9					0.4162 0.6591 0.7272 0.747 0.9318
^{236}Th			37.5 m	β^- /≈ 1.0					Pa k x-ray 0.1107
^{237}Th			5.0 m	β^-					
$_{91}$Pa									
^{215}Pa		215.0190	≈ 14. ms	α					
^{216}Pa		216.0190	0.19 s	α/8.010	7.720/ 7.820/ 7.920/				
^{217m}Pa			2. ms	α/	10.16/				
^{217}Pa		217.0183	5. ms	α/8.490	8.340(10)/				
^{218}Pa		218.0200	0.12 ms	α/	9.54/ 9.61/				
^{221}Pa		221.0219	6. μs	α	9.08(3)				

T A B L E O F T H E I S O T O P E S

Elem. or Isot.	Natural Abundance (%)	Atomic Mass or Weight	Half-Life	Decay Mode/Energy (/MeV)	Particle Energy /Intensity (MeV / %)	Spin (h/2π)	Nuclear Magnetic Mom.(nm)	Elect. Quadr. Mom. (b)	γ-ray Energy /Intensity (MeV / %)
^{222}Pa		222.0237	≈ 4.3 ms	α/8.700	8.180/50 8.330/20 8.540/30				
^{223}Pa		223.0240	≈ 6.5 ms	α/8.340	8.006(10)/55 8.196(10)/45				
^{224}Pa		224.0256	≈ 0.95 s	α/7.630	7.555(10)/75(3) 7.46(1)/25(3)				
^{225}Pa		225.0261	1.8 s	α/7.380	7.195(10)/30 7.245(10)/70				
^{226}Pa		226.02792	1.8 m	α/(74)/6.987 EC/(26)/2.83	6.728(10)/0.7 6.823(10)/35 6.863(10)/39				
^{227}Pa		227.02879	38.3 m	α/(85)/6.582 EC/(15)/1.02	6.357(4)/7 6.376(10)/2.2 6.401(4)/8 6.416(4)/13 6.423(10)/10 6.465(4)/43	(5/2-)			0.0649 0.0669 0.1100
^{228}Pa		228.03100	22. h	EC/(98)/2.111 α/(2)	5.779/0.23 5.805/0.15 6.078/0.4 6.105/0.25 6.118/0.22	(3+)	+3.5		Th k x-ray 0.20939 0.27026 0.28202 0.32767 0.32807 0.46310 0.91116 0.96464 0.96897 (0.1 - 1.96)
^{229}Pa		229.03209	1.5 d	EC/(99.8)/0.32 α/(0.2)/5.836	5.536(2)/0.02 5.579(2)/0.09 5.668(2)/0.05	(5/2)			0.04244 (0.024 - 0.18)
^{230}Pa		230.034532	17.4 d	EC/(90)/1.310 β^- /(10)/0.563	0.51/	(2-)	2.0		Th L x-ray Th k x-ray 0.4437 0.45477 0.89876 0.91856 0.95199 (0.053-1.07)
^{231}Pa		231.035878	3.25x10^4 y	α/5.148	4.6781(5)/1.5 4.7102(5)/1.0 4.7343(5)/8.4 4.8513(5)/1.4 4.9339(5)/3 4.9505(5)/22.8 4.9858(5)/1.4 5.0131(5)/25.4 5.0292(5)/20 5.0318(5)/2.5 5.0587(5)/11	3/2-	2.01	-1.	Ac L x-ray Ac k x-ray 0.01899 0.027396 0.03823 0.04639 0.25586 0.26029 0.28367 0.30007 0.30264 0.33007 (0.02 - 0.61)
^{232}Pa		232.03858	1.31 d	β^- /1.34		(2-)			U k x-ray 0.10900 0.15009 0.89439 0.96934 (0.10 - 1.17)
^{233}Pa		233.040239	27.0 d	β^- /0.571	0.15/40 0.256/60	3/2-	+4.0	-3.0	U L x-ray U k x-ray 0.30017 0.31201 0.34059
^{234m}Pa			1.17 m	β^- /(99.9)/2.29 I.T./(0.13)/		(0-)			U k x-ray 0.25818/0.07 0.76641/0.32 1.0009/0.85 (0.06 - 1.96)
^{234}Pa		234.043303	6.69 h	β^- /2.197	0.51/	(4+)			U L x-ray U k x-ray 0.1312/0.03 0.5695/0.02 0.9256/0.02 (0.02 - 1.99)
^{235}Pa		235.04544	24.4 m	β^- /1.41	1.4/97	(3/2-)			0.0308-0.65893

Elem. or Isot.	Natural Abundance (%)	Atomic Mass or Weight	Half-Life	Decay Mode/Energy (/MeV)	Particle Energy /Intensity (MeV / %)	Spin ($h/2\pi$)	Nuclear Magnetic Mom.(nm)	Elect. Quadr. Mom. (b)	γ-ray Energy /Intensity (MeV / %)
^{236}Pa		236.0487	9.1 m	β^- /2.9	1.1/40	(1-)			U k x-ray
					2.0/50				0.64235
					3.1/10				0.68759
									1.7630
									(0.04 - 2.18)
^{237}Pa		237.0511	8.7 m	β^- /2.3	1.1/60	(1/2+)			0.4986
					1.6/30				0.5293
					2.3/10				0.5407
									0.8536
									0.8650
									(0.04 - 1.4)
^{238}Pa		238.0545	2.3 m	β^- /3.5	1.2/	(3-)			0.10350
					1.7/				0.1785
									0.4484
									0.6350
									0.6800
									1.01446
									(0.04 - 2.5)
$_{92}$U		**238.0289(1)**							
^{218}U			≈0.002 s	α	8.63(3)/				
^{219}U			0.04 ms	α	9.68(4)/				
^{222}U		222.0261	≈ 1.μs	α					
^{223}U		223.0277	0.02 s	α/	8.78(4)/				
^{224}U		224.02759	≈ 1. ms	α/	8.46/100				
^{225}U		225.02938	0.09 s	α/	7.88/85				
					7.82/15				
^{226}U		226.02933	0.5 s	α/7.560	7.430/	0+			
^{227}U		227.03113	1.1 m	α/7.200	6.870/				
^{228}U		228.03137	9.1 m	α/6.803	6.404(6)/0.6	0+			0.095
					6.440(5)/0.7				0.152
					6.589(5)/29				0.187
					6.681(6)/70				0.246
^{229}U		229.03350	58. m	EC/(80)/1.31	6.223/3	(3/2+)			
				α/(20)/6.473	6.297(3)/11				
					6.332(3)/20				
					6.360(3)/64				
^{230}U		230.033927	20.8 d	α/5.992	5.5866(3)/0.01	0+			Th L x-ray
					5.6624(3)/0.26				0.07218
					5.6663(3)/0.38				0.15421
					5.8178(3)/32				0.23034
					5.8887(3)/67				
^{231}U		231.03626	4.2 d	EC/0.36		(5/2-)			Pa L x-ray
				α/(10^{-3})	5.46/1.6 x 10^{-3}				Pa k x-ray
					5.47/1.4 x 10^{-3}				0.02564
					5.40/1. x 10^{-3}				0.08420
^{232}U		232.037146	68.9 y	α/5.414	4.9979(1)/0.003	0+			
					5.1367(1)/0.3				
					5.2635(1)/31				
					5.3203(1)/69				
^{233}U		233.039627	1.59x10^5 y	α/4.909	4.7830(8)/13.2	5/2+	+0.59	3.66	Th L x-ray
					4.8247(8)/84.4				0.04244
					4.510-4.804				0.09714
									(0.0252-1.119)
^{234}U	0.0055(5)	234.040945	2.45x10^5 y	α/4.856	4.604(1)/0.24	0+			0.05323/0.156
					4.7231(1)/27.5				0.12091
					4.776(1)/72.5				
^{235m}U			26. m	I.T./0.0007		1/2+			
^{235}U	0.720(1)	235.043922	7.04x10^8 y	α/4.6793	4.1525(9)/0.9	7/2-	-0.38	4.6	Th L x-ray
					4.2157(9)/5.7				Th k x-ray
					4.3237(9)/4.6				0.10917
					4.3641(9)/11				0.14378
					4.370(4)/6				0.16338
					4.3952(9)/55				0.18574
					4.4144(9)/2.1				0.20213
					4.5025(9)/1.7				0.20533
					4.5558(9)/4.2				0.22140
					4.5970(9)/5.0				(0.03 - 0.79)
^{236}U		236.045561	2.34x10^7 y	α/4.569	4.332(8)/0.26	0+			Th L x-ray
					4.445(5)/26				0.04937
					4.494(3)/74				0.11275
^{237}U		237.048723	6.75 d	β^- /0.519	0.24/	1/2+			Np L x-ray
					0.25/				Np k x-ray
									0.05953
									0.20801

Elem. or Isot.	Natural Abundance (%)	Atomic Mass or Weight	Half-Life	Decay Mode/Energy (/MeV)	Particle Energy /Intensity (MeV / %)	Spin ($h/2\pi$)	Nuclear Magnetic Mom.(nm)	Elect. Quadr. Mom. (b)	γ-ray Energy /Intensity (MeV / %)
^{238}U	99.2745(15)	238.050784	4.46x10^9 y	α	4.039(5)/0.23 4.147(5)/23 4.196(5)/77	0+			Th L x-ray 0.04955/.06 0.1135/.01
^{239}U		239.054289	23.5 m	β^- /1.265	1.2/ 1.3/	5/2+			0.04354 0.07467
^{240}U		240.056585	14.1 h	β^- /0.39	0.36/	0+			Np L x-ray 0.04410 0.05558 0.06760
^{242}U			16.8 m	β^- /$\approx$ 1.2					
$_{93}$Np									
^{226}Np		226.0351	0.03 s	α/	8.04(2)/				
^{227}Np		227.0350	0.51 s	α/	7.65(2)/ 7.68(1)/				
^{228}Np		228.0362	61. s	EC/60(7)/ α/40(7)/,SF					
^{229}Np		229.0363	4.0 m	α/7.010	6.890(20)				
^{230}Np		230.0378	4.6 m	EC/97 /3.6 α/3	6.660(20)				
^{231}Np		231.03823	48.8 m	EC/98 /1.8 α/2 /6.368	6.280/2	5/2			0.2629 0.3475 0.3703
^{232}Np		232.0400	14.7 m	EC/99 /2.7		(4-)			U L x-ray U k x-ray 0.3268 0.81925 0.86683
^{233}Np		233.0410	36.2 m	EC/1.2		(5/2+)			U L x-ray U k x-ray 0.29887 0.31201
^{234}Np		234.04289	4.4 d	β^+ ,EC/1.81	0.79/	(0+)			U L x-ray U k x-ray 1.5272 1.5587 1.6022
^{235}Np		235.044055	1.085 y	EC/99.9 /0.124 α/0.001/5.191		5/2+			U k x-ray
^{236m}Np			22.5 h	EC/52 / β^- /48 /		(1-)			U L x-ray Pu L x-ray U k x-ray 0.64235 0.68759
^{236}Np		236.04657	1.55x10^5 y	EC/91 /0.94 β^- /9 /0.49		(6-)			U L x-ray U k x-ray 0.10423 0.16031
^{237}Np		237.048166	2.14x10^6 y	α/4.957	4.6395(5)/6.5 4.766(5)/9.7 4.7715(5)/22.7 4.7884(5)/47.8 4.558-4.873	5/2+	+3.14	+3.89	Pa L x-ray Pa k x-ray 0.029378/15 0.08653/12 (0.03-0.28)
^{238}Np		238.050940	2.117 d	β^- /1.292	1.2/	2+			Pu L x-ray Pu k x-ray 0.98447/25.2 1.02855/18.3 (.044-1.026)
^{239}Np		239.052931	2.355 d	β^- /0.722	0.341/30 0.438/48	5/2+			Pu L x-ray Pu k x-ray 0.10613 0.228186/11 0.27760/15 (0.04-0.50)
^{240m}Np			7.22 m	β^- /99.9 / I.T./0.1 /	2.18/	(1+)			0.25143 0.26333 0.55454 0.59735
^{240}Np		240.05617	1.032 h	β^- /2.20	0.89/	5+			0.1471/ 0.5664 0.6008
^{241}Np		241.0583	13.9 m	β^- /1.3	1.3/	5/2+			0.1330/ 0.1740 0.280
^{242m}Np			2.2 m	β^- /		(1+)			0.15910 0.2651/ 0.78570 0.9448/

Elem. or Isot.	Natural Abundance (%)	Atomic Mass or Weight	Half-Life	Decay Mode/Energy (/MeV)	Particle Energy /Intensity (MeV / %)	Spin (h/2π)	Nuclear Magnetic Mom.(nm)	Elect. Quadr. Mom. (b)	γ-ray Energy /Intensity (MeV / %)
^{242}Np		242.0616	5.5 m	β^- /2.7	2.7/	6+			0.6209 0.73620 0.78074 1.47340 (0.04-2.37)
$_{94}$Pu									
^{228}Pu				α/	7.81(2)/				
^{229}Pu				α/	7.46(3)/				
^{230}Pu		230.03964		α/	7.05/				
^{232}Pu		232.04118	34. m	EC/ > 80/1.1 α/ < 20/6.716	 6.542(10)/38 6.600(10)/62	0+			
^{233}Pu		233.04299	20.9 m	EC(99.9)/1.9 α/0.1 /6.416	 6.300(20)/0.1				0.1503 0.1804 0.2353 0.5002 0.5346/ 1.0352/
^{234}Pu		234.04331	8.8 h	EC/94 /0.39 α/6 /6.310	 6.035(3)/0.024 6.149(3)/1.9 6.200(3)/4.0	0+			
^{235}Pu		235.0453	25.3 m	EC/99+ /1.2 α/0.003/5.957	 5.850(20)/0.003	(5/2+)			
^{236}Pu		236.046048	2.87 y	α/5.867	5.611/0.21 5.7210/30.5 5.7677(1)/69.3	0+			0.0476/0.07 0.109/0.02 (0.17 - 0.97)
^{237}Pu		237.048403	45.7 d	EC/99.9 /0.220 α/0.003 /5.747	 5.334(4)/0.0015 5.356(4)/0.0006 5.650(4)/0.0007	7/2-			Np L x-ray Np k x-ray 0.026344 0.03319 0.05954 (0.03-0.5)
^{238}Pu		238.049553	87.74 y	α/5.593	5.3583(1)/0.10 5.465(1)/28.3 5.4992(1)/71.6	0+			U k x-ray 0.04347 (0.04-1.1)
^{239}Pu		239.052156	2.411 x 10^{4} y	α/5.244	5.055/0.047 5.076/0.078 5.106/11.9 5.144/17.1 5.157/70.8 (4.74 -5.03)	1/2+	+0.203		U k x-ray 0.05162 0.05682 0.12928 0.37502 0.41369
^{240}Pu		240.053807	6537. y	α/5.255	5.0212(1)/0.07 5.1237(1)/26.4 5.1681(1)/73.5	0+			U L x-ray 0.04524 0.10423 (0.04-0.97)
^{241}Pu		241.056844	14.4 y	β^- /99+ /0.0208 α/0.002 /5.139	4.853(7)/3x10^{-4} 4.8966(7)/0.002	5/2+	-0.683	+6.	0.14854
^{242}Pu		242.058736	3.76 x 10^{5} y	α/4.983	4.7546(7)/0.098 4.8564(7)/22.4 4.9006(7)/78	0+			U L x-ray 0.04491 0.10350
^{243}Pu		243.061996	4.956 h	β^- /0.582	0.49/21 0.58/60	7/2+			Am L x-ray 0.0417 0.0839
^{244}Pu		244.064197	8.2x10^{7} y	α/99.9/4.665 S.F./0.1 /	4.546(1)/19.4 4.589(1)/80.5	0+		U L x-ray	0.0439
^{245}Pu		245.06774	10.5 h	β^- /1.21	0.93/57 1.21/11	(9/2-)			Am L x-ray Am k x-ray [illegible] 0.30832 0.32752 0.56014 (0.03-1.2)
^{246}Pu		246.07020	10.85 d	β^- /0.40	0.150/85 0.35/10	0+			Am L x-ray Am k x-ray 0.04379 0.22371
$_{95}$Am									
^{232}Am			0.9 m	EC/ ≈ 5.0					
^{234}Am		234.0478	2.3 m	EC/4.2					
^{237}Am		237.0503	1.22 h	EC/99.98 /1.7 α/0.02 /6.20	 6.042(5)/0.02	(5/2-)			Pu k x-ray 0.14559 0.28026 0.43845

Elem. or Isot.	Natural Abundance (%)	Atomic Mass or Weight	Half-Life	Decay Mode/Energy (/MeV)	Particle Energy /Intensity (MeV / %)	Spin (h/2π)	Nuclear Magnetic Mom.(nm)	Elect. Quadr. Mom. (b)	γ-ray Energy /Intensity (MeV / %)
^{238}Am		238.05198	1.63 h	EC/2.26 α/0.0001 /6.04	 5.940/0.0001	1+			Pu L x-ray Pu k x-ray 0.91870 0.96278
^{239}Am		239.053018	11.9 h	EC/99.99 /0.803 a/0.01 /5.924	 5.734(2)/0.001 5.776(2)/0.008	5/2-			Pu L x-ray Pu k x-ray 0.18172 0.22818 0.27760
^{240}Am		240.05529	2.12 d	EC/1.38 α/5.592	 5.378(1)/16x10^{-4}	(3-)			Pu L x-ray Pu k x-ray 0.88878 0.98764 (0.1-1.3)
^{241}Am		241.056822	432.2 y	α/5.637	5.2443(1)/0.002 5.3221(1)/0.015 5.3884(1)/1.4 5.4431(1)/12.8 5.4857(1)/85.2 5.5116(1)/0.20 5.5442(1)/0.34	5/2-	+1.61	+4.	Np L x-ray 0.02634 0.033192 0.059536 (0.03-1.128)
^{242m}Am			141. y	I.T./99.5 /0.048 α/0.5 /5.62	 5.141(4)/0.026 5.2070(2)/0.4	5-	+1.0	+6.5	Am L x-ray 0.04863 0.08648 0.10944 0.16304
^{242}Am		242.059542	16.02 h	β^- /83 /0.665 EC/17 /0.750	0.63/46 0.67/37	1-	+0.388	-2.4	Pu L x-ray Cm L x-ray Pu k x-ray 0.0422 0.04453
^{243}Am		243.061372	7.37x10^3 y	α/5.438	5.1798(5)/1.1 5.2343(5)/11 5.2766(5)/88 5.394(5)/0.12 5.3500(5)/0.16	5/2-	+1.61	+4.	0.04354 0.07467 0.08657 0.11770 0.14197
^{244m}Am			≈ 26. m	β^- /1.498		(1-)			0.0429
^{244}Am		244.064279	10.1 h	β^- /1.428					Am L x-ray Cm k x-ray 0.7460 0.9000
^{245}Am		245.066444	2.05 h	β^- /0.894	0.65/19 0.90/77	(5/2+)			Cm L x-ray Cm k x-ray 0.25299
^{246m}Am			25.0 m	β^- /	1.3/79. 1.60/14 2.1/7	2-			Cm L x-ray Cm k x-ray 0.27002 0.79881 1.06201 1.07885 (0.04-2.29)
^{246}Am		246.06977	39. m	β^- /2.38	1.2/	(7-)			Cm L x-ray Cm k x-ray 0.1529 0.2046 0.6786
^{247}Am		247.0722	22. m	β^- /1.7					Cm L x-ray Cm k x-ray 0.2267 / 0.2853 /
$_{96}$Cm									
^{236}Cm		236.0514		EC/1.7					
^{237}Cm		237.0529		EC/2.5					
^{238}Cm		238.05302	2.4 h	EC/>90 /0.97 α/<10 /6.632	 6.520(50)/<10	0+			
^{239}Cm		239.0548	≈ 3. h	EC/1.7					0.0407 0.1466 0.1874
^{240}Cm		240.055519	27. d	α/6.397	5.989/0.014 6.147/0.05 6.2478(6) /28.8 6.2906(6) /70.6	0+			

Elem. or Isot.	Natural Abundance (%)	Atomic Mass or Weight	Half-Life	Decay Mode/Energy (/MeV)	Particle Energy /Intensity (MeV / %)	Spin (h/2π)	Nuclear Magnetic Mom.(nm)	Elect. Quadr. Mom. (b)	γ-ray Energy /Intensity (MeV / %)
^{241}Cm		241.057646	32.8 d	EC/99 /0.768 α/1 /6.184	 5.8842(4)/0.12 5.9291(4)/0.18 5.9389(4)/0.69	1/2+			Am k x-ray 0.13241 0.16505 0.18028 0.43063 0.47181
^{242}Cm		242.058828	162.8 d	α/6.216	5.9694(1)/0.035 6.069(1)/25 6.1129(1)/74	0+			Pu L x-ray 0.04408 0.10189 (0.04-1.2)
^{243}Cm		243.061381	28.5 y	α/6.167	5.6815(5) /0.2 5.6856(5)/1.6 5.7420(5)/10.6 5.7859(5)/73.3 5.9922(5)/6.5 6.0103(5)/1.0 6.0589(5)/5 6.0666(5)/1.5	5/2+	0.41		Pu L x-ray Pu k x-ray 0.10612 0.20975 0.22819 0.27760 0.28546 0.33431 (0.04-0.7)
^{244}Cm		244.062745	18.11 y	α/5.902	5.6656(1)/0.02 5.7528(1)/24 5.8050(1)/76	0+			Pu L x-ray 0.04282 0.09885 0.15262
^{245}Cm		245.065485	8.5x10^{3} y	α /5.623	5.235(10)/0.3 5.3038(10)/5.0 5.3620(7)/93 5.4927(11)/0.8 5.5331(11)/0.6	7/2+	0.5		Pu L x-ray Pu k x-ray 0.04195 0.13299 0.13606 0.17494
^{246}Cm		246.067217	4.78x10^{3} y	α/5.476	5.343(3)/21 5.386(3)/79	0+			Pu L x-ray 0.04453
^{247}Cm		247.070346	1.56x10^{7} y	α/5.352	4.818(4)/4.7 4.8690(20)/71 4.941(4)/1.6 4.9820(20)/2.0 5.1436(20)/1.2 5.2104(20)/5.7 5.2659(20)/13.8	9/2- 9/2-	0.37		Pu k x-ray 0.2792 0.2886 0.3471 0.4035
^{248}Cm		248.072341	3.4x10^{5} y	α/92 /5.162 S.F./8/	4.931(5)/0.07 5.0349(2)/16.5 5.0784(2)/(75)/1	0+			
^{249}Cm		249.075946	64.15 m	β^- /0.900	0.9/	1/2+			Bk k x-ray 0.56039 0.63431
^{250}Cm		250.07835	≈ 9.7x10^{3} y	S.F./ α/5.27		0+			
^{251}Cm		251.08228	16.8 m	β^- /1.42	0.90/16	(1/2+)			0.3896 / 0.5299 0.5425
$_{97}$Bk									
^{238}Bk		238.0583	2.4 m	EC/5.0					
^{242}Bk		242.0621	7.0 m	EC/3.0					
^{243}Bk		243.063001	4.5 h	EC/99.8 /1.508 α/0.15 /6.871	6.542(4)/0.03 6.5738(2)/0.04 6.7180(22)/0.02 6.7581(20)/0.02	(3/2-)			0.1466 0.1874 0.755 0.840 0.946
^{244}Bk		244.0652	4.4 h	EC/99.99 /2.26 α/0.01 /6.778	 6.625(4)/0.003 6.667(4)/0.003	(4-)			0.1445 0.1876 0.2176 0.9815 0.9215/
^{245}Bk		245.066355	4.94 d	EC/99.9 /0.810 α/0.1 /6.453	 5.8851(5)/0.03 6.1176(9)/0.01 6.1467(5)/0.02 6.3087(5)/0.014 6.3492(5)/0.018	3/2-			Cm L x-ray Cm k x-ray 0.25299 0.3809 0.3851
^{246}Bk		246.0687	1.80 d	EC/1.35		(2-)			Cm L x-ray Cm k x-ray 0.79881 1.08142

Elem. or Isot.	Natural Abundance (%)	Atomic Mass or Weight	Half-Life	Decay Mode/Energy (/MeV)	Particle Energy /Intensity (MeV / %)	Spin ($h/2\pi$)	Nuclear Magnetic Mom.(nm)	Elect. Quadr. Mom. (b)	γ-ray Energy /Intensity (MeV / %)
^{247}Bk		247.07030	1.4x10^{3} y	α/5.889	5.465(5)/1.5 5.501(5)/7 5.532(5)/45 5.6535(20)/5.5 5.678(2)/13 5.712(2)/17 5.753(2)/4.3 5.794(2)/5.5	(3/2-)			0.04175 0.0839 0.268
^{248}Bk		248.07311	23.7 h	β^- /70 /0.87 EC/30 /0.72	0.86/	(1-)			Cm L x-ray Cf L x-ray Cm k x-ray Cf k x-ray 0.5507
^{249}Bk		249.074980	320. d	β^- /0.125 α/0.001 /5.525	0.125/100 5.390(1)/0.0002 5.4174(6)/0.001	7/2+	2.0		0.327/10^{-5} 0.308/10^{-6}
^{250}Bk		250.078309	3.217 h	β^- /1.780	0.74/	2-			Cf L x-ray Cf k x-ray 0.98912 1.03184 (0.04-1.6)
^{251}Bk		251.08075	56. m	β^- /1.09		(3/2-)			0.02481 0.1528 0.1776
$_{98}$Cf									
^{239}Cf		239.0626	≈ 0.7 m	α					
^{240}Cf		240.0623	1.1 m	α/7.719	7.590(10)/	0+			
^{241}Cf		241.0637	4. m	EC/3.3 α/7.60	 7.335(5)/				
^{242}Cf		242.06369	3.5 m	α/7.509	7.351(6)/20 7.385(4)/80	0+			
^{243}Cf		243.0654	11. m	EC/86 /2.2 α/14 /7.40	7.060(6)/20 7.170/4	(1/2+)			
^{244}Cf		244.065990	20. m	α/7.328	7.168(5)/25 7.210(5)/75	0+			
^{245}Cf		245.068038	44. m	α/30 /7.255 EC/70 /1.569	6.886/ 6.983/ 7.036/ 7.084/ 7.137(2)/				
^{246}Cf		246.068798	1.49 h	α/6.869	6.6156(10)/0.18 6.7086(7)/21.8 6.7501(7)/78.0	0+			Cm L x-ray 0.04221 0.0945 0.147
^{247}Cf		247.07099	3.11 h	EC/99.96 /0.65 α/0.04 /6.55	 6.301(5)/	7/2+			Bk k x-ray 0.2941 0.4778
^{248}Cf		248.07218	334. d	α/6.369	6.220(5)/17 6.262(5)/83	0+			
^{249}Cf		249.074846	351. y	α/6.295	5.7582(2)/3.7 5.8119(2)/84 5.8488(2)/1.0 5.9029(2)/2.8 5.9451(2)/4.0 6.1401(2)/1.1 6.1940(2)/2.2	9/2-			Cm L x-ray Cm k x-ray 0.25299 0.33351 0.38832
^{250}Cf		250.076399	13.1 y	α/6.129	5.8913(4)/0.3 5.9889(4)/15 6.0310(4)/84.5	0+			Cm L x-ray 0.04285
^{251}Cf		251.079579	9.0x10^{2} y	α/6.172	5.56448(7)/1.5 5.632(1)/4.5 5.648(1)/3.5 5.6773(6)/35 5.762(3)/3.8 5.7937(7)/2.0 5.8124(8)/4.2 5.8514(6)/27 6.0140(7)/11.6 6.0744(7)/2.7	1/2+			
^{252}Cf		252.081619	2.65 y	α/96.9 /6.217 S.F./3.1/	5.7977(1)/0.23 6.0756(4)/15.2 6.1184(4)/81.6	0+			Cm L x-ray 0.04339 0.1002
^{253}Cf		253.08512	17.8 d	β^- /99.7 /0.29 α/0.3 /6.126	0.27/100 5.921(5)/0.02	(7/2+)			
^{254}Cf		254.08732	60.5 d	S.F./99.7/ α/0.3/5.930	 5.792(5)/0.05 5.834(5)/0.26	0+			

T A B L E   O F   T H E   I S O T O P E S

Elem. or Isot.	Natural Abundance (%)	Atomic Mass or Weight	Half-Life	Decay Mode/Energy (/MeV)	Particle Energy /Intensity (MeV / %)	Spin (h/2π)	Nuclear Magnetic Mom.(nm)	Elect. Quadr. Mom. (b)	γ-ray Energy /Intensity (MeV / %)
^{255}Cf		255.0910	1.4 h	β^- /0.7					
^{255}Cf			12. m	SF					
$_{99}$**Es**									
^{243}Es		243.0696	21. s	α/ > 30 /	7.89/ > 30				
				EC/ < 70 /4.0					
^{244}Es		244.0709	37. s	EC/76 /4.6					
				α/4 /	7.57/4				
^{245}Es		245.0713	1.3 m	α/40 /7.858	7.74				
				EC/60 /3.1					
^{246}Es		246.0730	7.7 m	EC/90 /3.9					
				α/10 /	7.35				
^{247}Es		247.07365	4.8 m	EC/93 /2.48					
				α/7 /	7.32				
^{248}Es		248.0755	26. m	EC/99.7 /3.1					
				α/0.3 /	6.87				
^{249}Es		249.07640	1.70 h	EC/99.4 /1.45		(7/2+)			0.3795
				α/0.6 /	6.77				0.8132
^{250m}Es			2.2 h	EC/		(1-)			Cf L x-ray
				β^+					Cf k x-ray
									0.9891
									1.0319
^{250}Es		250.0787	8.6 h	EC/2.1		(6+)			Cf L x-ray
									Cf k x-ray
									0.30339
									0.34948
									0.82883
^{251}Es		251.07998	1.38 d	EC/99.5 /0.38		(3/2-)			
				α/0.5 /	6.462/0.05				
					6.492/0.4				
^{252}Es		252.08297	1.29 y	α/76 /	6.632/61.0	(3-)			
				EC/24 /1.26	6.562/10.3				
^{253}Es		253.084818	20.47 d	α/	6.633/89.8	7/2+	+4.10	7.	0.04180
					6.5916/6.6				0.3892
^{254m}Es			1.64 d	β^- /99.6 /	0.475	2+	2.9	3.7	Fm L x-ray
				α/0.3 /6.67	6.382	2+			Fm k x-ray
									0.6488
									0.6938
^{254}Es		254.088017	276. d	α/	6.429	(7+)			0.064
^{255}Es		255.09027	40. d	β^- /92 /0.29		(7/2+)			
				α/8 /	6.26				
				S.F./0.004/	6.300				
^{256m}Es			7.6 h	β^- /		(8+)			0.218
									0.232
									0.862
^{256}Es		256.0936	25. m	β^- /1.7		(1+)			
$_{100}$**Fm**									
^{242}Fm			0.8 ms	SF					
^{243}Fm		243.0745	0.2 s	α/	8.55				
^{244}Fm		244.0741	3.7 ms	S.F./		0+			
^{245}Fm		245.0754	4. s	α/	8.15/				
^{246}Fm		246.07528	1.2 s	α/92 /	8.24/	0+			
				S.F./8/					
^{247m}Fm			9. s	α/	8.18/				
^{247}Fm		247.0768	35. s	α/8.20	7.87/70				
				EC/2.9	7.93/30				
^{248}Fm		248.07718	36. s	α/99.9 /8.001	7.83/20	0+			
				S.F./0.1/	7.87/80				
^{249}Fm		249.0790	3. m	EC/2.4		(7/2+)			
				α/	7.53				
^{250m}Fm			1.8 s	I.T./					
^{250}Fm		250.07951	30. m	α/	7.43/	0+			
				EC/0.8					
^{251}Fm		251.08157	5.3 h	EC/98 /1.47		(9/2-)			
				α/2 /	6.833				
^{252}Fm		252.08246	1.058 d	α/7.154	6.998/15	0+			
				SF	7.039/85				
^{253}Fm		253.085175	3.0 d	EC(88%)/0.333	6.676/	1/2+			Es k x-ray
				α/12 /	6.943/				0.2719
^{254}Fm		254.086847	3.240 h	α/	7.150	0+			
				S.F./0.06/	7.192				
^{255}Fm		255.089955	20.1 h	α/	6.9635(5)/5.0	7/2+			
					7.0225(5)/93.4				
^{256}Fm		256.09177	2.63 h	S.F./92/		0+			
				α/18 /	6.92/				

TABLE OF THE ISOTOPES

Elem. or Isot.	Natural Abundance (%)	Atomic Mass or Weight	Half-Life	Decay Mode/Energy (/MeV)	Particle Energy /Intensity (MeV / %)	Spin (h/2π)	Nuclear Magnetic Mom.(nm)	Elect. Quadr. Mom. (b)	γ-ray Energy /Intensity (MeV / %)
^{257}Fm		257.09510	100.5 d	α/99.8 /	6.519	(9/2+)			0.1794
				S.F./0.2					0.2410
^{258}Fm		258.0971	0.37 ms	S.F./					
^{259}Fm		259.1006	1.5 s	S.F./					
$_{101}$**Md**									
^{247}Md		247.0817	3. s	α	8.43				
^{248}Md		248.0828	7. s	EC/80 /5.3	8.32/15				
				α/20 /	8.36/5				
^{249}Md		249.0830	24. s	E.C>/<80 /3.7					
				α/>20 /8.46	8.030(20)/				
^{250}Md		250.0845	50. s	EC/94 /4.6	7.75/4				
				α/6 /8.25	7.83/2				
^{251}Md		251.0849	4.0 m	EC/>94 /3.1					
				α/<6 /	7.55/				
^{252}Md		252.0866	2. m	EC/>50 /3.9					
				α/<50 /	7.73/				
^{253}Md		253.0873	≈6 m	EC/2.0					
^{254m}Md			30. m	EC/					
^{254}Md		254.0897	10. m	EC/2.7					
^{255}Md		255.09108	27. m	EC/92 /1.04		(7/2-)			0.4531
				α/8 /	7.33				0.4056
^{256}Md		256.0941	1.30 h	EC/89 /2.13	7.22/47				
				α/11 /	7.155/21				Fm k x-ray
									(0.112-0.834)
^{257}Md		257.095535	5.5 h	EC/85 /0.41	7.074	(7/2-)			Fm k x-ray
				α/15 /, S.F.	7.014				(0.181-0.389)
^{258m}Md			57. m	EC/		(1-)			Fm k x-ray
^{258}Md		258.098427	51.5 d	α/7.40	6.718(2)/	(8-)			0.3678
					6.763(4)/				(0.057 - 0.448)
^{259}Md		259.1005	1.64 h	S.F./		7/2+			
				α/<1.3					
^{260}Md		260.104	32. d	S.F./					
$_{102}$**No**									
^{250}No			0.25 ms	S.F./		0+			
^{251}No		251.0889	0.8 s	α/	8.60/80				
					8.68/20				
^{252}No		252.08897	2.3 s	α/73 /8.551	8.42	0+			
				S.F./27/	8.37				
^{253}No		253.0907	1.7 m	α/	8.010(20)	(9/2-)			
				EC/3.2					
^{254m}No			0.28 s	I.T./					
^{254}No		254.09095	55. s	α/	8.09	0+			
				EC/1.1					
^{255}No		255.09323	3.1 m	α/62 /	8.12/	1/2+			0.187
				EC/38/2.01	7.93				
					8.08				
^{256}No		256.09428	2.9 s	α/	8.43	0+			
^{257}No		257.09685	25. s	α/	8.22	(7/2+)			
					8.27				
					8.32				
^{258}No		258.0983	≈ 1.2 ms	S.F./		0+			
^{259}No		259.1011	58. m	α/78 /7.794	7.52	(9/2+)			
				EC/22/0.5	7.55				
^{262}No		262.108	≈ 8. ms	S.F./					
$_{103}$**Lr**									
^{253}Lr		253.0953	1.3 s	α/	8.80				
					8.72				
^{254}Lr		254.0965	13. s	α/	8.45				
				EC/5.2					
^{255}Lr		255.0967	22. s	α/	8.37/60				
				EC/3.2	8.43/40				
^{256}Lr		256.0988	28. s	α/99.7 /8.554	8.43/				
				EC/4.2	8.39				
^{257}Lr		257.0996	0.65 s	α/	8.80	7/2+			
				EC/2.5	8.80				
^{258}Lr		258.1019	3.9 s	α/	8.60				
				EC/3.4	8.62				
					8.56				
^{259}Lr		259.1030	6.1 s	α/80	8.44(1)				
				SF/20					
^{260}Lr		260.1056	3. m	α/	8.03				
^{261}Lr		261.1069	40. m	SF					
^{262}Lr		262.110	3.6 h	EC/2.					

TABLE OF THE ISOTOPES

Elem. or Isot.	Natural Abundance (%)	Atomic Mass or Weight	Half-Life	Decay Mode/Energy (/MeV)	Particle Energy /Intensity (MeV / %)	Spin (h/2π)	Nuclear Magnetic Mom.(nm)	Elect. Quadr. Mom. (b)	γ-ray Energy /Intensity (MeV / %)
$_{104}$**104**									
^{253}104			≈ 1.8 s	SF					
^{254}104			0.5 ms	SF					
^{255}104		255.1015	1.7 s	SF					
^{256}104		256.10118	7. ms	SF,α/8.92	8.81				
^{257}104		257.1032	4.7 s	α/9.22	8.77				0.117
					9.01				
					8.95				
					8.62				
^{258}104		258.1035	12. ms	SF/					
^{259}104		259.1056	3.4 s.	α/9.09	8.77(2)/				
				SF/	8.86/				
^{260}104		260.1065	20. ms	SF/					
^{261}104		261.1089	1.1 m	α/8.78	8.28/				
^{262}104		262.1101	≈1.2 s	SF/					
^{263}104			<20. m	SF					
$_{105}$**105**									
^{255}105			≈ 1.5 s	SF					
^{257}105		257.1079	1.5 s	α/	8.97/				
					9.07/				
					9.16/				
^{258}105		258.1093	4.2 s	α/	9.17/				
				E.C/5.3	9.08/				
^{259}105		259.1097	≈ 1.2 s	S.F./					
^{260}105		260.1114	1.5 s	α/	9.05/				
				S.F./	9.08/				
					9.13/				
^{261}105		261.1121	1.8 s	α/	8.93/				
				S.F./					
^{262}105		262.1144	34. s	S.F./					
				α/	8.45/				
					8.53/				
					8.67/				
^{263}105		263.1153	≈0.45 m	S.F./57/, α/43/	8.35/43				
$_{106}$**106**									
^{259}106		259.1147	0.5 s	α,S.F./	9.62				
^{260}106		260.11444	4. ms	α,S.F./	9.77				
^{261}106		261.1164	0.3 s	α,S.F./	9.56				
^{263}106		263.1186	0.8 s	S.F.,α	9.06				
					9.25/				
^{265}106		265.121	<30. s	α/>50					
				SF/<50					
^{266}106		266.122	<30. s	α/>50	8.63(5)				
				SF/<50					
$_{107}$**107**									
^{261}107		261.1218	12. ms	α/,SF	10.40				
					10.10				
					10.03				
^{262m}107			8. ms	α/	10.37				
					10.24				
^{262}107		262.1231	0.10 s	α/	10.06				
					9.91				
					9.74				
^{264}107									
$_{108}$**108**									
^{263}108				α/					
^{264}108		264.1286	≈ 0.08 ms	α/,SF	11.0				
^{265}108		265.1306	2. ms	α/	10.36				
$_{109}$**109**									
^{266}109		266.1378	≈ 3.4 ms	α	11.10				
^{268}109									
$_{110}$**110**									
^{269}110			0.17 ms	α					
$_{111}$**111**									
^{272}111			1.5 ms	α					

NEUTRON SCATTERING AND ABSORPTION PROPERTIES*

(Revised 1993)

Norman E. Holden
High Flux Beam Reactor
Reactor Division
Brookhaven National Laboratory
Upton, New York 11973

This Table presents an evaluated set of values for the experimental quantities, which characterize the properties for scattering and absorption of neutrons. The neutron cross section is given for room temperature neutrons, 20.43° C, corresponds to a thermal neutron energy of 0.0253 electron volts (eV) or a neutron velocity of 2200 meters/second. The neutron resonance integral is defined over the energy range from 0.5 eV to 0.1 x 10^6 eV, or 0.1 MeV. A list of the major references used is given below. The literature cutoff date is October 1993. Uncertainties are given in parentheses. Parentheses with two or more numbers indicate values to the excited state(s) and to the ground state of the product nucleus.

TABLE LAYOUT

Column No.	Column title	Description
1	Isotope or Element	For elements, the atomic number and chemical symbol are listed. For nuclides, the mass number and chemical symbol are listed. Isomers are indicated by the addition of m, m1, or m2.
2	Isotopic Abundance	in atom percent.
3	Half-life	Half-life in decimal notation. μs = microseconds; ms = milliseconds; s = seconds; m = minutes; h = hours; d = days; and y = years.
4	Thermal Neutron Cross Section	Cross sections for neutron capture reactions in barns (10^{-24} cm^2) or in millibarns (mb). Proton, alpha production and fission reactions are designated by σ_p, σ_α, σ_f, respectively. Separate values are listed for isomeric production.
5	Neutron Resonance Integral	Resonance integrals for neutron capture reactions in barns (10^{-24} cm^2) or in millibarns (mb). Proton, alpha production and fission reactions are designated by RI_p, RI_α, RI_f, respectively. Separate values are listed for isomeric production.
6	Neutron Scattering Length	Bound Coherent scattering length for neutron scattering reactions in units of femtometer (fm), which is equal to fermis (10^{-13} cm).

GENERAL NUCLEAR DATA REFERENCES

The following references represent the major sources of the neutron data presented:

1. Mughabghab, S.F.; Divadeenam, M.; Holden, N.E.; Neutron Cross Sections, Vol. I *Neutron Resonance Parameters and Thermal Cross Sections* Part A Z=1–60. Academic Press Inc, New York, New York (1981); Mughabghab, S.F.; Part B, Z–61–100. Academic Press Inc, Orlando, Florida (1984).
2. Walker, F.W.; Parrington, J.R.; Feiner, F.; *Chart of the Nuclides, 14th Edition*, Knolls Atomic Power Laboratory (April 1988).
2. Holden, N.E.; *Total and Spontaneous Fission Half-lives for Uranium, Plutonium, Americium and Curium Nuclides*, Pure & Applied Chemistry 61, 1483 (1989).
3. Holden, N.E.; *Fifty Years with Nuclear Fission* Conference, Wash., D.C., Gaithersburg, Md. April 26 – 29, 1989, p.946. American Nuclear Society, LaGrange Park, Illinois (1989).
4. Holden, N.E.; *Review of Thermal Neutron Cross Sections and Isotopic Composition of the Elements*, BNL–NCS–42224 (March 1989) and Holden, N.E.; *Thermal Neutron Cross Sections for the 1991 Table of the Isotopes*, BNL–45255 (Feb. 1991).
5. Tuli, J.K.; *Nuclear Wallet Cards*, Brookhaven National Laboratory (July 1990).
6. Holden, N.E.; *Half-lives of Selected Nuclides*, Pure & Applied Chemistry 62, 941 (1990).
7. Koester, L.; Rauch, H.; Seymann, E.; *Neutron Scattering Lengths: A Survey of Experimental Data and Methods*, Atomic Data Nuclear Data Tables 49, 65 (1991).
8. Sears, V.F.; *Neutron Scattering Lengths and Cross Sections*, Neutron News 3, (3), 26 (1992).

*Research carried out under auspices of the US Department of Energy, Contract No. DE–AC02–76CH00016

NEUTRON SCATTERING AND ABSORPTION PROPERTIES (continued)

Elem. or Isot.	Natural Abundance (%)	Half-life	Thermal Neut. Cross-Section (barns)	Resonance Integral (barns)	Coh. Scat. Length (fermi)
$_1H$			0.332(2)	0.149(1)	– 3.739(1)
^1H	99.985(1)		0.332(2)	0.149(1)	– 3.741(1)
^2H	0.015(1)		0.52(1)mb	0.23(2) mb	6.671(4)
^3H		12.32 y	< 6. μb		4.79(3)
$_2He$			< 0.05		3.26(3)
^3He	1.37×10^{-6}		$\sigma_p = 5.33(1) \times 10^3$	$RI_p = 2.39(1) \times 10^3$	5.74(7)
			0.05(1) mb		
^4He	≈ 100.				3.26(3)
$_3Li$			71.(2)	32.(1)	– 1.90(2)
^6Li	7.5(2)		$\sigma_\alpha = 9.4(1) \times 10^2$	$RI_\alpha = 425.(4)$	2.0(1)
			39.(5) mb	18.(2) mb	
^7Li	92.5(2)		45.(5) mb	20.(2) mb	– 2.22(2)
$_4Be$			8.(1) mb	3.6(5) mb	7.79(1)
^7Be		53.28 d	$\sigma_p = 3.9(1) \times 10^4$	$RI_p = 1.75(5) \times 10^4$	
^9Be	100.		8.(1) mb	3.6(5) mb	7.79(1)
^{10}Be		1.52×10^6 y	< 1. mb		
$_5B$			$7.6(1) \times 10^2$	344.(5)	5.30(4)
^{10}B	19.9(2)		$\sigma_\alpha = 38.4(1) \times 10^2$	$RI_\alpha = 17.3(1) \times 10^2$	– 0.2(4)
			0.3(1)	0.13(4)	
			$\sigma_p = 7.(1)$ mb		
			$\sigma_t = 8.(2)$ mb		
^{11}B	80.1(2)		5.(3) mb	2.5(1) mb	6.65(4)
$_6C$			3.5(1) mb	1.6(1) mb	6.646(1)
^{12}C	98.89(1)		3.5(1) mb	1.6(1) mb	6.651(2)
^{13}C	1.11(1)		1.4(1) mb	1.7(2) mb	6.19(9)
^{14}C		5715. y	< 1. μb		
$_7N$			1.9(1)	0.90(5)	9.36(2)
^{14}N	99.634(9)		$\sigma_p = 1.8(1)$	$RI_p = 0.87(3)$	9.37(2)
			0.080(1)	0.030(1)	
^{15}N	0.366(9)		0.04(1) mb	34.(3) mb	6.44(3)
$_8O$			0.29(1) mb	0.26(4) mb	5.805(4)
^{16}O	99.76(1)		0.19(1) mb	0.21(4) mb	5.805(5)
^{17}O	0.04		$\sigma_\alpha = 0.24(1)$	0.11(1)	5.8(2)
			0.54(7) mb	0.39(5) mb	
^{18}O	0.20(1)		0.16(1) mb	0.81(4) mb	5.84(7)
$_9F$			9.5(7) mb	21.(3) mb	5.65(1)
^{19}F	100.		9.5(7) mb	21.(3) mb	5.65(1)
$_{10}Ne$			0.04(1)	0.02(1)	4.566(6)
^{20}Ne	90.48(3)		0.04(1)	0.02(1)	4.631(6)
^{21}Ne	0.27(1)		0.7(1)	0.3(1)	6.7(2)
^{22}Ne	9.25(3)		0.05(1)	0.02(1)	3.87(1)
$_{11}Na$			0.525(5)	0.32(2)	3.63(2)

Elem. or Isot.	Natural Abundance (%)	Half-life	Thermal Neut. Cross-Section (barns)	Resonance Integral (barns)	Coh.Scat. Length (fermi)
^{22}Na		2.605 y	σ_p=2.8(3)x10^4 σ_α=2.6(3)x10^2	< 2x10^5	
^{23}Na	100.		σ_m=0.43(3)	(0.28+0.04)	3.63(2)
$_{12}$Mg			63.(3) mb	38.(4) mb	5.37(1)
^{24}Mg	78.99(3)		0.053(6)	32.(4) mb	5.5(2)
^{25}Mg	10.00(1)		0.17(5)	98.(15) mb	3.6(1)
^{26}Mg	11.01(2)		0.037(2)	25.(2) mb	4.9(2)
^{27}Mg		9.45 m	0.07(2)	0.03(1)	
$_{13}$Al			0.230(2)	0.17(1)	3.45(1)
^{27}Al	100.		0.230(2)	0.17(1)	3.45(1)
$_{14}$Si			0.166(7)	0.09(2)	4.149(1)
^{28}Si	92.23(2)		0.17(1)	76.(5) mb	4.107(6)
^{29}Si	4.67(2)		0.12(1)	77.(15) mb	4.7(1)
^{30}Si	3.10(1)		0.107(4)	0.71(6)	4.58(8)
^{31}Si		2.62 h	73.(6) mb	33.(3) mb	
$_{15}$P			0.16(2)	85.(10) mb	5.13(1)
^{31}P	100.		0.16(2)	85.(10) mb	5.13(1)
$_{16}$S			0.54(2)	0.25(2)	2.847(1)
^{32}S	95.02(9)		0.55(5)	0.26	2.804(2)
^{33}S	0.75(4)		0.46(3)	0.23	4.7(2)
^{34}S	4.21(8)		0.29(6)	0.13	3.48(3)
^{36}S	0.02(1)		0.20(3)	0.17(4)	
$_{17}$Cl			33.6(3)	12.(2)	9.578(1)
^{35}Cl	75.77(7)		43.7(4);σ_p=0.5	15.(2);RI$_p$=0.6	11.70(9)
^{36}Cl		3.01x10^5 y	σ_p=46.(2) mb < 10.	RI$_p$=0.02	
^{37}Cl	24.23(7)		(0.05+0.37)	(0.04+0.26)	3.1(1)
$_{18}$Ar			0.66(3)	0.42(5)	1.909(6)
^{36}Ar	0.337(3)		5.6(5);σ_p<1.5 mb	2.(1)	24.90(7)
^{37}Ar		35.0 d	σ_α=2.0(3)x10^3		
^{38}Ar	0.063(1)		0.8(2)	0.4(1)	
^{39}Ar		268. y	6.(2)x10^2		
^{40}Ar	99.600(3)		0.64(3)	0.41(5)	1.7
^{41}Ar		1.82 h	0.5(1)	0.2(1)	
$_{19}$K			2.1(1)	1.0(1)	3.67(2)
^{39}K	93.2581(44)		2.1(2)	0.9(1)	3.74(2)
^{40}K	0.0117(1)	1.26x10^9 y	30.(8) σ_p=4.4(4)	13.(4)	
^{41}K	6.7302(44)		1.46(3)	1.4(2)	2.69(8)
$_{20}$Ca			0.43(2)	0.20(2)	4.70(2)

NEUTRON SCATTERING AND ABSORPTION PROPERTIES (continued)

Elem. or Isot.	Natural Abundance (%)	Half-life	Thermal Neut. Cross-Section (barns)	Resonance Integral (barns)	Coh. Scat. Length (fermi)
^{40}Ca	96.941(18)		0.41(3)	0.22(7)	4.80(5)
^{41}Ca		1.03x10^{5} y	≈ 4.		
^{42}Ca	0.647(9)		0.65(10)	0.39(4)	3.4(1)
^{43}Ca	0.135(6)		6.(1)	3.9(2)	- 1.56(9)
^{44}Ca	2.086(12)		0.8(2)	0.56(1)	1.42(6)
^{45}Ca		162.7 d	≈ 15.		
^{46}Ca	0.004(3)		0.70(2)	1.0(1)	
^{48}Ca	0.187(4)		1.1(1)	0.4(1)	0.39(9)
$_{21}$Sc			27.2(2)	12.(1)	12.3(1)
^{45}Sc	100.		(12.+15.)	(5.+7.)	12.3(1)
^{46}Sc		83.81 d	8.(1)	3.6(5)	
$_{22}$Ti			6.1(1)	2.8(2)	- 3.438(2)
^{46}Ti	8.25(3)		0.6(2)	0.4(1)	4.93(6)
^{47}Ti	7.44(2)		1.6(2)	1.6(2)	3.63(1)
^{48}Ti	73.72(3)		7.9(9)	3.6(2)	- 6.09(2)
^{49}Ti	5.41(2)		1.9(5)	1.2(2)	1.04(5)
^{50}Ti	5.18(2)		0.179(3)	0.12(2)	6.18(8)
$_{23}$V			5.0(1)	2.8(1)	- 0.382(1)
^{50}V	0.250(2)	>1.4x10^{17} y	40.(20)	60.(40)	7.6(6)
^{51}V	99.750(2)		4.9(1)	2.7(3)	- 0.402(2)
$_{24}$Cr			3.0(2)	1.7(1)	3.635(7)
^{50}Cr	4.345(13)		15.(1)	8.(1)	- 4.50(5)
^{51}Cr		27.70 d	< 10.		
^{52}Cr	83.79(2)		0.8(1)	0.6(2)	4.92(1)
^{53}Cr	9.50(2)		18.(2)	8.9(10)	- 4.20(3)
^{54}Cr	2.365(7)		0.36(4)	0.25(5)	4.6(1)
$_{25}$Mn			13.3(1)	14.0(3)	- 3.73(2)
^{53}Mn		3.7x10^{6} y	70.(10)	32.(5)	
^{54}Mn		312.2 d	< 10.		
^{55}Mn	100.		13.3(1)	14.0(3)	- 3.73(2)
$_{26}$Fe			2.60(4)	1.4(2)	9.45(2)
^{54}Fe	5.85(4)		2.7(5)	1.4(2)	4.2(1)
^{55}Fe		2.73 y	13.(2)	6.(1)	
^{56}Fe	91.75(4)		2.6([illegible])	1.1([illegible])	9.93(3)
^{57}Fe	2.12(1)		2.5(5)	1.6(3)	2.3(1)
^{58}Fe	0.28(1)		1.3(1)	1.2(2)	15.(7)
^{59}Fe		44.51 d	13.(3)	6.(1)	
$_{27}$Co			37.19(8)	74.(2)	2.49(2)
^{58m}Co		9.1 h	1.4(1)x10^{5}	2.5(10)x10^{5}	
^{58}Co		70.88 d	1.9(2)x10^{3}	7.(1)x10^{3}	
^{59}Co	100.		(20.7+16.5)	(39.+35.)	2.49(2)
^{60m}Co		10.47 m	58.(8)	230.(50)	
^{60}Co		5.271 y	2.0(2)	4.3(10)	
$_{28}$Ni			4.5(2)	2.3(2)	10.3(1)

Elem. or Isot.	Natural Abundance (%)	Half-life	Thermal Neut. Cross-Section (barns)	Resonance Integral (barns)	Coh.Scat. Length (fermi)
^{58}Ni	68.077(9)		4.6(4)	2.3(2)	14.4(1)
^{59}Ni		≈7.6x10^{4} y	92.(4)	138.(8)	
^{60}Ni	26.223(8)		2.9(3)	1.5(2)	2.8(1)
^{61}Ni	1.140(1)		2.5(5)	1.5(4)	7.6(6)
^{62}Ni	3.634(2)		15.(1)	6.8(3)	- 8.7(2)
^{63}Ni		100. y	24.(3)		
^{64}Ni	0.926(1)		1.8(1)	1.2(2)	- 0.38(7)
^{65}Ni		2.517 h	22.(2)	10.(1)	
$_{29}$Cu			3.8(1)	4.1(4)	7.718(4)
^{63}Cu	69.17(3)		4.5(2)	5.(1)	6.43(15)
^{64}Cu		12.701 h	≈ 270.		
^{65}Cu	30.83(3)		2.17(3)	2.2(1)	10.61(19)
^{66}Cu		5.10 m	1.4(1)x10^{2}	60.(20)	
$_{30}$Zn			1.1(2)	2.8(4)	5.680(5)
^{64}Zn	48.6(3)		0.74(5)	1.4(3)	5.22(4)
^{65}Zn		243.8 d	66.(8)		
^{66}Zn	27.9(2)		1.0(2)	1.8(2)	5.97(5)
^{67}Zn	4.1(1)		6.9(1.4)	25.(5)	7.56(8)
^{68}Zn	18.8(4)		(0.072+0.8)	(0.2+2.9)	6.03(3)
^{70}Zn	0.6(1)		(8.1+83.) mb	0.9(2)	
$_{31}$Ga			2.9(1)	22.(3)	7.288(2)
^{69}Ga	60.108(9)		1.68(7)	16.(2)	7.88(4)
^{71}Ga	39.892(9)		4.7(2);σ_m=0.15	31.(3)	6.40(3)
$_{32}$Ge			2.2(1)	6.(2)	8.19(2)
^{70}Ge	21.23(4)		(0.3+2.7)	2.3(1)	10.0(1)
^{72}Ge	27.66(3)		0.9(2)	0.8(3)	8.5(1)
^{73}Ge	7.73(1)		15.(1)	66.(20)	5.02(4)
^{74}Ge	35.94(2)		(0.14+0.28)	(0.4+0.5)	7.6(1)
^{76}Ge	7.44(2)		(0.09+0.06)	(1.3+0.6)	8.2(15)
$_{33}$As			4.0(4)	61.(5)	6.58(1)
^{75}As	100.		4.0(4)	61.(5)	6.58(1)
$_{34}$Se			12.(1)	14.(3)	7.970(9)
^{74}Se	0.89(2)		50.(4)	650.(50)	0.8(3)
^{75}Se		119.78 d	3.3(10)x10^{2}		
^{76}Se	9.36(11)		(22.+63.)	(6.+38.)	12.2(1)
^{77}Se	7.63(6)		42.(4)	30.(5)	8.25(8)
^{78}Se	23.78(9)		σ_m=0.38(2)	RI$_m$=3.7(5)	8.24(9)
^{80}Se	49.61(10)		(0.07+0.39)	(0.3+1.7)	7.48(3)
^{82}Se	8.73(6)		(39.+5.8) mb	0.04(1)	6.34(8)
$_{35}$Br			6.8(2)	92.(8)	6.79(2)
^{76}Br		16.0 h	224.(42)		
^{79}Br	50.69(7)		(2.5+8.3)	(36.+96.)	6.79(7)
^{81}Br	49.31(7)		(2.4+0.24)	51.(5)	6.78(7)

Elem. or Isot.	Natural Abundance (%)	Half-life	Thermal Neut. Cross-Section (barns)	Resonance Integral (barns)	Coh.Scat. Length (fermi)
$_{36}$Kr			24.(1)	39.(6)	7.81(2)
^{78}Kr	0.35(2)		(0.17+6.)	20.(2)	
^{80}Kr	2.25(2)		(4.6+7.)	57.(6)	
^{82}Kr	11.6(1)		(14.+7.)	130.(13)	
^{83}Kr	11.5(1)		183.(30)	183.(20)	
^{84}Kr	57.0(3)		σ_m=0.09;σ_{m+g}=0.11	2.4(3)	
^{85}Kr		10.73 y	1.7(2)	2.(1)	
^{86}Kr	17.3(2)		3.(2)mb	≈ 1. mb	8.1(3)
$_{37}$Rb			0.39(4)	6.(3)	7.08(2)
^{84}Rb		32.9 d	σ_p=12.(2)		
^{85}Rb	72.17(2)		(0.06+0.44)	(0.7+7.)	7.0(1)
^{86}Rb		18.65 d	< 20.		
^{87}Rb	27.83(2)	4.88x10^{10} y	0.10(1)	2.3(2)	7.3(1)
^{88}Rb		17.7 m	1.2(3)	0.5	
$_{38}$Sr			1.2(1)	7.(4)	7.02(2)
^{84}Sr	0.56(1)		(0.6+0.2)	(9.+1.)	5.(2)
^{86}Sr	9.86(1)		σ_m=0.81(4)	RI$_m$=4.(1)	5.68(5)
^{87}Sr	7.00(1)			118.(30)	7.41(7)
^{88}Sr	82.58(1)		5.8(4)mb	0.07(3)	7.16(6)
^{89}Sr		50.52 d	0.42(4)	0.2	
^{90}Sr		29.1 y	9.7(7)mb	4. mb	
$_{39}$Y			(0.001+1.25)	1.0(1)	7.75(2)
^{89}Y	100.		(0.001+1.25)	(0.006+1.0)	7.75(2)
^{90}Y		2.67 d	< 7.		
^{91}Y		58.5 d	1.4(3)	0.6	
$_{40}$Zr			0.19(1);σ_α<0.1mb	0.95(9)	7.16(3)
^{90}Zr	51.45(3)		≈0.014	0.2(1)	6.4(1)
^{91}Zr	11.22(4)		1.2(3)	5.(2)	8.8(1)
^{92}Zr	17.15(2)		0.2(1)	0.6(2)	7.5(2)
^{93}Zr		1.5x10^6 y	< 4.	16.(5)	
^{94}Zr	17.38(4)		0.049(6)	0.25(3)	8.3(2)
^{96}Zr	2.80(2)		0.020(3)	5.0(5)	5.5(1)
$_{41}$Nb			1.11(1);σ_α<0.1mb	8.5(6)	7.14(3)
^{93}Nb	100.		1.1;σ_m=0.86	(6.2+2.3)	7.14(3)
^{94}Nb		2.4x10^4 y	σ_{m+g}=15.(1) σ_m=0.6(1)	126.(13)	
^{95}Nb		34.97 d	< 7.	< 200.	
$_{42}$Mo			2.5(1)	26.(5)	6.72(2)
^{92}Mo	14.84(4)			≈ 0.8	6.93(8)
^{94}Mo	9.25(3)			≈ 0.8	6.82(7)
^{95}Mo	15.92(5)		13.4(5)	109.(5)	6.93(6)
^{96}Mo	16.68(5)			17.(3)	6.22(6)
^{97}Mo	9.55(3)		2.5(3)	14.(3)	7.26(8)
^{98}Mo	24.13(7)		0.14(1)	7.2(7)	6.60(7)
^{100}Mo	9.63(3)		0.19(1)	3.7(3)	6.75(7)

Elem. or Isot.	Natural Abundance (%)	Half-life	Thermal Neut. Cross-Section (barns)	Resonance Integral (barns)	Coh.Scat. Length (fermi)
$_{43}$Tc					
^{98}Tc		4.2x10^{6} y	σ_m=0.9		
^{99}Tc		2.13x10^{5} y	20.	340.(20)	6.8(3)
$_{44}$Ru			2.6 (1)	43.(5)	7.03(3)
^{96}Ru	5.52(6)		0.23(4)	7.(2)	
^{98}Ru	1.88(6)		< 8.		
^{99}Ru	12.7(1)		4.(1)	195.(20)	
^{100}Ru	12.6(1)		5.8(6)	11.(2)	
^{101}Ru	17.0(1)		5.(1)	80.(10)	
^{102}Ru	31.6(2)		1.2(1)	4.3(5)	
^{103}Ru		39.27 d	<20.	≈ 30.	
^{104}Ru	18.7(2)		0.49(2)	6.(2)	
^{105}Ru		4.44 h	0.30(3)		
^{106}Ru		1.020 y	0.15(4)	2.0(6)	
$_{45}$Rh			145.(2)	1.2(1)x10^{3}	5.88(4)
^{103}Rh	100.		(11.+134.)	(0.09+1.1)x10^{3}	5.88(4)
^{104m}Rh		4.36 m	800.(100)		
^{104}Rh		42.3 s	40.(30)		
^{105}Rh		35.4 h	1.1(3)x10^{4}	1.7(4)x10^{4}	
$_{46}$Pd			7.(1)	82.(8)	5.91(6)
^{102}Pd	1.02(1)		3.2(10)	10.(2)	
^{104}Pd	11.14(8)			16.(2)	
^{105}Pd	22.33(8)		22.(2)	60.(20)	5.5(3)
^{106}Pd	27.33(3)		(0.013+0.28)	(0.2+5.5)	6.4(4)
^{107}Pd		6.5x10^{6} y	1.8(2)	108.(4)	
^{108}Pd	26.46(9)		(0.19+8.5)	(2.+240.)	4.1(3)
^{110}Pd	11.72(9)		(0.033+0.7)	(0.7+8.)	
$_{47}$Ag			62.(1)	767.(60)	5.922(7)
^{107}Ag	51.839(7)		(0.37+35.)	(1.2+105.)	7.56(1)
^{109}Ag	48.161(7)		(4.2+87.)	(0.7+14.1)x10^{2}	4.17(1)
^{110m}Ag		249.8 d	82.(11)	20.(4)	
^{111}Ag		7.47 d	3.(2)	105.(20)	
$_{48}$Cd			2.52(5)x10^{3}	73.(8)	4.87(5)
^{106}Cd	1.25(4)		0.20(3)	4.(1)	5.4(1)
^{108}Cd	0.89(2)		1.	14.(3)	
^{109}Cd		462.0 d	≈180.	6.7(12)x10^{3}	
^{110}Cd	12.49(12)		(0.06+11.)	(6.+34.)	
^{111}Cd	12.80(8)		24.(3)	51.(6)	6.5(1)
^{112}Cd	24.13(14)		(0.012+2.2)	15.	6.4(1)
^{113}Cd	12.22(8)	9x10^{15} y	2.06(4)x10^{4}	390.(40)	- 8.0(2)
^{114}Cd	28.73(28)		(0.04+0.30)	16.(7)	7.5(1)
^{116}Cd	7.49(12)		(26.+52.)mb	1.2	6.3(1)
$_{49}$In			197.(4)	3.3(2)x10^{3}	4.07(2)
^{113}In	4.29(2)		(3.1+5.4+3.9)	(230.+80)	5.39(6)
^{115}In	95.71(2)	4.4x10^{14} y	(88.+73.+44.)	(1.5+1.2+0.7)x10^{3}	4.01(2)
$_{50}$Sn			0.61(3)	8.(2)	6.225(2)

Elem. or Isot.	Natural Abundance (%)	Half-life	Thermal Neut. Cross-Section (barns)	Resonance Integral (barns)	Coh.Scat. Length (fermi)
^{112}Sn	0.97(1)		(0.15+0.40)	(8.+19.)	
^{113}Sn		115.1 d	≈ 9.	210.(50)	
^{114}Sn	0.65(1)		≈ 0.12	5.(1)	6.2(3)
^{115}Sn	0.34(1)		σ_α=0.06 mb	29.(6)	
^{116}Sn	14.53(11)		(0.006+0.14)	(0.5+11.)	5.93(5)
^{117}Sn	7.68(7)		1.1(1)	16.(5)	6.48(5)
^{118}Sn	24.23(11)		σ_m=4. mb	4.7(5)	6.07(5)
^{119}Sn	8.59(4)		2.(1)	2.9(5)	6.12(5)
^{120}Sn	32.59(10)		(0.001+0.13)	1.2(3)	6.49(5)
^{122}Sn	4.63(3)		(0.15+0.001)	0.81(4)	5.74(5)
^{124}Sn	5.79(5)		(0.13+0.005)	RI$_m$=8.0(2)	5.97(5)
$_{51}$Sb			5.3(2)	169.(20)	5.57(3)
^{121}Sb	57.21(5)		(0.4+5.8)	(13.+192.)	5.71(6)
^{123}Sb	42.79(5)		(0.02+0.04+4.0)	(1.+119.)	5.38(7)
^{124}Sb		60.20 d	17.(3)	≈ 8.	
$_{52}$Te			4.2(1)	47.(3)	5.80(3)
^{120}Te	0.096(2)		(0.25+2.0)	≈ 1.	5.3(5)
^{122}Te	2.603(4)		2.4(5)	80.(20)	3.8(2)
^{123}Te	0.908(2)	1.3x10^{13} y	370.(40)	4.5(3)x10^3	- 0.05
^{124}Te	4.816(6)		(0.05+7.)	5.2(7)	8.0(1)
^{125}Te	7.139(6)		1.6(2)	21.(4)	5.02(8)
^{126}Te	18.952(11)		(0.12+0.8)	8.2(6)	5.56(7)
^{128}Te	31.687(11)		(0.016+0.20)	(0.08+1.5)	5.89(7)
^{130}Te	33.799(10)	2.5x10^{21} y	(0.03+0.20)	(0.08+0.34)	6.02(7)
$_{53}$I			6.15(10)	148.(7)	5.28(2)
^{125}I		59.4 d	900.(100)	1.4(2)x10^4	
^{127}I	100.		6.15(10)	148.(7)	5.28(2)
^{128}I		25.00 m	22.(4)	≈ 10.	
^{129}I		1.7x10^7 y	(20.7+10.3)	36.(4)	
^{130}I		12.36 h	18.(3)	≈ 8.	
^{131}I		8.040 d	≈ 0.7	8.(4)	
$_{54}$Xe			25.(1)	263.(50)	4.92(3)
^{124}Xe	0.10(1)		(28.+137.)	(0.6+3.0)x10^3	
^{125}Xe		17.1 h	σ_α≤0.03		
^{126}Xe	0.09(1)		(0.45+3.)	(8.+52.)	
^{127}Xe		36.4 d	σ_α≤0.01		
^{128}Xe	1.91(3)		σ_m=0.48	RI$_m$=38.(10)	
^{129}Xe	26.4(6)		22.(5)	250.(50)	
^{130}Xe	4.1(1)		σ_m=0.45	RI$_m$=16.(4)	
^{131}Xe	21.2(4)		90.(10)	8.(1)x10[illegible]	
^{132}Xe	26.9(5)		(0.05+0.4)	(0.9+3.7)	
^{133}Xe		5.243 d	190.(90)		
^{134}Xe	10.4(2)		(0.003+0.26)	0.40(4)	
^{135}Xe		9.10 h	2.65(11)x10^6	7.6(5)x10^3	
^{136}Xe	8.9(1)		0.26(2)	0.7(2)	
$_{55}$Cs			30.4(8)	422.(50)	5.42(2)
^{132}Cs		6.48 d	σ_α< 0.15		
^{133}Cs	100.		(2.7+27.7)	(32.+390.)	
^{134}Cs		2.065 y	140.(10)	54.(9)	
^{135}Cs		2.3x10^6 y	8.9(5)	90.(20)	
^{137}Cs		30.3 y	0.25(1)	0.36(7)	

NEUTRON SCATTERING AND ABSORPTION PROPERTIES (continued)

Elem. or Isot.	Natural Abundance (%)	Half-life	Thermal Neut. Cross-Section (barns)	Resonance Integral (barns)	Coh. Scat. Length (fermi)
$_{56}$Ba			1.3(2)	10.(2)	5.07(3)
^{130}Ba	0.106(2)		(1.+8.)	(25.+200.)	- 3.6(6)
^{132}Ba	0.101(2)		(0.84+9.7)	(4.7+24.)	7.8(3)
^{133}Ba		10.53 y	4.(1)	85.(30)	
^{134}Ba	2.417(27)		(0.1+1.3)	(5.6+18.)	5.7(1)
^{135}Ba	6.592(18)		(0.014+5.8)	(0.47+131.)	4.7(1)
^{136}Ba	7.854(36)		(0.010+0.44)	(0.1+1.5)	4.91(8)
^{137}Ba	11.23(4)		5.(1)	4.(1)	6.8(1)
^{138}Ba	71.70(7)		0.41(2)	0.4(1)	4.84(8)
^{139}Ba		1.396 h	5.(1)	2.2(5)	
^{140}Ba		12.75 d	1.6(3)	14.(1)	
$_{57}$La			9.2(2)	12.(1)	8.24(4)
^{138}La	0.0902(2)	1.06×10^{11} y	57.(6)	$4.1(9) \times 10^2$	
^{139}La	99.9098(2)		9.2(2)	12.(1)	8.24(4)
^{140}La		1.678 d	2.7(3)	69.(4)	
$_{58}$Ce			0.64(4)	0.71(6)	4.84(2)
^{136}Ce	0.19(1)		(1.0+6.5)	58.(12)	5.80(9)
^{138}Ce	0.25(1)		(0.018+1.)	(1.5+5.2)	6.70(9)
^{140}Ce	88.43(10)		0.58(4)	0.50(5)	4.84(9)
^{141}Ce		32.50 d	29.(3)	13.(2)	
^{142}Ce	11.13(10)		0.97(3)	1.3(3)	4.75(9)
^{143}Ce		1.38 d	6.1(7)	2.7(3)	
^{144}Ce		284.6 d	1.0(1)	2.6(3)	
$_{59}$Pr			11.5(4)	14.(3)	4.58(5)
^{141}Pr	100.		(4.+7.5)	14.(3)	4.58(5)
^{142}Pr		19.12 h	20.(3)	9.(1)	
^{143}Pr		13.57 d	90.(10)	190.(25)	
$_{60}$Nd			51.(2)	49.(5)	7.69(5)
^{142}Nd	27.13(12)		19.(1)	34.(11)	7.7(3)
^{143}Nd	12.18(6)		330.(10)	128.(30)	
^{144}Nd	23.80(12)	2.1×10^{15} y	3.6(3)	3.9(5)	2.8(3)
^{145}Nd	8.30(6)		47.(6)	260.(40)	
^{146}Nd	17.19(9)		1.5(2)	3.0(4)	8.7(2)
^{147}Nd		10.98 d	440.(150)	210.	
^{148}Nd	5.76(3)		2.4(1)	13.(2)	5.7(3)
^{150}Nd	5.64(3)		1.0(1)	14.(2)	5.3(2)
$_{61}$Pm					
^{146}Pm		5.53 y	$8.4(1.7) \times 10^3$		
^{147}Pm		2.6234 y	(84.+96.)	(1000.+1280.)	12.6(4)
^{148m}Pm		41.3 d	10600.(800)		
^{148}Pm		5.37 d	$\approx 10^3$	$3.6(2.4) \times 10^3$	
^{149}Pm		2.212 d	1400.(200)		
^{151}Pm		1.183 d	$\approx$ 150.		
$_{62}$Sm			$5.6(1) \times 10^3$	$1.4(2) \times 10^3$	
^{144}Sm	3.1(1)		1.6(1)	2.4(3)	
^{145}Sm		340. d	280.(20)	600.(90)	
^{147}Sm	15.0(2)	1.06×10^{11} y	56.(4)	710.(50)	14.(3)
^{148}Sm	11.3(1)	7×10^{15} y	2.4(6)	27.(14)	

NEUTRON SCATTERING AND ABSORPTION PROPERTIES (continued)

Elem. or Isot.	Natural Abundance (%)	Half-life	Thermal Neut. Cross-Section (barns)	Resonance Integral (barns)	Coh.Scat. Length (fermi)
^{149}Sm	13.8(1)	10^{16} y	40100.(600)	3100.(500)	
^{150}Sm	7.4(1)		102.(5)	290.(30)	14.(3)
^{151}Sm		90. y	15200.(300)	3520.(60)	
^{152}Sm	26.7(2)		206.(15)	3000.(300)	− 5.0(6)
^{153}Sm		1.929 d	420.(180)		
^{154}Sm	22.7(2)		7.5(3)	32.(6)	9.(1)
$_{63}$Eu			4570.(100)	3.8(5)x10^3	5.3(3)
^{151}Eu	47.8(5)		(4.+3150.+6000.)	(2000.+4000.)	
^{152m1}Eu		9.30 h	68000.(15000)	< 100000.	
^{152}Eu		13.48 y	11000.(2000)	1600.(200)	
^{153}Eu	52.2(5)		300.(20)	1800.(400)	8.2(1)
^{154}Eu		8.59 y	1500.(300)	1600.(200)	
^{155}Eu		4.71 y	3900.(200)	16000.(2000)	
$_{64}$Gd			48.8(6)x10^3	400.(10)	9.5(2)
^{152}Gd	0.20(1)		700.(200)	700.(200)	
^{153}Gd		241.6 d	20000.(10000)		
^{154}Gd	2.18(3)		(0.06+60.)	230.(50)	
^{155}Gd	14.80(5)		61000.(1000)	1540.(100)	
^{156}Gd	20.47(4)		≈2.0	104.(15)	6.3(4)
^{157}Gd	15.65(3)		2.54(3)x10^5	800.(100)	
^{158}Gd	[illegible](10)		2.3(3)	73.(7)	9.(2)
^{160}Gd	21.86(4)		1.5(7)	6.(1)	9.15(5)
^{161}Gd		3.66 m	2.0(6)x10^4		
$_{65}$Tb			23.2(5)	420.(50)	7.38(3)
^{159}Tb	100.		23.2(5)	420.(50)	7.38(3)
^{160}Tb		72.3 d	570.(110)		
$_{66}$Dy			9.5(2)x10^2	1.5(2)x10^3	16.9(3)
^{156}Dy	0.06(1)		33.(3)	1000.(100)	
^{158}Dy	0.10(1)		43.(6)	120.(10)	6.1(5)
^{159}Dy		144. d	8.(2)x10^3		
^{160}Dy	2.34(6)		60.(10)	1100.(200)	6.7(4)
^{161}Dy	18.9(2)		600.(50)	1100.(100)	10.3(4)
^{162}Dy	25.5(2)		170.(20)	2755.(300)	− 1.4(5)
^{163}Dy	24.9(2)		120.(10)	1600.(400)	5.0(4)
^{164}Dy	28.2(2)		(1.7+1.0)x10^3	(420.+100.)	49.4(2)
^{165m}Dy		1.26 m	2.0(6)x10^3		
^{165}Dy		2.33 h	3.5(3)x10^3	2.2(3)x10^4	
$_{67}$Ho			61.(2)	670.(40)	8.01(8)
^{165}Ho	100.		(3.1+58.)	(?+670.)	8.01(8)
^{166}Ho		1.2x10^3 y	9140.(650)	1140.(90)	8.01(8)
$_{68}$Er			169.(20)	730.(10)	7.79(2)
^{162}Er	0.14(1)		19.(3)	480.(50)	8.8(2)
^{164}Er	1.61(2)		13.(3)	105.(10)	8.2(2)
^{166}Er	33.6(2)		(15.+5.)	96.(12)	10.6(2)
^{167}Er	22.95(15)		700.(200)	2970.(70)	3.0(3)
^{168}Er	26.8(2)		2.0(6)	37.(5)	7.4(4)
^{170}Er	14.9(2)		6.(2)	26.(4)	9.6(5)
^{171}Er		7.52 h	370.(40)	170.(20)	

NEUTRON SCATTERING AND ABSORPTION PROPERTIES (continued)

Elem. or Isot.	Natural Abundance (%)	Half-life	Thermal Neut. Cross-Section (barns)	Resonance Integral (barns)	Coh.Scat. Length (fermi)
$_{69}$Tm			106.(5)	1.5(2)x10^3	7.07(3)
^{169}Tm	100		(8.+98.)	1.5(2)x10^3	7.07(3)
^{170}Tm		128.6 d	100.(20)	460.(50)	
^{171}Tm		1.92 y	≈ 160.	118.(6)	
$_{70}$Yb			52.(10)	1.7(2)x10^2	12.43(3)
^{168}Yb	0.13(1)		2.4(2)x10^3	2.0(5)x10^4	
^{169}Yb		32.03 d	3.6(3)x10^3	5200.(500)	
^{170}Yb	3.05(6)		12.(2)	320.(30)	6.8(1)
^{171}Yb	14.3(2)		50.(10)	315.(30)	9.7(1)
^{172}Yb	21.9(3)		≈ 1.3	25.(3)	9.4(1)
^{173}Yb	16.12(21)		16.(2)	380.(30)	9.56(7)
^{174}Yb	31.8(4)		(46.+74.)	(13.+47.)	19.3(1)
^{176}Yb	12.7(2)		3.1(2)	8.(2)	8.7(1)
$_{71}$Lu			78.(7)	8.3(7)x10^2	7.21(3)
^{175}Lu	97.41(2)		(16.+8.)	(550.+270.)	7.24(3)
^{176}Lu	2.59(2)	3.8x10^{10} y	(2.+2100.)	(3.+1100.)	6.1(2)
^{177m}Lu		160.7 d	3.2(3)	1.4(2)	
^{177}Lu		6.68 d	1000.(300)		
$_{72}$Hf			106.(3)	19.7(5)x10^2	7.8(1)
^{174}Hf	0.162(3)	2.0x10^{15} y	600.(50)	400.(50)	
^{176}Hf	5.206(5)		23.(4)	700.(100)	11.(1)
^{177}Hf	18.606(13)		(1.+375.)	7170.(200)	6.6(2)
^{178}Hf	27.297(4)		(53.+32.)	(1000.+850.)	5.9(2)
^{179}Hf	13.629(6)		(0.43+46.)	(6.8+620.)	7.5(2)
^{180}Hf	35.100(7)		13.(1)	32.(2)	13.2(3)
^{181}Hf		42.4 d	30.(25)		
$_{73}$Ta			20.(1)	650(20.)	6.91(7)
^{180m}Ta	0.012(2)	> 1.2x10^{15} y	≈ 560.	1350.(100)	
^{181}Ta	99.998(2)		(0.012+20.)	(0.4+650.)	6.91(7)
^{182}Ta		114.43 d	8200.(600)	900.(90)	
$_{74}$W			18.(1)	3.6(3)x10^2	4.86(2)
^{180}W	0.12(1)		≈ 4.	210.(30)	
^{182}W	26.50(3)		20.(1)	600.(90)	6.97(4)
^{183}W	14.31(1)		10.5(3)	340.(50)	6.53(4)
^{184}W	30.64(1)		(0.002+2.0)	15.(2)	7.48(6)
^{185}W		74.8 d	≈ 3.3	300.(50)	
^{186}W	28.43(4)		37.(2)	510.(50)	- 0.72(4)
^{187}W		23.9 h	70.(10)	2760.(550)	
$_{75}$Re			90.(4)	8.4(2)x10^2	9.2(3)
^{185}Re	37.40(2)		(0.33+110.)	1700.(50)	9.0(3)
^{187}Re	62.60(2)	4.2x10^{10} y	(2.+72.)	(9.+310.)	9.3(3)
$_{76}$Os			17.(1)	1.5(1)x10^2	10.7(2)
^{184}Os	0.020(3)		3.3(3)x10^3	660.(60)	
^{186}Os	1.58(2)	2.x10^{15} y	≈ 80.	280.(40)	12(2)
^{187}Os	1.6(4)		2.(1)x10^2	500.(70)	
^{188}Os	13.3(1)		≈ 5.	150.(20)	7.6(3)
^{189}Os	16.1(1)		(0.00026+40.)	(0.013+670.)	10.7(3)
^{190}Os	26.4(2)		(9.+4.)	(22.+8.)	11.0(3)

NEUTRON SCATTERING AND ABSORPTION PROPERTIES (continued)

Elem. or Isot.	Natural Abundance (%)	Half-life	Thermal Neut. Cross-Section (barns)	Resonance Integral (barns)	Coh.Scat. Length (fermi)
^{191}Os		15.4 d	$3.8(6)x10^2$		
^{192}Os	41.0(3)		3.(1)	6.(1)	11.5(4)
^{193}Os		30.5 h	40.(10)		
$_{77}Ir$			$4.2(1)x10^2$	$2.8(4)x10^3$	10.6(3)
^{191}Ir	37.27(9)		(660.+260.)	(1000.+4200.)	
^{192}Ir		73.83 d	1450.(250)	4800.(700)	
^{193}Ir	62.73(9)		(6.+110.)	1400.(200)	
^{194}Ir		19.3 h	$1.5(3)x10^3$		
$_{78}Pt$			10.(1)	$1.3(1)x10^2$	9.60(1)
^{190}Pt	0.01(1)	$6.5x10^{11}$ y	$1.5(1)x10^2$	70.(10)	9(1)
^{192}Pt	0.79(6)		(2.0+6.)	115.(20)	9.9(5)
^{194}Pt	32.9(6)		(0.1+1.1)	(4.+?)	10.55(8)
^{195}Pt	33.8(6)		28.(1)	365.(50)	8.8(1)
^{196}Pt	25.3(6)		(0.045+0.55)	7.(2)	9.89(8)
^{198}Pt	7.2(2)		(0.027+3.6)	(0.5+56.)	7.8(1)
^{199}Pt		30.8 m	≈ 16.		
$_{79}Au$			98.7(1)	$1.55(3)x10^3$	7.90(7)
^{197}Au	100.		00.7(1)	$1.66(3)x10^3$	7.90(7)
^{198}Au		2.694 d	$26.5(15)x10^3$	$≈ 4.x10^4$	
^{199}Au		3.14 d	≈ 30.		
$_{80}Hg$			$3.7(1)x10^2$	87.(5)	12.69(2)
^{196}Hg	0.15(1)		(105.+3000.)	(53.+410.)	30.(1)
^{198}Hg	9.97(8)		(0.017+2.)	(1.7+70.)	
^{199}Hg	16.87(10)		$2.1(2)x10^3$	435(20)	16.9(4)
^{200}Hg	23.10(16)		< 60.	2.1(5)	
^{201}Hg	13.18(8)		< 60.	30.(3)	
^{202}Hg	29.86(20)		4.9(5)	4.5(2)	
^{204}Hg	6.87(4)		0.4(1)	0.8(2)	
$_{81}Tl$			3.3(1)	12.5(8)	8.776(5)
^{203}Tl	29.52(1)		11.(1)	41.(2)	7.0(2)
^{204}Tl		3.78 y	22.(2)	90.(20)	
^{205}Tl	70.48(1)		0.11(2)	0.6(2)	9.52(7)
$_{82}Pb$			0.172(2)	0.14(4)	9.405(3)
^{204}Pb	1.4(1)		0.68(7)	2.0(2)	9.9(1)
^{205}Pb		$1.51x10^7$ y	≈ 5.		
^{206}Pb	24.1(1)		0.030(1)	0.10(1)	9.22(5)
^{207}Pb	22.1(1)		0.70(1)	0.38(1)	9.28(4)
^{208}Pb	52.4(1)		0.49(3) mb	2.0(2) mb	9.50(2)
^{210}Pb		22.6 y	< 0.5		
$_{83}Bi$			0.034(1)	0.19(2)	8.532(2)
^{209}Bi	100.		(11.+23.) mb	0.19(2)	8.532(2)
^{210m}Bi		$3.0x10^6$ y	54.(4) mb	0.20(3)	
$_{84}Po$					
^{210}Po		138.38 d	$\sigma_m < 0.5$ mb $\sigma_g < 30.$mb		

Elem. or Isot.	Natural Abundance (%)	Half-life	Thermal Neut. Cross-Section (barns)	Resonance Integral (barns)	Coh.Scat. Length (fermi)
$_{85}$At					
$_{86}$Rn					
^{220}Rn		55.6 s	< 0.2		
^{222}Rn		3.8235 d	0.74(5)		
$_{88}$Ra					
^{223}Ra		11.435	$1.3(2)\times10^{2}$		
^{224}Ra		3.66 d	12.0(5)		
^{226}Ra		1599. y	≈ 13.	280.(50)	10.(1)
^{228}Ra		5.76 y	36.(5)		
$_{89}$Ac					
^{227}Ac		21.77 y	$8.8(7)\times10^{2}$	$1.5(4)\times10^{3}$	
$_{90}$Th			7.4	85.(3)	10.31(3)
^{227}Th		18.72 d	$\sigma_f=2.0(2)\times10^{2}$		
^{228}Th		1.913 y	$1.2(2)\times10^{2}$	1014.(400)	
^{229}Th		7.9×10^{3} y	≈ 60.	$1.0(2)\times10^{3}$ $RI_f=466.(75)$	
^{230}Th		7.54×10^{4} y	23.4(5) $\sigma_f<$ 0.5mb	$1.0(1)\times10^{3}$	
^{232}Th	100.	1.4×10^{10}y	7.37(4) $\sigma_f=3.(1)$ μb $\sigma_\alpha<$ 1. mb	85.(3)	10.31(3)
^{233}Th		22.3 m	$1.5(1)\times10^{3}$ $\sigma_f=15.(2)$	$4.(1)\times10^{2}$	
^{234}Th		24.10 d	1.8(5) $\sigma_f<$ 0.01		
$_{91}$Pa					
^{230}Pa		17.4 d	$1.5(3)\times10^{3}$		
^{231}Pa		3.25×10^{4} y	$2.0(1)\times10^{2}$ $\sigma_f=20.(1)$ mb	750.(80) $RI_f=0.05(1)$	9.1(3)
^{232}Pa		1.31 d	$4.6(10)\times10^{2}$ $\sigma_f=7.(1)\times10^{2}$	300.(70)	
^{233}Pa		27.0 d	39.(2) $\sigma_m=20.(4)$ $\sigma_g=19.(3)$ $\sigma_f<$ 0.1	(460.+440.)	
$_{92}$U			3.4(3);$\sigma_f=4.2(1)$	2.8(2)	8.417(5)
^{230}U		20.8 d	$\sigma_f\approx$ 25.		
^{231}U		4.2 d	$\sigma_f\approx$ 250.		
^{232}U		68.9 y	73.(2) $\sigma_f=74.(8)$	280.(15) $RI_f=350.(30)$	
^{233}U		1.59×10^{5} y	47.(2) $\sigma_f=5.3(1)\times10^{2}$ $\sigma_\alpha<$ 0.2 mb	137.(6) $RI_f=760.(17)$	10.1(2)
^{234}U	0.0055(5)	2.45×10^{5} y	96.(2) $\sigma_f<$ 5. mb	660.(70) $RI_f=6.5$	12.(4)
^{235}U	0.720(1)	7.04×10^{8} y	95.(5) $\sigma_f=586.(2)$ $\sigma_\alpha<$ 0.1 mb	144.(6) $RI_f=275(5)$	10.47(4)

NEUTRON SCATTERING AND ABSORPTION PROPERTIES (continued)

Elem. or Isot.	Natural Abundance (%)	Half-life	Thermal Neut. Cross-Section (barns)	Resonance Integral (barns)	Coh.Scat. Length (fermi)
^{236}U		2.34x10^{7} y	5.1(3)	360.(15)	
			σ_f=0.04	RI$_f$=4.38(50)	
^{237}U		6.75 d	≈ 10^{2}	1200.(200)	
			σ_f< 0.35		
^{238}U	99.2745(15)	4.46x10^{9} y	2.7(1)	277.(3)	8.402(5)
			σ_f ≈ 3.μb	1.54(15) mb	
			σ_α=1.4(5) μb		
^{239}U		23.5 m	22.(2)		
			σ_f=15.(3)		
$_{93}$Np					
^{234}Np		4.4 d	σ_f=9.(3)x10^{2}		
^{235}Np		1.085 y	1.6(1)x10^{2}		
^{236m}Np		22.5 h	σ_f=2.7(2)x10^{3}	700.(400)	
^{236}Np		1.55x10^{5} y	σ_f=2.6(5)x10^{3}	1030.(100)	
^{237}Np		2.14x10^{6} y	180.	8.(2)x10^{2}	10.6(1)
			σ_f=20.(4)mb	RI$_f$=0.32	
^{238}Np		2.117 d	σ_f=2.1(1)x10^{3}	9.(1)x10^{2}	
^{239}Np		2.355 d	(32.+19.)		
			σ_f< 1.		
$_{94}$Pu					
^{236}Pu		2.87 y	σ_f=1.6(3)x10^{2}	1000.(60)	
^{237}Pu		45.2 d	σ_f=2.3(3)x10^{3}		
^{238}Pu		87.74 y	5.1(2)x10^{2}	1.6(2)x10^{2}	14.1(5)
			σ_f=17.(1)	RI$_f$=33.(5)	
^{239}Pu		2.411 x 10^{4}y	2.7(1)x10^{2}	2.0(2)x10^{2}	7.7(1)
			σ_f=752.(3)	3.0(1)x10^{2}	
			σ_α≤ 0.3 mb		
^{240}Pu		6537. y	2.9(1)x10^{2}	8.4(3)x10^{3}	3.5(1)
			σ_f ≈ 44.mb	RI$_f$=2.4	
^{241}Pu		14.4 y	3.7(1)x10^{2}	1.6(1)x10^{2}	
			σ_f=1.01(1)x10^{3}	5.7(4)x10^{2}	
^{242}Pu		3.76 x 10^{5} y	19.(1)	1.1(1)x10^{3}	8.1(1)
			σ_f<0.2	RI$_f$=0.23	
^{243}Pu		4.956 h	<100.		
			σ_f=2.0(2)x10^{2}		
^{244}Pu		8.2x10^{7} y	1.7(1)	41.(3)	
^{245}Pu		10.5 h	1.5(3)x10^{2}	220.(40)	
$_{95}$Am					
^{241}Am		432.2 y	(0.5+5.7)x10^{2}	(2.0+12.3)x10^{2}	
			σ_f=3.1(2)	14.(1)	
^{242m}Am		141. y	1.7(4)x10^{3}	≈ 200.	
			σ_f=7.0(3)x10^{3}	RI$_f$=1.8(1)x10^{3}	
^{242}Am		16.02 h	σ_f=2.1(2)x10^{3}	RI$_f$< 300.	
^{243}Am		7.37x10^{3}y	(75.+5.)	(1.0+17.1)x10^{2}	8.3(2)
			σ_f=74.(4) mb	RI$_f$=0.056	
^{244m}Am		≈ 26. m	σ_f=1.6(3)x10^{3}		
^{244}Am		10.1 h	σ_f=2.2(3)x10^{3}		
$_{96}$Cm					
^{242}Cm		162.8 d	≈ 20.	120.(50)	
			σ_f ≈ 5.		

NEUTRON SCATTERING AND ABSORPTION PROPERTIES (continued)

Elem. or Isot.	Natural Abundance (%)	Half-life	Thermal Neut. Cross-Section (barns)	Resonance Integral (barns)	Coh.Scat. Length (fermi)
^{243}Cm		28.5 y	$1.3(1)x10^2$	214.(20)	
			σ_f=6.2(2)x10^2	RI_f=1.6(1)x10^3	
^{244}Cm		18.11 y	15.(1)	640.(50)	9.5(3)
			σ_f=1.1(2)	RI_f=10.8(8)	
^{245}Cm		8.5x10^3 y	3.5(2)x10^2	110.(10)	
			σ_f=2.1(1)x10^3	RI_f=8.(1)x10^2	
^{246}Cm		4.78x10^3y	1.2(2)	120.(10)	9.3(2)
			σ_f=0.16(7)	13.(2)	
^{247}Cm		1.56x10^7y	60.(30)	5.(1)x10^2	
			σ_f=82.(5)	7.3(7)x10^2	
^{248}Cm		3.4x10^5 y	2.6(3)	270.(30)	7.7(2)
			σ_f=0.36(7)	13.(2)	
^{249}Cm		64.15 m	≈1.6		
^{250}Cm		≈ 9.7x10^3 y	≈80.		
$_{97}Bk$					
^{249}Bk		320. d	7.(1)x10^2	9.(1)x10^2	
			σ_f ≈ 0.1		
^{250}Bk		3.217 h	σ_f=1.0(2)x10^3		
$_{98}Cf$					
^{249}Cf		351. y	5.0(3)x10^2	7.7(4)x10^2	
			σ_f=1.7(1)x10^3	RI_f=2.1(3)x10^3	
^{250}Cf		13.1 y	2.0(2)x10^3	12.(2)x10^3	
			σ_f=110.(90)	RI_f=160.(40)	
^{251}Cf		9.0x10^2 y	2.9(2)x10^3	1.6(1)x10^3	
			σ_f=4.5(5)x10^3	RI_f=5.5(3)x10^3	
^{252}Cf		2.64 y	20.(2)	43.(3)	
			σ_f=32.(4)	RI_f=1.1(3)x10^2	
^{253}Cf		17.8 d	18.(2)	8.(1)	
			σ_f=1.3(2)x10^3		
^{254}Cf		60.5 d	4.5(10)	2.	
$_{99}Es$					
^{253}Es		20.47 d	(180.+5.8)	(37.5+1.1)x10^2	
^{254m}Es		1.64 d	σ_f=1.8(1)x10^3		
^{254}Es		276. d	28.(3)	18.(2)	
			σ_f=1.8(2)x10^3	RI_f=1.2(3)x10^3	
^{255}Es		40. d	≈ 55.		
$_{100}Fm$					
^{255}Fm		20.1 h	26.(3)	14.(2)	
			σ_f=3.3(2)x10^3		
^{257}Fm		100.5 d	σ_f=3.0(2)x10^3		

COSMIC RADIATION

A. Gregory and R. W. Clay

THE NATURE OF COSMIC RAYS

Primary cosmic radiation, in the form of high-energy nuclear particles, electrons, and photons, from outside the solar system and from the Sun, continually bombards our atmosphere.

Secondary radiation, resulting from the interaction of primary cosmic rays with atmospheric gas, is present at sea-level and throughout the atmosphere. The secondary radiation is collimated by absorption and scattering in the atmosphere and consists of a number of components associated with different particle species. High-energy primary particles can produce a large number of secondary particles forming an extensive air shower. Thus, a number of particles may be detected in coincidence at sea level. Primary particle energies range from $\sim 10^8$ eV to $\sim 10^{20}$ eV.[1,15]

PRIMARY PARTICLE ENERGY SPECTRUM

Figure 1 shows the spectrum of primary particle energies. In differential form it is roughly a power law (with an index of -3). There appears to be a knee (a steepening) at about 10^{15} eV and an ankle (a flattening) above $\sim 10^{18}$ eV.

At energies below $\sim 10^{13}$ eV, solar system magnetic fields and plasma can modulate the primary component and Figure 2 shows the extent of this modulation between solar maximum and minimum.[10,12]

PRIMARY PARTICLE ENERGY DENSITY

If the above spectrum is corrected for solar effects the energy density above a particle energy of 10^9 eV outside the solar system is found to be $\sim 5 \times 10^5$ eV m^{-3}. As the threshold energy is increased, the energy density decreases rapidly, being 2×10^4 eV m^{-3} above 10^{12} eV and 10^2 eV m^{-3} above 10^{15} eV.[16]

PRIMARY PARTICLE ISOTROPY

This is measured as an anisotropy $(I_{max} - I_{min})/(I_{max} + I_{min}) \times 100\%$, where I, the intensity ($m^{-2}s^{-1}sr^{-1}$), is usually measured with an angular resolution of a few degrees.

The anisotropy is small and energy dependent. It is roughly constant at between 0.05 and 0.1% for energies between 10^{11} eV and 10^{14} eV and increases at higher energies roughly as $0.4 \times (\text{energy (eV)}/10^{16})^{1/2}\%$ up to $\sim 10^{18}$ eV.[16]

PRIMARY PARTICLE COMPOSITION

The composition of low energy cosmic rays (it is uncertain at high energies) is close to universal abundances except where propagation effects are present, e.g. Li, Be, B which are spallation products, are over-abundant by ~ 6 orders of magnitude.

Composition at 10^{11} eV/Nucleus

Charge	1	2	(3—5)	(6—8)	(10—14)	(16—24)	(26—28)	≥30
% Composition	50	25	1	12	7	4	4	0.1

($\sim$10% uncertainty)

Cosmic ray composition at low energies is often quoted at a fixed *energy per nucleon.* When presented in this way, protons constitute roughly 90% of the flux, helium nuclei $\sim$10% and the remainder sum to a total of $\sim$1%.

Certain radioactive isotopic ratios show lifetime effects. The ratio of Be^{10}/Be^9 abundances is used to measure an 'age' of cosmic rays since Be^{10} is unstable (half life $= 1.6 \times 10^6$ years). A ratio of 0.6 would be expected in the absence of Be^{10} decay and a ratio of 0.2 is found experimentally.[10,13]

PRIMARY ELECTRONS

Primary electrons constitute $\sim$1% of the cosmic ray beam. The positron to electron ratio is about 10%.[14]

ANTIMATTER IN THE PRIMARY BEAM

The ratio of antiprotons to protons in the primary cosmic ray beam (at $\sim$100 GeV) is $\sim 5 \times 10^{-4}$.[11]

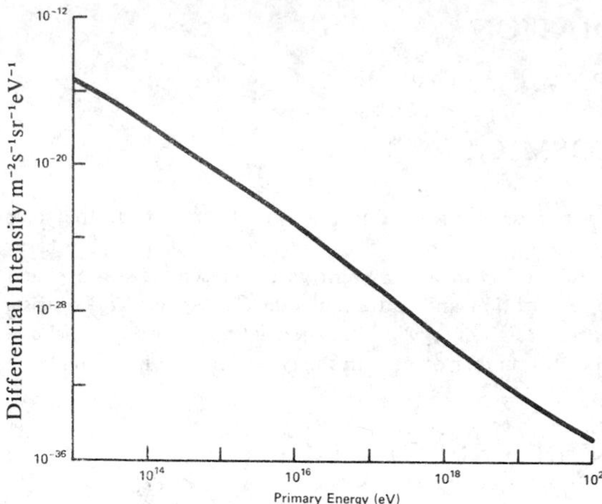

FIGURE 1. Energy Spectrum of Cosmic Ray Particles. This spectrum is of a differential form and can be converted to an integral spectrum by integrating over all energies above a required threshold (E).

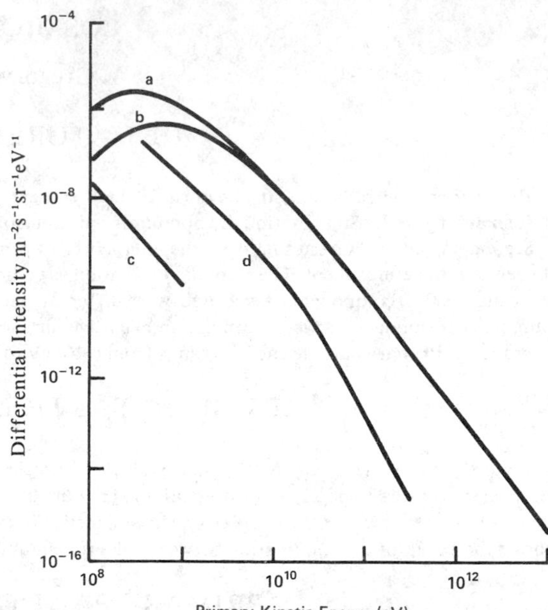

FIGURE 2. Energy Spectrum of Cosmic Ray Particles at Lower Energies. (a) Solar minimum proton energy spectrum, (b) solar maximum proton energy spectrum, (c) γ-ray energy spectrum, (d) local interstellar electron spectrum.

PRIMARY GAMMA RAYS

The flux of primary gamma rays is low at high energies. At 1 GeV the ratio of gamma rays to protons is about 10^{-6}. The arrival directions of these gamma rays are strongly concentrated in the plane of the Milky Way although there is also a diffuse near-isotropic background flux.

Since the absorption cross section for gamma rays above 100 MeV is approximately 20 mbarn/electron, less than 10% of gamma rays reach mountain altitudes.[16] Gamma rays from point sources have confidently been identified with energies up to 10^{16} eV.

SEA LEVEL COSMIC RADIATION

The sea level cosmic ray dose is 30 millirad. year^{-1} and the sea level ionisation is 2.22×10^6 ion pairs $m^{-1}s^{-1}$. The sea level flux has a soft component, which can be absorbed in ∿10cm of lead and a more penetrating (largely muon) hard component. The sea level radiation is a secondary component produced in the atmosphere and its flux is dependent somewhat on the solar cycle and the geomagnetic latitude of the observer.

Absolute Flux of Hard Component[2]

Vertical integral intensity	(I) ∼ 100 $m^{-2} \cdot s^{-1} \cdot sr^{-1}$
Angular dependence	$I(\theta) \sim I(O) \cos^2\theta$
Integrated intensity	∼ 200 $m^{-2} \cdot s^{-1}$

Flux of Soft Component

In free air the soft component comprises about one third of the total cosmic ray flux.

Latitude Effect

The geomagnetic field influences the trajectories of lower energy cosmic rays approaching the Earth. As a result, the background flux is reduced by about 7% at the geomagnetic equator. The effect decreases towards the poles and is negligible at latitudes above about 40°.

Flux of Protons

The proton component is strongly attenuated by the atmosphere with an attenuation length (reduction by a factor of e) of ∼ 120 $g \cdot cm^{-2}$, it constitutes about 1% of the total vertical sea level flux.

Absorption

The soft component is absorbed in about 100 g·cm^{-2} of lead. The hard component is absorbed much more slowly:

Absorption in lead	6%/100 g·cm^{-2}
Absorption in rock	8.5%/100 g·cm^{-2}
Absorption in water	10%/100 g·cm^{-2}

Absorption for depths less than 100 g·cm^{-2} is given by Greisen in Reference 7.

Altitude Dependence

Cosmic ray background in the atmosphere has a maximum intensity of about 15 times that at sea level at a depth of ~ 150 g·cm^{-2} (15 km altitude). At maximum intensity the soft and hard component contribute roughly equally but the hard component is then attenuated more slowly.[8]

COSMIC RAY SHOWERS

High energy cosmic rays (< 10^{13}eV) produce measurable cascades of secondary particles in the atmosphere. The primary particle progressively loses energy which is transferred through the production of successive generations of secondary particles to a cascade of hadrons, an electromagnetic shower component (electrons and gamma rays) and muons. The secondary particles are relativistic and all travel effectively at the speed of light. As a result, they reach sea level at approximately the same time, but, due to Coulomb scattering and production angles, are spread laterally into a disk-like shower front with a typical width of ~ 100 m and thickness ~ 2 to 3 m. The number of particles at sea level is roughly proportional to the primary energy.

Number of particles at sea level ~ 10^{-10} × energy (eV). At altitudes below a few kilometers, the number of particles in a shower attenuates with an *attenuation length* ~ 200 g·cm^{-2}, i.e., $N = N_o \times \exp(-\text{depth}/200)$. The rate of observation of showers of a given size at different depths of absorber attenuates with an *absorption length* of ~ 100 g·cm^{-2}.[15]

ATMOSPHERIC BACKGROUND LIGHT FROM COSMIC RAYS

Cosmic ray particles produce Cerenkov light in the atmosphere and produce fluorescent light through the excitation of atmospheric molecules.

Cerenkov Light

High energy charged particles will emit Cerenkov light in air if their energies are above ~30 MeV (electrons). This threshold is pressure (and hence altitude) dependent. A typical Cerenkov light pulse (at sea level, 100 m from core, 10^{16} eV primary energy, in the wavelength band 430 to 530 nm) has a width of a few nanoseconds and in this time has a flux of ~10^{14} photons m^{-2}s^{-1}. For comparison, the general night sky background flux is ~6 × 10^{11} photons m^{-2}·s^{-1}·sr^{-1} in the same wavelength band.

Fluorescent Light

Cosmic ray particles in the atmosphere excite atmospheric molecules which emit fluorescent light. This is weak compared to the Cerenkov component when viewed in the direction of the incident cosmic ray particle but is emitted isotropically. Typical pulse widths are expected to be longer than 50 ns and may be up to a few microseconds for distant large showers.[5,9]

CERENKOV EFFECTS IN TRANSPARENT MEDIA

Background cosmic ray particles will produce Cerenkov light in transparent material with a photon yield

$$\sim \frac{2\pi}{137} \sin^2 \theta_c \int_{\lambda_1}^{\lambda_2} \frac{d\lambda}{\lambda^2} \text{ photons (unit length)}^{-1}$$

where θ_c, the Cerenkov angle, = $\cos^{-1}$ (1/refractive index). This background light is known to affect sensitive light detectors, e.g. photomultipliers, and can be a major source of background noise.[6]

EFFECTS ON ELECTRONIC COMPONENTS

If background cosmic rays pass through electronic components, they may deposit sufficient energy to affect the state of, e.g. a transistor flip-flop. This effect may be significant where reliability is of great importance and the background flux is high. For instance, it has been estimated that in communication satellite operation an error rate of ~ 2 × 10^{-3} per transistor per year may be found. Permanent damage may also result. A significant error rate may be found even at sea level in large electronic memories. This error rate is dependent on the sensitivity of the component devices to the deposition of electrons in their sensitive volumes.[17]

BIOPHYSICAL SIGNIFICANCE

When cosmic rays interact with living tissue they produce radiation damage, the amount of damage depending on the total dose of radiation. Radiation doses are commonly measured in rads (radiation absorbed dose ≡ 100 ergs g^{-1}) or rems (radiation equivalent-man ≡ Quality factor × rad). The quality factor of radiation depends on the type of particle and its energy for most cosmic ray applications it will be ~1.

At sea level the cosmic radiation dose rate is small compared with doses from other sources, but both the quantity, and quality of the radiation change rapidly with altitude. Approximate dose rates under various conditions

	Cosmic rays			
Conditions	Sea level	10 km (subsonic jets)	18 km (supersonic transport)	Mean total dose rate at sea level
Dose (mrem $year^{-1}$)	30	2000	10,000	300

Astronauts would be subject to radiation from galactic (0.05 rads d^{-1}) and solar (~ few hundred rads per solar flare) cosmic rays as well as large fluxes of low energy radiation when passing through the Van Allen belts (~0.3 rads per traverse).

Both astronauts and S.S.T. travellers would be subject to a small flux of low energy heavy nuclei stopping in the body. Such particles are capable of destroying cell nuclei and could be particularly harmful in the early stages of development of an embryo. The rates of heavy nuclei stopping in tissue in supersonic transports and spacecraft are approximately as follows (from Reference 1 and 2)

Conditions (altitude)	16 km SST	20 km SST	Spacecraft
Stopping nuclei (cm^3 tissue)$^{-1}$ hr^{-1}	5.10^{-4}	5.10^{-3}	0.15

CARBON DATING

Radiocarbon is produced in the atmosphere due to the action of cosmic ray slow neutrons. Solar cycle modulation of the low energy cosmic rays causes an anticorrelation of the atmospheric ^{14}C activity with sunspot number with a mean amplitude of ~0.5%. In the long term, modulation of cosmic rays by a varying geomagnetic field may be important.[4]

PRACTICAL USES OF COSMIC RAYS

There are few direct practical uses of cosmic rays. Their attenuation in water and snow has however enabled automatic monitors of water and snow depth to be constructed, and a search for hidden cavities in pyramids has been carried out using a muon 'telescope'.

OTHER EFFECTS

Stellar X-rays have been observed to affect the transmission times of radio signals between distant stations by altering the depth of the ionospheric reflecting layer. It has also been suggested that variations in ionization of the atmosphere due to solar modulation may have observable effects on climatic conditions.

REFERENCES

1. **Allkofer, O. C.,** *Introduction to Cosmic Radiation,* Verlag Karl Thiemig, Munchen, Germany, 1975.
2. **Allkofer, O. C.,** *J. Phys.,* Sect. G, 1, L51, 1975.
3. **Allkofer, O. C. and Heinrich, W.,** *Health Phys.,* 27, 543, 1974.
4. **Burchuladze, A. A., Pagava, S. V., Povinec, P., Togonidze, G. I., and Usacev, S.,** Proc. 16th Int. Cosmic Ray Conf. Kyoto, 3, 201, Univ. of Tokyo, 3, 201, 1979.
5. **Cassiday, G. L., et al.,** Proc. 15th Int Cosmic Ray Conf., Plovdiv, Bulgarian Academy of Sciences, 8, 258, 1977.
6. **Clay, R. W. and Gregory, A. G.,** *J. Phys.,* Sect. A, 10, 135, 1977.
7. **Greisen, K.,** *Physical Rev.,* 63, 323, 1943.
8. **Hayakawa, S.,** *Cosmic Ray Physics,* Wiley-Interscience, New York, 1969.
9. **Jelley, J. V.,** *Prog. in Elementary Particle and Cosmic Ray Physics,* 9, 41, 1967.
10. **Juliusson, E.,** Proc. 14th Int. Cos Ray Conf. Munich, 8, 2689, Max-Planck Institute for Extraterrestriche Physik, Munchen, Germany, 1975.

11. Kiraly, P., Szabelski, J., Wdowczyk, J., and Wolfendale, A. W., *Nature,* 293, 120, 1981.
12. Linsley, J., *Origin of Cosmic Rays,* I.A.U. Symposium 94, 53, D Reidel Publishing, Dordrecht, Holland, 1981.
13. Meyer, P., *Origin of Cosmic Rays,* I.A.U. Symposium 94, 7, D. Reidel Publishing, Dordrecht, Holland, 1981.
14. Tan, L. C., and Ng, L. K., *J. Phys.,* Sect. G, 7, 1135, 1981.
15. Wilson, J. G., *Cosmic Rays,* Wykeham Publishing, London, 1976.
16. Wolfendale, A. W., *Pramana,* 12, 631, 1979.
17. Ziegler, J. F., *IEEE Trans. Electron Devices,* ED-28, 560, 1981.

Section 12
Properties of Solids

TECHNIQUES FOR MATERIALS CHARACTERIZATION

EXPERIMENTAL TECHNIQUES USED TO DETERMINE THE COMPOSITION, STRUCTURE, AND ENERGY STATES OF SOLIDS AND LIQUIDS

H.P.R.Frederikse

The many experimental methods, originally designed to study the chemical and physical behavior of solids and liquids, have grown into a new field known as Materials Characterization (or Materials Analysis). During the past 30 years a host of techniques aimed at the study of surfaces and thin films has been added to the many tools for the analysis of bulk samples. The field has benefitted particularly from the development of computers and microprocessors, which have vastly increased the speed and accuracy of the measuring devices and the recording of their output. Materials characterization was and is a very important tool in the search for new physical and chemical phenomena. It plays an essential role in new applications of solids and liquids in industry, communications, and medicine. Many of its techniques are used in quality control, in safety regulations, and in the fight against pollution.

In most Materials Characterization experiments the sample is subjected to some kind of radiation: electromagnetic, acoustic, thermal, or particles (electrons, ions, neutrons, etc.). The surface analysis techniques usually require a high vacuum. As a result of interactions between the solid (or liquid) and the incoming radiation a beam of a similar (or a different) nature will emerge from the sample. Measurement of the physical and/or chemical attributes of this emerging radiation will yield qualitative, and often quantitative, information about the composition and the properties of the material being probed.

The modern tendency of describing practically everything in this world by a combination of a few letters (acronyms) has also penetrated the field of Materials Characterization. The table below gives the meaning of the acronym for every technique listed, the form and size of the required sample (bulk, surface, film, liquid, powder, etc.), the nature of the incoming and of the emerging radiation, the depth and the lateral spatial resolution that can be probed, and the information obtained from the experiment. The last column lists one or two major references to the technique described.

Technique	Sample	In	Out	Depth	Lateral resolution	Information obtained	Ref.
OPTICAL AND MASS SPECTROSCOPIES FOR CHEMICAL ANALYSIS							
1. AAS Atomic Absorption Spectroscopy	Atomize (flame, electro, thermal, etc.)	Light e.g., glow discharge	Absorption spectrum	—	—	Concentration of atomic species (quantitative, using standards)	1,2
2. ICP-AES Induct. Coupled Plasma — Atomic Emission Spectroscopy	Atomize (flame, electro, thermal, ICP, etc.)	—	Emission spectrum	—	—	Concentration of atomic species (quantitative, using standards)	3
3. Dynamic SIMS Dynamic Secondary Ion Mass Spectroscopy	Surface	Ion beam (1–20 keV)	Secondary ions; analysis with mass spectrometer	2 nm–1 μm (or deeper: ion milling)	0.50 nm	Elemental and isotopic analysis; depth profile (all elements); detection limits: ppb-ppm	4
4. Static SIMS Static Secondary Ion Mass Spectroscopy	Surface	Ion beam (0.5–20 keV)	Secondary ions, analysis with mass spectrometer	0.1–0.5 nm	10 μm	Elemental analysis of surface layers; molecular analysis; detection limits: ppb-ppm	4
5. SNMS Sputtered Neutral Mass Spectroscopy	Surface, bulk	Plasma discharge; noble gases: 0.5–20 keV	Sputtered atoms ionized by atoms or electrons; then mass analyzed	0.1–0.5 nm (or deeper: ion milling)	1 cm	Elemental analysis $Z \geq 3$; depth profile; detection limit: ppm	4,6
6. SALI Surface Analysis by Laser Ionization	Surface	e-beam, ion-beam, or laser for sputtering	Sputtered atoms ionized by laser; then mass analyzed	0.1–0.5 nm up to 3 μm in milling mode	60 nm	Surface analysis; depth profiling	7
7. LIMS Laser Ionization Mass Spectroscopy	Surface, bulk	u.v. laser (ns pulses)	Ionized species; analyzed with mass spectrometer	50–150 nm	5 μm–1 mm	Elemental (micro)analysis; detection limits: 1–100 ppm	8
8. SSMS Spark Source Mass Spectroscopy	Sample in the form of two electrodes	High voltage R.F. spark produces ions	Ions — analyzed in mass spectrometer	1–5 μm	—	Survey of trace elements; detection limit: 0.01–0.05 ppm	9
9. GDMS Glow Discharge Mass Spectroscopy	Sample forms the cathode for a D.C. glow discharge	Sputtered atoms ionized in plasma	Ions — analyzed in mass spectrometer	0.1–100 μm	3–4 mm	(Bulk) trace element analysis; detection limit: sub-ppb	9,10
10. ICPMS Induct. Coupled Plasma Mass Spectroscopy	Liquid-dissolved sample carried by gas stream into R.F. induction coil	Ions produced in argon plasma	Ions — analyzed in quadrupole mass spectrometer	—	—	High sensitivity analysis of trace elements	11

TECHNIQUES FOR MATERIALS CHARACTERIZATION (continued)

Technique	Sample	In	Out	Depth	Lateral resolution	Information obtained	Ref.
			PHOTONS — ABSORPTION, REFLECTION AND ELECTRON EMISSION				
11. IRS Infrared Spectroscopy	Thin crystal, glass, liquid	I.R. light (W-filament, globar, Hg-arc)	I.R. spectrum	—	—	Electronic transitions (mainly in semiconductors and superconductors); vibrational modes (in crystals and molecules)	12,13, 14
12. FTIR Fourier Transform I.R. Spectroscopy	Solid, liquid; transmission or reflection	White light (all frequencies)	Fourier Transform of spectrum (interferometer)	—	—	Spectra obtained at higher speed and resolution	15
13. ATR Attenuated Total Reflection	Surface or thin crystal	—	—	μm's	—	Atomic or molecular spectra of surfaces and films	16
14. (μ)-RS (Micro-) Raman Spectroscopy	Solid, liquid (1 μm–1 cm)	Laser beam, e.g., Ar-line, YAG-line	Raman spectra	0.5 μm	0.5 μm	Molecular and crystal vibrations	12,14, 17
15. CARS Coherent Anti-Stokes Raman Spectroscopy	Solid, liquid (50 μm–3 cm)	Pump beam (ω_0)+ probe beam (ω_s)	Anti-Stokes spectrum	—	—	High resolution Raman spectra	14
16. Ellipsometry	Transparent films, crystals, adsorbed layers	Polarized light	Change in polarization	0.05 nm–5 μm	25 μm (or sample thickness)	Refractive index *and* absorption	18,19
17. UPS Ultraviolet Photo-electron Spectroscopy	Surfaces, adsorbed layers	u.v. light, 10–100 eV; 200 eV (synchrotron)	Electrons	0.2–10 nm	0.1–10 nm	Energies of electronic states of surfaces and free molecules	20,21
18. PSD Photon Stimulated Desorption	Surfaces with adsorbed species	Far u.v. light E > 10 eV	Ions — analyzed with mass spectrometer	0.1–2 nm	—	Structure and desorption kinetics of adsorbed atoms and molecules	22
			X-RAYS				
19. XRD X-Ray Diffraction	Single crystals, powders films	X-rays: λ = 0.05–0.2 nm (6–17 keV)	Diffracted X-ray beam	1–1000 μm	0.1–10 mm	Identification of crystallographic structures; all elements (low Z difficult)	23,24
20. XRF/EDS X-Ray Fluorescence/Energy Dispersive Spectroscopy	Thin films, single layer	Prim. X-ray beam λ = 0.02–0.1 nm 12–80 keV	Fluorescent X-rays	1–100 μm	10 mm	Elemental analysis; all elements except H, He, Li — (EDS also used in XRD, SEM, TEM and EPMA)	25,26
21. EXAFS Extended X-Ray Absorption Fine Structure	Films, foils	High intensity X-rays (synchrotron)	Spectrum near absorption edge	nm–μm	—	Local atomic structure: order/disorder in vicinity of absorbing atom	27
22. XPS/ESCA X-Ray Photo-electron Spectroscopy/Electron Spect. for Chemical Analysis	Surfaces, thin films (≈20 atomic layers)	Soft X-rays (1–20 keV)	Core electrons; valence electrons	0.5–10 nm	5 nm–50 μm	(Quantitative) identification of all elements in surface layer or film	28,29
			ELECTRONS				
23. CL Cathode Luminescence	Insulators, semiconductors	Electrons 5–50 keV	Photons 0.1–5 eV	1 nm–2 μn	1 or 2 μm	Energy levels of impurities and point defects	30
24. APS Appearance Potential Spectroscopy	Surface (≈20 atomic layers)	Electrons (energy scan) 50–2000 eV	X-rays to pinpoint electron energy threshold	—	—	Identification of surface species	21, see also C

TECHNIQUES FOR MATERIALS CHARACTERIZATION (continued)

Technique	Sample	In	Out	Depth	Lateral resolution	Information obtained	Ref.
25. AES Auger Electron Spectroscopy	Thin films, surfaces	Electrons 3–10 keV	Auger electrons 20–2000 eV	0.3–3 nm	≈30 nm	Elemental composition of surface (except H, He); detection limit 0.1–1%	28,29
26. EELS Electron Energy Loss Spectroscopy	Very thin samples (<200 nm)	Electrons (100–400 keV)	(Retarded) electrons; minus 1–1000 eV	<200 nm	1–100 nm	Local elemental concentration; electronic structure, chem. bonding; interatomic distances	31
27. EXELFS Extended Electron Energy Loss Fine Structure	Thin films	Electrons (100–400 keV)	Electrons energies 0–30 eV above edge	<200 nm	1–100 nm	Density of states of valence electrons (above Fermi level)	27,32
28. ESD Electron Stimulated Desorption	Adsorbed species	Electrons E > 10 eV	Ions — analyzed with mass spectrometer	—	—	Structure and desorption properties of adsorbed atoms and molecules	22
29. ESDIAD ESD-Ion Angular Distribution	(See ESD)	(See ESD)	Directional dependence of emitted ions	—	—	Geometries of adsorbed species (atoms or molecules)	22
30. EPMA Electron Probe (X-Ray) Micro Analysis	Solid conductors and insulators <1 cm thick	Electrons 5–30 keV	Characteristic X-ray 0.1–15 keV	100 nm–5 μm	1 μm	Elemental analysis, $Z \leq 4$, major, minor and trace amounts	33,34
31. LEED Low Energy Electron Diffraction	Surface	Mono-energetic electron beam 10–1000 eV	Diffracted electrons	0.4–2 nm	<5 μm	Crystallographic structure of surface; resolution: 0.01 nm	35
32. RHEED Reflection High Energy Electron Diffraction	Surface	Electron beam at grazing angle 5–50 keV	Reflected electrons	0.2–10 nm	<5 μm	Surface symmetry	36,37
33. SEM Scanning Electron Microscopy	Bulk, films (conducting)	High energy electrons usually ~30 keV	Secondary and backscattered electrons	1 nm–5 μm	1–20 nm	Surface image, defect structure; resolution 5–15 nm; magnification 300,000×	33,34
34. (S)TEM (Scanning) Transmission Electron Microscopy	Thin specimen — <200 nm	High energy electrons typically 300 keV	Transmitted and diffracted electrons	(Sample thickness)	2–20 nm	(Defect) structure of cryst. solids; microchemistry; high resol.: 0.2 nm	33
35. FEM Field Emission Microscopy	Metals, alloys (sharp point)	—	Electron emission (with appl. electric field — 50 kV)	≈0.5 nm	10–100 nm	Surface image, crystallographic structure	34
36. STM Scanning Tunneling Microscopy	Polished or cleaved surface (conducting)	Tunneling current controls distance between sample and very sharp tip		1–5 nm	2–10 nm	Atomic-scale relief map of surface; resolution: vert. 0.002 nm, hor. 0.2 nm	39
37. SPM Scanned Probe Microscopy	Very flat surface	Any field: e.g. mechan. vibration recorded with laser probe; same with magnetic, electric or thermal field		1–100 nm	1–100 nm	Surface-magnetic field, surface-thermal conductivity, etc.	39a
38. AFM Atomic Force Microscopy	Very flat surface	Similar to STM; force measured with cantilever spring		0.5–5 nm	0.2–130 nm	Surface topography with atomic resolution; interatomic force	40
			IONS AND NEUTRONS				
39. ISS (or LEIS) Ion Scattering Spectroscopy (Low Energy Ion Scattering)	Surface	Ion beam He^+ or Ne^+ <3 keV	Sputtered ions (energy analysis)	0.1–0.5 nm	1–100 μm	Elemental analysis (better for low Z) detection limits: 0.01–1%	41
40. FIM Field Ion Microscopy	Surface: metals, alloys; very sharp tip	(He gas above sample)	He ions + high electric field produce image	≈0.1 nm	0.1–2 nm	Atomic structure of surface	34,42
41. RBS Rutherford Back Scattering	Solids, thin films	Mono-energetic ions (H^+ or He^{++}) 0.5–3 MeV	Backscattered ions	10 nm–1 μm	1 mm	Element identification (Li to U) detection limit: 0.01–1%	46

TECHNIQUES FOR MATERIALS CHARACTERIZATION (continued)

Technique	Sample	In	Out	Depth	Lateral resolution	Information obtained	Ref.
42. NRA Nuclear Reaction Analysis	Solids, thin films	Mono-energetic ions (Li, Be, B, etc.) 200 keV–6 MeV	Protons, deuterons ^3He, α-particles, γ-rays	0.1–5 μm	10 μm–10 mm	Element identification (all) detection limit: 10^{-12}–10^{-2}	47
43. PIXE Particle Induced X-ray Emission	Thin films, surface layers	High energy ions (H^+ or He^{++})	Characteristic X-rays	<10 μm	1 μm–2 mm	Trace impurities: $Z>3$ detection limit: 0.1–100 ppm (depending on sample thickness)	48
44. INS Ion Neutralization Spectroscopy	Surface	He-ions (≈5 eV)	Electrons	—	—	Energies of valence electrons	49
45. NAA Neutron Activation Analysis	Bulk, >0.5 g	Thermal neutrons	Characteristic γ-rays, (≈1 MeV)	Bulk	—	Trace concentrations (of isotopes) of elements: trans. metals, Pt-group; detection limit: 10^8–10^{14} atoms/cm^3	43
46. N(P)D Neutron (Powder) Diffraction	Crystalline solids	Thermal neutrons E ≈0.0025 eV	Diffracted neutrons	Bulk	—	Crystallographic structure porosity, particle size	44
47. SANS Small Angle Neutron Scattering	Inhomogeneous solids; powders; porous samples	Thermal neutrons $2\theta = 10^{-2}$–10^{-4}	Scattered neutrons	1–25 mm	—	Average size of inhomogeneities; range: 1 nm–1 mm	45
			ACOUSTIC				
48. SLAM Scanning Laser Acoustic Microscopy	Bulk, film	Acoustic wave produced by laser 1 MHz–1 GHz	Reflected acoustic wave	μm–cm	0.1–20 mm	Defect structure; thickness measurement	50
			THERMAL				
49. DTA Differential Thermal Analysis	Specimen and reference sample	Uniform heating	Temperature difference	Bulk	—	Phase transitions, crystallization	51
50. DSC Differential Scanning Calorimetry	Specimen and ref. sample	Controlled heating	Measure heat required for equal temperature	Bulk	—	Phase transitions, crystallization; activation energies	51
51. TGA Thermo Gravimetric Analysis	Bulk, 1–100 g	Controlled heating	Weight as function of temperature (and time)	Bulk	—	Decomposition, non-stoichiometry, kinetics of reaction	52
			RESONANCE				
52. EPR (ESR) Electron Paramagnetic (Spin) Resonance	Paramagnetic solids or liquids	Microwave radiation in magnetic field 3–300 GHz; 1–100 kG	Microwave absorption (at resonance)	Bulk	—	Local environment of paramagnetic ion; concentration of paramagnetic, species; detection limit: 10^{11} spins/cm^3	53,54
53. ECR Electron Cyclotron Resonance	Semiconductors, metals; free electrons (low temperature)	Microwave radiation in magnetic field 10–30 GHz; 5–10 kG	Microwave absorption (at resonance)	Bulk	—	Electronic energy bands, effective masses	55
54. Mössbauer Effect	Source and absorber	Mono-energetic γ-rays: 5–100 keV	Mössbauer spectrum (Doppler shifted (lines)	50 m	1 cm	Interaction between nucleus and its environment (local electric, magnetic fields; bonds; valency; diffusion, etc.)	56
55. NMR (MRI) Nuclear Magnetic Resonance (Magnetic Resonance Imaging)	Solids, liquids	R.F. radiation + magnetic field; e.g. for protons: 60 MHz, 14 kG	R.F. absorption	<1 cm	1 cm	Quant. analysis; local magnetic environment; diffusion; imaging	58
56. ENDOR Electron Nuclear Double Resonance	Solids, liquids	R.F. + microwave radiation in magn. field.	Microwave absorption	—	—	Hyperfine interaction → local atomic structure	54

Technique	Sample	In	Out	Depth	Lateral resolution	Information obtained	Ref.
57. NQR Nuclear Quadrupole Resonance	Solids	R.F. radiation 0.5–1000 MHz	R.F. absorption	—	—	Asymmetry of the charge distribution at the nucleus	55,59
OTHER							
58. BET Brunauer-Emmett-Teller	(Large) surface area 1–20 m^2/g	Adsorbed gas (e.g., N_2 at low temp.) as function of pressure (monolayer coverage)		—	—	Surface area measurement	60

REFERENCES

General References

A. Wachtman, J. B., *Characterization of Materials*, Butterworth-Heinemann, Boston, 1993.
B. Brundle, C. R., Evans, C. A., and Wilson, S., Eds., *Encyclopedia of Materials*, Butterworth-Heinemann, Boston, 1992.
C. Woodruff, D. P. and Delchar, T. A., *Modern Techniques of Surface Science*, Cambridge University Press, Cambridge, 1986.
D. *Metals Handbook,* 9th Edition, Vol. 10, Materials Characterization, Whan, R. E., Coordinator, American Society for Metals, Metals Park, OH, 1986.

Specific References

1. Slavin, M., *Atomic Absorption Spectroscopy*, 2nd Edition, John Wiley & Sons, New York, 1978.
2. Schrenk, W. G., *Analytical Atomic Spectroscopy*, Plenum Press, New York, 1975.
3. Dean, J. A. and Rains, T. E., *Flame Emission and Atomic Absorption Spectroscopy*, Vols. 1—3, Marcel Dekker, New York, 1969.
4. Benninghoven, A., Rudenauer, F. G., and Werner, H. W., *Secondary Ion Mass Spectroscopy,* John Wiley & Sons, New York, 1987.
5. Bird, J. R. and Williams, J. S., Eds., in *Ion Beams for Materials Analysis*, Academic Press, New York, 1989, pp. 515—537.
6. Smith, G. C., *Quantitative Surface Analysis for Materials Science*, The Institute of Metals, London, 1991.
7. Becker, E. H., in *Ion Spectroscopies for Surface Analysis*, Czanderna, A. W. and Hercules, D. M., Eds., Plenum Press, New York, 1991, p. 273.
8. Simons, D. S., *Int. J. Mass Spectrometry and Ion Processes*. 55, 15, 1983.
9. White, F. A. and Wood, G. M., *Mass Spectrometry: Applications in Science and Engineering*, John Wiley & Sons, New York, 1986.
10. Harrison, W. W. and Bentz, B. L., *Prog. Anal. Spectrometry,* 11, 53, 1988.
11. Bowmans, P. W. J. M., *Inductively Coupled Plasma Emission Spectroscopy*, Parts I and II, John Wiley & Sons, New York, 1987.
12. Brame, Jr., E. G. and Grasselli, J., *Infrared and Raman Spectroscopy*, Practical Spectroscopy Series, Vol. I, Marcel Dekker, New York, 1976.
13. Hollas, J. M., *Modern Spectroscopy* John Wiley & Sons, New York, 1987.
14. Turrell, G., *Infrared and Raman Spectroscopy of Crystals*, Academic Press, New York and London, 1972.
15. Griffith, P. R. and Haseth, J. A., *Fourier Transform Infrared Spectroscopy*, John Wiley & Sons, New York, 1986.
16. Barnowski, M. K., *Fundamentals of Optical Fiber Communications*, Academic Press, New York, 1976.
17. Long, D. A., *Raman Spectroscopy*, McGraw-Hill, New York, 1977.
18. Azzam, R. M. A., *Ellipsometry and Polarized Light*, Elsevier-North Holland, Amsterdam, 1977.
19. Hecht, E., *Optics,* 2nd Edition, Addison-Wesley, Reading MA, 1987.
20. Brundle, C. R., in *Molecular Spectroscopy*, West, A. R., Ed., Heyden, London, 1976.
21. Park, R. L., in *Experimental Methods in Catalytic Research*, Vol. III, Anderson, R. B. and Dawson, P. T., Academic Press, New York, 1976, pp. 1—39.
22. Madey, T. E. and Stockbauer, R., in *Solid State Physics: Surfaces,* Vol. 22 of Methods of Experimental Physics, Park, R.L. and Lagally, M. G., Eds., Academic Press, New York, 1985.
23. Cullity, B. D., *Elements of X-Ray Diffraction*, 2nd Edition, Addison-Wesley, Reading, MA, 1978.
24. Schwartz, L. H. and Cohen, J. B., *Diffraction from Materials*, Springer Verlag, Berlin, 1987.

TECHNIQUES FOR MATERIALS CHARACTERIZATION (continued)

25. deBoer, D. K. G., in *Advances in X-Ray Analysis*, Vol. 34, Barrett, C. S. et. al., Eds., Plenum Press, New York, 1991.
26. Birks, L. S., *X-Ray Spectrochemical Analysis*, 2nd Edition, John Wiley & Sons, New York, 1969.
27. Bonnelle, C. and Mande, C., *Advances in X-Ray Spectroscopy*, Pergamon Press, Oxford, 1982.
28. *Practical Surface Analysis by Auger and X-Ray Photo-Electric Spectroscopy*, Briggs, D. and Seah, M. P., Eds., John Wiley & Sons, New York, 1983.
29. Powell, C. J. and Seah, M. P., *J. Vac. Sci. Technol. A*, Vol. 8, 735, 1990.
30. Yacobi, G. G. and Holt, D. B., *Cathodeluminescence Microscopy of Inorganic Solids*, Plenum Press, New York, 1990.
31. Egerton, R. F., *Electron Energy Loss Spectroscopy in the Electron Microscope*, Plenum Press, New York, 1986.
32. Disko, M. M., Krivanek, O. L., and Rez, P., *Phys. Rev.*, B25, 4252, 1982.
33. Goldstein, J. I., et. al., *Scanning Electron Microscopy and X-Ray Microanalysis*, 2nd Edition, Plenum Press, New York, 1986.
34. Murr, L. E., *Electron and Ion Microscopy and Microanalysis,* Marcel Dekker, New York, 1982.
35. Armstrong, R. A., in *Experimental Methods in Catalytic Research*, Vol. III, Anderson, R. B., and Dawson, P. T.,Eds., Academic Press, New York, 1976.
36. Dobson, P. J. et. al., *Vacuum*, 33, 593, 1983.
37. Rymer, T. B., *Electron Diffraction*, Methuen, London, 1970.
38. Reimer, L., *Transmission Election Microscopy*, Springer-Verlag, Berlin, 1984.
39. *Scanning Tunneling Microscopy and Related Methods*, Behm, R. J., Garcia, N., and Rohrer, H., Kluwer, Eds., Academic Publishers, Norwell, MA, 1990.

39a. Wikramasinghe, H.K., *Scientific American*, Vol. 261, No. 4, pp. 98—105, Oct. 1989.

40. Rugar, D. and Hansma, P., *Physics Today,* 43(10), pp. 23—30, 1990.
41. Feldman, C. C. and Mayer, J. W., *Fundamentals of Surface and Thin Film Analysis*, North-Holland, Amsterdam, 1986.
42. Muller, E. W. and Tsong, T. T., *Field Ion Microscopy*, Elsevier, Amsterdam, 1969.
43. Amiel, S., *Nondestructive Activation Analysis*, Elsevier, Amsterdam, 1981.
44. Bacon, G. E., *Neutron Diffraction*, 3rd Edition, Clarendon Press, Oxford, 1975.
45. Neutron Scattering, Part A., in *Methods of Experimental Physics*, Vol. 23, Skold, K. and Price, D. L., Eds., Academic Press, New York, 1986.
46. Chu, W. K., Mayer, J. W., and Nicolet, M. A., *Backscattering Spectroscopy*, Academic Press, New York, 1987.
47. Rickey, F. A., in *High Energy and Heavy Ion Beams in Materials Analysis*, Tesmer, J. R., et. al., Eds., MRS, 1990, pp. 3—26.
48. Johansson, S. A. E. and Campbell, J. L., *PIXE: A Novel Technique for Elemental Analysis*, John Wiley & Sons, New York, 1988.
49. Hagstrum, H. D., in *Inelastic Ion-Surface Collisions*, Tolk, N. H. et. al., Eds., Academic Press, New York, 1977, pp. 1—46.
50. Nikoonahad, M., in *Research Techniques in Nondestructive Testing*, Vol. VI, Sharpe, R.S., Ed., Academic Press, New York, 1984, pp. 217—257.
51. Gallagher, P. K., *Characterization of Materials by Thermoanalytical Techniques*, MRS - Bulletin, Vol. 13, No. 7, pp. 23—27, 1988.
52. Earnest, C. M., *Compositional Analysis by Thermogravimetry*, ASTM Special Technical Publication 997, 1988.
53. Poole, C. P., *Electron Spin Resonance — A Comprehensive Treatise on Experimental Techniques*, 2nd Edition, John Wiley & Sons, New York, 1983.
54. Atherton, N. M., *Principles of Electron Spin Resonance,* Ellis Horwood Ltd., Chichester, U.K., 1993.
55. Kittel, C., *Introduction to Solid State Physics*, 6th Edition, John Wiley & Sons, New York, 1986, p. 196.
56. Gibb, T. C., *Principles of Mössbauer Spectroscopy*, Chapman & Hall, London, 1976.
57. Slichter, C. P., *Principles of Magnetic Resonance*, 3rd Edition, Springer-Verlag, Berlin, 1990.
58. *NMR Spectroscopy Techniques*, Dybrowski, C. and Lichter, R. L., Eds., Marcel Dekker, New York, 1987.
59. Das, T. P. and Hahn, E. L., *Nuclear Quadrupole Resonance Spectroscopy*, Academic Press, New York, 1958.
60. Somorjai, G. A., *Principles of Surface Chemistry*, Prentice-Hall, Englewood Cliffs, NJ, 1972, p. 216.

SYMMETRY OF CRYSTALS

L. I. Berger

The ability of a body to coincide with itself in its different positions regarding a coordinate system is called its symmetry. This property reveals itself in iteration of the parts of the body in space. The iteration may be done by reflection in mirror planes, rotation about certain axes, inversions and translations. These actions are called the symmetry operations. The planes, axes, points, etc., are known as symmetry elements. Essentially, mirror reflection is the only truly primitive symmetry operation. All other operations may be done by a sequence of reflections in certain mirror planes. Hence, the mirror plane is the only true basic symmetry element. But for clarity, it is convenient to use the other symmetry operations, and accordingly, the other aforementioned symmetry elements. The symmetry elements and operations are presented in Table 1.

The entire set of symmetry elements of a body is called its symmetry class. There are thirty-two symmetry classes that describe all crystals which have ever been noted in mineralogy or been synthesized (more than 150,000). The denominations and symbols of the symmetry classes are presented in Table 2.

There are several known approaches to classification of individual crystals in accordance with their symmetry and crystallochemistry. The particles which form a crystal are distributed in certain points in space. These points are separated by certain distances (translations) equal to each other in any chosen direction in the crystal. Crystal lattice is a diagram that describes the location of particles (individual or groups) in a crystal. The lattice parameters are three non-coplanar translations that form the crystal lattice. Three basic translations form the unit cell of a crystal. August Bravais (1848) has shown that all possible crystal lattice structures belong to one or another of fourteen lattice types (Bravais lattices). The Bravais lattices, both primitive and non-primitive, are the contents of Table 3.

Among the three-dimensional figures, there is a group of polyhedrons that are called regular, which have all faces of the same shape and all edges of the same size (regular polygons). It has been shown that there are only five regular polyhedrons. Because of their importance in crystallography and solid state physics, a brief description of these polyhedrons is included in Table 4.

The systematic description of crystal structures is presented primarily in the well known *Structurbericht*. The classification of crystals by the Structurbericht does not reflect their crystal class, the Bravais lattice, but is based on the crystallochemical type. This makes it inconvenient to use the Structurbericht categories for comparison of some individual crystals. Thus, there have been several attempts to provide a more convenient classification of crystals. Table 5 presents a compilation of different classifications which allows the reader to correlate the Structurbericht type with the international and Schoenflies point and space groups and with Pearson's symbols, based on the Bravais lattice and chemical composition of the class prototype. The information included in Table 5 has been chosen as an introduction to a more detailed crystallophysical and crystallochemical description of solids.

TABLE 1
Symmetry Operations and Elements

	Symmetry element			Presentation on the stereographic projection	
		Symbol			
Symmetry operation	**Name**	**International (Hermann-Mauguin)**	**Schoenflies**	**Parallel**	**Perpendicular**
Reflection in a plane	Plane	m	C_s		
Rotation by angle $\alpha = 360°/n$ about an axis	Axis	n = 1, 2, 3, 4 or 6	C_n		
		n = 2	C_2		
		n = 3	C_3		
		n = 4	C_4		
		n = 6	C_6		
Rotation about an axis and inversion in a symmetry center lying on the axis	Inversion (improper) axis	$\bar{n} = \bar{3}, \bar{4}, \bar{6}$	C_{ni}		
		$\bar{n} = \bar{3}$	C_{3i}		
		$\bar{n} = \bar{4}$	C_{4i}		

TABLE 1
Symmetry Operations and Elements (continued)

Symmetry operation	Symmetry element: Name	Symbol: International (Hermann-Mauguin)	Symbol: Schoenflies	Presentation on the stereographic projection: Parallel	Presentation on the stereographic projection: Perpendicular
		$\bar{n} = \bar{6}$	C_{6i}		
Inversion in a point	Center	$\bar{1}$	C_i	● ○	×
Parallel translation	Translation vector $\vec{a}, \vec{b}, \vec{c}$				
Reflection in a plane and translation parallel to the plane	Glide–plane	a, b, c, n, d			
Rotation about an axis and translation parallel to the axis	Screw axis	n_m (m = 1, 2, .., n – 1)			
Rotation about an axis and reflection in a plane perpendicular to the axis	Rotatory-reflection axis	$\tilde{n}$ $\tilde{n} = \tilde{1}, \tilde{2}, \tilde{3}, \tilde{4}, \tilde{6}$	S_n		

TABLE 2
The Thirty-Two Symmetry Classes

Crystal symbol	Primitive Int	Primitive Sch	Central Int	Central Sch	Planal Int	Planal Sch	Axial Int	Axial Sch	Plane-axial Int	Plane-axial Sch	Inversion primitive Int	Inversion primitive Sch	Inversion-planal Int	Inversion-planal Sch
Triclinic	1	C_1	$\bar{1}$	C_i										
Monoclinic					m	C_s	2	C_2	2/m	C_{2h}				
Ortho-rhombic					mm2	C_{2v}	222	D_2	mmm	D_{2h}				
Trigonal	3	C_3	$\bar{3}$	C_{3i}	3m	C_{3v}	32	D_3	$\bar{3}$m	C_{3d}				
Tetragonal	4	C_4	4/m	C_{4h}	4mm	C_{4v}	422	D_4	4/mmm	D_{4h}	$\bar{4}$	S_4	$\bar{4}$2m	D_{2d}
Hexagonal	6	C_6	6/m	C_{6h}	6mm	C_{6v}	622	D_6	6/mmm	D_{6h}	$\bar{6}$	C_{3h}	$\bar{6}$m2	D_{3h}
Cubic	23	T	m3	T_h	$\bar{4}$3m	T_d	432	O	m3m	O_h				

Class name[a] and its symbol — International (Int) and Schoenflies (Sch)

[a] Per Fedorov Institute of Crystallography, USSR Academy of Sciences, nomenclature.

SYMMETRY OF CRYSTALS (continued)

TABLE 3
The Fourteen Possible Space Lattices (Bravais Lattices)

Crystal system	Metric category of the system	No. of different lattices in the system	Lattice type[a] (marked by +)					No. of identi-points per unit cell	Characteristic parameters (marked by +)						Description of characteristic parameters $a \subset X, b \subset Y, c \subset Z$ $\alpha \equiv (b,c), \beta \equiv (a,c), \gamma \equiv (a,b)$	Symmetry of the lattice	
			P	C	I	F	R		a	b	c	α	β	γ		Int	Sch
Triclinic	Trimetric	1	+					1	+	+	+	+	+	+	$a \neq b \neq c, \alpha \neq \beta \neq \gamma$	1	C
Monoclinic	Trimetric	2	+	+				1 or 2	+	+	+		+		$a \neq b \neq c, \alpha = \gamma = 90° \neq \beta$	2/m	C_{2h}
Orthorhombic	Trimetric	4	+	+	+	+		1, 2 or 4	+	+	+				$a \neq b \neq c, \alpha = \beta = \gamma = 90°$	mmm	D_{2h}
Trigonal (rhombohedral)	Dimetric	1					+	1	+			+			$a = b = c, 120° > \alpha = \beta = \gamma \neq 90°$	3m	D_{3d}
Tetragonal	Dimetric	2	+		+			1 or 2	+		+				$a = b \neq c, \alpha = \beta = \gamma = 90°$	4/mmm	D_{4h}
Hexagonal	Dimetric	1	+					1	+		+				$a = b \neq c, \alpha = \beta = 90°, \gamma = 120°$	6/mmm	D_{6h}
Isometric (cubic)	Monometric	3	+		+	+		1, 2 or 4	+						$a = b = c, \alpha = \beta = \gamma = 90°$	m3m	O_h

[a] Designations of the space-lattice types: P — primitive, C — side-centered (base-centered), I — body-centered, F — face-centered, R — rhombohedral.

TABLE 4
The Five Possible Regular Polyhedrons

Polyhedron	Symmetry (Schoenflies)		Form of faces	Number of[a]		
	Class	Elements		Faces (F)	Edges (E)	Vertices (V)
Tetrahedron	T	$4C_3 3C_2$	Equilateral triangle	4	6	4
Cube (hexahedron)	O	$3C_4 4C_3 6C_2$	Square	6	12	8
Octahedron	O	$3C_4 4C_3 6C_2$	Equilateral triangle	8	12	6
Pentagonal dodecahedron	J	$6C_5 10C_3 15C_2$	Regular pentagon	12	30	20
Icosahedron	J	$6C_5 10C_3 15C_2$	Equilateral triangle	20	30	12

[a] Per formula by Leonhard Euler: $F + V - E = 2$

TABLE 5
Classification of Crystals

Strukturbericht symbol 1	Structure name 2	Symmetry group		Pearson symbol[a] 5	Standard ASTM E157-82a symbol[b] 6
		International 3	Schoenflies 4		
A1	Cu	Fm3m	O^4_h	cF4	F
A2	W	Im3m	O^9_h	cI2	B
A3	Mg	$P6_3/mmc$	D^4_{6h}	hP2	H
A4	C	Fd3m	O^7_h	cF8	F
A5	Sn	If_1/amd	D^{19}_{4h}	tI4	U
A6	In	I4/mmm	D^{17}_{4h}	tI2	U
A7	As	$R\bar{3}m$	D^5_{3d}	hR2	R
A8	Se	$P3_121$ or $P3_221$	D^4_3 (D^6_3)	hP3	H
A10	Hg	$R\bar{3}m$	D^5_{3d}	hR1	R
A11	Ga	Cmca	D^{18}_{2h}	oC8	Q
A12	α-Mn	$I\bar{4}3m$	T^3_d	cI58	B
A13	β-Mn	$P4_132$	O^7	cP20	C
A15	OW_3	Pm3n	O^3_h	cP8	C
A20	α-U	Cmcm	D^{17}_{2h}	oC4	Q
B1	ClNa	Fm3m	O^5_h	cF8	F
B2	ClCs	Pm3m	O^1_h	cP2	C
B3	SZn	$F\bar{4}3m$	T^2_d	cF8	F
B4	SZn	$P6_3mc$	C^4_{6v}	hP4	H
$B8_1$	AsNi	$P6_3/mmc$	D^4_{6h}	hP4	H
$B8_2$	$InNi_2$	$P6_3/mmc$	D^4_{6h}	hP6	H
B9	HgS	$P3_121$ or $P3_221$	D^4_3 or D^6_3	hP6	H
B10	OPb	P4/nmm	D^7_{4h}	tP4	T
B11	γ-CuTi	P4/nmm	D^7_{4h}	tP4	T
B13	NiS	$R\bar{3}m$	D^5_{3d}	hR6	R
B16	GeS	Pnma	D^{16}_{2h}	oP8	O
B17	PtS	$P4_2/mmc$	D^9_{4h}	tP4	T
B18	CuS	$P6_3/mmc$	D^4_{6h}	hP12	H
B19	AuCd	Pmma	D^5_{2h}	oP4	O
B20	FeSi	$P2_13$	T^4	cP8	C

TABLE 5
Classification of Crystals (continued)

Strukturbericht symbol 1	Structure name 2	Symmetry group: International 3	Symmetry group: Schoenflies 4	Pearson symbol[a] 5	Standard ASTM E157-82a symbol[b] 6
B27	BFe	Pnma	D^{16}_{2h}	oP8	O
B31	MnP	Pnma	D^{16}_{2h}	oP8	O
B32	NaTl	Fd3m	O^{7}_{h}	cF16	F
B34	Pds	$P4_2/m$	C^{2}_{4h}	tP16	T
B35	CoSn	P6/mmm	D^{1}_{6h}	hP6	H
B37	SeTl	I4/mcm	D^{18}_{4h}	tI16	U
B_e	CdSb	Pbca	D^{15}_{2h}	oP16	O
B_f (B33)	ξ-BCr	Cmcm	D^{17}_{2h}	oC8	Q
B_g	BMo	$I4_1/amd$	D^{19}_{4h}	tI4	U
B_h	CW	P6m2	D^{1}_{3h}	hP2	H
B_i	γ-CMo (AsTi)	$P6_3/mmc$	D^{4}_{6h}	hP8	H
C1	CaF_2	Fm3m	O^{5}_{h}	cF12	F
$C1_b$	AgAsMg	$F\bar{4}3m$	T^{2}_{d}	cF12	F
C2	FeS_2	Pa3	T^{6}_{h}	cP12	C
C3	Cu_2O	Pn3m	O^{4}_{h}	cP6	C
C4	O_2Ti	$P4_2/mnm$	D^{14}_{4h}	tP6	T
C6	CdI_2	P3m1	D^{3}_{3d}	hP3	H
C7	MoS_2	$P6_3/mmc$	D^{4}_{6h}	hP6	H
$C11_a$	C_2Ca	I4/mmm	D^{17}_{4h}	tI6	U
$C11_b$	$MoSi_2$	I4/mmm	D^{17}_{4h}	tI6	U
C12	$CaSi_2$	$R\bar{3}m$	D^{5}_{3d}	hR6	R
C14	$MgZn_2$	$P6_3/mmc$	D^{4}_{6h}	hP12	H
C15	Cu_2Mg	Fd3m	O^{7}_{h}	cF24	F
$C15_b$	$AuBe_5$	$F\bar{4}3m$ or F23	T^{2}_{d} or T^{2}	cF24	F
C16	Al_2Cu	I4/mcm	D^{18}_{4h}	tI12	U
C18	FeS_2	Pnnm	D^{12}_{2h}	oP6	O
C19	$CdCl_2$	$R\bar{3}m$	D^{5}_{3d}	hR3	R
C22	Fe_2P	$P2\bar{6}m$	D^{1}_{3h}	hP9	H
C23	Cl_2Pb	Pnma	D^{16}_{2h}	oP12	O
C32	AlB_2	P6/mmm	D^{1}_{6h}	hP3	H
C33	Bi_2STe_2	$R\bar{3}m$	D^{5}_{3d}	hR5	R
C34	$AuTe_2$	C2/m (P2/m)	C^{3}_{2h} (C^{1}_{2h})	mC6	N
C36	$MgNi_2$	$P6_3/mmc$	D^{4}_{6h}	hP24	H
C38	Cu_2Sb	P4/nmm	D^{7}_{4h}	tP6	T
C40	$CrSi_2$	$P6_222$	D^{4}_{6}	hP9	H
C42	SiS_2	Ibam	D^{26}_{2h}	oI12	P
C44	GeS_2	Fdd2	C^{19}_{2v}	oF72	S
C46	$AuTe_2$	Pma2	C^{4}_{2v}	oP24	O
C49	Si_2Zr	Cmcm	D^{17}_{2h}	oC12	Q
C54	Si_2Ti	Fddd	D^{24}_{2h}	oF24	S
C_c	Si_2Th	$I4_1/amd$	D^{19}_{4h}	tI12	U
C_e	$CoGe_2$	Aba2	C^{17}_{2v}	oC23	Q
DO_2	As_3Co	Im3	T^{5}_{h}	cI32	B
DO_3	BiF_3	Fm3m	O^{5}_{h}	cF16	F
DO_9	O_3Re	Pm3m	O^{1}_{h}	cP4	C
DO_{11}	CFe_3	Pnma	D^{16}_{2h}	oP16	O
DO_{18}	$AsNa_3$	$P6_3/mmc$	D^{4}_{6h}	hP8	H
DO_{19}	Ni_3Sn	$P6_3/mmc$	D^{4}_{6h}	hP8	H
DO_{20}	Al_3Ni	Pnma	D^{16}_{2h}	oP16	O

TABLE 5
Classification of Crystals (continued)

Strukturbericht symbol	Structure name	Symmetry group		Pearson symbol[a]	Standard ASTM E157-82a symbol[b]
		International	Schoenflies		
1	2	3	4	5	6
DO_{21}	Cu_3P	$P\bar{3}c1$	$D^4{}_{3d}$	hP24	H
DO_{22}	Cu_3P	I4/mmm	$D^{17}{}_{4h}$	tI8	U
DO_{23}	Al_3Zr	I4/mmm	$D^{17}{}_{4h}$	tI16	U
DO_{24}	Ni_3Ti	$P6_3/mmc$	$D^4{}_{6h}$	hP16	H
DO_c	SiU_3	I4/mcm	$D^{18}{}_{4h}$	tI16	U
DO_e	Ni_3P	$I\bar{4}$	$S^2{}_4$	tI32	U
$D1_3$	Al_4Ba	I4/mmm	$D^{17}{}_{4h}$	tI10	U
$D1_a$	$MoNi_4$	I4/m	$C^5{}_{4h}$	tI10	U
$D1_b$	Al_4U	Imma	$D^{28}{}_{2h}$	oI20	P
$D1_c$	$PtSn_4$	Aba2	$C^{17}{}_{2v}$	oC20	Q
$D1_e$	B_4Th	P4/mbm	$D^5{}_{4h}$	tP20	T
$D1_f$	BMn_4	Fddd	$D^{24}{}_{2h}$	oF40	S
$D2_1$	B_6Ca	Pm3m	$O^1{}_h$	cP7	C
$D2_3$	$NaZn_{13}$	Fm3m	$O^5{}_h$	cF112	F
$D2_b$	$Mn_{12}Th$	I4/mmm	$D^{17}{}_{4h}$	tI26	U
$D2_c$	MnU_6	I4/mcm	$D^{18}{}_{4h}$	tI28	U
$D2_d$	$CaCu_5$	P6/mmm	$D^1{}_{6h}$	hP6	H
$D2_f$	$B_{12}U$	Fm3m	$O^5{}_h$	cF52	F
$D2_h$	Al_6Mn	Cmcm	$D^{17}{}_{2h}$	oC28	Q
$D5_1$	α-Al_2O_3	$R\bar{3}c$	$D^6{}_{3d}$	hR10	R
$D5_2$	La_2O_3	$P\bar{3}m1$	$D^3{}_{3d}$	hP5	H
$D5_3$	Mn_2O_3	Ia3	$T^7{}_h$	cI80	B
$D5_8$	S_3Sb_2	Pnma	$D^{16}{}_{2h}$	oP20	O
$D5_9$	P_2Zn_3	$P4_2/mmc$	$D^9{}_{4h}$	tP40	T
$D5_{10}$	C_2C_3	Pnma	$D^{16}{}_{2h}$	oP20	O
$D5_{13}$	Al_3Ni_2	$P\bar{3}m1$	$D^3{}_{3d}$	hP5	H
$D5_a$	Si_2U_3	P4/mbm	$D^5{}_{4h}$	tP10	T
$D5_c$	C_3Pu_2	$I\bar{4}3d$	$T^6{}_d$	cI40	B
$D7_1$	Al_4C_3	$R\bar{3}m$	$D^5{}_{3d}$	hR7	R
$D7_3$	P_4Th_3	$I\bar{4}3d$	$T^6{}_d$	cI28	B
$D7_b$	B_4Ta_3	Immm	$D^{25}{}_{2h}$	oI14	P
$D8_1$	Fe_3Zn_{10}	Im3m	$O^9{}_h$	cI52	B
$D8_2$	Cu_5Zn_8	$I\bar{4}3m$	$T^3{}_d$	cI52	B
$D8_3$	Al_4Cu_9	$P\bar{4}3m$	$T^1{}_d$	cP52	C
$D8_4$	C_6Cr23	Fm3m	$O^5{}_h$	cF116	F
$D8_5$	Fe_7W_6	$R\bar{3}m$	$D^5{}_{3d}$	hR13	R
$D8_6$	$Cu_{15}Si_4$	$I\bar{4}3m$	$T^3{}_d$	cI76	B
$D8_8$	Mn_5Si_3	$P6_3/mcm$	$D^3{}_{6h}$	hP16	H
$D8_9$	Co_9S_8	Fm3m	$O^5{}_h$	cF68	F
$D8_{10}$	Al_8Cr_5	R3m	$C^5{}_{3v}$	hR26	R
$D8_{11}$	Al_5Co_2	$P6_3/mcm$	$D^3{}_{6h}$	hP28	H
$D8_a$	$Mn_{23}Th_6$	Fm3m	$O^5{}_h$	cF116	F
$D8_b$	σ-phase of Cr-Fe	$p4_2/mnm$	$D^{14}{}_{4h}$	tP30	T
$D8_e$	$(Al,Zn)_{49}Mg_{32}$	Im3	$T^5{}_h$	cI162	B
$D8_f$	Ge_7Ir_3	Im3m	$O^9{}_h$	cI40	B
$D8_h$	B_5W_2	$P6_3/mmc$	$D^4{}_{6h}$	hP14	H
$D8_i$	B_5Mo_2	$R\bar{3}m$	$D^5{}_{3d}$	hR7	R
$D8_l$	B_3Cr_5	I4/mcm	$D^{18}{}_{4h}$	tI32	U
$D8_m$	Si_3W_5	I4/mcm	$D^{18}{}_{4h}$	tI32	U

TABLE 5
Classification of Crystals (continued)

Strukturbericht symbol 1	Structure name 2	Symmetry group International 3	Symmetry group Schoenflies 4	Pearson symbol[a] 5	Standard ASTM E157-82a symbol[b] 6
$D10_1$	C_3Cr_7	P31c	$C^4{}_{3v}$	hP80	H
$D10_2$	Fe_3Th_7	$P6_3mc$	$C^4{}_{6v}$	hP20	H
$E0_1$	ClFPb	P4/nmm	$D^7{}_{4h}$	tP6	T
$E1_1$	$CuFeS_2$	$I\bar{4}2d$	$D^{12}{}_{2d}$	tI16	U
$E2_1$	CaO_3Ti	Pm3m	$O^1{}_h$	cP5	C
$E2_4$	S_3Sn_2	Pnma	$D^{16}{}_{2h}$	oP20	O
E3	Al_2CdS_4	$I\bar{4}$	$S^2{}_4$	tI14	U
$E9_3$	$SiFe_3W_3$	Fd3m	$O^7{}_h$	cF112	F
$E9_a$	Al_7Cu_2Fe	P4/mnc	$D^6{}_{4h}$	tP40	T
$E9_b$	$AlLi_3N_2$	Ia3	$T^7{}_h$	cI96	B
$F0_1$	NiSSb	$P2_13$	T^4	cP12	C
$F5_1$	$CrNaS_2$	$R\bar{3}m$ or R32	$D^5{}_{3d}$ or $D^7{}_3$	hR4	R
$F5_6$	CuS_2Sb	Pnma	$D^{16}{}_{2h}$	oP16	O
$H1_1$	Al_2MgO_4	Fd3m	$O^7{}_h$	cF56	F
$H2_4$	Cu_3S_4V	$P\bar{4}3m$	$T^1{}_d$	cP8	C
$H2_5$	$AsCu_3S_4$	$Pmn2_1$	$C^7{}_{2v}$	oP16	O
$L1_0$	AuCu	P4/mmm	$D^1{}_{4h}$	tP4	T
$L1_2$	$AlCu_3$	Pm3m	$O^1{}_h$	cP4	C
$L2_1$	$AlCu_2Mn$	Fm3m	$O^5{}_h$	cF16	F
$L2_2$	Sb_2Tl_7	Im3m	$O^9{}_h$	cI54	B
$L'2_b$	H_2Th	I4/mmm	$D^{17}{}_{4h}$	tI6	U
L′3	Fe_2N	$P6_3/mmc$	$D^4{}_{6h}$	hP3	H
$L6_0$	$CuTi_3$	P4/mmm	$D^1{}_{4h}$	tP4	T

[a] The first letter denotes the crystal system: triclinic (a), monoclinic (m), orthorhombic (o), tetragonal (t), hexagonal (h) and cubic (c). Trigonal (rhombohedral) system is presented by combination hR. The second letter of Pearson's symbol denotes lattice type: primitive (P), edge-(base-) centered (C), body-centered (I) or face-centered (F). The following number denotes amount of atoms in the crystal unit cell.

[b] Standard ASTM E157-82a has the Bravais lattices designations as following: C — primitive cubic; B — body-centered cubic; F — face-centered cubic; T — primitive tetragonal; U — body-centered tetragonal; R — rhombohedral; H — hexagonal; O — primitive orthorhombic; P — body-centered orthorhombic; Q — base-centered orthorhombic; S — face-centered orthorhombic; M — primitive monoclinic; N — centered monoclinic; A — triclinic.

REFERENCES

1. A. Schoenflies, *Kristallsysteme und Kristallstructur*, Leipzig, 1891.
2. E. S. Fedorow, Zusammenstellung der kristallographischen Resultate, *Zs. Krist.*, 20, 1892.
3. P. Groth, *Elemente der physikalischen und chemischen Krystallographie*, R. Oldenbourg, München/Berlin, 1921.
4. N. V. Belov, *Class Method of Deriving Space Groups of Symmetry*, Trudy Instituta Kristallodraffi imeni Fedorova (Transactions of the Fedorov Inst. of Crystallography), 5, 25, 1951, in Russian.
5. W. B. Pearson, *Handbook of Lattice Spacings and Structures of Metals and Alloys*, Vol. 1, Pergamon Press, 1958; Vol. 2, 1967.
6. Ch. Kittel, *Introduction to Solid State Physics*, John Wiley & Sons, 1956.
7. G. S. Zhdanov, *Fizika Tverdogo Tela (Solid State Physics)*, Moscow University Press, 1962, in Russian.
8. M. J. Buerger, *Elementary Crystallography*, John Wiley & Sons, 1963.
9. F. D. Bloss, *Crystallography & Crystal Chemistry*, Holt, Rinehart & Winston, 1971.
10. T. Janssen, *Crystallographic Groups*, North-Holland/American Elsevier, 1973.
11. M. P. Shaskolskaya, *Kristallografiya (Crystallography)*, Vysshaya Shkola, Moscow, 1976, in Russian.
12. T. Hahn, Ed., Internat. *Tables for Crystallography*, Vol. A, D. Reidel Publishing, Boston, 1983.
13. Crystal Data. Determinative Tables, Volumes 1—6, 1966—1983, JCPDS-Intern Centre for Diffraction Data and U.S. Dept. of Commerce.

IONIC RADII IN CRYSTALS

Howard T. Evans, Jr.

This table lists ionic radii R_i in Ångstrom units corresponding to various coordination numbers *CN*. Values are based on $R_i(O^{-2})$ = 1.40 Å for *CN* = 6.

sq = square and py = pyramidal.

REFERENCE

Shannon, R. D., *Acta Crystallogr.*, A32, 751, 1974.

Ion	*CN*	R_i/Å
Anions		
F (–1)	6	1.33
Cl (–1)	6	1.81
Br (–1)	6	1.96
I (–1)	6	2.20
OH (–1)	6	1.37
O (–2)	3	1.36
	6	1.40
S (–2)	6	1.84
Se (–2)	6	1.98
Te (–2)	6	1.07
Cations (alphabetical)		
Ag (+1)	4	1.00
	6	1.15
Ag (+2)	4sq	0.79
	6	0.94
Al (+3)	4	0.39
	6	0.54
As (+3)	6	0.58
As (+5)	4	0.34
	6	0.46
Au (+1)	6	1.37
Au (+3)	4sq	0.64
	6	0.85
Ba (+2)	8	1.42
	12	1.61
Be (+2)	4	0.27
	6	0.45
Bi (+3)	6	1.03
Bi (+5)	6	0.76
Br (+5)	3py	0.31
Br (+7)	4	0.25
Ca (+2)	6	1.00
	8	1.12
Cd (+2)	4	0.78
	6	0.95
	8	1.10
Ce (+3)	8	1.14
Ce (+4)	6	0.87
	8	0.97
Cl (+5)	3py	0.12
Cl (+7)	4	0.08
Co (+2)	6	0.65
Co (+3)	6	0.55
Cr (+2)	6	0.73
Cr (+3)	6	0.62
Cr (+6)	4	0.26
Cs (+1)	8	1.74
	12	1.88
Cu (+1)	2	0.46
	4	0.60
Cu (+2)	4sq	0.57
	6	0.73
Dy (+3)	8	1.03
Er (+3)	8	1.00
Eu (+2)	8	1.25
Eu (+3)	8	1.07
Fe (+2)	6	0.61
Fe (+3)	4	0.49
	6	0.55
Ga (+3)	4	0.47
	6	0.62
Gd (+3)	8	1.05
Ge (+4)	4	0.39
	6	0.53
Hf (+4)	8	0.83
Hg (+1)	6	1.19
Hg (+2)	2	0.69
	6	1.02
I (+5)	3py	0.44
I (+7)	4	0.42
	6	0.53
In (+3)	4	0.62
	6	0.80
Ir (+3)	6	0.68
Ir (+4)	6	0.63
Ir (+5)	6	0.57
K (+1)	8	1.51
	12	1.64
La (+3)	8	1.16
Li (+1)	4	0.59
	6	0.76
Lu (+3)	8	0.98
Mg (+2)	6	0.72
Mn (+2)	6	0.67
Mn (+3)	6	0.58
Mn (+4)	4	0.39
	6	0.53
Mn (+5)	4	0.33
Mn (+6)	4	0.26
Mn (+7)	4	0.25
Mo (+3)	6	0.69
Mo (+4)	6	0.65
Mo (+5)	6	0.61
Mo (+6)	6	0.59
	7	0.73
Na (+1)	6	1.02
	9	1.24
Nb (+3)	6	0.72
Nb (+4)	6	0.68
Nb (+5)	6	0.64
Ni (+2)	4sq	0.44
	6	0.69
Ni (+3)	6	0.56
Os (+4)	6	0.63
Os (+5)	6	0.58
Os (+6)	6	0.55
Os (+8)	4	0.39
P (+5)	4	0.17
Pb (+2)	6	1.19
	10	1.40
Pb (+4)	4	0.65
	6	0.78
Pd (+2)	4sq	0.64
Pd (+3)	6	0.76
Pd (+4)	6	0.62
Pm (+3)	8	1.09
Pr (+3)	8	1.13
Pt (+2)	4sq	0.60
Pt (+4)	6	0.63
Ra (+2)	8	1.62
	12	1.84
Rb (+1)	8	1.61
	12	1.72
Re (+4)	6	0.63
Re (+5)	6	0.58
Re (+6)	6	0.55
Re (+7)	4	0.38
	6	0.53
Rh (+3)	6	0.67
Rh (+4)	6	0.60
Rh (+5)	6	0.55
Ru (+3)	6	0.68
Ru (+4)	6	0.62
Ru (+5)	6	0.57
Ru (+7)	4	0.38
Ru (+8)	4	0.36
S (+4)	6	0.37
S (+6)	4	0.37
	6	0.29

IONIC RADII IN CRYSTALS (CONTINUED)

Ion	*CN*	R_i/Å	Ion	*CN*	R_i/Å	Ion	*CN*	R_i/Å
Sb (+3)	6	0.76	Te (+4)	6	0.97	V (+3)	6	0.64
Sb (+5)	6	0.60	Te (+6)	6	0.56	V (+4)	5	0.53
Sc (+3)	6	0.75	Th (+4)	8	1.05		6	0.58
Se (+4)	6	0.50	Ti (+3)	6	0.67	V (+5)	5	0.46
Se (+6)	4	0.50	Ti (+4)	6	0.61		6	0.54
	6	0.42	Tl (+1)	8	1.59	W (+4)	6	0.66
Si (+4)	4	0.26		12	1.70	W (+5)	6	0.62
Sm (+3)	8	1.08	Tl (+3)	6	0.89	W (+6)	4	0.42
Sr (+2)	8	1.26	Tm (+3)	7	1.09		6	0.60
	12	1.44	U (+3)	6	1.03	Y (+3)	8	1.02
Ta (+3)	6	0.72	U (+4)	6	0.89	Yb (+2)	8	1.14
Ta (+4)	6	0.68	U (+5)	6	0.76	Yb (+3)	8	0.99
Ta (+5)	6	0.64	U (+6)	2	0.45	Zn (+2)	4	0.60
Tb (+3)	8	1.18		7	0.81		6	0.74
Tb (+4)	8	0.88	V (+2)	6	0.79	Zr (+4)	8	0.84

CRYSTAL STRUCTURES AND LATTICE PARAMETERS OF ALLOTROPES OF THE ELEMENTS

H. W. King

The crystal structures of the allotropic forms of the elements are presented in terms of the Pearson symbol, the Strukturbericht designation, and the prototype of the structure. The temperatures of the phase transformations are listed in degrees Celsius and the pressures are in GPa. A consistent nomenclature is used, whereby all allotropes are labeled by Greek letters. The lattice parameters of the unit cells are given in nanometers (nm) and are considered to be accurate to ±2 in the last reported digit.

This compilation is restricted to changes in crystal structure that occur as a result of a change in temperature or pressure. Low-temperature structures are included for the diatomic and rare gases, which show many similarities with respect to the metallic elements. The elements identified with an asterisk (*) have polymorphic structures based on different molecular configurations. The crystal data given for these elements refer to the most stable structure at room temperature.

Reprinted with the permission of ASM International from T. B. Massalski, Ed., Binary Alloy Phase Diagrams, ASM International, Metals Park, Ohio, 1986.

Element	Temperature, °C	Pressure, GPa	Pearson symbol	Space group	Strukturbericht designation	Prototype	Lattice parameters, nm *a*	*b*	*c*	Comment, *c/a* or α or β
Ac	25	atm	*cF4*	*Fm3m*	*A1*	Cu	0.5311	...	...	...
Ag	25	atm	*cF4*	*Fm3m*	*A1*	Cu	0.40857	...	...	...
αAl	25	atm	*cF4*	*Fm3m*	*A1*	Cu	0.40496	...	...	...
βAl	25	>20.5	*hP2*	$P6_3/mmc$	*A3*	Mg	0.2693	...	0.4398	1.6331
α′Am	25	atm	*hP4*	$P6_3/mmc$	*A3′*	αLa	0.34681	...	1.1241	2*1.621
αAm	>769	atm	*cF4*	*Fm3m*	*A1*	Cu	0.4894	...	...	...
βAm	>1074	atm	*cI2*	*Im3m*	*A2*	W	?	...	...	...
γAm	25	>15	*oC4*	*Cmcm*	*A20*	αU	0.3063	0.5968	0.5169	...
αAr	<−189.35	atm	*cF4*	*Fm3m*	*A1*	Cu	0.5316	...	...	...
(βAr)	<−189.40	atm	*hP2*	$P6_3/mmc$	*A3*	Mg	0.3760	...	0.6141	1.633
αAs	25	atm	*hR2*	$R\bar{3}m$	*A7*	αAs	0.41319	...	...	α = 54.12°
εAs	>448	atm	*oC8*	*Cmca*	...	P (black)	0.362	1.085	0.448	...
Au	25	atm	*cF4*	*Fm3m*	*A1*	Cu	0.40782	...	...	...
βB	25	atm	*hR105*	*R3m*	...	βB	1.017	...	...	α = 65.12°
αBa	25	atm	*cI2*	$Im\bar{3}m$	*A2*	W	0.50227	...	...	...
βBa	25	>5.33	*hP2*	$P6_3/mmc$	*A3*	Mg	0.3901	...	0.6154	1.5775
γBa	25	>23	?	?	...	...	...	...	...	...
αBe	25	atm	*hP2*	$P6_3/mmc$	*A3*	Mg	0.22859	...	0.35845	1.5681
βBe	>1270	atm	*cI2*	*Im3m*	*A2*	W	0.25515	...	...	...
γBe	25	>9.3	?	...	...	...	...	...	...	...
αBi	25	atm	*hR2*	$R\bar{3}m$	*A7*	αAs	0.47460	...	...	α = 57.23°
βBi	25	>2.6	*mC4*	*C2/m*	...	βBi	0.6674	0.6117	0.3304	β = 110.33°
γBi	25	>3.0	*mP3*	?	...	...	0.605	0.42	0.465	β = 85.33°
σBi	25	>4.3	?	?	...	...	...	...	...	...
εBi	25	>6.5	?	?	...	...	...	...	...	...
ζBi	25	>9.0	*cI2*	*Im3m*	*A2*	W	0.3800	...	...	...
αBk	25	atm	*hP4*	$P6_3/mmc$	*A3′*	αLa	0.3416	...	1.1069	2*1.620
βBk	>977	atm	*cF4*	*Fm3m*	*A1*	Cu	0.4997	...	...	...
Br	<7.25	atm	*oC8*	*Cmca*	...	Cl	0.668	0.449	0.874	...
C (graphite)	25	atm	*hP4*	$P6_3/mmc$	*A9*	C (graphite)	0.24612	...	0.6709	2.7258
C (diamond)	25	>60	*cF8*	*Fd3m*	*A4*	C (diamond)	0.35669	...	...	...
C (hd)	25	HP	*hP4*	$P6_3/mmc$	...	C (hd)	0.2522	...	0.4119	1.633
αCa	25	atm	*cF4*	*Fm3m*	*A1*	Cu	0.55884	...	...	...
βCa	>443	atm	*cI2*	*Im3m*	*A2*	W	0.4480	...	...	...
γCa	25	>1.5	?	...	...	...	...	...	...	...
Cd	25	atm	*hP2*	$P6_3/mmc$	*A3*	Mg	0.29793	...	0.56196	1.8862
αCe	<−177	atm	*cF4*	*Fm3m*	*A1*	Cu	0.485	...	...	...
βCe	25	atm	*hP4*	$P6_3/mmc$	*A3′*	αLa	0.36810	...	1.1857	2*1.611
γCe	>61	atm	*cF4*	*Fm3m*	*A1*	Cu	0.51610	...	...	...
σCe	>726	atm	*cI2*	*Im3m*	*A2*	W	0.412	...	...	...
α′Ce	25	>5.4	*oC4*	*Cmcm*	*A20*	αU	0.3049	0.5998	0.5215	...
αCf	25	atm	*hP4*	$P6_3/mmc$	*A3′*	αLa	0.339	...	1.1015	2*1.625
βCf	>590	atm	*cF4*	*Fm3m*	*A1*	Cu	?	...	...	...
Cl	25	atm	*oC8*	*Cmca*	...	Cl	0.624	0.448	0.826	...
αCm	25	atm	*hP4*	$P6_3/mmc$	*A3′*	αLa	0.3496	...	1.1331	2*1.621
βCm	>1277	atm	*cF4*	*Fm3m*	*A1*	Cu	0.4382	...	...	...
εCo	25	atm	*hP2*	$P6_3/mmc$	*A3*	Mg	0.25071	...	0.40686	1.6228
αCo	>422	atm	*cF4*	*Fm3m*	*A1*	Cu	0.35447	...	...	...
αCr	25	atm	*cI2*	*Im3m*	*A2*	W	0.38848	...	...	...
α′Cr	25	HP	*tI2*	*I4/mmm*	...	α′Cr	0.2882	...	0.2887	1.002
αCs	25	atm	*cI2*	*Im3m*	*A2*	W	0.6141	...	...	...
βCs	25	>2.37	*cF4*	*Fm3m*	*A1*	Cu	0.6465	...	...	...
β′Cs	25	>4.22	*cF4*	*Fm3m*	*A1*	Cu	0.5800	...	...	...
γCs	25	>4.27	?	...	...	...	...	...	...	...
Cu	25	atm	*cF4*	*Fm3m*	*A1*	Cu	0.36146	...	...	...
α′Dy	<−187	atm	*oC4*	*Cmcm*	...	α′Dy	0.3595	0.6184	0.5678	...
αDy	25	atm	*hP2*	$P6_3/mmc$	*A3*	Mg	0.35915	...	0.56501	1.5732

Element	Temperature, °C	Pressure, GPa	Pearson symbol	Space group	Strukturbericht designation	Prototype	Lattice parameters, nm a	b	c	Comment, c/a or α or β
βDy	>1381	atm	cI2	Im3m	A2	W	(0.398)	···	···	···
γDy	25	>7.5	hR3	$R\bar{3}m$	···	αSm	0.3436	···	2.483	4.5*1.606
Er	25	atm	hP2	$P6_3/mmc$	A3	Mg	0.35592	···	0.55850	1.5692
αEs	25	atm	hP4	$P6_3/mmc$	A3′	αLa	?	···	···	···
βEs	?	atm	cF4	Fm3m	A1	Cu	?	···	···	···
Eu	25	atm	cI2	Im3m	A2	W	0.45827	···	···	···
αF	<−227.6	atm	mC8	C2/c	···	αF	0.550	0.338	0.728	β = 102.17°
βF	<−219.67	atm	cP16	Pm3n	···	γO	0.667	···	···	···
αFe	25	atm	cI2	Im3m	A2	W	0.28665	···	···	···
γFe	>912	atm	cF4	Fm3m	A1	Cu	0.36467	···	···	···
σFe	>1394	atm	cI2	Im3m	A2	W	0.29315	···	···	···
εFe	25	>13	hP2	$P6_3/mmc$	A3	Mg	0.2468	···	0.396	1.603
αGa	25	atm	oC8	Cmca	A11	αGa	0.45186	0.76570	0.45258	···
βGa	25	>1.2	tI2	I4/mmm	A6	In	0.2808	···	0.4458	1.588
γGa	−53	>3.0	oC40	Cmcm	···	γGa	1.0593	1.3523	0.5203	···
αGd	25	atm	hP2	$P6_3/mmc$	A3	Mg	0.36336	···	0.57810	1.5910
βGd	>1235	atm	cI2	Im3m	A2	W	0.406	···	···	···
γGd	25	>3.0	hR3	R3m	···	αSm	0.361	···	2.603	4*1.60
αGe	25	atm	cF8	Fd3m	A4	C (diamond)	0.56574	···	···	···
βGe	25	>12	tI4	$I4_1/amd$	A5	βSn	0.4884	···	0.2692	0.551
γGe	25	>12 → atm	tP12	$P4_12_12$	···	σGe	0.593	···	0.698	1.18
σGe	LT	>12	cI16	Im3m	···	γSi	0.692	···	···	···
αH	<−271.9	atm	cF4	Fm3m	A1	Cu	0.5338	···	···	···
βH	<−259.34	atm	hP2	$P6_3/mmc$	A3	Mg	0.3776	···	0.6162	1.632
αHe	<−268.94	atm	hP2	$P6_3/mmc$	A3	Mg	0.3555	···	0.5798	1.631
βHe	>−258	0.125	cF4	Fm3m	A1	Cu	0.4240	···	···	···
γHe	<−271.47	0.03	cI2	Im3m	A2	W	0.4110	···	···	···
αHf	25	atm	hP2	$P6_3/mmc$	A3	Mg	0.31946	···	0.50510	1.5811
βHf	>1995	atm	cI2	Im3m	A2	W	0.3610	···	···	···
αHg	<−38.84	atm	hR1	$R\bar{3}m$	A10	αHg	0.3005	···	···	α = 70.53°
βHg	<−101	HP	tI2	I4/mmm	···	βHg	0.3995	···	0.2825	0.707
γHg	<−194	c.w.	hR1	?	···	···	···	···	···	···
αHo	25	atm	hP2	$P6_3/mmc$	A3	Mg	0.35778	···	0.56178	1.5702
βHo	25	>7.5	hR3	$R\bar{3}m$	···	αSm	0.334	···	2.45	4.5*1.63
I	25	atm	oC8	Cmca	···	Cl	0.72697	0.47903	0.97942	···
In	25	atm	tI2	I4/mmm	A6	In	0.3253	···	0.49470	1.5210
Ir	25	atm	cF4	Fm3m	A1	Cu	0.38392	···	···	···
K	25	atm	cI2	Im3m	A2	W	0.5321	···	···	···
Kr	<−157.00	atm	cF4	Fm3m	A1	Cu	0.5810	···	···	···
αLa	25	atm	hP4	$P6_3/mmc$	A3′	αLa	0.37740	···	1.2171	2*1.6125
βLa	>310	atm	cF4	Fm3m	A1	Cu	0.5303	···	···	···
γLa	>865	atm	cI2	Im3m	A2	W	0.426	···	···	···
β′La	25	>2.0	cF4	Fm3m	A1	Cu	0.517	···	···	···
αLi	<−193	atm	hP2	$P6_3/mmc$	A3	Mg	0.3111	···	0.5093	1.637
βLi	25	atm	cI2	Im3m	A2	W	0.35093	···	···	···
γLi	<−201	c.w.	cF4	Fm3m	A1	Cu	0.4388	···	···	···
Lu	25	atm	hP2	$P6_3/mmc$	A3	Mg	0.35052	···	0.55494	1.5832
Mg	25	atm	hP2	$P6_3/mmc$	A3	Mg	0.32094	···	0.52107	1.6236
αMn	25	atm	cI58	$I\bar{4}3m$	A12	αMn	0.89126	···	···	···
βMn	>710	atm	cP20	$P4_132$	A13	βMn	0.63152	···	···	···
γMn	>1079	atm	cF4	Fm3m	A1	Cu	0.3860	···	···	···
σMn	>1143	atm	cI2	Im3m	A2	W	0.3080	···	···	···
Mo	25	atm	cI2	Im3m	A2	W	0.31470	···	···	···
αN	<−237.6	atm	cP8	Pa3	···	αN	0.5661	···	···	···
βN	<−210.00	atm	hP4	$P6_3/mmc$	···	βN	0.4050	···	0.6604	1.631
γN	<−253	>3.3	tP4	$P4_2/mnm$	···	γN	0.3957	···	0.5109	1.291
αNa	<−233	atm	hP2	$P6_3/mmc$	A3	Mg	0.3767	···	0.6154	1.634
βNa	25	atm	cI2	Im3m	A2	W	0.42906	···	···	···
Nb	25	atm	cI2	Im3m	A2	W	0.33004	···	···	···
αNd	[illegible]	atm	hP4	[illegible]	[illegible]	[illegible]	0.36580	[illegible]	1.17966	2*1.6124
βNd	>863	atm	cI2	Im3m	A2	W	0.413	···	···	···
γNd	25	>5.0	cF4	Fm3m	A1	Cu	0.480	···	···	···
Ne	<−243.59	atm	cF4	Fm3m	A1	Cu	0.4462	···	···	···
Ni	25	atm	cF4	Fm3m	A1	Cu	0.35240	···	···	···
αNp	25	atm	oP8	Pnma	A_c	αNp	0.6663	0.4723	0.4887	···
βNp	>280	atm	tP4	$P42_12$	A_d	βNp	0.4883	···	0.3389	0.694
γNp	>576	atm	cI2	Im3m	A2	W	0.352	···	···	···
αO	<−243.3	atm	mC4	C2m	···	αO	0.5403	0.3429	0.5086	β = 132.53°
βO	<−229.6	atm	hR2	$R\bar{3}m$	···	βO	0.4210	···	···	α = 46.27°
γO	<−218.79	atm	cP16	Pm3n	···	γO	0.683	···	···	···
Os	25	atm	hP2	$P6_3/mmc$	A3	Mg	0.27341	···	0.43918	1.6063
P (black)	25	atm	oC8	Cmca	···	P (black)	0.33136	1.0478	0.43763	···
αPa	25	atm	tI2	I4/mmm	A_a	αPa	0.3921	···	0.3235	0.825
βPa	>1170	atm	cI2	Im3m	A2	W	0.381	···	···	···
αPb	25	atm	cF4	Fm3m	A1	Cu	0.49502	···	···	···
βPb	25	>10.3	hP2	$P6_3/mmc$	A3	Mg	0.3265	···	0.5387	1.650

CRYSTAL STRUCTURES AND LATTICE PARAMETERS OF ALLOTROPES OF THE ELEMENTS (continued)

Element	Temperature, °C	Pressure, GPa	Pearson symbol	Space group	Strukturbericht designation	Prototype	Lattice parameters, nm *a*	*b*	*c*	Comment, *c/a* or α or β
Pd	25	atm	*cF*4	*Fm*3*m*	*A*1	Cu	0.38903	···	···	···
αPm	25	atm	*hP*4	$P6_3/mmc$	*A*3′	αLa	0.365	···	1.165	2*1.60
βPm	>890	atm	*cI*2	*Im*3*m*	*A*2	W	?	···	···	···
αPo	25	atm	*cP*1	*Pm*3*m*	A_h	αPo	0.3366	···	···	···
βPo	>54	atm	*hR*1	*R*$\bar{3}$*m*	···	βPo	0.3373	···	···	α = 98.08°
αPr	25	atm	*hP*4	$P6_3/mmc$	*A*3′	αLa	0.36721	···	1.18326	2*1.6111
βPr	>795	atm	*cI*2	*Im*3*m*	*A*2	W	0.413	···	···	···
γPr	25	>4.0	*cF*4	*Fm*3*m*	*A*1	Cu	0.488	···	···	···
Pt	25	atm	*cF*4	*Fm*3*m*	*A*1	Cu	0.39236	···	···	···
αPu	25	atm	*mP*16	$P2_1/m$	···	αPu	0.6183	0.4822	1.0963	β = 101.97°
βPu	>125	atm	*mI*34	*I*2/*m*	···	βPu	0.9284	1.0463	0.7859	β = 92.13°
γPu	>215	atm	*oF*8	*Fddd*	···	γPu	0.31587	0.57682	1.0162	···
σPu	>320	atm	*cF*4	*Fm*3*m*	*A*1	Cu	0.46371	···	···	···
σ′Pu	>463	atm	*tI*2	*I*4/*mmm*	*A*6	In	0.33261	···	0.44630	1.3418
εPu	>483	atm	*cI*2	*Im*3*m*	*A*2	W	0.36343	···	···	···
Ra	25	atm	*cI*2	*Im*3*m*	*A*2	W	0.5148	···	···	···
αRb	25	atm	*cI*2	*Im*3*m*	*A*2	W	0.5705	···	···	···
βRb	25	>1.08	?	···	···	···	···	···	···	···
γRb	25	>2.05	?	···	···	···	···	···	···	···
Re	25	atm	*hP*2	$P6_3/mmc$	*A*3	Mg	0.27609	···	0.4458	1.6145
Rh	25	atm	*cF*4	*Fm*3*m*	*A*1	Cu	0.38032	···	···	···
Ru	25	atm	*hP*2	$P6_3/mmc$	*A*3	Mg	0.27058	···	0.42816	1.5824
αS	25	atm	*oF*128	*Fddd*	*A*16	αS	1.0464	1.28660	2.44860	···
αSb	25	atm	*hR*2	*R*$\bar{3}$*m*	*A*7	αAs	0.45067	···	···	α = 57.11°
βSb	25	>5.0	*cP*1	*Pm*3*m*	A_h	αPo	0.2992	···	···	···
γSb	25	>7.5	*hP*2	$P6_3/mmc$	*A*3	Mg	0.3376	···	0.5341	1.582
σSb	25	>14.0	*mP*3	?	···	···	0.556	0.404	0.422	β = 86.0°
αSc	25	atm	*hP*2	$P6_3/mmc$	*A*3	Mg	0.33088	···	0.52680	1.5921
βSc	>1337	atm	*cI*2	*Im*3*m*	*A*2	W	0.4541	···	···	···
γSe	25	atm	*hP*3	$P3_121$	*A*8	γSe	0.43659	···	0.49537	1.1346
αSi	25	atm	*cF*8	*Fd*3*m*	*A*4	C (diamond)	0.54306	···	···	···
βSi	25	>9.5	*tI*4	$I4_1/amd$	*A*5	βSn	0.4686	···	0.2585	0.552
γSi	25	>16.0	*cI*16	*Im*3*m*	···	γSi	0.6636	···	···	···
σSi	25	>16 → atm	*hP*4	$P6_3/mmc$	*A*3′	αLa	0.380	···	0.628	1.653
αSm	25	atm	*hR*3	*R*$\bar{3}$*m*	···	αSm	0.36290	···	2.6207	4*1.6048
βSm	>734	atm	*hP*2	$P6_3/mmc$	*A*3	Mg	0.36630	···	0.58448	1.5956
γ′Sm	>922	atm	*cI*2	*Im*3*m*	*A*2	W	?	···	···	···
σSm	25	>4.0	*hP*4	$P6_3/mmc$	*A*3′	αLa	0.3618	···	1.166	2*1.611
αSn	<13	atm	*cF*8	*Fd*3*m*	*A*4	C (diamond)	0.64892	···	···	···
βSn	25	atm	*tI*4	$I4_1/amd$	*A*5	βSn	0.58318	···	0.31818	0.5456
γSn	25	>9.0	*tI*2	?	···	γSn	0.370	···	0.337	0.91
αSr	25	atm	*cF*4	*Fm*3*m*	*A*1	Cu	0.6084	···	···	···
βSr	>547	atm	*cI*2	*Im*3*m*	*A*2	W	0.487	···	···	···
β′Sr	25	>3.5	*cI*2	*Im*3*m*	*A*2	W	0.4437	···	···	···
Ta	25	atm	*cI*2	*Im*3*m*	*A*2	W	0.33030	···	···	···
α′Tb	<−53	atm	*oC*4	*Cmcm*	···	α′Dy	0.3605	0.6244	0.5706	···
αTb	25	atm	*hP*2	$P6_3/mmc$	*A*3	Mg	0.36055	···	0.56966	1.5800
βTb	>1289	atm	*cI*2	*Im*3*m*	*A*2	W	(0.402)	···	···	···
γTb	25	>6.0	*hR*3	*R*$\bar{3}$*m*	···	αSm	0.341	···	2.45	4*1.60
Tc	25	atm	*hP*2	$P6_3/mmc$	*A*3	Mg	0.2738	···	0.4393	1.604
αTe	25	atm	*hP*3	$P3_121$	*A*8	γSe	0.44566	···	0.59264	1.3298
βTe	25	>2.0	*hR*2	*R*$\bar{3}$*m*	*A*7	αAs	0.469	···	···	α = 53.30°
γTe	25	>7.0	*hR*1	*R*$\bar{3}$*m*	···	βPo	0.3002	···	···	α = 103.3°
αTh	25	atm	*cF*4	*Fm*3*m*	*A*1	Cu	0.50842	···	···	···
βTh	>1360	atm	*cI*2	*Im*3*m*	*A*2	W	0.411	···	···	···
αTi	25	atm	*hP*2	$P6_3/mmc$	*A*3	Mg	0.29506	···	0.46835	1.5873
βTi	>882	atm	*cI*2	*Im*3*m*	*A*2	W	0.33065	···	···	···
ωTi	25	HP → atm	*hP*3	*P*6/*mmm*	···	ωTi	0.4625	···	0.2813	0.6082
αTl	25	atm	*hP*2	$P6_3/mmc$	*A*3	Mg	0.34566	···	0.55248	1.5983
βTl	>230	atm	*cI*2	*Im*3*m*	*A*2	W	0.3879	···	···	···
γTl	25	HP	*cF*4	*Fm*3*m*	*A*1	Cu	?	···	···	···
Tm	25	atm	*hP*2	$P6_3/mmc$	*A*3	Mg	0.35375	···	0.55540	1.5700
αU	25	atm	*oC*4	*Cmcm*	*A*20	αU	0.28537	0.58695	0.49548	···
βU	>668	atm	*tP*30	$P4_2/mnm$	A_b	βU	1.0759	···	0.5656	0.526
γU	>776	atm	*cI*2	*Im*3*m*	*A*2	W	0.3524	···	···	···
V	25	atm	*cI*2	*Im*3*m*	*A*2	W	0.30240	···	···	···
W	25	atm	*cI*2	*Im*3*m*	*A*2	W	0.31652	···	···	···
Xe	<−111.76	atm	*cF*4	*Fm*3*m*	*A*1	Cu	0.6350	···	···	···
αY	25	atm	*hP*2	$P6_3/mmc$	*A*3	Mg	0.36482	···	0.57318	1.5711
βY	>1478	atm	*cI*2	*Im*3*m*	*A*2	W	(0.407)	···	···	···
αYb	<−3	atm	*hP*2	$P6_3/mmc$	*A*3	Mg	0.38799	···	0.63859	1.6459
βYb	25	atm	*cF*4	*Fm*3*m*	*A*1	Cu	0.54848	···	···	···
γYb	>795	atm	*cI*2	*Im*3*m*	*A*2	W	0.444	···	···	···
Zn	25	atm	*hP*2	$P6_3/mmc$	*A*3	Mg	0.26650	···	0.49470	1.8563
αZr	25	atm	*hP*2	$P6_3/mmc$	*A*3	Mg	0.32316	···	0.51475	1.5929
βZr	>863	atm	*cI*2	*Im*3*m*	*A*2	W	0.36090	···	···	···
ωZr	25	HP → atm	*hP*2	*P*6/*mmm*	···	ωTi	0.5036	···	0.3109	0.617

LATTICE ENERGIES

H. D. B. Jenkins

Table 1 contains calculated values of the lattice energies (total lattice potential energies), U_{POT}, of crystalline salts, M_aX_b. U_{POT} is expressed in units of kilojoules per mole, kJ mol^{-1}. M and X can be either simple or complex ions.

Also listed in the table is the lattice energy, U_{POT}^{BFHC}, obtained from the application of the Born-Fajans-Haber cycle (BFHC) described below, using thermochemical data from Reference 1 plus certain other data which are given in Table 2 of this section (Reference 2).

The lattice enthalpy, ΔH_L, is given by the cycle:

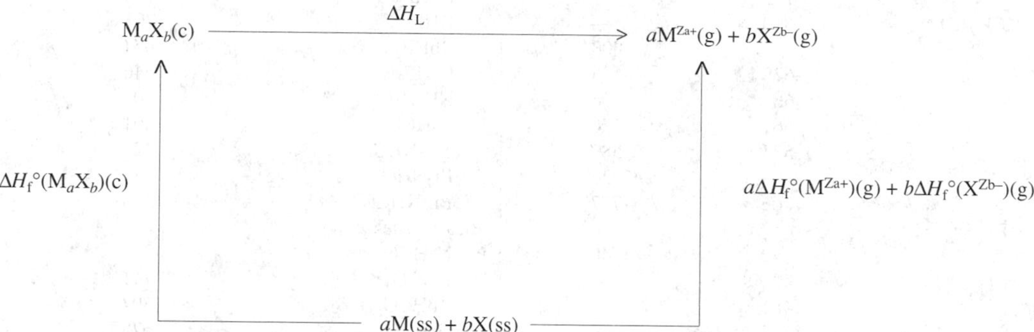

where (ss) is the standard state of the element concerned.

The lattice enthalpy, ΔH_L, is obtained using the equation:

$$\Delta H_L = a\Delta H_f°(M^{Za+})(g) + b\Delta H_f°(X^{Zb-})(g) - \Delta H_f°(M_aX_b)(c)$$

and is further related to the total lattice potential energy, U_{POT}^{BFHC}, by the relationship:

$$U_{POT}^{BFHC}(M_aX_b) = \Delta H_L - \left[a\left(\frac{n_{M^{Za+}}}{2} - 2\right) + b\left(\frac{n_{X^{Zb-}}}{2} - 2\right)\right]RT$$

where $n_{M^{Za+}}$ and $n_{X^{Zb-}}$ is equal to 3 for monatomic ions, 5 for linear polyatomic ions and 6 for polyatomic non-linear ions.

Substances are arranged by chemical class.

REFERENCES

1. Wagman, D. D., et al., *The NBS Tables of Chemical Thermodynamic Properties, J. Phys. Chem. Ref. Data*, 11, Suppl. 2, 1982.
2. *Adv. Inorg. Chem. Radiochem.*, 22, 1, 1978.

Table 1
LATTICE ENERGIES

Substance	Calc. U_{POT} (kJ mol^{-1})	U_{POT}^{BFHC} (kJ mol^{-1})
Acetates		
$Li(CH_3COO)$	—	881
$Na(CH_3COO)$	761	763
$K(CH_3COO)$	686	682
$Rb(CH_3COO)$	715	656
$Cs(CH_3COO)$	682	—
$NH_4(CH_3COO)$	725	695
$Ag(CH_3COO)$	—	863
$Tl(CH_3COO)$	—	750
$Ca(CH_3COO)_2$	2431	2294
$Sr(CH_3COO)_2$	2280	2166
$Ba(CH_3COO)_2$	2180	2033
$Mn(CH_3COO)_2$	2548	2616
$Cu(CH_3COO)_2$	—	2835
$Zn(CH_3COO)_2$	2615	2750
$Hg(CH_3COO)_2$	2368	2595
$Pb(CH_3COO)_2$	2247	2225
Acetylides		
CaC_2	2911	2904
SrC_2	2788	2784
BaC_2	2647	2654
Ammonium salts		
NH_4HF_2	705	705
NH_4HCO_3	—	741

Table 1 (continued)

Substance	Calc. U_{POT} (kJ mol^{-1})	U_{POT}^{BFHC} (kJ mol^{-1})
NH_4HSO_4	640	(645)
NH_4BF_4	582	—
NH_4NCO	724	—
NH_4CN	617	670
NH_4HCO_2	715	—
$NH_4(NH_2CH_2CO_2)$	650	—
NH_4IO_3	—	808
$NH_4IO_2F_2$	678	685
NH_4HS	661	666
NH_4NO_3	661	676
NH_4ClO_4	583	580
$(NH_4)_2S$	2008	(2026)
$(NH_4)_2SO_4$	1766	1777
$(NH_4)_2GeF_6$	1657	—
$(NH_4)_2IrCl_6$	1442	1440
$(NH_4)_2OsCl_6$	1433	—
$(NH_4)_2PdCl_6$	1481	—
$(NH_4)_2PbCl_6$	1355	—
$(NH_4)_2PbCl_6$	1468	—
$(NH_4)_2ReCl_6$	1402	1390
$(NH_4)_2SiF_6$	1657	1727
$(NH_4)_2SnCl_6$	1370	1334
$(NH_4)_2SnBr_6$	1319	—
$(NH_4)_2TeCl_6$	1318	—
$(NH_4)_2TeBr_6$	1294	—
$(NH_4)_2TiCl_6$	1413	—
NH_4CNS	605	611
$(NH_4)_2SeCl_6$	1420	—
$(NH_4)_2SeBr_6$	1380	—
Arsenates		
$Mg_3(AsO_4)_2$	10669	10716
$Ca_3(AsO_4)_2$	9749	9653
$Sr_3(AsO_4)_2$	9330	9266
$Ba_3(AsO_4)_2$	8870	8985
$AlAsO_4$	7255	—
$GaAsO_4$	7243	—
Astatides		
LiAt	720	—
NaAt	657	—
KAt	615	—
RbAt	594	—
CsAt	586	—
FrAt	573	—
Azides		
LiN_3	812	818
NaN_3	732	731
KN_3	659	658
RbN_3	637	632
CsN_3	612	604
AgN_3	854	—
TlN_3	689	686
$Ca(N_3)_2$	2186	2162
$Sr(N_3)_2$	2056	2066
$Ba(N_3)_2$	2021	1965
$Mn(N_3)_2$	2408	2416
$Cu(N_3)_2$	2730	2738
$Zn(N_3)_2$	2840	2848
$Cd(N_3)_2$	2446	2454
$Pb(N_3)_2$	—	2173
Bihalide salts		
$LiHF_2$	893	866
$NaHF_2$	788	788
KHF_2	703	698
$RbHF_2$	674	676
$CsHF_2$	646	628
NH_4HF_2	705	705
$CsHCl_2$	601	—
Me_4NHCl_2	427	—
Et_4NHCl_2	346	—
Bu_4NHCl_2	290	—
Bicarbonates		
$NaHCO_3$	820	818
$KHCO_3$	741	736
$RbHCO_3$	707	714
$CsHCO_3$	678	709
NH_4HCO_3	—	741
$Ca(HCO_3)_2$	2402	(2403)
$Sr(HCO_3)_2$	2255	(2272)
$Ba(HCO_3)_2$	2155	(2159)
Bisulphates		
NH_4HSO_4	640	(645)
Borides		
CaB_6	5146	—
SrB_6	5104	—
BaB_6	5021	—
YB_6	7447	—
LaB_6	7406	—
CeB_6	10083	—
PrB_6	7447	—
NdB_6	7447	—
PmB_6	7406	—
SmB_6	7447	—
EuB_6	5104	—
GdB_6	7489	—
TbB_6	7489	—
DyB_6	7489	—
HoB_6	7489	—
ErB_6	7489	—
TmB_6	7489	—
YbB_6	5146	—
LuB_6	7489	—
ThB_6	10167	—
Borohydrides		
$LiBH_4$	778	—
$NaBH_4$	703	—
KBH_4	665	—
$RbBH_4$	648	—
$CsBH_4$	628	—
Borohalides		
$LiBF_4$	699	—
$NaBF_4$	657	619
KBF_4	611	631
$RbBF_4$	577	605
$CsBF_4$	556	—

Table 1 (continued)

Substance	Calc. U_{POT} (kJ mol^{-1})	U_{POT}^{BFHC} (kJ mol^{-1})
NH_4BF_4	582	—
$Co(BF_4)_2$	2127	—
$Ni(BF_4)_2$	2136	—
$Zn(BF_4)_2$	2063	—
$Cd(BF_4)_2$	1937	—
$KBCl_4$	506	—
$RbBCl_4$	489	—
$CsBCl_4$	473	—
Carbonates		
Li_2CO_3	2523	2269
Na_2CO_3	2301	2030
K_2CO_3	2084	1858
Rb_2CO_3	2000	1795
Cs_2CO_3	1920	1702
$MgCO_3$	3180	3122
$CaCO_3$	2804	2810
$SrCO_3$	2720	2688
$BaCO_3$	2615	2554
$MnCO_3$	3046	3151
$FeCO_3$	3121	3171
$CoCO_3$	3443	3232
$NiCO_3$		3297
$CuCO_3$	3494	3327
$ZnCO_3$	3121	3273
$CdCO_3$	2929	3052
$SnCO_3$	2904	(2853)
$PbCO_3$	2728	2750
Cyanates		
LiNCO	770	—
NaNCO	736	—
KNCO	653	—
RbNCO	615	—
CsNCO	586	—
NH_4NCO	724	—
Cyanides		
LiCN	—	849
NaCN	738	739
KCN	674	669
RbCN	646	(640)
CsCN	612	602
NH_4CN	617	670
$Ca(CN)_2$	2268	2191
$Sr(CN)_2$	2138	(2076)
$Ba(CN)_2$	2046	1960
CuCN	—	1035
AgCN	741	914
$Zn(CN)_2$	2431	2768
$Cd(CN)_2$	2284	2542
Formates		
$Li(HCO_2)$	865	872
$Na(HCO_2)$	791	811
$K(HCO_2)$	713	729
$Rb(HCO_2)$	685	682
$Cs(HCO_2)$	651	644
$NH_4(HCO_2)$	715	—
$Mg(HCO_2)_2$	2674	—
$Ca(HCO_2)_2$	2360	2390
$Sr(HCO_2)_2$	2221	2261
$Ba(HCO_2)_2$	2092	2134
$Mn(HCO_2)_2$	2598	2701
$Co(HCO_2)_2$	—	2792
$Ni(HCO_2)_2$	—	2880
$Cu(HCO_2)_2$	2870	2913
$Zn(HCO_2)_2$	2791	2847
$Cd(HCO_2)_2$	2556	—
$Pb(HCO_2)_2$	2276	2330
Germanates		
Mg_2GeO_4	7991	—
Ca_2GeO_4	7301	7306
Sr_2GeO_4	6987	—
Ba_2GeO_4	6653	6643
Glycinates		
$Na(NH_2CH_2COO_2)$	739	—
$K(NH_2CH_2COO_2)$	668	—
$Rb(NH_2CH_2COO_2)$	648	—
$NH_4(NH_2CH_2COO_2)$	650	—
$Cu(NH_2CH_2CO_2)_2$	2694	—
Halates		
$LiBrO_3$	883	897
$NaBrO_3$	803	814
$KBrO_3$	740	745
$RbBrO_3$	720	742
$CsBrO_3$	694	681
$NaClO_3$	770	770
$KClO_3$	711	706
$RbClO_3$	690	687
$CsClO_3$	—	647
$LiIO_3$	975	—
$NaIO_3$	883	882
KIO_3	820	806
$RbIO_3$	791	—
$CsIO_3$	761	769
NH_4IO_3	—	808
$Mg(ClO_3)_2$	2535	(2475)
$Ca(ClO_3)_2$	2259	2286
$Sr(ClO_3)_2$	2138	2155
$Ba(ClO_3)_2$	2021	2027
Halides		
LiF	1030	1036
LiCl	834	853
LiBr	788	807
LiI	730	757
NaF	910	923
NaCl	769	786
NaBr	732	747
NaI	682	704
KF	808	821
KCl	701	715
KBr	671	682
KI	632	649
RbF	774	785
RbCl	680	689
RbBr	651	660
RbI	617	630

Table 1 (continued)

Substance	Calc. U_{POT} (kJ mol^{-1})	U_{POT}^{BFHC} (kJ mol^{-1})
CsF	744	740
CsCl	657	659
CsBr	632	631
CsI	600	604
FrF	715	—
FrCl	632	—
FrBr	611	—
FrI	582	—
CuCl	921	996
CuBr	879	979
CuI	835	966
AgF	953	967
AgCl	864	915
AgBr	830	904
AgI	808	889
AuCl	1013	1066
AuBr	1015	1061
AuI	1015	1070
InCl	—	763
InBr	—	767
InI	—	732
TlF	—	845
TlCl	782	751
TlBr	713	735
TlI	687	709
BeF_2	3150	3505
$BeCl_2$	3004	3020
$BeBr_2$	2950	2914
BeI_2	2653	2800
MgF_2	2913	2957
$MgCl_2$	2326	2526
$MgBr_2$	2097	2440
MgI_2	1944	2327
CaF_2	2609	2630
$CaCl_2$	2223	2258
$CaBr_2$	2132	2176
CaI_2	1905	2074
SrF_2	2476	2492
$SrCl_2$	2127	2156
SrI_2	2008	2075
SrI_2	1937	1963
BaF_2	2341	2352
$BaCl_2$	2033	2056
$BaBr_2$	1950	1985
BaI_2	1831	1877
RaF_2	2284	—
$RaCl_2$	2004	—
$RaBr_2$	1929	—
RaI_2	1803	—
$ScCl_2$	2380	—
$ScBr_2$	2291	—
ScI_2	2201	—
TiF_2	2724	—
$TiCl_2$	2431	2501
$TiBr_2$	2360	2419
TiI_2	2259	2329
VCl_2	2607	2579
VBr_2	—	2523
VI_2	—	2456
CrF_2	2778	2917
$CrCl_2$	2455	2586
$CrBr_2$	2377	2523
CrI_2	2269	2425
$MoCl_2$	2485	2733
$MoBr_2$	2448	2742
MoI_2	2422	2630
MnF_2	2644	—
$MnCl_2$	2368	2537
$MnBr_2$	2304	2471
MnI_2	2212	—
FeF_2	2769	—
$FeCl_2$	2525	2631
$FeBr_2$	2464	2569
FeI_2	2382	2480
CoF_2	2878	3018
$CoCl_2$	2709	2691
$CoBr_2$	2648	2629
CoI_2	2569	2545
NiF_2	2845	3066
$NiCl_2$	2753	2772
$NiBr_2$	2699	2709
NiI_2	2607	2623
$PdCl_2$	2766	2778
$PdBr_2$	—	2741
PdI_2	—	2748
CuF_2	3046	3082
$CuCl_2$	2774	2811
$CuBr_2$	2711	2763
CuI_2	2640	—
AgF_2	2919	2942
ZnF_2	2930	3032
$ZnCl_2$	2690	2734
$ZnBr_2$	2632	2678
ZnI_2	2549	2605
CdF_2	2740	2809
$CdCl_2$	2226	2552
$CdBr_2$	2468	2507
CdI_2	2406	2441
HgF_2	2757	(2798)
$HgCl_2$	2569	2651
$HgBr_2$	2598	2628
HgI_2	2569	2610
SnF_2	2551	—
$SnCl_2$	2276	2297
$SnBr_2$	2211	2245
SnI_2	2123	2193
PbF_2	2460	2522
$PbCl_2$	2229	2269
$PbBr_2$	2169	2219
PbI_2	2086	2163
ScF_3	5096	5492
$ScCl_3$	4874	4866
$ScBr_3$	4711	4729
ScI_3	4640	—

Table 1 (continued)

Substance	Calc. U_{POT} (kJ mol^{-1})	U_{POT}^{BFHC} (kJ mol^{-1})
YF_3	4983	—
YCl_3	4447	4506
YBr_3	4410	—
YI_3	4125	4240
TiF_3	5644	—
$TiCl_3$	5134	5134
$TiBr_3$	5012	5007
TiI_3	4845	—
ZrF_3	—	(5400)
$ZrCl_3$	—	4791
$ZrBr_3$	—	4758
ZrI_3	—	(4591)
VF_3	5895	—
VCl_3	5322	5315
BVr_3	5192	5214
VI_3	5058	5121
$NbCl_3$	5062	—
$NbBr_3$	4980	—
NbI_3	4860	—
CrF_3	5958	6033
$CrCl_3$	5473	5509
$CrBr_3$	5355	—
CrI_3	5201	5274
MoF_3	(6459)	5230
$MoCl_3$	—	5230
$MoBr_3$	—	5156
MoI_3	—	5073
MnF_3	6017	—
$MnCl_3$	5544	—
$MnBr_3$	5448	—
MnI_3	5330	—
$TcCl_3$	5270	—
$TcBr_3$	5215	—
TcI_3	5188	—
FeF_3	5870	—
$FeCl_3$	5364	5359
$FeBr_3$	5268	5333
FeI_3	5117	—
$RuCl_3$	—	5245
$RuBr_3$	—	5223
RuI_3	—	5222
CoF_3	5991	6118
$RhCl_3$	—	5641
IrF_3	(6112)	—
$IrBr_3$	(4794)	—
NiF3	(6111)	—
AuF_3	(5777)	—
$AuCl_3$	(4605)	—
$ZnCl_3$	5832	—
$ZnBr_3$	5732	—
ZnI_3	5636	—
AlF_3	5924	5215
$AlCl_3$	5376	5492
$AlBr_3$	5247	5361
AlI_3	5070	5218
GaF_3	—	6205
$GaCl_3$	5217	5645
$GaBr_3$	4966	5552
GaI_3	4611	5476
$InCl_3$	4736	5187
$InBr_3$	4535	5124
InI_3	4234	5005
TlF_3	5493	—
$TlCl_3$	5252	5258
$TlBr_3$	5171	—
TlI_3	5088	—
$AsBr_3$	—	5497
AsI_3	(3758)	4824
SbF_3	—	5295
$SbCl_3$	—	5032
$SbBr_3$	—	4954
SbI_3	—	4867
$BiCl_3$	—	4689
$BiBr_3$	—	—
BiI_3	(3774)	—
LaF_3	4682	—
$LaCl_3$	4343	4242
$LaBr_3$	4209	—
LaI_3	3916	3986
$CeCl_3$	4297	4284
CeI_3	—	4029
$PrCl_3$	4322	4326
PrI_3	—	4071
$NdCl_3$	4343	—
$SmCl_3$	4376	—
$EuCl_3$	4393	—
$GdCl_3$	4406	—
$DyCl_3$	4481	—
$HoCl_3$	4501	—
$ErCl_3$	4527	
$TmCl_3$	4548	4550
TmI_3	—	4314
$YbCl_3$	—	4546
$AcCl_3$	4096	—
UCl_3	4243	—
$NpCl_3$	4268	—
$PuCl_3$	4289	—
$PuBr_3$	(3959)	—
$AmCl_3$	4293	—
TiF_4	10012	9908
$TiCl_4$	9431	—
$TiBr_4$	9288	9039
TiI_4	9108	8893
ZrF_4	8853	8971
$ZrCl_4$	8096	8144
$ZrBr_4$	7916	7984
ZrI_4	7661	7801
MoF_4	8795	—
$MoCl_4$	8556	9573
$MoBr_4$	8510	9475
MoI_4	8427	—
$SnCl_4$	8355	—
$SnBr_4$	7970	8833
PbF_4	—	9461

Table 1 (continued)

Substance	Calc. U_{POT} (kJ mol^{-1})	U_{POT}^{BFHC} (kJ mol^{-1})
CrF_2Cl	5795	—
CrF_2Br	5753	—
CrF_2I	5669	—
$CrCl_2Br$	5448	—
$CrCl_2I$	5381	—
$CrBr_2I$	5330	—
CuFCl	2891	—
CuFBr	2853	—
CuFI	2803	—
CuClBr	2753	—
CuClI	2694	—
CuBrI	2669	—
FeF_2Cl	5711	—
FeF_2Br	5653	—
FeF_2I	5569	—
$FeCl_2Br$	5339	—
$FeCl_2I$	5272	—
FBr_2I	5209	—
$LiIO_2F_2$	845	—
$NaIO_2F_2$	766	764
KIO_2F_2	699	697
$RbIO_2F_2$	674	671
$CsIO_2F_2$	636	—
$NH_4IO_2F_2$	678	—
$AgIO_2F_2$	736	685
Hydrides		
LiH	858	920
NaH	782	808
KH	699	714
RbH	674	685
CsH	648	644
TiH	—	1407
ZrH	—	1590
VH	—	(1344)
NbH	—	(1633)
PdH	—	(1368)
CuH	—	1254
TiH	996	1407
ZrH	916	1590
HfH	904	—
LaH	828	—
VH	1184	(1344)
NbH	1163	(1633)
TaH	1021	—
CrH	1050	—
NiH	929	—
PdH	979	1368
PtH	937	—
CuH	828	1254
AgH	941	—
AuH	1033	—
TlH	745	—
GeH	950	—
PbH	778	—
BeH_2	3205	3295
MgH_2	2791	2706
CaH_2	2410	2394

Substance	Calc. U_{POT} (kJ mol^{-1})	U_{POT}^{BFHC} (kJ mol^{-1})
SrH_2	2250	2253
BaH_2	2121	2121
ScH_2	2711	2659
YH_2	(2598)	2670
LaH_2	2380	2522
CeH_2	2414	2484
PrH_2	2448	2398
NdH_2	2464	2367
PmH_2	2519	—
SmH_2	2510	2389
GdH_2	2494	2706
AcH_2	2372	—
ThH_2	2711	—
PuH_2	2519	—
AmH_2	2544	—
TiH_2	2866	2845
ZrH_2	2711	2999
CuH_2	2941	—
ZnH_2	2870	—
HgH_2	2707	—
AlH_3	5924	—
FeH_3	5724	—
ScH_3	5439	—
YH_3	5063	—
LaH_3	4895	4493
FeH_3	5724	—
GaH_3	5690	—
InH_3	5092	—
TlH_3	5092	—
Hydroselenides		
NaHse	703	—
KHSe	644	—
RbHSe	623	—
CsHSe	598	—
Hydrosulphides		
LiHS	759	821
NaHS	704	747
KHS	650	659
RbHS	623	637
CsHS	582	595
NH_4HS	661	666
$Ca(HS)_2$	2184	(2171)
$Sr(HS)_2$	2063	—
$Ba(HS)_2$	1979	(1956)
Hydroxides		
LiOH	1021	1039
NaOH	887	900
KOH	789	804
RbOH	766	773
CsOH	721	724
$Be(OH)_2$	3477	3629
$Mg(OH)_2$	2870	3006
$Ca(OH)_2$	2506	2645
$Sr(OH)_2$	2330	2483
$Ba(OH)_2$	2141	2339
$Ti(OH)_2$	—	2962
$Mn(OH)_2$	2909	3008

Table 1 (continued)

Substance	Calc. U_{POT} (kJ mol^{-1})	U_{POT}^{BFHC} (kJ mol^{-1})
$Fe(OH)_2$	2653	3055
$Co(OH)_2$	2786	3115
$Ni(OH)_2$	2832	3193
$Pd(OH)_2$	—	3175
$Cu(OH)_2$	2870	3237
CuOH	1006	—
AgOH	918	—
AuOH	1033	—
TlOH	705	—
$Zn(OH)_2$	2795	3158
$Cd(OH)_2$	2607	2918
$Hg(OH)_2$	2669	—
$Sn(OH)_2$	2489	2729
$Pb(OH)_2$	2376	2623
$Sc(OH)_3$	5063	—
$Y(OH)_3$	4707	—
$La(OH)_3$	4443	—
$Cr(OH)_3$	5556	—
$Mn(OH)_3$	6213	—
$Al(OH)_3$	5627	—
$Ga(OH)_3$	5732	—
$In(OH)_3$	5280	—
$Tl(OH)_3$	5314	—
$Ti(OH)_4$	9456	—
$Zr(OH)_4$	8619	—
$Mn(OH)_4$	10933	—
$Sn(OH)_4$	9188	—
Imides		
CaNH	3293	—
SrNH	3146	—
BaNH	2975	—
Metaniobates		
$NaNbO_3$	789	—
$Ca(NbO_3)_2$	2315	—
$Fe(NBO_3)_2$	2502	—
Metatantalates		
$NaTaO_3$	789	—
$Ca(TaO_3)_2$	2315	—
$Fe(TaO_3)_2$	2502	—
Metavanadates		
Li_3VO_4	3945	—
Na_3VO_4	3766	—
K_3VO_4	3376	—
Rb_3VO_4	3243	—
Cs_3VO_4	3137	—
Nitrates		
$LiNO_3$	848	848
$NaNO_3$	755	756
KNO_3	685	687
$RbNO_3$	662	658
$CsNO_3$	648	625
NH_4NO_3	661	676
$AgNO_3$	820	822
$TlNO_3$	690	700
$Mg(NO_3)_2$	2468	2503
$Ca(NO_3)_2$	2209	2228
$Sr(NO_3)_2$	2092	2132
$Ba(NO_3)_2$	1975	2016
$Mn(NO_3)_2$	2318	2519
$Fe(NO_3)_2$	—	(2563)
$CO(NO_3)_2$	2560	2626
$Ni(NO_3)_2$	—	2709
$Cu(NO_3)_2$	—	2720
$Zn(NO_3)_2$	2376	2628
$Cd(NO_3)_2$	2238	2443
$Ha(NO_3)_2$	2255	—
$Sn(NO_3)_2$	2155	2254
$Pb(NO_3)_2$	2067	2189
Nitrides		
ScN	7547	7506
LaN	6876	6793
TiN	8130	8033
ZrN	7633	7723
VN	8283	8233
NbN	7939	8022
CrN	8269	8358
Nitrites		
$NaNO_2$	774	748
KNO_2	660	664
$RbNO_2$	638	763
$CsNO_2$	598	—
$Ca(NO_2)_2$	2460	2225
$Sr(NO_2)_2$	2305	2111
$Ba(NO_2)_2$	2205	1987
Oxides		
Li_2O	2799	—
Na_2O	2481	—
K_2O	2238	—
Rb_2O	2163	—
Cu_2O	3273	—
Ag_2O	3002	
Tl_2O	2659	—
LiO_2	(878)	(872)
NaO_2	799	796
KO_2	741	725
RbO_2	706	695
CsO_2	679	668
Li_2O_2	2592	256
Na_2O_2	2309	2305
K_2O_2	2114	2078
Rb_2O_2	2025	2006
Cs_2O_2	1948	1901
MgO_2	3356	3526
CaO_2	3144	3133
SrO_2	3037	2849
KO_3	697	—
BeO	4293	4443
MgO	3795	3791
CaO	3414	3401
SrO	3217	3223
BaO	3029	3054
TiO	3832	3811
VO	3932	3863
MnO	3724	3745

Table 1 (continued)

Substance	Calc. U_{POT} (kJ mol^{-1})	U_{POT}^{BFHC} (kJ mol^{-1})
FeO	3795	3865
CoO	3837	3910
NiO	3908	4010
PdO	3736	—
CuO	4135	4050
ZnO	4142	3971
CdO	3806	—
HgO	3907	—
GeO	3919	—
SnO	3652	—
PbO	3520	—
Sc_2O_3	13557	13708
Y_2O_3	12705	—
La_2O_3	12452	—
Ce_2O_3	12661	—
Pr_2O_3	12703	—
Nd_2O_3	12736	—
Pm_2O_3	12811	—
Sm_2O_3	12878	—
Eu_2O_3	12945	—
Gd_2O_3	12996	—
Tb_2O_3	13071	—
Dy_2O_3	13138	—
Ho_2O_3	13180	—
Er_2O_3	13263	—
Tm_2O_3	13322	—
Yb_2O_3	13380	—
Lu_2O_3	13665	—
Ac_2O_3	12573	—
Ti_2O_3	—	14149
V_2O_3	15096	14520
Cr_2O_3	15276	14957
Mn_2O_3	15146	15035
Fe_2O_3	14309	14774
Al_2O_3	15916	—
Ga_2O_3	15590	15220
In_2O_3	13928	—
Ti_2O_3	14702	—
Pb_2O_3	(14841)	—
CeO_2	9627	—
ThO_2	10397	—
PaO_2	10573	—
VO_2	10644	—
NpO_2	10707	—
PuO_2	10786	—
AmO_2	10799	—
CmO_2	10832	—
TiO_2	12150	—
ZrO_2	11188	—
MoO_2	11648	—
MnO_2	12970	—
SiO_2	13125	—
GeO_2	12828	—
SnO_2	11807	—
PbO_2	11217	
Perchlorates		
$LiClO_4$	709	723
$NaClO_4$	643	648
$KClO_4$	599	602
$RbClO_4$	564	582
$CsClO_4$	636	(542)
NH_4ClO_4	583	580
$Ca(ClO_4)_2$	1958	1971
$Sr(ClO_4)_2$	1862	1862
$Ba(ClO_4)_2$	1795	1769
$NaMnO_4$	661	—
$KMnO_4$	607	—
$RbMnO_4$	586	—
$CsMnO_4$	565	—
$Ca(MnO_4)_2$	1937	—
$Sr(MnO_4)_2$	1845	—
$Ba(MnO_4)_2$	1778	—
Phosphates		
$Mg_4(PO_4)_2$	11632	11407
$Ca_3(PO_4)_2$	10602	10479
$Sr_3(PO_4)_2$	10125	10075
$Ba_3(PO_4)_3$	9652	9654
$MnPO_4$	7397	—
$FePO_4$	7251	7303
BPO_4	8201	—
$AlPO_4$	7427	7509
$GaPO_4$	7381	
Phosphonium salts		
PH_4Br	616	—
PH_4I	590	—
Selenides		
Li_2Se	2364	—
Na_2Se	2130	—
K_2Se	1933	—
Rb_2Se	1837	—
CS_2Se	1745	—
Ag_2Se	2686	—
Tl_2Se	2209	—
BeSe	3431	—
MgSe	3071	—
CaSe	2858	2862
SrSe	2736	—
BaSe	2611	—
MnSe	3176	3194
FeSe	3499	3396
CoSe	3554	3471
NiSe	3658	3558
CuSe	3736	3662
ZnSe	3502	3514
CdSe	3330	—
HgSe	3501	—
SnSe	3058	—
PbSe	3050	—
Selenites		
Li_2SeO_3	2171	—
Na_2SeO_3	1950	1931
K_2SeO_3	1774	—
Rb_2SeO_3	1715	1675
Cs_2SeO_3	1640	—

Table 1 (continued)

Substance	Calc. U_{POT} (kJ mol^{-1})	U_{POT}^{BFHC} (kJ mol^{-1})
Tl_2SeO_3	1879	—
Ag_2SeO_3	2113	—
$BeSeO_3$	3322	—
$MgSeO_3$	3012	2996
$CaSeO_3$	2732	—
$SrSeO_3$	2586	2586
$BaSeO_3$	2460	2448
$RaSeO_3$	2456	—
$MnSeO_3$	2975	—
$FeSeO_3$	2895	—
$CoSeO_3$	3155	—
$NiSeO_3$	2945	—
$CuSeO_3$	3209	—
$ZnSeO_3$	3167	—
$CdSeO_3$	2962	—
$PbSeO_3$	2669	—
Selenates		
Li_2SeO_4	2054	—
Na_2SeO_4	1879	—
K_2SeO_4	1732	—
Rb_2SeO_4	1686	—
Cs_2SeO_4	1615	—
Cu_2SeO_4	2201	—
Ag_2SeO_4	2033	—
Tl_2SeO_4	1766	—
Hg_2SeO_4	2163	—
$BeSeO_4$	3448	—
$MgSeO_4$	2895	—
$CaSeO_4$	2632	—
$SrSeO_4$	2489	—
$BaSeO_4$	2385	—
$RaSeO_4$	2364	—
$GdSeO_4$	2753	—
$MnSeO_4$	2837	—
$FeSeO_4$	3008	—
$CoSeO_4$	2912	—
$NiSeO_4$	3079	—
$CuSeO_4$	3104	—
$ZnSeO_4$	3021	—
$CdSeO_4$	2833	—
$HgSeO_4$	2845	—
$PbSeO_4$	2561	—
Sulphides		
Li_2S	2464	2472
Na_2S	2192	2203
K_2S	1979	(2052)
Rb_2S	1929	1949
Cs_2S	1892	1850
$(NH_4)_2S$	2008	(2026)
Cu_2S	2786	2865
Ag_2S	2606	2677
Au_2S	2908	—
Tl_2S	2298	2258
Sulphates		
Li_2SO_4	2229	2142
Na_2SO_4	1827	1938
K_2SO_4	1700	1796
Rb_2SO_4	1636	1748
Cs_2SO_4	1596	1658
$(NH_4)_2SO_4$	1766	1777
Cu_2SO_4	2276	2166
Ag_2SO_4	2104	1989
Tl_2SO_4	1828	1722
Hg_2SO_4	—	2127
$CaSO_4$	2489	2480
$SrSO_4$	2577	2484
$BaSO_4$	2469	2374
$MnSO_4$	2920	2825
$FeSO_4$	2983	2921
$CoSO_4$	3088	2917
$NiSO_4$	3167	3044
$CuSO_4$	3167	3066
$ZnSO_4$	3100	3006
$CdSO_4$	2891	—
$PbSO_4$	2635	2534
$Sn(SO_4)_2$	—	9616
Ternary salts		
Cs_2CuCl_4	1393	—
Rb_2ZnCl_4	1529	—
Cs_2ZnCl_4	1492	—
Rb_2ZnBr_4	1498	—
Cs_2ZnBr_4	1454	—
$(Me_4N)_2ZnBr_4$	1364	—
Cs_2ZnI_4	1386	—
$CsGaCl_4$	494	—
$NaAlCl_4$	556	—
$CsAlCl_4$	486	—
$NaFeCl_4$	492	—
Rb_2CoCl_4	1447	—
Cs_2CoCl_4	1391	—
K_2PtCl_4	1574	1550
Cs_2GeF_6	1573	—
$(NH_4)_2GeF_6$	1657	—
Cs_2GeCl_6	1375	1375
K_2HfCl_6	1345	1461
K_2IrCl_6	1442	1440
$(NH_4)_2IrCl_6$	1442	—
Na_2MoCl_6	1526	—
K_2MoCl_6	1418	1433
Rb_2MoCl_6	1399	1399
Cs_2MoCl_6	1347	1347
K_2NbCl_6	1375	1398
Rb_2NbCl_6	1371	1385
Cs_2NbCl_6	1381	1344
K_2OsCl_6	1447	1447
Cs_2OsCl_6	1409	—
$(NH_4)_2OsCl_6$	1433	—
K_2OsBr_6	1396	—
K_2PdCl_6	1450	1466
Rb_2PdCl_6	1449	—
Cs_2PdCl_6	1426	—
$(NH_4)_2PdCl_6$	1481	—
Rb_2PbCl_6	1343	1343
Cs_2PbCl_6	1344	—

Table 1 (continued)

Substance	Calc. U_{POT} (kJ mol^{-1})	U_{POT}^{BFHC} (kJ mol^{-1})
$(NH_4)_2PbCl_6$	1355	—
K_2PtCl_6	1468	—
Rb_2PtCl_6	1464	—
Cs_2PtCl_6	1444	—
$(NH_4)_2PtCl_6$	1468	—
Tl_2PtCl_6	1546	—
Ag_2PtCl_6	1773	—
$BaPtCl_6$	2047	—
K_2PtBr_6	1423	—
Ag_2PtBr_6	1791	—
K_2PtI_6	1421	—
K_2ReCl_6	1416	1442
Rb_2ReCl_6	1414	—
Cs_2ReCl_6	1398	—
$(NH_4)_2ReCl_6$	1402	1390
K_2ReBr_6	1375	1375
K_2SiF_6	1670	1628
Rb_2SiF_6	1639	1621
Cs_2SiF_6	1604	1498
Tl_2SiF_6	1675	—
$(NH_4)_2SiF_6$	1657	1727
K_2SnCl_6	1363	1390
Rb_2SnCl_6	1361	1363
Cs_2SnCl_6	1358	—
Tl_2SnCl_6	1437	—
$(NH_4)_2SnCl_6$	1370	1334
Rb_2SnBr_6	1309	—
Cl_2SnBr_6	1306	—
$(NH_4)_2SnBr_6$	1319	—
Rb_2SnI_6	1226	—
Cs_2SnI_6	1243	—
K_2TeCl_6	1318	1320
Rb_2TeCl_6	1321	—
Cs_2TeCl_6	1323	—
Tl_2TeCl_6	1392	—
$(NH_4)_2TeCl_6$	1318	—
K_2RuCl_6	1451	—
Rb_2CoF_6	1688	—
Cs_2CoF_6	1632	—
K_2NiF_6	1721	—
Rb_2NiF_6	1688	—
Rb_2SbCl_6	1357	—
Rb_2SeCl_6	1409	—
Cs_2SeCl_6	1397	—
$(NH_4)_2SeCl_6$	1420	—
K_2SeBr_6	1379	—
$(NH_4)_2SeBr_6$	1380	—
$(NH_4)_2PoCl_6$	1338	—
Cs_2PoBr_6	1286	—
$(NH_4)_2PoBr_6$	1292	—
Cs_2CrF_6	1603	—
Rb_2MnF_6	1688	—
Cs_2MnF_6	1620	—
K_2MnCl_6	1462	—
Rb_2MnCl_6	1451	—
$(NH_4)_2MnCl_6$	1464	—
Cs_2TeBr_6	1306	—
$(NH_4)_2TeBr_6$	1294	—
Cs_2TeI_6	1246	—
K_2TiCl_6	1412	1443
Rb_2TiCl_6	1415	1425
Cs_2TiCl_6	1402	1370
Tl_2TiCl_6	1560	1553
$(NH_4)_2TiCl_6$	1413	—
K_2TiBr_6	1379	1379
Rb_2TiBr_6	1341	1365
Cs_2TiBr_6	1339	1316
Na_2UBr_6	1504	—
K_2UBr_6	1484	—
Rb_2UBr_6	1473	—
Cs_2UBr_6	1459	—
K_2WCl_6	1398	1423
Rb_2WCl_6	1397	1434
Cs_2WCl_6	1392	1366
K_2WBr_6	1408	1408
Rb_2WBr_6	1361	1391
Cs_2WBr_6	1362	1332
K_2ZrCl_6	1339	1371
Rb_2ZrCl_6	1341	—
Cs_2ZrCl_6	1339	1308
Tetraalkyl ammonium salts		
Me_4NCl	566	—
Me_4NBr	553	—
Me_4NI	544	—
Tellurides		
Li_2Te	2212	—
Na_2Te	1997	2095
K_2Te	1830	—
Rb_2Te	1837	—
Cs_2Te	1745	
Cu_2Te	2706	2683
Ag_2Te	2607	2600
Tl_2Te	2084	2172
BeTe	3319	—
MgTe	2878	3081
CaTe	2721	—
SrTe	2599	—
BaTe	2473	—
RaTe	2481	—
MnTe	3041	—
FeTe	3399	—
CoTe	3429	—
NiTe	3534	—
CuTe	3639	—
ZnTe	3416	—
CdTe	3212	—
PbTe	2930	—
Thiocyanates		
LiCNS	764	(765)
NaCNS	682	682
KCNS	623	616
RbCNS	623	619
CsCNS	623	568
NH_4CNS	605	611
$Ca(CNS)_2$	2184	2118
$Sr(CNS)_2$	2063	1957

Table 1 (continued)

Substance	Calc. U_{POT} (kJ mol^{-1})	U_{POT}^{BFHC} (kJ mol^{-1})
$Ba(CNS)_2$	1979	1852
$Mn(CNS)_2$	2280	2351
$Zn(CNS)_2$	2335	2560
$Cd(CNS)_2$	2201	2374
$Hg(CNS)_2$	2146	2492
$Sn(CNS)_2$	2117	2142
$Pb(CNS)_2$	2058	—
Vanadates		
$LiVO_3$	810	—
$NaVO_3$	761	—
KVO_3	686	—
$RbVO_3$	657	—
$CsVO_3$	628	—

Table 2
ANCILLARY THERMODYNAMIC DATA

Substance	State	$\Delta H_f°$ (kJ mol^{-1})
Li^+	g	687.16
Na^+	g	609.84
K^+	g	514.19
Cs^+	g	452.3
O_2^-	g	–74
CN^-	g	41
NH_4^+	g	630.2
C_2^{-2}	g	918
O_2^{-2}	g	553
NO_3^-	g	–320.1
$IO_2F_2^-$	g	–693
IO_3^-	g	–208
ClO_3^-	g	–200
BrO_3^-	g	–145
HSO_4^-	g	(–1012)
HCO_3^-	g	–738
$NH_2CH_2CO_2^-$	g	–564
N^{-3}	g	(2588)
CO_3^{-2}	g	–321
PO_4^{-3}	g	(291)
AsO_4^{-3}	g	(289)
GeO_4^{-4}	g	(1460)
CH_3COO^-	g	–554
ClO_4^-	g	–344
NO_2^-	g	–219
SeO_3^{-2}	g	(–249)
HCO_2^-	g	–463
BH_4^-	g	–96
HS^-	g	–120
$TiCl_6^{-2}$	g	–1330
$TiBr_6^{-2}$	g	–1142
$ZrCl_6^{-2}$	g	–1526
$HfCl_6^{-2}$	g	–1640
$NbCl_6^{-2}$	g	–1224
$TaCl_6^{-2}$	g	–1275
$MoCl_6^{-2}$	g	–1070
WCl_6^{-2}	g	–985
WBr_6^{-2}	g	–705
$ReCl_6^{-2}$	g	–919
$ReBr_6^{-2}$	g	–689
$OsCl_6^{-2}$	g	–752
$IrCl_6^{-2}$	g	–785
$PdCl_6^{-2}$	g	–749
$PtCl_6^{-2}$	g	–774
$PtBr_6^{-2}$	g	–645
SiF_6^{-2}	g	–2207
$GeCl_6^{-2}$	g	–981
$SuCl_6^{-2}$	g	–1156
$PbCl_6^{-2}$	g	–940
$TeCl_6^{-2}$	g	902
NO_2^-	g	–202
LiO_2	c	–259.4
NaO_2	c	–260.7
KO_2	c	–284.5
CsO_2	c	–289.5
Li_2O_2	c	–632.6
Na_2O_2	c	–531.2
K_2O_2	c	–495.8
Cs_2O_2	c	–402.5
ThO_2	c	–1230.5
MgO_2	c	–623.0
$NaIO_2F_2$	c	–848.7
KIO_2F_2	c	–876.6
$RbIO_2F_2$	c	–875.3
$NH_4IO_2F_2$	c	–747.7
$LiBrO_3$	c	–356.4
$CsBrO_3$	c	–374.5
$CsHCO_3$	c	[illegible]
$Ca(HCO_3)_2$	c	(–1950)
$Sr(HCO_3)_2$	c	(–1954)
$Ba(HCO_3)_2$	c	(–1971)
CH_3COONH_4	c	–618.4
CH_3COOLi	c	–748.9
CH_3COORb	c	–720.9
$LiClO_4$	c	–380.7
$CsClO_4$	c	(–435.1)
LiH	c	–90.62
NaH	c	–56.4
KH	c	–57.8
CsH	c	–49.9

Table 2
ANCILLARY THERMODYNAMIC DATA (continued)

Substance	State	$\Delta H_f°$ (kJ mol^{-1})
LiN_3	c	10.8
NaN_3	c	21.25
KN_3	c	–1.38
RbN_3	c	–0.29
CsN_3	c	–9.92
TlN_3	c	233.4
AgN_3	c	310.3
CsHS	c	–263.2
$Ca(HS)_2$	c	(–481)
$Ba(HS)_2$	c	(–531.4)
LiHS	c	–254.4
K_2TiCl_6	c	–1747
Rb_2TiCl_6	c	–1767
Cs_2TiCl_6	c	–1797
Tl_2TiCl_6	c	–1330
K_2TiBr_6	c	–1493
Rb_2TiBr_6	c	–1517
Cs_2TiBr_6	c	–1553
K_2ZrCl_6	c	–1932
Cs_2ZrCl_6	c	–1992
K_2HfCl_6	c	–1957
K_2NbCl_6	c	–1594
Rb_2NbCl_6	c	–1619
Cs_2NbCl_6	c	–1663
K_2ReCl_6	c	–1333
$(NH_4)_2ReCl_6$	c	–1056
K_2ReBr_6	c	–1036
K_2OsCl_6	c	–1171
K_2IrCl_6	c	–1197
K_2SiF_6	c	–2807
Rb_2SiF_6	c	–2838
Cs_2SiF_6	c	–2801
$(NH_4)_2SiF_6$	c	–2681
Rb_2PbCl_6	c	–1293
$Mn(CN)_2$	c	60.7
RbCN	c	(–108.8)
CsCN	c	(–108.7)
LiCN	c	–121
$Sr(CN)_2$	c	(–205)
$LiNO_3$	c	–482.3
$RbNO_3$	c	–489.7
$CsNO_3$	c	–494.2
$Fe(NO_3)_2$	c	(–447.7)
$Sn(NO_3)_2$	c	–456.1
$NaBF_4$	c	–1774
$CsIO_3$	c	–526.3
$CsClO_3$	c	–395.8
$Mg(ClO_3)_2$	c	(–523)
$Ca(ClO_3)_2$	c	–757
$Sr(ClO_3)_2$	c	–766
ScN	c	–284.5
Li_2CO_3	c	–1216.0
Cs_2CO_3	c	–1118.8
$NiCO_3$	c	–689.1
$CuCO_3$	c	–595.0
$SnCO_3$	c	(–740.6)
HCO_2Li	c	–649.3
HCO_2Rb	c	–656.0
TiH	c	–130.5
ZrH	c	–173.5
VH	c	(–32)
NbH	c	(–100)
PdH	c	(–37)
CuH	c	(–21.4)
LaH_2	c	–230.1
CeH_2	c	–141.8
PrH_2	c	–200.0
NdH_2	c	–187.4
SmH_2	c	–187.4
GdH_2	c	–196.2
CsOH	c	–406.7
$Ti(OH)_2$	c	(–778.0)
LiOH	c	–487.2
K_2TaCl_6	c	–1648
Rb_2TaCl_6	c	–1669
Cs_2TaCl_6	c	–1711
Na_2MoCl_6	c	–1376
K_2MoCl_6	c	–1475
Rb_2MoCl_6	c	–1479
Cs_2MoCl_6	c	–1512
K_2WCl_6	c	–1380
Rb_2WCl_6	c	–1429
Cs_2WCl_6	c	–1446
K_2WBr_6	c	–1085
Rb_2WBr_6	c	–1106
Cs_2WBr_6	c	–1132
K_2PdCl_6	c	–1187
Ag_2PtCl_6	c	–527
$BaPtCl_6$	c	–1180
K_2PtBr_6	c	–1040
Ag_2PtBr_6	c	–398
Cs_2GeCl_6	c	–1451
K_2SnCl_6	c	–1518
Rb_2SnCl_6	c	–1529
$(NH_4)_2SnCl_6$	c	–1237
Rb_2TeCl_6	c	–1237

THE MADELUNG CONSTANT AND CRYSTAL LATTICE ENERGY

If U is the crystal lattice energy and M is the Madelung constant, then[a]

$$U = \frac{N M z_i z_j \, e^2}{r}(1 - 1/n)$$

Substance	Ion type	Crystal form[b]	M
Sodium chloride, NaCl	M^+, X^-	FCC	1.74756
Cesium chloride, CsCl	M^+, X^-	BCC	1.76267
Calcium chloride, $CaCl_2$	M^{++}, $2X^-$	Cubic	2.365
Calcium fluoride (fluorite), CaF_2	M^{++}, $2X^-$	Cubic	2.51939
Cadmium chloride, $CdCl_2$	M^{++}, $2X^-$	Hexagonal	2.244[c]
Cadmium iodide (α), CdI_2	M^{++}, $2X^-$	Hexagonal	2.355[c]
Magnesium fluoride, MgF_2	M^{++}, $2X^-$	Tetragonal	2.381[c]
Cuprous oxide (cuprite), Cu_2O	$2M^+$, X^{--}	Cubic	2.22124
Zinc oxide, ZnO	M^{++}, X^{--}	Hexagonal	1.4985[c]
Sphalerite (zinc blende), ZnS	M^{++}, X^{--}	FCC	1.63806
Wurtzite, ZnS	M^{++}, X^{--}	Hexagonal	1.64132[c]
Titanium dioxide (anatase), TiO_2	M^{4+}, $2X^{--}$	Tetragonal	2.400[c]
Titanium dioxide (rutile), TiO_2	M^{4+}, $2X^{--}$	Tetragonal	2.408[c]
β-Quartz, SiO_2	M^{4+}, $2X^{--}$	Hexagonal	2.2197[c]
Corundum, Al_2O_3	$2M^{3+}$, 3X	Rhombohedral	4.1719

[a] N is Avogadro's number, z_i and z_j are the integral charges on the ions(in units of e), and e is the charge on the electron in electrostatic units ($e = 4.803 \times 10^{-10}$ esu). r is the shortest distance between cation-anion pairs in centimeters. Then U is in ergs (1 erg = 10^{-7} J).

[b] FCC = face centered cubic; BCC = body centered cubic.

[c] For tetragonal and hexagonal crystals the value of M depends on the details of the lattice parameters.

The Born Exponent, n is:

Ion type	n
He, Li^+	5
Ne, Na^+, F^-	7
Ar, K^+, Cu^+, Cl^-	9
Kr, Rb^+, Ag^+, Br^-	10
Xe, Cs^+, Au^+, I^-	12

For a crystal with a mixed-ion type, an average of the values of n in this table is to be used (6 for LiF, for example).

LATTICE SPACING OF COMMON ANALYZING CRYSTALS

Crystal	Reflection plane	*d* Spacing (Å)
ADP[a]	101	5.31
ADP[a]	110	5.325
ADP[a]	200	3.75
Beryl	$10\bar{1}0$	7.98
Calcite	100	3.036
EDDT[b]	020	4.404
Germanium	111	3.265
Graphite	001	6.69
Gypsum	010	7.600
KAP[c]	001	13.32
KBr	200	3.29
KCl	200	3.14
Lead stearate		51
LiF	200	2.014
LiF	220	1.424
Mica	002	9.96
NaCl	200	2.820
Oxalic acid	001	5.85
PET[d]	002	4.371
Quartz	$10\bar{1}0$	4.225
Quartz	$10\bar{1}1$	3.343
Quartz	$11\bar{2}0$	2.456
Silicon	111	3.13
Topaz	303	1.356

While several of the above spacings have been measured to more than four significant figures, no more than four figures are given here because complications introduced by the index of refraction, anomalous dispersion, temperature coefficient of expansion, and crystal impurities must be considered before the additional figures are useful.

[a] Ammonium dihydrogen phosphate.
[b] Ethylenediamine *d*-tartrate.
[c] Potassium acid phthalate.
[d] Pentaerythritol.

LATTICE CONSTANTS FOR CUBIC CRYSTALS

From Volume 14, pages 689, 690, and 691 of the Analytical Edition of *Industrial and Engineering Chemistry*, with permission.

a/Å	Substance
A 4	
3.56	C (diamond)
5.42	Si
5.62	Ge
6.46	α-Sn
A 1	
3.517	Ni
3.554	α-Co
3.60	Taenite (57.7% Fe, 40.8% Ni, 0.5% P)
3.608	Cu
3.63	γ-Fe (1370 K.)
3.797	Rh
3.831	Ir
3.880	Pd
3.88—4.04	Pd-H
3.912	Pt
4.041	Al
4.070	Au
4.077	Ag
4.30	Co-N
4.40—4.46	Ti-H
4.52	Ne (4 K.)
4.66	Zr-H
4.84	β-Tl
4.939	Pb
5.08	Th
5.14	α-Ce
5.296	β-La
5.43	A (4 K.)
5.56	Ca
5.59	Kr (20 K.)
5.70	Kr (92 K.)
6.05	Sr
6.20	X (88 K.)
A 2	
2.861	α-Fe
2.873	α-Cr
2.90	β-Fe (1070 K.)
2.93	δ-Fe (1700 K.)
3.03	V
3.03—3.41	V-C
3.140	Mo
3.157	W
3.295	Cb
3.30	Ta
3.32	β-Ti (1200 K.)
3.46	Li (~80 K.)
3.50	Li
3.61	β-Zr (1120 K.)
4.24	Na (~80 K.)
4.29	Na
5.02	Ba
5.20	K (120 K.)
5.33	K
5.62	Rb (~80 K.)
6.05	Cs (~80 K.)
B 1	
4.018	LiF
4.065	LiD
4.08	VO
4.09	LiH
4.12	(Li_2TiO_3)
4.12—4.20	(Li_2TiO_3-MgO)
4.13	VN
4.14	CrN
4.14	VC
4.142	($.63Li_2Fe_2O_4 \cdot .37Li_2TiO_3$)
4.173	NiO
4.207	MgO
4.282	MgO (1570 K.)
4.225	TiN
4.235	TiO
4.24	80 TiN-20 TiC
4.27	CoO
4.28	V-N
4.283	FeO (160 K.)
4.290	FeO (299 K.)
4.30	VC (ε-phase)
4.315	TiC
4.40	CbC
4.41	CbN
4.426	MnO (117 K.)
4.436	MnO (299 K.)
4.44	ScN
4.446	TaC
4.458	HfC
4.615	NaF
4.62	ZrN
4.69	CdO
4.69	ZrC
4.80	CaO
4.82	(Na_2CeO_3)
4.84	(Na_2PrO_3)
4.88	NaH
4.92	AgF
5.006	CaNH
5.13	SrO
5.14	LiCl
5.14	NdN
5.19	MgS
5.192	MnS (130 K.)
5.210	MnS (299 K.)
5.33	KF
5.45	MgSe
5.45	MnSe
5.45	SrNH
5.49	LiBr
5.52	BaO
5.545	AgCl
5.55—5.76	AgCl-AgBr
5.627	NaCl
5.63	RbF
5.68	CaS
5.69	SnAs
5.70	KH
5.755	AgBr
5.76—5.92	AgBr-AgI
5.83	NdP
5.83	NaCN
5.84	BaNH
5.87	SrS
5.91	CaSe
5.94	PbS
5.95	NaBr
5.957	EuS
5.96	NdAs
6.00	PrAs
6.00	LiI
6.01	CsF
6.04	RbH
6.05	β-NaSH (>360 K.)
6.06	CeAs
6.13	LaAs
6.14	PbSe
6.23	SrSe
6.278	KCl
6.285	SnTe
6.31	NdSb
6.345	CaTe
6.35	PrSb
6.36	BaS
6.38	CsH
6.40	CeSb
6.44	PbTe
6.462	NaI
6.45	PrBi
6.48	LaSb
6.49	CeBi
6.53	KCN
6.53	NH_4Cl (>457 K.)
6.56	RbCl
6.57	LaBi
6.58	KBr
6.59	BaSe
6.60	β-KSH (>440 K.)
6.65	SrTe
6.82	RbCN
6.86	RbBr
6.90	NH_4Br (>411 K.)
6.93	β-RbSH (470 K.)
6.99	BaTe
7.052	KI
7.10	β-CsCl (>730 K.)
7.24	NH_4I (>255 K.)
7.325	RbI
H 0_5	
6.96 ± 0.04	$AgClO_4$ (453 ± 20 K.)

a/Å		Substance
7.16 ± 0.10		$NaClO_4$ (618 ± 35 K.)
7.49 ± 0.02		$KClO_4$ (598 ± 15 K.)
7.65 ± 0.05		$TlClO_4$(553 K.)
7.65 ± 0.02		NH_4ClO_4(528 ± 15 K.)
7.68 ± 0.03		$RbClO_4$(583 ± 10 K.)
7.97 ± 0.01		$CsClO_4$ (513 ± 10 K.)
	B 3	
4.255		CuF
4.36		CSi IV
4.855		BeS
5.10		BeSe
5.304		$(Cu, Fe, Mo, Sn)_4(S, As, Te)_2$, cousite
5.41		CuCl
5.425		β-ZnS
5.43		AlP
5.44		GaP
5.58		BeTe
5.60		MnS (red)
5.63		AlAs
5.635		GaAs
5.655		ZnSe
5.68		CuBr
5.82		β-CdS
5.84		HgS
5.86		InP
6.04		CdSe
6.04		InAs
6.05		CuI
6.07		HgSe
6.08		ZnTe
6.103		α-Cu_2HgI_4
6.12		AlSb
6.12		GaSb
6.13		SnSb
6.383		α-Ag_2HgI_4
6.40		HgTe
6.43		CdTe
6.45		InSb
6.48		AgI
	B 32	
6.195		LiGa
6.209		LiZn
6.36		LiAl
6.687		LiCd
6.786		LiIn
7.297		NaIn
7.373		$(CeMg_3)$
7.373		$(PrMg_3)$
7.473		NaTl
	B 20	
4.437		NiSi
4.438		FeSi
4.438		CoSi
4.548		MnSi
4.620		CrSi
	B 2	
2.603		NiBe
2.606		CoBe
2.69		CuBe
2.813		PdBe
2.82		AlNi
2.945		CuZn
2.989		CuPd
3.146		AuZn
3.156		AgZn
3.168		AgLi
3.259		AuMg
3.275		AgMg
3.287		HgLi
3.325		AgCd
3.34		AuCd (670 K.)
3.424		LiTl
3.442		HgMg
3.628		MgTl
3.67		PrZn
3.70		CeZn
3.73		AlNd
3.74		α-RbCl (83 K.)
3.75		LaZn
3.82		TlCn
3.82		PrCd
3.84		TlSb
3.835		TlCl
3.847		CaTl
3.86		NH_4Cl (<457 K.)
3.86		CeCd
3.88		MgPr
3.90		LaCd
3.97		TlBr
3.98		TlBi
4.024		SrTl
4.05		NH_4Br (<411 K.)
4.112		CsCl
4.20		TlI
4.20		CsCl (<720 K.)
4.25		CsCN
4.287		CsBr
4.29		CsSH
4.37		NH_4I (290 K.)
4.56		CsI
	D 2_1	
4.07		YB_6
4.07		ErB_6
4.10		NdB_6
4.12		GdB_6
4.12		PrB_6
4.13		CeB_6
4.13		YbB_6
4.14		CaB_6
4.15		LaB_6
4.15		ThB_6
4.19		SrB_6
4.33		BaB_6
	C 1	
4.33		Be_2C
4.619		Li_2O
5.06		$(3ZrO_2 \cdot MgO)$
5.07		ZrO_2
5.08		$(95ZrO_2 \cdot 5CeO_2)$
5.13		$(95HiO_2 \cdot 5CeO_2)$
5.38		PrO_2
5.40		CeO_2
5.40		CdF_2
5.406		CuF_2
5.45		CaF_2
5.47		UO_2
5.526		$(66CaF_2 \cdot 33YF_3)$
5.53		$(91CaF_2 \cdot 9ThF_4)$
5.54		HgF_2
5.55		Na_2O
5.58		ThO_2
5.59		Cu_2S
5.704		Li_2S
5.749		Cu_2Se
5.782		SrF_2
5.796		EuF_2
5.838		$(66SrF_2 \cdot 33LaF_3$
5.91		$PtAl_2$
5.91		$PtGa_2$
5.935		β-PbF_2(520 K.)
5.99		Al_2Au
6.005		Li_2Se
6.06		$AuGa_2$
6.19		BaF_2
6.34		Mg_2Si
6.35		$PtIn_2$
6.368		RaF_2
6.379		Mg_2Ge
6.436		K_2O
6.50		Li_2Te
6.50		$AuIn_2$
6.526		Na_2S
6.763		Mg_2Sn
6.809		Na_2Se
6.81		Mg_2Pb
6.98		$SrCl_2$
7.314		Na_2Te
7.38		K_2S
7.65		RbS_2
7.676		K_2Se
8.152		K_2Te
	C 15	
5.94		Be_2Cu
6.287		Be_2Ag
6.435		Be_2Ti
6.96		MgNiZn
7.03		Cu_2Mg
7.61		W_2Zr
7.79		Au_2Na

a/Å		Substance
7.91		Au_2Pb
7.94		Au_2Bi
8.02		Al_2Ca
8.04		Al_2Ce
8.16		Al_2La
9.50		Bi_2K
	C 2	
5.41		FeS_2
5.42		$(Fe, Ni)S_2$ (6.5% Ni)
5.57		RbS_2
5.57		RuS_2
5.57		Bravoite (53.8% NiS_2, 39.1% FeS_2, 7.1% CoS_2)
5.62		OsS_2
5.64		CoS_2
5.65		$(Cu, Ni, Co, Fe)(S, Se)_2$
5.68		PtP_2
5.74		NiS_2
5.85		$CoSe_2$
5.92		$RuSe_2$
5.93		$OsSe_2$
5.94		$PtAs_2$
5.97		$PdAs_2$
6.02		$NiSe_2$
6.096		MnS_2
6.36		$RuTe_2$
6.37		$OsTe_2$
6.43		$PtSb_2$
6.44		$PdSb_2$
6.64		$AuSb_2$
6.94		$MnTe_2$
	C 3	
4.25		Cu_2O
4.73		Ag_2O
	F 1	
5.55		CoAsS
5.68		NiAsS
5.90		NiSbS
		(Ni, Fe)AsS, plessite
		Ni(As, Sb)S, corynite
		Ni(Sb, Bi)S, kallilite
		(Co, Ni)SbS, willyamite
	D 5_3	
8.13		Be_3N_2
9.37		$(Mn, Fe)_2O_3$
9.42		Mn_2O_3
9.74		Zn_3N_2
9.79		Sc_2O_3
9.94		Mg_3N_2
10.12		In_2O_3
10.15		Be_3P_2
10.37		Lu_2O_3
10.39		Yb_2O_3
10.52		Tm_2O_3
10.54		Er_2O_3
10.57		Tl_2O_3
10.58		Ho_2O_3
10.60		Y_2O_3
10.63		Dy_2O_3
10.70		Tb_2O_3
10.79		Gd_2O_3
10.79		Cd_2N_2
10.84		Eu_2O_3
10.85		Sm_2O_3
11.05		Nd_2O_3
11.40		α-Ca_2N_2
12.02		Mg_3P_2
12.33		Mg_3As_2
	D 6_1	
11.05		As_4O_6
11.14		Sb_4O_6
	D 1_1	
10.32		$ZrCl_4$
11.25		$TiBr_4$
(11.34)		(CBr_4) (>320 K.)
(11.62)		(CI_4)
11.89		GeI_4
11.99		SiI_4
12.00		TiI_4
12.23		SnI_4
	E 2_1	
3.67		$YAlO_3$
3.75		$CdTiO_3$
3.78		$LaAlO_3$
3.80		$CaTiO_3$
3.83		$NaWO_3$
3.85		(Na, Ce, Ca)(Ti, Cb)O_3 Loparite
3.88		$NaTaO_3$
3.89		$LaGaO_3$
3.89		$NaCbO_3$
3.91		$SrTiO_3$
3.92		$CaSnO_3$
3.97		$BaTiO_3$
3.98		$KTaO_3$
3.99		$CaZrO_3$
4.00		$KMgF_3$
4.005		$KNiF_3$
4.01		$KCbO_3$
4.03		$SrSnO_3$
4.05		$KZnF_3$
4.07		$KCoF_3$
4.07		$SrHfO_3$
4.09		$SrZrO_3$
4.18		$BaZrO_3$
4.35		$BaPrO_3$
4.38		$BaCeO_3$
4.46		KIO_3
4.48		$BaThO_3$
4.5		NH_4IO_3
4.52		$RbIO_3$
4.66		$CsIO_3$
5.12		$MgZrO_3$
5.20		$CsCdCl_3$
5.33		$CsCdBr_3$
5.44		$CsHgCl_3$
5.77		$CsHgBr_3$
	G 0_3	
6.57		$NaClO_3$
6.71		$NaBrO_3$
	G 2_1	
7.60		$Ca(NO_3)_2$
7.81		$Sr(NO_3)_2$
7.84		$Pb(NO_3)_2$
8.11		$Ba(NO_3)_2$
	H 1_1	
8.045		$NiAl_2O_4$
8.07		$CuAl_2O_4$
8.07		$CoCo_2O_4$
8.07		$MgAl_2O_4$
8.08		$CoAl_2O_4$
8.08		$ZnAl_2O_4$
8.10		$FeAl_2O_4$
8.11		$(Ni, Co)(Co, Ni)_2O_4$
8.11		$(Zn, Co)Co_2O_4$
8.11		$MgCo_2O_4$
8.27		$MnAl_2O_4$
8.27		$(Mn, Co)(Co, Mn)_2O_4$
	H 1_1	
8.28		$MgGa_2O_4$
8.30		$NiCr_2O_4$
8.30		$MgCr_2O_4$
8.31		$ZnCr_2O_4$
8.32		$CoCr_2O_4$
8.32		$ZnGa_2O_4$
8.35		$NiFe_2O_4$
8.35		$Cu_2Cr_2O_4$
8.35		$FeCr_2O_4$
8.36		$MgFe_2O_4$
8.38		$CoFe_2O_4$
8.38		$NiMn_2O_4$
8.40		$ZnFe_2O_4$
8.40		$FeFe_2O_4$
8.42		$(Mn, Mg)Fe_2O_4$
8.42		$TiCo_2O_4$
8.43		$MnCr_2O_4$
8.43		$TiMg_2O_4$
8.43		$TiZn_2O_4$
8.44		$CuFe_2O_4$
8.47		FeV_2O_4
8.49		$MnCr_2O_4$
8.50		$TiFe_2O_4$
8.54		$MnFe_2O_4$
8.58		$CdCr_2O_4$
8.58		$SnMg_2O_4$
8.61		$SnCo_2O_4$
8.63		$SnZn_2O_4$
8.67		$CdFe_2O_4$
8.67		$TiMn_2O_4$
8.81		$MgIn_2O_4$
9.26		Ag_2MoO_4

LATTICE CONSTANTS FOR CUBIC CRYSTALS (continued)

a/Å	Substance
9.4	$CoCo_2S_4$
9.45	$(Co, Ni)_3S_4$
9.46	$CuCo_2O_4$
9.5	NiN_2S_4
9.92	$ZnCr_2S_4$
10.05	$MnCr_2S_4$
10.19	$CdCr_2S_4$
12.54	$K_2Zn(Cn)_4$
12.76	$K_2Hg(CN)_4$
12.84	$K_2Cd(CN)_4$
	$H\ 5_8$
10.08	$2Na_2SO_4 \cdot NaCl \cdot NaF$
	$H\ 4_{13}$
12.11	$KCr(SO_4)_2 \cdot 12H_2O$
12.12	$KAl(SO_4)_2 \cdot 12H_2O$
12.15	$NH_4Al(SO_4)_2 \cdot 12H_2O$
12.20	$RbAl(SO_4)_2 \cdot 12H_2O$
12.21	$TlAl(SO_4)_2 \cdot 12H_2O$
12.31	$CsAl(SO_4)_2 \cdot 12H_2O$
12.44	$NH_3 \cdot CH_3Al(SO_4)_2 \cdot 12H_2O$ (β-alum)
	Langbeinite
9.93	$K_2Mg(SO_4)_3$
10.2	$K_2(Ca_2Mg)SO_3)_3$
	$H\ 2_1$
6.00	Ag_3PO_4
6.120	Ag_3AsO_4(90 K)
6.130	Ag_3AsO_4 (380 K)
	$H\ 2_4$
5.37	Cu_3VS_4
	$J\ 1_1$
8.17	K_2SiF_6
8.35	$(NH_4)_2SiF_6$
8.38	$Rb_2CrF_5H_2O$
8.41	$Tl_2CrF_5H_2O$
8.42	$(NH_4)_2VF_5H_2O$
8.42	$Rb_2VF_5H_2O$
8.45	$Tl_2VF_5H_2O$
8.45	Rb_2SiF_6
8.58	Tl_2SiF_6
8.87	Cs_2SiF_6
8.99	Cs_2GeF_6
9.73	K_2PtCl_6
9.73	K_2OsCl_6
9.76	Tl_2PtCl_6
9.84	$(NH_4)_2PtCl_6$
9.86	K_2ReCl_6
9.88	Rb_2PtCl_6
9.92	Rb_2TiCl_6
9.94	$(NH_4)_2SeCl_6$
9.97	K_2SnCl_6
9.97	Tl_2SnCl_6
9.98	Rb_2SeCl_6
10.02	Rb_2PdBr_6
10.04	$(NH_4)SnCl_6$
10.08	$Ni(NH_3)_6Cl_2$
10.10	Rb_2SnCl_6
10.10	$Co(NH_3)_6Cl_2$
10.11	Tl_2TeCl_6
10.14	$(NH_4)_2PBCl_6$
10.14	K_2TeCl_6
10.15	$Fe(NH_2)_6Cl_2$
10.16	$Mg(NH_3)_6Cl_2$
10.17	Cs_2PtCl_6
	$J\ 1_1$
10.18	Rb_2ZrCl_6
10.18	$(NH_4)_2TeCl_6$
10.20	$Mn(NH_3)_6Cl_2$
10.20	Rb_2PbCl_6
10.22	Cs_2TiCl_6
10.23	Rb_2TeCl_6
10.25	$Zn(NH_2)_6(ClI_4)_3$
10.26	Cs_2SeCl_6
10.30	K_2OSBr_6
10.35	Cs_2SnCl_6
10.36	K_2SeBr_6
10.36	K_2PtBr_6
10.39	$Co(NH_3)_6Br_2$
10.4	$Ni(NH_2)_6Br_2$
10.41	Cs_2ZrCl_6
10.42	Cs_2PbCl_6
10.45	Cs_2TeCl_6
10.45	$Co(NH_2)H_2OSO_4Br$
10.46	$(NH_4)SeBr_6$
10.46	$Zn(NH_2)_6Br_2$
10.47	$Fe(NH_3)_6Br_2$

a/Å	Substance
10.47	$Mg(NH_3)_6Br_2$
10.48	K_2SnBr_6
10.51	$Co(NH_3)_6SO_4Br$
10.52	$Mn(NH_3)_6Br_2$
10.54	$Sr_2Ni(NO_2)_6$
10.55	$Pb_2Ni(NO_2)_6$
10.57	$(NH_4)SnBr_6$
10.58	Rb_2SnBr_6
10.62	$Co(NH_3)_5H_2OSO_4I$
10.63	$Co(NH_3)_6SeO_4Br$
10.67	$Ba_2Ni(NO_2)_6$
10.71	$Ca(NH_2)_6Br_2$
10.71	$Co(NH_3)_6SO_4I$
10.77	Cs_2SnBr_6
10.79	$Co(NH_3)_6SeO_4I$
10.9	$Ni(NH_3)_6I_2$
10.91	$Co(NH_3)_6I_2$
10.96	$Zn(NH_3)_6I_2$
10.97	$Fe(NH_3)_6I_2$
10.98	$Mg(NH_2)_6I_2$
11.04	$Mn(NH_3)_6I_2$
11.04	$Cd(NH_3I_2$
11.24	$Ca(NH_3)_6I_2$
11.27	$Ni(NH_3)_6(BF_4)_2$
11.3	$Co(NH_3)_6(BF_4)_2$
(11.3)	$Zn(NH_3)_6(ClO_4)_2$
11.34	$Mg(NH_3)_6(BF_4)_2$
11.34	$Fe(NH_3)_6(BF_4)_2$
11.37	$Mn(NH_3)_6(BF_4)_2$
11.38	$Cd(NH_3)_6(BF_4)_2$
11.41	$Ni(NH_3)_6(ClO_4)_2$
11.43	$Co(NH_3)_6(ClO_4)_2$
11.46	$Ni(NH_3)_6(SO_3F)_2$
11.49	$Co(NH_2)_6(SO_3F)_2$
11.52	$Fe(NH_3)_6(ClO_4)_2$
11.53	$Mg(NH_3)_6(ClO_4)_2$
11.54	$Cd(NH_3)_6Br_2$
11.54	$Fe(NH_3)_6(SO_2F)_2$
11.58	$Mn(NH_3)_6(ClO_4)_2$
11.59	$Cd(NH_3)_6(ClO_4)_2$
11.59	$Mn(NH_2)_6(SO_2F)_2$
11.62	$Cd(NH_2)_6(SO_3F)_2$
11.91	$Ni(NH_3)_6(PF_6)_2$
11.94	$Co(NH_2)_6(PF_6)_2$
12.03	$Ni(NH_2 \cdot CH_3)_6I_2$
12.05	$Co(NH_2 \cdot CH_3)_6I_2$
12.19	$\{NH(CH_3)_3\}_2SnCl_6$
12.41	$\{S(CH_3)_2\}_2SnCl_6$
12.65	$\{N(CH_3)_4\}_2PtCl_6$
12.80	$\{S(CH_3)_2C_2H_5\}_2SnCl_6$
12.87	$\{N(CH_3)_4\}_2SnCl_6$
13.17	$\{N(CH_3)_2CH_5\}_2SnCl_6$
13.51	$\{N(CH_3)(C_2H_5)_3\}_2SnCl_6$
13.93	$\{P(CH_3)(C_2H_5)_3\}_2SnCl_6$
	$J\ 2_1$ and Related Structures
8.88	Li_3FeF_6
8.90	$(NH_4)_3AlF_6$
9.01	$(NH_4)_3CrF_6$
9.04	$(NH_4)_3VF_6$
9.10	$(NH_4)_3FeF_6$
9.10	$(NH_4)MoO_3F_2$
9.26	Na_2FeF_6
9.93	K_2FeF_6
9.96	$CuLi_2Fe(CN)_6$
10.0	$CuR_2Fe(CN)_6R$ = Na, K, Rb, NH_4, Tl
10.15	$K_2CdFe(NO_2)_6$
10.17	$K_2CaCo(NO_2)_6$
10.19	$K_2CaFe(NO_2)_6$
10.2	$Fe^{III}RFe^{II}(CN)_6$ R = Na, K, Rb, NH_4
10.22	$K_2HgFe(NO_2)_6$
10.23	$K_2SrCo(NO_2)_6$
10.25	$(NH_4)_2CaFe(NO_2)_6$
10.25	$NaTl_2Co(NO_2)_6$
10.28	$K_2CdNi(NO_2)_6$
10.28	$(NH_4)_2CdFe(NO_2)_6$
10.29	$K_2HgNi(NO_2)_6$
10.30	$K_2SrFe(NO_2)_6$
10.30	$Tl_2CaFe(NO_2)_6$
10.31	$K_2PbFe(NO_2)_6$
10.32	$K_2CaNi(NO_2)_6$
10.34	$(NH_4)_2SrFe(NO_2)_6$
10.37	$(NH_4)_2PbFe(NO_2)_6$
10.37	$Tl_2CdNi(NO_2)_6$
10.39	$Tl_2PbFe(NO_2)_6$

a/Å	Substance
10.39	$NaRb_2Co(NO_2)_6$
10.40	$Tl_2SrFe(NO_2)_6$
10.4	$K_2PbCo(NO_2)_6$
10.41	$(NH_4)_2CdNi)NO_2)_6$
10.42	$Tl_2HgNi(NO_2)_6$
10.43	$K_2BaFe(NO_2)_6$
10.45	$K_2BaCo(NO_2)_6$
10.45	$K_2Co(NO_2)_6$
10.46	$(NH_4)_2HgNi(NO_2)_6$
10.47	$Rb_2HgNi(NO_2)_6$
10.49	$K_2SrNi(NO_2)_6$
10.49	$K_4Ni(NO_2)_6$
10.50	$(NH_4)_2BaFe(NO_2)_6$
10.54	$K_2LiBi(NO_2)_6$
10.55	$K_2PbNi(NO_2)_6$
10.55	$Tl_2BaFe(NO_2)_6$
10.58	Rb_2CdNi, $Cd(NO_2)_6$
10.58	$K_3Ir(NO_2)_6$
10.59	$Rb_2LiBi(NO_2)_6$
10.6	$K_2PbCu(NO_2)_6$
10.63	$K_3Rh(NO_2)_6$
10.63	$(NH_4)_2LiBi(NO_2)_6$
10.64	$Tl_2LiBi(NO_2)_6$
10.67	$K_2BaNi(NO_2)_6$
10.70	$NaCs_2Co(NO_2)_6$
10.70	$Ba_1\{Rh(NO_2)_6\}_2$
10.72	$Tl_3Co(NO_2)_6$
10.73	$Rb_3Co(NO_2)_6$
10.73	$(NH_4)_3Ir(NO_2)_6$
10.73	$Tl_3Ir(NO_2)_6$
10.77	$Rb_3Ir(NO_2)_6$
10.8	$(NH_4)_2Co(NO_2)_6$
10.81	$Cs_2Cd\{Ni, Cd(NO_2)_6\}$
10.82	$Co(NH_3)_5H_2OI_3$
10.83	$Rb_3Rh(NO_2)_6$
10.88	$K_2NaBi(NO_2)_6$
10.89	$Co(NH_3)_6I_3$
10.91	$Tl_3Rh(NO_2)_6$
10.91	$(NH_4)_3Rh\ (NO_2)_6$
10.94	$Cs_2LiBi(NO_2)_6$
10.95	$K_2AgBi(NO_2)_6$
10.98	$Rb_2NaBi(NO_2)_6$
10.99	$(NH_4)_2NaBi(NO_2)_6$
11.01	$Tl_2NaBi(NO_2)_6$
11.05	$Rb_2AgBi(NO_2)_6$
11.06	$Tl_2AgBi(NO_2)_6$
11.10	$(NH_4)_2AgBi(NO_2)_6$
11.15	$Cs_2NaBi(NO_2)_6$
11.15	$Cs_3Co(NO_2)_6$
11.17	$Cs_3Ir(NO_2)_6$
11.19	$Cs_3Bi(NO_2)_6$
11.19	$Cs_2AgBi(NO_2)_6$
11.21	$Co(NH_3)_6(BF_4)_3$
11.30	$Cs_3Rh(NO_2)_6$
11.32	$\{Co(NH_3)_6 \cdot H_2O\}(ClO_4)_3$
11.39	$Co(NH_3)_6(ClO_4)_3$
11.67	$Co(NH_3)_6(PF_6)$
K 6_1	
7.46	SiP_2O_7
7.80	TiP_2O_7
7.98	SnP_2O_7
8.18	HfP_2O_7
8.20	ZrP_2O_7
8.61	UP_2O_7
S 1_4	
11.51	$Al_2(Mg, Fe)_3(SiO_4)_3$, pyrope
11.51	$Al_2Fe_2(SiO_4)_3$, almandite
11.60	$Al_2Mn_3(SiO_4)_3$, spessartite
[illegible]	$Al_2Ca_3(SiO_4)_3$, grossularite
11.89	$(Al, Fe)_2Ca_3(SiO_4)_3$, hessonite
11.95	$Cr_2Ca_3(SiO_4)_3$, uvarovite
12.03	$Fe_2Ca_3(SiO_4)_3$, andradite
12.10	$(Na, Li)_3AlF_6$, cryolithionite
12.35—12.46	$(Mg, Mn)_2(Ca, Na)_3AsO_4)_3$, berzelite
S 6_1	
13.68	$NaAlSi_2O_6H_2O$
S 0_8	
13.82	$Al_{13}Si_5O_{20}(OH, F)_{18}Cl$, zunyite
S 6_2	
8.87	$Na_4(AlSiO_4)_3Cl$, sodalite
Tetrahedrite	
10.19	$(Cu, Fe)_{12}As_4S_{13}$, binnite
10.2—10.6	$(Cu, Ag)_{10}(Zn, Fe)_2(Sb, As)_4S_{13}$

ELASTIC CONSTANTS OF SINGLE CRYSTALS

H. P. R. Frederikse

This table gives selected values of elastic constants for single crystals. The values believed most reliable were selected from the original literature. The substances are arranged by crystal system and, within each system, alphabetically by name. A reference to the original literature is given for each value; a useful compilation of published values from many sources may be found in Reference 1 below.

Data are given for the single-crystal density and for the elastic constants c_{ij}, in units of 10^{11} N/m^2, which is equivalent to 10^{12} dyn/cm^2.

GENERAL REFERENCES

1. Simmons, G., and Wang, H., *Single Crystal Elastic Constants and Calculated Aggregate Properties: A Handbook, Second Edition,* The MIT Press, Cambridge, MA, 1971.
2. Gray, D.E., Ed., *American Institute of Physics Handbook, Third Edition*, McGraw-Hill, New York, 1972.

CUBIC CRYSTALS

Name	Formula	ρ/g cm^{-3}	*T*/K	Ref.	C_{11}	C_{12}	C_{44}
Aluminum	Al	2.6970	298	1	1.0675	0.6041	0.2834
Aluminum antimonide	AlSb	4.3600	300	2	0.8939	0.4427	0.4155
Ammonium bromide	NH_4Br	2.4314	300	3	0.3414	0.0782	0.0722
Ammonium chloride	NH_4Cl	1.5279	290	4	0.3814	0.0866	0.0903
Argon	Ar	1.7710	4.2	5	0.0529	0.0135	0.0159
Barium fluoride	BaF_2	4.8860	298	6	0.9199	0.4157	0.2568
Barium nitrate	$Ba(NO_3)_2$	3.2560	293	7	0.2925	0.2065	0.1277
Calcium fluoride	CaF_2	3.810	298	8	1.6420	0.4398	0.8406
Calcium telluride	CaTe	5.8544	298	9	0.5351	0.3681	0.1994
Cesium	Cs	1.9800	78	10	0.0247	0.0206	0.0148
Cesium bromide	CsBr	4.4560	298	11	0.3063	0.0807	0.0750
Cesium chloride	CsCl	3.9880	298	11	0.3644	0.0882	0.0804
Cesium iodide	CsI	4.5250	298	11	0.2446	0.0661	0.0629
Chromite	$FeCr_2O_4$	4.4500	RT	12	3.2250	1.4370	1.1670
Chromium	Cr	7.20	298	13	3.398	0.586	0.990
Cobalt oxide	CoO	6.44	298	14	2.6123	1.4699	0.8300
Cobalt zinc ferrite	$CoZnFeO_2$	5.43	303	12	2.660	1.530	0.780
Copper	Cu	8.932	298	15	1.683	1.221	0.757
Gallium antimonide	GaSb	5.6137	298	16	0.8839	0.4033	0.4316
Gallium arsenide	GaAs	5.3169	298	17	1.1877	0.5372	0.5944
Gallium phosphide	GaP	4.1297	300	18	1.4120	0.6253	0.7047
Garnet (yttrium-iron)	$Y_3Fe_2(FeO_4)_3$	5.17	298	19	2.680	1.106	0.766
Germanium	Ge	5.313	298	20	1.2835	0.4823	0.6666
Gold	Au	19.283	296.5	21	1.9244	1.6298	0.4200
Indium antimonide	InSb	5.7890	298	22	0.6720	0.3670	0.3020
Indium arsenide	InAs	5.6720	293	23	0.8329	0.4526	0.3959
Indium phosphide	InP	4.78	RT	24	1.0220	0.5760	0.4600
Iridium	Ir	22.52	300	25	5.80	2.42	2.56
Iron	Fe	7.8672	298	26	2.26	1.40	1.16
Lead	Pb	11.34	296	27	0.4966	0.4231	0.1498
Lead fluoride	PbF_2	7.79	300	28	0.8880	0.4720	0.2454
Lead nitrate	$Pb(NO_3)_2$	4.547	293	29	0.3729	0.2765	0.1347
Lead telluride	PbTe	8.2379	303.2	30	1.0795	0.0764	0.1343
Lithium	Li	0.5326	298	31	0.1350	0.1144	0.0878
Lithium bromide	LiBr	3.47	RT	32	0.3940	0.1880	0.1910
Lithium chloride	LiCl	2.068	295	33	0.4927	0.2310	0.2495
Lithium fluoride	LiF	2.638	RT	34	1.1397	0.4767	0.6364
Lithium iodide	LiI	4.061	RT	32	0.2850	0.1400	0.1350
Magnesium oxide	MgO	3.579	298	20	2.9708	0.9536	1.5613
Magnetite	Fe_3O_4	5.18	RT	32	2.730	1.060	0.971
Manganese oxide	MnO	5.39	298	35	2.23	1.20	0.79
Mercury telluride	HgTe	8.079	290	36	0.548	0.381	0.204
Molybdenum	Mo	10.2284	273	37	4.637	1.578	1.092

CUBIC CRYSTALS (continued)

Name	Formula	ρ/g cm^{-3}	*T*/K	Ref.	C_{11}	C_{12}	C_{44}
Nickel	Ni	8.91	298	15	2.481	1.549	1.242
Niobium	Nb	8.578	300	38	2.4650	1.3450	0.2873
Palladium	Pd	12.038	300	39	2.2710	1.7604	0.7173
Platinum	Pt	21.50	300	40	3.4670	2.5070	0.7650
Potassium	K	0.851	295	41	0.0370	0.0314	0.0188
Potassium bromide	KBr	2.740	298	11	0.3468	0.0580	0.0507
Potassium chloride	KCl	1.984	298	11	0.4069	0.0711	0.0631
Potassium cyanide	KCN	1.553	RT	32	0.1940	0.1180	0.0150
Potassium fluoride	KF	2.480	295	33	0.6490	0.1520	0.1232
Potassium iodide	KI	3.128	300	42	0.2710	0.0450	0.0364
Pyrite	FeS_2	5.016	RT	43	3.818	0.310	1.094
Rubidium	Rb	1.58	170	44	0.0296	0.0250	0.0171
Rubidium bromide	RbBr	3.350	300	45	0.3152	0.0500	0.0380
Rubidium chloride	RbCl	2.797	300	45	0.3624	0.0612	0.0468
Rubidium iodide	RbI	3.551	300	45	0.2556	0.0382	0.0278
Silicon	Si	2.331	298	46	1.6578	0.6394	0.7962
Silver	Ag	10.50	300	47	1.2399	0.9367	0.4612
Silver bromide	AgBr	5.585	300	48	0.5920	0.3640	0.0616
Sodium	Na	0.971	299	49	0.0739	0.0622	0.0419
Sodium bromate	$NaBrO_3$	3.339	RT	32	0.5450	0.1910	0.1500
Sodium bromide	NaBr	3.202	300	33	0.3970	0.1001	0.0998
Sodium chlorate	$NaClO_3$	2.485	RT	50	0.4920	0.1420	0.1160
Sodium chloride	NaCl	2.163	298	11	0.4947	0.1288	0.1287
Sodium fluoride	NaF	2.804	300	51	0.9700	0.2380	0.2822
Sodium iodide	NaI	3.6609	300	52	0.3007	0.0912	0.0733
Spinel	$MgAl_2O_4$	3.6193	298	53	2.9857	1.5372	1.5758
Strontium fluoride	SrF_2	4.277	300	54	1.2350	0.4305	0.3128
Strontium nitrate	$Sr(NO_3)_2$	2.989	293	29	0.4255	0.2921	0.1590
Strontium oxide	SrO	4.99	300	55	1.601	0.435	0.590
Strontium titanate	$SrTiO_3$	5.123	RT	56	3.4817	1.0064	4.5455
Tantalum	Ta	16.626	298	57	2.6023	1.5446	0.8255
Tantalum carbide	TaC	14.65	RT	58	5.05	0.73	0.79
Thallium bromide	TlBr	7.4529	298	59	0.3760	0.1458	0.0757
Thorium	Th	11.694	300	60	0.7530	0.4890	0.4780
Thorium oxide	ThO_2	9.991	298	61	3.670	1.060	0.797
Tin telluride	SnTe	6.445	300	62	1.1250	0.0750	0.1172
Titanium carbide	TiC	4.940	RT	107	5.00	1.13	1.75
Tungsten	W	19.257	297	64	5.2239	2.0437	1.6083
Uranium carbide	UC	13.63	300	65	3.200	0.850	0.647
Uranium dioxide	UO_2	10.97	298	66	3.960	1.210	0.641
Vanadium	V	6.022	300	67	2.287	1.190	0.432
Zinc selenide	ZnSe	5.262	298	68	0.8096	0.4881	0.4405
Zinc sulfide	ZnS	4.088	298	68	1.0462	0.6534	0.4613
Zinc telluride	ZnTe	5.636	298	68	0.7134	0.4078	0.3115
Zirconium carbide	ZrC	6.606	298	63	4.720	0.987	1.593

TETRAGONAL CRYSTALS

Name	Formula	ρ/g cm^{-3}	T/*K*	Ref.	C_{11}	C_{12}	C_{13}	C_{16}	C_{33}	C_{44}	C_{66}
Ammonium dihydrogen arsenate (ADA)	$NH_4H_2AsO_4$	2.3110	298	69	0.6747	-0.106	0.1652		0.3022	0.0685	0.0639
Ammonium dihydrogen phosphate (ADP)	$NH_4H_2PO_4$	1.8030	293	69	0.6200	-0.050	0.1400		0.3000	0.0910	0.0610
Barium titanate	$BaTiO_3$	5.9988	298	70	2.7512	1.7897	1.5156		1.6486	0.5435	1.1312
Calcium molybdate	$CaMoO_4$	4.255	298	79	1.447	0.664	0.466	0.134	1.265	0.369	0.451
Indium	In	7.300	RT	71	0.4450	0.3950	0.4050		0.4440	0.0655	0.1220
Magnesium fluoride	MgF_2	3.177	RT	72	1.237	0.732	0.536		1.770	0.552	0.978
Nickel sulfate hexahydrate	$NiSO_4 \cdot 6H_2O$	2.070	RT	73	0.3209	0.2315	0.0209		0.2931	0.1156	0.1779
Potassium dihydrogen arsenate (KDA)	KH_2AsO_4	2.867	RT	12	0.530	-0.060	-0.020		0.370	0.120	0.070
Potassium dihydrogen phosphate (KDP)	KH_2PO_4	2.388	RT	71	0.7140	-0.049	0.1290		0.5620	0.1270	0.0628
Rubidium dihydrogen phosphate (RDP)	RbH_2PO_4	2.800	298	74	0.5562	-0.064	0.0279		0.4398	0.1142	0.0350
Rutile	TiO_2	4.260	298	75	2.7143	1.7796	1.4957		4.8395	1.2443	1.9477
Tellurium oxide	TeO_2	5.99	RT	76	0.5320	0.4860	0.2120		1.0850	0.2440	0.5520
Tin (white)	Sn	7.29	288	77	0.7529	0.6156	0.4400		0.9552	0.2193	0.2336
Zircon	$ZrSiO_4$	4.70	RT	78	2.585	1.791	1.542		3.805	0.733	1.113

ORTHORHOMBIC CRYSTALS

Name	Formula	ρ/g cm^{-3}	*T*/K	Ref.	C_{11}	C_{12}	C_{13}	C_{22}	C_{23}	C_{33}	C_{44}	C_{55}	C_{66}
Acenaphthene	$C_{12}H_{10}$	1.220	293	80	0.1380	0.0210	0.0410	0.1262	0.0460	0.1117	0.0265	0.0290	0.0185
Ammonium sulfate	$(NH_4)_2SO_4$	1.774	293	81	0.3607	0.1651	0.1580	0.2981	0.1456	0.3534	0.1025	0.0717	0.0974
Aragonite	$CaCO_3$	2.93	RT	82	1.5958	0.3663	0.0197	0.8697	0.1597	0.8503	0.4132	0.2564	0.4274
Barite	$BaSO_4$	4.40	RT	82	0.8941	0.4614	0.2691	0.7842	0.2676	1.0548	0.1190	0.2874	0.2778
Benzene	C_6H_6	1.061	250	83	0.0614	0.0352	0.0401	0.0656	0.0390	0.0583	0.0197	0.0378	0.0153
Benzophenone	$(C_6H_5)_2CO$	1.219	RT	32	0.1070	0.0550	0.0169	0.1000	0.0321	0.0710	0.0203	0.0155	0.0353
Bronzite	$(MgFe)SiO_3$	3.38	RT	78	1.876	0.686	0.605	1.578	0.561	2.085	0.700	0.592	0.544
Calcium sulfate	$CaSO_4$	2.962	RT	84	0.9382	0.1650	0.1520	1.845	0.3173	1.1180	0.3247	0.2653	0.0926
Celestite	$SrSO_3$	3.96	RT	12	1.044	0.773	0.605	1.061	0.619	1.286	0.135	0.279	0.266
Cesium sulfate	Cs_2SO_4	4.243	293	81	0.4490	0.1958	0.1815	0.4283	0.1800	0.3785	0.1326	0.1319	0.1323
Fosterite	Mg_2SiO_4	3.224	298	85	3.2848	0.6390	0.6880	1.9980	0.7380	2.3530	0.6515	0.8120	0.8088
Iodic acid	HIO_3	4.630	RT	73	0.3030	0.1194	0.1169	0.5448	0.0548	0.4359	0.1835	0.2193	0.1736
Lithium ammonium tartrate	$LiNH_4C_4H_4O_6 \cdot 4H_2O$	1.71	RT	12	0.3864	0.1655	0.0875	0.5393	0.2007	0.3624	0.1190	0.0667	0.2326
Magnesium sulfate heptahydrate	$MgSO_4 \cdot 7H_2O$	1.68	RT	86	0.325	0.174	0.182	0.288	0.182	0.315	0.078	0.156	0.090
Natrolite	$(Na,Al)SiO_3$	2.25	RT	78	0.716	0.261	0.297	0.632	0.297	1.378	0.196	0.248	0.423
Nickel sulfate heptahydrate	$NiSO_4 \cdot 7H_2O$	1.948	RT	86	0.353	0.198	0.201	0.311	0.201	0.335	0.091	0.172	0.099
Olivine	$(MgFe)SiO_4$	3.324	RT	87	3.240	0.590	0.790	1.980	0.780	2.490	0.667	0.810	0.793
Potassium pentaborate	$KB_5O_8 \cdot 4H_2O$	1.74	RT	71	0.582	0.229	0.174	0.359	0.231	0.255	0.164	0.046	0.057
Potassium sulfate	K_2SO_4	2.665	293	81	0.5357	0.1999	0.2095	0.5653	0.1990	0.5523	0.195	0.1879	0.1424
Rochelle salt	$NaK(C_4H_4O_6) \cdot 4H_2O$	1.79	RT	71	0.255	0.141	0.116	0.381	0.146	0.371	0.134	0.032	0.098
Rubidium sulfate	Rb_2SO_4	3.621	293	81	0.5029	0.1965	0.1999	0.5098	0.1925	0.4761	0.1626	0.1589	0.1407
Sodium ammonium tartrate	$NaNH_4C_4H_4O_6 \cdot 4H_2O$	1.587	RT	12	0.3685	0.2725	0.3083	0.5092	0.3472	0.5541	0.1058	0.0303	0.0870
Sodium tartrate	$Na_2C_4H_4O_6 \cdot 2H_2O$	1.794	RT	12	0.461	0.286	0.320	0.547	0.352	0.665	0.124	0.031	0.098
Strontium formate dihydrate	$Sr(CHO_2)_2 \cdot 2H_2O$	2.25	RT	12	0.4391	0.1037	-0.149	0.3484	-0.014	0.3746	0.1538	0.1075	0.1724
Sulfur	S	2.07	RT	12	0.240	0.133	0.171	0.205	0.159	0.483	0.043	0.087	0.076
Thallium sulfate	$TlSO_4$	6.776	293	81	0.4106	0.2573	0.2288	0.3885	0.2174	0.4268	0.1125	0.1068	0.0751
Topaz	$Al_2SiO_3(OH,F)_2$	3.52	RT	82	2.8136	1.2582	0.8464	3.8495	0.8815	2.9452	1.0811	1.3298	1.3089
Uranium (alpha)	U	19.0453	293	88	2.1486	0.4622	0.2176	1.9983	1.0764	2.6763	1.2479	0.7379	0.7454
Zinc sulfate heptahydrate	$ZnSO_4 \cdot 7H_2O$	1.970	RT	86	0.3320	0.1720	0.2000	0.2930	0.1980	0.3200	0.0780	0.1530	0.0830

MONOCLINIC CRYSTALS

Name	Formula	ρ/g cm^{-3}	*T*/K	Ref.	C_{11}	C_{12}	C_{13}	C_{15}	C_{22}
Aegirine	$(NaFe)Si_2O_6$	3.50	RT	89	1.858	0.685	0.707	0.098	1.813
Anthracene	$C_{14}H_{10}$	1.258	RT	90	0.0852	0.0672	0.0590	-0.0192	0.1170
Cobalt sulfate heptahydrate	$CoSO_4{\cdot}7H_2O$	1.948	RT	86	0.335	0.205	0.158	0.016	0.378
Diopside	$(CaMg)Si_2O_6$	3.31	RT	91	2.040	0.884	0.0883	-0.193	1.750
Dipotassium tartrate	$KHC_4H_4O_6$	1.97	RT	12	0.4294	0.1399	0.3129	-0.0105	0.3460
Feldspar (microceine)	$KAlSi_3O_8$	2.56	RT	92	0.664	0.438	0.259	-0.033	1.710
Ferrous sulfate heptahydrate	$FeSO_4{\cdot}7H_2O$	1.898	RT	86	0.349	0.208	0.174	-0.020	0.376
Lithium sulfate monohydrate	$Li_2SO_4{\cdot}H_2O$	2.221	RT	32	0.5250	0.1715	0.1730	-0.0196	0.5060
Naphthalene	$C_{10}H_8$	1.127	RT	93	0.0780	0.0445	0.0340	-0.006	0.0990
Potassium tartrate	$K_2C_4H_4O_6$	1.987	RT	32	0.3110	0.1720	0.1690	0.0287	0.3900
Sodium thiosulfate	$Na_2S_2O_3$	1.7499	RT	12	0.3323	0.1814	0.1875	0.0225	0.2953
Stilbene	$(C_6H_5CH)_2$	1.60	RT	94	0.0930	0.0570	0.0670	-0.003	0.0920
Triglycine sulfate (TGS)	$(NH_2CH_2COOH)_3{\cdot}H_2SO_4$	1.68	RT	32	0.4550	0.1720	0.1980	-0.030	0.3210

Name	C_{23}	C_{25}	C_{33}	C_{35}	C_{44}	C_{46}	C_{55}	C_{66}
Aegirine	0.626	0.094	2.344	0.214	0.692	0.077	0.510	0.474
Anthracene	0.0375	-0.0170	0.1522	-0.0187	0.0272	0.0138	0.0242	0.0399
Cobalt sulfate heptahydrate	0.158	-0.018	0.371	-0.047	0.060	0.016	0.058	0.101
Diopside	0.482	-0.196	2.380	-0.336	0.675	-0.113	0.588	0.705
Dipotassium tartrate	0.1173	0.0176	0.6816	0.0294	0.0961	-0.0044	0.1270	0.0841
Feldspar (microceine)	0.192	-0.148	1.215	-0.131	0.143	-0.015	0.238	0.361
Ferrous sulfate heptahydrate	0.172	-0.019	0.360	-0.014	0.064	0.001	0.056	0.096
Lithium sulfate monohydrate	0.0368	0.0571	0.5400	-0.0254	0.1400	-0.0054	0.1565	0.2770
Naphthalene	0.0230	-0.0270	0.1190	0.0290	0.0330	-0.0050	0.0210	0.0415
Potassium tartrate	0.1330	0.0182	0.5540	0.0710	0.0870	0.0072	0.1040	0.0826
Sodium thiosulfate	0.1713	0.0983	0.4590	-0.0678	0.0569	-0.0268	0.1070	0.0598
Stilbene	0.0485	-0.005	0.0790	-0.005	0.0325	0.0050	0.0640	0.0245
Triglycine sulfate (TGS)	0.2080	-0.0036	0.2630	-0.0500	0.0950	-0.0026	0.1110	0.0620

ELASTIC CONSTANTS OF SINGLE CRYSTALS (continued)

HEXAGONAL CRYSTALS

Name	Formula	ρ/g cm^{-3}	*T*/K	Ref.	C_{11}	C_{12}	C_{13}	C_{33}	C_{55}
Apatite	$Ca_5(PO_4)_3(OH,F,Cl)$	3.218	RT	12	1.667	0.131	0.655	1.396	0.663
Beryl	$Be_3Al_2Si_6O_{18}$	2.68	RT	12	2.800	0.990	0.670	2.480	0.658
Beryllium	Be	1.8477	300	95	2.923	0.267	0.140	3.364	1.625
Beryllium oxide	BeO	3.01	RT	96	4.70	1.68	1.19	4.94	1.53
Cadmium	Cd	8.652	300	97	1.1450	0.3950	0.3990	0.5085	0.1985
Cadmium selenide	CdSe	5.655	298	68	0.7046	0.4516	0.3930	0.8355	0.1317
Cadmium sulfide	CdS	4.824	298	98	0.8431	0.5208	0.4567	0.9183	0.1458
Cobalt	Co	8.836	298	99	3.071	1.650	1.027	3.581	0.755
Dysprosium	Dy	8.560	298	100	0.7466	0.2616	0.2233	0.7871	0.2427
Erbium	Er	9.064	298	100	0.8634	0.3050	0.2270	0.8554	0.2809
Gadolinium	Gd	7.888	298	101	0.6667	0.2499	0.2132	0.7191	0.2089
Hafnium	Hf	12.727	298	102	1.881	0.772	0.661	1.969	0.557
Ice	H_2O(solid)	0.920	250	103	0.1410	0.0660	0.0624	0.1515	0.0288
Indium	In	7.2788	300	104	0.4535	0.4006	0.4151	0.4515	0.0651
Magnesium	Mg	1.7364	298	105	0.5950	0.2612	0.2180	0.6155	0.1635
Rhenium	Re	21.024	298	100	6.1820	2.7530	2.0780	6.8350	1.6060
Ruthenium	Ru	12.3615	298	100	5.6260	1.8780	1.6820	6.2420	1.8060
Thallium	Tl	11.560	300	106	0.4080	0.3540	0.2900	0.5280	0.0726
Titanium	Ti	4.5063	298	102	1.6240	0.9200	0.6900	1.8070	0.4670
Titanium diboride	TiB_2	4.95	RT	107	6.90	4.10	3.20	4.40	2.50
Yttrium	Y	4.472	300	108	0.7790	0.2850	0.2100	0.7690	0.2431
Zinc	Zn	7.134	295	109	1.6368	0.3640	0.5300	0.6347	0.3879
Zinc oxide	ZnO	5.6760	298	110	2.0970	1.2110	1.0510	2.1090	0.4247
Zinc sulfide	ZnS	4.089	298	96	1.2420	0.6015	0.4554	1.4000	0.2864
Zirconium	Zr	6.505	298	102	1.434	0.728	0.653	1.648	0.320

TRIGONAL CRYSTALS

Name	Formula	ρ/g cm^{-3}	*T*/K	Ref.	C_{11}	C_{12}	C_{13}	C_{14}	C_{33}	C_{44}
Aluminum oxide	Al_2O_3	3.986	300	111	4.9735	1.6397	1.1220	-0.2358	4.9911	1.4739
Aluminum phosphate	$AlPO_4$	2.556	RT	73	1.0503	0.2934	0.6927	0.1271	1.3353	0.2314
Antimony	Sb	6.70	295	112	1.0130	0.3450	0.2920	0.2090	0.4500	0.3930
Bismuth	Bi	9.80	295	112	0.6370	0.2490	0.2470	0.0717	0.3820	0.1123
Calcite	$CaCO_3$	2.712	300	113	1.4806	0.5578	0.5464	-0.2058	0.8557	0.3269
Hematite	Fe_2O_3	5.240	RT	82	2.4243	0.5464	0.1542	-0.1247	2.2734	0.8569
Lithium niobate	$LiNbO_3$	4.70	RT	114	2.030	0.530	0.750	0.090	2.450	0.600
Lithium tantalate	$LiTaO_3$	7.45	RT	114	2.330	0.470	0.800	-0.110	2.750	0.940
Quartz	SiO_2	2.6485	298	115	0.8680	0.0704	0.1191	-0.1804	1.0575	0.5820
Selenium	Se	4.838	300	116	0.1870	0.0710	0.2620	0.0620	0.7410	0.1490
Sodium nitrate	$NaNO_3$	2.27	RT	12	0.8670	0.1630	0.1600	0.0820	0.3740	0.2130
Tourmaline		3.05	RT	82	2.7066	0.6927	0.0872	-0.0774	1.6070	0.6682

REFERENCES

1. Thomas, J. F., *Phys. Rev.*, 175, 955-962, 1968.
2. Bolef, D. I. and M. Menes, *J. Appl. Phys.*, 31, 1426-1427, 1960.
3. Garland, C. W. and C. F. Yarnell, *J. Chem. Phys.*, 44, 1112-1120, 1966.
4. Garland, C. W. and R. Renard, *J. Chem. Phys.*, 44, 1130-1139, 1966.

5. Gsänger, M., H. Egger and E. Lüscher, *Phys. Letters*, 27A, 695-696, 1968.
6. Wong, C. and D. E. Schuele, *J. Phys. Chem. Solids*, 29, 1309-1330, 1968.
7. Haussühl, S., *Phys. Stat. Sol.*, 3, 1072-1076, 1963.
8. Wong, C. and D. E. Schuele, *J. Phys. Chem. Solids*, 28, 1225-1231, 1967.
9. McSkimin, H. J. and D. G. Thomas, *J. Appl. Phys.*, 33, 56-59, 1962.
10. Kollarits, F. J. and J. Trivisonno, *J. Phys. Chem. Solids*, 29, 2133-2139, 1968.
11. Slagle, D. D. and H. A. McKinstry, *J. Appl. Phys.*, 38, 446-458, 1967.
12. Hearmon, R. F. S., *Adv. Phys.*, 5, 323-382, 1956.
13. Sumer, A. and J. F. Smith, *J. Appl. Phys.*, 34, 2691-2694, 1963.
14. Alexandrov, K. S. et. al., *Sov. Phys. Sol. State*, 10, 1316-1321, 1968.
15. Epstein, S. G. and O. N. Carlson, *Acta Metal.*, 13, 487-491, 1965.
16. McSkimin, H. J., et. al., *J. Appl. Phys.*, 39, 4127-4128, 1968.
17. McSkimin, H. J., et. al., *J. Appl. Phys.*, 38, 2362-2364, 1967.
18. Weil, R. and W. O. Groves, *J. Appl. Phys.*, 39, 4049-4051, 1968.
19. Bateman, T. B., *J. Appl. Phys.*, 37, 2194-2195, 1966.
20. Bogardus, E. H., *J. Appl. Phys.*, 36, 2504-2513, 1965.
21. Golding, B., S. C. Moss and B. L. Averbach, *Phys. Rev.*, 158, 637-645, 1967.
22. Bateman, T. B., H. J. McSkimin and J. M. Whelan, *J. Appl. Phys.*, 30, 544-545, 1959.
23. Gerlich, D., *J. Appl. Phys.*, 35, 3062, 1964.
24. Hickernell, F. S. and W. R. Gayton, *J. Appl. Phys.*, 37, 462, 1966.
25. MacFarlane, R. E., et. al., *Phys. Letters*, 20, 234-235, 1966.
26. Leese, J. and A. E. Lord Jr., *J. Appl. Phys.*, 39, 3986-3988, 1968.
27. Miller, R. A. and D. E. Schuele, *J. Phys. Chem. Solids*, 30, 589-600, 1969.
28. Wasilik, J. H. and M. L. Wheat, *J. Appl. Phys.*, 36, 791-793, 1965.
29. Haussühl, S., *Phys. Stat. Sol.*, 3, 1072-1076, 1963.
30. Houston, B., et. al., *J. Appl. Phys.*, 39, 3913-3916, 1968.
31. Trivisonno, J. and C. S. Smith, *Acta Metal.*, 9, 1064-1071, 1961.
32. Alexandrov, K. S. and T. V. Ryzhova, *Sov. Phys. Cryst.*, 6, 228-252, 1961.
33. Lewis, J. T., A. Lehoczky and C. V. Briscoe, *Phys. Rev.*, 161, 877-887, 1967.
34. Drabble, J. R. and R. E. B. Strathen, *Proc. Phys. Soc.*, 92, 1090-1995, 1967.
35. Oliver, D. W., *J. Appl. Phys.*, 40, 893, 1969.
36. Alper, T., and G. A. Saunders, *J. Phys. Chem. Solids*, 28, 1637-1642, 1967.
37. Dickinson, J. M. and P. E. Armstrong, *J. Appl. Phys.*, 38, 602-606, 1967.
38. Bolef, D. I., *J. Appl. Phys.*, 32, 100-105, 1961.
39. Rayne, J. A., *Phys. Rev.*, 112, 1125-1130, 1958.
40. MacFarlane, R. E., et. al., *Phys. Letters*, 18, 91-92, 1965.
41. Smith, P. A. and C. S. Smith, *J. Phys. Chem. Solids*, 26, 279-289, 1965.
42. Norwood, M. H. and C. V. Briscoe, *Phys. Rev.*, 112, 45-48, 1958.
43. Simmons, G. and F. Birch, *J. Appl. Phys.*, 34, 2736-2738, 1963.
44. Gutman, E. J. and J. Trivisonno, *J. Phys. Chem. Sol.*, 28, 805-809, 1967.
45. Ghafelehbashi, M., et. al., *J. Appl. Phys.*, 41, 652-666, 1970.
46. McSkimin, H. J. and P. Andreatch Jr., *J. Appl. Phys.*, 35, 2161-2165, 1964.
47. Neighbours, J. R. and G. A. Alers, *Phys. Rev.*, 111, 707-712, 1958.
48. Hidshaw, W., J. T. Lewis, and C. V. Briscoe, *Phys. Rev.*, 163, 876-881, 1967.
49. Daniels, W. B., *Phys. Rev.*, 119, 1246-1252, 1960.
50. Viswanathan, R., *J. Appl. Phys.*, 37, 884-886, 1966.
51. Miller, R. A. and C. S. Smith, *J. Phys. Chem. Sol.*, 25, 1279-1292, 1964.
52. Claytor, R. N. and B. J. Marshall, *Phys. Rev.*, 120, 332-334, 1960.
53. Schreiber, E., *J. Appl. Phys.*, 38, 2508-2511, 1967.
54. Gerlich, D., *Phys. Rev.*, 136, A1366-A1368, 1964.
55. Johnston, D. L., P. H. Thrasher and R. J. Kearney, *J. Appl. Phys.*, 41, 427-428, 1970.
56. Poindexter, E. and A. A. Giardini, *Phys. Rev.*, 110, 1069, 1958.
57. Soga, N., *J. Appl. Phys.*, 37, 3416-3420, 1966.
58. Bartlett, R. W. and C. W. Smith, *J. Appl. Phys.*, 38, 5428-5429, 1967.
59. Morse, G. E. and A. W. Lawson, *J. Phys. Chem. Sol.*, 28, 939-950, 1967.
60. Armstrong, P. E., O. N. Carlson and J. F. Smith, *J. Appl. Phys.*, 30, 36-41, 1959.
61. Macedo, P. M., W. Capps and J. B. Wachtman, *J. Am. Cer. Soc.*, 47, 651, 1964.
62. Beattie, A. G., *J. Appl. Phys.*, 40, 4818-4821, 1969.
63. Chang, R. and L. J. Graham, *J. Appl Phys.*, 37, 3778-3783, 1966.
64. Lowrie, R. and A. M. Gonas, *J. Appl. Phys.*, 38, 4505-4509. 1967.
65. Graham, L. J., H. Nadler and R. Chang, *J. Appl. Phys.*, 34, 1572-1573, 1963.

66. Wachtman, J. B. Jr., et. al., *J. Nucl. Mat.*, 16, 39-41, 1965.
67. Bolef, D. I., *J. Appl. Phys.*, 32, 100-105, 1961.
68. Berlincourt, D., H. Jaffe and L. R. Shiozawa, *Phys. Rev.*, 129, 1009-1017, 1963.
69. Adhav. R. S. J. *Acoust. Soc. Am.*, 43, 835-838, 1968.
70. Berlincourt, D. and H. Jaffe, *Phys. Rev.*, 111, 143-148, 1958.
71. Huntington, H. B., in *Solid State Pysics*, *Vol. 7*, Seitz, F., and Turnbull, D., Ed., pp. 213-285; Academic Press, New York 1958.
72. Cutler, H. R., J. J. Gibson and K. A. McCarthy, *Sol. State Comm.*, 6, 431-433, 1968.
73. Mason, W. P., *Piezoelectric Crystals and Their Application to Ultrasonics*, D. Van Nostrand Co., Inc., New York, 1950.
74. Adhav, R. S., *J. Appl. Phys.*, 40, 2725-2727, 1969.
75. Manghnani, M. H., *J. Geophys. Res.*, 74, 4317-4328, 1969.
76. Uchida, N. and Y. Ohmachi, *J. Appl Phys.*, 40, 4692-4695, 1969.
77. House, D. G. and E. Y. Vernon, *Br. J. Appl. Phys.*, 11, 254-259, 1960.
78. Ryzhova, T. V., et. al., *Bull. Acad. Sci. USSR, Earth Phys. Ser.*, English Transl., no. 2, 111-113, 1966.
79. Alton, W. J. and A. J. Barlow, *J. Appl. Phys.*, 38, 3817-3820, 1967.
80. Michard, F., et. al., *C. R. Acad. Sci., Paris*, 265, 565-567, 1967.
81. Haussühl, S., *Acta Cryst.*, 18, 839-842, 1965.
82. Hearmon, R. F. S., *Rev. Mod. Phys.*, 18, 409-440, 1946.
83. Heseltine, J. C. W., D. W. Elliott and O. B. Wilson, *J. Chem. Phys.*, 40, 2584-2587, 1964.
84. Schwerdtner, W. M., et. al., *Canad. J. Earth Sci.*, 2, 673-683, 1965.
85. Kumazawa, M. and O. L. Anderson, *J. Geophys. Res.*, 74, 5961-5972, 1969 .
86. Alexandrov, K. S., et. al., *Sov. Phys. Cryst.*, 7, 753-755, 1963.
87. Verma, R. K., *J. Geophys. Soc.*, 65, 757-766, 1960.
88. McSkimin, H. J. and E. S. Fisher, *J. Appl. Phys.*, 31, 1627-1639, 1960.
89. Alexandrov, K. S. and T.V. Ryzhova, *Bull. Acad. Sci. USSR, Geophys. Ser.*, English Transl., no.8, 871-875, 1961.
90. Afanaseva, G. K., et. al, *Phys. Stat. Sol.*, 24, K61-K63, 1967.
91. Alexandrov, K. S., et. al., *Sov. Phys. Cryst.*, 8, 589-591, 1964.
92. Alexandrov, K. S. and T. V. Ryzhova, *Bull Acad. Sci. USSR, Geophys. Ser.*, English Transl., no.2, 129-131, 1962.
93. Alexandrov, K. S., et. al., *Sov. Phys. Cryst.*, 8, 164-166, 1963.
94. Teslenko, V. F., et. al., *Sov. Phys. Cryst.*, 10, 744-747, 1966.
95. Smith, J. F. and C. L. Arbogast, *J. Appl. Phys.*, 31, 99-102, 1960.
96. Cline, C. F., H. L. Dunegan and G. M. Henderson, *J. Appl. Phys.*, 38, 1944-1948, 1967.
97. Chang, Y. A. and L. Himmel, *J. Appl. Phys.*, 37, 3787-3790, 1966.
98. Gerlich, D., *J. Phys. Chem. Solids*, 28, 2575-2579, 1967.
99. McSkimin, H. J., *J. Appl. Phys.*, 26, 406-409, 1955.
100. Fisher, E. S. and D. Dever, *Trans. Met. Soc. AIME*, 239, 48-57, 1967.
101. Fisher, E. S. and D. Dever, *Proc. Conf. Rare Earth Res.*, 6th, Gatlinburg, Tenn., 522-533, 1967.
102. Fisher, E. S. and C. J. Renken, Phys. Rev., 135, A482-A494, 1964.
103. Proctor, T. M. Jr., *J. Acoust. Soc. Am.*, 39, 972-977, 1966.
104. Chandrasekhar, B. S. and J. A. Rayne, *Phys. Rev.*, 124, 1011-1041, 1961.
105. Wazzan, A. R. and L. B. Robinson, *Phys. Rev.*, 155, 586-594, 1967.
106. Ferris, R. W., et. al., *J. Appl. Phys.*, 34, 768-770, 1963.
107. Gilman, J. J. and B. W. Roberts, *J. Appl. Phys.*, 32, 1405, 1961.
108. Smith, J. F. and J. A. Gjevre, *J. Appl. Phys.*, 31, 645-647, 1960.
109. Alers, G. A. and J. R. Neighbours, *J. Phys. Chem. Solids*, 7, 58-64, 1908.
110. Bateman, T. B., *J. Appl. Phys.*, 33, 3309-3312, 1962.
111. Tefft, W. E., *J. Res. Natl. Bur. Stand.*, 70A, 277-280, 1966.
112. DeBretteville, Jr., A. et. al., *Phys. Rev.*, 148, 575-579, 1966.
113. Dandekar, D. P. and A. L. Ruoff, *J. Appl. Phys.*, 39, 6004-6009, 1968.
114. Warner, A. W., M. Onoe and G. A. Coquin, *J. Acoust. Soc. Am.*, 42, 1223-1231, 1967.
115. McSkimin, H. J., P. Andreatch and R. N. Thurston, *J. Appl. Phys.*, 36, 1624-1632, 1965.
116. Mort, J., *J. Appl. Phys.*, 38, 3414-3415, 1967.

ELECTRICAL RESISTIVITY OF PURE METALS

The first part of this table gives the electrical resistivity, in units of $10^{-8}\ \Omega$ m, for 28 common metallic elements as a function of temperature. The data refer to polycrystalline samples. The number of significant figures indicates the accuracy of the values. However, at low temperatures (especially below 50 K) the electrical resistivity is extremely sensitive to sample purity. Thus the low-temperature values refer to samples of specified purity and treatment. The references should be consulted for further information on this point, as well as for values at additional temperatures.

The second part of the table gives resistivity values in the neighborhood of room temperature for other metallic elements that have not been studied over an extended temperature range.

REFERENCES

1. C. Y. Ho, et al., *J. Phys. Chem. Ref. Data*, 12, 183—322, 1983; 13, 1069—1096, 1984; 13, 1097—1130, 1984, 13, 1131—1172, 1984.
2. R. A. Matula, *J. Phys Chem. Ref. Data*, 8, 1147—1298, 1979.
3. T. C. Chi, *J. Phys. Chem. Ref. Data*, 8, 339—438, 1979; 8, 439—498, 1979.
4. K. H. Helwege, Ed., *Landolt-Börnstein Numerical Data and Functional Relationships in Science and Technology*, Group III, Vol. 15, Subvolume a, Springer-Verlag, Heidelberg, 1982.
5. L. A. Hall, *Survey of Electrical Resistivity Measurements on 16 Pure Metals in the Temperature Range 0 to 273 K*, NBS Technical Note 365, U.S. Superintendent of Documents, 1968.

ELECTRICAL RESISTIVITY IN $10^{-8}\ \Omega$ m

T/K	Aluminum	Barium	Beryllium	Calcium	Cesium	Chromium	Copper
1	0.000100	0.081	0.0332	0.045	0.0026		0.00200
10	0.000193	0.189	0.0332	0.047	0.243		0.00202
20	0.000755	0.94	0.0336	0.060	0.86		0.00280
40	0.0181	2.91	0.0367	0.175	1.99		0.0239
60	0.0959	4.86	0.067	0.40	3.07		0.0971
80	0.245	6.83	0.075	0.65	4.16		0.215
100	0.442	8.85	0.133	0.91	5.28	1.6	0.348
150	1.006	14.3	0.510	1.56	8.43	4.5	0.699
200	1.587	20.2	1.29	2.19	12.2	7.7	1.046
273	2.417	30.2	3.02	3.11	18.7	11.8	1.543
293	2.650	33.2	3.56	3.36	20.5	12.5	1.678
298	2.709	34.0	3.70	3.42	20.8	12.6	1.712
300	2.733	34.3	3.76	3.45	21.0	12.7	1.725
400	3.87	51.4	6.76	4.7		15.8	2.402
500	4.99	72.4	9.9	6.0		20.1	3.090
600	6.13	98.2	13.2	7.3		24.7	3.792
700	7.35	130	16.5	8.7		29.5	4.514
800	8.70	168	20.0	10.0		34.6	5.262
900	10.18	216	23.7	11.4		39.9	6.041

T/K	Gold	Hafnium	Iron	Lead	Lithium	Magnesium	Manganese
1	0.0220	1.00	0.0225		0.007	0.0062	7.02
10	0.0226	1.00	0.0238		0.008	0.0069	18.9
20	0.035	1.11	0.0287		0.012	0.0123	54
40	0.141	2.52	0.0758		0.074	0.074	116
60	0.308	4.53	0.271		0.345	0.261	131
80	0.481	6.75	0.693	4.9	1.00	0.557	132
100	0.650	9.12	1.28	6.4	1.73	0.91	132
150	1.061	15.0	3.15	9.9	3.72	1.84	136
200	1.462	21.0	5.20	13.6	5.71	2.75	139
273	2.051	30.4	8.57	19.2	8.53	4.05	143
293	2.214	33.1	9.61	20.8	9.28	4.39	144
298	2.255	33.7	9.87	21.1	9.47	4.48	144
300	2.271	34.0	9.98	21.3	9.55	4.51	144
400	3.107	48.1	16.1	29.6	13.4	6.19	147
500	3.97	63.1	23.7	38.3		7.86	149

ELECTRICAL RESISTIVITY OF PURE METALS (continued)

T/K	Gold	Hafnium	Iron	Lead	Lithium	Magnesium	Manganese
600	4.87	78.5	32.9			9.52	151
700	5.82		44.0			11.2	152
800	6.81		57.1			12.8	
900	7.86					14.4	

T/K	Molybdenum	Nickel	Palladium	Platinum	Potassium	Rubidium	Silver
1	0.00070	0.0032	0.0200	0.002	0.0008	0.0131	0.00100
10	0.00089	0.0057	0.0242	0.0154	0.0160	0.109	0.00115
20	0.00261	0.0140	0.0563	0.0484	0.117	0.444	0.0042
40	0.0457	0.068	0.334	0.409	0.480	1.21	0.0539
60	0.206	0.242	0.938	1.107	0.90	1.94	0.162
80	0.482	0.545	1.75	1.922	1.34	2.65	0.289
100	0.858	0.96	2.62	2.755	1.79	3.36	0.418
150	1.99	2.21	4.80	4.76	2.99	5.27	0.726
200	3.13	3.67	6.88	6.77	4.26	7.49	1.029
273	4.85	6.16	9.78	9.6	6.49	11.5	1.467
293	5.34	6.93	10.54	10.5	7.20	12.8	1.587
298	5.47	7.12	10.73	10.7	7.39	13.1	1.617
300	5.52	7.20	10.80	10.8	7.47	13.3	1.629
400	8.02	11.8	14.48	14.6			2.241
500	10.6	17.7	17.94	18.3			2.87
600	13.1	25.5	21.2	21.9			3.53
700	15.8	32.1	24.2	25.4			4.21
800	18.4	35.5	27.1	28.7			4.91
900	21.2	38.6	29.4	32.0			5.61

T/K	Sodium	Strontium	Tantalum	Tungsten	Vanadium	Zinc	Zirconium
1	0.0009	0.80	0.10	0.000016		0.0100	0.250
10	0.0015	0.80	0.102	0.000137	0.0145	0.0112	0.253
20	0.016	0.92	0.146	0.00196	0.039	0.0387	0.357
40	0.172	1.70	0.751	0.0544	0.304	0.306	1.44
60	0.447	2.68	1.65	0.266	1.11	0.715	3.75
80	0.80	3.64	2.62	0.606	2.41	1.15	6.64
100	1.16	4.58	3.64	1.02	4.01	1.60	9.79
150	2.03	6.84	6.19	2.09	8.2	2.71	17.8
200	2.89	9.04	8.66	3.18	12.4	3.83	26.3
273	4.33	12.3	12.2	4.82	18.1	5.46	38.8
293	4.77	13.2	13.1	5.28	19.7	5.90	42.1
298	4.88	13.4	13.4	5.39	20.1	6.01	42.9
300	4.93	13.5	13.5	5.44	20.2	6.06	43.3
400		17.8	18.2	7.83	28.0	8.37	60.3
500		22.2	22.9	10.3	34.8	10.82	76.5
600		26.7	27.4	13.0	41.1	13.49	91.5
700		31.2	31.8	15.7	47.2		104.2
800		35.6	35.9	18.6	53.1		114.9
900			40.1	21.5	58.7		123.1

Element	T/K	Electrical resistivity $10^{-8}\ \Omega$ m
Antimony	273	39
Bismuth	273	107
Cadmium	273	6.8
Cerium	290—300	82.8
Cobalt	273	5.6
Dysprosium	290—300	92.6
Erbium	290—300	86.0
Europium	290—300	90.0
Gadolinium	290—300	131
Gallium	273	13.6
Holmium	290—300	81.4
Indium	273	8.0
Iridium	273	4.7
Lanthanum	290—300	61.5
Lutetium	290—300	58.2
Mercury	273	94.1
Neodymium	290—300	64.3
Niobium	273	15.2
Osmium	273	8.1
Polonium	273	40
Praseodymium	290—300	70.0
Promethium	290—300	75
Protactinium	273	17.7
Rhenium	273	17.2
Rhodium	273	4.3
Ruthenium	273	7.1
Samarium	290—300	94.0
Scandium	290—300	56.2
Terbium	290—300	115
Thallium	273	15
Thorium	273	14.7
Thulium	290—300	67.6
Tin	273	11.5
Titanium	273	39
Uranium	273	28
Ytterbium	290—300	25.0
Yttrium	290—300	59.6

ELECTRICAL RESISTIVITY OF SELECTED ALLOYS

Values of the resistivity are given in units of $10^{-8}\ \Omega$ m. General comments in the preceeding table for pure metals also apply here.

REFERENCE

C. Y. Ho, et al., *J. Phys. Chem. Ref. Data,* 12, 183—322, 1983.

Alloy Aluminum-Copper

Wt % Al	273 K	293 K	300 K	350 K	400 K
99[a]	2.51	2.74	2.82	3.38	3.95
95[a]	2.88	3.10	3.18	3.75	4.33
90[b]	3.36	3.59	3.67	4.25	4.86
85[b]	3.87	4.10	4.19	4.79	5.42
80[b]	4.33	4.58	4.67	5.31	5.99
70[b]	5.03	5.31	5.41	6.16	6.94
60[b]	5.56	5.88	5.99	6.77	7.63
50[b]	6.22	6.55	6.67	7.55	8.52
40[c]	7.57	7.96	8.10	9.12	10.2
30[c]	11.2	11.8	12.0	13.5	15.2
25[f]	16.3[aa]	17.2	17.6	19.8	22.2
15[h]	—	12.3	—	—	—
19[g]	10.8[aa]	11.0	11 1	11.7	12.3
5[e]	9.43	9.61	9.68	10.2	10.7
1[b]	4.46	4.60	4.65	5.00	5.37

Alloy—Aluminum-Magnesium

Wt % Al	273 K	293 K	300 K	350 K	400 K
99[c]	2.96	3.18	3.26	3.82	4.39
95[c]	5.05	5.28	5.36	5.93	6.51
90[e]	7.52	7.76	7.85	8.43	9.02
85	—	—	—	—	—
80	—	—	—	—	—
70	—	—	—	—	—
60	—	—	—	—	—
50	—	—	—	—	—
40	—	—	—	—	—
30	—	—	—	—	—
25	—	—	—	—	—
15	—	—	—	—	—
10[b]	17.1	17.4	17.6	18.4	19.2
5[b]	13.1	13.4	13.5	14.3	15.2
1[a]	5.92	6.25	6.37	7.20	8.03

Alloy—Copper-Gold

Wt % Cu	273 K	293 K	300 K	350 K	400 K
99[c]	1.73	1.86[aa]	1.91[aa]	2.24[aa]	2.58[aa]
95[c]	2.41	2.54[aa]	2.59[aa]	2.92[aa]	3.26[aa]
90[c]	3.29	4.42[aa]	3.46[aa]	3.79[aa]	4.12[aa]
85[c]	4.20	4.33	4.38[aa]	4.71[aa]	5.05[aa]
80[c]	5.15	5.28	5.32	5.65	5.99
70[c]	7.12	7.25	7.30	7.64	7.99
60[c]	9.18	9.13	9.36	9.70	10.05
50[c]	11.07	11.20	11.25	11.60	11.94
40[c]	12.70	[illegible]	[illegible]	13.27[aa]	13.65[aa]
30[c]	13.77	13.93	13.99[aa]	14.38[aa]	14.78[aa]
25[c]	13.93	14.09	14.14	14.54	14.94
15[c]	12.75	12.91	12.96[aa]	13.36[aa]	13.77
10[c]	10.70	10.86	10.91	11.31	11.72
5[c]	7.25	7.41[aa]	7.46	7.87	8.28
1[c]	3.40	3.57	3.62	4.03	4.45

Alloy—Copper-Nickel

Wt % Cu	273 K	293 K	300 K	350 K	400 K
99[c]	2.71	2.85	2.91	3.27	3.62
95[c]	7.60	7.71	7.82	8.22	8.62
90[c]	13.69	13.89	13.96	14.40	14.81
85[c]	19.63	19.83	19.90	2032	20.70
80[c]	25.46	25.66	25.72	26.12[aa]	26.44[aa]
70[i]	36.67	36.72	36.76	36.85	36.89
60[i]	45.43	45.38	45.35	45.20	45.01
50[i]	50.19	50.05	50.01	49.73	49.50
40[c]	47.42	47.73	47.82	48.28	48.49
30[i]	40.19	41.79	42.34	44.51	45.40
25[c]	33.46	35.11	35.69	39.67[aa]	42.81[aa]
15[c]	22.00	23.35	23.85	27.60	31.38
10[c]	16.65	17.82	18.26	21.51	25.19
5[c]	11.49	12.50	12.90	15.69	18.78
1[c]	7.23	8.08	8.37	10.63[aa]	13.18[aa]

Alloy—Copper-Palladium

Wt % Cu	273 K	293 K	300 K	350 K	400 K
99[c]	2.10	2.23	2.27	2.59	2.92
95[c]	4.21	4.35	4.40	4.74	5.08
90[c]	6.89	7.03	7.08	7.41	7.74
85[c]	9.48	9.61	9.66	10.01	10.36
80[c]	11.99	12.12	12.16	12.51[aa]	12.87
70[c]	16.87	17.01	17.06	17.41	17.78
60[c]	21.73	21.87	21.92	22.30	22.69
50[c]	27.62	27.79	27.86	28.25	28.64
40[c]	35.31	35.51	35.57	36.03	36.47
30[c]	46.50	46.66	46.71	47.11	47.47
25[c]	46.25	46.45	46.52	46.99[aa]	47.43[aa]
15[c]	36.52	36.99	37.16	38.28	39.35
10[c]	28.90	29.51	29.73	31.19[aa]	32.56[aa]
5[c]	20.00	20.75	21.02	22.84[aa]	24.54[aa]
1[c]	11.90	12.67	12.93[aa]	14.82[aa]	16.68[aa]

Alloy—Copper-Zinc

Wt % Cu	273 K	293 K	300 K	350 K	400 K
99[b]	1.84	1.97	2.02	2.36	2.71
95[b]	2.78	2.92	2.97	3.33	3.69
90[b]	3.66	3.81	3.86	4.25	4.63
85[b]	4.37	4.54	4.60	5.02	5.44
80[b]	5.01	5.19	5.26	5.71	6.17
70[b]	5.87	6.08	6.15	6.67	7.19
60	—	—	—	—	—
50	—	—	—	—	—
40	—	—	—	—	—
30	—	—	—	—	—
25	—	—	—	—	—
15	—	—	—	—	—
10	—	—	—	—	—
5	—	—	—	—	—
1	—	—	—	—	—

Alloy—Gold-Palladium

Wt % Au	273 K	293 K	300 K	350 K	400 K
99[c]	2.69	2.86	2.91	3.32	3.73
95[c]	5.21	5.35	5.41	5.79	6.17
90[i]	8.01	8.17	8.22	8.56	8.93
85[b]	10.50[aa]	10.66	10.72[aa]	11.100[aa]	11.48[aa]
80[b]	12.75	12.93	12.99	13.45	13.93
70[c]	18.23	18.46	18.54	19.10	19.67
60[b]	26.70	26.94	27.02	27.63[aa]	28.23[aa]
50[a]	27.23	27.63	27.76	28.64[aa]	29.42[aa]
40[a]	24.65	25.23	25.42	26.74	27.95
30[b]	20.82	21.49	21.72	23.35	24.92
25[b]	18.86	19.53	19.77	21.51	23.19
15[a]	15.08	15.77	16.01	17.80	19.61
10[a]	13.25	13.95	14.20[aa]	16.00[aa]	17.81[aa]
5[a]	11.49[aa]	12.21	12.46[aa]	14.26[aa]	16.07[aa]
1[a]	10.07	10.85[aa]	11.12[aa]	12.99[aa]	14.80[aa]

Alloy—Gold-Silver

Wt % Au	273 K	293 K	300 K	350 K	400 K
99[b]	2.58	2.75	2.80[aa]	3.22[aa]	3.63[aa]
95[a]	4.58	4.74	4.79	5.19	5.59
90[j]	6.57	6.73	6.78	7.19	7.58
85[j]	8.14	8.30	8.36[aa]	8.75	9.15
80[j]	9.34	9.50	9.55	9.94	10.33
70[j]	10.70	10.86	10.91	11.29	11.68[aa]
60[j]	10.92	11.07	11.12	11.50	11.87
50[j]	10.23	10.37	10.42	10.78	11.14
40[j]	8.92	9.06	9.11	9.46[aa]	9.81
30[a]	7.34	7.47	7.52	7.85	8.19
25[a]	6.46	6.59	6.63	6.96	7.30[aa]
15[a]	4.55	4.67	4.72	5.03	5.34
10[a]	3.54	3.66	3.71	4.00	4.31
5[i]	2.52	2.64[aa]	2.68[aa]	2.96[aa]	3.25[aa]
1[b]	1.69	1.80	1.84[aa]	2.12[aa]	2.42[aa]

Alloy—Iron-Nickel

Wt % Fe	273 K	293 K	300 K	350 K	400 K
99[a]	10.9	12.0	12.4	—	18.7
95[c]	18.7	19.9	20.2	—	26.8
90[c]	24.2	25.5	25.9	—	33.2
85[c]	27.8	29.2	29.7	—	37.3
80[c]	30.1	31.6	32.2	—	40.0
70[b]	32.3	33.9	34.4	—	42.4
60[c]	53.8	57.1	58.2	—	73.9
50[d]	28.4	30.6	31.4	—	43.7
40[d]	19.6	21.6	22.5	—	34.0
30[c]	15.3	17.1	17.7	—	27.4
25[b]	14.3	15.9	16.4	—	25.1
15[c]	12.6	13.8	14.2	—	21.1
10[c]	11.4	12.5	12.9	—	18.9
5[c]	9.66	10.6	10.9	—	16.1[aa]
1[b]	7.17	7.94	8.12	—	12.8

Alloy—Silver-Palladium

Wt % Ag	273 K	293 K	300 K	350 K	400 K
99[b]	1.891	2.007	2.049	2.35	2.66
95[b]	3.58	3.70	3.74	4.04	4.34
90[b]	5.82	5.94	5.98	6.28	6.59
85[k]	7.92[aa]	8.04[aa]	8.08	8.38[aa]	8.68[aa]
80[k]	10.01	10.13	10.17	10.47	10.78
70[k]	14.53	14.65	14.69	14.99	15.30
60[i]	20.9	21.1	21.2	21.6	22.0
50[k]	31.2	31.4	31.5	32.0	32.4
40[m]	42.2	42.2	42.2	42.3	42.3
30[b]	40.4	40.6	40.7	41.3	41.7
25[k]	36.67[aa]	37.06	37.19	38.1[aa]	38.8[aa]
15[i]	27.08[aa]	26.68[aa]	27.89[aa]	29.3[aa]	30.6[aa]
10[i]	21.69	22.39	22.63	24.3	25.9
5[b]	15.98	16.72	16.98	18.8[aa]	20.5[aa]
1[a]	11.06	11.82	12.08[aa]	13.92[aa]	15.70[aa]

[a] Uncertainty in resistivity is ± 2%.
[b] Uncertainty in resistivity is ± 3%.
[c] Uncertainty in resistivity is ± 5%.
[d] Uncertainty in resistivity is ± 7% below 300 K and ± 5% at 300 and 400 K.
[e] Uncertainty in resistivity is ± 7%.
[f] Uncertainty in resistivity is ± 8%.
[g] Uncertainty in resistivity is ± 10%.
[h] Uncertainty in resistivity is ± 12%.
[i] Uncertainty in resistivity is ± 4%.
[j] Uncertainty in resistivity is ± 1%.
[k] Uncertainty in resistivity is ± 3% up to 300 K and ± 4% above 300 K.
[m] Uncertainty in resistivity is ± 2% up to 300 K and ± 4% above 300 K.
[a] Crystal usually a mixture of α-hcp and fcc lattice.
[aa] In temperature range where no experimental data are available.

PERMITTIVITY (DIELECTRIC CONSTANT) OF INORGANIC SOLIDS

H. P. R. Frederikse

This table lists the permittivity ϵ, frequently called the dielectric constant, of a number of inorganic solids. When the material is not isotropic, the individual components of the permittivity are given. A superscript S indicates a measurement made under constant strain ("clamped" dielectric constant). If the constraint is removed, the measurement yields ϵ^T, the "unclamped" or free dielectric constant.

The temperature of the measurement is given when available; the symbol r.t. indicates a value at nominal room temperature. The frequency of the measurement is given in the last column (i.r. indicates a measurement in the infrared).

Substances are listed in alphabetical order by chemical formula.

REFERENCE

Young, K. F. and Frederikse, H. P. R., *J. Phys. Chem. Ref. Data,* 2, 313, 1973.

Formula	Name	ϵ_{ijk}	*T*/K	*ν*/Hz
Ag_3AsS_3	Silver thioarsenate (Proustite)	$\epsilon_{11}^{T} = 16.5$, $\epsilon_{11}^{S} = 14.5$	r.t.	2×10^7
		$\epsilon_{33}^{T} = 20.0$, $\epsilon_{33}^{S} = 18.0$	r.t.	2×10^7
AgBr	Silver bromide	12.50	r.t.	
AgCN	Silver cyanide	5.6	r.t.	10^6
AgCl	Silver chloride	11.15	r.t.	
$AgNO_3$	Silver nitrate	9.0	293	5×10^5
$AgNa(NO_2)_2$	Silver sodium nitrite	4.5 ± 0.5	r.t.	9.4×10^9
Ag_2O	Silver oxide	8.8	r.t.	
$(AlF)_2SiO_4$	Aluminum fluosilicate (topaz)	$\epsilon_{11} = 6.62$	297	7×10^3
		$\epsilon_{22} = 6.58$	297	7×10^3
		$\epsilon_{33} = 6.95$	297	7×10^3
Al_2O_3	Aluminum oxide (alumina)	$\epsilon_{11} = \epsilon_{22} = 9.34$	298	10^2 8×10^9
		$\epsilon_{33} = 11.54$	298	10^2—8×10^9
$AlPO_4$	Aluminum phosphate	$\epsilon_{11}^{T} = 6.05$	r.t.	10^3
AlSb	Aluminum antimonide	11.21	300	i.r.
AsF_3	Arsenic trifluoride	5.7	r.t.	
BN	Boron nitride	7.1	r.t.	i.r.
$BaCO_3$	Barium carbonate	8.53	291	2×10^5
$Ba(COOH)_2$	Barium formate	$\epsilon_{11} = 7.9$	r.t.	10^3
		$\epsilon_{22} = 5.9$	r.t.	10^3
		$\epsilon_{33} = 7.5$	r.t.	10^3
$BaCl_2$	Barium chloride	9.81	r.t.	
$BaCl_2 \cdot 2H_2O$	Barium chloride dihydrate	9.00	r.t.	10^3
BaF_2	Barium fluoride	7.32	292	5×10^2—10^{11}
$Ba(NO_3)_2$	Barium nitrate	4.95	292	2×10^5
$Ba_2NaNb_5O_{15}$	Barium sodium niobate ("Bananas")	$\epsilon_{11}^{S} = 222$, $\epsilon_{11}^{T} = 235$	296	10^4
		$\epsilon_{22}^{S} = 227$, $\epsilon_{22}^{T} = 247$	296	
		$\epsilon_{33}^{S} = 32$, $\epsilon_{33}^{T} = 51$	296	
BaO	Barium oxide (baria)	34 ± 1	248, 333	60×10^7
BaO_2	Barium peroxide	10.7	r.t.	2×10^6
BaS	Barium sulfide	19.23	r.t.	7.25×10^6
$BaSO_4$	Barium sulfate	11.4	288	10^8
$BaSnO_3$	Barium stannate	18	298	[illegible]
$BaTiO_3$	Barium titanate	$\epsilon_{11}^{T} = 3600$	298	10^5
		$\epsilon_{11}^{S} = 2300$	298	2.5×10^8
		$\epsilon_{33}^{T} = 150$	298	10^5
		$\epsilon_{33}^{S} = 80$	298	2.5×10^8
$Ba_6Ti_2Nb_8O_{30}$	Barium titanium niobate	$\epsilon_{11} = \epsilon_{22} \approx 190$	298	
		$\epsilon_{33} \approx 220$	298	
$BaWO_4$	Barium tungstate	$\epsilon_{11} = \epsilon_{22} = 35.5 \pm 0.2$	297.5	1.6×10^3
		$\epsilon_{33} = 37.2 \pm 0.2$	297.5	1.6×10^3
$BaZrO_3$	Barium zirconate	43	r.t.	

Formula	Name	ϵ_{ijk}	T/K	ν/Hz
$Be_3Al_2Si_6O_{18}$	Beryllium aluminum silicate (Beryl)	$\epsilon_{33} = 5.95$	297	7×10^3
		$\epsilon_{11} = \epsilon_{22} = 6.86$	297	7×10^3
$BeCO_3$	Beryllium carbonate	9.7	291	2×10^5
BeO	Beryllium oxide (beryllia)	7.35 ± 0.2	293	2×10^6
$BiFeO_3$	Bismuth iron oxide	40 ± 3	300	9.4×10^9
$Bi_{12}GeO_{20}$	Bismuth germanite	$\epsilon^S_{11} = 38$	r.t.	
$Bi(GeO_4)_3$	Bismuth germanate	16	293	
Bi_2O_3	Bismuth sesquioxide	18.2	r.t.	2×10^6
$Bi_4Ti_3O_{12}$	Bismuth titanate	112	r.t.	10^3
C	Diamond			
	Type I	5.87 ± 0.19	300	10^3
	Type IIa	5.66 ± 0.04	300	10^3
$C_4H_4O_6$	Tartaric acid	$\epsilon_{11} = \epsilon_{22} = 4.3$	298	
		$\epsilon_{33} = 4.5$	298	
		$\epsilon_{13} = 0.55$	298	
$C_6H_{14}N_2O_6$	Ethylene diamine tartrate (EDT)	$\epsilon^T_{11} = 5.0$	293	
		$\epsilon^T_{22} = 8.3$	293	
		$\epsilon^T_{33} = 6.0$	293	
		$\epsilon^T_{13} = 0.7$	293	
$C_6H_{12}O_6NaBr$	Dextrose sodium bromide	$\epsilon^T_{11} = 4.0$	r.t.	10^3
$(CH_3NH_3)Al(SO_4)_2 \cdot 2H_2O$	Methyl ammonium alum (MASD)	19	197	
$Ca_2B_6O_{11} \cdot 5H_2O$	Colemanite	$\epsilon_{11} = 20$	293	10^3
		$\epsilon_{33} = 25$	293	10^3
$CaCO_3$	Calcium carbonate	$\epsilon_{11} = 8.67$	r.t.	9.4×10^{10}
		$\epsilon_{22} = 8.69$	r.t.	9.4×10^{10}
		$\epsilon_{33} = 8.31$	r.t.	9.4×10^{10}
$CaCeO_3$	Calcium cerate	21	r.t.	
CaF_2	Calcium fluoride	6.81	300	5×10^2—10^{11}
$CaMoO_4$	Calcium molybdate	$\epsilon_{11} = \epsilon_{22} = 24.0 \pm 0.2$	297.5	<10
		$\epsilon_{33} = 20.0 \pm 0.2$	297.5	<10
$Ca(NO_3)_2$	Calcium nitrate	6.54	292	2×10^5
$CaNb_2O_6$	Calcium niobate	$\epsilon_{11} = 22.8 \pm 1.9$	r.t.	$(5—500) \times 10^3$
$Ca_2Nb_2O_7$	Calcium pyroniobate	~45	r.t.	5×10^7
CaO	Calcium oxide	11.8 ± 0.3	283	2×10^6
CaS	Calcium sulfide	6.699	r.t.	7.25×10^6
$CaSO_4 \cdot 2H_2O$	Calcium sulfate dihydrate	$\epsilon_{11} = 5.10$	r.t.	
		$\epsilon_{22} = 5.24$	r.t.	
		$\epsilon_{33} = 10.30$	r.t.	
$CaTiO_3$	Calcium titanate	165	r.t.	
$CaWO_4$	Calcium tungstate	$\epsilon_{11} = \epsilon_{22} = 11.7 \pm 0.1$	297.5	1.59×10^3
		$\epsilon_{33} = 9.5 \pm 0.2$	297.5	1.59×10^3
Cd_3As_2	Cadmium arsenide	$\epsilon_{33} = 18.5$	4	
$CdBr_2$	Cadmium bromide	8.6	293	5×10^5
CdF_2	Cadmium fluoride	8.33 ± 0.08	300	10^5—10^7
CdS	Cadmium sulfide	$\epsilon_{11} = \epsilon_{22} = 8.7$	300	i.r.
		$\epsilon_{33} = 9.25$	300	i.r.
		$\epsilon_{11} = \epsilon_{22} = 8.37$	8	i.r.
		$\epsilon_{33} = 9.00$	8	i.r.
		$\epsilon^T_{11} = 8.48$	77	10^4
		$\epsilon^T_{33} = 9.48$	77	10^4
		$\epsilon^S_{11} = 9.02$, $\epsilon^T_{11} = 9.35$	298	10^4
		$\epsilon^S_{33} = 9.53$, $\epsilon^T_{33} = 10.33$	298	10^4
CdSe	Cadmium selenide	$\epsilon^S_{11} = 9.53$, $\epsilon^T_{11} = 9.70$	298	10^4
		$\epsilon^S_{33} = 10.2$, $\epsilon^T_{33} = 10.65$	298	10^4
CdTe	Cadmium telluride	$\epsilon_{11} = \epsilon_{22} = 10.60 \pm 0.15$	297	i.r.
		$\epsilon_{33} = 7.05 \pm 0.05$	297	i.r.
$Cd_2Nb_2O_7$	Cadmium pyroniobate	500—580	293	10^3
CeO_2	Cerium oxide	7.0	r.t.	2×10^6

PERMITTIVITY (DIELECTRIC CONSTANT) OF INORGANIC SOLIDS (continued)

Formula	Name	ϵ_{ijk}	*T*/K	*ν*/Hz
$CoNb_2O_6$	Cobalt niobate	$\epsilon_{11} = 18.4 \pm 0.6$	r.t.	$(5—500) \times 10^3$
		$\epsilon_{22} = 21.4 \pm 1.1$	r.t.	$(5—500) \times 10^3$
		$\epsilon_{33} = 33.0 \pm 0.7$	r.t.	$(5—500) \times 10^3$
CoO	Cobalt oxide	12.9	298	$10^2—10^{10}$
Cr_2O_3	Chromic sesquioxide	$\epsilon_{11} = \epsilon_{22} = 13.3$	298.5	10^3
		$\epsilon_{33} = 11.9$	298.5	10^3
		8	315 (T_N)	6×10^{10}
$CsAl(SO_4)_2 \cdot 12H_2O$	Cesium alum	5.0	r.t.	$20—20 \times 10^3$
CsBr	Cesium bromide	6.38	298	1.6×10^3
Cs_2CO_3	Cesium carbonate	6.53	291	2×10^5
CsCl	Cesium chloride	7.2	298	
$Cs_2H_2AsO_4$	Cesium dihydrogen arsenate (CDA)	4.8	273	9.5×10^9
$Cs_2H_2PO_4$	Cesium dihydrogen phosphate (CDP)	6.15	285	9.5×10^9
$CsH_3(SeO_3)_2$	Cesium trihydrogen selenite	$\epsilon_{11} = 80$	273	10^5
		$\epsilon_{22} = 63$	273	10^5
		$\epsilon_{33} = 12$	273	10^5
CsI	Cesium iodide	6.31	298	1.6×10^3
$CsNO_3$	Cesium nitrate	$\epsilon_{11} = \epsilon_{22} = 9.4$	r.t.	5×10^5
		$\epsilon_{33} = 8.3$	r.t.	5×10^5
$CsPbCl_3$	Cesium lead chloride	14.37	300	$10^5—10^6$
CuBr	Cuprous bromide	8.0	293	5×10^5
CuCl	Cuprous chloride	9.8 ± 0.5	r.t.	10^3
CuO	Cupric oxide	18.1	r.t.	2×10^6
Cu_2O	Cuprous oxide (Cuprite)	7.60 ± 0.06	r.t.	10^5
$CuSO_4 \cdot 5H_2O$	Cupric sulfate pentahydrate	6.60	r.t.	
EuF_2	Europium fluoride	7.7 ± 0.2	298	$(1—300) \times 10^3$
$Eu_2(MoO_4)_3$	Europium molybdate	9.5	298	
EuS	Europium sulfide	13.10 ± 0.04	80	5×10^2 10^5
FeO	Ferrous oxide	14.2	r.t.	2×10^6
Fe_2O_3	Ferric sesquioxide	4.5	r.t.	$10^5—10^7$
Fe_2O_3-α	Ferric sesquioxide (hematite)	12		6×10^{10}
Fe_3O_4	Ferrosoferric oxide (magnetite)	20	r.t.	$10^5—10^7$
GaAs	Gallium arsenide	13.13	300	
		12.90	4	i.r.
GaP	Gallium phosphide	11.1	r.t.	
		10.75 ± 0.1	1.6	i.r.
GaSb	Gallium antimonide	15.69	r.t.	
		15.7	4	i.r.
$Gd_2(MoO_4)_3$	Gadolinium molybdate	$\epsilon^T = 10$	298	
		$\epsilon^S = 9.5$	298	10^3
Ge	Germanium	16.0 ± 0.3	4	9.2×10^9
		15.8 ± 0.2	r.t.	$500—3 \times 10^{10}$
GeO_2	Germanium dioxide	$\epsilon_{11} = \epsilon_{22} = 7.44$	r.t.	i.r.
HIO_3	Iodic acid	$\epsilon_{11} = 7.5$	r.t.	10^3
		$\epsilon_{22} = 12.4$	r.t.	10^3
		$\epsilon_{33} = 8.1$	r.t.	10^3
$HNH_4(ClCH_2COO)_2$	Hydrogen ammonium dichloroacetate	$\epsilon_{[102]} = 5.9$	r.t.	10^5
H_2O	Ice I (P = 0 kbar)	99	243	
	Ice III (P = 3 kbar)	117	243	
	Ice V(P = 5 kbar)	114	243	
	Ice VI (P = 8 kbar)	193	243	
HgCl	Mercurous chloride (Calumel)	$\epsilon_{11} = \epsilon_{22} = 14.0$	r.t.	10^{12}
$HgCl_2$	Mercuric chloride	6.5	r.t.	10^{12}
HgS	Mercurous sulfide (Cinnabar)	$\epsilon_{11} = \epsilon_{22} = 18.0$	r.t.	i.r.
		$\epsilon_{33} = 32.5$	r.t.	i.r.
HgSe	Mercurous selenide	25.6	r.t.	$10^4—10^6$
I_2	Iodine	$\epsilon_{11} = 6$	r.t.	$5 \times 10^4—10^7$
		$\epsilon_{22} = 3$	r.t.	$5 \times 10^4—10^7$
		$\epsilon_{33} = 40$	r.t.	$5 \times 10^4—10^7$

Formula	Name	ϵ_{ijk}	T/K	ν/Hz
InAs	Indium arsenide	14.55 ± 0.3	r.t.	i.r.
		15.15	4	i.r.
InP	Indium phosphide	12.61	r.t.	i.r.
InSb	Indium antimonide	17.88	4	i.r.
$KAl(SO_4)_2 \cdot 12H_2O$	Potassium alum	6.5	r.t.	$20—20 \times 10^3$
KBr	Potassium bromide	4.88	300	
		4.53	4.2	
$KBrO_3$	Potassium bromate	7.3	r.t.	2×10^6
KCN	Potassium cyanide	6.15	r.t.	2×10^6
K_2CO_3	Potassium carbonate	4.96	291	2×10^5
$K_2C_4H_4O_6 \cdot {}^1/_2\, H_2O$	Dipotassium tartrate (DKT)	$\epsilon_{11} = 6.44$	r.t.	
		$\epsilon_{22} = 5.80$	r.t.	
		$\epsilon_{33} = 6.49$	r.t.	
		$\epsilon_{13} = 0.005$	r.t.	
KCl	Potassium chloride	4.86 ± 0.02	r.t.	5×10^3
		4.50	4.2	
$KClO_3$	Potassium chlorate	5.1	r.t.	2×10^6
$KClO_4$	Potassium perchlorate	5.9	r.t.	2×10^6
K_2CrO_4	Potassium chromate	7.3	r.t.	6×10^7
$KCr(SO_4)_2 \cdot 12H_2O$	Potassium chrome alum	6.5	100—240	175×10^3
KD_2AsO_4	Potassium dideuterium arsenate (KDDA)	$\epsilon_{11} = 70$	298	
		$\epsilon_{33} = 31$	298	
KD_2PO_4	Potassium dideuterium phosphate (KDDP)	50 ± 2	297	10^3
KF	Potassium fluoride	6.05		2×10^6
KH_2AsO_4	Potassium dihydrogen arsenate (KDA)	$\epsilon_{11} = 60$	298	
		$\epsilon_{33} = 24$	298	
KH_2PO_4	Potassium dihydrogen phosphate (KDP)	46	298	10^3
		$\epsilon_{11} = 42$	r.t.	
		$\epsilon_{33} = 21$	r.t.	
K_2HPO_4	Dipotassium monohydrogen orthophosphate	9.05	r.t.	2×10^6
KI	Potassium iodide	5.00	r.t.	9.4×10^{10}
KIO_3	Potassium iodate	170	255	10^5
		10	293	10^5
		$\epsilon_{[101]} \approx 40{,}70$	r.t.	10^5
		16.85	r.t.	2×10^6
$(K,H)Al_3(SiO_4)_3$	Mica (muscovite)	5.4	299	$10^2—3 \times 10^9$
$(K,H)Mg_3Al(SiO_4)_3$	Mica (Canadian)	$\epsilon_{11} = \epsilon_{22} = 6.9$	298	$10^2—10^4$
		$\epsilon_{33} = 7.3$	298	10^4
KNO_2	Potassium nitrite	25	305	
KNO_3	Potassium nitrate	4.37	293	2×10^5
$KNbO_3$	Potassium niobate	700	r.t.	
K_3PO_4	Potassium orthophosphate	7.75	r.t.	2×10^6
KSCN	Potassium thiocyanate	7.9	r.t.	2×10^6
K_2SO_4	Potassium sulfate	6.4	r.t.	2×10^6
$K_2S_3O_6$	Potassium trithionate	5.7	293	1.8×10^6
$K_2S_4O_6$	Potassium tetrathionate	5.5	293	1.8×10^6
$K_2S_5O_6 \cdot H_2O$	Potassium pentathionate	7.8	293	1.8×10^6
$K_2S_6O_6$	Potassium hexathionate	7.8	293	1.8×10^6
K_2SeO_4	Potassium selenate	$\epsilon_{11} = 5.9$	r.t.	10^3
		$\epsilon_{22} = 7.7$	r.t.	10^3
$KSr_2Nb_5O_{15}$	Potassium strontium niobate	$\epsilon_{11} = \epsilon_{11} \approx 1200$	298	
		$\epsilon_{33} \approx 800$	298	
$KTaNbO_3$	Potassium tantalate niobate (KTN)	34,000	273	10^4
		6,000	293	10^4
$KTaO_3$	Potassium tantalate	242	298	2×10^5
$LaScO_3$	Lanthanum scandate	30	r.t.	
LiBr	Lithium bromide	12.1	r.t.	2×10^6
Li_2CO_3	Lithium carbonate	4.9	291	2×10^5

Formula	Name	ϵ_{ijk}	*T*/K	*ν*/Hz
LiCl	Lithium chloride	11.05	r.t.	2×10^6
LiD	Lithium deuteride	14.0 ± 0.5	r.t.	i.r.
LiF	Lithium fluoride	9.00	298	10^2—10^7
		9.11	353	10^2—10^7
$LiGaO_2$	Lithium metagallate	$\epsilon^T_{11} = 7.0$, $\epsilon^T_{22} = 6.0$	r.t.	
		$\epsilon^T_{33} = 9.5$	r.t.	
		$\epsilon^S_{11} = 6.8$, $\epsilon^S_{22} = 5.8$	r.t.	
Li^6H	Lithium-6 hydride	13.2 ± 0.5	r.t.	
Li^7H	Lithium-7 hydride	12.9 ± 0.5	r.t.	
$LiH_3(SeO_3)_2$	Lithium trihydrogen selenite	29	298	10^4
		$\epsilon_{11} = 13.0$	r.t.	
		$\epsilon_{22} = 12.9$	r.t.	
		$\epsilon_{33} = 46$	r.t.	
LiI	Lithium iodide	11.03	r.t.	2×10^6
$LiIO_3$	Lithium iodate	$\epsilon_{11} = \epsilon_{22} = 65$	294.5	10^3
		$\epsilon_{33} = 554$	298	
$LiNH_4C_4H_4O_6 \cdot H_2O$	Lithium ammonium tartrate (LAT)	$\epsilon^T_{11} = 7.2$	298	
		$\epsilon^T_{22} = 8.0$	298	
		$\epsilon^T_{33} = 6.9$	298	
$LiNa_3CrO_4 \cdot 6H_2O$	Lithium trisodium chromate	8.0	r.t.	10^3
$LiNa_3MoO_4 \cdot 6H_2O$	Lithium trisodium molybdate	$\epsilon_{11} = 6.7$	r.t.	10^3
		$\epsilon_{33} = 5.3$	r.t.	10^3
$LiNbO_3$	Lithium niobate	$\epsilon_{11} = \epsilon_{22} = 82$	298	10^5
		$\epsilon_{33} = 30$	298	10^5
$Li_2SO_4 \cdot H_2O$	Lithium sulfate monohydrate	$\epsilon_{11} = 5.6$	298	
		$\epsilon_{22} = 10.3$	298	
		$\epsilon_{33} = 6.5$	298	
		$\epsilon_{13} = 0.07$	298	
$LiTaO_3$	Lithium tantalate	$\epsilon_{11} = \epsilon_{22} = 53$	r.t.	10^5
		$\epsilon_{33} = 46$	r.t.	10^5
		$\epsilon^S_{11} = \epsilon^S_{22} = 41$	r.t.	
		$\epsilon^S_{33} = 43$	r.t.	
		$\epsilon^T_{11} = \epsilon^T_{22} = 51$	r.t.	
		$\epsilon^T_{33} = 45$	r.t.	
$LiTlC_4H_4O_6 \cdot H_2O$	Lithium thallium tartrate (LTT)	$\epsilon_{11} \approx 20$	80	
$Mg_3B_7O_{13}Cl$	Magnesium borate monochloride (boracite)	$\epsilon_{11} = 14.1$	r.t.	5×10^5
$MgCO_3$	Magnesium carbonate	8.1	291	2×10^5
$MgNb_2O_6$	Magnesium niobate	$\epsilon_{11} = 16.4 \pm 0.5$	r.t.	$(5$—$500) \times 10^3$
		$\epsilon_{22} = 20.9 \pm 0.5$	r.t.	$(5$—$500) \times 10^3$
		$\epsilon_{33} = 32.4 \pm 0.5$	r.t.	$(5$—$500) \times 10^3$
MgO	Magnesium oxide (Periclase)	9.65	298	10^2—10^8
$(MgO)_xAl_2O_3$	Spinel	8.6	r.t.	
$MgSO_4$	Magnesium sulfate	8.2	r.t.	
$MgSO_4 \cdot 7H_2O$	Magnesium sulfate septa hydrate	5.46	r.t.	
$MgTiO_3$	Magnesium titanate	17.5	r.t.	
$MgWO_4$	Magnesium tungstate	$\epsilon_{11} = 18.0 \pm 1$	r.t.	$(5$—$500) \times 10^3$
		$\epsilon_{22} = 18.0 \pm 1$	r.t.	$(5$—$500) \times 10^3$
$MnNb_2O_6$	Manganese niobate	$\epsilon_{11} = 17.4 \pm 2$	r.t.	$(5$—$500) \times 10^3$
		$\epsilon_{22} = 16.1 \pm 0.5$	r.t.	$(5$—$500) \times 10^3$
		$\epsilon_{33} = 30.7 \pm 1$	r.t.	$(5$—$500) \times 10^3$
MnO	Manganese oxide (Pyrolusite)	12.8	r.t.	6×10^{10}
MnO_2	Manganese dioxide	$\sim 10^4$	298	10^4
Mn_2O_3	Manganese sesquioxide	8	r.t.	6×10^{10}
$MnWO_4$	Manganese tungstate	$\epsilon_{11} = 19.3 \pm 1.3$	r.t.	$(5$—$500) \times 10^3$
		$\epsilon_{22} = 14.3 \pm 0.5$	r.t.	$(5$—$500) \times 10^3$
		$\epsilon_{33} = 16.5 \pm 1.1$	r.t.	$(5$—$500) \times 10^3$
$N(CH_3)_4HgBr_3$	Tetramethylammonium tribromo mercurate (TTM)	~ 10	233—373	

Formula	Name	ϵ_{ijk}	*T*/K	*ν*/Hz
$N(CH_3)_4HgI_3$	Tetramethylammonium triiodo mercurate (TTM)	~10	233—373	
$N_4(CH_2)_6$	Hexamethylene tetramine (HMTA)	2.6 ± 0.2	r.t.	10^9—10^{10}
$(ND_4)_2BeF_4$	Deuteroammonium fluoberyllate	$\epsilon_{11} = 10$	r.t.	
		$\epsilon_{22} = 9$	r.t.	
		$\epsilon_{33} = 9$	r.t.	
$(ND_4)_2SO_4$	Deuteroammonium sulfate	$\epsilon_{11} = 9$	r.t.	
		$\epsilon_{22} = 10$	r.t.	
		$\epsilon_{33} = 9$	r.t.	
$(NH_2 \cdot CH_2COOH)_3 \cdot H_2SO_4$	Triglycine sulfate (TGS)	$\epsilon_{11} = 9$	273	10^4
		$\epsilon_{22} = 30$	273	10^4
		$\epsilon_{33} = 6.5$	273	10^4
$(NH_2 \cdot CH_2COOH)_3 \cdot H_2SeO_4$	Triglycine selenate (TGSe)	200	293	1.6×10^3
$(NH_2 \cdot CH_2\ COOH)_3 \cdot H_2BeF_4$	Triglycine fluorberyllate (TGFB)	$\epsilon_{22} = 12$	273	10^4
$NH_4Al(SO_4)_2 \cdot 12H_2O$	Ammonium alum	6	r.t.	10^{12}
$(NH_4)_2BeF_4$	Ammonium fluorberyllate	$\epsilon_{11} = \epsilon_{22} = 7.8$	123	10^5
		$\epsilon_{33} = 7.1$	123	10^5
		$\epsilon_{11} = \epsilon_{22} = 8.8$	293	10^5
		$\epsilon_{33} = 9.2$	293	10^5
NH_4Br	Ammonium bromide	7.1	r.t.	7×10^5
NH_4I	Ammonium iodide	9.8	r.t.	
$(NH_4)_2C_2H_6O_6$	Ammonium tartrate	$\epsilon_{11} = 6.45$	r.t.	10^3
		$\epsilon_{22} = 6.8$	r.t.	10^3
		$\epsilon_{33} = 6.0$	r.t.	10^3
$(NH_4)_2Cd_2(SO_4)_3$	Ammonium cadmium sulfate	10.0	r.t.	10^4
NH_4Cl	Ammonium chloride	6.9	r.t.	7×10^5
$NH_4(ClCH_2COO)$	Ammonium monochloroacetate	5	r.t.	2×10^6
$NH_4Cr(SO_4)_2 \cdot 12H_2O$	Ammonium chrome alum	6.5	r.t.	175×10^3
NH_4HSO_4	Ammonium bisulfate	165	273	5×10^4
$NH_4H_2AsO_4$	Ammonium dihydrogen arsenate (ADA)	5.1	265	9.5×10^9
		$\epsilon_{11} = \epsilon_{22} = 85$	298	10^3
		$\epsilon_{33} = 22$	298	
$NH_4H_2PO_4$	Ammonium dihydrogen phosphate (ADP)	$\epsilon_{11} = \epsilon_{22} = 57.1 \pm 0.6$	294.5	10^5—35×10^9
		$\epsilon_{33} = 14.0 \pm 0.3$	294	10^5—36×10^9
$ND_4D_2PO_4$	Ammonium dideuterium phosphate (ADDP)	$\epsilon_{11} = \epsilon_{22} = 74$, $\epsilon_{33} = 24$	300	
NH_4NO_3	Ammonium nitrate	10.7	322	$(5$—$50) \times 10^3$
$(NH_4)_2SO_4$	Ammonium sulfate	$\epsilon_{11} = \epsilon_{22} = 8.0$	123	10^5
		$\epsilon_{33} = 6.3$	123	10^5
		$\epsilon_{11} = \epsilon_{22} = 10.0$	293	10^5
		$\epsilon_{33} = 9.3$	293	10^5
$(NH_4)_2UO_2(C_2O_4)_2$	Ammonium uranyl oxalate	8.03	r.t.	10^4—3.3×10^9
$(NH_4)_2UO_2(C_2O_4)_2 \cdot 3H_2O$	Ammonium uranyl oxalate trihydrate	6.06	r.t.	10^4—3.3×10^9
NaBr	Sodium bromide	6.44	298	1.6×10^3
$NaBrO_3$	Sodium bromate	$\epsilon^T_{11} = 5.70$	298	10^3
NaCN	Sodium cyanide	7.55	293	10^5
$NaCO_3$	Sodium carbonate	8.75	291	2×10^5
$NaCO_3 \cdot 10H_2O$	Sodium carbonate decahydrate	5.3	r.t.	6×10^7
NaCl	Sodium chloride	5.9	298	10^2—10^7
		5.45	4.2	
$NaClO_3$	Sodium chlorate	$\epsilon^T_{11} = 5.76$	301	10^3
		5.28	r.t.	10^3
$NaClO_4$	Sodium perchlorate	5.76	r.t.	10^3
NaF	Sodium fluoride	5.08 ± 0.02	r.t.	5×10^3
$NaH_3(SeO_3)_2$	Sodium trihydrogen selenite	$\epsilon_{11} \approx 75$	273	2×10^5
$NaD_3(SeO_3)_2$	Sodium trideuterium selenite	$\epsilon_{11} \approx 220$	273	2×10^5
NaI	Sodium iodide	7.28 ± 0.03	r.t.	
$NaK(C_4H_2D_2O_6) \cdot 4D_2O$	Sodium potassium tartrate tetradeutrate	$\epsilon_{11} = 70$	273	10^3
	(double deuterated Rochelle salt)	$\epsilon_{22} = 8.9$	273	10^3

PERMITTIVITY (DIELECTRIC CONSTANT) OF INORGANIC SOLIDS (continued)

Formula	Name	ϵ_{ijk}	*T*/K	ν/Hz
$NaK(C_4H_4O_6) \cdot 4H_2O$	Sodium potassium tartrate tetrahydrate (Rochelle salt)	$\epsilon_{11} = 170$	273	10^3
		$\epsilon_{22} = 9.1$	273	10^3
$NaNH_4(C_4H_4O_6) \cdot 4H_2O$	Sodium ammonium tartrate (Ammonium Rochelle salt)	$\epsilon_{11} = 8.4$	298	
		$\epsilon_{22} = 9.2$	298	
		$\epsilon_{33} = 9.5$	298	
$NaNbO_3$	Sodium niobate	$\epsilon_{33} = 670 \pm 13$	r.t.	
		$\epsilon_{11} = \epsilon_{22} = 76 \pm 2$	r.t.	
$NaNO_2$	Sodium nitrite	$\epsilon_{11} = 7.4$	r.t.	5×10^5
		$\epsilon_{22} = 5.5$	r.t.	5×10^5
		$\epsilon_{33} = 5.0$	r.t.	5×10^5
$NaNO_3$	Sodium nitrate	6.85	292	2×10^5
$NaSO_4$	Sodium sulfate	7.90	r.t.	
$NaSO_4 \cdot 10H_2O$	Sodium sulfate decahydrate	5.0	r.t.	
$Na_2S_2O_3 \cdot 5H_2O$	Sodium sulfate pentahydrate	7	250—290	$300—10^4$
$Na_2UO_2(C_2O_4)_2$	Sodium uranyl oxalate	5.18	r.t.	
$NdAlO_3$	Neodymium aluminate	17.5	r.t.	
$NdScO_3$	Neodymium scandate	27	r.t.	
$Ni_3B_7O_{13}I$	Nickel iodine boracite	$\epsilon_{11} = 14$	260	
$NiNb_2O_6$	Nickel niobate	$\epsilon_{11} = 16.0 \pm 0.5$	r.t.	$(5—500) \times 10^3$
		$\epsilon_{22} = 23.8 \pm 1.8$	r.t.	$(5—500) \times 10^3$
		$\epsilon_{33} = 31.3 \pm 2.5$	r.t.	$(5—500) \times 10^3$
NiO	Nickel oxide	11.9	298	10^5
$NiSO_4 \cdot 6H_2O$	Nickel sulfate hexahydrate	$\epsilon_{11} = 6.2$	r.t.	
		$\epsilon_{33} = 6.8$	r.t.	
$NiWO_4$	Nickel tungstate	$\epsilon_{11} = 17.4 \pm 2.4$	r.t.	$(5—500) \times 10^3$
		$\epsilon_{22} = 13.6 \pm 1.0$	r.t.	$(5—500) \times 10^3$
		$\epsilon_{33} = 19.7 \pm 0.6$	r.t.	$(5—500) \times 10^3$
P	Phosphorous (red)	4.1	r.t.	10^8
	Phosphorous (yellow)	3.6	r.t.	10^8
$[P(CH_3)_4]HgBr_3$	Tetramethylphosphonium tribromo mercurate (TTM)	~10	233—373	
$PbBr_2$	Lead bromide	>30	293	$(0.5—3) \times 10^6$
$PbCO_3$	Lead carbonate	18.6	288	10.8
$Pb(C_2H_3O_2)_2$	Lead acetate	2.6	290—295	10^6
$PbCl_2$	Lead chloride	33.5	273	$(0.5—3) \times 10^6$
Pb_2CoWO_6	Lead cobalt tungstate	~250	r.t.	
PbF_2	Lead fluoride	26.3	r.t.	
$PbHfO_3$	Lead hafnate	390	300	10^5
		185	400	
PbI_2	Lead iodide	20.8	293	$(0.5—3) \times 10^6$
$Pb_3MgNb_2O_9$	Lead magnesium niobate	10,000	297	
$PbMoO_4$	Lead molybdate	$\epsilon_{11} = 34.0 \pm 0.4$	297.5	1.6×10^3
		$\epsilon_{33} = 40.6 \pm 0.2$	297.5	1.6×10^3
$Pb(NO_3)_2$	Lead nitrate	16.8	r.t.	$(0.5—3) \times 10^6$
$PbNb_2O_6$	Lead niobate	$\epsilon^T_{33} = 180$	298	
PbO	Lead oxide	25.9	r.t.	2×10^6
PbS	Lead sulfide (Galena)	190	77	i.r.
		200 ± 35	r.t.	i.r.
$PbSO_4$	Lead sulfate	14.3	290—295	10^6
$PbSe$	Lead selenide	280	r.t.	i.r.
$PbTa_2O_6$	Lead metatantalate	$\epsilon_{11} = \epsilon_{22} \approx 300$	r.t.	10^4
		$\epsilon_{33} = 150$	r.t.	10^4
$PbTe$	Lead telluride	450	r.t.	i.r.
		40	77	$10^4—15 \times 10^4$
		430	4.2	$10^4—15 \times 10^4$
$PbTiO_3$	Lead titanate	~200	r.t.	10^3
$PbWO_4$	Lead tungstate	$\epsilon_{11} = \epsilon_{22} = 23.6 \pm 0.3$	297.5	1.59×10^3
		$\epsilon_{33} = 31.0 \pm 0.4$	297.5	1.59×10^3
$Pb(Zn_{1/3}Nb_{2/3})O_3$	Lead zinc niobate	7	300	10^3, 300×10^3
$PbZrO_3$	Lead zirconate	200	400	

Formula	Name	ϵ_{ijk}	T/K	ν/Hz
$RbAl(SO_4)_2 \cdot 12H_2O$	Rubidium alum	5.1	r.t.	10^{12}
RbBr	Rubidium bromide	4.83	300	
Rb_2CO_3	Rubidium carbonate	4.87 ± 0.02	r.t.	5×10^3
RbCl	Rubidium chloride	4.91 ± 0.02	r.t.	5×10^3
$RbCr(SO_4)_2 \cdot 12H_2O$	Rubidium chrome alum	5.0	r.t.	10^{12}
RbF	Rubidium fluoride	5.91	r.t.	2×10^6
$RbHSO_4$	Rubidium bisulfate	$\epsilon_{11} = 7$	r.t.	10^5
		$\epsilon_{22} = 8$	r.t.	10^5
		$\epsilon_{33} = 10$	r.t.	10^5
RbH_2AsO_4	Rubidium dihydrogen arsenate (RDA)	3.90	273	9.5×10^9
RbH_2PO_4	Rubidium dihydrogen phosphate (RDP)	6.15	285	9.5×10^9
RbI	Rubidium iodide	4.94 ± 0.02	r.t.	5×10^3
$RbInSO_4$	Rubidium indium sulfate	6.85	r.t.	
$RbNO_3$	Rubidium nitrate	20—380	433—488	10^6
		30	488—538	10^6
S	Sulfur	$\epsilon_{11} = 3.75$	298	10^2—10^3
		$\epsilon_{22} = 3.95$	298	10^2—10^3
		$\epsilon_{33} = 4.44$	298	10^2—10^3
	sublimed	3.69	298	10^2—10^3
$SC(NH_2)_2$	Thiourea	$\epsilon_{11} = \epsilon_{33} \approx 3$	77—300	10^3
		$\epsilon_{22} = 35$	300	10^3
Sb_2O_3	Antimonous sesquioxide	12.8	r.t.	$(1.5—2) \times 10^3$
Sb_2S_3	Antimonous sulfide (stibnite)	$\epsilon_{11} = \epsilon_{22} = 15$	r.t.	10^3
		$\epsilon_{33} = 180$	r.t.	10^3
Sb_2Se_3	Antimonous selenide	~110	r.t.	$(10—16.5) \times 10^9$
SbSI	Antimonous sulfide iodide	2000	273	10^5
		$\epsilon_{11} = \epsilon_{22} \approx 25$	r.t.	10^3—10^5
		$\epsilon_{33} \approx 5 \times 10^4$	295	10^3—10^5
Se	Selenium (monocrystal)	$\epsilon_{11} = \epsilon_{22} = 11$	300	24×10^9
		$\epsilon_{33} = 21$	300	24×10^9
	(amorphous)	6.0	298	10^2—10^{10}
Si	Silicon	12.1	4.2	10^7—10^9
SiC	Silicon carbide			
	cubic	9.72	r.t.	i.r.
	6H	$\epsilon_{11} = \epsilon_{22} = 9.66$	r.t.	i.r.
		$\epsilon_{33} = 10.03$	r.t.	i.r.
		9.7 ± 0.1	1.8	i.r.
Si_3N_4	Silicon nitride	4.2 (film)	r.t.	10^3
SiO	Silicon monoxide	5.8	r.t.	10^3
SiO_2	Silicon dioxide	$\epsilon_{11} = 4.42$	r.t.	9.4×10^{10}
		$\epsilon_{22} = 4.41$	r.t.	9.4×10^{10}
		$\epsilon_{33} = 4.60$	r.t.	9.4×10^{10}
$Sm_2(MoO_4)_3$	Samarium molybdate	12	298	
SnO_2	Stannic dioxide	$\epsilon_{11} = \epsilon_{22} = 14 \pm 2$	r.t.	10^4—10^{10}
		$\epsilon_{33} = 9.0 \pm 0.5$	r.t.	10^4—10^{10}
SnSb	Tin antimonide	147	r.t.	10^4—10^6
SnTe	Tin telluride	1770 ± 300	r.t.	i.r.
$Sr(COOH)_2 \cdot 2H_2O$	Strontium formate dihydrate	6.1	r.t.	10^3
$SrCO_3$	Strontium carbonate	8.85	298	2×10^5
$SrCl_2$	Strontium chloride	9.19	r.t.	
$Sr_4Cl_2 \cdot 6H_2O$	Strontium chloride hexahydrate	8.52	r.t.	
SrF_2	Strontium fluoride	6.50	300	5×10^2—10^{11}
$SrMoO_4$	Strontium molybdate	$\epsilon_{11} = \epsilon_{22} = 31.7 \pm 0.2$	297.5	1.59×10^3
		$\epsilon_{33} = 41.7 \pm 0.2$	297.5	1.59×10^3
$Sr(NO_3)_2$	Strontium nitrate	5.33	292	2×10^5
$Sr_2Nb_2O_7$	Strontium niobate	$\epsilon_{11} = 75$	r.t.	10^3
		$\epsilon_{22} = 46$	r.t.	10^3
		$\epsilon_{33} = 43$	r.t.	10^3

PERMITTIVITY (DIELECTRIC CONSTANT) OF INORGANIC SOLIDS (continued)

Formula	Name	ϵ_{ijk}	*T*/K	ν/Hz
SrO	Strontium oxide	13.3 ± 0.3	273	2×10^6
SrS	Strontium sulfide	11.3	r.t.	7.25×10^6
$SrSO_4$	Strontium sulfate	11.5	r.t.	
$SrTiO_3$	Strontium titanate	332	298	10^3
		2080	78	10^3
$SrWO_4$	Strontium tungstate	$\epsilon_{11} = \epsilon_{22} = 25.7 \pm 0.2$	297.5	1.6×10^3
		$\epsilon_{33} = 34.1 \pm 0.2$	297.5	1.6×10^3
Ta_2O_5	Tantalum pentoxide (tantala)			
	α phase	$\epsilon_{11} = \epsilon_{22} = 30$	77	10^3
		$\epsilon_{33} = 65$	77	10^3
	β phase	24	292	10^3
$Tb(MoO_4)_3$	Terbium molybdate	11	298	
		$\epsilon_{11} = \epsilon_{22} = 33$	100—200	9.4×10^9
		$\epsilon_{33} = 53$	100—200	9.4×10^9
Te	Tellurium	$\epsilon_{11} = \epsilon_{22} = 33$	r.t.	
		$\epsilon_{33} = 54$	r.t.	
	polycrystalline	27.5	r.t.	i.r.
	monocrystalline	28.0	r.t.	i.r.
ThO_2	Thorium dioxide	18.9 ± 0.4	r.t.	3×10^5
TiO_2	Titanium dioxide (rutile)	$\epsilon_{11} = \epsilon_{22} = 86$	300	10^4—10^6
		$\epsilon_{33} = 170$	300	10^4—10^6
Ti_2O_3	Titanium sesquioxide	30	77	6×10^{10}
TlBr	Thallium bromide	30	293	10^3—10^7
TlCl	Thallous chloride	32.2 ± 0.2	293	10^3—10^5
TlI	Thallous iodide (orthorhombic)	20.7 ± 0.2	293	10^4
		37.3	193	10^7
$TlNO_3$	Thallous nitrate	16.5	293	5×10^5
$TlSO_4$	Thallous sulfate	25.5	293	5×10^5
UO_2	Uranium dioxide	24	r.t.	3×10^5
WO_3	Tungsten trioxide	300		
$YMnO_3$	Yttrium manganate	20	r.t.	2×10^7
Y_2O_3	Yttrium sesquioxide	10	r.t.	10^6
$YbMnO_3$	Ytterbium manganate	20	r.t.	2×10^7
Yb_2O_3	Ytterbium sesquioxide	5.0 (film)	r.t.	10^3
ZnO	Zinc monoxide	$\epsilon^S_{11} = 8.33$	r.t.	
		$\epsilon^S_{33} = 8.84$	r.t.	
		$\epsilon^T_{11} = 9.26$	r.t.	
		$\epsilon^T_{33} = 11.0$	r.t.	
		$\epsilon_{11} = 9.26$	r.t.	
		$\epsilon_{33} = 8.2$	r.t.	
		8.15	r.t.	i.r.
ZnS	Zinc sulfide	$\epsilon^S_{11} = 8.08 \pm 2\%$	77	10^4
		$\epsilon^S_{11} = 8.32 \pm 2\%$	298	10^4
		$\epsilon^T_{11} = 8.14 \pm 2\%$	77	10^4
		$\epsilon^T_{11} = 8.37 \pm 2\%$	298	10^4
ZnSe	Zinc selenide	$\epsilon^T_{11} = \epsilon^S_{11} = 9.12 \pm 2\%$	298	10^4
ZnTe	Zinc telluride	$\epsilon^T_{11} = \epsilon^S_{11} = 10.10 \pm 2\%$	r.t.	
$ZnWO_4$	Zinc tungstate	$\epsilon_{22} = 16.1 \pm 0.5$	r.t.	$(5—500) \times 10^3$
ZrO_2	Zirconium dioxide (zirconia)	12.5	r.t.	2×10^6

CURIE TEMPERATURE OF SELECTED FERROELECTRIC CRYSTALS

H. P. R. Frederikse

The following table lists the major ferroelectric crystals and their Curie temperatures, T_C.

REFERENCE

Young, K. F. and Frederikse, H. P. R., *J. Phys. Chem. Ref. Data*, 2, 313, 1973.

Name or acronym	Formula	T_C/K
Potassium dihydrogen phosphate group		
KDP	KH_2PO_4	123
KDA	KH_2AsO_4	97
KDDP	KD_2PO_4	213
KDDA	KD_2AsO_4	162
RDP	RbH_2PO_4	146
RDA	RbH_2AsO_4	111
RDDP	RbD_2PO_4	218
RDDA	RbD_2AsO_4	178
CDP	CsH_2PO_4	159
CDA	CsH_2AsO_4	143
CDDA	CsD_2AsO_4	212
Rochelle salt group		
Rochelle salt	$NaKC_4H_4O_6{\cdot}4H_2O$	255—297
Deuterated Rochelle salt	$NaKC_4H_2D_2O_6{\cdot}4H_2O$	251—308
Ammonium Rochelle salt	$NaNH_4C_4H_4O_6{\cdot}4H_2O$	109
LAT	$LiNH_4C_4H_4O_6{\cdot}H_2O$	106
Triglycine sulfate group		
TGS	$(NH_2CH_2COOH)_3{\cdot}H_2SO_4$	322
TGSe	$(NH_2CH_2COOH)_3{\cdot}H_2SeO_4$	295
TGFB	$(NH_2CH_2COOH)_3{\cdot}H_2BeF_4$	346
AFB	$(NH_4)_2BeF_4$	176
HADA	$HNH_4(ClCH_2COO)_2$	128
Perovskites and related compounds		
Barium titanate	$BaTiO_3$	406, 278,193
Lead titanate	$PbTiO_3$	765
Potassium niobate	$KNbO_3$	712
Potassium tantalate niobate	$KTa_{2/3}Nb_{1/3}O_3$	271, 220, 170
Lithium niobate	$LiNbO_3$	1483
Lithium tantalate	$LiTaO_3$	891
Barium titanium niobate	$Ba_6Ti_2Nb_8O_{30}$	521
Ba-Na niobate ("Bananas")	$Ba_2NaNb_5O_{15}$	833
Potassium iodate	KIO_3	485, 343, 257—263, 83
Lithium iodate	$LiIO_3$	529
Potassium nitrate	KNO_3	397
Sodium nitrate	$NaNO_3$	548
Rubidium nitrate	$RbNO_3$	437—487
Miscellaneous compounds		
Cesium trihydrogen selenite	$CsH_3(SeO_3)_2$	143
Lithium trihydrogen selenite	$LiH_3(SeO_3)_2$	$T_C > T_{mp}$

Name or acronym	Formula	T_C/K
Potassium selenate	K_2SeO4	93
Methyl ammonium alum (MASD)	$CH_3NH_3Al(SO_4)_2 \cdot 12H_2O$	177
Ammonium cadmium sulfate	$(NH_4)_2Cd_2(SO_4)_3$	95
Ammonium bisulfate	$(NH_4)HSO_4$	271
Ammonium sulfate	$(NH_4)_2SO_4$	224
Ammonium nitrate	NH_4NO_3	398, 357, 305, 255
Colemanite	$CaB_3O_4(OH)_3 \cdot H_2O$	266
Cadmium pyroniobite	$Cd_2Nb_2O_7$	185
Gadolinium molybdate	$Gd_2(MoO_4)_3$	432

PROPERTIES OF ANTIFERROELECTRIC CRYSTALS

H. P. R. Frederikse

Some important antiferroelectric crystals are listed here with their Curie Temperatures T_C. The last column gives the constant T_0 which appears in the Curie-Weiss law describing the dielectric constant of these materials above the Curie Temperature:

$$\varepsilon = \text{const.}/(T - T_0)$$

Name or acronym	Formula	T_C/K	T_0/K
ADP	$NH_4H_2PO_4$	148	
ADA	$NH_4H_2AsO_4$	216	
ADDP	$NH_4D_2PO_4$	242, 245	
ADDA	$NH_4D_2AsO_4$	299	
A_dDDP	$ND_4D_2PO_4$	243	
A_dDDA	$ND_4D_2AsO_4$	304	
Sodium niobate	$NaNbO_3$	911, 793	
Lead hafnate	$PbHfO_3$	476	378
Lead zirconate	$PbZrO_3$	503	475
Lead metaniobate	$PbNb_2O_6$	843	530
Lead metatantalate	$PbTa_2O_6$	543	533
Tungsten trioxide	WO_3	1010	
Potassium strontium niobate	$KSr_2Nb_5O_{15}$	427	413
Sodium nitrite	$NaNO_2$	437	437
Sodium trihydrogen selenite	$NaH_3(SeO_3)_2$	193	192
Sodium trideuterium selenite	$NaD_3(SeO_3)_2$	271	245
Ammonium trihydrogen iodate	$(NH_4)_2H_3IO_3$	245	

DIELECTRIC CONSTANTS OF GLASSES

Type	Dielectric constant at 100 MHz (20°C)	Volume resistivity (350°C megohm-cm)	Loss factor[a]
Corning 0010	6.32	10	0.015
Corning 0080	6.75	0.13	0.058
Corning 0120	6.65	100	0.012
Pyrex 1710	6.00	2,500	0.025
Pyrex 3320	4.71	—	0.019
Pyrex 7040	4.65	80	0.013
Pyrex 7050	4.77	16	0.017
Pyrex 7052	5.07	25	0.019
Pyrex 7060	4.70	13	0.018
Pyrex 7070	4.00	1,300	0.0048
Vycor 7230	3.83	—	0.0061
Pyrex 7720	4.50	16	0.014
Pyrex 7740	5.00	4	0.040
Pyrex 7750	4.28	50	0.011
Pyrex 7760	4.50	50	0.0081
Vycor 7900	3.9	130	0.0023
Vycor 7910	3.8	1,600	0.00091
Vycor 7911	3.8	4,000	0.00072
Corning 8870	9.5	5,000	0.0085
G. E. Clear (silica glass)	3.81	4,000—30,000	0.00038
Quartz (fused)	3.75 4.1 (1 MHz)	—	0.0002 (1 MHz)

[a] Power factor × dielectric constant equals loss factor.

PROPERTIES OF SUPERCONDUCTORS

L. I. Berger and B. W. Roberts

The following tables include superconductive properties of selected elements, compounds, and alloys. Individual tables are given for thin films, elements at high pressures, superconductors with high critical magnetic fields, and high critical temperature superconductors.

The historically first observed and most distinctive property of a superconductive body is the near total loss of resistance at a critical temperature (T_c) that is characteristic of each material. Figure 1(a) below illustrates schematically two types of possible transitions. The sharp vertical discontinuity in resistance is indicative of that found for a single crystal of a very pure element or one of a few well annealed alloy compositions. The broad transition, illustrated by broken lines, suggests the transition shape seen for materials that are not homogeneous and contain unusual strain distributions. Careful testing of the resistivity limit for superconductors shows that it is less than 4×10^{-23} ohm cm, while the lowest resistivity observed in metals is of the order of 10^{-13} ohm cm. If one compares the resistivity of a superconductive body to that of copper at room temperature, the superconductive body is at least 10^{17} times less resistive.

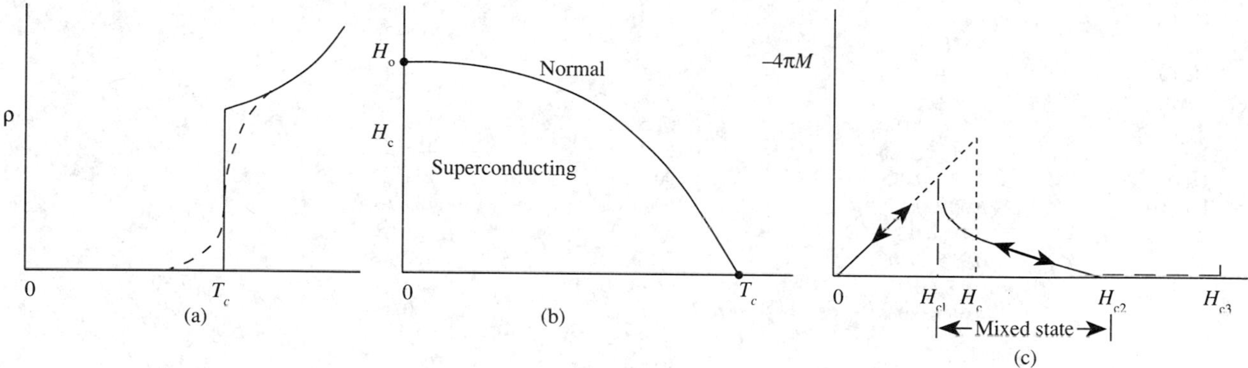

FIGURE 1. Physical properties of superconductors. (a) Resistivity vs. temperature for a pure and perfect lattice (solid line); impure and/or imperfect lattice (broken line). (b) Magnetic-field temperature dependence for Type-I or "soft" superconductors. (c) Schematic magnetization curve for "hard" or Type-II superconductors.

The temperature interval ΔT_c, over which the transition between the normal and superconductive states takes place, may be of the order of as little as 2×10^{-5} K or several K in width, depending on the material state. The narrow transition width was attained in 99.9999% pure gallium single crystals.

A Type-I superconductor below T_c, as exemplified by a pure metal, exhibits perfect diamagnetism and excludes a magnetic field up to some critical field H_c, whereupon it reverts to the normal state as shown in the *H-T* diagram of Figure 1(b).

The magnetization of a typical high-field superconductor is shown in Figure 1(c). The discovery of the large current-carrying capability of Nb_3Sn and other similar alloys has led to an extensive study of the physical properties of these alloys. In brief, a high-field superconductor, or Type-II superconductor, passes from the perfect diamagnetic state at low magnetic fields to a mixed state and finally to a sheathed state before attaining the normal resistive state of the metal. The magnetic field values separating the four stages are given as H_{c1}, H_{c2}, and H_{c3}. The superconductive state below H_{c1} is perfectly diamagnetic, identical to the state of most pure metals of the "soft" or Type-I superconductor. Between H_{c1} and H_{c2} a "mixed superconductive state" is found in which fluxons (a minimal unit of magnetic flux) create lines of normal flux in a superconductive matrix. The volume of the normal state is proportional to $-4\pi M$ in the "mixed state" region. Thus at H_{c2} the fluxon density has become so great as to drive the interior volume of the superconductive body completely normal. Between H_{c2} and H_{c3} the superconductor has a sheath of current-carrying superconductive material at the body surface, and above H_{c3} the normal state exists. With several types of careful measurement, it is possible to determine H_{c1}, H_{c2}, and H_{c3}. Table 6 contains some of the available data on high-field superconductive materials.

High-field superconductive phenomena are also related to specimen dimension and configuration. For example, the Type-I superconductor, Hg, has entirely different magnetization behavior in high magnetic fields when contained in the very fine sets of filamentary tunnels found in an unprocessed Vycor glass. The great majority of superconductive materials are Type-II. The elements in very pure form and a very few precisely stoichiometric and well annealed compounds are Type I with the possible exceptions of vanadium and niobium.

Metallurgical Aspects. The sensitivity of superconductive properties to the material state is most pronounced and has been used in a reverse sense to study and specify the detailed state of alloys. The mechanical state, the homogeneity, and the presence of impurity atoms and other electron-scattering centers are all capable of controlling the critical temperature and the current-carrying capabilities in high-magnetic fields. Well annealed specimens tend to show sharper transitions than those that are strained or inhomogeneous. This sensitivity to mechanical state underlines a general problem in the tabulation of properties for superconductive materials. The occasional divergent values of the critical temperature and of the critical fields quoted for a Type-II superconductor may lie in the variation in sample preparation. Critical temperatures of materials studied early in the history of superconductivity must be evaluated in light of the probable metallurgical state of the material, as well as the availability of less pure starting elements. It has been noted that recent work has given extended consideration to the metallurgical aspects of sample preparation.

Symbols in tables: T_c: Critical temperature; H_o: Critical magnetic field in the $T = 0$ limit; θ_D: Debye temperature; and γ: Electronic specific heat.

TABLE 1
Selective Properties of Superconductive Elements

Element	T_c(K)	H_o(oersted)	θ_D(K)	γ(mJ mol^{-1}K^{-1})
Al	1.175 ± 0.002	104.9 ± 0.3	420	1.35
Am* (α,?)	0.6			
Am* (β,?)	1.0			
Be	0.026			0.21
Cd	0.517 ± 0.002	28 ± 1	209	0.69
Gd	1.083 ± 0.001	58.3 ± 0.2	325	0.60
Gd (β)	5.9, 6.2	560		
Gd (γ)	7	950, HF[a]		
Gd (Δ)	7.85	815, HF		
Hf	0.128	12.7		2.21
Hg (α)	4.154 ± 0.001	411 ± 2	87, 71.9	1.81
Hg (β)	3.949	339	93	1.37
In	3.408 ± 0.001	281.5 ± 2	109	1.672
Ir	0.1125 ± 0.001	16 ± 0.05	425	3.19
La (α)	4.88 ± 0.02	800 ± 10	151	9.8
La (β)	6.00 ± 0.1	1096, 1600	139	11.3
Lu	0.1 ± 0.03	350 ± 50		
Mo	0.915 ± 0.005	96 ± 3	460	1.83
Nb	9.25 ± 0.02	2060 ± 50, HF	276	7.80
Os	0.66 ± 0.03	70	500	2.35
Pa	1.4			
Pb	7.196 ± 0.006	803 ± 1	96	3.1
Re	1.697 ± 0.006	200 ± 5	4.5	2.35
Ru	0.49 ± 0.015	69 ± 2	580	2.8
Sn	3.722 ± 0.001	305 ± 2	195	1.78
Ta	4.47 ± 0.04	829 ± 6	258	6.15
Tc	7.8 ± 0.1	1410, HF	411	6.28
Th	1.38 ± 0.02	1.60 ± 3	165	4.32
Ti	0.40 ± 0.04	56	415	3.3
Tl	2.38 ± 0.02	178 ± 2	78.5	1.47
U	0.2			
V	5.40 ± 0.05	1408	383	9.82
W	0.0154 ± 0.0005	1.15 ± 0.03	383	0.90
Zn	0.85 ± 0.01	54 ± 0.3	310	0.66
Zr	0.61 ± 0.15	47	290	2.77
Zr (ω)	0.65, 0.95			

TABLE 2
Range of Critical Temperatures Observed for Superconductive Elements in Thin Films Condensed Usually at Low Temperatures

Element	T_c Range (K)	Comments	Element	T_c Range (K)	Comments
Al	1.15—5.7	HF[a]	Nb	2.0—10.1	
Be	5—9.75	HF	Pb	1.8—7.5	
Bi	6.17—6.6		Re	1.7—7	
Cd			Sn	3.5—6	
(Disordered)	0.79—0.91		Ta	<1.7—4.51	HF[a]
(Ordered)	0.53—0.59		Tc	4.6—7.7	
Ga	2.5—8.5	HF	Ti	1.3 Max	
Hg	3.87—4.5		Tl	2.33—2.96	
In	2.2—5.6	HF	V	1.8—6.02	
La	3.55—6.74		W	<1.0—4.1	
Mo	3.3—8.0		Zn	0.77—1.9	

[a] HF denotes high magnetic field superconductive properties.

TABLE 3
Elements Exhibiting Superconductivity Under or After Application of High Pressure

Element	T_c Range (K)	Pressure (kbar)
Al	1.98—0.075	0—62
As	0.31—0.5	220—140
	0.2—0.25	140—100
Ba II	1—1.8	55—85
III	1.8—5	85—144
IV	4.5—5.4	144—190
Bi II	3.9	25—27
III	6.55—7.25	28—38
IV	7.0, 8.7—6.0	43, 43—62
V	6.7, 8.3	48—80
VI	8.55	90, 92—101
VII(?)	8.2	30
Ce (α)	0.020—0.045	20—35
Ce (α′)	1.9—1.3	45—125
Cs V	1.5	>125
Ga II	6.38	≥35
II′	7.5	≥35 then P removed
Ge	5.35	115
La	5.5—12.9	0—210
Lu	0.022—1.0	45—190
P	5.8	170
Pb II	3.55	160
Re II	2.3 Max.	"Plastic" compression
Sb (prepared 120 kbar, held below 77K)	2.6—2.7	
Sb II	3.55—3.40	85—150
Se II	6.75, 6.95	130
Si	6.7—7.1	120—130
Sn II	5.2—4.85	125—160
III	5.30	113
Te II	2.4—5.1	38—55
	4.1—4.2	53—62
IV	4.72—4	63—80
( )	3.3—2.8	100—260
Tl (cubic form)	1.45	35
(hexagonal form)	1.95	35
U	2.4—0.4	10—85
Y	1.7—2.5	110—160
Zr (omega form, metastable)	1—1.7	60—130

TABLE 4
Superconductive Compounds and Alloys

All compositions are denoted on an atomic basis, i.e., AB, AB_2, or AB_3 for compounds, unless noted. Solid solutions or odd compositions may be denoted as A_zB_{1-z} or A_zB. A series of three or more alloys is indicated as A_xB_{1-x} or by actual indication of the atomic fraction range, such as $A_{0-0.6}B_{1-0.4}$. The critical temperature of such a series of alloys is denoted by a range of values or possibly the maximum value.

The selection of the critical temperature from a transition in the effective permeability, or the change in resistance, or possibly the incremental changes in frequency observed by certain techniques is not often obvious from the literature. Most authors choose the mid-point of such curves as the probable critical temperature of the idealized material, while others will choose the highest temperature at which a deviation from the normal state property is observed. In view of the previous discussion concerning the variability of the superconductive properties as a function of purity and other metallurgical aspects, it is recommended that appropriate literature be checked to determine the most probable critical temperature or critical field of a given alloy.

A very limited amount of data on critical fields, H_o, is available for these compounds and alloys; these values are given at the end of the table.

A. SUPERCONDUCTORS WITH T_c < 10 K

Substance	T_c, K	Crystal structure type
$Ag_{3.3}Al$	0.34	A12-cI58 (Mn)
$Ag_xAl_yZn_{1-x-y}$	0.15	Cubic
$AgBi_2$	2.87—3.0	
$Ag_7F_{0.25}N_{0.75}O_{10.25}$	0.85—0.90	
Ag_2F	0.0.066	
Ag_7FO_8	0.3	Cubic
$Ag_{0.8-0.3}Ga_{0.2-0.7}$	6.5—8	
Ag_4Ge	0.85	Hex., c.p.
$Ag_{0.438}Hg_{0.562}$	0.64	$D8_2$
$AgIn_2$	~2.4	C16

TABLE 4
Superconductive Compounds and Alloys (continued)

Substance	T_c, K	Crystal structure type
$Ag_{0.1}In_{0.9}Te$ ($n = 1.4 \times 10^{22}$)*	1.2—1.89	B1
$Ag_{0.2}In_{0.8}Te$ ($n = 1.07 \times 10^{22}$)	0.77—1.00	B1
AgLa	0.94	B2-cP2 (CsCl)
AgLa (9.5 kbar)	1.2	B2
AgLu	0.33	B2-cP2
$AgMo_4S_5$	9.1	hR15 (Mo_6PbS_8)
$Ag_{1.2}Mo_6Se_8$	5.9	Same
Ag_7NO_{11}	1.04	Cubic
Ag_xPb_{1-x}	7.2 max.	
Ag_4Sn	0.1	h**
Ag_xSn_{1-x}	1.5—3.7	
Ag_xSn_{1-x} (film)	2.0—3.8	
$AgTe_3$	2.6	Cubic
AgTh	2.2	C16-tI12 (Al_2Cu)
$AgTh_2$	2.26	C16
$Ag_{0.03}Tl_{0.97}$	2.67	
$Ag_{0.94}Tl_{0.06}$	2.32	
AgY	0.33	B2-cP2 (CsCl)
Ag_xZn_{1-x}	0.5—0.845	
$AlAu_4$	0.4—0.7	Like A13
Al_2Au	0.1	C1-cF12 (CaF_2)
Al_2CMo_3	9.8—10.2	A13+trace 2nd. phase)
Al_2CaSi	5.8	
$Al_{0.131}Cr_{0.088}V_{0.781}$	1.46	Cubic
$AlGe_2$	1.75	
Al_2Ge_2U	1.6	LI_2-cP4 (Cu_3Au)
$AlLa_3$	5.57	DO_{19}
Al_2La	3.23	C15
Al_2Lu	1.02	C15-cF24 (Cu_2Mg)
Al_3Mg_2	0.84	F.C.C.
$AlMo_3$	0.58	A15
$AlMo_6Pd$	2.1	
AlN	1.55	B4
Al_2NNb_3	1.3	A13
Al_3Nb	0.64	tI8 (Al_3Ti)
AlOs	0.39	B2
Al_3Os	5.90	
AlPb (film)	1.2—7	
Al_2Pt	0.48—0.55	C1
Al_5Re_{24}	3.35	A12
AlSb	2.8	B4-tI4 (Sn)
Al_2Sc	1.02	C15-cF24 (Cu_2Mg)
Al_2Si_2U	1.34	LI_2-cP4 (Cu_3Au)
$AlTh_2$	0.1	C16-tI12 (Al_2Cu)
Al_3Th	0.75	DO_{19}
$Al_xTi_yV_{1-x-y}$	2.05—3.62	Cubic
$Al_{0.108}V_{0.892}$	1.82	Cubic
Al_2Y	0.35	C15-cF24 (Cu_2Mg)
Al_3Yb	0.94	LI_2-cP4 (Cu_3Au)
Al_xZn_{1-x}	0.5—0.845	
$AlZr_3$	0.73	LI_2
AsBiPb	9.0	
AsBiPbSb	9.0	
AsHfOs	3.2	C22-hP9 (Fe_2P)
AsHfRu	4.9	same

TABLE 4
Superconductive Compounds and Alloys (continued)

Substance	T_c, K	Crystal structure type
$As_{0.33}InTe_{0.67}$ (n = 1.24×10^{22})	0.85—1.15	B1
$As_{0.5}InTe_{0.5}$ (n = 0.97×10^{22})	0.44—0.62	B1
As_4La_3	0.6	cI28 (Th_3P_4)
$AsNb_3$	0.3	$L1_2$-tP32
$As_{0.50}Ni_{0.06}Pd_{0.44}$	1.39	C2
$AsNi_{0.25}Pd_{0.75}$	1.6	$B8_1$-hP4 (NiAs)
AsOsZr	8.0	C22-hP9 (Fe_2P)
AsPb	8.4	
$AsPd_2$ (low-temp. phase)	0.60	Hexagonal
$AsPd_2$ (high-temp. phase)	1.70	C22
$AsPd_5$	0.46	Complex
As_3Pd_5	1.9	
AsRh	0.58	B31
$AsRh_{1.4—1.6}$	< 0.03—0.56	Hexagonal
AsSn	4.10	
AsSn (n = 2.14×10^{22})	3.41—3.65	B1
$As_{\sim 2}Sn_{\sim 3}$	3.5—3.6; 1.21—1.17	
As_3Sn_4 (n = 0.56×10^{22})	1.16—1.19	Rhombohedral
AsV_3	0.20	A15-cP8 (Cr_3Si)
Au_5Ba	0.4—0.7	$D2_d$
AuBe	2.64	B20
Au_2Bi	1.80	C15
Au_5Ca	0.34—0.38	$C15_b$
$AuGa_2$	1.6	C1-cF12 (CaF_2)
AuGa	1.2	B31
$Au_{0.40—0.92}Ge_{0.60—0.08}$	<0.32—1.63	Complex
$AuIn_2$	0.2	C1-cF12
AuIn	0.4—0.6	Complex
AuLu	<0.35	B2
$AuNb_3$	1.2	A2
$AuPb_2$	3.15	
$AuPb_2$ (film)	4.3	
$AuPb_3$	4.40	
$AuPb_3$ (film)	4.25	
Au_2Pb	1.18; 6—7	C15
$AuSb_2$	0.58	C2
AuSn	1.25	$B8_1$
Au_xSn_{1-x} (film)	2.0—3.8	
Au_5Sn	0.7—1.1	A3
$AuTa_{4.3}$	0.55	A15-cP8 (Cr_3Si)
Au_3Te_5	1.62	Cubic
$AuTh_2$	3.08	C16
AuTl	1.92	
AuV_3	0.74	A15
Au_xZn_{1-x}	0.50—0.845	
$AuZn_3$	1.21	Cubic
Au_xZr_y	1.7—2.8	A3
$AuZr_3$	0.92	A15
$B_2Ba_{0.67}Pt_3$	5.60	hP12 (B_2BaPt_3)
$BCMo_2$	5.4	Orthorhombic
$BCMo_2$	5.3—7.0	Same
$B_2Ca_{0.67}Pt_3$	1.57	hP12
B_4ErIr_4	2.1	tP18 (B_4CeCo_4)
B_4ErRh_4	4.3	oC108 (B_4LuRh_4)
B_4ErRh_4	8.7	tP18 (B_4CeCo_4)

TABLE 4
Superconductive Compounds and Alloys (continued)

Substance	T_c, K	Crystal structure type
BHf	3.1	Cubic
B_4HoIr_4	2.0	tP18
B_4HoRh_4	1.4	oC108
B_2Ir_3La	1.65	hP6 ($CaCu_5$)
B_2Ir_3Th	2.09	Same
B_4Ir_4Tm	1.6	tP18
B_6La	5.7	
B_2LaRh_3	2.82	hP6
$B_{12}Lu$	0.48	
B_2LuOs	2.66	oP16 (B_2LuRu)
B_2LuOs_3	4.62	hP6
B_4LuRh_4	6.2	oC108
B_2LuRu	9.86	oP16
B_4LuRu_4	2.0	tI72 (B_4LuRu_4)
BMo	0.5 (extrapol.)	
BMo_2	4.74	C16
BNb	8.25	B_f
B_4NdRh_4	5.3	tP18
B_2OsSc	1.34	oP16
B_2OsY	2.22	oP16
$B_2Pt_3Sr_{0.67}$	2.78	hP12 (B_2BaPt_3)
BRe_2	2.80; 4.6	
$B_4Rh_{3.4}Ru_{0.6}$	8.38	tI72
B_4Rh_4Sm	2.7	tP18
B_4Rh_4Th	4.3	Same
B_4Rh_4Tm	9.8	Same
B_4Rh_4Tm	5.4	oC108
$B_{0.3}Ru_{0.7}$	2.58	$D10_2$
B_4Ru_4Sc	7.2	tI72
B_2Ru_3Th	1.79	hP6
B_2Ru_3Y	2.85	Same
$B_2Ru\ Y$	7.80	oP16
B_4Ru_4Y	1.4	tI72
$B_{12}Sc$	0.39	
BTa	4.0	B_f
BTa_2	3.12	C16-tI12 (Al_2Cu)
B_6Th	0.74	
BW_2	3.1	C16
B_6Y	6.5—7.1	
$B_{12}Y$	4.7	
BZr	3.4	Cubic
$B_{12}Zr$	5.82	
$BaBi_3$	5.69	Tetragonal
$Ba_2Mo_{15}Se_{19}$	2.75	hP15 (Mo_6PbS_8)
$Ba_xO_3Sr_{1-x}Ti$ ($n = 4.2 \times 10^{19}$)	<0.1—0.55	
$Ba_{0.13}O_3W$	1.9	Tetragonal
$Ba_{0.14}O_3W$	<1.25—2.2	Hexagonal
$BaRh_2$	6.0	C15
$Be_{22}Mo$	2.51	Cubic ($Be_{22}Re$)
$Be_8Nb_5Zr_2$	5.2	
$Be_{0.98—0.92}Re_{0.02—0.08}$ (quenched)	9.5—9.75	Cubic
$Be_{0.957}Re_{0.043}$	9.62	Cubic ($Be_{22}Re$)
BeTc	5.21	Cubic
$Be_{22}W$	4.12	Cubic ($Be_{22}Re$)

TABLE 4
Superconductive Compounds and Alloys (continued)

Substance	T_c, K	Crystal structure type
$Be_{13}W$	4.1	Tetragonal
Bi_3Ca	2.0	
$Bi_{0.5}Cd_{0.13}Pb_{0.25}Sn_{0.12}$ (weight fractions)	8.2	
BiCo	0.42—0.49	
Bi_2Cs	4.75	C15
Bi_xCu_{1-x} (electrodeposited)	2.2	
BiCu	1.33—1.40	
Bi_3Fe	1.0	m**
$Bi_{0.019}In_{0.981}$	3.86	
$Bi_{0.05}In_{0.95}$	4.65	α-phase
$Bi_{0.10}In_{0.90}$	5.05	Same
$Bi_{0.15—0.30}In_{0.85—0.70}$	5.3—5.4	α- and β-phases
$Bi_{0.34—0.48}In_{0.66—0.52}$	4.0—4.1	
Bi_3In_5	4.1	
$BiIn_2$	5.65	β-phase
Bi_2Ir	1.7—2.3	
Bi_2Ir (quenched)	3.0—3.96	
BiK	3.6	
Bi_2K	3.58	C15
BiLi	2.47	$L1_o$, α-phase
$Bi_{4—9}Mg$	0.7—~1.0	
Bi_3Mo	3—3.7	
BiNa	2.25	$L1_o$
$BiNb_3$	4.5	A15-cP8 (Cr_3Si)
$BiNb_3$ (high pressure and temperature)	3.05	A15
BiNi	4.25	$B8_1$
Bi_3Ni	4.06	Orthorhombic
$BiNi_{0.5}Rh_{0.5}$	3.0	$B8_1$-hP4 (AsNi)
$Bi_{0.5}NiSb_{0.5}$	2.0	Same
$Bi_{1-0}Pb_{0-1}$	7.26—9.14	
$Bi_{1-0}Pb_{0-1}$ (film)	7.25—8.67	
$Bi_{0.05—0.40}Pb_{0.95—0.60}$	7.35—8.4	H.C.P. to ε-phase
Bi_2Pb	4.25	t**
BiPbSb	8.9	
$Bi_{0.5}Pb_{0.31}Sn_{0.19}$ (weight fractions)	8.5	
$Bi_{0.5}Pb_{0.25}Sn_{0.25}$	8.5	
$BiPd_2$	4.0	
$Bi_{0.4}Pd_{0.6}$	3.7—4	Hexagonal, ordered
BiPd	3.7	Orthorhombic
Bi_2Pd	1.70	Monoclinic, α-phase
Bi_2Pd	4.25	Tetragonal, β-phase
BiPd[illegible]Pt[illegible]	3.7	$B8_1$-hP4 (NiAs)
BiPdSe	1.0	C2
BiPdTe	1.2	C2
BiPt	1.21	$B8_1$
$Bi_{0.1}PtSb_{0.9}$	2.05; 1.5	$B8_1$-hP4 (NiAs)
BiPtSe	1.45	C2
BiPtTe	1.15	C2
Bi_2Pt	0.155	Hexagonal
Bi_2Rb	4.25	C15
$BiRe_2$	1.9—2.2	
BiRh	2.06	$B8_1$

TABLE 4
Superconductive Compounds and Alloys (continued)

Substance	T_c, K	Crystal structure type
Bi_3Rh	3.2	Orthorhombic (NiB_3)
Bi_4Rh	2.7	Hexagonal
BiRu	5.7	m**
Bi_3Sn	3.6—3.8	
BiSn	3.8	
Bi_xSn_y	3.85—4.18	
Bi_3Sr	5.62	$L1_2$
Bi_3Te	0.75—1.0	
Bi_5Tl_3	6.4	
$Bi_{0.26}Tl_{0.74}$	4.4	Cubic, disordered
$Bi_{0.26}Tl_{0.74}$	4.15	$L1_2$, ordered (?)
Bi_2Y_3	2.25	
Bi_3Zn	0.8—0.9	
$Bi_{0.3}Zr_{0.7}$	1.51	
$BiZr_3$	2.4—2.8	
$BrMo_6Se_7$	7.1	hP15 (Mo_6PbS_8)
$Br_3Mo_6Se_5$	7.1	Same
CCs_x	0.020—0.135	Hexagonal
CFe_3	1.30	DO_{11}-oP16 (Fe_3C)
$CGaMo_2$	3.7—4.1	Hexagonal
$CHf_{0.5}Mo_{0.5}$	3.4	B1
$CHf_{0.3}Mo_{0.7}$	5.5	B1
$CHf_{0.25}Mo_{0.75}$	6.6	B1
$CHf_{0.7}Nb_{0.3}$	6.1	B1
$CHf_{0.6}Nb_{0.4}$	4.5	B1
$CHf_{0.5}Nb_{0.5}$	4.8	B1
$CHf_{0.4}Nb_{0.6}$	5.6	B1
$CHf_{0.25}Nb_{0.75}$	7.0	B1
$CHf_{0.2}Nb_{0.8}$	7.8	B1
$CHf_{0.9—0.1}Ta_{0.1—0.9}$	5.0—9.0	B1
CK (excess K)	0.55	Hexagonal
C_8K	0.39	Hexagonal
C_2La	1.66	tI6 (CaC_2)
C_2Lu	3.33	Same
$C_{0.40—0.44}Mo_{0.60—0.56}$	9—13	
C_3MoRe	3.8	B1-cF8
$C_{0.6}Mo_{4.8}Si_3$	7.6	$D8_8$
$CMo_{0.2}Ta_{0.8}$	7.5	B1
$CMo_{0.5}Ta_{0.5}$	7.7	B1
$CMo_{0.75}Ta_{0.25}$	8.5	B1
$CMo_{0.8}Ta_{0.2}$	8.7	B1
$CMo_{0.85}Ta_{0.15}$	8.9	B1
CMo_xV_{1-x}	2.9—9.3	B1
CMo_xZr_{1-x}	9.8	B1
$C_{0.984}Nb$	9.8	B1
CNb_2	9.1	
CNb_xTi_{1-x}	<4.2—8.8	B1
$CNb_{0.1—0.9}Zr_{0.9—0.1}$	4.2—8.4	B1
CRb_x (Au)	0.023—0.151	Hexagonal
$CRe_{0.06}W$	5.0	
CRu	2.00	hP2 (CW)
$C_{0.98}7Ta$	9.7	
$C_{0.848—0.987}$	2.04—9.7	
CTa (film)	5.09	B1
CTa_2	3.26	L'_3

TABLE 4
Superconductive Compounds and Alloys (continued)

Substance	T_c, K	Crystal structure type
$CTa_{0.4}Ti_{0.6}$	4.8	B1
$Cta_{1—0.4}W_{0—0.6}$	8.5—10.5	B1
$CTa_{0.2—0.9}Zr_{0.8—0.1}$	4.6—8.3	B1
CTc (excess C)	3.85	Cubic
$CTi_{0.5—0.7}W_{0.5—0.3}$	6.7—2.1	B1
CW	1.0	
CW_2	2.74	L'_3
CW_2	5.2	F.C.C.
C_2Y	3.88	tI6 (CaC_2)
$Ca_3Co_4Sn_{13}$	5.9	cP40 ($Pr_3Rh_2Sn_{13}$)
$Ca_3Ge_{13}Rh_4$	2.1	Same
CaHg	1.6	B2-cP2 (CsCl)
$CaHg_3$	1.6	hP8 (Ni_3Sn)
$CaIr_2$	6.15	C15
$Ca_3Ir_4Sn_{13}$	7.1	cP40
$Ca_xO_3Sr_{1-x}Ti$ ($n = 3.7—11 \times 10^{19}$)	<0.1—0.55	
$Ca_{0.1}O_3W$	1.4—3.4	Hexagonal
CaPb	7.0	
$CaRh_2$	6.40	C15
$CaRh_{1.2}Sn_{4.5}$	8.7	cP40
$CaTl_3$	2.0	B2-cP2
$Cd_{0.3—0.5}Hg_{0.7—0.5}$	1.70—1.92	
CdHg	1.77; 2.15	Tetragonal
$Cd_{0.0075—0.05}In_{0.9925—0.95}$	3.24—3.36	Tetragonal
$Cd_{0.97}Pb_{0.03}$	4.2	
CdSn	3.65	
$Cd_{0.17}Tl_{0.83}$	2.3	
$Cd_{0.18}Tl_{0.82}$	2.54	
$CeCo_2$	0.84	C15
$CeCo_{1.67}Ni_{0.33}$	0.46	C15
$CeCo_{1.67}Rh_{0.33}$	0.47	C15
$Ce_xGd_{1-x}Ru_2$	3.2—5.2	C15
$CeIr_3$	3.34	
$CeIr_5$	1.82	
$Ce_{0.005}La_{0.995}$	4.6	
Ce_xLa_{1-x}	1.3—6.3	
$Ce_xPr_{1-x}Ru_2$	1.4—5.3	C15
Ce_xPt_{1-x}	0.7—1.55	
$CeRu_2$	6.0	C15
$Ce_3Mo_6Se_5$	5.7	hR15 (Mo_6PbS_8)
$Ce_2Mo_6Te_6$	1.7	Same
$Co_xFe_{1-x}Si_2$	1.4 (max.)	C1
$CoHf_2$	0.56	$E9_3$
$CoLa_3$	4.28	
$Co_4La_3Sn_{13}$	2.8	cP40
$CoLu_3$	~0.35	
Co_xLuSn_y	1.5	cP40
$Co_{0—0.01}Mo_{0.8}Re_{0.2}$	2—10	
$Co_{0.02—0.10}Nb_3Rh_{0.98—0.90}$	2.28—1.90	A15
$Co_xNi_{1-x}Si_2$	1.4 (max.)	C1
$Co_{0.5}Rh_{0.5}Si_2$	2.5	
$Co_xRh_{1-x}Si_2$	3.65 (max.)	
$Co_{\sim0.3}So_{\sim0.7}$	~0.35	
$Co_4Sc_5Si_{10}$	5.0	tP38 ($Co_4Sc_5Si_{10}$)
$CoSi_2$	1.40; 1.22	C1

TABLE 4
Superconductive Compounds and Alloys (continued)

Substance	T_c, K	Crystal structure type
Co_xSn_yYb	2.5	cP40
Co_3Th_7	1.83	$D10_2$
Co_xTi_{1-x}	2.8 (max.)	Co in α-Ti
Co_xTi_{1-x}	3.8 (max.)	Co in β-Ti
$CoTi_2$	3.44	$E9_3$
CoTi	0.71	A2
CoU	1.7	B2, distorted
CoU_6	2.29	$D2_c$
$Co_{0.28}Y_{0.72}$	0.34	
CoY_3	<0.34	
$CoZr_2$	6.3	C16
$Co_{0.1}Zr_{0.9}$	3.9	A3
$Cr_{0.6}Ir_{0.4}$	0.4	H.C.P.
$Cr_{0.65}Ir_{0.35}$	0.59	H.C.P.
$Cr_{0.7}Ir_{0.3}$	0.76	H.C.P.
$Cr_{0.72}Ir_{0.28}$	0.83	
Cr_3Ir	0.45	A15
$Cr_{0—0.1}Nb_{1—0.9}$	4.6—9.2	A2
$Cr_{0.80}Os_{0.20}$	2.5	Cubic
Cr_3Os	4.68	A15-cP8 (Cr_3Si)
Cr_xRe_{1-x}	1.2—5.2	
$Cr_{0.4}Re_{0.6}$	2.15	$D8_b$
$Cr_{0.8—0.6}Rh_{0.2—0.4}$	0.5—1.10	A3
Cr_3Rh	0.3	A15-cP8
Cr_3Ru (annealed)	3.3	A15
Cr_2Ru	2.02	$D8_b$
Cr_3Ru_2	2.10	$D8_b$-tP30 (CrFe)
$Cr_{0.1—0.5}Ru_{0.9—0.5}$	0.34—1.65	A3
Cr_xTi_{1-x}	3.6 (max.)	Cr in α-Ti
Cr_xTi_{1-x}	4.2 (max.)	Cr in β-Ti
$Cr_{0.1}Ti_{0.3}V_{0.6}$	5.6	
$Cr_{0.0175}U_{0.9825}$	0.75	β-phase
$Cs_{0.32}O_3W$	1.12	Hexagonal
$Cu_{0.15}In_{0.85}$ (film)	3.75	
$Cu_{0.04—0.08}In_{0.94—0.92}$	4.4	
CuLa	5.85	
$Cu_2Mo_6O_2S_6$	9	hR15 (Mo_6PbS_8)
$Cu_2Mo_6Se_8$	5.9	Same
Cu_xPb_{1-x}	5.7—7.7	
CuS	1.62	B18
CuS_2	1.48—1.53	C18
CuSSe	1.5—2.0	C18
$CuSe_2$	2.3—2.43	C18
CuSeTe	1.6—2.0	C18
Cu_xSn_{1-x}	3.2—3.7	
Cu_xSn_{1-x} (film, made at 10K)	3.6—7	
Cu_xSn_{1-x} (film, made at 300K)	2.8—3.7	
$CuTe_2$	<1.25—1.3	C18
$CuTh_2$	3.49	C16
$Cu_{0—0.027}V$	3.9—5.3	A2
CuY	0.33	B2-cP2 (CsCl)
Cu_xZn_{1-x}	0.5—0.845	
$DyMo_6S_8$	2.1	hR15
Er_xLa_{1-x}	1.4—6.3	
$ErMo_6S_8$	2.2	hR15

TABLE 4
Superconductive Compounds and Alloys (continued)

Substance	T_c, K	Crystal structure type
$ErMo_6Se_8$	6.2	hR15
$Fe_3Lu_2Si_5$	6.1	tP40 ($Fe_3Sc_2Si_5$)
$Fe_{0-0.04}Mo_{0.8}Re_{0.2}$	1—10	
$Fe_{0.05}Ni_{0.05}Zr_{0.90}$	~3.9	
Fe_3Re_2	6.55	$D8_b$-tP30 (FeCr)
$Fe_3Sc_2Si_5$	4.52	tP40
Fe_3Si_5Tm	1.3	Same
$Fe_3Si_5Y_2$	2.4	Same
Fe_3Th_7	1.86	D10
Fe_xTi_{1-x}	3.2 (max.)	Fe in α-Ti
Fe_xTi_{1-x}	3.7 (max.)	Fe in β-Ti
$Fe_xTi_{0.6}V_{1-x}$	6.8 (max.)	
FeU_6	3.86	$D2_c$
$Fe_{0.1}Zr_{0.9}$	1.0	A3
$Ga_{0.5}Ge_{0.5}Nb_3$	7.3	A15
Ga_2Ge_2U	0.87	B2-cP2
$GaHf_2$	0.21	C16-tI12 (Al_2Cu)
$GaLa_3$	5.84	
Ga_3Lu	2.3	B2-cP2
Ga_2Mo	9.5	
$GaMo_3$	0.76	A15
GaN (black)	5.85	B4
$Ga_{0.7}Pt_{0.3}$	2.9	C1
GaPt	1.74	B20
GaSb (120kbar, 77K, annealed)	4.24	A5
GaSb (unannealed)	~5.9	
$Ga_{0-1}Sn_{1-0}$ (quenched)	3.47—4.18	
$Ga_{0-1}Sn_{1-0}$ (annealed)	2.6—3.85	
GaTe	0.17	mC24 (GaTe)
Ga_5V_2	3.55	Tetragonal (Mn_2Hg_5)
$GaV_{4.5}$	9.15	
Ga_3Zr	1.38	
Ga_3Zr_5	3.8	$D8_b$-hP16 (Mn_5Si_3)
Gd_xLa_{1-x}	< 1.0—5.5	
$GdMo_6S_8$	3.5	hR15
$GdMo_6Se_8$	5.6	hR15
$Gd_xOs_2Y_{1-x}$	1.4—4.7	
$Gd_xRu_2Th_{1-x}$	3.6 (max.)	C15
$Ge_{10}As_4Y_5$	9.06	tP38 ($C0_4Sc_5Si_{10}$)
GeIr	4.7	B31
GeIrLa	1.64	tI12 (LaPtSi)
$Ge_{10}Ir_4Lu_5$	2.60	tP38
$Ge_{10}Ir_4Y_5$	2.62	tP38
Ge_2La	1.49; 2.2	Orthorhombic, distorted (Mn_2Hg_5)
GeLaPt	3.53	tI12
$Ge_{13}Lu_3Os_4$	3.6	cP40 ($Pr_3Rh_2Sn_{13}$)
$Ge_{10}Lu_5Rh_4$	2.79	tP38
$Ge_{13}Lu_3Ru_4$	2.3	cP40
$GeMo_3$	1.43	A15
$GeNb_2$	1.9	
$Ge_{0.29}Nb_{0.71}$	6	A15
GePt	0.40	B31
Ge_3Rh_5	2.12	Orthorhombic, related to $InNi_2$

TABLE 4
Superconductive Compounds and Alloys (continued)

Substance	T_c, K	Crystal structure type
GeRh	0.96	B31-oP8 (MnP)
$Ge_{13}Rh_4Sc_3$	1.9	c P40
$Ge_{10}Rh_4Y_5$	1.35	tP38
$Ge_{13}Ru_4Y_3$	1.7	cP40
Ge_2So	1.3	
$GeTa_3$	8.0	A15-cP8 (Cr_3Si)
Ge_3Te_4 ($n = 1.06 \times 10^{22}$)	1.55—1.80	Rhombohedral
Ge_xTe_{1-x} ($n = 8.5—64 \times 10^{20}$)	0.07—0.41	R1
GeV_3	6.01	A15
Ge_2Y	3.80	C_c
$Ge_{1.62}Y$	2.4	
Ge_2Zr	0.30	oC12 ($ZrSi_2$)
$GeZr_3$	0.4	$L1_2$-tP32 (Ti_3P)
$H_{0.33}Nb_{0.67}$	7.28	B.C.C.
$H_{0.1}Nb_{0.9}$	7.38	Same
$H_{0.05}Nb_{0.95}$	7.83	Same
$H_{0.12}Ta_{0.88}$	2.81	B.C.C.
$H_{0.08}Ta_{0.92}$	3.26	Same
$H_{0.04}Ta_{0.96}$	3.62	Same
HfIrSi	3.50	C37-cP12 (Co_2Si)
$HfMo_2$	0.05	hP24 (Ni_2Mn)
$HfN_{0.989}$	6.6	B1
$Hf_{0—0.5}Nb_{1—0.5}$	8.3—9.5	A2
$Hf_{0.75}Nb_{0.25}$	>4.2	
$HfOs_2$	2.69	C14
HfOsP	6.1	C22-hP9 (Fe_2P)
HfPRu	9.9	Same
$HfRe_2$	4.80	C14
$Hf_{0.14}Re_{0.86}$	5.86	A12
$Hf_{0.99—0.96}Rh_{0.01—0.04}$	0.85—1.51	
$Hf_{0—0.55}Ta_{1—0.45}$	4.4—6.5	A2
HfV_2	8.9—9.6	C15
Hg_xIn_{1-x}	3.14—4.55	
HgIn	3.81	
Hg_2K	1.20	Orthorhombic
Hg_3K	3.18	
Hg_4K	3.27	
Hg_8K	3.42	
Hg_3Li	1.7	Hexagonal
$HgMg_3$	0.17	hP8 (Na_3As)
Hg_2Mg	4.0	tI6 ($MoSi_2$)
Hg_3Mg_5	0.48	$D8_b$-hP16 (Mn_5Si_3)
Hg_2Na	1.62	Hexagonal
Hg_4Na	3.05	
Hg_xPb_{1-x}	4.14—7.26	
HgSn	4.2	
Hg_xTl_{1-x}	2.30—4.19	
Hg_5Tl_2	3.86	
Ho_xLa_{1-x}	1.3—6.3	
$Ho_{1.2}Mo_6Se_8$	6.1	$D10_2$-hR12 (Be_3Nb)
$In_{1—0.86}Mg_{0—0.14}$	3.395—3.363	
$In_2Mo_6Te_6$	2.6	hR15 (Mo_6PbS_8)
$InNb_3$ (high pressure and temp.)	4—8; 9.2	A15
$In_{0.5}Nb_3Zr_{0.5}$	6.4	
$In_{0.11}O_3W$	< 1.25—2.8	Hexagonal

TABLE 4
Superconductive Compounds and Alloys (continued)

Substance	T_c, K	Crystal structure type
$In_{0.95—0.85}Pb_{0.05—0.15}$	3.6—5.05	
$In_{0.98—0.91}Pb_{0.02—0.09}$	3.45—4.2	
InPb	6.65	
InPd	0.7	B2
InSb (quenched from 170 kbar into liquid N_2)	4.8	Like A5
InSb	2.1	B4
$(InSb)_{0.95—0.10}Sn_{0.05—0.90}$ (various heat treatments)	3.8—5.1	
$(InSb)_{0—0.07}Sn_{1—0.93}$	3.67—3.74	
In_3Sn	~5.5	
In_xSn_{1-x}	3.4—7.3	
$In_{0.82—1}Te$ ($n = 0.83—1.71 \times 10^{22}$)	1.02—3.45	B1
$In_{1.000}Te_{1.002}$	3.5—3.7	B1
In_3Te_4 ($n = 4.7 \times 10^{21}$)	1.15—1.25	Rhombohedral
In_xTl_{1-x}	2.7—3.374	
$In_{0.8}Tl_{0.2}$	3.223	
$In_{0.62}Tl_{0.38}$	2.760	
$In_{0.78—0.69}Tl_{0.22—0.31}$	3.18—3.32	Tetragonal
$In_{0.69—0.62}Tl_{0.31—0.38}$	2.98—3.3	F.C.C.
Ir_2La	0.48	C15
Ir_3La	2.32	$D10_2$
Ir_3La_7	2.24	$D10_2$
Ir_5La	2.13	
$IrLaSi_2$	2.03	oC16 ($CeNiSi_2$)
$IrLaSi_3$	2.7	tI10 ($BaNiSn_3$)
Ir_2Lu	2.47	C15
Ir_3Lu	2.89	C15
$Ir_4Lu_5Si_{10}$	3.9	tP38 ($Co_4Sc_5Si_{10}$)
IrMo	< 1.0	A3
$IrMo_3$	9.6	A15
$IrMo_3$	6.8	$D8_b$
$IrNb_3$	1.9	A15
$Ir_{0.4}Nb_{0.6}$	9.8	$D8_b$
$Ir_{0.37}Nb_{0.63}$	2.32	$D8_b$
IrNb	7.9	$D8_b$
$Ir_{1.15}Nb_{0.85}$	4.6	oP12 (IrTa)
$Ir_{0.02}Nb_3Rh_{0.98}$	2.43	A15
$Ir_{0.05}Nb_3Rh_{0.95}$	2.38	A15
$Ir_{0.287}O_{0.14}Ti_{0.573}$	5.5	$E9_3$
$Ir_{0.265}O_{0.035}Ti_{0.65}$	2.30	$E9_3$
Ir_xOs_{1-x}	0.3—0.98	
$Ir_{1.5}Os_{0.5}$	2.4	C14
IrOsY	2.6	C15
IrSiY	2.70	C37-oP12 (Co_2Si)
IrSiZr	2.04	Same
Ir_2Sc	2.07	C15
$Ir_{2.5}Sc$	2.46	C15
$Ir_4Sc_5Si_{10}$	8.46	tP38
Ir_2Si_2Th	2.14	tI10
$IrSi_3Th$	1.75	tI10
IrSiTh	6.50	tI12 (LaPtSi)
Ir_2Si_2Y	2.60	tI10 (Al4Ba)
$Ir_4Si_{10}Y_5$	3.10	tP38
$Ir_3Si_5Y_2$	2.83	oI40

TABLE 4
Superconductive Compounds and Alloys (continued)

Substance	T_c, K	Crystal structure type
$IrSn_2$	0.65—0.78	C1
Ir_2Sr	5.70	C15
Ir_7Ta_{13}	1.2	$D8_b$-tP30 (FeCr)
$Ir_{0.5}Te_{0.5}$	~3	
$IrTe_3$	1.18	C2
IrTh	< 0.37	B_f
Ir_2Th	6.50	C15
Ir_3Th	4.71	
Ir_3Th_7	1.52	$D10_2$
Ir_5Th	3.93	$D2_d$
$IrTi_3$	5.40	A15
IrV_2	1.39	A15
IrW_3	3.82	
$Ir_{0.28}W_{0.72}$	4.49	
Ir_2Y	2.18; 1.38	C15
$Ir_{0.69}Y_{0.31}$	1.98; 1.44	C15
$Ir_{0.70}Y_{0.30}$	2.16	C15
Ir_2Y_3	1.61	
Ir_3Y	3.50	$D10_2$-hR13 (Be_3Nb)
Ir_xY_{1-x}	0.3—3.7	
Ir_2Zr	4.10	C15
$Ir_{0.1}Zr_{0.9}$	5.5	A3
$K_2Mo_{15}S_{19}$	3.32	hR15
$K_{0.27—0.31}O_3W$	0.50	Hexagonal
$K_{0.40—0.57}O_3W$	1.5	Tetragonal
$La_{0.55}Lu_{0.45}$	2.2	Hexagonal, La type
$La_{0.8}Lu_{0.2}$	3.4	Same
$LaMg_2$	1.05	C15
$LaMo_6S_8$	7.1	hR15
LaN	1.35	
$LaOs_2$	6.5	C15
$LaPt_2$	0.46	C15
$La_{0.28}Pt_{0.72}$	0.54	C15
LaPtSi	3.48	tI12
$LaRh_3$	2.60	
$LaRh_5$	1.62	
La_7Rh_3	2.58	$D10_2$
$LaRhSi_2$	3.42	oC16 ($CeNiSi_2$)
$La_2Rh_3Si_5$	4.45	oI40 ($Co_3Si_5U_2$)
$LaRhSi_3$	2.7	tI10 ($BaNiSn_3$)
$LaRh_2Si_2$	3.90	tI10 (Al_4Ba)
$LaRu_2$	1.63	C15
La_3S_4	6.5	$D7_3$
La_3Se_4	8.6	$D7_3$
$LaSi_2$	2.3	C_c
La_xY_{1-x}	1.7—5.4	
LaZn	1.04	B2
$Li_2Mo_6S_8$	4.2	hR15
LiPb	7.2	
$LuOs_2$	3.49	C14
$Lu_{0.275}Rh_{0.725}$	1.27	C15
$LuRh_5$	0.49	
$Lu_5Rh_4Si_{10}$	3.95	tP38 ($Co_4So_5Si_{10}$)
$LuRu_2$	0.86	C14
$Mg_{1.14}Mo_{6.6}S_8$	3.5	hR15

TABLE 4
Superconductive Compounds and Alloys (continued)

Substance	T_c, K	Crystal structure type
Mg2Nb	5.6	
$Mg_{\sim0.47}Tl_{\sim0.53}$	2.75	B2
MgZn	0.9	A3-oP4 (AuCd)
Mn_xTi_{1-x}	2.3 (max.)	Mn in -Ti
Mn_xTi_{1-x}	1.1—3.0	Mn in -Ti
MnU_6	2.32	$D2_c$
Mo_2N	5.0	F.C.C.
$Mo_6Na_2S_8$	8.6	hR15
Mo_xNb_{1-x}	0.016—9.2	
$Mo_{5.25}Nb_{0.75}Se_8$	6.2	hR15
Mo_6NdSa_8	8.2	hR15
Mo_3Os	7.2	A15
$Mo_{0.62}Cs_{0.38}$	5.65	$D8_b$
Mo_3P	5.31	DO_e
$Mo_6Pb_{1.2}Se_8$	6.75	hR15
$Mo_{0.5}Pd_{0.5}$	3.52	A3
Mo_6PrSe_8	9.2	hR15
MoRe	7.8	$D8_b$-tP30
$MoRe_3$	9.25; 9.89	A12
Mo_xRe_{1-x}	1.2—12.2	
$Mo_{0.42}Re_{0.58}$	6.35	$D8_b$
MoRh	1.97	A3
Mo_xRh_{1-x}	1.5—8.2	B.C.C.
MoRu	9.5—10.5	A3
$Mo_{0.61}Ru_{0.39}$	7.18	$D8_b$
$Mo_{0.2}Ru_{0.8}$	1.66	A3
Mo_3Ru_2	7.0	$D8_b$-tP30
$Mo_4Ru_2Te_8$	1.7	hR15
Mo_6S_8	1.85	hR15
Mo_6S_8Sc	3.6	hR15
$Mo_6S_8Sm_{1.2}$	2.9	hR15
Mo_6S_8Tb	2.0	hR15
Mo_6S_8Tl	8.7	hR15
$Mo_6S_8Tm_{1.2}$	2.1	hR15
$Mo_6S_8Y_{1.2}$	3.0	hR15
Mo_6S_8Yb	9.2	hR15
$Mo_{6.6}S_8Zn_{11}$	3.6	hR15
Mo_3Sb_4	2.1	
Mo_6Se_8	6.3	hR15
$Mo_6Se_8Sm_{1.2}$	6.8	hR15
$Mo_6Se_8Sn_{1.2}$	6.8	hR15
Mo_6Se_8Tb	5.7	hR15
Mo_3Se_3Tl	4.0	hP14
$Mo_6Se_8Tm_{1.2}$	6.3	hR15
Mo_6Se_8Yb	6.2	hR15
Mo_3Si	1.30	A15
$MoSi_{0.7}$	1.34	
Mo_xSiV_{3-x}	4.54—16.0	A15
$Mo_{5.25}Ta_{0.75}Te_8$	1.7	hR15
Mo_6Te_8	1.7	hR15
$Mo_{0.16}Ti_{0.84}$	4.18; 4.25	
$Mo_{0.913}Ti_{0.087}$	2.95	
$Mo_{0.04}Ti_{0.96}$	2.0	Cubic
$Mo_{0.025}Ti_{0.975}$	1.8	
Mo_xU_{1-x}	0.7—2.1	

TABLE 4
Superconductive Compounds and Alloys (continued)

Substance	T_c, K	Crystal structure type
Mo_xV_{1-x}	0—~5.3	
Mo_2Zr	4.25—4.75	C15
NNb (film)	6—9	B1
$N_xO_yTi_z$	2.9—5.6	Cubic
$N_xO_yV_z$	5.8—8.2	Cubic
$N_{0.34}Re$	4—5	F.C.C.
NTa (film)	4.84	B1
$N_{0.6—0.987}Ti$	<1.17—5.8	B1
$N_{0.82—0.99}V$	2.9—7.9	B1
NZr	9.8	B1
$N_{0.906—0.984}Zr$	3.0—9.5	B1
$Na_{0.28—0.35}O_3W$	0.56	Tetragonal
$Na_{0.28}Pb_{0.72}$	7.2	
NbO	1.25	
$NbOs_2$	2.52	A12
Nb_3Os	1.05	A15
$Nb_{0.6}Os_{0.4}$	1.89; 1.78	$D8_b$
$Nb_3Os_{0.02—0.10}Rh_{0.98—0.90}$	2.42—2.30	A15
Nb_3P	1.8	$L1_2$tP32 (Ti_3P)
NbPRh	4.08	C37-oP12 (Co_2Si)
$Nb_{0.6}Pd_{0.4}$	1.60	$D8_f$ plus cubic
$Nb_3Pd_{0.02—0.10}Rh_{0.92—0.90}$	2.49—2.55	A15
$Nb_{0.62}Pt_{0.38}$	4.21	$D8_b$
Nb_5Pt_3	3.73	$D8_b$
$Nb_3Pt_{0.02—0.98}Rh_{0.98—0.02}$	2.52—9.6	A15
$NbRe_3$	5.27	$D8_b$-tP30 (FeCr)
$Nb_{0.38—0.18}Re_{0.62—0.82}$	2.43—9.70	A15
NbRe	3.8	$D8_b$-tP30
NbReSi	5.1	oI36 (FeTiSi)
Nb_3Rh	2.64	A15
$Nb_{0.6}Rh_{0.40}$	4.21	$D8_b$ plus other
$Nb_{0.9}Rh_{1.1}$	3.07	A3-oP4 (AuCd)
$Nb_3Rh_{0.98—0.90}Ru_{0.02—0.10}$	2.42—2.44	A15
Nb_xRu_{1-x}	1.2—4.8	
NbRuSi	2.65	oI36
NbS_2	6.1—6.3	Hexagonal, $NbSe_2$ type
NbS_2	5.0—5.5	Hexagonal, three-layer type
Nb_3Sb	0.2	$L1_2$-tP32 (Ti_3P)
$Nb_3Sb_{0—0.7}Sn_{1—0.3}$	6.8—18	A15
$NbSe_2$	5.15—5.62	Hexagonal
$Nb_{1—1.05}Se_2$	2.2—7.0	Same
Nb_3Se_4	2.0	hP14
Nb_3Si	1.5	$L1_2$
Nb_3SiSnV_3	4.0	
$NbSn_2$	2.60	Orthorhombic
Nb_6Sn_5	2.8	oI44 (Sn_5Ti_6)
NbSnTaV	6.2	A15
$NbSnV_2$	5.5	A15
Nb_2SnV	9.8	A15
Nb_xTa_{1-x}	4.4—9.2	A2
Nb_3Te_4	1.8	hP14
Nb_xTi_{1-x}	0.6—9.8	
$Nb_{0.6}Ti_{0.4}$	9.8	
Nb_xU_{1-x}	1.95 (max.)	
$Nb_{0.88}V_{0.12}$	5.7	A2

TABLE 4
Superconductive Compounds and Alloys (continued)

Substance	T_c, K	Crystal structure type
$Nb_{0.5}V_{1.5}Zr$	4.3	C15-hP12 ($MgZn_2$)
$Ni_{0.3}Th_{0.7}$	1.98	$D10_2$
$NiZr_2$	1.52	
$Ni_{0.1}Zr_{0.9}$	1.5	A3
$O_3Rb_{0.27—0.29}W$	1.98	Hexagonal
OSn	3.81	tP4 (PbO)
O_3SrTi ($n = 1.7—12.0 \times 10^{19}$)	0.12—0.37	
O_3SrTi ($n = 10^{18}—10^{21}$)	0.05—0.47	
O_3SrTi ($n = 10^{20}$)	0.47	
$O_3Sr_{0.08}W$	2—4	Hexagonal
OTi	0.58	
$O_3Tl_{0.30}W$	2.0—2.14	Hexagonal
OV_3Zr_3	7.5	$E9_3$
OW_3 (film)	3.35; 1.1	A15
OsPti	1.2	C22-hP9 (Fe_2P)
OsPZr	7.4	Same
OsReY	2.0	C14
Os_2Sc	4.6	C14
OsTa	1.95	A12
Os_3Th_7	1.51	$D10_2$
Os_xW_{1-x}	0.9—4.1	
OsW_3	~3	
Os_2Y	4.7	C14
Os_2Zr	3.0	C14
Os_xZr_{1-x}	1.5—5.6	
PPb	7.8	
OsW_2	3.81	$D8_b$-tP30 (FeCr)
$PPd_{3.0—3.2}$	<0.35—0.7	DO_{11}
P_3Pd_7 (high temperature)	1.0	Rhombohedral
P_3Pd_7 (low temperature)	0.70	Complex
PRh	1.22	
PRh_2	1.3	C1
P_4Rh_5	1.22	oP28 ($CaFe_2O_4$)
PRhTa	4.41	C37-oP12 (Co_2Si)
PRhZr	1.55	Same
PRuTi	1.3	C22-hP9 (Fe_2P)
PRuZr	3.46	C37-oP12
PW_3	2.26	DO_e
Pb_2Pd	2.95	C16
Pb_4Pt	2.80	Related to C16
Pb_2Rh	2.66	C16
PbSb	6.6	
PbTe (plus 0.1 w/o Pb)	5.19	
PbTe (plus 0.1 w/o Te)	5.24—5.27	
$PbTl_{0.27}$	6.43	
$PbTl_{0.17}$	6.73	
$PbTl_{0.12}$	6.88	
$PbTl_{0.075}$	6.98	
$PbTl_{0.04}$	7.06	
$Pb_{1—0.26}Tl_{0—0.74}$	7.20—3.68	
$PbTl_2$	3.75—4.1	
Pb_3Zr_5	4.60	$D8_8$
$PbZr_3$	0.76	A15
$Pd_{0.9}Pt_{0.1}Te_2$	1.65	C6
$Pd_{0.05}Ru_{0.05}Zr_{0.9}$	~9	

TABLE 4
Superconductive Compounds and Alloys (continued)

Substance	T_c, K	Crystal structure type
$Pd_{2.2}S$ (quenched)	1.63	Cubic
$PdSb_2$	1.25	C2
PdSb	1.5	$B8_1$
PdSbSe	1.0	C2
PdSbTe	1.2	C2
Pd_4Se	0.42	Tetragonal
$Pd_{6–7}Se$	0.66	Like Pd_4Te
$Pd_{2.8}Se$	2.3	
Pd_xSe_{1-x}	2.5 (max.)	
PdSi	0.93	B31
PdSn	0.41	B31
$PdSn_2$	3.34	
Pd_2Sn	0.41	C37
Pd_3Sn	0.47—0.64	$B8_2$
Pd_2SnTm	1.77	DO_3-cF16 (BiF_3)
Pd_2SnY	4.92	Same
Pd_2SnYb	1.79	Same
PdTe	2.3; 3.85	$B8_1$
$PdTe_{1.02—1.08}$	2.56—1.88	$B8_1$
$PdTe_2$	1.69	C6
$PdTe_{2.1}$	1.89	C6
$PdTe_{2.3}$	1.85	C6
$Pd_{1.1}Te$	4.07	$B8_1$
Pd_3Te	0.76	cI2 (W)
$PdTh_2$	0.85	C16
$Pd_{0.1}Zr_{0.9}$	7.5	A3
PtSb	2.1	$B8_1$
PtSi	0.88	B31
PtSn	0.37	$B8_1$
$PtSn_4$	2.38	C16-oC20 ($PdSn_4$)
Pt_3Ta_7	1.5	$D8_b$-tP30
$PtTa_3$	0.4	A15-cP8 (Cr_3Si)
PtTe	0.59	Orthorhombic
PtTh	0.44	B_f
Pt_3Th_7	0.98	$D10_2$
Pt_5Th	3.13	
$PtTi_3$	0.58	A15
$Pt_{0.02}U_{0.98}$	0.87	β-phase
$PtV_{2.5}$	1.36	A15
PtV_3	2.87—3.20	A15
$PtV_{3.5}$	1.26	A15
$Pt_{0.5}W_{0.5}$	1.45	A1
Pt_xW_{1-x}	0.4—2.7	
Pt_2Y_3	0.90	
Pt_2Y	1.57; 1.70	C15
Pt_3Y_7	0.82	$D10_2$
PtZr	3.0	A3
Re_2Sc	4.2	C15-hP12 ($MgZn_2$)
$Re_{24}Sc_5$	2.2	A12-cI58 (Mg)
ReSiTa	4.4	oI36 (FeTiSi)
$Re_3Si_5Y_2$	1.76	tP40 ($Fe_3Sc_2Si_5$)
Re_3Ta_2	1.4	$D8_b$-tP30 (FeCr)
$Re_{0.64}Ta_{0.36}$	1.46	A12
Re_3Ta	6.78	A12-cI58 (Mn)
$Re_{24}Ti_5$	6.60	A12

TABLE 4
Superconductive Compounds and Alloys (continued)

Substance	T_c, K	Crystal structue type
Re_xTi_{1-x}	6.6 (max.)	
$Re_{0.76}V_{0.24}$	4.52	$D8_b$
Re_3V	6.26	$D8_b$-tP30
$Re_{0.92}V_{0.08}$	6.8	A3
$Re_{0.6}W_{0.4}$	6.0	
$Re_{0.5}W_{0.5}$	5.12	$D8_b$
$Re_{13}W_{12}$	5.2	$D8_b$-tP30
Re_3W	9.0	A12-cI58
Re_2Y	1.83	C14
Re_2Zr	5.9	C14
Re_3Zr	7.40	A12-cI58
Re_6Zr	7.40	Same
$Rh_{17}S_{15}$	5.8	Cubic
$Rh_{\sim 0.24}Sc_{\sim 0.76}$	0.88; 0.92	
$Rh_4Sc_5Si_{10}$	8.54	tP38
$Rh_4Sc_3Sn_{13}$	4.5	cP40
Rh_xSe_{1-x}	6.0 (max.)	
$RhSi_3Th$	1.76	tI10
$Rh_{0.86}Sc_{1.04}Th$	6.45	tI12
Rh_2Si_2Y	3.11	tI10
$Rh_3Si_5Y_2$	2.70	oI40
$Rh_4Sn_{13}Sr_3$	4.3	cP40
Rh_xSn_yTh	1.9	cI2 (W)
Rh_xSn_yTm	2.3	cP40
$Rh_4Sn_{13}Y_3$	3.2	cP40
Rh_2Sr	6.2	C15
$Rh_{0.4}Ta_{0.6}$	2.35	$D8_b$
$RhTe_2$	1.51	C2
$Rh_{0.67}Te_{0.33}$	0.49	
Rh_xTe_{1-x}	1.51 (max.)	
RhTh	0.36	B_1
Rh_3Th_7	2.15	$D10_2$
Rh_5Th	1.07	
Rh_xTi_{1-x}	2.25—3.95	
$Rh_{0.02}U_{0.98}$	0.96	
RhV_3	0.38	A15
RhW	~3.4	A3
RhY_3	0.65	
Rh_2Y_3	1.48	
Rh_3Y	1.07	C15
Rh_5Y	0.56	
Rh_3Y_7	0.32	hP20 (Fe_3Th_7)
$Rh_{0.005}Zr$ (annealed)	5.8	
$Rh_{0-0.45}Zr_{1-0.55}$	2.1—10.8	
$Rh_{0.1}Zr_{0.9}$	9.0	H.C.P.
Ru_2Sc	1.67	C14
RuSiTa	3.15	oI36
Ru_3Si_2Th	3.98	hP12
Ru_3Si_2Y	3.51	hP12
$Ru_{1.1}Sn_{3.1}Y$	1.3	cP40
Ru_2Th	3.56	C15
RuTi	1.07	B2
$Ru_{0.05}Ti_{0.95}$	2.5	
$Ru_{0.1}Ti_{0.9}$	3.5	
$Ru_xTi_{0.6}V_y$	6.6 (max.)	

TABLE 4
Superconductive Compounds and Alloys (continued)

Substance	T_c, K	Crystal structue type
Ru_3U	0.15	$L1_2$-cP4
$Ru_{0.45}V_{0.55}$	4.0	B2
RuW	7.5	A3
Ru_2Y	1.52	C14
Ru_2Zr	1.84	C14
$Ru_{0.1}Zr_{0.9}$	5.7	A3
STh	0.5	B1-cF8 (NaCl)
SbSn	1.30—1.42	B1 or distorted
$SbTa_3$	0.72	A15-cP8 (Cr_3Si)
$SbTi_3$	5.8	Same
Sb_2Ti_7	5.2	
$Sb_{0.01–0.03}V_{0.99–0.97}$	3.76—2.63	A2
SbV_3	0.80	A15
SeTh	1.7	B1-cF8
$SiMo_3$	1.4	A15-cP8
Si_2Th	3.2	C_c, α-phase
Si_2Th	2.4	C32, β-phase
$SiV_{2.7}Ru_{0.3}$	2.9	A15
Si_2W_3	2.8; 2.84	
$SiZr_3$	0.5	$L1_2$-tP32 (Ti_3P)
$Sn_{0.174–0.104}Ta_{0.826–0.896}$	6.5—< 4.2	A15
$SnTa_3$	8.35	A15, highly ordered
$SnTa_3$	6.2	A15, partially ordered
$SnTaV_2$	2.8	A15
$SnTa_2V$	3.7	A15
Sn_xTe_{1-x} (n = 10.5—20 × 10^{20})	0.07—0.22	B1
Sn_3Th	3.33	$L1_2$-cP4
$SnTi_3$	5.80	A15-cP8
Sn_xTl_{1-x}	2.37—5.2	
SnV_3	3.8	A15
$Sn_{0.02–0.057}V_{0.98–0.943}$	2.87—~1.6	A2
$SnZr_3$	0.92	A15-cP8
$Ta_{0.025}Ti_{0.975}$	1.3	Hexagonal
$Ta_{0.05}Ti_{0.95}$	2.9	Hexagonal
$Ta_{0.05–0.75}V_{0.95–0.25}$	4.30—2.65	A2
$Ta_{0.8–1}W_{0.2–0}$	1.2—4.4	A2
$Tc_{0.1–0.4}W_{0.9–0.6}$	1.25—7.18	Cubic
$Tc_{0.50}W_{0.50}$	7.52	α plus
$Tc_{0.60}W_{0.40}$	7.88	plus α
Tc_6Zr	9.7	A12
TeY	1.02	B1-cF8
$ThTl_3$	0.87	$L1_2$-cP4
$Th_{0–0.55}Y_{1–0.45}$	1.2—1.8	
$Ti_{0.70}V_{0.30}$	6.14	Cubic
Ti_xV_{1-x}	0.2—7.5	
$Ti_{0.5}Zr_{0.5}$ (annealed)	1.23	
$Ti_{0.5}Zr_{0.5}$ (quenched)	2.0	
Tl_3Y	1.52	$L1_2$-cP4
V_2Zr	8.80	C15
$V_{0.26}Zr_{0.74}$	5.9	
W_2Zr	2.16	C15
YZn	0.33	B2-cP2 (CsCl)

* n denotes current carriers concentration in cm^{-3}.

TABLE 4
Superconductive Compounds and Alloys (continued)

B. SUPERCONDUCTORS WITH T_c > 10K

Substance	T_c,K	Crystal structure type
Al_2CMo_3	10.0	A13
$Al_{0.5}Ge_{0.5}Nb$	12.6	A15
$Al_{\sim 0.8}Ge_{\sim 0.2}Nb_3$	20.7	A15
$AlNb_3$	18.0	A15 (Cr_3Si)
$AlNb_3$	12.0	(FeCr)
Al_xNb_{1-x}	<4.2—13.5	$D8_b$
Al_xNb_{1-x}	12—17.5	A15
$Al_{0.27}Nb_{0.73-0.48}V_{0-0.25}$	14.5—17.5	A15
$Al\ Nb_xV_{1-x}$	4.4—13.5	
$Al_{0.1}Si_{0.9}V_3$	14.05	
AlV_3	11.8	A15 (Cr_3Si)
$AuNb_3$	11.5	A15
$Au_{0-0.3}Nb_{1-0.7}$	1.1—11.0	
$Au_{0.02-0.98}Nb_3Rh_{0.98-0.02}$	2.53—10.9	A15
$AuNb_{3(1-x)}V_{3x}$	1.5—11.0	A15
$B_{0.03}C_{0.51}Mo_{0.47}$	12.5	
B_4LuRh_4	11.7	(B_4CeCo_4)
B_2LuRu	10	
B_4Rh_4Y	11.3	(B_4CeCo_4)
$B_{0.1}Si_{0.9}V_3$	15.8	A15
$BaBi_{0.2}O_3Pb_{0.8}$	13.2	
$Ba_2CaCu_2O_8Tl_2$	120	
$Ba_2Cu_3LaO_6$	80	
$Ba_2Cu_3O_7Tm$	101	
$Ba_2Cu_3O_7Y$	90	
$(Ba,La)_2CuO_4$	36	A15 (K_2NiF_4)
$Bi_2CaCu_2O_8Sr_2$	110	
$Br_2Mo_6S_6$	13.8	(Mo_6PbS_8)
C_3La	11.0	(C_3Pu_2)
CMo	14.3	B1 (NaCl)
CMo_2	12.2	o**
$C_{0.5}Mo_xNb_{1-x}$	10.8—12.5	B1
CMo_xTi_{1-x}	10.2(max)	B1
$CMo_{0.83}Ti_{0.17}$	10.2	B1
$C_{0-0.38}N_{1-0.62}Ta$	10.0—11.3	
CNb (whiskers)	7.5—10.5	
CNb	11.5	B1
$C_{0.7-1.0}Nb_{0.3-0}$	6—11	B1
CNb_xTa_{1-x}	8.2—13.9	
$CNb_{0.6-0.9}W_{0.4-0.1}$	12.5—11.6	B1
$C_{0.1}Si_{0.9}V_3$	16.4	A15
CTa	10.3	B1
$CTa_{1-0.4}W_{0-0.6}$	8.5—10.5	B1
$C_{0.66}Th_{0.13}Y_{0.21}$	17	(C_3Pu_2)
C_3Y_2	11.5	(C_3Pu_2)
CW	10	B1
$(Ca,La)_2CuO_4$	18	(K_2NiF_4)
$Cu(La,Sr)_2O_4$	39	
$Cu_{1.8}Mo_6S_8$	10.8	(Mo_6PbS_8)
$Cr_{0.3}SiV_{2.7}$	11.3	A15
$GaNb_3$	14.5	A15 (Cr_3Si)
$Ga_xNb_3Sn_{1-x}$	14—18.37	A15

TABLE 4
Superconductive Compounds and Alloys (continued)

Substance	T_c,K	Crystal structure type	
GaV_3	16.8	A15	
$GaV_{2.1—3.5}$	6.3—14.45	A15	
$GeNb_3$	23.2	A15	
$GeNb_3$ (quenched)	6—17	A15	
$Ge_xNb_3Sn_{1-x}$	17.6—18.0	A15	
$Ge_{0.5}Nb_3Sn_{0.5}$	11.3		
$Ge_{0.1}Si_{0.9}V_3$	14.0	A15	
GeV_3	11	A15	
$InLa_3$	9.83; 10.4	LI_2	($AuCu_3$)
$InLa_3$ (0—35 kbar)	9.75—10.55		
$In_{0—0.3}Nb_3Sn_{1—0.7}$	18.0—18.19	A15	
InV_3	13.9	A15	
$Ir_{0.4}Nb_{0.6}$	10		(FeCr)
$LaMo_6Se_8$	11.4		(Mo_6PbS_8)
LiO_4Ti_2	13.7		($A1_2MgO_4$)
MoN	12; 14.8		h*
Mo_3Os	12.7	A15	
$Mo_6Pb_{0.9}S_{7.5}$	15.2		(Mo_6PbS_8)
Mo_3Re	10.0; 15	A15	
Mo_xRe_{1-x}	1.2—12.2		
$Mo_{0.52}Re_{0.48}$	11.1		
$Mo_{0.57}Re_{0.43}$	14.0		
$Mo_{\sim 0.60}Re_{0.395}$	10.6		
MoRu	9.5—10.5	A3	
Mo_3Ru	10.6	A15	
Mo_6Se_8T1	12.2		(Mo_6PbS_8)
$Mo_{0.3}SiV_{2.7}$	11.7	A15	
Mn_3Si	12.5	A15	
Mo_3Tc	15	A15	
$Mo_{0.3}Tc_{0.7}$	12.0	A15	
Mo_xTc_{1-x}	10.8—15.8		
$MoTc_3$	15.8		
NNb (whiskers)	10—14.5		
NNb (diffusion wires)	16.10		
$N_{0.988}Nb$	14.9; 17.3	B1	
$N_{0.824—0.988}Nb$	14.4—15.3	B1	
$N_{0.7—0.795}Nb$	11.3—12.9		
NNb_xO_y	13.5—17.0	B1	
NNb_xO_y	6.0—11		
$N_{100—42w/o}Nb_{0—58w/o}Ti$	15—16.8		
$N_{100—75w/o}Nb_{0—25w/o}Zr$	12.5—16.35		
NNb_xZr_{1-x}	9.8—13.8	B1	
$N_{0.93}Nb_{0.85}Zr_{0.15}$	13.8	B1	
NTa	12—14	B1	
NZr	10.7	B1	
Nb_3Pt	10.9	A15	
$Nb_{0.18}Re_{0.82}$	10		(Mn)
Nb_3Si	19	A15	
$Nb_{0.3}SiV_{2.7}$	12.8	A15	
Nb_3Sn	18.05	A15	
$Nb_{0.8}Sn_{0.2}$	18.18; 18.5	A15	
Nb_xSn_{1-x} (film)	2.6—18.5		o*
Nb_3Sn_2	16.6		t*
$NbSnTa_2$	10.8	A15	
Nb_2SnTa	16.4	A15	

TABLE 4
Superconductive Compounds and Alloys (continued)

Substance	T_c,K	Crystal structure type
$Nb_{2.5}SnTa_{0.5}$	17.6	A15
$Nb_{2.75}SnTa_{0.25}$	17.8	A15
$Nb_{3x}SnTa_{3(1-x)}$	6.0—18.0	
$Nb_2SnTa_{0.5}V_{0.5}$	12.2	A15
$NbTc_3$	10.5	A12
$Nb_{0.75}Zr_{0.25}$	10.8	
$Nb_{0.66}Zr_{0.33}$	10.8	
$PbTa_3$	17	A15
$RhTa_3$	10	A15
$RhZr_2$	10.8; 11.3	C16 ($A1_2Cu$)
$Rh_{0-0.45}Zr_{1-0.55}$	2.1—10.8	
$SiTi_{0.3}V_{2.7}$	10.9	A15
SiV_3	17.1	A15
$SiV_{2.7}Zr_{0.3}$	13.2	A15

TABLE 5
Critical Field Data

Substance	H_o oersteds	Substance	H_o oersteds
Ag_2F	2.5	InSb	1100
Ag_7NO_{11}	57	In_xTl_{1-x}	252—284
Al_2CMo_3	1700	$In_{0.8}Tl_{0.2}$	252
$BaBi_3$	740	$Mg_{0.47}Tl_{0.53}$	220
Bi_2Pt	10	$Mo_{0.16}Ti_{0.84}$	<985
Bi_3Sr	530	$NbSn_2$	620
Bi_5Tl_3	>400	$PbTl_{0.27}$	756
CdSn	>266	$PbTl_{0.17}$	796
$CoSi_2$	105	$PbTl_{0.12}$	849
$Cr_{0.1}Ti_{0.3}V_{0.6}$	1360	$PbTl_{0.075}$	880
$In_{1-0.86}Mg_{0-0.14}$	272.4—259.2	$PbTl_{0.04}$	864

TABLE 6
High Critical Magnetic-Field Superconductive Compounds and Alloys

Substance	T_c, K	H_{c1}, kOe	H_{c2}, kOe	H_{c3}, kOe	T_{obs}, K[a]
Al_2CMo_3	9.8—10.2	0.091	156		1.2
$AlNb_3$		0.375			
$Ba_xO_3Sr_{1-x}Ti$	<0.1—0.55	0.0039 max.			
$Bi_{0.5}Cd_{0.1}Pb_{0.27}Sn_{0.13}$			>24		3.06
Bi_xPb_{1-x}	7.35—8.4	0.122 max.	30 max.		4.2
$Bi_{0.56}Pb_{0.44}$	8.8		15		4.2
$Bi_{7.5w/o}Pb_{92.5w/o}$[b]			2.32		
$Bi_{0.099}Pb_{0.901}$		0.29	2.8		
$Bi_{0.02}Pb_{0.98}$		0.46	0.73		
$Bi_{0.53}Pb_{0.32}Sn_{0.16}$			>25		3.06
$Bi_{1-0.93}Sn_{0-0.07}$			0—0.032		3.7
Bi_5Tl_3	6.4		>5.6		3.35
C_8K (excess K)	0.55		0.160 ($H\perp c$)		0.32
			0.730 ($H\|\|c$)		0.32

TABLE 6
High Critical Magnetic-Field Superconductive Compounds and Alloys (continued)

Substance	T_c, K	H_{c1}, kOe	H_{c2}, kOe	H_{c3}, kOe	T_{obs}, K[a]
C_8K	0.39		0.025 (H⊥c)		0.32
			0.250 (H‖c)		0.32
$C_{0.44}Mo_{0.56}$	12.5—13.5	0.087	98.5		1.2
CNb	8—10	0.12	16.9		4.2
$CNb_{0.4}Ta_{0.6}$	10—13.6	0.19	14.1		1.2
CTa	9—11.4	0.22	4.6		1.2
$Ca_xO_3Sr_{1-x}Ti$	<0.1—0.55	0.002—0.004			
$Cd_{0.1}Hg_{0.9}$ (by weight)		0.23	0.34		2.04
$Cd_{0.05}Hg_{0.95}$		0.28	0.31		2.16
$Cr_{0.10}Ti_{0.30}V_{0.60}$	5.6	0.071	84.4		0
GaN	5.85	0.725			4.2
Ga_xNb_{1-x}			>28		4.2
GaSb (annealed)	4.24		2.64		3.5
$GaV_{1.95}$	5.3		73[e]		
$GaV_{2.1-3.5}$	6.3—14.45		230—300[d]		0
GaV_3		0.4	350[e]		0
			500[d]		
$GaV_{4.5}$	9.15		121[c]		0
Hf_xNb_y			>52—>102		1.2
Hf_xTa_y			>28—>86		1.2
$Hg_{0.05}Pb_{0.95}$		0.235	2.3		
$Hg_{0.101}Pb_{0.899}$		0.23	4.3		4.2
$Hg_{0.15}Pb_{0.85}$	6.75		>13		2.93
$In_{0.98}Pb_{0.02}$	3.45	0.1		0.12	2.76
$In_{0.96}Pb_{0.04}$	3.68	0.1	0.12	0.25	2.94
$In_{0.94}Pb_{0.06}$	3.90	0.095	0.18	0.35	3.12
$In_{0.913}Pb_{0.087}$	4.2	~10.17	0.55	2.65	
$In_{0.316}Pb_{0.684}$		0.155	3.7		4.2
$In_{0.17}Pb_{0.83}$			2.8	5.5	4.2
$In_{1.000}Te_{1.002}$	3.5—3.7		1.2[c]		0
$In_{0.95}Tl_{0.05}$		0.263	0.263		3.3
$In_{0.90}Tl_{0.10}$		0.257	0.257		3.25
$In_{0.83}Tl_{0.17}$		0.242	0.39		3.21
$In_{0.75}Tl_{0.25}$		0.216	0.50		3.16
LaN	1.35	0.45			0.76
La_3S_4	6.5	≈0.15	>25		1.3
La_3Se_4	8.6	≈0.2	>25		1.25
$Mo_{0.52}Re_{0.48}$	11.1		14—21	22—33	4.2
			18—28	37—43	1.3
$Mo_{0.6}Re_{0.395}$	10.6		14—20	20—37	4.2
			19—26	26—37	1.3
$Mo_{0.5}Ti_{0.5}$			75[c]		0
$Mo_{0.16}Ti_{0.84}$	4.18	0.028	98.7[c]		0
			36—38		3.0
$Mo_{0.913}Ti_{0.087}$	2.95	0.060	15		4.2
$Mo_{0.1-0.3}U_{0.9-0.7}$	1.85—2.06		>25		
$Mo_{0.17}Zr_{0.83}$			30		
$N_{(12.8\ w/o)}Nb$	15.2		>9.5		13.2
NNb (wires)	16.1		153[c]		0
			132		4.2
			95		8
			53		12
NNb_xO_{1-x}	13.5—17.0		38		

TABLE 6
High Critical Magnetic-Field Superconductive Compounds and Alloys (continued)

Substance	T_c, K	H_{c1}, kOe	H_{c2}, kOe	H_{c3}, kOe	T_{obs}, K[a]
NNb_xZr_{1-x}	9.8—13.8		4- >130		4.2
$N_{0.93}Nb_{0.85}Zr_{0.15}$	13.8		>130		4.2
$Na_{0.086}Pb_{0.914}$		0.19	6.0		
$Na_{0.016}Pb_{0.984}$		0.28	2.05		
Nb	9.15		2.020		1.4
			1.710		4.2
Nb		0.4—1.1	3—5.5		4.2
Nb (unstrained)		1.1—1.8	3.40	6—9.1	4.2
Nb (strained)		1.25—1.92	3.44	6.0—8.7	4.2
Nb (cold-drawn wire)		2.48	4.10	≈10	4.2
Nb (film)			>25		4.2
NbSc			>30		
Nb_3Sn		0.170	221		4.2
			70		14.15
			54		15
			34		16
			17		17
$Nb_{0.1}Ta_{0.9}$		0.084	0.154		4.195
$Nb_{0.2}Ta_{0.8}$			10		4.2
$Nb_{0.65\text{-}0.73}Ta_{0.02\text{-}0.10}Zr_{0.25}$			>70—>90		4.2
Nb_xTi_{1-x}			148 max.		1.2
			120 max.		4.2
$Nb_{0.222}U_{0.778}$		1.98	23		1.2
Nb_xZr_{1-x}			127 max.		1.2
			94 max.		4.2
O_3SrTi	0.43	0.0049[c]	0.504[c]		0
O_3SrTi	0.33	0.00195[c]	0.420[c]		0
$PbSb_{1\ w/o}$(quenched)			>1.5		4.2
$PbSb_{1\ w/o}$(annealed)			>0.7		4.2
$PbSb_{2.8\ w/o}$(quenched)			>2.3		4.2
$PbSb_{2.8\ w/o}$(annealed)			>0.7		4.2
$Pb_{0.871}Sn_{0.129}$		0.45	1.1		
$Pb_{0.965}Sn_{0.035}$		0.53	0.56		
$Pb_{1\text{-}0.26}Tl_{0\text{-}0.74}$	7.20 3.68		2—6.9[c]		0
$PbTl_{0.17}$	6.73		4.5[c]		0
$Re_{0.26}W_{0.74}$			>30		
$Sb_{0.93}Sn_{0.07}$			0.12		3.7
SiV_3	17.0	0.55	156[e]		
Sn_xTe_{1-x}		0.00043—0.00236	0.005—0.0775		0.012—0.079
Ta (99.95%)		0.425	1.850		1.3
		0.325	1.425		2.27
		0.275	1.175		2.66
		0.090	0.375		3.72
$Ta_{0.5}Nb_{0.5}$			3.55		4.2
$Ta_{0.65\text{-}0}Ti_{0.35\text{-}1}$	4.4—7.8		>14—138		1.2
$Ta_{0.5}Ti_{0.5}$			138		1.2
Te	3.3	0.25[c]			0
Tc_xW_{1-x}	5.75—7.88		8—44		4.2
Ti				2.7	4.2
$Ti_{0.75}V_{0.25}$	5.3	0.029[c]	199[c]		0
$Ti_{0.775}V_{0.225}$	4.7	0.024[c]	172[c]		0
$Ti_{0.615}V_{0.385}$	7.07	0.050	34		4.2
$Ti_{0.516}V_{0.484}$	7.20	0.062	28		4.2
$Ti_{0.415}V_{0.585}$	7.49	0.078	25		4.2

TABLE 6
High Critical Magnetic-Field Superconductive Compounds and Alloys (continued)

Substance	T_c, K	H_{c1}, kOe	H_{c2}, kOe	H_{c3}, kOe	T_{obs}, K[a]
$Ti_{0.12}V_{0.88}$			17.3	28.1	4.2
$Ti_{0.09}V_{0.91}$			14.3	16.4	4.2
$Ti_{0.06}V_{0.94}$			8.2	12.7	4.2
$Ti_{0.03}V_{0.97}$			3.8	6.8	4.2
Ti_xV_{1-x}			108 max.		1.2
V	5.31	0.8	3.4		1.79
		0.75	3.15		2
		0.45	2.2		3
		0.30	1.2		4
$V_{0.26}Zr_{0.74}$	≈5.9	0.238			1.05
		0.227			1.78
		0.185			3.04
		0.165			3.5
W (film)	1.7—4.1		>34		1

[a] Temperature of critical field measurement.
[b] w/o denotes weight percent.
[c] Extrapolated.
[d] Linear extrapolation.
[e] Parabolic extrapolation.

REFERENCES

1. B. W. Roberts, in *Superconductive Materials and Some of Their Properties. Progress in Cryogenics*, Vol. IV, 1964, pp. 160—231.
2. B. W. Roberts, Superconductive Materials and Some of Their Properties, NBS Technical Notes 408 and 482, U.S. Government Printing Office, 1966 and 1969; B. W. Roberts, *J. Phys. Chem. Ref. Data*, 5, 581, 1976.
3. B. W. Roberts, Properties of Selected Superconductive Materials, 1978 Supplement, NBS Technical Note 983, 1978.
4. T. Claeson, *Phys. Rev.*, 147, 340, 1966.
5. C. J. Raub, W. H. Zachariasen, T. H. Geballe, and B. T. Matthias, *J. Phys. Chem. Solids*, 24, 1093, 1963.
6. T. H. Geballe, B. T. Matthias, V. B. Compton, E. Corenzwit, G. W. Hull, Jr., and L. D. Longinotti, *Phys. Rev.*, 1A, 119, 1965.
7. C. J. Raub, V. B. Compton, T. H. Geballe, B. T. Matthias, J. P. Maita, and G. W. Hull, Jr., *J. Phys. Chem. Solids*, 26, 2051, 1965.
8. R. D. Blaugher, J. K. Hulm, and P. N. Yocom, *J. Phys. Chem. Solids*, 26, 2037, 1965.
9. T. Claeson and H. L. Luo, *J. Phys. Chem. Solids*, 27, 1081, 1966.
10. S. C. Ng and B. N. Brockhouse, *Solid State Comm.*, 5, 79, 1967.
11. O. I. Shulishova and I. A. Shcherbak, *Izv. AN SSSR, Neorg. Materials*, 3, 1495, 1967.
12. T. F. Smith and H. L. Luo, *J. Phys. Chem. Solids*, 28, 569, 1967.
13. A. C. Lawson, *J. Less–Common Metals*, 23, 103, 1971.
14. R. Chevrel, M. Sergent, and J. Prigent, *J. Solid State Chem.*, 3, 515, 1971.
15. M. Marezio, P. D. Dernier, J. P. Remeika, and B. T. Matthias, *Mat. Res. Bull.*, 8, 657, 1973.
16. J. K. Hulm and R. D., *Blaugher in Superconductivity in d– and f–Band Metals*, D. H. Douglass,Ed., American Institute of Physics, 4, 1, 1972.
17. R. N. Shelton, A. C. Lawson, and D. C. Johnston, *Mat. Res. Bull.*, 10, 297, 1975.
18. H. D. Wiesinger, *Phys. Status Sol.*, 41A, 465, 1977.
19. O. Fisher, *Applied Phys.*, 16, 1, 1978.
20. D. C. Johnston, *Solid State Comm.*, 24, 699, 1977.
21. H. C. Ku and R. H. Shelton, *Mat. Res. Bull.*, 15, 1441, 1980.
22. H. Barz, *Mat. Res. Bull.*, 15, 1489, 1980.
23. G. P. Espinosa, A. S. Cooper, H. Barz, and J. P. Remeika, *Mat. Res. Bull.*, 15, 1635, 1980.
24. E. M. Savitskii, V. V. Baron, Yu. V. Efimov, M. I. Bychkova, and L. F. Myzenkova, in *Superconducting Materials*, Plenum Press, 1981, p. 107.
25. R. Fluckiger and R. Baillif, in Topics in *Current Physics*, O. Fischer and M. B. Maple, Eds., Springer Verlag, 34, 113, 1982.
26. R. N. Shelton, in *Superconductivity in d– and f–Band Metals*, W. Buckel and W. Weber, Eds., Kernforschungszentrum, Karlsruhe, 1982, p. 123.
27. D. C. Johnston and H. F. Braun, *Topics in Current Phys.*, 32, 11, 1982.
28. R. Chevrel and M. Sergent, *Topics in Current Phys.*, 32, 25, 1982.

29. G. P. Espinosa, A. S. Cooper, and H. Barz, *Mat. Res. Bull.*, 17, 963, 1982.
30. R. Muller, R. N. Shelton, J. W. Richardson, Jr., and R. A. Jacobson, *J. Less–Comm. Met.*, 92, 177, 1983.
31. You–Xian Zhao and Shou–An He, in *High Pressure in Science and Technology*, North Holland, 22, 51, 1983.
32. You–Xian Zhao and Shou–An He, *Solid State Comm.*, 24, 699, 1983.
33. G. P. Meisner and H. C. Ku, *Appl. Phys.*, A31, 201, 1983.
34. R. J. Cava, D. W. Murphy, and S. M. Zahurak, *J. Electrochem. Soc.*, 130, 2345, 1983.
35. R. N. Shelton, *J. Less–Comm. Met.*, 94, 69, 1983.
36. B. Chevalier, P. Lejay, B. Lloret, Wang Xian–Zhong, J. Etourneau, and P. Hagenmuller, *Annales de Chemie*, 9, 191, 1984.
37. G. Venturini, M. Meot–Meyer, E. McRae, J. F. Mareche, and B. Rogues, *Mat. Res. Bull.*, 19, 1647, 1984.
38. J. M. Tarascon, F. G. DiSalvo, D. W. Murphy, G. Hull, and J. V. Waszczak, *Phys. Rev.*, 29B, 172, 1984.
39. G. V. Subba and G. Balakrishnan, *Bull. Mat. Sci.*, 6, 283, 1984.
40. B. Batlog, *Physica*, 126B, 275, 1984.
41. M. J. Johnson, Ames Lab (USA) Report IS-T-1140, 1984.
42. I. M. Chapnik, *J. Mat. Sci. Lett.*, 4, 370, 1985.
43. W. Rong–Yao, L. Qi–Guang, and Z. Xiao, *Phys. Status Sol.*, 90A, 763, 1985.
44. W. Xian–Zhong, B. Chevalier, J. Etourneau, and P. Hagenmuller, *Mat. Res. Bull.*, 20, 517, 1985.
45. H. R. Ott, F. Hulliger, H. Rudigier, and Z. Fisk, *Phys. Rev.*, 31B, 1329, 1985.
46. P. Villars and L. D. Calver, *Pearson's Handbook of Crystallographic Data for Intermetallic Phases*, Vol. 1—3, ASM, 1985.
47. G. V. Subba Rao, K. Wagner, G. Balakhrishnan, J. Jakani, W. Paulus, and R. Scollhorn, *Bull. Mat. Sci.*, 7, 215, 1985.
48. J. G. Bednorz and K. A. Muller, *Zs. Physik*, B64, 189, 1986.
49. W. Rong–Yao, *Phys. Status Sol.*, 94A, 445, 1986.
50. H. D. Yang, R. N. Shelton, and H. F. Braun, *Phys. Rev.*, 33B, 5062, 1986.
51. G. Venturini, M. Kanta, E. McRae, J. F. Mareche, B. Malaman, and B. Roques, *Mat. Res. Bull.*, 21, 1203, 1986.
52. W. Rong–Yao, *J. Mat. Sci. Lett.*, 5, 87, 1986.
53. M. K. Wu, J. R. Ashburn, C. J. Torng, P. H. Hor, R. L. Meng, L. Gao, Z. J. Huang, Y. Q. Wang, and C. W. Chu, *Phys. Rev. Lett.*, 58, 908, 1987.
54. R. J. Cava, R. B. Van Dover, B. Batlog, and E. A. Rietman, *Phys. Rev. Lett.*, 58, 408, 1987.
55. L. C. Porter, T. J. Thorn, U. Geiser, A. Umezawa, H. H. Wang, W. K. Kwok, H–C. I. Kao, M. R. Monaghan, G. W. Crabtree, K. D. Carlson, and J. M. Williams, *Inorg. Chem.*, 26, 1645, 1987.
56. A. M. Kini, U. Geiser, H–C. I. Kao, K. D. Carlson, H. H. Wang, M. R. Monaghan, and K. M. Williams, *Inorg. Chem.*, 26, 1834, 1987.
57. T. Penney, S. von Molnar, D. Kaiser, F. Holtzberg, and A. W. Kleinsasser, *Phys. Rev.*, B38, 2918, 1988.
58. Y. K. Tao, J. S. Swinnea, A. Manthiram, J. S. Kim, J. B. Goodenoug, and H. Steinfink, *J. Mat. Res.*, 3, 248, 1988.
59. G. G. Peterson, B. R. Weinberger, L. Lynds, and H. A. Krasinski, *J. Mat. Res.*, 3, 605, 1988.
60. J. B. Torrance, Y. Tokura, A. Nazzai, and S. S. P. Parkin, *Phys. Rev. Lett.*, 60, 542, 1988.
61. K. Kourtakis, M. Robbins, P. K. Gallagher, and T. Teifel, *J. Mat. Res.*, 4, 1289, 1989.
62. J. C. Phillips, *Physics of High-T_c Superconductors*, Academic Press, 1989, p. 336.
63. Shui Wai Lin and L. I. Berger, *Rev. Sci. Instrum.*, 60, 507, 1989.

HIGH TEMPERATURE SUPERCONDUCTORS

C. N. R. Rao and A. K. Raychaudhuri

The following tables give properties of a number of high temperature superconductors. Table 1 lists the crystal structure (space group and lattice constants) and the critical transition temperature T_c for the more important high temperature superconductors so far studied. Table 2 gives energy gap, critical current density, and penetration depth in the superconducting state. Table 3 gives electrical and thermal properties of some of these materials in the normal state. The tables were prepared in November 1992 and updated in November 1994.

REFERENCES

1. Ginsburg, D.M., Ed., *Physical Properties of High-Temperature Superconductors*, Vols. I—III, World Scientific, Singapore, 1989—1992.
2. Rao, C.N.R., Ed., *Chemistry of High-Temperature Superconductors*, World Scientific, Singapore, 1991.
3. Shackelford, J.F., *The CRC Materials Science and Engineering Handbook*, CRC Press, Boca Raton, 1992, 98—99 and 122—123.
4. Kaldis, E., Ed., *Materials and Crystallographic Aspects of HT$_c$-Superconductivity*, Kluwer Academic Publ., Dordrecht, The Netherlands, 1992.
5. Malik, S.K. and Shah, S.S., Ed., *Physical and Material Properties of High Temperature Superconductors*, Nova Science Publ., Commack, N.Y., 1994.
6. Chmaissem, O. et. al., *Physica*, C230, 231—238, 1994.
7. Antipov, E.V. et. al., *Physica*, C215, 1—10, 1993.

Table 1
Structural Parameters and Approximate T_c Values of High-Temperature Superconductors

Material	Structure	T_c/K (maximum value)
$La_2CuO_{4+\delta}$	Bmab; a = 5.355, b = 5.401, c = 13.15 Å	39
$La_{2-x}Sr_x(Ba_x)CuO_4$	I4/mmm; a = 3.779, c = 13.23 Å	35
$La_2Ca_{1-x}Sr_xCu_2O_6$	I4/mmm; a = 3.825, c = 19.42 Å	60
$YBa_2Cu_3O_7$	Pmmm; a = 3.821, b = 3.885, c = 11.676 Å	93
$YBa_2Cu_4O_8$	Ammm; a = 3.84, b = 3.87, c = 27.24 Å	80
$Y_2Ba_4Cu_7O_{15}$	Ammm; a = 3.851, b = 3.869, c = 50.29 Å	93
$Bi_2Sr_2CuO_6$	Amaa; a = 5.362, b = 5.374, c = 24.622 Å	10
$Bi_2CaSr_2Cu_2O_8$	A$_2$aa; a = 5.409, b = 5.420, c = 30.93 Å	92
$Bi_2Ca_2Sr_2Cu_3O_{10}$	A$_2$aa; a = 5.39, b = 5.40, c = 37 Å	110
$Bi_2Sr_2(Ln_{1-x}Ce_x)_2Cu_2O_{10}$	P4/mmm; a = 3.888, c = 17.28 Å	25
$Tl_2Ba_2CuO_6$	A$_2$aa; a = 5.468, b = 5.472, c = 23.238 Å; I4/mmm; a = 3.866, c = 23.239 Å	92
$Tl_2CaBa_2Cu_2O_8$	I4/mmm; a = 3.855, c = 29.318 Å	119
$Tl_2Ca_2Ba_2Cu_3O_{10}$	I4/mmm; a = 3.85, c = 35.9 Å	128
$Tl(BaLa)CuO_5$	P4/mmm; a = 3.83, c = 9.55 Å	40
$Tl(SrLa)CuO_5$	P4/mmm; a = 3.7, c = 9 Å	40
$(Tl_{0.5}Pb_{0.5})Sr_2CuO_5$	P4/mmm; a = 3.738, c = 9.01 Å	40
$TlCaBa_2Cu_2O_7$	P4/mmm; a = 3.856, c = 12.754 Å	103
$(Tl_{0.5}Pb_{0.5})CaSr_2Cu_2O_7$	P4/mmm; a = 3.80, c = 12.05 Å	90
$TlSr_2Y_{0.5}Ca_{0.5}Cu_2O_7$	P4/mmm; a = 3.80, c = 12.10 Å	90
$TlCa_2Ba_2Cu_3O_8$	P4/mmm; a = 3.853, c = 15.913 Å	110
$(Tl_{0.5}Pb_{0.5})Sr_2Ca_2Cu_3O_9$	P4/mmm; a = 3.81, c = 15.23 Å	120
$TlBa_2(La_{1-x}Ce_x)_2Cu_2O_9$	I4/mmm; a = 3.8, c = 29.5 Å	40
$Pb_2Sr_2La_{0.5}Ca_{0.5}Cu_3O_8$	Cmmm; a = 5.435, b = 5.463, c = 15.817 Å	70
$Pb_2(Sr,La)_2Cu_2O_6$	P22$_1$2; a = 5.333, b = 5.421, c = 12.609 Å	32
$(Pb,Cu)Sr_2(La,Ca)Cu_2O_7$	P4/mmm; a = 3.820, c = 11.826 Å	50
$(Pb,Cu)(Sr,Eu)(Eu,Ce)Cu_2O_x$	I4/mmm; a = 3.837, c = 29.01 Å	25
$Nd_{2-x}Ce_xCuO_4$	I4/mmm; a = 3.95, c = 12.07 Å	30
$Ca_{1-x}Sr_xCuO_2$	P4/mmm; a = 3.902, c = 3.35 Å	110
$Sr_{1-x}Nd_xCuO_2$	P4/mmm; a = 3.942, c = 3.393 Å	40
$Ba_{0.6}K_{0.4}BiO_3$	Pm3m; a = 4.287 Å	31
Rb_2CsC_{60}	a = 14.493 Å	31
$NdBa_2Cu_3O_7$	Pmmm; a = 3.878, b = 3.913, c = 11.753	58

Table 1
Structural Parameters and Approximate T_c Values of High-Temperature Superconductors (continued)

Material	Structure	T_c/K (maximum value)
$SmBaSrCu_3O_7$	I4/mmm; $a = 3.854$, $c = 11.62$	84
$EuBaSrCu_3O_7$	I4/mmm; $a = 3.845$, $c = 11.59$	88
$GdBaSrCu_3O_7$	I4/mmm; $a = 3.849$, $c = 11.53$	86
$DyBaSrCu_3O_7$	Pmmm; $a = 3.802$, $b = 3.850$, $c = 11.56$	90
$HoBaSrCu_3O_7$	Pmmm; $a = 3.794$, $b = 3.849$, $c = 11.55$	87
$ErBaSrCu_3O_7$ (multiphase)	Pmmm; $a = 3.787$, $b = 3.846$, $c = 11.54$	82
$TmBaSrCu_3O_7$ (multiphase)	Pmmm; $a = 3.784$, $b = 3.849$, $c = 11.55$	88
$YBaSrCu_3O_7$	Pmmm; $a = 3.803$, $b = 3.842$, $c = 11.54$	84
$HgBa_2CuO_4$	I4/mmm; $a = 3.878$, $c = 9.507$	94
$HgBa_2CaCu_2O_6$ (annealed in O_2)	I4/mmm; $a = 3.862$, $c = 12.705$	127
$HgBa_2Ca_2Cu_3O_8$	Pmmm; $a = 3.85$, $c = 15.85$	133
$HgBa_2Ca_3Cu_4O_{10}$	Pmmm; $a = 3.854$, $c = 19.008$	126

Table 2
Superconducting Properties

J_c (0): Critical current density extrapolated to 0 K
λ_{ab}: Penetration depth in a-b plane
k_B: Boltzmann constant

Material	Form	Energy gap (Δ)		$10^{-6} \times J_c$ (0)/A cm^{-2}	λ_{ab}/Å
		$2\Delta_{pp}/k_BT_c$*	$2\Delta_{fit}/k_BT_c$†		
Y $Ba_2Cu_3O_7$	Single Crystal	5–6	4–5	30 (film)	1400
$Bi_2Sr_2CaCu_2O_8$	Single Crystal	8–9	5.5–6.5	2	2700
$Tl_2Ba_3CaCu_2O_8$	Ceramic	6–7	4–6	10 (film, 80 K)	2000
$La_{2-x}Sr_xCuO_4$, $x = 0.15$	Ceramic	7–9	4–6		
$Nd_{2-x}Ce_xCuO_4$	Ceramic	8	4–5	0.2 (film)	

* Obtained from peak to peak value.
† Obtained from fit to BCS-type relation.

Table 3
Normal State Properties

ρ_{ab}: Resistivity in the *a-b* plane
ρ_c: Resistivity along the *c* axis
+ve: ρ_c has positive temperature coefficient of resistivity
–ve: ρ_c has negative temperature coefficient of resistivity
n_H: Hall density
k: Thermal conductivity
in plane: Along *a-b* plane
out of plane: Perpendicular to *a-b* plane

Material	Form	$\rho_{ab}/\mu\Omega$ cm		ρ_c/mΩ cm	$d\rho_c/dT$	$10^{-21} \times n_H/cm^{-3}$		*k*/(mW/cm K) at 300 K	
		300 K	100 K	300 K		300 K	100 K	in plane	out of plane
$YBa_2Cu_3O_7$	Single Crystal	110	35	5	+ve	11–16	4–6	120	3
	film	200–300	60–100			5–9	2–3		
$YBa_2Cu_4O_8$	Single Crystal	75	20	10	–ve	14			
	film	100–200	20–50			22	17		
$Bi_2Sr_2CuO_6$	Single Crystal	300	150	5000	–ve	6	5		
$Bi_2Sr_2CaCu_2O_8$	Single Crystal	150	50	>1000	–ve	4	3	60	8
$Tl_2Ba_2CuO_6$	Single Crystal	300–400	50–75	200–300	+ve	3.1	2.5		
$Tl_2Ba_2Ca_2Cu_3O_{10}$	Ceramic	***	**				≈ 2*		
$La_{2-x}Sr_xCuO_4$, $x = 0.12$	Single Crystal	900	350	200	+ve for T >225 K	2.5			
$La_{2-x}Sr_xCuO_4$, $x = 0.20$	Single Crystal	400	200	80	+ve for T >150 K	10		50 (for $x = 0.04$)	20
	film	400	160			8.4	6.3		
$Nd_{2-x}Ce_xCuO_4$, $x = 0.17$	Single Crystal	500	275			53	17		
$x = 0.15$	film	140–180	35			32	11	250 (for $x = 0.15$)	

* At 200 K
** ρ ~0.4 mΩ cm at 120 K
*** ρ ~1.5 mΩ cm at 300 K

ORGANIC SUPERCONDUCTORS

H.P.R. Frederikse

Although the vast majority of organic compounds are insulating gases, liquids or solids, a small number of organic solids show a considerable amount of electrical conductivity. Some of these appear to be superconductors. The superconducting organics fall primarily into two groups: those containing fulvalenes (pentagonal rings containing sulfur or selenium) and fullerenes based on the nearly spherical cluster C_{60}.

The transition temperatures T_c of the fulvalene derivatives are shown in Table 1. The abbreviations of the various molecular groups are listed in Table 2 and their chemical structures are depicted in Figure 1. Most of the T_c's are between 1 and 12 K. Several of the compounds only show superconductivity under pressure.

The fullerenes are A_3C_{60} compounds, where A represents a single or a combination of alkali atoms. The C_{60} cluster is shown in Figure 2a, while Figure 2b illustrates how the alkali atoms fit into the A_3C_{60} molecule to form the A15 crystallographic structure. Their superconducting transition temperatures range from 8 to 31.3 K (see Table 3).

REFERENCES

1. Ishigura, T. and Yamaji, K., *Organic Superconductors*, Springer-Verlag, Berlin, 1990.
2. Williams, Jack M. et al., *Organic Superconductors (Including Fullerenes)*, Prentice Hall, Englewood Cliffs, N.J., 1992.
3. *The Fullerenes*, Ed.: Krato, H.W., Fisher, J.E., and Cox, D.E., Pergamon Press, Oxford, 1993.
4. Schluter, M. et al., in *The Fullerenes* (Ref. 3), p. 303.

Table 1
Critical Pressure and Maximum Critical Temperature of Organic Superconductors

Material	P_c/kbar	T_c/K	Material	P_c/kbar	T_c/K
$(TMTSF)_2PF_6$	6.5	1.2	β-$(ET)_2IBr_2$	0	2.8
$(TMTSF)_2AsF_6$	9	1.3	β-$(ET)_2AuI_2$	0	4.8
$(TMTSF)_2SbF_6$	11	0.4	$(ET)_4Hg_{2.89}Cl_8$	0	4.2
$(TMTSF)_2TaF_6$	12	1.4	$(ET)_4Hg_{2.89}Br_8$	12	1.8
$(TMTSF)_2ClO_4$	0	1.4	$(ET)_3Cl_2(H_2O)_2$	16	2
$(TMTSF)_2ReO_4$	9.5	1.3	κ-$(ET)_2Cu(NCS)_2$	0	10.4
$(TMTSF)_2FSO_3$	5	3	κ-(d-ET)$_2$Cu(NCS)$_2$	0	11.4
$(ET)_4(ReO_4)_2$	4.5	2	$(DMET)_2Au(CN)_2$	1.5	0.9
β_L-$(ET)_2I_3$	0	1.4	$(DMET)_2AuI_2$	5	0.6
β_H-$(ET)_2I_3$	0	8.1	$(DMET)_2AuBr_2$	0	1.9
γ-$(ET)_3I_{2.5}$	0	2.5	$(DMET)_2AuCl_2$	0	0.9
ε-$(ET)_2I_3(I_8)_{0.5}$	0	2.5	$(DMET)_2I_3$	0	0.6
α-$(ET)_2I_3I_2$-doped	0	3.3	$(DMET)_2IBr_2$	0	0.7
α_t-$(ET)_2I_3$	0	8	(MDT-TTF)$_2$AuI$_2$	0	3.5
ε→β-$(ET)_2I_3$[a]	0	6	TTF[Ni(dmit)$_2$]$_2$	2	1.6[b]
θ-$(ET)_2I_3$	0	3.6	TTF[Pd(dmit)$_2$]$_2$	20	6.5
κ-$(ET)_2I_3$	0	3.6	$(CH_3)_4$N[Ni(dmit)$_2$]$_2$	7	5

[a] Converted form ε-type to β-type by thermal treatment.
[b] For 7 kbar.

From Ishigura, T. and Yamaji, K., *Organic Superconductors*, Springer-Verlag, Berlin, 1990. With permission.

Table 2
List of Symbols and Abbreviations

TTF	tetrathiafulvalene
TMTSF	tetramethyltetraselenafulvalene
BEDT-TTF or "ET"	bis(ethylenedithio)tetrathiafulvalene
MDT-TTF	methylenedithiotetrathiafulvalene
DMET	[dimethyl(ethylenedithio)diselenadithiafulvalene]
dmit	4,5-dimercapto-1,3-dithiole-2-thione
T_c	transition temperature to superconducting state
P_c	minimum pressure required for superconducting transition

TMTSF

Tetramethyltetraselenafulvalene

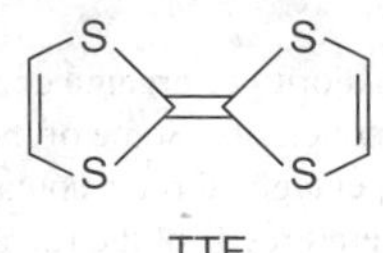

TTF

Tetrathiafulvalene

BEDT – TTF or ET

Bis(ethylenedithio)tetrathiafulvalene

DMET

Dimethyl(ethylenedithio)diselenadithiafulvalene

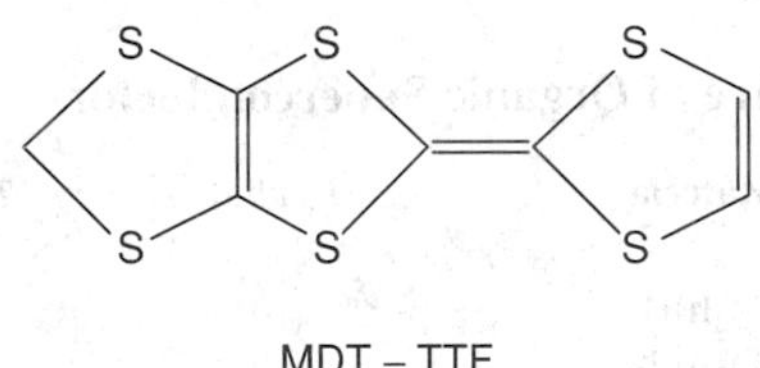

MDT – TTF

Methylenedithiotetrathiafulvalene

M=Ni, Pd, Pt

$M(dmit)_2^{2-}$

Ligand is 4,5-dimercapto-1.3-dithiole-2-thione

FIGURE 1. Structures of various donor molecules and acceptor species.

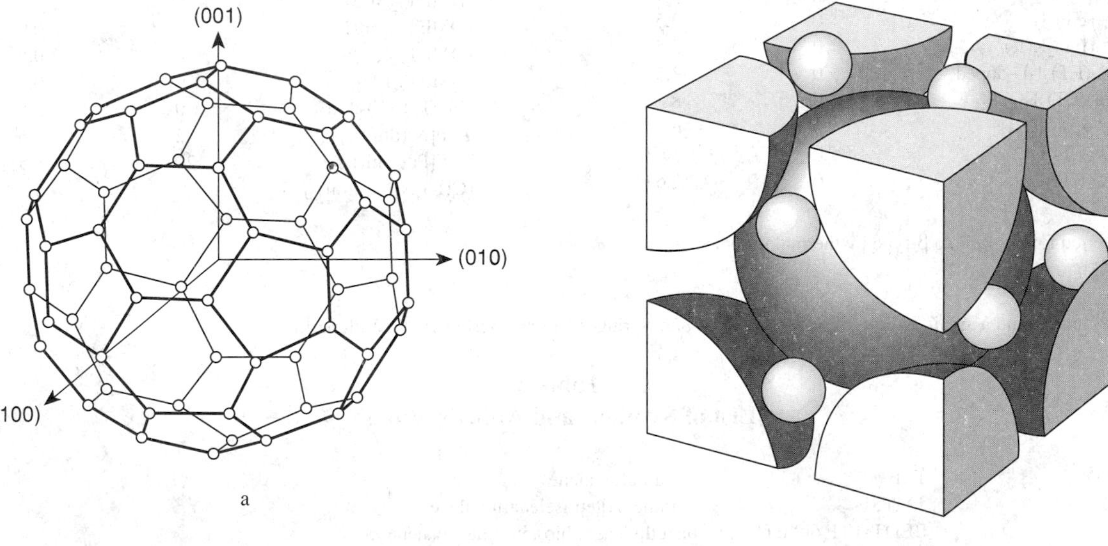

FIGURE 2. (a) C_{60} cluster placed in a fcc lattice. Each crystal axis crosses a double bond shared by two hexagons. (b) A hypothetical A_3C_{60} with the A15 structure. The structure can be seen to be an ordered defect structure of A_6C_{60}.

Table 3
Unit Cell and T_c for FCC-A_3 C_{60}

	Lattice parameter(s) (Å)	T_c/K
$Na_2Rb_{0.5}Cs_{0.5}C_{60}$	14.148(3)	8.0
Na_2CsC_{60} No. 1[a]	14.132(2)	10.5
Na_2CsC_{60} No. 2[a]	14.176(9)	14.0
K_3C_{60}	14.253(3)	19.3
K_2RbC_{60}	14.299(2)	21.8
Rb_2KC_{60} No. 1[a]	14.336(1)	24.4
Rb_2KC_{60} No. 2[a]	14.364(5)	26.4
Rb_3C_{60}	14.436(2)	29.4
Rb_2CsC_{60}	14.493(2)	31.3

[a] Samples labeled No. 1 and No. 2 have the same nominal composition.

From Schluter, M et. al., *The Fullerenes*, Ed.: Krato, H.W., Fisher, J.E., and Cox, D.E., Pergamon Press, Oxford, 1993. With permission.

PROPERTIES OF SEMICONDUCTORS

L. I. Berger and B. R. Pamplin

The term "semiconductor" is applied to a material in which electric current is carried by electrons or holes and whose electrical conductivity, when extremely pure, rises exponentially with temperature and may be increased from its low "intrinsic" value by many orders of magnitude by "doping" with electrically active impurities.

Semiconductors are characterized by an energy gap in the allowed energies of electrons in the material which separates the normally filled energy levels of the *valence band* (where "missing" electrons behave like positively charged current carriers "holes") and the *conduction band* (where electrons behave rather like a gas of free negatively charged carriers with an effective mass dependent on the material and the direction of the electrons' motion). This energy gap depends on the nature of the material and varies with direction in anisotropic crystals. It is slightly dependent on temperature and pressure, and this dependence is usually almost linear at normal temperatures and pressures.

Data are presented in three tables. Table I "General Properties of Semiconductors" lists the main crystallographic and semiconducting properties of a large number of semiconducting materials in three main categories: "Tetrahedral Semiconductors" in which every atom is tetrahedrally co-ordinated to four nearest neighbor atoms (or atomic sites) as for example in the diamond structure; "Octahedral Semiconductors" in which every atom is octahedrally co-ordinated to six nearest neighbor atoms — as for examples the halite structure; and "Other Semiconductors."

Table II gives more detailed information about some better known semiconductors, while Table III gives some information about the electronic energy band structure parameters of the best known materials.

Table I
PHYSICO-CHEMICAL PROPERTIES OF SEMICONDUCTORS (LISTED BY CRYSTAL STRUCTURE)

PART A. ADAMANTINE SEMICONDUCTORS

§A1. Diamond Structure Elements (Strukturbericht symbol A4, Space Group Fd3m-O_h^7)

Substance	Molecular mass	Average atomic mass	Lattice parameters (Å room temp.)	Density (g/cm³)	Melting point (K)	Microhardness, N/mm² (M-Mohs Scale)	Specific heat, J/kg·K (300K)	Debye temp. (K)	Coefficient of thermal linear expansion [10^{-6} K^{-1} (300K)]	Thermal conductivity [mW/cm·K (300K)]
C		12.01	3.56683	3.51	≈3850 Transition to graphite > 980	10(M)	471.5	2340	1.18	9900(I) 23200(IIA) 13600(IIB)
Si		28.09	5.43072	2.3283	1685 ± 2	11270	702	645	4.68	1240
Ge		72.59	5.65754	5.3234	1231	7644	321.9	374	6.1	640
α-Sn		118.69	6.4912	5.765 (281 K)	505.2 (Tr. 286.4)		213	230	5.4 (220 K)	

§A2. Sphalerite (Zinc Blende) Structure Compounds (Strukturbericht symbol B3 Space Group F$\bar{4}$3m-T_d^2)

Substance	Molecular mass	Average atomic mass	Lattice parameters (Å room temp.)	Density (g/cm³)	Melting point (K)	Microhardness, N/mm² (M-Mohs Scale)	Specific heat, J/kg·K (300K)	Debye temp. (K)	Coefficient of thermal linear expansion [10^{-6} K^{-1} (300K)]	Thermal conductivity [mW/cm·K (300K)]
I VII Compounds										
CuF	82.54	41.27	4.255		1181					
CuCl	98.99	49.49	5.4057	3.53	695	2.3 (M)	490	240	12.1	8.4
CuBr	143.36	71.73	5.6905	4.98	770	2.5(M)	381	207	15.4	12.5
CuI	190.46	95.23	6.60427	5.63	878	192	276	181	19.2	16.8
AgBr	187.78	93.89		6.473	>1570 (Tr. 410)	2.5(M)	270			
AgI	234.77	117.39	6.502	5.67	831	2.5(M)	232	134	−2.5	4.2
II VI Compounds										
BeS	41.08	20.54	4.865	2.36						
BeSe	87.97	43.99	5.139	4.315						
BeTe	136.61	68.31	5.626	5.090						
BePo	(218)	(109)	5.838	7.3						
ZnO	81.37	40.69	4.63	5.675	2248	5.0 (M)	494	416	2.9	234
ZnS	97.43	48.72	5.4093	4.079	2100 (Tr. 1295)	1780	472	530	6.36	251
ZnSe	144.34	72.17	5.6676	5.42	1790	1350	339	400	7.2	140
ZnTe	192.99	96.5	6.101	6.34	1568	900	264	223	8.19	108
ZnPo	(274)	(137)	6.309							
CdS	144.46	72.23	5.832	4.826	1750	1250	330	219	4.7	200
CdSe	191.36	95.68	6.05	5.674	1512	1300	255	181	3.8	90
CdTe	240.00	120.00	6.477	5.86	1365	600	205	200	4.9	58.5
CdPo	(321)	(161)	6.665							
HgS	232.65	116.33	5.8517	7.73	1820	3(M)	210			
HgSe	279.55	139.78	6.084	8.25	1070	2.5(M)	178	151	5.46	10
HgTe	328.19	164.10	6.4623	8.17	943	300	164	242	4.6	20
III V Compounds										
BN	24.82	12.41	3.615	3.49	≈3300	10(M)	793	≈1900		200
BP(L.T.)	41.78	20.87	4.538	2.9	≈2800	37000		≈980		
BAs	85.73	42.87	4.777		≈2300	19000		≈625		

Table I
PHYSICO-CHEMICAL PROPERTIES OF SEMICONDUCTORS (LISTED BY CRYSTAL STRUCTURE) (continued)

Substance	Molecular mass	Average atomic mass	Lattice parameters (Å room temp.)	Density (g/cm^3)	Melting point (K)	Microhardness, N/mm^2 (M-Mohs Scale)	Specific heat, J/kg·K (300 K)	Debye temp. (K)	Coefficient of thermal linear expansion [10^{-6} K^{-1} (300K)]	Thermal conductivity [mW/cm·K (300K)]
AlP	57.95	28.98	5.451	2.42	≈2100	5.5(M)		588		920
AlAs	101.90	50.95	5.6622	3.81	2013	5000		417	3.5	840
AlSb	148.73	74.37	6.1355	4.218	1330	4000		292	4.2	600
GaP	100.69	50.35	5.4505	4.13	1750	9450		446	5.3	752
GaAs	144.64	72.32	5.65315	5.316	1510	7500		344	5.4	560
GaSb	191.47	95.74	6.0954	5.619	980	4480	320	265	6.1	270
InP	145.79	72.90	5.86875	4.787	1330	4100		321	4.6	800
InAs	189.74	94.87	6.05838	5.66	1215	3300	268	249	4.7	290
InSb	236.57	118.29	6.47877	5.775	798	2200	144	202	4.7	160
Other sphalerite structure compounds										
MnS	87.0	43.5	5.011							
MnSe	133.9	66.95	5.82							
β-SiC	40.1	20.1	4.348	3.21	3070					
Ga_2Se_3	376.32	75.26	5.429	4.92	1020	3160			8.9	50
Ga_2Te_3	522.24	104.45	5.899	5.75	1063	2370				47
In_2Te_3(H.T.)	608.44	121.7	6.150	5.8	940	1660				69
$MgGeP_2$	158.84	39.71	5.652							
$ZnSnP_2$	246.00	61.5	5.65		1200					
$ZnSnAs_2$(H.T.)	333.90	82.38	5.851	5.53	1050					76
$ZnSnSb_2$	427.56	106.89	6.281	5.67	870	2500				76

§A3. Wurtzite (Zincite) Structure Compounds (Strukturbericht symbol B4, Space Group P 6_3mc-C^4_{6v})

Substance	Molecular mass	Average atomic mass	Lattice parameters (Å room temp.)	Density (g/cm^3)	Melting point (K)	Microhardness, N/mm^2 (M-Mohs Scale)	Specific heat, J/kg·K (300 K)	Debye temp. (K)	Coefficient of thermal linear expansion [10^{-6} K^{-1} (300K)]	Thermal conductivity [mW/cm·K (300K)]
I VII Compounds										
CuCl	99.0	49.5	3.91 6.42		T_c680K					
CuBr	143.46	71.73	4.06 6.66		T_c658K					
CuI	190.46	95.23	4.31 7.09							
AgI	234.80	117.40	4.580 7.494							
II VI Compounds										
BeO	25.01	12.51	2.698 4.380		2800					
MgTe	151.9	76.0	4.54 7.39	3.85	≈2800					
ZnO	81.37	40.69	3.24950 5.2069	5.66	2250					600
ZnS	97.43	48.72	3.8140 6.2576	4.1	2100					460
ZnTe	192.99	46.50	4.27 6.99		1568					
CdS	144.46	72.23	4.1348 6.7490	4.82	1748					401
CdSe	191.36	95.68	4.299 7.010	5.66	1512					316
CdTe	240.00	120.00	4.57 7.47							
III V Compounds										
BP(H.T.)	41.79	20.90	3.562 5.900							
AlN	40.99	20.50	3.111 4.978	3.26	≈2500					823
GaN	83.73	41.87	3.190 5.189	6.10	1500					656
InN	128.83	64.42	3.533 5.693	6.88	1200					556
Other wurtzite structure compounds										
MnS	87.0	43.5	3.985 6.45	3.248						
MnSe	133.9	66.95	4.12 6.72							
SiC	40.1	20.1	3.076 5.048							
MnTe	182.54	91.27	4.078 6.701							
Al_2S_3	150.14	30.03	3.579 5.829	2.55	1400					
Al_2Se_3	290.84	58.17	3.890 6.30	3.91	1250					

§A4. Chalcopyrite Structure Compounds (Strukturbericht symbol $E1_1$, Space Group I $\bar{4}$ 2d-D^{12}_{2d})

Substance	Molecular mass	Average atomic mass	Lattice parameters (Å room temp.)	Density (g/cm^3)	Melting point (K)	Microhardness, N/mm^2 (M-Mohs Scale)	Specific heat, J/kg·K (300 K)	Debye temp. (K)	Coefficient of thermal linear expansion [10^{-6} K^{-1} (300K)]	Thermal conductivity [mW/cm·K (300K)]
I III VI_2 Compounds										
$CuAlS_2$	154.65	38.66	5.323 10.44	3.47	[illegible]					
$CuAlSe_2$	248.45	62.11	5.617 10.92	4.70	2260					
$CuAlTe_2$	345.73	86.43	5.976 11.80	5.50	2550					
$CuGaS_2$	197.39	49.35	5.360 10.49	4.35	2300					
$CuGaSe_2$	291.19	72.80	5.618 11.01	5.56	1970	4200		275	5.4	42
$CuGaTe_2$	388.47	97.12	6.013 11.93	5.99	2400	3500			6.9	27
$CuInS_2$	242.49	60.62	5.528 11.08	4.75	1400	2550				
$CuInSe_2$	336.29	84.07	5.785 11.56	5.77	1600	2050			6.6	37
$CuInTe_2$	433.57	108.39	6.179 12.365	6.10	1660	400		195	7.1	49
$CuTlS_2$	332.05	83.01	5.580 11.17	6.32						
$CuTlSe_2$(L.T.)	425.85	106.46	5.844 11.65	7.11	900					
$CuFeS_2$	183.51	45.88	5.25 10.32	4.088						
$CuFeSe_2$	277.31	69.33			850					

Table I
PHYSICO-CHEMICAL PROPERTIES OF SEMICONDUCTORS (LISTED BY CRYSTAL STRUCTURE) (continued)

Substance	Molecular mass	Average atomic mass	Lattice parameters (Å room temp.)		Density (g/cm³)	Melting point (K)	Microhardness, N/mm² (M-Mohs Scale)	Specific heat, J/kg·K (300 K)	Debye temp. (K)	Coefficient of thermal linear expansion [10^{-6} K^{-1} (300K)]	Thermal conductivity [mW/cm·K (300K)]
$CuLaS_2$	266.58	66.65	5.65	10.86							
$AgAlS_2$	198.97	49.74	5.707	10.28	3.94						
$AgAlSe_2$	292.77	73.19	5.968	10.77	5.07	1220					
$AgAlTe_2$	390.05	97.51	6.309	11.85	6.18	1000					
$AgGaS_2$	241.71	60.43	5.755	10.28	4.72						
$AgGaSe_2$	335.51	83.88	5.985	10.90	5.84	1120	4400				
$AgGaTe_2$	432.79	108.2	6.301	11.96	6.05	990	1800		212		10
$AgInS_2$(L.T.)	286.87	71.70	5.828	11.19	5.00		2250				
$AgInSe_2$	380.61	95.15	6.102	11.69	5.81	1053	1850				30
$AgInTe_2$	477.89	119.47	6.42	12.59	6.12	965				9.49, 0.69	
$AgFeS_2$	227.83	56.96	5.66	10.30	4.53						
II IV V_2 Compounds											
$ZnSiP_2$	155.40	38.85	5.400	10.441	3.39	1640	11000				
$ZnGeP_2$	199.90	49.98	5.465	10.771	4.17	1295	8100				180
$ZnSnP_2$	246.00	61.5					6500				
$CdSiP_2$	202.43	50.61	5.678	10.431	4.00	≈1470	10500		282		
$CdGeP_2$	246.94	61.74	5.741	10.775	4.48	1049	5650				110
$CdSnP_2$	243.03	73.26	5.900	11.518			5000		195		140
$ZnSiAs_2$	242.20	60.55	5.61	10.88	4.70	1311	9200				
$ZnGeAs_2$	287.80	71.95	5.672	11.153	5.32	1150	6800		263		110
$ZnSnAs_2$	333.90	83.48	5.8515	11.704	5.53	1048	4550		271		150
$CdSiAs_2$	290.34	72.58	5.884	10.882			6850				
$CdGeAs_2$	334.83	83.71	5.9427	11.2172	5.60	938	4700				48
$CdSnAs_2$	380.93	95.23	6.0944	11.9182	5.72	880	3450				40

§A5. Other Ternary Semiconductors with Tetrahedral Coordination

Substance	Molecular mass	Average atomic mass	Lattice parameters (Å room temp.)		Density (g/cm³)	Melting point (K)	Microhardness, N/mm² (M-Mohs Scale)	Specific heat, J/kg·K (300 K)	Debye temp. (K)	Coefficient of thermal linear expansion [10^{-6} K^{-1} (300K)]	Thermal conductivity [mW/cm·K (300K)]
I_2 IV VI_3 Compounds											
Cu_2SiS_3(H.T.)	251.36	41.89	3.684	6.004	3.81	1200					23
Cu_2SiS_3(L.T.)			5.290	10.156	3.63						
Cu_2SiTe_3	537.98	89.66	5.93		5.47						
Cu_2GeS_3(H.T.)	295.88	49.31	5.317		4.45	1210	4550	510	254	7.2	12
Cu_2GeS_3(L.T.)			5.327	5.215	4.46						
Cu_2GeSe_3	436.56	72.76	5.589	5.485	5.57	1030	3840	340	168	8.4	24
Cu_2GeTe_3	582.51	97.09	5.958	5.935	5.92		2890				130
Cu_2SnS_3	341.98	57.00	5.436		5.02	1110	2770	440	214	7.8	28
$CuSnSe_3$	482.66	80.44	5.687		5.94	960	2510	310	148	8.9	35
Cu_2SnTe_3	628.61	104.77	6.048		6.51	680	1970				144
Ag_2GeSe_3	525.21	87.54									
Ag_2SnSe_3	571.31	95.22									
Ag_2GeTe_3	671.13	111.86									
Ag_2SnTe_3	717.23	119.54									
I_3 V VI_4 Compounds											
Cu_3PS_4	349.85	40.73	7.44	6.19							
Cu_3AsS_4	393.79	49.22	6.43	6.14	4.37					3.2	30.2
Cu_3AsSe_4	581.37	72.67	5.570	10.957	5.61				169	9.5	19
Cu_3SbS_4	440.64	55.08	5.38	16.76	4.90						
Cu_3SbSe_4	628.22	78.53	5.654	11.256	6.0				131	12.4	14.6
I IV_2 V_3 Compounds											
$CuSi_2P_3$	212.64	35.44	5.25								
$CuGe_2P_3$	301.65	50.28	5.375		4.318	1113	8500	429	8.21	37.6	
$AgGe_2P_3$	345.97	57.66				1015	6150				

§A6. "Defect Chalcopyrite" Structure Compounds (Strukturbericht symbol E3, Space Group I $\bar{4}$-S_4^2)

Substance	Molecular mass	Average atomic mass	Lattice parameters (Å room temp.)		Density (g/cm³)	Melting point (K)
$ZnAl_2Se_4$	435.18	62.17	5.503	10.90	4.37	
$ZnAl_2Te_4$(?)	629.74	84.96	5.904	12.05	4.95	
$ZnGa_2S_4$(?)	333.06	47.58	5.274	10.44	3.80	
$ZnGa_2Se_4$(?)	520.66	74.38	5.496	10.99	5.21	
$ZnGa_2Te_4$(?)	715.22	102.17	5.937	11.87	5.67	
$ZnIn_2Se_4$	610.86	87.27	5.711	11.42	5.44	1250
$ZnIn_2Te_4$	805.42	115.06	6.122	12.24	5.83	1075
$CdAl_2S_4$	294.61	42.09	5.564	10.32	3.06	
$CdAl_2Se_4$	482.21	68.89	5.747	10.68	4.54	
$CdAl_2Te_4$(?)	676.77	97.68	6.011	12.21	5.10	

Table I
PHYSICO-CHEMICAL PROPERTIES OF SEMICONDUCTORS (LISTED BY CRYSTAL STRUCTURE) (continued)

Substance	Molecular mass	Average atomic mass	Lattice parameters (Å room temp.)		Density (g/cm³)	Melting point (K)	Microhardness, N/mm² (M-Mohs Scale)	Specific heat, J/kg·K (300 K)	Debye temp. (K)	Coefficient of thermal linear expansion [10^{-6} K^{-1} (300K)]	Thermal conductivity [mW/cm·K (300K)]
$CdGa_2S_4$	380.09	54.30	5.577	10.08	4.03						
$CdGa_2Se_4$	567.69	81.10	5.743	10.73	5.32						
$CdGa_2Te_4$	762.25	108.89	6.093	11.81	5.77						
$CdIn_2Te_4$	852.45	121.78	6.205	12.41	5.9	1060					
$HgAl_2S_4$	382.79	54.68	5.488	10.26	4.11						
$HgAl_2Se_4$	570.39	82.48	5.708	10.74	5.05						
$HgAl_2Te_4$(?)	764.48	109.28	6.004	12.11	5.81						
$HgGa_2S_4$	468.27	66.90	5.507	10.23	5.00						
$HgGa_2Se_4$	655.87	93.70	5.715	10.78	6.18						
$HgIn_2Se_4$	746.07	106.58	5.764	11.80	6.3	1100					
$HgIn_2Te_4$(?)	940.63	134.38	6.186	12.37	6.3	980					

§A7. Other Adamantine Compounds

Substance	Molecular mass	Average atomic mass	Lattice parameters (Å room temp.)		Density (g/cm³)	Melting point (K)	Microhardness	Specific heat	Debye temp.	Coefficient of thermal linear expansion	Thermal conductivity
αSiC	40.1	20.1	3.0817 15.1183		3.21	3070					
$Hg_5Ga_2Te_8$	2163.19	144.21	6.235								
$Hg_5In_2Te_8$	2253.39	150.23	6.328								
$CdIn_2Se_4$	657.89	93.98	a = c − 5.823								

PART B. OCTAHEDRAL SEMICONDUCTORS

§B1. HALITE STRUCTURE SEMICONDUCTORS (Strukturbericht symbol B1, Space Group Fm3m-O_h^5)

Substance	Molecular mass	Average atomic mass	Lattice parameters (Å room temp.)		Density (g/cm³)	Melting point (K)	Microhardness	Specific heat	Debye temp.	Coefficient of thermal linear expansion	Thermal conductivity
GeTe	200.19	100.1	5.98		6.14						
SnSe	197.65	98.83	6.020			1133					
SnTe	246.29	123.15	6.313		6.45	1080 (max)					91
PbS	239.26	119.63	5.9362		7.61	1390					23
PbSe	286.16	143.08	6.1243		8.15	1340					17
PbTe	334.8	167.4	6.454		8.16	1180					23
Selected other binary halites											
BiSe	287.94	143.97	5.99		7.98	880					
BiTe	336.58	168.29	6.47								
EuSe	230.92	115.46	6.191			2300					2.4
GdSe	236.21	118.11	5.771			2400					
NiD	60.71	30.35	4.1684		6.6	2260					
CdO	128.41	64.21	4.6953			1700					7
SrSW	119.68	59.84	6.0199		3.643	3000					

PART C. OTHER SEMICONDUCTORS

§C1. Antifluorite Structure Compounds Fm3m (225)

Substance	Molecular mass	Average atomic mass	Lattice parameters (Å room temp.)		Density (g/cm³)	Melting point (K)	Microhardness	Specific heat	Debye temp.	Coefficient of thermal linear expansion	Thermal conductivity
Mg_2Si	76.70	25.57	6.338		1.88	1375					
Mg_2Ge	121.20	40.4	6.380		3.08	1388					
Mg_2Sn	167.3	55.77	6.765		3.53	1051					92
Mg_2Pb	225.81	85.27	6.836		5.1	823					

§C2. Tetradymite Structure Compounds ($\bar{R}3m - D_{3d}^5$)

Substance	Molecular mass	Average atomic mass	Lattice parameters (Å room temp.)		Density (g/cm³)	Melting point (K)	Microhardness	Specific heat	Debye temp.	Coefficient of thermal linear expansion	Thermal conductivity
Sb_2Te_3	626.3	125.26	4.25	30.3	6.44	895					
Bi_2Se_3	654.84	130.97	4.14	28.7	7.51	979	167				24
Bi_2Te_3	800.76	160.15	4.38	30.45	7.73	858	155	16			30

§C3. Selected Multinary Compounds

Substance	Molecular mass	Average atomic mass	Lattice parameters (Å room temp.)		Density (g/cm³)	Melting point (K)	Microhardness	Specific heat	Debye temp.	Coefficient of thermal linear expansion	Thermal conductivity
$AgSbSe_2$	387.54	96.88	5.786		6.60	910					10.5
$AgSbTe_2$ (or $Ag_{19}Sb_{29}Te_{52}$)	484.82	121.2	6.078		7.12	830					86, 0.3
$AgBiS_2$(H.T.)	380.97	95.24	5.648								
$AgBiSe_2$(H.T.)	474.77	118.69	5.82								
$AgBiTe_2$(H.T.)	572.05	143.01	6.155								
Cu_2CdSnS_4	486.43	60.80	5.586		10.83						

Table I
PHYSICO-CHEMICAL PROPERTIES OF SEMICONDUCTORS (LISTED BY CRYSTAL STRUCTURE) (continued)

Substance	Molecular mass	Average atomic mass	Lattice parameters (Å room temp.)	Density (g/cm^3)	Melting point (K)	Microhardness, N/mm^2 (M-Mohs Scale)	Specific heat, J/kg·K (300 K)	Debye temp. (K)	Coefficient of thermal linear expansion [10^{-6} K^{-1} (300K)]	Thermal conductivity [mW/cm·K (300K)]
§C4. Some Elemental Semiconductors										
B		10.81	4.91 12.6	2.34	2348	9.5(M)	1277	1370	8.3	600
Se(gray)		78.96	4.36 4.95	4.81	493	350	292.6		(‖C) 17.89 (⊥C) 74.09	(‖C) 45.2 (⊥C) 13.1
Te		127.6	4.45 5.91	6.23	723		196.5		16.8	(‖C) 33.8 (⊥C) 19.7

Table II
BASIC THERMODYNAMIC, ELECTRICAL, AND MAGNETIC PROPERTIES OF SEMICONDUCTORS (LISTED BY CRYSTAL STRUCTURE)

Substance	Heat of formation [kJ/mole (300K)]	Volume compressibility (10^{-10} m^2/N)	Static dielectric constant	Atomic magnetic susceptibility (10^{-6} CGS)	Index of refraction	Minimum room temperature energy gap (eV)	Mobility (Room temp.) ($cm^2/V·s$) Electrons	Holes	Optical transition	Remarks
PART A. ADAMANTINE SEMICONDUCTORS										
§A1. Diamond Structure Elements (Strukturbericht symbol A4, Space Group Fd3m-O_h^7)										
C	714.4	18	5.7	−5.88	2.419 (589 nm)	5.4	1800	1400	i*	
Si	324	0.306	11.8	−3.9	3.49 (589 nm)	1.107	1900	500	i	
Ge	291	0.768	16	−0.12	3.99 (589 nm)	0.67	3800	1820	i	
α-Sn	267.5		24		2.75 (589 nm)	0.0; 0.8	2500	2400		
§A2. Sphalerite (Zinc Blende) Structure Compounds (Strukturbericht symbol B3 Space Group F $\bar{4}$ 3m-T_d^2)										
I VII Compounds										
CuF										
CuCl	481	0.26	7.9		1.93	3.17			d	Nantokite
CuBr	481	0.26	7.9		2.12	2.91			d	
CuI	439	0.27	6.5		2.346	2.95			d	Marshite
AgBr	486		12.4		2.253	2.50	4000		i	Bromirite
AgI	389	0.41	10		2.22	2.22	30		d	Miersite
II VI Compounds										
BeS					4.17				i	
BeSe					3.61				i	
BeTe					1.45		20		d	
BePo										
ZnO										See A3
ZnS	477		8.9	−9.9	2.356	3.54	180	5(400°C)	d	See also A3
ZnSe	422		9.2		2.89	2.58	540	28	d	
ZnTe	376		10.4		3.56	2.26	340	100	d	
ZnPo										
CdS										See A3
CdSe										See A3
CdTe	339		7.2		2.50	1.44	1200	50	d	
CdPo										
HgS					2.85		250		d	Metacinnabarite
HgSe	247					2.10 (α)	20000	≈1.5	s	Tiemannite
HgTe	242					−0.06	25000	350	s	Coloradoite
III V Compounds										
BN	815					4.6				Borazone
BP(L.T.)						≈2.1	500	70		Ignites 470K
BAs						≈1.5				
AlP						2.45	80		i	
AlAs	627		10.9			2.16	1200	420	i	
AlSb	585	0.571	11		3.2	1.60	200—400	550	i	
GaP	635	0.110	11.1	−13.8	3.2	2.24	300	150	i	
GaAs	535	0.771	13.2	−16.2	3.30	1.35	8800	400	d	
GaSb	493	0.457	15.7	−14.2	3.8	0.67	4000	1400	d	
InP	560	0.735	12.4	−22.8	3.1	1.27	4600	150	d	
InAs	477	0.549	14.6	−27.7	3.5	0.36	33000	460	d	
InSb	447	0.442	17.7	−32.9	3.96	0.163	78000	750	d	

* i = indirect, d = direct, s = semimetal.

Table II
BASIC THERMODYNAMIC, ELECTRICAL, AND MAGNETIC PROPERTIES OF SEMICONDUCTORS (LISTED BY CRYSTAL STRUCTURE) (continued)

Substance	Heat of formation [kJ/mole (300K)]	Volume compressibility (10^{-10} m²/N)	Static dielectric constant	Atomic magnetic susceptibility (10^{-6} CGS)	Index of refraction	Minimum room temperature energy gap (eV)	Mobility (Room temp.) (cm²/V·s) Electrons	Holes	Optical transition	Remarks
Other sphalerite structure compounds										
MnS										See also §A3
MnSe										See also §A3
β-SiC					2.697	2.3	4000			
Ga_2Te_3	271			−13.5		1.35	50			
In_2Te_3(H.T.)	198			−13.6		1.04	50			
$MgGeP_2$										E1—T^{d12}
$ZnSnP_2$						2.1				Same
$ZnSnAs_2$(H.T.)						≈0.7				Same
$ZnSnSb_2$						0.4				Same

§A3. Wurtzite (Zincite) Structure Compounds (Strukturbericht symbol B4, Space Group P $6_3mc\text{-}C^4_{6v}$)

Substance	Heat of formation [kJ/mole (300K)]	Volume compressibility (10^{-10} m²/N)	Static dielectric constant	Atomic magnetic susceptibility (10^{-6} CGS)	Index of refraction	Minimum room temperature energy gap (eV)	Electrons	Holes	Optical transition	Remarks
I VII Compounds										
CuCl										
CuBr										
CuI										
AgI						2.63				Iodargirite
II VI Compounds										
BeO										
MgTe										
ZnO	−350					3.2	180			
ZnS	−206					3.67				
ZnTe	−163									
CdS			8.45; 9.12		2.32	2.42	350	40	d	Greenockide
CdSe						1.74	900	50	d	Cadmoselite
CdTe						1.50	650			
III V Compounds										
BP(H.T.)										
AIN						6.02				
GaN						3.34				
InN						2.0				
Other wurtzite structure compounds										
MnS										
MnSe										
SiC					2.654					
MnTe						≈1.0				
Al_2S_3	426					4.1				
Al_2Se_3	367					3.1				

§A4. Chalcopyrite Structure Compounds (Strukturbericht symbol $E1_1$, Space Group I $\bar{4}$ 2d-D^{12}_{2d})

Substance	Heat of formation [kJ/mole (300K)]	Volume compressibility (10^{-10} m²/N)	Static dielectric constant	Atomic magnetic susceptibility (10^{-6} CGS)	Index of refraction	Minimum room temperature energy gap (eV)	Electrons	Holes	Optical transition	Remarks
I III VI_2 Compounds										
$CuAlS_2$		0.106				2.5				
$CuAlSe_2$						1.1				
$CuAlTe_2$						0.88				
$CuGaS_2$		0.106				2.38				
$CuGaSe_2$		0.141				0.96, 1.63				
$CuGaTe_2$		0.227				0.82, 1.0				
$CuInS_2$		0.141				1.2				
$CuInSe_2$		0.187				0.86, 0.92				
$CuInTe_2$		0.278				0.95				
$CuTlS_2$										
$CuTlSe_2$(L.T.)						1.07				
$CuFeS_2$						0.53				Chalcopyrite
$CuFeSe_2$						0.16				
$CuLaS_2$										
$AgAlS_2$										
$AgAlSe_2$						0.7				
$AgAlTe_2$						0.56				
$AgGaS_2$		0.150				1.66				
$AgGaSe_2$		0.182				1.1				
$AgGaTe_2$		0.280				1.9				
$AgInS_2$(L.T.)		0.185				1.18				
$AgInSe_2$		0.238				0.96, 0.52				
$AgInTe_2$		0.338								
$AgFeS_2$										

Table II
BASIC THERMODYNAMIC, ELECTRICAL, AND MAGNETIC PROPERTIES OF SEMICONDUCTORS (LISTED BY CRYSTAL STRUCTURE) (continued)

Substance	Heat of formation [kJ/mole (300K)]	Volume compressibility (10^{-10} m²/N)	Static dielectric constant	Atomic magnetic susceptibility (10^{-6} CGS)	Index of refraction	Minimum room temperature energy gap (eV)	Mobility (Room temp.) (cm²/V·s)		Optical transition	Remarks
							Electrons	Holes		
$II\ IV\ V_2$ Compounds										
$ZnSiP_2$	312					2.3	1000			
$ZnGeP_2$	293					2.2				
$ZnSnP_2$	275					1.45				
$CdSiP_2$		0.103				2.2	1000			
$CdGeP_2$	289					1.8				
$CdSnP_2$	270					1.5				
$ZnSiAs_2$	290					1.7		50		
$ZnGeAs_2$	271			−14.4		0.85				
$ZnSnAs_2$	252			−18.4		0.65		300		Disorders at 910K
$CdSiAs_2$		0.143				1.6				
$CdGeAs_2$	266			−23.4		0.53	70	25		Disorders at 903
$CdSnAs_2$	247		13.7	−21.5		0.26	22000	250		

§A5. Other Ternary Semiconductors with Tetrahedral Coordination

Substance	Heat of formation [kJ/mole (300K)]	Volume compressibility (10^{-10} m²/N)	Static dielectric constant	Atomic magnetic susceptibility (10^{-6} CGS)	Index of refraction	Minimum room temperature energy gap (eV)	Electrons	Holes	Optical transition	Remarks
$I_2\ IV\ VI_3$ Compounds										
Cu_2SiS_3(H.T.)										Wurtzite
Cu_2SiS_3(L.T.)										Tetragonal
Cu_2SiTe_3										Cubic
Cu_2GeS_3(H.T.)				−18.7						Cubic
Cu_2GeS_3(L.T.)							360			Tetragonal
Cu_2GeSe_3	211.5			−21.3		0.94	238			Same
Cu_2GeTe_3	190.2			−23.4						Same
Cu_2SnS_3				−18.2		0.91	405			Cubic
$CuSnSe_3$				−21.0		0.66	870			Cubic
Cu_2SnTe_3				−28.4						Cubic
Ag_2GeSe_3				−29.6		0.91 (77K)				
Ag_2SnSe_3				−29.5		0.81				
Ag_2GeTe_3				−31.4		0.25				
Ag_2SnTe_3				−31.0		0.08				
$I_3\ V\ VI_4$ Compounds										
Cu_3PS_4										Enargite
Cu_3AsS_4	269.6			−15.8		1.24				
Cu_3AsSe_4	161.3			−13.1		0.88				Famatinite
Cu_3SbS_4				−8.3		0.74				Famatinite
Cu_3SbSe_4	127.1			−20.5		0.31				
$I\ IV_2\ V_3$ Compounds										
$CuSi_2P_3$										E1
$CuGe_2P_3$		0.12				0.9				E1
$AgGe_2P_3$										

§A6. "Defect Chalcopyrite" Structure Compounds (Strukturbericht symbol E3, Space Group I $\bar{4}$-S_4^2)

Substance	Heat of formation [kJ/mole (300K)]	Volume compressibility (10^{-10} m²/N)	Static dielectric constant	Atomic magnetic susceptibility (10^{-6} CGS)	Index of refraction	Minimum room temperature energy gap (eV)	Electrons	Holes	Optical transition	Remarks
$ZnAl_2Se_4$										
$ZnAl_2Te_4$(?)										
$ZnGa_2S_4$(?)						≈3.4				
$ZnGa_2Se_4$(?)						≈2.2				
$ZnGa_2Te_4$(?)						1.35				
$ZnIn_2Se_4$	206					1.82	35			
$ZnIn_2Te_4$	198					1.2				
$CdAl_2S_4$										
$CdAl_2Se_4$										
$CdAl_2Te_4$(?)										
$CdGa_2S_4$	256					3.44	60			
$CdGa_2Se_4$	216					2.43	33			
$CdGa_2Te_4$										
$CdIn_2Te_4$	195					(1.26 or 0.9)	4000			
$HgAl_2S_4$										
$HgAl_2Se_4$										
$HgAl_2Te_4$(?)										
$HgGa_2S_4$	249					2.84				
$HgGa_2Se_4$	204					1.95	400			
$HgIn_2Se_4$	196					0.6	290			
$HgIn_2Te_4$(?)	188					0.86	200			

Table II
BASIC THERMODYNAMIC, ELECTRICAL, AND MAGNETIC PROPERTIES OF SEMICONDUCTORS (LISTED BY CRYSTAL STRUCTURE) (continued)

Substance	Heat of formation [kJ/mole (300K)]	Volume compressibility (10^{-10} m²/N)	Static dielectric constant	Atomic magnetic susceptibility (10^{-6} CGS)	Index of refraction	Minimum room temperature energy gap (eV)	Mobility (Room temp.) (cm²/V·s)		Optical transition	Remarks
							Electrons	Holes		
§A7. Other Adamantine Compounds										
αSiC			10.2	−6.4	2.67	2.86	400			6H structure
$Hg_5Ga_2Te_8$										B3 with superlattice
$Hg_5In_2Te_8$						0.7	2000			B3 with superlattice
$CdIn_2Se_4$						1.55				
PART B. OCTAHEDRAL SEMICONDUCTORS										
§B1. HALITE STRUCTURE SEMICONDUCTORS (Strukturbericht symbol B1, Space Group Fm3m-O_h^5)										
GeTe										
SnSe										
SnTe										
PbS	435					0.5	600	600		
PbSe	393		161			0.37	1000	900		
PbTe	393		280			0.26	1600	600		Altaite
			360			0.25				
Selected other binary halites										
BiSe										
BiTe						0.4				
EuSe										
GdSe						1.8	4			
NiD						2.0 or 3.7	100			
CdO	531					2.5				
SrSW						4.1				
PART C. OTHER SEMICONDUCTORS										
§C1. Antifluorite Structure Compounds Fm3m (225)										
Mg_2Si						0.77	405	70		
Mg_2Ge						0.74	520	110		
Mg_2Sn						0.36	320	260		
Mg_2Pb						0.1				
§C2. Tetradymite Structure Compounds ($R\bar{3}m - D_{3d}^5$)										
Sb_2Te_3						0.3		360		
Bi_2Se_3						0.35	600			
Bi_2Te_3						0.21	1140	680		R3m (166)
§C3. Selected Multinary Compounds										
$AgSbSe_2$						0.58				
$AgSbTe_2$ (or $Ag_{19}Sb_{29}Te_{52}$)						0.7, 0.27				
$AgBiS_2$(H.T.)										
$AgBiSe_2$(H.T.)										
$AgBiTe_2$(H.T.)										
Cu_2CdSnS_4						1.16	<2			
§C4. Some Elemental Semiconductors										
B	397.1			−6.7	3.4	1.55	10			
Se(gray)			6.6 (0.1 GHz)	−22.1	2.5	1.5		5		$P3_121(152)$
Te				−39.5	3.3	0.33	1700	1200		Same

Table III
SEMICONDUCTING PROPERTIES OF SELECTED MATERIALS

Substance	Minimum Energy Gap (eV) R.T.	Minimum Energy Gap (eV) 0 K	$\frac{dE_g}{dT}$ x 10^4 eV/°C	$\frac{dE_g}{dP}$ x 10^6 eV·cm²/kg	Density of States Electron Effective Mass m_{d_n} (m_o)	Electron Mobility and Temperature Dependence μ_n cm²/V·s	Electron Mobility and Temperature Dependence −x	Density of States Hole Effective Mass m_{d_p} (m_o)	Hole Mobility and Temperature Dependence μ_p cm²/V·s	Hole Mobility and Temperature Dependence −x
Si	1.107	1.153	−2.3	−2.0	1.1	1,900	2.6	0.56	500	2.3
Ge	0.67	0.744	−3.7	+7.3	0.55	3,800	1.66	0.3	1,820	2.33
αSn	0.08	0.094	−0.5		0.02	2,500	1.65	0.3	2,400	2.0
Te	0.33				0.68	1,100		0.19	560	
III–V Compounds										
AlAs	2.2	2.3				1,200			420	
AlSb	1.6	1.7	−3.5	−1.6	0.09	200	1.5	0.4	500	1.8
GaP	2.24	2.40	−5.4	−1.7	0.35	300	1.5	0.5	150	1.5
GaAs	1.35	1.53	−5.0	+9.4	0.068	9,000	1.0	0.5	500	2.1
GaSb	0.67	0.78	−3.5	+12	0.050	5,000	2.0	0.23	1,400	0.9
InP	1.27	1.41	−4.6	+4.6	0.067	5,000	2.0		200	2.4
InAs	0.36	0.43	−2.8	+8	0.022	33,000	1.2	0.41	460	2.3
InSb	0.165	0.23	−2.8	+15	0.014	78,000	1.6	0.4	750	2.1
II–VI Compounds										
ZnO	3.2		−9.5	+0.6	0.38	180	1.5			
ZnS	3.54		−5.3	+5.7		180			5(400°C)	
ZnSe	2.58	2.80	−7.2	+6		540			28	
ZnTe	2.26			+6		340			100	
CdO	2.5 ± .1		−6		0.1	120				
CdS	2.42		−5	+3.3	0.165	400		0.8		
CdSe	1.74	1.85	−4.6		0.13	650	1.0	0.6		
CdTe	1.44	1.56	−4.1	+8	0.14	1,200		0.35	50	
HgSe	0.30				0.030	20,000	2.0			
HgTe	0.15		−1		0.017	25,000		0.5	350	
Halite Structure Compounds										
PbS	0.37	0.28	+4		0.16	800		0.1	1,000	2.2
PbSe	0.26	0.16	+4		0.3	1,500		0.34	1,500	2.2
PbTe	0.25	0.19	+4	−7	0.21	1,600		0.14	750	2.2
Others										
ZnSb	0.50	0.56			0.15	10				1.5
CdSb	0.45	0.57	−5.4		0.15	300			2,000	1.5
Bi_2S_3	1.3					200			1,100	
Bi_2Se_3	0.27					600			675	
Bi_2Te_3	0.13		−0.95		0.58	1,200	1.68	1.07	510	1.95
Mg_2Si		0.77	−6.4		0.46	400	2.5		70	
Mg_2Ge		0.74	−9			280	2		110	
Mg_2Sn	0.21	0.33	−3.5		0.37	320			260	
Mg_3Sb_2		0.32				20			82	
Zn_3As_2	0.93					10	1.1		10	
Cd_3As_2	0.55				0.046	100,000	0.88			
GaSe	2.05		3.8						20	
GaTe	1.66	1.80	−3.6			14	−5			
InSe	1.8					900				
TlSe	0.57		−3.9		0.3	30		0.6	20	1.5
$CdSnAs_2$	0.23				0.05	25,000	1.7			
Ga_2Te_3	1.1	1.55	−4.8							
α-In_2Te_3	1.1	1.2			0.7				50	1.1
β-In_2Te_3	1.0								5	
$Hg_5In_2Te_8$	0.5								11,000	
SnO_2									78	

Table IV
BAND PROPERTIES OF SEMICONDUCTORS

PART A. DATA ON VALENCE BANDS OF SEMICONDUCTORS (ROOM TEMPERATURES)

Substance	Band Curvature Effective Mass (Expressed as fraction of free electron mass): Heavy Holes	Light Holes	"Split-off" Band Holes	Energy Separation of "Split-off" Band (eV)	Measured (Light) Hole Mobility cm²/V·s
Semiconductors with Valence Band Maximum at the Center of the Brillouin Zone ('Γ')					
Si	0.52	0.16	0.25	0.044	500
Ge	0.34	0.043	0.08	0.3	1,820
Sn	0.3				2,400
AlAs					
AlSb	0.4			0.7	550
GaP				0.13	100
GaAs	0.8	0.12	0.20	0.34	400
GaSb	0.23	0.06		0.7	1,400
InP				0.21	150
InAs	0.41	0.025	0.083	0.43	460
InSb	0.4	0.015		0.85	750
CdTe	0.35				50
HgTe	0.5				350

Table IV
BAND PROPERTIES OF SEMICONDUCTORS (continued)

Semiconductors with Multiple Valence Band Maxima

Substance	Number of Equivalent Valleys and Direction	Band Curvature Effective Masses Longitudinal m_L	Transverse m_T	Anisotropy $K = m_L/m_T$	Measured (Light) Hole Mobility $cm^2/V{\cdot}s$
PbSe	4 "L" [111]	0.095	0.047	2.0	1,500
PbTe	4 "L" [111]	0.27	0.02	10	750
Bi_2Te_3	6	0.207	~0.045	4.5	515

PART B. DATA ON CONDUCTION BANDS OF SEMICONDUCTORS (Room Temperature Data)

Single Valley Semiconductors

Substance	Energy Gap (eV)	Effective Mass (m_o)	Mobility ($cm^2/V.s$)	Comments
GaAs	1.35	0.067	8,500	3(or 6?) equivalent [100] valleys 0.36 eV above this maximum with a mobility of ~50
InP	1.27	0.067	5,000	3(or 6?) equivalent [100] valleys 0.4 eV above this minimum.
InAs	0.36	0.022	33,000	equivalent valleys ~1.0 eV above this minimum.
InSb	0.165	0.014	78,000	
CdTe	1.44	0.11	1,000	4(or 8?) equivalent [111] valleys 0.51 eV above this minimum.

Multivalley Semiconductors

Substance	Energy Gap	Number of Equivalent Valleys and Direction	Band Curvature Effective Mass Longitudinal m_L	Transverse m_T	Anisotropy $K = m_L/m_T$	Comments
Si	1.107	6 in [100] "Δ"	0.90	0.192	4.7	
Ge	0.67	4 in [111] at "L"	1.588	0.0815	19.5	
GaSb	0.67	as Ge (?)	~1.0	~0.2	~5	
PbSe	0.26	4 in [111] at "L"	0.085	0.05	1.7	
PbTe	0.25	4 in [111] at "L"	0.21	0.029	5.5	
Bi_2Te_3	0.13	6			~0.05	

Table V
RESISTIVITY OF SEMICONDUCTING MINERALS

Mineral	ρ (ohm · m)
Diamond (C)	2.7
Sulfides	
Argentite, Ag_2S	1.5 to 2.0 × 10^{-3}
Bismuthinite, Bi_2S_3	3 to 570
Bornite, $Fe_2S_3 \cdot n\,Cu_2S$	1.6 to 6000 × 10^{-6}
Chalcocite, Cu_2S	80 to 100 × 10^{-6}
Chalcopyrite, $Fe_2S_3 \cdot Cu_2S$	150 to 9000 × 10^{-6}
Covellite, CuS	0.30 to 83 × 10^{-6}
Galena, PbS	6.8 × 10^{-6} to 9.0 × 10^{-2}
Haverite, MnS_2	10 to 20
Marcasite, FeS_2	1 to 150 × 10^{-3}
Metacinnabarite, 4HgS	2 × 10^{-6} to 1 × 10^{-3}
Millerite, NiS	2 to 4 × 10^{-7}
Molybdenite, MoS_2	0.12 to 7.5
Pentlandite, $(Fe, Ni)_9S_8$	1 to 11 × 10^{-6}
Pyrrhotite, Fe_7S_8	2 to 160 × 10^{-6}
Pyrite, FeS_2	1.2 to 600 × 10^{-3}
Sphalerite, ZnS	2.7 × 10^{-3} to 1.2 × 10^{4}
Antimony-sulfur compounds	
Berthierite, $FeSb_2S_4$	0.0083 to 2.0
Boulangerite, $Pb_5Sb_4S_{11}$	2 × 10^{3} to 4 × 10^{4}
Cylindrite, $Pb_3Sn_4Sb_2S_{14}$	2.5 to 60
Franckeite, $Pb_5Sn_3Sb_2S_{14}$	1.2 to 4
Hauchecornite, $Ni_9(Bi, Sb)_2S_8$	1 to 83 × 10^{-6}
Jamesonite, $Pb_4FeSb_6S_{14}$	0.020 to 0.15
Tetrahedrite, Cu_3SbS_3	0.30 to 30,000
Arsenic-sulfur compounds	
Arsenopyrite, FeAsS	20 to 300 × 10^{-6}
Cobaltite, CoAsS	6.5 to 130 × 10^{-3}
Enargite, Cu_3AsS_4	0.2 to 40 × 10^{-3}
Gersdorffite, NiAsS	1 to 160 × 10^{-6}
Glaucodote, (Co, Fe)AsS	5 to 100 × 10^{-6}
Antimonide	
Dyscrasite, Ag_3Sb	0.12 to 1.2 × 10^{-6}
Arsenides	
Allemonite, $SbAs_3$	70 to 60,000
Lollingite, $FeAs_2$	2 to 270 × 10^{-6}
Nicollite, NiAs	0.1 to 2 × 10^{-6}
Skutterudite, $CoAs_3$	1 to 400 × 10^{-6}
Smaltite, $CoAs_2$	1 to 12 × 10^{-6}
Tellurides	
Altaite, PbTe	20 to 200 × 10^{-6}
Calavarite, $AuTe_2$	6 to 12 × 10^{-6}
Coloradoite, HgTe	4 to 100 × 10^{-6}
Hessite, Ag_2Te	4 to 100 × 10^{-6}
Nagyagite, $Pb_6Au(S, Te)_{14}$	20 to 80 × 10^{-6}
Sylvanite, $AgAuTe_4$	4 to 20 × 10^{-6}
Oxides	
Braunite, Mn_2O_3	0.16 to 1.0
Cassiterite, SnO_2	4.5 × 10^{-4} to 10,000
Cuprite, Cu_2O	10 to 50
Hollandite, $(Ba, Na, K)Mn_8O_{16}$	2 to 100 × 10^{-3}
Ilmenite, $FeTiO_3$	0.001 to 4
Magnetite, Fe_3O_4	52 × 10^{-6}
Manganite, MnO · OH	0.018 to 0.5
Melaconite, CuO	6000
Psilomelane, $KMnO \cdot MnO_2 \cdot n\,H_2O$	0.04 to 6000
Pyrolusite, MnO_2	0.007 to 30
Rutile, TiO_2	29 to 910
Uraninite, UO	1.5 to 200

From Carmichael, R. S., ed., *Handbook of Physical Properties of Rocks,* Vol. I, CRC Press, 1982.

DIFFUSION DATA FOR SEMICONDUCTORS

B. L. Sharma

The diffusion coefficient D in many semiconductors may be expressed by an Arrhenius-type relation

$$D = D_o \exp(-Q/kT)$$

where D_o is a frequency factor, Q is the activation energy for diffusion, k is the Boltzmann constant, and T is the absolute temperature. This table lists D_o and Q for various diffusants in common semiconductors.

Abbreviations used in the table are

AES — Auger Electron Spectroscopy
DLTS — Deep Level Transient Spectroscopy
SEM — Scanning Electron Microscopy
SIMS — Secondary Ion Mass Spectrometry
D(c) — Concentration Dependent Diffusion Coefficient
D_{max} — Maximum Diffusion Coefficient

(f) — Fast Diffusion Component
(i) — Interstitial Diffusion Component
(s) — Slow Diffusion Component
(||) — Parallel to c Direction
(⊥) — Perpendicular to c Direction

Semiconductor	Diffusant	Frequency factor, D_o (cm²/s)	Activation energy, Q (eV)	Temperature range (°C)	Method of measurement	Ref.
Si	Li	2.5×10^{-3}	0.65	25—1350	Electrical	1
	Na	1.65×10^{-3}	0.77	800—1100	Electrical and flame photometry	2
	K	1.1×10^{-3}	0.76	800—1100	Electrical and flame photometry	2
	Cu	4×10^{-2}	1.0	800—1100	Radioactive	3
		4.7×10^{-3}	0.43 (i)	300—700	Radioactive	4
	Ag	2×10^{-3}	1.6	1100—1350	Radioactive	5
	Au	2.4×10^{-4}	0.39 (i)	700—1300	Radioactive	6
		2.75×10^{-3}	2.05 (s)			
	Be	($D \sim 10^{-7}$)	—	1050	Electrical	7
	Zn	1×10^{-1}	1.4	980—1270	Electrical	8
	B	2.46	3.59	1100—1250	Electrical	9
		2.4×10^{1}	3.87	840—1250	Electrical	10
	Al	1.38	3.41	1119—1390	Electrical	11
		1.8	3.2	1025—1175	Electrical	12
	Ga	3.74×10^{-1}	3.39	1143—1393	Electrical	11
		6×10^{1}	3.89	900—1050	Radioactive	13
	In	7.85×10^{-1}	3.63	1180—1389	Electrical	11
		1.94×10^{1}	3.86	1150—1242	Radioactive	14
	Tl	1.37	3.7	1244—1338	Electrical	11
		1.65×10^{1}	3.9	1105—1360	Electrical	15
	Ti	1.45×10^{-2}	1.79	950—1200	DLTS	16
	C	3.3×10^{-1}	2.92	1070—1400	Radioactive	17
	Si (self)	1.54×10^{2}	4.65	855—1175	SIMS	18
		1.6×10^{3}	4.77	1200—1400	Radioactive	19
	Ge	3.5×10^{-1}	3.92	855—1000	Radioactive	20
		2.5×10^{3}	4.97	1030—1302	Radioactive	20
		7.55×10^{3}	5.08	1100—1300	SIMS	21
	Sn	3.2×10^{1}	4.25	1050—1200	Neutron activation	22
	P	2.02×10^{1}	3.87	1100—1250	Electrical	9
		1.1	3.4	900—1200	Radioactive	23
		7.4×10^{-2}	3.3	1130—1405	Electrical	24
	As	6.0×10^{1}	4.2	950—1350	Radioactive	25
		6.55×10^{-2}	3.44	1167—1394	Electrical	26
		2.29×10^{1}	4.1	900—1250	Electrical	27
	Sb	1.29×10^{1}	3.98	1190—1398	Radioactive	28
		2.14×10^{-1}	3.65	1190—1405	Electrical	26
	Bi	1.03×10^{3}	4.64	1220—1380	Electrical	15
		1.08	3.85	1190—1394	Electrical	26

Semiconductor	Diffusant	Frequency factor, D_o (cm²/s)	Activation energy, Q (eV)	Temperature range (°C)	Method of measurement	Ref.
	Cr	1×10^{-2}	1	1100—1250	Radioactive	29
	O	7×10^{-2}	2.44	700—1250	SIMS	30
		1.4×10^{-1}	2.53	700—1160	SIMS	31
	S	5.95×10^{-3}	1.83	975—1200	Radioactive	32
	Se	9.5×10^{-1}	2.6	1050—1250	Electrical	33
	Te	($D < 10^{-11}$)	—	1050—1250	Electrical	33
	Mn	6.9×10^{-4}	0.63	900—1200	Radioactive	34
	Fe	1.3×10^{-3}	0.68	30—1250	Radioactive	35
	Co	2×10^{-3}	0.69	700—1300	Radioactive	36
	Ni	2×10^{-3}	0.47	800—1300	Radioactive	37
	Rh	($D \sim 10^{-6}$—10^{-4})	—	1000—1200	Electrical	38
	Os	($D \sim 2 \times 10^{-6}$)	—	1280	Electrical	39
	Ir	4.2×10^{-2}	1.3	950—1250	Electrical	40
	Pt	($D \sim 9.8 \times 10^{-16}$— 3.6×10^{-14})	—	700—800	DLTS	41
Ga	Li	1.3×10^{-3}	0.46	350—800	Electrical	42
		9.1×10^{-3}	0.57	800—500	Electrical	43
	Na	3.95×10^{-1}	2.03	700—850	Radioactive	44
	Cu	1.9×10^{-4}	0.18 (i)	750—900	Radioactive	45
		4×10^{-2}	0.99 (s)	600—700		
		4×10^{-3}	0.33 (i)	350—750	Radioactive	4
	Ag	4.4×10^{-2}	1.0 (i)	700—900	Radioactive	46, 47
		4×10^{-2}	2.23 (s)	800—900	Radioactive	48
	Au	2.25×10^{2}	2.5	600—900	Radioactive	49
	Be	5×10^{-1}	2.5	720—900	Electrical	50
	Zn	5	2.7	600—900	Radioactive and electrical	51
	Cd	1.75×10^{9}	4.4	760—915	Radioactive	52
	B	1.8×10^{9}	4.55	600—900	Electrical	51
	Al	1.0×10^{3}	3.45	554—905	SIMS	53
		$\sim 1.6 \times 10^{2}$	~3.24	750—850	Electrical	54
	Ga	1.4×10^{2}	3.35	554—916	SIMS	55
		3.4×10^{1}	3.1	600—900	Electrical	51
	In	1.8×10^{4}	3.67	554—919	SIMS	56
		3.3×10^{1}	3.02	700—855	Radioactive	57
	Tl	1.7×10^{3}	3.4	800—930	Radioactive	58
	Si	2.4×10^{-1}	2.9	650—900	(γ) resonance	59
	Ge (self)	2.48×10^{1}	3.14	549—891	Radioactive	60
		7.8	2.95	766—928	Radioactive	61
	Sn	1.7×10^{-2}	1.9	—	Radioactive	45
	P	3.3	2.5	600—900	Electrical	51
	As	2.1	2.39	700—900	Electrical	62
	Sb	3.2	2.41	700—855	Radioactive	57
		1.0×10^{1}	2.5	600—900	Radioactive and electrical	51
	Bi	3.3	2.57	650—850		63
	O	$4 \times 10^{[illegible]}$	2.08	—	Optical	64
	S	($D \sim 10^{-9}$)	—	920	—	65
	Se	($D \sim 10^{-10}$)	—	920	—	65
	Te	5.6	2.43	750—900	Radioactive	66
	Fe	1.3×10^{-1}	1.08	750—900	Radioactive	67
	Co	1.6×10^{-1}	1.12	750—850	Radioactive	47
	Ni	8×10^{-1}	0.9	670—900	Electrical	68
GaAs	Li	5.3×10^{-1}	1.0	250—500	Electrical and chemical	69
	Cu	3×10^{-2}	0.53	100—500	Radioactive	70

Semiconductor	Diffusant	Frequency factor, D_o (cm²/s)	Activation energy, Q (eV)	Temperature range (°C)	Method of measurement	Ref.
		6×10^{-2}	0.98	450—750	Ultrasonic	71
		1.5×10^{-3}	0.6	800—1000	Radioactive	72
	Ag	4×10^{-4}	0.8	500—1150	Radioactive	73
	Au	1×10^{-3}	1.0	740—1025	Radioactive	74
	Be	7.3×10^{-6}	1.2	800—990	Electrical	75
	Mg	4×10^{-5}	1.22	800—1200	Electrical	76
	Zn	1.5×10^{1}	2.49	600—980	Radioactive	77
		2.5×10^{-1}	3.0	750—1000	Radioactive	78
	Cd	1.3×10^{-3}	2.2	800—1100	Radioactive	79
		5×10^{-2}	2.43	868—1149	Radioactive	77
	Hg	($D \sim 5 \times 10^{-14}$)	—	1100	Radioactive	80
	Al	($D \sim 4 \times 10^{-18}$—10^{-14})	4.3	850—1100	AES	81
	Ga (self)	4×10^{-5}	2.6	1025—1100	Radioactive	82
		1×10^{7}	5.6	1125—1230	Radioactive	83
	Si	1.1×10^{-1}	2.5	850—1050	SIMS	84
	Ge	1.6×10^{-5}	2.06	650—850	SIMS	85
	Sn	6×10^{-4}	2.5	1060—1200	Radioactive	86
		1×10^{-5}	2	800—1000	Radioactive	87
	P	($D \sim 10^{-12}$—10^{-10})	2.9	800—1150	Reflectance measurements	88
	As (self)	7×10^{-1}	3.2	—	Radioactive	78
	Cr	2.04×10^{-6}	0.83 (f)	750—1000	SIMS	89
			1.7 (s)	700—900		
		7.9×10^{-3}	2.2	800—1100	Chemical analysis	90
	O	2×10^{-3}	1.1	700—900	Mass spectroscopy	91
	S	1.85×10^{-2}	2.6	1000—1300	Radioactive	92
		1.1×10^{1}	2.95	750—900	Electrical	93
	Se	3×10^{3}	4.16	1025—1200	Radioactive	83
	Te	1.5×10^{-1}	3.5	1000—1150	Radioactive	94
	Mn	6.5×10^{-1}	2.49	850—1100	Radioactive	95
	Fe	4.2×10^{-2}	1.8	850—1150	Radioactive	96
		2.2×10^{-3}	2.32	750—1050	Radioactive	97
	Co	5×10^{2}	2.5	800—1000	Radioactive	98
		1.2×10^{-1}	2.64	750—1050	Radioactive	97
	Tm	2.3×10^{-16}	1.0	800—1000	Radioactive	99
GaSb	Li	2.3×10^{-4}	1.9 (s)	527—657	Electrical and flame photometry	100
		1.2×10^{-1}	0.7 (f)	277—657		
	Cu	4.7×10^{-3}	0.9	470—650	Radioactive	101
	Zn	($D \sim 2 \times 10^{-13}$—1×10^{-11})	2	510—600	Radioactive	102
	Cd	1.5×10^{-6}	0.72	640—800	Electrical	103
	Ga (self)	3.2×10^{3}	3.15	658—700	Radioactive	104
	In	1.2×10^{-7}	0.53	320—650	Radioactive	105
	Sn	2.4×10^{-5}	0.8	320—650	Radioactive	106
		1.3×10^{-5}	1.1	500—650	Radioactive	107
	Sb (self)	3.4×10^{4}	3.45	658—700	Radioactive	104
	Se	($D \sim 2.4 \times 10^{-13}$—$1.37 \times 10^{-11}$)	—	400—500	Radioactive	106
	Te	3.8×10^{-4}	1.20	320—650	Radioactive	106
	Fe	5×10^{-2}	1.9 (I)	500—650	Radioactive	108
		5×10^{2}	2.3 (II)	500—650		
GaP	Ag	—	—	1000—1300	Radioactive	109
	Au	8	2.5 (I)	1050—1250	Radioactive	110
		20	2.4 (II)	1100—1250	Diffusion (I) A face and (II) B face	

DIFFUSION DATA FOR SEMICONDUCTORS (continued)

Semiconductor	Diffusant	Frequency factor, D_o (cm^2/s)	Activation energy, Q (eV)	Temperature range (°C)	Method of measurement	Ref.
	Be	($D_{max} \sim 2.4 \times 10^{-9}$— 8.5×10^{-8})	—	900—1000	Atomic absorption analysis	111
	Mg	5×10^{-5}	1.4	700—1050	Electrical	112
	Zn	1.0	2.1	700—1300	Radioactive	113
	Ge	—	—	900—1000	Radioactive	114
	Cr	6.2×10^{-4}	1.2	900—1130	Radioactive; ESR	115
	S	3.2×10^{3}	4.7	1120—1305	Radioactive	92
	Mn	2.1×10^{9}	4.7	$T < 950$	Radioactive; ESR	116
		1.1×10^{-6}	0.9	950—1130		
	Fe	1.6×10^{-1}	2.3	980—1180	Radioactive	117
	Co	2.8×10^{-3}	2.9	850—1100	Radioactive	118
InP	Cu	3.8×10^{-3}	0.69	600—900	Radioactive	119
	Ag	3.6×10^{-4}	0.59	500—900	Radioactive	120
	Au	1.32×10^{-5}	0.48	600—820	Radioactive	121
		1.37×10^{-4}	0.73	600—900	Radioactive	122
	Zn	1.6×10^{-8}	0.3	750—900	Electrical	123
		($D \sim 2 \times 10^{-9}$— 4×10^{-8})	—	700—900	Radioactive	124
	Cd	1.8	1.9	700—900	Radioactive	125
		1.1×10^{-7}	0.72	700—900	Electrical	123
		($D \sim 7 \times 10^{-13}$— 2×10^{-10})	—	450—650	Electrical	126
	In (self)	1×10^{5}	3.85	830—990	Radioactive	83
	Sn	($D \sim 3 \times 10^{-8}$)	—	550	Etching and cathodo-luminescence	127
	P (self)	7×10^{10}	5.65	900—1000	Radioactive	83
	Cr	—	—	600—900	Radioactive	128
	S	—	—	607—710	Electrical	129
	Mn	—	2.9	650—750	SIMS	130
	Fe	3	2	600—950	Radioactive	131
		6.8×10^{5}	3.4	600—700	SIMS	132
	Co	9×10^{-1}	1.8	600—950	Radioactive	131
InAs	Cu	3.6×10^{-3}	0.52	342—875	Radioactive	133
		2.2×10^{-2}	0.54	525—890	Radioactive	134
	Ag	7.3×10^{-4}	0.26	450—900	Radioactive	135
	Au	5.8×10^{-3}	0.65	600—900	Radioactive	136
	Mg	1.98×10^{-6}	1.17	600—900	Electrical	137
	Zn	4.2×10^{-3}	0.96	600—900	Radioactive	138
		3.11×10^{-3}	1.17	600—900	Electrical	137
	Cd	7.4×10^{-4}	1.15	650—900	Radioactive	139
	Hg	1.45×10^{-5}	1.32	650—850	Radioactive	140
	In (self)	6×10^{5}	4.0	740—900	Radioactive	141
	Ge	3.74×10^{-6}	1.17	600—900	Electrical	137
	Sn	1.49×10^{-6}	1.17	600—900	Electrical	137
	As (self)	3×10^{7}	4.45	740—900	Radioactive	141
	S	6.78	2.2	600—900	Electrical	137
	Se	12.6	2.2	600—900	Electrical	137
	Te	3.43×10^{-5}	1.28	600—900	Electrical	137
InSb	Li	7×10^{-4}	0.28	0—210	Electrical	142
	Cu	9×10^{-4}	1.08	200—500	Radioactive	143
		3×10^{-5}	0.37	230—490	Radioactive	144
	Ag	1×10^{-7}	0.25	440—510	Radioactive	145
	Au	7×10^{-4}	0.32	140—510	Radioactive	144
	Zn	5×10^{-1}	1.35	362—508	Radioactive	146
		—	1.5	355—455	SIMS	147
	Cd	1×10^{-5}	1.1	250—500	Radioactive	148

Semiconductor	Diffusant	Frequency factor, D_o (cm^2/s)	Activation energy, Q (eV)	Temperature range (°C)	Method of measurement	Ref.
		1.3×10^{-4}	1.2	360—500	Electrical	149
	Hg	4×10^{-6}	1.17	425—500	Radioactive	150
	In (self)	5×10^{-2}	1.82	478—520	Radioactive	104
		1.8×10^{13}	4.3	475—517	Radioactive	151
	Sn	5.5×10^{-8}	0.75	390—512	Radioactive	152
	Pb	($D \sim 2.7 \times 10^{-15}$)	—	500	Radioactive	153
	Sb (self)	5×10^{-2}	1.94	478—520	Radioactive	104
		3.1×10^{13}	4.3	475—517	Radioactive	151
	S	9×10^{-2}	1.4	360—500	Electrical	154
	Se	1.6	1.87	380—500	Electrical	155
	Te	1.7×10^{-7}	0.57	300—500	Radioactive	156
	Fe	1×10^{-7}	0.25	440—510	Radioactive	145
	Co	2.7×10^{-11}	0.39	420—500	Radioactive	157
AlAs	Ga	($D \sim 2 \times 10^{-18}$—10^{-15})	3.6	850—1100	AES	81
	Zn	($D \sim 9 \times 10^{-11}$)	—	557	SEM	158
AlSb	Cu	3.5×10^{-3}	0.36	150—500	Radioactive	159
	Zn	3.3×10^{-1}	1.93	660—860	Radioactive	160
	Cd	D(c) $\sim 4 \times 10^{-12}$—3×10^{-10}	—	900	Radioactive	161
	Al (self)	2	1.88	570—620	X-ray	162
	Sb (self)	1	1.7	570—620	X-ray	162
ZnS	Cu	2.6×10^{-3}	0.79	470—750	Radioactive	163
		4.3×10^{-4}	0.64	250—1200	Electroluminescence	164
		9.75×10^{-3}	1.04	400—800	Luminescence	165
	Au	1.75×10^{-4}	1.16	500—800	Radioactive	166
	Zn (self)	3×10^{-4}	1.5	$925<T<940$	Radioactive	167
		1.5×10^{4}	3.26	$940<T<1030$		
		1×10^{16}	6.5	$1030<T<1075$		
	Cd	($D \sim 10^{-10}$)	—	1100	Luminescence	168
	Al	5.69×10^{-4}	1.28	800—1000	Luminescence	167
	In	3×10^{1}	2.2	750—1000	Radioactive	163
	S (self)	2.16×10^{4}	3.15	600—800	Radioactive	166
		8×10^{-5}	2.2	740—1100	Radioactive	169
	Se	($D \sim 5 \times 10^{-13}$)	—	1070	X-ray microprobe	170
	Mn	2.3×10^{3}	2.46	500—800	Radioactive	166
ZnSe	Cu	1×10^{-4}	0.66	400—800	Luminescence	171
		1.7×10^{-5}	0.56	200—570	Radioactive	172
	Ag	2.2×10^{-2}	1.18	400—800	Luminescence	171
	Zn (self)	9.8	3.0	760—1150	Radioactive	173
	Cd	6.39×10^{-4}	1.87	700—950	Photoluminescence	174
	Al	2.3×10^{-2}	1.8	800—1100	Luminescence	171
	Ga	1.81×10^{2}	3.0	900—1100	Luminescence	165
		—	1.3	700—850	Electron probe	175
	In	($D \sim 2 \times 10^{-12}$)	—	940	—	176
	S	($D \sim 8 \times 10^{-12}$)	—	1060	X-ray microprobe	170
	Se (self)	1.3×10^{1}	2.5	860—1020	Radioactive	173
		2.3×10^{-1}	2.7	1000—1050	Radioactive	177
	Ni	($D \sim 1.5 \times 10^{-8}$—$1.7 \times 10^{-7}$)	—	740—910	Luminescence	165
ZnTe	Li	2.9×10^{-2}	1.22 (s)	400—700	Nuclear and chemical analysis	178
		1.7×10^{-4}	0.78 (f)			
	Zn (self)	2.34	2.56	760—860	Radioactive	179
		1.4×10^{1}	2.69	667—1077	Radioactive	180
	Al	—	2.0	700—1000	Electrical and optical	181

DIFFUSION DATA FOR SEMICONDUCTORS (continued)

Semiconductor	Diffusant	Frequency factor, D_o (cm²/s)	Activation energy, Q (eV)	Temperature range (°C)	Method of measurement	Ref.
	In	4	1.96	1100—1300	Radioactive	182
	Te (self)	2×10^{4}	3.8	727—977	Radioactive	180
CdS	Li	3×10^{-6}	0.68	610—960	Microhardness	183
	Na	($D \sim 3 \times 10^{-7}$)	—	800	Radioactive	184
	Cu	1.5×10^{-3}	0.76	400—700	Radioactive	185
		1.2×10^{-2}	1.05	300—700	Ultrasonic	186
		8×10^{-5}	0.72	20—200	Electrical	187
	Ag	2.5×10^{1}	1.2 (s)	300—500	Radioactive	188
		2.4×10^{-1}	0.8 (f)			
	Au	2×10^{2}	1.8	500—800	Radioactive	189
	Zn	1.27×10^{-9}	0.86 (s)	720—1000	Radioactive	190
		1.22×10^{-8}	0.66 (f)			
	Cd (self)	3.4	2.0	700—1100	Radioactive	191
	Ga	—	—	667—967	Optical and microprobe	192
	In	6×10^{1}	2.3 (∥)	650—930	Radioactive, optical and microprobe	193
		1×10^{1}	2.03 (⊥)			
	P	6.5×10^{-4}	1.6	800—1100	Radioactive	194
	S (self)	1.6×10^{-2}	2.05	800—900	Radioactive	195
		—	2.4	750—1050	Radioactive	196
	Se	($D \sim 1.2 \times 10^{-9}$)	—	900	Radioactive	177
	Te	1.3×10^{-7}	10.4	700—1000	Radioactive	197
	Cl	($D \sim 3 \times 10^{-10}$)	—	800	Electrical	184
	I	($D \sim 5 \times 10^{-12}$)	—	1000	Radioactive	184
	Ni	6.75×10^{-3}	10.9	570—900	Luminescence	165
	Yb	($D \sim 1.3 \times 10^{-9}$)	—	960	Photoluminescence	198
CdSe	Ag	2×10^{-4}	0.53	22—400	Ultrasonic	199
	Cd (self)	1.6×10^{-3}	1.5	700—1000	Radioactive	200
		6.3×10^{-2}	1.25 (I)	600—900	Radioactive;	201
		4.12×10^{-2}	2.18 (II)	600—900	(I) saturated Cd and (II) saturated Se pressure	
	P	($D \sim 5.3 \times 10^{-12}$— 6×10^{-11})	—	900—1000	Radioactive	202
	Se (self)	2.6×10^{3}	1.55	700—1000	Radioactive; saturated Se pressure	177
CdTe	Li	($D \sim 1.5 \times 10^{-10}$)	—	300	Ion microprobe	203
	Cu	3.7×10^{-4}	0.67	97—300	Radioactive	204
		8.2×10^{-8}	0.64	290—350	Ion backscattering	205
	Ag	—	—	700—800	Electrical and photo-luminescence	206
	Au	6.7×10^{1}	2.0	600—1000	Radioactive	207
	Cd (self)	1.26	2.07	700—1000	Radioactive	208
		3.26×10^{2}	2.67 (I)	650—900	Radioactive;	209
		1.58×10^{1}	2.44 (II)		(I) saturated Cd and (II) saturated Te pressure	
	In	8×10^{-2}	1.61	650—1000	Radioactive	210
		$1.17 \times {}^{2}$	2.21 (I)	500—850	Radioactive; (I) saturated Cd and (II) saturated Te pressure	211
		6.48×10^{-4}	1.15 (II)			
	Sn	8.3×10^{-2}	2.2	700—925	Radioactive	212
	P	($D \sim 1.2 \times 10^{-10}$)	—	900	Radioactive	202
	As	—	—	850	—	213
	O	5.6×10^{-9}	1.22	200—650	Mass spectrometry	214
		6.0×10^{-10}	0.29	650—900		
	Se	1.7×10^{-4}	1.35	700—1000	Radioactive	215

Semiconductor	Diffusant	Frequency factor, D_o (cm^2/s)	Activation energy, Q (eV)	Temperature range (°C)	Method of measurement	Ref.
	Te (self)	8.54×10^{-7}	1.42 (I)	600—900	Radioactive; (I) saturated Cd and (II) saturated Te pressure	209
		1.66×10^{-4}	1.38 (II)	500—800		
	Cl	7.1×10^{-2}	1.6	520—800	Radioactive	216
	Fe	$(D \sim 4 \times 10^{-8})$	0.77	900	Radioactive	217
HgSe	Sb	6.3×10^{-5}	0.85	540—630	Radioactive	218
	Se (self)	—	—	200—400	Radioactive	219
HgTe	Ag	6×10^{-4}	0.8	250—350	Radioactive	220
	Zn	5×10^{-8}	0.6	250—350	Radioactive	220
	Cd	3.1×10^{-4}	0.66	250—350	Radioactive	221
	Hg (self)	2×10^{-8}	0.6	200—350	Radioactive	220
	In	6×10^{-6}	0.9	200—300	Radioactive	220
	Sn	1.72×10^{-6}	0.66 (s)	200—300	Radioactive	222
		1.8×10^{-3}	0.80 (f)			
	Te (self)	10^{-6}	1.4	200—400	Radioactive	220
	Mn	1.5×10^{-4}	1.3	250—350	Radioactive	220
PbS	Cu	4.6×10^{-4}	0.36	150—450	Electrical	223
		5×10^{-3}	0.31	100—400	Electrical	224
	Pb (self)	8.6×10^{-5}	1.52	500—800	Radioactive	225
	S (self)	6.8×10^{-5}	1.38	500—750	Radioactive	226
	Ni	1.78×10^{1}	0.95	200—500	Electrical	227
PbSe	Na	1.5×10^{1}	1.74 (s)	400—850	Radioactive	228
		5.6×10^{-6}	0.4 (f)			
	Cu	2×10^{-5}	0.31	93—520	Radioactive	229
	Ag	7.4×10^{-4}	0.35	400—850	Radioactive	228
	Pb (self)	4.98×10^{-6}	0.83	400—800	Radioactive	230
	Sb	3.4×10^{-1}	2.0	650—850	Radioactive	231
	Se (self)	2.1×10^{-5}	1.2	650—850	Radioactive	231
	Cl	1.6×10^{-8}	0.45	400—850	Radioactive	232
	Ni	$(D \sim 1 \times 10^{-10})$	—	700	Radioactive	233
PbTe	Na	1.7×10^{-1}	1.91	600—850	Radioactive	234
	Sn	3.1×10^{-2}	1.56	500—800	Radioactive	231
	Pb (self)	2.9×10^{-5}	0.6	250—500	Radioactive	231
	Sb	4.9×10^{-2}	1.54	500—800	Radioactive	231
	Te	2.7×10^{-6}	0.75	500—800	Radioactive	231
	Cl	$(D > 2.3 \times 10^{-10})$	—	700	Radioactive	235
	Ni	$(D > 1 \times 10^{-6})$	—	700	Radioactive	235

REFERENCES

1. E. M. Pell, *Phys. Rev.*, 119, 1960; 119, 1014, 1960.
2. L. Svob, *Solid State Electron,* 10, 991, 1967.
3. B. I. Boltaks and I. I. Sosinov, *Zh. Tekh. Fiz.*, 28, 3, 1958.
4. R. N. Hall and J. N. Racette, *J. Appl. Phys.*, 35, 379, 1964.
5. B. I. Boltaks and Hsueh Shih-Yin, *Sov. Phys. Solid State*, 2, 2383, 1961.
6. W. R. Wilcox and T. J. LaChapelle, *J. Appl. Phys.*, 35, 240, 1964.
7. E. A. Taft and R. O. Carlson, *J. Electrochem. Soc.*, 117, 711, 1970.
8. R. Sh. Malkovich and N. A. Alimbarashvili, *Sov. Phys. Solid State,* 4, 1725, 1963.
9. R. N. Ghoshtagore, *Solid State Electron,* 15, 1113, 1972.
10. C. Hill, *Semiconductor Silicon 1981,* H. R. Huff, R. J. Kreiger, and Y. Takeishi, Eds., p. 988, *Electrochem. Soc.,* 1981.
11. R. N. Ghoshtagore, *Phys. Rev. B,* 3, 2507, 1971.
12. W. Rosnowski, *J. Electrochem. Soc.*, 125, 957, 1978.

13. J. S. Makris and B. J. Masters, *J. Appl. Phys.,* 42, 3750, 1971.
14. M. F. Millea, *J. Phys. Chem. Solids,* 27, 315, 1965 (refer Reference 2).
15. C. S. Fuller and J. A. Ditzenberger, *J. Appl. Phys.,* 27, 544, 1956.
16. S. Hocine and D. Mathiot, *Appl. Phys. Lett.,* 53, 1269, 1988.
17. R. C. Newman and J. Wakefield, *J. Phys. Chem. Solids,* 19, 230, 1961.
18. L. Kalinowski and R. Seguin, *Appl. Phys. Lett.,* 35, 211, 1979; *Appl. Phys. Lett.,* 36, 171, 1980.
19. R. F. Peart, *Phys. Stat. Sol.,* 15, K 119, 1966.
20. G. Hettich, H. Mehrer and K. Maler, *Inst. Phys. Conf. Ser.,* 46, 500, 1979.
21. M. Ogina, Y. Oana and M. Watanabe, *Phys. Stat. Sol. (a),* 72, 535, 1982.
22. T. H. Yeh, S. M. Hu, and R. H. Kastl, *Appl. Phys.,* 39, 4266, 1968.
23. I. Franz and W. Langheinrich, *Solid State Electron,* 14, 835, 1971.
24. R. N. Ghoshtagore, *Hys. Rev. B,* 3, 389, 1971.
25. B. J. Masters and J. M. Fairfield, *J. Appl. Phys.,* 40, 2390, 1969.
26. R. N. Goshtagore, *Phys. Rev. B,* 3, 397, 1971.
27. R. S. Fair and J. C. C. Tsai, *J. Electrochem. Soc.,* 122, 1689, 1975.
28. J. J. Rohan, N. E. Pickering and J. Kennedy, *J. Electrochem. Soc.,* 106, 705, 1969.
29. W. Wuerker, K. Roy, and J. Hesse, *Matsr. Res. Bull.,* 9, 971, 1974.
30. J. C. Mikkelsen, Jr., *Appl. Phys. Lett.,* 40, 336, 1982.
31. S. Tang Lee and D. Nicols, *Appl. Phys. Lett.,* 47, 1001, 1985.
32. P. L. Gruzin, S. V. Zemskii, A. D. Bullkin, and N. M. Makarov, *Sov. Phys. Sem.,* 7, 1241, 1974.
33. N. S. Zhdanovich and Yu. I. Kozlov, *Svoistva Legir, Poluprovodn.,* V. S. Zemskov, Ed., Nauka, Moscow, 1977, 115-120; *Fiz Tekh. Poluprovod.,* 9, 1594, 1975.
34. D. Gilles, W. Bergholze, and W. Schroeter, *J. Appl. Phys.,* 59, 3590, 1986.
35. E. R. Weber, *Appl. Phys. A,* 30, 1, 1983.
36. E. R. Weber, Properties of Silicon, EMIS Datareviews Ser. No. 4, INSPEC Publications, 1988, 409-451.
37. M. K. Bakhadyrkhanov, S. Zainabidinov, and A. Khamidov, *Sov. Phys. Sem.,* 14, 243, 1980.
38. S. A. Azimov, M. S. Yunosov, F. K. Khatamkulov, and G. Nasyrov, *Poluprovod.,* N. Kh. Abrikosov and V. S. Zemskov, Eds., Nauka, Moscow, 1975, 21-23.
39. S. A. Azimov, M. S. Yunosov, G. Nurkuziev, and F. R. Karimov, *Sov. Phys. Sem.,* 12, 981, 1978.
40. S. A. Azimov, B. V. Umarov, and M. S. Yunusov, *Sov. Phys. Sem.,* 10, 842, 1976.
41. S. Mantovani, F. Nava, C. Nobili, and G. Ottaviani, *Phys. Rev. B,* 33, 5536, 1986.
42. C. S. Fuller and J. A. Ditzenberger, *Phys. Rev.,* 91, 193, 1953.
43. B. Pratt and F. Friedman, *J. Appl. Phys.,* 37, 1893, 1966.
44. M. Stojic, V. Spiric, and D. Kostoski, *Inst. Phys. Conf. Ser.,* 31, 304, 1976.
45. B. I. Boltaks, *Diffusion in Semiconductors,* Inforsearch, London, 1963, 162.
46. A. A. Bugai, V. E. Kosenko, and E. G. Miselyuk, *Zh. Tekh. Fiz.,* 27, 67, 1957.
47. L. Y. Wei, *J. Phys. Chem. Solids,* 18, 162, 1961.
48. V. E. Kosenko, *Sov. Phys. Solid State,* 4, 42, 1962.
49. W. C. Dunlap, Jr., *Phys. Rev.,* 97, 614, 1955
50. Yu. I. Belyaev and V. A. Zhidkov, *Sov. Phys. Solid State,* 3, 133, 1961.
51. W. C. Dunlap, Jr. *Phys. Rev.,* 94, 1531, 1954.
52. V. E. Kosenko, *Sov. Phys. Solid State,* 1, 1481, 1960.
53. P. Dorner, W. Gust, A. Lodding, H. Odelius, B. Predel, and U. Roll, *Acta Metall.,* 30, 941, 1982.
54. W. Meer and D. Pommerrening, *Z. Agnew. Phys.,* 23, 369, 1967.
55. U. Sodervall, H. Odelius, A. Lodding, U. Roll, B. Predel, W. Gust, and P. Dorner, *Phil. Mag. A,* 54, 539, 1986.
56. P. Dorner, W. Gust, A. Lodding, H. Odelius, B. Predel, and U. Roll, *Z. Metalkd.,* 73, 325, 1982.
57. P. V. Pavlov, *Sov. Phys. Solid State,* 8, 2377, 1967.
58. V. I. Tagirov and A. A. Kuliev, *Sov. Phys. Solid State,* 4, 196, 1962.
59. J. Raisanen, J. Hirvonen, and A. Anttila, *Solid State Electron.,* 24, 333, 1981.
60. G. Vogel, G. Hettich, and H. Mehrer, *J. Phys. C,* 16, 6197, 1983.
61. H. Letaw, Jr., W. M. Portnoy, and L. Slifkin, *Phys. Rev.,* 102, 363, 1956.
62. W. Bosenberg, *Z. Naturforsch.,* 10a, 285, 1955.
63. V. M. Glazov and V. S. Zemskov, Physicochemical Principles of Semiconductor Doping, Israel Program for Scientific Translation, Jerusalem, 1968.
64. J. W. Corbett, R. S. McDonald, and G. D. Watkins, *J. Phys. Chem. Solids,* 25, 873, 1964.
65. W. W. Tyler, *J. Phys. Chem. Solids,* 8, 59, 1959.
66. V. D. Ignatkov and V. E. Kosenko, *Sov. Phys. Solid State,* 4, 1193, 1962.
67. A. A. Bugal, V. E. Kosenko, and E. G. Miseluk, *Zh. Tekh. Fiz.,* 27, 210, 1957.
68. F. van der Maesen and J. A. Brenkman, *Phillips Res. Rep.,* 9, 255, 1954.
69. C. S. Fuller and K. B. Wolfstrin, *J. Appl. Phys.,* 33, 2507, 1962.

70. R. N. Hall and J. H. Racette, *J. Appl. Phys.*, 25, 379, 1964.
72. V. S. Vasilev, I. N. Kamevoskii, and V. B. Osvenskii, *Sov. Phys. Sem.*, 2, 1495, 1969.
73. B. I. Boltaks and F. S. Shishiyanu, *Sov. Phys. Solid State*, 5, 1680, 1964.
74. V. I. Sokolov and F. S. Shishiyanu, *Sov. Phys. Solid State*, 6, 265, 1964.
75. E. A. Poltoratskii and V. M. Stuchebnikov, *Sov. Phys. Solid State*, 8, 770, 1966.
76. E. I. Andreivckii, S. V. Mashkin, and S. S. Zludkov, *Diffusion in Semiconductor*, (Russian), Gorkii, 1969.
77. B. Goldstein, *Phys. Rev.*, 118, 1024, 1960.
78. D. L. Kendall, *Semiconductors and Semimetals*, Vol. 4, R. K. Willardson and A. C. Beer, Eds., Academic, 163-259.
79. M. Fujimoto, K. Kudo, and N. Hishinuma, *Jpn. J. Appl. Phys.*, 8, 725, 1969.
80. J. A. Kanz, (p. 227 Reference 78).
81. L. L. Chang and A. Koma, *Appl. Physics Lett.*, 29, 138, 1976.
82. H. D. Palfrey, M. Brown, and A. F. W. Willoughby, *J. Electrochem. Soc.*, 128, 2224, 1981.
83. B. Goldstein, *Phys. Rev.*, 121, 1305, 1961.
84. M. E. Greiner and J. B. Gibbons, *Appl. Phys. Lett.*, 44, 750, 1984.
85. K. Sarma, R. Dalby, K. Rose, O. Aina, W. Katz, and N. Lewis, *J. Appl. Phys.*, 56, 2703, 1984.
86. S. Goldstein and H. Keller, *J. Appl. Phys.*, 32, 1180, 1961.
87. H. Yamazaki, Y. Kawaski, M. Futjimoto, and K. Kudo, *Jpn. J. Appl. Phys.*, 14, 717, 1975.
88. G. C. Jain, D. K. Sadana, and B. K. Das, *Solid State Electron.*, 19, 731, 1976.
89. M. D. Deal and D. A. Stevenson, *J. Appl. Phys.*, 59, 2398, 1986.
90. S. S. Khludkov, G. L., Prikhodko, and T. A. Karchina, *Izv. Akad. Nauk SSR, Heorg. Mater.*, 8, 1044, 1972.
91. J. Rachmann and R. Biermann, *Solid State Commun.*, 7, 1771, 1969.
92. A. S. Young and G. L. Pearson, *J. Phys. Chem. Solids*, 31, 517, 1970.
93. H. Matino, *Solid State Electron.*, 17, 35, 1974.
94. T. A. Karelina, T. T. Lavrischev, G. L. Prokhodko, and S. S. Khludkov, *Izv. Akad. Nauk SSSR, Neorg. Mater.*, 10, 228, 1974.
95. M. S. Seltzer, *J. Phys. Chem. Solids*, 26, 243, 1965.
96. B. I. Boltaks, G. S. Kulikov, I. N. Nikulista, and F. S. Shishiyanu, *Inorg. Mater.*, 11, 292, 1975.
97. V. A. Uskov and V. P. Sorvina, *Izv. Akad. Nauk SSSR, Neorg. Mater.*, 8, 758, 1972.
98. T. D. Dzhafarov, E. A. Skoryatina, E. S. Guds, and I. E. Moronchuk, *Phys. Stat. Sol. (a)*, 51, K221, 1979.
99. H. C. Casey, Jr. and G. L. Pearson, *J. Appl. Phys.*, 35, 3401, 1965.
100. M. H. van Maaren, *Phys. Stat. Sol.*, 24, K125, 1967.
101. A. I. Blashku and T. D. Dzhafarov, *Sov. Phys. Solid State*, 15, 536, 1973.
102. A. I. Blashku, B. I. Boltaks, I. I. Burdiyan, T. D. Dzhafarov, and M. A. Rzaev, *Sov. Phys. Sem.*, 6, 402, 1972.
103. J. Bougnot, L. Szepessy, and S. F. DaCunka, *Phys. Stat. Sol.*, 26, K127, 1968.
104. F. H. Eissen and C. E. Birchenall, *Acta Met.*, 5, 265, 1957.
105. R. K. Willardson, *Ultrapurification of Materials*, (Conf. Report, Boston), Macmillan, 1962.
106. B. I. Boltaks and Yu. A. Gutorov, *Sov. Phys. Solid State*, 1, 930, 1960.
107. V. A. Uskov, *Fiz. i Tekhn. Poluprov.*, 8, 2414, 1974.
108. F. S. Shishiyanu, *Diffusion and Degradation in Semiconductor Materials*, (Russian), Kishinev, 1978, 55.
109. S. V. Shudyakov, *Sov. Phys. Sem.*, 15, 4, 1981.
110. T. D. Dzhafarov, A. A. Litvin, and S. V. Khudyakov, *Sov. Phys. Solid State*, 20, 152, 1978.
111. M. Ilegems and W. C. O'Mara, *J. Appl. Phys.*, 43, 1190, 1972.
112. F. S. Shishiyanu, *Diffusion and Degradation in Semiconductor Materials*, (Russian), Kishinev, 1978, 78.
113. H. W. Allison, *J. Appl. Phys.*, 34, 231, 1963.
114. M. Schneider and E. Nebaur, *Phys. Stat. Sol. (a)*, 32, 333, 1975.
115. V. I. Kirillov, N. N. Pribylov, S. I. Rembeza, and A. I. Spirin, *Sov. Phys. Sem.*, 12, 1342, 1978.
116. V. I. Kirillov, N. N. Pribylov, S. I. Rembeza, and A. I. Spirin, *Sov. Phys. Solid State*, 22, 1945, 1980.
117. F. S. Shishiyanu and V. G. Georgiu, *Sov. Phys. Sem.*, 10, 1301, 1976.
118. T. D. Dzhafarov, Yu. P. Demakov, and N. N. Pribylov, *Fiz. Tverd. Tela*, 17, 3110, 1975.
119. K. A. Arseni, *Sov. Phys. Solids State*, 10, 2263, 1969.
120. K. A. Arseni and B. I. Boltaks, *Sov. Phys. Solid State*, 10, 2190, 1969.
121. S. I. Rembeza, *Sov. Phys. Sem.*, 3, 519, 1969.
122. K. A. Arseni, *Sov. Phys. Sem.*, 2, 1454, 1969.
123. R. M. Kundukhov, S. G. Metreveli, and N. V. Siukaev, *Sov. Phys. Sem.*, 1, 765, 1967.
124. L. L. Chang and H. Y. Casey, Jr., *Solid State Electron*, 7, 481, 1964.
125. K. A. Arseni, B. I. Boltaks, V. I. Gordin, and Ya. A. Ugai, *Izv. Akad. Nauk SSSR, Neorg. Mater.*, 3, 1679, 1967.
126. K. Ohtsuka, T. Nishino, and Y. Hamakawa, *Jpn. J. Appl. Phys.*, 21, 1170, 1982.
127. A. K. Chin, I. Camlibel, B. V. Dutt, V. Swaminathan, W. A. Bonner, and A. A. Ballman, *Appl. Phys. Lett.*, 42, 901, 1983.
128. M. R. Brozel, E. J. Foulkes, and B. Tuck, *Phys. Stat. Sol. (a)*, 72, K159, 1982; *Electron. Lett.*, 17, 532, 1981.
129. T. Kawakami and M. Okamura, *Electron. Lett.*, 15, 503, 1979.
130. R. Chaplain, M. Gaunear, H. H. L'Haridon, and R. Rupert, *J. Appl. Phys.*, 58, 1803, 1985.

131. F. S. Shishiyanu, V. Gh. Gheorgiu, and S. K. Palazov, *Phys. Stat. Sol. (a)*, 40, 29, 1977.
132. R. E. Holmes and R. G. Wilson, *J. Appl. Phys.*, 52, 3396, 1981.
133. C. S. Fuller and K. B. Wolfstirn, *J. Electrochem. Soc.*, 114, 856, 1967.
134. B. I. Boltaks, S. I. Rembeza, and M. K. Bakhadyrkhanov, *Sov. Phys. Solid State*, 10, 432, 1968.
135. B. I. Boltaks, S. I. Rembeza, and B. L. Sharma, *Sov. Phys. Sem.*, 1, 196, 1967.
136. S. I. Rembeza, *Sov. Phys. Sem.*, 1, 516, 1967.
137. E. Schillman, *Z. Naturforsch.*, 11a, 472, 1956.
138. B. I. Boltaks and S. I. Rembeza, *Sov. Phys. Solid State*, 8, 2177, 1967.
139. K. A. Arseni, B. I. Boltaks and S. I. Rembeza, *Sov. Phys. Solid State*, 8, 2248, 1967.
140. B. L. Sharma, R. K. Purohit, and S. N. Mukerjee, *J. Phys. Chem. Solids*, 32, 1397, 1971.
141. H. Kato, M. Yokozawa, R. Kohara, Y. Okabayashi, and S. Takayanagi, *Solid State Electron.*, 12, 137, 1969.
142. T. T. Takabatake, H. Ikari, and Y. Uyeda, *Jpn. J. Appl. Phys.*, 5, 839, 966.
143. H. J. Stocker, *Phys. Rev.*, 130, 2160, 1963.
144. B. I. Boltaks and V. I. Sokolov, *Sov. Phys. Solid State*, 6, 600, 1964.
145. L. A. K. Watt and W. S. Chen, *Bull. Am. Phys. Soc.*, 7, 89, 1962.
146. B. Goldstein, *Properties of Elemental and Compound Semiconductors*, Interscience, 1960, 155.
147. K. Nishitani, K. Hagahama, and T. Murotani, *Jpn. J. Appl. Phys.*, 22, 836, 1983.
148. B. I. Boltaks and V. I. Sokolov, *Sov. Phys. Solid State*, 5, 785, 1963.
149. R. B. Wilson and E. L. Heasell, *Proc. Soc. (London)*, 79, 403, 1962.
150. I. A. Gusev and A. N. Murin, *Sov. Phys. Solid State*, 6, 1229, 1964.
151. D. L. Kendall and R. A. Huggins, *J. Appl. Phys.*, 40, 2750, 1969.
152. S. M. Sze and L. Y. Wei, *Phys. Rev.*, 124, 84, 1961.
153. D. L. Kendall, *Semiconductors and Semimetals*, Vol. 4, R. K. Willardson and A. C. Beer, Eds., Academic, 1968, 255.
154. G. I. Rekalova, U. Kebe, and L. A. Mezrinea, *Sov. Phys. Sem.*, 5, 685, 1971.
155. G. I. Rekalova, A. A. Shakov, and V. V. Gaurushko, *Sov. Phys. Sem.*, 2, 1452, 1969.
156. B. I. Boltaks and G. S. Kulikov, *Sov. Phys. Tech. Phys.*, 2, 67, 1957.
157. I. A. Gusev and A. N. Nurin, *Sov. Phys. Solid State*, 6, 2274, 1965.
158. S. W. Kirchoefer, N. Holonyak, J. J. Coleman, and P. D. Dapkus, *J. Appl. Phys.*, 53, 766, 1982.
159. R. H. Wieber, H. C. Gorton, and C. S. Peet, *J. Appl. Phys.*, 31, 608, 1960.
160. D. Shaw, P. Joans, and D. Hazelby, *Proc. Phys. Soc. (London)*, 80, 167, 1962.
161. D. Shaw and S. R. Showan, *Phys. Stat. Sol.*, 34, 475, 1969.
162. B. Y. Pines and E. F. Shaikovskii, *Sov. Phys. Solid State*, 1, 864, 1959.
163. H. Nelkowski and G. Bollman, *Z. Naturforsch.*, A24, 1302, 1969.
164. V. M. Korsun and A. M. Nemchenko, *Sov. Phys. Solid State*, 8, 2988, 1967.
165. T. Lukaszewicz, *Phys. Stat. Sol. (a)*, 73, 611, 1982.
166. V. A. Williams, *J. Mat. Sci.*, 7, 807, 1972.
167. E. A. Secco, *J. Chem. Phys.*, 29, 406, 1958.
168. H. J. Biter and F. Williams, *J. Luminescence*, 3, 395, 1971.
169. H. Gobrecht, H. Nelkowski, J. W. Bears, and M. Weight, *Solid State Comm.*, 5, 777, 1967.
170. S. Asami, A. Ebina, and T. Takahashi, *Jpn. J. Appl. Phys.*, 17, 779, 1978.
171. T. Lukaszewicz and J. Zmija, *Phys. Stat. Sol. (A)*, 62, 695, 1980.
172. M. Aven and R. E. Halsted, *Phys. Rev.*, 137, A228, 1965.
173. M. M. Henneberg and D. A. Stevenson, *Phys. Stat. Sol. (b)*, 48, 255, 1971.
174. W. E. Martin, *J. Appl. Phys.*, 44, 5639, 1973.
175. T. Muranoi and M. Furukoshi, *Thin Solid Films*, 86, 307, 1981.
176. A. K. Kun and R. J. Robinson, *J. Electron. Mat.*, 5, 23, 1976.
177. H. H. Woodbury and R. B. Hall, *Phys. Rev.*, 157, 641, 1967.
178. P. Martin and A. Bontemps, *J. Phys. Chem. Solids*, 41, 1171, 1980.
179. E. A. Secco and S. C. Yeo, *Can. J. Chem.*, 49, 1953, 1971.
180. R. A. Reynold and D. A. Stevenson, *J. Phys. Chem. Solids*, 30, 139, 1969.
181. M. Aven and E. L. Kreiger, *J. Appl. Phys.*, 41, 1930, 1970.
182. M. Yokozawa, H. Kato, and S. Takayanagi, *Denki Kagaku*, 36, 282, 1968.
183. E. R. Dobrovinska, N. I. Krainyukov, Ya. A. Obukhouski, and L. A. Sysoev, *Ukr, Fiz. Zh.*, 13, 861, 1968.
184. H. H. Woodbury, *II-VI Semiconducting Compounds*, D. G. Thomas, Ed., Benjamin, NY, 1967, 244-276.
185. R. L. Clarke, *J. Appl. Phys.*, 30, 957, 1959.
186. J. L. Sullivan, *J. Phys. D: Appl. Phys.*, 6, 552, 1973.
187. B. Lepley, P. H. Nguyen, C. Boutrit, and S. Revelet, *J. Phys. D: Appl. Phys.*, 12, 145, 1979.
188. H. H. Woodbury, *J. Appl. Phys.*, 36, 2287, 1965.
189. E. Nebauer, *Phys. Stat. Sol.*, 29, 269, 1968.
190. J. Zimija and L. Sados, *Biul. Wojek, Akad. Tech.*, 20, No. 4, 105, 1971.

191. H. H. Woodbury, *Phys. Rev.*, 134, A492, 1964.
192. E. D. Jones and H. Mykura, *J. Phys. Chem. Solids*, 41, 1261, 1980.
193. E. D. Jones and H. Mykura, *J. Phys. Chem. Solids*, 39, 11, 1978.
194. E. Nebauer, *Phys. Stat. Sol. (b)*, 60, K57, 1973.
195. V. Kumar and F. A. Kroger, *J. Solid State Chem.*, 3, 387, 1971.
196. L. A. Sysoev, A. Ya, Gelfman, A. D. Kovaleva, and N. G. Kravchencko, *Izv. Akad. Nauk. SSSR., Neorg. Mater.*, 5, 2208, 1969.
197. E. Nebauer and J. Lautenbach, *Phys. Stat. Sol. (b)*, 48, 657, 1971.
198. D. G. Girton and W. E. Anderson, *Trans. Met. Soc. AIME*, 245, 465, 1969.
199. J. L. Sullivan, *Thin Solid Films*, 25, 245, 1975.
200. N. V. Dmitrieva, A. V. Vanyukov, and S. G. Yakovlev, *Elektron. Tekh. Nauk Tekh. Sb. Mater.*, No. 5, 150, 1970.
201. J. Zmija, *Acta Phys. Pol.*, A43, 345, 1973.
202. R. B. Hall and H. H. Woodbury, *J. Appl. Phys.*, 39, 5361, 1968.
203. L. Svob and C. Grattepain, *Calloq. Metall.*, 19, 725, 1976.
204. H. H. Woodbury and M. Aven, *J. Appl. Phys.*, 39, 5485, 1968.
205. H. Mann, G. Linker, and O. Meyer, *Solid State Comm.*, 11, 475, 1972.
206. J. P. Chamonal, E. Molva, J. L. Pautrat, and L. Revoil, *J. Cryst. Growth*, 59, 297, 1982.
207. I. Teramoto and S. Takayanagi, *J. Phys. Soc. Jpn.*, 17, 1137, 1982.
208. R. C. Whelan and D. Shaw, *II-IV Semiconducting Compounds*, D. C. Thomas, Ed., Benjamin, NY, 1967, 451.
209. P. M. Borsenberger and D. A. Stevenson, *J. Phys. Chem. Solids*, 29, 1277, 1968.
210. M. Yokozawa, H. Kato, and S. Takayanagi, *Denki Kagaku*, 34, 828, 1966.
211. E. Watson and D. Shaw, *J. Phys. C: Solid State Phys.*, 16, 515, 1983.
212. O. E. Panchuk, L. P. Shcherbak, P. I. Felchuk, and A. V. Savitskii, *Izv. Akad. Nauk SSSR, Neorg. Mater.*, 14, 54, 1978.
213. F. F. Morehead and G. Mandel, *Phys. Lett.*, 10, 5, 1964.
214. F. F. Vodovatov, G. V. Indenbaum, and A. V. Vanyukov, *Sov. Phys. Solid State*, 12, 17, 1970.
215. H. Kato, M. Yokozawa, and S. Takayanagi, *Jpn. J. Appl. Phys.*, 4, 1019, 1965.
216. D. Shaw and E. Watson, *J. Phys. C: Solid State Phys.*, 17, 4945, 1985.
217. O. E. Panchuk, R. N. Fesh, A. V. Savitskii, and L. P. Shcherbak, *Izv. Akad. Nauk SSSR, Neorg. Mater.*, 17, 1354, 1981.
218. B. I. Boltaks, *Diffusion in Semiconductors*, Inforsearch, 1963, 300.
219. F. F. Kharakhorin, *Izv. Akad. Nauk SSSR, Neorg. Mater.*, 5, 2212, 1969.
220. F. A. Zaitov, V. I. Stafeev, G. S., Khodakov, *Sov. Phys. Solid State*, 14, 2628, 1973.
221. V. V. Belov, F. A. Zaitov, and G. E. Popovyan, *Sov. Phys. Solid State*, 11, 1627, 1970.
222. F. A. Zaitov, *Sov. Phys. Solid State*, 13, 219, 1971.
223. V. S. Andramonov, N. S. Baryshev, and I. S. Aver'yanov, *Soviet Phys. Solid State*, 4, 1626, 1963.
224. J. Bloem and F. A. Kroger, *Philips Res. Rep.*, 12, 281, 1957.
225. G. Simkovich and J. B. Wagner, Jr., *J. Chem. Phys.*, 38, 1368, 1963.
226. M. S. Seltzer and J. B. Wagner, Jr., *J. Phys. Chem. Solids*, 26, 233, 1965.
227. J. Bloem and F. A. Kroger, Philips Res. Rep., 12, 303, 1957.
228. N. A. Fedorovich, *Soviet Phys. Solid State*, 7, 1289, 1965.
229. F. F. Kharakhorin, D. A. Gambarova, and V. V. Aksenov, *Soviet Phys. Solid State*, 7, 2813, 1966.
230. M. S. Seltzer and J. B. Wagner, Jr., *J. Chem. Phys.*, 36, 130, 1962.
231. B. I. Boltaks and Yu. N. Mokhov, *Zh. Tekh. Fiz.*, 28, 1046, 1958; *Zh. Tekh. Fiz.*, 26, 2448, 1956.
232. N. A. Fedorovich, *Soviet Phys. Solid State*, 7, 1291, 1965.
233. B. Swaroop and J. B. Wagner, Jr., *Appl. Phys. Lett.*, 12, 267, 1968.
234. A. J. Crocker and B. F. Dorning, *J. Phys. Chem. Solids*, 29, 155, 1968.
235. T. D. George and J. B. Wagner, Jr., *J. Electrochem. Soc.*, 115, 956, 1968; 116, 848, 1969.

VALUES FOR THE LANGEVIN FUNCTION £(*u*)

Compiled by Allen L. King

Because of random thermal rotations a dipole or dipole-like element ordinarily has an average dipole moment of zero. If it is placed in an orienting field F however it tends to align itself with the field so that the average component of the dipole moment parallel to the field equals p. Classically, if the system is in thermal equilibrium,

$$\bar{p} = p_o(\coth u - 1/u) = p_o£(u)$$

Here p_o is the permanent moment (electric or magnetic) of the dipole, and £(u) is the Langevin function of the argument u which equals Fp_o/kT.
The following table gives values of £ (u). Note that for $u \ll 1$, £ $(u) \approx u/3$; for $u \gg 1$, £ $(u) \approx (u - 1)/u$.

(*u*)	0	1	2	3	4	5	6	7	8	9	10	Diff.
0.0	.0000	.0033	.0066	.0100	.0133	.0166	.0200	.0233	.0267	.0300	.0333	33
0.1	.0333	.0366	.0400	.0433	.0466	.0499	.0532	.0566	.0599	.0632	.0665	33
0.2	.0665	.0698	.0731	.0764	.0797	.0830	.0863	.0896	.0928	.0961	.0994	33
0.3	.0994	.1027	.1059	.1092	.1125	.1157	.1190	.1222	.1255	.1287	.1319	32.5
0.4	.1319	.1352	.1384	.1416	.1448	.1480	.1512	.1544	.1576	.1608	.1640	32
0.5	.1640	.1671	.1703	.1734	.1766	.1797	.1829	.1860	.1891	.1922	.1954	31
0.6	.1954	.1985	.2016	.2046	.2077	.2108	.2139	.2169	.2200	.2230	.2261	31
0.7	.2261	.2291	.2321	.2351	.2381	.2411	.2441	.2471	.2500	.2530	.2559	30
0.8	.2559	.2589	.2618	.2647	.2677	.2706	.2735	.2763	.2792	.2821	.2850	29
0.9	.2850	.2878	.2906	.2935	.2963	.2991	.3019	.3047	.3075	.3103	.3130	28
1.0	.3130	.3158	.3185	.3213	.3240	.3267	.3294	.3321	.3348	.3375	.3401	27
1.1	.3401	.3428	.3454	.3481	.3507	.3533	.3559	.3585	.3611	.3636	.3662	26
1.2	.3662	.3688	.3713	.3738	.3763	.3789	.3813	.3838	.3863	.3888	.3912	25
1.3	.3912	.3937	.3961	.3985	.4010	.4034	.4057	.4081	.4105	.4129	.4152	24
1.4	.4152	.4175	.4199	.4222	.4245	.4268	.4291	.4314	.4336	.4359	.4381	23
1.5	.4381	.4404	.4426	.4448	.4470	.4492	.4514	.4535	.4557	.4578	.4600	22
1.6	.4600	.4621	.4642	.4663	.4684	.4705	.4726	.4747	.4767	.4788	.4808	21
1.7	.4808	.4828	.4849	.4869	.4889	.4908	.4928	.4948	.4967	.4987	.5006	20
1.8	.5006	.5026	.5045	.5064	.5083	.5102	.5120	.5139	.5158	.5176	.5194	19
1.9	.5194	.5213	.5231	.5249	.5267	.5285	.5303	.5321	.5338	.5356	.5373	18
2.0	.5373	.5390	.5408	.5425	.5442	.5459	.5476	.5493	.5509	.5526	.5543	17
2.1	.5543	.5559	.5575	.5592	.5608	.5624	.5640	.5656	.5672	.5687	.5703	16
2.2	.5703	.5719	.5734	.5750	.5765	.5780	.5795	.5810	.5825	.5840	.5855	15
2.3	.5855	.5870	.5885	.5899	.5914	.5928	.5943	.5957	.5971	.5985	.5999	14.5
2.4	.5999	.6013	.6027	.6041	.6055	.6068	.6082	.6096	.6109	.6122	.6136	14
2.5	.6136	.6149	.6162	.6175	.6188	.6201	.6214	.6227	.6240	.6252	.6265	13
2.6	.6265	.6277	.6290	.6302	.6314	.6327	.6339	.6351	.6363	.6375	.6387	12
2.7	.6387	.6399	.6411	.6422	.6434	.6446	.6457	.6469	.6480	.6492	.6503	12
2.8	.6503	.6514	.6525	.6536	.6547	.6558	.6569	.6580	.6591	.6602	.6612	11
2.9	.6612	.6623	.6634	.6644	.6655	.6665	.6675	.6686	.6696	.6706	.6716	10.5
3.0	.6716	.6726	.6736	.6746	.6756	.6766	.6776	.6786	.6796	.6805	.6815	10
3.1	.6815	.6824	.6834	.6843	.6853	.6862	.6872	.6881	.6890	.6899	.6908	9
3.2	.6908	.6917	.6926	.6935	.6944	.6953	.6962	.6971	.6980	.6988	.6997	9
3.3	.6997	.7006	.7014	.7023	.7031	.7040	.7048	.7056	.7065	.7073	.7081	8.5
3.4	.7081	.7089	.7097	.7106	.7114	.7122	.7130	.7138	.7145	.7153	.7161	8
3.5	.7161	.7169	.7177	.7184	.7192	.7200	.7207	.7215	.7222	.7230	.7237	8
3.6	.7237	.7245	.7252	.7259	.7267	.7274	.7281	.7288	.7295	.7302	.7310	7
3.7	.7310	.7317	.7324	.7331	.7337	.7344	.7351	.7358	.7365	.7372	.7378	7
3.8	.7378	.7385	.7392	.7398	.7405	.7412	.7418	.7425	.7431	.7438	.7444	7
3.9	.7444	.7450	.7457	.7463	.7469	.7476	.7482	.7488	.7494	.7501	.7507	6

PROPERTIES OF MAGNETIC ALLOYS

Name	Composition,* Weight percent					Remanence, B_r (Gauss)	Coercive force H_c (Oersteds)	Maximum energy product, $(BH)_{max}$ (Gauss-Oersteds $\times 10^{-6}$)
	Al	Ni	Co	Cu	Other			
Alnico I	12	20–22	5	....	......	7,100	440	1.4
Alnico II	10	17	12.5	6	......	7,200	540	1.6
Alnico III	12	24–26	....	3	......	6,900	470	1.35
Alnico IV	12	27–28	5	....	......	5,500	700	1.3
Alnico V†	8	14	24	3	......	12,500	600	5.0
Alnico V DG†	8	14	24	3	......	13,100	640	6.0
Alnico VI†	8	15	24	3	1.25 Ti	10,500	750	3.75
Alnico VII†	8.5	18	24	3	5 Ti	7,200	1,050	2.75
Alnico XII	6	18	35	....	8 Ti	5,800	950	1.6
Carbon steel	....	....	....	....	1 Mn 0.9 C	10,000	50	0.2
Chromium steel	....	....	....	....	3.5 Cr 0.9 C 0.3 Mn	9,700	65	0.3
Cobalt steel	....	....	17	....	2.5 Cr 8 W 0.75 C	9,500	150	0.65

Name	Composition,* Weight percent					Remanence, B_r (Gauss)	Coercive force H_c (Oersteds)	Maximum energy product, $(BH)_{max}$ (Gauss-Oersteds $\times 10^{-6}$)
	Al	Ni	Co	Cu	Other			
Cunico	....	21	29	50	.....	3,400	660	0.80
Cunife	....	20	....	60	.....	5,400	550	1.5
Ferroxdur 1	BaFe$_{12}$O$_{19}$					2,200	1,800	1.0
Ferroxdur 2	BaF$_{12}$O$_{19}$ (oriented)					3,840	2,000	3.5
Platinum-Cobalt	....	....	23	....	77 Pt	6,000	4,300	7.5
Remalloy	....	....	12	....	17 Mo	10,500	250	1.1
Silmanol	4.4	....	....	....	86.6 Ag 8.8 Mn	550	6,000	0.075
Tungsten steel	....	....	....	....	5 W 0.3 Mn 0.7 C	10,300	70	0.32
Vicalloy I	....	....	52	....	10 V	8,800	300	1.0
Vicalloy II (wire)	....	....	52	....	14 V	10,000	510	3.5
Germany								
Alni 90	12	21	....	....	.....	8,000	350	1.2
Alni 120	13	27	....	....	.....	6,000	570	1.2
Alnico 130	12	23	5	....	.....	6,300	620	1.4
Alnico 160	11	24	12	4	.....	6,200	700	1.6
Alnico 190	12	21	15	4	.....	7,000	700	1.8
Alnico 250	8	19	23	4	6 Ti	6,500	1,000	2.2
Alnico 400†	9	15	23	4	.....	12,000	650	4.8
Alnico 580† (semicolumnar)	9	15	23	4	.....	13,000	700	6.0
Oerstit 800	9	18	19	4	4 Ti	6,600	750	1.95
Great Britain								
Alcomax I	7.5	11	25	3	1.5 Ti	12,000	475	3.5
Alcomax II	8	11.5	24	4.5	.....	12,400	575	4.7
Alcomax IISC (semicolumnar)	8	11	22	4.5	.....	12,800	600	5.15
Alcomax III	8	13.5	24	3	0.8 Nb	12,500	670	5.10
Alcomax IIISC (semicolumnar)	8	13.5	24	3	0.8 Nb	13,000	700	5.80
Alcomax IV	8	13.5	24	3	2.5 Nb	11,200	750	4.30
Alcomax IVSC (semicolumnar)	8	13.5	24	3	2.5 Nb	11,700	780	5.10
Alni, high B_r	13	24	....	3.5	.....	6,200	480	1.25
Alni, normal	....	....	....	....	.....	5,600	580	1.25
Alni, high H_c	12	32	....	....	0–0.5 Ti	5,000	680	1.25
Alnico, high B_r	10	17	12	6	.....	8,000	500	1.70
Alnico, normal	....	....	....	....	.....	7,250	560	1.70
Alnico, high H_c	10	20	13.5	6	0.25 Ti	6,600	620	1.70
Columax (columnar)	similar to Alcomax III or IV					13,000–14,000	700–800	7.0–8.5
Hycomax	9	21	20	1.6	.....	9,500	830	3.3

* Remainder of unlisted composition is either iron or iron plus trace impurities
† Cast anisotropic. Unmarked ones are cast isotropic.

HIGH PERMEABILITY MAGNETIC ALLOYS

(See separate table for magnetic properties)

Name	Composition,* Weight percent	Sp. gr., g/cm^3	Tensile strength		Remark	Use
			kg/mm^2**	Form		
Silicon iron AISI M 15	Si 4	7.68–7.64	44.3	....	Annealed 4 hrs 802–1093°C	Low core losses
Silicon iron AISI M 8	Si 3	7.68–7.64	44.2	Grain oriented	Annealed 4 hrs 802–1204°C	
45 Permalloy	Ni 45; Mn 0.3	8.17	....	....	....	Audio transformer, coils, relays
Monimax	Ni 47; Mo 3	8.27	....	....	....	High frequency coils
4-79 Permalloy	Ni 79; Mo 4; Mn 0.3	8.74	55.4	....	H_2 annealed 1121°C	Audio coils, transformers, magnetic shields
Sinimax	Ni 43; Si 3	7.70	....	....	....	High frequency coils
Nu-metal	Ni 75; Cr 2; Cu 5	8.58	44.8	....	H_2 annealed 1221°C	Audio coils, magnetic shields, transformers
Supermalloy	Ni 79; Mo 5; Mn 0.3	8.77	....	....	....	Pulse transformers, magnetic amplifiers, coils
2-V Permendur	Co 40; V 2	8.15	46.3	....	....	D-c electromagnets, pole tips

* Iron is additional alloying metal.
** kg/mm^2 $\times$ 1422.33 = lbs/in.^{2}

PROPERTIES OF MAGNETIC ALLOYS (continued)

CAST PERMANENT MAGNETIC ALLOYS

(See separate table for magnetic properties)

Alloy name, country of manufacture*	Composition,** weight percent	Sp. gr., g/cm³	Thermal expansion		Tensile strength		Remark†	Use
			$\frac{Cm \times 10^{-6}}{cm \times °C}$	Between °C	***kg/mm²	Form		
Alnico I (USA)*	Al 12; Ni 20–22; Co 5	6.9	12.6	20–300	2.9	Cast	i.	Permanent magnets
Alnico II (USA)*	Al 10; Ni 17; Cu 6; Co 12.5	7.1	12.4	20–300	2.1 45.7	Cast Sintered	i.	Temperature controls, magnetic toys and novelties
Alnico III (USA)*	Al 12; Ni 24–26; Cu 3	6.9	12	20–300	8.5	Cast	i.	Tractor magnetos
Alnico IV (USA)*	Al 12; Ni 27–28; Co 5	7.0	13.1	20–300	6.3 42.1	Cast Sintered	i.	Application requiring high coercive force
Alnico V (USA)*	Al 8; Ni 14; Co 24; Cu 3	7.3	11.3	......	3.8 35	Cast Sintered	a.	Application requiring high energy
Alnico V DG (USA)*	Al 8; Ni 14; Co 24; Cu 3	7.3	11.3	......	....	....	a., c.	
Alnico VI (USA)*	Al 8; Ni 15; Co 24; Cu 3; Ti 1.25	7.3	11.4	......	16.1	Cast	a.	Application requiring high energy
Alnico VII (USA)*	Al 8.5; Ni 18; Cu 3; Co 24; Ti 5	7.17	11.4	......	....	....	a.	
Alnico XII (USA)*	Al 6; Ni 18; Co 35; Ti 8	7.2	11	20–300	....	....	.....	Permanent magnets
Comol (USA)*	Co 12; Mo 17	8.16	9.3	20–300	88.6	....	.....	Permanent magnets
Cunife (USA)*	Cu 60; Ni 20	8.52	....	.....	70.3	....	.....	Permanent magnets
Cunico (USA)*	Cu 50; Ni 21	8.31	....	......	70.3	....	.....	Permanent magnets
Barium ferrite Feroxdur (USA)*	Ba $Fe_{12}O_{19}$	4.7	10	......	70.3	....	....	Ceramics
Alcomax I (GB)*	Al 7.5; Ni 11; Co 25; Cu 3; Ti 1.5	....	....	......	....	....	a.	Permanent magnets
Alcomax II (GB)*	Al 8; Ni 11.5; Co 24; Cu 4.5	....	....	......	....	....	a.	Permanent magnets
Alcomax II SC(GB)*	Al 8; Ni 11; Co 22; Cu 4.5	7.3	....	......	....	....	a., sc.	
Alcomax III (GB)*	Al 8; Ni 13.5; Co 24; Nb 0.8	7.3	....	....		....	a.	Magnets for motors, loudspeakers
Alcomax IV (GB)*	Al 8; Ni 13.5; Cu 3; Co 24; Nb 2.5	....	....	......	....	....	.....	Magnets for cycle-dynamos
Columax (GB)*	Similar to Alcomax III or IV	....	....	......	....	....	a., sc.	Permanent magnets, heat treatable
Hycomax (GB)*	Al 9; Ni 21; Co 20; Cu 1.6	....	....	......	....	....	a.	Permanent magnets
Alnico (high H_c) (GB)*	Al 10; Ni 20; Co 13.5; Cu 6; Ti 0.25	7.3	....	......	....	....	i.	
Alnico (high B_r) (GB)*	Al 10; Ni 17; Co 12; Cu 6	7.3	....	......	....	....	i.	
Alni (high H_c) (GB)*	Al 12; Ni 32; Ti 0–0.5	6.9	....	......	....	....	i.	
Alni (high B_r) (GB)*	Al 13; Ni 24; Cu 3.5	....	....	......	....	....	i.	
Alnico 580 (Ger)*	Al 9; Ni 15; Co 23; Cu 4	....	....	......	....	....	i.	
Alnico 400 (Ger)*	Al 9; Ni 15; Co 23; Cu 4	....	....	......	....	....	a.	
Oerstit 800 (Ger)*	Al 9; Ni 18; Co 19; Cu 4; Ti 4	....	....	.....	....	....	i.	Permanent magnets
Alnico 250 (Ger)*	Al 8; Ni 19; Co 23; Cu 4; Ti 6	....	....	......	....	....	i.	
Alnico 190 (Ger)*	Al 12; Ni 21; Cu 4; Co 15	....	....	......	....	....	i.	
Alnico 160 (Austria)	Al 11; Ni 24; Co 12; Cu 4	....	....	......	....	....	i.	Permanent magnets, sintered
Alnico 130 (Ger)*	Al 12; Ni 23; Co 5	....	....	......	....	....	i.	
Alni 120 (Ger)*	Al 13; Ni 27	....	....	......	....	....	i.	
Alni 90 (Ger)*	Al 12; Ni 21	....	....	......	....	....	i.	

* USA—United States; GB—Great Britain; Ger—Germany.

** Iron is the additional alloying metal for each of the magnets listed.

*** kg/mm × 1422.33 = lbs/in.²

† i. = isotropic; a. = anisotropic; c. = columnar; sc. = semicolumnar.

PROPERTIES OF MAGNETIC ALLOYS (continued)

Properties of Antiferromagnetic Compounds

Compound	Crystal Symmetry	θ_N(K)	θ_P(K)	$(P_A)_{eff}$ μ_B	P_A μ_B
$CoCl_2$	Rhombohedral	25	−38.1	5.18	3.1 ± 0.6
CoF_2	Tetragonal	38	50	5.15	3.0
CoO	Tetragonal	291	330	5.1	3.8
Cr	Cubic	475			
Cr_2O_3	Rhombohedral	307	485	3.73	3.0
CrSb	Hexagonal	723	550	4.92	2.7
$CuBr_2$	Monoclinic	189	246	1.9	
$CuCl_2{\cdot}2H_2O$	Orthorhombic	4.3	4–5	1.9	
$CuCl_2$	Monoclinic	~70	109	2.08	
$FeCl_2$	Hexagonal	24	−48	5.38	4.4 ± 0.7
FeF_2	Tetragonal	79–90	117	5.56	4.64
FeO	Rhombohedral	198	507	7.06	3.32
α-Fe_2O_3	Rhombohedral	953	2940	6.4	5.0
α-Mn	Cubic	95			
$MnBr_2{\cdot}4H_2O$	Monoclinic	2.1	$\begin{Bmatrix}2.5\\1.3\end{Bmatrix}$	5.93	
$MnCl_2{\cdot}4H_2O$	Monoclinic	1.66	1.8	5.94	
MnF_2	Tetragonal	72–75	113.2	5.71	5
MnO	Rhombohedral	122	610	5.95	5.0
β-MnS	Cubic	160	982	5.82	5.0
MnSe	Cubic	~173	361	5.67	
MnTe	Hexagonal	310–323	690	6.07	5.0
$NiCl_2$	Hexagonal	50	−68	3.32	
NiF_2	Tetragonal	78.5–83	115.6	3.5	2.0
NiO	Rhombohedral	533–650	~2000	4.6	2.0
$TiCl_3$		100			
V_2O_3		170			

1. θ_N = Néel temperature, determined from susceptibility maxima or from the disappearance of magnetic scattering.
2. θ_P = a constant in the Curie-Weiss law written in the form $\chi_A = C_A/(T + \theta_P)$, which is valid for antiferromagnetic material for $T > \theta_N$.
3. $(P_A)_{eff}$ = effective moment per atom, derived from the atomic Curie constant $C_A = (P_A)^2_{eff}(N^2/3R)$ and expressed in units of the Bohr magneton, $\mu_B = 0.9273 \times 10^{-20}$ erg gauss^{-1}.
4. P_A = magnetic moment per atom, obtained from neutron diffraction measurements in the ordered state.

SATURATION CONSTANTS AND CURIE POINTS OF FERROMAGNETIC ELEMENTS

Element	σ_s (20°C)	M_s (20°C)	σ_s (0 K)	n_B	Curie Point (°C)
Fe	218.0	1,714	221.9	2.219	770
Co	161	1,422	162.5	1.715	1,131
Ni	54.39	484.1	57.50	0.604	358
Gd	0	0	253.5	7.12	16

σ_s = saturation magnetic moment/gram; M_s = saturation magnetic moment/cm^3, in cgs units. n_B = magnetic moment per atom in Bohr magnetons.

From American Institute of Physics Handbook, McGraw-Hill Company (1963) by permission.

MAGNETIC PROPERTIES OF TRANSFORMER STEELS

Ordinary Transformer Steel

B (Gauss)	H (Oersted)	Permeability = B/H
2,000	0.60	3,340
4,000	0.87	4,600
6,000	1.10	5,450
8,000	1.48	5,400
10,000	2.28	4,380
12.000	3.85	3,120
14,000	10.9	1,280
16,000	43.0	372
18,000	149	121

HIGH SILICON TRANSFORMER STEEL

B	H	Permeability
2,000	0.50	4,000
4,000	0.70	5,720
6,000	0.90	6,670
8,000	1.28	6,250
10,000	1.99	5,020
12,000	3.60	3,340
14,000	9.80	1,430
16,000	47.4	338
18,000	165	109

PROPERTIES OF MAGNETIC ALLOYS (continued)

SATURATION CONSTANTS FOR MAGNETIC SUBSTANCES

Substance	Field intensity	Induced magnetization	Substance	Field intensity	Induced magnetization
	(For saturation)			(For saturation)	
Cobalt	9000	1300	Nickel, hard	8000	400
Iron, wrought	2000	1700	annealed	7000	515
cast	4000	1200	Vicker's steel	15000	1600
Manganese steel	7000	200			

INITIAL PERMEABILITY OF HIGH PURITY IRON FOR VARIOUS TEMPERATURES

L. Alberts and B. J. Shepstone

Temperature °C	Permeability (gauss/oersted)
0	920
200	1040
400	1440
600	2550
700	3900
770	12580

MAGNETIC MATERIALS

High-permeability Materials

Material	Form	Approximate percent composition: Fe	Ni	Co	Mo	Other	Typical heat treatment °C	Permeability at $B = 20$ gausses	Maximum permeability	Saturation flux density B gausses	Hysteresis ‡ loss, W_h ergs/cm³	Coercive ‡ force H_c oersteds	Resistivity microhm cm	Density, g/cm³
Cold rolled steel	Sheet	98.5	—	—	—	—	950 Anneal	180	2,000	21,000	—	1.8	10	7.88
Iron	Sheet	99.91	—	—	—	—	950 Anneal	200	5,000	21,500	5,000	1.0	10	7.88
Purified iron	Sheet	99.95	—	—	—	—	1480 H_2 + 880	5,000	180,000	21,500	300	.05	10	7.88
4% Silicon-iron	Sheet	96	—	—	—	4 Si	800 Anneal	500	7,000	19,700	3,500	.5	60	7.65
Grain oriented*	Sheet	97	—	—	—	3 Si	800 Anneal	1,500	30,000	20,000	—	.15	47	7.67
45 Permalloy	Sheet	54.7	45	—	—	.3 Mn	1050 Anneal	2,500	25,000	16,000	1,200	.3	45	8.17
45 Permalloy †	Sheet	54.7	45	—	—	.3 Mn	1200 H_2 Anneal	4,000	50,000	16,000	—	.07	45	8.17
Hipernik	Sheet	50	50	—	—	—	1200 H_2 Anneal	4,500	70,000	16,000	220	.05	50	8.25
Monimax	Sheet	—	—	—	—	—	1125 H_2 Anneal	2,000	35,000	15,000	—	.1	80	8.27
Sinimax	Sheet	—	—	—	—	—	1125 H_2 Anneal	3,000	35,000	11,000	—	—	90	—
78 Permalloy	Sheet	21.2	78.5	—	—	.3 Mn	1050 + 600 Q§	8,000	100,000	10,700	200	.05	16	8.60
4-79 Permalloy	Sheet	16.7	79	—	4	.3 Mn	1100 + Q	20,000	100,000	8,700	200	.05	55	8.72
Mu metal	Sheet	18	75	—	—	2 Cr, 5 Cu	1175 H_2	20,000	100,000	6,500	—	.05	62	8.58
Supermalloy	Sheet	15.7	79	—	5	.3 Mn	1300 H_2 + Q	100,000	800,000	8,000		.002	60	8.77
Permendur	Sheet	49.7	—	50	—	.3 Mn	800 Anneal	800	5,000	24,500	12,000	2.0	7	8.3
2V Permendur	Sheet	49	—	49	—	2 V	800 Anneal	800	4,500	24,000	6,000	2.0	26	8.2
Hiperco	Sheet	64	—	34	—	Cr	850 Anneal	650	10,000	24,200	—	1.0	25	8.0
2-81 Permalloy	Insulated powder	17	81	—	2	—	650 Anneal	125	130	8,000	—	<1.0	10^6	7.8
Carbonyl iron	Insulated powder	99.9	—	—	—	—	—	55	132	—	—	—	—	7.86
Ferroxcube III	Sintered powder	$MnFe_2O_4$ + $ZnFe_2O_4$					—	1,000	1,500	2,500	—	.1	10^8	5.0

*Properties in direction of rolling.
† Similar properties for Nicaloi, 4750 alloy, Carpenter 49, Armco 48.
‡ At saturation.
§ Q, quench or controlled cooling.

PROPERTIES OF MAGNETIC ALLOYS (continued)

Permanent Magnet Alloys

Material	Percent composition (remainder Fe)	Heat treatment* (temperature, °C)	Magnetizing force $H_{max.}$ oersteds	Coercive force H_c oersteds	Residual induction B_r gausses	Energy product $BH_{max.}$ $\times 10^{-6}$	Method of fabrication†	Mechanical properties‡	Weight lb/in.³
Carbon steel	1 Mn, 0.9 C	Q 800	300	50	10,000	.20	HR, M, P	H, S	.280
Tungsten steel	5 W, 0.3 Mn, 0.7 C	Q 850	300	70	10,300	.32	HR, M, P	H, S	.292
Chromium steel	3.5 Cr, 0.9 C, 0.3 Mn	Q 830	300	65	9,700	.30	HR, M, P	H, S	.280
17% Cobalt steel	17 Co, 0.75 C, 2.5 Cr, 8 W	—	1,000	150	9,500	.65	HR, M, P	H, S	—
36% Cobalt steel	36 Co, 0.7 C, 4 Cr, 5 W	Q 950	1,000	240	9,500	.97	HR, M, P	H, S	.296
Remalloy or Comol	17 Mo, 12 Co	Q 1200, B 700	1,000	250	10,500	1.1	HR, M, P	H	.295
Alnico I	12 Al, 20 Ni, 5 Co	A 1200, B 700	2,000	440	7,200	1.4	C, G	H, B	.249
Alnico II	10 Al, 17 Ni, 2.5 Co, 6 Cu	A 1200, B 600	2,000	550	7,200	1.6	C, G	H, B	.256
Alnico II (sintered)	10 Al, 17 Ni, 2.5 Co, 6 Cu	A 1300	2,000	520	6,900	1.4	Sn, G	H	.249
Alnico IV	12 Al, 28 Ni, 5 Co	Q 1200, B 650	3,000	700	5,500	1.3	Sn, C, G	H	.253
Alnico V	8 Al, 14 Ni, 24 Co, 3 Cu	AF 1300, B 600	2,000	550	12,500	4.5	C, G	H, B	.264
Alnico VI	8 Al, 15 Ni, 24 Co, 3 Cu, 1 Ti	—	3,000	750	10,000	3.5	C, G	H, B	.268
Alnico XII	6 Al, 18 Ni, 35 Co, 8 Ti	—	3,000	950	5,800	1.5	C, G	H, B	.26
Vicalloy I	52 Co, 10 V	B 600	1,000	300	8,800	1.0	C, CR, M, P	D	.295
Vicalloy II (wire)	52 Co, 14 V	CW + B 600	2,000	510	10,000	3.5	C, CR, M, P	D	.292
Cunife (wire)	60 Cu, 20 Ni	CW + B 600	2,400	550	5,400	1.5	C, CR, M, P	D, M	.311
Cunico	50 Cu, 21 Ni, 29 Co	—	3,200	660	3,400	.80	C, CR, M, P	D, M	.300
Vectolite	30 Fe_2O_3, 44 Fe_3O_4, 26 C_2O_3	—	3,000	1,000	1,600	.60	Sn, G	W	.113
Silmanal	86.8 Ag, 8.8 Mn, 4.4 Al	—	20,000	6,000[a]	550	.075	C, CR, M, P	D, M	.325
Platinum-cobalt	77 Pt, 23 Co	Q 1200, B 650	15,000	3,600	5,900	6.5	C, CR, M	D	—
Hyflux	Fine powder	—	2,000	390	6,600	.97	—	—	.176

[a] Value given is intrinsic H_c.

* Q—Quenched in oil or water. A—Air cooled. B—Baked. F—Cooled in magnetic field. CW—Cold worked.

† HR—Hot rolled or forged. CR—Cold rolled or drawn. M—Machined. G—Must be ground. P—Punched. C—Cast. Sn—Sintered.

‡ H—Hard. B—Brittle. S—Strong. D—Ductile. M—Malleable. W—Weak.

ELECTRON WORK FUNCTIONS OF THE ELEMENTS

Compiled by Herbert B. Michaelson, 1977

The measured values cited for polycrystalline and single-crystal specimens are selected as being the best available data at this time. The selection is based on (1) The validity of the experimental technique (e.g., vacua of 10^{-9} or 10^{-10} Torr, clean surfaces, and identification of crystal-face distribution and other surface conditions), and (2) Best agreement with preferred values and theoretical values of the true work function (given variously by Fomenko,[1] Rivière,[2] Trasatti,[3] and Lang and Kohn[4]). Experimental data that are not well substantiated according to these criteria are listed in *italics*. Crystallographic directions for single-crystal data are indicated by parentheses.

Abbreviations apply to the experimental method: T, thermionic; P, photoelectric; CPD, contact potential difference; F, field emission. Important distinctions among such measurements are discussed in the Rivière[2] paper, pp. 180 to 198.

Element	Experimental value, ϕ (eV)	Experimental method	Ref.
Ag	*4.26*	P	5
	4.64 (100)	P	5
	4.52 (110)	P	5
	4.74 (111)	P	6
Al	4.28	P	7
	4.41 (100)	P	8
	4.06 (110)	P	7
	4.24 (111)	P	8
As	*3.75*	P	9
Au	5.1	P	10
	5.47 (100)	P	11
	5.37 (110)		11
	5.31 (111)		11
B	*4.45*	T	12
Ba	*2.7*	T	13
Be	4.98	P	14
Bi	*4.22*	P	15
C	*5.0*	CPD	16
Ca	2.87	P	17
Cd	*4.22*	CPD	18
Ce	2.9	P	10
Co	5.0	P	10
Cr	4.5	P	10
Cs	2.14	P	19
Cu	4.65	P	10
	4.59 (100)	P	20
	4.48 (110)	P	20
	4.94 (111)	P	20
	4.53 (112)	P	20
Eu	2.5	P	10
Fe	4.5	P	10
	4.67 (100)	P	21
	4.81α (111)	P	22
	4.70α	P	23
	4.62β	P	23
	4.68γ	P	23
Ga	4.2	CPD	24
Ge	5.0	CPD	25
	4.80 (111)	P	26
Gd	3.1	P	10
Hf	3.9	P	10
Hg	4.49	P	27
In	4.12	P	28
Ir	*5.27*	T	29
	5.42 (110)	F	30
	5.76 (111)	F	30
	5.67 (100)	F	31
	5.00 (210)	F	31
K	2.30	P	32
La	3.5	P	10
Li	*2.9*	F	33
Lu	*3.3*	CPD	34
Mg	*3.66*	P	35
Mn	4.1	P	10
Mo	4.6	P	10
	4.53 (100)	P	36
	4.95 (110)	P	36
	4.55 (111)	P	36
	4.36 (112)	P	36
	4.50 (114)	P	36
	4.55 (332)	P	36
Na	2.75	P	37
Nb	4.3	P	10
	4.02 (001)	T	38
	4.87 (110)	T	38

Element	Experimental value, ϕ (eV)	Experimental method	Ref.
	4.36 (111)	T	38
	4.63 (112)	T	38
	4.29 (113)	T	38
	3.95 (116)	T	38
	4.18 (310)	T	38
Nd	3.2	P	10
Ni	5.15	P	10
	5.22 (100)	P	39
	5.04 (110)	P	39
	5.35 (111)	P	39
Os	*4.83*	T	29
Pb	4.25	P	40
Pd	5.12	P	31
	5.6 (111)	P	41
Pt	5.65	P	10
	5.7 (111)	P	41
Rb	*2.16*	P	27
Re	*4.96*	T	29
	5.75 (1011)	F	33
Rh	4.98	P	31
Ru	4.71	P	31
Sb	4.55 (amorph.)	–	42
	4.7 (100)	–	43
Sc	3.5	P	10
Se	5.9	P	44
Si	4.85n	CPD	40
	*4.91*p (100)	CPD	45
	4.60p (111)	P	46
Sm	2.7	P	10
Sn	*4.42*	CPD	47
Sr	*2.59*	T	48
Ta	4.25	T	29
	4.15 (100)	T	49
	4.80 (110)	T	49
	4.00 (111)	T	49
Tb	3.0	P	50
Te	4.95	P	44
Th	3.4	T	51
Ti	4.33	P	10
Tl	*3.84*	CPD	52
U	3.63	P & CPD	53
	3.73 (100)	P & CPD	54
	3.90 (110)	P & CPD	54
	3.67 (113)	P & CPD	54
V	4.3	P	10
W	4.55	CPD	55
	4.63 (100)	F	30
	5.25 (110)	F	30
	4.47 (111)	F	30
	4.18 (113)	CPD	56
	4.30 (116)	T	57
Y	3.1	P	10
Zn	*4.33*	P	15
	4.9 (0001)	CPD	58
Zr	4.05	P	10

REFERENCES

1. Fomenko, V. S., *Emission Properties of Materials*, 3rd ed., Naukova Dumka, Kiev, 1970 (in Russian).
2. Rivière, J. C., *Solid State Surface Science*, Green, M., Ed., Vol. 1, Marcel Dekker, 1969, chap. 4.
3. Trasatti, S., *Chim. Ind.* (Milan), 53(6), 559, 1971.
4. Lang, N. D. and Kohn, W., *Phys. Rev. B*, 3(4), 1215, 1971.
5. Dweydari, A. W. and Mee, C. H. B., *Phys. Status Solidi A*, 27, 223, 1975.
6. Dweydari, A. W. and Mee, C. H. B., *Phys. Status Solidi A*, 17, 247, 1973.
7. Eastment, R. M. and Mee, C. H. B., *J. Phys. F*, 3, 1738, 1973.
8. Grepstad, J. K., Gartland, P. O., and Slagsvold, B. J., *Surf. Sci.*, 57, 348, 1976.
9. Raisin, C. and Pinchaux, R., *Solid State Commun.*, 16, 941, 1975.
10. Eastman, D. E., *Phys. Rev. Sect. B*, 2, 1, 1970.
11. Potter, H. C. and Blakeley, J. M., *J. Vac. Sci. Technol.*, 12, 635, 1975 and Potter, H. C., Ph.D. thesis Cornell University, Materials Science Center Rep. No. 1353, 1970.
12. Adirovich, E. I. and Gol'dshtein, L. M., *Fiz. Tverdogo Tela* (Leningrad), 9, 1258, 1967.
13. Bondarenko, B. V. and Makhov, V. I., *Sov. Phys. Solid State*, 12(7), 1522, 1971.
14. Gustafsson, Broden, and Nilsson, *J. Phys. F*, 4, 2351, 1974.
15. Suhrmann, R. and Wedler, G., *Z. Angew. Phys.*, 14, 70, 1962.
16. Robrieux, B., Faure, R., and Dussaulcy, J. P., *C. R. Acad. Sci. Ser. B*, 278(14), 659, 1974.
17. Gaudart, L. and Riviora, R., *Appl. Opt.*, 10, 2336, 1971.
18. Anderson, P. A., *Phys. Rev.*, 98, 1739, 1955.
19. Boutry, G. A. and Dormont, H., *Philips Tech. Rev.*, 30, 225, 1969.
20. Gartland, P. O., *Phys. Norv.*, 6(3,4), 201, 1972.
21. Ueda, K. and Shimizu, R., *Jp. J. Appl. Phys.*, 11(6), 916, 1972.
22. Kobayashi, H. and Kato, S., *Surf. Sci.*, 18(2), 341, 1969.
23. Cardwell, A., *Phys. Rev.*, 92, 554, 1953.
24. Osipova, E. V., Shurmovskaya, N. A., and Burshtein, R. Kh., *Elektrokhimiya*, 5(10), 1139, 1969 (in Russian).
25. Boiko, B. A., Gorodetskii, D. A., and Yas'ko, A. A., *Sov. Phys. Solid State*, 15(11), 2101, 1974.
26. Gobeli, G. W. and Allen, F. G., *Surf. Sci.*, 2, 402, 1964.
27. Lazarev, V. B. and Malov, Y. I., *Fiz. Met. Metalloved.*, 24(3), 565, 1967.
28. Peisner, J., Roboz, P., and Barna, P. B., *Phys. Stat. A*, 4, K187, 1971.
29. Wilson, R. G., *J. Appl. Phys.*, 37, 3170, 1966.
30. Strayer, R. W., Mackie, W., and Swanson, L. W., *Surf. Sci.*, 34, 225, 1973.
31. Nieuwenhuys, Bouwman, and Sachtler, *Thin Solid Films*, 21, 51, 1974.
32. Van Oirschot, Th. G. J., van den Brink, M., and Sachtler, W. H. M., *Surf. Sci.*, 29, 189, 1972.
33. Ovchinnikov, A. P. and Tsarev, B. M., *Sov. Phys. Solid State*, 9(12), 2766, 1968.
34. Bondarenko, B. V. and Makhov, V. I., *Sov. Phys. Solid State*, 12, 2986, 1971.
35. Garron, R., *C. R. Acad. Sci.*, 258, 1458, 1964.
36. Berge, Gartland, and Slagsvold, *Surf. Sci.*, 43, 275, 1974.
37. Whitefield, R. J. and Brady, J. J., *Phys. Rev. Lett.*, 26(7), 380, 1971.
38. Leblanc, R. P., Vanbrugghe, B. C., and Girouard, F. E., *Can. J. Phys.*, 52, 1589, 1974.
39. Baker, B. G., Johnson, E. B., and Maire, G. I. C., *Surf. Sci.*, 24, 572, 1971.
40. Thanailakis, A., *Inst. Phys. Conf. Ser.*, p. 59, 1974.
41. Demuth, J. E., *Chem. Phys. Lett.*, 45, 12, 1977.
42. Gorodetskii, D. A. and Yas'ko, A. A., *Sov. Phys. Solid State*, 13(11), 2928, 1972.
43. Gorodetskii, D. A. and Yas'ko, A. A., *Sov. Phys. Solid State*, 13(5), 1085, 1971.
44. Williams, R. H. and Polanco, J. I., *J. Phys. C*, 7, 2745, 1974.
45. Allen, F. G., *J. Phys. Chem. Solids*, 8, 119, 1959.
46. Allen, F. G. and Gobeli, G. W., *J. Appl. Phys.*, 35, 597, 1964.
47. Simmons, J. G., *Phys. Rev. Lett.*, 10, 10, 1963.
48. Alleau, T., *Surface Phenomena in Thermionic Emitters, Round Table Conf.*, Inst. Tech. Phys. Julich Nucl. Res. Establ., Julich, Germany, 1969, p. 54 (in English).
49. Protopopov, Mikheeva, Shreinberg, and Shuppe, *Fiz. Tverdogo Tela*, 8(4), 1140, 1966.
50. Nemchenok, R. L., Strakovskaya, S. E., and Titenskii, A. I., *Fiz. Tverdogo Tela* 11(9), 2692, 1969.

51. Estrup, P. J., Anderson, J. R., and Danforth, W. E., *Surf. Sci.*, 4, 286, 1966.
52. Klein, O. and Lange, E., *Z. Elektrochem.*, 44, 542, 1938.
53. Hopkins, B. J. and Sargood, A. J., *Nuovo Cimento*, 5, 459, 1967.
54. Lea, C. and Mee, C. H. B., *J. Appl. Phys.*, 39, 5890, 1968.
55. Hopkins, B. J. and Rivière, J. C., *Proc. Phys. Soc.* (London), 81, 590, 1963.
56. Love, H. M. and Dyer, G. L., *Can. J. Phys.*, 40, 1837, 1962.
57. Sultanov, V. M., *Radio Eng. Electron.*, 9, 252, 1964 (English translation).
58. Baker, J. M. and Blakeley, J. M., *Surf. Sci.*, 32, 45, 1972.

SECONDARY ELECTRON EMISSION

The secondary emission yield, or secondary emission ratio, δ, is the average number of secondary electrons emitted from a bombarded material for every incident primary electron. It is a function of the primary electron energy E_p. The maximum yield δ_{max} corresponds to a primary electron energy E_{pmax} (see figure). The two primary electron energies corresponding to a yield of unity are denoted the first and second crossovers (E_I and E_{II}). An insulating target, or a conducting target that is electrically floating, will charge positively or negatively depending on the primary electron energy. For $E_I < E_p < E_{II}$, $\delta > 1$ and the surface charges positively provided there is a collector present that is positive with respect to the target. For $E_p < E_I$ or $E_p > E_{II}$, $\delta < 1$, and the surface charges negatively with respect to the potential of the source of primary electrons.

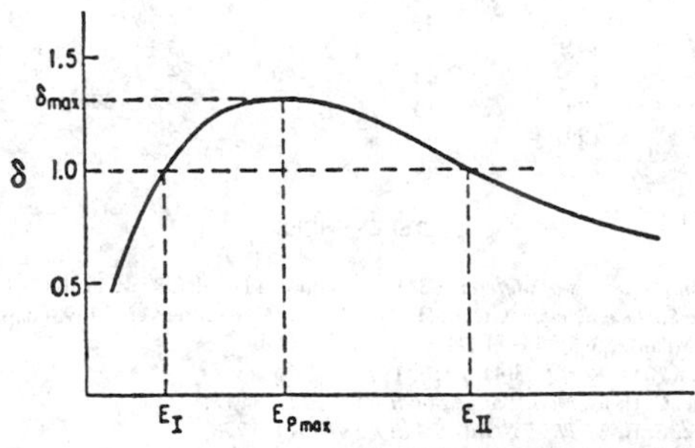

PRIMARY ELECTRON ENERGY (Ep)

The secondary emission yield is very sensitive to surface contamination, such as oxide films and carbon deposits. Whenever possible, yields believed to be most typical of clean surfaces have been selected. The yields are for measurements at room temperature and normal incidence of the primary electrons.

Element	δ_{max}	E_{pmax} (eV)	E_I (eV)	E_{II} (eV)	Element	δ_{max}	E_{pmax} (eV)	E_I (eV)	E_{II} (eV)
Ag	1.5	800	200	>2000	Li	0.5	85	None	None
Al	1.0	300	300	300	Mg	0.95	300	None	None
Au	1.4	800	150	>2000	Mo	1.25	375	150	1200
B	1.2	150	50	600	Na	0.82	300	None	None
Ba	0.8	400	None	None	Nb	1.2	375	150	1050
Bi	1.2	550	None	None	Ni	1.3	550	150	>1500
Be	0.5	200	None	None	Pb	1.1	500	250	1000
C (diamond)	2.8	750	None	>5000	Pd	>1.3	>250	120	None
C (graphite)	1.0	300	300	300	Pt	1.8	700	350	3000
C (soot)	0.45	500	None	None	Rb	0.9	350	None	None
Cd	1.1	450	300	700	Sb	1.3	600	250	2000
Co	1.2	600	200	None	Si	1.1	250	125	500
Cs	0.7	400	None	None	Sn	1.35	500	None	None
Cu	1.3	600	200	1500	Ta	1.3	600	250	>2000
Fe	1.3	400	120	1400	Th	1.1	800	None	None
Ga	1.55	500	75	None	Ti	0.9	280	None	None
Ge	1.15	500	150	900	Tl	1.7	650	70	>1500
Hg	1.3	600	350	>1200	W	1.4	650	250	>1500
K	0.7	200	None	None	Zr	1.1	350	None	None

SECONDARY ELECTRON EMISSION (continued)

Compound	δ_{max}	E_{pmax} (eV)
Alkali halides		
CsCl	6.5	
KBr (crystal)	14	1800
KCl (crystal)	12	1600
KCl (layer)	7.5	1200
KI (crystal)	10	1600
KI (layer)	5.6	
LiF (crystal)	8.5	
LiF (layer)	5.6	700
NaBr (crystal)	24	1800
NaBr (layer)	6.3	
NaCl (crystal)	14	1200
NaCl (layer)	6.8	600
NaF (crystal)	14	1200
NaF (layer)	5.7	
NaI (crystal)	19	1300
NaI (layer)	5.5	
RbCl (layer)	5.8	
Oxides		
Ag_2O	1.0	
Al_2O_3 (layer)	2—9	
BaO (layer)	2.3—4.8	400
BeO	3.4	2000
CaO	2.2	500
Cu_2O	1.2	400
MgO (crystal)	20—25	1500
MgO (layer)	3—15	400—1500
MoO_2	1.2	
SiO_2 (quartz)	2.1—4	400
SnO_2	3.2	640
Sulfides		
MoS_2	1.1	
PbS	1.2	500
WS_2	1.0	
ZnS	1.8	350
Others		
BaF_2 (layer)	4.5	
CaF_2 (layer)	3.2	
$BiCs_3$	6	1000
BiCs	1.9	1000
GeCs	7	700
Rb_3Sb	7.1	450
$SbCs_3$	6	700
Mica	2.4	350
Glasses	2—3	300—450

OPTICAL PROPERTIES OF METALS AND SEMICONDUCTORS

J. H. Weaver and H. P. R. Frederikse

These tables list the index of refraction n, the extinction coefficient k, and the normal incidence reflection R ($\phi = 0$) as a function of photon energy E, which is expressed in electron volts (eV). To convert the energy in eV to wavelength in µm, use $\lambda = 1.2398/E$. To compute the dielectric function $\tilde{\varepsilon} = \varepsilon_1 + i\varepsilon_2$ from the complex index of refraction $\tilde{N} = n + ik$, use $\varepsilon_1 = n^2 - k^2$ and $\varepsilon_2 = 2nk$.

The optical constants in these tables are abridged from three more extensive tabulations:

- *Optical Properties of Metals* (OPM), Volumes I and II, *Physics Data, Nr.* 18-1 and 18-2, J. H. Weaver, C. Krafka, D. W. Lynch, and E. E. Koch, Fachinformationzentrum, Karlsruhe, Germany.
- *Handbook of Optical Constants* (HOC), Vol. I, 1985, and Vol. II, 1991. Edited by E. D. Palik, published by Academic Press, Inc.
- *American Institute of Physics Handbook* (AIPH), 3rd Edition, Coord. Editor D. E. Gray, published by McGraw-Hill Book Co., New York, 1972.

The first two of these major sources provide detailed comparisons of all optical data available in the literature at the time of the compilation. For critical applications the reader should refer to the original work. References for individual metals and semiconductors are listed at the end of the tables. Generally, tabulated values for the optical properties are accurate to better than 10%. Data in parentheses are extrapolated or interpolated values. For most elements the spectral range covered is from the far infrared (0.010 or 0.10 eV) to the far ultraviolet (10, 30 or 300 eV). The intervals between successive energies in the tables are chosen in such a way that the major spectral features are preserved.

Very small values of k are expressed in exponential notation, e.g., 1.23E-5 means 1.23×10^{-5}.

The following table is convenient for identifying the energy entries in these tables with the corresponding wavelengths:

λ	*E*/eV	λ	*E*/eV
1 mm	0.00124	6000 Å	2.066
500 µm	0.00248	5000 Å	2.480
100 µm	0.01240	4000 Å	3.100
50 µm	0.02480	3000 Å	4.133
10 µm	0.12398	2000 Å	6.199
5 µm	0.24797	1000 Å	12.398
1 µm	1.240	400 Å	30.996

Energy (eV)	n	k	$R(\phi = 0)$
Aluminium[1]			
0.040	98.595	203.701	0.9923
0.050	74.997	172.199	0.9915
0.060	62.852	150.799	0.9906
0.070	53.790	135.500	0.9899
0.080	45.784	123.734	0.9895
0.090	39.651	114.102	0.9892
0.100	34.464	105.600	0.9889
0.125	24.965	89.250	0.9884
0.150	18.572	76.960	0.9882
0.175	14.274	66.930	0.9879
0.200	11.733	59.370	0.9873
0.250	8.586	48.235	0.9858
0.300	6.759	40.960	0.9844
0.350	5.438	35.599	0.9834
0.400	4.454	31.485	0.9826
0.500	3.072	25.581	0.9817
0.600	2.273	21.403	0.9806
0.700	1.770	18.328	0.9794
0.800	1.444	15.955	0.9778
0.900	1.264	14.021	0.9749
1.000	1.212	12.464	0.9697
1.100	1.201	11.181	0.9630
1.200	1.260	10.010	0.9521
1.300	1.468	8.949	0.9318
1.400	2.237	8.212	0.8852
1.500	2.745	8.309	0.8678
1.600	2.625	8.597	0.8794
1.700	2.143	8.573	0.8972
1.800	1.741	8.205	0.9069
1.900	1.488	7.821	0.9116
2.000	1.304	7.479	0.9148
2.200	1.018	6.846	0.9200
2.400	0.826	6.283	0.9228
2.600	0.695	5.800	0.9238
2.800	0.598	5.385	0.9242
3.000	0.523	5.024	0.9241
3.200	0.460	4.708	0.9243
3.400	0.407	4.426	0.9245
3.600	0.363	4.174	0.9246
3.800	0.326	3.946	0.9247
4.000	0.294	3.740	0.9248
4.200	0.267	3.552	0.9248
4.400	0.244	3.380	0.9249
4.600	0.223	3.222	0.9249
4.800	0.205	3.076	0.9249
5.000	0.190	2.942	0.9244
6.000	0.130	2.391	0.9257
6.500	0.110	2.173	0.9260
7.000	0.095	1.983	0.9262
7.500	0.082	1.814	0.9265
8.000	0.072	1.663	0.9269
8.500	0.063	1.527	0.9272
9.000	0.056	1.402	0.9277
9.500	0.049	1.286	0.9282
10.000	0.044	1.178	0.9286
10.500	0.040	1.076	0.9293
11.000	0.036	0.979	0.9298
11.500	0.033	0.883	0.9283
12.000	0.033	0.791	0.9224
12.500	0.034	0.700	0.9118
13.000	0.038	0.609	0.8960
13.500	0.041	0.517	0.8789
14.000	0.048	0.417	0.8486
14.200	0.053	0.373	0.8312
14.400	0.058	0.327	0.8102
14.600	0.067	0.273	0.7802
14.800	0.086	0.211	0.7202
15.000	0.125	0.153	0.6119
15.200	0.178	0.108	0.4903
15.400	0.234	0.184	0.3881
15.600	0.280	0.073	0.3182
15.800	0.318	0.065	0.2694
16.000	0.351	0.060	0.2326
16.200	0.380	0.055	0.2031
16.400	0.407	0.050	0.1789
16.750	0.448	0.045	0.1460
17.000	0.474	0.042	0.1278
17.250	0.498	0.040	0.1129
17.500	0.520	0.038	0.1005
17.750	0.540	0.036	0.0899
18.000	0.558	0.035	0.0809
18.500	0.591	0.032	0.0664
19.000	0.620	0.030	0.0554
19.500	0.646	0.028	0.0467
20.000	0.668	0.027	0.0398
20.500	0.689	0.025	0.0342
21.000	0.707	0.024	0.0296
21.500	0.724	0.023	0.0258
22.000	0.739	0.022	0.0226
22.500	0.753	0.021	0.0199
23.000	0.766	0.021	0.0177
23.500	0.778	0.020	0.0157
24.000	0.789	0.019	0.0140
24.500	0.799	0.018	0.0126
25.000	0.809	0.018	0.0113

Energy (eV)	n	k	$R(\phi = 0)$
25.500	0.817	0.017	0.0102
26.000	0.826	0.016	0.0092
27.000	0.840	0.015	0.0076
28.000	0.854	0.014	0.0063
29.000	0.865	0.014	0.0053
30.000	0.876	0.013	0.0044
35.000	0.915	0.010	0.0020
40.000	0.940	0.008	0.0010
45.000	0.957	0.007	0.0005
50.000	0.969	0.006	0.0003
55.000	0.979	0.005	0.0001
60.000	0.987	0.004	0.0000
65.000	0.995	0.004	0.0000
70.000	1.006	0.004	0.0000
72.500	1.025	0.004	0.0002
75.000	1.011	0.024	0.0002
77.500	1.008	0.025	0.0002
80.000	1.007	0.024	0.0002
85.000	1.007	0.028	0.0002
90.000	1.005	0.031	0.0002
95.000	0.999	0.036	0.0003
100.000	0.991	0.030	0.0002
110.000	0.994	0.025	0.0002
120.000	0.991	0.024	0.0002
130.000	0.987	0.021	0.0001
140.000	0.989	0.016	0.0001
150.000	0.990	0.015	0.0001
160.000	0.989	0.014	0.0001
170.000	0.989	0.011	0.0001
180.000	0.990	0.010	0.0000
190.000	0.990	0.009	0.0000
200.000	0.991	0.007	0.0000
220.000	0.992	0.006	0.0000
240.000	0.993	0.005	0.0000
260.000	0.993	0.004	0.0000
280.000	0.994	0.003	0.0000
300.000	0.995	0.002	0.0000

Carbon (diamond)[2]

Energy (eV)	n	k	$R(\phi = 0)$
0.06199	2.3741		0.166
0.06888	2.3741		0.166
0.07749	2.3745		0.166
0.08856	2.3750		0.166
0.1033	2.3757		0.166
0.1240	2.3765		0.166
0.1550	2.3772		0.166
0.1907		3.1 E-05	
0.2066	2.3779	5.7 E-05	0.166
0.22		1.21E-04	
0.23		2.36E-04	
0.24		3.82E-04	
0.25		5.21E-04	
0.26		2.96E-04	
0.27		4.39E-04	
0.28		2.75E-04	
0.29		7.82E-05	
0.30		1.32E-04	
0.31	2.3787	1.30E-04	0.167
0.32		1.11E-04	
0.33		2.99E-05	
0.34		1.89E-05	
0.35		2.11E-05	
0.36		2.47E-05	
0.37		2.80E-05	
0.38		3.11E-05	
0.39		3.67E-05	
0.40		3.58E-05	
0.41		3.25E-05	
0.4133	2.3795		0.167
0.42		2.94E-05	
0.43		2.87E-05	
0.44		3.14E-05	
0.45		3.62E-05	
0.46		3.22E-05	
0.47		1.57E-05	
0.48		6.17E-06	
0.4959	2.3801		0.167
0.6199	2.3813		0.167
0.8266	2.3837		0.167
1.240	2.3905		0.168
1.378	2.3934		0.169
1.459	2.3953		0.169
1.550	2.3975		0.169
1.653	2.4003		0.170
1.771	2.4036		0.170
1.889	2.4073		0.171
1.926	2.4084		0.171
2.066	2.4133		0.171
2.105	2.4147		0.172
2.271	2.4210		0.173
2.480	2.4299		0.174
2.650	2.4380		0.175
2.845		3.82E-07	
3.100	2.4627		0.178
3.131	2.4849		0.182
3.576	2.4955		0.183
3.961		8.97E-07	
4.160	2.5465		0.190
4.511		1.29E-06	
4.8187	2.6205	1.47E-06	0.200
5.00	2.6383		0.203
5.30		2.98E-06	
5.35		6.45E-06	
5.40		1.04E-05	
5.50		3.41E-05	
5.55		5.48E-04	
5.60	2.740	1.48E-03	0.216
5.80	2.780	5.02E-03	0.222
6.00	2.826	7.99E-03	0.228
6.10	2.852	8.62E-03	0.231
6.20	2.879	9.30E-03	0.235
6.30	2.910	9.74E-03	0.239
6.40	2.944	9.87E-03	0.243
6.50	2.985	1.10E-02	0.248
6.60	3.031	1.47E-02	0.254
6.70	3.085	2.20E-02	0.261
6.80	3.146	3.44E-02	0.268
6.90	3.220	5.24E-02	0.277
7.00	3.322	9.35E-02	0.289
7.10	3.444	0.210	0.304
7.15	3.464	0.307	0.308
7.20	3.437	0.388	0.307
7.30	3.376	0.473	0.303
7.40	3.335	0.515	0.300
7.50	3.321	0.533	0.299
7.60	3.306	0.592	0.300
7.80	3.276	0.659	0.300
8.00	3.251	0.712	0.300
8.25	3.232	0.765	0.301
8.50	3.228	0.806	0.303
8.75	3.247	0.855	0.308
9.00	3.272	0.910	0.314
9.25	3.308	0.978	0.322
9.50	3.348	1.055	0.331
9.75	3.398	1.147	0.342
10.00	3.453	1.258	0.355
10.25	3.514	1.403	0.371
10.50	3.565	1.581	0.389
10.75	3.600	1.813	0.411
11.00	3.582	2.078	0.434
11.25	3.507	2.380	0.460
11.50	3.346	2.693	0.488
11.75	3.090	2.986	0.518
12.00	2.736	3.228	0.551
12.20	2.383	3.354	0.580
12.40	1.983	3.382	0.610
12.60	1.532	3.265	0.641
12.80	1.312	2.953	0.627
13.00	1.223	2.722	0.604
13.50	1.129	2.379	0.557
14.00	1.070	2.178	0.526
14.50	1.018	2.034	0.504
15.00	0.972	1.929	0.489
15.50	0.917	1.845	0.482
16.00	0.861	1.767	0.477
16.50	0.805	1.692	0.474
17.00	0.753	1.619	0.471
17.50	0.707	1.546	0.467
18.00	0.665	1.476	0.463
18.50	0.626	1.408	0.459
19.00	0.589	1.341	0.455
19.50	0.557	1.273	0.449
20.00	0.527	1.203	0.442
21.00	0.487	1.052	0.413
22.00	0.518	0.888	0.330
23.00	0.597	0.850	0.270
24.00	0.586	0.829	0.268
25.00	0.562	0.787	0.265
26.00	0.538	0.736	0.260
27.00	0.516	0.679	0.252
28.00	0.501	0.616	0.239
29.00	0.494	0.552	0.221
30.00	0.493	0.490	0.201

Cesium (evaporated)[3]

Energy (eV)	n	k	$R(\phi = 0)$
2.145	0.264	1.123	0.631
2.271	0.278	0.950	0.561
2.845	0.425	0.438	0.235
3.064	0.540	0.320	0.127
3.397	0.671	0.233	0.057
3.966	0.827	0.174	0.018
4.889	0.916	0.143	0.007

Chromium[4]

Energy (eV)	n	k	$R(\phi = 0)$
0.06	21.19	42.00	0.962
0.10	11.81	29.76	0.955
0.14	15.31	26.36	0.936
0.18	8.73	25.37	0.53
0.22	5.30	20.62	0.954
0.26	3.91	17.12	0.951
0.30	3.15	14.28	0.943
0.42	3.47	8.97	0.862
0.54	3.92	7.06	0.788
0.66	3.96	5.95	0.736
0.78	4.13	5.03	0.680

Energy (eV)	n	k	$R(\phi = 0)$
0.90	4.43	4.60	0.650
1.00	4.47	4.43	0.639
1.12	4.53	4.31	0.631
1.24	4.50	4.28	0.629
1.36	4.42	4.30	0.631
1.46	4.31	4.32	0.632
1.77	3.84	4.37	0.639
2.00	3.48	4.36	0.644
2.20	3.18	4.41	0.656
2.40	2.75	4.46	0.677
2.60	2.22	4.36	0.698
2.80	1.80	4.06	0.703
3.00	1.54	3.71	0.695
3.20	1.44	3.40	0.670
3.40	1.39	3.24	0.657
3.60	1.26	3.12	0.661
3.80	1.12	2.95	0.660
4.00	1.02	2.76	0.651
4.20	0.94	2.58	0.639
4.40	0.90	2.42	0.620
4.50	0.89	2.35	0.607
4.60	0.88	2.28	0.598
4.70	0.86	2.21	0.586
4.80	0.86	2.13	0.572
4.90	0.86	2.07	0.557
5.00	0.85	2.01	0.542
5.10	0.86	1.94	0.523
5.20	0.87	1.87	0.503
5.40	0.93	1.80	0.466
5.60	0.95	1.74	0.443
5.80	0.97	1.74	0.437
6.00	0.94	1.73	0.444
6.20	0.89	1.69	0.446
6.40	0.85	1.66	0.447
6.60	0.80	1.59	0.444
6.80	0.75	1.51	0.439
7.00	0.74	1.45	0.425
7.20	0.71	1.39	0.414
7.40	0.69	1.33	0.404
7.60	0.66	1.23	0.378
7.80	0.67	1.15	0.347
8.00	0.68	1.07	0.315
8.20	0.71	1.00	0.278
8.50	0.74	0.92	0.235
9.0	0.83	0.81	0.170
9.50	0.92	0.74	0.132
10.00	0.98	0.73	0.120
10.50	1.01	0.72	0.112
11.00	1.05	0.69	0.103
11.50	1.09	0.69	0.100
12.00	1.13	0.70	0.101
12.50	1.15	0.73	0.108
13.00	1.15	0.77	0.119
13.50	1.12	0.80	0.128
14.00	1.09	0.82	0.135
14.50	1.03	0.82	0.142
15.00	1.00	0.82	0.143
15.50	0.96	0.80	0.141
16.00	0.92	0.77	0.139
16.50	0.31	0.75	0.134
17.00	0.90	0.73	0.132
17.50	0.88	0.72	0.130
18.00	0.87	0.70	0.129
18.50	0.84	0.69	0.130
19.00	0.82	0.68	0.131
20.00	0.77	0.64	0.130
20.5	0.76	0.63	0.129
21.0	0.74	0.58	0.121
21.5	0.72	0.55	0.116
22.0	0.71	0.52	0.112
22.5	0.70	0.50	0.109
23.0	0.69	0.48	0.105
23.5	0.68	0.45	0.101
24.0	0.68	0.43	0.096
24.5	0.67	0.39	0.089
25.0	0.68	0.36	0.080
25.5	0.68	0.33	0.072
26.0	0.70	0.31	0.063
26.5	0.71	0.28	0.055
27.0	0.72	0.26	0.048
27.5	0.73	0.25	0.043
28.0	0.75	0.23	0.037
29.0	0.77	0.22	0.032
30.0	0.78	0.21	0.030

Cobalt, single crystal, $\vec{E} \parallel \hat{c}$[5]

Energy (eV)	n	k	$R(\phi = 0)$
0.10	6.71	37.87	0.982
0.15	4.66	25.47	0.973
0.20	3.55	18.78	0.962
0.25	3.98	14.59	0.933
0.30	4.04	12.16	0.907
0.40	4.24	9.13	0.847
0.50	4.41	7.19	0.782
0.60	4.91	6.13	0.729
0.70	5.24	5.85	0.713
0.80	5.17	5.89	0.716
0.90	4.94	5.95	0.720
1.00	4.46	5.86	0.722
1.10	4.07	5.61	0.715
1.20	3.81	5.36	0.706
1.30	3.60	5.20	0.701
1.40	3.37	5.09	0.701
1.50	3.10	4.96	0.701
1.60	2.84	4.77	0.697
1.70	2.66	4.57	0.690
1.80	2.45	4.41	0.687
1.90	2.31	4.18	0.675
2.00	2.21	4.00	0.664
2.10	2.13	3.85	0.654
2.20	2.07	3.70	0.642
2.30	2.01	3.59	0.634
2.40	1.95	3.49	0.627
2.50	1.88	3.40	0.622
2.60	1.81	3.32	0.618
2.70	1.73	3.24	0.615
2.80	1.66	3.13	0.607
2.90	1.61	3.05	0.600
3.00	1.55	2.96	0.594
3.20	1.46	2.80	0.579
3.40	1.38	2.64	0.563
3.60	1.31	2.48	0.544
3.80	1.28	2.33	0.519
4.00	1.26	2.20	0.495
4.20	1.25	2.10	0.471
4.40	1.24	2.01	0.452
4.60	1.24	1.94	0.435
4.80	1.23	1.88	0.423
5.00	1.22	1.83	0.411
5.20	1.21	1.79	0.403
5.40	1.19	1.77	0.399
5.60	1.16	1.75	0.400
5.80	1.10	1.73	0.406
6.00	1.03	1.68	0.407
6.20	0.97	1.62	0.401
6.40	0.94	1.53	0.386
6.60	0.91	1.46	0.368
6.80	0.91	1.38	0.345
7.00	0.91	1.32	0.326
7.00	0.91	1.26	0.305
7.40	0.92	1.21	0.286
7.60	0.93	1.17	0.269
7.80	0.94	1.13	0.253
8.00	0.95	1.09	0.239

Cobalt, single crystal, $\vec{E} \perp \hat{c}$[5]

Energy (eV)	n	k	$R(\phi = 0)$
0.10	5.83	32.36	0.979
0.15	4.24	21.37	0.965
0.20	3.87	15.53	0.042
0.30	4.34	10.01	0.865
0.40	4.66	7.39	0.785
0.50	5.17	5.75	0.709
0.60	5.77	5.17	0.682
0.70	6.15	5.20	0.685
0.80	6.08	5.61	0.702
0.90	5.57	5.93	0.715
1.00	4.83	5.94	0.721
1.10	4.31	5.60	0.711
1.20	4.02	5.34	0.701
1.30	3.78	5.16	0.694
1.40	3.55	5.05	0.692
1.50	3.26	4.93	0.692
1.60	3.03	4.74	0.687
1.70	2.83	4.60	0.684
1.80	2.61	4.45	0.683
1.90	2.41	4.27	0.677
2.00	2.25	4.09	0.670
2.10	2.13	3.89	0.659
2.20	2.04	3.72	0.646
2.30	1.99	3.56	0.632
2.40	1.95	3.44	0.620
2.50	1.90	3.34	0.611
2.60	1.86	3.26	0.605
2.70	1.79	3.19	0.602
2.80	1.72	3.11	0.596
2.90	1.66	3.03	0.591
3.00	1.60	2.94	0.586
3.20	1.50	2.78	0.571
3.40	1.42	2.62	0.553
3.60	1.36	2.47	0.533
3.80	1.33	2.33	0.511
4.00	1.31	2.21	0.488
4.20	1.28	2.12	0.471
4.40	1.27	2.03	0.452
4.60	1.26	1.95	0.435
4.80	1.25	1.90	0.423
5.00	1.24	1.84	0.411
5.20	1.22	1.80	0.403
5.40	1.21	1.78	0.399
5.60	1.17	1.76	0.400
5.80	1.11	1.74	0.406
6.00	1.04	1.69	0.407
6.20	0.98	1.62	0.401
6.40	0.94	1.54	0.386
6.60	0.92	1.46	0.368
6.80	0.91	1.38	0.345

Energy (eV)	n	k	R(φ = 0)
7.00	0.91	1.32	0.326
7.20	0.91	1.26	0.305
7.40	0.92	1.21	0.285
7.60	0.93	1.17	0.269
7.80	0.94	1.13	0.253

Copper[6]

Energy (eV)	n	k	R(φ = 0)
0.10	29.69	71.57	0.980
0.50	1.71	17.63	0.979
1.00	0.44	8.48	0.976
1.50	0.26	5.26	0.965
1.70	0.22	4.43	0.958
1.75	0.21	4.25	0.956
1.80	0.21	4.04	0.952
1.85	0.22	3.85	0.947
1.90	0.21	3.67	0.943
2.00	0.27	3.24	0.910
2.10	0.47	2.81	0.814
2.20	0.83	2.60	0.673
2.30	1.04	2.59	0.618
2.40	1.12	2.60	0.602
2.60	1.15	2.50	0.577
2.80	1.17	2.36	0.545
3.00	1.18	2.21	0.509
3.20	1.23	2.07	0.468
3.40	1.27	1.95	0.434
3.60	1.31	1.87	0.407
3.80	1.34	1.81	0.387
4.00	1.34	1.72	0.364
4.20	1.42	1.64	0.336
4.40	1.49	1.64	0.329
4.60	1.52	1.67	0.334
4.80	1.53	1.71	0.345
5.00	1.47	1.78	0.366
5.20	1.38	1.80	0.380
5.40	1.28	1.78	0.389
5.60	1.18	1.74	0.391
5.80	1.10	1.67	0.389
6.00	1.04	1.59	0.380
6.50	0.96	1.37	0.329
7.00	0.97	1.20	0.271
7.50	1.00	1.09	0.230
8.00	1.03	1.03	0.206
8.50	1.03	0.98	0.189
9.00	1.03	0.92	0.171
9.50	1.03	0.87	0.154
10.00	1.04	0.82	0.139
11.00	1.07	0.75	0.118
12.00	1.09	0.73	0.111
13.00	1.08	0.72	0.109
14.00	1.06	0.72	0.111
14.50	1.03	0.72	0.111
15.00	1.01	0.71	0.111
15.50	0.98	0.69	0.109
16.00	0.95	0.67	0.106
17.00	0.91	0.62	0.097
18.00	0.89	0.56	0.084
19.00	0.88	0.51	0.071
20.00	0.88	0.45	0.059
21.00	0.90	0.41	0.048
22.00	0.92	0.38	0.040
23.00	0.94	0.37	0.035
24.00	0.96	0.37	0.035
25.00	0.96	0.40	0.040
26.00	0.92	0.40	0.044
27.00	0.88	0.38	0.043
28.00	0.86	0.35	0.039
29.00	0.85	0.30	0.032
30.00	0.86	0.26	0.025
31.00	0.88	0.24	0.020
32.00	0.89	0.22	0.017
33.00	0.90	0.21	0.015
34.00	0.91	0.20	0.014
35.00	0.92	0.20	0.013
36.00	0.92	0.19	0.012
37.00	0.92	0.19	0.011
38.00	0.93	0.18	0.010
39.00	0.93	0.17	0.009
40.00	0.93	0.17	0.009
41.00	0.94	0.16	0.008
42.00	0.94	0.16	0.007
43.00	0.94	0.15	0.007
44.00	0.95	0.15	0.007
45.00	0.95	0.15	0.006
46.00	0.95	0.15	0.006
47.00	0.95	0.14	0.006
48.00	0.95	0.14	0.006
49.00	0.95	0.14	0.005
50.00	0.95	0.13	0.005
51.00	0.95	0.13	0.005
52.00	0.95	0.13	0.005
53.00	0.96	0.12	0.004
54.00	0.96	0.12	0.004
55.00	0.96	0.12	0.004
56.00	0.96	0.11	0.004
57.00	0.96	0.11	0.004
58.00	0.96	0.11	0.004
59.00	0.97	0.11	0.003
60.00	0.97	0.11	0.003
61.00	0.97	0.11	0.003
62.00	0.97	0.11	0.003
63.00	0.96	0.10	0.003
64.00	0.96	0.10	0.003
65.00	0.97	0.10	0.003
66.00	0.97	0.10	0.003
67.00	0.97	0.09	0.003
68.00	0.97	0.09	0.002
69.00	0.97	0.09	0.002
70.00	0.97	0.09	0.002
75.00	0.98	0.09	0.002
80.00	0.98	0.09	0.002
85.00	0.97	0.09	0.002
90.00	0.96	0.08	0.002

Gallium (liquid)[7]

Energy (eV)	n	k	R(φ = 0)
1.425	2.40	9.20	0.900
1.550	2.09	8.50	0.898
1.771	1.65	7.60	0.898
2.066	1.25	6.60	0.897
2.480	0.89	5.60	0.898
3.100	0.59	4.50	0.896

Germanium, single crystal[8]

Energy (eV)	n	k	R(φ = 0)
0.01240	(4.0065)	3.00E-03	0.361
0.01364	4.0063	2.40E-03	0.361
0.01488	(4.0060)	1.70E-03	0.361
0.01612	(4.0060)	1.55E-03	0.361
0.01736	(4.0060)	1.50E-03	0.361
0.01860		1.50E-03	
0.01984		1.60E-03	
0.02108		1.60E-03	
0.02232		1.55E-03	
0.02356		1.53E-03	
0.02480		1.50E-03	
0.02604		1.25E-03	
0.02728		8.50E-04	
0.02852		6.50E-04	
0.02976		7.00E-04	
0.03100	3.9827	8.50E-04	0.358
0.03224		1.55E-03	
0.03348		2.75E-03	
0.03472		3.55E-03	
0.03596	(3.9900)	3.05E-03	0.359
0.03720		2.75E-03	
0.03844		2.70E-03	
0.03968	(3.9930)	2.90E-03	0.359
0.04092		2.95E-03	
0.04215		3.20E-03	
0.04339		6.30E-03	
0.04463		3.40E-03	
0.04587	(3.9955)	2.50E-03	0.360
0.04711		2.10E-03	
0.04835		2.00E-03	
0.04959		8.00E-04	
0.05083		1.40E-03	
0.05207		1.35E-03	
0.05331		1.10E-03	
0.05455		8.00E-04	
0.05579		6.00E-04	
0.05703		9.0 E-04	
0.05827		6.5 E-04	
0.05951		4.6 E-04	
0.06075		4.0 E-04	
0.06199	3.9992	3.98E-04	0.360
0.06323		4.0 E-04	
0.06447		4.3 E-04	
0.06571		4.4 E-04	
0.06695	(4.0000)	4.3 E-04	0.360
0.06819		3.1 E-04	
0.06943		3.3 E-04	
0.07067		3.8 E-04	
0.07191		3.3 E-04	
0.07315		2.5 E-04	
0.07439		1.9 E-04	
0.07514		1.58E-04	
0.07749	4.0009	9.55E-05	0.360
0.07999	4.0011	1.71E-04	0.360
0.08266	4.0013	9.78E-05	0.360
0.08551	4.0015	5.77E-05	0.360
0.08920		3.98E-05	
0.09460		4.59E-05	
0.09840		3.51E-05	
0.1	4.0063	3.70E-05	0.361
0.2	4.0108		0.361
0.3	4.0246		0.362
0.4	4.0429		0.364
0.5	(4.074)		0.367
0.6	(4.104)	6.58E-07	0.370
0.7	4.180	1.27E-04	0.377
0.8	4.275	5.67E-03	0.385
0.9	4.285	7.45E-02	0.386
1.0	4.325	8.09E-02	0.390
1.1	4.385	0.103	0.395
1.2	4.420	0.123	0.398
1.3	4.495	0.167	0.405

Energy (eV)	*n*	*k*	*R*(ϕ = 0)
1.4	4.560	0.190	0.411
1.5	4.635	0.298	0.418
1.6	4.763	0.345	0.428
1.7	4.897	0.401	0.439
1.8	5.067	0.500	0.453
1.9	5.380	0.540	0.475
2.0	5.588	0.933	0.495
2.1	5.748	1.634	0.523
2.2	5.283	2.049	0.516
2.3	5.062	2.318	0.519
2.4	4.610	2.455	0.508
2.5	4.340	2.384	0.492
2.6	4.180	2.309	0.480
2.7	4.082	2.240	0.471
2.8	4.035	2.181	0.464
2.9	4.037	2.140	0.461
3.0	4.082	2.145	0.463
3.1	4.141	2.215	0.471
3.2	4.157	2.340	0.482
3.3	4.128	2.469	0.490
3.4	4.070	2.579	0.497
3.5	4.020	2.667	0.502
3.6	3.985	2.759	0.509
3.7	3.958	2.863	0.517
3.8	3.936	2.986	0.527
3.9	3.920	3.137	0.539
4.0	3.905	3.336	0.556
4.1	3.869	3.614	0.579
4.2	3.745	4.009	0.612
4.3	3.338	4.507	0.659
4.4	2.516	4.669	0.705
4.5	1.953	4.297	0.713
4.6	1.720	3.960	0.702
4.7	1.586	3.709	0.690
4.8	1.498	3.509	0.677
4.9	1.435	3.342	0.664
5.0	1.394	3.197	0.650
5.1	1.370	3.073	0.636
5.2	1.364	2.973	0.622
5.3	1.371	2.897	0.609
5.4	1.383	2.854	0.600
5.5	1.380	2.842	0.598
5.6	1.360	2.846	0.602
5.7	1.293	2.163	0.479
5.8	1.209	2.873	0.632
5.9	1.108	2.813	0.641
6.0	1.30	2.34	0.517
6.5	1.10	2.05	0.489
7.0	1.00	1.80	0.448
7.5		1.60	
8.0	0.92	1.40	0.348
8.5	0.92	1.20	0.282
9.0	0.92	1.14	0.262
9.5		1.00	
10.0	0.93	0.86	0.167
20.0		0.237	
22.0		0.179	
24.0		0.144	
26.0		0.110	
28.0		0.0747	
30.0		0.1020	
32.0		0.0999	
34.0		0.0856	
36.0		0.0740	
38.0		0.0651	
40.0		0.0604	

Gold, electropolished, Au (110)[9]

Energy (eV)	*n*	*k*	*R*(ϕ = 0)
0.10	8.17	82.83	0.995
0.20	2.13	41.73	0.995
0.30	0.99	27.82	0.995
0.40	0.59	20.83	0.995
0.50	0.39	16.61	0.994
0.60	0.28	13.78	0.994
0.70	0.22	11.75	0.994
0.80	0.18	10.21	0.993
0.90	0.15	9.01	0.993
1.00	0.13	8.03	0.992
1.20	0.10	6.54	0.991
1.40	0.08	5.44	0.989
1.60	0.08	4.56	0.986
1.80	0.09	3.82	0.979
2.00	0.13	3.16	0.953
2.10	0.18	2.84	0.925
2.20	0.24	2.54	0.880
2.40	0.50	1.86	0.647
2.50	0.82	1.59	0.438
2.60	1.24	1.54	0.331
2.70	1.43	1.72	0.356
2.80	1.46	1.77	0.368
2.90	1.50	1.79	0.368
3.00	1.54	1.80	0.369
3.10	1.54	1.81	0.371
3.20	1.54	1.80	0.368
3.30	1.55	1.78	0.362
3.40	1.56	1.76	0.356
3.50	1.58	1.73	0.349
3.60	1.62	1.73	0.346
3.70	1.64	1.75	0.351
3.80	1.63	1.79	0.360
3.90	1.59	1.81	0.366
4.00	1.55	1.81	0.369
4.10	1.51	1.79	0.368
4.20	1.48	1.78	0.367
4.30	1.45	1.77	0.368
4.40	1.41	1.76	0.370
4.50	1.35	1.74	0.370
4.60	1.30	1.69	0.364
4.70	1.27	1.64	0.354
4.80	1.25	1.59	0.344
4.90	1.23	1.54	0.332
5.00	1.22	1.49	0.319
5.20	1.21	1.40	0.295
5.40	1.21	1.33	0.275
5.60	1.21	1.27	0.256
5.80	1.21	1.20	0.236
6.00	1.22	1.14	0.218
6.20	1.24	1.09	0.203
6.40	1.25	1.05	0.190
6.60	1.27	1.01	0.177
6.80	1.30	0.97	0.167
7.00	1.34	0.95	0.162
7.20	1.36	0.95	0.161
7.40	1.38	0.96	0.164
7.60	1.38	0.98	0.169
7.80	1.35	0.99	0.171
8.00	1.31	0.96	0.165
8.20	1.30	0.92	0.155
8.40	1.30	0.89	0.147
8.60	1.31	0.88	0.144
8.80	1.31	0.86	0.140
9.00	1.30	0.83	0.133
9.20	1.31	0.81	0.126
9.40	1.33	0.78	0.122
9.60	1.36	0.78	0.121
9.80	1.37	0.79	0.124
10.00	1.37	0.80	0.126
10.20	1.36	0.80	0.127
10.40	1.35	0.80	0.125
10.60	1.34	0.79	0.123
10.80	1.34	0.77	0.120
11.00	1.34	0.76	0.116
11.20	1.34	0.74	0.113
11.40	1.35	0.73	0.111
11.60	1.36	0.72	0.109
11.80	1.38	0.71	0.108
12.00	1.39	0.71	0.109
12.40	1.44	0.73	0.115
12.80	1.45	0.79	0.127
13.20	1.42	0.84	0.137
13.60	1.37	0.86	0.140
14.00	1.33	0.86	0.140
14.40	1.29	0.86	0.139
14.80	1.26	0.84	0.135
15.20	1.24	0.83	0.132
15.60	1.22	0.81	0.127
16.00	1.21	0.79	0.123
16.40	1.20	0.78	0.119
16.80	1.19	0.76	0.116
17.20	1.19	0.75	0.114
17.60	1.19	0.74	0.111
18.00	1.19	0.74	0.109
18.40	1.19	0.73	0.109
18.80	1.20	0.74	0.110
19.20	1.21	0.76	0.116
19.60	1.21	0.80	0.125
20.00	1.18	0.83	0.133
20.40	1.14	0.85	0.141
20.80	1.10	0.87	0.149
21.20	1.05	0.88	0.156
21.60	1.00	0.88	0.162
22.00	0.94	0.86	0.164
22.40	0.89	0.83	0.163
22.80	0.85	0.79	0.157
23.20	0.82	0.75	0.149
23.60	0.80	0.70	0.138
24.00	0.80	0.66	0.125
24.40	0.80	0.62	0.113
24.80	0.80	0.58	0.101
25.20	0.82	0.56	0.090
25.60	0.83	0.54	0.084
26.00	0.84	0.52	0.079
26.40	0.85	0.51	0.074
26.80	0.85	0.50	0.071
27.20	0.86	0.49	0.068
27.60	0.86	0.49	0.065
28.00	0.87	0.48	0.063
28.40	0.88	0.48	0.062
28.80	0.88	0.48	0.062
29.20	0.88	0.48	0.062
29.60	0.87	0.48	0.064
30.00	0.86	0.48	0.064

Hafnium, single crystal, $\vec{E} \parallel \hat{c}$[10]

Energy (eV)	*n*	*k*	*R*(ϕ = 0)
0.52	1.48	4.11	0.747

Energy (eV)	*n*	*k*	*R*(ϕ = 0)
0.56	1.84	3.29	0.615
0.60	2.34	2.62	0.486
0.66	3.21	2.13	0.428
0.70	3.70	2.03	0.441
0.76	4.31	2.10	0.476
0.80	4.61	2.31	0.504
0.86	4.71	2.70	0.533
0.90	4.64	2.85	0.541
0.95	4.54	2.96	0.545
1.00	4.45	3.00	0.545
1.10	4.28	3.08	0.547
1.20	4.08	3.10	0.544
1.30	3.87	3.04	0.536
1.40	3.72	2.95	0.525
1.50	3.60	2.85	0.514
1.60	3.52	2.73	0.500
1.70	3.52	2.61	0.488
1.80	3.57	2.56	0.485
1.90	3.63	2.59	0.489
2.00	3.65	2.67	0.498
2.10	3.64	2.81	0.511
2.20	3.53	2.99	0.526
2.30	3.34	3.09	0.534
2.40	3.15	3.11	0.537
2.50	2.99	3.13	0.540
2.60	2.83	3.12	0.542
2.70	2.68	3.10	0.542
2.80	2.54	3.08	0.543
2.90	2.40	3.04	0.544
3.00	2.27	3.00	0.544
3.10	2.14	2.95	0.544
3.20	2.00	2.89	0.544
3.30	1.87	2.79	0.538
3.40	1.78	2.68	0.528
3.50	1.71	2.58	0.517
3.60	1.66	2.48	0.503
3.70	1.63	2.40	0.491
3.80	1.60	2.33	0.481
3.90	1.56	2.27	0.473
4.00	1.52	2.21	0.466
4.10	1.48	2.14	0.455
4.20	1.45	2.07	0.442
4.30	1.43	2.01	0.431
4.40	1.41	1.95	0.420
4.50	1.39	1.89	0.407
4.60	1.39	1.83	0.394
4.70	1.39	1.79	0.382
4.80	1.38	1.75	0.373
4.90	1.38	1.71	0.364
5.00	1.37	1.68	0.356
5.20	1.36	1.61	0.341
5.40	1.35	1.55	0.324
5.60	1.35	1.51	0.314
5.80	1.32	1.48	0.308
6.00	1.28	1.41	0.295
6.20	1.26	1.35	0.278
6.40	1.26	1.28	0.258
6.60	1.27	1.22	0.240
6.80	1.28	1.16	0.224
7.00	1.31	1.13	0.212
7.20	1.33	1.10	0.204
7.40	1.34	1.07	0.197
7.60	1.36	1.05	0.191
7.80	1.37	1.02	0.183
8.00	1.40	1.01	0.179
8.20	1.43	1.01	0.178
8.40	1.45	1.01	0.180
8.60	1.47	1.02	0.183
8.80	1.48	1.04	0.186
9.00	1.49	1.07	0.193
9.20	1.50	1.10	0.201
9.40	1.48	1.14	0.211
9.60	1.46	1.18	0.222
9.80	1.41	1.21	0.230
10.00	1.36	1.22	0.235
10.20	1.32	1.22	0.238
10.40	1.28	1.22	0.240
10.60	1.24	1.21	0.241
10.80	1.20	1.20	0.242
11.00	1.16	1.19	0.242
11.20	1.13	1.17	0.241
11.40	1.10	1.16	0.241
11.60	1.07	1.14	0.239
11.80	1.04	1.12	0.238
12.00	1.02	1.10	0.236
12.40	0.96	1.06	0.232
12.80	0.92	1.01	0.225
13.20	0.88	0.96	0.218
13.60	0.84	0.90	0.205
14.00	0.83	0.83	0.186
14.40	0.83	0.80	0.172
14.80	0.81	0.76	0.167
15.20	0.79	0.70	0.153
15.60	0.79	0.64	0.132
16.00	0.83	0.60	0.111
16.40	0.81	0.60	0.114
16.80	0.79	0.55	0.105
17.20	0.79	0.50	0.089
17.60	0.80	0.46	0.077
18.00	0.81	0.42	0.064
18.40	0.84	0.38	0.051
18.80	0.87	0.34	0.040
19.00	0.89	0.33	0.036
19.60	0.93	0.32	0.030
20.00	0.94	0.31	0.027
20.60	0.97	0.30	0.023
21.00	0.99	0.29	0.022
21.60	1.01	0.28	0.020
22.00	1.03	0.28	0.020
22.60	1.06	0.28	0.020
23.00	1.07	0.28	0.021
23.60	1.09	0.29	0.022
24.00	1.09	0.30	0.023
24.60	1.10	0.31	0.024

Hafnium, single crystal, $\vec{E} \perp \hat{c}$[10]

Energy (eV)	*n*	*k*	*R*(ϕ = 0)
0.52	2.25	4.65	0.723
0.56	2.34	3.66	0.623
0.60	2.84	2.89	0.512
0.66	3.71	2.35	0.469
0.70	4.26	2.21	0.482
0.76	4.97	2.33	0.521
0.80	5.41	2.62	0.554
0.86	5.46	3.36	0.593
0.90	5.22	3.62	0.601
0.95	4.95	3.72	0.602
1.00	4.76	3.76	0.602
1.10	4.43	3.80	0.601
1.20	4.07	3.74	0.594
1.30	3.79	3.55	0.578
1.40	3.61	3.36	0.561
1.50	3.55	3.13	0.540
1.60	3.58	3.01	0.529
1.70	3.63	2.98	0.526
1.80	3.66	3.02	0.530
1.90	3.63	3.14	0.541
2.00	3.51	3.26	0.551
2.10	3.35	3.33	0.558
2.20	3.18	3.36	0.563
2.30	2.99	3.39	0.568
2.40	2.78	3.35	0.569
2.50	2.65	3.26	0.562
2.60	2.54	3.22	0.560
2.70	2.42	3.17	0.559
2.80	2.31	3.13	0.558
2.90	2.20	3.08	0.558
3.00	2.08	3.05	0.561
3.10	1.94	2.98	0.560
3.20	1.83	2.88	0.555
3.30	1.74	2.78	0.547
3.40	1.68	2.69	0.538
3.50	1.62	2.61	0.529
3.60	1.57	2.52	0.519
3.70	1.53	2.45	0.510
3.80	1.49	2.38	0.501
3.90	1.45	2.32	0.493
4.00	1.41	2.25	0.484
4.10	1.38	2.18	0.474
4.20	1.35	2.11	0.462
4.30	1.33	2.05	0.451
4.40	1.31	1.99	0.438
4.50	1.30	1.93	0.427
4.60	1.29	1.88	0.415
4.70	1.28	1.82	0.402
4.80	1.28	1.77	0.389
4.90	1.27	1.73	0.379
5.00	1.27	1.69	0.367
5.20	1.27	1.62	0.349
5.40	1.27	1.57	0.335
5.60	1.26	1.52	0.322
5.80	1.24	1.48	0.313
6.00	1.21	1.42	0.302
6.20	1.19	1.36	0.285
6.40	1.18	1.29	0.265
6.60	1.19	1.22	0.244
6.80	1.21	1.18	0.230
7.00	1.22	1.14	0.217
7.20	1.23	1.10	0.206
7.40	1.26	1.06	0.194
7.60	1.28	1.04	0.187
7.80	1.30	1.02	0.180
8.00	1.33	1.00	0.174
8.20	1.35	0.99	0.173
8.40	1.38	0.99	0.173
8.60	1.40	1.00	0.174
8.80	1.42	1.02	0.178
9.00	1.43	1.04	0.184
9.20	1.45	1.08	0.193
9.40	1.43	1.12	0.204
9.60	1.40	1.16	0.214
9.80	1.37	1.19	0.223
10.00	1.32	1.21	0.230
10.20	1.27	1.21	0.234
10.40	1.23	1.20	0.235
10.60	1.19	1.20	0.237

Energy (eV)	*n*	*k*	$R(\phi = 0)$
10.80	1.15	1.19	0.237
11.00	1.12	1.17	0.237
11.20	1.08	1.16	0.237
11.40	1.05	1.14	0.236
11.60	1.03	1.12	0.235
11.80	1.00	1.10	0.233
12.00	0.97	1.08	0.231
12.40	0.92	1.04	0.226
12.80	0.88	0.99	0.219
13.20	0.83	0.94	0.211
13.60	0.80	0.88	0.196
14.00	0.79	0.81	0.177
14.40	0.80	0.77	0.160
14.80	0.77	0.73	0.154
15.20	0.76	0.68	0.140
15.60	0.76	0.61	0.119
16.00	0.81	0.58	0.099
16.40	0.78	0.57	0.102
16.80	0.77	0.53	0.092
17.20	0.77	0.48	0.077
17.60	0.79	0.44	0.065
18.00	0.80	0.39	0.053
18.40	0.82	0.36	0.041
18.80	0.86	0.33	0.032
19.00	0.88	0.32	0.030
19.60	0.91	0.31	0.025
20.00	0.93	0.30	0.023
20.60	0.96	0.29	0.021
21.00	0.97	0.29	0.020
21.60	1.00	0.28	0.019
22.00	1.01	0.28	0.019
22.60	1.03	0.27	0.018
23.00	1.05	0.28	0.019
23.60	1.06	0.28	0.020
24.00	1.07	0.29	0.021
24.60	1.09	0.30	0.022

Iridium[11]

Energy (eV)	*n*	*k*	$R(\phi = 0)$
0.10	28.49	60.62	0.975
0.15	15.32	45.15	0.973
0.20	9.69	35.34	0.972
0.25	6.86	28.84	0.969
0.30	5.16	24.25	0.967
0.35	4.11	20.79	0.964
0.40	3.42	18.06	0.960
0.45	3.05	15.82	0.954
0.50	2.98	14.06	0.944
0.60	2.79	11.58	0.925
0.70	2.93	9.78	0.895
0.80	3.14	8.61	0.862
0.90	3.19	7.88	0.840
1.00	3.15	7.31	0.822
1.10	3.04	6.84	0.808
1.20	2.96	6.41	0.791
1.30	2.85	6.07	0.779
1.40	2.72	5.74	0.767
1.50	2.65	5.39	0.750
1.60	2.68	5.08	0.728
1.70	2.69	4.92	0.716
1.80	2.64	4.81	0.710
1.90	2.57	4.68	0.704
2.00	2.50	4.57	0.699
2.10	2.40	4.48	0.697
2.20	2.29	4.38	0.695
2.30	2.18	4.26	0.692
2.40	2.07	4.14	0.689
2.50	1.98	4.00	0.682
2.60	1.91	3.86	0.673
2.70	1.85	3.73	0.665
2.80	1.81	3.61	0.655
2.90	1.77	3.51	0.646
3.00	1.73	3.43	0.640
3.20	1.62	3.26	0.629
3.40	1.53	3.05	0.610
3.60	1.52	2.81	0.573
3.80	1.61	2.69	0.541
4.00	1.64	2.68	0.535
4.20	1.58	2.71	0.549
4.40	1.45	2.68	0.561
4.60	1.31	2.60	0.567
4.80	1.18	2.49	0.570
5.00	1.10	2.35	0.559
5.20	1.04	2.22	0.543
5.40	1.00	2.09	0.522
5.60	0.98	1.98	0.499
5.80	0.96	1.86	0.474
6.00	0.95	1.78	0.454
6.20	0.94	1.68	0.427
6.40	0.94	1.59	0.401
6.60	0.94	1.50	0.375
6.80	0.95	1.42	0.345
7.00	0.97	1.34	0.318
7.20	0.99	1.27	0.290
7.40	1.02	1.20	0.262
7.60	1.03	1.14	0.241
7.80	1.08	1.06	0.208
8.00	1.13	1.03	0.191
8.20	1.18	1.00	0.179
8.40	1.22	0.98	0.171
8.60	1.26	0.96	0.164
8.80	1.29	0.95	0.160
9.00	1.33	0.94	0.157
9.20	1.36	0.95	0.159
9.40	1.39	0.95	0.161
9.60	1.42	0.97	0.163
9.80	1.44	0.99	0.169
10.00	1.45	1.01	0.175
10.20	1.45	1.04	0.182
10.40	1.44	1.07	0.187
10.60	1.43	1.09	0.193
10.80	1.41	1.12	0.200
11.00	1.38	1.13	0.206
11.20	1.34	1.14	0.208
11.40	1.31	1.13	0.208
11.60	1.28	1.12	0.206
11.80	1.25	1.10	0.203
12.00	1.24	1.08	0.199
12.40	1.21	1.05	0.191
12.80	1.19	1.01	0.181
13.20	1.18	0.98	0.173
13.60	1.17	0.95	0.165
14.00	1.16	0.91	0.155
14.40	1.17	0.88	0.147
14.80	1.18	0.87	0.142
15.20	1.19	0.84	0.136
15.60	1.20	0.83	0.133
16.00	1.21	0.83	0.131
16.40	1.23	0.82	0.129
16.80	1.25	0.82	0.127
17.20	1.28	0.83	0.131
17.60	1.30	0.87	0.140
18.00	1.30	0.93	0.154
18.40	1.27	0.97	0.166
18.80	1.24	1.00	0.176
19.20	1.20	1.03	0.187
19.60	1.15	1.05	0.197
20.00	1.10	1.06	0.205
20.50	1.04	1.05	0.210
21.00	0.99	1.04	0.215
21.50	0.94	1.02	0.220
22.00	0.89	1.00	0.222
22.50	0.84	0.99	0.228
23.00	0.79	0.96	0.232
23.50	0.76	0.92	0.228
24.00	0.73	0.87	0.223
24.50	0.70	0.83	0.218
25.00	0.69	0.79	0.209
25.50	0.68	0.76	0.200
26.00	0.67	0.72	0.192
26.50	0.67	0.69	0.181
27.00	0.66	0.66	0.174
27.50	0.66	0.63	0.166
28.00	0.66	0.61	0.158
28.50	0.66	0.59	0.151
29.00	0.65	0.57	0.148
29.50	0.64	0.55	0.145
30.00	0.64	0.53	0.140
32.00	0.62	0.44	0.119
34.00	0.64	0.35	0.091
36.00	0.69	0.27	0.059
38.00	0.73	0.24	0.044
40.00	0.76	0.22	0.034

Iron[5]

Energy (eV)	*n*	*k*	$R(\phi = 0)$
0.10	6.41	33.07	0.978
0.15	6.26	22.82	0.956
0.20	3.68	18.23	0.958
0.26	4.98	13.68	0.911
0.30	4.87	12.05	0.892
0.36	4.68	10.44	0.867
0.40	4.42	9.75	0.858
0.50	4.14	8.02	0.817
0.60	3.93	6.95	0.783
0.70	3.78	6.17	0.752
0.80	3.65	5.60	0.725
0.90	3.52	5.16	0.700
1.00	3.43	4.79	0.678
1.10	3.33	4.52	0.660
1.20	3.24	4.26	0.641
1.30	3.16	4.07	0.626
1.40	3.12	3.87	0.609
1.50	3.05	3.77	0.601
1.60	3.00	3.60	0.585
1.70	2.98	3.52	0.577
1.80	2.92	3.46	0.573
1.90	2.89	3.37	0.563
2.00	2.85	3.36	0.563
2.10	2.80	3.34	0.562
2.20	2.74	3.33	0.563
2.30	2.65	3.34	0.567
2.40	2.56	3.31	0.567
2.50	2.46	3.31	0.570
2.60	2.34	3.30	0.576
2.70	2.23	3.25	0.575
2.80	2.12	3.23	0.580

Energy (eV)	n	k	R(φ = 0)
2.90	2.01	3.17	0.580
3.00	1.88	3.12	0.583
3.10	1.78	3.04	0.580
3.20	1.70	2.96	0.576
3.30	1.62	2.87	0.572
3.40	1.55	2.79	0.565
3.50	1.50	2.70	0.556
3.60	1.47	2.63	0.548
3.70	1.43	2.56	0.542
3.83	1.38	2.49	0.534
4.00	1.30	2.39	0.527
4.17	1.26	2.27	0.510
4.33	1.23	2.18	0.494
4.50	1.20	2.10	0.482
4.67	1.16	2.02	0.470
4.83	1.14	1.93	0.451
5.00	1.14	1.87	0.435
5.17	1.12	1.81	0.425
5.33	1.11	1.75	0.408
5.50	1.09	1.17	0.401
5.67	1.09	1.65	0.383
5.83	1.10	1.61	0.373
6.00	1.09	1.59	0.366
6.17	1.08	1.57	0.365
6.33	1.04	1.55	0.365
6.50	1.02	1.51	0.358
6.67	1.00	1.47	0.351
6.83	0.97	1.43	0.346
7.00	0.96	1.39	0.333
7.17	0.94	1.35	0.327
7.33	0.94	1.30	0.311
7.50	0.94	1.26	0.298
7.67	0.94	1.23	0.288
7.83	0.94	1.21	0.279
8.00	0.94	1.18	0.272
8.17	0.94	1.16	0.265
8.33	0.94	1.14	0.258
8.50	0.94	1.12	0.251
8.67	0.94	1.10	0.246
8.83	0.92	1.08	0.240
9.00	0.93	1.07	0.236
9.17	0.92	1.06	0.233
9.33	0.91	1.04	0.231
9.50	0.90	1.02	0.226
9.67	0.90	1.00	0.221
9.83	0.89	0.99	0.218
10.00	0.88	0.97	0.213
10.17	0.87	0.94	0.203
10.33	0.87	0.91	0.196
10.50	0.87	0.89	0.189
10.67	0.88	0.87	0.179
10.83	0.89	0.85	0.170
11.00	0.91	0.83	0.162
11.17	0.92	0.83	0.159
11.33	0.93	0.84	0.159
11.50	0.93	0.84	0.160
11.67	0.93	0.84	0.162
11.83	0.92	0.84	0.163
12.00	0.91	0.84	0.163
12.17	0.90	0.84	0.165
12.33	0.89	0.83	0.164
12.50	0.98	0.83	0.165
12.67	0.87	0.82	0.166
12.83	0.86	0.81	0.166
13.00	0.85	0.80	0.162
13.17	0.84	0.79	0.161
13.33	0.84	0.78	0.160
13.50	0.83	0.77	0.159
13.67	0.82	0.76	0.157
13.83	0.81	0.75	0.154
14.00	0.81	0.73	0.151
14.17	0.80	0.72	0.149
14.33	0.80	0.71	0.146
14.50	0.79	0.79	0.144
14.67	0.79	0.69	0.141
14.83	0.78	0.67	0.138
15.00	0.78	0.66	0.135
15.17	0.78	0.65	0.131
15.33	0.78	0.64	0.238
15.50	0.77	0.63	0.126
15.67	0.77	0.62	0.123
15.83	0.77	0.61	0.119
16.00	0.77	0.60	0.116
16.17	0.78	0.58	0.112
16.33	0.78	0.58	0.110
16.50	0.78	0.57	0.107
16.67	0.77	0.56	0.106
16.83	0.78	0.55	0.103
17.00	0.78	0.55	0.102
17.17	0.78	0.54	0.100
17.33	0.78	0.54	0.098
17.50	0.77	0.53	0.097
17.67	0.77	0.52	0.095
17.83	0.78	0.51	0.092
18.00	0.78	0.51	0.091
18.17	0.78	0.51	0.090
18.33	0.78	0.50	0.089
18.50	0.77	0.50	0.089
18.67	0.77	0.50	0.088
18.83	0.77	0.49	0.087
19.00	0.77	0.49	0.087
19.17	0.76	0.49	0.088
19.33	0.76	0.48	0.087
19.50	0.75	0.47	0.086
19.67	0.75	0.47	0.085
19.83	0.75	0.46	0.084
20.00	0.74	0.45	0.083
20.17	0.74	0.44	0.081
20.33	0.74	0.44	0.081
20.50	0.74	0.42	0.080
20.67	0.73	0.43	0.079
20.83	0.73	0.42	0.078
21.00	0.73	0.41	0.077
21.17	0.72	0.40	0.076
21.33	0.72	0.39	0.074
21.50	0.72	0.38	0.073
21.67	0.72	0.38	0.071
21.83	0.72	0.37	0.070
22.00	0.72	0.36	0.068
22.17	0.71	0.35	0.067
22.33	0.72	0.34	0.064
22.50	0.72	0.34	0.063
22.67	0.72	0.33	0.062
22.83	0.72	0.32	0.059
23.00	0.72	0.31	0.058
23.17	0.72	0.30	0.056
23.33	0.72	0.29	0.054
23.50	0.73	0.28	0.050
23.67	0.73	0.28	0.049
23.83	0.74	0.27	0.047
24.00	0.74	0.27	0.045
24.17	0.74	0.26	0.044
24.33	0.74	0.26	0.043
24.50	0.74	0.25	0.042
24.67	0.75	0.25	0.040
24.83	0.75	0.24	0.039
25.00	0.75	0.24	0.038
26.00	0.76	0.21	0.031
27.00	0.78	0.18	0.026
28.00	0.79	0.16	0.021
29.00	0.81	0.14	0.017
30.00	0.82	0.13	0.014
Lithium[12]			
0.14	0.659	38.0	0.998
0.54	0.661	12.6	0.984
0.75	0.561	7.68	0.963
1.05	0.448	5.58	0.946
1.35	0.338	4.36	0.935
1.65	0.265	3.55	0.925
1.95	0.221	2.94	0.913
2.25	0.206	2.48	0.892
2.55	0.217	2.11	0.854
2.85	0.247	1.82	0.797
3.15	0.304	1.60	0.715
3.45	0.334	1.45	0.656
3.75	0.345	1.32	0.611
4.05	0.346	1.21	0.578
4.35	0.333	1.11	0.557
4.65	0.317	1.01	0.540
4.95	0.302	0.906	0.520
5.25	0.299	0.795	0.484
5.55	0.310	0.688	0.434
5.85	0.342	0.594	0.365
6.15	0.376	0.522	0.306
6.45	0.408	0.460	0.256
6.75	0.440	0.407	0.214
7.05	0.466	0.364	0.183
7.35	0.492	0.320	0.155
7.65	0.517	0.282	0.131
7.95	0.545	0.246	0.109
8.25	0.572	0.214	0.091
8.55	0.601	0.189	0.075
8.85	0.624	0.163	0.063
9.15	0.657	0.144	0.050
9.45	0.680	0.130	0.042
9.75	0.708	0.119	0.034
10.1	0.726	0.108	0.029
10.4	0.743	0.102	0.025
10.6	0.753	0.080	0.022
Magnesium (evaporated)[13]			
2.145	0.48	3.71	0.880
2.270	0.57	3.47	0.843
2.522	0.53	2.92	0.805
2.845	0.52	2.65	0.777
3.064	0.52	2.05	0.681
5.167	0.10	1.60	0.894
5.636	0.15	1.50	0.832
6.200	0.20	1.40	0.765
6.889	0.25	1.30	0.693
7.750	0.20	1.20	0.722
8.857	0.15	0.95	0.730
10.335	0.25	0.40	0.419

Energy (eV)	*n*	*k*	$R(\phi = 0)$
Manganese[14]			
0.64	3.89	5.95	0.738
0.77	3.78	5.41	0.710
0.89	3.65	5.02	0.688
1.02	3.48	4.74	0.673
1.14	3.30	4.53	0.662
1.26	3.10	4.35	0.653
1.39	2.97	4.18	0.643
1.51	2.83	4.03	0.634
1.64	2.70	3.91	0.627
1.76	2.62	3.78	0.617
1.88	2.56	3.65	0.606
2.01	2.51	3.54	0.596
2.13	2.47	3.43	0.585
2.26	2.39	3.33	0.577
2.38	2.32	3.23	0.567
2.50	2.25	3.14	0.559
2.63	2.19	3.06	0.552
2.75	2.11	2.98	0.545
2.88	2.06	2.90	0.536
3.00	2.00	2.82	0.528
3.12	1.96	2.74	0.518
3.25	1.92	2.67	0.509
3.37	1.89	2.59	0.498
3.50	1.89	2.51	0.484
3.62	1.87	2.45	0.475
3.74	1.86	2.38	0.463
3.87	1.86	2.32	0.451
3.99	1.86	2.25	0.438
4.12	1.86	2.19	0.427
4.24	1.85	2.14	0.417
4.36	1.85	2.08	0.406
4.49	1.86	2.03	0.395
4.61	1.85	1.99	0.388
4.74	1.84	1.94	0.378
4.86	1.83	1.91	0.372
4.98	1.82	1.86	0.362
5.11	1.82	1.82	0.354
5.23	1.81	1.79	0.348
5.36	1.78	1.76	0.342
5.48	1.74	1.73	0.337
5.60	1.73	1.70	0.331
5.73	1.72	1.67	0.325
5.85	1.70	1.64	0.319
5.98	1.67	1.61	0.313
6.10	1.63	1.58	0.307
6.22	1.62	1.55	0.301
6.35	1.59	1.52	0.295
6.47	1.55	1.50	0.292
6.60	1.48	1.47	0.288
Mercury (liquid)[15]			
0.2	13.99	14.27	0.869
0.3	11.37	11.95	0.846
0.4	9.741	10.65	0.830
0.5	8.528	9.805	0.818
0.6	7.574	9.195	0.808
0.8	6.086	8.312	0.796
1.0	4.962	7.643	0.789
1.2	4.050	7.082	0.786
1.4	3.324	6.558	0.785
1.6	2.746	6.054	0.783
1.8	2.284	5.582	0.782
2.0	1.910	5.150	0.782
2.2	1.620	4.751	0.780
2.4	1.384	4.407	0.779
2.6	1.186	4.090	0.779
2.8	1.027	3.802	0.779
3.0	0.898	3.538	0.777
3.2	0.798	3.294	0.773
3.4	0.713	3.074	0.770
3.6	0.644	2.860	0.763
3.8	0.589	2.665	0.755
4.0	0.542	2.502	0.749
4.2	0.507	2.341	0.738
4.4	0.477	2.195	0.727
4.6	0.452	2.058	0.715
4.8	0.431	1.929	0.701
5.0	0.414	1.806	0.685
5.2	0.401	1.687	0.666
5.4	0.394	1.569	0.642
5.6	0.386	1.454	0.617
5.7	0.386	1.396	0.601
5.8	0.386	1.341	0.585
5.9	0.385	1.287	0.569
6.0	0.386	1.232	0.551
6.1	0.388	1.176	0.531
6.2	0.390	1.118	0.510
6.3	0.399	1.058	0.481
6.4	0.412	1.002	0.450
6.5	0.428	0.949	0.418
6.6	0.436	0.898	0.392
6.7	0.438	0.836	0.367
6.8	0.459	0.756	0.320
6.9	0.510	0.676	0.255
7.0	0.585	0.617	0.191
7.1	0.663	0.589	0.148
7.2	0.717	0.584	0.128
7.3	0.769	0.575	0.111
7.4	0.817	0.574	0.100
7.5	0.860	0.580	0.094
7.6	0.893	0.597	0.093
7.8	0.929	0.623	0.096
8.0	0.946	0.639	0.098
8.2	0.952	0.645	0.099
8.4	0.953	0.638	0.097
8.6	0.956	0.624	0.093
8.8	0.965	0.607	0.087
9.0	0.975	0.588	0.082
9.2	0.988	0.568	0.076
9.4	1.009	0.548	0.069
9.6	1.044	0.541	0.066
9.8	1.061	0.557	0.069
10.0	1.062	0.567	0.071
10.2	1.054	0.569	0.072
10.4	1.045	0.561	0.070
10.6	1.041	0.550	0.068
10.8	1.039	0.537	0.065
11.0	1.039	0.523	0.062
11.5	1.050	0.491	0.055
12.0	1.064	0.467	0.050
12.5	1.078	0.445	0.045
13.0	1.092	0.430	0.042
13.5	1.104	0.416	0.040
14.0	1.115	0.404	0.038
14.5	1.125	0.394	0.037
15.0	1.135	0.383	0.035
15.5	1.146	0.374	0.034
16.0	1.159	0.368	0.034
16.5	1.170	0.367	0.034
17.0	1.177	0.367	0.034
17.5	1.184	0.366	0.034
18.0	1.191	0.367	0.035
18.5	1.195	0.367	0.035
19.0	1.200	0.366	0.035
19.5	1.208	0.364	0.035
Molybdenum[16]			
0.10	18.53	68.51	0.985
0.15	8.78	47.54	0.985
0.20	5.10	35.99	0.985
0.25	3.36	28.75	0.984
0.30	2.44	23.80	0.983
0.34	2.00	20.84	0.982
0.38	1.70	18.44	0.980
0.42	1.57	16.50	0.978
0.46	1.46	14.91	0.975
0.50	1.37	13.55	0.971
0.54	1.35	12.36	0.966
0.58	1.34	11.34	0.960
0.62	1.38	10.44	0.952
0.66	1.43	9.67	0.942
0.70	1.48	8.99	0.932
0.74	1.51	8.38	0.921
0.78	1.60	7.83	0.906
0.82	1.64	7.35	0.892
0.86	1.70	6.89	0.876
9.90	1.74	6.48	0.859
1.00	1.94	5.58	0.805
1.10	2.15	4.85	0.743
1.20	2.44	4.22	0.671
1.30	2.77	3.74	0.608
1.40	3.15	3.40	0.562
1.50	3.53	3.30	0.550
1.60	3.77	3.41	0.562
1.70	3.84	3.51	0.570
1.80	3.81	3.58	0.576
1.90	3.74	3.58	0.576
2.00	3.68	3.52	0.571
2.10	3.68	3.45	0.565
2.20	3.76	3.41	0.562
2.30	3.79	3.61	0.578
2.40	3.59	3.78	0.594
2.50	3.36	3.73	0.591
2.60	3.22	3.61	0.582
2.70	3.13	3.51	0.573
2.80	3.08	3.42	0.565
2.90	3.05	3.33	0.566
3.00	3.04	3.27	0.550
3.10	3.03	3.21	0.544
3.20	3.05	3.18	0.540
3.30	3.06	3.18	0.540
3.40	3.06	3.19	0.541
3.50	3.06	3.21	0.543
3.60	3.05	3.23	0.546
3.70	3.04	3.27	0.550
3.80	3.04	3.31	0.554
3.90	3.04	3.40	0.564
4.00	3.01	3.51	0.576
4.20	2.77	3.77	0.610
4.40	2.39	3.88	0.640
4.60	2.06	3.84	0.658
4.80	1.75	3.76	0.678
5.00	1.46	3.62	0.695
5.20	1.22	3.42	0.706

Energy (eV)	n	k	$R(\phi = 0)$
5.40	1.07	3.20	0.706
5.60	0.96	2.99	0.700
5.80	0.89	2.80	0.688
6.00	0.85	2.64	0.674
6.20	0.81	2.50	0.660
6.40	0.79	2.36	0.641
6.60	0.78	2.24	0.619
6.80	0.78	2.13	0.592
7.00	0.80	2.04	0.568
7.20	0.81	1.98	0.548
7.40	0.81	1.95	0.542
7.60	0.75	1.90	0.552
7.80	0.71	1.81	0.542
8.00	0.69	1.73	0.530
8.20	0.67	1.65	0.512
8.40	0.66	1.57	0.495
8.60	0.65	1.49	0.475
8.80	0.65	1.41	0.450
9.00	0.65	1.33	0.420
9.20	0.67	1.25	0.385
9.40	0.69	1.19	0.355
9.60	0.71	1.12	0.320
9.80	0.74	1.05	0.285
10.00	0.77	0.99	0.250
10.20	0.81	0.93	0.217
10.40	0.86	0.88	0.188
10.60	0.91	0.83	0.162
10.80	0.98	0.79	0.138
11.00	1.05	0.77	0.125
11.20	1.12	0.78	0.123
11.40	1.18	0.80	0.125
11.60	1.23	0.85	0.135
11.80	1.25	0.89	0.145
12.00	1.26	0.92	0.154
12.40	1.25	0.98	0.168
12.80	1.23	1.00	0.178
13.20	1.20	1.02	0.185
13.60	1.17	1.02	0.187
14.00	1.15	1.01	0.185
14.40	1.13	1.00	0.182
14.80	1.13	0.99	0.179
15.00	1.14	0.99	0.179
15.60	1.15	1.01	0.184
16.00	1.14	1.04	0.194
16.60	1.10	1.10	0.216
17.00	1.04	1.12	0.233
17.60	0.94	1.14	0.257
18.00	0.87	1.12	0.270
18.60	0.77	1.08	0.283
19.00	0.71	1.02	0.284
19.60	0.66	0.94	0.275
20.00	0.64	0.89	0.264
20.60	0.62	0.81	0.245
21.00	0.61	0.77	0.234
21.60	0.61	0.71	0.215
22.00	0.60	0.69	0.207
22.60	0.59	0.63	0.195
23.00	0.58	0.60	0.185
23.60	0.58	0.53	0.166
24.00	0.58	0.49	0.151
24.60	0.60	0.43	0.124
25.00	0.62	0.39	0.106
25.60	0.66	0.35	0.085
26.00	0.68	0.33	0.072
26.50	0.71	0.31	0.060

Energy (eV)	n	k	$R(\phi = 0)$
27.00	0.73	0.29	0.050
27.50	0.76	0.28	0.041
28.00	0.79	0.27	0.036
28.50	0.81	0.26	0.031
29.00	0.83	0.26	0.028
29.50	0.86	0.26	0.025
30.00	0.88	0.26	0.023
31.00	0.92	0.29	0.024
32.00	0.92	0.32	0.030
33.00	0.90	0.33	0.032
34.00	0.91	0.34	0.034
35.00	0.87	0.37	0.043
36.00	0.82	0.34	0.043
37.00	0.81	0.30	0.038
38.00	0.81	0.27	0.033
39.00	0.82	0.25	0.029
40.00	0.83	0.23	0.025

Nickel[17]

Energy (eV)	n	k	$R(\phi = 0)$
0.10	9.54	45.82	0.983
0.15	5.45	30.56	0.978
0.20	4.12	22.48	0.969
0.25	4.25	17.68	0.950
0.30	4.19	15.05	0.934
0.35	4.03	13.05	0.918
0.40	3.84	11.43	0.900
0.50	4.03	9.64	0.864
0.60	3.84	8.35	0.835
0.70	3.59	7.48	0.813
0.80	3.38	6.82	0.794
0.90	3.18	6.23	0.774
1.00	3.06	5.74	0.753
1.10	2.97	5.38	0.734
1.20	2.85	5.10	0.721
1.30	2.74	4.85	0.708
1.40	2.65	4.63	0.695
1.50	2.53	4.47	0.688
1.60	2.43	4.31	0.679
1.70	2.28	4.18	0.677
1.80	2.14	4.01	0.670
1.90	2.02	3.82	0.659
2.00	1.92	3.65	0.649
2.10	1.85	3.48	0.634
2.20	1.80	3.33	0.620
2.30	1.75	3.19	0.605
2.40	1.71	3.06	0.590
2.50	1.67	2.93	0.575
2.60	1.65	2.81	0.557
2.70	1.64	2.71	0.542
2.80	1.63	2.61	0.525
2.90	1.62	2.52	0.509
3.00	1.61	2.44	0.495
3.10	1.61	2.36	0.480
3.20	1.61	2.30	0.467
3.30	1.61	2.23	0.454
3.40	1.62	2.17	0.441
3.50	1.63	2.11	0.428
3.60	1.64	2.07	0.416
3.70	1.66	2.02	0.405
3.80	1.69	1.99	0.397
3.90	1.72	1.98	0.393
4.00	1.73	1.98	0.392
4.20	1.74	2.01	0.396
4.40	1.71	2.06	0.409
4.60	1.63	2.09	0.421

Energy (eV)	n	k	$R(\phi = 0)$
4.80	1.53	2.11	0.435
5.00	1.40	2.10	0.449
5.20	1.27	2.04	0.454
5.40	1.16	1.94	0.449
5.60	1.09	1.83	0.435
5.80	1.04	1.73	0.417
6.20	1.00	1.54	0.371
6.40	1.01	1.46	0.345
6.60	1.01	1.40	0.325
6.80	1.02	1.35	0.308
7.00	1.03	1.30	0.291
7.20	1.03	1.27	0.282
7.40	1.03	1.24	0.273
7.60	1.02	1.22	0.265
7.80	1.01	1.18	0.256
8.00	1.01	1.15	0.248
8.20	1.00	1.13	0.242
8.40	0.99	1.11	0.235
8.60	0.98	1.08	0.228
8.80	0.97	1.05	0.220
9.00	0.97	1.01	0.211
9.20	0.96	0.99	0.203
9.40	0.95	0.96	0.194
9.60	0.95	0.93	0.185
9.80	0.95	0.89	0.175
10.00	0.95	0.87	0.166
10.20	0.95	0.83	0.155
10.40	0.95	0.80	0.145
10.60	0.97	0.76	0.129
10.80	0.99	0.75	0.123
11.00	1.01	0.73	0.115
11.25	1.04	0.72	0.111
11.50	1.05	0.71	0.109
11.75	1.07	0.71	0.108
12.00	1.07	0.71	0.108
12.25	1.07	0.71	0.107
12.50	1.08	0.71	0.106
12.75	1.08	0.71	0.106
13.00	1.08	0.71	0.105
13.25	1.08	0.71	0.105
13.50	1.07	0.70	0.105
13.75	1.07	0.70	0.105
14.00	1.07	0.71	0.106
14.25	1.06	0.70	0.106
14.50	1.05	0.70	0.106
14.75	1.04	0.70	0.107
15.00	1.03	0.70	0.107
15.25	1.02	0.69	0.106
15.50	1.01	0.69	0.105
15.75	1.00	0.68	0.104
16.00	0.99	0.67	0.103
16.50	0.98	0.66	0.101
17.00	0.96	0.64	0.098
17.50	[illegible]	0.63	0.096
18.00	0.92	0.61	0.092
18.50	0.91	0.58	0.087
19.00	0.90	0.56	0.082
19.50	0.90	0.54	0.077
20.00	0.89	0.51	0.071
20.50	0.89	0.49	0.066
21.00	0.90	0.47	0.061
21.50	0.91	0.46	0.057
22.00	0.91	0.45	0.055
22.50	0.91	0.44	0.053
23.00	0.92	0.44	0.051

Energy (eV)	*n*	*k*	$R(\phi = 0)$
23.50	0.91	0.44	0.052
24.00	0.90	0.43	0.051
24.50	0.90	0.43	0.051
25.00	0.89	0.42	0.050
26.00	0.88	0.39	0.046
27.00	0.87	0.37	0.042
28.00	0.87	0.35	0.040
29.00	0.86	0.34	0.037
30.00	0.86	0.32	0.034
35.00	0.86	0.24	0.022
40.00	0.87	0.18	0.014
45.00	0.88	0.13	0.008
50.00	0.92	0.10	0.004
60.00	0.96	0.08	0.002
65.00	0.98	0.09	0.002
68.00	0.96	0.12	0.004
70.00	0.94	0.11	0.004
75.00	0.94	0.09	0.003
80.00	0.94	0.07	0.002
90.00	0.94	0.06	0.002

Niobium[18]

Energy (eV)	*n*	*k*	$R(\phi = 0)$
0.12	15.99	53.20	0.979
0.20	7.25	34.14	0.976
0.24	5.47	28.88	0.975
0.28	4.26	24.95	0.974
0.35	3.11	20.03	0.970
0.45	2.28	15.58	0.964
0.55	1.83	12.67	0.956
0.65	1.57	10.59	0.947
0.75	1.41	9.00	0.935
0.85	1.35	7.74	0.918
0.95	1.35	6.70	0.893
1.05	1.44	5.86	0.857
1.15	1.55	5.18	0.814
1.25	1.65	4.63	0.768
1.35	1.76	4.13	0.715
1.45	1.95	3.68	0.650
1.55	2.15	3.37	0.595
1.65	2.36	3.13	0.552
1.75	2.54	2.99	0.527
1.85	2.69	2.89	0.510
1.95	2.82	2.86	0.505
2.05	2.89	2.87	0.505
2.15	2.92	2.87	0.505
2.25	2.93	2.87	0.505
2.35	2.92	2.88	0.506
2.45	2.89	2.90	0.509
2.55	2.83	2.92	0.512
2.65	2.74	2.90	0.511
2.75	2.66	2.86	0.507
2.85	2.58	2.80	0.500
3.00	2.51	2.68	0.485
3.10	2.48	2.60	0.475
3.20	2.45	2.53	0.465
3.30	2.44	2.45	0.453
3.40	2.46	2.38	0.442
3.50	2.48	2.33	0.435
3.60	2.52	2.29	0.428
3.70	2.56	2.27	0.426
3.80	2.59	2.28	0.427
3.90	2.62	2.29	0.429
4.00	2.64	2.33	0.434
4.20	2.64	2.42	0.447
4.40	2.53	2.56	0.467
4.60	2.39	2.56	0.470
4.80	2.32	2.52	0.465
5.00	2.26	2.57	0.475
5.20	2.16	2.62	0.487
5.40	2.00	2.68	0.505
5.60	1.81	2.67	0.518
5.80	1.63	2.60	0.522
6.00	1.49	2.49	0.520
6.20	1.38	2.38	0.512
6.40	1.31	2.25	0.496
6.60	1.26	2.14	0.480
6.80	1.24	2.04	0.460
7.00	1.23	1.96	0.441
7.20	1.22	1.91	0.430
7.40	1.20	1.88	0.427
7.60	1.14	1.85	0.430
7.80	1.07	1.78	0.428
8.00	1.02	1.69	0.412
8.20	1.00	1.60	0.390
8.40	0.99	1.51	0.365
8.60	0.99	1.43	0.340
8.70	0.99	1.39	0.328
8.80	1.00	1.36	0.315
9.00	1.01	1.29	0.290
9.20	1.04	1.22	0.265
9.40	1.07	1.18	0.245
9.60	1.10	1.13	0.227
9.80	1.13	1.09	0.209
10.00	1.18	1.05	0.194
10.20	1.23	1.04	0.187
10.40	1.27	1.04	0.185
10.60	1.30	1.06	0.190
10.80	1.32	1.08	0.195
11.00	1.32	1.10	0.200
11.20	1.31	1.12	0.204
11.40	1.30	1.13	0.207
11.60	1.28	1.13	0.209
11.80	1.27	1.13	0.210
12.00	1.25	1.12	0.209
12.40	1.24	1.10	0.204
12.80	1.24	1.09	0.200
13.20	1.24	1.09	0.201
13.60	1.23	1.12	0.208
14.00	1.20	1.13	0.216
14.40	1.16	1.15	0.225
14.80	1.11	1.16	0.234
15.00	1.08	1.16	0.238
15.60	0.99	1.14	0.247
16.00	0.92	1.11	0.250
16.60	0.85	1.04	0.245
17.00	0.80	0.99	0.240
17.20	0.79	0.96	0.236
17.40	0.77	0.93	0.230
17.80	0.75	0.87	0.217
18.00	0.74	0.85	0.209
18.60	0.73	0.77	0.185
19.00	0.72	0.72	0.170
19.60	0.72	0.66	0.150
20.00	0.72	0.62	0.137
20.60	0.71	0.55	0.119
21.00	0.72	0.50	0.100
21.60	0.75	0.43	0.075
22.00	0.78	0.40	0.063
22.60	0.82	0.35	0.045
23.00	0.85	0.33	0.038
23.60	0.88	0.30	0.029
24.00	0.91	0.29	0.025
24.60	0.94	0.28	0.022
25.00	0.96	0.27	0.020
25.60	0.99	0.26	0.018
26.00	1.00	0.26	0.017
26.60	1.03	0.25	0.016
27.00	1.04	0.25	0.015
27.60	1.06	0.25	0.015
28.00	1.08	0.24	0.015
28.60	1.11	0.24	0.016
29.00	1.13	0.25	0.017
29.60	1.16	0.26	0.020
30.00	1.18	0.28	0.023
31.00	1.18	0.31	0.026
32.00	1.20	0.34	0.031
33.00	1.21	0.38	0.038
34.00	1.20	0.42	0.044
35.20	1.17	0.47	0.051
36.00	1.15	0.50	0.056
37.50	1.07	0.53	0.064
39.50	0.95	0.50	0.063
40.50	0.92	0.47	0.059

Osmium (Polycrystalline)[9]

Energy (eV)	*n*	*k*	$R(\phi = 0)$
0.10	4.08	50.23	0.994
0.15	2.90	33.60	0.990
0.20	2.44	25.11	0.985
0.25	2.35	19.99	0.977
0.30	2.23	16.54	0.969
0.35	2.33	14.06	0.955
0.40	2.45	12.32	0.940
0.45	2.43	11.02	0.927
0.50	2.41	9.97	0.913
0.55	2.33	9.12	0.901
0.60	2.21	8.37	0.890
0.65	2.11	7.68	0.877
0.70	2.02	7.04	0.862
0.75	2.00	6.46	0.842
0.80	2.00	5.95	0.820
0.85	2.01	5.51	0.796
0.90	2.03	5.10	0.769
0.95	2.05	4.74	0.742
1.00	2.09	4.41	0.712
1.10	2.15	3.84	0.651
1.20	2.16	3.35	0.592
1.30	2.25	2.77	0.506
1.40	2.49	2.23	0.419
1.50	2.84	1.80	0.369
1.60	3.36	1.62	0.379
1.70	3.70	1.75	0.411
1.80	3.78	1.83	0.423
1.90	3.81	1.75	0.418
2.00	3.98	1.60	0.418
2.10	4.26	1.54	0.432
2.20	4.58	1.62	0.457
2.30	4.84	1.76	0.479
2.40	5.10	2.01	0.506
2.50	5.28	2.38	0.532
2.60	5.36	2.82	0.557
2.70	5.30	3.29	0.580
2.80	5.07	3.78	0.603
2.90	4.65	4.18	0.624
3.00	4.05	4.40	0.639
3.20	3.29	3.96	0.614

Energy (eV)	*n*	*k*	*R*(ϕ = 0)
3.40	2.93	3.79	0.607
3.60	2.75	3.45	0.577
3.80	2.73	3.32	0.562
4.00	2.71	3.34	0.565
4.20	2.53	3.44	0.584
4.40	2.24	3.44	0.599
4.60	2.01	3.31	0.598
4.80	1.88	3.19	0.592
5.00	1.74	3.12	0.596
5.20	1.58	3.00	0.597
5.40	1.46	2.88	0.593
5.60	1.36	2.77	0.589
5.80	1.27	2.65	0.582
6.00	1.20	2.54	0.575
6.20	1.13	2.44	0.571
6.40	1.06	2.33	0.562
6.60	1.01	2.21	0.548
6.80	0.97	2.11	0.532
7.00	0.95	2.00	0.514
7.20	0.92	1.91	0.497
7.40	0.91	1.81	0.476
7.60	0.90	1.72	0.451
7.80	0.90	1.63	0.426
8.00	0.91	1.55	0.400
8.20	0.91	1.48	0.375
8.40	0.94	1.40	0.344
8.60	0.96	1.34	0.319
8.80	0.98	1.29	0.296
9.00	1.01	1.24	0.274
9.20	1.04	1.19	0.255
9.40	1.08	1.16	0.238
9.60	1.10	1.14	0.229
9.80	1.13	1.11	0.217
10.00	1.16	1.10	0.209
10.20	1.19	1.08	0.203
10.30	1.20	1.08	0.201
10.40	1.22	1.08	0.200
10.50	1.23	1.09	0.201
10.60	1.24	1.10	0.203
10.80	1.25	1.11	0.206
11.00	1.24	1.13	0.213
11.20	1.23	1.14	0.217
11.40	1.19	1.15	0.223
11.60	1.17	1.12	0.216
11.80	1.16	1.10	0.211
12.00	1.15	1.08	0.205
12.40	1.14	1.03	0.191
12.80	1.15	1.01	0.183
13.20	1.16	0.98	0.174
13.60	1.17	0.97	0.170
14.00	1.17	0.96	0.169
14.40	1.16	0.94	0.165
14.80	1.16	0.91	0.156
15.20	1.17	0.89	0.148
15.60	1.20	0.86	0.140
16.00	1.25	0.87	0.140
16.40	1.28	0.90	0.147
16.80	1.28	0.94	0.157
17.20	1.27	0.97	0.167
17.60	1.26	1.01	0.178
18.00	1.23	1.04	0.189
18.40	1.19	1.08	0.200
18.80	1.14	1.10	0.210
19.20	1.10	1.10	0.219
19.60	1.05	1.11	0.227
20.00	0.96	1.10	0.239
20.40	0.93	1.09	0.240
20.80	0.89	1.05	0.240
21.20	0.86	1.02	0.237
21.60	0.83	0.99	0.235
22.00	0.80	0.96	0.230
22.40	0.78	0.93	0.226
22.80	0.77	0.90	0.220
23.20	0.75	0.88	0.217
23.60	0.75	0.86	0.211
24.00	0.73	0.84	0.209
24.40	0.72	0.82	0.207
24.80	0.70	0.80	0.205
25.20	0.69	0.77	0.202
25.60	0.67	0.75	0.199
26.00	0.66	0.72	0.195
26.40	0.65	0.69	0.189
26.80	0.63	0.66	0.183
27.20	0.65	0.62	0.165
28.00	0.64	0.59	0.156
28.40	0.64	0.57	0.148
28.80	0.65	0.55	0.140
29.20	0.65	0.53	0.134
29.60	0.65	0.51	0.128
30.00	0.65	0.49	0.121
31.00	0.65	0.45	0.111
32.00	0.66	0.41	0.095
33.00	0.68	0.37	0.079
34.00	0.70	0.34	0.068
35.00	0.72	0.31	0.057
36.00	0.74	0.29	0.048
37.00	0.77	0.27	0.040
38.00	0.79	0.26	0.035
39.00	0.81	0.26	0.031
40.00	0.84	0.26	0.026

Palladium[19]

Energy (eV)	*n*	*k*	*R*(ϕ = 0)
0.10	4.13	54.15	0.994
0.15	3.13	35.82	0.990
0.20	3.07	26.59	0.983
0.26	3.11	20.15	0.971
0.30	3.56	17.27	0.955
0.36	3.98	14.41	0.932
0.40	4.27	13.27	0.916
0.46	4.27	12.11	0.902
0.50	4.10	11.44	0.896
0.56	3.92	10.49	0.883
0.60	3.80	9.96	0.876
0.72	3.51	8.70	0.854
0.80	3.35	8.06	0.840
1.00	2.99	6.89	0.811
1.10	2.81	6.46	0.800
1.20	2.65	6.10	0.790
1.30	2.50	5.78	0.781
1.40	2.34	5.50	0.774
1.50	2.17	5.22	0.767
1.60	2.08	4.95	0.755
1.70	2.00	4.72	0.745
1.80	1.92	4.54	0.737
1.90	1.82	4.35	0.729
2.00	1.75	4.18	0.721
2.10	1.67	4.03	0.714
2.20	1.60	3.88	0.707
2.30	1.53	3.75	0.700
2.40	1.47	3.61	0.693
2.50	1.41	3.48	0.685
2.60	1.37	3.36	0.676
2.70	1.32	3.25	0.668
2.80	1.29	3.13	0.658
2.90	1.26	3.03	0.648
3.00	1.23	2.94	0.639
3.10	1.20	2.85	0.630
3.20	1.17	2.77	0.622
3.30	1.14	2.68	0.613
3.40	1.12	2.60	0.602
3.50	1.10	2.52	0.591
3.60	1.08	2.45	0.581
3.70	1.07	2.38	0.570
3.80	1.06	2.31	0.558
3.90	1.05	2.25	0.547
4.00	1.03	2.19	0.537
4.20	1.04	2.09	0.510
4.40	1.03	2.01	0.493
4.60	1.03	1.94	0.476
4.80	1.01	1.90	0.470
5.00	0.96	1.86	0.472
5.20	0.90	1.79	0.474
5.40	0.85	1.70	0.463
5.60	0.81	1.62	0.449
5.80	0.78	1.54	0.437
6.00	0.76	1.45	0.418
6.20	0.74	1.37	0.397
6.40	0.73	1.29	0.375
6.60	0.72	1.21	0.350
6.80	0.73	1.13	0.316
7.00	0.73	1.05	0.287
7.20	0.75	0.98	0.255
7.40	0.77	0.91	0.223
7.60	0.79	0.85	0.195
7.80	0.83	0.78	0.163
8.00	0.88	0.73	0.133
8.20	0.94	0.70	0.117
8.40	0.96	0.70	0.114
8.60	1.00	0.65	0.097
8.80	1.04	0.65	0.094
9.00	1.07	0.64	0.090
9.50	1.12	0.65	0.089
10.00	1.14	0.65	0.088
10.50	1.16	0.65	0.087
11.00	1.18	0.64	0.086
11.50	1.19	0.65	0.087
12.00	1.20	0.66	0.089
12.50	1.19	0.67	0.091
13.00	1.18	0.67	0.091
13.50	1.18	0.67	0.092
14.00	1.17	0.67	0.093
14.50	1.15	0.68	0.095
15.00	1.13	0.69	0.098
15.50	1.10	0.68	0.096
16.00	1.08	0.66	0.092
16.50	1.06	0.63	0.086
17.00	1.07	0.61	0.081
17.50	1.06	0.61	0.080
18.00	1.07	0.59	0.077
18.50	1.07	0.59	0.077
19.00	1.08	0.59	0.077
19.50	1.08	0.61	0.080
20.00	1.07	0.65	0.090
20.50	1.03	0.67	0.098
21.00	0.99	0.67	0.103

Energy (eV)	n	k	$R(\phi = 0)$
21.50	0.95	0.66	0.103
22.00	0.91	0.64	0.103
22.50	0.88	0.62	0.101
23.00	0.86	0.59	0.097
23.50	0.85	0.56	0.091
24.00	0.84	0.54	0.086
25.00	0.81	0.51	0.084
26.40	0.80	0.43	0.066
27.80	0.81	0.38	0.052
29.20	0.82	0.35	0.046
Platinum[20]			
0.10	13.21	44.72	0.976
0.15	8.18	31.16	0.969
0.20	5.90	23.95	0.962
0.25	4.70	19.40	0.954
0.30	3.92	16.16	0.945
0.35	3.28	13.66	0.936
0.40	2.81	11.38	0.922
0.45	3.03	9.31	0.882
0.50	3.91	7.71	0.813
0.55	4.58	7.14	0.777
0.60	5.13	6.75	0.753
0.65	5.52	6.66	0.746
0.70	5.71	6.83	0.751
0.75	5.57	7.02	0.759
0.80	5.31	7.04	0.762
0.85	5.05	6.98	0.763
0.90	4.77	6.91	0.765
0.95	4.50	6.77	0.763
1.00	4.25	6.62	0.762
1.10	3.86	6.24	0.753
1.20	3.55	5.92	0.746
1.30	3.29	5.61	0.736
1.40	3.10	5.32	0.725
1.50	2.92	5.07	0.716
1.60	2.76	4.84	0.706
1.70	2.63	4.64	0.697
1.80	2.51	4.43	0.686
1.90	2.38	4.26	0.678
2.00	2.30	4.07	0.664
2.10	2.23	3.92	0.654
2.20	2.17	3.77	0.642
2.30	2.10	3.67	0.636
2.40	2.03	3.54	0.626
2.50	1.96	3.42	0.616
2.60	1.91	3.30	0.605
2.70	1.87	3.20	0.595
2.80	1.83	3.10	0.585
2.90	1.79	3.01	0.575
3.00	1.75	2.92	0.565
3.20	1.68	2.76	0.546
3.40	1.63	2.62	0.527
3.60	1.58	2.48	0.507
3.80	1.53	2.37	0.491
4.00	1.49	2.25	0.472
4.20	1.45	2.14	0.452
4.40	1.43	2.04	0.432
4.60	1.39	1.95	0.415
4.80	1.38	1.85	0.392
5.00	1.36	1.76	0.372
5.20	1.36	1.67	0.350
5.40	1.36	1.61	0.332
5.60	1.36	1.54	0.315
5.80	1.36	1.47	0.295

Energy (eV)	n	k	$R(\phi = 0)$
6.00	1.38	1.40	0.276
6.20	1.39	1.35	0.261
6.40	1.42	1.29	0.246
6.60	1.45	1.26	0.236
6.80	1.48	1.24	0.231
7.00	1.50	1.24	0.230
7.20	1.50	1.25	0.231
7.40	1.49	1.23	0.228
7.60	1.48	1.22	0.225
7.80	1.48	1.20	0.221
8.00	1.47	1.18	0.216
8.20	1.47	1.17	0.212
8.40	1.47	1.15	0.209
8.60	1.47	1.14	0.205
8.80	1.47	1.13	0.202
9.00	1.48	1.12	0.200
9.20	1.49	1.11	0.198
9.40	1.49	1.12	0.200
9.60	1.49	1.13	0.203
9.80	1.48	1.15	0.207
10.00	1.46	1.15	0.209
10.20	1.43	1.16	0.211
10.40	1.40	1.15	0.210
10.60	1.37	1.14	0.207
10.80	1.35	1.12	0.203
11.00	1.33	1.10	0.199
11.20	1.31	1.08	0.194
11.40	1.30	1.06	0.188
11.60	1.29	1.04	0.183
11.80	1.29	1.01	0.177
12.00	1.29	1.00	0.173
12.40	1.29	0.97	0.165
12.80	1.29	0.94	0.158
13.20	1.31	0.93	0.155
13.60	1.31	0.93	0.155
14.00	1.31	0.93	0.155
14.40	1.30	0.93	0.156
14.80	1.27	0.93	0.157
15.20	1.27	0.93	0.155
15.60	1.25	0.92	0.151
16.00	1.24	0.89	0.146
16.50	1.24	0.87	0.142
17.00	1.25	0.86	0.138
17.50	1.27	0.85	0.135
18.00	1.31	0.88	0.142
18.50	1.30	0.94	0.157
19.00	1.28	0.99	0.171
19.50	1.23	1.03	0.184
20.00	1.18	1.06	0.197
20.50	1.11	1.09	0.212
21.00	1.03	1.10	0.226
21.50	0.94	1.08	0.238
22.00	0.87	1.04	0.240
22.50	0.81	0.98	0.235
23.00	0.77	0.92	0.226
23.50	0.75	0.87	0.213
24.00	0.74	0.82	0.201
24.50	0.73	0.77	0.187
25.00	0.73	0.73	0.174
25.50	0.73	0.70	0.162
26.00	0.74	0.67	0.150
26.50	0.74	0.65	0.142
27.00	0.74	0.63	0.136
27.50	0.74	0.62	0.130
28.00	0.75	0.60	0.125

Energy (eV)	n	k	$R(\phi = 0)$
28.50	0.75	0.59	0.121
29.00	0.75	0.58	0.118
29.50	0.74	0.58	0.120
30.00	0.73	0.58	0.124
Potassium[21]			
0.55	0.139	7.10	0.989
0.58	0.119	6.72	0.990
0.63	0.106	6.32	0.990
0.67	0.091	5.79	0.990
0.73	0.079	5.30	0.989
0.81	0.066	4.75	0.989
0.92	0.056	4.19	0.988
1.05	0.044	3.58	0.987
1.23	0.040	3.04	0.985
1.44	0.040	2.56	0.979
1.65	0.044	2.19	0.970
1.87	0.050	1.84	0.955
2.07	0.053	1.62	0.943
2.27	0.049	1.43	0.938
2.45	0.046	1.28	0.933
2.64	0.043	1.14	0.928
2.82	0.043	1.02	0.919
2.95	0.041	0.898	0.913
3.06	0.041	0.799	0.905
3.40	0.052	0.549	0.852
3.71	0.089	0.288	0.719
3.97	0.287	0.091	0.310
4.00	0.34	0.08	0.245
4.065	0.38	0.07	0.204
4.133	0.41	0.07	0.177
4.203	0.45	0.06	0.145
4.275	0.48	0.06	0.125
4.350	0.52	0.05	0.101
4.428	0.55	0.05	0.085
4.509	0.58	0.05	0.072
4.592	0.61	0.05	0.060
4.679	0.64	0.04	0.049
4.769	0.66	0.04	0.043
4.862	0.68	0.04	0.037
4.959	0.70	0.04	0.032
5.061	0.72	0.04	0.027
5.166	0.74	0.04	0.023
5.276	0.76	0.04	0.019
5.391	0.78	0.04	0.016
5.510	0.79	0.05	0.015
5.637	0.81	0.05	0.012
5.767	0.83	0.05	0.009
6.048	0.85	0.05	0.007
6.199	0.87	0.05	0.006
6.358	0.88	0.05	0.005
6.526	0.90	0.06	0.004
6.702	0.91	0.06	0.003
6.888	0.92	0.06	0.003
7.085	0.92	0.06	0.003
7.293	0.93	0.06	0.002
7.514	0.93	0.06	0.002
7.749	0.94	0.06	0.002
7.999	0.94	0.06	0.002
8.260	0.94	0.06	0.002
8.551	0.94	0.06	0.002
8.856	0.94	0.05	0.002
9.184	0.94	0.05	0.002
9.537	0.94	0.04	0.001
9.919	0.94	0.04	0.001

Energy (eV)	n	k	R(ϕ = 0)
10.33	0.94	0.03	0.001
11.0		0.03	
12.0		0.028	

Rhenium, single crystal, $\vec{E} \parallel \hat{c}$ [9]

Energy (eV)	n	k	R(ϕ = 0)
0.10	6.06	51.03	0.991
0.15	4.66	33.96	0.984
0.20	4.16	25.36	0.975
0.25	4.03	20.10	0.962
0.30	4.37	16.69	0.943
0.35	4.50	14.53	0.925
0.40	4.53	12.96	0.909
0.45	4.53	11.78	0.893
0.50	4.53	10.88	0.878
0.55	4.50	10.26	0.867
0.60	4.29	9.75	0.861
0.65	4.07	9.35	0.856
0.70	3.80	8.94	0.853
0.75	3.48	8.55	0.850
0.80	3.21	8.10	0.846
0.85	2.96	7.68	0.841
0.90	2.73	7.24	0.835
0.95	2.56	6.79	0.826
1.00	2.45	6.36	0.813
1.10	2.38	5.61	0.778
1.20	2.35	5.02	0.742
1.30	2.39	4.54	0.702
1.40	2.44	4.13	0.662
1.50	2.50	3.79	0.624
1.60	2.59	3.49	0.587
1.70	2.70	3.27	0.557
1.80	2.82	3.10	0.535
1.90	2.90	3.00	0.520
2.00	2.97	2.91	0.510
2.10	3.03	2.86	0.504
2.20	3.06	2.84	0.501
2.30	3.07	2.82	0.499
2.40	3.06	2.81	0.498
2.50	3.02	2.80	0.497
2.60	2.96	2.77	0.493
2.70	2.89	2.68	0.482
2.80	2.89	2.57	0.468
2.90	2.99	2.47	0.457
3.00	3.11	2.57	0.470
3.20	2.90	2.68	0.482
3.40	2.83	2.50	0.459
3.60	2.93	2.48	0.457
3.80	2.86	2.56	0.467
4.00	2.81	2.51	0.460
4.20	2.86	2.55	0.466
4.40	2.81	2.74	0.489
4.60	2.56	2.83	0.504
4.80	2.41	2.71	0.493
5.00	2.39	2.68	0.488
5.20	2.34	2.75	0.500
5.40	2.20	2.81	0.515
5.60	2.02	2.84	0.530
5.80	1.83	2.80	0.538
6.00	1.65	2.71	0.541
6.20	1.54	2.59	0.532
6.40	1.45	2.50	0.526
6.80	1.32	2.31	0.508
7.00	1.26	2.23	0.500
7.20	1.20	2.15	0.493
7.40	1.16	2.06	0.480
7.60	1.12	1.99	0.470
7.80	1.08	1.89	0.454
8.00	1.05	1.80	0.435
8.20	1.05	1.71	0.411
8.40	1.05	1.62	0.386
8.60	1.06	1.55	0.360
8.80	1.09	1.48	0.336
9.00	1.11	1.43	0.317
9.20	1.13	1.39	0.301
9.40	1.16	1.34	0.281
9.60	1.18	1.32	0.274
9.80	1.20	1.29	0.264
10.00	1.23	1.26	0.252
10.20	1.25	1.25	0.246
10.40	1.28	1.25	0.242
10.60	1.29	1.25	0.242
10.80	1.30	1.26	0.244
11.00	1.30	1.27	0.247
11.20	1.29	1.28	0.249
11.40	1.28	1.28	0.252
11.60	1.26	1.28	0.252
11.80	1.24	1.26	0.249
12.00	1.23	1.24	0.244
12.40	1.22	1.21	0.237
12.80	1.21	1.18	0.230
13.20	1.22	1.16	0.222
13.60	1.22	1.13	0.215
14.00	1.24	1.12	0.209
14.40	1.27	1.11	0.204
14.80	1.29	1.15	0.213
15.20	1.29	1.19	0.225
15.60	1.26	1.22	0.236
16.00	1.23	1.25	0.248
16.40	1.19	1.27	0.259
16.80	1.14	1.29	0.269
17.00	1.12	1.30	0.275
17.40	1.07	1.30	0.286
18.00	0.99	1.30	0.300
18.40	0.93	1.29	0.311
18.80	0.87	1.28	0.321
19.20	0.81	1.25	0.330
19.60	0.77	1.21	0.332
20.00	0.73	1.18	0.333
20.40	0.70	1.14	0.332
20.80	0.67	1.11	0.332
21.20	0.64	1.08	0.334
21.60	0.61	1.04	0.335
22.00	0.58	1.01	0.340
22.40	0.55	0.97	0.341
22.80	0.53	0.93	0.338
23.20	0.51	0.89	0.334
23.60	0.50	0.85	0.329
24.00	0.48	0.80	0.319
24.40	0.48	0.76	0.207
24.80	0.47	0.72	0.296
25.20	0.47	0.68	0.282
25.60	0.47	0.65	0.270
26.00	0.47	0.61	0.255
26.40	0.48	0.57	0.240
26.80	0.48	0.54	0.225
27.20	0.49	0.51	0.208
27.60	0.50	0.48	0.193
28.00	0.51	0.45	0.176
29.00	0.54	0.39	0.145
30.00	0.57	0.33	0.114
31.00	0.62	0.29	0.086
32.00	0.66	0.26	0.065
33.00	0.68	0.24	0.054
34.00	0.72	0.21	0.041
35.00	0.76	0.20	0.031
36.00	0.79	0.20	0.025
37.00	0.82	0.19	0.021
38.00	0.85	0.20	0.018
39.00	0.89	0.21	0.016
40.00	0.88	0.26	0.022
42.00	0.88	0.26	0.022
44.00	0.89	0.29	0.026
46.00	0.85	0.32	0.035
48.00	0.82	0.30	0.036
50.00	0.80	0.30	0.038
52.00	0.78	0.30	0.044
54.00	0.72	0.30	0.055
56.00	0.66	0.24	0.061
58.00	0.65	0.16	0.055

Rhenium, single crystal, $\vec{E} \perp \hat{c}$ [9]

Energy (eV)	n	k	R(ϕ = 0)
0.10	4.25	42.83	0.991
0.15	3.28	28.08	0.984
0.20	3.28	20.66	0.971
0.25	3.47	16.27	0.951
0.30	3.73	13.44	0.926
0.35	3.93	11.54	0.900
0.40	3.99	10.15	0.875
0.45	4.17	9.03	0.846
0.50	4.31	8.26	0.821
0.55	4.45	7.73	0.801
0.60	4.53	7.40	0.788
0.65	4.44	7.26	0.784
0.70	4.13	7.09	0.784
0.75	3.77	6.75	0.779
0.80	3.55	6.32	0.766
0.85	3.39	5.95	0.752
0.90	3.26	5.61	0.737
0.95	3.17	5.27	0.719
1.00	3.09	4.96	0.701
1.10	3.05	4.39	0.658
1.20	3.08	3.89	0.613
1.30	3.20	3.56	0.578
1.40	3.23	3.38	0.559
1.50	3.23	3.12	0.532
1.60	3.29	2.88	0.507
1.70	3.38	2.72	0.491
1.80	3.47	2.59	0.480
1.90	3.54	2.50	0.473
2.00	3.63	2.43	0.469
2.10	3.74	2.40	0.470
2.20	3.83	2.38	0.472
2.30	3.93	2.44	0.481
2.40	4.00	2.55	0.492
2.50	4.01	2.70	0.505
2.60	3.90	2.84	0.514
2.70	3.74	2.92	0.517
2.80	3.57	2.88	0.511
2.90	3.49	2.75	0.497
3.00	3.53	2.71	0.493
3.20	3.55	2.84	0.506
3.40	3.34	2.88	0.508
3.60	3.25	2.83	0.501
3.80	3.24	2.84	0.502
4.00	3.19	2.94	0.513

Energy (eV)	n	k	R(ϕ = 0)
4.20	3.05	3.06	0.526
4.40	2.88	3.15	0.539
4.60	2.67	3.18	0.548
4.80	2.44	3.17	0.554
5.00	2.25	3.12	0.556
5.20	2.10	3.04	0.555
5.40	1.96	2.96	0.553
5.60	1.84	2.88	0.551
5.80	1.73	2.81	0.549
6.00	1.61	2.74	0.549
6.20	1.51	2.64	0.545
6.40	1.42	2.56	0.541
6.80	1.28	2.37	0.526
7.00	1.22	2.28	0.517
7.20	1.16	2.19	0.508
7.40	1.12	2.08	0.493
7.60	1.12	1.98	0.468
7.80	1.08	1.93	0.463
8.00	1.05	1.83	0.443
8.20	1.05	1.74	0.418
8.40	1.05	1.66	0.397
8.60	1.06	1.58	0.372
8.80	1.07	1.52	0.351
9.00	1.09	1.46	0.327
9.20	1.11	1.41	0.309
9.40	1.14	1.36	0.290
9.60	1.17	1.31	0.273
9.80	1.20	1.27	0.258
10.00	1.24	1.24	0.244
10.20	1.29	1.22	0.234
10.40	1.33	1.23	0.233
10.60	1.36	1.25	0.238
10.80	1.38	1.28	0.245
11.00	1.37	1.31	0.253
11.20	1.36	1.33	0.259
11.40	1.33	1.34	0.264
11.60	1.31	1.34	0.266
11.80	1.28	1.33	0.266
12.00	1.26	1.32	0.264
12.40	1.23	1.29	0.257
12.80	1.22	1.26	0.251
13.20	1.20	1.23	0.245
13.60	1.19	1.20	0.236
14.00	1.20	1.16	0.225
14.40	1.22	1.13	0.214
14.80	1.27	1.12	0.207
15.20	1.31	1.17	0.218
15.60	1.31	1.23	0.234
16.00	1.28	1.28	0.251
16.40	1.24	1.33	0.270
16.80	1.17	1.37	0.288
17.00	1.14	1.38	0.297
17.40	1.06	1.39	0.314
18.00	0.95	1.38	0.334
18.40	0.88	1.36	0.346
18.80	0.82	1.33	0.355
19.20	0.76	1.29	0.360
19.60	0.72	1.25	0.363
20.00	0.67	1.21	0.369
20.40	0.64	1.15	0.364
20.80	0.61	1.10	0.357
21.20	0.60	1.06	0.349
21.60	0.58	1.02	0.342
22.00	0.57	0.98	0.336
22.40	0.56	0.95	0.328
22.80	0.55	0.92	0.325
23.20	0.53	0.89	0.322
23.60	0.52	0.85	0.317
24.00	0.50	0.82	0.314
24.40	0.49	0.79	0.309
24.80	0.48	0.75	0.303
25.20	0.47	0.72	0.295
25.60	0.47	0.68	0.286
26.00	0.46	0.64	0.276
26.40	0.46	0.61	0.263
26.80	0.46	0.57	0.249
27.20	0.47	0.53	0.231
27.60	0.48	0.50	0.216
28.00	0.49	0.47	0.198
29.00	0.51	0.41	0.164
30.00	0.55	0.34	0.129
31.00	0.59	0.29	0.097
32.00	0.64	0.26	0.072
33.00	0.67	0.24	0.060
34.00	0.70	0.22	0.047
35.00	0.74	0.20	0.036
36.00	0.77	0.19	0.029
37.00	0.80	0.19	0.023
38.00	0.84	0.19	0.018
39.00	0.88	0.21	0.016
40.00	0.87	0.25	0.023
42.00	0.87	0.25	0.023
44.00	0.88	0.28	0.026
46.00	0.84	0.31	0.035
48.00	0.82	0.30	0.036
50.00	0.80	0.30	0.039
52.00	0.77	0.30	0.044
54.00	0.71	0.29	0.055
56.00	0.66	0.23	0.061
58.00	0.64	0.16	0.055

Rhodium[11]

Energy (eV)	n	k	R(ϕ = 0)
0.10	18.48	69.43	0.986
0.20	8.66	37.46	0.977
0.30	5.85	25.94	0.967
0.40	4.74	19.80	0.955
0.50	4.20	16.07	0.941
0.60	3.87	13.51	0.925
0.70	3.67	11.72	0.908
0.80	3.63	10.34	0.887
0.90	3.62	9.36	0.867
1.00	3.71	8.67	0.848
1.10	3.67	8.26	0.837
1.20	3.51	7.94	0.832
1.30	3.26	7.63	0.829
1.40	3.01	7.31	0.827
1.50	2.78	6.97	0.823
1.60	2.60	6.64	0.818
1.70	2.42	6.33	0.813
1.80	2.30	6.02	0.805
1.90	2.20	5.76	0.798
2.00	2.12	5.51	0.789
2.10	2.05	5.30	0.780
2.20	2.00	5.11	0.772
2.30	1.94	4.94	0.765
2.40	1.90	4.78	0.756
2.50	1.88	4.65	0.748
2.60	1.85	4.55	0.743
2.70	1.80	4.49	0.742
2.90	1.63	4.36	0.748
3.00	1.53	4.29	0.753
3.10	1.41	4.20	0.760
3.20	1.30	4.09	0.764
3.30	1.20	3.97	0.767
3.40	1.11	3.84	0.769
3.50	1.04	3.71	0.768
3.60	0.99	3.58	0.764
3.70	0.95	3.45	0.759
3.80	0.91	3.34	0.753
3.90	0.88	3.23	0.747
4.00	0.86	3.12	0.739
4.20	0.83	2.94	0.722
4.40	0.80	2.76	0.706
4.60	0.78	2.60	0.684
4.80	0.79	2.46	0.659
5.00	0.79	2.34	0.635
5.20	0.79	2.23	0.613
5.40	0.80	2.14	0.591
5.60	0.80	2.06	0.573
5.80	0.79	2.00	0.561
6.00	0.76	1.93	0.556
6.20	0.73	1.85	0.544
6.40	0.70	1.77	0.534
6.60	0.68	1.69	0.518
6.80	0.67	1.60	0.498
7.00	0.66	1.52	0.476
7.20	0.66	1.43	0.452
7.40	0.66	1.35	0.423
7.60	0.67	1.27	0.394
7.80	0.68	1.20	0.363
8.00	0.69	1.12	0.329
8.20	0.71	1.04	0.288
8.40	0.74	0.97	0.252
8.60	0.78	0.89	0.212
8.80	0.83	0.83	0.179
9.00	0.88	0.77	0.148
9.20	0.95	0.73	0.125
9.40	1.01	0.71	0.110
9.60	1.07	0.69	0.102
9.80	1.12	0.69	0.098
10.00	1.17	0.69	0.098
10.60	1.26	0.73	0.106
11.00	1.29	0.76	0.113
11.60	1.32	0.80	0.124
12.00	1.32	0.82	0.127
12.60	1.32	0.82	0.129
13.00	1.32	0.83	0.131
13.60	1.32	0.85	0.134
14.00	1.32	0.86	0.138
14.60	1.30	0.89	0.144
15.00	1.28	0.90	0.147
15.60	1.25	0.90	0.147
16.00	1.24	0.89	0.147
16.50	1.23	0.88	0.145
17.00	1.22	0.88	0.144
17.50	1.22	0.87	0.143
18.00	1.23	0.88	0.145
18.50	1.25	0.92	0.155
19.00	1.24	0.98	0.172
19.50	1.18	1.05	0.193
20.00	1.10	1.09	0.213
20.50	1.00	1.09	0.230
21.00	0.91	1.05	0.234
21.50	0.86	1.00	0.228
22.00	0.83	0.95	0.219

Energy (eV)	n	k	R(ϕ = 0)
22.50	0.81	0.92	0.214
23.00	0.79	0.90	0.213
23.50	0.75	0.87	0.214
24.00	0.73	0.84	0.210
24.50	0.70	0.81	0.208
25.00	0.69	0.77	0.202
25.50	0.67	0.74	0.195
26.00	0.66	0.70	0.188
26.50	0.65	0.66	0.176
27.00	0.65	0.64	0.168
27.50	0.65	0.61	0.159
28.00	0.65	0.59	0.152
29.00	0.65	0.54	0.137
30.00	0.66	0.51	0.127
31.00	0.64	0.49	0.127
32.00	0.61	0.44	0.126
33.00	0.60	0.37	0.110
34.00	0.65	0.30	0.074
35.00	0.69	0.28	0.058
36.00	0.73	0.27	0.049
37.00	0.74	0.28	0.047
38.00	0.74	0.27	0.045
39.00	0.75	0.25	0.041
Ruthenium, single crystal, $\vec{E} \parallel \hat{c}$[9]			
0.10	11.50	51.38	0.984
0.20	5.93	27.14	0.970
0.30	4.33	18.50	0.953
0.40	3.60	13.97	0.933
0.50	3.18	11.04	0.909
0.60	3.28	8.89	0.865
0.70	3.62	7.73	0.822
0.80	3.42	7.02	0.801
0.90	3.25	6.12	0.766
1.00	3.39	5.33	0.715
1.10	3.66	4.83	0.675
1.20	3.84	4.57	0.654
1.30	3.94	4.38	0.638
1.40	4.02	4.19	0.624
1.50	4.16	4.07	0.614
1.60	4.33	4.08	0.615
1.70	4.42	4.21	0.624
1.80	4.40	4.38	0.636
1.90	4.29	4.61	0.651
2.00	4.04	4.81	0.667
2.10	3.69	4.90	0.679
2.20	3.35	4.82	0.683
2.30	3.09	4.70	0.681
2.40	2.89	4.55	0.677
2.50	2.74	4.40	0.671
2.60	2.64	4.25	0.663
2.70	2.58	4.14	0.656
2.80	2.54	4.05	0.650
2.90	2.48	4.03	0.650
3.00	2.38	4.03	0.656
3.10	2.26	4.00	0.661
3.20	2.13	3.96	0.666
3.30	2.00	3.91	0.671
3.40	1.87	3.83	0.673
3.50	1.76	3.74	0.674
3.60	1.66	3.65	0.675
3.70	1.57	3.55	0.673
3.80	1.49	3.45	0.672
3.90	1.42	3.35	0.668
4.00	1.37	3.24	0.661

Energy (eV)	n	k	R(ϕ = 0)
4.20	1.29	3.08	0.649
4.40	1.22	2.93	0.639
4.60	1.16	2.79	0.628
4.80	1.11	2.67	0.617
5.00	1.06	2.56	0.607
5.20	1.01	2.46	0.600
5.40	0.95	2.35	0.593
5.60	0.92	2.23	0.576
5.80	0.90	2.14	0.559
6.00	0.88	2.05	0.545
6.20	0.87	1.98	0.531
6.40	0.84	1.91	0.521
6.60	0.82	1.84	0.510
6.80	0.79	1.77	0.500
7.00	0.76	1.69	0.489
7.20	0.75	1.61	0.472
7.40	0.73	1.54	0.455
7.60	0.73	1.46	0.433
7.80	0.73	1.39	0.411
8.00	0.72	1.33	0.391
8.20	0.72	1.26	0.366
8.40	0.73	1.20	0.342
8.60	0.74	1.14	0.318
8.80	0.74	1.08	0.295
9.00	0.75	1.02	0.267
9.20	0.77	0.97	0.243
9.40	0.79	0.91	0.217
9.60	0.82	0.86	0.190
9.80	0.85	0.81	0.167
10.00	0.88	0.76	0.144
10.20	0.92	0.72	0.125
10.40	0.96	0.69	0.110
10.60	1.01	0.67	0.100
10.80	1.05	0.66	0.094
11.00	1.09	0.65	0.090
11.20	1.12	0.65	0.088
11.40	1.15	0.65	0.087
11.60	1.18	0.65	0.088
11.80	1.21	0.66	0.090
12.00	1.23	0.67	0.092
12.40	1.26	0.69	0.098
12.80	1.27	0.72	0.104
13.20	1.28	0.74	0.108
13.60	1.28	0.75	0.111
14.00	1.28	0.76	0.114
14.40	1.27	0.76	0.114
14.80	1.27	0.76	0.114
15.00	1.27	0.76	0.114
15.60	1.28	0.77	0.115
16.00	1.30	0.78	0.118
16.50	1.32	0.80	0.123
17.00	1.34	0.85	0.136
17.50	[illegible]	0.90	0.155
18.00	1.26	0.99	0.173
18.50	1.18	1.02	0.185
19.00	1.11	1.02	0.192
19.50	1.05	1.02	0.199
20.00	0.99	1.02	0.208
20.50	0.92	0.99	0.212
21.00	0.86	0.94	0.209
21.50	0.83	0.90	0.203
22.00	0.81	0.86	0.193
23.00	0.77	0.79	0.182
24.00	0.74	0.74	0.171
25.00	0.71	0.69	0.163

Energy (eV)	n	k	R(ϕ = 0)
26.00	0.68	0.63	0.154
27.00	0.67	0.57	0.140
28.00	0.66	0.51	0.124
29.00	0.67	0.46	0.107
30.00	0.67	0.43	0.097
31.00	0.67	0.37	0.084
32.00	0.69	0.33	0.070
33.00	0.71	0.30	0.058
34.00	0.73	0.27	0.048
35.00	0.75	0.25	0.039
36.00	0.77	0.24	0.035
37.00	0.79	0.23	0.039
38.00	0.80	0.22	0.027
39.00	0.82	0.22	0.024
40.00	0.83	0.22	0.022
Ruthenium, single crystal, $\vec{E} \perp \hat{c}$[5]			
0.10	11.85	50.81	0.983
0.20	6.68	27.18	0.966
0.30	4.94	18.92	0.950
0.40	3.90	14.51	0.933
0.50	3.27	11.63	0.915
0.60	2.98	9.54	0.888
0.70	2.82	7.99	0.856
0.80	2.73	6.71	0.815
0.90	2.82	5.54	0.751
1.00	3.17	4.59	0.670
1.10	3.69	3.91	0.604
1.20	4.28	3.66	0.585
1.30	4.66	3.72	0.593
1.40	4.86	3.79	0.601
1.50	4.99	3.89	0.609
1.60	5.08	4.03	0.618
1.70	5.12	4.22	0.629
1.80	5.10	4.45	0.642
1.90	4.96	4.78	0.660
2.00	4.61	5.06	0.677
2.10	4.21	5.09	0.682
2.20	3.94	5.00	0.681
2.30	3.69	4.97	0.684
2.40	3.44	4.88	0.684
2.50	3.27	4.77	0.681
2.60	3.14	4.66	0.677
2.70	3.06	4.59	0.674
2.80	2.99	4.59	0.676
2.90	2.87	4.64	0.686
3.00	2.64	4.69	0.701
3.10	2.40	4.64	0.710
3.20	2.18	4.55	0.717
3.30	2.00	4.43	0.721
3.40	1.84	4.30	0.723
3.50	1.71	4.16	0.723
3.60	1.60	4.02	0.722
3.70	1.50	3.90	0.721
3.80	1.41	3.77	0.718
3.90	1.35	3.64	0.713
4.00	1.29	3.53	0.707
4.20	1.21	3.31	0.694
4.40	1.16	3.13	0.679
4.60	1.13	2.97	0.662
4.80	1.09	2.86	0.652
5.00	1.03	2.75	0.648
5.20	0.97	2.64	0.643
5.40	0.91	2.52	0.635
5.60	0.88	2.40	0.622

Energy (eV)	n	k	R(ϕ = 0)
5.80	0.86	2.29	0.605
6.00	0.84	2.20	0.591
6.20	0.82	2.11	0.576
6.40	0.81	2.04	0.564
6.60	0.78	1.97	0.556
6.80	0.76	1.89	0.545
7.00	0.73	1.82	0.538
7.20	0.70	1.75	0.527
7.40	0.68	1.67	0.513
7.60	0.67	1.59	0.496
7.80	0.66	1.51	0.476
8.00	0.66	1.44	0.454
8.20	0.65	1.36	0.430
8.40	0.66	1.29	0.403
8.60	0.66	1.22	0.378
8.80	0.68	1.15	0.346
9.00	0.69	1.09	0.317
9.20	0.70	1.02	0.286
9.40	0.73	0.95	0.251
9.60	0.77	0.89	0.216
9.80	0.82	0.84	0.185
10.00	0.86	0.81	0.163
10.20	0.90	0.77	0.143
10.40	0.94	0.74	0.127
10.60	0.99	0.72	0.115
10.80	1.04	0.71	0.108
11.00	1.08	0.70	0.104
11.20	1.11	0.70	0.102
11.40	1.14	0.70	0.101
11.60	1.17	0.71	0.102
11.80	1.20	0.72	0.104
12.00	1.22	0.73	0.107
12.40	1.25	0.76	0.113
12.80	1.26	0.78	0.118
13.20	1.27	0.81	0.124
13.60	1.27	0.83	0.129
14.00	1.26	0.84	0.132
14.40	1.25	0.84	0.132
14.80	1.25	0.84	0.133
15.00	1.25	0.84	0.133
15.60	1.25	0.85	0.134
16.00	1.27	0.85	0.134
16.50	1.28	0.89	0.145
17.00	1.28	0.94	0.158
17.50	1.25	1.00	0.175
18.00	1.19	1.04	0.190
18.50	1.12	1.05	0.200
19.00	1.07	1.05	0.205
19.50	1.02	1.04	0.212
20.00	0.97	1.04	0.219
20.50	0.91	1.03	0.228
21.00	0.85	1.01	0.234
21.50	0.80	0.97	0.234
22.00	0.77	0.94	0.233
23.00	0.71	0.87	0.229
24.00	0.67	0.79	0.218
25.00	0.64	0.73	0.205
26.00	0.61	0.66	0.194
27.00	0.60	0.59	0.177
28.00	0.60	0.53	0.155
29.00	0.61	0.48	0.134
30.00	0.62	0.45	0.123
31.00	0.61	0.40	0.114
32.00	0.63	0.34	0.093
33.00	0.65	0.31	0.077
34.00	0.67	0.28	0.065
35.00	0.70	0.26	0.054
36.00	0.72	0.25	0.047
37.00	0.73	0.23	0.041
38.00	0.75	0.22	0.035
39.00	0.77	0.22	0.031
40.00	0.79	0.22	0.028

Selenium, single crystal, $\vec{E} \parallel \hat{c}$ [22]

Energy (eV)	n	k	R(ϕ = 0)
0.01364	2.914	0.248	0.242
0.01488	3.175	9.95E-02	0.272
0.01612	3.263	2.13E-03	0.282
0.01736	3.306	3.81E-02	0.287
0.01860	3.330	7.04E-03	0.290
0.01984	3.346	4.23E-02	0.291
0.02108	3.358	3.40E-03	0.293
0.02232	3.366	5.31E-02	0.294
0.02356	3.372	1.96E-03	0.294
0.02480	3.377	2.39E-02	0.295
0.02604	3.380		0.295
0.02728		1.16E-02	
0.02976		7.96E-03	
0.03224		8.57E-03	
0.03472		2.70E-02	
0.03720	3.397	1.72E-02	0.297
0.04463		1.13E-02	
0.04959	3.403	2.79E-03	0.298
0.05703		1.56E-03	
0.06199	3.405	1.35E-03	0.298
0.06819		5.79E-04	
0.07439	3.407	4.44E-04	0.298
0.08059		4.41E-04	
0.08679	3.408	4.32E-04	0.298
0.09299		2.44E-04	
0.09919	3.409	3.23E-04	0.299
0.1116	3.409	2.87E-04	0.299
0.1240	3.410	2.71E-04	0.299
0.2480	3.417	2.67E-04	0.299
0.3720	3.427	1.90E-04	0.301
0.4959	3.442	1.41E-04	0.302
0.6199	3.462	1.12E-04	0.304
0.7439	3.486	9.42E-05	0.307
0.8679	3.516	8.07E-05	0.310
0.9919	3.551	7.11E-05	0.314
1.116	3.592	6.37E-05	0.319
1.240	3.640	5.81E-05	0.324
1.50		1.33E-04	
1.60		1.59E-04	
1.70		6.27E-04	
1.80	4.46	2.20E-02	0.402
2.0	4.79	0.76	0.438
2.2	4.49	1.19	0.431
2.4	4.28	1.21	0.417
2.6	4.40	1.32	0.430
2.8	4.59	1.70	0.462
3.0	4.44	2.29	0.490
3.2	3.92	2.59	0.493
3.4	3.69	2.76	0.502
3.6	3.39	3.01	0.521
3.8	(3.00)		
4.0	(2.65)		
4.2	(2.30)		
4.5	1.92	2.78	0.528
5.0	1.50	2.31	0.482
6.0	1.57	1.49	0.288
7.0	1.84	1.45	0.276
8.0	1.35	1.68	0.353
9.0	1.35	1.64	0.342
10.0	0.92	1.07	0.238
12.0	1.00	1.10	0.232
14.0	0.81	0.91	0.211
16.0	0.65	0.61	0.160
18.0	0.65	0.48	0.120
20.0	0.69	0.36	0.076
22.0	0.81	0.25	0.030
24.0	0.91	0.18	0.011
26.0	0.86	0.15	0.012
28.0	0.85	0.13	0.011
30.0	0.87	0.11	0.008

Selenium, single crystal, $\vec{E} \perp \hat{c}$ [22]

Energy (eV)	n	k	R(ϕ = 0)
0.01364	2.854	0.0239	0.231
0.01488	2.932	0.0325	0.241
0.01612	3.140	0.1750	0.269
0.01736	2.959	1.3300	0.321
0.01860	2.111	0.2550	0.133
0.01984	2.356	0.0746	0.164
0.02108	2.462	0.0276	0.178
0.02232	2.502	0.0442	0.184
0.02356	2.543	0.0097	0.190
0.02480	2.550	0.0239	0.191
0.02604	2.582		0.195
0.02728	2.600	0.0101	0.198
0.02976	2.576	9.95E-03	0.194
0.03224	2.598	1.16E-02	0.197
0.03472	2.607	1.68E-02	0.199
0.03720	2.613	1.54E-02	0.199
0.04463		1.17E-02	
0.04959	2.627	3.58E-03	0.201
0.05703		8.65E-04	
0.06199	2.632	2.07E-03	0.202
0.06819		2.89E-04	
0.07439	2.635	1.59E-04	0.202
0.08059		1.35E-04	
0.08679	2.636	1.42E-04	0.202
0.09299		1.04E-04	
0.09919	2.637	8.95E-05	0.203
0.1116	2.638	8.84E-05	0.203
0.1240	2.639	8.51E-05	0.203
0.2480	2.645	5.97E-05	0.204
0.3720	2.652	5.44E-05	0.205
0.4959	2.654	4.58E-05	0.205
0.6199	2.675	3.82E-05	0.208
0.7439	2.692	3.32E-05	0.210
0.8679	2.713	2.96E-05	0.213
0.9919	2.739	2.69E-05	0.216
1.116	2.772	2.48E-05	0.221
1.240	2.816	2.31E-05	0.226
1.50		7.37E-05	
1.60		8.63E-05	
1.70		3.60E-04	
1.80	3.32	0.11	0.289
2.00	3.38	0.65	0.310
2.20	3.07	0.73	0.282
2.40	2.93	0.61	0.259
2.60	3.00	0.53	0.263
2.80	3.12	0.58	0.279
3.00	3.30	0.70	0.305
3.20	3.35	1.01	0.328
3.40	3.22	1.24	0.334

Energy (eV)	n	k	R(ϕ = 0)
3.60	3.06	1.47	0.344
3.80	2.84	1.66	0.351
4.00	2.51	1.81	0.356
4.20	2.18	1.83	0.352
4.50	1.75	1.94	0.382
5.00	1.25	1.50	0.316
6.00	1.32	0.73	0.107
7.00	1.62	0.61	0.105
8.00	1.81	0.69	0.135
9.00	1.66	1.02	0.182
10.00	1.72	0.95	0.171
12.00	1.25	1.02	0.181
14.00	0.98	0.92	0.178
16.00	0.68	0.96	0.274
18.00	0.61	0.65	0.191
20.00	0.73	0.48	0.094
22.00	0.78	0.39	0.060
24.00	0.78	0.32	0.046
26.00	0.78	0.26	0.036
28.00	0.80	0.19	0.023
30.00	0.79	0.14	0.020

Silicon, single crystal[23]

Energy (eV)	n	k	R(ϕ = 0)
0.01240	3.4185	2.90E-04	0.300
0.01488	3.4190	2.30E-04	0.300
0.01736	3.4192	1.90E-04	0.300
0.01984	3.4195	1.70E-04	0.300
0.02480	3.4197		0.300
0.03100	3.4199		0.300
0.04092	3.4200		0.300
0.04463		1.08E-04	
0.04959	3.4201	9.15E-05	0.300
0.05703		1.56E-04	
0.06199	3.4204	2.86E-04	0.300
0.06943		3.84E-04	
0.07439		7.16E-04	
0.08059	(3.4207)	1.52E-04	0.300
0.08679		1.02E-04	
0.09299		2.59E-04	
0.09919		1.77E-04	
0.1054		1.53E-04	
0.1116		2.02E-04	
0.1178		1.22E-04	
0.1240	3.4215	6.76E-05	0.300
0.1364		5.49E-05	
0.1488		2.41E-05	
0.1612		2.49E-05	
0.1736	(3.4230)	1.68E-05	0.300
0.1798		2.45E-05	
0.1860		2.66E-06	
0.1922		1.74E-06	
0.1984		8.46E-07	
0.2046		5.64E-07	
0.2108	(3.4244)	4.17E-07	0.300
0.2170		4.05E-07	
0.2232		3.94E-07	
0.2294		3.26E-07	
0.2356		2.97E-07	
0.2418		2.82E-07	
0.2480	3.4261	1.99E-07	0.300
0.3100	3.4294		0.301
0.3626	3.4327		0.301
0.4568	3.4393	2.50E-09	0.302
0.6199	3.4490		0.303
0.8093	3.4784		0.306
1.033	3.5193		0.311
1.1	(3.5341)	1.30E-05	0.312
1.2		1.80E-04	
1.3		2.26E-03	
1.4		7.75E-03	
1.5	3.673	5.00E-03	0.327
1.6	3.714	8.00E-03	0.331
1.7	3.752	1.00E-02	0.335
1.8	3.796	0.013	0.340
1.9	3.847	0.016	0.345
2.0	3.906	0.022	0.351
2.1	3.969	0.030	0.357
2.2	4.042	0.032	0.364
2.3	4.123	0.048	0.372
2.4	4.215	0.060	0.380
2.5	4.320	0.073	0.390
2.6	4.442	0.090	0.400
2.7	4.583	0.130	0.412
2.8	4.753	0.163	0.426
2.9	4.961	0.203	0.442
3.0	5.222	0.269	0.461
3.1	5.570	0.387	0.486
3.2	6.062	0.630	0.518
3.3	6.709	1.321	0.561
3.4	6.522	2.705	0.592
3.5	5.610	3.014	0.575
3.6	5.296	2.987	0.564
3.7	5.156	3.058	0.563
3.8	5.065	3.182	0.568
3.9	5.016	3.346	0.577
4.0	5.010	3.587	0.591
4.1	5.020	3.979	0.614
4.2	4.888	4.639	0.652
4.3	4.086	5.395	0.703
4.4	3.120	5.344	0.726
4.5	2.451	5.082	0.740
4.6	1.988	4.678	0.742
4.7	1.764	4.278	0.728
4.8	1.658	3.979	0.710
4.9	1.597	3.749	0.693
5.0	1.570	3.565	0.675
5.1	1.571	3.429	0.658
5.2	1.589	3.354	0.646
5.3	1.579	3.353	0.647
5.4	1.471	3.366	0.663
5.5	1.340	3.302	0.673
5.6	1.247	3.206	0.675
5.7	1.180	3.112	0.673
5.8	1.133	3.045	0.672
5.9	1.083	2.982	0.673
6.0	1.010	2.909	0.677
6.5	0.847	2.73	0.688
7.0	0.682	2.46	0.691
7.5	0.563	2.21	0.693
8.0	0.478	2.00	0.691
8.5	0.414	1.82	0.688
9.0	0.367	1.66	0.683
9.5	0.332	1.51	0.672
10.0	0.306	1.38	0.661
12.0	0.257	0.963	0.590
14.0	0.275	0.641	0.460
16.0	0.345	0.394	0.297
18.0	0.455	0.219	0.159
20.0	0.567	0.0835	0.079
22.14	0.675	0.0405	0.038
24.31	0.752	0.0243	0.020
26.38	0.803	0.0178	0.012
28.18	0.834	0.0152	0.008
30.24	0.860	0.0138	0.006
31.79	0.877	0.0132	0.004
34.44	0.899	0.0121	0.003
36.47	0.913	0.0113	0.002
38.75	0.925	0.0104	0.002
40.00	0.930	0.0100	0.001

Silver[6]

Energy (eV)	n	k	R(ϕ = 0)
0.10	9.91	90.27	0.995
0.20	2.84	45.70	0.995
0.30	1.41	30.51	0.994
0.40	0.91	22.89	0.993
0.50	0.67	18.32	0.992
1.00	0.28	9.03	0.987
1.50	0.27	5.79	0.969
2.00	0.27	4.18	0.944
2.50	0.24	3.09	0.914
3.00	0.23	2.27	0.864
3.25	0.23	1.86	0.816
3.50	0.21	1.42	0.756
3.60	0.23	1.13	0.671
3.70	0.30	0.77	0.475
3.77	0.53	0.40	0.154
3.80	0.73	0.30	0.053
3.90	1.30	0.36	0.040
4.00	1.61	0.60	0.103
4.10	1.73	0.85	0.153
4.20	1.75	1.06	0.194
4.30	1.73	1.13	0.208
4.50	1.69	1.28	0.238
4.75	1.61	1.34	0.252
5.00	1.55	1.36	0.257
5.50	1.45	1.34	0.257
6.00	1.34	1.28	0.246
6.50	1.25	1.18	0.225
7.00	1.18	1.06	0.196
7.50	1.14	0.91	0.157
8.00	1.16	0.75	0.114
9.00	1.33	0.56	0.074
10.00	1.46	0.56	0.082
11.00	1.52	0.56	0.088
12.00	1.61	0.59	0.100
13.00	1.66	0.64	0.112
14.00	1.72	0.78	0.141
14.50	1.64	0.88	0.152
15.00	1.56	0.92	0.156
16.00	1.42	0.91	0.151
17.00	1.33	0.86	0.139
18.00	1.28	0.80	0.124
19.00	1.07	[illegible]	0.111
20.00	1.29	0.71	0.103
21.00	1.35	0.75	0.112
21.50	1.37	0.80	0.124
22.00	1.34	0.87	0.141
22.50	1.26	0.93	0.157
23.00	1.17	0.94	0.163
23.50	1.10	0.93	0.165
24.00	1.04	0.90	0.165
24.50	0.99	0.87	0.160
25.00	0.95	0.83	0.154
25.50	0.91	0.78	0.144
26.00	0.90	0.74	0.133

Energy (eV)	*n*	*k*	$R(\phi = 0)$
26.50	0.89	0.69	0.121
27.00	0.89	0.65	0.109
27.50	0.89	0.62	0.099
28.00	0.90	0.59	0.090
28.50	0.91	0.57	0.084
29.00	0.92	0.56	0.079
30.00	0.93	0.54	0.074
31.00	0.93	0.53	0.072
32.00	0.92	0.53	0.072
33.00	0.90	0.51	0.071
34.00	0.88	0.49	0.067
35.00	0.86	0.45	0.061
36.00	0.89	0.44	0.055
38.00	0.89	0.39	0.043
40.00	0.90	0.37	0.039
42.00	0.90	0.35	0.036
44.00	0.90	0.33	0.033
46.00	0.90	0.32	0.031
48.00	0.89	0.31	0.030
50.00	0.88	0.29	0.027
52.00	0.89	0.28	0.024
54.00	0.88	0.17	0.024
56.00	0.87	0.26	0.024
58.00	0.87	0.24	0.021
60.00	0.87	0.22	0.018
62.00	0.88	0.21	0.016
64.00	0.88	0.21	0.016
66.00	0.88	0.21	0.016
68.00	0.87	0.21	0.017
70.00	0.83	0.20	0.021
72.00	0.85	0.18	0.016
74.00	0.85	0.17	0.014
76.00	0.85	0.16	0.013
78.00	0.85	0.15	0.013
80.00	0.85	0.14	0.012
85.00	0.85	0.11	0.011
90.00	0.85	0.08	0.009
95.00	0.86	0.06	0.007
100.00	0.87	0.04	0.005
Sodium[24]			
0.55	0.262	9.97	0.990
0.58	0.241	9.45	0.989
0.63	0.207	8.80	0.990
0.67	0.175	8.09	0.990
0.73	0.147	7.42	0.990
0.81	0.123	6.67	0.989
0.92	0.099	5.82	0.989
1.05	0.078	5.11	0.989
1.23	0.064	4.35	0.987
1.44	0.053	3.72	0.986
1.65	0.050	3.22	0.983
1.87	0.049	2.76	0.978
2.07	0.053	2.48	0.971
2.27	0.059	2.23	0.961
2.45	0.063	2.07	0.953
2.64	0.066	1.88	0.943
2.82	0.068	1.76	0.936
2.95	0.068	1.63	0.928
3.06	0.069	1.54	0.921
3.20	0.065	1.47	0.921
3.40	0.061	1.33	0.916
3.71	0.055	1.13	0.908
3.97	0.049	1.01	0.908
6.199	0.390		0.193
6.358	0.454		0.141
6.526	0.485		0.120
6.702	0.533		0.093
6.888	0.574		0.073
7.130	0.616		0.056
7.328	0.641		0.048
7.583	0.674		0.038
7.847	0.700		0.031
8.015	0.710		0.029
8.634	0.762		0.018
9.143	0.800		0.012
9.709	0.819		0.010
10.20	0.843		0.007
11.08	0.870		0.005
11.83	0.887		0.004
12.73	0.907		0.002
13.05	0.913		0.002
13.42	0.914		0.002
13.73	0.917		0.002
14.07	0.922		0.002
14.83	0.934		0.001
15.05	0.936		0.001
15.46	0.942		0.001
16.21	0.948		0.001
18.10	0.964		0.000
21.12	0.979		0.000
25.51	0.993		0.000
26.95	1.00		0.000
27.68	1.01		0.000
28.37	1.01		0.000
29.52	1.02		0.000
Tantalum[16]			
0.10	10.14	66.39	0.984
0.15	9.45	46.41	0.9834
0.20	5.77	35.46	0.982
0.26	3.67	27.53	0.981
0.30	2.87	23.90	0.980
0.38	2.03	18.87	0.978
0.50	1.37	14.26	0.974
0.58	1.15	12.19	0.970
0.70	0.96	9.92	0.962
0.78	0.89	8.77	0.956
0.90	0.84	7.38	0.942
1.00	0.89	6.47	0.992
1.10	0.93	5.75	0.899
1.20	0.98	5.14	0.872
1.30	1.00	4.62	0.842
1.40	1.04	4.15	0.805
1.50	1.09	3.73	0.762
1.60	1.15	3.33	0.707
1.70	1.24	2.95	0.640
1.80	1.35	2.60	0.560
1.90	1.57	2.24	0.460
2.00	1.83	1.99	0.388
2.10	2.10	1.84	0.354
2.20	2.36	1.81	0.351
2.30	2.56	1.86	0.365
2.40	2.68	1.92	0.378
2.50	2.75	1.98	0.388
2.60	2.80	2.02	0.395
2.70	2.84	2.08	0.405
2.80	2.85	2.14	0.412
2.90	2.84	2.20	0.420
3.00	2.81	2.24	0.425
3.20	2.73	2.31	0.432
3.40	2.61	2.33	0.435
3.60	2.49	2.30	0.430
3.80	2.40	2.22	0.418
4.00	2.36	2.14	0.406
4.20	2.35	2.06	0.392
4.40	2.39	2.01	0.384
4.60	2.45	2.00	0.384
4.80	2.53	2.06	0.394
5.00	2.58	2.20	0.416
5.20	2.52	2.44	0.450
5.40	2.31	2.61	0.480
5.60	2.06	2.67	0.501
5.80	1.83	2.63	0.510
6.00	1.63	2.56	0.515
6.20	1.48	2.45	0.512
6.40	1.37	2.33	0.504
6.60	1.29	2.22	0.492
6.80	1.23	2.11	0.478
7.00	1.18	2.01	0.462
7.20	1.15	1.91	0.445
7.40	1.13	1.82	0.425
7.60	1.12	1.75	0.406
7.80	1.11	1.68	0.390
8.00	1.11	1.61	0.370
8.20	1.12	1.55	0.350
8.40	1.13	1.50	0.332
8.60	1.14	1.45	0.317
8.80	1.17	1.41	0.301
9.00	1.19	1.40	0.294
9.20	1.21	1.38	0.289
9.40	1.21	1.38	0.287
9.60	1.21	1.38	0.285
9.80	1.21	1.37	0.285
10.00	1.20	1.37	0.286
10.20	1.19	1.37	0.286
10.40	1.18	1.37	0.287
10.60	1.16	1.36	0.288
10.80	1.15	1.36	0.289
11.00	1.13	1.35	0.290
11.20	1.11	1.35	0.292
11.40	1.09	1.34	0.293
11.60	1.07	1.33	0.294
11.80	1.05	1.32	0.295
12.00	1.02	1.31	0.296
12.20	1.00	1.29	0.295
12.40	0.98	1.28	0.294
12.60	0.96	1.26	0.292
12.80	0.94	1.24	0.289
13.00	0.93	1.22	0.286
13.60	0.91	1.16	0.272
14.00	0.90	1.15	0.272
14.60	0.85	1.15	0.285
15.00	0.80	1.13	0.293
15.60	0.72	1.08	0.301
16.00	0.68	1.04	0.304
16.60	0.63	0.97	0.301
17.00	0.60	0.92	0.296
17.60	0.60	0.92	0.296
18.00	0.55	0.79	0.274
18.60	0.53	0.71	0.254
19.00	0.53	0.65	0.236
19.60	0.53	0.57	0.207
20.00	0.54	0.52	0.185
20.60	0.55	0.44	0.153

Energy (eV)	n	k	$R(\phi = 0)$
21.00	0.57	0.39	0.127
21.60	0.64	0.34	0.089
22.00	0.64	0.32	0.081
22.60	0.69	0.27	0.058
23.00	0.73	0.24	0.043
23.60	0.80	0.26	0.033
24.00	0.80	0.26	0.034
24.60	0.82	0.25	0.029
25.00	0.83	0.25	0.026
25.60	0.86	0.24	0.022
26.00	0.88	0.25	0.022
26.60	0.87	0.26	0.023
27.00	0.87	0.25	0.022
27.60	0.89	0.23	0.019
28.00	0.90	0.23	0.017
28.60	0.91	0.22	0.015
29.00	0.92	0.22	0.014
29.60	0.94	0.22	0.014
30.00	0.95	0.22	0.014
31.00	0.97	0.23	0.014
32.00	0.98	0.24	0.015
33.00	0.98	0.25	0.015
34.00	0.99	0.25	0.016
35.00	0.99	0.26	0.017
36.00	0.99	0.27	0.018
37.00	0.99	0.28	0.019
38.00	0.98	0.28	0.021
39.00	0.97	0.29	0.022
40.00	0.95	0.29	0.023

Tellurium, $\vec{E} \parallel \hat{c}$[25]

Energy (eV)	n	k	$R(\phi = 0)$
0.01364	4.82	0.118	0.431
0.01488	5.26	0.0505	0.463
0.01612	5.47	0.0278	0.477
0.01736	5.59	0.0174	0.485
0.01860		0.0796	
0.01984		0.0696	
0.02108		0.0749	
0.02232		0.1900	
0.02356		0.2220	
0.02480		0.0716	
0.02604		0.0682	
0.02728		0.0832	
0.02976		0.0149	
0.03224		2.14E-03	
0.03472		1.71E-02	
0.03720	5.94	3.71E-03	0.507
0.03968		2.44E-03	
0.04339	5.96	1.59E-03	0.508
0.04711		7.85E-04	
0.05083		7.38E-04	
0.05579		3.89E-04	
0.06199	5.98	3.09E-04	0.509
0.07439		2.52E-04	
0.08679		2.96E-04	
0.09919		3.68E-04	
0.12400	6.246	3.34E-04	0.524
0.15500	6.253		0.525
0.20660	6.286		0.526
0.24800	6.316	7.48E-05	0.528
0.31	6.372	1.18E-05	0.531
0.35		4.93E-04	
0.41		6.74E-03	
0.5	6.53	2.30E-02	0.539
0.6	6.71	7.50E-02	0.549
0.7	7.00	0.24	0.563
0.8	7.23	0.48	0.574
0.9	7.48	0.94	0.589
1.0	7.70	1.56	0.606
1.2	6.99	2.22	0.593
1.4	7.11	2.46	0.604
1.6	6.75	2.91	0.606
1.8	6.89	3.70	0.637
2.0	4.67	4.67	0.654
2.2	4.94	5.16	0.681
2.4	3.94	5.08	0.686
2.6	3.25	4.77	0.681
2.8	2.73	4.42	0.674
3.0	2.30	4.16	0.674
3.5	1.69	3.44	0.646
4.0	1.33	2.64	0.571
4.5	1.32	1.96	0.428
5.0	1.63	1.60	0.312
5.5	1.72	1.57	0.302
6.0	1.73	1.45	0.276
6.5	1.78	1.36	0.257
7.0	1.83	1.36	0.257
7.5	1.72	1.51	0.289
8.0	1.54	1.37	0.260
8.5	1.55	1.23	0.226
9.0	0.99	0.93	0.179
9.5	1.47	1.25	0.233
10.0	0.86	0.86	0.181
11.0	0.80	0.77	0.165
12.0	0.79	0.76	0.164
14.0	0.67	0.59	0.146
16.0	0.59	0.49	0.147
18.0	0.48	0.31	0.160
20.0	0.74	0.20	0.035
22.0	0.83	0.18	0.018
24.0	0.85	0.15	0.013
26.0	0.87	0.12	0.009
28.0	0.89	0.090	0.006
30.0	0.90	0.045	0.003

Tellurium, $\vec{E} \perp \hat{c}$[25]

Energy (eV)	n	k	$R(\phi = 0)$
0.01364	2.61	0.2980	0.204
0.01488	3.65	0.0894	0.325
0.01612	4.10	0.0535	0.370
0.01736	4.63	0.4990	0.420
0.01860		0.1170	
0.01984		0.0343	
0.02108	(4.42)	0.0421	0.398
0.02232		0.1060	
0.02356		0.0880	
0.02480		0.0458	
0.02604		0.0928	
0.02728		0.0886	
0.02976		0.0232	
0.03224		3.06E-03	
0.03472		1.25E-02	
0.03720	4.71	2.65E-03	0.422
0.03968		1.89E-03	
0.04339	4.74	1.41E-03	0.425
0.04711		8.38E-04	
0.05083		6.79E-04	
0.05579		1.59E-04	
0.06199	4.77	1.16E-04	0.427
0.07439		7.23E-05	
0.08679		5.34E-05	
0.09919		4.28E-05	
0.1240	4.796	3.18E-05	0.429
0.1550	4.809		0.430
0.2066	4.838		0.432
0.2480	4.864	2.19E-05	0.434
0.31	4.929	3.18E-05	0.439
0.35		7.89E-02	
0.41		0.149	
0.5	4.90		0.437
0.6	4.93		0.439
0.7	4.95	0.11	0.441
0.8	5.10	0.13	0.452
0.9	5.22	0.22	0.461
1.0	5.35	0.45	0.472
1.2	5.17	0.63	0.462
1.4	5.56	0.63	0.488
1.6	5.88	1.15	0.517
1.8	6.10	1.80	0.545
2.0	5.94	2.69	0.571
2.2	5.10	3.61	0.594
2.4	4.24	3.77	0.593
2.6	3.57	3.75	0.591
2.8	3.03	3.63	0.588
3.0	2.51	3.39	0.578
3.5	1.72	2.70	0.532
4.0	1.32	2.01	0.440
4.5	1.28	1.28	0.251
5.0	1.47	0.82	0.132
5.5	1.74	0.51	0.104
6.0	1.94	0.39	0.118
6.5	2.19	0.32	0.148
7.0	2.48	0.40	0.192
7.5	2.60	0.69	0.226
8.0	2.59	0.91	0.245
8.5	2.39	1.00	0.235
9.0	1.11	1.24	0.259
9.5	2.08	1.11	0.224
10.0	0.99	1.04	0.215
11.0	0.84	1.01	0.237
12.0	0.87	0.87	0.182
14.0	0.59	0.87	0.282
16.0	0.64	0.55	0.144
18.0	0.52	0.41	0.161
20.0	0.50	0.38	0.165
22.0	0.56	0.29	0.110
24.0	0.54	0.25	0.113
26.0	0.50	0.20	0.127
28.0	0.48	0.17	0.135
30.0	0.46	0.088	0.140

Titanium (Polycrystalline)[14]

Energy (eV)	n	k	$R(\phi = 0)$
0.10	5.03	23.38	0.965
0.15	3.00	15.72	0.954
0.20	2.12	11.34	0.939
0.25	2.05	8.10	0.890
0.30	6.39	9.94	0.833
0.35	2.74	6.21	0.792
0.40	2.49	4.68	0.708
0.45	3.35	3.25	0.545
0.50	4.43	3.22	0.555
0.60	4.71	3.77	0.597
0.70	4.38	3.89	0.603
0.80	4.04	3.82	0.596
0.90	3.80	3.65	0.582
1.00	3.62	3.52	0.570

Energy (eV)	*n*	*k*	$R(\phi = 0)$	Energy (eV)	*n*	*k*	$R(\phi = 0)$	Energy (eV)	*n*	*k*	$R(\phi = 0)$
1.10	3.47	3.40	0.560	11.00	0.79	0.72	0.152	0.86	2.92	4.37	0.661
1.20	3.35	3.30	0.550	11.20	0.81	0.69	0.139	0.90	3.11	4.44	0.660
1.30	3.28	3.25	0.546	11.40	0.81	0.69	0.139	0.94	3.15	4.43	0.658
1.40	3.17	3.28	0.549	11.60	0.79	0.68	0.139	0.98	3.15	4.36	0.653
1.50	2.98	3.32	0.557	11.80	0.78	0.67	0.137	1.00	3.14	4.32	0.649
1.60	2.74	3.30	0.559	12.00	0.77	0.65	0.132	1.10	3.05	4.04	0.627
1.70	2.54	3.23	0.557	12.80	0.76	0.55	0.106	1.20	3.00	3.64	0.590
1.80	2.36	3.11	0.550	13.20	0.76	0.52	0.097	1.30	3.12	3.24	0.545
1.90	2.22	2.99	0.540	13.60	0.76	0.48	0.087	1.40	3.29	2.96	0.515
2.00	2.11	2.88	0.530	14.00	0.77	0.45	0.077	1.50	3.48	2.79	0.500
2.10	2.01	2.77	0.520	14.40	0.77	0.42	0.069	1.60	3.67	2.68	0.494
2.20	1.92	2.67	0.509	14.80	0.79	0.38	0.058	1.70	3.84	2.79	0.507
2.30	1.86	2.56	0.495	15.20	0.79	0.36	0.052	1.80	3.82	2.91	0.518
2.40	1.81	2.47	0.483	15.60	0.79	0.32	0.045	1.90	3.70	2.94	0.518
2.50	1.78	2.39	0.471	16.00	0.83	0.31	0.037	2.00	3.60	2.89	0.512
2.60	1.75	2.34	0.462	16.40	0.84	0.28	0.030	2.10	3.54	2.84	0.506
2.70	1.71	2.29	0.456	16.80	0.87	0.27	0.025	2.20	3.49	2.76	0.497
2.80	1.68	2.25	0.451	17.20	0.90	0.25	0.020	2.30	3.49	2.72	0.494
2.90	1.63	2.21	0.447	17.60	0.93	0.25	0.017	2.40	3.45	2.72	0.493
3.00	1.59	2.17	0.444	18.00	0.94	0.24	0.165	2.50	3.38	2.68	0.487
3.10	1.55	2.15	0.442	18.40	0.94	0.23	0.017	2.60	3.34	2.62	0.480
3.20	1.50	2.12	0.442	18.80	0.95	0.24	0.016	2.70	3.31	2.55	0.472
3.30	1.44	2.09	0.442	19.20	0.96	0.25	0.016	2.80	3.31	2.49	0.466
3.40	1.37	2.06	0.443	19.60	0.97	0.25	0.017	2.90	3.32	2.45	0.461
3.50	1.30	2.01	0.443	20.00	0.98	0.27	0.018	3.00	3.35	2.42	0.459
3.60	1.24	1.96	0.441	20.40	0.98	0.27	0.019	3.10	3.39	2.41	0.460
3.70	1.17	1.90	0.436	20.60	1.00	0.29	0.020	3.20	3.43	2.45	0.465
3.80	1.11	1.83	0.430	21.20	0.99	0.31	0.023	3.30	3.45	2.55	0.476
3.85	1.08	1.78	0.423	21.60	0.99	0.31	0.024	3.40	3.39	2.66	0.485
3.90	1.06	1.73	0.413	22.00	0.98	0.32	0.025	3.50	3.24	2.70	0.488
4.00	1.04	1.62	0.389	22.40	0.98	0.33	0.027	3.60	3.13	2.67	0.482
4.20	1.05	1.45	0.333	22.80	0.97	0.33	0.028	3.70	3.05	2.62	0.476
4.40	1.13	1.33	0.284	23.20	0.96	0.34	0.030	3.80	2.99	2.56	0.468
4.60	1.17	1.29	0.265	23.60	0.95	0.35	0.031	3.90	2.96	2.50	0.460
4.80	1.21	1.23	0.244	24.00	0.92	0.35	0.033	4.00	2.95	2.43	0.451
5.00	1.24	1.21	0.236	24.5	0.91	0.34	0.032	4.20	3.02	2.33	0.440
5.20	1.27	1.20	0.228	25.0	0.91	0.33	0.032	4.40	3.13	2.32	0.442
5.40	1.17	1.16	0.228	25.5	0.89	0.33	0.032	4.60	3.24	2.41	0.455
5.60	1.24	1.21	0.234	26.0	0.89	0.33	0.032	4.80	3.33	2.57	0.475
5.80	1.21	1.22	0.241	26.5	0.88	0.32	0.032	5.00	3.40	2.85	0.505
6.00	1.15	1.21	0.244	27.0	0.86	0.31	0.032	5.20	3.27	3.27	0.548
6.20	1.11	1.18	0.240	27.5	0.85	0.30	0.033	5.40	2.92	3.58	0.586
6.40	1.08	1.14	0.232	28.0	0.84	0.29	0.033	5.60	2.43	3.70	0.618
6.60	1.04	1.06	0.212	28.5	0.82	0.26	0.029	5.80	2.00	3.61	0.637
6.80	1.05	1.02	0.198	29.0	0.83	0.25	0.027	6.00	1.70	3.42	0.643
7.00	1.06	0.97	0.182	30.0	0.84	0.22	0.022	6.20	1.47	3.24	0.646
7.20	1.07	0.95	0.175					6.40	1.32	3.04	0.640
7.40	1.11	0.94	0.167	**Tungsten[27]**				6.60	1.21	2.87	0.631
7.60	1.09	0.92	0.165	0.10	14.06	54.71	0.983	6.80	1.12	2.70	0.619
7.80	1.11	0.93	0.165	0.20	3.87	28.30	0.981	7.00	1.06	2.56	0.607
8.00	1.10	0.94	0.169	0.25	2.56	22.44	0.980	7.20	1.01	2.43	0.593
8.20	1.10	0.95	0.171	0.30	1.83	18.32	0.979	7.40	0.98	2.30	0.573
8.40	1.08	0.95	0.175	0.34	1.71	15.71	0.973	7.60	0.95	2.18	0.556
8.60	1.04	0.96	0.181	0.38	1.86	13.88	0.963	7.80	0.93	2.06	0.533
8.80	1.02	0.95	0.181	0.42	1.92	12.63	0.954	8.00	0.94	1.95	0.505
9.00	1.00	0.94	0.182	0.46	1.69	11.59	0.952	8.20	0.94	1.86	0.481
9.20	0.97	0.93	0.182	0.50	1.40	10.52	0.952	8.40	0.96	1.76	0.449
9.40	0.95	0.91	0.181	0.54	1.23	9.45	0.948	8.60	0.99	1.70	0.422
9.60	0.94	0.90	0.179	0.58	1.17	8.44	0.938	8.80	1.01	1.65	0.401
9.80	0.91	0.88	0.179	0.62	1.28	7.52	0.917	9.00	1.01	1.60	0.388
10.00	0.89	0.88	0.180	0.66	1.45	6.78	0.888	9.20	1.02	1.55	0.369
10.20	0.86	0.85	0.178	0.70	1.59	6.13	0.856	9.40	1.03	1.50	0.352
10.40	0.85	0.83	0.175	0.74	1.83	5.52	0.810	9.60	1.05	1.44	0.329
10.60	0.81	0.79	0.167	0.78	2.12	5.00	0.759	9.80	1.09	1.38	0.307
10.80	0.80	0.76	0.162	0.82	2.36	4.61	0.710	10.00	1.13	1.34	0.287

Energy (eV)	n	k	$R(\phi = 0)$
10.20	1.19	1.33	0.274
10.40	1.24	1.34	0.270
10.60	1.27	1.36	0.274
10.80	1.29	1.39	0.282
11.00	1.28	1.42	0.290
11.20	1.27	1.44	0.297
11.40	1.25	1.46	0.305
11.60	1.22	1.48	0.313
11.80	1.20	1.48	0.318
12.00	1.16	1.48	0.323
12.40	1.10	1.47	0.329
12.80	1.04	1.44	0.333
13.20	0.98	1.40	0.332
13.60	0.94	1.35	0.325
14.00	0.91	1.28	0.312
14.40	0.90	1.23	0.296
14.80	0.90	1.17	0.276
15.20	0.93	1.13	0.255
15.60	0.97	1.12	0.246
16.00	0.98	1.14	0.249
16.40	0.97	1.17	0.260
16.80	0.94	1.19	0.273
17.20	0.90	1.21	0.289
17.60	0.85	1.21	0.304
18.00	0.80	1.20	0.317
18.40	0.74	1.18	0.330
18.80	0.69	1.15	0.340
19.20	0.64	1.11	0.347
19.60	0.60	1.07	0.353
20.00	0.56	1.02	0.354
20.40	0.54	0.97	0.350
20.80	0.52	0.92	0.342
21.20	0.50	0.87	0.331
21.60	0.50	0.82	0.318
22.00	0.49	0.77	0.303
22.40	0.49	0.73	0.287
22.80	0.49	0.69	0.272
23.20	0.49	0.66	0.263
23.60	0.48	0.62	0.252
24.00	0.49	0.57	0.234
24.40	0.50	0.53	0.213
24.80	0.51	0.49	0.191
25.20	0.53	0.46	0.171
25.60	0.55	0.43	0.150
26.00	0.57	0.40	0.132
26.40	0.59	0.38	0.117
26.80	0.61	0.37	0.105
27.00	0.62	0.36	0.099
27.50	0.64	0.34	0.085
28.00	0.67	0.32	0.073
28.50	0.69	0.31	0.065
29.00	0.71	0.30	0.057
29.50	0.73	0.30	0.052
30.00	0.75	0.29	0.047
31.00	0.78	0.29	0.042
32.00	0.79	0.29	0.040
33.00	0.82	0.28	0.033
34.00	0.84	0.29	0.032
35.00	0.85	0.31	0.033
36.00	0.85	0.32	0.036
37.00	0.84	0.33	0.039
38.00	0.83	0.33	0.040
39.00	0.81	0.33	0.042
40.00	0.80	0.33	0.045

Energy (eV)	n	k	$R(\phi = 0)$
Vanadium[9]			
0.10	12.83	45.89	0.978
0.20	3.90	24.30	0.975
0.28	2.13	17.35	0.973
0.36	1.54	13.32	0.966
0.44	1.28	10.74	0.957
0.52	1.16	8.93	0.945
0.60	1.10	7.59	0.929
0.68	1.07	6.54	0.909
0.76	1.08	5.67	0.882
0.80	1.10	5.30	0.864
0.90	1.18	4.50	0.811
1.00	1.34	3.80	0.730
1.10	1.60	3.26	0.632
1.20	1.93	2.88	0.543
1.30	2.25	2.71	0.498
1.40	2.48	2.72	0.491
1.50	2.57	2.79	0.499
1.60	2.57	2.84	0.507
1.70	2.52	2.88	0.512
1.80	2.45	2.88	0.515
1.90	2.36	2.85	0.514
2.00	2.34	2.81	0.509
2.10	2.31	2.78	0.506
2.20	2.28	2.80	0.510
2.30	2.23	2.83	0.516
2.40	2.15	2.88	0.528
2.50	2.02	2.91	0.540
2.60	1.89	2.92	0.552
2.70	1.74	2.89	0.561
2.80	1.61	2.85	0.569
2.90	1.48	2.80	0.577
3.00	1.36	2.73	0.582
3.20	1.16	2.55	0.585
3.40	0.99	2.37	0.586
3.60	0.87	2.17	0.575
3.80	0.80	1.96	0.547
4.00	0.78	1.76	0.503
4.20	0.80	1.60	0.449
4.40	0.83	1.47	0.400
4.60	0.87	1.38	0.355
4.80	0.90	1.31	0.326
5.00	0.91	1.26	0.304
5.25	0.93	1.18	0.271
5.50	0.94	1.14	0.258
5.75	0.96	1.09	0.235
6.00	0.98	1.06	0.223
6.25	0.97	1.02	0.212
6.50	0.97	0.98	0.199
6.75	0.97	0.94	0.185
7.00	0.98	0.91	0.175
7.33	0.97	0.89	0.170
7.66	0.98	0.87	0.162
8.00	0.98	0.85	0.155
8.33	0.98	0.81	0.146
8.66	0.98	0.81	0.145
9.00	0.96	0.79	0.142
9.50	0.94	0.77	0.136
10.00	0.91	0.74	0.133
10.50	0.89	0.71	0.126
11.00	0.87	0.65	0.112
11.50	0.88	0.58	0.091
12.00	0.90	0.58	0.089
12.50	0.89	0.57	0.086
13.00	0.88	0.55	0.082

Energy (eV)	n	k	$R(\phi = 0)$
13.50	0.87	0.53	0.079
14.00	0.86	0.51	0.075
14.50	0.86	0.49	0.070
15.00	0.86	0.47	0.065
15.50	0.86	0.46	0.062
16.00	0.85	0.45	0.061
16.50	0.84	0.43	0.059
17.00	0.84	0.41	0.056
17.50	0.83	0.40	0.054
18.00	0.82	0.38	0.051
18.50	0.82	0.37	0.048
19.00	0.82	0.35	0.045
19.50	0.82	0.34	0.043
20.00	0.81	0.32	0.041
20.50	0.81	0.31	0.038
21.00	0.81	0.29	0.036
21.50	0.81	0.28	0.033
22.00	0.81	0.27	0.032
22.50	0.81	0.25	0.029
23.00	0.82	0.24	0.027
23.50	0.82	0.23	0.025
24.00	0.82	0.22	0.024
24.50	0.83	0.21	0.022
25.00	0.83	0.20	0.020
25.50	0.83	0.19	0.019
26.00	0.83	0.18	0.018
26.50	0.84	0.17	0.016
27.00	0.84	0.16	0.015
27.50	0.85	0.16	0.014
28.00	0.85	0.15	0.013
28.50	0.86	0.14	0.012
29.00	0.86	0.14	0.011
29.50	0.86	0.13	0.010
30.00	0.87	0.13	0.009
31.00	0.88	0.12	0.008
32.00	0.90	0.11	0.007
33.00	0.90	0.10	0.005
34.00	0.91	0.10	0.005
35.00	0.92	0.09	0.004
36.00	0.94	0.10	0.004
37.00	0.94	0.10	0.004
38.00	0.95	0.11	0.004
39.00	0.95	0.12	0.004
40.00	0.95	0.13	0.005
Zinc, $\vec{E} \parallel \hat{c}$[28]			
0.7514	1.9241	7.5619	0.883
0.827	1.7921	6.9973	0.874
0.866	1.5571	6.7753	0.881
0.952	1.4824	6.2296	0.868
0.992	1.5762	5.8843	0.847
1.033	1.5407	5.3192	0.823
1.078	1.5853	4.9013	0.793
1.127	1.7768	4.5307	0.748
1.181	1.9808	4.2004	0.701
1.240	2.8821	3.4766	0.575
1.305	3.2039	3.0042	0.520
1.377	2.9459	3.5761	0.584
1.459	3.2523	4.2447	0.640
1.550	3.8086	4.6212	0.657
1.653	3.7577	4.6239	0.659
1.722	3.5908	4.4614	0.650
1.823	3.4234	4.3232	0.642
1.937	3.0132	3.9974	0.624
1.984	1.8562	3.9706	0.690

Energy (eV)	n	k	R(ϕ = 0)
2.066	1.4856	4.0555	0.737
2.094	1.2525	3.9961	0.762
2.119	1.0017	3.8683	0.789
2.275	0.7737	3.9129	0.832
2.445	0.6395	3.4013	0.821
2.666	0.4430	3.1379	0.851
2.917	0.3589	2.8140	0.853
3.220	0.3069	2.5088	0.847
3.594	0.2737	2.1737	0.828
4.065	0.2510	1.8528	0.799
4.678	0.2354	1.6357	0.776

Zinc, $\vec{E} \perp \hat{c}$[28]

Energy (eV)	n	k	R(ϕ = 0)
0.751	1.4469	7.4158	0.905
0.827	1.4744	6.9688	0.892
0.866	1.3628	6.6886	0.892
0.952	1.3165	6.2212	0.881
0.992	1.3835	5.8910	0.863
1.033	1.2889	5.4001	0.850
1.078	1.3095	4.9025	0.822
1.127	1.6897	4.4062	0.746
1.181	1.9701	4.0176	0.684
1.240	2.8717	3.2873	0.555
1.305	3.3991	2.7684	0.497
1.377	3.1807	3.4709	0.569
1.459	3.5064	4.1994	0.630
1.550	4.1241	4.7768	0.664
1.653	4.0269	4.8027	0.667
1.722	3.9369	4.6356	0.657
1.823	3.7549	4.3042	0.635
1.937	3.4512	4.1942	0.631
1.984	3.2515	4.2980	0.644
2.066	2.0802	4.7231	0.738
2.094	1.7084	4.7923	0.774
2.119	1.3329	4.4751	0.791
2.275	0.9725	4.2879	0.825
2.455	0.7568	3.7627	0.824
2.666	0.5470	3.4277	0.845
2.917	0.4774	3.0476	0.834
3.220	0.3911	2.7463	0.835
3.594	0.3147	2.3041	0.821
4.065	0.3013	2.0077	0.789
4.678	0.2806	1.7997	0.770

Zirconium (Polycrystalline)[28]

Energy (eV)	n	k	R(ϕ = 0)
0.10	6.18	1.76	0.300
0.15	3.37	1.30	0.123
0.20	2.34	1.08	0.058
0.26	2.24	1.06	0.052
0.30	2.59	1.14	0.073
0.36	3.17	1.26	0.110
0.40	3.09	1.24	0.105
0.46	3.36	1.30	0.123
0.50	4.13	1.44	0.175
0.56	5.01	1.58	0.231
0.60	5.18	1.61	0.242
0.70	4.54	1.51	0.202
0.80	4.03	1.42	0.168
0.90	3.74	1.37	0.149
0.96	3.69	1.36	0.145
1.00	3.66	1.35	0.143
1.10	3.65	1.35	0.142
1.20	3.53	1.33	0.134
1.30	3.25	1.27	0.116
1.40	3.10	1.25	0.106
1.50	3.02	1.23	0.100
1.60	2.88	1.20	0.091
1.70	2.68	1.16	0.078
1.80	2.49	1.12	0.067
2.00	2.14	1.03	0.047
2.10	1.99	1.00	0.040
2.20	1.87	0.97	0.034
2.30	1.78	0.94	0.030
2.40	1.71	0.92	0.027
2.50	1.62	0.90	0.024
2.60	1.54	0.88	0.022
2.70	1.46	0.86	0.019
2.80	1.40	0.84	0.018
2.90	1.34	0.82	0.016
3.00	0.30	0.81	0.016
3.10	1.26	0.80	0.015
3.30	1.19	0.77	0.014
3.40	1.16	0.76	0.013
3.50	1.13	0.75	0.013
3.60	1.10	0.74	0.013
3.70	1.07	0.73	0.013
3.80	1.04	0.72	0.012
3.90	1.01	0.71	0.012
4.00	0.98	0.70	0.012
4.20	0.94	0.68	0.013
4.40	0.89	0.67	0.013
4.60	0.85	0.65	0.014
4.80	0.81	0.64	0.014
5.00	0.78	0.63	0.015
5.20	0.77	0.62	0.016
5.40	0.77	0.62	0.016
5.60	0.80	0.63	0.014
5.80	0.87	0.66	0.013
6.00	1.00	0.71	0.012
6.20	1.11	0.75	0.013
6.40	1.23	0.78	0.014
6.60	1.33	0.81	0.016
6.80	1.42	0.84	0.018
7.00	1.49	0.86	0.020
7.20	1.54	0.88	0.022
7.40	1.58	0.89	0.023
7.60	1.61	0.90	0.024
7.80	1.63	0.90	0.025
8.00	1.66	0.91	0.026
8.20	0.67	0.91	0.026
8.40	1.68	0.92	0.026
8.60	1.68	0.92	0.026
8.80	1.66	0.91	0.026
9.00	1.65	0.91	0.025
9.20	1.63	0.90	0.025
9.40	1.60	0.89	0.024
9.60	1.57	0.89	0.023
9.80	1.52	0.87	0.021
10.00	1.47	0.86	0.020
10.20	1.42	0.84	0.018
10.40	1.35	0.82	0.016
10.50	1.32	0.81	0.016
10.60	1.28	0.80	0.015
10.80	1.23	0.78	0.014
11.00	1.19	0.77	0.014
11.20	1.16	0.76	0.013
11.40	1.13	0.75	0.013
11.60	1.11	0.74	0.013
11.80	1.09	0.74	0.013
12.00	1.08	0.73	0.013
12.40	1.05	0.72	0.012
12.80	1.01	0.71	0.012
13.20	0.98	0.70	0.012
13.60	0.95	0.69	0.013
14.00	0.92	0.68	0.013
14.40	0.89	0.67	0.013
14.80	0.90	0.67	0.013
15.20	0.92	0.68	0.013
15.60	0.95	0.69	0.013
16.00	0.98	0.70	0.012
16.40	1.01	0.71	0.012
16.80	1.04	0.72	0.012
17.20	1.09	0.74	0.013
17.60	1.13	0.75	0.013
18.00	1.17	0.76	0.014
18.40	1.21	0.78	0.014
18.80	1.24	0.79	0.014
19.20	1.27	0.80	0.015
19.60	1.29	0.80	0.015
20.00	1.30	0.81	0.015
20.60	1.29	0.80	0.015
21.00	1.27	0.80	0.015
21.60	1.23	0.78	0.014
22.00	1.20	0.77	0.014
22.60	1.15	0.76	0.013
23.00	1.12	0.75	0.013
23.60	1.08	0.73	0.013
24.00	1.05	0.73	0.013
24.60	1.02	0.71	0.012
25.00	1.00	0.71	0.012
25.60	0.97	0.69	0.012
26.00	0.95	0.69	0.013
26.60	0.91	0.67	0.013
27.00	0.88	0.66	0.013
27.60	0.84	0.65	0.014
28.00	0.83	0.64	0.014
28.60	0.82	0.64	0.014
29.00	0.81	0.64	0.014
29.60	0.82	0.64	0.014
30.00	0.82	0.64	0.014

REFERENCES

1. Shiles, E., Sasaki, T., Inokuti, M., and Smith, D. Y., *Phys. Rev. Sect. B*, 22, 1612, 1980.
2. Edwards, D. F., and Philipp, H. R., in *HOC-I*, p.665.
3. Ives, H. E., and Briggs, N. B., *J. Opt. Soc. Am.*, 27, 395, 1937.
4. Bos, L. W., and Lynch, D. W., *Phys. Rev. Sect. B*, 2, 4567, 1970.
5. Weaver, J. H., Colavita, E., Lynch, D. W., and Rosei, R., *Phys. Rev. Sect. B*, 19, 3850, 1979.
6. Hagemann, H. J., Gudat, W., and Kunz, C., *J. Opt. Soc. Am.*, 65, 742, 1975.
7. Schulz, L. G., *J. Opt. Soc. Am.*, 47, 64, 1957.
8. Potter, R. F., in *HOC-I*, p.465.
9. Olson, C. G., Lynch, D. W., and Weaver, J. H., unpublished.
10. Lynch, D. W., Olson, C. G., and Weaver, J. H., unpublished.
11. Weaver, J. H., Olson, C. G., and Lynch, D. W., *Phys. Rev. Sect. B*, 15, 4115, 1977.
12. Lynch, D. W., and Hunter, W. R., in *HOC-II*, p.345.
13. Priol, M. A., Daudé, A., and Robin, S., *Compt. Rend.*, 264, 935, 1967.
14. Johnson, P. B., and Christy, R. W., *Phys. Rev. Sect. B*, 9, 5056, 1974.
15. Arakawn, E. T., and Inagaki, T., in *HOC-II*, p.461.
16. Weaver, J. H., Lynch, D. W., and Olson, D. G., *Phys. Rev. Sect. B*, 10, 501, 1973.
17. Lynch, D. W., Rosei, R., and Weaver, J. H., *Solid State Commun.*, 9, 2195, 1971.
18. Weaver, J. H., Lynch, D. W., and Olson, C. G., *Phys. Rev. Sect. B*, 7, 4311, 1973.
19. Weaver, J. H., and Benbow, R. L., *Phys. Rev. Sect. B*, 12, 3509, 1975.
20. Weaver, J. H., *Phys. Rev.,Sect. B*, 11, 1416, 1975.
21. Lynch, D. W., and Hunter, W. R., in *HOC-II*, p.364.
22. Palik, E. D., in *HOC-II*, p. 691.
23. Edwards, D. F., in *HOC-I*, p. 547.
24. Lynch, D. W., and Hunter, W. R., in *HOC-II*, p.354.
25. Palik, E. D., in *HOC-II*, p. 709.
26. Lynch, D. W., Olson, C. G., and Weaver, J. H., *Phys. Rev. Sect. B*, 11, 3671, 1975.
27. Weaver, J. H., Lynch, D. W., and Olson, C. G., *Phys. Rev. Sect. B*, 12, 1293, 1975.
28. Lanham, A. P., and Terherne, D. M., *Proc. Phys. Soc.*, 83, 1059, 1964.

ELASTO-OPTIC, ELECTRO-OPTIC, AND MAGNETO-OPTIC CONSTANTS

When a crystal is subjected to a stress field, an electric field, or a magnetic field, the resulting optical effects are in general dependent on the orientation of these fields with respect to the crystal axes. It is useful, therefore, to express the optical properties in terms of the refractive index ellipsoid (or indicatrix):

$$\frac{x^2}{n_x^2}+\frac{y^2}{n_y^2}+\frac{z^2}{n_z^2}=1$$

or

$$\sum_{ij} B_{ij}x_i y_j = 1 \quad (i,j=1,2,3)$$

where

$$B_{ij}=\left[\frac{1}{\varepsilon}\right]_{ij}=\left[\frac{1}{n^2}\right]_{ij}$$

ε is the dielectric constant or permeability; the quantity B_{ij} has the name impermeability.

A crystal exposed to a *stress* **S** will show a change of its impermeability. The photo-elastic (or elasto-optic) constants, P_{ijkl}, are defined by

$$\Delta\left[\frac{1}{\varepsilon}\right]_{ij}=\Delta\left[\frac{1}{n^2}\right]_{ij}=\sum_{kl} P_{ijkl}\, S_{kl}$$

where n is the refractive index and S_{kl} are the strain tensor elements; the P_{ijkl} are the elements of a 4th rank tensor.

When a crystal is subjected to an *electric field* ***E*** two possible changes of the refractive index may occur depending on the symmetry of the crystal.

1. All materials, including isotropic solids and polar liquids, show an electro-optic birefringence (Kerr effect) which is proportional to the square of the electric field, ***E***:

$$\left[\frac{1}{n^2}\right]_{ij}=\sum_{k,l=1,2,3} K_{ijkl}E_kE_l=\sum_{k,l=1,2,3} g_{ijkl}P_kP_l$$

   where E_k and E_l are the components of the electric field and P_k and P_l the electric polarizations. The coefficients, K_{ijkl}, are the quadratic electro-optic coefficients, while the constants g_{ijkl} are known as the Kerr constants.

2. The other electro-optic effect only occurs in the 20 piezo-electric crystal classes (no center of symmetry). This effect is known as the Pockels effect. The optical impermeability changes linearly with the static field

$$\Delta\left[\frac{1}{n^2}\right]_{ij}=\sum_k r_{ij,k}E_k$$

The coefficients $r_{ij,k}$ have the name (linear) electro-optic coefficients.

The values of the electro-optic coefficients depend on the boundary conditions. If the superscripts T and S denote respectively the conditions of zero stress (free) and zero strain (clamped) one finds:

$$r_{ij}^{\mathrm{T}}=r_{ij}^{\mathrm{S}}+q_{ik}^{\mathrm{E}}e_{jk}=r_{ij}^{\mathrm{S}}+P_{ik}^{\mathrm{E}}d_{jk}$$

where $e_{jk}=(\partial T_k/\partial E_j)_S$ and $d_{jk}=(\partial S_k/\partial E_j)_T$ are the appropriate piezo-electric coefficients.

The interaction between a *magnetic field* and a light wave propagating in a solid or in a liquid gives rise to a rotation of the plane of polarization. This effect is known as *Faraday rotation*. It results from a difference in propagation velocity for left and right circular polarized light.

The Faraday rotation, θ_F, is linearly proportional to the magnetic field H:

$$\theta_F = VlH$$

where l is the light path length and V is the *Verdet* constant (minutes/oersted·cm).

For ferromagnetic, ferrimagnetic, and antiferromagnetic materials the magnetic field in the above expression is replaced by the magnetization M and the magneto-optic coefficient in this case is known as the Kund constant K:

Specific Faraday rotation $F = KM$

In the tables below the *Faraday rotation* is listed at the saturation magnetization per unit length, together with the absorption coefficient α, the temperature T, the critical temperature T_C (or T_N), and the wavelength of the measurement.

In the tables which follow, the properties are presented in groups:

- Elasto-optic coefficients (photoelastic constants)
- Linear electro-optic coefficients (Pockels constants)
- Quadratic electro-optic coefficients (Kerr constants)
- Magneto-optic coefficients:
  - Verdet constants
  - Faraday rotation parameters

Within each group, materials are classified by crystal system or physical state. References are given at the end of each group of tables.

ELASTO-OPTIC COEFFICIENTS (PHOTOELASTIC CONSTANTS)

Name Cubic (43m, 432, m3m)	Formula	λ/μm	p_{11}	p_{12}	p_{44}	p_{11}-p_{12}	Ref.
Sodium fluoride	NaF	0.633	0.08	0.20	-0.03	-0.12	1
Sodium chloride	NaCl	0.589	0.115	0.159	-0.011	-0.042	2
Sodium bromide	NaBr	0.589	0.148	0.184	-0.0036	-0.035	1
Sodium iodide	NaI	0.589	—	—	0.0048	-0.0141	3
Potassium fluoride	KF	0.546	0.26	0.20	-0.029	0.06	1
Potassium chloride	KCl	0.633	0.22	0.16	-0.025	0.06	4
Potassium bromide	KBr	0.589	0.212	0.165	-0.022	0.047	5
Potassium iodide	KI	0.590	0.212	0.171	—	0.041	6
Rubidium chloride	RbCl	0.589	0.288	0.172	-0.041	0.116	7,8
Rubidium bromide	RbBr	0.589	0.293	0.185	-0.034	0.108	7,8
Rubidium iodide	RbI	0.589	0.262	0.167	-0.023	0.095	7,8
Lithium fluoride	LiF	0.589	0.02	0.13	-0.045	-0.11	5
Lithium chloride	LiCl	0.589			-0.0177	-0.0407	3
Ammonium chloride	NH_4Cl	0.589	0.142	0.245	0.042	-0.103	9
Cadmium telluride	CdTe	1.06	-0.152	-0.017	-0.057	-0.135	10
Calcium fluoride	CaF_2	0.55-0.65	0.038	0.226	0.0254	-0.183	11
Copper chloride	CuCl	0.633	0.120	0.250	-0.082	-0.130	12
Copper bromide	CuBr	0.633	0.072	0.195	-0.083	-0.123	12
Copper iodide	CuI	0.633	0.032	0.151	-0.068	-0.119	12
Diamond	C	0.540-0.589	-0.278	0.123	-0.161	-0.385	13
Germanium	Ge	3.39	-0.151	-0.128	-0.072	-0.023	14
Gallium arsenide	GaAs	1.15	-0.165	-0.140	-0.072	-0.025	15
Gallium phosphide	GaP	0.633	-0.151	-0.082	-0.074	-0.069	15
Strontium fluoride	SrF_2	0.633	0.080	0.269	0.0185	-0.189	16
Strontium titanite	$SrTiO_3$	0.633	0.15	0.095	0.072	—	17
KRS-5	Tl(Br,I)	0.633	-0.140	0.149	-0.0725	-0.289	18,20
KRS-6	Tl(Br,Cl)	0.633	-0.451	-0.337	-0.164	-0.114	19,20
Zinc sulfide	Zn	0.633	0.091	-0.01	0.075	0.101	15

Rare Gases	Formula	λ/μm	p_{11}	p_{12}	p_{44}	p_{11}-p_{12}	Ref.
Neon (T = 24.3 K)	Ne	0.488	0.157	0.168	0.004	-0.011	21
Argon (T = 82.3 K)	Ar	0.488	0.256	0.302	0.015	-0.046	22
Krypton (T = 115.6 K)	Kr	0.488	0.34	0.34	0.037	0	21
Xenon (T = 160.5 K)	Xe	0.488	0.284	0.370	0.029	-0.086	22

Garnets	Formula	λ/μm	p_{11}	p_{12}	p_{44}	p_{11}-p_{12}	Ref.
GGG	$Gd_3Ga_5O_{12}$	0.514	-0.086	-0.027	-0.078	-0.059	23
YIG	$Y_3Fe_5O_{12}$	1.15	0.025	0.073	0.041	—	15
YGG	$Y_3Ga_5O_{12}$	0.633	0.091	0.019	0.079	—	17
YAG	$Y_3Al_5O_{12}$	0.633	-0.029	0.0091	-0.0615	-0.038	15

ELASTO-OPTIC COEFFICIENTS (PHOTOELASTIC CONSTANTS) continued)

Name Cubic (23, m3)	Formula	λ/μm	p_{11}	p_{12}	p_{44}	p_{13}	Ref.
Barium nitrate	Ba(NO3)2	0.589	—	p_{11}-p_{22} = 0.992	-0.0205	p_{11}-p_{13} = 0.713	13
Lead nitrate	$Pb(NO_3)_2$	0.589	0.162	0.24	-0.0198	0.20	24,25
Sodium bromate	$NaBrO_3$	0.589	0.185	0.218	-0.0139	0.213	26
Sodium chlorate	$NaClO_3$	0.589	0.162	0.24	-0.0198	0.20	26
Strontium nitrate	$Sr(NO_3)_2$	0.41	0.178	0.362	-0.014	0.316	27

Hexagonal (mmc,6mm)	Formula	λ/μm	p_{11}	p_{12}	p_{13}	p_{31}	p_{33}	p_{44}	Ref.
Beryl	$Be_3Al_2Si_6O_{18}$	0.589	0.0099	0.175	0.191	0.313	0.023	-0.152	28
Cadmium sulfide	CdS	0.633	-0.142	-0.066	-0.057	-0.041	-0.20	-0.099	15,2
Zinc oxide	ZnO	0.633	±0.222	±0.099	-0.111	±0.088	-0.235	0.0585	30
Zinc sulfide	ZnS	0.633	-0.115	0.017	0.025	0.0271	-0.13	-0.0627	31

Trigonal (3m,32,$\bar{3}$m)	Formula	λ/μm	p_{11}	p_{12}	p_{13}	p_{14}	p_{31}
Sapphire	Al_2O_3	0.644	-0.23	-0.03	0.02	0.00	-0.04
Calcite	$CaCO_3$	0.514	0.062	0.147	0.186	-0.011	0.241
Lithium niobate	$LiNbO_3$	0.633	±0.034	±0.072	±0.139	±0.066	±0.178
Lithium tantalate	$LiTaO_3$	0.633	-0.081	0.081	0.093	-0.026	0.089
Cinnabar	HgS	0.633			±0.445		
Quartz	SiO_2	0.589	0.16	0.27	0.27	-0.030	0.29
Proustite	Ag_3AsS_3	0.633	±0.10	±0.19	±0.22		±0.24
Sodium nitrite	$NaNO_3$	0.633		±0.21	±0.215	±0.027	±0.25
Tellurium	Te	10.6	0.155	0.130	—	—	—

Trigonal (3m,32,$\bar{3}$m) (continued)	p_{33}	p_{41}	p_{44}	Ref.
Sapphire	-0.20	0.01	-0.10	15,32
Calcite	0.139	-0.036	-0.058	33
Lithium niobate	+-0.060	±0.154	±0.300	15,34
Lithium tantalate	-0.044	-0.085	0.028	15,35
Cinnabar	+-0.115	—	—	36
Quartz	0.10	-0.047	-0.079	37
Proustite	+-0.20	—	—	38
Sodium nitrite		0.055	-0.06	39
Tellurium		—	—	15

Tetragonal (4/mmm,$\bar{4}$2m,422)	Formula	λ/μm	p_{11}	p_{12}	p_{13}	p_{31}
Ammonium dihydrogen phosphate	ADP	0.589	0.319	0.277	0.169	0.197
Barium titanate	$BaTiO_3$	0.633	0.425	—	—	—
Cesium dihydrogen arsenate	CDA	0.633	0.267	0.225	0.200	0.195
Magnesium fluoride	MgF_2	0.546	—	—	—	—
Calomel	Hg_2Cl_2	0.633	±0.551	±0.440	±0.256	±0.137
Potassium dihydrogen phosphate	KDP	0.589	0.287	0.282	0.174	0.241
Rubidium dihydrogen arsenate	RDA	0.633	0.227	0.239	0.200	0.205
Rubidium dihydrogen phosphate	RDP	0.633	0.273	0.240	0.218	0.210
Strontium barium niobate	$Sr_{0.75}Ba_{0.25}Nb_2O_6$	0.633	0.16	0.10	0.08	0.11
Strontium barium niobate	$Sr_{0.5}Ba_{0.5}Nb_2O_6$	0.633	0.06	0.08	0.17	0.09
Tellurium oxide	TeO_2	0.633	0.0074	0.187	0.340	0.090
Rutile	TiO_2	0.633	0.017	0.143	-0.139	-0.080

ELASTO-OPTIC COEFFICIENTS (PHOTOELASTIC CONSTANTS) (continued)

Name Tetragonal (4/mmm,$\bar{4}$2m,422) (continued)	p_{33}	p_{44}	p_{66}	Ref.
Ammonium dihydrogen phosphate	0.167	-0.058	-0.091	40
Barium titanate	—	—	—	41
Cesium dihydrogen arsenate	0.227	—	—	42
Magnesium fluoride	—	±0.0776	±0.0488	43
Calomel	±0.010	—	±0.047	44
Potassium dihydrogen phosphate	0.122	-0.019	-0.064	45
Rubidium dihydrogen arsenate	0.182	—	—	41
Rubidium dihydrogen phosphate	0.208	—	—	41
Strontium barium niobate	0.47	—	—	46
Strontium barium niobate	0.23	—	—	46
Tellurium oxide	0.240	-0.17	-0.046	47
Rutile	-0.057	-0.009	-0.060	48

Tetragonal (4,$\bar{4}$,4/m)	Formula	λ/μm	p_{11}	p_{12}	p_{13}	p_{16}	p_{31}
Cadmium molybdate	$CdMoO_4$	0.633	0.12	0.10	0.13	—	0.11
Lead molybdate	$PbMoO_4$	0.633	0.24	0.24	0.255	0.017	0.175
Sodium bismuth molybdate	$NaBi(MoO_4)_2$	0.633	0.243	0.205	0.25	—	0.21

Tetragonal (4,$\bar{4}$,4/m) (continued)	p_{33}	p_{44}	p_{45}	p_{61}	p_{66}	Ref.
Cadmium molybdate	0.18	—	—	—	—	49
Lead molybdate	0.300	0.067	-0.01	0.013	0.05	52
Sodium bismuth molybdate	0.29	—	—	—	—	—

Orthorhombic (222,m22,mmm)	Formula	λ/μm	p_{11}	p_{12}	p_{13}	p_{21}	p_{22}	p_{23}
Ammonium chlorate	NH_4ClO_4	0.633	—	0.24	0.18	0.23	—	0.20
Ammonium sulfate	$(NH_4)_2SO_4$	0.633	0.26	0.19	±0.260	±0.230	±0.27	±0.254
Rochelle salt	$NaKC_4H_4O_6$	0.589	0.35	0.41	0.42	0.37	0.28	0.34
Iodic acid (α)	HIO_3	0.633	0.302	0.496	0.339	0.263	0.412	0.304
Sulfur (α)	S	0.633	0.324	0.307	0.268	0.272	0.301	0.310
Barite	$BaSO_4$	0.589	0.21	0.25	0.16	0.34	0.24	0.19
Topaz	$Al_2SiO_4(OH,F)_2$	—	-0.085	0.069	0.052	0.095	-0.120	0.065

Orthorhombic (222,m22,mmm) (continued)	p_{31}	p_{32}	p_{33}	p_{44}	p_{55}	p_{66}	Ref.
Ammonium chlorate	0.19	0.18	±0.02	<±0.02	—	±0.04	51
Ammonium sulfate	0.20	±0.26	0.26	0.015	±0.0015	0.012	52
Rochelle salt	0.36	0.35	0.36	-0.030	0.0046	0.0[illegible]	53
Iodic Acid (α)	0.251	0.345	0.336	0.084	-0.030	0.098	54
Sulfur (α)	0.203	0.232	0.270	0.143	0.019	0.118	54
Barite	0.28	0.22	0.31	0.002	-0.012	0.037	55
Topaz	0.095	0.085	-0.083	-0.095	-0.031	0.098	28

Monoclinic (2,m,2/m)	Formula	λ/μm			
Taurine	$C_2H_7NO_3S$	0.589	p_{11} = 0.313 p_{12} = 0.251 p_{13} = 0.270 p_{15} = -0.10	p_{25} = -0.0025 p_{31} = 0.362 p_{32} = 0.275 p_{33} = 0.308	p_{51} = -0.014 p_{52} = 0.006 p_{53} = 0.0048 p_{55} = 0.047

ELASTO-OPTIC COEFFICIENTS (PHOTOELASTIC CONSTANTS) (continued)

Name Monoclinic (2,m,2/m)	Formula	λ/μm			
Taurine (continued)			p_{21} = 0.281 p_{22} = 0.252 p_{23} = 0.272	p_{35} = -0.003 p_{44} = 0.0025 p_{46} = -0.0056	p_{64} = 0.0024 p_{66} = 0.0028

Isotropic	Formula	λ/μm	p_{11}	p_{12}	p_{44}	Ref.
Fused silica	SiO_2	0.633	0.121	0.270	-0.075	15
Water	H_2O	0.633	±0.31	±0.31		15
Polystyrene		0.633	±0.30	±0.31		25
Lucite		0.633	±0.30	0.28		25
Orpiment	As_2S_3-glass	1.15	0.308	0.299	0.0045	15
Tellurium oxide	TeO_2-glass	0.633	0.257	0.241	0.0079	56
Laser glasses	LGS-247-2	0.488	±0.168	±0.230		57
	LGS-250-3		±0.135	±0.198		
	LGS-1		±0.214	±0.250		
	KGSS-1621		±0.205	±0.239		
Dense flint glasses	LaSF	0.633	0.088	0.147	-0.030	58
(examples)	SF_4		0.215	0.243	-0.014	
	U10502		0.172	0.179	-0.004	
	$TaFd_7$		0.099	0.138	-0.020	

REFERENCES

A. Narasimhamurty, T. S., *Photoelastic and Electro-Optic Properties of Crystals*, Plenum Press, New York, 1981; pp. 290-293.
B. Weber, M. J., Ed., *CRC Handbook of Laser Science and Technology*, Volume IV, Part 2, CRC Press, Boca Raton, FL, 1986; pp. 324-331.
1. Petterson, H. E., *J. Opt. Soc. Am.*, 63, 1243, 1973.
2. Burstein, E. and Smith, P. L., *Phys. Rev.*, 74, 229, 1948.
3. Pakhnev, A. V., et al., *Sov. Phys. J. (transl.)*, 18, 1662, 1975.
4. Feldman, A., Horovitz, D., and Waxler, R. M., *Appl. Opt.*, 16, 2925, 1977.
5. Iyengar, K. S., *Nature (London)*, 176, 1119, 1955.
6. Bansigir, K. G. and Iyengar, K. S., *Acta Crystallogr.*, 14, 727, 1961.
7. Pakhev, A. V., et al., *Sov. Phys. J. (transl.)*, 20, 648, 1975.
8. Bansigir, K. G., *Acta Crystallogr.*, 23, 505, 1967.
9. Krishna Rao, K. V. and Krishna Murty, V. G., *Ind. J. Phys.*, 41, 150, 1967.
10. Weil, R. and Sun, M. J., *Proc. Int. Symp. CdTe (Detectors)*, Strasbourg Centre de Rech. Nucl., 1971, XIX-1 to 6, 1972.
11. Schmidt, E. D. D. and Vedam, K., *J. Phys. Chem. Solids*, 27, 1563, 1966.
12. Biegelsen, D. K., et al., *Phys. Rev. B*, 14, 3578, 1976.
13. Helwege, K. H., *Landolt-Börnstein, New Series Group III*, Vol. II, Springer-Verlag Berlin, 1979.
14. Feldman, A., Waxler, R. M., and Horovitz, D., *J. Appl. Phys.*, 49, 2589, 1978.
15. Dixon, R. W., *J. Appl. Phys.*, 38, 5149, 1967.
16. Shabin, O. V., et al., *Sov. Phys. Sol. State (transl.)*, 13, 3141, 1972.
17. Reintjes, J. and Schultz, M. B., *J. Appl. Phys.*, 39, 5254, 1968.
18. Rivoallan, L. and Favre, F., *Opt. Commun.*, 8, 404, 1973.
19. Rivoallan, L. and Favre, F., *Opt. Commun.*, 11, 296, 1974.
20. Afanasev, I. I., et al., *Sov. J. Opt. Technol.*, 46, 663, 1979.
21. Rand, S. C., et al., *Phys. Rev. B*, 19, 4205, 1979.
22. Sipe, J. E., *Can J. Phys.*, 56, 199, 1978.
23. Christyi, I. L., et al., *Sov. Phys. Sol. State (transl.)*, 17, 922, 1975.
24. Narasimhamurty, T. S., *Curr. Sci. (India)*, 23, 149, 1954.
25. Smith, T. M. and Korpel, A., *IEEE J. Quant. Electron.*, QE-1, 283, 1965.
26. Narasimhamurty, T. S., *Proc. Ind. Acad. Sci.*, A40, 164, 1954.
27. Rabman, A., *Bhagarantam Commem. Vol.*, Bangalore Print. and Publ., 173, 1969.
28. Eppendahl, R., *Ann. Phys. (IV)*, 61, 591, 1920.
29. Laurenti, J. P. and Rouzeyre, M., *J. Appl. Phys.*, 52, 6484, 1981.
30. Sasaki, H., et al., *J. Appl. Phys.*, 47, 2046, 1976.
31. Uchida, N. and Saito, S., *J. Appl. Phys.*, 43, 971, 1972.

ELASTO-OPTIC COEFFICIENTS (PHOTOELASTIC CONSTANTS) (continued)

32. Waxler, R. M. and Farabaugh, E. M., *J. Res. Natl. Bur. Stand.*, A74, 215, 1970.
33. Nelson, D. F., Lazay, P. D., and Lax, M., *Phys. Rev.*, B6, 3109, 1972.
34. O'Brien, R. J., Rosasco, G. J., and Weber, A., *J. Opt. Soc. Am.*, 60, 716, 1970.
35. Avakyants, L. P., et al., *Sov. Phys.*, 18, 1242, 1976.
36. Sapriel, J., *Appl. Phys. Litt.*, 19, 533, 1971.
37. Narasimhamurty, T. S., *J. Opt. Soc. Am.*, 59, 682, 1969.
38. Zubrinov, I. I., et al., *Sov. Phys. Sol. State (transl.)*, 15, 1921, 1974.
39. Kachalov, O. V. and Shpilko, I. O., *Sov. Phys. JETP (transl.)*, 35, 957, 1972.
40. Narasimhamurty, T. S., et al., *J. Mater. Sci.*, 8, 577, 1973.
41. Tada, K. and Kikuchi, K., *Jpn. J. Appl. Phys.*, 19, 1311, 1980.
42. Aleksandrov, K. S., et al., *Sov. Phys. Sol. State (transl.)*, 19, 1090, 1977.
43. Afanasev, I. I., et al., *Sov. Phys. Sol. State (transl.)*, 17, 2006, 1975.
44. Silvestrova, I. M., et al., *Sov. Phys. Cryst. (transl.)*, 20, 649, 1975.
45. Veerabhadra Rao, K. and Narasimhamurty, T. S., *J. Mater. Sci.*, 10, 1019, 1975.
46. Venturini, E. L., et al., *J. Appl. Phys.*, 40, 1622, 1969.
47. Vehida, N. and Ohmachi, Y., *J. Appl. Phys.*, 40, 4692, 1969.
48. Grimsditch, M. H. and Ramdus, A. K., *Phys. Rev. B*, 22, 4094, 1980.
49. Schinke, D. P. and Viehman,W., unpublished Data.
50. Coquin, G. A., et al., *J. Appl. Phys.*, 42, 2162, 1971.
51. Vasquez, F., et al., *J. Phys. Chem. Solids*, 37, 451, 1976.
52. Luspin, Y. and Hauret, G., *C.R.Ac. Sci. Paris*, B274, 995 1972.
53. Narasimhamurty, T. S., *Phys. Rev.*, 186, 945, 1969.
54. Haussühl, S. and Weber, H. J., *Z. Kristall.*, 132, 266, 1970.
55. Vedam, K., *Proc. Ind. Ac. Sci.*, A34, 161, 1951.
56. Yano, T., Fukumoto, A., and Watanabe, A., *J. Appl. Phys.*, 42, 3674, 1971.
57. Manenkov, A. A. and Ritus, A. I., *Sov. J. Quant. Electr.*, 8, 78, 1978.
58. Eschler, H. and Weidinger, F., *J. Appl. Phys.*, 46, 65, 1975.

LINEAR ELECTRO-OPTIC COEFFICIENTS

Name

Cubic ($\bar{4}$3m)	**Formula**	**λ/μm**	**r_{41} pm/V**
Cuprous bromide	CuBr	0.525	0.85
Cuprous chloride	CuCl	0.633	3.6
Cuprous iodide	CuI	0.55	-5.0
Eulytite (BSO)	$Bi_4Si_3O_{12}$	0.63	0.54
Germanium eulytite (BGO)	$Bi_4Ge_3O_{12}$	0.63	1.0
Gallium arsenide	GaAs	10.6	1.6
Gallium phosphide	GaP	0.56	-1.07
Hexamethylenetetramine	$C_6H_{12}N_4$	0.633	0.78
Sphalerite	ZnS	0.65	2.1
Zinc selenide	ZnSe	0.546	2.0
Zinc telluride	ZnTe	3.41	4.2
Cadmium telluride	CdTe	3.39	6.8

Cubic (23)	**Formula**	**λ/μm**	**r_{41} pm/V**
Ammonium chloride (77 K)	NH_4Cl	—	1.5
Ammonium cadmium langbeinite	$(NH_4)_2Cd_2(SO_4)_3$	0.546	0.70
Ammonium manganese langbeinite	$(NH_4)_2Mn_2(SO_4)_3$	0.546	0.53
Thallium cadmium langbeinite	$Tl_2Cd_2(SO_4)_3$	0.546	0.37
Potassium magnesium langbeinite	$K_2Mg_2(SO_4)_3$	0.546	0.40
Bismuth monogermanate	$Bi_{12}GeO_{20}$	—	3.3
Bismuth monosilicate	$Bi_{12}SiO_{20}$	—	3.3
Sodium chlorate	$NaClO_3$	0.589	0.4
Sodium uranyl acetate	$NaUO_2(CH_3COO)_3$	0.546	0.87

LINEAR ELECTRO-OPTIC COEFFICIENTS (continued)

Name Cubic (23)	Formula	λ/μm	r_{41} pm/V
Trenhydrobromide	$N(CH_2CH_2NH_2)_3$ 3HBr	—	1.5
Trenhydrochloride	$N(CH_2CH_2NH_2)_3$ 3HCl	—	1.7

Tetragonal ($\bar{4}$2m)	Formula	T_{tran} K	r_{41} pm/V	r_{63} pm/V
Ammonium dihydrogen phosphate (ADP)	$NH_4H_2PO_4$	148	24.5	-8.5
Ammonium dideuterium phosphate (AD*P)	$NH_4D_2PO_4$	242	—	11.9
Ammonium dihydrogen arsenate (ADA)	$NH_4H_2AsO_4$	—	—	9.2
Cesium dihydrogen arsenate (CsDA)	CsH_2AsO_4	143	—	18.6
Cesium dideuterium arsenate (CsD*A)	CsD_2AsO_4	212	—	36.6
Potassium dihydrogen phosphate (KDP)	KH_2PO_4	123	8.6	-10.5
Potassium dideuterium phosphate (KD*P)	KD_2PO_4	222	8.8	23.8
Potassium dihydrogen arsenate (KDA)	KH_2AsO_4	97	12.5	10.9
Potassium dideuterium arsenate (KD*A)	KD_2AsO_4	162	—	18.2
Rubidium dihydrogen phosphate (RDP)	RbH_2PO_4	147	—	15.5
Rubidium dihydrogen arsenate (RDA)	RbH_2AsO_4	110	—	13.0
Rubidium dideuterium arsenate (RD*A)	RbD_2AsO_4	178	—	21.4

Tetragonal (4mm)	Formula	T_{tran} K	r_{13} pm/V	r_{33} pm/V	r_{51} pm/V
Barium titanate	$BaTiO_3$	406	8	28	—
Potassium lithium niobate	$K_3Li_2Nb_5O_{15}$	693	8.9	5.9	—
Lead titanate	$PbTiO_3$	765	13.8	5.9	—
Strontium barium niobate (SBN75)	$Sr_{0.75}Ba_{0.25}Nb_2O_6$	330	6.7	1340	42
Strontium barium niobate (SBN46)	$Sr_{0.46}Ba_{0.54}Nb_2O_6$	602	~180	35	—

Hexagonal (6mm)	Formula	r_{13} pm/V	r_{33} pm/V	r_{42} pm/V	r_{51} pm/V
Greenockite	CdS	3.1	2.9	2.0	3.7
Greenockite (const. strain)	CdS	1.1	2.4	—	—
Wurzite	ZnS	0.9	1.8	—	—
Zincite	ZnO	-1.4	+2.6	—	—

Hexagonal (6)	Formula	r_{13} pm/V	r_{33} pm/V	r_{42} pm/V	r_{51} pm/V
Lithium iodate	$LiIO_3$	4.1	6.4	1.4	3.3
Lithium potassium sulfate	$LiKSO_4$	r_{13}-r_{33} = 1.6	—	—	—

Trigonal (3m)	Formula	T_{tran} K	r_{13} pm/V	r_{22} pm/V	r_{33} pm/V	r_{42} pm/V
Cesium nitrate	$CsNO_3$	425	—	0.43	—	—
Lithium niobate	$LiNbO_3$	1483	8.6	7.0	30.8	28
Lithium tantalate	$LiTaO_3$	890	8.4	—	30.5	—
Lithium sodium sulfate	$LiNaSO_4$	—	—	<0.02	—	—
Tourmaline	—	—	—	0.3	—	—

Trigonal (32)	Formula	T_{tran} K	r_{11} pm/V	r_{41} pm/V
Cesium tartrate	$Cs_2C_4H_4O_6$	—	1.0	—
Cinnabar	HgS	659	3.1	1.5

LINEAR ELECTRO-OPTIC COEFFICIENTS (continued)

Name Trigonal (32)	Formula	T_{tran} K	r_{11} pm/V	r_{41} pm/V
Potassium dithionate	$K_2S_2O_6$	—	0.26	—
Strontium dithionate	$SrS_2O_6 \cdot 4H_2O$	—	0.1	—
Quartz	SiO_2	1140	-0.47	0.2
Selenium	Se	398	2.5	

Orthorhombic (222)	Formula	T_{tran} K	r_{41} pm/V	r_{52} pm/V	r_{63} pm/V
Ammonium oxalate	$(NH_4)_2C_2O_4 \cdot 4H_2O$	—	230	330	250
Rochelle salt	$KNaC_4H_4O_6 \cdot 4H_2O$	$T_u = 297$ $T_l = 255$	-2.0	-1.7	+0.32

Orthorhombic (mm2)	Formula	Ttrans K	r_{13} pm/V	r_{23} pm/V	r_{33} pm/V	r_{42} pm/V	r_{51} pm/V
Barium sodium niobate (BSN)	Ba_2NaNbO_{15}	833	15	13	48	92	90
Potassium niobate	$KNbO_3$	476	28	1.3	64	380	105

Monoclinic (2)	Formula	T_{trans} K	r_{22} pm/V	r_{32} pm/V
Calcium pyroniobate	$Ca_2Nb_2O_7$	—	0.33	13.7
Triglycine sulfate (TGS)	$(NH_2CH_2COOH)_3 \cdot H_2SO_4$	322	7.2	13.6

REFERENCES

1. Narasimhamurty, T. S., *Photoelastic and Electro-Optic Properties of Crystals*, Plenum Press, New York, 1981, pp. 405-407.
2. Weber, M. J., Ed., *CRC Handbook of Laser Science and Technology*, Vol. IV, CRC Press, Boca Raton, FL, 1986, pp. 258-278.

QUADRATIC ELECTRO-OPTIC COEFFICIENTS

Kerr Constants of Ferroelectric Crystals[1,2]

Name	Formula	T_{tran} K	λ µm	g_{11} 10^{10} esu	g_{12} 10^{10} esu	g_{11}-g_{12} 10^{10} esu	g_{44} 10^{10} esu
Barium titanate	$BaTiO_3$	406	0.633	1.33	-0.11	1.44	
Strontium titanate	$SrTiO_3$	—	0.633	—	—	1.56	—
Potassium tantalate niobate	$KTa_{0.65}Nb_{0.35}O_3$	330	0.633	1.50	-0.42	1.92	1.63
Potassium tantalate	$KTaO_3$	13	0.633	—	—	1.77	1.33
Lithium niobate	$LiNbO_3$	1483	—	0.94	0.25	0.7	0.6
Lithium tantalate	$LiTaO_3$	938	—	1.0	0.17	0.8	0.7
Barium sodium niobate (BSN)	$Ba_{0.8}Na_{0.4}Nb_2O_6$	833	—	1.55	0.44	1.11	

Kerr Constants of Selected Liquids[2]

K is the Kerr constant at a wavelength of 589 nm and at room temperature; ε is the static dielectric constant; T_m is the melting point; and T_b is the normal boiling point

Name	Molecular formula	K 10^{-7} esu	ε	T_m °C	T_b °C
Carbon disulfide	CS_2	+3.23	2.63	-111.5	+46.3
Acetone	C_3H_6O	+16.3	21.0	-94.8	+56.1
Methyl ethyl ketone	C_4H_8O	+13.6	18.56	-86.67	+79.6

QUADRATIC ELECTRO-OPTIC COEFFICIENTS (continued)

Kerr Constants of Selected Liquids (continued)[2]

Name	Molecular formula	K 10^{-7} esu	ε	T_m °C	T_b °C
Pyridine	C_5H_5N	+20.4	13.26	-42	+115.23
Ethyl cyanoacetate	$C_5H_7NO_2$	+38.8	31.6	-22.5	205
o-Dichlorobenzene	$C_6H_4Cl_2$	+42.6	10.12	-16.7	180
Benzenesulfonyl chloride	$C_6H_5ClO_2S$	+89.9	28.90	+14.5	247
Nitrobenzene	$C_6H_5NO_2$	+326	35.6	+5.7	210.8
Ethyl 3-aminocrotonate	$C_6H_{11}NO_2$	+31.0	—	+33.9	210
Paraldehyde	$C_6H_{12}O_3$	-23.0	14.7 12.0[a]	+12.6	124
Benzaldehyde	C_7H_6O	+80.8	17.85 14.1[a]	-26	179.05
p-Chlorotoluene	C_7H_7Cl	+23.0	6.25	+7.5	162.4
o-Nitrotoluene	$C_7H_7NO_2$	+174	26.26	-10	222.3
m-Nitrotoluene	$C_7H_7NO_2$	+177	24.95	+15.5	232
p-Nitrotoluene	$C_7H_7NO_2$	+222	22.2	+51.6	238.3
Benzyl alcohol	C_7H_8O	-15.4	11.92 10.8[a]	-15.3	205.8
m-Cresol	C_7H_8O	+21.2	12.44 5.0[a]	+11.8	202.27
m-Chloroacetophenone	C_8H_7ClO	+69.1			
Acetophenone	C_8H_8O	+66.6	17.44 15.8[a]	+19.7	202.3
Quinoline	C_9H_7N	+15.0	9.16	-14.78	237.16
Ethyl salicylate	$C_9H_{10}O_3$	+19.6	8.48	+1.3	231.5
Carvone	$C_{10}H_{14}O$	+23.6	11.2	<0	230
Ethyl benzoylacetate	$C_{11}H_{12}O_3$	+16.0	13.50	<0	270
Water	H_2O	+4.0	80.10	0.00	100.0

[a] Dielectric constant at radiofrequencies (10^8-10^9 Hz).

REFERENCES

1. Narasimhamurty, T. S., *Photoelastic and Electro-Optic Properties of Crystals*, Plenum Press, New York, 1981, p. 408.
2. Gray, D. E., Ed., *AIP Handbook of Physics*, McGraw Hill, New York, 1972, p. 6-241.

MAGNETO-OPTIC CONSTANTS

Verdet Constants of Non-Magnetic Crystals[1]

V is the Verdet constant; *n* is the refractive index; and λ is the wavelength

Material	*T* K	λ nm	*n*	*V* min/Oe cm
Al_2O_3	300	546.1	1.771	0.0240
	300	589.3	1.768	0.0210
$BaTaO_3$	403	427		0.95
	403	496		0.38
	403	620		0.18
	403	826		0.072
$Bi_4Ge_3O_{12}$	300	442	2.077	0.289
	300	632.8	2.048	0.099
	300	1064	2.031	0.026
C (diamond)	300	589.3	2.417	0.0233
$CaCO_3$	300	589.3	1.658	0.019
CaF_2	300	589.3	1.434	0.0088
$Cd_{0.55}Mn_{0.45}Te$	300	632.8		6.87

MAGNETO-OPTIC CONSTANTS (continued)

Verdet Constants of Non-Magnetic Crystals (continued)[1]

Material	*T* K	λ nm	*n*	*V* min/Oe cm
CuCl	300	546.1	1.93	0.20
GaSe	298	632.8		0.80
$KAl(SO_4)_2 \cdot 12H_2O$	300	589.3	1.456	0.0124
KBr	300	546.1	1.564	0.0500
	300	589.3	1.560	0.0425
KCl	300	589.3	1.490	0.0275
KI	300	546.1	1.673	0.083
	300	589.3	1.666	0.070
$KTaO_3$	296	352		0.44
	296	413		0.19
	296	496		0.096
	296	620		0.051
	296	826		0.022
LaF_3	300	325	1.639	0.054
(H‖c)	300	442	1.615	0.028
	300	632.8	1.601	0.012
	300	1064	1.592	0.006
$MgAl_2O_4$	300	589.3	1.718	0.021
$NH_4AlSO_4 \cdot 12H_2O$	300	589.3	1.459	0.0128
NH_4Br	300	589.3	1.711	0.0504
NH_4Cl	300	546.1		0.0410
	300	589.3	1.643	0.0362
NaBr	300	546.1		0.0621
NaCl	300	546.1		0.0410
	300	589.3	1.544	0.0345
$NaClO_3$	300	546.1		0.0105
	300	589.3	1.515	0.0081
$NiSO_4 \cdot 6H_2O$	297	546.1		0.0256
	297	589.3	1.511	0.0221
SiO_2	300	546.1	1.546	0.0195
	300	589.3	1.544	0.0166
$SrTiO_3$	298	413	2.627	0.78
	298	496		0.31
	298	620		0.14
	298	826		0.066
ZnS	300	546.1		0.287
	300	589.3	2.368	0.226
ZnSe	300	476	2.826	1.50
	300	496	2.759	1.04
	300	514	2.721	0.839
	300	587	2.627	0.529
	300	632.8	2.592	0.406

Verdet Constants of Rare-Earth Aluminum Garnets at Various Wavelengths[1]

The absorption coefficient α for these materials ranges from 0.2 to 0.6 cm^{-1} at 300 K

Material	T/K	V in min/Oe cm							
		λ = 405 nm	450 nm	480 nm	520 nm	546 nm	578 nm	635 nm	670 nm
$Tb_3Al_5O_{12}$	300	-2.266	-1.565	-1.290	-1.039	-0.912	-0.787	-0.620	-0.542
	77		-102.16	-83.45	-3.425	-3.051	-2.603	-2.008	-1.815
	4.2				-64.80	-58.35	-53.77	48.39	-45.15
	1.45		-200.95	-172.52	-139.28	-125.07	-111.27	97.47	-93.42
$Dy_3Al_5O_{12}$	300	-1.241	-0.942	-0.803	-0.667	-0.592	-0.518	-0.411	-0.359
$Ho_3Al_5O_{12}$	300	-0.709	-0.320	-0.260	-0.335	-0.304	-0.299		-0.206

MAGNETO-OPTIC CONSTANTS (continued)

Verdet Constants of Rare-Earth Aluminum Garnets at Various Wavelengths (continued)[1]

Material	T/K	V in min/Oe cm							
		λ = 405 nm	450 nm	480 nm	520 nm	546 nm	578 nm	635 nm	670 nm
$Er_3Al_5O_{12}$	300	-0.189	-0.240	-0.154	-0.162	-0.157	-0.145	-0.105	-0.089
$Tm_3Al_5O_{12}$	300	+0.151	+0.103	+0.093	0.076	0.069	+0.059	+0.048	
$Yb_3Al_5O_{12}$	298	0.287	0.215	0.186	0.140	0.133	0.116	0.094	
	77	0.718	0.540	0.481	0.393	0.342	0.302	0.239	

Verdet Constants for KDP-Type Crystals[1]

Measurements refer to T = 298 K and λ = 632.8 nm, with ***k*** ∥ [001]

Material	*V* min/Oe cm
KH_2PO_4 (KDP)	0.0124
$KH_{0.3}D_{1.7}PO_4$ (KD*P)	0.145
$NH_4H_2PO_4$ (ADP)	0.138
KH_2AsO_4 (KDA)	0.238
$KH_{0.1}D_{1.9}AsO_4$ (KD*A)	0.245
$NH_4H_2AsO_4$ (ADH)	0.244

Verdet Constants of Gases[2]

Values refer to *T* = 0°C and P = 101.325 kPa (760 mmHg); n_D is the refractive index at a wavelength of 589 nm

Gas	$(n_D - 1) \times 10^3$	$10^6 \times V$ min/Oe cm
He	0.036	+0.40
Ar	2.81	+9.36
H_2		+6.29
N_2	0.297	+6.46
O_2	0.272	+5.69
Air	0.293	+6.27
Cl_2	0.773	+31.9
HCl	0.447	+21.5
H_2S	0.63	+41.5
NH_3	0.376	+19.0
CO	0.34	+11.0
CO_2	0.45	+9.39
NO	0.297	-58
CH4	0.444	+17.4
n-C4H10		+44.0

Verdet Constants of Liquids[2]

n_D is the refractive index at a wavelength of 589 nm and a temperature of 20°C, unless otherwise indicated. *V* is the Verdet constant

Liquid	λ/nm	*T*/°C	$10^2 \times V$ min/Oe cm	n_D
P	589	33	+13.3	
S	589	114	+8.1	1.929 (110°C)
H_2O	589	20	+1.309	1.3328
D_2O	589	19.7	+1.257	1.3384
H_3PO_4	578	97.4	+1.35	
CS_2	589	20	+4.255	1.6255
CCl_4	578-589	25.1	+1.60	1.463 (15°C)
$SbCl_5$	578	18	+7.45	1.601 (14°C)
$TiCl_4$	578	17	-1.65	1.61
$TiBr_4$	578	46	-5.3	
Methanol	589	18.7	+0.958	1.3289
Acetone	578-589	20.0	+1.116	1.3585
Toluene	578-589	15.0	+2.71	1.4950
Benzene	578-589	15.0	+3.00	1.5005
Chlorobenzene	589	15	+2.92	1.5246
Nitrobenzene	589	15	+2.17	1.5523
Bromoform	589	17.9	+3.13	1.5960

MAGNETO-OPTIC CONSTANTS (continued)

Verdet Constants of Rare Earth Paramagnetic Crystals[1]

n is the refractive index, and *V* is the Verdet constant at the wavelength and temperature indicated

Rare Earth	Host	*T*/K	λ/nm	*n*	*V* min/Oe cm
Ce^{3+}(30%)	CaF_2	300	325	1.516	-0.956
		300	442	1.502	-0.297
		300	633	1.494	-0.111
		300	1064	1.489	-0.035
Ce^{3+}	CeF_3	300	442	1.613	-1.05
		300	633	1.598	-0.406
		77	633		-1.418
		300	1064	1.589	-0.113
Pr^{3+}(5%)	CaF_2	300	266	1.471	-0.172
		300	325	1.461	-0.0818
		300	442	1.451	-0.0089
		300	633	1.445	-0.0168
		300	1064	1.441	-0.0045
Nd^{3+}(2.9%)	CaF_2	4.2	426		-0.19
Nd^{3+}	NdF_3	300	442	1.60	-0.553
		290	633	1.59	-0.209
		77	633		-0.755
		300	1064	1.58	-0.097
Eu^{3+}(3%)	CaF_2	4.2	430		29
		4.2	440		22
Eu^{2+}	EuF_2	300	450		-4.5
		300	500		-2.6
		300	550		-1.6
		300	600		-1.1
		300	650		-0.8
		300	1064		-0.19
Tb^{3+}	KTb_3F_{10}	300	325	1.531	-2.174
		300	442	1.518	-0.933
		300	633	1.510	-0.386
		77	633		-1.94
		300	1064	1.505	-0.114
Tb^{3+}	$LiTbF_4$	300	325	1.493	-1.9
		300	442	1.481	-0.98
		300	633	1.473	-0.44
		300	1064	1.469	-0.13
Tb^{3+}	$Tb_3Ga_5O_{12}$	300	500	1.989	-0.749
		300	570	1.981	-0.581
		300	633	1.976	-0.461
		300	830	1.967	-0.21
		300	1060	1.954	-0.12

MAGNETO-OPTIC CONSTANTS (continued)

Verdet Constants of Paramagnetic Glasses[1]

The Verdet constant *V* is given at room temperature for the wavelengths indicated

Rare earth phosphate glasses of composition $R_2O_3{\cdot}xP_2O_5$, where *x* is given in the second column

		Verdet constant *V* in min/Oe cm									
R	*x*	**λ = 405 nm**	**λ = 436 nm**	**λ = 480 nm**	**λ = 500 nm**	**λ = 520 nm**	**λ= 546 nm**	**λ= 578 nm**	**λ= 600 nm**	**λ= 635 nm**	**λ= 670 nm**
La		0.037	0.030	0.024	0.022	0.020	0.018	0.015	-0.014	0.013	—
Ce	2.67	-0.672	0.510	-0.366	-0.326	-0.287	-0.253	-0.217	-0.197	-0.173	-0.150
Pr	3.09	-0.447	-0.332	-0.283	-0.261	-0.236	-0.208	-0.182	-0.170	-0.150	-0.132
Nd	2.92	-0.250	-0.209	-0.167	-0.155	-0.136	-0.134	-0.094	-0.080	-0.080	-0.071
Sm	2.87	0.026	0.024	0.020	0.020	0.017	0.015	0.014	0.012	0.011	0.010
Eu	2.93	-0.025	-0.017	-0.010	-0.006	-0.006	-0.005	-0.004	-0.003	-0.002	-0.002
Gd	3.01	0.018	0.015	0.014	0.012	0.012	0.011	0.011	0.010	0.009	0.009
Tb	2.94	-0.560	-0.458	-0.357	-0.323	-0.295	-0.261	-0.226	-0.206	-0.190	-0.164
Dy	2.51	-0.540	-0.453	-0.359	-0.331	-0.301	0.268	-0.237	-0.217	-0.197	-0.173
Ho	2.94	-0.299	-0.313	-0.156	-0.153	-0.138	-0.138	-0.119	-0.110	-0.098	-0.084
Er	3.01	-0.139	-0.121	-0.100	-0.111	-0.095	-0.062	-0.060	-0.057	-0.051	-0.044
Tm	2.79	0.019	0.013	0.012	0.009	0.008	0.006	0.005	0.004	0.004	0.007
Yb	3.01	0.087	0.072	0.056	0.050	0.045	0.041	0.036	0.032	0.029	0.024

The following are rare earth borate glasses with composition:
for La and Pr: $R_2O_3{\cdot}xP_2O_5$; for Tb-Pr and Dy-Pr: $R_2O_3{\cdot}xB_2O_3$; and for other elements: $R_2O_3{\cdot}0.85La_3O_3{\cdot}xB_2O_3$.

R	*x*	λ = 405 nm	λ = 436 nm	λ = 480 nm	λ = 500 nm	λ = 520 nm	λ= 546 nm	λ= 578 nm	λ= 600 nm	λ= 635 nm	λ= 670 nm
La	3.04	0.043	0.036	0.029	0.026	0.023	0.022	0.019	0.018	0.016	0.014
Pr-La	5.44	-0.380	-0.307	-0.230	-0.220	-0.201	-0.178	-0.153	-0.146	-0.128	-0.110
Nd-La	5.41	-0.180	-0.147	-0.120	-0.111	-0.096	-0.094	-0.100	-0.059	-0.056	-0.046
Sm-La	4.97	0.032	0.030	0.025	0.024	0.022	0.019	0.017	0.016	0.014	0.012
Eu-La	4.69	-0.081	-0.060	-0.038	-0.033	-0.029	-0.024	0.019	-0.016	0.014	-0.012
Gd-La	4.71	0.032	0.026	0.024	0.022	0.021	0.020	0.018	0.017	0.015	0.013
Tb-La	4.73	-0.512	-0.419	-0.319	-0.288	-0.262	-0.234	-0.205	-0.186	-0.167	-0.142
Dy-La	4.88	-0.436	-0.361	-0.299	-0.273	-0.246	-0.220	-0.193	-0.177	-0.159	-0.138
Ho-La	4.36	-0.269	-0.252	-0.123	-0.131	-0.112	-0.128	-0.104	-0.096	—	-0.074
Er-La	4.50	-0.093	-0.078	-0.068	-0.082	—	-0.045	-0.042	-0.040	-0.035	-0.034
Tm-La	4.75	0.060	0.046	0.039	0.034	0.031	0.026	0.023	0.021	0.018	0.016
Yb-La	8.58	0.115	0.094	0.073	0.066	0.060	0.054	0.046	0.043	0.037	0.033
Tb-Pr	4.99	-0.940	-0.786	-0.560	-0.536	-0.489	-0.436	-0.380	-0.348	-0.306	-0.265
Dy-Pr	4.63	-0.850	—	—	-0.497	-0.465	-0.413	-0.358	-0.332	-0.290	-0.252
Pr	2.56	-0.843	-0.646	-0.471	-0.480	-0.432	-0.390	-0.334	-0.317	-0.271	-0.243

MAGNETO-OPTIC CONSTANTS (continued)

Verdet Constants of Diamagnetic Glasses[1]

The Verdet constant V is given at room temperature for the wavelengths indicated

Glass type	Composition (wt. %)	Verdet constant V in min/Oe cm: λ = 325 nm	λ = 442 nm	λ = 633 nm	λ = 1064 nm
SiO_2	100% SiO_2			0.013	
B_2O_3	100% B_2O_3			0.010	
CdO	47.5% CdO, 52.5% P_2O_5	0.079	0.033	0.022	
ZnO	36.4% ZnO, 63.6% P_2O_5	0.072	0.044	0.020	
TeO_2	88.9% TeO_2, 11.1% P_2O_5		0.196	0.076	0.022
ZrF_4	63.1% ZrF_4, 14.9% BaF_2, 7.2% LaF_3, 1.9% AlF_3, 9.1% PbF_2, 3.8% LiF			0.011	

Glass type	Composition (wt. %)	λ = 700 nm	λ = 853 nm	λ = 1060 nm
Bi_2O_3	95% Bi_2O_3, 5% B_2O_3	0.086	0.051	0.033
PbO	95% PbO, 5% B_2O_3	0.093	0.061	0.031
	82% PbO, 18% SiO_2	0.077	0.045	0.027
	50% PbO, 15% K_2O, 35% SiO_2	0.032	0.020	0.011
Tl_2O	95% Tl_2O, 5% B_2O_3	0.092	0.061	0.032
	82% Tl_2O, 18% SiO_2	0.100	0.067	0.043
	50% Tl_2O, 15% K_2O, 35% SiO_2	0.036	0.022	0.012
SnO	76% SnO, 13% B_2O_3, 11% SiO_2	0.071	0.046	0.026
TeO_3	75% TeO_2, 25% Sb_2O_3	0.076	0.052	0.032
	80% TeO_2, 20% $ZnCl_2$	0.073	0.046	0.025
	84% TeO_2, 16% BaO	0.056	0.041	0.029
	70% TeO_2, 30% WO_3	0.052	0.035	0.022
	20% TeO_2, 80% PbO	0.128	0.075	0.048
Sb_2O_3	25% Sb_2O_3, 75% TeO_2	0.076	0.050	0.032
	75% Sb_2O_3, 75% Cs_2O, 5% Al_2O_3	0.074	0.044	0.025
	75% Sb_2O_3, 10% Cs_2O, 10% Rb_2O, 5% Al_2O_3	0.078	0.052	0.030

MAGNETO-OPTIC CONSTANTS (continued)

Verdet Constants of Commercial Glasses[1]

This table gives the density, ρ, refractive index at 589 nm, n_D, and Verdet constant, V, for the wavelengths indicated; the data refer to room temperature

Glass type	ρ g/cm³	n_D	V in min/Oe cm λ = 365.0 nm	λ = 404.7 nm	λ = 435.8 nm	λ = 546.1 nm	λ = 578.0 nm
BSC	2.49	1.5096	0.0499	0.0392	0.0333	0.02034	0.01798
HC	2.53	1.5189	0.0561	0.0440	0.0372	0.0225	0.01995
LBC	2.87	1.5406	0.0609	0.0477	0.0403	0.0245	0.0216
LF	3.23	1.5785	0.1143	0.0850	0.0693	0.0394	0.0344
BLF	3.48	1.6047	0.1112	0.0832	0.0685	0.0393	0.0344
DBC	3.56	1.6122	0.0662	0.0517	0.0435	0.0261	0.0231
DF	3.63	1.6203	0.1473	0.1076	0.0872	0.0485	0.0423
EDF	3.9	1.6533	0.1725	0.1248	0.1007	0.0556	0.0483

The composition of the glasses in weight percent is:

Glass type	SiO_2	B_2O_3	K_2O	CaO	Al_2O_3	As_2O_3	Na_2O	BaO	ZnO	PbO
BSC	69.6	6.7	20.5	2.9	0.3	0.1	—	—	—	—
HC	72.0	—	10.1	11.4	0.3	0.2	6.1	—	—	—
LBC	57.1	1.8	13.7	0.3	0.2	0.1	—	26.9	—	—
LF	52.5	—	9.5	0.3	0.2	0.1	—	—	—	37.6
BLF	45.2	—	7.8	—	—	0.4	—	16.0	8.3	22.2
DBC	36.2	7.7	0.2	0.2	3.5	0.7	—	44.6	6.7	—
DF	46.3	—	1.1	0.3	0.2	0.1	5.0	—	—	47.0
EDF	40.6	—	7.5	0.2	0.2	0.2	0.1	—	—	51.5

REFERENCES

1. Weber, M. J., *CRC Handbook of Laser Science and Technology*, Vol. IV, Part 2, CRC Press, Boca Raton, FL, 1988, p. 299-310.
2. Gray, D. E., Ed., *American Institute of Physics Handbook*, Third Edition, McGraw Hill, New York, 1972, p. 6-230.

FARADAY ROTATION

Ferro-, Ferri-, and Antiferromagnetic Solids

Material	T_c K	$4\pi M_s$ gauss	F deg/cm	α cm⁻¹	$2F/\alpha$	T K	λ nm
Fe	1043	21,800	4.4×10^5	6.5×10^5	1.4	300	500
			6.5×10^5	5.0×10^5	2.6	300	1000
			7×10^5	4.2×10^5	3.3	300	1500
			7×10^5	3.5×10^5	4.0	300	2000
Co	1390	18,200	2.9×10^5	—	—	300	500
			5.5×10^5	6.1×10^5	1.8	300	1000
			5.5×10^5	4.5×10^5	2.4	300	1500
			5.5×10^5	3.6×10^5	2.7	300	2000
Ni	633	6,400	0.8×10^5	—	—	300	500
			2.6×10^5	5.8×10^5	0.9	300	1000
			1.5×10^5	4.8×10^5	0.6	300	1500
			1×10^5	4.1×10^5	0.25	300	2000
Permalloy (Ni/Fe = 82/18)	803	10,700	1.2×10^5	6×10^5	0.4	300	500

FARADAY ROTATION (continued)

Ferro-, Ferri-, and Antiferromagnetic Solids (continued)

Material	T_c K	$4\pi M_s$ gauss	F deg/cm	α cm^{-1}	$2F/\alpha$	T K	λ nm
Ni/Fe = 100/0		6,000	1.2×10^5	7.05×10^5	0.34	300	632.8
Ni/Fe = 80/20		10,800	2.2×10^5	7.10×10^5	0.62	300	632.8
Ni/Fe = 60/40		14,900	2.9×10^5	7.54×10^5	0.77	300	632.8
Ni/Fe = 40/60		14,400	2.2×10^5	8.17×10^5	0.54	300	632.8
Ni/Fe = 20/80		19,400	3.3×10^5	8.10×10^5	0.81	300	632.8
Ni/Fe = 0/100	639	21,600	3.5×10^5	8.13×10^5	0.86	300	632.8
MnBi		7,700	4.2×10^5	6.1×10^5	1.4	300	450
			7.5×10^5	4.2×10^5	3.6	300	900
MnAs	313	—	0.44×10^5	5.0×10^5	0.174	300	500
			0.62×10^5	4.4×10^5	0.28	300	900
CrTe	334	1015	0.5×10^5	2.0×10^5	0.5	300	550
			0.4×10^5	1.2×10^5	0.7	300	900
FeRh	333	—	0.9×10^5	3.3×10^5	0.56	348	700
$Y_3Fe_5O_{12}$ (YIG)	560	2500	2400	1500	3.2	300	555
			1250	1400	1.8	300	625
			750	450	3.3	300	770
			175	<0.06	$>3 \times 10^3$	300	5000 to 1500
$Gd_3Fe_5O_{12}$ (GdIG)	Tn = 564	7300	-2000	6000	0.6	300	500
	T = 286		-1050	900	2.3	300	600
			-300	100	6.0	300	800
			-80	70	2.3	300	1000
$NiFe_2O_4$	858	3350	2.0×10^4	5.9×10^4	0.7	300	286
			-1.0×10^4	10×10^4	0.2	300	500
			-120	38	6	300	1500
			+75	15	10	300	3000
			+110	32	7	300	5000
$CoFe_2O_4$	793	4930	2.75×10^4	12×10^4	0.5	300	286
			3.6×10^4	17×10^4	0.4	300	400
			-2.5×10^4	6×10^4	0.8	300	660
$MgFe_2O_4$	593-713[e]	1450[e]	-60	100	1	300	2500
			0	12	0	300	4000
			+35	6	11	300	6000
$Li_{0.5}Fe_{2.5}O_4$	863-953[e]	3240[e]	-440	150	6	300	1500
		to 3900	+10	85	0.2	300	3000
			+110	44	5	300	5000
			+135	80	3	300	7000
$BaFe_{12}O_{19}$	723	—	-50	-38	3	300	2000
			+75	20	7.5	300	3000
			+150	20	15	300	5000
			+165	22	15	300	7000
$Ba_2Zn_2Fe_{12}O_{19}$	—	—	90	120	1.5	300	5000
			75	65	2.0	300	7000
$RbNiF_3$	220	1250	360	35	20	77	450[a]
			70	10	14	77	600[a]
			310	70	9	77	800[a]
			75	25	6	77	1000[a]
$RbNi_{0.75}Co_{0.25}F_3$	109	—	180	9	40	77	600[b]
$RbFeF_3$	102	—	3400	7	900	82	300[c]
			1600	3	1100	82	400[c]
			620	1.5	830	82	600[c]
			300	2.5	240	82	800[c]
FeF_3	365	40	670	14	95	300	349[d]
		at 300 K	180	4.4	82	300	522.5[d]
$CrCl_3$	16.8	3880	2000	200	20	1.5	410

FARADAY ROTATION (continued)

Ferro-, Ferri-, and Antiferromagnetic Solids (continued)

Material	T_c K	$4\pi M_s$ gauss	F deg/cm	α cm⁻¹	$2F/\alpha$	T K	λ nm
			-500	300	3	1.5	450
			-1000	70	30	1.5	590
$CrBr_3$	32.5	3390	3×10^5	3×10^3	200	1.5	478
			1.6×10^5	1.4×10^4	23	1.5	500
CrI_3	68	2690	1.1×10^5	6.3×10^3	35	1.5	970
			0.8×10^5	3×10^3	53	1.5	1000
$FeBO_3$	348	115	3200	140	45	300	500
		at 300 K	450	38	24	300	700
EuO	69	23700	-1.0×10^5	0.5×10^4	40	5	1100
			5×10^5	9.7×10^4	10	5	700
			0.5×10^5	7.8×10^4	1.3	5	500
			3×10^4	>0.5	~105	20	2500
			660	>1.0	1300	20	10600
EuS	16.3	—	-1.6×10^5	0	—	6	825
			-9.6×10^5	3.3×10^4	58	6	690
			$+5.5 \times 10^5$	1.2×10^5	9.2	6	563
EuSe	7.0	13,200	1.45×10^5	80	3600	4.2	750
			0.95×10^5	60	3170	4.2	800

[a] Measured along the C-axis (magnetic hard axis).
[b] Measured along the C-axis (magnetic easy axis).
[c] Measured along the C-axis ([100]-direction at room temperature).
[d] Strong natural birefringence interferes with the Faraday effect.
[e] Depends on heat treatment.

REFERENCE

1. Weber, M. J., Ed., *CRC Handbook of Laser Science and Technology,* Vol. IV, Part 2, CRC Press, Boca Raton, FL, 1988, pp. 288-296.

NONLINEAR OPTICAL CONSTANTS

H. P. R. Frederikse

The relation between the polarization density P of a dielectric medium and the electric field E is linear when E is small, but becomes nonlinear as E acquires values comparable with interatomic electric fields (10^5 to 10^8 V/cm). Under these conditions the relation between P and E can be expanded in a Taylor's series

$$P = \varepsilon_0 \chi^{(1)} E + 2\chi^{(2)} E^2 + 4\chi^{(3)} E^3 + \cdots \quad (1)$$

where ε_o is the permittivity of free space, while $\chi^{(1)}$ is the linear and $\chi^{(2)}$, $\chi^{(3)}$ etc. the nonlinear optical susceptibilities.

If we consider two optical fields, the first $E_j^{\omega_1}$ (along the j-direction at frequency ω_1) and the second $E_k^{\omega_2}$ (along the k-direction at frequency ω_2) one can write the second term of the Taylor's series as follows

$$P_i(\omega_1\omega_2) = 2\chi_{ijk}^{\omega_3=\omega_1\pm\omega_2} E_j^{\omega_1} E_k^{\omega_2}$$

When $\omega_1 \neq \omega_2$ the (parametric) mixing of the two fields gives rise to two new polarizations at the frequencies $\omega_3 = \omega_1 + \omega_2$ and $\omega_3' = \omega_1 - \omega_2$. When the two frequencies are equal, $\omega_1 = \omega_2 = \omega$, the result is Second Harmonic Generation (SHG) $\chi_{ijk}(2\omega, \omega, \omega)$, while equal and opposite frequencies, $\omega_1 = \omega$ and $\omega_2 = -\omega$ leads to Optical Rectification (OR): $\chi_{ijk}(0,\omega,-\omega)$. In the SHG case the following convention is adopted: the second order nonlinear coefficient d is equal to one half of the second order nonlinear susceptibility

$$d_{ijk} = 1/2\chi^{(2)}$$

Because of the symmetry of the indices j and k one can replace these two by a single index (subscript) m. Consequently the notation for the SHG nonlinear coefficient in reduced form is d_{im} where m takes the values 1 to 6. Only noncentrosymmetric crystals can possess a nonvanishing d_{ijk} tensor (third rank). The unit of the SHG coefficients is m/V (in the MKSQ/SI system).

In centrosymmetric media the dominant nonlinearity is of the third order. This effect is represented by the third term in the Taylor's series (Equation 1); it is the result of the interaction of a number of optical fields (one to three) producing a new frequency $\omega_4 = \omega_1 + \omega_2 + \omega_3$. The third order polarization is given by

$$P_j(\omega_1\omega_2\omega_3) = g_4 \chi_{jklm} E_k^{\omega_1} E_l^{\omega_2} E_m^{\omega_3}$$

Third Harmonic Generation (THG) is achieved when $\omega_1 = \omega_2 = \omega_3 = \omega$. In this case the constant $g_4 = 1/4$. The third order nonlinear coefficient C is related to the third order susceptibility as follows

$$C_{jklm} = 1/4\chi_{jklm}$$

This coefficient is a fourth rank tensor. In the THG case the matrices must be invariant under permutation of the indices k, l, and m; as a result the notation for the third order nonlinear coefficient can be simplified to C_{jn}. The unit of C_{jn} is $m^2 \cdot V^{-2}$ (in the MKSQ/SI system).

Applications of second order nonlinear optical materials include the generation of higher (up to sixth) optical harmonics, the mixing of monochromatic waves to generate sum or difference frequencies (frequency conversion), the use of two monochromatic waves to amplify a third wave (parametric amplification) and the addition of feedback to such an amplifier to create an oscillation (parametric oscillation).

Third order nonlinear optical materials are used for THG, self-focusing, four wave mixing, optical amplification, and optical conjugation. Many of these effects — as well as the variation and modulation of optical propagation caused by mechanical, electric, and magnetic fields (see the preceeding table on "Elasto-Optic, Electro-Optic, and Magneto-Optic Constants") are used in the areas of optical communication, optical computing, and optical imaging.

REFERENCES (NONLINEAR OPTICS)

1. *Handbook of Laser Science and Technology*, Vol. 111, Part 1; Ed.: Marvin J. Weber, Publ.: CRC Press, Inc., Boca Raton, FL, 1986.
2. Dmitriev, V.G., Gurzadyan, G.G., and Nikogosyan, D., *Handbook of Nonlinear Optical Crystals*, Springer-Verlag, Berlin, 1991.
3. Shen, Y.R., *The Principles of Nonlinear Optics*, John Wiley, New York, 1984.
4. Yariv, A., *Quantum Electronics*, 3rd edition, John Wiley, New York, 1988.
5. Bloembergen, N., *Nonlinear Optics,* W.A. Benjamin, New York, 1965.
6. Zernike F. and Midwinter, J.E., *Applied Nonlinear Optics*, John Wiley, New York, 1973.
7. Hopf, F.A. and Stegeman, G.I., *Applied Classical Electrodynamics*, Volume 2: Nonlinear Optics, John Wiley, New York, 1986.
8. *Nonlinear Optical Properties of Organic Molecules and Crystals,* Eds.: D.S. Chemla and J. Zyss, Publ.: Academic Press, Orlando, FL, 1987.
9. *Optical Phase Conjugation*, Ed.: R.A. Fisher, Publ.: Academic Press, New York, 1983.
10. Zyss, J., *Molecular Nonlinear Optics: Materials, Devices and Physics*, Academic Press, Boston, 1994.
11. Nonlinear Optics, 5 articles in *Physics Today, (Am. Inst. of Phys.)*, Vol. 47, No. 5, May, 1994.

Selected SHG Coefficients of NLO Crystals*

Material	Symmetry class	$d_{im} \times 10^{12}$ m/V	λ μm
GaAs	$\bar{4}3$ m	$d_{14} = 134.1 \pm 42$	10.6
GaP	$\bar{4}3$ m	$d_{14} = 71.8 \pm 12.3$	1.058
InAs	$\bar{4}3$ m	$d_{14} = 364 \pm 47$	1.058
		$d_{14} = 210$	10.6
ZnSe	$\bar{4}3$ m	$d_{14} = 78.4 \pm 29.3$	10.6
		$d_{36} = 26.6 \pm 1.7$	1.058
β-ZnS	$\bar{4}3$ m	$d_{14} = 30.6 \pm 8.4$	10.6
		$d_{36} = 20.7 \pm 1.3$	1.058
ZnTe	$\bar{4}3$ m	$d_{14} = 92.2 \pm 33.5$	10.6
		$d_{14} = 83.2 \pm 8.4$	1.058
		$d_{36} = 89.6 \pm 5.7$	1.058
CdTe	$\bar{4}3$ m	$d_{14} = 167.6 \pm 63$	10.6
Bi_4GeO_{12}	$\bar{4}3$ m	$d_{14} = 1.28$	1.064
$N_4(CH_2)_6$ (hexamine)	$\bar{4}3$ m	$d_{14} = 4.1$	1.06
$LiIO_3$	6	$d_{33} = -7.02$	1.06
		$d_{31} = -5.53 \pm 0.3$	1.064
ZnO	6 mm	$d_{33} = -5.86 \pm 0.16$	1.058
		$d_{31} = 1.76 \pm 0.16$	1.058
		$d_{15} = 1.93 \pm 0.16$	1.058
α-ZnS	6 mm	$d_{33} = 11.37 \pm 0.07$	1.058
		$d_{33} = 37.3 \pm 12.6$	10.6
		$d_{31} = -18.9 \pm 6.3$	10.6
		$d_{15} = 21.37 \pm 8.4$	10.6
CdS	6 mm	$d_{33} = 25.8 \pm 1.6$	1.058
		$d_{31} = -13.1 \pm 0.8$	1.058
		$d_{15} = 14.4 \pm 0.8$	1.058
CdSe	6 mm	$d_{33} = 54.5 \pm 12.6$	10.6
		$d_{31} = -26.8 \pm 2.7$	10.6
$BaTiO_3$	4 mm	$d_{33} = 6.8 \pm 1.0$	1.064
		$d_{31} = 15.7 \pm 1.8$	1.064
		$d_{15} = 17.0 \pm 1.8$	1.064
$PbTiO_3$	4 mm	$d_{33} = 7.5 \pm 1.2$	1.064
		$d_{31} = 37.6 \pm 5.6$	1.064
		$d_{15} = 33.3 \pm 5$	1.064
$K_3Li_2Nb_5O_{15}$	4 mm	$d_{33} = 11.2 \pm 1.6$	1.064
		$d_{31} = 6.18 \pm 1.28$	1.064
		$d_{15} = 5.45 \pm 0.54$	1.064
$K_{0.8}Na_{0.2}Ba_2Nb_5O_{15}$	4 mm	$d_{31} = 13.6 \pm 1.6$	1.064
$SrBaNb_5O_{15}$	4 mm	$d_{33} = 11.3 \pm 3.3$	1.064
		$d_{31} = 4.31 \pm 1.32$	1.064
		$d_{15} = 5.98 \pm 2$	1.064
$NH_4H_2PO_4$ (ADP)	$\bar{4}2$ m	$d_{36} = 0.53$	1.064
		$d_{36} = 0.85$	0.694
KH_2PO_4 (KDP)	$\bar{4}2$ m	$d_{36} = 0.44$	1.064
		$d_{36} = 0.47 \pm 0.07$	0.694
KD_2PO_4 (KD*P)	$\bar{4}2$ m	$d_{36} = 0.38 \pm 0.016$	1.058
		$d_{36} = 0.34 \pm 0.06$	0.694
		$d_{14} = 0.37$	1.058
KH_2AsO_4 (KDA)	$\bar{4}2$ m	$d_{36} = 0.43 \pm 0.025$	1.06
		$d_{36} = 0.39 \pm 0.4$	0.694
$CdGeAs_2$	$\bar{4}2$ m	$d_{36} = 351 \pm 105$	10.6
$AgGaS_2$	$\bar{4}2$ m	$d_{36} = 18 \pm 2.7$	10.6
$AgGaSe_2$	$\bar{4}2$ m	$d_{36} = 37.4 \pm 6.0$	10.6
$(NH_2)_2CO$ (urea)	$\bar{4}2$ m	$d_{36} = 1.3$	1.06
$AlPO_4$	32	$d_{11} = 0.35 \pm 0.03$	1.058
Se	32	$d_{11} = 97 \pm 25$	10.6

Selected SHG Coefficients of NLO Crystals (continued)*

Material	Symmetry class	$d_{im} \times 10^{12}$ m/V	λ µm
Te	32	$d_{11} = 650 \pm 30$	10.6
SiO_2 (quartz)	32	$d_{11} = 0.335$	1.064
HgS	32	$d_{11} = 50.3 \pm 17$	10.6
$(C_6H_5CO)_2$ [benzil]	32	$d_{11} = 3.6 \pm 0.5$	1.064
β-BaB_2O_4 [BBO]	3 m	$d_{22} = 2.22 \pm 0.09$	1.06
		$d_{31} = 0.16 \pm 0.08$	1.06
$LiNbO_3$	3 m	$d_{33} = 34.4$	1.06
		$d_{31} = -5.95$	1.06
		$d_{22} = 2.76$	1.06
$LiTaO_3$	3 m	$d_{33} = -16.4 \pm 2$	1.058
		$d_{31} = -1.07 \pm 0.2$	1.058
		$d_{22} = +1.76 \pm 0.2$	1.058
Ag_3AsS_3 [proustite]	3 m	$d_{31} = 11.3 \pm 2.5$	10.6
		$d_{22} = 18.0 \pm 2.5$	10.6
Ag_3SbS_3 [pyrargerite]	3m	$d_{31} = 12.6 \pm 4$	10.6
		$d_{22} = 13.4 \pm 4$	10.6
α-HIO_3	222	$d_{36} = 5.15 \pm 0.16$	1.064
$NO_2 \cdot CH_3NOC_5H_4 \cdot$ (POM)	222	$d_{36} = 6.4 \pm 1.0$	1.064
$Ba_2NaNb_5O_{15}$ [Banana]	mm 2	$d_{33} = -17.6 \pm 1.28$	1.064
		$d_{31} = -12.8 \pm 1.28$	1.064
$C_6H_4(NO_2)_2$ [MDB]	mm 2	$d_{33} = 0.74$	1.064
		$d_{32} = 2.7$	1.064
		$d_{31} = 1.78$	1.064
$Gd_2(MoO_4)_3$	mm 2	$d_{33} = -0.044 \pm 0.008$	1.064
		$d_{32} = +2.42 \pm 0.36$	1.064
		$d_{31} = -2.49 \pm 0.37$	1.064
$KNbO_3$	mm 2	$d_{33} = -19.58 \pm 1.03$	1.064
		$d_{32} = +11.34 \pm 1.03$	1.064
		$d_{31} = -12.88 \pm 1.03$	1.064
$KTiOPO_4$ [KTP]	mm 2	$d_{33} = 13.7$	1.06
		$d_{32} = \pm 5.0$	1.06
		$d_{31} = \pm 6.5$	1.06
$NO_2C_6H_4 \cdot NH_2$ [mNA]	mm 2	$d_{33} = 13.12 \pm 1.28$	1.064
		$d_{32} = 1.02 \pm 0.22$	1.064
		$d_{31} = 12.48 \pm 1.28$	1.064
$C_{10}H_{12}N_3O_6$ [MAP]	2	$d_{23} = 10.67 \pm 1.3$	1.064
		$d_{22} = 11.7 \pm 1.3$	1.064
		$d_{21} = 2.35 \pm 0.5$	1.064
		$d_{25} = -0.35 \pm 0.3$	1.064
$(NH_2CH_2COOH)_3H_2SO_4$ [TGS]	2	$d_{23} = 0.32$	0.694

* These data are taken from References 1 and 2.

Selected THG Coefficients of Some NLO Materials*

Material	NLO process	$C_{jn} \times 10^{20}$ m²/V⁻²	λ μm
$NH_4H_2PO_4$ [ADP]	(–3ω,ω,ω,ω)	$C_{11} = 0.0104$	1.06
		$C_{18} = 0.0098$	1.06
C_6H_6 [benzene]	(–3ω,ω,ω,ω)	$C_{11} = 0.0184 \pm 0.0042$	1.89
$CdGeAs_2$	(–3ω,ω,ω,ω)	$C_{11} = 182 \pm 84$	10.6
p-type: 5×10^{16} cm^{-3}		$C_{16} = 175$	10.6
		$C_{18} = -35$	10.6
$C_{40}H_{56}$ [β-carotene]	(–3ω,ω,ω,ω)	C_{11} 0.263 ± 0.08	1.89
GaAs	(–3ω,ω,ω,–ω)	$C_{11} = 62 \pm 31$	1.06
high-resistivity			
Ge	(–3ω,ω,ω,–ω)	$C_{11} = 23.5 \pm 12$	1.06
$LiIO_3$	(–3ω,ω,ω,–ω)	$C_{12} = 0.2285$	1.06
		$C_{35} = 6.66 \pm 1$	1.06
KBr	(–3ω,ω,ω,–ω)	$C_{11} = 0.0392$	1.06
		$C_{18}/C_{11} = 0.3667$	1.06
KCl	(–3ω,ω,ω,–ω)	$C_{11} = 0.0168$	1.06
		$C_{18}/C_{11} = 0.28$	1.06
KH_2PO_4 [KDP]	(–3ω,ω,ω,–ω)	$C_{11}–3C_{18} = 0.04$	1.06
Si	(–3ω,ω,ω,–ω)	$C_{11} = 82.8 \pm 25$	1.06
p-type: 10^{14} cm^{-3}			
NaCl	(–3,ω,ω,ω,–ω)	$C_{11} = 0.0168$	1.06
		$C_{18}/C_{11} = 0.4133$	1.06
NaF	(–3ω,ω,ω,–ω)	$C_{11} = 0.0035$	1.06

* These data are taken from Reference 1.

HEAT CAPACITY OF SELECTED SOLIDS

This table gives the molar heat capacity at constant pressure of representative metals, semiconductors, and other crystalline solids as a function of temperature in the range 200 to 600 K.

REFERENCES

1. Chase, M. W., et al., *JANAF Thermochemical Tables, 3rd ed., J. Phys. Chem. Ref. Data,* 14, Suppl. 1, 1985.
2. Garvin, D., Parker, V. B., and White, H. J., *CODATA Thermodynamic Tables,* Hemisphere Press, New York, 1987.
3. DIPPR Database of Pure Compound Properties, Design Institute for Physical Properties Data, American Institute of Chemical Engineers, New York, 1987.

Name	C_p in J/mol K						
	200 K	250 K	300 K	350 K	400 K	500 K	600 K
Aluminum	21.33	23.08	24.25	25.11	25.78	26.84	27.89
Aluminum oxide	51.12	67.05	79.45	88.91	96.14	106.17	112.55
Anthracene	138.6	173.9	210.7	248.8	288.4		
Benzoic acid	102.7	123.5	147.4	172.0			
Beryllium	9.98	13.58	16.46	18.53	19.95	21.94	23.34
Biphenyl	131.0	162.5	197.2				
Boron	5.99	8.82	11.40	13.65	15.69	18.72	20.78
Calcium	24.54	25.41	25.94	26.32	26.87	28.49	30.38
Calcium carbonate	66.50	75.66	83.82	91.51	96.97	104.52	109.86
Calcium oxide	33.64	38.59	42.18	45.07	46.98	49.33	50.72
Cesium chloride	50.13	51.34	52.48	53.58	54.68	56.90	59.10
Chromium	19.86	22.30	23.47	24.39	25.23	26.63	27.72
Cobalt	22.23	23.98	24.83	25.68	26.53	28.20	29.66
Copper	22.63	23.77	24.48	24.95	25.33	25.91	26.48
Copper oxide	34.80		42.41	44.95	46.78	49.19	50.83
Copper sulfate	77.01	89.25	99.25	107.65	114.93	127.19	136.31
Germanium			23.25	23.85	24.31	24.96	25.45
Gold			25.41	25.37	25.51	26.06	26.65
Graphite	5.01	6.82	8.58	10.24	11.81	14.62	16.84
Hexachlorobenzene	162.7	183.6	202.4				
Iodine	51.57	53.24	54.51	58.60			
Iron	21.59	23.74	25.15	26.28	27.39	29.70	32.05
Lead	25.87	26.36	26.85	27.30	27.72	28.55	29.40
Lithium	21.57	23.42	24.64	25.96	27.60	29.28	
Lithium chloride	43.35	46.08	48.10	49.66	50.97	53.34	55.59
Magnesium	22.72	24.02	24.90	25.57	26.14	27.17	28.18
Magnesium oxide			37.38	40.59	42.77	45.56	47.30
Manganese	23.05	24.95	26.35	27.52	28.53	30.29	31.90
Naphthalene	105.8	134.1	167.8	204.1			
Potassium	27.00	28.01	29.60				
Potassium chloride	48.44	50.10	51.37	52.31	53.08	54.71	56.35
Silicon	15.64	18.22	20.04	21.28	22.14	23.33	24.15
Silicon dioxide	32.64	39.21	44.77	49.47	53.43	59.64	64.42
Silver			25.36	25.55	25.79	26.36	26.99
Sodium	22.45	27.01	28.20	30.14			
Sodium chloride	46.89	48.85	50.21	51.25	52.14	53.96	55.81
Tantalum	24.08	[illegible]	25.31	25.60	25.84	26.35	26.84
Titanium	22.37	24.07	25.28	26.17	26.86	27.88	28.60
Tungsten	22.49	23.69	24.30	24.65	24.92	25.36	25.79
Vanadium	21.88	23.70	24.93	25.68	26.23	26.94	27.49
Zinc	24.05	25.02	25.45	25.88	26.35	27.39	28.59
Zirconium	23.87	24.69	25.22	25.61	25.93	26.56	27.28

THERMAL AND PHYSICAL PROPERTIES OF PURE METALS

This table gives the following properties for the metallic elements:

T_m: Melting point in °C
T_b: Normal boiling point in °C, at a pressure of 101.325 kPa (760 Torr)
$\Delta_{fus} H$: Enthalpy of fusion at the melting point in J/g
ρ_{25}: Density at 25°C in g/cm^3
α: Coefficient of linear expansion at 25°C in K^{-1} (the quantity listed is $10^6 \times \alpha$)
c_p: Specific heat capacity at constant pressure at 25°C in J/g K
λ: Thermal conductivity at 27°C in W/cm K

REFERENCES

1. Dinsdale, A. T., *CALPHAD*, 15, 317, 1991 (melting points, enthalpy of fusion).
2. Touloukian, Y. S., *Thermophysical Properties of Matter*, Vol. 12, Thermal Expansion, IFI/Plenum, New York, 1975 (coefficient of expansion, density).
3. Ho, C. Y., Powell, R. W., and Liley, P. E., *J. Phys. Chem. Ref. Data*, 3, Suppl. 1, 1974 (thermal conductivity).
4. Cox, J. D., Wagman, D. D., and Medvedev, V. A., *CODATA Key Values for Thermodynamics*, Hemisphere Publishing Corp., New York, 1989 (heat capacity).
5. Glushko, V. P., Ed., *Thermal Constants of Substances*, VINITI, Moscow, (enthalpy of fusion, heat capacity).
6. Wagman, D. D., et. al., *The NBS Tables of Chemical Thermodynamic Properties, J. Phys. Chem. Ref. Data*, 11, Suppl. 2, 1982 (heat capacity).
7. Chase, M. W., et. al., *JANAF Thermochemical Tables*, 3rd ed., J. Phys. Chem. Ref. Data, 14, Suppl. 1, 1985 (heat capacity, enthalpy of fusion).
8. Gschneidner, K. A., *Bull. Alloy Phase Diagrams*, 11, 216—224, 1990 (various properties of the rare earth metals).
9. Hellwege, K. H., Ed., *Landolt Börnstein, Numerical Values and Functions in Physics, Chemistry, Astronomy, Geophysics, and Technology*, Vol. 2, Part 1, Mechanical-Thermal Properties of State, 1971 (density).

Metal (symbol)	Atomic weight	T_m °C	T_b °C	$\Delta_{fus} H$ J/g	ρ_{25} g/cm^3	$\alpha \times 10^6$ K^{-1}	c_p J/g K	λ W/cm K
Actinium (Ac)		1051	3198		10			
Aluminum (Al)	26.98	660.32	2519	397	2.70	23.1	0.897	2.37
Antimony (Sb)	121.76	630.63	1587	163.2	6.68	11.0	0.207	0.243
Barium (Ba)	137.33	727	1897	52	3.62	20.6	0.204	0.184
Beryllium (Be)	9.01	1287	2471	877	1.85	11.3	1.825	2.00
Bismuth (Bi)	208.98	271.40	1564	54.1	9.79	13.4	0.122	0.0787
Cadmium (Cd)	112.41	321.07	767	55.1	8.69	30.8	0.232	0.968
Calcium (Ca)	40.08	842	1484	213	1.54	22.3	0.647	2.00
Cerium (Ce)	140.11	799	3424	39.0	8.16	5.2	0.192	0.113
Cesium (Cs)	132.91	28.44	671	15.8	1.93		0.242	0.359
Chromium (Cr)	52.00	1907	2671	404	7.15	4.9	0.449	0.937
Cobalt (Co)	58.93	1495	2927	275	8.86	13.0	0.421	1.00
Copper (Cu)	63.55	1084.62	2562	208.7	8.96	16.5	0.385	4.01
Dysprosium (Dy)	162.50	1411	2561	66.3	8.55	9.9	0.173	0.107
Erbium (Er)	167.26	1529	2862	119	9.07	12.2	0.168	0.145
Europium (Eu)	151.96	822	1596	60.6	5.24	35	0.182	0.14
Gadolinium (Gd)	157.25	1314	3264	62.4	7.90	9	0.236	0.105
Gallium (Ga)	69.72	29.76	2204	80.2	5.91		0.371	0.406
Gold (Au)	196.97	1064.18	2856	63.7	19.3	14.2	0.129	3.17
Hafnium (Hf)	178.49	2233	4603	152.4	13.3	5.9	0.144	0.230
Holmium (Ho)	164.93	1472	2694	71.3	8.80	11.2	0.165	0.162
Indium (In)	114.82	156.60	2072	28.6	7.31	32.1	0.233	0.816
Iridium (Ir)	192.22	2446	4428	213.9	22.5	6.4	0.131	1.47
Iron (Fe)	55.85	1538	2861	247.3	7.87	11.8	0.449	0.802
Lanthanum (La)	138.91	920	3455	44.6	6.15	12.1	0.195	0.134
Lead (Pb)	207.20	327.46	1749	23.0	11.3	28.9	0.129	0.353
Lithium (Li)	6.94	180.5	1342	432	0.534	46	3.582	0.847
Lutetium (Lu)	174.97	1663	3393	106.6	9.84	9.9	0.154	0.164
Magnesium (Mg)	24.30	650	1090	349	1.74	24.8	1.023	1.56

THERMAL AND PHYSICAL PROPERTIES OF PURE METALS (continued)

Metal (symbol)	Atomic weight	T_m °C	T_b °C	$\Delta_{fus}H$ J/g	ρ_{25} g/cm³	$\alpha \times 10^6$ K⁻¹	c_p J/g K	λ W/cm K
Manganese (Mn)	54.94	1246	2061	235.0	7.3	21.7	0.479	0.0782
Mercury (Hg)	200.59	-38.83	356.73	11.4	13.5336		0.140	0.0834
Molybdenum (Mo)	95.94	2623	4639	390.7	10.2	4.8	0.251	1.38
Neodymium (Nd)	144.24	1016	3066	49.5	7.01	9.6	0.190	0.165
Neptunium (Np)		644			20.2			0.063
Nickel (Ni)	58.69	1455	2913	298	8.90	13.4	0.444	0.907
Niobium (Nb)	92.91	2477	4744	323	8.57	7.3	0.265	0.537
Osmium (Os)	190.23	3033	5012	304.2	22.59	5.1	0.130	0.876
Palladium (Pd)	106.42	1554.9	2963	157.3	12.0	11.8	0.246	0.718
Platinum (Pt)	195.08	1768.4	3825	113.6	21.5	8.8	0.133	0.716
Plutonium (Pu)		640	3228		19.7	46.7		0.0674
Polonium (Po)		254	962		9.20			0.20
Potassium (K)	39.10	63.38	759	59.3	0.89		0.757	1.024
Praseodymium (Pr)	140.91	931	3510	48.9	6.77	6.7	0.193	0.125
Promethium (Pm)		1042	3000		7.26	11		0.15
Protactinium (Pa)	231.04	1572		53.4	15.4			
Radium (Ra)		700			5			
Rhenium (Re)	186.21	3186	5596	324.5	20.8	6.2	0.137	0.479
Rhodium (Rh)	102.91	1964	3695	258.4	12.4	8.2	0.243	1.50
Rubidium (Rb)	85.47	39.31	688	25.6	1.53		0.363	0.582
Ruthenium (Ru)	101.07	2334	4150	381.8	12.1	6.4	0.238	1.17
Samarium (Sm)	150.36	1072	1790	57.3	7.52	12.7	0.197	0.133
Scandium (Sc)	44.96	1541	2830	314	2.99	10.2	0.568	0.158
Silver (Ag)	107.87	961.78	2162	104.8	10.5	18.9	0.235	4.29
Sodium (Na)	22.99	97.72	883	113	0.97	71	1.228	1.41
Strontium (Sr)	87.62	777	1382	84.8	2.64	22.5	0.301	0.353
Tantalum (Ta)	180.95	3017	5458	202.1	16.4	6.3	0.140	0.575
Technetium (Tc)		2157	4265		11			0.506
Terbium (Tb)	158.93	1359	3221	63.9	8.23	10.3	0.182	0.111
Thallium (Tl)	204.38	304	1473	20.3	11.8	29.9	0.129	0.461
Thorium (Th)	232.04	1750	4788	59.5	11.7	11.0	0.113	0.540
Thulium (Tm)	168.93	1545	1946	99.7	9.32	13.3	0.160	0.169
Tin (Sn)	118.71	231.93	2602	59.2	7.26	22.0	0.228	0.666
Titanium (Ti)	47.88	1668	3287	295	4.51	8.6	0.523	0.219
Tungsten (W)	183.84	3422	5555	284.5	19.3	4.5	0.132	1.74
Uranium (U)	238.03	1135	4131	38.4	19.1	13.9	0.116	0.276
Vanadium (V)	50.94	1910	3407	422	6.0	8.4	0.489	0.307
Ytterbium (Yb)	173.04	824	1194	44.3	6.90	26.3	0.155	0.385
Yttrium (Y)	88.91	1526	3336	128.2	4.47	10.6	0.298	0.172
Zinc (Zn)	65.39	419.53	907	112	7.14	30.2	0.388	1.16
Zirconium (Zr)	91.22	1855	4409	230	6.52	5.7	0.278	0.227

THERMAL CONDUCTIVITY OF METALS AND SEMICONDUCTORS AS A FUNCTION OF TEMPERATURE

This table gives the temperature dependence of the thermal conductivity of several metals and of carbon, germanium, and silicon. For graphite, separate entries are given for the thermal conductivity parallel (||) and perpendicular (⊥) to the layer planes. The thermal conductivity of all these materials is very sensitive to impurities at low temperatures, especially below 100 K. Therefore, the values given here should be regarded as typical values for a highly purified specimen; the thermal conductivity of different specimens can vary by more than an order of magnitude in the low-temperature range. See Reference 2 for details.

REFERENCES

1. Ho, C. Y., Powell, R. W., and Liley, P. E., *J. Phys. Chem. Ref. Data*, 1, 279, 1972.
2. White, G. K., and Minges, M. L., *Thermophysical Properties of Some Key Solids*, CODATA Bulletin No. 59, 1985.

Thermal Conductivity in W/cm K

				Carbon (C)						
				Diamond (type)			Pyrolytic graphite			
T/K	Ag	Al	Au	I	IIa	IIb	∥	⊥	Cr	Cu
1	39.4	41.1	5.46						0.402*	42.2
2	78.3	81.8	10.9	0.0138*	0.033*	0.0200*			0.803	84.0
3	115	121	16.1	0.0461	0.111	0.0676			1.20	125
4	147	157	20.9	0.108	0.261	0.160			1.60	162
5	172	188	25.2	0.206	0.494	0.307			2.00	195
6	187	213	28.5	0.344	0.820	0.510			2.39	222
7	193	229	30.9	0.523	1.24	0.778			2.27	239
8	190	237	32.3	0.762	1.77	1.12			3.14	248
9	181	239	32.7	1.05	2.41	1.53			3.50	249
10	168	235	32.4	1.40	3.17	2.03	0.811	0.0116	3.85	243
15	96.0	176	24.6	3.96	8.65	5.66			5.24	171
20	51.0	117	15.8	7.87	16.8	11.2	4.20	0.0397	5.93	108
30	19.3	49.5	7.55	18.8	38.9	26.5	9.86	0.0786	5.49	44.5
40	10.5	24.0	5.15	29.4	65.9	44.0	16.4	0.120	4.25	21.7
50	7.0	13.5	4.21	35.3	92.1	59.1	23.1	0.152	3.17	12.5
60	5.5	8.5	3.74	37.4	112	67.5	29.8	0.173	2.48	8.29
70	4.97	5.85	3.48	36.9	119	69.1	36.6	0.181	2.07	6.47
80	4.71	4.32	3.32	35.1	117	65.7	42.8	0.181	1.84	5.57
90	4.60	3.42	3.28	32.7	109	60.0	47.5	0.176	1.69	5.08
100	4.50	3.02	3.27	30.0	100	54.2	49.7	0.168	1.59	4.82
150	4.32	2.48	3.25	19.5	60.2	32.5	45.1	0.125	1.29	4.29
200	4.30	2.37	3.23	14.1	40.3	22.6	32.3	0.0923	1.11	4.13
250	4.29	2.35	3.21	11.0	29.7	17.0	24.4	0.0711	1.00	4.06
300	4.29	2.37	3.17	8.95	23.0	13.5	19.5	0.0570	0.937	4.01
350	4.27	2.40	3.14	7.55*	18.5*	11.1*	16.2	0.0477	0.929	3.96
400	4.25	2.40	3.11	6.5*	15.4*	9.32*	13.9	0.0409	0.909	3.93
500	4.19	2.36	3.04				10.8	0.0322	0.860	3.86
600	4.12	2.31	2.98				8.92	0.0268	0.807	3.79
800	3.96	2.18	2.84				6.67	0.0201	0.713	3.66
1000	3.79		2.70				5.34	0.0160	0.654	3.52
1200	3.61*		2.55				4.48	0.0134	0.619	3.39
1400							3.84	0.0116	0.588	
1600							3.33	0.0100	0.556	
1800							2.93	0.00895	0.526*	
2000							2.62	0.00807	0.494*	

THERMAL CONDUCTIVITY OF METALS AND SEMICONDUCTORS AS A FUNCTION OF TEMPERATURE (continued)

T/K	Fe	Ge[a]	Mg	Ni	Pb	Pt	Si[a]	Sn	Ti	W
1	1.71	0.274	9.86	2.17	27.9	2.31	0.0693*	183	0.0144*	14.4
2	3.42	2.06	19.6	4.34	44.6	4.60	0.454	323	0.0288*	28.7
3	5.11	5.35	29.0	6.49	35.8	6.79	1.38	297	0.0432	42.8
4	6.77	8.77	37.6	8.59	22.2	8.8	2.97	181	0.0575	56.3
5	8.39	11.6	45.0	10.6	13.8	10.5	5.27	117	0.0719	68.7
6	9.93	13.9	50.8	12.5	8.10	11.8	8.23	76	0.0863	79.5
7	11.4	15.5	54.7	14.2	4.86	12.6	11.7	52	0.101	88.0
8	12.7	16.6	56.7	15.8	3.20	12.9	15.5	36	0.115	93.8
9	13.9	17.3	57.0	17.1	2.30	12.8	19.5	26	0.129	96.8
10	14.8	17.7	55.8	18.1	1.78	12.3	23.3	19.3	0.143	97.1
15	17.0	17.3	41.1	19.5	0.845	8.41	41.6	6.3	0.212	72.0
20	15.4	14.9	27.2	16.5	0.591	4.95	49.8	3.2	0.275	40.5
30	10.0	10.8	12.9	9.56	0.477	2.15	48.1	1.79	0.365	14.4
40	6.23	7.98	7.19	5.82	0.451	1.39	35.3	1.33	0.390	6.92
50	4.05	6.15	4.65	4.00	0.436	1.09	26.8	1.15	0.374	4.27
60	2.85	4.87	3.27	3.08	0.425	0.947	21.1	1.04	0.355	3.14
70	2.16	3.93	2.49	2.50	0.416	0.862	16.8	0.96	0.340	2.58
80	1.75	3.25	2.02	2.10	0.409	0.815	13.4	0.915	0.326	2.29
90	1.50	2.70	1.78	1.83	0.403	0.789	10.8	0.880	0.315	2.17
100	1.34	2.32	1.69	1.64	0.397	0.775	8.84	0.853	0.305	2.08
150	1.04	1.32	1.61	1.22	0.379	0.740	4.09	0.779	0.270	1.92
200	0.94	0.968	1.59	1.07	0.367	0.726	2.64	0.733	0.245	1.85
250	0.865	0.749	1.57	0.975	0.360	0.718	1.91	0.696	0.229	1.80
300	0.802	0.599	1.56	0.907	0.353	0.716	1.48	0.666	0.219	1.74
350	0.744	0.495	1.55	0.850	0.347	0.717	1.19	0.642	0.210	1.67
400	0.695	0.432	1.53	0.802	0.340	0.718	0.989	0.622	0.204	1.59
500	0.613	0.338	1.51	0.722	0.328	0.723	0.762	0.596	0.197	1.46
600	0.547	0.273	1.49	0.656	0.314	0.732	0.619		0.194	1.37
800	0.433	0.198	1.46*	0.676		0.756	0.422		0.197	1.25
1000	0.323	0.174		0.718		0.787	0.312		0.207	1.18
1200	0.283	0.174		0.762		0.826	0.257		0.220	1.12
1400	0.312			0.804		0.871	0.235		0.236	1.08
1600	0.330					0.919	0.221		0.253	1.04
1800	0.345*					0.961			0.270*	1.01
2000						0.994*				0.98

[a] Values below 300 K are typical values.

* Extrapolated.

THERMAL CONDUCTIVITY OF ALLOYS AS A FUNCTION OF TEMPERATURE

This table lists the thermal conductivity of selected alloys at various temperatures. The indicated compositions refer to weight percent. Since the thermal conductivity is sensitive to exact composition and processing history, especially at low temperatures, these values should be considered approximate.

REFERENCES

1. Powell, R. L., and Childs, G. E., in *American Institute of Physics Handbook, 3rd Edition*, Gray, D. E., Ed., McGraw-Hill, New York, 1972.
2. Ho, C. Y., et al., *J. Phys. Chem. Ref. Data*, 7, 959, 1978.

Thermal conductivity in W/m K

	Alloy	**4 K**	**20 K**	**77 K**	**194 K**	**273 K**	**373 K**	**573 K**	**973 K**
Aluminum:	1100	50	240	270	220	220			
	2024	3.2	17	56	95	130			
	3003	11	58	140	150	160			
	5052	4.8	25	77	120	140			
	5083, 5086	3	17	55	95	120			
	Duralumin	5.5	30	91	140	160	180		
Bismuth:	Rose metal		5.5	8.3	14	16			
	Wood's metal	4	17	23					
Copper:	electrolytic tough pitch	330	1300	550	400	390	380	370	350
	free cutting, leaded	200	800	460	380	380			
	phosphorus, deoxidized	7.5	42	120	190	220			
	brass, leaded	2.3	12	39	70	120			
	bronze, 68% Cu; 32% Zn	2.3	16	48	92	110			
	beryllium	2	17	36	70	90	113	172	
	german silver	0.75	7.5	17	20	23	25	30	40
	silicon bronze A		3.4	11	23	30			
	manganin	0.48	3.2	14	17	22			
	constantan	0.9	8.6	17	19	22			
Ferrous:	commercial pure iron	15	72	106	82	76	66	54	34
	plain carbon steel(AISI 1020)	13	20	58	65	65			
	plain carbon steel(AISI 1095)		8.5	31	41	45			
	3% Ni; 0.7% Cr; 0.6% Mo		6	22		33	35	36	30
	4% Si					20	24	28	26
	stainless steel	0.3	2	8	13	14	16	19	25
	27% Ni; 15% Cr		1.7	55		11	12	16	21
Gold:	colbalt thermocouple	1.2	8.6	20					
	65% Au; 35% Ag		12	24		61	89		
Indium:	85.5% In; 14.5% Pb	1.9	7.8	24	41				
Lead:	60% Pb; 40% Sn (soft solder)		28	44					
	64.35% Pb; 35.65% In	0.8	3.26	9.1		20.2			
Nickel:	80% Ni; 20% Cr					12	14	17	23
	contracid	0.2	2	7.3	9.5	13			
	inconel	0.5	4.2	12.5	13	15	16	19	26
	monel	0.9	7.1	15	20	21	24	30	43
Platinum:	90% Pt; 10% Ir					31	31.4		
	90% Pt; 10% Rh					30.1	30.5		
Silver:	silver solder		12	34	58				
	normal Ag thermocouple	48	230	310					
Tin:	60% Sn; 40% Pb	16	55	51					
Titanium:	5.5% Al; 2.5% Sn;0.2% Fe		1.8	4.3	6.4	7.8	8.4	10.8	
	4.7% Mn; 3.99% Al; 0.14% C		1.7	4.5	6.5	8.5			

THERMAL CONDUCTIVITY OF CRYSTALLINE DIELECTRICS

This table lists the thermal conductivity of a number of crystalline dielectrics, including some which find use as optical materials. Values are given at temperatures for which data are available.

REFERENCE

Powell, R. L., and Childs, G. E., in *American Institute of Physics Handbook, 3rd Edition*, Gray, D. E., Ed., McGraw-Hill, New York, 1972.

Material	*T*/K	Ther. cond. W/m K
AgCl	223	1.3
	273	1.2
	323	1.1
	373	1.1
Al,B silicate (tourmaline)	398	2.9
∥ to *c* axis	540	3.2
	723	3.5
Al,Be silicate (beryl)	315	6.4
Al,F silicate (topaz)	315	17.7
∥ to *c* axis	358	15.6
	417	13.3
Al,Fe silicate (garnet)	315	35.8
	358	35.4
	377	35.6
Al_2O_3 (sapphire):		
36° to *c* axis	4.2	110
	20	3500
	35	6000
	77	1100
∥ to *c* axis	373	2.6
	523	3.9
	773	5.8
Al_2O_3 (sintered)	4.2	0.5
	20	23
	77	150
	194	48
	273	35
	373	26
	973	8
Ar	8	6.0
	10	3.7
	20	1.4
	77	0.31
As_2S_3 (glass)	283	0.16
	323	0.21
	373	0.27
BN	1047	36.2
	1475	22.7
	1928	21.9
	2111	18.5
BaF_2	225	20
	260	13.4
	305	10.9
	370	10.5
$BaTiO_3$	5	4.2
	30	24.0
	40	25.0
	100	12.0
	250	4.8
	300	6.2

Material	*T*/K	Ther. cond. W/m K
BeO	4.2	0.3
	20	16
	77	270
	373	210
	573	120
	1273	29
Bi_2Te_3	80	6.4
	204	2.8
	303	3.6
	370	4.6
C (diamond)	4.2	13
type I	20	800
	77	3550
	194	1450
	273	1000
$CaCO_3$		
∥ to *c* axis	83	25
	273	5.5
⊥ to *c* axis	83	17
	194	6.5
	273	4.6
	373	3.6
CaF_2	83	39
	223	18
	273	10
	323	9.2
	373	9
$CaWO_4$ (scheelite)	422	11.3
CdTe	160	7.0
	297	3.6
	422	2.9
CsBr	223	1.2
	273	0.94
	323	0.81
	373	0.77
CsI	223	1.4
	273	1.2
	323	1
	373	0.95
Cu_2O (cuprite)	102	3.74
	163	7.76
	299	5.58
	360	4.86
Fe_3O_4 (magnetite)	4.5	27.4
	20.5	293.0
	126.5	7.4
	304	7.0
Glass:		
phoenix	4.2	0.095
	20	0.13
	77	0.37

Material	T/K	Ther. cond. W/m K
plastic perspex	4.2	0.058
	20	0.074
pyrex	77	0.44
	194	0.88
	273	1
H_2 (para + 0.5% ortho)	2.5	100
	3	150
	4	200
	6	30
	10	3
H_2O (ice)	173	3.5
	223	2.8
	273	2.2
He^3 (high pressure)	0.6	25
	1	2
	1.5	0.57
	2	0.21
He^4 (high pressure)	0.5	42
	0.8	120
	1	24
	2	0.18
I_2	300	0.45
	325	0.42
	350	0.4
KBr	2	150
	4.2	360
	100	12
	273	5
	323	4.8
	373	4.8
KCl	4.2	500
	25	140
	80	35
	194	10
	273	7.0
	323	6.5
	373	6.3
KI	4.2	700
	80	13
	194	4.6
	273	3.1
Kr	4.2	0.48
	10	1.7
	20	1.2
	77	0.36
LaF_3	78	7.8
	197	5.0
	274	5.4
LiF	4.2	620
	20	1800
	77	150
$MgO \cdot Al_2O_3$ (spinel)	373	13
	773	8.5
MnO	4.2	0.25
	40	55
	120	8
	573	3.5
NaCl	4.2	440
	20	300
	77	30
	273	6.4
	323	5.6
	373	5.4
NaF	5	1100
	50	250
	100	90
Ne	2	3.0
	3	4.6
	4.2	4.2
	10	0.8
	20	0.3
NH_4Cl	77	17
	194	23
	230	38
	273	27
$NH_4H_2PO_4$		
‖ to optic axis	315	0.71
	339	0.71
⊥ to optic axis	313	1.26
	342	1.34
NiO	4.2	5.9
	40	400
	194	82
SiO_2 (quartz)		
‖ to *c* axis	20	720
	194	20
	273	12
⊥ to *c* axis	20	370
	194	10
	273	6.8
SiO_2 (fused silica)	4.2	0.25
	20	0.7
	77	0.8
	194	1.2
	273	1.4
	373	1.6
	673	1.8
$SrTiO_3$	5	2.4
	30	21.0
	40	19.2
	100	18.5
	250	12.5
	300	11.2
TlBr	316	0.59
TlCl	311	0.75
TiO_2 (rutile)		
‖ to optic axis	4.2	200
	20	1000
	273	13
⊥ to optic axis	4.2	160
	20	690
	273	9

THERMAL CONDUCTIVITY OF CERAMICS AND OTHER INSULATING MATERIALS

Thermal conductivity values for ceramics, refractory oxides, and miscellaneous insulating materials are given here. The thermal conductivity refers to samples with density indicated in the second column. Since most of these materials are highly variable, the values should only be considered as a rough guide.

REFERENCES

1. Powell, R. L., and Childs, G. E., in *American Institute of Physics Handbook, 3rd Edition*, Gray, D. E., Ed., McGraw-Hill, New York, 1972.
2. Perry, R. H., and Green, D., *Perry's Chemical Engineers' Handbook, Sixth Edition*, McGraw-Hill, New York, 1984.

Material	Dens. g/cm³	*T* °C	Ther. cond. W/m K
Alumina (Al_2O_3)	3.8	100	30
		400	13
		1300	6
		1800	7.4
	3.5	100	17
		800	7.6
Al_2O_3 + MgO		100	15
		400	10
		1000	5.6
Asbestos	0.4	-100	0.07
		100	1
		0	0.09
Asbestos + 85% MgO	0.3	30	0.08
Asphalt	2.1	20	0.06
Beryllia (BeO)	2.8	100	210
		400	90
		1000	20
		1800	15
	1.85	50	64
		200	40
		600	23
Brick, dry	1.54	0	0.04
Brick, refractory:			
alosite		1000	1.3
aluminous	1.99	400	1.2
		1000	1.3
diatomaceous	0.77	100	0.2
		500	0.24
	0.4	100	0.08
		500	0.1
fireclay	2	400	1
		1000	1.2
silicon carbide	2	200	2
		600	2.4
vermiculite	0.77	200	0.26
		600	0.31
Calcium oxide		100	16
		400	9
		1000	7.5
Cement mortar	2	90	0.55
Charcoal	0.2	20	0.055
Coal	1.35	20	0.26
Concrete	1.6	0	0.8
Cork	0.05	0	0.03
		100	0.04
	0.35	0	0.06
		100	0.08
Cotton wool	0.08	30	0.04
Diatomite	0.2	0	0.05
		400	0.09
	0.5	0	0.09
		400	0.16
Ebonite	1.2	0	0.16
Felt, flax	0.2	30	0.05
	0.3	30	0.04
Fuller's earth	0.53	30	0.1
Glass wool	0.2	-200 to 20	0.005
		50	0.04
		100	0.05
		300	0.08
Graphite			
100 mesh	0.48	40	0.18
20-40 mesh	0.7	40	1.29
Linoleum cork	0.54	20	0.08
Magnesia (MgO)		100	36
		400	18
		1200	5.8
		1700	9.2
MgO + SiO_2		100	5.3
		400	3.5
		1500	2.3
Mica:			
muscovite		100	0.72
		300	0.65
		600	0.69
phlogopite		100	0.66
Canadian		300	0.19
		600	0.2
Micanite		30	0.3
Mineral wool	0.15	30	0.04
Perlite, expanded	0.1	-200 to 20	0.002
Plastics:			
bakelite	1.3	20	1.4
celluloid	1.4	30	0.02
polystyrene foam	0.05	-200 to 20	0.033
mylar foil	0.05	-200 to 20	0.0001
nylon		-253	0.10
		-193	0.23
		25	0.30
teflon		-253	1.43
		-193	2.15
		25	2.25
		230	2.5
urethane foam	0.07	20	0.06
Porcelain		90	1

Material	Dens. g/cm³	*T* °C	Ther. cond. W/m K
Rock:			
basalt		20	2
chalk		20	0.92
granite	2.8	20	2.2
limestone	2	20	1
sandstone	2.2	20	1.3
slate, ⊥		95	1.4
slate, ∥		95	2.5
Rubber:			
sponge	0.2	20	0.05
92 percent		25	0.16
Sand, dry	1.5	20	0.33
Sawdust	0.2	30	0.06
Shellac		20	0.23
Silica aerogel	0.1	-200 to 20	0.003
Snow	0.25	0	0.16
Steel wool	0.1	55	0.09
Thoria (ThO_2)		100	10
		400	5.8
		1500	2.4
Titanium dioxide		100	6.5
		400	3.8
		1200	3.3
Uranium dioxide		100	9.8
		400	5.5
		1000	3.4
Wood:			
balsa, ⊥	0.11	30	0.04
fir, ⊥	0.54	20	0.14
fir, ∥	0.54	20	0.35
oak		20	0.16
plywood		20	0.11
pine, ⊥	0.45	60	0.11
pine, ∥	0.45	60	0.26
walnut, ⊥	0.65	20	0.14
Wool	0.09	30	0.04
Zinc oxide		200	17
		800	5.3
Zirconia (ZrO_2)		100	2
		400	2
		1500	2.5
Zirconia + silica		200	5.6
		600	4.6
		1500	3.7

THERMAL CONDUCTIVITY OF GLASSES

This table gives the composition of various types of glasses and the thermal conductivity *k* as a function of temperature. Because of the variability of glasses, the data should be regarded as only approximate.

Type of glass	Composition: SiO_2 (wt%)	Composition: Other oxides (wt%)		*T* °C	*k* W/m K
Vitreous silica	100			–150	0.85
				–100	1.05
				–50	1.20
				0	1.30
				50	1.40
				100	1.50
Vycor glass	96	B_2O_3	3	–100	1.00
				0	1.25
				100	1.40
Pyrex type chemically-resistant borosilicate glasses	80–81	B_2O_3	12–13	–100	0.90
		Na_2O	4	0	1.10
		Al	2	100	1.25
Borosilicate crown glasses	60–65	B_2O_3	15–20	–100	0.65–0.75
				0	0.90–0.95
				100	1.00–1.05
	65–70	B_2O_3	10–15	100	0.75–0.80
				0	0.95–1.00
				100	1.05–1.15
	70–75	B_3O_3	5–10	–100	0.80–0.85
				0	1.05–1.10
				100	1.15–1.20
Zinc crown glasses (i)	55–65	ZnO	5–15	–100	0.88–0.92
		Remainder:		0	1.10–1.15
		B_2O_3, Al_2O_3		100	1.15–1.25
		ZnO	5–15	–100	0.60–0.70
		Remainder:		0	0.70–0.90
		Na_2O, K_2O		100	0.85–0.95
		ZnO	15–25	–100	0.88–0.92
		Remainder:		0	1.10–1.15
		B_2O_3, Al_2O_3		100	1.15–1.20
		ZnO	15–25	–100	0.65–0.80
		Remainder:		0	0.85–0.95
		Na_2O, K_2O		100	0.90–1.05
Zinc crown glasses (ii)	65–75	ZnO	5–15	–100	0.88–0.92
		Remainder:		0	1.15–1.15
		B_2O_3, Al_2O_3		100	1.20–1.30
		ZnO	5–15	–100	0.70–0.85
		Remainder:		0	0.90–1.05
		Na_2O, K_2O		100	1.00–1.15

Type of glass	Composition: SiO_2 (wt%)	Composition: Other oxides (wt%)		T °C	k W/m K
		ZnO	15–25	–100	0.90–0.95
		Remainder:		0	1.15–1.15
		B_2O_3, Al_2O_3		100	1.20–1.25
		ZnO	15–25	–100	0.65–0.85
		Remainder:		0	0.85–1.00
		Na_2O, K_2O		100	1.05–1.20
Barium crown glasses	31	B_2O_3	12	–100	0.55
		Al_2O_3	8	0	0.70
		BaO	48	100	0.80
	41	B_2O_3	6	–100	0.60
		Al_2O_3	2	0	0.75
		ZnO	8	100	0.85
		BaO	43		
	47	B_2O_3	4	–100	0.65
		Na_2O	1	0	0.75
		K_2O	7	100	0.90
		ZnO	8		
		BaO	32		
	65	B_2O_3	2	–100	0.70
		Na_2O	5	0	0.90
		K_2O	15	100	1.00
		ZnO	2		
		BaO	10		
Borate glasses					
Borate flint glass	9	B_2O_3	36	–100	0.55
		Na_2O	1	0	0.65
		K_2O	2	100	0.80
		PbO	36		
		Al_2O_3	10		
		ZnO	6		
Borate flint glass	0	B_2O_3	56	–100	0.50
		Al_2O_3	12	0	0.65
		PbO	32	100	0.85
Borate flint glass	0	B_2O_3	43	–100	0.40
		Al_2O_3	5	0	0.55
		PbO	52	100	0.70
Borate glass	4	B_2O_3	55	–100	0.65
		Al_2O_3	14	0	0.80
		PbO	11	100	0.90
		K_2O	4		
		ZnO	12		
Borate crown glass	0	B_2O_3	64	–100	0.50
		Na_2O	8	0	0.65
		K_2O	3	100	0.85
		BaO	4		
		PbO	3		
		Al_2O_3	18		

THERMAL CONDUCTIVITY OF GLASSES (continued)

Type of glass	Composition: SiO_2 (wt%)	Other oxides (wt%)		T °C	k W/m K
Light borate crown glass	0	B_2O_3	69	–100	0.55
		Na_2O	8	0	0.70
		BaO	5	100	0.90
		Al_2O_3	18		
Zinc borate glass	0	B_2O_3	40	–100	0.65
		ZnO	60	0	0.75
				100	0.85
Phosphate crown glasses					
Potash phosphate glass	0	P_2O_5	70	0	0.75
		B_2O_3	3	100	0.85
		K_2O	12		
		Al_2O_3	10		
		MgO	4		
Baryta phosphate glass	0	P_2O_5	60	45	0.75
		B_2O_3	3		
		Al_2O_3	8		
		BaO	28		
Soda-lime glasses	75	Na_2O	17	–100	0.75
		CaO	8	0	0.95
				100	1.10
	73	Na_2O	12	–100	0.90
		CaO	13	0	1.10
				100	1.15
	72	Na_2O	15	–100	0.80
		CaO	11	0	1.00
		Al_2O_3	2	100	1.15
	65	Na_2O	25	–100	0.65
		CaO	10	0	0.85
				100	0.95
	65	Na_2O	15	–100	0.85
		CaO	20	0	1.00
				100	1.10
	60	Na_2O	20	–100	0.75
		CaO	20	0	0.90
				100	1.00
Other crown glasses					
Crown glass	75	Na_2O	9	–100	0.80
		K_2O	11	0	1.00
		CaO	5	100	1.10
High dispersion crown glass	68	Na_2O	16	–100	0.65
		ZnO	3	0	0.85
		PbO	13	100	1.00

Type of glass	Composition			T	k
	SiO_2 (wt%)	Other oxides (wt%)		°C	W/m K
Miscellaneous flint glasses					
(i) Silicate flint glasses					
Light flint glasses	65	PbO	25	–100	0.65–0.70
		Others	10	0	0.88–0.92
				100	1.00–1.05
	55	PbO	35	–100	0.60–0.65
		Others	10	0	0.75–0.85
				100	0.88–0.92
Ordinary flint glass	45	PbO	45	–100	0.50–0.60
		Others	10	0	0.65–0.75
				100	0.80–0.85
Heavy flint glass	35	PbO	60	–100	0.45–0.50
		Others	5	0	0.60–0.65
				100	0.70–0.75
Very heavy flint glasses	25	PbO	73	–100	0.40–0.45
		Others	2	0	0.55–0.60
				100	0.63–0.67
	20	PbO	80	–100	0.40
				0	0.50
				100	0.60
(ii) Borosilicate flint glass	33	B_2O_3	31	–100	0.65
		PbO	25	0	0.85
		Al_2O_3	7	100	0.95
		K_2O	3		
		Na_2O	1		
(iii) Barium flint glass	50	BaO	24	–100	0.60
		PbO	6	0	0.70
		K_2O	8	100	0.85
		Na_2O	3		
		ZnO	8		
		Sb_2O_3	1		
Other glasses					
Potassium glass	59	K_2O	33	50	0.88–0.92
		CaO	8		
Iron glasses	63	Fe_2O_3	10	–100	0.80
		Na_2O	17	0	0.95
		MgO	4	100	1.05
		CaO	3		
		Al_2O_3	2		
	67	Fe_2O_3	15	0	0.88—0.92
		Na_2O_3	18	100	1.00—1.05
	62	Fe_2O_3	20	0	0.85—0.90
		Na_2O	18	100	0.95—1.00
Rock glasses					
Obsidian				0	1.35
Artificial diabase				100	1.25

REFRACTORY MATERIALS

Borides

Name	Formula	Molecular weight	Melting point, °C	Crystalline form	Lattice parameter, A	X-ray density, g/cm³
Chromium boride	CrB_2	73.65	1,850[29]a	Hexagonal AlB_2 type [C 32]	a = 2.969 c = 3.066[39]	5.16
Hafnium boride	HfB_2	200.14	3,100[31]	Hexagonal AlB_2 type [C 32]	a = 3.14[6] c = 3.47	10.5
Molybdenum boride	MoB	106.77	2,180[40]	Tetragonal	a = 3.110[41] c = 16.95	8.77
	MoB_2	117.59	2,100[40]	Hexagonal AlB_2 type [C 32]	a = 3.05 c = 3.113[42]	7.78
	Mo_2B	202.72	2,000[40] (decomposes)	Tetragonal $CuAl_2$ type [C 16]	a = 5.543[41] c = 4.735	9.31
Niobium boride	NbB	103.73	>0.2000[3]	Orthorhombic	a = 3.298 b = 8.724 c = 3.137[35]	
	NbB_2	114.55	2,900[36] (decomposes)	Hexagonal AlB_2 type [C 32]	a = 3.086[28] c = 3.306	7.21
Tantalum boride	TaB	191.77	>0.2000[3]	Orthorhombic	a = 3.276 b = 8.669 c = 3.157[37]	14.29
	TaB_2	202.59	3,000[6]	Hexagonal AlB_2 type [C 32]	a = 3.088[28] c = 3.241	12.60
Thorium boride	ThB_4	275.53	>0.2500[6]	Tetragonal D_{4h}^5-P4/mbm	a = 7.256[43] c = 4.113	8.45
Titanium boride	TiB_2	69.54	2,980[6]	Hexagonal AlB_2 type [C 32]	a = 3.028[28] c = 3.228	4.52
Tungsten boride	WB	194.68	2,860[29]	Tetragonal	a = 3.115[6] c = 16.92	16.0
	W_2B	378.54	2,770[29]	Tetragonal $CuAl_2$ type [C 16]	a = 5.564[41] c = 4.740	16.72
Uranium boride	UB_{12}	367.91	>0.1500[6] (decomposes)	Face-centered cubic	a = 7.473[44]	5.82
Vanadium boride	VB_2	72.59	2,100[29]	Hexagonal AlB_2 type [C 32]	a = 2.998[28] c = 3.057	5.10
Zirconium boride	ZrB_2	112.86	3,040 ± 50[30]	Hexagonal AlB_2 type [C 32]	a = 3.169[28] c = 3.530	6.09
	ZrB_{12}	221.06	2,680[30]	Face centered cubic	a = 7.408[33]	

Name	Thermal conductivity, cal-sec⁻¹-cm⁻²-cm-°C⁻¹	Electrical resistivity, microhm-cm	Ductility relative scale[b]	Resistance to oxidation[c]	Hardness[d]
Chromium boride		21 at room temperature[38]			1,800 kg/mm²[29]
Hafnium boride		10[31]	3[3]		
Molybdenum boride		α-MoB = 45 at room temperature[40] β-MoB = 23 at room temperature 45 at room temperature[40] 40 at room temperature[40]	3[3]	3[3]	8 Mohs[20] 1,570 kg/mm²[29] 1,280 kg/mm²[29] 8–9 Mohs[20]
Niobium boride	0.040 at 20°C[36]	6.45 at room temperature[36] 32.0 at room temperature[36]			>0.8 Mohs[36]
Tantalum boride	0.026 at 20°C[6]	100 at room temperature[38] 68 at room temperature[38]	3[3]	3[3]	
Thorium boride					
Titanium boride	0.0624 at 200°C[28] (15% porosity)	28.4[28]	3[3]	2–3[3]	3,400 kg/mm²[29]
Tungsten boride			3[3]	3[3]	9 Mohs[6]
Uranium boride					
Vanadium boride	16 at 20°C[34]				8–9 Mohs[6]
Zirconium boride	0.0550 at 200°C[28] 0.029 at room temperature[33]	9.2 at 20°C[31] 60–80 at room temperature[30]	2[3]	2–3[3]	8 Mohs[32]

Carbides

Name	Formula	Molecular weight	Melting point, °C	Crystalline form	Lattice parameter, A	X-ray density, g/cm³
Boron carbide	B_4C	55.29	2,450	Hexagonal	a_0 = 5.60[6] c_0 = 12.12	2.52
Chromium carbide	Cr_3C_2	180.05	1,895[15]	Orthorhombic ($D5_{10}$)	a = 2.82 b = 5.53 c = 11.47[16]	6.7
	Cr_7Cr_3	400.01	1,780[15]	Hexagonal [$C_{3v}{}^4$]	a = 14.01 c = 4.532[17]	6.92
Graphite	C	12.01	3,700 ± 100[3]	Hexagonal	Orthohexagonal axes a_0 = 2.46 b_0 = 4.28 c_0 = 6.71	2.25
Hafnium carbide	HfC	190.51	3,890[9]	Cubic NaCl type (B1)	a = 4.46[10]	12.7

REFRACTORY MATERIALS (Continued)

Carbides (Continued)

Name	Formula	Molecular weight	Melting point, °C	Crystalline form	Lattice parameter, A	X-ray density, g/cm^3
	MoC	107.96	2,695[9]	Face-centered cubic	a = 4.28[19]	
Molybdenum carbide	Mo_2C	203.91	2,690[6]	Hexagonal (L′3)	a = 3.002	9.2
					c = 4.724[18]	
Niobium carbide	NbC	104.92	3,500[9]	Cubic NaCl type (B1)	a = 4.461	7.85
Silicon carbide						
β	SiC	40.07	Trans. to α at 2,100	Face-centered cubic	a_o = 4.3590[6]	3.22
α	SiC		2,700[6]	Hexagonal (Wurtzite)	a_o = 3.081[6]	
					c_o = 5.0394	
Tantalum carbide	TaC	192.96	3,880[9], 4,730[4]	Cubic NaCl type (B1)	a = 4.455[14]	14.5
Thorium carbide	ThC	244.06	2,625[23]	Cubic NaCl type (B1)	a = 5.34[9]	10.67
	ThC_2	256.07	2,655[29]	Monoclinic [C 2/e]	a = 6.53[24]	9.6
					b = 4.24	
Titanium carbide	TiC	59.91	3,160[4]	Cubic NaCl type (B1)	a = 4.32[6]	4.938
Tungsten carbide	W_2C	379.73	2,730[21]	Hexagonal (L′3)	a = 2.98[6]	17.34
					c = 4.71	
	WC	195.87	2,630[21] (decomposes)	Hexagonal (L′3)	a = 2.900[22]	15.77
					c = 2.831	
Uranium carbide	UC	250.08	2,450–2,500[25]	Cubic NaCl type (B1)	a = 4.955[26]	13.6
	UC_2	262.09	2,350–2,400[25]	Body-centered tetragonal $CaCl_2$ type	a = 3.517[27]	11.68
					c = 5.987	
Vanadium carbide	VC	62.96	2,830[1]	Cubic NaCl type (B1)	a = 4.16[6]	5.8
Zirconium carbide	ZrC	103.23	3,030[4]	Cubic NaCl type (B1)	a = 4.689[5]	6.44

Name	Thermal conductivity, cal-sec^{-1}-cm^{-2}-cm-°C^{-1}	Electrical resistivity, microhm-cm	Ductility relative scale[b]	Resistance to oxidation[c]	Hardness[d]
Boron carbide	0.05 at 20–425°C[1]	0.30–0.80[1]	3[3]	3[3]	9.3 Mohs
Chromium carbide			3[3]	3[3]	1,300 kg/mm^2 [8]
			3[3]	3[3]	
Graphite	0.268–0.451 at room temperature	65 at room temperature[71]	3[3]	4[3]	
Hafnium carbide		109 at room temperature[1]	3[3]	3[3]	
Molybdenum carbide		97[6]	3[3]	5[3]	1.800 kg/mm^2 [6]
			3[3]	5[3]	>0.7 Mohs[20]
Niobium carbide	0.034 at room temperature[7]	74[6]	2–3[3]	3[3]	2,470 kg/mm^2 [13]
Silicon carbide					
β	0.10 at 20–425°C[1]	107–200 ohm cm[1] at room temperature	3[3]	2[3]	9.2 Mohs
Tantalum carbide	0.053 at room temperature[6]	30[6]	2–3[3]	3[3]	1,800 kg/mm^2 [8]
Thorium carbide			3[3]	3[3]	
Titanium carbide	0.049 at room temperature[6]	180–250[4]	3[3]	3[4]	3,200 kg/mm^2 [8]
Tungsten carbide		80[6]	3[3]	5[3]	3,000 kg/mm^2 [6]
		53[6]	3[3]	5[3]	2,400 kg/mm^2 [6]
Uranium carbide			3[3]		
				5[3]	
Vanadium carbide		150[6]	3[3]		2,800 kg/mm^2 [11]
					2,100 kg/mm^2 [12]
Zirconium carbide	0.049 at room temperature[6]	70 at room temperature[4]	3[3]	3[3]	2,600 kg/mm^2 [6]

Nitrides

Name	Formula	Molecular weight	Melting point, °C	Crystalline form	Lattice parameter, A	X-ray density, g/cm^3
Boron nitride	BN	24.83	2,730[6]	Hexagonal (B 12)	a = 2.51 ± .02	2.25
					c = 6.70 ± .04[52]	
Chromium nitride	CrN	66.02	1,500 (decomposes)	Cubic NaCl type (B1)	a = 4.140[6]	6.14
Hafnium nitride	HfN	192.60	3,310[6]			
Niobium nitride	NbN	106.92	2,050[46]	Cubic NaCl type (B1)	a = 4.41–4.375[6]	7.28
Tantalum nitride	Ta_2N	375.77	3,090[46]	Hexagonal	a = 3.05[48]	14.1
					c = 4.95[53]	
Thorium nitride	ThN	246.13	2,630[48]	Cubic NaCl type (B1)	a = 5.2[48]	
Titanium nitride	TiN	61.91	2,930 N_2 liberated on melting[46]	Cubic NaCl type (B1)	a = 4.23[53]	5.43
Uranium nitride	UN		2,650	Cubic NaCl type (B1)	a = 4.880[58]	14.32
Vanadium nitride	VN	64.96	2,050[46]	Cubic NaCl type (B1)	a = 4.129[6]	6.102
Zirconium nitride	ZrN	105.22	2,980[47]	Cubic NaCl type (B1)	a = 4.567[53]	7.349

Nitrides (Continued)

Name	Thermal conductivity, cal-sec^{-1}-cm^{-2}-cm-°C^{-1}	Electrical resistivity, microhm-cm	Ductility relative scale[b]	Resistance to oxidation[c]	Hardness[d]
Boron nitride		1,900 at 2,000°C[(1)]	3[(3)]	2[(3)]	2.0 Mohs
Chromium nitride					
Hafnium nitride					
Niobium nitride		200 at room temperature[(46)]	5[(3)]	3[(3)]	+8 Mohs[(46)]
Tantalum nitride		135 at room temperature[(47)]	3[(3)]	5[(3)]	+8
Thorium nitride					
Titanium nitride		21.7 at room temperature	3[(3)]	3[(3)]	Between 9 and 10 Mohs[(46)]
Uranium nitride					
Vanadium nitride		200 at room temperature[(46)]	3[(3)]		
Zirconium nitride		13.6 at room temperature[(46)]	3[(3)]	3[(3)]	+8 Mohs[(6)]

Oxides

Name	Formula	Molecular weight	Melting point, °C	Crystalline form	Lattice parameter, Å	X-ray density, g/cm³
Aluminum oxide	Al_2O_3	101.92	2,015[(1)]	Rhombohedral	a = 5.13 axial angle = 55° 6′	3.965
Beryllium oxide	BeO	25.02	2,550[(1)]	Hexagonal	a = 2.70 c = 4.39	3.03
Cerium oxide	CeO_2	172.3	2,600[(1)]	Face-centered cubic	a = 5.41	7.13
Chromic oxide	Cr_2O_3	152.02	2,265[(1)]	Rhombohedral	a = 5.38 axial angle = 54° 50′	5.21
Hafnium oxide	HfO_2	210.6	2,777[(1)]	Face-centered cubic	a = 5.11	9.68
Magnesium oxide	MgO	40.32	2,800[(1)]	Face-centered cubic	a = 4.20	3.58
Silicon oxide	SiO_2	60.06	1,728[(1)]	Hexagonal	a = 4.90 c = 5.39	2.32 (low cristobalite)
Thorium oxide	ThO_2	264.12	3,300[(1)]	Face-centered cubic	a = 5.59	9.69
Titanium oxide	TiO_2	79.90	1,840[(1)]	Tetragonal	a = 4.58 c = 2.98	4.24
Uranium oxide	UO_2	270.07	2,280[(1)]	Face-centered cubic	a = 5.47	10.96
Zirconium oxide	ZrO_2	123.22	2,677[(1)]	Monoclinic	a = 5.21 c = 5.37	5.56

Name	Thermal conductivity, cal-sec^{-1}-cm^{-2}-cm-°C^{-1}	Electrical resistivity, microhm-cm	Ductility relative scale[b]	Resistance to oxidation[c]	Hardness[d]
Aluminum oxide	0.0723 at 100°C[(2)]	1×10^{22} at 14°C[(1)] 3×10^{19} at 300°C[(1)] 3.5×10^{14} at 800°C[(1)]	3[(3)]	1[(3)]	9 Mohs[(1)]
Beryllium oxide	0.525 at 100°C[(2)]	4×10^{14} at 600°C[(1)] 5×10^{12} at 1,100°C[(1)] 8×10^{8} at 2,100°C[(1)]			9 Mohs[(1)]
Cerium oxide		6.5×10^{10} at 800°C[(1)] 3.4×10^{8} at 1,200°C[(1)]			6 Mohs[(1)]
Chromium oxide		1.3×10^{9} at 350°C[(1)] 2.3×10^{7} at 1,200°C[(1)] 6.8×10^{3} at 600°C[(1)] 4.5×10^{3} at 1,100°C[(1)]	3[(3)]	1[(3)]	
Hafnium oxide		5×10^{15} at 400°C[(1)] 1×10^{3} at 1,500°C[(1)]			
Magnesium oxide	0.0860 at 100°C[(2)]	2×10^{14} at 850°C[(1)] 3×10^{13} at 980°C[(1)] 4.5×10^{8} at 2,100°C[(1)]			6 Mohs[(1)]
Silicon oxide		1×10^{21} at 20°C[(1)] 7×10^{12} at 600°C[(1)] 4×10^{5} at 1,300°C[(1)] (vitreous)	3[(4)] (vitreous)	1[(4)] (vitreous)	6–7 Mohs[(1)] (cristobalite)
Thorium oxide	0.0243 at 100°C[(2)]	2.6×10^{13} at 550°C[(1)] 8×10^{11} at 800°C[(1)] 1.5×10^{10} at 1,200°C[(1)]			6.5 Mohs[(1)]
Titanium oxide	0.0156 at 100°C[(2)]	1.2×10^{10} at 800°C[(1)] 8.5×10^{6} at 1,200°C[(1)]			5.5–6.0 Mohs[(1)]
Uranium oxide	0.0234 at 100°C[(2)]	3.8×10^{10} at 20°C[(1)] 5×10^{8} at 500°C[(1)]			
Zirconium oxide	0.00466 at 100°C[(2)]	1×10^{12} at 385°C[(1)] 2.2×10^{10} at 700°C[(1)] 3.6×10^{8} at 1,200°C[(1)]	3[(3)]	1[(3)]	6.5 Mohs[(1)]

Silicides

Name	Formula	Molecular weight	Melting point, °C	Crystalline form	Lattice parameter, Å	X-ray density g/cm³
Chromium silicide	$CrSi_2$	108.13	1,570(51)	Hexagonal	a = 4.42 c = 6.35(51)	
	CrSi	80.07	1,870 (decomposes in presence of C)	Tetragonal (C 11b)	a = 3.20 c = 7.86(6)	
Molybdenum silicide	$MoSi_2$	152.07	1,870 (decomposes in presence of C)	Tetragonal (C 11b)	a = 3.20 c = 7.86(64)(51)	6.24
Niobium silicide	$NbSi_2$		1,950(50)	Hexagonal $CrSi_2$ type	a = 4.785 c = 6.576(51)	5.29
Tantalum silicide	$TaSi_2$	237.00	2,400(50)	Hexagonal $CrSi_2$ type	a = 4.773 c = 6.552(51)	8.83
Titanium silicide	$TiSi_2$	104.02	1,540(49)	Orthorhombic	a = 8.24 b = 4.79 c = 8.52(59)	4.13
	Ti_5Si_3	323.68	2,120(49)	Hexagonal	a = 7.465 c = 5.162(60)	4.32
Tungsten silicide	WSi_2	240.04	2,050(50)	Tetragonal $MoSi_2$ Structure (C 11b)	a = 3.21 c = 7.83(66)	9.3
Uranium silicide	βUSi_2	294.19	1,700(6)	Hexagonal	a = 3.85(6) c = 4.06	9.25
	U_3Si_2	770.33	1,665(6)	Tetragonal	a = 7.3151(6) c = 3.8925	12.20
Vanadium silicide	VSi_2	107.07	1,750(50)	Hexagonal $CrSi_2$ type	a = 4.562 c = 6.359(51)	4.71

Name	Thermal conductivity, cal-sec⁻¹-cm⁻²-cm-°C⁻¹	Electrical resistivity, microhm-cm	Ductility relative scale[b]	Resistance to oxidation[c]	Hardness[d]
Chromium silicide					1,150 kg/mm² (50)
Molybdenum silicide	0.075 at room temperature to 200°C(6)	21.5 at room temperature 18.9 at −80°C(69)	2(3)	1(3)	1,290 kg/mm² (50)
Niobium silicide		6.3(68)	2(3)	4(3)	1,050 kg/mm² (50)
Tantalum silicide		8.5(68)	2(3)	3(3)	1,560 kg/mm² (50)
Titanium silicide		123 (hot pressed)(1)	3(3) 3(3)	4(3) 4(3)	870 kg/mm² (70) 986 kg/mm² (49)
Tungsten silicide		33.4(68)		1–2(3)	1,310 kg/mm² (68) 1,090 kg/mm² (68)
Uranium silicide					
Vanadium silicide		9.5(68)			1.090 kg/mm² (50)

[a]Numbers in parentheses refer to references at end of table.
[b]Ductility – 1: capable of being severely drawn, rolled, or otherwise worked without failure; 2: capable of withstanding slight deformation, or consisting of individually ductile crystals fragilely bound together; 3: incapable of being worked, of glasslike brittleness.
[c]Resistance to oxidation – classed according to the temperature range in which the rate of attack by air would cause severe erosion or failure of the coated specimen within a few hours. 1: above 1,700°C; 2: 1,400 to 1,700°C; 1.100 to 1,400°C; 4: 800 to 1.000°C; 5: 500 to 800°C.
[d]Microhardness values taken with 100-g load.

REFERENCES

1. **Campbell, I. E.,** *High Temperature Technology,* Electrochemical Society, John Wiley & Sons New York, 1956.
2. **Kingery, W. D., Franch, J., Coble, R. L., and Vasilos, T.,** Thermal conductivity: X, data for several pure oxide materials corrected to zero porosity, *J. Am. Ceram. Soc.,* 37, 2, 107, 1954.
3. **Powel, C. F., Campbell, I. E., and Gonser, B. W.,** *Vapor Plating,* John Wiley & Sons, New York, 1955.
4. **Friederich, E. and Sittig, L.,** *Z. Anorg. Allg. Chem.,* 144, 169, 1925.
5. **Norton, J. T. and Merory, A. L.,** *Trans. Am. Inst. Min. Metall. Pet. Eng.,* 185, 133, 1949.
6. **Schwarzkopf, P. and Kieffer, R.,** *Refractory Hard Metals,* Macmillan, New York, 1953.
7. **Schwarzkopf, P. and Sindeband, S. J.,** Electrochemical Society, 97th Meeting, Cleveland, Ohio, 1950.
8. **Kieffer, R. and Kölbl, F.,** *Powder Metall. Bull.,* 4, 4, 1949.
9. **Agte, C. and Alterthum, H.,** *Z. Tech. Phys.,* 11, 182, 1930.
10. **Becker, K. and Ebert, F.,** *Z. Phys.,* 31, 268, 1925.
11. **Ruff, O. and Martin, W.,** *Z. Angew. Chem.,* 25, 49, 1912.
12. **Hinnuber, J.,** *Z. VDI,* 92, 111, 1950.
13. **Foster, L. S., Forbes, L. W., Jr., Friar, L. B., Moody, L. S., and Smith, W. H.,** *J. Am. Ceram. Soc.,* 33, 27, 1950.
14. **Ellinger, F. H.,** *Trans. Am. Soc. Met.,* 31, 89, 1943.
15. **Bloom, D. S. and Grant, N. J.,** *Trans. Am. Inst. Min. Metall. Pet. Eng.,* 188, 41, 1950.
16. **Hellstrom, K. and Westgren, A.,** *Kem. Tidskr.,* 45, 141, 1933.
17. **Westgren, A.,** *Jernkontorets Ann.,* 119, 231, 1935.
18. **Kuo, K. and Hagg, G.,** *Nature,* 170, 245, 1952.
19. **Nowotny, H. and Kieffer, R.,** *Z. Anorg. Allg. Chem.,* 267, 261, 1952.
20. **Weiss, G.,** *Ann. Chem.,* 1, 446, 1946.
21. **Brewer, L., Bromely, L. A., Gilles, P. W., and Lofgren, N. L.,** *The Chemistry and Metallurgy of Miscellaneous Materials: Thermodynamics,* McGraw-Hill, New York, 1950.

22. **Becker, K.,** *Z. Phys.,* 51, 481, 1928.
23. **Wilhelm, H. A. and Chiotti, P.,** *Trans. Am. Soc. Met.,* 42, 1295, 1950.
24. **Hunt, E. B. and Rundle, R. E.,** *J. Am. Chem. Soc.,* 73, 4777, 1951.
25. **Mallet, W., Gerds, A. F., and Nelson, H. R.,** *J. Electrochem. Soc.,* 99, 197, 1952.
26. **Litz, L. M., Gurrett, A. B., and Croxton, F. C.,** *J. Am. Chem. Soc.,* 70, 1718, 1948.
27. **Rundle, R. E., Baenziger, N. C., Wilson, A. S., and McDonald, R. A.,** *J. Am. Chem. Soc.,* 70, 99, 1948.
28. **Norton, J. T., Blumenthal, H., and Sindeband, S. J.,** *Trans. Am. Inst. Min. Metall. Pet. Eng.,* 185, 749, 1949.
29. **Honak, E. R.,** Thesis, Tech. Hochsch. Graz, 1951.
30. **Glaser, F. W. and Post, B.,** *J. Met.,* 1953.
31. **Moers, K.,** *Z. Anorg. Allg. Chem.,* 198, 262, 1931.
32. **Andrieux, L.,** *Rev. Mét.,* 45, 49, 1948.
33. **Post, B. and Glaser, F. W.,** *J. Met.,* 4, 631, 1952.
34. **Moers, K.,** *Z. Anorg. Allg. Chem.,* 198, 243, 1931.
35. **Anderson, L. H. and Kiessling, R.,** *Acta Chem. Scand.,* 4, 160, 209, 1950.
36. **Glaser, F. W.,** *J. Met.,* 4, 391, 1952.
37. **Kiessling, R.,** *Acta Chem. Scand.,* 3, 603, 1949.
38. **Moers, K.,** *Z. Anorg. Allg. Chem.,* 198, 262, 1931.
39. **Kiessling, R.,** *Acta Chem. Scand.,* 3, 595, 1949.
40. **Steintz, R.,** *J. Met.,* 4, 148, 1952.
41. **Kiessling, R.,** *Acta Chem. Scand.,* 1, 893, 1947.
42. **Bertaut, F. and Blum, P.,** *Acta Crystallogr.,* 4, 72, 1951.
43. **Zalkin, A. and Templeton, D. H.,** *J. Chem. Phys.,* 18, 391, 1950.
44. **Bertaut, F. and Blum, P.,** *Comptes rendus,* 229, 666, 1949.
45. **Baumann, H. N., Jr.,** Electrochemical Society, 99th Meeting, Washington, D.C., April 1951.
46. **Friederich, E. and Sittig, L.,** *Z. Anorg. Allg. Chem.,* 143, 293, 1925.
47. **Agte, C. and Moers, K.,** *Z. Anorg. Chem.,* 198, 233, 1931.
48. **Chiotti, P.,** *J. Am. Ceram. Soc.,* 35, 123, 1952.
49. **Hansen, M., Klasler, H. D., and McPherson, D. J.,** *Trans. Am. Soc. Met.,* 44, 518, 1952.
50. **Cerwenka, E.,** Thesis, Tech. Hochsch. Graz, 1951.
51. **Wallbaum, H. J.,** *Z. Metallkd.,* 33, 378, 1941.
52. **Pease, R. S.,** *Acta Crystallogr.,* 5, 356, 1952.
53. **van Arkel, A. E.,** *Physica,* 4, 296, 1924.
54. **Horn, F. H. and Ziegler, W. T.,** *J. Am. Chem. Soc.,* 69, 2762, 1947.
55. **Pauling, L., Killeffer, D. H., and Linz, A.,** *Molybdenum Compounds,* Interscience, New York, 1952.
56. **Hägg, G.,** *Z. Phys. Chem. Abt. B,* 7, 339, 1930.
57. **Kiessling, R. and Lier, Y. H.,** *J. Met.,* 3, 639, 1951.
58. **Rundle, R. E., Baenziger, N. C., Wilson, A. S., and McDonald, R. A.,** *J. Am. Chem. Soc.,* 70, 99, 1941.
59. **Laues, F. and Wallbaum, H. J.,** *Z. Kristallogr. A,* 101, 78, 1939.
60. **Pietrokowsky, P. and Duwez, P.,** *J. Met.,* 3, 772, 1951.
61. **Naray Szako, S. V.,** *Z. Kristallogr. A,* 97, 223, 1937.
62. **Lundin, C. E., McPherson, D. J., and Hansen, M.,** American Society for Metals, 34th Ann. Convention, Preprint No. 41, 1952.
63. **Wallbaum, H. J.,** *Z. Metallkd.,* 31, 362, 1939.
64. **Zachariasen, W. H.,** *Z. Phys. Chem.,* 128, 39, 1927.
65. **Templeton, D. H. and Dauben, C. H.,** *Acta Crystallogr.,* 3, 261, 1950.
66. **Nowotny, H., Kieffer, R., and Schachner, H.,** *Mh. Chem.,* 83, 1243, 1952.
67. **Brauer, G. and Mitices, A.,** *Z. Anorg. Allg. Chem.,* 249, 325, 1942.
68. **Gallistl, E.,** Thesis, Tech. Hochsch. Graz, 1951.
69. **Glaser, F. W.,** *J. Appl. Phys.,* 22, 103, 1951.
70. **Cerwenka, E.,** Thesis, Tech. Hochsch. Graz, 1951.
71. **Currie, L. M., Hamister, V. C., and MacPherson, H. G.,** paper presented at the United Nations International Conference on The Peaceful Uses of Atomic Energy, Geneva, 1955.

COMMERCIAL METALS AND ALLOYS

This table gives typical values of mechanical, thermal, and electrical properties of several common commercial metals and alloys. Values refer to ambient temperature (0 to 25°C). All values should be regarded as typical, since these properties are dependent on the particular type of alloy, heat treatment, and other factors. Values for individual specimens can vary widely.

REFERENCES

1. *ASM Metals Reference Book, Second Edition*, American Society for Metals, Metals Park, OH, 1983.
2. Lynch, C. T., *CRC Practical Handbook of Materials Science*, CRC Press, Boca Raton, FL, 1989.
3. Shackelford, J. F., and Alexander, W., *CRC Materials Science and Engineering Handbook*, CRC Press, Boca Raton, FL, 1991.

Common name	Thermal conductivity W/cm K	Density g/cm^3	Coeff. of linear expansion 10^{-16}/°C	Electrical resistivity μΩ cm	Modulus of elasticity GPa	Tensile strength MPa	Approx. melting point °C
Ingot iron	0.7	7.86	11.7	9.7	205	-	1540
Plain carbon steel AISI-SAE 1020	0.52	7.86	11.7	18	205	450	1515
Stainless steel type 304	0.15	7.9	17.3	72	195	550	1425
Cast gray iron	0.47	7.2	10.5	67	90	180	1175
Malleable iron		7.3	12	30	170	345	1230
Hastelloy C	0.12	8.94	11.3	125	200	780	1350
Inconel	0.15	8.25	11.5	103	200	800	1370
Aluminum alloy 3003, rolled	1.9	2.73	23.2	3.7	70	110	650
Aluminum alloy 2014, annealed	1.9	2.8	23.0	3.4	70	185	650
Aluminum alloy 360	1.5	2.64	21.0	7.5	70	325	565
Copper, electrolytic (ETP)	3.9	8.94	16.5	1.7	120	300	1080
Yellow brass (high brass)	1.2	8.47	20.3	6.4	100	300-800	930
Aluminum bronze	0.7	7.8	16.4	12	120	400-600	1050
Beryllium copper 25	0.8	8.23	17.8	7	130	500-1400	925
Cupronickel 30%	0.3	8.94	16.2		150	400-600	1200
Red brass, 85%	1.6	8.75	18.7	11	90	300-700	1000
Chemical lead	0.35	11.34	29.3	21	13	17	327
Antimonial lead (hard lead)	0.3	10.9	26.5	23	20	47	290
Solder 50-50	0.5	8.89	23.4	15	-	42	215
Magnesium alloy AZ31B	1.0	1.77	26	9	45	260	620
Monel	0.3	8.84	14.0	58	180	545	1330
Nickel (commercial)	0.9	8.89	13.3	10	200	460	1440
Cupronickel 55-45 (constantan)	0.2	8.9	18.8	49	160	-	1260
Titanium (commercial)	1.8	4.5	8.5	43	110	330-500	1670
Zinc (commercial)	1.1	7.14	32.5	6	-	130	419
Zirconium (commercial)	0.2	6.5	5.85	41	95	450	1855

HARDNESS OF MINERALS AND CERAMICS

There are several hardness scales for describing the resistance of a material to indentation or scratching. This table lists a number of common materials in order of increasing hardness. Values are given, when available, on three different hardness scales: the original Mohs Scale (range 1 to 10); the modified Mohs Scale (range 1 to 15), and the Knoop Hardness Scale. In the last case, a load of 100 g is assumed.

REFERENCE

Shackelford, J. F. and Alexander, W., *CRC Materials Science and Engineering Handbook*, CRC Press, Boca Raton, FL, 1991.

Material	Formula	Mohs	Modified mohs	Knoop
Graphite	C	0.5		
Talc	$3MgO \cdot 4SiO_2 \cdot H_2O$	1	1	
Alabaster	$CaSO_4 \cdot 2H_2O$	1.7		
Gypsum	$CaSO_4 \cdot 2H_2O$	2	2	32
Halite (rock salt)	NaCl	2		
Stibnite (antimonite)	Sb_2S_3	2.0		
Galena	PbS	2.5		
Mica		2.8		
Calcite	$CaCO_3$	3	3	135
Barite	$BaSO_4$	3.3		
Marble		3.5		
Aragonite	$CaCO_3$	3.5		
Dolomite	$CaMg(CO_3)_2$	3.5		
Fluorite	CaF_2	4	4	163
Magnesia	MgO	5		370
Apatite	$CaF_2 \cdot 3Ca_3(PO_4)_2$	5	5	430
Opal		5		
Feldspar (orthoclase)	$K_2O \cdot Al_2O \cdot 6SiO_2$	6	6	560
Augite		6		
Hematite	Fe_2O_3	6		750
Magnetite	Fe_3O_4	6		
Rutile	TiO_2	6.2		
Pyrite	FeS_2	6.3		
Agate	SiO_2	6.5		
Uranium dioxide	UO_2	6.7		600
Silica (fused)	SiO_2		7	
Quartz	SiO_2	7	8	820
Flint		7		
Silicon	Si	7		
Andalusite	Al_2OSiO_4	7.5		
Zircon	$ZrSiO_4$	7.5		
Zirconia	ZrO_2			1200
Aluminum nitride	AlN			1225
Beryl	$Be_3Al_2Si_6O_{18}$	7.8		
Beryllia	BeO			1300
Topaz	$Al_2SiO_4(OH,F)_2$	8	9	1340
Garnet	$Al_2O_3 \cdot 3FeO \cdot 3SiO_2$		10	1360
Emery	Al_2O_3 (impure)	8		
Zirconium nitride	ZrN	8+		1510
Zirconium boride	ZrB_2			1560
Titanium nitride	ZrN	9		1770
Zirconia (fused)	ZrO_2		11	
Tantalum carbide	TaC			1800
Tungsten carbide	WC			1880
Corundum (alumina)	Al_2O_3	9		2025
Zirconium carbide	ZrC			2150
Alumina (fused)	Al_2O_3		12	
Beryllium carbide	Be_2C			2400

HARDNESS OF MINERALS AND CERAMICS (continued)

Material	Formula	Mohs	Modified mohs	Knoop
Titanium carbide	TiC			2470
Carborundum (silicon carbide)	SiC	9.3	13	2500
Aluminum boride	AlB			2500
Tantalum boride	TaB_2			2600
Boron carbide	B_4C		14	2800
Boron	B	9.5		
Titanium boride	TiB_2			2850
Diamond	C	10	15	7000

PHYSICAL CONSTANTS OF CLEAR FUSED QUARTZ
Based on information contained in Fused Quartz Catalogue Q-7A General Electric Company

Property	Clear fused quartz
Density	2.2 g cm^{-3}
Hardness	4.9 (Mohs')
Tensile strength	7000 psi
Compressive strength	>160,000 psi
Bulk modulus	(approx.) 5.3×10^6 psi
Rigidity modulus	4.5×10^6 psi
Young's modulus	10.4×10^6 psi
Poisson's ratio	0.16
Coefficient of thermal expansion	(av.) 5.5×10^{-7} cm cm^{-1} °C^{-1} { 20°C 320°C
Thermal conductivity	0.0033 cal cm^{-1} s^{-1} °C^{-1} cm^{-1}
Specific heat	0.18 cal g^{-1}
Softening point	(approx.) 1665°C
Annealing point	(approx.) 1140°C
Strain point	1070°C
Electrical resistance	9.5 $\log_{10}$ R for cm^3 at 350°C
Dielectric constant	3.75 at 20°C. 1 MHz
Dielectric loss factor	less than 0.0004 at 20°C. 1 MHz
Dissipation factor	less than 0.0001 at 20°C. 1 MHz
Index of refraction	1.4585
Velocity of sound — shear wave	3.75×10^5 cm s^{-1}
Velocity of sound — compressional wave	5.90×10^5 cm s^{-1}
Sonic attenuation	less than 0.033 dB ft^{-1} MHz^{-1}

Section 13
Polymer Properties

NAMING ORGANIC POLYMERS

Robert B. Fox

REGULAR LINEAR POLYMERS

Both polymer science and its language have undergone extensive development in the last 20 years. Much of the effort in language has come from IUPAC Commission IV.1, on Macromolecular Nomenclature, founded in 1968. The following is an adaptation of the Commission's 1975 rules, "Nomenclature of Regular Single-Strand Organic Polymers", which are currently (1990) being updated and expanded. This nomenclature is also the basis for indexing polymers in Chemical Abstracts.

Traditionally, polymer names have been derived from their monomeric source: polystyrene is the polymer made from styrene, often indexed as "styrene, polymer of". Where the monomer name includes more than a single word, it should be enclosed in parentheses to avoid ambiguity: poly(bromostyrene) is the polymer made from bromostyrene (Br position unspecified) while polybromostyrene is styrene brominated at many unspecified positions. This method of naming polymers became increasingly inadequate as polymers became more complex. Logically, an organic polymer molecule ought to be named like an organic molecule, i. e. on the basis of its chemical structure; the rules of organic nomenclature (see Section 2) should be generally applicable.

The 1975 IUPAC rules are limited to regular single-strand polymer chains hypothetically free of branching, chain defects and impurities, and they do not cite stereoregularity. Monomer names are not involved. To the extent that a polymer chain can be fully described as a multiple of a repeating segment of the chain, the polymer could be named "poly(repeating unit)", where the repeating unit is an organic bivalent radical in most cases. However, different repeating units can often be written for a single chain. Thus, for a unique and unambiguous polymer name, it is necessary to define a starting point in the chain and a direction to move along the chain to the point of repetition. This repeating chain segment is called the Structural Repeating Unit or SRU (the 1975 rules use the term "constitutional repeating unit"). Once the SRU is defined, it can be named by the rules of organic nomenclature.

The steps to be taken, then, are: (1) write as much of the structure as necessary to include at least two repeating sequences; (2) select the SRU; (3) name the SRU. The name of the polymer is then poly(SRU).

SELECTION OF THE SRU

As a part of the main chain, every SRU is composed of one or more multivalent subunits such as a single atom (–O– or –N=), an atomic grouping (CH_2–CH_2–), or a ring system; subunits can have substituents. Polymers whose SRU consists of only one subunit are simply named poly(subunit) (or poly(SRU)). Any substituents in the SRU are given lowest locant numbers in the name as in poly(1-chloroethylene) for the structure $[-CHCl-CH_2-]_n$. Unsaturation is also given the lowest locant: $[-CH-CHCH_2CH_2-]_n$ is poly(1-butenylene). Free valences in the SRU are minimized: –CH=CH–, giving poly(vinylene), is preferred over =CH–CH=. Free valences in subunits composed of rings are given lowest locants consistent with fixed numbering as is used in heterocyclic rings.

SENIORITY AMONG SUBUNITS

Many polymer repeating units contain more than one subunit. In such structures, identification of the SRU depends upon a subunit seniority system. The SRU that is selected will begin with the most senior subunit and move in the direction of the closest (counting only atoms in the main chain, using the shortest path in rings) next senior subunit. For citation of the first subunit, the order of decreasing seniority is (1) heterocyclic rings, (2) chains containing hetero atoms, (3) carbocyclic rings, and, lowest, (4) chains containing only carbon atoms. This order is unaffected by the presence of substituents.

Within each major seniority group, there is a sub-order of seniority. Among heterocyclic rings, the descending order of seniority is (a) a ring system with nitrogen in the ring; (b) a ring system containing a hetero atom other than nitrogen as high as possible in the order of hetero atoms given below; (c) a ring system having the greatest number of rings; (d) a ring system with the largest individual ring; (e) a ring system with the largest number of hetero atoms; (f) a ring system with the greatest variety of hetero atoms; (g) the ring system with the greatest number of hetero atoms in the order of hetero atoms given below. When two ring systems are identical except for the locants for the hetero atoms, the one with lower locants is senior. When the systems are identical except for unsaturation, the senior subunit is that with the least hydrogenation. Among assemblies of identical rings, the ring of highest seniority is that with lowest numbers for the point of attachment between the rings. Further choice is based on the number and kind of substituents: (a) largest number of substituents; (b) substituents with lowest locants; (c) alphabetical order of substituents.

Among hetero atoms, the descending order of seniority is O, S, Se, N, P, As, Sb, Si, Ge, Sn, B, Hg with other atoms fitted in according to their place in the periodic table. Seniority among hetero atoms of the same kind falls to the one having (a) the largest number of substituents and (b) the substituent earliest in the alphabet.

Seniority among carbocycles is, in descending order, (a) the system with the largest number of rings; (b) the largest individual ring at the first point of difference; (c) the largest number of atoms common to the rings; (d) the lowest locant numbers at the first point of difference for ring junctions; and (e) the least hydrogenated ring. Further choice, based on substituents, is the same as with heterocyclic rings. Among acyclic chains, seniority falls to (a) the largest number of carbon atoms in the chain and (b) number and location of substituents.

DIRECTION OF PATH FOR CITATION OF SUBUNITS

To name the polymer, the SRU is written to read from left to right, with the most senior subunit at the left; the polymer name is simply the prefix poly, followed by citation of the subunits in the SRU in order as written. For example:

[– N –CH_2–O–]$_n$ Poly(3,5-pyridinediylmethyleneoxy-1,4-phenylene)

NAMING ORGANIC POLYMERS (continued)

Frequently, there will be a choice of path to be followed from the most senior subunit to the next senior subunit. Priority is given to the shortest path, and paths are unidirectional. Where the most senior subunit is repeated within the SRU, the path is the shortest one between them and thence to the next senior subunit. If the choices of paths between equal highest seniority subunits are the same length and different lower seniority subunits are traversed, the path selected is the one having the most senior traversed path. For example, if the descending order of seniority is ABCD, then the SRU –ABDADC– is preferred to –ADBACD– because B is closer to the first A in the former series.

SUBSTITUENTS

The largest possible subunits, based on main-chain atoms and rings, that have names approved in the organic rules are used: $–CO(CH_2)_6CO–$ is treated as a single subunit named adipoyl, as is $–(CH_3CH)–$, ethylidene, for example. Note, however, that more systematic names, such as hexanedioyl and 1,1-ethanediyl are gaining acceptance in the IUPAC rules. Substituents along the main chain other than those included in the name of a subunit are prefixed, using substitutive organic nomenclature, to the name of the subunit to which they are bound and the combination enclosed in parentheses. Lowest locants are used, consistent with other criteria for selecting the SRU.

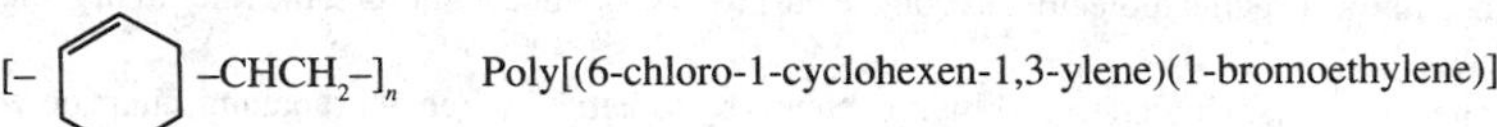

End groups are prefixed to the name of the polymer. They are denoted by α-, for the group attached to left side of the SRU, and ω- for the other group. The form is seen in a name such as α–(trichloromethyl)-ω-chloropoly(1,4-phenylenemethylene).

EXAMPLES

The reader is referred to the Table of Glass Transition Temperatures of Polymers for examples of both structure- and source-based polymer nomenclature. Additional examples as well as other details of the structure-based system can be found in the 1975 IUPAC Recommendations, *Pure Appl. Chem.*, 48, 373—385, 1976.

LINEAR COPOLYMERS

In the IUPAC document "Source-Based Nomenclature for Copolymers", Recommendations, 1985 (*Pure Appl. Chem.*, 57, 1427—1440, 1985), copolymer names take the form "poly(A-*connective*-B)", where A and B represent the names of the monomers (real or hypothetical) from which they are derived and sequence arrangements are denoted by the connectives *-co-* (unspecified), *-stat-* (statistical), *-ran-* (random), *-alt-* (alternating), *-per-* (periodic), *-block-* (block), and *-graft-* (graft). Examples are:

1. An unspecified copolymer of styrene and methyl methacrylate:

   Poly[styrene-*co*-(methyl methacrylate)]

2. A statistical copolymer (in which the sequential distribution of monomeric units obeys known statistical laws, as is the case with most simultaneous polymerizations of two or more monomers) of styrene, butadiene, and acrylonitrile:

   Poly(styrene-*stat*-butadiene-*stat*-acrylonitrile)

3. A random copolymer (a special case of statistical) of ethylene and vinyl acetate:

   Poly[ethylene-*ran*-(vinyl acetate)]

4. An alternating copolymer (the sequence of which is regular and therefore can be named on the basis of structure as well as source) of styrene and maleic anhydride:

   Poly[styrene-*alt*-(maleic anhydride)]

5. A periodic copolymer (one in which three or more monomeric units appear in ordered seqence) of formaldehyde and ethylene oxide:

   Poly[formaldehyde-*per*-(ethylene oxide)-*per*-(ethylene oxide)]

6. A block copolymer derived from a block of polystyrene and a block of polybutadiene (a) without specified junction unit; (b) joined through the bivalent radical $–(CH_3)_2Si–$; (c) with a repeated sequence of alternating blocks:

(a) Polystyrene-*block*-polybutadiene
(b) Polystyrene-*block*-dimethylsilylene-*block*-polybutadiene or Polystyrene—dimethylsilylene—polybutadiene
(c) Poly(polystyrene-*block*-polybutadiene)

6. A graft copolymer in which polystyrene is grafted to polyethylene:

Polyethylene-*graft*-polystyrene

Any of the connectives can be used in combination to describe more complex sequence arrangements; end groups may be denoted by α- and ω- as in structure-based nomenclature; and mass fractions, mole fractions, and degrees of polymerization may be specified in parentheses following the name of the copolymer. Further examples will be found in the 1985 Recommendations.

GLASS TRANSITION TEMPERATURE FOR SELECTED POLYMERS

Robert B. Fox

Polymer names are based on the IUPAC structure-based nomenclature system described in the table "Naming Organic Polymers". Within each category, names are listed in alphabetical order. Source-based and trivial names are also given (in italics) for the most common polymers. The table does not include polymers for which T_g is not clearly defined because of variability of structure or because of reactions taking place near the glass transition.

All values of T_g cited in this table have been determined by differential scanning calorimetry (DSC) except those values indicated by:

(D) dynamic method
(Dil) dilatometry
(M) mechanical method

Polymer name	Glass transition temperature (T_g/K)
ACYCLIC CARBON CHAINS	
Polyalkadienes	
Poly(alkenylene) *Polyalkadiene* –[CH=CHCH$_2$CH$_2$]–	
Poly(*cis*-1-butenylene) *cis-1,3-polybutadiene* [PBD]	171
Poly(*trans*-1-butenylene) *trans-1,3-polybutadiene* [PBD]	215
Poly(1-chloro-*cis*-1-butenylene) *cis-1,3-polychloroprene*	253
Poly(1-chloro-*trans*-1-butenylene) *trans-1,3-polychloroprene*	233
Poly(1-methyl-*cis*-1-butenylene) *cis-1,3-polyisoprene*	200
Poly(1-methyl-*trans*-1-butenylene) *trans-1,3-polyisoprene*	207
Poly(1,4,4-trifluoro-1-butenylene)	238
Polyalkenes	
Poly(alkylethylene) *Poly(alkylethylene)* -[RCHCH$_2$]-	
Poly(1-benzylethylene)	333
Poly(1-butylethylene)	223
Poly(1-cyclohexylethylene) (atactic)	393
Poly(1-cyclohexylethylene) (isotactic)	406 (D)
Poly(1,1-dimethylethylene) *Polyisobutylene* [PIB]	200
Poly(ethylene)	148
Poly(methylene)	155
Poly(1-phenethylethylene)	283
Poly(propylene) (isotactic)	272
Poly(propylene) (syndiotactic)	ca. 265
Poly[1-(2-pyridyl)ethylene]	377
Poly[1-(4-pyridyl)ethylene]	415
Poly(1-vinylethylene)	273
Polyacrylics	
Poly[1-(alkoxycarbonyl)ethylene] *Poly(alkyl acrylate)* –[(ROCO)CHCH$_2$]–	
Poly[1-(benzyloxycarbonyl)ethylene]	279
Poly[1-(butoxycarbonyl)ethylene] *Poly(butyl acrylate)* [PBA]	219 (M)

Polymer name	Glass transition temperature (T_g/K)
Poly[1-(*sec*-butoxycarbonyl)ethylene]	251
Poly[1-(butoxycarbonyl)-1-cyanoethylene]	358
Poly[1-(butylcarbamoyl)ethylene]	319 (M)
Poly(1-carbamoylethylene)	438
Polyacrylamide [PAM]	
Poly(1-carboxyethylene)	379
Poly(acrylic acid) [PAA]	
Poly[1-(2-chlorophenoxycarbonyl)ethylene]	326
Poly[1-(4-chlorophenoxycarbonyl)ethylene]	331
Poly[1-(4-cyanobenzyloxycarbonyl)ethylene]	317
Poly[1-(2-cyanoethoxycarbonyl)ethylene]	277
Poly[1-(cyanomethoxycarbonyl)ethylene)]	433 Dil
Poly[1-(4-cyanophenoxycarbonyl)ethylene]	363
Poly[1-(cyclohexyloxycarbonyl)ethylene]	292
Poly[1-(2,4-dichlorophenoxycarbonyl)ethylene]	333
Poly[1-(dimethylcarbamoyl)ethylene]	362
Poly[1-(ethoxycarbonyl)ethylene]	249
Poly(ethyl acrylate) [PEA]	
Poly[1-(ethoxycarbonyl)-1-fluoroethylene]	316
Poly[1-(2-ethoxycarbonylphenoxycarbonyl)ethylene]	303
Poly[1-(3-ethoxycarbonylphenoxycarbonyl)ethylene]	297
Poly[1-(4-ethoxycarbonylphenoxycarbonyl)ethylene]	310
Poly[1-(2-ethoxyethoxycarbonyl)ethylene]	223
Poly[1-(3-ethoxypropoxycarbonyl)ethylene]	218
Poly[1-(isopropoxycarbonyl)ethylene]	267-270
Poly[1-(methoxycarbonyl)ethylene]	283
Poly(methyl acrylate) [PMA]	
Poly[1-(2-methoxycarbonylphenoxycarbonyl)ethylene]	319
Poly[1-(3-methoxycarbonylphenoxycarbonyl)ethylene]	311
Poly[1-(4-methoxycarbonylphenoxycarbonyl)ethylene]	340
Poly[1-(2-methoxyethoxycarbonyl)ethylene]	223
Poly[1-(4-methoxyphenoxycarbonyl)ethylene]	324
Poly[1-(3-methoxypropoxycarbonyl)ethylene]	198
Poly[1-(2-naphthyloxycarbonyl)ethylene]	358
Poly[1-(pentachlorophenoxycarbonyl)ethylene]	420
Poly[1-(phenethoxycarbonyl)ethylene]	270
Poly[1-(phenoxycarbonyl)ethylene]	330
Poly[1-(*m*-tolyloxycarbonyl)ethylene]	298
Poly[1-(*o*-tolyloxycarbonyl)ethylene]	325
Poly[1-(*p*-tolyloxycarbonyl)ethylene]	316
Poly[1-(2,2,2-trifluorethoxycarbonyl)ethylene]	263
Polymethacrylics	
Poly[1-(alkoxycarbonyl)-1-methylethylene] *Poly(alkyl methacrylate)* –[(ROCO)(Me)CCH_2]–	
Poly[1-(benzyloxycarbonyl)-1-methylethylene]	327
Poly[1-(2-bromoethoxycarbonyl)-1-methylethylene]	325
Poly[1-(butoxycarbonyl)-1-methylethylene]	293
Poly(butyl methacrylate) [PBMA]	
Poly[1-(*sec*-butoxycarbonyl)-1-methylethylene]	333
Poly[1-(*tert*-butoxycarbonyl)-1-methylethylene)]	391
Poly[1-(2-chloroethoxycarbonyl)-1-methylethylene]	ca 315
Poly[1-(2-cyanoethoxycarbonyl)-1-methylethylene]	364
Poly[1-(4-cyanophenoxycarbonyl)-1-methylethylene]	428
Poly[1-(cyclohexyloxycarbonyl)-1-methylethylene] (atactic)	356
Poly[1-(cyclohexyloxycarbonyl)-1-methylethylene)] (isotactic)	324

Polymer name	Glass transition temperature (T_g/K)
Poly[1-(dimethylaminoethoxycarbonyl)-1-methylethylene]	292
Poly[1-(ethoxycarbonyl)-1-ethylethylene]	300
Poly[1-(ethoxycarbonyl)-1-methylethylene] (atactic) *Poly(ethyl methacrylate)* [PEMA]	338
Poly[1-(ethoxycarbonyl)-1-methylethylene] (isotactic)	285
Poly[1-(ethoxycarbonyl)-1-methylethylene)] (syndiotactic)	339
Poly[1-(hexyloxycarbonyl)-1-methylethylene]	268
Poly[1-(isobutoxycarbonyl)-1-methylethylene]	326
Poly[1-(isopropoxycarbonyl)-1-methylethylene]	354
Poly[1-(methoxycarbonyl)-1-methylethylene] (atactic) *Poly(methyl methacrylate)* [PMMA]	378
Poly[1-(methoxycarbonyl)-1-methylethylene)] (isotactic)	311
Poly[1-(methoxycarbonyl)-1-methylethylene)] (syndiotactic)	378
Poly[1-(4-methoxycarbonylphenoxy)-1-methylethylene]	379
Poly[1-(methoxycarbonyl)-1-phenylethylene)] (atactic)	391
Poly[1-(methoxycarbonyl)-1-phenylethylene)] (isotactic)	397
Poly[1-methyl-1-(phenethoxycarbonyl)ethylene]	299
Poly[1-methyl-1-(phenoxycarbonyl)ethylene]	383
Polyvinyl ethers, alcohols, and ketones	
Poly(1-alkoxyethylene) *Poly(alkyl vinyl ether)* –[$ROCHCH_2$]–	
Poly(1-hydroxyethylene) *Poly(vinyl alcohol)* –[$HOCHCH_2$]–	
Poly(1-alkanoylethylene) *Poly(alkyl vinyl ketone)* –[$RCOCHCH_2$]–	
Poly(1-butoxyethylene)	218
Poly(1-*sec*-butoxyethylene)	253
Poly(1-*tert*-butoxyethylene)	361
Poly[1-(butylthio)ethylene]	253
Poly(1-ethoxyethylene)	230
Poly[1-(4-ethylbenzoyl)ethylene]	325
Poly(1-hydroxyethylene) *Poly(vinyl alcohol)* [PVA]	358 (D)
Poly(hydroxymethylene)	407
Poly(1-isopropoxyethylene)	270
Poly[1-(4-methoxybenzoyl)ethylene]	319 (M)
Poly(1-methoxyethylene) *Poly(methyl vinyl ether)* [PMVE]	242
Poly[1-(methylthio)ethylene]	272
Poly(1-propoxyethylene)	224
Poly[1-(trifluoromethoxy)trifluoroethylene]	268
Polyvinyl halides and nitriles	
Poly(1-haloethylene) *Poly(vinyl halide)* –[$XCHCH_2$]–	
Poly(1-cyanoethylene) *Poly(acrylonitrile)* –[$NCCHCH_2$]–	
Poly(1-chloroethylene) *Poly(vinyl chloride)* [PVC]	354
Poly(chlorotrifluoroethylene)	373
Poly(1-cyanoethylene) *Polyacrylonitrile* [PAN]	370
Poly(1-cyano-1-methylethylene) *Polymethacrylonitrile*	393
Poly(1,1-dichloroethylene) *Poly(vinylidene chloride)*	255
Poly(1,1-difluoroethylene) *Poly(vinylidene fluoride)*	ca 233
Poly(1-fluoroethylene) *Poly(vinyl fluoride)*	314 (M)

Polymer name	Glass transition temperature (T_g/K)
Poly(1-hexafluoropropylene)	425
Poly[1-(2-iodoethyl)ethylene]	343
Poly(tetrafluoroethylene)	(160)
Poly[1-(trifluoromethyl)ethylene]	300
Polyvinyl esters	
Poly[1-(alkanoyloxy)ethylene] *Poly(vinyl alkanoate)* –[RCOOCHCH$_2$]–	
Poly(1-acetoxyethylene) *Poly(vinyl acetate)* [PVAc]	305
Poly[1-(benzoyloxy)ethylene]	344
Poly[1-(4-bromobenzoyloxy)ethylene]	365
Poly[1-(2-chlorobenzoyloxy)ethylene]	335
Poly[1-(3-chlorobenzoyloxy)ethylene]	338
Poly[1-(4-chlorobenzoyloxy)ethylene]	357
Poly[1-(cyclohexanoyloxy)ethylene]	349 (M)
Poly[1-(4-ethoxybenzoyloxy)ethylene]	343
Poly[1-(4-ethylbenzoyloxy)ethylene]	326
Poly[1-(4-isopropylbenzoyloxy)ethylene]	342
Poly[1-(2-methoxybenzoyloxy)ethylene]	338
Poly[1-(3-methoxybenzoyloxy)ethylene]	ca 317
Poly[1-(4-methoxybenzoyloxy)ethylene]	360
Poly[1-(4-methylbenzoyloxy)ethylene]	343
Poly[1-(4-nitrobenzoyloxy)ethylene]	395
Poly[1-(propionoyloxy)ethylene]	283 (M)
Polystyrenes	
Poly(1-phenylethylene) *Polystyrene* –[C$_6$H$_5$CHCH$_2$]–	
Poly[1-(4-acetylphenyl)ethylene]	389 (M)
Poly[1-(4-benzoylphenyl)ethylene]	371 (M)
Poly[1-(4-bromophenyl)ethylene]	391
Poly[1-(4-butoxyphenyl)ethylene]	ca 320 (M)
Poly[1-(4-butoxycarbonylphenyl)ethylene]	349 (M)
Pol[(1-(4-butylphenyl)ethylene]	279
Poly[1-(4-carboxyphenyl)ethylene]	386 (M)
Poly[1-(2-chlorophenyl)ethylene]	392
Poly[1-(3-chlorophenyl)ethylene]	363
Poly[1-(4-chlorophenyl)ethylene]	383
Poly[1-(2,4-dichlorophenyl)ethylene]	406
Poly[1-(2,5-dichlorophenyl)ethylene]	379
Poly[1-(2,6-dichlorophenyl)ethylene]	440
Poly[1-(3,4-dichlorophenyl)ethylene]	401
Poly[1-(2,4-dimethylphenyl)ethylene]	385
Poly[1-(4-(dimethylamino)phenyl)ethylene]	398 (M)
Poly[1-(4-ethoxyphenyl)ethylene]	ca 359 (M)
Poly[1-(4-ethoxycarbonylphenyl)ethylene]	367 (M)
Poly[1-(4-fluorophenyl)ethylene]	[illegible]
Poly[1-(4-iodophenyl)ethylene]	429
Poly[1-(4-methoxyphenyl)ethylene]	386
Poly[1-(4-methoxycarbonylphenyl)ethylene]	386 (M)
Poly(1-methyl-1-phenylethylene) *Poly(α-methylstyrene)*	373
Poly[1-(2-(methylamino)phenyl)ethylene]	462 (M)
Poly(1-phenylethylene) *Polystyrene* [PS]	373

Polymer name	Glass transition temperature (T_g/K)
Poly[1-(4-propoxyphenyl)ethylene]	343 (M)
Poly[1-(4-propoxycarbonylphenyl)ethylene]	365 (M)
Poly(1-*o*-tolylethylene)	409
CHAINS WITH CARBOCYCLIC UNITS	
Poly(arylenealkylene) –[–Ar–$(CH_2)_n$]–	
Poly[1-(2-bromo-1,4-phenylene)ethylene]	353 (M)
Poly[1-(2-chloro-1,4-phenylene)ethylene]	343 (M)
Poly[1-(2-cyano-1,4-phenylene)ethylene]	363 (M)
Poly[1-(2,5-dimethyl-1,4-phenylene)ethylene]	373 (M)
Poly[1-(2-ethyl-1,4-phenylene)ethylene]	298 (M)
Poly[1-(1,4-naphthylene)ethylene]	433 (M)
Poly[1-(1,4-phenylene)ethylene]	ca 353 (M)
CHAINS WITH HETEROATOM UNITS	
Main chain oxide units	
Poly(oxyalkylene) *Poly(alkylene oxide)* –[O$(CH_2)_n$]–	
Poly[oxy(1,1-bis(chloromethyl)trimethylene)]	265
Poly[oxy(1-(bromomethyl)ethylene)]	259
Poly[oxy(1-(butoxymethyl)ethylene)]	194
Poly[oxy(1-butylethylene)]	203
Poly[oxy(1-*tert*-butylethylene)]	308
Poly[oxy(1-(chloromethyl)ethylene)] *Poly(epichlorohydrin)*	251
Poly[oxy(2,6-dimethoxy-1,4-phenylene)]	440
Poly[oxy(1,1-dimethylethylene)]	264
Poly[oxy(2,6-dimethyl-1,4-phenylene)]	482
Poly[oxy(2,6-diphenyl-1,4-phenylene)]	493
Poly[oxy(1-ethylethylene)]	203
Poly(oxyethylidene) *Polyacetaldehyde*	243
Poly[oxy(1-(methoxymethyl)ethylene)]	211
Poly[oxy(2-methyl-6-phenyl-1,4-phenylene)]	428
Poly[oxy(1-methyltrimethylene)]	223 (D)
Poly[oxy(2-methyltrimethylene)]	218
Poly(oxy-1,4-phenylene) *Poly(phenylene oxide)* [PPO]	358
Poly[oxy(1-phenylethylene)]	313
Poly(oxytetramethylene) *Poly(tetrahydrofuran)* [PTMO]	189
Poly(oxytrimethylene)	195
Main-chain ester or anhydride units	
Poly(oxyalkyleneoxyalkanedioyl) *Poly(alkylene alkanedioate)*–[O$(CH_2)_m$OCO$(CH_2)_n$CO]–	
Poly(oxyadipoyloxydecamethylene)	217
Poly(oxyadipoyloxy-1,4-phenyleneisopropylidene-1,4-phenylene)	341
Poly(oxycarbonyloxy-1,4-phenylene-isopropylidene-1,4-phenylene) *Bisphenol A polycarbonate*	422
Poly(oxycarbonylpentamethylene)	213
Poly(oxycarbonyl-1,4-phenylenemethylene-1,4-phenylene)	395
Poly(oxycarbonyl-1,4-phenyleneisopropylidene-1,4-phenylene)	333

Polymer name	Glass transition temperature (T_g/K)
Poly[oxy(2,6-dimethyl-1,4-phenyleneisopropylidene-3,5-dimethyl-1,4-phenylene)oxysebacoyl]	318
Poly(oxyethylenecarbonyl-1,4-cyclohexylenecarbonyl) (trans)	291
Poly(oxyethyleneoxycarbonyl-1,4-naphthylenecarbonyl)	337
Poly(oxyethyleneoxycarbonyl-1,5-naphthylenecarbonyl)	344
Poly(oxyethyleneoxycarbonyl-2,6-naphthylenecarbonyl)	386
Poly(oxyethyleneoxycarbonyl-2,7-naphthylenecarbonyl)	392
Poly(oxyethyleneoxyterephthaloyl) *Poly(ethylene terephthalate)* [PET]	342
Poly(oxyisophthaloyl)	403 (D)
Poly(oxy(1-oxo-2,2-dimethyltrimethylene)) *Poly(pivalolactone)*	263
Poly(oxy-1,4-phenyleneisopropylidene-1,4-phenyleneoxysebacoyl)	280
Poly(oxy-1,4-phenyleneoxy-1,4-phenyleneoxy-carbonyl-1,4-phenylene) [PEEK]	416
Poly(oxypropyleneoxyterephthaloyl)	341
Poly[oxyterephthaloyloxy(2,6-dimethyl-1,4-phenyleneisopropylidene-3,5-dimethyl-1,4-(D)phenylene)]	498
Poly(oxyterephthaloyloxyoctamethylene)	318 (D)
Poly(oxyterephthaloyloxy-1,4-phenyleneisopropylidene-1,4-phenylene) *Poly(bisphenol A terephthalate)*	478
Poly(oxytetramethyleneoxyterephthaloyl) *Poly(butylene terephthalate)* [PBT]	323
Main-chain amide units	
Poly(iminoalkyleneiminoalkanedioyl) *Poly(alkylene alkanediamide)*–[$NH(CH_2)_mNHCO(CH_2)_nCO$]–	
Poly(iminoadipoyliminodecamethylene) *Nylon 10,6*	313
Poly(iminoadipoyliminohexamethylene) *Nylon 6,6*	ca 323
Poly(iminoadipoyliminooctamethylene) *Nylon 8,6*	318
Poly[iminoadipoyliminotrimethylene(methylimino)trimethylene]	278
Poly(iminocarbonyl-1,4-cyclohexylenemethylene)	466
Poly[iminocarbonyl-1,4-phenylene(2-oxoethylene)iminohexamethylene]	377
Poly(iminoethylene-1,4-phenyleneethyleneiminosebacoyl)	378 (D)
Poly(iminohexamethyleneiminoazelaoyl) *Nylon 6,9*	331
Poly(iminohexamethyleneiminododecanedioyl) *Nylon 6, 12*	319
Poly(iminohexamethyleneiminopimeloyl) *Nylon 6,7*	331
Poly(iminohexamethyleneiminosebacoyl) *Nylon 6,10*	323
Poly(iminohexamethyleneiminosuberoyl) *Nylon 6,8*	330
Poly(iminoisophthaloylimino-4,4′-biphenylylene)	558
Poly(iminoisophthaloyliminohexamethylene)	390
Poly(iminoisophthaloyliminomethylene-1,4-cyclohexylenemethylene)	481
Poly(iminoisophthaloyliminomethylene-1,3-phenylenemethylene)	438 (M)
Poly[iminomethylene(2,5-dimethyl-1,4-phenylene)methyleneiminosuberoyl]	351
Poly(imino-1,5-naphthyleneiminoisophthaloyl)	598
Poly(imino-1,5-naphthyleneiminoterephthaloyl)	578
Poly(iminooctamethyleneiminodecanedioyl) *Nylon 8,10*	333
Poly(iminooxalyliminohexamethylene) *Nylon 6,2*	430
Poly[imino(1-oxohexamethylene)] *Nylon 6*	326

Polymer name	Glass transition temperature (T_g/K)
Poly[imino(1-oxodecamethylene)]	315
Nylon 10	
Poly[imino(1-oxoheptamethylene)]	325
Nylon 7	
Poly[imino(1-oxo-3-methyltrimethylene]	369
Poly[imino(1-oxononamethylene)]	319
Nylon 9	
Poly[imino(1-oxooctamethylene)]	323
Nylon 8	
Poly[imino(1-oxotrimethylene)]	384
Nylon 3	
Poly(iminopentamethyleneiminoadipoyl)	318
Nylon 5,6	
Poly[iminopentamethyleneiminocarbonyl-1,4-phenylene(2-oxoethylene)]	376
Poly(imino-1,3-phenyleneiminoisophthaloyl)	553 (M)
Poly(imino-1,4-phenyleneiminoterephthaloyl)	618
Poly(iminopimeloyliminoheptamethylene)	328
Nylon 7,7	
Poly(iminoterephthaloylimino-4,4′-biphenylylene)	613
Poly(iminotetramethyleneiminoadipoyl)	316
Nylon 4,6	
Poly[iminotetramethyleneiminocarbonyl-1,4-phenylene(2-oxoethylene)]	357
Poly(iminotrimethyleneiminoadipoyliminotrimethylene)	307
Poly[iminotrimethyleneiminocarbonyl-1,4-phenylene(2-oxoethylene)]	382
Poly(oxy-1,4-phenyleneiminoterephthaloyl-imino-1,4-phenylene)	613
Poly(sulfonylimino-1,4-phenyleneiminoadipoylimino-1,4-phenylene)	467
Main-chain urethane units	
Poly(oxyalkyleneoxycarbonyliminoalkyleneiminocarbonyl)–[O(CH_2)$_m$OCONH(CH_2)$_n$NHCO]–	
Poly(oxyethyleneoxycarbonyliminohexamethyleneiminocarbonyl)	329
Poly[oxyethyleneoxycarbonylimino(6-methyl-1,3-phenylene)iminocarbonyl]	325
Poly(oxyethyleneoxycarbonylimino-1,4-phenylenemethylene-1,4-phenyleneiminocarbonyl)	412
Poly(oxyhexamethyleneoxycarbonyliminohexamethyleneiminocarbonyl)	332
Poly[oxyhexamethyleneoxycarbonylimino(6-methyl-1,3-phenylene)iminocarbonyl]	305
Poly(oxyhexamethyleneoxycarbonylimino-1,4-phenylenemethylene-1,4-phenyleneiminocarbonyl)	364
Poly(oxyoctamethyleneoxycarbonyliminohexamethyleneiminocarbonyl)	331
Poly[oxyoctamethyleneoxycarbonylimino(6-methyl-1,3-phenylene)iminocarbonyl]	337
Poly(oxyoctamethyleneoxycarbonylimino-1,4-phenylenemethylene-1,4-phenyleneiminocarbonyl)	352
Poly(oxytetramethyleneoxycarbonyliminohexamethyleneiminocarbonyl)	332
Poly[oxytetramethyleneoxycarbonylimino(6-methyl-1,3-phenylene)iminocarbonyl]	315
Poly(oxytetramethyleneoxycarbonylimino-1,4-phenylenemethylene-1,4-phenyleneiminocarbonyl)	382
Main-chain siloxanes	
Poly[oxy(dialkylsilylene)] *Poly(dialkylsiloxane)* –[O(R_2Si)]–	
Poly[oxy(dimethylsilylene)]	148
Poly(dimethylsiloxane) [PDMS]	
Poly[oxy(dimethylsilylene)oxy-1,4-phenylene]	363 (M)
Poly[oxy(dimethylsilylene)oxy-1,4-phenyleneisopropylidene-1,4-phenylene]	318 (M)
Poly[oxy(diphenylsilylene)]	238
Poly(diphenylsiloxane)	
Poly[oxy(diphenylsilylene)-1,3-phenylene]	ca 331
Poly[oxy((methyl)phenylsilylene)]	187
Poly[oxy((methyl)-3,3,3-trifluoropropylsilylene]	<193

Polymer name	Glass transition temperature (T_g/K)
Main-chain sulfur-containing units	
Poly(dithioethylene)	223
Poly(dithiomethylene-1,4-phenylenemethylene)	296
Poly(oxy-4,4′-biphenylylene-1,4-phenylenesulfonyl-1,4-phenylene)	503 (M)
Poly(oxycarbonyloxy-1,4-phenylenethio-1,4-phenylene)	ca 383
Poly(oxyethylenedithioethylene)	220 (M)
Poly[oxy(2-hydroxytrimethylene)oxy-1,4-phenylenesulfonyl-1,4-phenylene]	428
Poly(oxymethyleneoxyethylenedithioethylene)	214
Poly(oxy-1,4-phenylenesulfinyl-1,4-phenyleneoxy-1,4-phenylenecarbonyl-1,4-phenylene)	478 (M)
Poly(oxy-1,4-phenylenesulfinyl-1,4-phenyleneoxy-1,4-phenyleneisopropylidene-1,4-phenylene)	438 (M)
Poly(oxy-1,4-phenylenesulfonyl-1,4-phenylene)	487
Poly(oxy-1,4-phenylenesulfonyl-4,4′-biphenylylenesulfonyl-1,4-phenylene)	533
Poly[oxy-1,4-phenylenesulfonyl-1,4-phenyleneoxy(2,6-dimethyl-1,4-phenylene)isopropylidene (3,5-dimethyl-1,4-phenylene)]	508 (M)
Poly(oxy-1,4-phenylenesulfonyl-1,4-phenyleneoxy-1,4-phenylenecarbonyl-1,4-phenylene)	478 (M)
Poly[oxy-1,4-phenylenesulfonyl-1,4-phenyleneoxy-1,4-phenylene(hexafluoroisopropylidene)1,4-phenylene]	478 (M)
Poly(oxy-1,4-phenylenesulfonyl-1,4-phenyleneoxy-1,4-phenyleneisopropylidene-1,4-phenylene)	449
Poly(oxy-1,4-phenylenesulfonyl-1,4-phenyleneoxy-1.4-phenylenemethylene-1,4-phenylene)	453 (M)
Poly(oxy-1,4-phenylenesulfonyl-1,4-phenyleneoxy-1.4-phenylenethio-1,4-phenylene)	448 (M)
Poly(oxy-1,4-phenylenesulfonyl-1,4-phenyleneoxyterephthaloyl)	522
Poly(oxytetramethylenedithiotetramethylene)	197
Poly(sulfonyl-1,2-cyclohexylene)	401
Poly(sulfonyl-1,3-cyclohexylene)	381
Poly(sulfonyl-1,4-phenylenemethylene-1,4-phenylene)	497
Poly(thio-1,3-cyclohexylene)	221
Poly[thio(difluoromethylene)]	155
Poly(thioethylene)	223
Poly[thio(1-ethylethylene]	218
Poly[thio(1-methyl-3-oxotrimethylene)]	285
Poly[thio(1-methyltrimethylene)]	214
Pol[(thio(1-oxohexamethylene)]	292
Poly(thio-1,4-phenylene)	370
Poly(thiopropylene)	226
Main-chain heterocyclic units	
Poly(1,3-dioxa-4,6-cyclohexylenemethylene) *Poly(vinyl formal)*	378
Poly[(2,6-dioxopiperidine-1,4-diyl)trimethylene]	363
Poly[(2-methyl-1,3-dioxa-4,6-cyclohexylene)methylene] *Poly(vinyl acetal)*	355
Poly(1,4-piperazinediylcarbonyloxyethyleneoxycarbonyl)	333
Poly(1,4-piperazinediylisophthaloyl)	465 (M)
Poly[(2-propyl-1,3-dioxa-4,6-cyclohexylene)methylene] *Poly(vinyl butyral)*	322
Poly(3,6-pyridazinediyloxy-1,4-phenyleneisopropylidene-1,4-phenyleneoxy)	453 (M)
Poly(2,5-pyridinediylcarbonyliminohexamethyleneiminocarbonyl)	322

DIELECTRIC CONSTANTS OF SOME PLASTICS AND RUBBERS

Name	°C	Frequency (hertz) 1×10^3	1×10^6	1×10^8
Plastics				
Phenol-formaldehyde	25—27	5.15—8.61	4.45—5.05	4.1—4.5
	57	6.35	4.90	4.5
	88	8.5	5.2	4.7
Phenol-aniline-formaldehyde	25	4.50	4.31	4.11
	79	4.75	4.51	4.35
Melamine-formaldehyde	24—28	6.0—6.90	5.82—6.20	5.5—5.55
	57	6.95	5.40	4.90
	88	11.8	6.0	5.5
Urea-formaldehyde	24	6.7	6.0	5.2
	80	7.8	6.8	—
Polyamide resins				
Nylon 66	25	3.75	3.33	3.16
Nylon 610	25	3.50	3.14	3.0
	84	11.2	4.4	3.4
Cellulose acetate	26	3.50—4.48	3.28—3.90	3.05—3.40
Cellulose nitrate	27	8.4	6.6	5.2
	78	7.5	6.2	5.2
Methyl cellulose	22	6.8	5.7	4.3
Ethyl cellulose	25	3.09	3.01	2.90
Silicone resins	25	3.79—3.91	3.79—3.82	3.82
Polyethylene	−12	2.37	2.35	2.33
	23	2.26	2.26	2.26
Polyisobutylene	25	2.23	2.23	2.23
Vinylite QYNA	20	3.10	2.88	2.85
	76	3.83	3.0	2.8
	110	8.6		
Vinylite 5544	25	7.20	4.13	3.05
Vinylite 5901	25	5.5	3.4	3.0
Vinylite VU	24	5.65	3.30	2.80
	79	8.15	5.5	3.4
Vinylite VYHW	20	3.12	2.91	2.83
Vinylite VYNW	20	3.15	2.90	2.8
Polyvinyl chloride	25	4.55 (1×10^4)	3.3	—

Name	°C	Frequency (hertz) 1×10^3	1×10^6	1×10^8
Polyvinylidene and vinyl chloride	23	4.65	3.18	2.82
	84	4.94	4.40	3.2
Polychlorotrifluoroethylene	25	2.76	2.48	2.36
Polytetrafluoroethylene (Teflon)	22	2.1	2.1	2.1
	100	2.04	2.04	—
Polyvinylalcohol acetate	25	7.8	5.2	—
	85	100	10	—
Polyvinylacetals	26—27	3.02—3.12	2.86—2.92	2.67
	88	3.5	3.1	2.85
Polyacrylates				
Lucite	−12	2.9	2.63	2.50
	23	2.84	2.63	2.58
	81	3.45	2.72	2.59
Plexiglas	27	3.12	2.76	—
Polystyrene	25	2.54—2.56	2.54—2.56	2.55
	80	2.54	2.54	2.54
Styrene copolymers	25	2.55—2.95	2.55—2.80	2.55—2.77
Polyesters	25	3.22—4.3	3.12—4.0	2.94—2.98
Alkyd resins				
Alkyd isocyanate foam	25	1.223	1.218	1.20
Plaskon, clay filled	25	5.26	4.92	4.77
Plaskon, glass filled	25	5.04	4.73	4.50
Epoxy resins	25	3.63—3.67	3.52—3.62	3.32—3.35
Rubbers				
Hevea, vulcanized	27	2.94	2.74	2.42
Hevea compound	27	36	9	6.8
Gutta percha	25	2.60	2.53	2.47
Balata	25	2.50	2.50	2.42
Buna S	20	2.66	2.56	2.52
Butyl rubber compound	25	2.42	2.40	2.39
Neoprene	24	6.60	6.26	4.5
Silicon rubber	25	3.12—3.30	3.10—3.20	3.06—3.18

Section 14
Geophysics, Astronomy, and Acoustics

ASTRONOMICAL CONSTANTS

Victor Abalakin

The constants in this table are based primarilarly on the set of constants adopted by the International Astronomical Union (IAU) in 1976. Updates have been made when new data were available. All values are given in SI Units; thus masses are expressed in kilograms and distances in meters.

The astronomical unit of time is a time interval of one day (1 d) equal to 86400 s. An interval of 36525 d is one Julian century (1 cy).

REFERENCES

1. Seidelmann, P. K., *Explanatory Supplement to the Astronomical Almanac*, University Science Books, Mill Valley, CA, 1990.
2. Lang, K. R., *Astrophysical Data: Planets and Stars*, Springer-Verlag, New York, 1992.

Defining constants

Gaussian gravitational constant	$k = 0.01720209895$ m^3 kg^{-1} s^{-2}
Speed of light	$c = 299792458$ m s^{-1}

Primary constants

Light-time for unit distance (1 AU)	$\tau_A = 499.004782$ s
Equatorial radius of earth	$a_e = 6378140$ m
Equatorial radius of earth (IUGG value)	$a_e = 6378136$ m
Dynamical form-factor for earth	$J_2 = 0.001082626$
Geocentric gravitational constant	$GE = 3.986005 \times 10^{14}$ m^{3}s^{-2}
Constant of gravitation	$G = 6.672 \times 10^{-11}$ m^{3}kg^{-1}s^{-2}
Ratio of mass of moon to that of earth	$\mu = 0.01230002$
	$1/\mu = 81300.587$
General precession in longitude, per Julian century, at standard epoch J2000	$\rho = 5029''.0966$
Obliquity of the ecliptic at standard epoch J2000	$\varepsilon = 23°26'21''.448$

Derived constants

Constant of nutation at standard epoch J2000	$N = 9''.2025$
Unit distance (AU = $c\tau_A$)	AU $= 1.49597870 \times 10^{11}$ m
Solar parallax ($\pi_0 = \arcsin(a_e/\mathrm{AU})$)	$\pi_0 = 8''.794148$
Constant of aberration for standard epoch J2000	$\kappa = 20''.49552$
Flattening factor for the earth	$f = 1/298.257 = 0.00335281$
Heliocentric gravitational constant ($GS = A^3k^2/D^2$)	$GS = 1.32712438 \times 10^{20}$ m^3 s^{-2}
Ratio of mass of sun to that of the earth (S/E) = $(GS)/(GE)$)	$S/E = 332946.0$
Ratio of mass of sun to that of earth + moon	$(S/E)/(1 + \mu) = 328900.5$
Mass of the sun ($S = (GS)/G$)	$S = 1.9891 \times 10^{30}$ kg

Ratios of mass of sun to masses of the planets

Mercury	6023600
Venus	408523.5
Earth + moon	328900.5
Mars	3098710
Jupiter	1047.355
Saturn	3498.5
Uranus	22869
Neptune	19314
Pluto	3000000

PROPERTIES OF THE SOLAR SYSTEM

The following tables give various properties of the planets and characteristics of their orbits in the solar system. Certain properties of the sun and of the earth's moon are also included.

Explanations of the column headings:

- *Den.*: mean density in g/cm^3
- *Radius*: radius at the equator in km
- *Flattening*: degree of oblateness, defined as $(r_e - r_p)/r_e$, where r_e and r_p are the equatorial and polar radii, respectively
- *Potential coefficients*: coefficients in the spherical harmonic representation of the gravitational potential U by the equation

$$U(r,\phi) = (GM/r)\,[1 - \Sigma J_n(a/r)^n P_n(\sin\phi)]$$

where G is the gravitational constant, r the distance from the center of the planet, a the radius of the planet, M the mass, ϕ the latitude, and P_n the Legendre polynomial of degree n.
- *Gravity*: acceleration due to gravity at the surface
- *Escape velocity*: velocity needed at the surface of the planet to escape the gravitational pull
- *Dist. to sun*: semi-major axis of the elliptical orbit (1 AU = 1.496×10^8 km)
- ε: eccentricity of the orbit
- *Ecliptic angle*: angle between the planetary orbit and the plane of the earth's orbit around the sun
- *Inclin.*: angle between the equatorial plane and the plane of the planetary orbit
- *Rot. period*: period of rotation of the planet measured in earth days
- *Albedo*: ratio of the light reflected from the planet to the light incident on it
- T_{sur}: mean temperature at the surface
- P_{sur}: pressure of the atmosphere at the surface

The following general information on the solar system is of interest:

Mass of the earth = $M_e = 5.9742 \times 10^{24}$ kg
Total mass of planetary system = 2.669×10^{27} kg = 447 M_e
Total angular momentum of planetary system = 3.148×10^{43} kg m^2/s
Total kinetic energy of the planets = 1.99×10^{35} J
Total rotational energy of planets = 0.7×10^{35} J

Properties of the sun:

Mass = 1.9891×10^{30} kg = 332946.0 M_e
Radius = 6.9599×10^8 m
Surface area = 6.087×10^{18} m^2
Volume = 1.412×10^{27} m^3
Mean density = 1.409 g/cm^3
Gravity at surface = 27398 cm/s^2
Escape velocity at surface = 6.177×10^5 m/s
Effective temperature = 5780 K
Total radiant power emitted (luminosity) = 3.86×10^{26} W
Surface flux of radiant energy = 6.340×10^7 W/m^2
Flux of radiant energy at the earth (Solar Constant) = 1373 W/m^2

REFERENCES

1. Seidelmann, P. K., Editor, *Explanatory Supplement to the Astronomical Almanac*, University Science Books, Mill Valley, CA, 1992.
2. Lang, K. R., *Astrophysical Data: Planets and Stars*, Springer-Verlag, New York, 1992.
3. Allen, C. W., *Astrophysical Quantities, Third Edition*, Athlone Press, London, 1977.

PROPERTIES OF THE SOLAR SYSTEM (continued)

Planet	Mass 10^{24} kg	Den. g/cm^3	Radius km	Flattening	Potential coeffients $10^3 J_2$	$10^6 J_3$	$10^6 J_4$	Gravity cm/s^2	Escape vel. km/s
Mercury	0.33022	5.43	2439.7	0				370	4.25
Venus	4.8690	5.24	6051.9	0	0.027			887	10.4
Earth	5.9742	5.515	6378.140	0.00335364	1.08263	-2.54	-1.61	980	11.2
(Moon)	0.073483	3.34	1738	0	0.2027			162	2.37
Mars	0.64191	3.94	3397	0.00647630	1.964	36		371	5.02
Jupiter	1898.8	1.33	71492	0.0648744	14.75	-580		2312	59.6
Saturn	568.50	0.70	60268	0.0979624	16.45	-1000		896	35.5
Uranus	86.625	1.30	25559	0.0229273	12			777	21.3
Neptune	102.78	1.76	24764	0.0171	4			1100	23.3
Pluto	0.015	1.1	1151	0				72	1.1

Planet	Dist. to sun AU	ε	Ecliptic angle	Inclin.	Rot. period d	Albedo	No. of satellites
Mercury	0.38710	0.2056	7.00°	0°	58.6462	0.106	0
Venus	0.72333	0.0068	3.39°	177.3°	-243.01	0.65	0
Earth	1.00000	0.0167		23.45°	0.99726968	0.367	1
(Moon)				6.68°	27.321661	0.12	
Mars	1.52369	0.0933	1.85°	25.19°	1.02595675	0.150	2
Jupiter	5.20283	0.048	1.31°	3.12°	0.41354	0.52	16
Saturn	9.53876	0.056	2.49°	26.73°	0.4375	0.47	18
Uranus	19.19139	0.046	0.77°	97.86°	-0.65	0.51	15
Neptune	30.06107	0.010	1.77°	29.56°	0.768	0.41	8
Pluto	39.52940	0.248	17.15°	118°	-6.3867	0.3	1

Planet	T_{sur} K	P_{sur} bar	Atmospheric composition CO_2	N_2	O_2	H_2O	H_2	He	Ar	Ne	CO
Mercury	440	2×10^{-15}					2%	98%			
Venus	730	90	96.4%	3.4%	69 ppm	0.1%			4 ppm		20 ppm
Earth	288	1	0.03%	78.08%	20.95%	0 to 3%			0.93%	18 ppm	1 ppm
Mars	218	0.007	95.32%	2.7%	0.13%	0.03%			1.6%	3 ppm	0.07%
Jupiter	129						86.1%	13.8%			
Saturn	97						92.4%	7.4%			
Uranus	58						89%	11%			
Neptune	56						89%	11%			
Pluto	50	1×10^{-5}									

SATELLITES OF THE PLANETS

This table gives characteristics of the known satellites of the planets. The parameters covered are:

- Orbital period in units of earth days. An R following the value indicates a retrograde motion.
- Distance from the planet, as measured by the semi-major axis of the orbit.
- Eccentricity of the orbit.
- Inclination of the satellite orbit with respect to the equator of the planet.
- Mass of the satellite relative to the planet.
- Radius of the satellite in km.
- Mean density of the satellite.
- Geometric albedo, which is a measure of the fraction of incident sunlight reflected by the satellite.

REFERENCES

1. Seidelmann, P. K., Editor, *Explanatory Supplement to the Astronomical Almanac,* University Science Books, Mill Valley, CA, 1992.
2. Lang, K. R., *Astrophysical Data: Planets and Stars*, Springer-Verlag, New York, 1992.
3. Burns, J. A., and Matthews, M. S., Eds., *Satellites*, University of Arizona Press, Tucson, 1986.

Planet	Satellite		Orb. Period d	Distance 10^3 km	Eccentricity	Inclination	Rel. mass	Radius km	Den. g/cm^3	Albedo
Earth		Moon	27.321661	384.400	0.054900489	18.28-28.58°	0.01230002	1738	3.34	0.12
Mars	I	Phobos	0.31891023	9.378	0.015	1.0°	1.5×10^{-8}	13.5 × 10.8 × 9.4	<2	0.06
	II	Deimos	1.2624407	23.459	0.0005	0.9-2.7°	3×10^{-9}	7.5 × 6.1 × 5.5	<2	0.07
Jupiter	I	Io	1.769137786	422	0.004	0.04°	4.68×10^{-5}	1815	3.55	0.61
	II	Europa	3.551181041	671	0.009	0.47°	2.52×10^{-5}	1569	3.04	0.64
	III	Ganymede	7.15455296	1070	0.002	0.21°	7.80×10^{-5}	2631	1.93	0.42
	IV	Callisto	16.6890184	1883	0.007	0.51°	5.66×10^{-5}	2400	1.83	0.20
	V	Amalthea	0.49817905	181	0.003	0.40°	3.8×10^{-9}	135 × 83 × 75		0.05
	VI	Himalia	250.5662	11480	0.15798	27.63°	5.0×10^{-9}	93		0.03
	VII	Elara	259.6528	11737	0.20719	24.77°	4×10^{-10}	38		0.03
	VIII	Pasiphae	735 R	23500	0.378	145°	1×10^{-10}	25		
	IX	Sinope	758 R	23700	0.275	153°	0.4×10^{-10}	18		
	X	Lysithea	259.22	11720	0.107	29.02°	0.4×10^{-10}	18		
	XI	Carme	692 R	22600	0.20678	164°	0.5×10^{-10}	20		
	XII	Ananke	631 R	21200	0.16870	147°	0.2×10^{-10}	15		
	XIII	Leda	238.72	11094	0.14762	26.07°	0.03×10^{-10}	8		
	XIV	Thebe	0.6745	222	0.015	0.8°	4×10^{-10}	55 × 45		0.05
	XV	Adrastea	0.29826	129			0.1×10^{-10}	12.5 × 10 × 7.5		0.05
	XVI	Metis	0.294780	128			0.5×10^{-10}	20		0.05
Saturn	I	Mimas	0.942421813	185.52	0.0202	1.53°	8.0×10^{-8}	196	1.44	0.5
	II	Enceladus	1.370217855	238.02	0.00452	1.86°	1.3×10^{-7}	250	1.13	1.0
	III	Tethys	1.887802160	294.66	0.00000	1.86°	1.3×10^{-6}	530	1.20	0.9
	IV	Dione	2.736914742	377.40	0.002230	0.02°	1.85×10^{-6}	560	1.41	0.7
	V	Rhea	4.517500436	527.04	0.00100	0.35°	4.4×10^{-6}	765	1.33	0.7
	VI	Titan	15.94542068	1221.83	0.029192	0.33°	2.38×10^{-4}	2575	1.88	0.21
	VII	Hyperion	21.2766088	1481.1	0.104	0.43°	3×10^{-8}	205 × 130 × 110		0.3
	VIII	Iapetus	79.3301825	3561.3	0.02828	14.72°	3.3×10^{-6}	730	1.15	0.2
	IX	Phoebe	550.48 R	12952	0.16326	177°	7×10^{-10}	110		0.06

SATELLITES OF THE PLANETS (continued)

Planet	Satellite		Orb. Period d	Distance 10^3 km	Eccentricity	Inclination	Rel. mass	Radius km	Den. g/cm^3	Albedo
	X	Janus	0.6945	151.472	0.007	0.14°		110 × 100 × 80		0.8
	XI	Epimetheus	0.6942	151.422	0.009	0.34°		70 × 60 × 50		0.8
	XII	Helene	2.7369	377.40	0.005	0.0°		18 × 16 × 15		0.7
	XIII	Telesto	1.8878	294.66				17 × 14 × 13		0.5
	XIV	Calypso	1.8878	294.66				17 × 11 × 11		0.6
	XV	Atlas	0.6019	137.670	0.000	0.3°		20 × 10		0.9
	XVI	Prometheus	0.6130	139.353	0.003	0.0°		70 × 50 × 40		0.6
	XVII	Pandora	0.6285	141.700	0.004	0.0°		55 × 45 × 35		0.9
	XVIII	Pan	0.5750	133.583				10		0.5
Uranus	I	Ariel	2.52037935	191.02	0.0034	0.3°	1.56×10^{-5}	579	1.55	0.34
	II	Umbriel	4.1441772	266.30	0.0050	0.36°	1.35×10^{-5}	586	1.58	0.18
	III	Titania	8.7058717	435.91	0.0022	0.14°	4.06×10^{-5}	790	1.69	0.27
	IV	Oberon	13.4632389	583.52	0.0008	0.10°	3.47×10^{-5}	762	1.64	0.24
	V	Miranda	1.41347925	129.39	0.0027	4.2°	0.08×10^{-5}	240	1.25	0.27
	VI	Cordelia	0.335033	49.77	<0.001	0.1°		13		0.07
	VII	Ophelia	0.376409	53.79	0.010	0.1°		15		0.07
	VIII	Bianca	0.434577	59.17	<0.001	0.2°		21		0.07
	IX	Cressida	0.463570	61.78	<0.001	0.0°		31		0.07
	X	Desdemona	0.473651	62.68	<0.001	0.2°		27		0.07
	XI	Juliet	0.493066	64.35	<0.001	0.1°		42		0.07
	XII	Portia	0.513196	66.09	<0.001	0.1°		54		0.07
	XIII	Rosalind	0.558459	69.94	<0.001	0.3°		27		0.07
	XIV	Belinda	0.623525	75.26	<0.001	0.0°		33		0.07
	XV	Puck	0.761832	86.01	<0.001	0.31°		77		0.07
Neptune	I	Triton	5.8768541 R	354.76	0.000016	157.345°	2.09×10^{-4}	1353	2.05	0.7
	II	Nereid	360.13619	5513.4	0.7512	27.6°	2×10^{-7}	170		0.4
	III	Naiad	0.294396	117.6	<0.001	4.74°		29		0.06
	IV	Thalassa	0.311485	73.6	<0.001	0.21°		40		0.06
	V	Despina	0.334655	52.6	<0.001	0.07°		74		0.06
	VI	Galatea	0.428745	62.0	<0.001	0.05°		79		0.06
	VII	Larissa	0.554654	50.0	<0.0014	0.20°		104 × 89		0.06
	VIII	Proteus	1.122315	48.2	<0.001	0.55°		218 × 208 × 201		0.06
Pluto	I	Charon	6.38725	19.6	<0.001	99°	0.22	593		0.5

MASS, DIMENSIONS, AND OTHER PARAMETERS OF THE EARTH

This table is a collection of data on various properties of the Earth. Most of the values are given in SI units. Note that 1 AU (astronomical unit) = 149,597,870 km.

REFERENCES

1. Seidelmann, P. K., Editor, *Explanatory Supplement to the Astronomical Almanac*, University Science Books, Mill Valley, CA, 1992.
2. Lang, K. R., *Astrophysical Data: Planets and Stars*, Springer-Verlag, New York, 1992.

Quantity	Symbol	Value	Unit
Mass	M	$5.9742 \cdot 10^{27}$	g
Major orbital semi-axis	a_{orb}	1.000000	AU
		$1.4959787 \cdot 10^{8}$	km
Distance from sun at perihelion	r_π	0.9833	AU
Distance from sun at aphelion	r_α	1.0167	AU
Moment of perihelion passage	T_π	Jan. 2, 4 h 52 min	
Moment of aphelion passage	T_α	July 4, 5 h 05 min	
Siderial rotation period around sun	P_{orb}	$31.5581 \cdot 10^{6}$	s
		365.25636	d
Mean rotational velocity	U_{orb}	29.78	km/s
Mean equatorial radius	$\overline{a}$	6378.140	km
Mean polar compression (flattening factor)	α	1/298.257	
Difference in equatorial and polar semi-axes	$a - c$	21.385	km
Compression of meridian of major equatorial axis	α_a	1/295.2	
Compression of meridian of minor equatorial axis	α_b	1/298.0	
Equatorial compression	ε	1/30 000	
Difference in equatorial semi-axes	$a - b$	213	m
Difference in polar semi-axes	$c_N - c_S$	~70	m
Polar asymmetry	η	$\sim 1.10^{-5}$	
Mean acceleration of gravity at equator	g_e	9.78036	m/s^2
Mean acceleration of gravity at poles	g_p	9.83208	m/s^2
Difference in acceleration of gravity at pole and at equator	$g_p - g_e$	5.172	cm/s^2
Mean acceleration of gravity for entire surface of terrestrial ellipsoid	g	9.7978	m/s^2
Mean radius	R	6371.0	km
Area of surface	S	$5.10 \cdot 10^{8}$	km^2
Volume	V	$1.0832 \cdot 10^{12}$	km^3
Mean density	ρ	5.515	g/cm^3
Siderial rotational period	P	86,164.09	s
Rotational angular velocity	ω	$7.292116 \cdot 10^{-5}$	rad/s
Mean equatorial rotational velocity	v	0.46512	km/s
Rotational angular momentum	L	$5.861 \cdot 10^{33}$	J s
Rotational energy	E	$2.137 \cdot 10^{29}$	J
Ratio of centrifugal force to force of gravity at equator	q_c	0.0034677 = 1/288	
Moment of inertia	I	$8.070 \cdot 10^{37}$	kg m^2
Relative braking of earth's rotation due to tidal friction	$\Delta\omega_e/\omega$	$-4.2 \cdot 10^{-8}$	century^{-1}
Relative secular acceleration of earth's rotation	$\Delta\omega_i/\omega$	$+1.4 \cdot 10^{-8}$	century^{-1}
Not secular braking of earth's rotation	$\Delta\omega/\omega$	$-2.8 \cdot 10^{-8}$	century^{-1}
Probable value of total energy of tectonic deformation of earth	E_t	$\sim 1 \cdot 10^{23}$	J/century
Secular loss of heat of earth through radiation into space	$\Delta' E_k$	$1 \cdot 10^{23}$	J/century
Portion of earth's kinetic energy transformed into heat as a result of lunar and solar tides in the hydrosphere	$\Delta'' E_k$	$1.3 \cdot 10^{23}$	J/century

MASS, DIMENSIONS, AND OTHER PARAMETERS OF THE EARTH (continued)

Quantity	Symbol	Value	Unit
Differences in duration of days in March and August	ΔP	0.0025 (March-August)	s
Corresponding relative annual variation in earth's rotational velocity	$\Delta^*\omega/\omega$	$2.9 \cdot 10^{-8}$ (Aug.-March)	
Presumed variation in earth's radius between August and March	Δ^*R	–9.2 (Aug.-March)	cm
Annual variation in level of world ocean	Δh_o	~10 (Sept.-March)	cm
Area of continents	S_C	$1.49 \cdot 10^{8}$	km^2
		29.2	% of surface
Area of world ocean	S_o	$3.61 \cdot 10^{8}$	km^2
		70.8	% of surface
Mean height of continents above sea level	h_C	875	m
Mean depth of world ocean	h_o	3794	m
Mean thickness of lithosphere within the limits of the continents	$h_{c.l.}$	35	km
Mean thickness of lithosphere within the limits of the ocean	$h_{o.l.}$	4.7	km
Mean rate of thickening of continental lithosphere	$\Delta h/\Delta t$	10 - 40	$m/10^6$ y
Mean rate of horizontal extension of continental lithosphere	$\Delta l/\Delta t$	0.75 - 20	$km/10^6$ y
Mass of crust	m_l	$2.36 \cdot 10^{22}$	kg
Mass of mantle		$4.05 \cdot 10^{24}$	kg
Amount of water released from the mantle and core in the course of geological time		$3.40 \cdot 10^{21}$	kg
Total reserve of water in the mantle		$2 \cdot 10^{23}$	kg
Present content of free and bound water in the earth's lithosphere		$2.4 \cdot 10^{21}$	kg
Mass of hydrosphere	m_h	$1.664 \cdot 10^{21}$	kg
Amount of oxygen bound in the earth's crust		$1.300 \cdot 10^{21}$	kg
Amount of free oxygen		$1.5 \cdot 10^{18}$	kg
Mass of atmosphere	m_a	$5.136 \cdot 10^{18}$	kg
Mass of biosphere	m_b	$1.148 \cdot 10^{16}$	kg
Mass of living matter in the biosphere		$3.6 \cdot 10^{14}$	kg
Density of living matter on dry land		0.1	g/cm^2
Density of living matter in ocean		$15 \cdot 10^{-8}$	g/cm^3
Age of the earth		$4.55 \cdot 10^{9}$	y
Age of oldest rocks		$4.0 \cdot 10^{9}$	y
Age of most ancient fossils		$3.4 \cdot 10^{9}$	y

GEOLOGICAL TIME SCALE

Period or Epoch	Beginning and end, in 10^6 years	Key events
	Cenozoic era	
Quaternian		
Contemporary	0–10,000 y ± 2,000 y	
Pleistocene	10,000–1,000,000 y ± 50,000 y	Homo Erectus breakout
Tertiary		
Pliocene	1.8–5.3	Ape man fossils
Miocene	5–25	Origin of grass
Oligocene	25–37	Rise of cats, dogs, pigs
Eocene	37–55	Debut of hoofed mammals
Paleocene	55–67	Earliest primates
	Mesozoic era	
Cretaceous	67–138	Demise of dinosaurs
Jurassic	138–208	First birds
Triassic	208–245	Appearance of dinosaurs
	Paleozoic era	
Permian	245–290	Flowers, insect pollination
Carboniferous	290–360	First conifers
Devonian	360–410	First vertebrates ashore
Silurian	410–435	Spore-bearing plants
Ordovician	435–520	First animals ashore
Cambrian	520–570	Vertebrates appear
	Pre-Cambrian	
Pre-Cambrian III (Proterozoic)	570–2500	First plants, jellyfish
Pre-Cambrian II (Archean)	2500–3800	Photosynthetic bacteria
Pre-Cambrian I (Hadean)	3800–4450	Earth formed 4600 million years ago

Reference: Calder, N., *Timescale - An Atlas of the Fourth Dimension,* Viking Press, New York, 1983.

ACCELERATION DUE TO GRAVITY

The acceleration due to gravity is tabulated here as a function of latitude and height above the earth's surface. Values were calculated from the expression

$$g/(\mathrm{m/s^2}) = 9.780356\,(1 + 0.0052885\,\sin^2\phi - 0.0000059\,\sin^2 2\phi) - 0.003086\,H$$

where ϕ is the latitude and H is the height in kilometers.

REFERENCE

Jursa, A. S., Ed., *Handbook of Geophysics and the Space Environment,* 4th ed., Air Force Geophysics Laboratory, 1985, p. 14-17.

ϕ	$H = 0$	$H = 1$ km	$H = 5$ km	$H = 10$ km
0	9.78036	9.77727	9.76493	9.74950
5	9.78075	9.77766	9.76532	9.74989
10	9.78191	9.77882	9.76648	9.75105
15	9.78381	9.78072	9.76838	9.75295
20	9.78638	9.78330	9.77095	9.75552
25	9.78956	9.78647	9.77413	9.75870
30	9.79324	9.79016	9.77781	9.76238
35	9.79732	9.79424	9.78189	9.76646
40	9.80167	9.79858	9.78624	9.77081
45	9.80616	9.80307	9.79073	9.77530
50	9.81065	9.80757	9.79522	9.77979
55	9.81501	9.81193	9.79958	9.78415
60	9.81911	9.81602	9.80368	9.78825
65	9.82281	9.81972	9.80738	9.79195
70	9.82601	9.82292	9.81058	9.79515
75	9.82860	9.82551	9.81317	9.79774
80	9.83051	9.82743	9.81508	9.79965
85	9.83168	9.82860	9.81625	9.80082
90	9.83208	9.82899	9.81665	9.80122

DENSITY, PRESSURE, AND GRAVITY AS A FUNCTION OF DEPTH WITHIN THE EARTH

This table gives the density ρ, pressure p, and acceleration due to gravity g as a function of depth below the earth's surface, as calculated from the model of the structure of the earth in Reference 1. The model assumes a radius of 6371 km for the earth. The boundary between the crust and mantle (the Mohorovicic discontinuity) is taken as 21 km, while in reality it varies considerable with location.

REFERENCES

1. Anderson, D. L., and Hart, R. S., *J. Geophys. Res.*, 81, 1461, 1976.
2. Carmichael, R. S., *CRC Practical Handbook of Physical Properties of Rocks and Minerals*, p.467, CRC Press, Boca Raton, FL, 1989.

Depth km	ρ g/cm³	p kbar	g cm/s²
	Crust		
0	1.02	0	981
3	1.02	3	982
3	2.80	3	982
21	2.80	5	983
	Mantle (solid)		
21	3.49	5	983
41	3.51	12	983
61	3.52	19	984
81	3.48	26	984
101	3.44	33	984
121	3.40	39	985
171	3.37	56	987
221	3.34	73	989
271	3.37	89	991
321	3.47	106	993
371	3.59	124	994
571	3.95	199	999
871	4.54	328	997
1171	4.67	466	992
1471	4.81	607	991
1771	4.96	752	994
2071	5.12	903	1002
2371	5.31	1061	1017
2671	5.45	1227	1042
2886	5.53	1352	1069
	Outer core (liquid)		
2886	9.96	1352	1069
2971	10.09	1442	1050
3371	10.63	1858	953
3671	11.00	2154	874
4071	11.36	2520	760
4471	11.69	2844	641
4871	11.99	3116	517
5156	12.12	3281	427
	Inner core (solid)		
5156	12.30	3281	427
5371	12.48	3385	355
5771	12.52	3529	218
6071	12.53	3592	122
6371	12.58	3617	0

ABUNDANCE OF ELEMENTS IN THE EARTH'S CRUST AND IN THE SEA

This table gives the estimated abundance of the elements in the continental crust (in mg/kg, equivalent to parts per million by mass) and in seawater near the surface (in mg/L). Values represent the median of reported measurements. The concentrations of the less abundant elements may vary with location by several orders of magnitude.

REFERENCES

1. Carmichael, R. S., Ed., *CRC Practical Handbook of Physical Properties of Rocks and Minerals,* CRC Press, Boca Raton, FL, 1989.
2. Bodek,I., et al., *Environmental Inorganic Chemistry*, Pergamon Press, New York, 1988.
3. Ronov, A. B., and Yaroshevsky, A. A.,"Earth's Crust Geochemistry", in *Encyclopedia of Geochemistry and Environmental Sciences,* Fairbridge, R. W.,Ed., Van Nostrand, New York, 1969.

	Abundance			Abundance	
Element	Crust mg/kg	Sea mg/L	Element	Crust mg/kg	Sea mg/L
Ac	5.5×10^{-10}		N	1.9×10^{1}	5×10^{-1}
Ag	7.5×10^{-2}	4×10^{-5}	Na	2.36×10^{4}	1.08×10^{4}
Al	8.23×10^{4}	2×10^{-3}	Nb	2.0×10^{1}	1×10^{-5}
Ar	3.5	4.5×10^{-1}	Nd	4.15×10^{1}	2.8×10^{-6}
As	1.8	3.7×10^{-3}	Ne	5×10^{-3}	1.2×10^{-4}
Au	4×10^{-3}	4×10^{-6}	Ni	8.4×10^{1}	5.6×10^{-4}
B	1.0×10^{1}	4.44	O	4.61×10^{5}	8.57×10^{5}
Ba	4.25×10^{2}	1.3×10^{-2}	Os	1.5×10^{-3}	
Be	2.8	5.6×10^{-6}	P	1.05×10^{3}	6×10^{-2}
Bi	8.5×10^{-3}	2×10^{-5}	Pa	1.4×10^{-6}	5×10^{-11}
Br	2.4	6.73×10^{1}	Pb	1.4×10^{1}	3×10^{-5}
C	2.00×10^{2}	2.8×10^{1}	Pd	1.5×10^{-2}	
Ca	4.15×10^{4}	4.12×10^{2}	Po	2×10^{-10}	1.5×10^{-14}
Cd	1.5×10^{-1}	1.1×10^{-4}	Pr	9.2	6.4×10^{-7}
Ce	6.65×10^{1}	1.2×10^{-6}	Pt	5×10^{-3}	
Cl	1.45×10^{2}	1.94×10^{4}	Ra	9×10^{-7}	8.9×10^{-11}
Co	2.5×10^{1}	2×10^{-5}	Rb	9.0×10^{1}	1.2×10^{-1}
Cr	1.02×10^{2}	3×10^{-4}	Re	7×10^{-4}	4×10^{-6}
Cs	3	3×10^{-4}	Rh	1×10^{-3}	
Cu	6.0×10^{1}	2.5×10^{-4}	Rn	4×10^{-13}	6×10^{-16}
Dy	5.2	9.1×10^{-7}	Ru	1×10^{-3}	7×10^{-7}
Er	3.5	8.7×10^{-7}	S	3.50×10^{2}	9.05×10^{2}
Eu	2.0	1.3×10^{-7}	Sb	2×10^{-1}	2.4×10^{-4}
F	5.85×10^{2}	1.3	Sc	2.2×10^{1}	6×10^{-7}
Fe	5.63×10^{4}	2×10^{-3}	Se	5×10^{-2}	2×10^{-4}
Ga	1.9×10^{1}	3×10^{-5}	Si	2.82×10^{5}	2.2
Gd	6.2	7×10^{-7}	Sm	7.05	4.5×10^{-7}
Ge	1.5	5×10^{-5}	Sn	2.3	4×10^{-6}
H	1.40×10^{3}	1.08×10^{5}	Sr	3.70×10^{2}	7.9
He	8×10^{-3}	7×10^{-6}	Ta	2.0	2×10^{-6}
Hf	3.0	7×10^{-6}	Tb	1.2	1.4×10^{-7}
Hg	8.5×10^{-2}	3×10^{-5}	Te	1×10^{-3}	
Ho	1.3	2.2×10^{-7}	Th	9.6	1×10^{-6}
I	4.5×10^{-1}	6×10^{-2}	Ti	5.65×10^{3}	1×10^{-3}
In	2.5×10^{-1}	2×10^{-2}	Tl	8.5×10^{-1}	1.9×10^{-5}
Ir	1×10^{-3}		Tm	5.2×10^{-1}	1.7×10^{-7}
K	2.09×10^{4}	3.99×10^{2}	U	2.7	3.2×10^{-3}
Kr	1×10^{-4}	2.1×10^{-4}	V	1.20×10^{2}	2.5×10^{-3}
La	3.9×10^{1}	3.4×10^{-6}	W	1.25	1×10^{-4}
Li	2.0×10^{1}	1.8×10^{-1}	Xe	3×10^{-5}	5×10^{-5}
Lu	8×10^{-1}	1.5×10^{-7}	Y	3.3×10^{1}	1.3×10^{-5}
Mg	2.33×10^{4}	1.29×10^{3}	Yb	3.2	8.2×10^{-7}
Mn	9.50×10^{2}	2×10^{-4}	Zn	7.0×10^{1}	4.9×10^{-3}
Mo	1.2	1×10^{-2}	Zr	1.65×10^{2}	3×10^{-5}

SOLAR SPECTRAL IRRADIANCE

From NASA Technical Report R-351, "The Solar Constant and the Solar Spectrum Measured from a Research Aircraft". Report edited by Matthew P. Thekaekara, Goddard Space Flight Center, Greenbelt, Maryland 20771. Discussion of previously reported values and of measurements and calculations leading to the following table are contained in the NASA Technical Report R-351.

SOLAR SPECTRAL IRRADIANCE — PROPOSED STANDARD CURVE

λ — Wavelength in μm
P_λ — Solar spectral irradiance averaged over small bandwidth centered at λ, in W cm^{-2} μm^{-1}*
A_λ — Area under the solar spectral irradiance curve in the wavelength range 0 to λ, mW cm^{-2}
D_λ — Percentage of the solar constant associated with wavelengths shorter than λ
Solar Constant — 135.30 mW cm^{-2}, or 1.940 cal min^{-1} cm^{-2}**

λ	P_λ	A_λ	D_λ	λ	P_λ	A_λ	D_λ	λ	P_λ	A_λ	D_λ
0.120	0.000010	0.00059992	0.00044	0.475	0.2044	25.6001	18.921	2.4	0.0064	129.695	95.858
0.140	0.000003	0.00072999	0.00053	0.480	0.2074	26.6296	19.681	2.5	0.0054	130.285	96.294
0.150	0.000007	0.00077999	0.00057	0.485	0.1976	27.6421	20.430	2.6	0.0048	130.795	96.671
0.160	0.000023	0.00092999	0.00068	0.490	0.1950	28.6236	21.155	2.7	0.0043	131.250	97.007
0.170	0.000063	0.00135999	0.00100	0.495	0.1960	29.6011	21.878	2.8	0.00390	131.660	97.3103
0.180	0.000125	0.00229999	0.00169	0.500	0.1942	30.5766	22.599	2.9	0.00350	132.030	97.5838
0.190	0.000271	0.00427999	0.00316	0.505	0.1920	31.5421	23.312	3.0	0.00310	132.360	97.8277
0.200	0.00107	0.010984	0.0081	0.510	0.1882	32.4926	24.015	3.1	0.00260	132.645	98.0383
0.210	0.00229	0.027784	0.0205	0.515	0.1833	33.4214	24.701	3.2	0.00226	132.888	98.2179
0.220	0.00575	0.067984	0.0502	0.520	0.1833	34.3379	25.379	3.3	0.00192	133.097	98.3724
0.225	0.00649	0.098584	0.0728	0.525	0.1852	35.2591	26.059	3.4	0.00166	133.276	98.5047
0.230	0.00667	0.131484	0.0971	0.530	0.1842	36.1826	26.742	3.5	0.00146	133.432	98.6200
0.235	0.00593	0.162984	0.1204	0.535	0.1818	37.0976	27.418	3.6	0.00135	133.573	98.7238
0.240	0.00630	0.193559	0.1430	0.540	0.1783	37.9979	28.084	3.7	0.00123	133.702	98.8192
0.245	0.00723	0.227384	0.1680	0.545	0.1754	38.8821	28.737	3.8	0.00111	133.819	98.9056
0.250	0.00704	0.263059	0.1944	0.550	0.1725	39.7519	29.380	3.9	0.00103	133.926	98.9847
0.255	0.0104	0.306659	0.226	0.555	0.1720	40.6131	30.017	4.0	0.00095	134.025	99.0579
0.260	0.0130	0.365159	0.269	0.560	0.1695	41.4669	30.648	4.1	0.00087	134.116	99.1252
0.265	0.0185	0.443909	0.328	0.565	0.1705	42.3169	31.276	4.2	0.00078	134.198	99.1861
0.270	0.0232	0.548159	0.405	0.570	0.1712	43.1711	31.907	4.3	0.00071	134.273	99.2412
0.275	0.0204	0.657159	0.485	0.575	0.1719	44.0289	32.541	4.4	0.00065	134.341	99.2915
0.280	0.0222	0.763659	0.564	0.580	0.1715	44.8874	33.176	4.5	0.00059	134.403	99.3373
0.285	0.0315	0.897909	0.663	0.585	0.1712	45.7441	33.809	4.6	0.00053	134.459	99.3787
0.290	0.0482	1.09715	0.810	0.590	0.1700	46.5971	34.439	4.7	0.00048	134.509	99.4160
0.295	0.0584	1.36365	1.007	0.595	0.1682	47.4426	35.064	4.8	0.00045	134.556	99.4504
0.300	0.0514	1.63815	1.210	0.600	0.1666	48.2796	35.683	4.9	0.00041	134.599	99.482195
0.305	0.0603	1.91740	1.417	0.605	0.1647	49.1079	36.295	5.0	0.0003830	134.63905	99.511500
0.310	0.0689	2.24040	1.655	0.610	0.1635	49.9284	36.902	6.0	0.0001750	134.91805	99.717708
0.315	0.0764	2.60365	1.924	0.620	0.1602	51.5469	38.098	7.0	0.0000990	135.05505	99.818965
0.320	0.0830	3.00215	2.218	0.630	0.1570	53.1329	39.270	8.0	0.0000600	135.13455	99.877723
0.325	0.0975	3.45340	2.552	0.640	0.1544	54.6899	40.421	9.0	0.0000380	135.18355	99.913939
0.330	0.1059	3.96190	2.928	0.650	0.1511	56.2174	41.550	10.0	0.0000250	135.21505	99.937220
0.335	0.1081	4.49690	3.323	0.660	0.1486	57.7159	42.657	11.0	0.0000170	135.23605	99.952742
0.340	0.1074	5.03565	3.721	0.670	0.1456	59.1869	43.744	12.0	0.0000120	135.25055	99.963458
0.345	0.1069	5.57140	4.117	0.680	0.1427	60.6284	44.810	13.0	0.0000087	135.26090	99.971108
0.350	0.1093	6.11190	4.517	0.690	0.1402	62.0429	45.855	14.0	0.0000055	135.26800	99.976356
0.355	0.1083	6.65590	4.919	0.700	0.1369	63.4284	46.879	15.0	0.0000049	135.27320	99.980199
0.360	0.1068	7.19365	5.316	0.710	0.1344	64.7849	47.882	16.0	0.0000038	135.27755	99.983414
0.365	0.1132	7.74365	5.723	0.720	0.1314	66.1139	48.864	17.0	0.0000031	135.28100	99.985964
0.370	0.1181	8.32190	6.150	0.730	0.1290	67.4159	49.826	18.0	0.0000024	135.28375	99.987997
0.375	0.1157	8.90640	6.582	0.740	0.1260	68.6909	50.769	19.0	0.0000020	135.28595	99.989623
0.380	0.1120	9.47565	7.003	0.750	0.1235	69.9384	51.691	20.0	0.0000016	135.28775	99.990953
0.385	0.1098	10.0301	7.413	0.800	0.1107	75.7934	56.018	25.0	0.000000610	135.29328	99.995036
0.390	0.1098	10.5791	7.819	0.850	0.0988	81.0309	59.889	30.0	0.000000300	135.29555	99.996718
0.395	0.1189	11.1509	8.241	0.900	0.0889	85.7234	63.358	35.0	0.000000160	135.29670	99.997568
0.400	0.1429	11.8054	8.725	0.950	0.0835	90.0334	66.543	40.0	0.000000094	135.29734	99.998037
0.405	0.1644	12.5736	9.293	1.000	0.0746	93.9859	69.464	50.0	0.000000038	135.29800	99.998525
0.410	0.1751	13.4224	9.920	1.100	0.0592	100.675	74.409	60.0	0.000000019	135.29828	99.998736
0.415	0.1774	14.3036	10.571	1.200	0.0484	106.055	78.385	80.0	0.000000007	135.29854	99.998928
0.420	0.1747	15.1839	11.222	1.300	0.0396	110.455	81.637	100.0	0,000000003	135.29864	99.999002
0.425	0.1693	16.0439	11.858	1.400	0.0336	114.115	84.342	1000.0	0	135.30000	100.000000
0.430	0.1639	16.8769	12.473	1.500	0.0287	117.230	86.645				
0.435	0.1663	17.7024	13.083	1.600	0.0244	119.885	88.607				
0.440	0.1810	18.5706	13.725	1.700	0.0202	122.115	90.255				
0.445	0.1922	19.5036	14.415	1.800	0.0159	123.920	91.589				
0.450	0.2006	20.4856	15.140	1.900	0.0126	125.345	92.642				
0.455	0.2057	21.5014	15.891	2.000	0.0103	126.490	93.489				
0.460	0.2066	22.5321	16.653	2.100	0.0090	127.455	94.202				
0.465	0.2048	23.5606	17.413	2.200	0.0079	128.300	94.826				
0.470	0.2033	24.5809	18.167	2.300	0.0068	129.035	95.370				

* The variable $P_\lambda(\lambda)$, expressed in the units W cm^{-2} μm^{-1}, is related to $P_\nu(\nu)$, expressed in the units W cm^{-2} Hz^{-1}, through the relation $P_\nu(\nu) = (d\lambda/d\nu)P_\lambda(\lambda) = (c/\nu^2)P_\lambda(\lambda)$ where c is the speed of light (2.998×10^{14} μm/s) and ν is the frequency. More explicitly, $P_\nu(\nu) = (2.998 \times 10^{14})/\nu^2\ P_\lambda(c/\nu)$.
The conversion of $P_\nu(\nu)$ or $P_\lambda(\lambda)$ to *Spectral Energy Density* is given by $\mu(\nu)$ [joule cm^{-3} Hz^{-1}] = $(1/c)P_\nu(\nu)$ and $\mu(\lambda)$ [joule cm^{-3} μ^{-1}] = $(1/c)P_\lambda(\lambda)$ where c now has the value, 2.998×10^{10} cm s^{-1}.

** The best current value (1991) of the solar constant is 1373 W/m^2.

TOTAL MONTHLY SOLAR RADIATION IN A CLOUDLESS SKY

Total radiation is the sum of the direct and scattered radiation that strikes the earth's surface. It is influenced by the degree of cloudiness, atmospheric transparency, duration of sunshine, and height of elevation at which the measurements are taken. Deviations from values in this table are bound to occur; however, it is believed that the deviations do not exceed ±10% for the summer months and ±15% for the winter months. The radiation units in this table are kcal/cm².

$\varphi°$	January	February	March	April	May	June	July	August	September	October	November	December
90 N	0	0	0.1	10.0	21.9	26.0	23.8	12.9	2.4	0	0	0
85	0	0	0.7	10.2	21.8	25.8	23.4	13.1	3.0	0	0	0
80	0	0	2.4	10.8	21.4	25.2	23.0	13.4	4.3	0.5	0	0
75	0	0.5	4.0	11.7	21.0	24.5	22.2	13.8	5.8	1.3	0	0
70	0	1.6	6.0	13.1	20.5	23.6	21.2	14.6	7.5	2.7	0.5	0
65	0.7	2.8	8.0	14.5	20.1	22.8	21.0	15.6	9.5	4.3	1.4	0.2
60	1.8	4.3	9.9	16.0	20.8	22.9	21.4	16.7	11.3	6.1	2.6	1.1
55	3.1	6.2	11.7	17.3	21.4	23.4	21.9	17.9	12.9	7.8	4.0	2.3
50	4.8	8.2	13.3	18.5	22.2	23.7	22.6	19.1	14.4	9.7	5.8	3.9
45	6.7	10.3	14.8	19.5	22.6	23.9	23.2	20.1	15.8	11.5	7.8	5.9
40	8.8	12.2	16.4	20.3	23.0	24.0	23.4	20.9	17.0	13.2	9.7	7.7
35	10.7	14.0	17.6	21.0	23.0	24.0	23.6	21.6	18.1	14.7	11.4	9.7
30	12.5	15.5	18.6	21.4	23.0	23.8	23.4	21.8	19.1	16.1	13.1	11.5
25	14.1	16.8	19.5	21.6	23.0	23.4	23.1	21.8	19.8	17.4	14.6	13.1
20	15.5	17.9	20.2	21.6	22.5	22.8	22.6	21.6	20.4	18.5	16.1	14.7
15	16.9	19.0	20.8	21.4	21.9	22.0	21.9	21.2	20.9	19.3	17.4	16.1
10	18.1	19.8	21.1	21.2	21.2	21.0	21.1	20.6	21.2	20.1	18.5	17.5
5	19.3	20.4	21.4	21.0	20.2	19.9	20.0	20.0	21.3	20.6	19.5	18.8
0	20.2	20.9	21.5	20.4	19.3	18.8	19.1	19.3	21.2	21.2	20.4	19.2
5S	20.1	21.4	21.4	19.9	18.3	17.6	17.9	18.3	20.8	21.4	21.8	21.0
10	22.0	21.8	21.1	19.2	17.7	16.3	16.3	17.3	20.4	21.4	21.8	22.0
15	22.6	22.0	20.6	18.3	16.0	14.9	15.4	16.1	19.7	21.3	22.4	22.9
20	23.2	22.0	20.0	17.2	14.7	13.4	13.8	14.9	18.9	21.0	22.6	23.6
25	23.6	22.0	19.4	16.0	13.0	12.0	12.6	13.6	18.0	20.6	23.0	24.1
30	23.9	21.8	18.6	14.9	11.9	10.6	11.1	12.1	17.0	20.1	23.0	24.6
35	24.0	21.3	17.6	13.6	10.4	9.0	9.6	10.6	15.9	19.5	23.0	25.0
40	24.0	20.6	16.4	12.2	8.7	7.3	8.1	9.0	14.3	18.7	22.8	25.2
45	24.0	19.9	15.2	10.7	7.1	5.5	6.3	7.3	13.4	17.7	22.4	25.2
50	23.6	18.9	13.8	9.2	5.4	3.8	4.6	5.5	12.0	16.6	21.8	25.0
55	23.2	17.8	12.3	7.5	3.8	2.3	3.0	3.8	10.3	15.4	21.2	24.6
60	22.6	16.6	10.8	5.6	2.4	1.0	1.6	2.3	8.5	14.1	21.0	24.4
65	22.4	15.3	9.1	3.9	1.1	0.1	0.4	1.0	6.7	12.6	20.8	24.5
70	22.6	14.2	7.3	2.2	0.1	0	0	0	5.0	11.4	21.0	24.9
75	23.2	13.4	5.7	0.9	0	0	0	0	3.5	10.4	21.2	25.4
80	24.0	12.8	4.3	0	0	0	0	0	2.1	9.7	21.9	26.0
85	24.6	12.4	2.9	0	0	0	0	0	0.9	9.2	22.4	26.6
90	24.9	12.3	1.7	0	0	0	0	0	0	9.0	22.6	27.0

U.S. STANDARD ATMOSPHERE (1976)

From, "U.S. Standard Atmosphere, 1976", National Oceanic and Atmospheric Administration, National Aeronautics and Space Administration and the United States Air Force, 1976. The above referenced book is available from the Superintendent of Documents, U.S. Government Printing Office, Washington, D.C. 20402. The book contains considerably more extensive tables than those presented below plus the development of the equations used in the calculations of the tables as well as a discussion of the bases for selection of constants used in the equations.

The U.S. Standard Atmosphere, 1976 is an idealized, steady-state representation of the earth's atmosphere from the surface to 1000 km, as it is assumed to exist in a period of moderate solar activity. The air is assumed to be dry, and at heights sufficiently below 86 km, the atmosphere is assumed to be homogeneously mixed with a relative-volume composition leading to a mean molecular weight M. The molecular weights and assumed fractional-volume composition of sea-level dry air were

Gas species	Molecular weight M_i (kg/kmol)	Fractional volume F_i (dimensionless)
N_2	28.0134	0.78084
O_2	31.9988	0.209476
Ar	39.948	0.00934
CO_2	44.00995	0.000314
Ne	20.183	0.00001818
He	4.0026	0.00000524
Kr	83.80	0.00000114
Xe	131.30	0.000000087
CH_4	16.04303	0.000002
H_2	2.01594	0.0000005

SYMBOLS, ABBREVIATIONS, AND UNITS RELATING TO FOLLOWING TABLE

a	a coefficient, used in specifying the elliptical segment of the temperature-height profile, T (Z)
a_i	a set of species-dependent coefficients which, along with values of b_i are used in defining the set of height-dependent functions D_i
A	a coefficient used in specifying the elliptical segment of T (Z)
b	a dimensionless subscript designating a set of integers
b_i	a set of species-dependent exponents, which, along with values of a_i, are used to define the set of height-dependent functions D_i
C_s	the height-dependent speed of sound
D_i	the set of height-dependent, species-dependent, molecular-diffusion coefficients, for O, O_2, Ar, He, and H
$f(Z)$	the hydrostatic term in the height-dependent expression for n_i
F_i	the set of sea-level, fractional-volume concentrations, for each of the several atmospheric gas species
F_i'	the set of fractional-volume concentrations of the several atmospheric gas species adjusted for 86-km height to account for the dissociation of O_2
g	the height-dependent, 45-degree-latitude, acceleration of gravity
g_0'	the adopted constant, involved in the definition of the standard geopotential meter, and in the relationship between geopotential height and geometric height
H	geopotential height used as the argument for all tables up to 84.852 km′ (86.000 km)
H_P	the height-dependent, local, pressure scale height of the mixture of gases comprising the atmosphere
H_ρ	the height-dependent, local, density scale height of the mixture of gases comprising the atmosphere
i	a subscript designating the ith member of a set of gas species
k	the Boltzmann constant
k_t	the height-dependent coefficient of thermal conductivity
K	the height-dependent, eddy-diffusion (or turbulent-diffusion) coefficient
L	the height-dependent, mean free path
$L_{M,b}$	a set of gradients of T_M with respect to H
$L_{K,b}$	a set of gradients of T with respect to Z
M	the height-dependent, mean molecular weight of the mixture of gases constituting the atmosphere
M_i	the set of molecular weights of the several atmospheric gas species
N	the height-dependent, total, number density of the mixture of neutral atmospheric gas particles
n_i	the set of height-dependent, number densities of the several atmospheric gas species
N_A	the Avogadro constant
0	a subscript designating the sea-level value of the associated variable
P	the height-dependent, total atmospheric pressure
P_i	the partial pressure of the ith gas specie

q_i — one set of six adopted sets of species-dependent, constants, i.e., set q_i, set Q_i, set u_i, set U_i, set w_i, and set W_i, all used in an empirical species-dependent expression for the flux term $v_i/(D_i + K)$

Q_i — see q_i

r_0 — the adopted, effective earth's radius, 6356.766 km, used for computing g (Z) for 45-degree north latitude, and used for relating H and Z at that latitude

R^* — the universal gas constant

S — the Sutherland constant, used in computing μ

t — the height-dependent Celsius temperature

T — the height-dependent, Kelvin kinetic temperature, defined as a function of Z for all heights above 86 km and derived from T_M for heights below 86 km

T_c — a derived coefficient used in specifying an elliptical segment of $T(Z)$

T_M — the height-dependent, molecular-scale temperature, defined as a function of H for all heights from sea-level to 86 km

T_∞ — the exospheric temperature

u_i — see q_i

U_i — see q_i

v_i — the flow velocity of the ith gas species

v_m — the height-dependent mole volume

V — the height-dependent mean particle speed

w_i — see q_i

W_i — see q_i

Z — geometric height used as the argument of all tables at heights above 86 km

Z_c — the height coordinate of the center of the ellipse defining a portion of $T(Z)$

α_i — the set of species-dependent, thermal-diffusion coefficients

β — a constant used for computing μ

γ — a constant taken to represent the ratio of specific heat at constant pressure to the specific heat at constant volume, and used in defining C_s

Γ — the ratio g_0/g_0'

ε — a factor relating F_i to F_i'

η — the height-dependent kinematic viscosity

λ — a coefficient used in specifying the exponential expression defining a portion of $T(Z)$

μ — the height-dependent coefficient of dynamic viscosity

ν — the height-dependent mean collision frequency

ξ — a function of Z used in the exponential expression defining a portion of $T(Z)$

ρ — the height-dependent mass density of air

σ — the effective mean collision diameter used in defining L and ν

τ — a height-dependent coefficient representing the reduced height of the atomic hydrogen relative to a particular reference height and used in the computation of $n(\mathrm{H})$

ϕ — the vertical flux of atomic hydrogen

Φ_G — the potential energy per unit mass of gravitational attraction

Φ_C — the potential energy per unit mass associated with centrifugal force

Some equations useful for computing certain values of the Standard Atmosphere are:

$-1{,}524 \leqslant H < 11{,}000$ m:

$T = 288.15 - 0.006500\,H$

$P = 1013.25(288.15/T)^{-5.255877}$

$H = 44{,}331.514 - 11{,}880.516\,P^{0.1902632}$

$11{,}000 \leqslant H < 20{,}000$ m:

$T = 216.650$

$P = 226.32\,e^{-0.000156768832(H - 11{,}000.00)}$

$H = 45{,}383.967 - 6{,}341.6237\ln P$

$20{,}000 \leqslant H < 32{,}000$ m:

$T = 216.650 + 0.00100(H - 19{,}999.997)$

$P = 54.7487\,(216.650/T)^{34.16319}$

$H = -196{,}650.0 + 243{,}580.85/P^{0.0292713}$

$-1{,}524 \leqslant H < 32{,}000$ m:

$\varrho = 0.3483677\,(P/T)$

$H = rZ/(r + Z)$

$Z = rH/(r - H)$

Units are:

P : mb (1 mb = 100 Pa)

T : K

H : m

ρ : kg/m^3

Altitude		Acceleration due to gravity	Pressure scale height	Number density	Particle speed	Collision frequency	Mean free path	Molecular weight	Temperature	Pressure	Density	Sound speed	Dynamic viscosity	Kinematic viscosity	Thermal conductivity
													μ	η	10^{-3} k
Z(m)	H(m)	g(m/s²)	H_p (m)	n (m⁻³)	V (m/s)	ν (s⁻¹)	L (m)	M (kg/kmol)	T(K)	P(mb)	ϱ(kg/m³)	C_s(m/s)	(N·s/m²)	(m²/s)	(J/m·s·K)
−5000	−5004	9.8221	9371.8	4.0151 (+25)	484.15	1.1506 (+10)	4.2078 (−8)	28.964	320.676	1.7776 (+3)	1.9311 (+0)	358.99	1.9422 (−5)	1.0058 (−5)	2.7882 (−5)
−4500	−4503	9.8295	9278.2	3.8445 (+25)	481.69	1.0961 (+10)	4.3945 (−8)	28.964	317.421	1.6848 (+3)	1.8491 (+0)	357.16	1.9273 (−5)	1.0423 (−5)	2.7634 (−5)
−4000	−4003	9.8190	9184.5	3.6795 (+25)	479.22	1.0437 (+10)	4.5915 (−8)	28.964	314.166	1.5959 (+3)	1.7697 (+0)	355.32	1.9123 (−5)	1.0806 (−5)	2.7384 (−5)
−3500	−3502	9.8175	9090.8	3.5201 (+25)	476.73	9.9328 (+9)	4.7995 (−8)	28.964	310.913	1.5109 (+3)	1.6930 (+0)	353.48	1.8972 (−5)	1.1206 (−5)	2.7134 (−5)
−3000	−3001	9.8159	8997.1	3.3660 (+25)	474.23	9.4481 (+9)	5.0193 (−8)	28.964	307.659	1.4297 (+3)	1.6189 (+0)	351.63	1.8820 (−5)	1.1625 (−5)	2.6884 (−5)
−2500	−2501	9.8144	8903.4	3.2171 (+25)	471.71	8.9824 (+9)	5.2515 (−8)	28.964	304.406	1.3520 (+3)	1.5473 (+0)	349.76	1.8668 (−5)	1.2065 (−5)	2.6632 (−5)
−2000	−2001	9.8128	8809.6	3.1017 (+25)	469.69	8.6231 (+9)	5.4469 (−8)	28.964	301.154	1.2778 (+3)	1.4782 (+0)	347.89	1.8515 (−5)	1.2525 (−5)	2.6380 (−5)
−1500	−1500	9.8113	8715.9	2.9346 (+25)	466.65	8.1056 (+9)	5.7571 (−8)	28.964	297.902	1.2069 (+3)	1.4114 (+0)	346.00	1.8361 (−5)	1.3009 (−5)	2.6126 (−5)
−1000	−1000	9.8097	8622.1	2.8007 (+25)	464.09	7.6934 (+9)	6.0324 (−8)	28.964	294.651	1.1393 (+3)	1.3470 (+0)	344.11	1.8206 (−5)	1.3516 (−5)	2.5872 (−5)
−500	−500	9.8082	8528.3	2.6715 (+25)	461.53	7.2980 (+9)	6.3240 (−8)	28.964	291.400	1.0747 (+3)	1.2849 (+0)	342.21	1.8050 (−5)	1.4048 (−5)	2.5618 (−5)
0	0	9.8066	8434.5	2.5470 (+25)	458.94	6.9189 (+9)	6.6332 (−8)	28.964	288.150	1.01325 (+3)	1.2250 (+0)	340.29	1.7894 (−5)	1.4607 (−5)	2.5326 (−5)
500	500	9.8051	8340.7	2.4269 (+25)	456.35	6.5555 (+9)	6.9613 (−8)	28.964	284.900	9.5461 (+2)	1.1673 (+0)	338.37	1.7737 (−5)	1.5195 (−5)	2.5106 (−5)
1000	1000	9.8036	8246.9	2.3113 (+25)	453.74	6.2075 (+9)	7.3095 (−8)	28.964	281.651	8.9876 (+2)	1.1117 (+0)	336.43	1.7579 (−5)	1.5813 (−5)	2.4849 (−5)
1500	1500	9.8020	8153.0	2.2000 (+25)	451.12	5.8743 (+9)	7.6795 (−8)	28.964	278.402	8.4559 (+2)	1.0581 (+0)	334.49	1.7420 (−5)	1.6463 (−5)	2.4591 (−5)
2000	1999	9.8005	8059.2	2.0928 (+25)	448.48	5.5554 (+9)	8.0728 (−8)	28.964	275.154	7.9501 (+2)	1.0066 (+0)	332.53	1.7260 (−5)	1.7147 (−5)	2.4333 (−5)
2500	2499	9.7989	7965.3	1.9897 (+25)	445.82	5.2504 (+9)	8.4912 (−8)	28.964	271.906	7.4691 (+2)	9.5695 (−1)	330.56	1.7099 (−5)	1.7868 (−5)	2.4073 (−5)
3000	2999	9.7974	7871.4	1.8905 (+25)	443.15	4.9588 (+9)	8.9367 (−8)	28.964	268.659	7.0121 (+2)	9.0925 (−1)	328.58	1.6938 (−5)	1.8628 (−5)	2.3813 (−5)
3500	3498	9.7959	7777.5	1.7952 (+25)	440.47	4.6802 (+9)	9.4113 (−8)	28.964	265.413	6.5780 (+2)	8.6340 (−1)	326.59	1.6775 (−5)	1.9429 (−5)	2.3552 (−5)
4000	3997	9.7943	7683.6	1.7036 (+25)	437.76	4.4141 (+9)	9.9173 (−8)	28.964	262.166	6.1660 (+2)	8.1935 (−1)	324.59	1.6612 (−5)	2.0275 (−5)	2.3290 (−5)
4500	4497	9.7928	7589.7	1.6156 (+25)	435.05	4.1602 (+9)	1.0457 (−7)	28.964	258.921	5.7752 (+2)	7.7704 (−1)	322.57	1.6448 (−5)	2.1167 (−5)	2.3028 (−5)
5000	4996	9.7912	7495.7	1.5312 (+25)	432.31	3.9180 (+9)	1.1034 (−7)	28.964	255.676	5.4048 (+2)	7.3643 (−1)	320.55	1.6282 (−5)	2.2110 (−5)	2.2765 (−5)
5500	5495	9.7897	7401.8	1.4502 (+25)	429.56	3.6871 (+9)	1.1650 (−7)	28.964	252.431	5.0539 (+2)	6.9747 (−1)	318.50	1.6116 (−5)	2.3107 (−5)	2.2500 (−5)
6000	5994	9.7882	7307.8	1.3725 (+25)	426.79	3.4671 (+9)	1.2310 (−7)	28.964	249.187	4.7217 (+2)	6.6011 (−1)	316.45	1.5949 (−5)	2.4161 (−5)	2.2236 (−5)
6500	6493	9.7866	7213.8	1.2980 (+25)	424.00	3.2577 (+9)	1.3016 (−7)	28.964	245.943	4.4075 (+2)	6.6431 (−1)	314.39	1.5781 (−5)	2.5278 (−5)	2.1970 (−5)
7000	6992	9.7851	7119.8	1.2267 (+25)	421.20	3.0584 (+9)	1.3772 (−7)	28.964	242.700	4.1105 (+2)	5.9002 (−1)	312.31	1.5612 (−5)	2.6461 (−5)	2.1703 (−5)
7500	7491	9.7836	7025.8	1.1585 (+25)	418.37	2.8689 (+9)	1.4583 (−7)	28.964	239.457	3.8299 (+2)	5.5719 (−1)	310.21	1.5442 (−5)	2.7714 (−5)	2.1436 (−5)
8000	7990	9.7820	6931.7	1.0932 (+25)	415.53	2.6888 (+9)	1.5454 (−7)	28.964	236.215	3.5651 (+2)	5.2579 (−1)	308.11	1.5271 (−5)	2.9044 (−5)	2.1168 (−5)
8500	8489	9.7805	6837.7	1.0308 (+25)	412.67	2.5178 (+9)	1.6390 (−7)	28.964	232.974	3.3154 (+2)	4.9576 (−1)	305.98	1.5099 (−5)	3.0457 (−5)	2.0899 (−5)
9000	8987	9.7789	6743.6	9.7110 (+24)	409.79	2.3555 (+9)	1.7397 (−7)	28.964	229.733	3.0800 (+2)	4.6706 (−1)	303.85	1.4926 (−5)	3.1957 (−5)	2.0630 (−5)
9500	9486	9.7774	6649.5	9.1413 (+24)	406.89	2.2016 (+9)	1.8482 (−7)	28.964	226.492	2.8584 (+2)	4.3966 (−1)	301.70	1.4752 (−5)	3.3553 (−5)	2.0359 (−5)
10000	9984	9.7759	6555.4	8.5976 (+24)	403.97	2.0558 (+9)	1.9651 (−7)	28.964	223.252	2.6499 (+2)	4.1351 (−1)	299.53	1.4577 (−5)	3.5251 (−5)	2.0088 (−5)
10500	10483	9.7743	6461.3	8.0790 (+24)	401.03	1.9177 (+9)	2.0912 (−7)	28.964	220.013	2.4540 (+2)	3.8857 (−1)	297.35	1.4400 (−5)	3.7060 (−5)	1.9816 (−5)
11000	10981	9.7728	6367.2	7.5854 (+24)	398.07	1.7871 (+9)	2.2274 (−7)	28.964	216.774	2.2699 (+2)	3.6480 (−1)	295.15	1.4223 (−5)	3.8988 (−5)	1.9543 (−5)
11500	11479	9.7713	6364.6	7.0157 (+24)	397.95	1.6525 (+9)	2.4081 (−7)	28.964	216.650	2.0984 (+2)	3.3743 (−1)	295.07	1.4216 (−5)	4.2131 (−5)	1.9533 (−5)
12000	11977	9.7697	6365.6	6.4857 (+24)	397.95	1.5277 (+9)	2.6049 (−7)	28.964	216.650	1.9399 (+2)	3.1194 (−1)	295.07	1.4216 (−5)	4.5574 (−5)	1.9533 (−5)
12500	12475	9.7682	6366.6	5.9958 (+24)	397.95	1.4123 (+9)	2.8178 (−7)	28.964	216.650	1.7934 (+2)	2.8838 (−1)	295.07	1.4216 (−5)	4.9297 (−5)	1.9533 (−5)
13000	12973	9.7667	6367.6	5.5430 (+24)	397.95	1.3056 (+9)	3.0479 (−7)	28.964	216.650	1.6579 (+2)	2.6660 (−1)	295.07	1.4216 (−5	5.3325 (−5)	1.9533 (−5)
13500	13471	9.7651	6368.6	5.1244 (+24)	397.95	1.2070 (+9)	3.2969 (−7)	28.964	216.650	1.5327 (+2)	2.4646 (−1)	295.07	1.4216 (−5)	5.7680 (−5)	1.9533 (−5)
14000	13969	9.7636	6369.6	4.7375 (+24)	397.95	1.1159 (+9)	3.5662 (−7)	28.964	216.650	1.4170 (+2)	2.2786 (−1)	295.07	1.4216 (−5)	6.2391 (−5)	1.9533 (−5)
14500	14467	9.7621	6370.6	4.3799 (+24)	397.95	1.0317 (+9)	3.8574 (−7)	28.964	216.650	1.3100 (+2)	2.1066 (−1)	295.07	1.4216 (−5)	6.7485 (−5)	1.9533 (−5)
15000	14965	9.7605	6371.6	4.0493 (+24)	397.95	9.5380 (+8)	4.1723 (−7)	28.964	216.650	1.2111 (+2)	1.9476 (−1)	295.07	1.4216 (−5)	7.2995 (−5)	1.9533 (−5)
16000	15960	9.7575	6373.6	3.4612 (+24)	397.95	8.1528 (+8)	4.8812 (−7)	28.964	216.650	1.0352 (+2)	1.6647 (−1)	295.07	1.4216 (−5)	8.5397 (−5)	1.9533 (−5)
17000	16955	9.7544	6375.6	2.9587 (+24)	397.95	6.9691 (+8)	5.7102 (−7)	28.964	216.650	8.8497 (+1)	1.4230 (−1)	295.07	1.4216 (−5)	9.9901 (−5)	1.9533 (−5)
18000	17949	9.7513	6377.6	2.5292 (+24)	397.95	5.9576 (+8)	6.6797 (−7)	28.964	216.650	7.5652 (+1)	1.2165 (−1)	295.07	1.4216 (−5)	1.1686 (−4)	1.9533 (−5)
19000	18943	9.7483	6379.6	2.1622 (+24)	397.95	5.0931 (+8)	7.8135 (−7)	28.964	216.650	6.4674 (+1)	1.0400 (−1)	295.07	1.4216 (−5)	1.3670 (−4)	1.9533 (−5)
20000	19937	9.7452	6381.6	1.8486 (+24)	397.95	4.3543 (+8)	9.1393 (−7)	28.964	216.650	5.5293 (+1)	8.8910 (−2)	295.07	1.4216 (−5)	1.5989 (−4)	1.9533 (−5)
21000	20931	9.7422	6411.0	1.5742 (+24)	398.81	3.7161 (+8)	1.0732 (−6)	28.964	217.581	4.7289 (+1)	7.5715 (−2)	295.07	1.4267 (−5)	1.8843 (−4)	1.9611 (−5)
22000	21924	9.7391	6442.3	1.3413 (+24)	399.72	3.1733 (+8)	1.2596 (−6)	28.964	218.574	4.0475 (+1)	6.4510 (−2)	296.38	1.4322 (−5)	2.2201 (−4)	1.9695 (−5)
23000	22917	9.7361	6473.6	1.1437 (+24)	400.62	2.7119 (+8)	1.4772 (−6)	28.964	219.567	3.4668 (+1)	5.5006 (−2)	297.05	1.4376 (−5)	2.6135 (−4)	1.9778 (−5)
24000	23910	9.7330	6504.9	9.7591 (+23)	401.53	2.3194 (+8)	1.7312 (−6)	28.964	220.560	2.9717 (+1)	4.6938 (−2)	297.72	1.4430 (−5)	3.0743 (−4)	1.9862 (−5)
25000	24902	9.7300	6536.2	8.3341 (+23)	402.43	1.9852 (+8)	2.0272 (−6)	28.964	221.552	2.5492 (+1)	4.0084 (−2)	298.39	1.4484 (−5)	3.6135 (−4)	1.9945 (−5)
26000	25894	9.7269	6567.5	7.1225 (+23)	403.33	1.7004 (+8)	2.3720 (−6)	28.964	222.544	2.1883 (+1)	3.4257 (−2)	299.06	1.4538 (−5)	4.2439 (−4)	2.0029 (−5)
27000	26886	9.7239	6598.9	6.0916 (+23)	404.23	1.4575 (+8)	2.7734 (−6)	28.964	223.536	1.8799 (+1)	2.9298 (−2)	299.72	1.4592 (−5)	4.9805 (−4)	2.0112 (−5)
28000	27877	9.7208	6630.2	5.2138 (+23)	405.12	1.2502 (+8)	3.2404 (−6)	28.964	224.527	1.6161 (+1)	2.5076 (−2)	300.39	1.4646 (−5)	5.8405 (−4)	2.0195 (−5)
29000	28868	9.7178	6661.6	4.4657 (+23)	406.01	1.0732 (+8)	3.7832 (−6)	28.964	225.518	1.3904 (+1)	2.1478 (−2)	301.05	1.4699 (−5)	6.8437 (−4)	2.0278 (−5)
30000	29859	9.7147	6692.0	3.8278 (+23)	406.91	9.2192 (+8)	4.4137 (−6)	28.964	226.509	1.1970 (+1)	1.8410 (−2)	301.71	1.4753 (−5)	8.0134 (−4)	2.0361 (−5)
31000	30850	9.7117	6724.3	3.2833 (+23)	407.79	7.9251 (+8)	5.1456 (−6)	28.964	227.500	1.0312 (+1)	1.5792 (−2)	302.37	1.4806 (−5)	9.3759 (−4)	2.0443 (−5)
32000	31840	9.7087	6755.7	2.8133 (+23)	408.68	6.8175 (+8)	5.9945 (−6)	28.964	228.490	8.8906 (+0)	1.3555 (−2)	303.02	1.4859 (−5)	1.0962 (−3)	2.0526 (−5)
33000	32830	9.7056	6831.2	2.4062 (+23)	410.90	5.8522 (+8)	7.0212 (−6)	28.964	230.973	7.6730 (+0)	1.1573 (−2)	304.67	1.4992 (−5)	1.2955 (−3)	2.0733 (−5)
34000	33819	9.7026	6915.4	2.0558 (+23)	413.35	5.0297 (+8)	8.2182 (−6)	28.964	233.743	6.6341 (+0)	9.8874 (−3)	306.49	1.5140 (−5)	1.5312 (−3)	2.0963 (−5)
35000	34808	9.6995	6999.5	1.7597 (+23)	415.79	4.3307 (+8)	9.6010 (−6)	28.964	236.513	5.7459 (+0)	8.4634 (−3)	308.30	1.5287 (−5)	1.8062 (−3)	2.1193 (−5)
36000	35797	9.6965	7083.7	1.5090 (+23)	418.22	3.7356 (+8)	1.1196 (−5)	28.964	239.282	4.9852 (+0)	7.2579 (−3)	310.10	1.5433 (−5)	2.1264 (−3)	2.1422 (−5)

Altitude		Acceleration due to gravity	Pressure scale height	Number density	Particle speed	Collision frequency	Mean free path	Molecular weight	Temperature	Pressure	Density	Sound speed	Dynamic viscosity	Kinematic viscosity	Thermal conductivity
													μ	η	$10^{-3}\,k$
Z(m)	H(m)	g(m/s²)	H_p (m)	n (m⁻³)	V (m/s)	ν (s⁻¹)	L (m)	M (kg/kmol)	T(K)	P(mb)	ϱ(kg/m³)	C_s(m/s)	(N·s/m²)	(m²/s)	(J/m·s·K)
38000	37774	9.6904	7252.1	1.1158 (+23)	423.03	2.7939 (+7)	1.5141 (−5)	28.964	244.818	3.7713 (+0)	5.3666 (−3)	313.67	1.5723 (−5)	2.9297 (−3)	2.1878 (−5)
40000	39750	9.6844	7420.6	8.3077 (+22)	427.78	2.1036 (+7)	2.0336 (−5)	28.964	250.350	2.8714 (+0)	3.9957 (−3)	317.19	1.6009 (−5)	4.0066 (−3)	2.2331 (−5)
42000	41724	9.6783	7589.2	6.2266 (+22)	432.48	1.5939 (+7)	2.7133 (−5)	28.964	255.878	2.1996 (+0)	2.9948 (−3)	320.67	1.6293 (−5)	5.4404 (−3)	2.2781 (−5)
44000	43698	9.6723	7757.9	4.6965 (+22)	437.13	1.2152 (+7)	3.5973 (−5)	28.964	261.403	1.6949 (+0)	2.2589 (−3)	324.12	1.6573 (−5)	7.3371 (−3)	2.3229 (−5)
46000	45669	9.6662	7926.7	3.5640 (+22)	441.72	9.3182 (+6)	4.7404 (−5)	28.964	266.925	1.3134 (+0)	1.7142 (−3)	327.52	1.6851 (−5)	9.8305 (−3)	2.3764 (−5)
48000	47640	9.6602	8042.4	2.7376 (+22)	444.79	7.2075 (+6)	6.1713 (−5)	28.964	270.650	1.0229 (+0)	1.3167 (−3)	329.80	1.7037 (−5)	1.2939 (−2)	2.3973 (−5)
50000	49610	9.6542	8047.4	2.1351 (+22)	444.79	5.6201 (+6)	7.9130 (−5)	28.964	270.650	7.9779 (−1)	1.0269 (−3)	329.80	1.7037 (−5)	1.6591 (−2)	2.3973 (−5)
52000	51578	9.6482	8004.3	1.6750 (+22)	443.46	4.3966 (+6)	1.0086 (−4)	28.964	269.031	6.2214 (−1)	8.0562 (−3)	328.81	1.6956 (−5)	2.1047 (−2)	2.3843 (−5)
54000	53545	9.6421	7845.3	1.3286 (+22)	438.90	3.4515 (+6)	1.2716 (−4)	28.964	263.524	4.3337 (−1)	6.3901 (−3)	325.43	1.6600 (−5)	2.6104 (−2)	2.3400 (−5)
56000	55511	9.6361	7686.2	1.0488 (+22)	434.29	2.6961 (+6)	1.6108 (−4)	28.964	258.019	3.7362 (−1)	5.0445 (−4)	322.01	1.6402 (−5)	3.2514 (−2)	2.2955 (−5)
58000	57476	9.6316	7527.0	8.2390 (+21)	429.63	2.0952 (+6)	2.0506 (−4)	28.964	252.518	2.8723 (−1)	3.9627 (−4)	318.56	1.6121 (−5)	4.0682 (−2)	2.2508 (−5)
60000	59439	9.6241	7367.8	6.4387 (+21)	424.93	1.6195 (+6)	2.6239 (−4)	28.964	247.021	2.1958 (−1)	3.0968 (−4)	315.07	1.5837 (−5)	5.1141 (−2)	2.2058 (−5)
65000	64342	9.6091	6969.1	3.3934 (+21)	412.95	8.2945 (+5)	4.9787 (−4)	28.964	233.292	1.0929 (−1)	1.6321 (−4)	306.19	1.5116 (−5)	9.2617 (−2)	2.0926 (−5)
70000	69238	9.5942	6569.9	1.7222 (+21)	400.64	4.0839 (+5)	9.8104 (−4)	28.964	219.585	5.2209 (−2)	8.2829 (−5)	297.06	1.4377 (−5)	1.7357 (−1)	1.9780 (−5)
75000	74125	9.5793	6244.9	8.3003 (+20)	390.30	1.9175 (+5)	2.0354 (−3)	28.964	208.399	2.3881 (−2)	3.9921 (−5)	289.40	1.3759 (−5)	3.4465 (−1)	1.8834 (−5)
80000	79006	9.5644	5961.7	3.8378 (+20)	381.05	8.6559 (+5)	4.4022 (−3)	28.964	198.639	1.0524 (−3)	1.8458 (−5)	282.54	1.3208 (−5)	7.1557 (−1)	1.8001 (−5)
85000	83878	9.5496	5678.0	1.7090 (+20)	371.59	9.8858 (+5)	9.8858 (−3)	28.964	188.893	4.4568 (−3)	8.2196 (−6)	275.52	1.2647 (−5)	1.5386 (+0)	1.7162 (−5)
85500	84365	9.5481	5649.6	1.5727 (+20)	370.63	3.4501 (+5)	1.0743 (−2)	28.964	187.920	4.0802 (−3)	7.5641 (−6)	274.81	1.2590 (−5)	1.6645 (+0)	1.7078 (−5)
86000	84852	9.5466	5621	1.447 (+20)	369.7	3.17 (+4)	1.17 (−2)	28.95	186.87	3.7338 (−3)	6.958 (−6)	—	—	—	—
90000	88744	9.5348	5636	7.116 (+19)	369.9	1.56 (+4)	2.37 (−2)	28.91	186.87	1.8359	3.416	—	—	—	—
95000	93610	9.5200	5727	2.920 (+19)	372.6	6.44 (+4)	5.79 (−2)	28.73	188.42	7.5966 (−4)	1.393	—	—	—	—
100000	98451	9.5052	6009	1.189 (+19)	381.4	2.68 (+3)	1.42 (−1)	28.40	195.08	3.2011	5.604 (−7)	—	—	—	—
110000	108129	9.4759	7723	2.144 (+18)	431.7	5.48 (+2)	7.88 (−1)	27.27	240.00	7.1042 (−5)	9.708 (−8)	—	—	—	—
120000	117777	9.4466	12091	5.107 (+17)	539.3	1.63 (+2)	3.31 (+0)	26.20	360.00	2.5382	2.222	—	—	—	—
130000	127395	9.4175	16288	1.930 (+17)	625.0	7.1 (+1)	8.8 (+0)	25.44	469.27	1.2505	8.152 (−9)	—	—	—	—
140000	136983	9.3886	20025	9.322 (+16)	691.9	3.8 (+1)	1.8 (+1)	24.75	559.63	7.2028 (−6)	3.831	—	—	—	—
150000	146542	9.3597	23380	5.186 (+16)	746.5	2.3 (+1)	3.3 (+1)	24.10	634.39	4.5422	2.076	—	—	—	—
160000	156072	9.3310	26414	3.162 (+16)	792.2	1.5 (+1)	5.3 (+1)	23.49	696.29	3.0395	1.233	—	—	—	—
170000	165572	9.3024	29175	2.055 (+16)	831.3	1.0 (+1)	8.2 (+1)	22.90	747.57	2.1210	7.815 (−10)	—	—	—	—
180000	175043	9.2740	31703	1.400 (+16)	865.3	7.2 (+0)	1.2 (+2)	22.34	790.07	1.5271	5.194	—	—	—	—
190000	184486	9.2457	34030	9.887 (+15)	895.1	5.2 (+0)	1.7 (+2)	21.81	825.16	1.1266	3.581	—	—	—	—
200000	193899	9.2175	36183	7.182 (+15)	921.6	3.9 (+0)	2.4 (+2)	21.30	854.56	8.4736 (−7)	2.541	—	—	—	—
210000	—	—	—	—	—	—	—	—	878.84	6.4756	1.846	—	—	—	—
220000	212641	9.1615	40043	4.040 (+15)	966.5	2.3 (+0)	4.2 (+2)	20.37	899.01	5.0149	1.367	—	—	—	—
240000	231268	9.1061	43405	2.420 (+15)	1003.2	1.4 (+0)	7.0 (+2)	19.56	929.73	3.1059	7.858 (−11)	—	—	—	—
260000	249784	9.0511	46346	1.515 (+15)	1033.5	9.3 (−1)	1.1 (+3)	18.85	950.99	1.9894	4.742	—	—	—	—
280000	268187	8.9966	48925	9.807 (+14)	1058.7	6.1 (−1)	1.7 (+3)	18.24	965.75	1.3076	2.971	—	—	—	—
300000	286480	8.9427	51193	6.509 (+14)	1079.7	4.2 (−1)	2.6 (+3)	17.73	976.01	8.7704 (−8)	1.916	—	—	—	—
320000	304663	8.8892	53199	4.405 (+14)	1097.4	2.9 (−1)	3.8 (+3)	17.29	983.16	5.9796	1.264	—	—	—	—
340000	322738	8.8361	54996	3.029 (+14)	1112.4	2.0 (−1)	5.6 (+3)	16.91	988.15	4.1320	8.503 (−12)	—	—	—	—
360000	340705	8.7836	56637	2.109 (+14)	1125.5	1.4 (−1)	8.0 (+3)	16.57	991.65	2.8878 (−8)	5.805 (−12)	—	—	—	—
380000	358565	8.7315	58178	1.485 (+14)	1137.4	1.0 (−1)	1.1 (+4)	16.27	994.10	2.0384 (−8)	4.013 (−12)	—	—	—	—
400000	376320	8.6799	59678	1.056 (+14)	1148.5	7.2 (−2)	1.6 (+4)	15.98	995.83	1.4518 (−8)	2.803 (−12)	—	—	—	—
450000	420250	8.5529	63644	4.678 (+13)	1177.4	3.3 (−2)	3.6 (+4)	15.25	998.22	6.4468 (−9)	1.184 (−12)	—	—	—	—
500000	463540	8.4286	68785	2.192 (+13)	1215.0	1.6 (−2)	7.7 (+4)	14.33	999.24	3.0236 (−9)	5.215 (−13)	—	—	—	—
550000	506202	8.3070	76427	1.097 (+13)	1271.5	8.4 (−3)	1.5 (+5)	13.09	999.67	1.5137 (−9)	2.384 (−13)	—	—	—	—
600000	548252	8.1880	88244	5.950 (+12)	1356.4	4.8 (−3)	2.8 (+5)	11.51	999.85	8.2130 (−10)	1.137 (−13)	—	—	—	—
650000	589701	8.0716	105992	3.540 (+12)	1476.0	3.1 (−3)	4.8 (+5)	9.72	999.93	4.8865 (−10)	5.712 (−14)	—	—	—	—
700000	630563	7.9576	130630	2.311 (+12)	1627.0	2.2 (−3)	7.3 (+5)	8.00	999.97	3.1908 (−10)	3.070 (−14)	—	—	—	—
750000	670850	7.8460	161074	1.637 (+12)	1793.9	1.7 (−3)	1.0 (+6)	6.58	999.00	2.2599 (−10)	1.788 (−14)	—	—	—	—
800000	710574	7.7368	193862	1.234 (+12)	1954.3	1.4 (−3)	1.4 (+6)	5.54	999.99	1.7036 (−10)	1.136 (−14)	—	—	—	—
850000	749747	7.6298	224737	9.717 (+11)	2089.6	1.2 (−3)	1.7 (+6)	4.85	1000.00	1.3415 (−10)	7.824 (−15)	—	—	—	—
900000	788380	7.5250	250894	7.876 (+11)	2192.6	1.0 (−3)	2.1 (+6)	4.40	1000.00	1.0873 (−10)	5.759 (−15)	—	—	—	—
950000	826484	7.4224	271754	6.505 (+11)	2266.4	8.7 (−4)	2.6 (+6)	4.12	1000.00	8.9816 (−11)	4.453 (−15)	—	—	—	—
1000000	864071	7.3218	288203	5.442 (+11)	2318.1	7.5 (−4)	3.1 (+6)	3.94	1000.00	7.5138 (−11)	3.561 (−15)	—	—	—	—

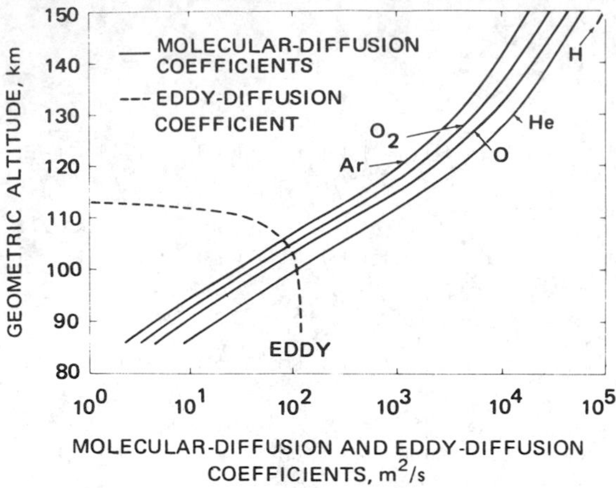

FIGURE 1. Molecular-diffusion and eddy-diffusion coefficients as a function of geometric altitude.

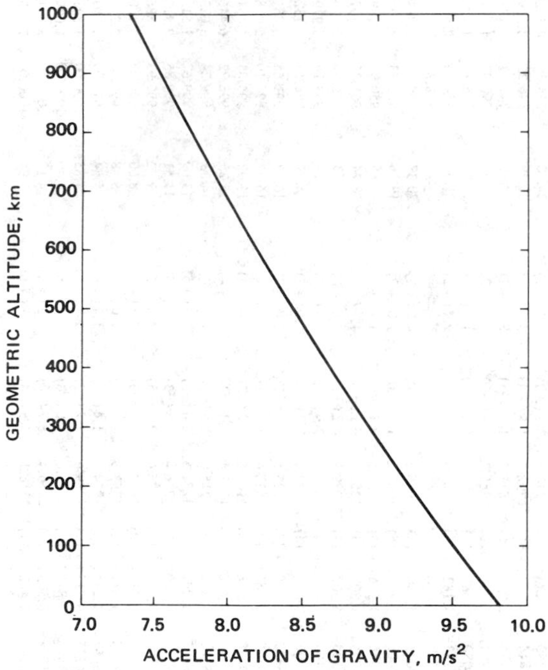

FIGURE 2. Acceleration of gravity as a function of geometric altitude.

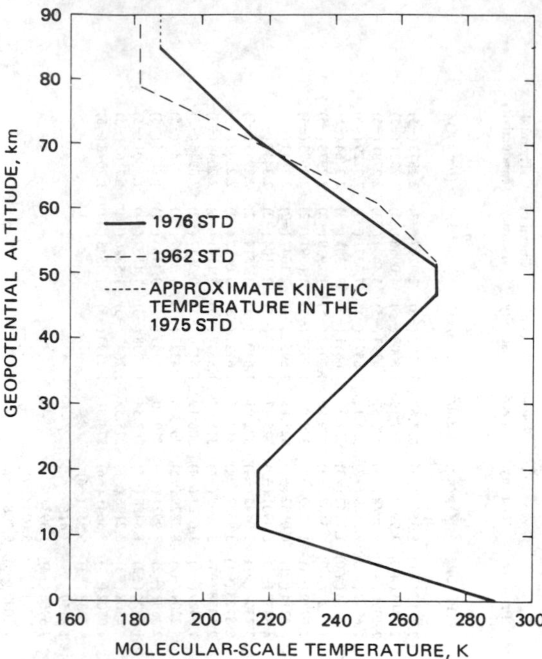

FIGURE 3. Molecular-scale temperature as a function of geopotential altitude.

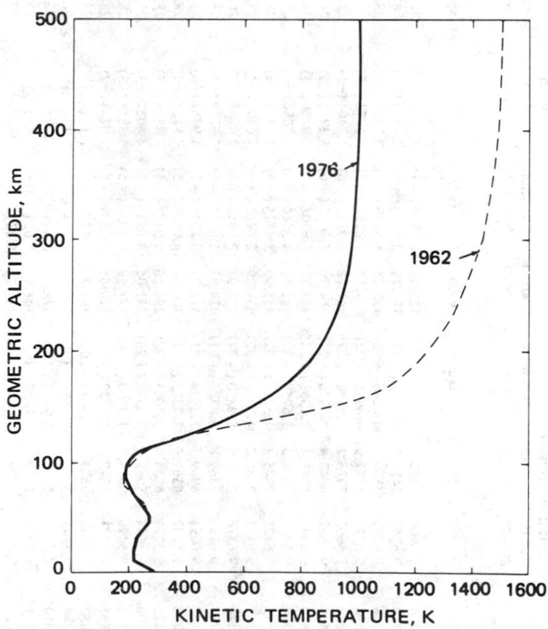

FIGURE 4. Kinetic temperature as a function of geometric altitude.

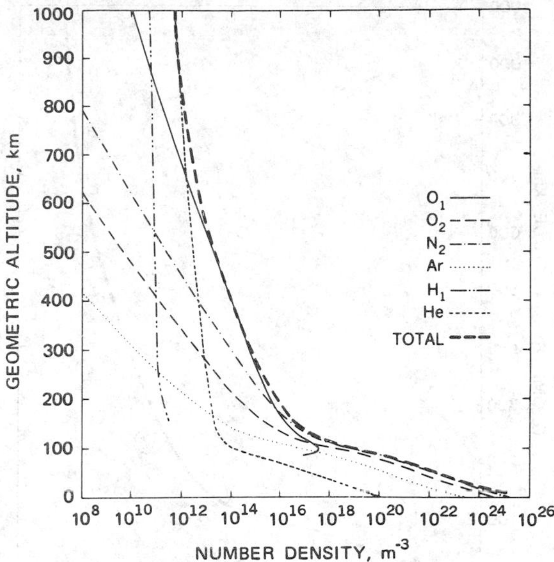

FIGURE 5. Number density of individual species and total number density as a function of geometric altitude.

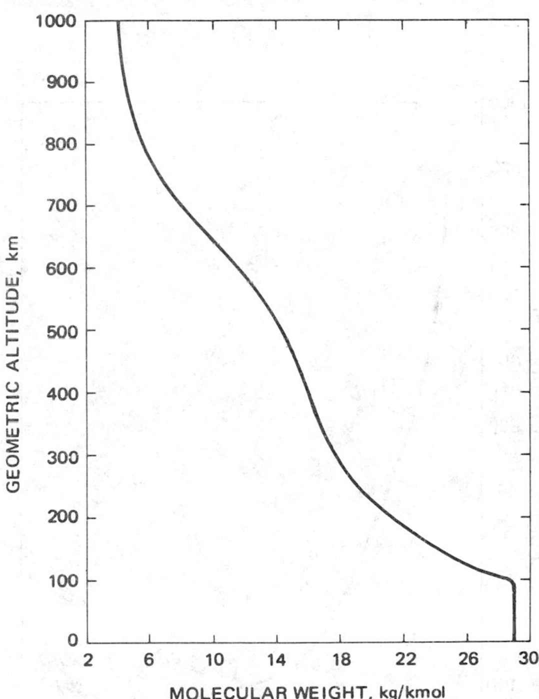

FIGURE 6. Mean molecular weight as a function of geometric altitude.

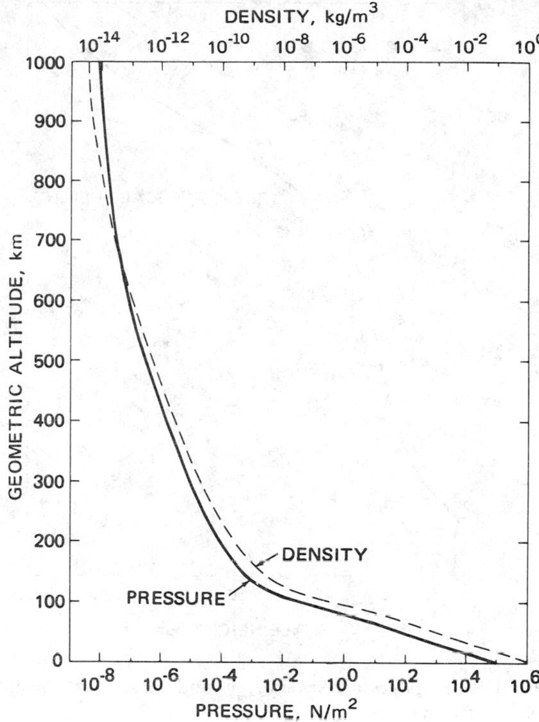

FIGURE 7. Total pressure and mass density as a function of geometric altitude.

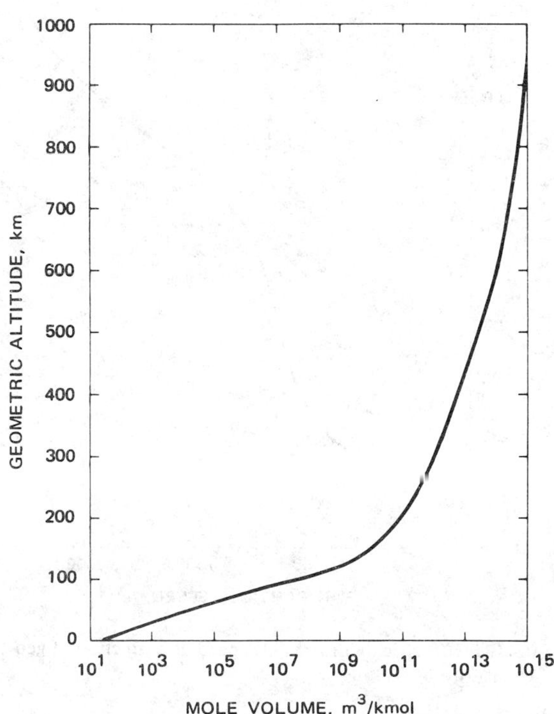

FIGURE 8. Mole volume as a function of geometric altitude.

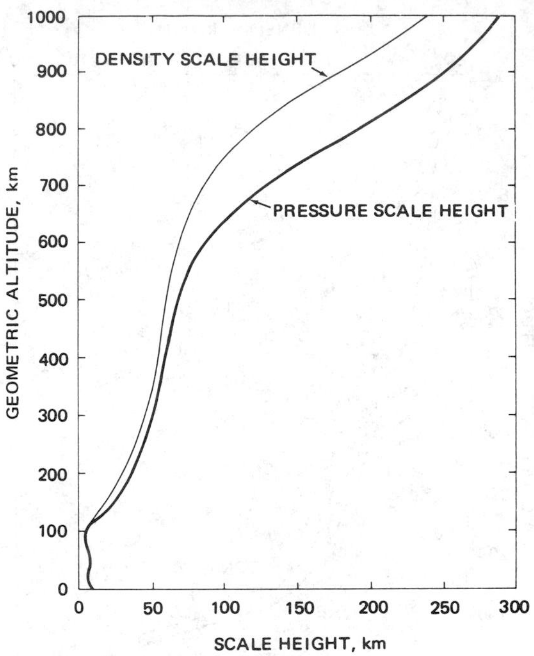

FIGURE 9. Pressure scale height and density scale height as a function of geometric altitude.

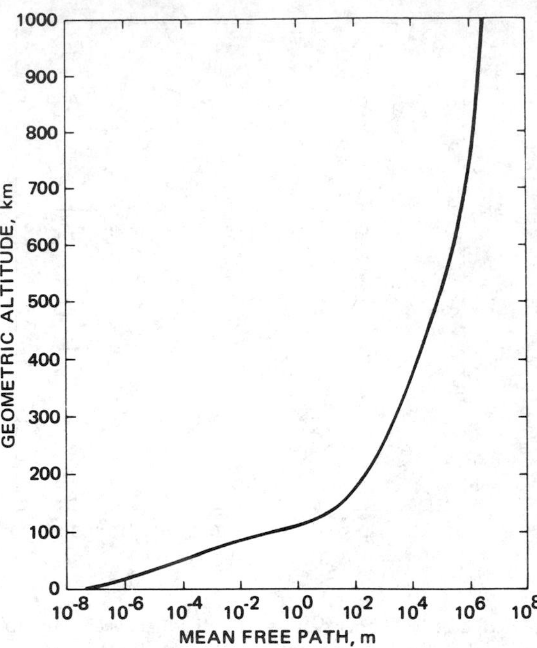

FIGURE 11. Mean free path as a function of geometric altitude.

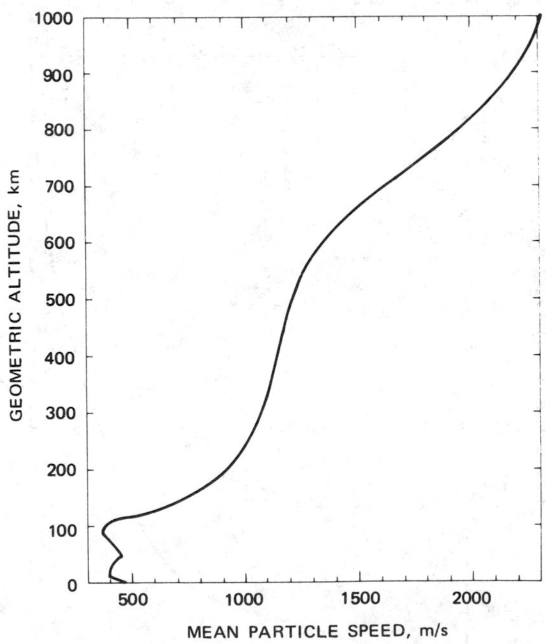

FIGURE 10. Mean air-particle speed as a function of geometric altitude.

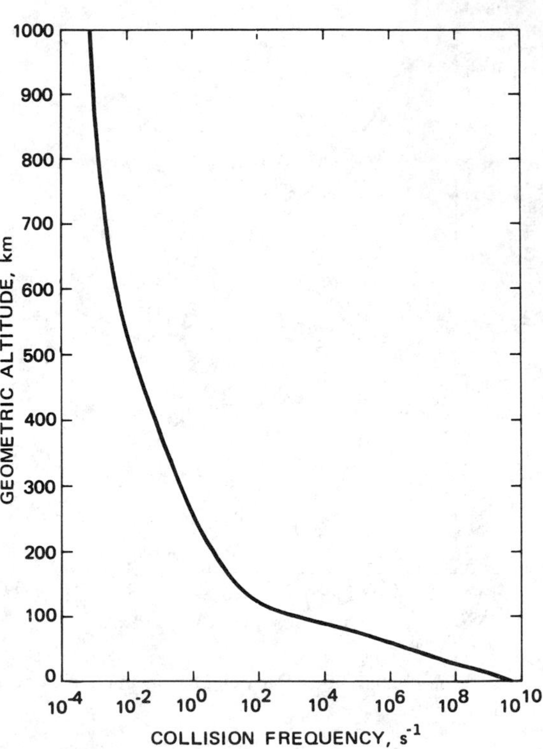

FIGURE 12. Collision frequency as a function of geometric altitude.

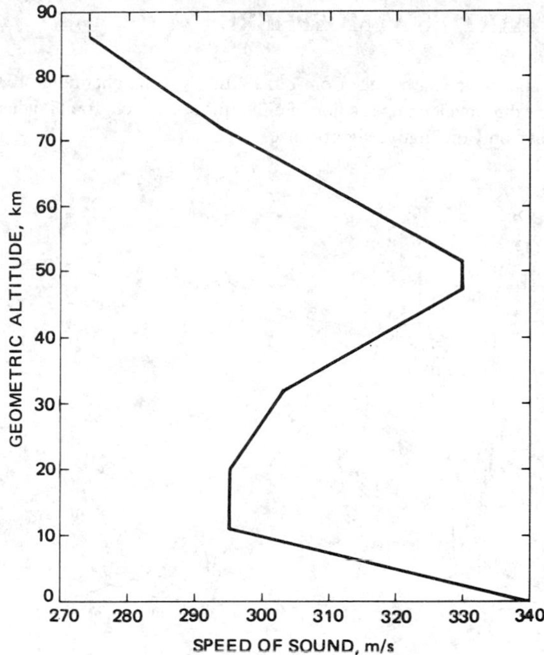

FIGURE 13. Speed of sound as a function of geometric altitude.

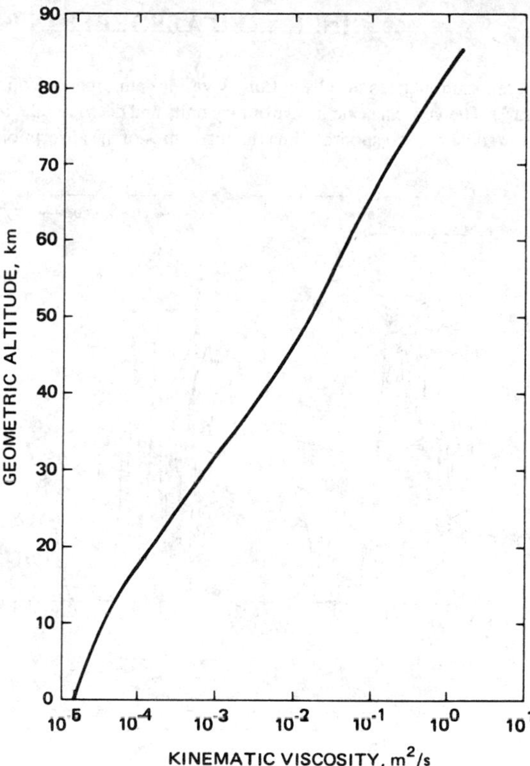

FIGURE 15. Kinematic viscosity as a function of geometric altitude.

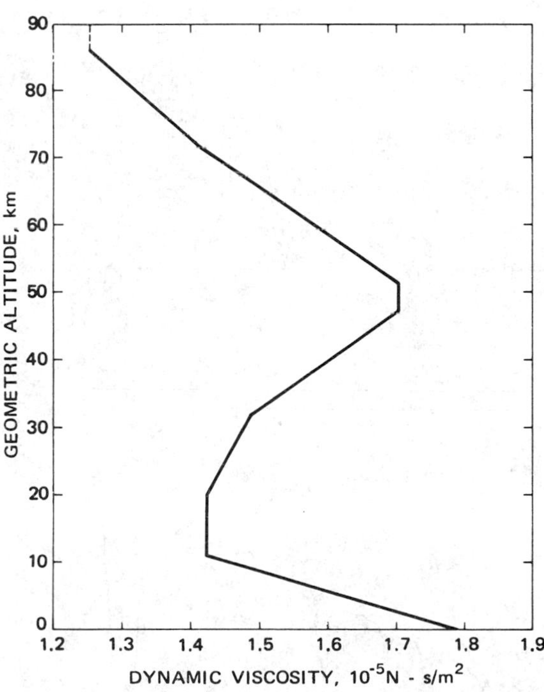

FIGURE 14. Dynamic viscosity as a function of geometric altitude.

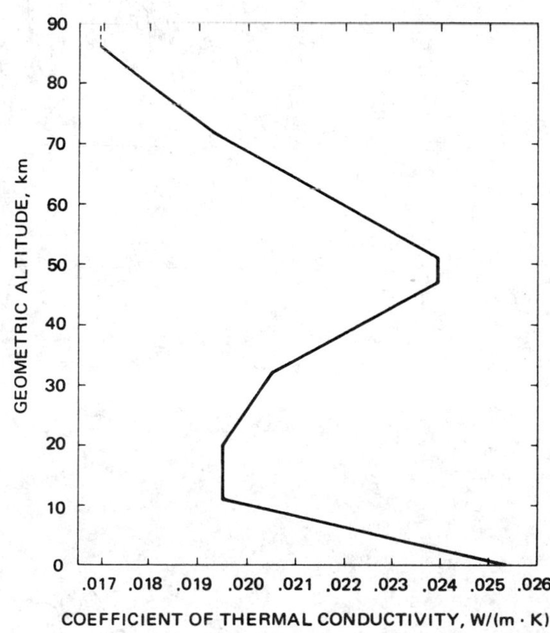

FIGURE 16. Coefficient of thermal conductivity as a function of geometric altitude.

INFRARED ABSORPTION BY THE EARTH'S ATMOSPHERE

This graph summarizes the absorption by various atmospheric constituents in the wavelength range from 1 to 13 μm (wavenumber range 10,000 to 770 cm^{-1}). The vertical scale is in arbitrary units and does not take into account the wavelength variation of either the solar background radiation or the infrared detector response. Thus the intensities of the absorption bands have only qualitative significance.

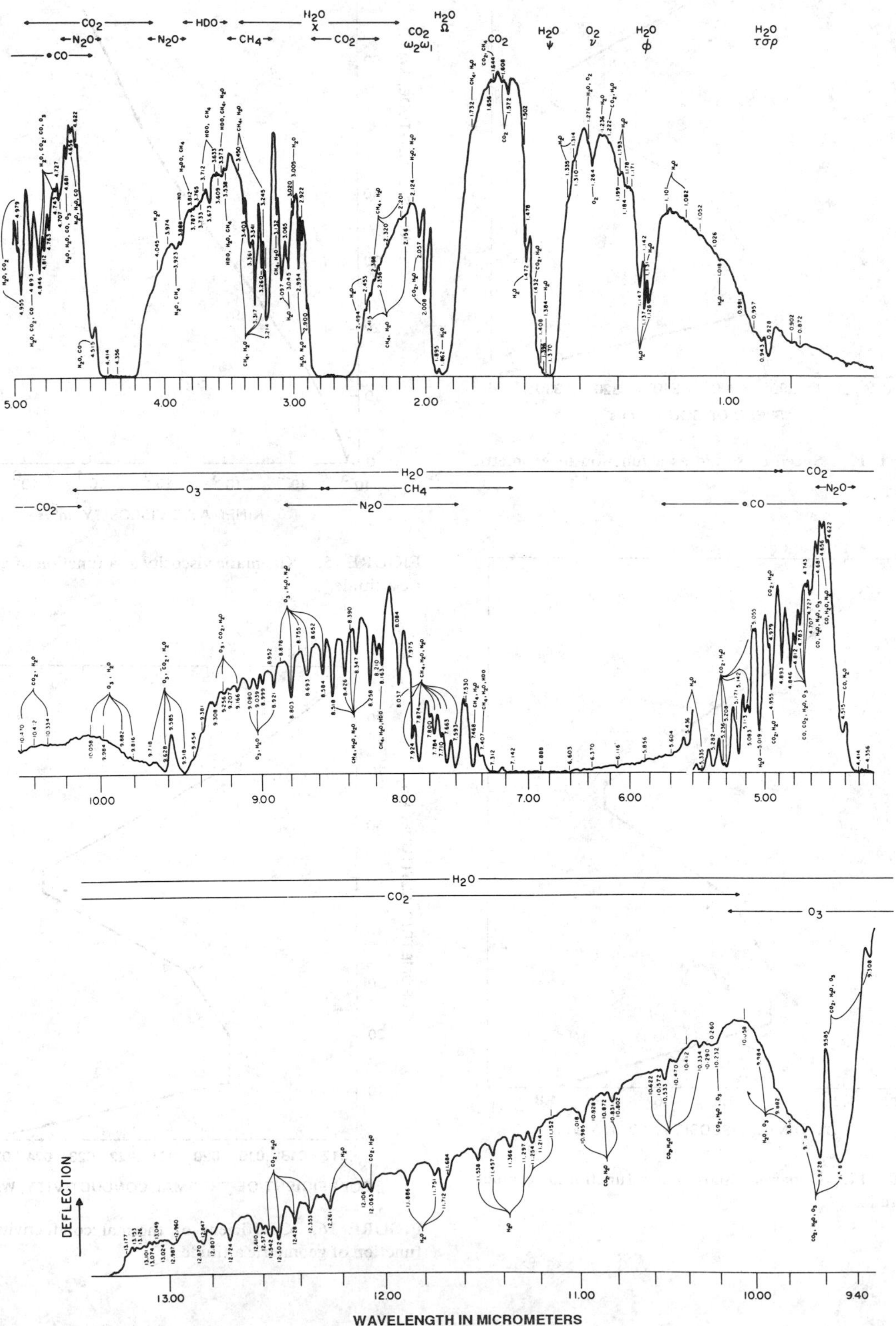

ATMOSPHERIC CONCENTRATION OF CARBON DIOXIDE, 1958—1990

The data were taken at Mauna Loa Observatory in Hawaii and represent averages adjusted to the 15th of each month. Concentration of CO_2 is given in parts per million by volume.

REFERENCE

TRENDS 90, A Compendium of Data on Global Change, Carbon Dioxide Information Analysis Center, Oak Ridge National Laboratory, P.O. Box 2008, Oak Ridge, TN 37831-6335.

CO_2 Concentrations (ppmv)

Year	Jan	Feb	Mar	April	May	June	July	Aug	Sept	Oct	Nov	Dec	Ann
1958			315.55	317.29	317.34		315.69	314.78	313.03		313.18	314.50	
1959	315.42	316.32	316.49	317.56	318.13	318.00	316.39	314.66	313.68	313.18	314.66	315.43	315.83
1960	316.27	316.81	317.42	318.87	319.87	319.43	318.01	315.75	314.00	313.68	314.84	316.03	316.75
1961	316.73	317.54	318.38	319.31	320.42	319.61	318.42	316.64	314.83	315.15	315.95	316.85	317.49
1962	317.78	318.40	319.53	320.41	320.85	320.45	319.44	317.25	316.12	315.27	316.53	317.53	318.30
1963	318.58	318.92	319.70	321.22	322.08	321.31	319.58	317.61	316.05	315.83	316.91	318.20	318.83
1964	319.41				322.06	321.73	320.27	318.54	316.54	316.71	317.53	318.55	
1965	319.27	320.28	320.73	321.97	322.00	321.71	321.05	318.71	317.65	317.14	318.71	319.25	319.87
1966	320.46	321.43	322.22	323.54	323.91	323.59	322.26	320.21	318.48	317.94	319.63	320.87	321.21
1967	322.17	322.34	322.88	324.25	324.83	323.93	322.39	320.76	319.10	319.23	320.56	321.80	322.02
1968	322.40	322.99	323.73	324.86	325.41	325.19	323.97	321.92	320.10	319.96	320.97	322.48	322.83
1969	323.52	323.89	325.04	326.01	326.67	325.96	325.13	322.90	321.61	321.01	322.08	323.37	323.93
1970	324.34	325.30	326.29	327.54	327.54	327.21	325.98	324.42	322.91	322.90	323.85	324.96	325.27
1971	326.01	326.51	327.01	327.62	328.76	328.40	327.20	325.28	323.20	323.40	324.64	325.85	326.16
1972	326.60	327.47	327.58	329.56	329.90	328.92	327.89	326.17	324.68	325.04	326.34	327.39	327.29
1973	328.37	329.40	330.14	331.33	332.31	331.90	330.70	329.15	327.34	327.02	327.99	328.48	329.51
1974	329.18	330.55	331.32	332.48	332.92	332.08	331.02	329.24	327.28	327.21	328.29	329.41	330.08
1975	330.23	331.24	331.87	333.14	333.80	333.42	331.73	329.90	328.40	328.17	329.32	330.59	330.99
1976	331.58	332.39	333.33	334.41	334.71	334.17	332.88	330.77	329.14	328.77	330.14	331.52	331.98
1977	332.75	333.25	334.53	335.90	336.57	336.10	334.76	332.59	331.41	330.98	332.24	333.68	333.73
1978	334.80	335.22	336.47	337.59	337.84	337.72	336.37	334.51	332.60	332.37	333.75	334.79	335.34
1979	336.05	336.59	337.79	338.71	339.30	339.12	337.56	335.92	333.74	333.70	335.13	336.56	336.68
1980	337.84	338.19	339.90	340.60	341.29	341.00	339.39	337.43	335.72	335.84	336.93	338.04	338.52
1981	339.06	340.30	341.21	342.33	342.74	342.07	340.32	338.27	336.52	336.68	338.19	339.44	339.76
1982	340.57	341.44	342.53	343.39	343.96	343.18	341.88	339.65	337.80	337.69	339.09	340.32	340.96
1983	341.20	342.35	342.93	344.77	345.58	345.14	343.81	342.22	339.69	339.82	340.98	342.82	342.61
1984	343.52	344.33	345.11	346.88	347.25	346.61	345.22	343.11	340.90	341.17	342.80	344.04	344.25
1985	344.79	345.82	347.25	348.17	348.75	348.07	346.38	344.52	342.92	342.63	344.06	345.38	345.73
1986	346.12	346.79	347.69	349.38	350.04	349.38	347.78	345.75	344.70	344.01	345.50	346.75	346.99
1987	347.86	348.32	349.26	350.84	351.70	351.11	349.37	347.97	346.31	346.22	347.68	348.82	348.79
1988	350.29	351.58	352.08	353.45	354.08	353.66	352.25	350.30	348.58	348.74	349.93	351.21	351.35
1989	352.62	352.93	353.54	355.27	355.52	354.97	353.74	351.51	349.63	349.82	351.12	352.35	352.75
1990	353.47	354.51	355.18	355.98	356.94	355.99	354.58	352.68	350.72	350.92	352.55	353.91	353.95

MEAN TEMPERATURES IN THE UNITED STATES, 1901—1987

Historical records of atmospheric temperatures in the U.S. have been analyzed to obtain mean temperatures in °C for 23 climatically-distinct regions for the period 1901—1987. This table gives the average of the values for these 23 regions covering the entire country. Data for the individual regions and for other parts of the world may be found in the reference.

REFERENCE

TRENDS 90, A Compendium on Global Change, Carbon Dioxide Information Analysis Center, ORNL, P.O. Box 2008, Oak Ridge, Tenn. 37831-6335.

Year	Ann*	Win	Spr	Sum	Fall	Ann†
1901	11.46	0.55	10.59	22.92	12.36	11.61
1902	11.33	–0.21	11.39	21.75	12.46	11.35
1903	10.79	–0.57	10.94	21.23	11.70	10.82
1904	11.03	–0.82	10.74	21.31	12.56	10.95
1905	11.03	–1.37	11.39	21.97	12.25	11.06
1906	11.44	0.88	10.27	21.77	12.15	11.27
1907	11.31	1.51	10.43	21.32	12.07	11.33
1908	11.59	1.20	11.56	21.62	12.21	11.65
1909	11.21	1.35	9.93	22.31	12.68	11.57
1910	11.78	–1.60	12.74	22.10	12.75	11.50
1911	11.65	1.02	11.53	22.35	11.60	11.63
1912	10.61	–1.09	10.02	21.40	11.96	10.57
1913	11.39	0.04	10.62	22.42	12.41	11.37
1914	11.55	0.83	11.17	22.51	12.91	11.85
1915	11.27	–0.39	10.47	20.98	12.88	10.98
1916	11.01	0.30	10.89	21.99	11.59	11.19
1917	10.43	–1.07	9.10	21.85	11.79	10.42
1918	11.55	–1.17	11.66	22.66	12.06	11.30
1919	11.38	1.14	10.99	22.51	12.00	11.66
1920	11.05	–0.28	9.88	21.58	12.26	10.86
1921	12.63	2.04	12.09	22.83	13.15	12.52
1922	11.64	0.22	10.93	22.58	13.12	11.71
1923	11.41	0.88	10.14	22.02	12.08	11.28
1924	10.80	0.83	9.69	21.98	12.32	11.21
1925	11.86	0.36	12.21	22.49	11.57	11.66
1926	11.48	1.16	10.61	22.14	12.17	11.52
1927	11.65	1.39	11.12	21.28	13.30	11.77
1928	11.51	0.27	10.97	21.70	12.28	11.31
1929	10.97	–1.24	11.26	22.14	11.52	10.92
1930	11.59	0.96	11.26	22.64	12.03	11.72
1931	12.52	1.63	10.58	23.12	13.95	12.32
1932	11.51	2.06	10.45	22.65	11.95	11.78
1933	12.24	0.33	10.91	23.14	13.36	11.93
1934	12.75	2.22	12.59	23.61	13.29	12.93
1935	11.56	1.30	10.70	22.61	11.99	11.65
1936	11.69	–1.78	11.87	23.72	12.22	11.51
1937	11.32	–0.26	10.66	22.98	12.30	11.42
1938	12.25	1.43	11.78	22.64	12.85	12.18
1939	12.29	0.96	11.74	22.68	13.15	12.13
1940	11.49	0.34	10.89	22.59	12.42	11.56
1941	12.06	1.61	11.28	22.33	12.96	12.04
1942	11.57	0.76	11.29	22.30	12.58	11.73
1943	11.65	1.08	10.78	22.87	11.98	11.68
1944	11.54	1.12	10.41	22.21	12.75	11.62

MEAN TEMPERATURES IN THE UNITED STATES, 1901—1987 (continued)

Year	Ann*	Win	Spr	Sum	Fall	Ann†
1945	11.46	0.53	11.30	21.69	12.60	11.53
1946	12.16	0.48	12.41	22.10	12.32	11.83
1947	11.55	1.21	10.46	22.21	12.98	11.72
1948	11.44	–0.12	11.01	22.31	12.52	11.43
1949	11.74	–0.13	11.41	22.62	12.92	11.71
1950	11.32	1.40	10.05	21.42	12.63	11.38
1951	11.14	0.88	10.38	21.90	11.63	11.19
1952	11.78	1.21	10.52	22.98	12.07	11.70
1953	12.33	2.28	10.88	22.71	13.42	12.32
1954	12.33	2.32	10.95	22.78	13.24	12.32
1955	11.43	0.16	11.41	22.39	12.17	11.53
1956	11.78	0.32	10.88	22.46	12.57	11.56
1957	11.65	1.64	10.82	22.43	11.58	11.62
1958	11.55	1.27	10.62	22.23	12.98	11.77
1959	11.69	0.21	11.32	22.81	11.71	11.51
1960	11.26	0.49	10.10	22.42	13.05	11.51
1961	11.47	0.71	10.65	22.42	11.95	11.43
1962	11.49	–0.11	10.91	21.88	12.91	11.40
1963	11.64	–0.43	11.74	22.28	14.05	11.91
1964	11.39	–0.57	10.83	22.09	12.25	11.15
1965	11.41	0.55	10.24	21.61	12.60	11.25
1966	11.29	0.05	11.08	22.24	12.37	11.44
1967	11.45	0.93	11.05	21.73	12.11	11.45
1968	11.24	0.22	11.00	21.99	12.24	11.36
1969	11.28	–0.21	10.60	22.24	11.96	11.14
1970	11.38	0.32	10.50	22.60	12.00	11.36
1971	11.42	0.54	10.21	22.26	12.56	11.39
1972	11.25	0.81	11.51	21.95	11.54	11.45
1973	11.75	–0.05	11.02	22.44	12.98	11.60
1974	11.72	1.02	11.77	22.07	12.01	11.72
1975	11.29	0.84	9.82	22.06	12.35	11.27
1976	11.23	1.72	11.15	21.82	10.84	11.38
1977	11.88	–0.83	12.29	22.79	12.76	11.76
1978	11.02	–1.29	10.91	22.42	12.64	11.16
1979	10.96	–2.47	10.86	21.92	12.48	10.69
1980	11.71	1.06	10.60	22.90	12.46	11.75
1981	12.12	1.77	11.81	22.65	12.60	12.20
1982	11.28	–0.25	10.96	21.87	12.06	11.16
1983	11.50	2.11	10.02	22.79	12.97	11.98
1984	11.57	–0.49	10.41	22.47	12.14	11.14
1985	11.21	–0.34	12.48	22.14	11.84	11.53
1986	12.30	0.76	12.40	22.72	12.21	12.03
1987	12.28	1.62	12.24	22.66	12.58	12.27

*Calendar year mean (Jan–Dec).

†Season year mean (Win = Dec–Feb; Spr = Mar–May; Sum = Jun–Aug; Fall = Sep–Nov).

ATMOSPHERIC ELECTRICITY

Hans Dolezalek, Hannes Tammet, John Latham, and Martin A. Uman

I. SURVEY AND GLOBAL CIRCUIT

Hans Dolezalek

The science of atmospheric electricity originated in 1752 by an experimental proof of a related earlier hypothesis (that lightning is an electrical event). In spite of a large effort, in part by such eminent physicists as Coulomb, Lord Kelvin, and many others, an overall, proven theory able to generate models with sufficient resolution is not yet available. Generally accepted and encompassing text books are now more than 20 years old. The voluminous proceedings of the, so far, nine international atmospheric electricity conferences (1954 to 1992) give much valuable detail and demonstrate impressive progress, as do a number of less comprehensive textbooks published in the last 20 years, but a general theory as indicated above is not yet created. Only now, certain related measuring techniques and mathematical possibilities are emerging.

Applications to practical purposes do exist in the field of lightning research (including the electromagnetic radiation emanating from lightning) by the establishment of lightning-location networks and by the now developing possibility to detect electrified clouds which pose hazards to aircraft. Application of atmospheric electricity to other parts of meteorology seems to be promising but so far has seldom been instituted. Because some atmospheric electric signals propagate around the earth and because of the existence of a global circuit, applications for the monitoring of global change processes and conditions are now being proposed. Significant secular changes in the global circuit would indicate a change in the global climate; the availability of many old data (about a span of 100 years) could help detect a long term trend.

The concept of the "global circuit" is based on the theory of the global spherical capacitor: both, the solid (and liquid) earth as one electrode, and the high atmospheric layers (about the ionosphere) as the other, are by orders of magnitude more electrically conductive than the atmosphere between them. According to the "classical picture of atmospheric electricity", this capacitor is continuously charged by the common action of all thunderstorms to a d.c. voltage difference of several hundred kilovolts, the earth being negative. The much smaller but still existing conductivity of the atmosphere allows a current flowing from the ionosphere to the ground, integrated for all sink areas of the whole earth, of the order of 1.5 kA. In this way, a global circuit is created with many generators and sink-areas both interspaced and distributed over the whole globe, all connected to two nodes: ionosphere and ground. Within the scope of the global circuit, for each location, the current density (order of several pA/m^2) is determined by the voltage difference between ionosphere and ground (which is the same for all locations but varying in time) and the columnar resistance reaching from the ground up to the ionosphere (in the order of 10^{17} Ω m^2).

Natural processes, especially meteorological processes and some human activity, which produce or move electric charges ("space charges") or affect the ion distribution, constitute local generators and thereby "local circuits", horziontally and/or in parallel or antiparallel to the local part of the global circuit. In many cases, the local currents are much stronger than the global ones, making the measurement of the global current at a given location and/or during a period of time very difficult or, often, impossible. The strongest local circuits usually occur with certain weather conditions (precipitation, fog, high wind, blown-up dust or snow, heavy cloudiness) which make measurement of the global circuit impossible everywhere; but even in their absence local generators exist in varying magnitudes and of different characters. The separation of the local and global shares in the measured values of current density is a central problem of the science of atmospheric electricity. Aerological measurements are of high value in this regard.

The above description is within the "classical picture" of atmospheric electricity, a group of hypotheses to explain the electricifation of the atmosphere. It is probably fundamentally correct but certainly not complete; it has not yet been confirmed by systems of measurements resulting in no inner contradictions. In particular, extraterrestrial influences must be permitted; their general significance is still under debate.

Within this "classical picture" a kind of electric standard atmosphere may be constructed as shown in Table 1.

Values with a star, *, are rough average values from measurement. A star in parentheses, (*), points to a typical value from one or a few measurements. All other values have been calculated from starred values, under the assumption that at 2 km 50% and at 12 km 90% of the columnar resistance is reached. Voltage drop along one of the partial columns can be calculated by subtracting the value for the lower column from that of the upper one. Columnar resistances, conductances, and capacitances are valid for that particular part of the column which is indicated at the left. Capacitances are calculated with the formula for plate capacitors, and this fact must be considered also for the time constants for columns.

According to measurements, U, the potential difference between 0 m and 65 km may vary by a factor of approximately 2. The total columnar resistance, R_c, is estimated to vary up to a factor of 3, the variation being due to either reduction of conductivity in the exchange layer (about lowest 2 km of this table) or to the presence of high mountains; in both cases the variation is caused in the troposphere. Smaller variations in the stratosphere and mesosphere are being discussed because of aerosols there. The air-earth current density in fair weather varies by a factor of 3 to 6 accordingly. Conductivity near the ground varies by a factor of about 3 but only decreasing; increase of conductivity due to extraordinary radioactivity is a singular event. The field strength near the ground varies as a consequence of variations of air-earth current density and conductivity from about 1/3 to about 10 times of the value quoted in the table. Conductivity near the ground shows a diurnal and an annual variation which depends strongly on the locality: air-earth current density shows a diurnal and annual variation because the earth-ionosphere potential difference undergoes such variations, and also because the columnar resistance is supposed to have a diurnal and probably an annual variation.

Conductivities and air-earth current densities on high mountains are greater than at sea level by factors of up to 10. Conductivity decreases when atmospheric humidity increases. Values for space charges are not quoted because measurements are too few to allow calculation of average values. Values of parameters over the oceans are still rather uncertain.

Theoretically, in fair-weather conditions, Ohm's law must be fulfilled for the electric field, the conduction current density, and the electrical conductivity of the atmosphere. Deviations point to shortcomings in the applied measuring techniques. Data which are representative for a large area (in the extreme, "globally representative data", i.e. data on the global circuit), can on the ground be obtained only by stations on an open plane and only if local generators are either small or constant or are independently measured. Certain measurements with instrumented aircraft provide globally representative information valid for the period of the actual measurement.

TABLE 1

Electrical Parameters of the Clear (Fair-Weather) Atmosphere, Pertinent to the Classical Picture of Atm. Electricity (Electric Standard Atmosphere)

Part of atmosphere for which the values are calculated (elements are in free, cloudless atmosphere)	Currents, I, in A; and current densities, i, in A/m²	Potential differences, U, in V; field strength E in V/m; U = 0 at sea level	Resistances, R, in Ω; columnar resistances, R_c, in Ω m² and resistivities, ρ, in Ω m	Conductances, G, in Ω^{-1}; columnar conductances G_c, in Ω^{-1} m^{-2}; total conductivities, γ, in Ω^{-1} m^{-1}	Capacitances, C, in F; columnar capacitances, C_c, in F m^{-2} and capacitivities, ε, in F m^{-1}	Time constants τ, in seconds
Volume element at about sea level, 1 m³	$i = 3 \times 10^{-12}$*	$E_0 = 1.2 \times 10^{2}$*	$\rho_0 = 4 \times 10^{13}$	$\gamma_0 = 2.5 \times 10^{-14}$	$\varepsilon_0 = 8.9 \times 10^{-12}$*	$\tau_0 = 3.6 \times 10^{2}$
Lower column of 1 m² cross section from sea level to 2 km height	$i = 3 \times 10^{-12}$	At upper end: $U_1 = 1.8 \times 10^{5}$	$R_{c1} = 6 \times 10^{16}$	$G_{c1} = 1.7 \times 10^{-17}$	$C_{c1} = 4.4 \times 10^{-15}$	$\tau_{c1} = 2.6 \times 10^{2}$
Volume element at about 2 km height, 1 m³	$i = 3 \times 10^{-12}$	$E_2 = 6.6 \times 10^{1}$	$\rho_2 = 2.2 \times 10^{13(*)}$	$\gamma_2 = 4.5 \times 10^{-14}$	$\varepsilon_2 = 8.9 \times 10^{-12}$	$\tau_2 = = 2 \times 10^{2}$
Center column of 1 m² cross section from 2 to 12 km	$i = 3 \times 10^{-12}$	At upper end: $U_m = 3.15 \times 10^{5}$	$R_{cm} = 4.5 \times 10^{16}$	$G_{cm} = 5 \times 10^{-17}$	$C_{cm} = 8.8 \times 10^{-16}$	$\tau_{cm} = 1.8 \times 10^{1}$
Volume element at about 12 km height, 1 m³	$i = 3 \times 10^{-12}$	$E_{12} = 3.9 \times 10^{0}$	$\rho_{12} = 1.3 \times 10^{12(*)}$	$\gamma_{12} = 7.7 \times 10^{-13}$	$\varepsilon_{12} = 8.9 \times 10^{-12}$	$\tau_{12} = 1.2 \times 10^{1}$
Upper column of 1 m² cross section from 12 to 65 km height	$i = 3 \times 10^{-12}$	At upper end: $U_u = 3.5 \times 10^{5}$	$R_{cu} = 1.5 \times 10^{16}$	$G_{cu} = 2.5 \times 10^{-17}$	$C_{cu} = 1.67 \times 10^{-16}$	$\tau_{cm} = 6.7 \times 10^{0}$
Whole column of 1 m² cross section from 0 to 65 km height	$i = 3 \times 10^{-12}$	At upper end: $U = 3.5 \times 10^{5}$	$R_c = 1.2 \times 10^{17}$	$G_c = 8.3 \times 10^{-18}$	$C_c = 1.36 \times 10^{-16}$	$\tau_c = 1.64 \times 10^{1}$
Total spherical capacitor area: 5×10^{14} m²	$1 = 1.5 \times 10^{3}$	$U = 3.5 \times 10^{5}$*	$R = 2.4 \times 10^{2}$	$G = 4.2 \times 10^{-3}$	$C = 6.8 \times 10^{-2}$	$\tau = 1.64 \times 10^{1}$

Note: All currents and fields listed are part of the global circuit, i.e., circuits of local generators are not included. Values are subject to variations due to latitude and altitude of the point of observation above sea level, locality with respect to sources of disturbances, meteorological and climatological factors, and man-made changes. For more explanations, see text.

II. AIR IONS

Hannes Tammet

The term " air ions" signifies all airborne particles which are the carriers of the electrical current in the air and have drift velocities determined by the electric field.

The probability of electrical dissociation of molecules in the atmospheric air under thermodynamic equilibrium is near to zero. The average ionization at the ground level over the ocean is $2 \cdot 10^6$ ion pairs $m^{-3}s^{-1}$. This ionization is produced mainly by cosmic rays. Over the continents the ionizing radiation from soil and from radioactive substances in the air each add about $4 \cdot 10^6$ $m^{-3}s^{-1}$. The total average ionization rate of 10^7 $m^{-3}s^{-1}$ is equivalent to 17 μR/h which is a customary expression of the background level of the ionizing radiations. The ionization rate over the ground varies in space due to the radioactivity of soil, and in time depending on the exchange of air between the atmosphere and radon-containing soil. Radioactive pollution increases the ionization rate. A temporary increase of about 10 times was registered in Sweden after the Chernobyl accident in 1986. The emission of Kr^{85} from nuclear power plants can noticeably increase the global ionization rate in the next century. The ionization rate decreases with altitude near the ground and increases at higher altitudes up to 15 km, where it has a maximum about $5 \cdot 10^7$ $m^{-3}s^{-1}$. Solar X-ray and extreme UV radiation cause a new increase at altitudes over 60 km.

Local sources of air ions are point discharges in strong electric fields, fluidization of charged drops from waves, etc.

The enhanced chemical activity of an ion results in a chain of ion-molecule reactions with the colliding neutrals, and, in the first microsecond of the life of an air ion, a charged molecular cluster called the *cluster ion* is formed. According to theoretical calculations in the air free from exotic trace gases the following cluster ions should be dominant:

$$NO_3^- \cdot (HNO_3) \cdot H_2O,\ NO_2^- \cdot (H_2O)_2,\ NO_3^- \cdot H_2O,\ O_2^- \cdot (H_2O)_4,\ O_2^- \cdot (H_2O)_5,$$
$$H_3O^+ \cdot (H_2O)_6,\ NH_4^+ \cdot (H_2O)_2,\ NH_4^+ \cdot (H_2O),\ H_3O^+ \cdot (H_2O)_5,\ NH_4^+ \cdot NH_3$$

A measurable parameter of air ions is the electrical mobility k, characterizing the drift velocity in the unit electric field. The mobility is inversely proportional to the density of air, and the results of measurements are as a rule reduced to normal conditions. According to mobility the air ions are called: fast or small or light ions with mobility $k > 5 \cdot 10^{-5}$ $m^2V^{-1}s^{-1}$, intermediate ions, and slow or large or heavy ions with mobility $k < 10^{-6}$ $m^2V^{-1}s^{-1}$. The boundary between intermediate and slow ions is conventional.

Cluster ions are fast ions. The masses of cluster ions may be measured with mass spectrometers, but the possible ion-molecule reactions during the passage of the air through nozzles to the vacuum chamber complicate the measurement. Mass and mobility of cluster ions are highly correlated. The experimental results[5] can be expressed by the empirical formula

$$m \approx \frac{850\ \text{u}}{\left[0.3 + k/\left(10^{-4}\,\text{m}^2\text{V}^{-1}\text{s}^{-1}\right)\right]^3}$$

where u is the unified atomic mass unit.

The value of the transport cross-section of a cluster ion is needed to calculate its mobility according to the kinetic theory of Chapman and Enskog. The theoretical estimation of transport cross-sections is rough and cannot be used to identify the chemical structures of cluster ions. Mass spectrometry is the main technique of identification of cluster ions.[2]

Märk and Castleman (1986) presented an overview of over 1000 publications on the experimental studies of cluster ions. Most of them present information about ions of millisecond age range. The low concentration makes it difficult to get detailed information about masses and mobilities of the natural atmospheric ions at ground level. The results of a 1-year continuous measurement[6] are as follows:

	+ ions	**- ions**	**unit**
Average mobility	1.36	1.56	10^{-4} $m^2V^{-1}s^{-1}$
The corresponding mass	190	130	u
The corresponding diameter	0.69	0.61	nm
The average concentration	400	360	10^6 m^{-3}
The corresponding conductivity	8.7	9.0	fS

The distribution of tropospheric cluster ions according to the mobility and estimated mass is depicted in Figure 1.

The problems and results of direct mass spectrometry of natural cluster ions are analysed by Eisele (1986) for ground level and by Meyerott, Reagan and Joiner (1980) for stratospheric measurements. Air ions in the high atmosphere are a subject of ionospheric physics.

During its lifetime (about 1 min), a cluster ion at ground level collides with nearly 10^{12} molecules. Thus the cluster ions are able to concentrate trace gases of very low concentration if they have an extra high electron or proton affinity. For example, Eisele (1986) demonstrated that a considerable fraction of positive atmospheric cluster ions in the unpolluted atmosphere at ground level probably consist of a molecule derived from pyridine. The concentration of these constituents is estimated to be about 10^{-12}. Therefore, air-ion mass and mobility spectrometry is considered as a promising technique for trace analysis in the air. Mass and mobility spectrometry of millisecond-age air ions has been developed as a technique of chemical analysis known as "plasma chromatography" (Carr, 1984). The sensitivity of the detection grows with the age of the cluster ions measured.

The mechanisms of annihilation of cluster ions are ion-ion recombination (on the average 3%) and sedimentation on aerosol particles (on the average 97% of cluster ions at ground level). The result of the combination of a cluster ion and neutral particle is a charged particle called an *aerosol ion*. In conditions of detailed thermodynamic equilibrium the probability that a spherical particle of diameter d carries q elementary charges is calculated from the Boltzmann distribution:

$$p_q(d) = (2\pi d/d_0)^{1/2} \exp(-q^2 d_0/2d)$$

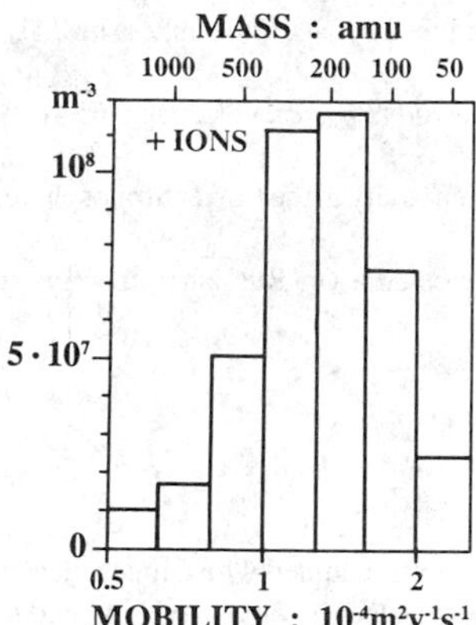

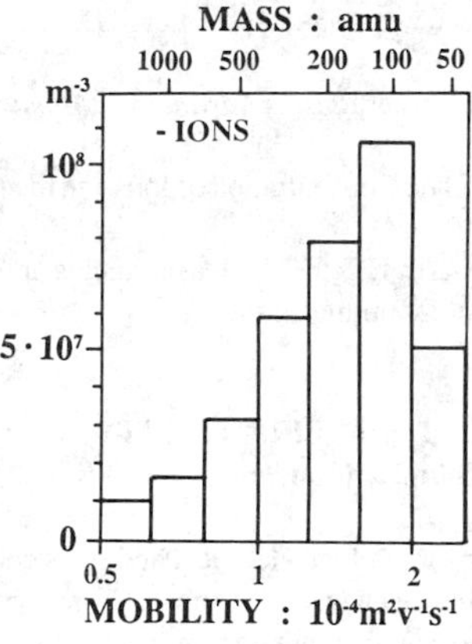

Figure 1. Average mobility and mass spectra of natural tropospheric cluster ions. Concentrations of the mobility fractions were measured in a rural site every 5 min over 1 year.[6] Ion mass is estimated according to the above empirical formula.

where d_0 = 115 nm (at 18°C). The supposition about the detailed equilibrium is an approximation and the formula is not valid for particles less than d_0. On the basis of numerical calculations by Hoppel and Frick (1990) the following charge probabilities can be derived:

d	3	10	30	100	300	1000	3000	nm
p_0	98	90	70	42	24	14	8	%
$p_{-1} + p_1$	2	10	30	48	41	25	15	%
$p_{-2} + p_2$	0	0	0	10	23	21	14	%
$p_{q>2}$	0	0	0	0	12	40	63	%
k_1	15000	1900	250	28	5.1	1.11	0.33	$10^{-9}m^2V^{-1}s^{-1}$

The last line of the table presents the mobility of a particle carrying one elementary charge. The distribution of the atmospheric aerosol ions over mobility is demonstrated in Figure 2.

Although the concentration of aerosol in continental air at ground level is an order of magnitude higher than the concentration of cluster ions, the mobilities of aerosol ions are so small that their percentage in air conductivity is less than 1%.

A specific class of aerosol ions are condensed aerosol ions produced as a result of the condensation of gaseous matter on the cluster ions. In aerosol physics the process is called ion-induced nucleation; it is considered as one among the processes of gas-to-particle conversion. The condensed aerosol ions have an inherent charge. Their sizes and mobilities are between the sizes and mobilities of cluster ions and of ordinary aerosol ions. Water and standard constituents of atmospheric air are not able to condense on the cluster ions in the real atmosphere. Thus the concentration of condensed aerosol ions depends on the trace constituents in the air and is very low in unpolluted air. Knowledge about condensed aerosol ions is poor because of measurement difficulties.

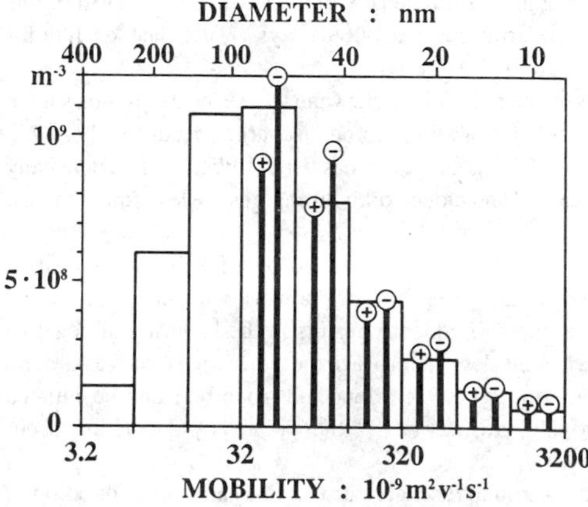

Figure 2. Mobility and size spectra of tropospheric aerosol ions.[6] The wide bars mark the fraction concentrations theoretically estimated on the basis of the standard size distribution of tropospheric aerosol. The pin bars with head + and - mark average values of positive and negative aerosol ion fraction concentrations measured in a rural site every 5 min during 4 months.

REFERENCES

1. Carr, T. W., Ed., *Plasma Chromatography*, Plenum Press, New York and London, XII + 259 pp., 1984.
2. Eisele, F. L., Identification of Tropospheric Ions, *J. Geophys. Res.*, vol. 91, no. D7, pp. 7897-7906, 1986.

3. Hoppel, W. A., and Frick, G. M., The Nonequilibrium Character of the Aerosol Charge Distributions Produced by Neutralizers, *Aerosol Sci. Technol.*, vol. 12, no. 3, pp. 471-496, 1990.
4. Mark, T. D., and Castleman, A. W., Experimental Studies on Cluster Ions, in *Advances in Atomic and Molecular Physics*, vol. 20, pp. 65.-172, Academic Press, 1985.
5. Meyerott, R. E., Reagan, J. B., and Joiner, R. G., The Mobility and Concentration of Ions and the Ionic Conductivity in the Lower Stratosphere, *J. Geophys. Res.*, vol. 85, no. A3, pp. 1273-1278, 1980.
6. Salm, J., Tammet, H., Iher, H., and Hörrak, U., Atmospheric Electrical Measurements in Tahkuse, Estonia (in Russian), in *Voprosy Atmosfernogo Elektrichestva*, pp. 168-175, Gidrometeoizdat, Leningrad, 1990.

III. THUNDERSTORM ELECTRICITY

John Latham

The development of improved radar techniques and instruments for in-cloud electrical and physical measurements, coupled with a much clearer recognition by the research community that establishment of the mechanism or mechanisms responsible for electric field development in thunderclouds, culminating in lightning, is inextricably linked to the concomitant dynamical and microphysical evolution of the clouds, has led to significant progress over the past decade.

Field studies indicate that in most thunderclouds the electrical development is associated with the process of glaciation, which can occur in a variety of incompletely understood ways. In the absence of ice, field growth is slow, individual hydrometeor charges are low, and lightning is produced only rarely. Precipitation — in the solid form, as graupel — also appears to be a necessary ingredient for significant electrification, as does significant convective activity and mixing between the clouds and their environments, via entrainment.

Increasingly, the view is being accepted that charge transfer leading to field-growth is largely a consequence of rebounding collisions between graupel pellets and smaller vapor-grown ice crystals, followed by the separation under gravity of these two types of hydrometeor. These collisions occur predominantly within the temperature range -15 to -30°C, and for significant charge transfer need to occur in the presence of supercooled cloud droplets.

The field evidence is inconsistent with an inductive mechanism, and extensive laboratory studies indicate that the principal charging mechanism is non-inductive and associated — in ways yet to be identified — with differences in surface characteristics of the interacting hydrometeors.

Laboratory studies indicate that the two most favored sites for corona emission leading to the lightning discharge are the tips of ephemeral liquid filaments, produced during the glancing collisions of supercooled raindrops, and protuberances on large ice crystals or graupel pellets. The relative importance of these alternatives will depend on the hydrometeor characteristics and the temperature in the regions of strongest fields; these features are themselves dependent on air-mass characteristics and climatological considerations.

A recently identified but unresolved question is why, in continental Northern Hemisphere thunderclouds at least, the sign of the charge brought to ground by lightning is predominantly negative in summer but more evenly balanced in winter.

IV. LIGHTNING

Martin A. Uman

From both ground-based weather-station data and satellite measurements, it has been estimated that there are about 100 lightning discharges, both cloud and ground flashes, over the whole earth each second; representing an average global lightning flash density of about 6 $km^{-2}yr^{-1}$. Most of this lightning occurs over the earth's land masses. For example, in central Florida, where thunderstorms occur about 90 days/yr, the flash density for discharges to earth is about 15 $km^{-2}yr^{-1}$. Some tropical areas of the earth have thunderstorms up to 300 days/yr.

Lightning can be defined as a transient, high-current electric discharge whose path length is measured in kilometers and whose most common source is the electric charge separated in the ordinary thunderstorm or cumulonimbus cloud. Well over half of all lightning discharges occur totally within individual thunderstorm clouds and are referred to as intracloud discharges. Cloud-to-ground lightning, however, has been studied more extensively than any other lightning form because of its visibility and its more practical interest. Cloud-to-cloud and cloud-to-air discharges are less common than intracloud or cloud-to-ground lightning.

Lightning between the cloud and earth can be categorized in terms of the direction of motion, upward or downward, and the sign of the charge, positive or negative, of the developing discharge (called a *leader*) which initiates the overall event. Over 90% of the worldwide cloud-to-ground discharges is initiated in the thundercloud by downward-moving negatively-charged leaders and subsequently results in the lowering of negative charge to earth. Cloud-to-ground lightning can also be initiated by downward-moving positive leaders, less than 10% of the worldwide cloud-to-ground lightning being of this type although the exact percentage is a function of season and latitude. Lightning between cloud and ground can also be initiated by leaders which develop upward from the earth. These upward-initiated discharges are relatively rare, may be of either polarity, and generally occur from mountaintops and tall man-made structures.

We discuss next the most common type of cloud-to-ground lightning. A negative cloud-to-ground discharge or *flash* has an overall duration of some tenths of a second and is made up of various components, among which are typically three or four high-current pulses called *strokes*. Each stroke lasts about a millisecond, the separation time between strokes being typically several tens of milliseconds. Such lightning often appears to "flicker" because the human eye can just resolve the individual light pulse associated with each stroke. A drawing of the components of a negative cloud-to-ground flash is found in Figure 3. Some values for salient parameters are found in Table 1. The negatively-charged *stepped leader* initiates the first stroke in a flash by propagating from cloud to ground through virgin air in a series of discrete steps. Photographically observed leader steps in clear

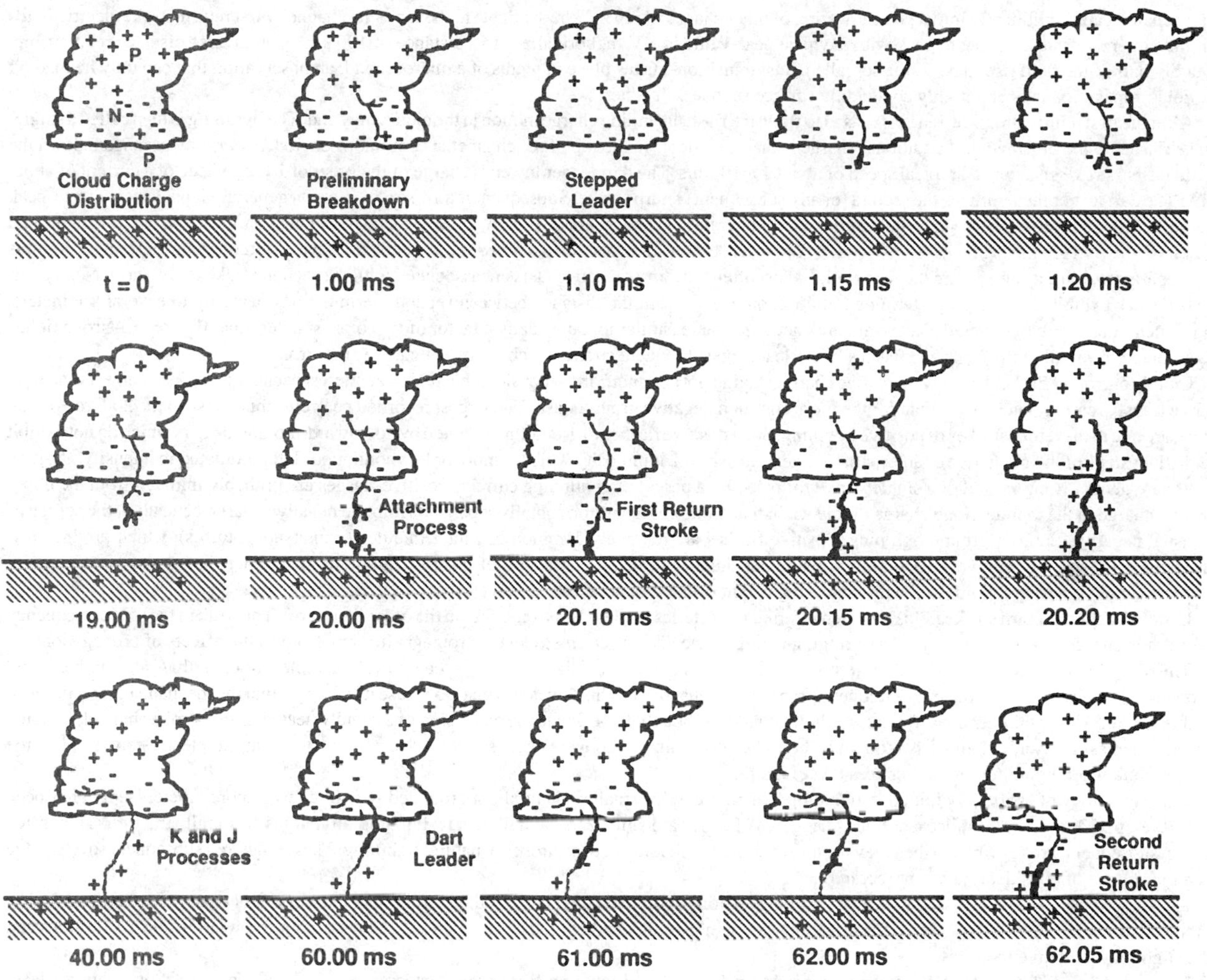

Figure 3. Sequence of steps in cloud-to-ground lightning.

air are typically 1 μs in duration and tens of meters in length, with a pause time between steps of about 50 μs. A fully developed stepped leader lowers up to 10 or more coulombs of negative cloud charge toward ground in tens of milliseconds with an average downward speed of about 2×10^5 m/s. The average leader current is in the 100 to 1000 A range. The steps have pulse currents of at least 1 kA. Associated with these currents are electric- and magnetic-field pulses with widths of about 1 μs or less and risetimes of about 0.1 μs or less. The stepped leader, during its trip toward ground, branches in a downward direction, resulting in the characteristic downward-branched geometrical structure commonly observed. The electric potential of the bottom of the negatively-charged leader channel with respect to ground has a magnitude in excess of 10^7 V. As the leader tip nears ground, the electric field at sharp objects on the ground or at irregularities of the ground itself exceeds the breakdown value of air, and one or more upward-moving discharges (often called upward leaders) are initiated from those points, thus beginning the *attachment process*. An understanding of the physics of the attachment process is central to an understanding of the operation of lightning protection of ground-based objects and the effects of lightning on humans and animals, since it is the attachment process that determines where the lightning connects to objects on the ground and the value of the early currents which flow. When one of the upward-moving discharges from the ground (or from a lightning rod or an individual) contacts the tip of the downward-moving stepped leader, typically some tens of meters above the ground, the leader tip is effectively connected to ground potential. The negatively-charged leader channel is then discharged to earth when a ground potential wave, referred to as the first *return stroke,* propagates continuously up the leader path. The upward speed of a return stroke near the ground is typically near one third the speed of light, and the speed decreases with height. The first return stroke produces a peak current near ground of typically 30 kA, with a time from zero to peak of a few microseconds. Currents measured at the ground fall to half of the peak value in about 50 μs, and currents of the order of hundreds of amperes may flow for times of a few milliseconds up to several hundred milliseconds. The longer-lasting currents are known as *continuing currents*. The rapid release of return stroke energy heats the leader channel to a temperature near 30,000 K and creates a high-pressure channel which expands and generates the shock waves that eventually become thunder, as further discussed later. The return stroke effectively lowers to ground the charge originally deposited onto the stepped-leader channel and additionally initiates the lowering of other charge which may be available to the top of its channel. First return-stroke electric fields exhibit a microsecond scale rise to peak with a typical peak value of 5 V/m, normalized to a distance of 100 km by an inverse distance relationship. Roughly half of the field rise to peak, the so-called “fast transition”, takes place in tenths of a microsecond, an observation that can only be made if the field propagation is over a highly conducting surface such as salt water.

After the first return-stroke current has ceased to flow, the flash, including charge motion in the cloud, may end. The lightning is then called a single-stroke flash. On the other hand, if additional charge is made available to the top of the channel, a continuous or *dart leader* may propagate down the residual first-stroke channel at a typical speed of about 1×10^7 m/s. The dart leader lowers a charge of the order of 1 C by virtue of a current of about 1 kA. The dart leader then initiates the second (or any subsequent) return stroke. Subsequent return-stroke currents generally have faster zero-to-peak rise times than do first-stroke currents, but similar maximum rates of change, about 100 kA/μs. Some leaders begin as dart leaders, but toward the end of their trip toward ground become stepped leaders. These leaders are known as *dart-stepped leaders* and may have different ground termination points (and separate upward leaders) from the first stroke. Most often the dart-stepped leaders are associated with the second stroke of the flash. Nearly half of all flashes exhibit more than one termination point on ground with the distance between separate terminations being up to several kilometers. Subsequent return-stroke radiated electric and magnetic fields are similar to, but usually a factor of two or so smaller, than first return-stroke fields. About one third of all multiple-stroke flashes has at least one subsequent stroke which is larger than the first stroke.

Cloud-to-ground flashes that lower positive charge, though not common, are of considerable practical interest because their peak currents and total charge transfer can be much larger than for the more common negative ground flash. The largest recorded peak currents, those in the 200- to 300-kA range, are due to the return strokes of positive lightning. Such positive flashes to ground are initiated by downward-moving leaders which do not exhibit the distinct steps of their negative counterparts. Rather, they show a luminosity which is more or less continuous but modulated in intensity. Positive flashes are generally composed of a single stroke followed by a period of continuing current. Positive flashes are probably initiated from the upper positive charge in the thundercloud charge dipole when that cloud charge is horizontally separated from the negative charge beneath it, the source of the usual negative cloud-to-ground lightning. Positive flashes are relatively common in winter thunderstorms (snow storms), which produce few flashes overall, and are relatively uncommon in summer thunderstorms. The fraction of positive lightning in summer thunderstorms apparently increases with increasing latitude and with increasing height of the ground above sea level.

Distant lightning return stroke fields are often referred to as sferics (called “atmospherics” in the older literature). The peak in the sferics frequency spectrum is near 5 kHz due to the bipolar or ringing nature of the distant return-stroke electromagnetic signal and to the effects of propagation.

Thunder, the acoustic radiation associated with lightning, is sometimes divided into the categories “audible”, sounds that one can hear, and “infrasonic”, below a few tens of hertz, a frequency range that is inaudible. This division is made because it is thought that the mechanisms that produce audible and infrasonic thunder are different. Audible thunder is thought to be due to the expansion of a rapidly heated return stroke channel, as noted earlier, whereas infrasonic thunder is thought to be associated with the conversion to sound of the energy stored in the electrostatic field of the thundercloud when lightning rapidly reduces that cloud field.

The technology of artificially initiating lightning by firing upward small rocket trailing grounded wire of a few hundred meters length has been well-developed during the past decade. Such “triggered” flashes are similar to natural upward-initiated discharges from tall structure. They often contain subsequent strokes which, when they occur, are similar to the subsequent strokes in natural lightning. These triggered subsequent strokes have been the subject of considerable recent research.

Also in the past 10 years or so sophisticated lightning locating equipment has been installed throughout the world. For example, all ground flashes in the U.S. are now centrally monitored for research, for better overall weather prediction, and for hazard warning for aviation, electric utilities and other lightning-sensitive facilities.

Information on lightning physics can be found in M. A. Uman, *The Lightning Discharge*, Academic Press, San Diego, 1987; on lightning death and injury in *Medical Aspects of Lightning Injury*, editors C. Andrews, M. A. Cooper, M. Darveniza, and D. Mackerras, CRC Press, 1992. Ground flash location information for the U.S., in real time or archived, is available from Geomet Data Service of Tucson, AZ, which is also a source of the names of providers of those data in other countries.

Table 2 has data for cloud-to-ground lightning discharges bringing negative charge to earth. The values listed are intended to convey a rough feeling for the various physical parameters of lightning. No great accuracy is claimed since the results of different investigators are often not in good agreement. These values may, in fact, depend on the particular environment in which the lightning discharge is generated. The choice of some of the entries in the table is arbitrary.

TABLE 2
Data for Cloud-to-Ground Lightning Discharges

	Minimum[a]	Representative values	Maximum[a]
Stepped leader			
Length of step, m	3	50	200
Time interval between steps, μs	30	50	125
Average speed of propagation of stepped leader, m/s[b]	1.0×10^5	2.0×10^5	3.0×10^6
Charged deposited on stepped-leader channel, coulombs	3	5	20
Dart leader			
Speed of propagation, m/s[b]	1.0×10^6	1.0×10^7	2.4×10^7
Charged deposited on dart-leader channel, coulombs	0.2	1	6
Return stroke[c]			
Speed of propagation, m/s[b]	2.0×10^7	1.0×10^8	2.0×10^8
Maximum current rate of increase, kA/μs	<1	100	400
Time to peak current, μs	<1	2	30
Peak current, kA	2	30	200
Time to half of peak current, μs	10	50	250
Charge transferred excluding continuing current, coulombs	0.02	3	20
Channel length, km	2	5	15
Lightning flash			
Number of strokes per flash	1	4	26
Time interval between strokes in absence of continuing current, ms	3	60	100
Time duration of flash, s	10^{-2}	0.5	2
Charge transferred including continuing current, coulombs	3	30	200

[a] The words maximum and minimum are used in the sense that most measured values fall between these limits.
[b] Speeds of propagation are generally determined from photographic data and are "two-dimensional". Since many lightning flashes are not vertical, values stated are probably slight underestimates of actual values.
[c] First return strokes have longer times to current peak and generally larger charge transfer than do subsequent return strokes.

Adapted from Uman, M. A., *Lightning*, Dover Paperbook, New York, 1986, and Uman, M. A., *The Lightning Discharge*, Academic Press, San Diego, 1987.

VELOCITY OF SOUND IN VARIOUS MEDIA

Compiled by Gordon E. Becker, Bell Telephone Laboratories

The data for the Velocity of Sound is in Various Materials were compiled from a variety of sources. For more extensive tables one is referred to the following books:

AIP Handbook, Smithsonian Tables.
Mason: Physical Acoustics and the Properties of Solids (1958).
Chalmers and Quarrell: Physical Examination of Metals (1960).
Mason: Piezoelectric Crystals and their Application to Ultrasonics (1950).
Bergmann: Der Ultraschall (Hirzel, 1954).

Definition of Terms: V_l = Velocity of plane longitudinal wave in bulk material
V_s = Velocity of plane transverse (shear) wave
V_{ext} = Velocity of longitudinal wave (extensional wave) in thin rods.

SOLIDS

Substance	Density g/cm^3	V_l m/s	V_s m/s	V_{ext} m/s
Metals				
Aluminum, rolled	2.7	6420	3040	5000
Berylium	1.87	12890	8880	12870
Brass (70 Cu, 30 Zn)	8.6	4700	2110	3480
Copper, annealed	8.93	4760	2325	3810
Copper, rolled	8.93	5010	2270	3750
Duralumin 17S	2.79	6320	3130	5150
Gold, hard-drawn	19.7	3240	1200	2030
Iron, electrolytic	7.9	5950	3240	5120
Iron, Armco	7.85	5960	3240	5200
Lead, annealed	11.4	2160	700	1190
Lead, rolled	11.4	1960	690	1210
Magnesium, drawn, annealed	1.74	5770	3050	4940
Molybdenum	10.1	6250	3350	5400
Monel metal	8.90	5350	2720	4400
Nickel (unmagnetized)	8.85	5480	2990	4800
Nickel	8.9	6040	3000	4900
Platinum	21.4	3260	1730	2800
Silver	10.4	3650	1610	2680
Steel, mild	7.85	5960	3235	5200
Steel, 347 Stainless	7.9	5790	3100	5000
Steel (1%C)	7.84	5940	3220	5180
Steel (1%C, hardened)	7.84	5854	3150	5070
Tin, rolled	7.3	3320	1670	2730
Titanium	4.5	6070	3125	5080
Tungsten, annealed	19.3	5220	2890	4620
Tungsten, drawn	19.3	5410	2640	4320
Tungsten Carbide	13.8	6655	3980	6220
Zinc, rolled	7.1	4210	2440	3850
Various				
Fused silica	2.2	5968	3764	5760
Glass, pyrex	2.32	5640	3280	5170
Glass, heavy silicate flint	3.88	3980	2380	3720
Glass, light borate crown	2.24	5100	2840	4540
Lucite	1.18	2680	1100	1840
Nylon 6-6	1.11	2620	1070	1800
Polyethylene	0.90	1950	540	920
Polystyrene	1.06	2350	1120	2240
Rubber, butyl	1.07	1830		
Rubber, gum	0.95	1550		
Rubber neoprene	1.33	1600		
Brick	1.8			3650
Clay rock	2.2			3480
Cork	0.25			500
Marble	2.6			3810
Paraffin	0.9			1300
Tallow				390
Woods				
Ash, along the fiber				4670
Ash, across the rings				1390
Ash, along the rings				1260
Beech, along the fiber				3340
Elm, along the fiber				4120
Maple, along the fiber				4110
Oak, along the fiber				3850

LIQUIDS

Substance	Formula	Density g/cm^3	Velocity at 25°C m/s	$-\delta v/\delta t$ m/sec °C
Acetone	C_3H_6O	0.79	1174	4.5
Benzene	C_6H_6	0.870	1295	4.65
Carbon disulphide	CS_2	1.26	1149	—
Carbon tetrachloride	CCl_4	1.595	926	2.7
Castor oil	$C_{11}H_{10}O_{10}$	0.969	1477	3.6
Chloroform	$CHCl_3$	1.49	987	3.4
Ethanoi	C_2H_6O	0.79	1207	4.0
Ethanol amide	C_2H_7NO	1.018	1724	3.4
Ethyl ether	$C_4H_{10}O$	0.713	985	4.87
Ethylene glycol	$C_2H_6O_2$	1.113	1658	2.1
Glycerol	$C_3H_8O_3$	1.26	1904	2.2
Kerosene	—	0.81	1324	3.6
Mercury	Hg	13.5	1450	—
Methanol	CH_4O	0.791	1103	3.2
Nitrobenzene	$C_6H_5NO_2$	1.20	1463	3.6
Turpentine	—	0.88	1255	—
Water (distilled)	H_2O	0.998	1496.7+ 2	−2.4
Water (sea)	—	1.025	1531	−2.4
Xylene hexafluoride	$C_8H_4F_6$	1.37	879	—

GASES AND VAPORS

Substance	Formula	Density g/l	Velocity m/s	$\delta v/\delta t$ m/sec °C
Gases (0°C)				
Air, dry		1.293	331.45	0.59
Ammonia	NH_3	0.771	415	
Argon	Ar	1.783	319*	0.56
Carbon monoxide	CO	1.25	338	0.6
Carbon dioxide	CO_2	1.977	259	0.4
Chlorine	CL_2	3.214	206	
Deuterium	D_2		890	1.6
Ethane (10°C)	C_2H_6	1.356	308	
Ethylene	C_2H_4	1.260	317	
Helium	He	0.178	965	0.8
Hydrogen	H_2	0.0899	1284	2.2
Hydrogen bromide	HBr	3.50	200	
Hydrogen chloride	HCl	1.639	296	
Hydrogen iodide	HI	5.66	157	
Hydrogen sulfide	H_2S	1.539	289	
Illuminating (Coal gas)			453	
Methane	CH_4	0.7168	430	
Neon	Ne	0.900	435	0.8
Nitric oxide (10°C)	NO	1.34	324	
Nitrogen	N_2	1.251	334	0.6
Nitrous oxide	N_2O	1.977	263	0.5
Oxygen	O_2	1.429	316	0.56
Sulfur dioxide	SO_2	2.927	213	0.47
Vapors (97.1°C)				
Acetone	C_3H_6O		239	0.32
Benzene	C_6H_6		202	0.3
Carbon tetrachloride	CCl_4		145	
Chloroform	$CHCl_3$		171	0.24
Ethanol	C_2H_6O		269	0.4
Ethyl ether	$C_4H_{10}O$		206	0.3
Methanol	CH_4O		335	0.46
Water vapor (134°C)	H_2O		494	

* At 20°C.

ABSORPTION AND VELOCITY OF SOUND IN STILL AIR

The following data refer only to the temperature 20°C (68°F). They were abstracted from an extensive compilation prepared by I. B. Evans and H. E. Bass. The entire report, Tables of Absorption and Velocity of Sound in Still Air at 68°F (20°C), AD-738 576 is available from National Technical Information Service, U.S. Department of Commerce, 5285 Port Royal Road, Springfield, Va. 22151.

Frequency (Hz)	Absorption (dB/1000 ft)	Absorption (dB/Km)	Absorption (dB/s)	Velocity (1000 ft/s)
Relative Humidity = 0%				
20.	0.154	0.51	0.174	1.126892
40.	0.327	1.07	0.368	1.127013
50.	0.384	1.26	0.433	1.127050
63.	0.436	1.43	0.491	1.127085
100.	0.509	1.67	0.573	1.127131
200.	0.560	1.84	0.631	1.127161
400.	0.596	1.96	0.672	1.127169
630.	0.645	2.11	0.727	1.127171
800.	0.692	2.27	0.780	1.127172
1250.	0.861	2.82	0.970	1.127172
2000.	1.262	4.14	1.423	1.127173
4000.	2.696	8.84	3.039	1.127178
6300.	4.541	14.89	5.118	1.127182
10000.	8.013	26.28	9.032	1.127184
12500.	10.918	35.81	12.306	1.127186
16000.	15.901	52.15	17.923	1.127187
20000.	22.978	75.37	25.901	1.127187
40000.	81.405	267.01	91.759	1.127188
63000.	196.544	644.66	221.542	1.127188
80000.	314.677	1032.14	354.700	1.127188
Relative Humidity = 5%				
20.	0.031	0.10	0.034	1.126973
40.	0.074	0.24	0.083	1.126996
50.	0.092	0.30	0.104	1.127004
63.	0.114	0.37	0.129	1.127009
100.	0.179	0.59	0.202	1.127019
200.	0.449	1.47	0.506	1.127028
400.	1.451	4.76	1.635	1.127043
630.	3.211	10.53	3.619	1.127067
800.	4.774	15.66	5.380	1.127088
1250.	9.164	30.06	10.329	1.127147
2000.	15.175	49.77	17.106	1.127228
4000.	22.685	74.41	25.573	1.127321
6300.	26.245	86.08	29.587	1.127352
10000.	30.781	100.96	34.701	1.127365
12500.	34.306	112.52	38.676	1.127369
16000.	40.263	132.06	45.391	1.127372
20000.	48.653	159.58	54.850	1.127373
40000.	115.903	380.16	130.666	1.127378
63000.	242.070	793.99	272.905	1.127381
80000.	367.063	1203.97	413.821	1.127383
Relative Humidity = 10%				
20.	0.021	0.07	0.024	1.127167
40.	0.064	0.21	0.072	1.127183
50.	0.084	0.28	0.095	1.127191
63.	0.108	0.35	0.122	1.127199
100.	0.161	0.53	0.181	1.127213
200.	0.289	0.95	0.326	1.127225
400.	0.706	2.32	0.796	1.127230
630.	1.501	4.92	1.692	1.127234
800.	2.297	7.54	2.590	1.127238
1250.	5.155	16.91	5.811	1.127254
2000.	11.658	38.24	13.141	1.127287
4000.	31.023	101.76	34.975	1.127386
6300.	47.085	154.44	53.087	1.127462
10000.	61.578	201.98	69.431	1.127522
12500.	68.146	223.52	76.837	1.127540
16000.	76.231	250.04	85.955	1.127555
20000.	85.605	280.78	96.525	1.127563
40000.	151.938	498.36	171.321	1.127576
63000.	277.191	909.19	312.555	1.127581
80000.	403.662	1324.01	455.162	1.127583

Frequency (Hz)	Absorption (dB/1000 ft)	Absorption (dB/Km)	Absorption (dB/s)	Velocity (1000 ft/s)
Relative Humidity = 20%				
20.	0.013	0.04	0.014	1.127568
40.	0.045	0.15	0.051	1.127577
50.	0.066	0.22	0.074	1.127582
63.	0.093	0.30	0.105	1.127587
100.	0.164	0.54	0.185	1.127603
200.	0.285	0.93	0.321	1.127624
400.	0.476	1.56	0.537	1.127633
630.	0.789	2.59	0.890	1.127636
800.	1.103	3.62	1.244	1.127637
1250.	2.277	7.47	2.568	1.127640
2000.	5.310	17.42	5.987	1.127645
4000.	18.991	62.29	21.416	1.127670
6300.	41.151	134.98	46.406	1.127710
10000.	79.657	261.28	89.836	1.127778
12500.	103.004	337.85	116.170	1.127817
16000.	130.465	427.93	147.146	1.127859
20000.	155.809	511.05	175.736	1.127893
40000.	251.952	826.40	284.192	1.127960
63000.	382.062	1253.16	430.957	1.127977
80000.	508.369	1667.45	573.431	1.127982
Relative Humidity = 30%				
20.	0.009	0.03	0.010	1.127976
40.	0.034	0.11	0.038	1.127980
50.	0.051	0.17	0.057	1.127984
63.	0.075	0.25	0.085	1.127987
100.	0.151	0.50	0.170	1.127999
200.	0.309	1.01	0.349	1.128023
400.	0.484	1.59	0.546	1.128037
630.	0.682	2.24	0.770	1.128041
800.	0.868	2.85	0.979	1.128044
1250.	1.552	5.09	1.751	1.128045
2000.	3.333	10.93	3.760	1.128047
4000.	11.856	38.89	13.374	1.128056
6300.	27.626	90.61	31.164	1.128072
10000.	62.493	204.98	70.498	1.128105
12500.	89.659	294.08	101.148	1.128131
16000.	128.814	422.51	145.324	1.128166
20000.	171.847	563.66	193.879	1.128204
40000.	338.710	1110.97	382.172	1.128316
63000.	499.838	1639.47	563.998	1.128361
80000.	635.085	2083.08	716.614	1.128375
Relative Humidity = 40%				
20.	0.007	0.02	0.008	1.128386
40.	0.027	0.09	0.030	1.128388
50.	0.041	0.13	0.046	1.128390
63.	0.062	0.20	0.070	1.128392
100.	0.134	0.44	0.151	1.128402
200.	0.318	1.04	0.359	1.128424
400.	0.524	1.72	0.592	1.128443
630.	0.692	2.27	0.781	1.128449
800.	0.829	2.72	0.935	1.128451
1250.	1.309	4.29	1.477	1.128453
2000.	2.544	8.34	2.870	1.128456
4000.	8.523	27.96	9.618	1.128460
6300.	19.995	65.58	22.564	1.128467
10000.	47.390	155.44	53.478	1.128484
12500.	70.857	232.41	79.962	1.128499
16000.	108.308	355.25	122.228	1.128521
20000.	154.838	507.87	174.742	1.128549
40000.	380.371	1247.62	429.310	1.128663
63000.	596.091	1955.18	672.827	1.128733
80000.	753.514	2471.53	850.536	1.128758
Relative Humidity = 50%				
20.	0.006	0.02	0.006	1.128795
40.	0.022	0.07	0.025	1.128797
50.	0.034	0.11	0.038	1.128798
63.	0.052	0.17	0.058	1.128800
100.	0.117	0.38	0.132	1.128807
200.	0.313	1.03	0.353	1.128826
400.	0.563	1.85	0.636	1.128849

Frequency (Hz)	Absorption (dB/1000 ft)	Absorption (dB/Km)	Absorption (dB/s)	Velocity (1000 ft/s)
630.	0.734	2.41	0.828	1.128858
800.	0.851	2.79	0.961	1.128860
1250.	1.231	4.04	1.390	1.128862
2000.	2.176	7.14	2.457	1.128865
4000.	6.752	22.15	7.622	1.128867
6300.	15.643	51.31	17.659	1.128871
10000.	37.521	123.07	42.357	1.128881
12500.	57.009	186.99	64.357	1.128890
16000.	89.594	293.87	101.143	1.128903
20000.	132.719	435.32	149.830	1.128922
40000.	382.722	1255.33	432.101	1.129020
63000.	654.606	2147.11	739.115	1.129099
80000.	844.024	2768.40	953.016	1.129133
Relative Humidity = 60%				
20.	0.005	0.02	0.005	1.129207
40.	0.018	0.06	0.021	1.129208
50.	0.029	0.09	0.032	1.129209
63.	0.044	0.15	0.050	1.129210
100.	0.103	0.34	0.117	1.129215
200.	0.301	0.99	0.339	1.129232
400.	0.593	1.94	0.669	1.129254
630.	0.782	2.57	0.884	1.129266
800.	0.896	2.94	1.012	1.129269
1250.	1.223	4.01	1.381	1.129273
2000.	1.997	6.55	2.256	1.129275
4000.	5.711	18.73	6.449	1.129277
6300.	12.962	42.51	14.638	1.129279
10000.	31.050	101.84	35.064	1.129286
12500.	47.462	155.67	53.598	1.129292
16000.	75.544	247.78	85.312	1.129300
20000.	113.958	373.78	128.694	1.129313
40000.	364.443	1195.37	411.598	1.129389
63000.	677.023	2220.64	764.675	1.129467
80000.	899.912	2951.71	1016.458	1.129508
Relative Humidity = 70%				
20.	0.004	0.01	0.005	1.129618
40.	0.016	0.05	0.018	1.129619
50.	0.025	0.08	0.028	1.129619
63.	0.039	0.13	0.044	1.129620
100.	0.092	0.30	0.104	1.129623
200.	0.284	0.93	0.321	1.129638
400.	0.611	2.01	0.691	1.129662
630.	0.829	2.72	0.937	1.129673
800.	0.947	3.11	1.070	1.129678
1250.	1.250	4.10	1.412	1.129683
2000.	1.915	6.28	2.163	1.129685
4000.	5.056	16.58	5.712	1.129687
6300.	11.197	36.73	12.649	1.129689
10000.	26.624	87.33	30.077	1.129693
12500.	40.758	133.69	46.044	1.129697
16000.	65.246	214.01	73.708	1.129704
20000.	99.365	325.92	112.254	1.129712
40000.	339.153	1112.42	383.165	1.129770
63000.	673.667	2209.63	761.137	1.129842
80000.	924.481	3032.30	1044.556	1.129884
Relative Humidity = 80%				
20.	0.004	0.01	0.004	1.130030
40.	0.014	0.05	0.016	1.130030
50.	0.022	0.07	0.025	1.130031
63.	0.034	0.11	0.039	1.130032
100.	0.082	0.27	0.093	1.130034
200.	0.267	0.88	0.302	1.130047
400.	0.620	2.03	0.701	1.130069
630.	0.870	2.85	0.983	1.130082
800.	0.998	3.27	1.128	1.130088
1250.	1.293	4.24	1.461	1.130094
2000.	1.887	6.19	2.132	1.130096
4000.	4.626	15.17	5.228	1.130098
6300.	9.976	32.72	11.274	1.130100
10000.	23.466	76.97	26.519	1.130102
12500.	35.896	117.74	40.566	1.130106
16000.	57.588	188.89	65.081	1.130110
20000.	88.144	289.11	99.612	1.130116

Frequency (Hz)	Absorption (dB/1000 ft)	Absorption (dB/Km)	Absorption (dB/s)	Velocity (1000 ft/s)
40000.	313.525	1028.36	354.334	1.130161
63000.	655.282	2149.33	740.615	1.130223
80000.	925.566	3035.86	1046.134	1.130264

Relative Humidity = 90%

Frequency (Hz)	Absorption (dB/1000 ft)	Absorption (dB/Km)	Absorption (dB/s)	Velocity (1000 ft/s)
20.	0.003	0.01	0.004	1.130443
40.	0.013	0.04	0.014	1.130443
50.	0.019	0.06	0.022	1.130444
63.	0.031	0.10	0.035	1.130444
100.	0.074	0.24	0.084	1.130446
200.	0.250	0.82	0.283	1.130457
400.	0.621	2.04	0.702	1.130478
630.	0.903	2.96	1.021	1.130493
800.	1.045	3.43	1.182	1.130498
1250.	1.344	4.41	1.520	1.130505
2000.	1.892	6.21	2.139	1.130508
4000.	4.337	14.22	4.903	1.130511
6300.	9.099	29.84	10.286	1.130512
10000.	21.133	69.32	23.891	1.130514
12500.	32.258	105.81	36.468	1.130516
16000.	51.762	169.78	58.518	1.130520
20000.	79.424	260.51	89.791	1.130526
40000.	290.117	951.58	327.995	1.130561
63000.	629.701	2065.42	711.948	1.130612
80000.	911.286	2989.02	1030.346	1.130651

Relative Humidity = 100%

Frequency (Hz)	Absorption (dB/1000 ft)	Absorption (dB/Km)	Absorption (dB/s)	Velocity (1000 ft/s)
20.	0.003	0.01	0.003	1.130856
40.	0.011	0.04	0.013	1.130856
50.	0.018	0.06	0.020	1.130857
63.	0.028	0.09	0.031	1.130857
100.	0.068	0.22	0.076	1.130858
200.	0.234	0.77	0.265	1.130868
400.	0.616	2.02	0.696	1.130888
630.	0.928	3.05	1.050	1.130902
800.	1.087	3.57	1.230	1.130909
1250.	1.399	4.59	1.582	1.130917
2000.	1.917	6.29	2.168	1.130921
4000.	4.140	13.58	4.682	1.130924
6300.	8.451	27.72	9.557	1.130925
10000.	19.357	63.49	21.891	1.130926
12500.	29.461	96.63	33.318	1.130929
16000.	47.227	154.90	53.410	1.130931
20000.	72.538	237.93	82.036	1.130935
40000.	269.598	884.28	304.905	1.130963
63000.	601.714	1973.62	680.543	1.131007
80000.	888.113	2913.01	1004.492	1.131042

VELOCITY OF SOUND IN DRY AIR

Data in this table apply only to dry air. These data have been calculated with air being treated as a perfect gas.

Temp °C	0 m/s	1 m/s	2 m/s	3 m/s	4 m/s	5 m/s	6 m/s	7 m/s	8 m/s	9 m/s
60	366.05	366.60	367.14	367.69	368.24	368.78	369.33	369.87	370.42	370.96
50	360.51	361.07	361.62	362.18	362.74	363.29	363.84	364.39	364.95	365.50
40	354.89	355.46	356.02	356.58	357.15	357.71	358.27	358.83	359.39	359.95
30	349.18	349.75	350.33	350.90	351.47	352.04	352.62	353.19	353.75	354.32
20	343.37	343.95	344.54	345.12	345.70	346.29	346.87	347.44	348.02	348.60
10	337.46	338.06	338.65	339.25	339.84	340.43	341.02	341.61	342.20	342.78
0	331.45	332.06	332.66	333.27	333.87	334.47	335.07	335.67	336.27	336.87
−10	325.33	324.71	324.09	323.47	322.84	322.22	321.60	320.97	320.34	319.72
−20	319.09	318.45	317.82	317.19	316.55	315.92	315.28	314.64	314.00	313.36
−30	312.72	312.08	311.43	310.78	310.14	309.49	308.84	308.19	307.53	306.88
−40	306.22	305.56	304.91	304.25	303.58	302.92	302.26	301.59	300.92	300.25
−50	299.58	298.91	298.24	297.56	296.89	296.21	295.53	294.85	294.16	293.48
−60	292.79	292.11	291.42	290.73	290.03	289.34	288.64	287.95	287.25	286.55
−70	285.84	285.14	284.43	283.73	283.02	282.30	281.59	280.88	280.16	279.44
−80	278.72	278.00	277.27	276.55	275.82	275.09	274.36	273.62	272.89	272.15
−90	271.41	270.67	269.92	269.18	268.43	267.68	266.93	266.17	265.42	264.66

MUSICAL SCALES

EQUAL TEMPERED CHROMATIC SCALE

$A_4 = 440$

American Standard pitch. Adopted by the American Standards Association in 1936

Note	Frequency	Note	Frequency	Note	Frequency	Note	Frequency
C_0	16.35	C_2	65.41	C_4	261.63	C_6	1046.50
$C\#_0$	17.32	$C\#_2$	69.30	$C\#_4$	277.18	$C\#_6$	1108.73
D_0	18.35	D_2	73.42	D_4	293.66	D_6	1174.66
$D\#_0$	19.45	$D\#_2$	77.78	$D\#_4$	311.13	$D\#_6$	1244.51
E_0	20.60	E_2	82.41	E_4	329.63	E_6	1318.51
F_0	21.83	F_2	87.31	F_4	349.23	F_6	1396.91
$F\#_0$	23.12	$F\#_2$	92.50	$F\#_4$	369.99	$F\#_6$	1479.98
G_0	24.50	G_2	98.00	G_4	392.00	G_6	1567.98
$G\#_0$	25.96	$G\#_2$	103.83	$G\#_4$	415.30	$G\#_6$	1661.22
A_0	27.50	A_2	110.00	A_4	440.00	A_6	1760.00
$A\#_0$	29.14	$A\#_2$	116.54	$A\#_4$	466.16	$A\#_6$	1864.66
B_0	30.87	B_2	123.47	B_4	493.88	B_6	1975.53
C_1	32.70	C_3	130.81	C_5	523.25	C_7	2093.00
$C\#_1$	34.65	$C\#_3$	138.59	$C\#_5$	554.37	$C\#_7$	2217.46
D_1	36.71	D_3	146.83	D_5	587.33	D_7	2349.32
$D\#_1$	38.89	$D\#_3$	155.56	$D\#_5$	622.25	$D\#_7$	2489.02
E_1	41.20	E_3	164.81	E_5	659.26	E_7	2637.02
F_1	43.65	F_3	174.61	F_5	698.46	F_7	2793.83
F#1	46.25	$F\#_3$	185.00	$F\#_5$	739.99	$F\#_7$	2959.96
G_1	49.00	G_3	196.00	G_5	783.99	G_7	3135.96
$G\#_1$	51.91	$G\#_3$	207.65	$G\#_5$	830.61	$G\#_7$	3322.44
A_1	55.00	A_3	220.00	A_5	880.00	A_7	3520.00
$A\#_1$	58.27	$A\#_3$	233.08	$A\#_5$	932.33	$A\#_7$	3729.31
B_1	61.74	B_3	246.94	B_5	987.77	B_7	3951.07
						C_8	4186.01

EQUAL TEMPERED CHROMATIC SCALE

$A_4 = 435$

International Pitch, adopted 1891

Note	Frequency	Note	Frequency	Note	Frequency	Note	Frequency
C_0	16.17	C_2	64.66	C_4	258.65	C_6	1034.61
$C\#_0$	17.13	$C\#_2$	68.51	$C\#_4$	274.03	$C\#_6$	1096.13
D_0	18.15	D_2	72.58	D_4	290.33	D_6	1161.31
$D\#_0$	19.22	$D\#_2$	76.90	$D\#_4$	307.59	$D\#_6$	1230.37
E_0	20.37	E_2	81.47	E_4	325.88	E_6	1303.53
F_0	21.58	F_2	86.31	F_4	345.26	F_6	1381.04
$F\#_0$	22.86	$F\#_2$	91.45	$F\#_4$	365.79	$F\#_6$	1463.16
G_0	24.22	G_2	96.89	G_4	387.54	G_6	1550.16
$G\#_0$	25.66	$G\#_2$	102.65	$G\#_4$	410.59	$G\#_6$	1642.34
A_0	27.19	A_2	108.75	A_4	435.00	A_6	1740.00
$A\#_0$	28.80	$A\#_2$	115.22	$A\#_4$	460.87	$A\#_6$	1843.47
B_0	30.52	B_2	122.07	B_4	488.27	B_6	1953.08
C_1	32.33	C_3	129.33	C_5	517.31	C_7	2069.22
$C\#_1$	34.25	$C\#_3$	137.02	$C\#_5$	548.07	$C\#_7$	2192.26
D_1	36.29	D_3	145.16	D_5	580.66	D_7	2322.62
$D\#_1$	38.45	$D\#_3$	153.80	$D\#_5$	615.18	$D\#_7$	2460.73
E_1	40.74	E_3	162.94	E_5	651.76	E_7	2607.05
F_1	43.16	F_3	172.63	F_5	690.52	F_7	2762.08
$F\#_1$	45.72	$F\#_3$	182.89	$F\#_5$	731.58	$F\#_7$	2926.32
G_1	48.44	G_3	193.77	G_5	775.08	G_7	3100.33
$G\#_1$	51.32	$G\#_3$	205.29	$G\#_5$	821.17	$G\#_7$	3284.68
A_1	54.38	A_3	217.50	A_5	870.00	A_7	3480.00
$A\#_1$	57.61	$A\#_3$	230.43	$A\#_5$	921.73	$A\#_7$	3686.93
B_1	61.03	B_3	244.14	B_5	976.54	B_7	3906.17
						C_8	4138.44

SCIENTIFIC OR JUST SCALE

$C_4 = 256$

Note	Frequency	Note	Frequency	Note	Frequency	Note	Frequency
C_0	16	C_2	64	C_4	256	C_6	1024
D_0	18	D_2	72	D_4	288	D_6	1152
E_0	20	E_2	80	E_4	320	E_6	1280
F_0	21.33	F_2	85.33	F_4	341.33	F_6	1365.33
G_0	24	G_2	96	G_4	384	G_6	1536
A_0	26.67	A_2	106.67	A_4	426.67	A_6	1706.67
B_0	30	B_2	120	B_4	480	B_6	1920
C_1	32	C_3	128	C_5	512	C_7	2048
D_1	36	D_3	144	D_5	576	D_7	2304
E_1	40	E_3	160	E_5	640	E_7	2560
F_1	42.67	F_3	170.67	F_5	682.67	F_7	2730.67
G_1	48	G_3	192	G_5	768	G_7	3072
A_1	53.33	A_3	213.33	A_5	853.33	A_7	3413.33
B_1	60	B_3	240	B_5	960	B_7	3840
						C_8	4096

Section 15
Practical Laboratory Data

STANDARD ITS-90 THERMOCOUPLE TABLES

The Instrument Society of America (ISA) has assigned standard letter designations to a number of thermocouple types having specified emf-temperature relations. These designations and the approximate metal compositions which meet the required relations, as well as the useful temperature ranges, are given below:

Type B	(Pt + 30% Rh) vs. (Pt + 6% Rh)	0 to 1820°C
Type E	(Ni + 10% Cr) vs. (Cu + 43% Ni)	-270 to 1000°C
Type J	Fe vs. (Cu + 43%Ni)	-210 to 1200°C
Type K	(Ni + 10% Cr) vs. (Ni + 2% Al + 2% Mn + 1% Si)	-270 to 1372°C
Type N	(Ni + 14% Cr + 1.5% Si) vs. (Ni + 4.5% Si + 0.1% Mg)	-270 to 1300°C
Type R	(Pt + 13% Rh) vs. Pt	-50 to 1768°C
Type S	(Pt + 10% Rh) vs. Pt	-50 to 1768°C
Type T	Cu vs. (Cu + 43% Ni)	-270 to 400°C

The compositions are given in weight percent, and the positive leg is listed first. It should be emphasized that the standard letter designations do not imply a precise composition but rather that the specified emf-temperature relation is satisfied.

The first set of tables below lists, for each thermocouple type, the emf as a function of temperature on the International Temperature Scale of 1990 (ITS-90). The coefficients in the equation used to generate the table are also given. The second set of tables gives the inverse relationships, i.e., the coefficients in the polynomial equation which expresses the temperature as a function of thermocouple emf. The accuracy of these equations is also stated.

Further details and tables at closer intervals may be found in Reference 1.

REFERENCES

1. Burns, G. W., Seroger, M. G., Strouse, G. F., Croarkin, M. C., and Guthrie, W. F., *Temperature-Electromotive Force Reference Functions and Tables for the Letter-Designated Thermocouple Types Based on the ITS-90*, Natl. Inst. Stand. Tech. (U.S.) Monogr. 175, 1993.
2. Schooley, J. F., *Thermometry*, CRC Press, Boca Raton, FL, 1986.

STANDARD ITS-90 THERMOCOUPLE TABLES (continued)

Type B thermocouples: emf-temperature (°C) reference table and equations.

Thermocouple emf as a Function of Temperature in Degrees Celsius (ITS-90)

emf in Millivolts — Reference Junctions at 0 °C

°C	0	10	20	30	40	50	60	70	80	90	100
0	0.000	−0.002	−0.003	−0.002	−0.000	0.002	0.006	0.011	0.017	0.025	0.033
100	0.033	0.043	0.053	0.065	0.078	0.092	0.107	0.123	0.141	0.159	0.178
200	0.178	0.199	0.220	0.243	0.267	0.291	0.317	0.344	0.372	0.401	0.431
300	0.431	0.462	0.494	0.527	0.561	0.596	0.632	0.669	0.707	0.746	0.787
400	0.787	0.828	0.870	0.913	0.957	1.002	1.048	1.095	1.143	1.192	1.242
500	1.242	1.293	1.344	1.397	1.451	1.505	1.561	1.617	1.675	1.733	1.792
600	1.792	1.852	1.913	1.975	2.037	2.101	2.165	2.230	2.296	2.363	2.431
700	2.431	2.499	2.569	2.639	2.710	2.782	2.854	2.928	3.002	3.078	3.154
800	3.154	3.230	3.308	3.386	3.466	3.546	3.626	3.708	3.790	3.873	3.957
900	3.957	4.041	4.127	4.213	4.299	4.387	4.475	4.564	4.653	4.743	4.834
1000	4.834	4.926	5.018	5.111	5.205	5.299	5.394	5.489	5.585	5.682	5.780
1100	5.780	5.878	5.976	6.075	6.175	6.276	6.377	6.478	6.580	6.683	6.786
1200	6.786	6.890	6.995	7.100	7.205	7.311	7.417	7.524	7.632	7.740	7.848
1300	7.848	7.957	8.066	8.176	8.286	8.397	8.508	8.620	8.731	8.844	8.956
1400	8.956	9.069	9.182	9.296	9.410	9.524	9.639	9.753	9.868	9.984	10.099
1500	10.099	10.215	10.331	10.447	10.563	10.679	10.796	10.913	11.029	11.146	11.263
1600	11.263	11.380	11.497	11.614	11.731	11.848	11.965	12.082	12.199	12.316	12.433
1700	12.433	12.549	12.666	12.782	12.898	13.014	13.130	13.246	13.361	13.476	13.591
1800	13.591	13.706	13.820								

Temperature Ranges and Coefficients of Equations Used to Compute the Above Table

The equations are of the form: $E = c_0 + c_1t + c_2t^2 + c_3t^3 + \ldots c_nt^n$, where E is the emf in millivolts, t is the temperature in degrees Celsius (ITS-90), and c_0, c_1, c_2, c_3, *etc.* are the coefficients. These coefficients are extracted from NIST Monograph 175.

		0 °C to 630.615 °C	630.615 °C to 1820 °C
c_0	=	0.000 000 000 0 . . .	−3.893 816 862 1 . . .
c_1	=	−2.465 081 834 6 X 10^{-4}	2.857 174 747 0 X 10^{-2}
c_2	=	5.904 042 117 1 X 10^{-6}	−8.488 510 478 5 X 10^{-5}
c_3	=	−1.325 793 163 6 X 10^{-9}	1.578 528 016 4 X 10^{-7}
c_4	=	1.566 829 190 1 X 10^{-12}	−1.683 534 486 4 X 10^{-10}
c_5	=	−1.694 452 924 0 X 10^{-15}	1.110 979 401 3 X 10^{-13}
c_6	=	6.299 034 709 4 X 10^{-19}	−4.451 543 103 3 X 10^{-17}
c_7	=	. . . .	9.897 564 082 1 X 10^{-21}
c_8	=	. . . .	−9.379 133 028 9 X 10^{-25}

Type E thermocouples: emf-temperature (°C) reference table and equations.

Thermocouple emf as a Function of Temperature in Degrees Celsius (ITS-90)

emf in Millivolts Reference Junctions at 0 °C

°C	0	−10	−20	−30	−40	−50	−60	−70	−80	−90	−100
−200	−8.825	−9.063	−9.274	−9.455	−9.604	−9.718	−9.797	−9.835			
−100	−5.237	−5.681	−6.107	−6.516	−6.907	−7.279	−7.632	−7.963	−8.273	−8.561	−8.825
0	0.000	−0.582	−1.152	−1.709	−2.255	−2.787	−3.306	−3.811	−4.302	−4.777	−5.237

°C	0	10	20	30	40	50	60	70	80	90	100
0	0.000	0.591	1.192	1.801	2.420	3.048	3.685	4.330	4.985	5.648	6.319
100	6.319	6.998	7.685	8.379	9.081	9.789	10.503	11.224	11.951	12.684	13.421
200	13.421	14.164	14.912	15.664	16.420	17.181	17.945	18.713	19.484	20.259	21.036
300	21.036	21.817	22.600	23.386	24.174	24.964	25.757	26.552	27.348	28.146	28.946
400	28.946	29.747	30.550	31.354	32.159	32.965	33.772	34.579	35.387	36.196	37.005
500	37.005	37.815	38.624	39.434	40.243	41.053	41.862	42.671	43.479	44.286	45.093
600	45.093	45.900	46.705	47.509	48.313	49.116	49.917	50.718	51.517	52.315	53.112
700	53.112	53.908	54.703	55.497	56.289	57.080	57.870	58.659	59.446	60.232	61.017
800	61.017	61.801	62.583	63.364	64.144	64.922	65.698	66.473	67.246	68.017	68.787
900	68.787	69.554	70.319	71.082	71.844	72.603	73.360	74.115	74.869	75.621	76.373
1000	76.373										

Temperature Ranges and Coefficients of Equations Used to Compute the Above Table

The equations are of the form: $E = c_0 + c_1t + c_2t^2 + c_3t^3 + \ldots c_nt^n$, where E is the emf in millivolts, t is the temperature in degrees Celsius (ITS-90), and c_0, c_1, c_2, c_3, *etc.* are the coefficients. These coefficients are extracted from NIST Monograph 175.

		−270 °C to 0 °C	0 °C to 1000 °C
c_0	=	0.000 000 000 0 . . .	0.000 000 000 0 . . .
c_1	=	$5.866\ 550\ 870\ 8 \times 10^{-2}$	$5.866\ 550\ 871\ 0 \times 10^{-2}$
c_2	=	$4.541\ 097\ 712\ 4 \times 10^{-5}$	$4.503\ 227\ 558\ 2 \times 10^{-5}$
c_3	=	$-7.799\ 804\ 868\ 6 \times 10^{-7}$	$2.890\ 840\ 721\ 2 \times 10^{-8}$
c_4	=	2.580 016 084 3 X 10[illegible]	−3.305 689 665 2 X 10[illegible]
c_5	=	$-5.945\ 258\ 305\ 7 \times 10^{-10}$	$6.502\ 440\ 327\ 0 \times 10^{-13}$
c_6	=	$-9.321\ 405\ 866\ 7 \times 10^{-12}$	$-1.919\ 749\ 550\ 4 \times 10^{-16}$
c_7	=	$-1.028\ 760\ 553\ 4 \times 10^{-13}$	$-1.253\ 660\ 049\ 7 \times 10^{-18}$
c_8	=	$-8.037\ 012\ 362\ 1 \times 10^{-16}$	$2.148\ 921\ 756\ 9 \times 10^{-21}$
c_9	=	$-4.397\ 949\ 739\ 1 \times 10^{-18}$	$-1.438\ 804\ 178\ 2 \times 10^{-24}$
c_{10}	=	$-1.641\ 477\ 635\ 5 \times 10^{-20}$	$3.596\ 089\ 948\ 1 \times 10^{-28}$
c_{11}	=	$-3.967\ 361\ 951\ 6 \times 10^{-23}$	. . . .
c_{12}	=	$-5.582\ 732\ 872\ 1 \times 10^{-26}$	. . . .
c_{13}	=	$-3.465\ 784\ 201\ 3 \times 10^{-29}$	. . . .

Type J thermocouples: emf-temperature (°C) reference table and equations.

Thermocouple emf as a Function of Temperature in Degrees Celsius (ITS-90)

emf in Millivolts — Reference Junctions at 0 °C

°C	0	−10	−20	−30	−40	−50	−60	−70	−80	−90	−100
−200	−7.890	−8.095									
−100	−4.633	−5.037	−5.426	−5.801	−6.159	−6.500	−6.821	−7.123	−7.403	−7.659	−7.890
0	0.000	−0.501	−0.995	−1.482	−1.961	−2.431	−2.893	−3.344	−3.786	−4.215	−4.633

°C	0	10	20	30	40	50	60	70	80	90	100
0	0.000	0.507	1.019	1.537	2.059	2.585	3.116	3.650	4.187	4.726	5.269
100	5.269	5.814	6.360	6.909	7.459	8.010	8.562	9.115	9.669	10.224	10.779
200	10.779	11.334	11.889	12.445	13.000	13.555	14.110	14.665	15.219	15.773	16.327
300	16.327	16.881	17.434	17.986	18.538	19.090	19.642	20.194	20.745	21.297	21.848
400	21.848	22.400	22.952	23.504	24.057	24.610	25.164	25.720	26.276	26.834	27.393
500	27.393	27.953	28.516	29.080	29.647	30.216	30.788	31.362	31.939	32.519	33.102
600	33.102	33.689	34.279	34.873	35.470	36.071	36.675	37.284	37.896	38.512	39.132
700	39.132	39.755	40.382	41.012	41.645	42.281	42.919	43.559	44.203	44.848	45.494
800	45.494	46.141	46.786	47.431	48.074	48.715	49.353	49.989	50.622	51.251	51.877
900	51.877	52.500	53.119	53.735	54.347	54.956	55.561	56.164	56.763	57.360	57.953
1000	57.953	58.545	59.134	59.721	60.307	60.890	61.473	62.054	62.634	63.214	63.792
1100	63.792	64.370	64.948	65.525	66.102	66.679	67.255	67.831	68.406	68.980	69.553
1200	69.553										

Temperature Ranges and Coefficients of Equations Used to Compute the Above Table

The equations are of the form: $E = c_0 + c_1t + c_2t^2 + c_3t^3 + \ldots c_nt^n$, where E is the emf in millivolts, t is the temperature in degrees Celsius (ITS-90), and c_0, c_1, c_2, c_3, *etc.* are the coefficients. These coefficients are extracted from NIST Monograph 175.

		−210 °C to 760 °C	760 °C to 1200 °C
c_0	=	0.000 000 000 0 . . .	$2.964\,562\,568\,1 \times 10^{2}$
c_1	=	$5.038\,118\,781\,5 \times 10^{-2}$	−1.497 612 778 6 . . .
c_2	=	$3.047\,583\,693\,0 \times 10^{-5}$	$3.178\,710\,392\,4 \times 10^{-3}$
c_3	=	$-8.568\,106\,572\,0 \times 10^{-8}$	$-3.184\,768\,670\,1 \times 10^{-6}$
c_4	=	$1.322\,819\,529\,5 \times 10^{-10}$	$1.572\,081\,900\,4 \times 10^{-9}$
c_5	=	$-1.705\,295\,833\,7 \times 10^{-13}$	$-3.069\,136\,905\,6 \times 10^{-13}$
c_6	=	$2.094\,809\,069\,7 \times 10^{-16}$	....
c_7	=	$-1.253\,839\,533\,6 \times 10^{-19}$	....
c_8	=	$1.563\,172\,569\,7 \times 10^{-23}$	....

Type K thermocouples: emf-temperature (°C) reference table and equations.

Thermocouple emf as a Function of Temperature in Degrees Celsius (ITS-90)

emf in Millivolts — Reference Junctions at 0 °C

°C	0	−10	−20	−30	−40	−50	−60	−70	−80	−90	−100
−200	−5.891	−6.035	−6.158	−6.262	−6.344	−6.404	−6.441	−6.458			
−100	−3.554	−3.852	−4.138	−4.411	−4.669	−4.913	−5.141	−5.354	−5.550	−5.730	−5.891
0	0.000	−0.392	−0.778	−1.156	−1.527	−1.889	−2.243	−2.587	−2.920	−3.243	−3.554

°C	0	10	20	30	40	50	60	70	80	90	100
0	0.000	0.397	0.798	1.203	1.612	2.023	2.436	2.851	3.267	3.682	4.096
100	4.096	4.509	4.920	5.328	5.735	6.138	6.540	6.941	7.340	7.739	8.138
200	8.138	8.539	8.940	9.343	9.747	10.153	10.561	10.971	11.382	11.795	12.209
300	12.209	12.624	13.040	13.457	13.874	14.293	14.713	15.133	15.554	15.975	16.397
400	16.397	16.820	17.243	17.667	18.091	18.516	18.941	19.366	19.792	20.218	20.644
500	20.644	21.071	21.497	21.924	22.350	22.776	23.203	23.629	24.055	24.480	24.905
600	24.905	25.330	25.755	26.179	26.602	27.025	27.447	27.869	28.289	28.710	29.129
700	29.129	29.548	29.965	30.382	30.798	31.213	31.628	32.041	32.453	32.865	33.275
800	33.275	33.685	34.093	34.501	34.908	35.313	35.718	36.121	36.524	36.925	37.326
900	37.326	37.725	38.124	38.522	38.918	39.314	39.708	40.101	40.494	40.885	41.276
1000	41.276	41.665	42.053	42.440	42.826	43.211	43.595	43.978	44.359	44.740	45.119
1100	45.119	45.497	45.873	46.249	46.623	46.995	47.367	47.737	48.105	48.473	48.838
1200	48.838	49.202	49.565	49.926	50.286	50.644	51.000	51.355	51.708	52.060	52.410
1300	52.410	52.759	53.106	53.451	53.795	54.138	54.479	54.819			

Temperature Ranges and Coefficients of Equations Used to Compute the Above Table

The equations are of the form: $E = c_0 + c_1t + c_2t^2 + c_3t^3 + \ldots c_nt^n$, where E is the emf in millivolts, t is the temperature in degrees Celsius (ITS-90), and c_0, c_1, c_2, c_3, *etc.* are the coefficients. In the 0 °C to 1372 °C range there is also an exponential term that must be evaluated and added to the equation. The exponential term is of the form: $c_0 e^{c_1 (t - 126.9686)^2}$, where t is the temperature in °C, e is the natural logarithm base, and c_0 and c_i are the coefficients. These coefficients are extracted from NIST Monograph 175.

	−270 °C to 0 °C	0 °C to 1372 °C	0 °C to 1372 °C (exponential term)
c_0 =	0.000 000 000 0 . . .	−1.760 041 368 6 X 10^{-2}	1.185 976 X 10^{-1}
c_1 =	3.945 012 802 5 X 10^{-2}	3.892 120 497 5 X 10^{-2}	−1.183 432 X 10^{-4}
c_2 =	2.362 237 359 8 X 10^{-5}	1.855 877 003 2 X 10^{-5}	....
c_3 =	−3.285 890 678 4 X 10^{-7}	−9.945 759 287 4 X 10^{-8}	....
c_4 =	−4.990 482 877 7 X 10^{-9}	3.184 094 571 9 X 10^{-10}	....
c_5 =	−6.750 905 917 3 X 10^{-11}	−5.607 284 488 9 X 10^{-13}	....
c_6 =	−5.741 032 742 8 X 10^{-13}	5.607 505 905 9 X 10^{-16}	....
c_7 =	−3.108 887 289 4 X 10^{-15}	−3.202 072 000 3 X 10^{-19}	....
c_8 =	−1.045 160 936 5 X 10^{-17}	9.715 114 715 2 X 10^{-23}	....
c_9 =	−1.988 926 687 8 X 10^{-20}	−1.210 472 127 5 X 10^{-26}	....
c_{10} =	−1.632 269 748 6 X 10^{-23}	....	....

Type N thermocouples: emf-temperature (°C) reference table and equations.

Thermocouple emf as a Function of Temperature in Degrees Celsius (ITS-90)

emf in Millivolts — Reference Junctions at 0 °C

°C	0	−10	−20	−30	−40	−50	−60	−70	−80	−90	−100
−200	−3.990	−4.083	−4.162	−4.226	−4.277	−4.313	−4.336	−4.345			
−100	−2.407	−2.612	−2.808	−2.994	−3.171	−3.336	−3.491	−3.634	−3.766	−3.884	−3.990
0	0.000	−0.260	−0.518	−0.772	−1.023	−1.269	−1.509	−1.744	−1.972	−2.193	−2.407

°C	0	10	20	30	40	50	60	70	80	90	100
0	0.000	0.261	0.525	0.793	1.065	1.340	1.619	1.902	2.189	2.480	2.774
100	2.774	3.072	3.374	3.680	3.989	4.302	4.618	4.937	5.259	5.585	5.913
200	5.913	6.245	6.579	6.916	7.255	7.597	7.941	8.288	8.637	8.988	9.341
300	9.341	9.696	10.054	10.413	10.774	11.136	11.501	11.867	12.234	12.603	12.974
400	12.974	13.346	13.719	14.094	14.469	14.846	15.225	15.604	15.984	16.366	16.748
500	16.748	17.131	17.515	17.900	18.286	18.672	19.059	19.447	19.835	20.224	20.613
600	20.613	21.003	21.393	21.784	22.175	22.566	22.958	23.350	23.742	24.134	24.527
700	24.527	24.919	25.312	25.705	26.098	26.491	26.883	27.276	27.669	28.062	28.455
800	28.455	28.847	29.239	29.632	30.024	30.416	30.807	31.199	31.590	31.981	32.371
900	32.371	32.761	33.151	33.541	33.930	34.319	34.707	35.095	35.482	35.869	36.256
1000	36.256	36.641	37.027	37.411	37.795	38.179	38.562	38.944	39.326	39.706	40.087
1100	40.087	40.466	40.845	41.223	41.600	41.976	42.352	42.727	43.101	43.474	43.846
1200	43.846	44.218	44.588	44.958	45.326	45.694	46.060	46.425	46.789	47.152	47.513
1300	47.513										

Temperature Ranges and Coefficients of Equations Used to Compute the Above Table

The equations are of the form: $E = c_0 + c_1t + c_2t^2 + c_3t^3 + \dots c_nt^n$, where E is the emf in millivolts, t is the temperature in degrees Celsius (ITS-90), and c_0, c_1, c_2, c_3, *etc.* are the coefficients. These coefficients are extracted from NIST Monograph 175.

		−270 °C to 0 °C	0 °C to 1300 °C
c_0	=	0.000 000 000 0 . . .	0.000 000 000 0 . . .
c_1	=	$2.615\,910\,596\,2 \times 10^{-2}$	$2.592\,939\,460\,1 \times 10^{-2}$
c_2	=	$1.095\,748\,422\,8 \times 10^{-5}$	$1.571\,014\,188\,0 \times 10^{-5}$
c_3	=	$-9.384\,111\,155\,4 \times 10^{-8}$	$4.382\,562\,723\,7 \times 10^{-8}$
c_4	=	$-4.641\,203\,975\,9 \times 10^{-11}$	$-2.526\,116\,979\,4 \times 10^{-10}$
c_5	=	$-2.630\,335\,771\,6 \times 10^{-12}$	$6.431\,181\,933\,9 \times 10^{-13}$
c_6	=	$-2.265\,343\,800\,3 \times 10^{-14}$	$-1.006\,347\,151\,9 \times 10^{-15}$
c_7	=	$-7.608\,930\,079\,1 \times 10^{-17}$	$9.974\,533\,899\,2 \times 10^{-19}$
c_8	=	$-9.341\,966\,783\,5 \times 10^{-20}$	$-6.086\,324\,560\,7 \times 10^{-22}$
c_9	=	. . . .	$2.084\,922\,933\,9 \times 10^{-26}$
c_{10}	=	. . . .	$-3.068\,219\,615\,1 \times 10^{-29}$

STANDARD ITS-90 THERMOCOUPLE TABLES (continued)

Type R thermocouples: emf-temperature (°C) reference table and equations.

Thermocouple emf as a Function of Temperature in Degrees Celsius (ITS-90)

emf in Millivolts — Reference Junctions at 0 °C

°C	0	−10	−20	−30	−40	−50	−60	−70	−80	−90	−100
0	0.000	−0.051	−0.100	−0.145	−0.188	−0.226					

°C	0	10	20	30	40	50	60	70	80	90	100
0	0.000	0.054	0.111	0.171	0.232	0.296	0.363	0.431	0.501	0.573	0.647
100	0.647	0.723	0.800	0.879	0.959	1.041	1.124	1.208	1.294	1.381	1.469
200	1.469	1.558	1.648	1.739	1.831	1.923	2.017	2.112	2.207	2.304	2.401
300	2.401	2.498	2.597	2.696	2.796	2.896	2.997	3.099	3.201	3.304	3.408
400	3.408	3.512	3.616	3.721	3.827	3.933	4.040	4.147	4.255	4.363	4.471
500	4.471	4.580	4.690	4.800	4.910	5.021	5.133	5.245	5.357	5.470	5.583
600	5.583	5.697	5.812	5.926	6.041	6.157	6.273	6.390	6.507	6.625	6.743
700	6.743	6.861	6.980	7.100	7.220	7.340	7.461	7.583	7.705	7.827	7.950
800	7.950	8.073	8.197	8.321	8.446	8.571	8.697	8.823	8.950	9.077	9.205
900	9.205	9.333	9.461	9.590	9.720	9.850	9.980	10.111	10.242	10.374	10.506
1000	10.506	10.638	10.771	10.905	11.039	11.173	11.307	11.442	11.578	11.714	11.850
1100	11.850	11.986	12.123	12.260	12.397	12.535	12.673	12.812	12.950	13.089	13.228
1200	13.228	13.367	13.507	13.646	13.786	13.926	14.066	14.207	14.347	14.488	14.629
1300	14.629	14.770	14.911	15.052	15.193	15.334	15.475	15.616	15.758	15.899	16.040
1400	16.040	16.181	16.323	16.464	16.605	16.746	16.887	17.028	17.169	17.310	17.451
1500	17.451	17.591	17.732	17.872	18.012	18.152	18.292	18.431	18.571	18.710	18.849
1600	18.849	18.988	19.126	19.264	19.402	19.540	19.677	19.814	19.951	20.087	20.222
1700	20.222	20.356	20.488	20.620	20.749	20.877	21.003				

Temperature Ranges and Coefficients of Equations Used to Compute the Above Table

The equations are of the form: $E = c_0 + c_1t + c_2t^2 + c_3t^3 + ... c_nt^n$, where E is the emf in millivolts, t is the temperature in degrees Celsius (ITS-90), and c_0, c_1, c_2, c_3, *etc.* are the coefficients. These coefficients are extracted from NIST Monograph 175.

	−50 °C to 1064.18 °C	1064.18 °C to 1664.5 °C	1664.5 °C to 1768.1 °C
c_0 =	0.000 000 000 00 . . .	2.951 579 253 16 . . .	1.522 321 182 09 X 10^{2}
c_1 =	5.289 617 297 65 X 10^{-3}	−2.520 612 513 32 X 10^{-3}	−2.688 198 885 45 X 10^{-1}
c_2 =	1.391 665 897 82 X 10^{-5}	1.595 645 018 65 X 10^{-5}	1.712 802 804 71 X 10^{-4}
c_3 =	−2.388 556 930 17 X 10^{-8}	−7.640 859 475 76 X 10^{-9}	−3.458 957 064 53 X 10^{-8}
c_4 =	3.569 160 010 63 X 10^{-11}	2.053 052 910 24 X 10^{-12}	−9.346 339 710 46 X 10^{-15}
c_5 =	−4.623 476 662 98 X 10^{-14}	−2.933 596 681 73 X 10^{-16}	....
c_6 =	5.007 774 410 34 X 10^{-17}	....	....
c_7 =	−3.731 058 861 91 X 10^{-20}	....	....
c_8 =	1.577 164 823 67 X 10^{-23}	....	....
c_9 =	−2.810 386 252 51 X 10^{-27}	....	....

Type S thermocouples: emf-temperature (°C) reference table and equations.

Thermocouple emf as a Function of Temperature in Degrees Celsius (ITS-90)

emf in Millivolts — Reference Junctions at 0 °C

°C	0	−10	−20	−30	−40	−50	−60	−70	−80	−90	−100
0	0.000	−0.053	−0.103	−0.150	−0.194	−0.236					

°C	0	10	20	30	40	50	60	70	80	90	100
0	0.000	0.055	0.113	0.173	0.235	0.299	0.365	0.433	0.502	0.573	0.646
100	0.646	0.720	0.795	0.872	0.950	1.029	1.110	1.191	1.273	1.357	1.441
200	1.441	1.526	1.612	1.698	1.786	1.874	1.962	2.052	2.141	2.232	2.323
300	2.323	2.415	2.507	2.599	2.692	2.786	2.880	2.974	3.069	3.164	3.259
400	3.259	3.355	3.451	3.548	3.645	3.742	3.840	3.938	4.036	4.134	4.233
500	4.233	4.332	4.432	4.532	4.632	4.732	4.833	4.934	5.035	5.137	5.239
600	5.239	5.341	5.443	5.546	5.649	5.753	5.857	5.961	6.065	6.170	6.275
700	6.275	6.381	6.486	6.593	6.699	6.806	6.913	7.020	7.128	7.236	7.345
800	7.345	7.454	7.563	7.673	7.783	7.893	8.003	8.114	8.226	8.337	8.449
900	8.449	8.562	8.674	8.787	8.900	9.014	9.128	9.242	9.357	9.472	9.587
1000	9.587	9.703	9.819	9.935	10.051	10.168	10.285	10.403	10.520	10.638	10.757
1100	10.757	10.875	10.994	11.113	11.232	11.351	11.471	11.590	11.710	11.830	11.951
1200	11.951	12.071	12.191	12.312	12.433	12.554	12.675	12.796	12.917	13.038	13.159
1300	13.159	13.280	13.402	13.523	13.644	13.766	13.887	14.009	14.130	14.251	14.373
1400	14.373	14.494	14.615	14.736	14.857	14.978	15.099	15.220	15.341	15.461	15.582
1500	15.582	15.702	15.822	15.942	16.062	16.182	16.301	16.420	16.539	16.658	16.777
1600	16.777	16.895	17.013	17.131	17.249	17.366	17.483	17.600	17.717	17.832	17.947
1700	17.947	18.061	18.174	18.285	18.395	18.503	18.609				

Temperature Ranges and Coefficients of Equations Used to Compute the Above Table

The equations are of the form: $E = c_0 + c_1t + c_2t^2 + c_3t^3 + \ldots c_nt^n$, where E is the emf in millivolts, t is the temperature in degrees Celsius (ITS-90), and c_0, c_1, c_2, c_3, *etc.* are the coefficients. These coefficients are extracted from NIST Monograph 175.

	−50 °C to 1064.18 °C	1064.18 °C to 1664.5 °C	1664.5 °C to 1768.1 °C
c_0 =	0.000 000 000 00 . . .	1.329 004 440 85 . . .	$1.466\ 282\ 326\ 36 \times 10^{2}$
c_1 =	$5.403\ 133\ 086\ 31 \times 10^{-3}$	$3.345\ 093\ 113\ 44 \times 10^{-3}$	$-2.584\ 305\ 167\ 52 \times 10^{-1}$
c_2 =	$1.259\ 342\ 897\ 40 \times 10^{-5}$	$6.548\ 051\ 928\ 18 \times 10^{-6}$	$1.636\ 935\ 746\ 41 \times 10^{-4}$
c_3 =	$-2.324\ 779\ 686\ 89 \times 10^{-8}$	$-1.648\ 562\ 592\ 09 \times 10^{-9}$	$-3.304\ 390\ 469\ 87 \times 10^{-8}$
c_4 =	$3.220\ 288\ 230\ 36 \times 10^{-11}$	$1.299\ 896\ 051\ 74 \times 10^{-14}$	$-9.432\ 236\ 906\ 12 \times 10^{-15}$
c_5 =	$-3.314\ 651\ 963\ 89 \times 10^{-14}$	....	....
c_6 =	$2.557\ 442\ 517\ 86 \times 10^{-17}$	....	....
c_7 =	$-1.250\ 688\ 713\ 93 \times 10^{-20}$	....	....
c_8 =	$2.714\ 431\ 761\ 45 \times 10^{-24}$	....	....

STANDARD ITS-90 THERMOCOUPLE TABLES (continued)

Type T thermocouples: emf-temperature (°C) reference table and equations.

Thermocouple emf as a Function of Temperature in Degrees Celsius (ITS-90)

emf in Millivolts — Reference Junctions at 0 °C

°C	0	−10	−20	−30	−40	−50	−60	−70	−80	−90	−100
−200	−5.603	−5.753	−5.888	−6.007	−6.105	−6.180	−6.232	−6.258			
−100	−3.379	−3.657	−3.923	−4.177	−4.419	−4.648	−4.865	−5.070	−5.261	−5.439	−5.603
0	0.000	−0.383	−0.757	−1.121	−1.475	−1.819	−2.153	−2.476	−2.788	−3.089	−3.379

°C	0	10	20	30	40	50	60	70	80	90	100
0	0.000	0.391	0.790	1.196	1.612	2.036	2.468	2.909	3.358	3.814	4.279
100	4.279	4.750	5.228	5.714	6.206	6.704	7.209	7.720	8.237	8.759	9.288
200	9.288	9.822	10.362	10.907	11.458	12.013	12.574	13.139	13.709	14.283	14.862
300	14.862	15.445	16.032	16.624	17.219	17.819	18.422	19.030	19.641	20.255	20.872
400	20.872										

Temperature Ranges and Coefficients of Equations Used to Compute the Above Table

The equations are of the form: $E = c_0 + c_1t + c_2t^2 + c_3t^3 + \ldots c_nt^n$, where E is the emf in millivolts, t is the temperature in degrees Celsius (ITS-90), and c_0, c_1, c_2, c_3, etc. are the coefficients. These coefficients are extracted from NIST Monograph 175.

		−270 °C to 0 °C	0 °C to 400 °C
c_0	=	0.000 000 000 0 . . .	0.000 000 000 0 . . .
c_1	=	$3.874\,810\,636\,4 \times 10^{-2}$	$3.874\,810\,636\,4 \times 10^{-2}$
c_2	=	$4.419\,443\,434\,7 \times 10^{-5}$	$3.329\,222\,788\,0 \times 10^{-5}$
c_3	=	$1.184\,432\,310\,5 \times 10^{-7}$	$2.061\,824\,340\,4 \times 10^{-7}$
c_4	=	$2.003\,297\,355\,4 \times 10^{-8}$	$-2.188\,225\,684\,6 \times 10^{-9}$
c_5	=	$9.013\,801\,955\,9 \times 10^{-10}$	$1.099\,688\,092\,8 \times 10^{-11}$
c_6	=	$2.265\,115\,659\,3 \times 10^{-11}$	$-3.081\,575\,877\,2 \times 10^{-14}$
c_7	=	$3.607\,115\,420\,5 \times 10^{-13}$	$4.547\,913\,529\,0 \times 10^{-17}$
c_8	=	$3.849\,393\,988\,3 \times 10^{-15}$	$-2.751\,290\,167\,3 \times 10^{-20}$
c_9	=	$2.821\,352\,192\,5 \times 10^{-17}$	. . . .
c_{10}	=	$1.425\,159\,477\,9 \times 10^{-19}$	. . . .
c_{11}	=	$4.876\,866\,228\,6 \times 10^{-22}$	. . . .
c_{12}	=	$1.079\,553\,927\,0 \times 10^{-24}$	. . . .
c_{13}	=	$1.394\,502\,706\,2 \times 10^{-27}$	. . . .
c_{14}	=	$7.979\,515\,392\,7 \times 10^{-31}$	. . . .

Type B thermocouples: coefficients (c_i) of polynomials for the computation of temperatures in °C as a function of the thermocouple emf in various temperature and emf ranges.

Temperature Range:		250 °C to 700 °C	700 °C to 1820 °C
emf Range:		0.291 mV to 2.431 mV	2.431 mV to 13.820 mV
c_0	=	$9.842\,332\,1 \times 10^{1}$	$2.131\,507\,1 \times 10^{2}$
c_1	=	$6.997\,150\,0 \times 10^{2}$	$2.851\,050\,4 \times 10^{2}$
c_2	=	$-8.476\,530\,4 \times 10^{2}$	$-5.274\,288\,7 \times 10^{1}$
c_3	=	$1.005\,264\,4 \times 10^{3}$.	9.916 080 4 . . .
c_4	=	$-8.334\,595\,2 \times 10^{2}$	−1.296 530 3 . . .
c_5	=	$4.550\,854\,2 \times 10^{2}$	$1.119\,587\,0 \times 10^{-1}$
c_6	=	$-1.552\,303\,7 \times 10^{2}$	$-6.062\,519\,9 \times 10^{-3}$
c_7	=	$2.988\,675\,0 \times 10^{1}$	$1.866\,169\,6 \times 10^{-4}$
c_8	=	−2.474 286 0 . . .	$-2.487\,858\,5 \times 10^{-6}$

NOTE—The above coefficients are extracted from NIST Monograph 175 and are for an expression of the form shown in Section 10.3.2. They yield approximate values of temperature that agree within ± 0.03 °C with the values given in Table 10.2.

Type E thermocouples: coefficients (c_i) of polynomials for the computation of temperatures in °C as a function of the thermocouple emf in various temperature and emf ranges.

Temperature Range:		−200 °C to 0 °C	0 °C to 1000 °C
emf Range:		−8.825 mV to 0.0 mV	0.0 mV to 76.373 mV
c_0	=	0.000 000 0 . . .	0.000 000 0 . . .
c_1	=	$1.697\,728\,8 \times 10^{1}$	$1.705\,703\,5 \times 10^{1}$
c_2	=	$-4.351\,497\,0 \times 10^{-1}$	$-2.330\,175\,9 \times 10^{-1}$
c_3	=	$-1.585\,969\,7 \times 10^{-1}$	$6.543\,558\,5 \times 10^{-3}$
c_4	=	$-9.250\,287\,1 \times 10^{-2}$	$-7.356\,274\,9 \times 10^{-5}$
c_5	=	$-2.608\,431\,4 \times 10^{-2}$	$-1.789\,600\,1 \times 10^{-6}$
c_6	=	$-4.136\,019\,9 \times 10^{-3}$	$8.403\,616\,5 \times 10^{-8}$
c_7	=	$-3.403\,403\,0 \times 10^{-4}$	$-1.373\,587\,9 \times 10^{-9}$
c_8	=	$-1.156\,489\,0 \times 10^{-5}$	$1.062\,982\,3 \times 10^{-11}$
c_9	=	. . . .	$-3.244\,708\,7 \times 10^{-14}$

NOTE—The above coefficients are extracted from NIST Monograph 175 and are for an expression of the form shown in Section 10.3.2. They yield approximate values of temperature that agree within ± 0.02 °C with the values given in Table 10.4.

Type J thermocouples: coefficients (c_i) of polynomials for the computation of temperatures in °C as a function of the thermocouple emf in various temperature and emf ranges.

Temperature Range:	−210 °C to 0 °C	0 °C to 760 °C	760 °C to 1200 °C
emf Range:	−8.095 mV to 0.0 mV	0.0 mV to 42.919 mV	42.919 mV to 69.553 mV
c_0 =	0.000 000 0 . . .	0.000 000 . . .	−3.113 581 87 X 10^{3}
c_1 =	1.952 826 8 X 10^{1}	1.978 425 X 10^{1}	3.005 436 84 X 10^{2}
c_2 =	−1.228 618 5 . . .	−2.001 204 X 10^{-1}	−9.947 732 30 . . .
c_3 =	−1.075 217 8 . . .	1.036 969 X 10^{-2}	1.702 766 30 X 10^{-1}
c_4 =	−5.908 693 3 X 10^{-1}	−2.549 687 X 10^{-4}	−1.430 334 68 X 10^{-3}
c_5 =	−1.725 671 3 X 10^{-1}	3.585 153 X 10^{-6}	4.738 860 84 X 10^{-6}
c_6 =	−2.813 151 3 X 10^{-2}	−5.344 285 X 10^{-8}	. . . .
c_7 =	−2.396 337 0 X 10^{-3}	5.099 890 X 10^{-10}	. . . .
c_8 =	−8.382 332 1 X 10^{-5}	. . . .	. . . .

NOTE—The above coefficients are extracted from NIST Monograph 175 and are for an expression of the form shown in Section 10.3.2. They yield approximate values of temperature that agree within ± 0.05 °C with the values given in Table 10.6.

Type K thermocouples: coefficients (c_i) of polynomials for the computation of temperatures in °C as a function of the thermocouple emf in various temperature and emf ranges.

Temperature Range:	−200 °C to 0 °C	0 °C to 500 °C	500 °C to 1372 °C
emf Range:	−5.891 mV to 0.0 mV	0.0 mV to 20.644 mV	20.644 mV to 54.886 mV
c_0 =	0.000 000 0 . . .	0.000 000 0 . . .	−1.318 058 X 10^{2}
c_1 =	2.517 346 2 X 10^{1}	2.508 355 X 10^{1}	4.830 222 X 10^{1}
c_2 =	−1.166 287 8 . . .	7.860 106 X 10^{-2}	−1.646 031 . . .
c_3 =	−1.083 363 8 . . .	−2.503 131 X 10^{-1}	5.464 731 X 10^{-2}
c_4 =	−8.977 354 0 X 10^{-1}	8.315 270 X 10^{-2}	−9.650 715 X 10^{-4}
c_5 =	−3.734 237 7 X 10^{-1}	−1.228 034 X 10^{-2}	8.802 193 X 10^{-6}
c_6 =	−8.663 264 3 X 10^{-2}	9.804 036 X 10^{-4}	−3.110 810 X 10^{-8}
c_7 =	−1.045 059 8 X 10^{-2}	−4.413 030 X 10^{-5}	. . . .
c_8 =	−5.192 057 7 X 10^{-4}	1.057 734 X 10^{-6}	. . . .
c_9 =	. . . .	−1.052 755 X 10^{-8}	. . . .

NOTE—The above coefficients are extracted from NIST Monograph 175 and are for an expression of the form shown in Section 10.3.2. They yield approximate values of temperature that agree within ± 0.05 °C with the values given in Table 10.8.

Type N thermocouples: coefficients (c_i) of polynomials for the computation of temperatures in °C as a function of the thermocouple emf in various temperature and emf ranges.

	−200 °C to 0 °C	0 °C to 600 °C	600 °C to 1300 °C
Temperature Range:	−200 °C to 0 °C	0 °C to 600 °C	600 °C to 1300 °C
emf Range:	−3.990 mV to 0.0 mV	0.0 mV to 20.613 mV	20.613 mV to 47.513 mV
c_0 =	0.000 000 0 . . .	0.000 00 . . .	1.972 485 X 10^1
c_1 =	3.843 684 7 X 10^1	3.868 96 X 10^1	3.300 943 X 10^1
c_2 =	1.101 048 5 . . .	−1.082 67 . . .	−3.915 159 X 10^{-1}
c_3 =	5.222 931 2 . . .	4.702 05 X 10^{-2}	9.855 391 X 10^{-3}
c_4 =	7.206 052 5 . . .	−2.121 69 X 10^{-6}	−1.274 371 X 10^{-4}
c_5 =	5.848 858 6 . . .	−1.172 72 X 10^{-4}	7.767 022 X 10^{-7}
c_6 =	2.775 491 6 . . .	5.392 80 X 10^{-6}	. . . .
c_7 =	7.707 516 6 X 10^{-1}	−7.981 56 X 10^{-8}	. . . .
c_8 =	1.158 266 5 X 10^{-1}	. . . .	. . . .
c_9 =	7.313 886 8 X 10^{-3}	. . . .	. . . .

NOTE—The above coefficients are extracted from NIST Monograph 175 and are for an expression of the form shown in Section 10.3.2. They yield approximate values of temperature that agree within ± 0.04 °C with the values given in Table 10.10.

Type R thermocouples: coefficients (c_i) of polynomials for the computation of temperatures in °C as a function of the thermocouple emf in various temperature and emf ranges.

	−50 °C to 250 °C	250 °C to 1200 °C	1064 °C to 1664.5 °C	1664.5 °C to 1768.1 °C
Temperature Range:	−50 °C to 250 °C	250 °C to 1200 °C	1064 °C to 1664.5 °C	1664.5 °C to 1768.1 °C
emf Range:	−0.226 mV to 1.923 mV	1.923 mV to 13.228 mV	11.361 mV to 19.739 mV	19.739 mV to 21.103 mV
c_0 =	0.000 000 0 . . .	1.334 584 505 X 10^1	−8.199 599 416 X 10^1	3.406 177 836 X 10^4
c_1 =	1.889 138 0 X 10^2	1.472 644 573 X 10^2	1.553 962 042 X 10^2	−7.023 729 171 X 10^3
c_2 =	−9.383 529 0 X 10^1	−1.844 024 844 X 10^1	−8.342 197 663 . . .	5.582 903 813 X 10^2
c_3 =	1.306 861 9 X 10^2	4.031 129 726 . . .	4.279 433 549 X 10^{-1}	−1.952 394 635 X 10^1
c_4 =	−2.270 358 0 X 10^2	−6.249 428 360 X 10^{-1}	−1.191 577 910 X 10^{-2}	2.560 740 231 X 10^{-1}
c_5 =	3.514 565 9 X 10^2	6.468 412 046 X 10^{-2}	1.492 290 091 X 10^{-4}	. . . .
c_6 =	−3.895 390 0 X 10^2	−4.458 750 426 X 10^{-3}	. . . .	. . . .
c_7 =	2.823 947 1 X 10^2	1.994 710 149 X 10^{-4}	. . . .	. . . .
c_8 =	−1.260 728 1 X 10^2	−5.313 401 790 X 10^{-6}	. . . .	. . . .
c_9 =	3.135 361 1 X 10^1	6.481 976 217 X 10^{-8}	. . . .	. . . .
c_{10} =	−3.318 776 9 . . .	. . . .	. . . .	. . . .

NOTE—The above coefficients are extracted from NIST Monograph 175 and are for an expression of the form shown in Section 10.3.2. They yield approximate values of temperature that agree within ± 0.02 °C with the values given in Table 10.12.

Type S thermocouples: coefficients (c_i) of polynomials for the computation of temperatures in °C as a function of the thermocouple emf in various temperature and emf ranges.

Temperature Range:	−50 °C to 250 °C	250 °C to 1200 °C	1064 °C to 1664.5 °C	1664.5 °C to 1768.1 °C
emf Range:	−0.235 mV to 1.874 mV	1.874 mV to 11.950 mV	10.332 mV to 17.536 mV	17.536 mV to 18.693 mV
c_0 =	0.000 000 00 . . .	1.291 507 177 X 10^{1}	−8.087 801 117 X 10^{1}	5.333 875 126 X 10^{4}
c_1 =	1.849 494 60 X 10^{2}	1.466 298 863 X 10^{2}	1.621 573 104 X 10^{2}	−1.235 892 298 X 10^{4}
c_2 =	−8.005 040 62 X 10^{1}	−1.534 713 402 X 10^{1}	−8.536 869 453 . . .	1.092 657 613 X 10^{3}
c_3 =	1.022 374 30 X 10^{2}	3.145 945 973 . . .	4.719 686 976 X 10^{-1}	−4.265 693 686 X 10^{1}
c_4 =	−1.522 485 92 X 10^{2}	−4.163 257 839 X 10^{-1}	−1.441 693 666 X 10^{-2}	6.247 205 420 X 10^{-1}
c_5 =	1.888 213 43 X 10^{2}	3.187 963 771 X 10^{-2}	2.081 618 890 X 10^{-4}	. . . .
c_6 =	−1.590 859 41 X 10^{2}	−1.291 637 500 X 10^{-3}	. . . .	. . . .
c_7 =	8.230 278 80 X 10^{1}	2.183 475 087 X 10^{-5}	. . . .	. . . .
c_8 =	−2.341 819 44 X 10^{1}	−1.447 379 511 X 10^{-7}	. . . .	. . . .
c_9 =	2.797 862 60 . . .	8.211 272 125 X 10^{-9}	. . . .	. . . .

NOTE—The above coefficients are extracted from NIST Monograph 175 and are for an expression of the form shown in Section 10.3.2. They yield approximate values of temperature that agree within ± 0.02 °C with the values given in Table 10.14.

Type T thermocouples: coefficients (c_i) of polynomials for the computation of temperatures in °C as a function of the thermocouple emf in various temperature and emf ranges.

Temperature Range:	−200 °C to 0 °C	0 °C to 400 °C
emf Range:	−5.603 mV to 0.0 mV	0.0 mV to 20.872 mV
c_0 =	0.000 000 0 . . .	0.000 000 . . .
c_1 =	2.594 919 2 X 10^{1}	2.592 800 X 10^{1}
c_2 =	−2.131 696 7 X 10^{-1}	−7.602 961 X 10^{-1}
c_3 =	7.901 869 2 X 10^{-1}	4.637 791 X 10^{-2}
c_4 =	4.252 777 7 X 10^{-1}	−2.165 394 X 10^{-3}
c_5 =	1.330 447 3 X 10^{-1}	6.048 144 X 10^{-5}
c_6 =	2.024 144 6 X 10^{-2}	−7.293 422 X 10^{-7}
c_7 =	1.266 817 1 X 10^{-3}	. . . .

NOTE—The above coefficients are extracted from NIST Monograph 175 and are for an expression of the form shown in Section 10.3.2. They yield approximate values of temperature that agree within ± 0.04 °C with the values given in Table 10.16.

PROPERTIES OF COMMON LABORATORY SOLVENTS

This table give properties of 200 organic solvents which are frequently used in laboratory and industrial applications. Compounds are listed in alphabetical order by the most common name; synonyms are given in some cases. The properties tabulated are:

MF: Molecular formula
CAS RN: Chemical Abstracts Service Registry Number
M_r: Molecular weight
T_m: Melting point in °C
T_b: Normal boiling point in °C
ρ: Density in g/cm^3 at the temperature in °C indicated by the superscript
c_p: Specific heat capacity of the liquid at constant pressure at 25°C in J/g K
vp: Vapor pressure at 25°C in kPa (1 kPa = 7.50 mmHg)
μ: Electric dipole moment in debye units. Values in parentheses are measurements on the pure liquid or in solution; these are less reliable than the other values, which were obtained in the gas phase.
FP: Flash point temperature in °C
Fl. Lim.: Flammable (explosive) range in air in percent by volume
Ign. Temp.: Autoignition temperature in °C
TLV: Threshold limit for allowable airborne concentration, given in parts per million by volume at 25°C and atmospheric pressure (see table "Threshold Limit Values for Airborne Contaminants" in Section 16)

REFERENCES

1. Lide, D.R., *Handbook of Organic Solvents*, CRC Press, Boca Raton, FL, 1994.
2. Lide, D.R., and Kehiaian, H.V., *CRC Handbook of Thermophysical and Thermochemical Data*, CRC Press, Boca Raton, FL, 1994.
3. Riddick, J.A., Bunger, W.B., and Sakano, T.K., *Organic Solvents, Fourth Edition*, John Wiley & Sons, New York, 1986.
4. *Fire Protection Guide to Hazardous Materials*, 10th Edition , National Fire Protection Association, Quincy, MA, 1991.

Name	MF	CAS RN	M_r	T_m/°C	T_b/°C	ρ/g cm^{-3}	c_p/J g^{-1}K^{-1}	vp/kPa	μ/D	FP/°C	Fl. Lim.	Ign. Temp./°C	TLV
Acetal (1,1-Diethoxyethane)	$C_6H_{14}O_2$	105-57-7	118.18	-100	102.2	0.8254^{20}	2.01	3.68	(1.4)	-21	2-10%	230	
Acetic acid	$C_2H_4O_2$	64-19-7	60.05	17	118	1.0492^{20}	2.06	2.07	1.70	39	4-20%	463	10
Acetone	C_3H_6O	67-64-1	58.08	-95	56	0.7899^{20}	2.18	30.8	2.88	-20	3-13%	465	750
Acetonitrile	C_2H_3N	75-05-8	41.05	-44	82	0.7857^{20}	2.23	11.8	3.92	6	3-16%	524	40
Acetylacetone	$C_5H_8O_2$	123-54-6	100.12	-23	138	0.9721^{25}	2.08	1.02	(2.8)	34		340	
Acrylonitrile	C_3H_3N	107-13-1	53.06	-83.5	77.3	0.8060^{20}	2.05	14.1	3.87	0	3-17%	481	2
Adiponitrile	$C_6H_8N_2$	111-69-3	108.14	1	295	0.9676^{20}	1.19	<0.01		93	2-5%	550	2
Allyl alcohol	C_3H_6O	107-18-6	58.08	-129	97.0	0.8540^{20}	2.39	3.14	1.60	21	3-18%	378	2
Allylamine	C_3H_7N	107-11-9	57.10	-88.2	53.3	0.758^{20}		33.1	1.2	-29	2-22%	374	
2-Aminoisobutanol	$C_4H_{11}NO$	124-68-5	89.14	25.5	165.5	0.934^{20}				67			
Benzal chloride	$C_7H_6Cl_2$	98-87-3	161.03	-17	205	1.26^{25}		0.06	(2.1)				
Benzaldehyde	C_7H_6O	100-52-7	106.12	-26	179.0	1.0415^{10}	1.62	0.17	(3.0)	63		192	
Benzene	C_6H_6	71-43-2	78.11	6	80	0.8765^{20}	1.74	12.7	0	-11	1-8%	498	10
Benzonitrile	C_7H_5N	100-47-0	103.12	-12.7	191.1	1.0093^{15}	1.60	0.11	4.18				
Benzyl chloride	C_7H_7Cl	100-44-7	126.59	-45	179	1.1004^{20}	1.44	0.16	(1.8)	67	1%-	585	1
Bromochloromethane	CH_2BrCl	74-97-5	129.38	-87.9	68.0	1.9344^{20}	0.41	19.5	(1.7)				200
Bromoform (Tribromomethane)	$CHBr_3$	75-25-2	252.73	8.0	149	2.899^{15}	0.52	0.73	0.99	83			0.5
Butyl acetate	$C_6H_{12}O_2$	123-86-4	116.16	-78	126	0.8825^{20}	1.96	1.66	(1.9)	22	2-8%	425	150

PROPERTIES OF COMMON LABORATORY SOLVENTS (continued)

Name	MF	CAS RN	M_r	T_m/°C	T_b/°C	ρ/g cm^{-3}	c_p/J g^{-1}K^{-1}	vp/kPa	μ/D	FP/°C	Fl. Lim.	Ign. Temp./°C	TLV
Butyl alcohol	$C_4H_{10}O$	71-36-3	74.12	-90	118	0.8098^{20}	2.39	0.86	1.66	37	1-11%	343	50
sec-Butyl alcohol	$C_4H_{10}O$	78-92-2	74.12	-114.7	99.5	0.8063^{20}	2.66	2.32	(1.8)	24	2-10%	405	100
tert-Butyl alcohol	$C_4H_{10}O$	75-65-0	74.12	26	82	0.7887^{20}	2.97	5.52	(1.7)	11	2-8%	478	100
Butylamine	$C_4H_{11}N$	109-73-9	73.14	-49	77	0.7414^{20}	2.45	12.2	1.0	-12	2-10%	312	5
tert-Butylamine	$C_4H_{11}N$	75-64-9	73.14	-67	44	0.6958^{20}	2.63	48.4	(1.3)	-9	2-9%	380	
Butyl methyl ketone	$C_6H_{12}O$	591-78-6	100.16	-56	128	0.8113^{20}	2.13	1.54	(2.7)	25	1-8%	423	5
p-tert-Butyltoluene	$C_{11}H_{16}$	98-51-1	148.25	-52	190	0.8612^{20}		0.09	≈0	68			10
γ-Butyrolactone	$C_4H_6O_2$	96-48-0	86.09	-43.3	204	1.1284^{16}	1.64	0.43	4.27	98			
Caprolactam	$C_6H_{11}NO$	105-60-2	113.16	69	270		1.38	<0.01	(3.9)	125			5
Carbon disulfide	CS_2	75-15-0	76.14	-112	46	1.2632^{20}	1.00	48.2	0	-30	1-50%	90	10
Carbon tetrachloride	CCl_4	56-23-5	153.82	-23	77	1.5940^{20}	0.85	15.2	0				5
1-Chloro-1,1-difluoroethane	$C_2H_3ClF_2$	75-68-3	100.50	-131	-10	1.107^{25}	1.30	351	2.14				
Chlorobenzene	C_6H_5Cl	108-90-7	112.56	-45	132	1.1058^{20}	1.33	1.6	1.69	28	1-10%	593	10
Chloroform	$CHCl_3$	67-66-3	119.38	-64	61	1.4832^{20}	0.96	26.2	1.04				10
Chloropentafluoroethane	C_2ClF_5	76-15-3	154.47	-99	-38	1.5673^{-42}	1.19	912	0.52				1000
Cumene (Isopropylbenzene)	C_9H_{12}	98-82-8	120.19	-96.0	152	0.8618^{20}	1.75	0.61	0.79	36	1-7%	424	50
Cyclohexane	C_6H_{12}	110-82-7	84.16	7	81	0.7785^{20}	1.84	13.0	≈0	-20	1-8%	245	300
Cyclohexanol	$C_6H_{12}O$	108-93-0	100.16	25	161	0.9624^{20}	2.08	0.10		68	1-9%	300	50
Cyclohexanone	$C_6H_{10}O$	108-94-1	98.14	-31	155	0.9478^{20}	1.86	0.53	2.87	44	1-9%	420	25
Cyclohexylamine	$C_6H_{13}N$	108-91-8	99.18	-18	134	0.8191^{20}		1.20	(1.3)	31	1-9%	293	10
Cyclopentane	C_5H_{10}	287-92-3	70.13	-93.8	49.3	0.7457^{20}	1.84	42.3	≈0	<-7	2%-	361	600
Cyclopentanone	C_5H_8O	120-92-3	84.12	-51.3	130.5	0.9487^{20}	1.84	1.55	3.3	26			
p-Cymene	$C_{10}H_{14}$	99-87-6	134.22	-69	177	0.8573^{20}	1.76	0.19	≈0	47	1-6%	436	
cis-Decalin	$C_{10}H_{18}$	493-01-6	138.25	-42.9	195.8	0.8965^{20}	1.68	0.10	≈0				
trans-Decalin	$C_{10}H_{18}$	493-02-7	138.25	-30.3	187.3	0.8699^{20}	1.65	0.16	≈0	54	1-5%	255	
Diacetone alcohol	$C_6H_{12}O_2$	123-42-2	116.16	-44	168	0.9387^{20}	1.91	0.22	(3.2)	58	2-7%	643	50
1,2-Dibromoethane	$C_2H_4Br_2$	106-93-4	187.86	9.9	131.6	2.1791^{20}	0.72	1.55	(1.2)				
Dibromofluoromethane	$CHBr_2F$	1868-53-7	191.83	-78	64.9	2.421^{20}							
Dibromomethane	CH_2Br_2	74-95-3	173.83	-52.5	97	2.4969^{20}	0.61	6.12	1.43				
1,2-Dibromotetrafluoroethane	$C_2Br_2F_4$	124-73-2	259.82	-110.4	47.3	2.149^{25}	0.69	43.4					
Dibutylamine	$C_8H_{19}N$	111-92-2	129.25	-62	160	0.7670^{20}	2.27	0.34	(1.0)	47	1-6%		
o-Dichlorobenzene	$C_6H_4Cl_2$	95-50-1	147.00	-17	180	1.3059^{20}	1.10	0.18	2.50	66	2-9%	648	25
1,1-Dichloroethane	$C_2H_4Cl_2$	75-34-3	98.96	-97	57	1.1757^{20}	1.28	30.5	2.06	-17	5-11%	458	100
1,2-Dichloroethane	$C_2H_4Cl_2$	107-06-2	98.96	-36	84	1.2351^{20}	1.30	10.6	(1.8)	13	6-16%	413	10
1,1-Dichloroethylene	$C_2H_2Cl_2$	75-35-4	96.94	-122.5	31.6	1.213^{20}	1.15	80.0	1.34	-28	7-16%	570	5
cis-1,2-Dichloroethylene	$C_2H_2Cl_2$	156-59-2	96.94	-80	60	1.2837^{20}	1.20	26.8	1.90	2	3-15%	460	200
trans-1,2-Dichloroethylene	$C_2H_2Cl_2$	156-60-5	96.94	-50	49	1.2565^{20}	1.20	44.2	0	2	6-13%	460	200
Dichloroethyl ether	$C_4H_8Cl_2O$	111-44-4	143.01	-52	179	1.22^{20}	1.54	0.14	(2.6)	55	3%-	369	5
Dichloromethane	CH_2Cl_2	75-09-2	84.93	-95	40	1.3266^{20}	1.19	58.2	1.60		13-23%	556	50
1,2-Dichloropropane	$C_3H_6Cl_2$	78-87-5	112.99	-100	96	1.1560^{20}	1.32	6.62	(1.8)	16	3-15%	557	75
1,2-Dichlorotetrafluoroethane	$C_2Cl_2F_4$	76-14-2	170.92	-94	4	1.518^{4}	0.96	215	0.5				1000
Diethanolamine	$C_4H_{11}NO_2$	111-42-2	105.14	28	269	1.0966^{20}	2.22	<0.01	(2.8)	172	2-13%	662	0.46
Diethylamine	$C_4H_{11}N$	109-89-7	73.14	-50	55	0.7056^{20}	2.31	30.1	0.92	-23	2-10%	312	5
Diethyl carbonate	$C_5H_{10}O_3$	105-58-8	118.13	-43	126	0.9752^{20}	1.80	1.63	1.10	25			
Diethylene glycol	$C_4H_{10}O_3$	111-46-6	106.12	-10	246	1.1197^{15}	2.31	<0.01	(2.3)	124	2-17%	224	
Diethylene glycol dimethyl ether	$C_6H_{14}O_3$	111-96-6	134.18	-68	162	0.9434^{20}	2.04	0.31	(2.0)	67			
Diethylene glycol monoethyl ether	$C_6H_{14}O_3$	111-90-0	134.18		196	0.9885^{20}	2.24	0.02	(1.6)	96			
Diethylene glycol monoethyl ether acetate	$C_8H_{16}O_4$	112-15-2	176.21	-25	218.5	1.0096^{20}		0.03	(1.8)	110		425	

PROPERTIES OF COMMON LABORATORY SOLVENTS (continued)

Name	MF	CAS RN	M_r	T_m/°C	T_b/°C	ρ/g cm⁻³	c_p/J g⁻¹K⁻¹	vp/kPa	μ/D	FP/°C	Fl. Lim.	Ign. Temp./°C	TLV
Diethylene glycol monomethyl ether	$C_5H_{12}O_3$	111-77-3	120.15		193	1.035^{20}	2.26	0.02	(1.6)	96	1-23%	240	
Diethylenetriamine	$C_4H_{13}N_3$	111-40-0	103.17	-39	207	0.9569^{20}	2.46	0.03	(1.9)	98	2-7%	358	1
Diethyl ether	$C_4H_{10}O$	60-29-7	74.12	-116	34	0.7138^{20}	2.33	71.7	1.15	-45	2-36%	180	400
Diisobutyl ketone	$C_9H_{18}O$	108-83-8	142.24	-42	169	0.8062^{20}	2.09	0.23	(2.7)	49	1-7%	396	25
Diisopropyl ether	$C_6H_{14}O$	108-20-3	102.18	-87	69	0.7241^{20}	2.12	19.9	1.13	-28	1-8%	443	250
N,N-Dimethylacetamide	C_4H_9NO	127-19-5	87.12	-20	165	0.9366^{25}	2.02	0.07	(3.7)	70	2-12%	490	10
Dimethylamine	C_2H_7N	124-40-3	45.08	-92	7	0.6804^{0}	3.05	203	1.01	20	3-14%	400	5
Dimethyl disulfide	$C_2H_6S_2$	624-92-0	94.20	-85	109.8	1.0625^{20}	1.55	3.82	(1.8)	24			
N,N-Dimethylformamide	C_3H_7NO	68-12-2	73.09	-60	153	0.944^{25}	2.06	0.44	3.82	58	2-15%	445	10
Dimethyl sulfoxide	C_2H_6OS	67-68-5	78.14	19	189	1.1014^{20}	1.96	0.08	3.96	95	3-42%	215	
1,4-Dioxane	$C_4H_8O_2$	123-91-1	88.11	12	101	1.0337^{20}	1.74	4.95	0	12	2-22%	180	25
1,3-Dioxolane	$C_3H_6O_2$	646-06-0	74.08	-95	78	1.060^{20}	1.59	14.6	1.19	2			
Dipentene	$C_{10}H_{16}$	7705-14-8	136.24	-95.5	178	0.8402^{21}	1.83	0.26		45		237	
Epichlorohydrin	C_3H_5ClO	106-89-8	92.52	-26	116	1.1812^{20}	1.42	2.2	(1.8)	31	4-21%	411	2
Ethanolamine (Glycinol)	C_2H_7NO	141-43-5	61.08	11	171	1.0180^{20}	3.20	0.05	(2.3)	86	3-24%	410	3
Ethyl acetate	$C_4H_8O_2$	141-78-6	88.11	-84	77	0.9003^{20}	1.94	12.6	1.78	-4	2-12%	426	400
Ethyl acetoacetate	$C_6H_{10}O_3$	141-97-9	130.14	-45	180.8	1.0368^{10}	1.91	0.09		57	1-10%	295	
Ethyl alcohol	C_2H_6O	64-17-5	46.07	-114	78	0.7893^{20}	2.44	7.87	1.69	13	3-19%	363	1000
Ethylamine	C_2H_7N	75-04-7	45.08	-81	17	0.686^{17}	2.88	142	1.22	<-18	4-14%	385	5
Ethylbenzene	C_8H_{10}	100-41-4	106.17	-95	136	0.8670^{20}	1.73	1.28	0.59	21	1-7%	432	100
Ethyl bromide	C_2H_5Br	74-96-4	108.97	-118.6	38.5	1.4604^{20}	0.93	62.5	2.03		7-8%	511	5
Ethyl chloride	C_2H_5Cl	75-00-3	64.51	-139	12	0.909^{12}	1.62	160	2.05	-50	4-15%	519	1000
Ethylene carbonate	$C_3H_4O_3$	96-49-1	88.06	36.4	248	1.3214^{39}	1.52	<0.01	(4.9)	143			
Ethylenediamine	$C_2H_8N_2$	107-15-3	60.10	11	117	0.8979^{20}	2.87	1.62	1.99	40	3-12%	385	10
Ethylene glycol	$C_2H_6O_2$	107-21-1	62.07	-13	197	1.1088^{20}	2.41	0.01	2.28	111	3-22%	398	50
Ethylene glycol diethyl ether	$C_6H_{14}O_2$	629-14-1	118.18	-74	119.4	0.8484^{20}	2.19	4.33		35			
Ethylene glycol dimethyl ether	$C_4H_{10}O_2$	110-71-4	90.12	-58	85	0.8691^{20}	2.14	9.93		-2		202	
Ethylene glycol monobutyl ether	$C_6H_{14}O_2$	111-76-2	118.18	-75	168	0.9015^{20}	2.38	0.15	(2.1)	69	4-13%	238	25
Ethylene glycol monoethyl ether	$C_4H_{10}O_2$	110-80-5	90.12	-70	135	0.9297^{20}	2.34	0.71	(2.1)	43	3-18%	235	5
Ethylene glycol ethyl ether acetate	$C_6H_{12}O_3$	111-15-9	132.16	-62	156	0.9740^{20}	2.85	0.24	(2.2)	56	2-8%	379	5
Ethylene glycol monomethyl ether	$C_3H_8O_2$	109-86-4	76.10	-85	124	0.9647^{20}	2.25	1.31	2.36	39	2-14%	285	5
Ethylene glycol momomethyl ether acetate	$C_5H_{10}O_3$	110-49-6	118.13	-70	143	1.0074^{19}	2.62	0.67	(2.1)	49	2-12%	392	5
Ethyl formate	$C_3H_6O_2$	109-94-4	74.08	-80	54	0.9168^{20}	2.02	32.3	1.9	-20	3-16%	455	100
Furan	C_4H_4O	110-00-9	68.08	-86	31	0.9514^{20}	1.69	80.0	0.66	<0	2-14%		
Furfural	$C_5H_4O_2$	98-01-1	96.09	-37	162	1.1594^{20}	1.70	0.29	(3.5)	60	2-19%	316	2
Furfuryl alcohol	$C_5H_6O_2$	98-00-0	98.10	-31	171	1.1296^{20}	2.08	0.10	(1.9)	75	2-16%	491	10
Glycerol	$C_3H_8O_3$	56-81-5	92.09	18	290	1.2613^{20}	2.38	<0.01	(2.6)	199	3-19%	370	
Heptane	C_7H_{16}	142-82-5	100.20	-91	98	0.6837^{20}	2.24	6.09	≈0	-4	1-7%	204	400
1-Heptanol	$C_7H_{16}O$	111-70-6	116.20	-34	176.4	0.8219^{20}	2.34						
Hexane	C_6H_{14}	110-54-3	86.18	-95	69	0.6548^{25}	2.27	20.2	≈0	-22	1-8%	225	50
1-Hexanol (Caproyl alcohol)	$C_6H_{14}O$	111-27-3	102.18	-44.6	157.6	0.8136^{20}	2.35	0.11		63			
Hexylene glycol	$C_6H_{14}O_2$	107-41-5	118.18	-50	197	0.923^{15}	2.84	<0.01	(2.9)	102	1-9%	306	25
Hexyl methyl ketone	$C_8H_{16}O$	111-13-7	128.21	-16	172.5	0.820^{20}	2.13		(2.7)	52			
Isobutyl acetate	$C_6H_{12}O_2$	110-19-0	116.16	-99	117	0.8712^{20}	2.01	2.39	(1.9)	18	1-11%	421	150
Isobutyl alcohol	$C_4H_{10}O$	78-83-1	74.12	-108	108	0.8018^{20}	2.44	1.39	1.64	28	2-11%	415	50
Isobutylamine	$C_4H_{11}N$	78-81-9	73.14	-87	68	0.724^{25}	2.50	19.0	(1.3)	-9	2-12%	378	
Isopentyl acetate	$C_7H_{14}O_2$	123-92-2	130.19	-79	143	0.876^{15}	1.91	0.73	(1.9)	25	1-8%	360	100

PROPERTIES OF COMMON LABORATORY SOLVENTS (continued)

Name	MF	CAS RN	M_r	T_m/°C	T_b/°C	ρ/g cm^{-3}	c_p/J g^{-1}K^{-1}	vp/kPa	μ/D	FP/°C	Fl. Lim.	Ign. Temp./°C	TLV
Isophorone	$C_9H_{14}O$	78-59-1	138.21	-8	215	0.9255^{20}	1.83	0.06		84	1-4%	460	5
Isopropyl acetate	$C_5H_{10}O_2$	108-21-4	102.13	-73	89	0.8713^{20}	1.95	8.1		2	2-8%	460	250
Isopropyl alcohol	C_3H_8O	67-63-0	60.10	-90	82	0.7855^{20}	2.58	6.02	1.56	12	2-13%	399	400
Isoquinoline	C_9H_7N	119-65-3	129.16	26.47	243.2	1.0910^{30}	1.52		2.73				
d-Limonene (Citrene)	$C_{10}H_{16}$	5989-27-5	136.24	-97	178	0.8411^{20}	1.83	0.28		49			
2,6-Lutidine	C_7H_9N	108-48-5	107.16	-6.1	144.1	0.9226^{20}	1.73	0.75	(1.7)				
Mesitylene	C_9H_{12}	108-67-8	120.19	-45	165	0.8652^{20}	1.74	0.33	0	50	1-5%	559	25
Mesityl oxide	$C_6H_{10}O$	141-79-7	98.14	-59	130	0.8653^{20}	2.17	1.47	(2.8)	31	1-7%	344	15
Methyl acetate	$C_3H_6O_2$	79-20-9	74.08	-98	57	0.9342^{20}	1.92	28.8	1.72	-10	3-16%	454	200
Methylal	$C_3H_8O_2$	109-87-5	76.10	-105	42	0.8593^{20}	2.12	53.1	(0.7)	-32	2-14%	237	1000
Methyl alcohol	CH_4O	67-56-1	32.04	-98	65	0.7914^{20}	2.53	16.9	1.70	11	6-36%	464	200
Methylamine	CH_5N	74-89-5	31.06	-93	-6	0.656^{25}	3.29	353	1.31	0	5-21%	430	5
Methyl benzoate	$C_8H_8O_2$	93-58-3	136.15	-15	199	1.0933^{15}	1.63	0.05	(1.9)	83			
Methylcyclohexane	C_7H_{14}	108-87-2	98.19	-127	101	0.7694^{20}	1.88	6.18	≈0	-4	1-7%	250	400
Methyl ethyl ketone	C_4H_8O	78-93-3	72.11	-87	80	0.8054^{20}	2.20	12.6	2.78	-9	1-11%	404	200
N-Methylformamide	C_2H_5NO	123-39-7	59.07	-3.8	199.5	1.011^{19}	2.10		3.83				
Methyl formate	$C_2H_4O_2$	107-31-3	60.05	-99	32	0.9742^{20}	1.98	78.1	1.77	-19	5-23%	449	100
Methyl iodide	CH_3I	74-88-4	141.94	-66.4	42.5	2.279^{20}	0.89	53.9	1.62				2
Methyl isobutyl ketone	$C_6H_{12}O$	108-10-1	100.16	-84	116	0.7978^{20}	2.13	2.64		18	1-8%	448	50
Methyl isopentyl ketone	$C_7H_{14}O$	110-12-3	114.19		144	0.888^{20}		0.69		36	1-8%	191	50
2-Methylpentane	C_6H_{14}	107-83-5	86.18	-153.7	60.2	0.653^{25}	2.25	28.2	≈0	<-29	1-7%	264	
4-Methyl-2-pentanol	$C_6H_{14}O$	108-11-2	102.18	-90	132	0.8075^{20}	2.67	0.70		41	1-6%		25
Methyl pentyl ketone	$C_7H_{14}O$	110-43-0	114.19	-35	151	0.8111^{20}	2.04	0.49	(2.6)	39	1-8%	393	50
Methyl propyl ketone	$C_5H_{10}O$	107-87-9	86.13	-77	102	0.809^{20}	2.14	4.97	(2.7)	7	2-8%	452	200
N-Methyl-2-pyrrolidone	C_5H_9NO	872-50-4	99.13	-24	202	1.0230^{25}	3.11	0.04	(4.1)	96	1-10%	346	
Morpholine	C_4H_9NO	110-91-8	87.12	-5	128	1.0005^{20}	1.89	1.34	1.55	37	1-11%	290	20
Nitrobenzene	$C_6H_5NO_2$	98-95-3	123.11	6	211	1.2037^{20}	1.51	0.03	4.22	88	2-9%	482	1
Nitroethane	$C_2H_5NO_2$	79-24-3	75.07	-90	114	1.0448^{25}	1.79	2.79	3.23	28	3-17%	414	100
Nitromethane	CH_3NO_2	75-52-5	61.04	-29	101	1.1371^{20}	1.75	4.79	3.46	35	7-22%	418	20
1-Nitropropane	$C_3H_7NO_2$	108-03-2	89.09	-108	131.1	0.9961^{25}	1.97	1.36	3.66	36	2%-	421	25
2-Nitropropane	$C_3H_7NO_2$	79-46-9	89.09	-91	120	0.9821^{25}	1.91	2.3	3.73	24	3-11%	428	10
Octane	C_8H_{18}	111-65-9	114.23	-57	126	0.6986^{25}	2.23	1.86	≈0	13	1-7%	206	300
1-Octanol	$C_8H_{18}O$	111-87-5	130.23	-15.5	195.1	0.8262^{25}	2.34	0.01	(1.8)	81			
Pentachloroethane	C_2HCl_5	76-01-7	202.29	-29	160	1.6796^{20}	0.86	0.48	0.92				
Pentamethylene glycol	$C_5H_{12}O_2$	111-29-5	104.15	-18	239	0.9914^{20}	3.08		(2.5)	129		335	
Pentane	C_5H_{12}	109-66-0	72.15	-130	36	0.6262^{20}	2.32	68.3	≈0	<-40	2-8%	260	600
1-Pentanol	$C_5H_{12}O$	71-41-0	88.15	-79	138	0.8144^{20}	2.36	0.26	(1.7)	33	1-10%	300	
Pentyl acetate	$C_7H_{14}O_2$	628-63-7	130.19	-71	149	0.8755^{20}	2.00	0.60	1.75	16	1-8%	360	100
2-Picoline	C_6H_7N	109-06-8	93.13	-67	129	0.9443^{20}	1.70	1.5	1.85	39		538	
α-Pinene	$C_{10}H_{16}$	80-56-8	136.24	-64	156	0.8539^{25}		0.64		35		275	
β-Pinene	$C_{10}H_{16}$	127-91-3	136.24	-61.5	166	0.860^{25}		0.61		38		275	
Piperidine	$C_5H_{11}N$	110-89-4	85.15	-11	106	0.8606^{20}	2.11	4.28	(1.2)	16	1-10%		
Propanenitrile	C_3H_5N	107-12-0	55.08	-93	97	0.7818^{20}	2.17	6.14	4.05	2	3-14%	512	
Propyl acetate	$C_5H_{10}O_2$	109-60-4	102.13	-93	102	0.8878^{20}	1.92	4.49	(1.8)	13	2-8%	450	200
Propyl alcohol	C_3H_8O	71-23-8	60.10	-126	97	0.8035^{20}	2.39	2.76	1.55	23	2-14%	412	200
Propylamine	C_3H_9N	107-10-8	59.11	-83	47	0.7173^{20}	2.75	42.1	1.17	-37	2-10%	318	
Propylbenzene	C_9H_{12}	103-65-1	120.19	-99.5	159.2	0.8620^{20}	1.79		≈0	30	1-6%	450	
Propylene glycol	$C_3H_8O_2$	57-55-6	76.10	-60	188	1.0361^{20}	2.51	0.02	(2.2)	99	3-13%	371	
Pseudocumene	C_9H_{12}	95-63-6	120.19	-44	169	0.8758^{20}	1.79	0.30	≈0	44	1-6%	500	25

PROPERTIES OF COMMON LABORATORY SOLVENTS (continued)

Name	MF	CAS RN	M_r	T_m/°C	T_b/°C	ρ/g cm^{-3}	c_p/J g^{-1}K^{-1}	vp/kPa	μ/D	FP/°C	Fl. Lim.	Ign. Temp./°C	TLV
Pyridine	C_5H_5N	110-86-1	79.10	-42	115	0.9819^{20}	1.68	2.76	2.21	20	2-12%	482	5
Pyrrole	C_4H_5N	109-97-7	67.09	-23.4	129.7	0.9698^{20}	1.90	1.10	1.74	39			
Pyrrolidine	C_4H_9N	123-75-1	71.12	-57.8	86.5	0.8586^{20}	2.20	8.40	(1.6)	3			
2-Pyrrolidone	C_4H_7NO	616-45-5	85.11	25	251	1.120^{20}	1.99		(3.5)	129			
Quinoline	C_9H_7N	91-22-5	129.16	-14.78	237.1	1.0977^{15}	1.51		2.29			480	
Styrene	C_8H_8	100-42-5	104.15	-31	145	0.9060^{20}	1.75	0.81		31	1-7%	490	50
Sulfolane	$C_4H_8O_2S$	126-33-0	120.17	28	287	1.2723^{18}	1.50	<0.01	(4.8)	177			
α-Terpinene	$C_{10}H_{16}$	99-86-5	136.24		174	0.8375^{19}							
1,1,1,2-Tetrachloro-2,2-difluoroethane	$C_2Cl_4F_2$	76-11-9	203.83	40.6	91.5	1.649^{25}		7.36					500
1,1,2,2-Tetrachloro-1,2-difluoroethane	$C_2Cl_4F_2$	76-12-0	203.83	26	93	1.6447^{25}	0.85	7.51					500
1,1,1,2-Tetrachloroethane	$C_2H_2Cl_4$	630-20-6	167.85	-70	131	1.5406^{20}	0.92	1.6		47	5-12%		
1,1,2,2-Tetrachloroethane	$C_2H_2Cl_4$	79-34-5	167.85	-44	146	1.5953^{20}	0.97	0.62	1.32	62	20-54%		1
Tetrachloroethylene	C_2Cl_4	127-18-4	165.83	-22	121	1.6227^{20}	0.86	2.42	0	45			50
Tetraethylene glycol	$C_8H_{18}O_5$	112-60-7	194.23	-6.2	328	1.1285^{15}	2.21			182			
Tetrahydrofuran	C_4H_8O	109-99-9	72.11	-108	65	0.8892^{20}	1.72	21.6	1.75	-14	2-12%	321	200
1,2,3,4-Tetrahydronaphthalene	$C_{10}H_{12}$	119-64-2	132.21	-36	208	0.9660^{25}	1.65	0.05	≈0	71	1-5%	385	
Tetrahydropyran	$C_5H_{10}O$	142-68-7	86.13	-45	88	0.8814^{20}	1.82	9.54	1.74	-20			
Tetramethylsilane	$C_4H_{12}Si$	75-76-3	88.22	-99.0	26.6	0.648^{19}	2.31	94.2	0				
Toluene	C_7H_8	108-88-3	92.14	-95	111	0.8669^{20}	1.70	3.79	0.37	4	1-7%	480	50
o-Toluidine	C_7H_9N	95-53-4	107.16	-16.3	200.3	0.9984^{20}	1.96	0.04	(1.6)	85		482	2
Triacetin	$C_9H_{14}O_6$	102-76-1	218.21	-78	259	1.1583^{20}	1.76	<0.01		138	1%-	433	
Tributylamine	$C_{12}H_{27}N$	102-82-9	185.35	-70	217	0.7770^{20}		0.01	(0.8)	86	1-5%		
1,1,1-Trichloroethane	$C_2H_3Cl_3$	71-55-6	133.40	-30	74	1.3390^{20}	1.08	16.5	1.76	-1	8-13%	537	350
1,1,2-Trichloroethane	$C_2H_3Cl_3$	79-00-5	133.40	-37	114	1.4397^{20}	1.13	3.1	(1.4)	32	6-28%	460	10
Trichloroethylene	C_2HCl_3	79-01-6	131.39	-85	87	1.4642^{20}	0.95	9.91	(0.8)	32	8-11%	420	50
Trichlorofluoromethane	CCl_3F	75-69-4	137.37	-111	24	1.478^{24}	0.89	106	0.46				1000
1,1,2-Trichlorotrifluoroethane	$C_2Cl_3F_3$	76-13-1	187.38	-35	48	1.5635^{25}	0.91	44.8					1000
Triethanolamine	$C_6H_{15}NO_3$	102-71-6	149.19	21	335	1.1242^{20}	2.61	<0.01	(3.6)	179	1-10%		0.5
Triethylamine	$C_{10}H_{22}O_2$	121-44-8	101.19	-115	89	0.7275^{20}	2.17	7.70	0.66	-7	1-8%	249	1
Triethylene glycol	$C_6H_{14}O_4$	112-27-6	150.17	-7	285	1.1274^{15}	2.18			177	1-9%	371	
Triethyl phosphate	$C_6H_{15}O_4P$	78-40-0	182.16	-56.4	215.5	1.0695^{20}			(3.1)	115		454	
Trimethylamine	C_3H_9N	75-50-3	59.11	-117	3	0.627^{25}	2.33	215	0.61	-7	2-12%	190	5
Trimethylene glycol	$C_3H_8O_2$	504-63-2	76.10	-26.7	214.4	1.0538^{20}			(2.5)			400	
Trimethyl phosphate	$C_3H_9O_4P$	512-56-1	140.08	-46	197.2	1.2144^{20}		0.11	(3.2)	107			
Veratrole	$C_8H_{10}O_2$	91-16-7	138.17	22.5	206	1.0810^{25}			(1.3)				
o-Xylene	C_8H_{10}	95-47-6	106.17	-25	144	0.8802^{10}	1.75	0.88	0.64	32	1-7%	463	100
m-Xylene	C_8H_{10}	108-38-3	106.17	-48	139	0.8642^{20}	1.72	1.13	≈0	27	1-7%	527	100
p-Xylene	C_8H_{10}	106-42-3	106.17	13	138	0.8611^{20}	1.71	1.19	0	27	1-7%	528	100

DEPENDENCE OF BOILING POINT ON PRESSURE

The normal boiling point T_b of a liquid is defined as the temperature at which the vapor pressure reaches standard atmospheric pressure, 101.325 kPa. The change in boiling point with pressure may be calculated from the representation of the vapor pressure by the Antoine Equation,

$$\ln p = A_1 - A_2/(T + A_3)$$

where p is the vapor pressure, T the absolute temperature, and A_1, A_2, and A_3 are constants. This table, which has been calculated using the Antoine constants in Reference 1, gives values of $\Delta T/\Delta p$ for a number of liquids, in units of both °C/kPa and °C/mmHg. The correction is generally accurate to 0.1 to 0.2 °C as long as the pressure is within 10% of standard atmospheric pressure.

A slightly less accurate estimate of $\Delta T/\Delta p$ may be obtained from the Claussius-Clapeyron equation, with the assumption that the change in volume upon vaporization equals the ideal-gas volume of the vapor. This leads to the equation

$$\Delta T/\Delta p = RT_b^2/p_0\,\Delta_{vap}H(T_b)$$

where R is the molar gas constant, p_0 is 101.325 kPa, and $\Delta_{vap}H(T_b)$ is the molar enthalpy of vaporization at the normal boiling point. Values of the last quantity may be obtained from the table "Enthalpy of Vaporization" in Section 6.

REFERENCE

1. Lide, D.R., and Kehiaian, H.V., *CRC Handbook of Thermophysical and Thermochemical Data*, CRC Press, Boca Raton, FL, 1994, pp. 49-59.

Compound	T_b °C	$\Delta T/\Delta p$ °C/kPa	$\Delta T/\Delta p$ °C/mmHg
Acetaldehyde	20.1	0.261	0.0348
Acetic acid	117.9	0.324	0.0432
Acetone	56.0	0.289	0.0385
Acetonitrile	81.6	0.316	0.0421
Ammonia	-33.33	0.198	0.0264
Aniline	184.1	0.378	0.0504
Anisole	153.7	0.367	0.0489
Benzaldehyde	179.0	0.392	0.0523
Benzene	80.0	0.321	0.0428
Bromine	58.8	0.300	0.0400
Butane	-0.5	0.267	0.0356
1-Butanol	117.7	0.278	0.0371
Carbon disulfide	46.2	0.304	0.0405
Chlorine	-34.04	0.224	0.0299
Chlorobenzene	131.7	0.365	0.0487
1-Chlorobutane	78.6	0.321	0.0428
Chloroethane	12.3	0.262	0.0349
Chloroethylene	-13.3	0.241	0.0321
Cyclohexane	80.7	0.328	0.0437
Cyclohexanol	160.8	0.344	0.0459
Cyclohexanone	155.4	0.382	0.0509
Decane	174.1	0.388	0.0517
Dibutyl ether	140.2	0.363	0.0484
Dichloromethane	39.6	0.276	0.0368
Diethyl ether	34.5	0.278	0.0371
Dimethyl sulfoxide	189.0	0.379	0.0505
1,4-Dioxane	101.5	0.321	0.0428
Dipropyl ether	90.0	0.326	0.0435
Ethanol	78.2	0.249	0.0332
Ethyl acetate	77.1	0.300	0.0400
Ethylene glycol	197.3	0.331	0.0441
Heptane	98.5	0.336	0.0448
Hexafluorobenzene	80.2	0.305	0.0407
Hexane	68.7	0.314	0.0419
1-Hexanol	157.6	0.318	0.0424
Hydrogen fluoride	20.1	0.276	0.0368
Iodomethane	42.5	0.291	0.0388
Isobutane	-11.7	0.254	0.0339
Methanol	64.6	0.251	0.0335
Methyl acetate	56.8	0.282	0.0376
Methyl formate	31.7	0.582	0.0776
N-Methylaniline	196.2	0.396	0.0528
N-Methylformamide	199.5	0.371	0.0495
Nitrobenzene	210.8	0.418	0.0557
Nitromethane	101.1	0.320	0.0427
1-Octanol	195.1	0.360	0.0480
Pentane	36.0	0.289	0.0385
1-Pentanol	137.9	0.296	0.0395
Phenol	181.8	0.349	0.0465
Propane	-42.1	0.224	0.0299
1-Propanol	97.2	0.261	0.0348
2-Propanol	82.3	0.247	0.0329
Pyridine	115.2	0.340	0.0453
Pyrrole	129.7	0.330	0.0440
Pyrrolidine	86.5	0.309	0.0412
Styrene	145.1	0.369	0.0492
Sulfur dioxide	-10.05	0.221	0.0295
Tetrachloroethylene	121.3	0.354	0.0472
Tetrachloromethane	76.8	0.325	0.0433
Toluene	110.6	0.353	0.0471
Trichloroethylene	87.2	0.330	0.0440
Trichloromethane	61.1	0.302	0.0403
Trimethylamine	2.8	0.248	0.0331
Water	100.0	0.276	0.0368
o-Xylene	144.5	0.373	0.0497
m-Xylene	139.1	0.368	0.0491
p-Xylene	138.3	0.369	0.0492

EBULLIOSCOPIC CONSTANTS FOR CALCULATION OF BOILING POINT ELEVATION

The boiling point T_b of a dilute solution of a non-volatile, non-dissociating solute is elevated relative to that of the pure solvent. If the solution is ideal (i.e., follows Raoult's Law), the amount of elevation depends only on the number of particles of solute present. Hence the change in boiling point ΔT_b can be expressed as

$$\Delta T_b = E_b m_2$$

where m_2 is the molality (moles of solute per kilogram of solvent) and E_b is the Ebullioscopic Constant, a characteristic property of the solvent. The Ebullioscopic Constant may be calculated from the relation

$$E_b = R T_b^2 M/\Delta_{vap}H$$

where R is the molar gas constant, M the molar mass of the solvent, and $\Delta_{vap}H$ the molar enthalpy (heat) of vaporization of the solvent.

This table lists E_b values for some common solvents, as calculated from data in the table "Enthalpy of Vaporization" in Section 6.

Compound	E_b/°C kg mol^{-1}
Acetic acid	3.22
Acetone	1.80
Acetonitrile	1.44
Aniline	3.82
Anisole	4.20
Benzaldehyde	4.24
Benzene	2.64
1-Butanol	2.17
Carbon disulfide	2.42
Chlorobenzene	4.36
1-Chlorobutane	3.13
Cyclohexane	2.92
Cyclohexanol	3.5
Decane	6.10
Dichloromethane	2.42
Diethyl ether	2.20
Dimethyl sulfoxide	3.22
1,4-Dioxane	3.01
Ethanol	1.23
Ethyl acetate	2.82
Ethylene glycol	2.26
Heptane	3.62
Hexane	2.90
Iodomethane	4.31
Methanol	0.86
Methyl acetate	2.21
N-Methylaniline	4.3
N-Methylformamide	2.2
Nitrobenzene	5.2
Nitromethane	2.09
1-Octanol	5.06
Phenol	3.54
1-Propanol	1.66
2-Propanol	1.58
Pyridine	2.83
Pyrrole	2.33
Pyrrolidine	2.32
Tetrachloroethylene	6.18
Tetrachloromethane	5.26
Toluene	3.40
Trichloroethylene	4.52
Trichloromethane	3.80
Water	0.513
o-Xylene	4.25

CRYOSCOPIC CONSTANTS FOR CALCULATION OF FREEZING POINT DEPRESSION

The freezing point T_f of a dilute solution of a non-volatile, non-dissociating solute is depressed relative to that of the pure solvent. If the solution is ideal (i.e., follows Raoult's Law), this lowering is a function only of the number of particles of solute present. Thus the absolute value of the lowering of freezing point ΔT_f can be expressed as

$$\Delta T_f = E_f m_2$$

where m_2 is the molality (moles of solute per kilogram of solvent) and E_f is the Cryoscopic Constant, a characteristic property of the solvent. The Cryoscopic Constant may be calculated from the relation

$$E_f = R\, T_f^2\, M/\Delta_{fus}H$$

where R is the molar gas constant, M the molar mass of the solvent, and $\Delta_{fus}H$ the molar enthalpy (heat) of vaporization of the solvent.

This table lists cryscopic constants for selected substances, as calculated from data in the table "Enthalpy of Fusion" in Section 6.

Compound	E_f/°C kg mol^{-1}
Acetamide	3.92
Acetic acid	3.63
Acetophenone	5.16
Aniline	5.23
Benzene	5.07
Benzonitrile	5.35
Benzophenone	8.58
(+)-Camphor	37.8
1-Chloronaphthalene	7.68
o-Cresol	5.92
m-Cresol	7.76
p-Cresol	7.20
Cyclohexane	20.8
Cyclohexanol	42.2
cis-Decahydronaphthalene	6.42
trans-Decahydronaphthalene	4.70
Dibenzyl ether	6.17
p-Dichlorobenzene	7.57
Diethanolamine	3.16
Dimethyl sulfoxide	3.85
1,4-Dioxane	4.63
Diphenylamine	8.38
Ethylene glycol	3.11
Formamide	4.25
Formic acid	2.38
Glycerol	3.56
Methylcyclohexane	2.60
Naphthalene	7.45
Nitrobenzene	6.87
Phenol	6.84
Pyridine	4.26
Quinoline	6.73
Succinonitrile	19.3
1,1,2,2-Tetrabromoethane	21.4
1,1,2,2-Tetrachloro-1,2-difluoroethane	41.0
Toluene	3.55
p-Toluidine	4.91
Tribromomethane	15.0
Water	1.86
p-Xylene	4.31

RELATIVE HUMIDITY AND DEW-POINT

The table gives the relative humidity of the air for temperature t and dew-point d.

(From Smithsonian Meterological Tables)

Depression of dew-point t-d, °C	Dew-Point (d)				
	−10	0	+10	+20	+30
0.0	100%	100%	100%	100%	100%
0.2	98	99	99	99	99
0.4	97	97	97	98	98
0.6	95	96	96	96	97
0.8	94	94	95	95	96
1.0	92	93	94	94	94
1.2	91	92	92	93	93
1.4	90	90	91	92	92
1.6	88	89	90	91	91
1.8	87	88	89	90	90
2.0	86	87	88	88	89
2.2	84	85	86	87	88
2.4	83	84	85	86	87
2.6	82	83	84	85	86
2.8	80	82	83	84	85
3.0	79	81	82	83	84
3.2	78	80	81	82	83
3.4	77	79	80	81	82
3.6	76	77	79	80	82
3.8	75	76	78	79	81
4.0	73	75	77	78	80
4.2	72	74	76	77	79
4.4	71	73	75	77	78
4.6	70	72	74	76	77
4.8	69	71	73	75	76
5.0	68	70	72	74	75
5.2	67	69	71	73	75

Depression of dew-point t-d, °C	Dew-Point (d)				
	−10	0	+10	+20	+30
5.4	66	68	70	72	74
5.6	65	67	69	71	73
5.8	64	66	69	70	72
6.0	63	66	68	70	71
6.2	62	65	67	69	71
6.4	61	64	66	68	70
6.6	60	63	65	67	69
6.8	60	62	64	66	68
7.0	59	61	63	66	68
7.2	58	60	63	65	67
7.4	57	60	62	64	66
7.6	56	59	61	63	65
7.8	55	58	60	63	65
8.0	54	57	60	62	64
8.2	54	56	59	61	63
8.4	53	56	58	60	63
8.6	52	55	57	60	62
8.8	51	54	57	59	61
9.0	51	53	56	58	61
9.2	50	53	55	58	60
9.4	49	52	55	57	59
9.6	48	51	54	56	59
9.8	48	51	53	56	58
10.0	47	50	53	55	57
10.5	45	48	51	54	—
11.0	44	47	49	52	—
11.5	42	45	48	51	—

Depression of dew-point t-d, °C	Dew-Point (d)				
	−10	0	+10	+20	+30
12.0	41	44	47	49	—
12.5	39	42	45	48	—
13.0	38	41	44	46	—
13.5	37	40	43	45	—
14.0	35	38	41	44	—
14.5	34	37	40	43	—
15.0	33	36	39	42	—
15.5	32	35	38	40	—
16.0	31	34	37	39	—
16.5	30	33	36	38	—
17.0	29	32	35	37	—
17.5	28	31	34	36	—
18.0	27	30	33	35	—
18.5	26	29	32	34	—
19.0	25	28	31	33	—
19.5	24	27	30	33	—
20.0	24	26	29	32	—
21.0	22	25	27	—	—
22.0	21	23	26	—	—
23.0	19	22	24	—	—
24.0	18	21	23	—	—
25.0	17	19	22	—	—
26.0	16	18	21	—	—
27.0	15	17	20	—	—
28.0	14	16	19	—	—
29.0	13	15	18	—	—
30.0	12	14	17	—	—

RELATIVE HUMIDITY FROM WET AND DRY BULB THERMOMETER

This table gives the approximate relative humidity directly from the reading of the air temperature (dry bulb) (t °C) and the wet bulb (t °C). It is computed for a barometric pressure of 74.27 cm Hg. Errors resulting from the use of this table for air temperatures above −10°C and between 77.5 and 71 cm Hg will usually be within the errors of observation.

Condensed from Bulletin of the U.S. Weather Bureau No. 1071

t-t′t	0.2	0.4	0.6	0.8	1.0	1.2	1.4	1.6	1.8	2.0	2.2	2.4	2.6	2.8	3.0	3.2	3.4	3.6	3.8	4.0	4.5	5.0	5.5	6.0	6.5	7.0	7.5	8.0		9.0	9.5	10.0	10.5	11.0
−10	93	87	80	74	67	61	54	48	41	35	28	22	16	9																				
−9	94	88	81	75	69	63	57	51	45	39	33	27	21	15	9																			
−8	94	88	83	77	71	65	60	54	48	43	37	32	26	20	15	10																		
−7	95	89	84	78	73	67	62	57	52	46	41	36	31	25	20	15	10	5																
−6	95	90	85	79	74	69	64	59	54	49	45	40	35	30	25	20	15	11	6															
−5	95	90	86	81	76	71	66	62	57	52	48	43	39	34	29	25	20	16	11	7														
−4	95	91	86	82	77	73	68	64	59	55	51	46	42	38	33	29	25	21	17	12														
−3	96	91	87	82	78	74	70	66	62	57	53	49	45	41	37	33	29	25	21	17	8													
−2	96	92	88	84	79	75	71	68	64	60	56	52	48	44	40	37	33	29	25	22	12													
−1	96	92	88	84	81	77	73	69	66	62	58	54	51	47	43	40	36	33	29	26	17	8												
0	96	93	89	85	81	78	74	71	67	64	60	57	53	50	46	43	40	36	33	29	21	13	5											
1	97	93	90	86	83	80	76	73	70	66	63	59	56	53	49	46	43	40	36	33	25	17	10											
2	97	93	90	87	84	81	78	74	71	68	65	62	59	55	52	49	46	43	40	37	29	22	14	7										
3	97	94	91	88	84	82	78	76	72	70	67	64	61	58	55	52	49	46	43	40	33	26	19	12	5									
4	97	94	91	88	85	82	79	77	74	71	68	65	62	60	57	54	51	48	46	43	36	29	22	16	9									
5	97	94	91	88	86	83	80	77	75	72	69	67	64	61	58	56	53	51	48	45	39	33	26	20	13	7								
6	97	94	92	89	86	84	81	78	76	73	70	68	65	63	60	58	55	53	50	48	41	35	29	24	17	11	5							
7	97	95	92	89	87	84	82	79	77	74	72	69	67	64	62	59	57	54	52	50	44	38	32	26	21	15	10							
8	97	95	92	90	87	85	82	80	77	75	73	70	68	65	63	61	58	56	54	51	46	40	35	29	24	19	14	8						
9	98	95	93	90	88	85	83	81	78	76	74	71	69	67	64	62	60	58	55	53	48	42	37	32	27	22	17	12	7					
10	98	95	93	90	88	86	83	81	79	77	74	72	70	68	66	63	61	59	57	55	50	44	39	34	29	24	20	15	10	6				
11	98	95	93	91	89	86	84	82	80	78	75	73	71	69	67	65	62	60	58	56	51	46	41	36	32	27	22	18	13	9	5			
12	98	96	93	91	89	87	85	82	80	78	76	74	72	70	68	66	64	62	60	58	53	48	43	39	34	29	25	21	16	12	8			
13	98	96	93	91	89	87	85	83	81	79	77	75	73	71	69	67	65	63	61	59	54	50	45	41	36	32	28	23	19	15	11	7		
14	98	96	94	92	90	88	86	84	82	79	78	76	74	72	70	68	66	64	62	60	56	51	47	42	38	34	30	26	22	18	14	10	6	
15	98	96	94	92	90	88	86	84	82	80	78	76	74	73	71	69	67	65	63	61	57	53	48	44	40	36	32	27	24	20	16	13	9	6

	0.5	1.0	1.5	2.0	2.5	3.0	3.5	4.0	4.5	5.0	5.5	6.0	6.5	7.0	7.5	8.0	8.5	9.0	9.5	10.0	10.5	11.0	11.5	12.0	12.5	13.0	13.5	14.0	14.5	15.0	16.0	17.0	18.0	19.0	20.0
16	95	90	85	81	76	71	67	63	58	54	50	46	42	38	34	30	26	23	19	15	12	8	5												
17	95	90	86	81	76	72	68	64	60	55	51	47	43	40	36	32	28	25	21	18	14	11	8												
18	95	91	86	82	77	73	69	65	61	57	53	49	45	41	38	34	30	27	23	20	17	14	10	7											
19	95	91	87	82	78	74	70	65	62	58	54	50	46	43	39	36	32	29	26	22	19	16	13	10	7										
20	96	91	87	83	78	74	70	66	63	59	55	51	48	44	41	37	34	31	28	24	21	18	15	12	9	6									
21	96	91	87	83	79	75	71	67	64	60	56	53	49	46	42	39	36	32	29	26	23	20	17	14	12	9	6								
22	96	92	87	83	80	76	72	68	64	61	57	54	50	47	44	40	37	34	31	28	25	22	19	17	14	11	8	6							
23	96	92	88	84	80	76	72	69	65	62	58	55	52	48	45	42	39	36	33	30	27	24	21	19	16	13	11	8	6						
24	96	92	88	84	80	77	73	69	66	62	59	56	53	49	46	43	40	37	34	31	29	26	23	20	18	15	13	10	8	5					
25	96	92	88	84	81	77	74	70	67	63	60	57	54	50	47	44	41	39	36	33	30	28	25	22	20	17	15	12	10	8					
26	96	92	88	85	81	78	74	71	67	64	61	58	54	51	49	46	43	40	37	34	32	29	26	24	21	19	17	14	12	10	5				
27	96	92	89	85	82	78	75	71	68	65	62	58	56	52	50	47	44	41	38	36	33	31	28	26	23	21	18	16	14	12	7				
28	96	93	89	85	82	78	75	72	69	65	62	59	56	53	51	48	45	42	40	37	34	32	29	27	25	22	20	18	16	13	9	5			
29	96	93	89	86	82	79	76	72	69	66	63	60	57	54	52	49	46	43	41	38	36	33	31	28	26	24	22	19	17	15	11	7			
30	96	93	89	86	83	79	76	73	70	67	64	61	58	55	52	50	47	44	42	39	37	35	32	30	28	25	23	21	19	17	13	9	5		
31	96	93	90	86	83	80	77	73	70	67	64	61	59	56	53	51	48	45	43	40	38	36	33	31	29	27	25	22	20	18	14	11	7		
32	96	93	90	86	83	80	77	74	71	68	65	62	60	57	54	51	49	46	44	41	39	37	35	32	30	28	26	24	22	20	16	12	9	5	
33	97	93	90	87	83	80	77	74	71	68	66	63	60	57	55	52	50	47	45	42	40	38	36	33	31	29	27	25	23	21	17	14	10	7	
34	97	93	90	87	84	81	78	75	72	69	66	63	61	58	56	53	51	48	46	43	41	39	37	35	32	30	28	26	24	23	19	15	12	8	5
35	97	94	90	87	84	81	78	75	72	69	67	64	61	59	56	54	51	49	47	44	42	40	38	36	34	32	30	28	26	24	20	17	13	10	7
36	97	94	90	87	84	81	78	75	73	70	67	64	62	59	57	54	52	50	48	45	43	41	39	37	35	33	31	29	27	25	21	18	15	11	8
37	97	94	91	87	84	82	79	76	73	70	68	65	63	60	58	55	53	51	48	46	44	42	40	38	36	34	32	30	28	26	23	19	16	13	10
38	97	94	91	88	84	82	79	76	74	71	68	66	63	61	58	56	54	51	49	47	45	43	41	39	37	35	33	31	29	27	24	20	17	14	11
39	97	94	91	88	85	82	79	77	74	71	69	66	64	61	59	57	54	52	50	48	46	43	42	39	38	36	34	32	30	28	25	22	18	15	12
40	97	94	91	88	85	82	80	77	74	72	69	67	64	62	59	57	54	53	51	48	46	44	42	40	38	36	35	33	31	29	26	23	20	16	14

CONSTANT HUMIDITY SOLUTIONS

Anthony Wexler

An excess of a water soluble salt in contact with its saturated solution and contained within an enclosed space produces a constant relative humidity and water vapor pressure according to

$$RH = A\exp(B/T)$$

where RH is the percent relative humidity (generally accurate to ±2 %), T is the temperature in kelvin, and the constants A and B and the range of valid temperatures are given in the table below. The vapor pressure, p, can be calculated from

$$p = (RH/100) \times p_0$$

where p_0 is the vapor pressure of pure water at temperature T as given in the table in Section 6 titled "Vapor Pressure of Water from 0 to 370°C".

REFERENCES

1. Wexler, A. S. and Seinfeld, J. H., *Atmospheric Environment*, 25A, 2731, 1991.
2. Greenspan, L., *J. Res. National Bureau of Standards*, 81A, 89, 1977.
3. Broul, et al., *Solubility of Inorganic Two-Component Systems*, Elsevier, New York, 1981.
4. Wagman, D. D. et al., *J. Phys. Chem. Ref. Data*, Vol. 11, Suppl. 2, 1982.

Compound	Temperature range (°C)	*RH* 25°C	*A*	*B*
$NaOH \cdot H_2O$	15—60	6	5.48	27
$LiBr \cdot 2H_2O$	10—30	6	0.23	996
$ZnBr_2 \cdot 2H_2O$	5—30	8	1.69	455
$KOH \cdot 2H_2O$	5—30	9	0.014	1924
$LiCl \cdot H_2O$	20—65	11	14.53	–75
$CaBr_2 \cdot 6H_2O$	11—22	16	0.17	1360
$LiI \cdot 3H_2O$	15—65	18	0.15	1424
$CaCl_2 \cdot 6H_2O$	15—25	29	0.11	1653
$MgCl_2 \cdot 6H_2O$	5—45	33	29.26	34
$NaI \cdot 2H_2O$	5—45	38	3.62	702
$Ca(NO_3)_2 \cdot 4H_2O$	10—30	51	1.89	981
$Mg(NO_3)_2 \cdot 6H_2O$	5—35	53	25.28	220
$NaBr \cdot 2H_2O$	0—35	58	20.49	308
NH_4NO_3	10—40	62	3.54	853
KI	5—30	69	29.35	254
$SrCl_2 \cdot 6H_2O$	5—30	71	31.58	241
$NaNO_3$	10—40	74	26.94	302
NaCl	10—40	75	69.20	25
NH_4Cl	10—40	79	35.67	235
KBr	5—25	81	40.98	203
$(NH_4)_2SO_4$	10—40	81	62.06	79
KCl	5—25	84	49.38	159
$Sr(NO_3)_2 \cdot 4H_2O$	5—25	85	28.34	328
$BaCl_2 \cdot 2H_2O$	5—25	90	69.99	75
CsI	5—25	91	70.77	75
KNO_3	0—50	92	43.22	225
K_2SO_4	10—50	97	86.75	34

EFFICIENCY OF DRYING AGENTS

Compiled by John H. Yoe

A. Drying agents depending upon chemical action (absorption) for their efficiency:*

Substance	Residual water, mg per liter of dry air**	Reference
P_2O_5	<1 mg in 40,000 liters	Morley, Am. J. Sci., 34, 199 (1887); J.A.C.S., 26, 1171 (1904).
$Mg(ClO_4)_2$ anhyd.	"Unweighable" in 210 liters	Willard and Smith, J.A.C.S., 44, 2255 (1922).
BaO	0.00065	Bower, Bur. Std. J. Res., 12, 241 (1934).
KOH fused	0.002	Baxter and Starkweather, J.A.C.S., 38, 2038 (1916).
CaO	0.003	Bower, loc. cit.
H_2SO_4	0.003	Baxter and Starkweather, loc. cit.
$CaSO_4$ anhyd.	0.005	Bower loc. cit.
Al_2O_3	0.005	Ibid.
KOH sticks	0.014	Ibid.
NaOH fused	0.16	Baxter and Starkweather, loc. cit.
$CaBr_2$	0.18	Baxter and Warren, J.A.C.S., 33, 340 (1911).
$CaCl_2$ fused	0.34	Baxter and Starkweather, loc. cit.
NaOH sticks	0.80	Bower loc. cit.
$Ba(ClO_4)_2$	0.82	Ibid.
$ZnCl_2$	0.85	Baxter and Warren, loc. cit.
$ZnBr_2$	1.16	Ibid.
$CaCl_2$ granular	1.5	Bower, loc. cit.
$CuSO_4$ anhyd.	2.8	Ibid.

B. Drying agents depending upon physical action (adsorption) for their efficiency:* Alumina (low temperature fired), asbestos, charcoal, clay and porcelain (low temperature fired), glass wool, kieselguhr, silica gel, refrigeration.

* It should be noted that the efficiency of some drying agents (e.g. Al_2O_3 and anhydrous $CaCl_2$, and probably also BaO, anhydrous $Mg(ClO_4)_2$, $Mg(ClO_4)_2 \cdot 3H_2O$, anhydrous $Ba(ClO_4)_2$, and $CaSO_4$) depends upon both adsorption and absorption.

** 30°C. for Bower's values; others 25°C. or room temp.

LOW TEMPERATURE LIQUID BATHS

Liquid thermostat baths suitable for many physical measurements can be produced by using a stirred solid–liquid mixture at its melting point. Dry-ice or liquid air can be used to produce the solid. A Dewar flask is preferable as a container and, with adequate insulation or immersion in another somewhat colder bath, good temperature constancy can be maintained over several hours. Such baths are especially useful over the temperature range between dry-ice (−78C) and liquid air (−190). The following table gives the melting and normal boiling points of some readily available organic liquids suitable for this purpose. The compounds are listed in order of their increasing melting points. Temperatures are in degrees Celcius.

Compound	M.P.	B.P.	Compound	M.P.	B.P.
Isopentane (2-Methyl butane)	−159.9	27.85	Ethyl acetate	− 84	77
Methyl cyclopentane	−142.4	71.8	(Dry-ice + acetone)	− 78	——
Allyl chloride	−134.5	45	*p*-Cymene	− 67.9	177.1
n-Pentane	−129.7	36.1	Chloroform	− 63.5	61.7
Allyl alcohol	−129	97	N-Methyl aniline	− 57	196
Ethyl alcohol	−117.3	78.5	Chlorobenzene	− 45.6	132
Carbon disulfide	−110.8	46.3	Anisole	− 37.5	155
Isobutyl alcohol	−108	108.1	Bromobenzene	− 30.8	156
Acetone	− 95.4	56.2	Carbon tetrachloride	− 23	76.5
Toluene	− 95	110.6	Benzonitrile	− 13	205

WIRE TABLES

The resistance per unit length of wires of various metals is tabulated here. Values were calculated from resistivity values in the tables "Electrical Resistivity of Pure Metals" and "Electrical Resistivity of Selected Alloys", which appear in Section 12. In practice, resistance may vary because of differing heat treatments and metal composition. The values in the table refer to 20°C, but values at other temperatures may be calculated from the following resistivity data:

Metal	Resistivity in 10^{-8} Ω m at temperature			
	0°C	20°C	25°C	100°C
Aluminum	2.417	2.650	2.709	3.56
Brass (70% Cu, 30% Zn)	5.87	6.08	6.13	6.91
Constantan (60% Cu, 40% Ni)	45.43	45.38	45.35	45.11
Copper	1.543	1.678	1.712	2.22
Nichrome (79% Ni, 21% Cr)	107.3	107.5	107.6	108.3
Platinum	9.6	10.5	10.7	13.6
Silver	1.467	1.587	1.617	2.07
Tungsten	4.82	5.28	5.39	7.18

Resistance per unit length at 20°C in Ω/m

B & S Gauge	Diameter (mm)	Aluminum	Brass	Constantan	Copper	Nichrome	Platinum	Silver	Tungsten
0	8.252	0.000495	0.00114	0.00848	0.000314	0.0201	0.00196	0.000297	0.00099
2	6.543	0.000788	0.00181	0.0135	0.000499	0.0320	0.00312	0.000472	0.00157
4	5.189	0.00125	0.00287	0.0214	0.000793	0.0508	0.00496	0.000750	0.00250
6	4.115	0.00199	0.00457	0.0341	0.00126	0.0808	0.00789	0.00119	0.00397
8	3.264	0.00317	0.00727	0.0542	0.00200	0.128	0.0125	0.00190	0.00631
10	2.588	0.00504	0.0115	0.0863	0.00319	0.204	0.0200	0.00302	0.0100
12	2.053	0.00800	0.0184	0.137	0.00507	0.325	0.0317	0.00479	0.0159
14	1.628	0.0127	0.0292	0.218	0.00806	0.516	0.0504	0.00762	0.0254
16	1.291	0.0202	0.0464	0.347	0.0128	0.821	0.0802	0.0121	0.0403
18	1.024	0.0322	0.0738	0.551	0.0204	1.30	0.127	0.0193	0.0641
20	0.8118	0.0512	0.117	0.877	0.0324	2.08	0.203	0.0307	0.102
22	0.6439	0.0814	0.187	1.39	0.0515	3.30	0.322	0.0487	0.162
24	0.5105	0.129	0.297	2.22	0.0820	5.25	0.513	0.0775	0.258
26	0.4049	0.206	0.472	3.52	0.130	8.35	0.815	0.123	0.410
28	0.3211	0.327	0.751	5.60	0.207	13.3	1.30	0.196	0.652
30	0.2548	0.520	1.19	8.90	0.329	21.1	2.06	0.311	1.03
32	0.2019	0.828	1.90	14.2	0.524	33.6	3.28	0.496	1.65
34	0.1601	1.32	3.02	22.5	0.833	53.4	5.22	0.788	2.62
36	0.1270	2.09	4.80	35.8	1.32	84.9	8.29	1.25	4.17
38	0.1007	3.33	7.63	57.0	2.11	135	13.2	1.99	6.63
40	0.07988	5.29	12.1	90.5	3.35	214	20.9	3.17	10.5

CHARACTERISTICS OF PARTICLES AND PARTICLE DISPERSOIDS

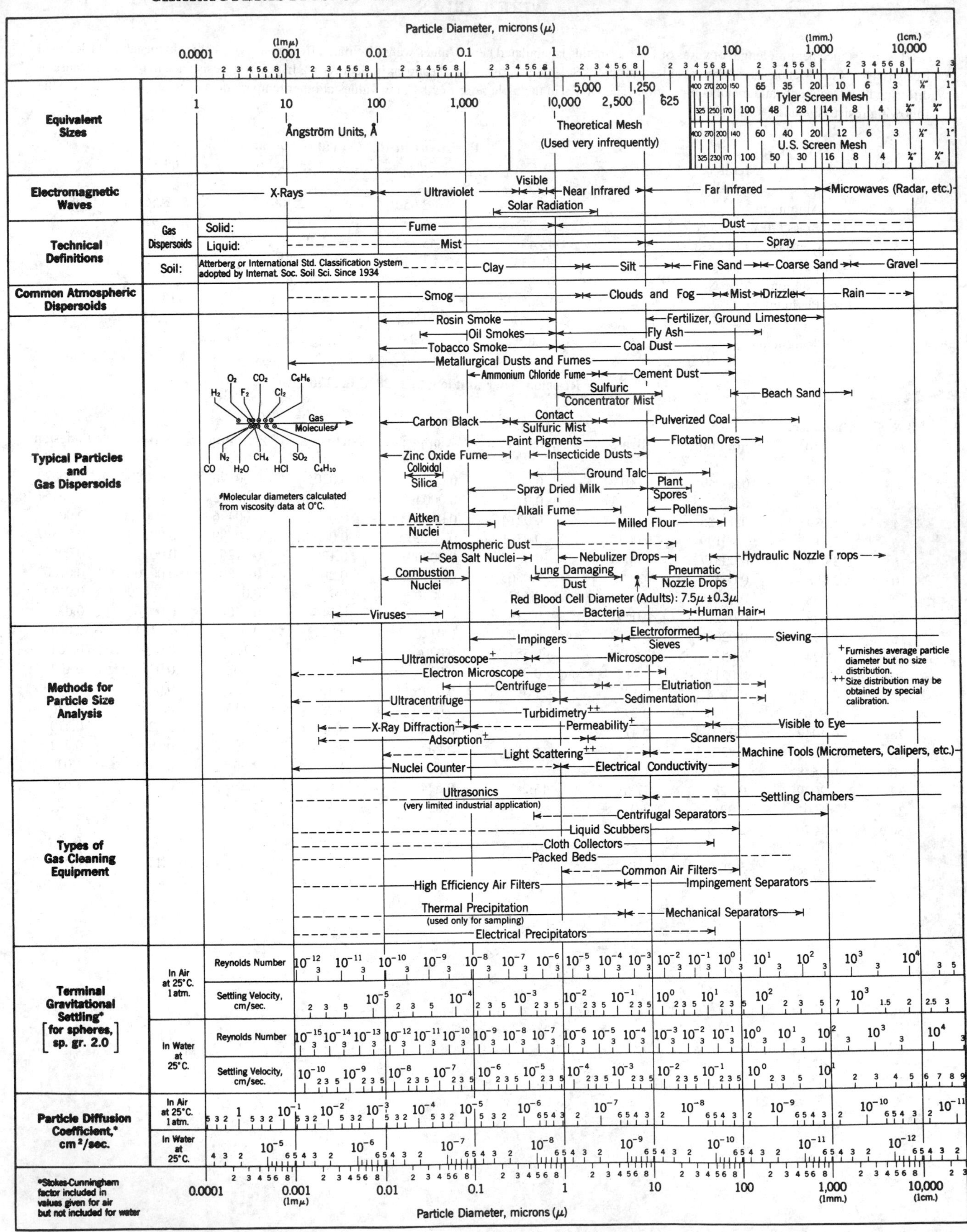

C. E. Lapple, Stanford Research Institute Journal, Vol. 5, p.95 (Third Quarter, 1961)

DENSITY OF VARIOUS SOLIDS

The approximate density of various solids at ordinary atmospheric temperature.
In the case of substances with voids such as paper or leather the bulk density is indicated rather than the density of the solid portion.

(Selected principally from the Smithsonian Tables.)

Substance	Grams per cu cm	Pounds per cu ft
Agate	2.5–2.7	156–168
Alabaster, carbon-		
ate	2.69–2.78	168–173
sulfate	2.26–2.32	141–145
Albite	2.62–2.65	163–165
Amber	1.06–1.11	66–69
Amphiboles	2.9–3.2	180–200
Anorthite	2.74–2.76	171–172
Asbestos	2.0–2.8	125–175
Asbestos slate	1.8	112
Asphalt	1.1–1.5	69–94
Basalt	2.4–3.1	150–190
Beeswax	0.96–0.97	60–61
Beryl	2.69–2.7	168–169
Biotite	2.7–3.1	170–190
Bone	1.7–2.0	106–125
Brick	1.4–2.2	87–137
Butter	0.86–0.87	53–54
Calamine	4.1–4.5	255–280
Calcspar	2.6–2.8	162–175
Camphor	0.99	62
Caoutchouc	0.92–0.99	57–62
Cardboard	0.69	43
Celluloid	1.4	87
Cement, set	2.7–3.0	170.190
Chalk	1.9–2.8	118–175
Charcoal, oak	0.57	35
pine	0.28–0.44	18–28
Cinnabar	8.12	507
Clay	1.8–2.6	112–162
Coal, anthracite	1.4–1.8	87–112
bituminous	1.2–1.5	75–94
Cocoa butter	0.89–0.91	56–57
Coke	1.0–1.7	62–105
Copal	1.04–1.14	65–71
Cork	0.22–0.26	14–16
Cork linoleum	0.54	34
Corundum	3.9–4.0	245–250
Diamond	3.01–3.52	188–220
Dolomite	2.84	177
Ebonite	1.15	72
Emery	4.0	250
Epidote	3.25–3.50	203–218
Feldspar	2.55–2.75	159–172
Flint	2.63	164
Fluorite	3.18	198
Galena	7.3–7.6	460–470
Gamboge	1.2	75
Garnet	3.15–4.3	197–268
Gas carbon	1.88	117
Gelatin	1.27	79

Substance	Grams per cu cm	Pound per cu ft
Glass, common	2.4–2.8	150–175
flint	2.9–5.9	180–370
Glue	1.27	79
Granite	2.64–2.76	165–172
Graphite*	2.30–2.72	144–170
Gum arabic	1.3–1.4	81–87
Gypsum	2.31–2.33	144–145
Hematite	4.9–5.3	306–330
Hornblende	3.0	187
Ice	0.917	57.2
Ivory	1.83–1.92	114–120
Leather, dry	0.86	54
Lime, slaked	1.3–1.4	81–87
Limestone	2.68–2.76	167–171
Linoleum	1.18	74
Magnetite	4.9–5.2	306–324
Malachite	3.7–4.1	231–256
Marble	2.6–2.84	160–177
Meerschaum	0.99–1.28	62–80
Mica	2.6–3.2	165–200
Muscovite	2.76–3.00	172–187
Ochre	3.5	218
Opal	2.2	137
Paper	0.7–1.15	44–72
Paraffin	0.87–0.91	54–57
Peat blocks	0.84	52
Pitch	1.07	67
Porcelain	2.3–2.5	143–156
Porphyry	2.6–2.9	162–181
Pressed wood		
pulp board	0.19	12
Pyrite	4.95–5.1	309–318
Quartz	2.65	165
Resin	1.07	67
Rock salt	2.18	136
Rubber, hard	1.19	74
Rubber, soft		
commercial	1.1	69
pure gum	0.91–0.93	57–58
Sandstone	2.14–2.36	134–147
Serpentine	2.50–2.65	156–165
Silica, fused trans-		
parent	2.21	138
translucent	2.07	129
Slag	2.0–3.9	125–240
Slate	2.6–3.3	162–205
Soapstone	2.6–2.8	162–175
Spermacéti	0.95	59
Starch	1.53	95
Sugar	1.59	99
Talc	2.7–2.8	168–174

Substance	Grams per cu cm	Pounds per cu ft
Tallow, beef	0.94	59
mutton	0.94	59
Tar	1.02	66
Topaz	3.5–3.6	219–223
Tourmaline	3.0–3.2	190–200
Wax, sealing	1.8	112
Wood (seasoned)		
alder	0.42–0.68	26–42
apple	0.66–0.84	41–52
ash	0.65–0.85	40–53
balsa	0.11–0.14	7–9
bamboo	0.31–0.40	19–25
basswood	0.32–0.59	20–37
beech	0.70–0.90	43–56
birch	0.51–0.77	32–48
blue gum	1.00	62
box	0.95–1.16	59–72
butternut	0.38	24
cedar	0.49–0.57	30–35
cherry	0.70–0.90	43–56
dogwood	0.76	47
ebony	1.11–1.33	69–83
elm	0.54–0.60	34–37
hickory	0.60–0.93	37–58
holly	0.76	47
juniper	0.56	35
larch	0.50–0.56	31–35
lignum vitae	1.17–1.33	73–83
locust	0.67–0.71	42–44
logwood	0.91	57
mahogany		
Honduras	0.66	41
Spanish	0.85	53
maple	0.62–0.75	39–47
oak	0.60–0.90	37–56
pear	0.61–0.73	38–45
pine, pitch	0.83–0.85	52–53
white	0.35–0.50	22–31
yellow	0.37–0.60	23–37
plum	0.66–0.78	41–49
poplar	0.35–0.5	22–31
satinwood	0.95	59
spruce	0.48–0.70	30–44
sycamore	0.40–0.60	24–37
teak, Indian	0.66–0.88	41–55
African	0.98	61
walnut	0.64–0.70	40–43
water gum	1.00	62
willow	0.40–0.60	24–37

* Some values reported as low as 1.6.

COEFFICIENT OF FRICTION

Compiled by Harold Minshall

The coefficient of friction between two surfaces is the ratio of the force required to move one over the other to the total force pressing the two together. If F is the force required to move one surface over another and W, the force pressing the surfaces together, the coefficient of friction,

$$\mu = \frac{F}{W}$$

Materials	Condition	Temperature °C	μ (Static)
A. STATIC FRICTION			
Non Metals			
Glass on glass	clean	—	0.9–1.0
" " "	lubricated with paraffin oil	—	0.5–0.6
" " "	" " liquid fatty acids	—	0.3–0.6
" " "	" " solid hydrocarbons, alcohols or fatty acids	—	0.1
" " metal	clean	—	0.5–0.7
" " "	lubricated	—	0.2–0.3
Diamond on diamond	clean	—	0.1
" " "	lubricated	—	0.05–0.1
" " metal	clean	—	0.1–0.15
" " "	lubricated	—	0.1
Sapphire on sapphire	clean or lubricated	—	0.2
" " steel	" " "	—	0.15
Hard carbon on carbon	clean	—	0.16
" " " "	lubricated	—	0.12–0.14
Graphite on graphite	clean or lubricated	—	0.1
" " "	outgassed	—	0.5–0.8
" " steel	clean or lubricated	—	0.1
Mica on mica	freshly cleaved	—	1.0
" " "	contaminated	—	0.2–0.4
Crystals of $NaNO_3$, KNO_3, NH_4Cl on self	clean	—	0.5
" " " " " " "	lubricated with long chain polar compounds	—	0.12
Tungsten carbide on tungsten carbide	clean	room	0.17
Tungsten carbide on tungsten carbide	outgassed	room	0.58
" " " " "	clean	820	0.35
" " " " "	"	970	0.40
" " " " "	"	1010	0.45
" " " " "	"	1160	0.5
" " " " "	"	1220	0.7
" " " " "	"	1440	1.2
" " " " "	"	1600	1.8
" " " graphite	outgassed	room	0.62
" " " "	clean	"	0.15
" " " "	"	800	0.32
" " " "	"	910	0.30
" " " "	"	1000	0.25
" " " "	"	1120	0.29
" " " "	"	1220	0.26
" " " "	"	1300	0.25
" " " "	"	1410	0.25
" " " "	"	1800	0.24
" " " "	"	2030	0.25
" " " steel	"	—	0.4–0.6
" " " "	lubricated	—	0.1–0.2
Polymethyl methacrylate on self	clean	—	0.8
" " " steel	"	—	0.4–0.5
Polystyrene on self	"	—	0.5
" " steel	"	—	0.3–0.35
Polyethylene on self	"	—	0.2
" " steel	"	—	0.2
Polytetrafluoroethylene on self	"	—	0.04
" " steel	"	—	0.04
Nylon on nylon[1]	"	—	0.15–0.25
Silk on silk	commercially clean	—	0.2–0.3
Cotton on cotton (thread)	" "	—	0.3
" " " (from cotton wool)	" "	—	0.6
Rubber on solids	" "	—	1–4
Wood on wood	" " and dry	—	0.25–0.5
" " "	" " " wet	—	0.2
" " metals	" " " dry	—	0.2–0.6
" " "	" " " wet	—	0.2
" " brick	" "	—	0.6
" " leather	" "	—	0.3–0.4
Leather on metal	" "	—	0.6
" " "	" " and wet	—	0.4
" " "	greasy	—	0.2
Brake material on cast iron	commercially clean	—	0.4
" " " " "	" " and wet	—	0.2
" " " " "	lubricated with mineral oil	—	0.1
Wool fiber on horn	clean (against scales)	—	0.8–1.0
" " " "	" (with scales)	—	0.4–0.6
" " " "	greasy (against scales)	—	0.5–0.8
" " " "	" (with scales)	—	0.3–0.4
Metals			
Steel on steel	clean	20	0.58
" " "	vegetable oil lubricant (a) castor oil	20	0.095
		100	0.105
" " "	(b) rape	20	0.105
		100	0.105
" " "	(c) olive	20	0.105
		100	0.105
" " "	(d) coconut	20	0.08
		100	0.08

[1] Registered trade name.

Materials	Condition	Temperature °C	μ (Static)
A. STATIC FRICTION (Cont.)			
Metals (Cont.)			
Steel on steel	Animal oil lubricant		
" " "	(a) sperm	20	0.10
		100	0.10
" " "	(b) pale whale	20	0.095
		100	0.095
" " "	(c) neatsfoot	20	0.095
		100	0.095
" " "	(d) lard	20	0.085
		100	0.085
" " "	Mineral oil lubricant		
" " "	(a) light machine	20	0.16
		100	0.19
" " "	(b) thick gear	20	0.125
		100	0.15
" " "	(c) solvent refined	20	0.15
		100	0.20
" " "	(d) heavy motor	20	0.195
		100	0.205
" " "	(e) extreme pressure	20	0.09–0.1
		100	0.09–0.1
" " "	(f) graphited oil	20	0.13
		100	0.15
" " "	(g) B.P. Paraffin	20	0.18
		100	0.22
" " "	lubricated with trichloroethylene	20	0.33
" " "	" " benzene	20	0.48
" " "	" " glycerol	20	0.2
" " "	" " ethyl alcohol	20	0.43
" " "	" " butyl alcohol	room	0.3
" " "	" " octyl	"	0.23
" " "	" " decyl	"	0.16
" " "	" " cetyl	"	0.10
" " "	lubricated with nonane	room	0.26
" " "	" " decane	"	0.23
" " "	" " acetic acid	"	0.5
" " "	" " proprionic acid	"	0.4
" " "	" " valeric acid	"	0.17
" " "	" " caproic acid	"	0.12
" " "	" " pelargonic acid	"	0.11
" " "	" " capric acid	"	0.11
" " "	" " lauric acid	"	0.11
" " "	" " myristic acid	"	0.11
" " "	" " oleic acid	20–100	0.08
" " "	" " palmitic acid	room	0.11
" " "	" " stearic acid	"	0.10
" " hard steel	" " rape oil	—	0.14
" " " "	" " castor oil	—	0,12
" " " "	" " mineral oil	—	0.16
" " " "	" " long chain fatty acid	—	0.09
" " cast iron	" " rape oil	—	0.11
" " " "	" " castor oil	—	0.15
" " " "	" " mineral oil	—	0.21
" " " "	clean	—	0.4
" " gun metal	lubricated with rape oil	—	0.15
" " " "	" " castor oil	—	0.16
" " " "	" " mineral oil	—	0.21
" " bronze	" " rape oil	—	0.12
" " "	" " caster oil	—	0.12
" " "	" " mineral oil	—	0.16
" " lead	" " " "	—	0.5
" " "	" " long chain fatty acid	—	0.22
" " base white metal	" " mineral oil	—	0.1
" " " " "	" " long chain fatty acid	—	0.08
" " "	clean	—	0.55
" " tin	lubricated with mineral oil	—	0.6
" " "	" " long chain fatty acid	—	0.21
" " white metal, tin base	" " mineral oil	—	0.1
" " " " " "	" " long chain fatty acid	—	0.07
" " " " " "	clean	—	0.8
" " sintered bronze	lubricated with mineral oil	—	0.13
" " brass	" " " "	—	0.19
" " "	" " castor oil	—	0.11
" " "	" " long chain fatty acid	—	0.13
" " "	clean	—	0.35
" " copper–lead alloy	"	—	0.22
" " Wood's alloy	"	—	0.7
" " phosphor bronze	"	—	0.35
" " aluminum bronze	"	—	0.45
" " constantan	"	—	0.4
[2] " " indium film deposited on steel	4 kg load, clean	—	0.08
[2] " " " " " " "	8 kg " "	—	0.04
[2] " " " " " " " silver	4 kg " "	—	0.1
[2] " " " " " " " "	8 kg " "	—	0.07
[2] " " lead film deposited on copper	4 kg " "	—	0.18
[2] " " " " " " "	8 kg " "	—	0.12
[2] " " copper film deposited on steel	4 kg " "	—	0.3
[2] " " " " " " "	8 kg " "	—	0.2
[2]Al on Al	in air or O_2	—	1.9
" " "	" H_2O vapor	—	1.1
[3]Cu on Cu	" H_2 or N_2	—	4.0
" " "	" air or O_2	—	1.6

[2] Hemispherical steel slider having 0.6 cm. diameter. The thin, 10^{-3} to 10^{-4} cm., thin metallic films were deposited on various substrates as indicated. Amonton's Law is not obeyed in this case.
[3] The metals which were spectroscopically pure were outgassed in a vacuum prior to other gases being admitted. When clean and in vacuum there is gross seizure.

Materials	Condition	Temperature °C	μ (Static)
A. STATIC FRICTION (Cont.)			
Metals (Cont.)			
[3]Au on Au	in H_2 or N_2	—	4.0
,, ,, ,,	,, air or O_2	—	2.8
,, ,, ,,	,, H_2O vapor	—	2.5
[3]Fe on Fe	,, in air or O_2	—	1.2
,, ,, ,,	,, H_2O vapor	—	1.2
[3]Mo on Mo	,, air or O_2	—	0.8
,, ,, ,,	,, H_2O vapor	—	0.8
[3]Ni on Ni	,, H_2 or N_2	—	5.0
,, ,, ,,	,, air or O_2	—	3.0
,, ,, ,,	,, H_2O vapor	—	1.6
[3]Pt on Pt	,, air or O_2	—	3.0
,, ,, ,,	,, H_2O vapor	—	3.0
[3]Ag on Ag	,, air or O_2	—	1.5
,, ,, ,,	,, H_2O vapor	—	1.5
Various Materials on Snow and Ice			
Ice on ice	clean	0	0.05–0.15
,, ,, ,,	,,	−12	0.3
,, ,, ,,	,,	−71	0.5
,, ,, ,,	,,	−82	0.5
,, ,, ,,	,,	−110	0.5
Polymethylmethylacrylate	on wet snow	0	0.5
,,	,, dry ,,	0	0.3
,,	,, ,, ,,	−10	0.34
,,	,, ,, ,,	−32	0.4
Polyester of teraphthalic acid and ethylene glycol	,, wet ,,	0	0.5
Polyester of teraphthalic acid and ethylene glycol	,, dry ,,	0	0.35
Polyester of teraphthalic acid and ethylene glycol	,, ,, ,,	−10	0.38
[1]Nylon	on wet snow	0	0.4
,,	,, dry ,,	0	0.3
,,	,, ,, ,,	−10	0.3
Polytetrafluoroethylene	,, wet ,,	0	0.05
,,	,, dry ,,	0	0.02
,,	,, ,, ,,	−10	0.08
,,	,, ,, ,,	−32	0.1
Paraffin wax	,, wet ,,	0	0.06
,, ,,	,, dry ,,	0	0.06
,, ,,	,, ,, ,,	−10	0.35
,, ,,	,, ,, ,,	−32	0.4
Swiss wax	,, wet ,,	0	0.05
,, ,,	,, dry ,,	0	0.03
,, ,,	,, ,, ,,	−10	0.2
,, ,,	,, ,, ,,	−32	0.2
Ski wax	,, wet ,,	0	0.1
,, ,,	,, dry ,,	0	0.04
,, ,,	,, ,, ,,	−10	0.2
,, ,,	,, ,, ,,	−32	0.2
,, laquer	,, wet ,,	0	0.2
,, ,,	,, dry ,,	0	0.1
,, ,,	,, ,, ,,	−10	0.4
,, ,,	,, ,, ,,	−32	0.4
Aluminum	,, wet ,,	0	0.4
,,	,, dry ,,	0	0.35
,,	,, ,, ,,	−10	0.38

Materials	Condition	Temperature °C	μ (Kinetic)
B. KINETIC FRICTION Various Materials			
Unwaxed hickory	4 m/sec on dry snow	−3	0.08
Waxed ,,	0.1 m/sec ,, wet ,,	0	0.14
,, ,,	0.1 m/sec ,, dry ,,	0	0.04
,, ,,	0.1 m/sec ,, ,, ,,	−3	0.09
,, ,,	4 m/sec ,, ,, ,,	−3	0.03
Waxed hickory	0.1 m/sec on dry snow	−10	0.18
,, ,,	0.1 m/sec ,, ,, ,,	−40	0.4
Ice on ice	4m/sec, clean	0	0.02
,, ,, ,,	,, , ,,	−10	0.035
,, ,, ,,	,, , ,,	−20	0.050
,, ,, ,,	,, , ,,	−40	0.075
,, ,, ,,	,, , ,,	−60	0.085
,, ,, ,,	,, , ,,	−80	0.09
Ebonite	4m/sec on ice	0	0.02
,,	,, ,, ,,	−10	0.05
,,	,, ,, ,,	−20	0.065
,,	,, ,, ,,	−40	0.085
,,	,, ,, ,,	−60	0.10
,,	,, ,, ,,	−80	0.11
Brass	4m/sec on ice	0	0.02
,,	,, ,, ,,	−10	0.075
,,	,, ,, ,,	−20	0.085
,,	,, ,, ,,	−40	0.115
,,	,, ,, ,,	−60	0.14
,,	,, ,, ,,	−80	0.15
Natural rubber, vulcanized	100m/min on ground glass, clean	—	1.07
,, ,, , ,,	100m/min ,, ,, ,, , wetted with water	—	0.94
,, ,, , ,,	100m/min on concrete, clean	—	1.02
,, ,, , ,,	100m/min ,, ,, , wetted with water	—	0.97
,, ,, , ,,	100m/min on bitumen, clean	—	1.07
,, ,, , ,,	100m/min ,, ,, , wetted with water	—	0.95
,, ,, , ,,	100m/min on rubber flooring or rubber tread vulcanisate, clean	—	1.16
,, ,, , ,,	100m/min on bitumen containing rubber powder, clean	—	1.15 (Varies with quantity of powder)
,, ,, , ,,	100m/min on bitumen containing rubber powder, wetted with water	—	1.03

[1]Registered trade name. [3] The metals which were spectroscopically pure were outgassed in a vacuum prior to other gases being admitted. When clean and in vacuum there is gross seizure.

Section 16
Health and Safety Information

HANDLING AND DISPOSAL OF CHEMICALS IN LABORATORIES

Robert Joyce and Blaine C. McKusick

The following material has been extracted from two books prepared under the auspices of the Committee on Hazardous Substances in the Laboratory of the National Academy of Sciences — National Research Council. Readers are referred to these books for full details:

Prudent Practices for Handling Hazardous Chemicals in Laboratories, National Academy Press, Washington, 1981.
Prudent Practices for Disposal of Chemicals from Laboratories, National Academy Press, Washington, 1983.

The permission of the National Academy Press to use these extracts is gratefully acknowledged.

INCOMPATIBLE CHEMICALS

The term "incompatible chemicals" refers to chemicals that can react with each other

- Violently
- With evolution of substantial heat
- To produce flammable products
- To produce toxic products

Good laboratory safety practice requires that incompatible chemicals be stored, transported, and disposed of in ways that will prevent their coming together in the event of an accident. Tables 1 and 2 give some basic guidelines for the safe handling of acids, bases, reactive metals, and other chemicals. Neither of these tables is exhaustive, and additional information on incompatible chemicals can be found in the following references.

1. L. Bretherick, *Handbook of Reactive Chemical Hazards*, 3rd ed., Butterworths, London–Boston, 1985.
2. L. Bretherick, Ed., *Hazards in the Chemical Laboratory*, 3rd ed., Royal Society of Chemistry, London, 1981.
3. *Manual of Hazardous Chemical Reactions, A Compilation of Chemical Reactions Reported to be Potentially Hazardous,* National Fire Protection Association, NFPA 491M, 1975, NFPA, 470 Atlantic Avenue, Boston, MA 02210.

TABLE 1
General Classes of Incompatible Chemicals

A	B
Acids	Bases, reactive metals
Oxidizing agents[a]	Reducing agents[a]
Chlorates	Ammonia, anhydrous and aqueous
Chromates	Carbon
Chromium trioxide	Metals
Dichromates	Metal hydrides
Halogens	Nitrites
Halogenating agents	Organic compounds
Hydrogen peroxide	Phosphorus
Nitric acid	Silicon
Nitrates	Sulfur
Perchlorates	
Peroxides	
Permanganates	
Persulfates	

[a] The examples of oxidizing and reducing agents are illustrative of common laboratory chemicals; they are not intended to be exhaustive.

TABLE 2
Examples of Incompatible Chemicals

Chemical	Is incompatible with
Acetic acid	Chromic acid, nitric acid, hydroxyl compounds, ethylene glycol, perchloric acid, peroxides, permanaganates
Acetylene	Chlorine, bromine, copper, fluorine, silver, mercury
Acetone	Concentrated nitric and sulfuric acid mixtures
Alkali and alkaline earth metals (such as powdered aluminum or magnesium, calcium, lithium, sodium, potassium)	Water, carbon tetrachloride or other chlorinated hydrocarbons, carbon dioxide, halogens
Ammonia (anhydrous)	Mercury (in manometers, for example), chlorine, calcium hypochlorite, iodine, bromine, hydrofluoric acid (anhydrous)
Ammonium nitrate	Acids, powdered metals, flammable liquids, chlorates, nitrites, sulfur, finely divided organic or combustible materials
Aniline	Nitric acid, hydrogen peroxide
Arsenical materials	Any reducing agent
Azides	Acids
Bromine	See Chlorine
Calcium oxide	Water
Carbon (activated)	Calcium hypochlorite, all oxidizing agents
Carbon tetrachloride	Sodium
Chlorates	Ammonium salts, acids, powdered metals, sulfur, finely divided organic or combustible materials
Chromic acid and chromium troixide	Acetic acid, naphthalene, camphor, glycerol, alcohol, flammable liquids in general
Chlorine	Ammonia, acetylene, butadiene, butane, methane, propane (or other petroleum gases), hydrogen, sodium carbide, benzene, finely divided metals, turpentine
Chlorine dioxide	Ammonia, methane, phosphine, hydrogen sulfide
Copper	Acetylene, hydrogen peroxide
Cumene hydroperoxide	Acids (organic or inorganic)
Cyanides	Acids
Flammable liquids	Ammonium nitrate, chromic acid, hydrogen peroxide, nitric acid, sodium peroxide, halogens
Fluorine	Everything
Hydrocarbons (such as butane, propane, benzene)	Fluorine, chlorine, bromine, chromic acid, sodium peroxide
Hydrocyanic acid	Nitric acid, alkali
Hydrofluoric acid (anhydrous)	Ammonia (aqueous or anhydrous)
Hydrogen peroxide	Copper, chromium, iron, most metals or their salts, alcohols, acetone, organic materials, aniline, nitromethane, combustible matierals
Hydrogen sulfide	Fuming nitric acid, oxidizing gases
Hypochlorites	Acids, activated carbon
Iodine	Acetylene, ammonia (aqueous or anhydrous), hydrogen
Mercury	Acetylene, fulminic acid, ammonia
Nitrates	Sulfuric acid
Nitric acid (concentrated)	Acetic acid, aniline, chromic acid, hydrocyanic acid, hydrogen sulfide, flammable liquids, flammable gases, copper, brass, any heavy metals
Nitrites	Acids
Nitroparaffins	Inorganic bases, amines
Oxalic acid	Silver, mercury

TABLE 2
Examples of Incompatible Chemicals (continued)

Chemical	Is incompatible with
Oxygen	Oils, grease, hydrogen, flammable liquids, solids, or gases
Perchloric acid	Acetic anhydride, bismuth and its alloys, alcohol, paper, wood, grease, oils
Peroxides, organic	Acids (organic or mineral), avoid friction, store cold
Phosphorus (white)	Air, oxygen, alkalis, reducing agents
Potassium	Carbon tetrachloride, carbon dioxide, water
Potassium chlorate	Sulfuric and other acids
Potassium perchlorate (see also chlorates)	Sulfuric and other acids
Potassium permanganate	Glycerol, ethylene glycol, benzaldehyde, sulfuric acid
Selenides	Reducing agents
Silver	Acetylene, oxalic acid, tartartic acid, ammonium compounds, fulminic acid
Sodium	Carbon tetrachloride, carbon dioxide, water
Sodium nitrite	Ammonium nitrate and other ammonium salts
Sodium peroxide	Ethyl or methyl alcohol, glacial acetic acid, acetic anhydride, benzaldehyde, carbon disulfide, glycerin, ethylene glycol, ethyl acetate, methyl acetate, furfural
Sulfides	Acids
Sulfuric acid	Potassium chlorate, potassium perchlorate, potassium permanganate (similar compounds of light metals, such as sodium, lithium)
Tellurides	Reducing agents

EXPLOSION HAZARDS

Table 3 lists some common classes of laboratory chemicals that have potential for producing a violent explosion when subjected to shock or friction. These chemicals should never be disposed of as such, but should be handled by procedures given in *Prudent Practices for Disposal of Chemicals from Laboratories*, National Academy Press, 1983, chapters 6 and 7. Additional information on these, as well as on some less common classes of explosives, can be found in L. Bretherick, *Handbook of Reactive Chemical Hazards,* 3rd ed., Butterworths, London–Boston, 1985.

Table 4 lists some illustrative combinations of common laboratory reagents that can produce explosions when they are brought together or that form reaction products that can explode without any apparent external initiating action. This list is not exhaustive, and additional information on potentially explosive reagent combinations can be found in Manual of Hazardous Chemical Reactions, A Compilation of Chemical Reactions Reported to be Potentially Hazardous, National Fire Protection Association, NFPA 491M, 1975, NFPA, 470 Atlantic Avenue, Boston, MA 02210.

WATER-REACTIVE CHEMICALS

Table 5 lists some common laboratory chemicals that react violently with water and that should always be stored and handled so that they do not come into contact with liquid water or water vapor. Procedures for decomposing laboratory quantities are given in Prudent Practices for Disposal of Chemicals from Laboratories, chapter 6; the pertinent section of that chapter is given in parentheses.

PYROPHORIC CHEMICALS

Many members of the classes of readily oxidized, common laboratory chemicals listed in Table 6 ignite spontaneously in air. A more extensive list can be found in L. Bretherick, *Handbook of Reactive Chemical Hazards*, 3rd ed., Butterworths, London-Boston, 1985. Pyrophoric chemicals should be stored in tightly closed containers under an inert atmosphere (or, for some, an inert liquid), and all transfers and manipulations of them must be carried out under an inert atmosphere or liquid. Suggested procedures for decomposing them are given in Prudent Practices for Disposal of Chemicals from Laboratories, chapter 6; the pertinent section of that chapter is given in parentheses.

TABLE 3
Shock–Sensitive Compounds

Acetylenic compounds, especially polyacetylenes, haloacetylenes, and heavy metal salts of acetylenes (copper, silver, and mercury salts are particularly sensitive)
Acyl nitrates
Alkyl nitrates, particularly polyol nitrates such as nitrocellulose and nitroglycerine
Alkyl and acyl nitrites
Alkyl perchlorates
Amminemetal oxosalts: metal compounds with coordinated ammonia, hydrazine, or similar nitrogenous donors and ionic perchlorate, nitrate, permanganate, or other oxidizing group
Azides, including metal, nonmetal, and organic azides
Chlorite salts of metals, such as $AgClO_2$ and $Hg(ClO_2)_2$
Diazo compounds such as CH_2N_2
Diazonium slats, when dry
Fulminates (silver fulminate, AgCNO, can form in the reaction mixture from the Tollens' test for aldehydes if it is allowed to stand for some time; this can be prevented by adding dilute nitric acid to the test mixture as soon as the test has been completed)
Hydrogen peroxide becomes increasingly treacherous as the concentration rises above 30%, forming explosive mixtures with organic materials and decomposing violently in the presence of traces of transition metals
N–Halogen compounds such as difluoroamino compounds and halogen azides
N–Nitro compounds such as *N*–nitromethylamine, nitrourea, nitroguanidine, and nitric amide
Oxo salts of nitrogenous bases: perchlorates, dichromates, nitrates, iodates, chlorites, chlorates, and permanganates of ammonia, amines, hydroxylamine, guanidine, etc.
Perchlorate salts. Most metal, nonmetal, and amine perchlorates can be detonated and may undergo violent reaction in contact with combustible materials
Peroxides and hydroperoxides, organic (see Chapter 6, Section II.P)
Peroxides (solid) that crystallize from or are left from evaporation of peroxidizable solvents (see Chapter 6 and Appendix I)
Peroxides, transition–metal salts
Picrates, especially salts of transition and heavy metals, such as Ni, Pb, Hg, Cu, and Zn; picric acid is explosive but is less sensitive to shock or friction than its metal salts and is relatively safe as a water–wet paste (see Chapter 7)
Polynitroalkyl compounds such as tetranitromethane and dinitroacetonitrile
Polynitroaromatic compounds, especially polynitro hydrocarbons, phenols, and amines

TABLE 4
Potentially Explosive Combinations of Some Common Reagents

Acetone + chloroform in the presence of base
Acetylene + copper, silver, mercury, or their salts
Ammonia (including aqueous solutions) + Cl_2, Br_2, or I_2
Carbon disulfide + sodium azide
Chlorine + an alcohol
Chloroform or carbon tetrachloride + powdered Al or Mg
Decolorizing carbon + an oxidizing agent
Diethyl ether + chlorine (including a chlorine atmosphere)
Dimethyl sulfoxide + an acyl halide, $SOCl_2$, or $POCl_3$
Dimethyl sulfoxide + CrO_3
Ethanol + calcium hypochlorite
Ethanol + silver nitrate
Nitric acid + acetic anhydride or acetic acid
Picric acid + a heavy–metal salt, such as of Pb, Hg, or Ag
Silver oxide + ammonia + ethanol
Sodium + a cholrinated hydrocarbon
Sodium hypochlorite + an amine

TABLE 5
Water–Reactive Chemicals

Alkali metals (III.D)
Alkali metal hydrides (III.C.2)
Alkali metal amides (III.C.7)
Metal alkyls, such as lithium alkyls and aluminum alkyls (IV.A)
Grignard reagents (IV.A)
Halides of nonmetals, such as BCl_3, BF_3, PCl_3, PCl_5, $SiCl_4$, S_2Cl_2 (III.F)
Inorganic acid halides, such as $POCl_3$, $SOCl_2$, SO_2Cl_2 (III.F)
Anhydrous metal halides, such as $AlCl_3$, $TiCl_4$, $ZrCl_4$, $SnCl_4$ (III.E)
Phosphorus pentoxide (III.I)
Calcium carbide (IV.E)
Organic acid halides and anhydrides of low molecular weight (II.J)

TABLE 6
Classes of Pyrophoric Chemicals

Grignard reagents, RMgX (IV.A)
Metal alkyls and aryls, such as RLi, RNa, R_3Al, R_2Zn (IV.A)
Metal carbonyls, such as Ni $(CO)_4$, $Fe(CO)_5$, $Co_2(CO)_8$ (IV.B)
Alkali metals such as Na, K (III.D.l)
Metal powders, such as Al, Co, Fe, Mg, Mn, Pd, Pt, Ti, Sn, Zn, Zr (III.D.2)
Metal hydrides, such as NaH, $LiAlH_4$ (IV.C.2)
Nonmetal hydrides, such as B_2H_6 and other boranes, PH_3, AsH_3 (III.G)
Nonmetal alkyls, such as R_3B, R_3P, R_3As (IV.C)
Phosphorus (white) (III.H)

HAZARDS FROM PEROXIDE FORMATION

Many common laboratory chemicals can form peroxides when allowed access to air over a period of time. A single opening of a container to remove some of the contents can introduce enough air for peroxide formation to occur. Some types of compounds form peroxides that are treacherously and violently explosive in concentrated solution or as solids. Accordingly, peroxide–containing liquids should never be evaporated near to or to dryness. Peroxide formation can also occur in many polymerizable unsaturated compounds, and these peroxides can initiate a runaway, sometimes explosive, polymerization reaction. Procedures for testing for peroxides and for removing small amounts from laboratory chemicals are given in Prudent Practices for Disposal of Chemicals from Laboratories, chapter 6, Section II.P.

Table 7 provides a list of structural characteristics in organic compounds that can peroxidize. These structures are listed in approximate order of decreasing hazard. Reports of serious incidents involving the last five structural types are extremely rare, but these structures are listed because laboratory workers should be aware that they can form peroxides that can influence the course of experiments in which they are used.

Table 8 gives examples of common laboratory chemicals that are prone to form peroxides on exposure to air. The lists are not exhaustive, and analogous organic compounds that have any of the structural features given in Table 7 should be tested for peroxides before being used as solvents or reagents, or before being distilled. The recommended retention times begin with the date of synthesis or of opening the original container.

DISPOSAL OF TOXIC CHEMICALS

It is often desirable to precipitate toxic cations or hazardous anions from solution to facilitate recovery or disposal. Table 9 lists precipitants for many common cations, and Table 10 gives precipitants for some hazardous anions. Many cations can be precipitated as sulfides by adding sodium sulfide solution (preferable to the highly toxic hydrogen sulfide) to a neutral solution of the cation (Table 11). Control of pH is important because some sulfides will redissolve in excess sulfide ion. After precipitation, excess sulfide can be destroyed by addition of hypochlorite.

Most metal cations are precipitated as hydroxides or oxides at high pH. Since many of these precipitates will redissolve in excess base, it is often necessary to control pH. Table 12 shows the recommended pH range for precipitating many cations in their most common oxidation state. The notation "1 N" in the right–hand column indicates that the precipitate will not dissolve in 1 N sodium hydroxide (pH 14).

The distinctions between high and low toxicity or hazard are based on toxicological and other data, and are relative. There is no implication of a sharp distinction between high and low, or that any cations or anions are totally without hazard.

TABLE 7
Types of Chemicals That Are Prone to Form Peroxides

A. Organic structures (in approximate order of decreasing hazard)

1.	$>\mathrm{C(H)-O-}$	Ethers and acetals with α hydrogen atoms
2.	$>\mathrm{C{=}C(-)-C(H)}<$	Olefins with allylic hydrogen atoms
3.	$>\mathrm{C{=}C(X)-}$	Chloroolefins and fluoroolefins
4.	$\mathrm{CH_2{=}C}<$	Vinyl halides, esters, and ethers
5.	$>\mathrm{C{=}C(-)-C(-){=}C}<$	Dienes
6.	$>\mathrm{C{=}C(H)-C{\equiv}CH}$	Vinylacetylenes with α hydrogen atoms
7.	$>\mathrm{C(H)-C{\equiv}CH}$	Alkylacetylenes with α hydrogen atoms
8.	$>\mathrm{C(H)-Ar}$	Alkylarenes that contain tertiary hydrogen atoms
9.	$-\mathrm{C(-)(-)-H}$	Alkanes and cycloalkanes that contain tertiary hydrogen atoms
10.	$>\mathrm{C{=}C(-)-CO_2R}$	Acrylates and methacrylates
11.	$>\mathrm{C(H)-OH}$	Secondary alcohols
12.	$-\mathrm{C({=}O)-C(H)}<$	Ketones that contain α hydrogen atoms
13.	$-\mathrm{C(H){=}O}$	Aldehydes
14.	$-\mathrm{C({=}O)-N(H)-C(H)}<$	Ureas, amides, and lactams that have a hydrogen atom on a carbon atom attached to nitrogen

TABLE 7
Types of Chemicals That Are Prone to Form Peroxides (continued)

B. Inorganic substances

1. Alkali metals, especially potassium, rubidium, and cesium (see Chapter 6, Section III.D)
2. Metal amides (see Chapter 6, Section III.C.7)
3. Organometallic compounds with a metal atom bonded to carbon (see Chapter 6, Section IV)
4. Metal alkoxides

TABLE 8
Common Peroxide-Forming Chemicals

LIST A

Severe Peroxide Hazard on Storage with Exposure to Air

Discard within 3 months

- Diisopropyl ether (isopropyl ether)
- Divinylacetylene (DVA)[a]
- Potassium metal
- Potassium amide
- Sodium amide (sodamide)
- Vinylidene chloride (1,1–dichloroethylene)[a]

LIST B

Peroxide Hazard on Concentration; Do Not Distill or Evaporate Without First Testing for the Presence of Peroxides

Discard or test for peroxides after 6 months

- Acetaldehyde diethyl acetal (acetal)
- Cumene (isopropylbenzene)
- Cyclohexene
- Cyclopentene
- Decalin (decahydronaphthalene)
- Diacetylene
- Dicyclopentadiene
- Diethyl ether (ether)
- Diethylene glycol dimethyl ether (diglyme)
- Dioxane
- Ethylene glycol dimethyl ether (glyme)
- Ethylene glycol ether acetates
- Ethylene glycol monoethers (cellosolves)
- Furan
- Methylacetylene
- Methylcyclopentane
- Methyl isobutyl ketone
- Tetrahydrofuran (THF)
- Tetralin (tetrahydronaphthalene)
- Vinyl ethers[a]

LIST C

Hazard of Rapid Polymerization Initiated by Internally Formed Peroxides[a]

a. Normal Liquids; Discard or test for peroxides after 6 months[b]

- Chloroprene (2–chloro–1,3–butadiene)[c]
- Styrene
- Vinyl acetate
- Vinylpyridine

b. Normal Gases; Discard after 12 months[d]

- Butadiene[c]
- Tetrafluoroethylene (TFE)[c]
- Vinylacetylene (MVA)[c]
- Vinyl chloride

[a] Polymerizable monomers should be stored with a polymerization inhibitor from which the monomer can be separated by distillation just before use.

[b] Although common acrylic monomers such as acrylonitrile, acrylic acid, ethyl acrylate, and methyl methacrylate can form peroxides, they have not been reported to develop hazardous levels in normal use and storage.

[c] The hazard from peroxides in these compounds is substantially greater when they are stored in the liquid phase, and if so stored without an inhibitor they should be considered as in LIST A.

TABLE 8
Common Peroxide-Forming Chemicals (continued)

[d] Although air will not enter a gas cylinder in which gases are stored under pressure, these gases are sometimes transferred from the original cylinder to another in the laboratory, and it is difficult to be sure that there is no residual air in the receiving cylinder. An inhibitor should be put into any such secondary cylinder before one of these gases is transferred into it; the supplier can suggest inhibitors to be used. The hazard posed by these gases is much greater if there is a liquid phase in such a secondary container, and even inhibited gases that have been put into a secondary container under conditions that create a liquid phase should be discarded within 12 months.

Note: Laboratory workers should label all containers of peroxidizable solvents or reagents with one of the following:

[LIST A]

Peroxidizable compound

	Received	Opened
Date	________	______

Discard 3 months after opening

[LISTS B AND C]

Peroxidizable compound

	Received	Opened
Date	________	_______

Discard or test for peroxides
6 months after opening

TABLE 9
Relative Toxicity of Cations

High toxic hazard	Precipitant[a]	Low toxic hazard	Precipitant[a]
Antimony	OH^-, S^{2-}	Aluminum	OH^-
Arsenic	S^{2-}	Bismuth	OH^-, S^{2-}
Barium	SO_4^{2-}, CO_3^{2-}	Calcium	SO_4^{2-}, CO_3^{2-}
Beryllium	OH^-	Cerium	OH^-
Cadmium	OH^-, S^{2-}	Cesium	
Chromium (III)[b]	OH^-	Copper[c]	OH^-, S^{2-}
Cobalt (II)[b]	OH^-, S^{2-}	Gold	OH^-, S^{2-}
Gallium	OH^-	Iron[c]	OH^-, S^{2-}
Germanium	OH^-, S^{2-}	Lanthanides	OH^-
Hafnium	OH^-	Lithium	
Indium	OH^-, S^{2-}	Magnesium	OH^-
Iridium	OH^-, S^{2-}	Molybdenum (VI)[b,d]	
Lead	OH^-, S^{2-}	Niobium (V)	OH^-
Manganese (II)[b]	OH^-, S^{2-}	Palladium	OH^-, S^{2-}
Mercury	OH^-, S^{2-}	Potassium	
Nickel	OH^-, S^{2-}	Rubidium	
Osmium (IV)[b,e]	OH^-, S^{2-}	Scandium	OH^-
Platinum (II)[b]	OH^-, S^{2-}	Sodium	
Rhenium (VII)[b]	S^{2-}	Strontium	SO_4^{2-}, CO_3^{2-}
Rhodium (III)[b]	OH^-, S^{2-}	Tantalum	OH^-
Ruthenium (III)[b]	OH^-, S^{2-}	Tin	OH^-, S^{2-}
Selenium	S^{2-}	Titanium	OH^-
Silver	Cl^-, OH^-, S^{2-}	Yttrium	OH^-
Tellurium	S^{2-}	Zinc[c]	OH^-, S^{2-}
Thallium	OH^-, S^{2-}	Zirconium	OH^-
Tungsten (VI)[b,d]			
Vanadium	OH^-, S^{2-}		

TABLE 9
Relative Toxicity of Cations (continued)

[a] Precipitants are listed in order of preference:
OH^- = base (sodium hydroxide or sodium carbonate)
S^{2-} = sulfide
Cl^- = chloride
SO_4^{2-} = sulfate
CO_3^{2-} = carbonate

[b] The precipitant is for the indicated valence state.

[c] Maximum tolerance levels have been set for these low–toxicity ions by the U.S. Public Health Service, and large amounts should not be put into public sewer systems. The small amounts typically used in laboratories will not normally affect water supplies.

[d] These ions are best precipitated as calcium molybdate or calcium tungstate.

[e] CAUTION: OsO_4, a volatile, extremely poisonous substance, is formed from almost any osmium compound under acid conditions in the presence of air.

TABLE 10
Relative Hazard of Anions

High–hazard anions			
Ion	Hazard type[a]	Precipitant	Low-hazard anions
Aluminum hydride, AlH_4	F	—	Bisulfite, HSO_3^-
Amide, NH_2^-	F, E[b]	—	Borate, BO_3^{3-}, $B_4O_7^{2-}$
Arsenate, AsO_3^-, AsO_4^{3-}	T	Cu^{2+}, Fe^{2+}	Bromide, Br^-
Arsenite, AsO_2^-, AsO_3^{3-}	T	Pb^{2+}	Carbonate, CO_3^{2-}
Azide, N_3^-	E, T	—	Chloride, Cl^-
Borohydride, BH_4^-	F	—	Cyanate, OCN^-
Bromate, BrO_3^-	O, E	—	Hydroxide, OH^-
Chlorate, ClO_3^-	O, E	—	Iodide, I^-
Chromate, CrO_4^{2-}, $Cr_2O_7^{2-}$	T, O	[c]	Oxide, O^{2-}
Cyanide, CN^-	T	—	Phosphate, PO_4^{3-}
Ferricyanide, $Fe(CN)_6^{3-}$	T	Fe^{2+}	Sulfate, SO_4^{2-}
Ferrocyanide, $Fe(CN)_6^{4-}$	T	Fe^{3+}	Sulfite, SO_3^{2-}
Fluoride, F^-	T	Ca^{2+}	Thiocyanate, SCN^-
Hydride, H^-	F	—	
Hydroperoxide, O_2H^-	O, E	—	
Hydrosulfide, SH^-	T	—	
Hypochlorite, OCl^-	O	—	
Iodate, IO_3^-	O, E	—	
Nitrate, NO_3^-	O	—	
Nitrite, NO_2^-	T, O	—	
Perchlorate, ClO_4^-	O, E	—	
Permanganate, MnO_4^-	T, O	[d]	
Peroxide, O_2^{2-}	O, E	—	
Persulfate, $S_2O_8^{2-}$	O	—	
Selenate, SeO_4^{2-}	T	Pb^{2+}	
Selenide, Se^{2-}	T	Cu^{2+}	
Sulfide, S^{2-}	T	[e]	

[a] Toxic, T: oxidant, O; flammable, F; explosive, E.

[b] Metal amides readily form explosive peroxides on exposure to air.

[c] Reduce and precipitate as Cr(III); see Table 9.

[d] Reduce and precipitate as Mn(II); see Table 9.

[e] See Table 11.

TABLE 11
Precipitation of Sulfides

Precipitated at pH 7	Not precipitated at low pH	Forms a soluble complex at high pH
Ag^{+}		
As^{3+}[a]		X
Au^{+}[a]		X
Bi^{3+}		
Cd^{2+}		
Co^{2+}	X	
Cr^{3+}[a]		
Cu^{2+}		
Fe^{2+}[a]	X	
Ge^{2+}		X
Hg^{2+}		X
In^{3+}	X	
Ir^{4+}		X
Mn^{2+}[a]	X	
Mo^{3+}		X
Ni^{2+}	X	
Os^{4+}		
Pb^{2+}		
Pd^{2+}[a]		
Pt^{2+}[a]		X
Re^{4+}		
Rh^{2+}[a]		
Ru^{4+}		
Sb^{3+}[a]		X
Se^{2+}		X
Sn^{2+}		X
Te^{4+}		X
Tl^{+}[a]	X	
V^{4+}[a]		
Zn^{2+}	X	

[a] Higher oxidation states of this ion are reduced by sulfide ion and precipitated as this sulfide.

TABLE 12
pH Range for Precipitation of Metal Hydroxides and Oxides

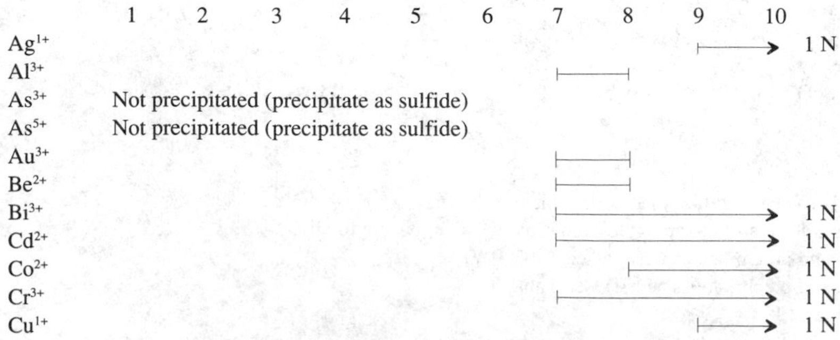

TABLE 12
pH Range for Precipitation of Metal Hydroxides and Oxides (continued)

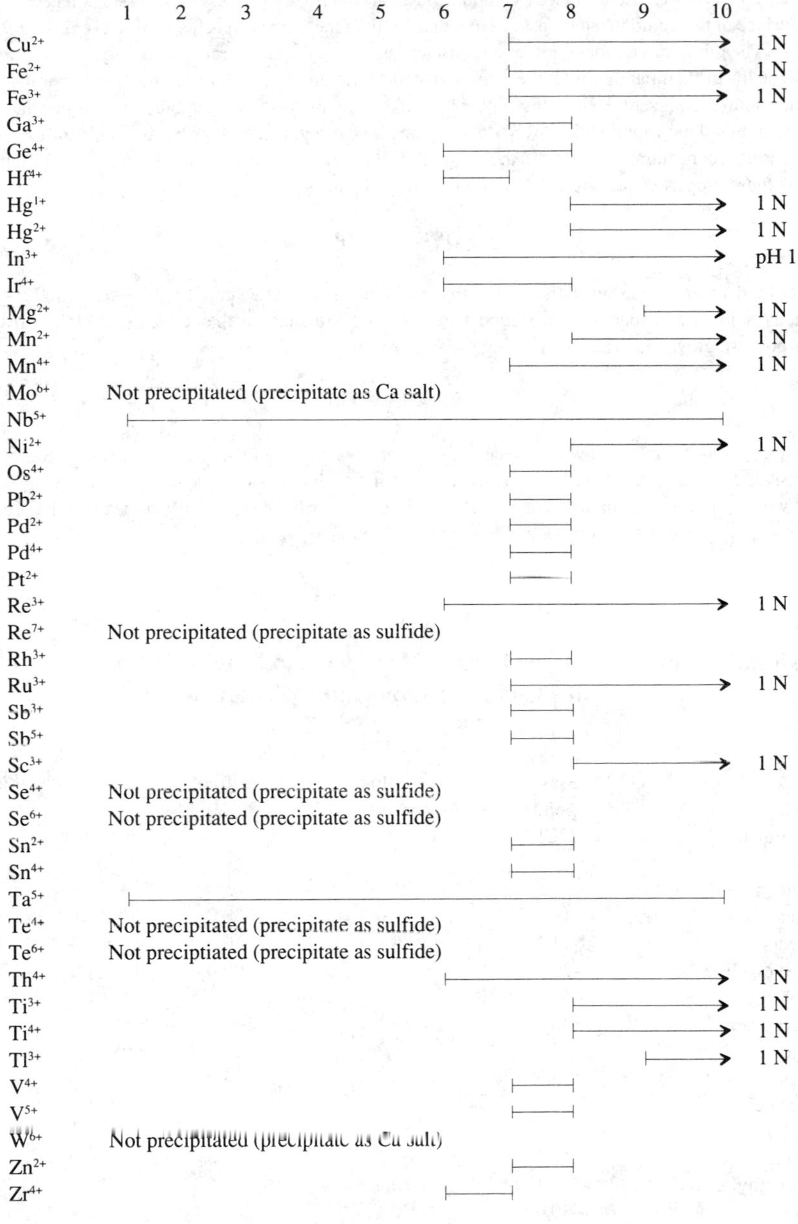

REFERENCES

L. Erdey, *Gravimetric Analysis,* Part II, Pergamon Press, New York, 1965.

D. T. Burns, A. Towsend, and A. H. Carter, *Inorganic Reaction Chemistry,* Vol. 2, Ellis Horwood, New York, 1981.

FIRE HAZARDS

Flammable solvents are a common source of laboratory fires. The relative ease with which some common laboratory solvents can be ignited is indicated by the following properties.

Flash Point — The lowest temperature, as determined by standard tests, at which a liquid emits vapor in sufficient concentration to form an ignitable mixture with air near the surface of the liquid in a test vessel. Note that many of these common chemicals have flash points below room temperature.

Ignition Temperature — The minimum temperature required to initiate self–sustained combustion, regardless of the heat source.

Flammable Limits — The lower flammable limit is the minimum concentration (percent by volume) of a vapor in air below which a flame is not propagated when an ignition source is present. Below this concentration the mixture is too lean to burn. The upper flammable limit is the maximum concentration (percent by volume) of the vapor in air above which a flame is not propagated. Above this concentration the mixture is too rich to burn. The flammable range comprises all concentrations between these two limits. This range becomes wider with increasing temperature and in oxygen–rich atmospheres. Table 13 lists these properties for a few common laboratory chemicals.

GLOVE MATERIALS

It is good safety practice (and mandated in some laboratories) to wear rubber gloves while handling chemicals that can cause injury when in contact with, or absorbed through, the skin. The various common rubbers are not equally resistant to all chemicals. Table 14 provides guidelines for selecting the best, and avoiding the poorest, glove material for handling a given chemical.

RESPIRATORS

In the event of a laboratory accident or spill, it will be necessary for someone to enter the contaminated area for cleanup. If significant quantities of a chemical are spilled, or even minor quantities of a known toxic material, it is essential to wear the correct kind of respirator equipment when entering the area. If it is not known whether the contamination is of a chemical "Immediately dangerous to life or health", the prudent course is to assume that it is, and to use the corresponding type of respirator. Guidelines are presented in Table 15.

TABLE 13
Flash Points, Boiling Points, Ignition Temperatures, and Flammable Limits of Some Common Laboratory Chemicals

Chemical	Flash point (°C)	Boiling point (°C)	Ignition temp. (°C)	Flammable limit (percent by volume in air)	
				Lower	Upper
Acetaldehyde	–37.8	21.1	175.0	4.0	60.0
Acetone	–19.0	56.0	538.0	2.6	12.8
Benzene	–11.1	80.1	560.0	1.4	8.0
Carbon disulfide	–30.0	45.8	90.0	1.0	44.0
Cyclohexane	–18.0	80.7	260.0	1.3	8.0
Diethyl ether	–45.0	34.4	160.0	1.8	48.0
Ethanol	12.0	78.3	363.0	3.3	19.0
n–Heptane	–3.9	98.4	204.0	1.0	6.7
n–Hexane	–21.7	68.7	223.0	1.2	7.5
Isopropyl alcohol	11.7	82.2	398.9	2.0	12.0
Methanol	11.1	64.5	385.0	6.0	36.5
Methyl ethyl ketone	–6.1	79.6	515.6	1.9	11.0
Pentane	–40.0	36.1	260.0	1.4	7.8
Styrene	31.0	145.0	490.0	1.1	6.1
Toluene	4.4	110.6	530.0	1.3	7.0
p–Xylene	25.0	132.4	529.0	1.1	7.0

Note: For a more extensive listing, see the table "Properties of Common Solvents" in Section 15.

TABLE 14
Resistance to Chemicals of Common Glove Materials
(E = Excellent, G = Good, F = Fair, P = Poor)

Chemical	Natural rubber	Neoprene	Nitrile	Vinyl
Acetaldehyde	G	G	E	G
Acetic acid	E	E	E	E
Acetone	G	G	G	F
Acrylonitrile	P	G	—	F
Ammonium hydroxide (sat)	G	E	E	E
Aniline	F	G	E	G
Benzaldehyde	F	F	E	G
Benzene[a]	P	F	G	F
Benzyl chloride[a]	F	P	G	P
Bromine	G	G	—	G
Butane	P	E	—	P
Butyraldehyde	P	G	—	G
Calcium hypochlorite	P	G	G	G
Carbon disulfide	P	P	G	F
Carbon tetrachloride[a]	P	F	G	F
Chlorine	G	G	—	G
Chloroacetone	F	E	—	P
Chloroform[a]	P	F	G	P
Chromic acid	P	F	F	E
Cyclohexane	F	E	—	P
Dibenzyl ether	F	G	—	P
Dibutyl phthalate	F	G	—	P
Diethanolamine	F	E	—	E
Diethyl ether	F	G	E	P
Dimethyl sulfoxide[b]	—	—	—	—
Ethyl acetate	F	G	G	F
Ethylene dichloride[a]	P	F	G	
Ethylene glycol	G	G	E	E
Ethylene trichloride[a]	P	P	—	P
Fluorine	G	G	—	G
Formaldehyde	G	E	E	E
Formic acid	G	E	E	E
Glycerol	G	G	E	E
Hexane	P	E	—	P
Hydrobromic acid (40%)	G	E	—	E
Hydrochloric acid (conc)	G	G	G	E
Hydrofluoric acid (30%)	G	G	G	E
Hydrogen peroxide	G	G	G	E
Iodine	G	G	—	G
Methylamine	G	G	E	E
Methyl cellosolve	F	E	—	P
Methyl chloride[a]	P	E	—	P
Methyl ethyl ketone	F	G	G	P
Methylene chloride[a]	F	F	G	F
Monoethanolamine	F	E	—	E
Morpholine	F	E	—	E
Naphthalene[a]	G	G	E	G
Nitric acid (conc)	P	P	P	G
Perchloric acid	F	G	F	E
Phenol	G	E	—	E
Phosphoric acid	G	E	—	E
Potassium hydroxide (sat)	G	G	G	E

TABLE 14
Resistance to Chemicals of Common Glove Materials
(E = Excellent, G = Good, F = Fair, P = Poor) (continued)

Chemical	Natural rubber	Neoprene	Nitrile	Vinyl
Propylene dichloride[a]	P	F	—	P
Sodium hydroxide	G	G	G	E
Sodium hypochlorite	G	P	F	G
Sulfuric acid (conc)	G	G	F	G
Toluene[a]	P	F	G	F
Trichloroethylene[a]	P	F	G	F
Tricresyl phosphate	P	F	—	F
Triethanolamine	F	E	E	E
Trinitrotoluene	P	E	—	P

[a] Aromatic and halogenated hydrocarbons will attack all types of natural and synthetic glove materials. Should swelling occur, the user should change to fresh gloves and allow the swollen gloves to dry and return to normal.

[b] No data on the resistance to dimethyl sulfoxide of natural rubber, neoprene, nitrile rubber, or vinyl materials are available; the manufacturer of the substance recommends the use of butyl rubber gloves.

TABLE 15
Guide for Selection of Respirators

Type of hazard	Type of respirator
Oxygen deficiency	Self–contained breathing apparatus Hose mask with blower Combination of air–line respirator and auxiliary self–contained air supply or air–storage receiver with alarm
Gas and vapor contaminants Immediately dangerous to life or health	Self–contained breathing apparatus Hose mask with blower Air–purifying full–facepiece respirator with chemical canister (gas mask) Self–rescue mouthpiece respirator (for escape only) Combination of air–line respirator and auxiliary self–contained air supply or air–storage receiver with alarm
Not immediately dangerous to life or health	Air–line respirator Hose mask with blower Air–purifying half–mask or mouthpiece respirator with chemical cartridge
Particulate Contaminants Immediately dangerous to life or health	Self–contained breathing apparatus Hose mask with blower Air–purifying full–facepiece respirator with appropriate filter Self-rescue mouthpiece respirator (for escape only) Combination of air–line respirator and auxiliary self–contained air supply or air–storage receiver with alarm

TABLE 15
Guide for Selection of Respirators (continued)

Type of hazard	Type of respirator
Not immediately dangerous to life or health	Air–purifying half–mask or mouthpiece respirator with filter pad or cartridge Air–line respirator Air–line abrasive–blasting respirator Hose mask with blower
Combination of gas, vapor, and particulate contaminants Immediately dangerous to life or health	Self–contained breathing apparatus Hose mask with blower Air–purifying full–facepiece respirator with chemical canister and appropriate filter (gas mask with filter) Self–rescue mouthpiece respirator (for escape only) Combination of air–line respirator and auxiliary self–contained air supply or air–storage receiver with alarm
Not immediately dangerous to life or health	Air–line respirator Hose mask without blower Air–purifying half–mask or mouthpiece respirator with chemical cartridge and appropriate filter

Source: ANSI Standard Z88.2 (1969).

THRESHOLD LIMIT VALUES FOR AIRBORNE CONTAMINANTS

This table gives threshold limits for selected airborne contaminants in the workplace. The data are derived from a more extensive list in the publication:

Threshold Limit Values and Biological Exposure Indices for 1994-1995
American Conference of Governmental Industrial Hygienists
6500 Glenway Ave., Bldg. D-7
Cincinnati, OH 45211-4438

Threshold limit values are airborne concentrations under which it is believed that nearly all workers may be repeatedly exposed day after day without adverse effect. However, it should be noted that a small percentage of individuals may be affected at concentrations at or below these limits because of unusual susceptibility or pre-existing conditions. The publication cited above should be consulted for full details.

In this table the threshold limit valve (TLV) is given in parts per million (ppm) by volume in air at 25°C and 101.325 kPa (1 atmosphere). For most substances the quantity tabulated is the time weighted average (TWA) for a normal 8-hr workday and 40-hr workweek. However, a C following a TLV entry indicates a ceiling valve, which should not be exceeded even for short periods.

Explanation of abbreviations and notes:

Carcinogen: indicates that the compound is a suspected or confirmed human carcinogen

Asphyxiant: indicates a compound that may not have significant physiological effects but is an asphyxiant when present in high concentrations. Fire and explosion hazards may also exist for these compounds.

CAS RN: Chemical Abstracts Serviced Registry Number.

C: indicates a ceiling value.

Compounds are listed by systematic name. A synonym index follows the table

Substance	Molecular formula	CAS RN	TLV-TWA ppm		Note
Acetaldehyde	C_2H_4O	75-07-0	100		carcinogen
Acetic acid	$C_2H_4O_2$	64-19-7	10		
Acetic anhydride	$C_4H_6O_3$	108-24-7	5		
Acetone	C_3H_6O	67-64-1	750		
Acetonitrile	C_2H_3N	75-05-8	40		
Acetylene	C_2H_2	74-86-2			asphyxiant
Allyl alcohol	C_3H_6O	107-18-6	2		
Ammonia	H_3N	7664-41-7	25		
Anisidine (*o*, *p*)	C_7H_9NO	29191-52-4	0.1		carcinogen
Aniline	C_6H_7N	62-53-3	2		carcinogen
Argon	Ar	7440-37-1			asphyxiant
Arsine	AsH_3	7784-42-1	0.05		
Benzene	C_6H_6	71-43-2	10		carcinogen
Benzenethiol	C_6H_6S	108-98-5	0.5		
p-Benzoquinone	$C_6H_4O_2$	106-51-4	0.1		
Benzyl chloride	C_7H_7Cl	100-44-7	1		carcinogen
Biphenyl	$C_{12}H_{10}$	92-52-4	0.2		
Bis(2-amimoethyl)amine	$C_4H_{13}N_3$	111-40-0	1		
Bis(2-chloroethyl) ether	$C_4H_8Cl_2O$	111-44-4	5		
Bis(chloromethyl) ether	$C_2H_4Cl_2O$	542-88-1	0.001		carcinogen
Boron tribromide	BBr_3	10294-33-4	1	C	
Boron trifluoride	BF_3	7637-07-2	1	C	
Bromine	Br_2	7726-95-6	0.1		
Bromine pentafluoride	BrF_5	7789-30-2	0.1		
Bromochloromethane	CH_2BrCl	74-97-5	200		
Bromoethane	C_2H_5Br	74-96-4	5		carcinogen
Bromomethane	CH_3Br	74-83-9	5		carcinogen
1,3-Butadiene	C_4H_6	106-99-0	2		carcinogen
Butane	C_4H_{10}	106-97-8	800		
1-Butanethiol	$C_4H_{10}S$	109-79-5	0.5		
1-Butanol	$C_4H_{10}O$	71-36-3	50	C	
2-Butanol	$C_4H_{10}O$	78-92-2	100		
2-Butanone	C_4H_8O	78-93-3	200		
2-Butoxyethanol	$C_6H_{14}O_2$	111-76-2	25		

Substance	Molecular formula	CAS RN	TLV-TWA ppm		Note
Butyl acetate	$C_6H_{12}O_2$	123-86-4	150		
sec-Butyl acetate	$C_6H_{12}O_2$	105-46-4	200		
tert-Butyl acetate	$C_6H_{12}O_2$	540-88-5	200		
Butyl acrylate	$C_7H_{12}O_2$	141-32-2	10		
Butylamine	$C_4H_{11}N$	109-73-9	5	C	
p-tert-Butyltoluene	$C_{11}H_{16}$	98-51-1	10		
Camphor	$C_{10}H_{16}O$	76-22-2	2		
Caprolactam	$C_6H_{11}NO$	105-60-2	5		
Carbon dioxide	CO_2	124-38-9	5000		
Carbon disulfide	CS_2	75-15-0	10		
Carbon monoxide	CO	630-08-0	25		
Carbonyl chloride	CCl_2O	75-44-5	0.1		
Carbonyl fluoride	CF_2O	353-50-4	2		
Chlorine	Cl_2	7782-50-5	0.5		
Chlorine dioxide	ClO_2	10049-04-4	0.1		
Chlorine trifluoride	ClF_3	7790-91-2	0.1	C	
Chloroacetaldehyde	C_2H_3ClO	107-20-0	1	C	
Chloroacetyl chloride	$C_2H_2Cl_2O$	79-04-9	0.05		
Chlorobenzene	C_6H_5Cl	108-90-7	10		
2-Chloro-1,3-butadiene	C_4H_5Cl	126-99-8	10		carcinogen
Chlorodifluoromethane	$CHClF_2$	75-45-6	1000		
Chloroethane	C_2H_5Cl	75-00-3	1000		
2-Chloroethanol	C_2H_5ClO	107-07-3	1	C	
Chloroethylene	C_2H_3Cl	75-01-4	5		carcinogen
Chloromethane	CH_3Cl	74-87-3	50		carcinogen
Chloropentafluoroethane	C_2ClF_5	76-15-3	1000		
3-Chloropropene	C_3H_5Cl	107-05-1	1		carcinogen
o-Chlorotoluene	C_7H_7Cl	95-49-8	50		
Chromyl chloride (CrO_2Cl_2)	Cl_2CrO_2	14977-61-8	0.025		carcinogen
Chrysene	$C_{18}H_{12}$	218-01-9			carcinogen
o-Cresol	C_7H_8O	95-48-7	5		
m-Cresol	C_7H_8O	108-39-4	5		
p-Cresol	C_7H_8O	106-44-5	5		
Cumene	C_9H_{12}	98-82-8	50		
Cyanogen	C_2N_2	460-19-5	10		
Cyanogen chloride	$CClN$	506-77-4	0.3	C	
Cyclohexane	C_6H_{12}	110-82-7	300		
Cyclohexanol	$C_6H_{12}O$	108-93-0	50		
Cyclohexanone	$C_6H_{10}O$	108-94-1	25		
Cyclohexene	C_6H_{10}	110-83-8	300		
Cyclohexylamine	$C_6H_{13}N$	108-91-8	10		
1,3-Cyclopentadiene	C_5H_6	542-92-7	75		
Cyclopentane	C_5H_{10}	287-92-3	600		
Diacetone alcohol	$C_6H_{12}O_2$	123-42-2	50		
Diazomethane	CH_2N_2	334-88-3	0.2		carcinogen
Diborane	B_2H_6	19287-45-7	0.1		
Dibromodifluoromethane	CBr_2F_2	75-61-6	100		
1,2-Dibromoethane	$C_2H_4Br_2$	106-93-4			carcinogen
Dichloroacetylene	C_2Cl_2	7572-29-4	0.1	C	carcinogen
o-Dichlorobenzene	$C_6H_4Cl_2$	95-50-1	25	C	
p-Dichlorobenzene	$C_6H_4Cl_2$	106-46-7	10		carcinogen
Dichlorodifluoromethane	CCl_2F_2	75-71-8	1000		
1,1-Dichloroethane	$C_2H_4Cl_2$	75-34-3	100		
1,2-Dichloroethane	$C_2H_4Cl_2$	107-06-2	10		carcinogen
1,1-Dichloroethylene	$C_2H_2Cl_2$	75-35-4	5		carcinogen
cis-1,2-Dichloroethylene	$C_2H_2Cl_2$	156-59-2	200		
trans-1,2-Dichloroethylene	$C_2H_2Cl_2$	156-60-5	200		

Substance	Molecular formula	CAS RN	TLV-TWA ppm		Note
Dichlorofluoromethane	$CHCl_2F$	75-43-4	10		
Dichloromethane	CH_2Cl_2	75-09-2	50		carcinogen
1,2-Dichloropropane	$C_3H_6Cl_2$	78-87-5	75		carcinogen
1,3-Dichloropropene	$C_3H_4Cl_2$	542-75-6	1		carcinogen
1,2-Dichlorotetrafluoroethane	$C_2Cl_2F_4$	76-14-2	1000		
Diethanolamine	$C_4H_{11}NO_2$	111-42-2	0.46		
Diethylamine	$C_4H_{11}N$	109-89-7	5		
Diethyl ether	$C_4H_{10}O$	60-29-7	400		
Diisopropylamine	$C_6H_{15}N$	108-18-9	5		
Dimethoxymethane	$C_3H_8O_2$	109-87-5	1000		
N,*N*-Dimethylacetamide	C_4H_9NO	127-19-5	10		
Dimethylamine	C_2H_7N	124-40-3	5		
N,*N*-Dimethylaniline	$C_8H_{11}N$	121-69-7	5		
N,*N*-Dimethylformamide	C_3H_7NO	68-12-2	10		
2,6-Dimethyl-4-heptanone	$C_9H_{18}O$	108-83-8	25		
1,1-Dimethylhydrazine	$C_2H_8N_2$	57-14-7	0.5		carcinogen
1,4-Dioxane	$C_4H_8O_2$	123-91-1	25		carcinogen
Epichlorohydrin	C_3H_5ClO	106-89-8	2		carcinogen
Ethane	C_2H_6	74-84-0			asphyxiant
1,2-Ethanediamine	$C_2H_8N_2$	107-15-3	10		
Ethanethiol	C_2H_6S	75-08-1	0.5		
Ethanol	C_2H_6O	64-17-5	1000		
Ethanolamine	C_2H_7NO	141-43-5	3		
2-Ethoxyethanol	$C_4H_{10}O_2$	110-80-5	5		
Ethyl acetate	$C_4H_8O_2$	141-78-6	400		
Ethyl acrylate	$C_5H_8O_2$	140-88-5	5		carcinogen
Ethylamine	C_2H_7N	75-04-7	5		
Ethylbenzene	C_8H_{10}	100-41-4	100		
Ethylene	C_2H_4	74-85-1			asphyxiant
Ethylene glycol	$C_2H_6O_2$	107-21-1	50	C	
Ethyleneimine	C_2H_5N	151-56-4	0.05		carcinogen
Ethylene oxide	C_2H_4O	75-21-8	1		carcinogen
Ethyl formate	$C_3H_6O_2$	109-94-4	100		
Fluorine	F_2	7782-41-4	1		
Formaldehyde	CH_2O	50-00-0	0.3	C	carcinogen
Formamide	CH_3NO	75-12-7	10		
Formic acid	CH_2O_2	64-18-6	5		
Furfural	$C_5H_4O_2$	98-01-1	2		
Furfuryl alcohol	$C_5H_6O_2$	98-00-0	10		
Heptane	C_7H_{16}	142-82-5	400		
2-Heptanone	$C_7H_{14}O$	110-43-0	50		
Hexachloro-1,3-butadiene	C_4Cl_6	87-68-3	0.02		carcinogen
Hexachloroethane	C_2Cl_6	67-72-1	1		carcinogen
Hexane	C_6H_{14}	110-54-3	50		
2-Hexanone	$C_6H_{12}O$	591-78-6	5		
Hexylene glycol	$C_6H_{14}O_2$	107-41-5	25		
Hydrazine	H_4N_2	302-01-2	0.01		carcinogen
Hydrogen bromide	BrH	10035-10-6	3	C	
Hydrogen chloride	ClH	7647-01-0	5	C	
Hydrogen cyanide	CHN	74-90-8	4.7	C	
Hydrogen fluoride	FH	7664-39-3	3	C	
Hydrogen peroxide	H_2O_2	7722-84-1	1		
Hydrogen selenide	H_2Se	7783-07-5	0.05		
Hydrogen sulfide	H_2S	7783-06-4	10		
Indene	C_9H_8	95-13-6	10		
Iodine	I_2	7553-56-2	0.1	C	
Iodomethane	CH_3I	74-88-4	2		carcinogen

Substance	Molecular formula	CAS RN	TLV-TWA ppm		Note
Isobutyl acetate	$C_6H_{12}O_2$	110-19-0	150		
Isopentyl acetate	$C_7H_{14}O_2$	123-92-2	100		
Isophorone	$C_9H_{14}O$	78-59-1	5		
Isopropyl acetate	$C_5H_{10}O_2$	108-21-4	250		
Isopropylamine	C_3H_9N	75-31-0	5		
Ketene	C_2H_2O	463-51-4	0.5		
Maleic anhydride	$C_4H_2O_3$	108-31-6	0.25		
Mesitylene	C_9H_{12}	108-67-8	25		
Mesityl oxide	$C_6H_{10}O$	141-79-7	15		
Methacrylic acid	$C_4H_6O_2$	79-41-4	20		
Methane	CH_4	74-82-8			asphyxiant
Methanethiol	CH_4S	74-93-1	0.5		
Methanol	CH_4O	67-56-1	200		
2-Methoxyethanol	$C_3H_8O_2$	109-86-4	5		
2-Methoxyethyl acetate	$C_5H_{10}O_3$	110-49-6	5		
Methyl acetate	$C_3H_6O_2$	79-20-9	200		
Methyl acrylate	$C_4H_6O_2$	96-33-3	10		
Methylacrylonitrile	C_4H_5N	126-98-7	1		
Methylamine	CH_5N	74-89-5	5		
o-Methylaniline	C_7H_9N	95-53-4	2		carcinogen
m-Methylaniline	C_7H_9N	108-44-1	2		
p-Methylaniline	C_7H_9N	106-49-0	2		carcinogen
N-Methylaniline	C_7H_9N	100-61-8	0.5		
Methylcyclohexane	C_7H_{14}	108-87-2	400		
Methyl formate	$C_2H_4O_2$	107-31-3	100		
Methylhydrazine	CH_6N_2	60-34-4	0.2	C	carcinogen
Methylisocyanate	C_2H_3NO	624-83-9	0.02		
Methyl methacrylate	$C_5H_8O_2$	80-62-6	100		
Methyloxirane	C_3H_6O	75-56-9	20		carcinogen
4-Methyl-2-pentanone	$C_6H_{12}O$	108-10-1	50		
2-Methyl-2-propanol	$C_4H_{10}O$	75-65-0	100		
2-Methyl-1-propanol	$C_4H_{10}O$	78-83-1	50		
o-Methylstyrene	C_9H_{10}	611-15-4	50		
m-Methylstyrene	C_9H_{10}	100-80-1	50		
p-Methylstyrene	C_9H_{10}	622-97-9	50		
Morpholine	C_4H_9NO	110-91-8	20		
Naphthalene	$C_{10}H_8$	91-20-3	10		
Nickel carbonyl ($Ni(CO)_4$)	C_4NiO_4	13463-39-3	0.05		carcinogen
Nitric oxide	NO	10102-43-9	25		
Nitric acid	HNO_3	7697-37-2	2		
Nitrobenzene	$C_6H_5NO_2$	98-95-3	1		
Nitroethane	$C_2H_5NO_2$	79-24-3	100		
Nitrogen dioxide	NO_2	10102-44-0	3		
Nitrogen trifluoride	F_3N	7783-54-2	10		
Nitromethane	CH_3NO_2	75-52-5	20		
2-Nitropropane	$C_3H_7NO_2$	79-46-9	10		carcinogen
o-Nitrotoluene	$C_7H_7NO_2$	88-72-2	2		
m-Nitrotoluene	$C_7H_7NO_2$	99-08-1	2		
p-Nitrotoluene	$C_7H_7NO_2$	99-99-0	2		
Nonane	C_9H_{20}	111-84-2	200		
Octane	C_8H_{18}	111-65-9	300		
2-Oxetanone	$C_3H_4O_2$	57-57-8	0.05		carcinogen
Oxygen difluoride	F_2O	7783-41-7	0.05	C	
Ozone	O_3	10028-15-6	0.1	C	
Pentaborane	B_5H_9	19624-22-7	0.005		
Pentanal	$C_5H_{10}O$	110-62-3	50		
Pentane	C_5H_{12}	109-66-0	600		
2-Pentanone	$C_5H_{10}O$	107-87-9	200		
3-Pentanone	$C_5H_{10}O$	96-22-0	200		

THRESHOLD LIMIT VALUES FOR AIRBORNE CONTAMINANTS (continued)

Substance	Molecular formula	CAS RN	TLV-TWA ppm		Note
Pentyl acetate	$C_7H_{14}O_2$	628-63-7	100		
Perchloryl fluoride	$ClFO_3$	7616-94-6	3		
Perfluoroacetone	C_3F_6O	684-16-2	0.1		
Phenol	C_6H_6O	108-95-2	5		
Phenylhydrazine	$C_6H_8N_2$	100-63-0	0.1		carcinogen
Phosphine	H_3P	7803-51-2	0.3		
Phosphorus pentachloride	Cl_5P	10026-13-8	0.1		
Phosphorus trichloride	Cl_3P	7719-12-2	0.2		
Phosphoryl chloride	Cl_3OP	10025-87-3	0.1		
Phthalic anhydride	$C_8H_4O_3$	85-44-9	1		
Propane	C_3H_8	74-98-6			asphyxiant
Propanoic acid	$C_3H_6O_2$	79-09-4	10		
1-Propanol	C_3H_8O	71-23-8	200		
2-Propanol	C_3H_8O	67-63-0	400		
Propargyl alcohol	C_3H_4O	107-19-7	1		
Propenal	C_3H_4O	107-02-8	0.1		
Propenamide	C_3H_5NO	79-06-1			carcinogen
Propenenitrile	C_3H_3N	107-13-1	2		carcinogen
Propenoic acid	$C_3H_4O_2$	79-10-7	2		
Propyl acetate	$C_5H_{10}O_2$	109-60-4	200		
Propyleneimine	C_3H_7N	75-55-8	2		carcinogen
Propyne	C_3H_4	74-99-7	1000		
Pyridine	C_5H_5N	110-86-1	5		
Selenium fluoride (SeF_6)	F_6Se	7783-79-1	0.05		
Silane	H_4Si	7803-62-5	5		
Stibine	H_3Sb	7803-52-3	0.1		
Styrene	C_8H_8	100-42-5	50		carcinogen
Sulfur dioxide	O_2S	7446-09-5	2		
Sulfur fluoride (SF_4)	F_4S	7783-60-0	0.1	C	
Sulfur fluoride (SF_5)	F_5S	10546-01-7	0.01	C	
Sulfur fluoride (SF_6)	F_6S	2551-62-4	1000		
Sulfuryl fluoride	F_2O_2S	2699-79-8	5		
Tetrabromomethane	CBr_4	558-13-4	0.1		
Tetrachloro-1,2-difluoroethane	$C_2Cl_4F_2$	76-12-0	500		
1,1,2,2-Tetrachloroethane	$C_2H_2Cl_4$	79-34-5	1		carcinogen
Tetrachloroethylene	C_2Cl_4	127-18-4	50		carcinogen
Tetrachloromethane	CCl_4	56-23-5	5		carcinogen
Tetrahydrofuran	C_4H_8O	109-99-9	200		
Thionyl chloride	Cl_2OS	7719-09-7	1	C	
Toluene	C_7H_8	108-88-3	50		
Toluene-2,4-diisocyanate	$C_9H_6N_2O_2$	584-84-9	0.005		carcinogen
Tribromomethane	$CHBr_3$	75-25-2	0.5		
1,2,4-Trichlorobenzene	$C_6H_3Cl_3$	120-82-1	5	C	
1,1,1-Trichloroethane	$C_2H_3Cl_3$	71-55-6	350		
1,1,2-Trichloroethane	$C_2H_3Cl_3$	79-00-5	10		carcinogen
Trichloroethylene	C_2HCl_3	79-01-6	50		carcinogen
Trichlorofluoromethane	CCl_3F	75-69-4	1000	C	
Trichloromethane	$CHCl_3$	67-66-3	10		carcinogen
1,1,2-Trichlorotrifluoroethane	$C_2Cl_3F_3$	76-13-1	1000		
Triethanolamine	$C_6H_{15}NO_3$	102-71-6	0.5		
Triethylamine	$C_6H_{15}N$	121-44-8	1		
Triiodomethane	CHI_3	75-47-8	0.6		
Trimethylamine	C_3H_9N	75-50-3	5		
1,2,4-Trimethylbenzene	C_9H_{12}	95-63-6	25		
Vinyl acetate	$C_4H_6O_2$	108-05-4	10		carcinogen
o-Xylene	C_8H_{10}	95-47-6	100		
m-Xylene	C_8H_{10}	108-38-3	100		
p-Xylene	C_8H_{10}	106-42-3	100		

THRESHOLD LIMIT VALUES FOR AIRBORNE CONTAMINANTS (continued)

SYNONYM INDEX

Acetyl acetate: see Acetic anhydride
Acetylene tetrachloride: see 1,1,2,2-Tetrachloroethane
Acrolein: see Propenal
Acrylamide: see Propenamide
Acrylic acid: see Propenoic acid
Acrylonitrile: see Propenenitrile
Allyl chloride: see 3-Chloropropene
2-Aminoethanol: see Ethanolamine
2-Aminopropane: see Isopropylamine
Amyl acetate: see Pentyl acetate
Azine: see Pyridine
Aziridine: see Ethyleneimine
Benzenamine: see Aniline
1,2-Benzenedicarboxylic anhydride: see Phthalic anhydride
1,2-Benzophenanthrene: see Chrysene
Bis(2-hydroxyethyl)amine: see Diethanolamine
Bromoform: see Tribromomethane
cis-Butenedioic anhydride: see Maleic anhydride
Butyl alcohol: see 1-Butanol
sec-Butyl alcohol: see 2-Butanol
tert-Butyl alcohol: see 2-Methyl-2-propanol
Butyl mercaptan: see 1-Butanethiol
Butyl methyl ketone: see 2-Hexanone
Butyl-2-propenoate: see Butyl acrylate
Carbon tetrabromide: see Tetrabromomethane
Carbon tetrachloride: see Tetrachloromethane
Chloroform: see Trichloromethane
(Chloromethyl)benzene: see Benzyl chloride
(Chloromethyl)oxirane: see Epichlorohydrin
Cyclohexyl alcohol: see Cyclohexanol
Dichloroethyl ether: see Bis(2-chloroethyl) ether
Diethylenetriamine: see Bis(2-amimoethyl)amine
Diethyl ketone: see 3-Pentanone
Diisobytyl ketone: see 2,6-Dimethyl-4-heptanone
2,4-Diisocyanatotoluene: see Toluene-2,4-diisocyanate
1,2-Dimethylbenzene: see *o*-Xylene
1,3-Dimethylbenzene: see *m*-Xylene
1,4-Dimethylbenzene: see *p*-Xylene
Diphenyl: see Biphenyl
Ethanal: see Acetaldehyde
Ethanedinitrile: see Cyanogen
1,2-Ethanediol: see Ethylene glycol
Ethanenitrile: see Acetonitrile
Ethene: see Ethylene
Ethenone: see Ketene
Ethyl alcohol: see Ethanol
Ethyl bromide: see Bromoethane
Ethyl chloride: see Chloroethane
Ethylene chlorohydrin: see 2-Chloroethanol
Ethylenediamine: see 1,2-Ethanediamine
Ethylene dibromide: see 1,2-Dibromoethane
Ethylene dichloride: see 1,2-Dichloroethane
Ethylene glycol monobutyl ether: see 2-Butoxyethanol
Ethylene glycol monoethyl ether: see 2-Ethoxyethanol
Ethylene glycol monomethyl ether: see 2-Methoxyethanol
Ethyl ether: see Diethyl ether
Ethylidene dichloride: see 1,1-Dichloroethane
Ethyl mercaptan: see Ethanethiol
Ethyl propenoate: see Ethyl acrylate
Ethyne: see Acetylene
Fluorine oxide (F_2O): see Oxygen difluoride
2-Furaldehyde: see Furfural
2-Furanmethanol: see Furfuryl alcohol
Hexafluoro-2-propanone: see Perfluoroacetone
Hexahydro-2-azepinone: see Caprolactam
Hydrobromic acid: see Hydrogen bromide
Hydrochloric acid: see Hydrogen chloride
Hydrocyanic acid: see Hydrogen cyanide
Hydrofluoric acid: see Hydrogen fluoride
Hydroxybenzene: see Phenol
4-Hydroxy-4-methyl-2-pentanone: see Diacetone alcohol
Iodoform: see Triiodomethane
Isoamyl acetate: see Isopentyl acetate
Isobutyl alcohol: see 2-Methyl-1-propanol
Isobutyl methyl ketone: see 4-Methyl-2-pentanone
Isopropyl alcohol: see 2-Propanol
Isopropylbenzene: see Cumene
Methanal: see Formaldehyde
Methanamide: see Formamide
2-Methoxyethyl ethanoate: see 2-Methoxyethyl acetate
Methylacetylene: see Propyne
Methylal: see Dimethoxymethane
Methyl alcohol: see Methanol
2-Methylaziridine: see Propyleneimine
Methylbenzene: see Toluene
Methyl bromide: see Bromomethane
Methyl chloride: see Chloromethane
Methyl chloroform: see 1,1,1-Trichloroethane
Methylene chloride: see Dichloromethane
Methyl ethyl ketone: see 2-Butanone
Methyl iodide: see Iodomethane
Methyl mercaptan: see Methanethiol
Methyl 2-methylpropenoate: see Methyl methacrylate
2-Methyl-2,4-pentanediol: see Hexylene glycol
4-Methyl-3-penten-2-one: see Mesityl oxide
Methyl pentyl ketone: see 2-Heptanone
2-Methylphenol: see *o*-Cresol
3-Methylphenol: see *m*-Cresol
4-Methylphenol: see *p*-Cresol
2-Methylpropenenitrile: see Methylacrylonitrile
Methyl propenoate: see Methyl acrylate
2-Methylpropenoic acid: see Methacrylic acid
Methyl propyl ketone: see 2-Pentanone
Nitrogen monoxide: see Nitric oxide
Oxirane: see Ethylene oxide
Oxolane: see Tetrahydrofuran
Phenyl chloride: see Chlorobenzene
Phenyl mercaptan: see Benzenethiol
Phosgene: see Carbonyl chloride
Phosphorus oxychloride: see Phosphoryl chloride
2-Propanone: see Acetone
1-Propen-3-ol: see Allyl alcohol
β-Propiolactone: see 2-Oxetanone
Propionic acid: see Propanoic acid
Propyl alcohol: see 1-Propanol
Propylene dichloride: see 1,2-Dichloropropane

SYNONYM INDEX (continued)

1,2-Propylene oxide: see Methyloxirane
1-Propyne-3-ol: see Propargyl alcohol
Pseudocumene: see 1,2,4-Trimethylbenzene
Sulfinyl dichloride: see Thionyl chloride
Sulfonyl difluoride: see Sulfuryl fluoride
Sulfur hexafluoride: see Sulfur fluoride (SF_6)
Tetrahydro-1,4-oxazine: see Morpholine
o-Toluidine: see *o*-Methyl aniline
m-Toluidine: see *m*-Methyl aniline
p-Toluidine: see *p*-Methylaniline
o-Tolyl chloride: see *o*-Chlorotoluene
Trichloroethene: see Trichloroethylene
1,3,5-Trimethylbenzene: see Mesitylene
3,3,5-Trimethyl-2-cyclohexen-1-one: see Isophorone
Tris(2-hydroxyethyl)amine: see Triethanolamine
Valeraldehyde: see Pentanal
Vinylbenzene: see Styrene
Vinyl chloride: see Chloroethylene
Vinylidene chloride: see 1,1-Dichloroethylene
2-Vinyltoluene: see *o*-Methylstyrene
3-Vinyltoluene: see *m*-Methylstyrene
4-Vinyltoluene: see *p*-Methylstyrene

UNDERWRITERS' LABORATORIES' CLASSIFICATION OF COMPARATIVE LIFEHAZARD OF GASES AND VAPORS

(Group number definition)

Group	Definition	Examples
1	Gases or vapors which in concentrations of the order of 1/2 to 1% for durations of exposure of the order of 5 min are lethal or produce serious injury.	Sulfur dioxide
2	Gases or vapors which in concentrations of the order of 1/2 to 1% for duration of exposure of the order of 1/2 h are lethal or produce serious injury	Ammonia, methyl bromide
3	Gases or vapors which in concentations of the order of 2 to 2 1/2% for durations of exposure of the order of 1 hr are lethal or produce serious injury.	Bromochloromethane carbon tetrachloride, chloroform, methyl formate
4	Gases or vapors which in concentrations of the order of 2 to 2 1/2% for durations of exposure of the order of 2 h are lethal or produce seirous injury	Dichloroethylene methyl chloride, ethylbromide
Between 4 and 5	Appear to classify as somewhat less toxic than Group 4.	Methylene chloride, ethyl chloride.
	Much less toxic than group 4 but somewhat more toxic than Group 5.	Refrigerant 112[a] Refrigerant 113 Refrigerant 21
5a	Gases or vapors much less toxic than Group 4 but more toxic than Group 6.	Refrigerant 11 Refrigerant 22 Refrigerant 114B2 Refrigerant 502 Carbon dioxide
5b	Gases or vapors which available data indicate would classify as either Group 5a or Group 6.	Ethane, propane, butane
6	Gases or vapors which in concentrations up to at least about 20% by volume for duration of exposure of the order of 2 h do not appear to produce injury.	Refrigerant 13B1 Refrigerant 12 Refrigerant 114 Refrigerant 115 Refrigerant 13[a] Refrigerant 14[a] Refrigerant 23[a] Refrigerant 116[a] Refrigerant C318[a]

[a] Not tested by U.L. but estimated to belong in group indicated.

PROPERTIES OF LARGE PRODUCTION AND PRIORITY ORGANIC POLLUTANTS

This table gives properties of environmental interest for 89 important organic compounds. The properties included are pertinent to prediction of the environmental fate and exposure hazard of these substances. The properties given are:

T_m: melting point
T_b: boiling point
log P: logarithm of octanol-water partition coefficient
pK_a: negative of the logarithm (base 10) of the dissociation constant in aqueous solution (for acids, phenols, amines)
Solubility: in water, at the indicated temperature
Vapor pressure: at 0 and 25° C. An entry of 0 indicates a vapor pressure less than 0.1 kPa. Note that 1 kPa = 0.01 bar ≈ 7.5 mmHg
Henry's Law constant: Ratio of vapor pressure to solubility; therefore, a measure of the partition of the compound between air and water (7E–3 means 7×10^{-3})
Compounds are listed by molecular formula in Hill order.

Molecular formula	Name	CAS RN	Mol. wt.	T_m/°C	T_b/°C	log P	pK_a	Solubility		Vap. Press.(kPa)		Henry's Const. (bar m³/mol)
								Value	T/°C	0° C	25° C	
CCl_4	Carbon tetrachloride	56-23-5	153.92	–23	77	2.83		1.2 g/L	25	4.5	15.2	
CS_2	Carbon disulfide	75-15-0	76.14	–112	46			2.1 g/L	20	16.8	48.0	
$CHCl_3$	Chloroform	67-66-3	119.38	–64	61	1.97		10 g/L	15	7.9	26.3	
CH_2O	Formaldehyde	50-00-0	30.03	–92	–19	0.35		misc.		221.0	519.0	3.3E–7
CH_3Br	Methyl bromide	74-83-9	94.94	–94	4	1.19		17 g/L	20	88.2	218.0	6.3E–3
CH_3Cl	Methyl chloride	74-87-3	50.49	–98	–24	0.91				260.0	575.0	2.4E–2
CH_3I	Methyl iodide	74-88-4	141.94	–66	42	1.91		14 g/L	20	18.6	54.0	
C_2Cl_4	Tetrachloroethylene	127-18-4	165.83	–22	121			0.15 g/L	25	0.6	2.5	
C_2Cl_6	Hexachloroethane	67-72-1	236.74	187	187	3.82		0.05 g/L	22	0.0	0.0	2.8E–3
C_2HCl_3	Trichloroethylene	79-01-6	131.39	–85	87	2.29				2.7	9.8	
$C_2H_2Cl_2$	1,1-Dichloroethylene	75-35-4	96.94	–122	32	2.13		2.4 g/L	25	28.6	80.1	3.0E–2
$C_2H_2Cl_4$	1,1,2,2-Tetrachloroethane	79-34-5	167.85	–43	146	2.39		1.9 g/L	20	0.1	0.7	
C_2H_3Cl	Vinyl chloride	75-01-4	62.50	–154	–13	1.38		2.8 g/L	25	175.0	398.0	1.1E–2
C_2H_3ClO	Acetyl chloride	75-36-5	78.50	–113	51			misc.		12.1	38.4	
$C_2H_3Cl_3$	1,1,1-Trichloroethane	71-55-6	133.40	–30	74	2.49		4.4 g/L	20	4.8	16.5	
$C_2H_4Br_2$	1,2-Dibromoethane	106-93-4	187.86	10	131			4 g/L	20	0.0	1.8	
$C_2H_4Cl_2$	1,2-Dichloroethane	107-06-2	98.96	–36	83	1.48		8.7 g/L	20	2.8	10.6	
$C_2H_4Cl_2O$	Bis(chloromethyl) ether	542-88-1	114.96	–41	106			hydrol.				
C_2H_4O	Acetaldehyde	75-07-0	44.05	–123	20	0.45		misc.		44.3	120.0	
C_2H_4O	Ethylene oxide	75-21-8	44.06	–112	11	–0.30		misc.		66.4	176.0	1.2E–4
C_3H_3N	Acrylonitrile	107-13-1	53.06	–83	77	0.25		75 g/L	25	4.2	14.5	1.1E–4
C_3H_4O	Acrolein	107-02-8	56.06	–88	53	–0.01		208 g/L	20	11.9	36.6	4.4E-06
$C_3H_4O_2$	Acrylic acid	79-10-7	72.06	12	141	0.16	4.25	misc.		0.0	0.5	3.2E-07
C_3H_5ClO	Epichlorohydrin	106-89-8	92.53	–57	116	0.30		66 g/L	20	0.5	2.2	
C_3H_5NO	Acrylamide	79-06-1	71.08	84	193	–0.78		2.1 kg/L	30	0.0	0.0	3.2E-10
C_3H_6O	Allyl alcohol	107-18-6	58.08	–129	97	0.17		misc.		0.7	3.5	4.9E-06
C_3H_6O	1,2-Propylene oxide	75-56-9	58.08	–112	35	0.03		480 g/L	25	24.5	71.2	8.5E–5
C_3H_7NO	*N,N*-Dimethylformamide	68-12-2	73.10	–60	153	–1.01		misc.		0.1	0.5	
C_3H_9N	Trimethylamine	75-50-3	59.11	–117	3	0.16		400 g/L	20	90.1	215.3	
$C_4H_2O_3$	Maleic anhydride	108-31-6	98.06	53	202			hydrol.		0.0	0.0	

PROPERTIES OF LARGE PRODUCTION AND PRIORITY ORGANIC POLLUTANTS (continued)

Molecular formula	Name	CAS RN	Mol. wt.	T_m/°C	T_b/°C	log *P*	pK_a	Solubility Value	Solubility *T*/°C	Vap. Press.(kPa) 0° C	Vap. Press.(kPa) 25° C	Henry's Const. (bar m^3/mol)
$C_4H_4O_4$	Maleic acid	10-16-7	116.07	130		–0.50		788 g/L	25	0.0	0.0	7E-14
C_4H_6	1,3-Butadiene	06-99-0	54.09	–109	–4	1.99		0.73 g/L	25	120.5	281.0	2.57
$C_4H_6O_2$	Vinyl acetate	08-05-3	86.09	–93	72	0.73		20 g/L	20	4.2	15.3	4.9E–4
$C_4H_8Cl_2O$	*Bis*(2-chloroethyl) ether	11-44-4	143.01	–52	178	1.29		1.02 g/L	20	0.0	0.1	2.9E–4
C_4H_8O	Tetrahydrofuran	09-99-9	72.11	–108	65	0.82		300 g/L	25	6.4	21.6	
$C_4H_8O_2$	1,4-Dioxane	23-91-1	88.11	11	101	–0.27		misc.		0.0	5.1	
$C_5H_8O_2$	Methyl methacrylate	80-62-6	100.12	–48	100	1.38		16 g/L	20	1.1	5.1	3.3E–4
C_5H_{12}	*n*-Pentane	09-66-0	72.15	–130	36	3.45		42 mg/L	25	24.3	68.6	
C_6Cl_6	Hexachlorobenzene	18-74-1	284.78	232	325	5.31		5.0 g/L	25	0.0	0.0	1.3E–3
$C_6H_3Cl_3$	1,2,4-Trichlorobenzene	20-82-1	181.45	17	213	4.02		38 mg/L	25	0.0	0.1	1.4E–3
$C_6H_3Cl_3$	1,3,5-Trichlorobenzene	08-70-3	181.45	63	208	4.49		6.0 mg/L	25	0.0	0.0	1.0E–3
$C_6H_3Cl_3O$	2,4,6-Trichlorophenol	88-06-2	197.45	69	246	3.69	7.42	0.9 g/L	25	0.0	0.0	6.2E–8
$C_6H_4ClNO_2$	*o*-Chloronitrobenzene	88-73-3	157.56	32	245	2.24		0.20 g/L	25	0.0	0.0	3.6E–5
$C_6H_4ClNO_2$	*p*-Chloronitrobenzene	00-00-5	157.56	83	242	2.39		0.22 g/L	20	0.0	0.0	3.6E–5
$C_6H_4Cl_2$	*o*-Dichlorobenzene	95-50-1	147.00	–17	180	3.38		0.15 g/L	25	0.0	0.2	1.2E–3
$C_6H_4Cl_2$	*m*-Dichlorobenzene	41-73-1	147.00	–25	173	3.60		0.11 g/L	25	0.0	0.3	1.8E–3
$C_6H_4Cl_2$	*p*-Dichlorobenzene	06-46-7	147.00	53	174	3.52		0.08 g/L	25	0.0	0.0	1.5E–3
C_6H_5Cl	Chlorobenzene	08-90-7	112.56	–45	132	2.84		0.49 g/L	25	0.4	1.6	3.5E–3
C_6H_5ClO	*o*-Chlorophenol	95-57-8	128.56	9	175	2.15	8.52	22 g/L	25	0.0	0.3	5.7E–7
C_6H_5ClO	*m*-Chlorophenol	08-43-0	128.56	33	214	2.50	9.12	22 g/L	25	0.0	0.0	5.7E—7
C_6H_5ClO	*p*-Chlorophenol	06-48-9	128.56	43	220	2.39	9.41	27 g/L	25	0.0	0.0	5.7E–7
$C_6H_5NO_2$	Nitrobenzene	98-95-3	123.11	6	211	1.85		1.9 g/L	20	0.0	0.0	2.4E–5
$C_6H_5NO_3$	*o*-Nitrophenol	88-75-5	139.11	45	216	1.79	7.23	1.1 g/L	20	0.0	0.0	3.5E–6
$C_6H_5NO_3$	*m*-Nitrophenol	54-84-7	139.11	97		2.00	8.36	13 g/L	25	0.0	0.0	1.0E–5
$C_6H_5NO_3$	*p*-Nitrophenol	00-02-7	139.11	114		1.91	7.16	11 g/L	20	0.0	0.0	3.3E–8
C_6H_6	Benzene	71-43-2	78.11	5	80	2.13		1.77 g/L	25	0.0	12.7	
C_6H_6ClN	*o*-Chloroaniline	95-51-2	127.57	–14	209	1.90	2.66	3.8 g/L	20	0.0	0.0	7.5E–6
C_6H_6ClN	*p*-Chloroaniline	06-47-8	127.57	72	232	1.83	3.98	3.9 g/L	20	0.0	0.0	1.1E–5
C_6H_6O	Phenol	08-95-2	94.11	41	182	1.50	9.99	87 g/L	25	0.0	0.1	4.0E–7
C_6H_7N	Aniline	62-53-3	93.13	–6	185	0.90	4.60	36 g/L	25	0.0	0.1	1.4E–1
$C_6H_{10}O_4$	Adipic acid	24-04-9	146.14	153	337	0.08	4.44	15 g/L	15	0.0	0.0	9.5E–7
$C_6H_{11}NO$	Caprolactam	05-60-2	113.16	69	139	–0.19		5.2 kg/L	25	0.0	0.0	
C_6H_{12}	Cyclohexane	10-82-7	84.16	7	81	3.44		58 mg/L	25	0.0	13.1	
C_6H_{14}	*n*-Hexane	10-54-3	86.18	–95	69	4.00		11 mg/L	25	6.1	20.4	
$C_7H_5Cl_3$	Trichloromethylbenzene	98-07-7	195.48	–5	221	2.92		hydrol.		0.0	0.0	
$C_7H_6O_2$	Benzoic acid	65-85-0	122.12	122	249	1.87	4.20	2.7 g/L	18	0.0	0.0	7.1E–8
C_7H_7Cl	Chloromethylbenzene	00-44-7	126.59	–45	179	2.30		0.49 g/L	20	0.0	0.2	3.5E–4
$C_7H_7NO_2$	*o*-Nitrotoluene	88-72-2	137.14	–10	222	2.30		0.65 g/L	30	0.0	0.0	5.6E–5
$C_7H_7NO_2$	*p*-Nitrotoluene	99-99-0	137.14	53	238	2.42		0.44 g/L	30	0.0	0.0	5E–5
C_7H_8	Toluene	08-88-3	92.14	–95	111	2.73		0.53 g/L	25	0.9	3.8	
C_7H_8O	*o*-Cresol	95-48-7	108.14	30	191	1.98	10.30	31 g/L	40	0.0	0.0	1.6E–6
C_7H_8O	*m*-Cresol	08-39-4	108.14	12	202	1.98	10.09	23 g/L	25	0.0	0.0	8.8E–7

PROPERTIES OF LARGE PRODUCTION AND PRIORITY ORGANIC POLLUTANTS (continued)

Molecular formula	Name	CAS RN	Mol. wt.	T_m/°C	T_b/°C	log P	pK_a	Solubility		Vap. Press.(kPa)		Henry's Const. (bar m³/mol)
								Value	T/°C	0° C	25° C	
C_7H_8O	*p*-Cresol	106-44-5	108.14	36	202	1.97	10.26	23 g/L	40	0.0	0.0	9.7E–7
$C_7H_{10}N_2$	Toluene-2,4-diamine	95-80-7	122.17	99	292	0.34				0.0	0.0	
C_7H_{16}	*n*-Heptane	142-82-5	100.20	–91	98	4.50		2.3 mg/L	25	1.5	6.1	
$C_8H_4O_3$	Phthalic anhydride	85-44-9	148.12	131	295			hydrol.		0.0	0.0	6.2E–9
C_8H_8	Styrene	100-42-5	104.15	–31	145	3.05		0.25 g/L	25	0.1	0.8	2.8E–3
C_8H_{10}	Ethylbenzene	100-41-1	106.17	–95	136	3.15		0.16 g/L	25	0.3	1.3	8.5E–3
C_8H_{10}	*o*-Xylene	95-47-6	106.17	–25	144	3.12		0.17 g/L	25	0.2	0.9	
C_8H_{10}	*m*-Xylene	108-38-3	106.17	–48	139	3.20		0.16 g/L	25	0.2	1.1	
C_8H_{10}	*p*-Xylene	106-42-3	106.17	13	138	3.15		0.18 g/L	25	0.0	1.2	
$C_8H_{10}O$	2,4-Xylenol	105-67-9	122.17	25	211	2.35	10.63	6.2 g/L	25	0.0	0.0	6.4E–7
$C_9H_6N_2O_2$	Toluene-2,4-diisocyanate	584-84-9	174.16	21	251					0.0	0.0	
$C_9H_6N_2O_2$	Toluene-2,6-diisocyanate	91-08-7	174.16	0	131					0.0	0.0	
$C_{10}H_8$	Naphthalene	91-20-3	128.17	78	218	3.35		0.03 g/L	25	0.0	0.0	4.9E–4
$C_{10}H_{10}O_4$	Dimethyl phthalate	131-11-3	194.19	5	284	1.56		4.0 g/L	25	0.0	0.0	1.1E–7
$C_{12}H_{12}N_2$	Benzidine	92-87-5	184.24	120		1.34	4.66	0.52 g/L	25	0.0	0.0	3.9E–11
$C_{12}H_{14}O_4$	Diethyl phthalate	84-66-2	222.23	–40	295	2.47		1.1 g/L	25	0.0	0.0	4.8E–7
$C_{16}H_{22}O_4$	Dibutyl phthalate	84-74-2	278.35	–35	340	4.72		11 mg/L	25	0.0	0.0	4.6E–7

REFERENCES

General

Hazardous Substances Data Bank (TOXNET, National Library of Medicine)

P. H. Howard, *Handbook of Environmental Fate and Exposure Data*, Vol. I, Lewis Publishers, Chelsea, Michigan, 1989.

N. I. Sax and R. J. Lewis, *Hazardous Chemicals Desk Reference*, Van Nostrand Reinhold Co., New York, 1987.

Chemical Safety Data Sheets, Vol. I, Royal Society of Chemistry, Cambridge, U.K., 1989.

Solubility

IUPAC Solubility Data Series, Pergamon Press, Oxford.

log P

J. Sangster, Octanol-water partition coefficients of simple organic compounds, *J. Phys. Chem. Ref. Data*, 18, 1111, 1989.

Vapor pressure

DIPPR Data Compilation of Pure Compound Properties, American Institute of Chemical Engineers, New York, 1987.

TRC Thermodynamic Tables, Thermodynamics Research Center, Texas A & M University, College Station, Texas, 1985.

PERMISSIBLE QUARTERLY INTAKE OF RADIONUCLIDES

Karl Z. Morgan

The following table lists the permissible quarterly intakes (oral and inhalation) of radionuclides for critical body organs as recommended by NCPR for occupational exposure.

Radionuclide	Radioactive half-life days	Solubility state	Oral intake: Critical organ	Oral intake: Permissible quarterly intake microcuries	Intake by inhalation: Critical organ	Intake by inhalation: Permissible quarterly intake microcuries
3-H	4.50E 03	Sol.	Body tissue	6.4E 03(2)	Body tissue	3.1E 03(2)
7-BE	5.36E 01	Sol.	GI LLI	3.6E 03	Tot. body	3.5E 03
		Insol.	GI LLI	3.6E 03	Lungs	7.5E 02
14-C	2.00E 06	Sol.	Fat	1.6E 03	Fat	2.2E 03
18-F	7.80E-02	Sol.	GI SI	1.7E 03	GI SI	3.3E 03
		Insol.	GI ULI	1.0E 03	GI ULI	1.6E 03
22-NA	9.50E 02	Sol.	Tot. body	8.0E 01	Tot. body	1.1E 02
		Insol.	GI LLI	6.0E 01	Lungs	5.3E 00
24-NA	6.30E-01	Sol.	GI SI	3.8E 02	GI SI	7.6E 02
		Insol.	GI LLI	5.6E 01	GI LLI	8.9E 01
31-SI	1.10E-01	Sol.	GI S	1.8E 03	GI S	3.5E 03
		Insol.	GI ULI	3.8E 02	GI ULI	6.2E 02
32-P	1.43E 01	Sol.	Bone	3.8E 01	Bone	4.4E 01
		Insol.	GI LLI	4.6E 01	Lungs	4.9E 01
35-S	8.71E 01	Sol.	Testes	1.3E 02	Testes	1.7E 02
		Insol	GI LLI	5.4E 02	Lungs	1.6E 02
36-CL	1.20E 08	Sol.	Tot. body	1.6E 02	Tot. body	2.2E 02
		Insol.	GI LLI	1.2E 02	Lungs	1.4E 01
38-CL	2.60E-02	Sol.	GI S	8.0E 02	GI S	1.6E 03
		Insol	GI S	8.0E 02	GI S	1.3E 03
42-K	5.20E-01	Sol.	GI S	6.2E 02	GI-S	1.2E 03
		Insol.	GI LLI	4.0E 01	GI LLI	6.7E 01
45-CA	1.64E 02	Sol.	Bone	1.8E 01	Bone	2.0E 01
		Insol.	GI LLI	3.6E 02	Lungs	7.5E 01
47-CA	4.90E 00	Sol.	Bone	1.0E 02	Bone	1.1E 02
		Insol.	GI LLI	6.6E 01	GI LLI	1.1E 02
					Lungs	1.2E 02
46-SC	8.50E 01	Sol.	GI LLI	7.6E 01	Liver	1.5E 02
					GI LLI	1.5E 02
		Insol.	GI LLI	7.6E 01	Lungs	1.4E 01
47-SC	3.43E 00	Sol.	GI LLI	1.8E 02	GI LLI	3.6E 02
		Insol.	GI LLI	1.8E 02	GI LLI	2.9E 02
48-SC	1.83E 00	Sol.	GI LLI	5.4E 01	GI LLI	1.1E 02
		Insol.	GI LLI	5.4E 01	GI LLI	8.7E 01
48-V	1.61E 01	Sol.	GI LLI	5.8E 01	GI LLI	1.2E 02
		Insol.	GI LLI	5.8E 01	Lungs	3.5E 01
51-CR	2.78E 01	Sol.	GI LLI	3.2E 03	GI LLI	6.4E 03
					Tot. body	6.7E 03
		Insol.	GI LLI	3.0E 03	Lungs	1.4E 03
52-MN	5.55E 00	Sol.	GI LLI	6.6E 01	GI LLI	1.3E 02
		Insol.	GI LLI	6.0E 01	Lungs	8.7E 01
					GI LLI	9.5E 01
54-MN	3.00E 02	Sol.	GI LLI	2.6E 02	Liver	2.4E 02
		Insol.	GI LLI	2.4E 02	Lungs	2.2E 01
56-MN	1.10E-01	Sol.	GI LLI	2.4E 02	GI LLI	4.7E 02
		Insol.	GI LLI	2.0E 02	GI LLI	3.3E 02
55-FE	1.10E 03	Sol.	Spleen	1.6E 03	Spleen	5.3E 02
		Insol.	GI LLI	4.6E 03	Lungs	6.4E 02
59-FE	4.51E 01	Sol.	GI LLI	1.2E 02	Spleen	9.3E 01
		Insol.	GI LLI	1.1E 02	Lungs	3.3E 01
57-CO	2.70E 02	Sol.	GI LLI	1.1E 03	GI LLI	2.2E 03
		Insol.	GI LLI	7.6E 02	Lungs	1.0E 02
58M-CO	3.80E-01	Sol.	GI LLI	5.6E 03	GI LLI	1.1E 04
		Insol.	GI LLI	4.0E 03	Lungs	5.5E 03
58-CO	7.20E 01	Sol	GI LLI	2.6E 02	GI LLI	5.3E 02
					Tot. body	6.0E 02
		Insol	GI LLI	1.8E 02	Lungs	3.5E 01
60-CO	1.90E 03	Sol	GI LLI	9.8E 01	GI LLI	2.0E 02
					Tot. body	1.1E 01
		Insol.	GI LLI	7.0E 01	Lungs	5.5E 00
59-NI	2.90E 07	Sol.	Bone	4.0E 02	Bone	2.9E 02
		Insol.	GI LLI	4.0E 03	Lungs	4.7E 02
63-NI	2.90E 04	Sol	Bone	5.4E 01	Bone	4.0E 01
		Insol	GI LLI	1.4E 03	Lungs	1.7E 02
65-NI	1.10E-01	Sol.	GI ULI	2.8E 02	GI ULI	5.6E 02
		Insol.	GI ULI	2.0E 02	GI ULI	3.3E 02
64-CU	5.30E-01	Sol.	GI LLI	6.6E 02	GI LLI	1.3E 03
		Insol.	GI LLI	4.2E 02	GI LLI	6.6E 02
65-ZN	2.45E 02	Sol.	Tot. body	2.0E 02	Tot. body	6.6E 01
			Prostate	2.4E 02	Prostate	8.0E 01
			Liver	2.6E 02		
		Insol.	GI LLI	3.6E 02	Lungs	3.6E 01
69M-ZN	5.80E-01	Sol.	GI LLI	1.4E 02	Prostate	2.4E 02
		Insol.	GI LLI	1.2E 02	GI LLI	2.0E 02
69-ZN	3.60E-02	Sol.	GI S	3.6E 03	Prostate	4.4E 03
		Insol.	GI S	3.6E 03	GI S	5.8E 03
72-GA	5.90E-01	Sol.	GI LLI	7.4E 01	GI LLI	1.5E 02
		Insol.	GI LLI	7.4E 01	GI LLI	1.2E 02
71-GE	1.20E 01	Sol.	GI LLI	3.2E 03	GI LLI	6.4E 03
		Insol.	GI LLI	3.2E 03	Lungs	4.0E 03

PERMISSIBLE QUARTERLY INTAKE OF RADIONUCLIDES (continued)

Radionuclide	Radioactive half-life days	Solubility state	Oral intake: Critical organ	Oral intake: Permissible quarterly intake microcuries	Intake by inhalation: Critical organ	Intake by inhalation: Permissible quarterly intake microcuries
73-AS	7.60E 01	Sol.	GI LLI	9.6E 02	Tot. body	1.3E 03
		Insol.	GI LLI	9.2E 02	Lungs	2.4E 02
74-AS	1.75E 01	Sol.	GI LLI	1.1E 02	GI LLI	2.2E 02
		Insol.	GI LLI	1.0E 02	Lungs	7.8E 01
76-AS	1.10E 00	Sol.	GI LLI	4.0E 01	GI LLI	8.0E 01
		Insol.	GI LLI	3.8E 01	GI LLI	6.2E 01
77-AS	1.60E 00	Sol.	GI LLI	1.6E 02	GI LLI	3.3E 02
		Insol.	GI LLI	1.6E 02	GI LLI	2.5E 02
75-SE	1.27E 02	Sol.	Kidneys	6.0E 02	Kidneys	7.8E 02
			Tot. body	6.8E 02		
		Insol.	GI LLI	5.4E 02	Lungs	7.6E 01
82-BR	1.50E 00	Sol.	Tot. body	5.2E 02	Tot. body	7.1E 02
			GI SI	5.6E 02		
		Insol.	GI LLI	7.4E 01	GI LLI	1.2E 02
86-RB	1.86E 01	Sol.	Tot. body	1.3E 02	Tot. body	1.8E 02
			Pancreas	1.3E 02	Pancreas	1.8E 02
					Liver	2.5E 02
		Insol.	GI LLI	4.8E 01	Lungs	4.2E 01
87-RB	1.80E 13	Sol.	Pancreas	2.2E 02	Pancreas	3.1E 02
					Tot. body	4.0E 02
					Liver	4.2E 02
		Insol.	GI LLI	3.4E 02	Lungs	4.0E 01
85M-SR	4.90E-02	Sol.	GI SI	1.3E 04	GI SI	2.5E 04
		Insol.	GI SI	1.3E 04	GI SI	2.2E 04
85-SR	6.50E 01	Sol.	Tot. body	1.9E 02	Tot. body	1.4E 02
		Insol.	GI LLI	3.4E 02	Lungs	6.6E 01
89-SR	5.05E 01	Sol.	Bone	2.4E 01	Bone	1.7E 01
		Insol.	GI LLI	5.6E 01	Lungs	2.2E 01
90-SR	1.00E 04	Sol.	Bone	8.0E-01	Bone	7.3E-01
		Insol.	GI LLI	7.0E 01	Lungs	3.5E 00
91-SR	4.00E-01	Sol.	GI LLI	1.4E 02	GI LLI	2.7E 02
		Insol.	GI LLI	9.8E 01	GI LLI	1.6E 02
92-SR	1.10E-01	Sol.	GI ULI	1.4E 02	GI ULI	2.7E 02
		Insol.	GI ULI	1.2E 02	GI ULI	1.8E 02
90-Y	2.68E 00	Sol.	GI LLI	4.0E 01	GI LLI	8.0E 01
		Insol.	GI LLI	4.0E 01	GI LLI	6.4E 01
91M-Y	3.50E-02	Sol.	GI SI	6.8E 03	GI SI	1.4E 04
		Insol.	GI SI	6.8E 03	GI SI	1.1E 04
91-Y	5.80E 01	Sol.	GI LLI	5.2E 01	Bone	2.2E 01
		Insol.	GI LLI	5.2E 01	Lungs	2.0E 01
92-Y	1.50E-01	Sol.	GI ULI	1.2E 02	GI ULI	2.4E 02
		Insol.	GI ULI	1.2E 02	GI ULI	1.8E 02
93-Y	4.20E-01	Sol.	GI LLI	5.4E 01	GI LLI	1.1E 02
		Insol.	GI LLI	5.4E 01	GI LLI	8.6E 01
93-ZR	4.00E 08	Sol.	GI LLI	1.6E 03	Bone	8.0E 01
		Insol.	GI LLI	1.6E 03	Lungs	2.0E 02
95-ZR	6.33E 01	Sol.	GI LLI	1.3E 02	Tot. body	8.0E 01
		Insol.	GI LLI	1.3E 02	Lungs	2.0E 01
97-ZR	7.10E-01	Sol.	GI LLI	3.6E 01	GI LLI	7.3E 01
		Insol.	GI LLI	3.6E 01	GI LLI	5.6E 01
93M-NB	3.70E 03	Sol.	GI LLI	8.0E 02	Bone	7.6E 01
		Insol.	GI LLI	8.0E 02	Lungs	1.0E 02
95-NB	3.50E 01	Sol.	GI LLI	1.9E 02	Tot. body	2.9E 02
					GI LLI	3.8E 02
		Insol.	GI LLI	1.9E 02	Lungs	6.2E 01
97-NB	5.10E-02	Sol.	GI ULI	1.9E 03	GI ULI	3.8E 03
		Insol.	GI ULI	1.9E 03	GI ULI	2.9E 03
99-MO	2.79E 00	Sol.	Kidneys	3.6E 02	Kidneys	4.5E 02
			GI LLI	4.8E 02		
		Insol.	GI LLI	7.8E 01	GI LLI	1.3E 02
96M-TC	3.60E-02	Sol.	GI LLI	2.4E 04	GI LLI	4.7E 04
		Insol.	GI LLI	2.0E 04	Lungs	1.8E 04
96-TC	4.30E 00	Sol.	GI LLI	2.0E 02	GI LLI	4.0E 02
		Insol.	GI LLI	9.4E 01	GI LLI	1.5E 02
97M-TC	9.20E 01	Sol.	GI LLI	7.2E 02	GI LLI	1.4E 03
		Insol.	GI LLI	3.4E 02	Lungs	9.5E 01
97-TC	3.70E 06	Sol.	GI LLI	3.4E 03	GI LLI	6.7E 03
					Kidneys	8.0E 03
		Insol.	GI LLI	1.6E 03	Lungs	1.8E 02
99M-TC	2.50E-01	Sol.	GI ULI	1.1E 04	GI ULI	2.4E 04
		Insol.	GI ULI	5.6E 03	GI ULI	8.7E 03
99-TC	7.70E 07	Sol.	GI LLI	6.6E 02	GI LLI	1.3E 03
		Insol.	GI LLI	3.2E 02	Lungs	3.8E 01
97-RU	2.80E 00	Sol.	GI LLI	7.2E 02	GI LLI	1.4E 03
		Insol.	GI LLI	7.0E 02	GI LLI	1.1E 03
103-RU	4.10E 01	Sol.	GI LLI	1.6E 02	GI LLI	3.3E 02
		Insol.	GI LLI	1.6E 02	Lungs	5.3E 01
105-RU	1.90E-01	Sol.	GI ULI	2.2E 02	GI ULI	4.4E 02
		Insol.	GI ULI	2.0E 02	GI ULI	3.3E 02
106-RU	3.65E 02	Sol.	GI LLI	2.4E 01	GI LLI	4.7E 01
		Insol.	GI LLI	2.4E 01	Lungs	3.5E 00
103M-RH	3.80E-02	Sol.	GI S	2.4E 04	GI S	4.7E 04
		Insol.	GI S	2.4E 04	GI S	3.8E 04
105-RH	1.52E 00	Sol.	GI LLI	2.6E 02	GI LLI	5.3E 02
		Insol.	GI LLI	2.0E 02	GI LLI	3.3E 02
103-PD	1.70E 01	Sol.	GI LLI	6.8E 02	Kidneys	8.4E 02
		Insol.	GI LLI	5.4E 02	Lungs	4.7E 02
109-PD	5.70E-01	Sol.	GI LLI	1.8E 02	GI LLI	3.5E 02
		Insol.	GI LLI	1.4E 02	GI LLI	2.2E 02

PERMISSIBLE QUARTERLY INTAKE OF RADIONUCLIDES (continued)

Radionuclide	Radioactive half-life days	Solubility state	Oral intake		Intake by inhalation	
			Critical organ	Permissible quarterly intake microcuries	Critical organ	Permissible quarterly intake microcuries
105-AG	4.00E 01	Sol.	GI LLI	1.9E 02	GI LLI	3.8E 02
		Insol.	GI LLI	1.9E 02	Lungs	5.1E 01
110M-AG	2.70E 02	Sol.	GI LLI	6.0E 01	GI LLI	1.2E 02
		Insol.	GI LLI	6.0E 01	Lungs	6.4E 00
111-AG	7.50E 00	Sol.	GI LLI	8.8E 01	GI LLI	1.8E 02
		Insol.	GI LLI	8.6E 01	GI LLI	1.4E 02
109-CD	4.75E 02	Sol.	GI LLI	3.4E 02	Liver	3.3E 01
					Kidneys	3.5E 01
		Insol.	GI LLI	3.4E 02	Lungs	4.5E 01
115M-CD	4.30E 01	Sol.	GI LLI	5.0E 01	Liver	2.2E 01
		Insol.	GI LLI	5.0E 01	Lungs	2.2E 01
115-CD	2.20E 00	Sol.	GI LLI	6.8E 01	GI LLI	1.4E 02
		Insol.	GI LLI	7.2E 01	GI LLI	1.1E 02
134-CS	8.40E 02	Sol.	Tot. body	1.7E 01	Tot. body	2.4E 01
		Insol.	GI LLI	8.0E 01	Lungs	8.0E 00
135-CS	1.10E 09	Sol.	Liver	2.2E 02	Liver	2.9E 02
			Spleen	2.4E 02	Spleen	3.3E 02
			Tot. body	2.6E 02	Tot. body	3.6E 02
		Insol.	GI LLI	4.6E 02	Lungs	5.6E 01
136-CS	1.30E 01	Sol.	Tot. body	1.7E 02	Tot. body	2.4E 02
		Insol.	GI LLI	1.3E 02	Lungs	1.1E 02
137-CS	1.10E 04	Sol.	Tot. body	3.0E 01	Tot. body	4.0E 01
			Liver	3.6E 01		
			Spleen	4.4E 01		
			Muscle	4.8E 01		
		Insol.	GI LLI	8.8E 01	Lungs	9.1E 00
131-BA	1.16E 01	Sol.	GI LLI	3.6E 02	GI LLI	7.3E 02
		Insol.	GI LLI	3.4E 02	Lungs	2.2E 02
140-BA	1.28E 01	Sol.	GI LLI	5.2E 01	Bone	8.0E 01
		Insol.	GI LLI	5.0E 01	Lungs	2.7E 01
140-LA	1.68E 00	Sol.	GI LLI	4.8E 01	GI LLI	9.6E 01
		Insol.	GI LLI	4.8E 01	GI LLI	7.6E 01
141-CE	3.20E 01	Sol.	GI LLI	1.8E 02	Liver	2.7E 02
					GI LLI	3.6E 02
					Bone	3.6E 02
		Insol.	GI LLI	1.8E 02	Lungs	9.8E 01
143-CE	1.33E 00	Sol.	GI LLI	8.0E 01	GI LLI	1.6E 02
		Insol.	GI LLI	8.0E 01	GI LLI	1.3E 02
144-CE	2.90E 02	Sol.	GI LLI	2.4E 01	Bone	6.0E 00
		Insol.	GI LLI	2.4E 01	Lungs	4.0E 00
142-PR	8.00E-01	Sol.	GI LLI	6.0E 01	GI LLI	1.2E 02
		Insol.	GI LLI	6.0E 01	GI LLI	9.8E 01
143-PR	1.37E 01	Sol.	GI LLI	9.8E 01	GI LLI	2.0E 02
		Insol.	GI LLI	9.8E 01	Lungs	1.1E 02
144-ND	7.30E 17	Sol.	Bone	1.3E 02	Bone	5.1E-02
		Insol.	GI LLI	1.6E 02	Lungs	1.8E-01
147-ND	1.13E 01	Sol.	GI LLI	1.2E 02	Liver	2.2E 02
					GI LLI	2.4E 02
		Insol.	GI LLI	1.2E 02	Lungs	1.4E 02
149-ND	8.30E-02	Sol.	GI LLI	5.6E 02	GI LLI	1.1E 03
		Insol.	GI ULI	5.6E 02	GI ULI	9.1E 02
147-PM	9.20E 02	Sol.	GI LLI	4.4E 02	Bone	4.0E 01
		Insol.	GI LLI	4.4E 02	Lungs	6.0E 01
149-PM	2.20E 00	Sol.	GI LLI	8.8E 01	GI LLI	1.8E 02
		Insol.	GI LLI	8.8E 01	GI LLI	1.4E 02
147-SM	4.80E 13	Sol.	Bone	1.1E 02	Bone	4.4E-02
		Insol.	GI LLI	1.4E 02	Lungs	1.6E-01
151-SM	3.70E 04	Sol.	GI LLI	7.4E 02	Bone	4.0E 01
		Insol.	GI LLI	7.4E 02	Lungs	8.7E 01
153-SM	1.96E 00	Sol.	GI LLI	1.5E 02	GI LLI	3.1E 02
		Insol.	GI LLI	1.5E 02	GI LLI	2.5E 02
152M-EU	3.80E-01	Sol.	GI LLI	1.3E 02	GI LLI	2.5E 02
		Insol.	GI LLI	1.3E 02	GI LLI	2.0E 02
152-EU	4.70E 03	Sol.	GI LLI	1.5E 02	Kidneys	7.6E 00
		Insol.	GI LLI	1.5E 02	Lungs	1.1E 01
154-EU	5.80E 03	Sol.	GI LLI	4.4E 01	Kidneys	2.4E 00
					Bone	2.4E 00
		Insol.	GI LLI	4.4E 01	Lungs	4.4E 00
155-EU	6.21E 02	Sol.	GI LLI	4.0E 02	Kidneys	5.8E 01
					Bone	6.2E 01
		Insol.	GI LLI	4.0E 02	Lungs	4.5E 01
153-GD	2.36E 02	Sol.	GI LLI	4.2E 02	Bone	1.4E 02
		Insol.	GI LLI	4.2E 02	Lungs	5.6E 01
159-GD	7.50E-01	Sol.	GI LLI	1.6E 02	GI LLI	3.1E 02
		Insol.	GI LLI	1.6E 02	GI LLI	2.5E 02
113M-IN	7.30E-02	Sol.	GI ULI	2.6E 03	GI ULI	5.3E 03
		Insol.	GI ULI	2.6E 03	GI ULI	4.2E 03
114M-IN	4.90E 01	Sol.	GI LLI	3.4E 01	Kidneys	6.4E 01
					GI LLI	6.7E 01
					Spleen	7.1E 01
		Insol.	GI LLI	3.4E 01	Lungs	1.3E 01
115M-IN	1.90E-01	Sol.	GI ULI	7.4E 02	GI ULI	1.5E 03
		Insol.	GI ULI	7.4E 02	GI ULI	1.2E 03
115-IN	2.20E 17	Sol.	GI LLI	1.8E 02	Kidneys	1.5E 02
		Insol.	GI LLI	1.8E 02	Lungs	2.2E 01
113-SN	1.12E 02	Sol.	GI LLI	1.7E 02	Bone	2.2E 02
		Insol.	GI LLI	1.6E 02	Lungs	3.3E 01

PERMISSIBLE QUARTERLY INTAKE OF RADIONUCLIDES (continued)

Radionuclide	Radioactive half-life days	Solubility state	Oral intake		Intake by inhalation	
			Critical organ	Permissible quarterly intake microcuries	Critical organ	Permissible quarterly intake microcuries
125-SN	9.50E 00	Sol.	GI LLI	3.6E 01	GI LLI	7.3E 01
		Insol.	GI LLI	3.4E 01	Lungs	5.3E 01
					GI LLI	5.5E 01
122-SB	2.80E 00	Sol.	GI LLI	5.8E 01	GI LLI	1.2E 02
		Insol.	GI LLI	5.8E 01	GI LLI	9.1E 01
124-SB	6.00E 01	Sol.	GI LLI	4.6E 01	GI LLI	9.3E 01
		Insol.	GI LLI	4.6E 01	Lungs	1.2E 01
125-SB	8.77E 02	Sol.	GI LLI	2.0E 02	Lungs	3.3E 02
					Tot. body	3.6E 02
					GI LLI	4.0E 02
					Bone	4.5E 02
		Insol.	GI LLI	2.0E 02	Lungs	1.7E 01
125M-TE	5.80E 01	Sol.	Kidneys	3.2E 02	Kidneys	2.2E 02
			GI LLI	3.6E 02		
			Testes	4.4E 02		
		Insol.	GI LLI	2.4E 02	Lungs	8.0E 01
127M-TE	1.05E 02	Sol.	Kidneys	1.2E 02	Kidneys	8.2E 01
					Testes	8.7E 01
		Insol.	GI LLI	1.1E 02	Lungs	2.5E 01
127-TE	3.90E-01	Sol.	GI LLI	5.2E 02	GI LLI	1.0E 03
		Insol.	GI LLI	3.4E 02	GI LLI	5.3E 02
129M-TE	3.30E 01	Sol.	GI LLI	6.6E 01	Kidneys	4.9E 01
					Testes	5.6E 01
		Insol.	GI LLI	4.0E 01	Lungs	2.0E 01
129-TE	5.10E-02	Sol.	GI S	1.6E 03	GI S	3.3E 03
		Insol.	GI ULI	1.6E 03	GI ULI	2.5E 03
131M-TE	1.25E 00	Sol.	GI LLI	1.2E 02	GI LLI	2.4E 02
		Insol.	GI LLI	7.4E 01	GI LLI	1.2E 02
132-TE	3.20E 00	Sol.	GI LLI	6.4E 01	GI LLI	1.3E 02
		Insol.	GI LLI	4.2E 01	GI LLI	6.6E 01
126-I	1.33E 01	Sol.	Thyroid	3.4E 00	Thyroid	4.5E 00
		Insol.	GI LLI	1.8E 02	Lungs	2.0E 02
129-I	6.30E 09	Sol.	Thyroid	7.6E-01	Thyroid	1.0E 00
		Insol.	GI LLI	4.2E 02	Lungs	4.5E 01
131-I	8.00E 00	Sol.	Thyroid	4.0E 00	Thyroid	5.3E 00
		Insol.	GI LLI	1.3E 02	GI LLI	2.0E 02
					Lungs	2.0E 02
132-I	9.70E-02	Sol.	Thyroid	1.1E 02	Thyroid	1.5E 02
		Insol.	GI ULI	3.6E 02	GI ULI	5.8E 02
133-I	8.70E-01	Sol.	Thyroid	1.5E 01	Thyroid	2.0E 01
		Insol.	GI LLI	8.2E 01	GI LLI	1.3E 02
134-I	3.60E-02	Sol.	Thyroid	2.4E 02	Thyroid	3.1E 02
		Insol.	GI S	1.2E 03	GI S	2.0E 03
135-I	2.80E-01	Sol.	Thyroid	4.8E 01	Thyroid	6.4E 01
		Insol.	GI LLI	1.4E 02	GI LLI	2.2E 02
131-CS	1.00E 01	Sol.	Tot. body	4.8E 03	Tot. body	6.6E 03
					Liver	8.0E 03
		Insol.	GI LLI	1.9E 03	Lungs	2.0E 03
134M-CS	1.30E-01	Sol.	GI S	1.1E 04	GI S	2.2E 04
		Insol.	GI ULI	2.2E 03	GI ULI	3.6E 03
160-TB	7.30E 01	Sol.	GI LLI	8.8E 01	Bone	6.2E 01
		Insol.	GI LLI	9.0E 01	Lungs	2.0E 01
165-DY	9.70E-02	Sol.	GI ULI	8.0E 02	GI ULI	1.6E 03
		Insol.	GI ULI	8.0E 02	GI ULI	1.3E 03
166-DY	3.40E 00	Sol.	GI LLI	7.6E 01	GI LLI	1.5E 02
		Insol.	GI LLI	7.6E 01	GI LLI	1.2E 02
166-HO	1.10E 00	Sol.	GI LLI	6.2E 01	GI LLI	1.2E 02
		Insol.	GI LLI	6.2E 01	GI LLI	1.0E 02
169-ER	9.40E 00	Sol.	GI LLI	1.9E 02	GI LLI	3.8E 02
		Insol.	GI LLI	1.9E 02	Lungs	2.4E 02
171-ER	3.10E-01	Sol.	GI ULI	2.2E 02	GI ULI	4.4E 02
		Insol.	GI ULI	2.2E 02	GI ULI	3.6E 02
170-TM	1.27E 02	Sol.	GI LLI	9.2E 01	Bone	2.2E 01
		Insol.	GI LLI	9.2E 01	Lungs	2.2E 01
171-TM	6.94E 02	Sol.	GI LLI	1.0E 03	Bone	6.9E 01
		Insol.	GI LLI	1.0E 03	Lungs	1.4E 02
175-YB	4.10E 00	Sol.	GI LLI	2.2E 02	GI LLI	4.4E 02
		Insol.	GI LLI	2.2E 02	GI LLI	3.6E 02
177-LU	6.80E 00	Sol.	GI LLI	2.0E 02	GI LLI	4.0E 02
		Insol.	GI LLI	2.0E 02	GI LLI	3.3E 02
					Lungs	4.4E 02
181-HF	4.60E 01	Sol.	GI LLI	1.4E 02	Spleen	2.4E 01
		Insol.	GI LLI	1.4E 02	Lungs	4.5E 01
182-TA	1.12E 02	Sol.	GI LLI	8.0E 01	Liver	2.4E 01
		Insol.	GI LLI	8.0E 01	Lungs	1.4E 01
181-W	1.40E 02	Sol.	GI LLI	7.2E 02	GI LLI	1.4E 03
		Insol.	GI LLI	6.4E 02	Lungs	7.8E 01
185-W	7.40E 01	Sol.	GI LLI	2.4E 02	GI LLI	4.7E 02
		Insol.	GI LLI	2.2E 02	Lungs	6.9E 01
187-W	1.00E 00	Sol.	GI LLI	1.4E 02	GI LLI	2.7E 02
		Insol.	GI LLI	1.2E 02	GI LLI	2.0E 02
183-RE	7.30E 01	Sol.	GI LLI	1.1E 03	Tot. body	1.6E 03
		Insol.	GI LLI	5.6E 02	Lungs	9.8E 01
186-RE	3.79E 00	Sol.	GI LLI	1.9E 02	GI LLI	3.8E 02
		Insol.	GI LLI	9.4E 01	GI LLI	1.5E 02
187-RE	1.80E 13	Sol.	GI LLI	5.0E 03	Skin	5.6E 03
			Skin	5.6E 03		
		Insol.	GI LLI	5.0E 03	Lungs	3.1E 02

PERMISSIBLE QUARTERLY INTAKE OF RADIONUCLIDES (continued)

Radionuclide	Radioactive half-life days	Solubility state	Oral intake: Critical organ	Oral intake: Permissible quarterly intake microcuries	Intake by inhalation: Critical organ	Intake by inhalation: Permissible quarterly intake microcuries
188-RE	7.10E-01	Sol.	GI LLI	1.3E 02	GI LLI	2.5E 02
		Insol.	GI LLI	6.2E 01	GI LLI	1.0E 02
185-OS	9.50E 01	Sol.	GI LLI	1.5E 02	GI LLI	2.9E 02
		Insol.	GI LLI	1.3E 02	Lungs	2.9E 01
191M-OS	5.80E-01	Sol.	GI LLI	5.0E 03	GI LLI	1.0E 04
		Insol.	GI LLI	4.8E 03	Lungs	5.8E 03
191-OS	1.60E 01	Sol.	GI LLI	3.4E 02	GI LLI	6.7E 02
		Insol.	GI LLI	3.2E 02	Lungs	2.5E 02
193-OS	1.30E 00	Sol.	GI LLI	1.2E 02	GI LLI	2.4E 02
		Insol.	GI LLI	1.1E 02	GI LLI	1.7E 02
190-IR	1.20E 01	Sol.	GI LLI	4.0E 02	GI LLI	8.0E 02
		Insol.	GI LLI	3.6E 02	Lungs	2.5E 02
192-IR	7.45E 01	Sol.	GI LLI	8.0E 01	Kidneys	7.8E 01
		Insol.	GI LLI	7.2E 01	Lungs	1.6E 01
194-IR	7.90E-01	Sol.	GI LLI	6.8E 01	GI LLI	1.4E 02
		Insol.	GI LLI	6.0E 01	GI LLI	9.6E 01
191-PT	3.00E 00	Sol.	GI LLI	2.4E 02	GI LLI	4.7E 02
		Insol.	GI LLI	2.2E 02	GI LLI	3.5E 02
193M-PT	3.40E 00	Sol.	GI LLI	2.2E 03	GI LLI	4.4E 03
		Insol.	GI LLI	2.0E 03	GI LLI	3.3E 03
					Lungs	4.0E 03
193-PT	1.80E 05	Sol.	Kidneys	1.9E 03	Kidneys	6.4E 02
		Insol.	GI LLI	3.0E 03	Lungs	2.0E 02
197M-PT	5.60E-02	Sol.	GI ULI	2.0E 03	GI ULI	4.0E 03
		Insol.	GI ULI	1.9E 03	GI ULI	2.9E 03
197-PT	7.50E-01	Sol.	GI LLI	2.4E 02	GI LLI	4.7E 02
		Insol.	GI LLI	2.2E 02	GI LLI	3.5E 02
196-AU	5.60E 00	Sol.	GI LLI	3.2E 02	GI LLI	6.4E 02
		Insol.	GI LLI	3.0E 02	Lungs	3.8E 02
198-AU	2.70E 00	Sol.	GI LLI	1.0E 02	GI LLI	2.0E 02
		Insol.	GI LLI	9.2E 01	GI LLI	1.5E 02
199-AU	3.15E 00	Sol.	GI LLI	3.4E 02	GI LLI	6.7E 02
		Insol.	GI LLI	3.2E 02	GI LLI	5.1E 02
197M-HG	1.00E 00	Sol.	Kidneys	3.8E 02	Kidneys	4.5E 02
		Insol.	GI LLI	3.4E 02	GI LLI	5.3E 02
197-HG	2.70E 00	Sol.	Kidneys	6.0E 02	Kidneys	7.3E 02
		Insol.	GI LLI	9.8E 02	GI LLI	1.5E 03
203-HG	4.58E 01	Sol.	Kidneys	3.6E 01	Kidneys	4.4E 01
		Insol.	GI LLI	2.2E 02	Lungs	7.8E 01
200-TL	1.13E 00	Sol.	GI LLI	8.8E 02	GI LLI	1.6E 03
		Insol.	GI LLI	4.4E 02	GI LLI	7.1E 02
201-TL	3.00E 00	Sol.	GI LLI	6.2E 02	GI LLI	1.2E 03
		Insol.	GI LLI	3.4E 02	GI LLI	5.5E 02
202-TL	1.20E 01	Sol.	GI LLI	2.4E 02	GI LLI	4.7E 02
		Insol.	GI LLI	1.4E 02	Lungs	1.5E 02
204-TL	1.10E 03	Sol.	GI LLI	2.2E 02	Kidneys	3.8E 02
					GI LLI	4.4E 02
		Insol.	GI LLI	1.2E 02	Lungs	1.6E 01
203-PB	2.17E 00	Sol.	GI LLI	7.8E 02	GI LLI	1.6E 03
		Insol.	GI LLI	7.0E 02	GI LLI	1.1E 03
210-PB	7.10E 03	Sol.	Tot. body	2.4E-01	Kidneys	7.8E-02
			Kidneys	2.8E-01		
		Insol.	GI LLI	3.6E 02	Lungs	1.5E-01
212-PB	4.40E-01	Sol.	GI LLI	3.8E 01	Kidneys	1.1E 01
			Kidneys	4.0E 01		
		Insol.	GI LLI	3.4E 01	Lungs	1.2E 01
206-BI	6.40E 00	Sol.	GI LLI	7.6E 01	Kidneys	1.2E 02
		Insol.	GI LLI	7.6E 01	Lungs	9.1E 01
207-BI	2.90E 03	Sol.	GI LLI	1.3E 02	Kidneys	1.1E 02
		Insol.	GI LLI	1.3E 02	Lungs	8.6E 00
210-BI	5.00E 00	Sol.	GI LLI	8.2E 01	Kidneys	4.0E 00
		Insol.	GI LLI	8.2E 01	Lungs	3.6E 00
212-BI	4.20E-02	Sol.	GI S	7.0E 02	Kidneys	6.0E 01
		Insol.	GI S	7.0E 02	Lungs	1.3E 02
210-PO	1.38E 02	Sol.	Spleen	1.4E 00	Spleen	3.1E-01
					Kidneys	3.1E-01
		Insol.	GI LLI	5.8E 01	Lungs	1.3E-01
211-AT	3.00E-01	Sol.	Thyroid	3.4E 00	Thyroid	4.4E 00
			Ovaries	3.4E 00		
		Insol.	GI ULI	1.5E 02	Lungs	2.2E 01
220-RN[3]	6.40E-04				Lungs	1.8E 02
222-RN[3]	3.82E 00				Lungs	1.8E 01
223-RA	1.17E 01	Sol.	Bone	1.4E 00	Bone	1.1E 00
		Insol.	GI LLI	8.4E 00	Lungs	1.5E-01
224-RA	3.64E 00	Sol.	Bone	4.6E 00	Bone	3.5E 00
		Insol.	GI LLI	1.1E 01	Lungs	4.5E-01
226-RA	5.90E 05	Sol.	Bone	2.4E-02	Bone	1.8E-02
		Insol.	GI LLI	6.4E 01	Lungs	3.3E-02
228-RA	2.40E 03	Sol.	Bone	5.6E-02	Bone	4.2E-02
		Insol.	GI LLI	5.0E 01	Lungs	2.4E-02
227-AC	8.00E 03	Sol.	Bone	3.8E 00	Bone	1.5E-03
		Insol.	GI LLI	6.0E 02	Lungs	1.6E-02
228-AC	2.60E-01	Sol.	GI ULI	1.7E 02	Liver	4.7E 01
					Bone	5.3E 01
		Insol.	GI ULI	1.7E 02	Lungs	1.1E 01
227-TH	1.84E 01	Sol.	GI LLI	3.6E 01	Bone	2.2E-01
		Insol.	GI LLI	3.6E 01	Lungs	1.1E-01
228-TH	7.00E 02	Sol.	Bone	1.5E 01	Bone	5.6E-03
		Insol.	GI LLI	2.6E 01	Lungs	3.8E-03

PERMISSIBLE QUARTERLY INTAKE OF RADIONUCLIDES (continued)

Radionuclide	Radioactive half-life days	Solubility state	Oral intake: Critical organ	Oral intake: Permissible quarterly intake microcuries	Intake by inhalation: Critical organ	Intake by inhalation: Permissible quarterly intake microcuries
230-TH	2.90E 07	Sol.	Bone	3.4E 00	Bone	1.4E-03
		Insol.	GI LLI	6.4E 01	Lungs	6.4E-03
231-TH	1.07E 00	Sol.	GI LLI	4.6E 02	GI LLI	9.3E 02
		Insol.	GI LLI	4.6E 02	GI LLI	7.5E 02
232-TH	5.10E 12	Sol.	Bone	3.0E 00	Bone	1.2E-03[4]
		Insol.	GI LLI	7.6E 01	Lungs	7.3E-03
234-TH	2.41E 01	Sol.	GI LLI	3.4E 01	Bone	3.6E 01
		Insol.	GI LLI	3.4E 01	Lungs	2.0E 01
NAT-TH[5]		Sol.	Bone	2.6E 00	Bone	1.0E-03[4]
		Insol.	GI LLI	1.9E 01	Lungs	2.5E-03
230-PA	1.77E 01	Sol.	GI LLI	4.8E 02	Bone	1.1E 00
		Insol.	GI LLI	5.0E 02	Lungs	4.9E-01
231-PA	1.30E 07	Sol.	Bone	1.7E 00	Bone	7.1E-04
		Insol.	GI LLI	5.4E 01	Lungs	6.7E-02
233-PA	2.74E 01	Sol.	GI LLI	2.4E 02	Kidneys	3.8E 02
		Insol.	GI LLI	2.4E 02	Lungs	1.1E 02
230-U	2.08E 01	Sol.	Kidneys	4.8E 00	Kidneys	1.8E-01
		Insol.	GI LLI	9.2E 00	Lungs	7.1E-02
232-U	2.70E 04	Sol.	Bone	1.7E 00	Bone	6.4E-02
		Insol.	GI LLI	5.8E 01	Lungs	1.7E-02
233-U	5.90E 07	Sol.	Bone	8.4E 00	Bone	3.3E-01
		Insol.	GI LLI	6.4E 01	Lungs	7.5E-02
234-U	9.10E 07	Sol.	Bone	8.6E 00	Bone	3.5E-01
		Insol.	GI LLI	6.4E 01	Lungs	7.5E-02
235-U[5]	2.60E 11	Sol.	Kidneys	7.4F 00	Kidneys	2.9E-01
					Bone	3.6E-01
		Insol.	GI LLI	5.6E 01	Lungs	8.0E-02
236-U[5]	8.70E 09	Sol.	Bone	9.0E 00	Bone	3.6E-01
		Insol.	GI LLI	6.8E 01	Lungs	7.8E-02
238-U[5]	1.60E 12	Sol.	Kidneys	1.2E 00	Kidneys	4.5E-02
		Insol.	GI LLI	7.0E 01	Lungs	8.6E-02
240-U	5.88E-01	Sol.	GI LLI	6.8E 01	GI LLI	1.4E 02
		Insol.	GI LLI	6.8E 01	GI LLI	1.1E 02
NAT-U[5]		Sol.	Kidneys	1.2E 00	Kidneys	4.5E-02
		Insol.	GI LLI	3.2E 01	Lungs	4.0E-02
237-NP	8.00E 08	Sol.	Bone	6.2E 00	Bone	2.5E-03
		Insol.	GI LLI	7.0E 01	Lungs	7.5E-02
239-NP	2.33E 00	Sol.	GI LLI	2.6E 02	GI LLI	5.3E 02
		Insol.	GI LLI	2.6E 02	GI LLI	4.2E 02
238-PU	3.30E 04	Sol.	Bone	1.0E 01	Bone	1.2E-03
		Insol.	GI LLI	5.6E 01	Lungs	2.2E-02
239-PU	8.90E 06	Sol.	Bone	9.0E 00	Bone	1.1E-03
		Insol.	GI LLI	5.8E 01	Lungs	2.4E-02
240-PU	2.40E 06	Sol.	Bone	9.0E 00	Bone	1.1E-03
		Insol.	GI LLI	5.8E 01	Lungs	2.4E-02
241-PU	4.80E 03	Sol.	Bone	4.6E 02	Bone	5.6E-02
		Insol.	GI LLI	2.6E 03	Lungs	2.4E 01
242-PU	1.40E 08	Sol.	Bone	9.4E 00	Bone	1.1E-03
		Insol.	GI LLI	6.2E 01	Lungs	2.4E-02
243-PU	2.08E-01	Sol.	GI ULI	6.8E 02	GI ULI	1.1E 03
		Insol.	GI ULI	6.8E 02	GI ULI	1.4E 03
244-PU	2.80E 10	Sol.	Bone	8.6E 00	Bone	1.0E-03
		Insol.	GI LLI	2.2E 01	Lungs	2.0E-02
241-AM	1.70E 05	Sol.	Kidneys	7.6E 00	Bone	3.6E-03
					Kidneys	3.8E-03
		Insol.	GI LLI	5.4E 01	Lungs	6.4E-02
242M-AM	5.60E 04	Sol.	Bone	8.8E 00	Bone	3.5E-03
		Insol.	GI LLI	1.8E 02	Lungs	1.6E-01
242-AM	6.77E-01	Sol.	GI LLI	2.6E 02	Liver	2.4E 01
		Insol.	GI LLI	2.6E 02	Lungs	2.9E 01
243-AM	2.90E 06	Sol.	Bone	8.8E 00	Bone	3.5E-03
					Kidneys	3.8E-03
		Insol.	GI LLI	5.6E 01	Lungs	6.7E-02
244-AM	1.81E-02	Sol.	GI SI	9.6E 03	Bone	2.5E 03
					Kidneys	2.7E 03
		Insol.	GI SI	9.6E 03	Lungs	1.5E 04
					GI(SI)	1.5E 04
242-CM	1.62E 02	Sol.	GI LLI	4.8E 01	Liver	7.5E-02
		Insol.	GI LLI	5.0E 01	Lungs	1.0E-01
243-CM	1.30E 04	Sol.	Bone	1.0E 01	Bone	4.0E-03
		Insol.	GI LLI	5.0E 01	Lungs	6.2E-02
244-CM	6.70E 03	Sol.	Bone	1.4E 01	Bone	5.6E-03
		Insol.	GI LLI	5.2E 01	Lungs	6.2E-02
245-CM	7.30E 06	Sol.	Bone	7.0E 00	Bone	2.9E-03
		Insol.	GI LLI	5.6E 01	Lungs	6.7E-02
246-CM	2.40E 06	Sol.	Bone	7.2E 00	Bone	2.9E-03
		Insol.	GI LLI	5.6E 01	Lungs	6.6E-02
247-CM	3.30E 10	Sol.	Bone	7.2E 00	Bone	2.9E-03
		Insol.	GI LLI	4.4E 01	Lungs	6.7E-02
248-CM	1.70E 08	Sol.	Bone	8.8E-01	Bone	3.6E-04
		Insol.	GI LLI	2.6E 00	Lungs	8.2E-03
249-CM	4.40E-02	Sol.	GI S	4.4E 03	Bone	7.8E 03
		Insol.	GI S	4.4E 03	GI S	7.1E 03
249-BK	2.90E 02	Sol.	GI LLI	1.2E 03	Bone	5.8E-01
		Insol.	GI LLI	1.2E 03	Lungs	7.5E 01
250-BK	1.34E-01	Sol.	GI ULI	4.4E 02	Bone	9.1E 01
		Insol.	GI ULI	4.4E 02	GI ULI	7.1E 02
249-CF	1.70E 05	Sol.	Bone	8.2E 00	Bone	9.8E-04
		Insol.	GI LLI	4.8E 01	Lungs	6.2E-02

PERMISSIBLE QUARTERLY INTAKE OF RADIONUCLIDES (continued)

Radionuclide	Radioactive half-life days	Solubility state	Oral intake: Critical organ	Oral intake: Permissible quarterly intake microcuries	Intake by inhalation: Critical organ	Intake by inhalation: Permissible quarterly intake microcuries
250-CF	3.70E 03	Sol.	Bone	2.6E 01	Bone	3.1E-03
		Insol.	GI LLI	5.0E 01	Lungs	6.2E-02
251-CF	2.90E 05	Sol.	Bone	8.6E 00	Bone	1.0E-03
		Insol.	GI LLI	5.2E 01	Lungs	6.2E-02
252-CF	8.03E 02	Sol.	GI LLI	1.5E 01	Bone	4.0E-03
		Insol.	GI LLI	1.5E 01	Lungs	2.0E-02
253-CF	1.80E 01	Sol.	GI LLI	2.8E 02	Bone	5.3E-01
		Insol.	GI LLI	2.8E 02	Lungs	4.7E-01
254-CF	5.60E 01	Sol.	GI LLI	2.4E-01	Bone	3.3E-03
		Insol.	GI LLI	2.4E-01	Lungs	3.1E-03
253-ES	2.00E 01	Sol.	GI LLI	4.6E 01	Bone	4.7E-01
		Insol.	GI LLI	4.6E 01	Lungs	3.8E-01
254M-ES	1.60E 00	Sol.	GI LLI	3.8E 01	Bone	3.3E 00
		Insol.	GI LLI	3.8E 01	Lungs	3.6E 00
254-ES	4.80E 02	Sol.	GI LLI	2.8E 01	Bone	1.2E-02
		Insol.	GI LLI	2.8E 01	Lungs	6.7E-02
255-ES	3.00E 01	Sol.	GI LLI	5.6E 01	Bone	3.1E-01
		Insol.	GI LLI	5.6E 01	Lungs	2.5E-01
254-FM	1.35E-01	Sol.	GI ULI	2.4E 02	Bone	4.0E 01
		Insol.	GI ULI	2.4E 02	Lungs	4.4E 01
255-FM	8.96E-01	Sol.	GI LLI	6.6E 01	Bone	1.0E 01
		Insol.	GI LLI	6.6E 01	Lungs	6.7E 00
256-FM	1.11E-01	Sol.	GI ULI	1.8E 00	Bone	1.7E 00
		Insol.	GI ULI	1.8E 00	Lungs	1.1E 00

[1] These quarterly intakes are calculated from values of $(MPC)_a$ and $(MPC)_w$ recommended by the NCRP for occupational exposure (see NBS Handbook 69 and NCRP Report No. 32)* except for uranium, ^{90}Sr, and certain transuranic isotopes given by the ICRP (see ICRP Publications 2 and 6).** Specifically, except where indicated in the footnotes, the quarterly intakes are calculated to two-digit accuracy as $91 \times 2 \times 10^7 \times (MPC)_a$ and $91 \times 2200 \times (MPC)_w$ where 91 ~ days/quarter, 2×10^7 ~ cc of air breathed/day, 2200 ~ cc of water intake/day, and $(MPC)_a$, $(MPC)_w$ are μCi/cc of air and water recommended as limits for occupational exposure for the 168-hour week. These quarterly intakes are calculated for standard adult man and would be expected to deliver a total dose (= dose commitment) to the critical organ of 1/4 the maximum permissible annual dose (MPD) recommended as limiting by the NCRP for occupational exposure. These limiting MPD are the following: Bone, skin, thyroid—MPD = 30 rem; blood-forming organs, lenses of the eyes, gonads—MPD = 5 rem/yr; other organs—MPD = 15 rem/yr. These limits are not appropriate for exposure of the population or, specifically, for persons below the age of 18 years. For more detailed information on appropriate exposure limits recommended by the NCRP, see NBS Handbook 69 and NCRP Report No. 32.*

The abbreviations GI, S, SI, ULI, and LLI refer to gastrointestinal tract, stomach, small intestine, upper large intestine, and lower large intestine, respectively.

With few exceptions, the above intakes do not take account of chemical toxicity which should be considered separately.

[2] This intake includes absorption through the skin (see ICRP Publication 2, p. 22).**

[3] The daughter isotopes of ^{220}Rn and ^{222}Rn are assumed present to the extent they occur in unfiltered air (see NBS Handbook 69, p. 12).* For all other isotopes, the daughter elements are not considered as part of the intake and if present must be considered on the basis of the rules for mixtures.

[4] These are provisional values for ^{232}Th and nat-Th (see footnote, p. 83, NBS Handbook 69).*

[5] A curie of natural uranium is here considered to correspond to 2.94 g and a curie of natural thorium to 9.0 g of these materials (see NBS Handbook 69, p. 14).* For soluble nat-U, ^{238}U, ^{236}U, and ^{235}U these limits are based on considerations of chemical toxicity (see ICRP Publication 6, p. 13).**

* NBS Handbook 69, *Maximum Permissible Body Burdens and Maximum Permissible Concentrations of Radionuclides in Air and Water for Occupational Exposure*, 1959 (available from the Superintendent of Documents, U.S. Government Printing Office, Washington, D.C. 20402). NCRP Report No. 32, *Radiation Protection in Educational Institutions*, 1966 (NCRP Publications, P.O. Box 4867, Washington, D.C. 20008).

** *Recommendations of the International Commission on Radiological Protection*, ICRP Publication 2 (Pergamon Press, London, 1959) or with the bibliography, *Healthy Phys.* 3, 1–380 (June 1960); ICRP Publication 6 (A Pergamon Press Book, 1964, distributed by The Macmillan Co., New York).

CHEMICAL CARCINOGENS

The following substances are listed in the *Seventh Annual Report on Carcinogens: Summary 1994*, released by the National Institute of Environmental Health Sciences (NIEHS) under the National Toxicology Program (NTP). Substances are grouped into three classes:

1. Known to be carcinogens from studies in humans
2. Reasonably anticipated to be carcinogens, based on limited human studies or studies on experimental animals
3. Associated with technological processes that are known to be carcinogenic

The table includes the name used by the NTP, the Chemical Abstracts Service Registry Number (CAS RN) as given in the NTP report, and, where possible, the systematic name of the substance (generally the CAS index name).

REFERENCE

Seventh Annual Report on Carcinogens: Summary 1994, Public Health Service, National Toxicology Program, Central Data Management, P.O. Box 12233, MD A0-01, Research Triangle Park, NC 27709.

NTP name	CAS RN	Systematic name
SUBSTANCES KNOWN TO BE CARCINOGENIC		
Aflatoxins	1402-68-2	
Aminobiphenyl	92-67-1	[1,1´-Biphenyl]-4-amine
Analgesic mixtures containing phenacetin		Acetamide, N-(4-ethoxyphenyl)- (in analgesic mixtures)
Arsenic	7440-38-2	Arsenic
Certain arsenic compounds		
Asbestos	1332-21-4	Asbestos
Azathioprine	446-86-6	Purine, 6-((1-methyl-4-nitroimidazol-5-yl)thio)
Benzene	71-43-2	Benzene
Benzidine	92-87-5	[1,1´-Biphenyl]-4,4´-diamine
Bis(chloromethyl) ether	542-88-1	Methane, oxybis[chloro-
Chloromethyl methyl ether (techical grade)	107-30-2	Methane, chloromethoxy-
1,4-Butanediol dimethyl-sulfonate (Myleran)	55-98-1	1,4-Butanediol, dimethanesulfonate
Chlorambucil	305-03-3	Benzenebutanoic acid, 4-[bis(2-chloroethyl)amino]-
(2-Chloroethyl)-3-(4-methylcyclohexyl)-1-nitrosourea (MeCCNU)	13909-09-6	Urea, 1-(2-chloroethyl)-3-(4-methylcyclohexyl)-1-nitroso-, (*E*)
Chromium	7440-47-3	Chromium
Certain chromium compounds		
Conjugated estrogens		
Cyclophosphamide	50-18-0	2H-1,3,2-Oxazaphosphorine, 2(bis(2-chloroethyl)amino)-tetrahydro, 2-oxide
Diethylstilbestrol	56-53-1	Phenol, 4,4´-(1,2-diethyl-1,2-ethenediyl)bis-, (*E*)-
Erionite	66733-21-9	Erionite (Calcium sodium aluminum silicate)
Melphalan	148-82-3	Alanine, 3-(*p*-(bis(2-chloroethyl)amino)phenyl)-, *L*
Methoxsalen with ultraviolet A therapy (PUVA)		
Mustard gas	505-60-2	Ethane, 1,1´-thiobis[2-chloro-
2-Naphthylamine	91-59-8	2-Naphthalenamine
Radon	10043-92-2	Radon
Thorium dioxide	1314-20-1	Thorium(IV) oxide
Vinyl chloride	75-01-4	Ethene, chloro-
SUBSTANCES ANTICIPATED TO BE CARCINOGENIC		
Acetaldehyde	75-07-0	Acetaldehyde
2-Acetylaminofluorene	53-96-3	Acetamide, *N*-9H-fluoren-2-yl-
Acrylamide	79-06-1	2-Propenamide
Acrylonitrile	107-13-1	2-Propenenitrile
2-Aminoanthraquinone	117-79-3	9,10-Anthracenedione, 2-amino-
o-Aminoazotoluene	97-56-3	Benzenamine, 2-methyl-4-[(2-methylphenyl)azo]-
1-Amino-2-methylanthraquinone	82-28-0	9,10-Anthracenedione, 1-amino-2-methyl-
Amitrole	61-82-5	1H-1,2,4-Triazol-3-amine
o-Anisidine hydrochloride	134-29-2	Benzenamine, 2-methoxy-, hydrochloride
Benzotrichloride	98-07-7	Benzene, (trichloromethyl)-

CHEMICAL CARCINOGENS (continued)

NTP name	CAS RN	Systematic name
Benzotrichloride	98 07 7	Benzene, (trichloromethyl)-
Beryllium	7440-41-7	Beryllium
Certain beryllium compounds		
Bischloroethyl nitrosourea	154-93-8	Urea, *N,N*′-bis(2-chloroethyl)-*N*-nitroso-
Bromodichloromethane	75-27-4	Methane, bromodichloro-
1,3-Butadiene	106-99-0	1,3-Butadiene
Butylated hydroxyanisole	25013-16-5	Phenol, (1,1-dimethylethyl)-4-methoxy-
Cadmium	7440-43-9	Cadmium
Certain cadmium compounds		
Carbon tetrachloride	56-23-5	Methane, tetrachloro-
Ceramic fibers (respirable size)		
Chlorendic acid	115-28-6	5-Norbornene-2,3-dicarboxylic acid, 1,4,5,6,7,7-hexachloro-
Chlorinated paraffins (C_{12}, 60% chlorine)	108171-26-2	
1-(2-Chloroethyl)-3-cyclohexyl-1-nitrosourea (CCNU)	13010-47-4	Urea, 1-(2-chloroethyl)-3-cyclohexyl-1-nitroso-
Chloroform	67-66-3	Methane, trichloro-
3-Chloro-2-methylpropene	563-47-3	1-Propene, 3-chloro-2-methyl-
4-Chloro-*o*-phenylenediamine	95-83-0	1,2-Benzenediamine, 4-chloro-
C.I. Basic Red 9 Monohydrochloride	569-61-9	Benzenamine, 4-[(4-aminophenyl)(4-imino-2,5-cyclohexadien-1- ylidene)methyl]-, monohydrochloride
Cisplatin	15663 27 1	Platinum(II), diamminedichloro-, *cis*
p-Cresidine	120-71-8	Benzenamine, 2-methoxy-5-methyl-
Cupferron	135-20-6	Benzenamine, *N*-hydroxy-*N*-nitroso-, ammonium salt
DDT	50-29-3	Benzene, 1,1′-(2,2,2-trichloroethylidene)bis[4-chloro-
2,4-Diaminoanisole sulfate	39156-41-7	1,3-Benzenediamine, 4-methoxy-, sulfate
2,4-Diaminotoluene	95-80-7	1,3-Benzenediamine, 4-methyl-
1,2-Dibromo-3-chloropropane	96-12-8	Propane, 1,2-dibromo-3-chloro-
1,2-Dibromoethane (Ethylene dibromide)	106-93-4	Ethane, 1,2-dibromo-
1,4-Dichlorobenzene	106-46-7	Benzene, 1,4-dichloro-
3,3′-Dichlorobenzidine	91-94-1	[1,1′-Biphenyl]-4,4′-diamine, 3,3′-dichloro-
3,3′-Dichlorobenzidine 2HCl	612-83-9	[1,1′-Biphenyl]-4,4′-diamine, 3,3′-dichloro-, dihydrochloride
1,2-Dichloroethane	107-06-2	Ethane, 1,2-dichloro-
Dichloromethane (Methylene chloride)	75-09-2	Methane, dichloro-
1,3-Dichloropropene (technical grade)	542-75-6	1-Propene, 1,3-dichloro- (mixed isomers)
Diepoxybutane	1464-53-5	2,2′-Bioxirane
Di(2-ethylhexyl) phthalate	117 81 7	1,2-Benzenedicarboxylic acid, bis(2-ethylhexyl) ester
Diethyl sulfate	64-67-5	Sulfuric acid, diethyl ester
Diglycidyl resorcinol ether	101-90-6	Oxirane, 2,2′-[1,3-phenylenebis(oxymethylene)]bis-
3,3′-Dimethoxybenzidine	119-90-4	[1,1′-Biphenyl]-4,4′-diamine, 3,3′-dimethoxy-
3,3′-Dimethoxybenzidine 2HCl	20325-40-0	[1,1′-Biphenyl]-4,4′-diamine, 3,3′-dimethoxy-, dihydrochloride
4-Dimethylamino-azobenzene	60-11-7	Benzenamine, *N,N*-dimethyl-4-(phenylazo)-
3,3′-Dimethylbenzidine	119-93-7	[1,1′-Biphenyl]-4,4′-diamine, 3,3′-dimethyl-
Dimethylcarbamoyl chloride	79-44-7	Carbamic chloride, dimethyl-
1,1-Dimethylhydrazine	57-14-7	Hydrazine, 1,1-dimethyl-
Dimethyl sulfate	77-78-1	Sulfuric acid, dimethyl ester
Dimethylvinyl chloride	513-37-1	1-Propene, 1-chloro-2-methyl-
1,4-Dioxane	123-91-1	1,4-Dioxane
Direct Black 38	1937-37-7	2,7-Naphthalenedisulfonic acid, 4-amino-3-((4′-((2,4-diaminophenyl)azo)(1,1′-biphenyl)-4-yl)azo)-5-hydroxy-6-(phenylazo)-, disodium salt
Direct Blue 6	2602-46-2	2,7-Naphthalenedisulfonic acid, 3,3′-((4,4′-biphenylylene)bis(azo))bis(5-amino-4-hydroxy-, tetrasodium salt
Epichlorohydrin	106-89-8	Propane, 1-chloro-2,3-epoxy-
Estrogens (not conjugated): Estradiol-17beta	50-28-2	Estra-1,3,5(10)-triene-3,17-diol (17ß)-
Estrogens (not conjugated): Estrone	53-16-7	Estra-1,3,5(10)-trien-17-one, 3-hydroxy-
Estrogens (not conjugated): Ethinylestradiol	57-63-6	19-Norpregna-1,3,5(10)-trien-20-yne-3,17-diol, (17α)-
Estrogens (not conjugated): Mestranol	72-33-3	17-α-19-Norpregna-1,3,5(10)-trien-20-yn-17-ol, 3-methoxy
Ethyl acrylate	140-88-5	2-Propenoic acid, ethyl ester
Ethylene oxide	75-21-8	Oxirane

NTP name	CAS RN	Systematic name
Ethylene thiourea	96-45-7	2-Imidazolidinethione
Ethyl methanesulfonate	62-50-0	Methanesulfonic acid, ethyl ester
Formaldehyde (gas)	50-00-0	Formaldehyde
Glasswool (respirable size)		
Glycidol	556-52-5	1-Propanol, 2,3-epoxy
Hexachlorobenzene	118-74-1	Benzene, hexachloro-
Hexachloroethane	67-72-1	Ethane, hexachloro-
Hexamethyl-phosphoramide	680-31-9	Phosphoric triamide, hexamethyl-
Hydrazine	302-01-2	Hydrazine
Hydrazine sulfate	10034-93-2	Hydrazine sulfate
Hydrazobenzene	122-66-7	Hydrazine, 1,2-diphenyl-
Iron dextran complex	9004-66-4	Imferon
Kepone (chlordecone)	143-50-0	1,3,4-Metheno-2H-cyclobuta[cd]pentalen-2-one, 1,1a,3,3a,4,5,5,5a,5b,6-decachlorooctahydro-
Lead acetate	301-04-2	Lead(II) acetate
Lead phosphate	7446-27-7	Lead(II) phosphate
Lindane and other hexachlorocyclohexane isomers		Cyclohexane, 1,2,3,4,5,6-hexachloro- (mixed isomers)
2-Methylaziridine (Propyleneimine)	75-55-8	Aziridine, 2-methyl-
4,4´-Methylenebis(2-chloroaniline) (MBOCA)	101-14-4	Benzenamine, 4,4´-methylenebis[2-chloro-
4,4´-Methylenebis(*N*,*N*-dimethylbenzenamine)	101-61-1	Benzenamine, 4,4´-methylenebis[*N*,*N*-dimethyl-
4,4´-Methylenedianiline	101-77-9	Benzenamine, 4,4´-methylenebis-
4,4´-Methylenedianiline 2HCl	13552-44-8	Aniline, 4,4´-methylenedi-, dihydrochloride
Methyl methanesulfonate	66-27-3	Methanesulfonic acid, methyl ester
N-Methyl-*N*-nitro-*N*-nitrosoguanidine	70-25-7	Guanidine, *N*-methyl-*N*´-nitro-*N*-nitroso-
Metronidazole	443-48-1	1H-Imidazole-1-ethanol, 2-methyl-5-nitro-
Michler´s ketone	90-94-8	Methanone, bis[4-(dimethylamino)phenyl]-
Mirex	2385-85-5	1,3,4-Metheno-1H-cyclobuta[cd]pentalene, 1,1a,2,2,3,3a,4,5,5,5a,5b,6-dodecachlorooctahydro-
Nickel	7440-02-0	Nickel
Certain nickel compounds		
Nitrilotriacetic acid	139-13-9	Glycine, *N*,*N*-bis(carboxymethyl)-
Nitrofen	1836-75-5	Benzene, 2,4-dichloro-1-(4-nitrophenoxy)-
Nitrogen mustard hydrochloride	55-86-7	Ethanamine, 2-chloro-*N*-(2-chloroethyl)-*N*-methyl-, hydrochloride
2-Nitropropane	79-46-9	Propane, 2-nitro-
N-Nitrosodi-*n*-butylamine	924-16-3	1-Butanamine, *N*-butyl-*N*-nitroso-
N-Nitrosodiethanolamine	1116-54-7	Ethanol, *N*-nitrosoiminodi-
N-Nitrosodiethylamine	55-18-5	Ethanamine, *N*-ethyl-*N*-nitroso-
N-Nitrosodimethylamine	62-75-9	Methanamine, *N*-methyl-*N*-nitroso-
N-Nitrosodi-*n*-propylamine	621-64-7	1-Propanamine, *N*-nitroso-*N*-propyl-
N-Nitroso-*n*-ethylurea	759-73-9	Urea, *N*-ethyl-*N*-nitroso-
4-(*N*-Nitrosomethyl-amino)-1-(3-pyridyl)-1-butanone (NNK)	64091-91-4	Ketone, 3-pyridyl-3-(*N*-methyl-*N*-nitrosamino)propyl-
N-Nitroso-*n*-methylurea	684-93-5	Urea, *N*-methyl-*N*-nitroso-
N-Nitrosomethyl-vinylamine	4549-40-0	Ethenylamine, *N*-methyl-*N*-nitroso-
N-Nitrosomorpholine	59-89-2	Morpholine, 4-nitroso-
N-Nitrosonornicotine	16543-55-8	Pyridine, 3-(1-nitroso-2-pyrrolidinyl)-, (*S*)-
N-Nitrosopiperidine	100-75-4	Piperidine, 1-nitroso-
N-Nitrosopyrrolidine	930-55-2	Pyrrolidine, 1-nitroso-
N-Nitrososarcosine	13256-22-9	Glycine, *N*-methyl-*N*-nitroso-
Norethisterone	68-22-4	19-Nor-17-α-pregn-4-en-20-yn-3-one, 7-hydroxy
Ochratoxin A	303-47-9	Alanine, *N*-((5-chloro-8-hydroxy-3-methyl-1-oxo-7-isochromanyl)carbonyl)-3-phenyl-, (-)
4,4´-Oxydianiline	101-80-4	Benzenamine, 4,4´-oxybis-
Oxymetholone	434-07-1	5-α,17-ß-Androstan-3-one, 17-hydroxy-2-(hydroxymethylene)-17-methyl-
Phenacetin	62-44-2	Acetamide, *N*-(4-ethoxyphenyl)-
Phenazopyridine hydrochloride	136-40-3	Pyridine, 2,6-diamino-3-(phenylazo), monohydrochloride
Phenoxybenzamine hydrochloride	63-92-3	Benzenemethanamine, *N*-(2-chloroethyl)-*N*-(1-methyl-2-phenoxyethyl)-, hydrochloride

CHEMICAL CARCINOGENS (continued)

NTP name	CAS RN	Systematic name
Phenytoin	57-41-0	2,4-Imidazolidinedione, 5,5-diphenyl-
Polybrominated biphenyls		
Polychlorinated biphenyls	1336-36-3	
Polycyclic aromatic hydrocarbons, 15 listings		
Procarbazine hydrochloride	366-70-1	*p*-Toluamide, *N*-isopropyl-α-(2-methylhydrazino)-, monohydrochloride
Progesterone	57-83-0	Pregn-4-ene-3,20-dione
1,3-Propane sultone	1120-71-4	1,2-Oxathiolane, 2,2-dioxide
beta-Propiolactone	57-57-8	2-Oxetanone
Propylene oxide	75-56-9	Propane, 1,2-epoxy
Propylthiouracil	51-52-5	4(1H)-Pyrimidinone, 2,3-dihydro-6-propyl-2-thioxo-
Reserpine	50-55-5	Yohimban-16-carboxylic acid, 11,17-dimethoxy-18-[(3,4,5-trimethoxybenzoyl)oxy]-, methyl ester, (3ß,16ß,17α18ß,20α)-
Saccharin	128-44-9	1,2-Benzisothiazol-3(2H)-one, 1,1-dioxide, sodium salt
Safrole	94-59-7	1,3-Benzodioxole, 5-(2-propenyl)-
Selenium sulfide	7446-34-6	Selenium sulfide (SSe)
Silica, crystalline (respirable size)		Silicon dioxide
Streptozotocin	18883-66-4	Glucopyranose, 2-deoxy-2-(3-methyl-3-nitrosoureido)-, *D*
Sulfallate	95-06-7	Carbamodithioic acid, diethyl-, 2-chloro-2-propenyl ester
2,3,7,8-Tetrachlorodibenzo-*p*-dioxin (TCDD)	1746-01-6	Dibenzo-*p*-dioxin, 2,3,7,8-tetrachloro-
Tetrachloroethylene (perchloroethylene)	127-18-4	Ethene, tetrachloro-
Tetranitromethane	509-14-8	Methane, tetranitro-
Thioacetamide	62-55-5	Ethanethioamide
Thiourea	62-56-6	Thiourea
Toluene diisocyanate	26471-62-5	Benzene, 1,3-diisocyanato-methyl- (mixed isomers)
o-Toluidine	95-53-4	Benzenamine, 2-methyl-
o-Toluidine hydrochloride	636-21-5	Benzenamine, 2-methyl-, hydrochloride
Toxaphene	8001-35-2	Camphene, polychlorinated
2,4,6-Trichlorophenol	88-06-2	Phenol, 2,4,6-trichloro-
Tris(1-aziridinyl)phosphine sulfide (ThioTEPA)	52-24-4	Phosphine sulfide, tris(1-aziridinyl)
Tris(2,3-dibromopropyl) phosphate	126-72-7	1-Propanol, 2,3-dibromo-, phosphate (3:1)
Urethane	51-79-6	Carbamic acid, ethyl ester
4-Vinyl-1-cyclohexene diepoxide	106-87-6	7-Oxabicyclo[4.1.0]heptane, 3-oxiranyl-

SUBSTANCES ASSOCIATED WITH TECHNOLOGICAL PROCESSES KNOWN TO BE CARCINOGENIC

Coke oven emissions
Soots, tars, and mineral oils

Appendix A
Mathematical Tables

MISCELLANEOUS MATHEMATICAL CONSTANTS

π CONSTANTS

π = 3.14159 26535 89793 23846 26433 83279 50288 41971 69399 37511
$1/\pi$ = 0.31830 98861 83790 67153 77675 26745 02872 40689 19291 48091
π^2 = 9.86960 44010 89358 61883 44909 99876 15113 53136 99407 24079
$\log_e \pi$ = 1.14472 98858 49400 17414 34273 51353 05871 16472 94812 91531
$\log_{10} \pi$ = 0.49714 98726 94133 85435 12682 88290 89887 36516 78324 38044
$\log_{10} \sqrt{2\pi}$ = 0.39908 99341 79057 52478 25035 91507 69595 02099 34102 92128

CONSTANTS INVOLVING e

e = 2.71828 18284 59045 23536 02874 71352 66249 77572 47093 69996
$1/e$ = 0.36787 94411 71442 32159 55237 70161 46086 74458 11131 03177
e^2 = 7.38905 60989 30650 22723 04274 60575 00781 31803 15570 55185
$M = \log_{10} e$ = 0.43429 44819 03251 82765 11289 18916 60508 22943 97005 80367
$1/M = \log_e 10$ = 2.30258 50929 94045 68401 79914 54684 36420 76011 01488 62877
$\log_{10} M$ = 9.63778 43113 00536 78912 29674 98645 −10

π^e AND e^π CONSTANTS

π^e = 22.45915 77183 61045 47342 71522
e^π = 23.14069 26327 79269 00572 90864
$e^{-\pi}$ = 0.04321 39182 63772 24977 44177
$e^{1/2\pi}$ = 4.81047 73809 65351 65547 30357
$i^i = e^{-1/2\pi}$ = 0.20787 95763 50761 90854 69556

NUMERICAL CONSTANTS

$\sqrt{2}$ = 1.41421 35623 73095 04880 16887 24209 69807 85696 71875 37695
$\sqrt[3]{2}$ = 1.25992 10498 94873 16476 72106 07278 22835 05702 51464 70151
$\log_e 2$ = 0.69314 71805 59945 30941 72321 21458 17656 80755 00134 36026
$\log_{10} 2$ = 0.30102 99956 63981 19521 37388 94724 49302 67881 89881 46211
$\sqrt{3}$ = 1.73205 08075 68877 29352 74463 41505 87236 69428 05253 81039
$\sqrt[3]{3}$ = 1.44224 95703 07408 38232 16383 10780 10958 83918 69253 49935
$\log_e 3$ = 1.09861 22886 68109 69139 52452 36922 52570 46474 90557 82275
$\log_{10} 3$ = 0.47712 12547 19662 43729 50279 03255 11530 92001 28864 19070

OTHER CONSTANTS

Euler's Constant γ = 0.57721 56649 01532 86061
$\log_e \gamma$ = −0.54953 93129 81644 82234
Golden Ratio ϕ = 1.61803 39887 49894 84820 45868 34365 63811 77203 09180

EXPONENTIAL AND HYPERBOLIC FUNCTIONS AND THEIR COMMON LOGARITHMS

	e^x		e^{-x}	sinh x		cosh x		tanh x
x	Value	$\log^{10}$	(value)	Value	$\log^{10}$	Value	$\log^{10}$	(value)
0.00	1.0000	0.00000	1.00000	0.0000	$-\infty$	1.0000	0.00000	0.00000
0.01	1.0101	.00434	0.99005	.0100	$\bar{2}.00001$	1.0001	.00002	.01000
0.02	1.0202	.00869	.98020	.0200	$\bar{2}.30106$	1.0002	.00009	.02000
0.03	1.0305	.01303	.97045	.0300	$\bar{2}.47719$	1.0005	.00020	.02999
0.04	1.0408	.01737	.96079	.0400	$\bar{2}.60218$	1.0008	.00035	.03998
0.05	1.0513	.02171	.95123	.0500	$\bar{2}.69915$	1.0013	.00054	.04996
0.06	1.0618	.02606	.94176	.0600	$\bar{2}.77841$	1.0018	.00078	.05993
0.07	1.0725	.03040	.93239	.0701	$\bar{2}.84545$	1.0025	.00106	.06989
0.08	1.0833	.03474	.92312	.0801	$\bar{2}.90355$	1.0032	.00139	.07983
0.09	1.0942	.03909	.91393	.0901	$\bar{2}.95483$	1.0041	.00176	.08976
0.10	1.1052	.04343	.90484	.1002	$\bar{1}.00072$	1.0050	.00217	.09967
0.11	1.1163	.04777	.89583	1102	$\bar{1}.04227$	1.0061	.00262	.10956
0.12	1.1275	.05212	.88692	.1203	$\bar{1}.08022$	1.0072	.00312	.11943
0.13	1.1388	.05646	.87809	.1304	$\bar{1}.11517$	1.0085	.00366	.12927
0.14	1.1503	.06080	.86936	.1405	$\bar{1}.14755$	1.0098	.00424	.13909
0.15	1.1618	.06514	.86071	.1506	$\bar{1}.17772$	1.0113	.00487	.14889
0.16	1.1735	.06949	.85214	.1607	$\bar{1}.20597$	1.0128	.00554	.15865
0.17	1.1853	.07383	.84366	.1708	$\bar{1}.23254$	1.0145	.00625	.16838
0.18	1.1972	.07817	.83527	.1810	$\bar{1}.25762$	1.0162	.00700	.17808
0.19	1.2092	.08252	.82696	.1911	$\bar{1}.28136$	1.0181	.00779	.18775
0.20	1.2214	.08686	.81873	.2013	$\bar{1}.30392$	1.0201	.00863	.19738
0.21	1.2337	.09120	.81058	.2115	$\bar{1}.32541$	1.0221	.00951	.20697
0.22	1.2461	.09554	.80252	.2218	$\bar{1}.34592$	1.0243	.01043	.21652
0.23	1.2586	.09989	.79453	.2320	$\bar{1}.36555$	1.0266	.01139	.22603
0.24	1.2712	.10423	.78663	.2423	$\bar{1}.38437$	1.0289	.01239	.23550
0.25	1.2840	.10857	.77880	.2526	$\bar{1}.40245$	1.0314	.01343	.24492
0.26	1.2969	.11292	.77105	.2629	$\bar{1}.41986$	1.0340	.01452	.25430
0.27	1.3100	.11726	.76338	.2733	$\bar{1}.43663$	1.0367	.01564	.26362
0.28	1.3231	12160	.75578	.2837	$\bar{1}.45282$	1.0395	.01681	.27291
0.29	1.3364	.12595	.74826	.2941	$\bar{1}.46847$	1.0423	.01801	.28213
0.30	1.3499	.13029	.74082	.3045	$\bar{1}.48362$	1.0453	.01926	.29131
0.31	1.3634	.13463	.73345	.3150	$\bar{1}.49830$	1.0484	.02054	.30044
0.32	1.3771	.13897	.72615	.3255	$\bar{1}.51254$	1.0516	.02187	.30951
0.33	1.3910	.14332	.71892	.3360	$\bar{1}.52637$	1.0549	.02323	.31852
0.34	1.4049	.14766	.71177	.3466	$\bar{1}.53981$	1.0584	.02463	.32748
0.35	1.4191	.15200	.70469	.3572	$\bar{1}.55290$	1.0619	.02607	.33638
0.36	1.4333	.15635	.69768	.3678	$\bar{1}.56564$	1.0655	.02755	.34521
0.37	1.4477	.16069	.69073	.3785	$\bar{1}.57807$	1.0692	.02907	.35399
0.38	1.4623	.16503	.68386	.3892	$\bar{1}.59019$	1.0731	.03063	.36271
0.39	1.4770	.16937	.67706	.4000	$\bar{1}.60202$	1.0770	.03222	.37136
0.40	1.4918	.17372	.67032	.4108	$\bar{1}.61358$	1.0811	.03385	.37995
0.41	1.5063	.17806	.66365	.4216	$\bar{1}.62488$	1.0852	.03552	.33847
0.42	1.5220	.18240	.65705	.4325	$\bar{1}.63594$	1.0895	.03723	.39693
0.43	1.5373	.18675	.65051	.4434	$\bar{1}.64677$	1.0939	.03897	.40532
0.44	1.5527	.19109	.64404	.4543	$\bar{1}.65738$	1.0984	.04075	.41364
0.45	1.5683	.19543	.63763	.4653	$\bar{1}.66777$	1.1030	.04256	.42190
0.46	1.5841	.19978	.63128	.4764	$\bar{1}.67797$	1.1077	.04441	.43008
0.47	1.6000	.20412	.62500	.4875	$\bar{1}.68797$	1.1125	.04630	.43820
0.48	1.6161	.20846	.61878	.4986	$\bar{1}.69779$	1.1174	.04822	.44624
0.49	1.6323	.21280	.61263	.5098	$\bar{1}.70744$	1.1225	.05018	.45422
0.50	1.6487	.21715	.60653	.5211	$\bar{1}.71692$	1.1276	.05217	.46212
0.51	1.6653	.22149	.60050	.5324	$\bar{1}.72624$	1.1329	.05419	.46995
0.52	1.6820	.22583	.59452	.5438	$\bar{1}.73540$	1.1383	.05625	.47770
0.53	1.6989	.23018	.58860	.5552	$\bar{1}.74442$	1.1438	.05834	.48538
0.54	1.7160	.23452	.58275	.5666	$\bar{1}.75330$	1.1494	.06046	.49299
0.55	1.7333	.23886	.57695	.5782	$\bar{1}.76204$	1.1551	.06262	.50052
0.56	1.7507	.24320	.57121	.5897	$\bar{1}.77065$	1.1609	.06481	.50798
0.57	1.7683	.24755	.56553	.6014	$\bar{1}.77914$	1.1669	.06703	.51536
0.58	1.7860	.25189	.55990	.6131	$\bar{1}.78751$	1.1730	.06929	.52267
0.59	1.8040	.25623	.55433	.6248	$\bar{1}.79576$	1.1792	.07157	.52990
0.60	1.8221	.26058	.54881	.6367	$\bar{1}.80390$	1.1855	.07389	.53705
0.61	1.8404	.26492	.54335	.6485	$\bar{1}.81194$	1.1919	.07624	.54413
0.62	1.8589	.26926	.53794	.6605	$\bar{1}.81987$	1.1984	.07861	.55113
0.63	1.8776	.27361	.53259	.6725	$\bar{1}.82770$	1.2051	.08102	.55805
0.64	1.8965	.27795	.52729	.6846	$\bar{1}.83543$	1.2119	.08346	.56490
0.65	1.9155	.28229	.52205	.6967	$\bar{1}.84308$	1.2188	.08593	.57167
0.66	1.9348	.28664	.51685	.7090	$\bar{1}.85063$	1.2258	.08843	.57836
0.67	1.9542	.29098	.51171	.7213	$\bar{1}.85809$	1.2330	.09095	.58498
0.68	1.9739	.29532	.50662	.7336	$\bar{1}.86548$	1.2402	.09351	.59152
0.69	1.9937	.29966	.50158	.7461	$\bar{1}.87278$	1.2476	.09609	.59798
0.70	2.0138	.30401	.49659	.7586	$\bar{1}.88000$	1.2552	.09870	.60437
0.71	2.0340	.30835	.49164	.7712	$\bar{1}.88715$	1.2628	.10134	.61068
0.72	2.0544	.31269	.48675	.7838	$\bar{1}.89423$	1.2706	.10401	.61691

EXPONENTIAL AND HYPERBOLIC FUNCTIONS AND THEIR COMMON LOGARITHMS (continued)

x	e^x Value	e^x $\log^{10}$	e^{-x} (value)	sinh x Value	sinh x $\log^{10}$	cosh x Value	cosh x $\log^{10}$	tanh x (value)
0.73	2.0751	.31703	.48191	.7966	$\bar{1}$.90123	1.2785	.10670	.62307
0.74	2.0959	.32138	.47711	.8094	$\bar{1}$.90817	1.2865	.10942	.62915
0.75	2.1170	.32572	.47237	.8223	$\bar{1}$.91504	1.2947	.11216	.63515
0.76	2.1383	.33006	.46767	.8353	$\bar{1}$.92185	1.3030	.11493	.64108
0.77	2.1598	.33441	.46301	.8484	$\bar{1}$.92859	1.3114	.11773	.64693
0.78	2.1815	.33875	.45841	.8615	$\bar{1}$.93527	1.3199	.12055	.65721
0.79	2.2034	.34309	.45384	.8748	$\bar{1}$.94190	1.3286	.12340	.65841
0.80	2.2255	.34744	.44933	.8881	$\bar{1}$.94846	1.3374	.12627	.66404
0.81	2.2479	.35178	.44486	.9015	$\bar{1}$.95498	1.3464	.12917	.66959
0.82	2.2705	.35612	.44043	.9150	$\bar{1}$.96144	1.3555	.13209	.67507
0.83	2.2933	.36046	.43605	.9286	$\bar{1}$.96784	1.3647	.13503	.68048
0.84	2.3164	.36481	.43171	.9423	$\bar{1}$.97420	1.3740	.13800	.68581
0.85	2.3396	.36915	.42741	.9561	$\bar{1}$.98051	1.3835	.14099	.69107
0.86	2.3632	.37349	.42316	.9700	$\bar{1}$.98677	1.3932	.14400	.69626
0.87	2.3869	.37784	.41895	.9840	$\bar{1}$.99299	1.4029	.14704	.70137
0.88	2.4100	.38218	.41478	.9981	1.99916	1.4128	.15009	.70642
0.89	2.4351	.38652	.41066	1.0122	0.00528	1.4229	.15317	.71139
0.90	2.4596	.39087	.40657	1.0265	.01137	1.4331	.15627	.21630
0.91	2.4843	39521	.40242	1.0409	.01741	1.4434	.15939	.72113
0.92	2.5093	.39955	.39852	1.0554	.02341	1.4539	.16254	.72590
0.93	2.5345	.40389	.39455	1.0700	.02937	1.4645	.16570	.73059
0.94	2.5600	.40824	.39063	1.0847	.03530	1.4753	.16888	.73522
0.95	2.5857	.41258	.38674	1.0995	.04119	1.4862	.17208	.73978
0.96	2.6117	.41692	.38289	1.1144	.04704	1.4973	.17531	.74428
0.97	2.6379	.42127	.37908	1.1294	.05286	1.5085	.17855	.74870
0.98	2.6645	.42561	.37531	1.1446	.05864	1.5199	.18181	.75307
0.99	2.6912	.42995	.37158	1.1598	.06439	1.5314	.18509	.75736
1.00	2.7183	.43429	.36788	1.1752	.07011	1.5431	.18839	.76159
1.01	2.7456	.43864	.36422	1.1907	.07580	1.5549	.19171	.76576
1.02	2.7732	.44298	.36060	1.2063	.06146	1.5669	.19504	.76987
1.03	2.8011	.44732	.35701	1.2220	.08708	1.5790	.19839	.77391
1.04	2.8292	.45167	.35345	1.2379	.09268	1.5913	.20176	.77789
1.05	2.8577	.45601	.34994	1.2539	.09825	1.6038	.20515	.78181
1.06	2.8864	.46035	.34646	1.2700	.10379	1.6164	.20855	.78566
1.07	2.9154	.46470	.34301	1.2862	.10930	1.6292	.21197	.78946
1.08	2.9447	.46904	.33960	1.3025	.11479	1.6421	.21541	.79320
1.09	2.9743	.47338	.33622	1.3190	.12025	1.6552	.21886	.79688
1.10	3.0042	.47772	.33287	1.3356	.12569	1.6685	.22233	.80050
1.11	3.0344	.48207	.32956	1.3524	.13111	1.6820	.22582	.80406
1.12	3.0659	.48641	.32628	1.3693	.13649	1.6956	.22931	.80757
1.13	3.0957	.49075	.32303	1.3863	.14186	1.7083	.23283	.81102
1.14	3.1268	.49510	.31982	1.4035	.14720	1.7233	.23636	.81441
1.15	3.1582	.49944	.31644	1.4208	.15253	1.7374	.23990	.81775
1.16	3.1899	.50378	.31349	1.4382	.15783	1.7517	.24346	.82104
1.17	3.2220	.50812	.31037	1.4558	.16311	1.7662	.24703	.82427
1.18	3.2544	.51247	.30728	1.4735	.16836	1.7808	.25062	.82745
1.19	3.2871	.51681	.30422	1.4914	.17360	1.7957	.25422	.83058
1.20	3.3201	.52115	.30119	1.5095	.17882	1.8107	.25784	.83365
1.21	3.3535	.52550	.29820	1.5276	.18402	1.8258	.26146	.83668
1.22	3.3872	.52984	.29523	1.5460	.18920	1.8412	.26510	.83965
1.23	3.4212	.53418	.29229	1.5645	.19437	1.8568	.26876	.84258
1.24	3.4556	.53853	.28938	1.5831	.19951	1.8725	.27242	.83546
1.25	3.4903	.54287	.28650	1.6019	.20464	1.8884	.27610	.84828
1.26	3.5254	.54721	.28365	1.6209	.20975	1.9045	.27979	.85106
1.27	3.5609	.55155	.28083	1.6400	.21485	1.9208	.28349	.85380
1.28	3.5996	.55590	.27804	1.6593	.21993	1.9373	.28721	.85648
1.29	3.6328	.56024	.27527	1.6788	.22499	1.9540	.29093	.85913
1.30	3.6693	.56458	.27253	1.6984	.23004	1.9709	.29467	.86172
1.31	3.7062	.56893	.26982	1.7182	.23507	1.9880	.29842	.86428
1.32	3.7434	.57327	.26714	1.7381	.24009	2.0053	.30217	.86678
1.33	3.7810	.57761	.26448	1.7583	.24509	2.0228	.30594	.86925
1.34	3.8190	.58195	.26185	1.7786	.25008	2.0404	.30972	.87167
1.35	3.8574	.58630	.25924	1.7991	.25505	2.0583	.31352	.87405
1.36	3.8962	.59064	.25666	1.8198	.26002	2.0764	.31732	.87639
1.37	3.9354	.59498	.25411	1.8406	.26496	2.0947	.32113	.87869
1.38	3.9749	.59933	.25158	1.8617	.26990	2.1132	.32495	.88095
1.39	4.0149	.60367	.24908	1.8829	.27482	2.1320	.32878	.88317
1.40	4.0552	.60801	.24660	1.9043	.27974	2.1509	.33262	.88535
1.41	4.0960	.61236	.24414	1.9259	.28464	2.1700	.33647	.88749
1.42	4.1371	.61670	.24171	1.9477	.28952	2.1894	.34033	.88960
1.43	4.1787	.62104	.23931	1.9697	.29440	2.2090	.34420	.89167

EXPONENTIAL AND HYPERBOLIC FUNCTIONS AND THEIR COMMON LOGARITHMS (continued)

	e^x		e^{-x}	sinh x		cosh x		tanh x
x	Value	$\log_{10}$	(value)	Value	$\log_{10}$	Value	$\log_{10}$	(value)
1.44	4.2207	.62538	.23693	1.9919	.29926	2.2288	.34807	.89370
1.45	4.2631	.62973	.23457	2.0143	.30412	2.2488	.35196	.89569
1.46	4.3060	.63407	.23224	2.0369	.30896	2.2691	.35585	.89765
1.47	4.3492	.63841	.22993	2.0597	.31379	2.2896	.35976	.89958
1.48	4.3929	.64276	.22764	2.0827	.31862	2.3103	.36367	.90147
1.49	4.4371	.64710	.22537	2.1059	.32343	2.3312	.36759	.90332
1.50	4.4817	.65144	.22313	2.1293	.32823	2.3524	.37151	.90515
1.51	4.5267	.65578	.22091	2.1529	.33303	2.3738	.37545	.90694
1.52	4.5722	.66013	.21871	2.1768	.33781	2.3955	.37939	.90870
1.53	4.6182	.66447	.21654	2.2008	.34258	2.4174	.38334	.91042
1.54	4.6646	.66881	.21438	2.2251	.34735	2.4395	.38730	.91212
1.55	4.7115	.67316	.21225	2.2496	.35211	2.4619	.39126	.91379
1.56	4.7588	.67750	.21014	2.2743	.35686	2.4845	.39524	.91542
1.57	4.8066	.68184	.20805	2.2993	.36160	2.5073	.39921	.91703
1.58	4.8550	.68619	.20598	2.3245	.36633	2.5305	.40320	.91860
1.59	4.9037	.69053	.20393	2.3499	.37105	2.5538	.40719	.92015
1.60	4.9530	.69487	.20190	2.3756	.37577	2.5775	.41119	.92167
1.61	5.0028	.69921	.19989	2.4015	.38048	2.6013	.41520	.92316
1.62	5.0531	.70356	.19790	2.4276	.38518	2.6255	.41921	.92462
1.63	5.1039	.70790	.19593	2.4540	.38987	2.6499	.42323	.92606
1.64	5.1552	.71224	.19398	2.4806	.39456	2.6746	.42725	.92747
1.65	5.2070	.71659	.19205	2.5075	.39923	2.6995	.43129	.92886
1.66	5.2593	.72093	.19014	2.5346	.40391	2.7247	.43532	.93022
1.67	5.3122	.72527	.18825	2.5620	.40857	2.7502	.43937	.93155
1.68	5.3656	.72961	.18637	2.5896	.41323	2.7760	.44341	.93286
1.69	5.4195	.73396	.18452	2.6175	.41788	2.8020	.44747	.93415
1.70	5.4739	.73830	.18268	2.6456	.42253	2.8283	.45153	.93541
1.71	5.5290	.74264	.18087	2.6740	.42717	2.8549	.45559	.93665
1.72	5.5845	.74699	.17907	2.7027	.43180	2.8818	.45966	.93786
1.73	5.6407	.75133	.17728	2.7317	.43643	2.9090	.46374	.93906
1.74	5.6973	.75567	.17552	2.7609	.44105	2.9364	.46782	.94023
1.75	5.7546	.76002	.17377	2.7904	.44567	2.9642	.47191	.94138
1.76	5.8124	.76436	.17204	2.8202	.45028	2.9922	.47600	.94250
1.77	5.8709	.76870	.17033	2.8503	.45488	3.0206	.48009	.94361
1.78	5.9299	.77304	.16864	2.8806	.45948	3.0492	.48419	.94470
1.79	5.9895	.77739	.16696	2.9112	.46408	3.0782	.48830	.94576
1.80	6.0496	.78173	.16530	2.9422	.46867	3.1075	.49241	.94681
1.81	6.1104	.78607	.16365	2.9734	.47325	3.1371	.49652	.94783
1.82	6.1719	.79042	.16203	3.0049	.47783	3.1669	.50064	.94884
1.83	6.2339	.79476	.16041	3.0367	.48241	3.1972	.50476	.94983
1.84	6.2965	.79910	.15882	3.0689	.48698	3.2277	.50889	.95080
1.85	6.3598	.80344	.15724	3.1013	.49154	3.2585	.51302	.95175
1.86	6.4237	.80779	.15567	3.1340	.49610	3.2897	.51716	.95268
1.87	6.4383	.81213	.15412	3.1671	.50066	3.3212	.52130	.95359
1.88	6.5535	.81647	.15259	3.2005	.50521	3.3530	.52544	.95449
1.89	6.6194	.82082	.15107	3.2341	.50976	3.3852	.52959	.95537
1.90	6.6859	.82516	.14957	3.2682	.51430	3.4177	.53374	.95624
1.91	6.7531	.82950	.14808	3.3025	.51884	3.4506	.53789	.95709
1.92	6.8210	.83385	.14661	3.3372	.52338	3.4838	.54205	.95792
1.93	6.8895	.83819	.14515	3.3722	.52791	3.5173	.54621	.95873
1.94	6.9588	.84253	.14370	3.4075	.53244	3.5512	.55038	.95953
1.95	7.0287	.84687	.14227	3.4432	.53696	3.5855	.55455	.96032
1.96	7.0993	.85122	.14086	3.4792	.54148	3.6201	.55872	.96109
1.97	7.1707	.85556	.13946	3.5156	.54600	3.6551	.56290	96185
1.98	7.2427	.85990	.13807	3.5923	.55051	3.6904	.56707	.96259
1.99	7.3155	.86425	.13670	3.5894	.55502	3.7261	.57126	.96331
2.00	7.3891	.86859	.13534	3.6269	.55953	3.7622	.57544	.96403
2.01	7.4633	.87293	.13399	3.6647	.56403	3.7987	.57963	.96473
2.02	7.5383	.87727	.13266	3.7028	.56853	3.8335	.58382	.96541
2.03	7.6141	.88162	.13134	3.7414	.57303	3.8727	.58802	.96609
2.04	7.6906	.88596	.13003	3.7803	.57753	3.9103	.59221	.96675
2.05	7.7679	.89030	.12873	3.8196	.58202	3.9483	.59641	.96740
2.06	7.8460	.89465	.12745	3.8593	.58650	3.9867	.60061	.96803
2.07	7.9248	.89899	.12619	3.8993	.59099	4.0255	.60482	.96865
2.08	8.0045	.90333	.12493	3.9398	.59547	4.0647	.60903	.96926
2.09	8.0849	.90768	.12369	3.9806	.59995	4.1043	.61324	.96986
2.10	8.1662	.91202	.12246	4.0219	.60443	4.1443	.61745	.97045
2.11	8.2482	.91636	.12124	4.0635	.60890	4.1847	.62167	.97103
2.12	8.3311	.92070	.12003	4.1056	.61337	4.2256	.62589	.97159
2.13	8.4149	.92505	.11884	4.1480	.61784	4.2669	.63011	.97215
2.14	8.4994	.92939	.11765	4.1909	.62231	4.3085	.63433	.97269
2.15	8.5849	.93373	.11648	4.2342	62677	4.3507	.63856	.97323

EXPONENTIAL AND HYPERBOLIC FUNCTIONS AND THEIR COMMON LOGARITHMS (continued)

x	e^x Value	e^x $\log_{10}$	e^{-x} (value)	sinh x Value	sinh x $\log_{10}$	cosh x Value	cosh x $\log_{10}$	tanh x (value)
2.16	8.6711	.93808	.11533	4.2779	.63123	4.3932	.64278	.97375
2.17	8.7583	.94242	.11418	4.3221	.63569	4.4362	.64701	.97426
2.18	8.8463	.94676	.11304	4.3666	.64015	4.4797	.65125	.97477
2.19	8.9352	.95110	.11192	4.4116	.64460	4.5236	.65548	.97526
2.20	9.0250	.95545	.11080	4.4571	.64905	4.5679	.65972	.97574
2.21	9.1157	.95979	.10970	4.5030	.65350	4.6127	.66396	.97622
2.22	9.2073	.96413	.10861	4.5494	.65795	4.6580	.66820	.97668
2.23	9.2999	.96848	.10753	4.5962	.66240	4.7037	.67244	.97714
2.24	9.3933	.97282	.10646	4.6434	.66684	4.7499	.67668	.97759
2.25	9.4877	.97716	.10540	4.6912	.67128	4.7966	.68093	.97803
2.26	9.5831	.98151	.10435	4.7394	.67572	4.8437	.68518	.97846
2.27	9.6794	.98585	.10331	4.7880	.68016	4.8914	.68943	.97888
2.28	9.7767	.99019	.10228	4.8372	.68459	4.9395	.69368	.97929
7.29	9.8749	.99453	.10127	4.8868	.68903	4.9881	.69794	97970
2.30	9.9742	.99888	.10026	4.9370	.69346	5.0372	.70219	.98010
2.31	10.074	1.00322	.09926	4.9876	.69789	5.0868	.70645	.98049
2.32	10.176	1.00756	.09827	5.0387	.70232	5.1370	.71071	.98087
2.33	10.278	1.01191	.09730	5.0903	.70675	5.1876	.71497	.98124
2.34	10.381	1.01625	.09633	5.1425	.71117	5.2388	.71923	.98161
2.35	10.486	1.02059	.09537	5.1951	.71559	5.2905	.72349	.98197
2.36	10.591	1.02493	.09442	5.2483	.72002	5.3427	.72776	.98233
2.37	10.697	1.02928	.09348	5.3020	.72444	5.3954	.73203	.98267
2.38	10.805	1.03362	.09255	5.3562	.72885	5.4487	.73630	.98301
2.39	10.913	1.03796	.09163	5.4109	.73327	5.5026	.74056	.98335
2.40	11.023	1.04231	.09072	5.4662	.73769	5.5569	.74484	.98367
2.41	11.134	1.04665	.08982	5.5221	.74210	5.6119	.74911	.98400
2.42	11.246	1.05099	.08892	5.5785	.74652	5.6674	.75338	.98431
2.43	11.359	1.05534	08804	5.6354	.75093	5.7235	.75766	.98462
2.44	11.473	1.05968	08716	5.6929	.75534	5.7801	.76194	.98492
2.45	11.588	1.06402	.08629	5.7510	.75975	5.8373	.76621	.98522
2.46	11.705	1.06836	.08543	5.8097	.76415	5.8951	.77049	.98551
2.47	11.822	1.07271	.08458	5.8689	.76856	5.9535	.77477	.98579
2.48	11.941	1.07705	.08374	5.9288	.77296	6.0125	.77906	.98607
2.49	12.061	1.08139	.08291	5.9892	.77737	6.0721	.78334	.98635
2.50	12.182	1.08574	.08208	6.0502	.78177	6.1323	.78762	.98661
2.51	12.305	1.09008	.08127	6.1118	.78617	6.1931	.79191	.98688
2.52	12.429	1.09442	.08046	6.1741	.79057	6.2545	.79619	.98714
2.53	12.554	1.09877	.07966	6.2369	.79497	6.3166	.80048	.98739
2.54	12.680	1.10311	.07887	6.3004	.79937	6.3793	.80477	.98764
2.55	12.807	1.10745	.07808	6.3645	.80377	6.4426	.80906	.98788
2.56	12.936	1.11179	.07730	6.4293	.80816	6.5066	.81335	.98812
2.57	13.066	1.11614	.07654	6.4946	.81256	6.5712	.81764	.98835
2.58	13.197	1.12048	.07577	6.5607	.81695	6.6365	.82194	.98858
2.59	13.330	1.12482	.07502	6.6274	.82134	6.7024	.82623	.98881
2.60	13.464	1.12917	.07427	6.6947	.82573	6.7690	.83052	.98903
2.61	13.599	1.13351	.07353	6.7628	.83012	6.8363	.83482	.98924
2.62	13.736	1.13785	.07280	6.8315	.83451	6.9043	.83912	.98946
2.63	13.874	1.14219	.07208	6.9008	.83890	6.9729	.84341	.98966
2.64	14.013	1.14654	.07136	6.9709	.84329	7.0423	.84771	98987
2.65	14.154	1.15008	.07065	7.0417	.84768	7.1123	.85201	.99007
2.66	14.296	1.15522	.06995	7.1132	.85206	7.1831	.85631	.99026
2.67	14.440	1.15957	.06925	7.1854	.85645	7.2546	.86061	.99045
2.68	14.585	1.16391	.06856	7.2583	.86083	7.3268	.86492	.99064
2.69	14.732	1.16825	.06788	7.3319	.86522	7.3998	.86922	.99083
2.70	14.880	1.17260	.06721	7.4063	.86960	7.4735	.87352	.99101
2.71	15.029	1.17694	.06654	7.4814	.87398	7.5479	.87783	.99118
2.72	15.180	1.18128	.06587	7.5572	.87836	7.6231	.88213	.99136
2.73	15.333	1.18562	.06522	7.6338	.88274	7.6991	.89644	.99153
2.74	15.487	1.18997	.06457	7.7112	.88712	7.7758	.89074	.99170
2.75	15.643	1.19431	06393	7.7894	.89150	7.8533	.89505	.99186
2.76	15.800	1.19865	.06329	7.8683	.89588	7.9316	.89936	.99202
2.77	15.959	1.20300	.06266	7.9480	.90026	8.0106	.90367	.99218
2.78	16.119	1.20734	.06204	8.0285	.90463	8.0905	.90798	.99233
2.79	16.281	1.21168	.06142	8.1098	.90901	8.1712	.91229	.99248
2.80	16.445	1.21602	.06081	8.1919	.91339	8.2527	.91660	.99263
2.81	16.610	1.22037	.06020	8.2749	.91776	8.3351	92091	.99278
2.82	16.777	1.22471	.05961	8.3586	.92213	8.4182	.92522	.99292
2.83	16.945	1.22905	.05901	8.4432	.92651	8.5022	.92953	.99306
2.84	17.116	1.23340	.05843	8.5287	.93088	8.5871	.93385	.99320
2.85	17.288	1.23774	.05784	8.6150	.93525	8.6728	.93816	.99333
2.86	17.462	1.24208	.05727	8.7021	.93963	8.7594	.94247	.99346

EXPONENTIAL AND HYPERBOLIC FUNCTIONS AND THEIR COMMON LOGARITHMS (continued)

x	e^x Value	e^x $\log^{10}$	e^{-x} (value)	sinh x Value	sinh x $\log^{10}$	cosh x Value	cosh x $\log^{10}$	tanh x (value)
2.87	17.637	1.24643	.05670	8.7902	.94400	8.8469	.94679	.99359
2.88	17.814	1.25077	.05613	8.8791	.94837	8.9352	95110	.99372
2.89	17.993	1.25511	.05558	8.9689	.95274	9.0244	.95542	.99384
2.90	18.174	1.25945	.05502	9.0596	.95711	9.1146	.95974	.99396
2.91	18.357	1.26380	.05448	9.1512	.96148	9.2056	.96405	.99408
2.92	18.541	1.26814	.05393	9.2437	.96584	9.2976	.96837	.99420
2.93	18.728	1.27248	.05340	9.3371	.97021	9.3905	.97269	.99431
2.94	18.916	1.27683	.05287	9.4315	.97458	9.4844	.97701	.99443
2.95	19.106	1.28117	.05234	9.5268	.97895	9.5791	.98133	.99454
2.96	19.298	1.28551	.05182	9.6231	.98331	9.6749	.98565	.99464
2.97	19.492	1.28985	.05130	9.7203	.98768	9.7716	.98997	.99475
2.98	19.688	1.29420	.05079	9.8185	.99205	9.8693	.99429	.99485
2.99	19.886	1.29854	.05029	9.9177	.99641	9.9680	.99861	.99496
3.00	20.086	1.30288	.04979	10.018	1.00078	10.068	1.00293	0.99505
3.05	21.115	1.32460	.04736	10.534	1.02259	10.581	1.02454	0.99552
3.10	22.198	1.34631	.04505	11.076	1.04440	11.122	1.04616	0.99595
3.15	23.336	1.36803	.04285	11.647	1.06620	11.690	1.06779	0.99633
3.20	24.533	1.38974	.04076	12.246	1.08799	12.287	1.08943	0.99668
3.25	25.790	1.41146	.03877	12.876	1.10977	12.915	1.11108	0.99700
3.30	27.113	1.43317	.03688	13.538	1.13155	13.575	1.13273	0.99728
3.35	28.503	1.45489	.03508	14.234	1.15332	14.269	1.15439	0.99754
3.40	29.964	1.47660	.03337	14.965	1.17509	14.999	1.17605	0.99777
3.45	31.500	1.49832	.03175	15.734	1.19685	15.766	1.19772	0.99799
3.50	33.115	1.52003	.03020	16.543	1.21860	16.573	1.21940	0.99818
3.55	34.813	1.54175	.02872	17.392	1.24036	17.421	1.24107	0.99835
3.60	36.598	1.56346	.02732	18.286	1.26211	18.313	1.26275	0.99851
3.65	38.475	1.58517	02599	19.224	1.28385	19.250	1.28444	9.99865
3.70	40.447	1.60689	.02472	20.211	1.30559	20.236	1.30612	0.99878
3.75	42.521	1.62860	.02352	21.249	1.32733	21.272	1.32781	0.99889
3.80	44.701	1.65032	.02237	22.339	1.34907	22.362	1.34951	0.99900
3.85	46.993	1.67203	.02128	23.486	1.37081	23.507	1.37120	0.99909
3.90	49.402	1.69375	.02024	24.691	1.39254	24.711	1.39290	0.99918
3.95	51.935	1.71546	.01925	25.958	1.41427	25.977	1.41459	0.09926
4.00	54.598	1.73718	.01832	27.290	1.43600	27.308	1.43629	0.99933
4.10	60.340	1.78061	.01657	30.162	1.47946	30.178	1.47970	0.99945
4.20	66.686	1.82404	.01500	33.336	1.52291	33.351	1.52310	0.99955
4.30	73.700	1.86747	.01357	36.843	1.56636	36.857	1.56652	0.99963
4.40	81.451	1.91090	.01227	40.719	1.60980	40.732	1.60993	0.99970
4.50	90.017	1.95433	.01111	45.003	1.65324	45.014	1.65335	0.99975
4.60	99.484	1.99775	.01005	49.737	1.69668	49.747	1.69677	0.99980
4.70	109.95	2.04118	.00910	54.969	1.74012	54.978	1.74019	0.99983
4.80	121.51	2.08461	.00823	60.751	1.78355	60.759	1.78361	0.99986
4.90	134.29	2.12804	.00745	67.141	1.82699	67.149	1.82704	0.99989
5.00	148.41	2.17147	.00674	74.203	1.87042	74.210	1.87046	0.99991
5.10	164.02	2.21490	.00610	82.008	1.91389	82.014	1.91389	0.99993
5.20	181.27	2.25833	.00552	90.633	1.95729	90.639	1.95731	0.99994
5.30	200.34	2.30176	.00499	100.17	2.00074	100.17	2.00074	0.99995
5.40	221.41	2.34519	.00452	110.70	2.04415	110.71	2.04417	0.99996
5.50	244.69	2.38862	.00409	122.34	2.08758	122.35	2.08760	0.99997
5.60	270.43	2.43205	.00370	135.21	2.13101	135.22	2.13103	0.99997
5.70	298.87	2.47548	.00335	149.43	2.17444	149.44	2.17445	0.99998
5.80	330.30	2.51891	.00303	165.15	2.21787	165.15	2.21788	0.99998
5.90	365.04	2.56234	.00274	182.52	2.26130	182.52	2.26131	0.99998
6.00	403.43	2.60577	.00248	201.71	2.30473	201.72	2.30474	0.99999
6.25	518.01	2.71434	.00193	259.01	2.41331	259.01	2.41331	0.99999
6.50	665.14	2.82291	.00150	332.57	2.52188	332.57	2.52189	1.00000
6.75	854.06	2.93149	.00117	427.03	2.63046	427.03	2.63046	1.00000
7.00	1096.6	3.04006	.00091	548.32	2.73904	548.32	2.73903	1.00000
7.50	1808.0	3.25721	.00055	904.02	2.95618	904.02	2.95618	1.00000
8.00	2981.0	3.47436	.00034	1490.5	3.17333	1490.5	3.17333	1.00000
8.50	4914.8	3.69150	.00020	2457.4	3.39047	2457.4	3.39047	1.00000
9.00	8103.1	3.90865	.00012	4051.5	3.60762	4051.5	3.60762	1.00000
9.50	13360.	4,12580	.00007	6679.9	3.82477	6679.9	3.82477	1.00000
10.00	22026.	4.34294	.00005	11013.	4.04191	11013.	4.04191	1.00000

NATURAL TRIGONOMETRIC FUNCTIONS TO FOUR PLACES

X radians	X degrees	sin x	cos x	tan x	cot x	sec x	csc x		
.0000	**0° 00′**	.0000	1.0000	.0000	—	1.000	—	**90° 00′**	1.5708
.0029	10	.0029	1.0000	.0029	343.8	1.000	343.8	50	1.5679
.0058	20	.0058	1.0000	.0058	171.9	1.000	171.9	40	1.5650
.0087	30	.0087	1.0000	.0087	114.6	1.000	114.6	30	1.5621
.0116	40	.0116	.9999	.0116	85.94	1.000	85.95	20	1.5592
.0145	50	.0145	.9999	.0145	68.75	1.000	68.76	10	1.5563
.0175	**1° 00′**	.0175	.9998	.0175	57.29	1.000	57.30	**89° 00′**	1.5533
.0204	10	.0204	.9998	.0204	49.10	1.000	49.11	50	1.5504
.0233	20	.0233	.9997	.0233	42.96	1.000	42.98	40	1.5475
.0262	30	.0262	.9997	.0262	38.19	1.000	38.20	30	1.5446
.0291	40	.0291	.9996	.0291	34.37	1.000	34.38	20	1.5417
.0320	50	.0320	.9995	.0320	31.24	1.001	31.26	10	1.5388
.0349	**2° 00′**	.0349	.9994	.0349	28.64	1.001	28.65	**88° 00′**	1.5359
.0378	10	.0378	.9993	.0378	26.43	1.001	26.45	50	1.5330
.0407	20	.0407	.9992	.0407	24.54	1.001	24.56	40	1.5301
.0436	30	.0436	.9990	.0437	22.90	1.001	22.93	30	1.5272
.0465	40	.0465	.9989	.0466	21.47	1.001	21.49	20	1.5243
.0495	50	.0494	.9988	.0495	20.21	1.001	20.23	10	1.5213
.0524	**3° 00′**	.0523	.9986	.0524	19.08	1.001	19.11	**87° 00′**	1.5184
.0553	10	.0552	.9985	.0553	18.07	1.002	18.10	50	1.5155
.0582	20	.0581	.9983	.0582	17.17	1.002	17.20	40	1.5126
.0611	30	.0610	.9981	.0612	16.35	1.002	16.38	30	1.5097
.0640	40	.0640	.9980	.0641	15.60	1.002	15.64	20	1.5068
.0669	50	.0669	.9978	.0670	14.92	1.002	14.96	10	1.5039
.0698	**4° 00′**	.0698	.9976	.0699	14.30	1.002	14.34	**86° 00′**	1.5010
.0727	10	.0727	.9974	.0729	13.73	1.003	13.76	50	1.4981
.0756	20	.0756	.9971	.0758	13.20	1.003	13.23	40	1.4952
.0785	30	.0785	.9969	.0787	12.71	1.003	12.75	30	1.4923
.0814	40	.0814	.9967	.0816	12.25	1.003	12.29	20	1.4893
.0844	50	.0843	.9964	.0846	11.83	1.004	11.87	10	1.4864
.0873	**5° 00′**	.0872	.9962	.0875	11.43	1.004	11.47	**85° 00′**	1.4835
.0902	10	.0901	.9959	.0904	11.06	1.004	11.10	50	1.4806
.0931	20	.0929	.9957	.0934	10.71	1.004	10.76	40	1.4777
.0960	30	.0958	.9954	.0963	10.39	1.005	10.43	30	1.4748
.0989	40	.0987	.9951	.0992	10.08	1.005	10.13	20	1.4719
.1018	50	.1016	.9948	.1022	9.788	1.005	9.839	10	1.4690
.1047	**6° 00′**	.1045	.9945	.1051	9.514	1.006	9.567	**84° 00′**	1.4661
.1076	10	.1074	.9942	.1080	9.255	1.006	9.309	50	1.4632
.1105	20	.1103	.9939	.1110	9.010	1.006	9.065	40	1.4603
.1134	30	.1132	.9936	.1139	8.777	1.006	8.834	30	1.4573
.1164	40	.1161	.9932	.1169	8.556	1.007	8.614	20	1.4544
.1193	.50	.1190	.9929	.1198	8.345	1.007	8.405	10	1.4515
.1222	**7° 00′**	.1219	.9925	.1228	8.144	1.008	8.206	**83° 00′**	1.4486
.1251	10	.1248	.9922	.1257	7.953	1.008	8.016	50	1.4457
.1280	20	.1276	.9918	.1287	7.770	1.008	7.834	40	1.4428
.1309	30	.1305	.9914	.1317	7.596	1.009	7.661	30	1.4399
.1338	40	.1334	.9911	.1346	7.429	1.009	7.496	20	1.4370
.1367	50	.1363	.9907	.1376	7.269	1.009	7.337	10	1.4341
.1396	**8° 00′**	.1392	.9903	.1405	7.115	1.010	7.185	**82° 00′**	1.4312
.1425	10	.1421	.9899	.1435	6.968	1.010	7.040	50	1.4283
.1454	20	.1449	.9894	.1465	6.827	1.011	6.900	40	1.4254
.1484	30	.1478	.9890	.1495	6.691	1.011	6.765	30	1.4224
.1513	40	.1507	.9886	.1524	6.561	1.012	6.636	20	1.4195
.1542	50	.1536	.9881	.1554	6.435	1.012	6.512	10	1.4166
.1571	**9° 00′**	.1564	.9877	.1584	[illegible]	1.012	6.392	**81° 00′**	1.4137
.1600	10	.1593	.9872	.1614	6.197	1.013	6.277	50	1.4108
.1629	20	.1622	.9868	.1644	6.084	1.013	6.166	40	1.4079
.1658	30	.1650	.9863	.1673	5.976	1.014	6.059	30	1.4050
.1687	40	.1679	.9858	.1703	5.871	1.014	5.955	20	1.4021
.1716	50	.1708	.9853	.1733	5.769	1.015	5.855	10	1.3992
.1745	**10° 00′**	.1736	.9848	.1763	5.671	1.015	5.759	**80° 00′**	1.3963
.1774	10	.1765	.9843	.1793	5.576	1.016	5.665	50	1.3934
.1804	20	.1794	.9838	.1823	5.485	1.016	5.575	40	1.3904
.1833	30	.1822	.9833	.1853	5.396	1.017	5.487	30	1.3875
.1862	40	.1851	.9827	.1883	5.309	1.018	5.403	20	1.3846
.1891	50	.1880	.9822	.1914	5.226	1.018	5.320	10	1.3817
.1920	**11° 00′**	.1908	.9816	.1944	5.145	1.019	5.241	**79° 00′**	1.3788
.1949	10	.1937	.9811	.1974	5.066	1.019	5.164	50	1.3759
.1978	20	.1965	.9805	.2004	4.989	1.020	5.089	40	1.3730
.2007	30	.1994	.9799	.2035	4.915	1.020	5.016	30	1.3701
		cos x	**sin x**	**cot x**	**tan x**	**csc x**	**sec x**	**x degrees**	**x radians**

NATURAL TRIGONOMETRIC FUNCTIONS TO FOUR PLACES (continued)

x radians	x degrees	sin x	cos x	tan x	cot x	sec x	csc x		
.2036	40	.2022	.9793	.2065	3.843	1.021	4.945	20	1.3672
.2065	50	.2051	.9787	.2095	4.773	1.022	4.876	10	1.3643
.2094	**12° 00′**	.2079	.9781	.2126	4.705	1.022	4.810	**78° 00′**	1.3614
.2123	10	.2108	.9775	.2156	4.638	1.023	4.745	50	1.3584
.2153	20	.2136	.9769	.2186	4.574	1.024	4.682	40	1.3555
.2182	30	.2164	.9763	.2217	4.511	1.025	4.620	30	1.3526
.2211	40	.2193	.9757	.2247	4.449	1.025	4.560	20	1.3497
.2240	50	.2221	.9750	.2278	4.390	1.026	4.502	10	1.3468
.2269	**13° 00′**	.2250	.9744	.2309	4.331	1.026	4.445	**77° 00′**	1.3439
.2298	10	.2278	.9737	.2339	4.275	1.027	4.390	50	1.3410
.2327	20	.2306	.9730	.2370	4.219	1.028	4.336	40	1.3381
.2356	30	.2334	.9724	.2401	4.165	1.028	4.284	30	1.3352
.2385	40	.2363	.9717	.2432	4.113	1.029	4.232	20	1.3323
.2414	50	.2391	.9710	.2462	4.061	1.030	4.182	10	1.3294
.2443	**14° 00′**	.2419	.9703	.2493	4.011	1.031	4.134	**76° 00′**	1.3265
.2473	10	.2447	.9696	.2524	3.962	1.031	4.086	50	1.3235
.2502	20	.2476	.9689	.2555	3.914	1.032	4.039	40	1.3206
.2531	30	.2404	.9681	.2586	3.867	1.033	3.994	30	1.3177
.2560	40	.2532	.9674	.2617	3.821	1.034	3.950	20	1.3148
.2589	50	.2560	.9667	.2648	3.776	1.034	3.906	10	1.3119
.2618	**15° 00′**	.2588	.9659	.2679	3.732	1.035	3.864	**75° 00′**	1.3090
.2647	10	.2616	.9652	.2711	3.689	1.036	3.822	50	1.3061
.2676	20	.2644	.9644	.2732	3.647	1.037	3.782	40	1.3032
.2705	30	.2672	.9636	.2773	3.606	1.038	3.742	30	1.3003
.2734	40	.2700	.9628	.2805	3.566	1.039	3.703	20	1.2974
.2763	50	.2728	.9621	2836	3.526	1.039	3.665	10	1.2945
.2793	**16° 00′**	.2756	.9613	.2867	3.487	1.040	3.628	**74° 00′**	1.2915
.2822	10	.2784	.9605	.2899	3.450	1.041	3.592	50	1.2886
.2851	20	.2812	.9596	.2931	3.412	1.042	3.556	40	1.2857
.2880	30	.2840	.9588	.2962	3.376	1.043	3.521	30	1.2828
.2909	40	.2868	.9580	.2994	3.340	1.044	3.487	20	1.2799
.2938	50	.2896	.9572	.3026	3.305	1.045	3.453	10	1.2770
.2967	**17° 00′**	.2924	.9563	.3057	3.271	1.046	3.420	**73° 00′**	1.2741
.2996	10	.2952	.9555	.3089	3.237	1.047	3.388	50	1.2712
.3025	20	.2979	.9546	.3121	4.204	1.048	3.356	40	1.2683
.3054	30	.3007	.9537	.3153	3.172	1.049	3.326	30	1.2654
.3083	40	.3035	.9528	.3185	3.140	1.049	3.295	20	1.2625
.3113	50	.3062	.9520	.3217	3.108	1.050	3.265	10	1.2595
.3142	**18° 00′**	.3090	.9511	.3249	3.078	1.051	3.236	**72° 00′**	1.2566
.3171	10	.3118	.9502	.3281	3.047	1.052	3.207	50	1.2537
.3200	20	.3145	.9492	.3314	3.018	1.053	3.179	40	1.2508
.3229	30	.3173	.9483	.3346	2.989	1.054	3.152	30	1.2479
.3258	40	.3201	.9474	.3378	2.960	1.056	3.124	20	1.2450
.3287	50	.3228	.9465	.3411	2.932	1.057	3.098	10	1.2421
.3316	**19° 00′**	.3256	.9455	.3443	2.904	1.058	3.072	**71° 00′**	1.2392
.3345	10	.3283	.9446	.3476	2.877	1.059	3.046	50	1.2363
.3374	20	.3311	.9436	.3508	2.850	1.060	3.021	40	1.2334
.3403	30	.3338	.9426	.3541	2.824	1.061	2.996	30	1.2305
.3432	40	.3365	.9417	.3574	2.798	1.062	2.971	20	1.2275
.3462	50	.3393	.9407	.3607	2.773	1.063	2.947	10	1.2246
.3491	**20° 00′**	.3420	.9397	[illegible]	2.747	1.064	2.924	**70° 00′**	1.2217
.3520	10	.3448	.9387	.3673	2.723	1.065	2.901	50	1.2188
.3599	20	.3475	.9377	.3706	2.699	1.066	2.878	40	1.2159
.3578	30	.3502	.9367	.3739	2.675	1.068	2.855	30	1.2130
.3607	40	.3529	.9356	.3772	2.651	1.069	2.833	20	1.2101
.3636	50	.3557	.9346	.3805	2.628	1.070	2.812	10	1.2072
.3665	**21° 00′**	.3584	.9336	.3839	2.605	1.071	2.790	**69° 00′**	1.2043
.3694	10	.3611	.9325	.3872	2.583	1.072	2.769	50	1.2014
.3723	20	.3638	.9315	.3906	2.560	1.074	2.749	40	1.1985
.3752	30	.3665	.9304	.3939	2.539	1.075	2.729	30	1.1956
.3782	40	.3692	.9293	.3973	2.517	1.076	2.709	20	1.1926
.3811	50	.3719	.9283	.4006	2.496	1.077	2.689	10	1.1897
.3840	**22° 00′**	.3746	.9272	.4040	2.475	1.079	2.669	**68° 00′**	1.1868
.3869	10	.3773	.9261	.4074	2.455	1.080	2.650	50	1.1839
.3898	20	.3800	.9250	.4108	2.434	1.081	2.632	40	1.1810
.3927	30	.3827	.9239	.4142	2.414	1.082	2.613	30	1.1781
.3956	40	.3854	.9228	.4176	2.394	1.084	2.595	20	1.1752
.3985	50	.3881	.9216	.4210	2.375	1.085	2.577	10	1.1723
.4014	**23° 00′**	.3907	.9205	.4245	2.356	1.086	2.559	**67° 00′**	1.1694
.4043	10	.3934	.9194	.4279	2.337	1.088	2.542	50	1.1665
		cos x	sin x	cot x	tan x	csc x	sec x	x degrees	x radians

NATURAL TRIGONOMETRIC FUNCTIONS TO FOUR PLACES (continued)

x radians	x degrees	sin x	cos x	tan x	cot x	sec x	csc x		
.4072	20	.3961	.9182	.4314	2.318	1.089	2.525	40	1.1636
.4102	30	.3987	.9171	.4348	2.300	1.090	2.508	30	1.1606
4131	40	.4014	.9159	.4383	2.282	1.092	2.491	20	1.1577
.4160	50	4041	.9147	.4417	2.264	1.093	2.475	10	1.1548
.4189	**24° 00′**	.4067	.9135	.4452	2.246	1.095	2.459	**66° 00′**	1.1519
.4218	10	.4094	.9124	.4487	2.229	1.096	2.443	50	1.1490
.4247	20	.4120	.9112	.4522	2.211	1.097	2.427	40	1.1461
.4276	30	.4147	.9100	.4557	2.194	1.099	2.411	30	1.1432
.4305	40	.4173	.9088	.4592	2.177	1.100	2.396	20	1.1403
.4334	50	.4200	.9075	.4628	2.161	1.102	2.381	10	1.1374
.4363	**25° 00′**	.4226	.9063	.4663	2.145	1.103	2.366	**65° 00′**	1.1345
.4392	10	.4253	.9051	.4699	2.128	1.105	2.352	50	1.1316
.4422	20	.4279	.9038	.4734	2.112	1.106	2.337	40	1.1286
.4451	30	.4305	.9026	.4770	2.097	1.108	2.323	30	1.1257
.4480	40	.4331	.9013	.4806	2.081	1.109	2.309	20	1.1228
.4509	50	.4358	.9001	.4841	2.066	1.111	2.295	10	1.1199
.4538	**26° 00′**	.4384	.8988	.4877	2.050	1.113	2.281	**64° 00′**	1.1170
.4567	10	.4410	.8975	.4913	2.035	1.114	2.268	50	1.1141
.4596	20	.4436	.8962	.4950	2.020	1.116	2.254	40	1.1112
.4625	30	.4462	.8949	.4986	2.006	1.117	2.241	30	1.1083
.4654	40	.4488	.8936	.5022	1.991	1.119	2.228	20	1.1054
.4683	50	.4514	.8923	.5059	1.977	1.121	2.215	10	1.1025
.4712	**27° 00′**	.4540	.8910	.5095	1.963	1.122	2.203	**63° 00′**	1.0996
.4741	10	.4566	.8897	.5132	1.949	1.124	2.190	50	1.0966
.4771	20	.4592	.8884	.5169	1.935	1.126	2.178	40	1.0937
.4800	30	.4617	.8870	.5206	1.921	1.127	2.166	30	1.0908
.4829	40	.4643	.8857	.5243	1.907	1.129	2.154	20	1.0879
.4858	50	.4669	.8843	.5280	1.894	1.131	2.142	10	1.0850
.4887	**28° 00′**	.4695	.8829	.5317	1.881	1.133	2.130	**62° 00′**	1.0821
.4916	10	.4720	.8816	.5354	1.868	1.134	2.118	50	1.0792
.4945	20	.4746	.8802	.5392	1.855	1.136	2.107	40	1.0763
.4974	30	.4772	.8788	.5430	1.842	1.138	2.096	30	1.0734
.5003	40	.4797	.8774	.5467	1.829	1.140	2.085	20	1.0705
.5032	50	.4823	.8760	.5505	1.816	1.142	2.074	10	1.0676
.5061	**29° 00′**	.4848	.8746	.5543	1.804	1.143	2.063	**61° 00′**	1.0647
.5091	10	.4874	.8732	.5581	1.792	1.145	2.052	50	1.0617
.5120	20	.4899	.8718	.5619	1.780	1.147	2.041	40	1.0588
.5149	30	.4924	.8704	.5658	1.767	1.149	2.031	30	1.0559
.5178	40	.4950	.8689	.5696	1.756	1.151	2.020	20	1.0530
.5207	50	.4975	.8675	.5735	1.744	1.153	2.010	10	1.0501
.5236	**30° 00′**	.5000	.8660	.5774	1.732	1.155	2.000	**60° 00′**	1.0472
.5265	10	.5025	.8646	.5812	1.720	1.157	1.990	50	1.0443
.5294	20	.5050	.8631	.5851	1.709	1.159	1.980	40	1.0414
.5323	30	.5075	.8616	.5890	1.698	1.161	1.970	30	1.0385
.5352	40	.5100	.8601	.5930	1.686	1.163	1.961	20	1.0356
.5381	50	.5125	.8587	.5969	1.675	1.165	1.951	10	1.0327
.5411	**31° 00′**	.5150	.8572	.6009	1.664	1.167	1.942	**59° 00′**	1.0297
.5440	10	.5175	.8557	.6048	1.653	1.169	1.932	50	1.0268
.5469	20	.5200	.8542	.6088	1.643	1.171	1.923	40	1.0239
.5498	30	.5225	.8526	.6128	1.632	1.173	1.914	30	1.0210
.5527	40	.5250	.8511	.6168	1.621	1.175	1.905	20	1.0181
.5556	50	.5275	.8496	.6208	1.611	1.177	1.896	10	1.0152
.5585	**32° 00′**	.5299	8480	.6249	1.600	1.179	1.887	**58° 00′**	1.0123
.5614	10	.5324	.8465	.6289	1.590	1.181	1.878	50	1.0094
.5643	20	.5348	.8450	.6330	1.580	1.184	1.870	40	1.0065
.5672	30	.5373	.8434	.6371	1.570	1.186	1.861	30	1.0036
.5701	40	.5398	.8418	.6412	1.560	1.188	1.853	20	1.0007
.5730	50	.5422	.8403	.6453	1.550	1.190	1.844	10	.9977
.5760	**33° 00′**	.5446	.8397	.6494	1.540	1.192	1.836	**57° 00′**	.9948
.5789	10	.5471	.8371	.6536	1.530	1.195	1.828	50	.9919
.5818	20	.5495	.8355	.6577	1.520	1.197	1.820	40	.9890
.5847	30	.5519	.8339	.6619	1.511	1.199	1.812	30	.9861
.5876	40	.5544	.8323	.6661	1.501	1.202	1.804	20	.9832
.5905	50	.5568	.8307	.6703	1.492	1.204	1.796	10	.9803
.5934	**34° 00′**	.5592	.8290	.6745	1.483	1.206	1.788	**56° 00′**	.9774
.5963	10	.5616	.8274	.6787	1.473	1.209	1.781	50	.9745
.5992	20	.5640	.8258	.6830	1.464	1.211	1.773	40	.9716
.6021	30	.5664	.8241	.6873	1.455	1.213	1.766	30	.9687
.6050	40	.5688	.8225	.6916	1.446	1.216	1.758	20	.9657
.6080	50	.5712	.8208	.6959	1.437	1.218	1.751	10	.9628
		cos x	sin x	cot x	tan x	csc x	sec x	x degrees	x radians

NATURAL TRIGONOMETRIC FUNCTIONS TO FOUR PLACES (continued)

x radians	x degrees	sin x	cos x	tan x	cot x	sec x	csc x		
.6109	**35° 00′**	.5736	.8192	.7002	1.428	1.221	1.743	**55° 00′**	.9599
.6138	10	.5760	.8175	.7046	1.419	1.223	1.736	50	.9570
.6167	20	.5783	.8158	.7089	1.411	1.226	1.729	40	.9541
.6196	30	.5807	.8141	.7133	1.402	1.228	1.722	30	.9512
.6225	40	.5831	.8124	.7177	1.393	1.231	1.715	20	.9483
.6254	50	.5854	.8107	.7221	1.385	1.233	1.708	10	.9454
.6283	**36° 00′**	.5878	.8090	.7265	1.376	1.236	1.701	**54° 00′**	.9425
.6312	10	.5901	.8073	.7310	1.368	1.239	1.695	50	.9396
.6341	20	.5925	.8056	.7355	1.360	1.241	1.688	40	.9367
.6370	30	.5948	.8039	.7400	1.351	1.244	1.681	30	.9338
.6400	40	.5972	.8021	.7445	1.343	1.247	1.675	20	.9308
.6429	50	.5995	.8004	.7490	1.335	1.249	1.668	10	.9279
.6458	**37° 00′**	.6018	.7986	.7536	1.327	1.252	1.662	**53° 00′**	.9250
.6487	10	.6041	.7969	.7581	1.319	1.255	1.655	50	.9221
.6516	20	.6065	.7951	.7627	1.311	1.258	1.649	40	.9192
.6545	30	.6088	.7934	.7673	1.303	1.260	1.643	30	.9163
.6574	40	.6111	.7916	.7720	1.295	1.263	1.636	20	.9134
.6603	50	.6134	.7898	.7766	1.288	1.266	1.630	10	.9105
.6632	**38° 00′**	.6157	.7880	.7813	1.280	1.269	1.624	**52° 00′**	.9076
.6661	10	.6180	.7862	.7860	1.272	1.272	1.618	50	.9047
.6690	20	.6202	.7844	.7907	1.265	1.275	1.612	40	.9018
.6720	30	.6225	.7826	.7954	1.257	1.278	1.606	30	.8988
.6749	40	.6248	.7808	.8002	1.250	1.281	1.601	20	.8959
.6778	50	.6271	.7790	.8050	1.242	1.284	1.595	10	.8930
.6807	**39° 00′**	.6293	.7771	.8098	1.235	1.287	1.589	**51° 00′**	.8901
.6836	10	.6316	.7753	.8146	1.228	1.290	1.583	50	.8872
.6865	20	.6338	.7735	.8195	1.220	1.293	1.578	40	.8843
.6894	30	.6361	.7716	.8243	1.213	1.296	1.572	30	.8814
.6923	40	.6383	.7698	.8292	1.206	1.299	1.567	20	.8785
.6952	50	.6406	.7679	.8342	1.199	1.302	1.561	10	.8756
.6981	**40° 00′**	.6428	.7660	.8391	1.192	1.305	1.556	**50° 00′**	.8727
.7010	10	.6450	.7642	.8441	1.185	1.309	1.550	50	.8698
.7039	20	.6472	.7623	.8491	1.178	1.312	1.545	40	.8668
.7069	30	.6494	.7604	.8541	1.171	1.315	1.540	30	.8639
.7098	40	.6517	.7585	.8591	1.164	1.318	1.535	20	.8610
.7127	50	.6539	.7566	.8642	1.157	1.322	1.529	10	.8581
.7156	**41° 00′**	.6561	.7547	.8693	1.150	1.325	1.524	**49° 00′**	.8552
.7185	10	.6583	.7528	.8744	1.144	1.328	1.519	50	.8523
.7214	20	.6604	.7509	.8796	1.137	1.332	1.514	40	.8494
.7243	30	.6626	.7490	.8847	1.130	1.335	1.509	30	.8465
.7272	40	.6648	.7470	.8899	1.124	1.339	1.504	20	.8436
.7301	50	.6670	.7451	.8952	1.117	1.342	1.499	10	.8407
.7330	**42° 00′**	.6691	.7431	.9004	1.111	1.346	1.494	**48° 00′**	.8378
.7359	10	.6713	.7412	.9057	1.104	1.349	1.490	50	.8348
7389	20	.6734	.7392	.9110	1.098	1.353	1.485	40	.8319
.7418	30	.6756	.7373	.9163	1.091	1.356	1.480	30	.8290
.7447	40	.6777	.7353	.9217	1.085	1.360	1.476	20	.8261
.7476	50	.6799	.7333	.9271	1.079	1.364	1.471	10	.8232
.7505	**43° 00′**	.6820	.7314	.9325	1.072	1.367	1.466	**47° 00′**	.8203
.7534	10	.6841	.7294	.9380	1.066	1.371	1.462	50	.8174
.7563	20	.6862	.7274	.9435	1.060	1.375	1.457	40	.8145
.7592	30	.6884	.7254	.9490	1.054	1.379	1.453	30	.8116
.7621	40	.6905	.7234	.9545	1.048	1.382	1.448	20	.8087
.7650	50	.6926	.7214	.9601	1.042	1.386	1.444	10	.8058
.7679	**44° 00′**	.6947	.7193	.9657	1.036	1.390	1.440	**46° 00′**	.8029
.7709	10	.6967	.7173	.9713	1.030	1.394	1.435	50	.7999
.7738	20	.6988	.7153	.9770	1.024	1.398	1.431	40	.7970
.7767	30	.7009	.7133	.9827	1.018	1.402	1.427	30	.7941
.7796	40	.7030	.7112	.9884	1.012	1.406	1.423	20	.7912
.7825	50	.7050	.7092	.9942	1.006	1.410	1.418	10	.7883
.7854	**45° 00′**	.7071	.7071	1.0000	1.0000	1.414	1.414	**45° 00′**	.7854
		cos x	sin x	cot x	tan x	csc x	sec x	x degrees	x radians

RELATION OF ANGULAR FUNCTIONS IN TERMS OF ONE ANOTHER

TRIGONOMETRIC FUNCTIONS

Function	sin α	cos α	tan α	cot α	sec α	csc α
$\sin\alpha$	$\sin\alpha$	$\pm\sqrt{1-\cos^2\alpha}$	$\dfrac{\tan\alpha}{\pm\sqrt{1+\tan^2\alpha}}$	$\dfrac{1}{\pm\sqrt{1+\cot^2\alpha}}$	$\dfrac{\pm\sqrt{\sec^2\alpha-1}}{\sec\alpha}$	$\dfrac{1}{\csc\alpha}$
$\cos\alpha$	$\pm\sqrt{1-\sin^2\alpha}$	$\cos\alpha$	$\dfrac{1}{\pm\sqrt{1+\tan^2\alpha}}$	$\dfrac{\cot\alpha}{\pm\sqrt{1+\cot^2\alpha}}$	$\dfrac{1}{\sec\alpha}$	$\dfrac{\pm\sqrt{\csc^2\alpha-1}}{\csc\alpha}$
$\tan\alpha$	$\dfrac{\sin\alpha}{\pm\sqrt{1-\sin^2\alpha}}$	$\dfrac{\pm\sqrt{1-\cos^2\alpha}}{\cos\alpha}$	$\tan\alpha$	$\dfrac{1}{\cot\alpha}$	$\pm\sqrt{\sec^2\alpha-1}$	$\dfrac{1}{\pm\sqrt{\csc^2\alpha-1}}$
$\cot\alpha$	$\dfrac{\pm\sqrt{1-\sin^2\alpha}}{\sin\alpha}$	$\dfrac{\cos\alpha}{\pm\sqrt{1-\cos^2\alpha}}$	$\dfrac{1}{\tan\alpha}$	$\cot\alpha$	$\dfrac{1}{\pm\sqrt{\sec^2\alpha-1}}$	$\pm\sqrt{\csc^2\alpha-1}$
$\sec\alpha$	$\dfrac{1}{\pm\sqrt{1-\sin^2\alpha}}$	$\dfrac{1}{\cos\alpha}$	$\pm\sqrt{1+\tan^2\alpha}$	$\dfrac{\pm\sqrt{1+\cot^2\alpha}}{\cot\alpha}$	$\sec\alpha$	$\dfrac{\csc\alpha}{\pm\sqrt{\csc^2\alpha-1}}$
$\csc\alpha$	$\dfrac{1}{\sin\alpha}$	$\dfrac{1}{\pm\sqrt{1-\cos^2\alpha}}$	$\dfrac{\pm\sqrt{1+\tan^2\alpha}}{\tan\alpha}$	$\pm\sqrt{1+\cot^2\alpha}$	$\dfrac{\sec\alpha}{\pm\sqrt{\sec^2\alpha-1}}$	$\csc\alpha$

Note: The choice of sign depends upon the quadrant in which the angle terminates.

HYPERBOLIC FUNCTIONS

Function	sinh x	cosh x	tanh x
$\sinh x =$	$\sinh x$	$+\sqrt{\cosh^2 x-1}$	$\dfrac{\tanh x}{\sqrt{1-\tanh^2 x}}$
$\cosh x =$	$\sqrt{1+\sinh^2 x}$	$\cosh x$	$\dfrac{1}{\sqrt{1-\tanh^2 x}}$
$\tanh x =$	$\dfrac{\sinh x}{\sqrt{1+\sinh^2 x}}$	$\pm\dfrac{\sqrt{\cosh^2 x-1}}{\cosh x}$	$\tanh x$
$\text{cosech}\, x =$	$\dfrac{1}{\sinh x}$	$\pm\dfrac{1}{\sqrt{\cosh^2 x-1}}$	$\dfrac{\sqrt{1-\tanh^2 x}}{\tanh x}$
$\text{sech}\, x =$	$\dfrac{1}{\sqrt{1+\sinh^2 x}}$	$\dfrac{1}{\cosh x}$	$\sqrt{1-\tanh^2 x}$
$\coth x =$	$\dfrac{\sqrt{1+\sinh^2 x}}{\sinh x}$	$\dfrac{\pm\cosh x}{\sqrt{\cosh^2 x-1}}$	$\dfrac{1}{\tanh x}$

Function	cosech x	sech x	coth x
$\sinh x =$	$\dfrac{1}{\text{cosech}\, x}$	$\pm\dfrac{\sqrt{1-\text{sech}^2 x}}{\text{sech}\, x}$	$\dfrac{\pm 1}{\sqrt{\coth^2 x-1}}$
$\cosh x =$	$\pm\dfrac{\sqrt{\text{cosech}^2 x+1}}{\text{cosech}\, x}$	$\dfrac{1}{\text{sech}\, x}$	$\pm\dfrac{\coth x}{\sqrt{\coth^2 x-1}}$
$\tanh x =$	$\dfrac{1}{\sqrt{\text{cosech}^1 x+1}}$	$\pm\sqrt{1-\text{sech}^2 x}$	$\dfrac{1}{\coth x}$
$\text{cosech}\, x =$	$\text{cosech}\, x$	$\pm\dfrac{\text{sech}\, x}{\sqrt{1-\text{sech}^1 zsx}}$	$\pm\dfrac{\sqrt{\coth^2 x-1}}{1}$
$\text{sech}\, x =$	$\pm\dfrac{\text{cosec}\, x}{\sqrt{\text{cosech}^2 x+1}}$	$\text{sech}\, x$	$\pm\dfrac{\sqrt{\coth^2 x-1}}{\coth x}$
$\coth x =$	$\sqrt{\text{cosech}^2 x+1}$	$\pm\dfrac{1}{\sqrt{1-\text{sech}^2 x}}$	$\coth x$

Whenever two signs are shown, choose + sign if x is positive, − sign if x is negative.

Derivatives*

In the following formulas u, v, w represent functions of x, while a, c, n represent fixed real numbers. All arguments in the trigonometric functions are measured in radians, and all inverse trigonometric and hyperbolic functions represent principal values.

1. $\frac{d}{dx}(a) = 0$

2. $\frac{d}{dx}(x) = 1$

3. $\frac{d}{dx}(au) = a\frac{du}{dx}$

4. $\frac{d}{dx}(u + v - w) = \frac{du}{dx} + \frac{dv}{dx} - \frac{dw}{dx}$

5. $\frac{d}{dx}(uv) = u\frac{dv}{dx} + v\frac{du}{dx}$

6. $\frac{d}{dx}(uvw) = uv\frac{dw}{dx} + vw\frac{du}{dx} + uw\frac{dv}{dx}$

7. $\frac{d}{dx}\left(\frac{u}{v}\right) = \frac{v\frac{du}{dx} - u\frac{dv}{dx}}{v^2} = \frac{1}{v}\frac{du}{dx} - \frac{u}{v^2}\frac{dv}{dx}$

8. $\frac{d}{dx}(u^n) = nu^{n-1}\frac{du}{dx}$

9. $\frac{d}{dx}(\sqrt{u}) = \frac{1}{2\sqrt{u}}\frac{du}{dx}$

10. $\frac{d}{dx}\left(\frac{1}{u}\right) = -\frac{1}{u^2}\frac{du}{dx}$

11. $\frac{d}{dx}\left(\frac{1}{u^n}\right) = -\frac{n}{u^{n+1}}\frac{du}{dx}$

12. $\frac{d}{dx}\left(\frac{u^n}{v^m}\right) = \frac{u^{n-1}}{v^{m+1}}\left(nv\frac{du}{dx} - mu\frac{dv}{dx}\right)$

13. $\frac{d}{dx}(u^n v^m) = u^{n-1}v^{m-1}\left(nv\frac{du}{dx} + mu\frac{dv}{dx}\right)$

14. $\frac{d}{dx}[f(u)] = \frac{d}{du}[f(u)] \cdot \frac{du}{dx}$

* Let $y = f(x)$ and $\frac{dy}{dx} = \frac{d[f(x)]}{dx} = f'(x)$ define respectively a function and its derivative for any value x in their common domain. The differential for the function at such a value x is accordingly defined as

$$dy = d[f(x)] = \frac{dy}{dx}dx = \frac{d[f(x)]}{dx}dx = f'(x)\,dx$$

Each derivative formula has an associated differential formula. For example, formula 6 above has the differential formula

$$d(uvw) = uv\,dw + vw\,du + uw\,dv$$

15. $\frac{d^2}{dx^2}[f(u)] = \frac{df(u)}{du} \cdot \frac{d^2u}{dx^2} + \frac{d^2f(u)}{du^2} \cdot \left(\frac{du}{dx}\right)^2$

16. $\frac{d^n}{dx^n}[uv] = \binom{n}{0} v \frac{d^n u}{dx^n} + \binom{n}{1} \frac{dv}{dx} \frac{d^{n-1}u}{dx^{n-1}} + \binom{n}{2} \frac{d^2 v}{dx^2} \frac{d^{n-2}u}{dx^{n-2}}$

$+ \cdots + \binom{n}{k} \frac{d^k v}{dx^k} \frac{d^{n-k}u}{dx^{n-k}} + \cdots + \binom{n}{n} u \frac{d^n v}{dx^n}$

where $\binom{n}{r} = \frac{n!}{r!(n-r)!}$ the binomial coefficient, n non-negative integer and $\binom{n}{0} = 1$.

17. $\frac{du}{dx} = \frac{1}{\frac{dx}{du}}$ if $\frac{dx}{du} \neq 0$

18. $\frac{d}{dx}(\log_a u) = (\log_a e)\frac{1}{u}\frac{du}{dx}$

19. $\frac{d}{dx}(\log_e u) = \frac{1}{u}\frac{du}{dx}$

20. $\frac{d}{dx}(a^u) = a^u(\log_e a)\frac{du}{dx}$

21. $\frac{d}{dx}(e^u) = e^u\frac{du}{dx}$

22. $\frac{d}{dx}(u^v) = vu^{v-1}\frac{du}{dx} + (\log_e u)u^v\frac{dv}{dx}$

23. $\frac{d}{dx}(\sin u) = \frac{du}{dx}(\cos u)$

24. $\frac{d}{dx}(\cos u) = -\frac{du}{dx}(\sin u)$

25. $\frac{d}{dx}(\tan u) = \frac{du}{dx}(\sec^2 u)$

26. $\frac{d}{dx}(\cot u) = -\frac{du}{dx}(\csc^2 u)$

27. $\frac{d}{dx}(\sec u) = \frac{du}{dx}\sec u \cdot \tan u$

28. $\frac{d}{dx}(\csc u) = -\frac{du}{dx}\csc u \cdot \cot u$

29. $\frac{d}{dx}(\text{vers } u) = \frac{du}{dx}\sin u$

30. $\frac{d}{dx}(\arcsin u) = \frac{1}{\sqrt{1-u^2}}\frac{du}{dx}, \quad \left(-\frac{\pi}{2} \leq \arcsin u \leq \frac{\pi}{2}\right)$

31. $\frac{d}{dx}(\text{arc cos } u) = -\frac{1}{\sqrt{1-u^2}}\frac{du}{dx}, \qquad (0 \le \text{arc cos } u \le \pi)$

32. $\frac{d}{dx}(\text{arc tan } u) = \frac{1}{1+u^2}\frac{du}{dx}, \qquad \left(-\frac{\pi}{2} < \text{arc tan } u < \frac{\pi}{2}\right)$

33. $\frac{d}{dx}(\text{arc cot } u) = -\frac{1}{1+u^2}\frac{du}{dx}, \qquad (0 \le \text{arc cot } u \le \pi)$

34. $\frac{d}{dx}(\text{arc sec } u) = \frac{1}{u\sqrt{u^2-1}}\frac{du}{dx}, \qquad \left(0 \le \text{arc sec } u < \frac{\pi}{2}, -\pi \le \text{arc sec } u < -\frac{\pi}{2}\right)$

35. $\frac{d}{dx}(\text{arc csc } u) = -\frac{1}{u\sqrt{u^2-1}}\frac{du}{dx}, \qquad \left(0 < \text{arc csc } u \le \frac{\pi}{2}, -\pi < \text{arc csc } u \le -\frac{\pi}{2}\right)$

36. $\frac{d}{dx}(\text{arc vers } u) = \frac{1}{\sqrt{2u-u^2}}\frac{du}{dx}, \qquad (0 \le \text{arc vers } u \le \pi)$

37. $\frac{d}{dx}(\sinh u) = \frac{du}{dx}(\cosh u)$

38. $\frac{d}{dx}(\cosh u) = \frac{du}{dx}(\sinh u)$

39. $\frac{d}{dx}(\tanh u) = \frac{du}{dx}(\text{sech}^2 u)$

40. $\frac{d}{dx}(\coth u) = -\frac{du}{dx}(\text{csch}^2 u)$

41. $\frac{d}{dx}(\text{sech } u) = -\frac{du}{dx}(\text{sech } u \cdot \tanh u)$

42. $\frac{d}{dx}(\text{csch } u) = -\frac{du}{dx}(\text{csch } u \cdot \coth u)$

43. $\frac{d}{dx}(\sinh^{-1} u) = \frac{d}{dx}[\log(u + \sqrt{u^2+1})] = \frac{1}{\sqrt{u^2+1}}\frac{du}{dx}$

44. $\frac{d}{dx}(\cosh^{-1} u) = \frac{d}{dx}[\log(u + \sqrt{u^2-1})] = \frac{1}{\sqrt{u^2-1}}\frac{du}{dx}, \qquad (u > 1, \cosh^{-1} u > 0)$

45. $\frac{d}{dx}(\tanh^{-1} u) = \frac{d}{dx}\left[\frac{1}{2}\log\frac{1+u}{1-u}\right] = \frac{1}{1-u^2}\frac{du}{dx}, \qquad (u^2 < 1)$

46. $\frac{d}{dx}(\coth^{-1} u) = \frac{d}{dx}\left[\frac{1}{2}\log\frac{u+1}{u-1}\right] = \frac{1}{1-u^2}\frac{du}{dx}, \qquad (u^2 > 1)$

47. $\frac{d}{dx}(\text{sech}^{-1} u) = \frac{d}{dx}\left[\log\frac{1+\sqrt{1-u^2}}{u}\right] = -\frac{1}{u\sqrt{1-u^2}}\frac{du}{dx}, \qquad (0 < u < 1, \text{sech}^{-1} u > 0)$

48. $\frac{d}{dx}(\text{csch}^{-1} u) = \frac{d}{dx}\left[\log\frac{1+\sqrt{1+u^2}}{u}\right] = -\frac{1}{|u|\sqrt{1+u^2}}\frac{du}{dx}$

49. $\dfrac{d}{dq}\displaystyle\int_p^q f(x)\,dx = f(q)$, [$p$ constant]

50. $\dfrac{d}{dp}\displaystyle\int_p^q f(x)\,dx = -f(p)$, [$q$ constant]

51. $\dfrac{d}{da}\displaystyle\int_p^q f(x, a)\,dx = \int_p^q \frac{\partial}{\partial a}[f(x, a)]\,dx + f(q, a)\frac{dq}{da} - f(p, a)\frac{dp}{da}$

INTEGRATION

The following is a brief discussion of some integration techniques. A more complete discussion can be found in a number of good text books. However, the purpose of this introduction is simply to discuss a few of the important techniques which may be used, in conjunction with the integral table which follows, to integrate particular functions.

No matter how extensive the integral table, it is a fairly uncommon occurrence to find in the table the exact integral desired. Usually some form of transformation will have to be made. The simplest type of transformation, and yet the most general, is substitution. Simple forms of substitution, such as $y = ax$, are employed almost unconsciously by experienced users of integral tables. Other substitutions may require more thought. In some sections of the tables, appropriate substitutions are suggested for integrals which are similar to, but not exactly like, integrals in the table. Finding the right substitution is largely a matter of intuition and experience.

Several precautions must be observed when using substitutions:

1. Be sure to make the substitution in the dx term, as well as everywhere else in the integral.
2. Be sure that the function substituted is one-to-one and continuous. If this is not the case, the integral must be restricted in such a way as to make it true. See the example following.
3. With definite integrals, the limits should also be expressed in terms of the new dependent variable. With indefinite integrals, it is necessary to perform the reverse substitution to obtain the answer in terms of the original independent variable. This may also be done for definite integrals, but it is usually easier to change the limits.

Example:

$$\int \frac{x^4}{\sqrt{a^2 - x^2}}\,dx$$

Here we make the substitution $x = |a| \sin \theta$. Then $dx = |a| \cos \theta \, d\theta$, and

$$\sqrt{a^2 - x^2} = \sqrt{a^2 - a^2 \sin^2 \theta} = |a|\sqrt{1 - \sin^2 \theta} = |a \cos \theta|$$

Notice the absolute value signs. It is very important to keep in mind that a square root radical always denotes the positive square root, and to assure the sign is always kept positive. Thus $\sqrt{x^2} = |x|$. Failure to observe this is a common cause of errors in integration.

Notice also that the indicated substitution is not a one-to-one function, that is, it does not have a unique inverse. Thus we must restrict the range of θ in such a way as to make the function one-to-one. Fortunately, this is easily done by solving for θ

$$\theta = \sin^{-1} \frac{x}{|a|}$$

and restricting the inverse sine to the principal values, $-\dfrac{\pi}{2} \le \theta \le \dfrac{\pi}{2}$.

Thus the integral becomes

$$\int \frac{a^4 \sin^4 \theta |a| \cos \theta \, d\theta}{|a| |\cos \theta|}$$

Now, however, in the range of values chosen for θ, $\cos \theta$ is always positive. Thus we may remove the absolute value signs from $\cos \theta$ in the denominator. (This is one of the reasons that the principal values of the inverse trigonometric functions are defined as they are.)

Then the $\cos \theta$ terms cancel, and the integral becomes

$$a^4 \int \sin^4 \theta \, d\theta$$

By application of integral formulas 299 and 296, we integrate this to

$$-a^4 \frac{\sin^3 \theta \cos \theta}{4} - \frac{3a^4}{8} \cos \theta \sin \theta + \frac{3a^4}{8} \theta + C$$

We now must perform the inverse substitution to get the result in terms of x. We have

$$\theta = \sin^{-1} \frac{x}{|a|}$$

$$\sin \theta = \frac{x}{|a|}$$

Then

$$\cos \theta = \pm \sqrt{1 - \sin^2 \theta} = \pm \sqrt{1 - \frac{x^2}{a^2}} = \pm \frac{\sqrt{a^2 - x^2}}{|a|}.$$

Because of the previously mentioned fact that $\cos \theta$ is positive, we may omit the $\pm$ sign. The reverse substitution then produces the final answer

$$\int \frac{x^4}{\sqrt{a^2 - x^2}} dx = -\tfrac{1}{4} x^3 \sqrt{a^2 - x^2} - \tfrac{3}{8} a^2 x \sqrt{a^2 - x^2} + \frac{3a^4}{8} \sin^{-1} \frac{x}{|a|} + C.$$

Any rational function of x may be integrated, if the denominator is factored into linear and irreducible quadratic factors. The function may then be broken into partial fractions, and the individual partial fractions integrated by use of the appropriate formula from the integral table. See the section on partial fractions for further information.

Many integrals may be reduced to rational functions by proper substitutions. For example,

$$z = \tan \frac{x}{2}$$

will reduce any rational function of the six trigonometric functions of x to a rational function of z. (Frequently there are other substitutions which are simpler to use, but this one will always work. See integral formula number 484.)

Any rational function of x and $\sqrt{ax + b}$ may be reduced to a rational function of z by making the substitution

$$z = \sqrt{ax + b}.$$

Other likely substitutions will be suggested by looking at the form of the integrand.

The other main method of transforming integrals is integration by parts. This involves applying formula number 5 or 6 in the accompanying integral table. The critical factor in this method is the choice of the functions u and v. In order for the method to be successful, $v = \int dv$ and $\int v \, du$ must be easier to integrate than the original integral. Again, this choice is largely a matter of intuition and experience.

Example:

$$\int x \sin x \, dx$$

Two obvious choices are $u = x$, $dv = \sin x \, dx$, or $u = \sin x$, $dv = x \, dx$. Since a preliminary mental calculation indicates that $\int v \, du$ in the second choice would be more, rather than less, complicated than the original integral (it would contain x^2), we use the first choice.

$$u = x \qquad\qquad du = dx$$

$$dv = \sin x \, dx \qquad\qquad v = -\cos x$$

$$\int x \sin x \, dx = \int u \, dv = uv - \int v \, du = -x \cos x + \int \cos x \, dx$$

$$= \sin x - x \cos x$$

Of course, this result could have been obtained directly from the integral table, but it provides a simple example of the method. In more complicated examples the choice of u and v may not be so obvious, and several different choices may have to be tried. Of course, there is no guarantee that any of them will work.

Integration by parts may be applied more than once, or combined with substitution. A fairly common case is illustrated by the following example.

Example:

$$\int e^x \sin x \, dx$$

Let

$$u = e^x \qquad \text{Then} \quad du = e^x \, dx$$

$$dv = \sin x \, dx \qquad\qquad v = -\cos x$$

$$\int e^x \sin x \, dx = \int u \, dv = uv - \int v \, du = -e^x \cos x + \int e^x \cos x \, dx$$

In this latter integral,

$$\text{let} \quad u = e^x \qquad \text{Then} \quad du = e^x \, dx$$

$$dv = \cos x \, dx \qquad\qquad v = \sin x$$

$$\int e^x \sin x \, dx = -e^x \cos x + \int e^x \cos x \, dx = -e^x \cos x + \int u \, dv$$

$$= -e^x \cos x + uv - \int v \, du$$

$$= -e^x \cos x + e^x \sin x - \int e^x \sin x \, dx$$

This looks as if a circular transformation has taken place, since we are back at the same integral we started from. However, the above equation can be solved algebraically for the required integral:

$$\int e^x \sin x \, dx = \tfrac{1}{2}(e^x \sin x - e^x \cos x)$$

In the second integration by parts, if the parts had been chosen as $u = \cos x$, $dv = e^x \, dx$, we would indeed have made a circular transformation, and returned to the starting place.

In general, when doing repeated integration by parts, one should never choose the function u at any stage to be the same as the function v at the previous stage, or a constant times the previous v.

The following rule is called the extended rule for integration by parts. It is the result of $n + 1$ successive applications of integration by parts.

If

$$g_1(x) = \int g(x)\,dx, \qquad g_2(x) = \int g_1(x)\,dx,$$

$$g_3(x) = \int g_2(x)\,dx, \ldots, g_m(x) = \int g_{m-1}(x)\,dx, \ldots,$$

then

$$\int f(x)\cdot g(x)\,dx = f(x)\cdot g_1(x) - f'(x)\cdot g_2(x) + f''(x)\cdot g_3(x) - + \cdots + (-1)^n f^{(n)}(x) g_{n+1}(x) + (-1)^{n+1}\int f^{(n+1)}(x) g_{n+1}(x)\,dx.$$

A useful special case of the above rule is when $f(x)$ is a polynomial of degree n. Then $f^{(n+1)}(x) = 0$, and

$$\int f(x)\cdot g(x)\,dx = f(x)\cdot g_1(x) - f'(x)\cdot g_2(x) + f''(x)\cdot g_3(x) - + \cdots + (-1)^n f^{(n)}(x) g_{n+1}(x) + C$$

Example:

If $f(x) = x^2$, $g(x) = \sin x$

$$\int x^2 \sin x\,dx = -x^2 \cos x + 2x \sin x + 2 \cos x + C$$

Another application of this formula occurs if

$$f''(x) = af(x) \quad \text{and} \quad g''(x) = bg(x),$$

where a and b are unequal constants. In this case, by a process similar to that used in the above example for $\int e^x \sin x\,dx$, we get the formula

$$\int f(x)g(x)\,dx = \frac{f(x)\cdot g'(x) - f'(x)\cdot g(x)}{b - a} + C$$

This formula could have been used in the example mentioned. Here is another example.

Example:

If $f(x) = e^{2x}$, $g(x) = \sin 3x$, then $a = 4$, $b = -9$, and

$$\int e^{2x} \sin 3x\,dx = \frac{3\,e^{2x}\cos 3x - 2\,e^{2x}\sin 3x}{-9 - 4} + C = \frac{e^{2x}}{13}(2 \sin 3x - 3 \cos 3x) + C$$

The following additional points should be observed when using this table.

1. A constant of integration is to be supplied with the answers for indefinite integrals.
2. Logarithmic expressions are to base $e = 2.71828\ldots$, unless otherwise specified, and are to be evaluated for the absolute value of the arguments involved therein.
3. All angles are measured in radians, and inverse trigonometric and hyperbolic functions represent principal values, unless otherwise indicated.
4. If the application of a formula produces either a zero denominator or the square root of a negative number in the result, there is usually available another form of the answer which

avoids this difficulty. In many of the results, the excluded values are specified, but when such are omitted it is presumed that one can tell what these should be, especially when difficulties of the type herein mentioned are obtained.

5. When inverse trigonometric functions occur in the integrals, be sure that any replacements made for them are strictly in accordance with the rules for such functions. This causes little difficulty when the argument of the inverse trigonometric function is positive, since then all angles involved are in the first quadrant. However, if the argument is negative, special care must be used. Thus if $u > 0$,

$$\sin^{-1} u = \cos^{-1}\sqrt{1-u^2} = \csc^{-1}\frac{1}{u}, \text{ etc.}$$

However, if $u < 0$,

$$\sin^{-1} u = -\cos^{-1}\sqrt{1-u^2} = -\pi - \csc^{-1}\frac{1}{u}, \text{ etc.}$$

See the section on inverse trigonometric functions for a full treatment of the allowable substitutions.

6. In integrals 340–345 and some others, the right side includes expressions of the form

$$A \tan^{-1}[B + C \tan f(x)].$$

In these formulas, the $\tan^{-1}$ does not necessarily represent the principal value. Instead of always employing the principal branch of the inverse tangent function, one must instead use that branch of the inverse tangent function upon which $f(x)$ lies for any particular choice of x.

Example:

$$\int_0^{4\pi} \frac{dx}{2+\sin x} = \frac{2}{\sqrt{3}}\tan^{-1}\frac{2\tan\frac{x}{2}+1}{\sqrt{3}}\Bigg]_0^{4\pi}$$

$$= \frac{2}{\sqrt{3}}\left[\tan^{-1}\frac{2\tan 2\pi + 1}{\sqrt{3}} - \tan^{-1}\frac{2\tan 0 + 1}{\sqrt{3}}\right]$$

$$= \frac{2}{\sqrt{3}}\left[\frac{13\pi}{6} - \frac{\pi}{6}\right] = \frac{4\pi}{\sqrt{3}} = \frac{4\sqrt{3}\pi}{3}$$

Here

$$\tan^{-1}\frac{2\tan 2\pi + 1}{\sqrt{3}} = \tan^{-1}\frac{1}{\sqrt{3}} = \frac{13\pi}{6},$$

since $f(x) = 2\pi$; and

$$\tan^{-1}\frac{2\tan 0 + 1}{\sqrt{3}} = \tan^{-1}\frac{1}{\sqrt{3}} = \frac{\pi}{6},$$

since $f(x) = 0$.

7. B_n and E_n where used in Integrals represents the Bernoulli and Euler numbers as defined in tables of Bernoulli and Euler polynomials contained in certain mathematics reference and handbooks, as for example, Beyer, W. H., *Handbook of Mathematical Sciences,* 5th ed., CRC Press, Inc., West Palm Beach 1978, 577–583.

ELEMENTARY FORMS

1. $\int a\,dx = ax$

2. $\int a \cdot f(x)\,dx = a\int f(x)\,dx$

3. $\int \phi(y)\,dx = \int \frac{\phi(y)}{y'}\,dy, \qquad \text{where } y' = \frac{dy}{dx}$

4. $\int (u + v)\,dx = \int u\,dx + \int v\,dx, \qquad$ where u and v are any functions of x

5. $\int u\,dv = u\int dv - \int v\,du = uv - \int v\,du$

6. $\int u\frac{dv}{dx}dx = uv - \int v\frac{du}{dx}dx$

7. $\int x^n\,dx = \frac{x^{n+1}}{n+1}, \qquad \text{except } n = -1$

8. $\int \frac{f'(x)\,dx}{f(x)} = \log f(x), \qquad (df(x) = f'(x)\,dx)$

9. $\int \frac{dx}{x} = \log x$

10. $\int \frac{f'(x)\,dx}{2\sqrt{f(x)}} = \sqrt{f(x)}, \qquad (df(x) = f'(x)\,dx)$

11. $\int e^x\,dx = e^x$

12. $\int e^{ax}\,dx = e^{ax}/a$

13. $\int b^{ax}\,dx = \frac{b^{ax}}{a\log b}, \qquad (b > 0)$

14. $\int \log x\,dx = x\log x - x$

15. $\int a^x \log a\,dx = a^x, \qquad (a > 0)$

16. $\int \frac{dx}{a^2 + x^2} = \frac{1}{a}\tan^{-1}\frac{x}{a}$

17. $\int \frac{dx}{a^2 - x^2} = \begin{cases} \frac{1}{a}\tanh^{-1}\frac{x}{a} \\ \text{or} \\ \frac{1}{2a}\log\frac{a+x}{a-x}, \qquad (a^2 > x^2) \end{cases}$

18. $\int \frac{dx}{x^2 - a^2} = \begin{cases} -\frac{1}{a}\coth^{-1}\frac{x}{a} \\ \text{or} \\ \frac{1}{2a}\log\frac{x-a}{x+a}, \qquad (x^2 > a^2) \end{cases}$

19. $$\int \frac{dx}{\sqrt{a^2 - x^2}} = \begin{cases} \sin^{-1} \dfrac{x}{|a|} \\ \text{or} \\ -\cos^{-1} \dfrac{x}{|a|}, \quad (a^2 > x^2) \end{cases}$$

20. $$\int \frac{dx}{\sqrt{x^2 \pm a^2}} = \log (x + \sqrt{x^2 \pm a^2})$$

21. $$\int \frac{dx}{x\sqrt{x^2 - a^2}} = \frac{1}{|a|} \sec^{-1} \frac{x}{a}$$

22. $$\int \frac{dx}{x\sqrt{a^2 \pm x^2}} = -\frac{1}{a} \log \left(\frac{a + \sqrt{a^2 \pm x^2}}{x} \right)$$

FORMS CONTAINING $(a + bx)$

For forms containing $a + bx$, but not listed in the table, the substitution $u = \dfrac{a + bx}{x}$ may prove helpful.

23. $$\int (a + bx)^n \, dx = \frac{(a + bx)^{n+1}}{(n + 1)b}, \quad (n \neq -1)$$

24. $$\int x(a + bx)^n \, dx$$

$$= \frac{1}{b^2(n + 2)}(a + bx)^{n+2} - \frac{a}{b^2(n + 1)}(a + bx)^{n+1}, \quad (n \neq -1, -2)$$

25. $$\int x^2(a + bx)^n \, dx = \frac{1}{b^3}\left[\frac{(a + bx)^{n+3}}{n + 3} - 2a\frac{(a + bx)^{n+2}}{n + 2} + a^2\frac{(a + bx)^{n+1}}{n + 1}\right]$$

26. $$\int x^m(a + bx)^n \, dx = \begin{cases} \dfrac{x^{m+1}(a + bx)^n}{m + n + 1} + \dfrac{an}{m + n + 1}\displaystyle\int x^m(a + bx)^{n-1} \, dx \\ \text{or} \\ \dfrac{1}{a(n + 1)}\left[-x^{m+1}(a + bx)^{n+1} \right. \\ \qquad \left. + (m + n + 2)\displaystyle\int x^m(a + bx)^{n+1} \, dx \right] \\ \text{or} \\ \dfrac{1}{b(m + n + 1)}\left[x^m(a + bx)^{n+1} - ma\displaystyle\int x^{m-1}(a + bx)^n \, dx \right] \end{cases}$$

27. $$\int \frac{dx}{a + bx} = \frac{1}{b} \log (a + bx)$$

28. $$\int \frac{dx}{(a + bx)^2} = -\frac{1}{b(a + bx)}$$

29. $$\int \frac{dx}{(a + bx)^3} = -\frac{1}{2b(a + bx)^2}$$

30. $$\int \frac{x \, dx}{a + bx} = \begin{cases} \dfrac{1}{b^2}[a + bx - a \log (a + bx)] \\ \text{or} \\ \dfrac{x}{b} - \dfrac{a}{b^2} \log (a + bx) \end{cases}$$

31. $\int \frac{x\,dx}{(a+bx)^2} = \frac{1}{b^2}\left[\log(a+bx) + \frac{a}{a+bx}\right]$

32. $\int \frac{x\,dx}{(a+bx)^n} = \frac{1}{b^2}\left[\frac{-1}{(n-2)(a+bx)^{n-2}} + \frac{a}{(n-1)(a+bx)^{n-1}}\right], \qquad n \neq 1, 2$

33. $\int \frac{x^2\,dx}{a+bx} = \frac{1}{b^3}\left[\frac{1}{2}(a+bx)^2 - 2a(a+bx) + a^2\log(a+bx)\right]$

34. $\int \frac{x^2\,dx}{(a+bx)^2} = \frac{1}{b^3}\left[a + bx - 2a\log(a+bx) - \frac{a^2}{a+bx}\right]$

35. $\int \frac{x^2\,dx}{(a+bx)^3} = \frac{1}{b^3}\left[\log(a+bx) + \frac{2a}{a+bx} - \frac{a^2}{2(a+bx)^2}\right]$

36. $\int \frac{x^2\,dx}{(a+bx)^n} = \frac{1}{b^3}\left[\frac{-1}{(n-3)(a+bx)^{n-3}} + \frac{2a}{(n-2)(a+bx)^{n-2}} - \frac{a^2}{(n-1)(a+bx)^{n-1}}\right], \qquad n \neq 1, 2, 3$

37. $\int \frac{dx}{x(a+bx)} = -\frac{1}{a}\log\frac{a+bx}{x}$

38. $\int \frac{dx}{x(a+bx)^2} = \frac{1}{a(a+bx)} - \frac{1}{a^2}\log\frac{a+bx}{x}$

39. $\int \frac{dx}{x(a+bx)^3} = \frac{1}{a^3}\left[\frac{1}{2}\left(\frac{2a+bx}{a+bx}\right)^2 + \log\frac{x}{a+bx}\right]$

40. $\int \frac{dx}{x^2(a+bx)} = -\frac{1}{ax} + \frac{b}{a^2}\log\frac{a+bx}{x}$

41. $\int \frac{dx}{x^3(a+bx)} = \frac{2bx-a}{2a^2x^2} + \frac{b^2}{a^3}\log\frac{x}{a+bx}$

42. $\int \frac{dx}{x^2(a+bx)^2} = -\frac{a+2bx}{a^2x(a+bx)} + \frac{2b}{a^3}\log\frac{a+bx}{x}$

FORMS CONTAINING $c^2 \pm x^2$, $x^2 - c^2$

43. $\int \frac{dx}{c^2+x^2} = \frac{1}{c}\tan^{-1}\frac{x}{c}$

44. $\int \frac{dx}{c^2-x^2} = \frac{1}{2c}\log\frac{c+x}{c-x}, \qquad (c^2 > x^2)$

45. $\int \frac{dx}{x^2-c^2} = \frac{1}{2c}\log\frac{x-c}{x+c}, \qquad (x^2 > c^2)$

46. $\int \frac{x\,dx}{c^2 \pm x^2} = \pm\frac{1}{2}\log(c^2 \pm x^2)$

47. $\int \frac{x\,dx}{(c^2 \pm x^2)^{n+1}} = \mp\frac{1}{2n(c^2 \pm x^2)^n}$

48. $\int \frac{dx}{(c^2 \pm x^2)^n} = \frac{1}{2c^2(n-1)}\left[\frac{x}{(c^2 \pm x^2)^{n-1}} + (2n-3)\int\frac{dx}{(c^2 \pm x^2)^{n-1}}\right]$

49. $\int \frac{dx}{(x^2 - c^2)^n} = \frac{1}{2c^2(n-1)}\left[-\frac{x}{(x^2 - c^2)^{n-1}} - (2n-3)\int\frac{dx}{(x^2 - c^2)^{n-1}}\right]$

50. $\int \frac{x\,dx}{x^2-c^2} = \frac{1}{2}\log(x^2 - c^2)$

51. $\int \frac{x\,dx}{(x^2 - c^2)^{n+1}} = -\frac{1}{2n(x^2 - c^2)^n}$

FORMS CONTAINING $a + bx$ and $c + dx$

$$u = a + bx, \qquad v = c + dx, \qquad k = ad - bc$$

If $k = 0$, then $v = \frac{c}{a}u$

52. $\int \frac{dx}{u \cdot v} = \frac{1}{k} \cdot \log\left(\frac{v}{u}\right)$

53. $\int \frac{x\,dx}{u \cdot v} = \frac{1}{k}\left[\frac{a}{b}\log(u) - \frac{c}{d}\log(v)\right]$

54. $\int \frac{dx}{u^2 \cdot v} = \frac{1}{k}\left(\frac{1}{u} + \frac{d}{k}\log\frac{v}{u}\right)$

55. $\int \frac{x\,dx}{u^2 \cdot v} = \frac{-a}{bku} - \frac{c}{k^2}\log\frac{v}{u}$

56. $\int \frac{x^2\,dx}{u^2 \cdot v} = \frac{a^2}{b^2ku} + \frac{1}{k^2}\left[\frac{c^2}{d}\log(v) + \frac{a(k - bc)}{b^2}\log(u)\right]$

57. $\int \frac{dx}{u^n \cdot v^m} = \frac{1}{k(m - 1)}\left[\frac{-1}{u^{n-1} \cdot v^{m-1}} - (m + n - 2)b\int \frac{dx}{u^n \cdot v^{m-1}}\right]$

58. $\int \frac{u}{v}\,dx = \frac{bx}{d} + \frac{k}{d^2}\log(v)$

59. $$\int \frac{u^m\,dx}{v^n} = \begin{cases} \dfrac{-1}{k(n - 1)}\left[\dfrac{u^{m+1}}{v^{n-1}} + b(n - m - 2)\displaystyle\int \dfrac{u^m}{v^{n-1}}\,dx\right] \\ \text{or} \\ \dfrac{-1}{d(n - m - 1)}\left[\dfrac{u^m}{v^{n-1}} + mk\displaystyle\int \dfrac{u^{m-1}}{v^n}\,dx\right] \\ \text{or} \\ \dfrac{-1}{d(n - 1)}\left[\dfrac{u^m}{v^{n-1}} - mb\displaystyle\int \dfrac{u^{m-1}}{v^{n-1}}\,dx\right] \end{cases}$$

FORMS CONTAINING $(a + bx^n)$

60. $\int \frac{dx}{a + bx^2} = \frac{1}{\sqrt{ab}}\tan^{-1}\frac{x\sqrt{ab}}{a}, \qquad (ab > 0)$

61. $$\int \frac{dx}{a + bx^2} = \begin{cases} \dfrac{1}{2\sqrt{-ab}}\log\dfrac{a + x\sqrt{-ab}}{a - x\sqrt{-ab}}, & (ab < 0) \\ \text{or} \\ \dfrac{1}{\sqrt{-ab}}\tanh^{-1}\dfrac{x\sqrt{-ab}}{a}, & (ab < 0) \end{cases}$$

62. $\int \frac{dx}{a^2 + b^2x^2} = \frac{1}{ab}\tan^{-1}\frac{bx}{a}$

63. $\int \frac{x\,dx}{a + bx^2} = \frac{1}{2b}\log(a + bx^2)$

64. $\int \frac{x^2\,dx}{a + bx^2} = \frac{x}{b} - \frac{a}{b}\int \frac{dx}{a + bx^2}$

65. $\int \frac{dx}{(a + bx^2)^2} = \frac{x}{2a(a + bx^2)} + \frac{1}{2a}\int \frac{dx}{a + bx^2}$

66. $$\int \frac{dx}{a^2 - b^2x^2} = \frac{1}{2ab} \log \frac{a + bx}{a - bx}$$

67. $$\int \frac{dx}{(a + bx^2)^{m+1}} = \begin{cases} \dfrac{1}{2ma} \dfrac{x}{(a + bx^2)^m} + \dfrac{2m - 1}{2ma} \displaystyle\int \frac{dx}{(a + bx^2)^m} \\ \text{or} \\ \dfrac{(2m)!}{(m!)^2}\left[\dfrac{x}{2a} \displaystyle\sum_{r=1}^{m} \frac{r!(r-1)!}{(4a)^{m-r}(2r)!(a + bx^2)^r} + \frac{1}{(4a)^m} \int \frac{dx}{a + bx^2}\right] \end{cases}$$

68. $$\int \frac{x\,dx}{(a + bx^2)^{m+1}} = -\frac{1}{2bm(a + bx^2)^m}$$

69. $$\int \frac{x^2\,dx}{(a + bx^2)^{m+1}} = \frac{-x}{2mb(a + bx^2)^m} + \frac{1}{2mb} \int \frac{dx}{(a + bx^2)^m}$$

70. $$\int \frac{dx}{x(a + bx^2)} = \frac{1}{2a} \log \frac{x^2}{a + bx^2}$$

71. $$\int \frac{dx}{x^2(a + bx^2)} = -\frac{1}{ax} - \frac{b}{a} \int \frac{dx}{a + bx^2}$$

72. $$\int \frac{dx}{x(a + bx^2)^{m+1}} = \begin{cases} \dfrac{1}{2am(a + bx^2)^m} + \dfrac{1}{a} \displaystyle\int \frac{dx}{x(a + bx^2)^m} \\ \text{or} \\ \dfrac{1}{2a^{m+1}}\left[\displaystyle\sum_{r=1}^{m} \frac{a^r}{r(a + bx^2)^r} + \log \frac{x^2}{a + bx^2}\right] \end{cases}$$

73. $$\int \frac{dx}{x^2(a + bx^2)^{m+1}} = \frac{1}{a} \int \frac{dx}{x^2(a + bx^2)^m} - \frac{b}{a} \int \frac{dx}{(a + bx^2)^{m+1}}$$

74. $$\int \frac{dx}{a + bx^3} = \frac{k}{3a}\left[\frac{1}{2} \log \frac{(k + x)^3}{a + bx^3} + \sqrt{3} \tan^{-1} \frac{2x - k}{k\sqrt{3}}\right], \qquad \left(k = \sqrt[3]{\frac{a}{b}}\right)$$

75. $$\int \frac{x\,dx}{a + bx^3} = \frac{1}{3bk}\left[\frac{1}{2} \log \frac{a + bx^3}{(k + x)^3} + \sqrt{3} \tan^{-1} \frac{2x - k}{k\sqrt{3}}\right], \qquad \left(k = \sqrt[3]{\frac{a}{b}}\right)$$

76. $$\int \frac{x^2\,dx}{a + bx^3} = \frac{1}{3b} \log (a + bx^3)$$

77. $$\int \frac{dx}{a + bx^4} = \frac{k}{2a}\left[\frac{1}{2} \log \frac{x^2 + 2kx + 2k^2}{x^2 - 2kx + 2k^2} + \tan^{-1} \frac{2kx}{2k^2 - x^2}\right], \qquad \left(ab > 0, k = \sqrt[4]{\frac{a}{4b}}\right)$$

78. $$\int \frac{dx}{a + bx^4} = \frac{k}{2a}\left[\frac{1}{2} \log \frac{x + k}{x - k} + \tan^{-1} \frac{x}{k}\right], \qquad \left(ab < 0, k = \sqrt[4]{-\frac{a}{b}}\right)$$

79. $$\int \frac{x\,dx}{a + bx^4} = \frac{1}{2bk} \tan^{-1} \frac{x^2}{k}, \qquad \left(ab > 0, k = \sqrt{\frac{a}{b}}\right)$$

80. $$\int \frac{x\,dx}{a + bx^4} = \frac{1}{4bk} \log \frac{x^2 - k}{x^2 + k}, \qquad \left(ab < 0, k = \sqrt{-\frac{a}{b}}\right)$$

81. $$\int \frac{x^2\,dx}{a + bx^4} = \frac{1}{4bk}\left[\frac{1}{2} \log \frac{x^2 - 2kx + 2k^2}{x^2 + 2kx + 2k^2} + \tan^{-1} \frac{2kx}{2k^2 - x^2}\right], \qquad \left(ab > 0, k = \sqrt[4]{\frac{a}{4b}}\right)$$

82. $$\int \frac{x^2\,dx}{a+bx^4} = \frac{1}{4bk}\left[\log\frac{x-k}{x+k} + 2\tan^{-1}\frac{x}{k}\right], \qquad \left(ab<0,\ k=\sqrt[4]{-\frac{a}{b}}\right)$$

83. $$\int \frac{x^3\,dx}{a+bx^4} = \frac{1}{4b}\log(a+bx^4)$$

84. $$\int \frac{dx}{x(a+bx^n)} = \frac{1}{an}\log\frac{x^n}{a+bx^n}$$

85. $$\int \frac{dx}{(a+bx^n)^{m+1}} = \frac{1}{a}\int\frac{dx}{(a+bx^n)^m} - \frac{b}{a}\int\frac{x^n\,dx}{(a+bx^n)^{m+1}}$$

86. $$\int \frac{x^m\,dx}{(a+bx^n)^{p+1}} = \frac{1}{b}\int\frac{x^{m-n}\,dx}{(a+bx^n)^p} - \frac{a}{b}\int\frac{x^{m-n}\,dx}{(a+bx^n)^{p+1}}$$

87. $$\int \frac{dx}{x^m(a+bx^n)^{p+1}} = \frac{1}{a}\int\frac{dx}{x^m(a+bx^n)^p} - \frac{b}{a}\int\frac{dx}{x^{m-n}(a+bx^n)^{p+1}}$$

88. $$\int x^m(a+bx^n)^p\,dx = \begin{cases} \dfrac{1}{b(np+m+1)}\left[x^{m-n+1}(a+bx^n)^{p+1} - a(m-n+1)\displaystyle\int x^{m-n}(a+bx^n)^p\,dx\right] \\ \text{or} \\ \dfrac{1}{np+m+1}\left[x^{m+1}(a+bx^n)^p + anp\displaystyle\int x^m(a+bx^n)^{p-1}\,dx\right] \\ \text{or} \\ \dfrac{1}{a(m+1)}\left[x^{m+1}(a+bx^n)^{p+1} - (m+1+np+n)b\displaystyle\int x^{m+n}(a+bx^n)^p\,dx\right] \\ \text{or} \\ \dfrac{1}{an(p+1)}\left[-x^{m+1}(a+bx^n)^{p+1} + (m+1+np+n)\displaystyle\int x^m(a+bx^n)^{p+1}\,dx\right] \end{cases}$$

FORMS CONTAINING $c^3 \pm x^3$

89. $$\int \frac{dx}{c^3 \pm x^3} = \pm\frac{1}{6c^2}\log\frac{(c\pm x)^3}{c^3\pm x^3} + \frac{1}{c^2\sqrt{3}}\tan^{-1}\frac{2x\mp c}{c\sqrt{3}}$$

90. $$\int \frac{dx}{(c^3\pm x^3)^2} = \frac{x}{3c^3(c^3\pm x^3)} + \frac{2}{3c^3}\int\frac{dx}{c^3\pm x^3}$$

91. $$\int \frac{dx}{(c^3\pm x^3)^{n+1}} = \frac{1}{3nc^3}\left[\frac{x}{(c^3\pm x^3)^n} + (3n-1)\int\frac{dx}{(c^3\pm x^3)^n}\right]$$

92. $$\int \frac{x\,dx}{c^3\pm x^3} = \frac{1}{6c}\log\frac{c^3\pm x^3}{(c\pm x)^3} \pm \frac{1}{c\sqrt{3}}\tan^{-1}\frac{2x\mp c}{c\sqrt{3}}$$

93. $$\int \frac{x\,dx}{(c^3\pm x^3)^2} = \frac{x^2}{3c^3(c^3\pm x^3)} + \frac{1}{3c^3}\int\frac{x\,dx}{c^3\pm x^3}$$

94. $\int \frac{x\,dx}{(c^3 \pm x^3)^{n+1}} = \frac{1}{3nc^3}\left[\frac{x^2}{(c^3 \pm x^3)^n} + (3n-2)\int \frac{x\,dx}{(c^3 \pm x^3)^n}\right]$

95. $\int \frac{x^2\,dx}{c^3 \pm x^3} = \pm\frac{1}{3}\log(c^3 \pm x^3)$

96. $\int \frac{x^2\,dx}{(c^3 \pm x^3)^{n+1}} = \mp\frac{1}{3n(c^3 \pm x^3)^n}$

97. $\int \frac{dx}{x(c^3 \pm x^3)} = \frac{1}{3c^3}\log\frac{x^3}{c^3 \pm x^3}$

98. $\int \frac{dx}{x(c^3 \pm x^3)^2} = \frac{1}{3c^3(c^3 \pm x^3)} + \frac{1}{3c^6}\log\frac{x^3}{c^3 \pm x^3}$

99. $\int \frac{dx}{x(c^3 \pm x^3)^{n+1}} = \frac{1}{3nc^3(c^3 \pm x^3)^n} + \frac{1}{c^3}\int \frac{dx}{x(c^3 \pm x^3)^n}$

100. $\int \frac{dx}{x^2(c^3 \pm x^3)} = -\frac{1}{c^3x} \mp \frac{1}{c^3}\int \frac{x\,dx}{c^3 \pm x^3}$

101. $\int \frac{dx}{x^2(c^3 \pm x^3)^{n+1}} = \frac{1}{c^3}\int \frac{dx}{x^2(c^3 \pm x^3)^n} \mp \frac{1}{c^3}\int \frac{x\,dx}{(c^3 \pm x^3)^{n+1}}$

FORMS CONTAINING $c^4 \pm x^4$

102. $\int \frac{dx}{c^4 + x^4} = \frac{1}{2c^3\sqrt{2}}\left[\frac{1}{2}\log\frac{x^2 + cx\sqrt{2} + c^2}{x^2 - cx\sqrt{2} + c^2} + \tan^{-1}\frac{cx\sqrt{2}}{c^2 - x^2}\right]$

103. $\int \frac{dx}{c^4 - x^4} = \frac{1}{2c^3}\left[\frac{1}{2}\log\frac{c+x}{c-x} + \tan^{-1}\frac{x}{c}\right]$

104. $\int \frac{x\,dx}{c^4 + x^4} = \frac{1}{2c^2}\tan^{-1}\frac{x^2}{c^2}$

105. $\int \frac{x\,dx}{c^4 - x^4} = \frac{1}{4c^2}\log\frac{c^2 + x^2}{c^2 - x^2}$

106. $\int \frac{x^2\,dx}{c^4 + x^4} = \frac{1}{2c\sqrt{2}}\left[\frac{1}{2}\log\frac{x^2 - cx\sqrt{2} + c^2}{x^2 + cx\sqrt{2} + c^2} + \tan^{-1}\frac{cx\sqrt{2}}{c^2 - x^2}\right]$

107. $\int \frac{x^2\,dx}{c^4 - x^4} = \frac{1}{2c}\left[\frac{1}{2}\log\frac{c+x}{c-x} - \tan^{-1}\frac{x}{c}\right]$

108. $\int \frac{x^3\,dx}{c^4 \pm x^4} = \pm\frac{1}{4}\log(c^4 \pm x^4)$

FORMS CONTAINING $(a + bx + cx^2)$

$X = a + bx + cx^2$ and $q = 4ac - b^2$

If $q = 0$, then $X = c\left(x + \frac{b}{2c}\right)^2$, and formulas starting with 23 should be used in place of these.

109. $\int \frac{dx}{X} = \frac{2}{\sqrt{q}}\tan^{-1}\frac{2cx + b}{\sqrt{q}}, \qquad (q > 0)$

110. $$\int \frac{dx}{X} = \begin{cases} \frac{-2}{\sqrt{-q}}\tanh^{-1}\frac{2cx+b}{\sqrt{-q}} \\ \text{or} \\ \frac{1}{\sqrt{-q}}\log\frac{2cx + b - \sqrt{-q}}{2cx + b + \sqrt{-q}}, \qquad (q < 0) \end{cases}$$

111. $\int \frac{dx}{X^2} = \frac{2cx + b}{qX} + \frac{2c}{q}\int \frac{dx}{X}$

112. $\int \frac{dx}{X^3} = \frac{2cx+b}{q}\left(\frac{1}{2X^2} + \frac{3c}{qX}\right) + \frac{6c^2}{q^2}\int \frac{dx}{X}$

113. $\int \frac{dx}{X^{n+1}} = \begin{cases} \frac{2cx+b}{nqX^n} + \frac{2(2n-1)c}{qn}\int \frac{dx}{X^n} \\ \text{or} \\ \frac{(2n)!}{(n!)^2}\left(\frac{c}{q}\right)^n\left[\frac{2cx+b}{q}\sum_{r=1}^{n}\left(\frac{q}{cX}\right)^r\left(\frac{(r-1)!r!}{(2r)!}\right) + \int \frac{dx}{X}\right] \end{cases}$

114. $\int \frac{x\,dx}{X} = \frac{1}{2c}\log X - \frac{b}{2c}\int \frac{dx}{X}$

115. $\int \frac{x\,dx}{X^2} = -\frac{bx+2a}{qX} - \frac{b}{q}\int \frac{dx}{X}$

116. $\int \frac{x\,dx}{X^{n+1}} = -\frac{2a+bx}{nqX^n} - \frac{b(2n-1)}{nq}\int \frac{dx}{X^n}$

117. $\int \frac{x^2}{X}\,dx = \frac{x}{c} - \frac{b}{2c^2}\log X + \frac{b^2-2ac}{2c^2}\int \frac{dx}{X}$

118. $\int \frac{x^2}{X^2}\,dx = \frac{(b^2-2ac)x+ab}{cqX} + \frac{2a}{q}\int \frac{dx}{X}$

119. $\int \frac{x^m\,dx}{X^{n+1}} = -\frac{x^{m-1}}{(2n-m+1)cX^n} - \frac{n-m+1}{2n-m+1}\cdot\frac{b}{c}\int \frac{x^{m-1}\,dx}{X^{n+1}} + \frac{m-1}{2n-m+1}\cdot\frac{a}{c}\int \frac{x^{m-2}\,dx}{X^{n+1}}$

120. $\int \frac{dx}{xX} = \frac{1}{2a}\log\frac{x^2}{X} - \frac{b}{2a}\int \frac{dx}{X}$

121. $\int \frac{dx}{x^2X} = \frac{b}{2a^2}\log\frac{X}{x^2} - \frac{1}{ax} + \left(\frac{b^2}{2a^2} - \frac{c}{a}\right)\int \frac{dx}{X}$

122. $\int \frac{dx}{xX^n} = \frac{1}{2a(n-1)X^{n-1}} - \frac{b}{2a}\int \frac{dx}{X^n} + \frac{1}{a}\int \frac{dx}{xX^{n-1}}$

123. $\int \frac{dx}{x^mX^{n+1}} = -\frac{1}{(m-1)ax^{m-1}X^n} - \frac{n+m-1}{m-1}\cdot\frac{b}{a}\int \frac{dx}{x^{m-1}X^{n+1}} - \frac{2n+m-1}{m-1}\cdot\frac{c}{a}\int \frac{dx}{x^{m-2}X^{n+1}}$

FORMS CONTAINING $\sqrt{a+bx}$

124. $\int \sqrt{a+bx}\,dx = \frac{2}{3b}\sqrt{(a+bx)^3}$

125. $\int x\sqrt{a+bx}\,dx = -\frac{2(2a-3bx)\sqrt{(a+bx)^3}}{15b^2}$

126. $\int x^2\sqrt{a+bx}\,dx = \frac{2(8a^2-12abx+15b^2x^2)\sqrt{(a+bx)^3}}{105b^3}$

127. $\int x^m\sqrt{a+bx}\,dx = \begin{cases} \frac{2}{b(2m+3)}\left[x^m\sqrt{(a+bx)^3} - ma\int x^{m-1}\sqrt{a+bx}\,dx\right] \\ \text{or} \\ \frac{2}{b^{m+1}}\sqrt{a+bx}\sum_{r=0}^{m}\frac{m!(-a)^{m-r}}{r!(m-r)!(2r+3)}(a+bx)^{r+1} \end{cases}$

128. $\int \frac{\sqrt{a+bx}}{x}dx = 2\sqrt{a+bx} + a\int \frac{dx}{x\sqrt{a+bx}}$

129. $\int \frac{\sqrt{a+bx}}{x^2}dx = -\frac{\sqrt{a+bx}}{x} + \frac{b}{2}\int \frac{dx}{x\sqrt{a+bx}}$

130. $\int \frac{\sqrt{a+bx}}{x^m}dx = -\frac{1}{(m-1)a}\left[\frac{\sqrt{(a+bx)^3}}{x^{m-1}} + \frac{(2m-5)b}{2}\int \frac{\sqrt{a+bx}}{x^{m-1}}dx\right]$

131. $\int \frac{dx}{\sqrt{a+bx}} = \frac{2\sqrt{a+bx}}{b}$

132. $\int \frac{x\,dx}{\sqrt{a+bx}} = -\frac{2(2a-bx)}{3b^2}\sqrt{a+bx}$

133. $\int \frac{x^2\,dx}{\sqrt{a+bx}} = \frac{2(8a^2-4abx+3b^2x^2)}{15b^3}\sqrt{a+bx}$

134. $$\int \frac{x^m\,dx}{\sqrt{a+bx}} = \begin{cases} \frac{2}{(2m+1)b}\left[x^m\sqrt{a+bx} - ma\int \frac{x^{m-1}\,dx}{\sqrt{a+bx}}\right] \\ \text{or} \\ \frac{2(-a)^m\sqrt{a+bx}}{b^{m+1}}\sum_{r=0}^{m}\frac{(-1)^r m!(a+bx)^r}{(2r+1)r!(m-r)!a^r} \end{cases}$$

135. $\int \frac{dx}{x\sqrt{a+bx}} = \frac{1}{\sqrt{a}}\log\left(\frac{\sqrt{a+bx}-\sqrt{a}}{\sqrt{a+bx}+\sqrt{a}}\right), \quad (a > 0)$

136. $\int \frac{dx}{x\sqrt{a+bx}} = \frac{2}{\sqrt{-a}}\tan^{-1}\sqrt{\frac{a+bx}{-a}}, \quad (a < 0)$

137. $\int \frac{dx}{x^2\sqrt{a+bx}} = -\frac{\sqrt{a+bx}}{ax} - \frac{b}{2a}\int \frac{dx}{x\sqrt{a+bx}}$

138. $$\int \frac{dx}{x^n\sqrt{a+bx}} = \begin{cases} -\frac{\sqrt{a+bx}}{(n-1)ax^{n-1}} - \frac{(2n-3)b}{(2n-2)a}\int \frac{dx}{x^{n-1}\sqrt{a+bx}} \\ \text{or} \\ \frac{(2n-2)!}{[(n-1)!]^2}\left[-\frac{\sqrt{a+bx}}{a}\sum_{r=1}^{n-1}\frac{r!(r-1)!}{x^r(2r)!}\left(-\frac{b}{4a}\right)^{n-r-1} + \left(-\frac{b}{4a}\right)^{n-1}\int \frac{dx}{x\sqrt{a+bx}}\right] \end{cases}$$

139. $\int (a+bx)^{\pm\frac{n}{2}}\,dx = \frac{2(a+bx)^{\frac{2\pm n}{2}}}{b(2\pm n)}$

140. $\int x(a+bx)^{\pm\frac{n}{2}}\,dx = \frac{2}{b^2}\left[\frac{(a+bx)^{\frac{4\pm n}{2}}}{4\pm n} - \frac{a(a+bx)^{\frac{2\pm n}{2}}}{2\pm n}\right]$

141. $\int \frac{dx}{x(a+bx)^{\frac{m}{2}}} = \frac{1}{a}\int \frac{dx}{x(a+bx)^{\frac{m-2}{2}}} - \frac{b}{a}\int \frac{dx}{(a+bx)^{\frac{m}{2}}}$

142. $\int \frac{(a+bx)^{\frac{n}{2}}\,dx}{x} = b\int (a+bx)^{\frac{n-2}{2}}\,dx + a\int \frac{(a+bx)^{\frac{n-2}{2}}}{x}\,dx$

143. $\int f(x, \sqrt{a+bx})\,dx = \frac{2}{b}\int f\left(\frac{z^2-a}{b}, z\right) z\,dz, \qquad (z = \sqrt{a+bx})$

FORMS CONTAINING $\sqrt{a+bx}$ and $\sqrt{c+dx}$

$u = a + bx \qquad v = c + dx \qquad k = ad - bc$

If $k = 0$, then $v = \frac{c}{a}u$, and formulas starting with 124 should be used in place of these.

144. $$\int \frac{dx}{\sqrt{uv}} = \begin{cases} \frac{2}{\sqrt{bd}} \tanh^{-1} \frac{\sqrt{bduv}}{bv}, & bd > o,\ k < o \\ \text{or} \\ \frac{2}{\sqrt{bd}} \tanh^{-1} \frac{\sqrt{bduv}}{du}, & bd > o,\ k > o. \\ \text{or} \\ \frac{1}{\sqrt{bd}} \log \frac{(bv + \sqrt{bduv})^2}{v}, & (bd > 0) \end{cases}$$

145. $$\int \frac{dx}{\sqrt{uv}} = \begin{cases} \frac{2}{\sqrt{-bd}} \tan^{-1} \frac{\sqrt{-bduv}}{bv} \\ \text{or} \\ -\frac{1}{\sqrt{-bd}} \sin^{-1}\left(\frac{2bdx + ad + bc}{|k|}\right), & (bd < 0) \end{cases}$$

146. $\int \sqrt{uv}\,dx = \frac{k + 2bv}{4bd}\sqrt{uv} - \frac{k^2}{8bd}\int \frac{dx}{\sqrt{uv}}$

147. $$\int \frac{dx}{v\sqrt{u}} = \begin{cases} \frac{1}{\sqrt{kd}} \log \frac{d\sqrt{u} - \sqrt{kd}}{d\sqrt{u} + \sqrt{kd}} \\ \text{or} \\ \frac{1}{\sqrt{kd}} \log \frac{(d\sqrt{u} - \sqrt{kd})^2}{v}, & (kd > 0) \end{cases}$$

148. $\int \frac{dx}{v\sqrt{u}} = \frac{2}{\sqrt{-kd}} \tan^{-1} \frac{d\sqrt{u}}{\sqrt{-kd}}, \qquad (kd < 0)$

149. $\int \frac{x\,dx}{\sqrt{uv}} = \frac{\sqrt{uv}}{bd} - \frac{ad + bc}{2bd}\int \frac{dx}{\sqrt{uv}}$

150. $\int \frac{dx}{v\sqrt{uv}} = \frac{-2\sqrt{uv}}{kv}$

151. $\int \frac{v\,dx}{\sqrt{uv}} = \frac{\sqrt{uv}}{b} - \frac{k}{2b}\int \frac{dx}{\sqrt{uv}}$

152. $\int \sqrt{\frac{v}{u}}\,dx = \frac{v}{|v|}\int \frac{v\,dx}{\sqrt{uv}}$

153. $\int v^m \sqrt{u}\,dx = \frac{1}{(2m+3)d}\left(2v^{m+1}\sqrt{u} + k\int \frac{v^m\,dx}{\sqrt{u}}\right)$

154. $$\int \frac{dx}{v^m\sqrt{u}} = -\frac{1}{(m-1)k}\left(\frac{\sqrt{u}}{v^{m-1}} + \left(m - \frac{3}{2}\right)b\int \frac{dx}{v^{m-1}\sqrt{u}}\right)$$

155. $$\int \frac{v^m\,dx}{\sqrt{u}} = \begin{cases} \dfrac{2}{b(2m+1)}\left[v^m\sqrt{u} - mk\displaystyle\int \frac{v^{m-1}}{\sqrt{u}}\,dx\right] \\ \text{or} \\ \dfrac{2(m!)^2\sqrt{u}}{b(2m+1)!}\displaystyle\sum_{r=0}^{m}\left(-\frac{4k}{b}\right)^{m-r}\frac{(2r)!}{(r!)^2}v^r \end{cases}$$

FORMS CONTAINING $\sqrt{x^2 \pm a^2}$

156. $$\int \sqrt{x^2 \pm a^2}\,dx = \tfrac{1}{2}[x\sqrt{x^2 \pm a^2} \pm a^2 \log(x + \sqrt{x^2 \pm a^2})]$$

157. $$\int \frac{dx}{\sqrt{x^2 \pm a^2}} = \log(x + \sqrt{x^2 \pm a^2})$$

158. $$\int \frac{dx}{x\sqrt{x^2 - a^2}} = \frac{1}{|a|}\sec^{-1}\frac{x}{a}$$

159. $$\int \frac{dx}{x\sqrt{x^2 + a^2}} = -\frac{1}{a}\log\left(\frac{a + \sqrt{x^2 + a^2}}{x}\right)$$

160. $$\int \frac{\sqrt{x^2 + a^2}}{x}\,dx = \sqrt{x^2 + a^2} - a\log\left(\frac{a + \sqrt{x^2 + a^2}}{x}\right)$$

161. $$\int \frac{\sqrt{x^2 - a^2}}{x}\,dx = \sqrt{x^2 - a^2} - |a|\sec^{-1}\frac{x}{a}$$

162. $$\int \frac{x\,dx}{\sqrt{x^2 \pm a^2}} = \sqrt{x^2 \pm a^2}$$

163. $$\int x\sqrt{x^2 \pm a^2}\,dx = \tfrac{1}{3}\sqrt{(x^2 \pm a^2)^3}$$

164. $$\int \sqrt{(x^2 \pm a^2)^3}\,dx = \frac{1}{4}\left[x\sqrt{(x^2 \pm a^2)^3} \pm \frac{3a^2x}{2}\sqrt{x^2 \pm a^2} + \frac{3a^4}{2}\log(x + \sqrt{x^2 \pm a^2})\right]$$

165. $$\int \frac{dx}{\sqrt{(x^2 \pm a^2)^3}} = \frac{\pm x}{a^2\sqrt{x^2 \pm a^2}}$$

166. $$\int \frac{x\,dx}{\sqrt{(x^2 \pm a^2)^3}} = \frac{-1}{\sqrt{x^2 \pm a^2}}$$

167. $$\int x\sqrt{(x^2 \pm a^2)^3}\,dx = \tfrac{1}{5}\sqrt{(x^2 \pm a^2)^5}$$

168. $$\int x^2\sqrt{x^2 \pm a^2}\,dx = \frac{x}{4}\sqrt{(x^2 \pm a^2)^3} \mp \frac{a^2}{8}x\sqrt{x^2 \pm a^2} - \frac{a^4}{8}\log(x + \sqrt{x^2 \pm a^2})$$

169. $$\int x^3\sqrt{x^2 + a^2}\,dx = (\tfrac{1}{5}x^2 - \tfrac{2}{15}a^2)\sqrt{(a^2 + x^2)^3}$$

170. $$\int x^3\sqrt{x^2 - a^2}\,dx = \frac{1}{5}\sqrt{(x^2 - a^2)^5} + \frac{a^2}{3}\sqrt{(x^2 - a^2)^3}$$

171. $$\int \frac{x^2\,dx}{\sqrt{x^2 \pm a^2}} = \frac{x}{2}\sqrt{x^2 \pm a^2} \mp \frac{a^2}{2}\log(x + \sqrt{x^2 \pm a^2})$$

172. $\int \frac{x^3\,dx}{\sqrt{x^2 \pm a^2}} = \frac{1}{3}\sqrt{(x^2 \pm a^2)^3} \mp a^2\sqrt{x^2 \pm a^2}$

173. $\int \frac{dx}{x^2\sqrt{x^2 \pm a^2}} = \mp\frac{\sqrt{x^2 \pm a^2}}{a^2 x}$

174. $\int \frac{dx}{x^3\sqrt{x^2 + a^2}} = -\frac{\sqrt{x^2 + a^2}}{2a^2x^2} + \frac{1}{2a^3}\log\frac{a + \sqrt{x^2 + a^2}}{x}$

175. $\int \frac{dx}{x^3\sqrt{x^2 - a^2}} = \frac{\sqrt{x^2 - a^2}}{2a^2x^2} + \frac{1}{2|a^3|}\sec^{-1}\frac{x}{a}$

176. $\int x^2\sqrt{(x^2 \pm a^2)^3}\,dx = \frac{x}{6}\sqrt{(x^2 \pm a^2)^5} \mp \frac{a^2x}{24}\sqrt{(x^2 \pm a^2)^3} - \frac{a^4x}{16}\sqrt{x^2 \pm a^2}$

$$\mp \frac{a^6}{16}\log(x + \sqrt{x^2 \pm a^2})$$

177. $\int x^3\sqrt{(x^2 \pm a^2)^3}\,dx = \frac{1}{7}\sqrt{(x^2 \pm a^2)^7} \mp \frac{a^2}{5}\sqrt{(x^2 \pm a^2)^5}$

178. $\int \frac{\sqrt{x^2 \pm a^2}\,dx}{x^2} = -\frac{\sqrt{x^2 \pm a^2}}{x} + \log(x + \sqrt{x^2 \pm a^2})$

179. $\int \frac{\sqrt{x^2 + a^2}}{x^3}\,dx = -\frac{\sqrt{x^2 + a^2}}{2x^2} - \frac{1}{2a}\log\frac{a + \sqrt{x^2 + a^2}}{x}$

180. $\int \frac{\sqrt{x^2 - a^2}}{x^3}\,dx = -\frac{\sqrt{x^2 - a^2}}{2x^2} + \frac{1}{2|a|}\sec^{-1}\frac{x}{a}$

181. $\int \frac{\sqrt{x^2 \pm a^2}}{x^4}\,dx = \mp\frac{\sqrt{(x^2 \pm a^2)^3}}{3a^2x^3}$

182. $\int \frac{x^2\,dx}{\sqrt{(x^2 \pm a^2)^3}} = \frac{-x}{\sqrt{x^2 \pm a^2}} + \log(x + \sqrt{x^2 \pm a^2})$

183. $\int \frac{x^3\,dx}{\sqrt{(x^2 \pm a^2)^3}} = \sqrt{x^2 \pm a^2} \pm \frac{a^2}{\sqrt{x^2 \pm a^2}}$

184. $\int \frac{dx}{x\sqrt{(x^2 + a^2)^3}} = \frac{1}{a^2\sqrt{x^2 + a^2}} - \frac{1}{a^3}\log\frac{a + \sqrt{x^2 + a^2}}{x}$

185. $\int \frac{dx}{x\sqrt{(x^2 - a^2)^3}} = -\frac{1}{a^2\sqrt{x^2 - a^2}} - \frac{1}{|a^3|}\sec^{-1}\frac{x}{a}$

186. $\int \frac{dx}{x^2\sqrt{(x^2 \pm a^2)^3}} = -\frac{1}{a^4}\left[\frac{\sqrt{x^2 \pm a^2}}{x} + \frac{x}{\sqrt{x^2 \pm a^2}}\right]$

187. $\int \frac{dx}{x^3\sqrt{(x^2 + a^2)^3}} = -\frac{1}{2a^2x^2\sqrt{x^2 + a^2}} - \frac{3}{2a^4\sqrt{x^2 + a^2}}$

$$+ \frac{3}{2a^5}\log\frac{a + \sqrt{x^2 + a^2}}{x}$$

188. $\int \frac{dx}{x^3\sqrt{(x^2 - a^2)^3}} = \frac{1}{2a^2x^2\sqrt{x^2 - a^2}} - \frac{3}{2a^4\sqrt{x^2 - a^2}} - \frac{3}{2|a^5|}\sec^{-1}\frac{x}{a}$

189. $\int \frac{x^m}{\sqrt{x^2 \pm a^2}}\,dx = \frac{1}{m}x^{m-1}\sqrt{x^2 \pm a^2} \mp \frac{m-1}{m}a^2\int \frac{x^{m-2}}{\sqrt{x^2 \pm a^2}}\,dx$

190. $$\int \frac{x^{2m}}{\sqrt{x^2 \pm a^2}}\,dx = \frac{(2m)!}{2^{2m}(m!)^2}\left[\sqrt{x^2 \pm a^2}\sum_{r=1}^{m}\frac{r!(r-1)!}{(2r)!}(\mp a^2)^{m-r}(2x)^{2r-1} + (\mp a^2)^m \log\left(x + \sqrt{x^2 \pm a^2}\right)\right]$$

191. $$\int \frac{x^{2m+1}}{\sqrt{x^2 \pm a^2}}\,dx = \sqrt{x^2 \pm a^2}\sum_{r=0}^{m}\frac{(2r)!(m!)^2}{(2m+1)!(r!)^2}(\mp 4a^2)^{m-r}x^{2r}$$

192. $$\int \frac{dx}{x^m\sqrt{x^2 \pm a^2}} = \mp\frac{\sqrt{x^2 \pm a^2}}{(m-1)a^2x^{m-1}} \mp \frac{(m-2)}{(m-1)a^2}\int \frac{dx}{x^{m-2}\sqrt{x^2 \pm a^2}}$$

193. $$\int \frac{dx}{x^{2m}\sqrt{x^2 \pm a^2}} = \sqrt{x^2 \pm a^2}\sum_{r=0}^{m-1}\frac{(m-1)!m!(2r)!2^{2m-2r-1}}{(r!)^2(2m)!(\mp a^2)^{m-r}x^{2r+1}}$$

194. $$\int \frac{dx}{x^{2m+1}\sqrt{x^2 + a^2}} = \frac{(2m)!}{(m!)^2}\left[\frac{\sqrt{x^2 + a^2}}{a^2}\sum_{r=1}^{m}(-1)^{m-r+1}\frac{r!(r-1)!}{2(2r)!(4a^2)^{m-r}x^{2r}} + \frac{(-1)^{m+1}}{2^{2m}a^{2m+1}}\log\frac{\sqrt{x^2 + a^2} + a}{x}\right]$$

195. $$\int \frac{dx}{x^{2m+1}\sqrt{x^2 - a^2}} = \frac{(2m)!}{(m!)^2}\left[\frac{\sqrt{x^2 - a^2}}{a^2}\sum_{r=1}^{m}\frac{r!(r-1)!}{2(2r)!(4a^2)^{m-r}x^{2r}} + \frac{1}{2^{2m}|a|^{2m+1}}\sec^{-1}\frac{x}{a}\right]$$

196. $$\int \frac{dx}{(x-a)\sqrt{x^2 - a^2}} = -\frac{\sqrt{x^2 - a^2}}{a(x-a)}$$

197. $$\int \frac{dx}{(x+a)\sqrt{x^2 - a^2}} = \frac{\sqrt{x^2 - a^2}}{a(x+a)}$$

198. $$\int f(x, \sqrt{x^2 + a^2})\,dx = a\int f(a\tan u, a\sec u)\sec^2 u\,du, \qquad \left(u = \tan^{-1}\frac{x}{a}, a > 0\right)$$

199. $$\int f(x, \sqrt{x^2 - a^2})\,dx = a\int f(a\sec u, a\tan u)\sec u\tan u\,du, \qquad \left(u = \sec^{-1}\frac{x}{a}, a > 0\right)$$

FORMS CONTAINING $\sqrt{a^2 - x^2}$

200. $$\int \sqrt{a^2 - x^2}\,dx = \frac{1}{2}\left[x\sqrt{a^2 - x^2} + a^2\sin^{-1}\frac{x}{|a|}\right]$$

201. $$\int \frac{dx}{\sqrt{a^2 - x^2}} = \begin{cases}\sin^{-1}\dfrac{x}{|a|}\\ \text{or}\\ -\cos^{-1}\dfrac{x}{|a|}\end{cases}$$

202. $$\int \frac{dx}{x\sqrt{a^2 - x^2}} = -\frac{1}{a}\log\left(\frac{a + \sqrt{a^2 - x^2}}{x}\right)$$

203. $$\int \frac{\sqrt{a^2 - x^2}}{x}\,dx = \sqrt{a^2 - x^2} - a\log\left(\frac{a + \sqrt{a^2 - x^2}}{x}\right)$$

204. $$\int \frac{x\,dx}{\sqrt{a^2 - x^2}} = -\sqrt{a^2 - x^2}$$

205. $$\int x\sqrt{a^2 - x^2}\,dx = -\tfrac{1}{3}\sqrt{(a^2 - x^2)^3}$$

206. $\int \sqrt{(a^2 - x^2)^3}\, dx = \frac{1}{4}\left[x\sqrt{(a^2 - x^2)^3} + \frac{3a^2x}{2}\sqrt{a^2 - x^2} + \frac{3a^4}{2}\sin^{-1}\frac{x}{|a|}\right]$

207. $\int \frac{dx}{\sqrt{(a^2 - x^2)^3}} = \frac{x}{a^2\sqrt{a^2 - x^2}}$

208. $\int \frac{x\, dx}{\sqrt{(a^2 - x^2)^3}} = \frac{1}{\sqrt{a^2 - x^2}}$

209. $\int x\sqrt{(a^2 - x^2)^3}\, dx = -\frac{1}{5}\sqrt{(a^2 - x^2)^5}$

210. $\int x^2\sqrt{a^2 - x^2}\, dx = -\frac{x}{4}\sqrt{(a^2 - x^2)^3} + \frac{a^2}{8}\left(x\sqrt{a^2 - x^2} + a^2\sin^{-1}\frac{x}{|a|}\right)$

211. $\int x^3\sqrt{a^2 - x^2}\, dx = (-\frac{1}{5}x^2 - \frac{2}{15}a^2)\sqrt{(a^2 - x^2)^3}$

212. $\int x^2\sqrt{(a^2 - x^2)^3}\, dx = -\frac{1}{6}x\sqrt{(a^2 - x^2)^5} + \frac{a^2x}{24}\sqrt{(a^2 - x^2)^3}$

$$+ \frac{a^4x}{16}\sqrt{a^2 - x^2} + \frac{a^6}{16}\sin^{-1}\frac{x}{|a|}$$

213. $\int x^3\sqrt{(a^2 - x^2)^3}\, dx = \frac{1}{7}\sqrt{(a^2 - x^2)^7} - \frac{a^2}{5}\sqrt{(a^2 - x^2)^5}$

214. $\int \frac{x^2\, dx}{\sqrt{a^2 - x^2}} = -\frac{x}{2}\sqrt{a^2 - x^2} + \frac{a^2}{2}\sin^{-1}\frac{x}{|a|}$

215. $\int \frac{dx}{x^2\sqrt{a^2 - x^2}} = -\frac{\sqrt{a^2 - x^2}}{a^2x}$

216. $\int \frac{\sqrt{a^2 - x^2}}{x^2}\, dx = -\frac{\sqrt{a^2 - x^2}}{x} - \sin^{-1}\frac{x}{|a|}$

217. $\int \frac{\sqrt{a^2 - x^2}}{x^3}\, dx = -\frac{\sqrt{a^2 - x^2}}{2x^2} + \frac{1}{2a}\log\frac{a + \sqrt{a^2 - x^2}}{x}$

218. $\int \frac{\sqrt{a^2 - x^2}}{x^4}\, dx = -\frac{\sqrt{(a^2 - x^2)^3}}{3a^2x^3}$

219. $\int \frac{x^2\, dx}{\sqrt{(a^2 - x^2)^3}} = \frac{x}{\sqrt{a^2 - x^2}} - \sin^{-1}\frac{x}{|a|}$

220. $\int \frac{x^3\, dx}{\sqrt{a^2 - x^2}} = -\frac{2}{3}(a^2 - x^2)^{\frac{3}{2}} - x^2(a^2 - x^2)^{\frac{1}{2}} = -\frac{1}{3}\sqrt{a^2 - x^2}(x^2 + 2a^2)$

221. $\int \frac{x^3\, dx}{\sqrt{(a^2 - x^2)^3}} = 2(a^2 - x^2)^{\frac{1}{2}} + \frac{x^2}{(a^2 - x^2)^{\frac{1}{2}}} = \frac{a^2}{\sqrt{a^2 - x^2}} + \sqrt{a^2 - x^2}$

222. $\int \frac{dx}{x^3\sqrt{a^2 - x^2}} = -\frac{\sqrt{a^2 - x^2}}{2a^2x^2} - \frac{1}{2a^3}\log\frac{a + \sqrt{a^2 - x^2}}{x}$

223. $\int \frac{dx}{x\sqrt{(a^2 - x^2)^3}} = \frac{1}{a^2\sqrt{a^2 - x^2}} - \frac{1}{a^3}\log\frac{a + \sqrt{a^2 - x^2}}{x}$

224. $\int \frac{dx}{x^2\sqrt{(a^2 - x^2)^3}} = \frac{1}{a^4}\left[-\frac{\sqrt{a^2 - x^2}}{x} + \frac{x}{\sqrt{a^2 - x^2}}\right]$

225. $\int \frac{dx}{x^3\sqrt{(a^2 - x^2)^3}} = -\frac{1}{2a^2x^2\sqrt{a^2 - x^2}} + \frac{3}{2a^4\sqrt{a^2 - x^2}}$

$$- \frac{3}{2a^5}\log\frac{a + \sqrt{a^2 - x^2}}{x}$$

226. $$\int \frac{x^m}{\sqrt{a^2-x^2}}\,dx = -\frac{x^{m-1}\sqrt{a^2-x^2}}{m} + \frac{(m-1)a^2}{m}\int \frac{x^{m-2}}{\sqrt{a^2-x^2}}\,dx$$

227. $$\int \frac{x^{2m}}{\sqrt{a^2-x^2}}\,dx = \frac{(2m)!}{(m!)^2}\left[-\sqrt{a^2-x^2}\sum_{r=1}^{m}\frac{r!(r-1)!}{2^{2m-2r+1}(2r)!}a^{2m-2r}x^{2r-1} + \frac{a^{2m}}{2^{2m}}\sin^{-1}\frac{x}{|a|}\right]$$

228. $$\int \frac{x^{2m+1}}{\sqrt{a^2-x^2}}\,dx = -\sqrt{a^2-x^2}\sum_{r=0}^{m}\frac{(2r)!(m!)^2}{(2m+1)!(r!)^2}(4a^2)^{m-r}x^{2r}$$

229. $$\int \frac{dx}{x^m\sqrt{a^2-x^2}} = -\frac{\sqrt{a^2-x^2}}{(m-1)a^2x^{m-1}} + \frac{m-2}{(m-1)a^2}\int \frac{dx}{x^{m-2}\sqrt{a^2-x^2}}$$

230. $$\int \frac{dx}{x^{2m}\sqrt{a^2-x^2}} = -\sqrt{a^2-x^2}\sum_{r=0}^{m-1}\frac{(m-1)!m!(2r)!2^{2m-2r-1}}{(r!)^2(2m)!a^{2m-2r}x^{2r+1}}$$

231. $$\int \frac{dx}{x^{2m+1}\sqrt{a^2-x^2}} = \frac{(2m)!}{(m!)^2}\left[-\frac{\sqrt{a^2-x^2}}{a^2}\sum_{r=1}^{m}\frac{r!(r-1)!}{2(2r)!(4a^2)^{m-r}x^{2r}} + \frac{1}{2^{2m}a^{2m+1}}\log\frac{a-\sqrt{a^2-x^2}}{x}\right]$$

232. $$\int \frac{dx}{(b^2-x^2)\sqrt{a^2-x^2}} = \frac{1}{2b\sqrt{a^2-b^2}}\log\frac{(b\sqrt{a^2-x^2}+x\sqrt{a^2-b^2})^2}{b^2-x^2}, \qquad (a^2>b^2)$$

233. $$\int \frac{dx}{(b^2-x^2)\sqrt{a^2-x^2}} = \frac{1}{b\sqrt{b^2-a^2}}\tan^{-1}\frac{x\sqrt{b^2-a^2}}{b\sqrt{a^2-x^2}}, \qquad (b^2>a^2)$$

234. $$\int \frac{dx}{(b^2+x^2)\sqrt{a^2-x^2}} = \frac{1}{b\sqrt{a^2+b^2}}\tan^{-1}\frac{x\sqrt{a^2+b^2}}{b\sqrt{a^2-x^2}}$$

235. $$\int \frac{\sqrt{a^2-x^2}}{b^2+x^2}\,dx = \frac{\sqrt{a^2+b^2}}{|b|}\sin^{-1}\frac{x\sqrt{a^2+b^2}}{|a|\sqrt{x^2+b^2}} - \sin^{-1}\frac{x}{|a|}$$

236. $$\int f(x,\sqrt{a^2-x^2})\,dx = a\int f(a\sin u, a\cos u)\cos u\,du, \qquad \left(u=\sin^{-1}\frac{x}{a},\ a>0\right)$$

FORMS CONTAINING $\sqrt{a+bx+cx^2}$

$$X = a+bx+cx^2,\ q = 4ac-b^2, \text{ and } k = \frac{4c}{q}$$

If $q=0$, then $\sqrt{X} = \sqrt{c}\left|x+\frac{b}{2c}\right|$

237. $$\int \frac{dx}{\sqrt{X}} = \begin{cases} \frac{1}{\sqrt{c}}\log(2\sqrt{cX}+2cx+b) \\ \text{or} \\ \frac{1}{\sqrt{c}}\sinh^{-1}\frac{2cx+b}{\sqrt{q}}, \qquad (c>0) \end{cases}$$

238. $$\int \frac{dx}{\sqrt{X}} = -\frac{1}{\sqrt{-c}}\sin^{-1}\frac{2cx+b}{\sqrt{-q}}, \qquad (c<0)$$

239. $$\int \frac{dx}{X\sqrt{X}} = \frac{2(2cx+b)}{q\sqrt{X}}$$

240. $$\int \frac{dx}{X^2\sqrt{X}} = \frac{2(2cx+b)}{3q\sqrt{X}}\left(\frac{1}{X} + 2k\right)$$

241. $$\int \frac{dx}{X^n\sqrt{X}} = \begin{cases} \dfrac{2(2cx+b)\sqrt{X}}{(2n-1)qX^n} + \dfrac{2k(n-1)}{2n-1}\displaystyle\int \frac{dx}{X^{n-1}\sqrt{X}} \\ \text{or} \\ \dfrac{(2cx+b)(n!)(n-1)!4^n k^{n-1}}{q[(2n)!]\sqrt{X}} \displaystyle\sum_{r=0}^{n-1} \frac{(2r)!}{(4kX)^r(r!)^2} \end{cases}$$

242. $$\int \sqrt{X}\,dx = \frac{(2cx+b)\sqrt{X}}{4c} + \frac{1}{2k}\int \frac{dx}{\sqrt{X}}$$

243. $$\int X\sqrt{X}\,dx = \frac{(2cx+b)\sqrt{X}}{8c}\left(X + \frac{3}{2k}\right) + \frac{3}{8k^2}\int \frac{dx}{\sqrt{X}}$$

244. $$\int X^2\sqrt{X}\,dx = \frac{(2cx+b)\sqrt{X}}{12c}\left(X^2 + \frac{5X}{4k} + \frac{15}{8k^2}\right) + \frac{5}{16k^3}\int \frac{dx}{\sqrt{X}}$$

245. $$\int X^n\sqrt{X}\,dx = \begin{cases} \dfrac{(2cx+b)X^n\sqrt{X}}{4(n+1)c} + \dfrac{2n+1}{2(n+1)k}\displaystyle\int X^{n-1}\sqrt{X}\,dx \\ \text{or} \\ \dfrac{(2n+2)!}{[(n+1)!]^2(4k)^{n+1}}\left[\dfrac{k(2cx+b)\sqrt{X}}{c}\displaystyle\sum_{r=0}^{n} \frac{r!(r+1)!(4kX)^r}{(2r+2)!} + \int \frac{dx}{\sqrt{X}}\right] \end{cases}$$

246. $$\int \frac{x\,dx}{\sqrt{X}} = \frac{\sqrt{X}}{c} - \frac{b}{2c}\int \frac{dx}{\sqrt{X}}$$

247. $$\int \frac{x\,dx}{X\sqrt{X}} = -\frac{2(bx+2a)}{q\sqrt{X}}$$

248. $$\int \frac{x\,dx}{X^n\sqrt{X}} = -\frac{\sqrt{X}}{(2n-1)cX^n} - \frac{b}{2c}\int \frac{dx}{X^n\sqrt{X}}$$

249. $$\int \frac{x^2\,dx}{\sqrt{X}} = \left(\frac{x}{2c} - \frac{3b}{4c^2}\right)\sqrt{X} + \frac{3b^2-4ac}{8c^2}\int \frac{dx}{\sqrt{X}}$$

250. $$\int \frac{x^2\,dx}{X\sqrt{X}} = \frac{(2b^2-4ac)x + 2ab}{cq\sqrt{X}} + \frac{1}{c}\int \frac{dx}{\sqrt{X}}$$

251. $$\int \frac{x^2\,dx}{X^n\sqrt{X}} = \frac{(2b^2-4ac)x + 2ab}{(2n-1)cqX^{n-1}\sqrt{X}} + \frac{4ac+(2n-3)b^2}{(2n-1)cq}\int \frac{dx}{X^{n-1}\sqrt{X}}$$

252. $$\int \frac{x^3\,dx}{\sqrt{X}} = \left(\frac{x^2}{3c} - \frac{5bx}{12c^2} + \frac{5b^2}{8c^3} - \frac{2a}{3c^2}\right)\sqrt{X} + \left(\frac{3ab}{4c^2} - \frac{5b^3}{16c^3}\right)\int \frac{dx}{\sqrt{X}}$$

253. $$\int \frac{x^n\,dx}{\sqrt{X}} = \frac{1}{nc}x^{n-1}\sqrt{X} - \frac{(2n-1)b}{2nc}\int \frac{x^{n-1}\,dx}{\sqrt{X}} - \frac{(n-1)a}{nc}\int \frac{x^{n-2}\,dx}{\sqrt{X}}$$

254. $$\int x\sqrt{X}\,dx = \frac{X\sqrt{X}}{3c} - \frac{b(2cx+b)}{8c^2}\sqrt{X} - \frac{b}{4ck}\int \frac{dx}{\sqrt{X}}$$

255. $$\int xX\sqrt{X}\,dx = \frac{X^2\sqrt{X}}{5c} - \frac{b}{2c}\int X\sqrt{X}\,dx$$

256. $\int xX^n\sqrt{X}\,dx = \frac{X^{n+1}\sqrt{X}}{(2n+3)c} - \frac{b}{2c}\int X^n\sqrt{X}\,dx$

257. $\int x^2\sqrt{X}\,dx = \left(x - \frac{5b}{6c}\right)\frac{X\sqrt{X}}{4c} + \frac{5b^2 - 4ac}{16c^2}\int \sqrt{X}\,dx$

258. $\int \frac{dx}{x\sqrt{X}} = -\frac{1}{\sqrt{a}}\log\frac{2\sqrt{aX} + bx + 2a}{x}, \qquad (a > 0)$

259. $\int \frac{dx}{x\sqrt{X}} = \frac{1}{\sqrt{-a}}\sin^{-1}\left(\frac{bx + 2a}{|x|\sqrt{-q}}\right), \qquad (a < 0)$

260. $\int \frac{dx}{x\sqrt{X}} = -\frac{2\sqrt{X}}{bx}, \qquad (a = 0)$

261. $\int \frac{dx}{x^2\sqrt{X}} = -\frac{\sqrt{X}}{ax} - \frac{b}{2a}\int \frac{dx}{x\sqrt{X}}$

262. $\int \frac{\sqrt{X}\,dx}{x} = \sqrt{X} + \frac{b}{2}\int \frac{dx}{\sqrt{X}} + a\int \frac{dx}{x\sqrt{X}}$

263. $\int \frac{\sqrt{X}\,dx}{x^2} = -\frac{\sqrt{X}}{x} + \frac{b}{2}\int \frac{dx}{x\sqrt{X}} + c\int \frac{dx}{\sqrt{X}}$

FORMS INVOLVING $\sqrt{2ax - x^2}$

264. $\int \sqrt{2ax - x^2}\,dx = \frac{1}{2}\left[(x - a)\sqrt{2ax - x^2} + a^2\sin^{-1}\frac{x - a}{|a|}\right]$

265. $\int \frac{dx}{\sqrt{2ax - x^2}} = \begin{cases} \cos^{-1}\frac{a - x}{|a|} \\ \text{or} \\ \sin^{-1}\frac{x - a}{|a|} \end{cases}$

266. $\int x^n\sqrt{2ax - x^2}\,dx = \begin{cases} -\frac{x^{n-1}(2ax - x^2)^{\frac{3}{2}}}{n + 2} + \frac{(2n + 1)a}{n + 2}\int x^{n-1}\sqrt{2ax - x^2}\,dx \\ \text{or} \\ \sqrt{2ax - x^2}\left[\frac{x^{n+1}}{n + 2} - \sum_{r=0}^{n}\frac{(2n + 1)!(r!)^2a^{n-r+1}}{2^{n-r}(2r + 1)!(n + 2)!n!}x^r\right] \\ \qquad + \frac{(2n + 1)!a^{n+2}}{2^n n!(n + 2)!}\sin^{-1}\frac{x - a}{|a|} \end{cases}$

267. $\int \frac{\sqrt{2ax - x^2}}{x^n}\,dx = \frac{(2ax - x^2)^{\frac{3}{2}}}{(3 - 2n)ax^n} + \frac{n - 3}{(2n - 3)a}\int \frac{\sqrt{2ax - x^2}}{x^{n-1}}\,dx$

268. $\int \frac{x^n\,dx}{\sqrt{2ax - x^2}} = \begin{cases} \frac{-x^{n-1}\sqrt{2ax - x^2}}{n} + \frac{a(2n - 1)}{n}\int \frac{x^{n-1}}{\sqrt{2ax - x^2}}\,dx \\ \text{or} \\ -\sqrt{2ax - x^2}\sum_{r=1}^{n}\frac{(2n)!r!(r - 1)!a^{n-r}}{2^{n-r}(2r)!(n!)^2}x^{r-1} \\ \qquad + \frac{(2n)!a^n}{2^n(n!)^2}\sin^{-1}\frac{x - a}{|a|} \end{cases}$

269. $$\int \frac{dx}{x^n\sqrt{2ax-x^2}} = \begin{cases} \dfrac{\sqrt{2ax-x^2}}{a(1-2n)x^n} + \dfrac{n-1}{(2n-1)a}\displaystyle\int \frac{dx}{x^{n-1}\sqrt{2ax-x^2}} \\ \text{or} \\ -\sqrt{2ax-x^2}\displaystyle\sum_{r=0}^{n-1} \frac{2^{n-r}(n-1)!n!(2r)!}{(2n)!(r!)^2a^{n-r}x^{r+1}} \end{cases}$$

270. $$\int \frac{dx}{(2ax-x^2)^{\frac{3}{2}}} = \frac{x-a}{a^2\sqrt{2ax-x^2}}$$

271. $$\int \frac{x\,dx}{(2ax-x^2)^{\frac{3}{2}}} = \frac{x}{a\sqrt{2ax-x^2}}$$

MISCELLANEOUS ALGEBRAIC FORMS

272. $$\int \frac{dx}{\sqrt{2ax+x^2}} = \log(x+a+\sqrt{2ax+x^2})$$

273. $$\int \sqrt{ax^2+c}\,dx = \frac{x}{2}\sqrt{ax^2+c} + \frac{c}{2\sqrt{a}}\log(x\sqrt{a}+\sqrt{ax^2+c}), \qquad (a>0)$$

274. $$\int \sqrt{ax^2+c}\,dx = \frac{x}{2}\sqrt{ax^2+c} + \frac{c}{2\sqrt{-a}}\sin^{-1}\left(x\sqrt{-\frac{a}{c}}\right), \qquad (a<0)$$

275. $$\int \sqrt{\frac{1+x}{1-x}}\,dx = \sin^{-1}x - \sqrt{1-x^2}$$

276. $$\int \frac{dx}{x\sqrt{ax^n+c}} = \begin{cases} \dfrac{1}{n\sqrt{c}}\log\dfrac{\sqrt{ax^n+c}-\sqrt{c}}{\sqrt{ax^n+c}+\sqrt{c}} \\ \text{or} \\ \dfrac{2}{n\sqrt{c}}\log\dfrac{\sqrt{ax^n+c}-\sqrt{c}}{\sqrt{x^n}}, \qquad (c>0) \end{cases}$$

277. $$\int \frac{dx}{x\sqrt{ax^n+c}} = \frac{2}{n\sqrt{-c}}\sec^{-1}\sqrt{-\frac{ax^n}{c}}, \qquad (c<0)$$

278. $$\int \frac{dx}{\sqrt{ax^2+c}} = \frac{1}{\sqrt{a}}\log(x\sqrt{a}+\sqrt{ax^2+c}), \qquad (a>0)$$

279. $$\int \frac{dx}{\sqrt{ax^2+c}} = \frac{1}{\sqrt{-a}}\sin^{-1}\left(x\sqrt{-\frac{a}{c}}\right), \qquad (a<0)$$

280. $$\int (ax^2+c)^{m+\frac{1}{2}}\,dx = \begin{cases} \dfrac{x(ax^2+c)^{m+\frac{1}{2}}}{2(m+1)} + \dfrac{(2m+1)c}{2(m+1)}\displaystyle\int (ax^2+c)^{m-\frac{1}{2}}\,dx \\ \text{or} \\ x\sqrt{ax^2+c}\displaystyle\sum_{r=0}^{m} \frac{(2m+1)!(r!)^2c^{m-r}}{2^{2m-2r+1}m!(m+1)!(2r+1)!}(ax^2+c)^r \\ \quad + \dfrac{(2m+1)!c^{m+1}}{2^{2m+1}m!(m+1)!}\displaystyle\int \frac{dx}{\sqrt{ax^2+c}} \end{cases}$$

281. $$\int x(ax^2+c)^{m+\frac{1}{2}}\,dx = \frac{(ax^2+c)^{m+\frac{3}{2}}}{(2m+3)a}$$

282. $\displaystyle\int \frac{(ax^2+c)^{m+\frac{1}{2}}}{x}\,dx = \begin{cases} \dfrac{(ax^2+c)^{m+\frac{1}{2}}}{2m+1} + c\displaystyle\int \frac{(ax^2+c)^{m-\frac{1}{2}}}{x}\,dx \\ \text{or} \\ \sqrt{ax^2+c}\displaystyle\sum_{r=0}^{m} \frac{c^{m-r}(ax^2+c)^r}{2r+1} + c^{m+1}\displaystyle\int \frac{dx}{x\sqrt{ax^2+c}} \end{cases}$

283. $\displaystyle\int \frac{dx}{(ax^2+c)^{m+\frac{1}{2}}} = \begin{cases} \dfrac{x}{(2m-1)c(ax^2+c)^{m-\frac{1}{2}}} + \dfrac{2m-2}{(2m-1)c}\displaystyle\int \frac{dx}{(ax^2+c)^{m-\frac{1}{2}}} \\ \text{or} \\ \dfrac{x}{\sqrt{ax^2+c}}\displaystyle\sum_{r=0}^{m-1} \frac{2^{2m-2r-1}(m-1)!m!(2r)!}{(2m)!(r!)^2c^{m-r}(ax^2+c)^r} \end{cases}$

284. $\displaystyle\int \frac{dx}{x^m\sqrt{ax^2+c}} = -\frac{\sqrt{ax^2+c}}{(m-1)cx^{m-1}} - \frac{(m-2)a}{(m-1)c}\int \frac{dx}{x^{m-2}\sqrt{ax^2+c}}$

285. $\displaystyle\int \frac{1+x^2}{(1-x^2)\sqrt{1+x^4}}\,dx = \frac{1}{\sqrt{2}}\log\frac{x\sqrt{2}+\sqrt{1+x^4}}{1-x^2}$

286. $\displaystyle\int \frac{1-x^2}{(1+x^2)\sqrt{1+x^4}}\,dx = \frac{1}{\sqrt{2}}\tan^{-1}\frac{x\sqrt{2}}{\sqrt{1+x^4}}$

287. $\displaystyle\int \frac{dx}{x\sqrt{x^n+a^2}} = -\frac{2}{na}\log\frac{a+\sqrt{x^n+a^2}}{\sqrt{x^n}}$

288. $\displaystyle\int \frac{dx}{x\sqrt{x^n-a^2}} = -\frac{2}{na}\sin^{-1}\frac{a}{\sqrt{x^n}}$

289. $\displaystyle\int \sqrt{\frac{x}{a^3-x^3}}\,dx = \frac{2}{3}\sin^{-1}\left(\frac{x}{a}\right)^{\frac{3}{2}}$

FORMS INVOLVING TRIGONOMETRIC FUNCTIONS

290. $\displaystyle\int (\sin ax)\,dx = -\frac{1}{a}\cos ax$

291. $\displaystyle\int (\cos ax)\,dx = \frac{1}{a}\sin ax$

292. $\displaystyle\int (\tan ax)\,dx = -\frac{1}{a}\log\cos ax = \frac{1}{a}\log\sec ax$

293. $\displaystyle\int (\cot ax)\,dx = \frac{1}{a}\log\sin ax = -\frac{1}{a}\log\csc ax$

294. $\displaystyle\int (\sec ax)\,dx = \frac{1}{a}\log(\sec ax + \tan ax) = \frac{1}{a}\log\tan\left(\frac{\pi}{4}+\frac{ax}{2}\right)$

295. $\displaystyle\int (\csc ax)\,dx = \frac{1}{a}\log(\csc ax - \cot ax) = \frac{1}{a}\log\tan\frac{ax}{2}$

296. $\displaystyle\int (\sin^2 ax)\,dx = -\frac{1}{2a}\cos ax\sin ax + \frac{1}{2}x = \frac{1}{2}x - \frac{1}{4a}\sin 2ax$

297. $\displaystyle\int (\sin^3 ax)\,dx = -\frac{1}{3a}(\cos ax)(\sin^2 ax + 2)$

298. $\displaystyle\int (\sin^4 ax)\,dx = \frac{3x}{8} - \frac{\sin 2ax}{4a} + \frac{\sin 4ax}{32a}$

299. $\displaystyle\int (\sin^n ax)\,dx = -\frac{\sin^{n-1}ax\cos ax}{na} + \frac{n-1}{n}\int (\sin^{n-2} ax)\,dx$

300. $$\int (\sin^{2m} ax)\, dx = -\frac{\cos ax}{a} \sum_{r=0}^{m-1} \frac{(2m)!(r!)^2}{2^{2m-2r}(2r+1)!(m!)^2} \sin^{2r+1} ax + \frac{(2m)!}{2^{2m}(m!)^2} x$$

301. $$\int (\sin^{2m+1} ax)\, dx = -\frac{\cos ax}{a} \sum_{r=0}^{m} \frac{2^{2m-2r}(m!)^2(2r)!}{(2m+1)!(r!)^2} \sin^{2r} ax$$

302. $$\int (\cos^2 ax)\, dx = \frac{1}{2a} \sin ax \cos ax + \frac{1}{2}x = \frac{1}{2}x + \frac{1}{4a}\sin 2ax$$

303. $$\int (\cos^3 ax)\, dx = \frac{1}{3a}(\sin ax)(\cos^2 ax + 2)$$

304. $$\int (\cos^4 ax)\, dx = \frac{3x}{8} + \frac{\sin 2ax}{4a} + \frac{\sin 4ax}{32a}$$

305. $$\int (\cos^n ax)\, dx = \frac{1}{na}\cos^{n-1} ax \sin ax + \frac{n-1}{n}\int (\cos^{n-2} ax)\, dx$$

306. $$\int (\cos^{2m} ax)\, dx = \frac{\sin ax}{a} \sum_{r=0}^{m-1} \frac{(2m)!(r!)^2}{2^{2m-2r}(2r+1)!(m!)^2} \cos^{2r+1} ax + \frac{(2m)!}{2^{2m}(m!)^2} x$$

307. $$\int (\cos^{2m+1} ax)\, dx = \frac{\sin ax}{a} \sum_{r=0}^{m} \frac{2^{2m-2r}(m!)^2(2r)!}{(2m+1)!(r!)^2} \cos^{2r} ax$$

308. $$\int \frac{dx}{\sin^2 ax} = \int (\csc^2 ax)\, dx = -\frac{1}{a}\cot ax$$

309. $$\int \frac{dx}{\sin^m ax} = \int (\csc^m ax)\, dx = -\frac{1}{(m-1)a} \cdot \frac{\cos ax}{\sin^{m-1} ax} + \frac{m-2}{m-1}\int \frac{dx}{\sin^{m-2} ax}$$

310. $$\int \frac{dx}{\sin^{2m} ax} = \int (\csc^{2m} ax)\, dx = -\frac{1}{a}\cos ax \sum_{r=0}^{m-1} \frac{2^{2m-2r-1}(m-1)!m!(2r)!}{(2m)!(r!)^2 \sin^{2r+1} ax}$$

311. $$\int \frac{dx}{\sin^{2m+1} ax} = \int (\csc^{2m+1} ax)\, dx =$$

$$-\frac{1}{a}\cos ax \sum_{r=0}^{m-1} \frac{(2m)!(r!)^2}{2^{2m-2r}(m!)^2(2r+1)!\sin^{2r+2} ax} + \frac{1}{a}\cdot\frac{(2m)!}{2^{2m}(m!)^2}\log\tan\frac{ax}{2}$$

312. $$\int \frac{dx}{\cos^2 ax} = \int (\sec^2 ax)\, dx = \frac{1}{a}\tan ax$$

313. $$\int \frac{dx}{\cos^n ax} = \int (\sec^n ax)\, dx = \frac{1}{(n-1)a}\cdot\frac{\sin ax}{\cos^{n-1} ax} + \frac{n-2}{n-1}\int \frac{dx}{\cos^{n-2} ax}$$

314. $$\int \frac{dx}{\cos^{2m} ax} = \int (\sec^{2m} ax)\, dx = \frac{1}{a}\sin ax \sum_{r=0}^{m-1} \frac{2^{2m-2r-1}(m-1)!m!(2r)!}{(2m)!(r!)^2 \cos^{2r+1} ax}$$

315. $$\int \frac{dx}{\cos^{2m+1} ax} = \int (\sec^{2m+1} ax)\, dx =$$

$$\frac{1}{a}\sin ax \sum_{r=0}^{m-1} \frac{(2m)!(r!)^2}{2^{2m-2r}(m!)^2(2r+1)!\cos^{2r+2} ax}$$

$$+ \frac{1}{a}\cdot\frac{(2m)!}{2^{2m}(m!)^2}\log(\sec ax + \tan ax)$$

316. $$\int (\sin mx)(\sin nx)\, dx = \frac{\sin(m-n)x}{2(m-n)} - \frac{\sin(m+n)x}{2(m+n)}, \qquad (m^2 \neq n^2)$$

317. $$\int (\cos mx)(\cos nx)\, dx = \frac{\sin(m-n)x}{2(m-n)} + \frac{\sin(m+n)x}{2(m+n)}, \qquad (m^2 \neq n^2)$$

318. $$\int (\sin ax)(\cos ax)\, dx = \frac{1}{2a}\sin^2 ax$$

319. $\int (\sin mx)(\cos nx)\,dx = -\dfrac{\cos(m-n)x}{2(m-n)} - \dfrac{\cos(m+n)x}{2(m+n)}, \qquad (m^2 \neq n^2)$

320. $\int (\sin^2 ax)(\cos^2 ax)\,dx = -\dfrac{1}{32a}\sin 4ax + \dfrac{x}{8}$

321. $\int (\sin ax)(\cos^m ax)\,dx = -\dfrac{\cos^{m+1} ax}{(m+1)a}$

322. $\int (\sin^m ax)(\cos ax)\,dx = \dfrac{\sin^{m+1} ax}{(m+1)a}$

323. $$\int (\cos^m ax)(\sin^n ax)\,dx = \begin{cases} \dfrac{\cos^{m-1} ax \sin^{n+1} ax}{(m+n)a} + \dfrac{m-1}{m+n}\displaystyle\int (\cos^{m-2} ax)(\sin^n ax)\,dx \\ \text{or} \\ -\dfrac{\sin^{n-1} ax \cos^{m+1} ax}{(m+n)a} + \dfrac{n-1}{m+n}\displaystyle\int (\cos^m ax)(\sin^{n-2} ax)\,dx \end{cases}$$

324. $$\int \frac{\cos^m ax}{\sin^n ax}\,dx = \begin{cases} -\dfrac{\cos^{m+1} ax}{(n-1)a\sin^{n-1} ax} - \dfrac{m-n+2}{n-1}\displaystyle\int \frac{\cos^m ax}{\sin^{n-2} ax}\,dx \\ \text{or} \\ \dfrac{\cos^{m-1} ax}{a(m-n)\sin^{n-1} ax} + \dfrac{m-1}{m-n}\displaystyle\int \frac{\cos^{m-2} ax}{\sin^n ax}\,dx \end{cases}$$

325. $$\int \frac{\sin^m ax}{\cos^n ax}\,dx = \begin{cases} \dfrac{\sin^{m+1} ax}{a(n-1)\cos^{n-1} ax} - \dfrac{m-n+2}{n-1}\displaystyle\int \frac{\sin^m ax}{\cos^{n-2} ax}\,dx \\ \text{or} \\ -\dfrac{\sin^{m-1} ax}{a(m-n)\cos^{n-1} ax} + \dfrac{m-1}{m-n}\displaystyle\int \frac{\sin^{m-2} ax}{\cos^n ax}\,dx \end{cases}$$

326. $\int \dfrac{\sin ax}{\cos^2 ax}\,dx = \dfrac{1}{a\cos ax} = \dfrac{\sec ax}{a}$

327. $\int \dfrac{\sin^2 ax}{\cos ax}\,dx = -\dfrac{1}{a}\sin ax + \dfrac{1}{a}\log\tan\left(\dfrac{\pi}{4} + \dfrac{ax}{2}\right)$

328. $\int \dfrac{\cos ax}{\sin^2 ax}\,dx = -\dfrac{1}{a\sin ax} = -\dfrac{\csc ax}{a}$

329. $\int \dfrac{dx}{(\sin ax)(\cos ax)} = \dfrac{1}{a}\log\tan ax$

330. $\int \dfrac{dx}{(\sin ax)(\cos^2 ax)} = \dfrac{1}{a}\left(\sec ax + \log\tan\dfrac{ax}{2}\right)$

331. $\int \dfrac{dx}{(\sin ax)(\cos^n ax)} = \dfrac{1}{a(n-1)\cos^{n-1} ax} + \int \dfrac{dx}{(\sin ax)(\cos^{n-2} ax)}$

332. $\int \dfrac{dx}{(\sin^2 ax)(\cos ax)} = -\dfrac{1}{a}\csc ax + \dfrac{1}{a}\log\tan\left(\dfrac{\pi}{4} + \dfrac{ax}{2}\right)$

333. $\int \dfrac{dx}{(\sin^2 ax)(\cos^2 ax)} = -\dfrac{2}{a}\cot 2ax$

334. $$\int \frac{dx}{\sin^m ax \cos^n ax} = \begin{cases} -\dfrac{1}{a(m-1)(\sin^{m-1} ax)(\cos^{n-1} ax)} \\ \qquad\qquad + \dfrac{m+n-2}{m-1}\displaystyle\int \frac{dx}{(\sin^{m-2} ax)(\cos^n ax)} \\ \text{or} \\ \dfrac{1}{a(n-1)\sin^{m-1} ax \cos^{n-1} ax} \\ \qquad\qquad - \dfrac{m+n-2}{n-1}\displaystyle\int \frac{dx}{\sin^m ax \cos^{n-2} ax} \end{cases}$$

335. $$\int \sin(a+bx)\,dx = -\frac{1}{b}\cos(a+bx)$$

336. $$\int \cos(a+bx)\,dx = \frac{1}{b}\sin(a+bx)$$

337. $$\int \frac{dx}{1 \pm \sin ax} = \mp \frac{1}{a}\tan\left(\frac{\pi}{4} \mp \frac{ax}{2}\right)$$

338. $$\int \frac{dx}{1+\cos ax} = \frac{1}{a}\tan\frac{ax}{2}$$

339. $$\int \frac{dx}{1-\cos ax} = -\frac{1}{a}\cot\frac{ax}{2}$$

***340.** $$\int \frac{dx}{a+b\sin x} = \begin{cases} \dfrac{2}{\sqrt{a^2-b^2}}\tan^{-1}\dfrac{a\tan\frac{x}{2}+b}{\sqrt{a^2-b^2}} \\ \text{or} \\ \dfrac{1}{\sqrt{b^2-a^2}}\log\dfrac{a\tan\frac{x}{2}+b-\sqrt{b^2-a^2}}{a\tan\frac{x}{2}+b+\sqrt{b^2-a^2}} \end{cases}$$

***341.** $$\int \frac{dx}{a+b\cos x} = \begin{cases} \dfrac{2}{\sqrt{a^2-b^2}}\tan^{-1}\dfrac{\sqrt{a^2-b^2}\tan\frac{x}{2}}{a+b} \\ \text{or} \\ \dfrac{1}{\sqrt{b^2-a^2}}\log\left(\dfrac{\sqrt{b^2-a^2}\tan\frac{x}{2}+a+b}{\sqrt{b^2-a^2}\tan\frac{x}{2}-a-b}\right) \end{cases}$$

***342.** $$\int \frac{dx}{a+b\sin x+c\cos x}$$

$$= \begin{cases} \dfrac{1}{\sqrt{b^2+c^2-a^2}}\log\dfrac{b-\sqrt{b^2+c^2-a^2}+(a-c)\tan\frac{x}{2}}{b+\sqrt{b^2+c^2-a^2}+(a-c)\tan\frac{x}{2}}, & \text{if } a^2 < b^2+c^2, a \neq c \\ \text{or} \\ \dfrac{2}{\sqrt{a^2-b^2-c^2}}\tan^{-1}\dfrac{b+(a-c)\tan\frac{x}{2}}{\sqrt{a^2-b^2-c^2}}, & \text{if } a^2 > b^2+c^2 \\ \text{or} \\ \dfrac{1}{a}\left[\dfrac{a-(b+c)\cos x-(b-c)\sin x}{a-(b-c)\cos x+(b+c)\sin x}\right], & \text{if } a^2 = b^2+c^2, a \neq c. \end{cases}$$

*See note 6 on page A-19.

*343. $\int \frac{\sin^2 x\,dx}{a+b\cos^2 x} = \frac{1}{b}\sqrt{\frac{a+b}{a}}\tan^{-1}\left(\sqrt{\frac{a}{a+b}}\tan x\right) - \frac{x}{b}, \qquad (ab > 0, \text{ or } |a| > |b|)$

*344. $\int \frac{dx}{a^2\cos^2 x + b^2\sin^2 x} = \frac{1}{ab}\tan^{-1}\left(\frac{b\tan x}{a}\right)$

*345. $\int \frac{\cos^2 cx}{a^2 + b^2\sin^2 cx}dx = \frac{\sqrt{a^2+b^2}}{ab^2c}\tan^{-1}\frac{\sqrt{a^2+b^2}\tan cx}{a} - \frac{x}{b^2}$

346. $\int \frac{\sin cx\cos cx}{a\cos^2 cx + b\sin^2 cx}dx = \frac{1}{2c(b-a)}\log(a\cos^2 cx + b\sin^2 cx)$

347. $\int \frac{\cos cx}{a\cos cx + b\sin cx}dx = \int \frac{dx}{a + b\tan cx} =$

$\frac{1}{c(a^2+b^2)}[acx + b\log(a\cos cx + b\text{ si}$

348. $\int \frac{\sin cx}{a\sin cx + b\cos cx}dx = \int \frac{dx}{a + b\cot cx} =$

$\frac{1}{c(a^2+b^2)}[acx - b\log(a\sin cx + b\cos x)]$

*349. $\int \frac{dx}{a\cos^2 x + 2b\cos x\sin x + c\sin^2 x} = \begin{cases} \frac{1}{2\sqrt{b^2-ac}}\log\frac{c\tan x + b - \sqrt{b^2-ac}}{c\tan x + b + \sqrt{b^2-ac}} \qquad (b^2 > ac) \\ \text{or} \\ \frac{1}{\sqrt{ac-b^2}}\tan^{-1}\frac{c\tan x + b}{\sqrt{ac-b^2}}, \qquad (b_2 < ac) \\ \text{or} \\ -\frac{1}{c\tan x + b}, \qquad (b^2 = ac) \end{cases}$

350. $\int \frac{\sin ax}{1 \pm \sin ax}dx = \pm x + \frac{1}{a}\tan\left(\frac{\pi}{4} \mp \frac{ax}{2}\right)$

351. $\int \frac{dx}{(\sin ax)(1 \pm \sin ax)} = \frac{1}{a}\tan\left(\frac{\pi}{4} \mp \frac{ax}{2}\right) + \frac{1}{a}\log\tan\frac{ax}{2}$

352. $\int \frac{dx}{(1+\sin ax)^2} = -\frac{1}{2a}\tan\left(\frac{\pi}{4} - \frac{ax}{2}\right) - \frac{1}{6a}\tan^3\left(\frac{\pi}{4} - \frac{ax}{2}\right)$

353. $\int \frac{dx}{(1-\sin ax)^2} = \frac{1}{2a}\cot\left(\frac{\pi}{4} - \frac{ax}{2}\right) + \frac{1}{6a}\cot^3\left(\frac{\pi}{4} - \frac{ax}{2}\right)$

354. $\int \frac{\sin ax}{(1+\sin ax)^2}dx = -\frac{1}{2a}\tan\left(\frac{\pi}{4} - \frac{ax}{2}\right) + \frac{1}{6a}\tan^3\left(\frac{\pi}{4} - \frac{ax}{2}\right)$

355. $\int \frac{\sin ax}{(1-\sin ax)^2}dx = -\frac{1}{2a}\cot\left(\frac{\pi}{4} - \frac{ax}{2}\right) + \frac{1}{6a}\cot^3\left(\frac{\pi}{4} - \frac{ax}{2}\right)$

356. $\int \frac{\sin x\,dx}{a + b\sin x} = \frac{x}{b} - \frac{a}{b}\int \frac{dx}{a + b\sin x}$

357. $\int \frac{dx}{(\sin x)(a + b\sin x)} = \frac{1}{a}\log\tan\frac{x}{2} - \frac{b}{a}\int \frac{dx}{a + b\sin x}$

358. $\int \frac{dx}{(a + b\sin x)^2} = \frac{b\cos x}{(a^2-b^2)(a + b\sin x)} + \frac{a}{a^2-b^2}\int \frac{dx}{a + b\sin x}$

*See note 6 on page A-19.

359. $$\int \frac{\sin x\,dx}{(a+b\sin x)^2} = \frac{a\cos x}{(b^2-a^2)(a+b\sin x)} + \frac{b}{b^2-a^2}\int \frac{dx}{a+b\sin x}$$

***360.** $$\int \frac{dx}{a^2+b^2\sin^2 cx} = \frac{1}{ac\sqrt{a^2+b^2}}\tan^{-1}\frac{\sqrt{a^2+b^2}\tan cx}{a}$$

***361.** $$\int \frac{dx}{a^2-b^2\sin^2 cx} = \begin{cases} \dfrac{1}{ac\sqrt{a^2-b^2}}\tan^{-1}\dfrac{\sqrt{a^2-b^2}\tan cx}{a}, & (a^2>b^2) \\ \text{or} & \\ \dfrac{1}{2ac\sqrt{b^2-a^2}}\log\dfrac{\sqrt{b^2-a^2}\tan cx + a}{\sqrt{b^2-a^2}\tan cx - a}, & (a^2<b^2) \end{cases}$$

362. $$\int \frac{\cos ax}{1+\cos ax}\,dx = x - \frac{1}{a}\tan\frac{ax}{2}$$

363. $$\int \frac{\cos ax}{1-\cos ax}\,dx = -x - \frac{1}{a}\cot\frac{ax}{2}$$

364. $$\int \frac{dx}{(\cos ax)(1+\cos ax)} = \frac{1}{a}\log\tan\left(\frac{\pi}{4}+\frac{ax}{2}\right) - \frac{1}{a}\tan\frac{ax}{2}$$

365. $$\int \frac{dx}{(\cos ax)(1-\cos ax)} = \frac{1}{a}\log\tan\left(\frac{\pi}{4}+\frac{ax}{2}\right) - \frac{1}{a}\cot\frac{ax}{2}$$

366. $$\int \frac{dx}{(1+\cos ax)^2} = \frac{1}{2a}\tan\frac{ax}{2} + \frac{1}{6a}\tan^3\frac{ax}{2}$$

367. $$\int \frac{dx}{(1-\cos ax)^2} = -\frac{1}{2a}\cot\frac{ax}{2} - \frac{1}{6a}\cot^3\frac{ax}{2}$$

368. $$\int \frac{\cos ax}{(1+\cos ax)^2}\,dx = \frac{1}{2a}\tan\frac{ax}{2} - \frac{1}{6a}\tan^3\frac{ax}{2}$$

369. $$\int \frac{\cos ax}{(1-\cos ax)^2}\,dx = \frac{1}{2a}\cot\frac{ax}{2} - \frac{1}{6a}\cot^3\frac{ax}{2}$$

370. $$\int \frac{\cos x\,dx}{a+b\cos x} = \frac{x}{b} - \frac{a}{b}\int \frac{dx}{a+b\cos x}$$

371. $$\int \frac{dx}{(\cos x)(a+b\cos x)} = \frac{1}{a}\log\tan\left(\frac{x}{2}+\frac{\pi}{4}\right) - \frac{b}{a}\int \frac{dx}{a+b\cos x}$$

372. $$\int \frac{dx}{(a+b\cos x)^2} = \frac{b\sin x}{(b^2-a^2)(a+b\cos x)} - \frac{a}{b^2-a^2}\int \frac{dx}{a+b\cos x}$$

373. $$\int \frac{\cos x}{(a+b\cos x)^2}\,dx = \frac{a\sin x}{(a^2-b^2)(a+b\cos x)} - \frac{b}{a^2-b^2}\int \frac{dx}{a+b\cos x}$$

***374.** $$\int \frac{dx}{a^2+b^2-2ab\cos cx} = \frac{2}{c(a^2-b^2)}\tan^{-1}\left(\frac{a+b}{a-b}\tan\frac{cx}{2}\right)$$

375. $$\int \frac{dx}{a^2+b^2\cos^2 cx} = \frac{1}{ac\sqrt{a^2+b^2}}\tan^{-1}\frac{a\tan cx}{\sqrt{a^2+b^2}}$$

***376.** $$\int \frac{dx}{a^2-b^2\cos^2 cx} = \begin{cases} \dfrac{1}{ac\sqrt{a^2-b^2}}\tan^{-1}\dfrac{a\tan cx}{\sqrt{a^2-b^2}}, & (a^2>b^2) \\ \text{or} & \\ \dfrac{1}{2ac\sqrt{b^2-a^2}}\log\dfrac{a\tan cx - \sqrt{b^2-a^2}}{a\tan cx + \sqrt{b^2-a^2}}, & (b^2>a^2) \end{cases}$$

377. $$\int \frac{\sin ax}{1\pm\cos ax}\,dx = \mp\frac{1}{a}\log(1\pm\cos ax)$$

*See note 6 on page A-19.

378. $\displaystyle\int \frac{\cos ax}{1 \pm \sin ax}\,dx = \pm \frac{1}{a}\log(1 \pm \sin ax)$

379. $\displaystyle\int \frac{dx}{(\sin ax)(1 \pm \cos ax)} = \pm \frac{1}{2a(1 \pm \cos ax)} + \frac{1}{2a}\log\tan\frac{ax}{2}$

380. $\displaystyle\int \frac{dx}{(\cos ax)(1 \pm \sin ax)} = \mp \frac{1}{2a(1 \pm \sin ax)} + \frac{1}{2a}\log\tan\left(\frac{\pi}{4} + \frac{ax}{2}\right)$

381. $\displaystyle\int \frac{\sin ax}{(\cos ax)(1 \pm \cos ax)}\,dx = \frac{1}{a}\log(\sec ax \pm 1)$

382. $\displaystyle\int \frac{\cos ax}{(\sin ax)(1 \pm \sin ax)}\,dx = -\frac{1}{a}\log(\csc ax \pm 1)$

383. $\displaystyle\int \frac{\sin ax}{(\cos ax)(1 \pm \sin ax)}\,dx = \frac{1}{2a(1 \pm \sin ax)} \pm \frac{1}{2a}\log\tan\left(\frac{\pi}{4} + \frac{ax}{2}\right)$

384. $\displaystyle\int \frac{\cos ax}{(\sin ax)(1 \pm \cos ax)}\,dx = -\frac{1}{2a(1 \pm \cos ax)} \pm \frac{1}{2a}\log\tan\frac{ax}{2}$

385. $\displaystyle\int \frac{dx}{\sin ax \pm \cos ax} = \frac{1}{a\sqrt{2}}\log\tan\left(\frac{ax}{2} \pm \frac{\pi}{8}\right)$

386. $\displaystyle\int \frac{dx}{(\sin ax \pm \cos ax)^2} = \frac{1}{2a}\tan\left(ax \mp \frac{\pi}{4}\right)$

387. $\displaystyle\int \frac{dx}{1 + \cos ax \pm \sin ax} = \pm \frac{1}{a}\log\left(1 \pm \tan\frac{ax}{2}\right)$

388. $\displaystyle\int \frac{dx}{a^2\cos^2 cx - b^2\sin^2 cx} = \frac{1}{2abc}\log\frac{b\tan cx + a}{b\tan cx - a}$

389. $\displaystyle\int x(\sin ax)\,dx = \frac{1}{a^2}\sin ax - \frac{x}{a}\cos ax$

390. $\displaystyle\int x^2(\sin ax)\,dx = \frac{2x}{a^2}\sin ax - \frac{a^2x^2 - 2}{a^3}\cos ax$

391. $\displaystyle\int x^3(\sin ax)\,dx = \frac{3a^2x^2 - 6}{a^4}\sin ax - \frac{a^2x^3 - 6x}{a^3}\cos ax$

392. $$\int x^m \sin ax\,dx = \begin{cases} -\dfrac{1}{a}x^m\cos ax + \dfrac{m}{a}\displaystyle\int x^{m-1}\cos ax\,dx \\ \text{or} \\ \cos ax \displaystyle\sum_{r=0}^{\left[\frac{m}{2}\right]} (-1)^{r+1}\frac{m!}{(m-2r)!}\cdot\frac{x^{m-2r}}{a^{2r+1}} \\ +\sin ax \displaystyle\sum_{r=0}^{\left[\frac{m-1}{2}\right]} (-1)^r \frac{m!}{(m-2r-1)!}\cdot\frac{x^{m-2r-1}}{a^{2r+2}} \end{cases}$$

Note: $[s]$ means greatest integer $\le s$; $[3\frac{1}{2}] = 3$, $[\frac{1}{2}] = 0$, etc.

393. $\displaystyle\int x(\cos ax)\,dx = \frac{1}{a^2}\cos ax + \frac{x}{a}\sin ax$

394. $\displaystyle\int x^2(\cos ax)\,dx = \frac{2x\cos ax}{a^2} + \frac{a^2x^2 - 2}{a^3}\sin ax$

395. $\displaystyle\int x^3(\cos ax)\,dx = \frac{3a^2x^2 - 6}{a^4}\cos ax + \frac{a^2x^3 - 6x}{a^3}\sin ax$

396. $$\int x^m(\cos ax)\,dx = \begin{cases} \dfrac{x^m \sin ax}{a} - \dfrac{m}{a}\displaystyle\int x^{m-1} \sin ax\,dx \\ \text{or} \\ \sin ax \displaystyle\sum_{r=0}^{\left[\frac{m}{2}\right]} (-1)^r \frac{m!}{(m-2r)!} \cdot \frac{x^{m-2r}}{a^{2r+1}} \\ \quad + \cos ax \displaystyle\sum_{r=0}^{\left[\frac{m-1}{2}\right]} (-1)^r \frac{m!}{(m-2r-1)!} \cdot \frac{x^{m-2r-1}}{a^{2r+2}} \end{cases}$$

See note integral 392.

397. $$\int \frac{\sin ax}{x}\,dx = \sum_{n=0}^{\infty} (-1)^n \frac{(ax)^{2n+1}}{(2n+1)(2n+1)!}$$

398. $$\int \frac{\cos ax}{x}\,dx = \log x + \sum_{n=1}^{\infty} (-1)^n \frac{(ax)^{2n}}{2n(2n)!}$$

399. $$\int x(\sin^2 ax)\,dx = \frac{x^2}{4} - \frac{x \sin 2ax}{4a} - \frac{\cos 2ax}{8a^2}$$

400. $$\int x^2(\sin^2 ax)\,dx = \frac{x^3}{6} - \left(\frac{x^2}{4a} - \frac{1}{8a^3}\right) \sin 2ax - \frac{x \cos 2ax}{4a^2}$$

401. $$\int x(\sin^3 ax)\,dx = \frac{x \cos 3ax}{12a} - \frac{\sin 3ax}{36a^2} - \frac{3x \cos ax}{4a} + \frac{3 \sin ax}{4a^2}$$

402. $$\int x(\cos^2 ax)\,dx = \frac{x^2}{4} + \frac{x \sin 2ax}{4a} + \frac{\cos 2ax}{8a^2}$$

403. $$\int x^2(\cos^2 ax)\,dx = \frac{x^3}{6} + \left(\frac{x^2}{4a} - \frac{1}{8a^3}\right) \sin 2ax + \frac{x \cos 2ax}{4a^2}$$

404. $$\int x(\cos^3 ax)\,dx = \frac{x \sin 3ax}{12a} + \frac{\cos 3ax}{36a^2} + \frac{3x \sin ax}{4a} + \frac{3 \cos ax}{4a^2}$$

405. $$\int \frac{\sin ax}{x^m}\,dx = -\frac{\sin ax}{(m-1)x^{m-1}} + \frac{a}{m-1}\int \frac{\cos ax}{x^{m-1}}\,dx$$

406. $$\int \frac{\cos ax}{x^m}\,dx = -\frac{\cos ax}{(m-1)x^{m-1}} - \frac{a}{m-1}\int \frac{\sin ax}{x^{m-1}}\,dx$$

407. $$\int \frac{x}{1 \pm \sin ax}\,dx = \mp \frac{x \cos ax}{a(1 \pm \sin ax)} + \frac{1}{a^2} \log(1 \pm \sin ax)$$

408. $$\int \frac{x}{1 + \cos ax}\,dx = \frac{x}{a} \tan\frac{ax}{2} + \frac{2}{a^2} \log \cos\frac{ax}{2}$$

409. $$\int \frac{x}{1 - \cos ax}\,dx = -\frac{x}{a} \cot\frac{ax}{2} + \frac{2}{a^2} \log \sin\frac{ax}{2}$$

410. $$\int \frac{x + \sin x}{1 + \cos x}\,dx = x \tan\frac{x}{2}$$

411. $$\int \frac{x - \sin x}{1 - \cos x}\,dx = -x \cot\frac{x}{2}$$

412. $$\int \sqrt{1 - \cos ax}\,dx = -\frac{2 \sin ax}{a\sqrt{1 - \cos ax}} = -\frac{2\sqrt{2}}{a} \cos\left(\frac{ax}{2}\right)$$

413. $$\int \sqrt{1 + \cos ax}\,dx = \frac{2 \sin ax}{a\sqrt{1 + \cos ax}} = \frac{2\sqrt{2}}{a} \sin\left(\frac{ax}{2}\right)$$

414. $$\int \sqrt{1 + \sin x}\,dx = \pm 2\left(\sin\frac{x}{2} - \cos\frac{x}{2}\right),$$

[use + if $(8k-1)\frac{\pi}{2} < x \le (8k+3)\frac{\pi}{2}$, otherwise $-$; k an integer]

415. $\int \sqrt{1 - \sin x}\, dx = \pm 2\left(\sin\frac{x}{2} + \cos\frac{x}{2}\right),$

[use + if $(8k - 3)\frac{\pi}{2} < x \le (8k + 1)\frac{\pi}{2}$, otherwise − ; k an integer]

416. $\int \frac{dx}{\sqrt{1 - \cos x}} = \pm\sqrt{2}\log\tan\frac{x}{4},$

[use + if $4k\pi < x < (4k + 2)\pi$, otherwise − ; k an integer]

417. $\int \frac{dx}{\sqrt{1 + \cos x}} = \pm\sqrt{2}\log\tan\left(\frac{x + \pi}{4}\right),$

[use + if $(4k - 1)\pi < x < (4k + 1)\pi$, otherwise − ; k an integer]

418. $\int \frac{dx}{\sqrt{1 - \sin x}} = \pm\sqrt{2}\log\tan\left(\frac{x}{4} - \frac{\pi}{8}\right),$

[use + if $(8k + 1)\frac{\pi}{2} < x < (8k + 5)\frac{\pi}{2}$, otherwise − ; k an integer]

419. $\int \frac{dx}{\sqrt{1 + \sin x}} = \pm\sqrt{2}\log\tan\left(\frac{x}{4} + \frac{\pi}{8}\right),$

[use + if $(8k - 1)\frac{\pi}{2} < x < (8k + 3)\frac{\pi}{2}$, otherwise − ; k an integer]

420. $\int (\tan^2 ax)\, dx = \frac{1}{a}\tan ax - x$

421. $\int (\tan^3 ax)\, dx = \frac{1}{2a}\tan^2 ax + \frac{1}{a}\log\cos ax$

422. $\int (\tan^4 ax)\, dx = \frac{\tan^3 ax}{3a} - \frac{1}{a}\tan x + x$

423. $\int (\tan^n ax)\, dx = \frac{\tan^{n-1} ax}{a(n - 1)} - \int (\tan^{n-2} ax)\, dx$

424. $\int (\cot^2 ax)\, dx = -\frac{1}{a}\cot ax - x$

425. $\int (\cot^3 ax)\, dx = -\frac{1}{2a}\cot^2 ax - \frac{1}{a}\log\sin ax$

426. $\int (\cot^4 ax)\, dx = -\frac{1}{3a}\cot^3 ax + \frac{1}{a}\cot ax + x$

427. $\int (\cot^n ax)\, dx = -\frac{\cot^{n-1} ax}{a(n - 1)} - \int (\cot^{n-2} ax)\, dx$

428. $\int \frac{x}{\sin^2 ax}\, dx = \int x(\csc^2 ax)\, dx = -\frac{x\cot ax}{a} + \frac{1}{a^2}\log\sin ax$

429. $\int \frac{x}{\sin^n ax}\, dx = \int x(\csc^n ax)\, dx = -\frac{x\cos ax}{a(n - 1)\sin^{n-1} ax}$

$$- \frac{1}{a^2(n - 1)(n - 2)\sin^{n-2} ax} + \frac{(n - 2)}{(n - 1)}\int \frac{x}{\sin^{n-2} ax}\, dx$$

430. $\int \frac{x}{\cos^2 ax}\, dx = \int x(\sec^2 ax)\, dx = \frac{1}{a}x\tan ax + \frac{1}{a^2}\log\cos ax$

431. $\int \frac{x}{\cos^n ax}\, dx = \int x(\sec^n ax)\, dx = \frac{x\sin ax}{a(n - 1)\cos^{n-1} ax}$

$$- \frac{1}{a^2(n - 1)(n - 2)\cos^{n-2} ax} + \frac{n - 2}{n - 1}\int \frac{x}{\cos^{n-2} ax}\, dx$$

432. $$\int \frac{\sin ax}{\sqrt{1+b^2\sin^2 ax}}\,dx = -\frac{1}{ab}\sin^{-1}\frac{b\cos ax}{\sqrt{1+b^2}}$$

433. $$\int \frac{\sin ax}{\sqrt{1-b^2\sin^2 ax}}\,dx = -\frac{1}{ab}\log\left(b\cos ax + \sqrt{1-b^2\sin^2 ax}\right)$$

434. $$\int (\sin ax)\sqrt{1+b^2\sin^2 ax}\,dx = -\frac{\cos ax}{2a}\sqrt{1+b^2\sin^2 ax} - \frac{1+b^2}{2ab}\sin^{-1}\frac{b\cos ax}{\sqrt{1+b^2}}$$

435. $$\int (\sin ax)\sqrt{1-b^2\sin^2 ax}\,dx = -\frac{\cos ax}{2a}\sqrt{1-b^2\sin^2 ax} - \frac{1-b^2}{2ab}\log\left(b\cos ax + \sqrt{1-b^2\sin^2 ax}\right)$$

436. $$\int \frac{\cos ax}{\sqrt{1+b^2\sin^2 ax}}\,dx = \frac{1}{ab}\log\left(b\sin ax + \sqrt{1+b^2\sin^2 ax}\right)$$

437. $$\int \frac{\cos ax}{\sqrt{1-b^2\sin^2 ax}}\,dx = \frac{1}{ab}\sin^{-1}(b\sin ax)$$

438. $$\int (\cos ax)\sqrt{1+b^2\sin^2 ax}\,dx = \frac{\sin ax}{2a}\sqrt{1+b^2\sin^2 ax} + \frac{1}{2ab}\log\left(b\sin ax + \sqrt{1+b^2\sin^2 ax}\right)$$

439. $$\int (\cos ax)\sqrt{1-b^2\sin^2 ax}\,dx = \frac{\sin ax}{2a}\sqrt{1-b^2\sin^2 ax} + \frac{1}{2ab}\sin^{-1}(b\sin ax)$$

440. $$\int \frac{dx}{\sqrt{a+b\tan^2 cx}} = \frac{\pm 1}{c\sqrt{a-b}}\sin^{-1}\left(\sqrt{\frac{a-b}{a}}\sin cx\right), \qquad (a > |b|)$$

[use + if $(2k-1)\frac{\pi}{2} < x \leq (2k+1)\frac{\pi}{2}$, otherwise −; k an integer]

FORMS INVOLVING INVERSE TRIGONOMETRIC FUNCTIONS

441. $$\int (\sin^{-1} ax)\,dx = x\sin^{-1} ax + \frac{\sqrt{1-a^2x^2}}{a}$$

442. $$\int (\cos^{-1} ax)\,dx = x\cos^{-1} ax - \frac{\sqrt{1-a^2x^2}}{a}$$

443. $$\int (\tan^{-1} ax)\,dx = x\tan^{-1} ax - \frac{1}{2a}\log(1+a^2x^2)$$

444. $$\int (\cot^{-1} ax)\,dx = x\cot^{-1} ax + \frac{1}{2a}\log(1+a^2x^2)$$

445. $$\int (\sec^{-1} ax)\,dx = x\sec^{-1} ax - \frac{1}{a}\log\left(ax + \sqrt{a^2x^2-1}\right)$$

446. $$\int (\csc^{-1} ax)\,dx = x\csc^{-1} ax + \frac{1}{a}\log\left(ax + \sqrt{a^2x^2-1}\right)$$

447. $$\int \left(\sin^{-1}\frac{x}{a}\right)dx = x\sin^{-1}\frac{x}{a} + \sqrt{a^2-x^2}, \qquad (a > 0)$$

448. $$\int \left(\cos^{-1}\frac{x}{a}\right)dx = x\cos^{-1}\frac{x}{a} - \sqrt{a^2-x^2}, \qquad (a > 0)$$

449. $\int \left(\tan^{-1}\frac{x}{a}\right) dx = x\tan^{-1}\frac{x}{a} - \frac{a}{2}\log(a^2 + x^2)$

450. $\int \left(\cot^{-1}\frac{x}{a}\right) dx = x\cot^{-1}\frac{x}{a} + \frac{a}{2}\log(a^2 + x^2)$

451. $\int x[\sin^{-1}(ax)]\, dx = \frac{1}{4a^2}[(2a^2x^2 - 1)\sin^{-1}(ax) + ax\sqrt{1 - a^2x^2}]$

452. $\int x[\cos^{-1}(ax)]\, dx = \frac{1}{4a^2}[(2a^2x^2 - 1)\cos^{-1}(ax) - ax\sqrt{1 - a^2x^2}]$

453. $\int x^n[\sin^{-1}(ax)]\, dx = \frac{x^{n+1}}{n+1}\sin^{-1}(ax) - \frac{a}{n+1}\int \frac{x^{n+1}\, dx}{\sqrt{1 - a^2x^2}}, \quad (n \neq -1)$

454. $\int x^n[\cos^{-1}(ax)]\, dx = \frac{x^{n+1}}{n+1}\cos^{-1}(ax) + \frac{a}{n+1}\int \frac{x^{n+1}\, dx}{\sqrt{1 - a^2x^2}}, \quad (n \neq -1)$

455. $\int x(\tan^{-1} ax)\, dx = \frac{1 + a^2x^2}{2a^2}\tan^{-1} ax - \frac{x}{2a}$

456. $\int x^n(\tan^{-1} ax)\, dx = \frac{x^{n+1}}{n+1}\tan^{-1} ax - \frac{a}{n+1}\int \frac{x^{n+1}}{1 + a^2x^2}\, dx$

457. $\int x(\cot^{-1} ax)\, dx = \frac{1 + a^2x^2}{2a^2}\cot^{-1} ax + \frac{x}{2a}$

458. $\int x^n(\cot^{-1} ax)\, dx = \frac{x^{n+1}}{n+1}\cot^{-1} ax + \frac{a}{n+1}\int \frac{x^{n+1}}{1 + a^2x^2}\, dx$

459. $\int \frac{\sin^{-1}(ax)}{x^2}\, dx = a\log\left(\frac{1 - \sqrt{1 - a^2x^2}}{x}\right) - \frac{\sin^{-1}(ax)}{x}$

460. $\int \frac{\cos^{-1}(ax)\, dx}{x^2} = -\frac{1}{x}\cos^{-1}(ax) + a\log\frac{1 + \sqrt{1 - a^2x^2}}{x}$

461. $\int \frac{\tan^{-1}(ax)\, dx}{x^2} = -\frac{1}{x}\tan^{-1}(ax) - \frac{a}{2}\log\frac{1 + a^2x^2}{x^2}$

462. $\int \frac{\cot^{-1} ax}{x^2}\, dx = -\frac{1}{x}\cot^{-1} ax - \frac{a}{2}\log\frac{x^2}{a^2x^2 + 1}$

463. $\int (\sin^{-1} ax)^2\, dx = x(\sin^{-1} ax)^2 - 2x + \frac{2\sqrt{1 - a^2x^2}}{a}\sin^{-1} ax$

464. $\int (\cos^{-1} ax)^2\, dx = x(\cos^{-1} ax)^2 - 2x - \frac{2\sqrt{1 - a^2x^2}}{a}\cos^{-1} ax$

465. $$\int (\sin^{-1} ax)^n\, dx = \begin{cases} x(\sin^{-1} ax)^n + \frac{n\sqrt{1 - a^2x^2}}{a}(\sin^{-1} ax)^{n-1} \\ \qquad - n(n-1)\int (\sin^{-1} ax)^{n-2}\, dx \\ \text{or} \\ \sum_{r=0}^{\left[\frac{n}{2}\right]} (-1)^r \frac{n!}{(n-2r)!} x(\sin^{-1} ax)^{n-2r} \\ \qquad + \sum_{r=0}^{\left[\frac{n-1}{2}\right]} (-1)^r \frac{n!\sqrt{1 - a^2x^2}}{(n-2r-1)!a}(\sin^{-1} ax)^{n-2r-1} \end{cases}$$

Note: $[s]$ means greatest integer $\leq s$. Thus $[3.5]$ means 3; $[5] = 5$, $[\frac{1}{2}] = 0$.

466. $$\int (\cos^{-1} ax)^n \, dx = \begin{cases} x(\cos^{-1} ax)^n - \dfrac{n\sqrt{1-a^2x^2}}{a}(\cos^{-1} ax)^{n-1} \\ \qquad\qquad -n(n-1)\displaystyle\int (\cos^{-1} ax)^{n-2}\, dx \\ \quad \text{or} \\ \displaystyle\sum_{r=0}^{\left[\frac{n}{2}\right]} (-1)^r \frac{n!}{(n-2r)!} x(\cos^{-1} ax)^{n-2r} \\ \qquad - \displaystyle\sum_{r=0}^{\left[\frac{n-1}{2}\right]} (-1)^r \frac{n!\sqrt{1-a^2x^2}}{(n-2r-1)!a}(\cos^{-1} ax)^{n-2r-1} \end{cases}$$

467. $$\int \frac{1}{\sqrt{1-a^2x^2}}(\sin^{-1} ax)\, dx = \frac{1}{2a}(\sin^{-1} ax)^2$$

468. $$\int \frac{x^n}{\sqrt{1-a^2x^2}}(\sin^{-1} ax)\, dx = -\frac{x^{n-1}}{na^2}\sqrt{1-a^2x^2}\sin^{-1} ax + \frac{x^n}{n^2a} + \frac{n-1}{na^2}\int \frac{x^{n-2}}{\sqrt{1-a^2x^2}}\sin^{-1} ax\, dx$$

469. $$\int \frac{1}{\sqrt{1-a^2x^2}}(\cos^{-1} ax)\, dx = -\frac{1}{2a}(\cos^{-1} ax)^2$$

470. $$\int \frac{x^n}{\sqrt{1-a^2x^2}}(\cos^{-1} ax)\, dx = -\frac{x^{n-1}}{na^2}\sqrt{1-a^2x^2}\cos^{-1} ax - \frac{x^n}{n^2a} + \frac{n-1}{na^2}\int \frac{x^{n-2}}{\sqrt{1-a^2x^2}}\cos^{-1} ax\, dx$$

471. $$\int \frac{\tan^{-1} ax}{a^2x^2+1}\, dx = \frac{1}{2a}(\tan^{-1} ax)^2$$

472. $$\int \frac{\cot^{-1} ax}{a^2x^2+1}\, dx = -\frac{1}{2a}(\cot^{-1} ax)^2$$

473. $$\int x \sec^{-1} ax\, dx = \frac{x^2}{2}\sec^{-1} ax - \frac{1}{2a^2}\sqrt{a^2x^2-1}$$

474. $$\int x^n \sec^{-1} ax\, dx = \frac{x^{n+1}}{n+1}\sec^{-1} ax - \frac{1}{n+1}\int \frac{x^n\, dx}{\sqrt{a^2x^2-1}}$$

475. $$\int \frac{\sec^{-1} ax}{x^2}\, dx = -\frac{\sec^{-1} ax}{x} + \frac{\sqrt{a^2x^2-1}}{x}$$

476. $$\int x \csc^{-1} ax\, dx = \frac{x^2}{2}\csc^{-1} ax + \frac{1}{2a^2}\sqrt{a^2x^2-1}$$

477. $$\int x^n \csc^{-1} ax\, dx = \frac{x^{n+1}}{n+1}\csc^{-1} ax + \frac{1}{n+1}\int \frac{x^n\, dx}{\sqrt{a^2x^2-1}}$$

478. $$\int \frac{\csc^{-1} ax}{x^2}\, dx = -\frac{\csc^{-1} ax}{x} - \frac{\sqrt{a^2x^2-1}}{x}$$

FORMS INVOLVING TRIGONOMETRIC SUBSTITUTIONS

479. $$\int f(\sin x)\, dx = 2\int f\left(\frac{2z}{1+z^2}\right)\frac{dz}{1+z^2}, \qquad \left(z = \tan\frac{x}{2}\right)$$

480. $$\int f(\cos x)\, dx = 2\int f\left(\frac{1-z^2}{1+z^2}\right)\frac{dz}{1+z^2}, \qquad \left(z = \tan\frac{x}{2}\right)$$

*481. $\int f(\sin x)\,dx = \int f(u)\frac{du}{\sqrt{1-u^2}}, \qquad (u = \sin x)$

*482. $\int f(\cos x)\,dx = -\int f(u)\frac{du}{\sqrt{1-u^2}}, \qquad (u = \cos x)$

*483. $\int f(\sin x, \cos x)\,dx = \int f(u, \sqrt{1-u^2})\frac{du}{\sqrt{1-u^2}}, \qquad (u = \sin x)$

484. $\int f(\sin x, \cos x)\,dx = 2\int f\left(\frac{2z}{1+z^2}, \frac{1-z^2}{1+z^2}\right)\frac{dz}{1+z^2}, \qquad \left(z = \tan\frac{x}{2}\right)$

LOGARITHMIC FORMS

485. $\int (\log x)\,dx = x\log x - x$

486. $\int x(\log x)\,dx = \frac{x^2}{2}\log x - \frac{x^2}{4}$

487. $\int x^2(\log x)\,dx = \frac{x^3}{3}\log x - \frac{x^3}{9}$

488. $\int x^n(\log ax)\,dx = \frac{x^{n+1}}{n+1}\log ax - \frac{x^{n+1}}{(n+1)^2}$

489. $\int (\log x)^2\,dx = x(\log x)^2 - 2x\log x + 2x$

490. $$\int (\log x)^n\,dx = \begin{cases} x(\log x)^n - n\int(\log x)^{n-1}\,dx, & (n \neq -1) \\ \text{or} \\ (-1)^n n!\,x\sum_{r=0}^{n}\frac{(-\log x)^r}{r!} \end{cases}$$

491. $\int \frac{(\log x)^n}{x}\,dx = \frac{1}{n+1}(\log x)^{n+1}$

492. $\int \frac{dx}{\log x} = \log(\log x) + \log x + \frac{(\log x)^2}{2\cdot 2!} + \frac{(\log x)^3}{3\cdot 3!} + \cdots$

493. $\int \frac{dx}{x\log x} = \log(\log x)$

494. $\int \frac{dx}{x(\log x)^n} = -\frac{1}{(n-1)(\log x)^{n-1}}$

495. $\int \frac{x^m\,dx}{(\log x)^n} = -\frac{x^{m+1}}{(n-1)(\log x)^{n-1}} + \frac{m+1}{n-1}\int\frac{x^m\,dx}{(\log x)^{n-1}}$

496. $$\int x^m(\log x)^n\,dx = \begin{cases} \frac{x^{m+1}(\log x)^n}{m+1} - \frac{n}{m+1}\int x^m(\log x)^{n-1}\,dx \\ \text{or} \\ (-1)^n\frac{n!}{m+1}x^{m+1}\sum_{r=0}^{n}\frac{(-\log x)^r}{r!(m+1)^{n-r}} \end{cases}$$

497. $\int x^p\cos(b\ln x)\,dx = \frac{x^{p+1}}{(p+1)^2+b^2}\cdot[b\sin(b\ln x) + (p+1)\cos(b\ln x)] + c$

498. $\int x^p\sin(b\ln x)\,dx = \frac{x^{p+1}}{(p+1)^2+b^2}\cdot[(p+1)\sin(b\ln x) - b\cos(b\ln x)] + c$

499. $\int [\log(ax+b)]\,dx = \frac{ax+b}{a}\log(ax+b) - x$

*The square roots appearing in these formulas may be plus or minus, depending on the quadrant of x. Care must be used to give them the proper sign.

500. $$\int \frac{\log(ax+b)}{x^2}\,dx = \frac{a}{b}\log x - \frac{ax+b}{bx}\log(ax+b)$$

501. $$\int x^m[\log(ax+b)]\,dx = \frac{1}{m+1}\left[x^{m+1} - \left(-\frac{b}{a}\right)^{m+1}\right]\log(ax+b) - \frac{1}{m+1}\left(-\frac{b}{a}\right)^{m+1}\sum_{r=1}^{m+1}\frac{1}{r}\left(-\frac{ax}{b}\right)^r$$

502. $$\int \frac{\log(ax+b)}{x^m}\,dx = -\frac{1}{m-1}\frac{\log(ax+b)}{x^{m-1}} + \frac{1}{m-1}\left(-\frac{a}{b}\right)^{m-1}\log\frac{ax+b}{x} + \frac{1}{m-1}\left(-\frac{a}{b}\right)^{m-1}\sum_{r=1}^{m-2}\frac{1}{r}\left(-\frac{b}{ax}\right)^r, \quad (m>2)$$

503. $$\int \left[\log\frac{x+a}{x-a}\right]dx = (x+a)\log(x+a) - (x-a)\log(x-a)$$

504. $$\int x^m\left[\log\frac{x+a}{x-a}\right]dx = \frac{x^{m+1}-(-a)^{m+1}}{m+1}\log(x+a) - \frac{x^{m+1}-a^{m+1}}{m+1}\log(x-a) + \frac{2a^{m+1}}{m+1}\sum_{r=1}^{\left[\frac{m+1}{2}\right]}\frac{1}{m-2r+2}\left(\frac{x}{a}\right)^{m-2r+2}$$

See note integral 392.

505. $$\int \frac{1}{x^2}\left[\log\frac{x+a}{x-a}\right]dx = \frac{1}{x}\log\frac{x-a}{x+a} - \frac{1}{a}\log\frac{x^2-a^2}{x^2}$$

506. $$\int(\log X)\,dx = \begin{cases} \left(x+\dfrac{b}{2c}\right)\log X - 2x + \dfrac{\sqrt{4ac-b^2}}{c}\tan^{-1}\dfrac{2cx+b}{\sqrt{4ac-b^2}}, & (b^2-4ac<0) \\ \text{or} & \\ \left(x+\dfrac{b}{2c}\right)\log X - 2x + \dfrac{\sqrt{b^2-4ac}}{c}\tanh^{-1}\dfrac{2cx+b}{\sqrt{b^2-4ac}}, & (b^2-4ac>0) \\ \text{where} & \\ X = a+bx+cx^2 & \end{cases}$$

507. $$\int x^n(\log X)\,dx = \frac{x^{n+1}}{n+1}\log X - \frac{2c}{n+1}\int\frac{x^{n+2}}{X}\,dx - \frac{b}{n+1}\int\frac{x^{n+1}}{X}\,dx$$

where $X = a + bx + cx^2$

508. $$\int[\log(x^2+a^2)]\,dx = x\log(x^2+a^2) - 2x + 2a\tan^{-1}\frac{x}{a}$$

509. $$\int[\log(x^2-a^2)]\,dx = x\log(x^2-a^2) - 2x + a\log\frac{x+a}{x-a}$$

510. $$\int x[\log(x^2\pm a^2)]\,dx = \tfrac{1}{2}(x^2\pm a^2)\log(x^2\pm a^2) - \tfrac{1}{2}x^2$$

511. $$\int[\log(x+\sqrt{x^2\pm a^2})]\,dx = x\log(x+\sqrt{x^2\pm a^2}) - \sqrt{x^2\pm a^2}$$

512. $$\int x[\log(x+\sqrt{x^2\pm a^2})]\,dx = \left(\frac{x^2}{2}\pm\frac{a^2}{4}\right)\log(x+\sqrt{x^2\pm a^2}) - \frac{x\sqrt{x^2\pm a^2}}{4}$$

513. $$\int x^m[\log(x+\sqrt{x^2\pm a^2})]\,dx = \frac{x^{m+1}}{m+1}\log(x+\sqrt{x^2\pm a^2}) - \frac{1}{m+1}\int\frac{x^{m+1}}{\sqrt{x^2\pm a^2}}\,dx$$

514. $$\int \frac{\log(x+\sqrt{x^2+a^2})}{x^2}\,dx = -\frac{\log(x+\sqrt{x^2+a^2})}{x} - \frac{1}{a}\log\frac{a+\sqrt{x^2+a^2}}{x}$$

515. $$\int \frac{\log(x+\sqrt{x^2-a^2})}{x^2}\,dx = -\frac{\log(x+\sqrt{x^2-a^2})}{x} + \frac{1}{|a|}\sec^{-1}\frac{x}{a}$$

516. $$\int x^n \log(x^2-a^2)\,dx = \frac{1}{n+1}\left[x^{n+1}\log(x^2-a^2) - a^{n+1}\log(x-a) - (-a)^{n+1}\log(x+a) - 2\sum_{r=0}^{\left[\frac{n}{2}\right]}\frac{a^{2r}x^{n-2r+1}}{n-2r+1}\right]$$

See note integral 392.

EXPONENTIAL FORMS

517. $$\int e^x\,dx = e^x$$

518. $$\int e^{-x}\,dx = -e^{-x}$$

519. $$\int e^{ax}\,dx = \frac{e^{ax}}{a}$$

520. $$\int x\,e^{ax}\,dx = \frac{e^{ax}}{a^2}(ax-1)$$

521. $$\int x^m e^{ax}\,dx = \begin{cases} \dfrac{x^m e^{ax}}{a} - \dfrac{m}{a}\displaystyle\int x^{m-1}e^{ax}\,dx \\ \text{or} \\ e^{ax}\displaystyle\sum_{r=0}^{m}(-1)^r\frac{m!\,x^{m-r}}{(m-r)!\,a^{r+1}} \end{cases}$$

522. $$\int \frac{e^{ax}\,dx}{x} = \log x + \frac{ax}{1!} + \frac{a^2x^2}{2\cdot 2!} + \frac{a^3x^3}{3\cdot 3!} + \cdots$$

523. $$\int \frac{e^{ax}}{x^m}\,dx = -\frac{1}{m-1}\frac{e^{ax}}{x^{m-1}} + \frac{a}{m-1}\int \frac{e^{ax}}{x^{m-1}}\,dx$$

524. $$\int e^{ax}\log x\,dx = \frac{e^{ax}\log x}{a} - \frac{1}{a}\int \frac{e^{ax}}{x}\,dx$$

525. $$\int \frac{dx}{1+e^x} = x - \log(1+e^x) = \log\frac{e^x}{1+e^x}$$

526. $$\int \frac{dx}{a+be^{px}} = \frac{x}{a} - \frac{1}{ap}\log(a+be^{px})$$

527. $$\int \frac{dx}{ae^{mx}+be^{-mx}} = \frac{1}{m\sqrt{ab}}\tan^{-1}\left(e^{mx}\sqrt{\frac{a}{b}}\right), \qquad (a>0, b>0)$$

528. $$\int \frac{dx}{ae^{mx}-be^{-mx}} = \begin{cases} \dfrac{1}{2m\sqrt{ab}}\log\dfrac{\sqrt{a}\,e^{mx}-\sqrt{b}}{\sqrt{a}\,e^{mx}+\sqrt{b}} \\ \text{or} \\ \dfrac{-1}{m\sqrt{ab}}\tanh^{-1}\left(\sqrt{\dfrac{a}{b}}\,e^{mx}\right), \qquad (a>0, b>0) \end{cases}$$

529. $$\int (a^x - a^{-x})\,dx = \frac{a^x + a^{-x}}{\log a}$$

530. $$\int \frac{e^{ax}}{b+ce^{ax}}\,dx = \frac{1}{ac}\log(b+ce^{ax})$$

531. $$\int \frac{x\,e^{ax}}{(1+ax)^2}\,dx = \frac{e^{ax}}{a^2(1+ax)}$$

532. $\int x e^{-x^2} dx = -\frac{1}{2} e^{-x^2}$

533. $\int e^{ax} [\sin (bx)] \, dx = \dfrac{e^{ax}[a \sin (bx) - b \cos (bx)]}{a^2 + b^2}$

534. $$\int e^{ax} [\sin (bx)][\sin (cx)] \, dx = \frac{e^{ax}[(b - c) \sin (b - c)x + a \cos (b - c)x]}{2[a^2 + (b - c)^2]} - \frac{e^{ax}[(b + c) \sin (b + c)x + a \cos (b + c)x]}{2[a^2 + (b + c)^2]}$$

535. $$\int e^{ax}[\sin (bx)][\cos (cx)] \, dx = \begin{cases} \dfrac{e^{ax}[a \sin (b - c)x - (b - c) \cos (b - c)x]}{2[a^2 + (b - c)^2]} + \dfrac{e^{ax}[a \sin (b + c)x - (b + c) \cos (b + c)x]}{2[a^2 + (b + c)^2]} \\ \text{or} \\ \dfrac{e^{ax}}{\rho}[(a \sin bx - b \cos bx)[\cos (cx - \alpha)] - c(\sin bx) \sin (cx - \alpha)] \\ \text{where} \\ \rho = \sqrt{(a^2 + b^2 - c^2)^2 + 4a^2c^2}, \\ \rho \cos \alpha = a^2 + b^2 - c^2, \qquad \rho \sin \alpha = 2ac \end{cases}$$

536. $$\int e^{ax}[\sin (bx)][\sin (bx + c)] \, dx = \frac{e^{ax} \cos c}{2a} - \frac{e^{ax}[a \cos (2bx + c) + 2b \sin (2bx + c)]}{2(a^2 + 4b^2)}$$

537. $$\int e^{ax}[\sin (bx)][\cos (bx + c)] \, dx = \frac{-e^{ax} \sin c}{2a} + \frac{e^{ax}[a \sin (2bx + c) - 2b \cos (2bx + c)]}{2(a^2 + 4b^2)}$$

538. $\int e^{ax}[\cos (bx)] \, dx = \dfrac{e^{ax}}{a^2 + b^2}[a \cos (bx) + b \sin (bx)]$

539. $$\int e^{ax}[\cos (bx)][\cos (cx)] \, dx = \frac{e^{ax}[(b - c) \sin (b - c)x + a \cos (b - c)x]}{2[a^2 + (b - c)^2]} + \frac{e^{ax}[(b + c) \sin (b + c)x + a \cos (b + c)x]}{2[a^2 + (b + c)^2]}$$

540. $$\int e^{ax}[\cos (bx)][\cos (bx + c)] \, dx = \frac{e^{ax} \cos c}{2a} + \frac{e^{ax}[a \cos (2bx + c) + 2b \sin (2bx + c)]}{2(a^2 + 4b^2)}$$

541. $$\int e^{ax}[\cos (bx)][\sin (bx + c)] \, dx = \frac{e^{ax} \sin c}{2a} + \frac{e^{ax}[a \sin (2bx + c) - 2b \cos (2bx + c)]}{2(a^2 + 4b^2)}$$

542. $$\int e^{ax}[\sin^n bx] \, dx = \frac{1}{a^2 + n^2b^2}\left[(a \sin bx - nb \cos bx) e^{ax} \sin^{n-1} bx + n(n - 1)b^2 \int e^{ax}[\sin^{n-2} bx] \, dx\right]$$

543. $$\int e^{ax}[\cos^n bx] \, dx = \frac{1}{a^2 + n^2b^2}\left[(a \cos bx + nb \sin bx) e^{ax} \cos^{n-1} bx + n(n - 1)b^2 \int e^{ax}[\cos^{n-2} bx] \, dx\right]$$

544. $$\int x^m e^x \sin x \, dx = \frac{1}{2} x^m e^x(\sin x - \cos x) - \frac{m}{2}\int x^{m-1} e^x \sin x \, dx + \frac{m}{2}\int x^{m-1} e^x \cos x \, dx$$

545. $$\int x^m e^{ax}[\sin bx] \, dx = \begin{cases} x^m e^{ax} \dfrac{a \sin bx - b \cos bx}{a^2 + b^2} - \dfrac{m}{a^2 + b^2}\displaystyle\int x^{m-1} e^{ax}(a \sin bx - b \cos bx) \, dx \\ \text{or} \\ e^{ax} \displaystyle\sum_{r=0}^{m} \frac{(-1)^r m! x^{m-r}}{\rho^{r+1}(m-r)!} \sin [bx - (r+1)\alpha] \\ \text{where} \\ \rho = \sqrt{a^2 + b^2}, \quad \rho \cos \alpha = a, \quad \rho \sin \alpha = b \end{cases}$$

546. $$\int x^m e^x \cos x \, dx = \frac{1}{2} x^m e^x(\sin x + \cos x) - \frac{m}{2}\int x^{m-1} e^x \sin x \, dx - \frac{m}{2}\int x^{m-1} e^x \cos x \, dx$$

547. $$\int x^m e^{ax} \cos bx \, dx = \begin{cases} x^m e^{ax} \dfrac{a \cos bx + b \sin bx}{a^2 + b^2} - \dfrac{m}{a^2 + b^2}\displaystyle\int x^{m-1} e^{ax}(a \cos bx + b \sin bx) \, dx \\ \text{or} \\ e^{ax} \displaystyle\sum_{r=0}^{m} \frac{(-1)^r m! x^{m-r}}{\rho^{r+1}(m-r)!} \cos [bx - (r+1)\alpha] \\ \text{where} \\ \rho = \sqrt{a^2 + b^2}, \quad \rho \cos \alpha = a, \quad \rho \sin \alpha = b \end{cases}$$

548. $$\int e^{ax}(\cos^m x)(\sin^n x)\,dx = \begin{cases} \dfrac{e^{ax}\cos^{m-1} x \sin^n x[a\cos x + (m+n)\sin x]}{(m+n)^2 + a^2} \\ \quad -\dfrac{na}{(m+n)^2+a^2}\displaystyle\int e^{ax}(\cos^{m-1} x)(\sin^{n-1} x)\,dx \\ \quad +\dfrac{(m-1)(m+n)}{(m+n)^2+a^2}\displaystyle\int e^{ax}(\cos^{m-2} x)(\sin^n x)\,dx \\ \quad \text{or} \\ \dfrac{e^{ax}\cos^m x \sin^{n-1} x[a\sin x - (m+n)\cos x]}{(m+n)^2 + a^2} \\ \quad +\dfrac{ma}{(m+n)^2+a^2}\displaystyle\int e^{ax}(\cos^{m-1} x)(\sin^{n-1} x)\,dx \\ \quad +\dfrac{(n-1)(m+n)}{(m+n)^2+a^2}\displaystyle\int e^{ax}(\cos^m x)(\sin^{n-2} x)\,dx \\ \quad \text{or} \\ \dfrac{e^{ax}(\cos^{m-1} x)(\sin^{n-1} x)(a\sin x\cos x + m\sin^2 x - n\cos^2 x)}{(m+n)^2 + a^2} \\ \quad +\dfrac{m(m-1)}{(m+n)^2+a^2}\displaystyle\int e^{ax}(\cos^{m-2} x)(\sin^n x)\,dx \\ \quad +\dfrac{n(n-1)}{(m+n)^2+a^2}\displaystyle\int e^{ax}(\cos^m x)(\sin^{n-2} x)\,dx \\ \quad \text{or} \\ \dfrac{e^{ax}(\cos^{m-1} x)(\sin^{n-1} x)(a\cos x\sin x + m\sin^2 x - n\cos^2 x)}{(m+n)^2 + a^2} \\ \quad +\dfrac{m(m-1)}{(m+n)^2+a^2}\displaystyle\int e^{ax}(\cos^{m-2} x)(\sin^{n-2} x)\,dx \\ \quad +\dfrac{(n-m)(n+m-1)}{(m+n)^2+a^2}\displaystyle\int e^{ax}(\cos^m x)(\sin^{n-2} x)\,dx \end{cases}$$

549. $$\int x\, e^{ax}(\sin bx)\,dx = \frac{x\, e^{ax}}{a^2+b^2}(a\sin bx - b\cos bx) - \frac{e^{ax}}{(a^2+b^2)^2}[(a^2-b^2)\sin bx - 2ab\cos bx]$$

550. $$\int x\, e^{ax}(\cos bx)\,dx = \frac{x\, e^{ax}}{a^2+b^2}(a\cos bx + b\sin bx) - \frac{e^{ax}}{(a^2+b^2)^2}[(a^2-b^2)\cos bx + 2ab\sin bx]$$

551. $$\int \frac{e^{ax}}{\sin^n x}\,dx = -\frac{e^{ax}[a\sin x + (n-2)\cos x]}{(n-1)(n-2)\sin^{n-1} x} + \frac{a^2+(n-2)^2}{(n-1)(n-2)}\int \frac{e^{ax}}{\sin^{n-2} x}\,dx$$

552. $$\int \frac{e^{ax}}{\cos^n x}\,dx = -\frac{e^{ax}[a\cos x - (n-2)\sin x]}{(n-1)(n-2)\cos^{n-1} x} + \frac{a^2+(n-2)^2}{(n-1)(n-2)}\int \frac{e^{ax}}{\cos^{n-2} x}\,dx$$

553. $$\int e^{ax}\tan^n x\,dx = e^{ax}\frac{\tan^{n-1} x}{n-1} - \frac{a}{n-1}\int e^{ax}\tan^{n-1} x\,dx - \int e^{ax}\tan^{n-2} x\,dx$$

HYPERBOLIC FORMS

554. $$\int (\sinh x)\,dx = \cosh x$$

555. $$\int (\cosh x)\,dx = \sinh x$$

556. $\int (\tanh x)\, dx = \log \cosh x$

557. $\int (\coth x)\, dx = \log \sinh x$

558. $\int (\operatorname{sech} x)\, dx = \tan^{-1}(\sinh x)$

559. $\int \operatorname{csch} x\, dx = \log \tanh\left(\frac{x}{2}\right)$

560. $\int x(\sinh x)\, dx = x \cosh x - \sinh x$

561. $\int x^n(\sinh x)\, dx = x^n \cosh x - n \int x^{n-1}(\cosh x)\, dx$

562. $\int x(\cosh x)\, dx = x \sinh x - \cosh x$

563. $\int x^n(\cosh x)\, dx = x^n \sinh x - n \int x^{n-1}(\sinh x)\, dx$

564. $\int (\operatorname{sech} x)(\tanh x)\, dx = -\operatorname{sech} x$

565. $\int (\operatorname{csch} x)(\coth x)\, dx = -\operatorname{csch} x$

566. $\int (\sinh^2 x)\, dx = \frac{\sinh 2x}{4} - \frac{x}{2}$

567. $$\int (\sinh^m x)(\cosh^n x)\, dx = \begin{cases} \frac{1}{m+n}(\sinh^{m+1} x)(\cosh^{n-1} x) \\ \qquad + \frac{n-1}{m+n}\int (\sinh^m x)(\cosh^{n-2} x)\, dx \\ \text{or} \\ \frac{1}{m+n}\sinh^{m-1} x \cosh^{n+1} x \\ - \frac{m-1}{m+n}\int (\sinh^{m-2} x)(\cosh^n x)\, dx, \qquad (m+n \neq 0) \end{cases}$$

568. $$\int \frac{dx}{(\sinh^m x)(\cosh^n x)} = \begin{cases} -\frac{1}{(m-1)(\sinh^{m-1} x)(\cosh^{n-1} x)} \\ \qquad - \frac{m+n-2}{m-1}\int \frac{dx}{(\sinh^{m-2} x)(\cosh^n x)}, \qquad (m \neq 1) \\ \text{or} \\ \frac{1}{(n-1)\sinh^{m-1} x \cosh^{n-1} x} \\ \qquad + \frac{m+n-2}{n-1}\int \frac{dx}{(\sinh^m x)(\cosh^{n-2} x)}, \qquad (n \neq 1) \end{cases}$$

569. $\int (\tanh^2 x)\, dx = x - \tanh x$

570. $\int (\tanh^n x)\, dx = -\frac{\tanh^{n-1} x}{n-1} + \int (\tanh^{n-2} x)\, dx, \qquad (n \neq 1)$

571. $\int (\operatorname{sech}^2 x)\, dx = \tanh x$

572. $\int (\cosh^2 x)\, dx = \frac{\sinh 2x}{4} + \frac{x}{2}$

573. $\int (\coth^2 x)\,dx = x - \coth x$

574. $\int (\coth^n x)\,dx = -\frac{\coth^{n-1} x}{n-1} + \int \coth^{n-2} x\,dx, \qquad (n \neq 1)$

575. $\int (\operatorname{csch}^2 x)\,dx = -\operatorname{ctnh} x$

576. $\int (\sinh mx)(\sinh nx)\,dx = \frac{\sinh (m+n)x}{2(m+n)} - \frac{\sinh (m-n)x}{2(m-n)}, \qquad (m^2 \neq n^2)$

577. $\int (\cosh mx)(\cosh nx)\,dx = \frac{\sinh (m+n)x}{2(m+n)} + \frac{\sinh (m-n)x}{2(m-n)}, \qquad (m^2 \neq n^2)$

578. $\int (\sinh mx)(\cosh nx)\,dx = \frac{\cosh (m+n)x}{2(m+n)} + \frac{\cosh (m-n)x}{2(m-n)}, \qquad (m^2 \neq n^2)$

579. $\int \left(\sinh^{-1}\frac{x}{a}\right) dx = x \sinh^{-1}\frac{x}{a} - \sqrt{x^2 + a^2}, \qquad (a > 0)$

580. $\int x\left(\sinh^{-1}\frac{x}{a}\right) dx = \left(\frac{x^2}{2} + \frac{a^2}{4}\right)\sinh^{-1}\frac{x}{a} - \frac{x}{4}\sqrt{x^2 + a^2}, \qquad (a > 0)$

581. $\int x^n(\sinh^{-1} x)\,dx = \frac{x^{n+1}}{n+1}\sinh^{-1} x - \frac{1}{n+1}\int \frac{x^{n+1}}{(1+x^2)^{\frac{1}{2}}}\,dx, \qquad (n \neq -1)$

582. $\int \left(\cosh^{-1}\frac{x}{a}\right) dx = \begin{cases} x\cosh^{-1}\frac{x}{a} - \sqrt{x^2 - a^2}, & \left(\cosh^{-1}\frac{x}{a} > 0\right) \\ \text{or} & \\ x\cosh^{-1}\frac{x}{a} + \sqrt{x^2 - a^2}, & \left(\cosh^{-1}\frac{x}{a} < 0\right), \quad (a > 0) \end{cases}$

583. $\int x\left(\cosh^{-1}\frac{x}{a}\right) dx = \frac{2x^2 - a^2}{4}\cosh^{-1}\frac{x}{a} - \frac{x}{4}(x^2 - a^2)^{\frac{1}{2}}$

584. $\int x^n(\cosh^{-1} x)\,dx = \frac{x^{n+1}}{n+1}\cosh^{-1} x - \frac{1}{n+1}\int \frac{x^{n+1}}{(x^2 - 1)^{\frac{1}{2}}}\,dx, \qquad (n \neq -1)$

585. $\int \left(\tanh^{-1}\frac{x}{a}\right) dx = x\tanh^{-1}\frac{x}{a} + \frac{a}{2}\log(a^2 - x^2), \qquad \left(\left|\frac{x}{a}\right| < 1\right)$

586. $\int \left(\coth^{-1}\frac{x}{a}\right) dx = x\coth^{-1}\frac{x}{a} + \frac{a}{2}\log(x^2 - a^2), \qquad \left(\left|\frac{x}{a}\right| > 1\right)$

587. $\int x\left(\tanh^{-1}\frac{x}{a}\right) dx = \frac{x^2 - a^2}{2}\tanh^{-1}\frac{x}{a} + \frac{ax}{2}, \qquad \left(\left|\frac{x}{a}\right| < 1\right)$

588. $\int x^n\left(\tanh^{-1} x\right) dx = \frac{x^{n+1}}{n+1}\tanh^{-1} x - \frac{1}{n+1}\int \frac{x^{n+1}}{1 - x^2}\,dx, \qquad (n \neq -1)$

589. $\int x\left(\coth^{-1}\frac{x}{a}\right) dx = \frac{x^2 - a^2}{2}\coth^{-1}\frac{x}{a} + \frac{ax}{2}, \qquad \left(\left|\frac{x}{a}\right| > 1\right)$

590. $\int x^n(\coth^{-1} x)\,dx = \frac{x^{n+1}}{n+1}\coth^{-1} x + \frac{1}{n+1}\int \frac{x^{n+1}}{x^2 - 1}\,dx, \qquad (n \neq -1)$

591. $\int (\operatorname{sech}^{-1} x)\,dx = x\operatorname{sech}^{-1} x + \sin^{-1} x$

592. $\int x\operatorname{sech}^{-1} x\,dx = \frac{x^2}{2}\operatorname{sech}^{-1} x - \frac{1}{2}\sqrt{1 - x^2}$

593. $\int x^n\operatorname{sech}^{-1} x\,dx = \frac{x^{n+1}}{n+1}\operatorname{sech}^{-1} x + \frac{1}{n+1}\int \frac{x^n}{(1 - x^2)^{\frac{1}{2}}}\,dx, \qquad (n \neq -1)$

594. $\int \operatorname{csch}^{-1} x\,dx = x\operatorname{csch}^{-1} x + \frac{x}{|x|}\sinh^{-1} x$

595. $\int x \operatorname{csch}^{-1} x \, dx = \frac{x^2}{2} \operatorname{csch}^{-1} x + \frac{1}{2} \frac{x}{|x|} \sqrt{1 + x^2}$

596. $\int x^n \operatorname{csch}^{-1} x \, dx = \frac{x^{n+1}}{n+1} \operatorname{csch}^{-1} x + \frac{1}{n+1} \frac{x}{|x|} \int \frac{x^n}{(x^2+1)^{\frac{1}{2}}} dx, \qquad (n \neq -1)$

DEFINITE INTEGRALS

597. $\int_0^\infty x^{n-1} e^{-x} dx = \int_0^1 \left(\log \frac{1}{x}\right)^{n-1} dx = \frac{1}{n} \prod_{m=1}^{\infty} \frac{\left(1 + \frac{1}{m}\right)^n}{1 + \frac{n}{m}}$

$= \Gamma(n), n \neq 0, -1, -2, -3, \ldots$ (Gamma Function)

598. $\int_0^\infty t^n p^{-t} dt = \frac{n!}{(\log p)^{n+1}}, \qquad (n = 0, 1, 2, 3, \ldots \text{ and } p > 0)$

599. $\int_0^\infty t^{n-1} e^{-(a+1)t} dt = \frac{\Gamma(n)}{(a+1)^n}, \qquad (n > 0, a > -1)$

600. $\int_0^1 x^m \left(\log \frac{1}{x}\right)^n dx = \frac{\Gamma(n+1)}{(m+1)^{n+1}}, \qquad (m > -1, n > -1)$

601. $\Gamma(n)$ is finite if $n > 0$, $\Gamma(n+1) = n\Gamma(n)$

602. $\Gamma(n) \cdot \Gamma(1-n) = \frac{\pi}{\sin n\pi}$

603. $\Gamma(n) = (n-1)!$ if $n =$ integer > 0

604. $\Gamma(\frac{1}{2}) = 2 \int_0^\infty e^{-t^2} dt = \sqrt{\pi} = 1.7724538509 \cdots = (-\frac{1}{2})!$

605. $\Gamma(n + \frac{1}{2}) = \frac{1 \cdot 3 \cdot 5 \ldots (2n-1)}{2^n} \sqrt{\pi} \qquad n = 1, 2, 3, \ldots$

606. $\Gamma(-n + \frac{1}{2}) = \frac{(-1)^n 2^n \sqrt{\pi}}{1 \cdot 3 \cdot 5 \ldots (2n-1)} \qquad n = 1, 2, 3, \ldots$

607. $\int_0^1 x^{m-1}(1-x)^{n-1} dx = \int_0^\infty \frac{x^{m-1}}{(1+x)^{m+n}} dx = \frac{\Gamma(m)\Gamma(n)}{\Gamma(m+n)} = B(m, n)$

(Beta function)

608. $B(m, n) = B(n, m) = \frac{\Gamma(m)\Gamma(n)}{\Gamma(m+n)}$, where m and n are any positive real numbers.

609. $\int_a^b (x-a)^m (b-x)^n dx = (b-a)^{m+n+1} \frac{\Gamma(m+1) \cdot \Gamma(n+1)}{\Gamma(m+n+2)},$

$(m > -1, n > -1, b > a)$

610. $\int_1^\infty \frac{dx}{x^m} = \frac{1}{m-1}, \qquad [m > 1]$

611. $\int_0^\infty \frac{dx}{(1+x)x^p} = \pi \csc p\pi, \qquad [p < 1]$

612. $\int_0^\infty \frac{dx}{(1-x)x^p} = -\pi \cot p\pi, \qquad [p < 1]$

613. $\int_0^\infty \frac{x^{p-1} dx}{1+x} = \frac{\pi}{\sin p\pi}$

$= B(p, 1-p) = \Gamma(p)\Gamma(1-p), \qquad [0 < p < 1]$

614. $\int_0^\infty \frac{x^{m-1} dx}{1+x^n} = \frac{\pi}{n \sin \frac{m\pi}{n}}, \qquad [0 < m < n]$

615. $\displaystyle\int_0^\infty \frac{x^a\,dx}{(m+x^b)^c} = \frac{m^{\frac{a+1-bc}{b}}}{b}\left[\frac{\Gamma\left(\frac{a+1}{b}\right)\Gamma\left(c-\frac{a+1}{b}\right)}{\Gamma(c)}\right]$ $\left(a > -1, b > 0, m > 0, c > \frac{a+1}{b}\right)$

616. $\displaystyle\int_0^\infty \frac{dx}{(1+x)\sqrt{x}} = \pi$

617. $\displaystyle\int_0^\infty \frac{a\,dx}{a^2+x^2} = \frac{\pi}{2}$, if $a > 0$; 0, if $a = 0$; $-\frac{\pi}{2}$, if $a < 0$

618. $\displaystyle\int_0^a (a^2-x^2)^{\frac{n}{2}}\,dx = \frac{1}{2}\int_{-a}^a (a^2-x^2)^{\frac{n}{2}}\,dx = \frac{1\cdot 3\cdot 5\ldots n}{2\cdot 4\cdot 6\ldots(n+1)}\cdot\frac{\pi}{2}\cdot a^{n+1}$ **(n odd)**

619. $\displaystyle\int_0^a x^m(a^2-x^2)^{\frac{n}{2}}\,dx = \begin{cases} \frac{1}{2}a^{m+n+1}B\left(\frac{m+1}{2},\frac{n+2}{2}\right) \\ \text{or} \\ \frac{1}{2}a^{m+n+1}\dfrac{\Gamma\left(\frac{m+1}{2}\right)\Gamma\left(\frac{n+2}{2}\right)}{\Gamma\left(\frac{m+n+3}{2}\right)} \end{cases}$

620. $\displaystyle\int_0^{\pi/2} (\sin^n x)\,dx = \begin{cases} \displaystyle\int_0^{\pi/2} (\cos^n x)\,dx \\ \text{or} \\ \dfrac{1\cdot 3\cdot 5\cdot 7\ldots(n-1)}{2\cdot 4\cdot 6\cdot 8\ldots(n)}\dfrac{\pi}{2}, & (n \text{ an even integer}, n \neq 0) \\ \text{or} \\ \dfrac{2\cdot 4\cdot 6\cdot 8\ldots(n-1)}{1\cdot 3\cdot 5\cdot 7\ldots(n)}, & (n \text{ an odd integer}, n \neq 1) \\ \text{or} \\ \dfrac{\sqrt{\pi}}{2}\dfrac{\Gamma\left(\frac{n+1}{2}\right)}{\Gamma\left(\frac{n}{2}+1\right)}, & (n > -1) \end{cases}$

621. $\displaystyle\int_0^\infty \frac{\sin mx\,dx}{x} = \frac{\pi}{2}$, if $m > 0$; 0, if $m = 0$; $-\frac{\pi}{2}$, if $m < 0$

622. $\displaystyle\int_0^\infty \frac{\cos x\,dx}{x} = \infty$

623. $\displaystyle\int_0^\infty \frac{\tan x\,dx}{x} = \frac{\pi}{2}$

624. $\displaystyle\int_0^\pi \sin ax\cdot\sin bx\,dx = \int_0^\pi \cos ax\cdot\cos bx\,dx = 0,$ ($a \neq b$; a, b integers)

625. $\displaystyle\int_0^{\pi/a} [\sin(ax)][\cos(ax)]\,dx = \int_0^\pi [\sin(ax)][\cos(ax)]\,dx = 0$

626. $\displaystyle\int_0^\pi [\sin(ax)][\cos(bx)]\,dx = \frac{2a}{a^2-b^2}$, if $a-b$ is odd, or 0 if $a-b$ is even

627. $\displaystyle\int_0^\infty \frac{\sin x\cos mx\,dx}{x}$

$= 0$, if $m < -1$ or $m > 1$; $\frac{\pi}{4}$, if $m = \pm 1$; $\frac{\pi}{2}$, if $m^2 < 1$

628. $\int_0^\infty \frac{\sin ax \sin bx}{x^2} dx = \frac{\pi a}{2}, \quad (a \le b)$

629. $\int_0^\pi \sin^2 mx\, dx = \int_0^\pi \cos^2 mx\, dx = \frac{\pi}{2}$

630. $\int_0^\infty \frac{\sin^2 (px)}{x^2} dx = \frac{\pi p}{2}$

631. $\int_0^\infty \frac{\sin x}{x^p} dx = \frac{\pi}{2\Gamma(p)\sin(p\pi/2)}, \quad 0 < p < 1$

632. $\int_0^\infty \frac{\cos x}{x^p} dx = \frac{\pi}{2\Gamma(p)\cos(p\pi/2)}, \quad 0 < p < 1$

633. $\int_0^\infty \frac{1 - \cos px}{x^2} dx = \frac{\pi p}{2}$

634. $\int_0^\infty \frac{\sin px \cos qx}{x} dx = \left\{0, \quad q > p > 0; \quad \frac{\pi}{2}, \quad p > q > 0; \quad \frac{\pi}{4}, \quad p = q > 0\right\}$

635. $\int_0^\infty \frac{\cos (mx)}{x^2 + a^2} dx = \frac{\pi}{2|a|} e^{-|ma|}$

636. $\int_0^\infty \cos (x^2)\, dx = \int_0^\infty \sin (x^2)\, dx = \frac{1}{2}\sqrt{\frac{\pi}{2}}$

637. $\int_0^\infty \sin ax^n\, dx = \frac{1}{na^{1/n}} \Gamma(1/n) \sin \frac{\pi}{2n}, \quad n > 1$

638. $\int_0^\infty \cos ax^n\, dx = \frac{1}{na^{1/n}} \Gamma(1/n) \cos \frac{\pi}{2n}, \quad n > 1$

639. $\int_0^\infty \frac{\sin x}{\sqrt{x}} dx = \int_0^\infty \frac{\cos x}{\sqrt{x}} dx = \sqrt{\frac{\pi}{2}}$

640. (a) $\int_0^\infty \frac{\sin^3 x}{x} dx = \frac{\pi}{4}$ (b) $\int_0^\infty \frac{\sin^3 x}{x^2} dx \frac{3}{4} \log 3$

641. $\int_0^\infty \frac{\sin^3 x}{x^3} dx = \frac{3\pi}{8}$

642. $\int_0^\infty \frac{\sin^4 x}{x^4} dx = \frac{\pi}{3}$

643. $\int_0^{\pi/2} \frac{dx}{1 + a \cos x} = \frac{\cos^{-1} a}{\sqrt{1 - a^2}}, \quad (a < 1)$

644. $\int_0^\pi \frac{dx}{a + b \cos x} = \frac{\pi}{\sqrt{a^2 - b^2}}, \quad (a > b \ge 0)$

645. $\int_0^{2\pi} \frac{dx}{1 + a \cos x} = \frac{2\pi}{\sqrt{1 - a^2}}, \quad (a^2 < 1)$

646. $\int_0^\infty \frac{\cos ax - \cos bx}{x} dx = \log \frac{b}{a}$

647. $\int_0^{\pi/2} \frac{dx}{a^2 \sin^2 x + b^2 \cos^2 x} = \frac{\pi}{2ab}$

648. $\int_0^{\pi/2} \frac{dx}{(a^2 \sin^2 x + b^2 \cos^2 x)^2} = \frac{\pi(a^2 + b^2)}{4a^3 b^3}, \quad (a, b > 0)$

649. $\int_0^{\pi/2} \sin^{n-1} x \cos^{m-1} x\, dx = \frac{1}{2} B\left(\frac{n}{2}, \frac{m}{2}\right)$, m and n positive integers

650. $\displaystyle\int_0^{\pi/2} (\sin^{2n+1}\theta)\,d\theta = \frac{2\cdot 4\cdot 6\ldots(2n)}{1\cdot 3\cdot 5\ldots(2n+1)}, \qquad (n = 1, 2, 3\ldots)$

651. $\displaystyle\int_0^{\pi/2} (\sin^{2n}\theta)\,d\theta = \frac{1\cdot 3\cdot 5\ldots(2n-1)}{2\cdot 4\ldots(2n)}\left(\frac{\pi}{2}\right), \qquad (n = 1, 2, 3\ldots)$

652. $\displaystyle\int_0^{\pi/2} \frac{x}{\sin x}\,dx = 2\left\{\frac{1}{1^2} - \frac{1}{3^2} + \frac{1}{5^2} - \frac{1}{7^2} + \cdots\right\}$

653. $\displaystyle\int_0^{\pi/2} \frac{dx}{1+\tan^m x} = \frac{\pi}{4}$

654. $\displaystyle\int_0^{\pi/2} \sqrt{\cos\theta}\,d\theta = \frac{(2\pi)^{\frac{3}{2}}}{[\Gamma(\frac{1}{4})]^2}$

655. $\displaystyle\int_0^{\pi/2} (\tan^h\theta)\,d\theta = \frac{\pi}{2\cos\left(\dfrac{h\pi}{2}\right)}, \qquad (0 < h < 1)$

656. $\displaystyle\int_0^{\infty} \frac{\tan^{-1}(ax) - \tan^{-1}(bx)}{x}\,dx = \frac{\pi}{2}\log\frac{a}{b}, \qquad (a, b > 0)$

657. The area enclosed by a curve defined through the equation $x^{\frac{b}{c}} + y^{\frac{b}{c}} = a^{\frac{b}{c}}$ where $a > 0$, c a positive odd integer and b a positive even integer is given by

$$\frac{\left[\Gamma\left(\dfrac{c}{b}\right)\right]^2}{\Gamma\left(\dfrac{2c}{b}\right)}\left(\frac{2ca^2}{b}\right)$$

658. $I = \displaystyle\iiint_R x^{h-1}y^{m-1}z^{n-1}\,dv$, where R denotes the region of space bounded by the co-ordinate planes and that portion of the surface $\left(\dfrac{x}{a}\right)^p + \left(\dfrac{y}{b}\right)^q + \left(\dfrac{z}{c}\right)^k = 1$, which lies in the first octant and where $h, m, n, p, q, k, a, b, c$, denote positive real numbers is given by

$$\int_0^a x^{h-1}\,dx \int_0^{b\left[1-\left(\frac{x}{a}\right)^p\right]^{\frac{1}{q}}} y^m\,dy \int_0^{c\left[1-\left(\frac{x}{a}\right)^p-\left(\frac{y}{b}\right)^q\right]^{\frac{1}{k}}} z^{n-1}\,dz$$

$$= \frac{a^h b^m c^n}{pqk}\,\frac{\Gamma\left(\dfrac{h}{p}\right)\Gamma\left(\dfrac{m}{q}\right)\Gamma\left(\dfrac{n}{k}\right)}{\Gamma\left(\dfrac{h}{p}+\dfrac{m}{q}+\dfrac{n}{k}+1\right)}$$

659. $\displaystyle\int_0^{\infty} e^{-ax}\,dx = \frac{1}{a}, \qquad (a > 0)$

660. $\displaystyle\int_0^{\infty} \frac{e^{-ax} - e^{-bx}}{x}\,dx = \log\frac{b}{a}, \qquad (a, b > 0)$

661. $\displaystyle\int_0^{\infty} x^n e^{-ax}\,dx = \begin{cases} \dfrac{\Gamma(n+1)}{a^{n+1}}, & (n > -1, a > 0) \\ \quad\text{or} & \\ \dfrac{n!}{a^{n+1}}, & (a > 0, n \text{ positive integer}) \end{cases}$

662. $\displaystyle\int_0^{\infty} x^n \exp(-ax^p)\,dx = \frac{\Gamma(k)}{pa^k}, \qquad \left(n > -1, p > 0, a > 0, k = \frac{n+1}{p}\right)$

663. $\displaystyle\int_0^{\infty} e^{-a^2x^2}\,dx = \frac{1}{2a}\sqrt{\pi} = \frac{1}{2a}\Gamma\left(\frac{1}{2}\right), \qquad (a > 0)$

664. $\displaystyle\int_0^{\infty} x\,e^{-x^2}\,dx = \frac{1}{2}$

665. $\int_0^\infty x^2 e^{-x^2}\,dx = \frac{\sqrt{\pi}}{4}$

666. $\int_0^\infty x^{2n} e^{-ax^2}\,dx = \frac{1\cdot 3\cdot 5\ldots(2n-1)}{2^{n+1}a^n}\sqrt{\frac{\pi}{a}}$

667. $\int_0^\infty x^{2n+1} e^{-ax^2}\,dx = \frac{n!}{2a^{n+1}}, \qquad (a > 0)$

668. $\int_0^1 x^m e^{-ax}\,dx = \frac{m!}{a^{m+1}}\left[1 - e^{-a}\sum_{r=0}^{m}\frac{a^r}{r!}\right]$

669. $\int_0^\infty e^{\left(-x^2-\frac{a^2}{x^2}\right)}\,dx = \frac{e^{-2a}\sqrt{\pi}}{2}, \qquad (a \geq 0)$

670. $\int_0^\infty e^{-nx}\sqrt{x}\,dx = \frac{1}{2n}\sqrt{\frac{\pi}{n}}$

671. $\int_0^\infty \frac{e^{-nx}}{\sqrt{x}}\,dx = \sqrt{\frac{\pi}{n}}$

672. $\int_0^\infty e^{-ax}(\cos mx)\,dx = \frac{a}{a^2+m^2}, \qquad (a > 0)$

673. $\int_0^\infty e^{-ax}(\sin mx)\,dx = \frac{m}{a^2+m^2}, \qquad (a > 0)$

674. $\int_0^\infty x e^{-ax}[\sin(bx)]\,dx = \frac{2ab}{(a^2+b^2)^2}, \qquad (a > 0)$

675. $\int_0^\infty x e^{-ax}[\cos(bx)]\,dx = \frac{a^2-b^2}{(a^2+b^2)^2}, \qquad (a > 0)$

676. $\int_0^\infty x^n e^{-ax}[\sin(bx)]\,dx = \frac{n![(a+ib)^{n+1}-(a-ib)^{n+1}]}{2i(a^2+b^2)^{n+1}}, \qquad (i^2 = -1, a > 0)$

677. $\int_0^\infty x^n e^{-ax}[\cos(bx)]\,dx = \frac{n![(a-ib)^{n+1}+(a+ib)^{n+1}]}{2(a^2+b^2)^{n+1}}, \qquad (i^2 = -1, a > 0)$

678. $\int_0^\infty \frac{e^{-ax}\sin x}{x}\,dx = \cot^{-1} a, \qquad (a > 0)$

679. $\int_0^\infty e^{-a^2x^2}\cos bx\,dx = \frac{\sqrt{\pi}}{2a}\exp\left(-\frac{b^2}{4a^2}\right), \qquad (ab \neq 0)$

680. $\int_0^\infty e^{-t\cos\phi}\,t^{b-1}\sin(t\sin\phi)\,dt = [\Gamma(b)]\sin(b\phi), \qquad \left(b > 0, -\frac{\pi}{2} < \phi < \frac{\pi}{2}\right)$

681. $\int_0^\infty e^{-t\cos\phi}\,t^{b-1}[\cos(t\sin\phi)]\,dt = [\Gamma(b)]\cos(b\phi), \qquad \left(b > 0, -\frac{\pi}{2} < \phi < \frac{\pi}{2}\right)$

682. $\int_0^\infty t^{b-1}\cos t\,dt = [\Gamma(b)]\cos\left(\frac{b\pi}{2}\right), \qquad (0 < b < 1)$

683. $\int_0^\infty t^{b-1}(\sin t)\,dt = [\Gamma(b)]\sin\left(\frac{b\pi}{2}\right), \qquad (0 < b < 1)$

684. $\int_0^1 (\log x)^n\,dx = (-1)^n\cdot n!$

685. $\int_0^1 \left(\log\frac{1}{x}\right)^{\frac{1}{2}} dx = \frac{\sqrt{\pi}}{2}$

686. $\int_0^1 \left(\log\frac{1}{x}\right)^{-\frac{1}{2}} dx = \sqrt{\pi}$

687. $$\int_0^1 \left(\log \frac{1}{x}\right)^n dx = n!$$

688. $$\int_0^1 x \log (1 - x)\, dx = -\tfrac{3}{4}$$

689. $$\int_0^1 x \log (1 + x)\, dx = \tfrac{1}{4}$$

690. $$\int_0^1 x^m (\log x)^n\, dx = \frac{(-1)^n n!}{(m + 1)^{n+1}}, \qquad m > -1, n = 0, 1, 2, \ldots$$

If $n \neq 0, 1, 2, \ldots$ replace $n!$ by $\Gamma(n + 1)$.

691. $$\int_0^1 \frac{\log x}{1 + x}\, dx = -\frac{\pi^2}{12}$$

692. $$\int_0^1 \frac{\log x}{1 - x}\, dx = -\frac{\pi^2}{6}$$

693. $$\int_0^1 \frac{\log (1 + x)}{x}\, dx = \frac{\pi^2}{12}$$

694. $$\int_0^1 \frac{\log (1 - x)}{x}\, dx = -\frac{\pi^2}{6}$$

695. $$\int_0^1 (\log x)[\log (1 + x)]\, dx = 2 - 2 \log 2 - \frac{\pi^2}{12}$$

696. $$\int_0^1 (\log x)[\log (1 - x)]\, dx = 2 - \frac{\pi^2}{6}$$

697. $$\int_0^1 \frac{\log x}{1 - x^2}\, dx = -\frac{\pi^2}{8}$$

698. $$\int_0^1 \log \left(\frac{1 + x}{1 - x}\right) \cdot \frac{dx}{x} = \frac{\pi^2}{4}$$

699. $$\int_0^1 \frac{\log x\, dx}{\sqrt{1 - x^2}} = -\frac{\pi}{2} \log 2$$

700. $$\int_0^1 x^m \left[\log \left(\frac{1}{x}\right)\right]^n dx = \frac{\Gamma(n + 1)}{(m + 1)^{n+1}}, \qquad \text{if } m + 1 > 0, n + 1 > 0$$

701. $$\int_0^1 \frac{(x^p - x^q)\, dx}{\log x} = \log \left(\frac{p + 1}{q + 1}\right), \qquad (p + 1 > 0, q + 1 > 0)$$

702. $$\int_0^1 \frac{dx}{\sqrt{\log \left(\frac{1}{x}\right)}} = \sqrt{\pi}, \text{ (same as integral 686)}$$

703. $$\int_0^\infty \log \left(\frac{e^x + 1}{e^x - 1}\right) dx = \frac{\pi^2}{4}$$

704. $$\int_0^{\pi/2} (\log \sin x)\, dx = \int_0^{\pi/2} \log \cos x\, dx = -\frac{\pi}{2} \log 2$$

705. $$\int_0^{\pi/2} (\log \sec x)\, dx = \int_0^{\pi/2} \log \csc x\, dx = \frac{\pi}{2} \log 2$$

706. $$\int_0^\pi x (\log \sin x)\, dx = -\frac{\pi^2}{2} \log 2$$

707. $$\int_0^{\pi/2} (\sin x)(\log \sin x)\, dx = \log 2 - 1$$

708. $\displaystyle\int_0^{\pi/2} (\log \tan x)\, dx = 0$

709. $\displaystyle\int_0^{\pi} \log (a \pm b \cos x)\, dx = \pi \log \left(\frac{a + \sqrt{a^2 - b^2}}{2}\right), \qquad (a \geq b)$

710. $\displaystyle\int_0^{\pi} \log (a^2 - 2ab \cos x + b^2)\, dx = \begin{cases} 2\pi \log a, & a \geq b > 0 \\ 2\pi \log b, & b \geq a > 0 \end{cases}$

711. $\displaystyle\int_0^{\infty} \frac{\sin ax}{\sinh bx}\, dx = \frac{\pi}{2b} \tanh \frac{a\pi}{2b}$

712. $\displaystyle\int_0^{\infty} \frac{\cos ax}{\cosh bx}\, dx = \frac{\pi}{2b} \operatorname{sech} \frac{a\pi}{2b}$

713. $\displaystyle\int_0^{\infty} \frac{dx}{\cosh ax} = \frac{\pi}{2a}$

714. $\displaystyle\int_0^{\infty} \frac{x\, dx}{\sinh ax} = \frac{\pi^2}{4a^2}$

715. $\displaystyle\int_0^{\infty} e^{-ax}(\cosh bx)\, dx = \frac{a}{a^2 - b^2}, \qquad (0 \leq |b| < a)$

716. $\displaystyle\int_0^{\infty} e^{-ax}(\sinh bx)\, dx = \frac{b}{a^2 - b^2}, \qquad (0 \leq |b| < a)$

717. $\displaystyle\int_0^{\infty} \frac{\sinh ax}{e^{bx} + 1}\, dx = \frac{\pi}{2b} \csc \frac{a\pi}{b} - \frac{1}{2a}$

718. $\displaystyle\int_0^{\infty} \frac{\sinh ax}{e^{bx} - 1}\, dx = \frac{1}{2a} - \frac{\pi}{2b} \cot \frac{a\pi}{b}$

719. $$\int_0^{\pi/2} \frac{dx}{\sqrt{1 - k^2 \sin^2 x}} = \frac{\pi}{2}\left[1 + \left(\frac{1}{2}\right)^2 k^2 + \left(\frac{1 \cdot 3}{2 \cdot 4}\right)^2 k^4 + \left(\frac{1 \cdot 3 \cdot 5}{2 \cdot 4 \cdot 6}\right)^2 k^6 + \cdots\right], \text{ if } k^2 < 1$$

720. $$\int_0^{\pi/2} \sqrt{1 - k^2 \sin^2 x}\, dx = \frac{\pi}{2}\left[1 - \left(\frac{1}{2}\right)^2 k^2 - \left(\frac{1 \cdot 3}{2 \cdot 4}\right)^2 \frac{k^4}{3} - \left(\frac{1 \cdot 3 \cdot 5}{2 \cdot 4 \cdot 6}\right)^2 \frac{k^6}{5} - \cdots\right], \text{ if } k^2 < 1$$

721. $\displaystyle\int_0^{\infty} e^{-x} \log x\, dx = -\gamma = -0.5772157\ldots$

722. $\displaystyle\int_0^{\infty} e^{-x^2} \log x\, dx = -\frac{\sqrt{\pi}}{4}(\gamma + 2 \log 2)$

723. $\displaystyle\int_0^{\infty} \left(\frac{1}{1 - e^{-x}} - \frac{1}{x}\right) e^{-x}\, dx = \gamma = 0.5772157\ldots$ [Euler's Constant]

724. $\displaystyle\int_0^{\infty} \frac{1}{x}\left(\frac{1}{1 + x} - e^{-x}\right) dx = \gamma = 0.5772157\ldots$

For n even:

725. $$\int \cos^n x\, dx = \frac{1}{2^{n-1}} \sum_{k=0}^{\frac{n}{2}-1} \binom{n}{k} \frac{\sin (n - 2k) x}{(n - 2k)} + \frac{1}{2^n}\binom{n}{\frac{n}{2}} x$$

726. $\int \sin^n x\,dx = \frac{1}{2^{n-1}} \sum_{k=0}^{\frac{n}{2}-1} \binom{n}{k} \frac{\sin\left[(n-2k)\left(\frac{\pi}{2}-x\right)\right]}{2k-n} + \frac{1}{2^n}\binom{n}{\frac{n}{2}} x$

For n odd:

727. $\int \cos^n x\,dx = \frac{1}{2^{n-1}} \sum_{k=0}^{\frac{n-1}{2}} \binom{n}{k} \frac{\sin (n-2k)\,x}{(n-2k)}$

728. $\int \sin^n x\,dx = \frac{1}{2^{n-1}} \sum_{k=0}^{\frac{n-1}{2}} \binom{n}{k} \frac{\sin\left[(n-2k)\left(\frac{\pi}{2}-x\right)\right]}{2k-n}$

DIFFERENTIAL EQUATIONS

SPECIAL FORMULAS

Certain types of differential equations occur sufficiently often to justify the use of formulas for the corresponding particular solutions. The following set of tables 1 to XIV covers all first, second, and *n*th order ordinary linear differential equations with constant coefficients for which the right members are of the form $P(x)e^{rx}\sin sx$ or $P(x)e^{rx}\cos sx$, where r and s are constants and $P(x)$, is a polynomial of degree n.

When the right member of a reducible linear partial differential equation with constant coefficients is not zero, particular solutions for certain types of right members are contained in tables XV to XXI. In these tables both F and P are used to denote polynomials, and it is assumed that no denominator is zero. In any formula the roles of x and y may be reversed throughout, changing a formula in which x dominates to one in which y dominates. Tables XIX, XX, XXI are applicable whether the equations are reducible or not. The symbol $\binom{m}{n}$ stands for $\frac{m!}{(m-n)!n!}$ and is the $n+1$ st coefficient in the expansion of $(a+b)^m$. Also $0! = 1$ by definition.

The tables as herewith given are those contained in the text *Differential Equations* by Ginn and Company (1955) and are published with their kind permission and that of the author, Professor Frederick H. Steen.

Solution of Linear Differential Equations with Constant Coefficients

Any linear differential equation with constant coefficients may be written in the form

$$p(D)y = R(x)$$

where D is the differential operation

$$Dy = \frac{dy}{dx}$$

$p(D)$ is a polynomial in D,
y is the dependent variable,
x is the independent variable,
$R(x)$ is an arbitrary function of x.

A power of D represents repeated differentiation, that is

$$D^n y = \frac{d^n y}{dx^n}$$

For such an equation, the general solution may be written in the form

$$y = y_c + y_p$$

where y_p is any particular solution, and y_c is called the *complementary function*. This complementary function is defined as the general solution of the *homogeneous equation*, which is the original differential equation with the right side replaced by zero, i.e.

$$p(D)y = 0$$

The complementary function y_c may be determined as follows:

1. Factor the polynomial $p(D)$ into real and complex linear factors, just as if D were a variable instead of an operator.
2. For each nonrepeated linear factor of the form $(D - a)$, where a is real, write down a term of the form

   $$ce^{ax}$$

   where c is an arbitrary constant.
3. For each repeated real linear factor of the form $(D - a)^n$, write down n terms of the form

$$c_1e^{ax} + c_2xe^{ax} + c_3x^2e^{ax} + \cdots + c_nx^{n-1}e^{ax}$$

where the c_i's are arbitrary constants.

4. For each non-repeated conjugate complex pair of factors of the form $(D - a + ib)(D - a - ib)$, write down 2 terms of the form

$$c_1e^{ax}\cos bx + c_2e^{ax}\sin bx$$

5. For each repeated conjugate complex pair of factors of the form $(D - a + ib)^n(D - a - ib)^n$, write down $2n$ terms of the form

$$c_1e^{ax}\cos bx + c_2e^{ax}\sin bx + c_3xe^{ax}\cos bx + c_4xe^{ax}\sin bx + \cdots$$
$$+ c_{2n-1}x^{n-1}e^{ax}\cos bx + c_{2n}x^{n-1}e^{ax}\sin bx$$

6. The sum of all the terms thus written down is the complementary function y_c.

To find the particular solution y_p, use the following tables, as shown in the examples. For cases not shown in the tables, there are various methods of finding y_p. The most general method is called *variation of parameters*. The following example illustrates the method:

Find y_p for $(D^2 - 4)\,y = e^x$.

This example can be solved most easily by use of equation 63 in the tables following. However it is given here as an example of the method of variation of parameters.

The complementary function is

$$y_c = c_1e^{2x} + c_2e^{-2x}$$

To find y_p, replace the constants in the complementary function with unknown functions,

$$y_p = ue^{2x} + ve^{-2x}$$

We now prepare to substitute this assumed solution into the original equation. We begin by taking all the necessary dervatives:

$$y_p = ue^{2x} + ve^{-2x}$$
$$y_p' = 2ue^{2x} - 2ve^{-2x} + u'e^{2x} + v'e^{-2x}$$

For each derivative of y_p except the highest, we set the sum of all the terms containing u' and v' to 0. Thus the above equation becomes

$$u'e^{2x} + v'e^{-2x} = 0 \quad \text{and} \quad y_p' = 2ue^{2x} - 2ve^{-2x}$$

Continuing to differentiate, we have

$$y_p'' = 4ue^{2x} + 4ve^{-2x} + 2u'e^{2x} - 2v'e^{-2x}$$

When we substitute into the original equation, all the terms not containing u' or v' cancel out. This is a consequence of the method by which y_p was set up.

Thus all that is necessary is to write down the terms containing u' or v' in the highest order derivative of y_p, multiply by the constant coefficient of the highest power of D in $p(D)$, and set it equal to $R(x)$. Together with the previous terms in u' and v' which were set equal to 0, this gives us as many linear equations in the first derivatives of the unknown functions as there are unknown functions. The first derivatives may then be solved for by algebra, and the unknown functions found by integration. In the present example, this becomes

$$u'e^{2x} + v'e^{-2x} = 0$$
$$2u'e^{2x} - 2v'e^{-2x} = e^x$$

We eliminate v' and u' separately, getting

$$4u'e^{2x} = e^x$$
$$4v'e^{-2x} = -e^x$$

Thus

$$u' = \tfrac{1}{4}e^{-x}$$
$$v' = -\tfrac{1}{4}e^{3x}$$

Therefore, by integrating

$$u = -\tfrac{1}{4}e^{-x}$$
$$v = -\tfrac{1}{12}e^{3x}$$

A constant of integration is not needed, since we need only one particular solution. Thus

$$y_p = ue^{2x} + ve^{-2x} = -\tfrac{1}{4}e^{-x}e^{2x} - \tfrac{1}{12}e^{3x}e^{-2x}$$
$$= -\tfrac{1}{4}e^x - \tfrac{1}{12}e^x = -\tfrac{1}{3}e^x$$

and the general solution is

$$y = y_c + y_p = c_1e^{2x} + c_2e^{-2x} - \tfrac{1}{3}e^x$$

The following examples illustrate the use of the tables.
Example 1. Solve $(D^2 - 4)y = \sin 3x$.
Substitution of $q = -4$, $s = 3$ in formula 24 gives

$$y_p = \frac{\sin 3x}{-9 - 4}$$

wherefore the general solution is

$$y = c_1 e^{2x} + c_2 e^{-2x} - \frac{\sin 3x}{13}$$

Example 2. Obtain a particular solution of $(D^2 - 4D + 5)y = x^2 e^{3x} \sin x$.
Applying formula 40 with $a = 2, b = 1, r = 3, s = 1$, $P(x) = x^2$, $s + b = 2$, $s - b = 0$, $a - r = -1$, $(a - r)^2 + (s + b)^2 = 5$, $(a - r)^2 + (s - b)^2 = 1$, we have

$$\begin{aligned} y_p = {} & \frac{e^{3x}\sin x}{2}\left[\left(\frac{2}{5} - \frac{0}{1}\right)x^2 + \left(\frac{2(-1)2}{25} - \frac{2(-1)0}{1}\right)2x \right. \\ & \left. + \left(\frac{3\cdot 1\cdot 2 - 2^3}{125} - \frac{3\cdot 1\cdot 0 - 0}{1}\right)2\right] \\ & - \frac{e^{3x}\cos x}{2}\left[\left(\frac{-1}{5} - \frac{-1}{1}\right)x^2 + \left(\frac{1 - 4}{25} - \frac{1 - 0}{1}\right)2x \right. \\ & \left. + \left(\frac{-1 - 3(-1)4}{125} - \frac{-1 - 3(-1)0}{1}\right)2\right] \\ = {} & \left(\tfrac{1}{5}x^2 - \tfrac{4}{25}x - \tfrac{2}{125}\right)e^{3x}\sin x + \left(-\tfrac{2}{5}x^2 + \tfrac{28}{25}x - \tfrac{136}{125}\right)e^{3x}\cos x \end{aligned}$$

The special formulas effect a very considerable saving of time in problems of this type.
Example 3. Obtain a particular solution of $(D^2 - 4D + 5)y = x^2 e^{2x} \cos x$. (Compare with Example 2.)
Formula 40 is not applicable here since for this equation $r = a$, $s = b$, wherefore the denominator $(a - r)^2 + (s - b)^2 = 0$. We turn instead to formula 44. Substituting $a = 2$, $b = 1$, $P(x) = x^2$ and replacing sin by cos, cos by $-\sin$, we obtain

$$\begin{aligned} y_p &= \frac{e^{2x}\cos x}{4}\left(x^2 - \tfrac{2}{4}\right) + \frac{e^{2x}\sin x}{2}\int\left(x^2 - \tfrac{1}{2}\right)dx \\ &= \left(\frac{x^2}{4} - \frac{1}{8}\right)e^{2x}\cos x + \left(\frac{x^3}{6} - \frac{x}{4}\right)e^{2x}\sin x \end{aligned}$$

which is the required solution.
Example 4. Find z_p for $(D_x - 3D_y)z = \ln(y + 3x)$.
Referring to Table XV we note that formula 69 (not 68) is applicable. This gives

$$z_p = x \ln(y + 3x)$$

It is easily seen that $-y/3 \ln(y + 3x)$ would serve equally well.
Example 5. Solve $(D_x + 2D_y - 4)z = y\cos(y - 2x)$.
Since R in formula 76 contains a polynomial in x, not y, we rewrite the given equation in the form

$$\left(D_y + \tfrac{1}{2}D_x - 2\right)z = \tfrac{1}{2}y\cos(y - 2x)$$

Then

$$z_c = e^{2y}F\left(x - \tfrac{1}{2}y\right) = e^{2y}f(2x - y)$$

and by the formula

$$\begin{aligned} z_p &= -\tfrac{1}{2}\cos(y - 2x)\cdot\left(\frac{y}{2} + \frac{\frac{1}{2}}{2}\right) \\ &= -\tfrac{1}{8}(2y + 1)\cos(y - 2x) \end{aligned}$$

Example 6. Find z_p for $(D_x + 4D_y)^3 z = (2x - y)^2$.
Using formula 79, we obtain

$$z_p = \frac{\iiint u^2 du^3}{[2 + 4(-1)]^3} = \frac{u^5}{5\cdot 4\cdot 3\cdot(-8)} = -\frac{(2x - y)^5}{480}$$

Example 7. Find z_p for $(D_x^3 + 5D_x^2 D_y - 7D_x + 4)z = e^{2x+3y}$.
By formula 87

$$z_p = \frac{e^{2x+3y}}{2^3 + 5\cdot 2^2\cdot 3 - 7\cdot 2 + 4} = \frac{e^{2x+3y}}{58}$$

Example 8. Find z_p for

$$(D_x^4 + 6D_x^3 D_y + D_x D_y + D_y^2 + 9)z = \sin(3x + 4y)$$

Since every term in the left member is of even degree in the two operators D_x and D_y, formula 90 is applicable. It gives

$$z_p = \frac{\sin(3x + 4y)}{(-9)^2 + 6(-9)(-12) + (-12) + (-16) + 9}$$

$$= \frac{\sin(3x + 4y)}{710}$$

TABLE I: $(D - a)y = R$

R	y_p
1. e^{rx}	$\dfrac{e^{rx}}{r-a}$
2. $\sin sx$*	$-\dfrac{a\sin sx + s\cos sx}{a^2+s^2} = -\dfrac{1}{\sqrt{a^2+s^2}}\sin\left(sx + \tan^{-1}\dfrac{s}{a}\right)$
3. $P(x)$	$-\dfrac{1}{a}\left[P(x) + \dfrac{P'(x)}{a} + \dfrac{P''(x)}{a^2} + \cdots + \dfrac{P^{(n)}(x)}{a^n}\right]$
4. $e^{rx}\sin sx$*	Replace a by $a - r$ in formula 2 and multiply by e^{rx}.
5. $P(x)e^{rx}$	Replace a by $a - r$ in formula 3 and multiply by e^{rx}.
6. $P(x)\sin sx$*	$-\sin sx\left[\dfrac{a}{a^2+s^2}P(x) + \dfrac{a^2-s^2}{(a^2+s^2)^2}P'(x) + \dfrac{a^3-3as^2}{(a^2+s^2)^3}P''(x) + \cdots + \dfrac{a^k - \binom{k}{2}a^{k-2}s^2 + \binom{k}{4}a^{k-4}s^4 - \cdots}{(a^2+s^2)^k}P^{(k-1)}(x) + \cdots\right]$ $-\cos sx\left[\dfrac{s}{a^2+s^2}P(x) + \dfrac{2as}{(a^2+s^2)^2}P'(x) + \dfrac{3a^2s - s^3}{(a^2+s^2)^3}P''(x) + \cdots + \dfrac{\binom{k}{1}a^{k-1}s - \binom{k}{3}a^{k-3}s^3 + \cdots}{(a^2+s^2)^k}P^{(k-1)}(x) + \cdots\right]$
7. $P(x)e^{rx}\sin sx$*	Replace a by $a - r$ in formula 6 and multiply by e^{rx}.
8. e^{ax}	xe^{ax}
9. $e^{ax}\sin sx$*	$-\dfrac{e^{ax}\cos sx}{s}$
10. $P(x)e^{ax}$	$e^{ax}\displaystyle\int P(x)\,dx$
11. $P(x)e^{ax}\sin sx$*	$\dfrac{e^{ax}\sin sx}{s}\left[\dfrac{P'(x)}{s} - \dfrac{P'''(x)}{s^3} + \dfrac{P^{v}(x)}{s^5} - \cdots\right] - \dfrac{e^{ax}\cos sx}{s}\left[P(x) - \dfrac{P''(x)}{s^2} + \dfrac{P^{iv}(x)}{s^4} - \cdots\right]$

* For $\cos sx$ in R replace "sin" by "cos" and "cos" by "$-$ sin" in y_p.

$$D^n = \frac{d^n}{dx^n} \qquad \binom{m}{n} = \frac{m!}{(m-n)!n!} \qquad 0! = 1$$

TABLE II: $(D - a)^2y = R$

R	y_p
12. e^{rx}	$\dfrac{e^{rx}}{(r-a)^2}$
13. $\sin sx$*	$\dfrac{1}{(a^2+s^2)^2}[(a^2-s^2)\sin sx + 2as\cos sx] = \dfrac{1}{a^2+s^2}\sin\left(sx + \tan^{-1}\dfrac{2as}{a^2-s^2}\right)$
14. $P(x)$	$\dfrac{1}{a^2}\left[P(x) + \dfrac{2P'(x)}{a} + \dfrac{3P''(x)}{a^2} + \cdots + \dfrac{(n+1)P^{(n)}(x)}{a^n}\right]$
15. $e^{rx}\sin sx$*	Replace a by $a - r$ in formula 13 and multiply by e^{rx}.
16. $P(x)e^{rx}$	Replace a by $a - r$ in formula 14 and multiply by e^{rx}.
17. $P(x)\sin sx$*	$\sin sx\left[\dfrac{a^2-s^2}{(a^2+s^2)^2}P(x) + 2\dfrac{a^3-3as^2}{(a^2+s^2)^3}P'(x) + 3\dfrac{a^4-6a^2s^2+s^4}{(a^2+s^2)^4}P''(x) + \cdots + (k-1)\dfrac{a^k - \binom{k}{2}a^{k-2}s^2 + \binom{k}{4}a^{k-4}s^4 - \cdots}{(a^2+s^2)^k}P^{(k-2)}(x) + \cdots\right]$ $+\cos sx\left[\dfrac{2as}{(a^2+s^2)^2}P(x) + 2\dfrac{3a^2s - s^3}{(a^2+s^2)^3}P'(x) + 3\dfrac{4a^3s - 4as^3}{(a^2+s^2)^4}P''(x) + \cdots + (k-1)\dfrac{\binom{k}{1}a^{k-1}s - \binom{k}{3}a^{k-3}s^3 + \cdots}{(a^2+s^2)^k}P^{(k-2)}(x) + \cdots\right]$
18. $P(x)e^{rx}\sin sx$*	Replace a by $a - r$ in formula 17 and multiply by e^{rx}.
19. e^{ax}	$\frac{1}{2}x^2e^{ax}$
20. $e^{ax}\sin sx$*	$-\dfrac{e^{ax}\sin sx}{s^2}$
21. $P(x)e^{ax}$	$e^{ax}\displaystyle\iint P(x)\,dx\,dx$
22. $P(x)e^{ax}\sin sx$*	$-\dfrac{e^{ax}\sin sx}{s^2}\left[P(x) - \dfrac{3P''(x)}{s^2} + \dfrac{5P^{iv}(x)}{s^4} - \dfrac{7P^{vi}(x)}{s^6} + \cdots\right] - \dfrac{e^{ax}\cos sx}{s^2}\left[\dfrac{2P'(x)}{s} - \dfrac{4P'''(x)}{s^3} + \dfrac{6P^{v}(x)}{s^5} - \cdots\right]$

*** For $\cos sx$ in R replace "sin" by "cos" and "cos" by "$-$ sin" in y_p.**

TABLE III: $(D^2 + q)y = R$

R	y_p
23. e^{rx}	$\dfrac{e^{rx}}{r^2+q}$
24. $\sin sx$*	$\dfrac{\sin sx}{-s^2+q}$
25. $P(x)$	$\dfrac{1}{q}\left[P(x) - \dfrac{P''(x)}{q} + \dfrac{P^{\text{iv}}(x)}{q^2} - \cdots + (-1)^k \dfrac{P^{(2k)}(x)}{q^k} \cdots\right]$
26. $e^{rx}\sin sx$*	$\dfrac{(r^2-s^2+q)e^{rx}\sin sx - 2rse^{rx}\cos sx}{(r^2-s^2+q)^2+(2rs)^2} = \dfrac{e^{rx}}{\sqrt{(r^2-s^2+q)^2+(2rs)^2}}\sin\left[sx - \tan^{-1}\dfrac{2rs}{r^2-s^2+q}\right]$
27. $P(x)e^{rx}$	$\dfrac{e^{rx}}{r^2+q}\left[P(x) - \dfrac{2r}{r^2+q}P'(x) + \dfrac{3r^2-q}{(r^2+q)^2}P''(x) - \dfrac{4r^3-4qr}{(r^2+q)^3}P'''(x) + \cdots + (-1)^{k-1}\dfrac{\binom{k}{1}r^{k-1} - \binom{k}{3}r^{k-3}q + \binom{k}{5}r^{k-5}q^2 - \cdots}{(r^2+q)^{k-1}}P^{(k-1)}(x) + \cdots\right]$
28. $P(x)\sin sx$*	$\dfrac{\sin sx}{(-s^2+q)}\left[P(x) - \dfrac{3s^2+q}{(-s^2+q)^2}P''(x) + \dfrac{5s^4+10s^2q+q^2}{(-s^2+q)^4}P^{\text{iv}}(x) + \cdots + (-1)^k\dfrac{\binom{2k+1}{1}s^{2k} + \binom{2k+1}{3}s^{2k-2}q + \binom{2k+1}{5}s^{2k-4}q^2 + \cdots}{(-s^2+q)^{2k}}P^{(2k)}(x) + \cdots\right]$ $- \dfrac{s\cos sx}{(-s^2+q)}\left[\dfrac{2P'(x)}{(-s^2+q)} - \dfrac{4s^2+4q}{(-s^2+q)^3}P'''(x) + \cdots + (-1)^{k+1}\dfrac{\binom{2k}{1}s^{2k-2} + \binom{2k}{3}s^{2k-4}q + \cdots}{(-s^2+q)^{2k-1}}P^{(2k-1)}(x) + \cdots\right]$

TABLE IV: $(D^2 + b^2)y = R$

R	y_p
29. $\sin bx$*	$-\dfrac{x\cos bx}{2b}$
30. $P(x)\sin bx$*	$\dfrac{\sin bx}{(2b)^2}\left[P(x) - \dfrac{P''(x)}{(2b)^2} + \dfrac{P^{\text{iv}}(x)}{(2b)^4} - \cdots\right] - \dfrac{\cos bx}{2b}\displaystyle\int\left[P(x) - \dfrac{P''(x)}{(2b)^2} + \cdots\right]dx$

* For cos sx in R replace "sin" by "cos" and "cos" by " – sin" in y_p.

TABLE V: $(D^2 + pD + q)y = R$

R	y_p
31. e^{rx}	$\dfrac{e^{rx}}{r^2+pr+q}$
32. $\sin sx$*	$\dfrac{(q-s^2)\sin sx - ps\cos sx}{(q-s^2)^2+(ps)^2} = \dfrac{1}{\sqrt{(q-s^2)^2+(ps)^2}}\sin\left(sx - \tan^{-1}\dfrac{ps}{q-s^2}\right)$
33. $P(x)$	$\dfrac{1}{q}\left[P(x) - \dfrac{p}{q}P'(x) + \dfrac{p^2-q}{q^2}P''(x) - \dfrac{p^3-2pq}{q^3}P'''(x) + \cdots + (-1)^n\dfrac{p^n - \binom{n-1}{1}p^{n-2}q + \binom{n-2}{2}p^{n-4}q^2 - \cdots}{q^n}P^{(n)}(x)\right]$
34. $e^{rx}\sin sx$*	Replace p by $p + 2r$, q by $q + pr + r^2$ in formula 32 and multiply by e^{rx}.
35. $P(x)e^{rx}$	Replace p by $p + 2r$, q by $q + pr + r^2$ in formula 33 and multiply by e^{rx}.

TABLE VI: $(D - b)(D - a)y = R$

R	y_p
36. $P(x)\sin sx$*	$\dfrac{\sin sx}{b-a}\left[\left(\dfrac{a}{a^2+s^2} - \dfrac{b}{b^2+s^2}\right)P(x) + \left(\dfrac{a^2-s^2}{(a^2+s^2)^2} - \dfrac{b^2-s^2}{(b^2+s^2)^2}\right)P'(x) + \left(\dfrac{a^3-3as^2}{(a^2+s^2)^3} - \dfrac{b^3-3bs^2}{(b^2+s^2)^3}\right)P''(x) + \cdots\right]$ $+ \dfrac{\cos sx}{b-a}\left[\left(\dfrac{s}{a^2+s^2} - \dfrac{s}{b^2+s^2}\right)P(x) + \left(\dfrac{2as}{(a^2+s^2)^2} - \dfrac{2bs}{(b^2+s^2)^2}\right)P'(x) + \left(\dfrac{3a^2s-s^3}{(a^2+s^2)^3} - \dfrac{3b^2s-s^3}{(b^2+s^2)^3}\right)P''(x) + \cdots\right]$†
37. $P(x)e^{rx}\sin sx$*	Replace a by $a - r$, b by $b - r$ in formula 36 and multiply by e^{rx}.
38. $P(x)e^{ax}$	$\dfrac{e^{ax}}{a-b}\left[\displaystyle\int P(x)dx + \dfrac{P(x)}{(b-a)} + \dfrac{P'(x)}{(b-a)^2} + \dfrac{P''(x)}{(b-a)^3} + \cdots + \dfrac{P^{(n)}(x)}{(b-a)^{n+1}}\right]$

* For cos sx in R replace "sin" by "cos" and "cos" by " – sin" in y_p.

† For additional terms, compare with formula 6.

TABLE VII: $(D^2 - 2aD + a^2 + b^2)y = R$

R	y_p
39. $P(x)\sin sx$*	$\frac{\sin sx}{2b}\left[\left(\frac{s+b}{a^2+(s+b)^2} - \frac{s-b}{a^2+(s-b)^2}\right)P(x) + \left(\frac{2a(s+b)}{[a^2+(s+b)^2]^2} - \frac{2a(s-b)}{[a^2+(s-b)^2]^2}\right)P'(x) + \left(\frac{3a^2(s+b)-(s+b)^3}{[a^2+(s+b)^2]^3} - \frac{3a^2(s-b)-(s-b)^3}{[a^2+(s-b)^2]^3}\right)P''(x) + \cdots\right] - \frac{\cos sx}{2b}\left[\left(\frac{a}{a^2+(s+b)^2} - \frac{a}{a^2+(s-b)^2}\right)P(x) + \left(\frac{a^2-(s+b)^2}{[a^2+(s+b)^2]^2} - \frac{a^2-(s-b)^2}{[a^2+(s-b)^2]^2}\right)P'(x) + \left(\frac{a^3-3a(s+b)^2}{[a^2+(s+b)^2]^3} - \frac{a^3-3a(s-b)^2}{[a^2+(s-b)^2]^3}\right)P''(x) + \cdots\right]$†
40. $P(x)e^{rx}\sin sx$*	Replace a by $a - r$ in formula 39 and multiply by e^{rx}.
41. $P(x)e^{ax}$	$\frac{e^{ax}}{b^2}\left[P(x) - \frac{P''(x)}{b^2} + \frac{P^{iv}(x)}{b^4} - \cdots\right]$
42. $e^{ax}\sin sx$*	$\frac{e^{ax}\sin sx}{-s^2+b^2}$
43. $e^{ax}\sin bx$*	$-\frac{xe^{ax}\cos bx}{2b}$
44. $P(x)e^{ax}\sin bx$*	$\frac{e^{ax}\sin bx}{(2b)^2}\left[P(x) - \frac{P''(x)}{(2b)^2} + \frac{P^{iv}(x)}{(2b)^4} - \cdots\right] - \frac{e^{ax}\cos bx}{2b}\int\left[P(x) - \frac{P''(x)}{(2b)^2} + \frac{P^{iv}(x)}{(2b)^4} - \cdots\right]dx$

* For $\cos sx$ in R replace "sin" by "cos" and "cos" by "− sin" in y_p.
† For additional terms, compare with formula 6.

TABLE VIII: $f(D)y = [D^n + a_{n-1}D^{n-1} + \cdots + a_1D + a_0]y = R$

R	y_p
45. e^{rx}	$\frac{e^{rx}}{f(r)}$
46. $\sin sx$*	$\frac{[a_0 - a_2s^2 + a_4s^4 - \cdots]\sin sx - [a_1s - a_3s^3 + a_5s^5 + \cdots]\cos sx}{[a_0 - a_2s^2 + a_4s^4 - \cdots]^2 + [a_1s - a_3s^3 + a_5s^5 - \cdots]^2}$

TABLE IX: $f(D^2)y = R$

R	y_p
47. $\sin sx$*	$\frac{\sin sx}{f(-s^2)} = \frac{\sin sx}{a_0 - a_2s^2 + \cdots \pm s^{2n}}$

TABLE X: $(D - a)^n y = R$

R	y_p
48. e^{rx}	$\frac{e^{rx}}{(r-a)^n}$
49. $\sin sx$*	$\frac{(-1)^n}{(a^2+s^2)^n}\{[a^n - \binom{n}{2}a^{n-2}s^2 + \binom{n}{4}a^{n-4}s^4 - \cdots]\sin sx + [\binom{n}{1}a^{n-1}s - \binom{n}{3}a^{n-3}s^3 + \cdots]\cos sx\}$
50. $P(x)$	$\frac{(-1)^n}{a^n}\left[P(x) + \binom{n}{1}\frac{P'(x)}{a} + \binom{n+1}{2}\frac{P''(x)}{a^2} + \binom{n+2}{3}\frac{P'''(x)}{a^3} + \cdots\right]$
51. $e^{rx}\sin sx$*	Replace a by $a - r$ in formula 49 and multiply by e^{rx}.
52. $e^{rx}P(x)$	Replace a by $a - r$ in formula 50 and multiply by e^{rx}.
53. $P(x)\sin sx$*	$(-1)^n\sin sx[A_nP(x) + \binom{n}{1}A_{n+1}P'(x) + \binom{n+1}{2}A_{n+2}P''(x) + \binom{n+2}{3}A_{n+3}P'''(x) + \cdots] + (-1)^n\cos sx[B_nP(x) + \binom{n}{1}B_{n+1}P'(x) + \binom{n+1}{2}B_{n+2}P''(x) + \binom{n+2}{3}B_{n+3}P'''(x) + \cdots]$; $A_1 = \frac{a}{a^2+s^2},\ A_2 = \frac{a^2-s^2}{(a^2+s^2)^2},\ \ldots,\ A_k = \frac{a^k - \binom{k}{2}a^{k-2}s^2 + \binom{k}{4}a^{k-4}s^4 - \cdots}{(a^2+s^2)^k}$. $B_1 = \frac{s}{a^2+s^2},\ B_2 = \frac{2as}{(a^2+s^2)^2},\ \ldots,\ B_k = \frac{\binom{k}{1}a^{k-1}s - \binom{k}{3}a^{k-3}s^3 + \cdots}{(a^2+s^2)^k}$.
54. $P(x)e^{rx}\sin sx$*	Replace a by $a - r$ in formula 53 and multiply by e^{rx}.

55. $e^{ax}P(x)$ $\qquad e^{ax}\int\int\cdots\int P(x)dx^n$

56. $P(x)e^{ax}\sin sx$* $\qquad \dfrac{(-1)^{\frac{n-1}{2}} e^{ax}\sin sx}{s^n}\left[\binom{n}{n-1}\dfrac{P'(x)}{s} - \binom{n+2}{n-1}\dfrac{P'''(x)}{s^3} + \binom{n+4}{n-1}\dfrac{P^{v}(x)}{s^5} - \cdots\right]$

$$+ \frac{(-1)^{\frac{n+1}{2}} e^{ax}\cos sx}{s^n}\left[\binom{n-1}{n-1}P(x) - \binom{n+1}{n-1}\frac{P''(x)}{s^2} + \binom{n+3}{n-1}\frac{P^{iv}(x)}{s^4} - \cdots\right] \text{ (}n\text{ odd)}$$

$$\frac{(-1)^{\frac{n}{2}} e^{ax}\sin sx}{s^n}\left[\binom{n-1}{n-1}P(x) - \binom{n+1}{n-1}\frac{P''(x)}{s^2} + \binom{n+3}{n-1}\frac{P^{iv}(x)}{s^4} - \cdots\right]$$

$$+ \frac{(-1)^{\frac{n}{2}} e^{ax}\cos sx}{s^n}\left[\binom{n}{n-1}\frac{P'(x)}{s} - \binom{n+2}{n-1}\frac{P'''(x)}{s^3} + \binom{n+4}{n-1}\frac{P^{v}(x)}{s^5} - \cdots\right] \text{ (}n\text{ even)}$$

* For cos sx in R replace "sin" by "cos" and "cos" by "– sin" in y_p.

TABLE XI: $(D - a)^n f(D)y = R$

57. e^{ax} $\qquad \dfrac{x^n}{n!}\cdot\dfrac{e^{ax}}{f(a)}$

* For cos sx in R replace "sin" by "cos" and "cos" by "– sin" in y_p.

TABLE XII: $(D^2 + q)^n y = R$

R	y_p
58. e^{rx}	$e^{rx}/(r^2+q)^n$
59. $\sin sx$*	$\sin sx/(q-s^2)^n$
60. $P(x)$	$\dfrac{1}{q^n}\left[P(x) - \binom{n}{1}\dfrac{P''(x)}{q} + \binom{n+1}{2}\dfrac{P^{iv}(x)}{q^2} - \binom{n+2}{3}\dfrac{P^{vi}(x)}{q^3} + \cdots\right]$
61. $e^{rx}\sin sx$*	$\dfrac{e^{rx}}{(A^2+B^2)^n}\{[A^n - \binom{n}{2}A^{n-2}B^2 + \binom{n}{4}A^{n-4}B^4 - \cdots]\sin sx - [\binom{n}{1}A^{n-1}B - \binom{n}{3}A^{n-3}B^3 + \cdots]\cos sx\}$ $A = r^2 - s^2 + q, \quad B = 2rs$

TABLE XIII: $(D^2 + b^2)^n y = R$

62. $\sin bx$* $\qquad (-1)^{\frac{n+1}{2}}\dfrac{x^n\cos bx}{n!(2b)^n}$ (n odd), $\qquad (-1)^{\frac{n}{2}}\dfrac{x^n\sin bx}{n!(2b)^n}$ (n even)

TABLE XIV: $(D^n - q)y = R$

63. e^{rx} $\qquad e^{rx}/(r^n - q)$

64. $P(x)$ $\qquad -\dfrac{1}{q}\left[P(x) + \dfrac{P^{(n)}(x)}{q} + \dfrac{P^{(2n)}(x)}{q^2} + \cdots\right]$

65. $\sin sx$* $\qquad -\dfrac{q\sin sx + (-1)^{\frac{n-1}{2}} s^n\cos sx}{q^2 + s^{2n}}$ (n odd), $\qquad \dfrac{\sin sx}{(-s^2)^{n/2} - q}$ (n even)

66. $e^{rx}\sin sx$* $\qquad \dfrac{Ae^{rx}\sin sr - Be^{rx}\cos sx}{A^2+B^2} = \dfrac{e^{rx}}{\sqrt{A^2+B^2}}\sin\left(sx - \tan^{-1}\dfrac{B}{A}\right)$

$$A = [r^n - \binom{n}{2}r^{n-2}s^2 + \binom{n}{4}r^{n-4}s^4 - \cdots] - q, \qquad B = [\binom{n}{1}r^{n-1}s - \binom{n}{3}r^{n-3}s^3 + \cdots]$$

* For cos sx in R replace "sin" by "cos" and "cos" by "– sin" in y_p.

TABLE XV: $(D_x + mD_y)z = R$

R	z_p
67 e^{ax+by}	$\dfrac{e^{ax+by}}{a+mb}$
68. $f(ax+by)$	$\dfrac{\int f(u)du}{a+mb}, \; u = ax+by$
69. $f(y-mx)$	$xf(y-mx)$
70. $\phi(x,y)f(y-mx)$	$f(y-mx)\int\phi(x, a+mx)dx$ ($a = y - mx$ after integration)

TABLE XVI: $(D_x + mD_y - k)z = R$

71. e^{ax+by}	$\dfrac{e^{ax+by}}{a+mb-k}$
72. $\sin(ax+by)$*	$-\dfrac{(a+bm)\cos(ax+by)+k\sin(ax+by)}{(a+bm)^2+k^2}$
73. $e^{\alpha x+\beta y}\sin(ax+by)$*	Replace k in 72 by $k-\alpha-m\beta$ and multiply by $e^{\alpha x+\beta y}$
74. $e^{xk}f(ax+by)$	$\dfrac{e^{kx}\int f(u)du}{a+mb}$, $u = ax+by$
75. $f(y-mx)$	$-\dfrac{f(y-mx)}{k}$
76. $P(x)f(y-mx)$	$-\dfrac{1}{k}f(y-mx)\left[P(x)+\dfrac{P'(x)}{k}+\dfrac{P''(x)}{k^2}+\cdots+\dfrac{P^{(n)}(x)}{k^n}\right]$
77. $e^{kx}f(y-mx)$	$xe^{kx}f(y-mx)$

* For cos $(ax+by)$ replace "sin" by "cos," and "cos" by "$-$sin" in z_p.

$$D_x = \frac{\partial}{\partial x};\quad D_y = \frac{\partial}{\partial y};\quad D_x{}^k D_y{}^r = \frac{\partial^{k+r}}{\partial_x{}^k\partial_y{}^r}$$

TABLE XVII: $(D_x + mD_y)^n z = R$

R	z_p
78. e^{ax+by}	$\dfrac{e^{ax+by}}{(a+mb)^n}$
79. $f(ax+by)$	$\dfrac{\int\int\cdots\int f(u)du^n}{(a+mb)^n}$, $u = ax+by$
80. $f(y-mx)$	$\dfrac{x^n}{n!}f(y-mx)$
81. $\phi(x,y)f(y+mx)$	$f(y-mx)\int\int\cdots\int\phi(x,a+mx)dx^n$ ($a = y-mx$ after integration)

TABLE XVIII: $(D_x + mD_y - k)^n z = R$

82. e^{ax+by}	$\dfrac{e^{ax+by}}{(a+mb-k)^n}$
83. $f(y-mx)$	$\dfrac{(-1)^n f(y-mx)}{k^n}$
84. $P(x)f(y-mx)$	$\dfrac{(-1)^n}{k^n}f(y-mx)\left[P(x)+\binom{n}{1}\dfrac{P'(x)}{k}+\binom{n+1}{2}\dfrac{P''(x)}{k^2}+\binom{n+2}{3}\dfrac{P'''(x)}{k^3}+\cdots\right]$
85. $e^{kx}f(ax+by)$	$\dfrac{e^{kx}\int\int\cdots\int f(u)du^n}{(a+mb)^n}$, $u = ax+by$
86. $e^{kx}f(y-mx)$	$\dfrac{x^n}{n!}e^{kx}f(y-mx)$

TABLE XIX: $[D_x{}^n + a_1D_x{}^{n-1}D_y + a_2D_x{}^{n-2}D_y{}^2 + \cdots + a^nD_y{}^n]z = R$

87. e^{ax+by}	$\dfrac{e^{ax+by}}{a+a_1a^{n-1}b+a_2a^{n-2}b^2+\cdots+a_nb^n}$
88. $f(ax+by)$	$\dfrac{\int\int\cdots\int f(u)du^n}{a^n+a_1a^{n-1}b+a_2a^{n-2}b^2+\cdots+a^nb^n}$, $(u = ax+by)$

TABLE XX: $F(D_x, D_y)z = R$

89. e^{ax+by}	$\dfrac{e^{ax+by}}{F(a,b)}$

TABLE XXI: $F(D_x{}^2, D_xD_y, D_y{}^2)z = R$

90. $\sin(ax+by)$*	$\dfrac{\sin(ax+by)}{F(-a^2,-ab,-b^2)}$

* For cos $(ax+by)$ replace "sin" by "cos", and "cos" by "$-$sin" in z_p.

DIFFERENTIAL EQUATIONS

Differential equation	Method of solution
Separation of variables $f_1(x)\,g_1(y)\,dx + f_2(x)\,g_2(y)\,dy = 0$	$\int \frac{f_1(x)}{f_2(x)}\,dx + \int \frac{g_2(y)}{g_1(y)}\,dy = c$
Exact equation $M(x,y)dx + N(x,y)dy = 0$ where $\partial M/\partial y = \partial N/\partial x$	$\int M\,\partial x + \int \left(N - \frac{\partial}{\partial y}\int M\,\partial x\right) dy = c$ where ∂x indicates that the integration is to be performed with respect to x keeping y constant.
Linear first order equation $\frac{dy}{dx} + P(x)y = Q(x)$	$ye^{\int P dx} = \int Qe^{\int P dx}\,dx + c$
Bernoulli's equation $\frac{dy}{dx} + P(x)y = Q(x)y^n$	$ve^{(1-n)\int P dx} = (1-n)\int Qe^{(1-n)\int P dx}\,dx + c$ where $v = y^{1-n}$. If $n = 1$, the solution is $\ln y = \int (Q - P)\,dx + c$
Homogeneous equation $\frac{dy}{dx} = F\left(\frac{y}{x}\right)$	$\ln x = \int \frac{dv}{F(v) - v} + c$ where $v = y/x$. If $F(v) = v$, the solution is $y = cx$
Reducible to homogeneous $(a_1x + b_1y + c_1)\,dx + (a_2x + b_2y + c_2)\,dy = 0$ $\frac{a_1}{a_2} \neq \frac{b_1}{b_2}$	Set $u = a_1x + b_1y + c_1$ $v = a_2x + b_2y + c_2$ Eliminate x and y and the equation becomes homogenous
Reducible to separable $(a_1x + b_1y + c_1)\,dx + (a_2x + b_2y + c_2)\,dy = 0$ $\frac{a_1}{a_2} = \frac{b_1}{b_2}$	Set $u = a_1x + b_1y$ Eliminate x or y and equation becomes separable

$y\,F(xy)\,dx + x\,G(xy)dy = 0$	$\ln x = \int \frac{G(v)\,dv}{v\{G(v) - F(v)\}} + c$ where $v = xy$. If $G(v) = F(v)$, the solution is $xy = c$.
Linear, homogeneous second order equation $\frac{d^2y}{dx^2} + b\frac{dy}{dx} + cy = 0$ b,c are real constants	Let m_1, m_2 be the roots of $m^2 + bm + c = 0$. Then there are 3 cases: Case 1. m_1, m_2 real and distinct: $y = c_1 e^{m_1 x} + c_2 e^{m_2 x}$ Case 2. m_1, m_2 real and equal: $y = c_1 e^{m_1 x} + c_2 x e^{m_1 x}$ Case 3. $m_1 = p + qi$, $m_2 = p - qi$: $y = e^{px}(c_1 \cos qx + c_2 \sin qx)$ where $p = -b/2$, $q = \sqrt{4c - b^2}/2$
Linear, nonhomogeneous second order equation $\frac{d^2y}{dx^2} + b\frac{dy}{dx} + cy = R(x)$ b, c are real constants	There are 3 cases corresponding to those immediately above: Case 1. $y = c_1 e^{m_1 x} + c_2 e^{m_2 x} + \frac{e^{m_1 x}}{m_1 - m_2}\int e^{-m_1 x} R(x)\,dx + \frac{e^{m_2 x}}{m_2 - m_1}\int e^{-m_2 x} R(x)\,dx$ Case 2. $y = c_1 e^{m_1 x} + c_2 x e^{m_1 x} + x e^{m_1 x}\int e^{-m_1 x} R(x)\,dx - e^{m_1 x}\int x e^{-m_1 x} R(x)\,dx$ Case 3. $y = e^{px}(c_1 \cos qx + c_2 \sin qx) + \frac{e^{px}\sin qx}{q}\int e^{-px} R(x)\cos qx\,dx - \frac{e^{px}\cos qx}{q}\int e^{-px} R(x)\sin qx\,dx$

Euler or Cauchy equation $x^2 \dfrac{d^2y}{dx^2} + bx\dfrac{dy}{dx} + cy = S(x)$	Putting $x = e^t$, the equation becomes $\dfrac{d^2y}{dt^2} + (b-1)\dfrac{dy}{dt} + cy = S(e^t)$ and can then be solved as a linear second order equation.
Bessel's equation $x^2\dfrac{d^2y}{dx^2} + x\dfrac{dy}{dx} + (\lambda^2x^2 - n^2)y = 0$	$y = c_1 J_n(\lambda x) + c_2 Y_n(\lambda x)$
Transformed Bessel's equation $x^2\dfrac{d^2y}{dx^2} + (2p+1)x\dfrac{dy}{dx} + (\alpha^2x^{2r} + \beta^2)y = 0$	$y = x^{-p}\left\{c_1 J_{q/r}\left(\frac{\alpha}{r}x^r\right) + c_2 Y_{q/r}\left(\frac{\alpha}{r}x^r\right)\right\}$ where $q = \sqrt{p^2 - \beta^2}$.
Legendre's equation $(1-x^2)\dfrac{d^2y}{dx^2} - 2x\dfrac{dy}{dx} + n(n+1)y = 0$	$y = c_1 P_n(x) + c_2 Q_n(x)$

FOURIER SERIES

1. If $f(x)$ is a bounded periodic function of period 2L (i.e. $f x + 2L) = f(x)$, and satisfies the *Dirichlet conditions:*

   A. In any period $f(x)$ is continuous, except possibly for a finite number of jump discontinuities.
   B. In any period $f(x)$ has only a finite number of maxima and minima.

   Then $f(x)$ may be represented by the *Fourier series*

$$\frac{a_0}{2} + \sum_{n=1}^{\infty}\left(a_n \cos\frac{n\pi x}{L} + b_n \sin\frac{n\pi x}{L}\right)$$

where a_n and b_n are as determined below. This series will converge to $f(x)$ at every point where $f(x)$ is continuous, and to

$$\frac{f(x^+) + f(x^-)}{2}$$

(i.e. the average of the left-hand and right-hand limits) at every point where $f(x)$ has a jump discontinuity.

$$a_n = \frac{1}{L}\int_{-L}^{L} f(x)\cos\frac{n\pi x}{L}\,dx, \quad n = 0, 1, 2, 3, \ldots;$$

$$b_n = \frac{1}{L}\int_{-L}^{L} f(x)\sin\frac{n\pi x}{L}\,dx, \quad n = 1, 2, 3, \ldots$$

We may also write

$$a_n = \frac{1}{L}\int_{\alpha}^{\alpha+2L} f(x)\cos\frac{n\pi x}{L}\,dx \text{ and } b_n = \frac{1}{L}\int_{\alpha}^{\alpha+2L} f(x)\sin\frac{n\pi x}{L}\,dx$$

where α is any real number. Thus if $\alpha = 0$,

$$a_n = \frac{1}{L}\int_{0}^{2L} f(x)\cos\frac{n\pi x}{L}\,dx, \quad n = 0, 1, 2, 3, \ldots;$$

$$b_n = \frac{1}{L}\int_{0}^{2L} f(x)\sin\frac{n\pi x}{L}\,dx, \quad n = 1, 2, 3, \ldots$$

2. If in addition to the above restrictions, $f(x)$), is even (i.e., $f(-x) = f(x)$ the Fourier series reduces to

$$\frac{a_0}{2} + \sum_{n=1}^{\infty} a_n \cos \frac{n\pi x}{L}$$

That is, $b_n = 0$. In this case, a simpler formula for a_n is

$$a_n = \frac{2}{L} \int_0^L f(x) \cos \frac{n\pi x}{L}\, dx, \quad n = 0, 1, 2, 3, \ldots$$

3. If in addition to the restrictions in (1), $f(x)$ is an odd function (i.e., $f(-x) = -f(x)$), then the Fourier series reduces to

$$\sum_{n=1}^{\infty} b_n \sin \frac{n\pi x}{L}$$

That is, $a_n = 0$. In this case, a simpler formula for the b_n is

$$b_n = \frac{2}{L} \int_0^L f(x) \sin \frac{n\pi x}{L}\, dx, \quad n = 1, 2, 3, \ldots$$

4. If in addition to the restrictions in (2) above, $f(x) = -f(L - x)$, then a_n will be 0 for all even values of n, including $n = 0$. Thus in this case, the expansion reduces to

$$\sum_{m=1}^{\infty} a_{2m-1} \cos \frac{(2m-1)\pi x}{L}$$

5. If in addition to the restrictions in (3) above, $f(x), = f(L - x)$, then b_n will be 0 for all even values of n. Thus in this case, the expansion reduces to

$$\sum_{m=1}^{\infty} b_{2m-1} \sin \frac{(2m-1)\pi x}{L}$$

(The series in (4) and (5) are known as *odd-harmonic series,* since only the odd harmonics appear. Similar rules may be stated for even-harmonic series, but when a series appears in the even-harmonic form, it means that $2L$ has not been taken as the smallest period of $f(x)$. Since any integral multiple of a period is also a period, series obtained in this way will also work, but in general computation is simplified if $2L$ is taken to be the smallest period.)

6. If we write the Euler definitions for cos θ and sin θ, we obtain the complex form of the Fourier Series known either as the "Complex Fourier Series" or the "Exponential Fourier Series" of $f(x)$. It is represented as

$$f(x) = \frac{1}{2} \sum_{n=-\infty}^{n=+\infty} c_n e^{i\omega_n x}$$

where

$$c_n = \frac{1}{L} \int_{-L}^{L} f(x) e^{-i\omega_n x}\, dx, \quad n = 0, \pm 1, \pm 2, \pm 3, \ldots$$

with $\omega_n = \frac{n\pi}{L}, \quad n = 0, \pm 1, \pm 2, \ldots$

The set of coefficients $\{c_n\}$ is often referred to as the Fourier spectrum.

7. If both sine and cosine terms are present and if $f(x)$ is of period $2L$ and expandable by a Fourier series, it can be represented as

$$f(x) = \frac{a_0}{2} + \sum_{n=1}^{\infty} c_n \sin\left(\frac{n\pi x}{L} + \phi_n\right), \text{ where } a_n = c_n \sin \phi_n,$$

$$b_n = c_n \cos \phi_n, \quad c_n = \sqrt{a_n^2 + b_n^2}, \quad \phi_n = \arctan\left(\frac{a_n}{b_n}\right)$$

It can also be represented as

$$f(x) = \frac{a_0}{2} + \sum_{n=1}^{\infty} c_n \cos\left(\frac{n\pi x}{L} + \phi_n\right), \text{ where } a_n = c_n \cos \phi_n,$$

$$b_n = -c_n \sin \phi_n, \quad c_n = \sqrt{a_n^2 + b_n^2}, \quad \phi_n = \arctan\left(-\frac{b_n}{a_n}\right)$$

where ϕ_n is chosen so as to make a_n, b_n, and c_n hold.

8. The following table of trigonometric identities should be helpful for developing Fourier Series.

	n	n even	n odd	$n/2$ odd	$n/2$ even
$\sin n\pi$	0	0	0	0	0
$\cos n\pi$	$(-1)^n$	$+1$	-1	$+1$	$+1$
$^*\sin \frac{n\pi}{2}$		0	$(-1)^{(n-1)/2}$	0	0
$^*\cos \frac{n\pi}{2}$		$(-1)^{n/2}$	0	-1	$+1$
$\sin \frac{n\pi}{4}$			$\frac{\sqrt{2}}{2}(-1)^{(n^2+4n+11)/8}$	$(-1)^{(n-2)/4}$	0

*A useful formula for $\sin \frac{n\pi}{2}$ and $\cos \frac{n\pi}{2}$ is given by

$$\sin \frac{n\pi}{2} = \frac{(i)^{n+1}}{2}[(-1)^n - 1] \text{ and } \cos \frac{n\pi}{2} = \frac{(i)^n}{2}[(-1)^n + 1], \text{ where } i^2 = -1.$$

AUXILIARY FORMULAS FOR FOURIER SERIES

$$1 = \frac{4}{\pi}\left[\sin \frac{\pi x}{k} + \frac{1}{3}\sin \frac{3\pi x}{k} + \frac{1}{5}\sin \frac{5\pi x}{k} + \cdots\right] \qquad [0 < x < k]$$

$$x = \frac{2k}{\pi}\left[\sin \frac{\pi x}{k} - \frac{1}{2}\sin \frac{2\pi x}{k} + \frac{1}{3}\sin \frac{3\pi x}{k} - \cdots\right] \qquad [-k < x < k]$$

$$x = \frac{k}{2} - \frac{4k}{\pi^2}\left[\cos \frac{\pi x}{k} + \frac{1}{3^2}\cos \frac{3\pi x}{k} + \frac{1}{5^2}\cos \frac{5\pi x}{k} + \cdots\right] \qquad [0 < x < k]$$

$$x^2 = \frac{2k^2}{\pi^3}\left[\left(\frac{\pi^2}{1} - \frac{4}{1}\right)\sin \frac{\pi x}{k} - \frac{\pi^2}{2}\sin \frac{2\pi x}{k} + \left(\frac{\pi^2}{3} - \frac{4}{3^3}\right)\sin \frac{3\pi x}{k}\right.$$
$$\left. - \frac{\pi^2}{4}\sin \frac{4\pi x}{k} + \left(\frac{\pi^2}{5} - \frac{4}{5^3}\right)\sin \frac{5\pi x}{k} + \cdots\right] \qquad [0 < x < k]$$

$$x^2 = \frac{k^2}{3} - \frac{4k^2}{\pi^2}\left[\cos \frac{\pi x}{k} - \frac{1}{2^2}\cos \frac{2\pi x}{k} + \frac{1}{3^2}\cos \frac{3\pi x}{k} - \frac{1}{4^2}\cos \frac{4\pi x}{k} + \cdots\right] \qquad [-k < x < k]$$

$$1 - \frac{1}{3} + \frac{1}{5} - \frac{1}{7} + \cdots = \frac{\pi}{4}$$

$$1 + \frac{1}{2^2} + \frac{1}{3^2} + \frac{1}{4^2} + \cdots = \frac{\pi^2}{6}$$

$$1 - \frac{1}{2^2} + \frac{1}{3^2} - \frac{1}{4^2} + \cdots = \frac{\pi^2}{12}$$

$$1 + \frac{1}{3^2} + \frac{1}{5^2} + \frac{1}{7^2} + \cdots = \frac{\pi^2}{8}$$

$$\frac{1}{2^2} + \frac{1}{4^2} + \frac{1}{6^2} + \frac{1}{8^2} + \cdots = \frac{\pi^2}{24}$$

FOURIER EXPANSIONS FOR BASIC PERIODIC FUNCTIONS

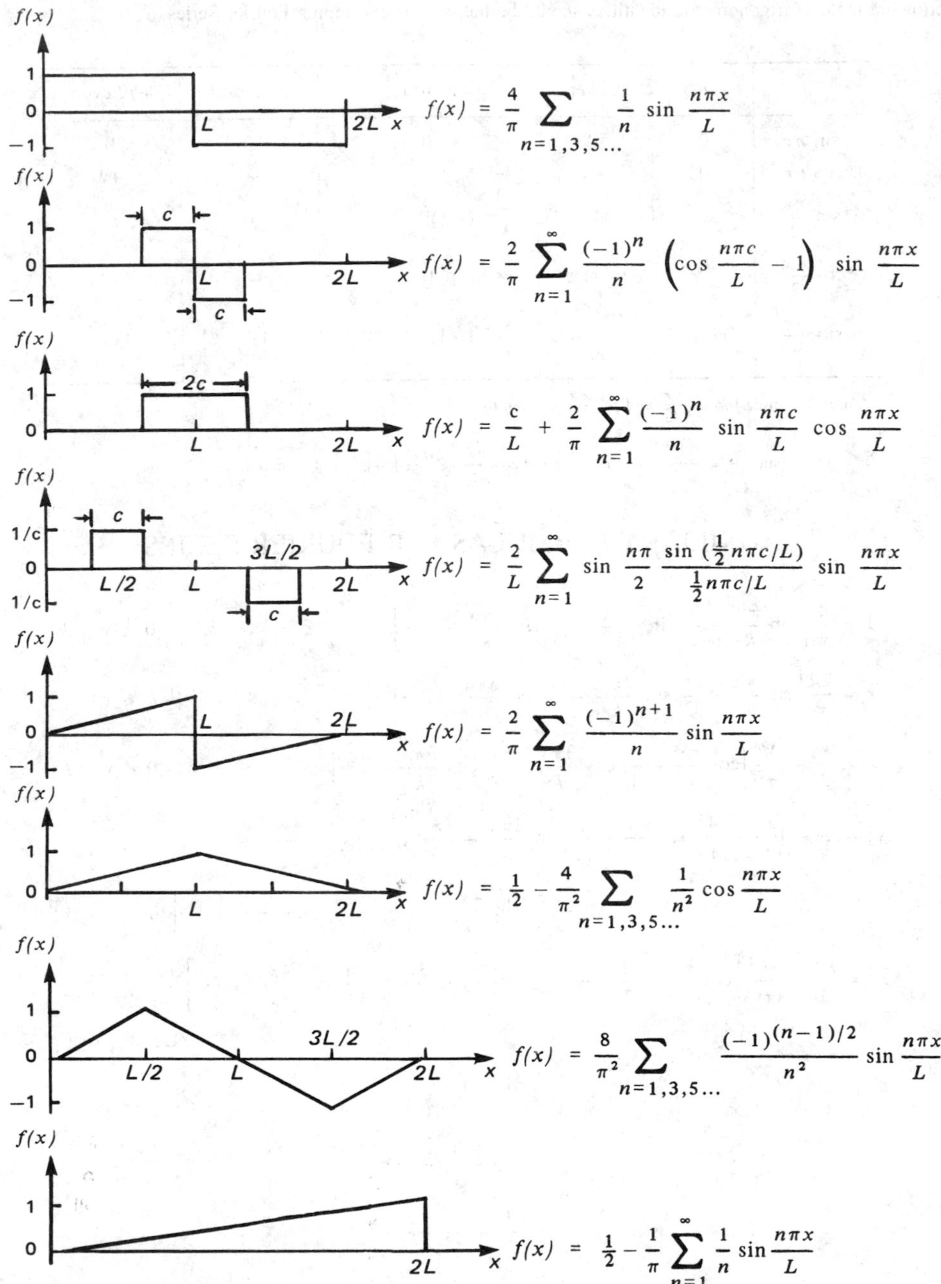

$$f(x) = \frac{4}{\pi} \sum_{n=1,3,5\ldots} \frac{1}{n} \sin \frac{n\pi x}{L}$$

$$f(x) = \frac{2}{\pi} \sum_{n=1}^{\infty} \frac{(-1)^n}{n} \left(\cos \frac{n\pi c}{L} - 1 \right) \sin \frac{n\pi x}{L}$$

$$f(x) = \frac{c}{L} + \frac{2}{\pi} \sum_{n=1}^{\infty} \frac{(-1)^n}{n} \sin \frac{n\pi c}{L} \cos \frac{n\pi x}{L}$$

$$f(x) = \frac{2}{L} \sum_{n=1}^{\infty} \sin \frac{n\pi}{2} \frac{\sin (\frac{1}{2} n\pi c/L)}{\frac{1}{2} n\pi c/L} \sin \frac{n\pi x}{L}$$

$$f(x) = \frac{2}{\pi} \sum_{n=1}^{\infty} \frac{(-1)^{n+1}}{n} \sin \frac{n\pi x}{L}$$

$$f(x) = \frac{1}{2} - \frac{4}{\pi^2} \sum_{n=1,3,5\ldots} \frac{1}{n^2} \cos \frac{n\pi x}{L}$$

$$f(x) = \frac{8}{\pi^2} \sum_{n=1,3,5\ldots} \frac{(-1)^{(n-1)/2}}{n^2} \sin \frac{n\pi x}{L}$$

$$f(x) = \frac{1}{2} - \frac{1}{\pi} \sum_{n=1}^{\infty} \frac{1}{n} \sin \frac{n\pi x}{L}$$

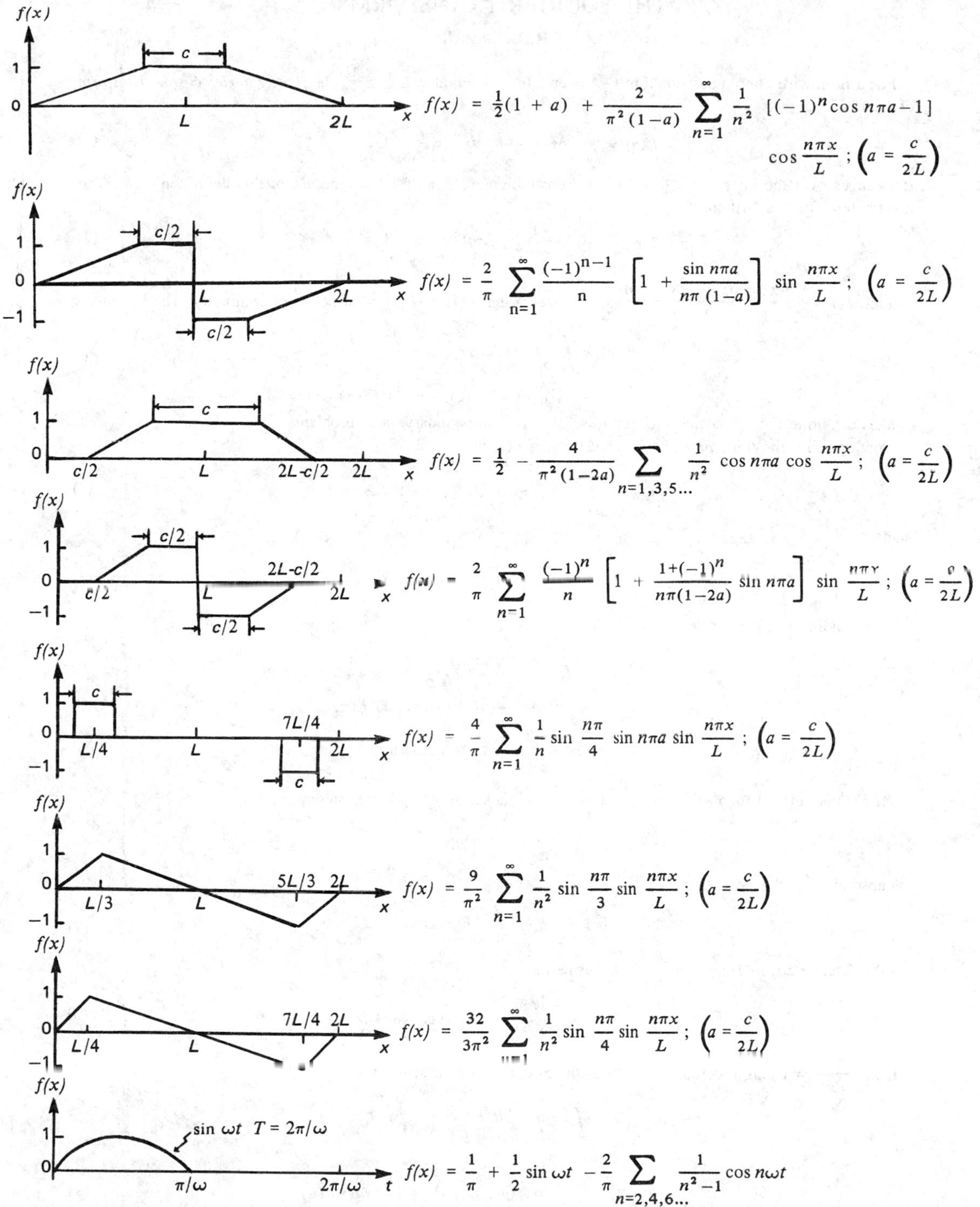

$$f(x) = \frac{1}{2}(1+a) + \frac{2}{\pi^2(1-a)}\sum_{n=1}^{\infty}\frac{1}{n^2}\left[(-1)^n\cos n\pi a - 1\right]\cos\frac{n\pi x}{L};\quad \left(a=\frac{c}{2L}\right)$$

$$f(x) = \frac{2}{\pi}\sum_{n=1}^{\infty}\frac{(-1)^{n-1}}{n}\left[1+\frac{\sin n\pi a}{n\pi(1-a)}\right]\sin\frac{n\pi x}{L};\quad \left(a=\frac{c}{2L}\right)$$

$$f(x) = \frac{1}{2} - \frac{4}{\pi^2(1-2a)}\sum_{n=1,3,5\ldots}\frac{1}{n^2}\cos n\pi a\cos\frac{n\pi x}{L};\quad \left(a=\frac{c}{2L}\right)$$

$$f(x) = \frac{2}{\pi}\sum_{n=1}^{\infty}\frac{(-1)^n}{n}\left[1+\frac{1+(-1)^n}{n\pi(1-2a)}\sin n\pi a\right]\sin\frac{n\pi x}{L};\quad \left(a=\frac{c}{2L}\right)$$

$$f(x) = \frac{4}{\pi}\sum_{n=1}^{\infty}\frac{1}{n}\sin\frac{n\pi}{4}\sin n\pi a\sin\frac{n\pi x}{L};\quad \left(a=\frac{c}{2L}\right)$$

$$f(x) = \frac{9}{\pi^2}\sum_{n=1}^{\infty}\frac{1}{n^2}\sin\frac{n\pi}{3}\sin\frac{n\pi x}{L};\quad \left(a=\frac{c}{2L}\right)$$

$$f(x) = \frac{32}{3\pi^2}\sum_{n=1}^{\infty}\frac{1}{n^2}\sin\frac{n\pi}{4}\sin\frac{n\pi x}{L};\quad \left(a=\frac{c}{2L}\right)$$

$$f(x) = \frac{1}{\pi} + \frac{1}{2}\sin\omega t - \frac{2}{\pi}\sum_{n=2,4,6\ldots}\frac{1}{n^2-1}\cos n\omega t$$

Extracted from graphs and formulas, pages 372, 373, Differential Equations in Engineering Problems, Salvadori and Schwarz, published by Prentice-Hall, Inc., 1954.

THE FOURIER TRANSFORMS*

R. E. Gaskell

For a piecewise continuous function $F(x)$ over a finite interval $0 \leqq x \leqq \pi$, the *finite Fourier cosine transform* of $F(x)$ is

$$f_c(n) = \int_0^\pi F(x)\cos nx\,dx \quad (n = 0, 1, 2, \ldots) \tag{1}$$

If x ranges over the interval $0 \leqq x \leqq L$, the substitution $x' = \pi x/L$ allows the use of this definition, also. The inverse transform is written.

$$\bar{F}(x) = \frac{1}{\pi} f_c(0) - \frac{2}{\pi}\sum_{n=1}^{\infty} f_c(n)\cos nx \quad (0 < x < \pi) \tag{2}$$

where $F(x) = \dfrac{[F(x + o) + F(x - o)]}{2}$. We observe that $F(x) = F(x) =$ at points of continuity. The formula

$$\begin{aligned} f_c^{(2)}(n) &= \int_0^\pi F''(x)\cos nx\,dx \\ &= -n^2 f_c(n) - F'(0) + (-1)^n F'(\pi) \end{aligned} \tag{3}$$

makes the finite Fourier cosine transform useful in certain boundary value problems.

Analogously, the *finite Fourier sine transform* of $F(x)$ is

$$f_s(n) = \int_0^\pi F(x)\sin nx\,dx \quad (n = 1, 2, 3, \ldots) \tag{4}$$

and

$$\bar{F}(x) = \frac{2}{\pi}\sum_{n=1}^{\infty} f_s(n)\sin nx \quad (0 < x < \pi) \tag{5}$$

Corresponding to (3) we have

$$\begin{aligned} f_s^{(2)}(n) &= \int_0^\pi F''(x)\sin nx\,dx \\ &= -n^2 f_s(n) - nF(0) - n(-1)^n F(\pi) \end{aligned} \tag{6}$$

Fourier Transforms

If $F(x)$ is defined for $x \geqq 0$ and is piecewise continuous over any finite interval, and if

$$\int_0^\infty F(x)\,dx$$

is absolutely convergent, then

$$f_c(\alpha) = \sqrt{\frac{2}{\pi}}\int_0^\infty F(x)\cos(\alpha x)\,dx \tag{7}$$

is the *Fourier cosine transform* of $F(x)$. Furthermore.

$$\bar{F}(x) = \sqrt{\frac{2}{\pi}}\int_0^\infty f_c(\alpha)\cos(\alpha x)\,d\alpha \tag{8}$$

If $\lim\limits_{x\to\infty} \dfrac{d^n F}{dx^n} = 0$, an important property of the Fourier cosine transform

$$\begin{aligned} f_c^{(2r)}(\alpha) &= \sqrt{\frac{2}{\pi}}\int_0^\infty \left(\frac{d^{2r}F}{dx^{2r}}\right)\cos(\alpha x)\,dx \\ &= -\sqrt{\frac{2}{\pi}}\sum_{n=0}^{r-1}(-1)^n a_{2r-2n-1}\alpha^{2n} + (-1)^r \alpha^{2r} f_c(\alpha) \end{aligned} \tag{9}$$

where $\lim\limits_{x\to\infty} \dfrac{d^r F}{dx^r} = a_r$, makes it useful in the solution of many problems.

Under the same conditions.

$$f_s(\alpha) = \sqrt{\frac{2}{\pi}}\int_0^\infty F(x)\sin(\alpha x)\,dx \tag{10}$$

* From Beyer, W. H., Ed., *CRC Handbook of Mathematical Sciences*, 5th ed., CRC Press, Boca Raton, 1978, 592—598. With permission.

defines the *Fourier sine transform* of $F(x)$, and

$$\bar{F}(x) = \sqrt{\frac{2}{\pi}} \int_0^\infty f_s(\alpha) \sin(\alpha x)\, d\alpha \tag{11}$$

Corresponding to (9) we have

$$f_s^{(2r)}(\alpha) = \sqrt{\frac{2}{\pi}} \int_0^\infty \frac{d^{2r} F}{dx^{2r}} \sin(\alpha x)\, dx$$

$$= -\sqrt{\frac{2}{\pi}} \sum_{n=1}^{r} (-1)^n \alpha^{2n-1} a_{2r-2n} + (-1)^{r-1} \alpha^{2r} f_s(\alpha) \tag{12}$$

Similarly, if $F(x)$ is defined for $-\infty < x < \infty$, and if $\int_{-\infty}^{\infty} F(x)\, dx$ is absolutely convergent, then

$$f(\alpha) = \frac{1}{\sqrt{2\pi}} \int_{-\infty}^{\infty} F(x) e^{i\alpha x}\, dx \tag{13}$$

is the *Fourier transform* of $F(x)$, and

$$\bar{F}(x) = \frac{1}{\sqrt{2\pi}} \int_{-\infty}^{\infty} f(\alpha) e^{-i\alpha x}\, d\alpha \tag{14}$$

Also, if

$$\lim_{|x| \to \infty} \left| \frac{d^n F}{dx^n} \right| = 0 \quad (n = 1, 2, \ldots, r-1)$$

then

$$f^{(r)}(\alpha) = \frac{1}{\sqrt{2\pi}} \int_{-\infty}^{\infty} F^{(r)}(x) e^{i\alpha x}\, dx = (-i\alpha)^r f(\alpha) \tag{15}$$

Finite Sine Transforms

	$f_s(n)$	$F(x)$
1	$f_s(n) = \int_0^\pi F(x) \sin nx\, dx \; (n = 1, 2, \cdots)$	$F(x)$
2	$(-1)^{n+1} f_s(n)$	$F(\pi - x)$
3	$\frac{1}{n}$	$\frac{\pi - x}{\pi}$
4	$\frac{(-1)^{n+1}}{n}$	$\frac{x}{\pi}$
5	$\frac{1 - (-1)^n}{n}$	1
6	$\frac{2}{n^2} \sin \frac{n\pi}{2}$	$\begin{cases} x & \text{when } 0 < x < \pi/2 \\ \pi - x & \text{when } \pi/2 < x < \pi \end{cases}$
7	$\frac{(-1)^{n+1}}{n^3}$	$\frac{x(\pi^2 - x^2)}{6\pi}$
8	$\frac{1 - (-1)^n}{n^3}$	$\frac{x(\pi - x)}{2}$
9	$\frac{\pi^2 (-1)^{n-1}}{n} - \frac{2[1 - (-1)^n]}{n^3}$	x^2
10	$\pi(-1)^n \left(\frac{6}{n^3} - \frac{\pi^2}{n} \right)$	x^3
11	$\frac{n}{n^2 + c^2} [1 - (-1)^n e^{c\pi}]$	e^{cx}
12	$\frac{n}{n^2 + c^2}$	$\frac{\sinh c(\pi - x)}{\sinh c\pi}$
13	$\frac{n}{n^2 - k^2} \; (k \neq 0, 1, 2, \cdots)$	$\frac{\sin k(\pi - x)}{\sin k\pi}$
14	$\begin{cases} \frac{\pi}{2} & \text{when } n = m \\ 0 & \text{when } n \neq m \end{cases} \; (m = 1, 2, \cdots)$	$\sin mx$

	$f_s(n)$	$F(x)$
15	$\frac{n}{n^2-k^2}[1-(-1)^n \cos k\pi]$ $(k \neq 1, 2, \cdots)$	$\cos kx$
16	$\begin{cases} \frac{n}{n^2-m^2}[1-(-1)^{n+m}] & \text{when } n \neq m = 1, 2, \cdots \\ 0 & \text{when } n = m \end{cases}$	$\cos mx$
17	$\frac{n}{(n^2-k^2)^2}$ $(k \neq 0, 1, 2, \cdots)$	$\frac{\pi \sin kx}{2k \sin^2 k\pi} - \frac{x \cos k(\pi - x)}{2k \sin k\pi}$
18	$\frac{b^n}{n}$ $(\lvert b \rvert \leq 1)$	$\frac{2}{\pi} \arctan \frac{b \sin x}{1 - b \cos x}$
19	$\frac{1-(-1)^n}{n} b^n$ $(\lvert b \rvert \leq 1)$	$\frac{2}{\pi} \arctan \frac{2b \sin x}{1 - b^2}$

Finite Cosine Transforms

	$f_c(n)$	$F(x)$
1	$f_c(n) = \int_0^\pi F(x) \cos nx \, dx$ $(n = 0, 1, 2, \cdots)$	$F(x)$
2	$(-1)^n f_c(n)$	$F(\pi - x)$
3	0 when $n = 1, 2, \cdots;\ f_c(0) = \pi$	1
4	$\frac{2}{n} \sin \frac{n\pi}{2};\ f_c(0) = 0$	$\begin{cases} 1 \text{ when } 0 < x < \pi/2 \\ -1 \text{ when } \pi/2 < x < \pi \end{cases}$
5	$-\frac{1-(-1)^n}{n^2};\ f_c(0) = \frac{\pi^2}{2}$	x
6	$\frac{(-1)^n}{n^2};\ f_c(0) = \frac{\pi^2}{6}$	$\frac{x^2}{2\pi}$
7	$\frac{1}{n^2};\ f_c(0) = 0$	$\frac{(\pi - x)^2}{2\pi} - \frac{\pi}{6}$
8	$3\pi^2 \frac{(-1)^n}{n^2} - 6\frac{1-(-1)^n}{n^4};\ f_c(0) = \frac{\pi^4}{4}$	x^3
9	$\frac{(-1)^n e^c \pi - 1}{n^2 + c^2}$	$\frac{1}{c} e^{cx}$
10	$\frac{1}{n^2 + c^2}$	$\frac{\cosh c(\pi - x)}{c \sinh c\pi}$
11	$\frac{k}{n^2 - k^2}[(-1)^n \cos \pi k - 1]$ $(k \neq 0, 1, 2, \cdots)$	$\sin kx$
12	$\frac{(-1)^{n+m} - 1}{n^2 - m^2};\ f_c(m) = 0$ $(m = 1, 2, \cdots)$	$\frac{1}{m} \sin mx$
13	$\frac{1}{n^2 - k^2}$ $(k \neq 0, 1, 2, \cdots)$	$-\frac{\cos k(\pi - x)}{k \sin k\pi}$
14	0 when $n = 1, 2, \cdots;$ $f_c(m) = \frac{\pi}{2}$ $(m = 1, 2, \cdots)$	$\cos mx$

Fourier Sine Transforms*

	$F(x)$	$f_s(\alpha)$
1	$\begin{cases} 1 & (0 < x < a) \\ 0 & (x > a) \end{cases}$	$\sqrt{\frac{2}{\pi}}\left[\frac{1 - \cos \alpha}{\alpha}\right]$
2	x^{p-1} $(0 < p < 1)$	$\sqrt{\frac{2}{\pi}} \frac{\Gamma(p)}{\alpha^p} \sin \frac{p\pi}{2}$
3	$\begin{cases} \sin x & (0 < x < a) \\ 0 & (x > a) \end{cases}$	$\frac{1}{\sqrt{2\pi}}\left[\frac{\sin[a(1-\alpha)]}{1-\alpha} - \frac{\sin[a(1+\alpha)]}{1+\alpha}\right]$
4	e^{-x}	$\sqrt{\frac{2}{\pi}}\left[\frac{\alpha}{1+\alpha^2}\right]$
5	$xe^{-x^2/2}$	$\alpha e^{-\alpha^2/2}$
6	$\cos \frac{x^2}{2}$	$\sqrt{2}\left[\sin \frac{\alpha^2}{2} C\left(\frac{\alpha^2}{2}\right) - \cos \frac{\alpha^2}{2} S\left(\frac{\alpha^2}{2}\right)\right]^*$

	$F(x)$	$f_s(\alpha)$
7	$\sin \frac{x^2}{2}$	$\sqrt{2}\left[\cos \frac{\alpha^2}{2}\, C\left(\frac{\alpha^2}{2}\right) + \sin \frac{\alpha^2}{2}\, S\left(\frac{\alpha^2}{2}\right)\right]^*$

* $C(y)$ and $S(y)$ are the Fresnel integrals

$$C(y) = \frac{1}{\sqrt{2\pi}} \int_0^y \frac{1}{\sqrt{t}} \cos t \, dt,$$

$$S(y) = \frac{1}{\sqrt{2\pi}} \int_0^y \frac{1}{\sqrt{t}} \sin t \, dt$$

* More extensive tables of the Fourier sine and cosine transforms can be found in Fritz Oberhettinger, *Tabellen zur-Fourier Transformation*, Springer, 1957.

Fourier Cosine Transforms

	$F(x)$	$f_c(\alpha)$
1	$\begin{cases} 1 & (0 < x < a) \\ 0 & (x > a) \end{cases}$	$\sqrt{\frac{2}{\pi}} \frac{\sin a\alpha}{\alpha}$
2	$x^{p-1} \quad (0 < p < 1)$	$\sqrt{\frac{2}{\pi}} \frac{\Gamma(p)}{\alpha^p} \cos \frac{p\pi}{2}$
3	$\begin{cases} \cos x & (0 < x < a) \\ 0 & (x > a) \end{cases}$	$\frac{1}{\sqrt{2\pi}}\left[\frac{\sin[a(1-\alpha)]}{1-\alpha} + \frac{\sin[a(1+\alpha)]}{1+\alpha}\right]$
4	e^{-x}	$\sqrt{\frac{2}{\pi}}\left(\frac{1}{1+\alpha^2}\right)$
5	$e^{-x^2/2}$	$e^{-\alpha^2/2}$
6	$\cos \frac{x^2}{2}$	$\cos\left(\frac{\alpha^2}{2} - \frac{\pi}{4}\right)$
7	$\sin \frac{x^2}{2}$	$\cos\left(\frac{\alpha^2}{2} + \frac{\pi}{4}\right)$

Fourier Transforms*

	$F(x)$	$f(\alpha)$
1	$\frac{\sin ax}{x}$	$\begin{cases} \sqrt{\frac{\pi}{2}} & \lvert\alpha\rvert < a \\ 0 & \lvert\alpha\rvert > a \end{cases}$
2	$\begin{cases} e^{iwx} & (p < x < q) \\ 0 & (x < p,\ x > q) \end{cases}$	$\frac{i}{\sqrt{2\pi}} \frac{e^{ip(w+\alpha)} - e^{iq(w+\alpha)}}{(w+\alpha)}$
3	$\begin{cases} e^{-cx+iwx} & (x > 0) \\ 0 & (x < 0) \end{cases} \quad (c > 0)$	$\frac{i}{\sqrt{2\pi}(w + \alpha + ic)}$
4	$e^{-px^2} \quad R(p) > 0$	$\frac{1}{\sqrt{2p}} e^{-\alpha^2/4p}$
5	$\cos px^2$	$\frac{1}{\sqrt{2p}} \cos\left[\frac{\alpha^2}{4p} - \frac{\pi}{4}\right]$
6	$\sin px^2$	$\frac{1}{\sqrt{2p}} \cos\left[\frac{\alpha^2}{4p} + \frac{\pi}{4}\right]$
7	$\lvert x\rvert^{-p} \quad (0 < p < 1)$	$\sqrt{\frac{2}{\pi}} \frac{\Gamma(1-p)\sin\frac{p\pi}{2}}{\lvert\alpha\rvert^{(1-p)}}$
8	$\frac{e^{-a\lvert x\rvert}}{\sqrt{\lvert x\rvert}}$	$\frac{\sqrt{\sqrt{(a^2+\alpha^2)} + a}}{\sqrt{a^2+\alpha^2}}$
9	$\frac{\cosh ax}{\cosh \pi x} \quad (-\pi < a < \pi)$	$\sqrt{\frac{2}{\pi}} \frac{\cos\frac{a}{2}\cosh\frac{\alpha}{2}}{\cosh\alpha + \cos a}$
10	$\frac{\sinh ax}{\sinh \pi x} \quad (-\pi < a < \pi)$	$\frac{1}{\sqrt{2\pi}} \frac{\sin a}{\cosh\alpha + \cos a}$
11	$\begin{cases} \frac{1}{\sqrt{a^2 - x^2}} & (\lvert x\rvert < a) \\ 0 & (\lvert x\rvert > a) \end{cases}$	$\sqrt{\frac{\pi}{2}} J_0(a\alpha)$

	$F(x)$	$f(\alpha)$
12	$\dfrac{\sin[b\sqrt{a^2+x^2}]}{\sqrt{a^2+x^2}}$	$\begin{cases} 0 & (\lvert\alpha\rvert > b) \\ \sqrt{\frac{\pi}{2}}\, J_0(a\sqrt{b^2-\alpha^2}) & (\lvert\alpha\rvert < b) \end{cases}$
13	$\begin{cases} P_n(x) & (\lvert x\rvert < 1) \\ 0 & (\lvert x\rvert > 1) \end{cases}$	$\dfrac{i^n}{\sqrt{\alpha}} J_{n+\frac{1}{2}}(\alpha)$
14	$\begin{cases} \dfrac{\cos[b\sqrt{a^2-x^2}]}{\sqrt{a^2-x^2}} & (\lvert x\rvert < a) \\ 0 & (\lvert x\rvert > a) \end{cases}$	$\sqrt{\frac{\pi}{2}}\, J_0(a\sqrt{a^2+b^2})$
15	$\begin{cases} \dfrac{\cosh[b\sqrt{a^2-x^2}]}{\sqrt{a^2-x^2}} & (\lvert x\rvert < a) \\ 0 & (\lvert x\rvert > a) \end{cases}$	$\sqrt{\frac{\pi}{2}}\, J_0(a\sqrt{\alpha^2-b^2})$

* More extensive tables of Fourier transforms can be found in W. Magnus and F. Oberhettinger, *Formulas and Theorems of the Special Functions of Mathematical Physics.* Chelsea, 1949, 116—120.

The following functions appear among the entries of the tables on transforms.

Function	Definition	Name
$Ei(x)$	$\int_{-\infty}^{x} \frac{e^v}{v}\,dv$; or sometimes defined as $-Ei(-x) = \int_x^{\infty} \frac{e^{-v}}{v}\,dv$	Sine, Cosine, and Exponential Integral tables pages 548–556
$Si(x)$	$\int_0^x \frac{\sin v}{v}\,dv$	Sine, Cosine, and Exponential Integral tables pages 548–556
$Ci(x)$	$\int_{\infty}^{x} \frac{\cos v}{v}\,dv$; or sometimes defined as negative of this integral	Sine, Cosine, and Exponential Integral tables pages 548–546
$erf(x)$	$\frac{2}{\sqrt{\pi}} \int_0^x e^{-v^2}\,dv$	Error function
$erfc(x)$	$1 - erf(x) = \frac{2}{\sqrt{\pi}} \int_x^{\infty} e^{-v^2}\,dv$	Complementary function to error function
$L_n(x)$	$\frac{e^x}{n!}\frac{d^n}{dx^n}(x^n e^{-x}),\ n = 0, 1, \cdots$	Laguerre polynomial of degree n

SERIES EXPANSION

The expression in parentheses following certain of the series indicates the region of convergence. If not otherwise indicated it is to be understood that the series converges for all finite values of x.

BINOMIAL

$$(x+y)^n = x^n + nx^{n-1}y + \frac{n(n-1)}{2!}x^{n-2}y^2 + \frac{n(n-1)(n-2)}{3!}x^{n-3}y^3 + \cdots \quad (y^2 < x^2)$$

$$(1 \pm x)^n = 1 \pm nx + \frac{n(n-1)x^2}{2!} \pm \frac{n(n-1)(n-2)x^3}{3!} + \cdots \quad (x^2 < 1)$$

$$(1 \pm x)^{-n} = 1 \mp nx + \frac{n(n+1)x^2}{2!} \mp \frac{n(n+1)(n+2)x^3}{3!} + \cdots \quad (x^2 < 1)$$

$$(1 \pm x)^{-1} = 1 \mp x + x^2 \mp x^3 + x^4 \mp x^5 + \cdots \quad (x^2 < 1)$$
$$(1 \pm x)^{-2} = 1 \mp 2x + 3x^2 \mp 4x^3 + 5x^4 \mp 6x^5 + \cdots \quad (x^2 < 1)$$

REVERSION OF SERIES

Let a series be represented by

$$y = a_1x + a_2x^2 + a_3x^3 + a_4x^4 + a_5x^5 + a_6x^6 + \cdots \quad (a_1 \neq 0)$$

to find the coefficients of the series

$$x = A_1y + A_2y^2 + A_3y^3 + A_4y^4 + \cdots$$

$$A_1 = \frac{1}{a_1} \qquad A_2 = -\frac{a_2}{a_1^3} \qquad A_3 = \frac{1}{a_1^5}(2a_2^2 - a_1a_3)$$

$$A_4 = \frac{1}{a_1^7}(5a_1a_2a_3 - a_1^2a_4 - 5a_2^3)$$

$$A_5 = \frac{1}{a_1^9}(6a_1^2a_2a_4 + 3a_1^2a_3^2 + 14a_2^4 - a_1^3a_5 - 21a_1a_2^2a_3)$$

$$A_6 = \frac{1}{a_1^{11}}(7a_1^3a_2a_5 + 7a_1^3a_3a_4 + 84a_1a_2^3a_3 - a_1^4a_6 - 28a_1^2a_2^2a_4 - 28a_1^2a_2a_3^2 - 42a_2^5)$$

$$A_7 = \frac{1}{a_1^{13}}(8a_1^4a_2a_6 + 8a_1^4a_3a_5 + 4a_1^4a_4^2 + 120a_1^2a_2^3a_4 + 180a_1^2a_2^2a_3^2 + 132a_2^6 - a_1^5a_7 - 36a_1^3a_2^2a_5 - 72a_1^3a_2a_3a_4 - 12a_1^3a_3^3 - 330a_1a_2^4a_3)$$

TAYLOR

1. $$f(x) = f(a) + (x-a)f'(a) + \frac{(x-a)^2}{2!}f''(a) + \frac{(x-a)^3}{3!}f'''(a) + \cdots + \frac{(x-a)^n}{n!}f^{(n)}(a) + \cdots \text{ (Taylor's Series)}$$

(Increment form)

2. $$f(x+h) = f(x) + hf'(x) + \frac{h^2}{2!}f''(x) + \frac{h^3}{3!}f'''(x) + \cdots$$
$$= f(h) + xf'(h) + \frac{x^2}{2!}f''(h) + \frac{x^3}{3!}f'''(h) + \cdots$$

3. If $f(x)$ is a function possessing derivatives of all orders throughout the interval $a \leqq x \leqq b$, then there is a value X, with $a < X < b$, such that

$$f(b) = f(a) + (b-a)f'(a) + \frac{(b-a)^2}{2!}f''(a) + \cdots + \frac{(b-a)^{n-1}}{(n-1)!}f^{(n-1)}(a) + \frac{(b-a)^n}{n!}f^{(n)}(X)$$

$$f(a+h) = f(a) + hf'(a) + \frac{h^2}{2!}f''(a) + \cdots + \frac{h^{n-1}}{(n-1)!}f^{(n-1)}(a) + \frac{h^n}{n!}f^{(n)}(a+\theta h), \quad b = a+h,\ 0 < \theta < 1.$$

or

$$f(x) = f(a) + (x-a)f'(a) + \frac{(x-a)^2}{2!}f''(a) + \cdots + (x-a)^{n-1}\frac{f^{(n-1)}(a)}{(n-1)!} + R_n,$$

where

$$R_n = \frac{f^{(n)}[a + \theta\cdot(x-a)]}{n!}(x-a)^n, \quad 0 < \theta < 1.$$

The above forms are known as Taylor's series with the remainder term.

4. *Taylor's series for a function of two variables*

If $\left(h\frac{\partial}{\partial x} + k\frac{\partial}{\partial y}\right)f(x,y) = h\frac{\partial f(x,y)}{\partial x} + k\frac{\partial f(x,y)}{\partial y}$;

$$\left(h\frac{\partial}{\partial x} + k\frac{\partial}{\partial y}\right)^2 f(x,y) = h^2\frac{\partial^2 f(x,y)}{\partial x^2} + 2hk\frac{\partial^2 f(x,y)}{\partial x\,\partial y} + k^2\frac{\partial^2 f(x,y)}{\partial y^2}$$

etc., and if $\left(h\frac{\partial}{\partial x} + k\frac{\partial}{\partial y}\right)^n f(x,y)\Big|_{\substack{x=a\\y=b}}$ with the bar and subscripts means that after differentiation we are to replace x by a and y by b,

$$f(a+h, b+k) = f(a,b) + \left(h\frac{\partial}{\partial x} + k\frac{\partial}{\partial y}\right)f(x,y)\Big|_{\substack{x=a\\y=b}} + \cdots + \frac{1}{n!}\left(h\frac{\partial}{\partial x} + k\frac{\partial}{\partial y}\right)^n f(x,y)\Big|_{\substack{x=a\\y=b}} + \cdots$$

MACLAURIN

$$f(x) = f(0) + xf'(0) + \frac{x^2}{2!}f''(0) + \frac{x^3}{3!}f'''(0) + \cdots + x^{n-1}\frac{f^{(n-1)}(0)}{(n-1)!} + R_n,$$

where

$$R_n = \frac{x^n f^{(n)}(\theta x)}{n!}, \quad 0 < \theta < 1.$$

EXPONENTIAL

$$e = 1 + \frac{1}{1!} + \frac{1}{2!} + \frac{1}{3!} + \frac{1}{4!} + \cdots$$

$$e^x = 1 + x + \frac{x^2}{2!} + \frac{x^3}{3!} + \frac{x^4}{4!} + \cdots \qquad \text{(all real values of } x\text{)}$$

$$a^x = 1 + x \log_e a + \frac{(x \log_e a)^2}{2!} + \frac{(x \log_e a)^3}{3!} + \cdots$$

$$e^x = e^a\left[1 + (x - a) + \frac{(x - a)^2}{2!} + \frac{(x - a)^3}{3!} + \cdots\right]$$

LOGARITHMIC

$$\log_e x = \frac{x-1}{x} + \frac{1}{2}\left(\frac{x-1}{x}\right)^2 + \frac{1}{3}\left(\frac{x-1}{x}\right)^3 + \cdots \qquad (x > \tfrac{1}{2})$$

$$\log_e x = (x - 1) - \tfrac{1}{2}(x - 1)^2 + \tfrac{1}{3}(x - 1)^3 - \cdots \qquad (2 \geq x > 0)$$

$$\log_e x = 2\left[\frac{x-1}{x+1} + \frac{1}{3}\left(\frac{x-1}{x+1}\right)^3 + \frac{1}{5}\left(\frac{x-1}{x+1}\right)^5 + \cdots\right] \qquad (x > 0)$$

$$\log_e(1 + x) = x - \tfrac{1}{2}x^2 + \tfrac{1}{3}x^3 - \tfrac{1}{4}x^4 + \cdots \qquad (-1 < x \leq 1)$$

$$\log_e(n + 1) - \log_e(n - 1) = 2\left[\frac{1}{n} + \frac{1}{3n^3} + \frac{1}{5n^5} + \cdots\right]$$

$$\log_e(a + x) = \log_e a + 2\left[\frac{x}{2a + x} + \frac{1}{3}\left(\frac{x}{2a + x}\right)^3 + \frac{1}{5}\left(\frac{x}{2a + x}\right)^5 + \cdots\right]$$

$$(a > 0, -a < x < +\infty)$$

$$\log_e \frac{1 + x}{1 - x} = 2\left[x + \frac{x^3}{3} + \frac{x^5}{5} + \cdots + \frac{x^{2n-1}}{2n - 1} + \cdots\right], \qquad -1 < x < 1$$

$$\log_e x = \log_e a + \frac{(x - a)}{a} - \frac{(x - a)^2}{2a^2} + \frac{(x - a)^3}{3a^3} - + \cdots, \qquad 0 < x \leqq 2a$$

TRIGONOMETRIC

$$\sin x = x - \frac{x^3}{3!} + \frac{x^5}{5!} - \frac{x^7}{7!} + \cdots \qquad \text{(all real values of } x\text{)}$$

$$\cos x = 1 - \frac{x^2}{2!} + \frac{x^4}{4!} - \frac{x^6}{6!} + \cdots \qquad \text{(all real values of } x\text{)}$$

$$\tan x = x + \frac{x^3}{3} + \frac{2x^5}{15} + \frac{17x^7}{315} + \frac{62x^9}{2835} + \cdots + \frac{(-1)^{n-1}\, 2^{2n}\,(2^{2n} - 1)\, B_{2n}}{(2n)!} x^{2n-1} + \cdots,$$

$$\left[x^2 < \frac{\pi^2}{4}, \text{ and } B_n \text{ represents the } n\text{th Bernoulli number.}\right]$$

$$\cot x = \frac{1}{x} - \frac{x}{3} - \frac{x^3}{45} - \frac{2x^5}{945} - \frac{x^7}{4725} - \cdots - \frac{(-1)^{n+1}2^{2n}}{(2n)!} B_{2n}x^{2n-1} - \cdots,$$

$$[x^2 < \pi^2, \text{ and } B_n \text{ represents the } n\text{th Bernoulli number.}]$$

$$\sec x = 1 + \frac{x^2}{2} + \frac{5}{24}x^4 + \frac{61}{720}x^6 + \frac{277}{8064}x^8 + \cdots + \frac{(-1)^n}{(2n)!} E_{2n}x^{2n} + \cdots,$$

$$\left[x^2 < \frac{\pi^2}{4}, \text{ and } E_n \text{ represents the } n\text{th Euler number.}\right]$$

$$\csc x = \frac{1}{x} + \frac{x}{6} + \frac{7}{360}x^3 + \frac{31}{15{,}120}x^5 + \frac{127}{604{,}800}x^7 + \cdots$$

$$+ \frac{(-1)^{n+1}2(2^{2n-1} - 1)}{(2n)!} B_{2n}x^{2n-1} + \cdots,$$

$$[x^2 < \pi^2, \text{ and } B_n \text{ represents } n\text{th Bernoulli number.}]$$

$$\sin x = x\left(1 - \frac{x^2}{\pi^2}\right)\left(1 - \frac{x^2}{2^2\pi^2}\right)\left(1 - \frac{x^2}{3^2\pi^2}\right)\cdots \qquad (x^2 < \infty)$$

$$\cos x = \left(1 - \frac{4x^2}{\pi^2}\right)\left(1 - \frac{4x^2}{3^2\pi^2}\right)\left(1 - \frac{4x^2}{5^2\pi^2}\right)\cdots \qquad (x^2 < \infty)$$

$$\sin^{-1} x = x + \frac{x^3}{2\cdot 3} + \frac{1\cdot 3}{2\cdot 4\cdot 5}x^5 + \frac{1\cdot 3\cdot 5}{2\cdot 4\cdot 6\cdot 7}x^7 + \cdots \qquad \left(x^2 < 1, -\frac{\pi}{2} < \sin^{-1} x < \frac{\pi}{2}\right)$$

$$\cos^{-1} x = \frac{\pi}{2} - \left(x + \frac{x^3}{2\cdot 3} + \frac{1\cdot 3}{2\cdot 4\cdot 5}x^5 + \frac{1\cdot 3\cdot 5x^7}{2\cdot 4\cdot 6\cdot 7} + \cdots\right) \qquad (x^2 < 1, 0 < \cos^{-1} x < \pi)$$

$$\tan^{-1} x = x - \frac{x^3}{3} + \frac{x^5}{5} - \frac{x^7}{7} + \cdots \qquad (x^2 < 1)$$

$$\tan^{-1} x = \frac{\pi}{2} - \frac{1}{x} + \frac{1}{3x^3} - \frac{1}{5x^5} + \frac{1}{7x^7} - \cdots \qquad (x > 1)$$

$$\tan^{-1} x = -\frac{\pi}{2} - \frac{1}{x} + \frac{1}{3x^3} - \frac{1}{5x^5} + \frac{1}{7x^7} - \cdots \qquad (x < -1)$$

$$\cot^{-1} x = \frac{\pi}{2} - x + \frac{x^3}{3} - \frac{x^5}{5} + \frac{x^7}{7} - \cdots \qquad (x^2 < 1)$$

$$\log_e \sin x = \log_e x - \frac{x^2}{6} - \frac{x^4}{180} - \frac{x^6}{2835} - \cdots \qquad (x^2 < \pi^2)$$

$$\log_e \cos x = -\frac{x^2}{2} - \frac{x^4}{12} - \frac{x^6}{45} - \frac{17x^8}{2520} - \cdots \qquad \left(x^2 < \frac{\pi^2}{4}\right)$$

$$\log_e \tan x = \log_e x + \frac{x^2}{3} + \frac{7x^4}{90} + \frac{62x^6}{2835} + \cdots \qquad \left(x^2 < \frac{\pi^2}{4}\right)$$

$$e^{\sin x} = 1 + x + \frac{x^2}{2!} - \frac{3x^4}{4!} - \frac{8x^5}{5!} - \frac{3x^6}{6!} + \frac{56x^7}{7!} + \cdots$$

$$e^{\cos x} = e\left(1 - \frac{x^2}{2!} + \frac{4x^4}{4!} - \frac{31x^6}{6!} + \cdots\right)$$

$$e^{\tan x} = 1 + x + \frac{x^2}{2!} + \frac{3x^3}{3!} + \frac{9x^4}{4!} + \frac{37x^5}{5!} + \cdots \qquad \left(x^2 < \frac{\pi^2}{4}\right)$$

$$\sin x = \sin a + (x - a)\cos a - \frac{(x-a)^2}{2!}\sin a - \frac{(x-a)^3}{3!}\cos a + \frac{(x-a)^4}{4!}\sin a + \cdots$$

VECTOR ANALYSIS

Definitions

Any quantity which is completely determined by its magnitude is called a *scalar*. Examples of such are mass, density, temperature, etc. Any quantity which is completely determined by its magnitude and direction is called a *vector*. Examples of such are velocity, acceleration, force, etc. A vector quantity is represented by a directed line segment, the length of which represents the magnitude of the vector. A vector quantity is usually represented by a boldfaced letter such as **V**. Two vectors $\mathbf{V}_1$ and $\mathbf{V}_2$ are equal to one another if they have equal magnitudes and are acting in the same directions. A negative vector, written as $-\mathbf{V}$, is one which acts in the opposite direction to **V**, but is of equal magnitude to it. If we represent the magnitude of **V** by v, we write $|\mathbf{V}| = v$. A vector parallel to **V**, but equal to the reciprocal of its magnitude is written as $\mathbf{V}^{-1}$ or as $\frac{1}{\mathbf{V}}$.

The *unit vector* $\frac{\mathbf{V}}{v}$ $(v \neq 0)$ is that vector which has the same direction as **V**, but has a magnitude of unity (sometimes represented as $\mathbf{V}_0$ or $\hat{\mathbf{v}}$).

Vector Algebra

The vector sum of $\mathbf{V}_1$ and $\mathbf{V}_2$ is represented by $\mathbf{V}_1 + \mathbf{V}_2$. The vector sum of $\mathbf{V}_1$ and $-\mathbf{V}_2$, or the difference of the vector $\mathbf{V}_2$ from $\mathbf{V}_1$ is represented by $\mathbf{V}_1 - \mathbf{V}_2$.

If r is a scalar, then $r\mathbf{V} = \mathbf{V}r$, and represents a vector r times the magnitude of **V**, in the same direction as **V** if r is positive, and in the opposite direction if r is negative. If r and s are scalars, $\mathbf{V}_1$, $\mathbf{V}_2$, $\mathbf{V}_3$, vectors, then the following rules of scalars and vectors hold:

$$\mathbf{V}_1 + \mathbf{V}_2 = \mathbf{V}_2 + \mathbf{V}_1$$

$$(r + s)\mathbf{V}_1 = r\mathbf{V}_1 + s\mathbf{V}_1; \qquad r(\mathbf{V}_1 + \mathbf{V}_2) = r\mathbf{V}_1 + r\mathbf{V}_2$$

$$\mathbf{V}_1 + (\mathbf{V}_2 + \mathbf{V}_3) = (\mathbf{V}_1 + \mathbf{V}_2) + \mathbf{V}_3 = \mathbf{V}_1 + \mathbf{V}_2 + \mathbf{V}_3$$

Vectors in Space

A plane is described by two distinct vectors $\mathbf{V}_1$ and $\mathbf{V}_2$. Should these vectors not intersect each other, [illegible] is displaced parallel to itself until they do (fig. 1.) Any other vector **V** lying in this plane is given by

$$\mathbf{V} = r\mathbf{V}_1 + s\mathbf{V}_2$$

A *position vector* specifies the position in space of a point relative to a fixed origin. If therefore $\mathbf{V}_1$ and $\mathbf{V}_2$ are the position vectors of the points A and B, relative to the origin O, then any point P on the line AB has a position vector **V** given by

$$\mathbf{V} = r\mathbf{V}_1 + (1 - r)\mathbf{V}_2$$

The scalar "r" can be taken as the parametric representation of P since $r = 0$ implies $P = B$ and $r = 1$ implies $P = A$. (fig. 2). If P divides the line AB in the ratio $r{:}s$ then

$$\mathbf{V} = \left(\frac{r}{r+s}\right)\mathbf{V}_1 + \left(\frac{s}{r+s}\right)\mathbf{V}_2$$

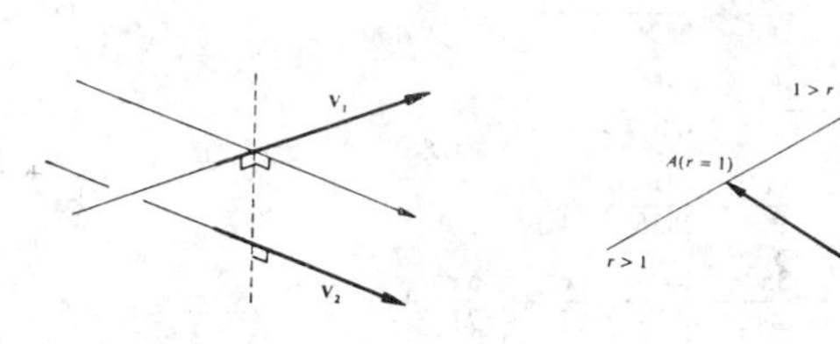

Fig. 1.

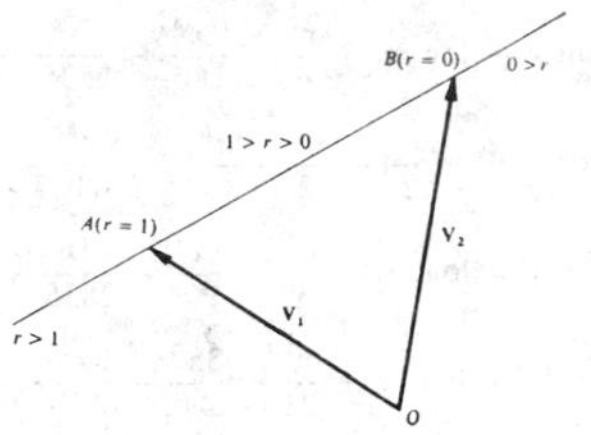

Fig. 2.

The vectors $\mathbf{V}_1, \mathbf{V}_2, \mathbf{V}_3, \ldots, \mathbf{V}_n$ are said to be *linearly dependent* if there exist scalars $r_1, r_2, r_3, \ldots, r_n$, not all zero, such that

$$r_1\mathbf{V}_1 + r_2\mathbf{V}_2 + \cdots + r_n\mathbf{V}_n = 0$$

A vector $\mathbf{V}$ is linearly dependent upon the set of vectors $\mathbf{V}_1, \mathbf{V}_2, \mathbf{V}_3, \ldots, \mathbf{V}_n$ if

$$\mathbf{V} = r_1\mathbf{V}_1 + r_2\mathbf{V}_2 + r_3\mathbf{V}_3 + \cdots + r_n\mathbf{V}_n$$

Three vectors are linearly dependent if and only if they are co-planar.

All points in space can be uniquely determined by linear dependence upon three *base vectors* i.e. three vectors any one of which is linearly independent of the other two. The simplest set of base vectors are the unit vectors along the coordinate Ox, Oy and Oz axes. These are usually designated by **i**, **j** and **k** respectively.

If **V** is a vector in space, and a, b and c are the respective magnitudes of the projections of the vector along the axes then

$$\mathbf{V} = a\mathbf{i} + b\mathbf{j} + c\mathbf{k}$$

and

$$v = \sqrt{a^2 + b^2 + c^2}$$

and the direction cosines of **V** are

$$\cos\alpha = a/v, \quad \cos\beta = b/v, \quad \cos\gamma = c/v.$$

The law of addition yields

$$\mathbf{V}_1 + \mathbf{V}_2 = (a_1 + a_2)\mathbf{i} + (b_1 + b_2)\mathbf{j} + (c_1 + c_2)\mathbf{k}$$

The Scalar, Dot, or Inner Product of Two Vectors $\mathbf{V}_1$ and $\mathbf{V}_2$

This product is represented as $\mathbf{V}_1 \cdot \mathbf{V}_2$ and is defined to be equal to $v_1v_2\cos\theta$, where θ is the angle from $\mathbf{V}_1$ to $\mathbf{V}_2$, i.e.,

$$\mathbf{V}_1 \cdot \mathbf{V}_2 = v_1v_2\cos\theta$$

The following rules apply for this product:

$$\mathbf{V}_1 \cdot \mathbf{V}_2 = a_1a_2 + b_1b_2 + c_1c_2 = \mathbf{V}_2 \cdot \mathbf{V}_1$$

It should be noted that this verifies that scalar multiplication is commutative.

$$(\mathbf{V}_1 + \mathbf{V}_2) \cdot \mathbf{V}_3 = \mathbf{V}_1 \cdot \mathbf{V}_3 + \mathbf{V}_2 \cdot \mathbf{V}_3$$
$$\mathbf{V}_1 \cdot (\mathbf{V}_2 + \mathbf{V}_3) = \mathbf{V}_1 \cdot \mathbf{V}_2 + \mathbf{V}_1 \cdot \mathbf{V}_3$$

If $\mathbf{V}_1$ is perpendicular to $\mathbf{V}_2$ then $\mathbf{V}_1 \cdot \mathbf{V}_2 = 0$, and if $\mathbf{V}_1$ is parallel to $\mathbf{V}_2$ then $\mathbf{V}_1 \cdot \mathbf{V}_2 = v_1v_2 = rw_1^2$

In particular

$$\mathbf{i} \cdot \mathbf{i} = \mathbf{j} \cdot \mathbf{j} = \mathbf{k} \cdot \mathbf{k} = 1,$$

and

$$\mathbf{i} \cdot \mathbf{j} = \mathbf{j} \cdot \mathbf{k} = \mathbf{k} \cdot \mathbf{i} = 0$$

The Vector or Cross Product of Vectors $\mathbf{V}_1$ and $\mathbf{V}_2$

This product is represented as $\mathbf{V}_1 \times \mathbf{V}_2$ and is defined to be equal to $v_1v_2(\sin\theta)\mathbf{1}$, where θ is the angle from $\mathbf{V}_1$ to $\mathbf{V}_2$ and **1** is a unit vector perpendicular to the plane of $\mathbf{V}_1$ and $\mathbf{V}_2$ and so directed that a right-handed screw driven in the direction of **1** would carry $\mathbf{V}_1$ into $\mathbf{V}_2$, i.e.,

$$\mathbf{V}_1 \times \mathbf{V}_2 = v_1v_2(\sin\theta)\mathbf{1}$$

and

$$\tan\theta = \frac{|\mathbf{V}_1 \times \mathbf{V}_2|}{\mathbf{V}_1 \cdot \mathbf{V}_2}$$

The following rules apply for vector products:

$$\mathbf{V}_1 \times \mathbf{V}_2 = -\mathbf{V}_2 \times \mathbf{V}_1$$
$$\mathbf{V}_1 \times (\mathbf{V}_2 + \mathbf{V}_3) = \mathbf{V}_1 \times \mathbf{V}_2 + \mathbf{V}_1 \times \mathbf{V}_3$$
$$(\mathbf{V}_1 + \mathbf{V}_2) \times \mathbf{V}_3 = \mathbf{V}_1 \times \mathbf{V}_3 + \mathbf{V}_2 \times \mathbf{V}_3$$
$$\mathbf{V}_1 \times (\mathbf{V}_2 \times \mathbf{V}_3) = \mathbf{V}_2(\mathbf{V}_3 \cdot \mathbf{V}_1 - \mathbf{V}_3(\mathbf{V}_1 \cdot \mathbf{V}_2)$$
$$\mathbf{i} \times \mathbf{i} = \mathbf{j} \times \mathbf{j} = \mathbf{k} \times \mathbf{k} = 0.\mathbf{1} \text{ (zero vector)}$$
$$= \mathbf{0}$$

$$\mathbf{i} \times \mathbf{j} = \mathbf{k}, \qquad \mathbf{j} \times \mathbf{k} = \mathbf{i}, \qquad \mathbf{k} \times \mathbf{i} = \mathbf{j}$$

If $\mathbf{V}_1 = a_1\mathbf{i} + b_1\mathbf{j} + c_1\mathbf{k}, \qquad \mathbf{V}_2 = a_2\mathbf{i} + b_2\mathbf{j} + c_2\mathbf{k}, \qquad \mathbf{V}_3 = a_3\mathbf{i} + b_3\mathbf{j} + c_3\mathbf{k},$

then

$$\mathbf{V}_1 \times \mathbf{V}_2 = \begin{vmatrix} \mathbf{i} & \mathbf{j} & \mathbf{k} \\ a_1 & b_1 & c_1 \\ a_2 & b_2 & c_2 \end{vmatrix} = (b_1c_2 - b_2c_1)\mathbf{i} + (c_1a_2 - c_2a_1)\mathbf{j} + (a_1b_2 - a_2b_1)\mathbf{k}$$

It should be noted that, since $\mathbf{V}_1 \times \mathbf{V}_2 = -\mathbf{V}_2 \times \mathbf{V}_1$, the vector product is not commutative.

Scalar Triple Product

There is only one possible interpretation of the expression $\mathbf{V}_1 \cdot \mathbf{V}_2 \times \mathbf{V}_3$ and that is $\mathbf{V}_1 \cdot (\mathbf{V}_2 \times \mathbf{V}_3)$ which is obviously a scalar.

Further $\mathbf{V}_1 \cdot (\mathbf{V}_2 \times \mathbf{V}_3) = (\mathbf{V}_1 \times \mathbf{V}_2) \cdot \mathbf{V}_3 = \mathbf{V}_2 \cdot (\mathbf{V}_3 \times \mathbf{V}_1)$

$$= \begin{vmatrix} a_1 & b_1 & c_1 \\ a_2 & b_2 & c_2 \\ a_3 & b_3 & c_3 \end{vmatrix}$$

$$= v_1v_2v_3 \cos\phi \sin\theta,$$

Where θ is the angle between $\mathbf{V}_2$ and $\mathbf{V}_3$ and ϕ is the angle between $\mathbf{V}_1$ and the normal to the plane of $\mathbf{V}_2$ and $\mathbf{V}_3$.

This product is called the *scalar triple product* and is written as $[\mathbf{V}_1\mathbf{V}_2\mathbf{V}_3]$.

The determinant indicates that it can be considered as the volume of the parallelepiped whose three determining edges are $\mathbf{V}_1$, $\mathbf{V}_2$ and $\mathbf{V}_3$.

It also follows that cyclic permutation of the subscripts does not change the value of the scalar triple product so that

$$[\mathbf{V}_1\mathbf{V}_2\mathbf{V}_3] = [\mathbf{V}_2\mathbf{V}_3\mathbf{V}_1] = [\mathbf{V}_3\mathbf{V}_1\mathbf{V}_2]$$

$$\text{but } [\mathbf{V}_1\mathbf{V}_2\mathbf{V}_3] = -[\mathbf{V}_2\mathbf{V}_1\mathbf{V}_3] \text{ etc.} \qquad \text{and } [\mathbf{V}_1\mathbf{V}_1\mathbf{V}_2] \equiv 0 \text{ etc.}$$

Given three non-coplanar reference vectors $\mathbf{V}_1$, $\mathbf{V}_2$ and $\mathbf{V}_3$, the *reciprocal system is* given by $\mathbf{V}_1^*$, $\mathbf{V}_2^*$ and $\mathbf{V}_3^*$, where

$$1 = v_1v_1^* = v_2v_2^* = v_3v_3^*$$

$$0 = v_1v_2^* = v_1v_3^* = v_2v_1^* \quad \text{etc.}$$

$$\mathbf{V}_1^* = \frac{\mathbf{V}_2 \times \mathbf{V}_3}{[\mathbf{V}_1\mathbf{V}_2\mathbf{V}_3]}, \qquad \mathbf{V}_2^* = \frac{\mathbf{V}_3 \times \mathbf{V}_1}{[\mathbf{V}_1\mathbf{V}_2\mathbf{V}_3]}, \qquad \mathbf{V}_3^* = \frac{\mathbf{V}_1 \times \mathbf{V}_2}{[\mathbf{V}_1\mathbf{V}_2\mathbf{V}_3]}$$

The system $\mathbf{i}, \mathbf{j}, \mathbf{k}$ is its own reciprocal.

Vector Triple Product

The product $\mathbf{V}_1 \times (\mathbf{V}_2 \times \mathbf{V}_3)$ defines the *vector triple product*. Obviously, in this case, the brackets are vital to the definition.

$$\mathbf{V}_1 \times (\mathbf{V}_2 \times \mathbf{V}_3) = (\mathbf{V}_1 \cdot \mathbf{V}_3)\mathbf{V}_2 - (\mathbf{V}_1 \cdot \mathbf{V}_2)\mathbf{V}_3$$

$$= \begin{vmatrix} \mathbf{i} & \mathbf{j} & \mathbf{k} \\ a_1 & b_1 & c_1 \\ \begin{vmatrix} b_2 & c_2 \\ b_3 & c_3 \end{vmatrix} & \begin{vmatrix} c_2 & a_2 \\ c_3 & a_3 \end{vmatrix} & \begin{vmatrix} a_2 & b_2 \\ a_3 & b_3 \end{vmatrix} \end{vmatrix}$$

i.e. it is a vector, perpendicular to $\mathbf{V}_1$, lying in the plane of $\mathbf{V}_2$, $\mathbf{V}_3$.

Similarly

$$(\mathbf{V}_1 \times \mathbf{V}_2) \times \mathbf{V}_3 = \begin{vmatrix} \mathbf{i} & \mathbf{j} & \mathbf{k} \\ \begin{vmatrix} b_1 & c_1 \\ b_2 & c_2 \end{vmatrix} & \begin{vmatrix} c_1 & a_1 \\ c_2 & a_2 \end{vmatrix} & \begin{vmatrix} a_1 & b_1 \\ a_2 & b_2 \end{vmatrix} \\ a_3 & b_3 & c_3 \end{vmatrix}$$

$$\mathbf{V}_1 \times (\mathbf{V}_2 \times \mathbf{V}_3) + \mathbf{V}_2 \times (\mathbf{V}_3 \times \mathbf{V}_1) + \mathbf{V}_3 \times (\mathbf{V}_1 \times \mathbf{V}_2) \equiv 0$$

If $\mathbf{V}_1 \times (\mathbf{V}_2 \times \mathbf{V}_3) = (\mathbf{V}_1 \times \mathbf{V}_2) \times \mathbf{V}_3$ then $\mathbf{V}_1$, $\mathbf{V}_2$, $\mathbf{V}_3$ form an *orthogonal set.* Thus $\mathbf{i}, \mathbf{j}, \mathbf{k}$ form an orthogonal set.

Geometry of the Plane, Straight Line and Sphere

The position vectors of the fixed points A, B, C, D relative to O are $\mathbf{V}_1$, $\mathbf{V}_2$, $\mathbf{V}_3$, $\mathbf{V}_4$ and the position vector of the variable point P is $\mathbf{V}$.

The vector form of the equation of the straight line through A parallel to $\mathbf{V}_2$ is

$$\mathbf{V} = \mathbf{V}_1 + r\mathbf{V}_2$$

$$\text{or} \quad (\mathbf{V} - \mathbf{V}_1) = r\mathbf{V}_2$$

$$\text{or} \quad (\mathbf{V} - \mathbf{V}_1) \times \mathbf{V}_2 = 0$$

while that of the plane through A perpendicular to $\mathbf{V}_2$ is

$$(\mathbf{V} - \mathbf{V}_1) \cdot \mathbf{V}_2 = 0$$

The equation of the line AB is

$$\mathbf{V} = r\mathbf{V}_1 + (1 - r)\mathbf{V}_2$$

and those of the bisectors of the angles between $\mathbf{V}_1$ and $\mathbf{V}_2$ are

$$\mathbf{V} = r\left(\frac{\mathbf{V}_1}{v} \pm \frac{\mathbf{V}_2}{v_2}\right)$$

$$\text{or} \qquad \mathbf{V} = r(\hat{\mathbf{v}}_1 \pm \hat{\mathbf{v}}_2)$$

The perpendicular from C to the line through A parallel to $\mathbf{V}_2$ has as its equation

$$\mathbf{V} = \mathbf{V}_1 - \mathbf{V}_3 - \hat{\mathbf{v}}_2 \cdot (\mathbf{V}_1 - \mathbf{V}_3)\hat{\mathbf{v}}_2.$$

The condition for the intersection of the two lines,

$$\mathbf{V} = \mathbf{V}_1 + r\mathbf{V}_3$$

$$\text{and} \qquad \mathbf{V} = \mathbf{V}_2 + s\mathbf{V}_4$$

$$\text{is} \qquad [(\mathbf{V}_1 - \mathbf{V}_2)\mathbf{V}_3\mathbf{V}_4] = 0.$$

The common perpendicular to the above two lines is the line of intersection of the two planes

$$[(\mathbf{V} - \mathbf{V}_1)\mathbf{V}_3(\mathbf{V}_3 \times \mathbf{V}_4)] = 0$$

$$\text{and} \qquad [(\mathbf{V} - \mathbf{V}_2)\mathbf{V}_4(\mathbf{V}_3 \times \mathbf{V}_4)] = 0$$

and the length of this perpendicular is

$$\frac{[(\mathbf{V}_1 - \mathbf{V}_2)\mathbf{V}_3\mathbf{V}_4]}{|\mathbf{V}_3 \times \mathbf{V}_4|}.$$

The equation of the line perpendicular to the plane ABC is

$$\mathbf{V} = \mathbf{V}_1 \times \mathbf{V}_2 + \mathbf{V}_2 \times \mathbf{V}_3 + \mathbf{V}_3 \times \mathbf{V}_1$$

and the distance of the plane from the origin is

$$\frac{[\mathbf{V}_1\mathbf{V}_2\mathbf{V}_3]}{|(\mathbf{V}_2 - \mathbf{V}_1) \times (\mathbf{V}_3 - \mathbf{V}_1)|}.$$

In general the vector equation

$$\mathbf{V} \cdot \mathbf{V}_2 = r$$

defines the plane which is perpendicular to $\mathbf{V}_2$, and the perpendicular distance from A to this plane is

$$\frac{r - \mathbf{V}_1 \cdot \mathbf{V}_2}{v_2}.$$

The distance from A, measured along a line parallel to $\mathbf{V}_3$, is

$$\frac{r - \mathbf{V}_1 \cdot \mathbf{V}_2}{\mathbf{V}_2 \cdot \hat{\mathbf{v}}_3} \quad \text{or} \quad \frac{r - \mathbf{V}_1 \cdot \mathbf{V}_2}{v_2 \cos\theta}$$

where θ is the angle beween $\mathbf{V}_2$ and $\mathbf{V}_3$.
(If this plane contains the point C then $r = \mathbf{V}_3 \cdot \mathbf{V}_2$ and if it passes through the origin then $r = 0$.)

Given two planes

$$\mathbf{V} \cdot \mathbf{V}_1 = r$$

$$\mathbf{V} \cdot \mathbf{V}_2 = s$$

then any plane through the line of intersection of these two planes is given by

$$\mathbf{V} \cdot (\mathbf{V}_1 + \lambda\mathbf{V}_2) = r + \lambda s$$

where λ is a scalar parameter. In particular $\lambda = \pm v_1/v_2$ yields the equation of the two planes bisecting the angle between the given planes.

The plane through A parallel to the plane of $\mathbf{V}_2$, $\mathbf{V}_3$ is

$$\mathbf{V} = \mathbf{V}_1 + r\mathbf{V}_2 + s\mathbf{V}_3$$

$$\text{or} \quad (\mathbf{V} - \mathbf{V}_1) \cdot \mathbf{V}_2 \times \mathbf{V}_3 = 0$$

$$\text{or} \quad [\mathbf{V}\mathbf{V}_2\mathbf{V}_3] - [\mathbf{V}_1\mathbf{V}_2\mathbf{V}_3] = 0$$

so that the expansion in rectangular Cartesian coordinates yields

$$\begin{vmatrix} (x - a_1) & (y - b_1) & (z - c_1) \\ a_2 & b_2 & c_2 \\ a_3 & b_3 & c_3 \end{vmatrix} = 0 \qquad (\mathbf{V} \equiv x\mathbf{i} + y\mathbf{j} + z\mathbf{k})$$

which is obviously the usual linear equation in x, y and z.

The plane through AB parallel to $\mathbf{V}_3$ is given by

$$[(\mathbf{V} - \mathbf{V}_1)(\mathbf{V}_1 - \mathbf{V}_2)\mathbf{V}_3] = 0$$

$$\text{or} \quad [\mathbf{V}\mathbf{V}_2\mathbf{V}_3] - [\mathbf{V}\mathbf{V}_1\mathbf{V}_3] - [\mathbf{V}_1\mathbf{V}_2\mathbf{V}_3] = 0$$

The plane through the three points A, B and C is

$$\mathbf{V} = \mathbf{V}_1 + s(\mathbf{V}_2 - \mathbf{V}_1) + t(\mathbf{V}_3 - \mathbf{V}_1)$$

or $\mathbf{V} = r\mathbf{V}_1 + s\mathbf{V}_2 + t\mathbf{V}_3 \qquad (r + s + t = 1)$

or $[(\mathbf{V} - \mathbf{V}_1)(\mathbf{V}_1 - \mathbf{V}_2)(\mathbf{V}_2 - \mathbf{V}_3)] = 0$

or $[\mathbf{V}\mathbf{V}_1\mathbf{V}_2] + [\mathbf{V}\mathbf{V}_2\mathbf{V}_3] + [\mathbf{V}\mathbf{V}_3\mathbf{V}_1] - [\mathbf{V}_1\mathbf{V}_2\mathbf{V}_3] = 0$

For four points A, B, C, D to be coplanar, then

$$r\mathbf{V}_1 + s\mathbf{V}_2 + t\mathbf{V}_3 + u\mathbf{V}_4 \equiv 0 \equiv r + s + t + u$$

The following formulae relate to a sphere when the vectors are taken to lie in three dimensional space and to a circle when the space is two dimensional. For a circle in three dimensions take the intersection of the sphere with a plane.

The equation of a sphere with center O and radius OA is

$$\mathbf{V} \cdot \mathbf{V} = v_1^2 \qquad (\text{not } \mathbf{V} = \mathbf{V}_1)$$

or $(\mathbf{V} - \mathbf{V}_1) \cdot (\mathbf{V} + \mathbf{V}_1) = 0$

while that of a sphere with center B radius v_1 is

$$(\mathbf{V} - \mathbf{V}_2) \cdot (\mathbf{V} - \mathbf{V}_2) = v_1^2$$

or $\mathbf{V} \cdot (\mathbf{V} - 2\mathbf{V}_2) = v_1^2 - v_2^2$

If the above sphere passes through the origin then

$$\mathbf{V} \cdot (\mathbf{V} - 2\mathbf{V}_2) = 0$$

(note that in two dimensional polar coordinates this is simply)

$$r = 2a \cdot \cos\theta$$

while in three dimensional Cartesian coordinates it is

$$x^2 + y^2 + z^2 - 2(a_2x + b_2y + c_2x) = 0.$$

The equation of a sphere having the points A and B as the extremities of a diameter is

$$(\mathbf{V} - \mathbf{V}_1) \cdot (\mathbf{V} - \mathbf{V}_1) = 0$$

The square of the length of the tangent from C to the sphere with center B and radius v_1 is given by

$$(\mathbf{V}_3 - \mathbf{V}_2) \cdot (\mathbf{V}_3 - \mathbf{V}_2) = v_1^2$$

The condition that the plane $\mathbf{V} \cdot \mathbf{V}_3 = s$ is tangential to the sphere $(\mathbf{V} - \mathbf{V}_2) \cdot (\mathbf{V} - \mathbf{V}_2) = v_1^2$ is

$$(s - \mathbf{V}_3 \cdot \mathbf{V}_2) \cdot (s - \mathbf{V}_3 \cdot \mathbf{V}_2) = v_1^2 v_3^2.$$

The equation of the tangent plane at D, on the surface of sphere $(\mathbf{V} - \mathbf{V}_2) \cdot (\mathbf{V} - \mathbf{V}_2) = v_1^2$, is

$$(\mathbf{V} - \mathbf{V}_4) \cdot (\mathbf{V}_4 - \mathbf{V}_2) = 0$$

or $\mathbf{V} \cdot \mathbf{V}_4 - \mathbf{V}_2 \cdot (\mathbf{V} + \mathbf{V}_4) = v_1^2 - v_2^2$

The condition that the two circles $(\mathbf{V} - \mathbf{V}_2) \cdot (\mathbf{V} - \mathbf{V}_2) = v_1^2$ and $(\mathbf{V} - \mathbf{V}_4) \cdot (\mathbf{V} - \mathbf{V}_4) = v_3^2$ intersect orthogonally is clearly

$$(\mathbf{V}_2 - \mathbf{V}_4) \cdot (\mathbf{V}_2 - \mathbf{V}_4) = v_1^2 + v_3^2$$

The polar plane of D with respect to the circle

$$(\mathbf{V} - \mathbf{V}_2) \cdot (\mathbf{V} - \mathbf{V}_2) = v_1^2 \quad \text{is}$$

$$\mathbf{V} \cdot \mathbf{V}_4 - \mathbf{V}_2 \cdot (\mathbf{V} + \mathbf{V}_4) = v_1^2 - v_2^2$$

Any sphere through the intersection of the two spheres $(\mathbf{V} - \mathbf{V}_2) \cdot (\mathbf{V} - \mathbf{V}_2) = v_1^2$ and $(\mathbf{V} - \mathbf{V}_4) \cdot (\mathbf{V} - \mathbf{V}_4) = v_3^2$ is given by

$$(\mathbf{V} - \mathbf{V}_2) \cdot (\mathbf{V} - \mathbf{V}_2) + \lambda(\mathbf{V} - \mathbf{V}_4) \cdot (\mathbf{V} - \mathbf{V}_4) = v_1 + \lambda v_3$$

while the radical plane of two such spheres is

$$\mathbf{V} \cdot (\mathbf{V}_2 - \mathbf{V}_4) = -\tfrac{1}{2}(v_1^2 - v_2^2 - v_3^2 + v_4^2)$$

Differentiation of Vectors

If $\mathbf{V}_1 = a_1\mathbf{i} + b_1\mathbf{j} + c_1\mathbf{k}$, and $\mathbf{V}_2 = a_2\mathbf{i} + b_2\mathbf{j} + c_2\mathbf{k}$, and if $\mathbf{V}_1$ and $\mathbf{V}_2$ are functions of the scalar t, then

$$\frac{d}{dt}(\mathbf{V}_1 + \mathbf{V}_2 + \cdots) = \frac{d\mathbf{V}_1}{dt} + \frac{d\mathbf{V}_2}{dt} + \cdots,$$

where $\dfrac{d\mathbf{V}_1}{dt} = \dfrac{da_1}{dt}\mathbf{i} + \dfrac{db_1}{dt}\mathbf{j} + \dfrac{dc_1}{dt}\mathbf{k}$, etc.

$$\frac{d}{dt}(\mathbf{V}_1 \cdot \mathbf{V}_2) = \frac{d\mathbf{V}_1}{dt} \cdot \mathbf{V}_2 + \mathbf{V}_1 \cdot \frac{d\mathbf{V}_2}{dt}$$

$$\frac{d}{dt}(\mathbf{V}_1 \times \mathbf{V}_2) = \frac{d\mathbf{V}_1}{dt} \times \mathbf{V}_2 + \mathbf{V}_1 \times \frac{d\mathbf{V}_2}{dt}$$

$$\mathbf{V} \cdot \frac{d\mathbf{V}}{dt} = v \cdot \frac{dv}{dt}$$

In particular, if $\mathbf{V}$ is a vector of constant length then the right hand side of the last equation is identically zero showing that $\mathbf{V}$ is perpendicular to its derivative.

The derivatives of the triple products are

$$\frac{d}{dt}[\mathbf{V}_1\mathbf{V}_2\mathbf{V}_3] = \left[\left(\frac{d\mathbf{V}_1}{dt}\right)\mathbf{V}_2\mathbf{V}_3\right] + \left[\mathbf{V}_1\left(\frac{d\mathbf{V}_2}{dt}\right)\mathbf{V}_3\right] + \left[\mathbf{V}_1\mathbf{V}_2\left(\frac{d\mathbf{V}_3}{dt}\right)\right]$$

and

$$\frac{d}{dt}\left\{\mathbf{V}_1 \times (\mathbf{V}_2 \times \mathbf{V}_3)\right\} = \left(\frac{d\mathbf{V}_1}{dt}\right) \times (\mathbf{V}_2 \times \mathbf{V}_3) + \mathbf{V}_1 \times \left(\left(\frac{d\mathbf{V}_2}{dt}\right) \times \mathbf{V}_3\right) + \mathbf{V}_1 \times \left(\mathbf{V}_2 \times \left(\frac{d\mathbf{V}_3}{dt}\right)\right)$$

Geometry of Curves in Space

s = the *length of arc*, measured from some fixed point on the curve (fig. 3).

$\mathbf{V}_1$ = the position vector of the point A on the curve

$\mathbf{V}_1 + \delta\mathbf{V}_1$ = the position vector of the point P in the neighborhood of A

$\hat{\mathbf{t}}$ = the *unit tangent* to the curve at the point A, measured in the direction of s increasing.

The *normal plane* is that plane which is perpendicular to the unit tangent. The principal normal is defined as the intersection of the normal plane with the plane defined by $\mathbf{V}_1$ and $\mathbf{V}_1 + \delta\mathbf{V}_1$ in the limit as $\delta\mathbf{V}_1 - 0$.

$\hat{\mathbf{n}}$ = the *unit normal* (principal) at the point A. The plane defined by $\hat{\mathbf{t}}$ and $\hat{\mathbf{n}}$ is called the *osculating plane* (alternatively plane of curvature or local plane).

ρ = the radius of curvature at A

$\delta\theta$ = the angle subtended at the origin by $\delta\mathbf{V}_1$.

$$\kappa = \frac{d\theta}{ds} = \frac{1}{\rho}$$

$\hat{\mathbf{b}}$ = the *unit binormal* i.e. the unit vector which is parallel to $\hat{\mathbf{t}} \times \hat{\mathbf{n}}$ at the point A:

λ = the *torsion* of the curve at A

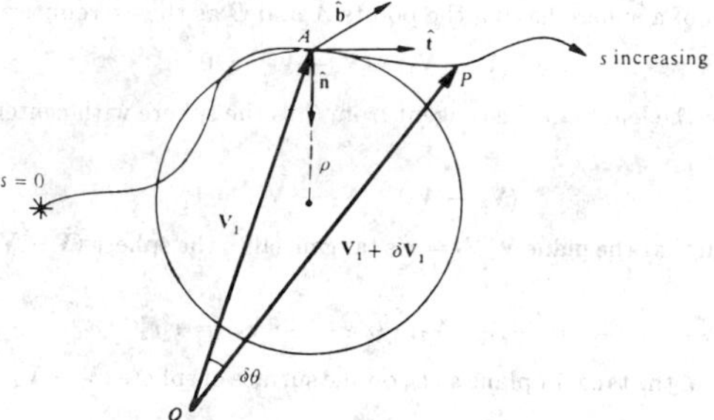

Figure 3.

Frenet's Formulae:

$$\frac{d\hat{\mathbf{t}}}{ds} = \kappa\hat{\mathbf{n}}$$

$$\frac{d\hat{\mathbf{n}}}{ds} = -\kappa\hat{\mathbf{t}} + \lambda\hat{\mathbf{b}}$$

$$\frac{d\hat{\mathbf{b}}}{ds} = -\lambda\hat{\mathbf{n}}$$

The following formulae are also applicable:

Unit tangent $$\hat{\mathbf{t}} = \frac{d\mathbf{V}_1}{ds}$$

Equation of the tangent $$(\mathbf{V} - \mathbf{V}_1) \times \hat{\mathbf{t}} = 0$$
or $$\mathbf{V} = \mathbf{V}_1 + q\hat{\mathbf{t}}$$

Unit normal $$\hat{\mathbf{n}} = \frac{1 d^2\mathbf{V}_1}{\kappa ds^2}$$

Equation of the normal plane $$(\mathbf{V} - \mathbf{V}_1) \cdot \hat{\mathbf{t}} = 0$$

Equation of the normal $$(\mathbf{V} - \mathbf{V}_1) \times \hat{\mathbf{n}} = 0$$
or $$\mathbf{V} = \mathbf{V}_1 + r\hat{\mathbf{n}}$$

Unit binormal $$\hat{\mathbf{b}} = \hat{\mathbf{t}} \times \hat{\mathbf{n}}$$

Equation of the binormal $$(\mathbf{V} - \mathbf{V}_1) \times \hat{\mathbf{b}} = 0$$
or $$\mathbf{V} = \mathbf{V}_1 + u\hat{\mathbf{b}}$$
or $$\mathbf{V} = \mathbf{V}_1 + w\frac{d\mathbf{V}_1}{ds} \times \frac{d^2\mathbf{V}_1}{ds^2}$$

Equation of the osculating plane:

$$[(\mathbf{V} - \mathbf{V}_1)\hat{\mathbf{t}}\hat{\mathbf{n}}] = 0$$

or

$$\left[(\mathbf{V} - \mathbf{V}_1)\left(\frac{d\mathbf{V}_1}{ds}\right)\left(\frac{d^2\mathbf{V}_1}{ds^2}\right)\right] = 0$$

A *geodetic line* on a surface is a curve, the osculating plane of which is everywhere normal to the surface.

The differential equation of the geodetic is

$$[\hat{\mathbf{n}} d\mathbf{V}_1\, d^2\mathbf{V}_1] = 0$$

Differential Operators—Rectangular Coordinates

$$dS = \frac{\partial S}{\partial x} \cdot dx + \frac{\partial S}{\partial y} \cdot dy + \frac{\partial S}{\partial z} \cdot dz$$

By definition

$$\nabla \equiv \text{del} \equiv \mathbf{i}\frac{\partial}{\partial x} + \mathbf{j}\frac{\partial}{\partial y} + \mathbf{k}\frac{\partial}{\partial z}$$

$$\nabla^2 \equiv \text{Laplacian} \equiv \frac{\partial^2}{\partial x^2} + \frac{\partial^2}{\partial y^2} + \frac{\partial^2}{\partial z^2}$$

If S is a scalar function, then

$$\nabla S \equiv \text{grad } S \equiv \frac{\partial S}{\partial x}\mathbf{i} + \frac{\partial S}{\partial y}\mathbf{j} + \frac{\partial S}{\partial z}\mathbf{k}$$

Grad S defines both the direction and magnitude of the maximum rate of increase of S at any point. Hence the name *gradient* and also its vectorial nature. ∇S is independent of the choice of rectangular coordinates.

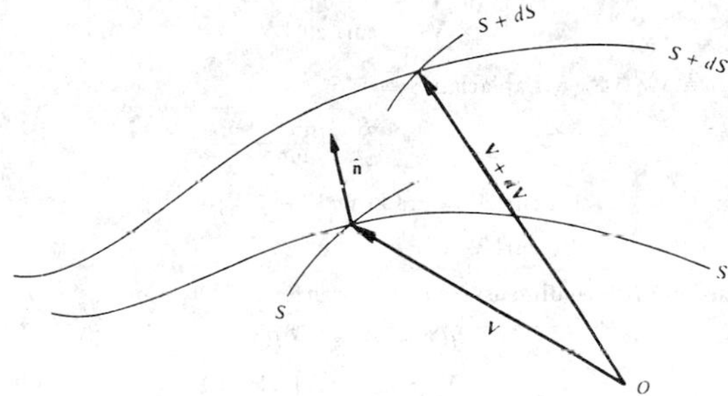

Figure 4

$$\nabla S = \frac{\partial S}{\partial n}\hat{\mathbf{n}}$$

where $\hat{\mathbf{n}}$ is the unit normal to the surface S = constant, in the direction of S increasing. The total derivative of S at a point having the position vector $\mathbf{V}$ is given by (fig. 4)

$$dS = \frac{\partial S}{\partial n}\hat{\mathbf{n}} \cdot d\mathbf{V}$$
$$= d\mathbf{V} \cdot \nabla S$$

and the directional derivative of S in the direction of $\mathbf{U}$ is

$$\mathbf{U} \cdot \nabla S = \mathbf{U} \cdot (\nabla S) = (\mathbf{U} \cdot \nabla)S$$

Similarly the directional derivative of the vector $\mathbf{V}$ in the direction of $\mathbf{U}$ is

$$(\mathbf{U} \cdot \nabla)\mathbf{V}$$

The *distributive* law holds for finding a gradient. Thus if S and T are scalar functions

$$\nabla(S + T) = \nabla S + \nabla T$$

The *associative* law becomes the rule for differentiating a product:

$$\nabla(ST) = S\nabla T + T\nabla S$$

If $\mathbf{V}$ is a vector function with the magnitudes of the components parallel to the three coordinate axes V_x, V_y, V_z, then

$$\nabla \cdot \mathbf{V} \equiv \text{div } \mathbf{V} \equiv \frac{\partial V_x}{\partial x} + \frac{\partial V_y}{\partial y} + \frac{\partial V_z}{\partial z}$$

The divergence obeys the distributive law. Thus, if $\mathbf{V}$ and $\mathbf{U}$ are vectors functions, then

$$\nabla \cdot (\mathbf{V} + \mathbf{U}) = \nabla \cdot \mathbf{V} + \nabla \cdot \mathbf{U}$$
$$\nabla \cdot (S\mathbf{V}) = (\nabla S) \cdot \mathbf{V} + S(\nabla \cdot \mathbf{V})$$
$$\nabla \cdot (\mathbf{U} \times \mathbf{V}) = \mathbf{V} \cdot (\nabla \times \mathbf{U}) - \mathbf{U} \cdot (\nabla \times \mathbf{V})$$

As with the gradient of a scalar, the divergence of a vector is invariant under a transformation from one set of rectangular coordinates to another.

$$\nabla \times \mathbf{V} \equiv \text{curl } \mathbf{V} \text{ (sometimes } \nabla \wedge \mathbf{V} \text{ or rot } \mathbf{V})$$

$$\equiv \left(\frac{\partial \mathbf{V}_z}{\partial y} - \frac{\partial \mathbf{V}_y}{\partial z}\right)\mathbf{i} + \left(\frac{\partial \mathbf{V}_x}{\partial z} - \frac{\partial \mathbf{V}_z}{\partial x}\right)\mathbf{j} + \left(\frac{\partial \mathbf{V}_y}{\partial x} - \frac{\partial \mathbf{V}_x}{\partial y}\right)\mathbf{k}$$

$$= \begin{vmatrix} \mathbf{i} & \mathbf{j} & \mathbf{k} \\ \frac{\partial}{\partial x} & \frac{\partial}{\partial y} & \frac{\partial}{\partial z} \\ V_x & V_y & V_z \end{vmatrix}$$

The *curl* (or *rotation*) of a vector is a vector which is invariant under a transformation from one set of rectangular coordinates to another.

$$\nabla \times (\mathbf{U} + \mathbf{V}) = \nabla \times \mathbf{U} + \nabla \times \mathbf{V}$$
$$\nabla \times (S\mathbf{V}) = (\nabla S) \times \mathbf{V} + S(\nabla \times \mathbf{V})$$
$$\nabla \times (\mathbf{U} \times \mathbf{V}) = (\mathbf{V} \cdot \nabla)\mathbf{U} - (\mathbf{U} \cdot \nabla)\mathbf{V} + \mathbf{U}(\nabla \cdot \mathbf{V}) - \mathbf{V}(\nabla \cdot \mathbf{U})$$

$$\begin{aligned}\text{grad } (\mathbf{U} \cdot \mathbf{V}) &= \nabla(\mathbf{U} \cdot \mathbf{V}) \\ &= (\mathbf{V} \cdot \nabla)\mathbf{U} + (\mathbf{U} \cdot \nabla)\mathbf{V} + \mathbf{V} \times (\nabla \times \mathbf{U}) + \mathbf{U} \times (\nabla \times \mathbf{V})\end{aligned}$$

If
$$\mathbf{V} = V_x\mathbf{i} + V_y\mathbf{j} + V_z\mathbf{k}$$
$$\nabla \cdot \mathbf{V} = \nabla V_x \cdot \mathbf{i} + \nabla V_y \cdot \mathbf{j} + \nabla V_z \cdot \mathbf{k}$$
and
$$\nabla \times \mathbf{V} = \nabla V_x \times \mathbf{i} + \nabla V_y \times \mathbf{j} + \nabla V_z \times \mathbf{k}$$

The operator ∇ can be used more than once. The number of possibilities where ∇ is used twice are

$$\begin{aligned}\nabla \cdot (\nabla\theta) &\equiv \text{div grad } \theta \\ \nabla \times (\nabla\theta) &\equiv \text{curl grad } \theta \\ \nabla(\nabla \cdot \mathbf{V}) &\equiv \text{grad div } \mathbf{V} \\ \nabla \cdot (\nabla \times \mathbf{V}) &\equiv \text{div curl } \mathbf{V} \\ \nabla \times (\nabla \times \mathbf{V}) &\equiv \text{curl curl } \mathbf{V}\end{aligned}$$

Thus: $\text{div grad } S \equiv \nabla \cdot (\nabla S) \equiv \text{Laplacian } S \equiv \nabla^2 S$

$$\equiv \frac{\partial^2 S}{\partial x^2} + \frac{\partial^2 S}{\partial y^2} + \frac{\partial^2 S}{\partial z^2}$$

$\text{curl grad } S \equiv 0;$ $\qquad \text{curl curl } \mathbf{V} \equiv \text{grad div } \mathbf{V} - \nabla^2\mathbf{V};$

$$\text{div curl } \mathbf{V} \equiv 0$$

Taylor's expansion in three dimensions can be written

$$f(\mathbf{V} + \boldsymbol{\varepsilon}) = e^{\boldsymbol{\varepsilon} \cdot \nabla} f(\mathbf{V})$$

where $\mathbf{V} = x\mathbf{i} + y\mathbf{j} + z\mathbf{k}$

and $\boldsymbol{\varepsilon} = h\mathbf{i} + l\mathbf{j} + m\mathbf{k}$

(note the analogy with $f_p = e^{phD} f_0$ in finite difference methods).

Orthogonal Curvilinear Coordinates

If at a point P there exist three uniform point functions u, v and w so that the surfaces u = const., v = const., and w = const., intersect in three distinct curves through P then the surfaces are called the *coordinate surfaces* through P. The three lines of intersection are referred to as the *coordinate lines* and their tangents a, b, and c as the *coordinate axes*. When the coordinate axes form an orthogonal set the system is said to define *orthogonal curvilinear coordinates* at P.

Consider an infinitesimal volume enclosed by the surfaces u, v, w, $u + du$, $v + dv$, and $w + dw$ (fig. 5).

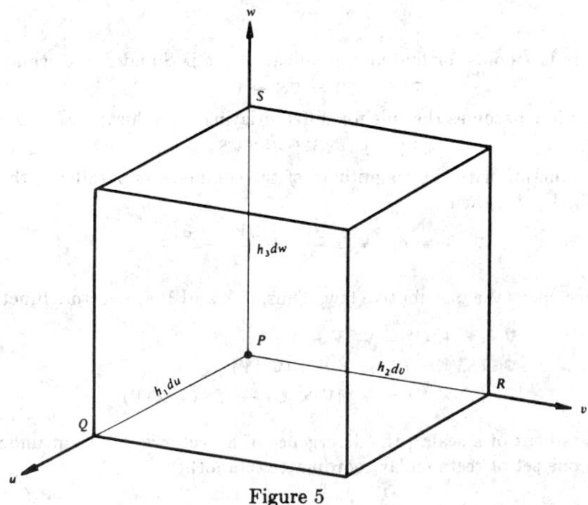

Figure 5

The surface $PRS \equiv u$ = const., and the face of the curvilinear figure immediately opposite this is $u + du$ = const. etc.

In terms of these surface constants

$$P = P(u,v,w)$$
$$Q = Q(u + du,v,w) \qquad \text{and} \qquad PQ = h_1 du$$

$$R = R(u,v + dv,w) \qquad PR = h_2 dv$$
$$S = S(u,v,w + dw) \qquad PS = h_3 dw$$

where h_1, h_2, and h_3 are functions of u, v, and w.

In rectangular Cartesians **i**, **j**, **k**

$$h_1 = 1, \qquad h_2 = 1, \qquad h_3 = 1.$$

$$\frac{\hat{\mathbf{a}}}{h_1}\frac{\partial}{\partial u} = \mathbf{i}\frac{\partial}{\partial x}, \qquad \frac{\hat{\mathbf{b}}}{h_2}\frac{\partial}{\partial v} = \mathbf{j}\frac{\partial}{\partial y}, \qquad \frac{\hat{\mathbf{c}}}{h_3}\frac{\partial}{\partial w} = \mathbf{k}\frac{\partial}{\partial z}.$$

In cylindrical coordinates $\hat{\mathbf{r}}$, $\hat{\boldsymbol{\phi}}$, $\hat{\mathbf{k}}$

$$h_1 = 1, \qquad h_2 = r \qquad h_3 = 1.$$

$$\frac{\hat{\mathbf{a}}}{h_1}\frac{\partial}{\partial u} = \hat{\mathbf{r}}\frac{\partial}{\partial r}, \qquad \frac{\hat{\mathbf{b}}}{h_2}\frac{\partial}{\partial v} = \frac{\hat{\boldsymbol{\phi}}}{r}\frac{\partial}{\partial \phi} \qquad \frac{\hat{\mathbf{c}}}{h_3}\frac{\partial}{\partial w} = \hat{\mathbf{k}}\frac{\partial}{\partial z}$$

In spherical coordinates $\hat{\mathbf{r}}$, $\hat{\boldsymbol{\theta}}$, $\hat{\boldsymbol{\phi}}$

$$h_1 = 1, \qquad h_2 = r, \qquad h_3 = r\sin\theta$$

$$\frac{\hat{\mathbf{a}}}{h_1}\frac{\partial}{\partial u} = \hat{\mathbf{r}}\frac{\partial}{\partial r}, \qquad \frac{\mathbf{b}}{h_2}\frac{\partial}{\partial v} = \frac{\hat{\boldsymbol{\phi}}}{r}\frac{\partial}{\partial \theta}, \qquad \frac{\hat{\mathbf{c}}}{h_3}\frac{\partial}{\partial w} = \frac{\hat{\boldsymbol{\phi}}}{r\sin\theta}\frac{\partial}{\partial \phi}$$

The general expressions for grad, div and curl together with those for ∇^2 and the directional derivative are, in orthogonal curvilinear coordinates, given by

$$\nabla S = \frac{\hat{\mathbf{a}}}{h_1}\frac{\partial S}{\partial u} + \frac{\hat{\mathbf{b}}}{h_2}\frac{\partial S}{\partial v} + \frac{\hat{\mathbf{c}}}{h_3}\frac{\partial S}{\partial w}$$

$$(\mathbf{V}\cdot\nabla)S = \frac{V_1}{h_1}\frac{\partial S}{\partial u} + \frac{V_2}{h_2}\frac{\partial S}{\partial v} + \frac{V_3}{h_3}\frac{\partial S}{\partial w}$$

$$\nabla\cdot\mathbf{V} = \frac{1}{h_1h_2h_3}\left\{\frac{\partial}{\partial u}(h_2h_3V_1) + \frac{\partial}{\partial v}(h_3h_1V_2) + \frac{\partial}{\partial w}(h_1h_2V_3)\right\}$$

$$\nabla\times\mathbf{V} = \frac{\hat{\mathbf{a}}}{h_2h_3}\left\{\frac{\partial}{\partial v}(h_3V_3) - \frac{\partial}{\partial w}(h_2V_2)\right\} + \frac{\hat{\mathbf{b}}}{h_3h_1}\left\{\frac{\partial}{\partial w}(h_1V_1) - \frac{\partial}{\partial u}(h_3V_3)\right\} + \frac{\hat{\mathbf{c}}}{h_1h_2}\left\{\frac{\partial}{\partial u}(h_2V_2) - \frac{\partial}{\partial v}(h_1V_1)\right\}$$

$$\nabla^2 S = \frac{1}{h_1h_2h_3}\left\{\frac{\partial}{\partial u}\left(\frac{h_2h_3}{h_1}\frac{\partial S}{\partial u}\right) + \frac{\partial}{\partial v}\left(\frac{h_3h_1}{h_2}\frac{\partial S}{\partial v}\right) + \frac{\partial}{\partial w}\left(\frac{h_1h_2}{h_3}\frac{\partial S}{\partial w}\right)\right\}$$

FORMULAS OF VECTOR ANALYSIS

	Rectangular coordinates	Cylindrical coordinates	Spherical coordinates
Conversion to rectangular coordinates		$x = r\cos\varphi \quad y = r\sin\varphi \quad z = z$	$x = r\cos\varphi\sin\theta \quad y = r\sin\varphi\sin\theta$ $z = r\cos\theta$
Gradient.	$\nabla\phi = \frac{\partial\phi}{\partial x}\mathbf{i} + \frac{\partial\phi}{\partial y}\mathbf{j} + \frac{\partial\phi}{\partial z}\mathbf{k}$	$\nabla\phi = \frac{\partial\phi}{\partial r}\mathbf{r} + \frac{1}{r}\frac{\partial\phi}{\partial\varphi}\boldsymbol{\phi} + \frac{\partial\phi}{\partial z}\mathbf{k}$	$\nabla\phi = \frac{\partial\phi}{\partial r}\mathbf{r} + \frac{1}{r}\frac{\partial\phi}{\partial\theta}\boldsymbol{\theta} + \frac{1}{r\sin\theta}\frac{\partial\phi}{\partial\varphi}\boldsymbol{\phi}$
Divergence	$\nabla\cdot\mathbf{A} = \frac{\partial A_x}{\partial x} + \frac{\partial A_y}{\partial y} + \frac{\partial A_z}{\partial z}$	$\nabla\cdot\mathbf{A} = \frac{1}{r}\frac{\partial(rA_r)}{\partial r} + \frac{1}{r}\frac{\partial A\varphi}{\partial\varphi} + \frac{\partial A_z}{\partial z}$	$\nabla\cdot\mathbf{A} = \frac{1}{r^2}\frac{\partial(r^2A_r)}{\partial r} + \frac{1}{r\sin\theta}\frac{\partial(A_\theta\sin\theta)}{\partial\theta} + \frac{1}{r\sin\theta}\frac{\partial A_\varphi}{\partial\varphi}$
Curl.	$\nabla\times\mathbf{A} = \begin{vmatrix} \mathbf{i} & \mathbf{j} & \mathbf{k} \\ \frac{\partial}{\partial x} & \frac{\partial}{\partial y} & \frac{\partial}{\partial z} \\ A_x & A_y & A_z \end{vmatrix}$	$\nabla\times\mathbf{A} = \begin{vmatrix} \frac{1}{r}\mathbf{r} & \boldsymbol{\phi} & \frac{1}{r}\mathbf{k} \\ \frac{\partial}{\partial r} & \frac{\partial}{\partial\varphi} & \frac{\partial}{\partial z} \\ A_r & rA_\varphi & A_z \end{vmatrix}$	$\nabla\times\mathbf{A} = \begin{vmatrix} \frac{\mathbf{r}}{r^2\sin\theta} & \frac{\boldsymbol{\theta}}{r\sin\theta} & \frac{\boldsymbol{\phi}}{r} \\ \frac{\partial}{\partial r} & \frac{\partial}{\partial\theta} & \frac{\partial}{\partial\varphi} \\ A_r & rA_\theta & rA_\varphi\sin\theta \end{vmatrix}$
Laplacian.	$\nabla^2\phi = \frac{\partial^2\phi}{\partial x^2} + \frac{\partial^2\phi}{\partial y^2} + \frac{\partial^2\phi}{\partial z^2}$	$\nabla^2\phi = \frac{1}{r}\frac{\partial}{\partial r}\left(r\frac{\partial\phi}{\partial r}\right) + \frac{1}{r^2}\frac{\partial^2\phi}{\partial\varphi^2} + \frac{\partial^2\phi}{\partial z^2}$	$\nabla^2\phi = \frac{1}{r^2}\frac{\partial}{\partial r}\left(r^2\frac{\partial\phi}{\partial r}\right) + \frac{1}{r^2\sin\theta}\frac{\partial}{\partial\theta}\left(\sin\theta\frac{\partial\phi}{\partial\theta}\right) + \frac{1}{r^2\sin^2\theta}\frac{\partial^2\phi}{\partial\varphi^2}$

Transformation of Integrals

s = the distance along some curve "C" in space and is measured from some fixed point.
S = a surface area
V = a volume contained by a specified surface
$\hat{\mathbf{t}}$ = the unit tangent to C at the point P
$\hat{\mathbf{n}}$ = the unit outward pointing normal
F = some vector function
ds = the vector element of curve (= $\hat{\mathbf{t}}\,ds$)
dS = the vector element of surface (= $\hat{\mathbf{n}}\,dS$)

Then
$$\int_{(c)} \mathbf{F} \cdot \hat{\mathbf{t}}\, ds = \int_{(c)} \mathbf{F} \cdot d\mathbf{s}$$

and when
$$\mathbf{F} = \nabla\phi$$

$$\int_{(c)} (\nabla\phi) \cdot \hat{\mathbf{t}}\, ds = \int_{(c)} d\phi$$

Gauss' Theorem (Green's Theorem)

When S defines a closed region having a volume V

$$\iiint_{(v)} (\nabla \cdot \mathbf{F})\, dV = \iint_{(s)} (\mathbf{F} \cdot \hat{\mathbf{n}})\, dS = \iint_{(s)} \mathbf{F} \cdot d\mathbf{S}$$

also
$$\iiint_{(v)} (\nabla\phi)\, dV = \iint_{(s)} \phi\hat{\mathbf{n}}\, dS$$

and
$$\iiint_{(v)} (\nabla \times \mathbf{F})\, dV = \iint_{(s)} (\hat{\mathbf{n}} \times \mathbf{F})\, dS$$

Stokes' Theorem

When C is closed and bounds the open surface S.

$$\iint_{(s)} \hat{\mathbf{n}} \cdot (\nabla \times \mathbf{F})\, dS = \int_{(c)} \mathbf{F} \cdot d\mathbf{s}$$

also
$$\iint_{(s)} (\hat{\mathbf{n}} \times \nabla\phi)\, dS = \int_{(c)} \phi\, d\mathbf{s}$$

Green's Theorem

$$\iint_{(s)} (\nabla\phi \cdot \nabla\theta)\, dS = \iint_{(s)} \phi\hat{\mathbf{n}} \cdot (\nabla\theta)\, dS = \iiint_{(v)} \phi(\nabla^2\theta)\, dV$$
$$= \iint_{(s)} \theta \cdot \hat{\mathbf{n}}(\nabla\phi)\, dS = \iiint_{(v)} \theta(\nabla^2\phi)\, dV$$

MOMENT OF INERTIA FOR VARIOUS BODIES OF MASS

The mass of the body is indicated by m

Body	Axis	Moment of inertia
Uniform thin rod	Normal to the length, at one end	$m\frac{l^2}{3}$
Uniform thin rod	Normal to the length, at the center	$m\frac{l^2}{12}$
Thin rectangular sheet, sides a and b	Through the center parallel to b	$m\frac{a^2}{12}$
Thin rectangular sheet, sides a and b	Through the center perpendicular to the sheet	$m\frac{a^2+b^2}{12}$
Thin circular sheet of radius r	Normal to the plate through the center	$m\frac{r^2}{2}$
Thin circular sheet of radius r	Along any diameter	$m\frac{r^2}{4}$
Thin circular ring. Radii r_1 and r_2	Through center normal to plane of ring	$m\frac{r_1^2+r_2^2}{2}$
Thin circular ring. Radii r_1 and r_2	Any diameter	$m\frac{r_1^2+r_2^2}{4}$
Rectangular parallelopiped, edges a, b, and c	Through center perpendicular to face ab, (parallel to edge c)	$m\frac{a^2+b^2}{12}$
Sphere, radius r	Any diameter	$m\frac{2}{5}r^2$
Spherical shell, external radius, r_1, internal radius r_2	Any diameter	$m\frac{2}{5}\frac{(r_1^5-r_2^5)}{(r_1^3-r_2^3)}$
Spherical shell, very thin, mean radius, r	Any diameter	$m\frac{2}{3}r^2$
Right circular cylinder of radius r, length l	The longitudinal axis of the solid	$m\frac{r^2}{2}$
Right circular cylinder of radius r, length l	Transverse diameter	$m\left(\frac{r^2}{4}+\frac{l^2}{12}\right)$
Hollow circular cylinder, length l, radii r_1 and r_2	The longitudinal axis of the figure	$m\frac{(r_1^2+r_2^2)}{2}$
Thin cylindrical shell, length l, mean radius, r	The longitudinal axis of the figure	mr^2
Hollow circular cylinder, length l, radii r_1 and r_2	Transverse diameter	$m\left[\frac{r_1^2+r_2^2}{4}+\frac{l^2}{12}\right]$
Hollow circular cylinder, length l, very thin, mean radius	Transverse diameter	$m\left(\frac{r^2}{2}+\frac{l^2}{12}\right)$
Elliptic cylinder, length l, transverse semiaxes a and b	Longitudinal axis	$m\left(\frac{a^2+b^2}{4}\right)$
Right cone, altitude h, radius of base r	Axis of the figure	$m\frac{3}{10}r^2$
Spheroid of revolution, equatorial radius r	Polar axis	$m\frac{2r^2}{5}$
Ellipsoid, axes $2a$, $2b$, $2c$	Axis $2a$	$m\frac{(b^2+c^2)}{5}$

Bessel Functions*

1. Bessel's differential equation for a real variable x is

$$x^2\frac{d^2y}{dx^2} + x\frac{dy}{dx} + (x^2 - n^2)y = 0$$

* From Beyer, W. H., Ed., *CRC Handbook of Mathematical Sciences*, 5th ed., CRC Press, Boca Raton, 1978, 500—503. With permission.

2. When n is not an integer, two independent solutions of the equation are $J_n(x)$, $J_{-n}(x)$, where

$$J_n(x) = \sum_{k=0}^{\infty} \frac{(-1)^k}{k!\,\Gamma(n+k+1)} \left(\frac{x}{2}\right)^{n+2k}$$

3. If n is an integer $J_n(x) = (-1)^n J_n(x)$, where

$$J_n(x) = \frac{x^n}{2^n n!} \left\{ 1 - \frac{x^2}{2^2 \cdot 1!(n+1)} + \frac{x^4}{2^4 \cdot 2!(n+1)(n+2)} - \frac{x^6}{2^6 \cdot 3!(n+1)(n+2)(n+3)} + \cdots \right\}$$

4. For $n = 0$ and $n = 1$, this formula becomes

$$J_0(x) = 1 - \frac{x^2}{2^2(1!)^2} + \frac{x^4}{2^4(2!)^2} - \frac{x^6}{2^6(3!)^2} + \frac{x^8}{2^8(4!)^2} - \cdots$$

$$J_1(x) = \frac{x}{2} - \frac{x^3}{2^3 \cdot 1!2!} + \frac{x^5}{2^5 \cdot 2!3!} - \frac{x^7}{2^7 \cdot 3!4!} + \frac{x^9}{2^9 \cdot 4!5!} - \cdots$$

5. When x is large and positive, the following asymptotic series may be used

$$J_0(x) = \left(\frac{2}{\pi x}\right)^{\frac{1}{2}} \left\{ P_0(x) \cos\left(x - \frac{\pi}{4}\right) - Q_0(x) \sin\left(x - \frac{\pi}{4}\right) \right\}$$

$$J_1(x) = \left(\frac{2}{\pi x}\right)^{\frac{1}{2}} \left\{ P_1(x) \cos\left(x - \frac{3\pi}{4}\right) - Q_1(x) \sin\left(x - \frac{3\pi}{4}\right) \right\}$$

where

$$P_0(x) \sim 1 - \frac{1^2 \cdot 3^2}{2!(8x)^2} + \frac{1^2 \cdot 3^2 \cdot 5^2 \cdot 7^2}{4!(8x)^4} - \frac{1^2 \cdot 3^2 \cdot 5^2 \cdot 7^2 \cdot 9^2 \cdot 11^2}{6!(8x)^6} + \cdots$$

$$Q_0(x) \sim -\frac{1^2}{1!8x} + \frac{1^2 \cdot 3^2 \cdot 5^2}{3!(8x)^3} - \frac{1^2 \cdot 3^2 \cdot 5^2 \cdot 7^2 \cdot 9^2}{5!(8x)^5} + - \cdots$$

$$P_1(x) \sim 1 + \frac{1^2 \cdot 3 \cdot 5}{2!(8x)^2} - \frac{1^2 \cdot 3^2 \cdot 5^2 \cdot 7 \cdot 9}{4!(8x)^4} + \frac{1^2 \cdot 3^2 \cdot 5^2 \cdot 7^2 \cdot 9^2 \cdot 11 \cdot 13}{6!(8x)^6} - + \cdots$$

$$Q_1(x) \sim \frac{1 \cdot 3}{1!8x} - \frac{1^2 \cdot 3^2 \cdot 5 \cdot 7}{3!(8x)^3} + \frac{1^2 \cdot 3^2 \cdot 5^2 \cdot 7^2 \cdot 9 \cdot 11}{5!(8x)^5} - \cdots$$

[In $P_1(x)$ the signs alternate from + to − after the first term]

6. If $x > 25$, it is convenient to use the formulas

$$J_0(x) = A_0(x) \sin x + B_0(x) \cos x$$
$$J_1(x) = B_1(x) \sin x - A_1(x) \cos x$$

where

$$A_0(x) = \frac{P_0(x) - Q_0(x)}{(\pi x)^{\frac{1}{2}}} \quad \text{and} \quad A_1(x) = \frac{P_1(x) - Q_1(x)}{(\pi x)^{\frac{1}{2}}}$$

$$B_0(x) = \frac{P_0(x) + Q_0(x)}{(\pi x)^{\frac{1}{2}}} \quad \text{and} \quad B_1(x) = \frac{P_1(x) + Q_1(x)}{(\pi x)^{\frac{1}{2}}}$$

7. The zeros of $J_0(x)$ and $J_1(x)$

If j_{0s} and j_{1s} are the sth zeros of $J_0(x)$ and $J_1(x)$ respectively, and if $a = 4_s - 1$, $b = 4_s + 1$

$$j_{0,s} \sim \frac{1}{4}\pi a \left\{ 1 + \frac{2}{\pi^2 a^2} - \frac{62}{3\pi^4 a^4} + \frac{15{,}116}{15\pi^6 a^6} - \frac{12{,}554{,}474}{105\pi^8 a^8} + \frac{8{,}368{,}654{,}292}{315\pi^{10} a^{10}} - + \cdots \right\}$$

$$j_{1,s} \sim \frac{1}{4}\pi b \left\{ 1 - \frac{6}{\pi^2 b^2} + \frac{6}{\pi^4 b^4} - \frac{4716}{5\pi^6 b^6} + \frac{3{,}902{,}418}{35\pi^8 b^8} - \frac{895{,}167{,}324}{35\pi^{10} b^{10}} + \cdots \right\}$$

$$J_1(j_{0,s}) \sim \frac{(-1)^{s+1} 2^{\frac{1}{2}}}{\pi a^{\frac{1}{2}}} \left\{ 1 - \frac{56}{3\pi^4 a^4} + \frac{9664}{5\pi^6 a^6} - \frac{7{,}381{,}280}{21\pi^8 a^8} + \cdots \right\}$$

$$J_0(j_{1,s}) \sim \frac{(-1)^s 2^{\frac{1}{2}}}{\pi b^{\frac{1}{2}}} \left\{ 1 + \frac{24}{\pi^4 b^4} - \frac{19{,}584}{10\pi^6 b^6} + \frac{2{,}466{,}720}{7\pi^8 b^8} - \cdots \right\}$$

8. Table of zeros for $J_0(x)$ and $J_1(x)$

$J_1(\alpha_n) = 0 \qquad J_0(\beta_n) = 0$

Roots α_n	$J_1(\alpha_n)$	Roots β_n	$J_0(\beta_n)$
2.4048	0.5191	0.0000	1.0000
5.5201	−0.3403	3.8317	−0.4028
8.6537	0.2715	7.0156	0.3001
11.7915	−0.2325	10.1735	−0.2497
14.9309	0.2065	13.3237	0.2184
18.0711	−0.1877	16.4706	−0.1965
21.2116	0.1733	19.6159	0.1801

9. Recurrence formulas

$$J_{n-1}(x) + J_{n+1}(x) = \frac{2n}{x} J_n(x) \qquad nJ_n(x) + xJ_n'(x) = xJ_{n-1}(x)$$

$$J_{n-1}(x) - J_{n+1}(x) = 2J_n'(x) \qquad nJ_n(x) - xJ_n'(x) = xJ_{n+1}(x)$$

10. If J_n is written for $J_n(x)$ and $J_n^{(k)}$ is written for $\frac{d^k}{dx^k}\{J_n(x)\}$, then the following derivative relationships are important

$$J_0^{(r)} = -J_1^{(r-1)}$$

$$J_0^{(2)} = -J_0 + \frac{1}{x} J_1 = \frac{1}{2}(J_2 - J_0)$$

$$J_0^{(3)} = \frac{1}{x} J_0 + \left(1 - \frac{2}{x^2}\right) J_1 = \frac{1}{4}(-J_3 + 3J_1)$$

$$J_0^{(4)} = \left(1 - \frac{3}{x^2}\right) J_0 - \left(\frac{2}{x} - \frac{6}{x^3}\right) J_1 = \frac{1}{8}(J_4 - 4J_2 + 3J_0), \text{ etc.}$$

11. Half order Bessel functions

$$J_{\frac{1}{2}}(x) = \sqrt{\frac{2}{\pi x}} \sin x$$

$$J_{-\frac{1}{2}}(x) = \sqrt{\frac{2}{\pi x}} \cos x$$

$$J_{n+\frac{1}{2}}(x) = -x^{n+\frac{1}{2}} \frac{d}{dx}\{x^{-(n+\frac{1}{2})} J_{n+\frac{1}{2}}(x)\}$$

$$J_{n-\frac{1}{2}}(x) = x^{-(n+\frac{1}{2})} \frac{d}{dx}\{x^{n+\frac{1}{2}} J_{n+\frac{1}{2}}(x)\}$$

n	$\left(\frac{\pi x}{2}\right)^{\frac{1}{2}} J_{n+\frac{1}{2}}(x)$	$\left(\frac{\pi x}{2}\right)^{\frac{1}{2}} J_{-(n+\frac{1}{2})}(x)$
0	$\sin x$	$\cos x$
1	$\frac{\sin x}{x} - \cos x$	$-\frac{\cos x}{x} - \sin x$
2	$\left(\frac{3}{x^2} - 1\right)\sin x - \frac{3}{x}\cos x$	$\left(\frac{3}{x^2} - 1\right)\cos x + \frac{3}{x}\sin x$
3	$\left(\frac{15}{x^3} - \frac{6}{x}\right)\sin x - \left(\frac{15}{x^2} - 1\right)\cos x$	$-\left(\frac{15}{x^3} - \frac{6}{x}\right)\cos x - \left(\frac{15}{x^2} - 1\right)\sin x$
	etc.	

12. Additional solutions to Bessel's equation are

$Y_n(x)$ (also called Weber's function, and sometimes denoted by $N_n(x)$)
$H_n^{(1)}(x)$ and $H_n^{(2)}(x)$ (also called Hankel functions)

These solutions are defined as follows

$$Y_n(x) = \begin{cases} \dfrac{J_n(x)\cos(n\pi) - J_{-n}(x)}{\sin(n\pi)} & n \text{ not an integer} \\ \lim\limits_{\nu \to n} \dfrac{J_\nu(x)\cos(\nu\pi) - J_{-\nu}(x)}{\sin(\nu\pi)} & n \text{ an integer} \end{cases}$$

$$H_n^{(1)}(x) = J_n(x) + iY_n(x)$$
$$H_n^{(2)}(x) = J_n(x) - iY_n(x)$$

The additional properties of these functions may all be derived from the above relations and the known properties of $J_n(x)$.

13. Complete solutions to Bessel's equation may be written as

$$c_1 J_n(x) + c_2 J_{-n}(x) \qquad \text{if } n \text{ is not an integer}$$

or

$$\left.\begin{array}{l} c_1 J_n(x) + c_2 Y_n(x) \\ \\ c_1 H_n^{(1)}(x) + c_2 H_n^{(2)}(x) \end{array}\right\} \text{for any value of } n$$

or

14. The modified (or hyperbolic) Bessel's differential equation is

$$x^2 \frac{d^2y}{dx^2} + x\frac{dy}{dx} - (x^2 + n^2)y = 0$$

15. When n is not an integer, two independent solutions of the equation are $I_n(x)$ and $I_{-n}(x)$, where

$$I_n(x) = \sum_{k=0}^{\infty} \frac{1}{k!\Gamma(n + k + 1)}\left(\frac{x}{2}\right)^{n+2k}$$

16. If n is an integer,

$$I_n(x) = I_{-n}(x) = \frac{x^n}{2^n n!}\left\{1 + \frac{x^2}{2^2 \cdot 1!(n + 1)} + \frac{x^4}{2^4 \cdot 2!(n + 1)(n + 2)} + \frac{x^6}{2^6 \cdot 3!(n + 1)(n + 2)(n + 3)} + \cdots\right\}$$

17. For $N = 0$ and $n = 1$, this formula becomes

$$I_0(x) = 1 + \frac{x^2}{2^2(1!)^2} + \frac{x^4}{2^4(2!)^2} + \frac{x^6}{2^6(3!)^2} + \frac{x^8}{2^8(4!)^2} + \cdots$$

$$I_1(x) = \frac{x}{2} + \frac{x^3}{2^3 \cdot 1!2!} + \frac{x^5}{2^5 \cdot 2!3!} + \frac{x^7}{2^7 \cdot 3!4!} + \frac{x^9}{2^9 \cdot 4!5!} + \cdots$$

18. Another solution to the modified Bessel's equation is

$$K_n(x) = \begin{cases} \dfrac{1}{2}\pi \dfrac{I_{-n}(x) - I_n(x)}{\sin(n\pi)} & n \text{ not an integer} \\ \lim\limits_{\nu \to n} \dfrac{1}{2}\pi \dfrac{I_{-\nu}(x) - I_\nu(x)}{\sin(\nu\pi)} & n \text{ an integer} \end{cases}$$

This function is linearly independent of $I_n(x)$ for all values of n. Thus the complete solution to the modified Bessel's equation may be written as

$$c_1 I_n(x) + c_2 I_{-n}(x) \qquad n \text{ not an integer}$$

or

$$c_1 I_n(x) + c_2 K_n(x) \qquad \text{any } n$$

19. The following relations hold among the various Bessel functions:

$$I_n(z) = i^{-m} J_m(iz)$$

$$Y_n(iz) = (i)^{n+1} I_n(z) - \frac{2}{\pi} i^{-n} K_n(z)$$

Most of the properties of the modified Bessel function may be deduced from the known properties of $J_n(x)$ by use of these relations and those previously given.

20. Recurrence formulas

$$I_{n-1}(x) - I_{n+1}(x) = \frac{2n}{x} I_n(x) \qquad I_{n-1}(x) + I_{n+1}(x) = 2I_n'(x)$$

$$I_{n-1}(x) - \frac{n}{x} I_n(x) = I_n'(x) \qquad I_n'(x) = I_{n+1}(x) + \frac{n}{x} I_n(z)$$

The Gamma Function*

Definition: $\Gamma (n = \int_0^\infty t^{n-1} e^{-t}\, dt \; n > 0$

Recursion Formula: $\Gamma(n + 1 = n\Gamma(n)$

$\Gamma(n) + 1) = n!$ if $n = 0,1,2, \ldots$ where $0! = 1$

For $n < 0$ the gamma function can be defined by using

* From Beyer, W. H., Ed., *CRC Handbook of Mathematical Sciences,* 5th ed., CRC Press, Boca Raton, 1978, 484—485. With permision.

$$\Gamma(n) = \frac{\Gamma(n+1)}{n}$$

Graph:

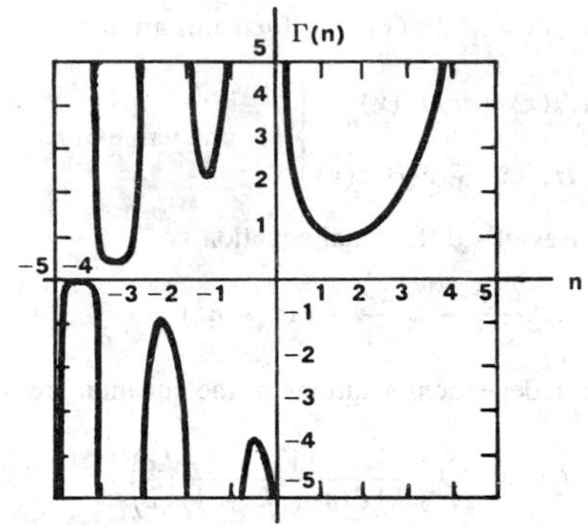

Special Values:

$$\Gamma(\tfrac{1}{2}) = \sqrt{\pi}$$

$$\Gamma(m + \tfrac{1}{2}) = \frac{1\cdot 3\cdot 5\cdots(2m-1)}{2^m}\sqrt{\pi} \qquad m = 1,2,3,\ldots$$

$$\Gamma(-m + \tfrac{1}{2}) = \frac{(-1)^m 2^m \sqrt{\pi}}{1\cdot 3\cdot 5\cdots(2m-1)} \qquad m = 1,2,3,\ldots$$

Definition:

$$\Gamma(x+1) = \lim_{k\to\infty} \frac{1\cdot 2\cdot 3\cdots k}{(x+1)(x+2)\cdots(x+k)} k^x$$

$$\frac{1}{\Gamma(x)} = xe^{\gamma x} \prod_{m=1}^{\infty} \left\{\left(1 + \frac{x}{m}\right) e^{-x/m}\right\}$$

This is an infinite product representation for the gamma function where γ is Euler's constant.

Properties:

$$\Gamma'(1) = \int_0^\infty e^{-x} \ln x \, dx = -\gamma$$

$$\frac{\Gamma'(x)}{\Gamma(x)} = -\gamma + \left(\frac{1}{1} - \frac{1}{x}\right) + \left(\frac{1}{2} - \frac{1}{x+1}\right) + \ldots + \left(\frac{1}{n} - \frac{1}{x+n-1}\right) + \ldots$$

$$\Gamma(x+1) = \sqrt{2\pi x}\, x^x e^{-x} \left\{1 + \frac{1}{12x} + \frac{1}{288x^2} - \frac{139}{51{,}840x^3} + \ldots\right\}$$

This is called *Stirling's asymptotic series.*

If we let $x = n$ a positive integer, then a useful approximation for $n!$ where n is large (e.g., $n > 10$) is given by *Stirling's formula*

$$n! \approx \sqrt{2\pi n}\, n^n e^{-n}$$

The Gamma Function*

Values of $\Gamma(n) = \int_0^\infty e^{-x} x^{n-1}\, dx;\ \Gamma(n+1) = n\Gamma(n)$

n	$\Gamma(n)$	n	$\Gamma(n)$	n	$\Gamma(n)$	n	$\Gamma(n)$
1.00	1.00000	1.25	.90640	1.50	.88623	1.75	.91906
1.01	.99433	1.26	.90440	1.51	.88659	1.76	.92137
1.02	.98884	1.27	.90250	1.52	.88704	1.77	.92376
1.03	.98355	1.28	.90072	1.53	.88757	1.78	.92623
1.04	.97844	1.29	.89904	1.54	.88818	1.79	.92877
1.05	.97350	1.30	.89747	1.55	.88887	1.80	.93138
1.06	.96874	1.31	.89600	1.56	.88964	1.81	.93408
1.07	.96415	1.32	.89464	1.57	.89049	1.82	.93685
1.08	.95973	1.33	.89338	1.58	.89142	1.83	.93969
1.09	.95546	1.34	.89222	1.59	.89243	1.84	.94261
1.10	.95135	1.35	.89115	1.60	.89352	1.85	.94561
1.11	.94740	1.36	.89018	1.61	.89468	1.86	.94869
1.12	.94359	1.37	.88931	1.62	.89592	1.87	.95184
1.13	.93993	1.38	.88854	1.63	.89724	1.88	.95507
1.14	.93642	1.39	.88785	1.64	.89864	1.89	.95838
1.15	.93304	1.40	.88726	1.65	.90012	1.90	.96177
1.16	.92980	1.41	.88676	1.66	.90167	1.91	.96523
1.17	.92670	1.42	.88636	1.67	.90330	1.92	.96877
1.18	.92373	1.43	.88604	1.68	.90500	1.93	.97240
1.19	.92089	1.44	.88581	1.69	.90678	1.94	.97610
1.20	.91817	1.45	.88566	1.70	.90864	1.95	.97988
1.21	.91558	1.46	.88560	1.71	.91057	1.96	.98374
1.22	.91311	1.47	.88563	1.72	.91258	1.97	.98768
1.23	.91075	1.48	.88575	1.73	.91466	1.98	.99171
1.24	.90852	1.49	.88595	1.74	.91683	1.99	.99581
						2.00	1.00000

* For large positive values of x, $\Gamma(x)$ approximates Stirling's asymptotic series

$$x^x e^{-x}\sqrt{\frac{2\pi}{x}}\left[1 + \frac{1}{12x} + \frac{1}{288x^2} - \frac{139}{51840x^3} - \frac{571}{2488320x^4} + \cdots\right]$$

The Beta Function*

Definition: $B(m,n) = \int_0^1 t^{m-1}(1-t)^{n-1}\, dt \qquad m > 0,\ n > 0$

Relationship with Gamma Function: $B(m,n) = \dfrac{\Gamma(m)\Gamma(n)}{\Gamma(m+n)}$

Properties:

$$B(m,n) = B(n,m)$$

$$B(m,n) = 2\int_0^{\pi/2} \sin^{2m-1}\theta \cos^{2n-1}\theta\, d\theta$$

$$B(m,n) = \int_0^\infty \frac{t^{m-1}}{(1+t)^{m+n}}\, dt$$

$$B(m,n) = r^n(r+1)^m \int_0^1 \frac{t^{m-1}(1-t)^{n-1}}{(r+t)^{m+n}}\, dt$$

The Error Function

Definition: $\operatorname{erf} x = \dfrac{2}{\sqrt{\pi}}\int_0^x e^{-t^2}\, dt$

Series: $\operatorname{erf} x = \dfrac{2}{\sqrt{\pi}}\left(x - \dfrac{x^3}{3} + \dfrac{1}{2!}\dfrac{x^5}{5} - \dfrac{1}{3!}\dfrac{x^7}{7} + \ldots\right)$

Property: $\operatorname{erf} x = -\operatorname{erf}(-x)$

Relationship with Normal Probability Function f(t): $\int_0^x f(t)\, dt = \frac{1}{2}\operatorname{erf}\left(\frac{x}{\sqrt{2}}\right)$

To evaluate *erf* (2.3), one proceeds as follows: Since $\frac{x}{\sqrt{2}} = 2.3$, one finds $x = (2.3)(\sqrt{2}) = 3.25$. In the normal probability function table (page A-), one finds the entry 0.4994 opposite the value 3.25. Thus *erf* (2.3) = 2(0.4994) = 0.9988.

* From Beyer, W. H., Ed., *CRC Handbook of Mathematical Sciences*, 5th ed., CRC Press, Boca Raton, 1978, 499. With permission.

$$erfc\ z = 1 - erf\ z = \frac{2}{\sqrt{\pi}} \int_z^{\infty} e^{-t^2}\, dt$$

is known as the complementary error function.

Orthogonal Polynomials*

I

Name: Legendre *Symbol:* $P_n(x)$ *Interval:* $[-1,1]$
Differential Equation: $(1 - x^2)\, y'' - 2\, xy' + n(n+1)\, y = 0$

$$y = P_n(x)$$

Explicit Expression: $P_n(x) = \frac{1}{2^n} \sum_{m=0}^{[n/2]} (-1)^m \binom{n}{m} \binom{2n - 2m}{n} x^{n-2m}$

Recurrence Relation: $(n+1)\, P_{n+1}(x) = (2n+1)\, x\, P_n(x) - nP_{n-1}(x)$
Weight: 1 *Standardization:* $P_n(1) = 1$

Norm: $\int_{-1}^{+1} [P_n(x)]^2\, dx = \frac{2}{2n+1}$

Rodrigues' Formula: $P_n(x) = \frac{(-1)^n}{2^n n!} \frac{d^n}{dx^n} \{(1 - x^2)^n\}$

Generating Function: $R^{-1} = \sum_{n=0}^{\infty} P_n(x) z^n; \; -1 < x < 1, \; |z| < 1,$

$$R = \sqrt{1 - 2xz + z^2}$$

Inequality: $|P_n(x)| \leq 1, \; -1 \leq x \leq 1.$

II

Name: Tschebysheff, First Kind *Symbol:* $T_n(x)$ *Interval:* $[-1,1]$
Differential Equation: $(1 - x^2)y - xy' + n^2y = 0$

$$y = T_n(x)$$

Explicit Expression: $\frac{n}{2} \sum_{m=0}^{[n/2]} (-1)^m \frac{(n-m-1)!}{m!(n-2m)!} (2x)^{n-2m} = \cos(n \arccos x) = T_n(x)$

Recurrence Relation: $T_{n+1}(x) = 2xT_n(x) - T_{n-1}(x)$
Weight: $(1 - x^2)^{-1/2}$ *Standardization:* $T_n(1) = 1$

Norm: $\int_{-1}^{+1} (1 - x^2)^{-1/2} [T_n(x)]^2\, dx = \begin{cases} \pi/2, & n \neq 0 \\ \pi, & n = 0 \end{cases}$

Rodrigues' Formula: $\frac{(-1)^n (1 - x^2)^{1/2} \sqrt{\pi}}{2^{n+1} \Gamma(n + \frac{1}{2})} \frac{d^n}{dx^n} \{(1 - x^2)^{n-(1/2)}\} = T_n(x)$

Generating Function: $\frac{1 - xz}{1 - 2xz + z^2} = \sum_{n=0}^{\infty} T_n(x) z^n, \; -1 < x < 1, \; |z| < 1$

Inequality: $|T_n(x)| \leq 1, \; -1 \leq x \leq 1.$

III

Name: Tschebysheff, Second Kind *Symbol:* $U_n(x)$ *Interval:* $[-1,1]$
Differential Equation: $(1 - x^2)\, y'' - 3\, xy' + n(n+2)y = 0$

$$y = U_n(x)$$

Explicit Expression: $U_n(x) = \sum_{m=0}^{[n/2]} (-1)^m \frac{(m-n)!}{m!(n-2m)!} (2x)^{n-2m}$

$$U_n(\cos\theta) = \frac{\sin[(n+1)\theta]}{\sin\theta}$$

* From Beyer, W. H., Ed., *CRC Handbook of Mathematical Sciences*, 5th ed., CRC Press, Boca Raton, 1978, 557—560. With permission.

Recurrence Relation: $U_{n+1}(x) = 2xU_n(x) - U_{n-1}(x)$
Weight: $(1 - x^2)^{1/2}$ *Standardization:* $U_n(1) = n + 1$

Norm: $\int_{-1}^{+1} (1 - x^2)^{1/2}[U_n(x)]^2\,dx = \frac{\pi}{2}$

Rodrigues' Formula: $U_n(x) = \frac{(-1)^n(n+1)\sqrt{\pi}}{(1-x^2)^{1/2}2^{n+1}\Gamma(n+\frac{3}{2})}\frac{d^n}{dx^n}\{(1-x^2)^{n+(1/2)}\}$

Generating Function: $\frac{1}{1 - 2xz + z^2} = \sum_{n=0}^{\infty} U_n(x)z^n, \ -1 < x < 1, \ |z| < 1$

Inequality: $|U_n(x)| \leq n + 1, \ -1 \leq x \leq 1.$

IV

Name: Jacobi *Symbol:* $P_n^{(\alpha,\beta)}(x)$ *Interval:* $[-1,1]$
Differential Equation:

$$(1 - x^2)y'' + [\beta - \alpha - (\alpha + \beta + 2)x]y' + n(n + \alpha + \beta + 1)y = 0$$
$$y = P_n^{(\alpha,\beta)}(x)$$

Explicit Expression: $P_n^{(\alpha,\beta)}(x) = \frac{1}{2^n}\sum_{m=0}^{n}\binom{n+\alpha}{m}\binom{n+\beta}{n-m}(x-1)^{n-m}(x+1)^m$

Recurrence Relation:
$$2(n+1)(n+\alpha+\beta+1)(2n+\alpha+\beta)P_{n+1}^{(\alpha,\beta)}(x) = (2n+\alpha+\beta+1)[(\alpha^2-\beta^2)+(2n+\alpha+\beta+2) \times (2n+\alpha+\beta)x]P_n^{(\alpha,\beta)}(x) - 2(n+\alpha)(n+\beta)(2n+\alpha+\beta+2)P_{n-1}^{(\alpha,\beta)}(x)$$

Weight: $(1 - x)^\alpha(1 + x)^\beta$; $\alpha,\beta > 1$ *Standardization:* $P_n^{(\alpha,\beta)}(x) = \binom{n+\alpha}{n}$

Norm: $\int_{-1}^{+1}(1-x)^\alpha(1+x)^\beta[P_n^{(\alpha,\beta)}(x)]^2\,dx = \frac{2^{\alpha+\beta+1}\Gamma(n+\alpha+1)\Gamma(n+\beta+1)}{(2n+\alpha+\beta+1)n!\,\Gamma(n+\alpha+\beta+1)}$

Rodrigues' Formula: $P_n^{(\alpha,\beta)}(x) = \frac{(-1)^n}{2^n n!(1-x)^\alpha(1+x)^\beta}\frac{d^n}{dx^n}\{(1-x)^{n+\alpha}(1+x)^{n+\beta}\}$

Generating Function: $R^{-1}(1 - z + R)^{-\alpha}(1 + z + R)^{-\beta} = \sum_{n=0}^{\infty} 2^{-\alpha-\beta}P_n^{(\alpha,\beta)}(x)z^n,$

$$R = \sqrt{1 - 2xz + z^2},\ |z| < 1$$

Inequality:
$$\max_{-1\leq x\leq 1}|P_n^{(\alpha,\beta)}(x)| = \begin{cases} \binom{n+q}{n} \sim n^q \text{ if } q = \max(\alpha,\beta) \geq -\frac{1}{2} \\ |P_n^{(\alpha,\beta)}(x')| \sim n^{-1/2} \text{ if } q < -\frac{1}{2} \\ x' \text{ is one of the two maximum points nearest} \\ \frac{\beta-\alpha}{\alpha+\beta+1} \end{cases}$$

V

Name: Generalized Laguerre *Symbol:* $L_n^{(\alpha)}(x)$ *Interval:* $[0, \infty]$
Differential Equation: $xy'' + (\alpha + 1 - x)y' + ny = 0$
$$y = L_n^{(\alpha)}(x)$$

Explicit Expression: $L_n^{(\alpha)}(x) = \sum_{m=0}^{n}(-1)^m\binom{n+\alpha}{n-m}\frac{1}{m!}x^m$

Recurrence Relation: $(n + 1)\,L_{n+1}^{(\alpha)}(x) = [(2n + \alpha + 1) - x]\,L_n^{(\alpha)}(x) - (n + \alpha)\,L_{n-1}^{(\alpha)}(x)$

Weight: $x^\alpha e^{-x}$, $\alpha > -1$ *Standardization:* $L_n^{(\alpha)}(x) = \frac{(-1)^n}{n!}x^n + \cdots$

Norm: $\int_0^\infty x^\alpha e^{-x}[L_n^{(\alpha)}(x)]^2\,dx = \frac{\Gamma(n+\alpha+1)}{n!}$

Rodrigues' Formula: $L_n^{(\alpha)}(x) = \frac{1}{n!\,x^\alpha e^{-x}}\frac{d^n}{dx^n}\{x^{n+\alpha}e^{-x}\}$

Generating Function: $(1 - z)^{-\alpha-1} \exp\left(\frac{xz}{z-1}\right) = \sum_{n=0}^{\infty} L_n^{(\alpha)}(x)\, z^n$

Inequality: $|L_n^{(\alpha)}(x)| \leq \frac{\Gamma(n+\alpha+1)}{n!\,\Gamma(\alpha+1)} e^{x/2}; \quad \begin{matrix} x \geq 0 \\ \alpha > 0 \end{matrix}$

$$|L_n^{(\alpha)}(x)| \leq \left[2 - \frac{\Gamma(\alpha+n+1)}{n!\,\Gamma(\alpha+1)}\right] e^{x/2}; \quad \begin{matrix} x \geq 0 \\ -1 < \alpha < 0 \end{matrix}$$

Orthogonal Polynomials

Name: Hermite *Symbol:* $H_n(x)$ *Interval:* $[-\infty, \infty]$
Differential Equation: $y'' - 2xy' + 2ny = 0$

Explicit Expression: $H_n(x) = \sum_{m=0}^{[n/2]} \frac{(-1)^m\, n!\, (2x)^{n-2m}}{m!\,(n-2m)!}$

Recurrence Relation: $H_{n+1}(x) = 2x\,H_n(x) - 2nH_{n-1}(x)$
Weight: e^{-x^2} *Standardization:* $H_n(1) = 2^n x^n + \ldots$

Norm: $\int_{-\infty}^{\infty} e^{-x^2} [H_n(x)]^2\, dx = 2^n\, n!\, \sqrt{\pi}$

Rodriques' Formula: $H_n(x) = (-1)^n e^{x^2} \frac{d^n}{dx^n}(e^{-x^2})$

Generating Function: $e^{-z^2 + 2zx} = \sum_{n=0}^{\infty} H_n(x) \frac{z^n}{n!}$

Inequality: $|H_n(x)| < e^{\frac{x^2}{2}}\, k\, 2^{n/2} \sqrt{n!} \quad k \approx 1.086435$

NORMAL PROBABILITY FUNCTION

Areas under the Standard Normal Curve from 0 to z

z	0	1	2	3	4	5	6	7	8	9
0.0	.0000	.0040	.0080	.0120	.0160	.0199	.0239	.0279	.0319	.0359
0.1	.0398	.0438	.0478	.0517	.0557	.0596	.0636	.0675	.0714	.0754
0.2	.0793	.0832	.0871	.0910	.0948	.0987	.1026	.1064	.1103	.1141
0.3	.1179	.1217	.1255	.1293	.1331	.1368	.1406	.1443	.1480	.1517
0.4	.1554	.1591	.1628	.1664	.1700	.1736	.1772	.1808	.1844	.1879
0.5	.1915	.1950	.1985	.2019	.2054	.2088	.2123	.2157	.2190	.2224
0.6	.2258	.2291	.2324	.2357	.2389	.2422	.2454	.2486	.2518	.2549
0.7	.2580	.2612	.2652	.2673	.2704	.2734	.2764	.2794	.2823	.2852
0.8	.2881	.2910	.2939	.2967	.2996	.3023	.3051	.3078	.3106	.3133
0.9	.3159	.3186	.3212	.3238	.3264	.3289	.3315	.3340	.3365	.3389
1.0	.3413	.3438	.3461	.3485	.3508	.3531	.3554	.3577	.3599	.3621
1.1	.3643	.3665	.3686	.3708	.3729	.3749	.3770	.3790	.3810	.3830
1.2	.3849	.3869	.3888	.3907	.3925	.3944	.3962	.3980	.3997	.4015
1.3	.4032	.4049	.4066	.4082	.4099	.4115	.4131	.4147	.4162	.4177
1.4	.4192	.4207	.4222	.4236	.4251	.4265	.4279	.4292	.4306	.4319
1.5	.4332	.4345	.4357	.4370	.4382	.4394	.4406	.4418	.4429	.4441
1.6	.4452	.4463	.4474	.4484	.4495	.4505	.4515	.4525	.4535	.4545
1.7	.4554	.4564	.4573	.4582	.4591	.4599	.4608	.4616	.4625	.4633
1.8	.4641	.4649	.4656	.4664	.4671	.4678	.4686	.4693	.4699	.4706
1.9	.4713	.4719	.4726	.4732	.4738	.4744	.4750	.4756	.4761	.4767
2.0	.4772	.4778	.4783	.4788	.4793	.4798	.4803	.4808	.4812	.4817
2.1	.4821	.4826	.4830	.4834	.4838	.4842	.4846	.4850	.4854	.4857
2.2	.4861	.4864	.4868	.4871	.4875	.4878	.4881	.4884	.4887	.4890
2.3	.4893	.4896	.4898	.4901	.4904	.4906	.4909	.4911	.4913	.4916
2.4	.4918	.4920	.4922	.4925	.4927	.4929	.4931	.4932	.4934	.4936
2.5	.4938	.4940	.4941	.4943	.4945	.4946	.4948	.4949	.4951	.4952
2.6	.4953	.4955	.4956	.4957	.4959	.4960	.4961	.4962	.4963	.4964
2.7	.4965	.4966	.4967	.4968	.4969	.4970	.4971	.4972	.4973	.4974
2.8	.4974	.4975	.4976	.4977	.4977	.4978	.4979	.4979	.4980	.4981
2.9	.4981	.4982	.4982	.4983	.4984	.4984	.4985	.4985	.4986	.4986
3.0	.4987	.4987	.4987	.4988	.4988	.4989	.4989	.4989	.4990	.4990
3.1	.4990	.4991	.4991	.4991	.4992	.4992	.4992	.4992	.4993	.4993
3.2	.4993	.4993	.4994	.4994	.4994	.4994	.4994	.4995	.4995	.4995
3.3	.4995	.4995	.4995	.4996	.4996	.4996	.4996	.4996	.4996	.4997
3.4	.4997	.4997	.4997	.4997	.4997	.4997	.4997	.4997	.4997	.4998
3.5	.4998	.4998	.4998	.4998	.4998	.4998	.4998	.4998	.4998	.4998
3.6	.4998	.4998	.4999	.4999	.4999	.4999	.4999	.4999	.4999	.4999
3.7	.4999	.4999	.4999	.4999	.4999	.4999	.4999	.4999	.4999	.4999
3.8	.4999	.4999	.4999	.4999	.4999	.4999	.4999	.4999	.4999	.4999
3.9	.5000	.5000	.5000	.5000	.5000	.5000	.5000	.5000	.5000	.5000

F(z) below refers to area under Standard Normal Curve from $-\infty$ to z

z	1.282	1.645	1.960	2.326	2.576	3.090
F(z)	.90	.95	.975	.99	.995	.999
2[1 − F(z)]	.20	.10	.05	.02	.01	.002

PERCENTAGE POINTS, STUDENT'S *t*-DISTRIBUTION

This table gives values of t such that

$$F(t) = \int_{-\infty}^{t} \frac{\Gamma\left(\frac{n+1}{2}\right)}{\sqrt{n\pi}\,\Gamma\left(\frac{n}{2}\right)} \left(1 + \frac{x^2}{n}\right)^{-\frac{n+1}{2}} dx$$

for n, the number of degrees of freedom, equal to 1, 2, . . ., 30, 40, 60, 120, ∞; and for $F(t) = 0.60, 0.75, 0.90, 0.95, 0.975, 0.99, 0.995,$ and 0.9995. The *t* distribution is symmetrical, so that $F(-t) = 1 - F(t)$

n \ F	.60	.75	.90	.95	.975	.99	.995	.9995
1	.325	1.000	3.078	6.314	12.706	31.821	63.657	636.619
2	.289	.816	1.886	2.920	4.303	6.965	9.925	31.598
3	.277	.765	1.638	2.353	3.182	4.541	5.841	12.924
4	.271	.741	1.533	2.132	2.776	3.747	4.604	8.610
5	.267	.727	1.476	2.015	2.571	3.365	4.032	6.869
6	.265	.718	1.440	1.943	2.447	3.143	3.707	5.959
7	.263	.711	1.415	1.895	2.365	2.998	3.499	5.408
8	.262	.706	1.397	1.860	2.306	2.896	3.355	5.041
9	.261	.703	1.383	1.833	2.262	2.821	3.250	4.781
10	.260	.700	1.372	1.812	2.228	2.764	3.169	4.587
11	.260	.697	1.363	1.796	2.201	2.718	3.106	4.437
12	.259	.695	1.356	1.782	2.179	2.681	3.055	4.318
13	.259	.694	1.350	1.771	2.160	2.650	3.012	4.221
14	.258	.692	1.345	1.761	2.145	2.624	2.977	4.140
15	.258	.691	1.341	1.753	2.131	2.602	2.947	4.073
16	.258	.690	1.337	1.746	2.120	2.583	2.921	4.015
17	.257	.689	1.333	1.740	2.110	2.567	2.898	3.965
18	.257	.688	1.330	1.734	2.101	2.552	2.878	3.922
19	.257	.688	1.328	1.729	2.093	2.539	2.861	3.883
20	.257	.687	1.325	1.725	2.086	2.528	2.845	3.850
21	.257	.686	1.323	1.721	2.080	2.518	2.831	3.819
22	.256	.686	1.321	1.717	2.074	2.508	2.819	3.792
23	.256	.685	1.319	1.714	2.069	2.500	2.807	3.767
24	.256	.685	1.318	1.711	2.064	2.492	2.797	3.745
25	.256	.684	1.316	1.708	2.060	2.485	2.787	3.725
26	.256	.684	1.315	1.706	2.056	2.479	2.779	3.707
27	.256	.684	1.314	1.703	2.052	2.473	2.771	3.690
28	.256	.683	1.313	1.701	2.048	2.467	2.763	3.674
29	.256	.683	1.311	1.699	2.045	2.462	2.756	3.659
30	.256	.683	1.310	1.697	2.042	2.457	2.750	3.646
40	.255	.681	1.303	1.684	2.021	2.423	2.704	3.551
60	.254	.679	1.296	1.671	2.000	2.390	2.660	3.460
120	.254	.677	1.289	1.658	1.980	2.358	2.617	3.373
∞	.253	.674	1.282	1.645	1.960	2.326	2.576	3.291

* This table is abridged from the "Statistical Tables" of R. A. Fisher and Frank Yates published by Oliver & Boyd. Ltd., Edinburgh and London, 1938. It is here published with the kind permission of the authors and their publishers.

PERCENTAGE POINTS, CHI-SQUARE DISTRIBUTION

This table gives values of χ^2 such that

$$F(\chi^2) = \int_0^{\chi^2} \frac{1}{2^{\frac{n}{2}}\,\Gamma\left(\frac{n}{2}\right)} x^{\frac{n-2}{2}} e^{-\frac{x}{2}}\,dx$$

for n, the number of degrees of freedom, equal to 1, 2, . . . , 30. For $n > 30$, a normal approximation is quite accurate. The expression $\sqrt{2\chi^2} - \sqrt{2n-1}$ is approximately normally distributed as the standard normal distribution. Thus χ^2_α, the α-point of the distribution, may be computed by the formula

$$\chi^2_\alpha = \tfrac{1}{2}[x_\alpha + \sqrt{2n-1}]^2,$$

where x_α is the α-point of the cumulative normal distribution. For even values of n, $F(\chi^2)$ can be written as

$$1 - F(\chi^2) = \sum_{x=0}^{x'-1} \frac{e^{-\lambda}\lambda^x}{x!}$$

with $\lambda = \frac{1}{2}\chi^2$ and $x' = \frac{1}{2}n$. Thus the cumulative Chi-Square distribution is related to the cumulative Poisson distribution.

Another approximate formula for large n

$$\chi^2_\alpha = n\left(1 - \frac{2}{9n} + z_\alpha\sqrt{\frac{2}{9n}}\right)^3$$

n = degrees of freedom
z_α = the normal deviate, (the value of x for which $F(x)$ = the desired percentile).

x	1.282	1.645	1.960	2.326	2.576	3.090
$F(x)$	.90	.95	.975	.99	.995	.999

$\chi^2_{.99} = 60[1 - 0.00370 + 2.326(0.06086)]^3 = 88.4$ is the 99th percentile for 60 degrees of freedom.

$$F(\chi^2) = \int_0^{\chi^2} \frac{1}{2^{\frac{n}{2}}\Gamma\left(\frac{n}{2}\right)} x^{\frac{n-2}{2}} e^{-\frac{x}{2}}\, dx$$

n \ F	.005	.010	.025	.050	.100	.250	.500	.750	.900	.950	.975	.990	.995
1	.0000393	.000157	.000982	.00393	.0158	.102	.455	1.32	2.71	3.84	5.02	6.63	7.88
2	.0100	.0201	.0506	.103	.211	.575	1.39	2.77	4.61	5.99	7.38	9.21	10.6
3	.0717	.115	.216	.352	.584	1.21	2.37	4.11	6.25	7.81	9.35	11.3	12.8
4	.207	.297	.484	.711	1.06	1.92	3.36	5.39	7.78	9.49	11.1	13.3	14.9
5	.412	.554	.831	1.15	1.61	2.67	4.35	6.63	9.24	11.1	12.8	1[illegible].1	16.7
6	.676	.872	1.24	1.64	2.20	3.45	5.35	7.84	10.6	12.6	14.4	16.8	18.5
7	.989	1.24	1.69	2.17	2.83	4.25	6.35	9.04	12.0	14.1	16.0	18.5	20.3
8	1.34	1.65	2.18	2.73	3.49	5.07	7.34	10.2	13.4	15.5	17.5	20.1	22.0
9	1.73	2.09	2.70	3.33	4.17	5.90	8.34	11.4	14.7	16.9	19.0	21.7	23.6
10	2.16	2.56	3.25	3.94	4.87	6.74	9.34	12.5	16.0	18.3	20.5	23.2	25.2
11	2.60	3.05	3.82	4.57	5.58	7.58	10.3	13.7	17.3	19.7	21.9	24.7	26.8
12	3.07	3.57	4.40	5.23	6.30	8.44	11.3	14.8	18.5	21.0	23.3	26.2	28.3
13	3.57	4.11	5.01	5.89	7.04	9.30	12.3	16.0	19.8	22.4	24.7	27.7	29.8
14	4.07	4.66	5.63	6.57	7.79	10.2	13.3	17.1	21.1	23.7	26.1	29.1	31.3
15	4.60	5.23	6.26	7.26	8.55	11.0	.4.3	18.2	22.3	25.0	27.5	30.6	32.8
16	5.14	5.81	6.91	7.96	9.31	11.9	15.3	19.4	23.5	26.3	28.8	32.0	34.3
17	5.70	6.41	7.56	8.67	10.1	12.8	16.3	20.5	24.8	27.6	30.2	33.4	35.7
18	6.26	7.01	8.23	9.39	10.9	13.7	17.3	21.6	26.0	28.9	31.5	34.8	37.2
19	6.84	7.63	8.91	10.1	11.7	14.6	18.3	22.7	27.2	30.1	32.9	36.2	38.6
20	7.43	8.26	9.59	10.9	12.4	15.5	19.3	23.8	28.4	31.4	34.2	37.6	40.0
21	8.03	8.90	10.3	11.6	13.2	16.3	20.3	24.9	29.6	32.7	35.5	38.9	41.4
22	8.64	9.54	11.0	12.3	14.0	17.2	21.3	26.0	30.8	33.9	36.8	40.3	42.8
23	9.26	10.2	11.7	13.1	14.8	18.1	22.3	27.1	32.0	35.2	38.1	41.6	44.2
24	9.89	10.9	12.4	13.8	15.7	19.0	23.3	28.2	33.2	36.4	39.4	43.0	45.6
25	10.5	11.5	13.1	14.6	16.5	19.9	24.3	29.3	34.4	37.7	40.6	44.3	46.9
26	11.2	12.2	13.8	15.4	17.3	20.8	25.3	30.4	35.6	38.9	41.9	45.6	48.3
27	11.8	12.9	14.6	16.2	18.1	21.7	26.3	31.5	36.7	40.1	43.2	47.0	49.6
28	12.5	13.6	15.3	16.9	18.9	22.7	27.3	32.6	37.9	41.3	44.5	48.3	51.0
29	13.1	14.3	16.0	17.7	19.8	23.6	28.3	33.7	39.1	42.6	45.7	49.6	52.3
30	13.8	15.0	16.8	18.5	20.6	24.5	29.3	34.8	40.3	43.8	47.0	50.9	53.7

PERCENTAGE POINTS, *F*-DISTRIBUTION

This table gives values of F such that

$$F(F) = \int_0^F \frac{\Gamma\left(\frac{m+n}{2}\right)}{\Gamma\left(\frac{m}{2}\right)\Gamma\left(\frac{n}{2}\right)} m^{\frac{m}{2}} n^{\frac{n}{2}} x^{\frac{m-2}{2}} (n + mx)^{-\frac{m+n}{2}}\, dx$$

for selected values of m, the number of degrees of freedom of the numerator of F; and for selected values of n, the number of degrees of freedom of the denominator of F. The table also provides values corresponding to $F(F)$ = .10, .05, .025, .01, .005, .001 since $F_{1-\alpha}$ for m and n degrees of freedom is the reciprocal of F_α for n and m degrees of freedom. Thus

$$F_{.05}(4, 7) = \frac{1}{F_{.95}(7, 4)} = \frac{1}{6.09} = .164\ .$$

$$F(F) = \int_0^F \frac{\Gamma\left(\frac{m+n}{2}\right)}{\Gamma\left(\frac{m}{2}\right)\Gamma\left(\frac{n}{2}\right)} m^{\frac{m}{2}} n^{\frac{n}{2}} x^{\frac{m}{2}-1} (n+mx)^{-\frac{m+n}{2}} dx = .90$$

n \ m	1	2	3	4	5	6	7	8	9	10	12	15	20	24	30	40	60	120	∞
1	39.86	49.50	53.59	55.83	57.24	58.20	58.91	59.44	59.86	60.19	60.71	61.22	61.74	62.00	62.26	62.53	62.79	63.06	63.33
2	8.53	9.00	9.16	9.24	9.29	9.33	9.35	9.37	9.38	9.39	9.41	9.42	9.44	9.45	9.46	9.47	9.47	9.48	9.49
3	5.54	5,46	5.39	5.34	5.31	5.28	5.27	5.25	5.24	5.23	5.22	5.20	5.18	5.18	5.17	5.16	5.15	5.14	5.13
4	4.54	4.32	4.19	4.11	4.05	4.01	3.98	3.95	3.94	3.92	3.90	3.87	3.84	3.83	3.82	3.80	3.79	3.78	3.76
5	4.06	3.78	3.62	3.52	3.45	3.40	3.37	3.34	3.32	3.30	3.27	3.24	3.21	3.19	3.17	3.16	3.14	3.12	3.10
6	3.78	3.46	3.29	3.18	3.11	3.05	3.01	2.98	2.96	2.94	2.90	2.87	2.84	2.82	2.80	2.78	2.76	2.74	2.72
7	3.59	3.26	3.07	2.96	2.88	2.83	2.78	2.75	2.72	2.70	2.67	2.63	2.59	2.58	2.56	2.54	2.51	2.49	2.47
8	3.46	3.11	2.92	2.81	2.73	2.67	2.62	2.59	2.56	2.54	2.50	2.46	2.42	2.40	2.38	2.36	2.34	2.32	2.29
9	3.36	3.01	2.81	2.69	2.61	2.55	2.51	2.47	2.44	2.42	2.38	2.34	2.30	2.28	2.25	2.23	2.21	2.18	2.16
10	3.29	2.92	2.73	2.61	2.52	2.46	2.41	2.38	2.35	2.32	2.28	2.24	2.20	2.18	2.16	2.13	2.11	2.08	2.06
11	3.23	2.86	2.66	2.54	2.45	2.39	2.34	2.30	2.27	2.25	2.21	2.17	2.12	2.10	2.08	2.05	2.03	2.00	1.97
12	3.18	2.81	2.61	2.48	2.39	2.33	2.28	2.24	2.21	2.19	2.15	2.10	2.06	2.04	2.01	1.99	1.96	1.93	1.90
13	3.14	2.76	2.56	2.43	2.35	2.28	2.23	2.20	2.16	2.14	2.10	2.05	2.01	1.98	1.96	1.93	1.90	1.88	1.85
14	3.10	2.73	2.52	2.39	2.31	2.24	2.19	2.15	2.12	2.10	2.05	2.01	1.96	1.94	1.91	1.89	1.86	1.83	1.80
15	3.07	2.70	2.49	2.36	2.27	2.21	2.16	2.12	2.09	2.06	2.02	1.97	1.92	1.90	1.87	1.85	1.82	1.79	1.76
16	3.05	2.67	2.46	2.33	2.24	2.18	2.13	2.09	2.06	2.03	1.99	1.94	1.89	1.87	1.84	1.81	1.78	1.75	1.72
17	3.03	2.64	2.44	2.31	2.22	2.15	2.10	2.06	2.03	2.00	1.96	1.91	1.86	1.84	1.81	1.78	1.75	1.72	1.69
18	3.01	2.62	2.42	2.29	2.20	2.13	2.08	2.04	2.00	1.98	1.93	1.89	1.84	1.81	1.78	1.75	1.72	1.69	1.66
19	2.99	2.61	2.40	2.27	2.18	2.11	2.06	2.02	1.98	1.96	1.91	1.86	1.81	1.79	1.76	1.73	1.70	1.67	1.63
20	2.97	2.59	2.38	2.25	2.16	2.09	2.04	2.00	1.96	1.94	1.89	1.84	1.79	1.77	1.74	1.71	1.68	1.64	1.61
21	2.96	2.57	2.36	2.23	2.14	2.08	2.02	1.98	1.95	1.92	1.87	1.83	1.78	1.75	1.72	1.69	1.66	1.62	1.59
22	2.95	2.56	2.35	2.22	2.13	2.06	2.01	1.97	1.93	1.90	1.86	1.81	1.76	1.73	1.70	1.67	1.64	1.60	1.57
23	2.94	2.55	2.34	2.21	2.11	2.05	1.99	1.95	1.92	1.89	1.84	1.80	1.74	1.72	1.69	1.66	1.62	1.59	1.55
24	2.93	2.54	2.33	2.19	2.10	2.04	1.98	1.94	1.91	1.88	1.83	1.78	1.73	1.70	1.67	1.64	1.61	1.57	1.53
25	2.92	2.53	2.32	2.18	2.09	2.02	1.97	1.93	1.89	1.87	1.82	1.77	1.72	1.69	1.66	1.63	1.59	1.56	1.62
26	2.91	2.52	2.31	2.17	2.08	2.01	1.96	1.92	1.88	1.86	1.81	1.76	1.71	1.68	1.65	1.61	1.58	1.54	1.50
27	2.90	2.51	2.30	2.17	2.07	2.00	1.95	1.91	1.87	1.85	1.80	1.75	1.70	1.67	1.64	1.60	1.57	1.53	1.49
28	2.89	2.50	2.29	2.16	2.06	2.00	1.94	1.90	1.87	1.84	1.79	1.74	1.69	1.66	1.63	1.59	1.56	1.52	1.48
29	2.89	2.50	2.28	2.15	2.06	1.99	1.93	1.89	1.86	1.83	1.78	1.73	1.68	1.65	1.62	1.58	1.55	1.51	1.47
30	2.88	2.49	2.28	2.14	2.05	1.98	1.93	1.88	1.85	1.82	1.77	1.72	1.67	1.64	1.61	1.57	1.54	1.50	1.46
40	2.84	2.44	2.23	2.09	2.00	1.93	1.87	1.83	1.79	1.76	1.71	1.66	1.61	1.57	1.54	1.51	1.47	1.42	1.38
60	2.79	2.39	2.18	2.04	1.95	1.87	1.82	1.77	1.74	1.71	1.66	1.60	1.54	1.51	1.48	1.44	1.40	1.35	1.29
120	2.75	2.35	2.13	1.99	1.90	1.82	1.77	1.72	1.68	1.65	1.60	1.55	1.48	1.45	1.41	1.37	1.32	1.26	1.19
∞	2.71	2.30	2.08	1.94	1.85	1.77	1.72	1.67	1.63	1.60	1.55	1.49	1.42	1.38	1.34	1.30	1.24	1.17	1.00

$F = \frac{s_1^2}{s_2^2} = \frac{S_1}{m} \Big/ \frac{S_2}{n}$, where $s_1^2 = S_1/m$ and $s_2^2 = S_2/n$ are independent mean squares estimating a common variance σ^2 and based on m and n degrees of freedom, respectively.

$$F(F) = \int_0^F \frac{\Gamma\left(\frac{m+n}{2}\right)}{\Gamma\left(\frac{m}{2}\right)\Gamma\left(\frac{n}{2}\right)} m^{\frac{m}{2}} n^{\frac{n}{2}} x^{\frac{m}{2}-1} (n+mx)^{-\frac{m+n}{2}} dx = .95$$

n \ m	1	2	3	4	5	6	7	8	9	10	12	15	20	24	30	40	60	120	∞
1	161.4	199.5	215.7	224.6	230.2	234.0	236.8	238.9	240.5	241.9	243.9	245.9	248.0	249.1	250.1	251.1	252.2	253.3	254.3
2	18.51	19.00	19.16	19.25	19.30	19.33	19.35	19.37	19.38	19.40	19.41	19.43	19.45	19.45	19.46	19.47	19.48	19.49	19.50
3	10.13	9.55	9.28	9.12	9.01	8.94	8.89	8.85	8.81	8.79	8.74	8.70	8.66	8.64	8.62	8.59	8.57	8.55	8.53
4	7.71	6.94	6.59	6.39	6.26	6.16	6.09	6.04	6.00	5.96	5.91	5.86	5.80	5.77	5.75	5.72	5.69	5.66	5.63
5	6.61	5.79	5.41	5.19	5.05	4.95	4.88	4.82	4.77	4.74	4.68	4.62	4.56	4.53	4.50	4.46	4.43	4.40	4.36
6	5.99	5.14	4.76	4.53	4.39	4.28	4.21	4.15	4.10	4.06	4.00	3.94	3.87	3.84	3.81	3.77	3.74	3.70	3.67
7	5.59	4.74	4.35	4.12	3.97	3.87	3.79	3.73	3.68	3.64	3.57	3.51	3.44	3.41	3.38	3.34	3.30	3.27	3.23
8	5.32	4.46	4.07	3.84	3.69	3.58	3.50	3.44	3.39	3.35	3.28	3.22	3.15	3.12	3.08	3.04	3.01	2.97	2.93
9	5.12	4.26	3.86	3.63	3.48	3.37	3.29	3.23	3.18	3.14	3.07	3.01	2.94	2.90	2.86	2.83	2.79	2.75	2.71
10	4.96	4.10	3.71	3.48	3.33	3.22	3.14	3.07	3.02	2.98	2.91	2.85	2.77	2.74	2.70	2.66	2.62	2.58	2.54
11	4.84	3.98	3.59	3.36	3.20	3.09	3.01	2.95	2.90	2.85	2.79	2.72	2.65	2.61	2.57	2.53	2.49	2.45	2.40
12	4.75	3.89	3.49	3.26	3.11	3.00	2.91	2.85	2.80	2.75	2.69	2.62	2.54	2.51	2.47	2.43	2.38	2.34	2.30
13	4.67	3.81	3.41	3.18	3.03	2.92	2.83	2.77	2.71	2.67	2.60	2.53	2.46	2.42	2.38	2.34	2.30	2.25	2.21
14	4.60	3.74	3.34	3.11	2.96	2.85	2.76	2.70	2.65	2.60	2.53	2.46	2.39	2.35	2.31	2.27	2.22	2.18	2.13
15	4.54	3.68	3.29	3.06	2.90	2.79	2.71	2.64	2.59	2.54	2.48	2.40	2.33	2.29	2.25	2.20	2.16	2.11	2.07
16	4.49	3.63	3.24	3.01	2.85	2.74	2.66	2.59	2.54	2.49	2.42	2.35	2.28	2.24	2.19	2.15	2.11	2.06	2.01
17	4.45	3.59	3.20	2.96	2.81	2.70	2.61	2.55	2.49	2.45	2.38	2.31	2.23	2.19	2.15	2.10	2.06	2.01	1.96
18	4.41	3.55	3.16	2.93	2.77	2.66	2.58	2.51	2.46	2.41	2.34	2.27	2.19	2.15	2.11	2.06	2.02	1.97	1.92
19	4.38	3.52	3.13	2.90	2.74	2.63	2.54	2.48	2.42	2.38	2.31	2.23	2.16	2.11	2.07	2.03	1.98	1.93	1.88
20	4.35	3.49	3.10	2.87	2.71	2.60	2.51	2.45	2.39	2.35	2.28	2.20	2.12	2.08	2.04	1.99	1.95	1.90	1.84
21	4.32	3.47	3.07	2.84	2.68	2.57	2.49	2.42	2.37	2.32	2.25	2.18	2.10	2.05	2.01	1.96	1.92	1.87	1.81
22	4.30	3.44	3.05	2.82	2.66	2.55	2.46	2.40	2.34	2.30	2.23	2.15	2.07	2.03	1.98	1.94	1.89	1.84	1.78
23	4.28	3.42	3.03	2.80	2.64	2.53	2.44	2.37	2.32	2.27	2.20	2.13	2.05	2.01	1.96	1.91	1.86	1.81	1.76
24	4.26	3.40	3.01	2.78	2.62	2.51	2.42	2.36	2.30	2.25	2.18	2.11	2.03	1.98	1.94	1.89	1.84	1.79	1.73
25	4.24	3.39	2.99	2.76	2.60	2.49	2.40	2.34	2.28	2.24	2.16	2.09	2.01	1.96	1.92	1.87	1.82	1.77	1.71
26	4.23	3.37	2.98	2.74	2.59	2.47	2.39	2.32	2.27	2.22	2.15	2.07	1.99	1.95	1.90	1.85	1.80	1.75	1.69
27	4.21	3.35	2.96	2.73	2.57	2.46	2.37	2.31	2.25	2.20	2.13	2.06	1.97	1.93	1.88	1.84	1.79	1.73	1.67
28	4.20	3.34	2.95	2.71	2.56	2.45	2.36	2.29	2.24	2.19	2.12	2.04	1.96	1.91	1.87	1.82	1.77	1.71	1.65
29	4.18	3.33	2.93	2.70	2.55	2.43	2.35	2.28	2.22	2.18	2.10	2.03	1.94	1.90	1.85	1.81	1.75	1.70	1.64
30	4.17	3.32	2.92	2.69	2.53	2.42	2.33	2.27	2.21	2.16	2.09	2.01	1.93	1.89	1.84	1.79	1.74	1.68	1.62
40	4.08	3.23	2.84	2.61	2.45	2.34	2.25	2.18	2.12	2.08	2.00	1.92	1.84	1.79	1.74	1.69	1.64	1.58	1.51
60	4.00	3.15	2.76	2.53	2.37	2.25	2.17	2.10	2.04	1.99	1.92	1.84	1.75	1.70	1.65	1.59	1.53	1.47	1.39
120	3.92	3.07	2.68	2.45	2.29	2.17	2.09	2.02	1.96	1.91	1.83	1.75	1.66	1.61	1.55	1.50	1.43	1.35	1.25
∞	3.84	3.00	2.60	2.37	2.21	2.10	2.01	1.94	1.88	1.83	1.75	1.67	1.57	1.52	1.46	1.39	1.32	1.22	1.00

$F = \frac{s_1^2}{s_2^2} = \frac{S_1}{m} \Big/ \frac{S_2}{n}$, where $s_1^2 = S_1/m$ and $s_2^2 = S_2/n$ are independent mean squares estimating a common variance σ^2 and based on m and n degrees of freedom, respectively.

$$F(F) = \int_0^F \frac{\Gamma\left(\frac{m+n}{2}\right)}{\Gamma\left(\frac{m}{2}\right)\Gamma\left(\frac{n}{2}\right)} m^{\frac{m}{2}} n^{\frac{n}{2}} x^{\frac{m}{2}-1} (n+mx)^{-\frac{m+n}{2}} \, dx = .975$$

n \ m	1	2	3	4	5	6	7	8	9	10	12	15	20	24	30	40	60	120	∞
1	647.8	799.5	864.2	899.6	921.8	937.1	948.2	956.7	963.3	968.6	976.7	984.9	993.1	997.2	1001	1006	1010	1014	1018
2	38.51	39.00	39.17	39.25	39.30	39.33	39.36	39.37	39.39	39.40	39.41	39.43	39.45	39.46	39.46	39.47	39.48	39.49	39.50
3	17.44	16.04	15.44	15.10	14.88	14.73	14.62	14.54	14.47	14.42	14.34	14.25	14.17	14.12	14.08	14.04	13.99	13.95	13.90
4	12.22	10.65	9.98	9.60	9.36	9.20	9.07	8.98	8.90	8.84	8.75	8.66	8.56	8.51	8.46	8.41	8.36	8.31	8.26
5	10.01	8.43	7.76	7.39	7.15	6.98	6.85	6.76	6.68	6.62	6.52	6.43	6.33	6.28	6.23	6.18	6.12	6.07	6.02
6	8.81	7.26	6.60	6.23	5.99	5.82	5.70	5.60	5.52	5.46	5.37	5.27	5.17	5.12	5.07	5.01	4.96	4.90	4.85
7	8.07	6.54	5.89	5.52	5.29	5.12	4.99	4.90	4.82	4.76	4.67	4.57	4.47	4.42	4.36	4.31	4.25	4.20	4.14
8	7.57	6.06	5.42	5.05	4.82	4.65	4.53	4.43	4.36	4.30	4.20	4.10	4.00	3.95	3.89	3.84	3.78	3.73	3.67
9	7.21	5.71	5.08	4.72	4.48	4.32	4.20	4.10	4.03	3.96	3.87	3.77	3.67	3.61	3.56	3.51	3.45	3.39	3.33
10	6.94	5.46	4.83	4.47	4.24	4.07	3.95	3.85	3.78	3.72	3.62	3.52	3.42	3.37	3.31	3.26	3.20	3.14	3.08
11	6.72	5.26	4.63	4.28	4.04	3.88	3.76	3.66	3.59	3.53	3.43	3.33	3.23	3.17	3.12	3.06	3.00	2.94	2.88
12	6.55	5.10	4.47	4.12	3.89	3.73	3.61	3.51	3.44	3.37	3.28	3.18	3.07	3.02	2.96	2.91	2.85	2.79	2.72
13	6.41	4.97	4.35	4.00	3.77	3.60	3.48	3.39	3.31	3.25	3.15	3.05	2.95	2.89	2.84	2.78	2.72	2.66	2.60
14	6.30	4.86	4.24	3.89	3.66	3.50	3.38	3.29	3.21	3.15	3.05	2.95	2.84	2.79	2.73	2.67	2.61	2.55	2.49
15	6.20	4.77	4.15	3.80	3.58	3.41	3.29	3.20	3.12	3.06	2.96	2.86	2.76	2.70	2.64	2.59	2.52	2.46	2.40
16	6.12	4.69	4.08	3.73	3.50	3.34	3.22	3.12	3.05	2.99	2.89	2.79	2.68	2.63	2.57	2.51	2.45	2.38	2.32
17	6.04	4.62	4.01	3.66	3.44	3.28	3.16	3.06	2.98	2.92	2.82	2.72	2.62	2.56	2.50	2.44	2.38	2.32	2.25
18	5.98	4.56	3.95	3.61	3.38	3.22	3.10	3.01	2.93	2.87	2.77	2.67	2.56	2.50	2.44	2.38	2.32	2.26	2.19
19	5.92	4.51	3.90	3.56	3.33	3.17	3.05	2.96	2.88	2.82	2.72	2.62	2.51	2.45	2.39	2.33	2.27	2.20	2.13
20	5.87	4.46	3.86	3.51	3.29	3.13	3.01	2.91	2.84	2.77	2.68	2.57	2.46	2.41	2.35	2.29	2.22	2.16	2.09
21	5.83	4.42	3.82	3.48	3.25	3.09	2.97	2.87	2.80	2.73	2.64	2.53	2.42	2.37	2.31	2.25	2.18	2.11	2.04
22	5.79	4.38	3.78	3.44	3.22	3.05	2.93	2.84	2.76	2.70	2.60	2.50	2.39	2.33	2.27	2.21	2.14	2.08	2.00
23	5.75	4.35	3.75	3.41	3.18	3.02	2.90	2.81	2.73	2.67	2.57	2.47	2.36	2.30	2.24	2.18	2.11	2.04	1.97
24	5.72	4.32	3.72	3.38	3.15	2.99	2.87	2.78	2.70	2.64	2.54	2.44	2.33	2.27	2.21	2.15	2.08	2.01	1.94
25	5.69	4.29	3.69	3.35	3.13	2.97	2.85	2.75	2.68	2.61	2.51	2.41	2.30	2.24	2.18	2.12	2.05	1.98	1.91
26	5.66	4.27	3.67	3.33	3.10	2.94	2.82	2.73	2.65	2.59	2.49	2.39	2.28	2.22	2.16	2.09	2.03	1.95	1.88
27	5.63	4.24	3.65	3.31	3.08	2.92	2.80	2.71	2.63	2.57	2.47	2.36	2.25	2.19	2.13	2.07	2.00	1.93	1.85
28	5.61	4.22	3.63	3.29	3.06	2.90	2.78	2.69	2.61	2.55	2.45	2.34	2.23	2.17	2.11	2.05	1.98	1.91	1.83
29	5.59	4.20	3.61	3.27	3.04	2.88	2.76	2.67	2.59	2.53	2.43	2.32	2.21	2.15	2.09	2.03	1.96	1.89	1.81
30	5.57	4.18	3.59	3.25	3.03	2.87	2.75	2.65	2.57	2.51	2.41	2.31	2.20	2.14	2.07	2.01	1.94	1.87	1.79
40	5.42	4.05	3.46	3.13	2.90	2.74	2.62	2.53	2.45	2.39	2.29	2.18	2.07	2.01	1.94	1.88	1.80	1.72	1.64
60	5.29	3.93	3.34	3.01	2.79	2.63	2.51	2.41	2.33	2.27	2.17	2.06	1.94	1.88	1.82	1.74	1.67	1.58	1.48
120	5.15	3.80	3.23	2.89	2.67	2.52	2.39	2.30	2.22	2.16	2.05	1.94	1.82	1.76	1.69	1.61	1.53	1.43	1.31
∞	5.02	3.69	3.12	2.79	2.57	2.41	2.29	2.19	2.11	2.05	1.94	1.83	1.71	1.64	1.57	1.48	1.39	1.27	1.00

$F = \frac{s_1^2}{s_2^2} = \frac{S_1}{m} \Big/ \frac{S_2}{n}$, where $s_1^2 = S_1/m$ and $s_2^2 = S_2/n$ are independent mean squares estimating a common variance σ^2 and based on m and n degrees of freedom, respectively.

$$F(F) = \int_0^F \frac{\Gamma\left(\frac{m+n}{2}\right)}{\Gamma\left(\frac{m}{2}\right)\Gamma\left(\frac{n}{2}\right)} m^{\frac{m}{2}} n^{\frac{n}{2}} x^{\frac{m}{2}-1} (n+mx)^{-\frac{m+n}{2}} \, dx = .99$$

n \ m	1	2	3	4	5	6	7	8	9	10	12	15	20	24	30	40	60	120	∞
1	4052	4999.5	5403	5625	5764	5859	5928	5982	6022	6056	6106	6157	6209	6235	6261	6287	6313	6339	6366
2	98.50	99.00	99.17	99.25	99.30	99.33	99.36	99.37	99.39	99.40	99.42	99.43	99.45	99.46	99.47	99.47	99.48	99.49	99.50
3	34.12	30.82	29.46	28.71	28.24	27.91	27.67	27.49	27.35	27.23	27.05	26.87	26.69	26.60	26.50	26.41	26.32	26.22	26.13
4	21.20	18.00	16.69	15.98	15.52	15.21	14.98	14.80	14.66	14.55	14.37	14.20	14.02	13.93	13.84	13.75	13.65	13.56	13.46
5	16.26	13.27	12.06	11.39	10.97	10.67	10.46	10.29	10.16	10.05	9.89	9.72	9.55	9.47	9.38	9.29	9.20	9.11	9.02
6	13.75	10.92	9.78	9.15	8.75	8.47	8.26	8.10	7.98	7.87	7.72	7.56	7.40	7.31	7.23	7.14	7.06	6.97	6.88
7	12.25	9.55	8.45	7.85	7.46	7.19	6.99	6.84	6.72	6.62	6.47	6.31	6.16	6.07	5.99	5.91	5.82	5.74	5.65
8	11.26	8.65	7.59	7.01	6.63	6.37	6.18	6.03	5.91	5.81	5.67	5.52	5.36	5.28	5.20	5.12	5.03	4.95	4.86
9	10.56	8.02	6.99	6.42	6.06	5.80	5.61	5.47	5.35	5.26	5.11	4.96	4.81	4.73	4.65	4.57	4.48	4.40	4.31
10	10.04	7.56	6.55	5.99	5.64	5.39	5.20	5.06	4.94	4.85	4.71	4.56	4.41	4.33	4.25	4.17	4.08	4.00	3.91
11	9.65	7.21	6.22	5.67	5.32	5.07	4.89	4.74	4.63	4.54	4.40	4.25	4.10	4.02	3.94	3.86	3.78	3.69	3.60
12	9.33	6.93	5.95	5.41	5.06	4.82	4.64	4.50	4.39	4.30	4.16	4.01	3.86	3.78	3.70	3.62	3.54	3.45	3.36
13	9.07	6.70	5.74	5.21	4.86	4.62	4.44	4.30	4.19	4.10	3.96	3.82	3.66	3.59	3.51	3.43	3.34	3.25	3.17
14	8.86	6.51	5.56	5.04	4.69	4.46	4.28	4.14	4.03	3.94	3.80	3.66	3.51	3.43	3.35	3.27	3.18	3.09	3.00
15	8.68	6.36	5.42	4.89	4.56	4.32	4.14	4.00	3.89	3.80	3.67	3.52	3.37	3.29	3.21	3.13	3.05	2.96	2.87
16	8.53	6.23	5.29	4.77	4.44	4.20	4.03	3.89	3.78	3.69	3.55	3.41	3.26	3.18	3.10	3.02	2.93	2.84	2.75
17	8.40	6.11	5.18	4.67	4.34	4.10	3.93	3.79	3.68	3.59	3.46	3.31	3.16	3.08	3.00	2.92	2.83	2.75	2.65
18	8.29	6.01	5.09	4.58	4.25	4.01	3.84	3.71	3.60	3.51	3.37	3.23	3.08	3.00	2.92	2.84	2.75	2.66	2.57
19	8.18	5.93	5.01	4.50	4.17	3.94	3.77	3.63	3.52	3.43	3.30	3.15	3.00	2.92	2.84	2.76	2.67	2.58	2.49
20	8.10	5.85	4.94	4.43	4.10	3.87	3.70	3.56	3.46	3.37	3.23	3.09	2.94	2.86	2.78	2.69	2.61	2.52	2.42
21	8.02	5.78	4.87	4.37	4.04	3.81	3.64	3.51	3.40	3.31	3.17	3.03	2.88	2.80	2.72	2.64	2.55	2.46	2.36
22	7.95	5.72	4.82	4.31	3.99	3.76	3.59	3.45	3.35	3.26	3.12	2.98	2.83	2.75	2.67	2.58	2.50	2.40	2.31
23	7.88	5.66	4.76	4.26	3.94	3.71	3.54	3.41	3.30	3.21	3.07	2.93	2.78	2.70	2.62	2.54	2.45	2.35	2.26
24	7.82	5.61	4.72	4.22	3.90	3.67	3.50	3.36	3.26	3.17	3.03	2.89	2.74	2.66	2.58	2.49	2.40	2.31	2.21
25	7.77	5.57	4.68	4.18	3.85	3.63	3.46	3.32	3.22	3.13	2.99	2.85	2.70	2.62	2.54	2.45	2.36	2.27	2.17
26	7.72	5.53	4.64	4.14	3.82	3.59	3.42	3.29	3.18	3.09	2.96	2.81	2.66	2.58	2.50	2.42	2.33	2.23	2.13
27	7.68	5.49	4.60	4.11	3.78	3.56	3.39	3.26	3.15	3.06	2.93	2.78	2.63	2.55	2.47	2.38	2.29	2.20	2.10
28	7.64	5.45	4.57	4.07	3.75	3.53	3.36	3.23	3.12	3.03	2.90	2.75	2.60	2.52	2.44	2.35	2.26	2.17	2.06
29	7.60	5.42	4.54	4.04	3.73	3.50	3.33	3.20	3.09	3.00	2.87	2.73	2.57	2.49	2.41	2.33	2.23	2.14	2.03
30	7.56	5.39	4.51	4.02	3.70	3.47	3.30	3.17	3.07	2.98	2.84	2.70	2.55	2.47	2.39	2.30	2.21	2.11	2.01
40	7.31	5.18	4.31	3.83	3.51	3.29	3.12	2.99	2.89	2.80	2.66	2.52	2.37	2.29	2.20	2.11	2.02	1.92	1.80
60	7.08	4.98	4.13	3.65	3.34	3.12	2.95	2.82	2.72	2.63	2.50	2.35	2.20	2.12	2.03	1.94	1.84	1.73	1.60
120	6.85	4.79	3.95	3.48	3.17	2.96	2.79	2.66	2.56	2.47	2.34	2.19	2.03	1.95	1.86	1.76	1.66	1.53	1.38
∞	6.63	4.61	3.78	3.32	3.02	2.80	2.64	2.51	2.41	2.32	2.18	2.04	1.88	1.79	1.70	1.59	1.47	1.32	1.00

$F = \frac{s_1^2}{s_2^2} = \frac{S_1}{m} \Big/ \frac{S_2}{n}$, where $s_1^2 = S_1/m$ and $s_2^2 = S_2/n$ are independent mean squares estimating a common variance σ^2 and based on m and n degrees of freedom, respectively.

$$F(F) = \int_0^F \frac{\Gamma\left(\frac{m+n}{2}\right)}{\Gamma\left(\frac{m}{2}\right)\Gamma\left(\frac{n}{2}\right)} m^{\frac{m}{2}} n^{\frac{n}{2}} x^{\frac{m}{2}-1} (n+mx)^{-\frac{m+n}{2}} dx = .995$$

n \ m	1	2	3	4	5	6	7	8	9	10	12	15	20	24	30	40	60	120	∞
1	16211	20000	21615	22500	23056	23437	23715	23925	24091	24224	24426	24630	24836	24940	25044	25148	25253	25359	25465
2	198.5	199.0	199.2	199.2	199.3	199.3	199.4	199.4	199.4	199.4	199.4	199.4	199.4	199.5	199.5	199.5	199.5	199.5	199.5
3	55.55	49.80	47.47	46.19	45.39	44.84	44.43	44.13	43.88	43.69	43.39	43.08	42.78	42.62	42.47	42.31	42.15	41.99	41.83
4	31.33	26.28	24.26	23.15	22.46	21.97	21.62	21.35	21.14	20.97	20.70	20.44	20.17	20.03	19.89	19.75	19.61	19.47	19.32
5	22.78	18.31	16.53	15.56	14.94	14.51	14.20	13.96	13.77	13.62	13.38	13.15	12.90	12.78	12.66	12.53	12.40	12.27	12.14
6	18.63	14.54	12.92	12.03	11.46	11.07	10.79	10.57	10.39	10.25	10.03	9.81	9.59	9.47	9.36	9.24	9.12	9.00	8.88
7	16.24	12.40	10.88	10.05	9.52	9.16	8.89	8.68	8.51	8.38	8.18	7.97	7.75	7.65	7.53	7.42	7.31	7.19	7.08
8	14.69	11.04	9.60	8.81	8.30	7.95	7.69	7.50	7.34	7.21	7.01	6.81	6.61	6.50	6.40	6.29	6.18	6.06	5.95
9	13.61	10.11	8.72	7.96	7.47	7.13	6.88	6.69	6.54	6.42	6.23	6.03	5.83	5.73	5.62	5.52	5.41	5.30	5.19
10	12.83	9.43	8.08	7.34	6.87	6.54	6.30	6.12	5.97	5.85	5.66	5.47	5.27	5.17	5.07	4.97	4.86	4.75	4.64
11	12.23	8.91	7.60	6.88	6.42	6.10	5.86	5.68	5.54	5.42	5.24	5.05	4.86	4.76	4.65	4.55	4.44	4.34	4.23
12	11.75	8.51	7.23	6.52	6.07	5.76	5.52	5.35	5.20	5.09	4.91	4.72	4.53	4.43	4.33	4.23	4.12	4.01	3.90
13	11.37	8.19	6.93	6.23	5.79	5.48	5.25	5.08	4.94	4.82	4.64	4.46	4.27	4.17	4.07	3.97	3.87	3.76	3.65
14	11.06	7.92	6.68	6.00	5.56	5.26	5.03	4.86	4.72	4.60	4.43	4.25	4.06	3.96	3.86	3.76	3.66	3.55	3.44
15	10.80	7.70	6.48	5.80	5.37	5.07	4.85	4.67	4.54	4.42	4.25	4.07	3.88	3.79	3.69	3.58	3.48	3.37	3.26
16	10.58	7.51	6.30	5.64	5.21	4.91	4.60	4.52	4.38	4.27	4.10	3.92	3.73	3.64	3.54	3.44	3.33	3.22	3.11
17	10.38	7.35	6.16	5.50	5.07	4.78	4.56	4.39	4.25	4.14	3.97	3.79	3.61	3.51	3.41	3.31	3.21	3.10	2.98
18	10.22	7.21	6.03	5.37	4.96	4.66	4.44	4.28	4.14	4.03	3.86	3.68	3.50	3.40	3.30	3.20	3.10	2.99	2.87
19	10.07	7.09	5.92	5.27	4.85	4.56	4.34	4.18	4.04	3.93	3.76	3.59	3.40	3.31	3.21	3.11	3.00	2.89	2.78
20	9.94	6.99	5.82	5.17	4.76	4.47	4.26	4.09	3.96	3.85	3.68	3.50	3.32	3.22	3.12	3.02	2.92	2.81	2.69
21	9.83	6.89	5.73	5.09	4.68	4.39	4.18	4.01	3.88	3.77	3.60	3.43	3.24	3.15	3.05	2.95	2.84	2.73	2.61
22	9.73	6.81	5.65	5.02	4.61	4.32	4.11	3.94	3.81	3.70	3.54	3.36	3.18	3.08	2.98	2.88	2.77	2.66	2.55
23	9.63	6.73	5.58	4.95	4.54	4.26	4.05	3.88	3.75	3.64	3.47	3.30	3.12	3.02	2.92	2.82	2.71	2.60	2.48
24	9.55	6.66	5.52	4.89	4.49	4.20	3.99	3.83	3.69	3.59	3.42	3.25	3.06	2.97	2.87	2.77	2.66	2.55	2.43
25	9.48	6.60	5.46	4.84	4.43	4.15	3.94	3.78	3.64	3.54	3.37	3.20	3.01	2.92	2.82	2.72	2.61	2.50	2.38
26	9.41	6.54	5.41	4.79	4.38	4.10	3.89	3.73	3.60	3.49	3.33	3.15	2.97	2.87	2.77	2.67	2.56	2.45	2.33
27	9.34	6.49	5.36	4.74	4.34	4.06	3.85	3.69	3.56	3.45	3.28	3.11	2.93	2.83	2.73	2.63	2.52	2.41	2.25
28	9.28	6.44	5.32	4.70	4.30	4.02	3.81	3.65	3.52	3.41	3.25	3.07	2.89	2.79	2.69	2.59	2.48	2.37	2.29
29	9.23	6.40	5.28	4.66	4.26	3.98	3.77	3.61	3.48	3.38	3.21	3.04	2.86	2.76	2.66	2.56	2.45	2.33	2.24
30	9.18	6.35	5.24	4.62	4.23	3.95	3.74	3.58	3.45	3.34	3.18	3.01	2.82	2.73	2.63	2.52	2.42	2.30	2.18
40	8.83	6.07	4.98	4.37	3.99	3.71	3.51	3.35	3.22	3.12	2.95	2.78	2.60	2.50	2.40	2.30	2.18	2.06	1.93
60	8.49	5.79	4.73	4.14	3.76	3.49	3.29	3.13	3.01	2.90	2.74	2.57	2.39	2.29	2.19	2.08	1.96	1.83	1.69
120	8.18	5.54	4.50	3.92	3.55	3.28	3.09	2.93	2.81	2.71	2.54	2.37	2.19	2.09	1.98	1.87	1.75	1.61	1.43
∞	7.88	5.30	4.28	3.72	3.35	3.09	2.90	2.74	2.62	2.52	2.36	2.19	2.00	1.90	1.79	1.67	1.53	1.36	1.00

$F = \frac{s_1^2}{s_2^2} = \frac{S_1}{m} \bigg/ \frac{S_2}{n}$, where $s_1^2 = S_1/m$ and $s_2^2 = S_2/n$ are independent mean squares estimating a common variance σ^2 and based on m and n degrees of freedom, respectively.

$$F(F) = \int_0^F \frac{\Gamma\left(\frac{m+n}{2}\right)}{\Gamma\left(\frac{m}{2}\right)\Gamma\left(\frac{n}{2}\right)} m^{\frac{m}{2}} n^{\frac{n}{2}} x^{\frac{m}{2}-1} (n+mx)^{-\frac{m+n}{2}} dx = .999$$

n \ m	1	2	3	4	5	6	7	8	9	10	12	15	20	24	30	40	60	120	∞
1	4053*	5000*	5404*	5625*	5764*	5859*	5929*	5981*	6023*	6056*	6107*	6158*	6209*	6235*	6261*	6287*	6313*	6340*	6366*
2	998.5	999.0	999.2	999.2	999.3	999.3	999.4	999.4	999.4	999.4	999.4	999.4	999.4	999.5	999.5	999.5	999.5	999.5	999.5
3	167.0	148.5	141.1	137.1	134.6	132.8	131.6	130.6	129.9	129.2	128.3	127.4	126.4	125.9	125.4	125.0	124.5	124.0	123.5
4	74.14	61.25	56.18	53.44	51.71	50.53	49.66	49.00	48.47	48.05	47.41	46.76	46.10	45.77	45.43	45.09	44.75	44.40	44.05
5	47.18	37.12	33.20	31.09	29.75	28.84	28.16	27.64	27.24	26.92	26.42	25.91	25.39	25.14	24.87	24.60	24.33	24.06	23.79
6	35.51	27.00	23.70	21.92	20.81	20.03	19.46	19.03	18.69	18.41	17.99	17.56	17.12	16.89	16.67	16.44	16.21	15.99	15.75
7	29.25	21.69	18.77	17.19	16.21	15.52	15.02	14.63	14.33	14.08	13.71	13.32	12.93	12.73	12.53	12.33	12.12	11.91	11.70
8	25.42	18.49	15.83	14.39	13.49	12.86	12.40	12.04	11.77	11.54	11.19	10.84	10.48	10.30	10.11	9.92	9.73	9.53	9.33
9	22.86	16.39	13.90	12.56	11.71	11.13	10.70	10.37	10.11	9.89	9.57	9.24	8.90	8.72	8.55	8.37	8.19	8.00	7.81
10	21.04	14.91	12.55	11.28	10.48	9.92	9.52	9.20	8.96	8.75	8.45	8.13	7.80	7.64	7.47	7.30	7.12	6.94	6.76
11	19.69	13.81	11.56	10.35	9.58	9.05	8.66	8.35	8.12	7.92	7.63	7.32	7.01	6.85	6.68	6.52	6.35	6.17	6.00
12	18.64	12.97	10.80	9.63	8.89	8.38	8.00	7.71	7.48	7.29	7.00	6.71	6.40	6.25	6.09	5.93	5.76	5.59	5.42
13	17.81	12.31	10.21	9.07	8.35	7.86	7.49	7.21	6.98	6.80	6.52	6.23	5.93	5.78	5.63	5.47	5.30	5.14	4.97
14	17.14	11.78	9.73	8.62	7.92	7.43	7.08	6.80	6.58	6.40	6.13	5.85	5.56	5.41	5.25	5.10	4.94	4.77	4.60
15	16.59	11.34	9.34	8.25	7.57	7.09	6.74	6.47	6.26	6.08	5.81	5.54	5.25	5.10	4.95	4.80	4.64	4.47	4.31
16	16.12	10.97	9.00	7.94	7.27	6.81	6.46	6.19	5.98	5.81	5.55	5.27	4.99	4.85	4.70	4.54	4.39	4.23	4.06
17	15.72	10.66	8.73	7.68	7.02	6.56	6.22	5.96	5.75	5.58	5.32	5.05	4.78	4.63	4.48	4.33	4.18	4.02	3.85
18	15.38	10.39	8.49	7.46	6.81	6.35	6.02	5.76	5.56	5.39	5.13	4.87	4.59	4.45	4.30	4.15	4.00	3.84	3.67
19	15.08	10.16	8.28	7.26	6.62	6.18	5.85	5.59	5.39	5.22	4.97	4.70	4.43	4.29	4.14	[illegible]	3.84	[illegible]	[illegible]
[illegible]	[illegible]	[illegible]	8.10	7.10	6.46	6.02	5.69	5.44	5.24	5.08	4.82	4.56	4.29	4.15	4.00	3.86	3.70	3.54	3.38
21	14.59	9.77	7.94	6.95	6.32	5.88	5.56	5.31	5.11	4.95	4.70	4.44	4.17	4.03	3.88	3.74	3.58	3.42	3.26
22	14.38	9.61	7.80	6.81	6.19	5.76	5.44	5.19	4.99	4.83	4.58	4.33	4.06	3.92	3.78	3.63	3.48	3.32	3.15
23	14.19	9.47	7.67	6.69	6.08	5.65	5.33	5.09	4.89	4.73	4.48	4.23	3.96	3.82	3.68	3.53	3.38	3.22	3.05
24	14.03	9.34	7.55	6.59	5.98	5.55	5.23	4.99	4.80	4.64	4.39	4.14	3.87	3.74	3.59	3.45	3.29	3.14	2.97
25	13.88	9.22	7.45	6.49	5.88	5.46	5.15	4.91	4.71	4.56	4.31	4.06	3.79	3.66	3.52	3.37	3.22	3.06	2.89
26	13.74	9.12	7.36	6.41	5.80	5.38	5.07	4.83	4.64	4.48	4.24	3.99	3.72	3.59	3.44	3.30	3.15	2.99	2.82
27	13.61	9.02	7.27	6.33	5.73	5.31	5.00	4.76	4.57	4.41	4.17	3.92	3.66	3.52	3.38	3.23	3.08	2.92	2.75
28	13.50	8.93	7.19	6.25	5.66	5.24	4.93	4.69	4.50	4.35	4.11	3.86	3.60	3.46	3.32	3.18	3.02	2.86	2.69
29	13.39	8.85	7.12	6.19	5.59	5.18	4.87	4.64	4.45	4.29	4.05	3.80	3.54	3.41	3.27	3.12	2.97	2.81	2.64
30	13.29	8.77	7.05	6.12	5.53	5.12	4.82	4.58	4.39	4.24	4.00	3.75	3.49	3.36	3.22	3.07	2.92	2.76	2.59
40	12.61	8.25	6.60	5.70	5.13	4.73	4.44	4.21	4.02	3.87	3.64	3.40	3.15	3.01	2.87	2.73	2.57	2.41	2.23
60	11.97	7.76	6.17	5.31	4.76	4.37	4.09	3.87	3.69	3.54	3.31	3.08	2.83	2.69	2.55	2.41	2.25	2.08	1.89
120	11.38	7.32	5.79	4.95	4.42	4.04	3.77	3.55	3.38	3.24	3.02	2.78	2.53	2.40	2.26	2.11	1.95	1.76	1.54
∞	10.83	6.91	5.42	4.62	4.10	3.74	3.47	3.27	3.10	2.96	2.74	2.51	2.27	2.13	1.99	1.84	1.66	1.45	1.00

* Multiply these entries by 100.

Index

INDEX

The most efficient way to use this index is to look for the pertinent *property* (e.g., vapor pressure, entropy), *process* (e.g., disposal of chemicals, calibration), or *general concept* (e.g., units, radiation). Most primary entries are subdivided into several secondary entries, e.g., under heat capacity there are 24 secondary entries such as acids, air, hydrocarbons, water, etc. Primary entries will be found for certain *classes of substances*, such as alloys, elements, organic compounds, refrigerants, semiconductors, etc. Primary entries are also given for the individual chemical elements and for about 175 common compounds. However, only the most important tables are listed under these common substances. Therefore, the user will find in most cases that it is best to look first for the property of interest, then examine the table or tables that are referenced.

The reference given for each index term is the inclusive pages of the pertinent table (e.g., **3**-168 to 183). The introduction to each table describes the method of ordering the substances within that table.

The Editor would be grateful for comments and suggestions on this index.

INDEX

A

B

D

F

I

J

K

L

M

N

O

P

Q

R

S

U

V

W

X

STANDARD ATOMIC WEIGHTS (1993)

At. no.	Name	Symbol	Atomic weight	At. no.	Name	Symbol	Atomic weight
1	Hydrogen	H	1.00794(7)	54	Xenon	Xe	131.29(2)
2	Helium	He	4.002602(2)	55	Cesium	Cs	132.90543(5)
3	Lithium	Li	6.941(2)	56	Barium	Ba	137.327(7)
4	Beryllium	Be	9.012182(3)	57	Lanthanum	La	138.9055(2)
5	Boron	B	10.811(5)	58	Cerium	Ce	140.115(4)
6	Carbon	C	12.011(1)	59	Praseodymium	Pr	140.90765(3)
7	Nitrogen	N	14.00674(7)	60	Neodymium	Nd	144.24(3)
8	Oxygen	O	15.9994(3)	61	Promethium	Pm	[145]
9	Fluorine	F	18.9984032(9)	62	Samarium	Sm	150.36(3)
10	Neon	Ne	20.1797(6)	63	Europium	Eu	151.965(9)
11	Sodium	Na	22.989768(6)	64	Gadolinium	Gd	157.25(3)
12	Magnesium	Mg	24.3050(6)	65	Terbium	Tb	158.92534(3)
13	Aluminum	Al	26.981539(5)	66	Dysprosium	Dy	162.50(3)
14	Silicon	Si	28.0855(3)	67	Holmium	Ho	164.93032(3)
15	Phosphorous	P	30.973762(4)	68	Erbium	Er	167.26(3)
16	Sulfur	S	32.066(6)	69	Thulium	Tm	168.93421(3)
17	Chlorine	Cl	35.4527(9)	70	Ytterbium	Yb	173.04(3)
18	Argon	Ar	39.948(1)	71	Lutetium	Lu	174.967(1)
19	Potassium	K	39.0983(1)	72	Hafnium	Hf	178.49(2)
20	Calcium	Ca	40.078(4)	73	Tantalum	Ta	180.9479(1)
21	Scandium	Sc	44.955910(9)	74	Tungsten	W	183.84(1)
22	Titanium	Ti	47.867(1)	75	Rhenium	Re	186.207(1)
23	Vanadium	V	50.9415(1)	76	Osmium	Os	190.23(3)
24	Chromium	Cr	51.9961(6)	77	Iridium	Ir	192.217(3)
25	Manganese	Mn	54.93805(1)	78	Platinum	Pt	195.08(3)
26	Iron	Fe	55.845(2)	79	Gold	Au	196.96654(3)
27	Cobalt	Co	58.93320(1)	80	Mercury	Hg	200.59(2)
28	Nickel	Ni	58.6934(2)	81	Thallium	Tl	204.3833(2)
29	Copper	Cu	63.546(3)	82	Lead	Pb	207.2(1)
30	Zinc	Zn	65.39(2)	83	Bismuth	Bi	208.98037(3)
31	Gallium	Ga	69.723(1)	84	Polonium	Po	[209]
32	Germanium	Ge	72.61(2)	85	Astatine	At	[210]
33	Arsenic	As	74.92159(2)	86	Radon	Rn	[222]
34	Selenium	Se	78.96(3)	87	Francium	Fr	[223]
35	Bromine	Br	79.904(1)	88	Radium	Ra	[226]
36	Krypton	Kr	83.80(1)	89	Actinium	Ac	[227]
37	Rubidium	Rb	85.4678(3)	90	Thorium	Th	232.0381(1)
38	Strontium	Sr	87.62(1)	91	Protactinium	Pa	231.03588(2)
39	Yttrium	Y	88.90585(2)	92	Uranium	U	238.0289(1)
40	Zirconium	Zr	91.224(2)	93	Neptunium	Np	[237]
41	Niobium	Nb	92.90638(2)	94	Plutonium	Pu	[244]
42	Molybdenum	Mo	95.94(1)	95	Americium	Am	[243]
43	Technetium	Tc	[98]	96	Curium	Cm	[247]
44	Ruthenium	Ru	101.07(2)	97	Berkelium	Bk	[247]
45	Rhodium	Rh	102.90550(3)	98	Californium	Cf	[251]
46	Palladium	Pd	106.42(1)	99	Einsteinium	Es	[252]
47	Silver	Ag	107.8682(2)	100	Fermium	Fm	[257]
48	Cadmium	Cd	112.411(8)	101	Mendelevium	Md	[258]
49	Indium	In	114.818(3)	102	Nobelium	No	[259]
50	Tin	Sn	118.710(7)	103	Lawrencium	Lr	[262]
51	Antimony	Sb	121.760(1)	104	Rutherfordium	Rf	[261]
52	Tellurium	Te	127.60(3)	105	Hahnium	Ha	[262]
53	Iodine	I	126.90447(3)				

Numbers in brackets are the mass numbers of the longest-lived isotope of elements for which a standard atomic weight cannot be defined.